THE
ENCYCLOPEDIA
OF
FANTASY

THE
ENCYCLOPEDIA
OF
FANTASY

EDITED BY
JOHN CLUTE AND JOHN GRANT

THE
ENCYCLOPEDIA
— OF —
FANTASY

EDITED BY
JOHN CLUTE AND JOHN GRANT

Contributing editors
MIKE ASHLEY
ROZ KAVENEY
DAVID LANGFORD
RON TINER

Consultant editors
DAVID G. HARTWELL
GARY WESTFAHL

St Martin's Press
New York

St Martin's Press
New York

THE ENCYCLOPEDIA OF FANTASY. Copyright © 1997 by John Clute and John Grant. All rights reserved. Printed in Great Britain. No part of this book may be used or reproduced in any manner whatsoever without written permission except in the case of brief quotations embodied in critical articles or reviews. For information, address St. Martin's Press, 175 Fifth Avenue, New York, N.Y. 10010.

ISBN 0-312-15897-1

First published in Great Britain by Orbit

First U.S. Edition: May 1997

10 9 8 7 6 5 4 3 2 1

Library of Congress Cataloging-in-Publication Data

Clute, John, 1940–
    The encyclopedia of fantasy / John Clute, John Grant.
      p.   cm.
   ISBN 0-312-15897-1
   1. Fantastic fiction—Encyclopedias.   I. Grant, John,
1949–   . II. Title.
PN3435.C55 1997
809.3'8766'03—dc21
                                      96-37472
                                       CIP

CONTENTS

INTRODUCTION

In 1993 there was published a book called *The Encyclopedia of Science Fiction*. Even before they had finished compiling that book, various members of its editorial team realized they had only begun to dip their toes into the waters at the edge of the ocean that is the literature of the fantastic. The prospect of exploring this vast ocean was irresistible, which is why the current volume has come into being.

Although the two are entirely independent, this encyclopedia can be regarded as a sibling volume to *The Encyclopedia of Science Fiction*. As with many siblings there are similarities but also many profound contrasts, primarily because fantasy is a field of literature radically different from science fiction. Its roots go much deeper into history, and its concerns are more archetypal. Unlike science fiction, it is a literature which is remarkably hard to define (we here use the term "literature" to cover all the modes – text, cinema, comics, art, etc. – in which fantasy is expressed, because the field is remarkably integrated). We recognized all of this before we decided to embark on the enterprise: what we did not realize was quite *how* different the new book would be.

Some similarities between the two volumes are very obvious, notably the ascription practice – the format we use in giving titles, dates and so forth – although even here we found we had to make minor changes. On occasion we have felt it would be helpful to cross-refer readers to entries in *The Encyclopedia of Science Fiction* (◊ *SFE*), usually to show that authors who have done significant work in both areas are discussed in both books. The breakdown of entry-types also looked superficially similar: in our huge draft list there were entries on authors, movies, recurring themes within the literature, etc. But the similarity is deceptive. The draft list also included a large number of entries, intended as a whole to cast a net over the field of fantasy, and whose topics were such that they could hardly be described as "themes"; we called them "motifs", although we were still not entirely happy with that terminology. Some of them (like ANCESTRAL MEMORIES and COMMEDIA DELL'ARTE) were known terms; some (like PERCEPTION and REALITY) were existing terms but with implications for fantasy that had not occurred to us before; and some (like CROSSHATCH and POLDER and WAINSCOT) were tools of literary analysis which we had found it necessary to create.

As the implications of the "motif tree" dawned on us, it became very evident that our map of fantasy was going to differ very substantially from the map of science fiction adumbrated by Peter Nicholls as long ago as 1975 when he was conceiving the first edition of *The Encyclopedia of Science Fiction*.

And so it proved. The entry on FANTASY and some of the further entries referred to there encapsulate much of our sense of the distinction both between fantasy and science fiction and between the structures of description we have used to cover the two fields. The editorial team of *The Encyclopedia of Science Fiction* was able to treat sf as a field with definable boundaries, to parse that field in various clearcut ways, and to hope to cover *everything* within those boundaries (obviously a doomed proposition, but the goal was clear). In *The Encyclopedia of Fantasy*, on the other hand, we have confronted a different challenge. The term "fantasy" is used to cover a very wide range of texts, movies, visual presentations and so on. Tales involving DREAMS and VISIONS, ALLEGORY and ROMANCE, SURREALISM and MAGIC REALISM, SATIRE and WONDERLAND, SUPERNATURAL FICTION, DARK FANTASY, WEIRD FICTION and HORROR – all of these and more besides, sometimes expressing conflicting understandings of the nature of fantasy, were theoretically within our remit. Clearly we could not cover everything anybody had at one time or another thought of as "fantasy". Although it was going to be impossible to establish fixed limits, we had in *some* way to define our field.

The critic Brian ATTEBERY has spoken of fantasy in language which we feel well describes our final sense of the way in which this book has been constructed. Fantasy, he has said, is a "fuzzy set". By this he means a set which cannot be defined by its boundaries but which *can* be understood through significant examples of what best represents it. The "fuzzy set" model is, therefore, both exploratory and prescriptive. The exemplary writers and motifs making up the set of significant examples are like spotlights shining on a very complex world: they illuminate paths through that world and they help define our space, but they cannot shine on *everything*. The boundaries of fantasy fade into WATER MARGINS in every direction. (This might sound the same as the classic simplistic definition: "You know it's a fantasy when you see it." But there is a difference. Much of the literature discussed in the pages of this book does *not* initially look like fantasy.)

At the centre of all the fuzzy sets is a rough definition of what we mean by fantasy: a fantasy text is a self-coherent narrative which, when set in our REALITY, tells a story which is impossible in the world as we perceive it (◊ PERCEPTION); when set in an OTHERWORLD or SECONDARY WORLD, that otherworld will be impossible, but stories set there will be possible *in the otherworld's terms*. An associated point, hinted at here, is that at the core of fantasy is STORY. Even the most surrealist of fantasies tells a tale.

This notion extends to all aspects of the field, not just printed texts. Two of our editorial team (Grant and Tiner) have argued extensively elsewhere that FANTASY ART is, at its heart, a narrative form: the "fuzzy set" model may cast some light on the regions of Surrealism and the Abstract, but fantasy-art images proper depict a moment which has both a "before" and an "after"; viewers have to construct the surrounding story for themselves and will obviously come up with many different versions, but all of those tales are inherent in the image. It is easier to recognize the narrative aspect in most other modes of fantasy – CINEMA, COMICS, OPERA, SONG, etc. – except possibly MUSIC, although it is hard to listen to Berlioz's *Symphonie Fantastique* or Stravinsky's *The Rite of Spring* without realizing that one is being told a story of some sort, even if one hasn't read the programme notes and doesn't know what the story *is*.

But to return to the central theme. As far as our coverage of texts, authors and movies goes, therefore, the heart of this enterprise is the kind of fantasy that evolved from a few decades before the beginning of the 19th century – through the elaborate fictions of writers like E.T.A. HOFFMANN and, somewhat later, Edgar Allan POE – in the work of George MACDONALD, William MORRIS, Lewis CARROLL, Abraham MERRITT, E.R. EDDISON, Robert E. HOWARD, J.R.R. TOLKIEN, C.S. LEWIS, L. Sprague DE CAMP and Fletcher PRATT, Fritz LEIBER . . . and so on down to the moderns who have woven worlds either out of these examples or anew. Important writers like Sheridan LE FANU, H.P. LOVECRAFT and M.R. JAMES are given extensive entries, but are deemed more significant as authors of SUPERNATURAL FICTION – an area which we cover in considerable depth – or HORROR, which we touch on very much more lightly.

Before the 19th century, we tread somewhat more carefully. We have included full entries on such topics as FAIRYTALES, FOLKLORE and MYTHS, along with articles concerning relevant writers or compilers like the GRIMM BROTHERS and Charles PERRAULT. Entries on the literature of the fantastic produced by different countries and cultures also cover the years before 1800. Authors of what we call TAPROOT TEXTS – those works which are both central to Western literature as a whole and have contributed to the CAULDRON OF STORY from which modern fantasy authors ladle many of their basic ideas – are widely included: among them are APULEIUS, BOCCACCIO, BUNYAN, CERVANTES, CHAUCER, DANTE, MALORY, MILTON, RABELAIS and SHAKESPEARE. We have delved back into ancient history in entries like MESOPOTAMIAN EPIC and SANSKRIT LITERATURE, and even have a brief note on the BIBLE – which, regarded as history or otherwise, is certainly an important taproot text. But, to repeat, our sense is that fantasy as a form of literature began in the decades preceding the start of the 19th century, and the weight of our entries reflects this.

Given our prescriptive though far-ranging model of fantasy literature, and given our sense of the almost infinite extent of the water margins that surround that model, we have not attempted to write entries for every author who might potentially be included in the book, or for every movie or tv programme with some fantasy content. At the same time, we have tried to include every author (or movie, or whatever) of importance to the fantasy field. Obviously we are fallible! Some of our inclusions and omissions may be contentious; some omissions, in particular, may be simply the product of human error.

There were also areas which we firmly decided to leave out. (a) Our coverage up to the end of 1995 is as complete as we can make it; coverage of items that appeared in 1996 is incomplete, and in particular books which we have not seen are indicated in the style «Title» (1996) to show that they are projected titles. (b) Some HORROR authors are of definite fantasy interest, because their work is full of fantasy (however dark); other horror authors seldom if ever stray into the fantasy field. To give a couple of random examples, it would have been wrong to omit Stephen KING, but there was no justification for including the (likewise bestselling) Richard Laymon. (c) The same went for the movies. Quite a number of the 500 or so movies treated in these pages are HORROR MOVIES, but they have been selected because of their fantasy content and their influence on the field. Some of these movies we do not *like*, but it would have been wrong to exclude them. Also, we decided only to cover movies which we had watched (or, often, re-watched) specifically to write the articles about them: too many of the secondary sources on the cinema are wildly inaccurate, and memory is always fallible. In the end, in a few cases – notably within long series of movies – we had to compromise on this because we simply could not track down a viewing copy. Such instances are indicated *in situ*. (d) An exactly analogous argument to that concerning horror can be extended to the field of SUPERNATURAL FICTION. (e) On the other hand, where we have come across a book or a movie or any other fantasy artefact, however obscure, that we felt should be brought to the attention of those interested in the field, we have included it.

The conception and writing of the actual text constituted a complex task, as is inevitable in any enterprise of this size and scope. Various editors named on the title page, plus colleagues whose assistance was less formal, made their impact at various stages of the process. It is difficult to distinguish just who fulfilled which function, because the enterprise has been collegiate throughout, but:

In particular John Clute and Roz Kaveney spent many hours working out a rough entry structure and generating a preliminary map of the interconnections between those entries; they were helped, out of sheer goodwill, by Chris Bell and Diana Wynne Jones – and countless others, including, of course, the remaining named editors. Clute also wrote the bulk of the "motif" and author entries. In particular John Grant and David Langford shaped oceans of copy into a text that made sense, complete with a reliable network of entries and cross-references. Grant also, with a deal of assistance from Langford and further assistance from Lydia Darbyshire, copyedited the entire text and himself wrote virtually every CINEMA entry. Langford wrote many "motif" and author entries, and also supplied a not inconsiderable amount of computer expertise. Mike Ashley wrote very many entries and shaped various parts of the book, notably those concerning ARTHUR and the MATTER of Britain, SUPERNATURAL FICTION, ANTHOLOGIES, MAGAZINES, FAIRYTALES and FOLKLORE. Ron Tiner did the same shaping task in the areas of COMICS, GRAPHIC NOVELS and ILLUSTRATION. David G. Hartwell and Gary Westfahl made stringent comments about the draft text; Westfahl also helped in the shaping of the initial entry list. But, as will be obvious to readers who look at the initials at the ends of articles, it was all rather more complicated than that: many of the articles in this book are truly collegiate, and few have not been altered in some way by an editor – perhaps several editors – other than the one whose initials appear. In addition, Ashley, Darbyshire and Langford proofread the entire text.

Rather more than half the text was written between them by Clute (about 400,000 words) and Grant (about 250,000 words). Ashley contributed about 200,000 words, Langford about 80,000, Tiner about 60,000, Brian Stableford about 50,000, Kaveney about 25,000 and Bill Cotter – who shaped the tv sections and wrote almost all the relevant entries – about 20,000. All of these word-counts are suspect, once again because of the collegiate nature of the enterprise: we have been unable to keep any track (and anyway it would have been a fool's venture) of the sentences and paragraphs the editors have inserted into each other's entries. Other contributors – among whom Gregory Feeley, who also recruited a bank of further fine writers, deserves special mention – wrote varying amounts.

The Encyclopedia of Fantasy has been a job for many people. It has also been a very considerable job. While we have made every effort to attain accuracy, certainly there must be errors. We invite readers to inform us of any they discover by writing, with chapter and verse, to one or other of us via John Clute, 221B Camden High Street, London NW1 7BU, UK (e-mail jclute@cix.co.uk). John Grant was, at the time of going to press, shortly to move house, but notes addressed to him c/o that same address will reach him.

John Clute
John Grant
September 1996

CONTRIBUTORS

IA	Ivan Adamovič	MH	Mirta Hillen
MA	Mike ASHLEY	SH	Steve Holland
BA	Brian ATTEBERY	JHH	John-Henri Holmberg
DVB	David V. Barrett	DWJ	Diana Wynne JONES
AB	Alberto Becattini	SJ	Steve JONES
CB	Chris Bell	RK	Roz KAVENEY
IB	Imants Belogrivs	DK	Don Keller
GB	Gediminas Beresnevičius	PK	Paul Kincaid
JB	Joe Bernstein	DRL	David LANGFORD
JCB	John C. Bunnell	GL	Gwendolyn Layne
EFB	E.F. BLEILER	LL	Lynne Lundquist
RB	Richard Bleiler	RL	Robert Latham
DB	David S. Bratman	SJL	Sam J. LUNDWALL
RB	Romualdas Buivydas	PJM	Paul J. McAuley
JC	John CLUTE	SM	Sean McMullen
JRC	John Robert COLOMBO	ShM	Shinji Maki
BC	Bill Cotter	PM	Patrick Marcel
RD	Richard DALBY	KLM	K.L. Maund
HD	Hugh Davies	RM	Richard Middleton
NdG	Norman Thomas di Giovanni	BM	Bob Morrish
SD	Stefan DZIEMIANOWICZ	CM	Caroline Mullan
GF	Gregory Feeley	EN	Egon Niczky
JF	Jo Fletcher	JO	Jaroslav Olša Jr
NG	Neil GAIMAN	CP	Carlo Pagetti
JG	John GRANT	SP	Steve Paulsen
LH	Larry Hammer	DP	David PRINGLE
JH	Judith Hanna	JR	Jilly Reed
DMH	Donald M. Hassler	DR	David Roache

MSR	Michael Scott ROHAN	MS	Marcial SOUTO
FR	Franz Rottensteiner	BS	Brian STABLEFORD
MR	Marcus Rowland	BT	Braulio Tavares
AFR	André-François Ruaud	JT	Jonathan Tickner
FS	Fay SAMPSON	RT	Ron TINER
AS	Andy Sawyer	LT	Lisa TUTTLE
WKS	William K. Schafer	HW	Henry Wessells
CS	Cyril Simsa	GW	Gary Westfahl
DRS	David R. Smith	CW	Colin WILSON
KS	Krzysztof Sokołowski	GKW	Gary K. Wolfe

ACKNOWLEDGEMENTS

We would not have been able to compile this book without the gratefully received help of the following:

Dominick Abel; Brian Ameringen; David Baldwin; Adrian Belcher; Chris Bell; E.F. Bleiler; Richard Bleiler; Charles N. Brown; Lucie M. Chin; William Contento; John Dallman; Lydia Darbyshire; Keith R.A. DeCandido; Robert Devereaux; Norman Thomas di Giovanni; Jeff Fisher; John Foyster; Kirsten Gong-Wong; Colin Greenland; John Gullidge; Paul Hamilton; Peter Heck; Tim Holman; David Hutchinson; Melissa Hyland; John Jarrold; Diana Wynne Jones; Paul Kincaid; Mary Kirchoff; Rosemary Kirstein; Hazel Langford; Fred Lerner; Paula Mardo (Exeter Central Library); Beth Meacham; Miwa Messer; Richard Middleton; Pam Miller; Julia Millette; Caroline Mullan; Colin Murray; Freida A. Murray; Valerie Paine; Karen Pender-Gunn; Dominic Petitfaux; Ann Polis; Andrew Porter; Christopher Priest; David Pringle; Susan Protter; Haydn Rawlinson; Jilly Reed; Robert Reginald; Roger Robinson; Alan Rodgers; Michael Scott Rohan; Yvonne Rousseau; Sharon Sbarsky; Corin See; Donal Sheridan (Bloomsbury Theatre, London); David R. Smith; Mark Stackpole; Bruce Sterling; Alan Stewart; Gordon Van Gelder; Al von Ruff; Wandering Jew (Ahasuerus the); Lawrence Watt-Evans; Bridget Wilkinson; Matt Williams; M.K. Wren.

In addition, acknowledgement should be made to the editors and authors of the wide range of scholarly and reference works now available to students of fantasy in all its forms. Ev Bleiler's pioneering scholarship and fatherly presence irradiated our words throughout. Two recent books that were particularly valuable were *The Supernatural Index* (**1995**) ed Mike Ashley and William Contento and *The St James Guide to Fantasy Writers* (**1996**) ed David Pringle, who kindly made available to us an advance copy of the text.

    Personal thanks are also due. This encyclopedia took a long time to write, and generated personality traits in at least one of its editors. John Clute would like to thank Judith Clute, Helen Nicholls and Anna Russell, who did not hit back too often. John Grant would like to thank Catherine Barnett, Jane Barnett, Fionna O'Sullivan, Muriel Barnett and Margaret Stewart, who tolerated his absences for long hours at the computer and gave unstinting friendship and support.

John Clute
John Grant
September 1996

ABBREVIATIONS AND SYMBOLS

*	applies to a title tied to an original movie, game, etc.
◊	see (i.e., a cross-reference)
◊◊	see also (i.e., an ancillary cross-reference)
«Title»	indicates a projected title; such titles may in fact have been published before the text went to press, but have not been seen by the editors

aka	also known as
anim dir	animation director(s)
anon	anonymous(ly)
anth	anthology
assoc pr	associate producer(s)
b/w	black-and-white
chap	chapbook (here defined as a book under 100 pages long)
coll	collection
cut	abridged, by either author or publisher
dir	director(s)
dos	bound back-to-back alongside an independent text
ed	edited by
edn	edition
exec pr	executive producer(s)
exp	expanded
F&SF	*The Magazine of Fantasy and Science Fiction*
fixup	collection of (usually revised) stories presented as a novel
fx	effects
graph	graphic book
illus	illustrated by
LOTR	*The Lord of the Rings*
mufx	makeup effects
OED	*Oxford English Dictionary*

omni	omnibus edition
ot	original title
pr	producer(s)
rev	revised
RPG	role-playing game
S&S	sword and sorcery
sf	science fiction
SFE	*The Encyclopedia of Science Fiction* (2nd edn)
spfx	special effects
sv	supervisor(s)
trans	translated by
tv	television
tvm	movie made for tv
var mags	published in various magazines
vfx	visual effects
vol	volume
vt	variant title
WT	*Weird Tales*
WWI	World War I
WWII	World War II
WWIII	World War III
YA	Fiction marketed for a young adult audience

Dates given in **bold** are those of first publication
Dates given in ***bold italics*** are those of first movie release

AAMODT, DONALD (1935-   ) US writer whose **Zarathandra** sequence of comic fantasies – *A Name to Conjure With* (**1989**) and *A Troubling Along the Border* (**1991**) – features the adventures of Sandy McGregor, a human called in error to an ALTERNATE WORLD by a SORCERER in search of a DEMON to help him; here he learns the rules of the game, gains COMPANIONS and becomes himself a sorcerer to reckon with, though his discovery in the first volume – that all his actions have been moves in a cynical GODGAME – is chastening. The second volume, a conventional QUEST tale, is better told but less fun.   [JC]

ABBEY, LYNN Working name of US writer Marilyn Lorraine Abbey (1948-   ), best known for co-creating and co-editing with Robert Lynn ASPRIN (*whom see for details*) the **Thieves' World** sequence of SHARED-WORLD anthologies, some of them worked into BRAIDS; the sequence is now complete in 12 vols. LA's first solo work, the **Rifkind** sequence – whose heroine is a WITCH devoted uncomfortably to serving the GODDESS – comprises *Daughter of the Bright Moon* (**1979**) and *The Black Flame* (**1980**). The **Unicorn and Dragon** series – *Unicorn & Dragon* (**1987**) and *Conquest* (**1988**; vt *The Green Man* 1989 UK) – revisits a similar venue, in this case 11th-century Britain, and again features a young woman whose TALENTS make her eligible to become a priestess for the Goddess. The main conflict – Anglo-Saxons, devoted to the Celtic religion, versus invading Normans; Wicca versus patriarchal eastern sorcery – is of no historical validity, but does generate hints of THINNING throughout, providing a sense of the possible course of future volumes. The **Ultima Saga** books – *The Forge of Virtue* * (**1991**) and *The Temper of Wisdom* * (**1992**) – novelize the "Ultima Saga" GAME. LA has a serviceable style and a knack for historical fantasy, and her protagonists display an unusual complexity and grittiness, but her action sequences are sometimes tangled.   [JC]
Other works: *The Guardians* (**1982**); *Phule's Company* * (**1990**), part of the **Phule** shared-world sequence of sf comedies; *The Wooden Sword* (**1991**) and its sequel, *Beneath the Web* (**1994**), which features a fetch with AMNESIA; *The Brazen Gambit* * (**1994**) and *Cinnabar Shadows* * (**1995**), contributions to the **Dark Sun** series of novels based on the game.

ABBOTT AND COSTELLO MEET DR JEKYLL AND MR HYDE US movie (***1953***). ◊ JEKYLL AND HYDE MOVIES.
ABBOTT AND COSTELLO MEET FRANKENSTEIN US movie (***1948***). ◊ FRANKENSTEIN MOVIES.
ABBOTT AND COSTELLO MEET THE INVISIBLE MAN US movie (***1951***). ◊ *The* INVISIBLE MAN (***1933***).
ABDULLAH, ACHMED Working name of Russian-born UK writer Alexander Nicholayevitch Romanoff (1881-1945), who adopted the name Achmed Abdullah Nadir Khan el-Durani el Iddrissyeh after his parents divorced; he wrote also as A.A. Nadir and John Hamilton. He was educated in India, the UK and France, served in the British Army, and became a resident of the USA, where most of his work first appeared. He is best-known for his novelization of *The* THIEF OF BAGDAD (***1924***), *The Thief of Bagdad* * (**1924**) – featuring a wraparound dustwrapper by Willy POGANY – which he freely adapted from the original script by Elton Thomas and Lotta Woods. This rousing ARABIAN FANTASY is not typical of his work. Most of his stories and novels appeared first in the US pulp MAGAZINES, where he rapidly established a wide readership for his contemporary Oriental and near-eastern adventure stories. All have good characterization and local colour, but there is usually only a hint of the fantastic, often involving REINCARNATION and FATE, as in the stories in *Alien Souls* (coll **1922**). A few are more overtly supernatural (◊ SUPERNATURAL FICTION), as in *Wings: Tales of the Psychic* (coll **1920**), where evil SPIRITS and astral projection feature. These same themes emerge in his novels – *The Mating of the Blades* (**1920**) and *The Flower of the Gods* (**1936**), the latter written with Fulton Oursler (1893-1952). Few of AA's other novels are fantastic, though his early series **The God of the Invincibly Strong Arms** – three serials with that title (*All-Story Weekly* 1915-16), of which only two appeared in book-form, as *The Red Stain* (**1915**) and *The Blue-Eyed Manchu* (**1916**) – about a fanatical cult of Kali worshippers, is redolent of Sax ROHMER's **Fu Manchu** series. AA's peak was in the post-WWI decade; thereafter he concentrated more on romances and plays.   [MA]
Other works: *Mysteries of Asia* (coll **1935** UK).
As editor: *Fifty Enthralling Stories of the Middle East* (anth **1937** UK).

Further reading: *The Cat Had Nine Lives* (**1933**; vt *My Nine Lives* 1934 UK), autobiography.

ABÉ, KOBO (1924-1993) Japanese writer, some of whose later novels – notably *Suna no Onna* (**1962**; trans E. Dale Saunders as *Woman in the Dunes* **1964** US) – are famous for their inspired absorption and transformation of Western renderings of alienation and the Absurd, including specifically the work of Franz KAFKA and Samuel Beckett (1906-1989). The stories assembled as *Beyond the Curve* (coll trans Juliet Winters Carpenter **1991** US), first published 1949-66, show how early and how intensely KA had translated these influences into an original idiom, marked by narrative urgency, urban bleakness and an occasionally convulsive wit. In "Dendorokakariya" (1949), trans here as "Dendrocacalia", the TRANSFORMATION of an ordinary man into a plant reads simultaneously as a TWICE-TOLD version of the myth of Daphne, a POSTHUMOUS FANTASY and an acidulous example of the comedy of the Absurd. The other tales in *Beyond the Curve*, though very various in content, share a similar driving wit. KA's novels in translation – including *Moetsukita Chizu* (**1967**; trans E. Dale Saunders as *The Ruined Map* **1969** US), *Tanin no Kao* (**1964**; trans E. Dale Saunders as *The Face of Another* **1966** US) and *Hakobune Sakura Maru* (**1984**; trans Juliet Winters Carpenter as *The Ark Sakura* **1988** US) – are less mobile in their handling of the metaphysics of solitudinous alienation. *Dai-Yon Kampyoki* (**1959**; trans E. Dale Saunders as *Inter Ice Age 4* **1970** US) is sf. [JC]

À BECKETT, ARTHUR WILLIAM (1844-1899) UK author, a son of Gilbert Abbott À BECKETT. His *The Member for Wrottenborough: Passages from his Life in Parliament* (**1892**) satirizes, among other targets, a matriarchy purporting to govern the ISLAND of Eden-minus-Adamia. [JC]

À BECKETT, GILBERT ABBOTT (1811-1856) UK author of about 50 plays including *Peter Wilkins* (**1846** chap) with Mark Lemon (1809-1870), which dramatizes *The Life and Adventures of Peter Wilkins* (**1751**) by Robert Paltock (1697-1767), a popular FANTASTIC VOYAGE. [JC]

À BECKETT, GILBERT ARTHUR (1837-1891) UK writer, a son of Gilbert Abbott À BECKETT. He collaborated with W.S. GILBERT (*whom see for details*) on *The Happy Land: A Burlesque Version of "The Wicked World"* (**1873** chap). [JC]

ab HUGH, DAFYDD (1960-   ) US writer (whose Welsh name has been legalized) perhaps best-known for his novella, "The Coon Rolled Down and Ruptured his Larinks, a Squeezed Novel by Mr Skunk" (1990 *IASFM*). His first novel, *Heroing, or How He Wound Down the World* (**1987**), and its sequel, *Warriorwards* (**1990**), are SWORD AND SORCERY tales starring a female mercenary (◊ GENDER) named Jiana. In the first tale she aids (and later thwarts) a prince in his QUEST for the wish-granting "World's Dream", while herself examining the nature of heroism; in the second, she becomes (willingly) a slave in order ultimately to free a young woman, and subsequently to train her as a swordswoman. As REVISIONIST FANTASY the sequence is of some interest, though the painful "richness" of DaH's style distracts from the argument. The **Arthur War Lord** sequence – *Arthur War Lord* (**1994**) and *Far Beyond the Wave* (**1994**) – is a RATIONALIZED FANTASY of TIME TRAVEL, in which a man is sent back to the time of ARTHUR where, as LANCELOT, he must pursue a female IRA agent who is attempting to change history by assassinating the king. [JC]

Other works: *Star Trek: Deep Space Nine: Fallen Heroes* * (**1994**) and *Star Trek: The Next Generation: Balance of Power* * (**1994**); two contributions to the **Doom** series of GAME ties, being *Knee-Deep in the Dead* * (**1995**) and *Hell on Earth* * (**1995**), both ascribed in some sources as anon collaborations with Brad Linaweaver (1952-   ; ◊ *SFE*).

ABSENT MINDED PROFESSOR, THE US movie (*1961*). ◊ DISNEY.

ABSURDIST FANTASY Given the definition of FANTASY used in this encyclopedia – a central thread of which is the argument that fantasy narratives tend to be STORIES which end with a sense that a meaningful tale has been both told and *understood* – this term will not often be found.

The Absurd, as a critical concept, was generated as a response to the works of Albert Camus (1913-1960), who used the term to describe the gap between the need for order – for stories which tell the tale of the world and are true and humanly meaningful – and a world which does not supply that need. Although Camus himself came to a philosophical position which trumped the Absurd through a grimly exultant stoicism about the human condition, the term itself, as normally used, refuses any such transcendence. The plays of Samuel Beckett (1906-1989) and Eugene IONESCO have frequently been described as defining a "Theatre of the Absurd": their eschewing of consistent character development and of linear plots – their refusal, in other words, to fabricate story – represents an antithesis to fantasy, though the works may be FANTASTIC. [JC]

ACCURSED WANDERERS The figure of the AW goes back as far as Cain and the medieval legends of the WANDERING JEW, both of whom give many features to the character: AWs are taboo-breaking or blasphemous, they may bear a distinctive mark, and they enjoy supernatural protection against mundane threats. But the legend did not take on classic form until the early days of ROMANTICISM, when Lord BYRON and Charles MATURIN in particular not only reinvigorated the archetype in their works but also tried to live it out in their own lives. In doing so, they added some important details – an extension of the taboo-breaking element to cover disapproved sexual practices and the implication that the curse is infectious: the eponymous AW featured in Maturin's *Melmoth the Wanderer* (**1820**) can free himself from a PACT WITH THE DEVIL only by persuading someone else to take it on, and therefore continually puts people in situations where they might be tempted to do so. Byron, in *Childe Harold's Pilgrimage* (**1812-18**), attempted (with only partial success) to create an AW who is also a CHILDE; and in *Cain* (1821; **1822**) creates a protagonist who realizes that his sins are minuscule in comparison with those of the God who has condemned him. Here the AW is an antinomian figure – that is, one of the blessed elect in spite of or even because of his vile acts – who perceives the WRONGNESS of things in a way not open to those with calmer lives.

Sometimes the AW comes to see his curse as a NIGHT JOURNEY in which he has been instructed in virtue; Samuel Taylor COLERIDGE's Ancient Mariner, in "The Rime of the Ancient Mariner" (1798), is a person who LEARNS BETTER. Something of this also affects the eventual fate of GOETHE's FAUST, whose continual damaging of other people's lives nonetheless redeems him when it leads him to break the CONDITIONS of his pact in a moment of high-minded and forgetful altruism; this sort of QUIBBLE has as its descendant the redemption by true love which Richard WAGNER added to the tale of the

FLYING DUTCHMAN (◊ OPERA). Many others have used this device to soften, sentimentalize and revise LEGENDS.

The AW is often also an OBSESSED SEEKER; HIDDEN MONARCHS may be AWs, or at least closely resemble them when first met – in J.R.R. TOLKIEN's *The Lord of the Rings* (**1954-5**), Aragorn in his role as Strider often resembles the archetype. AWs may have much to offer humanity yet be cursed with an affliction so disgusting that humanity rejects their aid. Some AWs are SHAPESHIFTERS and most of fantasy's VAMPIRES are AWs, not least because one of the clinching adaptations of the vampire myth into Western European art was John POLIDORI's melodramatic lampoon of Byron in *The Vampyre: A Tale* (**1819** chap). Other immortals (◊ IMMORTALITY) are often AWs, simply because of the drying of their affections by long life; even SATAN can appear as an AW – the 19th-century Romantic misreading of John MILTON is crucial to this particular form of the archetype. The AW is frequently an AVATAR of some earlier figure – in which case the curse may be simply a matter of leftover bad karma, as in Robert JORDAN's **Wheel of Time** sequence (**1990** onwards).

Michael MOORCOCK's **Elric** is the classic type from HIGH FANTASY; Moorcock clearly intended to give him almost every conceivable feature of the archetype. Elric is incestuous, a kinslayer and a traitor; he is marked out both by his status as prince of an ELDER RACE and by his albinism; he is almost inevitably fatal to everyone who comes near him, particularly his lovers and friends; he gradually learns he has been manipulated by the Lords of CHAOS and Order, and that any crime he may have committed is trivial by comparison; he is effectively a vampire as a result of his dependency on lives taken for him by his SWORD; he is on several occasions redeemed, temporarily, by the love of a good woman.

The AW is a useful protagonist in REVISIONIST FANTASY like Moorcock's because his position is readily seen as an injustice and therefore an indictment, from a radical, atheist or anarchist position, of a THEODICY like Tolkien's. Sometimes the AW is a figure of revisionism from a standpoint more rigorously orthodox than Tolkien's Catholic pietism. The protagonist of Stephen R. DONALDSON's **Thomas Covenant** sequence has many characteristics of the AW because the element of theodicy in the two trilogies' resolution is specifically an antirationalist one, in which grace comes capriciously through wild MAGIC rather than because Covenant has done the right thing throughout. Because he is an embodiment of grace, Covenant's worst acts turn out to have been for the good.

In HEROIC FANTASY curses are likely to be a generator of plot, either because they cause things to happen or because the AW's attempt to free himself of the curse (or of the compulsion to wander) provides the driving force of a TEMPLATE. As a general rule, heroic fantasy avoids the deeper implications of the archetype in favour of its narrative conveniences. An example of this in action is the curse on Brion Rouwen in Robert VARDEMAN's *The Accursed* (**1994**), which causes him constantly to be instrumental in the deaths of allies when he comes to care for them, and accordingly forces him continually into alliances with people he despises and distrusts. In Bernard KING's *Starkadder* (**1985**) the eponymous character (drawn directly from Norse mythology) is cursed by the GODS to live three lives consecutively, the end of each being marked by his betrayal of someone to whom he has sworn loyalty; he must reconcile his wish finally to die with his reluctance to do so in dishonour.

[RK]

See also: BELATEDNESS.

ACHERON There is a tendency in modern fantasy to treat the geography of the UNDERWORLD with some looseness. But the River Acheron – Acheron originally being the Titans' watercarrier, whom ZEUS transformed into a river as an eternal punishment (◊ BONDAGE) – is properly described as its central RIVER. Four tributaries flow into the Acheron: the Styx, Phlegethon, Cocytus and Lethe. The passage of the dead across all five streams is full of anguish; only after Lethe causes AMNESIA is there any peace.

The CITY of Acheron, in Robert E. HOWARD's *Conan the Conqueror* (**1935-6** *WT* as "The Hour of the Dragon"; **1950**), has nothing to do with the river. [JC]

ACHILLEOS, CHRIS (1947-   ) Working name of fantasy illustrator Christos Achilleos, born in Cyprus, in the UK since 1960. He is widely renowned for his meticulously painted fantasy erotica, done in an elegant, polished style. He works in cell paint with brush and airbrush on illustration board.

CA studied scientific and technical illustration at Hornsey College of Art, but quickly turned to fantasy and sf book covers. He has produced cover paintings for editions of Edgar Rice BURROUGHS's **Pellucidar** books, Robert E. HOWARD titles, John NORMAN's **Gor** books, as well as for editions of books by H.G. WELLS, Michael MOORCOCK and Target Books' **Dr Who** series (1975-7). In 1974 he began the series of erotic paintings for Paul Raymond's girlie magazine *Mayfair* that brought him worldwide recognition. He continues to produce fantasy book covers and movie posters, including Corgi Books' **Star Trek** series (1983-4) and *Heavy Metal Movie* (1980). [RT]

Other works: *Beauty and the Beast* (graph **1978**); *Sirens* (graph **1986**); *Medusa* (graph **1988**).

ACHILLES Born of a mortal and Thetis, a nymph, this great though bad-tempered warrior, who dies young, is one of the Greek HEROES who besiege Troy and is a central figure in HOMER's *Iliad* (*c*8th century BC). Fantasy tends to make limited use of Achilles as a generalized UNDERLIER figure. Brian M. STABLEFORD's **Dies Irae** trilogy (**1971**) is a comparatively rare example in which Achilles is central. [JC]

See also: ACHILLES' HEEL.

ACHILLES' HEEL One of the LEGENDS surrounding ACHILLES that post-dates HOMER's *Iliad* (*c*8th century BC) tells how his mother took her infant son to the Styx, where she bathed him to make him invulnerable. Unfortunately, she held him by one heel which, not having been immersed, remained vulnerable. Achilles was later killed by an arrow to the heel. In modern HIGH FANTASY, most characters of any significance have an analogue of an Achilles' Heel, and many plots turn on discovering it. In J.R.R. TOLKIEN's *The Lord of the Rings* (**1954-5**), to give a single instance, the Nazgûl king cannot be slain by any man, so a woman is given the task. [JC]

ACKROYD, PETER (1949-   ) UK writer whose early surrealistic poetry was assembled in volumes like *London Lickpenny* (coll **1973** chap) and *Country Life* (coll **1978** chap), and whose nonfiction *Dressing Up: Transvestism and Drag, the History of an Obsession* (**1979**) early manifested his interest in double or multiple selves. Several of his titles signal PA's deep involvement with LONDON, and it is as an author of complex URBAN FANTASY that he is of greatest genre interest. Within this urban frame, his novels almost invariably conflate present and past through patterns of association and POSSESSION, and his contemporary protagonists sometimes seem little more than impersonating MASKS of figures sunk deep in time. Sometimes the impersonation is less a ravishment of these protagonists than a masquerade: PA's interest in music hall and travesty is marked, especially in the nonfantasy *Last Testament of Oscar Wilde* (**1983**), which pastiches the great wit, and *Dan Leno and the Limehouse Golem* (**1994**; vt *The Trial of Elizabeth Cree: A Novel of the Limehouse Murders* 1995 US), a nonfantasy GASLIGHT ROMANCE set in late-19th-century London. Elements of pastiche occur throughout PA's fiction, particularly in his quasi-fantasies.

In his first novel, *The Great Fire of London* (**1982**), the makers of a movie of Charles DICKENS's *Little Dorrit* (**1857**) are haunted by the original characters, who are themselves haunted by the great city in whose coils they are caught. The eponymous contemporary detective in *Hawksmoor* (**1985**) is increasingly obsessed by an 18th-century architect – based on the historical Nicholas Hawksmoor (1661-1736) – who is himself haunted by a numerology-based geography of London, a system of correspondences through which occult patterns can be descried and the future governed. Similarly, in *The House of Doctor Dee* (**1993**) a contemporary protagonist finds himself tracing out, through the London of the MAGUS John DEE, some sense of his own shrunken self; for it proves he may be nothing more than an impersonating SHADOW, the reborn (or recast) image of the HOMUNCULUS which Dee has been attempting to create.

Other novels of interest include: *First Light* (**1989**), in which much is made of the fact that, as a result of precession, the night sky over a neolithic grave being excavated duplicates that of 26,000 years ago, when the monument was built; and *English Music* (**1992**), a nonfantasy paean to the ISLAND domain. PA is one of the central fabulists of the CITY.   [JC]

ACOLYTE, THE ◊ MAGAZINES.

ADAIR, GILBERT (?   -   ) UK writer and journalist whose first novels – *Alice Through the Needle's Eye: A Third Adventure for Lewis Carroll's "Alice"* (**1984**) and *Peter Pan and the Only Children* (**1987**) – are SEQUELS BY OTHER HANDS to, respectively, Lewis CARROLL's **Alice** books and J.M. BARRIE's **Peter Pan** works. "The Giant Rat of Sumatra" (1988 *Observer*) features SHERLOCK HOLMES. The first novel is perhaps the greatest interest, taking Alice into a WONDER-LAND where her adventures are (it eventually proves) governed by the letters of the alphabet, in correct order (she initially falls into an A-stack, is bothered by spelling bees, makes her way to the sea, etc.). Of his later novels, *The Death of the Author* (**1992**) alertly parodies the voice of Vladimir NABOKOV, is textually haunted by the ghost of Paul de Man (1919-1984), and concludes with an intricate play on the conventions of the POSTHUMOUS FANTASY. In *The Postmodernist Always Rings Twice: Reflections on Culture in the 90s* (coll **1992**) he subjects various icons, from BATMAN to Umberto ECO, to an analysis based on the argument that the media through which culture is transmitted are themselves a form of product. He has also translated *La Disparition* (**1969**) by Georges Perec (1936-1982) as *A Void* (**1994**); both original and translation avoid using the letter "e".
                                                                                            [JC]

ADAM ADAMANT LIVES! UK tv series (1966-7). BBC. **Pr** Verity Lambert. **Dir** Moira Armstrong, Laurence Bourne, Anthea Browne-Wilkinson, Philip Dudley, Leonard Lewis, Ridley Scott, William Slater and others. **Writers** Brian Clemens and many others. **Created by**

Donald Cotton, Harris. **Starring** Peter Ducrow (The Face), Juliet Harmer (Georgina Jones), Gerald Harper (Adam Llewellyn de Vere Adamant), Jack May (William E. Simms). 29 50min episodes in 2 series. B/w.

In 1902 the master-villain The Face embedded his adversary Adamant in a block of ice; when this was thawed in 1966 the world gasped, for Adamant still lived. Catapulted into London's Swingin' Sixties, he gathered sidekicks Jones (an independent-minded mod beauty) and Simms (a valet) as COMPANIONS in his fight against crime and corruption – and eventually encountered The Face once more. *AA* is often regarded as a sort of offshoot of *The* AVENGERS (1961-9) but – after a desperately creaky episode #1 – it established its own style, centring on the quasi-quixotic valour of a man stranded out of his own time and combating modern intriguers while himself armed only with a swordstick. Although very much of its era, *AA* still conjures more than nostalgic delights.   [JG]

ADAM AND EVE As key players in the Judeo-Christian CREATION MYTH, Adam and Eve are allegorical figures of considerable significance, and their SERPENT-assisted Original Sin and subsequent expulsion from EDEN have been examined from every possible philosophical angle in literary works of all kinds. Problematic aspects of the story's moral are clear even in such apologetic analyses as John MILTON's *Paradise Lost* (**1667**), but it was not until the late 19th century that a more robust and challenging scepticism emerged in such works as Rémy de GOURMONT's *Lilith* (**1892**), which may have provided an exemplar for George MACDONALD's *Lilith* (**1895**). The addition of Lilith – Adam's alleged first wife, not recorded in *Genesis* but acknowledged in some apocryphal versions – gives an extra twist to the dynamics of the Fall that was further exploited in John ERSKINE's SATIRE on sexual mores *Adam and Eve* (**1927**), which extended a series of US burlesques begun with Nathaniel HAWTHORNE's "The New Adam and Eve" (**1843**) and Mark TWAIN's *Extracts from Adam's Diary* (1893; **1904** chap) and *Eve's Diary* (**1905** chap). UK sarcastic fantasies of a broadly similar kind include the first part of George Bernard SHAW's *Back to Methuselah* (**1921**) and Rudyard KIPLING's "The Enemies to Each Other" (1924). More contemplative deployments of the myth can be found in C.L. MOORE's "Fruit of Knowledge" (1940), *The Cool of the Evening* (**1942**) by Horace Horsnell (1882-1949), C.S. LEWIS's *Perelandra* (**1943**), *The Windfall: A Fable* (**1985** chap) by Christopher Milne (1920-1996) and John CROWLEY's "The Nightingale Sings at Night" (1989). Stories in which the situation is metaphorically recreated are numerous. Many modern commentators on the myth have observed that part of its function is to put the blame for the iniquities of the human condition on female shoulders.
                                                                                            [BS]

ADAMS, NEAL (1941-   ) Award-winning, prolific and gifted US COMIC-book artist, with a strong, dynamic, sophisticated line style, who has breathed new life into a number of established comics characters. The great vitality of his cover designs brought him particular acclaim in the 1970s, and he has worked in movie-poster design and advertising, but his main area of creative activity remains the comics.

NA worked for Archie Comics before establishing himself in syndicated newspaper strips with a strip version of the tv series *Ben Casey* (1962-6). His first work in comic books was on DC COMICS's DEADMAN. He worked also on The SPECTRE,

BATMAN and many others, bringing to all a fresh and exciting new dynamism. Among his most successful characters in this regard were **Green Lantern** and **Green Arrow**, whose team-up series included the drugs storyline (in "Snowbirds Don't Fly" [*Green Lantern-Green Arrow #85*] and "They Say It'll Kill Me But They Don't Say When" [*#86*]) that won him and writer Denny O'Niell an Academy of Comic Book Arts Award in 1972.

His output continued to be remarkably consistent and prolific, and in 1987 he formed his own publishing company, Continuity Comics. [RT]

ADAMS, NICHOLAS ◊ Debra DOYLE.

ADAMS, RICHARD (GEORGE) (1920-) UK writer who became instantly famous with his first novel, *Watership Down* (**1972**), a long, grave, well crafted ANIMAL FANTASY, written ostensibly for children but clearly accessible to adults as well; it was followed by a pendant title, *The Watership Down Film Picture Book* * (**1978**), which retells the same story (◊ WATERSHIP DOWN [*1978*]). The very considerable surge in the publication of animal fantasies after 1972 – especially in the UK – was certainly stimulated by the success of this book. The story represents a sustained attempt to render in PASTORAL terms something of the real existence and the mythopoetic implications inherent in rabbit society as a model for survival in a threatening world, and as an extended ALLEGORY on the nature of just kingship. The basic plot can be described as a search for a POLDER: warned by premonitions on the part of Fiver (a rabbit Cassandra whose prophecies are usually ignored) that their original warren, whose king has become aged and indolent, is under threat, three rabbit COMPANIONS (more join quickly) travel in search of a safe haven, finding it at last in Watership Down. But, before they can reconstitute themselves as a society, they must find females, and in so doing they encounter two warrens (one tame, one tyrannized) which generate a powerful sense of a BAD PLACE and of unjust rule. Finally – their morale lifted in the meanwhile by the telling of several inserted stories which represent the rabbit hero-god, El-Ahrairah, as a TRICKSTER figure – they achieve their goal. They make Watership Down into a safe polder – a haven of ecological sanity – in a world constantly darkening under the "husbandry" of *Homo sapiens*.

The first of the **Beklan Empire** sequence – which comprises *Shardik* (**1974**) and *Maia* (**1984**) – followed. None of RA's subsequent work has enjoyed anything like the huge success of his first novel, but his further accomplishments have been considerable. *Shardik* is set in a LAND – the Beklan Empire may be intended to represent a pre-Christian hegemony on Earth, but the geography is deliberately left so vague that the landscapes depicted could easily be understood as those of a SECONDARY WORLD – riven by cruelties and a bad war. Shardik himself – an enormous bear with parnormal powers – may be a messenger of God, or even a MESSIAH-figure; but the cruel ambivalence of his behaviour makes him into a kind of black, worldly PARODY of figures like C.S. LEWIS's Aslan. In *Maia*, set somewhat later in the same territory, the eponymous heroine, a slave girl profoundly attractive to almost everyone she meets, goes through a long sequence of adventures; the style of storytelling has been likened to that of Jane GASKELL's **Atlan** sequence, but the tale itself does not come to any transforming or catastrophic climax.

The Plague Dogs (**1977**) might be described as a kind of TECHNOFANTASY in that the dogs of the title are both sentient, as in any animal fantasy, and severely affected by animal experimentation: the protagonist has the gift of PROPHECY, but only because part of his brain has been excised. The story itself is thesis-ridden (RA has no sympathy for any sort of animal experimentation) and arousing, as the dogs escape the farm, learn how to live in the wild, and find their human master in the end. *The Iron Wolf and Other Stories* (coll **1980**; vt *The Unbroken Web: Stories and Fables* 1980 US) assembles short work supplementary to the main novels. *The Girl in a Swing* (**1980**) is a SUPERNATURAL FICTION which leaves unsettled the status – GHOST or DEMON – of the girl. *Traveler* (**1988** US) tells the story of the AMERICAN CIVIL WAR through the eyes of General Robert E. Lee's horse, Traveller: there are moments when the surreal obscenity of mass slaughter comes across with extrardinary vividness.

The impulse to fantasy is not powerful in RA's work, but it is clear that he has understood the useful potency of the structure of the genre in his attempts to convey a variety of messages about the world. His concerns – ecology, religion, a peculiarly feudal concept of honour as a binding force in the just society, the enslavement of nonhuman species – sometimes dominate his tales but, when a balance is achieved, the power of his sustained anger and disillusion profoundly transforms the modes he has chosen. [JC]

Other works: *Sinister and Unnatural Stories* (anth **1978**); a version of *Grimm's Fairy Tales* (anth **1981**); *Richard Adams's Favourite Animal Stories* (anth **1981**); *The Legend of Te Tuna* (**1982** chap US), which retells a South Seas legend; *The Bureaucats* (**1985** chap), a fantasy for children; *The Day Gone By: An Autobiography* (**1991**).

ADAMS, (FRANKLIN) ROBERT (1932-1990) US soldier and writer who remains best-known for the **Horseclans** sequence of post-HOLOCAUST adventures which, though its protagonists behave for pages on end like SWORD-AND-SORCERY heroes, maintains a consistent sf rationale. The **Castaways in Time** sequence was also sf. The incomplete **Stairway to Forever** sequence – *The Stairway to Forever* (**1988**) and *Monsters and Magicians* (**1988**) – is CONTEMPORARY FANTASY, its initial setting being the US South, though it is clear that the hero's access via a PORTAL to a SECONDARY WORLD would increasingly have directed further volumes of his adventures, had RA lived to write them.

RA edited, in collaboration, a variety of anthologies, most of fantasy interest: *Barbarians* (anth **1985**) and *Barbarians II* (anth **1988**), both with Harry Martin Greenberg and Charles G. Waugh; *Magic in Ithkar* (anth **1985**), *#2* (anth **1985**), *#3* (anth **1986**) and *#4* (anth **1987**), all with Andre NORTON; *Hunger for Horror* (anth **1988**) with Pamela Crippen Adams (1961-) and Greenberg; *Phantom Regiments* (anth **1990**) with Adams and Greenberg, featuring GHOST soldiers. [JC]

ADDAMS, CHARLES (1912-1988) US cartoonist, a prolific contributor to the magazine *The New Yorker* from 1935; he had a macabre sense of humour and a firm brush-line and wash style. He is best-known for his creation of the grotesque Addams Family, whom he first introduced in *The New Yorker*; these characters have been widely exploited on tv and in the movies (*see subsequent entries*).

Collections of his work include *Drawn and Quartered* (graph coll **1943**), *Addams and Evil* (graph coll **1947**), *Afternoon in the Attic: Pictures by Charles Addams* (graph coll **1950**), *Monster Rally* (graph coll **1951**), *Charles Addams' Homebodies Collection* (graph coll **1954**), *Dear Dead Days* (graph coll **1954**), *Night Crawlers* (graph coll **1957**), *Black Maria* (graph coll **1960**), *The Groaning Board* (graph coll **1964**), *The Charles Addams Mother Goose* (graph coll **1967**), *My Crowd* (graph coll **1971**), *Favourite Haunts* (graph coll **1977**), *Creature Comforts* (graph coll **1981**), *The Addams Family Album* (graph coll **1991**) and *The World of Charles Addams* (graph coll **1991**), the last of which won a Hugo AWARD. [RT]

ADDAMS FAMILY, THE Two tv series have been based on the creations of Charles ADDAMS.

1. US tv series (1964-6). Filmways/NBC. **Pr** Nat Perrin. **Exec pr** David Levy. **Dir** Arthur Hiller, Jerry Hopper, Sidney Lanfield, Nat Perrin, Sidney Salkow and many others. **Developed for tv** by David Levy. **Writers** Hannibal Coons, Carol Henning, Preston Wood and many others. **Novelization** *The Addams Family* * (**1965**) by Jack Sharkey. **Starring** John Astin (Gomez Addams), Ted Cassidy (Lurch/Thing), Jackie Coogan (Uncle Fester), Carolyn Jones (Morticia Frump Addams/Ophelia Frump), Lisa Loring (Wednesday Addams), Blossom Rock (Grandmama Addams), Felix Silla (Cousin Itt), Ken Weatherwax (Pugsley Addams). 64 30min episodes. B/w.

For many years Addams had entertained magazine readers with a running series of sketches featuring the family, but now they were given names and distinct personalities. Gomez, the head of the family, hails from Spain; he is totally unaware he is unlike the rest of the world. Some of his more memorable moments come when his wife, Morticia, says anything in French, for this drives him to the heights of amorous ecstasy. Morticia, dressed in a slinky black gown so tight she can barely walk, enjoys working in her garden of carnivorous plants. The couple have two children: Pugsley, aged nine, and Wednesday, seven. The household also includes Uncle Fester, who loves explosives, sleeps on a bed of nails and puts his head in a vice to relieve headaches, and Lurch, the 6ft 9in butler, whose vocabulary consists almost exclusively of a sepulchral "You rang?". The family is completed by Thing, a disembodied hand.

In each episode some poor "regular" person would be drawn briefly into their madness. For example, one episode dealt with a visit from the children's teacher, who was concerned about their schoolwork; a short while in the mansion quickly convinced her never to worry about them again. This was one of two series debuting that year that dealt with unusual families, the other being *The* MUNSTERS (1964-6). [BC]

2. *The Addams Family* US animated tv series (1973-5). HANNA-BARBERA/NBC. **Pr** Iwao Takamoto. **Exec pr** William Hanna, Joseph Barbera. **Dir** Charles A. Nichols. **Comics adaptation** *The Addams Family* * (3 issues 1974-5). **Voice actors** Ted Cassidy (Lurch), Jackie Coogan (Uncle Fester), Jodie Foster (Pugsley Addams), Cindy Henderson (Wednesday Addams), Janet Waldo (Morticia Addams/Grandmama Addams), Lennie Weinrib (Gomez Addams). 37 30min episodes. Colour.

This animated version put the Addams family in a series of new settings, for they converted their two-storey mansion into a motor home and set out on the road. Each episode featured their misadventures as they travelled across the USA. Original cast members Jackie Coogan and Ted Cassidy reprised their roles. One intriguing (with hindsight) bit of casting was of the young Jodie Foster as Pugsley. The humour was rudimentary. [BC]

ADDAMS FAMILY MOVIES Three movies have been based on the creations of Charles ADDAMS. The later series (**2-3**) has probably ended, Raul Julia (1940-1995) having died.

1. *Halloween with the New Addams Family* (vt *The Addams Family*) US movie (**1977** tvm). NBC. **Pr** David Levy. **Exec pr** Charles Fries. **Writer** George Tibbles. **Dir** Dennis Steinmetz. **Starring** John Astin (Gomez Addams), Ted Cassidy (Lurch), Jackie Coogan (Uncle Fester), Henry Darrow (Pancho Addams), Carolyn Jones (Morticia Addams), Lisa Loring (Wednesday Addams), Kenneth Marquis (Pugsley Addams Jr), Jane Rose (Grandmama Addams), Felix Silla (Cousin Itt), Jennifer Surprenant (Wednesday Addams Jr), Ken Weatherwax (Pugsley Addams). 90 mins. Colour.

Ten years after tv's *The* ADDAMS FAMILY (1964-6) left the air, the cast re-united. The family gather to celebrate the biggest holiday of their year, HALLOWEEN. Proceedings are enlivened by two robbers who break into the mansion, for the Addamses think they are there for the party. Soon the criminals' biggest problem is escaping the very unwanted hospitality of their would-be victims. A minor work. [BC]

2. *The Addams Family* US movie (**1991**). Columbia TriStar/Paramount/Orion. **Pr** Scott Rudin. **Exec pr** Graham Place. **Dir** Barry Sonnenfeld. **Vfx** Alan Munro. **Screenplay** Caroline Thompson, Larry Wilson. **Novelization** *The Addams Family* * (**1991**) by Elizabeth Faucher. **Starring** Dan Hedaya (Tully Alford), Anjelica Huston (Morticia Addams), Raul Julia (Gomez Addams), Christopher Lloyd (Gordon Craven/Fester Addams), Judith Malina (Granny Addams), Christina Ricci (Wednesday Addams), Carel Struycken (Lurch), Elizabeth Wilson (Abigail Craven), Jimmy Workman (Pugsley Addams). 99 mins. Colour.

Loan shark Abigail Craven disguises her psychopathic son Gordon as Gomez's long-lost brother Fester and foists him onto the Addams family in hope of gaining their fortune. The plot succeeds, and the Addamses are cast out into the world to make their own way – which they do very badly. But Gordon has come to love the Addamses and enjoys being Fester, and ensures the reestablishment of the status quo. Spurred by inventiveness rather than narrative drive, *TAF* is full of throwaway pieces of fantastication – e.g., opening a copy of *Gone with the Wind* stirs up a gale (◊ BOOKS); as Gomez plays with his toy trains, a sudden viewpoint shift shows us a commuter staring through a train window at Gomez's gigantic face. [JG]

3. *Addams Family Values* US movie (**1993**). Paramount. **Pr** Scott Rudin. **Exec pr** David Nicksay. **Dir** Barry Sonnenfeld. **Vfx** Alan Munro. **Spfx** Albert Delgado. **Screenplay** Paul Rudnick. **Novelization** *Addams Family Values* * (**1993**) by Todd Strasser (1950-   ). **Starring** Joan Cusack (Debbie Jellinsky), Anjelica Huston (Morticia Addams), Raul Julia (Gomez Addams), Carol Kane (Granny Addams), David Krumholtz (Joel Glicker), Christopher Lloyd (Fester Addams), Christina Ricci (Wednesday Addams), Carel Struycken (Lurch), Jimmy Workman (Pugsley Addams). **Voice actor** Cheryl Chase (Pubert Addams). 94 mins. Colour.

Baby Pubert is born to Morticia and Gomez. The nanny they hire, Debbie, is a SERIAL KILLER, the Black Widow, who marries and murders ugly rich men (a surprisingly direct borrowing from the movie *Black Widow* [**1987**]); she weds Fester, severing him from the other Addamses, but finds him virtually unkillable. Pubert, emotionally disturbed by all this, becomes possessed (◊ POSSESSION) by Virtue and is transformed from a mustachioed little monster into a cute, blonde, tousled advertisement-style babe. Thing rescues Fester from Debbie's clutches, but it requires an "exorcised" Pubert to stop her assassinating the entire family. A funnier but less fantasticated movie than its predecessor, this contains many recursive references to classic HORROR MOVIES involving CHILDREN – ROSEMARY'S BABY (*1968*) and *Child's Play* (*1988*) are two. [JG]

ADELER, MAX Principal pseudonym of US writer and businessman Charles Heber Clark (1841-1915), of genre interest primarily for the influence upon Mark TWAIN's *A Connecticut Yankee in King Arthur's Court* (**1889**) of his TIME-TRAVEL tale, "Professor Baffin's Island" (1880; vt "The Fortunate Island" 1882), which appears in *The Old Fogey and Other Stories* (coll **1881** UK; cut vt *The Fortunate Island and Other Stories* 1882 US); the title story of the UK edition is a TIMESLIP fantasy. Other fantasies are assembled in *Random Shots* (coll **1878** UK), which includes "Mr Skinner's Night in the Underworld", *Transformations, Containing Mrs Shelmire's Djinn and A Desperate Adventure* (coll **1883** UK), and *By a Bend of the River* (coll **1914**). [JC]

ADKINS, PATRICK H. (1948-   ) US writer whose first work of genre interest was *Edgar Rice Burroughs: Bibliography and Price Guide* (**1974** chap), but who is of main interest for his **Titans** sequence – *Lord of the Crooked Paths* (**1987**), *Master of the Fearful Depths* (**1989**) and *Sons of the Titans* (**1990**) – which retells Greek MYTHS, the first two volumes recounting stories of the Titan Kronos and his allies and enemies, none of whom pay much attention to the newly created human race; the third volume takes up the story of the young ZEUS through a plotline that casts him as a HIDDEN MONARCH whose identity is concealed from his dread father, Kronos, and who is therefore not eaten alive. [JC]

ADONIS There are two stories about the god or GODS of this name. As the child god whom APHRODITE loves he is redolent of Eros, and is gored to death by a boar (at the behest of the GODDESS) while hunting; it is this story that William SHAKESPEARE tells in his *Venus and Adonis* (**1593**). The other (or continuing) story has perhaps more resonance within modern fantasy: as the god whom Persephone loves, Adonis is directed by ZEUS to spend half (or a third of) the year above-ground and the remainder in HADES. In this aspect he is one of several gods of vegetation, and RITUALS dedicated to Adonis – they occur at the SEASON of the harvest – annually mourn his death, which is the death of the consort of the Goddess. His story is very close to that of the Mesopotamian god Tammuz (◊ MESOPOTAMIAN EPIC).

In *From Ritual to Romance* (**1920**) Jessie L. WESTON identifies Adonis with the FISHER KING, whose HEALING through the finding of the GRAIL represents (in her view) the seasonal rebirth of the land. Weston's ideas are no longer taken very seriously by scholars. [JC]

ADULT FANTASY This term has two meanings.

1. At the end of the 1960s, when most fantasy – whatever its nature – was marketed for children, Lin CARTER began editing the BALLANTINE ADULT FANTASY series of reprints, including work by Lord DUNSANY, George MACDONALD and others, most of whom he claimed to have "rediscovered", and all of whom he claimed (rightly) were authors of adult fiction. In this encyclopedia the simple term FANTASY describes work for adults; we differentiate CHILDREN'S FANTASY.

2. The term AF has been used to describe erotic or pornographic texts. [JC]

ADVENTURE & MYSTERY STORY MAGAZINE ◊ HUTCHINSON'S MAGAZINES.

ADVENTURER, THE ◊ MAGAZINES.

ADVENTURER FANTASY GENRE FANTASY features many stories whose protagonists are ordinary or extraordinary people trying to make a living in FANTASYLAND. This large body of work, at its best often intriguingly PICARESQUE – and at its not infrequent worst merely a revision of standard TEMPLATE wish-fulfilment with fantasy tropes – can appropriately be called AF (a term devised by Eric Korn in 1995) rather than HEROIC FANTASY, which implies that larger-than-life QUESTS are essential to the form, or SWORD AND SORCERY, which term has a derogatory sting – as though the **Fafhrd and Gray Mouser** tales of Fritz LEIBER were intrinsically less significant than EPIC FANTASIES by the not necessarily very interesting writers who dominate the popular market. But the term AF, though most closely describing the central characters and plots of the form, has yet to attain any widespread use. A full discussion of the form indicated appears under SWORD AND SORCERY.

[JC/RK]

ADVENTURES OF ALICE UK movie (**1960** tvm). ◊ ALICE IN WONDERLAND (*1951*).

ADVENTURES OF BARON MUNCHAUSEN UK/West German movie (*1989*). Prominent/Laura Film/Columbia-TriStar. **Pr** Thomas Schühly. **Exec pr** Jack Eberts. **Dir** Terry GILLIAM. **Spfx** Richard Conway. **Screenplay** Gilliam, Charles McKeown, published as *The Adventures of Baron Munchausen* (**1989** US) by Gilliam and McKeown. **Based on** the works of Rudolph Erich RASPE. **Novelization** *The Adventures of Baron Munchausen* * (**1989** UK) by Gilliam and McKeown. **Starring** Valentina Cortese (Ariadne, Queen of the Moon), Winston Dennis (Albrecht), Eric Idle (Berthold), Peter Jeffrey (Sultan), McKeown (Adolphus), John Neville (Baron Munchausen), Sarah Polley (Sally), Jonathan Pryce (Horatio Jackson), Jack Purvis (Gustavus), Oliver Reed (Vulcan), Uma Thurman (Venus), Robin Williams (credited as Ray D. Tutto; Roger, King of the Moon). 126 mins. Colour.

AOBM, a financial disaster, is generally regarded as a movie of parts whose sum makes not much of a whole. Repeated viewing reveals it as far better than that. Certainly it is episodic, its nested-story narrative (◊ STORY CYCLE) more a PICARESQUE than a fully developed plot, with each episode being of independent fantasy interest. It starts in a CITY besieged by the Turk in an unspecified past age; a bad performance of *The Adventures of Baron Munchausen* is interrupted by the arrival of the Baron himself, who proceeds to tell the "truth". Thereafter there is constant manipulation of our PERCEPTIONS of what is going on, with characters from the "real world" (notably an actor's daughter, Sally) intruding into the Baron's stories, and vice versa, the two REALITIES merging. The adventures include a FANTASTIC VOYAGE to the MOON, whose King and Queen have heads existing independently of their bodies. Elsewhere the friends parlay with Vulcan and the Cyclopes, who, though human-sized, believe themselves GIANTS. Later the party is swallowed by a SEA MONSTER, inside which card-playing mariners argue as to whether they are in HEAVEN or HELL. By now it might be fair to feel that the tale is going out of control, but a secure sense of FANTASY is restored when the Valkyrie-like DEATH descends to seize Munchausen and

Sally intervenes to say he cannot yet die, since the STORY is unfinished. At the end of the recounted fables the Baron and his fantastic COMPANIONS (e.g., the DWARF Gustavus, who can blow up a hurricane) rout the Turks from around the city.

The main focus – as in John BARTH's analogous *The Last Voyage of Somebody the Sailor* (**1991**) – is indeed the power of Story: "I must inform you, my liege," remarks Munchausen acidly to the King of the Moon, "that without my adventures you wouldn't be here." Later, in mundane reality, Jackson spits at Munchausen: "There are certain rules as to the proper conduct of living. We cannot fly to the Moon. We cannot defy Death. We must face the facts, not the folly of fantasists like you, who do not live in the real world." But in this instance it is the fantasy that is driving the mundane reality, as is proven when the city's obscuring gates are opened and rationalism's blinkers removed from the eyes of all, for the Turk has indeed disappeared.   [JG]

Further reading: *Losing the Light: Terry Gilliam and the Munchausen Saga* (**1993**) by Andrew Yule.

See also: MÜNCHHAUSEN.

ADVENTURES OF BARON MUNCHHAUSEN, THE vt of *Münchhausen* (**1943**). ◊ MÜNCHHAUSEN.

ADVENTURES OF BATMAN, THE US tv series (1969-70). ◊ BATMAN.

ADVENTURES OF BATMAN AND ROBIN, THE (ot *Batman: The Animated Series*) US syndicated tv series (1992-4). Warner Bros. **Pr** Alan Burnett, Eric Radomski, Bruce W. Timm. **Exec pr** Jean MacCurdy, Tom Rueger. **Dir** Kevin Altieri and many others. **Based on** the COMIC-book characters created by Bob KANE. **Voice actors** Michael Ansara (Mr Freeze), Ed Asner (Roland Daggett), Adrienne Barbeau (Selina Kyle/Catwoman), Jeff Bennett (HARDAC), Kevin Conroy (Batman/Bruce Wayne), John Glover (The Riddler), Mark Hamill (The Joker), Bob Hastings (Commissioner Gordon), Aron Kincaid (Killer Croc), Loren Lester (Dick Grayson/Robin), Roddy McDowall (The Mad Hatter), Richard Moll (Harvey Dent/Two-Face), Ron Perlman (Matthew Hagen/Clayface), Diane Pershing (Poison Ivy), Henry Polic (The Scarecrow), Helen Slater (Talia), Arleen Sorkin (Harley Quinn), John Vernon (Rupert Thorne), David Warner (Ra's Al Ghul), Paul Williams (The Penguin), Efrem Zimbalist Jr (Alfred Pennyworth). 65 30min episodes. Colour.

This came in response to Tim BURTON's successful series of BATMAN MOVIES. Though using many of the characters first seen in the COMIC books, the series had a notably different style and look: more angular in nature than the comics, it also made effective use of shadows and perspective, setting it apart from both the comics and other, competitor animated shows. Batman and Robin battle the criminal element of Gotham City, though in a different continuity from the original comic books. The continued high ratings led to the ANIMATED MOVIE *Batman: Mark of the Phantasm* (**1993**) (◊ BATMAN MOVIES). The series generated a successful spinoff series of comic books (October 1992-current).

There had been several prior animated versions, all for a more juvenile audience; e.g., *Super Powers* (1985) (pr Norm Prescott, Lou Scheimer; dir Hal Sutherland). Warner Bros. is reported to be repackaging these for future release.   [BC]

ADVENTURES OF CAPTAIN MARVEL US serial movie (**1941**). ◊ CAPTAIN MARVEL.

ADVENTURES OF ICHABOD AND MR TOAD, THE US movie (**1949**). ◊ *The* WIND IN THE WILLOWS.

ADVENTURES OF LOIS AND CLARK, THE (vt *Lois and Clark: The Adventures of Superman*) US tv series (1993-current). Warner Bros./ABC. **Pr** Mel Efros, John MacNamara, Jim Michaels, Thania St John, Phillip J. Sgriccia. **Exec Pr** Robert Butler, James A. Contner, David Jacobs, Deborah Joy LeVine, Robert Singer. **Dir** Felix Enriquez Alcala and many others. **Writers** Hilary Bader and many others. **Based on** characters created by Jerry Siegel (1914-1996) and Joe Shuster (1914-1992). **Developed by** LeVine. **Starring** Dean Cain (Clark Kent/Superman), K Callan (Martha Kent), Teri Hatcher (Lois Lane), Eddie Jones (Jonathan Kent), Michael Landes (Jimmy Olsen 1993-4), Tracy Scoggins (Cat Grant 1993-4), John Shea (Lex Luthor), Lane Smith (Perry White), Justin Whalin (Jimmy Olsen 1994 onwards). 1 2hr and 42 1hr episodes to end of 1994-5 season. Colour.

Several decades after *The* ADVENTURES OF SUPERMAN (1951-7) and several years after SUPERBOY (1988-92), SUPERMAN returned with, this time, the focus on the complex love triangle between Lois, Clark and Superman. Though not quite as hare-brained as in earlier tv incarnations, Lois still finds ample opportunity to place herself in jeopardy in pursuit of a story. Overall, the approach worked reasonably well in the first season. Sensitive to criticism that the series was long on talk but short on action, the producers changed direction slightly for the next year, introducing more action elements. Lex Luthor, pivotal in the first series, was killed fleeing arrest, and Superman had to battle more villains who relied on muscles rather than brains. At the end of the second season Clark decided to reveal his secret to Lois and propose marriage, but the outcome has yet (March 1996) to be resolved.   [BC]

ADVENTURES OF ROBIN HOOD, THE UK tv series (1955-9). ITV/Sapphire Films. **Exec pr** Hannah Weinstein. **Dir** Lindsay Anderson, Terence Fisher and many others. **Writers** Eric Heath, Ian McLellan Hunter (pseudonymous), Ring Lardner Jr (pseudonymous), Anne Rodney, Smart, Palmer Thompson and many others. **Starring** Richard Coleman (Alan-a-Dale), Rufus Cruikshank (Little John in later episodes), Patricia Driscoll (Maid Marian in later episodes), Archie Duncan (Little John in early episodes), Paul Eddington (Will Scarlett in later episodes), Alexander Gauge (Friar Tuck), Richard Greene (Robin Hood), Ronald Howard (Will Scarlett in early episodes), Ian Hunter (King Richard), Bernadette O'Farrell (Maid Marian in early episodes), Donald Pleasance (Prince John), Alan Wheatley (Sheriff of Nottingham). 143 26min episodes. B/w.

One of TELEVISION's most successful fantasy series, this nevertheless did its best to suppress all potential fantasy elements: it was basically an adventure series. It was required viewing for children – and many adults – on both sides of the Atlantic, being screened in the USA by CBS and through syndication; in 1995 it remained in active syndication. The series' scenario stuck closely to a consensus version of the ROBIN HOOD legend; each self-contained episode formed a soon-familiar TEMPLATE. The success of *TAORH* gave rise to two direct imitators: *The* ADVENTURES OF SIR LANCELOT (1956-7), which had a very similar production and writing team and was filmed on an adjacent lot; and *The Adventures of William Tell* (1958-9), which again featured many of the people responsible for *TAORH*.   [JG]

ADVENTURES OF SINBAD, THE Japanese ANIMATED MOVIE (**1975**). ◊ SINBAD MOVIES.

ADVENTURES OF SINBAD, THE UK ANIMATED MOVIE (*1979*). ◊ SINBAD MOVIES.

ADVENTURES OF SIR LANCELOT, THE UK tv series (1956-7). ITV/Sapphire Films. **Pr** Dallas Bower, Sidney Cole, Bernard Knowles. **Exec pr** Hannah Weinstein. **Dir** Arthur Crabtree, Knowles, Ralph Smart, Anthony Squire and others. **Writers** Ian McLellan Hunter (pseudonymous), Ring Lardner Jr (pseudonymous), Leslie Poynton, Leighton Reynolds, Smart, and others. **Starring** Jane Hylton (Guinevere), Ronald Leigh-Hunt (Arthur in later episodes), William Russell (Sir Lancelot du Lac), Robert Scroggins (Brian), Bruce Seton (Arthur in early episodes), Cyril Smith (Merlin). 30 25min episodes. B/w.

The ADVENTURES OF ROBIN HOOD (1956-60) had been a great hit for ITV/Sapphire, and so they followed it up with *TAOSL*, shot on a next-door set at Nettlefold Studios. Like ITV's later *The Adventures of William Tell* (1958-9), this was essentially a rehash of the ROBIN HOOD series – and with much the same production team – but set in a different milieu. While the stories were not distinguished, the period reconstruction was, being based on research commissioned from the University of Oxford. [JG]

ADVENTURES OF SUPERMAN, THE US syndicated tv series (1951-7). Superman Inc/Lippert Productions. **Pr** Whitney Ellsworth, Bernard Lubar, Robert Maxwell. **Dir** Bob Barnes and many others. **Spfx** Danny Hayes, Thol Simonson. **Writers** Robert Leslie Bellem and many others. **Based on** characters created by Jerry Siegel (1914-1996) and Joe Shuster (1914-1992). **Starring** Phyllis Coates (Lois Lane 1951), John Hamilton (Perry White), Jack Larson (Jimmy Olsen), Noel Neill (Lois Lane 1953-7), George Reeves (Clark Kent/Superman), Robert Shayne (Inspector Bill Henderson). 104 30min episodes. B/w and colour.

One of the most popular TECHNOFANTASY series of all time. Directly based on the star of the SUPERMAN series of COMIC books, it followed the adventures of Superman, the sole survivor of the planet Krypton. It had its genesis in *Superman and the Mole Men* (*1951*) (◊ SUPERMAN MOVIES), a feature movie which saw the first appearance of George Reeves as the Man of Steel. This served as a pilot for the series, which during the first season featured a number of gritty and menacing episodes. A change of producers led to a more lighthearted approach; as time passed this became almost comedic. While many episodes dealt with the mundane criminal element of Metropolis, others included the use of mind-controlling drugs and devices, VOODOO, a crook who could read minds (◊ TALENTS) and a sacred jewel bearing a deadly CURSE. Further episodes dealt with sf tropes.

Unfortunately, all ended with the apparent suicide of Reeves. The producers tried their hands at another version, this one for a younger audience. The proposed series, *Superpup*, featured humans in dog costumes, and would have followed the adventures of one Bark Bent, ace reporter working for editor Terry Bite. A pilot episode was filmed in 1957 on the same sets used for *TAOS*, but mercifully failed to sell. In 1961 they tried again, this time with Johnny Rockwell in the title role of *Superboy*. This, too, failed to sell, and it was over 20 years before another version of the Man of Steel appeared on tv (◊ SUPERBOY [1988-92]). *The* ADVENTURES OF LOIS AND CLARK (1993-current), continues the tradition today. [BC]

ADVENTURES OF TARZAN, THE US movie (*1921*). ◊ TARZAN MOVIES.

ADVENTURES OF WILLIAM TELL, THE UK tv series (1958-9). ◊ *The* ADVENTURES OF ROBIN HOOD (1955-9); *The* ADVENTURES OF SIR LANCELOT (1956-7).

ADVENTURES OF WONDER WOMAN, THE US tv series (1976-9). Douglas S. Cramer Company/Bruce Lansbury Productions/Warner Bros./ABC, CBS. **Pr** Wilfred Lloyd Baumes, Charles B. Fitzsimons, Mark Rodgers. **Sv pr** Bruce Lansbury. **Exec pr** Baumes, Douglas S. Cramer. **Dir** Jack Arnold, Bruce Bilson, Michael Caffey, Barry Crane, Leonard J. Horn and many others. **Writers** Allan BRENNERT and many others. **Based on** the COMIC-book characters created by Charles Moulton. **Starring** Ed Begley Jr (Harold Farnum), Lynda Carter (Diana Prince/Wonder Woman), Carolyn Jones (Queen 1976-7), Cloris Leachman (Queen 1976-7), Beatrice Straight (Queen 1977-9), Lyle Waggoner (Major Steve Trevor/Steve Trevor Jr), Debra Winger (Drusilla/Wonder Girl 1976-7). 59 60min episodes. Colour.

The series began by continuing the successful formula of the second pilot movie – *The* NEW, ORIGINAL WONDER WOMAN (*1975*) – and is initially set during WORLD WAR II. Most of the early plots centre on battles against the Nazis – who in one episode create their own Wonder Woman – and against saboteurs and spies. Following 13 episodes on ABC, the series moved to CBS, where it was re-titled *The New Adventures of Wonder Woman*. With the change of networks, the setting became the present. The new story begins with Steve Trevor's son, Steve Jr (both conveniently played by Lyle Waggoner), crashlanding on Paradise Island. Wonder Woman returned there after WWII, but Steve brings her back to fight a new set of villains.

Once again Wonder Woman adopts the disguise of Diana Prince. Both she and Steve work as agents of IADC, the Inter-Agency Defense Council, fighting a wide variety of problems. Several stories deal with mind-enslaving aliens; others see a telekinetic Japanese soldier who does not know WWII is over, political blackmailers, a mad scientist who can create volcanoes, and a time traveller taking advantage of Wonder Woman's knowledge of the past. Perhaps the most unusual episode finds her helping a leprechaun recover his stolen gold. This new version proved so popular that the comic book was also updated to reflect the modern timeline. [BC]

ADVENTURE-STORY MAGAZINE ◊ HUTCHINSON'S MAGAZINES; MAGAZINES.

ADVERTISEMENTS ◊ COMMERCIALS.

A.E. or Æ Pseudonym of Irish writer and painter George William Russell (1867-1935), a colleague of William Butler YEATS, with whom he helped found the Dublin Lodge of the Theosophical Society (◊ THEOSOPHY). On the title page of his first book, *Homeward* (coll **1894**), Russell attempted to have printed the word AEON, the name of an eternal entity in the PANTHEON of Gnosticism; by printer's error this was rendered Æ. The tales assembled in *The Mask of Apollo, and Other Stories* (coll **1904**) are mystical in effect, but as watery as his sketches and paintings, which often explicitly depict the Sidhe (◊ FAIRIES). At the same time, his best work as a writer and painter conveyed a sense of some imminent revelation of Being that is characteristic of the Gnostic imagination; it is a feeling akin to SEHNSUCHT, and in its expression – however fleeting – A.E. reached its apogee as an artist. His novels – *The Interpreters* (**1922**) and *The Avatars: A Futurist Fantasy* (**1933**) – can with some difficulty be read as philosophical sf (◊ SFE), but his essential bent of mind was far from materialist. Although *The Avatars* is set in the future, its speculative content is governed by, not any analysis of the course of history, but a deep (though mildly expressed) longing for a world which has awoken into Being. The GODS who inhabit his work are inhabitants of that world.

The long introduction by Monk Gibbon to *The Living Torch* (coll **1937**), a posthumous volume of essays, is adulatory but informed. Along with his own works, AE deserves recognition for his influence on other writers: he discovered James STEPHENS, and James JOYCE acknowledged his influence by using him as a character in *Ulysses* (**1921**).   [JC]

AESIR Collective name for the PANTHEON of ELDER GODS of Northern Europe, including Ymir, father of the giants, and ODIN, Vili and Ve, who are sidebar descendants of the first GOD, and who defeat him and his offspring. They then create the Cosmos out of Ymir's corpse, and later take on normal pantheon duties, operating from their HEAVEN, Asgard. Asgard is also home to later gods – up to 42 of them, according to some totals – including Thor, Balder and LOKI. The Aesir inhabit the background of much NORDIC FANTASY as well as much RATIONALIZED FANTASY in the mode developed for UNKNOWN. Many can be found in *The Incomplete Enchanter* (**1941**) by L. Sprague DE CAMP and Fletcher PRATT. They appear also in much SWORD AND SORCERY, though normally along the fringes of the central action, as in Robert E. HOWARD's CONAN tale "Gods of the North" (1934 *Fantasy Fan*). *Job: A Comedy of Justice* (**1984**) by Robert A. HEINLEIN features the Aesir, mainly in the person of Loki (who stands in for Lucifer); members of the pantheon appear in Sterling E. Lanier's "The Kings of the Sea" (1968), and in disguise in Diana Wynne JONES's *Eight Days of Luke* (**1975**). The fascination exercised by the Aesir on commercial fantasy writers like POUL ANDERSON has primarily to do with their status as doomed within historical time and according to their own legends; they are thus characters apt for tales which emphasize THINNING and APOCALYPSE as well as for a more general atmosphere of melancholia, doom and BELATEDNESS. This and further aspects of the Aesir are guyed by Tom HOLT in *Expecting Someone Taller* (**1987**) and other titles.   [JC]

AESOP ◊ AESOPIAN FANTASY; FABLE.

AESOPIAN FANTASY According to some accounts, Aesop was a slave who wrote for masters; he was possibly deformed. Such accounts add up to the portrait of a man who might have had some reason to respond to the world with SATIRE, and also to utter ALLEGORY in code. Whatever the truth of Aesop's life, the BEAST FABLES assembled as *Aesop's Fables* were so written as to avoid any direct indictments, to obscure any references that might have been made to the Phrygian politicians of 600BC; hence the coinage by Russian critic Mikhail Saltykov of the term "aesopic language" to designate tales written in code under repressive circumstances.

An AF can be defined as a tale which tends to utilize TALKING ANIMALS to convey points about human nature, though stories featuring humans (or beasts) in moralistically construed OTHERWORLDS also clearly belong in this category, and a collection like *The Day of the Women and the Night of the Men* (coll **1977**) by Wolf MANKOWITZ, a set of fables about GODS and humans, may well explicitly evoke Aesop. For tales to be understood as Aesopian it is necessary only that their moral intention *seems to be concealed*. Didactic Victorian FAIRYTALES, which explicitly convey moral

lessons, are not AFs, but George ORWELL's *Animal Farm* (**1945** chap) is a classic example of the coded beast fable, even though its relevance to the world of 1945 only *seems* to be concealed through its author's obedience to the rules; and explicitly Aesopian works, like the verse *Beastly Tales from Here and There* (coll **1991**) by Vikram Seth (1952-   ), clearly fit the model. Of the innumerable PARODIES which use Lewis CARROLL's WONDERLAND, some – like Huang Chun-Sin's *Alice in Manialand* (**1959** chap Hong Kong), which can be readily be decipherable as an attack upon Communist China – are obviously Aesopian. Likewise, countless animated shorts (◊ ANIMATED MOVIES) deploy talking animals as surrogate humans with the full intention that audiences will understand straightway the moral, philosophical or political thrust.

The publication of an AF need not entail political risk to its author: many examples have been published in countries which do not normally punish coded satire, especially when it is mildly put. *Cadwallader: A Diversion* (**1959**) by Russell Lynes (1910-1991), starring rats, and *Ticker Khan* (**1974** chap) by Bamber Gascoigne (1935-   ), starring pheasants, both couch lessons about humanity in terms whose concealment is safely penetrable. On the other hand, the political and religious traumas of modern Eire and Ulster have tended to generate highly loaded fables, such as Flann O'BRIEN's long-unpublished dramatic fantasia on the ČAPEK brothers' *The Insect Play* (**1921**), *Rhapsody in Stephen's Green* (produced 1943; **1994**). Also densely coded are tales like "All in a Day's Donkey-Work" and "Two Windows and a Watertank" in *Lipstick on the Host* (coll **1992**) by Aidan Mathews (1956-   ). The situation in Ireland resembles conditions in earlier centuries: Benjamin DISRAELI's sharp-tongued "The Modern Aesop" (1826 *The Star Chamber*) was published anonymously, for instance, never acknowledged during his lifetime, and not released in his name until the publication of *The Dunciad of To-Day, and The Modern Aesop* (coll **1928** chap). And in other parts of the world coding has been essential through much of the 20th century. Writers in most of the countries dominated until 1990 by the USSR wrote prolifically in Aesopian terms; as a single example of self-censorship in Russia, Mikhail BULGAKOV's *Sobacheye Serdste* (written 1925; trans Michael Glenny from manuscript as *Heart of a Dog* **1968** UK) uses AF means to depict – with some ambivalence – human resistance to totalitarian personality control. During WWII French authors tended to create beast fables whose messages were clearly coded.   [JC]

AFFINITY GROUPS Writers in any genre literature tend to form AGs. This is true in two broad senses: conventions and language are shared; and lives are shared.

An AG, in the second sense, is a communication matrix linking people of like mind who wish to do or to appreciate similar things. The group of writers and readers who focused on and corresponded with H.P. LOVECRAFT is clearly one, though most of them did not meet physically. Meeting regularly to drink and read aloud to one another, C.S. LEWIS, J.R.R. TOLKIEN and Charles WILLIAMS informally declared themselves members of an AG they called the INKLINGS. The SOCIETY FOR CREATIVE ANACHRONISM informally assembles writers and readers to act out medieval fantasy routines. The JAMES GANG is a post-facto description of many writers who share an interest in the GHOST STORIES of M.R. JAMES, and who often consciously aspire to write like him. In MINNEAPOLIS, USA, a group of writers formed itself into a social and creative scrum they called the SCRIBBLIES; in Toronto, Canada, a similar group called itself the Bunch of Seven (◊ CANADA). Both groups have generated SHARED-WORLD tales or sequences (though it stretches the term beyond usefulness to describe a shared-world series as an AG enterprise). In Oxford, UK, the Pieria group included Robert HOLDSTOCK, Garry KILWORTH, David LANGFORD, Michael Scott ROHAN and Allan SCOTT.

Writing workshops could also be described as AGs: certainly the writers who have been involved in the Clarion Science Fiction Writers' Workshop and the US or UK version of the Milford Science Fiction Writers' Conference are members of formidable AGs. [JC]

AFTERLIFE The notion that death is not the end is common to most religions and mythologies; some have opted for REINCARNATION as a putative mode of extending the span of the individual SOUL, some look instead to GHOSTS, some have posited a continued beingness on an ASTRAL PLANE of some sort, and some claim the existence of what are in effect SECONDARY WORLDS – be they HEAVEN, HELL or VALHALLA – where the virtuous (depending upon how that term is defined) or the lucky enjoy a second, eternal "life". With such a rich array of options available from traditional sources, it is not surprising that the POSTHUMOUS FANTASY has proved such a popular subgenre, though modern taste tends not to find attractive the late-19th- and early-20th-century tendency to interpret afterlife experiences in terms of SPIRITUALISM. All too often the afterlife is experienced, as in Marie CORELLI's *The Soul of Lilith* (1892), through a state of enchanted sleep; or, as in R.H. Wright's *The Outer Darkness* (1906), is mostly submerged in pious ALLEGORY, though this tale, by treating PURGATORY in LOST-RACE or lost-world terms, does hint at developments later in this century (see below). A late Spiritualist experience of the afterlife appears in *Time Must Have a Stop* (1944) by Aldous Huxley (1894-1963), where a character dies, travels through the afterlife, and reports back on events (including the forthcoming WORLD WAR II) to his associates.

Depictions of the afterlife as a place (rather than as a condition) are reasonably common in both written and cinematic fantasy. *Job: A Comedy of Justice* (1984) by Robert A. HEINLEIN briefly portrays a peculiarly mundane, bureaucratic HEAVEN and a slightly more interesting HELL. Earlier, in *Inferno* (1975), Larry NIVEN and Jerry Pournelle reworked DANTE, sticking fairly closely to Dante's geography (although assigning contemporary political enemies, like environmentalists, to Dantean punishments they considered "appropriate"). It is unsurprising that these quasi-rationalizations have been produced by sf rather than fantasy writers.

Much more interesting from a fantasy viewpoint is the afterlife portrayed by John Kendrick BANGS in *A House-Boat on the Styx: Being Some Account of the Divers Doings of the Associated Shades* (1896) – sequelled by *The Pursuit of the House-Boat* (1897) – in which historical noteworthies are gathered in the eponymous venue; Frederic Arnold KUMMER's very similar afterlife stories include *Ladies in Hades* (1928) and *Gentlemen in Hades* (1930). A thematic descendant is Philip José FARMER's extended TECHNOFANTASY, the **Riverworld** series, in which a resurrected humanity is placed along the banks of a river many millions of miles long. Indeed, RIVERS are popular symbols in this context: Sheri S. TEPPER's *The Awakeners* (1987), for example, which blends sf and fantasy themes, again centres a large human

population along the banks of a hugely long river. The dead rise in countless local "afterlives", which have physical existence. Return of the dead to the areas occupied by the living is TABOO, as, for the most part, is the incursion of the living into the "afterlives". In John GRANT's *The World* (1992) multitudinous afterlives are equated with the territories of DREAMS, which are themselves seen as ALTERNATE REALITIES, so that the continued existence of the individual is in effect a form of REINCARNATION, but each time into a different venue.

Fantasy writers have also used the afterlife as a place for living protagonists to visit, usually via PORTAL, as with L. Sprague DE CAMP's and Fletcher PRATT's *The Incomplete Enchanter* (1941), whose protagonist participates in events – the travails of the Norse AESIR as FIMBULWINTER deepens – that mortals are meant to see only posthumously. Indeed, any fantasy visit to the abode of the GODS implies an afterlife context: for the gods may visit us whenever they wish, but we normally visit them only after we have "passed over" into their sphere.

A cluster of movies have depicted the afterlife, or more especially LIMBO. Limbo is portrayed in HERE COMES MR JORDAN (*1941*), which was sequelled by *Down to Earth* (*1947*) and remade as HEAVEN CAN WAIT (*1978*). A more extravagant version of Limbo appeared in *A* MATTER OF LIFE AND DEATH (*1946*), and the notion of Limbo as a sort of vast bureaucracy is perpetuated in movies like BEETLEJUICE (*1988*) and especially DEFENDING YOUR LIFE (*1991*). In OUTWARD BOUND (*1930*) and its remake, *Between Two Worlds* (*1944*), by contrast, Limbo is a steamship making its way from our world to the hereafter, at the end of which voyage the passengers shall be judged; suicides are doomed to repeat the journey endlessly. The Limbo of ALWAYS (*1989*) is a tract of green countryside, deserted save for the protagonist and an instructing ANGEL. Very few movies go beyond Limbo in their explorations of the afterlife: a couple of recent examples are Don BLUTH's *All Dogs Go to Heaven* (*1989*), which depicts Heaven, canine-style, and BILL & TED'S BOGUS JOURNEY (*1991*), which memorably portrays both Heaven and Hell. [JG/JC]

See also: GHOST STORIES.

AGINS, PHYLLIS CAROL (1947-   ) US writer whose first novel, *Suisan* (1992), renders in TWICE-TOLD terms the SNOW WHITE tale from the viewpoint of the DWARFS. Their confused and at times anguished BELATEDNESS, and the darkness of the outcome of their ambivalent playing of the game of the STORY, make *Suisan* into a representative REVISIONIST FANTASY, reminiscent at times of Donald BARTHELME's *Snow White* (1967). [JC]

AGON When Johann Huizinga (1872-1945) first published *Homo Ludens: Proeve Eener Bepaling van Het Spel-Element der Cultuur* (1938; trans R.F.C. Hull from the 1944 German edition, and from Huizinga's English notes as *Homo Ludens: A Study of the Play-Element in Culture* 1949 UK), he made little mention of agon; and the additional notes he wrote for the English translation are not apparently very clear. What is clear, however, is that he intended the term "agon" (originally a Greek word denoting a verbal contest between two characters in a play) to describe cultural situations in which contest cannot be distinguished from play – which he defines as "a voluntary activity or occupation executed within certain fixed limits of time and place, according to rules freely accepted but absolutely binding, having its aim in itself and accompanied by a feeling of tension, joy and the consciousness that it is 'different' from 'ordinary life'". An agon is, therefore, a contest conducted in accordance with artistic rules, "freely accepted but absolutely binding". Any complex society can be described – with some metaphorical latitude – as containing imagined or actual arenas where conflicts are ritually decided, and where RITES OF PASSAGE are concluded, according to rules which reconfirm the nature of that society through being obeyed and which convey a kind of joy; thus Huizinga's description of humanity as *Homo ludens*, man the player.

The beheading match in the anonymous 14th-century *Sir Gawain and the Green Knight* is a form of agon, and the essential health of ARTHUR and his ROUND TABLE at this stage of their long story is demonstrated by Gawain's absolute acceptance of the rules of the game he has embarked upon. Modern HIGH FANTASIES similarly tend to demonstrate the health of their worlds by depicting agons in working order. The sense of WRONGNESS that marks the threat or presence of disease in the fantasy world may similarly be marked by rigged agons, agons which PARODY rather than confirm the central rule-structure through which the world can be understood: in E.R. EDDISON's *The Worm Ouroboros* (**1922**) a wrestling match is arranged between Gorice and Goldry Bluszco, according to the way of things, in order to solve a political dispute; but Gorice cheats and, although Goldry still wins the match, the sense of wrongness generated by the violation of agon clearly ordains that grim and distorting conflicts must take place before any HEALING climax. In much SWORD AND SORCERY, that climax will take the shape of a resumed agon after the HERO triumphs and the LAND is saved. Most tales set in FANTASYLAND tend to feature agons which are acted and re-enacted within the iron cliché of the fantasyland premise.

It can be argued that secret histories (◊ FANTASIES OF HISTORY) tend to treat mundane REALITY as a kind of playground for the enactment of inner dramas perceivable only to elites capable of understanding the *rules of the game*. THEOSOPHY hovers perilously close to treating the material world as a kind of arena; occult faiths and programmes in general tend to treat the exoteric world as both blind to the truths on the other side of the MIRROR and fair game for the Elect to play with (◊ LIFESTYLE FANTASY; GODGAME). Michael MOORCOCK has used his concept of the MULTIVERSE to treat the mundane worlds under its sway as a PALIMPSEST array of agons.   [JC]

AHAB Master of the *Pequod* in Herman MELVILLE's *Moby-Dick* (**1851**). ◊ CHILDE; OBSESSED SEEKER.

AHASUERUS ◊ WANDERING JEW.

AICKMAN, ROBERT (FORDYCE) (1914-1981) UK writer whose passionately engaged life – as opera critic and founder of the Inland Waterways Association – fed into his fiction, which he began to write about 1949. He contributed to *We Are for the Dark: Six Ghost Stories* (coll **1951**) with Elizabeth Jane HOWARD. RA's stories in this volume – "The Trains", "The Insufficient Answer" and "The View" – are the work of a mature writer, and, although his most profound tales date from his last years, he never worked out an apprenticeship in public: he seemed fully formed – and fully armed – from the first, an impression strengthened by his autobiographies, *The Attempted Rescue* (**1966**) and *The River Runs Uphill: A Story of Success and Failure* (**1986**).

RA was perhaps the finest writer of the GHOST STORY in the second half of the 20th century, and although little of his work is pure fantasy – only a few of his 45 or so stories, and

neither of his novels, actually take leave of the mundane world in any sense, except as a prelude to death – he is of absorbing interest in our context because he demonstrates the range of meaning that may be extracted from the devices of any form of fantastic literature, when those devices are treated gravely by a writer of high quality. Most of RA's work is technically SUPERNATURAL FICTION, about which he comments interestingly in his introductions to his **Fontana Book of Great Ghost Stories** series of ANTHOLOGIES. He commonly focuses not upon the entity (usually a GHOST) which violates the fabric of REALITY but upon his richly conceived protagonists as they thrust towards, and sometimes across, thresholds (◊ PORTALS) they do not know how to "read". They cannot understand the ghost that faces them because that ghost, in RA's most typical stories – like "Compulsory Games" (1976 in *Frights* ed Kirby McCauley) – is a manifestation, a psychic portrait, of their failure to understand their own lives. Having failed to know themselves, his protagonists become frightened unto death by the fragmented images they glimpse across the UNCANNY threshold.

There is, therefore, much WRONGNESS in RA's work, and very little HEALING; an almost solitary exception is the very late "The Stains" (1980 in *New Terrors* ed Ramsey CAMPBELL) – winner of a 1981 BRITISH FANTASY AWARD – whose protagonist comes to terms with the fact that he is dying and that the nymph he meets is escorting him to the grave (RA's protagonists seldom sufficiently sort themselves out to come to a conclusion). In the end, the nymph takes him down into the coffin-like dwelling of her father, who is the natural world, and the protagonist realizes that he is happy and for a time "count[s] the good things only, as does a sundial".

After *We Are for the Dark* RA's stories appeared in various solo collections, including *Dark Entries* (coll **1964**), *Powers of Darkness* (coll **1966**), *Sub Rosa: Strange Tales* (coll **1968**), *Cold Hand in Mine: Eight Strange Stories* (coll dated 1975 but **1976**), *Tales of Love and Death* (coll **1977**), *Intrusions: Strange Stories* (coll **1980**) and *Night Voices* (coll **1985**). US collections included *Painted Devils: Strange Stories* (coll **1979** US) and *The Wine-Dark Sea* (coll **1988** US; cut 1990 UK), both selected from previous UK volumes; *The Unsettled Dust* (coll **1990**) is also a compilation. *The Late Breakfasters* (**1964**) and arguably *The Model* (**1987** US) are less successful than the shorter work.

Stories of note include: "Ringing the Changes" (1955 in *The Third Ghost Book* ed Cynthia ASQUITH), in which the risen dead paralyse a dangerously moribund marriage; "The Visiting Star" in *Powers of Darkness*; "Into the Wood" (◊ INTO THE WOODS) in *Sub Rosa*; "The Hospice" and "Pages From a Young Girl's Journal", which won a 1975 WORLD FANTASY AWARD, in *Cold Hand in Mine*; "The Fetch" in *Intrusions*; and "Mark Ingestre – The Customer's Tale" (1980 in *Dark Forces* ed Kirby McCauley). But there are few weak stories. RA can and should be read entire.   [JC]

As editor: *The Fontana Book of Great Ghost Stories* (anth **1964**); *The Second Fontana Book of Great Ghost Stories* (anth **1966**); *#3* (anth **1977**); *#4* (anth **1978**); *#5* (anth **1969**); *#6* (anth **1970**); *#7* (anth **1971**); *#8* (anth **1972**).

AIKEN, CONRAD (1889-1973) US poet, father of Joan AIKEN and John Aiken. Some of his narrative poems – like "John Deth" in *John Deth, A Metaphysical Legend, and Other Poems* (coll **1930**) – use ballad-like idiom and supernatural content to articulate their author's traumatized psyche (when he was very young his father killed his mother and committed suicide) and fluent Modernist instincts. For half a century after his first volume of poems, *Earth Triumphant* (coll **1914**), he was very prolific, much of his work evoking figures of MYTH and seasonal climaxes (◊ SEASONS), as in "Hallowe'en" (in *Skylight One* [coll **1949**]), a powerful evocation of Samhain (◊ HALLOWEEN).

Most of his fiction appeared during the interbellum. He may be longest remembered for novels like *King Coffin* (**1935**). His most famous single tale (out of a total of about 40) is "Mr Arcularis" (**1931**), a POSTHUMOUS FANTASY whose protagonist hears his death in the throbbing of the engine of a great ocean liner; CA later dramatized it as *Mr Arcularis: A Play* (**1957** chap). Other stories, like "Silent Snow, Secret Snow" (**1932**), perhaps more clearly express CA's abiding concern, the war between a domineering, private, hysterical psyche and the fantastical, luminescent world of the senses. Collections include *Bring! Bring!* (coll **1925**), *Costumes by Eros* (coll **1928**), *Gehenna* (coll **1930**) and *Among the Lost People* (coll **1934**); CA's short fiction was later assembled as *The Collected Stories of Conrad Aiken* (coll **1960**). [JC]

AIKEN, JOAN (DELANO) (1924-   ) UK writer, daughter of Conrad AIKEN, brother of John Aiken and stepdaughter of Martin ARMSTRONG. Most of her many adult novels have no explicit fantasy content, though some have supernatural implications. Of these titles, *Castle Barebane* (**1976** US) is a streamlined Gothic and *The Haunting of Lamb House* (coll of linked stories **1991**), set in the actual Lamb House in Rye, expose two of its real-life inhabitants – Henry JAMES and E.F. BENSON – to the GHOST of an 18th-century child who has had to endure the breakup of his family.

She has also written many CHILDREN'S FANTASIES for younger readers, characterized by the unusual combination of surreal invention and elegance of plotting; but of greatest genre interest are her books for older children, the most famous being the sequence – variously referred to as the **Willoughby Chase**, **Alternate England**, **Dido Twite** or **James III** series – set mostly in an ALTERNATE-WORLD version of England, with a Stuart dynasty (James III has acceded in 1832) and Hanoverian plotters. The series' internal chronology, with one exception matching the sequence of publication, begins with *The Wolves of Willoughby Chase* (**1962**) – filmed (rather flatly) as *The Wolves of Willoughby Chase* (*1988*) – and *Black Hearts in Battersea* (**1964** US): the youthful protagonists escape from an evil governess; in escaping they meet young Simon (a HIDDEN MONARCH) and the remarkable Dido Twite, whose pragmatic but fantastical personality is perhaps JA's most sustained creation. In *Nightbirds on Nantucket* (**1966**), *The Stolen Lake* (**1981**) and *The Cuckoo Tree* (**1971**) the focus shifts to Dido and her elaborate adventures at sea, in a LOST LAND in South America (where the ONCE AND FUTURE KING is encountered, along with Queen Ginevra, a figure of distorted BELATEDNESS, for she has been awaiting her husband's return for a millennium; ◊ ARTHUR), and returning to England on a mission vital to the monarchy, in opposition to her own father's Hanoverian plotting. After *Dido and Pa* (**1986**), the sequence darkens in *Is* (**1992**; vt *Is Underground* 1992 US), set in a muted STEAMPUNK industrial hell – though at book's end Simon is due to become king. *Cold Shoulder Road* (**1995**) focuses in a lighter vein on other members of the Twite family.

Similarly located, the **Felix** series of adventures – *Go Saddle the Sea* (**1977** US), *Bridle the Wind* (**1983**) and *The*

Teeth of the Gale (**1988**) – follows the travels of its young protagonists through magically heightened LANDSCAPES; it is an oddity of the sequence that Felix fails to realize that his companion Juan is in fact a girl (◊ DUOS; GENDER DISGUISE). Singletons of interest include *The Kingdom and the Cave* (**1960**), *The Whispering Mountain* (**1968**), *Midnight is a Place* (**1974**) and *The Shadow Guests* (**1980**), an extremely effective RITE-OF-PASSAGE tale whose protagonist must come to terms with the messages conveyed by several family GHOSTS. One play of exceptional interest – *Winterthing: A Child's Play* (**1972** chap US), assembled with *The Mooncusser's Daughter* (**1973** chap US) as *Winterthing, and The Mooncusser's Daughter* (omni **1973**) – compactly and gravely encompasses much of JA's strengths in its story of four unrelated children who have been stolen by the kleptomaniac "Auntie" and taken to Winter Island (◊ ISLANDS), which disappears every seventh winter (◊ TIME IN FAERIE). There they undergo intense rites of passage, are succoured by a sea-goddess, learn wisdom in the magic winter and prepare to re-enter the world the following spring (◊ SEASONS).

JA's short stories have been assembled in several volumes – confusingly, US and UK editions often vary remarkably in the tales included – and comprise an extremely impressive range of work, a gamut which incorporates delicately exemplary FAIRYTALES, nonfantasy and fantasy stories of rural and urban life (sometimes constructed with an effect of easeful MAGIC REALISM), GHOST STORIES and HORROR. Titles include *A Necklace of Raindrops* (coll **1968**), *A Small Pinch of Weather* (coll **1969**), *The Windscreen Weepers and Other Tales of Horror and Suspense* (coll **1969**; vt with differing contents *The Green Flash and Other Tales of Horror, Suspense, and Fantasy* 1971 US), *Smoke from Cromwell's Time* (coll **1970** US), *The Kingdom Under the Sea* (coll **1971**), *A Harp of Fishbones* (coll **1972**), *All But a Few* (coll **1974**), *Not What You Expected* (coll **1974** US), *A Bundle of Nerves* (coll **1976**; vt with differing contents *The Far Forests: Tales of Romance, Fantasy and Suspense* 1977 US), *The Faithless Lollybird* (coll **1977**; with differing contents 1978 US), *Tale of a One-Way Street* (coll **1978**), *A Whisper in the Night* (coll **1982**; with differing contents 1984 US), *Fog Hounds, Wind Cats, Sea Mice* (coll **1984**), *Up the Chimney Down* (coll **1984**), *The Last Slice of Rainbow* (coll **1985**), *Past Eight O'Clock* (coll **1986**), *A Goose on Your Grave* (coll **1987**), *Give Yourself a Fright* (coll **1989** US), *A Foot in the Grave* (coll **1989**), *A Fit of Shivers* (coll **1990**) and *A Creepy Company* (coll **1993**; rev 1995 US).

Throughout her career JA has generated work of an almost relentless fertility. There is a passionate knowingness in her invention of small details that is clearly a matter of her own satisfaction – few young readers would know, for instance, that the "hobey" played by Dido Twite's father is an oboe, the French *hautbois* having, in this world, been differently Englished. Her feverishness may derive to some degree from the example of her father; the loving urgency of her depiction of character and landscape and plot seems an intrinsic gift. [JC]

Other works:

Adult novels (many associational): *The Silence of Herondale* (**1964** US); *The Fortune Hunters* (**1965** US); *Trouble With Product X* (**1966**; vt *Beware of the Bouquet* 1966 US); *Hate Begins at Home* (**1967**; vt *Dark Interval* 1967 US); *The Ribs of Death* (**1967**; vt *The Crystal Cow* 1968 US) and its sequel *Foul Matter* (**1983**); *Night Fall* (**1969**); *The Embroidered Sunset* (**1970**); *Died on a Rainy Sunday* (**1972**); *The Butterfly Picnic* (**1972**; vt *A Cluster of Separate Sparks*

1972 US); *Voices in an Empty House* (**1975**); *Last Movement* (**1977**); *The Five-Minute Marriage* (**1977**); *The Smile of the Stranger* (**1978**) and its sequel *The Weeping Ash* (**1980** US; vt *The Lightning Tree* 1980 UK); *The Young Lady from Paris* (**1982**; vt *The Girl from Paris* 1982 US); *Mansfield Revisited* (**1984**); *Deception* (**1987**; vt *If I Were You* 1987 US); *Blackground* (**1989**); *Jane Fairfax* (**1990**); *Morningquest* (**1992**); *Eliza's Daughter* (**1994**); *The Winter Sleepwalker* (**1994**).

For younger children: *All You've Ever Wanted* (coll **1953**) and *More Than You Bargained For* (coll **1955**), both assembled as *All and More* (omni **1971**; cut vt *All But a Few* 1974); *Armitage, Armitage, Fly Away Home* (coll **1968** US); *Mice and Mendelson* (**1978** chap); the **Arabel and Mortimer** sequence about a young girl and her raven, comprising *Arabel's Raven* (**1972** chap), *The Escaped Black Mamba* (**1973** chap; vt *Arabel and the Escaped Black Mamba* 1984 chap) and *The Bread Bin* (**1974** chap) – all three assembled as *Tales of Arabel's Raven* (omni **1974**) – *Mortimer's Tie* (**1976** chap), *The Spiral Stair* (**1979** chap) and *Mortimer and the Sword Excalibur* (**1979** chap) – all three assembled as *Arabel and Mortimer* (omni **1980**) – *Mortimer's Portrait on Glass* (**1981** chap) and *The Mystery of Mr Jones's Disappearing Taxi* (**1982** chap) – both assembled with "Mortimer's Cross" as *Mortimer's Cross* (omni **1983**) – *The Kitchen Warriors* (**1983** chap) and *Mortimer Says Nothing, and Other Stories* (coll **1985**); *The Shoemaker's Boy* (**1991** chap US); *The Midnight Moropus* (**1993** chap US); *Hatching Trouble* (**1993** chap).

For older children: *The Skin Spinners* (coll **1976** US), poetry; *The Moon's Revenge* (**1987** US); *The Erl King's Daughter* (**1988**); *Voices* (**1988**; vt *Return to Harken House* 1990 US).

Nonfiction: *The Way to Write for Children* (**1982**); *Conrad Aiken, Our Father* (**1989**) with Jane Aiken Hodge (1917-) and John Aiken.

AIKEN, JOHN (KEMPTON) (1913-1990) US-born UK writer, son of Conrad AIKEN and brother of Joan AIKEN.

AINSWORTH, WILLIAM HARRISON (1805-1882) UK writer, primarily of historical romances. WHA also wrote as Will Brown and Cheviot Ticheburn. His style favoured sensationalism, mixing the Gothic fiction of Ann RADCLIFFE with the ROMANTICISM of Walter SCOTT, and paved the way for the sensationalistic novels and penny dreadfuls of the 1840s. The basis of his popularity was his heroic treatment of common folk; though, when this extended to convicted criminals – like Dick Turpin, whose myth he created in his first novel, *Rookwood* (**1834**) – WHA was accused of condoning crime, a charge levelled at Edward BULWER-LYTTON and at many horror/thriller writers since. His most popular works aside from *Rookwood* included *Jack Sheppard* (**1839**) – about another highwayman – *The Tower of London* (**1840**) and *Old St Paul's* (**1841**). It was only in his later works, as his popularity was waning, that WHA turned to LEGENDS and SUPERNATURAL FICTION. *Windsor Castle* (**1843**), set in the reign of Henry VIII, utilizes the FOLKTALE of HERNE THE HUNTER to pit paganism against Henry's new Church; *The Lancashire Witches* (**1849**) is based on the 17th-century trials of the WITCHES of Pendle Forest, and is a seminal novel in its field; *Auriol, or The Elixir of Life* (1844-5 *Ainsworth's Magazine*; exp in *The Works of William Harrison Ainsworth Vol 12* coll **1850**), develops its alchemical base into a story of ROSICRUCIANISM and DEVIL-worship. WHA was briefly editor of *Bentley's Miscellany* (1839-41) and less briefly proprietor and editor of *Ainsworth's Magazine* (1842-54)

and *The New Monthly Magazine* (1845-70), where many of his works appeared. Their effect was often enhanced by the superb artwork of George CRUIKSHANK, who later claimed to have provided WHA with some of his plots. WHA's work was much imitated, especially by the melodramatists G.P.R. James (1799-1860), Thomas Peckett Prest (1810-1859) and James Malcolm RYMER. [MA]

Further reading: *William Harrison Ainsworth and His Friends* (**1911**) by Stewart M. Ellis; *A Bibliographical Catalogue of the Published Novels and Ballads of William Harrison Ainsworth* (**1925**) by Harold Locke; *A Tale of Three Authors* (**1940**) by Leo Mason, which links WHA with Charles DICKENS and Edgar Allan POE.

AJVAZ, MICHAL (1949-) Czech writer. ◊ CZECH REPUBLIC.

AKERS, ALAN BURT ◊ Kenneth BULMER.

AKI, TANUKI [s] ◊ Charles DE LINT.

AKIRA Japanese ANIMATED MOVIE (*1987*). Akira Committee/Asahi/ICA Projects. **Pr** Shunzo Kato, Ryohei Suzuki. **Exec pr** Sawako Noma. **Dir** Katsuhiro OTOMO. **Anim dir** Hiroaki Sato, Yoshio Takeuchi. **Screenplay** Izo Hashimoto, Otomo. **Based on** the ongoing graphic series by Otomo. 124 mins. Colour.

21st-century Tokyo is nearly destroyed when the misguided efforts of the military and various psi-powered CHILDREN (◊ TALENTS) help unleash the nuclear and metaphysical threat of the Akira Project, which is, in effect, both FRANKENSTEIN monster and a created GOD. The first piece of ANIME to make a significant impact on the West, where it begat a continuing cult following of anime and MANGA, *A* is a long and complex work of TECHNOFANTASY that does not submit easily to analysis. As a work of SCIENCE FICTION it shows the thematic influence of the cyberpunk subgenre, while visually it is indebted to arcade games, to movies like *Blade Runner* (*1982*) and even, in part, *Star Wars* (*1977*), with its greatest debt being to COMICS art – indeed, *A* views like a graphic novel come to life; the US comic *Akira* (1988-current) continues the story. Its fantasy elements are metamorphic, contriving to produce an original vision from a combination of stock elements. [JG]

AKUTAGAWA, RYUNOSUKE (1892-1927) Japanese writer whose first literary work was a 1914 translation of Anatole FRANCE's fantasy *Balthasar* (**1889**) and who remains best-known for his third story, "Rashomon" (1915), a SUPERNATURAL FICTION which appears in *Rashomon and Other Stories* (coll trans Takashi Kojima **1951** chap US) and served as a base for Akira KUROSAWA's movie *Rashomon* (*1951*). Other short fiction, much of it based on old FOLK-TALES, is assembled in *Tales Grotesque and Curious* (coll trans G.W. Shaw **1930**). The protagonist of *Kappa* (**1927**; trans S. Shiojiri **1947**; new trans Geoffrey Bownas **1971** US) is propelled by the eponymous water spirit to an UNDERGROUND country whose culture is a TOPSY-TURVY version of Japanese life. The protagonist's disgust on returning to the surface has been taken by critics as reflecting *Gulliver's Travels* (**1726**) by Jonathan SWIFT; but RA's own emotional state resembled his protagonist's, and he committed suicide soon after completing the text. [JC]

ALADDIN US ANIMATED MOVIE (*1992*). Disney. **Pr** Ron Clements, John Musker. **Dir** Clements, Musker. **Spfx** Don Paul. **Screenplay** Clements, Ted Elliott, Musker, Terry Rossio. **Based on** the traditional tale (◊ ARABIAN FANTASY). **Voice actors** Aaron Blaise (Rajah), Jonathan Freeman (Jafar), Gilbert Gottfried (Iago), Brad Kane (Aladdin singing), Linda Larkin (Jasmine speaking), Lea Salonga

(Jasmine singing), Douglas Seale (Sultan), Scott Weinger (Aladdin speaking), Frank Welker (Abu), Robin Williams (Genie). 90 mins. Colour.

Wicked vizier Jafar hopes to depose the Sultan of Agrabah; to do so he requires the aid of the GENIE from a magic lamp in the Cave of Wonders, a vast UNDERGROUND treasure trove in the desert into which only the pure of heart can enter. He sends orphan Aladdin, imprisoned for having rescued Princess Jasmine from embarrassment in the souk, into the cave; there Aladdin and MONKEY sidekick Abu ally with the Genie and a magic carpet; thereafter the tale concerns Jafar's thwarted attempts to expunge Aladdin, enslave the Genie, marry Jasmine and oust the Sultan. In an effective denouement, the TRICKSTER Aladdin persuades Jafar to order the Genie to make him, Jafar, likewise a genie, with all a genie's magical powers. But genie-hood brings also bonds (◊ BONDAGE), and thus Jafar is pent inside the lamp.

A is set in "Hollywood Arabia", a LAND OF FABLE that bears little resemblance to the original. Jafar's face was modelled partly on that of Nancy Reagan, but the exceptional characterization is that of the Genie, with Robin Williams's manic vocal portrayal being superbly matched visually by the animation. Elsewhere, considerable debts are owed in terms of plot and characterization to both *The* THIEF OF BAGDAD (*1924*) and *The* THIEF OF BAGHDAD (*1940*). Most notable, though, is *A*'s overpowering visual effect: not since the early days have the Disney animators allowed themselves so lavishly imaginative and so surreal a visual interpretation of a plot, with the foreground ever and again erupting into a kaleidoscope of brilliant colour while the background fades to matte.

The sequel was *The* RETURN OF JAFAR (*1994*).   [JG]

ALAN, A.J. Working name and radio persona of UK broadcaster and civil servant Leslie Harrison Lambert (1883-1941), the first nationally popular radio storyteller. Between 1924 and 1940 he broadcast over 40 15min stories, many collected as *Good Evening, Everyone!* (coll **1928**) and *A.J. Alan's Second Book* (coll **1932**), the former containing most of his weird stories. Generally AJA told TALL TALES, with enough conviction that many listeners believed him. His weird tales tended to be GHOST STORIES, although he loved to make the absurd commonplace, as in "Percy the Prawn" (1934), which has a fortune-telling shellfish. AJA firmly established radio as a medium for storytelling, and thus prepared the way for his successor, Algernon BLACKWOOD.   [MA]
Other works: *The Best of A.J. Alan* (coll **1954**) ed Kenelm Foss.
Further reading: "A.J. Alan: The First Broadcaster" by Jack Adrian in *Million #8*, March/April 1992.

ALARCÓN, PEDRO A(NTONIO) DE (1833-1891) Spanish author of *El Sombrero de tres picos* (**1874**; trans as *The Three-Cornered Hat* **1891** UK), a FABLE-like tale which became instantly famous. The earlier *El amigo de la muerte* (**1852**; trans Mrs Francis J.A. Darr as *The Strange Friend of Tito Gil* **1890** US; new trans Mary J. Serrano as *The Friend of Death: A Fantastic Tale* **1891** UK) is a POSTHUMOUS FANTASY in which a suicide is gulled by DEATH into thinking he has not yet killed himself, but eventually awakens – after an illusory life in which he marries his beloved – into Death's kingdom. The tale climaxes in an APOCALYPSE set in the 24th century.   [JC]

ALASTAIR Working name of German-born writer and illustrator, pianist, dancer and mime artist Hans Henning Voigt (1887-1969), who claimed to be the illegitimate offspring of a hot-headed Bavarian prince and a pretty Irish girl. This account of his origins is as dubious as his mysteriously acquired title of "Baron", since, of all the Decadents (◊ DECADENCE), it is probably of him that it could most truly be said that he "created himself".

He was largely self-taught. His drawing style was heavily influenced by that of Aubrey BEARDSLEY. Pictures usually comprised a delicately outlined single figure or couple, the details filled in with solid areas and fine, lacelike patterning. In much of his work there is an intense eroticism, redolent with sadomasochistic elements. The faces are like MASKS, with large expressive eyes. He anonymously illustrated several privately printed books and pamphlets 1904-14 before being befriended by John Lane of the Bodley Head, who published *Forty Three Drawings by Alastair (With a Note of Exclamation by Robert Ross)* (graph **1914** UK). Ten years later, an entirely new selection of illustrations was published under the title *Fifty Drawings by Alastair* (graph **1925** US). A's other work includes ILLUSTRATIONS for books of fiction, poetry and drama, and translations of both prose and poetry from French and English into German, among them works by Charles BAUDELAIRE, Gustave FLAUBERT, William Makepeace THACKERAY, John KEATS and Oscar WILDE. He illustrated *The Sphinx* (1920 UK) by Wilde, *Carmen* (1920 Germany) by Prosper Mérimée (1803-1870), *Die Buchse der Pandora* ["Pandora's Box"] (1921) by Frank Wedekind (1864-1918), *Erdgeist* (1921) by Wedekind, *Sebastian von Storck* (1924 Austria) by Walter Pater (1695-1736), *Salome* (**1925** France) by Wilde, *The Fall of the House of Usher* (1928 France) by Edgar Allan POE, *L'Anniversaire de l'Enfante* (1928 France) by Wilde, *Manon Lescaut* (1928 UK/US) by Abbé Prévost (1697-1763) and *Les Liaisons Dangereuses* (2 vols 1929 France) by Pierre de Laclos (1741-1803). He also illustrated a small number of privately printed books of poetry and wrote *Das Flammende Tal* ["The Flaming Valley"] (**1920**), a book of poetry.   [RT]
Further reading: *Alastair: Illustrator of Decadence* (graph **1979**) by Victor Arwas.

ALCALA, ALFREDO P. (1925-   ) Filipino COMIC-book artist with an intricate pen-line style reminiscent of the Victorian steel engravers, but with a boldness and vitality rarely achieved by them. His work is also remarkable for its inventiveness and ingenuity in creating dramatic atmospheres. His early influences were US comics artists Harold FOSTER, Alex RAYMOND and Lou Fine (1914-1971); later he was influenced by the US classical illustrators Howard PYLE, N.C. WYETH and J.C. Leyendecker and the UK mural painter Frank Brangwyn (1867-1956).

APA's first job was as a sign-painter; he then worked with a manufacturer of wrought ironwork. His first comics work appeared in *Bituin Komiks* ["Star Comics"] in 1948. He proved a fast and reliable all-purpose artist, and frequently worked on eight comic books per month, drawing SUPERHERO tales, comedy, melodrama and fantasy with equal facility; he is credited with such enormous energy that he can work for up to 96 hours non-stop. His most important work was the epic **Voltar** (*Alcala Fight Komix* from 1963).

In the early 1970s he began working for US comic-book publishers, drawing mainly short horror and mystery tales for DC COMICS and MARVEL COMICS. The work that brought him the widest recognition in the US was on **Savage Sword of Conan** (beginning with #2, 1974), in which he inked and embellished the layouts of John Buscema (1927-   ); his work lent the stories a moody, magical antiquity. His other work includes **Kong the Untamed** (*#1-#5* 1975-6). [RT]

ALCHEMY The name given in Western Europe from the 12th century to various kinds of mystical proto-chemistry. Chinese alchemy is relatively well documented, but the evidence about ancient Greek alchemy is confined to a single manuscript, and the Arabic tradition is known only by allusion. From late medieval times onwards SCHOLARLY FANTASY moved to fill this historical void with copious invention.

Alchemy is traditionally associated with two particular quests: for the ELIXIR OF LIFE and the PHILOSOPHER'S STONE – i.e., the secret of transmuting "base metal" into gold. More recent revisionist historians tend to argue that these objectives ought to be construed metaphorically, and that alchemy is better regarded as a quest for spiritual enlightenment. The fact that medieval and Renaissance treatises were written in a bizarre symbolic code assists and encourages this kind of saving move.

Early literary works featuring alchemists tend to regard them as petty confidence tricksters, after the fashion of Geoffrey CHAUCER in "The Canon's Yeoman's Tale" (*c*1390) and Ben Jonson (1572-1639) in *The Alchemist* (**1610**). William GODWIN's *St Leon* (**1799**) suggests that success in the traditional alchemical quests would bring no joy, while Honoré de BALZAC's *La recherche de l'absolu* (**1834**) represents the alchemical quest as an archetypal exercise in futility. The occult revival of the late 19th century, however, brought forth a new wave of scholarly fantasies, usually associated with the various Rosicrucian societies (◊ ROSICRUCIANISM) inspired (at least in part) by Johann Andreae's *Chemical Wedding of Christian Rosenkreutz* (**1616**). Edward BULWER-LYTTON's *A Strange Story* (**1861**) and Joris-Karl HUYSMANS's *Là-Bas* (**1891**) are the best-known literary works giving evidence of the new credulity, but some Rosicrucian lifestyle fantasists – notably the Frenchman Joséphin Péladan – were prolific writers of bad fiction. Arthur MACHEN's involvement in such mysticism inevitably infected his work, his most elaborate alchemical tale being "The Spagyric Quest of Beroaldus Cosmopolita" (1923). Gustav MEYRINK's obsession with these materials gave extravagant inspiration to perhaps his finest novel, *The Angel of the West Window* (**1927**). Alchemical mysticism also plays a muted role in the metaphysical fantasies of Charles WILLIAMS, most notably in *Many Dimensions* (**1931**). Literary performances by contemporary scholarly fantasists – who usually take their inspiration from such works as Mircea ELIADE's *The Forge and the Crucible* (**1956**) and Frances Yates's *The Rosicrucian Enlightenment* (**1972**), but abandoning the conscience and sense of proportion carefully maintained by those authors – tend to be too earnest and/or too *recherché* for their own good.

Alchemy's intrusions into modern fantasy are not extensive, although alchemists feature in bit roles in many HIGH FANTASIES and sorcerers' lairs are often kitted out with the impedimenta of the alchemical laboratory. In Lyndon HARDY's *Master of the Five Magics* (**1980**) alchemy is discussed as one of five branches in a pseudo-taxonomy of MAGIC. It crops up in a number of notable historical fantasies, including Alexander DE COMEAU's *Monk's Magic* (**1931**), Avram DAVIDSON's *The Phoenix and the Mirror* (**1969**), John CROWLEY's *Aegypt* (**1987**), Patrick HARPUR's *Mercurius, or The Marriage of Heaven & Earth* (**1990**) and Derek Beaven's *Newton's Niece* (**1994**), and in a few exercises

in exoticism – including Frank OWEN's "Dr Shen Fu" (1938) and its sequels, which feature a Chinese alchemist – but is shown off to much better advantage in borderline-sf stories. Ian WATSON's *The Gardens of Delight* (**1980**), which interprets Hieronymus BOSCH's famous painting in terms of alchemical symbolism and actualizes its imagery, is the most ambitious exercise to date along these lines. [BS]

ALCINA In the **Orlando** sequences by Matteo BOIRDO and Lodovico ARIOSTO, Alcina is the sister of Morgana (◊ MORGAN LE FAY), and personifies the intoxication of the senses. She governs a magic ISLAND to which men are lured; once they are in her power, she METAMORPHOSES them into beasts or trees or rocks. Their BONDAGE seems irreversible until a young woman, Melissa, breaks the spell. Alcina is the subject of an OPERA by George Friedrich Handel, *Alcina* (**1735**), with libretto by Antonio Marchi. [JC]

ALDISS, BRIAN W(ILSON) (1925-   ) UK writer, anthologist and critic, best-known as a dominant figure in SCIENCE FICTION. He began his fiction career with "Criminal Record" for *Science Fantasy* in 1954, and after 40 years of constant activity has published well over 300 stories – several of them fantasy – and many novels, the most famous of which are sf; they include *Non-Stop* (**1958**), *The Long Afternoon of Earth* (**1962** US; exp vt *Hothouse* 1962 UK), *Barefoot in the Head* (**1969**), *Frankenstein Unbound* (**1973**) – which was filmed as *Frankenstein Unbound* (**1990**; ◊ FRANKENSTEIN MOVIES) – and the **Helliconia** sequence – *Helliconia Spring* (**1982**), *Helliconia Summer* (**1983**) and *Helliconia Winter* (**1985**). *Seasons in Flight* (coll **1984**) contains some fantasies later assembled – along with most of BWA's other short fiction of similar interest – as *A Romance of the Equator: Best Fantasy Stories* (coll **1989**). His later work, including stories assembled in *A Tupolev Too Far and Other Stories* (coll **1993**) and *The Secret of This Book* (coll **1995**), tends increasingly to subject the contemporary world to magic-realist transformations (◊ MAGIC REALISM) and to extract ironies – some savage – from the interplay; BWA has acknowledged the specific influence of Jorge Luis BORGES, but diverges from that model through the fury that governs his vision of the political and cultural cruelties which stain the late 20th century. To somewhat similar effect, a note of elegy countervails against the soothing, arabesque intricacies of the CITY that dominates *The Malacia Tapestry* (**1976**), BWA's only fantasy novel, a note which specifically contradicts the THEODICY implied by the title. Although set in a land with SECONDARY-WORLD elements (including DRAGONS, who may be the ancestral stock from which the humans of this world have evolved, and folk with WINGS), Malacia is governed, as fully as any sf world BWA has created, by entropy. It is an ANACHRONISM, deliberately held still against decay; and the novel's plot, unsurprisingly, deals with the threat of change or of absolute stasis. A condensed and rather surly version of Malacia figures in *Pile: Petals from St Klaed's Computer* (**1979** chap) illustrated by Mike WILKS, a long poem about an EDIFICE-like city which has lost its colour and is ruled by a computer built of stone and wood. [JC]

Further reading: *Apertures: A Study of the Writings of Brian Aldiss* (**1984**) by Brian Griffin and David Wingrove; *Brian W. Aldiss* (**1986**) by M.R. Collings; *Brian Wilson Aldiss: A Working Bibliography* (**1988** chap) by Phil Stephensen-Payne; *The Work of Brian W. Aldiss: An Annotated Bibliography and Guide* (**1992**) by Margaret Aldiss; *SFE*.

ALDRIDGE, ALAN (1943-   ) UK artist and writer,

active from about 1963. *The Penguin Book of Comics: A Slight History* (**1967**; rev 1971) with George Perry is a useful survey; *Phantasia of Dockland, Rockland, and Dodos* (**1981** chap US) comprises text and illustrations devoted to the CITY and other venues, with some fiction added. AA's fiction includes *The Gnole* (**1991**) with Stephen R. BOYETT, an ambitiously illustrated ecological FABLE in which the planet is seen to be at risk. AA's work as an illustrator has been in general extremely influential (for good and ill) on fantasy art at its most purely decorative, ornate and sentimental. [JC]

ALEXANDER, LLOYD (CHUDLEY) (1924-   ) US author best-known for his **Chronicles of Prydain** which, as with the works of J.R.R. TOLKIEN, progresses from a relatively lightweight CHILDREN'S FANTASY to deal with altogether higher, deeper and darker matters. The central series comprises *The Book of Three* (**1964**), *The Black Cauldron* (**1965**), *The Castle of Llyr* (**1966**), *Taran Wanderer* (**1967**) and *The High King* (**1968**). The LAND-OF-FABLE setting approximates to the mythic Wales of the MABINOGION, portions of which LA has freely and sometimes playfully adapted. This entails simplification: the *Mabinogion*'s engagingly amoral Gwydion and grim but honourable Arawn (a king of the UNDERWORLD) become respectively a spotless if unshowy hero and a routinely evil DARK LORD.

The Book of Three is something of a romp, whose villain the Horned King is routinely overcome, but introduces continuing characters: Gwydion (as above); the oracular pig Hen Wen; Taran, the rootless, aspiring (and often comically overreaching) young series hero, UGLY DUCKLING and Assistant Pig-Keeper; his mentor Dallben, recalling the MERLIN of T.H. WHITE; sharp-tongued Princess Eilonwy; the king and BARD Fflewddur Fflam, whose MAGIC harp's strings snap at his least untruth; and Taran's hairy, greedy, smelly and loyal pet/COMPANION (of uncertain species), Gurgi.

The sequence darkens almost at once in *The Black Cauldron*, where admirable characters can die, "good" men can turn traitor or evince major flaws (notably the corroding pride of one Prince Ellidyr), and real SACRIFICES are demanded. The Cauldron or Crochan, borrowed from the tragic Second Branch of the *Mabinogion*, restores the dead as ZOMBIE slave-warriors and requires a life as the price of its destruction . . . a price which is paid. Ambiguous light relief comes from the Crochan's guardians, the sinisterly comic hags Orwen, Orddu and Orgoch – who, despite lip-smacking eagerness to turn people into toads, later emerge as more than they seem (◊ FATES). *The Castle of Llyr* is quieter and focuses more on character: the amiable but colossally inept Prince Rhun acquires a certain stature almost despite Taran's impatience, Eilonwy has to make her own first significant sacrifice, Glew the self-made alchemical GIANT is a genuinely pitiable minor villain, and Queen Achren (the Enemy of this episode, vaguely reminiscent of the *Mabinogion*'s Arianrhod) also finally commands pity.

Taran Wanderer takes Taran on a bleaker, stonier road unfamiliar in HEROIC FANTASY: a QUEST for his roots among the neglected ordinary folk of the Free Commots, whose weaving, farming, pottery and SMITH skills supply the nobility of Prydain. Hoping to find himself well born, Taran suffers painful humiliation before settling to learn what he can of common crafts and emerging, slightly to his own surprise, as a kind of leader after all. This quality is tested almost to destruction in the final war of *The High King* – a Newbery Medal winner – which is full of betrayal, hard-

fought defeats and journeys without hope. Here we are shocked by deaths of characters who, in the standard grammar of CHILDREN'S FANTASY, seemed unkillable. As in Tolkien's Middle Earth, victory over Arawn ends the age of enchantment (◊ THINNING), and the conventionally noble or magical folk sail off to eternal bliss in the "Summer Country" . . . leaving the now predictable new High King and consort to the long job of HEALING Prydain's brutal scars.

Pendants to the main books are *Coll and His White Pig* (**1965**), *The Truthful Harp* (**1967**) and *The Foundling and Other Tales of Prydain* (coll **1973**). The movie *The* BLACK CAULDRON (*1985*) takes its story chiefly from *The Book of Three* and *The Black Cauldron*.

The five novels about Taran and his companions show Prydain and LA's talent at their best, and unlike his other work are constantly reprinted. [DRL]

Other works: The **Westmark** Graustarkian adventures (◊ RURITANIA), comprising *Westmark* (**1981**), *The Kestrel* (**1982**) and *The Beggar Queen* (**1984**); the **Vesper Holly** YA novels of an alternative Victorian era, comprising *The Illyrian Adventure* (**1986**), *The El Dorado Adventure* (**1987**), *The Drackenberg Adventure* (**1988**), *The Jedera Adventure* (**1989**) and *The Philadelphia Adventure* (**1990**); *The Remarkable Journey of Prince Jen* (**1991**), an ORIENTAL FANTASY; *The House Gobbaleen* (graph **1995**), for younger children; *The Arkadians* (**1995**), set in an ALTERNATE-WORLD ancient Greece.

ALEXANDER, MARC (ELWARD) (1929-   ) New Zealand writer who produces HORROR as by Mark Ronson and who began to publish fantasy with the unremarkable *The Mist Lizard* (**1977**). His work of greatest interest is the **Wells of Ythan** sequence – *Ancient Dreams* (**1988**), *Magic Casements* (**1989**), *Shadow's Realm* (**1991**) and *Enchantment's End* (**1992**) – set in a FANTASYLAND long ago subject to a THINNING when Princess Livia (whose UNDERLIER is SLEEPING BEAUTY) is cast into an enchanted sleep. The protagonists, often helped on their way by PLOT COUPONS (like The Esav, through which it is possible – using other coupons *en passant* – to locate the slumbering Livia), variously QUEST for the princess in order that the HEALING may begin; as they proceed, they tend to re-enact traditional tales in a series of linked plots. [JC]

Other works: *Not After Nightfall: Thirteen Stories to Haunt Your Dreams* (coll **1985**).

As Mark Ronson: *Bloodthirst* (**1979**); *Ghoul* (**1980**); *Ogre* (**1980**); *Plague Pit* (**1981**); *Grimalkin's Tales: Strange and Wonderful Cat Stories* (coll **1983**) with Stella Whitelaw and Judy Gardiner; *The Dark Domain* (**1984**); *Whispering Corner* (**1989**).

ALF MOVIES ◊ W.A. DARLINGTON.

ALI BABA ◊ ARABIAN FANTASY.

ALICE UK movie (**1965** tvm). ◊ ALICE IN WONDERLAND (*1951*).

ALICE (ot *Neco z Alenky*) Swiss/UK/West German live-action/stop-motion-ANIMATED MOVIE (*1988*). Condor/Film Four International/Hessischer Rundfunk. **Pr** Peter-Christian Fueter. **Exec pr** Keith Griffiths, Michael Havas. **Dir** Jan Svankmajer. **Anim dir** Bedřich Glaser. **Screenplay** Svankmajer, trans Alice Turner, Gerry Turner. **Based on** *Alice's Adventures in Wonderland* (**1865**) by Lewis CARROLL. **Starring** Kristýna Kohoutová (Alice). **Voice actors** Kohoutová (Alice), Camilla Power (Alice in English). 85 mins. Colour.

In many ways *A* is a very faithful adaptation of Carroll's

book – depicting all the important scenes, and representing accurately that reading which interprets the book as a nightmare – yet *A* has an affect that is entirely its own: it is *Alice's Adventures in Wonderland* for grown-ups (◊ WONDERLAND). There is no dialogue, only Alice's occasional narration; otherwise the soundtrack is composed of exaggerated gratings, scratchings, etc. The White Rabbit is initially a glass-cased stuffed rabbit in Alice's bedroom – leaking sawdust, it must frequently re-stuff itself – and vanishes not down a hole but into the drawer of a desk set alone in a windswept waste; the images of the desk and of Alice's ripping away the knob as she tries to open the drawer recur obsessively throughout the movie; others include scissors, clocks, keys and locks, cakes . . . The most nightmarish sequence is the Mad Hatter's Tea Party, in which the insanity is played for real to produce what is in effect a PSYCHOLOGICAL THRILLER in miniature. Everywhere domestic objects and Alice's TOYS are deployed; e.g., the Caterpillar is a living sock perched atop a darning-mushroom, while Alice, on shrinking, becomes one of her own DOLLS (on one occasion the child-Alice claws her way out of the doll, as if being SKINNED). When Alice wakes in her bedroom she is surrounded by the scattered ingredients of her DREAM – but the rabbit's case is missing, and the rabbit itself gone. Investigating, she finds a vertical drawer in the case's base; opening it, she discovers a pair of scissors . . .   [JG]

ALICE IN WONDERLAND US movie (*1933*). ◊ ALICE IN WONDERLAND (*1951*).

ALICE IN WONDERLAND US ANIMATED MOVIE (*1951*). Disney. **Pr sv** Ben Sharpsteen. **Dir** Clyde Geronimi, Wilfred Jackson, Hamilton Luske. **Special processes** Ub IWERKS. **Based on** *Alice's Adventures in Wonderland* (*1865*) by Lewis CARROLL. **Voice actors** Kathryn Beaumont (Alice), Jerry Colonna (March Hare), Verna Felton (Queen of Hearts), Sterling Holloway (Cheshire Cat), Joseph Kearns (Doorknob), Jim Macdonald (Dormouse), J. Pat O'Malley (Walrus, Carpenter, Tweedledum, Tweedledee), Bill Thompson (White Rabbit, Dodo), Ed Wynn (Mad Hatter). 75 mins. Colour.

Cast in the form of a series of DREAM incidents, most based on Carroll's book, *AIW* is among the least-liked of the DISNEY animated features. The mismatch between Disney's style and Carroll's might have been rendered less jarring had Walt DISNEY not attempted to ride roughshod over the original (the introduction of a major new character, a wacky Doorknob, is disastrous), chosen to regard Carroll's book as merely a children's story, and selected some of the voices with astonishing insensitivity. Isolated moments of fine fantasy are drowned in the general mess. As with FANTASIA (*1940*), *AIW*'s SURREALISM led to a surge in its popularity during the 1960s, when hippies interpreted the Caterpillar and its toadstool in terms of HALLUCINATION-inducing recreational DRUGS.

Decades earlier, Walt Disney and Iwerks had borrowed Carroll's notion for almost their first excursion into ANIMATED MOVIES: the series of 57 b/w shorts called the **Alice Comedies** (*1924-7*; *Alice's Wonderland* made 1923 but not properly released). In these a live-action girl ventured into an animated land of fantasy.

Other movies based on Carroll's novel include: *Alice in Wonderland* (*1933*), a live-action version in which the cast wore MASKS based on the TENNIEL illustrations; *Alice in Wonderland* (*1951*) dir Dallas Bower and Lou Bunin, a live-action/puppet version that was completely eclipsed by the Disney feature, but which is now becoming more widely appreciated thanks to video release; *The Adventures of Alice* (*1960* tvm), a live-action version; *Alice* (*1965* tvm), which was Dennis Potter's first interpretation and was concerned more with the relationship between Carroll/Dodgson and Alice Liddell (◊ DREAMCHILD [*1985*]); *Alice in Wonderland* (*1966* tvm) dir Jonathan Miller, a hauntingly effective live-action version; *Alice's Adventures in Wonderland* (*1972*), a rather dull live-action version with a plethora of star actors; *Alice in Wonderland* (*1985* tvm), an epic based on both the **Alice** books; and ALICE (*1988*). *Alice in Wonderland* (*1976*) is non-fantasy soft porn; *Through the Looking Glass* (*1976*) is also soft porn but is a fantasy, although unconnected with Carroll's (◊ PORNOGRAPHIC FANTASY MOVIES).   [JG]

See also: ALICE THROUGH THE LOOKING GLASS (*1960* tvm).

ALICE IN WONDERLAND (vt *Alice au Pays de Merveilles*) French/US/UK live-action/puppet movie (*1951*). ◊ ALICE IN WONDERLAND (*1951*).

ALICE IN WONDERLAND UK movie (*1966* tvm). ◊ ALICE IN WONDERLAND (*1951*).

ALICE IN WONDERLAND US movie (*1985* tvm). ◊ ALICE IN WONDERLAND (*1951*).

ALICE'S ADVENTURES IN WONDERLAND UK movie (*1972*). ◊ ALICE IN WONDERLAND (*1951*).

ALICE THROUGH THE LOOKING GLASS US movie (*1960* tvm). Alwynn/Dum & Dee. **Pr** Alan Handley, Bob Wynn. **Dir** Handley. **Music** Moose Charlap. **Lyrics** Elsie Simmons. **Screenplay** Albert Simmons. **Based loosely on** *Alice's Adventures in Wonderland* (*1865*) and *Through the Looking-glass* (*1871*) by Lewis CARROLL, with elements drawn from *The Wonderful Wizard of Oz* (*1900*) by L. Frank BAUM. **Starring** Roy Castle (Lester the Jester), Robert Coote (Red King), Jimmy Durante (Humpty Dumpty), Nanette Fabray (White Queen), Ricardo Montalban (White King), Agnes Moorehead (Red Queen), Jack Palance (Jabberwock), Judy Rolin (Alice), Smothers Brothers (Tweedledum and Tweedledee). *c*90 mins. Colour.

Having little of interest except that it exists, this tvm, performed as a stage musical, takes a few of the bits of the two **Alice** books and throws them together with elements from OZ – notably the Yellow Brick Road. Alice climbs through a MIRROR, finds a SECONDARY WORLD partly inhabited by CHESS pieces, is threatened by the Jabberwock, becomes a queen, and at last discovers everything has been a DREAM.
                                                [JG]

ALIEN MASSACRE vt of *The Wizard of Mars* (*1964*). ◊ *The* WIZARD OF OZ (*1939*).

ALL DOGS GO TO HEAVEN Irish ANIMATED MOVIE (*1989*). ◊ AFTERLIFE; Don BLUTH.

ALLEGORY A story is an allegory when, as Northrop FRYE puts it, "the events of a narrative obviously and continuously refer to another simultaneous structure of events or ideas". It differs from symbolism (◊ SYMBOLS AND SYMBOLISM) in that it defines the structure of the entire text. In allegory, the different levels of meaning relate to one another in a continuous and architectural manner. Allegory is perhaps unpopular in 20th-century literature and criticism because it posits a relative fixity of interaction; i.e., allegory insists not only that literature *means* something (means, in fact, something *else*) but that that meaning can be determined in advance. Allegory is a form of instruction.

Given the prevalence of allegory in the literature of previous centuries, it is unsurprising that many of the TAPROOT TEXTS which have fertilized the field of fantasy are allegories. Much of the BIBLE has been interpreted as typological allegory with, for instance, the Queen of Sheba in *The Song of Solomon* being a "type" of the Christian Church. The Myth of the Cave in PLATO's *Republic* (written *c*360BC) is an allegory. Other examples include: *Scipio's Dream* (written *c*50BC) by Marcus Tullius Cicero (106-43BC); OVID's *Art of Love* (written 3BC); the CUPID AND PSYCHE fable inserted into Lucius APULEIUS's *The Golden Ass* (written *c*AD155); the part of *Romance of the Rose* (written *c*1230) that was by Guillaume de Lorris; DANTE's *The Divine Comedy* (written *c*1320); *Jerusalem Delivered* (**1581**) by Torquato Tasso (1544-1595); Edmund SPENSER's *The Faerie Queene* (**1589**); John BUNYAN's *The Pilgrim's Progress* (**1678**); Jonathan SWIFT's *Gulliver's Travels* (**1726**); and various prophetic books by William BLAKE.

As the 19th century progressed, the form became less frequent, though short stories by writers like Hans Christian ANDERSEN and Nathaniel HAWTHORNE were often allegories; but, as the STORY told in the typical fantasy text was increasingly conveyed with a sense of coherent dense REALITY, fantasies were less and less likely to be mistaken for allegories in sheep's clothing. Even so, Edward BULWER-LYTTON's *Zanoni* (**1842**) is an allegory, as is Jean INGELOW's *Mopsa the Fairy* (**1869**). Although George MACDONALD's SECONDARY WORLDS verge on allegory, they do so with some unease, with some sense that what initially has all the timbre of an allegory straight from Christian homiletics tends to turn into a congeries of loaded symbols. PARODIES of Lewis CARROLL's **Alice** books set in WONDERLANDS tend to use the allegorical impulse to assault a wide range of targets.

In the 20th century, allegory is even less easy to identify. Walter DE LA MARE's *Henry Brocken: His Travels and Adventures in the Rich, Strange, Scarce-Imaginable Regions of Romance* (**1904**), David LINDSAY's *A Voyage to Arcturus* (**1920**), David GARNETT's *Lady into Fox* (**1922** chap) and Rex WARNER's *The Aerodrome* (**1941**) are UK examples; US examples are rare. Though C.S. LEWIS wrote allegories – like *The Pilgrim's Regress* (**1933**) and *The Great Divorce* (**1945** chap) – he argued that his **Ransom** and **Narnia** sequences were not allegorical, even though CHRISTIAN FANTASY as a whole tends to slip into allegorical exegetics. J.R.R. TOLKIEN's *The Lord of the Rings* (**1954-5**) is often treated as allegorical, against his denials; nevertheless, in "Concerning Tolkien" (1991) Isaac ASIMOV suggests that in the book the RING of power is allegorical of "industrial technology".

POSTHUMOUS FANTASIES and BEAST FABLES are often allegories. Almost every example of the first is by definition so, and in the second allegorical readings have an inevitable tendency to harden into the full structure, as in Eden PHILLPOTTS's *The Apes* (**1927**). George ORWELL's *Animal Farm* (**1945** chap) is generally understood as an allegory; and Richard ADAMS's *Watership Down* (**1972**) has, less convincingly, been offered as another example. [JC]

ALLEGRO NON TROPPO Italian live-action/ANIMATED MOVIE (1976). Essential/Specialty. **Pr** Bruno Bozzetto. **Dir** Bozzetto. **Spfx** Luciano Marzetti. **Anim dir** Bozzetto. **Screenplay** Bozzetto, Guido Manuli, Maurizio NICHETTI. **Starring** Nestor Garay (Conductor), Maria Luisa Giovannini (Stagehand), Maurizio Micheli (MC), Nichetti (Animator). 74 mins, restored to 85 mins. Colour and tinted b/w.

The live-action sequences in this PARODY of DISNEY's FANTASIA (*1940*) are of little interest, but the animated sequences are exquisite: *ANT* is arguably superior to its original.

In the *Prelude à l'Après-midi d'un Faun* sequence (music by Debussy), set in an Arcadian paradise, an ageing SATYR attempts unsuccessfully to relive his youth among the nymphs. The sequence accompanying Dvořák's *Slavonic Dance #7* portrays the metronomic rise of militarism through the human need to imitate. Ravel's *Bolero* accompanies a truly astonishing piece of animation, which both parodies *Fantasia*'s *Rite of Spring* sequence and wildly surpasses it, depicting the marching, rhythmic evolution of life from its lowly origins as a splash spilt from a Coke bottle. The sequence corresponding to Sibelius's *Valse Triste* is tragic: a half-starved stray cat, haunting a derelict house, sees there the GHOSTS of its onetime occupants before fading to become a ghost itself. Vivaldi's *Concerto in C Minor* is matched to a simple but very funny tale of an epicurean bee trying to pollinate in a field terrorized (in bee terms) by a pair of roiling lovers. And the sequence set to an extract from Stravinsky's *Firebird* retells the ADAM AND EVE legend, with the Serpent the one to eat EDEN's apple: his indigestion from so doing brings him apocalyptic VISIONS of a technological future. The overt subtext of this final sequence is that, although the Serpent is originally evil, his is mischievous EVIL, and thus innocent: the eating of the apple destroys that innocence. This revisionist view of the nature of primordial evil is among the most interesting in extant fantasy CINEMA. [JG]

ALLEN, F.M. Pseudonym of Irish writer Edmund Downey (1856-1937), author of some sf (◊ *SFE*) and fantasy, the latter including: *A House of Tears* (**1888** US) as by ED; *Brayhard: The Strange Adventures of One Ass and Seven Champions* (**1890**), which is humorous; and *The Little Green Man* (**1895**), in which an exported leprechaun starts the Gold Rush in California. FMA was an author of whimsy, sometimes with bite. [JC]

ALLEN, (CHARLES) GRANT (BLAIRFINDIE) (1848-1899) Canadian-born UK writer, scientist and educationalist. After a thwarted career in academia in the UK and Jamaica, GA turned in 1876 to writing, mostly on scientific and philosophical subjects. Seeking a more profitable sideline, he began to sell short fiction under the name J. Arbuthnot Wilson (he later used the pseudonyms Cecil Power, Olive Pratt Rayner and Martin Leach Warborough). His early stories were often thinly disguised lectures, and this professorial tone remained evident in much of his work. GA's sf and SUPERNATURAL FICTION, although a small part of his output, was significant. An evolutionist and emancipationist, he held strong views on the moral and social matters of the day, and he liked to challenge taboos, as demonstrated in his feminist novel *The Woman Who Did* (**1895**) (◊ FEMINISM). He used sf as a vehicle to comment upon Victorian society in *The British Barbarians* (**1895**), and challenged the concept of GOD in *The Evolution of the Idea of God* (**1897**). He sought to rationalize some aspects of the supernatural through his knowledge of FOLKLORE, ANTHROPOLOGY and RELIGION, and thereby helped the supernatural story advance into the 20th century.

His West Indian experience provided him with material for an early VOODOO story, "The Beckoning Hand" (1884), and other tales feature REVENANTS from CRYPTS, MUMMIES' TOMBS or burial mounds. His WEIRD FICTION is found alongside nonfantastic stories in *Strange Stories* (coll **1884**), *The Beckoning Hand* (coll **1887**), *Ivan Greet's Masterpiece* (coll **1893**), *The Desire of the Eyes* (coll **1895**) and – his selection of his best – *Twelve Tales* (coll **1899**). Quasi-supernaturalism creeps into a few of his novels, especially *Kalee's Shrine* (**1886**; vt *The Indian Mystery* 1902 US) with May Cotes, about the Thuggee cult and MESMERISM. [MA]

Other works: *The Devil's Die* (**1888**); *The White Man's Foot* (**1888**); *The Great Taboo* (**1890**); *The Tents of Shem* (coll **1891**); *Michael's Crag* (**1893**).

ALLENDE, ISABEL (1942-   ) South American writer – born in Peru, raised in Chile, exiled to Venezuela after the 1973 assassination of her uncle, Salvadore Allende (1908-1973), and eventually resident in the USA – who began writing plays and sketches around 1964 but came to international notice only with *La Casa de los espíritus* (**1982** Spain; trans Magda Bogin as *The House of the Spirits* **1985** US), a magic-realist (◊ MAGIC REALISM) confabulation of political SATIRE and fantasy in which the secular tragedies of contemporary Chilean life are illuminated against the story of a family with TALENTS. In their complex richness these talents represent the multifariousness of a threatened culture which ultimately suffers profound damage as the 20th century progresses. Critics have likened the book to Gabriel GARCIA MARQUÉZ's *Cien anos de soledad* (**1967**; trans as *One Hundred Years of Solitude* **1970** US). IA's third novel, *Eva Luna* (**1987**; trans Margaret Sayers Peden **1988** US), is structured as a series of stories that its protagonist – a SCHEHEREZADE-like figure – tells (self-creatingly) about herself. [JC]

ALL HALLOW'S EVE ◊ HALLOWE'EN.

ALLINGHAM, WILLIAM (1824-1889) UK writer. ◊ Richard DOYLE; Andrew LANG.

ALL OF ME US movie (**1984**). Kings Road. **Pr** Stephen Friedman. **Dir** Carl Reiner. **Spfx** Bruce Steinhemmer, Optical House. **Screenplay** Henry Olek, Phil Alden Robinson. **Based on** *Me Two* by Ed Davis. **Starring** Jason Bernard (Tyrone Wattell), Eric Christmas (Fred Hoskins), Dana Elcar (Burton Schuyler), Richard Libertini (Prahka Lasa), Steve Martin (Roger Cobb), Madolyn Smith (Peggy Schuyler), Victoria Tennant (Terry Hoskins), Lily Tomlin (Edwina Cutwater). 93 mins. Colour.

Wealthy dying heiress Cutwater hires oriental mystic Prahka Lasa to arrange the transmigration of her SOUL into the body of the stableman's beautiful daughter (and crook) Terry, the latter having agreed that her own soul will be liberated to "become one with the Universe". Hardbitten civil lawyer and part-time jazz guitarist Cobb is called in to help draw up the will leaving Cutwater's millions to "Terry". He protests, but Cutwater dies suddenly and the bowl in which Lasa has temporarily housed Cutwater's soul is dropped on Cobb's head, so that half his body is possessed (◊ POSSESSION) by Cutwater's independent soul. At last all is sorted satisfactorily.

AOM has many weaknesses, notably Libertini's patronizing "goodness gracious me" turn as Prahka Lasa. It also suffers a crisis much like that of the possessed Cobb: presumably intended to be a Steve Martin vehicle, the movie is stolen by Tomlin (and, among the lesser parts, Bernard). Nevertheless, this is Martin's best movie. The moral – Cobb realizes his life as a lawyer has been every bit as sterile as Cutwater's life chained to a wheelchair and the family millions – while not novel, is lovingly handled; and the movie as a whole leaves a good taste. [JG]

See also: IDENTITY EXCHANGE.

ALLONBY, EDITH (?   -before 1905) UK writer whose anonymously published **Lucifram** novels – *Jewel Sowers* (**1903**) and *Marigold* (**1905**) – are mild SATIRES set on the imaginary planet Lucifram, a name whose derivation from Lucifer is obvious. *The Fulfillment* (**1905**) is a POSTHUMOUS FANTASY set in HEAVEN and HELL. [JC]

ALL SOUL'S DAY ◊ HALLOWE'EN.

ALL-STORY ◊ MAGAZINES.

ALL THAT MONEY CAN BUY vt of *The* DEVIL AND DANIEL WEBSTER (*1941*).

ALL THE YEAR ROUND ◊ MAGAZINES.

ALMOST AN ANGEL US movie (*1990*). Paramount/ Ironbark. **Pr** John Cornell. **Exec pr** Paul Hogan. **Dir** Cornell. **Screenplay** Hogan. **Starring** Charlton Heston (uncredited; God), Hogan (Terry Dean), Elias Koteas (Steve Garner), Linda Kozlowski (Rose Garner), Doreen Lang (Mrs Garner). 95 mins. Colour.

Small-time criminal Terry is injured saving a child's life; unconscious in hospital, he DREAMS he is dead but GOD is returning him to Earth as an ANGEL. Coincidences conspire to convince him (but not us) this is more than a dream, and he amateurishly performs good deeds, some involving faked MIRACLES. Yet in the movie's denouement we discover this is the reverse of a RATIONALIZED FANTASY: he does indeed have supernatural attributes and abilities. *AAA* is a slight, wry, sentimental comedy, but in its modest way an extremely effective – and affecting – piece of fantasy. [JG]

ALTERITY In some critical discourses on the nature of FANTASY, this term is used to designate the sense of "otherness" consequent upon the fact that fantasy texts – as Gary K. Wolfe makes clear in *Critical Terms for Science Fiction and Fantasy* (**1986**) – violate the "ground rules" of the mundane world, becoming "dependent on the internal consistency of their own ground rules". But the term describes that otherness of the text from a cognitive remove; it does not describe the experience of reading a fantasy text. In a properly engrossing example, alterity may be sensed as a threshold experience, but not as an ongoing response to innovations within that text: for those innovations will be internally consistent, and when understood will be *recognized*. In SUPERNATURAL FICTION there is no alterity, because those elements in the tale which violate the ground rules of mundane REALITY remain contingent upon those rules. [JC]

ALTERNATE REALITIES There can be confusion in the usages of the terms "alternate worlds" and "alternate realities"; in essence, the latter term embraces the former, which is more a product of sf than of fantasy. An ALTERNATE WORLD may be looked upon as a form of AR, but there are other alternatives to our mundane REALITY than parallel Earths, many rooted in mystical ideas, like those of THEOSOPHY or the notion of the ASTRAL PLANE, rather than in science. A classic example of an AR is the Australian Aboriginal DREAMTIME, which not only was a period in the past but is still contiguous and isomorphic with the mundane world.

While certainly the PLAYGROUND of quantum physics has been rich territory for sf writers in search of alternate-world ideas, fantasy writers have played in it too, producing tales that draw on the alternate-world imagery yet envisage a quite different relationship between the other reality (or realities) and our own. For example, in Gene WOLFE's *There Are Doors* (**1988**), what can be regarded in the most superficial sense as an alternate world – a grim place into and from which the narrator stumbles repeatedly and uncontrollably – is in fact a place without any real past; whether it is a reality born of the narrator's deteriorating PERCEPTION of the world about him, or whether it has come into existence in response to the demands of his bitter unhappiness with his life – or even whether it has always existed, history-less – is a matter left open to the reader's interpretation; what is certain is that it is a reality other than our own which cannot be related to it by any simple, any quasi-scientific, explanation. The tale thus uses, in short, the imagery but not the mechanism, and thus conveys a completely different impression of what is actually the circumstance.

Most tales of ARs are, for obvious reasons, CROSS-HATCHES: most often the focus of the tale is some form of interaction between the alternate reality and our own, even if only to the extent that a protagonist may skip between the two. This is a clear difference between alternate-reality and alternate-world stories, in that many distinguished examples of the latter exist in which there is no such cross-linkage – as in *SS-GB* (**1978**) by Len Deighton and *Fatherland* (**1992**) by Robert Harris, where the Nazis won WORLD WAR II, or *Promises to Keep* (**1988**) by George Bernau, where Kennedy (or a President very like him) survives the Dallas assassination. Otherwise, an AR that is quite unlike our own and has no cross-linkage with it is almost certainly not an AR at all but a SECONDARY WORLD.

ARs pervade many novels whose concerns are about other things besides, as in Mark HELPRIN's *Winter's Tale* (**1983**). Yet some focus on the topic. The protagonist of Lisa TUTTLE's *Lost Futures* (**1992**) strays through a diversity of existences, enduring all sorts of "wrong" lives, before discovering that the true reality – the *real* world – is not the one she seemed to start from, which was ours. In *Dream Science* (**1990**) by Thomas PALMER ARs come about because reality is a far more complex construct than we perceive: the fullness of reality is made up of a Russian-doll-like set of concentric shells, each of which is a reality that has the same status as our own. In *Daughter of the Bear King* by Eleanor ARNASON a woman discovers that, at first in her DREAMS and then in her waking life, she dwells in an entrancing AR. In Ken GRIMWOOD's *Replay* (**1986**) the diverse realities, each sparked by changes of history, are sequential rather than concurrent, so that the tale thereby once more depends on a view of reality (specifically, in this instance, the nature of TIME) that is far more complex than what we perceive.

One special category of AR tale is where the other reality is a *created* one; that is, a character in one reality has created a SECONDARY WORLD of some type, and that secondary world comes to have a reality of its own, and possibly also an accompanying history that stretches back further than the moment of creation. Paul Kearney's *The Way to Babylon* (**1992**) is an example of a basic version of this theme: a fantasy writer discovers the existence, in some skewed direction, of the FANTASYLAND which he has created as the setting for his novels. A similar effect is brought about by a haunted room on a failed detective-story writer in *The Further Adventures of Captain Gregory Dangerfield* (**1973**) by Jeremy Lloyd: each different adventure-land created by the room's previous resident, a more successful writer, becomes, for a while, a reality experienced by the protagonist. Andrew M. Greeley's *God Game* (**1986**) sophisticates such ideas a little: the protagonist has created his world through the medium of a marvellously clever interactive computer GAME, whose semi-free-willed denizens recognize their own origins and his – for he is, though unworshipped, quite literally their GOD. In Ralph BAKSHI's movie COOL WORLD (*1992*) the created reality is the somewhat arbitrarily con-

structed world of a COMICS artist; its previous history not only predates the act of creation but has affected our world in that earlier time. John GRANT's *The World* (**1992**) takes a metafictional view: the created reality in question is the setting of his earlier novel, *Albion* (**1991**), and has come to have existence through the publication of that book. The reality within PICTURES (which serve as a PORTAL to it) has been a frequent source of fascination: "The Hall Bedroom" (1905) by Mary Wilkins FREEMAN is one example; in James Branch CABELL's "The Delta of Radegonde" (1924) a lover enters by MAGIC a picture of the long-dead Queen Radegonde, and is welcomed by her; and entering pictures is the premise of Ian WATSON's anthology *Pictures at an Exhibition* (anth **1981**), which thus features AR stories like David LANGFORD's "Transcends All Wit", set in Dürer's etching *Melencolia*; the protagonist of Stephen KING's *Rose Madder* (**1995**) discovers an AR (and herself) through venturing into a picture. Movies like *The* PURPLE ROSE OF CAIRO (*1984*) and LAST ACTION HERO (*1993*) credit the world behind the cinema screen as being a created AR. ADVENTURES OF BARON MUNCHAUSEN (*1989*) provides examples of scenes presented on the theatrical stage becoming full-blown realities, at least for a while.

There are also, of course, innumerable tales – from Lewis CARROLL's *Through the Looking-glass* (**1871**) onwards – of the worlds that exist on the far side of MIRRORS. These can be viewed as secondary-world stories except (normally) where the mirror-reality begins to interact with our own. Also, almost all POSTHUMOUS FANTASIES are by definition tales of ARs, in that they posit the existence of some unseen plane of reality. But by far the majority of AR stories do not require physiological or physical PORTALS, like death or the surface of a mirror: if there are portals, they generally lie in the mind.   [JG]

ALTERNATE WORLDS Clearly any SECONDARY WORLD or OTHERWORLD or WONDERLAND can be thought of as an alternate world. By definition, none of these regions are of *this* world, nor are they *arguable* extrapolations of the history of this world. They are *other*. Upon entering them, readers experience a clear sense of ALTERITY. The realities – the ground rules – that govern them are alternate to those of the real world. At the same time, to treat any fantasy set wholly or partially in such a world is effectively to treat most fantasy and the AW story as synonymous. In this encyclopedia the term is used in a more restricted sense, one that works in tandem with Brian STABLEFORD's definition of it in an sf context: "An alternate world" – as he phrases it in *SFE* – "is an account of Earth as it might have become in consequence of some hypothetical alteration in history." In fantasy, an alternate world is an account of our world as it might otherwise have been.

The difference is simple, but crucial. If a story presents the alteration of some specific event as a premise from which to argue a new version of history – favourite "branch points" include the victory of the Spanish Armada in 1588, the victory of the South in the AMERICAN CIVIL WAR, and HITLER WINS scenarios – then that story is likely to be sf. If, however, a story presents a different version of the history of Earth *without arguing the difference* – favourite differences include the significant, history-changing presence of MAGIC, or of actively participating GODS, or of ATLANTIS or other LOST LANDS, or of CROSSHATCHES with OTHERWORLDS – then that story is likely to be fantasy.

There is one obvious exception to this pattern of distinctions: the story in which a specific *but not arguable* event occurs to change the course of history. An example is the survival of Count DRACULA, and his subsequent marriage to Queen Victoria, in Kim NEWMAN's *Anno Dracula* (1992). As VAMPIRES do not exist, the history-changing triumph of the Count does not constitute an alteration in the outcome of a specific historical event. It is, rather, an *intervention* into world history, and because it violates history a tale like Newman's is more likely to be a SUPERNATURAL FICTION than a fantasy proper.

FANTASIES OF HISTORY – in which, typically, SECRET MASTERS manage the world – might seem to constitute a further exception, but in fact do not. The fantasy of history does not present an AW: it argues that it is *this* world which is different from what we think, and that the difference has been concealed from us. To treat such stories as AW tales is to deprive them of their paranoid bite.

There is a subtle distinction between the AW tale and the story of ALTERNATE REALITIES. The AW tale depicts a world that is related to our own, but different for one reason or another. The alternate-reality tale conceives that there are other REALITIES which may or may not be accessible from or interact with our own (although they usually are and/or do, because that is generally the crux of the tale) but which have no historical dependence on it and may be totally dissimilar from it. Such realities may occur because "total reality" is a far more complex construct than we are aware, as in Thomas PALMER's *Dream Science* (1990), or they may be realities of the mind rather than having physical existence (although who is to decide their physical status?), as in Gene WOLFE's *There Are Doors* (1988). Either way, such stories are likely to involve PERCEPTION, whereas in AW tales there is no mysterious other layer of existence to be unravelled, and characters' perceptions of the hovering presence of an AW, if they are perceiving both this world and the AW simultaneously, are likely to be definable as TROMPE L'OEIL.

One further type of AW story (although it can be read also as an ALTERNATE-REALITY story) can be exemplified by Ken GRIMWOOD's *Replay* (1986), in which the central characters re-experience the same chunk of history several times over, changing it through their actions both in detail or in the large. The rationale that emerges is that TIME itself has "knots", and that they are the only ones who can perceive this particular "knot cluster". The "knotting" of time has thus created a series of AWs, but these are, if properly perceived, *sequential* rather than *parallel*, as in the more usual AW story. (The movie GROUNDHOG DAY [*1993*] depicts a similar recycling through time, but it is not clear if this is due to time "repeating" or to the protagonist in effect timeslipping.)

AW fantasy texts include: Joan AIKEN's **Willoughby Chase** sequence; Brian W. ALDISS's *The Malacia Tapestry* (1976); Poul ANDERSON's *Operation Chaos* (coll of linked stories 1971) and *A Midsummer Tempest* (1974); Avram DAVIDSON's **Vergil** sequence (1969-87), **Peregrine** sequence (1971-81) and *The Enquiries of Dr Eszterhazy* (coll 1978); Peter DICKINSON's *The Blue Hawk* (1976) and *King and Joker* (1976), with the latter's sequel *Skeleton-in-Waiting* (1989); Robert A. HEINLEIN's "Magic, Inc" (1942 *Unknown* as "The Devil Makes the Law"); Robert E. HOWARD's **Conan** books (in book form from 1950); Diana Wynne JONES's **Chrestomanci** sequence (1977-88); Katherine KURTZ's **Deryni** books (from 1970); Ursula K. LE GUIN's *Orsinian Tales* (coll 1976) and *Malafrena* (1979); *Last Letters from Hav* (1985) by Jan Morris (1926-   ); *A Time to*

Choose (1973) by Richard Parker (1915-   ), whose protagonists inhabit two versions of the world; Milorad PAVIĆ's *Hazarski recnik* (1988; trans as *Dictionary of the Khazars* 1988); and Roger ZELAZNY's **Amber** sequence (from 1970). [JC]

ALVARENGA, LUCAS JOSÉ D' (1768-1831) Brazilian writer. ◊ BRAZIL.

ALWAYS US movie (*1989*). Universal/United Artists/Amblin. **Pr** Kathleen Kennedy, Frank Marshall, Steven SPIELBERG. **Dir** Spielberg. **Vfx sv** Bruce Nicholson. **Screenplay** Jerry Belson. **Based on** screenplay by Dalton Trumbo for *A Guy Named Joe* (*1944*). **Starring** Richard Dreyfuss (Pete), John Goodman (Al), Audrey Hepburn (Hap), Holly Hunter (Dorinda), Brad Johnson (Ted). 123 mins. Colour.

A POSTHUMOUS FANTASY set among firefighting pilots. Pete dies in a crash and finds himself in a sort of New Age AFTERLIFE where the ANGEL Hap explains to him that new pilots are always trained by the GHOSTS of dead ones. He is assigned to train Ted, but finds Ted is also taking over the affections of Pete's bereaved lover Dorinda. Only when Pete realizes he must release his emotional grip on Dorinda is he fit to ascend to a higher spiritual plane. The tale is slight and trite; the movie is long. [JG]

AMADO, JORGE (1912-   ) Brazilian writer who from his first novel, *O País do Carnaval* ["Carnival Country"] (1931), demonstrated a leftwing response to the inequities of Brazilian life and a capacity to create what read (at least to others) as myths for the people; he is the best-known Brazilian writer of the 20th century. Many of his books are naturalistic, but those most popular in translation combine TALL-TALE exuberance, a sophisticated use of folk material, sexual gaieties, elaborate double- and treble-layered plots, and a rhetorical insistence on the life-affirming virtues of the common people. Translated titles include: *Gabriela, Cravo e Canela* (1958; trans James L. Taylor and William Grossman as *Gabriela: Close and Cinnamon* 1962 US); *Os Velhos Marinheiros* (coll 1961; in 2 vols, vol 1 trans Harriet de Onís as *Home is the Sailor* 1964 US, vol 2 trans Barbara Shelby as *The Two Deaths of Quincas Wateryell* 1965 US); *Os Pastores da Noite* (1964; trans Harriet de Onís as *Shepherds of the Night* 1966 US); *Dona Flor e Seus Dois Maridos* (1966; trans Harriet de Onís as *Dona Flor and her Two Husbands* 1979 US), the first husband being a nagging GHOST for most of the tale; and *O Gato Malhado e a Andorinha Sinhá* (1976; trans Barbara Shelby as *The Swallow and the Tomcat* 1982 chap), a BEAST FABLE. [JC]

AMAZING ADVENTURES OF BARON MUNCH-HAUSEN, THE vt of *Münchhausen* (*1943*). ◊ MUNCHHAUSEN.

AMAZING SPIDER-MAN, THE US tv series (1978-9). Danchuk Productions/Charles Fries Productions/CBS. **Pr** Robert Janes, Ron Satlof, Lionel E. Siegel. **Exec pr** Charles Fries, Daniel R. Goodman. **Dir** Tom Blank and many others. **Writers** Alvin Boretz and many others. **Based on** the COMIC-book characters created by Stan LEE. **Starring** Ellen Bry (Julie), Chip Fields (Rita Conway), Nicholas Hammond (Peter Parker/Spider-Man), Michael Pataki (Captain Barbera), Robert F. Simon (J. Jonah Jameson), Irene Tedrow (Aunt May Parker). 12 60min episodes. Colour.

Spider-Man was the first comic-book SUPERHERO to have *problems*. While the comics began with Peter Parker (bitten by a radioactive SPIDER and thus himself possessed of arachnoid

TALENTS) in high school, the tv version started with him as a graduate student at Empire State University. His job as a photographer at the *Daily Bugle* brings him into contact with a number of crimes, where his special abilities as Spider-Man come into play. Chief among these are superstrength, for he is blessed with "the proportional strength of a spider", and "spider sense", an ESP-like TALENT that lets him detect danger. Able to stick to any surface, Spider-Man swings across the city using webs created by special web-shooters he has invented. This impressive arsenal is brought to bear against a number of mundane criminals, but serves Spider-Man especially well when he has to battle a clone and an Egyptian CURSE. The series lasted only one season; several episodes were later edited into feature-length tvms. [BC] See also: SPIDER-MAN (*1977* tvm).

AMAZING STORIES (magazine) ◊ FANTASTIC; FANTASTIC ADVENTURES; MAGAZINES; *SFE*.

AMAZING STORIES US tv series (1985-7). Amblin Entertainment/Universal/NBC. **Pr** Steven SPIELBERG, David E. Vogel. **Exec pr** Spielberg. **Created by** Joshua Brand, John Falsey, Spielberg. **Dir** Joe Dante (◊ *SFE*), Danny DeVito, Clint Eastwood, Paul Michael Glaser, Tobe Hooper, Peter Hyams, Burt Reynolds, Martin Scorsese, Spielberg, Robert ZEMECKIS and many others. **Writers** Jack FINNEY, Richard MATHESON, Spielberg and many others. **Stars included** Loni Anderson, Milton Berle, Beau Bridges, David Carradine, Kevin Costner, Dom DeLuise, Charles Durning, Stan Freberg, Mark Hamill, Gregory Hines, Harvey Keitel, John Lithgow, Christopher Lloyd, Richard Masur, Bronson Pinchot, Charlie Sheen, Kiefer Sutherland, Patrick Swayze, M. Emmett Walsh, Ray Walston, Sam Waterson. 2 2hr episodes and 43 30min episodes. Colour.

Spielberg's amazing theatrical success led NBC to sign him up to create a new series for the network for 1984; in an unprecedented show of faith, they committed to two seasons before a single episode had been filmed. Spielberg, who outlined most of the stories for the series, set about assembling an impressive group of directors and performers for what many assumed would be one of tv's finest moments. They were quickly disappointed. But, while the talent involved was excellent, the stories generally proved rather predictable (the same charge was levelled at *The* NEW TWILIGHT ZONE [1985-8], which debuted the same year). For example, in the first, "Ghost Train", it was obvious that a house built on the route of a now-vanished railroad would soon have a spectral visitor. "The Mission", starring Kevin Costner, was a dramatic portrayal of a WWII bombing mission – until the end, which featured a very out-of-place cartoon effect. Other stories featured TIMESLIPS, a demonic RING, an actor in an sf movie who is mistaken for a real MUMMY, and a lonely man overly attached to a DOLL. While NBC continued to voice support for Spielberg, work was halted before the entire set had been completed.

Spielberg attempted to create a series based on "Family Dog", an animated episode from this series, only to discover that producing a weekly animated series was far more complex than originally thought. After several embarrassing delays *Family Dog* (1993) finally made it to the air, but it vanished after a month.

Two book adaptations were produced: *Steven Spielberg's Amazing Stories* * (coll **1986**) and *Volume II of Steven Spielberg's Amazing Stories* * (coll **1986**), both by Steven BAUER. [BC]

AMAZONS In Greek mythology, a matriarchal race of warrior women who worship the MOON. The right breasts of Amazon girls are removed at puberty, according to some accounts, so they can continue to draw bowstrings. They usually reside in the south of Turkey, as in *The Green Scamander* (**1933**) by Maude Meagher (?1895-?1977), or in South America, though they have been granted more romantic homelands like ATLANTIS in novels like Ivor Bannet's *The Amazons* (**1948**). They figure in HOMER's *Iliad* (*c***800**BC), and their ultimately tragic role in the Trojan War has inspired various works, including *The Green Scamander* and at least one OPERA, *Penthesilea* (performed 1927) by Othmar Schoek (1878-1957). In the 19th and early 20th centuries they are sometimes discovered in LOST-WORLD venues, as in *The Amazonian Republic: Recently Discovered in the Interior of Peru* (**1842**) by Timothy Savage, *A Parisian Sultana* (coll trans H. Mainwaring Dunstan **1879** UK) by Adolphe Belot (1829-1890), *An Amazonian Queen, or Nick Carter Becomes a Gladiator* (**1907** chap) by Frederic Van Rensselaer Dey (1865-1922) writing as as Chickering Carter, Harold Mercer's *Amazon Island* (**1933**), where the lost race of women is engaging in a UTOPIAN experiment, and *Tarzan and the Amazons* (**1945**; ◊ TARZAN MOVIES). The most famous 20th-century use of the image is of course WONDER WOMAN, created in 1941.

More recently, the figure of the Amazon has become common in SWORD AND SORCERY as an icon of female autonomy, appearing in series like Sharon GREEN's **Amazon Warrior** sequence (**1982-6**) and in singletons like Megan LINDHOLM's *Harpy's Flight* (**1983**) and Evangeline WALTON's *The Sword is Forged* (**1983**), the latter presenting, through the figure of Theseus, a patriarchal challenge to the Amazon in terms which allow some inspired debate. Anthologies include Marion Zimmer BRADLEY's *Free Amazons of Darkover* (anth **1985**) and Jessica Amanda SALMONSON's *Amazons!* (anth **1979**) and *Amazons II* (anth **1982**); Salmonson's guide to the subject, *The Encyclopedia of Amazons: Women Warriors from Antiquity to the Modern Era* (**1991**), is extremely thorough. The Amazon continues to be used as a model for feminist speculations (◊ FEMINISM), one of the more recent examples of the subgenre being *Amazon* (**1992**) by Barbara G. Walker (1930-   ), a TIMESLIP fantasy in which an ancient Amazon is befriended by a contemporary women named Diana, and they write a book together which is critical of the patriarchal USA. [JC]

AMERICAN CIVIL WAR In sf, the American Civil War is a central focus for alternate-history texts (◊ ALTERNATE WORLDS); most of these posit various linchpin moments when – if things had gone differently – the South might have won. Because fantasy texts written in the 20th century tend to use TIMESLIP devices – which frequently impose a degree of constraint upon the traveller – to get their protagonists into venues like this, stories set during the ACW often differ from sf in that they tend to be tales of BONDAGE, like Connie WILLIS's *Lincoln's Dreams* (**1987**), where the timeslip connection between a contemporary woman and General Robert E. Lee evokes hypnotically distressing images of the seemingly unending war. The elderly ex-soldier protagonist of Dan SIMMONS's "Iverson's Pits" (1988) timeslips from 1913 to the Battle of Gettysberg in a doomed attempt to revenge himself on a destructively incompetent colonel. And, although not in a timeslip tale, the horse protagonist of Richard ADAMS's *Traveler* (1988) is a creature inherently bound to General Lee, and to the ensuing horrors.

But the pre-eminent author of ACW fantasies was himself a survivor of the conflict. Several volumes by Ambrose

BIERCE – *Tales of Soldiers and Civilians* (coll **1891**), *Can Such Things Be?* (coll **1893**) and volume 3 of *The Collected Works of Ambrose Bierce* (coll **1909**) – contain stories of great note, the single most famous being "An Occurrence at Owl Creek Bridge" (◊ POSTHUMOUS FANTASY); others include "A Tough Tussle", "A Cold Night" – in which a laid-out corpse is found, after a bitterly cold night, to have assumed a foetal position (a strikingly powerful presentation of the image of BONDAGE) – "The Other Lodgers" and "Two Military Executions".

Some of Bierce's stories involve GHOSTS, and there are many tales by other hands that attempt to capture the horrific nature of the ACW through the devices of SUPERNATURAL FICTION. In Manly Wade WELLMAN's "His Name on a Bullet" (1940) a bullet charmed by a WITCH keeps a soldier from harm, but causes the deaths of his mates. Among the best ACW GHOST STORIES is "Stonewall Jackson's Wife" from *Herself in Love* (coll **1987**) by Marianne Wiggins (1947-   ). The eponymous black ghost in *Beloved* (**1987**) by Toni Morrison (1931-   ) haunts her mother, who has killed her to prevent her being returned to slavery. An anthology of this material is *Confederacy of the Dead* (anth **1993**) ed Richard Gilliam, Martin H. GREENBERG and Edward E. Kramer.   [JC]

AMERICAN GOTHIC A term used generally to describe writers of the US South, like William Faulkner (1897-1962) and Truman Capote (1924-1984). However universal the implications of their work, AG authors tend to be thought of as regional and their works tend to give off a sense of BELATEDNESS. In their case, however, the anxiety of influence seems to relate more strongly to the BONDAGE of the soil than to the fatherly burden of prior writers. Later figures, like James Purdy (1923-   ) and Joyce Carol OATES, though not from the South, generate some similar effects, but AG in fantasy is generally associated with Southern writers like Manly Wade WELLMAN and Sharyn MCCRUMB. The influence of AG is strongly felt in SUPERNATURAL FICTION, and specifically in HORROR. [JC]

AMERICAN TAIL, AN US/Irish ANIMATED MOVIE (*1986*). Universal/Steven SPIELBERG/Amblin. **Pr** Don BLUTH, Gary Goldman, John Pomeroy. **Exec pr** Kathleen Kennedy, David Kirschner, Frank Marshall, Spielberg. **Dir** Bluth. **Spfx dir anim** Dorse A. Lanpher. **Screenplay** Judy Freudberg, Tony Geiss. **Based on** story by Kirschner. **Voice actors** Cathianne Blore (Bridget), Betsy Cathcart (Tanya Mousekewitz singing), Dom DeLuise (Tiger), John Finneggan (Warren T. Rat), Phillip Glasser (Feivel/Fievel Mousekewitz), Amy Green (Tanya Mousekewitz), Madeline Kahn (Gussie Mausheimer), Pat Musick (Tony Toponi), Christopher Plummer (Henri), Neil Ross (Honest John), Will Ryan (Digit). 80 mins. Colour.

In 1885 the Mousekewitz family, being Jewish mice, flee Russia for the USA, because "there are no cats in America, and the streets are paved with cheese"; this FAIRYTALE is compared with the legend of the Giant Mouse of Minsk, who terrified all the cats there. A storm at sea separates young Fievel from his parents and sister; by the time they are reunited they, along with countless other mice plus the friendly cat Tiger, must thwart Warren T. Rat, a cat disguised as a rat, who plots to enslave all immigrant mice; eventually Fievel succeeds through drawing upon the LEGEND of the Giant Mouse of Minsk. The ALLEGORY with the plight of 19th-century Europeans fleeing to the USA, and of their exploitation there, is bitterly worked out: *AAT*, in its

SATIRE of capitalist exploitation, is for the most part a very disillusioned-seeming movie.

This was Bluth's second animated feature after he had led a walkout from DISNEY in 1979 (the credits are littered with ex-Disney artists) and is his masterpiece. The animators play interesting TROMPE L'OEIL games with PERCEPTION (the storm-waves at sea are seen by Fievel as water-MONSTERS or -DEVILS) and SURREALISM (notably in several of the sung set-pieces). However, the allegory trips itself up in places: the "cats", of course, were *not* all driven out of the USA; and the prevailing ethic within the immigrant mouse community, that mousehood is more important than national origin, suggests the very racism the moviemakers were trying so hard to negate. Yet with *AAT* Bluth put commercial animation back on centre stage as a legitimate vehicle of serious moviemaking.   [JG]

AMERICAN TAIL, AN: FIEVEL GOES WEST (vt *Fievel Goes West*) US movie (*1991*). Universal/Amblin. **Pr** Steven SPIELBERG, Robert Watts. **Exec pr** Kathleen Kennedy, David Kirschner, Frank Marshall. **Dir** Phil Nibbelink, Simon Wells. **Spfx sv** Scott Santoro. **Screenplay** Flint Dille, Charles Swenson. **Voice actors** Cathy Cavadini (Tanya), John Cleese (Cat R. Waul), Dom DeLuise (Tiger), Phillip Glasser (Fievel), Amy Irving (Miss Kitty), James Stewart (Wylie Burp). 75 mins. Colour.

This sequel to *An* AMERICAN TAIL was made without Don BLUTH's participation; it came from the newly formed Amblimation, a London branch of Spielberg's empire. Feline shyster Cat R. Waul cons hordes of New York mice to rustic Green River with the promise that, out West, cats and mice are friends; the Mousekewitzes are ensnared. To defeat Waul, Fievel must enlist his hero, the now-decrepit canine Sheriff Wylie Burp, and himself become a Western HERO. Though faster and better plotted than its predecessor, this lacks the subtextual depth.   [JG]

AMERICAN WEREWOLF IN LONDON, AN US movie (*1981*). PolyGram/Lycanthrope. **Pr** George Folsey Jr. **Exec pr** Peter Guber, Jon Peters. **Dir** John Landis. **Spfx** Rick Baker, Effects Associates. **Screenplay** Landis. **Starring** Jenny Agutter (Alex Price), Griffin Dunne (Jack Goodman), David Naughton (David Kessler), John Woodvine (Dr Hirsch). 97 mins. Colour.

Hitch-hiking in the north of England, two US teenagers, Jack and David, are attacked on a lonely moor by a WERE-WOLF; Jack dies but David, surviving, is taken to a London hospital. There he is visited by the ZOMBIE-like REVENANT of Jack, who explains David is now also a werewolf. At the next full MOON David suffers radical transformation, goes on the rampage, slaughters the innocent and wakes in nude ignorance the following morning in the wolf enclosure of London Zoo. He tries to give himself up as a werewolf, but the authorities are incredulous. On his next rampage he is shot despite attempts by Alex, the nurse who has become his lover, to awaken the human within the MONSTER. Mixing HUMOUR with the conventions of the HORROR MOVIE – and with spectacular spfx – *AAWIL* is an uneasy movie. Interesting, though, is its depiction of the transformed werewolf not as a sleek, erotic beast but as a heavy, swift-lumbering brute – far removed from the sensitive, soft-eyed David.   [JG]

A. MERRITT'S FANTASY MAGAZINE ◊ FAMOUS FANTASTIC MYSTERIES.

AMERY, L(EOPOLD CHARLES MORRIS) S(TENNETT) (1873-1955) UK writer whose *The Stranger of the*

Ulysses (**1934**) features the incursion of a pagan GOD into life aboard a contemporary ship, *The Ulysses*. [JC]

AMES, MILDRED (1919-   ) US writer for older children whose work includes some sf as well as one fantasy of interest, *The Silver Link, The Silken Tie* (**1984**), in which two CHILDREN, haunted by traumas in their respective pasts, heal themselves through the TALENTS they discover; they also help fight off a national conspiracy against youth. *Conjuring Summer In* (**1986**) is a SUPERNATURAL FICTION. [JC]

AMIS, KINGSLEY (WILLIAM) (1922-1995) UK novelist, poet, editor and critic. KA was noted for numerous, often satirical, novels of social comedy, beginning with *Lucky Jim* (**1954**), which despite its geniality caused him to be journalistically labelled – with Colin WILSON and others – as an "Angry Young Man". His distinctions include the Booker Prize for *The Old Devils* (**1986**) and a knighthood (1990). An unsnobbish affection for genre fiction is shown by contributions to crime, fantasy and sf.

His one major fantasy, *The Green Man* (**1969**), is a fine novel by any standards, and among KA's best. Its disturbing supernatural-HORROR events are usefully told through an entertainingly flawed and very Amisian ("hero as shit") narrator, afflicted with alcoholism, family troubles and middle-aged lust. A 17th-century MAGICIAN, who lived in what is now the protagonist's eponymous coaching inn, once animated a tree-man (◊ GREEN MAN) to do unpleasant business. The magician's GHOST tempts the protagonist with power and IMMORTALITY (in fact intending POSSESSION) – against which the protagonist's self-loathing is a shield – and harries him through minutely described ILLUSIONS all too reminiscent of delirium tremens. There is a remarkable appearance of GOD, manifesting in the privacy of TIME stasis (◊ TIME IN FAERIE) as a rather unpleasant young man whose self-imposed CONDITIONS of operation require that the WRONGNESS of magician and tree-man be neutralized not directly but by a suitable catspaw. The reluctant hero is given the task, along with a TALISMAN cross and some chilling hints about the AFTERLIFE; an EXORCISM results. A BBC TV miniseries adaptation was broadcast in 1991.

"Who or What Was It?" (1972 *Playboy*) is a TALL TALE, originally a radio broadcast, with KA himself stumbling into the situation of *The Green Man* (◊ RECURSIVE FANTASY). "To See the Sun", first published in *Collected Short Stories* (coll **1980**; exp 1987) is an epistolary VAMPIRE novelette set in 1925 Dacia, containing some irony as its sceptical English researcher informs an alluring vampire that her family legend arises from ergot-induced HALLUCINATION. [DRL]

Other work: *The Alteration* (**1976**), ALTERNATE WORLD story.

Further reading: There is some relevant criticism in *The Amis Collection: Selected Non-Fiction 1954-1990* (coll **1990**); *Kingsley Amis in Life and Letters* (anth **1990**) ed Dale Salwak (1947-   ); *Memoirs* (**1991**), autobiographical; *Kingsley Amis: Modern Novelist* (**1992**) by Dale Salwak, biography.

AMIS, MARTIN (LOUIS) (1949-   ) UK writer, son of Kingsley AMIS. A surprisingly high proportion of MA's work is sf, though he has avoided any literary stigma for contributing to that genre. *Other People: A Mystery Story* (**1981**) is a kind of POSTHUMOUS FANTASY whose protagonist – she is ambiguously treated as being either dead or a victim of profound AMNESIA, or both – undertakes a Rake's Progress through a savagely depicted LONDON. Sometimes called sf, *Time's Arrow* (**1991**) sees, to simplify, a man possessed (◊

POSSESSION) by a spirit that is living backwards in TIME towards crimes committed during WORLD WAR II. [JC]

AMITYVILLE MOVIES ◊ HAUNTED DWELLINGS.

AMNESIA The amnesia of the protagonist, which he must struggle to penetrate, is a PLOT DEVICE common to most genres of popular literature. It has been used on innumerable occasions throughout the 20th century. In detective novels or thrillers – an example being Margery Allingham's *Traitor's Purse* (**1941**) – amnesia causes difficulties when the protagonist is accused of murder (or espionage) and must somehow penetrate the veil (or, in the case of hoax amnesia, pretend to) in order for the truth to be discovered; the trope is carried through to the cinema in the form of movies like *The Morning After* (**1986**). In sf, the amnesia of the protagonist will frequently conceal his true nature – superior mutation, android, robot, alien – or ultimate role: ruler of the Sevagram, perhaps.

In fantasy, amnesia – which may be a medical condition, or sometimes nothing more diagnosable than an absence of knowledge – is generally a form of BONDAGE which conceals from the protagonist her or his true role in the STORY whose unfolding may well unveil an ancient reality, revealing that he is the AVATAR of an ELDER GOD, or the god himself, or a HERO, or a HIDDEN MONARCH, or a CHANGELING or a FAIRY; or that he embodies the soul of a MALIGN SLEEPER who now awakens (◊ SHADOW); or that he is the REINCARNATION of an ancient MAGUS or other figure of power or knowledge (or, when the amnesiac is female, of SHE). Amnesia may have been imposed upon the protagonist as a protection or as a CURSE. It may be lifted through introspective study (this is not common), or through INITIATION, or through the re-enactment of events anciently significant, or through tracing a LABYRINTH, or through passing a THRESHOLD, or through the action of MAGIC, often made effective by reading a RUNE or by touching an AMULET or some other TALISMAN.

A typical amnesiac is the mortal soldier in A.E. VAN VOGT's *The Book of Ptath* (**1947**), who gradually discovers that he is the eponymous father of the gods. It is central to the SPELL imprisoning Chessie, the enchanted purple Labrador in Nancy KRESS's *The Prince of Morning Bells* (**1981**), that he not remember his name or life as a man. In Tim POWERS's *The Drawing of the Dark* (**1979**), Brian Duffy cannot remember that he is in truth ARTHUR. As a punishment for having committed some sin he cannot now remember against a goddess whose name he does not know, the protagonist of Gene WOLFE's **Latro** sequence cannot remember anything for more than 24 hours: he writes his own story in short sequences, as an aide memoire. More typical examples (out of many) are the hero of Garfield and Judith REEVES-STEVENS's **Chronicles of Galen Sword** sequence (1990-   ), a human on Earth haunted by flashes of memory of his high status in an ALTERNATE REALITY; and the heroic warrior in Lisa GOLDSTEIN's *Summer King, Winter Fool* (**1994**) who, on being asked how he has escaped the sorcery-imposed bondage of amnesia, says simply, "I remembered who I was." The affliction is extremely common.

Fantasy amnesia – unless it is imposed at the end of a tale in order to protect the protagonist, or the world, or the god – exists in order to be removed. Amnesia, in other words, is almost invariably a form of suspense. [JC]

AMULETS An amulet is traditionally a CHARM worn or carried as protection against hostile MAGIC or specific ills. The scabbard of ARTHUR's sword Excalibur has traditional amulet-like properties. Lord DUNSANY's *The Travel Tales of Mr Joseph Jorkens* (coll **1931**) features an amulet preventing death from thirst: its wearer fearlessly travels the desert and is drowned in a flash flood; he should have READ THE SMALL PRINT. Angharad's bracelet in Alan GARNER's *The Weirdstone of Brisingamen* (**1960**) protects its child wearer against a TROLL-like creature's attack. The six components of the Circle of Signs in Susan COOPER's *The Dark is Rising* (**1973**) are individually amulet-like and warn of, or repel, the Dark. The early SANDMAN episodes by Neil GAIMAN feature a demonic amulet of universal protection. One character in Terry PRATCHETT's *Guards! Guards!* (**1989**) has an amulet against the remote eventuality of crocodile bites, and upon losing it is promptly bitten by a crocodile.

Less conventionally, an amulet may be as versatile as a RING: the half-amulet in E. NESBIT's *The Story of the Amulet* (**1906**) functions as a PORTAL through TIME; the amulet of DOCTOR STRANGE has various mystic powers depending on the needs of COMIC-BOOK plotting; that of Slaye in Jack VANCE's *The Eyes of the Overworld* (**1966**) commands 30 DEMONS; the eponymous PLOT COUPON of Michael MOORCOCK's *The Mad God's Amulet* (**1968**) confers superhuman fighting ability but in bad hands can enslave (◊ BONDAGE); the villain of Piers ANTHONY's *Blue Adept* (**1981**) is a prolific creator of offensive amulets that grow and animate as demons; the DRAGON amulet in Barry Hughart's *Bridge of Birds* (**1984**) is nonmagical but serves as the MAP of a LABYRINTH, and also a key. [DRL]

See also: TALISMANS.

ANACHRONISM This dislocation applies most obviously to TIMESLIP fantasies. Deliberate anachronism may help establish timelessness, as with MERLIN in T.H. WHITE's *The Sword in the Stone* (**1938**), who lives backwards through time and whose wizardly paraphernalia thus includes cigarette cards and the *Encyclopedia Britannica*. Milieux are more subtly mixed for timeless effect in M. John HARRISON's **Viriconium** series.

Anachronism is an obvious source of HUMOUR: Duke Astolph, drawn from ARIOSTO, boasts of his Winchester school scarf and membership of London's "Sphinx Club" in *The Castle of Iron* (**1941**) by L. Sprague DE CAMP and Fletcher PRATT; *Bored of the Rings* (**1969**), Henry N. Beard's and Douglas C. Kennedy's *Harvard Lampoon* PARODY of J.R.R. TOLKIEN, relies heavily on thrusting anachronisms into a travestied MIDDLE-EARTH, US brand names being regarded as particularly hilarious (one character is the Jolly Green Giant); Robert ASPRIN's *Myth Conceptions* (**1980**) painfully describes a fantasy INN which is all too evidently a modern fast-food outlet; Terry PRATCHETT has systematized anachronism through a joke theory of "morphic resonance" between Earth and his **Discworld**, with DRUIDS rebooting their 66-megalith stone-circle computers, armoured guardsmen echoing tough lines from tv police procedurals, playwrights recreating William SHAKESPEARE's *Macbeth* (**1623**) in the manner of Samuel Beckett's *Waiting for Godot* (**1955**), etc.; and Esther M. FRIESNER's **Gnome** series likewise uses much anachronistic humour.

Inadvertent anachronism may or may not harm a book. J.R.R. TOLKIEN's *The Lord of the Rings* (**1954-5**) has potatoes in what is stated to be prehistoric Europe. Ursula K. LE GUIN's acute ear for DICTION led her to skewer Katherine KURTZ's CELTIC FANTASY *Deryni Rising* (**1970**) in "From Elfland to Poughkeepsie" (1973) for its too-modern, too-US dialogue; similarly, in the medieval setting of Peter MOR-

WOOD's *The Dragon Lord* (**1986**), words like "scenario" and "paranoia" can jar. Even the glaring PREHISTORIC-FANTASY anachronism of humans coexisting with DINOSAURS has gained a kind of common-law acceptance through sheer familiarity. Some anachronisms are debatable: was Guy Gavriel KAY unwise to make the aviational term "Aileron" a major character's name in **The Fionavar Tapestry**? Some are publisher-enforced: Garry KILWORTH's *The Drowners* (**1991**) uses metres and litres in 19th-century England, its young audience being deemed incapable of understanding yards and gallons.

In general, anachronism involves perceived intrusions from the future; similar contact with the past is less troubling, unless the gap of years compels the jolting recognition of a TIME ABYSS. An exception is the curious GHOST STORY *Time Out of Mind* (**1986**) by John R. MAXIM, in which it is through flashes of 19th-century scenes and events that the present-day characters first realize the WRONGNESS of the situation. [DRL]

ANAHITA ◊ GODDESS.

ANANSI or ANANSE A Ghanaian and West Indian TRICKSTER god in the shape of a SPIDER. Most of the material incorporating Anansi comprises FOLKTALES retold or newly imagined. *West African Trickster Tales* (coll **1994**) as retold by Martin Bennett, a useful compendium, includes several Ashanti tales from Ghana that deal with Ananse. The best-known West Indian version is probably the Anancy who figures in the REVISIONIST-FANTASY sequence by Andrew Salkey (1928-1995) – *Anancy's Score* (coll **1973**), *One* (coll **1985**) and *Anancy, Traveller* (coll **1992**) – where the trickster spider becomes involved in Third World struggles. A similar figure appears in *Anancy-Spiderman* (**1989**; vt *Spiderman-Anancy* 1989 US) by James Berry (1925-   ). [JC]

ANCESTRAL MEMORIES In SUPERNATURAL FICTION AMs – which may speak through a REVENANT or a PICTURE or a BAD PLACE or an object which has been infected – are almost certainly invasive, and represent an infliction of BONDAGE upon the protagonist; they are attempts to *ingest* the present. Examples are very numerous; most confirm a sense that the AM in supernatural stories is likely to be inimical, properly resisted by the healthy protagonist, and – if the ending is to be happy – properly laid to rest.

In TIMESLIP fantasies, the AM will quite possibly represent the call of the earlier self to the current self (◊ SHADOW); and it is likely that – rather than ingest the present self – the caller invokes that present self's need to inhabit the earlier version. E. NESBIT's *The House of Arden* (**1908**), its sequel *Harding's Luck* (**1909**) and Lucy BOSTON's **Green Knowe** books (from **1954**) all combine elements of the timeslip with an evocation of the AM as a form of the *genius loci*, the genius of place.

In most fantasies set in SECONDARY WORLDS, the AM serves as a reminder of the central STORY which is being revealed to the protagonists, who may well be AVATARS of a founding GOD or monarch. AMs, in this context, are messages and adjurations: they remind the protagonist of the purity of being which may have been lost through THINNING, of the restorative task that is therefore laid upon the protagonist, and of the story that must be continued to the end; and they often convey specific instructions. Where in supernatural fictions the AM is an enemy, in fantasy it is a gift. [JC]

ANDERSEN, HANS CHRISTIAN (1805-1875) Danish writer, father of the modern FAIRYTALE. Whereas the

GRIMM BROTHERS, Charles PERRAULT and to some extent Madame D'AULNOY had drawn heavily upon the oral tradition, HCA based only 12 of his 156 known fairy stories on FOLKTALES.

HCA came from a poor family, and the early death of his father meant he had to find work, which set back his education. In later years he pursued his love for the THEATRE, and it was here that his flights of fancy began. His first novel, *Improvisatoren* ["The Improvisator"] (**1835**), was an autobiographical exploration of a poor boy's integration into society. HCA was ever determined to overcome his impoverished roots, and he returned to the theme often in his stories, as in "The Ugly Duckling" (1845), a universally understood FABLE of self-discovery (◊ UGLY DUCKLING).

HCA's first fairytale was "The Spectre" (1829), later rewritten as "The Travelling Companion" (1836). It was soon after completing this story that he began to realize the potential of this new form. His first four stories were published as *Eventyr, fortalte for Børn* ["Fairytales, Told for Children"] (coll **1835** chap), and they were immediately popular. Thereafter, on an almost annual basis, Andersen issued a booklet of stories, each bearing the same title. The first three were assembled as *Eventyr, fortalte for Børn* (omni **1837**), and the same happened with the next three (omni **1842**). By 1845 he wanted to emphasize that his stories were accessible on more than one level, and should not be restricted to children, so he titled his new collection *Nye Eventyr* ["New Fairytales"] (coll **1845**), and that title remained with each successive volume. He achieved his desire for the wider acceptance of his stories in Denmark, but in English-speaking countries – where a torrent of translations (often bad) began to appear from 1846 – the prevailing Victorian attitude to folktales and fairytales placed them firmly in the children's domain. This rejection of his views annoyed HCA, who claimed he was a "poet for all ages". From 1852 he titled his regular collections **Eventyr og Historier** to distinguish between fairytales (*eventyr*) and stories (*historier*); though his *historier* were more philosophical and tended toward ALLEGORY, the supernatural often pervaded both.

Most of HCA's fairytales are more appreciable as WONDER TALES, as they open the imagination to aspects of life beyond the normal human experience. Some of the allegories have become bywords for the human condition – such as "The Emperor's New Clothes" (1837), about human vanity and snobbery, and "The Little Mermaid" (1837), a symbol of SACRIFICE. Some are deeply philosophical, like "On the Last Day" (1852) and "The Old Oak Tree's Last Dream" (1857), both parables of death. HCA frequently used DREAMS to transcend the humdrum – though sometimes this might mean death, as in "The Rose Elf" (1842) and (best-known of all) "The Little Match Girl" (1848). While humour pervades many of the stories, the later ones tend to be sombre to the extent they can be classed as DARK FANTASY. He was not afraid to frighten, as in his last story, "Auntie Toothache" (1872), in which the pain of a toothache takes on physical form: the devastating Madame Toothache.

Almost every fantasy theme and PLOT DEVICE appears in HCA's stories, which are often archetypes of the genre. Among stories of his that stand in their own right as fantasies are "The Nightingale" (1845) – portraying the power of beauty over death – "The Snow Queen" (1846) – with its transformation of the world into WRONGNESS – "The Elf

Hill" (1847) – one of his few stories to be set entirely in a SECONDARY WORLD of FAERIE – "The Ice Maiden" (1861) – exploring humankind's relationship with FATE – and two stories which can scarcely at all be regarded as CHILDREN'S FANTASY: "The Story of a Mother" (1848), wherein a recently bereaved mother meets the personification of DEATH as a gardener tending plants that are humankind, and "The Shadow" (1847), an early JEKYLL-AND-HYDE study of a man haunted by his old SHADOW.

Throughout HCA's fiction there is a message of hope over adversity: there is beauty and goodness in the world if one is prepared to look, even if it means self-sacrifice. This view has given rise to an oversentimental perception of HCA's fiction, when in fact he used a mixture of humour and pathos to re-focus, through the imagery of fantasy, his audience's minds on reality.

HCA's stories were translated throughout Europe, with four editions appearing in the UK in 1846 alone, starting with *Wonderful Stories for Children* (coll **1846**) ed Mary Howitt (1799-1888). One of his annual booklets was published first in English – HCA was on a tour of the UK at the time – as *A Christmas Greeting to My English Friends* (coll **1847**). His work influenced, among many others, Charles DICKENS, William THACKERAY and Oscar WILDE, and was the final impetus that inspired an upsurge in children's fairytales in Victorian England. His influence continues to this day, many of his stories serving as models for REVISIONIST FANTASIES. His effect on the fantasy genre has been immeasurable. HCA was the world's first great fantasy storyteller. [MA]

Other works: Over 400 different selections and editions of HCA's stories have appeared in the English language alone. The first complete edition outside Denmark of all his writings was **Hans Andersen Library** (**1869-87** 20 vols UK). Among the many beautifully illustrated editions are *Danish Fairy Tales and Legends* (coll **1897** UK, rev from *Danish Fairy Legends and Tales* trans Caroline Peachey 1846 UK) illustrated by W. Heath ROBINSON, *Hans Andersen's Fairy Tales* (coll **1910** US) illustrated by Frank C. PAPÉ, *Stories from Hans Andersen* (coll **1911** UK) illustrated by Edmund DULAC, *Fairy Tales* (coll **1913** UK) illustrated by Mabel Lucie Attwell (1879-1964) and *Fairy Tales* (coll **1932** UK) illustrated by Arthur RACKHAM. Translations of interest include *Hans Andersen's Fairy Tales* (coll **1930** UK) by M.R. JAMES, which is a selection, and the complete *Fairy Tales and Stories* (coll **1974** UK) by Erik Haugaard (1923-   ).

Further reading: HCA's autobiography is *Mit Livs Eventyr* (**1855**; trans Horace E. Scudder and updated by HCA as *The Story of My Life* **1871** US), although this is generally regarded as somewhat sanitized. Worthwhile biographies are *Hans Christian Andersen* by Svend Larsen (**1953** Denmark but in English trans Mabel Dyrup) and *Hans Christian Andersen: The Story of His Life and Work 1805-75* by Elias Bredsdorff (**1975** UK).

ANDERSON, DANA (1952-   ) US writer. ◊ Charles DE LINT.

ANDERSON, KAREN (1932-   ) US author and editor. ◊ Poul ANDERSON.

ANDERSON, MARGARET J(EAN) (1931-   ) US writer for young adults. Her books are mostly TIMESLIP adventures, including her first, *To Nowhere and Back* (**1975**), in which a young girl travels a century into the past, where she has another name. MJA is best-known for her **Time Trilogy** – *In the Keep of Time* (**1977**), *In the Circle of Time*

(**1979**) and *The Mists of Time* (**1984**) – which provides intensified interest because of the intricacy of its switches from time past to time future; its Scottish protagonists, in the 14th and a barbaric 21st century, fight to defend their native land. *The Druid's Gift* (**1989**) also involves timeslips as its young protagonist learns from the future how to build a life that may have some chance of continuing. *The Ghost Inside the Monitor* (**1990**), about a computer monitor which contains a GHOST, is a TECHNOFANTASY. [JC]

ANDERSON, POUL (WILLIAM) (1926-   ) US writer, mostly of sf, beginning with "Tomorrow's Children" with F.N. Waldrop for *Astounding* in 1947. A Nordic-twilight hue permeates much of PA's work; moreover, he has generated throughout his career a sense that heroic actions – though necessary and much to be praised – are essentially providential and that HEROES enact stories already told (◊ TWICE-TOLD). This sense of BELATEDNESS is not a note often struck in sf, and as a consequence much of PA's work has been understood as fantasy-tinged. In fact, his most significant work – novels like *Brain Wave* (**1954**) and *Tau Zero* (1967 *Galaxy* as "To Outlive Eternity"; exp **1970**) – has been hard sf, as have been his main series, and much of his seeming fantasy turns out in the end to be sf, carefully rationalized (◊ RATIONALIZED FANTASY), but retaining accord with, in W.H. Auden's term, "the Northern thing". Parallel to, or resisting, this melancholy is PA's adherence to the can-do optimism of the 20th-century USA and technology that we associate with John W. Campbell Jr and his magazines – though this adherence had soured by the 1990s. If at times his fantasies slip into glib resolutions, it is because, even in the world of MAGIC, PA wishes reason and special knowledge to be useful tools; when in *Three Hearts and Three Lions* (1953 *F&SF*; exp **1961**) a troll turns to stone, the hero instantly deduces that the troll's curse is in fact radioactive residue from TRANSMUTATION. This sort of thing is a staple of rationalized fantasy, but in PA's work it is, sometimes literally, a bulwark against CHAOS. In the short story "Pact" (1959) as by Winston P. Sanders, it is reason and the pursuit of knowledge with which the demon protagonist is punished by the human he has summoned and made a PACT with.

PA has been an intensely prolific author for nearly half a century, and several fantasies of strong interest have appeared. *The Broken Sword* (**1954**; rev 1971) is – especially in the original version – a violent, sometimes confused, plot-dominated tale set in a Dark-Age England occupied by humans and WAINSCOT societies of ELVES and TROLLS. Other residents of FAERIE – the Sidhe (◊ FAIRIES) have an ambiguous role – also intervene in the human world, the affairs of which are additionally affected by the AESIR who, though neutral in the war between elves and trolls which propels the plot, have their own agenda. The human protagonist is kidnapped as a baby – an evil CHANGELING being substituted – and becomes a HERO of sorts, though the cursed, eponymous SWORD requires him to kill each time he draws it. In the end, he plays a significant role in the cruel war (as does his SHADOW self), falling in love with his own sister *en passant*, but loses all and dies (as does his cruel elf shadow). A later work, *Hrolf Kraki's Saga* (**1973**), which reworks traditional Icelandic/Western Isles saga material, is told with a similar full-blooded bleakness; it won a 1975 BRITISH FANTASY AWARD.

Published in short form about the same time as *The Broken Sword*, though considerably smoother in effect, *Three Hearts*

and Three Lions is a more conventional exercise in rationalized fantasy, being clearly reminiscent of the **Incomplete Enchanter** stories published a decade earlier by L. Sprague DE CAMP and Fletcher PRATT. PA's protagonist – conveyed to a medieval world replete with WITCHES, DRAGONS and erotic danger in the person of MORGAN LE FAY – emigrates across a second THRESHOLD into Faerie, returning to the medieval venue to discover himself the current AVATAR of a Carolingian ETERNAL CHAMPION, destined to fight on the side of Law against the forces of CHAOS. The tale is brightly peopled – though PA's eclectic cast and ROMANCE setting became seriously overused in later decades. His most original stroke may have been his portrait of elves as remote, aristocratic, hauntingly seductive to humans.

The thematically linked *A Midsummer Tempest* (**1974**) – set in an ALTERNATE WORLD where all SHAKESPEARE's plays, particularly *A Midsummer Night's Dream* (performed *c*1595; **1600**) and *The Tempest* (performed *c*1611; **1623**), are chronicles of fact – romantically reverses the 16th-century THINNING of Merrie England: OBERON, PUCK and ARIEL – and ARTHUR and Charlemagne too – band together to protect Charles I and Faerie from Cromwell and his Puritans, who are defeated at Glastonbury Tor. The anachronisms in Shakespeare are taken as licence for a different level of technology – the Puritans in this reality are having an Industrial Revolution and Prince Rupert escapes by steam train. During the tale the protagonists arrive at a CLUB-STORY venue, the Phoenix Tavern (a paradigm version of the INN), where characters from divers PA novels meet and debate the virtues of various parallel worlds in a variety of DICTIONS – ordinary folk speak prose; those of high birth speak in blank verse. In *The Merman's Children* (fixup **1979**), on the other hand, there is no rescue of Faerie from the Thinning of a world in which Christianity triumphs; the story, which complexly interweaves several QUESTS in search of the lost domain, is movingly elegiac.

Also closely tied to the kind of fantasy espoused by John W. Campbell Jr for the magazine UNKNOWN, *Operation Chaos* (coll of linked stories **1971**) is set in an ALTERNATE WORLD where MAGIC works through the discovery and application of pragmatic laws. The cast (one of whom enters the debate in the Phoenix Tavern) is again populous and colourfully conceived, and includes afreets, various familiars and DEMONS (one of whom is an alternate version of Adolf Hitler); the male protagonist is a WEREWOLF and his female counterpart a virgin WITCH (◊ VIRGINITY). Again the war is between Law and Chaos; here, as in *A Midsummer Tempest*, the ending is not grim.

PA's later fantasies are perhaps less engaging, though *The Devil's Game* (**1980**), which features an apparent PACT WITH THE DEVIL, is interestingly cast as a suspense thriller. His most important later work, written in collaboration with his wife, Karen Anderson (1932-   ), is almost certainly the **King of Ys** sequence: *The King of Ys #1: Roma Mater* (**1986**), *#2: Gallicenae* (**1987**), *#3: Dahut* (**1988**) and *#4: The Dog and the Wolf* (**1988**), all assembled as *The King of Ys* (omni **1988** 2 vols). CELTIC-FANTASY elements intermix with RECURSIVE-FANTASY references to the cast of Rudyard KIPLING's *Puck of Pook's Hill* (coll of linked stories **1906**) in the long story of the city of Ys – a Graeco-Punic outpost in Brittany – and its effects on late-Roman politics. In general, however, the most significant achievements of PA's later career have been sf. Most of his fantasies have been cast as requiems, though sometimes disguised as romps; and perhaps PA now feels that the Northern way of life they mourn has – as he has told us more than once – indeed passed away. [JC]

Other works: *The Fox, the Dog and the Griffin: A Tale Adapted from the Danish of C. Molbech* (**1966**), for children; *The Demon of Scattery* (**1979**) with Mildred Downey BROXON; the **Last Viking** sequence of historical tales, comprising *The Golden Horn* (**1980**), *The Road of the Sea Horse* (**1980**) and *The Sign of the Raven* (**1980**); *Fantasy* (coll **1981**); *The Unicorn Trade* (coll **1984**) with Karen Anderson; *The Night Fantastic* (anth **1991**) ed with Karen Anderson, stories about DREAMS; *Loser's Night* (**1991** chap); *The Armies of Elfland* (coll **1992**).

ANDOM, R. Pseudonym of UK writer Alfred Walter Barrett (1869-1920), whose career was exclusively as an author of light fiction designed for the large markets that opened up in the 1890s, and for boys'-fiction journals like *Captain Library*, *Nuggets* and *Scraps* (which he edited); long before his death he had become a figure of the past. *We Three and Troddles: A Tale of London Life* (**1894**) and its several sequels – one of which, *In Fear of a Throne* (**1911**), is a RURITANIAN fantasy – display the domesticated, deflating inconsequentiality demanded by its readership, a style only a few writers, like H.G. WELLS and (less obviously) F. ANSTEY, seemed able to subvert or transcend. The title story of *The Strange Adventure of Roger Wilkins, and Other Stories* (coll **1895**) is an IDENTITY-EXCHANGE tale involving Wilkins and his office boy; *The Identity Exchange: A Story of Some Odd Transformations* (**1902**; vt *The Marvellous Adventures of Me* 1903) is similar, though more diffuse. *The Enchanted Ship: A Story of Mystery with a Lot of Imagination* (**1908**) is a GHOST STORY set on a PIRATE ship. The first tale in *The Magic Bowl, and The Blue-Stone Ring: Oriental Tales with Occi (or Acci) dental Fittings* (coll **1909**) features a GENIE who makes it possible for two young men to transform Russia into a UTOPIA, while "The Blue-Stone Ring" is a further identity-exchange story. [JC]

ANDRADE, MÁRIO DE (1893-1945) Brazilian writer. ◊ BRAZIL.

ANDREAE, JOHANN (VALENTIN) (1586-1654) German mystic, author of *The Chemical Wedding of Christian Rosenkreutz* (**1616**) and *Christianopolis* (**1619**). ◊ ALCHEMY; ROSICRUCIANISM.

ANDREWS, ALLEN (1913-) UK writer whose fantasy has been restricted to the **Plantagenet** ANIMAL-FANTASY sequence – *The Pig Plantagenet* (**1980**) and *Castle Crespin* (**1982**) – featuring a pig named Plantagenet. The setting is 13th-century France. The forest is endangered by the lord of the château, who wishes to massacre the local family of boars. Plantagenet saves them; but they must now QUEST for a safe place. [JC]

ANDREWS, VIRGINIA C(LEO) (1933-1986) US writer and painter, author of several bestselling PSYCHOLOGICAL THRILLERS. There is a sense of *almost* supernatural BONDAGE in the primal setting that defines her best-known sequence, the **Dollanganger Children** series – *Flowers in the Attic* (**1979**), *Petals on the Wind* (**1980**), *If There Be Thorns* (**1981**), *Seeds of Yesterday* (**1984**) and *Garden of Shadows* (**1987**), the last being completed by Andrew Neiderman (1940-). The children, imprisoned in an attic by their evil mother, create a private universe in their heated imaginations. Neiderman, himself a noted HORROR writer, has either completed from VCA's apparently copious notes or written solo a number of further books, the more recent of them being signed "The New Virginia Andrews". [JC]

ANDREYEV, LEONID (NIKOLAEVICH) (1871-1919) Russian writer and lawyer. His neurotic, pessimistic and suicidal nature provided sufficient brooding melancholy to develop a corpus of horror and macabre stories that made him one of Russia's most popular writers in the period 1898-1910. Most of his early fiction is straight HORROR with a few elements of SUPERNATURAL FICTION, but he later shifted toward the symbolic, seeking to evoke metaphysical horror, though more often attaining only cheap sensationalism. LA frequently used Biblical imagery to convey his message, and nowhere more powerfully than in "Lazarus" (1906; in *Lazarus, and The Man from San Francisco* anth trans Abraham Yarmolinsky **1918** US), a form of POSTHUMOUS FANTASY – but one set in the real world, where Lazarus is resurrected to a life that he finds too burdensome, and sucks all whom he encounters into eternal misery. "The Abyss" (1909) portrays a similar mental decline, and is best compared to various works by Guy DE MAUPASSANT.

LA's fictions are depressing but powerful. His *Complete Works* (coll **1913** 8 vols) have been published in Russia, but only selections have appeared in translation, of which *Silence* (coll trans **1910** UK), *The Little Angel* (coll trans **1915** UK) and *The Crushed Flower* (coll trans **1917** UK) are early examples, and *Selected Stories* (coll **1969** UK) the most recent. [MA]

Other works: *An Abyss* (**1909** UK).

ANDY WARHOL'S DRACULA vt of *Blood for Dracula* (*1974*). ◊ DRACULA MOVIES.

ANDY WARHOL'S FRANKENSTEIN (*1973*) ◊ FRANKENSTEIN MOVIES.

ANGELS The word "angel" (derived from the Latin *angelus*) is consistently used by translators of the Bible to render Hebrew and Greek words for "messenger". And it is as a messenger – usually a messenger of GOD – that the angel most often appears, almost invariably embodying a sense that something is being made visible that is otherwise ineffable, or of too heightened a nature for human PERCEPTION to bear. In modern fantasy the function of the angel as a LIMINAL BEING may well be detached from more traditional uses; but the angel, whenever encountered in its pure form, has a tendency to bear news.

The hierarchical division of angels into several courtly ranks – in descending order, Seraphim, Cherubim, Thrones, Dominions, Virtues, Powers, Principalities, Archangels, Angels – dates from the very earliest days of Christianity, though it was first articulated in the 5th century AD by Dionysus the pseudo-Areopagite. The Church did not officially renounce the hierarchy until well into the 15th century.

Perhaps because of their powerful association with a living religion, angels do not proliferate in SECONDARY WORLDS, being much more commonly encountered in SUPERNATURAL FICTION, where their liminality often registers as a beacon, signalling the presence of a higher state of being. Angels (as opposed to DEVILS, who are, like SATAN, fallen angels) appear frequently in the later 19th century, less so through the 20th century, until recent years. There are, though, many examples. The hero of Marie CORELLI's *Ardath: The Story of a Dead Self* (**1889**) falls in love with one in contemporary Babylon, then TIMESLIPS into an ancient world where he becomes worthy of her. In H.G. WELLS's *The Wonderful Visit* (**1895**) – the Arthur RACKHAM cameo on the cover is of a winged CUPID – an angel visits Earth and is immediately shot. Angels bedevil the protagonist of Mark

TWAIN's *Extracts from Captain Stormfield's Visit to Heaven* (**1909**). Anatole FRANCE's *La Révolte des anges* (**1914**; trans as *The Revolt of the Angels* **1914** UK) is one of the very few SATIRES in which angels do not seem mawkish. An angel surfaces through the ET IN ARCADIA EGO obfuscations of Barry PAIN's *Going Home* (**1921**). They appear in *The High Place* (**1923**) and *The Devil's Own Dear Son* (**1949**) by James Branch CABELL; they accompany Mr Weston (who is God) in T.F. POWYS's *Mr Weston's Good Wine* (**1927**); in Robert NATHAN's *The Bishop's Wife* (**1928**) an angel helps a bishop in a campaign to fund cathedral repairs; one becomes mortal (a theme typical of much SUPERNATURAL FICTION) in Helen BEAUCLERK's *The Love of the Foolish Angel* (**1929**); a multivalent angel performs many acts simultaneously in J.B.S. Haldane's *My Friend Mr Leakey* (**1937**); they have an intrinsic role in a Christian sequence like the **Ransom** trilogy (**1938-45**) by C.S. LEWIS; the Archangel Michael warns the protagonist of T.H. WHITE's *The Elephant and the Kangaroo* (**1947** US) of a second FLOOD; and it is made clear in J.R.R. TOLKIEN's posthumous *The Silmarillion* (**1977**) that Gandalf and the other wizards are indeed angels, probably Seraphim, though they bring news of a prior (rather than a higher) reality. Angels can be seen in Shamus FRAZER's *Blow, Blow Your Trumpets* (**1945**), in Taylor Caldwell's *Dialogues with the Devil* (**1967**), and elsewhere. The eponymous protagonist of Mervyn PEAKE's *Mr Pye* (**1953**) actually becomes one.

More recent genre examples include Manuel MUJICA LAINEZ's *El unicornio* (**1965**; trans Mary Fitton as *The Wandering Unicorn* **1982** Canada), in whose depiction of the Middle Ages FAIRIES and angels CROSSHATCH with cool equanimity; the "cavern angel" who monitors – in the form of the itinerant WillyBoy – the life of the Eternal Champion who fights the eponymous DARK LORD in Roderick MACLEISH's *Prince Ombra* (**1982**); R.A. MACAVOY's **Damiano** sequence (**1983-1984**), in which the archangel Raphael loses his angelic status when he becomes too much involved; Richard Condon's *Money is Love* (**1975**); Stephen BRUST's *To Reign in Hell* (**1984**), where they are central; Nancy WILLARD's *Things Invisible to See* (**1985**); Madeleine L'ENGLE's *Many Waters* (**1986**); *The Urth of the New Sun* (**1987** UK) by Gene WOLFE; Brian STABLEFORD's **Werewolves of London** sequence (**1990-94**), where they are again central; Neil GAIMAN's and Terry PRATCHETT's *Good Omens* (**1990**); Gary KILWORTH's **Angel** sequence (from **1993**); Elizabeth HAND's *Waking the Moon* (**1994**); and Sean STEWART's *Resurrection Man* (**1995**). In nongenre books, angels continue to have a metaphorical presence; an angel participates, for instance, in the CARNIVAL revolt of *The Milagro Beanfield War* (**1974**) by John Nichols (**1940-**); and the protagonist of his *American Blood* (**1987**) believes himself to be one. There is a tendency for recent angels to be androgynes, and evocative therefore of earlier religions than Christianity, specifically those faiths built around the worship of the GODDESS.

Movies in which angels feature are numerous, though are rarely very serious in their attempts to render angels in visual terms. Those in which angels wear wings are generally comic in intent, if not in accomplishment; while those featuring actors in togas often achieve their comic effects inadvertently. The most successful renderings of angels tend to be those in which they appear in mortal guise, as in Frank CAPRA's IT'S A WONDERFUL LIFE (**1946**) and Paul Hogan's ALMOST AN ANGEL (**1990**). In WINGS OF DESIRE

(**1987**) two angels discover the temptations of mortality in modern Berlin; although they appear human on the screen, it is understood that this is merely an appearance they have adopted, their true appearance being incomprehensible to we mortals. The possibility of angels physically taking on human form is, however, explicit: part of the plot is that the actor Peter Falk, playing himself, is in reality an angel who some years ago chose to become human. Other movies in which angels make themselves known include: FAUST: EINE DEUTSCHE VOLKSSAGE (**1926**); *The* PASSING OF THE THIRD FLOOR BACK (**1935**), debatably; HERE COMES MR JORDAN (**1941**) and its remake HEAVEN CAN WAIT (**1978**), in which angels are shown as having the same inefficient bureacracy as we mortals, much as they do in A MATTER OF LIFE AND DEATH (**1946**); ANGELS IN THE OUTFIELD (**1952**); BARBARELLA (**1967**), where the angel is a physical rather than a spiritual being; ALWAYS (**1989**), rather unimaginatively; *Angel on My Shoulder* (**1946**); FIELD OF DREAMS (**1989**), arguably (the mystic guiding voice is never formally identified); ORLANDO (**1992**); and even TWIN PEAKS: FIRE WALK WITH ME (**1992**). Angels of death appear in ORPHÉE (**1949**) and *Raiders of the Lost Ark* (**1981**; ◊ INDIANA JONES). Gabriel in person, although never seen, plays a part in GABRIEL OVER THE WHITE HOUSE (**1933**). Fallen angels – SATAN excepted – appear in BORN OF FIRE (**1986** tvm) and to especially chilling effect in *The* LAST TEMPTATION OF CHRIST (**1988**), where the angel/temptress, an emissary of Satan, is depicted as a young, seemingly innocent girl. The Devil attempts to regain readmission to God's chosen circle of angels in BEDAZZLED (**1967**). [JC/JG]

See also: AFTERLIFE.

ANGELS AND THE PIRATES, THE ◊ ANGELS IN THE OUTFIELD (**1952**).

ANGELS IN THE OUTFIELD (vt *Angels and the Pirates* UK). US movie (**1952**). MGM. **Pr** Clarence Brown. **Dir** Brown. **Spfx** Peter Ballbusch, A. Arnold Gillespie, Warren Newcombe. **Screenplay** Richard Conlin, Dorothy Kingsley, George Wells. **Starring** Ty Cobb (himself), Donna Corcoran (Bridget White), Joe DiMaggio (himself), Paul Douglas (Aloysius "Guffy" McGovern), Janet Leigh (Jennifer Paige). 99 mins. B/w.

McGovern, aggressive and foulmouthed, manages the Pittsburgh Pirates, a losing BASEBALL team. One night he is confronted by an invisible ANGEL, who tells him that, if he will simply reform, HEAVEN will chip in on his side with some MIRACLES. He obeys. Promptly the Pirates start to win games. 8-year-old orphan Bridget is the only person who can see that, during the Pirates' crucial innings, a celestial baseball team – a team of angels – is standing behind and inspiring the mortal players. Paige, a cub reporter assigned to cover the Pirates' season for the *Pittsburgh Messenger*, initially loathes McGovern but in due course, predictably, falls in love with him, the couple adopting Bridget. Before that, in scenes reminiscent of MIRACLE ON 34TH STREET (**1947**), McGovern is hauled up before a judge to examine the case for the existence or otherwise of angels.

In terms of the history of the fantasy CINEMA, this is one of the later examples of the feelgood-fantasy strand that included movies like IT'S A WONDERFUL LIFE (**1946**) and *Miracle on 34th Street*. It was remade under the original title in 1994 by DISNEY. [JG]

ANGELS OF MONS In "The Bowmen" (**1914** *Evening News*) Arthur MACHEN created one of the enduring LEGENDS of WORLD WAR I. The story – explicitly fictional, though this was widely ignored – is set at the Battle of Mons, whose outcome is in doubt until phantom bowmen, from the days of England's glory, come to the aid of the British Expeditionary Force. The story was included in Machen's *The Angels of Mons, The Bowmen, and Other Legends of the War* (coll **1915**; exp 1915). [JC]

ANIMA A term from JUNGIAN PSYCHOLOGY referring to that ARCHETYPE within the male collective unconscious which represents the feminine aspect. The nature of a man's individual anima is discovered through its projections into the outer world, where its incarnations include the male's own mother and (certainly in fantasy texts) literal manifestations of the GODDESS – as LAMIA, as MUSE and in her own right. The term "animus" was used by Jung to refer correspondingly to the woman's masculine side, but this term had narrower application because women were not deemed to have the same capacity of projection or the same creative needs. [JC]

ANIMAL FANTASY Three closely associated but distinct terms are used in this encyclopedia to cover some of the innumerable uses to which animals are put in fantasy. Two of these terms – AF and BEAST FABLE – describe categories of story; the third, TALKING ANIMALS, identifies such an animal when it (or she, or he) plays a part, often that of COMPANION, in a tale whose protagonists are human.

A pure AF is a tale which features sentient animals who almost certainly talk to one another and to other animal species, though not to humans, and who are described in terms which emphasize both their animal nature and the characteristic nature of the species to which they belong. A pure AF will almost certainly be set in the real world, and will usually teach its readers some natural history; the most extreme examples are little more than fictionalized BIOLOGY lessons.

A beast fable, on the other hand, is a tale whose animal protagonists are described in terms which permit a SATIRE- or ALLEGORY-based comparison of their behaviour and nature with that of humans; these comparisons are normally made without much attention being paid to real animal behaviour. Beast fables are sometimes set in a WONDERLAND or OTHERWORLD venue, but they are just as frequently set in a fantasticated version of the real world, where their protagonists act as humans and frequently interrelate with humans.

Examples of the AF include Richard ADAMS's *Watership Down* (**1972**), about rabbits, and Garry KILWORTH's *Hunter's Moon* (**1989**), about foxes, neither text allowing any communication between animals and humans. Many tales hover between the two extremes (*Watership Down* itself has been referred to by some as a beast fable). Kenneth GRAHAME's *The Wind in the Willows* (**1908**), Walter DE LA MARE's *The Three Mulla-Mulgars* (**1910**), George ORWELL's *Animal Farm* (**1945** chap) and William HORWOOD's **Duncton Chronicles** all feature protagonists who seem sometimes animal and sometimes to PARODY human foibles. Unlike the pure AF, where the real world constantly and intimately interacts with the animal community, a boundary does exist between the central venue and the real world in these books.

In the pure AF the initiating fantasy premise tends to dissolve into a narrative which heeds the laws of the world. Because they exist in the world, and because the communities they depict are subject to the laws of nature, AFs tend to end in tragedy. To tell a pure AF is, ultimately, to depart from fantasy. [JC]

ANIMAL FARM UK ANIMATED MOVIE (**1955**). Louis de Rochemont/Halas-Batchelor. **Pr** Joy Batchelor, John Halas. **Dir** Joy Batchelor, John Halas. **Anim dir** John F. Reed. **Screenplay** Joy Batchelor, John Halas, Borden Mace, Philip Stapp, Luther Wolff. **Based on** *Animal Farm: A Fairy Story* (**1945** chap) by George ORWELL. **Voice actors** Maurice Denham (all except Narrator), Gordon Heath (Narrator). 75 mins. Colour.

A fairly faithful rendition of Orwell's original, although with a slightly more upbeat ending and a minor shift in the slant of the SATIRE, so that it excoriates the totalitarian corruption of Communism rather than Communism as a notion (indeed, much of the movie, and in particular its final moments, seems to present a socialist viewpoint; it is not that the system is especially corruptible, but that individuals will corrupt it if the rest of society lets them), *AF* has been widely underestimated: its animation – especially of the pig Napoleon – is of DISNEY quality (albeit not the very *best* Disney), and the tale, mainly narrated but with some dialogue, is grippingly told in terms of both words and visuals. [JG]

ANIMAL MAN ◊ Jamie DELANO; Grant MORRISON.

ANIMALS UNKNOWN TO SCIENCE Fabulous animals play a significant role in MYTH and LEGEND, to the extent that many were included in Pliny the Elder's pioneering *Natural History* and in medieval BESTIARIES. Many were incorporated into heraldic devices, which conserved their familiarity and their symbolic authority, and no TRAVELLERS' TALE in any era could be deemed truly satisfying unless it included at least one such creature.

The boundary between animals "known to science" and those which are not has never been made entirely clear; a subspecies of fantasy, close to the borderline of SCIENCE FICTION, deals with ambiguous cases, as exemplified in Robert W. CHAMBERS's *In Search of the Unknown* (coll **1904**). Modern fantasy has been remarkably promiscuous in deploying fabulous animals, for exactly the same purposes as did ancient MYTHS – mostly for use as "straw monsters" to be slain (or possibly tamed) by ambitious heroes. GENRE FANTASY makes particularly abundant use of DRAGONS and UNICORNS, having reformulated their typical roles – and hence their potential symbolism – in the direction of benignity. Other generic types sufficiently important to warrant individual consideration are SEA MONSTERS and winged horses.

Most fabulous animals are compounded out of ill-matched parts of real animals, thus belonging to the class of chimeras, named after the Chimera, mentioned in HOMER's *Iliad*, which had the foreparts of a lion, the middle of a goat and the hindparts of a snake. The gryphon (griffin) has an eagle's head and wings, the rest of its body being a lion's. The sphinx sets a humanoid head on a lion's body (adding wings in the Greek version); the manticore adds a scorpion's tail and exotic teeth to a similar assembly. A modified griffin, called a hippogriff or "griffin horse", was invented by ARIOSTO for *Orlando Furioso* (**1516**). The other common strategy employed in the making of MONSTERS is, of course, giantism, but it is doubtful whether the results really qualify as "unknown to science" except in cases where obvious extremism is combined with an element of exotic symbolism, as in Jeremias Gotthelf's *The Black Spider* (**1842**) and the movie KING KONG (**1933**).

Jorge Luis BORGES's useful guide to fantastic BIOLOGY, *The Book of Imaginary Beings* (**1967**), includes descriptions of all the above-mentioned species plus the Amphisbaena (a snake with a head at each end), the Basilisk or Cockatrice (a snake or lizard whose gaze is credited with the power to kill), Harpies (vultures with human heads), the Hydra (a multi-headed creature which grows two new heads whenever one is cut off), the Salamander (a lizard-like creature able to live in fire) and many others of more idiosyncratic prove-nance and more limited use. A useful bibliography of references to traditional fantastic creatures is Margaret W. Robinson's *Fictitious Beasts* (**1961** chap).

Modern fantasy writers have found difficulty making sig-nificant additions to the range of fabulous creatures (◊ IMAGINARY ANIMALS *for discussion*), occasionally retreating into unrepentant vagueness after the fashion of Lewis CARROLL, who never specified what snarks and jabberwocks looked like. The It of Theodore STURGEON's "It" (1940) and Angela CARTER's Odd in *Miss Z, The Dark Young Lady* (**1970**) are modern examples. Nasty creatures are usually ugly in name as well as manner, as per H.P. LOVECRAFT's shoggoths, J.R.R. TOLKIEN's orcs (a frequently borrowed term which, confus-ingly, was originally used to describe a kind of sea monster) and Stephen R. DONALDSON's urviles.   [BS]

See also: MYTHICAL CREATURES.

ANIMATED MOVIES The nature of the contribution of animation to the CINEMA of the fantastic is enigmatic: at one and the same time it (a) cannot be overestimated and (b) is far too easy to overestimate. In the most pedantic sense, very few animated movies are not fantasy, in that a huge per-centage of them involve TALKING ANIMALS like MICKEY MOUSE, BUGS BUNNY and the rest. But this alone hardly qualifies them to be considered as full fantasies, for their events might otherwise be reasonably mundane: in some cases the whole point of an animated movie is that the cir-cumstances may be fantasticated solely as a means of putting into context the foibles of people in the real world. In this sense many animated movies are related closely to the BEAST FABLES of, say, Aesop (◊ AESOPIAN FANTASY).

If we eliminate such movies from consideration, we still find ourselves with a huge number of animated movies that are most certainly fantasy. In addition, an ever-increasing number of movies combine – thanks to technological advances – live-action with animation, and these too are preponderantly fantasies.

The history of animation really began with such artifices as the flickerbook and zoetrope in the 19th century – unless one considers the cave paintings at Altamira, Lascaux, etc., as animations, in that flickering firelight seems to give them motion. However, it is hard to pin down the very first ani-mated *movie*; often this is listed as Winsor MCCAY's *Little Nemo* (**1911**), but certainly the French animator Emile Cohl had produced some short movies several years earlier. However, *Little Nemo*, which McCay not only animated but hand-coloured, opened up to others the possibility of the animated movie as an artform; indeed, it is better viewed as a demonstration of possibilities than as a movie, because it made no attempt to tell a story. McCay's nine other shorts are described in the entry on him; here we should note that *The Story of a Mosquito* (**1912**) made the breakthrough into storytelling, while with *Gertie, the Dinosaur* (**1914**) he made the discovery that animations were best performed by more than one hand; in this short he himself executed all the drawings of Gertie, while his assistant John Fitzsimmons provided the backgrounds, based on McCay's originals. As examination of the credits of any more recent AM will con-firm, this division of labour became a persistent – and nec-essary – feature of animation.

Others, like John Randolph Bray, Raoul Barré, Otto Messmer and Walt DISNEY, had far more ambitious ideas. Bray is best remembered for having produced the series of shorts based on a MUNCHHAUSEN-like character, **Colonel Heeza Liar**; the first of these shorts, *Colonel Heeza Liar in Africa* (**1913**), was probably the first AM to be issued to cinemas rather than featuring as part of some other perfor-mance. Bray came up with various labour-saving ideas that were extended by Earl Hurd, who had the notion that much effort would be avoided if the moving part of the animation were drawn on transparent cels (sheets of celluloid) that could be overlaid on a painted background; through the use of several cels at once, even more time could be saved in the drawing, since not all parts of a moving figure are moving at any one moment.

Barré established various technical advances, but is best remembered for having founded the first animation studio, in 1914; his sidekick was Bill Nolan, another technical inno-vator. With animators Gregory LaCava and Frank Moser, they produced the **Animated Grouch Chasers** series of shorts, many of which were apparently fantasies. In 1916, though, the entrepreneur William Randolph Hearst (1863-1951) hired Barré's collaborators to found International Film Service, which produced a number of largely forgotten cartoons as well as the screen versions of **The Katzenjammer Kids** and **Krazy Kat** (although from 1919 much better versions of these cartoons were produced by Bray's studios under licence from Hearst). Barré went on to work with Charles Bowers, who had acquired movie rights to the **Mutt and Jeff** COMIC strip – although the credits on these animated shorts acknowledged only the strip's cre-ator, Bud Fisher. The next few years saw many changes in responsibility for the production of the **Mutt and Jeff** shorts, but they served as an instructional hotbed for new animators.

In 1919 Otto Messmer, then working for Pat Sullivan's studio, created Felix the Cat, and the first **Felix the Cat** car-toon appeared later that year. (Although Sullivan claimed the credit for these cartoons and their associated comic strips, in fact Messmer and a team under him – which even-tually included Barré – were entirely responsible.) **Felix** was still going strong while Walt DISNEY was producing his early cartoons, the **Laugh-O-grams**, **Alice Comedies** and **Oswald the Lucky Rabbit**, and it was Disney who forced animation to take its next great leap forward when he pro-duced *Steamboat Willie* (**1928**), the first sound cartoon. Starring MICKEY MOUSE, this has a soundtrack that is more whistles and yelps than anything else, but the noises were sufficiently well synchronized with the on-screen action to create the illusion of being integral to it.

The next few decades' progress in animation do not easily submit to an historical account because they saw great diversification of aims and techniques; they are too often seen as being a question of "Disney, and the rest". This was far from the case (although Disney's contribution should not be underestimated; ◊ DISNEY; Walt DISNEY). Of considerable importance in this period were, in their different ways, Tex AVERY, Max FLEISCHER, Chuck JONES, Walter LANTZ and Charles MINTZ – not to mention Disney's collaborator, Ub IWERKS. Nevertheless, it was Walt Disney who realized that colour would be animation's next big milestone, signing a contract with Technicolor that allowed him to produce the first cartoon in full colour (there had been earlier dead-end experiments with limited colour), *Flowers and Trees* (**1932**), and it was he who created the first full-length animated feature, SNOW WHITE AND THE SEVEN DWARFS (**1937**). Disney's pre-eminence also established a trend that continues to this day: the use of classic stories – especially WONDER TALES – as the bases for animated features. The first serious competitor to Disney in this field, Max Fleischer's GULLIVER'S TRAVELS (**1939**), conforms to this pattern, while Disney – and later the DISNEY studio – have over the decades produced PINOCCHIO (**1940**), *The Adventures of Ichabod and Mr Toad* (**1949**) – based on Washington IRVING's "The Legend of Sleepy Hollow" (*c*1819) and Kenneth GRAHAME's *The Wind in the Willows* (**1908**) (◊ *The* WIND IN THE WILLOWS) – CINDERELLA (**1950**), ALICE IN WONDERLAND (**1951**), PETER PAN (**1953**), SLEEPING BEAUTY (**1959**), *The Sword in the Stone* (**1963**) – a treatment of the ARTHUR legend, loosely based on T.H. WHITE's *The Sword in the Stone* (**1939**) – *The Jungle Book* (**1967**) – loosely based on Rudyard KIPLING's *The Jungle Book* (coll **1894**) – *Robin Hood* (**1973**), MICKEY'S CHRISTMAS CAROL (**1983**), *Oliver & Company* (**1988**) – very loosely based on Charles DICKENS's *Oliver Twist* (**1838**) – *The* LITTLE MERMAID (**1989**), *The Prince and the Pauper* (**1990**), BEAUTY AND THE BEAST (**1991**) and ALADDIN (**1992**). Warner Brothers have raided a similar stockpot with movies like *Treasure Island* (**1972**), *Oliver Twist* (**1974**) and *The Nutcracker Prince* (**1990**); Richard WILLIAMS has dipped his spoon with *Raggedy Ann and Andy* (**1977**), based on stories and characters created by Johnny Gruelle (1880-1938), and CHARLES DICKENS' A CHRISTMAS CAROL, BEING A GHOST STORY OF CHRISTMAS (**1971** tvm); Rankin-Bass have done more than that with movies like *Return to Oz* (**1964** tvm), *The Cricket on the Hearth* (**1967** tvm), *The Wacky World of Mother Goose* (**1967** tvm), *A* CHRISTMAS CAROL (**1970** tvm), *The Emperor's New Clothes* (**1972** tvm), *The Hobbit* (**1977** tvm) and *The* WIND IN THE WILLOWS (**1987** tvm). HANNA-BARBERA have been there with *Jack and the Beanstalk* (**1967** tvm), *Gulliver's Travels* (**1979** tvm) and *The Count of Monte Cristo* (**1973** tvm). Others of interest in this context include: ANIMAL FARM (**1955**); *Journey Back to Oz* (**1974**) by Hal Sutherland; *The* LORD OF THE RINGS (**1978**) by Ralph BAKSHI; *The* WATER BABIES (**1979**) by Lionel Jeffries; *The* LION, THE WITCH AND THE WARDROBE (**1979** tvm); *A* CHRISTMAS CAROL (**1984**) from Burbank Films; *Adventures of Sinbad* (**1975**), *The Adventures of Sinbad the Sailor* (**1975**) and *The Adventures of Sinbad* (**1979** tvm) (◊ SINBAD MOVIES); *Dr Jekyll and Mr Hyde* (**1986** tvm) (◊ JEKYLL AND HYDE MOVIES); *The New Gulliver* (**1933**), *Gulliver's Travels Beyond the Moon* (**1966**), *Gulliver's Travels* (**1976**) and *Gulliver's Travels Part 2* (**1983**) (◊ GULLIVER MOVIES); *The* PHANTOM OF THE OPERA (**1987**); and *The* PRINCESS AND THE GOBLIN (**1992**). A recent example of note is Don BLUTH's THUMBELINA (**1994**).

It might be easy, looking at the length of this list, to get the impression that animated movies are nothing but retellings of the classics. But this is certainly not true. Particularly outside the USA, commercial moviemakers have recognized that animation can be used to create new tales, and to challenge rather than merely entertain. Nowhere is this more obvious than in the career of the Czech animator Jan SVANKMAJER. He has produced only two features, both using a combination of live action, stop-motion animation and puppetry, and both might seem, to judge only by their titles, to fall into the pattern of animated classics: ALICE (**1988**) and

FAUST (**1994**). In fact both are extremely disturbing, extremely complex works in which the originals are mere springboards for tales that could probably be told in no other way: they are adult fare, not children's entertainments. With much lighter intent, but again with great ambition, the Italian Maurizio NICHETTI has made live-action/animated movies that are open to deeper interpretation than their wildly entertaining exteriors might suggest: ALLEGRO NON TROPPO (**1976**), which he cowrote, and VOLERE VOLARE (**1991**), which he cowrote and codirected. The US emigré Don BLUTH has produced various animated movies that are superficially entertainments but in fact pack a message, notably *An* AMERICAN TAIL (**1986**), while the Australian FERNGULLY: THE LAST RAINFOREST (**1992**) and the UK WATERSHIP DOWN (**1978**) likewise sugarcoat their different pills – as, indeed, did Disney's FANTASIA (**1940**), whose agenda was the popularization of classical music. In the USA, it is hard to think of a commercial animator who has consistently picked up the glove offered by the medium except Ralph BAKSHI, who has – not always with success – uncompromisingly produced adult material, including the fantasies WIZARDS (**1977**), *The* LORD OF THE RINGS (**1978**), FIRE AND ICE (**1982**) and the live-action/animated COOL WORLD (**1992**).

This last is an excellent example of the use of the mixture of live-action and animation to generate what is in effect a new breed of fantasy creature: the TOON, the animated character which has to cope with a physical, human-populated world. Other relevant movies include DUNDERKLUMPEN! (**1974**), PETE'S DRAGON (**1977**), VOLERE VOLARE (**1991**), *The* SECRET ADVENTURES OF TOM THUMB (**1993**), LAST ACTION HERO (**1993**) – to a much lesser extent than the others – and of course the magnificent WHO FRAMED ROGER RABBIT (**1988**). TRON (**1982**) reverses the process, making humans vulnerable in a toon environment.

In the 1990s animation has achieved a healthier state than it has enjoyed for decades, and is now regarded as an important component of the commercial CINEMA, with each new Disney feature regularly being among the top grossing movies of its year and with, for example, Tim BURTON's stop-motion *The* NIGHTMARE BEFORE CHRISTMAS (**1993**) being an astonishing international success. True, there are still some very poor – *patronizingly* poor – animated movies being released, but this is countered by the fact that people like Bakshi, Bluth, Nichetti and Švankmajer (not to mention DISNEY, of course) are still at work. Adding vigour to the scene is the fact that the COMICS and animation have – thanks probably to the huge popularity of Japanese ANIME – resumed the mutually beneficial relationship which they enjoyed in the earliest days, with very creditable attempts being made to transfer comics material to the screen – as in BATMAN – MASK OF THE PHANTASM (**1993**). In the same year, almost entirely ignored, came LITTLE NEMO: ADVENTURES IN SLUMBERLAND (**1993**), based on the Winsor MCCAY strip and with an astonishingly impressive credits list. And the future looks exciting, too, especially since reasonably priced animation software packages have become available for use on home computers. In spring 1996, moreover, the huge success of a string of recent Disney animated features brought many other studios, established or specifically set up, into the game. [JG]

Further reading: *The American Animated Cartoon: A Critical Anthology* (**1980**) ed Danny Peary and Gerald Peary, recommended; *Of Mice and Magic: A History of American*

Animated Cartoons (**1980**; rev 1987) by Leonard Maltin, recommended; *World Encyclopedia of Cartoons* (**1980**) by Maurice Horn; *Before Mickey: The Animated Film 1898-1928* (**1982**; rev 1993) by Donald Crafton, which also has a companion video compilation, *Before Mickey: An Animated Anthology* (**1993**), containing such shorts as *Fantasmagorie* (**1908**), *Gertie* (**1914**), *Out of the Inkwell: Perpetual Motion* (**1920**), *Alice's Mysterious Mystery* (**1926**), *The Lunch Hound* (**1927**) and *Felix the Cat in The Oily Bird* (**1928**); *The Great Cartoon Directors* (**1983**) by Jeff Lenburg; *That's All Folks!: The Art of Warner Bros. Animation* (**1988**) by Steve Schneider, which sadly lacks an index; *Tom and Jerry: Fifty Years of Cat and Mouse* (**1991**) by T.R. Adams, an exceptionally useful book covering Tom & Jerry in their various incarnations, containing also good filmographies (◊ HANNA-BARBERA) plus a lot about MGM animation; *The Encyclopedia of Animated Cartoons* (**1991**) by Jeff Lenburg (based in part on *The Encyclopedia of Animated Cartoon Series* [*c***1981**] by Lenburg), inaccurate but useful; *Animating Culture: Hollywood Cartoons from the Sound Era* (**1993**) by Eric Smoodin, concentrating on shorts; *Cartoons: One Hundred Years of Cinema Animation* (trans Anna Taraboletti-Segre **1994**) by Giannalberto Bendazzi. For books on the Disney output ◊ DISNEY.

See also: AKIRA (**1987**); *The* BLACK CAULDRON (**1985**); *The* BRAVE LITTLE TOASTER (**1987**); DICK TRACY (**1990**); DUCKTALES: THE MOVIE – TREASURE OF THE LOST LAMP (**1990**); DUMBO (**1941**); *The* LAST UNICORN (**1982**); MARY POPPINS (**1964**); *The* PAGEMASTER (**1994**); *The* RETURN OF JAFAR (**1994**); SUPERMAN MOVIES; YELLOW SUBMARINE (**1968**).

ANIMATE/INANIMATE The gap between the animate and the inanimate is, in fantasy, charged. The most famous example of the transition is the story of PYGMALION. His tale appears early in the history of Western literature. As OVID tells the story in *Metamorphoses* (written *c*AD1-8), Pygmalion is a king of Cyprus who is also an artist, and who falls in love with the STATUE of APHRODITE he has carved. He prays to the GODDESS that he may have a wife as beautiful as this work of art, and She gives life to the statue, which/whom Pygmalion then marries. Artists have for centuries portrayed the cusp or TROMPE L'OEIL moment of META-MORPHOSIS, when inanimate stone becomes animate human flesh, doing so in a manner that makes it impossible to tell whether what is seen is animate or inanimate, or magically both.   [JC]

See also: FRANKENSTEIN; GOLEM; REAL BOY; TOYS.

ANIME Japanese movie, tv and video animation. Experiments with animation by MANGA (comics) artists began in about 1914. Progress was slow, with isolated achievements being the first sound short, *Chikaro To Onna No Yononaka* ["The World of Power and Women"] (**1932**) dir Masaoka Kenzo, the first colour feature, *Hakujaden* ["The White Serpent"] (**1958**), 78 min, dir Taiji Yabushita and Toei Doga, and the first tv series, *Manga Calendar* ["Manga Calendar"] (1962; 54 episodes; Otagi). The modern anime industry began a year later with the tv series *Tetsuwan Atom* ["Astro Boy"] (1963 onwards; 193 episodes; Mushi) dir Osamu Tezuka (1929-1989), adapted by Tezuka from his bestselling manga *Atom Taishi* ["Ambassador Atom"] (1951-68 *Shonen Magazine*). *Tetsuwan Atom*'s success in Japan and the USA stimulated an enormous increase in tv anime production. Many of these early shows, like *Tetsujin 28-GO* ["Iron Man #28"] (1963 onwards; 96 episodes; TCJ) dir Tetsuo Imazawa, continued *Tetsuwan Atom*'s theme of

the robot saviour. From the 1970s onwards there was an increasing variety of shows, usually with sf or fantasy elements: among the more popular were *Lupin Sansei* ["Lupin III"] (1971 onwards; 23 episodes; Monkey Punch/TMS) dir M. Okuma, Isao Takahata and Hayao Miyazaki, the comedy adventures of a gentleman thief; *Devilman* (1972 onwards; 39 episodes; Go Nagai/Dynamic Productions) dir Go Nagai, a HORROR fantasy; and *Uchu Senkan Yamoto* ["Space Cruiser Yamoto"] (1974 onwards; 26 episodes; Westcape Corporation) dir Reiji Matsumoto and Yoshinobu Nishizaki, a space opera. The popularity of these led to further tv series and to big-screen treatments: *Uchu Senkan Yamoto* ["Space Cruiser Yamoto"] (**1977**; 130 min; Westcape Corporation) dir Reiji Matsumoto and Yoshinobu Nishizaki; *Lupin Sansei – Mamo Karano Chousen* ["Lupin III – Secret of Mamo"] (**1978**; 100 mins; Monkey Punch/TMS) dir Soji Yoshikawa; and others.

Original feature movies re-emerged in the 1980s. Hayao Miyazaki (1941-   ), the "Japanese DISNEY", directed some of the most popular and accomplished anime movies, notably the eco-fantasy *Kaze No Tani No Nausicaa* ["Nausicaa of the Valley of the Wind"] (**1983**; 118 mins; Nibariki/Tokuma Shoten/Hakuhodo/Toei) and the enchanting PASTORAL *Tonari No Totoro* ["My Neighbour Totoro"] (**1988**; 85 mins; Nibariki/Tokuma Shoten). Katsuhiro OTOMO directed the influential and acclaimed cyber-psychic movie AKIRA (**1987**).

A new format, the OAV (Original Animation Video) appeared in 1983, allowing high-quality animation to be produced and released on video at a fraction of the cost of movie and tv anime. This further increased the diversity of anime, which by now was firmly established as an integral part of Japanese popular culture. OAV particularly enabled the growth of sexually explicit anime like *Chojin Densetsu Urotsukidoji* ["The Wandering Kid"] (1987; 3 episodes; Maeda/Javn) dir Hideki Takayama, which mixed sex, violence and fantasy. Nudity and sexual subjects are common in anime in general, even in family tv series like the popular *Ranma Nibun No Ichi* ["Ranma 1/2"] (1990; Takahashi/Kitty Film) dir Rumiko Takahashi (1957-   ). Takahashi is a rare female creator in a male-dominated industry; her strong female characters are, however, not the exception, as women are generally well represented in anime, with all-female teams like those in the tv series *Dirty Pair* (1985 onwards; 24 episodes; Studio Nue/Sunrise/NTV) dir Haruki Takachiho and the OAV series *Bubblegum Crisis* (1987 onwards; 8 episodes to date; AIC/Artmic) dir Akiyama Katsuhito presenting images of independent, capable women.

The influence of anime on Japanese society and culture extends far beyond the movies, tv shows and OAVs; and, as the strong visual identities of anime characters are eminently marketable, merchandizing covers everything from soundtrack CDs to stationery, model kits and junk food. In the West, anime has become particularly popular in the form of videos, *Akira* probably marking the point where it emerged from obscure cult status, and there are now flourishing monthly magazines covering the subgenre.   [RM]

ANIMUS ◊ ANIMA.

ANMAR, FRANK ◊ Charles BEAUMONT; William F. NOLAN.

ANNUALS Annuals occasionally appeared in the summer but were more often a phenomenon of winter, owing something of their origins to the tradition of the almanac; of particular significance here was the *Vox Stellarum* of the astrologer Francis Moore (1657-1714), which first appeared in 1700 and is still an annual tradition as *Old Moore's Almanac*. The association with the dark evenings and particularly with Christmas (◊ CHRISTMAS BOOKS) meant that annuals frequently featured GHOST STORIES. As a result of both these factors annuals are closely linked to the field of SUPERNATURAL FICTION.

The first genuine annuals were highly illustrated and, with the new technique of steel engraving, artwork was much featured (◊ ILLUSTRATION). This made them more expensive than normal books or magazines, so they were issued as giftbooks. The format was set by the lithographer Rudolph Ackermann (1764-1834), who brought the idea from Germany. Already established in London from 1795 with his fine-art printing, he issued the first giftbook, **Forget-Me-Not** in 1823 (**1823-47** 25 vols ed Frederic Shoberl [1775-1853]). This rapidly begat imitations, of which the most successful was **The Keepsake** (**1828-57** 30 vols); the first volume ed William Harrison AINSWORTH, who was succeeded 1829-35 by Frederic M. Reynolds (?1800-1850). **The Keepsake** published several supernatural stories by Sir Walter SCOTT and Mary SHELLEY. Its US equivalents were **The Token** (**1828-42** 15 vols), notable for running many stories by Nathaniel HAWTHORNE, and **The Gift** (**1840-45** 4 vols), which published several HORROR stories by Edgar Allan POE, including "The Pit and the Pendulum" (1844). The expense of the annuals led to their downfall in the late 1840s.

By then, particularly in the UK, the publication of CHRISTMAS BOOKS had become increasingly popular. This was especially so after the success of the **Christmas Stories** by Charles DICKENS, which started with *A Christmas Carol* (**1843**) and continued with a new Christmas book each year until *The Haunted Man* (**1848**). The medium was continued by William Makepeace THACKERAY, whose FAIRYTALE *The Rose and the Ring* (**1855**) as by M.A. Titmarsh was the last of his Christmas stories. By this time the literary and popular MAGAZINES had established themselves, and they began to publish special Christmas issues, which during the 1850s began to supersede the literary annuals, though they were not so attractive. The leader was Dickens, who issued special Christmas issues of his magazine *Household Words* (1850-59) and its successor *All the Year Round* (1859-95); these frequently sold in excess of 250,000 copies. Some of the most popular of all Victorian GHOST STORIES appeared in these Christmas issues, including work by Wilkie COLLINS, Amelia B. Edwards (1831-1892), Elizabeth Gaskell (1810-1865) and J. Sheridan LE FANU.

The Christmas annual became firmly established with **Beeton's Christmas Annual** (39 vols **1860-98**), the best-remembered of the Victorian annuals because it published the first SHERLOCK HOLMES story, *A Study in Scarlet* (**1887**). A few volumes of **Beeton's** emphasized the unusual, like *Bach-O-Bahar* (anth **1871**), which used the ARABIAN FANTASY as a vehicle for SATIRE, *The Fortunate Island* (anth **1880**), which used a remote ISLAND in its FRAME STORY by Max ADELER, and *A Dead Town* (anth **1883**).

The early Christmas annuals of *Tinsley's Magazine*, during the editorship of Edmund Yates (1831-1894), are particularly interesting. The first three – *Storm-bound* (anth **1867**), *A Stable for Nightmares* (anth **1868**; rev 1896 US) and *Thirteen at Table* (anth **1869**) – had a frame device whereby people who had gathered together (either voluntarily or by

mishap) told tales – frequently ghost stories – to pass the time. *A Stable for Nightmares* is devoted entirely to the supernatural. All these annuals featured either complete novels or collections of stories often linked by a frame story. They are difficult to distinguish from ANTHOLOGIES, which were also emerging during this period.

The **Belgravia Annual** (**1867-96** 30 vols), a special additional Christmas volume of *Belgravia*, frequently carried ghost stories, including those by its initial editor, M.E. BRADDON. Other significant annuals in our context include the 1883 **Bow Bells Annual** – entitled *Stories with a Vengeance* – which contained tales of vengeful SPIRITS, and two volumes of **Unwin's Christmas Annual**, ed Sir Henry Norman (1858-1939). The first of these latter, *The Broken Shaft* (**1886**), set aboard a stranded liner, included the original publications of "The Upper Berth" by F. Marion CRAWFORD and "Markheim" by Robert Louis STEVENSON; the second, *The Witching Time* (**1887**), featured "By the Waters of Paradise" by Crawford and "A Mystery of the Campagna" by his sister Anne Crawford, Baroness Von Rabe (1846-?   ). **Arrowsmith's Christmas Annual** began in 1881 with the imitative frame story "Thirteen at Dinner" and included a few supernatural stories, but the main success of this series was the 1883 volume, which featured the novel *Called Back* (**1883** chap) by Hugh Conway, a murder mystery including telepathy (◊ TALENTS), which in this and subsequent editions sold almost half a million copies within three years.

Annuals continued to be a feature of magazine publishing until WWI, but thereafter they were almost wholly superseded by books. Some of these ANTHOLOGIES were annuals in terms of frequency of appearance (e.g., the NOT AT NIGHT series and the PAN BOOK OF HORROR STORIES), but they lacked the atmosphere or variety of the Victorian annuals.

The annual format switched to children's books. Children's annuals had been around almost as long as their adult equivalent, starting with **The Christmas Box** (**1828-9** 2 vols) ed T. Crofton Croker (1798-1854) and **The Juvenile Forget-Me-Not** (**1829-37** 9 vols) ed Anna Maria Hall (1800-1881). Children's annuals likewise soon became associated with magazines, but the tradition of the annual remained stronger in children's publishing, and a number of annuals appeared not directly associated with magazines; these included **Blackie's Children's Annual** (**1904-40** 37 vols) and **Joy Street Annual** (**1923-36** 14 vols; *#13* was numbered *#12a*). The latter included many fantasies by Algernon BLACKWOOD, Laurence HOUSMAN and Compton Mackenzie (1883-1972), plus poetry by Lord DUNSANY. Particularly influential in the fantasy world has been the **Rupert Annual** (◊ RUPERT THE BEAR).

The closest the fantasy genre has to annuals today are the annual selection of the year's best stories. The latest such selection is **The Year's Best Fantasy and Horror** (first issued as *The Year's Best Fantasy* anth **1988**; retitled *The Year's Best Fantasy and Horror* from *#3* anth **1990**) ed Ellen DATLOW and Terri WINDLING, and **Best New Horror** (*#1* anth **1990**) ed Stephen JONES and Ramsey CAMPBELL (ed Jones alone from *#6* anth **1995**). [MA]

Further reading: *Literary Annuals and Gift-Books* (**1912**) by F.W. Faxon; *An Index to the Annuals 1820-1850* (**1967**) by Andrew Boyle.

ANNWN ◊ AVALON.

ANSTEY, F. Working name of UK writer, humorist and long-term contributor to *Punch* Thomas Anstey Guthrie

(1856-1934); his intended pseudonym was T. Anstey, the "F" stemming from a misprint. He is best-known for humorous fantasies introducing TOPSY-TURVY elements into Victorian life through some arbitrary, capricious magical intrusion.

His debut novel, *Vice Versâ, or A Lesson to Fathers* (**1882**; rev 1883) (the circumflex is *sic*), was highly successful. A magic "Garudâ Stone" grants the insincere WISH of a stern Victorian father to be a boy again: his METAMORPHOSIS encourages his school-hating son to wish for adult stature and freedom, resulting in IDENTITY EXCHANGE. Unable to shed pompous speech-patterns, the father suffers condignly at boarding-school and, as a man who LEARNS BETTER, enjoys improved relations with his son when the SPELL is undone. The book has been twice filmed as VICE VERSA.

In *The Tinted Venus* (**1885**) a young man commits the classic error of jokingly placing his engagement RING on a park's statue of APHRODITE – which the GODDESS animates (◊ PYGMALION), loftily accepting his presumed love and causing predictable fiancée trouble. (Tim POWERS's *The Stress of Her Regard* [**1989**] reprised the notion.) This disruptive pattern continues with *A Fallen Idol* (**1886**), where an artist's life is afflicted by his innocent possession of a malevolent Indian (Jain) idol, which destroys a loved dog by toppling on it, ruins the painter's reputation by making unflattering changes to his portraits, and even renders him temporarily colour-blind (at least in ILLUSION).

FA's fantasy ideas are often imperfectly followed through: after *Vice Versâ* he seemed too ready to eke out situational humour with conventionally snobbish gibes at the lower middle classes (*The Tinted Venus*) or policemen and foreigners with "comic" accents (*A Fallen Idol*). Nor are his resolutions hugely ingenious, a low point being the otherwise entertaining *Tourmalin's Time Cheques (A Farcical Extravagance)* (**1885**; vt *The Time Bargain, or Tourmalin's Cheque Book* 1905). This TIME FANTASY supposes that idle hours – here, on a steamship voyage – may be deposited in the Time Bank at compound interest, and drawn out when required. When after landfall the hero wearies of LONDON and cashes a cheque for some idyllic sea-travel time, the restored hours arrive out of sequence and prove unexpectedly eventful: time paradoxes confuse his romantic entanglements, building to such delirious complexity that FA resorts to making the whole farrago a deckchair DREAM.

The Brass Bottle (**1900**) offers a fruitfully funny situation as a grateful released GENIE showers its benefactor with increasingly embarrassing boons – the transformation of his cheap London lodgings into a palace of more than Oriental splendour convinces staid prospective in-laws that he may lack thrift, and a belly-dancer is even worse received. Eventually the genie is neutralized by resourceful bluffing. A stage version appeared as *The Brass Bottle: A Farcical Fantastic Play* (**1911**); this was FA's nearest return to the early success of *Vice Versâ*.

In *Brief Authority* (**1915**) departs from FA's usual template, with a stiff Victorian lady becoming queen over the FOLKTALE country of the GRIMM BROTHERS (◊ RURITANIA); this was his last novel. Shorter stories and squibs appear in *The Black Poodle and Other Tales* (coll **1884**), *The Talking Horse* (coll **1891** US) – whose title story prefigures the masterful TALKING ANIMAL and submissive rider of C.S. LEWIS's *The Horse and His Boy* (**1954**) – *Salted Almonds* (coll **1906**), *Percy and Others* (coll **1915**) – the first five stories relating the adventures of a bee – and *The Last Load* (coll **1925**). These apply routine humour to animated TOYS, ANIMAL FANTASY, FAIRYTALES, family CURSES, GHOSTS and (a rare serious story) SIRENS. *Humour and Fantasy* (omni **1931**) assembles *Vice Versâ*, *The Tinted Venus*, *A Fallen Idol*, *The Talking Horse*, *Salted Almonds* and *The Brass Bottle*; FA's new introduction shows a certain resignation at never having outdone the then 49-year-old *Vice Versâ*. [DRL]

Other works: *The Giant's Robe* (**1884**) and *The Pariah* (**1889**), both associational; *The Statement of Stella Maberley, Written by Herself* (**1896**), published anon; *Paleface and Redskin, and Other Stories for Girls and Boys* (coll **1898**); *Only Toys!* (**1903**), for children.

Further reading: *A Long Retrospect* (**1936**), autobiography.

ANSWERED PRAYERS SENECA advised: "Do not ask for what you will wish you had not got." Common in most folk literatures (as in the *locus classicus*, the GRIMM BROTHERS' "The Three Wishes"), the motif of the answered prayer responds to a sense of the enormous *dangerousness* (◊ CONDITIONS; PROHIBITIONS; QUIBBLES) of making a WISH under conditions which seem to compel obedience, of attempting to trick the cosmos and the GODS into granting a desired outcome. The AP may be defined as a prayer which, when answered literally, brings unsought but terrible consequences; and it is in this sense that Charles DICKENS's *The Haunted Man and the Ghost's Bargain* (**1848**) can be seen as an effective early use of the AP as a literary device: the haunted man agrees with his DOPPELGÄNGER to have all memory of "sorrow, wrong, and trouble" expunged from his mind; and finds that he has transformed himself into a hollow man, an artificial creature divested of that which made him human.

The most famous AP story is almost certainly "The Monkey's Paw" (1902) by W.W. Jacobs (1863-1943). The eponymous TALISMAN has been cursed by an Indian fakir who "wanted to show that fate ruled people's lives, and that those who interfered with it did so to their sorrow". It will grant THREE WISHES. The first brings an old couple £200 – which turns out to be compensation for the death of their son. The couple then wish for his return. At the last moment, their third wish protects them from the horror that scrabbles at the door. In Graham Masterton's "The Taking of Mr Bill" (1993), a GASLIGHT-ROMANCE version of this tale, a horrific PETER PAN figure kills the SOULS of young boys; the mother of one dead lad calls him back, though without his used soul. In *Pet Sematary* (**1983**) by Stephen KING, filmed as PET SEMATARY (*1989*), the bereaved can bury their dead in an ancient cursed ground and know that so doing guarantees physical return of the loved ones – but with souls that are murderously different.

Answered Prayers (**1987**), a fragmentary nonfantasy novel by Truman Capote (1924-1984), examines people whom success has tricked; as does *Into the Woods* (**1987**) (◊ INTO THE WOODS) by Stephen Sondheim (1930-). When Pierce Moffatt, in John Crowley's *Aegypt* (**1987**), elaborately considers how to couch in safe terms the three wishes he longs to have, he is attempting in his imagination to avoid an AP. Jonathan CARROLL, in his sequence starting with *Bones of the Moon* (**1987** UK), has constructed a series of AP meditations on the costs of success in life, in art, in negotiations with FATE. Stephen King, in *Needful Things* (**1991**), combines the motif with that of the Little SHOP to demonstrate that the urge to get what one wants most is one that easily becomes damnable. [JC]

ANTHOLOGIES An anthology is a collection of stories by various authors, either new works or reprints, assembled by an editor and published in book form. It is usually distinguished from a MAGAZINE by being a one-off publication (although anthology series exist) and its lack of editorial features and departments (especially letter columns and book reviews). That said, since the emergence of mass-market paperback publishing in the 1940s, the distinction between anthologies and magazines has become increasingly blurred (◊ AVON FANTASY READER *for an example*).

The early tendency to publish works anonymously makes it difficult to ascertain whether many volumes of stories published before the 19th century are anthologies or single-author collections. Quite commonly writers or translators would copy or adapt the works of others, sometimes without acknowledgement. More frequently, where stories were from the oral tradition, no original author was known, so the credited author tends to be the person who first recorded the tale in writing. This is true of the earliest surviving anthologies, like the *Iliad* and the *Odyssey*, usually attributed to HOMER. It is likely that even earlier collections of tales and legends existed, as attested by a surviving fragment of the *Westcar Papyrus* (*c*2000 BC) from ancient Egypt.

In the Greek world, the best-known anthologists of MYTHS and LEGENDS included Hesiod (8th century BC), Herodotus (484-425BC) and Plato (428-347BC), but they were far exceeded by the Romans, who delighted in gathering tales. Those most linked with FANTASY and the supernatural were VIRGIL with *The Aeneid* (?**20**BC), OVID with *Metamorphoses* (AD**2-8**) and Gaius Petronius (? -AD65) with *The Satyricon* (AD**60**).

Such volumes recounting earlier tradition are common in all ancient cultures. The best-known is the BIBLE, the books of which were written down between about 1500BC and 440BC (Old Testament) and between AD40 and AD100 (New Testament), although the version we know today did not come into being until compiled by Jerome (*c*342-420) in the 4th century. Of equal size and scale are the great Indian epic poems the *Rámáyana* and the *Máhábharata*, steadily amassed between the 5th century BC and the 4th century AD, and the Persian *Panchatantra*, which dates back as far as the 2nd century BC, but which was written down only in the 8th century AD by Ibn al-Muqaffa.

These aggregations of tales and histories, with nonfiction often indistinguishable from fiction, continued throughout the centuries (◊ TAPROOT TEXTS); others included the Icelandic *Prose Edda* (**1222**) by Snorri Sturluson (1179-1241), the *Historia Regum Britanniae* (**1136**) by GEOFFREY OF MONMOUTH, and the *Gesta Romanorum* (?**1300**), of unknown authorship, which is one of the foundation anthologies of FAIRYTALES and was the first popular volume to establish the format of separate stories within a FRAME STORY. This device was used to considerable effect by Giovanni BOCCACCIO in his *Decameron* (**1358**), and was taken up by Geoffrey CHAUCER in *The Canterbury Tales* (?**1387**) and Giambattista Basile (1575-1632) in *The Pentamerone* (*Lo Cunto de li Cunti*) (**1634**). Such volumes are thus the precursors of modern anthologies. They all feature the TWICE-TOLD tale, with the concentration on STORY, which is symptomatic of fantasy. As other examples we can mention: *Liao-Chai chih i* (**1679**; part trans by Herbert A. Giles as *Strange Stories from a Chinese Studio* **1913**) compiled by P'u Sung-Ling (1640-1716); *The Arabian Nights* (◊ ARABIAN FANTASY) – some tales of which date back at least as far as the Persian book of fairytales *Hazar Afsanah* ["The Thousand Legends"] *c*850) – and which first appeared in

Europe as *Les Mille et Une Nuits* (**1704-17**) compiled by Antoine Galland (1646-1715); and of course the many volumes of fairytales produced by the GRIMM BROTHERS, beginning with *Die Kinder- und Hausmärchen* (**1812**).

When more formal anthologies emerged they likewise assembled stories from anonymous or attributable sources which were themselves based on FOLKTALES. During the 18th century, literary collections, often called miscellanies, began to appear with increasing frequency, some – such as *The Ladies Tale* (**1714**) ed anon and *Winter Evening Tales* (**1731**) ed anon – utilizing the oft-repeated frame device of stories told by each person among an assembled group. Several of these stories, though not all, might be macabre, with some venturing into the territory of SUPERNATURAL FICTION.

The first English-language anthology of fantastic tales was *Tales of the East* (anth **1812**) compiled for Walter SCOTT by his literary assistant, Henry William Weber (1783-1818). It is a massive collection of ORIENTAL FANTASIES. Scott had long been interested in Oriental and Gothic fiction, and had planned to assemble an anthology of the latter with Matthew Gregory LEWIS, but financial straits precluded it. Instead he issued *Tales of Terror: An Apology* (anth **1799** chap), a privately printed limited-edition booklet of 12 macabre poems by himself, Lewis and Robert SOUTHEY. Lewis, with Scott's help, did assemble an anthology of HORROR verse, *Tales of Wonder* (anth **1801**). Between them, Scott, Lewis and Weber laid the basis for the interest in the selection and promotion of Gothic and Oriental fiction and verse in anthology form.

At the same time appeared possibly the most influential anthology of Gothic fiction, *Tales of the Dead* (anth **1813**), translated and compiled by Sarah Utterson (?1782-1851). This book was read in its French version by Lord BYRON, John POLIDORI, Percy Bysshe SHELLEY and Mary SHELLEY during their holiday at the Villa Diodati, and inspired them to turn their minds to writing GHOST STORIES; the major result was Mary Shelley's *Frankenstein, or The Modern Prometheus* (**1818**).

Since then over 2000 anthologies of supernatural, fantasy and horror fiction have been published. These are discussed separately (with some unavoidable overlap) in the sections below: Fantasy Anthologies, Anthologies of Supernatural Fiction, and Horror Anthologies. [MA]

FANTASY ANTHOLOGIES

Distinguishing between FANTASY and the mundane was not an issue to early writers, by whom the existence of GODS and the supernatural was taken for granted. Similarly, until fantasy became a marketable product in the 1970s, little attempt was made to define it as distinct from other genres. This discussion recognizes fantasy as both (a) that field of literature that has MYTH and FOLKLORE as its roots and (b) a latter-day marketing niche.

The earliest compilations of traditional tales often included, alongside more mundane tales, accounts of the fantastic, with such events treated as everyday occurrences. Stories in which the fantastic was distinguished from mundane reality did not start to emerge until the 18th century. The first popular anthology of the fantastic was the series of ARABIAN FANTASIES *Les Mille et Une Nuits* (anth **1704-17**), compiled by Antoine Galland (1646-1715). These and other ORIENTAL FANTASIES dominated popular fiction of the century, and led eventually to the first comprehensive anthology of such stories, *Tales of the East* (anth **1812**) by

Henry William Weber (1783-1816). This included not only Galland's *Arabian Nights' Entertainments* but other Arabian, Persian, Indian and Mogul tales, plus the novels *The History of Nourjahad* (**1767**) by Frances Sheridan (1724-1766) and *The History of Abdalla* (**1729**) by Jean-Paul Bignon (1662-1743). Weber also edited a volume of FANTASTIC VOYAGES, *Popular Romances* (anth **1812**). These volumes emphasize Weber's pioneering role as collector and anthologist.

By the early 19th century the Oriental tale was being challenged for popularity by the Gothic tale (◊ GOTHIC FANTASY), which in turn fed into the development of GHOST STORIES and HORROR stories. During this transition, some Gothic anthologies continued to reflect traditional FOLK-TALES, such as those collected in *Interesting Tales* (anth **1796**) ed anon or more influentially in *Popular Tales and Romances of the Northern Nations* (anth **1823**), which included stories from GERMANY by Johann August APEL, Friedric FOUQUÉ, Johann Karl MUSÄUS and Ludwig TIECK. Their popularity was further enhanced by Thomas CARLYLE's translation of selected *Volksmärchen* in *German Romance* (anth **1827** 4 vols), which focused particularly on Musäus and Tieck but also included representative examples by E.T.A. HOFFMANN and GOETHE. The work of these writers was a significant influence on UK writers and on Edgar Allan POE, but more specifically on the development of SUPERNATURAL FICTION. The fantasy elements became increasingly consigned to the world of the FAIRYTALE, even though Tieck and Musäus mined the same motherlode as the GRIMM BROTHERS, who contemporaneously were assembling their *Die Kinder- und Hausmärchen* (anth **1812-15**). Fairytales, folktales, the wider world of myths and LEGENDS and the emerging world of the LITERARY FAIRYTALE thus dominated fantastic literature in the 19th century, with much of the work increasingly aimed either at younger readers or at folklorists and ethnologists, with little serious attempt to publish anthologies of fantasy for general adult readers. Anthologies at this time consisted almost wholly of horror and ghost stories, and even *The Garden of Romance* (anth **1897**) by Ernest Rhys included stories by Poe, Washington IRVING and Nathaniel HAWTHORNE in its wider selection of fantasy. Isolated but serious treatment came from Frederick de Berard (1853-1927), who compiled **Famous Tales** (anth **1899** 17 vols); this contained volumes devoted to *Famous Tales of Wonder*, including the early German Romantics (◊ ROMANTICISM), *Famous Tales of Fairyland and Fancy*, which focused on fairytales, *Famous Tales of the Orient*, *Famous Tales of Gods and Heroes* and, of more importance, *Famous Tales of Enchantment*, which included early WONDER TALES – by Herman MELVILLE, John RUSKIN, William MORRIS and Robert Louis STEVENSON – which were thus being distinguished from fairytales, a distinction rare at this time. More typical was *Tales of Fantasy* (anth **1902**) ed Tudor Jenks (1857-1922); this reprinted similar folktales and legends, but packaged them for young readers. Significant in this area were the tales retold by Andrew LANG in his variously coloured **Fairy Books**. Despite the publication at this stage of the works of William Morris and Lord DUNSANY, fantasy seemed inextricably linked to the fairytale, and almost all anthologies were so tailored.

One rare exception was *Through the Forbidden Gates* (anth **1903**) ed anon by Herman Umbstaetter (1851-1913), a selection of strange and unusual stories from the magazine *The Black Cat*. These stories were often on the borderline of the fantastic, and are really early examples of SLICK FAN-TASY. The creation of this genre occurred predominantly in

the MAGAZINES, and followed in book form only when anthologists mined magazines for appropriate stories. The pioneering anthologist Joseph L. French (1858-1936) assembled several such volumes, in particular *The Best Psychic Stories* (anth **1920**), which looked at the wider world of the supernatural beyond GHOSTS and SPIRITS, and his four-volume *Masterpieces of Mystery* (anth **1920**), which included several stories of the inexplicable. Although the occasional fantasy story continued to appear, ghost and horror stories dominated anthologies until after WWII.

The pioneering anthology of modern fantasy was *Pause to Wonder* (anth **1944**) ed Marjorie Fischer (1903-1961) and Rolfe Humphries (1894-1969). This compilation was a deliberate move against the HORROR tradition to include stories that were more lighthearted and/or uplifting while remaining "miraculous and marvellous"; by the time of this publication a sufficient body of fantastic literature had emerged in books and magazines to warrant a solidly representative selection. Although among the 80 or so items was material from Roman and medieval times, the bulk came from the previous 50 years, and included tales and poems by Max BEERBOHM, Ambrose BIERCE, John BUCHAN, G.K. CHESTERTON, John COLLIER, Walter DE LA MARE, F. Scott FITZGERALD, E.M. FORSTER, David GARNETT, Robert GRAVES, W.W. JACOBS, Henry JAMES, D.H. LAWRENCE, Arthur MACHEN, Liam O'FLAHERTY, John Steinbeck, Frank R. STOCKTON, James THURBER, Sylvia Townsend WARNER, H.G. WELLS, Oscar WILDE, Virginia WOOLF and W.B. YEATS. The tales were predominantly slick fantasy, but there were also Absurdist (◊ ABSURDIST FANTASY) and Surreal (◊ SURREALISM) fictions. Fischer and Humphries extended their selection in *Strange To Tell* (anth **1946**), with stories from around Europe. This volume showed that fantastic literature was more prevalent on the Continent than in the UK. The international status of fantasy had in fact been recognized in Argentina where *Antología de la literatura fantastica* (anth **1940**) had been compiled by Jorge Luis BORGES, Silvina Ocampo (1906-   ) and Adolfo BIOY CASARES, but it was not until its English translation as *The Book of Fantasy* in 1976 that the English-speaking world could fully appreciate the extent to which non-English writers had long adopted fantasy – particularly Absurdist, Surreal and INSTAURATION FANTASY – as a standard part of literature. Two other wartime anthologies did attempt to acknowledge the development of fantasy outside the UK and USA: *Angels and Beasts* (anth **1947**) ed Denis Saurat (1890-1958), which selected the latest strange stories from France, and *A Night with Jupiter and Other Fantastic Stories* (anth **1945**) ed Charles Henri Ford (1913-   ). Both contained stories of the Surreal, and the latter was one of the first in English to anthologize Latin American stories of MAGIC REALISM.

Encouraged by the public acceptance of these volumes, publishers became bolder. Of particular importance in the UK was a trio of anthologies assembled by Kay DICK (two as by Jeremy Scott). In *The Mandrake Root: An Anthology of Fantastic Tales* (anth **1946**) her final introductory note emphasized that the stories focused on the inexplicable in the known, or the intrusion of the unreal into the real, which we have defined in this encyclopedia to be the province of SUPERNATURAL FICTION. In the second volume, *At Close of Eve* (anth **1947**), the introduction was handed over to Daniel George (1890-1967) who, after listing many categories of fiction that fall under the heading of fantasy, defined the genre as "any piece of fiction in which the action transgresses what in our opinion is natural law". Dick's third volume, *The Uncertain Element* (anth **1950**), stopped trying to define the field and instead presented mostly new stories and factual accounts of the bizarre and macabre. These were ostensibly supernatural anthologies, but were among the first to take the traditional limit of the supernatural beyond the run-of-the-mill ghost story into more imaginative extravagances. In so doing they widened the public's awareness of the field.

Meanwhile, in the USA, paperback publishing was emerging from the pulp MAGAZINE field. Seeking to capture the magic of pulp fantasy, Donald A. WOLLHEIM began the AVON FANTASY READER in 1949. He did not attempt to define fantasy, other than by example, and his selections covered the whole spectrum of fantastic literature, including sf, although this was kept to a minimum. The emphasis was on stories from WEIRD TALES or from UK writers, and included the exotic fantasies of Clark Ashton SMITH and Lord DUNSANY as well as the HEROIC FANTASY of Robert E. HOWARD. At this time the *Avon Fantasy Reader* had a stronger influence on the magazine field than on books. Even the more literary fantasies appearing in magazines were generally ignored in the USA; although *The Saturday Evening Post Fantasy Stories* (anth **1951**) ed Barthold Fles (1902-) gave a nod to some stories of the supernatural, its contents were mostly sf, as were those of *The Post Reader of Fantasy & Science Fiction* (anth **1963**) ed anon. The most literary of fantasy anthologies in the 1950s came from Ray BRADBURY: both *Timeless Stories for Today and Tomorrow* (anth **1952**) and *The Circus of Dr Lao and Other Improbable Stories* (anth **1956**) selected slick fantasies with a Surreal and Absurdist edge. These were the first US mass-circulation paperback anthologies to select literary fantasies, and were important in laying the groundwork for the new generation. It was not until the appearance of *Best Fantasy Stories* (anth **1962**) ed Brian W. ALDISS that a similar anthology would take the next step, into the 1960s.

Two other anthologies in what was otherwise a wasteland of fantastic fiction in the 1950s also put down some foundations. *Witches Three* (anth **1952**) ed anon by Fletcher PRATT included Pratt's own novel *The Blue Star* (exp **1969**), the first HIGH FANTASY to appear as an original in an anthology. *Shanadu* (anth **1953**) ed Robert E. Briney (1933-) was not only the first SMALL-PRESS original anthology of any scale but had the first SHARED-WORLD setting in GENRE FANTASY.

Aside from Aldiss's volume, a few anthologies in the early 1960s began to break away from the usual run of horror and supernatural: they included *The Dream Adventure* (anth **1963**) ed Roger Caillois (1913-1978), which brought together several centuries of DREAM fantasies, and *Hell Hath Fury* (anth **1963**) ed George Hay (1922-), selecting from UNKNOWN, but the most significant step was that made by L. Sprague DE CAMP with *Swords & Sorcery* (anth **1963**). The growing need for marketing niches in the paperback world had led to the creation of a new subgenre of HEROIC FANTASY, dubbed SWORD AND SORCERY by Fritz LEIBER. These were fantasies set in SECONDARY WORLDS or OTHERWORLDS whose plots were reliant on mighty warriors battling evil sorcerers, as typified by Robert E. Howard's CONAN stories. De Camp's anthology, with some of the best fantasies of Poul ANDERSON, Dunsany, Howard, Henry KUTTNER, Leiber, C.L. MOORE, H.P. LOVECRAFT and Clark Ashton Smith, all effectively illustrated by Virgil FINLAY, was very popular and resulted in several sequels: *The Spell of Seven* (anth **1965**), *The Fantastic Swordsmen* (anth **1967**) and *Warlocks and Warriors* (anth **1970**). This last, as an interesting example of marketing synchronicity, appeared within months of the UK anthology *Warlocks and Warriors* (anth **1971**) ed Douglas Hill (1935-). By then the combined success of Howard's **Conan** books and J.R.R. TOLKIEN's *LOTR* in paperback had resulted in unprecedented interest in heroic and HIGH FANTASY. Before the field began to settle down, a run of imitative S&S anthologies appeared, including: *The Mighty Barbarians* (anth **1969**) and *The Mighty Swordsmen* (anth **1970**), both ed Hans Stefan Santesson (1914-1975); *Swords Against Tomorrow* (anth **1970**) ed Robert Hoskins (1933-); *Kingdoms of Sorcery* (anth **1976**) and *Realms of Wizardry* (anth **1976**) both ed Lin CARTER; and *Savage Heroes* (anth **1977**) ed Eric Pendragon (real name Michel Parry [1947-]). There were also two series of original anthologies, **Flashing Swords!** (**1973-81** 5 vols) ed Carter and **Swords Against Darkness** (**1977-9** 5 vols) ed Andrew J. OFFUTT. None of these improved upon de Camp's anthologies, all settling for the basic warrior-vs-wizard scenario. This was also true of the **Year's Best Fantasy Stories** series (**1975-88** 14 vols) first 6 vols ed Carter, last 8 vols ed Arthur W. Saha, which under Carter printed solely heroic fantasy. A more representative selection was made in various series edited by Terry CARR, starting with **New Worlds of Fantasy** (**1967-71** 3 vols). This deliberately excluded heroic fantasy in favour of stories at "the borders of man's imagination" and, in lieu of a definition, sought to establish the extremes of fantasy in the real world. Carr explored this further in *#2*, stating that "fantasy is the literary equivalent of dreams", by which he meant that fantasy was plumbing the depths of the psyche. This was the first paperback anthology series to give consideration to the wider world of fantasy. Carr later explored this on an annual basis in **The Year's Finest Fantasy** (**1978-9** 2 vols; vt in 3 vols **Fantasy Annual** 1981-2), although it was not until the leviathan efforts of Ellen DATLOW and Terri WINDLING in their annual **Year's Best Fantasy and Horror** (from **1988**; first 2 vols vt **Demons and Dreams** 1989-90 UK) that the field secured its Egon Ronay.

The 1970s saw fantasy seeking to define itself. It was given some history and structure by Lin Carter in the volumes he assembled for the BALLANTINE ADULT FANTASY series, especially the anthologies *The Young Magicians* (anth **1969**), *Dragons, Elves, and Heroes* (anth **1969**), *Golden Cities, Far* (anth **1970**), *New Worlds for Old* (anth **1971**), *Discoveries in Fantasy* (anth **1972**) and *Great Short Novels of Adult Fantasy* (anth **1972-3** 2 vols).

Unlike the case in sf, academics have found it difficult to grapple with fantasy. Attempts were made to analyse the field by example in *Fantasy: The Literature of the Marvelous* (anth **1974**) ed Leo P. Kelley (1928-) and more especially in the anthologies ed Robert H. Boyer and Kenneth J. Zahorski: these included *Dark Imaginings: A Collection of Gothic Fantasy* (anth **1978**), whose subtitle sought to exemplify the roots of the fantastic; *The Fantastic Imagination* (anth **1977-8** 2 vols), which considered high fantasy; *The Phoenix Tree* (anth **1980**), which explored MYTH fantasy; and *Visions of Wonder* (anth **1981**), which sought to explore CHRISTIAN FANTASY. The first academic breakthrough, however, came with *Fantastic Worlds* (anth **1979**) ed Eric S. RABKIN, which sought to codify and establish a pedigree for fantastic literature.

Other publishers recognized this distinctive marketing niche for anthologies that covered the full extent of fantasy. The main run of subsequent anthologies, however, limited themselves almost solely to MAGAZINE fantasy and are far less representative than their titles might suggest. They include: *A Treasury of Modern Fantasy* (anth **1981**; cut vt *Masters of Fantasy* 1992) ed Terry Carr and Martin H. GREENBERG; *The Fantasy Hall of Fame* (anth **1983**; vt *The Mammoth Book of Fantasy All-Time Greats* 1988 UK) ed Robert Silverberg and Greenberg, with stories voted by members of the World Fantasy Convention; and *The Oxford Book of Fantasy Stories* (anth **1994**) ed Tom SHIPPEY.

The more rewarding anthologies are those which seek to re-evaluate the borders of fantasy and, in so doing, reabsorb the literary fairytale which decades earlier had been exiled to the sphere of CHILDREN'S FANTASY, despite the fact that many were written on several levels. This had been explored by Jonathan Cott in *Beyond the Looking Glass* (anth **1971**), and was more fully exposed by Terri Windling and Mark Alan Arnold in their series **Elsewhere** (anth **1981-4** 3 vols) and by David G. HARTWELL in *Masterpieces of Fantasy and Enchantment* (anth **1988**) and *Masterpieces of Fantasy and Wonder* (anth **1989**). The most complete anthologies of the literary fairytale are *Spells of Enchantment* (anth **1991**; vt *The Penguin Book of Western Fairy Tales* 1993) ed Jack ZIPES and *The Oxford Book of Modern Fairy Tales* (anth **1993**) ed Alison Lurie.

As noted, the English-language edition (1976) of the Borges, Ocampo and Bioy Casares anthology directed attention towards the international dimension of fantasy. Franz Rottensteiner (1942-   ) further explored this area in his all too constrained *The Slaying of the Dragon* (anth **1984**), but its full potential has been best realized by Alberto MANGUEL in *Black Water* (anth **1983**) and *Black Water 2* (anth **1990**; vt *White Fire* 1991 UK); these two are the most complete modern examples of the spectrum of contemporary fantasy, although they include some older work and concentrate mostly on SUPERNATURAL FICTION, with some MAGIC REALISM.

Though the scope for such omnifarious compendia has been recognized by the widening audience for fantasy, the straightforward heroic-fantasy anthology has remained fundamental since de Camp's breakthrough in 1963. Recent examples include: the reprint series **Echoes of Valor** (**1987-91** 3 vols) ed Karl Edward WAGNER; the original series SWORD AND SORCERESS (from **1984**) ed Marion Zimmer BRADLEY; and **Xanadu** (from **1992**) ed Jane YOLEN. With the possible exceptions of *Lands of Never* (anth **1983**) and *Beyond Lands of Never* (anth **1984**), both ed Maxim Jakubowski (1944-   ), and **Other Edens** (**1987-9** 3 vols) ed Christopher Evans and Robert HOLDSTOCK, there has been no original anthology series that has sought to develop fantasy in the same way that many such anthologies developed sf during the 1960s and 1970s. Fantasy as a genre has found itself becalmed by the vastness of its own ocean, and, although the anthologies of the past 30 years have striven to give it an identity, they are only now establishing sufficient momentum to give it a direction.   [MA]

ANTHOLOGIES OF SUPERNATURAL FICTION
The more traditional anthology of SUPERNATURAL FICTION, containing stories of GHOSTS, VAMPIRES, WEREWOLVES and the occult, accounts for the largest body of books within the overall spectrum of anthologies of the FANTASTIC. The earliest supernatural anthologies grew out of the 18th-century

fascination in Western Europe for Gothic fiction (◊ GOTHIC FANTASY). Several anonymous collections were made of likewise anonymous German tales, so it is often difficult to distinguish anthologies from single-author collections. Thus the wholly anonymous *Popular Tales of the Germans* (coll trans **1791**), which has all the appearance of a selective anthology – the translation of which has been attributed to William BECKFORD – is actually a condensation of *Volksmärchen der deutschen* (auth **1782-6** 5 vols) by Johann Karl MUSÄUS. However, *Tales of the Wild and Wonderful* (anth **1825**), which has been mistakenly credited to George Borrow (1803-1881), is a genuine anthology of various European tales, recalled by the editor from his childhood and freely retold to fit the literary fashion in the wake of Byronic ROMANTICISM.

The book which to all intents started the movement was *Tales of the Dead* (anth **1813**). The stories had been translated by Sarah Utterson (?1782-1851) – who added material of her own – from the French volume *Fantasmagoriana* (anth trans **1812**), itself a translation by Jean Baptiste Eyriès (1767-1846) of the first two volumes of the German **Gespensterbuch** (coll **1811-15** 5 vols), a collection of embellished folklore assembled by Johann APEL and Friedrich Laun (real name Friedrich Schulze; 1770-1849). **Gespensterbuch** contained darker and more sinister FOLKTALES than those being collected at the same time by the GRIMM BROTHERS – indeed, the tales were deliberately revised to reflect the dark side of nature. The French translation, as noted, led to Mary SHELLEY's *Frankenstein, or The Modern Prometheus* (**1818**) and indirectly inspired John POLIDORI's *The Vampyre: A Tale* (**1819** chap). This anthology must thus be seen as seminal. The Gothic and Romantic movements were at their height, and many volumes sought to cash in on the market. Among them were *Popular Tales and Romances of the Northern Nations* (anth **1823** 3 vols) ed anon, *German Stories* (anth **1826** 3 vols) ed and trans R.P. Gillies (1788-1858), *German Romance* (anth **1827** 4 vols) ed and trans Thomas CARLYLE, *Tales of Terror* (anth **1835** 2 vols; rev vt *Evening Tales for the Winter* 1856 3 vols) ed Henry St Clair, *Miniature Romances from the German* (anth **1841**) attributed to Thomas Tracy (1781-1872), and *Tales from the German* (anth **1844**) trans John Oxenford (1812-1877) and C.A. Feiling. Coming at the end of the popular movement, this last volume offered new translations of more recent and lesser-known stories.

But already the UK was growing tired of Gothic fiction and, in particular, the Gothic GHOST STORY – so much so that, as early as 1823, Rudolph Ackermann (1764-1823) published *Ghost Stories Collected with a Particular View to Counteract the Vulgar Belief in Ghosts and Apparitions* (anth **1823**), a volume which sought to rationalize and de-sensationalize. Nevertheless, the ghost story *per se*, moralized and internalized by writers like J. Sheridan LE FANU and Charles DICKENS, continued to thrive and prosper during the Victorian era.

The main source for supernatural short fiction during this period was the popular MAGAZINES. Dickens, for example, built upon the tradition for winter fireside tales by starting the seasonal Christmas issues of his magazine *Household Words* (1850-59) (◊ CHRISTMAS BOOKS). These special issues frequently contained ghost stories, and increasingly they became distinct from the magazine proper: usually in paper covers but sometimes hardbound, they lie in the borderland between anthologies and magazines and are dealt with in more detail under ANNUALS.

Annuals continued to be a feature of magazine publishing until WWI but, as early as the late Victorian period, anthologies in their own right had begun to reappear. Significant in this respect was the 12-vol US **Little Classics** series (all anths **1875**) ed Rossiter Johnson (1840-1931). At least three of the volumes were composed predominantly of supernatural fiction – although the terms "supernatural" and "ghost story" were deliberately avoided in order to subdue accusations of sensationalism; the relevant volumes were thus called *Stories of Intellect* (anth **1875**), *Stories of Mystery* (anth **1875**) and *Stories of Tragedy* (anth **1875**), all including stories by Edward BULWER-LYTTON, Edgar Allan POE, Charles DICKENS, Catherine Crowe (1790-1872) and Nathaniel HAWTHORNE. A significant publication in the UK was *Dreamland and Ghostland* (anth **1887** 3 vols) ed anon and published by George Redway; drawing heavily on newspapers like *London Society*, it was the first book to anthologize several of Arthur Conan DOYLE's stories.

The 1890s saw a flurry of interest in the occult, but the improving quality of related fiction was reflected more in novels and magazines than in anthologies. A few series recognized the international status of supernatural fiction, though neither **Nuggets for Travellers** (**1888** 12 vols) – of which five volumes covered WEIRD FICTION from the USA, England, Germany, Ireland and Scotland – or **Terrible Tales** (**1891** 4 vols) – selecting stories from France, Germany, Italy and Spain – contained anything modern or original.

The very title of *Modern Ghosts* (anth **1890**) – #15 in the **Continental Classics** series assembled by the editorial staff at Harper's – showed an attempt to move away from the traditional ghost story. The stories, particularly those by Guy DE MAUPASSANT, showed the supernatural as a manifestation as much of the mind as of spirits. This anthology was well regarded because of the prestige of its publisher; it was later mined by Farnsworth WRIGHT, who reprinted most of the stories in WEIRD TALES.

Magazines ruled the first 20 years of the 20th century, and few significant supernatural anthologies were published. Two worth noting were *Twenty-Five Ghost Stories* (anth **1904**; vt *Twenty-Five Great Ghost Stories* 1943; vt *The Permabook of Ghost Stories* 1950; included in *The Haunted Hotel and 25 Other Ghost Stories* omni **1941**; cut vt *20 Great Ghost Stories* 1955) ed W. Bob Holland (1868-1932), which drew upon a store of anonymous and traditional UK and US tales, and *Shapes that Haunt the Dusk* (anth **1907**) ed William Dean HOWELLS and Henry Mills Alden (1836-1919), which selected good material from *Harper's Monthly*. Other than *Uncanny Stories* (anth **1916**) and *More Uncanny Stories* (anth **1918**), both ed anon and selecting stories from *The Novel Magazine*, there was no other supernatural anthology until the post-WWI revival of SPIRITUALISM caused a sudden resurgence of the subgenre.

The first swathe of new anthologies came from journalists who also worked as editors for mass-market publishers, and they selected fairly obvious Victorian and Edwardian stories. In the USA Joseph L. French (1858-1936) published a series starting with *Great Ghost Stories* (anth **1918**; rebound with *Ghosts Grim and Gentle* [anth **1926**] as *The Ghost Story Omnibus* omni **1933**). These clearly drew from *Modern Ghosts* and the Victorian anthologies, but they also included modern material by Algernon BLACKWOOD, M.R. JAMES and Richard MIDDLETON. French's most distinctive anthology was *The Best Psychic Stories* (anth **1920**), which, in seeking to explore Spiritualism, selected more modern and original

material, including some nonfiction. His other anthologies of interest were *Masterpieces of Mystery* (anth **1920** 4 vols; vol 2 titled *Ghost Stories*) and *Tales of Terror* (anth **1925**). French's rival at the publisher Thomas Crowell was Joseph W. McSpadden (1874-1960), who compiled the similar anthologies *Famous Ghost Stories* (anth **1918**) – almost a copy of French's first volume, even though compiled contemporaneously – and *Famous Psychic Stories* (anth **1920**), both assembled as *Famous Psychic and Ghost Stories* omni **1938**); these were less original than French's selections.

In the UK supernatural fiction was given the stamp of respectability by the editor of **Everyman's Library**, Ernest Rhys, who produced *The Haunted and the Haunters* (anth **1921**) which, despite the inclusion of Edward BULWER-LYTTON's title story, was an eclectic selection, relying on stories and factual accounts that drew upon folklore and tradition. Another popular title of the period was *A Muster of Ghosts* (anth **1924**; vt *The Best Ghost Stories* 1924 US) ed Bohun Lynch (1884-1928), which selected lesser-known but more commercial fiction, and is notable for being the first book to anthologize "Thurnley Abbey" (1908) by Perceval Landon (1869-1927).

The field was rapidly explored by educationalists. The pioneer was the US academic Dorothy Scarborough (1878-1935). Having already published her seminal work, *The Supernatural in Modern English Fiction* (**1917**), she compiled two representative anthologies – *Famous Modern Ghost Stories* (anth **1921**) and *Humorous Ghost Stories* (anth **1921**) – which were the first important modern GHOST-STORY anthologies. The depth of her research is evidenced by the inclusion of a much wider range of material, including several stories which remain uncommon to this day. Her work was soon matched in the UK by *Ghosts and Marvels: A Selection of Uncanny Tales from Daniel Defoe to Algernon Blackwood* (anth **1924**) ed Vere H. Collins, whose selection of stories is relatively traditional and predictable, although it was the first to gather "The Lifted Veil" by George Eliot (1819-1880) and "The Moon-Slave" by Barry PAIN. Its greatest importance lies in its introduction, by M.R. JAMES, who sets down his views on and rules for the ghost story. The anthology's success led to *More Ghosts and Marvels: A Selection of Uncanny Tales from Sir Walter Scott to Michael Arlen* (anth **1927**) ed Collins, which has a wider and more original selection of stories.

A significant development in the 1920s was the publication of an anthology of mainly original ghost stories. This was *The Ghost Book* (anth **1926**) ed Cynthia ASQUITH, which much later, from 1952, led to a long-running series (◊ *The* GHOST BOOK). Asquith was able to acquire new stories from major writers and secure reprint rights to material from lesser-known literary (rather than popular-fiction) magazines. These stories of precognition, fate, revenge and mental disintegration – as well as more conventional HAUNTINGS – brought the ghost story firmly into the 20th century, and marked the start of a Golden Age. Asquith produced other original anthologies, including *The Black Cap* (anth **1927**), *Shudders* (anth **1929**) and *When Churchyards Yawn* (anth **1931**), which contained a reliable mixture of supernatural and macabre fiction. These volumes were later gathered in a single volume as *A Century of Creepy Stories* (cut omni **1934**).

This period saw a vogue for huge anthologies. The trend began with the first anthology by Dorothy L. Sayers (1893-1957): *Great Short Stories of Detection, Mystery and Horror* (anth **1928**; rev vt *The Omnibus of Crime* 1929 US; vt in 2 vols *Great Short Stories of Detection, Mystery and Horror: Part I* 1939 and *Part II* 1939; cut vt in 3 vols *Tales of Detection and Mystery* anth 1962 US, *Stories of the Supernatural* anth 1963 US and *Human and Inhuman Stories* anth 1963 US). At over 1100 pages, it caught the imagination of devotees of both mystery and supernatural fiction, and sales were aided by Sayers's reputation. Her introduction, though concentrating on the detective story, showed a grasp and understanding of the supernatural story; and the selections, while feasting upon standard repast, had a few rarer aperitifs. Sayers compiled two further bumper volumes: *Great Short Stories of Detection, Mystery and Horror Vol 2* (anth **1931**; rev vt *The Second Omnibus of Crime* 1932 US; vt *The World's Great Crime Stories* 1939 US; vt in 2 vols *Great Short Stories of Detection, Mystery and Horror: Part III* 1939 and *Part IV* 1939) and *Great Short Stories of Detection, Mystery and Horror Vol 3* (anth 1934; rev vt *The Second Omnibus of Crime* 1942 US; vt in 2 vols *Great Short Stories of Detection, Mystery and Horror: Part V* 1939 and *Part VI* 1939).

The 1930s began with some studious books. *Great Ghost Stories* (anth **1930**) and *More Great Ghost Stories* (anth **1932**), both ed Harrison Dale (1885-1969), are more useful for their historical introductions than for their fairly conventional selection. *They Walk Again* (anth **1931**; vt *The Ghost Book*, 1932) ed Colin de la Mare (1906-   ) is notable not only for its revealing introduction by Walter DE LA MARE but for being the volume that rediscovered William Hope HODGSON. But the major anthologist of supernatural and Gothic literature was Montague Summers (1880-1948); he produced *The Supernatural Omnibus* (anth **1931**; omitting 9 stories and adding 6 1932 US; reissued in 2 vols 1967). Although not as massive as Sayers's tomes, this idiosyncratic selection, with its lengthy and detailed historical evaluation of the field, presented a considerable range of new material and has seldom been out of print. After Scarborough's compilation, it set the standard for supernatural anthologies; along with the Sayers books it established the parameters of the medium, with its subdivisions: hauntings, POSSESSION, diabolism and BLACK MAGIC. Summers produced two other worthy anthologies, *The Grimoire and Other Supernatural Stories* (anth 1936) and *Victorian Ghost Stories* (anth 1936); the latter is again valued for its detailed (if unbalanced) introduction.

Up to now, few anthologies had specialized in subgenres of the supernatural, but *Devil Stories* (anth **1921**) ed Maximilian Rudwin (1885-1946) and *The Devil in Scotland* (anth **1934**) ed Douglas Percy Bliss (1900-   ), both intended as literary studies, were early examples.

In the USA there was no equivalent of these anthologies, although some worthwhile volumes were appearing. C. Armitage Harper sought to establish a national pedigree for the field in *American Ghost Stories* (anth **1928**) – which contained some surprises – but the two most popular volumes drew their material from the magazines, especially the pulps. First was *Beware After Dark* (anth **1929**) ed T. Everett Harré (1884-1948), which mixed supernatural and horror stories, including selections from WEIRD TALES. Then came *Creeps by Night* (anth **1931**; cut vt *Modern Tales of Horror* 1932 UK) ed Dashiell Hammett (1894-1961); this, apart from a few SLICK FANTASIES, selected material from a wide range of pulps and literary magazines, including WT.

The 1930s, the Golden Age of supernatural fiction in the UK, saw scores of bumper anthologies that often mixed mystery and horror with supernatural fiction, with the latter usually taking pride of place. These books, mostly edited anonymously by publishers' in-house staff, include: *A Century of Thrillers from Poe to Arlen* (anth **1934**) and *A Century of Thrillers: Second Series* (anth **1935**; cut vt *Thrillers* 1994); *The Evening Standard Book of Strange Stories* (anth **1934**) and *The Evening Standard Second Book of Strange Stories* (anth **1937**); and *50 Years of Ghost Stories* (anth **1935**; exp vt *A Century of Ghost Stories*, **1936**; cut in 2 vols vt *Let's Talk of Graves* 1970 and *Walk in Dread* 1970) ed anon but in fact by Dorothy M. Tomlinson. Others, with editors credited, included: *The Mystery Book* (anth **1934**) and *The Great Book of Thrillers* (anth **1935**; rev 1937; cut vt *Great Tales of Terror* 1991) both ed H. Douglas Thomson (1905-1975); *A Century of Horror* (anth **1935**) ed Dennis WHEATLEY; *The Mammoth Book of Thrillers, Ghosts & Mysteries* (anth **1936**) ed J.M. Parrish and John R. Crossland (1892-   ); and *A Second Century of Creepy Stories* (anth **1937**) ed Hugh Walpole (1884-1941). This was also the period of the NOT AT NIGHT series and CREEPS LIBRARY.

The only US parallel was *The Haunted Omnibus* (anth **1937**; cut vt *Great Ghost Stories of the World* 1941), which reprinted familiar stories alongside lesser-known regional tales and ORIENTAL FANTASIES. Later, however, while WWII blighted UK publishing (although mini-anthologies, composed mostly of common reprints, continued to appear), the USA maintained the production of supernatural-fiction anthologies, particularly toward the end of the war. Some of these, like *Famous Ghost Stories* (anth **1944**) ed Bennet Cerf (1899-1971), were aimed more towards prestige and immediate popularity than originality, but others, like *Six Novels of the Supernatural* (anth **1944**) ed Edward WAGENKNECHT, are more interesting; the Wagenknecht book is a remarkable selection of lesser-known works from the previous 60 years. One of the most significant books of the period was *Great Tales of Terror and the Supernatural* (anth **1944**; cut vt *Tales of Terror and the Supernatural* 1994 as ed anon) ed Herbert A. Wise (1890-1961) and Phyllis A. Fraser (1915-   ). Although containing nothing new, the book seemed to capture the public's imagination, and has remained almost constantly in print; it has often been cited as an inspiration for the next generation of writers.

After this flurry, however, the number of supernatural-fiction anthologies decreased rapidly, as the post-WWII technology boom swept ghosts and phantoms into the corners. The only editor who sought to produce consistently high-quality anthologies was August W. DERLETH. Starting with the successful *Sleep No More* (anth **1944**), he assembled a series of anthologies that were resourceful and exciting. Although he to a fair extent relied on WEIRD TALES, this source did not dominate the anthologies; and his selections covered UK and regional writers, plus the occasional unknown. Derleth sustained the series through *Who Knocks?* (anth **1946**), *The Night Side* (anth **1947**) and *The Sleeping and the Dead* (anth **1947**) before the market began to fail. All are consistent in quality and coverage, and this remains an impressive body of work. He managed to issue one further anthology, *Night's Yawning Peal* (anth **1952**), through his own ARKHAM HOUSE, but by then the market really had folded. Apart from *The Supernatural Reader* (anth **1953**) ed Groff and Lucy Conklin, which like Derleth's selections came predominantly from the better-quality pulps and UK writers (and was mainly SLICK FANTASY), no other significant supernatural anthology appeared in the USA until the 1960s.

In the UK the next revival of interest started earlier. This was due almost solely to Herbert van Thal (1904-1983), who, through the company he worked for (Arthur Barker) and various publishing connections, encouraged others to edit anthologies and then did so himself. His earliest anthologies, *Told in the Dark* (anth **1950**) and *A Book of Strange Stories* (anth **1954**), were a mixture of well known and obscure items. Through Barker's he published R.C. Bull's highly original anthology *Perturbed Spirits* (anth **1954**), which concentrated almost wholly on rare and forgotten fiction and is significant for resurrecting stories by Grant ALLEN, Dick Donovan (real name J.E. Preston-Muddock; 1842-1934), William Hope HODGSON and Hume NISBET. The book was an "unqualified success", leading to a successor, *Snapdragon* (anth **1955**) ed Mervyn Savill, which contains more fantasy and horror than supernatural fiction, and has some rare selections by European writers. But it lacked the popular appeal of its predecessor, and consequently van Thal's proposal for an annual Christmas anthology was shelved. He returned to the idea with *The Pan Book of Horror Stories* (anth **1959**), whose success inaugurated the longest-running annual series of anthologies in the world (◊ PAN BOOK OF HORROR STORIES). The early volumes contained some supernatural fiction, but it soon began to focus almost solely on physical horror. The success of the series, and the resurgence of interest in HORROR MOVIES, began to revive the field, but most of the anthologies published in the early 1960s, in both the UK and the USA, carried lacklustre retreads. Only Derleth showed any enterprise, producing an original anthology of WEIRD FICTION, *Dark Mind, Dark Heart* (anth **1962**) – best-known now for introducing Ramsey CAMPBELL. Although Derleth's new anthologies (and reprints of his earlier ones) provided a foundation of quality, little else emerged until the mid-1960s, when the field was again revived.

The first hope came from Robert AICKMAN, who began *The* FONTANA BOOK OF GREAT GHOST STORIES series in 1964. Aickman was methodical in his choice of fiction, and never settled for second-best; he also endeavoured to have as much variety as possible. The success of the first volume inaugurated another long-running series. In the later volumes, edited by R. CHETWYND-HAYES, the quality dropped but the diversity remained, and the series opened up a market for new stories as well as reprints.

This was a characteristic of the next phase, too. In 1952 Cynthia ASQUITH had edited her *The Second Ghost Book* (anth **1952**), followed by *The Third Ghost Book* (anth **1955**). These anthologies had contained new or almost new stories by many of the writers from her original *The Ghost Book* (anth **1926**) plus the latest literary generation (◊ *The* GHOST BOOK). These stories brought a quiet, craftsmanlike quality to the ghost story, focusing on the subtle and psychological aspects rather than the overtly supernatural. James Turner (1909-1975) revived the series with *The Fourth Ghost Book* (anth **1965**), then the series became a regular annual under the control of Rosemary Timperley (1920-1988) from *The Fifth Ghost Book* (anth **1969**). With #4 the series had become a market for new stories, and thus encouraged tales from Joan AIKEN, Denys Val BAKER, George Mackay BROWN, D.G. Compton (1930-   ), Elizabeth Fancett, Jean Stubbs (1926-   ), Paul Tabori (1908-1974), William Trevor (1928-   ) and Fred Urquhart (1912-1995), plus the steadfast L.P. HARTLEY. The series continued under other editors – Aidan CHAMBERS and James Hale – and these editors in turn produced further anthologies of similar mood. A few other anthologies of the period shared the same minimalist approach to the supernatural – and often the same authors. These included: *Tales of Unease* (anth **1966**), *More Tales of Unease* (anth **1969**) and *New Tales of Unease* (anth **1976**), all ed John F. Burke (1922-   ); and *Prevailing Spirits* (anth **1976**) and *A Book of Contemporary Nightmares* (anth **1977**), both ed Giles Gordon (1940-   ) – though the latter volume ventured more into CONTEMPORARY FANTASY and horror. The real spiritual descendants of the **Ghost Book** series were the anthologies published by William Kimber, in particular the series ed Denys Val BAKER and Amy Myers (1938-   ). Baker's series started with *Haunted Cornwall* (anth **1973**) and then moved through a roughly thematic sequence with *Stories of the Night* (anth **1976**), *Stories of the Macabre* (anth **1976**) and others, but later it concentrated mostly on ghost stories, especially in rural settings: *Cornish Ghost Stories* (anth **1981**), *Ghosts in Country Houses* (anth **1981**), *Ghosts in Country Villages* (anth **1983**) and others. Myers instigated the **After Midnight Stories** series (*After Midnight Stories* anth **1985**; *The Second Book of After Midnight Stories* anth **1986**; *The Third Book* anth **1987**; *The Fourth Book* anth **1988**, and *The Fifth Book* anth **1991**), which contained mostly new material. All of these sustained, for over 20 years, a continuity of new and original writing applying both conventional and innovative treatment.

The third factor influencing anthologies in the mid-1960s was the work of the UK editor Peter HAINING. During a period of over 30 years Haining has become the most prolific solo anthologist in the field and, although his anthologies are predominantly of reprinted stories, he has striven to include rarer material among the better-known. Haining was there when the UK paperback market began to rediscover supernatural and horror fiction, and he was instrumental in the creation of the notion of the thematic anthology. Because he had come to the field through his interest in BLACK MAGIC and OCCULTISM, his books focused more on these aspects; although earlier anthologies had included such material, few had been devoted to these themes. (John Keir CROSS's *Best Black Magic Stories* [anth **1960**] and Frederick Pickersgill's *No Such Thing as a Vampire* [anth **1964**] were two exceptions.) From the mid-1960s, however, thematic anthologies began to erupt in both the UK and the USA, with Haining invariably leading the way. His anthologies tended to focus on the EVIL within humankind rather than in nature or beyond; consequently, many are more of horror than of the supernatural, certainly in intent if not in content. Those more overtly supernatural include three VAMPIRE anthologies – *The Midnight People* (anth **1968**; vt *Vampires at Midnight* 1970 US), *Shades of Dracula* (anth **1982**) and *Vampire* (anth **1985**) – three witchcraft (◊ WITCHES) anthologies – *The Witchcraft Reader* (anth **1969**), *A Circle of Witches* (anth **1971**) and *Hallowe'en Hauntings* (anth **1984**) – and in particular four anthologies of black magic and diabolism – *The Satanists* (anth **1969**), *The Necromancers* (anth **1971**), *The Magicians* (anth **1972**) and *The Black Magic Omnibus* (anth **1976**). Haining's main contributions to the development of supernatural-fiction anthologies were to broaden the market from straight ghost stories, which he felt had become passé, and to package them with a spicier image, thus attracting the more liberated readers of the late 1960s and 1970s. By the end of the 1970s the supernatural anthology field as a whole had shifted significantly away from ghosts toward the occult, while increasing numbers of anthologies were devoted solely to stories of vampires, werebeasts (◊ WEREWOLVES), ZOMBIES and other MONSTERS, not to mention the CTHULHU MYTHOS.

One other characteristic became evident in the 1960s and 1970s: the increasing number of supernatural-fiction anthologies designed for younger readers. Wilhelmina Harper (1884-1973) had assembled *Ghosts and Goblins* (anth **1936**; rev 1964), but this contained predominantly FOLK-TALES. The enterprising Franklin Watts had issued *Ghosts, Ghosts, Ghosts* (anth **1952**) ed Phyllis R. Fenner (1899-1982) which, while still relying on quaint old tales, did include more modern material. The time did not seem ripe, though, until the mid-1960s when Watts published *Spooks, Spooks, Spooks* (anth **1966**) ed Helen Hoke (1903-1990). This developed into a regular series, most volumes being identified by the trifold title. Titles of relevant interest include *Weirdies* (anth **1973**), *Ghosts & Ghastlies* (anth **1976**), *Ghostly, Grim and Gruesome* (anth **1976**), *Creepies, Creepies, Creepies* (anth **1977**; vt *Creepies: A Covey of Quiver-and-Quaver Tales* 1977 UK), *Eerie, Weird and Wicked* (anth **1977**), *Haunts! Haunts! Haunts!* (anth **1977**; vt *Spectres, Spooks & Shuddery Shades* 1977 UK); *Fear! Fear! Fear!* (anth **1980**), *More Ghosts, Ghosts, Ghosts* (anth **1981**), *Sinister, Strange and Supernatural* (anth **1981**), *Tales of Fear and Frightening Phenomena* (anth **1982**), *Ghostly, Ghoulish, Gripping Tales* (anth **1983**), *Uncanny Tales of Unearthly & Unexpected Horrors* (anth **1983**), *Spirits, Spooks, and Other Sinister Creatures* (anth **1984**), and *Horrifying and Hideous Hauntings* (anth **1986**) with Franklin Watts. In the UK several paperback series emerged. Christine Bernard (1926-   ) began the *The Armada Ghost Book* (anth **1967**), continued after *#2* by Mary Danby (1941-   ) and which reached *#15* (◊ FONTANA BOOK OF GREAT GHOST STORIES); Aidan CHAMBERS edited *Ghosts* (anth **1969**) with Nancy Chambers and *Ghosts 2* (anth **1972**), though the latter contained stories mostly by Chambers himself; Richard Davis (1945-   ) edited the **Spectre** series (*Spectre 1* anth **1973**; *#2* anth **1975**; *#3* anth **1976**; *#4* anth **1977**) and also edited *Animal Ghosts* (anth **1980**); and following *Haunting Tales* (anth **1973**), Barbara Ireson (1927-   ) edited **Spooky Stories** (*Spooky Stories* anth **1975**; *#2* anth **1979**; *#3* anth **1981**; *#4* anth **1982**; *#5* anth **1983**, and *#6* anth **1984**) plus other notable volumes including *Ghostly and Ghastly* (anth **1977**), *Creepy Creatures* (anth **1978**), *Ghostly Laughter* (anth **1981**) and *Fearfully Frightening* (anth **1984**). There were also such singleton volumes as *The House of the Nightmare* (anth **1967**) and *The Haunted and the Haunters* (anth **1977**) both ed Kathleen Lines (1902-1988), and *The Restless Ghost* (anth **1970**; vt *The Usurping Ghost* 1971 US; UK paperback cut vt in 2 vols *Ghostly Experiences* anth 1973 and *Ghostly Encounters* anth 1973) ed Susan Dickinson (1931-   ). More recently there have been *Ghost Stories* (anth **1988**) ed Robert WESTALL, *The Walker Book of Ghost Stories* (anth **1990**) ed Susan HILL, *The Young Oxford Book of Ghost Stories* (anth **1994**) ed Dennis Pepper and *Dread and Delight: A Century of Children's Ghost Stories* (anth **1995**) ed Philippa PEARCE. (◊ GHOST STORIES FOR CHILDREN.)

The popular anthologies of the 1960s led, in the 1970s, to an increase in attention from more serious and academic publishers. Everett F. BLEILER produced several anthologies of rare Victorian material, including *Five Victorian Ghost Novels* (anth **1971**), *Three Supernatural Novels of the Victorian Period* (anth **1975**) and *A Treasury of Victorian Ghost Stories* (anth **1981**). Sustaining the interest in

Victorian and rare early fiction were the many anthologies edited by Hugh Lamb (1946-   ), Richard DALBY and Jessica Amanda SALMONSON, plus *Victorian Ghost Stories* (anth **1991**) ed Michael Cox (1948-   ) and R.A. Gilbert (1942-   ). Further examples of academic interest in the 1970s were *Edges of Reality* (anth **1972**) ed Leo B. Kneer and Ruth S. Cohen, *The Supernatural in Fiction* (anth **1973**) ed Leo P. Kelley (1928-   ) and *Lost Souls* (anth **1983**) ed Jack Sullivan (1946-   ). The re-emergence of the ghost story received a stamp of approval from the London *Times* through a ghost-story competition that resulted in *The Times Anthology of Ghost Stories* (anth **1975**).

By the 1980s publishers were confident enough to return to the bumper-anthology style of the 1930s. In the UK, Hamlyn led the way with *The Best Ghost Stories* (anth **1977**; cut by 16 stories 1990 US) ed anon. The UK publisher Anthony Cheetham commissioned US literary agent Kirby McCauley (1941-   ) to produce his landmark anthology of supernatural horror *Dark Forces* (anth **1980**), which sought to set new parameters for the subgenre. Thereafter the field took off. In the USA Marvin KAYE assembled *Ghosts: A Treasury of Chilling Tales Old & New* (anth **1981**; cut by 14 stories and all appendices vt *A Classic Collection of Haunting Ghost Stories* 1993 UK). Kaye has since compiled several authoritative anthologies, including *Masterpieces of Terror and the Supernatural* (anth **1985**), *Devils and Demons* (anth **1987**), *Witches and Warlocks* (anth **1990**), *Haunted America* (anth **1991**) and *Masterpieces of Terror and the Unknown* (anth **1993**). In the UK there have been prestigious and high-profile anthologies like *Roald Dahl's Book of Ghost Stories* (anth **1983**) ed Roald DAHL, *Ghost Stories* (anth **1983**) ed Susan HILL – neither of which contained much that was new, but both of which gave the field added respectability – *The Penguin Book of Ghost Stories* (anth **1984**) ed J.A. CUDDON, *The Oxford Book of English Ghost Stories* (anth **1986**) ed Michael Cox and R.A. Gilbert, and *The Chatto Book of Ghosts* (anth **1994**) ed Jenny Uglow.

In the USA similar prestige came with *The Oxford Book of Canadian Ghost Stories* (anth **1990**) ed Alberto MANGUEL, *The Literary Ghost* (anth **1991**) ed Larry Dark (1959-   ) and *The New Gothic* (anth **1991**) ed Bradford Morrow (1951-   ) and Patrick McGrath (1950-   ). In addition there have been many supernatural anthologies compiled by Martin H. GREENBERG in collaboration with various co-editors, especially his series of regional ghost-story anthologies assembled with Frank D. McSherry (1927-   ) and Charles G. Waugh (1943-   ) – which culminated in *Great American Ghost Stories* (anth **1991**; cut 2 vols 1992, 1993). There were also the SHADOWS series (ed Charles L. GRANT) and the NIGHT VISIONS series (various editors), both of which ushered the traditional supernatural anthology into the domain of horror, a metamorphosis that was explored in *The Dark Descent* (anth **1987**) ed David G. HARTWELL.

Although some may regard the ghost story as a dying form, that view is belied by the continued – and increasing – appearance of ghost-story and supernatural-fiction anthologies. Over half the anthologies in this field have appeared in the past 20 years, many of a high standard and featuring new stories. In the absence of MAGAZINES carrying ghost and supernatural tales, the anthology has become the main proving ground for new authors . . . and the main salvation for vintage material.   [MA]

HORROR ANTHOLOGIES

Most early writings of GOTHIC FANTASY and SUPERNATURAL

FICTION were intended to induce emotions of horror, but this was more an issue of impact on and interaction with the reader than necessarily of content. HORROR fiction that contains no (or only ersatz) supernatural content – that is, fiction that locates EVIL in society and humankind (or individual humans) rather than in of the spirit world – is beyond the purview of this encyclopedia. It is distinct from FANTASY (notably, in this context, DARK FANTASY) and supernatural fiction in that it focuses upon (and generally offers no HEALING solution to) a breakdown or disintegration of the world order. This section of this article also explores those areas where authors go beyond the cultural conventions of the day to induce horror and revulsion while remaining within the realms of the fantastic.

Horror ran in tandem with supernatural fiction for most of the Victorian period. Although its parameters had been explored by Edgar Allan POE and others, it was not until Robert Louis STEVENSON's TECHNOFANTASY *The Strange Case of Dr Jekyll and Mr Hyde* (**1888**) that the true nature of the individual's potential for horror and evil emerged (at least, in fiction). At about the same time, the French writers Guy de MAUPASSANT and VILLIERS DE L'ISLE ADAM were developing the medium known as CONTES CRUELS; paralleling the French theatrical movement of GRAND GUIGNOL, this medium established a cultural acceptance of the treatment of physical horror. Nevertheless, the intent to explore this in fiction was limited and anyway most anthologies of the time were of reprinted material. No original anthology emerged until Frederick Stuart Greene (1870-1939) boldly came up with *The Grim 13* (anth **1917**), which collected together fiction that had been rejected by magazines as too horrible or unconventional. This anthology remains little known, despite the stature of its contributors, who included Stacy Aumonier (1887-1928), Dana Burnet (1888-1962), Will Levington Comfort (1878-1932), Marie Belloc Lowndes (1868-1947), Ethel Watts Mumford (1878-1940) and Vincent O'SULLIVAN. Few of its stories involve the supernatural, but all reflect an inherent understanding of cultural transgression.

WEIRD TALES (founded 1923) was the first magazine regularly to publish stories of the bizarre and unusual. Although most of the early stories were standard tales of HAUNTINGS or MONSTERS, there were also more gruesome stories of humankind's inhumanity. It was this more lurid context that appealed to Christine Campbell Thomson (1897-1985) when she assembled *Not at Night* (anth **1925**), which drew its contents wholly from the magazine. The success of this book led to a series, NOT AT NIGHT, which by #4 included mostly original material from UK writers. Among the better horror stories were "The Horror at Red Hook" and "Pickman's Model" by H.P. LOVECRAFT (this series was the first to anthologize Lovecraft's stories), "The Dead Woman" by David H. KELLER, "The Chain" by H. Warner MUNN, "The Copper Bowl" by George Fielding Eliot (1894-1971) and stories by Campbell Thomson herself (as Flavia Richardson) and her husband Oscar Cook (1888-1952). By the time the series concluded with the retrospective *Not at Night Omnibus* (omni **1937**), a similar UK series had emerged from publisher Philip Allan (1884-1973) as part of his CREEPS LIBRARY. A series of anthologies, compiled by Allan's house editor Charles BIRKIN – who delighted in the *conte cruel* – began with *Creeps* (anth **1932**). A typical horror story in the first volume is "The Charnel House" by Allan himself under his pseudonym Philip

Murray, about an anatomist who prepares bodies for study and, after his death, is aware of the preparation of his own body. Birkin's own first stories (as Charles Lloyd) appeared in this series. (Both the **Creeps** and **Not at Night** series were key sources when Herbert van Thal began compiling his PAN BOOK OF HORROR STORIES series 30 years later.)

Far superior were the occasional anthologies edited by T.I.F. Armstrong (John GAWSWORTH), usually anonymously, starting with *Strange Assembly* (anth **1932**) and including several bumper volumes: *New Tales of Horror* (anth **1934**), *Crimes, Creeps and Thrills* (anth **1936**), *Masterpiece of Thrills* (anth **1936**) and *Thrills* (anth **1936**). Gawsworth drew upon his friendships with Oswell Blakeston (1907-1985), Edgar JEPSON, Arthur MACHEN, E.H.W. Meyerstein (1889-1952), M.P. SHIEL, E.H. VISIAK, and other littérateurs of the 1930s to compile relatively sophisticated commercial anthologies. Many of the stories are nonsupernatural, but Gawsworth encouraged tales occupying the borderland between madness and the *outré*, which epitomized horror fiction of this period. It was in the 1930s that the word "horror" once more became acceptable in book titles, a standard anthology being *A Century of Horror* (anth **1935**) ed Dennis WHEATLEY.

The interest in horror faded during WWII and, despite their titles, anthologies like *Tales of Terror* (anth **1943**) ed Boris Karloff (real name William Henry Pratt; 1887-1969) and *Terror at Night* (anth **1947**) ed Herbert Williams (1914-) concentrated more on SUPERNATURAL FICTION. The only significant horror anthology of the 1950s was *Terror in the Modern Vein* (anth **1955**) ed Donald A. WOLLHEIM, which sought to demonstrate the effectiveness with which the supernatural could be deployed in modern surroundings and circumstances. Following the success of HAMMER movies like *The Curse of Frankenstein* (**1957**) and *The Horror of Dracula* (**1958**), the UK publisher Pan Books issued *The Pan Book of Horror Stories* (anth **1959**) ed Herbert VAN THAL. The volume was instantly successful, and spawned a series that ran for 30 vols (**1959-89**), the last six ed Clarence Paget (1909-1991); the series was succeeded by **Dark Voices**, beginning with *Dark Voices* (anth **1990**) ed Stephen JONES and Paget. Van Thal's series drew heavily upon the **Not at Night** and **Creeps** volumes, and increasingly focused on physical horror and graphic violence and depravity rather than supernatural horror, thus serving as forerunner to the Splatterpunk movement of the 1980s.

The success of the **Pan Book of Horror Stories** series revived publishers' interests in the medium in both the UK and the USA. The **Fontana Book of Great Horror Stories** series (anth **1966-84** 17 vols) ed Christine Bernard (1926-) for #1-#4 and from #5 by Mary Danby (◊ FONTANA BOOK OF GREAT GHOST STORIES); was more selective, and the stories were generally of a higher quality. Other anthologies at this time were: *The Tandem Book of Horror Stories* (anth **1965**) ed Charles Birkin, which initiated a short (retitled) series (*Tandem Horror 2* [anth **1968**] and *Tandem Horror 3* [anth **1969**] both ed Davis); *Best Tales of Terror* (anth **1962**) and *Best Tales of Terror 2* (anth **1965**) both ed Edmund Crispin (real name Bruce Montgomery; 1921-1978); *Best Horror Stories* (anth **1956**) and *Best Horror Stories 2* (anth **1965**), both ed John Keir CROSS with *Best Horror Stories 3* (anth **1972**) ed Alex Hamilton (1930-); *Taboo* (anth **1964**) ed anon Paul Neimark, which sought to test the boundaries of sexuality; and a steady flow of anthologies from Peter HAINING.

Haining's first anthology, *The Hell of Mirrors* (anth **1965**) explored the horror of DREAMS and nightmares, a theme he returned to in *The Nightmare Reader* (anth **1973**), but his excursion into horror became most noticeable with *The Evil People* (anth **1968**) and *The Unspeakable People* (anth **1969**), the latter noted for including stories considered outrageous or banned at the time of their original publication.

But it was really not until the popularity of the movie *The* EXORCIST (*1973*) and the novels of Stephen KING that horror reawakened as a marketable commodity. This became evident in the cluster of anthologies ed Michel Parry (1947-) in the mid-1970s, the Victorian and other selections ed Hugh Lamb (1946-) – which focused on horror imagery – the first anthology ed Ramsey CAMPBELL – *Superhorror* (**1976**; vt *The Far Reaches of Fear* 1980) – and *Frights* (anth **1976**; as 2 vols *Frights 1* and *Frights 2* 1979 UK) ed Kirby McCauley. It was Campbell and McCauley who established the landmark anthologies of horror: Campbell with *New Terrors* (anth **1980** 2 vols) and McCauley with *Dark Forces* (anth **1980**). Both these assemblages brought horror up to date by emphasizing the disintegration of society under supernatural or spiritual affliction.

In 1980 Karl Edward WAGNER took over for 3 vols the **Year's Best Horror Stories** series started by Richard Davis in 1971, being followed by Gerald W. Page (1939-) for 4 vols. Along with the original anthologies ed Charles L. GRANT, especially the SHADOWS series, these set the markers for horror fiction in the 1980s which became increasingly dominated by physical and erotic horror and the Splatterpunk movement. This field has been dominated by anthologies of original fiction, with contributors charged to push back the barriers. Prominent among these have been: the **Masques** series (**1984-91**) ed J.N. WILLIAMSON; *Cutting Edge* (anth **1986**) and *Metahorror* (anth **1992**) both ed Dennis ETCHISON; *Prime Evil* (anth **1988**) ed Douglas E. Winter (1950-); *Splatterpunks: Extreme Horror* (anth **1990**) ed Paul M. Sammon (1949-); the **Borderlands** series (**1990-93**) ed Thomas F. MONTELEONE; the **Dark Voices** series ed Stephen JONES and David A. Sutton (1947-); and the **Narrow Houses** series (**1992-4**) ed Peter CROWTHER, which concentrates on SUPERSTITIONS.

Attempts to define the field through anthologies were made in: *The Penguin Book of Horror Stories* (anth **1984**) ed J.A. CUDDON; *A Treasury of American Horror Stories* (anth **1985**) ed Frank D. McSherry (1927-), Charles G. Waugh (1943-) and Martin H. GREENBERG; the **Masters of Darkness** series ed Dennis ETCHISON; and two giant anthologies ed David G. HARTWELL, *The Dark Descent* (anth **1987**) and *Foundations of Fear* (anth **1992**) – although these latter two volumes consider horror for its effect rather than in the thematic sense, and thus include many SUPERNATURAL FICTIONS. Two annual series have been **The Year's Best Fantasy and Horror** (from **1988**) ed Ellen DATLOW and Terri WINDLING and **Best New Horror** (from **1990**) ed Ramsey Campbell and Stephen Jones. Both give an exemplary coverage of the field.

Horror for younger readers (◊ CHILDREN'S FANTASY) did not really emerge until the last decade, although Helen Hoke (1903-1990) edited a number of anthologies starting with *Witches, Witches, Witches* (anth **1958**); most are supernatural, and draw on folklore and children's tales. Ramsey Campbell sought to share his own childhood nightmares in *The Gruesome Book* (anth **1983**), and Mary Danby (1941-)

produced the short **Nightmares** series (**1983-5**) of original anthologies. But it was not until the young-horror boom of the 1990s that such anthologies became more graphic: examples are *Thirteen* (anth **1991**; vt *Thirteen Tales of Terror* 1992 UK) ed Tonya Pines, *Thirteen More Tales of Horror* (anth **1994**) ed A(nne) Finnis and *13 Again* (anth **1995**) ed Finnis.

The re-emergence of horror fiction is a reflection of society's increasing obsession with and terror of violence, and of the pressures of the late 20th century, when fear of the world about us has replaced our fear of the supernatural – although both fears are fuelled by the same imaginative powers. The modern horror anthology has become the home for stories of explicit violence and eroticism; in intent, at least, little has changed since the Gothic and supernatural anthologies published 200 years earlier. [MA]

Further reading: *The Supernatural Index* (**1995**) by Mike ASHLEY and William G. Contento indexes over 2200 anthologies.

ANTHONY, PIERS Working name of US writer Piers Anthony Dillingham Jacob (1934-) for all his published work, most of it sf in the first years of his career, when he released what remain his two most highly respected novels, *Chthon* (**1967**) and *Macroscope* (**1969**; cut 1972 UK). He began publishing work of genre interest with "Possible to Rue" for *Fantastic* in 1963, and assembled most of his nonseries short fiction as *Anthonology* (coll **1985**) and *Alien Plot* (coll **1992**). Since about 1970 PA has become exceedingly popular for his fantasy series.

The first and longest of these, the ongoing **Xanth** sequence, comprises *A Spell for Chameleon* (**1977**), which won the 1978 BRITISH FANTASY AWARD, *The Source of Magic* (**1979**) and *Castle Roogna* (**1979**) – all assembled as *The Magic of Xanth* (omni **1981**; vt *Three Complete Xanth Novels* 1994 – *Centaur Isle* (**1982**), *Ogre, Ogre* (**1982**), *Night Mare* (**1982**), *Dragon on a Pedestal* (**1983**), *Crewel Lye: A Caustic Yarn* (**1984**), *Golem in the Gears* (**1986**), *Vale of the Vole* (**1987**), *Heaven Cent* (**1988**), *Man from Mundania* (**1989**), *Isle of View* (**1990**), *Question Quest* (**1991**), *The Color of her Panties* (**1992**), *Demons Don't Dream* (**1993**), *Harpy Thyme* (**1993**), *Geis of the Gargoyle* (**1994**) and *Roc and a Hard Place* (**1995**). Surrounded by sea and by the nonfantasy lands collectively called Mundania, Xanth itself, a vast POLDER much resembling Florida (where PA lives), is a land permeated by ecologically balanced MAGIC, whose laws can be worked out pragmatically (◊ RATIONALIZED FANTASY), and where TIME moves independently of the outside world (◊ TIME IN FAERIE); words and puns (hence the titles of individual volumes) can operate in a literal fashion (◊ TRUE NAME), so that (for instance) cherry bombs grow on cherry trees and shoes on shoe trees. Most (if not all) of the **Xanth** books are constructed as QUEST tales, and the earlier titles in particular tend to engage young men and women in RITE-OF-PASSAGE storylines leading to a modest maturity and marriage with a suitable partner; some early protagonists, Bink in particular, have struck some critics as unattractively sexist (◊ GENDER; SEX), but later figures are subtler in their choices. Very few characters or MOTIFS common to GENRE FANTASY have failed to appear in the **Xanth** books, which in their playfulness, escapism, geography and dependence on word-play are strongly reminiscent of the Oz books by L. Frank BAUM and his successors. *Piers Anthony's Visual Guide to Xanth* (**1989**) with Jody Lynn NYE (*whom see for* **Crossroads Adventure** *ties set in the* **Xanth** *universe*) is a nonfiction survey of the series.

PA's second large sequence, the **Apprentice Adept** series – *Split Infinity* (**1980**), *Blue Adept* (**1981**) and *Juxtaposition* (**1982**), all assembled as *Double Exposure* (omni **1982**), plus *Out of Phaze* (**1987**), *Robot Adept* (**1988**), *Unicorn Point* (**1989**) and *Phaze Doubt* (**1990**) – juxtaposes fantasy and sf worlds whose interactions and conflicts constantly threaten an ecological balance between magic and science; the tales are lightly told, and late volumes are comparatively flimsy.

His third long sequence, the **Incarnations of Immortality** series – *On a Pale Horse* (**1983**), *Bearing an Hourglass* (**1984**), *With a Tangled Skein* (**1985**), *Wielding a Red Sword* (**1986**), *Being a Green Mother* (**1987**), *For Love of Evil* (**1988**) and *And Eternity* (**1990**) – more ambitiously presents an ALTER-NATE WORLD in which mortal humans are selected to become incarnations (◊ AVATARS) of figures like Death, Nature, Time, God and so forth, and to wage battle on humanity's behalf against SATAN. If there is an imbalance between the sweep of the concept and the intermittent flippancy of the telling, it is an imbalance clearly of PA's choice.

The **Dragon** series, with Robert Margroff (1930-   ) – *Dragon's Gold* (**1987**), *Serpent's Silver* (**1988**) and *Chaera's Copper* (**1990**), all three assembled as *The Adventures of Kelvin of Rud: Across the Frames* (omni **1992**; vt *Three Complete Novels* 1994), plus *Orc's Opal* (**1990**) and *Mouvar's Magic* (**1992**), both assembled as *The Adventures of Kelvin of Rud: Final Magic* (omni **1992**) – is of less interest. The ongoing **Mode** sequence – *Virtual Mode* (**1991**), *Fractal Mode* (**1992**) and *Chaos Mode* (**1993**) – is a love story involving parallel worlds. Fantasy singletons include: *Hasan* (1969 *Fantastic*; **1977**), an ARABIAN FANTASY; *Pretender* (**1979**) with Frances Hall (1914-   ); *Shade of the Tree* (**1986**), a tale of paranormal HORROR in which, typically of PA's work, various REALITIES intersect; *Ghost* (1966 *If* as "The Ghost Galaxies"; exp **1986**), in which the world is haunted by the Seven Deadly Sins; *Pornucopia* (**1989**), which tells of a man's search through fantasy environments for his lost penis, aided by an adaptive prosthetic tool; *Firefly* (**1990**), horror; and *The Tatham Mound* (**1991**), an ambitious RITE-OF-PASSAGE tale whose Native American protagonist must not only work out his own mature being but attempt to save his people.

Even more than in his sf, the fantasy work of PA constantly enthralls with its scope and frustrates through the pun-ridden, excessive facility of its telling. It sometimes seems difficult for PA to find worlds of the imagination that are sufficiently gritty to engage his full attention. When his imagination is properly involved, however, his work is explosive. [JC]

Other works: *Tales from the Great Turtle* (anth **1994**) with Richard Gilliam (1950-   ).

ANTHROPOLOGY The scientific study of humankind. The founders of the science, including Edward Tylor (1832-1917) and Sir James FRAZER, made the dubious assumption that, by studying the diversity of contemporary societies and describing a hierarchy extending from the most "primitive" to the most "highly developed", they could discover a single evolutionary pattern. Frazer described this imagined pattern very elaborately in the various editions of *The Golden Bough* (**1890**; rev 1900; exp rev 1911-15), which contends that the beliefs on which culture is founded inevitably evolve from the proto-theories of "sympathetic MAGIC" and animistic RELIGION through a series of more complex religious ideologies – beginning with FERTILITY cults and climaxing in monotheism – to the development of

scientific understanding. This SCHOLARLY FANTASY was parent to many others far wilder – including the notorious contentions of Margaret A. Murray (1863-1963) concerning WITCHCRAFT and *The White Goddess* (1947 US) by Robert Graves – and thus became the direct or indirect inspiration of a great deal of literary fantasy.

Scholarly fantasies of this evolutionistic stripe appeal to writers of PREHISTORIC-FANTASY and LOST-RACE stories because they provide useful schematics assisting the design of hypothetical primitive societies. Other unorthodox theories of aquasi-anthropological nature have exported materials into modern fantasy which have become imaginative staples of the genre, their origins obscured by endless repetition; Ignatius Donnelly's attempts to locate ATLANTIS within the theories of the "cultural diffusionists" have been a prolific inspiration (◊ SFE), and Lewis SPENCE's speculative reconstruction of *The Mysteries of Britain* (**1928**) has made a substantial contribution to the ideological background of CELTIC FANTASY. The endeavours of anthropological folklorists have, of course, provided modern fantasy with a veritable treasury of themes.

Frazer produced only one brief work of fiction, but his contemporary Andrew LANG used his own anthropological expertise to good effect in *In the Wrong Paradise and Other Stories* (coll **1886**) and his collaboration with H. Rider HAGGARD, *The World's Desire* (**1890**). Grant ALLEN also wrote a Frazerian fantasy, *The Great Taboo* (**1891**), and brought an anthropologist from the future to study tribalism and TABOOS in Victorian society in *The British Barbarians* (**1895**). 19th-century UK literary depictions of primitive or ancient society are, however, distinctly pedestrian when placed beside contemporaneous French works, where anthropological data were copiously imported into a framework of ideas owing much to Jean-Jacques Rousseau's notion of the "noble savage". A fascination with primitive and exotic societies – particularly the societies of the Orient – infects much French fiction from Chateaubriand's *Atala* (**1801**) through Judith Gautier's *The Imperial Dragon* (**1868**) to the FIN-DE-SIÈCLE works of Pierre Loti (1850-1923) and Claude FARRERE and the prehistoric fantasies of J.H. Rosny aîné (1856-1940).

US fiction of this period tended to demonize those indigenous tribal populations which had not yet been annihilated or pacified, and thus tended to avoid more objective anthropological perspectives – a fact which has persuaded many contemporary US writers to extravagant expressions of compensatory guilt. Modern UK writers occasionally indulge in similar guilt-trips regarding the Imperial past, but many writers from the one-time colonies feel this is an aspect of conscience that needs to be far more sharply pricked, and literary fantasy is one of the instruments sometimes employed to this end.

Any list of fantasies partly inspired and fuelled by anthropological theories and speculations is bound to be highly selective, but notable examples include: J. Leslie MITCHELL's Rousseauesque *Three Go Back* (**1932**) and *The Lost Trumpet* (**1932**); Louis Herrman's late addition to the canon of Gulliveriana, *In the Sealed Cave* (**1935**); *The Lord of the Flies* (**1954**) by William GOLDING, which has twice been filmed (◊ LORD OF THE FLIES); Henry TREECE's historical fantasies, especially *The Golden Strangers* (**1956**) and *The Green Man* (**1966**); the romance of ancient exploration *Eaters of the Dead* (**1976**) by Michael Crichton (1942-   ); Jessica Amanda SALMONSON's Japanese fantasies, including the trilogy begun with *Tomoe Gozen* (**1981**); Martin H. Brice's prehistoric fantasy *The Witch in the Cave* (**1986**); Pat MURPHY's Mayan romance *The Falling Woman* (**1986**); Michaela ROESSNER's account of the Australian Aboriginal DREAMTIME, *Walkabout Woman* (**1988**); Margaret BALL's Hindu Kush-set *Flameweaver* (**1991**); and Kenneth MORRIS's Toltec romance *The Chalchiuhite Dragon* (written *c*1930; **1992**). An interesting set of stories set in a world where Frazer's supposed "laws of magic" (◊ RATIONALIZED FANTASY) are real is Randall GARRETT's **Lord Darcy** series.   [BS]

ANTICHRIST In traditional Christian ESCHATOLOGY and legend, based loosely on *Revelation* and the gospels, the Antichrist is the false prophet mentioned by CHRIST whose appearance will be one of the signals of the END OF THE WORLD. His appearance is therefore a necessary precondition for APOCALYPSE. He will deceive most of humanity into adhering to the cause of EVIL, an adherence they will advertise by wearing the Number of the Beast on their foreheads. Eventually these events will provoke ARMAGEDDON, in which the Antichrist will be defeated. In Islamic tradition, he will be slain by the returned Christ at the gates of the Church of Lydda. The MYTH was given new life in the late Middle Ages when the label was hung on various oppressors. Many have been associated with the Number of the Beast – 666 (◊ GREAT BEAST) – through the application of NUMEROLOGY to their names. Martin Luther (1483-1546) stigmatized the papacy itself as the Antichrist.

According to tradition, the Antichrist was to be born the son of an INCUBUS and a nun, making the end of time the result of female inchastity. A corollary of this tradition was that the product of such a union, caught early and baptized, would be a powerful force for good, albeit of a wild and uncontrollable nature. In many versions of the ARTHUR legend, MERLIN was such an offspring.

Among fictional treatments, the most notable literary account of the Antichrist's career is *A Short Tale of the Antichrist* (**1900** Russia) by Vladimir Soloviev (1853-1900). Robert Hugh BENSON's *Lord of the World* (**1907**) is easily the most extravagant of the devout prophetic novels produced around the end of the last century, and is far more readable than such absurdities as Joseph Burroughs's *Titan, Son of Saturn: The Coming World Emperor* (**1905**) and Sydney Watson's *The Mark of the Beast* (**1911**). *Caesar Antichrist* (**1895**) by Alfred Jarry (1873-1907) and *Antichrist* (1905) by James Huneker (1860-1921), by contrast, treat the notion with ironic contempt. Charles WILLIAMS made more subtle and sophisticated use of it in *Shadow of Ecstasy* (**1933**)

The definitive horror treatment comes with the OMEN movies – *Omen IV: The Awakening* (**1991** tvm) offered a female Antichrist – and these and others are parodied in *Good Omens; The Nice and Accurate Prophecies of Agnes Nutter, Witch* (**1990**; rev 1991 US) by Neil GAIMAN and Terry PRATCHETT. Further treatments of the Antichrist in contemporary HORROR include an essentially thriller treatment in *The Number of the Beast* (**1992**) by Daniel Easterman (1949-   ). A more colourful account is offered in Jack CHALKER's sf/fantasy hybrid *The Messiah Choice* (**1985**), while the **Nightworld** sequence by F. Paul WILSON provides humanity with a similar adversary largely divorced from a specifically Christian eschatology.

The principal importance of the Antichrist tradition to GENRE FANTASY has been through its influence on the figure of the DARK LORD in J.R.R. TOLKIEN and elsewhere. [BS/RK]

ANTI-FANTASY A term used to make a theoretical distinction in the literatures of the FANTASTIC between FANTASY – which (it is argued) presents a world that requires an act of belief from the reader – and certain genres, like the FAIRYTALE, where "Enchantment exists of itself, and the reader's participation in it is scarcely intellectual", as W.R. IRWIN puts it in *The Game of the Impossible: a Rhetoric of Fantasy* (**1976**).   [JC]

ANTIHERO During the last century and a half the mimetic novel has increasingly concentrated on protagonists who have nothing in particular of the heroic about them – though they may be admirable in other more complex ways. In the mimetic novel an antihero is a character who behaves badly (not necessarily unheroically), who (if he tells his own story) may be an unreliable narrator, who embodies and profits from the advent of social change, and who cannot easily be identified with by the reader.

In fantasy, however, an unsympathetic protagonist is likely to be described in terms which contrast him with a HERO. Because the TAPROOT TEXTS from which much fantasy derives deal either directly or at second hand with the heroic tradition, antiheroes in fantasy are very frequently defined as refusers of heroic commitments, and tend to appear in REVISIONIST-FANTASY or PARODY contexts. (ACCURSED WANDERERS and OBSESSED SEEKERS may have some characteristics of the antihero, as do some of the more ruthless protagonists of HEROIC FANTASY and MILITARY FANTASY, but in most such cases this is intrinsic to the legendary material on which they draw.)

In fantasy, moreover, protagonists who seem initially to be antiheroes very frequently turn out to LEARN BETTER – Eustace Stubbs in C.S. LEWIS's *The Voyage of the "Dawn Treader"* (**1952**) is typical of many, in the way his original obnoxiousness is modulated by instruction and adversity into heroism. Conversions of this sort are central to fantasy: many important stories turn on the protagonist's RECOGNITION of his or her true self, which may well have been cloaked in antihero clothing. Tales in which that vital turn towards the happy ending is ultimately frustrated are told through the viewpoint of an unrepentant antihero; e.g., *Gormenghast* (**1950**), the central volume of Mervyn PEAKE's **Titus Groan** sequence, centres on the career of Steerpike, and his total failure to undergo any form of TRANSFORMATION into a less worldly person signals the sequence's overall refusal of a happy ending. Texts closer to SUPERNATURAL FICTION or HORROR may also focus on an antihero; the protagonist of Jack WILLIAMSON's *Darker Than You Think* (**1948**) starts as a likable scholar and becomes an amoral killer, the MESSIAH of a race of SHAPESHIFTERS.

The antihero finds a natural place in those sorts of fantasy, usually short stories, in which he or she can serve as an awful warning. SLICK FANTASY often has protagonists unlikable to the point of being antiheroes so that the eventual comeuppance is dramatically satisfying.   [RK]

ANTILLIA ◊ LOST LANDS AND CONTINENTS.

ANTRIM, DONALD (ELDRIDGE) (1958-   ) US writer, author of *Elect Mr Robinson for a Better World* (**1993**), a satiric FABULATION set in a US suburb where neighbours battle with semiautomatic weaponry in the local park and surround their houses with trenches and fortifications; additionally, the narrator's wife possesses a TALENT that enables her to "access memory at the cellular level and ride the DNA chain like a wave back into prehistory", thus achieving rapport with extinct sealife. The tension between the narrator's affectless voice and the grim goings-on tightens the narrative thread to a darkly hilarious pitch. [GF]

ANUBIS A GOD of ancient EGYPT, involved in the central Isis and Osiris MYTH. He resides in the UNDERWORLD, and as the god of cemeteries is a more than normally noticeable LIMINAL BEING, whose image – he has the head of a jackal – has been familiar for many centuries. He appears with some frequency in fantasy, especially DARK FANTASY. He can be found, for example, in Roger ZELAZNY's *Creatures of Light and Darkness* (**1969**) and in Tim POWERS's *The Anubis Gates* (**1983**). [JC]

APARTHUR, C. [s] ◊ Kenneth MORRIS.

APEL, JOHANN AUGUST (1771-1816) German lawyer, playwright and, later, librarian who became an authority on German FOLKTALES, especially GHOST STORIES. Working with Laun (real name Freidrich Schulze; 1770-1849), JAA produced five volumes of such tales, each called *Gespensterbuch* (anths **1811**, **1812**, **1813**, **1814** and **1815**; Laun alone may have produced further vols in 1816 and 1817), drawn as much from the oral as from the written tradition, like the contemporary work of the GRIMM BROTHERS – though JAA and Laun concentrated more on Gothic and occult aspects. JAA's most popular contribution to the series was "Der Freischütz" (1811; trans as "The Fatal Marksman" 1824 UK; vt "The Freeshot"; vt "The Magic Balls"), a PACT WITH THE DEVIL story which formed the basis of the OPERA *Der Freischütz* (1821) by Carl Weber (1786-1826). Selections from the first two volumes were translated into French by J.B.B. Eyriès as *Fantasmagoriana* ["Phantasmagoria"] (anth **1812** France); this was the book which (◊ ANTHOLOGIES), read by Lord BYRON and his house party, eventually inspired Mary SHELLEY's *Frankenstein, or The Modern Prometheus* (**1818**). It saw a UK translation as *Tales of the Dead* (anth **1813**), but the rest of the series remains untranslated. Apart from "The Boarwolf" (1812; vt "The Demon's Victim"), an early WEREWOLF story which also involves a Faustian pact, none of JAA's fiction has been credited in translation, and much of his work is therefore regarded as "anon". JAA's seminal work had considerable impact on the development of SUPERNATURAL FICTION. [MA]

APES In sf, apes (and cavemen) have long served as emblems of evolution, and offer through their clearly evident kinship with *Homo sapiens* a series of lessons and *arguments* about time, progress and the human condition. Fantasy, where evolution tends to be understood in terms of METAMORPHOSIS, does not normally provide that sort of perspective, and fantasy apes tend consequently not to represent arguments about the human condition but ALLEGORIES of it. A partial exception to this principle would be the PREHISTORIC FANTASY, but most examples of the form – which is almost entirely restricted to tales written after Charles Darwin (1809-1882) published *Origin of Species* (**1859**) – simply demonstrate how close to sf (or indeed to mundane fiction about the deep past) this subgenre is.

Some pre-Darwin tales involving apes invoke early versions of ANTHROPOLOGY in a fantasticated fashion, and can be thought of as satirical fantasies. Thomas Love PEACOCK's *Melincourt, or Sir Oran Haut-on* (**1817**) features an educated orang-utan who becomes a Member of Parliament. The LOST-RACE ape culture featured in *The Monikins* (**1835**) by James Fenimore Cooper (1789-1851) is similarly fantasticated. *Les Emotions de Polydore Marasquin* (**1857**; trans anon as *The Man Among the Monkeys, or Ninety Days in Apeland. To which is Added the Philosopher and his Monkeys, the Professor and the Crocodile, and Other Strange Stories of Men and Animals*

1873 UK; vt *The Emotions of Polydore Marasquin* 1888 UK; vt *Monkey Island* 1888 UK) by Léon Gozlan (1806-1866), with illustrations by Gustave DORÉ, comes close to fantasy in telling the story of a man who becomes king of the monkeys on an unknown ISLAND.

After Darwin, the ape is mostly found playing two roles. He (or she) may represent the "ape of God", and in this guise embody a PARODY of humankind or deity. Examples include: the humans (who are seen as apes in a game of PERCEPTION) in Wyndham LEWIS's *The Apes of God* (**1930**); the chimp in John COLLIER's *His Monkey Wife* (**1930**), eventually preferred to a human woman; the ape who dresses in lions' skins the donkey who plays ANTICHRIST in C.S. LEWIS's *The Last Battle* (**1956**); the man perceived as an ape in Herbert ROSENDORFER's *German Suite* (**1972**) because his mother had slept with one before she became pregnant; and the ape of God who is the final teller of the ARABIAN-NIGHTMARE tales in Robert IRWIN's *The Arabian Nightmare* (**1983**). The ape may represent the face of innocence in a nightmarish human world, usually in tales which can be described as BEAST FABLES; examples include: Wilhlem HAUFF's "The Monkey as a Man" (1827), which Hans Werner Henze (1926-) adapted into an OPERA, *Der Junge Lord* ["The Young Lord"] (**1965**), where he is specifically an ape; Jenny DISKI's *Monkey's Uncle* (**1994**), which features a talking orang-utan; and Scott BRADFIELD's *Animal Planet* (**1995**), one of whose protagonists is an ape who becomes a sexually exploited au pair in NEW YORK.

Almost certainly the most famous ape in fantasy is the Librarian of the Unseen University in Terry PRATCHETT's **Discworld** novels. [JC]

See also: BEAUTY AND THE BEAST.

APHRODITE The Greek GODDESS of LOVE and beauty, called Venus by the Romans; the equivalent of the Phoenician Astarte. Legend credits her with a key role in starting the Trojan War when she offered Helen to Paris as a bribe. She betrayed her husband Hephaestus with Ares. Her symbolic presence dominates *Aphrodite* (**1896**) by Pierre LOUŸS and "The Disinterment of Venus" (1934) by Clark Ashton SMITH and is of considerable importance in *Mistress of Mistresses* (**1935**) and its sequels by E.R. EDDISON. She is the central character of "Mrs Hephaestus" (1887) by George A. Baker (1849-1918) and *Venus the Lonely Goddess* (**1949**) by John ERSKINE. Other relevant mythological fantasies include "Mr Skinner's Night in the Underworld" (1878) by Max ADELER, *The World's Desire* (**1890**) by H. Rider HAGGARD and Andrew LANG, *The Night Life of the Gods* (**1931**) by Thorne SMITH and *Pagan Passions* (**1959**) by Randall GARRETT and Larry M. Harris.

Notable fantasies based on a tale reproduced in Robert Burton's *The Anatomy of Melancholy* (**1621**), in which a STATUE comes to life after a RING is placed on its finger, include "The Venus of Ille" (1837) by Prosper MÉRIMÉE, *The Tinted Venus* (**1885**) by F. ANSTEY, "St Eudaemon and his Orange Tree" (1907) by Vernon LEE, "The Bride" (1931) by Rosalie Muspratt (1906-1976) and *The Eve of Saint Venus* (**1964**) by Anthony BURGESS (◊ PYGMALION). Significant fantasies based on the story of the German knight Tannhäuser, who consorted with the love goddess in a cave-palace within the Venusberg, include "The Faithful Eckhart and the Tannenhaeuser" (1799) by Ludwig TIECK, the incomplete "Under the Hill" (1896) by Aubrey BEARDSLEY, completed by John Glassco as *Under the Hill* (**1959**), and "The Gods and Ritter Tanhuser" (1913) by Vernon Lee. A fantasy featuring Aphrodite's magic girdle, which made everyone who wore it the object of irresistible desire, is "The Girdle of Venus" (1947) by Harold Lawlor. [BS]

APOCALYPSE Apocalyptic literature was produced in abundance 200BC-AD200 as the Jews responded to political persecution and cultural upheaval, describing in calculatedly enigmatic terms a world-ending intervention of God on behalf of His chosen people. Parts of this tradition were taken up by the Christian Mythos, entering the New Testament in *Revelation*, whose imagery has (within the Western tradition) become perforce definitive. The term is nowadays employed more broadly to refer to any abrupt END OF THE WORLD.

The modern subgenre of apocalyptic fiction belongs more to sf than to fantasy, although the imagery of *Revelation* resounds continually within it. There was a brief fashion for such works in the UK at the beginning of the 19th century, sparked off by the anonymous *The Last Man, or Omegarus and Syderia: A Romance in Futurity* (**1806**; actually a trans from the French of Jean Cousin de Grainville's *Le Dernier Homme* [**1805**]), and apocalyptic imagery is also very evident in the paintings of John MARTIN. Edgar Quinet's *Ahasvérus* (**1833**) redesigns the Christian apocalypse according to a new allegorical logic. A second wave of apocalyptic fantasies followed around the end of the century, but those which were religiously inspired – including several novels by Sydney Watson – are mostly devoid of literary interest, and such sarcastic fantasies as H.G. WELLS's "A Vision of Judgment" (1899) treat the notion with contempt. In modern fantasy *Revelation*'s imagery is frequently invoked in a metaphorical fashion; particularly prolific use has been made of the enigmatic GREAT BEAST and the FOUR HORSE-MEN. The horsemen appear in person in a number of comedies and ironic fantasies, sometimes in light disguise, as in "The Big Flash" (1969) by Norman Spinrad (1940-   ), but also in more serious works like Sheri S. TEPPER's TECHNOFANTASY *A Plague of Angels* (**1993**). A sweeping comic version of *Revelation* can be found in *Good Omens* (**1990**) by Neil GAIMAN and Terry PRATCHETT. A distinctively modern version of *Revelation* is given rather fragmentary literal expression in the movie *The Rapture* (**1991**), but by far the most effective 20th-century account of a literal apocalypse solidly based in the Judeo-Christian Mythos is James BLISH's *Black Easter* (**1968**) and *The Day After Judgment* (**1972**).

The SECONDARY WORLDS of GENRE FANTASY are very often threatened with apocalyptic termination, but the formulaic plots of such works almost invariably require that disaster be averted in the nick of time. In INSTAURATION FANTASIES, however, such reprieves are not always granted, and in non-genre works the urge to wrathful destruction may be given free expression, as in Gustav MEYRINK's *Das grüne Gesicht* (**1916**; trans as *The Green Face* **1992** UK). [BS]
See also: ANTICHRIST.

APOLLINAIRE, GUILLAUME Pseudonym of Italian-born French writer Wilhelm de Kostrowitzky (1880-1918), initially important as an instigator and main literary promoter of movements like Fauvism and Cubism, and of painters like Georges Braque (1882-1963) and Pablo Picasso (1881-1973), during the first years of the 20th century. Towards the end of his life GA invented the term SURREALISM to describe the harlequinade-dominated set (◊ COMMEDIA DELL'ARTE) Picasso designed for Jean COCTEAU's ballet *Parade*, produced 1917 by Sergei Diaghilev (1872-

1929); he also used the term to describe his own play, *Les Mamelles de Tirésias* (produced 1917; trans as "The Breasts of Tiresias" *Odyssey* 1961). He wrote considerable fiction, some pornographic, like *Les Onze Mille Verges* (**1907**; trans as *Les Onze Mille Verges, or The Amorous Adventures of Prince Mony Vibescu* **1979** UK), whose bizarrenesses are almost supernaturally intense; but is best-known for his poetry.

The long title story of *L'Enchanteur pourrissant* ["The Rotting Magician"] (1904 *Le Festin d'Esope*; coll **1909**) constitutes a kind of promenade in front of the tomb of MERLIN, who cannot be awoken from his SPELL by the adjurations of the female narrator. "Onirocritique" ["Dream Criticism"], also in this volume, describes the Arthurian CITY of Orqueneseles. *L'Hérésiarque et Cie* (coll **1910**; trans Remy Inglis Hall as *The Heresiarch and Co* **1965** US; vt *The Wandering Jew and Other Stories* 1965 UK) contains a variety of tales whose dilettantish ornateness tends to hide some sharp ironies, as in "The Passerby of Prague", which features a robustly amoral WANDERING JEW. In "The Disappearance of Honoré Subrac" a man is so cowardly he turns into a WAINSCOT creature, invisible to others. In the **False Amphion** sequence a CONFIDENCE MAN projects varying images of himself across Paris, via a hologram-like device. *Le Poète Assassiné* (**1916**; trans Ron Padgett as *The Poet Assassinated* **1968** UK) is a surrealistic autobiographical tale which conveys a phantasmagorical sense of life in pre-War PARIS (◊ URBAN FANTASY); it also appears in *The Poet Assassinated and Other Stories* (coll trans **1984** US), a volume which includes "The Blue Eye", about an apparition which arouses young girls approaching menarche, and "Arthur, the Once and Future King", set in AD2105, when the monarch reawakens. [JC]

APOLLO A principal GOD of Greek and Roman mythology. Arrows fired from his bow were a significant instrument of divine punishment, but he functioned also as a deliverer and as a patron of ORACLES. His place in modern fantasy is more associated with his minor role as the god of MUSIC. He was also the god to whom SACRIFICES were offered when new towns were founded (thereby tacitly appointing him "god of civilization"); for this reason he was deployed as a crucial symbol by Friedrich Nietzsche (1844-1900) in *The Birth of Tragedy* (**1872**), which used the terms "Apollonian" and "Dionysian" (◊ DIONYSUS) to compare and contrast ascetic and hedonistic cultural phases. Nietzsche's notion of Apollonian ideals is neatly unpacked in *Evander* (**1919**) by Eden PHILLPOTTS. Apollo's symbolic presence as Sun-god is vital to *The Arrows of the Sun* (**1949**) by Ivor Bannet. He plays a leading role in "The Substitute for Apollo" (1833) by John STERLING, "The Dumb Oracle" (1878) and "The Poet of Panopolis" (1888) by Richard GARNETT, "The Gods and Ritter Tanhuser" (1913) by Vernon LEE, "Under the Sun" (1923) by R. Ellis Roberts (1879-1953) and *The Mask of Circe* (1948; **1975**) by Henry KUTTNER. [BS]

APPRENTICE ◊ SORCERER'S APPRENTICE.

APULEIUS (*c*AD 125-?   ) Latin writer, often credited with the name of Lucius (of which there is, in fact, no record), born in North Africa, active in Carthage, Athens, Rome and elsewhere; he remains best-known for his *Metamorphoseon sue de Asino Aureo Libri XI* (written *c*AD165; trans William Adlington as *The xi Bookes of the Golden Asse, Containing the Metamorphoses of Lucius Apuleius* **1566** UK; many trans since; good modern trans by Robert Graves **1950** and by P.G. Walsh **1994**), normally known as *The Golden Ass*. Within an elaborate FRAME STORY various episodes and tales are recounted, the most famous being that of CUPID AND PSYCHE, which is the first LITERARY FAIRY-TALE. The frame story itself – which may be in part borrowed from LUCIAN's *Lucius, or The Ass* (written *c*AD150) – is also of considerable interest. It is narrated by Lucius, a Greek who on a visit to Thessaly stays in the house of a sorcerer, whose wife Pamphile is able to metamorphose into a BIRD. Avid to learn about MAGIC and disregarding clear warnings (contained in stories he has heard) about the dangers of uncontrollable curiosity, Lucius asks the servant he has been sleeping with to give him the ointment necessary for the feat; but she gives him the wrong ointment and he is transformed into an ASS. Only when he manages to eat a rose will he become human again; but before he can find one he is stolen by thieves, and his PICARESQUE adventures begin in earnest. During their course he hears the story of "Cupid and Psyche", enters a CIRCUS, and has SEX with a woman. Eventually he escapes his various captors and begs the aid of Isis (◊ GODDESS), who initiates him into the MYSTERY RELIGION devoted to her. He is then transformed back into a man, Lucius Apuleius.

Almost inevitably, a tale so subversive and licentious at any literal level has been treated as ALLEGORY; early Christian exegetes – despite A's dislike of Christianity – tended to conceive of it as an allegorical justification of Christian morality, and the "Cupid and Psyche" tale (more plausibly) a description of the testing and growth of the SOUL (◊ BEAUTY AND THE BEAST). The frame story is, certainly, a FABLE, and the narrator does grow spiritually through his taxing NIGHT JOURNEY; but the equipoise and fixity of allegory is missing. Inventive, bawdy, hilarious and terrifying, *The Golden Ass* – the only Roman novel to survive intact – is an essential TAPROOT TEXT for fantasy. [JC]

ARABIAN FANTASY In *Arabesques: More Tales of the Arabian Nights* (anth **1988**), Susan SHWARTZ suggests that the stories making up *The Arabian Nights*, along with various tales associated with the *Nights*, should be described as the Matter of Araby – by analogy with the MATTER of Britain or France.

Some of the necessary conditions exist. AF as a whole inhabits a LAND-OF-FABLE environment whose deserts and oases, bazaars and slums, jewelled caverns and minaret-topped EDIFICES are immediately recognizable. The CITIES central to it – CAIRO and BAGHDAD – are classic venues for URBAN FANTASY. The cast – beggars, houris, eunuchs, caliphs, viziers, adventurers, GENIES and orcs – are also familiar MOTIFS; as are magic carpets and other appurtenances. Moreover, AF is almost always told – *Arabesques* is only one recent imitation – as a STORY CYCLE, the form in which *The Arabian Nights* has always been transmitted to the West. But certain elements are lacking. AF, as understood in the West, lacks centrally any epic version of the founding or defence of a domain, nation or culture. There is also a problem of geography. ARTHUR's Britain, Roland's France and others can be located in the mind's eye, and in fact upon the map. Araby, on the other hand, is a mirage. In *Arabesques* Shwartz provides a MAP of "The World of the Arabian Nights": it features locations from northern Africa through Asia Minor to China, but Arabia itself (being mostly desert) is almost empty.

This lack of focus – as far as Western users of the tales are concerned – makes it difficult to think of AF as dealing with the matter of Araby; but there is a more fundamental difficulty. Nowhere *within* the various cultures which originally contributed to the making of the *Nights* is that story cycle deemed to constitute a national epic. The matter of Araby, so far as it can be thought to exist at all, is a Western creation, as is AF as a whole.

Even the *Nights* themselves, as they have come down to us, are partly a Western fabrication. The original *Alf Layla wa-Layla* (literally "One Thousand Nights and a Night") began to take shape in about the 10th century; something like the essential cycle, assembled from Egyptian, Syrian and Persian sources, seems to have been in existence by about the 14th century. The FRAME STORY seems always to have shaped the cycle: the Sultan, Shahriyar, betrayed by an earlier wife, proclaims his decision to wed a virgin (◊ VIRGINITY) nightly and to behead her in the morning; SHE-HEREZADE volunteers for the role, saving her life for 1001 nights by telling the Sultan part of a STORY each night, breaking off at dawn before the tale can be concluded, so that the intrigued Shahriyar postpones her beheading 1001 times. But the actual number and sorting of the stories she tells have varied over the centuries, and several of the best-known tales in English have no known Arabic manuscript precursors.

The first translation into a Western language was by Antoine Galland (1646-1715) into French as *Les Mille et une nuits* (**1704-17**), based on (but not restricting itself to) a manuscript dating from the 14th/15th century. This manuscript was published as *Alf Layla wa-Layla* (**1984**) in a critical edition by Muhsin Mahdin, and is the version translated by Husain Haddawy (see below). It contains only 35 stories, however; there are hundreds more. Among the tales which seem either to have been invented by Galland or to derive from some now untraceable manuscript or oral transmission are "Aladdin", "Ali Baba", "The Ebony Horse" and "Prince Ahmed and his Two Sisters". (Arabic manuscripts of any of these tales postdate Galland's versions.) Also, though it is based on UNDERLIER material from numerous sources, and though Sinbad himself has served as an underlier for centuries, the manuscript Galland used does not include "The Seven Voyages of Sinbad".

From about 1706, English translations of selections from Galland began to appear. From the early 19th century, translations into English have depended upon variously trustworthy printed editions – all compiled or invented according to various criteria – of the *Nights* in Arabic. Robert IRWIN's *The Arabian Nights: A Companion* (**1994**) – from which this entry, with his permission, takes much data concerning the *Nights* – analyses at considerable length the nature of the texts used and the quality of the various recent translations, beginning with *The Thousand and One Nights* (**1838-41**) by Edward William Lane (1801-1876) and continuing with versions by John Payne (1842-1906) and Sir Richard BURTON, whose *A Plain and Literal Translation of the Arabian Nights Entertainments, Now Entitled the Book of the Thousand Nights and a Night* (**1885**), plus *Supplemental Nights to the Book of the Thousand Nights and a Night* (**1886-8**), though awkward, contains the most complete assembly of *Nights* and *Nights*-related tales. A translation in French by Joseph Charles Mardrus as *Le Livre des Mille et une nuits* (**1899-1904**), which is more an adaptation than a translation, was itself adapted into English by E. Powys Mathers as *The Book of the Thousand Nights and One Night* (**1923** 4 vols). Recent translations of merit include *Tales from the Thousand and One Nights* (**1973**) by N.J. Dawood, which takes a broad-church attitude toward the inclusion of late and/or apocryphal tales, though it is less omnivorous than Burton, and *The Arabian Nights* (**1990**) by Husain Haddawy, which restricts itself to tales found in the oldest extant manuscript.

Even before Galland's edition reached its posthumous conclusion (he was not responsible for some of the insertions in the final volume), writers in France and the UK had begun to mine the treasure trove. Given the state of textual criticism in the early 18th century – even now a secure pre-14th-century STEMMA for the *Nights* remains a scholarly dream – it is not surprising that a very wide range of original stories, rough adaptations and translations became associated with the *Nights*, and that the *Nights* came to stand for anything that might be identified as an AF, whenever or wherever compiled. Of these preceding and surrounding cycles, the most famous is probably the SANSKRIT *Panchatantra* ["Five Books"] (*c* 6th century AD), which was incorporated into the *Katha Sarit Sagara* (*c* 11th century AD; trans C.H. Tawney as *Katha Saritsagara* **1880-4**; this trans ed Norman PENZER vt *The Ocean of Story* **1924-8**). The Sanskrit compilation and the *Nights* have many motifs and tales in common.

This situation is of greater import to scholars than to modern fantasy writers, who know they are fabricating an "Araby" and a *Nights* that never existed, except in the West after 1700, and make frequent and increasingly sophisticated use not only of the *Nights* but of the OCEAN OF STORY in which the *Nights* remains only the most famous assemblage. In the UK, the first writer of interest to fabricate AF is probably Anthony Hamilton (?1646-1720), who wrote in French; his *Histoire de Fleur d'Épine* (**1949**; trans as *History of May-Flower* **1793** UK) is a deliberate PARODY of the newly translated *Nights*. Over the course of the 18th century, French writers frequently used AF models to generate SATIRE directed at French society, even though the government censors proved unkindly adept at penetrating this form of Aesopian language. Claude-Prosper CRÉBILLON fils was imprisoned for insufficient obscurity, though he escaped retribution for *The Sofa* (**1740**). Other tales using the *Nights* in this manner included *Les bijoux indiscrets* (**1748**) by Denis Diderot (1713-1784) and VOLTAIRE's *Zadig* (**1748**). Jacques CAZOTTE's *The Devil in Love* (**1772**) echoes the *Nights*. And Count Jan POTOCKI, who though Polish wrote in French, borrowed the complex frame-story structure of *The Manuscript Found at Saragossa* (written from 1797; **1847**) from the *Nights*. There were hundreds more.

At first the picture in the UK was less interesting, and it remained for some time sullied by an English assumption that the *Nights* were immoral. The first notable AF in English may be *Rasselas* (**1759**) by Samuel Johnson (1709-1784). Later works of interest include *Nourjahad* (**1767**) by Frances Sheridan (1724-1766), *Hieroglyphic Tales* (coll **1785** chap) by Horace WALPOLE, *The History of Charob, Queen of Aegypt* (**1785**) by Clara REEVE, Robert SOUTHEY's *Thalaba the Destroyer* (**1800**) and, pre-eminently, William BECKFORD's *Vathek* (**1786**), arguably the first significant DARK FANTASY and a book which has been influential since its publication. Many 19th-century works – often eschewing any element of the FANTASTIC to concentrate on travel and adventure – continued to show the influence, though often by using AF as a matrix through which to mock the imaginary habits of the imaginary "Arabians", or to PARODY contemporary life; Frederick MARRYAT's *The Pacha of Many Tales* (coll of linked stories **1835**) and William Makepeace

THACKERAY's *John Bull and his Wonderful Lamp* (**1849** chap) are typical. The most interesting fantasy tale was probably *The Shaving of Shagpat: An Arabian Entertainment* (**1855**) by George Meredith (1828-1909), which uneasily combines pastiche and ALLEGORY; and the most interesting collection was perhaps Robert Louis STEVENSON's *New Arabian Nights* (coll **1882**) which – with its sequel, *More New Arabian Nights: The Dynamiter* (coll **1885**) with Fanny de Grifft Stevenson, and the unrelated *Island Nights' Entertainments* (coll **1893**) – provided a model whereby the ambience of the *Nights* could help establish the sootier mood of the GASLIGHT ROMANCE.

In the 20th century, the influence of the *Nights* extends from Modernist literature – James JOYCE's *Ulysses* (**1922**) contains a layer of reference to Sinbad the Sailor, and Jorge Luis BORGES made constant reference to the *Nights* – on to Postmodernists like John BARTH, several of whose tales and novels – most obviously *The Last Voyage of Somebody the Sailor* (**1991**) – are explicit fantasias upon the characters and structure of the cycle, and Salman RUSHDIE, whose works also make a constant conversation with the *Nights*. Rather more tentatively than in the 19th century, popular authors – like James Elroy FLECKER, whose *The King of Alsander* (**1914**) and *Hassan* (**1922**), are both set in land-of-fable Arabias – continued to use AF material both to reflect comments on the contemporary world and as a romantic alternative to that world. More recent fantasy works which use the matrix – not all of them comic – include Ian DENNIS's **The Prince of Stars in the Cavern of Time** sequence, Seamus CULLEN's *A Noose of Light* (**1986** UK) and its sequel, *The Sultan's Turret* (**1986** UK), Esther FRIESNER's **Chronicles of the Twelve Kingdoms** sequence, Tom HOLT's *Djinn Rummy* (**1995**), L. Ron HUBBARD's *Typewriter in the Sky* (1940 *Unknown*; **1995**) and many others, Robert IRWIN's *The Arabian Nightmare* (**1983**) (◊ ARABIAN NIGHTMARE), Elizabeth Ann SCARBOROUGH's *The Harem of Aman Akbar, or The Djinn Decanted* (**1984**), Craig Shaw GARDNER's **Arabian Nights** sequence, and Tad WILLIAMS's and Nina Kiriki HOFFMAN's *Child of an Ancient City* (**1992** chap). Susan Shwartz's *Arabesques: More Tales of the Arabian Nights* (anth **1988**), discussed above, and *Arabesques II* (anth **1989**), provide a useful assembly of the range of AF-derived fantasy by contemporary writers. [JC]

ARABIAN NIGHTMARE Robert IRWIN's first novel, *The Arabian Nightmare* (**1983**), provides a clear model – and a convenient name – for the AN, a tale or DREAM in which other tales or dreams are embedded in a process with no clear outcome. It is a narrative structure which has existed as long as has fantasy as a conscious genre, though it takes its name, and many ICONS and narrative elements, from the ARABIAN NIGHTS. In Irwin's novel a man called Balian arrives in 14th-century Cairo on a complicated mission, settles into the local caravanserai, falls asleep, and enters through a PORTAL of dreams into a world of profound slumber. He soon dreams that he has awoken, but the dream continues. In the dream, he dreams again. He awakens (or dreams that he has awoken). He is suffering from the AN, a condition which may be described as that of inhabiting in dreams a STORY the only exit from which is a further and deeper dream; in Irwin's novel the AN is furthermore a condition which in dreams subjects one to suffering without end, and which no one can remember having experienced after awakening (but if the dreamer awakens from a nightmare which he takes solace in remembering, he may simply be dreaming, from within the ongoing AN, that he has awoken). Moreover, and even more nightmarishly, none of the stories he is trapped in ever seems to finish, so that Balian (or the story of Balian) is never complete. He has become the PARODY of a SOUL, and his course downwards through one truncated story after another constitutes a parodic mockery of the impulse of Story to come to a resolution.

Irwin's novel is a late and extremely sophisticated DARK-FANTASY rendering of the nightmare of the story that fails to lead its bearers back into the world, and is very literal in that the nightmare is in fact induced by an actual dream; but various realizations of the apprehension that underlies it have appeared in the literature of fantasy. They include the first "ten days" of Count POTOCKI's *The Saragossa Manuscript* (**1804-5**), Charles NODIER's *Smarra* (**1821**), L. Ron HUBBARD's *Fear* (**1940** *Unknown*; **1957**), Herbert ROSENDORFER's *The Architect of Ruins* (**1969**), Gene WOLFE's *Peace* (**1975**), William KOTZWINKLE's *Fata Morgana* (**1977**), Michael ENDE's *Mirror in the Mirror* (coll of linked stories **1984**), Lisa GOLDSTEIN's *Tourists* (**1989**), John FULLER's *Look Twice: An Entertainment* (**1991**), Scott BAKER's "Virus Dreams" (1993) and *Flesh & Blood* (**1994**) by Michele Roberts (1949-). [JC]

ARABIAN NIGHTS ◊ ARABIAN FANTASY.

ARBES, JAKUB (1840-1914) Czech writer. ◊ CZECH REPUBLIC.

ARCADIA The province of Arcadia or Arcady, in the Greek Peloponnesus, was never hospitable to husbandry or much else; but its function as a locus for nostalgia is important. The popular dream of Arcadia derives from VIRGIL, who conceived of it as a happy land in which pastoral virtues triumphed, men and women lived in harmony, and death did not seem to intrude.

The only version available until the 20th century of the prose romance *Arcadia* (rev 1593) by Sir Philip Sidney (1554-1586) is, though not fantastic, an important TAPROOT TEXT because it presents with great and mannered thoroughness a nostalgic vision of a secular but blessed GOLDEN AGE. Its plot is extremely complicated and artificial, and Sidney never completed the much-elaborated second version of the tale which he worked on until his death; consequently Sidney's unfinished narrative has never served as more than an abstract model for later writers to build upon. But Arcadia itself, a POLDER hovering at the edge of the fantastic, has always been an important locus for that nostalgia urban sophisticates feel for enclaves safe from the historical processes which give them the opportunity to relish their emotions (◊ ET IN ARCADIA EGO; GOLDEN AGE). [JC]

ARCHETYPES Term derived from a Greek word meaning an original pattern or template, and adapted into various metaphysical theses. Platonist philosophers used it to refer to the primal "ideas", or transcendent essences, of which existent things are inferior reproductions; scholastic philosophers used it to refer to the divine ideas determining the forms of Creation; John Locke (1632-1704) used it to refer to the external REALITIES to which our ideas and impressions correspond. Its modern usage is dominated by the psychological theories of Carl Jung (◊ JUNGIAN PSYCHOLOGY), who postulated that "the archetypes of the collective unconscious" are innate features of the human psyche which direct all fantasy activity, thus giving rise to a series of "archetypal images" which constantly resurface in MYTH, FOLKLORE, literature (in its "visionary" mode) and DREAMS. Jung's most significant essays on the topic 1934-55 can be found in Part 1 of volume 9 (**1959**) of his *Collected Works*. Jung never attempted a comprehensive list but he referred, among others, to the Mother (◊ GODDESS), REBIRTH, the Spirit and the TRICKSTER; his notions of the ANIMA and *animus* are affiliates of the same theoretical framework. Jungian folklorists and literary theorists have been more prolific than consistent in identifying archetypal images; those most frequently employed include the Earth Mother, the Divine Child, the Unwilling HERO, the Wise Old Man and the Enchanted Prince. A rare pre-Jungian use of archetypes in fantasy can be found in "The Seven Geases" (1934) by Clark Ashton SMITH. [BS]

ARCHIPELAGO There are numerous exceptions, but the landscapes depicted in SECONDARY-WORLD tales seem generally of two types: a LAND, consisting of an assortment of countries which frequently surround a central inland sea; and a world-straddling oceanscape, featuring archipelagos. Archipelagos are also found in OTHERWORLD and LAND-OF-FABLE fantasies and in PLANETARY ROMANCES, because the islands making up an archipelago tend naturally to differ very widely from one another. Whereas ISLAND stories – like T.H. WHITE's *The Master* (**1958**) – tend to house tyrants and UTOPIAS, archipelagos are useful whenever a range of possible societies is to be contrastingly depicted, as in Jonathan SWIFT's *Gulliver's Travels* (**1726**), Herman MELVILLE's *Mardia and a Voyage Thither* (**1849**), Clark Ashton SMITH's "The Voyage of King Euvoran" (1933) – whose protagonist discovers various eccentric domains in "the archipelagos of morning" eastward from Zothique – Salman RUSHDIE's *Haroun and the Sea of Stories* (**1990**) – in which the great city of Gup, choked with a WONDERLAND assortment of inhabitants, is built on an archipelago in the Ocean of the Streams of Story – Ursula K. LE GUIN's **Earthsea** quartet (**1968-90**), or the Efica island group in Peter CAREY's *The Unusual Life of Tristan Smith* (**1994**); they are also useful, as in William MORRIS's *The Water of the Wondrous Isles* (**1897**), when an author wishes to dramatize the flight from an unwanted circumstance to a longed-for destination. [JC]

ARCIMBOLDO, GIUSEPPE (1527-1593) Milanese painter who became Court Painter at Prague (1562-87) for two Holy Roman Emperors, Maximilian II and his son Rudolph II; it was at Prague that GA composed his series of FOLIATE HEADS composed of whole or fragmented vegetables, flowers, BIRDS and animals (heraldic or otherwise) and portions of LANDSCAPE. These series were generally made up of four portraits, and several of them depict the four SEASONS; the allegorical intent (◊ ALLEGORY) of these paintings is very clear, and their TROMPE L'OEIL effects may have been secondary to GA and to his patrons. To a late-20th-century viewer, GA's portraits seem to hover on the cusp of METAMORPHOSIS, and presage the dream-effects of SURREALISM, some of whose advocates claimed GA as an honoured ancestor. The conflation of world and countenance in GA can be understood as analogous to the intimate linkage between SECONDARY WORLD and protagonist in full-fantasy narratives. [JC]

See also: FACE OF GLORY.

Further reading: *Giuseppe Arcimboldo* (**1993**) by Werner Kriegeskorte.

ARDA ◊ FANTASYLAND; MIDDLE-EARTH; J.R.R. TOLKIEN.

ARDIZZONE, EDWARD (JEFFREY IRVING) (1900-1979) Chinese-born painter, illustrator and writer, in the UK from 1905. He began to work as an illustrator around 1926. He illustrated an edition of J. Sheridan LE FANU's *In a*

Glass Darkly in 1929; this was the first of over 170 books he illustrated. He started to write stories for younger children with *Little Tim and the Brave Sea Captain* (**1936**; illus redrawn 1955), beginning a series of loosely linked tales that mix the mundane and the fantastical in a manner typical of tales written for that age group. *Ardizzone's Hans Andersen* (coll **1978**) and *Ardizzone's English Fairy Tales* (coll **1980**) are also of interest. His best-known illustrations of fantasy subjects are those for various books by James REEVES, beginning with *The Blackbird in the Lilac* (**1952**), and Eleanor FARJEON, beginning with *The Little Bookroom* (**1955**). A collaboration with Reeves, *Prefabulous Animiles* (**1957**), generated numerous drawings of fabulated creatures (◊ BESTIARIES).

Other texts of interest with EA illustrations include an edition of John BUNYAN's *The Pilgrim's Progress* (1947), several books by Walter DE LA MARE, including the first post-WWII edition of *Peacock Pie* (coll 1946), Nicholas Stuart GRAY's *Down in the Cellar* (**1961**), editions of *The Little Fire Engine* (1973) by Graham Greene (1904-1991) and Greene's other CHILDREN'S FANTASIES, the late revision of Noel LANGLEY's *The Land of Green Ginger* (1966), a new edition of the TWICE-TOLD tale *The Dragon* (1966 chap) by Archibald Marshall (1866-1934), Naomi MITCHISON's *The Rib of the Green Umbrella* (**1960** chap), several children's books by John SYMONDS including *Lottie* (**1956** chap), and T.H. WHITE's *The Godstone and the Blackymor* (**1959**).

EA's style was seemingly casual, artfully simple, undemanding, unchallenging and unmistakable. [JC]

ARENAS, REINALDO (1943-1991) Cuban writer, in the USA from 1980, whose *El mundo alucinante* (**1968** France as *Le monde hallucinant*; first Spanish-language version **1969** Mexico; trans Gordon Brotherson as *Hallucinations: Being an Account of the Life and Adventures of Friar Servado Teresa de Mier* **1971** US; new trans Andrew Hurley, with RA, as *The Ill-Fated Peregrinations of Fray Servando* **1987** US) is a MAGIC-REALISM fabulation in which – in contradistinction to the imprisoned life RA was leading in Cuba – everything is alive: the words of the book, the protagonist, the thriving world. The **Pentagonía** – *Celestino antes del alba* (**1967**; rev vt *Cantando en el pozo* 1982 Spain; trans Hurley as *Singing from the Well* **1987** US), *El palacio de las blanquísimas mofetas* (**1975** France as *Le palais de très blanches mouffettes*; first Spanish-language version **1980** Venezuela; trans Hurley as *The Palace of the White Skunks* **1990** US), *Otra vez el mar* (**1982** Spain; trans Hurley as *Farewell to the Sea* **1986** US) and *El Asalto* (**1990** Spain; trans Hurley as *The Assault* **1994** US) – is a series of quasi-autobiographical narratives, told in differing voices, about a protagonist who is reincarnated (◊ REINCARNATION) into different lives, sometimes male, sometimes female, as the narratives succeed one another. [JC]

ARGOS US digest MAGAZINE, 3 issues, quarterly, Winter 1988-Summer 1988, published by Penrhyn Publishing, Renton, Washington; ed Ross Emry.

In its brief existence *A* was unable to carve itself a niche. It invoked some wrath in *#1* by suggesting that fantasy was a superior literary form to sf. It sought to publish both, but the emphasis was on LOW FANTASY, especially the intrusion or re-enactment of LEGEND. It attracted good fiction by John BRUNNER, Ru EMERSON, Janet MORRIS, Larry NIVEN, Mike RESNICK, Elizabeth SCARBOROUGH, Nancy SPRINGER and Keith TAYLOR. [MA]

ARGOSY, THE ◊ MAGAZINES.

ARIEL A name from the Hebrew, meaning "lion of God",

and hence an ANGEL name in various texts, including John MILTON's *Paradise Lost* (**1667**). Much the best-known appearance of Ariel is as the SHAPESHIFTER spirit of the air in SHAKESPEARE's *The Tempest* (performed *c*1611; **1623**) who was imprisoned by the WITCH Sycorax in a TREE (for which he has become an emblem of the kind of metamorphic BONDAGE characteristic of fantasy) and subsequently enslaved by CALIBAN. He is released by PROSPERO on condition that he serve until no longer needed. At the end of the play he is freed to the elements. As an important figure in a central TAPROOT TEXT, Ariel can be detected as an UNDERLIER figure in many tales. [JC]

ARIEL: THE BOOK OF FANTASY US large-format (9in × 12in) MAGAZINE, 4 issues, irregular (Autumn 1976, 1977, April and October 1978), published by Morning Star Press, Kansas; ed Thomas Durwood.

A:TBOF started as a magazine but ended as an artbook, distributed via Ballantine Books. A lavish, high-quality production, it featured a wide range of fantastic fiction, including COMICS, with an emphasis on HEROIC FANTASY. The series remains more collectible for its art, including works by Richard CORBEN, Frank FRAZETTA, Burne HOGARTH and Jeff JONES, than for its fiction, though it included short stories by Harlan ELLISON, Michael MOORCOCK, Larry NIVEN and Roger ZELAZNY. [MA]

ARIOSTO, LODOVICO (1474-1533) Italian poet whose *Orlando Furioso* (**1516**; exp 1532; trans Sir John Harington **1591** UK) is a direct continuation of *Orlando Innamorato* (**1487**) by Matteo Boiardo (1434-1494), taking up the tale at the point where the earlier poem stops short, due to Boiardo's death. Orlando (or Roland) and the other knights of Charlemagne's France remain variously ensorcelled through the agency of the pagan princess, Angelica, and others in the service of the Saracens; they spend much of their time – conveniently for the development of the main strands of the plot – in the coils of an illusory, sorcerer-spawned EDIFICE. LA's continuation emphasizes the PICARESQUE and the MARVELLOUS, incorporating battles and jousts and adventures in a fashion which makes the epic a fitting TAPROOT TEXT for 20th-century ADVENTURER FANTASY; in the most famous single episode a central character falls under the spell of ALCINA on her magic island. Orlando's madness (the title, literally translated, is *Crazy Roland*) visits him after a romantic betrayal, and causes his wits to fly to the Moon, where a colleague travels to recapture them.

Chelsea Quinn YARBRO's *Ariosto: Ariosto Furioso, a Romance for an Alternate Renaissance* (**1980**) places the real author in an alternate version of Italian history, one in which Lorenzo the Magnificent lives 30 more years and in which Ariosto's dreams of a fantasy America – described at length in terms which replicate *Orlando Furioso* – seem briefly to augur an Italian conquest of the new continent. [JC]

ARIS, ERNEST ALFRED (1882-1963) UK artist. ◊ ILLUSTRATION.

ARISTOPHANES (*c*445BC-*c*385BC) Greek dramatist who wrote at least 40 comedies. The 11 that survive are the sole extant examples of their extraordinary genre, Athenian "Old Comedy", which in general seems to have disregarded concern for narrative logic and demanded absurd ideas, so that most books on Aristophanes include a chapter on "fantasy"; this is certainly what his contemporaries saw his plays as. In relation to modern fantasy, however, his work falls into three groups: relatively realistic topical SATIRES; PARODIES of

MYTHS; and plays introducing mythic elements into the present day, of which four survive. During 427-421BC he presented 10 plays, nine of them mostly mundane, but *Peace* (**421**BC) depicts a flight to HEAVEN on a dung-beetle (parodying EURIPIDES' lost *Bellerophon*) to recover a person from imprisonment there. About 20 plays seem to date 417-405BC. *The Birds* (**414**BC; good trans William Arrowsmith **1961**) brilliantly parodies the theology and history of the Athenian Empire with the rise and ideology of its Cloud-cuckoo Land (a kind of WONDERLAND), TOPSY-TURVY and dystopia (◊ UTOPIAS). *The Frogs* (**405**BC; good trans Richmond Lattimore **1962**) more tamely shows DIONYSUS searching the UNDERWORLD for a good tragedian. *Plutus* (**388**BC; vt *Wealth*) is a fairytale with little of his vicious glitter. In the Middle Ages his most popular work, it is today the least.

Self-referential, obscene, relentlessly punning and allusive, prejudiced, slanderous, eloquent, musical . . . Aristophanes' work both demands and defies translation. Within a few decades of his death, it seems clear that fantasy was abandoned in Greek comedy (except in myth-parody). He matters to the history of fantasy as the sole example of what fantasy meant to the Greeks (◊ GREEK AND LATIN CLASSICS), while his importance to today's fantasy, or more broadly the FANTASTIC, lies in his early and extreme illustration of the power of absurdity (◊ ABSURDIST FANTASY).

Aristophanes appears as a prominent character in PLATO's *Symposium* and in Tom HOLT's **Walled Orchard** books, which travesty his career slightly less viciously than he travestied others'. [JB]

Other works: *The Acharnians* (**425**BC; good trans Douglass Parker **1961**); *The Knights* (**424**BC); *The Clouds* (**423**BC; original version lost; rev *c*417BC; good trans William Arrowsmith **1962**); *The Wasps* (**422**BC; good trans Douglass Parker **1962**); *Thesmophoriazusae* (**411**BC; vt *The Poet and the Women*); *Lysistrata* (**411**BC; good trans Douglass Parker **1964**); *Ecclesiazusae* (*c*392BC; good trans Douglass Parker as *The Congresswomen* **1967**; new trans [see below] vt *The Assemblywomen*). Additional recommended trans are *The Wasps/The Poet and the Women/The Frogs* (coll trans David Barrett **1964**) and *The Knights, Peace, The Birds, The Assemblywomen, Wealth* (coll trans David Barrett and Alan H. Sommerstein **1978**).

Further reading: *Aristophanic Comedy* (**1972**) by K.J. Dover.

ARKHAM COLLECTOR, THE ◊ WHISPERS.

ARKHAM HOUSE US SMALL PRESS which has been, over nearly 60 years, the most successful house specializing in the fantastic. It was founded – in Sauk City, Wisconsin – by August DERLETH and Donald WANDREI after they had failed in their attempts to interest mainstream firms in a volume of H.P. LOVECRAFT's stories; they had no intention, at first, to continue the firm. But *The Outsider and Others* (coll **1939**) – a vast, seminal, well produced volume – was initially a financial failure, and so they had to publish further books to service the debts incurred. Derleth bought out Wandrei in 1943, and the firm gradually became viable. Over the years it broadened its range from HORROR and SUPERNATURAL FICTION to all kinds of FANTASY and sf. Lovecraft associates published by AH include Frank Belknap LONG and Clark Ashton SMITH, as well as Derleth and Wandrei themselves; also from AH came the first significant US publication of William Hope HODGSON. AH published several first books: Robert BLOCH's *The Opener of the Way* (coll **1945**), A.E.

VAN VOGT's *Slan* (**1946**), Ray BRADBURY's *Dark Carnival* (coll **1947**) and Fritz LEIBER's *Night's Black Agents* (coll **1947**), Ramsey CAMPBELL's *The Inhabitant of the Lake and Less Welcome Tenants* (coll **1964**), Brian LUMLEY's *The Caller of the Black* (coll **1971**) and Phyllis EISENSTEIN's *Born to Exile* (coll **1978**).

After Derleth's death in 1971, editorial control passed (at the behest of Derleth's family and colleagues) to James Turner (1945-   ), who continues to publish and republish the canon centring on and around Lovecraft but also venturing with considerable success into new regions. Over the past decade or so AH has published a number of critically and commercially successful collections by writers never previously associated with the firm: these include Michael BISHOP's *Blooded on Arachne* (coll **1982**), Greg BEAR's *The Wind from a Burning Woman* (coll **1983**), *The Zanzibar Cat* (coll **1983**) by Joanna Russ (1937-   ), Tanith LEE's *Dreams of Dark and Light: The Great Short Fiction of Tanith Lee* (coll **1986**), Lucius SHEPARD's *The Jaguar Hunter* (coll **1987**), J.G. BALLARD's *Memories of the Space Age* (coll **1988**), *Crystal Express* (coll **1989**) by Bruce Sterling (1954-   ; ◊ SFE), *Her Smoke Rose Up Forever* (coll **1990**) by James Tiptree Jr (1915-1987), Michael SWANWICK's *Gravity's Angels* (coll **1991**), *Meeting at Infinity* (coll **1992**) by John Kessel (1950-   ), Nancy KRESS's *The Aliens of Earth* (coll **1993**) and *The Breath of Suspension* (coll **1994**) by Alexander Jablokov (1956-   ). Significantly, several of these were either the only or the most definitive collections of the authors in question.

AH maintains the editorial savvy and the production values it has always boasted, and remains commercially viable. [JC]

ARKHAM SAMPLER, THE ◊ MAGAZINES.

ARLEN, MICHAEL (1895-1956) UK-Armenian writer, born Dikran Kouyoumidjian, best-known for sentimentalized satires of London life like *The Green Hat* (**1924**) and for sf like *Man's Mortality* (**1933**), a tale of the *pax aeronautica* frequently prophesied by writers of the early 20th century.

His SUPERNATURAL FICTION appears in *These Charming People* (coll of linked stories **1923**) and *May Fair* (coll of linked stories **1924**), where it accompanies non-fantasy stories also dealing with the adventures of Lord Tarlyon and his cronies; it is reassembled alone in *Ghost Stories* (coll **1927**). Most of these tales involve supernatural incursions into contemporary life, as in "The Ancient Sin" (1923 *Pan*), in which Tarlyon comes across a father and son engaged upon so ancient a family quarrel that, after the son kills the father, the venue turns to ruins, for it is a drama that has been repeated from the beginning of time. *Hell! Said the Duchess* (**1934**) combines sf and fantasy elements: set in a Fascist near future, it concentrates on a man (in fact a DEMON) who transforms himself into the shape of a woman and murders men by throat-cutting after SEX. [JC]

ARMAGEDDON Literally "Megiddo's Mountain", the site at which – according to *Revelation* – the kings of the Earth will be drawn "to the battle of that great day of God Almighty". Like APOCALYPSE, the term is extensively used in fantastic literature to signify a violent END OF THE WORLD, LAST JUDGEMENT or confrontation with the ANTICHRIST. Examples include *Megiddo's Ridge* (**1937**) by S. Fowler Wright (1874-1965), "Armageddon" (1941) by Fredric BROWN, *The Armageddon Rag* (**1983**) by George R.R. MARTIN and the OMEN movies. [BS]

ARMSTRONG, ANTHONY Working name of UK author

and journalist George Anthony Armstrong Willis (1897-1976), one of the *Punch* magazine stable of humorists. His historical fantasies – *Lure of the Past* (**1920**) and *The Love of Prince Raameses* (**1921**) – used the then still fashionable REINCARNATION theme; *Wine of Death* (**1925**) is a violent LOST-LAND adventure about a surviving community from ATLANTIS.

AA is best remembered for humorous fantasy in which standard FAIRY-TALE settings are, with some SATIRE, updated to house contemporary comedy stereotypes: monarchs are tipsy and henpecked, testy FAIRIES are a recurring menace at christenings, and younger folk resemble the Bright Young Things chronicled in *Punch* by A.A. MILNE – the women outrageous modern flirts, the men wriggling through CONTRACTS and CONDITIONS. These stories are collected in *The Prince who Hiccupped and Other Tales* (coll **1932**) and *The Pack of Pieces* (coll **1942**; vt *The Naughty Princess* 1945). Fairytale motifs parodied by AA include CINDERELLA, QUESTS, WISHES ("I wish for three more wishes"), and, repeatedly, the FROG PRINCE. One metamorphosed prince (◊ METAMORPHOSES) becomes a traditional magic MIRROR on a bedroom wall where he enjoys unparalleled views of the princess dressing and undressing . . . summing up the tales' period sauciness. [DRL]

Other works: *When the Bells Rang* (**1943**) with Bruce Graeme, alternate history about a Nazi invasion of the UK; *The Strange Case of Mr Pelham* (**1957**).

ARMSTRONG, FYTTON ◊ John GAWSWORTH.

ARMSTRONG, MARTIN (DONISTHORPE) (1882-1974) UK writer, poet, critic and broadcaster; stepfather of Joan AIKEN. MA was a consummate stylist and a lover of the English language. His stories are polished and professional. Raised on the border of Northumbria with Scotland, in a region rife with LEGENDS, MA was fascinated by MYTH and fantasy while remaining a sceptic and rationalist – even though he once lived in a haunted house (◊ HAUNTED DWELLINGS). A later house provided the setting for his ingenious GHOST STORY "The Pipe Smoker". Probably his best story is "Presence of Mind", in which a man with an overzealous imagination progressively creates a fantasy world about him. Both these stories come from *General Buntop's Miracle* (coll **1934**), probably MA's most rewarding collection, but other good fantasies can be found in *The Bazaar* (coll **1924**), *The Fiery Dive* (coll **1929**) and *A Case of Conscience* (coll **1937**). This last includes "Mr Porter Transported" an amusing story about a man's sudden ability to fly (◊ FLYING) and is reminiscent of Mervyn PEAKE's *Mr Pye*. Generally his stories deal with a sudden and uncontrollable sense of WRONGNESS, though the rational mind never submits. [MA]

ARMSTRONG, T(ERENCE) I(AN) F(YTTON) ◊ John GAWSWORTH.

ARNASON, ELEANOR (ATWOOD) (1942-   ) US writer widely noticed as the author of complexly ambitious sf novels like *Ring of Swords* (**1993**), but whose early work was either fantasy or a mixture of the two genres. She began publishing with "A Clear Day in the Motor City" for *New Worlds Quarterly* #6 (anth **1973**) ed Michael MOORCOCK and Charles Platt (1945-   ). Her first novel, *The Sword Smith* (**1978**), carries its protagonist, a young SMITH reared by DRAGONS, through a modest RITE OF PASSAGE in an austere HIGH-FANTASY venue; almost completely lacking in heroics, the tale stays vividly in memory as a hard-crafted miniature. *To the Resurrection Station* (**1986**) is by contrast a

highly baroque SCIENCE FANTASY in which a robot butler may be the GHOST of a revered ancestor of the young protagonist, who flees with it (or him – or, ultimately, her) to Earth in search of the eponymous salvation; some FOLKTALES are embedded into the narrative, and some surreal special effects. *Daughter of the Bear King* (**1987**) is a CROSS-HATCH whose protagonist, a Minneapolis housewife, finds first her DREAMS and then her world invaded by visions of a life as a heroine in an OTHERWORLD increasingly marred by WRONGNESS, here embodied as a kind of this-worldly shoddiness. Faced with various challenges, she makes her escape by SHAPESHIFTING into a bear, a recourse understandable through the tale's constant crosshatching as a wry comment on the lengths to which it is sometimes necessary for women to transform themselves in order to cope with a patriarchal world.

EA's work – much of which acutely and variously tests GENDER roles – is muscular, sharp-witted, concise and fully packed. [JC]

ARMIDA A sorceress in *Jerusalem Delivered* (**1581**) by Torquato Tasso (1544-1595), a TAPROOT TEXT of some importance for its influence on Edmund SPENSER and others. She enchants Rinaldo into SEX with her, but when he comes to his senses she – loathing the separation – sets her palace afire and is slain in combat. At least 15 OPERAS of interest feature her. [JC]

See also: QUEEN OF AIR AND DARKNESS.

ARNIM, ACHIM VON (1781-1831) German folklorist. ◊ FOLKLORE.

ARNOLD, [Sir] EDWIN (1832-1904) UK poet, translator (mainly from Sanskrit) and journalist, editor of the *Daily Telegraph* from 1873; father of Edwin Lester ARNOLD. His blank-verse epic *The Light of Asia, or The Great Renunciation* (**1879**), an examination of the Buddha's life and philosophy, played a considerable role in popularizing such ideas. *The Light of the World, or The Great Consummation* (**1891**) is a similar account of the Christian Mythos. *The Voyage of Ithobal* (**1901**) is an ALLEGORY. [BS]

ARNOLD, EDWIN LESTER (1857-1935) UK writer, son of Sir Edwin ARNOLD. He derived from his father a strong interest in the ideas of REINCARNATION and *karma*; his fantasies are "karmic romances" modelled on the work of H. Rider HAGGARD. The hero of *The Wonderful Adventures of Phra the Phoenician* (**1891**; vt *Phra the Phoenician* 1910) recalls a series of past "awakenings" in different eras of British history – the withdrawal of the Romans, the Norman invasion of 1066, the Battle of Crécy, and Elizabethan intrigues – and continually meets women reminiscent of the WITCH-wife whose MAGIC was the source of his problematic IMMORTALITY and with whom he yearns to secure a more permanent reunion beyond the Earth. "Rutherford the Twice-Born" (1892), the only fantasy in *The Story of Ulla, and Other Tales* (coll **1895**), makes more formal use of the notion of *karma*, recounting the tribulations of a man who comes into a tainted inheritance but is absolved from guilt by a convenient VISION. *Lepidus the Centurion: A Roman of To-day* (**1901**) is a psychological melodrama in which a young Victorian and a virile Roman brought back from the dead are fragmentary aspects of a single SOUL and inevitably come into conflict over a woman.

Arnold's last fantasy novel, *Lieut. Gullivar Jones: His Vacation* (**1905**; vt *Gulliver of Mars* 1964 US), has similarities with the Martian fantasies of Edgar Rice BURROUGHS, although this is more likely due to the common influence of

Haggard's *She* (**1887**) than to any direct imitation. Jones, an impoverished US Naval officer, is whisked away by magic carpet to Mars, where decadent inheritors of an ancient civilization are trying to defend the remnants of their culture against marauding barbarians.

Arnold's fantasies are essentially playful, employing their philosophical devices as props to support amusing flights of romantic fancy, and do not seem to be weighed down by sincere belief, as Haggard's later reincarnation romances were. They do, however, echo a restless sense of frustration and dissatisfaction. [BS]

ARNOLD, MARK ALAN (1951-?   ) US editor. ◊ Midori SNYDER; Terri WINDLING.

ARSCOTT, DAVID (?   -   ) UK author, collaborator with David J. Marl (?   -   ) on two linked novels set in a nonmagical ALTERNATE WORLD which has a 19th-century flavour. In *The Frozen City* (**1984**) some symbolism is deployed, the nameless CITY being politically frozen under the unpleasant regime of the Red Blade. It is also a LABYRINTH, with sections accessible only by rooftop or secret doorway; art and beauty are exiled to a literally UNDERGROUND (but stagnant) Other City. The secret of the Blade's frozen permanence carries a George ORWELL-like conviction: it perpetuates itself through codes, regulations and a maze-like command chain which simulates tyrannical leadership, although the Leader who planned this is long dead. The book ends with a warm plea against hereditary kingship. *A Flight of Bright Birds* (**1985**), the colourful but rambling sequel, concludes with the heirs to the city's crown deviously contriving not to commit themselves to either refusing or accepting it. Fantasy's political default of monarchy is interestingly questioned by DA and Marl, but to ultimately inconsequential effect. [DRL]

ART ◊ FANTASY ART; ILLUSTRATION.

ARTEMIS ◊ GODDESS.

ARTHUR Quasihistorical British king or war-leader whose mythical adventures have formed the basis for the largest single subcategory of fantastic literature. Here we treat the Arthurian cycle under four heads: **King Arthur**, which sets the historical background; **Arthurian legend**, which considers the LEGEND and MOTIFS; **Arthurian romance**, which surveys the development of the legend through the medieval texts; and **Arthurian fiction**, which considers the treatment of the legend over the last century. As might be expected, there are crossovers between the various sections.

KING ARTHUR

Almost nothing is known about the real King Arthur. There are no contemporary references, and later annals make only scant record. The most reliable contemporary record, *De Excidio Britanniae* ["The Ruin of Britain"] (?**540**) by Gildas (?495-?570) makes no mention of Arthur, but generally sets the historical scene of the British and Romano-British being rallied by a High King and standing firm against the Saxon onslaught, achieving a decisive victory over the enemy at Badon. The earliest reference to Arthur by name is in the poem *Y Gododdin* (?**600**) attributed to Aneirin, a British bard of the early 7th century. It unfavourably compares a warrior of the Votadini with the strength and might of Arthur of old. This episode has been recreated in the historical CHILDREN'S FANTASY *The Shining Company* (**1990**) by Rosemary SUTCLIFF. The only other historical reference to Arthur is in the *Annales Cambriae* ["Welsh Annals"] (10th century), which date the death of Arthur at the Battle of Camlann in 537. Historians of the period drew upon either older records, now lost, or on oral tradition, and thus give an incomplete and at times misleading account of the period. The real Arthur was in all likelihood an army general, quite probably born and raised in a Roman or Romano-British family, whose skill and prowess as a battle leader (or *dux bellorum*, as he is described in the *Historia Brittonum* [**830**], attributed to Nennius [?   -809]), resulted in a decisive defeat of the Saxons at Badon in about 495. Thereafter Arthur was able to establish a period of relative prosperity and peace in Britain until his death at Camlann 40 years later. The extent of Arthur's kingdom is unclear, but the base of it was almost certainly the area of the Gododdin, a grouping of Celtic tribes between the Hadrian and Antonine Walls in the Scottish borders. He probably had overlordship of other minor British kingdoms, especially in central Britain and the Welsh borders. This may be all we can say with any reasonable accuracy about Arthur. It bears no relationship at all to the legend, which was given impetus by GEOFFREY OF MONMOUTH in his *Historia Regum Britanniae* ["History of the Kings of Britain"] (**1136**), which created a totally false history of King Arthur – while at the same time establishing the popularity of the character. [MA]

ARTHURIAN LEGEND

The accepted legend of King Arthur, generally now drawn from *Le Morte Darthur* (**1485**) by Sir Thomas MALORY, is a compendium of oral tradition and romantic invention covering a much deeper well of memory and myth that fuses tales from the MATTER of Britain, the MABINOGION and the Norman-French and Breton lays and romances, to form a of tales overlaying possible historical events with the motifs of the supernatural. As the legends became established they attracted other tales and legends about, for example, the GRAIL and LYONESSE, which have since become inextricably linked. Although the tales have as their core the character of Arthur, they rely heavily on other major characters who serve as ARCHETYPES for much later HEROIC FANTASY, especially the KNIGHTS of the ROUND TABLE, in particular LANCELOT, GAWAIN and PERCEVAL, the magus MERLIN and the enchantresses MORGAN LE FAY and the LADY OF THE LAKE. Around the characters and their adventures are deeper beliefs about death and RESURRECTION as exemplified by the BEHEADING MATCH between Gawain and the GREEN KNIGHT, the DOLOROUS STROKE leading to the WASTE LAND and recovery through the Cauldron of Plenty or the Grail, and the overarching legend that Arthur will come again in national time of need is identified by the phrase the ONCE AND FUTURE KING. All of these have their origins deep in Celtic legend and history and only came to be associated with the exploits of Arthur over a period of years.

The historicity of King ARTHUR is irrelevant to the legend other than in depicting a valiant hero protecting Britain from invasion, decline and ruin. The legends, drawing from their variant sources, offer different perspectives of Arthur, and in the three centuries from Geoffrey to Malory, Arthur moves from the heroic centre stage to become a tragic figure and victim of FATE. As the legends developed, particularly through the French romancers. Arthur's enemies became those within his kingdom, not external. Largely a reflection of 12th-century court intrigue and the civil wars within the Angevin Empire, the Arthurian legends became a catalogue of deceit, betrayal and treachery, against which Arthur increasingly relied on supernatural help: first from MERLIN and then from the Lady of the Lake.

The legends themselves, especially as consolidated by Malory, are too complex to detail here: a brief overview must suffice. In one respect all legends agree: Arthur is the son of Uther Pendragon and Ygraine, following the deceit worked upon Ygraine by Merlin. Merlin raises Arthur in secret as the foster son of Sir Ector, so that, when Arthur establishes his right to the throne by extracting the sword from the anvil (not a stone), he emerges as a HIDDEN MONARCH.

The deceit established by Merlin continues to weave its web. Not all of Arthur's vassal kings accept his claim. A rebellion is led by King Lot of Orkney which Arthur is forced to quell. Lot sends his wife Morgause/Morgawse (◊ MORGAN LE FAY), in reality Arthur's half-sister, to spy on the King. Arthur, unaware of his true parentage and relatives, beds Morgause and conceives Mordred (Modred), who will later prove his downfall. This web of deceit is paralleled by a vision Arthur has of the WASTE LAND. While on a hunt, Arthur sees a strange animal, the offspring of a woman who has slept with the DEVIL: it has the body of a leopard, the hindquarters of a lion, the feet of a deer and the head of a SERPENT, while from its stomach comes the sound of hounds baying – hence its name, the Questing Beast. When Merlin subsequently reveals that Arthur has slept with his half-sister, the parallel disturbs Arthur. Merlin adds to his disquiet by prophesying that a child born this coming Mayday will be Arthur's ruin, and that all children born that day should be killed. Arthur cannot bring himself to do this, but declares that all children of noble birth born in the past two months should be cast adrift in a ship. The ship founders on rocks and many of the infants perish, but among those saved is Mordred.

After many battles in which Arthur not only defeats his rebellious vassals but curbs the Saxons, he establishes a period of peace throughout Britain. Settled, he marries GUINEVERE, the most beautiful lady in the land. As part of the marriage arrangement, Arthur acquires the ROUND TABLE, made by Merlin, at which he commands his noblest knights to sit in his court at CAMELOT. To earn that right the knights must undertake deeds of considerable valour. One seat at the table remains empty, the Siege Perilous, which can be occupied only by the most holy and perfect of knights. The Round Table itself stands for purity and unity. The need for knights to prove their worth is a significant part of the Arthurian legend and the image of questing knights (◊ QUESTS) is often at the heart of people's ideas of the tales, though in truth the knights are usually more bloodthirsty than chivalrous.

After losing his original SWORD in a fight with King Pellinore, Arthur is taken by Merlin to a nearby lake where he meets the Lady of the Lake. She reveals to Arthur EXCALIBUR, held in a hand rising from the middle of the lake. Arthur claims the sword for his own. When Merlin asks Arthur which he feels is more precious, the sword or its scabbard, Arthur replies that it is the sword, but Merlin reveals that it is the scabbard that will protect him from harm (◊ AMULET). Arthur entrusts Excalibur to his half-sister, MORGAN LE FAY, unaware that she is plotting against him. Morgan creates a false Excalibur and gives the real sword to her champion, Sir Accolon, who fights and would defeat Arthur were it not for the intervention of the Lady of the Lake. Morgan attempts to steal the sword again, but succeeds in taking only the scabbard – which Arthur never recovers. From that day, Arthur's star begins to fall.

LANCELOT appears on the scene; he falls in LOVE with Guinevere and an adulterous affair begins. When Guinevere is abducted, it is Lancelot, not Arthur, who defies the very portals of death to rescue her.

The peace of the kingdom is now under threat. Sir Balin attacks Sir Pellam with the Lance of Longinus and delivers the DOLOROUS STROKE; famine and pestilence fall over the kingdom. (A historical plague did indeed sweep through Europe in the middle of the 6th century.) As the LAND darkens Arthur looks for guidance. At the Feast of Pentecost he and his knights have a vision of the GRAIL. The result is a series of QUESTS to attain the Grail and restore the LAND.

With the Grail quests underway we find Arthur's kingdom in disarray; his knights are far afield, while at home the rift between Arthur and Guinevere is widening. Eventually Guinevere's adultery is revealed and war breaks out between Arthur and Lancelot. The Order of the Round Table is sundered, as knights take sides. Lancelot, still wishing to remain loyal to Arthur, leaves Britain for France and is pursued by Arthur. In Arthur's absence, Mordred seizes the throne and seduces the Queen. Arthur returns in haste to Britain, and eventually meets Mordred at the battle of Camlann, where Arthur is mortally wounded. Arthur orders Bedivere to return Excalibur to the Lady of the Lake. The knight twice refuses, hiding it in the bushes, but Arthur knows he has lied and the third time Bedivere throws the sword far into the lake. A hand rises from the water, catches the sword by the hilt, brandishes it three times and withdraws. Arthur commands Bedivere to carry him down to the lake, and there a boat, carrying Morgan le Fay, bears Arthur's body away to the Isle of AVALON, there to heal. Britain awaits his return in times of peril.

There are many variations and additions to this basic tale, especially in the adventures of the individual knights – many of whom, such as GAWAIN, PERCEVAL and LANCELOT, have STORY CYCLES of their own. Although it has a natural appeal at a superficial level in the heroic adventures of Arthur and his knights, it has more appeal at its deeper levels. The mystical interpretations of the CYCLE of birth and death bring comparisons with the Earth's life cycle (◊ GREEN MAN); the parallel cycles of love, deceit and betrayal arising from either supernatural intervention or human failing, set against the search for perfection, underscore human nature; and the overriding appeal of the HERO who is not dead but only sleeping awaiting the call to save us (◊ SLEEPER UNDER THE HILL) is a final salve to us all. These core themes have entranced storytellers and their audiences for over 1000 years.   [MA]

ARTHURIAN ROMANCE

Although tales were almost certainly told of Arthur's exploits within a generation or two of his death – i.e., in the 6th and 7th centuries – no written accounts of these stories exist until much later, when the flowering of Welsh and Anglo-Norman literature resulted in the first industry of Arthurian fiction. The Welsh Triads, by which the bards recalled heroic events in sequences of three, sustained Arthurian exploits in simple summary form, relying on oral tradition to expand the narrative. These narratives were not written down until the 12th century, most especially in the MABINOGION, the oldest surviving texts of which are *The White Book of Rhydderch* (?**1325**) and *The Red Book of Hergest* (?**1400**), compiled by unknown hands. The oldest Arthurian story within this is *Culhwch and Olwen*, which possibly dates from the 11th century. The story is an archetype of a hero having to perform a series of almost impossible tasks in order to earn his reward. In this story Culhwch, a cousin of Arthur's, is cursed (◊ CURSES) with being allowed to marry only Olwen, the daughter of the GIANT Ysbaddaden. With Arthur's help, Culhwch finds Ysbaddaden but, before he can marry Olwen, the giant charges him with SEVEN impossible tasks. It is Arthur and seven of his court who achieve the tasks – which include a journey into the UNDERWORLD to gain the Cauldron of Plenty. This last adventure was drawn from an earlier poem, *Preiddeu Annwfyn* ["The Spoils of Annwn"] (?**900**). The other major Welsh poem featuring Arthur is *Breudwyt Rhonabwy* ["The Dream of Rhonabwy"] (?**1210**); this is really a SATIRE on power politics in 12th-century Wales; Rhonabwy – one of a band of soldiers seeking an outlaw – dreams of an Arthurian GOLDEN AGE full of heroes. Three other Celtic tales of knightly quests are drawn from the same source as romances by CHRÉTIEN DE TROYES – *Owain* – known also as *The Lady of the Fountain* (?**1250**) – *Geraint and Enid* (?**1250**) and, of special interest, *Peredur* (?**1250**) (◊ PERCEVAL), an amalgam of legends tracing the development of a lowly boy who is protected from society but who yearns for adventure. His later exploits, many of which parallel Arthur's in portraying a coming of age and RITE OF PASSAGE, also bear the kernel of the GRAIL legend. This motif forms the basis of much late 20th-century HIGH FANTASY (by, for example, Lloyd ALEXANDER, David EDDINGS and Tad WILLIAMS).

While the many Celtic legends are the sole source of the Arthurian tales, those not perpetuated in Celtic manuscripts almost certainly survived in oral tradition to surface in later stories as the tales grew in popularity and spread across Europe. Many incidents presented unadorned in the Celtic texts reappear grossly embellished in later French and German ROMANCES. It is ironic that Arthur's literary success was as a result of writings by those who conquered the Celts rather than by the Celts themselves.

As the Normans established themselves across England in the century after the Battle of Hastings (1066), their greatest threat came not from the vanquished Saxons but from the Celts, particularly the Welsh. Instead of quashing the Celtic legends, the Normans decided to adopt them, thus making the Welsh heroes their own and stemming the rising tide of Welsh nationalism. GEOFFREY OF MONMOUTH was commissioned to adapt an ancient Welsh book (now lost, if it existed at all) into Latin, and this emerged as the *Historia Regum Britanniae* ["History of the Kings of Britain"] (**1136**). This volume is predominantly a fanciful embellishment of oral tradition, tracing the Kings of Britain from Brutus, the son of Aeneas (◊ FANTASIES OF HISTORY; MATTER) to Cadwalladr, a 7th-century king of Gwynedd. Much of it is given over to the exploits of Arthur, presaged by the emergence and prophecies of MERLIN. Geoffrey's tale was the first extensive story of Arthur's life, and the first to appear in Latin, and the book became the medieval equivalent of a bestseller, firmly establishing King Arthur as a HERO. Apart from the involvement of Merlin, Geoffrey's account is generally devoid of the supernatural. It tells of how, through the use of Merlin's drugs, Uther Pendragon is able to transform himself into the likeness of Ygraine's husband, Gorlois, and thus how Arthur's conception arises from deception. It goes on to recount how Arthur's true background is later revealed, how he is made king, defeats the Saxons in a series of battles culminating in that at Bath (Badon), and enters into a halcyon reign. After the first 12

years, when he has established his court at Caerleon and developed his retinue of knights, Arthur embarks on a conquest of Europe. Challenged by Rome, Arthur marches on the Eternal City, but before he can conquer it learns of a revolt in Britain and returns to face the traitor Mordred. This leads to his final battle, where he is mortally wounded and taken away to AVALON. Geoffrey thus does not record Arthur's death, leaving open the possibility of a return and thereby establishing a basis for the eternal legend.

The interest in Arthur's exploits now led to an industry, encouraged during the expansion of empire in the reign of Henry II (reigned 1154-89). First came the *Roman de Brut* (**1155**) by the Norman monk Wace (?**1110**-?**1175**), a translation of Geoffrey's *Historia* into Norman French but with its own embellishments, in particular the introduction of the ROUND TABLE as the Order of Knighthood. This same work was then adapted into English by Layamon (?**1140**-?**1210**) as *Brut* (?**1200**), a rather bloodthirsty rendition which reintroduced episodes from Celtic tradition, including the role of MORGAN LE FAY as Queen of the ELVES.

Meanwhile CHRÉTIEN DE TROYES was converting the Arthurian quasi-history into romantic adventure. It was Chrétien who introduced the character of LANCELOT in *Lancelot* or *Le Chevalier de la Charrete* ["The Knight of the Cart"] (?**1177**), which explored Lancelot's love for Guinevere. Chrétien contrasted this relationship in *Le Chevalier au Lion* (**1177**; vt *Yvain*), and it was through this work, and his earlier romances – *Érec et Énide* (?**1170**) and *Cligés* (?**1176**) – that the knightly world of Arthurian chivalry emerged. Chrétien's unfinished final work, *Perceval, ou Le Conte de Graal* (begun ?**1182**), brought the GRAIL legend into the Arthurian canon. Chrétien does not identify the Grail but he does develop its mysteries, including the link with the Bleeding Lance. It was down to Chrétien's successor, Robert de Boron (? -1212) to link the Grail with CHRIST's chalice at the Last Supper, in his *Joseph d'Arimathie* (?**1200**; vt *Le Roman de l'Estoire dou Graal*). It was also Robert, in *Merlin* (?**1200**), who introduced the concept of the SWORD in the Stone, and it may also have been Boron who wrote the anonymous *Mort Artu* the second part of a prose rendition of the Arthurian tales which, along with *Queste del Saint Graal* and *Lancelot*, forms the basis of what is known as the **Vulgate Cycle** of Arthurian stories brought together by unknown writers during the period 1215-35. Between them Chrétien and Robert developed the primary Arthurian STORY CYCLE from which most other tales developed. Not only did their work feed the **Vulgate Cycle**, they also inspired the romancers Hartmann von Aue (?**1160**-?**1215**), Wolfram von Eschenbach (?**1170**-?**1220**) and Gottfried von Strassburg (? -?**1210**), who developed the Yvain, Perceval and Tristan tales in German in the early years of the 13th century.

There were scores of other romances written throughout Europe during this period, particularly the stories about Tristan and Isolde that were embellished by the Anglo-Norman poets and romancers Béroul (late 12th century), Thomas d'Angleterre (*fl*1170-75) and Marie de France (late 12th century). The other major and independent Arthurian medieval romance is the anonymous 14th-century *Sir Gawain and the Green Knight* (◊ GAWAIN).

The majority of the Arthurian legends had been captured in print, translated, retold, reinvented and consolidated by the end of the 14th century. It was these that were then used by Malory to produce the definitive *Le Morte Darthur*

(**1485**). Thereafter *Le Morte Darthur* (often referred to as *Le Morte D'Arthur*) has been treated as the standard Arthurian text from which, until recently, all Arthurian stories have been derived.   [MA]

ARTHURIAN FICTION

Arthurian imagery and motifs have remained strong in literature since the success of *Le Mort Darthur*, although during the Reformation they were comparatively dormant, kept alive through oral tradition and FOLKTALES like the story of TOM THUMB, which is set in Arthur's court. The regeneration of interest in legend and folktales that followed the rediscovery of FAIRYTALES in the UK in the second half of the 18th century had been presaged to some extent by the modernization of old texts and sagas by Robert SOUTHEY, who contributed his own translation of *Le Roman de Merlin* to an 1817 edition of *Le Morte Darthur*, Thomas Love Peacock (1785-1866), who adapted and reworked Welsh texts as *The Misfortunes of Elphin* (**1829**), and Lady Charlotte Guest (1812-1895), who produced the definitive Victorian edition of the MABINOGION (**1849**). These inspired and encouraged Alfred, Lord Tennyson (1809-1892) who, starting with "The Lady of Shalott" (**1832**; rev in *Poems* coll **1842**), developed a large body of Arthurian poetry, culminating in his sequence *Idylls of the King* (**1859**; exp 1870, 1873, 1886). Tennyson's popularity brought an immediate revival of interest in the Arthurian legend, with a new edition of *Le Morte D'Arthur* (**1862**) by James Knowles (1831-1908), while the whole Arthurian theme was championed by the PRE-RAPHAELITES, who took GALAHAD as their patron. Especially strong in this movement was William MORRIS, who early in his career produced his poem *The Defence of Guinevere* (**1858**) and in turn influenced Algernon Swinburne (1837-1909), whose work included *Tristram of Lyonesse* (**1882**) and *The Tale of Balen* (**1896**).

Arthurian imagery has continued to inspire poetry to this day; special reference must be made to *Lays of the Round Table* (**1905**) by Ernest Rhys, the sequence *Merlin* (**1917**), *Lancelot* (**1920**) and *Tristram* (**1927**) by Edwin Arlington Robinson (1869-1935), and *The Waste Land* (1922 *Criterion*; in *Poems: 1909-1925* coll **1925**) by T.S. Eliot (1888-1965).

By the close of the 19th century, however, interest in the Arthurian tales was moving away from poetry toward fiction. This initially took the form of adapting the work for children (◊ CHILDREN'S FANTASY), extolling the chivalric and manly virtues of the knights. Works including *The Boy's King Arthur* (**1880**) by Sidney Lanier (1842-1881), the beautifully illustrated adaptations by Howard PYLE – in *The Story of King Arthur and His Knights* (**1903**), *The Story of the Champions of the Round Table* (**1905**), *The Story of Sir Lancelot and His Companions* (**1907**) and *The Story of the Grail and the Passing of Arthur* (**1910**) – and many illustrated adaptations of Malory including *Le Morte D'Arthur* (1893) done by Aubrey BEARDSLEY, *Tales of King Arthur* (**1905**) retold by Andrew LANG and illustrated by Henry Ford (1860-1940), *King Arthur's Knights* (**1911-15**) retold by Henry Gilbert and illustrated by Walter CRANE, and *The Romance of King Arthur and his Knights of the Round Table* (**1917**) retold by Alfred W. Pollard and illustrated by Arthur RACKHAM. All of these contributed towards the 20th-century vision and interpretation of the Arthurian tales. It was in this environment that Mark TWAIN wrote *A Connecticut Yankee in King Arthur's Court* (**1889**), which contrasted US and Arthurian values and acknowledged that values had not necessarily

improved and might even have declined (◊ *A* CONNECTICUT YANKEE).

This regeneration of interest in Arthurian fiction took a number of shapes. Although the whole basis of Arthurian fiction is fantasy, the purely fantastic was among the latest forms to develop. Initially authors sought either to recreate the Arthurian world with as much historical exactitude as was possible or to use Arthurian imagery in a contemporary setting. These two categories are outside the strict purview of this encyclopedia, but their impact cannot be ignored in the development of the genre, and thus some of the key works are covered below.

Early fictional treatments of the tales remained virtuous and overly sentimental; examples are *Cian of the Chariots* (**1898**) by William H. Babcock (1849-1922), *Uther and Igraine* (**1903**) by Warwick DEEPING, *A Lady of King Arthur's Court* (**1907**) by Sara Hawks Sterling and *The Clutch of Circumstance* (**1908**) by Dorothy Senior. The 1920s brought the inevitable romantic interpretation, with undertones of SATIRE on contemporary society as portrayed in *Galahad* (**1926**) and *Tristan and Isolde* (**1932**), both by John Erskine (1879-1951), *Launcelot* (**1926**) by Ernest Hamilton (1858-1939), *Launcelot and the Ladies* (**1927**) by Will Bradley (1868-1962), *Pendragon* (**1930**) by W. Barnard Faraday (1874-1953) and *The Little Wench* (**1935**) by Philip Lindsay (1906-1958). These stories use the traditional Arthurian setting to explore various romantic entanglements, some lightheartedly, others (specially Lindsay's) with deeper interpretation. Further lighthearted society spoofs appearing at this time used the Arthurian world to satirize the aristocracy; these include "Sir Agravaine" (1912 *Colliers Weekly*) in *The Man Upstairs* (coll **1914**) by P.G. Wodehouse (1881-1975), a series of **Sir Archibald** stories (1922-5) by A.M. BURRAGE and "Sir Borlays and the Dark Knight" (1933) by Anthony ARMSTRONG.

When T.H. WHITE produced *The Sword in the Stone* (**1938**) Arthurian fiction took a giant step forward. White ignored the traditional Arthurian storyline, using stock characters and circumstances only when he needed them, and developed an Arthurian world of his own. The tale still reflects the contemporary mores of the 1920s and 1930s as used by Armstrong and Burrage, and is wonderfully anachronistic in its creation of the Arthurian world, using modern images and inventions when convenient. In so doing White depicts an acceptable Arthurian existence, a world beyond time where Merlin has taken young Wart (Arthur) for his training and preparedness for the world. White's book was the first in a series of novels that became darker, more subtle and less whimsical, partly because of the onset of WWII but also because of the increasing tragedy of the subject matter. *The Sword in the Stone*, *The Witch in the Wood* (**1939**) and *The Ill-Made Knight* (**1940**) were later revised and enlarged, with the addition of "The Candle in the Wind", as *The Once and Future King* (**1958**), to many still the definitive Arthurian volume of the 20th century. The book served as the basis for the stage musical *Camelot* (1960; movie *1967*) and DISNEY's ANIMATED MOVIE *The Sword in the Stone* (*1963*). *The Book of Merlyn* (**1977**), originally written as part of the sequence, is less successful.

The only other pre-WWII work of any note is *King of the World's Edge* (1939 *WT*; **1966** dos) by H. Warner MUNN, which postulates that surviving remnants of Arthur's army, under Merlin and Ventidius Varro, a Roman centurion, sail west and discover the Americas. Munn was fascinated by his post-Arthurian world and returned to it again in *The Ship from Atlantis* (**1967**; fixup with "King of the World's Edge" as *Merlin's Godson* **1976**) and *Merlin's Ring* (**1974**), though these books are more about other lands contemporary with and after Arthur.

The WWII years brought a new perspective to the Arthurian condition: the Battle of Britain and the Dunkirk spirit made the besieged island fortress a reality, and the Arthurian world took on a greater meaning. The result was a move away from fantasy toward a mixture of historical realism and mystical interpretation. John Cowper POWYS, who had used Arthurian themes in two earlier contemporary novels, depicted a harsh Arthurian world in *Porius* (**1951**). This harshness and sense of alienation is repeated in other novels of the 1950s, a decade which also brought a more perceptive delineation of the characters. This is especially true of the trilogies by Dorothy James Roberts (1903-1990) – *The Enchanted Cup* (**1953**), *Launcelot, My Brother* (**1954**) and *Kinsmen of the Grail* (**1963**) – which uses a more traditional setting, and by Henry TREECE – *The Eagles Have Flown* (**1954**), *The Great Captains* (**1956**) and *The Green Man* (**1966**) – which depicts a bitterly bleak Celtic world.

The 1950s also saw new attempts to use a fantastic treatment of the Arthurian legend to satirize contemporary society, much in the style of Twain; examples are *To the Chapel Perilous* (**1955**) by Naomi MITCHISON and *The Quest of Excalibur* (**1959**) by Leonard Wibberley (1915-1983), the first bringing a modern perspective to the past and the second bringing Arthurian values to the present. Less successful in their day were *The Queen's Knight* (**1955**) by Marvin Borowsky, which treats Arthur as a puppet king, and *The Pagan King* (**1959**) by Edison Marshall (1894-1967), which rationalizes legend but depicts Arthur creating his own myth as a BARD. *The Fair* (**1964**) by Robert NATHAN is a lighthearted fantasy more in keeping with Twain and White than contemporary works.

By the 1960s Arthurian fiction was establishing itself firmly into a number of subdivisions. At one level, more realism was being brought to an interpretation of the historical Arthurian world. The lead in this direction had been taken by Rosemary SUTCLIFF with *Sword at Sunset* (**1963**), which endeavoured to present a rational historical perspective. Other examples of note include: *The Duke of War* (1966) by Walter O'Meara (1897-1989) and *Twilight Province* (**1967** Australia; vt *Watch Fires to the North* 1968 US) by George Finkel (1909-1975), both exploring the political chaos of the period; the **Crimson Chalice** sequence by Victor Canning (1911-1986) – *The Crimson Chalice* (**1976**), *The Circle of the Gods* (**1977**) and *The Immortal Wound* (**1978**), assembled as *The Crimson Chalice* (omni **1980**) – which portrays a realistic Romano-British world with Arthur (Arturo) as an outcast driven by what he believes are the wishes of the gods; *Pendragon* (**1977**) by Douglas Carmichael (1923-), which considers Arthur's early life before kingship; *The Road to Avalon* (**1988**) by Joan Wolf; the **Dream of Eagles** sequence by Jack Whyte (1940-) – *The Skystone* (**1992**) and *The Singing Sword* (**1993**) – which traces the historical break-up of Roman Britain and the establishment of the Arthurian kingship (this is currently being expanded into the 6-vol **Camulod Chronicles**), and **The Warlord Chronicles** by Bernard Cornwell – comprising *The Winter King* (**1995**), «The Enemy of God» and «The Warlord» – which considers the political and military aspects of the history.

At the next level, the traditional Arthurian world provides a strong background for exploring the romantic conflict. Most of these books use the anachronistic world of Malory to make the Arthurian world a sort of alternate SECONDARY WORLD. The best example in this category is *Lionors* (**1975**) by Barbara Ferry Johnson (1923-), about the relationship between Arthur and a young girl that results in the birth of a blind daughter. Also in this category are the **Guinevere** sequence by Sharan NEWMAN, *The Sword and the Flame* (**1978**; vt *The Pendragon* 1979 US) by Catherine Christian (1901-) and *The Enchanter* (**1990**) by Christina Hamlett.

Linked to this category are those books which also portray the traditional Arthurian world but largely replace the supernatural elements by rationalized religious or mystical interpretations and present the stories as straight fiction, avoiding romantic or sentimental overtones. These include some of the best of all Arthurian fictions, including the sequence by Mary STEWART – *The Crystal Cave* (**1970**), *The Hollow Hills* (**1973**), *The Last Enchantment* (**1979**), *The Wicked Day* (**1984**) and *The Prince and the Pilgrim* (**1995**) – and Marion Zimmer BRADLEY's *The Mists of Avalon* (**1982**). Bradley's novel bridges that divide between the traditional Arthurian world and the more realistic historical world. Such RATIONALIZED FANTASIES, today forming the biggest category of new fiction about Arthur, emerged in force in the 1970s and 1980s. They sometimes take the historical reality of a post-Roman Britain, although some prefer the timeless quality of an Arthurian alter-world. Both weave fantastic and supernatural elements into the story, usually rationalized but sometimes taken as a natural part of that world. Examples are: *Drustan the Wanderer* (**1971**) by Anna Taylor (1944-), an authentic recreation of the Tristan legend; **The Three Damosels** sequence by Vera CHAPMAN; the **Parsival** series (**1977-80**) by Richard MONACO; *Percival and the Presence of God* (**1978**) by Jim Hunter (1939-), a strongly introspective novel on the nature of the Grail; *Arthur Rex* (**1978**) by Thomas BERGER, a rather more bawdy treatment of Arthur that acts as a counterpoint to the rising tide of feminist treatment and thus matches well with Robert NYE's erotic *Merlin* (**1978**) and Nicholas Seare's *Rude Tales and Glorious* (coll **1983**); *The Dragon Lord* (**1979**) by David DRAKE; the **Gwalchmai** sequence by Gillian BRADSHAW; *Firelord* (**1980**) and *Beloved Exile* (**1984**) by Parke GODWIN, which contrast the lives of Arthur and Guinevere; the **Bard** sequence by Keith TAYLOR; *The Idylls of the Queen* (**1982**) by Phyllis Ann KARR, an ingenious Arthurian murder mystery; *The Lady of the Fountain* (**1982**) by Kathleen Herbert, which retells the story of Linet; *The Last Knight of Albion* (**1987**) and *The Book of Mordred* (**1988**) by Peter Hanratty; the **Pendragon Cycle** by Stephen LAWHEAD; the **Guinevere** trilogy by Persia WOOLLEY; *Ghost King* (**1988**) and *Last Sword of Power* (**1988**) by David GEMMELL, set in the time of Uther Pendragon; *The White Raven* (**1988**) by Diana L. PAXSON, retelling the Tristan and Iseult romance; the **Daughter of Tintagel** sequence by Fay SAMPSON; the **Dragon's Heirs** trilogy by Courtway Jones (real name John Alan Jones; 1923-), being *In the Shadow of the Oak King* (**1991**), *Witch of the North* (**1993**) and *A Prince in Camelot* (**1995**); *The Dragon and Unicorn* (**1994**) and *Arthor* (**1995**) by A.A. ATTANASIO, where the author makes a characteristically more mystical and obtuse interpretation of the legend; the **Pendragon's Banner** trilogy by Helen Hollick, beginning with *The Kingmaking* (**1995**); and **The Mordred Cycle** by Haydn Middleton (1955-), beginning with *The King's Evil* (**1995**).

All these novels seek to recreate an Arthurian world, and demonstrate the diversity and scope of Arthurian literature. A smaller but rapidly growing body of work considers Arthurian relics or personalities surviving beyond their time. This was a natural extension of the ONCE AND FUTURE KING motif, but was as applicable to other Arthurian images (◊ EXCALIBUR; GRAIL; MERLIN). C.S. LEWIS brought Merlin and the FISHER KING into the nuclear age in *That Hideous Strength* (**1945**) in order to restore the world after a HOLO-CAUST. Leonard Wibberley brought Arthur into a contemporary UK in *The Quest of Excalibur* (1959) to con-front socialist ideals. Roger ZELAZNY brought LANCELOT into the modern day to defend the UK against Merlin in "The Last Defender of Camelot" (1979); Welwyn Wilton Katz tracked the fate of Arthur through the centuries in *The Third Magic* (**1990**). The movie KNIGHTRIDERS (*1981*) dir George A. Romero (1940-) quite faithfully retells the Arthur saga in terms of modern-day bikers trying to create a better USA. *The Lost History of Redwyn* (**1992**) by William Jay is set in the 14th-century ruins of Camelot, where Merlin seeks to combat the Black Death. Simon HAWKE reawakens Merlin in *The Wizard of Camelot* (**1993**), while Lawrence WATT-EVANS tries something similar in *The Rebirth of Wonder* (**1992**).

The immediate post-Arthurian world is the setting for: the **Dark Ages** trilogy – *Queen of the Lightning* (**1983**), *Ghost in the Sunlight* (**1986**) and *Bride of the Spear* (**1988**) – by Kathleen Herbert, which depicts the strife that riddles 6th-century Britain; *The Last Rainbow* (**1985**) by Parke GODWIN, which extends his Arthurian sequence to explore the life of St Patrick (5th century); *The Shining Company* (**1990**) by Rosemary SUTCLIFF, which recreates the Battle of the Gododdin; and *Druid Sacrifice* (**1993**) by Nigel Tranter (1909-), which links the life of Arthur with St Mungo (c518-603).

The concept of the Arthurian world reawakening has resulted in many excellent CHILDREN'S FANTASIES. These include: *The Weirdstone of Brisingamen* (**1960**) and *The Moon of Gomrath* (**1963**) by Alan GARNER; the **Dark is Rising** sequence by Susan COOPER; *Earthfasts* (**1966**) by William MAYNE, which resurrects a fearsome Arthur; *The Sleepers* (**1968**) by Jane CURRY, which follows traditional lines; *Excalibur* (**1973**) by Sanders Anne LAUBENTHAL, with chil-dren seeking the sword in the USA; *On All Hollows' Eve* (**1984**) and *Out of the Dark World* (**1985**) by Grace CHETWIN, wherein a young girl encounters MORGAN LE FAY; *The World of Amber* (**1985**) and *In the Ice King's Palace* (**1986**) by A(lice) Orr (1950-); and *The Pendragon Caper* (**1991**) by Richard H.R. Smithies, where a modern town holds an Arthurian spear of legend.

Tim POWERS originally planned a series of novels depict-ing the incarnation of the FISHER KING in different historical periods, but only *The Drawing of the Dark* (**1979**), set in the 16th century, appeared in that form. Arthurian worlds cre-ated in other times or ALTERNATE REALITIES may be found in *Father of Lies* (1962; rev **1968** dos US) by John BRUNNER, in which a child has such strong psychic powers he recreates an Arthurian world; *Fang, the Gnome* (**1988**) and *King of the Scepter'd Isle* (**1989**) by Michael G. Coney (1932-), in which Nyneve, wife of Merlin, creates her own alternate Arthurian world; and three novels by Andre NORTON – *Steel Magic* (1965; vt *Grey Magic* 1967), in which three children find their way into Avalon, *Here Abide Monsters* (**1973**), in which an alternate fantasy world includes Arthurian

characters, and *Merlin's Mirror* (**1975**), which recreates the Arthurian world as sf. Patricia KENNEALY-MORRISON recre-ates Arthurian characters in an ALTERNATE WORLD in her **Keltiad** sequence and its successor, *The Hawk's Gray Feather* (**1990**). In *The Last of Danu's Children* (**1981**) by Alison RUSH the Celtic world of legend, including Merlin, begins to intrude upon our own.

The above listing almost entirely excludes the many Arthurian short stories written over the past century. A sampling of these, plus new material, can be found in *The Pendragon Chronicles* (anth **1990**), *The Camelot Chronicles* (anth **1992**), *The Merlin Chronicles* (anth **1995**) and «The Chronicles of the Holy Grail» (anth 1996), all ed Mike ASH-LEY. Only new stories were included in *Invitation to Camelot* (anth **1988**) ed Parke Godwin and *Excalibur* (anth **1995**) ed Richard Gilliam (1950-), Martin H. GREENBERG and Edward Kramer, while *Camelot* (anth **1995**) ed Jane YOLEN is aimed at a children's/YA audience.

Many younger readers may have first encountered the Arthurian stories in COMICS. In most cases publishers would create a new character and incorporate him in the Arthurian world. The best-known are PRINCE VALIANT, created by Hal Foster (1892-1982), and **The Black Knight**, created by Stan LEE, which had its own short-lived magazine *The Black Knight* (5 issues May 1955-April 1956). More recently, in the 12-issue sequence *Camelot 3000* (1982-4), Arthur was res-urrected in AD3000 to combat an alien invasion led by Morgan Le Fay, while *Camelot Eternal* (1990) tracks an alternate world where Arthur was victorious at Camlann.

The Arthurian stories have not lent themselves well to filming. The emphasis has almost always been either on action and adventure or on spoof and comedy (the latter also lending itself to musical adaptation). In the first category are the earliest known Arthurian movies, *Lancelot and Elaine* (*1910*) and *Adventures of Sir Galahad* (*1919*), but it was not until the 1950s that big budgets were brought to produce *Knights of the Round Table* (*1953*), starring Robert Taylor and Ava Gardner, *The Black Knight* (*1954*), starring Alan Ladd, and *Lancelot and Guinevere* (*1963*), produced, dir and starring Cornel Wilde. The lighthearted strand of Arthurian moviemaking derives from the first version of *A Connecticut Yankee in King Arthur's Court* (*1920*), variously remade (◊ *A* CONNECTICUT YANKEE). Also in this mould are the Disney *The Sword in the Stone* (*1963*) and the screen version of the stage musical *Camelot* (*1967*), both adapted from T.H. White. Humour was at its best in MONTY PYTHON AND THE HOLY GRAIL (*1975*), where an insightful knowledge of the Arthurian motifs allowed for an intelligent satire.

More serious movie treatment of the tales came initially from French directors who, rather than taking the Arthurian world as a whole, selected specific aspects – such as *Tristan et Iseult* (*1972*) dir Yvan Lagrange, *Lancelot du Lac* (*1974*) dir Robert Bresson and *Perceval* (*1978*) dir Eric Rohmer – and gave them a more studied and intense inter-pretation. It was not until John Boorman directed EXCALIBUR (*1981*) that English-speaking audiences had a serious and more mystical adaptation of the main motifs (*Knightriders*, noted above, was released in the same week as Boorman's movie, and initially flopped, but, like *Excalibur*, is now much more highly regarded than at first). *First Knight* (*1995*), which focused on the love between Lancelot and Guinevere, marked a return to the big budget/minimal interpretation approach of the 1950s.

The legend has also inspired a number of musical

adaptations from *Lohengrin* (1850), *Tristan and Isolde* (1865) and *Parsifal* (1882) by Richard WAGNER through to *The Myths and Legends of King Arthur and The Knights of the Round Table* (1975) by Rick Wakeman (1949-   ).

There is not a single aspect of the artistic media that the Arthurian legends have not, at some time, inspired. Most recently a series of GAMES has been developed by Greg Stafford for Chaosium, set in the Arthurian Celtic world of which *The Boy King* and *Pendragon* focus on the adventures of Arthur.   [MA]

Further reading: The Arthurian field is overwhelmed with reference works, most studying the origins of King Arthur or the various interpretations of the legend. The core texts which identify and discuss Arthurian literature, including coverage of fantasy volumes, are *King Arthur Today: The Arthurian Legend in English and American Literature 1901-1953* (**1954**) by Nathan Comfort Starr; *Arthurian Literature in the Middle Ages* (**1959**) ed Roger Sherman Loomis; *The Development of Arthurian Romance* (**1963**) by Loomis; *The Mystery of King Arthur* (**1975**) by Elizabeth Jenkins; *The Arthurian Bibliography* (**1981-8** 3 vols) by C.E. Pickford and R.W. Last; *The Return of King Arthur: British and American Arthurian Literature Since 1800* (**1983**) by Beverly Taylor and Elisabeth Brewer; *Arthurian Legend and Literature: An Annotated Bibliography* (1984) ed Edmund Reiss, Louise H. Reiss and Beverly Taylor; *The Return from Avalon: A Study of the Arthurian Legend in Modern Fiction* (**1985**) by Raymond H. Thompson; *The Arthurian Encyclopedia* (**1986**) ed Norris J. Lacy (rev vt *The New Arthurian Encyclopedia* 1994); and *The Magical Quest: The Use of Magic in Arthurian Romance* (**1988**) by Anne Wilson. "Camelot in Four Colors" (*Amazing Heroes #55* September 15 1984) by Alan Stewart is a detailed study of the Arthurian legend in comic books. Two selections of early Arthurian literature and scholarship are *The Arthurian Legends* (anth **1979**) ed Richard Barber and *An Arthurian Reader* (anth **1988**) ed John Matthews.

ARTHUR, ROBERT (JAY) (1909-1969) US prose author and writer for tv and radio, not to be confused with the US movie producer Robert Arthur Feder (1909-1986), who worked under his given names. RA was a prolific contributor to the pulp MAGAZINES from 1931, most of his output being crime and mystery fiction, although by the late 1930s he had shifted more to fantasy; he joined MGM as a staff writer in 1937. Much of his work was SLICK FANTASY, often derivative from Lord DUNSANY and Stephen Vincent BENÉT; most of it was routine, but he scored high on creating an atmosphere of WRONGNESS. His **Murchison Morks** stories (*Argosy* 1940-41) are CLUB STORIES; some were collected as *Ghosts and More Ghosts* (coll **1963**). His best work – e.g., "Satan and Sam Shay" (1942 *Elk's Magazine*), a PACT WITH THE DEVIL story – remains uncollected.

In the late 1940s and early 1950s RA worked more for radio, scripting and directing a number of series, most importantly **The Mysterious Traveler** (1943-52), which was notable for its surprise endings and included many GHOST STORIES. He also edited *The Mysterious Traveler Magazine* (5 issues, November 51-Fall 52), which published several of his own stories, some under the pseudonyms Andrew Fell, Anthony Morton, Jay Norman, John West and Mark Williams. RA later worked closely with Alfred HITCHCOCK, ghost-editing several of Hitchcock's anthologies – titles include: *Stories for Late at Night* (anth **1961**; vt in 2 vols *12 Stories for Late at Night* 1962 US and *More*

Stories for Late at Night 1962 US; vt in 2 vols *Stories for Late at Night, Part One* 1964 UK and *Part Two* 1965 UK); *Alfred Hitchcock's Haunted Houseful* (anth **1961**); *Alfred Hitchcock's Ghostly Gallery* (anth **1962**); *Stories My Mother Never Told Me* (anth **1963**; vt in 2 vols as *Stories My Mother Never Told Me* 1965 US and *More Stories My Mother Never Told Me* 1965 US; vt in 2 vols *Stories My Mother Never Told Me, Part I* 1966 UK and *Part II* 1967 UK); *Alfred Hitchcock's Monster Museum* (anth **1965**; cut 1973 UK; rev 1982); *Alfred Hitchcock's Witches' Brew* (anth **1965**; vt *Alfred Hitchcock's Witch's Brew* 1977); *Stories Not for the Nervous* (anth **1965**; vt in 2 vols *Stories Not for the Nervous* 1966 US and *More Stories Not for the Nervous* 1966 US; vt in 2 vols *Stories Not for the Nervous, Book One* 1968 UK and *Book Two* 1969 UK); *Stories that Scared Even Me* (anth **1967**; vt in 2 vols *Scream Along With Me* 1970 US and *Slay Ride* 1971 US; vt in 2 vols *Stories That Scared Even Me, Part One* 1970 UK and *Part Two* 1970 UK); *Spellbinders in Suspense* (anth **1967**; rev 1982); *A Month of Mystery* (anth **1969**; vt in 2 vols *Dates With Death* 1972 US and *Terror Time* 1972 US; vt in 2 vols *A Month of Mystery, Book One* 1972 UK and *Book Two* 1972 UK); probably others. RA also edited his own juvenile anthologies of ghost and supernatural stories: *Davy Jones's Haunted Locker* (anth **1965**), *Monster Mix* (anth **1968**) and *Thrillers and More Thrillers* (anth **1968**).   [MA]

ARTHUR, RUTH M. Working name of UK writer Ruth Mabel Arthur Higgins (1905-1979), long active as a children's author, her career beginning with *Friendly Stories* (coll **1932**). Most of her early work, like the **Brownie** sequence – *The Crooked Brownie* (**1936**), *The Crooked Brownie in Town* (**1942**) and *The Crooked Brownie at the Seaside* (**1942**) – is for younger children, but with *Dragon Summer* (**1962**) and *A Candle in her Room* (**1966**) she began to produce the TIMESLIP romances for which she became best-known. They typically feature a teenage girl on the verge of adolescence, a crisis dramatically resolved through her absorption in an earlier, exemplary life-situation. In *Requiem for a Princess* (**1967**) an adopted girl timeslips to 16th-century Spain and into the SOUL of a girl destined to drown: a fate similar to but instructively more severe than being adopted. *On the Wasteland* (**1975**) sends an orphan back to Viking times, where she becomes the betrothed daughter of an important chief. RMA also wrote some GHOST STORIES, like *The Autumn People* (**1973**; vt *The Autumn Ghosts* 1976) and *Miss Ghost* (**1979**), in which the process is reversed: GHOSTS visit girls in trouble and offer solutions.   [JC]

Other work: *Mother Goose Stories* (coll **1938**); *The Whistling Boy* (**1969**); *The Saracen Lamp* (**1970**); *An Old Magic* (**1977**).

ARTHUR, WALSHINGHAM [s] ◊ Kenneth MORRIS.

ARTHUR C. CLARKE AWARD Given since 1987 for the best SCIENCE-FICTION novel published during the course of the previous calendar year in the UK. Although it is theoretically restricted to sf, more than one winning novel has been outright fantasy while several have strayed very far from the kind of sf written (and by implication espoused) by Clarke himself. The panel of judges is overseen by the Science Fiction Foundation, of which Clarke is a Patron, the British Science Fiction Association and the International Science Policy Foundation, each supplying two jurors on a rota basis. Clarke has underwritten the costs of administering the award, and the annual prize of £1000 is donated by him. In the list below dates are year of award rather than of publication; US novels in particular may have been published more than a year earlier.

Winners:

1987: Margaret Atwood, *The Handmaid's Tale*
1988: George Turner, *The Sea and Summer*
1989: Rachel POLLACK, *Unquenchable Fire*
1990: Geoff RYMAN, *The Child Garden*
1991: Colin GREENLAND, *Take Back Plenty*
1992: Pat Cadigan, *Synners*
1993: Marge Piercy, *Body of Glass* (ot *He, She and It*)
1994: Jeff Noon, *Vurt*
1995: Pat Cadigan, *Fools* [JC]

ARTMANN, H.C. (1921-   ) Austrian writer. ◊ AUSTRIA.

AS ABOVE, SO BELOW Earth as mirror of HEAVEN can provide justification for theocracy, for the Divine Right of Kings, for the Panglossian philosophy that things are best as they are, EVIL included. But mirroring works both ways, allowing Heaven and HELL to be slyly modelled on earthly hierarchies. Thus Rudyard KIPLING's "On the Gate" (**1926**) shows an AFTERLIFE run like a benevolent, paternalistic army or civil service, finding technical loopholes in the law so that the most hopeless cases can be admitted to Heaven; James Branch CABELL's *Jurgen* (**1919**) presents Hell as an enlightened democracy, proud not to be a dictatorship as above (though elections have admittedly been suspended during the war with Heaven); C.S. LEWIS's *The Great Divorce* (**1945** chap) models Hell on grey London suburbs; the ten-lane Heaven Expressway in Michael FRAYN's *Sweet Dreams* (**1973**) leads to the perfect reward for yuppies, a social round exactly like mortal life yet better; Robert A. HEINLEIN's *Job* (**1984**) updates Heaven with necessary data-processing, crowd-handling and transport systems, while Piers ANTHONY's **Incarnations of Immortality** sequence describes a similar high-tech SOUL-handling operation in PURGATORY; Terry PRATCHETT makes Hell a dreary hotel filled with middle-management types in *Eric* (**1990**); Tom HOLT's *Here Comes the Sun* (**1993**) has the entire Universe supernaturally run with the majestic inefficiency of UK heavy industry. Further examples abound.

The physical structures of many fantasy EDIFICES, and of the CITIES portrayed in much URBAN FANTASY, also mirror the principle. [DRL]

See also: GREAT AND SMALL; LITTLE BIG.

ASGARD ◊ AESIR; NORDIC FANTASY.

ASH, CONSTANCE (LEE) (1950-   ) US writer whose fantasy novels about **Glennys the Stallion Queen** – *The Horsegirl* (**1988**), *The Stalking Horse* (**1990**) and *The Stallion Queen* (**1992**) – describe her heroine's escape from BONDAGE to her psychic rapport (◊ TALENTS) with horses. The bondage is onerous, for the FANTASYLAND of the sequence is riven by sectarian hatred; any follower of the GODDESS – here identified as Eve, the First Mother – is adamantly opposed by fundamentalist patriarchs who espouse the primal guilt of women, in terms familiar to students of fundamentalist Christianity and Islam (◊ GENDER), and who despoil the LAND, justifying their behaviour in terms which are also familiar. Glennys rises in this world by dint of infrequent but extended sexual alliances with sympathetic male figures, and by the end of her tale – having become Stallion Queen of a newly founded domain, and after much bloodshed typical of DYNASTIC FANTASY – has attained much security and saved the land from RELIGION. Though hampered by the strictures of her choice of venue and plot-structure, CA demonstrates how thoroughly such venues and plots can serve more complex ends. [JC]

ASHLEY, MIKE Working name of UK editor and researcher Michael Raymond Donald Ashley (1948-   ), whose special area of expertise is the history and development of SCIENCE FICTION, FANTASY and SUPERNATURAL FICTION, and the relationship between these genres and LEGENDS. This has led to a number of ambitious reference works, in particular *Who's Who in Horror and Fantasy Fiction* (**1977**) – a seminal but all too brief coverage of some 400 writers – *Monthly Terrors: An Index to the Weird Fantasy Magazines Published in the United States and Great Britain* (**1985** US) with Frank H. Parnell (1916-   ), which covers many rare and small-circulation MAGAZINES, *Science Fiction, Fantasy and Weird Fiction Magazines* (**1985** US) with Marshall B. Tymn (1937-   ), which remains the most comprehensive study of this protean field, and *The Supernatural Index* (**1995** US) with William G. Contento (1947-   ), a painstaking index to over 2200 ANTHOLOGIES of supernatural, fantasy and HORROR fiction.

MA has long been working on a biography of Algernon BLACKWOOD; spinoffs *en route* have included *Algernon Blackwood: A Bio-Bibliography* (**1987** US) and two collections of Blackwood's stories: *Tales of the Supernatural* (coll **1983**) and *The Magic Mirror: Lost Tales and Mysteries* (coll **1989**). In the same vein MA has compiled *The Fantasy Readers' Guide to Ramsey Campbell* (**1980** chap) and the collections *Mrs Gaskell's Tales of Mystery and Horror* (coll **1978**) and *Robert E. Howard's World of Heroes* (coll **1989**) (◊ Robert E. HOWARD).

As an anthologist MA has striven to produce original work representative of its subject matter, as demonstrated in a series of Arthurian collections: *The Pendragon Chronicles* (anth **1990**), *The Camelot Chronicles* (anth **1992**), *The Merlin Chronicles* (anth **1995**) and «The Chronicles of the Holy Grail» (anth 1996) (◊ ARTHUR), plus the mini-guide *The Life and Times of King Arthur* (**1996** chap). Other anthologies of relevance are *Weird Legacies* (anth **1977**), with stories drawn from WEIRD TALES, *Jewels of Wonder* (anth **1981**), reprinting longer stories of HEROIC FANTASY, *The Mammoth Book of Short Horror Novels* (anth **1988**), *The Giant Book of Myths and Legends* (anth **1995**) and the children's anthologies «Ghost Stories» (anth 1996) and «Fantasy Stories» (anth 1996), containing stories of QUESTS and discovery. (◊ SFE for further bibliography.) [MA]

ASHTON, CHARLES (?   -   ) UK writer whose **Smoke and Dragon** YA fantasy sequence – *Jet Smoke and Dragon Fire* (**1991**), *Into the Spiral* (**1992**) and *The Singing Bridge* (**1993**) – is set in some isolated villages whose residents are kept from the rest of the world by a DRAGON; whether they inhabit a prison or a POLDER is a question which cannot be simply answered. [JC]

Other works: *Billy's Drift* (**1994**); *The Giant's Boot* (**1995**).

ASHTORETH ◊ GODDESS.

ASIMOV, ISAAC (1920-1993) US writer, of great importance to sf (◊ SFE) but of minor interest to fantasy. Two collections – *Azazel* (coll of linked stories **1988**) and «Magic: The Final Fantasy Collection» (coll 1996) – contain most of his fantasy of any interest. Many of these – rather in the mode of the **Mulliner** CLUB-STORY sequence by P.G. Wodehouse (1881-1975) – are TALL TALES told to an Asimov-like listener by George Butternut, whose COMPANION, an "extraterrestrial" DEMON named Azazel, fulfils WISHES and does other things to help George help others. Most of Azazel's feats backfire; the effect is broad, and occasionally amusing. The second volume also includes some essays on FANTASY. [JC]

Other works (as editor): *100 Great Fantasy Short Short Stories* (anth **1984**) with Terry CARR and Martin H. GREENBERG; *Isaac Asimov Presents the Best Fantasy of the 19th Century* (anth **1982**) and *Isaac Asimov Presents the Best Horror and Supernatural of the 19th Century* (anth **1983**), both with Greenberg and Charles G. Waugh (1943-   ); *Baker's Dozen: 13 Short Fantasy Novels* (anth **1985**; vt *The Mammoth Book of Short Fantasy Novels* 1988 UK); *Visions of Fantasy: Tales from the Masters* (anth **1989**).

Edited series: The **Magical Worlds of Fantasy**, all with Greenberg and Waugh, comprising *Isaac Asimov's Magical Worlds of Fantasy #1: Wizards* (anth **1983**); *#2: Witches* (anth **1984**); *#3: Cosmic Knights* (anth **1985**); *#4: Spells* (anth **1985**); *#5: Giants* (anth **1985**); *#6: Mythical Beasties* (anth **1986**; vt *Mythic Beasts* 1988 UK); *#7: Magical Wishes* (anth **1986**); *#8: Devils* (anth **1987**); *#9: Atlantis* (anth **1987**); *#10: Ghosts* (anth **1988**); *#11: Curses* (anth **1989**) and *#12: Faeries* (anth **1991**); *#1* and *#2* were assembled as *Isaac Asimov's Magical Worlds of Fantasy: Witches & Wizards* (omni **1985**).

ASIMOV'S SCIENCE FICTION MAGAZINE US digest MAGAZINE, originally quarterly, bimonthly from January/February 1978, monthly from January 1979, four-weekly since January 1981, Spring 1977-current, published by Bantam Doubleday Dell Magazines, New York; ed George H. Scithers (1929-   ) Spring 1977-February 1982, Kathleen Moloney March 1982-December 1982, Shawna McCarthy (1954-   ) January 1983-February 1986, Gardner Dozois (1947-   ) since March 1986.

Originally entitled *Isaac Asimov's Science Fiction Magazine*, when founded by Davis Magazines, the title changed in November 1992 with the magazine's sale to its new publisher. Despite the inclusion of "science fiction" in the title, and the association with Isaac ASIMOV, *ASFM* published fantasy stories from the outset, though these maintained some core trappings of sf, as in Roger ZELAZNY's award-winning "Unicorn Variations" (1981). Even Asimov's own continuation of his **Azazel** series, originally about a friendly DEMON, turned Azazel into an extraterrestrial for *ASFM*. Under the editorship of McCarthy, fantasy became an increasing part of the content, and Asimov's own February 1984 editorial made it explicit that in future more fantasy would be included. The main contributors to the change were Tanith LEE, whose REVISIONIST FANTASY "La Reine Blanche" (1983) produced the first fantasy-orientated cover, and Lucius SHEPARD, whose strong GHOST STORY "The Storming of Annie Kinsale" (1984) and powerful "How the Wind Spoke at Madaket" (1985) set cross-generic standards. Asimov himself hastened to assure readers that basic SWORD AND SORCERY would not appear in *ASFM*, and the closest it came to HEROIC FANTASY was the **Gilgamesh** series (1986-8) by Robert Silverberg. Most *ASFM* SECONDARY-WORLD fantasies are fashioned as FAIRYTALES. It also regularly publishes GHOST STORIES, plus many light fantasies in the UNKNOWN style; these have been more evident during Dozois's tenure. Overall, though, *ASFM* comes closest to *The* MAGAZINE OF FANTASY AND SCIENCE FICTION in its approach to a crossover of sf and fantasy, and could now be claimed as the major publisher of SLICK FANTASY. Main contributors include Michael BISHOP, Octavia E. Butler (1947-   ), Suzy McKee CHARNAS, John CROWLEY, Jack DANN, Avram DAVIDSON, Charles DE LINT, Gregory FROST, Alex Jablokov (1956-   ), Janet Kagan (1945-   ), Nancy KRESS, Bruce Sterling (1954-   ), S.P.

SOMTOW, Michael SWANWICK, Connie WILLIS and Jane YOLEN. In recent years more stories from *ASFM* – including a number of fantasies – have won AWARDS than from any other magazine; Dozois has won the Hugo for Best Professional Editor seven times since 1988, including an unbroken run in 1988-93. Many anthologies have been drawn from *ASFM*, but the most representative of its fantastic fiction are *Isaac Asimov's Fantasy* (anth **1985**; cut 1990) ed McCarthy, *Transcendental Tales from Isaac Asimov's Science Fiction Magazine* (anth **1989**) ed Dozois and *Isaac Asimov's Ghosts* (anth **1995**) ed Dozois and Sheila Williams.   [MA]

ASPRIN, ROBERT LYNN (1946-   ) US writer and anthologist who began publishing sf with *The Cold Cash War* (**1977**) and several further sf novels; he has become much better known, however, for fantasy, especially for the **Myth** series of ARABIAN-FANTASY romps, his most accomplished work: *Another Fine Myth…* (**1978**), *Myth Conceptions* (**1980**), *Myth Directions* (**1982**), *Hit or Myth* (dated 1983 but **1984**), *Myth-ing Persons* (**1984**), *Little Myth Marker* (**1985**), *M.Y.T.H. Inc Link* (**1986**), *Myth-Nomers and Im-pervections* (**1987**), *M.Y.T.H. Inc in Action* (**1990**) and *Sweet Myth-tery of Life* (**1994**). A US omnibus, *Myth Adventures* (omni **1984**), assembled the first four titles; two UK omnibuses, *The Myth-ing Omnibus* (omni **1992** UK) and *The Second Myth-ing Omnibus* (omni **1992** UK), assembled the first six titles. Two COMICS collections – *Myth Adventures One* (graph coll **1985**) and *Myth Adventures Two* (graph coll **1986**), both with artist Phil Foglio – were based on the first novel in the series; further comic work, not in book form, is *Myth Adventures #9-#12* (all 1986) and *Myth Conceptions #1-#8* (1985-7). According to RA, the series, which features the adventures of a wacky DUO of magician and DEMON, is based on the "ROAD TO" MOVIES featuring Bing Crosby and Bob Hope, and treats its large array of fantasy figures in an appropriately slapstick manner somewhat after the fashion of L. Sprague DE CAMP. The **Duncan and Mallory** series of GRAPHIC NOVELS – *Duncan and Mallory* (graph **1986**), *The Bar-None Ranch* (graph **1987**) and *The Raiders* (graph **1988**), all with artist Mel White – features a KNIGHT with a dragon COMPANION. Other fantasy of interest includes *Catwoman* * (**1992**; vt *Catwoman: Tiger Hunt* 1993 UK) with Lynn ABBEY, tied to BATMAN RETURNS (*1992*).

It is, however, in the creation and editorial supervision of the **Thieves' World** sequence of SHARED-WORLD anthologies, some worked up into novel-like BRAIDS, that RLA – in collaboration with Abbey – has done his most original work. The sequence comprises 12 volumes: *Thieves' World* * (anth **1979**), *Thieves World #2: Tales from the Vulgar Unicorn* * (anth **1980**) and *#3: Shadows of Sanctuary* * (anth **1981**), all three assembled as *Sanctuary* * (omni **1982**); *#4: Storm Season* * (anth **1982**), *#5: The Face of Chaos* * (anth **1983**) – from this volume on, Abbey is listed as co-editor – and *#6: Wings of Omen* * (anth **1984**), all three assembled as *Cross-Currents* * (omni **1985**); *#7: The Dead of Winter* * (anth **1985**), *#8: Soul of the City* * (**anth** 1985) and *#9: Blood Ties* * (anth **1986**), all three assembled as *The Shattered Sphere* * (omni **1986**); and *#10: Aftermath* * (anth **1987**), *#11: Uneasy Alliances* * (anth **1988**) and *#12: Stealer's Sky* * (anth **1989**), all three assembled as *The Price of Victory* * (omni **1990**). Graphic-novel versions of material from the sequence have been published as *Thieves' World Graphics #1* * (graph **1985**), *#2* * (graph **1986**), *#3* * (graph **1986**), *#4:* * (graph **1987**), *#5* * graph **1987**) and *#6* * (graph **1988**), all with artist Tim Sale. Set in the city of Sanctuary, in an unremarkable

FANTASYLAND, the sequence traces the adventures of various characters out of the repertory of SWORD AND SORCERY as they cope with their congenially criminous lives, and as Sanctuary itself must respond to various internal and external threats. Each main character has been used primarily by the author originally responsible for him/her, though each tends to appear in minor roles in others' stories as the sequence winds onward. The impressive list of authors who have contributed to **Thieves' World** includes both editors, Poul ANDERSON, Marion Zimmer BRADLEY, John BRUNNER, C.J. CHERRYH, David DRAKE, Diane DUANE, Philip José FARMER, Joe Haldeman, Vonda N. McIntyre, Janet MORRIS, Diana L. PAXSON and A.E. VAN VOGT. Associated novels are *Shadowspawn* * (**1987**) by Andrew J. OFFUTT, *Dagger* * (**1988**) by David DRAKE, *City at the Edge of Time* * (**1988**) by Chris and Janet MORRIS, and the **Tempus** series by Janet Morris.

RLA has also co-edited some volumes in the **Winds of Change** shared-world sequence run by Richard PINI.

As a writer of comic fantasy, RLA gives pleasure though his clarity and a good sense of timing; his language, however, is not resourceful. [JC]

Other works: *Mirror Friend, Mirror Foe* (**1979**) with George Takei; *The Bug Wars* (**1979**); *Tambu* (**1979**); the **Phule's Company** sequence of sf adventures comprising *Phule's Company* (**1990**) and *Phule's Paradise* (**1992**).

ASQUITH, [Lady] CYNTHIA (MARY EVELYN) (1887-1960) UK socialite and, from 1918 to 1937, private secretary to J.M. BARRIE. She developed a flair as an essayist and anthologist; her own fiction is fluid and graceful. *This Mortal Coil* (coll **1947** US; vt with 2 stories omitted and 1 added *What Dreams May Come* 1951 UK) features mostly GHOST STORIES of POSSESSION, with the SPIRITS seeking retribution. But CA's major contribution to SUPERNATURAL FICTION was as an anthologist. *The Ghost Book* (anth **1926**) was the first serious 20th-century attempt to produce an anthology of modern ghost stories written unsensationally and by the literary establishment (◊ *The* GHOST BOOK). CA produced two further anthologies of equal value: *Shudders* (anth **1929**) and *When Churchyards Yawn* (anth **1931**) – these two assembled with *The Black Cap* (anth **1927**), a volume of murder stories, as *A Century of Creepy Stories* (cut omni **1934**). She later returned to the **Ghost Book** series with *The Second Ghost Book* (anth **1952**; vt *A Book of Modern Ghosts* 1953 US) and *The Third Ghost Book* (anth **1955**). In all of these volumes the HAUNTINGS remain equivocal: they might be either physical or mental manifestations (◊ PERCEPTION). The books established a standard for ghost-story ANTHOLOGIES that few have bettered. CA applied the same exacting standards to her anthologies for children, which contain many magical fantasies. [MA]

As editor for children: *The Flying Carpet* (anth **1925**), *The Treasure Ship* (anth **1926**), *Sails of Gold* (anth **1927**), *The Treasure Cave* (anth **1928**), *The Funny Bone* (anth **1928**), *The Children's Cargo* (anth **1930**), *The Silver Ship* (anth **1932**), *The Children's Ship* (anth **1950**). *My Grimmest Nightmare* (anth **1935**) is often attributed to CA but was in fact ed anon Cecil Madden (1902-1987).

Further reading: *Cynthia Asquith* (**1984**) by Nicola Beauman.

ASS As one of the most stubborn of beasts, and one of the most cruelly exploited, the ass (or donkey) has been an object of general abuse for as long as there have been stories; probably for this reason it has been used – in FAIRYTALES,

BEAST FABLES and SATIRES – as an emblem of humble wisdom capable of conveying sharp moral lessons to the arrogant; and in stories of METAMORPHOSIS, in which concupiscent or greedy or simply thoughtless humans are comically transfigured.

The first Western version of the tale of the owner who learns from his ass is the story of Balaam (from the BIBLE's *Numbers*). Balaam is a SORCERER caught in a cleft stick. Balak, King of Moab, has called on him to come and curse the Jews; Yahweh warns him not to. Balak becomes more insistent. Yahweh tells Balaam to get on his ass and go, but to follow instructions. En route the ass sees an ANGEL with a drawn sword barring the way, and refuses to advance. Balaam then beats the ass, which Yahweh gives the gift of speech (◊ TALKING ANIMALS) so it can ask him why it's being so maltreated. But Balaam baulks at his ass's simple wisdom until Yahweh opens his eyes, and he too sees the angel. Christian iconography made considerable use of this and similar asses, all of them humble, sometimes tending to unction; but novels like Jean MORRIS's *The Donkey's Crusade* (**1983**), though set within a Christian context, complexify the portrait of the wise talking animal, who in this particular text becomes a genuine COMPANION and mentor.

Related tales concern talking mules. The best-known 20th-century example is the **Francis the Talking Mule** series by David J. STERN, which formed the basis of the sequence of FRANCIS MOVIES and the tv series MISTER ED.

Of greater fantasy interest are the tales of the man who is turned into an ass. These tales are of significance because they deal not only with TRANSFORMATION, a central motif in most fantasy, but with subsequent BONDAGE. The earliest tales of donkey-metamorphosis are at least as old as writing. The first written version of note, which is partially dependent on oral tradition, is LUCIAN's *Lucius, or The Ass* (written *c*150), a complex tale whose protagonist travels to a land of women and spies on one of them as she anoints her body and becomes a BIRD. Anxious to fly as well, he anoints himself – but with the wrong salve – and is transformed into an ass (◊ PARODY). His subsequent adventures are terrifying, though he also does duty with various women attracted by his sexual prowess; and he regains human form only when he eats a rose (which is sacred to Venus). Composed just afterwards, *The Golden Ass* (written *c*165) by APULEIUS tells a similar story, though the work as a whole is more complicated, the ass tale being a FRAME STORY within which (among other matter) the first LITERARY FAIRYTALE, "Cupid and Psyche", is narrated.

The exemplary transformation of a man into a donkey soon became a central TOPOS of popular and written literature. It surfaces in SHAKESPEARE's *A Midsummer Night's Dream* (performed *c*1595; **1600**), in Carlo COLLODI's *Pinocchio* (**1883**) and in Ann LAWRENCE's *Tom Ass, or The Second Gift* (**1972**). There are many more. [JC]

ASTARTE ◊ GODDESS.

ASTERIX Eponymous hero of the French COMIC strip by GOSCINNY AND UDERZO. The stories concern the inhabitants of a small village in Roman-occupied Gaul who are able to hold out against the invaders through the efficacy of a magic POTION brewed by their local druid, Panoramix, which temporarily endows all who consume it with incredible strength. The great success of the strip lay in its exaggerated praise of all things French and its SATIRE of the characteristics of all other nationalities; also important was Uderzo's clean and finely detailed artwork. The adventures of Asterix and his slow-witted gluttonous sidekick Obelix as they travel throughout Roman-occupied Europe were reprinted in book form from 1964 and ran to some 30 volumes, with translations in almost as many languages.

The UK translations by Anthea Bell and David Hockridge are witty and inventive, giving the characters even more amusing names and idiosyncrasies than those in the original French editions. The druid Panoramix is renamed Getafix; the inept village musician becomes Cacophonix; Romans include Christmus Bonus and Raucus Hallelujachorus. Asterix's great popularity in the UK was not matched in the USA.

Asterix has also starred in a series of ANIMATED MOVIES (◊ ASTERIX. [RT]

ASTERIX MOVIES A number of ANIMATED MOVIES, none particularly distinguished, have been made about ASTERIX.

1. *Asterix the Gaul* Belgian/French movie (ot *Astérix le Gaulois*) (*1967*). Dargaud/Belvision. **Dir** Rene Goscinny, Albert Uderzo. **Anim dir** Willy Lateste. **Screenplay** Lateste. **Based on** the COMICS by GOSCINNY AND UDERZO. 67 mins. Colour.

Caesar has conquered all Gaul except one small village, which resists the Romans thanks to the magic strength-giving POTION brewed by the druid Panoramix (UK readers of the comics will discover several of the main characters' names differ in this US dubbing). The local Roman commander sends his dopiest soldier, Caligula Minus, into the village to uncover this secret, then captures Panoramix and tortures him hoping to gain the formula. Asterix, the most resourceful in the village, comes to Panoramix's rescue, and the two fool the Romans into drinking a different potion, one that makes hair grow uncontrollably. This is, at least in the Amerenglish-language version, a surprisingly lacklustre adaptation, with a particularly rebarbative musical score. It is, however, probably the best of the series, later members of which are listed below.

2. *Asterix and Cleopatra* Belgian/French movie (*1968*). **Dir** Rene Goscinny, Albert Uderzo. 70 mins. Colour.

3. *The Twelve Tasks of Asterix* (ot *Les Douze Travaux d'Astérix*; vt *Asterix the Gaul 2: The Twelve Tasks of Asterix*) French movie (*1975*). Idéfix. **Dir** Rene Goscinny, Matt McCarthy, Albert Uderzo. 82 mins. Colour.

4. *Asterix in Britain* (ot *Astérix Chez les Brétons*; vt *Asterix in Britain: The Movie*) Danish/French movie (*1986*). Gaumont/Dargaud. **Pr** Yannik Piel. **Dir** Pino Van Lamsweerde. **Screenplay** Pierre Tchernia. 89 mins. Colour.

5. *Asterix and the Big Fight* (ot *Le Coup de Menhir*) French/West German movie (*1989*). Palace/Gaumont/Extrafilm. **Pr** Nicolas Pesques. **Dir** Philippe Grimond. **Screenplay** George Roubicek. 81 mins. Colour. [JG]

ASTRAL BODY An occult term used in SUPERNATURAL FICTION to describe a detachable aspect of the SOUL or spirit. The AB is more or less identical in "substance" to the AURA, except for the circumstance that it is often found – in fiction at least – to project itself, or to be projected (◊ ASTRAL PLANE). Many people who have undergone near-death experiences have described themselves as floating in an AB over their corpses, to which they have still been joined by a kind of psychic rope that has, in due course, enabled their AB to haul itself back "into" their physical body. The image is a potent one, and turns up fairly often in fiction. [JC]

ASTRAL PLANE An occult term used in SUPERNATURAL FICTION to describe where an ASTRAL BODY may be found; by extension, it has been used to describe a more generalized region which the astral body may jointly inhabit with other astral bodies, perhaps after the death of the worldly body (◊ AFTERLIFE). More broadly, the AP can be considered as an ALTERNATE REALITY. [JC]

ASTROLOGY Astrological beliefs, widespread in the ancient world, became briefly fashionable again during the Renaissance and are so yet again today. Nevertheless – and notwithstanding the fact that at least one notable fantasy writer, Jane GASKELL, is a practising astrologer – astrology plays a very minor role in modern fantasy; it is far less popular as a basis for stories than the TAROT.

Astrologers crop up as minor characters in many historical novels and Gothic tales, but are usually represented as charlatans; those in John Galt's "The Black Ferry" (*c*1820) and Washington IRVING's "Legend of the Arabian Astrologer" (1832) are exceptions – likewise Rudyard KIPLING's ironic "A Doctor of Medicine", whose hero infers from astrological portents that plague may be fought by killing rats. Kipling's "Children of the Zodiac" (1891) is an offbeat ALLEGORY, while *Children of the Zodiac* (**1929**) by Alice M. Williamson (1869-1933), who wrote as Mrs C.N. Williamson, is a curious exercise in literary game-playing. Two novels based on the premise that some bold pioneer might one day make astrology into a exact science, Edward S. Hyams's *The Astrologer* (**1950**) and John Cameron's *The Astrologer* (**1972**), both appreciate the irony inherent in what is essentially a *reductio ad absurdum*, as had Alan GRIFFITHS's *The Passionate Astrologer* (**1936**), which features an even less likely route to predictive certainty. Clark Ashton SMITH's "The Last Hieroglyph" (1935) is also ironic after its own gaudy fashion, but E. Hoffmann PRICE's "The Shadow of Saturn" (1950) appears to take a more reverent view of the astrologer's art. Piers ANTHONY's *Macroscope* (**1969**) deploys the signs of the ZODIAC in a ponderously symbolic manner in the interests of cultivating an eccentric bizarrerie. Credulity is rarely an asset in the composition of literary fantasies – a fact made painfully evident in the novels of sometime astrologer and general practitioner of the magical arts Dion FORTUNE – but *The Finger and the Moon* (**1973**), an eclectic magical romance by Geoffrey Ashe (1923-   ), is more graceful in its execution than most romances of its earnest stripe. [BS]

ASTURIAS, MIGUEL (ANGEL) (1899-1974) Guatemalan writer, winner of the 1967 Nobel Prize for Literature, ambassador to France in his later years, and a central figure in the evolution of Latin American literature, most significantly perhaps through the workings-out of his concept of "myth creation", a term he used to describe the writer's central role in the creation of structures of meaning relevant to a world in which SURREALISM was not a mode but a PERCEPTION of circumambient and pressing REALITY. Latin America could also be described as a world whose primal engendering stories (preserved sacred-book compendiums like the Mayan *Popol Vuh*, which MA trans **1927** from the French) could not perhaps be used holus-bolus, but served as necessary paradigms for new myths. The connections between "myth creation" and MAGIC REALISM are multifarious and intimate.

MA's first novel – which he part-drafted as early as 1922 – was *El Señor Presidente* (**1946** Mexico; trans Frances Partridge as *The President* **1963** UK; vt *El Señor Presidente* 1964 US), an extremely complex narrative set mainly in a nightmarish CITY dominated by a GOD-like all-seeing dictator who is opposed by his favourite "angel" – his name is Miguel Angel, MA's given names – whose rebellion extends only as far as the President allows it to. This PARODY of human autonomy, whose effects extend throughout the text, contributes centrally to the effect the novel generates of combined mythic richness and claustrophobia.

MA's second novel, *Hombres de maíz* (**1949** Argentina; trans Gerald Martin as *Men of Maize* **1975** US), is more difficult – combining at least six interwoven episodes involving protagonists who are each other's REINCARNA-TIONS, and more visibly written according to the precepts of myth creation. The story, insofar as it can be reduced to one central tale, is that of the nation (Guatemala) itself, rendered through PALIMPSESTS of exemplary MYTH. The novel contains deaths, resurrections, ACCURSED WANDERERS, SHAMANS, excursions to the UNDERWORLD, seasonal rites (◊ SEASONS) and, at the end, a glimpse of EDEN.

MA's last novel of fantasy importance, *Mulata de tal* (**1963** Argentina; trans Gregory Rabassa as *Mulata* **1967** US; vt *The Mulatta and Mr. Fly* 1967 UK), involves a kind of pact with the eponymous DEVIL (◊ PACTS WITH THE DEVIL), who takes the form of the mulatto woman the protagonist eventually marries (only to immure her); the novel then traces the ultimate devastation imposed upon the LAND by the consequences of the pact.

Other works of interest include the **Banana Trilogy** – *Viento fuerte* (**1950** Argentina; trans Darwin Flakoll and Claribel Alegría as *The Cyclone* **1967** UK), *El papa verde* (**1954** Argentina; trans Gregory Rabassa as *The Green Pope* **1971** US) and *Los ojos de los enterrados* (**1960** Argentina; trans Gregory Rabassa as *The Eyes of the Interred* **1973** US) – which incorporates some supernatural elements. MA, along with Jorge Luis BORGES, is regarded as the most important Latin American writer of the generation preceding Gabriel GARCIA MARQUÉZ. [JC]

ATAVISM The recurrence in an organism or part of an organism of a form typical of its remote ancestors. In fantasy, trivial atavistic symptoms are frequently associated with CURSES, but more extreme reversions to ancestral type are featured in stories of atavistic METAMORPHOSIS, many of which qualify as borderline SCIENCE FICTION. Max BRAND's oft-reprinted "That Receding Brow" (1919) is perhaps the best-known fantasy of atavism; an earlier example is Jack LONDON's "When the World Was Young" (1910). Atavisms which reveal unorthodox ancestries crop up in Royal W. Jimerson's "Medusa" (1928) and H.P. LOVECRAFT's *The Shadow over Innsmouth* (**1936**). Stories of induced atavism, involving some kind of "ancestral memory", include Leonard Cline's *The Dark Chamber* (**1927**), Alan Sullivan's *In the Beginning* (**1927**), Edison Marshall's *Ogden's Strange Story* (1928; **1934**), Norman Springer's *The Dark River* (**1928**), Neil BELL's *The Disturbing Affair of Noel Blake* (**1932**) and Paddy Chayevsky's *Altered States* (**1978**). Also, there are stories in which psi powers are an atavism, the thesis being that once all human beings had such TALENTS but these were extinguished by the process of civilization, yet can re-emerge atavistically in rare individuals. [BS]

ATHENE ◊ GODDESS.

ATKEY, BERTRAM (1880-1952) UK writer, best remembered today for his Cockney crime stories about **Smiler Bunn**, but who also produced several sf and fantasy stories for the popular MAGAZINES; only a small portion have been collected. The **Hobart Honey** series has the trappings of sf (travel to different periods of history) but the modus – pills

provided by a Tibetan lama – shifts them into fantasy; the tone is humorous. Two known novels in the series are "The Backslidings of Mr Hobart Honey" (1916 *Red Magazine*) and *The Escapes of Mr Honey* (**1944**). In the same tone BA updated the HERCULES legend, converting the 12 labours into a series of sports in *Hercules – Sportsman* (1917 *Red Magazine*; fixup **1922**). He also delighted in nature fantasies. *Folk of the Wild* (**1907**) looked at creatures of MYTH surviving until today, a theme recurring in more humorous vein in the series **Unnatural Nature Stories** (1918-19 *Red Magazine*), which involved DINOSAURS, ape-men (◊ APES), MUMMIES, centaurs and mermen (◊ MERMAIDS). [MA]

ATLANTIC MONTHLY ◊ MAGAZINES.

ATLANTIS Of all the mythic statements that have come down to us from the Classical Greek world only one remains vital – and, indeed, may number more believers now than it did two and a half millennia ago. This is the story of Atlantis, as narrated by Plato (*c*429-347BC) in his *Timaeus* and *Critias*.

According to Plato, the Athenian lawgiver Solon (*c*640-559BC), on a visit to Egypt, heard the story of Atlantis from an Egyptian priest, who used it as an illustration of CYCLES in the history of the world. Founded by Poseidon, Atlantis was a small continent in the Atlantic Ocean beyond the Gates of Hercules. Originally a land of just rule, it degenerated into an Oriental-type empire extending far into the Mediterranean. At the height of its power it was waging war against the men of the Eastern Mediterranean, who were led by Ur-Athens, when the GODS (led by ZEUS), outraged by Atlantean hubris and decadence, intervened, sinking Atlantis beneath the sea and incidentally destroying ancient Athens and its army. Since then the Atlantic Ocean had been unnavigable. All this happened about 9000 years ago and is recorded on Egyptian monuments, said the Egyptian priest, said Solon, said Critias, said Plato.

The ancient world was perplexed as to the interpretation of Plato's literary myth, since there was no other account of Atlantis. Was it a historical narrative or a flight of fancy? Aristotle (384-322BC), according to Strabo (*c*60BC-AD20), seems to have rejected historicity, while the members of the Platonic succession, for the most part, allegorized the story in various ways. Proclus (*c*410-485), in his *Commentary on the Timaeus*, mentions various interpretations: as a metaphor for the battle of GOOD AND EVIL stars, the conflict of the fixed stars with the moving planets, the battles of HEROES and DEMONS, and so on. Proclus and his master Iamblichus, however, took the story of Atlantis as a philosophical statement in which contraries would be destroyed to create a new unity. This seems to have been the general interpretation of Atlantis in the Neoplatonic stream (◊ NEOPLATONISM).

Today, our interpretation of Plato's account is closely linked to our knowledge of his sources. Modern rational conclusions about Plato's sources (and meaning) fall into several groups:

(a) Solon/Critias/Plato were honestly reporting a perhaps greatly garbled legend of Egyptian origin. Against this is that nothing substantiating Atlantis has been found in Egyptian documents and that the Egyptians were anyway largely indifferent to foreign history. A recent, not very convincing, version of the Egyptian theory interprets Plato's account as a distorted reminiscence of Troy and the Trojan War.

(b) The story of Atlantis, what with the thalassocracy and the catastrophism, is a folk memory of Minoan civilization, particularly the seismic destruction of Thēra. This is attractive, but negates almost all of Plato's account.

(c) Plato's narrative is a construct, a deliberately concocted mythic history incorporating echoes of the Persian Wars, elements of Greek FOLKLORE, TRAVELLERS' TALES and traditional MYTH, probably with a philosophical intention. This has been the overwhelming position among modern Classical scholars and historians.

As the story of Atlantis moved down from the Graeco-Roman world, it underwent changes. In the Middle Ages and early Renaissance the geographic component of Atlantis – an island in the Atlantic – was often taken literally, and maps up until the Renaissance occasionally show Atlantis along with Hy-Brasil, St Brendan's Island, the Hesperides and similar mythical ISLANDS (◊ LOST LANDS AND CONTINENTS). As geographical knowledge increased, however, the mode of interpretation changed again: it was not the "island-ness" of Atlantis that was accepted but a vague historicity. Thus geographers and speculative thinkers, both reasonable and crackpot (◊ SCHOLARLY FANTASY), often relocated Atlantis, since the Atlantic was somewhat unsuitable. For example, the 17th-century Swedish botanist Olof Rudbeck (1630-1702) wrote a huge work proving Atlantis was really pre-Christian Sweden, while the late 17th-century German Classical scholar Johann Albert Fabricius (1668-1736) set Atlantis in Palestine. At an earlier period, Francis Bacon (1561-1626), in his *The New Atlantis* (**1629**), identified Plato's Atlantis with the Americas, describing a Mexican raid on Europe and a Peruvian raid on Bensalem; Atlantis-America was then devastated by the Biblical FLOOD. It is not known how seriously Bacon took this history. This translocation of Atlantis has continued to the present, with Atlantean revisionists seriously relocating Plato's Atlantis almost everywhere in the world, from the Arctic to the Antarctic, Central America to Central Africa and Central Asia.

The bibliography of Atlantean studies is enormous, with hundreds of documents from the Middle Ages to the 19th century. Of all these, by far the most influential has been *Atlantis, the Antediluvian World* (**1882**) by Ignatius Donnelly (1831-1901). Donnelly, a US Populist politician and part-time scholar, set out to prove that Plato's account (surreptitiously dropping Zeus) was literally correct: Atlantis was an Atlantic island that sank very suddenly. He carried matters a step farther: drawing on contemporary archaeology, ANTHROPOLOGY, geology and philology, he claimed that Bronze-Age Atlantis was the mother of ancient civilizations from Peru and Mexico to Europe and Egypt, perhaps even China. He also believed that the fall of Atlantis should be a lesson to his contemporary world. A sidebranch to Donnelly was provided by Lewis SPENCE in *The History of Atlantis* (**1926**). Spence considered Atlantis the home and origin of Upper Palaeolithic Cro-Magnon Man, who came to Europe just before the subsidence.

Opposed to Donnelly's rational, if wrong, reconstruction of Atlantis is that of the occultists, the most notable of whom was H.P. BLAVATSKY, the founder of the Theosophical Society (◊ THEOSOPHY). According to *The Secret Doctrine* (**1888**), Atlantis marked an evolutionary step in the development of modern humanity. The Atlanteans were GIANTS, perhaps 25ft tall, who had a superscience based on occult powers. At first a moral people, the Atlanteans became "black with sin" and destroyed themselves and their continent by their obsession with BLACK MAGIC. This destruction was hundreds of thousands of years ago. Many more recent occultists, like the groups emergent from the Order of the GOLDEN DAWN, accept a later Atlantis, peopled by modern humanity, but one that was likewise concerned with magical operations that caused the destruction of the land. In the Society of the Inner Light, for example, Dion FORTUNE attempted a recreation of the magic practised before the Submersion, as described in part in her *The Sea Priestess* (**1938**).

Thanks to the renewed interest in Atlantis created by Donnelly and Blavatsky, it became a PLAYGROUND. A major research project would be required to identify all Atlantean fictions, but a safe estimate for the English language alone would range in the upper hundreds. Unfortunately, in this plethora of writing, there is very little of literary value.

As described by Plato, Atlantis was a rich land, fertilely set amid sheltering mountains, productive of tropical fruits, and for long the child of a GOD; among the exotic fauna were elephants. (Such richness and exoticism may have been familiar to Plato from his residence in Sicily, which would have offered contrast to the already overcultivated small lands of Greece.) This double characterization of Atlantis – as glamour place (like AVALON or EDEN) and as minatory victim (in its fall) – continues in other fiction, with proportions variously changed.

Three early novelistic treatments of Atlantis convey different images of the lost continent. *Atla: The Story of the Lost Island* (**1886**) by Mrs J. Gregory Smith (1818-1905) follows Donnelly's Bronze-Age transatlantic Atlantis closely, presenting, in Wardour Street language, a sentimental LOVE story. *Atlantis* (**1895** as by André Laurie; trans as *The Crystal City under the Sea* **1896**) by Paschal Grousset (1845-1909; ◊ SFE) is essentially a boys' costume romance with some attempt at character humour. It established the image of Atlantis surviving deep UNDER THE SEA under glass, a situation probably suggested by the powerful Victorian institution of the stove greenhouse. Earlier than this, Grousset's occasional collaborator Jules Verne, in *Twenty Thousand Leagues under the Sea* (**1869-70**), offered the wonderful promenade of Nemo and Arronax among the flooded statuary and temples of Atlantis. Popular in its day, *The Lost Continent* (**1900**) by C.J. Cutcliffe Hyne (1865-1944), heavily based on Donnelly and the novels of H. Rider HAGGARD, presented a stodgy dynastic romance that is now occasionally laughable. *The Scarlet Empire* (1906) by David M. Parry (1852-1915), President of the National Association of Manufacturers, used Atlantis as a situation for a bitterly anti-labour, anti-socialist *Tendenzfiktion*. His (perhaps ghosted) work describes a crystal-dome Atlantis that is totally regimented, corrupt, cynical and inhuman.

In the 1920s *L'Atlantide* (1919; translated variously as *Atlantida* and *The Queen of Atlantis* 1920) by Pierre Benoît (1886-1962) offered new elements: Atlantis in the Sahara, isolated by the disappearance of an ancient sea; a scholarly treatment of Roman remains in the area; and erotica. It centred on a feminist FEMME FATALE who avenged wrongs committed against Oriental women by killing her European lovers and copperplating their corpses. Benoît's novel has been filmed several times: *L'Atlantide* (**1921**); *Die Herrin von Atlantis* (**1932** vt *L'Atlantide*; vt *Lost Atlantis*; vt *The Mistress of Atlantis*; vt *Queen of Atlantis*); *Siren of Atlantis* (**1948**); and *L'Atlantide* (**1961**); the most significant of these is *Siren of Atlantis*. All in all, Benoît's story is the most readable of the early renderings of Atlantis.

Cultural forces played a part in the sinking of several Atlantises. "The Veiled Feminists of Atlantis" (1926) by Booth Tarkington (1869-1946) linked aggressive, unreasonable FEMINISM to destruction of values and black MAGIC to the Submersion, while "The Fall of Atlantis" (in *Recreations of a Psychologist* coll **1920**) by G(ranville) Stanley Hall (1844-1924) postulated a superscientific Atlantis that fell and was destroyed because of occupational syndicalism, a metaphor for our own times. The earlier *The Sin of Atlantis* (**1900**) by Roy Horniman (1874-1930), although largely a modern society novel, invokes an Atlantis that fell because of truck with evil. GOOD AND EVIL, in the form of hypostatizations, are associated with a surviving glass-bowl Atlantis in Arthur Conan DOYLE's "The Maracot Deep" (1927-8 *The Strand*).

Other popular novels dealing with Atlantis are *They Found Atlantis* (**1936**) by Dennis WHEATLEY, which ineptly combines a crime thriller with an underground Atlantis inhabited by only 12 men and women who enjoy sexual communism, astral wanderings around the world and highminded pursuits. *Three Go Back* (**1932**) by James Leslie Mitchell (1901-1935), superior novelistically to Wheatley's work, is based on Spence's concept of Atlantis as bearing an upper Palaeolithic culture. Three people are precipitated by a timewarp into a primitive land that contains both Cro-Magnon and Neanderthal humans. *The Ultimate Island* (1925) by Lance Sieveking (1896-1972) portrays a small surviving Atlantean island as a hermit land with a culture based on obsessive social integration and minimalism. The culture is interesting, but the plotting is badly handled sensationalism.

More recent novelistic treatments include: *Alas, That Great City* (**1948**) by Francis Ashton (1904-   ), a somewhat plodding adventure novel based on archetypal psychology and crank astronomy; Jane GASKELL's **Cija** series (**1963-77**), full of colourful, overwritten battles of SEX and power that might have taken place in any other FANTASYLAND; *The Romance of Atlantis* (**1975**) by Taylor Caldwell (1900-1985) with Jess Stearn, a piece of Caldwell's juvenilia – she was 12 when she first wrote it – that perhaps should have remained unpublished; Marion Zimmer BRADLEY's **Atlantean Chronicles** (**1982-4**) and Poul ANDERSON's *The Dancer from Atlantis* (**1971**), which edges into Minoan history.

The US pulp efflorescence of the first half of this century has been rich in Atlantean pryings, although there is little work of quality among the scores of such stories. Trivial Atlantean material appears in several **Tarzan** novels by Edgar Rice BURROUGHS – *The Return of Tarzan* (**1915**), *Tarzan and the Jewels of Opar* (**1918**), *Tarzan and the Golden Lion* (**1924**) and *Tarzan the Invincible* (**1931**). Adventure fantasies with ancient Atlantean settings are found also in the work of Henry KUTTNER and others of the pulp HEROIC-FANTASY school.

Descendants of Atlantis are found in the most varied places: in the Sahara, as in "The City of Glass" (1927 *WT*) by Joel Martin Nicholls; in the Arctic, in "Phalanxes of Atlans" (1931 *Astounding Stories*) by F. Van Wyck Mason (1901-1978); in Colorado, in "The Moth Message" (1934 *Wonder Stories*) by Laurence Manning (1899-1972); on Venus, in "The Swordsman of Sarvon" (1932 *Amazing Stories*) by Charles Cloukey; on Ganymede, in "What the Sodium Lines Revealed" (1929 *Amazing Stories Quarterly*) by L. Taylor Hansen (1899-1985); in other dimensions, as in "The Heads of Apex" (1931 *Astounding Stories*) by Francis Flagg (George Henry Weiss; 1898-1946); underneath New Jersey, in "Silver Dome" (1931 *Astounding Stories*) by Harl Vincent (1893-1968); and in gigantic mile-across cubes that float in the upper atmosphere, as in "The Empire in the Sky" (1930 *Wonder Stories*) by Ralph Wilkins. Such modern Atlanteans are usually, though not always, possessed of science much superior to our own.

In most instances Atlantis in the pulps stresses Plato's original notion of a glamour place, but occasionally it is a metaphor for social situations, sometimes minatory or premonitory. Thus "The Sunken World" (1928 *Amazing Stories Quarterly*) by Stanton A. Coblentz (1896-1982) depicted a glass-bowl Atlantis that embodied socialist perfection but was brought to ruin by cultural stasis and the human frailty of boredom. Laurence Manning's "Voice of Atlantis" (1934 *Wonder Stories*) was an antimechanistic plea for natural conservation and GOLDEN AGE values. The pacifist "The Third Vibrator" (1933 *Wonder Stories*) by John Beynon Harris (better known as John Wyndham; 1903-1969) describes cyclical cultural destructions, including Atlantis, as the result of the development of superweapons.

On occasion Atlantis cannot be kept down: it arises once again, with varied impact. In Jules Verne's "L'éternel Adam" (1910) the re-emergence of drowned Atlantis and the discovery of a dead high civilization shatter faith in progress and demonstrate both a cyclical history of mankind and the impermanence of civilization. In *The Crystal Button* (**1891**) by Chauncey Thomas (1822-1898) the rising of Atlantis, making available enormous supplies of gold, destroys monetary standards and helps usher in a new age. And Ursula K. LE GUIN's "The New Atlantis" (in *The New Atlantis* anth **1975** ed Robert Silverberg) shows Atlantis as a metaphor for a grinding totalitarianism.

Particularly strong in fiction is the emphasis on the magical aspect of Atlantis, both as a site of MAGIC and as itself a source of magic. Atlantis, as a glamour place, can become hope, perfection, aspiration, and promise. For example, in E.T.A. HOFFMANN's "Der goldne Topf" (1814) Atlantis is a wonderful dreamland, the home of ecstasy, the place of poesy, where a young would-be poet ventures spiritually. Socially, too, Atlantis has become the dream of something better. In *Young West* (**1894**) by Solomon Schindler (1842-1915), one of the SEQUELS BY OTHER HANDS to *Looking Backward, 2000-1887* (**1888**) by Edward Bellamy (1850-1898), the new Communist Boston is called Atlantis. Similarly, in *A.D. 2050, Electrical Development of Atlantis* (**1890**) by John Bachelder, another Bellamy epigone, a utopian community called Atlantis is established off the coast of California.

Despite its enormous significance in fantastic fiction, Atlantis has inspired relatively little in other arts. In music, Manuel de Falla (1876-1946) left an unfinished *Atlantida* for chorus and orchestra, and Darius Milhaud (1892-1974) composed the music for a ballet called *Atlantis* (performed 1964). There has been no good Atlantis movie; most are of the poor standard of *Atlantis, The Lost Continent* (**1961**) or *Warlords of Atlantis* (**1978**); the US tv series *Man from Atlantis* (1977-8) was similarly uninspiring.

Closely related to Atlantis, but without either its impressive history or its emotional appeal, are two poor relations, Lemuria and Mu (◊ LOST LANDS) that are likewise said to have been inundated. [EFB]

Further reading: *Lost Continents* (**1954**; rev 1970) by L.

Sprague de Camp; *The Search for Lost Worlds* (**1975**) by James Wellard; *Atlantis: Fact or Fiction?* (anth **1978**) ed Edwin S. Ramage; *The Guide to Supernatural Fiction* (**1983**), *Science-Fiction: The Early Years* (**1990**) and «Science-Fiction: The Gernsback Years» by Everett F. BLEILER; the video *Cousteau Odyssey 8: Calypso's Search for Atlantis* (*1978*) by Jacques Cousteau (1910-   ) has interesting sections on the excavation of Thēra.

ATOM MAN VS SUPERMAN (*1950*) US movie. ◊ SUPERMAN MOVIES.

ATOM TAISHI Japanese MANGA. ◊ ANIME.

ATTANASIO, A(LFRED) A(NGELO) (1951-   ) US writer, most noted for his sf, which he began to publish with his first story, "Once More, the Dream" as aa attanasio for *New Worlds Quarterly #7* (anth **1974**) ed Hilary Bailey and Charles Platt. His novels are generally sf – he is best-known for the **Radix Tetrad** sequence beginning with *Radix* (**1981**) – but the exorbitant intensity of his rendering of the visible world brings them at times close to MAGIC REALISM. Set mostly in the South Pacific in the early 17th century, *Wyvern* (**1988**), which is fantasy, follows the RITE OF PASSAGE into adulthood of a blond TARZAN-like *enfant sauvage* whom the "natives" of Borneo think is a DEMON and who is raised by a soul-catcher sorcerer to fulfil prophecies; in the event he becomes a PIRATE, shifting *en passant* from the realm of fantasy to the realm of historical romance. *Hunting the Ghost Dancer* (**1991**) is a PREHISTORIC FANTASY. *Kingdom of the Grail* (**1992**) treats the MATTER of Britain. *The Moon's Wife: A Hystery* (**1993**), a heavily cross-hatched CONTEMPORARY FANTASY (◊ CROSSHATCH) in which the masculine MOON woos a young woman, examines SEX, GENDER roles, and other preconceptions; her inability to determine whether or not she is hallucinating (◊ HALLUCINATION) her PERCEPTION of the Moon's courtship neatly exemplifies any definition of FANTASY which concentrates upon the UNCANNY. With *The Dragon and the Unicorn* (**1994** UK) and *Arthor* (**1995**) AAA began an ambitious Arthurian sequence (◊ ARTHUR), not visibly connected to *Kingdom of the Grail*. The first volume presents a complex cosmogony in which the UNICORN – a noncorporeal creature of the SUN – takes on bodily form only when caught in BONDAGE to the Earth. The figure of MERLIN gradually emerges as an AVATAR of a spirit or demon immemorially appalled (◊ TIME ABYSS) by matter. The second volume, which focuses on "Arthor", is bound more explicitly to traditional Arthurian material, though Arthor himself is an unusually brutal figure. Morgeu (i.e., MORGAN LE FAY) enters the picture, and there is much adversarial MAGIC, with Merlin opposing her.

A florid QUEST-like urgency permeates all of AAA's work; indeed, most of his protagonists are themselves driven by quest imperatives, usually the need to discover their own identities. The wide range of his work may have kept him from wide fame within the field, as may an erratic style (eloquence and bathos frequently abut within single paragraphs); there remains, however, a sense that this rich variousness will, sooner or later, become fully recognized; *The Dark Shore* (**1996**), set in a full-scale SECONDARY WORLD which rather resembles that of E.R. EDDISON's **Zimiamvia** books, is a move towards that recognition. [JC]

ATTEBERY, BRIAN (LEONARD) (1951-   ) US academic and writer whose first analysis of FANTASY, *The Fantasy Tradition in American Literature from Irving to Le Guin* (**1980**), is a sharply intelligent survey of the slow development of the form in the USA until the Frontier closed, the American Dream turned back on itself, and L. Frank BAUM created, in OZ, the first livable US OTHERWORLD. In *Strategies of Fantasy* (**1992**) BA suggests a description of fantasy central to some of the strategies employed in this encyclopedia. By describing fantasy as a "fuzzy set" – i.e., a grouping defined not by boundaries but by central examples – he identifies a significant difference between fantasy and sf. Sf can be treated as a field with boundaries, but fantasy moves outwards from its central examples into WATER MARGINS where clear boundaries do not exist. The critical acumen displayed in the book – both theoretically and in practical readings of authors like John CROWLEY and J.R.R. TOLKIEN – is very considerable. BA was given a Distinguished Scholarship Award at the 1991 International Conference on the Fantastic in the Arts (◊ INTERNATIONAL ASSOCIATION FOR THE FANTASTIC IN THE ARTS). [JC]

ATTWELL, MABEL LUCY (1879-1964) UK illustrator. ◊ ILLUSTRATION.

AUDREY ROSE US movie (*1977*). United Artists/Robert Wise. **Pr** Frank de FELITTA and Joe Wizan. **Dir** Wise. **Spfx** Henry Millar Jr. **Screenplay** de Felitta. **Based on** *Audrey Rose* (**1975**) by de Felitta. **Starring** John Beck (Bill Templeton), Anthony Hopkins (Elliot Hoover), Marsha Mason (Janice Templeton), Susan Swift (Ivy Templeton). 113 mins. Colour.

11 years ago Elliot Hoover's daughter Audrey Rose was burned alive in a car crash; now he believes her REINCARNATION is the Templetons' 11-year-old daughter Ivy, or at least that Ivy has been possessed by Audrey Rose's spirit (◊ POSSESSION), which has returned to corporeality too soon to have healed itself of the traumas of the agonizing death. Ivy is subject to nightmares and wild grief around each birthday; Hoover's presence calms her, but the attacks increase in severity and start to involve POLTERGEIST effects. A court case ensues; Ivy is subjected to regression hypnosis and, through reliving Audrey Rose's last moments, dies.

Often slated as an insufficiently frightening bandwagon HORROR MOVIE – it came after *The* EXORCIST (*1973*) and *The* OMEN (*1976*) – *AR* is in fact an excellent SUPERNATURAL FICTION; its portrait of supposed reincarnation cases is not especially sensationalized. Some ancillary elements are embarrassingly '70s, and Ivy's father, representing rationalism, is made too stupid to act as a proper counterargument, but overall this is a very rewarding movie. One brief sequence, as a child flees through the rain, is strongly reminiscent of DON'T LOOK NOW (*1973*). [JG]

AUEL, JEAN M(ARIE) (1936-   ) US writer of the PREHISTORIC-FANTASY **Earth's Children** sequence: *The Clan of the Cave Bear* (**1980**), *The Valley of Horses* (**1982**), *The Mammoth Hunters* (**1985**) and *The Plains of Passage* (**1990**). These long novels – telling the tale of a Cro-Magnon woman reared in a Neanderthal community – though rather flatly told, have been hugely popular. Part of the saga was filmed as *The* CLAN OF THE CAVE BEAR (*1985*). [JG]

AUGURS ◊ OMENS.

AULNOY, MADAME D' The usual appellation given to Marie-Cathérine le Jumel (?1650-1705) who, after her marriage in 1665 to François de la Motte, Baron d'Aulnoy (?1621-1700), became Comtesse d'Aulnoy. It was an unhappy marriage, and ended with the comtesse being imprisoned for plotting against her husband, but she escaped and led an adventurous life around Europe before settling in Paris in 1685. She established a literary *salon* that became all the rage, where women (and occasionally men) gathered for the discussion or art, science and politics. Often this discussion took the form of stories, using fiction as a form to satirize events at the court of Louis XIV. She herself incorporated elements from old FOLKTALES – some adapted from versions first incorporated into Giovanni Basile's *Pentamerone* (coll **1634-6**) – into what were ostensibly new stories, updating them so the characters and events related to current circumstances. Unlike her contemporary, Charles PERRAULT (a frequent visitor to her *salons*), who only occasionally used his FAIRYTALES for purposes of SATIRE, she made that her prime motive, with the result that, unlike Perrault's, her tales were composed primarily for adults – and were thus among the first LITERARY FAIRYTALES.

Her first was "L'île de la Félicité" ["The Isle of Happiness"] (incorporated in *L'Histoire d'Hippolyte, comte de Douglas* **1690**). This contains many of the standard motifs of FANTASY: a prince who is lost in a FOREST, but who is conveyed by Zephir to the ISLAND of PARADISE where he falls in love with a FAIRY. Time passes, and he finally discovers that the three months he has spent there have in fact been three centuries (◊ TIME IN FAERIE); at last, caught by Father Time, he dies. The story ends with a statement of two morals: "There is no avoiding Father Time; neither is there perfect happiness."

She eventually composed 25 such tales. These appeared in two books, each of four volumes: *Les contes de fées* ["Tales of the Fairies"] (vols 1-3 coll **1696-7**; vol 4 coll **1698**) and *Contes nouveaux, ou les fées à la mode* ["More Popular Fairy Tales"] (coll **1698**). Most stories are set either in a recognizable Middle Ages on which a supernatural world impinges/overlaps (◊ CROSSHATCH) or almost wholly in a SECONDARY WORLD of FAERIE. The tales are very strong on STORY, with a TWICE-TOLD element of timeless distance.

Although less known now, her stories contain all the basic PLOT DEVICES of FANTASY, and were highly influential in their day. "L'oiseau bleu" (1697; trans anon as "The Blue Bird" 1699) includes a wicked STEPMOTHER and tells of a handsome prince (in the English-language translation this saw the first use of the name "Prince Charming") who is transformed (◊ TRANSFORMATIONS) into a blue bird because of his love for the stepdaughter. "Le naine jaune" (1698; trans anon as "The Yellow Dwarf" 1721) tells of a princess who is betrothed to a hideous DWARF. Her beloved King of the Golden Mines, with the aid of a magic SWORD, undertakes a series of challenges to rescue her. He succeeds, but the dwarf gains the sword and kills the king, whereupon the princess dies of grief. "La chatte blanche" (1698; trans anon as "The White Cat" 1721) concerns a king, fearful he will be deposed by his sons, who sets them a series of tasks; the youngest son is always triumphant. The story involves various ENCHANTMENTS, including one of INVISIBILITY and the transformation of a princess into the eponymous white CAT. "Finette Cendron" (1698; trans anon as "Finetta the Cinder-Girl" 1721) is MD'A's version of CINDERELLA.

Her stories are usually much longer narrative constructions than the fairytales by Perrault or the GRIMM BROTHERS, and this fact has made them less memorable, despite their position among the earliest original fantasies. It was, however, with the translation of a few of her stories into English as *Tales of the Fairys* (coll trans **1699** UK), that the term "fairytales" passed into the language.   [MA]

AURA Central to occult beliefs is the doctrine that every natural thing emanates a kind of sphere of light-emitting "vital energy", normally called the aura. The doctrine takes some of its potency from the fact that, as long ago as in ancient EGYPT, the existence of something that might have been similar to what modern occultists call the aura had been speculated upon. In 15th-century Europe PARACELSUS brought the subject up. Towards the end of the 19th century, it became a popular manifestation in SUPERNATURAL FICTION, and was readily perceivable by OCCULT DETECTIVES. The aura should be distinguished from the apparent energy field – possibly an illusion – perceived by Kirlian photographs as surrounding any living tissue.   [JC]

See also: ASTRAL BODY.

AUSTIN, WILLIAM (1788-1841) US attorney and writer, remembered almost exclusively for one story, *Peter Rugg, the Missing Man* (1824 *New England Galaxy*; exp **1882** chap). Years before the narrator of the tale begins to speak of his experiences with the eponymous ACCURSED WANDERER, Rugg forswears friendship and hubristically defies the elements by declaring that "I will see home to-night, in spite of the last tempest, or may I never see home!" Half a century later (as the narrator records in the complex web of letters which makes up the text) Peter Rugg is still trying to reach Boston, in vain. When finally allowed to reach his destination, he discovers how the years have passed as though he had been enchanted (◊ TIME IN FAERIE), and that he is cursed to wander forever. As a presentation of the sense of BELATEDNESS so characteristic of US fiction in general, and the AMERICAN GOTHIC strand in particular, the tale is exemplary. WA wrote other short stories – all assembled in *Literary Papers of William Austin* (coll **1890**) – of which "The Man with the Cloaks: A Vermont Legend" (1836 *American Monthly Magazine*) is of interest for its TALL-TALE portrayal of a miser forced, because of a lack of charity, to put an extra cloak on every day to fight the supernatural chill. Finally, he is allowed to begin to remove them, constraining spirits evaporate from BONDAGE, and the miser becomes truly human at last.   [JC]

AUSTRALIA Australian 20th-century fantasy can be divided into two broad categories: stories of LOST RACES and worlds, and HEROIC FANTASY, with additional contributions from such fields as CHILDREN'S FANTASY and TALL TALES.

While many Australian lost-race tales of the 19th and early 20th centuries were moderately successful, they generally contained little that was original. Even when original works did begin to appear in the 20th century, Australia lacked a sufficiently large and sympathetic publishing industry to do them justice. "The Social Code" (1909 *The Lone Hand*), a short romance by Erle Cox (1873-1950), described a long-distance affair between a Martian girl and an astronomer from Earth; appearing three years before Edgar Rice BURROUGHS's first **Barsoom** novel, and with a similar feel, this might have achieved the same worldwide success had it been launched in a publishing industry as vital as that of the USA. Cox later wrote an early classic of Australian sf, *Out of the Silence* (1919 *The Argus*; **1925**; rev 1947), but he had to publish the first book edition himself.

Australian children's fantasy had no such problems at this time, although it integrated local settings and fauna so heavily that it cannot be described as classic fantasy. Two of the best-known examples, Norman LINDSAY's *The Magic Pudding* (**1918**) and *Snugglepot and Cuddlepie* (**1918**) by May Gibbs (1877-1969) were published just before *Out of the Silence*, but they have more in common with Beatrix POTTER than T.H. WHITE. Throughout the interbellum years the more specifically adult lost-race fantasy adventure remained dominant in Australia, eventually fading from the scene by the 1960s.

Phil Collas (real name Felix Edward Collas; 1907-1989) bypassed the problem of the local market, publishing *The Inner Domain* (1935 *Amazing*; **1989** chap) in the USA. The setting is initially local, featuring contemporary explorers in central Australia who are teleported into an Aboriginal civilization many miles underground. "The Reign of the Reptiles" (1935 *Wonder Stories*) by the Sydney teenager Allan Connell (1916-) was another early US publication by an Australian. Here explorers go back by time machine to when intelligent reptiles succoured an advanced level of civilization.

Five years after these stories Australia was at war, and subject to shipping restrictions. Government import bans included overseas magazines, so sales of locally written fiction boomed. Currawong Publishing accepted a three-part series of novellas written by Connell in the mid-1930s, the **Serpent Land** sequence – *Lords of Serpent Land* (**1945** chap), *Prisoners in Serpent Land* (**1945** chap), whose cover reads "Prisoners of Serpent Land", and *Warriors of Serpent Land* (**1945** chap). These were adventure fantasies set in an unexplored part of South America. While there was nothing very original about them, this was quite reasonable escapism for older teenagers.

Following WWII the great genre artist Stanley Pitt and the writer Frank Ashley invented a TARZAN-style character named **Yarmak**, who featured in what was probably the country's first fantasy COMIC series. Soon after, the sf author Eric North (real name Charles Bernard Cronin; 1884-1968) had *The Ant Men* (**1955** US) published in the USA; as with Connell's first story, explorers are sent back in time to a civilization in Central Australia where, in this case, a frog-monarch presides over giant ants. Although this was republished several times, the Australian lost-race subgenre was in decline, and the success was not repeated.

The 1970s were the great watershed for all of Australian fantasy's other strands. Mainstream author Peter CAREY brought the TALL TALE to a highpoint, with his part-fantasy collections winning mainstream AWARDS. Patricia WRIGHTSON's Australian-based fantasies for young readers won awards and inspired other authors to follow her lead. Even more significantly, from the mid-1970s Australian HEROIC FANTASY finally appeared in a clearly recognizable form. In 1974, cartoonist Roger Fletcher began the series **Orn**, about a warrior who rode an eagle, and in June 1976 this became **Torkan** (*Sunday Telegraph*); Torkan, very much in the CONAN mould, proved popular and ran for years. In 1979 the local magazine *Crux* began publishing Jay Hoffman's satirical Conan-in-space comic strip **Horg**, which was nominated for Australia's sf/fantasy genre AWARD, the Ditmar, in 1981. Better described as a fictional historical textbook than as a novel, *Australian Gnomes* (**1979**) by Robert Ingpen (1936-), won the Ditmar in 1980.

In 1975 Keith TAYLOR's first story, "Fugitives in Winter" as by Dennis More, appeared in *Fantastic Stories* (◊ FANTASTIC). This was the first of four tales set in Dark-Age Britain. Taylor dropped his pseudonym when his story "When Silence Rules" was published in *Distant Worlds* (anth **1981**) ed Paul Collins. This story later won a Ditmar. His "Dennis More" stories were later published by Ace Books as *Bard* (fixup **1981** US); this novel was nominated for the Ditmar in 1982 and subsequently grew into a five-part series, the third

in the series winning a Ditmar in 1987 and thereby becoming the first and so far only heroic fantasy to do so.

Paul Collins (1954-   ) – author, publisher and editor – created the first Australian market for heroic fantasy through the various incarnations of his publishing company Void Press. He began his own writing career with heroic fantasy, and his first two stories appeared in the US magazine WEIRDBOOK in 1977. Apart from publishing and editing the landmark **Worlds** series of anthologies, he was the first to publish heroic-fantasy novels locally – Taylor's second novel, *Lances of Nengesdul* (**1982**), and *The Tempting of the Witch King* (**1983**) by Russell Blackford (1954-   ), both of which were Ditmar nominees. In the early 1980s Australian heroic fantasy was gaining popularity. Of the Ditmar nominations for fantasy and sf during 1981-90, a quarter of the novels nominated were identifiably fantasy, and a fifth of these nominees subsequently won. Thus Australian genre fans apparently enjoyed local fantasy, even though local commercial publishers still ignored the field. Meanwhile other Australian heroic-fantasy authors were making progress. After winning the 1984 Ditmar for short fiction with "Above Atlas his Shoulders", Andrew Whitmore (1955-   ) saw his *The Fortress of Eternity* (**1990** US) published by Avon, and this work gained him another Ditmar nomination. The successful books of Terry Dowling (1947-   ) often read as marginal fantasy but are substantially sf (◊ *SFE*). Dowling has populated his futuristic inland Australia with an exotic people and fauna, playing on the real Australian landscape's striking similarities with the old-time genre image of Mars.

In 1990 Pan Australia published *Circle of Light* (**1990**) by the then unknown Martin Middleton (1954-   ). This boasted good genre cover art, and the cover blurb sold it unambiguously as heroic fantasy. As fantasy it was not groundbreaking, but its theme and style were of a type already proven successful in the Australian market by overseas authors. Within three years its sales had shattered existing local genre records. A sequel in his **Chronicles of the Custodians** sequence, *Sphere of Influence* (**1992**), soon appeared. Local critics were stunned by his success. The venture showed that Australians would buy local fantasy just as readily as that from overseas. Subsequent epics from Pan Macmillan have enjoyed similar levels of success: they include the **Andrakis** sequence (**1992-3**) by Tony Shillitoe (1955-   ) and *Zenith* (**1993**) by Dirk Strasser (1959-   ).

The Ditmar nominations during 1991-4, however, showed some reversals for fantasy. In short fiction there were no fantasy nominations, and just 15% of the nominated novels were fantasy (with no winners); this ratio is down on the 1980s, possibly because the fantasy publishing boom of the early 1990s coincided with a spectacular increase in both the quality and the quantity of Australian sf.

Australian fantasy is not confined to books. Strategic Studies Group, a Sydney-based company, has sold over 50,000 copies each of its computer strategy GAMES *Warlords* and *Warlords 2*, both of which have fantasy scenarios. In dollar terms this is the most successful and profitable fantasy venture ever to come out of Australia.

CHILDREN'S FANTASY experienced a boom in Australia far earlier than its adult counterpart. Wrightson began writing in the mid-1960s, had developed a formidable reputation by the mid-1970s, and has since consolidated her position as one of the top Australian authors in this area. Her **Wirrun** sequence (**1977-81**) has been among her most successful, the first title, *The Ice is Coming* (**1977** US), winning the Children's Book of the Year Award. *The Nargun and the Stars* (**1973**) was dramatized for radio and filmed; it, too, won the Children's Book of the Year Award, and a special edition won two Hans Christian Andersen Medals in 1986. Victor KELLEHER has written several successful fantasy novels for young readers and won several awards. *Master of the Grove* (**1982** UK) won the Children's Book of the Year Award, and *Brother Night* (**1990** UK) won a Children's Book of the Year Honour Award. Kelleher's fantasy generally has a less Australian flavour than Wrightson's, but it contains powerful moral and social messages.

Isobelle Carmody (1958-   ) is one of the newer fantasy authors for the YA readership. Her **Obernewtyn Chronicles** sequence (**1987-90**) is of interest, and her fantasy novel *The Gathering* (**1993**) tied as winner of the Children's Book Council Book of the Year Award.

In the 1990s, women finally made it into the ranks of the adult fantasy novelists on a significant scale. Carmody is now aiming her very popular novels at the adult market, even though it is still among the YA readership that most of her audience is concentrated. Bendigo author Sara Douglass (real name Sara Warneke; 1958-   ) has published *BattleAxe* (**1995**) and *Enchanter* (**1996**). In Perth, Shannah Jay (real name Sherry-Anne Jacobs; 1941-   ) has now had four fantasy novels published by Pan Macmillan.

Australian fantasy's greatest triumph has been to prove that local genre literature – aside from crime and romance – could make a comfortable profit; locally written sf and horror are in consequence also benefiting. The blockbuster mode of fantasy novel that first appeared in 1990 is now taken for granted, and readers expect that at least a dozen adult fantasy novels will be published in any given year.

The first high-fantasy anthology by Australian authors, «Dreamweavers» (anth 1996) ed Paul Collins, possibly heralds a resurgence in local short fantasy. Australia's first award specifically for fantasy – in the form of a "Best Fantasy" category, instigated 1995, of the Aurealis AWARDS – resulted in a Best Novel award to *Sabriel* (**1995**) by Garth Nix (1963-   ); the Best Short Fiction award went to "Harvest Bay" (1995 *Eidolon 19*) by Karen Attard.

Australian fantasy's final, and so far unattained, frontier is recognition among the big international awards; so far it has failed to achieve the nominations and wins that local sf has brought home over the past decade. Given the strength and potential for growth of Australian fantasy, however, this situation may be rectified soon.   [SM/SP]

AUSTRALIAN HORROR AND FANTASY MAGAZINE, THE ◊ MAGAZINES.

AUSTRIA Austrian fantasy, like Austrian literature in general, is inextricably linked to German fantasy (◊ GERMANY). Nevertheless, there has been no lack of attempts to define a specific "Austrianness"; in fantasy it might be said to manifest itself in a predominance of the baroque, grotesque, love of paradox, imaginary states, morbid LOVE, decay and death.

In the 18th century there was a spate of popular ghost novels (◊ GHOST STORIES), much as in the German states – some were imitations of the popular German writer Christian Heinrich Spiess (1755-1799). The 19th century saw the successful Viennese magic plays of Ferdinand Raimund (1790-1836) – e.g., *Der Diamant Des Geisterkönigs* ["The Diamond of the Spirit King"] (**1824**) – and Johann Nestroy (1801-1862) – e.g., *Lumpazivagabundus* (**1833**). Ghostly incidents and episodes are also to be found in the works of

Austria's national icon, the dramatist Franz Grillparzer (1791-1872). An isolated writer of often debunking humorous GHOST STORIES was Carl Stugau (1816-1890), with *Unbegreifliche Geschichten* ["Incomprehensible Tales"] (**1863-4**), but Austrian fantasy proper originated with the FIN DE SIÈCLE flowering of the arts and continued to develop until about 1930, when the rise of Nazism stifled it.

During these years fantasy became an important branch of Austrian literature – especially if one includes the German-language writers from Bohemia and Moravia. Of these writers of fantastic literature, Franz KAFKA is the only one firmly to have entered the mainstream of world literature; at the time, his voice was one among many. Unlike other writers of his day he had no love for the occult or supernatural; instead, he transformed the everyday world into an icy place ruled by anonymous bureacratic forces. Only a few stories – including the famous *Die Verwandlung* (1916; trans as "Metamorphosis") – appeared during his lifetime. His approximate contemporary, Gustav MEYRINK, became famous with *Der Golem* ["The Golem"] (**1915**), the Number One bestseller in Germany during WWI (◊ GOLEM). One of the reasons for Meyrink's continuing popularity, apart from his literary competence, is that many see him less as a fantasy writer than as a teacher of arcane wisdom. In many of his later novels the occult became more pronounced, but his complex *Der Engel vom Westlichen Fenster* (**1927**; trans as *The Angel of the West Window* **1991**) is centred on the Elizabethan magician John DEE.

A writer in a similar occult vein was Franz Spunda (1890-1963), who wrote theosophic novels (◊ THEOSOPHY) charged with a rather crude sexuality: *Devachan* (**1921**), *Der gelbe und der weiße Papst* ["The Yellow and the White Pope"] (**1923**), *Das ägyptische Totenbuch* ["The Egyptian Book of the Dead"] (**1924**) and *Baphomet* (**1928**); the last of these is about a scion of the last Emperor of Byzantium fighting satanic Knights Templar.

One of the "Three Musketeers" of the German fantastic revival after 1900 – the other two were Meyrink and Hanns Heinz EWERS – was Karl Hans Strobl (1877-1946), a rabid Czech-hater and German nationalist from Moravia. In later years he became a high official in the Nazi writers' organization, but earlier he did much for fantasy as a writer, editor and reviewer. In 1919-21 he edited the beautifully illustrated German fantasy magazine *Der Orchideengarten* ["The Orchid Garden"], the first such MAGAZINE in the world. His novels, such as *Eleagabal Kuperus* (**1910**), are marred by his openly propagandistic dislike for materialism and socialism, but in his shorter work he showed himself an able writer and he covered the whole spectrum of fantastic themes. His stories are collected in *Die Eingebungen des Arphaxat* ["The Inspirations of Arphaxat"] (coll **1904**), *Die knöcherne Hand* ["The Bone Hand"] (coll **1911**), *Lemuria* (coll **1917**) and others.

Alfred Kubin (1877-1959) was the leading German artist and illustrator of HORROR-fantasy, but he also wrote a novel that, in its sexual symbolism, catastrophism and exotic imagery, came to be seen as prophetic of the horrors to come: *Die andere Seite* (**1909**; trans Denver Lindley as *The Other Side* **1967** US). The narrator, the author's *alter ego*, visits an eccentric UTOPIA, the Traumreich ("Dream Realm"), founded somewhere in Asia by a friend of his youth, and witnesses the downfall and phantasmagoric destruction of it and its capital.

Quite the opposite of such writers as Meyrink and Spunda

was Leo PERUTZ, a Jewish writer from Prague; in the last decade his work has enjoyed a revival. Influenced by him was the aristocratic Alexander Lernet-Holenia (1897-1976), whose masterpiece is a fantasy about death, the novella "Der Baron Bagge" (1936; trans Jane B. Greene 1956); in translation it was collected in *Count Luna: Two Tales of the Real and Unreal* (coll **1956**) along with "Graf Luna" (1955; trans Richard and Clara Winston 1956).

Perutz's contemporary and perhaps rival Otto Soyka (1882-1955), once a bestseller, is now completely forgotten. His theme was power – the power of individuals over others through the use of psychic forces, suggestion or chemistry; examples are *Die Traumpeitsche* ["The Dream-Whip"] (**1921**) and *Der Seelenschmied* ["The Soul-Smith"] (**1931**). In *Eva Morsini, die Frau, die war* ["Eva Morsini – The Woman that Was"] (**1923**) a modern woman is apparently possessed (◊ POSSESSION) by the spirit of Catherine the Great. Paul Busson (1873-1924) wrote an excellent historical novel of REINCARNATION in *Die Wiedergeburt des Melchior Dronte* (**1921**; trans as *The Man who was Born Again* **1927**) and some collections of "curious tales", the best of which is *Seltsame Geschichten* ["Strange Tales"] (**1925**).

Perhaps the most typically Austrian fantasy writer was the scurrilous Fritz von Herzmanovsky-Orlando (1877-1954). Almost unpublished during his lifetime, his work achieved literary success in 1958-63, when the eminent critic Friedrich Torberg (1908-1979) brought out a heavily edited four-volume collection of his works, stressing the comic element and editing out the esoteric and mythological bits. Since then Herzmanovsky-Orlando's collected works have been released in 10 volumes, with the original texts restored. The only novel by Herzmanovsky-Orlando to be published during his lifetime, *Der Gaulschreck im Rosennetz* ["A Scare-nag in the Rosy Net"] (**1928**), tells of a lowly bureaucrat who wants to give his emperor as a birthday present a necklace made out of the milk teeth of virgins. Echoes of Herzmanovsky-Orlando are to be found in the works of Peter Marginter (1934-   ) – *Der Baron und die Fische* ["The Baron and the Fishes"] (**1966**; rev 1980) and many others – and Herbert Rosendorfer (1934-   ), especially in his first novel, *Der Ruinenbaumeister* (**1969**; trans Mike Mitchell as *The Architect of Ruins* **1992** UK), a convoluted work made up of stories within stories in the manner of Count Jan POTOCKI's *The Saragossa Manuscript* (**1804** and **1805**; trans **1960**).

Two singular and forgotten novels are the huge *Freinacht* ["Extra Night"] (**1935**) by Julius Pupp (1886-1974), a journey into a fantasy world of SPIRITS where the hero, accompanied by Cyrano de Bergerac (1619-1655), has adventures in many historical societies from Ancient Egypt to Babylon; and *Hans Adam Löwenmacht* (**1938**) by Rudolf Slawitschek (1880-1945), a panoramic journey through a baroque Bohemia, full of MAGIC lore and a plethora of fantastic beings – a wonderfully inventive romp.

Many other stories of the fantastic were contributed by Austrian literary figures such as Franz Werfel (1890-1945), Arthur Schnitzler (1862-1931) and Max Brod (1884-1968). Sometimes these tales were directly influenced by psychoanalysis; in turn, the early Viennese psychoanalysts – like Wilhelm Reich (1897-1957) – often commented favourably on fantasy writers. A GENRE-FANTASY writer much valued by some connoisseurs was the now almost completely forgotten Leonhard Stein.

Aside from Peter Marginter, several other writers on the current Austrian literary scene are fantasists. Daniel Wolfkind (real name Peter Vujica; 1937-   ) has written two volumes of surrealist (◊ SURREALISM) short stories, the first being *Mondnacht* ["Lunar Night"] (coll **1972**). Barbara Büchner (1950-   ) has written some material of note. The dark, brooding and sublimely erotic short stories of Barbara Neuwirth (1958-   ) have been published in *In den Gärten der Nacht* ["In the Gardens of Night"] (coll **1990**) and *Dunkler Fluß des Lebens* ["Dark River of Life"] (coll **1992**). Christoph RANSMAYR (1954-   ) wrote what proved an international bestseller: his Ovidean fantasy *Die letzte Welt* (**1988**; trans as *The Last World* **1990** US). The most consistently fantastic has been H.C. Artmann (1921-   ), a poet, playwright, author and translator who likes to play with language and elements of genre literature like the old Dime Novels. His forays into fantasy include PARODIES of DRACULA – *dracula, dracula* (**1966**) – and FRANKENSTEIN – *Frankenstein in Sussex* (**1969**); his *tök ph'rong süleng* is a WEREWOLF story.

Representative anthologies are Jean Goyory's *Phantastisches Österreich* ["Fantastic Austria"] (anth **1976**), a somewhat different version of an anthology first published in French in Belgium as *L'Autriche fantastique avant et après Kafka* (anth **1976**), and *The Dedalus/Ariadne Book of Austrian Fantasy: The Meyrink Years 1890-1930* (**1992** UK/US) ed and trans Mike Mitchell; the latter includes representative works by Meyrink, Strobl, Herzmanovsky-Orlando, Perutz, Schnitzler, Busson and many others. [FR]

AUTOMATA If an automaton is defined as a clockwork creature that lacks self-motivation or free will, we can understand the paucity of examples in fantasy. But the giant metal figure in L. Frank BAUM's *Ozma of Oz* (**1907**), whose axe crushes anyone attempting to gain access to the UNDERGROUND kingdom of the Nome King, is a fantasy automaton. E.T.A. HOFFMANN's "The Automatons" (1817) and "The Sandman" (1817), which inspired scenes in the OPERA *Tales of Hoffmann* (**1881**) by Jacques Offenbach (1819-1880), both deal with automata, and several of the protagonists of Robert AICKMAN's later stories suffer such BONDAGE – particularly perhaps in "Meeting Mr Millar" (1972). A modern cognate of the automaton is the TOON. [JC]

AUTOMOBILES ◊ CARS.

AUTUMN ◊ SEASONS.

AVALON The Isle of Avalon was the final resting place of ARTHUR, to which he was taken to recover after the Battle of Camlann. The etymology of the name is curious, and may result from confusion on the part of GEOFFREY OF MONMOUTH, who refers to the island in his *Historia Regum Britanniae* (**1136**) and *Vita Merlini* (?**1155**) variously as Insula Pomorum or the "Isle of Apples (also called Fortunate)" and Ynys Avallach or Insula Avallonis. There are many tales of ISLANDS to the west of Europe believed to be like Paradise, and in Geoffrey's time the adventures of the Irish St Brendan (?486-?575) as recorded in the ROMANCE *Navagatio Sancti Brendani Abbatis* ["The Voyage of St Brendan"] (before 10th century) were highly popular. Geoffrey may have linked this to the Celtic king Avalloch (known as Evelake in the French romances) – who ruled an island to the west and was, according to some versions, the father of MORGAN LE FAY – as he makes Avalon the home of Morgan. Morgan is regarded here as a beneficent priestess, head of a sisterhood of nine virgins (◊ VIRGINITY), and indeed there was such a sisterhood on the Île de Seine, off

Brittany, in the 1st century AD. Avalon has become closely associated with Glastonbury – the Tor was a marsh-surrounded island in Arthur's day – but it has also been linked with the Isle of Arran, the Isle of Man, Ireland, various islands off the Irish coast, and even with Sicily. Avalon is also linked with the Annwn of Celtic legend, particularly in the poems "*Immram Bran*" ["Voyage of Bran"] (?8th century) and "*Preiddeu Annwfyn*" ["The Spoils of Annwn"] (?**900**AD), where it is an OTHERWORLD which drifts in and out of our REALITY. Avalon thereby came to represent a HEAVEN or Paradise, but one from which heroes could return after being healed. Ogier the Dane is taken to Avalon by Morgan in the anonymous 13th-century *chanson de geste* called *Ogier le Danois*. By the 19th century, with the flowering of the English romantics, Avalon had come to be regarded as an Earthly Paradise and a place of escape from an increasingly polluted and industrialized UK. This was how it was depicted by William MORRIS in his sequence of poems *The Earthly Paradise* (**1868-70** 3 vols).

The name Avalon projects a timeless image of an existence on Earth but not of this world. It is in this sense that it often appears as the title of Arthurian books, especially *Avalon* (**1965**) by Anya Seton (1916-   ), where descendants of Arthur bring peace and tranquility to 10th-century Cornwall, and *The Mists of Avalon* (**1982**) by Marion Zimmer BRADLEY, where the word is also representative of the earthly Glastonbury. Few writers have attempted to explore the mythical Avalon, although Andre NORTON takes three children there in *Steel Magic* (**1965**; vt *Grey Magic* 1967), which is based more on the Celtic Annwn.   [MA]
See also: LOST LANDS AND CONTINENTS.

AVATAR Term used in the Vedic myths incorporated into Hindu religion to denote the various incarnations of the gods, particularly Brahma and Vishnu. It is sometimes generalized by commentators to refer to any temporary incarnation of a deity, particularly in the context of syncretizing theories which regard all similar myth-figures as aspects of the same root-image. Such theories underlie a good deal of fantasy, resulting in such "avatar stories" as Edgar JEPSON's *The Horned Shepherd* (**1904**). The term's use in "Avatar" (1856) by Théophile GAUTIER and *The Avatars: A Futurist Fantasy* (**1933**) by A.E. is more obviously metaphorical.   [BS]

AVENGERS, THE UK tv series (1961-9). ABC. **Pr** John Bryce, Brian Clemens, Albert Fennell, Leonard White, Julian Wintle. **Exec pr** Fennell, Gordon L.T. Scott, Wintle. **Dir** Robert Asher, Ray Austen, Bill Bain, Roy Baker, Charles Crichton, Robert Day, Peter DuFell, Gordon Flemyng, Cyril Frankel, Robert Fuest, Richmond Harding, Sidney Hayers, James Hill, John Hough, Roger Jenkins, Quentin Lawrence, Don Leaver, John Moxey, Leslie Norman, Gerry O'Hara, Cliff Owen, Peter Scott, Don Sharp, Peter Sykes. **Writers** Geoffrey Bellman, Clemens, Terence Dicks, Terence Feely, Malcolm Hulke, James Mitchell, Terry Nation, Dennis Spooner, John Whitney, Martin Woodhouse and many others. **Created by** Sydney Newman, Leonard White. **Novelizations** *The Avengers* * (**1963**) by Douglas Enefer; *The Avengers: Deadline* * (**1965**) and *The Avengers: Dead Duck* * (**1966**) both by Patrick Macnee and Peter Leslie; *The Avengers #1: The Floating Game* * (**1967**), *#2: The Laugh Was on Lazarus* * (**1967**), *#3: The Passing of Gloria Munday* * (**1967**) and *#4* * (**1967**), all four by John Garforth, *#5: The Afrit Affair* * (**1968**), *#6: The Drowned Queen* * (**1968**) and *#7: The Gold*

Bomb (**1968**), all three by Keith Laumer, *#8: The Magnetic Man* * (**1968**) and *#9: Moon Express* * (**1969**), both by Norman Daniels; *The Saga of Happy Valley* * (**1980**) by Geoff Barlow (an unauthorized but tolerated novel, with names changed to Steade and Peale); *John Steed – An Authorized Biography: Volume 1, Jealous in Honour* * (**1977**) by Tim Heald (first volume of an immediately aborted series); *Chapeau Melon et Bottes de Cuir* * (graph **1990** France) by Alain Carrazé and Jean-Luc Putheaud; *Too Many Targets* * (**1993**) by John Peel and Dave Rogers; *The New Avengers #1: House of Cards* * (**1976**) by Peter Cave, *#2: The Eagle's Nest* * (**1976**) by John Carter, *#3: To Catch a Rat* * (**1976**) by Walter Harris, *#4: Fighting Man* * (**1977**) by Justin Cartwright, *#5: Last of the Cybernauts* * (**1977**) by Cave and *#6: Hostage* * (**1977**) by Cave. **Starring** Honor Blackman (Cathy [Catherine] Gale), Ian Hendry (Dr David Keel), Gareth Hunt (Mike Gambit), Joanna Lumley (Purdey), Patrick Macnee (John [Jonathan] Steed), Patrick Newell (Mother), Diana Rigg (Emma Peel), Julie Stevens (Venus Smith), Linda Thorson (Tara King). 161 50min episodes in 6 seasons (season 1 26 episodes, season 2 26 episodes, season 3 26 episodes, season 4 26 episodes, season 5 25 episodes, season 6 32 episodes) plus 26 50min episodes of *The New Avengers*. B/w (seasons 1-4) and colour (seasons 5-6 and *The New Avengers*).

One of the most successful tv fantasy series of all time, this had its origins in the perfectly mundane crime series *Police Surgeon* (1960), which starred Ian Hendry as police surgeon cum detective Dr David Keel. The first series of *TA* was a direct continuation of this but with the format altered: episodes were now 50 mins long (as opposed to *Police Surgeon*'s 25min episodes) and a new character was introduced: secret agent John Steed. With Steed came the first whiff of fantasy and, when Hendry left, Macnee took over as the central character, being given an attractive – but deadly – female sidekick in the shape of widow Cathy Gale. (In a few early episodes Gale was one of two alternating sidekicks, the other being Venus Smith.) Gale departed at the end of season 3 and was replaced by Emma Peel, another widow. Season 6 – the last season of *TA* proper – saw the replacement of Peel (her husband had been discovered miraculously alive) by Tara King, a far younger woman than her predecessors, and not a widow. This change in nature of Steed's sidekick destroyed an integral part of *TA*'s appeal – that the sidekick was a female who was in no way subservient to Steed – and the show finished. A revival was attempted in 1976-7 with the UK/Canadian/French coproduction *The New Avengers* (pr Clemens and Fennell), with Macnee effectively backstaged – he was now in his mid-50s – to look on paternally as a younger duo, Purdey and Mike Gambit, performed most of the action. It would be tempting to say that the revival failed because *TA* was a product of the 1960s ill suited to the 1970s, but this is belied by the continuing popularity of repeat screenings of the earlier series. The truth is that *TA*'s success – both critically and popularly – relied on four factors: its imaginative SURREALISM (i.e., its fantasy); the style of Steed; the style of Gale and later, most especially, Peel; and the frisson of sexuality attached to the relationship between Steed and, in succession, these two highly independent women. Once any one of these qualities was removed, the rest foundered.

It is, therefore, reasonable to divide *TA*'s history into four parts: the pre-Gale years, which have no real fantasy interest; the Gale Years; the Peel years; and afterwards. It is no coincidence that the Peel years saw *TA*'s greatest popularity – when "Avengers" outfits, in imitation of Peel's, were manufactured and eagerly bought – for this was the period when the show's fantasy was at its height.

Increasingly during the Gale years the plots became more fantasticated – even surreal. Often the fantasy involved was the kind of TECHNOFANTASY that had been made popular by the **James Bond** movies and the tv series *The Man from U.N.C.L.E.* (1964-8): there is an evil conspiracy to destroy the British Way of Life, if not the world, using some implausible and despicable scientific device. This trend reached its peak during the Peel years. But the fantasy involved not just the plotting: far more impressive – again, especially during the Peel years – was the ambience, for *TA* became a series that was no longer set in the real world, and quite often stressed the artificiality of the locale where its events were taking place (e.g., by using overt stage flats in place of real buildings). In this OTHERWORLD there was no general population – merely characters important in some way to the plot – no grime, no bloodshed (indeed, no real violence), and in the end no one who was not as uppercrust as Steed and Peel themselves. In such a milieu it seemed by no means preposterous that anything and everything could happen to the plot – and generally did: one's disbelief had already been suspended.

The final series of *TA* never achieved this. Because of the balance of seniority/juniority between the characters of Steed and King, their relationship had inevitably to become a real one rather than the fantasy of that between Steed and Peel. This umbilicus to the real world led to a shattering of the illusion. A new character, Mother (Patrick Newell), Steed's improbably fat, wheelchair-bound boss, was introduced in an effort to recapture some sense of unlikelihood, but giving the mysterious organization for which Steel worked any face at all solidified what had been best left misty. Some desperate plotting attempts were made to fantasticate events as compensation for the loss of fantasticated environment (e.g., a briefing with Mother might be, for no conceivable reason, in a mini-submarine beneath the Thames), but such effects – which might have appeared reasonable in earlier years – now seemed just silly. Although *TA* was still doing moderately well in the ratings, it was wisely decided to let it close.   [JG]

Further reading: *The Complete Avengers* * (**1989**; exp vt *The Ultimate Avengers* **1995**) by Dave Rogers; *The Avengers Program Guide* * (**1994**) by Paul Cornell, Martin Day and Keith Topping.

See also: ADAM ADAMANT LIVES! (1966-7).

AVERY, TEX Working name from 1941 of US animator (and occasional voice actor) Frederick Bean Avery (1907-1980); earlier he was credited as Fred Avery. TA is best-known for his formative work on BUGS BUNNY (whom he also gave the catchphrase "What's up, Doc?"); other characters he created or radically developed include Chilly Willy, Daffy Duck, Droopy, Screwy Squirrel (aka Screwball Squirrel) and, based on the characters George and Lenny in John Steinbeck's *Of Mice and Men* (**1937**), George & Junior. Most of his significant career was over by 1956; thereafter he spent over two decades making commercials before spending the last year or so of his life with HANNA-BARBERA. But his earlier career – at Universal/Walter LANTZ (1930-35), Warner Bros. (1936-41), Paramount (1942) and most notably MGM (1942-55) – glittered. More than any other animator he was responsible for that stream of ANIMATED

MOVIES which glorified in eschewing realism for the sake of SURREALISM; because his approach did not require painstaking detail-work and hence colossal financial resources, it is arguable he had a far greater influence on the subsequent development of the animated short than, for example, the DISNEY studio. Frequently, too, his gag-packed cartoons contained jokes designed specifically for adults: the title of *Red Hot Riding Hood* (1943) is sufficient exemplification (the wolf in this short is briefly homaged in *The* MASK [1994]). Above all, he championed animation as a medium in itself rather than an illusion of life: in his cartoons events generally occurred within a LANDSCAPE bounded by four straight lines, the characters being often perfectly aware they were animators' creations.   [JG]

AVI Pseudonym of US writer Edward Irving Wortis (1937-), whose fantasy and HORROR tales, mostly for older children, include: *No More Magic* (**1975**); *Emily Upham's Revenge, or How Deadwood Dick Saved the Banker's Niece: A Massachusetts Adventure* (**1978**), which is humorous; *Devil's Race* (**1984**), a tale of POSSESSION; *Bright Shadow* (**1985**); *Something Upstairs* (**1988**), a GHOST STORY; and *The Man who was Poe* (**1989**), which features Edgar Allan POE's encounter – in the role of detective – with a cemetery ghost.   [JC]

AVON FANTASY READER US digest ANTHOLOGY, 18 vols, quarterly, 1947-52, published by Avon Publishing, New York; ed Donald A. WOLLHEIM.

AFR was the first regular paperback reprint fantasy anthology series, though its digest format has often caused it to be classified as a MAGAZINE. Wollheim selected primarily from WEIRD TALES, emphasizing the works of H.P. LOVECRAFT and Clark Ashton SMITH, and working in tandem with August W. DERLETH at ARKHAM HOUSE to promote the best of *WT*'s authors. Wollheim also reprinted stories from STRANGE TALES, *Adventure* and *Blue Book*, selecting fiction otherwise hard to find, but did not confine himself to the pulps, presenting a wide range of fiction from books, especially by UK authors like Algernon BLACKWOOD, Lord DUNSANY and William Hope HODGSON. Although *AFR* published sf, most of its content was SUPERNATURAL FICTION, plus some SLICK FANTASY and HEROIC FANTASY. The quality was high throughout.

AFR had a short-lived companion, *Avon Science-Fiction Reader* (3 issues, 1951-2), which published some SCIENCE FANTASY. The two titles were subsequently merged to create a new, genuine magazine, *Avon Science Fiction and Fantasy Reader* (2 issues January-April 1953, ed Sol Cohen), which included some SLICK FANTASY and HORROR.   [MA]

AVON SCIENCE FICTION AND FANTASY READER ◊ AVON FANTASY READER; MAGAZINES.

AVON SCIENCE-FICTION READER ◊ AVON FANTASY READER.

AWARDS Fantasy is a set of loose fields, and this may in part explain the fact that, by comparison with sf, relatively few awards focus on it. *Reginald's Science Fiction and Fantasy Awards: A Comprehensive Guide to the Awards and their Winners* (current ed **1993**) by Daryl F. Mallett and Robert Reginald lists 140 genre awards, and is not complete; only a small proportion of these concentrate, even partially, on fantasy. Of this total, 10 English-language awards receive individual entries here: the ARTHUR C. CLARKE AWARD, the BALROG AWARD, the BRAM STOKER AWARD, the BRITISH FANTASY AWARD, the GANDALF – which is given within the frame of the Hugo Awards – the INTERNATIONAL FANTASY AWARD, the LOCUS AWARDS – which recognize some fantasy titles –

the MYTHOPOEIC AWARD, the PHILIP K. DICK AWARD, the WILLIAM L. CRAWFORD AWARD and the WORLD FANTASY AWARD. [JC]

AWLINSON, RICHARD ◊ Scott CIENCIN; James LOWDER.

AXA Sexy, blonde, eponymous COMICS star of a SWORD-AND-SORCERY strip set in a FAR-FUTURE post-holocaust Earth. Created and drawn by erstwhile **Modesty Blaise** artist, Enrique Romero (1930-   ), and written by Donne Avenell (1925-   ), the strip was originally published in the UK newspaper *The Sun* (1978-85), then reprinted in the USA by Eclipse as *Axa #1: The Beginning* (graph coll **1982**), *#2: The Desired* (graph coll **1982**), *#3: The Brave* (graph coll **1983**), *#4: The Earthbound* (graph coll **1983**), *#5: The Eager* (graph coll **1983**) and *#6: The Dwarfed* (graph coll **1984**), *#7: The Mobile and The Unmasked* (graph coll **1985**), *#8: The Castaway and The Seeker* (graph coll 1986), *#9: The Escapist, The Starstruck and Betrayed* (graph coll **1988**) and one original graphic album, *Axa Adult Fantasy Colour Album* (graph **1985**). It proved popular enough to generate a short-lived comic-book series drawn by Romero and written by Chuck Dixon (1954-   ), also published by Eclipse: **Axa #1** (1987) and **Axa #2** (1987). [RT]

AXIS MUNDI ◊ WORLD-TREE.

AYESHA ◊ SHE.

AYMÉ, MARCEL (ANDRÉ) (1902-1967) French writer. His novels of French provincial life occasionally include Rabelaisian touches of fantasy. *La Jument verte* (**1933**; trans anon as *The Green Mare* 1938 UK; better trans Norman Denny 1955 UK) is about the complication of a 19th-century family feud by the eponymous supernatural visitor. The similarly supernatural figure of *La Vouivre* (**1943**; trans Eric Sutton as *The Fable and the Flesh* 1949 UK) is "the Lady of the Serpents", who leaves her jewels to tempt unwary thieves while she bathes. *Uranus* (**1948**; trans Denny as *Fanfare in Blémont* 1950 UK; vt *The Barkeep of Blémont* US) includes surreal passages in which one character believes he is being absorbed into the "negative being" of the eponymous planet. *La Belle image* (**1941**; trans Denny as *The Second Face* 1951 UK) is a more contemplative moral fable about a man whose face is remoulded in a more flattering way.

MA's plays, too, include several fantasies. *Clérambard* (**1950**; trans Denny **1952** UK; vt *The Count of Clerambard* US) involves a vision of St Francis of Assisi. *Les Oiseaux de lune* (**1956**) features a character who can turn humans into BIRDS. The title piece of *Le Minotaure* (coll **1967**) is a tale of atavistic TRANSFORMATION, while "Le Convention Belzébir", first performed a year earlier, is a futuristic SATIRE.

The FAIRYTALES in *Les contes du chat perché* (coll **1939**; trans Denny as *The Wonderful Farm* 1951 UK) and *Autres contes du chat perché* (coll **1954** chap; trans Denny as *Return to the Wonderful Farm* 1954 UK; vt *The Magic Pictures: More About the Wonderful Farm* US) are not intended solely as CHILDREN'S FANTASY. Stories selected from six MA collections – *Le Puits aux images* (coll **1932**), *Le Nain* (coll **1934**), *Derrière chez Martin* (coll **1938**), *Le Passe-Muraille* (coll **1943**), *Le Vin de Paris* (coll **1947**) and *En arrière* (coll **1950** France) – make up *Across Paris and Other Stories* (coll trans Denny **1950** UK; vt *The Walker Through Walls* 1962 US) and *The Proverb and Other Stories* (coll trans Denny **1961** UK); a selection from these two translated collections, with one other story, was issued as *The Walker-Through-Walls and Other Stories* (coll **1972** UK). These include numerous fantasies. [BS]

Further reading: *The Short Stories of Marcel Aymé* (**1980**) by Graham Lord; *Marcel Aymé* (**1987**) by Lord.

AYRTON, MICHAEL (1921-1975) UK artist, illustrator, stage designer and writer; many of his book ILLUSTRATIONS were for works of fantasy, and much of his painting focused on material – most significantly the myth of DAEDALUS and the visual and philosophical enigmas of the LABYRINTH – that was also of fantasy interest. Among the titles for which MA did typically dark-toned, highly expressive dustwrapper illustrations, in a Neo-Romantic style typical of many 1940s artists, are the UK edition of Ray BRADBURY's *Dark Carnival* (coll **1947**), a 1948 reprinting of Oscar WILDE's *The Picture of Dorian Gray*, *Here Are Ghosts and Witches* (coll **1954**) by J(ames) Wentworth Day (1899-?1983), the two volumes of Wyndham LEWIS's *The Human Age* (**1955** and **1956**), Marcel AYMÉ's *Across Paris* (trans **1957**), *The Blood Rushed to my Pockets* (**1957**) by George Foa and a 1957 reprinting of Edgar Allan POE's *Tales of Mystery and Imagination*; MA also did the dustwrappers for all his own books. His views on modern art were aggressively conservative (he took a jaundiced view of Pablo Picasso [1881-1973] in his *roman à clef*, *The Midas Consequence* [**1974**], and in essays, but he admired Wyndham LEWIS), and the estranged provinciality of his stance may not have much helped his development as an artist; his work became increasingly stiff. His costume designs for the 1947 stage production of *The Fairy Queen* (**1692**) by Henry Purcell (◊ OPERA) were, however, fluent and eloquent; they appear in *Purcell's The Fairy Queen as Presented by The Sadler's Wells Ballet and the Covent Garden Opera* (**1948**).

MA's first work of genre interest, *Tittivulus, or The Verbiage Collector* (**1953**), is a humorous, heavily illustrated SATIRE featuring the eponymous fiend (named after a medieval DEMON who had the job of collecting the words omitted from the Mass by slipshod, hasty and careless monks) whose 20th-century task is to collect unnecessary language emitted by human beings; he exploits his role to become a Stalin-like dictator of HELL. *The Testament of Daedalus* (**1962** chap) tells the story, accompanied by several intense illustrations, of ICARUS's flight and passion for APOLLO in the form of a philosophical meditation by DAEDALUS. *The Maze Maker* (**1967**) presents similar material in the form of a fully developed novel. Of the essays and tales assembled in *Fabrications* (coll **1972**), some are of genre interest, including "Tenebroso", a POSTHUMOUS FANTASY narrated by the dead Michelangelo. As with his pictorial work, MA's fiction was muscular but stiff, although at times it came vividly to life. [JC]

Other works: *Golden Sections* (coll **1957**), essays; *Maze and Minotaur: An Exhibition of Work on the Theme* (**1973**).

AZEVEDO, ALVARES DE (1831-1852) Brazilian writer. ◊ BRAZIL.

"B" ◊ A.C. BENSON.

BABBITT, LUCY CULLYFORD (1960-   ) US writer whose **Melde** sequence – *The Oval Amulet* (**1985**) and *Children of the Maker* (**1989**) – follows the travels of a girl who, disguised as a boy (◊ GENDER DISGUISE), undertakes a QUEST to find the woman who long before gave her an AMULET.   [JC]

BABBITT, NATALIE (MOORE) (1932-   ) US poet, writer and illustrator whose first book was *Dicke Foote and the Shark* (coll **1967**), poetry, and whose first novel, *The Search for Delicious* (**1969**), describes the adventures of a young man in search of a food that might properly represent the term "delicious" in the kingdom's new dictionary (◊ TRUE NAME). The QUEST in *Kneeknock Rise* (**1970**) has a similarly didactic tinge, though the stories in *The Devil's Storybook* (coll **1974**) and *The Devil's Other Storybook* (coll **1987**) are more straightforwardly humorous SUPERNATURAL FICTIONS, featuring a rather thick-witted DEVIL who trys to make trouble. NB's most impressive single work is *Tuck Everlasting* (**1975**), an eloquently subtle tale in which the human consequences of IMMORTALITY are examined through the eyes of a young girl as she gets to know the Tuck family, four people who drank at a magic spring and are now everlasting (◊ FOUNTAIN OF YOUTH). NB's work is ostensibly for children, but appeals strongly to adults.   [JC]

Other works: *The Something* (**1970**); *Goody Hall* (**1974**); *The Eyes of the Amaryllis* (**1977**).

BABEL In *Genesis xi* a tower is built with the intention of reaching HEAVEN; GOD halts the project and punishes its builders' hubris by fragmenting their language so they can no longer communicate – a MYTH OF ORIGIN for the world's diverse languages. John MILTON's *Paradise Lost* (**1667**) describes the incident approvingly. Jorge Luis BORGES elaborates the theme of noncommunication in "The Library of Babel" (1941), whose exhaustive LIBRARY contains all knowledge in every language, but lost amid all possible lies and gibberish: the hubris of attempted all-embracingness automatically brings confusion. C.S. LEWIS's *That Hideous Strength* (**1945**) echoes the description of Babel in the poem "The Monarchie" (1554) by Sir David Lindsay (1486-1555) – "The Shadow of that hyddeous strength/Sax myle and more it is of length" – and inflicts the CURSE of scrambled language on builders of a metaphorical tower, the hateful ideological structure of the N.I.C.E. (which is already mired in obfuscatory jargon). "The End of the Axletree" (1983) by Alasdair Gray (1934-   ) is a REVISIONIST-FANTASY retelling whose tower-builders pierce the physical roof of the sky and unleash a vast FLOOD.

The portrait of the Tower of Babel by Pieter BRUEGEL almost certainly shapes Western images of this first EDIFICE, and has been an UNDERLIER for fantasy edifices ever since.   [DRL]

BACCHUS ◊ DIONYSUS.

BACH, RICHARD (DAVID) (1936-   ) US writer who began publishing work of genre interest with "Cat" in 1962, collected in *A Gift of Wings* (coll **1974**). He is known mostly for *Jonathan Livingston Seagull* (**1970**), filmed as *Jonathan Livingston Seagull* (*1973*), an ANIMAL FANTASY about a philosophical gull who is profoundly affected by FLYING, but who demands too much of his community and is cast out by it. He becomes an extremely well behaved ACCURSED WANDERER, then dies, and in POSTHUMOUS-FANTASY sequences – though he is too wise really to question the fact of death, and too calmly confident to have doubts about his continuing upward mobility – he learns greater wisdom. Back on Earth, he continues to preach and heal and finally returns to HEAVEN, where he belongs. *There's No Such Place as Far Away* (**1979**), even more tendentiously sentimental than its predecessor, offers lessons in living from BIRDS to the narrator of the tale, who then gives a young girl, on her birthday, a RING which will enable her to receive the same wisdom. Of more interest to adults may be a thematic trilogy – *Illusions: The Adventures of a Reluctant Messiah* (**1977**), *The Bridge Across Forever: A Lovestory* (**1984**) and *One* (**1988**). The first recounts the frustrations of a human ORACLE who – like the famous gull – has difficulty in conveying his message; the second, cast in the form of a fictionalized autobiography, contains factual-seeming references to material exposited as fantasy in the first and third volumes; and *One* carries its protagonists in a flight across a vast sea where versions of the main characters inhabit points in a huge ARCHIPELAGO of ALTERNATE REALITIES. RB's style has become less choked, and his range has broadened; but an almost solipsistic concentration on individual salvation continues to attenuate the impact of his best work.   [JC]

BACON, MARTHA Working name of US writer Martha Sherman Bacon Oliver-Smith (1917-1981), whose *The Third Road* (**1971**) is a TIMESLIP fantasy in which three contemporary children are trapped in 17th-century Spain, having travelled there on a UNICORN's back. *Moth Manor: A Gothic Tale* (**1978**) grippingly describes the efforts of a child to save a DOLL-house and its animate inhabitants from her elders.   [JC]

BAD PLACE Any place with a powerful flavour of WRONGNESS and rottenness: HAUNTED DWELLINGS, evil woods, DARK TOWERS, sometimes even WASTE LANDS. What makes a BP may be the psychic taint of EVIL doings, as in the eponymous abandoned CARNIVAL ground in Dean R. KOONTZ's *The Bad Place* (**1990**), whose perceived wrongness stems from the activities of a SERIAL KILLER; or deluded PERCEPTION, as with the "somehow abominable" countryside in "The Bad Lands" (in *The Smoking Leg* coll **1925**) by John Metcalfe (1891-1965); or some RITUAL OF DESECRATION, as with Sauron's foul industrial slag-heaps outside the gates of Mordor in J.R.R. TOLKIEN's *The Lord of the Rings* (**1954-5**); or contamination following major misuse of MAGIC (often in long-past wars), as with Methwold FOREST in Brian STABLEFORD's *The Last Days of the Edge of the World* (**1978**); or a CURSE, as with the house in Diana Wynne JONES's *Power of Three* (**1976**), whose sickening waves of wrongness emanate from a cursed AMULET; or some SOUL in BONDAGE, as in Rudyard KIPLING's "The House Surgeon" (1909) and William Hope HODGSON's "The Whistling Room" (in *Carnacki the Ghost-Finder* coll **1910**), which room's physical fabric is hideously animated by the malevolent SPIRIT; or associations (perhaps *via* PORTALS) with dread inhuman entities, as with the risen ISLAND in H.P. LOVECRAFT's "The Call of Cthulhu" (1928), whose wrongness is additionally signalled by architecture with impossible geometries. Although a BP can and often does contain physical threats, its essence is a deeper spiritual discomfort or danger.

[DRL]

BAGHDAD Baghdad is the central CITY in the LAND-OF-FABLE Araby where ARABIAN FANTASIES are set. Many of these tales mention Baghdad, or use it as a minaret-bestrewn, particoloured backdrop. Relatively few tales – a rare example is Benjamin DISRAELI's *The Wondrous Tale of Alroy: The Rise of Iskander* (**1833**) – make any serious attempt to envision Baghdad as a physical place.   [JC]

See also: *The* THIEF OF BAGDAD.

BAGPUSS UK tv series (1974). ◊ *The* CLANGERS (1969-73).

BAILEY, HILARY (1936-   ) UK writer and editor. ◊ A.A. ATTANASIO; *SFE*.

BAILEY, ROBIN WAYNE Working name of US writer Robert Wayne Bailey (1952-   ), who began writing fantasy novels with the **Frost** sequence of tales about a warrior WITCH: *Frost* (**1983**), *Skull Gate* (**1985**) and *Bloodsongs* (**1986**). The combination of FEMINISM and SWORD AND SORCERY is intermittently effective, especially at points when Frost, the heroine, must struggle with guilt (she is a patricide) and with her son (inspired by a WIZARD, he threatens the LAND with WRONGNESS). A second series, **Brothers of the Dragon** – *Brothers of the Dragon* (**1992**), *Straight on Till Mourning* (**1993**) and *Flames of the Dragon* (**1994**) – carries its protagonists, who are brothers, through a PORTAL into a SECONDARY WORLD where they are treated as HEROES in an

eons-long war against a sorceress, who rules the Dark Lands, and a sorcerer who rules the Kingdoms of Night. The DRAGONS of the title are warrior dragons, telepathically linked to their riders; the UNICORNS who take a dominant role are Nazgûl-like ghouls (◊ J.R.R.TOLKIEN). Other echoes – specifically of Stephen R. DONALDSON's first **Covenant** sequence (**1977**) – attend the gradual unfolding of the brothers' redemptive role in the land and their troubles back in this world.   [JC]

Other works: *Enchanter* * (**1989**), tied to a computer GAME; *Philip José Farmer's The Dungeon #4: The Lake of Fire* * (**1989**); *Night Watch* (**1990**); *The Lost City of Zork* * (**1991**), tied to a game.

BAIN, F(RANCIS) W(ILLIAM) (1863-1940) UK writer long resident in India, who published a series of novella-length stories based on Hindu mythology: *A Digit of the Moon* (coll of linked stories **1899**), *The Descent of the Sun* (**1903**), *A Heifer of the Dawn* (**1904** chap) and *In the Great God's Hair* (**1904** chap), all four assembled as *A Digit of the Moon and Other Love Stories from the Hindoo* (omni **1910** US), plus *A Draught of the Blue* (**1905** chap), *An Essence of the Dusk* (**1906** chap), assembled as *A Draught of the Blue Together with An Essence of the Dusk* (omni **1906** US), plus *An Incarnation of the Snow* (**1908** chap), *A Mine of Faults* (**1909**), *The Ashes of a God* (**1911**), *Bubbles of the Foam* (**1912**), *A Syrup of the Bees* (**1914**), *The Livery of Eve* (**1917**) and *The Substance of a Dream* (**1919**), the last being assembled with *Bubbles of the Foam* as *The Substance of a Dream Together with Bubbles of the Foam* (omni **1919** US). In *A Digit of the Moon* FWB represents his tales as translations from a Sanskrit manuscript, *The Churning of the Ocean of Time*, a title clearly intended to echo the actual Sanskrit OCEAN OF STORY; by *The Substance of a Dream* he had admitted that he had composed them himself.

Each story has a ruminative preface explaining its meaning according to Hindu philosophy (often with an anecdote from FWB's life in India), as well as an apparatus of footnotes (prefiguring Jack VANCE's similar strategy) to explain the Sanskrit words and Hindu concepts (such as REINCARNATION) that lie thick upon the page. The stories are told in a cool but rapturous DICTION that owes something to Victorian ORIENTAL FANTASY and something perhaps to William MORRIS. *A Digit of the Moon* is a series of FABLES constituting a RIDDLE for a king trying to win the hand of a princess of great wisdom; most of the rest have FRAME STORIES told by the great god Maheshwara, including *The Substance of a Dream*, probably the finest: it is unusual in being told in the first person, by a young musician erotically obsessed with a sequestered queen he first saw in a DREAM. While remaining, like the rest, curiously chaste, it still attains a remarkable intensity of emotion. It seems a pity that, apart from a few excerpts, FWB's body of work has remained out of print for half a century.   [DK]

Other works: *Christina, Queen of Sweden* (**1890**), *The Bullion Report, and The Foundation of the Gold Standard* (privately printed **1896** chap), *The Corner in Gold: Its History and Theory* (privately printed **1893** chap).

Further reading: *Francis William Bain* (**1963**) by Keshav Mutalik.

BAKER, DENYS VAL (1917-1984) UK writer noted for his long series of Cornish novels and reminiscences. After an early career as a reporter, he turned to writing in 1941. His first story collection, *Worlds Without End* (coll **1945**), reflects the loss of identity and displacement-of-self experienced as a result of WORLD WAR II; its horrors are as much psychological as supernatural, with a delight in the inexplicable. In later years DVB's fiction mellowed, focusing on more traditional GHOST STORIES, usually with a Cornish setting. His SUPERNATURAL FICTIONS, generally dark and introspective, are scattered throughout his UK story collections; August W. DERLETH assembled a selection of them for ARKHAM HOUSE, *The Face in the Mirror* (coll 1971 US). DVB also established himself as an anthologist, producing a regular series of homely ghost stories.   [MA]

Other works: *The Secret Place* (coll 1977), mostly nonfantasy but with a strong evocation of place.

As editor: *One and All* (anth 1951), *Haunted Cornwall* (anth 1973), *Stories of the Night* (anth 1976), *Stories of the Macabre* (anth 1976), *Stories of Horror and Suspense* (anth 1977), *Stories of the Occult* (anth 1976), *Stories of the Supernatural* (anth 1979), *Stories of Fear* (anth 1980), *Cornish Ghost Stories* (anth 1981), *Ghosts in Country Houses* (anth 1981), *When Churchyards Yawn* (anth 1982), *Ghosts in Country Villages* (anth 1983), *Stories of Haunted Inns* (anth 1983), *Phantom Lovers* (anth 1984), *Haunted Travellers* (anth 1985).

BAKER, FRANK (1908-1983) UK writer and occasional actor whose fantasy novel *The Birds* (1936), which explores the avian manifestation of humankind's personal devils, was overshadowed by Daphne DU MAURIER's novella with the same title; Alfred HITCHCOCK's movie *The* BIRDS (*1963*) can be seen as based on an amalgam of the two (◊◊ BIRDS). FB's work contains some of the more memorable treatments and interpretations of common supernatural themes (◊ SUPERNATURAL FICTION). His most popular work during his lifetime was *Miss Hargreaves* (1940), in which an imaginary poetess comes into being. In such early works he considered the projection of the psyche; he firmly believed in the supernatural, and had direct experience of a HAUNTED DWELLING, but in his later books he chose to treat the supernatural within a religious/metaphysical framework. Although neither *Mr Allenby Loses the Way* (1945) nor *Before I Go Hence* (1946) is a TIME FANTASY in the style of J.B. PRIESTLEY, both explore the continuity of self through TIME as a possible separate existence to one's mortal self. Both, too, can be interpreted at a nonsupernatural level, even though the first seems provoked by the incursion of FAERIE and the second considers the nature of eternity. *Sweet Chariot* (1942) depicts an IDENTITY EXCHANGE between a man and an ANGEL, and somewhat incoherently explores the relationship between mortal and SPIRIT.

Time is also central to FB's few short stories, the best of which were collected as *Stories Strange and Sinister* (coll 1983). Rather more mystical elements of the supernatural emerge in *The Downs So Free* (coll 1948) and *Talk of the Devil* (coll 1956), both of which include traditional HAUNTINGS, though with untraditional explanations.   [MA]

Further reading: *I Follow by Myself* (1968), a revealing autobiography.

BAKER, GEORGE A(UGUSTUS) (1849-1906) US lawyer and writer whose *Mrs Hephaestus and Other Short Stories, Together with West Point, A Comedy in Three Acts* (coll 1887) contains some work of interest. The protagonist of "Mrs Hephaestus" is a seductress with a bronze complexion, and it may be assumed from her name that she has been forged by the Greek god of fire and SMITHS. In "The Spirit of the Age" a property sale forces a transference of FAERIE to Central Park.   [JC]

BAKER, G(EORGE) P(HILIP) (1879-1951) UK author of

the **Greenwood** sequence: *The Magic Tale of Harvanger and Yolande* (1914) and *The Romance of Palombris and Pallogris (The Second Magic Tale)* (1915). As Arthurian pastiches they are of some interest, though suburban in tone.   [JC]

BAKER, (ROBERT) MICHAEL (GRAHAM) (1938-   ) UK solicitor and writer, whose *The Mountain and the Summer Stars: An Old Tale Newly Minted* (1968 chap) carries its CHANGELING protagonist under the mountains (◊ UNDERGROUND) in search of his FAIRY mother. This recasting of the FOLKTALE "The Fairy Wife" is accomplished with grace.   [JC]

BAKER, SCOTT (MacMARTIN) (1947-   ) US-born writer, long resident in France, whose first novel, *Symbiote's Crown* (1978), is sf but most of whose work is DARK FANTASY or HORROR – including his best-known title, *Nightchild* (1979; rev 1983). The protagonist of this SUPERNATURAL FICTION undergoes a RITE OF PASSAGE into a VAMPIRE sect, for which it is intended he will serve as provender. But he proves a vampire himself, albeit one with good intentions: he helps recast vampires – an alien race – and humanity into a new harmony. In its revised form, this novel forms the first volume of the loose **Ashlu Cycle**, which continues with *Firedance* (1986) and *Drink the Fire from the Flames* (1987); the later volumes depict the complex interrelationship between two SHAMAN protagonists, the elder of whom is training the younger to replace him by means of a ritual AGON. *Dhampire* (1982; much rev vt *Ancestral Hungers* 1995) and *Webs* (1989) are both horror, the latter an ambitiously constructed tale involving multiple entanglements, from supernatural SPIDERS to engulfing SEX.   [JC]

BAKSHI, RALPH (1938-   ) Palestinian-born US animator and director of ANIMATED MOVIES. He first came to notice – indeed, to notoriety – with *Fritz the Cat* (*1972*), adapted from the COMICS of Robert H. Crumb (1943-   ). His first movie of fantasy interest was WIZARDS (*1977*), a tale of MAGIC and of the conflict between GOOD AND EVIL two million years after a world HOLOCAUST. This movie also saw RB pioneering the technique of rotoscoping, whereby scenes are shot in live action and then traced to produce the animated result. Rotoscoping does not, however, impose greater realism; its somewhat uncanny effect is not easy to describe. Rotoscoping featured also in his next movie, *The* LORD OF THE RINGS (*1978*), based on the J.R.R. TOLKIEN cycle: although the venture was commendably ambitious, the movie was a disappointing mishmash, crippled by lack of funds at crucial stages. Some nonfantasy work followed, and then FIRE AND ICE (*1983*) saw him return to the fantasy genre. In 1987-9 his *Mighty Mouse: The New Adventures* was screened on US Saturday-morning tv. It was with COOL WORLD (*1992*) that he achieved his most distinguished fantasy movie to date: mixing live action with animation (◊ TOONS), this is a sophisticated exploration of ALTERNATE REALITIES, one subjective (and created) and the other objective, and of the interplay between the two. It is often adversely compared with the more polished WHO FRAMED ROGER RABBIT (*1988*), yet *Cool World* shows the better grasp of fantasy. RB's career has been patchy, including some very bad movies, but at his best he has probably been animation's single most important contributor to the genre.   [JG]

Other movies: *Heavy Traffic* (*1973*); *Coonskin* (*1975*); *American Pop* (*1981*); *Hey, Good Lookin'* (*1982*); animated sequences in *Cannonball Run II* (*1983*).

BALANCE Traditionally Balance registers or influences the state of play between above and below (◊ AS ABOVE, SO

BELOW), GOOD AND EVIL, Law/Order/Stasis and CHAOS, male and female (◊ GENDER) or YIN AND YANG. Laws of MAGIC generally acknowledge Balance – in that action and reaction are equal – and the greatest of Ursula LE GUIN's vaguely Taoist WIZARDS in her **Earthsea** sequence avoids all but urgently needed magic since there must always be consequences. Often the world or universal Balance needs to be restored when upset by WRONGNESS: Michael MOORCOCK's *Stormbringer* (**1965**) has a concluding VISION of a literal Cosmic Balance righting itself with the defeat of Chaos. More liberatingly, his *The King of the Swords* (**1972**) finally sees this Balance swinging free with nothing to weigh, all the law and Chaos GODS having been joyously slain. Gordon R. DICKSON's *The Dragon and the George* (**1976**) explicitly acknowledges that one may need to fight for Chaos to prevent the Balance tilting to the stasis of perfect order; but the latter is the precise goal of John BRUNNER's eponymous *Traveler in Black* (coll **1971** US), whose purging of irrational magic from the Universe will ultimately bring TIME to a stop. R.A. LAFFERTY's *Aurelia* (**1982**) unusually suggests that one should be absolute for good or evil, and portrays the yin-yang symbol of Balance as a vicious weapon. Piers ANTHONY's *On a Pale Horse* (**1983**) makes much play with individuals' personal good/evil Balance, which if exact consigns them to PURGATORY rather than HEAVEN or HELL. Patricia C. WREDE's *The Seven Towers* (**1984**) pivots on the good/evil Balance in magic, which must not be wholly one or the other; Diana Wynne JONES's *Black Maria* (**1991**) explores the wrongness resulting from lack of male/female Balance in, again, magic. L.E. MODESITT's RATIONALIZED-FANTASY **Recluce** books make Balance a natural conservation law as for electrical charge: creating new Order automatically produces counterbalancing Chaos energies. Louise COOPER's **Time Master** sequence (**1985-7**) shows the lords of Chaos and Order as morally equivalent, and the temporary victory of Chaos as a necessity.

Occasionally, Balance itself is used as a PLOT DEVICE, as in Moorcock's *The Queen of the Swords* (**1971**), where the eponymous Chaos Queen is lured into violating a CONDITION of the Cosmic Balance, which then destroys her.   [DRL]

BALIOL, ALEXANDER (DE) (1953-   ) UK writer whose **Amulets of Darkness** cycle – *The Magefire* (**1990**) and *The Tears of Ginara* (**1992**) – follows the QUEST of a persecuted healer who must recover manuscripts which will prevent the election of a false High King. As he travels, he gathers around him a SEVEN SAMURAI-style cadre of COMPANIONS with various skills.   [JC]

BALL, MARGARET (ELIZABETH) (1947-   ) US writer who has written historical novels as Catherine Lyndell. Her first fantasy novel, *The Shadow Gate* (**1991**), carries an inhabitant of FAERIE into this world; here she makes friends who later, back in Faerie, become vital to her attempts at HEALING her domain. MB's **Tamai** sequence – *Flameweaver* (**1991**) and *Changeweaver* (**1993**) – is set in an alternate 19th century (◊ ALTERNATE WORLDS) in which China continues to exclude the imperialist West from its borders, partly through the use of DEMONS. Tamai, who has MAGIC powers, undergoes various learning experiences with her skill, then accompanies a UK explorer, Charles Carrington, on various adventures. By the end of the second volume they still have not reached the heart of China, so sequels may appear. MB has also co-written with Anne MCCAFFREY an sf novel, *PartnerShip* (**1992**).   [JC]

BALLANTINE ADULT FANTASY SERIES In the wake of the success of the US paperback editions of J.R.R. TOLKIEN's *The Hobbit* (**1937**) and *The Lord of the Rings* (**1954-5**), Ballantine Books began in 1966 to look for further works to feed the demand for material "in the tradition of Tolkien". During 1967-8 they reprinted E.R. EDDISON's *The Worm Ouroboros* (**1922**) and **Zimiamvian** trilogy (**1935-58**), Mervyn PEAKE's **Gormenghast** trilogy (**1946-59**) and David LINDSAY's *A Voyage to Arcturus* (**1920**); Peter S. BEAGLE's *A Fine and Private Place* (**1960**) and *The Last Unicorn* (**1968**) followed in early 1969. These all sold well, so Betty Ballantine and Lin CARTER conceived a regular series to be published "under the sign [logo] of the Unicorn's Head". Carter's nonfiction *Tolkien: A Look Behind "The Lord of the Rings"* (**1969**) was a book-length introduction to the tradition of fantasy the series intended to cover; as with the introductions he contributed to each volume, Carter's enthusiasm was infectious but his scholarship suspect.

Beginning in May 1969, with Fletcher PRATT's *The Blue Star* (1952; **1969**), the series brought into print about 60 volumes, many of them reprints of older fantasies, including several each by James Branch CABELL, William MORRIS and Lord DUNSANY. There were also some original works, most notably Joy CHANT's *Red Moon and Black Mountain* (**1970**) and Katherine KURTZ's *Deryni Rising* (**1970**). One of the main achievements of the series was that, after the reprint of her early novel *The Virgin and the Swine* (**1936**; vt [in BAF series] *The Island of the Mighty* 1970), Evangeline WALTON was coaxed into finishing the MABINOGION tetralogy – with *The Children of Llyr* (**1971**), *The Song of Rhiannon* (**1972**) and *Prince of Annwn* (**1974**) – which the first volume had all along been intended to initiate.

Carter edited a number of anthologies, with running commentary – beginning with *Dragons, Elves and Heroes* (anth **1969**) and *The Young Magicians* (anth **1969**) – as well as several collections by, among others, Clark Ashton SMITH, beginning with *Zothique* (coll of linked stories **1970**). He also wrote another full-length critical book, *Imaginary Worlds* (**1973**).

As the series wore on, sales fell off, and it was closed about five years after it began. (Some of the very last volumes, like *Prince of Annwn*, though acquired for the series were published without the logo or Carter's introduction, so a clear closure-date is hard to determine.) Just a few years later Del Rey Books initiated its own fantasy programme; only a very few of the BAF titles reappeared under the Del Rey imprint.   [DGK]

BALLARD, J(AMES) G(RAHAM) (1930-   ) UK writer, born and raised in Shanghai and interned in a Japanese civilian POW camp during WWII, experiences recalled in his autobiographical *Empire of the Sun* (**1984**), filmed by Steven SPIELBERG as *Empire of the Sun* (*1987*). Generally regarded as a writer of SCIENCE FICTION, and more recently accepted by the mainstream, JGB produces work that is *sui generis*, occasionally drifting into the realms of the FANTASTIC. The setting of his **Vermilion Sands** stories, which include his first story sale, "Prima Belladonna" (1956) for SCIENCE FANTASY, depicts the DECADENCE of a future resort for creative has-beens amid the atmosphere of a timeless AFTERLIFE. These stories were collected as *Vermilion Sands* (coll **1971** US; with 1 story added 1973 UK). A similar mood of desolation pervades many of JGB's other stories, such as "The Terminal Beach" (1964 *New Worlds*), and this tone was ideally suited for his disaster novels of the 1960s;

these began with *The Wind from Nowhere* (1961 *New Worlds* as "Storm-Wind"; **1962** US) but became increasingly more surreal by *The Crystal World* (fixup **1966**), which is almost an INSTAURATION FANTASY. This whole PERCEPTION of changing awareness took JGB toward fantasies of death and entombment, as conceptualized in *Crash* (**1973**), *Concrete Island* (**1974**) and *High-Rise* (**1975**) – which, while not URBAN FANTASIES, nevertheless used the urban image with considerable potency – and then inevitably to the POSTHUMOUS FANTASY *The Unlimited Dream Company* (**1979**), an instauration fantasy deploying the motifs of the urban FAIRYTALE. All of JGB's works focus on perception and change. While most are not intrinsically fantastic, neither are they rooted in reality.   [MA]

Other works: *The Drowned World* (**1962** US); *The Voices of Time and Other Stories* (coll **1962** US); *Billenium* (coll **1962** US); *The Voices of Time, And Other Stories* (coll **1962** US); *The Four-Dimensional Nightmare* (coll **1963**; rev 1974; vt *The Voices of Time* 1985); *Passport to Eternity* (coll **1963** US); *The Terminal Beach* (coll **1964** UK), not to be confused with *Terminal Beach* (coll **1964** US); *The Burning World* (**1964** US; rev vt *The Drought* 1965 UK); *The Drowned World and The Wind from Nowhere* (omni **1965** US); *The Impossible Man* (coll **1966** US); *By Day Fantastic Birds Flew through the Petrified Forest* (**1967** wall-poster), incorporating text from *The Crystal World*; *The Disaster Area* (coll **1967**); *The Day of Forever* (coll **1967**; rev 1971); *The Overloaded Man* (coll **1967**; rev vt *The Venus Hunters* 1980); *Why I Want to Fuck Ronald Reagan* (**1968** chap); *The Atrocity Exhibition* (coll **1970**; vt *Love and Napalm: Export USA* 1972 US; rev 1990 US); *Chronopolis and Other Stories* (coll **1971** US); *Low-Flying Aircraft* (coll **1976**); *The Best of J.G. Ballard* (coll **1977**); *The Best Short Stories of J.G. Ballard* (coll **1978** US); *Hello America* (**1981**); *Myths of the Near Future* (coll **1982**); *News from the Sun* (**1982** chap); *The Day of Creation* (**1987**); *Memories of the Space Age* (coll **1988** US); *Running Wild* (**1988** chap); *War Fever* (coll **1990**); *The Kindness of Women* (**1991**); *The Crystal World; Crash; Concrete Island* (omni **1991** US); *Rushing to Paradise* (**1994**).

BALROG AWARD US award given annually 1979-85 for work in fantasy, named after the demon servants of Melkor the Dark Enemy, who are referred to mainly in the backstory to J.R.R. TOLKIEN's *The Lord of the Rings* (**1954-5**), though one MALIGN SLEEPER Balrog does survive long enough to be defeated by Gandalf. There were many categories, some granted only intermittent recognition; only the more important are listed below. The award year is one year later than that of publication.   [JC]

Best novel

1979: *Blind Voices* by Tom REAMY
1980: *Dragondrums* by Anne MCCAFFREY
1981: *The Wounded Land* by Stephen R. DONALDSON
1982: *Camber the Heretic* by Katherine KURTZ
1983: *The One Tree* by Stephen R. DONALDSON
1984: *Armageddon Rag* by George R.R. MARTIN
1985: *The Practice Effect* by David Brin

Short fiction

1979: "Death from Exposure" by Pat Cadigan
1980: "The Last Defender of Camelot" by Roger ZELAZNY
1981: "The Web of the Magi" by Richard Cowper
1982: "A Thief in Korianth" by C.J. CHERRYH
1983: "All of us are Dying" by George Clayton Johnson
1984: "Wizard Goes A-Courtin'" by John MORRESSY
1985: "A Troll and Two Roses" by Patricia MCKILLIP

Best collection/anthology

1979: *Born to Exile* by Phyllis EISENSTEIN
1980: *Night Shift* by Stephen KING
1981: *Unfinished Tales* by J.R.R. TOLKIEN
1982: *Shadows of Sanctuary* ed Robert ASPRIN
1983: *Storm Season* ed Robert ASPRIN
1984: *Unicorn Variations* by Roger ZELAZNY
1985: *Daughter of Regals and Other Tales* by Stephen R. DONALDSON

BALZAC, HONORÉ de (1799-1850) French writer who, from about 1835, began to think of his mature work as comprising a vast series, **La comédie humaine** ["The Human Comedy"]; in the 15 years before his death, almost everything he wrote was fitted into the ever-expanding (and, realistically, never-completable) sequence. Most of his SUPERNATURAL FICTION, however, precedes the **Comedy**, and generally lacks the interconnected, astonishingly complex bustle of the later work. The most successful of these early, usually pseudonymous works – *Le Centenaire, ou le deux Behringeld* (**1822** as by Horace de Saint-Aubin; vt *Le Sorcier* 1837; trans George Edgar Slusser as *The Centenarian, or The Two Behringelds* 1976 US) – deals, in terms of demonic temptation (◊ DEMONS) and the search for IMMORTALITY, with one of HDB's lifelong obsessions: the idea that individuals have a fixed amount of energy, a kind of *élan vital* bank account which is drawn on until exhausted. Other relevant early work includes: "L'élixir de longue vie" ["The Elixir of Life"] (in *Romans et contes philosophiques* coll **1830**; trans as *Don Juan* chap US; vt *Elixir of Life* 1901 US; also in *The Unknown Masterpiece*, see below) (◊ DON JUAN); "Le recherche de l'absolu" (in *Études de moeurs au XIXve siècle* coll **1834**; trans as *The Philosopher's Stone* 1844 chap US; vt *Balthazar* 1859 UK; new trans Ellen Marriage as *The Quest of the Absolute* 1990 UK); and the novel-length "Séraphita" (in *Le Livre mystique* coll **1835** with "Louis Lambert"; trans anon in coll *Louis Lambert; Séraphita* coll **1889** 2 vols US; new trans Clara Bell in *Séraphita (and Louis Lambert & The Exiles)* coll **1989** UK), an occult romance (◊ OCCULT FANTASY) about an ANGEL.

Of much greater interest is *La Peau de chagrin* (**1831**; trans as *Luck and Leather: A Parisian Romance* 1842 US; vt *The Wild Ass's Skin* 1888; new trans Katharine Prescott Wormeley as *The Magic Skin* 1888 US), which HDB incorporated into the **Comedy**, and which, far more than earlier work, exemplifies his presentation of the CITY as inextricably complex, Gothic, tentacled, fog-ridden, haunting and haunted (◊◊ URBAN FANTASY). Suicidally desperate in the coils of Paris, the protagonist of the tale enters a magic SHOP. Its owner gives him a present: the skin of a wild ass. When rubbed, this skin will grant him any WISH he might make; however, in strict accordance with HDB's obsession about energy, the skin will shrink each time a wish is fulfilled, and when it finally disappears its owner will die. Cursed by this PACT WITH THE DEVIL, the protagonist attempts to eke out a life devoid of wishes, so that he can remain among the living. In the end, of course, he fails.

Of almost equal melodramatic impact is "Melmoth réconcilié" (in *Études philosophiques* coll **1835**; trans Ellen Marriage as "Melmoth Reconciled" in *The Unknown Masterpiece* coll **1896** UK), which continues (◊ SEQUELS BY OTHER HANDS) Charles MATURIN's *Melmoth the Wanderer* (**1820**) – as does, less explicitly, *Le Centenaire* (noted above). The venal young protagonist of "Melmoth réconcilié" is approached by a strange Englishman who turns out to be

Melmoth ($\lozenge$ ACCURSED WANDERER), and who offers to exchange SOULS with him; he accepts, gains unhuman powers, despairs, passes on his demonic "gift" to another, and dies. Again, the tale is driven by HDB's obsession with the lifeforce as a substance to be bartered and spent. Other germane work includes: "Massimilla Doni" (in *Une fille d'Eve* coll **1839**; trans in *The Human Comedy* omnis **1895-8**), made into an OPERA in 1935 by Othmar Schoeck; and *L'Histoire des Treize* (**1834-5**; part trans as *The Mystery of the Rue Soly* **1894** UK). These works take a peripheral place in HDB's overall oeuvre, but can be said to illuminate the shadows of that extraordinarily comprehensive world.

HDB published his titles in various forms, at various dates, in various states of completion, under various titles and overall rubrics; English-language translations, many pirated, are almost equally difficult to trace or ascribe with accuracy. [JC]

BANAT, D.R. [s] $\lozenge$ Ray BRADBURY.

BANGS, JOHN KENDRICK (1862-1922) US writer and editor, much of whose work was published under pseudonyms, some still unknown. He began to publish humorous sketches and stories while still in college, early material appearing in collections like *New Waggings of Old Tales* (coll **1888**) with Frank Dempster Sherman, writing together as Two Wags, and *Tiddlywink Tales* (coll **1891**). His first novel of interest, *Roger Camerden: A Strange Story* (**1887**), published anon, is a tale of supernatural dementia and, almost uniquely in his work, lacks any attempt at HUMOUR. More typical was *Toppleton's Client, or A Spirit in Exile* (**1893**), an overcomplicated GHOST STORY in which HORROR and humour intermix. In novels like this – and in collections like *The Water Ghost and Others* (coll **1894**), which contains his best work, *Ghosts I Have Met and Some Others* (coll **1898**) and *Over the Plum Pudding* (coll **1901**) – JKB attempted to master a comically debunking style of fantasy which was like that of Mark TWAIN in its occasional savagery (though lacking Twain's profundity or wit) and like that of L. Frank BAUM in the spoof literalism of some of its plots (though never to magic effect, as in OZ). For the most part, JKB's work has failed to survive.

A partial exception is the **Houseboat on the Styx** sequence: *A Houseboat on the Styx: Being Some Account of the Divers Doings of the Associated Shades* (**1895**), *The Pursuit of the Houseboat: Being Some Further Account of the Divers Doings of the Associated Shades, Under the Leadership of Sherlock Holmes* (**1897**) and *The Enchanted Type-Writer* (coll of linked stories **1899**). Though the idea that the dead might be imagined as conversing together in HELL has a provenance extending back at least as far as LUCIAN's *Dialogues of the Dead* (*c*AD150), what became known as the Bangsian fantasy ($\lozenge\lozenge$ AFTERLIFE; POSTHUMOUS FANTASY) established an amiable CLUB-STORY atmosphere and implicated its eminent dead souls – some historical, some fictional ($\lozenge$ RECURSIVE FANTASY) – in prankish escapades. Philip José FARMER's **Riverworld** series descends directly from KKB's sequence.

Further attempts at spoof fantasy – like *The Autobiography of Methuselah* (**1909**), which burlesques Noah and his contemporaries – have been forgotten. The last years of JKB's career were spent on the lecture circuit, where he enjoyed a deserved success. [JC]

Other works: *Mr Bonaparte of Corsica* (**1895**), anon; *The Idiot* (**1895**) and *The Inventions of an Idiot* (coll **1904**); *The Bicyclers, and Three Other Farces* (coll **1896**); *The Rebellious Heroine* (**1896**); *The Dreamers: A Club* (coll **1899**); *Mr Munchausen* (coll of linked stories **1901**) ($\lozenge$ MUNCHHAUSEN); *Bikey the Skicycle, & Other Tales of Jimmieboy* (coll **1902**); *Emblemland* (**1902**) with Charles Raymond Macauley, whose *Fantasma Land* (**1904**) is a Bangsian spoof; *Mollie and the Unwiseman* (**1902**); *Olympian Nights* (coll of linked stories **1902**); *The Worsted Man: A Musical Play for Amateurs* (**1905** chap); *Alice in Blunderland: An Iridescent Dream* (**1907**); *Jack and the Check Book* (**1911**); *Shylock Homes: His Posthumous Memoirs* (coll **1973**).

BANGSIAN FANTASY $\lozenge$ AFTERLIFE; John Kendrick BANGS; POSTHUMOUS FANTASY; RIVERS.

BANKS, LYNNE REID $\lozenge$ Lynne REID BANKS.

BANTOCK, NICK (1949-) UK artist and writer, now resident in Canada, who entered the fantasy field with the **Griffin & Sabine** sequence of GRAPHIC NOVELS: *Griffin & Sabine: An Extraordinary Correspondence* (graph **1991** US), *Sabine's Notebook* (graph **1992**) and *The Golden Mean: In Which the Extraordinary Correspondence of Griffin & Sabine Concludes* (graph **1993**). Griffin Moss is a lonely graphic artist and postcard designer in London; Sabine Strohem is a woman able, via some TALENT, to see through Griffin's eyes as he paints. As the sequence develops, and the two begin to flee from or search for one another, their mutual mystery begins to unpack in terms of JUNGIAN PSYCHOLOGY: she seems to be his MUSE, and they are each other's SHADOW. *The Egyptian Jukebox* (graph **1993**) presents picture clues to the disappearance of its protagonist; and *Kublai Khan* (graph **1994**) accompanies Samuel Taylor COLERIDGE's poem with pop-up illustrations. [JC]

BANVILLE, JOHN (1945-) Irish writer and journalist. *Long Lankin* (coll of linked stories **1970** UK; with 2 stories omitted and 1 story added rev 1984), though nonfantasy, displays features that persist through JB's later fiction: a governing (usually eponymous) metaphor drawn from supernatural lore; a broodingly GOTHIC manner; and a persistent tendency towards imagery that evokes the FANTASTIC. *Nightspawn* (**1971** US), *Birchwood* (**1973** UK) and *Mefisto* (**1986** UK) similarly employ emblems and images of fantasy to tell interior dramas.

Doctor Copernicus (**1976** UK) and *Kepler* (**1981** UK) dramatize the psyches of the eponymous scientists in the anguish of seeing their discoveries displace models of the Universe that had afforded a sense of spiritual place; the contemporary short novel *The Newton Letter* (**1982** UK) – whose protagonist is writing a monograph on Isaac Newton – deals (at one remove) with similar matters. In *Ghosts* (**1993** UK) the unnamed narrator – who is plainly the protagonist of *The Book of Evidence* (**1989** UK), now released from prison – is a man unmoored from ontological or moral benchmarks; he is in a sense the inheritor of the legacy created by JB's earlier scientist protagonists. The GHOSTS of the title are simply the hauntings of his past life; JB has throughout abjured actual fantasy. *Athena* (**1995** UK) is a sequel, in which the HAUNTINGS continue, and deepen. [GF]

BARBARELLA French/Italian movie (*1967*). Marianne/De Laurentiis. **Pr** Dino De Laurentiis. **Dir** Roger Vadim. **Spfx** August Lohman. **Sp anim fx** Gerard Cogan, Thierry Vincens-Fargo. **Screenplay** Vittorio Bonicelli, Claude Brûle, Brian Degas, Jean-Claude FOREST, Tudor Gates, Terry Southern, Vadim, Clement Biddle Wood. **Based on** the comic strip created by Jean-Claude FOREST. **Starring** Jane Fonda (Barbarella), David Hemmings (Dildano), John Phillip Law (Pigar), Milo O'Shea (Duran Duran), Anita Pallenberg (Black Queen). 98 mins. Colour.

A soft-porn SCIENCE FANTASY that, in a PICARESQUE plot, mixes SEX with an astonishing number of fantasy archetypes. Barbarella must rescue rogue scientist Duran Duran from the Tau Ceti system's evil city Sogo, where the court is run like that in an ARABIAN FANTASY. Along the way she is tormented by voracious AUTOMATA and teams up with, *inter alia*, a blind, flightless ANGEL, Pigar, who is later punished by crucifixion. [JG]

BARBAROSSA ◊ ONCE AND FUTURE KING.

BARCLAY, BILL ◊ Michael MOORCOCK.

BARDS Bards do not perform simply to entertain. The songs they sing, and the sagas they recite, comprise versions of the MATTER of the land to which they belong. In fantasy tales the versions of STORY generated by a bard may well have MAGIC force, confirming what has occurred or prophesying what is to come. [JC]

See also: MINSTRELS; SONG.

BARHAM, RICHARD HARRIS (1788-1845) UK curate and writer, creator of the **Ingoldsby Legends** – *The Ingoldsby Legends, or Mirths and Marvels: First Series* (coll **1840**), *Second Series* (coll **1842**), *Third Series* (coll **1847**) – ostensibly narrated by Thomas Ingoldsby Esquire; earlier serial publication, from 1837, was in *Bentley's Miscellany*, except for the first story, "A Singular Passage in the Life of the Late Henry Harris", which appeared in BLACKWOOD'S MAGAZINE in 1831 as by Jasper Ingoldsby. The stories and poems either retold local Kentish LEGENDS or unabashedly created new ones. Often humorous, many involved supernatural manifestations, usually GHOSTS, WITCHES or DEVILS. They were immensely popular, partly because they poked fun at the creaking horrors of Gothic fiction (◊ GOTHIC FANTASY) and thus established a basis for less tenebrous short SUPERNATURAL FICTION. RHB paved the way for J. Sheridan LE FANU and Charles DICKENS. [MA]

BARING, MAURICE (1874-1945) UK diplomat and writer who became an expert on Russia and its literature. His first writings were poetry and a play, *The Black Prince* (**1902**); his short fiction began to appear in *The Morning Post* while he was its foreign correspondent. His stories, often enigmatic, show a predilection for disassociation of mind – as in his best tales: "The Shadow of a Midnight" (1908), about precognition (◊ TALENTS), and "Venus" (1909), featuring astral projection (◊ ASTRAL BODY) to the eponymous planet. MB's stories were collected in *Orpheus in Mayfair* (coll **1909**), whose title story is a tale of translated MYTH, a theme MB further explored in *The Glass-Mender* (coll **1910**). A revised selection of MB's stories was assembled as *Half-a-Minute's Silence* (coll **1925**). Vernon LEE's collection *For Maurice* (coll **1927**) is dedicated to MB. [MA]

BARING-GOULD, S(ABINE) (1834-1924) UK clergyman, country squire and writer, mostly of nonfiction. He produced over 100 books on theology, topography (especially of Dartmoor), history, FOLKLORE and MYTH; he also wrote the hymn "Onward, Christian Soldiers" (1865). He began writing in 1857, but became most prolific when he inherited the family estate at Lew Trenchard, Devon, in 1881. An antiquarian with a fascination for local LEGENDS and beliefs, he issued several important early studies, including *The Book of Were-wolves* (**1865**), *Curious Myths of the Middle Ages Series I* (**1866**) and *Series II* (**1868**), *Curiosities of the Olden Times* (**1869**) and *A Book of Folklore* (**1913**). This interest also caused him to write a number of short stories relating to legends, WITCHES, GHOSTS and the paranormal, collected from over 50 years' output as *A Book*

of Ghosts (coll **1904**). SB-G taught himself Icelandic and Danish in order to undertake a modern adaptation of the *Grettir Saga* – *Grettir the Outlaw* (**1889**) – while an interest in FAIRYTALES resulted in *A Book of Fairy Tales Retold* (coll **1894**), *The Crock of Gold* (**1899**), both assembling tales of a TWICE-TOLD character, and a translation, *Fairy Tales from Grimm* (coll **1895**). His diverse works make SB-G something of a forerunner to writers as various as M.R. JAMES, Andrew LANG, William MORRIS, Eden PHILLPOTTS and Montague Summers (1880-1948). [MA]

Further reading: *Onward Christian Soldier* by W. Purcell (**1957**).

BARKER, CICELY MARY (1895-1973) UK children's book illustrator best-known for her **Flower Fairies** books, featuring appealingly drawn children in picturesque botanical costumes; these have hardly been out of print since first published in seven volumes 1923-35.

CMB briefly studied part-time at Croydon School of Art, but was largely self-taught. She came to be recognized as one of the most accomplished children's illustrators of her period. [RT]

BARKER, CLIVE (1952-   ) UK writer, producer, director and artist who made a ground-breaking literary debut in the HORROR field with the simultaneous publication of the first three volumes of **Clive Barker's Books of Blood** (coll **1984** 3 vols). Born and raised in Liverpool, he started writing plays and stories in his teens. At age 16 he was inspired when local author Ramsey CAMPBELL was invited to his school to talk to the pupils about writing horror. At age 21, after several years' toiling in Liverpool's fringe theatre, CB moved to London and spent the next eight years on welfare. He wrote and painted every day for his own enjoyment, being dismissed by one DHSS official as "unemployable" when he listed his occupation as "writer". He began to gain a measure of success writing plays with members of his theatre group, The Dog Company, and enjoyed two very successful seasons at the Edinburgh Festival with his production *The History of the Devil* (1981) and a short West End run in 1986 with *The Secret Life of Cartoons* (1983). He wrote the first three **Books of Blood** over a period of eight months, during the evenings and at weekends. He followed this auspicious debut with an ambitious Faustian novel (◊ FAUST), *The Damnation Game* (**1985**), and a further **Books of Blood** trio (coll **1985** 3 vols). During this period he also wrote the screenplays for two low-budget UK movies, *Underworld* (**1985** US; vt *Transmutations*) and *Rawhead Rex* (**1986**), the latter based on his own published story. Publicly dissatisfied with the results, he approached New World Pictures asking to write and direct HELLRAISER (**1987**), an audacious blend of sexual subtext and explicit gore with a sadomasochistic slant, loosely based on his novella "The Hell Bound Heart" (in *Night Visions 3* **1986** ed George R.R. MARTIN). The movie was a surprise hit, especially on video, creating a genuine horror ICON in the character of the demonic Pinhead and spawning three sequels to date, each scripted by long-time schoolfriend and associate Peter Atkins. Barker went on to write and direct NIGHTBREED (**1990**), based on *Cabal* (**1988**), and *Lord of Illusions* (**1995**), inspired by his story "The Last Illusion" (1985), both with higher budgets but less success, and executive-produce the two CANDYMAN films, based on his short story "The Forbidden" (1985). His subsequent novels, such as *Weaveworld* (**1987**), *Cabal, The Great and Secret Show: The First Book of The Art* (**1989**), *Imajica* (**1991**), *Everville: The*

Second Book of the Art (**1994**) and *Sacrament* (**1996**), have been EPIC FANTASIES, usually with pronounced horror elements, of ALTERNATE WORLDS or OTHERWORLDS, resolutely moving away from trappings of horror towards what the author likes to term the "fantastique". CB also wrote and illustrated the YA fantasy *The Thief of Always: A Fable* (**1992**). Much of his later work reflects his fascination with such recurring themes as hidden dimensions and physical TRANSFORMATION, and as a writer and moviemaker he continues to redefine the boundaries of the genre. Since 1991 he has lived and worked in Beverly Hills.　　　　　　[SJ]

Other work: *Incarnations: Three Plays* (coll **1995**).

As writer/director/producer/actor: *Salome* (made 1973; *1995*), 8mm experimental short film; *The Forbidden* (made 1975-8; *1995*), 16mm experimental short film; *Tales from the Darkside: The Yattering and Jack* (1987), scripted for tv from own story; *Hellbound: Hellraiser II* (*1988*), exec pr; CANDY-MAN (*1992*), exec pr; *Hellraiser III: Hell on Earth* (*1992*), exec pr; *Sleepwalkers* (*1992*), actor; *Candyman 2: Farewell to the Flesh* (*1995*), exec pr; *Hellraiser: Bloodline* (*1995*), exec pr; «Clive Barker's A-Z of Horror» (1996), as tv host; «The Thief of Always» (1996), musical ANIMATED MOVIE, exec pr.

Further reading: *Clive Barker's The Nightbreed Chronicles* * (coll **1990**) ed Stephen JONES; *Clive Barker's Nightbreed: The Making of the Film* * (**1990**) by Mark Salisbury and John Gilbert; *Clive Barker Illustrator* (graph coll **1990**) ed Steve Niles; *Clive Barker's Shadows in Eden* (anth **1991**) ed Jones; *Pandemonium* (anth **1991**) ed Michael Brown; *The Hellraiser Chronicles* * (coll **1992**) ed Jones; *Illustrator II: The Art of Clive Barker* (graph coll **1993**) by Fred Burke; *Clive Barker: Mythmaker for the Millennium* (**1994** chap) by Suzanne J. Barbieri; *Clive Barker's Short Stories: Imagination as Metaphor in the Books of Blood and Other Works* (**1994**) by Gary Hoppenstand; *Clive Barker's A-Z of Horror* (coll **1996**) ed Jones; *Clive Barker A-Z* (**1996**) by Burke.

BARKS, CARL (1901-　　) US artist and writer, the outstanding creator of COMICS stories featuring the DISNEY characters. CB developed an interest in cartooning as a child; in 1928-9 he sold his first gag cartoons to magazines like *Judge* and *The Calgary Eye-Opener*, whose staff he later joined. In 1935 he sent a few sample drawings to the Disney Studio in Hollywood, and was hired as an animation in-betweener, soon moving to the story department, where he started scripting and storyboarding **Donald Duck** and **Pluto** shorts. In 1942 he drew his first comic-book story, "Donald Duck Finds Pirate Gold" for *Four Color Comics #9*, modifying an already written script; in November that year he left Disney for Western Publishing, working for this firm exclusively for the rest of his career, contributing to at least 45 different comics titles released under various Western labels (these included Dell and Gold Key), along with some non-comics work for subsidiary firms (including Whitman and Golden Press). Some of the 45 titles are one-offs, or contain miscellaneous contributions. Three are of central importance. CB began to write and draw 10-page DONALD DUCK stories for the monthly *Walt Disney's Comics and Stories* in 1943, beginning with *#31* (though all CB stories from the mid-1960s on are reprints). Soon he was also creating longer stories (24-32 pages) starring Donald and nephews Huey, Dewey and Louie; these ran in *Four Color Comics* (from 1943). On screen and in the syndicated strips drawn by Al Taliaferro, Donald had until then been rather one-dimensional, but CB's Donald was a multifaceted character with all the qualities and faults of a normal human

being, an ANTIHERO with whom the reader could easily sympathize. The nephews were transformed from mischievous little ducklings into, as it were, their uncle's "triple good conscience". During 1943-52 CB wrote and drew for *Four Color Comics* a succession of unforgettable, ambitious stories, often set in exotic locales; they include "The Mummy's Ring" (1943), "Volcano Valley" (1947), "Lost in the Andes" (1949), "Ancient Persia" (1950) and "Old California" (1951). At the same time CB gave life to a number of other characters. Scrooge McDuck, Donald's stingy multimillionaire uncle, first appeared in "Christmas on Bear Mountain" (1947) in a **Donald Duck** one-shot; from 1952 Scrooge starred alongside Donald and the nephews in the third essential CB comics title, *Uncle Scrooge*, which he wrote and drew until the mid 1960s; stories included "Back to the Klondike" (1953), "Land Beneath the Ground!" (1956), "Island in the Sky" (1959) – CB's own favourite story – and "The Phantom of Notre Duck" (1965). Donald's implausibly lucky cousin, Gladstone Gander, has pestered him ever since his debut in 1948 in *Walt Disney's Comics #88*; the Beagle Boys, a gang of witless burglars, started trying to rob Scrooge's "Money Bin" in 1951 in *Walt Disney's Comics #134*; later came Gyro Gearloose (1952) in *Walt Disney's Comics #140*, a madcap "inventor of everything", and Magica DeSpell (1961) in *Uncle Scrooge #36*, a Neapolitan duck sorceress whose *idée fixe* is to steal Scrooge's "Old Number One", the first dime he ever earned.

When CB retired in June 1966 he had written over 500 stories, including a few starring MICKEY MOUSE, Grandma Duck and Gus Goose (whom CB had created in 1939 for a **Donald Duck** animated short) as well as Warner Bros's Porky Pig, Walter LANTZ's Andy Panda and MGM's Barney Bear and Benny Burro. Until 1973 he continued to contribute scripts (in layout form) to the *Uncle Scrooge*, *Donald Duck* and *Junior Woodchucks* comics. In 1971 CB started recreating the best scenes from his Duck comics as a series of oil paintings that now fetch high prices; he was permitted to continue this trade only after legal wrangling with Disney.

Although CB, like all Disney comics artists, worked anonymously, his name became familiar to fans as early as the 1960s, and he is now recognized worldwide. The monumental **Carl Barks Library**, published by Another Rainbow, has collected all his Disney stories. Western Publishing, Disney and Gladstone have constantly reprinted his stories in their comic books. In 1983 the asteroid (2730) Barks was named in his honour.　　　[AB]

Further reading: *Carl Barks and the Art of the Comic Book* (**1981**) by Michael Barrier.

BARNES, ARTHUR K. (1911-1969) US writer. ◊ Henry KUTTNER.

BARNETT, H.C. [s] ◊ Hugh B. CAVE.

BARNETT, LISA A(NNE) (1958-　　) US writer. ◊ Melissa SCOTT.

BARNETT, PAUL ◊ John GRANT.

BARON MÜNCHHAUSEN ◊ MÜNCHHAUSEN.

BARON MÜNCHHAUSEN Czech movie (*1962*). ◊ MÜNCHHAUSEN.

BARR, GEORGE (1937-　　) US illustrator. ◊ Virgil FINLAY.

BARR, KEN(NETH JOHN) (1933-　　) Scottish-born US artist and illustrator of fantasy and sf subjects who uses a wide range of very bright colours and whose compositions and subject matter have been heavily influenced by Frank FRAZETTA; his work shows also the influence of COMICS artists like Alex RAYMOND and Burne HOGARTH. Born in Glasgow, KB was apprenticed as a sign painter and then, after military service, worked in a commercial art studio in London. His first sf work was a cover for *Nebula* magazine in 1958. In 1968 he emigrated to the USA, where he continues to publish book and magazine covers and advertising material. [RT]

BARR, ROBERT (1850-1912) Scottish-born writer, raised in Canada, in the UK from 1881; of his 40 or more titles, some are sf and some are fantasy or SUPERNATURAL FICTION, like *From Whose Bourne* (1893), a POSTHUMOUS FANTASY in which a man "awakens" to find himself dead, and persuades a likewise deceased detective to solve his murder and thereby clear his wife of the crime. There are similar AFTERLIFE rectifications in "The Vengeance of the Dead", which appears in *Revenge!* (coll 1896). [JC]
Other works: *In a Steamer Chair and Other Shipboard Stories* (coll 1892); *The Justification of Andrew Lebrun* (1894); *The Adventures of Sherlaw Kombs* (1892; 1979 chap).

BARRETT, WILSON (1846-1904) UK writer, actor and theatre manager. ◊ Robert HICHENS.

BARRIE, [Sir] J(AMES) M(ATTHEW) (1860-1937) Scottish dramatist and novelist who received the Order of Merit and a baronetcy. His first foray into fantasy was the semi-autobiographical *The Little White Bird* (1902), in which a lonely, timid and melancholy man befriends a child in Kensington Gardens and makes up stories for him, including one about a magical boy who can fly and never grows older; the tale was later published separately as *Peter Pan in Kensington Gardens* (1906), with fine illustrations by Arthur RACKHAM. A modified version of this character became the hero of the play *Peter Pan, or The Boy who Would not Grow Up* (produced 1904; rev 1905; rev 1928) and the novel based on it, *Peter and Wendy* (1911; vt *Peter Pan and Wendy* 1921; vt *Peter Pan* 1951). The play established the tale as one of the most significant modern myths. In the earlier story, Peter lives on an island in Kensington Gardens; in the later one he lives in the Never Land (◊ NEVER-NEVER LAND: in different texts JMB spelled this ISLAND's name variously), is utterly committed to a life of adventure which revolves around his ongoing war with Captain Hook's PIRATES, and is steadfastly resistant to the innocently seductive appeal of Wendy Darling. Later JMB produced an epilogue to the story, *When Wendy Grew Up: An Afterthought* (produced 1908; 1957 chap); years have passed, and Wendy – a mother now, and no longer "innocent" – cannot fly back to Never-Never Land with Peter, so he takes her children instead, with her blessing. The underlying darkness of JMB's conception is signalled here by Peter Pan's inability – for he is locked in a kind of BONDAGE – even to remember Captain Hook. In 1920, JMB also wrote a Peter Pan screenplay, which was not, however, used in the movie *Peter Pan* (1924); it is printed in full in *Fifty Years of Peter Pan* (1954) by Roger Lancelyn GREEN. Other movies include DISNEY's animated PETER PAN (1953) and Steven SPIELBERG's HOOK (1991), the latter having something in common with *When Wendy Grew Up*.

Many versions of the story have appeared in various forms (◊ SEQUELS BY OTHER HANDS), the first of them probably being *The Peter Pan Picture Book* (1907) by Daniel O'Connor, illustrated by Alice B. Woodward (1862-1911). *Dear Brutus* (1917) is a play which sends a group of people

INTO THE WOODS so that they may briefly experience the lives they would have led had they made crucial choices differently; most discover they are happier as they are. *A Kiss for Cinderella: A Comedy* (produced 1916; 1920), a sarcastic farce updating the traditional CINDERELLA tale, includes a delusional sequence. *Mary Rose* (produced 1920; 1924) is a brilliantly poignant TIMESLIP romance whose heroine vanishes on a desolate islet, returning after many years without having aged (◊ TIME IN FAERIE). After her death she becomes a GHOST, unable to let go of her past even when confronting the grown son she left behind. Barrie's preoccupation with this theme was of considerable personal significance: he felt he had been alienated from his own mother's affections by her grief over the childhood death of his brother, who remained the same age in her memory; a tiny man himself, JMB offered his wife as sole explanation for his failure to consummate their marriage: "Boys can't love." His novella *Farewell, Miss Julie Logan: A Wintry Tale* (1931 chap) first appeared as a Christmas supplement to *The Times* in memory of Charles DICKENS's CHRISTMAS BOOKS; it is a lacklustre tale of a brief relationship between a young minister and a female ghost. [BS/JC]
Further reading: *Barrie: The Story of a Genius* (1929) by J.A. Hammerton; *J.M. Barrie: The Man Behind the Image* (1970) by J. Dunbar; *J.M. Barrie and the Lost Boys* (1979) by Andrew Birkin.

BARRINGER, LESLIE (1895-1968) UK editor and writer whose **Neustria Cycle** – *Gerfalcon* (1927), *Joris of the Rock* (1928) and *Shy Leopardess* (1948) – depicts events in a LAND OF FABLE, an imaginary French province in an ALTERNATE WORLD. The basic premise, vaguely presented, is that the Merovingian Dynasty does not split apart cAD750; instead, Neustria survives, and at the time of the three tales (c1400) is still thriving. Each tale follows a young man or woman through physical and moral perils into adulthood (◊ RITE OF PASSAGE). Of the three protagonists, Yolande of Baraine – the Shy Leopardess of the third novel – is perhaps the most interesting, as she successfully gambles her life (her "virtue" does not last the course) to gain autonomy in a male-dominated world. The sequence's alternate-world displacement serves not as an opening for magic but as a freeing of LB's imagination; the **Neustria Cycle** is far more intense and eloquent than his more straightforward historical novels. [JC]

BARRINGTON, E. Pseudonym of L. Adams BECK.

BARRINGTON, MICHAEL [s] ◊ Michael MOORCOCK.

BARTH, JOHN (1930-   ) US writer whose works generally display and advocate a playful attitude towards STORY that makes almost anything he has written understandable within a broad understanding of the FANTASTIC. The small craft that conveys the protagonist and his wife through *Sabbatical* (1982) is called *Story*, and the protagonist makes sense of his life only through beginning to tell stories, to re-create the mundane world. At the same time, only some of JB's novels and tales can be described as FANTASY in a more usual sense. Neither *The Sot-Weed Factor* (1960; rev 1967), despite its extraordinary elaborations of the mode of the PICARESQUE, nor *Giles Goat-Boy, or The Revised New Syllabus* (1966), which is a FABLE dressed as SCIENCE FICTION, can be so described. But some of the tales assembled in *Lost in the Funhouse: Fiction for Print, Tape, Live Voice* (coll 1968; exp 1969), like "Night-Sea Journey", are fantasy; and *Chimera* (coll of linked stories 1972) is full-blown fantasy. The first story here assembled, "Dunyazadiad", is an ARABIAN FANTASY centred on SCHEHERAZADE, who makes sense of her life (and, of course, literally saves it) through the telling of stories, which have been passed back through time by Barth himself when she cannot think of any, yet which are the stories she originally told. The second, "Perseid", takes the HERO into the confused world that comes after the great adventure has been accomplished, and in which he takes the form of a storyteller recounting his own past. The third, "Bellerophoniad", follows Bellerophon into preordained disaster – he plummets with Pegasus into the sea, and becomes not only a storyteller but, having undergone SHAPESHIFTING at the hands of a minor deity, an actual text of Story: a bundle of papers.

One of the epistolary contributors to *Letters* (**1979**) is part-insect, and hopes to breed with human females. *The Tidewater Tales* (coll of linked stories **1987**) features appearances, highly hypothetical though fabulous, of characters like Scheherazade (who reappears frequently in JB's work). *The Last Voyage of Somebody the Sailor* (**1992**) is constructed as a series of nested stories, in the coils of which are figures like Sinbad and, once more, Scheherazade. *Once Upon a Time* (**1994**) again provides a structure within which a Barth-like author creates, through MIRRORS and fabulated TIMESLIPS, a story which will serve to fabricate an author named Barth.

JB's *oeuvre* as a whole comprises a sometimes marvellous series of internal recursions, each tale commenting upon and re-interpreting its siblings, so that, in the end, everything he writes turns out to be the story of what he has written. [JC]

BARTHELME, DONALD (1931-1989) US writer whose short stories, for which he remains best known, are unduly constricted – and indeed betrayed – if treated as FANTASY. His short work was first assembled in several individual collections: *Come Back, Dr Caligari* (coll **1964**); *Unspeakable Practices, Unnatural Acts* (coll **1968**); *City Life* (coll **1970**); *Sadness* (coll **1972**); *Guilty Pleasures* (coll **1974**), published as nonfiction; *Amateurs* (coll **1976**); *Great Days* (coll **1979**); and *Overnight to Many Distant Cities* (coll **1983**). Two compilations, each containing some new stories, were *Sixty Stories* (coll **1981**) and *Forty Stories* (coll **1987**). *The Teachings of Don B.* (coll **1992**) ed Kim Herzinger assembles mostly uncollected material.

Of more direct fantasy interest are three of DB's four novels. *Snow White* (**1967**) is a REVISIONIST FANTASY in which the SNOW WHITE tale is subjected to a ruthless (and hilarious) secularization, with absurdist TIMESLIPS working to pile ANACHRONISM upon anachronism. *The Dead Father* (**1975**) gravely parodies the QUESTS common to fantasy narratives, as those who survive the death of the vast eponymous patriarch struggle to deliver his bier to a resting place; a fictional BOOK, *A Manual for Sons*, is inserted *passim*. *The King* (**1990**) places various actors out of the MATTER of Britain into the heart of WORLD WAR II, where they behave with exemplary (though dizzied) rectitude.

DB's value as a fantasy writer is greatest as a marker of boundaries. [JC]

Other works: *The Slightly Irregular Fire Engine* (**1971** chap), for younger children; *Paradise* (**1986**), associational.

BARTON FINK US movie (*1991*). Circle/20th Century-Fox. **Pr** Ethan Coen. **Exec pr** Ben Barenholtz, Bill Durkin, Ted and Jim Pedas. **Dir** Joel Coen. **Spfx** Stetson Visual Services. **Screenplay** Ethan and Joel Coen. **Starring** Judy Davis (Audrey Taylor), John Goodman (Charlie Meadows), Michael Lerner (Jack Lipnick), John Mahoney (Bill Mayhew), John Turturro (Barton Fink). 116 mins. Colour.

Firebrand playwright Fink is drawn in 1941 to Hollywood's Capitol Pictures, where his dreams of socialist movies for the common man are soon shattered. Lodged in the seedy Hotel Earle – where his sole MUSE is a peeling calendar bathing-belle picture on the damp wall – his writer's block is cemented by the intrusions of his neighbour, Meadows, a *real* common man rather than the ideal of Fink's earlier writing. Seeking inspiration from once-great writer Mayhew, now a drunk, Fink inadvertently beds Mayhew's mistress and ghost-screenwriter Taylor, waking next morning to find her murdered. From here the plunge into madness is irrevocable: the remainder of *BF* is, or is not, a fantasy of PERCEPTION. Fink's madness interacts synergistically with that of Meadows, who proves a SERIAL KILLER; only with Taylor's severed head as an AMULET in a box by his typewriter can Fink write – she is indeed a ghostwriter. But the SURREALISM extends beyond the plot that leads inevitably to Meadows's, and in a way Fink's, nemesis, for Hollywood itself is depicted as an ABSURDIST FANTASY, almost as an ALTERNATE REALITY. The triumph of *BF* is that its false realities become realer than our own.

Drawing in part on such sources as cinema clichés and URBAN LEGENDS, *BF* is also more specifically RECURSIVE: Capitol boss Lipnick is identifiable with Louis B. Mayer (1885-1957), Fink himself with Clifford Odets (1906-1963) and Mayhew with William Faulkner (1897-1962). [JG]

BASEBALL Baseball has had a powerful hold on the US imagination: "Casey at the Bat" and "A Visit from St Nicolas" are two of the best-known US poems. Because of the game's broad appeal and idyllic aura, baseball has frequently been a subject in US fantasy.

First, while the actual origins of baseball are unclear, writers have devised imaginary histories (indeed, the lie that Civil War general Abner Doubleday invented the game itself qualifies as a kind of MYTH OF ORIGIN). In Mark TWAIN's "Papers from the Adam Family" (written ?1870-1906; **1962**) Methuselah grumpily observes some compatriots playing that new fad, baseball, well before the Flood; and in W.P. KINSELLA's *The Iowa Baseball Confederacy* (**1986**) LEONARDO DA VINCI flies over a baseball field in a balloon and claims he invented the game.

Next, baseball frequently attracts divine and supernatural interest or intervention, a recurring theme of Kinsella, the writer who has most often produced baseball fantasies. In *Shoeless Joe* (**1982**) – filmed as FIELD OF DREAMS (*1989*) – a disembodied Voice commands a farmer to build a baseball field, where dead players gather to play their game once again. In *The Iowa Baseball Confederacy*, a dead Native American returns to observe an exhibition game lasting thousands of innings, whose outcome he believes will enable him to reunite with his lost love. In "The Last Pennant before Armageddon" (**1985**) the manager of the Chicago Cubs learns that GOD will end the world when the Cubs win another pennant, so he deliberately loses a key game. In "The Night Manny Mota Tied the Record" (**1985**) a spectator learns that a recently killed baseball player will experience RESURRECTION if he agrees to give up his own life.

Other stories that mingle baseball and divine beings include *The Great American Novel* (**1973**) by Philip Roth (1933-), narrated by the spirit of writing (called Word Smith), and Douglass WALLOP's *The Year the Yankees Lost the Pennant* (**1954**), about a frustrated Washington Senators fan who sells his soul to the DEVIL so he can become a star player for his team and help them win the World Series; the story later became a Broadway musical and was filmed as *Damn Yankees* (**1958**). In ANGELS IN THE OUTFIELD (**1952**; remade **1994**) a baseball team has ANGELS helping them win their games; a friendly GHOST helps a children's baseball team in David A. Adler's juvenile novel *Jeffrey's Ghost and the Leftover Baseball Team* (**1984**); and a frustrated ex-player is visited by the ghost of his catcher friend in the movie *Cooperstown* (**1992**). An aura of the supernatural permeates *The Natural* (**1952**) by Bernard Malamud (1914-1986), filmed as *The Natural* (**1984**). The protagonist of Robert COOVER's *The Universal Baseball Association, J. Henry Waugh, Prop.* (**1968**) makes up his own imaginary players and, the final scene indicates, also brings them to life in some alternate plane of REALITY. Leonard P. Kessler's juvenile *Old Turtle's Baseball Stories* (**1982**) adopts the form of the **Uncle Remus** tale (◊ Joel Chandler HARRIS) to tell about a ball-playing octopus, moose, kangaroo and squirrel. Michael BISHOP's fantasy *Brittle Innings* (**1994**) poetically mingles baseball with the FRANKENSTEIN mythos. [GW]

BATMAN Fantasy COMIC-book crimefighter, the archetypal MASKED AVENGER, tagged "The Caped Crusader" and "The Dark Knight", who has become a 20th-century ICON. He is not a SUPERHERO as the term is generally used, since he is represented as having no superhuman abilities.

Created by artist Bob KANE and writer Bill Finger, Batman is the bizarre *alter ego* of millionaire socialite Bruce Wayne. His adventures take place in Gotham City, a version of NEW YORK with Gothic overtones. Batman first appeared in *Detective Comics #27* (May 1939) in a six-page story subtitled *The Case of the Chemical Syndicate*; an introduction read: "The 'Bat-Man', a mysterious and adventurous figure fighting for righteousness and apprehending the wrong-doer, in his lone battle against the evil forces of society . . . his identity remains unknown." In the opening scene, Police Commissioner Gordon, a character who has remained a key figure throughout Batman's long history, is seen chatting with "his young socialite friend Bruce Wayne" when a telephone call informs him of a murder. Subsequent scenes introduce the Bat-Man, who dispatches one criminal by pitching him off a roof and another by dumping him in a vat of acid – "A fitting end to his kind," Batman opines dourly. The story concludes with Commissioner Gordon recounting the events to Wayne, whom he privately views as "a nice young chap – but he must lead a boring life". The last frame reveals that Wayne is the Bat-Man.

The character was an instant success, and further stories were published in *Detective Comics* and in a new quarterly comic book, *Batman* (#1 Spring 1940), with Batman dispassionately killing criminals in divers ways. When in 1941 the USA began to take part in WWII, however, Batman's publishers, DC National (later DC COMICS), became uncomfortable with the violence, and Batman became a more pacifist figure: "The Batman never carries or kills with a gun" (*Batman #4* 1941).

A two-page feature in *Detective Comics #33* (1940) told how a 12-year old Bruce Wayne witnessed the shooting of his parents by a petty criminal and vowed ". . . to avenge their deaths by spending my life warring on all criminals". A chance sighting of a bat as he ruminates on his need for a disguise leads to the birth of "The Batman". This MYTH OF ORIGIN has been retold and embellished many times.

Detective Comics #38 (1940) introduced young Dick Grayson, son of husband-and-wife trapeze act The Flying Graysons, who had been killed by a gang boss. Dick, recruited as Batman's aide, became Robin, billed from the beginning as "The Boy Wonder"; the pair were later dubbed "The Dynamic Duo". *Batman #16* introduced the chubby Englishman, Alfred Beagle, a Shakespearean actor and brilliant detective, who was to become the butler at Wayne Manor. In *Detective #83* (1944) Alfred became tall and thin, and in *Batman #214* (1969) he was renamed Alfred Pennyworth. Batwoman (wealthy heiress Kathy Kane) was introduced in *Detective Comics #233* (1956), and her niece Betty made a few appearances as Batgirl in 1961, wearing a flamboyant MASK and party frock; in 1966 Commissioner Gordon's librarian daughter donned a very sexy skintight costume to become a more readily recognizable Batgirl (*Detective Comics #359*).

With the publication of an increasing number of stories featuring Batman, the original idea was embellished with the introduction of the Batcave, a technology-packed cavern beneath Wayne Manor, accessed via a grandfather clock in the study; here were parked a specially built high-powered auto – first introduced in *Detective Comics #30* (1939) but not referred to as the Batmobile until *#48* (1941) – the Batgyro (*Detective Comics #31* 1939; later the Batcopter), the Batplane (*Batman #4* 1941), etc. All Batman's vehicles, along with the gadgets contained in his utility belt, have been frequently enhanced and updated.

Batman by now faced a gallery of bizarre villains – a convention already partly established by Chester Gould (1900-1985) in his long-running **Dick Tracy** newspaper strip. The first was Batman's most enduring adversary, The Joker (*Batman #1* 1940); others have been The Cat (*Batman #1*; renamed Catwoman in *#2*), The Penguin (*Detective #58* 1941), Two-Face – Harvey Kent, later renamed Harvey Dent (*Detective Comics #66* 1942) – and The Riddler (*Detective #140* 1948).

The character of Batman and the stories in which he features have undergone many changes since his first appearance. The "eerie figure of the night" image created in 1939 softened considerably after the creation of Robin: during WWII he swung in on his Batrope chirping wittily, "Mind if I join the *party*? So sorry to *drop in* so *unexpectedly* this way!" before socking wrongdoers cleanly on the jaw, wisecracking tediously the while, and taking advantage of every opportunity to exhort the reader to "buy war bonds and stamps".

The Kane/Finger creative team was extended to include artists Jerry Robinson (1922-   ), Sheldon Moldoff (1920-   ), Dick Sprang (1915-   ), George Roussos, Jack Burnley (1911-   ) and others. Finger remained the chief writer.

In the late 1940s and early 1950s there was increasing US public concern over the moral climate. During the McCarthy era (1950-54) this crusade intensified, focusing on popular literature, and many comics creators reacted by stressing the moral uprightness of their characters. Batman became, in the words of one commentator "a kind of benign scoutmaster" – who on one occasion actually gave his readers a full-page lecture on sportsmanship and racial equality (*Batman #57* 1950). Nevertheless, the main theme of the stories remained the catching of criminals – albeit some rather strange ones at times – through detection, Batman's own brand of forensic science, and the strength and agility of The Dynamic Duo. Nor did the tradition of Batman operating at night entirely disappear.

The publication of *Seduction of the Innocent* (**1954**) by Frederick Wertham caused a revolution in the COMICS publishing industry. Wertham claimed to present scientific proof of a significant causal connection between comics-reading and criminal behaviour; he also saw evidence of homosexuality in the relationship between Batman and Robin. DC National reacted by making the stories more fanciful. Batman began to have conflicts with aliens from distant planets and other dimensions, became involved in TIME TRAVEL, and encountered ferocious MONSTERS – indeed, actually became monstrous himself from time to time, turning on one occasion into a human fish (*Batman #118* 1958).

In 1964 a major revamp of the character – drawn by Carmine Infantino (1925-   ) – was undertaken in an attempt to return to the eerie and threatening figure of the night; this attempt was undermined when the tv series started in 1966. Batman now became a very hot property for DC: within months of the first broadcast, about 1000 **Batman** products had been licensed, including models, mask-and-cape sets and a vast range of paraphernalia carrying the Batman logo. Many of the comic-book stories aped the Pop Art camp of the tv shows.

This influence was short-lived. In 1969 new writers and artists were brought in to reassess thoroughly the original image of the "grim crimefighter driven by an obsession born of tragedy". The fantasy elements associated with the bat costume were retained, but the camp silliness and cod sf were jettisoned. Dick Grayson (Robin) went to college (*Batman #217* 1969) and was thus effectively discarded; although he turned up again from time to time and featured in a number of solo adventures, his association with Batman was eventually terminated.

Neal ADAMS, with writer Denny O'Neill (1939-   ), enhanced Infantino's concept of Batman. In Adams's hands – and those of artists Jim Aparo (1932-   ), Dick Giordano (1932-   ) and Irv Novick – the character underwent a remarkable metamorphosis. The visual image became darker and intensely dramatic. Now Batman was a realistically drawn fantasy figure of the night; the new stories were character-led crime thrillers or PSYCHOLOGICAL THRILLERS spiced with elements of SUPERNATURAL FICTION. The Joker, Two-Face and other weird villains were presented as psychotic criminals with appropriate psychological profiles. Batman had a number of romantic involvements with female characters – most memorably during this period with one Talia, the daughter of an interesting new villain, Ra's al Ghul; in Adams's memorable depiction of their first kiss a shirtless Batman, still wearing the batcowl, utility belt and tight black knickers, and with blood seeping from claw-wounds across his chest, clasps the yielding beauty in a scene redolent of sadomasochism (*Batman #244* 1972). Their relationship was actually consummated on-page in the one-off comic book *Batman: Son of the Demon* (1987), as a result of which, unknown to Batman, she bore him a son.

The gallery of villains continued to expand with the introduction in 1970 of a particularly appropriate and enduring new adversary created by Frank Robbins (1917-1994) and Adams – the Man-Bat, otherwise Professor Kirk Langstrom of Gotham's Museum of Natural History, who, after injecting himself with bat-serum, mutates into a very interesting variation on the bat theme (*Detective Comics #400*).

Dick Grayson was replaced as Robin by Jason Todd in *Batman #366* (1983), but readers were unenthusiastic and, when offered an opportunity to vote on his destiny in the three-part story *A Death in the Family* (*Batman #426-#429*), opted for his demise. Frank MILLER introduced a short, chunky, 13-year-old female Robin, Carrie Kelly, in *Batman: The Dark Knight Returns* (books 1-4 1986; graph coll **1986**) and another new Robin, Timothy Drake, was introduced in *Batman #442* 1989.

Miller brought his revolutionary storytelling techniques to bear in *The Dark Knight Returns*. Set in an unspecified future, this deals with an ageing, cynical and disillusioned Batman coming out of a 10-year retirement to deal with a murderous cult known as The Mutants. Commissioner Gordon, now 70, is forced to retire, and is succeeded by a woman, Ellen Yindel. The batmobile has developed into a tank, bristling with weapons and techno-wizardry, and the batcopter has become a similarly armoured juggernaut. Aided by Robin/Carrie, Batman defeats the leader of The Mutants, but his violence leads to his being labelled a criminal – and to the formation of another cult calling themselves Sons of the Batman. The Joker is released from Gotham's mental institution, Arkham Asylum, and immediately indulges in an orgy of mass-murder; he dies in a final confrontation with Batman, who then is himself hunted as a murderer. SUPERMAN is despatched to bring him to justice. In the showdown Batman humiliates Superman and then apparently dies of a heart-attack – to be revived after the burial by Robin. Batman enlists the remnants of The Mutants and Sons of the Batman to plan a new crimefighting strategy in the extensive caves under a now destroyed Wayne Manor.

The publication of *The Dark Knight Returns* signalled DC's more openminded policy regarding experimentation with the character, and a number of distinguished comics creators took a renewed interest. Alan MOORE and Brian Bolland embellished the now well established Batman-versus-Joker ethos with the remarkable *The Killing Joke* (1988), and Grant MORRISON and Dave MCKEAN looked deeper into the psychoses of some of Batman's bizarre adversaries in *Arkham Asylum* (1989). In *Batman, Year One* (*Batman #404-#407* 1987; graph coll **1988**), beautifully and realistically drawn by David Mazzuchelli (1960-   ), Miller went on to update the Batman MYTH OF ORIGIN, setting it in a 1980s Gotham. Other creators have set Batman in the 1880s for a JACK THE RIPPER story – *Gotham by Gaslight* (1989) by Brian Augustyn and Mike MIGNOLA, which is full of GASLIGHT-ROMANCE effects – and in a FAR FUTURE, when all that remains of Batman is a crimefighting computer which he spent his final years designing and constructing – *Digital Justice* (1990) by Jose Luis ["Pepe"] Moreno, in which all the artwork was computer-generated.

Other imaginative new treatments of Batman have continued to appear in the 1990s, leading to a chaotic proliferation of incompatible Batman stories running concurrently in various DC titles: it remains to be seen if any will have a lasting effect. For example, in *Brotherhood of the Bat* (1995) Batman is replaced by a strange league of new Batmen, all wearing slightly differing versions of the Batman costume. Batman currently (1995) features regularly in the following titles: *Batman, Detective, Shadow of the Bat, Legends of the Dark Knight, DC Showcase, The Batman Chronicles* and *The Batman and Robin Adventures* (this last being based on the tv cartoon series), as well as in a sporadic flow of one-offs and special editions.

Batman has also starred in newspaper strips: a daily and Sunday series (1943-7) written primarily by Alvin Schwartz and drawn by Kane and Burnley; a shortlived series (1953); and dailies (1966-72) and Sundays (1966-9) written by Whitney Ellsworth and E. Nelson Bridwell, and drawn by Infantino, Moldoff, Giella, Plastino and others, then (1989-current) written by Max Allan Collins (1948-) and William Messner Loebs (1949-) and drawn by Infantino and Marshall Rogers. [RT/DR]

Further reading: *Batman from the Thirties to the Seventies* (**1979**) reprints milestone stories from that period; *Tales of the Dark Knight* (**1989**) by Mark Cotta Vaz; *Batman and Me* (**1989**) by Bob Kane with Tim Andrae.

BATMAN US tv series (1966-8). Greenway Productions/20th Century-Fox/ABC. **Pr** Howie Horwitz. **Exec pr** William Dozier. **Dir** Robert Butler and many others **Writers** Lorenzo Semple Jr, Henry Slesar and many others. **Based on** the COMIC-book characters created by Bob KANE. **Starring** John Astin (The Riddler), Tallulah Bankhead (The Black Widow), Anne Baxter (Zelda), Madge Blake (Aunt Harriet Cooper), Milton Berle (Louie the Lilac), Victor Buono (King Tut), Art Carney (The Archer), Joan Collins (The Siren), Yvonne Craig (Barbara Gordon/Batgirl 1967-8), Howard Duff (Cabala), Maurice Evans (The Puzzler), Zsa Zsa Gabor (Minerva), Frank Gorshin (The Riddler), Neil Hamilton (Commissioner Gordon), Van Johnson (The Minstrel), Carolyn Jones (Marsha), Eartha Kitt (Catwoman), Bruce Lee (Kato), Liberace (Chandell), Ida Lupino (Dr Cassandra), Roddy McDowall (The Bookworm), Burgess Meredith (The Penguin), Alan Napier (Alfred Pennyworth), Julie Newmar (Catwoman), Otto Preminger (Mr Freeze), Vincent Price (Egghead), Michael Rennie (The Sandman), Stafford Repp (Chief O'Hara), Cliff Robertson (Shane), Cesar Romero (The Joker), Barbara Rush (Nora Clavicle), George Sanders (Mr Freeze), Walter Slezak (The Clock King), Malachi Throne (Falseface), Rudi Vallee (Lord Phogg), Eli Wallach (Mr Freeze), Burt Ward (Dick Grayson/Robin), David Wayne (The Mad Hatter), Adam West (Bruce Wayne/Batman), Van Williams (The Green Hornet), Shelley Winters (Ma Parker). 120 30min episodes. Colour.

One of the biggest tv hits of the 1960s, this series portrayed BATMAN strictly for laughs. While the basic setting remained the same, with Batman and Robin striking out against criminals from their Batcave, this version made a deliberate effort to be different – and succeeded admirably. Though the actors played their roles with apparent seriousness, the series went for a deliberately campy style, as evidenced in the plots, sets and dialogue.

Unusually, two episodes were aired each week: the first always ended with the heroes seemingly in deadly peril, and the second got them out of it. Only one person (a villain in the pilot episode, played by Jill St John) was ever killed or maimed; instead, the many fist fights were punctuated by unusual camera angles and giant "Splat!", "Pow!" and other comics-style visual gimmicks. Another trademark was the "Bat" labelling of everything in the Batcave, from the Batcomputer to the Batpoles that led down from stately Wayne Manor. The biggest eyecatcher was the atomic-powered Batmobile.

The series was an instant hit, and the USA was swept by Batmania. Batman costumes were the biggest sellers at Hallowe'en, and the studio was hounded by actors wanting to be guest VILLAINS; the producers cast many of them in cameo roles. As well as using villains straight out of the comic books, the series introduced several new ones. By far the most memorable VILLAINS were The Penguin, The Riddler, The Joker and Cat Woman (several actors appeared in some of these roles); *Batman* (**1986**) (◊ BATMAN MOVIES), which sprang from the series, featured all four.

Oddly, though, after just two seasons the ratings began to fall. In a move to save the show, the producers introduced a new character, Batgirl, and cut back to one episode per week. Played by Yvonne Craig in a skintight costume, Batgirl rode into battle on her customized Batcycle, a powerful motorcycle. Nevertheless, the Caped Crusader was vanquished by tv's deadliest foe – the Nielsen ratings. Stars West and Ward were later reunited to provide the voices of Batman and Robin in an animated series, *Batman and the Super Seven* (1980-81). [BC]

BATMAN AND THE SUPER SEVEN US animated tv series (1980-81). ◊ BATMAN (1966-8).

BATMAN MOVIES A number of movies have been based on DC COMICS' BATMAN.

1. *Batman* US serial movie (**1943**). Columbia. **Dir** Lambert Hillyer. **Starring** Douglas Croft (Robin), Lewis Wilson (Batman). 1 22min episode plus 14 16min episodes. B/w.

Batman and Robin, after various shenanigans, succeed in defeating a conspiracy headed by master-criminal Doctor Daka. We have been unable to obtain a viewing copy of this movie. [JG]

2. *Batman and Robin* US serial movie (**1949**). Columbia. **Pr** Sam Katzman. **Dir** Spencer Gordon Bennet. **Screenplay** Royal K. Cole, George H. Plympton, Joseph F. Poland. **Starring** Jane Adams (Vicki Vale), John Duncan (Robin/Dick Grayson), William Fawcett (Professor Hammil), Robert Lowery (Batman/Bruce Wayne), Leonard Penn (Carter/The Wizard), Lyle Talbot (Commissioner Gordon), Rick Vallin (Barry Brown). 1 22min episode plus 14 16min episodes. B/w.

A masked master-criminal, The Wizard, steals Professor Hammil's new invention, the Remote Control Machine, which gives him the power to take over control of any moving vehicle within a 50-mile radius. Batman and Robin set out to right matters, but The Wizard proves resourceful, eventually developing the technology for personal INVISIBILITY. This leaden TECHNOFANTASY features car chases and fisticuffs galore, but little real excitement. [JG]

3. *Batman* (vt *Batman – The Movie*; vt *Batman '66*) US movie (**1966**). 20th Century-Fox/Greenlawn/National Periodical Publications. **Pr** William Dozier. **Dir** Leslie H. Martinson. **Spfx** L.B. Abbott. **Mufx** Ben Nye. **Screenplay** Lorenzo Semple Jr. **Novelization** *Batman vs. The Fearsome Foursome* * (**1966**) by Winston Lyon. **Starring** Frank Gorshin (Riddler), Neil Hamilton (Commissioner Gordon), Burgess Meredith (Penguin), Lee Meriwether (Catwoman), Alan Napier (Alfred), Stafford Repp (Chief O'Hara), Cesar Romero (Joker), Burt Ward (Robin/Dick Grayson), Adam West (Batman/Bruce Wayne). 105 mins. Colour.

A camped-up tale based on the tv series BATMAN (1966-8) and guying, with cheerful unsubtlety, SUPERHERO conventions and COMIC-book pretensions (notably the heroes' ultra-wholesomeness). Credibility is thrown to the wind in a monumentally tortuous plot in which four prime VILLAINS – Catwoman, The Joker, The Riddler and The Penguin – unite in a world-domination attempt. *B* flags about the halfway mark, but is still a better movie than generally granted – and certainly a pleasingly lighthearted contrast to **4** and **5**. [JG]

4. *Batman* US movie (**1989**). Warner. **Pr** Peter Guber, Jon Peters. **Exec pr** Benjamin Melniker, Michael E. Uslan. **Dir** Tim BURTON. **Vfx** Derek Meddings. **Screenplay** Sam Hamm, Warren Skaaren. **Novelization** *Batman* * (**1989**) by Craig Shaw GARDNER. **Starring** Kim Basinger (Vicky/Vicki Vale), Michael Gough (Alfred), Jerry Hall (Alicia), Pat Hingle (Commissioner Gordon), Michael Keaton (Batman/Bruce Wayne), Jack Nicholson (Jack Napier/Joker), Jack Palance (Carl Grissom), Robert Wuhl (Alexander Knox). 126 mins. Colour.

Gotham City has come to regard Batman as merely an URBAN LEGEND, but then he apprehends Napier, treacherous sidekick of gang boss Grissom, who ends up in a vat of chemicals. Plastic surgery leaves Napier's face a ghastly MASK crossed by a rictus grin – and so he recreates himself as The Joker, gathering CIRCUS-clown thugs around him and aiming to become criminal master of Gotham City. Journalists Knox and Vale, working on the Batman story, trip over reclusive millionaire Bruce Wayne, whom Vale beds; later she has rough sex with Batman, yet fails to identify the two. The Joker, too, lusts for Vale, and she becomes the real focus of the struggle between the VILLAIN and Batman – which struggle Batman wins.

B owes much more to the dark, tormented soul portrayed in Frank MILLER's *Batman: The Dark Knight Returns* (1986; graph coll **1986**) than to the simpler SUPERHERO of the earlier COMICS, the tv series and **3**. It is dour – often visually splendid – with pretensions to psychological depth. Through music, lighting, timing and surreal cityscapes Burton tries to invest his URBAN FANTASY with the weight of MYTH, yet the burden seems too great for the subject matter easily to bear; occasional flashes of realism make the rest seem suddenly kitsch. Yet the visual ponderousness of *B* is, undeniably, impressive. [JG]

5. *Batman Returns* US movie (**1992**). Warner. **Pr** Tim BURTON, Denise Di Novi. **Exec pr** Peter Guber, Benjamin Melniker, Jon Peters, Michael Uslan. **Dir** Burton. **Spfx** Michael Fink, Chuck Gaspar. **Screenplay** Wesley Strick, Daniel Waters. **Novelizations** *Batman Returns* * (**1992**) by Craig Shaw GARDNER; *Catwoman* * (**1992**) by Lynn ABBEY and Robert ASPRIN. **Starring** Danny DeVito (Penguin/Oswald Cobblepot), Michael Keaton (Batman/Bruce Wayne), Michelle Pfeiffer (Catwoman/Selina Kyle), Christopher Walken (Max Shreck). 126 mins. Colour.

The sequel to **4**. Thrown into the sewers by his parents – revolted by their half-bird, half-man offspring – Oswald Cobblepot is now the Penguin, whose thugs, the Red Triangle Gang, terrorize Gotham City. Kyle, secretary to corrupt businessman Shreck, discovers his crookedness and is defenestrated by him; she survives the fall, becoming the villainous Catwoman by night while retaining her mundane personality by day. Kyle and Wayne fall in love even as Catwoman, allied to the Penguin, is at war with Batman. But her real loathing is for Shreck, and finally Catwoman and Batman together destroy the VILLAINS. Script and direction are leaden, and Pfeiffer is incapable of projecting the feline sexuality required of Catwoman (Lee Meriwether did a much finer job in **3**); since the Catwoman/Kyle dichotomy should be the movie's spine, the rest falls apart. [JG]

6. *Batman – Mask of the Phantasm* US ANIMATED MOVIE (**1993**). Warner Bros. **Pr** Benjamin Melniker, Michael

Uslan. **Exec pr** Tom Ruegger. **Dir** Eric Radomski, Bruce W. Timm. **Screenplay** Alan Burnett, Paul Dini, Martin Pasko, Michael Reaves. **Voice actors** Hart Bochner (Arthur Reeves), Kevin Conroy (Batman/Bruce Wayne), Dana Delany (Andrea Beaumont), Mark Hamill (Joker), Stacy Keach Jr (Carl Beaumont/Phantasm), Dick Miller (Chuckie Sol), John P. Ryan (Buzz Bronski), Abe Vigoda (Salvatore Valestra), Efrem Zimbalist Jr (Alfred). 73 mins. Colour.

Borrowing much of its visual style (notably the range of colour values) and musical ambience from **4** and **5**, this animated feature is probably better than both. Much of the action happens in flashbacks to a decade ago when Wayne's parents had not long died, and he resolved to become a vigilante crimefighter. Around then he met and became affianced to Andrea Beaumont, the one woman he has ever loved; but she and father Carl fled the country when Carl's criminal associates turned the heat on him. In the present, a sinister, fog-enshrouded, masked DEATH figure, The Phantasm (the MASK and husking voice are like Darth Vader's), is killing those associates, and Batman is blamed; also, Andrea is back, and she and Wayne attempt to resume their former relationship. Enter The Joker: ten years ago he was the least of the gang threatening Carl; now he is a criminal mastermind intent on destroying both The Phantasm (before The Phantasm destroys him) and Batman. In due course The Phantasm, assumed to be Carl, proves to be Andrea. She and Batman seemingly kill The Joker, and Batman assumes that she, too, dies in the process; in fact she escapes to leave his life again, this time forever.

Much of the animation is technically not sophisticated, yet the limitations are cleverly exploited to contribute to an immense stylishness – to create the effect of an excellent COMIC book brought to life. Many of the camera-angles and sequence-constructions owe more to live-action direction than to traditional animation. *B – MOTP* generates the thrill of believed-in, and hence somehow credible, modern MYTH. [JG]

7. *Batman Forever* US movie (**1995**). Warner Bros. **Pr** Tim BURTON, Peter MacGregor. **Exec pr** Benjamin Melniker, Michael E. Ulsan. **Dir** Joel Schumacher. **Vfx sv** John Dykstra. **Screenplay** Janet Scott Batchelor, Lee Batchelor, Akiva Goldsman. **Starring** Drew Barrymore (Sugar), Jim Carrey (Edward Nygma/Riddler), Michael Gough (Alfred Pennyworth), Pat Hingle (Commissioner Gordon), Tommy Lee Jones (Harvey Dent/Harvey Two-Face), Nicole Kidman (Dr Chase Meridian), Val Kilmer (Bruce Wayne/Batman), Debi Mazar (Spice), Chris O'Donnell (Dick Grayson/Robin). 122 mins. Colour.

The sequel to **4** and **5**. A new director pitches a new Batman against a new pair of supervillains. Perhaps truer to the earlier comics and tv series, it sees the arrival of Dick Grayson/Robin. The foes are Harvey Two-Face and The Riddler (a role originally intended for Robin Williams). Two-Face is an ex-DA turned schizoid; always in two minds, he believes the only true justice is luck and so flips a coin to decide whether his victims live or die – shades of Luke Rhinehart's *The Dice Man* (**1971**; rev 1983). He has two molls (the refined, virginal Sugar and the decadent, kinky Spice) and his abode is split into two opposing styles ("Heavy Metal meets Homes and Gardens"). The Riddler is mad scientist Edward Nygma, whose invention of an implement to utilize neural energy is dismissed by Wayne. As revenge The Riddler joins Two-Face and forms NygmaTech, selling a product to the public that animates their fantasies, projecting them onto a tv screen, but also taps into their brainwaves and feeds information to Nygma. The love interest is Meridian, a criminal psychologist who becomes obsessed with Batman but falls for Wayne. Robin becomes Batman's partner after his parents are killed by Two-Face, thus giving the two a common cause.

Schumacher's vision is not so much brighter than Burton's, but is strewn with neon strips and fluorescent paints among Gotham's lowlife and supervillains; he combines the use of colour and the Gothic darkness effectively. The plot is confusing and manically directed. Batman is overshadowed by the leading villains, particularly Carrey. [JT]

BATMAN: THE ANIMATED SERIES ◊ ot of *The* ADVENTURES OF BATMAN AND ROBIN (1992-4).

BATTLE OF THE ASTROS vt of *Kaiju Daisenso* (*1965*). ◊ GODZILLA MOVIES.

BAUDELAIRE, CHARLES (PIERRE) (1821-1867) French poet. Théophile GAUTIER, who provided CB's funeral oration and wrote his biography, identified his work as the definitive example of the "Decadent literary style" (◊ DECADENCE). His worldview – which arose out of a curious combination of a dissolute lifestyle, the interrupted voyage to the Orient on which he was sent in the hope of curing his bad habits, and the fact that his one major commercial project was the translation into French of the works of Edgar Allan POE – attained multifaceted expression in the classic poetry collection *Les Fleurs du Mal* (coll **1857**; exp 1861; part trans Richard Herne Shepherd in *Translations from Charles Baudelaire with a Few Original Poems* coll **1869** UK; full trans Cyril Scott as *The Flowers of Evil* coll **1909** US), which was sufficiently shocking to attract prosecution for obscenity. The contents included: CB's crucial contribution to the tradition of Literary Satanism (◊ SATAN) "Les Litanies de Satan"; two poems eroticizing the VAMPIRE motif, "Le Vampire" and "Les Métamorphoses du vampire"; and four poems entitled "Spleen" which delimited one of the notions central to the aesthetics of Decadence.

CB went on to establish the prose-poem as a significant form in Decadent art, his work in that vein including such hymns to ESCAPISM as "Anywhere out of the World" (1857), whose title is a quotation from Thomas Hood, "L'Invitation au voyage" (1862) and "La Chambre double" (1862), and such sarcastic fantasies as "Les Dons des fées" (1862) and "Les Tentations, ou Éros, Plutus et la Gloire" (1863). These were intended for publication together as «Le spleen de Paris», but were eventually issued in volume IV (coll **1869**) of *Oeuvres complètes* (coll **1868-70** 7 vols) along with the clinical study *Les paradis artificiels, opium et haschisch* (**1860**), which takes a jaundiced view of the potential of drug-assisted escapism. They are also in vol IV of the definitive *Oeuvres complètes* (coll **1923-65** 19 vols) ed Jacques Crépet, along with the novella "Le Jeune enchanteur; histoire tirée d'un palimpseste de Pompeïa" (1846 as by Baudelaire Dufays). Translations of the prose-poems can be found in many collections. [BS]
Further reading: *Charles Baudelaire: His Life* by *Théophile Gautier, Translated into English, with Selections from his Poems, "Little Poems in Prose" and Letters to Sainte-Beuve and Flaubert, and An Essay on his Influence* (coll **1915**) by Guy Thorne (1876-1923); *Baudelaire* (**1957**) by Enid Starkie; *Baudelaire* (**1994**) by Joanna Richardson.

BAUDINO, GAEL (1955-) US writer and harpist, formerly a priestess of Dianic Wicca. Her music and religious background inform several of her novels, particularly *Gossamer Axe* (**1990**), in which a 6th-century Celtic harpist flung into the 20th-century USA uses heavy rock to rescue her lover from her captivity by the Sidhe (◊ FAIRIES). GB's tetralogy about the last of the ELVES is remarkable for its lack of whimsy, with elves, heretics and followers of the Old Religion hunted down equally by the Inquisition. The series comprises *Strands of Starlight* (**1989**), *Maze of Moonlight* (**1993**) and *Shroud of Shadow* (**1993**), all set in a fictional country in a well drawn medieval Europe, plus *Strands of Sunlight* (**1994**), where the elf-blood reawakens in modern Denver. GB's **Dragon** trilogy – *Dragonsword* (**1988**), *Duel of Dragons* (**1991**) and *Dragon Death* (**1992**) – is more usual fantasy fare: a dumpy lecturer suddenly finds herself a stunningly beautiful sword-wielding dragon-rider in a standard FANTASYLAND which proves the mental creation of her professor. These books are characterized by a dark harshness that jars with the flip telling. [DVB]

BAUER, STEVEN (1940-) US writer in whose *Satyrday* (**1980**), a fantasy SATIRE, an owl kidnaps the Moon, among many other high-jinks. His versions of tales from AMAZING STORIES – *Steven Spielberg's Amazing Stories* * (coll **1986**) and *Volume II of Steven Spielberg's Amazing Stories* * (coll **1986**) – are competent. [JC]

BAUM, L(YMAN) FRANK (1856-1919) US writer whose immense importance to the field of fantasy rests upon his creation of the land of OZ, the first successful OTHERWORLD in US literature. He began his writing career in the 1880s with nonfiction, and over his life published some adult novels, including *The Last Egyptian: A Romance of the Nile* (**1908**), released anon; but he is of interest almost exclusively for his CHILDREN'S FANTASY, beginning with some of the stories assembled in *The Purple Dragon and Other Fantasies* (coll **1976**), which date from as early as 1897. Other early texts are *Mother Goose in Prose* (coll **1899**), remarkable mostly for the illustrations by Maxfield PARRISH, and *A New Wonderland* (**1900**; vt *The Surprising Adventures of the Magical Monarch of Mo* 1903).

In the same year as *The New Wonderland*, LFB published the first volume in the **Oz** sequence, *The Wonderful Wizard of Oz* (**1900**; vt *The New Wizard of Oz* 1903), upon which was based the most famous of all fantasy movies, *The* WIZARD OF OZ (*1939*). The sequence continued with *The Marvelous Land of Oz* (**1904**; vt *The Land of Oz* 1914), *Ozma of Oz* (**1907**; vt *Princess Ozma of Oz* 1942 UK), *Dorothy and the Wizard of Oz* (**1908**), *The Road to Oz* (**1909**), *The Emerald City of Oz* (**1910**), *The Patchwork Girl of Oz* (**1913**), *Tik-Tok of Oz* * (**1914**), *The Scarecrow of Oz* (**1915**), *Rinkitink in Oz* (**1916**), *The Lost Princess of Oz* (**1917**), *The Tin Woodman of Oz* (**1918**) – featuring a human woodsman who has suffered a gradual TRANSFORMATION into two successive versions of his Tin Woodman identity, problems of identity arising when the first version has a perplexed conversation with his former head and later, with the second version, encounters a patchwork character constructed from their former "meat" bodies – *The Magic of Oz* (**1919**) and *Glinda of Oz* (**1920**). Associated plays include *The Wizard* (produced 1902), *The Woggle-Bug* (produced 1905) – which generated *The Woggle-Bug Book* * (**1905**) – and *The Tik-Tok Man of Oz* (produced 1913). Six volumes of tales for younger children – including *Jack Pumpkinhead and the Sawhorse* (**1914** chap) – were published in 1914 as the **Little Wizard Series**; they were reissued in one volume as *Little Wizard Stories of Oz* (omni **1939** UK). It is generally agreed that the later **Oz** volumes represent a falling-off from the standards of the earlier titles; the sequence was so popular, however, that other writers, including Ruth Plumly Thompson (1891-1976), continued to generate new stories (◊ SEQUELS BY OTHER HANDS). LFB's other work is competent, but without genius.

OZ is a POLDER which is magically proximate to the surrounding desert of the Western USA; characters constantly make their way from Kansas across various THRESHOLDS into this magic land of their DREAMS – at least until Oz is rendered finally invisible . . . although reader protests forced a reappearance. At the same time Oz represents a contradiction of the "real" world (hence, perhaps, the distaste US educationists and librarians have notoriously felt for the series). It is a kind of EDEN inhabited by unfallen creatures, and stands in radical contrast to the implacable BONDAGE of impoverished dust-bowl Kansas, a state whose reality seems inescapable – hence, perhaps, some of the poignancy of Geoff RYMAN's "Was . . ." (**1992**; vt *Was* 1992 US), which also treats the historical Kansas as essentially inescapable.

The Wonderful Wizard of Oz is a classic FAIRYTALE in which a young orphan travels into FAERIE, where she is set a task and safely accomplishes it. Dorothy's QUEST for a way home – in the company of three magic COMPANIONS (a scarecrow, a TALKING ANIMAL – the Cowardly Lion who seems at first to be nothing but a Miles Gloriosus [◊ COMMEDIA DELL'ARTE] – and the woodsman) – is superbly balanced and told. Further volumes offer numerous additions to the population of Oz, some highly bizarre and accompanied by explicit pleas for tolerance of odd-looking outsiders. The sequence as a whole is important for Oz itself, for the huge inventiveness that makes it seem so remarkably populous, and for the sense of liberation from bondage it continues to convey. [JC]

Other works: *Dot and Tot in Fairyland* (**1901** chap); *American Fairy Tales* (coll **1901**; exp 1908); *The Master Key: An Electrical Fairy Tale* (**1901**); *The Life and Adventures of Santa Claus* (**1902**) and *A Kidnapped Santa Claus* (**1904**; **1969** chap) (◊ SANTA CLAUS); *The Enchanted Island of Yew* (**1903**); *Queen Zixi of Ix, or The Story of the Magic Cloak* (**1905**); *John Dough and the Cherub* (**1906**); *The Sea Fairies* (**1911**) and its sequel *Sky Island* (**1912**); *Jaglon and the Tiger Fairies* (**1953**); *Animal Fairy Tales* (coll **1989**), stories first published 1905.

See also: *The* WIZARD OF OZ for discussion of the movies based on LFB's creation.

BEAGLE, PETER S(OYER) (1939-   ) US writer who has published only four novels (plus a novella later issued in book form) in 35 years, establishing a reputation as craftsman and innovator with his first in 1960; his star dipped in the 1980s, but his skill and vitality were reaffirmed in the 1990s.

His beginnings were precocious. After a nongenre story – "Telephone Call" for *Seventeen* in 1956 – he published his first and by many still most-loved novel, *A Fine and Private Place* (**1960**), before he was 21. In its grave and controlled polish, it does not seem the work of a young man. Though the tale's warm intricacies – and PSB's occasional flights of sentimentality – tend to obscure the fact, *A Fine and Private Place* is a SUPERNATURAL FICTION in chamber-opera form, whose small cast of sharply realized characters play out their destinies within a narrow compass, without any genuine exit available into a fantasy HEALING: their coming to terms with life and death is, in the end, this-worldly. In a Bronx

graveyard lives Jonathan Rebeck, a recluse from life's challenges who is kept alive, in part, by a talking raven, and who has survived here for almost two decades (there is an echo of this in Anne RICE's *Interview with the Vampire* [**1976**], whose Lestat dwells long years hidden in a cemetery). He is surrounded by GHOSTS of those buried there, who seem to remain sentient – after the manner of the dead in the style of classic POSTHUMOUS FANTASY – until they have come to terms with the true shapes of the lives they had in the flesh. Rebeck, a live companion and two ghosts all duly discover that the stories they've been telling themselves about themselves are more or less false, and that they must live (or die) the true story; the novel closes with the two ghosts lying embraced in a single grave, while the two live characters go off to have breakfast. The influence of Robert NATHAN upon this novel has been acknowledged by PSB, who has said it is "a direct descendant" of Nathan's *One More Spring* (**1933**).

After another short story – "Come, Lady Death" (1963), in which DEATH is invited to a party and (as usual in FABLES of this sort) arraigns the hostess and guests – PSB published his most significant single work to date, *The Last Unicorn* (**1968**), a highly self-conscious, sophisticated, classic fantasy. The main characters are the eponymous UNICORN, who learns – in a manner which perfectly demonstrates that the PERCEPTION of WRONGNESS can easily be defined as a realization that a THINNING of the world has already occurred – that she is the last unicorn still at liberty; and the incompetent magician, Schmendrick, who is doomed never to age (◊ BELATEDNESS; BONDAGE) until he grows into his full power as MAGUS and artist (◊ KENOSIS). As the unicorn begins her QUEST to find her fellows, she becomes involved in a CIRCUS, where she meets some of the numerous UNDERLIER characters who fill the text with cunningly presented echoes of traditional fantasy narratives; and subsequently she gathers COMPANIONS (including Schmendrick) around her. As they continue, it becomes apparent that they know they are in a STORY, one which is full to the point of pixilation with figures and icons and PLOT COUPONS and PLOT DEVICES from the CAULDRON OF STORY; and every new underlier they meet (the band of outlaws who re-enact their leader's dream of being a new ROBIN HOOD; Molly Grue, the crone who rather unseriously takes on the role of perpetual virgin [◊ VIRGINITY]; the DRAGON-fighting prince) further deepens their sense that somehow they are going to have to live through – to tell by their own actions the predetermined story of – the events to come.

The quest ends in the WASTE LAND created by King Haggard, whose dark self (the Red Bull) has imprisoned all the other unicorns of this SECONDARY WORLD in the surf that pounds the beach that borders the EDIFICE where he lives in a state of self-knowing refusal of rebirth (he is a classic KNIGHT OF THE DOLEFUL COUNTENANCE). The unicorn disguises herself from Haggard by permitting Schmendrick to inflict upon her a thinning METAMORPHOSIS into human form as Lady Amalthea, and passes through the NIGHT JOURNEY of this bondage into the climax of the tale, the point at which every significant member of the cast (including the unicorn in her true form again) comes together on the beach, in a moment of tension before the story unfolds that we can characterize as a RECOGNITION of passage: "For Molly Grue, the world hung motionless in that glass moment. As though she were standing on a higher tower than King Haggard's, she looked down on a pale paring of land where a toy man and woman stared with their knitted eyes at a clay bull and a tiny ivory unicorn. Abandoned playthings – there was another doll, too, half-buried; and a sandcastle with a stick king propped up in one tilted turret. The tide would take it all in a moment, and nothing would be left but the flaccid birds of the beach, hopping in circles."

After this TROMPE L'OEIL moment the story then unfolds in the traditional, destined way. The unicorn defeats the Red Bull; the whitecaps reassemble themselves into unicorns; the edifice of King Haggard topples into the sea; and the waste land undergoes an almost instant HEALING, with green sprouting everywhere. Schmendrick and Molly Grue depart together. There are tears and laughter – for much of the book is hilarious. The story has been told.

PSB himself scripted the rather flat and disappointing ANIMATED-MOVIE version of the novel, *The* LAST UNICORN (*1982*); what comes across as charm on the printed page seems instead, when rendered as a screenplay, somewhat overladen with tweeness, on occasion excruciatingly so – a fault carried through to the animation and some of the voice acting.

The novella *Lila, the Werewolf* (1969 *Guabi* as "Farrell and Lila the Werewolf"; **1974** chap) is a very early example of the kind of CONTEMPORARY FANTASY on which many fantasy writers – like Charles DE LINT and most of the SCRIBBLIES – have concentrated since about 1980: URBAN FANTASY told in a colloquial tone of voice, and tending to treat CROSSHATCH moments as though they redeemed the modern world city. In this story the narrator's girlfriend turns out to be a WERE-WOLF whose transformations are tied to her periods; the tale is told with a kind of cool humorousness which has seemed offensive to some. Joe Farrell (the narrator) reappears in PSB's next and least esteemed novel, *The Folk of the Air* (**1986**), which amiably, but lacking some of his previous mythopoeic intensity, depicts a California cast, most of whom belong to the League for Archaic Pleasures (based on the SOCIETY FOR CREATIVE ANACHRONISM) and who are variously involved in activities which invoke a genuine crosshatching reaction from other realms in response to their games. The course of the tale gives PSB room to create a complex vision of the GODDESS, whose BONDAGE to a THINNING world – along with that of her son and others – becomes an intermittently powerful account of how, in a mature fantasy narrative, improvident role-playing – most of the characters are bound to UNDERLIER figures they have played – can generate arguments about the relationship of PARODY to authentic being. These arguments, which are relevant to any analysis of the RECURSIVE nature of much modern fantasy (and indeed, much modern literature in general), are clearly evoked here. But there is a lack of bite – which, along with PSB's seemingly easy recourse to sentimentalized character portrayals, gave rise to fears about his future importance.

As with Joe Farrell in *The Folk of the Air*, one of the protagonists of *The Innkeeper's Song* (**1993**) – an unnamed MAGUS who has called previous students to his side to help him fight off another one-time student who needs to send his SOUL to the UNDERWORLD to honour a bargain – is also from an earlier work; in this instance it is Schmendrick from *The Last Unicorn*. But here the presence of Schmendrick, identifiable only by inference, does not point to a softening of the texture of the new tale; and *The Innkeeper's Song* is not in any proper sense a sequel. Nor is it easily amenable itself to sequelizing: as the reader discovers only late in the text,

the events described have become, over the decades since they occurred, a familiar and oft-told STORY (◊◊ TWICE-TOLD), which ends. The plot is not particularly complex, for the old magus is saved in the end from the demons of the underworld when one of his young helpers acts as a psychopomp (in a reversal of the ORPHEUS myth) and guides him back to the surface of the world, a vaguely Oriental LAND OF FABLE. The complexity lies in the telling, as it is recounted from the perspective of a large number of first-person narrators, some important to the tale, others peripheral. As these narrators are all reliable at least insofar as their knowledge goes, the effect is not *Rashomon*-like but, rather, is of facets of story.

Throughout this glittering presentation of a loved story, certain motifs appear and reappear: for example, one of the magus's old students is a woman who seems a man, not by means of disguise but through a constant process of META-MORPHOSIS, so that it is at times impossible to understand what one is perceiving; and a SHAPESHIFTER man (he becomes a fox on occasion) turns out to be a fox who becomes a man. In short, the story is irradiated with transformations and passages. Although it takes place almost entirely in a rural inn – the main excursion being a useless attempt by the magus's students to find and defeat his opponent – *The Innkeeper's Song* encompasses much of the range available to modern fantasy. It demonstrates once again PSB's remarkable and exemplary grasp of his chosen genre.

[JC]

Other works: *The Fantasy Worlds of Peter Beagle* (omni **1978**; vt *The Fantasy World of Peter S. Beagle* 1980 UK), assembling *A Fine and Private Place*, *The Last Unicorn*, *Lila, the Werewolf* and "Come, Lady Death"; *The* LORD OF THE RINGS (*1978*), screenplay; *The Garden of Earthly Delights* (**1981**), a study of the painting by Hieronymus BOSCH; *The Last Unicorn*; *A Fine and Private Place* (omni **1991**); *Peter S. Beagle's The Immortal Unicorn* (anth **1995**) ed PSB, Janet Berliner and (anon) Martin H. GREENBERG.

BEAN, NORMAN ◊ Edgar Rice BURROUGHS.

BEAR, GREG(ORY DALE) (1951-   ) US author whose contributions to SCIENCE FICTION, especially hard sf, are extensive and important (◊ *SFE*). Occasionally his large-scale sf devices can be decoded as RATIONALIZED FANTASY – thus *Blood Music* (**1985**) suggests that a sufficient density of microscopic consciousnesses could adjust quantum REAL-ITY to block nuclear chain reactions by, effectively, a communal WISH.

Psychlone (**1979**) is a somewhat stumbling HORROR novel whose premise is that nuclear explosions smash victims' SPIRITS into electromagnetic fragments which, unable to depart the Earth, combine into potent ELEMENTAL-like forces. The eponymous psychic horror results from the WORLD WAR II bombings of Japan, and takes revenge on US towns via murderous POSSESSION. It is ultimately nullified by particle-beam weapons, leaving a stench of hubris and untranscendable WRONGNESS since the US Army now owns (and cannot resist testing) the means to destroy SOULS.

GB's prime work of fantasy is the **Michael Perrin** or **Songs of Earth and Power** diptych, comprising *The Infinity Concerto* (**1984**) and *The Serpent Mage* (**1986**) and assembled as *Songs of Earth and Power* (omni **1992** UK; rev 1994 US – the first UK edn claims revisions not in fact incorporated until the US edn). The story involves PORTAL crossings between Earth and a beautiful but harsh SEC-ONDARY WORLD, the "Realm", whose stylized FANTASYLAND qualities are deliberate: it is the imperfect creation of a self-elevated GOD, a place of exile for the haughty Sidhe (◊ ELVES), stray humans, and half-breeds ("Breeds"), following millennia-past wars. CHILDREN born there lack SOULS and become abominations; there are no DREAMS; unpleasant METAMORPHOSES abound.

The hero Michael crosses over (evading a vampiric LIMINAL BEING) and receives magical training from THREE ancient Breeds. This includes the notion of "casting a SHADOW" – sacrificing unwanted portions of one's own personality to make a tangible decoy or WEAPON. Magic also operates through "songs of power" in words, MUSIC, dance or even flavour (portals may be opened by one special wine): S.T. COLERIDGE's "Kubla Khan" (1816) is inevitably an unfinished SPELL, and a dangerous human magician is discomfited by Michael's completion of the poem in words that bring its pleasure dome to destruction. Now the BALANCE is awry, and the flawed realm's fatal THINNING accelerates; its inhabitants return disruptively to Earth. After consulting the Serpent Mage – a shape-cursed casualty of that prehistoric war, now confined to Loch Ness – Michael vies with the Sidhe adept who would at best be able to create another marred realm, and himself achieves a new creation which merges healingly with Earth. There is renewal, for this is an INSTAURATION FANTASY.

Songs of Earth and Power has many striking concepts, sometimes confusingly presented. It suffers from inner cross-purposes, with aspirations towards mysticism and myth undermined by the explanatory reasonableness of sf. GB is a welcome visitor to fantasy, rather than a denizen.
[DRL]

BEARDSLEY, AUBREY (VINCENT) (1872-1898) UK illustrator and writer, a central graphic artist of the late 19th century. Delicate, daring, perverse and profoundly erotic, his drawings remain instantly recognizable. He worked almost exclusively in black-and-white, his extremely elegant scraped Indian-ink line laying down and surrounding black patterns with a sense of ease, simplicity and effortless arabesque; his late work more and more hinted at obscure and liquid METAMORPHOSES, giving his last images a COMMEDIA DELL'ARTE atmosphere. An example of this is his illustration to Ernest Dowson's *The Pierrot of the Minute* (play **1897**).

Much of this work appeared in *The Yellow Book* (◊ MAGAZINES) – which he cofounded, and for which he was appointed art editor in 1894, losing the position after the Oscar WILDE scandal in 1895 – and *The Savoy*, which he also cofounded, in 1896. The first book he illustrated – an edition of Sir Thomas MALORY's *Le Morte Darthur* (**1894**) – portrayed GUINEVERE as a *belle dame sans merci* (◊ LAMIA), and his females almost invariably entwine his men (who themselves notoriously tend to boast huge phalluses). Other works of interest illustrated by AB include editions of Wilde's *Salome* (1894), Aristophanes' *Lysistrata* (1896), and LUCIAN's *True History*. His fiction of note is restricted to *The Story of Venus and Tannhauser* (cut version 1894 *The Yellow Book* as "Under the Hill"; **1907**), a highly erotized rendering of the tale; the bowdlerized *Yellow Book* version was published in *Under the Hill and Other Essays in Prose and Verse* (coll **1904**).

AB's life – he is reputed to have slept with only one person (his sister) and he died a Roman Catholic convert – and work have come to stand for subsequent generations as an essential definition of the FIN DE SIÈCLE. [JC]

BEAST FABLE Three closely associated but distinct terms are used in this encyclopedia to cover some of the innumerable uses to which animals are put in fantasy. As the entry on ANIMAL FANTASY argues, two of these terms – "beast fable" and "animal fantasy" – can overlap; the third, TALKING ANIMALS, describes the behaviour of animals in fantasy narratives whose protagonists are not themselves animals.

The beast fable features animal protagonists whose behaviour and nature can be compared with that of humans, usually in terms of SATIRE or ALLEGORY; these comparisons are normally made without much attention being paid to real animal behaviour, though authors of the beast fable tend to draw on FOLKLORE and FAIRYTALE, and upon our conventional sense of the way various animals act (◊◊ COLOUR CODING; ESTATES SATIRE), in order to provide useful caricatures: the cunning CAT, the treacherous SERPENT, the TRICKSTER fox, the idiot-savant ASS, and so on. Beast fables can be in a WONDERLAND or OTHERWORLD, but they are as often set in a fantasticated version of the real world, where their protagonists act as humans and frequently interrelate with them – as in PUSS IN BOOTS. The protagonist of a pure animal fantasy, by contrast, will almost certainly talk to others of its species, and to members of different species as well, but not to humans; and its home territory will be part of the dangerous real world.

The talking animals in Rudyard KIPLING's **Jungle Books** (**1894-5**) take on beast-fable characteristics whenever they are engaged in educating young Mowgli. Similarly, Kenneth GRAHAME's *The Wind in the Willows* (**1908**), set in an Arcadian LANDSCAPE whose connection to the mundane world is at times tenuous, is close to the pure beast fable, though with hints of animal fantasy when Mole or Ratty (but never Toad) are described in naturalistic terms. Other tales which share characteristics of both beast fable and animal fantasy include: Walter DE LA MARE's *The Three Mulla-Mulgars* (**1910**); A.A.MILNE's *Winnie-the-Pooh* (**1926**); George ORWELL's *Animal Farm* (**1945** chap); Ann LAWRENCE's *Tom Ass, or The Second Gift* (**1972**); Walter WANGERIN's *The Book of the Dun Cow* (**1978**); William HORWOOD's **Duncton Chronicles** (**1980-93**); Brian JACQUES's **Redwall** sequence (**1980**-current); and Mary STANTON's *The Heavenly Horse from the Outermost West* (**1988**). In these texts the boundary between the central action and the real world is porous and variable. All feature protagonists who seem sometimes animal and sometimes a PARODY of human foibles, and in all of them it can be difficult to recollect whether their featured animals do or do not wear clothes (animals in pure animal fantasies are, of course, never dressed).

In the pure beast fable, the real world – which may be portrayed in extremely exaggerated terms – tends to fade into a backdrop for the exemplary FABLE being told stage-front; and the behaviour of the animal characters who convey the story is sometimes very remote from what would seem characteristic of their true natures. This has always been so; the TAPROOT sources of the beast fable date from antiquity – *Mr Monkey and Other Sumerian Fables* (anth **1995**) adapted by Jessica Amanda SALMONSON (◊◊ MESOPOTAMIAN EPIC) presents some extremely early examples – and are very extensive. Every traditional literature in the world offers examples of the beast fable; and from Aesop (◊ AESOPIAN FANTASY) through the 12th-century fox tales assembled by Pierre de Saint-Cloud as *Roman de Renard* (partial trans Patricia Terry as *Romance of Reynard* **1983**; no complete trans into English) and the MONKEY epic (partial trans Arthur Waley as *Monkey* **1942**) by Wu Ch'êng-ên (1500-1582) to Jean de La Fontaine, it has served as a favourite literary device for propaganda and satire. Some examples from the literature of fantasy include: Joel Chandler HARRIS's **Uncle Remus** (from **1881**); *The Cunning Little Vixen* (**1920**) by Rudolf Tesnohlidek (1882-1928); Christopher MORLEY's *Where the Blue Begins* (**1922**); the education sequences in T.H. WHITE's *The Sword in the Stone* (**1938**); C.S. LEWIS's **Narnia** sequence (**1950-56**); Almet JENKS's *The Huntsman at the Gate* (**1952**); David GARNETT's *The Master Cat: The True and Unexpurgated Story of Puss in Boots* (**1974**); Günter GRASS's *The Flounder* **1977**) and *The Rat* **1986**); David Henry WILSON's *Coachman Rat* (**1985**); Deborah GRABIEN's *Plainsong* (**1990**); and *Beastly Tales: From Here and There* (coll **1991**) by Vikram Seth (1952-   ). In the field of ANIMATED MOVIES there are so many beast fables that it would be pointless to try to list them. Many PANTOMIMES, children's plays and children's tv series are likewise beast fables.

If the animal fantasy has an inherent bent towards the real world, the beast fable moves in another direction. It hovers at the threshold of full fantasy, and sometimes crosses over.
[JC]

BEAST WITH FIVE FINGERS, THE US movie (**1946**). Warner. **Pr** William Jacobs. **Exec pr** Jack L. Warner. **Dir** Robert Florey. **Spfx** H. Koenekamp, William McGann. **Screenplay** Curt Siodmak. **Based on** "The Beast with Five Fingers", in *The Beast with Five Fingers and Other Stories* (coll **1928**) by William F(ryer) HARVEY. **Starring** Robert Alda (Bruce Conrad [listed in credits as Conrad Ryler]), Victor Francen (Francis Ingram), Andrea King (Julie Holden), Peter Lorre (Hillary Cummins), J. Carrol Naish (Castanio). 88 mins. B/w.

Wheelchair-bound pianist Ingram, pillar of the small US community in an Italian fishing village, is murdered, and later his left hand vanishes from his coffin. Ingram's will, leaving all to his lovely young nurse, Holden, is disputed by greedy relatives and a crooked lawyer. The latter is strangled and one of the relatives attacked, both seemingly by the autonomous hand. So far this could be a PSYCHOLOGICAL THRILLER, but now we see the hand flaunting itself in front of Ingram's secretary (and murderer) Cummins; at last, as he tries to burn it, it crawls from the flames to choke him. The conclusion attempts to persuade us we have merely seen Cummins's crazed HALLUCINATIONS; but this rationalization (◊ RATIONALIZED FANTASY) does not explain all. At the very least *TBWFF* is a fantasy of PERCEPTION; it reads better as a DARK FANTASY. [JG]

BEATLES, THE UK musicians George Harrison (1943-   ), John Lennon (1940-1980), Paul McCartney (1942-   ) and Ringo Starr (real name Richard Starkey; 1940-   ). Their contributions to written fantasy were exclusively Lennon's books *In His Own Write* (**1964**) and *A Spaniard in the Works* (**1965**), collections of stories and sketches filled with *non sequiturs*, puns and other wordplay. These schoolboy entertainments have not aged well.

In their later years as a band, working primarily in the studio, the Beatles wrote and performed a number of "psychedelic" songs with fantastic or surrealistic language and sound effects. Notable examples include: McCartney's and Lennon's "Yellow Submarine" (1966), a children's song; Lennon's oracular "Tomorrow Never Knows" (1966);

Harrison's sermon on Hindu mysticism, "Within You Without You" (1967); McCartney's haunting "The Fool on the Hill" (1968), which some thought was a song about CHRIST; Lennon's chaotic and apocalyptic sound collage "Revolution #9" (1968); Lennon's "Cry Baby Cry" (1968), which mimics the language of nursery rhymes; and Starr's undersea fantasy "Octopus's Garden" (1969). Two of Lennon's songs stand out: "Lucy in the Sky with Diamonds" (1967), surely designed to convey the feeling of an LSD trip (though Lennon insisted the song was inspired by one of his son's drawings), and "I Am the Walrus" (1967), a grimmer journey through similar territory with references to Lewis CARROLL and Edgar Allan POE.

As moviemakers, TB were solely responsible for MAGICAL MYSTERY TOUR (*1967* tvm), generally regarded as not good. They had little control over their other movies, two of which are of fantasy interest: *Help!* (**1965**) is a spy spoof with sf elements, and YELLOW SUBMARINE (*1968*) is an ANIMATED MOVIE suggested by their songs. Years after their break-up TB inspired a less worthy live-action fantasy movie, *Sergeant Pepper's Lonely Hearts Club Band* (*1977*), with the Bee Gees and Peter Frampton as hapless stand-ins. A US tv animated series, *The Beatles* (**1964-6**), also with no Beatles participation, had some fantasy elements. There was also a movie version of the popular stage show *Beatlemania* (*1980*), with Beatles impersonators performing songs backed by spectacular visual effects.

The Beatles' solo careers after 1970 displayed much less interest in fantasy – though one can mention Starr's roles as MERLIN in *Son of Dracula* (*1974*), which he also produced (◊ DRACULA MOVIES), as the eponymous *Caveman* (*1981*), and as the miniature conductor in the first seasons of the children's tv series *Shining Time Station*, based on the Reverend W. Awdry's **Thomas the Tank Engine** stories; and Harrison served as executive producer and theme-song writer on Terry GILLIAM's TIME BANDITS (*1981*).

The Beatles entered the realm of urban FOLKLORE in 1969, when a fan observed that the mumbled words at the end of "Strawberry Fields Forever" (1967), played backwards, were "I buried Paul" (though Lennon claimed he had actually said either "Strawberry jam" or "I'm very bored"). Similar backwards messages were discovered in other songs – "Paul is gone, man, miss him, miss him" at the end of "I'm So Tired" (1968), and "Turn me on, dead man" in "Revolution #9" – which inspired the story that Paul had died in a late-1966 car accident ("He blew his mind out in a car" says "A Day in the Life" [1967]) but had been secretly replaced by a Scottish lookalike named William Campbell. A variant was that Paul had staged his own death so as to go live on a remote Greek island with other supposedly dead rock stars like Brian Jones, Janis Joplin, Jimi Hendrix and Jim Morrison; deciphering all the album clues would yield a phone number which one could dial to get a plane ticket to the island hideaway. [GW]

BEATLES, THE (tv series) ◊ The BEATLES; YELLOW SUBMARINE (*1968*).

BEAUCLERK, HELEN (1892-1969) UK writer and translator, in France much of her life; from 1923 until his death in 1953 she lived with Edmund DULAC, who illustrated her first two novels. Her fantasy novels transpose materials derived from the rich tradition of French fantastic fiction into an English mode. *The Green Lacquer Pavilion* (**1926**) celebrates, tongue-in-cheek, the enduring French fascination with various facets of the legendary Orient, and *The Love of* *the Foolish Angel* (**1929**) seeks to capture the spirit of the heretical fantasies of Anatole FRANCE. In the former novel an Oriental screen becomes a PORTAL whereby eight people gathered for an English country-house party in 1710 are conveyed into a vividly ornate SECONDARY WORLD. The latter tells of Tamael, who unwittingly attaches himself to the cause of rebellious LUCIFER and becomes a fallen ANGEL by accident; a blatant misfit in HELL, he is dispatched to Earth to tempt human beings to sin, but falls in love with the lovely Basilea and becomes her protector. Both novels are toned down by comparison with their French models, as might be expected in work published in English by a female author, but the resultant delicacy is not to their disadvantage. *The Mountain and the Tree* (**1936**) contains an opening sequence set in the Stone Age and draws upon Frazerian theories (◊ ANTHROPOLOGY) in constructing an account of the changing role of GENDER women in early human societies. [BS]

BEAUMONT, CHARLES Working (later legalized) name of US author and scriptwriter Charles Leroy Nutt (1929-1967), who produced artwork as E.T. Beaumont and Charles McNutt, and also wrote as C.B. Lovehill, Michael Phillips and Frank Anmar (the last two pseudonyms were sometimes used jointly with William F. NOLAN). An early sf enthusiast, CB produced his own fanzine, *Utopia*, in 1945, and supplied illustrations to sf MAGAZINES. After an abortive stint as an actor at Universal Studios he worked in MGM's art department, then turned to writing with "The Devil, You Say" (1951 *Amazing Stories*). His output soon shifted from sf to mystery and HORROR, often with a vein of the unreal in the style of Ray BRADBURY. CB favoured psychological horror over supernatural, although into his best work, which is SLICK FANTASY, he often wove the inexplicable. His short fiction was originally collected as *The Hunger* (coll **1957**; vt with title story cut *Shadow Play* 1964 UK), *Yonder* (coll **1958**), *Night Ride* (coll **1960**) and *The Magic Man* (coll **1965**) – the latter title has become identified with CB himself. His approach was ideally suited to the tv series *The* TWILIGHT ZONE, for which he produced over 20 scripts 1959-63. Fantasy and sf movies scripted or coscripted by CB include *Queen of Outer Space* (**1958**), *The Wonderful World of the Brothers Grimm* (**1962**), BURN WITCH BURN (**1961**; ot *The Night of the Eagle*) – based on *Conjure Wife* (**1953**) by Fritz LEIBER – *The Premature Burial* (**1961**) – one of Roger CORMAN's Edgar Allan POE cycle – *The Haunted Palace* (**1963**) – loosely based on *The Case of Charles Dexter Ward* (**1952** UK) by H.P. LOVECRAFT – and 7 FACES OF DR LAO (**1964**) – based on *The Circus of Dr Lao* (**1935**) by Charles G. FINNEY. One of his last movie scripts, for *Brain Dead* (**1989**), written in 1963 about a neurosurgeon who loses his identity, tragically reflects CB's own demise: he succumbed to Alzheimer's disease and died aged only 38. [MA]

Other works: *The Edge* (coll **1966** UK), selected from *Yonder* and *Night Ride*; *Best of Beaumont* (coll **1982**) ed Christopher Beaumont; *Charles Beaumont: Selected Stories* (coll **1988**; vt *The Howling Man* 1992) ed and with biography by Roger Anker, which won the BRAM STOKER AWARD.

As editor: *The Fiend in You* (anth **1962**), mostly psychological horror.

Further reading: *The Work of Charles Beaumont: An Annotated Bibliography and Guide* (**1986** chap; rev 1990) by William F. NOLAN.

BEAUMONT, E.T. ◊ Charles BEAUMONT.

BEAUMONT, MARIE LEPRINCE DE ◊ Marie LEPRINCE DE BEAUMONT.

BEAUTÉ DU DIABLE, LA Italian/French movie (**1949**). ◊ René CLAIR.

BEAUTIES OF THE NIGHT vt of *Les* BELLES DE NUIT (**1952**).

BEAUTY AND THE BEAST One of the most durable of all WONDER TALES, this contains as its core situation that is an UNDERLIER for many later stories. Versions vary, but in the best-known – that by Marie LEPRINCE DE BEAUMONT (1756; trans 1783) – the basic tale is: A bankrupted merchant, with three daughters of marriagable age, the youngest called Beauty, travels in hope of restoring his fortune. Before departing he asks the daughters what presents they would like him to bring back; while the elder sisters respond covetously (there is more than a shade of the CINDERELLA tale here), Beauty asks merely for a red rose. The merchant's journey is in vain. As he returns he is lost in a blizzard. He discovers a great house, inside which is warmth, a table set for dinner, and a good bed. In desperation, he eventually helps himself to food and sleeps in the bed. Next morning he finds in the castle grounds a rosebush, and takes a branch of it. At once a MONSTER appears and says that, while he had willingly given the merchant a night's lodging, the merchant must die for having rewarded his generosity with theft. The merchant pleads for mercy, and explains why he stole the roses. Beast (it is a proper name) allows the merchant three months to tidy up his affairs and gives him a chest of treasure, also saying that, if he can persuade one of his daughters to come to the castle and forfeit her life in his stead, he may consider the account settled. The merchant goes home, explains all, and Beauty insists she should take his place. At the castle, though, Beast spares her life, explaining he would rather wed her; she slowly becomes fond of him, despite revulsion at his terrifying exterior, but cannot consent to marriage. SCRYING in a MIRROR, she discovers her father is near death, and Beast permits her to make one last visit home, on her promise to return. A while later, in a DREAM, she sees that Beast is pining unto death for her; she realizes she loves him, and is delighted to awaken back in the castle, transported there by a FAIRY GODMOTHER figure. She finds Beast apparently dead; discovering him alive, she declares her LOVE and at once a long-ago CURSE is lifted from him and he becomes a handsome prince; they are happy ever after, while her proud and greedy sisters are condemned to lives of misery.

Leprince de Beaumont's version is based on that in *Les contes marins ou la jeune Américaine* (**1740**) by Gabrielle-Suzanne de Villeneuve (1695-1755); a translation of her version appears in *Beauties, Beasts and Enchantments* (anth **1989**) ed Jack ZIPES. The essence of the tale can in fact be traced back as far as that of CUPID AND PSYCHE (◊◊ Lucius APULEIUS); the story of the FROG PRINCE can be seen as a cognate. *The Wedding of Sir Gawen and Dame Ragnell* (?**1450**; vt "Sir Gawain and the Loathly Lady") is an early version from the English tradition (◊ GAWAIN); in this the sexes are reversed, Gawain saving ARTHUR's life by agreeing to marry a hag, who proves in fact a beautiful woman. The relationship between CALIBAN and Miranda in William SHAKESPEARE's *The Tempest* (performed *c*1611; **1623**) might be viewed as a skewed version; Tad WILLIAMS's *Caliban's Hour* (**1994**) exploits this possibility, with Miranda's daughter accepting Caliban in order both to save her mother and avoid an arranged marriage. There have been numerous retellings since Leprince de Beaumont, of which we can note: *Beauty and the Beast, or A Rough Outside with a Gentle*

Heart: A Poetical Version of an Ancient Tale (**1811**) attributed to Charles LAMB; *Beauty and the Beast* (play **1841**) by J.R. PLANCHÉ; *Beauty and the Beast* (graph **1875**) by Walter CRANE; "Beauty and the Beast" (in *The Blue Fairy Book* anth **1889**) ed Andrew LANG; "Beauty and the Beast" (in *The Sleeping Beauty and Other Tales from the Old French* coll **1910**) by Sir Arthur QUILLER-COUCH, illustrated by Edmund DULAC; *Beauty and the Beast* (play **1951**) by Nicholas Stuart GRAY; *Beauty: A Retelling of the Story of Beauty and the Beast* (**1978**) by Robin MCKINLEY, a recursive version (◊ RECURSIVE FANTASY) that is at the same time very faithful to the original; *Beauty and the Beast* (graph **1978**) by Chris ACHILLEOS, also recursive; "Beauty and the Beast" (in *Sleeping Beauty and Other Favourite Fairy Tales* **1982** coll) trans Angela CARTER, illustrated by Michael FOREMAN (also of interest is Carter's "The Courtship of Mr Lyon" [1979], which transposes the tale into modern times); and "Beauty" (in *Red as Blood, or Tales from the Sisters Grimmer* coll **1983**) by Tanith LEE. Jean Cocteau filmed the tale memorably as *La* BELLE ET LA BÊTE (*1946*); DISNEY made an ANIMATED MOVIE of it as BEAUTY AND THE BEAST (*1991*); there have been other movie versions.

As noted, the tale is of considerable interest as an UNDERLIER. In at least one instance this is overt: the tv series BEAUTY AND THE BEAST (1987-90), in which a beautiful woman befriends a mutant who dwells in NEW YORK's sewers. Gaston LEROUX's *The Phantom of the Opera* (**1910**) plays with the fascination a "beast" may exert over a beauty; all the movie versions (◊ PHANTOM OF THE OPERA) exploit this, with differing degrees of overtness. More often, though, the treatment of the BATB theme is covert, and sometimes the sexes may be reversed, as in *His Monkey Wife, or Married to a Chimp* (**1930**) by John COLLIER; by contrast, in Michael BISHOP's "The Monkey's Bride", a story inset in *Who Made Stevie Crye* (**1984**), it is the husband who is the "monkey" and the wife who accepts him for what he is, herself starting to grow fur. Cija's relationship with the reptilian Zerd in Jane GASKELL's **Atlan** sequence presents several repetitions of the BATB motif; furthermore, in the series' third volume, *The City* (**1966**), Cija conquers revulsion to take on an APE as a lover, seeing his innate goodness. KING KONG (*1933*) is a classic CINEMA hijacking of the theme. This underlier role extends far outside fantasy: Charlotte BRONTE's *Jane Eyre* (**1847**) is a noteworthy example, with Jane discovering that the unhandsome, elderly Rochester has inner virtue that inspires her love; the crime novel *Shadows on the Mirror* (**1989**) by Frances Fyfield, in which a beautiful harlot changes her ways on identifying the goodness within a grossly fat man, is a recent example plucked from many. [JG]

Further reading: *Beauty and the Beast: Visions and Revisions of an Old Tale* (**1989**) by Betsy Hearne is an exemplary discussion, containing also considerable bibliographical information.

BEAUTY AND THE BEAST US ANIMATED MOVIE (*1991*). Disney. **Pr** Don Hahn. **Exec pr** Howard Ashman. **Dir** Gary Trousdale, Kirk Wise. **Screenplay** Linda Wolverton. **Based loosely on** the version of the tale in *Contes Marins* (**1740-41**) by Gabrielle-Suzanne de Villeneuve (◊ BEAUTY AND THE BEAST). **Voice actors** Robby Benson (Beast), Jesse Corti (Le Fou), Rex Everhart (Maurice), Angela Lansbury (Mrs Potts), Paige O'Hara (Belle), Jerry Orbach (Lumiere), Bradley Michael Pierce (Chip), Hal Smith (Philippe), David Ogden Stiers (Cogsworth, Narrator), Richard White (Gaston). 84 mins. Colour.

A fairly standard retelling, *BATB* has much that is excellent, in particular the depiction of Beast – a far more complex character than one expects in an animated movie. Six character animators, led by Glen Keane, produced a MONSTER that is part lion, part bear, part buffalo, part gorilla, part wolf . . . and, despite its lumbering animality, very human; it is no surprise that Belle (i.e., Beauty) has a moment of doubt when this splendid creature is transformed into a rather bland, standard-issue Handsome Prince. Other characters are well formed – notably Belle, Lumiere (a living candle-holder) and Mrs Potts (a living teapot) – but the movie suffers a severe compression of timescale: the Beast's wooing and winning of Belle have to be crammed into about 24 hours to match the progress of events back in the village. *BATB* was DISNEY's fifth full-length animated FAIRYTALE, and was widely described on release as the best ANIMATED MOVIE of all time – a spurious claim. Nor does it have, amid the customary Disney fun and hijinks, anything of the affect of, say, Jean COCTEAU's *La* BELLE ET LA BÊTE (*1946*), from which it has borrowed some moments. But, on the level of simple entertainment, it is superb. [JG]

BEAUTY AND THE BEAST US tv series (1987-90). Witt-Thomas Productions/CBS. **Pr** Ron Koslow, Tony Thomas, Paul Junger Witt. **Dir** Virginia Attias and many others. **Created by** Ron Koslow. **Writers** George R.R. MARTIN and many others. **Starring** Jay Acovone (Deputy District Attorney Joe Maxwell), Jo Anderson (Diana Bennett), Cory Danziger (Kipper), David Greenlee (Mouse), Linda Hamilton (Catherine Chandler), Roy Dotrice (Father), Ron Perlman (Vincent), Ren Woods (Eddie). 55 60min episodes. Colour.

Desperate to escape pursuing robbers, Catherine Chandler escapes into a strange maze of tunnels under NEW YORK, and is saved by Vincent, a strange half-man, half-lion creature whose fierce looks belie a tender personality. He was abandoned as a baby in the tunnels, there being raised by Father, an enigmatic figure who led a tribe of other unfortunates. Chandler, a lawyer at the DA's office, finds herself relying on the help of Vincent and his fellow tunnel dwellers in her attempts to stop crime. As time passes, the two are drawn to each other, often reciting poetry. Though the soap-opera-like story romantically linked the two, when Linda Hamilton decided she wanted to give up the role of Chandler, her pregnant character was kidnapped and murdered. Vincent became a relentless vigilante, a move that quickly spelled doom for the series. The main attraction for some fans was the love story; for others it was the enigmatic and dimly lit sets of the tunnel world.

Novelization were *Beauty and the Beast* * (**1989**) by Barbara HAMBLY; *Beauty and the Beast: Song of Orpheus* * (**1990**) by Hambly, and *Beauty and the Beast: Masques* (**1990**) by Ru EMERSON. [BC]

BECK, C.C. (1910-1989) US COMICS artist. ◊ CAPTAIN MARVEL.

BECK, L(ILY) ADAMS Working name of UK writer and mystic Eliza Louisa Moresby (?1862-1931). She accompanied her father, John Moresby (1830-1922), on his expeditions to the East Indies in the 1870s; she later assisted him in his autobiography, *Two Admirals* (**1909**). She visited India, Ceylon, the East Indies, Burma and Japan, and these experiences influenced her life and writings. Her work, which drew upon tales and beliefs of the Orient and especially on stories of REINCARNATION, caught the public

interest in SPIRITUALISM that developed after WWI. Her ORIENTAL FANTASIES, which began to appear in magazines in 1919, were collected as *The Ninth Vibration, and Other Stories* (coll **1922**), *The Perfume of the Rainbow, and Other Stories* (coll **1923**) and *Dreams and Delights* (coll **1926**). These stories use old legends and fables much in the style of Lafcadio HEARN, but have a stronger mood of LOVE and romance, sometimes to the point of oversentimentality. Her most powerful stories appear in *The Openers of the Gate* (coll **1930**), which details the cases of a doctor who, like Algernon BLACKWOOD's John Silence, is something of an OCCULT DETECTIVE. LAB's novels allow her more scope, and are particularly strong in development of character and the Oriental background. The most complete are *The Way of Stars* (**1925**), where the powers of ancient EGYPT are reawakened leading to the fulfilment of a PROPHECY, and *The House of Fulfilment* (**1927**), a theosophical romance (◊ THEOSOPHY). Both these, and to a lesser extent *The Treasure of Ho* (**1924**) and *The Glory of Egypt* (**1926** UK) – the latter originally published as by Louis Moresby – include mysterious figures with TALENTS, especially clairvoyance, astral travel, telepathy and levitation. The novels can be likened to the work of Talbot MUNDY. LAB's nonfiction of interest includes *The Story of Oriental Philosophy* (**1923**) and two books about Buddha: *The Splendour of Asia* (**1927**), concerning his teachings, and *The Life of the Buddha* (**1939** UK). [MA]
Other works: *The Key of Dreams* (**1922**) and *The Garden of Vision* (**1929**), both on spiritual awakening. LAB also wrote historical romances as E. Barrington, the name under which she was better known in the UK.

BECKETT, SAMUEL (1906-1989) Irish writer and playwright, in France from 1937. Although rarely considered a writer of fantasy – when *En Attandant Godot* (**1952** France; trans SB as *Waiting for Godot: A Tragicomedy* **1954** UK) was first performed it was explicated in terms of existentialism or the "theatre of the Absurd" – SB has been an author of FABULATIONS throughout his long career. *Murphy* (**1938** UK) has no fantasy elements, but its antirealistic features marked SB as closer to the fantasia of James JOYCE and William Butler YEATS than to the concerns of 1930s social realism. *Watt* (written 1943-5; **1953** France), funnier and bleaker than *Murphy*, offers a protagonist who seeks, in Raymon Federman's words, to enter "a zone of creative consciousness where he would no longer be bound by temporal and spatial dimensions"; and the trilogy of novels that SB wrote in French – *Molloy* (**1951** France; trans SB and Paul Bowles **1955** US), *Malone meurt* (**1951** France; trans SB as *Malone Dies* **1956** US) and *L'Innommable* (**1953** France; trans SB as *The Unnamable* **1958** US) – assembled as *Three Novels* (omni **1965** US) – whose protagonists lose ontological fixity and seem to blur identities (*The Unnamable* can be read as a POSTHUMOUS FANTASY), are in no way realistic narratives.

SB's plays after *Godot* are more overtly fantastic than his prose. *Fin de partie: Act sans paroles* (produced 1957; coll **1957** France; trans SB as *Endgame, followed by Act without Words* coll **1958**) is sf, set in the aftermath of a worldwide HOLOCAUST. *Happy Days* (produced 1961; **1962**) features a protagonist embedded in a mound of earth. *Play* (produced 1963 as *Spiel* trans Erika and Elmar Tophoven; produced in English 1964; in *Play and Two Pieces for Radio* coll **1964**) is an AFTERLIFE fantasy, set in a harrowing PURGATORY that recalls DANTE. The tv play *Eh Joe* (produced 1966; in *Eh Joe and Other Writings* coll **1967**) tells of a HAUNTING. SB's last works of prose trace a similar path away from narrative

reality. *Comment C'est* (**1961** France; trans SB as *How It Is* **1964** US) is another fantasy of Purgatory, as are *Le Depeupleur* (**1971** chap France; trans SB as *The Lost Ones* **1972** chap UK) and the subsequent, still shorter prose pieces that SB produced towards the end of his life.

SB's influence on post-WWII literature has been enormous; most fabulists, from Donald BARTHELME and Thomas BERGER in English to Peter Handke and Thomas Bernhard in German, have created oeuvres recognizably shaped by his presence. Often seen as a follower of James Joyce but later realized to be closer to Franz KAFKA, SB incomparably dramatizes, as John UPDIKE has noted, "the guttering spiritual condition of human life amid our century's material blaze". [GF]
Other works: *More Pricks than Kicks* (coll **1934** UK); *Krapp's Last Tape and Other Dramatic Pieces* (coll **1960** UK); *Ghost Trio* (televised 1976 UK); *Worstward Ho* (**1984** chap US).

BECKFORD, WILLIAM (1760-1844) English eccentric and art connoisseur, and author of *Vathek* (ot *An Arabian Tale, from an Unpublished Manuscript: With Notes Critical and Explanatory* trans Samuel Henley from author's original French **1786**; reconstructed French text **1787** Switzerland; rev vt *Vathek, Conte Arab* 1787 France; final rev 1815 France; English edn vt *Vathek* 1816; vt *The History of the Caliph Vathek* 1868). *Vathek* is an ORIENTAL FANTASY that utilizes the settings made popular by the *Arabian Nights* (◊ ARABIAN FANTASY) and infuses them with the atmosphere of GOTHIC FANTASY, then gaining in popularity. It tells of an 8th-century caliph who renounces Islam and partakes of vile and depraved practices in order to obtain supernatural TALENTS. After years of corruption, during which he constructs a mighty pleasure tower (◊ EDIFICE), he descends into HELL to inherit a kingdom, but is rewarded instead with eternal damnation. The book served to consolidate a reputation that WB was already earning through his sexual philandering and diabolic habits as a man of EVIL. Like Matthew LEWIS, whose *The Monk* (**1796**) achieved equal notoriety a decade later, WB became the subject of speculation and hostility – a position he seemed to encourage through his outlandish lifestyle. The richest commoner in England, WB financed the construction of Fonthill Abbey, then the tallest private structure in the UK, where he lived as something of a recluse among his phenomenal art collection for almost 20 years.

WB completed *Vathek* in French by early 1783 (taking a year to write it, not the three days he later claimed) and then arranged for Samuel Henley (1740-1815), his cousins' tutor, to translate it into English. WB delayed publication by writing some new "episodes" which take place in Hell; this delay caused Henley to bring his own translation into print in England, thus forcing WB to publish a reconstructed French text in Lausanne (the original had been lost), still minus the "episodes".

WB never did completed the "episodes" to his own satisfaction, and they remained unpublished until found by Lewis Melville (real name Lewis S. Benjamin; 1874-1932) in 1909. Two of them, "Histoire de la Princesse Zulkaïs et de Prince Kalilah" (unfinished; completed by Clark Ashton SMITH as "The Third Episode of Vathek" *Leaves* 1937; in *The Abominations of Yondo* [coll **1960**]) and "Histoire du Prince Alasi et de la Princesse Firouzka", were serialized in their original French in *The English Review* 1909-10. They were then assembled with "The Story of Prince Barkiarokh" as *The Episodes of Vathek* (coll trans Sir Frank Marzials [1840-1912] **1912**). These stories relate the degradation of other princes who misuse the black arts in order to achieve power and wealth, and lack the impact of the original *Vathek* – indeed they are only more of the same – and are too self-indulgent. The first edition to combine both texts was *Vathek, with The Episodes of Vathek* (coll **1929**) ed Guy Chapman (1889-1972).

WB wasted a talent as a writer and artist. In his youth he was taught music by Wolfgang Amadeus Mozart (1756-1791) and became an acquaintance of Voltaire, on whose work he styled *Vathek*. WB was also a fluent linguist, but apart from one other novel, the satirical *Azemia* (**1797**) as Jacquetta Jenks, his remaining work published during his lifetime related to his European travels. He has been credited with the anonymous *Popular Tales of the Germans* (coll **1791**), a translation of the stories of Johann Karl MUSÄEUS, but there is no evidence to support his assertion.   [MA]
Other works: *Dreams, Waking Thoughts and Incidents* (coll **1783**), a reflection on life.
Further reading: *The Life and Letters of William Beckford* (**1910**) by Lewis Melville; *The Life of William Beckford* (**1932**) by John W. Oliver; *William Beckford* (**1937**; rev 1952) by Guy Chapman; *The Caliph of Fonthill* (**1956**) by H.A.N. Brockman; *England's Wealthiest Son: A Study of William Beckford* (**1962**) by Boyd Alexander; *William Beckford* (**1976**) by James Lees-Milne; *William Beckford* (**1977** US) by Robert J. Gemmett; *Beckford of Fonthill* (**1979**) by Brian Fothergill.

BÉCQUER, GUSTAVO ADOLFO (?   -   ) Spanish writer. ◊ SPAIN.

BEDAZZLED UK movie (*1967*). 20th Century-Fox. **Pr** Stanley Donen. **Dir** Donen. **Screenplay** Peter Cook. **Novelization** *Bedazzled* * (**1967**) by Michael J. Bird. **Starring** Michael Bates (Inspector Reg Clarke), Eleanor Bron (Margaret Spencer), Peter Cook (The Devil/George Spiggott), Barry Humphries (Envy), Dudley Moore (Stanley Moon), Robert Russell (Anger), Michael Trubshawe (Lord Dowdy), Raquel Welch (Lilian Lust), Lockwood West (St Peter). 96 mins. Colour.

Stanley, cook in a Wimpy Bar, yearns for waitress Margaret but is shy. On the verge of suicide he is approached by the DEVIL, in the guise of nightclub-owner Spiggott (misspelt "Spiggot" in the credits), who gives him seven WISHES in return for his SOUL (◊ PACT WITH THE DEVIL). Six wishes, all but one directed towards the acquisition of Margaret's love, are presented in the form of ALTERNATE REALITIES, with putative Stanleys and Margarets living the consequences of the wish; the results are inevitably, due to the Devil's QUIBBLES, disastrous. In between wish-fulfilment scenarios, the Devil explains that his position as instigator of EVIL was born out of a BET with GOD as to which could first claim 100 billion souls; he is close to winning, and thereby reattaining God's innermost circle of ANGELS. Stanley's final REALITY is as a fellow-nun alongside Margaret in an Order of Silence; he pleads with the Devil for mercy, and the latter returns Stanley's soul as his first benevolent act as an angel. Ascending to HEAVEN the Devil is rejected by God. Returned to Earth, the Devil approaches Stanley, now a Wimpy cook again, in hopes of striking up a new CONTRACT; but Stanley, having dared to ask Margaret for a date, decides, even though rejected, in turn to reject the Devil.

This reinterpretation of the FAUST legend in terms of the Swinging Sixties is . . . very Sixties. It has splendid moments, but in between there is much, perhaps too much, that is insufficiently sharp.   [JG]

BEDFORD-JONES, H(ENRY JAMES O'BRIEN) (1887-1949) Canadian-born US writer who from 1908 produced over 1000 stories for the US pulp MAGAZINES under his own name and at least 19 pseudonyms, of which Michael Gallister, Allan Hawkwood and Gordon Keyne were the most prolific and most associated with FANTASY. HB-J's stories were mainly historical adventure and Westerns, his best-known series in book form featuring **John Solomon**, a Bulldog Drummond-style Cockney crimefighter whose adventures occasionally drift into the fantastic, usually involving lost worlds and ancient, quasi-magical artefacts. The best in this vein was *The Seal of John Solomon* (1915 *Argosy*; **1924** UK as Allan Hawkwood). Other LOST-RACE works include "Khmer the Mysterious" (1919 *People's Magazine* as Allan Hawkwood) and "The Golden Woman of Khmer" (1919 *People's Magazine* as Allan Hawkwood), "The Brazen Peacock" (1920 *Blue Book*), and *The Temple of the Ten* (1921 *Adventure*; **1973**) with W.C. Robertson.

Of his many magazine series, **Trumpets from Oblivion** (1938-9 *Blue Book*) uses the sf device of time-viewing to explore ancient LEGENDS including those concerning AMAZONS, UNICORNS, WEREWOLVES and PRESTER JOHN; HB-J re-employed this gimmick in his **Counterclockwise** series (1943-4 *Blue Book*), which is of less fantasy interest. Other series of note include **The Adventures of a Professional Corpse** (1940-41 *WT*), about a man who could feign death, and **The Sphinx Emerald** (1946-7 *Blue Book*), about an accursed jewel. HB-J produced many singleton stories for the pulps, the best appearing in *Blue Book*; of these, "From Out the Dark Water" (1940 *Blue Book* as Michael Gallister) involves leprechauns repelling a Nazi invasion.   [MA]
Other works: *The Star Woman* (**1924**); *The Wizard of the Atlas* (**1928** UK as Allan Hawkwood). The other **John Solomon** stories in book form, all as by Hawkwood but not all fantastic, are *Solomon's Quest* (1915 *People's Magazine*; **1924** UK), *John Solomon, Supercargo* (1914 *Argosy*; **1924** UK), *Solomon's Carpet* (1915 *People's Magazine*; **1925** UK), *John Solomon, Incognito* (1921 *People's Magazine*; **1925** UK), *Gentleman Solomon* (1915 *People's Magazine*; **1925** UK) and *The Shawl of Solomon* (1917 *People's Magazine*; **1925** UK).

BEDKNOBS AND BROOMSTICKS US live-action/ANI-MATED MOVIE (*1971*). ◊ DISNEY.

BEERBOHM, (HENRY) MAX(IMILIAN) (1872-1956) UK author, essayist, dramatic critic and caricaturist. His skill at PARODY and pastiche shaped his fantasy excursions: thus *The Happy Hypocrite* (1896 *The Yellow Book* as "The Happy Hypocrite: A Fairy Tale for Tired Men"; **1896** chap) imitates the sophisticated FAIRYTALES of Oscar WILDE, with a sideswipe at *The Picture of Dorian Gray* (**1891**). Wicked Lord George Hell conceals his debauched features behind a saintly wax MASK, ultimately to find that virtuous living has changed his real face to match the false one. This seems a conventionally moral fable, but MB's tongue-in-cheek narration undermines the sentimentality – as when Lord George marries "according to the simple rites of a dear little registry-office in Covent Garden".

MB's acclaimed prose parodies in *A Christmas Garland* (coll **1912**) dip into fantasy when mocking Rudyard KIPLING (SANTA CLAUS is arrested by a brutish policeman while a sycophantic Kipling-narrator shouts encouragement) and the spiritual chorus of Thomas Hardy's *The Dynasts* (**1904-8**). In his classic if flawed humorous fantasia of Oxford, *Zuleika Dobson, or An Oxford Love Story* (**1911**), MB used various fantasy elements: a special dispensation from Clio, MUSE of history, allowing him to report characters' thoughts; pearls which change colour to signify their wearers' love or otherwise; and PORTENTS like the black owls foreshadowing the Duke of Dorset's end. The climax sees virtually all the undergraduates of Oxford, hearts broken by vapid heroine Zuleika's beauty, fling themselves after the Duke into the river Isis . . . leading to the black humour of a High Table dinner at which unconcerned dons fail to notice the students' absence. This was grimly (albeit unwittingly) prophetic: similar scenes shortly occurred during WWI.

Seven Men (coll **1919**; exp vt *Seven Men and Two Others* 1950) contains the fine PACT WITH THE DEVIL story "Enoch Soames", whose eponymous would-be writer is the shabbiest of figures in the lovingly pictured 1890s literary scene, but has faith that his name will endure. The Devil arranges TIME TRAVEL to 1997, where Soames puzzles out the future's phonetic spelling to find himself known only as protagonist of a "sumwot labud sattire" by MB. Other stories in *Seven Men* include SLICK FANTASIES playing with psychic projection and predestination; though the latter tale, "A.V. Laider", is ultimately a RATIONALIZED FANTASY, its story of imminent multiple death predicted through palmistry has the compelling quality of URBAN LEGEND – reflecting Laider's compulsion to invent and tell it.

As a writer MB was a fine miniaturist who pretended to world-weariness and dilettantism. The pose is deceptive; his polished prose and wit retain an enduring toughness beneath the glitter. [DRL]

Other works: *The Dreadful Dragon of Hay Hill* (**1928** chap); "Yai and the Moon" in *A Variety of Things* (coll **1928**); "Ten Years Ago . . ." in *A Peep into the Past* (coll **1972**).

Further reading: *Max: A Biography* (**1964**) by David Cecil.

BEETLEJUICE US movie (**1988**). Warner Bros/Geffen. **Pr** Michael Bender, Richard Hashimoto, Larry Wilson. **Dir** Tim BURTON. **Spfx** Chuck Gaspar, Robert Short. **Screenplay** Michael McDowell, Warren Skaaren. **Starring** Alec Baldwin (Adam Maitland), Geena Davis (Barbara Maitland), Jeffrey Jones (Charles Deetz), Michael Keaton (Betelgeuse), Catherine O'Hara (Delia Deetz), Winona Ryder (Lydia Deetz), Glenn Shadix (Otho), Sylvia Sidney (Juno). 92 mins. Colour.

After dying in an accident a young couple, the Maitlands, find themselves bound (◊ BONDAGE) to their much-loved home as GHOSTS: through its doors, for them, lie the desert wastes of Saturn, where WORMS roam and TIME plays tricks. When the appalling Deetzes buy the house and start "improving" it the Maitlands try HAUNTING it hideously, but only the daughter, Lydia, can see them. They seek aid in the AFTERLIFE, a sort of civil-service office; crusty case-worker Juno tells them they must fend for themselves for the first 125 years, and warns them against hiring the services of TRICKSTER Betelgeuse, whose ostensible trade is EXORCISM of the living. But they fall into the trap; only the guile and courage of the Maitlands and Lydia avert terminal disaster. In the new *status quo* all coexist, the ghosts becoming Lydia's surrogate parents and the Deetzes, in effect, the house's ghosts.

Superb spfx and brilliant performances by Davis, O'Hara and Ryder cannot disguise the fact that this comedy, although it touches almost all the right bases, finally fails to satisfy. There is too much of a presumption that Keaton's firecracker, scatological Betelgeuse (with a smidgen of Professor Marvel in *The* WIZARD OF OZ [*1939*]) will carry the movie. Burton might have been wise not to focus so much attention on the model village built by Adam Maitland, thereby emphasizing the fact that here, as in most of his movies – e.g., BATMAN (*1989*) and EDWARD SCISSORHANDS (*1990*) – there is an uncomfortable feeling that he is directing model people in model villages. [JG]

BEGBIE, (EDWARD) HAROLD (1871-1929) UK writer who collaborated with J. Stafford Ransome and M.H. Temple, all writing as Caroline Lewis, in two **Alice** PARODIES: *Alice in Blunderland* (**1902**) and *Lost in Blunderland: The Further Adventures of Clara* (**1903**). *The Day that Changed the World* (**1912** as by The Man Who Was Warned; 1914 as HB) can be read as sf, but is perhaps better thought of as CHRISTIAN FANTASY. *On the Side of the Angels* (**1915**) is a subfusc response to Arthur MACHEN's *The Bowmen* (coll **1915**). [JC]

BELASCO, DAVID (1859-1931) US dramatist and theatrical producer who wrote some unremarkable fantasy in *Fairy Tales Told by Seven Travellers at the Red Lion Inn* (coll **1906**). He also novelized, as *The Return of Peter Grimm** (**1912**), his own play of the previous year, in which the eponymous elderly bachelor dies and becomes a REVENANT to save his fosterdaughter from an unsuitable marriage. Other plays of interest include *The Darling of the Gods* (**1902**) with John Luther Long which, though not specifically supernatural, created a Japan so heightened from reality that its example helped inspire Lord DUNSANY to begin his own writing career. [JC]

BELATEDNESS There is a common reading of US literature which sees it as permeated by the sense that the present is already too late, that the Wilderness is past, that to perceive a Frontier (◊◊ AMERICAN GOTHIC) is to find it dust, that EDEN cannot be recaptured, that people are never born soon enough to live their true lives. The QUEST for a truly virgin frontier, which is also a quest for the ideal CITY, is by this reading a quest whose true bent is return. US literature, if read along these lines, sounds very much like any definition of FANTASY that treats the form as being haunted by precedents.

The usefulness of "belatedness" as a term of practical analysis is minor unless it is restricted to particular applications. It seems best to restrict the term to that cultural complex of disappointments and loss which is intrinsic to US literature – whether fantasy or otherwise.

Early US literature is characteristically peopled by protagonists who are attempting to recapture what has irretrievably gone. Along with Washington IRVING's Rip Van Winkle and most of the regret-drenched creations of Edgar Allan POE, the eponymous ACCURSED WANDERER of William AUSTIN's *Peter Rugg, the Missing Man* (1824 *New England Galaxy*; 1882 chap), one of the central early US fantasy texts, is constantly encountered trying to reach the CITY to which he longs to return but can never approach, for the faster he rides the more distant it seems. US texts seemingly influenced by this tale's haunting fixedness include Herman MELVILLE's "Bartleby the Scrivener" (1853) and "The Lightning-Rod Man" (1856). More generally, the FLYING DUTCHMAN was peculiarly suited to the 19th-century US imagination; when he does triumph, his victory tends to be as fatal as Captain Ahab's (in Melville's *Moby-Dick*

[**1851**]), and it has become a cliché that a grandiose Ahab-like futility underlies the QUEST structure of much US literature. Mark TWAIN's *The Adventures of Huckleberry Finn* (**1884**) owes some of its exhilarating effect to the fact that, when Huck decides to "light out for the Territory ahead of the rest . . .", he does so on the last page: only later will he become a typical adult American, according to the lines laid down by US writers in their most classic works. For belatedness is not only a relationship to the past: it is an experience of inherent disappointment.

That US fantasy can be seen as a set of strategies to counter disappointedness is an obvious – though extreme – reading of a complex set of texts; but certain characteristic modes of contemporary US fantasy – particularly DARK FANTASY – clearly represent a range of confrontations with the longing to return. Many of the novels of Stephen KING (and his imitators) are permeated by a feeling that all the HORRORS are occurring in an environment – usually a small town – that *should have been* still like it was before, in which case the horrors might not have happened. Stories like "The Search" (1981), *It* (**1986**), *The Tommyknockers* (**1987**) and *Needful Things* (**1991**) convey a frozen belatedness. Charles L. GRANT's **Oxrun Station** sequence evinces a similar effect, as do some of Gene WOLFE's confessional tales – *Peace* (**1975**) in particular. Most tales of THINNING are necessarily also tales of belatedness, as in the movie DRAGONSLAYER (*1981*), whose SORCERER'S APPRENTICE hero is too late in his inheritance of his master's MAGIC because the world is now turning towards a new form of "magic", Christianity. (DYING-EARTH stories are of course explicitly nostalgic about the deep past, as are many PLANETARY ROMANCES, but it would be difficult to argue that they represent in any engaged sense a wrestling with the sensation of having just missed the boat.) In all this it should not be ignored that, underlying the disappointedness that fuels so much US fantasy (here we ignore GENRE FANTASY), there is also a sense that recuperation may be possible; and that, even if failure is inevitable, the search for meaning is more than a gesture.

Perhaps the most original strategy devised by US writers to cope with belatedness can be found in the GNOSTIC FANTASIES of writers like John CROWLEY, whose continuing **Aegypt** sequence complexly conflates its protagonist's search for a true STORY, one that will make sense of his life, with a Gnostic explanatory structure by virtue of which he may come to understand that the contemporary world is a shadow of the *real* world (◊ REALITY), and that the real world is not irretrievably lost (i.e., inherently belated) but simply has not yet been *told*. [JC]

See also: KNIGHT OF THE DOLEFUL COUNTENANCE; LIMINAL BEINGS; OBSESSED SEEKER; RECURSIVE FANTASY; REVENANTS; TWICE-TOLD; WAINSCOTS.

BELL, CLARE (LOUISE) (1952-) UK-born writer, in the USA from age 5, who remains best-known for her sf; she lives with M. Coleman EASTON, writing sf with him under the name of Clare Coleman. The **Ratha** sequence of YA tales – *Ratha's Creature* (**1983**), *Clan Ground* (**1984**), *Ratha and Thistle-Chaser* (**1990**), *The Jaguar Princess* (**1993**) and *Ratha's Challenge* (**1995**) – can be read as ALTERNATE-WORLD sf where evolution has favoured the CAT, but the long saga of Ratha's growth as leader of her tribe of cougar-like talking felines, and the interplay of genetic and DYNASTIC FANTASY-like narrative strands, is really fantasy. [JC]

BELL, DOUGLAS (? -) US writer whose first novel,

Mojo and the Pickle Jar (**1991**), transforms the modern USA into a LAND OF FABLE – in a fashion similar to Terry BISSON's transformative technique in *Talking Man* (**1986**) – and sets in motion a QUEST plot involving the rescue of the heart of the world, which has been trapped in a pickle jar. The protagonist and his various COMPANIONS encounter many wonders, including a mafia of DEMONS, *en route* to a successful HEALING of the cancerous world. [JC]

BELL, JULIE (1958-) US artist. ◊ OLIVIA.

BELL, NEIL Best-known pseudonym of UK writer Stephen Southwold (1887-1964), who adopted that name legally although born Stephen Henry Critten; he used also the pseudonyms S.H. Lambert and Paul Martens on work of fantasy interest. He produced numerous collections of children's tales under his legal name, most of which contain fantasy stories. *Mixed Pickles: Short Stories* (coll **1935**) includes the allegorical "The Earthly Paradise", the delusional fantasy "The Mirror" and an effective sardonic novelette about a man who discovers that he can levitate and aspires to become a MESSIAH, "The Facts about Benjamin Crede". The latter is also in *Ten Short Stories* (coll **1948**). "The Queer Affair at Yattenbridge" in *The Smallways Rub Along* (coll **1938**) is a calculated trivialization of a similar theme. *Precious Porcelain* (**1931**) and *The Disturbing Affair of Noel Blake* (**1932**) are tales of artificially disordered personality on the border of fantasy and sf. *Death Rocks the Cradle* (**1933** as by Martens) is a VISIONARY FANTASY about a community of covert sadists. *Portrait of Gideon Power* (**1944** as by Lambert; 1962 as by Bell) is a psychological case-study in the form of evidence offered at the Seat of Judgement (◊ LAST JUDGEMENT). The fantasies in *Alpha and Omega* (coll **1946**) include "Strange Encounter", an interesting tale of REINCARNATION. *Three Pairs of Heels* (coll **1951**) and *Forty Stories* (coll **1958**) include a few trivial fantasies and weird tales. *The Secret Life of Miss Lottinger: A Novella and Twenty Short Stories* (coll **1953**) includes "Mr Albert Finkelman", a sarcastic account of the fate of the ELIXIR OF LIFE. *Village Casanova* (coll **1961**) contains an earnest SPIRITUALISM fantasy, "Mary Birch". His other collections include a smattering of weird fiction and the occasional futuristic SATIRE. [BS]

Other works (as Stephen Southwold): *In Between Stories* (coll **1923**); *Twilight Tales* (coll **1925**); *Old Gold: A Book of Fables and Parables* (coll **1926**); *Listen Children! Stories for Spare Moments* (coll **1926**); *Once Upon a Time Stories* (coll **1927**); *Ten-Minute Tales* (coll **1927**); *Listen Again, Children!* (coll **1928**); *Happy Families* (coll **1929**); *Hey Diddle Diddle* (coll **1930**); *Tales Quaint and Queer* (coll **1930**); *Tick Tock Tales* (coll **1930** chap); *Forty Tales* (coll **1931**); *Forty More Tales* (coll **1931**); *The Tales of Joe Egg* (coll **1936**); *Tell Me Another* (coll **1938**).

BELLAIRS, JOHN (1938-1991) US writer, mostly of YA books, but whose fantasy novels appeal to all ages. He began publishing work of interest with *St Fidgeta and Other Parodies* (coll **1966**), which PARODY fantasy devices in their mild spoofing of religion, but only established himself with *The Face in the Frost* (**1969**), a humorous fantasy. Two MAGI – Roger Bacon (*c*1214-1292) and PROSPERO – have to discover who is behind a series of supernatural attacks and then defeat him. Since the solution requires an artifice from our own REALITY (which is perceived as a fantasy world by Prospero and Bacon), the story is also a GNOSTIC FANTASY. The humour and characterization make this at times reminiscent of the work of Mervyn PEAKE, and there are also comparisons with Roger CORMAN's *The Raven* (*1963*), but JB never lets the slapstick get out of control, and on occasion the tale descends into moments of pure HORROR. The novel remains a unique classic.

JB's earlier works only hinted at his talent, though his first novel, *The Pedant and the Shuffly* (1968) – about an evil magician thwarted by a kindly old man – bears similarities with *The Face in the Frost*. The GOOD AND EVIL motif returned in *The House with a Clock in its Walls* (1973), JB's first book for children and the first of the **Chubby Lewis** series, which is set in JB's own childhood period, the late 1940s. Young Lewis, recently orphaned, lives with his Uncle Jonathan, whose house is full of junk. In this first novel Lewis and Jonathan have to defeat the legacy of an evil WIZARD who once lived in the house. In *The Figure in the Shadows* (1975) Lewis finds an AMULET that revives the shade of an old farmer who practised witchcraft before the Civil War. A third novel in this series, *The Letter, the Witch and the Ring* (1976), places more emphasis on Lewis's friends Rose Rita Pottinger and the WITCH Mrs Zimmerman.

With *The Treasure of Alpheus Winterborn* (1978) JB turned to a new character, **Anthony Monday**, who was more contemporary than Lewis, though the adventures concerned similar discoveries of and the combating of ancient evil; further novels in this series were *The Dark Secret of Weatherend* (1984), *The Lamp from the Warlock's Tomb* (1988) and *The Mansion in the Mist* (1992). JB produced an almost indistinguishable series featuring the adventures of **Johnny Dixon and the Professor**: *The Curse of the Blue Figurine* (1983), *The Mummy, the Will and the Crypt* (1983), *The Spell of the Sorcerer's Skull* (1984), *The Revenge of the Wizard's Ghost* (1985), *The Eyes of the Killer Robot* (1986), *The Trolly to Yesterday* (1989), *The Chessmen of Doom* (1989) and *The Secret of the Underground Room* (1990). Although these last two series, illustrated by Edward GOREY, are strong on atmosphere, the format became repetitive.

After JB's death Brad STRICKLAND completed further JB manuscripts, including the **Chubby Lewis** books *The Ghost in the Mirror* (1993), *The Vengeance of the Witchfinder* (1993) and *The Doom of the Haunted Opera* (1995) and the **Johnny Dixon** books *The Drum, the Doll and the Zombie* (1994) and *Johnny Dixon in the Hands of the Necromancer* (1996). [MA]

BELLAMY, FRANK (ALFRED) (1917-1976) UK artist of COMIC strips. Highly talented, FB had a polished line style and refined colour sense, and his bold, forceful treatment of the human figure has influenced many prominent artists in the medium. His first job on leaving grammar school was in a local art studio, painting cinema posters. Following wartime service in the Royal Artillery (1939-45) he did similar jobs before turning to comic strips, working for *Mickey Mouse Weekly* drawing DISNEY's *Living Desert* and *Monty Carstairs* (1953-4). He went on to draw adaptations of *King Arthur and his Knights* (1955) and *Robin Hood* (1956-7), among others, for the children's comic *Swift*, then took over the full-colour biographical strips on the back page of the boys' comic *Eagle* (1957-9) and, on that comic's front page, **Dan Dare** (1959-60), in part later collected as *The Terra Nova Trilogy* (graph coll **1994**) and *Project Nimbus and Other Stories* (graph coll **1994**). FB's other work of fantasy and sf interest included HEROS THE SPARTAN in *Eagle*, where he and Luis BERMEJO alternated stories 1962-5, **The Ghost World** (based on Harry Harrison's *Deathworld* [**1960**]) in *Boys' World* (1963-4), **Thunderbirds** for *TV Century 21*

(1966-9), a short adaptation of Edgar Allan POE's "The Pit and the Pendulum" (1843) for *Sunday Citizen* (1966), a number of individual illustrations in *Radio Times* for various episodes of *Dr Who* (1971-6; vt *Dr, Who: Timeview* graph coll **1985**) and the daily newspaper strip GARTH (1971-6) in the *Daily Mirror*, reprinted in *The Daily Mirror Book of Garth 1975* (graph coll **1975**), *The Daily Mirror Book of Garth 1976* (graph coll **1976**), *The Doomsmen* (graph **1981**), *Bride of Jenghiz Khan* (graph **1985**), *The Cloud of Balthus* (graph coll **1985**) and *The Women of Galba* (graph coll **1985**). He was involved in the production of one episode of *The* AVENGERS, "The Winged Avenger" (1967), for which he both designed the Winged Avenger's costume and drew some comic-book pages that appear in the episode, which tells of a comics artist who insanely comes to believe he is the avenging HERO he has created.

FB was a Fellow of the Society of Industrial Artists and received the Best Foreign Artist Award from the American Academy of Comic Book Arts (1972). In 1982 The UK Society of Strip Illustration (now the Comics Creators' Guild) instituted the Annual Frank Bellamy Award for Lifetime Achievement in Comics.

Other published collections of his work include *High Command* (graph coll **1981**) and *Fraser of Africa* (graph coll **1990**). [RT]

Further reading: Long interview in *Fantasy Advertiser* Vol 3 *#50* (1973).

BELLE, PAMELA (1952-   ) UK writer, at first of historical novels and more recently of fantasy. The **Silver City** sequence – *The Silver City* (**1994**) and *The Wolf Within* (**1995**), with a further volume projected – is a DYNASTIC FANTASY set in a robustly described FANTASYLAND over which wars rage. The small illegitimate son of the man who becomes king (in the first volume) undergoes (in the second) a RITE OF PASSAGE into adulthood and the exercise of his full powers of SORCERY. Romantic relationships, barbarians and genocide feature. [JC]

BELLE DAME SANS MERCI ◊ LAMIA.

BELLE ET LA BÊTE, LA (vt *Beauty and the Beast*) French movie (*1946*). Discina. **Pr** André Paulvé. **Dir** Jean COCTEAU. **Spfx** René Clément. **Screenplay** Cocteau. **Based on** the version of the tale by Jeanne-Marie LEPRINCE DE BEAUMONT. **Starring** Marcel André (Merchant), Michel Auclair (Ludovic), Josette Day (Belle), Nane Germon (Adelaide), Jean Marais (Avenant/La Bête/Prince), Raoul Marco (Moneylender), Mila Parély (Félicie). 100 mins (cut to 96 and 92 mins UK, 87 mins US). B/w.

Superficially *LBELB* seems a straightforward retelling of the BEAUTY AND THE BEAST legend, and a surprisingly pacy one, yet Cocteau is at pains to draw our attention to the subtext. It is a movie that pulsates with sex: La Bête's (Beast's) lust is explicit, and Belle's becomes almost as obvious. The moral spelled out by La Bête at the close ("Love may make a beast of man – love may also make an ugly man handsome") is trite enough, yet the physical exchange of forms between himself and Avenant (a handsome wastrel whom Belle once loved) adds a new level of meaning: Avenant and La Bête are, refracted through Belle's PERCEPTION, identified; both are Man rather than individual men. The kingdom to which La Bête flies Belle after his TRANSFOR-MATION is a kingdom of the heart.

In 1994 the composer Philip Glass presented an operatic version of *LBELB*, in which live musicians and singers perform in synchrony with the projected movie. Glass had earlier, in 1993, staged an opera using as libretto the script of Cocteau's other great masterpiece, ORPHÉE (*1949*), and planned to present a balletic interpretation of *Les Enfants Terribles* (*1950*) in 1995. [JG]

BELLES DE NUIT, LES (vt *Beauties of the Night*) French/Italian movie (*1952*). Franco London/Angelo Rizzoli. **Dir** René CLAIR. **Screenplay** Clair, with Pierre Barillet, Jean-Pierre Gredy. **Starring** Martine Carol (Edmé), Gina Lollobrigida (Leila), Gérard Philipe (Claude), Magali Vendeuil (Suzanne). 89 mins. B/w.

The vivid DREAMS of frustrated pianist Claude combine to create a set of ALTERNATE REALITIES in which he is an astonishingly successful composer pursued by beautiful women in three different historical eras, each regarded by its successors as a GOLDEN AGE: pre-Revolutionary France, 19th-century Algeria and the FIN DE SIÈCLE. Soon, however, events in his mundane reality start to impinge on the alternate realities and *vice versa*, while the alternate realities start leaking into each other. At last Claude, aided by mundane-reality friends, flees the vengeful husbands, brothers and fathers of his various amours in an astonishing car chase through history, from Neanderthal times, the FLOOD and the Roman era to the present – where it does not stop merely because it has obtruded into today. Steeped in MUSIC, this rich fantasy caused some sensation on release – if you peer closely you can just see Lollobrigida's left buttock – but until recently was largely forgotten. [JG]

BEMMANN, HANS (1922-   ) German writer, journalist and academic who taught at Bonn University 1971-83. After some pseudonymous works in the 1960s, he became a full-time writer with the publication of *Stein und Flöte* (**1983**; trans Anthea Bell as *The Stone and the Flute* **1986** US), a long fantasy novel set in a LAND OF FABLE resembling Northern Europe that depicts the life of its protagonist from childhood to old age and death. There are element of the *Bildungsroman* in the narrative, which is illuminated throughout by exemplary stirrings from fantasy's CAULDRON OF STORY. These half-lit, half-told stories – at times abbreviated to the concision of the motif – generate a sense of folkloric wisdom in a long tale that some critics found tediously lacking in event. In fact, the eventfulness in the book lies inward. Moreover, befitting HB's Catholicism, the novel is a drama of the maturing of the protagonist's SOUL, which evolves in a linear fashion beyond the compass of the text, and which sloughs off the world, joyously, at the end of things.

HB's second fantasy novel, as *The Broken Goddess* (**1990**; trans under that title Anthea Bell **1993** UK), similarly conveys its protagonist through a fantasy realm whose main function seems to be to impart lessons; but this is a more complex work of fiction, despite its comparative brevity. A callow young scholar of FOLKLORE, arriving at a conference, catches sight of a mutilated statue of a goddess – in fact, it proves, *the* GODDESS – and identifies her with the woman (never fully named) to whom he addresses his confessional narrative. He soon finds himself slipping across a THRESHOLD into an OTHERWORLD that resembles FAERIE. Here a ferryman takes him (for a fee he should not have paid) across a river to an ISLAND, in the centre of which a DARK TOWER beckons. On the island he is subjected to constant TRANSFORMATIONS. Moreover, the island is LITTLE BIG: the Tower recedes indefinitely, opening to him various adventures in the lands that lie between it and him. It is only on his third immersion into this otherworld that he finally behaves

with sufficient decorum and maturity to reach his destination, where the GODGAME to which he has been subjected comes to a climax. Within the Tower – which is also a LABYRINTH, a LIBRARY, and a RECOGNITION of passage, "the point where two REALITIES intersected, one of them forcing stored memories of my past into the other, which until now I had thought was imaginary" – he finally learns how properly to address the lioness in whom he recognizes both the Goddess and the woman within whom the Goddess resides in the mortal world. When he comes to maturity, the Tower dissolves (HB, it should be remembered, is a Christian), and the protagonist – grown, ready to LOVE – returns irrevocably to the world. [JC]

BENÉT, STEPHEN VINCENT (1898-1943) US poet, short-story writer and novelist, brother of William Rose BENÉT; he was primarily known in the interbellum for his poetry, including the book-length US epic *John Brown's Body* (**1928**) as well as individual poems like "American Lines", whose closing words – "Bury my heart at Wounded Knee" – would have a relevance to the Native American rights movement he could hardly have guessed. He was, in fact, deeply involved in the MATTER of America, and much of his fiction reflects that involvement, though often in SLICK-FANTASY terms – his best-known stories first appeared in the *Saturday Evening Post*. Of about 120 short stories – many never collected – a dozen or so are of fantasy interest. "The Barefoot Saint" (1929), whose mildly ironic cadences reflect the influence of James Branch CABELL, early demonstrated SVB's companionable smoothness of style, but is too visibly written as a parable to engage the imagination. It was not until the publication of his single most famous tale – *The Devil and Daniel Webster* (1936 *Saturday Evening Post*; **1937** chap), soon transformed into an OPERA (**1939** chap), a play (**1939** chap) and a movie, *The* DEVIL AND DANIEL WEBSTER (*1941*) – that he was able to deploy that smoothness in the telling of an effectively resonant TALL TALE. With impressive eloquence, the story fabulizes US history through the defence by Daniel Webster (1782-1852) of a farmer who has made an injudicious PACT WITH THE DEVIL; with Webster's victory the Frontier is metaphorically declared open for business. Two sequels, "Daniel Webster and the Sea Serpent" (1937) and "Daniel Webster and the Ides of March" (1939), are of lesser interest.

In SVB's second famous fantasy tale, *Johnny Pye & the Fool-Killer* (1937 *Saturday Evening Post*; **1938** chap), Pye outwits DEATH the Fool-Killer by refusing to accept an offer of IMMORTALITY. The two famous stories, plus several others of fantasy interest, were assembled in *Thirteen O'Clock: Stories of Several Worlds* (coll **1937**) and *Tales Before Midnight* (coll **1939**), both volumes themselves assembled as *Twenty-five Stories* (omni **1944**); a posthumous collection, *The Last Circle* (coll **1946**), includes less interesting material, though "The Land Where There is No Death" (1942) is a moving ALLEGORY in which a man's lifelong search for meaning transforms him into a teller of immortal tales – but, as is typical of SVB, the allegorical element tends to denature the fantasy. In his work the tall tale tends, in the end diminishingly, to be self-regarding. [JC]

Other works: *Selected Works of Stephen Vincent Benét* (coll **1942** 2 vols; cut vt *The Stephen Vincent Benét Pocket Book* 1946); *From the Earth to the Moon* (written 1935; **1958** chap); *The Devil and Daniel Webster and Other Stories* (coll **1967**).

See also: Archibald MACLEISH.

BENÉT, WILLIAM ROSE (1886-1950) US poet, editor and novelist, brother of Stephen Vincent BENÉT, married to Elinor WYLIE; he is best remembered as the editor of *The Reader's Encyclopedia* (**1948**), an extremely successful and accomplished guide. Of fantasy interest is his only novel for children, *The Flying King of Kurio* (**1926**), featuring an OZ-like land reachable through a magic PORTAL. The prose style is uneasy, but some of the interpolated poems are remarkable examples of assured NONSENSE verse. [JC]

BENNETT, ANNA ELIZABETH (1914-) UK writer. ◊ CHILDREN'S FANTASY.

BENNETT, (ENOCH) ARNOLD (1867-1931) UK writer, famous for *The Old Wives' Tale* (**1908**) and other naturalistic novels, but also the author – in a consciously more commercial vein – of some SUPERNATURAL FICTION. *The Ghost: A Fantasia on Modern Times* (**1907**), which appeared very early in AB's career as "For Love and Life" in an 1890s magazine, features a GHOST whose attempts to keep a living woman in his thrall cause disaster to her various suitors. *The Glimpse: An Adventure of the Soul* (**1909**), a POSTHUMOUS FANTASY, follows an insensitive, selfish husband who has died of a heart attack into a series of AFTERLIFE experiences which teach him wisdom; he is then brought back to life only to find his wife has committed suicide out of remorse because she had taunted him with a false admission of infidelity. [JC]

BENNETT, JOHN (1865-1956) US folklorist. ◊ FOLK-TALES.

BENSON, A(RTHUR) C(HRISTOPHER) (1862-1925) UK essayist, journal-keeper and writer of some fiction; elder brother of E.F. BENSON and Robert Hugh BENSON. Much closer to M.R. JAMES than to either brother, ACB can be considered a member of the JAMES GANG because of the academic, antiquarian tone of his stories. The most notable are collected in *The Hill of Trouble and Other Stories* (coll **1903**) and *The Isles of Sunset* (coll **1904**), both assembled as *Paul the Minstrel and Other Stories* (omni **1911**); it is also possible that ACB was the "B" who wrote several GHOST STORIES in the same mode, set in typical college surroundings; the best are assembled as *When the Door is Shut* (coll **1986** chap). The two novellas under his own name collected in *Basil Netherby* (coll **1926**) display a similar aspiring wistfulness, a precarious indecision about the advisability of penetrating THRESHOLDS into the antique-visaged unknown. *The Child of the Dawn* (**1912**) is a mild-mannered UTOPIA. [JC]

BENSON, E(DWARD) F(REDERICK) (1867-1940) UK writer, brother of A.C. BENSON and Robert Hugh BENSON, best-known during his lifetime for his first novel, *Dodo* (**1893**), but subsequently more famous for his nonfantasy **Mapp and Lucia** sequence (the mid-1980s sequels by Tom HOLT likewise have no fantasy content) and for his short SUPERNATURAL FICTIONS. His supernatural novels, like his short stories, tend to treat the supernatural as invasive (◊ HORROR). The first, *The Luck of the Vails* (**1901**), is a murder mystery in which a goblet embodies a CURSE upon its owners. In *The Image in the Sand* (**1905**) an occultist releases an evil SPIRIT from BONDAGE. *The Angel of Pain* (**1905** US) also features occult investigations that go desperately wrong. *The House of Defense* (**1906** Canada) is a horror tale. *David Blaize and the Blue Door* (**1918**) sentimentally conveys its young hero – who appears in two other (nonfantasy) novels, *David Blaize* (**1916**) and *David of King's* (**1924**) – into a WONDERLAND based fairly closely on Lewis CARROLL's. *Across the Stream* (**1919**) returns to the admonitory mode of earlier work in the story of the DEBASEMENT of its protagonist through the influence of a DEMON that haunts him in the shape of his dead brother. *Colin: A Novel* (**1923**) and *Colin II* (**1925**) together tell the story of a PACT WITH THE DEVIL whose consequences extend from Elizabethan times to the present, once more focusing with palpable fascination upon the EVIL that emanates from young men who embrace dissolute lives. Two of EFB's late novels are of interest: *The Inheritor* (**1930**) again features a family curse – engendering the birth of MONSTERS – and a protagonist tempted by perversion; *Raven's Brood* (**1934**) carries into a rural setting some of the same concern about the embracing of intrusive evil, with a wife who turns out to be a WITCH.

Perhaps more interesting to contemporary readers, EFB's short stories – the later ones are slimmer, less portentous and more effective than the earlier – traverse much the same territory. The most important collections are *The Room in the Tower* (coll **1912**) – a typical story in which is "The Shootings of Achnaleish" (1906), about a Scottish village whose inhabitants can shapeshift into hares – *Visible and Invisible* (coll **1923**), which includes the remarkable "The Outcast" (1922), whose outwardly beautiful but inwardly loathsome protagonist may well be a REINCARNATION of Judas Iscariot; *Spook Stories* (coll **1928**), which includes "The Temple" (1924), about a house built within a stone circle where ancient sacrifices are re-enacted; and *More Spook Stories* (coll **1934**). This last includes "Monkeys" (1933), in which an ancient curse causes the death of a surgeon who rifled an Egyptian tomb of broken vertebrae because he thought the technique used to heal them might be of use to 20th-century humanity, and "Pirates" (1928), which expresses nostalgia for an EDEN where, as a child with the ghosts of his siblings, the narrator can once again be a brig- and terrorizing chaste lawns. Posthumous collections emphasizing stories of fantasy interest include *The Tale of the Empty House and Other Ghost Stories* (coll **1986**) ed Cynthia Reavell, *The Flint Knife* (coll **1988**), *Desirable Residences* (coll **1991**) and *Fine Feathers* (coll **1994**), the last three ed Jack Adrian and including much previously uncollected material; *The Collected Ghost Stories* (coll **1992**) ed Richard DALBY, despite its title, omits most of the contents of these two volumes.

Although EFB was a friend of M.R. JAMES, he only occasionally wrote stories under James's overt influence (◊ JAMES GANG). He was a more versatile writer than James, and his sense of the allure of EVIL was far more urbane, even though his favourite locales were rural enclaves typically subjected to visits by city gentlemen who then LEARNED BETTER. EFB has been neglected for most of the past half-century but is now enjoying a modest revival. [JC]

Other works: *The Valkyries: A Romance Founded on Wagner's Opera* * (**1903**); *The Countess of Lowndes Square* (coll **1920**), containing three stories of genre interest; "*And the Dead Spake –* " and *The Horror-Horn* (coll **1923** chap US); *The Myth of Robert Louis Stevenson* (1925; **1992** chap); *The Horror Horn* (coll **1974**) ed Alexis Lykiard, mostly HORROR; *Demoniacal Possession* (**1992** chap), a talk; *The Man who Went Too Far* (1904; **1992** chap); *The Technique of the Ghost Story and Three Short Stories* (coll **1993** chap); *The Heart of India* (coll **1994** chap) ed Jack Adrian; *Weepies* (coll **1995** chap) ed Jack Adrian.

BENSON, ROBERT HUGH (1871-1914) UK writer, younger brother of A.C. BENSON and E.F. BENSON, and the most overtly religious of the three – their father, Edward White Benson (1829-1896) was Archbishop of Canterbury. The stories assembled in *The Light Invisible* (coll **1903**), published in the year RHB converted to Roman Catholicism, utilize supernatural devices for homiletic purposes, as does the sequence of CLUB STORIES assembled in *A Mirror of Shalott, Composed of Tales Told at a Symposium* (coll **1907**; cut 1907 US). More impressively, *The Necromancers* (**1909**) treats SPIRITUALISM as a form of NECROMANCY through the story of a young Roman Catholic convert who almost loses his SOUL when he communicates with the SPIRIT of his dead fiancée. [JC]

Other works: *The Lord of the World* (**1907**), in which the ANTICHRIST gains a brief victory, and its sequel, *The Dawn of All* (**1911**).

BENSON, STELLA (1892-1933) UK author who eventually settled in China. Her only fantasy novel, *Living Alone* (**1919**), is set during WORLD WAR I. Its dispirited heroine encounters a WITCH and follows her to the exotic "House of Living Alone", where she meets other individuals with magical TALENTS and her life is decisively changed. After a series of strange encounters and vivid adventures – including an aerial combat involving two broomstick-riding witches and an air raid whose bombs provoke a few of the dead to rise from their graves in deluded expectation of the LAST JUDGEMENT – she sets sail for a new life in the USA. One of the finest fantasies of the period, it is a deeply personal book but, in common with many other fantastically transfigured spiritual autobiographies, has considerable power to move readers capable of empathizing with the existential plight of its heroine.

All but one of SB's briefer fantasies – the exception is the ORIENTAL FANTASY *Kwan-yin* (**1922** chap) – can be found in *Collected Short Stories* (coll **1936**), although most had earlier appeared as pamphlets from various private presses. *The Awakening* (**1925** chap) is an allegory of divine underachievement akin to those of Laurence HOUSMAN. *The Man who Missed the 'Bus* (**1928** chap) is a darkly surreal and nightmarish story diametrically opposed in tone and content to *Living Alone*. *Christmas Formula* (**1932** chap), also reprinted in *Christmas Formula and Other Stories* (coll **1932**), is a satirical extrapolation of the "gospel of commercial intimacy" allegedly preached by the new advertising-supported magazines of its day. "A Dream" is an allegorical VISIONARY FANTASY. [BS]

BEOWULF Epic Anglo-Saxon poem, probably composed in one of the newly Christian English kingdoms in the 8th century. The HERO's feats take place in older, heathen settings in southern Scandinavia, from which the Angles came to Britain.

The poem begins with the ship-funeral of King Scyld. His descendant Hrothgar builds a feasting-hall, Heorot. Jealous of the sound of revelry, the man-eating DEMON Grendel raids Heorot and carries off those who sleep there. After 12 years, Beowulf wrestles with him bare-handed and wrenches off the monster's arm. Grendel crawls back to his swamp to die. The ensuing celebrations are devastated when Grendel's mother arrives by night to exact vengeance. She seizes Hrothgar's counsellor. Beowulf's party track her to a bloodstained mere and find the victim's head. Beowulf dives deep into the mere and fights her. The SWORD he was given fails. In a dry cavern he finds another and strikes the MONSTER dead, then discovers and decapitates Grendel's body. The blood consumes the blade. He returns to

rejoicing and is laden with gifts. In old age Beowulf is now King of the Geats. An outlaw finds a hoard of treasure in a barrow, guarded by a DRAGON. He steals a cup. The dragon burns up the countryside. Beowulf and his men are led to the barrow. Beowulf fights the dragon but is burned by its fire. All but one of his companions flee. Beowulf kills the dragon but dies himself. The treasure is loaded on his funeral-pyre and a barrow is raised as a beacon for ships.

Within this story are fragments of other dragon-slayings, hero-tales and treacheries.

Beowulf is a rare survival of early Anglo-Saxon STORY. Like the Celtic ARTHUR, Beowulf may be an historical figure who attracted legendary exploits to his name; "Beowulf", like "Arthur", may mean "bear". But, unlike Arthur's, Beowulf's story has not been much built upon by later writers – John GARDNER's *Grendel* (**1971**) shifts the focus to the adversary's viewpoint. *Beowulf* extols a hero-code of boasting, single combat, generous gift-giving (especially of gold) and loyalty to one's lord. The style shows a sober absence of exaggeration, other than the presence of monsters. Beowulf is saved by armour made by WELAND SMITH, but other WEAPONS may be less trustworthy, as friends may prove. There is a shadow of doom over the poem, akin to NORDIC FANTASY. [FS]

BERBERICK, NANCY VARIAN (1951-   ) US writer who began publishing fantasy with "The Merlin's Gift" for *Beyond* in 1986, and who has – after a **Dragonlance** tie, *Stormblade: Dragonlance Saga Heroes 2* * (**1988**) – published two series of moderate interest. The **Elvish** sequence – *The Jewels of Elvish* (**1989**), an independent title, even though published by TSR, and *A Child of Elvish* (**1992**) – is set in a FANTASYLAND where a number of races (including the human race) intersect, vie and mate; the plot of the first volume involves the alliance of two of these races – ELVES and humans – against barbarian invasions. The second is dominated by a search for lost jewels of power (◊ PLOT COUPONS). A second sequence – *Shadow of the Seventh Moon* (**1991**) and *The Panther's Hoard* (**1994**) – set in a LAND-OF-FABLE Britain after the fall of CAMELOT, is about the THINNING of the world, as told through Wotan's (◊ ODIN) campaign to exterminate all. [JC]

BERESFORD, ELISABETH (1926-   ) UK writer, daughter of J.D. Beresford (1873-1947) and creator of **The Wombles**; although she has written adult romances, most of her books are for children. Early work for tv led to her first book, *The Television Mystery* (**1957**), followed by *The Flying Doctor* (**1958**), a rather unconventional novel which included all manner of Australian effects, including an Aboriginal GHOST. For a while her children's stories were conventional mysteries and adventures, until *Awkward Magic* (**1964**; vt *The Magic World* 1965 US), in which a child befriends a griffin in search of its lost treasure, introduced her to CHILDREN'S FANTASY. Written in the vein of E. NESBIT, this triggered an occasional series of books in which children encounter PLOT COUPONS in the form of magical creatures or artifacts. The best of the series is *Dangerous Magic* (**1972**), involving cosmic pyrotechnics in the style of Susan COOPER. Other titles are *Travelling Magic* (**1965**; vt *The Vanishing Garden* 1967 US), *Sea-Green Magic* (**1968**), *Vanishing Magic* (**1970**), *Invisible Magic* (**1974**), *Secret Magic* (**1978**), *Curious Magic* (**1980**) and *Strange Magic* (**1986**). Less interesting are several books she has written involving friendly GHOSTS, including *The Happy Ghost* (**1979**), *The Ghosts of Lupus Street School* (**1986**) and *Emily and the Haunted Castle* (**1987**).

All these books were popular, but they have been overshadowed by the success of **The Wombles**. EB's WAINSCOT society of eco-friendly quasi-human creatures, who resemble eccentric teddy-bears, live UNDERGROUND and at night collect rubbish with the aim of transforming it into something useful, first appeared in *The Wombles* (**1968**). Their popularity resulted in a long series of books, a tv series, records and other paraphernalia. The **Wombles** stories demonstrate, if nothing else, the ability of fantasy animals to deliver a simple message effectively to children.   [MA]

The Wombles series: *The Wandering Wombles* (**1970**); *The Invisible Womble and Other Stories* (coll **1973**); *The Wombles in Danger* (**1973**); *The Wombles at Work* (**1973**); *The Wombles Go to the Seaside* (**1974**); *The Snow Womble* (**1975**); *Tomsk and the Tired Tree* (**1975**); *Wellington and the Blue Balloon* (**1975**); *Orinoco Runs Away* (**1975**); *The Wombles Make a Clean Sweep* (**1975**); *The Wombles to the Rescue* (**1975**); *The MacWombles Pipe Band* (**1976**); *Madame Cholet's Picnic Party* (**1976**); *Bungo Knows Best* (**1976**); *Tobermory's Big Surprise* (**1976**); *The Wombles Go Round the World* (**1976**); *The World of the Wombles* (**1976**) and *Wombling Free* (**1978**); plus *The Wombles Annual 1975-1978* (**1974-7** 4 vols) and *The Wombles Gift Book* (coll **1975**).

Other work: *Jack and the Magic Stove* (**1982**), FOLKTALE adaptation.

BERGER, THOMAS (LOUIS) (1924-   )US writer best-known as author of several novels telling the story of a representative 20th-century man named Reinhart – beginning with *Crazy in Berlin* (**1958**) – and for *Little Big Man* (**1964**). The latter, which takes the format of the Western as a kind of TEMPLATE upon which to inscribe an experiment in SATIRE, prefigures his fantasy, all of which similarly estranges the reader from the devices and venues being exploited. *Regiment of Women* (**1973**), because its role-reversal premise receives no real explanation, could be read as a fantasy exploration of a world in which male and female roles are reversed; but the tenor of the tale is sf-like. *Arthur Rex: A Legendary Novel* (**1978**) initially gives the impression that it comprises a fairly straightforward rendering of the MATTER of Britain and its protagonists in an undemanding TWICE-TOLD style, but the text as a whole – with its deadpan reiterations of traditional material that are simultaneously devastating and full of desiderium – is rather like Tennyson's *Idylls of the King* (**1870**) recited word-for-word by Woody Allen. *Nowhere* (**1985**) similarly deconstructs RURITANIA by transforming a typical venue into a WONDERLAND utopia; but the modest ambition of the original Ruritanian mode generates a strangely bland text. *Being Invisible* (**1987**) is similarly lacking in underlying bite, though some sharp ironies do attend the newly invisible (◊ INVISIBILITY) protagonist's attempts either to exploit his condition (which is that of Everyman in chinks of the world machine) or to do good. Of greater interest is *Changing the Past* (**1989**), whose protagonist is allowed to return through TIME and to gain various goals he longed for; inevitably, each ANSWERED PRAYER proves double-edged, for the weight of the world redresses any lost BALANCE. No matter how ingenious the WISH, mortals remain in BONDAGE, a fate which also afflicts the protagonists of *Granted Wishes* (coll **1984** chap), a collection of answered-prayer tales with some fantasy elements.   [JC]

BERGMAN, INGMAR (1918-   ) Swedish movie director, renowned as one of the world's leading moviemakers; he is also a distinguished theatrical producer and screenwriter.

Outside Sweden, IB is chiefly known for his psychological dramas in which he – like Rainer Werner Fassbinder (1945-1982) – creates a world of his own in the microcosmos of the film studio. But he has also made a number of light-hearted comedies dealing with the impossible complexities of LOVE.

Most of IB's movies contain obvious fantasy elements, from the internationally acclaimed *The SEVENTH SEAL* (*1957*; ot *Sjunde inseglet*), based on a medieval church painting of DEATH playing CHESS with a knight for the lives of a family of innocent jesters, to his celebrated family drama *Fanny and Alexander* (*1982*; ot *Fanny och Alexander*), with its overwhelmingly rich world, filled with Central European imagery – where horror marches side-by-side with happiness, where the dead stay with their families, helping them along, and where MAGIC is a part of everyday life. Magic, and its ways and means, play an important part in IB's comedy *The Devil's Eye* (*1960*; ot *Djävulens öga*), where the DEVIL, in a manner reminiscent of a FOLKTALE or perhaps a Mikhail BULGAKOV story, sends DON JUAN up to Earth in order to seduce an innocent young girl – who instead seduces Don Juan. Another well known fantasy by IB is the comedy *Wild Strawberries* (*1957*; ot *Smultronstället*), where an old man – played by none other than the great Swedish director Victor Sjöström (1879-1960), who himself directed several fantasy movies – dreams of his forthcoming death in a sequence that nods to Carl Th DREYER's famous VAMPIRE movie, VAMPYR (*1932*). *The Hour of the Wolf* (*1968*; ot *Vargtimmen*) is, by contrast, a terrifying movie, in which horrific nightmares (◊ DREAMS) seem to be reified. PERSONA (*1966*), perhaps the weirdest of the many IDENTITY-EXCHANGE movies that have been made, is often hailed as IB's masterpiece.

Yet IB's most interesting fantasy work might be his filmed theatre production of Mozart's *The Magic Flute* (*1975*; ot *Trollflöjten*; ◊ OPERA); a fantasy opera to start with, this is one of IB's most complex works as he develops the original into a brilliant trip into an archetypal world of dreams and startling images, sometimes reminiscent of August STRINDBERG's *A Dream Play* (*1902*; ot *Ett drömspel*).   [SJL]

BERMEJO, LUIS Working name of prolific Spanish COMIC-strip artist Luis Bermejo Rojo (1932-   ), whose stylish pen-and-brush drawing for UK comics became increasingly attractive and sophisticated throughout the 1960s; thereafter he transferred to Spanish comics, US comics, and Spanish comics again. His earliest work of note was the series *Aventuras del FBI* ["Adventures of the FBI"] (1954-5). In 1959 he formed a close working association with José ORTIZ, taking on a number of young and talented assistants, including, briefly, Esteban MAROTO.

The greater part of his work in the 1960s was in the UK for Fleetway Publications' digest-size comic books of WWII stories. His artwork developed markedly in dramatic quality and sensitivity in the two-year period during which he worked on **John Steel Casebooks**, a series of distinctive private-eye stories which he drew in a dramatic *film-noir* style for the companion title *Thriller Picture Library* (7 titles 1962-3).

LB's first work of fantasy interest was **Pike Mason** for *Boys' World*, a series of three stories about a freelance deep-sea salvage expert – *The Curse of Zentaca* (1963), *The Monster of Duncrana* (1963) and *The Sea Ape* (1963-4) – which he imbued with a shadowy, mysterious quality and rendered in line and wash. Immediately after this he worked on HEROS THE SPARTAN (1964-6), alternating stories with Frank BELLAMY. This was followed by the creation of one of the very few entirely UK SUPERHEROES, **Johnny Future** (1967-8), which LB sensitively rendered in b/w for the MARVEL COMICS reprint title *Fantastic*. He was now at the height of his creative powers, his refined use of fine line and black areas showing a remarkable crispness and sophistication. During 1969-75 he drew a number of children's FAIRYTALES and fantasy strips for *Once Upon a Time* and *Treasure*, and a strip adaptation of the Song of Roland for *Look and Learn* (1971).

LB drew a number of fantasy, sf and horror stories for WARREN PUBLISHING, for whom he also created the time-travelling character **The Rook**. One issue of CREEPY was devoted entirely to him – *Creepy #71* (1975). But LB's work suffered a subtle decline in quality, and his 1979 strip adaptation of J.R.R. TOLKIEN's *The Lord of the Rings* (1954-5), *El Señor de los Anelos* (graph **1979** 6 vols), scripted by Nicola Cuti, showed drawing which, although obviously accomplished, had lost some of its sensitivity and clarity of imaginative vision.

LB remains a leading artist of Spanish comics, with a very long list of fantasy, sf and Western strips to his credit. He drew the 48-page colour comic-strip album *Los 8 Anillos de Elibarin* (graph **1981**) and the fifth volume of the 25-volume **Relatos del Nuevo Mundo** ["Stories of the New World"] (graph **1992**) – a series in celebration of the 500th anniversary of Columbus's discovery of the Americas. His work has been published in Italy, Germany, France, Portugal, Scandinavia and Japan. He continues to draw a wide range of strips for regular publication throughout mainland Europe, including the two long-running Westerns **El Mestizo** ["The Halfbreed"] (1985-   ) and, in continued collaboration with Ortiz, the stories about the Native American **Jon Khe** (1986-current). [RT]

BERNANOS, GEORGES (1888-1948) French writer. ◊ Michel BERNANOS.

BERNANOS, MICHEL (1924-1964) French writer, son of the polemical Catholic writer Georges Bernanos. In the latter's allegorical novel *Sous le Soleil de Satan* (**1936**; trans Veronica Lucas as *Star of Satan* **1927** UK) an obsessive priest struggles against the odds to oppose the DEVIL, to whose cause the great majority of people unwittingly deliver themselves. MB's own *À Montagne morte de la vie* (**1967**; trans Elaine P. Halperin as *The Other Side of the Mountain* **1968** US), a more elaborate and far more enigmatic ALLEGORY embodying a similar worldview, invites consideration as one of the most phantasmagoric examples of POSTHUMOUS FANTASY. [BS]

BERTRAM, NOEL Pseudonym of UK writer Joseph Noel Thomas Boston (1910-1966), Vicar of Dereham and antiquarian, whose occasional GHOST STORIES, in the style of M.R. JAMES, were privately published as *Yesterday Knocks* (coll **1953**) as by Noel Boston. Most are set around the fictional cathedral of Losingham. Some were reprinted (along with new stories) in the **Supernatural Stories** series 1960-62 as by Noel Bertram. This came about through the intervention of R. Lionel Fanthorpe (1935-   ; ◊ SFE) who, as a favour, placed the stories with the publisher as if they were his own. [MA]

Further reading: "Noel Boston: Master Antiquarian" (*Ghosts & Scholars #5* 1983) by Mike ASHLEY.

BESANT, SIR WALTER (1836-1901) UK writer and social reformer. He wrote several bestselling novels in collaboration with James Rice (1844-1882), with whom he also produced – at first anonymously – *The Case of Mr Lucraft and*

Other Tales (coll **1876**). In the striking title story a young man leases his healthy appetite to an aged hedonist, accepting the unpleasant side-effects of the old man's sybaritic indulgence. Also included are four light-hearted GHOST STORIES and the didactic novella "Titania's Farewell", in which the FAIRIES leave England in protest against modern social trends. WB collaborated with Walter Herries POLLOCK on the novella "Sir Jocelyn's Cap" (1884-5; reprinted in *Uncle Jack, etc.* coll **1886** by WB), whose method and style anticipate the works of F. ANSTEY. His most important solo fantasy was *The Doubts of Dives* (**1889**; reprinted in *Verbena Camellia Stephanotis* coll **1892**), in which a bored socialite exchanges bodies (◊ IDENTITY EXCHANGE) with a poor friend who thinks that wealth will allow his literary talent to blossom; their respective fiancées find the swap disconcerting. Besant's interest in such experiments was further extrapolated in the dual-personality novel *The Ivory Gate* (**1892**). [BS]

Other works: *The Charm and Other Drawing-Room Plays* (coll **1896**) with Pollock.

BESTIARIES The classic medieval bestiary assembles notes on animals and interprets their habits as pious ALLEGORY; MYTHICAL CREATURES may appear owing to unreliable or mistranslated sources, with monkeys becoming SATYRS while TRAVELLERS' TALES of rhinoceros and narwhal generate the UNICORN. T.H. WHITE's *The Book of Beasts: Being a Translation from a Latin Bestiary of the Twelfth Century* (**1956**) adds learned and witty commentary on these matters, without scoffing at monastic scribes for lacking modern data on BIOLOGY. Generally, a bestiary can be any not purely factual collection of creature descriptions. Hilaire Belloc (1870-1953) rhymed with fantastic humour about real animals in *A Bad Child's Book of Beasts* (**1896**) and *More Beasts for Worse Children* (**1897**). S.H. SIME drew his own IMAGINARY ANIMALS, with accompanying verses, in *Bogey Beasts* (**1923**). Edward ARDIZZONE and James REEVES produced *Prefabulous Animiles* (**1957**). *The Book of Imaginary Beings* by Jorge Luis BORGES (**1967**) is a fine if eccentric bestiary of mythic and fictional creatures. Margaret W. Robinson's *Fictitious Beasts: A Bibliography* (**1961**) is self-explanatory. *The Encyclopaedia of Things that Never Were* (**1985**) by Michael Page (1922-   ) and Robert Ingpen (1936-   ) includes many fantasy-bestiary entries. Bestiary descriptions of gnomes and DRAGONS are expanded mock-seriously to book length in *Gnomes* (**1976**) by Wil Huygen (1922-   ) and Rien Poortvliet (?1933-   ) and in Peter DICKINSON's *The Flight of Dragons* (**1979**), both also vehicles for FANTASY ART. Some commercial fantasy bestiaries are GAME adjuncts describing and providing fighting statistics for MONSTERS in games like *Dungeons & Dragons* and *Call of Cthulhu* (◊◊ CTHULHU MYTHOS). [DRL]

See also: IMAGINARY ANIMALS.

BETANCOURT, JOHN GREGORY (1960-   ) US writer. Much of his work has been sf, though his first story of genre interest – "Vernon's Dragon" for *100 Great Fantasy Short-Short Stories* (anth **1984**) ed Isaac ASIMOV, Terry CARR and Martin H. GREENBERG – was fantasy. *Rogue Pirate* (**1987**) is a fantasy adventure. *The Blind Archer* (**1988**), in which a young mortal must learn to do MAGIC according to the strict precepts of the eponymous GOD, is set in the same fantasy CITY featured in *Slab's Tavern and Other Uncanny Places* (coll **1991** chap), a volume containing some CLUB STORIES. In 1988 JGB co-edited with George H. Scithers (1929-   ) and Darrell SCHWEITZER several issues of a revived WEIRD TALES. He also edited *The Ultimate Frankenstein* (anth **1991**) with Byron PREISS.   [JC]

BETHANCOURT, T. ERNESTO Working name of US writer Tomàs Ernesto Bethancourt Passailaigue (1932-), mostly known for borderline tales like *The Dog Days of Arthur Cane* (**1976**) and *Nightmare Town* (**1979**), and for the **Instruments** YA sf series: *The Mortal Instruments* (**1977**) and *Instruments of Darkness* (**1979**). A fantasy sequence – *Tune in Yesterday* (**1978**) and *The Tomorrow Connection* (**1984**) – is concerned with messages through TIME.   [JC]

BETWEEN TWO WORLDS (*1944*) US movie (*1944*). ◊ OUTWARD BOUND (*1930*).

BEWITCHED US movie (*1945*). MGM. **Pr** Jerry Bresler. **Dir** Arch Oboler. **Screenplay** Oboler. **Starring** Henry H. Daniels Jr (Bob Arnold), Edmund Gwenn (Dr Bergson), Horace McNally (Eric Russell), Phyllis Thaxter (Joan Alris Ellis). 65 mins. B/w.

Wealthy daughter Joan is haunted by her *alter ego* Karen, who is sexually precocious where Joan is prim, and who eventually murders Joan's quondam fiancé Bob. Lawyer Eric, for whom both Joan and Karen have fallen, successfully defends Joan at the trial; but she, discovering the truth, confesses guilt before the verdict is given. Condemned, she is spared at the last moment when Dr Bergson hypnotizes both TWINS individually, persuading Karen to die and thereby fulfilling the death sentence. A trim PSYCHOLOGICAL THRILLER, *B*, made at a time when the condition now called MULTIPLE PERSONALITY belonged more to the occult than to science, is a fantasy that would, some decades later, have been recast as sf.   [JG]

BEWITCHED US tv series (1964-72). ABC/Screen Gems. **Pr** William Asher. **Exec pr** Harry Ackerman. **Dir** Asher, Ida Lupino, William D. Russell and others. **Spfx** Dick Albain. **Writers** Sol Saks and many others. **Created by** Saks. **Starring** Maurice Evans (Maurice), Alice Ghostley (Esmeralda), Marion Lorne (Aunt Clara), Paul Lynde (Uncle Arthur), Elizabeth Montgomery (Serena/Samantha Stephens), Agnes Moorehead (Endora), Dick Sargent (Darrin Stephens 1969-72), Dick York (Darrin Stephens (1964-9). 254 25min episodes. Colour.

Samantha marries ad-agency man Darrin, confessing she is a WITCH only after the wedding; he entreats her never to use her MAGIC powers again, and she acquiesces. For almost all of an astonishing 254 episodes of this sitcom the same basic formula was repeated: early in the episode Samantha succumbs to temptation, often with the encouragement of her similarly talented relatives (◊ TALENTS), and for the remainder of the time she and Darrin attempt to cover up the truth behind the consequences. What made this immensely popular show so much better than it sounds was the inventiveness with which its huge array of screenwriters (upwards of 40, including the young Larry Cohen) manipulated the formula.

B was shamelessly imitated by NBC with I DREAM OF JEANNIE (1965-70); a further offshoot was *Tabitha* (1977-8), a short-lived series starring Lisa Hartman as the Stephens's daughter, now adult and working for a tv station. The movie BEWITCHED (*1945*) is quite unrelated.   [JG]

BEYOND FANTASY FICTION US digest MAGAZINE, 10 issues, bimonthly, July 1953-[January] 1955, published by Galaxy Publishing, New York; ed Horace L. GOLD.

A companion to *Galaxy Science Fiction*, *BFF* sought to bring the same sophistication to fantasy as *Galaxy* had to sf.

It succeeded to a large extent, and is generally acknowledged as the natural successor to UNKNOWN though, because it was shorter-lived and had a lower circulation, it was less influential. Along with FANTASY MAGAZINE and the early issues of FANTASTIC, it encouraged a brief but abortive revival of interest in fantasy, bringing a level of originality and quality to the field that would not be equalled for the next 25 years. *BFF*'s Surrealist covers sought to rival the slick magazines on the stalls, and most of the fiction had pretensions to be SLICK FANTASY. Among its 77 stories are such classics as ". . . And My Fear is Great" and "Talent" by Theodore STURGEON, "Babel II" by Damon Knight (1922-), "The Wall Around the World" by Theodore Cogswell (1918-1987), "Can Such Beauty Be?" by Jerome BIXBY, "The King of the Elves" by Philip K. Dick (1928-1982), "The Real People" by Algis Budrys (1931-), "Sorry, Right Number" by Richard MATHESON, "The God Business" by Philip José FARMER and "Sine of the Magus" (exp vt *The Magicians* 1976) by James E. Gunn (1923-), a diversity which shows that the magazine was prepared to publish a range of stories of FANTASY, HORROR and SUPERNATURAL FICTION, including DARK FANTASY and LOW FANTASY – although the closest it ventured to HEROIC FANTASY was the **Harold Shea** story "The Green Magician" by L. Sprague DE CAMP and Fletcher PRATT. Today the magazine is remembered more for what it stood for than for its individual contents, and there is much excellent material still to be rediscovered. Only one anthology has been chosen exclusively from *BFF*'s pages: *Beyond* (anth **1963**) ed (uncredited) Thomas A. Dardis (1926-). The first four issues were reprinted in cut form in the UK, November 1953-May 1954 (undated).   [MA]

BEYOND THE FIELDS WE KNOW ◊ MAGAZINES.

BIBIANA, JEAN GALLI DE (?1710-?1780) French writer. ◊ DOLLS.

BIBLE It is not the purpose of this encyclopedia to treat the Bible, a significant TAPROOT TEXT, as an anthology of LEGENDS, FABLES and ALLEGORIES. Nor is the rightness or wrongness of Harold BLOOM's argument – that the "best" tales in the Bible were written by one person of considerable literary sophistication, and that this person was a woman (named J) – a debate to be more than referred to here. Entries in which the Bible is considered in context include CHRISTIAN FANTASY and JEWISH RELIGIOUS LITERATURE. Other entries of interest include APOCALYPSE, BABEL, CHRIST, FLOOD and THEODICY.   [JC]

BIERCE, AMBROSE (GWINETT) (1842-*c*1914) US writer and journalist, of intense interest for his work in all areas of the fantastic. He began publishing work of genre interest with "The Haunted Valley" for the *Overland Monthly* in 1871, several years after he had left the US Army, having served throughout the Civil War – an experience that marked him for life. Indeed, throughout his career – which was marked by much travel, including a three-year stay in London – he gazed upon the USA with a born exile's cold and solitary eye, through fiction, essays and the black aphorisms published in newspapers over the years and assembled late in his life as *The Cynic's Word Book* (coll **1906**; exp vt *The Devil's Dictionary* 1911; exp ed Ernest Jerome Hopkins vt *The Enlarged Devil's Dictionary* 1967).

It is perhaps not surprising that his greatest stories involve solitary figures *in extremis* or that, when the supernatural intervenes, it does so with an effect of venomous irony. His best work – HORROR stories like "An Occurrence at Owl

Creek Bridge", "An Inhabitant of Carcosa" (1886) (*for both* ◊ POSTHUMOUS FANTASY) and "The Death of Halpin Frayser" – appears in two volumes of stories, *Tales of Soldiers and Civilians* (coll dated 1891 but **1892**; vt *In the Midst of Life* 1892 UK; exp under original title 1898 US) and *Can Such Things Be?* (coll **1893**; exp 1923), which also contain a number of sardonic GHOST STORIES, of which the best is "The Damned Thing". Some additional stories appear in various collected editions. AB's fantasy is almost completely restricted to two volumes: *Cobwebs from an Empty Skull* (coll dated 1874 but **1873** UK; vt *Cobwebs: Being Fables of Zambri and Parsee* c1874 UK) as by Dod Grile, which contains humorous satirical sketches in FABLE form; and *Fantastic Fables* (coll **1899**; cut vt *Fantastic Debunking Fables* c1926), some of whose contents are little more than aphorisms expanded into anecdotes. But even in his most minor work an astonishingly sharp mind, never at rest, makes itself felt.

[JC]

Other works: *The Fiend's Delight* (coll c1872 UK) and *Nuggets and Dust, Panned out of California* (coll c1872 UK), both as by Grile; *The Dance of Death* (**1877**) with Thomas A. Harcourt, AB writing as William Herman; *The Collected Works of Ambrose Bierce* (coll **1909-12** 12 vols); *Ten Tales* (coll **1925** UK); *Tales of Ghouls and Ghosts* (coll c1926 chap); *Tales of Haunted Houses* (coll c1926 chap); *Eyes of the Panther* (coll **1928** UK); *The Collected Writings of Ambrose Bierce* (coll **1946**; vt *The Best of Ambrose Bierce* ?1984); *Ghost and Horror Stories of Ambrose Bierce* (coll **1964**) ed E.F. BLEILER; *The Complete Short Stories of Ambrose Bierce* (coll **1970**); *The Devil's Advocate: An Ambrose Bierce Reader* (coll **1987**; vt *In the Midst of Life* UK); *An Occurrence at Owl Creek Bridge and Other Stories* (coll **1995** chap UK).

Further reading: *Ambrose Bierce: A Biography* (**1968**) by Richard O'Connor.

BIG (vt *Big!*) US movie (**1988**). 20th Century-Fox/Gracie Films/American Entertainment. **Pr** James L. Brooks, Robert Greenhut. **Co-pr** Gary Ross, Anne Spielberg. **Dir** Penny Marshall. **Screenplay** Ross, Spielberg. **Novelization** *Big! * (**1988**) by B.B. Hiller and Neil W. Hiller. **Starring** Tom Hanks (Josh Baskin), John Heard (Paul Davenport), Robert Loggia (MacMillan), David Moscow (Young Josh), Elizabeth Perkins (Susan Lawrence), Jared Rushton (Billy Kopeche). 102 mins. Colour.

Although involving only one person rather than two, *B* has an essentially IDENTITY-EXCHANGE plot. 12-year-old computer nerd Josh wants to be bigger so that he has a chance with school bimbette Cynthia; the latest humiliation comes when, in front of her, he is barred from a CARNIVAL ride. Nearby is electrical wishing-machine Zoltar; although unplugged, it lights up to grant him his WISH. Next morning he wakes to find himself adult; the carnival has gone, and his mother believes him an intruder. Only best friend Billy will believe what has happened; he helps Josh get to NEW YORK, install himself in a flophouse and start hunting the Zoltar machine; meantime Josh gets a job as a computer operator in MacMillan Toys. A chance encounter in a toyshop with MacMillan himself leads Josh, with his childish understanding of TOYS, to a Vice-Presidency in the company. It also leads him to Susan, who has been sleeping her way up the corporate structure; expecting merely sex, she finds instead love – and in due course sex. This consummation decides Josh to opt for adulthood, but at last, finding the Zoltar machine, he wishes himself back to childhood, bidding Susan and the adult world a tearful farewell.

B is largely held together by supporting actors Loggia and especially Rushton; its central romance, despite would-be sophisticated ironies, has a slightly off-colour feel. [JG]

BIG DUEL IN THE NORTH SEA vt of *Ebirah, Terror of the Deep* (**1966**). ◊ GODZILLA MOVIES.

BIGGEST BATTLE ON EARTH, THE vt of *Ghidora, The Three-Headed Monster* (**1965**). ◊ GODZILLA MOVIES.

BIGGEST FIGHT ON EARTH, THE vt of *Ghidora, The Three-Headed Monster* (**1965**). ◊ GODZILLA MOVIES.

BIGGLES (vt *Biggles: Adventures in Time*; vt *Biggles Gets Off the Ground*) UK movie (**1986**). Universal/Compact Yellowbill/Tambarle. **Pr** Pom Oliver, Kent Walwin. **Exec pr** Adrian Scrope. **Dir** John Hough. **Spfx** David Harris. **Screenplay** John Groves, Walwin. **Based on** characters created by W.E. Johns (1893-1968). **Novelizations** *Biggles: The Untold Story* * (**1986**) by Peter James; *Biggles: The Movie* * (**1986**) by Larry Milne. **Starring** Peter Cushing (Commodore William Raymond), Neil Dickson (Biggles), Daniel Flynn (Ginger), Marcus Gilbert (Erich von Stalhein), Francesca Gonshaw (Marie), William Hootkins (Chuck), Fiona Hutchinson (Debbie), Alex Hyde-White (Jim Ferguson), James Saxon (Bertie), Michael Siberry (Algy). 92 mins. Colour.

Present-day fast-foods entrepreneur Ferguson is a time-TWIN of the WORLD WAR I fighter ace Biggles, as he discovers through a series of TIMESLIPS. He is prepared to dismiss this as HALLUCINATION until ancient RAF Commodore Raymond persuades him that the aid he gives/gave to Biggles in destroying a German secret weapon is/was important if the present is to be preserved. At last Biggles and the weapon are brought into the present, where the weapon is destroyed. *B* is rather flatly scripted and feels like SPIELBERG-on-the-cheap, yet has interest because of its notion of travel not only through TIME (◊◊ TIME TRAVEL) but from the mundane into a created ALTERNATE REALITY, and *vice versa*. [JG]

BIG HEART, THE vt of MIRACLE ON 34TH STREET (**1947**).

BIG TROUBLE IN LITTLE CHINA US movie (**1986**). 20th Century-Fox/Taft-Barish-Monash. **Pr** Larry J. Franco. **Exec pr** Keith Barish, Paul Monash. **Dir** John Carpenter. **Vfx** Richard Edlund. **Screenplay** Gary Goldman, David Z. Weinstein, adapted by W.D. Richter. **Starring** Kate Burton (Margo), Kim Cattrall (Gracie Law), Dennis Dun (Wang Chi), James Hong (Lo Pan), Peter Kwong (Rain), Donald Li (Eddie Lee), Suzee Pai (Miao Yin), James Pax (Lightning), Kurt Russell (Jack Burton), Carter Wong (Thunder), Victor Wong (Egg Shen). 99 mins. Colour.

An Americanized CHINOISERIE plunging trucker Burton into Chinatown feuds involving three MAGIC fighters – the Storms (Lightning, Rain and Thunder) – and their ruler, the immortal sorcerer Lo Pan. Friend Lee's fiancée Miao Yin is sought by Lo Pan because she has green eyes, rare among Chinese women; through wedding and vampirizing such a woman he may incarnate himself as a youth rather than, as at present, an ancient. Lee's martial arts and Burton's no-nonsense redneck fists thwart the plan, and Burton is able to kill Lo Pan in the instant during which he is a vulnerable mortal. *BTILC* has all the requisite pyrotechnics, but lacks charm; there is no attempt to view events through Chinese eyes, because all the good Chinese involved are Chinese-*Americans*; and, since the lovely Miao Yin is deployed merely as a MCGUFFIN, there is little emotional interest in the outcome. [JG]

BILL & TED'S BOGUS JOURNEY US movie (*1991*). Columbia TriStar/Orion/Nelson Entertainment/Interscope Communications. **Pr** Scott Kroopf. **Exec pr** Robert W. Cort, Ted Field, Rick Finkelstein, Barry Spikings. **Sv pr** Neil Machlis. **Dir** Pete Hewitt. **Vfx** Gregory L. McMurry, Richard Yuricich. **Creature/mufx** Kevin Yagher. **Screenplay** Chris MATHESON, Ed Solomon. **Novelization** *Bill & Ted's Bogus Journey* * (**1991**) by Robert Tine. **Starring** Joss Ackland (De Momolos), George Carlin (Rufus), Keanu Reeves (Ted), William Sadler (Death/Grim Reaper), Alex Winter (Bill). **Voice actors** Neil Ross, Frank Welker. 93 mins. Colour.

The more interesting follow-up to *Bill & Ted's Excellent Adventure* (*1989*; ◊ SFE), in which we learnt that a future society depends for its existence on the success today of the rock group Wild Stallyns, fronted by Californian layabouts Bill and Ted. Now, in that future, martinet De Momolos wants to change its past in order to create a more disciplined society; he sends back android lookalikes of Bill and Ted to kill the originals. They succeed. The dead dudes venture into the AFTERLIFE, flee the hooded figure of DEATH (a direct descendant of that in Bergman's The SEVENTH SEAL [*1956*], with accent to match), try as GHOSTS and through POSSESSION to warn the living, confront the DEVIL, are cast into a HELL comprising their own worst memories of childhood, defeat Death at various games (e.g., Cluedo, Battleships & Cruisers, Twister), travel with him to HEAVEN where GOD grants them a return to life, enlist the help of Station, the greatest scientist in the Universe – who happens to be an alien – and, through a series of TIME paradoxes, destroy the evil androids, thwart De Momolos and, with Death on double bass and Station on congas, drive Wild Stallyns to interplanetary stardom. As one might expect from its scriptwriters, *B&TBJ* shows, amid the mirth, a firm and knowledgeable grasp of the fantasy genre; the depictions of Heaven and Hell, especially the latter, are stunning, with sophisticated use of TECHNOFANTASY tropes. [JG]

BILL & TED'S EXCELLENT ADVENTURE US movie (*1988*). ◊ BILL & TED'S BOGUS JOURNEY (*1991*); SFE.

BILLY THE KID William H. Bonney (1859-1881) was a thief who shot other men in the back. He was also involved in range wars, and possibly originally cast in heroic roles because set in situations which called for the HERO he was not. Even before his early death he had become legendary as Billy the Kid through dime novels written about his imaginary exploits; this relationship between life and LEGEND has itself proved to be of "mythic" import. He is, in his legendary form, a significant component of the consensual HEROIC-FANTASY hero, and the ambivalence between truth and legend is a useful distancing device within that form.

His story is told in Jorgé Luis BORGES's *Historia universal de la infamia* (coll **1954**; trans **1972**). David Thomson's nonfantasy *Silver Light* (**1990**) copes with the intractability in real life of Billy the Kid as ICON by revealing only his shadow in the tale itself, which may be treated as a rumination upon the thesis that Billy the Kid created Hollywood rather than *vice versa*. In fantasy terms, the Kid is a CHILDE, a solitary figure driven woundedly to QUEST for an outcome he is unlikely to survive. He appears in Samuel R. DELANY's *The Einstein Intersection* (**1967**) as Kid Death (and a version, with the same soubriquet, appears in Simon R. GREEN's **Deathstalker** sequence of space operas), in *A Captive in Time* (**1990**) by Sarah Dreher (1937-   ), in *The Ancient Child* (**1989**) by N. Scott Momaday (1934-   ), in Rebecca

ORE's sf novel *The Illegal Rebirth of Billy the Kid* (**1991**), in John JAKES's sf novel *Six-Gun Planet* (**1970**), and in *Blood Meridian* (**1985**) by Cormac McCarthy (1933-   ), in which the Kid is complicit with a gang run by DEATH. There are many movies, of which at least one, *Billy the Kid versus Dracula* (**1966**) (◊ DRACULA MOVIES), is SUPERNATURAL FICTION; *Billy the Kid and the Green Baize Vampire* (**1966**), although a supernatural movie, is about not the Kid but a contemporary snooker player. [JC]

BILLY THE KID VERSUS DRACULA (*1965*) ◊ DRACULA MOVIES.

BIOLOGY Fantasy biology has three main threads. On the one hand, there is a rich mythology of fabulous plants and animals, some of which were once believed to exist and thus figured in such classic protoscientific works as the *Historia Naturalis* (*c60AD*) of Pliny the Elder (23-79). Many are given separate consideration in these pages; others may be found under such headings as ANIMALS UNKNOWN TO SCIENCE, MYTHICAL CREATURES and BIRDS. The BESTIARIES popular in the 11th-13th centuries attributed complex symbolic meanings to both real and unreal animals – a device echoed in modern fantasy by Charles WILLIAMS's *The Place of the Lion* (**1931**) and Jorge Luis BORGES's *The Book of Imaginary Beings* (**1967**; trans **1969**). On the other hand, there is a tradition of thought which descends from the animistic assumption that all things are endowed with controlling SPIRITS, which equips plants with such associative entities as DRYADS and flower FAIRIES and which sometimes functions as an "explanation" of such phenomena as THERIOMORPHY. Such notions are tacitly echoed in traditional BEAST FABLES and in the modern literary convention which permits anthropomorphization of animal consciousness for narrative purposes (◊ ANIMAL FANTASY; TALKING ANIMALS). The third thread comprises IMAGINARY ANIMALS, devised by fantasy writers to populate their SECONDARY WORLDS. A sidebar concerns the biology of VAMPIRES, WEREWOLVES and the other denizens of SUPERNATURAL FICTION.

It is unusual to find much use being made of such exploded biological theories as the notion of spontaneous generation. Real biological data are, however, often used to add realism to anthropomorphized animal fantasies, as in Richard ADAMS's *Watership Down* (**1972**) and William HORWOOD's *Duncton Wood* (**1980**). Fantasies which allegorize the process of evolution include Charles KINGSLEY's *The Water Babies* (**1863**) and Gerald HEARD's *Gabriel and the Creatures* (**1952**). The extrapolation of ideas drawn from ecology into a form of "ecological mysticism" which supernaturalizes the supposed "balance of nature", pioneered by W.H. HUDSON in *A Crystal Age* (**1887**), is a significant strand in much modern fantasy; striking examples include *Cobwebwalking* (**1986**) by Sara Banerjii (1932-   ) and "Buffalo Gals, Won't You Come out Tonight" (1987) by Ursula K. LE GUIN. Notable phantasmagoric ecosystems which belong more to fantasy than to SCIENCE FICTION are featured in the fungal forest sequence of *Etidorhpa* (**1895**; rev 1901) by John Uri Lloyd (1849-1936), William Hope HODGSON's *The Night Land* (**1912**), David LINDSAY's *A Voyage to Arcturus* (**1920**) and *Inrock* (**1983**) by Desmond Morris (1928-   ). [BS]

BIOY CASARES, ADOLFO (1914-   ) Argentine writer, noted from his first book, *Prólogo* ["Prologue"] (**1929**) – which he later disavowed – for an adventurous SURREALISM and for a fascination (which he shared with his older contemporary and frequent collaborator Jorge Luis BORGES)

with various forms of popular literature, in particular the sf and detective models. Through these models both authors were able to counter, though in a manner profoundly parodic, their profound sense that the LABYRINTH of the world is a BONDAGE from which there is no ultimate escape. Under the shared pseudonym H. Bustos Domecq they published in *Seis problemas para Don Isidro Parodi* (coll **1942**; trans Norman Thomas di Giovanni as *Six Problems for Don Isidro Parodi* **1981** US) a set of witty but sad detections. ABC has always been more interested than Borges in riding sf in this fashion. *La invención de Morel* (**1940**; trans Ruth I.C. Simms in *The Invention of Morel and Other Stories* coll **1964** US) mocks sf more than fantasy in its recounting of a successful search for IMMORTALITY; it was filmed as *L'Invenzione di Morel* (**1974** Italy). *Dormir al sol* (**1973**; trans Suzanne Jill Levine as *Asleep in the Sun* **1978** US), comically hyperbolic sf, suggests that psychosurgery and totalitarianism are natural twins.

In the fantasy *El sueño del los héroes* (**1954**; trans Diana Thorold as *The Dream of the Heroes* **1987** US) a workman's life is saved by a mysterious figure who resembles the charismatic soul-fathers who appear so frequently in the literature of LATIN AMERICA, though ABC treats this material with deconstructive wryness. Many early short stories, often utilizing fantasy motifs, were assembled as *Selected Stories* (coll trans Suzanne Jill Levine **1994** US), based in part on *Guirnalda con amores* ["Garland of Loves"] (coll **1959**) and in part on more recent work. *Una muñeca rusa* (coll **1991** Spain; trans Suzanne Jill Levine as *A Russian Doll and Other Stories* **1992** US) assembles later material.

With his wife Silvina Ocampo and Borges, ABC edited a seminal anthology of the literature of the FANTASTIC: *Antología de la Literatura Fantástica* (anth **1940**; rev 1976; rev ed trans as *The Book of Fantasy* **1976** US). It contains little full fantasy, but its huge range of tales demonstrates the enormousness and centrality of the fantastic in world literatures other than English (◊ ANTHOLOGIES). [JC]

BIRDS Birds play significant roles in fantasy for several reasons. Their mastery of FLYING makes their condition seem desirable, and the metaphorical meanings which can be attached to the idea "flight" ensure that fantasies in which humans borrow the attributes of birds are not without a certain *gravitas*. Birds' mastery of song is sometimes given similar symbolic treatment. One of the most frequent FOLKTALE motifs is that of the bird-bride – often a swan-maiden – who must be captured by stealth and might be lost again through careless neglect. Popular superstition has long sought hidden meanings in the appearance and movement of certain bird species; birds of ill-omen include owls and ravens, while swallows and storks are generally considered harbingers of good fortune. The fact that some birds – notably parrots and ravens – can imitate human voices lends credence to fantasies about talking birds. The most famous fabulous birds are the PHOENIX and the gargantuan roc featured in the *Arabian Nights* (◊ ARABIAN FANTASY).

Notable fantasies in which birds symbolize human dreams of flight or "flight" include "The Eccentricity of Simon Parnacute" (1910) and *The Promise of Air* (**1918**) by Algernon BLACKWOOD, *Going Home* (**1921**) by Barry PAIN, *They Chose to be Birds* (**1935**) by Geoffrey Dearmer (1893-1993), *The Summer Birds* (**1962**) by Penelope FARMER, *Jonathan Livingston Seagull* (**1970** chap) by Richard BACH, *The Unlimited Dream Company* (**1979**) by J.G. BALLARD and *Birdy* (**1979**) by William WHARTON. Notable works in

which birdsong performs a similarly metaphorical role include *Luscignole* (**1892** France; trans 1928) by Catulle Mendès (1841-1909), "The Nightingale and the Rose" (1888) by Oscar WILDE and "My Lady Sweet, Arise" (1983) by Frank BAKER. Another significant ALLEGORY involving birds is *Sparrow Farm* (**1935** Germany; trans **1937**) by Hans Fallada (1893-1947). Effective "bird-bride" stories include Tchaivoksky's ballet *Swan Lake* (1877) and the proto-feminist allegory *Angel Island* (**1914**) by Inez Haynes Gillmore (1873-1970). A traditional swan-maiden is ironically featured in James Branch CABELL's *Figures of Earth* (**1921**); a non-sexist version of the theme is Nicholas GRAY's *The Seventh Swan* (**1962**). Birds of ill-omen are memorably featured in Samuel Taylor COLERIDGE's "The Rime of the Ancient Mariner" (1798), Edgar Allan POE's "The Raven" (1845), W.H. HUDSON's "Marta Riquelme" (1902), Walter DE LA MARE's "The Bird of Passage" (1923) and John COLLIER's "Bird of Prey" (1941). Among many tales of talking birds gifted with intelligence are *Dudley and Gilderoy* (**1929**) by Algernon BLACKWOOD, *Clovis* (**1948**) by Michael FESSIER and *A Fine and Private Place* (**1960**) by Peter S. BEAGLE. Birds are not often featured in ANIMAL FANTASIES, but two exceptions are *Satyrday* (**1980**) by Steven BAUER and *Callanish* (**1984**) by William HORWOOD.

Fabulous birds invented by modern fantasists include the Feng in Helen BEAUCLERK's *The Green Lacquer Pavilion* (**1925**) and the gazolba in Clark Ashton SMITH's "The Voyage of King Euvoran" (1933). Significant tales involving avian THERIOMORPHY include the allegorical *Lilith* (**1895**) by George MACDONALD, "The Albatross" (1931) by Hector Bolitho (1897-1974), the parodic *Gentleman into Goose* (**1924**) by Christopher WARD, *When the Birds fly South* (**1945**) by Stanton A. Coblentz (1896-1982) and (in its final cruel twist) *The Lost Traveller* (**1943**) by Ruthven Todd (1914-1978). A chimerical birdman is featured in *Blaedud the Birdman* (**1978**) by Vera CHAPMAN. There are several old literary works in which a man stands trial for the crimes of his species before a court of birds, the best-known being *States and Empires of the Sun* (**1662** France; trans 1687) by Cyrano de Bergerac (1619-1655); Cock Robin is the likely underlier. The motif is occasionally echoed in modern fantasy, but it is more common nowadays to find birds playing a prominent part in rebellion-of-nature stories, as in Frank BAKER's *The Birds* (**1936**) and Daphne DU MAURIER's "The Birds" (1952), both of which are bases for Alfred HITCHCOCK's movie *The* BIRDS (*1963*). [BS]

BIRDS, THE US movie (*1963*). Universal. **Pr** Alfred Hitchcock. **Dir** Hitchcock. **Spfx** Lawrence A. Hampton. **Special photographic advisor** Ub IWERKS. **Screenplay** Evan Hunter (◊ Ed MCBAIN). **Inspired by** "The Birds" by Daphne DU MAURIER in *The Apple Tree: A Short Novel and Some Stories* (coll **1952**); other contributory influences are *The Revolt of the Birds* (**1927**) by Melville Davisson Post (1871-1930) and *The Birds* (**1936**) by Frank BAKER. **Starring** Veronica Cartwright (Cathy Brenner), Ethel Griffies (Mrs Bundy), Tippi Hedren (Melanie Daniels), Suzanne Pleshette (Annie Hayworth), Jessica Tandy (Lydia Brenner), Rod Taylor (Mitchell Brenner). 119 mins. Colour.

Emotionally retarded socialite Daniels comes to Bodega Bay in romantic pursuit of chic lawyer Brenner; simultaneously, the birds of the resort begin to attack humans, with Daniels – at least initially – an apparent focus. The attacks build up in numbers and intensity – some of the set-pieces have become almost ICONS of the cinema – until, with the birds in complete control, they abruptly halt, the few human survivors being permitted to escape from their POLDER into the outside world . . . although reports are coming from there, too, of unprovoked avian attacks.

TB seems less a plotted tale than a string of incidents, each expertly tightening the terror, leading up to a resolution that never comes; there is no real development, and no explanation. Yet the movie has astonishing power, and has drawn countless interpretations, most rooted in psychoanalysis, precisely because of its lack of surface rationale – which lack might be taken as one marker of the borderline between fantasy and SCIENCE FICTION. One tempting interpretation is that *TB* is a fantasy of PERCEPTION: we are seeing everything through Daniels's eyes, her rejection by the birds being a parallel of her rejection as unsuitable prospective daughter-in-law by Brenner's mother Lydia; only in the final moments, with Daniels crazed through fear, do we see matters with the eyes of an outsider.

Bizarrely, *TB* was dismissed as froth by contemporary critics, who thought it was a HORROR MOVIE. [JG]

BIRKIN, CHARLES (LLOYD) (1907-1985) UK writer, editor and businessman; from 1942 5th Baronet Birkin. He became, in 1932, editor for the publisher Philip Allan and worked on CREEPS LIBRARY (1932-6), for which he compiled a series of anthologies – to which he also contributed stories (as Charles Lloyd). **Creeps Library** included his own first collection, *Devil's Spawn* (coll **1936**). His work is routine but intense, with the emphasis on physical horror and the CONTE CRUEL. CB stopped writing in 1939 until interest in his work was revived in the 1960s by Dennis WHEATLEY and Herbert van Thal; he then produced a stream of collections and was for a period very popular, his tales being likened to those of Edgar Allan POE and Roald DAHL. His best stories were assembled in *The Kiss of Death* (coll **1964**) and *The Smell of Evil* (coll 1965) – both books include tales of spirit POSSESSION and occultism. Later collections are *Where Terror Stalked* (coll **1966**), *My Name is Death* (coll **1966**), *Dark Menace* (coll **1968**), *So Pale, So Cold* (coll **1970**) and *Spawn of Satan* (coll **1971** US). CB edited *The Tandem Book of Ghost Stories* (anth **1965**; vt *The Haunted Dancers* 1967 US) and *The Tandem Book of Horror Stories* (anth **1965**; vt *The Witch Baiter* 1967 US). [MA]

BISHOP, MICHAEL (1945-　　) US writer of sf and some fantasy. Several stories in *One Winter in Eden* (coll **1984**) are AMERICAN GOTHIC fantasies; they include the title tale, "One Winter in Eden" (*Dragons of Light* anth **1980** ed Orson Scott CARD), in which a DRAGON disguised as a schoolteacher flares up at racial injustice, and must flee to a new hideaway, and *The Quickening* (in *Universe 11* ed Terry CARR anth **1981**; **1991** chap), in which the world's population is translated into a testing new environment which could be an EDEN, although already spoiled; the latter story won a Nebula AWARD. *Who Made Stevie Crye?* (**1984**) is a HORROR novel as well as a RECURSIVE meditation on fiction and dreaming (◊ DREAMS). *Unicorn Mountain* (**1988**) is an uneasily powerful fantasy (with some interesting TECHNO-FANTASY elements) which equates the arrival in this sphere of sick UNICORNS from the realm of DEATH with the passing over of AIDS victims into unicorn country. The moral and metaphysical transactions engaged upon are complexly argued. [JC]

Works edited: *Changes: Stories of Metamorphosis* (anth **1983**) with Ian WATSON; *Light Years and Dark: Science*

Fiction and Fantasy of and for Our Time (anth **1984**); *Nebula Awards 23* (anth **1989**); *Nebula Awards 24* (anth **1990**); *Nebula Awards 25* (anth **1991**).

BISLEY, SIMON (*c*1965-　　) UK COMICS artist whose distinctive brand of ultra-tough fantasy characters has spawned an army of imitators. SB's first published work, on **ABC Warriors** in *2,000 A.D.* (*#555-#558* 1988), done in pen-and-ink line, was impressive for its vitality and clarity of imaginative vision, but his huge popularity originated with the publication of the first episodes of **Slaine: The Horned God** (from 1989) written by Pat MILLS. SB's SLAINE was painted in full colour, and as the series progressed his work became dramatically looser and more accomplished, while there became increasingly evident a thread of self-mocking humour that remained a significant factor in SB's rapid rise to stardom. His use of rich but subtly muted colour in representing his very personal, massively muscled grotesques and his uninhibited inventiveness in decorating them with metal chains, bolts and motifs of death and destruction have made him perhaps the most sought-after artist in the comics field.

SB's other published work has included: *Lobo: The Last Czarnian* (in *Lobo #1-#4* 1989-90), *The Lobo Xmas Sanction* (1991) and *Lobo's Back* (*#1-#3* 1992; graph coll **1993**); several stories featuring **Judge Dredd** (in *Judge Dredd: The Megazine #14*, *#16* and *#19* 1991-2 and *Heavy Metal Dredd* [graph **1993**]); the **Batman/Dredd** teamup *Judgement on Gotham* (graph **1991**); and *Frank Frazetta's Death Dealer* * (graph **1995**). [RT]

BISSON, TERRY (BALLANTINE) (1942-　　) US writer increasingly perceived as a significant fabulator of sf, though much of his work has in fact been fantasy. His first novel, *Wyrldmaker: A Heroic Romance* (**1981**), combines genres: for most of its length it works as a vivid, telegraphically told SWORD AND SORCERY tale set in a complex of pocket-universe like LANDS, each being a wyrld. The eponymous SWORD – like many of its kind – loves to kill, and sings while doing so; and at one point, after its owner Kemen has followed his lover into an UNDERWORLD, serves as ORPHEUS – again like many of its kind. Then, at the climax of Kemen's QUEST for the goddess Noese, it is revealed that the wyrlds are in fact a kind of generation starship, and that Noese is the human-shaped projection of a consciousness responsible for guiding the wyrlds to a new star, and seeding it. At twice the length *Wyrldmaker* might have had sufficient gravity to sustain the very numerous twists of its plotting.

TB's second novel, *Talking Man* (**1986**), rather more successfully uses a CONTEMPORARY-FANTASY setting for an epic TALL-TALE trip across a USA that undergoes compressions and TRANSFORMATIONS not entirely dissimilar to those experienced by Kemen in the first novel. The eponymous WIZARD or SHAMAN – who operates a junkyard in rural Kentucky, does not talk, increasingly takes on TRICKSTER characteristics, and is in effect (along with his dark SHADOW sister), a SECRET MASTER and guardian of the cosmic BALANCE – leads his kinfolk to an Arctic THRESHOLD, beyond which a new USA unfolds. This new land is identical to the UTOPIA, located in an ALTERNATE WORLD, that is described in sf terms in TB's next novel, *Fire on the Mountain* (**1988**); but within the terms of *Talking Man* itself this transformation gives an initially modest fable some of the contours of INSTAURATION FANTASY.

Most of the tales assembled in *Bears Discover Fire* (coll **1993**) are also typical of TB in that they tend to offer multiple readings, and never settle into comfortable genre locations. The main device – it appears in the title story and in, for instance, "England Underway" – is the introduction of a logically absurd premise, which is then accepted into a narrative that treats it as a given, thus generating fantasy tales out of material that would more normally lead into ABSURDIST FANTASY or WONDERLAND games. TB is an exploratory writer, a tester of boundaries; the value of his work is considerable. [JC]

BIXBY, (DREXEL) JEROME (LEWIS) (1923-   ) US writer and editor; a prolific author of Westerns, mysteries and erotica, he occasionally strayed into SCIENCE FICTION, HORROR and FANTASY. His writing career began in 1949, and he worked as editor and assistant editor on a number of pulp-MAGAZINE titles. His sf is little remembered today, but some of his SUPERNATURAL FICTIONS are reprinted fairly regularly, especially "It's a *Good* Life" (in *Star SF Stories #2* anth **1953**), about an evil CHILD with psi TALENTS. His stories in BEYOND FANTASY FICTION and FANTASTIC all consider the clash between GOOD AND EVIL. *The Devil's Scrapbook* (coll **1964**; vt *Call for an Exorcist* **1974**) blends BLACK MAGIC with erotica. JB's involvement with the field since has related mostly to his contributions to **Star Trek**, and he has written little fantasy. [MA]

BIZARRE FANTASY TALES ◊ MAGAZINES; *The* MAGAZINE OF HORROR.

BLACK, ROBERT ◊ Robert P. HOLDSTOCK.

BLACK AFRICAN FANTASY Classical African art and oral tradition mix realistic, mystic and fantastic elements – and this influences modern fiction as well. Fantastic motifs can be found in a wide range of literary works, but only a few are fantastic as a whole.

One of the first examples of fantasy in modern African fiction is a part of *Gandoki* (**1934**) written in Hausa by the Nigerian A. Bello Kagara (1890-?   ). The main part of this novel is set in late-19th-century Nigeria, as British rule widened into the central part of the country and local kings lost their powers. In the second half of the novel the fictitious king Gandoki loses his war with the British and escapes with a small group of allies to the POLDER Salayana, somewhere in India, where he spends a few years among GHOSTS and other supernatural beings. When at last he returns to his native land the novel reverts to realism.

Various religious fantasies, the main aim of which was to support Christianity against animist beliefs, are on the borderline of fantastic literature. Such works often combine Christian ideas with local settings – various sinners encounter SATAN in *Sekoting sa lihele* ["In the Depths of Hell"] (**1956**) written in Sotho by D.P. Lebakeng, from Lesotho. A description of the AFTERLIFE can be found in the novelette "Maphunye" (in coll *Sebogoli sa Ntsoana-Tsatsi* ["The Watcher of Ntsoana-Tsatsi"] **1943**) by the Lesotho writer C.R. Moikangoa and in many other works. A more up-to-date variation is *Mission to Gehenna* (**1989**) by the Kenyan writer Karanja wa Kang'ethe.

Many fantastic fictions are ANIMAL FANTASIES. In traditional style is the collection *À la belle étoile* ["Under the Sky"] (coll **1962**) written in French by Beniamin Matip, from Cameroon. An Orwellian approach was taken by Mallane Libakeng Maile from Lesotho in *Pitso ea liphoofolo tsa hae* ["The Meeting of the Domestic Animals"] (**1956**), a SATIRE in which animals discuss their bad treatment by humans. Animals acted as a major force in the fight against South Africa's aggression in Angola in *E nas Florestas os*

Bichos Falaram ["And the Animals Started to Speak in the Forests"] (*c***1979**), which was written in Portuguese by Maria Eugenia Neto, the wife of a former Angolan president.

As supernatural elements still form an integral part of today's African perception of REALITY, they can also be found in mainly realistic works by major mainstream writers. A lake-GOD interferes with the life of a small traditional village in *Le Chant du Lac* ["The Song of a Lake"] (**1965**), written in French by Olympe Bhely-Quénum from Benin. A story of a nymph become human and revenging herself on her spirit husband, the Sea King, is *The Concubine* (**1966**), written in English by the Nigerian Elechi Amadi. Two ancestral spirits visit a small village to show how the world of the past looked in *A Dance of the Forests* (**1963**) by the Nobel laureate Wole Soyinka (1934-   ), from Nigeria. A few fantastic stories were written by other mainstream writers, including Camara Laye (1924-   ) from Guinea, Samuel Asare Konadu (1932-   ) from Ghana, Abioseh Nicole (1924-   ) from Sierra Leone and James T. Ngugi (known also as Ngugi wa Thiongo; 1938-   ) from Kenya.

There is a lot of fantastic writing in the form of modern adaptations and reinterpretations of classical LEGENDS and MYTHS; an early example is *L'Arbre fétiche* ["Tree Fetish"] (**1963**) and *Kondo le Requin* ["Shark Named Kondo"] (**1965**), both written in French by Jean Pliya (1931-   ) from Benin. A modern version of a classic tale of two brothers and their fight with a MONSTER is *Tsoana-makhulo* by L.E. Mahloane from Lesotho. One recent example of such works is an interpretation of Luo myth about a visit of a mysterious woman from an unknown OTHERWORLD to a small Kenyan village, *Miaha* (**1983**; trans Okoth Okombo as *The Strange Bride* **1989**), written in Dholuo by the Kenyan Grace Ogot (1930-   ).

The so-called "market literature" – fiction published mainly in pamphlet form and sold in East and West African marketplaces since the 1940s – is rich in fantastic elements. These adventures are a weird mixture of romance, thriller, spy and detective fiction, plus various types of fantasy. The influence of Western cinema is overt: local James Bond- and SUPERMAN-like characters are highly popular. A story where various supernatural characters interfere with the hero's life is *Adili na nduguze* ["Adili and His Brothers"], written in Swahili by Sh. Robert (probably Tanzanian). *Ndoa ya Mzimuni* ["Marriage in the Other World"] (**1974**), written in Swahili by Saidi M. Nurru (probably a Kenyan), is a romance about a girl who kills her fiancé, who becomes a SPIRIT. More ambitious is *Mfu aliyefufuka* ["A Dead Man who Resuscitated"] (**1974**) by one of the best Swahili stylists, the Kenyan H.C.M. Mbelwa.

Similar literary methods and themes to those found in market literature are visible in modern African adventure fictions. Although most of these are set in a real Africa – with its crime waves, political instability and wars – there are also supernatural motifs. A typical example is *The Instrument* (**1980**) by the Nigerian Victor Thorpe (1919-   ), which is a political thriller of a modern or near-future Nigeria torn with crime and terrorism, against which backdrop fight GOOD AND EVIL supernatural powers.

Two works interestingly combine various subgenres of fantastic literature. The Nigerian Umaru A. Dembo's *Tauraruwa mai wutsiya* ["The Comet"] (**1969**), written in Hausa, seems at first to be sf – an extraterrestrial takes a small boy into space in his UFO – but most of their subsequent adventures are supernatural, with motifs drawn from the traditional beliefs and tales of the Hausa and also from the *Arabian Nights* (◊ ARABIAN FANTASY). Fantastic HORROR is the theme of the collection *Silence, cimetière!* ["Silence, Cemetery!"] (coll **1979**), written in French by the Senegalese Nabil Ali Haidar, an admirer of Edgar Allan POE.

Only a few works of modern African fiction are fully fantasy, and all have been influenced by the Nigerian D.O. FAGUNWA, author of five fantasy novels written in Yoruba. The direct influence of Fagunwa's style and themes is visible in *Kórimále Nínú Igbó Adimúla* ["Korimale in Adimula's Forest"] (**1967**), written in Yoruba by D.I. Fatanmi (1938-). The main character is much like one of Fagunwa's – a brave hunter travelling through a mysterious FOREST full of supernatural beings. The popularity and success of Fagunwa stimulated other Yoruba writers – e.g., J. Ogunsina Ogundele (1926-) and J. Folahana Odunjo (1904-?) – to create their own fantasies. Ogundele's *Ibú Olókun* ["The Deeps where Olokun Reigns Supreme"] (**1956**) has as a main character a man who has TALENTS and is therefore involved in various FANTASTIC VOYAGES – from Earth to Heaven, to the depths of the ocean, etc.

Fagunwa directly influenced also the best-known African fantasy writer, Amos TUTUOLA, author of eight novels and a story collection. Although a Yoruba, Tutuola writes in a special kind of basic English, and his works have become more popular outside Africa than at home. [JO]

BLACK CAT, THE ◊ MAGAZINES.

BLACK CAULDRON, THE US ANIMATED MOVIE (**1985**). DISNEY. **Pr** Joe Hale. **Exec pr** Ron Miller. **Dir** Ted Berman, Richard Rich. **Spfx** Bill Kilduff, Phillip Meador, Ron Osenbaugh. **Based on** the **Chronicles of Prydain** series by Lloyd ALEXANDER. **Voice actors** Grant Bardsley (Taran), John Byner (Gurgi, Doli), Phil Fondacaro (Creeper), Nigel Hawthorne (Fflewddur Fflam), John Hurt (Horned King), John Huston (Narrator), Freddie Jones (Dallben), Susan Sheridan (Eilonwy). 80 mins. Colour.

Pigkeeper Taran, apprenticed to sorcerer Dallben (◊ SORCERER'S APPRENTICE), tries to shield the oracular pig Hen Wen from the Horned King, who seeks to discover through the animal the location of the Black Cauldron, into which, eons ago, the GODS cast the SPIRIT of the most EVIL king the world has known. From the Cauldron the Horned King hopes to release a host of deathless warriors, the Cauldron Born, and with them conquer the world. Aided by the waspish Eilonwy and various WITCHES and FAIRIES, Taran averts the threat. Although disliked by the critics and reputedly by Lloyd ALEXANDER himself (his several submitted treatments were rejected by Disney), *TBC* can be seen as an ambitious and largely successful attempt to bring HIGH FANTASY within the domain of the ANIMATED MOVIE. Previous essays in this field were Ralph BAKSHI's *The* LORD OF THE RINGS (*1978*) and Jim HENSON's *The* DARK CRYSTAL (*1982*). [JG]

BLACKENSTEIN (*1973*) ◊ FRANKENSTEIN MOVIES.

BLACKLIN, MALCOLM Pseudonym of UK writer and editor Aidan CHAMBERS.

BLACK MAGIC Magic worked with evil intent; the term is virtually synonymous with SORCERY. The latter is the term more often used in fantasy, while the former is more frequent in occult fiction. Marjorie BOWEN's ironic melodrama *Black Magic* (**1909**) is on the borderline between the two genres. BM stories might have become a subgenre when

Dennis WHEATLEY achieved bestseller status with *The Devil Rides Out* (**1935**) and its sequels, but no one else made significant inroads into his marketplace niche. John Keir CROSS's *Best Black Magic Stories* (anth **1960**) was a belated afterthought. [BS]

BLACK MASS Blasphemous rite allegedly carried out by Satanists (◊ SATANISM). The BM was distinguished from the SABBATS supposedly celebrated by WITCHES as a result of the "Chambre Ardente" affair of 1679, when Louis XIV's mistress Madame de Montespan was alleged to have been a client of Satanists involved in such ceremonies. It was adopted for dramatic purposes by the Hell-Fire Clubs of the 18th century. Whether such rites actually figured in the repertoire of the French lifestyle fantasists who dabbled in occultism in the late 19th century – as vividly depicted in J.K. HUYSMANS's quasi-documentary novel *Là-Bas* (**1891**) – is unclear, but that novel set an important precedent enthusiastically followed by many 20th-century writers wishing to cash in on the innate melodrama of the notion. Examples can be found in E.F. BENSON's *Colin II* (**1925**), Dion FORTUNE's *The Winged Bull* (**1935**) and several novels by Dennis WHEATLEY. The most effective use of the motif is perhaps "The Earlier Service" (1935) by Margaret IRWIN. A SCHOLARLY FANTASY detailing the "history" of the black mass is *The Satanic Mass* (**1954**) by H.T.F. Rhodes. [BS]

BLACKWOOD, ALGERNON (HENRY) (1869-1951) UK writer and broadcaster, a leading exponent of SUPERNATURAL FICTION in the early 20th century; now best remembered for his GHOST STORIES, in his day he was recognized also for his pantheistic fantasies and his books for children. Most of his fiction was derived from personal or mystical experience. AB's knowledge of OCCULTISM came from several years as an active member of the GOLDEN DAWN (to which he was introduced by W.B. YEATS), prior to which he had studied Eastern wisdoms (he was a Buddhist in early life) and THEOSOPHY. But his greatest inspiration – he said – came from a proximity to nature which imbued him with an enhanced consciousness, heightening his awareness of preternatural existence. Widely travelled, AB had his most inspirational experiences in the forests of northern Canada – hence his classic encounter with the call-of-the-wild personified, "The Wendigo" (1910 in *The Lost Valley*; ◊ WENDIGO) – and the remote mountains of the Caucasus, which experiences he combined with his studies of the works of Gustav Fechner (1801-1887), who postulated a sentient Mother Earth, to create *The Centaur* (**1911**), which portrays an encounter with the primeval spirits of the planet, resulting in a spiritual METAMORPHOSIS.

AB's early writings were largely theosophical, although his first story – *A Mysterious House* (1889 *Belgravia*; **1987** chap ed Richard DALBY) – arose from his fascination for psychic research; this tale of a traditional HAUNTED DWELLING is spoiled by its DREAM ending.

A sensitive and gullible youth, AB later chronicled his early days in Canada and New York (1890-99) in *Episodes Before Thirty* (**1923**; rev vt *Adventures Before Thirty* 1934); he had many trials and privations before making good as a private secretary. His horror of New York spilled out as, essentially, emotional therapy in his first two collections, *The Empty House and Other Ghost Stories* (coll **1906**) and *The Listener and Other Stories* (coll **1907**). The lesser stories betray a residue of Gothicism, but the best develop themes of psychic and occult intensity. "The Listener" is particularly potent: it tells of a room haunted by the GHOST of a leper. In the same collection "The Willows", inspired by a journey AB made down the Danube, depicts a vortex at the THRESHOLD of spiritual forces. This was one of H.P. LOVE-CRAFT's favourite stories and may have had some influence on *The Colour Out of Space* (1927; **1982** chap). Lovecraft regarded AB as *the* leading writer of supernatural fiction.

AB's third book, *John Silence, Physician Extraordinary* (coll **1908**), was his most successful, and the income from it allowed him to become a full-time writer. It introduced the eponymous OCCULT DETECTIVE, and both popularized and gave credentials to that theme. **John Silence** was a doctor of the spirit who had studied for years to master the paranormal. The stories are mostly about POSSESSION, of either person or place. One series story omitted from the collection, "A Victim of Higher Space" (1914 *Occult Review*), was included in *Day and Night Stories* (coll **1917**); it showed AB's interest in other dimensions.

During 1908-14, when AB lived in Switzerland, his work was at its most intense, and he moved from traditional ghost and occult themes towards a more mystical level. This became evident in *The Lost Valley and Other Stories* (coll **1910**) but most so in *Pan's Garden: A Volume of Nature Stories* (coll **1912**) and *Incredible Adventures* (coll **1914**); stories in the latter two books seek to explore the power of nature and its impact upon spiritual consciousness, dealing primarily with humanity's heightened PERCEPTION of the SPIRIT world. Spirit forces may rejuvenate, as in "The Regeneration of Lord Ernie" (in *Incredible Adventures*), bring spiritual union, as in "The Man Whom the Trees Loved" (1912 *London Magazine*), challenge the unwary, as in "Sand" (in *Pan's Garden*), or drain the vitality of the soul, as in "A Descent Into Egypt" (in *Incredible Adventures*).

AB had earlier merged his mystical beliefs with the occult in *The Human Chord* (**1910**), which considers the mastery of tonal vibrations in order to summon the power of Jehovah. He firmly believed in REINCARNATION, and this theme – which he had explored in stories like "The Insanity of Jones" (in *The Listener*) and "Old Clothes" (in *The Lost Valley*) – was further imbued with the occult in *Julius Le Vallon* (**1916**; written 1912), where an eternal spirit seeks to remedy problems unresolved in a former life. The experiment fails, and an elemental spirit enters an unborn child. That child's story is told in *The Bright Messenger* (**1921**), which considers the dilemma of a freeborn spirit trapped in a human frame (◊ BONDAGE). AB also explored this notion in *The Promise of Air* (**1918**) – indeed, it was a fundamental idea to AB, who believed he himself was a trapped spirit, as was his "soul mate", Maya Stuart-King (?1880-1945), the "M.S.-K." of his dedications.

After Maya had married (in fact, remarried) in 1916, AB's work lost a vital spark, and his inspiration was further stifled by WWI, in which he served as an intelligence agent and a Red Cross worker. The net result was a series of uninspired works of maudlin sentimentality, including *Karma: A Reincarnation Play* (play **1918**) with Violet Pearn (1880-1947) and *The Garden of Survival* (**1918**), composed after the death of his brother and seeking to portray a link with the AFTERLIFE. This mood was only slightly lightened by *The Wolves of God, and Other Fey Stories* (coll **1921**), the best of his latter-day collections, based on ideas by Wilfred Wilson (1875-1957). *Tongues of Fire and Other Sketches* (coll **1924**) is more ephemeral, the better stories – like "The Man who was Milligan" (1923 *Pearson's Magazine*) – showing AB's continued interest in TIME and the fourth dimension. This vein

was mined more diligently in his last significant collection, *Shocks* (coll **1935**), particularly in "Elsewhere and Otherwise", *Full Circle* (1925 *English Review*; **1929** chap) and "The Man who Lived Backwards" (1930 *World Radio*); it has to be said that this collection also shows an unhealthy preoccupation with suicide.

AB's other main works were either for children or, more enticingly, about them. His first completed novel, *Jimbo* (written 1900; **1909**), graphically envisions the DREAMS of a child in a coma who must learn to fly (◊ FLYING) to escape from an old house; the link with the idea of the spirit seeking to flee its corporeal prison is evident. *The Education of Uncle Paul* (**1909**) is a novel for adults about CHILDREN; it explores the land of lost childhood on the threshold between today and tomorrow. This novel was later adapted for the stage by Violet Pearn as *Through the Crack* (produced 1920; **1925**), and the theme was reworked by AB himself in *The Extra Day* (**1915**) and *The Fruit Stoners* (**1934**), books that became increasingly for children rather than about childhood. He had hoped that the play *The Starlight Express* (produced 1916), adapted by Pearn from *A Prisoner in Fairyland* (**1913**) – dealing with the astral projections (◊ ASTRAL PLANE) of children's spirits, released during sleep and spreading good thoughts in the form of stardust – would be as popular as J.M. BARRIE's *Peter Pan* (play **1904**), which it often imitates, but the production failed. During the 1920s AB wrote increasingly for children, with *Sambo and Snitch* (**1927**), *Mr Cupboard* (1927 *Joy Street Annual*; **1928** chap) and others. *Dudley & Gilderoy: A Nonsense* (**1929**; considerably cut by Marion Cothren as *The Adventures of Dudley and Gilderoy* 1941 chap US), although ostensibly a children's story about the adventures of a parrot and a cat, was too philosophical for children, and is really more a critique on 1920s society.

During the 1930s AB became popular as a radio broadcaster and storyteller (following the trend established by A.J. ALAN), and he further developed this career in the 1940s, extending it to tv (he received a Television Society Medal), but by now he was recycling old material, and found it difficult to produce anything new; it was only through the intervention of August W. DERLETH that his final original collection, *The Doll, and One Other* (coll **1946**) was published. His tv popularity led to the release of an omnibus of his stories, *Tales of the Uncanny and Supernatural* (coll **1949**; assembled with *Tales of the Mysterious and Macabre* [coll **1967**] as *Tales of Terror and Darkness* omni **1977**). This has become the standard volume of his work, although more representative selections are *Strange Stories* (coll **1929** UK; cut vt *The Best Supernatural Tales of Algernon Blackwood* 1973 US) and *The Tales of Algernon Blackwood* (coll **1938**) ed L.M. Lamont.

AB received a CBE in 1949. He suffered a stroke in his last year, although he continued to write and broadcast until his death.

AB was a prolific writer – over 200 stories and a dozen novels – and in his chosen field he has never been rivalled and seldom imitated. Even at its best, his work struggles to express abstract mystical experience in lay terminology; this limited its overall potential and restricted its commerciality. As a result, AB's contribution to SUPERNATURAL FICTION has generally not been fully appreciated; instead, he tends to have been measured by the radio soubriquet that made him famous but which he so disliked: the "Ghost Man". [MA]
Other works: *Ten Minute Stories* (coll **1914**); *Ancient*

Sorceries and Other Tales (coll **1927**); *The Dance of Death and Other Tales* (coll **1927**); *Short Stories of To-day & Yesterday* (coll **1930**) ed F.H. Pritchard; *The Willows and Other Queer Tales* (coll **1932**); *Selected Tales* (coll **1942**); *Selected Short Stories* (coll **1945**); *In the Realm of Terror* (coll **1957** US); *Selected Tales* (coll **1964**); *Ancient Sorceries and Other Stories* (coll **1968**); *Best Ghost Stories* (coll **1973** US) ed E.F. BLEILER; *Tales of the Supernatural* (coll **1983**) ed Mike ASHLEY; *The Magic Mirror: Lost Tales and Mysteries* (coll **1989**) ed Ashley.

Other works for children: *By Underground* (1929 *Joy Street Annual*; **1930** chap); *The Parrot and the – Cat* (1930 *Joy Street Annual*; **1931** chap); *The Italian Conjurer* (1931 *Joy Street Annual*; **1932** chap); *Maria (of England) in the Rain* (1932 *Joy Street Annual*; **1933** chap); *Sergeant Poppett and Policeman James* (1933 *Joy Street Annual*; **1934** chap); *The Fruit Stoners* (1934 *Joy Street Annual*; **1935** chap) which is not the same as the 1934 novel; and *How the Circus Came to Tea* (1935 *Joy Street Annual*; **1936** chap).

Further reading: *Algernon Blackwood: A Bio-Bibliography* (**1987**) by Mike ASHLEY.

BLACKWOOD'S MAGAZINE Scottish literary MAGAZINE, 1980 issues, monthly, April 1817-September 1980, published by Blackwood & Sons, Edinburgh; titled *The Edinburgh Monthly Magazine* April-September 1817, *Blackwood's Edinburgh Magazine* October 1817-December 1905, and thereafter *Blackwood's Magazine* (popularly known as the "Maga" almost since foundation). Founded by William Blackwood (1776-1834), it had its heyday under his son, John Blackwood (1818-1879), who was editor from 1845.

Although started as a political review, *BM* nearly folded after six issues in its failed rivalry with *The Edinburgh Review*. It was revamped, with the emphasis on literature. *BM* rapidly became the pre-eminent literary magazine of the first part of the 19th century, its format encouraging popular contributions from the literati of the day. Its early issues, especially during the heyday of editorial assistants John Lockhart (1794-1854), James HOGG and John Wilson (1785-1854), established a vibrant cauldron of debate and discussion. This included an early example of the literary hoax – the fictional BOOK the *Chaldee MS*, introduced by Hogg in 1819, purporting to be a lost biblical text. *BM* developed the habit of publishing stories that took the form of mock recollections, often introducing the supernatural via local LEGENDS and FOLKTALES. Among these were Hogg's *The Shepherd's Calendar* (1827-8; **1828**), Richard BARHAM's **Ingoldsby Legends** (which began in *BM* in 1831 before continuing in *Bentley's Miscellany*), and **The Diary of a Late Physician** (1830-37; **1831** US; rev 1832; rev 3 vols 1838) by Samuel Warren (1807-1877). Wilson and Lockhart championed German literature in *BM*, which regularly featured Gothic and Romantic fiction, and was instrumental in popularizing GOETHE and Schiller in the UK. Edgar Allan POE admired *BM*'s approach to sensational fiction, and one anonymous story in *BM*, "Who is the Murderer?" (1842), is attributed to him. Poe may have been influenced by several HORROR stories in *BM*, including "The Man in the Bell" (1821) by William Maginn (1793-1842), a tale of suffocation, and "The Iron Shroud" (1830) by William Mudford (1792-1848), of which traces can be found in Poe's "The Pit and the Pendulum" (1843). A representative selection of early fiction will be found in *Tales of Terror from Blackwood's Magazine* (anth **1995**) ed Robert Morrison and Chris Baldick (1954-   ).

Of special import was "The Haunted and the Haunters" (1859) by Edward BULWER-LYTTON, the most reprinted of all Victorian GHOST STORIES. The popularity of this and "The Lifted Veil" (1859) by George Eliot (1819-1880) made *BM* the primary mid-Victorian magazine for the sophisticated ghost story, and later contributions came from such writers as John BUCHAN, Hugh CONWAY, Frank Cowper (1849-1930), Thomas Hardy (1840-1928) and Margaret OLIPHANT (who published her best stories there over a 40-year period).

By the 1890s *BM* had lost its pre-eminence to magazines like *The Cornhill*. It suffered further against popular magazines like *The Strand* and *Pall Mall*, though it did well in presenting some of the early, challenging stories of Joseph CONRAD. It kept its traditional format and content, always sustaining a core of readers, though after WWII its feature stories concerning the outposts of Empire were becoming dated. The publishers made regular selections from the magazine for a long series of anthologies, including several that were thematic: *Strange Tales from Blackwood* (anth **1950**) and *Ghost Tales from Blackwood* (anth **1969**) both contain stories of the supernatural. The magazine published less WEIRD FICTION towards the end of its life, although Sheila Hodgson and Fred Urquhart (1912-1995) contributed ghost stories to its final issues.   [MA]

Further reading: *Annals of a Publishing House* (3 vols **1897-98**) by Margaret Oliphant and Mrs Gerald Porter.

BLACULA (*1972*) ◊ DRACULA MOVIES.

BLADE, ALEXANDER [hn] ◊ John W. JAKES.

BLAKE, WILLIAM (1757-1827) English writer and painter, the most visionary and genuinely revolutionary of the great Romantic poets. WB's vision of the human spirit freed of the shackles of institutional orthodoxy is essentially religious, and the vivid evocations in his lyrics, narrative poems and engravings of cosmic energies and animist-like powers (such as that of "the Tyger") must be understood in terms of the Christianity of his Dissenting background, whatever the temptation to impute a more modern sensibility. The *Songs of Innocence* (coll **1787** chap) and *Songs of Experience* (coll **1794** chap) have been immensely influential upon 20th-century literature, although probably only in the sense that the critic Harold BLOOM would call a "weak misreading": the WB – it is almost invariably early WB – who supplies 20th-century authors, genre ones especially, with lines for the titles and epigraphs of their books is an author whom fashion can adopt, not an author valued for his capacity to disrupt and surprise.

If WB's early verse has found an audience in recent decades, his later, more complex works may have to wait longer, for they are rarely read save for academic studies. His series of cosmological yet highly personal narrative poems, culminating in *Milton* (dated 1804 but not published until much later; probably composed 1803-1808, with the final plates etched and printed – both by WB – *c***1815**) and *Jerusalem* (dated 1804 but written and etched much later; the earliest copies issued by WB are on paper with watermark 1820), are of sufficient singularity and difficulty that as late as 1910 the *Encyclopaedia Britannica* (11th edn) was willing to question WB's sanity. Certainly their difficulty is compounded by a discursiveness at variance with the extreme compression of the *Songs* and *The Mental Traveller* (written *c*1803; **1905**); they are not made more accessible by the fact that many of WB's solemn myth-figures have names (Oothoon, Golgonooza) that strike modern ears as risible, nor the fact that appreciation of the poems requires study of the accompanying drawings (100 engraved plates accompany *Jerusalem's* more than 4000 lines). WB's more daunting late works, hortatory and forbidding as they are, reward persistence without requiring gloss; their vision of horrific struggle and agon possesses a dramatic force that is genuinely harrowing.

The unworldly, rather bardic WB of popular imagination (a compound of pre-Victorian William MORRIS craftsman and proto-hippie) will remain an anthology favourite, the savagery of his lyrics dimly apprehended; his engravings, familiar enough from postcards and posters, give a better sense of his imaginative power. Writing in *Jerusalem*, WB declares, "I must Create a System, or be enslav'd by another Man's"; and the intricacy and comprehensiveness of his system, viewed as an act of world-building, is rivalled only by that of J.R.R. TOLKIEN.

Surprisingly (in light of GENRE FANTASY's predilection for dramatizing the lives of the Romantic poets), fiction concerning WB is scarce. The wanderer called Taleswapper in Orson Scott CARD'S *Seventh Son* (**1987**) is plainly the WB of the early lyrics (which he quotes). Ray Nelson's *Blake's Progress* (**1975** Canada; rev vt *Timequest* 1985 US) is a TIME-TRAVEL tale of WB and his wife; it is an imaginative portrait, with the emphasis on WB the painter. The Jim Jarmusch movie «Dead Man» (1996) features a protagonist believed to be WB reborn, and *The Pit* * (**1993**) by Neil Penswick, in the **New Doctor Who Adventures**, has WB as a protagonist. Fantasies that make use of Blakean creations are rarer still. Blakean names (like Urizen and Theotormon) appear in Philip José FARMER's **World of Tiers** novels, but these are merely flavourings for the author's cosmological gallimaufry, which is essentially Jungian. The mountain range that is the work of WB has been thickly settled in its foothills, but its peaks remain all but unscaled. [GF]

BLAMIRES, HARRY (1916-   ) UK author of a reverent but heavily ironic Divine Comedy – in the tradition of C.S. LEWIS's *The Screwtape Letters* (**1942**) – begun with *The Devil's Hunting Grounds* (**1954**), an account of PURGATORY, and continued in *Cold War in Hell* (**1955**) and *Blessing Unbounded* (**1955**). Modern politics (including Church politics) and social trends are sharply satirized, but the third volume exercises cautious discretion in describing the road to HEAVEN without offering any elaborate account of that destination. [BS]

BLAMPIED, EDMUND (1886-1966) Acclaimed UK painter, draughtsman, etcher and lithographer, born on Jersey. His fantasy book ILLUSTRATIONS were usually drawn in a fluent, controlled line in pencil, pen-and-ink or Chinese brush style. He painted in oils and watercolours, designed stamps during the WWII German occupation of the Channel Islands, and was an accomplished clay sculptor. Books illustrated by him include *The Edmund Blampied Edition of Peter Pan* (**1939**) by J. M. BARRIE, *The Phoenix and the Carpet* (**1904**) by E. NESBIT and *Black Beauty* (1920) by Anna Sewell (1820-1978). [RT]

BLATTY, WILLIAM PETER (1928-   ) US author, former PR officer and screenwriter best known for *The Exorcist* (**1971**), which he adapted and produced as the movie *The* EXORCIST (**1973**). Based loosely on a true incident, the novel is an intelligent exploration of POSSESSION and EXORCISM; in the battle between GOOD AND EVIL, humanity is but a pawn. It is an intense and dark THEODICY.

The screen version, with its realistic and shocking effects, transformed the HORROR MOVIE and revived the moribund HORROR fiction market just at the time that Stephen KING emerged.

WPB was not involved with the movie's first sequel, but he did write a sequel to the *Exorcist*, *Legion* (**1983**; vt *Exorcist III: Legion* 1990), which again explores the world's inherent evil; it was filmed as *The Exorcist III* (**1990**; ◊ *The* EXORCIST). [MA]

Other works: *I, Billy Shakespeare* (**1965**).

BLAVATSKY, [Madame] H(ELENA) P(ETROVNA) (1831-1891) Russian-born founder (with others) of the Theosophical Society (◊ THEOSOPHY) in 1875 in the USA, and author of texts designed to promulgate the new occult faith. If the diffuse SPIRITUALISM so popular at this time can be seen to have its fictional analogue (and explication) in various forms of SUPERNATURAL FICTION, then the LIFESTYLE FANTASY of Theosophy can be seen as receiving its proper fictional due (and advocacy) in FANTASY itself; and HPB's two main philosophical works – *Isis Unveiled* (**1877** 2 vols) and *The Secret Doctrine* (**1888** 2 vols) – can arguably be seen as providing much raw material for creators of fantasy worlds from the time of William MORRIS onwards. *Key to Theosophy* (**1889**) is less important; her *Collected Works* (omni **1933-6** 3 vols) was ed A. Trevor Barker. In HPB's hands, Theosophy supplies models for the secret history of the world; for SECRET MASTERS; for a PARIAH ELITE; for an exceedingly varied cast of extras ontologically bound (like creatures of fantasy) to fulfil their preordained role in a great drama (◊ THEODICY); for a sense that the history of the Universe is a STORY that each of us, wittingly or unwittingly, helps to tell; for visions of TIME ABYSS, in which vast epochs dominated by LEMURIA or ATLANTIS are mere episodes in the long story, which spirals upwards in great CYCLES; for a geography of the world featuring POLDERS inhabited by the elect, a geography that supports the abiding sense that there are places – which one cannot quite call SECONDARY WORLDS – safely elsewhere from the mundane world; and for a yearning sense that the drama of cosmology leads in the end to EUCATASTROPHE. As far as writers of fiction are concerned, she had a remarkably useful imagination. The entry on THEOSOPHY expands upon these points.

HPB herself was a most extraordinary person, and might almost have served as a model for the TEMPORAL ADVENTURESS as created by Michael MOORCOCK. She left her husband, Nikifor Blavatsky, Vice-Governor of the province of Erivan in the Ukraine, at the age of 18; arrived in Constantinople soon after; led an adventurous life from that point – there are hints that she was once a snake-charmer in Cairo – until 1873, when she landed in the USA, a stocky, sultry chain-smoker much addicted to slapstick, and certainly very familiar with books like Edward BULWER-LYTTON's *Zanoni* (**1842**), itself an *omnium gatherum* of occult doctrines and influences, mainly from Eliphas Lévi (1810-1875); Zanoni himself is a Secret Master nonpareil. After the founding of the Theosophical Society, HPB took her entourage to India in 1879, where the full-blown Theosophy of her final book was evolved. She returned to Europe in 1884, where she faced down scandalous revelations about the authenticity of her mystical powers; met William Butler YEATS, who became a Theosophist, and Aleister CROWLEY; and died in time to avoid further embarrassment. She is as much a figure of the GASLIGHT ROMANCE as any fictional character. Her own fiction, most of which was assembled as *Nightmare Tales* (coll **1892**), is unimportant. Later fiction in which she appears in her own right includes Mark FROST's *The List of Seven* (**1993**).   [JC]

Further reading: *Madame Blavatsky's Baboon: Theosophy and the Emergence of the Western Guru* (**1993**) by Peter Washington.

BLAYLOCK, JAMES P. (1950-   ) US writer. His first published story was "The Red Planet" (1977) for *Unearth*. His initial books were the first two fantasies in his **Elfin** series, *The Elfin Ship* (**1982**) and *The Disappearing Dwarf* (**1983**). The series was subsequently expanded to contain a rather darker prequel, *The Stone Giant* (**1989**). All three novels are set in a FANTASYLAND which includes elves, dwarves and goblins; if they are influenced by J.R.R. TOLKIEN it is for the most part by the Shire chapters of *The Hobbit* (**1937**) and *The Lord of the Rings* (**1954-5**) in their combination of a society of small towns and isolated villages in which there is no heavy industry but a fair amount of user-friendly technology – in JPB's instance, barometers, airships, etc. The books' emphasis on a male collegiality full of shared meals and practical jokes echoes *The Wind in the Willows* (**1908**) by Kenneth GRAHAME. Grahame is also a possible source for the books' opposition between the RIVER, presented here as a potentially dangerous environment made manageable by human conviviality and cooperation, and the uncontrollable dangers which arise when characters venture INTO THE WOODS. All three books deal with the more or less genial suppression of an aspiring DARK LORD – in the first two, the dwarf Selznak, and in the prequel the rather similar Helstrom. The motives of the protagonists are never especially highflown: Jonathan Bing, protagonist of the first two books, is a Master Cheesemaker who takes action because worried about the restraint of trade the dwarf's incursions are causing. Selznak's magic watch, which stops time (◊ TIME FANTASY) is the first of the magical MCGUFFINS which dominate JPB's later books.

The other definable series in his work, **St Ives**, is a quintessential contribution to STEAMPUNK, combining as it does a vision of 19th-century LONDON as the archetypal CITY with endless proliferations of ALCHEMY and weird science and a constant atmosphere of conspiracy (◊ FANTASIES OF HISTORY) and SECRET GUARDIANS. The first of these two books, *Homunculus* (**1986**), won the PHILIP K. DICK AWARD for best paperback original, and deals in part with the pursuit of IMMORTALITY through essences drawn from carp; the machinations of the hunchback Narbondo, a MAGUS and Napoleon of Crime, are defeated by a band of adventurers which includes St Ives. (The name of JPB's hero is possibly a homage to Robert Louis STEVENSON's *St Ives* [**1897**]; Stevenson's *New Arabian Nights* [coll **1882**] is as important a source of JPB's vision of London in these books as the more often cited Charles DICKENS.) St Ives's colleagues are less important in the second book, *Lord Kelvin's Machine* (**1992**), which reads like a fixup. Narbondo's murder of St Ives's wife leads to an attempt to blackmail the governments of the world with a threat to turn off its magnetic field and to St Ives's going back in time to rid the world of his enemy in childhood. It is typical of JPB that St Ives's eventual deed is to give alms to Narbondo's mother rather than to commit infanticide.

Sometimes seen as Steampunk, *The Digging Leviathan* (**1984**) is the first of JPB's novels of modern California.

Another version of Narbondo is but one of the eccentrics and true believers and dastards who cluster round the boy Giles, born with webbed fingers and gill slits; as with the other Californian novels, the story is less a conventional plot than a device to interrupt the display of characters by periodic flurries of action; again like the other Californian books, it is characterized by an interest in weird science like the Dean Drive and the Hieronymus Machine, and in SCHOLARLY FANTASIES like the HOLLOW EARTH.

The remaining Californian novels, most set in northern California rather than the LOS ANGELES of *The Digging Leviathan*, are: *Land of Dreams* (**1987**), which includes a sinister travelling CARNIVAL (rather in the style of Ray BRADBURY) and a PORTAL; *The Last Coin* (**1988**), the characters of which include a travelling salesman who turns out to be the WANDERING JEW and which has as its McGuffin the 30 pieces of silver paid to Judas for betraying CHRIST; *The Paper Grail* (**1991**), which features a lost drawing by Hokusai, the bones of Joseph of Arimathea, a FISHER KING and echoes of the PRE-RAPHAELITES; *Night Relics* (**1994**), concerned with GHOSTS; and *All the Bells on Earth* (**1995**), where a small Californian town becomes a spiritual battlefield for the fight between GOOD AND EVIL. The juvenile *The Magic Spectacles* (**1991** UK), by contrast, is set in an OTHERWORLD, accessible only through use of the title's spectacles.

JPB has not published much short fiction, but his *Paper Dragons* (1985 in *Imaginary Lands* ed Robin MCKINLEY **1992** chap) won the WORLD FANTASY AWARD. It shares the northern Californian background of the later novels.

JPB was part of the AFFINITY GROUP that centred on Philip K. DICK in his latter years; so too were his friends K.W. Jeter and Tim POWERS, with the latter of whom JPB created the imaginary 19th-century poet Ashbless, a cognate of whom appears in *The Digging Leviathan* as well as in Powers's *The Anubis Gates* (**1983**). Powers and JPB have collaborated on minor works.   [RK]

Other works: *Twelve Hours of the Night* (**1985** chap) with Powers, together as William Ashbless; *The Pink of Fading Neon* (**1986** chap dos); *The Shadow on the Doorstep* (1986 *IASF*; **1987** chap dos with short stories by Edward BRYANT); *A Short Poem* (**1987** chap) as by William Ashbless; *Two Views of a Cave Painting* (coll **1987** chap dos); *A Postscript to Homunculus* (**1988** chap) as by William Hastings, a reprint of the last few pages of the limited edition of *Homunculus*; *Doughnuts* (**1994** chap).

BLAYRE, CHRISTOPHER Pseudonym of UK academic and writer Edward Heron-Allen (1861-1943). His early works, published under his own name, include *The Princess Daphne* (**1885**), a novel involving psychic vampirism (◊ VAMPIRES) written with Selina Delaro (uncredited) and *A Fatal Fiddle: The Commonplace Tragedy of a Snob* (**1890**). His academic works include the study *Barnacles in Nature and in Myth* (**1928**).

The items signed CB derive from *The Purple Sapphire* (coll **1921**), which offers a series of tongue-in-cheek weird tales and sf stories "selected from the unofficial records of the University of Cosmpoli". The ninth and longest, *The Cheetah-Girl* (**1923** chap), was excised by the publisher on grounds of obscenity and released in a very limited privately printed edition; four further stories, including the cream of CB's slyly sophisticated parodies, were also privately printed as *Some Women of the University* (coll **1932**). *The Strange Papers of Dr Blayre* (coll **1932**) augmented the contents of *The Purple Sapphire* with four new but relatively trivial items. [BS]

Other works: *Kisses of Fate: A Study of Mere Human Nature* (coll **1888**) as by EH-A.

BLEILER, E(VERETT) F(RANKLIN) (1920-) US editor and bibliographer whose work has contributed to a wider appreciation of and access to the fields of SUPERNATURAL FICTION and SCIENCE FICTION. EFB's seminal work, *The Checklist of Fantastic Literature: A Bibliography of Fantasy, Weird and Science Fiction Books Published in the English Language* (**1948**; rev vt *The Checklist of Science-Fiction and Supernatural Fiction* 1978) provided a grounding in sf and fantasy bibliography which remained unsurpassed until the works of Donald H. Tuck (1922-) and Robert Reginald (real name Michael R. Burgess; 1948-), although EFB also took his research forward with *The Guide to Supernatural Fiction* (**1983**), which provided plot summaries and analyses of 1775 books dated 1750-1960 and included further groundbreaking work in its phenomenology and motif index. EFB undertook a similar analysis of proto-sf in *Science Fiction: The Early Years* (**1991**), compiled with his son Richard Bleiler, and the impending «Science Fiction: The Gernsback Years», and also edited two significant studies of the respective fields in *Science Fiction Writers: Critical Studies of the Major Authors from the early Nineteenth Century to the Present Day* (anth **1982**) and *Supernatural Fiction Writers: Fantasy and Horror* (anth **1985** 2 vols). With Thaddeus E. Dikty, EFB co-edited a series of sf ANTHOLOGIES 1949-54. In 1955 he joined Dover Publications, becoming Executive Vice-President in 1967 and retiring in 1977. For Dover he compiled many books on a wide range of subjects, including many translations. His particular talent was in identifying rare and often forgotten Victorian works of singular quality. Volumes compiled by EFB of genre interest, most with detailed and exemplary introductions, include *Three Gothic Novels* (anth **1960**), *Ghost and Horror Stories of Ambrose Bierce* (coll **1964**), *Best Ghost Stories of J.S. Le Fanu* (coll **1964** US) and its companion *Ghost Stories and Mysteries* by J. Sheridan LE FANU (coll **1975** US), *The Best Tales of Hoffmann* (coll **1967**) by E.T.A. HOFFMANN, *The King in Yellow and Other Horror Stories* by Robert W. CHAMBERS (coll **1970**), *Five Victorian Ghost Novels* (anth **1971**), *Gods, Men and Ghosts* (coll **1972** US) by Lord DUNSANY, *Best Ghost Stories of Algernon Blackwood* (coll **1973** US), *Classic Ghost Stories by Charles Dickens and Others* (anth **1975**) ed anon, *Three Supernatural Novels of the Victorian Period* (anth **1975**), *The Golem and The Man who was Born Again* (omni **1976**), *The Collected Ghost Stories of Mrs J.H. Riddell* (coll **1977** US) by Mrs RIDDELL, *The Best Supernatural Stories of Arthur Conan Doyle* (coll **1979** US) by Arthur Conan DOYLE, and *Prophecies and Enigmas of Nostradamus* (trans **1979** US) (◊ NOSTRADAMUS). For Scribners' EFB compiled *A Treasury of Victorian Ghost Stories* (anth **1981**). For his distinguished contributions to the field of research EFB received the Pilgrim AWARD in 1984. [MA]

BLEILER, RICHARD (JAMES) (1959-) US bibliographer. ◊ E.F. BLEILER.

BLISH, JAMES (BENJAMIN) (1921-1975) US writer, in the UK from 1969. The bulk of his varied and uneven output is sf – including the bravura ESCHATOLOGY of *The Triumph of Time* (**1958**; vt *A Clash of Cymbals* UK), which ends his **Cities in Flight** sf sequence with the destruction of our Universe and one character's imposition of his own will

on the following cycle of CREATION. The minor *The Warriors of Day* (**1951** *Two Complete Science Adventure Books* as "Sword of Xota"; **1953**) offers a rationalized GOD who is a collective planetary consciousness, and who after using and absorbing his human AVATAR is finally, ironically, seen shaping a worthier tool: a bear.

JB's prime fantasy contribution is the patchwork sequence **After Such Knowledge**, comprising the sf-religious (◊ RELIGION) *A Case of Conscience* (part 1 in *If* 1953; **1958**), the fictional biography *Doctor Mirabilis* (**1964** UK; rev 1971 US), the fantasy *Black Easter, or Faust Aleph-Null* (**1968**) and *The Day After Judgment* (**1971**). The last two were regarded by JB as one novel and were republished as *Black Easter and The Day After Judgment* (omni **1980** US; vt *The Devil's Day* 1990 US); the full set was assembled as *After Such Knowledge* (omni **1991** UK). Taking its overall title from T.S. Eliot's line "After such knowledge, what forgiveness?", the sequence is supposedly linked by an old question of theology (◊ RELIGION), expressed in *The Day After Judgment* as "[whether] the possession and use of secular knowledge – or even the desire for it – is in itself evil".

This seems least relevant in *A Case of Conscience*, except in that the secular physics of spaceflight brings humanity into contact with Lithia, an alien EDEN whose "unfallen" inhabitants have adopted a CHRISTIAN ethos without any concept of GOD. The Jesuit priest-hero reasons that this must be a trap created by SATAN – which is heresy, since Satan cannot create. As an alien ANTICHRIST figure disrupts Earth society, the priest finds an orthodox explanation: that Lithia is a diabolical ILLUSION. When he pronounces a formal EXORCISM the planet is destroyed, ambiguously, through the "accidental" detonation of a thermonuclear bomb factory. This book won the Hugo AWARD.

Roger Bacon (*c*1214-1292), whose life is persuasively imagined in *Doctor Mirabilis*, engages more directly with the question. The part of his nature which dreams of universal Scientific Method appears as a SECRET SHARER, whispering ideas at odds with Church discipline. In a delirious, ALCHEMY-inspired DREAM – his own monkish figure opposes the Antichrist's forces with spyglasses and explosions – Bacon hears himself pronounce the formula for gunpowder as an anagrammatic puzzle which he must then decipher. The Church does not treat him kindly.

Black Easter offers two versions of the dangerous seeker. Theron Ware, a black magician, has entered into numerous PACTS with DEMONS whose powers – along with money derived from hiring out their services – he employs to gain knowledge of such scientific enigmas as quasars. Baines, an armaments tycoon, is bored with secular wars and commissions Ware to release 48 major demons from HELL for one night, merely to see what will happen. Incidental effects include TRANSMUTATION, supernatural killing and a SUCCUBUS. JB's bleak, spare descriptions convincingly establish BLACK MAGIC as a scholium having the same unforgiving precision as physics: the clinical tone heightens the horror of the long conjuration RITUAL in which the demons are summoned. The results escalate through nuclear war to ARMAGEDDON, with Hell victorious since one factor even Ware has tacitly relied upon seems absent: as the Sabbath Goat remarks at the finale, GOD is dead. The less potent continuation, *The Day After Judgment*, cannot top this, instead making some satirical play with US military efforts to analyse and counter a presumed invasion, including a wonderfully self-indulgent moment of RECURSIVE FANTASY as a US general identifies a glimpsed demon as Sax ROHMER's FU MANCHU; but the CITY of Dis, now manifest in Death Valley, proves resistant to multiple nuclear strikes and is defended by the head of Medusa. In verse which is a pastiche of John MILTON's, it emerges that the death or withdrawal of God has forced Satan to take his place and rule benignly.

JB was a professed agnostic who enjoyed applying his formidable intellect to theology, which he treated as a PLAYGROUND. A similar scholarly interest drew him to the language-puzzles of James JOYCE's *Finnegans Wake* (**1939**) – a theological poser from which he examined in *A Case of Conscience* – and to the erudite allusiveness of James Branch CABELL: JB was one editor of the Cabell Society magazine *Kalki*. His best fantasies have a chilly, compulsive fascination. [DRL]

Other works: "Cathedrals in Space" in *The Issue At Hand* (coll **1964**) discusses religion in sf/fantasy; some of the poems in *With All of Love: Selected Poems* (coll **1995**) are fantasy, including "Scenario: The Edifice", which neatly defines EDIFICE as that term is used in this encyclopedia.

See also: BOOKS.

BLISHEN, EDWARD (1920-   ) UK writer. ◊ Leon GARFIELD.

BLISS Australian movie (*1985*). ◊ Peter CAREY.

BLITHE SPIRIT UK movie (*1945*). Two Cities/Noël Coward-Cineguild. **Pr** Noël Coward (1899-1973), Anthony Havelock-Allan. **Dir** David Lean. **Spfx** Tom Howard. **Screenplay** adapted by Havelock-Allan, Lean, Ronald Neame from the play *Blithe Spirit* (**1941**) by Coward. **Starring** Constance Cummings (Ruth Condomine), Kay Hammond (Elvira Condomine), Rex Harrison (Charles Condomine), Margaret Rutherford (Mme Arcati). 96 mins. Colour.

Novelist Charles, married to Ruth after the death of Elvira, hires batty medium Mme Arcati to hold a SEANCE, expecting a fraud. To his horror, the GHOST of Elvira arrives and refuses to leave; there develops a bizarre love-triangle, with Ruth seeking EXORCISM to banish Elvira and Elvira seeking Charles's death so they can be reunited. But the car crash Elvira rigs kills Ruth instead, so now *two* shades covet the luckless writer . . . *BS*, though based quite closely on Coward's stage play (in which both Hammond and Harrison had starred), is very cinematic, thanks to fine camerawork and spfx. The dialogue is as barbed as one could wish. [JG]

BLIXEN, KAREN (1883-1962) Danish author who writes most often as Isak DINESEN.

BLOCH, ROBERT (1917-1994) US writer, best-known for psychological and supernatural HORROR, much of which has been dramatized for radio, cinema and tv. His earliest work was heavily influenced by the H.P. LOVECRAFT school, and he wrote a number of ornate CONTES CRUELS which qualify as fantasies, including "Black Lotus" (1935), "The Mandarin's Canaries" (1938) and – in collaboration with Henry KUTTNER – "The Black Kiss" (1937); the last was later retitled as the title story of *Sea-Kissed* (coll **1945** chap). Many of the stories he published in *Strange Stories* in 1939 as Tarleton Fiske are similar horror/fantasy hybrids. All of his pure fantasies are comedies.

RB modelled "A Good Knight's Work" (1942) and its sequel "The Eager Dragon" (1943) on the work of Damon Runyon (1884-1946), and "Nursemaid to Nightmares" (1942) and its sequel "Black Barter" (1943) on the works of Thorne SMITH. These four stories were collected in *Dragons and Nightmares* (coll **1969**), the second pair having earlier been run together as "Mr Margate's Mermaid" (1955) in *Imaginative Tales*. *Imaginative Tales* also reprinted "The Devil with You" (*Fantastic Adventures* 1950) as "Black Magic Holiday" (1955) and published two further novellas in the Thorne Smith mould: "The Miracle of Roland Weems" (1955) and "The Big Binge" (1955; reprinted as *It's All in Your Mind* **1971**); its companion MAGAZINE, *Imagination* had earlier published the similar "Hell's Angel" (1951). A pun-laden series of anecdotal short stories (*Fantastic Adventures* 1942-6) was collected as *Lost in Time and Space with Lefty Feep* (coll **1987**). This work proved far less commercial than RB's horror (the books cited were all issued by small presses), and his writings in this ebullient vein became increasingly infrequent, although he won a Hugo AWARD for "That Hell-Bound Train" (1958) not long before his thriller *Psycho* (**1959**) secured his reputation *via* Alfred HITCHCOCK's celebrated movie adaptation *Psycho* (*1960*).

Later weird stories by RB which embody significant fantasy elements – sometimes casually mingled with sf elements – include "All on a Golden Afternoon" (1956), "The Funnel of God" (1960) and "The World-Timer" (1960). The Lovecraftian stories collected in *The Mysteries of the Worm: All the Cthulhu Mythos Stories* (coll **1981**; rev exp 1993) are mostly rather slapdash shockers, but *Strange Eons* (**1979**) rounds off the series with a more wholehearted futuristic fantasy. *The Jekyll Legacy* (**1990**) with Andre NORTON is a sequel to Robert Louis STEVENSON's classic moral fantasy. A useful collection of RB's early fiction is *The Early Fears* (omni **1994**), which combines the contents of *The Opener of the Way* (coll **1945**; vt in 2 vols as *The Opener of the Way* and *House of the Hatchet* 1976 UK) and *Pleasant Dreams – Nightmares* (coll **1960**; cut vt *Nightmares* 1961; rev *Pleasant Dreams* 1979) with three new stories and an introduction by the author. [BS]

Other works: *Terror in the Night and Other Stories* (coll **1958**); *Blood Runs Cold* (coll **1961**; cut 1963 UK); *Yours Truly, Jack the Ripper* (coll **1962**; vt *The House of the Hatchet and Other Tales of Horror* 1965 UK); *More Nightmares* (coll **1962**); *Horror-7* (coll **1963**); *Bogey Men* (coll **1963**); *Tales in a Jugular Vein* (coll **1965**); *The Skull of the Marquis de Sade* (coll **1965**), whose title story was filmed as *The Skull* (*1965* UK); *Chamber of Horrors* (coll **1966**); *The Living Demons* (coll **1967**); *Fear Today, Gone Tomorrow* (coll **1971**); *The King of Terrors* (coll **1977**); *Cold Chills* (coll **1977**); *The Best of Robert Bloch* (coll **1977**); *Out of the Mouths of Graves* (coll **1978**); *Such Stuff as Screams are Made Of* (coll **1979**); *Out of my Head* (coll **1986**); *Midnight Pleasures* (coll **1987**); *The Selected Stories of Robert Bloch* (coll **1988** 3 vols: *#1 Final Reckonings* vt *The Complete Stories of Robert Bloch Vol 1: Final Reckonings* 1990; *#2 Bitter Ends*; *#3 Last Rites*); *Fear and Trembling* (coll **1989**).

Further reading: *Once Around the Bloch* (**1993**), autobiography.

BLOCK, FRANCESCA LIA (1962-   ) US writer and journalist whose first appearance in print was *Weetzie Bat* (**1989**), a short YA novel; in the same sequence are *Witch Baby* (**1991**), *Cherokee Bat and the Goat Guys* (**1992**), *Missing Angel Juan* (**1993**) and *Baby Be-Bop* (**1995**). The lives of a very odd extended family as its various members struggle with adolescence are told in a slangy, highly contemporary style and based in the multicultural street life of Los Angeles, striking a note of wonder which may merit the tag

MAGIC REALISM. *Ecstasia* (**1993**) and *Primavera* (**1994**) form a two-book sequence which is more conventional. Set in a possibly post-APOCALYPSE desert landscape, where there is only one CITY, it describes a rock'n'roll-dominated society where those who grow old must leave the city and go to the UNDERWORLD. *Ecstasia* retells the ORPHEUS myth (with points of resemblance to Samuel R. DELANY's *The Einstein Intersection* [**1967**]), while *Primavera* follows the myth of Persephone.

The Hanged Man (**1995**), again set in contemporary California, describes a teenager's attempts to deal with real life: her father is dead, her mother may be a WITCH, the wild boy she falls in with may be a VAMPIRE (the work is less surreal than FLB's earlier work and the fantasy elements can be interpreted as drug HALLUCINATIONS or fugues of insanity; ◊ RATIONALIZED FANTASY). The style is crisper and more pointed than before, and focuses tightly on the principle of a TAROT layout. It may be FLB's most fully realized work. [DGK]

BLOGGSLEIGH, AUBREY TYNDALL [s] ◊ Kenneth MORRIS.

BLOOD FOR DRACULA (*1974*; vt *Andy Warhol's Dracula*) ◊ DRACULA MOVIES.

BLOOD OF DRACULA (*1957*) ◊ DRACULA MOVIES.

BLOOD OF DRACULA'S CASTLE (*1969*) ◊ DRACULA MOVIES.

BLOOM, HAROLD (1930-   ) US critic, anthologist and writer, most famous in the literary world for his theory – first developed in *The Anxiety of Influence* (**1973**) – that, over the past several centuries in the Western world, strong writers have created their work in part through transgressive responses to their predecessors. Within the network of publicly shared assumptions, PLOT DEVICES and vocabulary that marks genre literatures, it is perhaps less easy to demonstrate this principle, though it could certainly be argued that Stephen R. DONALDSON's response to J.R.R. TOLKIEN shows all the signs of a strong writer whose response to the father is (to put it mildly) revisionist. *Kabbalah and Criticism* (**1975**) extends the arguments of HB's previous book in terms relevant to scholars of the Cabbala and gnostic literature in general. Some of the arguments in *Agon: Towards a Theory of Revisionism* (**1982**) provide a framework for beginning to understand REVISIONIST FANTASY. *The Western Canon: The Books and School of the Ages* (**1994**) is invaluable – in terms of fantasy – for its ample discussion of TAPROOT-TEXT writers like CHAUCER, SHAKESPEARE and GOETHE; it is also informative about Jorge Luis BORGES and Franz KAFKA.

In his long introduction to *The Book of J* (texts trans from Hebrew by David Rosenberg **1990**) HB argues – his conclusions have been vigorously disputed – that most of the great stories in the Old Testament were not only the work of one person (identified as J) but that this person was a woman. If he is correct, J is the first great writer of fantasy (◊ JEWISH RELIGIOUS LITERATURE). HB's only work of fiction, *The Flight to Lucifer: A Gnostic Fantasy* (**1979**), is an over-dense presentation in story form of the central gnostic QUEST for some opening out of the BONDAGE of unknowing into the light of REALITY; it shows the influence of David LINDSAY, for whom HB has expressed extravagant admiration.

HB is overall editor of the **Chelsea House Modern Critical Views** sequence of anthologies, projected to run to hundreds of volumes. Those of genre interest so far issued include *Mary Shelley* (anth **1985**), *Edgar Allan Poe* (anth **1985**), *Ursula K. Le Guin* (anth **1986**) and *Ursula K. Le Guin's The Left Hand of Darkness* (anth **1987**), *Doris Lessing* (anth **1986**), *George Orwell* (anth **1987**) and *George Orwell's 1984* (anth **1987**), *Classic Horror Writers* (anth **1993**), *Classic Fantasy Writers* (anth **1995**), *Classic Science Fiction Writers* (anth **1995**), *Science Fiction Writers of the Golden Age* (anth **1995**), *Modern Fantasy Writers* (anth **1995**) and *Modern Horror Writers* (anth **1995**). [JC]

BLUE BIRD, THE Two movies have been based on *L'Oiseau bleu* (play **1908**) by Maurice Maeterlinck (1862-1949).

1. US movie (*1940*). 20th Century-Fox. **Pr** Darryl F. Zanuck. **Assoc pr** Gene Markey. **Dir** Walter Lang. **Spfx** Fred Sersen. **Screenplay** Walter Bullock, Ernest Pascal. **Starring** Eddie Collins (Tylo), Helen Ericson (Light), Sybil Jason (Angela Berlingot), Jessie Ralph (Fairy Berylune), Johnny Russell (Tyltyl), Gale Sondergaard (Tylette), Shirley Temple (Mytyl). 98 mins. Colour and b/w.

Just before CHRISTMAS, peasant child Mytyl and her younger brother Tyltyl catch a bird in the woods. On the way home Mytyl refuses to trade it with sick child Angela. That night Mytyl complains of the misery of her lot. Approaching midnight, Mytyl and Tyltyl wake (and the movie shifts from b/w into Technicolor). The Fairy Berylune arrives and tells the children they must find the Blue Bird of Happiness, which could be anywhere; as guide she gives them Light (incarnated as a beautiful woman) and for COMPANIONS she transmogrifies into human form Tylo the dog (thick but loyal) and Tylette the cat (intelligent but treacherous, realizing that if the children's QUEST founders she will be freed from the BONDAGE of cat-hood). The children search the World of the Past, the World of Luxury and the World of the Future, which they find populated by all the children who have yet to be born; birth comes when Father Time calls the roll and the chosen children travel to Earth in a silver-sailed ship. Still there is no Blue Bird and, disappointed, Mytyl and Tyltyl return home . . . to wake in their beds realizing how lucky they are. Joyously, Mytyl discovers the bird she caught yesterday is now blue, and at once takes it to Angela. The bird escapes; Mytyl consoles Angela with the knowledge that one can *always* find the Blue Bird.

Much was expected of *TBB*, with Temple commercially a strong lead, but while it was in production *The* WIZARD OF OZ (*1939*) was released, and *TBB* more or less vanished without trace. The remake, **2**, although itself disastrous, gave new life to the original. This is no *Wizard of Oz* – Temple's tweeness cannot match Garland's acting – but neither should it have been forgotten. [JG]

2. US/USSR movie (*1976*). 20th Century-Fox/Lenfilm/Sovinfilm/Tower. **Pr** Paul Maslansky. **Exec pr** Edward Lewis. **Dir** George Cukor. **Spfx** Lev Cholmov, Wayne Fitzgerald/Pacific Title, Roy Field, Leonid Kajukov, Boris Michailov, Georgi Senotov, Aleksandr Zavtalov. **Screenplay** Alfred Hayes, Hugh Whitemore. **Starring** Harry Andrews (Oak), George Cole (Tylo), Jane Fonda (Night), Ava Gardner (Luxury), Patsy Kensit (Mytyl), Todd Lookinland (Tyltyl), Pheona McLellan (Sick Girl), Robert Morley (Time), Elizabeth Taylor (Light/Maternal Love/Mother/Witch), Cicely Tyson (Tylette). 83 mins. Colour.

This was intended as a gesture of West/East friendship, so there are Russian actors in minor roles (their lines usually dubbed, and clumsily plus interpolated fits of Russian ballet; the Russians contributed more impressively on the technical side. To the scenario of **1** is added a venture into the Palace of Night, where are found numerous blue birds; but they die with dawn's coming. Only Tyson seems able to bring any conviction to this excruciatingly bad movie; Taylor is wooden in four roles. The whole is redolent of a school play. [JG]

BLUE FAIRY ◊ PINOCCHIO.

BLUTH, DON (1938-   ) US animator and director of ANIMATED MOVIES, resident in Eire since 1979. He rose to prominence at DISNEY, working on such movies as *Robin Hood* (**1973**), PETE'S DRAGON (**1977**) and *The Rescuers* (**1977**) before becoming concerned that Disney feature animation was losing its way (which, at the time, it probably was). Attracted by the generous tax concessions offered by Eire to creative artists, he set up his own studio there, and since then has created a number of ANIMATED MOVIES in what he regards as the true spirit of Disney: their animation is certainly on a par, but their scripting is less assured and they can lack vivacity. All are, however, of fantasy interest; and all are worth watching.

The Secret of NIMH (**1982**), based on the Newbery-winning *Mrs Frisby and the Rats of NIMH* (**1971**) by Robert C. O'Brien (1922-1973), is a TECHNOFANTASY concerning intelligent laboratory rats escaping from a research establishment. *An* AMERICAN TAIL (**1986**) is DB's masterpiece to date, and represents an important advance in commercial animation: it takes on adult issues (such as antisemitism and exploitation) and mixes them adroitly with more conventional material to produce a movie that is both fun and thought-provoking for all the family. *The Land Before Time* (**1988**) is a fantasy adventure set among intelligent DINOSAURS: highly original in concept and excellently animated, it was generally disliked by the critics. *All Dogs Go to Heaven* (**1989**) is another extremely interesting movie, a canine POSTHUMOUS FANTASY in which a petty-criminal dog returns to Earth for vengeance on the canine gang boss who had him murdered. There are intriguing scenes in the AFTERLIFE. *Rock-a-Doodle* (**1990**) sees a farmhand suffering TRANSFORMATION by an owl into a cat; he gets help from an urban rock-star cockerel. THUMBELINA (**1994**) was the first DB movie to feature humans rather than TALKING ANIMALS – and also to challenge Disney's "Classic Fairytales" features head-on. The tale is nicely told, but still lacks the punch of the best of Disney's work in this field.

It could be argued that Disney's current output of animated features is as good as it is because of the incentive to surpass DB. Although there have been superb one-offs from other animation studios – FAI/Youngheart's FERNGULLY: THE LAST RAINFOREST (**1992**), for example – DB's work, taken as a whole, is quite exceptional. [JG]

BLYTH, JAMES (1864-1933) UK writer. ◊ Barry PAIN.

BLYTON, ENID (1897-1968) UK writer. ◊ CHILDREN'S FANTASY.

BOCCACCIO, GIOVANNI (1313-1375) Italian writer, a central figure in the development of vernacular Western literature and in the evolution of STORY out of exemplary FABLE and FAIRYTALE. *The Decameron* (coll *c*1350), a central TAPROOT TEXT, may be the first STORY CYCLE in the West whose contents are selfconsciously designed as literary entities, and the first since Roman times to be designedly secular in its impact. A CLUB-STORY-like structure works to free individual tales from the straitjacket of having to teach something "worthwhile". GB – as did CHAUCER a little later – opened the door for the free story; it has not been shut again. [JC]

BOHNHOFF, MAYA KAATHRYN (1954-   ) US author of the **Meri** sequence – *The Meri* (**1992**), *Taminy* (**1993**) and *The Crystal Rose* (**1995**) – about the interactions between humans and a race of mer-creatures (◊ MERMAIDS). [JC]

BOIARDO, MATTEO MARIO, COUNT OF SCANDIANO (1434-1494) Italian poet. ◊ Lodovico ARIOSTO; MORGAN LE FAY; TAPROOT TEXTS.

BOK, HANNES (VAJN) Working name of US artist, writer and astrologer Wayne Woodard (1914-1964). Inspired in his artwork by Maxfield PARRISH and in his writing by A. MERRITT, HB produced work that was flamboyant and exotic. His illustrations began to appear in the fan press in 1934. Through his friendship with Ray BRADBURY he secured work with WEIRD TALES, which published many of his covers and interior illustrations 1939-51. He also appeared frequently in *Astonishing Stories* and FAMOUS FANTASTIC MYSTERIES, and supplied nearly 30 covers for specialist SMALL-PRESS publishers 1942-55. His style came closest to Edd Cartier's in terms of his ability to create unthreatening aliens. His black-and-white illustrations succeeded in conveying both the dark menace as well as the allure of otherworldliness. Although HB jointly (with Ed Emshwiller [1925-1990]) won the first Best Cover Artist Hugo AWARD in 1953, he more or less abandoned the fantasy field thereafter, instead drifting towards mysticism and ASTROLOGY and leading a bohemian lifestyle. After his death, his friend Emil Petaja (1915-   ) created the Bokanalia Foundation to promote HB's work, and has issued a number of portfolios including *A Memorial Portfolio* (portfolio **1970**) and *The Hannes Bok Memorial Showcase of Fantasy Art* (portfolio **1974**). Gerry de la Ree (1924-1993) compiled *A Hannes Bok Sketchbook* (graph **1976**) and *Beauty and the Beasts* (graph **1978**), and more recently Stephen D. Korshak has assembled *A Hannes Bok Treasury* (graph **1994**) and *A Hannes Bok Showcase* (graph **1995**).

As a writer of exotic fantasy and sf, HB produced only six short stories and three solo novels. In all three novels, "Starstone World" (1942 *Science Fiction Quarterly*), *The Sorcerer's Ship* (1942 *Unknown*; **1969**) and *Beyond the Golden Stair* (1948 *Startling Stories* as "The Blue Flamingo"; rev **1970**) the hero is transported via a PORTAL to an OTHERWORLD where MAGIC works. These tales are stronger in imagery and wordplay than in plot or characterization. HB also developed two fragments left by Abraham MERRITT into novels: *The Fox Woman and the Blue Pagoda* (**1946**) and *The Black Wheel* (**1947**), both involving spirit POSSESSION and the fulfilment of a QUEST. HB's poetry, equally exotic, has been collected by Petaja as *Spinner of Silver and Thistle* (coll **1972**). [MA]

Further reading: *And Flights of Angels* (**1968**) by Emil Petaja; *Bok – A Tribute to the Late Fantasy Artist on the 60th Anniversary of his Birth and the 10th Anniversary of his Death* (anth **1974**) ed Gerry de la Ree.

BOLTON, JOHN (1951-   ) UK-born COMICS artist whose careful, polished style has matured steadily through the last two decades. JB trained as an engineer but turned to illustration in 1971, drawing spot illustrations before producing horror-movie adaptations for the magazine *House of Hammer*; these included "Dracula, Prince of Darkness" (1976), "Curse of the Leopard Men" (1976), "Curse of the Werewolf" (1977) and "One Million Years BC" (1977). Then came a strip version of the tv series *The Bionic Woman* (*Look In Magazine* 1978-9).

JB began to work for MARVEL COMICS UK, drawing **The**

Incredible Hulk (1979), **Nightraven** (1979), **Kull** (*Bizarre Adventures #26* 1980 and *Kull the Conqueror #2-#4* 1983-4). He meticulously painted **Marada the She-Wolf**, scripted by Chris Claremont (*Epic Magazine #10-#12* 1982 and *#22-#23* 1984). His work in the fantasy vein has continued to appear regularly in a wide range of US comic books, including *Twisted Tales* (*#4-#7* 1983-4), *Epic* (*#7, #15, #18, #24-#25* 1981-4), *Alien Worlds* (*#5* 1983), *Classic X-Men* (*#1-#28* and *#30-#35* 1986-9), *A1* (*#1-#4* 1989-90), *Hellraiser* (*#1* and *#7* 1990-91), *Tapping the Vein* (*#2* 1990) and *The Books of Magic* (*#1* 1990). His GRAPHIC NOVELS are *Someplace Strange* (graph **1988**), written by Ann Nocenti, and Clive BARKER's *The Yattering and Jack* (graph **1992**), adapted by Steve Niles. JB also drew a comic-book adaptation of the movie *Army of Darkness* (graph **1992-3** 3 vols). A series of paintings of naked female VAMPIRES (*Glamour International #16* 1991) has enhanced his reputation in Italy.   [RT]
Further reading: *Unmasked* (*#0-#1* 1995) features JB articles and sketchbook work.

BOMBA MOVIES A series of 12 JUNGLE MOVIES starring Johnny Sheffield, who had played, alongside Johnny Weissmuller as Tarzan, the part of Tarzan's adopted son Boy in eight of the TARZAN MOVIES; the character came from the **Bomba the Jungle Boy** series of 20 books published 1926-38 under the housename Roy Rockwood. The movies were a rather too obvious cut-price attempt to capitalize on the popularity of the **Tarzan** series, and have more or less been forgotten: they are *Bomba the Jungle Boy* (**1948**), *Bomba on Panther Island* (**1948**), *Bomba and the Hidden City* (**1949**), *Bomba and the Lost Volcano* (**1949**), *Bomba and the Elephant Stampede* (**1950**), *Bomba and the African Treasure* (**1951**), *Bomba and the Jungle Girl* (**1951**), *Bomba and the Lion Hunters* (**1951**), *Safari Drums* (**1953**), *The Golden Idol* (**1954**), *Killer Leopard* (**1954**) and *D of the Jungle* (**1955**).   [JG]

BOND, MICHAEL (1926-   ) UK writer. ◊ CHILDREN'S FANTASY.

BOND, NANCY (BARBARA) (1945-   ) US writer best-known for her first novel, *A String in the Harp* (**1976**), in which connections between contemporary life in Wales and a fantasy world derived from CELTIC FANTASY and the Arthurian cycle (◊ ARTHUR) are beguilingly woven. *The Voyage Begun* (**1981**) is YA sf. In *Another Shore* (**1989**) a young woman finds herself in a TIME-TRAVEL trap, locked into 18th-century conflicts in pre-Revolutionary America along with other trapped travellers.   [JC]

BOND, NELSON S(LADE) (1908-   ) US writer, rare-book dealer and philatelist. Although NSB has written little since 1952, his work remains fondly remembered; he was one of the most competent US writers of light and original SUPERNATURAL FICTION. Starting in 1937, he wrote for pulp and other MAGAZINES, especially WEIRD TALES, FANTASTIC ADVENTURES and *Blue Book*, where he produced a series of humorous – often Runyonesque – SLICK FANTASIES, ideally suited to the WWII years. NSB began with "Mr Mergenthwirker's Lobblies" (1937 *Scribner's Magazine*), about a man befriended by invisible creatures (◊ INVISIBLE COMPANIONS); this developed into a radio series, a story series in *Argosy* (1941-2) and a tv play, and formed the basis of his first collection, *Mr Mergenthwirker's Lobblies and Other Fantastic Tales* (coll **1946**). NSB delighted in stories where everyday folk create wonderful inventions either by luck or by mistake – as in the **Pat Pending** series in *Blue Book* (1942-8) – or are given miraculous powers by some ancient or enchanted object, as in "The Ring" (1943 *Blue Book* as "The Ring of Iscariot"), "Saint Mulligan" (1943 *Fantastic Adventures*), and other stories in *The Thirty-First of February* (coll 1949). In "The Bookshop" (1941 *Blue Book*) he created the collectors' dream of a bookshop (◊ SHOP) full of fictional BOOKS.

The best of NSB's later works, including the **Squaredeal Sam** series of TALL TALES, were all published in *Blue Book*, many being collected in *Nightmares and Daydreams* (coll **1968**).

A friend and correspondent of James Branch CABELL, NSB has compiled *James Branch Cabell: A Complete Bibliography* (**1974**).   [MA]
Other works: *Exiles of Time* (1940 *Blue Book*; **1949**); *Lancelot Biggs: Spaceman* (coll **1950**); *The Monster* (coll **1953** chap Australia); *No Time Like the Future* (coll **1954**); *Animal Farm: A Fable in Two Acts* (play **1964** chap) based on George ORWELL's novel.

BONDAGE A term of central importance to the understanding of the nature of STORY in the literatures of FANTASY when used to describe not (as in general parlance) an act of tying up but a state of being contained or trapped in a particular place, time, physical shape or moral condition. At first glance, a definition so broad might be assumed to apply not only to fantasy but to all narrative art. In almost any tale, for something to happen there must be something to happen *against*. There must be a given state or problem, an inertia, a resistance; and there must be some sort of process of change which acts upon that condition. The resistance could be called bondage; the process of change could be called Story.

In fantasy, however, this statement about the nature of story is true in ways which make the term "bondage" peculiarly appropriate. In mimetic tales the constraints against which stories are acted out may be no more (or less) than the author's understanding and presentation of the world as it is given, the world understood as a mortal coil. But in a fantasy the nature of the world is *not* a given. Even if most GENRE FANTASIES show little sign that their authors have made much creative use of this fact, it is still true that, in fantasy, the world – and the means within the story by which that world can be made to work and change – have at least theoretically been *chosen*. The consequence is a heightening of *all* the elements which in a mundane story would be understood as part of the consensual background. (This can also be seen as a demarcation between the full fantasy and the FANTASYLAND fantasy, because Fantasyland is likewise a consensual background whose details do not need to be spelled out: a wizard is a wizard is a wizard.) Humans die of old age in a mundane world; in a world of fantasy they die because, perhaps, the author has chosen not to give them IMMORTALITY. A storm in mundane London may be no more than weather; a pea-soup fog which derails a QUEST in a GASLIGHT ROMANCE set in LONDON is almost certainly meaningful (and may well be an actual agent in the tale). In fantasy, in other words, death and weather (and all sorts of other things) are not necessarily just part of the daily world; they may be forms of bondage.

In fantasy, then, death and weather – to stick with these examples – are peculiarly subject to interrogation and challenge. When they are easily defeated, a fantasy tale may be open to the (frequent) accusation that it is mere ESCAPISM. But when the world itself, or all the dramatic forms of bondage that can be found in fantasy, cannot glibly be transcended, then the "escape from prison" (as J.R.R. TOLKIEN regarded escapism in fantasy) can be understood as one of the movements of Story that we, as readers, respond to most deeply and most legitimately. Fantasy – more clearly than other forms of literature – seems well designed to satisfy this inherent human need for tales of escape (◊ FANTASY; METAMORPHOSIS; RECOGNITION; STORY).

There are reasons. Fantasy, as noted, tends to address bondage directly. And, because fantasy is a story-telling genre, it tends to treat bondage as a condition to escape from – what could be less story-like than a state of immobility, whatever its cause? Bondage is stasis. Stories *move*. They *tell* bondage away. The fully structured fantasy is a tale of escape. (Contrariwise, the triumph of bondage in fantasy is likely to be announced precisely through stories which do not end: the ARABIAN NIGHTMARE, in which nests of stories almost invariably fail to reach a conclusion, has as its main theme the deepening of bondage; and any POSTHUMOUS FANTASY – like Flann O'BRIEN's *The Third Policeman* (**1967**) or Gene WOLFE's *Peace* (**1975**) – whose protagonists fail to integrate their past lives will tend to convey that deepening of bondage through stories that stop before they end.)

Bondage/escape is such a pervasive pattern in fantasy that examples could be provided indefinitely. Almost any tale of METAMORPHOSIS, for instance, normally involves either an initial escape from the immobility of bondage (ARIEL liberated from the cloven pine in William SHAKESPEARE's *The Tempest* [performed *c*1611; **1623**]) or an entry into it (Lucius's TRANSFORMATION into an ASS in APULEIUS's *The Golden Ass* [*c***165**]), though only very rarely will the condition be permanent – significantly, Lucius's bondage in *The Golden Ass* is told in a kind of elaborated FRAME STORY; and much of the novel is taken up with stories which are told to completion in his hearing.

DOUBLES are in bondage to one another. MIRRORS transfix. Almost any fantasy QUEST can be understood to have liberation from bondage as its ultimate goal. Bondage may be personal (the protagonist may be bound into AMNESIA, may be under a SPELL which s/he must decipher, may be an UGLY DUCKLING in search of a true role, may be a PUPPET or some other form of constricted being who longs to be a REAL BOY, etc.) or it may be larger in scale (the LAND may have lost its communal memory, MAGIC may have disappeared to leave the world rigid and sere [◊ THINNING], the monarch of the land may have have become a FISHER KING, a DARK LORD may have desiccated the land through the bondage of his PARODY of just rule, etc.).

Indeed, the quest itself can be thought of a bondage to be transcended. In Nancy KRESS's *The Prince of Morning Bells* (**1981**) the quest is defined – through a succession of terms which could apply to all forms of fantasy bondage – as a form of "gyve" or shackle, which locks HEROES AND HEROINES into "Enchantment. Cloistering. Caesurea. Captivity. Arrestment. Often accompanied by bewitchment. You can't walk away from that. It holds you immobile. You turn into something you're not." [JC]

See also: ACCURSED WANDERER; OBSESSED SEEKER.

BONFIGLIOLI, KYRIL (1928-1985) UK editor. ◊ SCIENCE FANTASY (magazine).

BONTLY, THOMAS (1939-   ) US academic and author. His third novel, *Celestial Chess* (**1979**), concerns a scholar who investigates a 12th-century manuscript in Cambridge, embroiling himself eventually in SATANISM, a CURSE laid upon a colleague's family, and the fate of the SOUL of the manuscript's author, who entered into a PACT WITH THE DEVIL. [GF]

BOOKS Max BEERBOHM devoted an essay in his *And Even Now* (coll **1921**) to "Books Within Books", those fascinating volumes that exist only as titles and perhaps brief extracts within stories. Fantasy contains many – in Peter GREENAWAY's movie PROSPERO'S BOOKS (***1991***) they even become a more central "character" than PROSPERO himself.

James Branch CABELL presents, in *Beyond Life* (**1919**), a whole LIBRARY of such books (e.g., *The Complete Works of David Copperfield*), of real authors' unwritten projects (e.g., MILTON's verse *King Arthur*) and of actual books not as published but as their writers intended them. Neil GAIMAN's DREAM library in *Sandman #22* (**1990**) pays skewed homage to Cabell's notion with such imagined volumes as *The Man who Was October* by G.K. CHESTERTON. Nelson S. BOND's "The Bookshop" (1941) presents the bibliophile's dream: a magic SHOP full of fictional books. Nonexistent works of real authors are a frequent fictional MCGUFFIN: Umberto ECO's *The Name of the Rose* (**1980**; trans **1983**) uses the lost second book of Aristotle's *Poetics* in this way, and rediscovered SHAKESPEARE manuscripts often feature in crime fiction – as do lost gospels, especially ones written by Judas (confusingly, *The Judas Testament* [**1994**] by Daniel Easterman concerns the discovery of a manuscript written by CHRIST); or even the entire BIBLE, the intolerable "true" version of which is unearthed in Edward WHITTEMORE's *Sinai Tapestry* (**1979**). In *O 31!o! Peregrino* ["The 31st Pilgrim"] (**1993**) Rubens Teixeira Scavone (1925-   ) supplies an "undiscovered" tale from CHAUCER's *Canterbury Tales*.

Books naturally play a considerable role in fantasy as grimoires, but they sometimes function as magical devices in their own right. Books which exert a subtly malign influence over their readers are fairly common, the most famous being the verse playscript which gives Robert W. CHAMBERS's *The King in Yellow* (coll **1895**) its title; it reappears in more elaborate form in "More Light" (1970) by James BLISH, whose *Black Easter* (**1968**) discusses "real" grimoires and also features the adept Theron Ware's grim volume of diabolic PACTS. (Blish had earlier produced a neatly recomplicated version of the oft-told story of the book of biographies which absorbs its reader in "The Book of Your Life" [1955], and Jonathan CARROLL later employed a wholesale variant of the same theme in *The Land of Laughs* [**1980**].) Other examples include Tod ROBBINS's "For Art's Sake" (1920; exp as *The Master of Murder* 1933), Margaret IRWIN's "The Book" (1935) and Stephen Vincent BENÉT's "The Minister's Books" (1942). Chambers's stories helped to inspire H.P. LOVECRAFT to create the *Necronomicon*, which in turn inspired the members of his circle to create a series of similarly sinister tomes, various CTHULHU MYTHOS "titles" including the *Book of Eibon*, *Unaussprechlichen Kulten* by "von Junzt", *Cultes des Goules* by "the Comte d'Erlette", *De Vermis Mysteriis* by "Ludvig Prinn", the *Pnakotic Manuscripts* . . . A late invention in this line plays a leading role in Fritz LEIBER's *Our Lady of Darkness* (**1977**). A VAMPIRE book is featured in Michael HARRISON's "Where Thy Heart Is" (1926). Multi-volume sets lend themselves to PLOT-COUPON accumulation, as in Kenneth BULMER's *Kandar* (**1969**), where the fictional *Thaumalogicon*, *Ochre Scroll* and *Umbre Testament* each hold one-third of a vital spell. An endearingly polyphonic library of magical books plays a significant role in several of Terry PRATCHETT's **Discworld** novels: his *Necrotelecomnicon* or *Liber Paginarum*

Fulvarum gestures to both H.P. LOVECRAFT and to the *Yellow Pages*. Benign books which enhance their reader's lives are much rarer; the one in Henry KUTTNER's "Compliments of the Author" (1942) flatters only to deceive; and the volumes featured in Michael Kandel's *In Between Dragons* (1990) also have their problematic aspects.

In fantasy, opening a book may lead to unusual experiences. George MACDONALD's *Phantastes* (1858) has books that project one's viewpoint into the story; *The Book of Gramarye* in Susan COOPER's *The Dark is Rising* (1973) similarly offers engulfing multimedia tuition; the book within Michael ENDE's *The Neverending Story* (1979) is additionally a PORTAL into the land of Fantastica which it describes. In John GRANT's "Banedon's Telling" (in *The Tellings* * coll 1993) the protagonist reads a book that narrates the adventures he is currently, somewhere else, undergoing. Similarly, the real book and the imagined eponymous book within the real book interact upon protagonists in Charles DE LINT's *The Little Country* (1991); in Darrell SCHWEITZER's *The Mask of the Sorcerer* (1995), the fictional memoirs of Tannivar the Parricide shape the protagonist's life; in William Browning SPENCER's *Résumé With Monsters* (1995) the protagonist has written a Lovecraftian book called *The Despicable Quest*, and in his *Zod Wallop* (1995) the eponymous children's story designates the nature of various characters' experiences, shaping the outcome of the book for good or for ill, depending upon which draft of the original tale is dominant. The eponymous volume of Jorge Luis BORGES's "The Book of Sand" has infinitely many pages, so the chance of finding a given passage twice is infinitesimal. In *The* ADDAMS FAMILY (*1991*) books can behave like their titles, so that, say, opening a copy of *Gone With the Wind* (1936) brings a gale into the room.

Not all books are books. The calling-charm to ensnare an intelligent child in Diana Wynne JONES's *The Magicians of Caprona* (1980) manifests as a hypnotically readable book; a mysteriously missing corpse in Piers ANTHONY's **Xanth** sequence proves to have been topologically metamorphosed (◊ METAMORPHOSES) into book shape as *The Skeleton in the Closet*; one nameless pseudo-book in Gene WOLFE's *The Book of the New Sun* (1980-83) holds an angelic image of such power that its observer literally sweats blood (◊ FACE OF GLORY).

Books of record, often immutable, are frequently encountered: the book of Judgement, of FATE, of Destiny (e.g., in Gaiman's *Sandman*), or of the Norns (◊ FATES) – subject to a QUIBBLE in Cabell's *The Music from Behind the Moon* (1926) where, although no one "may alter any word", the hero rewrites history by inserting a decimal point. Lloyd ALEXANDER's *Book of Three* foretells the future, which by the end of his **Prydain** series has become history: the series itself. DEATH's biographies in Pratchett's **Discworld** continuously write themselves to record every person's own story.

Closely related are holy books and books of PROPHECY, regarded by believers as containing absolute truths which may be cryptically phrased (like the prophecies of NOSTRADAMUS). One such, to be studied only after ritual purification, hints at the way to William MORRIS's eponymous *The Well at the World's End* (1896); more profane is the infallible *The Nice and Accurate Prophecies of Agnes Nutter, Witch* in *Good Omens* (1990) by Neil GAIMAN and Terry PRATCHETT.

Fictional-historical sourcebooks are endemic, and often

recount the basic STORY that the narrative (and its protagonists) are in the process of uncovering; a fictional book is in this sense a natural device in many fantasy tales, where the act of RECOGNITION of that which has been lost (and is now found) is so central: thus Cabell grounds *Jurgen* (1919) in the invented medieval epic cycle *La Haulte Histoire de Jurgen*. J.R.R. TOLKIEN's *The Lord of the Rings* (1954-5) purports to be based on *The Red Book of Westmarch*, a personal record by its characters Bilbo and Frodo Baggins. William GOLDMAN's *The Princess Bride* (1973) purports to abridge a book called *The Princess Bride* by S. Morgenstern, a similarly shortened form of which was ostensibly read aloud to Goldman as a child. *The Book of True and Cruel* is periodically examined by protagonists of Tom DE HAVEN's **Chronicles of the King's Tramp** sequence (1990-92), and tells them the meaning of the events that are unfolding around them.

Works of fictional authors may significantly reflect or interact with their surrounding narratives. The character Sister Theodora, writing the book which is Italo CALVINO's *The Non-Existent Knight* (1959; trans 1962), finally reveals herself to be the AMAZON knight Bradamante (◊ Edmund SPENSER) from the story she tells. In Flann O'BRIEN's *The Third Policeman* (1967) the eccentric natural philosopher De Selby – who never appears – dominates much of the text via footnotes, quotations and his role as the narrator's MAGGOT; Vladimir NABOKOV used a similar device in *Pale Fire* (1962). Other examples are: the shockingly deceptive diary in Christopher PRIEST's *The Affirmation* (1981); *The Book of the Wonders of Urth and Sky* in Gene Wolfe's *The Book of the New Sun* (1980-83), providing inset stories which highlight the remoteness of this FAR FUTURE by showing the absorption of such fictions as Rudyard KIPLING's MOWGLI into anonymous, syncretic myths; and Fellowes Kraft's historical fantasy about Giordano Bruno (1548-1600), which complexly nests within or contains John CROWLEY's *Aegypt* (1987)

Books, at least in writers' minds, are all too natural a subject for books. [DRL/BS/JC]

BOOTHBY, GUY (NEWELL) (1867-1905) Australian-born writer, in the UK from 1894. The **Dr Nikola** series – *A Bid for Fortune* (1895; rev vt *Dr Nikola's Vendetta* 1908 US), *Doctor Nikola* (1896), *Dr Nikola's Experiment* (1899) and "*Farewell, Nikola*" (1901) – are thrillers in which the sinister, mysteriously gifted occultist Nikola searches for the secret of IMMORTALITY, giving short shrift to those who get in his way; he also plays a peripheral role in *The Lust of Hate* (1898). *Pharos the Egyptian: A Romance* (1899) is a more exotic thriller whose similarly enigmatic central character seemingly brings off the difficult trick of achieving immortality while also existing as a MUMMY. "A Professor of Egyptology" in *The Lady of the Island* (coll 1903) is a REINCARNATION romance; the other weird tales in that volume, like those in *Bushigrams* (coll 1897) and *Uncle Joe's Legacy and Other Stories* (coll 1902), are trivial. GB very obviously made his novel plots up as he went along – which is why the most interesting of them, *The Curse of the Snake* (1902), concludes with a woefully inadequate explanation of its marvellously creepy opening sequence. [BS]

Other works: *The Phantom Stockman* (1897 chap).

BORDERLAND ◊ MAGAZINES.

BORDERLANDS The difference between borderlands and WATER MARGINS is a question of edges. A water margin exists at the edge of, or may fully surround, a central POLDER or LAND or empire; but it has no further edge.

Characters cannot pass through water margins into other regions: if they remain close to the edge they may re-enter the central territory, but if they stray too deep they simply fade from view. A borderland, on the other hand, always boasts at least two edges, and generally serves as a marker, resting place or toll-gate between two differing kinds of REALITY. A borderland like that featured in the **Borderland** sequence of SHARED-WORLD anthologies ed Mark Alan Arnold and Terri WINDLING will be a hotbed of CROSS-HATCH activities, and may readily serve as a TEMPLATE venue for various sorts of EPIC FANTASY. [JC]

BORGES, JORGE LUIS (1899-1986) Argentine poet, librarian, essayist and short-story writer, a central figure in the explosive growth of Latin American literature; he was also central – since the 1960s, when he began to appear widely in translation – to 20th-century literature as a whole. His importance to fantasy is seminal, though he wrote no novels and relatively few of his short stories settle themselves with any security of tenure in anything remotely resembling the SECONDARY WORLDS or CROSSHATCHES common to modern fantasy. His huge influence on writers of fantasy and sf lies, like Gene WOLFE's, in his deeply inventive and suggestive manipulation of certain SYMBOLS, which include the fictional BOOK, the DOUBLE, the protagonist who is the DREAM of another, the LIBRARY, the IMAGINARY LAND, the LABYRINTH, the MIRROR . . . But his profound sense of BELATEDNESS – his sense that everything he wrote was a plagiarism or PARODY of earlier texts, and that no STORY could be told again, merely aped – led him never to tell any tale in which the entrapments and labyrinthine illusions of REALITY are genuinely penetrated by the HERO. In JLB's work, there are no genuine PORTALS out of the maze of the world. In this he reflects a deeply gnostic sense (◊ GNOSTIC FANTASY) that to exist in the world is to have fallen from the Pleroma, or Primordial Being. All JLB's symbols represent BONDAGE, for not one of them can truly guide any protagonist out of *here*.

JLB was a precocious child, publishing scattered poems and stories by the teens of the century. His first published book – *Fervor de Buenos Aires* ["Passion for Buenos Aires"] (coll **1923**) – was poetry; his last containing original material – *La cifra* ["The Cipher"] (coll **1981**) – was likewise. In his prose pieces the line between fiction and nonfiction early began to dissolve, and it is sometimes difficult without extra-textual perspectives to determine whether an essay (complete with scholarly apparatus) is only *seeming* to describe a real book, a real person, a real country, or whatever. But the essays assembled in *Inquisiciones* ["Inquisitions"] (coll **1925**) and *Otras inquisiciones (1937-1952)* (coll **1952**; exp 1960; trans Ruth L.C. Simms as *Other Inquisitions 1937-1952* **1964** US) are impressively acute studies, several of which concern the English-language writers – like G.K. CHESTERTON and H.G. WELLS – for whom JLB felt the greatest admiration, though he could not imitate the first's faith or the latter's attempts to bestride the world.

Historia universal de la infamia (coll **1935**; rev 1954; trans Norman Thomas di Giovanni as *A Universal History of Infamy* **1972** US) hovers between "genuine" nonfiction and essays on figures like BILLY THE KID – "El asesino desinteresado Bill Harrigan" (here trans as "The Disinterested Killer Bill Harrigan") – which imperiously mythologize their subjects (◊ MAGIC REALISM). The book also contains "Hombre de la esquina rosada" (1933; here trans as "Streetcorner Man"), JLB's first adult tale. His greatest fic-

tions followed soon, eight of them being assembled in *El jardin de senderos que se bifurcan* ["The Garden of Forking Paths"] (coll **1942**), which became Part One of *Ficciones* (coll **1944**; exp 1956; trans Anthony Kerrigan *et al* **1962** US/UK). The 1944 edition of *Ficciones* (the 1956 edition is little changed) is one of the most important books of short stories published this century. Part One contains *Tlön, Uqbar, Orbis Tertius* (1941 *Sur*; trans Alastair Reed; new trans James E. Irby **1983** chap Canada), "El acercamiento a Almotásim" (1935; trans Kerrigan as "The Approach to al-Mu'tasim"), "Pierre Menard, autor del Quixote" (1939; trans Anthony Bonner as "Pierre Menard, Author of Don Quixote"), "Las ruinas circulares" (1940; trans Bonner as "The Circular Ruins"), "La Lotería en Babilonia" (1941; trans Kerrigan as "The Babylon Lottery"), "Examen de la obra de Herbert Quain" (1941; trans Kerrigan as "An Examination of the Work of Herbert Quain"), "La biblioteca de Babel" (1941; trans Kerrigan as "The Library of Babel") and "El jardin de senderos que se bifurcan" (1941; trans Helen Temple and Ruthven Todd as "The Garden of Forking Paths"). In addition to some essays, Part Two (augmented in 1956) contains "Funes el memorioso" (1941; trans Kerrigan as "Funes the Memorious"), "La forma de la espada" (1941; trans Kerrigan as "The Form of the Sword"), "La muerte y la brújula" (1941; trans Kerrigan as "Death and the Compass") and "El Sur" (1953; trans Kerrigan as "The South").

Tlön, Uqbar, Orbis Tertius, which may be JLB's most famous single story, adduces apocryphal evidence from Johann Valentin Andreae (1586-1654; ◊ ROSICRUCIANISM) and others (including Adolfo BIOY CASARES, who is purported to discover a mysterious encyclopedia) to argue for the existence of Uqbar, an IMAGINARY LAND (somewhere towards the western side of Asia) whose literature in turn refers only to further imaginary lands like Tlön. Later, volume 11 of a second encyclopedia is discovered: *A First Encyclopedia of Tlön*, 1001 pages long (◊ ARABIAN FANTASY), describes an entire planet called Tlön (Uqbar is never mentioned again) for whose inhabitants ideas – metaphysics is a branch of fantastic literature – create REALITY, "which is a kind of amazement". Meanwhile, it turns out that a society of SECRET MASTERS on Earth not only generated the entire *Encyclopedia* (through which Tlön has been created) but is now preparing a new edition, written in a language of Tlön and tentatively called *Orbis Tertius*, whose writers (or "demiurges") will by this means transform Earth (◊ FANTASIES OF HISTORY) into Tlön, which is "a labyrinth plotted by men, a labyrinth destined to be deciphered by men". For there is no God; we are here, in a circular ruin, dreaming one another.

The other tales in the volume have a similar self-reflective density and air of entrapment. "Pierre Menard, Author of Don Quixote" describes the creation of a text identical to Miguel de CERVANTES's but which (being written at a different point, even by a man totally immersed in his version of Cervantes's life and world) cannot be the same book. In "The Circular Ruins", a WIZARD dreams a man into REALITY only to discover that he, too, is being dreamed. "The Library of Babel" describes a LIBRARY which contains all possible combinations of words, but which is beyond human access. "Death and the Compass" depicts the ratiocination of a detective as a predetermined threading of a LABYRINTH at whose heart awaits his DOUBLE and inevitable murderer. "Funes the Memorious" describes a man who, incapable of forgetting anything, can experience no new thing.

The stories assembled in JLB's next collection, *El Aleph* ["The Aleph"] (coll **1949**; exp 1952), have been translated in various volumes, including *Labyrinths: Selected Stories and Other Writings* (coll trans various **1962** US; rev 1964) ed Donald A. Yates and James E. Irby, which also includes much of *Ficciones*, and *The Aleph and Other Stories 1933-1969* (coll trans JLB and Norman Thomas di Giovanni **1970** US). "The Aleph" and "El Zahir" (here trans as "The Zahir") both deal with MAGIC objects that contain the whole Universe within them (◊ LITTLE BIG). "La casa de Asterión" (1947; here trans as "The House of Asterion") tells the story of the MINOTAUR from his own viewpoint. "El immortal" (1949; here trans as "The Immortal") complexly unveils a figure who may be HOMER as well as other writers: a centurion who is on a QUEST for the labyrinthine "City of the Immortals" (◊ CITY) which "contaminates" the future, and who is encased in IMMORTALITY, a terrible form of BONDAGE.

Later volumes – like *El hacedor* (coll **1960**; trans Mildred Boyer and Harold Morland as *Dreamtigers* 1964 US), *El Informe de Brodie* (coll **1970**; trans Norman Thomas di Giovanni as *Doctor Brodie's Report* 1972 US) and *El Libro de Arena* (coll **1975**; trans Norman Thomas di Giovanni as *The Book of Sand* 1977 US) – tend to reiterate, sometimes in a more "realistic" vein, the pressing, subtle, pessimistic gnosticism of earlier work. A new note is perhaps struck in "Utopía de un hombre que está cansado" (trans in *The Book of Sand* as "Utopia of a Tired Man"), in which an old man (like JLB then) TIMESLIPS into a desolate future where the will to make up realities has somehow desiccated, and then returns to the 20th century with a painting so faint it can hardly be deciphered.

With Adolfo Bioy Casares and Silvina Ocampo, JLB edited *Antología de la literatura fantástica* (anth **1940**; rev 1965; again rev 1976; trans as *The Book of Fantasy* **1988** UK), whose 81 stories and fragments demonstrate the cosmopolitanism of the three editors, who were at the centre of Argentine literary life for many years.

JLB himself spoke to the century.   [JC]

Other works (selected): *Seis problemas para don Isidro Parodi* (coll **1942**; trans Norman Thomas di Giovanni as *Six Problems for Don Isidro Parodi* **1981** US) with Bioy Casares; *Manual de zoología fantástica* (**1957**; exp vt *El libro de los seres imaginarios* 1967; trans di Giovanni of exp, plus further revs, as *The Book of Imaginary Beings* **1969** US) (◊ BESTIARIES); *Antología personal* (coll **1961**; trans Anthony Kerrigan as *A Personal Anthology* **1967** US); *Crónicas de Bustos Domecq* (coll **1967**; trans di Giovanni as *Chronicles of Bustos Domecq* **1976**) with Bioy Casares; *Borges: A Reader* (trans coll **1981** US); *Atlas* (coll **1984**; trans Kerrigan **1985** US).

BORN OF FIRE UK movie (tvm **1986**). Dehlavi/Film Four International. **Pr** Jamil Dehlavi, Thérése Pickard. **Dir** Dehlavi. **Spfx** Special Effects Universal Ltd. **Mufx** Sula Loizou. **Screenplay** Raficq Abdulla, Dehlavi. **Starring** Suzan Crowley (The Woman), Peter Firth (Paul Bergson), Stefan Kalipha (Bilal), Oh-Tee (Master Musician), Nabil Shaban (Silent One). 90 mins. Colour.

Probably the most conceptually complex fantasy ever filmed, *BOF* is concerned with the legend of the djinn Iblis (◊ GENIES), the only ANGEL who, born of fire, defied Allah's instruction to prostrate himself before Adam, born of clay, and was thus cast out. Ethereal MUSIC is heard by both a flautist (Bergson) during a recital and an astronomer (The

Woman) during a solar eclipse; after they meet she is plagued by apparitions and sees her DOPPELGÄNGER drowned in the bath. Bergson travels to Cappadocia, in Turkey, where years ago his father went to consult the Master Musician, an experience that killed the music in him. The Woman follows him . . . and there ensues the weaving of an intricate tapestry of SHAPESHIFTING and TIMESLIPS that defies summarization; dialogue, performances, cinematography, music and the spectacular geomorphology of the setting are further threads woven so that they become meaningless in isolation from each other. *BOF* is less a linear tale than a single event in which notions of causality and the Arrow of TIME must be abandoned. The power of the movie is undeniable.   [JG]

BORON, ROBERT DE (?   -1212) French chronicler. ◊ ARTHUR; GRAIL; MERLIN; PERCEVAL.

BORROWERS, THE US movie (*1973* tvm). Walt DeFaria Productions/Charles M. Schultz Creative Associates/Foote, Cone and Belding Productions/20th Century-Fox. **Pr** Walt DeFaria, Warren L. Lockhart. **Exec pr** Duane C. Bogie. **Dir** Walter C. Miller. **Screenplay** Jay Presson Allen. **Based on** *The Borrowers* (1952) by Mary NORTON. **Starring** Eddie Albert (Pod), Judith Anderson (Great-Aunt Sophy), Tammy Grimes, Barnard Hughes, Dennis Larson (Boy), Karen Pearson, Beatrice Straight (movie lacks proper credits). 78 mins. Colour.

The Boy comes to convalesce in the country mansion of his hard-drinking, bedridden Great-Aunt Sophy. She is aware, although she thinks they are ILLUSIONS born from her decanter, that the house has nonhuman occupants; her handyman and housekeeper, Mr and the spiteful Mrs Crampfurl, are not. Beneath the grandfather clock lives a family of tiny people, the Clocks (Homily, husband Pod and daughter Arrietty), who survive by "borrowing" small objects. One night Pod is seen by the Boy, who assumes he is a FAIRY; the Boy soon finds other evidence of the Clocks' presence and then discovers and befriends Arrietty. One night the Crampfurls, too, find the Clocks' home, and set a ferret after them; but, aided by the Boy, they escape to join the Hendready family of Borrowers living in a nearby field.

This is a somewhat stolid adaptation, resited to New England, of Norton's definitive WAINSCOT story. Budgetary considerations are evident in the script, with some events being described rather than portrayed, and the performances – the children emphatically excepted – are a little by-numbers, but the spfx are surprisingly convincing and the whole has a pleasing ambience.   [JG]

BOSCH, HIERONYMUS (*c*1450-1516) Dutch painter who may have been born in the town of 's Hertogenbosch, where he lived most of his life and died; hence perhaps his surname. He is an extremely important figure in art history, and a TAPROOT source for many painters and illustrators of fantasy material. The nightmarish exorbitance of his work, the surreal intensity of the LANDSCAPES of HELL which he appears to create out of whole cloth, all tend to make HB into a figure who seems inexplicable to later generations, but, while there is no questioning the intensity or the reach of his imagination, his great paintings were created in accordance with the precepts of ALLEGORY as understood by Christian artists at the end of the Middle Ages. They were perhaps heterodox in execution, but their intended meanings were at the time both manifest and unimpeachable.

HB's first extant painting probably dates from *c*1475. In possible chronological order his works of greatest fantasy interest include "The Seven Deadly Sins", "The Haywain", "The Last Judgement", "The World After the Flood"/"The Wicked World", "The Ship of Fools", "The Temptation of Saint Anthony" and "The Garden of Earthly Delights", the last being the most influential of all. Peter S. BEAGLE's nonfiction study, *The Garden of Earthly Delights* (**1981**), is an intense response to this panoramic triptych, whose CARNIVAL implications seem endless. Ian WATSON's quasi-sf *The Gardens of Delight* (**1980**) is probably the most important fictional response; Tom HOLT's romp *Faust Among Equals* (**1994**) sees HB employed as a landscape architect employed to redesign HELL as a theme park along the lines of this picture.

The picture's most famous single image is probably that of the "human tree" which dominates the right panel: the legs are tree-trunks, each bedded in a skiff afloat in obscure waters; the torso is hollow, contains humans and other figures; the head faces us, ruefully, from underneath a kind of vast plate/hat, around which a bevy of unhuman figures dance to music emitted by a great reddish bladder. Interpretations of the human TREE are exceedingly numerous: in fantasy terms alone, the interaction between BONDAGE and METAMORPHOSIS in the overall image is extremely powerful.

For the 20th century HB is inexhaustible, like Pieter BRUEGEL after him. Especially in the darkness of the dawn before the EUCATASTROPHE, his panoramas underlie the LANDSCAPES of fantasy. [JC]

BOSTON, LUCY M(ARIA WOOD) (1892-1990) UK writer for children and adults. Almost all her significant work is set in and about a single house (in reality, the Manor House at Hemingford Grey, north of Cambridge, UK), called Yew Hall or Green Knowe. The real-life house and its sheltering grounds can be regarded as a POLDER. Its use as a focusing agent for LMB's fiction – much of it involving TIMESLIP episodes whose participants are variously connected to the dwelling – helps give that fiction a texture of deeply imagined beingness, a sense that all the times of the house are embraced simultaneously within its arms.

LMB's first novel, *Yew Hall* (**1954**), for adults, contains no fantasy elements, though saturated with a sense of potential not easily described in terms of the mimetic tradition. But the **Green Knowe** sequence – *The Children of Green Knowe* (**1954**), *The Chimneys of Green Knowe* (**1958**; vt *Treasure of Green Knowe* 1958 US), *The Castle of Yew* (**1958**), *The River at Green Knowe* (**1959**), *A Stranger at Green Knowe* (**1961**), *An Enemy at Green Knowe* (**1964**) and *The Stones of Green Knowe* (**1976**) – is fantasy. The first two volumes introduce a young relative of Mrs Olknow (clearly a version of LMB herself) to the house of Green Knowe, and to the various GHOSTS who conjoin past and present within the sanction of the house. Beyond the walls – as most of the remaining volumes demonstrate – a more threatening world constantly impinges, either in the form of modern life itself (Hemingford Grey has not escaped unscathed from "development"), or through malign activities, like those of the WITCH who features in *An Enemy at Green Knowe*. In the final volume, a 12th-century inhabitant of the house is enabled, by the eponymous MAGIC stones, to trace down the centuries the fate of Green Knowe and of those who live there; unusually for a book written ostensibly for children, this concluding tale is dedicated almost exclusively to visions of THINNING.

LMB's refusal of the modern world deeply graces her work as a whole, and seems anything but sentimental. It may be that the unusually close relationship between REALITY and fantasy vision, which LMB conveyed with such exactitude, enabled her to couch her refusal in such unanswerable terms. [JC]

Other works: *Persephone* (**1969**; vt *Strongholds* 1969 US), associational; *The Horned Man, or Whom Will You Send to Fetch Her Away?* (**1970**), a play; *Memory in a House* (**1973**) and *Perverse and Foolish: A Memoir of Childhood and Youth* (**1979**), both autobiographies.

Other works (for younger children): *The Sea Egg* (**1967**); *The House that Grew* (**1969**); *Nothing Said* (**1971**); *The Guardians of the House* (**1974**); *The Fossil Snake* (**1975**).

BOSTON, NOEL ◊ Noel BERTRAM.

BOTTO, JÁN (1829-1881) Slovak poet. ◊ SLOVAKIA.

BOUCHER, ANTHONY Pseudonym of US editor and writer William Anthony Parker White (1911-1968), who, after a piece of juvenilia, "Ye Goode Olde Ghoste Storie" (1927 *WT*), began his career proper with a number of comic fantasies in UNKNOWN, including "Snulbug" (1941), "The Compleat Werewolf" (1942) and "Sriberdegibit" (1943). The first and third are reprinted in *Far and Away* (coll **1955**); the first and second appear in *The Compleat Werewolf and Other Stories of Fantasy and Science Fiction* (coll **1969**). The latter collection includes also the moralistic "We Print the Truth" (1943), perhaps AB's finest work. He wrote some notable sf and detective fiction, but his contribution to the fantasy genre was largely restricted to his service as co-editor (1949-54) and then sole editor (1954-58) of *The* MAGAZINE OF FANTASY & SCIENCE FICTION. He edited an annual selection of stories drawn from the magazine: *The Best from Fantasy and Science Fiction* (anth **1952**), *The Best from Fantasy and Science Fiction, Second Series* (anth **1953**) and *Third Series* (anth **1954**), both with J. Francis McComas (1911-1978), his co-editor on *F&SF*, plus, solo, *Fourth Series* (anth **1955**), *Fifth Series* (anth **1956**), *Sixth Series* (anth **1957**), *Seventh Series* (anth **1958**) and *Eighth Series* (anth **1959**). [BS]

BOXING HELENA US movie (*1993*). Main Line. **Pr** Philippe Caland, Carl Mazzocone. **Exec pr** James R. Schaeffer, Larry Sugar. **Dir** Jennifer Chambers Lynch. **Spfx** Bill "Splat" Johnson (prosthetics), Bob Shelley's Special Effects. **Screenplay** Lynch. **Based on** story by Caland. **Starring** Matt Berry (Young Nick Cavanaugh), Betsy Clark (Anne Garrett), Sherilyn Fenn (Helena), Bill Paxton (Ray O'Malley), Meg Register (Marion Cavanaugh), Julian Sands (Dr Nick Cavanaugh), Nicolette Scorsese (Fantasy Lover/ Nurse). 101 mins. Colour.

Cavanaugh, a brilliant surgeon obsessed by the promiscuous Helena, is tormented by his own guilty sexuality. When a car accident mangles her legs he conducts an emergency amputation at his home and secretes her there, then abandons his job to worship her. When Helena eventually strikes out at him he amputates also her arms, so that she is now merely a head and torso kept in a display box. But then, despite herself, she begins to educate him about SEX, about the sexual needs of women; through him she is able herself orgasmically to experience vicarious sex. By the time her macho ex-lover O'Malley tracks her down, her allegiances lie with Cavanaugh – though this does not stop O'Malley brutally beating him, to the point that . . . immediately after the original car accident Cavanaugh summoned an ambulance and shepherded her through the operation to save her life. The "boxing" of Helena has been either an instantaneous DREAM or an alternate past.

BH has been the victim of its own sensationalism – Kim Basinger celebratedly abandoned the title role because of the excessive nudity – and this has obscured the fact that this is a fascinating movie. Throughout, REALITY is aslide and untrustworthy, each moment being proffered conditionally and the integration of the whole being moulded by Cavanaugh's and our own PERCEPTIONS. The copious sex is viewed analytically and yet not entirely dispassionately; it is as if we, the audience, were like Helena boxed, immobile, distanced, lacking the necessary organs and restricted to voyeurism, yet not totally dissociated from the carnal activities. The influence of director Lynch's father, David, is obvious, yet her view is clearer and considerably less self-indulgent; it is tempting to say that *BH* is the best David LYNCH film to date, although that would undervalue Jennifer Lynch's own originality of vision. [JG]

BOWEN, MARJORIE Best-known pseudonym of UK novelist and biographer Gabrielle Margaret Campbell (1886-1952), who became a prolific writer from 1906 onwards in order to support her extravagant mother and sister and later her growing family (from two marriages). She produced 156 books under many pseudonyms, including George R. Preedy, Joseph Shearing, John Winch and Robert Paye. The speed of production meant that some of her works were slight, but this should not be regarded as detracting from the atmosphere or originality of her best. She drew heavily upon reported mysteries and historical events, as well as upon her own childhood nightmares and later experiences when she lived in a HAUNTED DWELLING in London. An early success was *Black Magic* (**1909**), a rather shallow novel exploring the influence of SATANISM on the Papacy in the Middle Ages (◊ FANTASIES OF HISTORY). Her best work comprises her GHOST STORIES, all of which have a brooding sense of EVIL, even when the GHOSTS themselves are harmless. Most of these, many with historical settings or PLOT DEVICES, are assembled in *The Last Bouquet* (coll **1933**), *The Bishop of Hell* (coll **1949**) and *Kecksies* (coll **1976** US), this last from ARKHAM HOUSE. Others are scattered through the collections listed below or tucked away in old magazines, to which she was a prolific contributor. Her best novella-length ghost story is *The Devil Snar'd* (**1932** dos) as Preedy; in more traditional vein is *The Fetch* (**1942** as Shearing; vt *The Spectral Bride* 1942 US).

MB edited two anthologies of supernatural HORROR, *Great Tales of Horror* (anth **1933**) – adapted from *The Great Weird Stories* (anth **1929** US) ed Arthur Neale – and *More Great Tales of Horror* (anth **1935**), to which she contributed several translations of early Gothic horrors. [MA]

Other works: *Curious Happenings* (coll **1917**); *The Haunted Vintage* (**1921**); *Dark Ann and Other Stories* (coll **1927**); *"Five Winds"* (**1927**); *The Knot Garden* (coll **1933**) as Preedy; *Julia Roseing Rave* (**1933**) as Paye; *Orange Blossoms* (coll **1938**) as Shearing.

Further reading: *The Debate Continues* (**1939**), autobiography, as Margaret Campbell; "Mistress of the Macabre" by Richard DALBY in *Fantasy Macabre* #7 (1985).

BOWKETT, STEPHEN (1953-) UK writer who has also published as Louis P. Garou and has written HORROR novels as Ben Leech; these latter include *The Community* (**1993**), *The Bidden* (**1994**) and *A Rare Breed* (**1996**). As SB he has concentrated on YA work, in various genres. His first novel, *Spellbinder* (**1985** chap), plays with MAGIC and the OCCULT; in *Gameplayers* (**1986**), which is TECHNOFANTASY,

a boy becomes implicated in a fantasy role-playing GAME. Other novels, like *Dualists* (**1987**) and *Frontiersville High* (coll of linked stories **1990**), are sf. [JC]

Other work: *Catch and Other Stories* (coll **1988**), horror and sf.

BOYER, ELIZABETH H. (?   -   ) US writer known almost exclusively for a succession of novels, some only loosely linked and some making up a coherent self-contained sequence, all set in the FANTASYLAND world of **Alfar**, a venue dominated by two categories of ELVES, the dark and the light: *The Sword and the Satchel* (**1980**), *The Elves and the Otterskin* (**1981**), *The Thrall and the Dragon's Heart* (**1982**), *The Wizard and the Warlord* (**1983**), plus the **Wizard's War** sequence – *The Troll's Grindstone* (**1986**), *The Curse of Slagfid* (**1989**), *The Dragon's Carbuncle* (**1990**) and *The Lord of Chaos* (**1991**) – and the **Skyla** sequence – *The Clan of the Warlord* (**1992**), *The Black Lynx* (**1993**) and *Keeper of Cats* (**1995**). Her storylines are not unconventional: young HEROES AND HEROINES undertake QUESTS, generally with COMPANIONS and in order to find themselves and/or to counter the machinations of evil WIZARDS. The range of action seldom extends beyond the normal limits of HEROIC FANTASY, though the **Wizard's War** sequence does climax in a conflict which, if the forces of GOOD are defeated, will result in a kind of FIMBULWINTER. EHB is not normally, however, that ambitious. [JC]

BOYETT, STEPHEN R. (1960-) US writer whose first novel, *Ariel* (**1983**), is a fantasy set – unusually – in a version of our world which experienced a CROSSHATCH incursion from FAERIE before the tale began, generating a Balkanized LANDSCAPE similar to that more often found in post-HOLOCAUST sf. The eponymous UNICORN and his human COMPANION undergo various experiences together which culminate in a transfigured NEW YORK, after which Ariel's friend falls in love and loses both his VIRGINITY and his ability to converse with unicorns.

SRB's second novel, *The Architect of Sleep* (**1986**), can easily be read as sf: the hero plunges through a fantasy PORTAL into an ALTERNATE-WORLD USA, described in the language of RATIONALIZED FANTASY, occupied by sentient, human-sized raccoons whose leaders – known as Architects of Sleep – are capable of precognitive DREAMS. Though engrossingly told, the tale ends with its plot strands only beginning to be knitted, leading to the disappointed assumption that a series may have been aborted; no further volumes had appeared by the end of 1995. His third novel, *The Gnole* (**1991** UK) with Alan ALDRIDGE, is an ambitiously illustrated ecological fable in which the planet is seen to be at risk. [JC]

BOZIC, STRETAN (1932-) Serbian-born writer who later wrote as B. WONGAR.

BRACKETT, LEIGH (DOUGLASS) (1915-1978) US writer, married to sf writer Edmond Hamilton (1904-1977) and herself an influential author of space opera and PLANETARY ROMANCE. Her work was strongly influenced by Edgar Rice BURROUGHS and, to a lesser extent, Robert E. HOWARD, though the development of SWORD AND SORCERY as a subgenre is as attributable to LB as to Howard. Her most fantastic works feature the character of **Eric John Stark**, whose adventures, originally set on Mars, were published in the pulp MAGAZINE *Planet Stories*, starting with *The Secret of Sinharat* (1949 as "Queen of the Martian Catacombs"; exp **1964** dos) and *People of the Talisman* (1951 as "Black Amazon of Mars"; exp **1964** dos), both apparently expanded for book form by Hamilton. They were later assembled as *Eric John Stark: Outlaw of Mars* (omni **1982**); by then LB's protagonist was an anachronistic though still vibrant figure. The **Stark** stories were revived in the 1970s, now set on the distant world of Skaith, in *The Ginger Star* (**1974**), *The Hounds of Skaith* (**1974**) and *The Reavers of Skaith* (**1976**), all assembled as *The Book of Skaith* (omni **1976**). Perhaps her most effective Martian vision, however, is in *The Sword of Rhiannon* (1949 *Thrilling Wonder Stories* as "Sea-Kings of Mars"; **1953** dos) and its allied story, "Sorcerer of Rhiannon" (1942 *Astounding*), which transfers some of the DYING-EARTH imagery also portrayed in works by Clark Ashton SMITH, Jack VANCE and Michael MOOR-COCK to a decadent Mars. LB was one of the most influential women writers for the pulps (along with C.L. MOORE), her work being an inspiration to Marion Zimmer BRADLEY, Lin CARTER, Jo CLAYTON, Moorcock and many others. Her movie credits include co-scripting *The Empire Strikes Back* (**1980**). [MA]

BRADBURY, EDWARD P. ◊ Michael MOORCOCK.

BRADBURY, RAY(MOND DOUGLAS) (1920-   ) US writer. RB discovered sf through the magazine *Amazing Stories* in 1928, and nurtured a childhood interest in **Buck Rogers** comic strips and the novels of Edgar Rice BUR-ROUGHS, but his imagination was also strongly shaped by his rural Midwest upbringing, nostalgic reflections of which colour much of his writing. He moved with his family to Los Angeles in 1934, and became involved in sf fandom in 1937. By 1938, he was publishing in fan magazines and the following year editing his own short-lived periodical, *Future Fantasia*. Henry KUTTNER, Leigh BRACKETT, Robert A. HEINLEIN, Hannes BOK and other members of the California sf community became friends and mentors.

RB's first professional sale, "Pendulum" (1941), written with Henry Hasse (1913–1977), appeared in *Super Science Stories*. It was followed by "The Candle" (1941), the first of 25 sales to WEIRD TALES, where he honed his style 1941-8 and emerged as the most distinguished talent in the MAGA-ZINE during its declining years. Evocative, poetic and suffused with youthful wonder, RB's tales broke with pulp conventions in their style and approach to the fantastic. Some, such as "The Scythe" (1943) and "Skeleton" (1945), refurbish Gothic clichés (◊ GOTHIC FANTASY) for use in modern situations. Others, including "The Lake" (1944), use their horrific elements to evoke sympathy and longing. A number are macabre with only hints of the supernatural: "The Jar" (1944), "Reunion" (1944) and "The Night" (1946) deploy images of loss to study the emotional lives of their characters. The majority were collected in his first book, *Dark Carnival* (coll **1947**; cut 1948 UK; cut vt *The Small Assassin* 1962 UK), where they mesh to form a small-town landscape in which the magic possibilities of ordinary life and the banality of the fantastic are indistinguishable from one another. RB's depiction of fantasy as an inextrica-ble element of daily life, which anticipated the contemporary DARK FANTASY movement, became more pro-nounced when the contents of *Dark Carnival* were modified for publication as *The October Country* (**1955**), with more than half the *WT* selections dropped in favour of stories from the slick magazines featuring grotesques and normal characters caught up in the dark side of human experience (◊ SLICK FANTASY). RB approached the same idea from an opposite tack in a contemporaneously written quartet of stories that domesticate and demystify the macabre: "The Traveller" (1946), "Homecoming" (1946), "Uncle Einar"

(1947) and "The April Witch" (1952), all featuring an extended family of VAMPIRES, WEREWOLVES and other supernatural beings who display emotional needs and moti-vations no different from those of mortals, with whom they secretly coexist.

In the mid-1940s RB expanded his creative reach through contributions to a wide variety of markets. Under his own name and the pseudonym D.R. Banat he published a hand-ful of crime stories, many concocted from bizarre and macabre premises, later collected in *A Memory of Murder* (**1984**). He also appeared with increasing frequency in *The American Mercury*, *Mademoiselle*, *The Saturday Evening Post* and other mainstream magazines. It was, though, in the sf pulps, which published the bulk of his stories written 1946-50, that he achieved his greatest renown, although these stories do not conform the prevailing attitudes and interests of Golden Age sf and are more fantasy than sf.

The Martian Chronicles (coll of linked stories **1950**; with "Usher II" cut and "The Fire Balloons" added, rev vt *The Silver Locusts* 1951 UK; with "The Wilderness" added as well, rev 1953 UK), which splices together a number of self-contained stories from this period concerned with Earth's colonization and eventual abandonment of Mars, is typical of Bradbury's use of sf tropes in the service of the same ideas that inform his fantasy. The planet's habitable terrain is not scientific, and the Earth civilization that settles it is locked into a mid-20th-century middle-American mindset seemingly untouched by technological advance. The book's portrait of Martian culture – recently vanished in some stories, long dead in others, varies to suit the needs of the moment. Several of the stories are weird tales tricked out as sf: "The Third Expedition" (vt "Mars is Heaven") tells of a Martian town that uses mind control to engineer a cruel fate for an expeditionary team, and "Usher II" (vt "Carnival of Madness") of a magnate whose mansion on Mars dis-penses Poe-esque deaths to his persecutors. The collection succeeds marvelously as a fantasia on the immutability of human nature, its individual episodes being linked by recur-ring images of SACRIFICE and loneliness. The arc of the collection's events parallels the transition from childhood wonder to adult disillusionment that underlies much of RB's fantasy. Although its mood is elegiac, the collection, quietly reassures that, even though humanity can never completely overcome its worst tendencies, those traits that ennoble the species are indomitable. The book's obvious appeal to universal meanings that transcend the genre explains in part its enduring popularity with general audi-ences, who recognize in Bradbury's work more than that of any other writer of fantasy and sf of the day his use of genre tropes as tools for probing the human condition.

Most of RB's other sf stories from this period show the same disregard as his Mars tales for science in their fantas-tic elaborations of technologically sophisticated but soulless futures. In "The Exiles" (1949; vt "The Mad Wizards of Mars") an interplanetary expedition to Mars finds that the planet has magically become a refuge for writers of classic SUPERNATURAL FICTION whose work has been banned by a clinically scientific Earth culture. In "Pillar of Fire" (1948) a miraculously resurrected troublemaker from the past finds it easy to get away with murder in a technologically advanced future where civilization has been disinfected of instinctive human fears and superstitions. RB has been criticized in the sf community for the apparent antitechno-logical bias that empowers such stories, although this strain in his fiction can be viewed simply as a different expression of the theme of lost innocence around which much of his fantasy coheres. His dystopic novel *Fahrenheit 451* (1951 *Galaxy* as "The Fireman"; with two short stories as coll **1953**; most later editions omit the short stories; rev 1979 with coda; rev 1982 with afterword) crystallizes this theme in its extrapolation of a future where BOOKS are burned to prevent dissemination of their ideas.

RB's weird tales and sf proved to be preparation for an informal series of semi-autobiographical novels that inventively recycle his trademark themes and represent his most distinguished work since the 1950s. Laced with fantasy, these novels can be read as the saga of a single character loosely based on RB himself, who grows to maturity but is sustained by the power of his youthful imagination. *Dandelion Wine* (1950-57 various mags; fixup **1957**) is set in Green Town, Illinois, and features Douglas Spaulding, a 12-year-old boy experiencing a transitional summer in which the loss of a friend, an acquaintance with death and his own changing outlook on life irrevocably close the door of childhood behind him and set him on the path of adulthood. Where the fantasy of this novel is limited almost exclusively to Douglas's imaginative TRANSFORMATIONS (◊◊ PERCEPTION) of the town and its people, its follow-up, *Something Wicked This Way Comes* (**1962**), is more overtly Gothic and supernatural. Here the initiation into adulthood is tainted with the potential for EVIL as a travelling CARNIVAL that steals the SOULS of its victims tempts the novel's two 14-year-old protagonists to embrace the adult entitlements its sideshows promise. Dismissed by many for its heavy-handed allegorizing (◊ ALLEGORY), the novel represents RB's most poignant evocation of the hopes and frustration of small-town life. Nearly a quarter-century separates it from his next novel for adults, *Death is a Lonely Business* (**1985**), which is noticeably a *roman à clef* based on RB's years as a pulp writer and founded on a premise that seems all the more provocative considering his success: its writer protagonist's literary ambitions provide him with hope and a zest for life that save him from a blight of death striking down his more discouraged friends. RB folds the tale's existential speculations into a hardboiled-detective narrative, a feat he duplicates in *A Graveyard for Lunatics* (**1990**), a murder mystery set in a Hollywood backlot where a screenwriter's efforts to solve a crime MIRROR his personal quest for meaning in a world of false surfaces and unreliable façades. RB's travel to Ireland to script John Huston's film of *Moby Dick* (**1956**) is the basis of *Green Shadows, White Whale* (1958-92, various mags; fixup **1992**), in which a screenwriter's hope to understand and integrate himself into the life of a small Irish town introduces him to a cast of eccentrics as inscrutable – and often as uncanny – as the Martians of his sf tales.

Notwithstanding his increasing stature as a novelist, RB's greatest renown is for his short stories, whose focused imagery and controlled prose usually prevent the descent into bathos that mars some of his longer work. Most of his best fiction from the 1940s and 1950s has been collected in: *The Illustrated Man* (coll **1951**; with 2 stories added and 4 deleted, rev 1952 UK), a fixup whose insubstantial FRAME STORY – the stories are living tattoos on the skin of a sideshow freak (◊ CARNIVAL) – does not diminish their power; *The Golden Apples of the Sun* (coll **1953**; with 2 stories deleted 1953 UK), and *A Medicine for Melancholy* (coll **1959**; vt with 4 stories removed and 5 added *The Day It Rained*

Forever 1959 UK). By comparison, the contents of *The Machineries of Joy* (coll **1964**; with 1 story cut 1964 UK), *I Sing the Body Electric* (coll **1969**), and *Long After Midnight* (coll **1976**) are drawn primarily from his later work for mainstream periodicals and lack the invention of those stories in which he manipulates the tropes of fantasy and sf to serve his ends. The definitive *Stories of Ray Bradbury* (coll **1980**; UK paperback in 2 vols 1983), however, brings together works from all of these books and affords a unique context for appreciating the different permutations of his favourite themes over time, as well as the maturation of his style.

RB's achievement as a fiction writer who bridged the gap between GENRE FANTASY and nongenre literature often obscures his other literary and extraliterary contributions, which are considerable. He has edited two anthologies, *Timeless Stories for Today and Tomorrow* (anth **1952**) and *The Circus of Dr Lao and Other Improbable Stories* (anth **1956**), comprising modern fantasy stories published in nongenre venues. He helped adapt a number of his stories for EC COMICS, reprinted as *The Autumn People* (graph coll **1965**) and *Tomorrow Midnight* (graph coll **1966**), and oversaw work on the seven volumes of graphic renderings of his work, compiled as *The Ray Bradbury Chronicles* (graph colls **1993-4**). He has written the juvenile novels *Switch on the Night* (**1955**) and *The Halloween Tree* (**1972**) and lent his imprimatur to the *Ray Bradbury Presents* novels (**1993-5**) for younger readers, based on his DINOSAUR tales. He has written several volumes of poetry which, though undistinguished, captures his unquenchable exuberance. The plays collected in his *On Stage: A Chrestomathy of His Plays* (coll **1991**) feature both original theatre pieces and adaptations that are interesting for their distillations of his essential themes for another medium. Reflections on the creative process form the basis of several essays in his nonfiction collections *Zen and the Art of Writing* (coll **1973**) and *Yestermorrow* (**1991**). Many of his best-known books have been adapted for CINEMA and tv, including *Fahrenheit 451* (**1966**), *The Illustrated Man* (**1968**), SOMETHING WICKED THIS WAY COMES (**1983**) and the tv miniseries *The Martian Chronicles* (1980), but none of these screen versions faithfully captures the spirit behind the texts. The movies *The Beast from Twenty Thousand Fathoms* (**1953**) and *It Came from Outer Space* (**1953**) are both based loosely on RB short stories. His own screenwriting credits range widely, from his work on *Moby Dick* (**1956**) and *Picasso Summer* (**1972** tvm), the latter based on his short story "In a Season of Calm Weather" (screenplay credited to "Douglas Spaulding") to the Academy-Award nominated short *Icarus Montgolfier Wright* (**1962**) and eight episodes of the highly regarded cable-tv series *Ray Bradbury Theatre* (1985-96).

RB has been honoured with a 1977 WORLD FANTASY AWARD and a 1989 Nebula Grandmaster AWARD, both for Lifetime Achievement. The tribute volume *The Bradbury Chronicles: Stories in Honor of Ray Bradbury* (**1991**), ed William F. NOLAN and Martin H. GREENBERG, features stories by writers influenced by RB's work. It is telling that the majority are dark fantasy. [SD]

Other works: *Sun and Shadow* (1953 *Reporter*; **1957** chap); *The Essence of Creative Writing* (**1962**), nonfiction; *R is For Rocket* (coll **1962**), all but two stories having appeared in earlier collections; *The Anthem Sprinters, and Other Antics* (coll **1963**), short plays; *The Pedestrian* (1952 *F&SF*; **1964** chap); *The Vintage Bradbury* (coll **1965**); *The Day It Rained Forever: A Comedy in One Act* (**1966**), play; *The Pedestrian: A Fantasy* *in One Act* (**1966**), a play; *S is for Space* (coll **1966**), all but 4 stories having appeared in earlier collections; *Twice 22* (omni **1966**, collecting the contents of *The Golden Apples of the Sun* and *A Medicine for Melancholy*); *Bloch and Bradbury* (anth **1969**; vt *Fever Dream and Other Fantasies* 1970 UK), collecting stories by RB and Robert BLOCH; *Old Ahab's Friend, and a Friend to Noah, Speak His Peace* (**1971**), verse; *The Wonderful Ice Cream Suit and Other Plays* (coll **1972**); *Madrigals for the Space Age* (coll **1972**), words with music by Lalo Schifrin; *When Elephants Last in the Dooryard Bloomed* (coll **1973**), verse; *Ray Bradbury* (coll **1975** UK), retrospective collection; *Pillar of Fire and Other Plays for Today, Tomorrow and Beyond Tomorrow* (coll **1975**), plays; *Where Robot Mice and Robot Men Run Round in Robot Towns* (coll **1977**), verse; *The Mummies of Guanajuato* (**1978**), illustrated version with photos by Archie Lieberman of "The Next in Line" (1947); *The Ghosts of Forever* (coll **1981**), verse; *The Complete Poems of Ray Bradbury* (coll **1982**); *Dinosaur Tales* (coll **1983**); *Fahrenheit 451/The Illustrated Man/Dandelion Wine/The Golden Apples of the Sun/The Martian Chronicles* (omni **1987** UK); *Fever Dream* (1948 *WT*; **1987** chap), juvenile illustrated by Darrel Anderson; *The Toynbee Convector* (coll **1988**); *Classic Stories 1* (coll **1990**), reprint anthology containing all but 5 stories from *The Golden Apples of the Sun* and *R is for Rocket*; *Classic Stories 2* (coll **1990**), reprinting most of *A Medicine for Melancholy* and *S is for Space*, with 4 of the 5 stories omitted from *Classic Stories 1*.

About the author: *The Ray Bradbury Companion: A Life and Career History, Photolog, and Comprehensive Checklist of Writings* (**1975**) by William F. Nolan, supplemented by *Bradbury Bits & Pieces: The Ray Bradbury Bibliography: 1974-1988* (**1991**) by Donn Albright; *The Bradbury Chronicles* (**1977** chap) by George Edgar Slusser; *The Drama of Ray Bradbury* (**1977**; rev vt *Ray Bradbury: Dramatist* 1989) by Ben Indick; *Ray Bradbury* (**1980**) by Wayne L. Johnson; *Ray Bradbury* (anth **1980**) ed Martin H. Greenberg and J.D. Olander; *Ray Bradbury and the Poetics of Reverie* (**1984**) and *Ray Bradbury* (**1989**) both by William F. Touponce; *Ray Bradbury* (**1986**) by David Mogen.

BRADDON, MARY E(LIZABETH) (1837-1915) Prolific Victorian novelist, notorious both for her private life – before their marriage in 1874 she lived with publisher John Maxwell (1824-1895) and bore two of his children while his first wife was still alive in an insane asylum – and for her sensationalist novels, of which the most successful was *Lady Audley's Secret* (**1862**), about bigamy and murder. MEB often flavoured her more sensational work with the supernatural – usually in terms of VENGEANCE, as in *Gerard, or The World, the Flesh and the Devil* (**1891**; vt *The World, the Flesh and the Devil* 1891 US), a version of the FAUST legend, and *The Conflict* (**1903**). Her more direct use of the supernatural was more effective in her shorter work – mostly GHOST STORIES in which the HAUNTING usually presages death, either through its being a PORTENT, as in "At Chrighton Abbey" (1871 *Belgravia*), or by inflicting guilt that leads to suicide, as in "The Cold Embrace" (1860 *The Welcome Guest*). Most of MEB's stories appeared anonymously or pseudonymously (as Babbington White) either in Maxwell's MAGAZINES – including *Temple Bar* (prior to its sale in 1866) and *Belgravia*, which she edited 1866-76 – or in other Victorian periodicals, and it is likely that much of her short fiction has yet to be discovered. No single collection brings together all her ghost stories, but several are in *Ralph the Bailiff and Other Tales* (coll **1862**), *Weavers and Weft*

(coll **1877**) and *My Sister's Confession* (coll **1879** US).

MEB was the mother of the novelist William B. Maxwell (1866-1938), who wrote the ghost story "The Last Man In" for Cynthia ASQUITH's *Shudders* (anth **1929**). [MA]

Further reading: *Time Gathered* (**1938**) by W.B. Maxwell.

BRADFIELD, SCOTT (MICHAEL) (1955-) US writer and academic, connected with the University of Connecticut since 1989. His first fiction of genre interest, "What Makes a Cage? Jamie Knows" in *Protostars* (anth *1971*) ed David Gerrold, was sf, and gave little sign of the dangerous hilarity of his mature FABULATIONS, most of which press the envelope of the mundane but take strength through not quite splitting from the world into supernatural explanations. His best short stories appear in *The Secret Life of Houses* (coll **1988** UK; with 4 stories added exp vt *Dream of the Wolf* 1990 US; further exp vt *Greetings From Earth: New and Collected Stories* 1993 UK), most of which are set in SB's home state of California, a venue whose "mundanity" is itself (◊ LOS ANGELES) irradiated with the surreal. His first two novels, *The History of Luminous Motion* (**1989** UK) and *What's Wrong with America* (**1994** UK), inhabit the same world. SB is of fantasy interest primarily for his third novel, *Animal Planet* (**1995**), a tale which explicitly harks back to George ORWELL's *Animal Farm* (**1945** chap) but which radically differs in technique. Although SB's SATIRE of the contemporary world contains – as does Orwell's – elements of ALLEGORY, his tale settles only intermittently into that fixed relationship between meanings that is necessary to allegory. Moments of fixity constantly dissolve into overheated, angry, extremely funny BEAST FABLE. The various revolutionary animals' relationship to humans is immensely complex, and constantly shifting: animals can be mere animals, TALKING ANIMALS, peons, underprivileged classes or media stars. [JC]

BRADLEY, MARION ZIMMER (1930-) US sf and fantasy writer who initially established her reputation with the **Darkover** series of novels, primarily sf, though later novels in the series are more in the realm of PLANETARY ROMANCE. She achieved bestseller status with her long CELTIC FANTASY *The Mists of Avalon* (**1982**), which retold the legend of ARTHUR through the eyes of MORGAN LE FAY, emphasizing the mystical aspects as well as the conflict between Christianity and the older RELIGIONS.

MZB made her first professional sale in 1953 with "Women Only" for *Vortex Science Fiction #2*. Although her early work was almost all sf, there was ever a flavour of the fantastic, inspired by her own delight in the works of C.L. MOORE and Henry KUTTNER, whose novel *The Dark World* (1946 *Startling Stories*) MZB pastiched in her first novel *Falcons of Narabedla* (1957 *Other Worlds*; **1964** dos), which presaged the **Darkover** series.

Darkover proper began with "The Planet Savers" (1958 *Amazing Stories*; **1962** dos). The series – which explores the fate of the descendants of human colonists on the planet Darkover, where the culture functions by psychic rather than technological power – shifted towards fantasy with *The Spell Sword* (**1974**); later **Darkover** novels, including the successor anthologies, are closer to HEROIC FANTASY than to sf, and since the mid-1970s almost all of her work has been marketed as fantasy.

Her fantasies fall into three categories. She is most prolific in the field of basic SWORD AND SORCERY, starting with the J.R.R. TOLKIEN-inspired stories *The Jewel of Arwen* (**1974** chap) and *The Parting of Arwen* (**1974** chap), set in Middle-

Earth, and continuing through the **Atlantis Chronicles** – *Web of Light* (**1982**) and *Web of Darkness* (**1984**), assembled as *Web of Darkness* (omni **1985** UK; vt *The Fall of Atlantis* 1987 US) – to reach a nadir with *Warrior Woman* (1985), an intended feminist rebuttal of John NORMAN's **Gor** series which failed to rise above even Norman's level. Her LOW-FANTASY work also includes the continuing story series **Lythande**, about a female MINSTREL magician, which developed from the **Thieves' World** SHARED WORLD but now has a life of its own. Early tales in the series are collected as *Lythande* (coll **1986**), which has one story by Vonda N. McIntyre (1948-   ). Since 1982 MZB has also encouraged new writers into the field, firstly by letting them share her **Darkover** world and then by inviting them to contribute to a series of ANTHOLOGIES that began with *Greyhaven* (anth **1983**) and extended into an annual series, SWORD AND SORCERESS, plus the singleton *Spells of Wonder* (anth **1989**). In 1988 she began to publish MARION ZIMMER BRADLEY'S FANTASY MAGAZINE, which furthers much of the same, though with a wider remit in terms of fantasy styles. This magazine has spawned its own ANTHOLOGY, *The Best of Marion Zimmer Bradley's Fantasy Magazine* (anth **1994**).

At the next level, MZB has published works of fantasy that explore more challenging concepts. These began with *The House Between the Worlds* (**1980**; rev 1981), a fairly routine novel of a PORTAL between two worlds, though it has a compelling development of character and motive. Others in this mode include *Night's Daughter* (**1985**) – based on Emanuel Schikaneder's libretto for Mozart's OPERA *The Magic Flute* (1791) – and *Ghostlight* (**1995**). MZB also contributed to the SCIENCE-FANTASY **Trillium** series, writing *Black Trillium* (**1990**) with Andre NORTON and Julian MAY and then writing solo the fourth volume, *Lady of the Trillium* (**1995**). This series has shallower characterization, but the imagery of the **Trillium** world is intense.

Thirdly there are MZB's more literary works, stories crafted, in the wake of *The Mists of Avalon*, to appeal to a wider readership than that for straightforward GENRE FANTASY. These books focus on strong lead characters in mytho-historic settings, and have only a fine thread of the fantastic. *The Firebrand* (**1987**) explores the character and fate of Cassandra during and after the seige of Troy. *The Forest House* (**1993** UK) relates to the MATTER of Britain at the time of the Roman settlement, and probes the clash between the two cultures. These works display MZB's talent for plot, character, vision and fine storytelling.   [MA]

Other works:
The Darkover sequence: *The Sword of Aldones* (**1962** dos) and *The Planet Savers* (**1962** dos; plus "The Waterfall" coll **1976**) – these two assembled as *The Planet Savers/The Sword of Aldones* (omni **1980**) – *The Bloody Sun* (**1964**; rev plus "To Keep the Oath" coll **1979**); *Star of Danger* (**1965**), *The Winds of Darkover* (**1970**); *The World Wreckers* (**1971**); *Darkover Landfall* (**1972**); *The Spell Sword* (**1974**); *The Heritage of Hastur* (**1975**); *The Shattered Chain* (**1976**); *The Forbidden Tower* (**1976**); *Stormqueen!* (**1978**); *Two to Conquer* (**1980**); *Sharra's Exile* (fixup **1981**) – which considerably revises and expands *The Sword of Aldones* – *Hawkmistress!* (**1982**); *Thendara House* (**1983**); *City of Sorcery* (**1984**); *The Heirs of Hammerfell* (**1989**); *Rediscovery* (**1993**) – with Mercedes LACKEY – and *Marion Zimmer Bradley's Darkover* (coll **1993**); plus the **Darkover** anthologies ed MZB listed below.
The Survivors sequence: *Hunters of the Red Moon* (**1973**) and *The Survivors* (**1979**), both with Paul Edwin ZIMMER (though was uncredited for the first).

Other sf/fantasy/Gothic: *The Door Through Space* (**1961**); *Seven from the Stars* (**1962**); *The Colors of Space* (**1963**; text restored 1983); *The Dark Intruder and Other Stories* (coll **1964**); *Castle Terror* (**1965**); *Souvenir of Monique* (**1967**); *Bluebeard's Daughter* (**1968**); *The Brass Dragon* (**1969**); *Dark Satanic* (**1972**) and its sequels *The Inheritor* (**1984**) and *Witch Hill* (**1990**); *In the Steps of the Master* * (**1973**), tie to *The Sixth Sense* tv series; *Can Ellen be Saved?* (**1975**); *Endless Voyage* (**1975**; exp vt *Endless Universe* 1979); *Drums of Darkness* (**1976**); *The Ruins of Isis* (**1978**); *The Mäenåds* (**1978** chap), poem based on the Greek legend; *Survey Ship* (**1980**); *The Best of Marion Zimmer Bradley* (coll **1985**; rev 1988; exp with new story "Jamie", vt *Jamie and Other Stories* 1993) ed Martin H. GREENBERG; *Tiger Burning Bright* (**1995**) with Andre NORTON and Mercedes LACKEY.

Nonfiction: *Men, Halflings and Hero-Worship* (**1973** chap); *The Necessity for Beauty: Robert W. Chambers and the Romantic Tradition* (**1974** chap) (◊ Robert W. CHAMBERS).

As editor (all in the Darkover series): *The Keeper's Price* (anth **1980**); *Sword of Chaos* (anth **1982**); *Free Amazons of Darkover* (anth **1985**); *The Other Side of the Mirror* (anth **1987**); *Red Sun of Darkover* (anth **1987**); *Four Moons of Darkover* (anth **1988**); *Domains of Darkover* (anth **1990**); *Renunciates of Darkover* (anth **1991**); *Leroni of Darkover* (anth **1991**); *Towers of Darkover* (anth **1993**); *Snows of Darkover* (anth **1994**).

Further reading: *Marion Zimmer Bradley* (**1985**) by Rosemarie Arbur; *Marion Zimmer Bradley, Mistress of Magic: A Working Bibliography* (**1991** chap) by Gordon Benson Jr and Phil Stephensen-Payne; "Marion Zimmer Bradley" by Wendy Bradley, *Interzone #42* (1990).

BRADSHAW, GILLIAN (MARUCHA) (1956-   ) US writer whose reputation in the fantasy field rests mainly on her first work, the **Gawain** trilogy: *Hawk of May* (**1980**), *Kingdom of Summer* (**1981**) and *In Winter's Shadow* (**1982**), assembled as *Down the Long Wind: The Magical Trilogy of Arthurian Britain* (omni **1988** UK). The venue is a LAND-OF-FABLE post-Roman Britain, skewed from mundane history mainly through the presence of patterns of MAGIC normal to Arthurian tales, and through a sense – central to Arthurian work of any seriousness – that the MATTER of Britain is being addressed. GAWAIN himself is a man of decent (though not irreproachable) instincts, caught up in the dynastic and cosmological war between ARTHUR, representative of the Sun GOD on Earth, and Gawain's own mother, Morgause, who is identified as a WITCH and as the QUEEN OF AIR AND DARKNESS. Any Christian element imposed upon later versions of the Arthurian Cycle is treated here with reserve. GB's style is literate and plain-spoken, though frequently eloquent; and her sense of Britain nicely balances LEGEND and historical verisimilitude.

Much of her subsequent work – like *Horses of Heaven* (**1991**), which has some fantastication – lies in the field of historical fiction. Some of her later tales – like *The Dragon and the Thief* (**1991**), *The Land of Gold* (**1992**) and *Beyond the North Wind* (**1993**) – are historical YA fantasies.   [JC]

BRAGGADOCIO ◊ COMMEDIA DELL'ARTE.

BRAMAH, ERNEST Working name, for all his fiction, of UK author Ernest Bramah Smith (1868-1942). His most influential fantasy creation is the impossibly romanticized Old China narrated and inhabited by professional storyteller **Kai Lung**. This series runs: *The Wallet of Kai Lung* (coll **1900**), whose lead story appeared separately as *The Transmutation of Ling* (**1911** chap); *Kai Lung's Golden Hours* (coll **1922**); *Kai Lung Unrolls His Mat* (coll **1928**), of which two stories were republished as *The Story of Wan and the Remarkable Shrub and The Story of Ching-Kwei and the Destinies* (coll **1927** chap US) and another as *Kin Weng and the Miraculous Tusk* (**1941** chap); *The Moon of Much Gladness* (**1932** vt *The Return of Kai Lung* 1937 US), a parodic detective story "related by" but not featuring Kai Lung; *Kai Lung Beneath the Mulberry Tree* (coll **1940**); and *Kai Lung: Six* (coll **1974** US chap), a posthumous round-up of uncollected 1940-41 tales from *Punch* magazine. *The Kai Lung Omnibus* (omni **1936**) assembles the first three books; a selection appeared as *The Celestial Omnibus* (coll **1963**).

Kai Lung undergoes various improbable though nonfantastic adventures in this milieu, and tells many included tales – sometimes, like SCHEHERAZADE, to delay his own execution (thus the frame story of *Kai Lung's Golden Hours*, leading some to consider this a novel). Often these inner stories feature tropes of fantasy CHINOISERIE, including DRAGONS, DEMONS, capricious and fallible GODS, IDENTITY EXCHANGE, METAMORPHOSIS and TRANSMUTATION. Several are humorous MYTHS OF ORIGIN – e.g., for bamboo, CHESS, tea and willow-pattern crockery. EB's achievement was to relate the whole sequence with unwavering mock-Chinese floweriness and conversational politesse, buttressed by a remarkable vocabulary and phrase-coining skill, all to considerable ironic effect.

EB's aphorisms are still much quoted by connoisseurs, and Kai Lung once enjoyed quasi-cult status. But the stories' constant irony and stylistic density may repel casual or lazy readers, and later visitors to this LAND-OF-FABLE China, such as Barry HUGHART, have presented it in diluted form. EB is not easily imitated. [DRL]

Other works: *The Mirror of Kong Ho* (**1905**), a comic novel about a stereotyped Chinaman in London; *What Might Have Been* (**1907** anon; vt *The Secret of the League* 1909 as by EB), sf; associational stories about blind detective **Max Carrados** in *Max Carrados* (coll **1914**), *The Eyes of Max Carrados* (coll **1923**), *Max Carrados Mysteries* (coll **1927**) and *The Bravo of London* (**1934**); *The Specimen Case* (coll **1924**), assembling tales from as early as 1894; *Ernest Bramah* (coll **1929**), reprinting four **Kai Lung** stories amid other material.

BRAM STOKER AWARD AWARD given annually by The Horror Writers of America for work in DARK FANTASY and HORROR. There are several categories, and multiple awards have not been uncommon. The awards are given for books and other material published during the previous year. Only book awards and life-achievement awards are listed here. [JC]

Novel
1988: *Misery* by Stephen KING and *Swan Song* by Robert McCammon
1989: *The Silence of the Lambs* by Thomas Harris.
1990: *Carrion Comfort* by Dan SIMMONS
1991: *Mine* by Robert McCammon
1992: *Boy's Life* by Robert McCammon
1993: *The Blood of the Lamb* by Thomas E. MONTELEONE
1994: *The Throat* by Peter STRAUB
First Novel
1988: *The Manse* by Lisa Cantrell
1989: *The Suiting* by Kelley Wilde

1990: *Sunglasses After Dark* by Nancy A. COLLINS
1991: *The Revelation* by Bentley LITTLE
1992: *The Cipher* by Kathe Koja and *Prodigal* by Melanie Tem
1993: *Sineater* by Elizabeth Massie
1994: *The Thread that Binds the Bones* by Nina Kiriki HOFFMAN
Collection
1988: *The Essential Ellison* by Harlan ELLISON
1989: *Charles Beaumont: Selected Stories* by Charles BEAUMONT
1990: *Richard Matheson: Collected Stories* by Richard MATHESON
1991: *Four Past Midnight* by Stephen KING
1992: *Prayers to Broken Stones* by Dan SIMMONS
1993: *Mr Fox* by Norman Partridge
1994: *Alone with the Horrors* by Ramsey CAMPBELL
Life Achievement
1988: Fritz LEIBER; Frank Belknap LONG; Clifford D. SIMAK
1989: Ray BRADBURY; R. CHETWYND-HAYES
1990: no award
1991: Hugh B. CAVE; Richard MATHESON
1992: Gahan WILSON
1993: Ray Russell
1994: Joyce Carol OATES (Special Trustees Award)

BRAM STOKER'S DRACULA (*1993*) ◊ DRACULA MOVIES.

BRAND, MAX Pseudonym of US writer Frederick (Schiller) Faust (1892-1944). ◊ ATAVISM.

BRANDNER, GARY (1933-) US author. ◊ CAT PEOPLE (*1982*); *The* HOWLING (*1980*).

BRAUTIGAN, RICHARD (GARY) (1933-1984) US writer now so inescapably identified with the proto-New Age culture of 1960s hippie California that he is rarely read seriously. From *A Confederate General from Big Sur* (**1964**) onwards RB's novels tended to make use of fantasy elements in a roseate, expressionist fashion not generally tied to any central story, and (perhaps deliberately) contradictive of any sense of self-contained narrative coherence. Some of the effects so engendered are moving, some attain MAGIC-REALISM intensity. Various of RB's novels – none of them dull – sufficiently trespass beyond the surreal or the adventitious to warrant notice in the fantasy context. *The Hawkline Monster: A Gothic Western* (**1974**) mixes the genres designated in its subtitle, and incorporates MONSTER and METAMORPHOSIS devices, though the conclusion – in which TRANSFORMATIONS and other sorts of BONDAGE are escaped through the discovery that the inimical principle is a kind of "light", which can be smothered – is perhaps oversanguine. *In Watermelon Sugar* (**1968**) is as much sf as fantasy. [JC]

BRAVE LITTLE TAILOR The title of this famous FAIRYTALE, collected by the GRIMM BROTHERS, was appropriated by Robert A. HEINLEIN as a widely applicable plot description (◊◊ LEARNS BETTER). The original tailor brags of his murderous prowess – "Seven at one blow", alarming those unaware that the "seven" were flies – and is thus expected to undertake awesome tasks like killing GIANTS, which he in fact achieves by cunning and trickery (◊◊ JACK; TRICKSTER). The BLT typifies the unheroic HERO (or seeming ANTI-HERO) who adopts or is thrust into a role initially far too large for him, and successfully grows to be worthy of it. J.R.R. TOLKIEN shows several such careers: Bilbo reluctantly agrees to become a burglar in *The Hobbit* (**1937**) and succeeds far beyond his own expectations; Frodo and Sam in turn accept the burden of the RING in *The Lord of the Rings*

(1954-5), and gain different kinds of heroic stature. Further examples are very numerous, from Ralph in William MOR-RIS's *The Well at the World's End* (**1896**) to Simon the kitchen-boy in Tad WILLIAMS's **Memory, Sorrow and Thorn** to MICKEY MOUSE in the animated short *Brave Little Tailor* (*1938*), which is a more straightforward retelling. Diana Wynne JONES adds a slight twist in *Castle in the Air* (**1990**), whose brave little carpet-merchant's adventures are partly choreographed by a GENIE so that he undergoes in reality all the least comfortable episodes from his boasts and daydreams. Barbara HAMBLY's "The Little Tailor and the Elves" (1994) inverts the character as a decidedly unheroic, unsuccessful, wife-beating tailor (◊ PARODY).   [DRL]

BRAVE LITTLE TOASTER, THE US ANIMATED MOVIE (*1987*). ITC/Hyperion/Kushner-Locke. **Pr** Donald Kushner, Thomas L. Wilhite. **Exec pr** Willard Carroll, Peter Locke. **Dir** Jerry Rees. **Screenplay** Joe Ranft, Rees. **Based on** *The Brave Little Toaster* (1981 *Fantasy Annual IV*; **1986** chap) by Thomas M. DISCH. **Voice actors** Timothy E. Day (Blanky), Phil Hartman (Air Conditioner), Wayne Kaatz (Bob), Jon Lovitz (Radio), Deanna Oliver (Toaster), Thurl Ravenscroft (Kirby), Tim Stack (Lampy). 90 mins. Colour.

The electrical appliances left behind in a summer cabin set off in QUEST of the lost master they recall, not realizing the years have turned their ONCE AND FUTURE KING from a little boy into a man. At last, after various alarms and excursions, they reach the CITY, where the Toaster saves the master's life and all are happily reunited. This is a delightfully simple FABLE, translating the tropes and ARCHETYPES of HIGH FAN-TASY into a quaintly outmoded TECHNOFANTASY context – for example, the Toaster himself is a JACK-type HERO. Interestingly, *TBLT* is confident enough in the strength of its STORY almost to abjure two standard tools of ANIMATED MOVIES: jokes and "business".   [JG]

BRAZIL The label "fantasy" is not as common in Brazil as in English-speaking countries. In Brazil, the word *fantasia* (which has strong connotations of "fancy") is often associ-ated with children's fiction; it is not much applied to serious fantastic literature.

The Romantic Movement in Europe had a strong influ-ence on Brazilian literature, and Brazilian writers became fond of two Romantic gateways to the fantastic: the UTOPIA, and the Hallucinatory Tale. An early example of the former is the short novel *Statira, e Zoroastes* ["Statira, and Zoroastes"] (**1826**) by Lucas José d'Alvarenga (1768-1831), a tale about an imaginary Oriental kingdom where men become idle and decadent and women have to take arms to stop a conquering invasion: after that, women rule the country. Published only five years after Independence, the book was dedicated to Brazil's Empress. Examples of the Hallucinatory Tale abound: early and illustrious are the stories in *Noite na Taverna* ["A Night in the Tavern"] (coll **1878**) by Alvares de Azevedo (1831-1852), a collection of Byronic tales of crime, lust and madness by an author who died very young; the book is still in print today.

Feminist utopias (◊ FEMINISM) returned with *A Rainha do Ignoto* ["The Queen of the Unknown"] (**1899**) by Emilia de Freitas (1855-1908), which tells of a mysterious "Island of the Mist" on the coast of Ceará, northeastern Brazil; the ISLAND hides a secret society of women, whose queen is described as being "a Spiritualist, an Abolitionist and a Republican". Not much later, Godofredo E. Barnsley (1874-?   ) published *São Paulo no Ano 2000* ["São Paulo in

the Year 2000"] (**1909**), a typical utopian descriptive tale, in which a future society is exhaustively described by a guide to a contemporary man who fell asleep in a park. Another typ-ical feminist utopia of the time was *Sua Excelência, a Presidente da República no Ano 2500* ["Her Excellency, the President of the Republic in the Year 2500"] (**1929**) by Adalzira Bittencourt (1904-1976), whose heroine falls in love with a crippled artist due to be eliminated by a gov-ernment eugenics programme. *O Reino de Kiato, ou No País da Verdade* ["The Kingdom of Kiato, or In the Country of Truth"] (**1922**) by de Rodolpho Teophilo (1853-1922) again depicts an island where everybody lives in harmony.

More recent examples of the Hallucinatory Tale are *Noite* (1954; trans L.L. Barrett as *Night* **1956** US/UK) by Erico Verissimo (1905-1975), one of the foremost Brazilian novel-ists; *Valete de Espadas* ["Jack of Swords"] (**1960**) by Gerardo Mello Mourão (1917-   ); and several novels by Campos de Carvalho (1916-   ), like *A Lua vem da Asia* ["The Moon comes from Asia"] (**1956**), *Vaca de Nariz Sutil* ["A Cow with a Subtle Nose"] (**1961**) and *A Chuva Imóvel* ["The Immobile Rain"] (**1963**).

HEROIC FANTASY has appeared only rarely in Brazilian lit-erature. A typical work in this vein is *Imortalidade* ["Immortality"] (**1925**) by Coelho Netto (1864-1934): in a LAND-OF-FABLE medieval Europe, Everardo, Lord of Crève-Coeur, becomes immortal after drinking an alchemist's POTION. Netto also wrote *Esfinge* ["Sphinx"] (**1908**), where a magician transplants a woman's head onto a man's body. Recently, though, due to the growth of UK/US GENRE FAN-TASY in both books and the movies, heroic-fantasy themes have become very popular among Brazilian readers. Fantasy GAMES also play an important role in this fast-expanding market, and maybe this is one reason many writers shy away from it: they believe that fantasy is just a set of fairytales for adolescents. They also point out (understandably) that, yes, stories based on Celtic mythology and the legends of ARTHUR can be enjoyed in Brazil, but should they be *written* by Brazilians?

There is a narrow but strong subgenre in Brazilian fantasy that could be called "Tales of the Mythical Brazilian". Brazil is a young nation, and a melting pot of European, Indian and African cultures, so building up a "typical" Brazilian hero is a literary puzzle that many writers have tried to solve. In such books, which use plot structures and narrative voices very close to oral literature, the hero is a Brazilian, usually poor (but also resourceful) and naive (but also streetwise); he sets out on a QUEST during which he will face dangers and temptations and will run into many fantastic creatures.

In *Macunaíma* (**1928**; trans E.A. Goodland as *Macunaíma* **1984** US/UK) by Mário de Andrade (1893-1945) the eponymous hero, an Indian born in the jungle, has his TAL-ISMAN stolen by a GIANT, whom he follows to the CITY of São Paulo. The story draws on the author's vast knowledge of Brazilian folklore and Indian myths. Other tales of mytho-logical HEROES are *Manuscrito Holandês, ou A Peleja do Caboclo Mitavaí contra o Monstro Macobeba* ["A Dutch Manuscript, or The Fight of the Caboclo Mitavaí against the Monster Macobeba"] (**1960**) by M. Cavalcanti Proença (1905-1966), and *As Pelejas de Ojuara* ["The Fights of Ojuara"] (**1986**) by Neil de Castro (1940-   ). A further important work is *A Pedra do Reino* ["The Stone of the Kingdom"] (**1971**) by poet and playwright Ariano Suassuna (1927-   ), which tells of a man who believes himself to be the heir of a kingdom in the backlands of the Brazilian northeast; this huge novel interweaves themes from Classical literature, Brazilian history and MYTHICAL CREATURES from the "literatura de cordel" (narrative poems, often fantastic, printed as booklets and sold to poor people in the Brazilian northeast).

Books about esoterism and the Occult have been selling more and more in Brazil in recent years, and at least one Brazilian writer has been basing a literary career on those themes. Paulo COELHO is today Brazil's biggest seller: his four OCCULT FANTASIES have together reportedly sold over three million copies, although this kind of fiction has not attained the same prestige that MAGIC REALISM still enjoys among many Brazilian critics. The boom in magical realism occurred at a time when Brazil was under a military dictatorship, and many mainstream writers turned to the genre (or to what they supposed it to be) in order to write books that would be basically ALLEGORIES, dealing with the theme of individual freedom. Erico Verissimo published *Incidente em Antares* ["Incident at Antares"] (**1971**), in which seven dead men refuse to be buried and, claiming justice, defy the population of a city. Singer/songwriter Chico Buarque de Hollanda (1944-   ) wrote an Orwellian BEAST FABLE in *Fazenda Modelo* ["Model Farm"] (**1974**), where cattle behave like people. Writers that became very popular at this time were José J. VEIGA and Murilo RUBIÃO, who are among the very few Brazilian writers whose names are primarily associated with the fantastic; but Moacyr SCLIAR, Osman LINS and João Guimarães Rosa (1908-1967) are authors who weave fantasies all of their own. Outstanding works by other writers are *Dona Flor e Seus Dois Maridos* (**1966**; trans Harriet de Onís as *Dona Flor and her Two Husbands* 1969 US/UK) by Jorge AMADO, a funny and erotic GHOST STORY; *O Beijo Antes do Sono* ["The Kiss before Sleep"] (**1974**) by Fausto Cunha (1923-   ), who is also a front-rank sf writer and critic; *Imperatriz no Fim do Mundo* ["Empress in the End of the World"] (**1992**) by Ivanir Calado (1953-   ), in which the GHOST of a 19th-century Brazilian Empress researches her own life in the National Library; *O 31!o! Peregrino* ["The 31st Pilgrim"] (**1993**) by Rubens Teixeira Scavone (1925-   ), a haunting and chilling "missing story" from Geoffrey CHAUCER's *The Canterbury Tales*.   [BT]

BRAZIL UK movie (**1985**). Embassy. **Pr** Arnon Milchan. **Dir** Terry GILLIAM. **Spfx** Richard Conway, George Gibbs. **Screenplay** Gilliam, Charles McKeown, Tom Stoppard. **Starring** Robert De Niro (Archibald "Harry" Tuttle), Kim Greist (Jill Layton), Ian Holm (Mr Kurtzmann), Katherine Helmond (Ida Lowry), Bob Hoskins (Spoor), Michael Palin (Jack Lint), Jonathan Pryce (Sam Lowry), Peter Vaughan (Mr Helpmann). 142 mins. Colour.

A TECHNOFANTASY set in an ALTERNATE REALITY whose technology is sometimes reminiscent of our gadgetry of the 1950s and sometimes futuristic. The part of the world in which the story is set is undefined, though it would seem to be an alternate London, capital of a bureaucracy-ridden, somewhat paranoid England.

Lowry is a humble clerk in the Department of Records. In his DREAM life, however, he is something quite other – a FLYING man, physically like the ANGEL in BARBARELLA (**1967**) – and must save a flaxen-haired beauty from threats that are sometimes realized, sometimes only sensed. In the daytime world, a machine malfunction causes an innocent man to be arrested as a terrorist; he dies in custody and, when Lowry

visits the widow, he discovers that the upstairs neighbour, Layton, is the woman from his dream – albeit crop-haired. However, having made the mistake of complaining about her neighbour's erroneous arrest, she is now herself classified as a terrorist, and goes on the lam. Lowry takes promotion from Records to Information Retrieval and abuses the bureaucratic system to track her down. When at last he persuades her to trust him they fall in love; he tampers with the Information Retrieval databank to persuade it that she is dead, but even so the department's agents discover the lovers after their single night of passion and Layton dies in a hail of bullets.

A strange mix of Orson Welles's version of *The* TRIAL (*1962*), of George ORWELL's *Nineteen Eighty-four* (**1949**), James THURBER's `The Secret Life of Walter Mitty' (1939), Lewis CARROLL's *Alice in Wonderland* (**1865**) and too many other CROSSHATCH sources to identify, let alone mention, *B* is as imaginative a movie as Gilliam's TIME BANDITS (*1981*) and *The* ADVENTURES OF BARON MUNCHAUSEN (*1989*), and among the most mature fantasies the CINEMA has produced.   [JG]

BREAK THE NEWS UK movie (*1938*) ◊ René CLAIR.
BREAKTHROUGH, THE Canadian/UK movie (*1993*). Lillian Gallo Entertainment/Filmline/Screen Partners/ Astral/World International Network. **Pr** Nicolas Clermont. **Dir** Piers Haggard. **Spfx** Ryal Cosgrove, Brock Jolliffe. **Screenplay** Mike Hodges, Gerard MacDonald. **Based on** "The Breakthrough" (1966) by Daphne DU MAURIER. **Starring** Mimi Kuzyk (Jessica Saunders), Corin Nemec (Ken Ryan), Hayley Reynolds (Niki Janus), Michael Rudder (Woody Gifford), Donald Sutherland (Dr Mac Maclean), Vlasta Vrana (Dr Robbie Allman). 88 mins. Colour.

Flagging CIA agent Saunders, racked by the accidental drowning of her young son, is sent to a governmental research station in Newfoundland to check what the team under Maclean are up to. She finds they are tapping and trapping the psychic energy given off by human beings at times of great stress; to this end the team includes a young man, Ryan, dying of leukemia, and a little girl, Niki, who because of the death of her twin sister is autistic yet has TALENTS. Saunders recommends the experiment be terminated forthwith, but her CIA boss suddenly sees the potential for power if life and death can indeed be controlled. As Ryan dies the team's computer, Charon, records the death experience, continuing well past the stage where physical life has departed. Ryan's SOUL, trapped in the project's hardware, is allowed at last its freedom.

This – though hampered by a clumsy screenplay – is a very intelligent TECHNOFANTASY. Much of its territory has been well tramped over before – by, e.g., FIRESTARTER [*1984*] and *The* FURY [*1978*] – but it manages to preserve an air of freshness, probably because, unlike those movies, it eschews violence and sensation, concentrating instead on the essential fascination of its AFTERLIFE theme.   [JG]

BRECCIA, ALBERTO (1919-1993) Uruguay-born Argentinian COMICS artist. AB has applied a wide range of drawing, painting and printmaking techniques in his very personal approach to comic strips, and has been quoted as a major influence by almost every leading practitioner in the field. He published his first drawings at age 17. His first strip of note was **Mu Fa, un Detective Oriental** ["Mu Fa, An Oriental Detective"] (1939), a humorous parody of FU MANCHU. Around this time he joined the Argentine Realist School of José Luis Salinas (1908-1987), along with two other young artists who were to gain worldwide acclaim, Arturo del Castillo (1925-1992) and Hugo Pratt (1927-1995); their book on the art of comic-strip drawing was *Tecnica de la Historieta* (**1966** Brazil).

During 1940-47 AB produced a number of short stories and adaptations of popular novels in the sf and horror vein for the magazines *El Gorion* and *Rataplan*, including *La Mano que Aprieta* ["The Gripping Hand"] (1940), *El Jorobado* ["The Hunchback"] (1942) and *La Hosteria Solitaria* ["The Lonely Inn"] (1943-4). He then embarked on the newspaper strip **Vito Nervio** (1947-58), a series with sf and fantasy elements about a secret-service agent. Thereafter he formed a lasting partnership with the writer Hector Oesterheld (? -1977), producing sf and horror series such as **Sherlock Time** (1958) and his great masterpiece, **Mort Cinder** (1962-3), for the Argentine magazine *Misterix*.

AB travelled to the UK in 1960 to work for Fleetway Publications, where he drew a number of 68-page Western and spy comic books in a bold, loose brush line style. He returned to Argentina in 1962.

In 1968 he briefly took over the long-running series **El Eternauta** ["The Eternaut"], and then put his own life in danger by creating his very personal *El Vida del Che* ["The Life of Che"] (graph **1986**). He buried the proofs of this series in his garden when the original art was confiscated and the printer killed by the military authorities: it was eventually published, years later, in Spain. AB then began adapting into strips H.P. LOVECRAFT's CTHULHU MYTHOS stories as **Los Mitos de Cthulhu** ["The Cthulhu Mythos"] (1973-4; graph coll **1985**), in which the terrifying subterranean creatures stood as an ALLEGORY for the grip of terror imposed by the Argentine secret police. In 1977 Oesterheld and his entire family were murdered by them.

AB began experimenting with a wide range of drawing and painting techniques including collage, frottage, resist and monotype, and his work became increasingly moody and expressive. In this way he rendered **Un Tal Daneri** ["A Man Called Daneri"] (1974 *Mengano*; graph coll **1977**), a moody monochromatic series of short stories about a contract killer, and brightly coloured, humorous adult versions of FAIRYTALES, **Cuentos de los Hermanos Grimm** ["Tales from the Brothers Grimm"] (1980-1; vt *Che ha Paura delle Fiabe?* ["Who's Afraid of Fairy Tales?"] graph coll **1981** Italy).

On the fall of Argentina's military dictators after the Falklands War (1982), AB and writer Juan Sasturain embarked on the biting political satire *Perramus* (graph **1984-6** 4 vols; vol 1 trans 1991-2), in which his close friend Jorge Luis BORGES features among many other famous faces. He followed this with a very personal version of *Dracula* (graph **1991**).

The publication of AB's fantasy work in English has been desultory at best, aside from *Perramus*. In HEAVY METAL have appeared *The Dunwich Horror* (1979), *Mister Valdemar* (1982), *Poe* (1985), *To Draw or Not . . .* (1988) and *Dracula* (1992).

AB's son Enrique (1944-) is also a considerable talent in South American comics. [RT]

BREITMANN, HANS ◊ Charles Godfrey LELAND.

BRENNAN, JOSEPH PAYNE (1918-1990) US writer, poet and librarian, one of the last memorable names to emerge from WEIRD TALES before its first demise in 1954. He started with "The Green Parrot" (1952), a routine

GHOST STORY. An early classic was "Slime" (1953 *WT*), featuring a blob-like entity. JPB's HORROR and ghost stories are often derivative but are always strong on setting and atmosphere and hence are much beloved by traditionalists; they often share a common locale, **Juniper Hill**. They are collected as *9 Horrors and a Dream* (coll **1959**), *The Dark Returners* (coll **1959**), *Scream at Midnight* (coll **1963**), *Stories of Darkness and Dread* (coll **1973**), *The Shapes of Midnight* (coll **1980**) and *The Borders Just Beyond* (coll **1986**).

In 1957 JPB began his own small-press MAGAZINE as a successor to *WT*: *Macabre* (23 issues 1957-76). It was in this magazine that he introduced his stories featuring **Lucius Leffing**, a small-town OCCULT DETECTIVE with a penchant for the Victorian age. Leffing's investigations (not all supernatural) are collected as *The Casebook of Lucius Leffing* (coll **1973**), *The Chronicles of Lucius Leffing* (coll **1977**), *The Adventures of Lucius Leffing* (coll **1990**) and the RECURSIVE FANTASY *Act of Providence* (**1979**) with Donald Grant (1927-). JPB produced another mystery novel with horror overtones, *Evil Always Ends* (**1982**).

A long-time devotee of WEIRD FICTION, JPB privately published *A Select Bibliography of H.P. Lovecraft* (**1952** chap; exp vt *H.P. Lovecraft, A Bibliography* 1952 chap) and *H.P. Lovecraft, An Evaluation* (**1955** chap). JPB's volumes of poetry include *Heart of Earth* (coll **1950** chap), *The Humming Stair* (coll **1953** chap), *The Wind of Time* (coll **1961** chap), *Nightmare Need* (coll **1964**), *Edges of Night* (coll **1974** chap), *Webs of Time* (coll **1979**), *Creep to Death* (coll **1981**), *Sixty Selected Poems* (coll **1985**) and *Look Back on Laurel Hills* (coll **1989**). [MA]

BRENNER, (MAYER) ALAN (1956-) US writer. His **Dance of the Gods** sequence – *Catastrophe's Spell* (**1989**), *Spell of Intrigue* (**1990**), *Spell of Fate* (**1992**) and *Spell of Apocalypse* (**1994**) – is an amiable recycling of some standard HEROIC-FANTASY material: a conflict between a DIRTY DOZEN of assorted WIZARDS and warriors and a pantheon of GODS who gradually turn out to be merely more powerful MAGIC-wielders. AB is entertaining and has a nice feel for acrobatic swashbuckling. His intermittently intelligent use of such tropes as IDENTITY EXCHANGE and TIME ABYSS – which latter reveals this to be a post-HOLOCAUST future – informs us that he has the capacity to be slightly more than he has so far shown. [RK]

BRENNERT, ALAN (MICHAEL) (1954-) US tv producer, writer and author, most of whose work has been DARK FANTASY and HORROR, beginning with "Nostalgia Tripping" in *Infinity 5* (anth **1973**) ed Robert Hoskins. His first novel, *City of Masques* (**1978**), is sf. *Kindred Spirits* (**1984**), marketed as YA, and *Time and Chance* (**1990**) are both dark fantasies in which two male protagonists SHADOW one another through life crises. The former offers a relatively simple communion of SOULS between two suicidal teenagers; in the latter a man finds himself slipping into an ALTERNATE REALITY where another version of him lives, and the two swap worlds.

The tales assembled in *Her Pilgrim Soul, and Other Stories* (coll **1990**) and *Ma Qui, and Other Phantoms* (coll **1991**) are also set mainly in the real world: some enlist fantasy or SUPERNATURAL-FICTION devices to illuminate RITES OF PASSAGE in their (not always adolescent) protagonists; others, like the title story (1990) of the second volume, are powerful GHOST STORIES – powerful because AB's GHOSTS almost invariably cry out to the real world for release from their BONDAGE . . . and do not necessarily gain it. *Weird Romance:*

Two One-Act Musicals of Speculative Fiction (coll **1993**) – with book and libretto by David Spencer and music by Alan Menken – is a CD presentation of two stories, the adaptation of AB's "Her Pilgrim Soul" being fantasy.

Much of AB's fantasy work is in his tv scriptwriting. He was Executive Story Consultant to *The* NEW TWILIGHT ZONE (1985-8), writing several of the scripts, and before that wrote for *The* ADVENTURES OF WONDER WOMAN (1978-9) and the space opera *Buck Rogers in the 25th Century* (1979-81). He edited *The New Twilight Zone* * (anth **1991**), comprising stories either written especially for the series or on which episodes had been based; his long introduction gives interesting background information about the series.

AB's COMICS work has mainly been scripts for BATMAN, some assembled as *The Greatest Batman Stories Ever Told* (graph coll **1989**). [JC]

BRENTANO, CLEMENS (1778-1842) German folklorist. ◊ Adelbert von CHAMISSO; FOLKLORE; SONG.

BRER RABBIT ◊ Joel Chandler HARRIS; TRICKSTER.

BRER RABBIT'S CHRISTMAS CAROL US movie (*1992* tvm). ◊ *A* CHRISTMAS CAROL.

BRIAR ROSE ◊ SLEEPING BEAUTY.

BRICE, MARTIN H(UBERT) (1935-   ) UK writer. ◊ ANTHROPOLOGY; WITCHCRAFT.

BRIDE, THE US/UK movie (*1985*). ◊ FRANKENSTEIN MOVIES.

BRIDE OF FRANKENSTEIN, THE US movie (*1935*). ◊ FRANKENSTEIN MOVIES.

BRIDES OF DRACULA UK movie (*1960*). ◊ DRACULA MOVIES.

BRIGADOON ◊ BRIGADOON (*1954*); Alan Jay LERNER; POLDER; TIME IN FAERIE.

BRIGADOON US movie (*1954*). MGM. **Pr** Arthur Freed. **Dir** Vincente Minnelli. **Spfx** Warren Newcombe. **Screenplay** Alan Jay LERNER. **Based on** *Brigadoon* (stage musical 1947) by Lerner and Frederick Loewe. **Starring** Virginia Bosler (Jean Campbell), Cyd Charisse (Fiona Campbell), Van Johnson (Jeff Douglas), Barry Jones (Mr Lundie), Gene Kelly (Tommy Albright), Hugh Laing (Harry Beaton), Jimmy Thompson (Charles Chisholm Dalrymple). 108 mins. Colour.

On a shooting holiday in Scotland two young Americans, Albright and Douglas, stray into the hamlet Brigadoon, which exists in our REALITY only one day every 100 years, "slumbering" in the interim: in Brigadoon the year is still 1754. This condition was brought about by its Minister, Mr Forsyth, who bartered his own life with God for a MIRACLE that would preserve the hamlet from the prevalent evil of witchcraft (◊ WITCHES). Should any of its inhabitants cross Brigadoon's boundaries, its next night will be everlasting – but this is exactly what thwarted lover Beaton proposes to do until shot dead by Douglas. The Americans flee, although Albright has so fallen in love with a Brigadoon girl, Fiona, that after a few months he rejects his NEW YORK life and returns to Scotland, the strength of his love being sufficient to conjure Brigadoon long enough from the mists that he may enter it.

B is very stagy – it was shot not on location but on the MGM lot – and the standard trappings of the musical (e.g., interminable dance routines) almost smother its central premise, drawn from Scottish folklore. Also interesting is its portrayal of 18th-century Scotland as a LAND OF FABLE – a Scotland which, once sampled, is worth giving everything to regain. [JG]

BRIGGS, K(ATHARINE) M(ARY) (1898-1980) UK folklorist and writer who began her career with *The Legend of Maiden-Hair* (**1915**), a tale for children published by a vanity press, and with plays like *The Garrulous Lady* (**1931**). Only after WWII did she publish *The Personnel of Fairyland: A Short Account of the Fairy People of Great Britain for Those Who Tell Stories to Children* (**1953**), the first of many studies and surveys, the most important almost certainly being *A Dictionary of Fairies: Hobgoblins, Brownies, Bogies and Other Supernatural Creatures* (**1976**; vt *An Encyclopedia of Fairies* 1976 US; cut vts *A Sampler of British Folk-tales* 1977 UK and *British Folk-tales* 1977 US). In this and other works KMB significantly helped accelerate the slow growth of understanding, during the 20th century, of the cultural importance and intrinsic narrative interest of the oral FOLKLORE tradition. Along with the UK scholars of the preceding generation, who had concentrated on folksong, her work came just as the final authentic bearers of folk traditions were reaching their last years.

Her first novel, *Hobberdy Dick* (**1955**), is historical fantasy set in the 17th century: it complexly opposes the traditional world of FAERIE to the THINNING influence of Puritan Christianity, as seen through a complicated triangular conflict between a Puritan paterfamilias, a WITCH, and the eponymous brownie. KMB's only other novel, *Kate Crackernuts* (**1963**; rev 1979), in which a wicked STEPMOTHER hires a witch to give her stepdaughter the head of a sheep, is a psychological drama; the intensity of the RITE OF PASSAGE undergone by the protagonists can easily be understood as translating the fantasy elements into adolescent psychosis.

KMB's fiction, though highly sensitive to the folkloric material it presents, seems never to have captured her full interest. She is now best remembered for the seminal importance, and exemplary clarity, of her scholarly work. [JC]

Other works (for younger children): *The Witches' Ride* (**1937**); *Stories Arranged for Mime* (coll **1937**), plays; *The Prince, the Fox, and the Dragon* (**1938**).

Other works (nonfiction): *The Anatomy of Puck: An Examination of Fairy Beliefs Among Shakespeare's Contemporaries and Successors* (**1959**); *Pale Hecate's Team: An Examination of the Beliefs on Witchcraft and Magic Among Shakespeare's Contemporaries and His Immediate Successors* (**1962**); *The Fairies in Tradition and Literature* (**1967**; vt *The Fairies in English Tradition and Literature* 1967 US); *A Dictionary of British Folk-tales in the English Language: Folk Narratives* (**1970**; vt *Folk-tales* 1970 US) both in 2 vols; *Folk Legends* (**1971**) in 2 vols; *The Vanishing People: A Study of Traditional Fairy Beliefs* (**1977**); *Abbey Lubbers, Banshees and Boggarts: A Who's Who of Fairies* (**1979**).

BRIGGS, RAYMOND (REDVERS) (1934-   ) UK illustrator and writer, initially popular for his children's books, mostly in formats that utilize the visual conventions of COMICS. His most notable single work for a younger audience is debatably *The Snowman* (graph **1978**), which was adapted as a 25min ANIMATED MOVIE, *The Snowman* (*1982* tvm); this movie has proved perennially popular, both in video form and when shown on tv (in the UK at least annually, at CHRISTMAS). The eponymous hero of RB's textless narrative lives in blissful ignorance that it is his fate eventually to melt. A later companion was *The Snowman Pop-Up Book* * (graph **1986**). Though much of his earlier output is specifically directed towards children, it is clear that mature concerns permeate his entire oeuvre, all of which is of exceptional interest. Throughout there is a sense of the fragility of contentment, of the soul-oppressing weight of prejudices, and of the profound importance of remaining true to oneself, despite the distorting demands of the outer world.

RB's most remarkable fantasy, however, is *Fungus the Bogeyman* (graph **1977**), which creates a WAINSCOT society of deliciously loathsome creatures whose TOPSY-TURVY life is a comic BONDAGE: their imposed duty of frightening people is endured, not enjoyed. Beneath the schoolboy filth-jokes, the hero is a KNIGHT OF THE DOLEFUL COUNTENANCE, glumly questioning the rituals of his absurd existence. A companion volume was *The Fungus the Bogeyman Plop-Up Book* * (graph **1982**).

RB's most famous work for adults is *When the Wind Blows* (graph **1982**), a dreadful-warning sf tale about nuclear holocaust, also in comics form; it was also published as an unillustrated play (**1982**), broadcast as an acclaimed radio play, and made into a cumulatively powerful animated movie, *When the Wind Blows* (**1986**). *The Tin Pot General and the Old Iron Woman* (graph **1984**) too closely anatomized the 1982 Falklands War to ingratiate itself with devoteees of UK Prime Minister Margaret Thatcher. In *The Man* (graph **1992**) a young boy encounters a diminutive man, member of another WAINSCOT society (◊ GREAT AND SMALL), who insists upon his autonomy (as well as his need for special food) while staying with the lad for three days. The melancholy which attends their inevitable separation is very effectively conveyed, as is the moral intensity of their conversational clashes. In RB's works costs are always registered, as in a parallel tale, *The Bear* (graph **1994**), in which a large polar bear visits a similar home. [JC]

Other works: *Midnight Adventure* (graph **1961**); *The Strange House* (graph **1961**); *Ring-a-Ring o' Roses* (anth **1962**), verse; *Sledges to the Rescue* (graph **1963**); *The White Land* (anth **1963**); *Fee Fi Fo Fum: A Picture Book of Nursery Rhymes* (anth **1964**); *The Elephant and the Bad Baby* (graph **1969**) with Elfrida Vipont (1902-); *Jim and the Beanstalk* (graph **1970**); *Father Christmas* (graph **1973**) and its sequel, *Father Christmas Goes on Holiday* (graph **1975**); *Gentleman Jim* (graph **1980**); *Unlucky Wally* (graph **1987**) and *Unlucky Wally Twenty Years On* (graph **1989**).

Selected works illustrated by RB: *Peter and the Piskies: Cornish Folk and Fairy Tales* (coll **1958**) by Ruth Manning-Sanders (1888-1988); *Look at Castles* (**1960**) by Alfred Duggan (1903-1964); *William's Wild Day Out* (**1963**) by Meriol Trevor; *Whistling Rufus* (**1964**) by William MAYNE; *Mother Goose Treasury* (anth **1966**); *The Hamish Hamilton Book of Giants* (anth **1968**) ed Mayne; *Richthofen the Red Baron* (**1968**) by Nicholas Fisk; *Fairy Tale Treasury* (anth **1972**) ed Virginia Haviland (1911-1988); *The Forbidden Forest, and Other Stories* (coll **1972**) by James REEVES.

BRIGGS, STEPHEN (1951-) UK writer/illustrator. ◊ Terry PRATCHETT.

BRIGHELLA ◊ COMMEDIA DELL'ARTE.

BRITE, POPPY Z. (1967-) US writer. PZB is known primarily as an author of HORROR/SUPERNATURAL-FICTION novels and stories – with many of the latter being collected in the excellent *Swamp Foetus* (coll **1993**). Here PZB's lush, atmospheric prose is shown to best effect in tales like "His Mouth Will Taste of Wormwood" and "Calcutta, Lord of Nerves". Rather less successful was *Lost Souls* (**1992**), a VAMPIRE novel, in which the plotting was occasionally indistinct.

Her second novel, *Drawing Blood* (**1993**), featured computer hacking, underground COMICS and Birdland, a SECONDARY WORLD suffused with *noir*; it boasts some of PZB's most inspired, fevered writing. PZB also co-edited (with Martin H. GREENBERG) *Love in Vein* (anth **1994**), a well received anthology of erotic vampire stories. She won a 1994 BRITISH FANTASY AWARD as Best Newcomer. [WKS]

Other works: *His Mouth Will Taste of Wormwood* (coll **1995** chap UK), 4 stories from *Swamp Foetus*.

BRITISH FANTASY AWARD Annual set of awards, in various categories, given by the British Fantasy Society since 1971. In honour of August DERLETH, the best novel award bears his name. Reflecting a Derlethian sense of the nature of fantasy, all awards consist of a statuette of the god Cthulhu (◊ CTHULHU MYTHOS; H.P. LOVECRAFT); over the years, DARK FANTASY and HORROR have increasingly dominated proceedings. Only the more important awards (categories have fluctuated) are listed here. The award year is one year later than that of publication. [JC]

August Derleth Award

1973: *The Knight of the Swords* by Michael MOORCOCK
1974: *The King of the Swords* by Michael MOORCOCK
1975: *Hrolf Kraki's Saga* by Poul ANDERSON
1976: *The Hollow Lands* by Michael MOORCOCK
1977: *The Dragon and the George* by Gordon R. DICKSON
1978: *A Spell for Chameleon* by Piers ANTHONY
1979: The Chronicles of Thomas Covenant the Unbeliever by Stephen R. DONALDSON
1980: *Death's Master* by Tanith LEE
1981: *To Wake the Dead* by Ramsey CAMPBELL
1982: *Cujo* by Stephen KING
1983: *The Sword of the Lictor* by Gene WOLFE
1984: *Floating Dragon* by Peter STRAUB
1985: *The Ceremonies* by T.E.D. KLEIN
1986: *The Ragged Astronauts* by Bob Shaw
1987: *It* by Stephen KING
1988: *The Influence* by Ramsey CAMPBELL
1990: *Carrion Comfort* by Dan SIMMONS
1991: *Midnight Sun* by Ramsey CAMPBELL
1992: *Outside the Dog Museum* by Jonathan CARROLL
1993: *Dark Sister* by Graham JOYCE
1994: *The Long Lost* by Ramsey CAMPBELL
1995: *Only Forward* by Michael Marshall Smith

Short Story

1975: "Sticks" by Karl Edward WAGNER
1977: "Two Suns Setting" by Karl Edward WAGNER
1978: "In the Bag" by Ramsey CAMPBELL
1979: "Jeffty is Five" by Harlan ELLISON
1980: "The Button Molder" by Fritz LEIBER
1981: "Stains" by Robert AICKMAN
1982: "The Dark Country" by Dennis ETCHISON
1983: "Breathing Method" by Stephen KING
1984: "Neither Brute Nor Human" by Karl Edward WAGNER
1985: "The Forbidden" by Clive BARKER
1986: "Kaeti and the Hangman" by Keith ROBERTS
1987: "The Olympic Runner" by Dennis ETCHISON
1988: "Leaks" by Steve Rasnic Tem
1989: "Fruiting Bodies" by Brian LUMLEY
1990: "On the Far Side of the Cadillac Desert with Dead Folks" by Joe R. Lansdale
1991: "The Man Who Drew Cats" by Michael Marshall Smith
1992: "The Dark Land" by Michael Marshall Smith
1993: "Night Shift Sister" by Nicholas Royle

1994: "The Dog Park" by Dennis ETCHISON
1995: "The Temptation of Dr Stein" by Paul J. McAuley
Best Film
1975: *The* EXORCIST
1977: *The* OMEN
1979: *Close Encounters of the Third Kind*
1980: *Alien*
1981: *The Empire Strikes Back*
1982: *Raiders of the Lost Ark* (◊ INDIANA JONES)
1983: *Blade Runner*
1984: *Videodrome*
1985: *A* NIGHTMARE ON ELM STREET
1987: *Aliens*
1988: HELLRAISER
1990: *Indiana Jones and the Last Crusade* (◊ INDIANA JONES)

BRITTAIN, C. DALE (1948-) US writer whose **Daimbert** sequence – *A Bad Spell in Yurt* (**1991**), *The Wood Nymph and the Cranky Sain* (**1993**), *Mage Quest* (**1993**), *Voima* (**1995**) and *The Witch and the Cathedral* (**1995**) – follows, with a light touch, the career in FANTASYLAND of the royal wizard of Yurt. [JC]

BROCK, C(HARLES) E(DMUND) (1870-1938) Very prolific and popular UK book illustrator with a strong, realistic pen-and-ink line style. He was a highly competent draughtsman, and his work's clarity and sureness of touch have ensured that it has not become dated.

CEB was the elder brother of H.M. BROCK, with whom he worked very closely. He studied under the sculptor Henry Wiles, and by the age of 21 was an established book and magazine illustrator. He and his brother amassed, in the studio they shared, a collection of antique furniture and costume for use as props in the authentic period ILLUSTRATIONS that constituted a large proportion of their output. [RT]

BROCK, H(ENRY) M(ATTHEW) (1875-1960) Younger of the two prolific illustrator brothers (◊ C.E. BROCK) who worked in a shared studio in Cambridge on a vast range of books and magazines in a strongly competent pen-line style. HMB was the more fanciful of the two, although his drawings did not until later in his career have a clarity comparable to that of his elder brother; by this time it was difficult to distinguish their work. He and his brother illustrated almost every leading author of their period. [RT]

BROD, MAX (1884-1968) Austrian writer. ◊ Franz KAFKA.

BRONTË, CHARLOTTE (1816-1855) English writer, author of *Jane Eyre* (**1847**) and others, and elder sister to Branwell (1817-1848), Emily (1818-1848) and Anne Brontë (1820-1849). CB's earliest writings, most unpublished until after her death, centred on the creation of IMAGINARY LANDS. In 1826, after the death of two elder sisters, CB and her surviving sisters and brother sought escape from their unhappy childhood in the detailed creation of an imaginary country called the Glass Town Confederacy, set in West Africa, and built around the characters of Branwell's toy soldiers. As they grew older, this closed universe became more sophisticated. Emily and Anne founded the neighbouring kingdom of Gondal, while Charlotte and Branwell developed the breakaway lands of Angria, Zamorna and Northangerland. Angria, a FANTASYLAND of wish-fulfilment, was sustained in CB's booklets, written in microscopic handwriting, until 1839, with Emily and Anne continuing the story of Gondal in similar fashion until 1845. The tales of Gondal and Angria may owe inspiration to Jonathan SWIFT's *Gulliver's Travels* (**1726**), but the lands

became considerably more developed, resulting in a proto-form of DYNASTIC FANTASY. One of the most robust stories is *The Spell* (written 1834; **1931** chap), set in Zamorna, which utilizes the DOPPELGÄNGER theme in portraying ensorcelled TWINS who will die if seen together.

The stories were collected initially as *The Twelve Adventurers and Other Stories* (coll **1925**) ed Clement Shorter, and then more comprehensively as *Legends of Angria* (coll **1933** US) ed Fannie Ratchford and W.C. de Vane. A selection was issued as *Tales from Angria* (coll **1954**) ed Phyllis Bentley. The most complete compilation has been the 3-volume *Early Writings of Charlotte Brontë, 1826-1832* (coll **1987**), *1883-1834* (coll **1991**) and *1834-1835* (coll **1991**), all ed Christine Alexander. [MA]

Further reading: *The Brontës' Web of Childhood* (**1941** US) by Fannie Ratchford; *The Bewitched Parsonage* (**1950**) by William S. Braithwaite; *Weaver of Dreams: Charlotte Brontë* (**1966**) by Elfrida Vipont; *The Early Writings of Charlotte Brontë* (**1983**) by Christine Alexander. Also of interest is *Gondal's Queen* by Emily Brontë (**1955** US) ed Ratchford.

BROOKE, WILLIAM HENRY (1772-1860) UK illustrator. ◊ ILLUSTRATION.

BROOKE, WILLIAM J. (1946-) US writer who has specialized almost exclusively in the production of retold versions of FAIRYTALES for YA audiences, in the mode of the REVISIONIST FANTASY or the TWICE-TOLD tale, or combinations of both, sometimes with a distorting effect on the material. His collections are *A Telling of the Tales* (coll **1990**), *Untold Tales* (coll **1992**), and *Teller of Tales* (coll **1994**). *A Brush with Magic* (**1993**), which retells a Chinese story, exhibits the same faults and virtues in presenting the tale of a young Chinese orphan capable of bringing things to life by painting them (◊◊ CHINOISERIE). [JC]

BROOKS, TERRY (1944-) Working name of US lawyer and writer Terence Dean Brooks, whose first novel, *The Sword of Shannara* (**1977**), was deliberately modelled on parts of *The Lord of the Rings* (**1954-5**), and who has always acknowledged his indebtedness to the work of J.R.R. TOLKIEN. With the publication of *The Sword of Shannara*, TB became the first modern fantasy author to appear on the *New York Times* bestseller list, a triumph anticipated by his editor at Ballantine Books, Lester DEL REY, who had been searching for a marketable successor to Tolkien. He welcomed TB's easy, open style and adventure-oriented narrative, and used the book to launch the Ballantine Del Rey imprint. Along with Tolkien, TB's main influences are writers like Alexandre DUMAS, which may account for the ease with which he translated the complex CHRISTIAN FANTASY of *LOTR*, and the SECONDARY WORLD in which it takes place, into a series of morally transparent GENRE FANTASY adventures set in an apparent FANTASYLAND.

The Sword of Shannara is the first volume in the **Shannara** sequence, which continues with *The Elfstones of Shannara* (**1982**) and *The Wishsong of Shannara* (**1985**). **The Heritage of Shannara** – *The Scions of Shannara* (**1990**), *The Druid of Shannara* (**1991**), *The Elf Queen of Shannara* (**1992**) and *The Talismans of Shannara* (**1993**) – is set 300 years after the **Shannara** books; and *The First King of Shannara* (**1996**) is a prequel to the other seven novels. The initial tale is simple, but told with happy clarity. A young Hobbit-like UGLY DUCKLING is told by a Gandalf-like WIZARD that he is a HIDDEN MONARCH, last of the Shannara line; and must now undertake an arduous QUEST, with suitable COMPANIONS, for the eponymous SWORD. This MAGIC sword had been crafted centuries earlier by an associate of the current DARK LORD, a REVENANT who threatens to effect a terminal THINNING of the world – which is, in fact, post-APOCYLYPSE Earth, a venue whose potential for change substantially undercuts the fantasyland surface of the tales. The second and third instalments of **Shannara** present similar quests, though starring different members of the central Ohmsford clan; at points, the dependence upon a familiar cast and standardized MOTIFS becomes mechanical. The **Heritage of Shannara**, which comprises one sustained tale, carries later Ohmsfords through a quest to restore MAGIC to the now-thinned world.

TB's second squence – the **Kingdom of Landover** series comprising *Magic Kingdom for Sale – Sold!* (**1986**), *The Black Unicorn* (**1987**), *Wizard at Large* (**1988**), *The Tangle Box* (**1994**), a title which refers to a form of magical BONDAGE, and *Witches' Brew: A Magic Kingdom of Landover Novel* (**1995**) – has been less popular than the best-selling **Shannara** books; but although TB is clearly an uneasy humourist, the underlying premise of the sequence has some interest. A man of this world – a deeply depressed, stale lawyer – responds to an ad offering for sale the throne of Landover, laying down a million dollars for the privilege of running a magic kingdom. Not unexpectedly, READ THE SMALL PRINT complications ensue, more or less endlessly, and there is an ongoing rumination on the costs of wish fulfilment. The series continues, and may darken.

TB has also novelized the Steven SPIELBERG movie HOOK (*1991*) as *Hook* * (**1991**). [JC/JHH]

BROOKS, WALTER R(OLLIN) (1886-1958) US journalist and writer, active for many years but now remembered almost exlusively for the **Freddy the Pig** series of ANIMAL FANTASIES; his only other work of fantasy interest, *Ernestine Takes Over* (**1935**), is a mild tale much under the influence of Thorne SMITH. The sequence featuring Freddy and his animal associates is: *To and Again* (**1927**; vt *Freddy Goes to Florida* 1949; vt *Freddy's First Adventure* 1949 UK), *More to and Again* (**1930**; vt *Freddy the Explorer* 1949 UK; vt *Freddy Goes to the North Pole* 1951 US), *Freddy the Detective* (**1932**), *The Story of Freginald* (**1936**; vt *Freddy and Freginald* 1952 UK), *The Clockwork Twin* (**1937**), *Wiggins for President* (**1939**; vt *Freddy the Politician* 1948), *Freddy's Cousin Weedly* (**1940**), *Freddy and the Ignoramus* (**1941**), *Freddy and the Perilous Adventure* (**1942**), *Freddy and the Bean Home News* (**1943**), *Freddy and Mr Camphor* (**1944**), *Freddy and the Popinjay* (**1945**), *Freddy and the Pied Piper* (**1946**), *Freddy the Magician* (**1947**), *Freddy Goes Camping* (**1948**), *Freddy Plays Football* (**1949**), *Freddy the Cowboy* (**1950**), *Freddy Rides Again* (**1951**), *Freddy the Pilot* (**1952**), *Freddy and the Space Ship* (**1953**), *The Collected Poems of Freddy the Pig* (coll **1953**) – this latter assembled from previous volumes – *Freddy and the Men from Mars* (**1954**), *Freddy and the Baseball Team from Mars* (**1955**), *Freddy and Simon the Dictator* (**1956**), *Freddy and the Flying Saucer Plans* (**1957**) and *Freddy and the Dragon* (**1958**). Later volumes are somewhat thin, lacking the linguistic inventiveness and ingenious plotting of the earlier ones; those early tales, though ostensibly written as CHILDREN'S FANTASY, are full-length adventures, featuring a richly various cast – animals and humans alike. The sequence continues to warrant attention. [JC]

Other works: *Jenny and the King of Smithia* (**1947**); *Henry's Dog Henry* (**1965**); *Jimmy Takes Vanishing Lessons* (**1965**).

BROPHY, BRIGID (ANTONIA) (1929-1995) UK writer and critic whose first novel, *Hackenfeller's Ape* (**1953**),

savages any exculpatory implications of the fantastic in its treatment of an intelligent but doomed chimpanzee (◊ APES). *The Adventures of God in his Search for the Black Girl* (coll **1973**) contains some surreal tales (◊ SURREALISM); the title story neatly reverses the thrust of George Bernard Shaw's *The Adventures of the Black Girl in her Search for God* (**1932** chap). The land of Evarchia, where the nonfantasy *Palace Without Chairs: A Baroque Novel* (**1978**) takes place, is a RURITANIA. [JC]

BROSTER, D(OROTHY) K(ATHLEEN) (1877-1950) Prolific UK writer. Her SUPERNATURAL FICTION is restricted to two books, *A Fire of Driftwood* (coll **1932**) and *Couching at the Door* (coll **1942**). As with the famous title story of the second collection – in which a decadent artist (◊ DECADENCE) is haunted by a familiar he has himself unleashed – her work in this area generally involves violations of the natural world, and retributions which uneasily seal things together again. [JC]

BROWN, CHARLES BROCKDEN (1771-1810) The first US professional author. His most famous work, *Wieland, or The Transformation: An American Tale* (**1798**), like all his other Gothic romances (◊ GOTHIC FANTASY), is a RATIONALIZED FANTASY: it offers a naturalistic explanation for events that seemed supernatural; e.g., the eponymous Wieland, obsessed by voices which instruct him ritually to exterminate his family, may have in fact been deluded by a ventriloquist (◊ VENTRILOQUISM). CBB is perhaps of more importance for his transmutation of the US wilderness, in a novel like *Arthur Mervyn* (**1798-9**), into a hallucinated LABYRINTH-like LANDSCAPE; and for his extremely early use of a CITY like Philadelphia, in the same novel, in passages that convey a sense of genuine URBAN FANTASY. His influence on 19th-century writers like Nathaniel HAWTHORNE and Edgar Allan POE was very considerable. [JC]

Other works: *Alcuin: A Dialogue* (**1798**); *Edgar Huntly, or Memoirs of a Sleepwalker* (**1799**); *Ormond, or The Secret Witness* (**1799**).

BROWN, FREDRIC (WILLIAM) (1906-1972) US writer, mostly of crime and mystery fiction and sf. He wrote less SUPERNATURAL FICTION, but it is of high quality. FB began by selling humorous stories to the technical trade press in 1936 before graduating to the detective pulps in 1938 and sf and fantasy in 1941, including sales to UNKNOWN and WEIRD TALES. The best of these early fantasies, some clearly inspired by the works of Charles Fort (1874-1932), were collected in *Angels and Spaceships* (coll **1954**) and show FB's abilities to depict the Universe out of joint as in "The Angelic Angleworm" (1943 *Unknown*; vt "The Angelic Earthworm" 1954), where an error in the universal linotype machine starts life malfunctioning. The collection included also several very short humorous stories, often no more than extended puns, at which FB became skilled. These were further developed in *Honeymoon in Hell* (coll **1958**) and *Nightmares and Geezenstacks* (coll **1961**) – both assembled as *And the Gods Laughed* (omni **1987**) – and show a particular fascination with BLACK MAGIC and VOODOO. "The Geezenstacks" (1943 *WT*) is about DOLLS which preordain events. FB also delighted in twists on FAIRYTALES, particularly involving METAMORPHOSES, as in "Bear Possibility" (1960 *Dude*), "Fish Story" (in *Nightmares and Geezenstacks*) and "Eine Kleine Nachtmusik" (1965 *F&SF*) with Carl Onspaugh, his last completed story. FB's humour makes him comparable to Robert BLOCH, though his solipsist view of the Universe, as shown in "Solipsist" (in *Angels and*

Spaceships coll **1954**) and "It Didn't Happen" (1963 *Playboy*), betrays more than a hint of Theodore STURGEON.   [MA]

Further reading: *Martians and Misplaced Clues: The Life and Work of Fredric Brown* (**1993**) by Jack Seabrook.

BROWN, GEORGE MACKAY (1921-1996) Scottish writer who was born in and rarely left the Orkney Islands; he visited England only once, in 1989. The Orkneys were the setting for his first poem – "Prayer to Magnus" for *The New Shetlander* in 1947 – and for almost all his subsequent poetry and fiction. Much of the latter is influenced by the world-view of the 13th-century *Orkneyinga Saga*, and his plots often explicitly rework these LEGENDS. Stories of genre interest appear in *A Calendar of Love* (coll **1957**), *A Time to Keep* (coll **1969**) – whose contents differ from the compilation *A Time to Keep* (coll **1986** US) – *Hawkfall* (coll **1974**), *The Sun's Net* (coll **1976**), *Witch* (coll **1977**), *Andrina* (coll **1983**), *Christmas Stories* (coll **1985**), *The Golden Bird: Two Orkney Stories* (coll **1987**), *The Masked Fisherman* (coll **1989**) and *The Sea-King's Daughter/Eureka!* (coll **1991** chap). Frequently a thread of fantasy is conveyed through plots subtly based on ballads. Of his novels, *Magnus* (**1973**) treats its historical protagonist in mythic fashion, and *Time in a Red Coat* (**1984**) is a TIME-TRAVEL fantasy in which a young girl's birthgift sends her through various epochs whose horrors gradually stain her clothing the colour of blood. *Beside the Ocean of Time* (**1994**) recounts the history of a northern ISLAND – including SELKIES – through the DREAMS of its protagonist, who finally represents the power of STORY to redeem the lost world. Mermen (◊ MERMAIDS) appear in the children's stories assembled in *The Two Fiddlers* (coll **1974**). Several compositions by Sir Peter Maxwell Davies (1934-   ), also resident in the Orkneys, are based on GMB's work, most notably perhaps *The Martyrdom of St Magnus* (1977), a fantasy OPERA based on *Magnus*.   [JC]

Works for children: *Pictures in the Cave* (**1977** chap); *Six Lives of Fankle the Cat* (**1980** chap); *Keepers of the House* (**1986** chap); *Letters to Gypsy* (**1990** chap).

Plays: *A Spell for Green Corn* (**1970**); *Three Plays* (coll **1984**), including "The Voyage of Saint Brandan".

BROWN, MARGARET WISE (1910-1952) US author who wrote also as Timothy Hay, Golden MacDonald and Juniper Sage. Although she worked in a frequently ignored area of literature – picture books for young children – MWB has slowly gained recognition as a major writer for *The Runaway Bunny* (**1942** chap) and *Goodnight Moon* (**1947** chap). *The Runaway Bunny*, an ANIMAL FANTASY, might serve as a textbook illustration of the purposes of fantasy. A boy rabbit envisions how he might transform himself and travel through various places to escape his mother, who responds to each suggestion by describing how she would transform herself to follow him. Finally realizing he can never be separated from his mother, he agrees to stay at home with her. Here, fantasy initially functions as a means of ESCAPISM but ultimately serves to reconcile the individual to the real world. MWB's masterpiece was *Goodnight Moon*, where a similar bunny sits on his bed in his room and slowly says goodnight to every item in his little world. His refusal to distinguish between ANIMATE/INANIMATE – his insistence that all the objects he sees merit recognition – reaffirms a childlike form of animism and, in effect, transforms a mundane world into a fantasy world.

Discovering MWB is one of the secret joys of parenthood.   [LL]

Further reading: Over 100 picture books for small children

are listed in *Margaret Wise Brown: Awakened by the Moon* (**1992**) by Leonard S. Marcus.

BROWN, MARY (1929-   ) UK writer whose first novel of fantasy interest, *The Unlikely Ones* (**1986**), refreshingly subjects a QUEST plot – involving a SEVEN-SAMURAI assortment of COMPANIONS (bound by a CURSE) including a CAT, a crow, a girl who believes herself ugly (having been deceived by a WITCH's MIRROR) and who therefore wears a MASK, a KNIGHT whose disability can be cured only if he marries the ugliest girl in the land, a UNICORN and others – to an acute though good-humoured interrogation. MB's second novel of interest, *Pigs Don't Fly* (**1994**), replicates the pattern of the first, down to the number of companions (seven) and the handicapped young heroine, but sustains most of the *joie de vivre*.   [JC]

BROWN, WILL ◊ W. Harrison AINSWORTH.

BROWNIES ◊ ELVES.

BROWNING, ROBERT (1812-1889) UK poet, remembered as much for his association and marriage with Elizabeth Barrett Browning (1806-1881) as for his dramatic and lyrical poetry. From the outset RB's writing was dark and sinister, as betrayed in his first work, *Pauline* (**1833** chap). Much of his later work, though lightened of its morbidity, gave serious vent to RB's awe of the divine and the growth and development of the SOUL, a theme explored in *Paracelsus* (**1835** chap), *Sordello* (**1840** chap) and *Luria and A Soul's Tragedy* (**1846** chap). *Dramatis Personae* (coll **1864**) allowed for an even greater mystical, psychological and spiritual insight into human and religious experience, including RB's views on IMMORTALITY. Perhaps his most enigmatic poem in this vein is "'Childe Roland to the Dark Tower Came'" (1855 in *Men and Women*), in which a world-weary KNIGHT faces his final challenge (◊ CHILDE; DARK TOWER). In only slightly lighter vein was his ever popular "The Pied Piper of Hamelin" (1842 in *Dramatic Lyrics*), probably the best known interpretation of the old FOLKTALE (◊ PIED PIPER). RB's masterwork, *The Ring and the Book* (**1868-9** 4 parts), based on a notorious 17th-century murder case, is not fantasy but is deep in its interpretation of human psychology.   [MA]

BROXON, MILDRED DOWNEY (1944-   ) US writer, raised in Brazil, who began publishing work of fantasy interest with "Asclepius Has Paws" for *Clarion III* (anth **1973**) ed Robin Scott Wilson. Her first novel, *Eric Brighteyes 2: A Witch's Welcome* * (**1979**) as by Sigfridur Skaldaspillir, is a sequel (◊ SEQUELS BY OTHER HANDS) to H. Rider HAGGARD's *Eric Brighteyes* (**1891**), and carries on from the doom-laden concluding passages of that book: the WITCH who has caused the deaths of the cast carries them off in a longboat towards outcomes MDB saw a chance to explicate. *The Demon of Scattery* (**1979**) with Poul ANDERSON is minor – a DRAGON is called upon to save 9th-century Ireland from Vikings – but *Too Long a Sacrifice* (**1981**) more ambitiously attempts to universalize the conflict between Protestant and Catholic in Ireland in a tale whose 6th-century protagonists have been transformed into SLEEPERS UNDER THE HILL, and who awaken in Ireland in the current time of Troubles. They have TALENTS, and are AVATARS of figures from Celtic mythology (◊ CELTIC FANTASY), the woman "representing" the GODDESS of summer and the man HERNE THE HUNTER. The present age is shown in terms of THINNING; and the female protagonist becomes an agent of the HEALING necessary for the world to survive.

MDB rarely fails of seriousness of purpose, but has

demonstrated only a modest capacity at the crafting of storylines capable of bearing her intentions. [JC]

BRUEGEL, PIETER (*c*1525-1569) Dutch painter, much influenced in his early work by Hieronymus BOSCH, though his later paintings were increasingly grounded in landscape and social observation. The grotesqueries depicted in this late work almost always derive from exaggerations of what PB actually observed; "The Parable of the Blind" (1568), for instance, though giving a sense that we are witnessing a supernatural DANCE OF DEATH, contains nothing which could not have been assembled from life. This is not true of earlier work, of which paintings of fantasy interest include "Landscape with the Fall of Icarus" (*c*1555), "The Fight Between Carnival and Lent" (1559), "The Triumph of Death" (*c*1562), which presents a full-blown Dance of Death, "The Fall of the Rebel Angels" (*c*1562), "'Dulle Griet' (Mad Meg)" (1562), the last of PB's work in which relatively free-floating fantastic SYMBOLS of human states (like Gluttony and Sin) are to be found, and "The Tower of Babel" (1566), the most famous single representation of BABEL.

It is perhaps marginally safer to try to understand PB in modern terms than to attempt the same exercise for Bosch; the half-century or so between the two artists saw profound changes in consciousness and in the use of religious material. Bosch was an allegorist, whose fecundity of image illuminated a way of understanding the sense of the world and the spirit that we can no longer easily apprehend. PB used some of the same language but with a looseness that makes him seem much more contemporary. Bosch is a foreign country, and haunts the dreams of modern fantasists with images he may not himself have found haunting; when PB haunts us with images, we tend to feel that he, too, was haunted.

PB is often referred to as "the Elder" to distinguish him from his son, Pieter Bruegel II (*c*1564-1638). [JC]

BRUNNER, JOHN (KILIAN HOUSTON) (1934-1995) UK writer more closely identified with sf, in which field he was a prolific writer from his first book, *Galactic Storm* (**1951**) as Gill Hunt (a Curtis Warren housename). It tends to be overlooked that he wrote a deal of FANTASY – some of which is alternatively classifiable as sf, like *Catch a Falling Star* (1958 *Science Fantasy* as "Earth is But a Star"; rev vt *The 100th Millenium* **1959** US; rev 1968 US), a TECHNOFANTASY set in the FAR FUTURE where science is indistinguishable from MAGIC, and *Father of Lies* (1962 *Science Fantasy*; rev **1968** dos US), in which a child's TALENTS are so strong he can recreate his own world of MYTH.

JB's best-known fantasy is the apocalyptic *The Traveler in Black* (coll of linked stories **1971** US; rev exp [1 new story] vt *The Compleat Traveler in Black* 1987 US), in which the Traveller is the guardian of humanity (◊ FISHER KING) who travels the world across eons of time in a ceaseless battle against entropy, endeavouring to protect humanity against its own ignorance.

The initial **Traveller in Black** stories appeared first in SCIENCE FANTASY, which published most of JB's early WEIRD FICTION under variations of his own name and the pseudonyms Keith Woodcott and Trevor Staines. These stories covered most of the standard themes of SUPERNATURAL FICTION, but always with an original twist. "No Future In It" (1956) blends ALCHEMY and TIME TRAVEL; "Proof Negative" (1956) is a charming tale about SANTA CLAUS; in "The Kingdoms of the World" (1957) creatures of an ancient race

wait to reclaim the world (◊ MALIGN SLEEPER; WAINSCOTS); the protagonist of "All the Devils in Hell" (1960) enters a PACT WITH THE DEVIL to create a FEMME FATALE; "Oeuf du Coq" (1962) is one of the few tales to include a cockatrice. Some of these stories are included in *Out of My Mind* (coll **1967** US; variant UK edn 1968). JB expanded others to novel length, as with *The Gaudy Shadows* (1960 *Science Fantasy*; exp **1970**), where drugs create a tangible nightmare world (◊ DREAMS), and "This Rough Magic" (1956 *Science Fantasy*; exp vt *Black is the Color* **1969** US), about a BLACK-MAGIC coven.

In many of his weird stories JB used the device of a narrator to give the tale the feel of a CLUB STORY. This is particularly true of the **Tommy Caxton** series, five stories which span his career – "The Man who Played the Blues" (1956 *Science Fantasy*), "When Gabriel . . ." (1956 *Science Fantasy*), "Whirligig" (1967 *Beyond Infinity*), "Djinn Bottle Blues" (1972 *Fantastic*) and the posthumous "The Drummer and the Skins" (1995 *Interzone*) – and which verge on the TALL TALE, and the **Mr Secrett** stories – nine stories, mostly in *F&SF* (1977-92) – which are sinister exercises in PERCEPTION. Almost all of JB's fantasies challenge REALITY, urging us to reconsider the nature of things about us. [MA]
Other works: *No Future In It* (coll **1962**); *The Devil's Work* (**1970**), a PSYCHOLOGICAL THRILLER; *Time-Jump* (coll **1973** US); *The Best of John Brunner* (coll **1988** US); *A Case of Painter's Ear* (in *Tales from the Forbidden Planet* ed Roz KAVENEY 1987; **1991** chap US); much sf.
Further reading: *The Happening Worlds of John Brunner* (critical anth **1975**) ed Joseph W. de Bolt; *John Brunner, Shockwave Rider: A Working Bibliography* (latest edn **1989** chap) by Gordon Benson Jr and Phil Stephensen-Payne; "Behind the Realities: The Fantasies of John Brunner" by Mike ASHLEY in *WT* for Spring 1992, a special JB issue; *SFE*.

BRUST, STEVEN (KARL ZOLTAN) (1955-   ) US writer; a member of the SCRIBBLIES. SB's work mostly stands at something of an angle to the CONTEMPORARY-FANTASY preoccupations of that group, while sharing their sense of aggressive pleasure in creating what they most want to read or hear. This glory in competent performance is one of the traits of Vlad Taltos, ANTIHERO of the sequence which dominates SB's work; there is also a contemporary-fantasy feel to the specifics of corruption and gang warfare in the CITY of Adrilankha that acts as Taltos's backdrop.

The **Taltos** sequence comprises *Jhereg* (**1983**), *Yendi* (**1984**) and *Teckla* (**1986**) – all three assembled as *Taltos the Assassin* (omni **1991** UK) – plus *Taltos* (**1988**; vt *Taltos and the Paths of the Dead* 1991 UK), *Phoenix* (**1990**) and *Athyra* (**1993**). Related is the **Khaavren** sequence: *The Phoenix Guard* (**1991**), *Five Hundred Years After* (**1994**) and «The Viscount of Adrilankha». *Brokedown Palace* (**1986**) contains associated material.

Taltos is one of the large racial minority of humans, apparently from Eastern Europe (a legendary version of their crossing the borders between worlds is given in *Brokedown Palace*) in an ALTERNATE WORLD dominated by tall, magically gifted nonhumans, the Dragaerans (i.e., ELVES). The Dragaerans appear to have been genetically manipulated by supernatural or alien beings – the same ones who inflicted humanity on them – so that each of their clans (effectively subspecies) has characteristics of its totemic animal.

The **Khaavren** books are "historical novels" written shortly before Taltos's time by a Dragaeran equivalent of

Alexandre Dumas père; various characters from them recur, later in their vast lifespans, as important characters in the **Taltos** sequence. Not without their sinister side, these are the jauntiest and most likeable of SB's books.

Taltos is a typical figure of a less innocent time: an assassin, tavern-keeper and pimp, he has risen to the status of minor functionary in his adopted Dragaeran clan, the Jhereg – the Mafia, in effect. He has a COMPANION in the shape of a small poisonous reptilian flier, a teckla, and some ability in the human varieties of MAGIC. He finds himself caught up in the court and metaphysical intrigues of the Dragaeran aristocracy and their GODS, not least because he is an AVATAR of an ancient Dragaeran folk-HERO. Latterly, he is compromised by his wife's involvement in a revolutionary movement, apparently imported from Earth; in *Athyra*, the only **Taltos** book not told in the first person, he is on the run in the pastoral Dragaeran hinterland. The **Taltos** books offer unusually sophisticated versions of standard HEROIC-FANTASY tropes; SB explores aspects of crime and punishment, politics and history not usually given this much weight in the form. Above all, these books deal unusually acutely with racism.

Of his nonseries work, *To Reign in Hell* (**1984**) is a dreamlike account of the intrigues whereby Jehovah ($\lozenge$ GOD), originally one of several equals, politically discredits, isolates and exiles his principal rivals, who become the DEVILS of Christian myth; it parallels, rather than imitates, Anatole FRANCE's *La Révolte des Anges* (**1914**). *Agyar* (**1993**) is a VAMPIRE novel which imitates Gene WOLFE's *Peace* (**1975**) in the refusal of its hero to talk about his situation. SB brings to *Gypsy* (**1992**) with Megan LINDHOLM an amoral sense of Hungarian gypsy FOLKLORE that blends interestingly with Lindholm's streetwise pieties. [RK]

Other works: *The Sun, the Moon, and the Stars* (**1987**), fairytale retelling; *Cowboy Feng's Space Bar and Grille* (**1990**), sf.

BRYAN, MICHAEL $\lozenge$ Brian MOORE.

BRYANT, EDWARD (WINSLOW Jr) (1945-   ) US writer who began publishing work of genre interest with "They Come Only in Dreams" for *Adam* in 1970. His early work is best understood as sf, though the DYING EARTH presented in *Cinnabar* (coll of linked stories **1976**) has the fantasy feel usual to FAR-FUTURE tales, and Cinnabar itself displays many of the lineaments of a fantasy CITY. EB has become less prolific in the 1980s-90s, and most of his later non-sf veers into HORROR, one exception being *The Cutter* (in *Silver Scream* anth **1988**; **1991** chap), which is fantasy. [JC]

BUBBLEGUM CRISIS Japanese video series (1987 onwards). $\lozenge$ ANIME.

BUCHAN, JOHN (1875-1940) UK writer, barrister and politician, Governor General of Canada from 1935 – when he was created 1st Baron Tweedsmuir of Elsfield – until his death. His over 80 books include histories, biographies, belles lettres and considerable fiction, some of it SUPERNATURAL FICTION. He began to publish work of genre interest with "The Keeper of Cademuir" for the *Glasgow University Magazine* in 1894, and fantasy stories – often featuring an incursion of the past (or other form of HAUNTING) imperilling the well-being of an outwardly successful contemporary man – appeared in various collections, including *Grey Weather: Moorland Tales of My Own People* (coll **1899**), *The Watcher by the Threshold* (coll **1902**; exp 1916 US), *The Moon Endureth: Tales and Fancies* (coll **1912**; cut 1912 US) and *The Runagates Club* (coll **1928**), the last being a volume of CLUB STORIES typical of that subgenre.

Throughout his career, tales like "The Watcher by the Threshold" (1900 *Blackwood's Magazine*), "The Grove of Ashtoreth" (1910 *Blackwood's Magazine*) and "The Wind in the Portico" (1928 *Pall Mall Magazine*) tended to reiterate a sense of the fragility of modern people when confronted with the inchoate past – with the uncontrollable pagan allure of old GODS.

Prester John (**1910**), a lost-world ($\lozenge$ LOST RACES) tale, is JB's first novel of genre interest. Within a specious EDEN hidden deep in Ethiopia, the White protagonist finds a Black religious leader, Laputa, acting out a threatening (and potentially revolutionary) revival of the life of the 15th-century African monarch PRESTER JOHN; all ends safely in Laputa's defeat and death – but not before the narrator recognizes that Laputa is a finer man than himself, and than most White men: a daring observation in its day. The latent unease expressed in *Prester John* about the security of White civilization was fully justified by the outbreak of WORLD WAR I, which affected JB profoundly. In his long and informative introduction to the 1987 reprint of *These for Remembrance: Memoirs of 6 Friends Killed in the Great War* (**1919** chap), Peter VANSITTART suggests that what JB most "loathed and feared was 'de-civilization': civilized people losing their way and returning to barbarism". For JB, WWI opened an abyss into "a great emptiness", a sickness of spirit his fiction tends to reflect through protagonists who mask their despair with a forced stoicism, and who behave heroically according to precepts which have become hollow.

This thinness of being is most evident in JB's many nonfantastic novels, but it also marks his supernatural work. In *The Dancing Floor* (**1926**) a nightmare-ridden young man must protect an English girl from Greeks who wish to sacrifice her to ancient gods of pillage and dread; the minister protagonist of *Witch Wood* (**1927**), set in 17th-century Scotland, must attempt to confront the destabilizing EVIL inherent in the ancient forest which surrounds his village, but discovers that no one can go INTO THE WOODS without suffering some form of transformation; *The Gap in the Curtain* (**1932**) deploys the TIME theories of J.W. Dunne (1875-1949) to fix its protagonists' lives within an ominously pre-told future; and in *The Long Traverse* (**1941**; vt *The Lake of Gold* 1941 US) an Indian inflicts occult visions of Canada's history on a young lad.

JB's short fiction of genre interest has been assembled in *The Far Islands, and Other Tales of Fantasy* (coll **1982** US) ed John Bell and *The Best Supernatural Stories of John Buchan* (coll **1991**) ed Peter HAINING. *The Magic Walking-Stick* (**1932**) is a CHILDREN'S FANTASY. *The Watcher by the Threshold* and *The Moon Endureth*, plus *The Thirty-Nine Steps* (**1915**) and *The Power-House* (**1916**), are to be found in the paradoxically titled *Four Tales* (omni **1936**). [JC]

Other works: *The Courts of the Morning* (**1929**); *The Causal and the Casual in History* (**1929** chap) discusses ALTERNATE WORLDS, giving five theoretical examples; *The Blanket of the Dark* (**1931**).

Further reading: *Memory Hold-The-Door* (**1940**), autobiography.

BÜCHNER, GEORG (1813-1837) German writer, physician and revolutionary who, despite his death at age 23, has been one of the most powerfully influential figures in European drama. His first play, *Danton's Tod* (bowdlerized **1835**; restored 1850), shows GB's ferocious opposition to the ROMANTICISM of the immediately preceding generation. *Leonce und Lena: Ein Lustspiel* (**1850**), a comedy set in two imaginary kingdoms (called Popo and Pipi, for scatological reasons), makes use of FAIRYTALE conventions to fiercely satirical effect (◊ SATIRE). GB might be considered one of the most impassioned anti-fantasists of the 19th century were it not that, by a tremendous irony, his fragment *Woyzeck* (**1850** as *Wozzeck*) has become, partly because of the 1921 OPERA by Alban Berg (1885-1935), an important forerunner of much 20th-century drama, exerting an unmistakable influence on the work of such fabulists as Samuel BECKETT, Bertolt Brecht (1898-1956) and Günter GRASS. The grandmother's tale recounted in the play evokes familiar FOLKLORE tropes to devastating effect. The three plays are translated as *Danton's Death, Leonce and Lena* and *Woyzeck* in *The Plays* (omni trans Geoffrey Dunlop **1927** UK). [GF]

BUFFY THE VAMPIRE SLAYER US movie (*1992*). 20th Century-Fox/Kuzui/Sandollar. **Pr** Kaz Kuzui. Howard Rosenman. **Exec pr** Carol Baum, Sandy Gallin, Fran Rubel Kuzui. **Dir** Fran Rubel Kuzui. **Spfx** Joseph Mercurio. **Mufx** William Forsche, Mark Maitre. **Screenplay** Joss Whedon. **Starring** Rutger Hauer (Lothos), Luke Perry (Pike), Paul Reubens (Amilyn), Donald Sutherland (Merrick), Kristy Swanson (Buffy). 94 mins. Colour.

Through various REINCARNATIONS Merrick has been training a succession of likewise reincarnated VAMPIRE slayers – all young women skilled in the martial arts. The latest is the Los Angeles high-school airhead Buffy, who initially resents the intrusion into her brattish life. In due course she becomes enamoured of vampire slaying, and rids the world of numerous, including the grand master Lothos; in the later stages she is aided by the youth Pike.

Amiable nonsense, this perhaps suffers from the fact that its SATIRE is too accurate: the bimbo attitudes of Buffy (on ecology, "What about the ozone layer?/Yeah, we gotta get rid of that") are so rebarbative that one feels *BTVS* is kicking those who are already down. Where it is effective, though, is in showing how Buffy's type are effectively social vampires, taking and – except finally in Buffy's case – never giving. [JG]

BUGS BUNNY Warner Bros.' flagship character and, after MICKEY MOUSE and DONALD DUCK, almost certainly the most internationally recognized star of animated shorts, although BB, like Tom & Jerry and Woody Woodpecker, has never become a cultural ICON in quite the same fashion as the other two. His first (anonymous) appearance was in *Porky's Hare Hunt* (*1938*), where he reprised a Daffy Duck role from the previous year's *Porky's Duck Hunt* (*1937*). This movie was a great hit, and a few shorts later BB was established as a star. (His name came from the nickname of his first director, Ben "Bugs" Hardaway.) Tex AVERY, Chuck JONES and the great voice artist Mel Blanc (1908-1989) all played their parts in BB's early development, with Avery providing, in *A Wild Hare* (*1940*), the final definition of the character whom, at least 160 shorts later (not to mention compilation features), we know today: the wisecracking, inevitably superior TRICKSTER bunny, ever ready with a disingenuous "What's up, Doc?" or a sympathy-squeezing faked-death scene.

BB had a small role in WHO FRAMED ROGER RABBIT (*1988*), being voiced by Mel Blanc – as in his very first short. [JG]

Further reading: *That's All Folks!: The Art of Warner Bros. Animation* (**1988**) by Steve Schneider; *Bugs Bunny: Fifty Years and Only One Gray Hare* (**1990**) by Joe Adamson.

BUILDINGS Various entries in this encyclopedia deal with buildings, either individually (e.g., BABEL) or in terms of categories relevant to fantasy (◊ EDIFICE; HAUNTED DWELLINGS; INNS; LABYRINTHS). [JC]

BULFINCH, THOMAS (1796-1867) US recounter of MYTHS and LEGENDS, a son of the architect Charles Bulfinch (1763-1844). He is best-known as the author of *The Age of Fable* (coll **1855**), retelling Greek, Roman, Celtic, Scandinavian and Oriental tales. His *The Age of Chivalry* (coll **1858**) recounts tales of ARTHUR and from the MABINOGION, while *Legends of Charlemagne* (coll **1863**) repeats the exercise for that HERO. TB was a somewhat slipshod and certainly bowdlerizing writer; his importance is that he brought such material to the attention of general readers. [JG]

Other works: *Hebrew Lyrical History* (coll **1853**); *Poetry of the Age of Fable* (**1863**).

BULGAKOV, MIKHAIL (1891-1940) Soviet dramatist and novelist whose brilliant but risk-taking use of Aesopian language (◊ AESOPIAN FANTASY) in his attempts to anatomize the transmogrification of Bolshevism into the Communism of the corporate state has made him a symbol of the freedom of the imagination. As he was only partially published during his lifetime, his fame in the West is mainly posthumous, and has grown as his works have been translated. Some – like *Belaya gvardiya* (**1925**; trans Michael Glenny as *The White Guard* **1971** UK) and *Cherny sneg* (written late 1930s; trans Michael Glenny as *Black Snow* **1967** UK) – carry their messages without recourse to FABULATION; but most of the stories assembled in *Dyaboliada* (coll **1925**; trans Carl R. Proffer as *Diaboliad and Other Stories* **1972** US) are either sf or fantasy. They include "The Crimson Island: A Novel by Comrade Jules Verne Translated from the French into the Aesopian" (**1924** Germany), a short story which dangerously transgresses true Aesopian tactics by admitting to their use; it was made into a play in 1928. Other plays – all fantastically exaggerating sf material – include *Adam and Eve* (written 1931; trans Carl R. Proffer and Ellendea Proffer in *The Early Plays* coll **1972** US), in which the capitalist West destroys Russia before her citizens can be made chemically immune to aggression, and *Bliss* (written 1934; trans in *The Early Plays*; new trans Mirra Ginsberg in *Flight & Bliss: Two Plays* coll **1985** US), a TIME-TRAVEL tale in which 1930s Soviet apparatchiks effortlessly corrupt the future, a lesson taught in reverse in *Ivan Vasilievich* (written 1935; trans in *The Early Plays*; new trans Laurence Senelick in *Russian Satiric Comedy* anth **1983** US), in which a similar figure trades place with Ivan the Terrible, and becomes him.

The same transfiguring sense of mockery shapes *Sobacheye Serdste* (written 1925; trans Michael Glenny from the manuscript as *Heart of a Dog* **1968** UK and by Mirra Ginsburg **1968** US), in which a scientist transforms a dog into a quasi-man incapable of acting like an urban citizen; this also appears as the title story in *The Heart of a Dog and Other Stories* (coll trans Kathleen Cook-Horujy and Avril Pyman **1990** Russia). *Master i Margarita* (written 1938; **1966-7** US; complete text trans Michael Glenny as *The Master and Margarita* **1967** UK; cut text trans Mirra Ginsburg **1967** US) is MB's only full-scale fantasy: the DEVIL comes to Moscow, and Christ is recrucified. This astonishing work of the imagination – replete with wit and a certain sly eroticism – was filmed rather pedestrianly as *Majstori i Margarita* (*1972* Yugoslavia/Italy; vt *The Master and Margarita*) dir Aleksandar Petrović. [JC]

BULL, EMMA (1954-   ) US writer, married to Will SHETTERLY, and with him among the founders of the SCRIBBLIES. She began publishing with "Rending Dark" in *Sword and Sorceress* (anth **1984**) ed Marion Zimmer BRADLEY, and much of her infrequent short fiction has been assembled as *Double Feature* (coll **1994**) with Shetterly. Her first novel, *War for the Oaks* (**1987**), is an important and shaping contribution to the CONTEMPORARY-FANTASY subgenre, and has done much to establish MINNEAPOLIS as a minor but thriving venue for URBAN FANTASY. A young female rock musician becomes suddenly aware that Minneapolis is crosshatched with FAERIE, and that a grave conflict between the Seelie and the Unseelie factions within Faerie has begun. Contacted by a phouka – who manifests either as a man or as a dog – she reluctantly accepts her central role in the upcoming war. At points, an ELF aristo who rather resembles David Bowie (1947-   ) makes appearances; elf rockers have since become a central feature of GENRE FANTASIES with CROSSHATCH venues.

EB's second full fantasy, *Finder: A Novel of the Borderlands* * (**1994**), set in the **Borderland** SHARED-WORLD sequence created by Terri WINDLING and Mark Arnold, exhibits few of the flaws common to such enterprises. The tone is dark; the plot and writing are tight and identifiably akin to EB's other work. Bordertown, which exists along the boundary between this world and Faerie and is home to individuals estranged from either or both, is a venue of crosshatch and a THRESHOLD. Within this venue a man named Orient – he is the eponymous Finder (◊ TALENTS) – becomes involved in a murder investigation whose *noir* tone conflates with a growing sense that thresholds not only are escape hatches but can also ship contagion across the border. In *The Princess and the Lord of Night* (**1994**), an illustrated FAIRYTALE for children, a princess must QUIBBLE with a CURSE in order to save her parents from death. It is neatly told, with a tinge of the darkness typical of EB's work.

EB is as well known for sf as for fantasy; both *Falcon* (**1988**) and *Bone Dance: A Fantasy for Technophiles* (**1991**) – which is sf – are of considerable interest. With Shetterly she created and edited the **Liavek** shared world, a FANTASYLAND in which the eponymous city of thieves serves as a convenient venue for ADVENTURER FANTASY: *Liavek* * (anth **1985**), *The Players of Luck* * (anth **1986**), *Wizard's Row* * (anth **1987**), *Spells of Binding* * (anth **1988**) and *Festival Week* * (anth **1990**).

EB is not prolific, and the inherent darkness of her vision sometimes wars with an occasionally feelgood surface manner; but a sharp intelligence is constantly evident.   [JC]

BULL, RENÉ (1872-1942) Irish-born UK illustrator. One of the UK's most talented and prolific war artists, he first covered the Armenian massacres in 1896, then accompanied every UK campaign for the next 10 years. His graphic illustrations were published widely in national magazines and newspapers.

Like Edmund DULAC and Warwick GOBLE, RB was greatly influenced by Oriental art. His expert knowledge of Arab customs and costume led to his most admired book ILLUSTRATIONS, for editions of *The Arabian Nights* (1912) and *The Rubáiyát of Omar Khayyám* (1913). The latter work has remained in print for over 80 years.

RB illustrated several other fantasy classics, including editions of Jean de la Fontaine's *Fables* (1905), Joel Chandler HARRIS's *Uncle Remus* (1906), Hans ANDERSEN's *Fairy Tales* (1926) and Jonathan SWIFT's *Gulliver's Travels* (1928).   [RD]

BULLETT, GERALD (1893-1958) UK author. In his only fully fledged fantasy novel, *Mr Godly Beside Himself* (**1924**), a businessman whose work and marriage have lost their flavour becomes involved, through his new secretary, with various strange characters and surreal situations, culminating in his entering FAERIE – from which she hails – and exchanging places with his fairy DOUBLE Godelik. Godly finds Faerie in political turmoil, and is distressed to realize a revolution is imminent; meanwhile, the naïve Godelik has problems adapting to the complicated demands and conventions of modern life. Like several other notable fantasies of the period, the novel pleads eloquently for a healthy reconciliation of reason and imagination, lest modern human life become utterly arid.

An everpresent surreal element in GB's work gives many of his short fantasies a nightmarish edge. A brief preparatory sketch of *Mr Godly Beside Himself* is "The Enchanted Moment" (in *The Street of the Eye and Nine Other Tales* coll **1923**), whose protagonist is snatched into a parallel world where DIONYSUS contemptuously cuts him down (literally) to a much smaller size. The collection's title story is a curious metaphysical fantasy whose paranoid central character suspects he is being closely monitored by an unkind deity; also included is the subtle theriomorphic (◊ SHAPESHIFTERS) fantasy "Dearth's Farm". The grotesque "The Dark House" and the POSTHUMOUS FANTASIES "Queer's Rival" and "Last Days of Binnacle" in *The Baker's Cart and Other Tales* (coll **1925**) are among the darkest of GB's fantasies, contrasting sharply with the sentimental posthumous fantasy "The Grasshopper" in *The World in Bud and Other Tales* (coll **1928**). The title story of *Helen's Lovers and Other Tales* (coll **1932**) is a notable TIMESLIP romance, while "Fiddler's Luck" and "Tangent in Trouble" are ironic tales involving human males with supernatural females in a modern context; "Three Men at Thark" is a disturbing tale of a POLTERGEIST, and the ghostly CONTE CRUEL "The Elder" is perhaps Bullett's best HORROR story. The brief fantasies in *Ten Minute Tales and Some Others* (coll **1960**) are slight, having been written to a precise length, but the best are elegant. *Twenty Four Tales* (coll **1938**) reprints stories from the earlier collections alongside one new fantasy, "Dr Jannock's Chair"; an earlier selection of reprints was *Short Stories of To-day and Yesterday* (coll **1929**).

Some of GB's later novels have marginal fantasy elements: *Marden Fee* (**1931**) juxtaposes eras of prehistory and history in order to construct a melancholy romance of eternal recurrence; *Cricket in Heaven* (**1949**) reworks the Classical story of Alcestis in a modern setting.   [BS]

BULMER, (HENRY) KENNETH (1921-   ) UK writer whose first professionally published works of interest were two sf novels, *Space Treason* (**1952**) and *Cybernetic Controller* (**1952**), both with A.V. Clarke (1922-   ). In KB's total bibliography, which runs to over 100 titles, sf does not bulk hugely, but it does represent his most significant work in the field of the fantastic. His fantasy titles are in fact more numerous, due to one series, the **Dray Prescot** SCIENCEFANTASY sequence set in the interstellar framework common to space opera and comprising an extended set of PLANETARY ROMANCES. The series includes 37 titles, all as by Alan Burt Akers, or as told to Alan Burt Akers by Dray Prescot (the central figure of the tales), or as by Dray Prescot: *Transit to Scorpio* (**1972** US), *The Suns of Scorpio* (**1973** US), *Warrior of Scorpio* (**1973** US), *Swordships of Scorpio* (**1973** US), *Prince of Scorpio* (**1974** US), *Manhounds of Antares* (**1975** US), *Arena* *of Antares* (**1974** US), *Fliers of Antares* (**1975** US), *Bladesman of Antares* (**1975** US), *Avenger of Antares* (**1975** US), *Armada of Antares* (**1976** US), *The Tides of Kregen* (**1976** US), *Renegades of Kregen* (**1976** US), *Krozair of Kregen* (**1977** US), *Secret Scorpio* (**1977** US), *Savage Scorpio* (**1978** US), *Captive Scorpio* (**1978** US), *Golden Scorpio* (**1978** US), *A Life for Kregen* (**1979** US), *A Sword for Kregen* (**1979** US), *A Fortune for Kregen* (**1979** US), *A Victory for Kregen* (**1980** US), *Beasts of Antares* (**1980** US), *Rebel of Antares* (**1980** US), *Legions of Antares* (**1981** US), *Allies of Antares* (**1981** US), *Mazes of Scorpio* (**1982** US), *Delia of Vallia* (**1982** US), *Fires of Scorpio* (**1983** US), *Talons of Scorpio* (**1983** US), *Masks of Scorpio* (**1984** US), *Seg the Bowman* (**1984** US), *Werewolves of Kregen* (**1985** US), *Witch of Kregen* (**1985** US), *Storm over Vallia* (**1985** US), *Omens of Kregen* (**1985** US) and *Warlords of Antares* (**1988** US). As SWORD AND SORCERY derived from Edgar Rice BURROUGHS, the **Dray Prescot** books are efficient. Similar titles include *Kandar* (**1969** US), which features a scholar HERO who takes up a SWORD, and *Swords of the Barbarians* (**1970** US). The **Odan the Half-God** sequence, as by Manning Norvil – *Dream Chariots* (**1977**), *Whetted Bronze* (**1978** US) and *Crown of the Sword God* (**1980** US) – conveys a Nordic flavour. [JC]

BULWER-LYTTON, (Sir) EDWARD (GEORGE EARLE LYTTON), FIRST BARON LYTTON (1803-1873) UK writer and politician, usually referred to as Lord Lytton; more properly, he was Edward Bulwer until 1838, then Sir Edward Bulwer until 1843, when he succeeded to his mother's estate at Knebworth and became Sir Edward Bulwer-Lytton, and finally he became 1st Baron Lytton of Knebworth in 1866.

EB-L is probably best-known in fantasy for his GHOST STORY *The Haunted and the Haunters, or The House and the Brain* (1859 *Blackwood's Magazine*; **1905** chap), an archetypal Victorian HAUNTED-DWELLING story – though it is profound in its assessment and investigation of the HAUNTING, and remarkably effective in its creation of atmosphere.

EB-L's writing career began with the publication, when he was still in his teens, of *Ismael: An Oriental Tale* (**1820**); at the time he was heavily under the influence of Lord BYRON. A scandalous affair and a disastrous marriage in 1827 left him penniless – his mother cut off his funds – and something of an outcast. This forced him to write for a living, and his interests turned to other outcasts: criminals and occultists. After his bestseller *Pelham* (**1828**) brought him back into society, EB-L remained a critic and observer. He weathered the literary scandals which arose from *Paul Clifford* (**1830**) and *Eugene Aram* (**1832**), later dubbed "Newgate novels" because they seemed to glorify crime and exonerate the criminal – a charge also levelled against W. Harrison AINSWORTH.

Although EB-L's fame came with his historical novels, like *The Last Days of Pompeii* (**1834**), he remained deeply interested in ROSICRUCIANISM and OCCULTISM, which inspired a series of works starting with *Asmodeus at Large* (1832-3 *New Monthly Magazine*; **1833** US), a post-Gothic extravaganza concerning the narrator's exploits with a DEMON.

In 1835, while researching books on ASTROLOGY and not long before his separation from his wife, EB-L had a DREAM in which the fabric of a novel came to him. He produced an incomplete version as "Zicci" (1838 *Monthly Chronicle*; in *Critical and Miscellaneous Writings* **1841**) and then reworked it substantially as *Zanoni* (**1842**). Bearing some semblance to

Charles MATURIN's *Melmoth the Wanderer* (**1820**), this was intended to be an ALLEGORY on the human condition, but is too philosophical to work effectively at that level; as a story, however, the tale of an immortal (◊ IMMORTALITY) adept and his SACRIFICE for LOVE became one of the classic works of Victorian supernaturalism. In *Zanoni* EB-L created the image of the "dweller on the threshold", a phrase beloved by writers of WEIRD FICTION ever since. The QUEST for an ELIXIR OF LIFE continued in EB-L's *A Strange Story* (1861-2 *All the Year Round*; **1862**), which further reworked *Zanoni*'s theme, though shifting more toward the spiritual than the diabolic.

EB-L's own outcast status, and particularly the perished relationship with his estranged wife, continued to fuel his imagination, so that now he himself became an OBSESSED SEEKER. This culminated in the nonsupernatural but highly charged *Kenelm Chillingly* (**1873**). EB-L discovered, through his fiction, that the QUEST is often more satisfying than the reward: in the proto-sf novel *The Coming Race* (**1871**; vt *Vril: The Power of the Coming Race* 1972 US) his protagonist realizes that the utopian life he discovers among a scientifically advanced subterranean civilization is . . . boring.

EB-L's work was sensationally popular in his day and had a strong influence on other writers. His occult works, along with those of J. Sheridan LE FANU, form the basis of modern SUPERNATURAL FICTION.

EB-L's son, the diplomat Robert Bulwer-Lytton (1831-1891), was also a poet and novelist, mostly as Owen Meredith. This pseudonymity has at times caused confusion, especially over the novel *The Ring of Amasis* (**1863**), which is sometimes mistakenly attributed to the father, even though EB-L regarded it as immature. The story uses the supernatural (in the form of a MUMMY's RING) to highlight guilt. [MA]

Other works: *Godolphin* (**1833**); *The Pilgrims of the Rhine* (coll of linked stories **1834**); *The Student* (coll **1835**); *King Arthur* (**1848**), a book-length narrative poem (◊ ARTHUR).

Further reading: *The Life of Edward Bulwer, First Lord Lytton* (**1913**) by the 2nd Earl of Lytton; "Lytton the Mystic" by Harold Armitage in *The Haunted and the Haunters* (**1925**) ed Armitage, which also reprints EB-L's story and provides analyses on the original location of the haunting; *Strange Stories, and Other Explorations in Victorian Fiction* (**1971**) by Robert Lee Wolff; *Gothic Immortals: The Fiction of the Brotherhood of the Rosy Cross* (**1990**) by Marie Roberts.

BULWER-LYTTON, ROBERT ◊ Edward BULWER-LYTTON.

BUNCH, CHRIS Working name of US scriptwriter – for such series as *The Rockford Files* – and author Christopher R. Bunch (1943-   ), all of whose work has been done in collaboration with Allan COLE. [JC]

BUNCH OF SEVEN Canadian AFFINITY GROUP. ◊ CANADA.

BUÑUEL, LUIS (1900-1983) Spanish-born movie director who during a restless life worked in Spain, France, the USA and Mexico. Of the 34 feature movies he directed – many of which he also either wrote or co-wrote – most are of fantasy interest if only in that LB, an early convert to SURREALISM, never lost his Surrealist eye: even his cheaply and quickly produced Mexican movies have the "feel" of fantasy.

Having served as assistant director to Jean Epstein (1897-1953), LB first made his mark with the notorious Surrealist short *Un Chien Andalou* (**1928**), which opened with the sight of a woman's eyeball being sliced with a razor. Much more significant than such juvenilia was *L'Age d'Or* (*1930* vt *The Golden Age*), his first feature movie, co-written with Salvador DALI: filled with Surrealist imagery, this was essentially a movie about the difficulties of LOVE, and caused a riot during an early screening – after which it was for a while banned. A somewhat patchy career over the next years saw him doing such chores as, briefly, working in Hollywood to produce Spanish-language versions of Warner Bros. movies. He went to Mexico in 1946, remaining there until 1961 although also making a number of movies in France. During this period, LB's movies of fantasy interest included *The Adventures of Robinson Crusoe* (*1952*; ot *Las Aventuras de Robinson Crusoe*), which mesmerizingly portrays no character at all except Crusoe (Dan O'Herlihy) until close to its end, when Friday (Jaime Fernandez) appears. A decade later there appeared *The Exterminating Angel* (*1962*; ot *El Angel Exterminador*), an intriguing work of fantasy in that, while we know the supernatural elements are there, they are kept off-screen. Like *L'Age d'Or* and others since, it sees LB ruthlessly assault middle-class values: a group of people, somehow unable to leave a room, soon descend into barbarism; when at last they escape they go to a cathedral, where the process repeats itself. *Belle de Jour* (*1966*) continued the assault, although rather more gently and wittily, in its tale of a beautiful woman (Catherine Deneuve), married to a successful surgeon, who feels she must spend her afternoons as a call-girl: we are never quite sure if the movie's surface events are erotic DREAM or reality. *The* DISCREET CHARM OF THE BOURGEOISIE (*1972*; ot *Le Charme Discret de la Bourgeoisie*) uses similar means to those of *The Exterminating Angel* to show corruption, as a group of "respectable" crooks and their entourage try to have a quiet meal. *The Phantom of Liberty* (*1974*; ot *Le Fantôme de la liberté*) resembles a collection of sketches rather than a feature; very funny, it belongs more to the FANTASTIC than to fantasy. *That Obscure Object of Desire* (*1977*; ot *Cet obscur objet du desir*) tells of a puritanical man's obsessive love for a beautiful younger woman, which obsession he is barred from consummating with her even after bedding her; the obscurity of the "object" is enhanced by the casting of two actresses (Carole Bouquet, Angela Molina) in the single role.

It is hard to locate LB within fantasy; it is equally hard to omit him from any consideration of fantasy. Such difficulties are a common hallmark of the most significant fantasy creators. [JG]

BUNYAN, JOHN (1628-1688) A tinker of Bedfordshire, England, who served under Cromwell in the Civil War. After years of spiritual anguish he experienced a religious conversion and became a Calvinist preacher. He was arrested for preaching without a licence and spent 12 years in jail. Here he wrote *Grace Abounding* (**1666**), his spiritual autobiography; it teems with imagery of fiends and dangerous journeying. He transformed this personal experience into an ALLEGORY others could recognize with the first part of *The Pilgrim's Progress from This World to That Which is to Come* (**1678-9**; Part II **1684**). (◊◊ PILGRIM'S PROGRESS.)

Christian, with a burden on his back, sets out from the City of Destruction. His wife and children refuse to accompany him. He has to cross the Slough of Despond and pass through the Wicket-Gate. He reaches the House of the Interpreter, full of emblematic scenes, comes to the Cross, where his burden rolls away into an open sepulchre, fights the fiend Apollyon, and passes through the Valley of the Shadow of Death. At Vanity Fair, he and his COMPANION Faithful refuse to buy and are arrested. Faithful is condemned to death; Christian escapes. With Hopeful he goes astray and they almost die in Doubting Castle, prisoners of Giant Despair. From the Delectable Mountains they see the Celestial City. To reach it, they must cross a deep river. Shining Ones welcome them on the other side. Many of these images have become common currency even among those who have never read the book.

The characters Christian meets are virtues and vices personified. Yet Bunyan brings them alive in recognizably 17th-century landscape and society. His is a *pilgrimage*, rather then a QUEST with a task to be accomplished no matter where it leads: the destination and the journey are what matter. Other writers had used the spiritual-pilgrimage form before, but Bunyan's sources are likely to have been FOLKLORE and medieval pulpit illustrations.

In *Mr Badman* (**1680**), the story of a dissolute businessman, he foreshadowed the biographical novel. JB returned to allegory in *The Holy War* (**1682**), using his military experience. The city of Mansoul is captured by Diabolus and has to be rescued for the king by his son. In 1684 came the second part of *The Pilgrim's Progress*. Christian's wife and children, with Mercy, decide to follow him. After being attacked, they are provided with a champion, Mr Greatheart. Their journey is thereafter easier than Christian's. Bunyan was reflecting his own experience as pastor of a congregation. [FS]

BURDEKIN, KATHARINE (PENELOPE) (1896-1963) UK writer who published her first novels – in the 1920s – under her own name, but who became better known in the 1930s as Murray Constantine. *Swastika Night* (**1937**), as by Constantine, remains her most famous single work; it is a HITLER WINS tale, almost certainly the first written, and is a vigorously feminist sf analysis of fascism. KB's first novel of fantasy interest, *The Burning Ring* (**1927**), intricately presents a young man's RITE OF PASSAGE into full manhood in the form of a tripartite NIGHT JOURNEY; through use of a magic RING he finds himself able to engage in TIME TRAVEL, and has maturing experiences in three different epochs. *The Children's Country* (**1929** US) as by Kay Burdekin is a CHILDREN'S FANTASY set in a land free of GENDER distortions. *The Devil, Poor Devil!* (**1934**), as by Constantine, is a SUPERNATURAL FICTION in which the DEVIL fatally experiences the THINNING of belief in his reality. *The Rebel Passion* (**1929**), *Proud Man* (**1934**) as by Constantine and *The End of This Day's Business* (**1990**) are all sf. [JC]

BURDEKIN, KAY ◊ Katharine BURDEKIN.

BURFORD, LOLAH (MARY) (1931-   ) US writer. In her TIMESLIP fantasy *The Vision of Stephen: An Elegy* (**1972**) the movement is unusually forward in time. The protagonist, the son of an Anglo-Saxon princeling, is treated brutally by his father, and escapes – though at first only by means of mental contact – to early-19th-century England, where he eventually finds genuine refuge. [JC]

BURGESS, ANTHONY Working name of UK writer and composer John Anthony Burgess Wilson (1917-1993), whose most famous single work, *A Clockwork Orange* (**1962**; cut 1963 US), is sf. Though much of his fiction is exuberantly experimental, displaying a tropical intensity of language which generates a magic-realist sense of transfigured scapes (◊ MAGIC REALISM), AB wrote relatively little outright fantasy. In *The Eve of Saint Venus* (**1964**) a RING brings the GODDESS to life; *Beard's Roman Women* (**1976**

US) is a GHOST STORY as well as a lament for lost LOVE; *A Long Trip to Teatime* (**19776**) is a surreal FAIRYTALE for older children; *Any Old Iron* (**1989**), whose title describes EXCALIBUR, posits its survival into the 20th century, with satirical consequences; and *Enderby's Dark Lady, or No End to Enderby* (**1985**) is a fantasy reflecting the life of SHAKESPEARE. Some of the stories assembled in *The Devil's Mode* (coll **1989**) are fantasy. Of all these titles, *Beard's Roman Women* is the most successful; the others tend to unease, perhaps generated by AB's overall distaste for genre literatures. [JC]

BURGESS, (FRANK) GELETT (1866-1951) US writer who created various noteworthy neologisms, the most famous being "bromide" and "blurb". *The White Cat* (**1907**), in which an evil doctor hypnotizes (◊ MESMERISM) a female patient and evokes her evil *alter ego* or DOPPELGÄNGER, owes much to Robert Louis STEVENSON, as does *Lady Mechante, or Life as it Should be: Being Divers Precious Episodes in the Life of a Naughty Nonpareille: A Farce in Filigree* (**1909**), whose spoof episodes reflect the influence of Stevenson's *New Arabian Nights* (coll **1882**), but which do not extend beyond hoax. He was noted for several humorous GHOST STORIES, especially "The Ghost-Extinguisher" (1905), where science finds a way to petrify ghosts. [JC]

BURNE-JONES, [Sir] EDWARD (COLEY) (1833-1898) UK painter. Along with Dante Gabriel ROSSETTI and William MORRIS, he was part of the second generation of the PRERAPHAELITES. He produced many tapestry and stained-glass designs for Morris, and these, together with his paintings, evoke a dreamy, romantic, medieval, literary never-never land in the style of Filippo Lippi (*c*1406-1469) and Sandro Botticelli (1444-1510). Among his greatest book ILLUSTRATIONS are the 87 plates for Morris's Kelmscott Press edition of the works of Geoffrey CHAUCER (1897). Typical among his many weird fantasy subjects was the only painting EB-J ever contributed to the Royal Academy Exhibition, "The Depths of the Sea" (1886), which shows a powerful MERMAID hugging a naked man around the waist, and pulling him down to his death.

EB-J's son, Philip Burne-Jones (1861-1926), was a painter in the same tradition, best-remembered for his portrait of "The Vampire" (1897), which coincided with the publication of Bram STOKER's *Dracula* (**1897**). This shows a woman with flowing black hair and long sharp teeth, clad in a clinging nightdress, astride a contented young man. The catalogue was enlivened with a poem called "The Vampire" by the artist's cousin, Rudyard KIPLING. [RD]

BURNETT, FRANCES (ELIZA) HODGSON (1849-1924) UK-born writer, in the USA from 1865, and a working author from 1868 to 1922, by which time she had published at least 70 books, few now remembered. Her most famous titles are *Little Lord Fauntleroy* (**1886**) and *The Secret Garden* (**1911**), an important early-20th-century children's book whose title itself has become a common term (◊ SECRET GARDEN). The supernatural content of the tale is slight – a touch of pantheism with reference to the life-affirming ambience of the garden itself, which cures two children of their worldviews; and a call from his dead wife to a bereft widower – but the book as a whole, like the best fantasies, generates a sense of earned TRANSFORMATION. (Fantasy is entirely absent from the two significant movie adaptations, *The Secret Garden* [**1949**] and *The Secret Garden* [**1993**], although the former lovingly stresses its Gothic ambience.) Generally, like many 19th-century writers, FHB tended to eschew outright fantasy, which at the time

continued to lack even the precarious adult cachet accorded to SUPERNATURAL FICTION. So a tale like "Behind the White Brick" (1879 *St Nicholas Magazine*) – later assembled in *Little St Elizabeth and Other Stories* (coll **1890**) – stands out. Unjustly punished, the child heroine, reversing Alice's descent (◊ Lewis CARROLL), falls up a chimney into Chimneyland, a LITTLE BIG paradise with clear WONDERLAND characteristics, where she meets fictional characters, SANTA CLAUS, a talking DOLL and others. The tale has a swing and a drive that make one regret that FHB did not write full-length fantasies. Of her other books of interest, *The White People* (**1917**) is about a young girl with second sight (◊ TALENTS) who sees GHOSTS and the eponymous folk. [JC]

Other works: *In the Closed Room* (**1904**); *The Land of the Blue Flower* (**1909** chap).

BURGESS, THORNTON W. (1874-1965) US children's writer. ◊ John CROWLEY.

BURNS, JIM (1948-   ) Welsh painter and illustrator. JB works in oils, and favours a colour scheme featuring muted browns in a composition consisting typically of static three-quarter-length standing figures against a background of space hardware and alien architecture. His early influences were Frank Hampson (1917-1985) and Frank BELLAMY. In recent years he has begun to use a fresher colour range and his compositions have become more varied and subtle.

On leaving St Martin's School of Art, London, JB joined the agency Young Artists and worked on book covers, inclining more and more towards sf. He worked on the aborted project to bring **Dan Dare** to tv, painting large extraterrestrial backgrounds. He did 30 large oil paintings to illustrate the novella *Planet Story* (graph 1979) by Harry Harrison. He worked briefly on Ridley Scott's ill-fated «Dune» project, and did design paintings for *Blade Runner* (**1982**). He also worked on the design stage of Sir Clive Sinclair's C5 miniature electric vehicle project. He did a series of b/w illustrations for *Eye* (coll **1985**) by Frank Herbert.

JB won the Hugo AWARD as Best Professional Artist in 1987; he was the first non-US artist to do so. He won it again in 1995. [RT]

Other work: *Lightship* (graph coll **1986**), with text by Chris Evans (1951-   ).

BURNS, JOHN M. (1935-   ) Prolific and very accomplished UK COMIC-strip artist, with a bold, slick and sophisticated line style; he is possibly the finest draughtsman in the comics medium. His firm control of composition and skilled command of narrative make him one of the most sought-after comics artists in both the UK and mainland Europe. However, little of his very substantial output has been of lasting value, since much has been in the form of jokey newspaper strips with a soft-porn slant or ephemeral children's comics featuring strip versions of tv series. His drawing is always arresting, and some of his newspaper strips have been republished in collected form, including DANIELLE, *The Seekers* (1966-71 *Daily Sketch*, reprinted in issues of *Menomonee Falls Gazette* and continued 1979-84 in *The Comic Reader* US) and *Modesty Blaise* (1978-9 *London Evening News*).

On leaving school at 16, JMB took an apprenticeship with Link Studios, London. He began to undertake freelance work on his own account in 1954. His remarkable drawing skill first became evident in WRATH OF THE GODS, a series set among the GODS of ancient Greece. He went on to draw **Kelpie**, a SORCERER'S-APPRENTICE tale for the children's weekly *Wham*. Further sf-oriented strips followed, including several for comics based on Gerry Anderson tv series.

In 1982 **Ertha**, written by Donne Avenell, began in full colour in the supplement to the Sunday newspaper *News of the World*. JMB resuscitated **Jane** for the *Daily Mirror*, produced the album series **Zetari** and did a brief run on **Dan Dare** for *Eagle*. His work for US comic books includes *Espers: Assassins* (1987), "Wild Cards" in *Epic #4* (1992) and *A Silent Armageddon* (graph **1993**) for the Dark Horse **James Bond** series. He continues to draw and paint **El Capitan Trueno** for the Spanish market and **Judge Dredd** in the UK. [RT]

BURNS, RICHARD (1958-1992) UK writer whose six books included two full fantasies, and who killed himself – because, it has been suggested, of the financial stress of freelance work – before fully establishing himself in any mode. *A Dance for the Moon* (**1986**) is not fantasy, but exhibits an intensity of language and longing for elsewhere that serves as a prelude to his later work, all of which is notable for a sense that its protagonists are estranged from normal life. *Khalindaine* (**1986**) is constructed around the search for the UGLY-DUCKLING heir to the throne of the eponymous FANTASYLAND kingdom (◊ HIDDEN MONARCH); a good deal of suspense is generated by the problem of deciding who is the true heir. In the sequel, *Troubadour* (**1988**), the new Emperor is taxed beyond endurance by a puritan Brotherhood which is incensed because he refuses to inflict upon himself a year-king mutilation (◊ GOLDEN BOUGH); civil war ensues and the dynasty threatens to topple. *Fond and Foolish Lovers* (**1990**), essentially a this-worldly meditation on death, contains passages that evoke the tone of POSTHUMOUS FANTASY; the reality of this level of interpretation is, however, insecure. [JC]

BURN WITCH BURN (ot *Night of the Eagle*) UK movie (**1961**). Independent Artists/Anglo Amalgamated. **Pr** Albert Fennell. **Exec pr** Leslie Parkyn, Julian Wintle. **Dir** Sidney Hayers. **Photography** Reginald Wyer. **Screenplay** George Baxt, Charles BEAUMONT, Richard MATHESON. **Based on** *Conjure Wife* (1943 *Unknown*; 1953) by Fritz LEIBER. **Starring** Janet Blair (Tansy Taylor), Kathleen Byron (Evelyn Sawtelle), Colin Gordon (Lindsay Carr), Margaret Johnston (Flora Carr), Peter Wyngarde (Norman Taylor). 87 mins. B/w.

The second movie based on the Leiber novel; the first was the very minor *Weird Woman* (**1944**). High-flying psychology don Norman discovers his wife Tansy is a WITCH; an arch-rationalist, he believes her talk of other faculty members using BLACK MAGIC against him is hogwash, and makes her burn all her paraphernalia. But from the next day his career starts to disintegrate, and at last he himself is driven to use a magical RITUAL in order to save her life. At the university he confronts a colleague, Flora Carr, who confesses to her own witchcraft and proceeds to burn a house of TAROT cards, saying that, through sympathetic MAGIC, this is also the Taylors' house, in which Tansy lies sleeping. He flees, but she plays over the public-address system a hexed tape that warps his (and our) PERCEPTIONS so that he believes he is being attacked by a statuary eagle. When Carr's husband innocently turns off the tape the SPELL is broken, and Tansy is saved from the flames. Foreshadowing ROSEMARY'S BABY (*1968*), BWB is a wry SATIRE of infradepartmental politicking but also a tightly constructed DARK FANTASY whose auctorial and directorial brio renders its impossible events quite plausible. [JG]

BURRAGE, A(LFRED) M(cLELLAND) (1889-1956) UK short-story writer who published a vast amount of fiction in popular MAGAZINES, including boys' magazines, for over 40 years, starting in 1905. AMB's total output has never been fully assessed, and more may be hidden under pseudonyms other than the known Frank Lelland and Ex-Private X. He wrote in all genres, but is remembered today for his GHOST STORIES, some of which are among the most effective ever written. AMB's GHOSTS, frequently menacing but not always EVIL, often serve as PORTENTS or warnings. With fecund originality, he frequently mixed his ghost stories with elements of TIMESLIP or switched the perspective to the ghost's, so as to allow alternate interpretations (◊ PERCEPTION). These tales have been collected as *Some Ghost Stories* (coll **1927**), *Someone in the Room* (coll **1931**) as by Ex-Private X, and the posthumous *Between the Minute and the Hour* (coll **1967**). Jack Adrian (1945-) has since assembled further collections from old magazines: *Warning Whispers* (coll **1988**, «rev» 1997), *Intruders* (coll **1995**) and «The Occult Files of Francis Chard» (coll 1996).

AMB also produced a BLACK-MAGIC thriller, *Seeker to the Dead* (**1942**), which was almost certainly based on the occult practices of Aleister CROWLEY. Not published in book form but typical of AMB's lighter moments was a series of amusing romps, in the style of P.G. Wodehouse (1881-1975), set in the time of ARTHUR and featuring **Sir Archibald**; these stories are SATIRES on 1920s society. Titles were "The Knightly Adventures of Sir Archibald" (1922 *Yellow Magazine*), "The Further Adventures of Sir Archibald" (1923 *Yellow Magazine*) and "The Further Adventures of Sir Archibald and the Knights of the Round Table" (1925 *Yellow Magazine*). [MA]

BURROUGHS, EDGAR RICE (1875-1950) US writer whose various series – **Tarzan**, the **Barsoom** sequence (featuring John Carter on the planet Mars), the **Venus** sequence, and the **Pellucidar** sequence (set within a HOLLOW EARTH) – have all had a profound effect on the development of both genre sf and GENRE FANTASY. A good case could be made for calling almost all of ERB's many titles both essentially sf and essentially fantasy. Within the definition of FANTASY followed in general throughout this encyclopedia, however, ERB falls marginally into the realm of sf, if for no other reason than that – sometimes only by implication – his venues and marvels are justified by some argument, however tenuous, of an sf nature. There is nothing deliberately impossible in his work; though, perhaps confusingly, a sense of pubescent wish-fulfilment can be detected throughout. The blissful ease with which TARZAN – who is both apeman and aristo, SHADOW and constitutional monarch of his SOUL – "solves" or soothes the 19th-century obsession with the dark DOUBLE is nothing but dreamily adolescent. But to identify Tarzan's LAND-OF-FABLE Africa or John Carter's Mars (◊ SCIENCE FANTASY) as arenas in which devoutly desired actions can freely be countenanced is not to identify these regions of the heart, or the actions which fill them, as fantasy. [JC]

Barsoom: *A Princess of Mars* (1912 *All-Story Magazine* as "Under the Moons of Mars" as by Norman Bean; **1917**); *The Gods of Mars* (1913 *All-Story*; **1918**); *The Warlord of Mars* (1913-14 *All-Story* **1919**); *Thuvia, Maid of Mars* (1916 *All-Story Weekly* **1920**), *The Chessmen of Mars* (1922), *The Master Mind of Mars* (**1928**), *A Fighting Man of Mars* (**1931**), *Swords of Mars* (**1936**), *Synthetic Men of Mars* (**1940**), *Llana of Gathol* (1941 *Amazing Science Fiction*; fixup **1948**) and

John Carter of Mars (1941-3 *Amazing Science Fiction*; coll **1964**). "John Carter and the Giant of Mars", in the last volume, was originally written as a juvenile tale with ERB's son, John Coleman Burroughs (1913-1979), and was later expanded by ERB.

Tarzan: *Tarzan of the Apes* (1912 *All-Story* **1914**); *The Return of Tarzan* (1913 *New Story*; **1915**); *The Beasts of Tarzan* (1914 *All-Story Cavalier*; **1916**); *The Son of Tarzan* (1915 *All-Story Cavalier*; **1917**); *Tarzan and the Jewels of Opar* (1916 *All-Story Cavalier*; **1918**); *Jungle Tales of Tarzan* (coll **1919**; vt *Tarzan's Jungle Tales* 1961 UK); *Tarzan the Untamed* (coll of linked stories **1920**); *Tarzan the Terrible* (**1921**); *Tarzan and the Golden Lion* (**1923**); *Tarzan and the Ant Men* (**1924**; rev 1924); *Tarzan, Lord of the Jungle* (**1928**); *Tarzan and the Lost Empire* (**1929**); *Tarzan at the Earth's Core* (**1930**); *Tarzan the Invincible* (**1931**); *Tarzan Triumphant* (**1932**); *Tarzan and the City of Gold* (1931 *Argosy*; **1933**; cut 1952); *Tarzan and the Lion Man* (**1934**); *Tarzan and the Leopard Men* (**1935**); *Tarzan's Quest* (**1936**); *Tarzan and the Forbidden City* (**1938**; cut vt *Tarzan in the Forbidden City* 1940); *Tarzan the Magnificent* (fixup **1939**); *Tarzan and the Foreign Legion* (**1947**); *Tarzan and the Madman* (**1964**); *Tarzan and the Castaways* (1939-41 various mags; coll **1965**). *The Tarzan Twins* (**1927**; cut 1935; rev by other hands vt *Tarzan and the Tarzan Twins in the Jungle* 1938) and its sequel, *Tarzan and the Tarzan Twins with Jad-Bal-Ja, the Golden Lion* (**1936**), both assembled as *Tarzan and the Tarzan Twins* (omni **1963**), are associated titles. (◊◊ TARZAN MOVIES.)

Pellucidar: *At the Earth's Core* (1914 *All-Story Weekly*; **1922**); *Pellucidar* (1915 *All-Story*; **1923**); *Tanar of Pellucidar* (**1930**); *Tarzan at the Earth's Core* (properly also a **Tarzan** title, see above); *Back to the Stone Age* (**1937**); *Land of Terror* (**1944**); *Savage Pellucidar* (1942 *Amazing Science Fiction*; fixup **1963**).

Venus: *Pirates of Venus* (1932 *Argosy*; **1934**); *Lost on Venus* (**1935**); *Carson of Venus* (**1939**); *Escape on Venus* (1941-2 *Fantastic Adventures*; fixup **1946**); *The Wizard of Venus* (coll **1970**; vt *The Wizard of Venus and Pirate Blood* 1984).

Miscellaneous: *The Land that Time Forgot* (1918 *Blue Book* in 3 parts; fixup **1924**; vt in 3 vols under original part-titles: *The Land that Time Forgot* 1982, *The People that Time Forgot* **1982** and *Out of Time's Abyss* 1982); *The Eternal Lover* (1914-15 *All-Story Weekly*; fixup **1925**; vt *The Eternal Savage* 1963); *The Cave Girl* (1913-17 *All-Story Weekly*; fixup **1925**); *The Moon Maid* (1923-5 *Argosy All-Story Weekly* as "The Moon Maid", "The Moon Men" and "The Red Hawk"; cut fixup **1926**; vt *The Moon Men* 1962; vt in 2 vols and with text restored as *The Moon Maid* 1962 and *The Moon Men* 1962); *The Mad King* (1914-15 *All-Story Weekly*; fixup **1926**); *The Monster Men* (1913 *All-Story* as "A Man without a Soul"; **1929**); *Jungle Girl* (1932; vt *Land of Hidden Men* 1963); *Beyond Thirty* (1916 *All Around Magazine*; *c*1955 chap; vt *The Lost Continent* 1963) and *The Man-Eater* (*c*1955 chap), both assembled as *Beyond Thirty and the Man-Eater* (omni **1957**); *Tales of Three Planets* (coll **1964**; cut vt *Beyond the Farthest Star* 1965).

BURROUGHS, JOSEPH B. (? -?) US writer. ◊ ANTICHRIST.

BURROUGHS, WILLIAM S(EWARD) (1914-) US writer whose early life as a heroin addict in corners of the Third World is reflected in all his work, beginning with *Junky* (**1953** as by William Lee; rev vt *Junkie* as by WSB 1977) and *Queer* (written 1950s; **1985**), where his

homosexuality is also exposed to light and laceration. His first fully characteristic text, *The Naked Lunch* (**1959** France; vt *Naked Lunch* 1962 US), like almost all his later work, is far more effective when its phantasmagoric language and imagery is understood within an sf frame. When WSB is treated as an author of fantasy he can be misunderstood as an aesthete of the Dark Side who *projects* nightmares into fictional form; in truth, his tales are meant to describe *this* world, and to serve as arguments about its infinitely corroded nature. This sense of his work makes the movie NAKED LUNCH (*1992*) a relatively successful presentation of fantasy-like distortions in PERCEPTION within an "objective" frame; for WSB, in his writings, the world *is* a vast REALITY-ensnaring conspiracy, and his works are reportage. Other texts of interest include *The Soft Machine* (**1961** France; rev 1966 US), *The Ticket that Exploded* (**1962** France; rev 1967 US), *Nova Express* (**1964**), *The Wild Boys: A Book of the Dead* (**1971**; rev 1979 UK), *Exterminator!* (**1973**), *Port of Saints* (**1973** Switzerland; rev 1980 US), *Cities of the Red Night* (**1981**), *The Place of Dead Roads* (**1984**) and *Interzone* (coll **1989**). [JC]

BURTON, [Sir] RICHARD (FRANCIS) (1821-1890) UK orientalist, traveller and translator. ◊ ARABIAN FANTASY.

BURTON, TIM (1960-) US movie director, animator, writer and producer, currently rivalled only by Steven SPIELBERG as the creator of sure-fire hit fantasy movies (◊ CINEMA); almost all of his feature movies are discussed individually in this encyclopedia. Those he has directed include *Pee-Wee's Big Adventure* (*1985*), BEETLEJUICE (*1988*), EDWARD SCISSORHANDS (*1991*) and *Ed Wood* (*1994*), the last being a biography of the "world's worst movie director", Edward D. Wood Jr (1922-1978), responsible for such genre favourites as *Plan 9 From Outer Space* (*1958*). TB revived the COMICS hero BATMAN with *Batman* (*1989*) and *Batman Returns* (*1992*), both of which he directed, and *Batman Forever* (*1995*), which he co-produced (◊ BATMAN MOVIES). He also produced *The* NIGHTMARE BEFORE CHRISTMAS (*1993*), based on his own story. There can be little doubting the power of TB's directorial vision, yet there is a distancing effect in many of his movies – notably *Edward Scissorhands* and his two **Batman** outings: it is as if TB were trying to emphasize that the events are occurring in an artificial POLDER rather than in any version of reality, so that his characters seem not real people coping with real emotions but merely roles being enacted at a director's behest; one cares for them as much or as little as one would care for a wooden puppet. TB's timing is also suspect: his **Batman** movies are slow, and even *Nightmare*, otherwise brilliant, ends in a rush, as if TB were eager to wrap the project up quick. Perhaps his most fascinating piece of animation is *Vincent* (*1982*), a short he made during his apprenticeship at DISNEY. [JG]

BUSHYAGER, LINDA E(YSTER) (1947-) US writer whose two fantasy novels reveal through their titles – *Master of Hawks* (**1979**) and *The Spellstone of Shaltus* (**1980**) – an element of homage to Andre NORTON in the first instance and Marion Zimmer BRADLEY in the second. Set in a PLANETARY-ROMANCE venue, and dealing with a prolonged conflict between an inimical Empire and the independent states which surround it, these tales come close to RATIONALIZED FANTASY in their use of telepathy and precognition (◊ TALENTS), and MAGIC whose effects are almost mechanically calculable. In the end, the Empire is defeated, and (in the second volume) an attractively presented heroine (◊

HEROES AND HEROINES) helps gain a better world for all. LEB has since fallen silent. [JC]

BYATT, A(NTONIA) S(USAN) (1936-   ) UK writer whose most recent work has eschewed any use of the FANTASTIC, but whose *Possession: A Romance* (**1990**), though nothing supernatural actually occurs, makes intense play with DOUBLES and creates a version of literary life in the 19th century that has all the enfolding antiquarian passion of the best FANTASIES OF HISTORY. There is nothing supernatural, either, in *Angels & Insects* (coll **1992**), but its vision of the 19th century is similarly hallucinatory. *The Djinn in the Nightingale's Eye* (coll **1994**) is a set of FAIRYTALES told in REVISIONIST-FANTASY mode; the long title story in particular provides an eloquently sinuous intertwining of STORY and world. [JC]

"BYRON" ◊ Ron EMBLETON.

BYRON, LORD (GEORGE GORDON) (1788-1824) English poet, a leading figure in the English Romantic movement (◊ ROMANTICISM). Descended from an aristocratic family (he became 6th Baron Byron in 1798), Byron inherited many of their traits (his forebears all seem to have been insane, profligate or rascals; Lady Caroline Lamb later called Byron "mad, bad and dangerous to know"). He needed the energy of Europe for inspiration. Although, after the swift success of *Childe Harold's Pilgrimage* (**1812** cantos I and II; **1816** canto III; **1818** canto IV), he was lionized by society, he despised England, and, though shocked at the reaction to his scandalous morality, did not regret his ostracism from England from 1816 on. Always an outsider, he became the very symbol of the ACCURSED WANDERER, a figure doomed never to rest, an image he had himself portrayed in *Childe Harold's Pilgrimage* and which has become regarded as quintessentially Byronic.

Following his departure from the UK Byron settled briefly with Mary SHELLEY and Percy Bysshe SHELLEY in Switzerland in 1816. It was after reading a collection of GHOST STORIES that Byron suggested they each write one. This contest resulted in Mary Shelley's *Frankenstein* (**1818**); Byron's physician, John POLIDORI, unable to finish his own story, borrowed an idea from Byron for "The Vampyre" (1819 *New Monthly Magazine*), first published as by Byron – though Polidori seemingly had assumed publication would be anonymous. Byron was forced to submit his own incomplete "Fragment of a Novel" to his publisher, John Murray, who published it as a supplement to *Mazeppa* (**1819**). There is more than a little of Byron in Polidori's VAMPIRE, Lord Ruthven, the name which had already been used by Caroline Lamb to disguise (thinly) Byron in her novel *Glenarvon* (**1816**). Polidori also drew upon the imagery used by Byron in the latter's poem *The Giaour* (**1813**), which includes reference to a vampire-like spectre.

Byron never completed his "Fragment". More important was *Manfred* (**1817**), written in the wake of the contest and inspired by the immensity of the Swiss scenery and Byron's own anguish as an OBSESSED SEEKER; it also owes something to his youthful love of Matthew LEWIS's *The Monk* (**1796**). The poem, the epitome of Byronic tragedy, brought him to the attention of GOETHE (there were those who claimed Byron had based *Manfred* on FAUST), who honoured him by including him, in the character of Euphorion, in the second part of **Faust** – *Der Tragödie zweiter Teil* (**1832**).

Although Byron often used the power and imagery of the supernatural in his poetry he never returned to it with such extensive intensity as in *Manfred*. His death from rheumatic fever while fighting with the Greeks in their war of independence marked the ultimate fate of the Byronic hero. [MA]

Further reading: *Byron and the Romantics in Switzerland, 1816* (**1978**) by Elma Dangerfield; *Byron* (**1982**) by Frederic Raphael; *Byron and the Eye of Appetite* (**1986**) by Mark Storey; *Byron's Travels* (**1988**) by Allan Massie; *The Politics of Paradise* (**1988**) by Michael Foot.

CABBALA Spelled also Cabala, Kabbala or Qabbalah, a set of esoteric doctrines which claim to represent in diagrammatical form the manner in which the ultimate and unknowable God (a *deus absconditus* or hidden God who much resembles the God of Gnosticism) unfolds his full nature (which may be called the pleroma) into the Universe, which is the shape of this unfolding. The central cabbalistic image of this process is that of a Tree of Life, which comprises 10 interlinked spheres, the Sephiroth (singular: Sephirah), each of which is an aspect of God, and each of which interacts with all the other spheres in order – and according to rules whose complexity is staggering – to generate and maintain the Universe.

Cabbalism, as a genuine form of mysticism, is less important in the history of fantasy than in that of SUPERNATURAL FICTION. Algernon BLACKWOOD's *The Human Chord* (**1910**); several of the novels of Gustav MEYRINK, in particular *The Golem* (**1915**), Charles WILLIAMS's *Many Dimensions* (**1931**) and Harold BLOOM's *The Flight to Lucifer* (**1979**) all learnedly engage with aspects of the Cabbala (much of Bloom's critical work has also been devoted to it). In a far more general sense, the cabbalistic Tree of Life – which in some of its diagrammatic representations has an uncanny resemblance to the kind of fantasy MAP that joins up PLOT COUPONS – may be one of the underlying images responsible for the large number of Stones of Power to be found in modern tales, and which must be linked together to generate (or to preserve) the whole of REALITY. "Cabbalistic" patterns found on the robes or in the tomes of WIZARDS may or may not be relevant. [JC]

CABELL, JAMES BRANCH (1879-1958) US (emphatically *Southern* US) writer of some of this century's finest ironical fantasies, and also of lesser work, descending in his last years to mannered weariness. Amid wildly fantastic surroundings his characters exemplify a fixed, chiefly bleak view of unchanging human nature. JBC categorized his male HEROES as ruled by CHIVALRY, Gallantry (philandering) or Poetry; women are either acidulous but ultimately comfortable wives or the elusive, unattainable WITCH-women which, according to JBC, every man desires (◊ SEHNSUCHT).

JBC's best-known works lie within the lengthy sequence **Biography of the Life of Manuel** – not a tautology, as the "life" of the hero Manuel, supposed redeemer of the IMAGINARY LAND of POICTESME, is traced through descendants to the then contemporary USA. All JBC's works before 1930 (including nonfantasies) were forcibly assimilated into the **Biography**. The official ordering, broadly chronological except for the first and last volumes (serving as prologue and epilogue), runs: *Beyond Life: Dizain des Démiurges* (**1919**), dramatized essays outlining JBC's theories of literature in the setting of a fantasy LIBRARY; *Figures of Earth: A Comedy of Appearances* (**1921**); *The Silver Stallion: A Comedy of Redemption* (coll of linked stories **1926**); the **Witch-Woman** sequence, being *The Music from Behind the Moon: Epitome of a Poet* (**1926**), *The White Robe: A Saint's Summary* (**1928**) – an unusually black WEREWOLF story – and *The Way of Ecben: A Comedietta Involving a Gentleman* (**1929**), assembled as *The Witch-Woman: A Trilogy About Her* (omni **1948**); *The Soul of Melicent* (**1913**; rev vt *Domnei: A Comedy of Woman-Worship* 1920); *Chivalry: Dizain des Reines* (coll of linked stories **1909**; rev 1921); *Jurgen: A Comedy of Justice* (**1919**; rev 1921 UK); *The Line of Love: Dizain des Mariages* (coll of linked stories **1905**; rev 1921); *The High Place: A Comedy of Disenchantment* (**1923**); *Gallantry: Dizain des Fêtes Galantes* (coll of linked stories **1907**; rev 1922); *Something about Eve: A Comedy of Fig-Leaves* (**1927**); *The Certain Hour: Dizain des Poètes* (coll 1916); *The Cords of Vanity: A Comedy of Shirking* (**1909**; rev 1920); *From the Hidden Way* (**1916** chap; rev 1924; rev vt *Ballades from the Hidden Way* 1928), verse; *The Jewel Merchants* (**1921** chap), a play; *The Rivet in Grandfather's Neck: A Comedy of Limitations* (**1915**); *The Eagle's Shadow: A Comedy of Purse-Strings* (**1904**; rev 1923); *The Cream of the Jest: A Comedy of Evasion* (**1917** rev 1922); *The Lineage of Lichfield: Another Comedy of Evasion* (**1922**), fictional genealogy; and *Straws and Prayer-Books: Dizain des Diversions* (**1924**), essays plus two allegorical pendants to *The Silver Stallion*. The 1927-30 "Storisende edition" of the **Biography** (rev throughout) is footnoted in *Townsend of Lichfield: Dizain des Adieux* (omni **1930**), which includes: *The White Robe, The Way of Ecben; Taboo* (**1921** chap), a satirical commentary on *Jurgen*'s legal toils; *Sonnets from*

Antan (dated 1929 but **1930** chap), verse; *Jurgen and the Law* (**1922** chap) ed Guy Holt; and miscellanea. *Preface to the Past* (coll **1931**) assembles JBC's prefaces to the deluxe **Storisende** edition.

Jurgen is a central JBC work, partly as perhaps his most fully achieved blend of characteristic irony, allusion, erudition, censor-teasing and creative joy, but also for its effect on his career. Its 1920 banning, instigated by the New York Society for the Suppression of Vice, was ostensibly for "obscene" double-entendres (the hero's SWORD, staff, sceptre, etc., are much admired by ladies in equivocal situations) but, as emerged in court, was also a reaction to SATIRE on RELIGION. After years of being admired by a select audience that had included Mark TWAIN, Cabell became and remained the highly notorious "author of *Jurgen*". (The obscenity case was tried and the publishers acquitted in 1922.)

Jurgen himself is a middle-aged poet-turned-pawnbroker who, having capriciously praised EVIL, is rewarded by Koshchei (maker of this Universe) with the disappearance of his harridan wife. Reluctant search for her leads Jurgen on a year-long adventure: first on centaur-back to the Garden Between Dawn and Sunrise where imaginary beings live, including all lovers' versions of their objects of desire; and onward to encounter the many-named goddess Sereda, who is also Cybele (◊ GODDESS), and who, outrageously flattered by Jurgen, gives him renewed youth together with a SHADOW – her own – which takes notes on his doings. These involve much dalliance: with his first teenage love on a past day revisited by TIME TRAVEL; with GUINEVERE shortly before she marries ARTHUR (Jurgen is also a previous user of Caliburn/EXCALIBUR); with his own step-grand-mother's GHOST; with the LADY OF THE LAKE, here identified as Anaïtis, goddess of erotic diversions, with whom he travels to Cocaigne; with a DRYAD in Leukê, land of Greek/Roman MYTHOLOGY; and very nearly with Helen of Troy, who is however so unsurpassably desirable that even Jurgen refrains. Leukê is conquered by the greyly reasonable forces of Philistia (America) and Jurgen condemned: later editions here incorporate *The Judging of Jurgen* (**1920** chap), satirizing JBC's legal tormentor as a dung-beetle who indicates WEAPONS held by pageboys and declares, "You are offensive because this page has a sword which I choose to say is not a sword . . ." (◊ RECURSIVE FANTASY). Having been deemed a solar myth by the Master Philologist, Jurgen must naturally winter underground in HELL (to the undoing of SATAN's wife and an attractive VAMPIRE). All along he has been promoting himself – duke, prince, king, emperor – and now tricks his way up Jacob's ladder into HEAVEN in the guise of Pope. GOD, as he suspected, is an ILLUSION (invented by Koshchei to please Jurgen's insistently pious grandmother): but even this illusion of perfect love overwhelms Jurgen. Still he ascends the throne of Heaven, but wearily renounces it; tricks Sereda into withdrawing his unnatural youth; and finally meets Koshchei, who like other potent beings encountered in his travels proves less intelligent than that "monstrous clever fellow", Jurgen himself. Though again offered Guinevere and Anaïtis and Helen, he chooses to have back his wife. The mercy of Koshchei is that the year's upsets become, retrospectively, a DREAM; and like the SUN to whose myth Jurgen has been relegated, the tale comes full circle to its starting place.

The **Biography**'s MULTIVERSE extends in various directions from *Jurgen*. *Figures of Earth* tells of the previous generation's Manuel the Redeemer, who may be very shrewd or very obtuse. When he sells ordinary goose feathers to kings who believe them magical – and to whom, through belief, they *are* magical, like the gifts of L. Frank BAUM's *The Wonderful Wizard of Oz* (**1900**) – Manuel's role in the bargain is passive, for the motto of Poictesme is *Mundus vult decipi*: "The world wishes to be deceived." We are not admitted to any of his thoughts, giving the book a chilly aspect despite fine poetic passages; his monomaniac desire to create and animate clay images is absurdly based on a maternal instruction to make himself a figure in the world. Manuel does so, betraying virtually everyone he deals with; even his reconquest of Poictesme is carried out for him by superannuated deities invoked by a WIZARD, and he holds the land only as a fiefdom under the mysterious demiurge Horvendile. When Grandfather DEATH comes for Dom Manuel, time bends as in E.R. EDDISON's *The Worm Ouroboros* (**1922**) and the Count of Poictesme is once again a young swineherd shaping figures futilely from mud (◊ TIME FANTASIES).

The Silver Stallion, another joyous book, relates the bizarre adventures of Manuel's barons after his death, and the growth of a sentimental, women-fostered and wholly untrue LEGEND of Manuel the Redeemer as quasi-Christian saviour and saint. One baron re-enacts, on a large and witty scale, the FAIRYTALE of THREE WISHES where the last wish must undo all the damage done – here to Koshchei's entire Universe. Another, dying in battle, is inadvertently carried off by the Valkyrie sent for his heathen foe, and adopted into a NORDIC-FANTASY pantheon of which Koshchei is but a minor member. Others meet fates almost as strange.

In the sulphurous *The High Place*, the amoral hero Florian enters the SLEEPING-BEAUTY story and (unlike Jurgen with Helen) does not draw back at the sight of excessive beauty. Complications ensue: Beauty is realistically diminished during pregnancy, the first-born child is forfeit to Satan under the PACT that guaranteed Florian's success, and an irascible saint is eager to call down holy fire on transgressors. Florian treads close to damnation and is saved only when Satan and the ANGEL Michael conspire to let recent events become, again, a dream: he has a rare second chance and LEARNS BETTER. *Something About Eve*, with a stronger than usual flavour of ALLEGORY, shows its non-hero feebly intending to gain promised glory awaiting in the land of "Antan" but forever delayed on Mispec Moor (anagram: "compromise"), wearing literal rose-coloured spectacles and beguiled by the woman Maya, while bolder folk like Solomon and ODYSSEUS pass by on the road to Antan.

The tortuousness of the Cabellian cosmology emerges in *The Cream of the Jest*, whose author hero Kennaston enjoys fantastic DREAMS thanks to a sigil (actually the broken top of a modern cosmetics bottle), and flits through history as Horvendile – who has power over all these fictions. But Kennaston is himself documented as a descendant of Manuel, and the FRAME STORY complicates matters further. Meanwhile Horvendile's eternal doomed pursuit of Ettarre (sister of Jurgen's first love Dorothy, and of Melicent, the object of long-frustrated, absurdly chivalric love in *Domnei*) is prefigured in *The Music from Behind the Moon*. These connections can be traced endlessly, though many links seem merely decorative – like the **Biography**'s repeated mentions of a sinister, undescribed RITUAL involving a MIRROR and two white pigeons. Such patterns, together with buried anagrams, verse passages disguised as prose, and JBC's great variety of historic, linguistic and mythic allusions, have been explored in the Cabell Society's magazine *Kalki*. Several "historical" **Biography** volumes (including *Chivalry, Gallantry* and *The Line of Love*) were initially illustrated by Howard PYLE; most of the major fantasies have appeared in ornate editions illustrated and decorated by Frank PAPÉ.

Once the **Manuel/Poictesme** sequence was complete, in 1932, JBC marked this career watershed by writing as Branch Cabell, though in 1946 he resumed his full name. His most substantial post-**Biography** fantasy was **The Nightmare Has Triplets**, a sequence comprising *Smirt: An Urbane Nightmare* (**1934**), *Smith: A Sylvan Interlude* (**1935**) and *Smire: An Acceptance in the Third Person* (**1937**). This explicitly emulates the logic and geography of DREAMS. The dreamer, another JBC-like author of romances, is triply eponymous – first as Smirt, a figure of Jurgen-like self-esteem who patronizes God and Satan, instructs the Stewards of Heaven in how to improve the world (by modelling it on Smirt's books), and sets up house with Arachne the SPIDER-woman; then as Smith, a retiring woodland deity presiding over inset tales of Smirt's dream-begotten sons; and finally, after further THINNING, as the dwindled wanderer Smire. Choruses of banal newspaper headlines and conversational platitudes give warning whenever wakefulness comes too near. It is indeed a successfully misty and dreamlike (if slightly inconsequential) work. The pamphlet *The Nightmare Has Triplets* (**1937** chap) discusses the trilogy.

JBC's avowed aim was "to write perfectly of beautiful happenings". Readers taking this at face value were shocked by bitter ironies, cynicism and determined avoidance of wish-fulfilment. Indeed JBC strove to write beautifully (though sometimes affectedly) and, in his best work, of morally realistic happenings, stressing that prices must be paid and time's inevitable ravages endured. Comforting illusions are necessary, and are accepted with full knowledge of their illusory nature. This toughness always underlies the surface frivolity, wit, erudition and spiciness. JBC can still take us by surprise. [DRL]

Other works: *Between Dawn and Sunrise: Selections from the Writings of James Branch Cabell* (anth **1930** coll) ed John Macy; *Some of Us: An Essay in Epitaphs* (coll **1930**), literary commentary; the **It Happened In Florida** sequence, being *There Were Two Pirates: A Comedy of Division* (**1946**) and *The Devil's Own Dear Son: A Comedy of the Fatted Calf* (**1949**); *The Nightmare Has Triplets: Smirt, Smith and Smire* (omni **1972**).

As Branch Cabell: Their Lives and Letters, a series comprising essays and imaginary letters, being *These Restless Heads: A Trilogy of Romantics* (**1932**), *Special Delivery: A Packet of Replies* (**1933**) and *Ladies and Gentlemen: A Parcel of Reconsiderations* (**1943**); **Heirs and Assigns**, associational historical fiction, being *The King was in His Counting House: A Comedy of Common-Sense* (**1938**), *Hamlet Had an Uncle: A Comedy of Honour* (**1940**) and *The First Gentleman of America: A Comedy of Conquest* (**1942**; vt *The First American Gentleman* 1942 UK); the first **It Happened In Florida** title, *The St Johns: A Parade of Diversities* (**1943**) with A.J. Hanna, history/topography of the St Johns river, Florida (trio continued as by JBC alone).

Further reading: *James Branch Cabell* (**1925** rev 1932) by Carl van Doren, including JBC's MAP of POICTESME; *Cabellian Harmonics* (**1928**) by Warren A. McNeill; the **Virginians Are Various** volumes, including JBC

autobiography and reminiscences, being *Let Me Lie* (coll **1947**), *Quiet, Please* (**1952**) and *As I Remember It: Some Epilogues in Recollection* (**1955**); *Between Friends: Letters of James Branch Cabell and Others* (**1962**) ed Padraic Colum and Margaret Freeman Cabell; "The James Branch Cabell Case Reopened", in *The Bit Between My Teeth* (coll **1965**) by Edmund Wilson; *James Branch Cabell: The Dream and the Reality* (**1967**) by Desmond Tarrant; *The Letters of James Branch Cabell* (**1975**) ed Edward WAGENKNECHT.

See also: BOOKS.

CABINET OF CALIGARI, THE US movie (*1962*). ◊ *The* CABINET OF DR CALIGARI (*1919*).

CABINET OF DR CALIGARI, THE (ot *Das Kabinett des Dr Caligari*) German movie (*1919*). Decla-Bioscop. **Pr** Erich Pommer. **Dir** Robert Wiene. **Screenplay** Hans Janowitz, Karl Mayer. **Starring** Lil Dagover (Jane Olsen, the Girl), Hans Feher (Francis, the Student), Werner Krauss (Dr Caligari), Hans V. Twardowsky (Alan, the Student's Friend), Conrad Veidt (Cesare, the Somnambulist). *c*70 mins. B/w, silent.

One day – young Francis tells his companion in a tranquil garden – the CARNIVAL came to his hometown of Holstenwall, and with it the sinister Dr Caligari with his somnambulist, Cesare, who had slept in a coffin for 25 years. Still sleeping, Cesare was lured nightly to perform feats of PROPHECY for the paying customers – and also, secretly, to commit murders at Caligari's behest. Francis at last cottoned on to what was happening, and chased Caligari to a lunatic asylum, where he and an initially incredulous staff discovered the awful truth about its Director's double life. "He is now a madman, chained to his cell," Francis tells his companion in the garden, but it emerges instead that it is Francis who is the madman, and that the characters in his deluded tale are fellow-inmates of the asylum. Or is that entirely the case? As the camera gives us a last close-up of the Director we see a glint in his eye . . .

A fantasy of PERCEPTION *par excellence* – we are left to choose which of the two REALITIES is valid – *TCODC* is also a brilliant piece of moviemaking. All the settings are unashamedly stage sets, their angles distorted bizarrely; frequently Wiene masks the lens to alter the frame-shape in order to heighten the sense of unreality. The figure of the somnambulist seems a direct ancestor of the cinematic FRANKENSTEIN's MONSTER. Most available versions of the movie are substantially cut, some by as much as 30 mins.

The Cabinet of Caligari (**1962**), often listed as a remake, resembles *TCODC* only insofar as there is a character of that name and the tale proves to be the misperception of a mad person, in this case an elderly woman being treated by a psychiatrist in an attempt to cure her of sexual fantasies. [JG]

Further reading: *The Cabinet of Dr Caligari: Texts, Contexts, Histories* (anth **1990**) by Mike Budd.

CABRERA INFANTE, GUILLERMO (1929-   ) Cuban-born UK writer, journalist and movie critic, in England since 1966; he also writes as G. Cain. He is best-known for those works translated into English, many of which are FABULATIONS that challenge our PERCEPTIONS of history to show the potential for past ALTERNATE REALITIES (◊◊ FANTASIES OF HISTORY). This is especially true of *Tres tristes tigres* (**1965** Spain; trans Donald Gardner and Suzanne Jill Levine as *Three Trapped Tigers* **1971** US), which portrays different interpretations of historical events, and the enigmatic *Vista del amanecer en el trópico* (coll of linked stories

1974 Spain; trans Levine as *A View of Dawn in the Tropics* **1978** US) – the closest GCI's work comes to MAGIC REALISM – which depicts events through a series of alternate sketches. GCI challenges even his own life in *La Habaña para un infante difunto* (**1979**; trans Levine and GCI as *Infante's Inferno* **1984** US) which, while purporting to be autobiographical, uses different perceptual techniques to provide a multilayered yet obfuscated interpretation of life. [MA]

Other works: *Asi en la paz como en la guerra* ["In Peace as in War"] (coll **1960**); screenplays for *Wonderwall* (**1968**) and *Vanishing Point* (**1970**).

CACOTOPIAS Synonym for dystopias. ◊ UTOPIAS.

CAIN, G. Pseudonym of Guilermo CABRERA INFANTE.

CAIRO Cairo is a venue for URBAN FANTASY. As such, it tends to be used in tales whose attitude towards the LAND OF FABLE of ARABIAN FANTASY is ironized (◊ REVISIONIST FANTASY) or negative. Robert IRWIN's *The Arabian Nightmare* (**1983**), for instance, treats Cairo as part of the nightmare, not as a backdrop. FANTASIES OF HISTORY – like Edward WHITTEMORE's **Jerusalem Quartet**, and *Name of the Beast* (**1992**) by Daniel Easterman (1949-   ) – routinely evoke Cairo as a place of allure and suffocation. [JC]

See also: BAGHDAD; EGYPT.

CALADO, IVANIR (1953-   ) Brazilian writer. ◊ BRAZIL.

CALDECOTT, MOYRA (1927-   ) Pseudonym of South African-born writer Olivia Brown, in the UK from 1951, who began publishing original work of fantasy interest with the **Tall Stones** sequence – *The Tall Stones* (**1977**), *The Temple of the Sun* (**1977**) and *Shadow on the Stones* (**1978**), assembled as *Guardians of the Tall Stones* (omni **1986**), plus *The Silver Vortex* (**1987**). Like most of her work, the sequence takes place in a LAND OF FABLE, in this case Bronze-Age Britain, and rather emolliently describes a conflict between pacific worshippers – the focus of their RELIGION being the stone circles that dominate their villages – and invaders whose religion involves sacrificial offerings. But the threatened THINNING – typical of novels which contrast the old faith and the new Christianity – does not actually occur; it might be noted that the invidious consequences of the REINCARNATIONS which proliferate throughout MC's work do not consort easily with the cyclic religions to which she grants her novelistic adherence. The **Ancient Egypt** sequence – *Son of the Sun* (**1986**), *Daughter of Amun* (**1989**) and *Daughter of Ra* (**1990**) – focuses to similar effect on the period dominated by Akhnaten. Singletons include *The Lily and the Bull* (**1979**), in which a mortal AVATAR of the GODDESS resists patriarchy in ancient Crete, and *The Tower and the Emerald* (**1985**), which mixes REINCARNATION motifs in an Arthurian setting (◊ ARTHUR).

Other works: Several TWICE-TOLD versions of old material, including *Weapons of the Wolfhound* (coll **1976**), *Twins of the Twylwyth Teg* (coll **1983**), *Taliesin and Avagddu* (coll **1984**), *Bran, Son of Llyr* (coll **1984**) and *The Green Lady and the King of Shadows* (**1989**), about Glastonbury Tor; *Adventures by Leaflight* (coll **1978** US), children's tales; *Child of the Dark Star* (**1984**), set in the future; *Etheldreda* (**1987**); *Women in Celtic Myth* (**1988**), nonfiction; *Crystal Legends* (**1990**), nonfiction; *Myths of the Sacred Tree* (**1993** US); *The Winged Man* (**1993**).

CALDECOTT, RANDOLPH (1846-1886) UK illustrator, cartoonist and watercolourist, active from 1868 in various journals, including *Will O' the Wisp*, *London Society* and *Punch*. From 1875, with his illustrated version of Washington IRVING's *Old Christmas* (**1875**), he became well known for comic tales, done in colour, for the CHRISTMAS issues of the *Graphic*; much of this work was later assembled in book form, beginning with *Randolph Caldecott's "Graphic" Pictures* (graph **1883**). It generally depicts rural life with a modicum of fantasy. The concurrent **Randolph Caldecott's Picture Books** sequence of 16 books for children (**1878-85**) mostly illustrates famous verses, like the first, *John Gilpin* (graph **1878**); the sequence was subsequently republished in various formats, culminating in *The Complete Collection of Randolph Caldecott's Pictures and Songs* (graph omni **1887**). Other illustrated books of interest include *Some of Aesop's Fables with Modern Influences* (graph **1883**) and *Lob Lie-by-the-Fire* (graph **1885**). His work, whose subjects are often caricatured with cartoon-like transparency, is delicately coloured, and almost unfailingly unveils an underlying kindliness and nostalgia for country values. [JC]

CALENDAR The yearly cycle of fantasy-TIME comes in three main flavours. Simplistic calendars note the SEASONS (often in paraphrase: the Time of Thaw and Sowing), with an optional midwinter or year-end festival. Others aim for historical resonance by borrowing ancient systems; e.g., the Year of the Fire Dragon, the Moon of Almost Anything, or most often a jumble of Celtic and pagan feasts like Imbolc (1 February), Beltane (Mayday and/or Mayday Eve, which is Walpurgisnacht), Lugnasad (1 August), Samhain (◊ HALLOWE'EN) and Yule. Such precise dating as "1 February", though not unknown in CONTEMPORARY FANTASY, is hazardous owing to calendar slippage and reform. Finally, some authors invent their own detailed calendars: J.R.R. TOLKIEN did so for the Shire in Middle-Earth (using a 365-day year with regular leap days, since this is prehistoric Earth), and likewise Terry PRATCHETT for **Discworld** (with 13 months totalling approximately 800 days). [DRL]

See also: MYTHOLOGY; SUN.

CALHOUN, MARY (1926-   ) US writer. ◊ CHILDREN'S FANTASY.

CALIBAN The "savage and deformed slave" of William SHAKESPEARE's *The Tempest* (produced *c*1611; **1623**), described as son of the witch Sycorax and tamed into servitude by PROSPERO. His name is a part-anagram of "cannibal". As well as being an emblem of the untamed races of the New World and of the perennial question of whether savagery is innate or engendered by oppression (he remarks that teaching him language has taught him how to curse), Caliban is a personification of the wild semi-human who occupies an important place in much fantasy. We can, for instance, see him in J.R.R. TOLKIEN's Gollum, whose RING (like Caliban's ISLAND, his by right of possession) is taken from him by superior beings and who is forced to serve his subjugator. Also like Gollum, Caliban fawns upon and worships and plots against his "masters" – and half-loves them.

Caliban is a rich enough figure to be the subject of reinterpretation of his role. Robert BROWNING shows him meditating on natural religion in his poem "Caliban upon Setebos". Tad WILLIAMS looks at the relationship between Caliban and Miranda in *Caliban's Hour* (**1995**), his "sequel" to *The Tempest*, told from Caliban's viewpoint. [AS]

See also: BONDAGE.

CALLANDER, DON (1930-   ) US writer who began publishing fantasy with the **Brightglade** sequence – *Pyromancer* (**1992**), *Aquamancer* (**1993**) and *Geomancer*

(1994) – in which the young WIZARD Brightglade and his partner (and lover) Myrn must master various MAGICS (those of Fire, Water and Earth to date, with Air presumably to come) in order to fend off threats to their world and their persons. The tone is light and reasonably assured. A new series seems to be underway with *Dragon Companion* (1995), whose librarian HERO finds himself hired out as COMPANION to a DRAGON. [JC]

CALTHROP, DION (WILLIAM PALGRAVE) CLAYTON (1878-1937) UK writer and illustrator whose *The Guide to Fairyland* (1906), a Cooks' Tour (◊ PLOT DEVICES) – for which he did both text and extensive illustration – treats his chosen venue in WONDERLAND terms; from the moment the Charon-like Captain Bullfinch whisks the narrator away, the dialogue tends to whimsical logic-chopping in a manner perhaps intended to evoke memories of Lewis CARROLL. Some of the interpolated FAIRYTALES show imagination, as do some of the sentimentalized visions of harlequinade (◊ COMMEDIA DELL'ARTE) expressed in *Rouge, Brown* (coll 1906) with Haldane MacFall (1860-1928) and *The Harlequin Set* (1911). *The Harlequinade: An Excursion* (1918) with Harley Granville-Barker (1877-1946) is a play with fantasy elements. In *Hyacinth: An Excursion* (1927) the immortal Hyacinth visits LONDON, spreading contentment. [JC]

CALVINO, ITALO (1923-1985) Italian novelist, born in Cuba, who began publishing soon after WWII, his experience of which (as a Partisan in the Ligurian mountains) appears in some of his early fiction. Although his earliest work was written in the prevailing climate of the post-WWII Italian Neorealists, his collection *Ultimo viene il corvo* (coll 1949; trans A. Colquhoun and P. Wright with some stories dropped and 1 added as *Adam, One Afternoon* 1957 UK), while containing no fantastic elements – the English translation adds the later fantasy "La Formica Argentina" (1952; trans as "The Argentine Ant") – possesses an Arcadian sunniness and sense of numinous wonder that already signal an essential difference in temperament from the *neorealisma* of, say, Alberto Moravia (1907-1990).

The two short fantasy novels that IC published in the 1950s, *Il Visconte dimezzato* (1952) and *Il Cavaliere inesistante* (1959) – assembled as *The Non-Existent Knight and The Cloven Viscount* (omni trans A. Colquhoun 1962 UK) – marked a shift to the kind of fiction for which he became famous: FABULATIONS, often set between the Middle Ages and the Enlightenment, that deal wittily and exuberantly with questions of knowledge, formalism and ontology. *Il Barone rampante* (1957; trans A. Colquhoun as *The Baron in the Trees* 1959 UK) – in which an 18th-century baron's son climbs a tree in an act of defiance and ends up spending his life in various treetops – displays the narrative extravagance and intellectual playfulness that characterize IC's more overt fantasies. All three volumes were assembled as *I nostri antenati* (omni 1960; trans as *Our Ancestors* 1980 US).

Le Cosmicomiche (coll of linked stories 1963; trans William Weaver as *Cosmicomics* 1968 US) and *Ti con zero* (coll of linked stories 1967; trans Weaver as *t zero* 1969 US; vt *Time and the Hunter* 1970 UK) are, with their cosmological ruminations (by an entity as old as the Universe), closer to sf than to fantasy. *Le città invisibili* (1972; trans Weaver as *Invisible Cities* 1974 US) and *Il Castello dei Destini incrociati* (coll of linked stories 1973; trans Weaver as *The Castle of Crossed Destinies* 1977 US) adopt the form that IC would employ for the rest of his life: meditations, with elaborate

variations and framing devices, on matters involving the interaction between humanity and the Universe. The former, comprising Marco Polo's discourses to Kublai Khan on 55 (imaginary) cities he has visited, and the latter, which organizes a series of stories through the complex rules set forth by a TAROT reading, are arranged with a formal rigour that shares much with the aesthetic of Oulipo and very little with the narrative-driven linearity of most GENRE FANTASY.

IC's urbane geniality and gift for giving airy expression to weighty themes are evident despite the (sometimes fevered) virtuosity of his formal ingenuity, which reaches its high point in *Se una notte d'inverno un viaggiatore* (1979; trans Weaver as *If on a Winter's Night a Traveler* 1981 US). The novel alternates the opening chapters of 10 different novels with instalments of a long discourse on the experience of reading that frequently segues into a fantastical narrative concerning the intrigues of a literary translator, who may be responsible for the 10 opening chapters. This most ambitious of IC's productions may, like *The Castle of Crossed Destinies*, have been influenced by the numerous interwoven narratives in his massive *Fiabe italiane* (anth 1956; trans George Martin as *Italian Folktales* 1980 US). IC's last novel, *Palomar* (1983; trans Weaver as *Mister Palomar* 1985 US), not fantasy, abjures the intensive virtuosity of the previous two novels for a lighter mode, but possesses a characteristically complex structure.

IC created a body of work that seems at home with the fabulations of Umberto ECO, Vladimir NABOKOV and Günter GRASS. His was a fancifully imagined and geometrically exacting country that genre fantasy has yet to discover. [GF]

Other works: *The Watcher and Other Stories* (coll trans Weaver 1971 US), three novellas, one sf and two fantasy; *Sotto il sole giaguaro* (coll 1986; trans Weaver as *Under the Jaguar Sun* 1988 US), three fabulations about the senses of taste, hearing and smell, part of an unfinished cycle of five; *Six Memos for the Next Millennium* (trans Patrick Creagh 1988 US), five lectures (of a projected six) IC wrote in the last year of his life for the Charles Eliot Norton Lectures at Harvard; *Primal che tu Dicta "Pronto"* (coll 1993; trans Tim Parks as *Numbers in the Dark and Other Stories* 1995 UK).

CAMELOT King ARTHUR's most famous court. The name was first used by CHRÉTIEN DE TROYES at the opening of *Lancelot* (?1177), where it is referred to as "magnificent". Chrétien probably took the name from an older text, now lost. GEOFFREY OF MONMOUTH located Arthur's capital at Caerleon in South Wales, based on its earlier description as the City of the Legions, an old Roman title which applies equally to Chester and Carlisle; but Chrétien's reference clearly distinguishes it from Caerleon. Others have argued in favour of Colchester (Britain's oldest city, which the Romans called Camulodunum), Winchester (the former Saxon capital of England) and Stirling (site of the original capital of the Gododdin), while the Stone-Age Cadbury Castle also has much support. Camelot became Arthur's main capital in the later ROMANCES and came to represent the centre of the Arthurian GOLDEN AGE. The image of the Arthurian world is often represented by the portrayal of Camelot as a many-towered castle, as described by Alfred, Lord Tennyson (1809-1892) in *Idylls of the King* (1859; exp 1870, 1873, 1886).

But Camelot is rarely explored in detail in Arthurian fiction, its use being symbolic – as in Alan Jay LERNER's musical *Camelot* (1960; movie **1967**) and in the anthologies *Invitation to Camelot* (anth **1988**) ed Parke GODWIN, *The Camelot Chronicles* (anth **1992**) ed Mike ASHLEY and *Camelot* (anth **1995**) ed Jane YOLEN. The most extensive descriptions appear in *A Connecticut Yankee in King Arthur's Court* (**1889**) by Mark TWAIN, *Galahad* (**1926**) by John Erskine (1879-1951) and *A Prince in Camelot* (**1995**) by Courtway Jones (real name John Alan Jones; 1923-   ).   [MA]

CAMELOT US movie (**1967**). ◊ Alan Jay LERNER.

CAMERON, ELEANOR (BUTLER) (1912-   ) Canadian-born US writer for a YA audience, best-known for her **Mushroom Planet** sf sequence: *The Wonderful Flight to the Mushroom Planet* (**1954**), *Stowaway to the Mushroom Planet* (**1956**), *Mr Bass's Planetoid* (**1958**), *A Mystery for Mr Bass* (**1960**) and *Time and Mr Bass* (**1967**). Her finest fantasies are *The Court of the Stone Children* (**1973**) and its sequel *To the Green Mountains* (**1975**), although *Beyond Silence* (**1980**), a TIMESLIP fantasy set in Scotland, has great elegance.   [JC]

Other works: *The Terrible Churnadryne* (**1959**); *The Mysterious Christmas Shell* (**1961**); *The Beast with the Magical Horn* (**1963**); *A Spell is Cast* (**1964**).

CAMERON, JOHN (1927-   ) US writer. ◊ ASTROLOGY.

CAMPBELL, ANGUS Pseudonym of R. CHETWYND-HAYES.

CAMPBELL, EDDIE (1956-   ) UK-born COMIC-strip artist and writer, resident in Australia since 1986, most notable for his semi-autobiographical strip **Alec** in *Escape* (1980-84), collected as *Alec: Episodes in the Life of Alec McGarry* (graph coll **1984** UK), *Love and Beerglasses* (graph coll **1985** UK), *Doggie in the Window* (graph coll **1986**), *The Complete Alec* (graph omni with new material **1990** UK), *Alec in the Dance of Lifey Death* graph coll **1994** US) and *Graphitti Kitchen* (graph coll **1995** US), inspired by the writings of Jack Kerouac (1922-1969) and Henry Miller (1891-1980).

EC attended art-school foundation but failed after the first year. However, he created a number of features for the music magazines *Sounds* and *New Musical Express*: **In the Days of the Ace Rock and Roll Club**; two semi-HORROR strips with Phil Elliott, together as Charlie Trumper – **Rodney: The Premonition** (1985) and **The Horrors of the Mammy** (1985-6) – and horror reportage with **The Pyjama Girl**, based on a 1935 murder in Sydney, Australia.

EC's longest-running series has been **Deadface**, stories about characters from Greek MYTHOLOGY who survive to the present day; these sharply observed retellings spotlight individuals like Bacchus (◊ DIONYSUS). Also of note is **The Eyeball Kid**. Stories from both series appear in various magazines and anthologies, the bulk being found in *Deadface* (#1-#8 1987-8), *Bacchus* (#1-#2 1988), *Immortality Isn't for Ever* (1991), *Deadface: Doing the Islands with Bacchus* (#1-#3 1991 US), *The Eyeball Kid* (#1-#3 1992 US) and *Deadface: Earth, Air, Fire and Water* (#1-#4 1992 US).

EC's controversial artwork for Alan MOORE's JACK THE RIPPER series **From Hell** (projected 16 vols, of which 7 published by end 1995) has attracted criticism for its draughtsmanship. He has also produced *The Dead Muse* (graph coll **1990** US), *Little Italy* (graph **1991** US), *The Cheque Mate* (graph **1992**) and *Hermes Versus the Eyeball Kid* (#1-#3 1994-5 US).   [RT/SH]

CAMPBELL, RAMSEY (1946-   ) UK writer, primarily of HORROR, whose work often melds influences from many genres to produce an amalgam of images of EVIL and alienation. RC's most notable primary influence was H.P.

LOVECRAFT. August DERLETH, RC's mentor in his formative years, encouraged him to establish his own milieu for his Lovecraftian stories. This led to RC's first professionally published story "The Church in High Street" (in *Dark Mind, Dark Heart* anth **1962** US ed Derleth) and his first book, *The Inhabitant of the Lake and Less Welcome Tenants* (coll **1964** US). Highly imitative, these stories, most fitting loosely into the CTHULHU MYTHOS, are glutinous with atmosphere but weak on character. By the time the book appeared RC was already moving away from Lovecraft towards deeper roots. His juvenilia, which he has no qualms over sharing with his readers, show the much stronger influence of writers like M.R. JAMES and Algernon BLACKWOOD, plus imagery from COMICS and HORROR MOVIES, particularly the *film noir* and the work of the early German directors. Movie imagery is evident in many of RC's stories, and is the basis of *Ancient Images* (**1989**). He also novelized three 1930s movies – *The Bride of Frankenstein* * (**1977** US; text restored 1978), *The Wolfman* * (**1977** US) and *Dracula's Daughter* * (**1977** US); the US editions all bore the pseudonym Carl Dreadstone, as did the first book in the UK, but the last two UK editions carried the name E.K. Leyton. RC has also written as Montgomery Comfort (on short fiction) and Jay Ramsay.

RC's earliest stories, dating from 1958-63, were collected in two special editions of *The* CRYPT OF CTHULHU: *The Tomb-Herd and Others* (coll **1986** US) and *Ghostly Tales* (coll **1987** US); he later compiled a retrospective of all his Lovecraftian fiction, *Cold Print* (coll **1985**; exp 1993). But the mid- to late 1960s saw his work markedly metamorphosing as he came to terms with the Lovecraftian elements and began to control his influences, placing greater emphasis on character, particularly the psychology of his protagonists. RC's next collections, *Demons by Daylight* (coll **1973** US) and *The Height of the Scream* (coll **1976** US), trace this catharsis. His stories now focused on the EVIL in humankind, an evil that may arise from supernatural origins but is as likely to be inherent. This new RC emerged with shocking violence in his first novel, *The Doll who Ate his Mother* (**1976** US; rev 1985 US), where dabblings in SATANISM result in the birth of an evil child (◊ CHILDREN). The events themselves are only by implication supernatural, underlining the dichotomy RC has liked to explore in his later work between whether evil is of supernatural or human origin. This malevolence is more profound in his second novel, *The Face that Must Die* (cut **1979**; restored 1983 US), a PSYCHOLOGICAL THRILLER of mental decline. RC also utilized the latent malevolence of the cityscape (◊ URBAN FANTASY) to heighten this sense of alienation. Many of his subsequent novels – *To Wake the Dead* (**1980**; rev vt *The Parasite* 1980 US); *The Nameless* (**1981**; rev 1985 US); *The Claw* (**1983** as Jay Ramsay; vt *Night of the Claw* 1983 US; as by RC 1992 US); *Incarnate* (**1983** US; rev 1990); *Obsession* (**1985** US); *The Influence* (**1988** US) and *The Count of Eleven* (**1991**) – follow this development, deploying concepts of the supernatural, especially satanic cults, as only a possible explanation (or excuse) for human failings and degradation. These novels are thus stories of POSSESSION, whether by human, supernatural or psychological intervention, often triggered by a dominant precursive malign figure, which rules through the SPIRIT (control may often manifest via DREAMS or VISIONS). All are interpretations of madness. *The One Safe Place* (**1995**) utilizes this development to produce a strongly anti-censorship nonfantastic novel that explores how social deprivation is the root cause of most corruption.

RC's shorter stories remain more firmly in the realms of SUPERNATURAL FICTION, and include some of the best GHOST STORIES of the 1970s and 1980s – he received the WORLD FANTASY AWARD for "The Chimney" (1977 US in *Whispers* ed Stuart David Schiff) and "Mackintosh Willy" (1979 US in *Shadows 2* ed Charles L. GRANT). The best of these are collected as *Dark Companions* (coll **1982** US) and *Dark Feasts* (coll **1987**); the two volumes overlap extensively. RC also produced some SWORD-AND-SORCERY stories while struggling to establish himself as a freelance writer; these include the four **Ryre the Warrior** tales, starting with "The Sustenance of Hoak" (1977), which appeared in the **Swords Against Darkness** series ed Andrew J. OFFUTT: none has been included in RC's story collections. He also completed some of Robert E. HOWARD's **Solomon Kane** stories, included in *Solomon Kane 2: The Hills of the Dead* (coll **1979** US) by Howard. At this time RC produced a series of erotic horror stories (◊ SEX), mostly for ANTHOLOGIES ed Michel Parry (1947-   ), and these have been collected as *Scared Stiff: Tales of Sex and Death* (coll **1986** US).

RC's later short stories show a stronger affinity with his novels, placing greater emphasis on psychological degradation, alienation and distortions of REALITY. This emerges most potently in his novella *Needing Ghosts* (**1990** chap), a story of lost identity. Collections of later material are *Waking Nightmares* (coll **1991** US) and *Strange Things and Stranger Places* (coll **1993** US), plus a comprehensive retrospective *Alone With the Horrors: The Great Short Fiction of Ramsey Campbell 1961-1991* (coll **1993** US).

The later novels have shifted slightly from the emphasis on the evil of humanity to the possession of place, exploring the implications of residual influences caused either by humanity's spiritual or artistic passions – as in *The Hungry Moon* (**1986** US) and *Ancient Images* (**1989**) – or by the spirit of place (◊◊ PAN), as in *Midnight Sun* (**1990**) and *The Long Lost* (**1993**). The last two, in particular, are rare among late-20th-century SUPERNATURAL FICTION in the complexity of their exploration of humanity's relationship with its surroundings.

[MA]

Other works: *L'Homme du souterrain* (coll **1979** France) ed Richard D. Nolane; *Watch the Birdie* (**1984** chap); *Through the Walls* (**1985** chap); *Slow* (**1986** chap US); *Medusa* (**1987** chap US); *Two Obscure Tales* (coll **1993** chap US); *The Guide/Der Reiseführer* (coll **1994** chap Germany).

As editor: *Superhorror* (anth **1976**; vt *The Far Reaches of Fear* 1980); *New Tales of the Cthulhu Mythos* (anth **1980** US); *New Terrors 1* (anth **1980**; cut vt *New Terrors* 1982 US) and *New Terrors 2* (anth **1980**; cut vt *New Terrors II* 1984 US) – assembled as *Omnibus of New Terrors* (omni **1985**); *The Gruesome Book* (anth **1983**); *Fine Frights: Stories that Scared Me* (anth **1988**); **Best New Horror** series co-ed with Stephen JONES, being *Best New Horror* (anth **1990**), *Best New Horror 2* (anth **1991**), *#3* (anth **1992**), *#4* (anth **1993**), *#5* (anth **1994**), with a selection from first three issued as *The Giant Book of Best New Horror* (anth **1993**); *Uncanny Banquet* (anth **1992**); *Deathport* (anth **1993**) with Martin H. GREENBERG.

Further reading: *The Fantasy Readers' Guide to Ramsey Campbell* (**1980**) by Mike ASHLEY; *Ramsey Campbell* (**1988**) by Gary W. Crawford; *The Core of Ramsey Campbell: A Bibliography and Reader's Guide* (**1995** US) by RC with Stefan DZIEMIANOWICZ and S.T. JOSHI.

CANADA Canadian fantasy divides into two quite disparate strands, English and French. Canada's first fantastic tales were recited in any of the 13 native languages, including Inuktitut (spoken by the Inuit or Eskimo) and Algonkian (the most widely spoken Native American language); only later were the traditional stories told and written in French or English, the country's official languages. These native tales assumed the presence of TRICKSTERS, SHAPESHIFTERS, culture HEROES, SHAMANS, WITCHES, oracular shaking tents, malevolent windigos (◊ WENDIGO), benevolent thunderbirds, DWARFS, GIANTS and TRANSFORMATIONS.

1. English In the 16th-19th centuries explorers, colonists and settlers from France, the UK and other countries established notable regional traditions of storytelling, often fantastic in nature, including the "habitant" legends of Québec (which influenced the province's characteristic *contes*), the supernatural folklore of the Ottawa Valley, the TALL TALES characteristic of the Prairies and the West, and fancies and fantasies concerning influences that emanated from the North Pole and North Magnetic Pole. Since Confederation (1867) four factors have bedevilled Canada's imaginative writers: the geographical size of the country, with its high travel costs; the weakness of the local book- and magazine-publishing industry; the ready availability of imported publications from the UK and USA; and an otherwise preoccupied reading public. In consequence many writers have emigrated, sought publication abroad or set their fiction elsewhere.

The quintessential Canadian contribution to world literature is the animal story. The drama of a wild animal in a natural setting, written to satisfy the conflicting demands of naturalist and litterateur, owes its existence to Sir Charles G.D. Roberts (1860-1943) and its popularity to Ernest Thompson Seton (1886-1946). Roberts wrote the earliest realistic story of this type: "Do Seek Their Meat from God" (*Harpers Monthly* 1892; in *Earth's Enigmas* coll **1896**). But the vogue for the genre was sparked by the success of *Wild Animals I Have Known* (coll **1898**), which Seton wrote and illustrated. In these stories the animals do not talk (◊ TALKING ANIMALS), although quite often human qualities other than speech are ascribed to them – rudimentary reasoning, sympathetic feelings and so on.

The Canadian woods are a dramatic backdrop for six of the compelling tales of Algernon BLACKWOOD, who lived for two formative years in Toronto and Muskoka. In stories like "A Haunted Island" (in *The Empty House* coll **1906**) and "The Wendigo" (in *The Lost Valley* coll **1910**) the fauna and flora of the backwoods are seen as effectively alien – the counter-convention of the animal story.

The first Canadian ANTHOLOGY of HORROR was *Not to Be Taken at Night* (anth **1981**) ed John Robert COLOMBO and Michael Richardson (1946-   ). It was followed by *13 Canadian Ghost Stories* (anth **1988**) ed Ted Stone (1947-   ) and *Shivers: An Anthology of Canadian Ghost Stories* (anth **1989**) ed Greg Ioannou (1953-   ) and Lynne Missen. The prolific anthologist Alberto MANGUEL compiled *The Oxford Book of Canadian Ghost Stories* (anth **1990**), which offers 26 GHOST STORIES that range from French-Canadian legends through Anglo-Canadian stories to Postmodern fictions in which the GHOSTS are more apparent by their absence than their presence. Don Hutchison is the editor of the anthology series that begins with *Northern Frights* (anth **1992**) and continues with *Northern Frights 2* (anth **1994**).

Canadian writers' contributions to pulp MAGAZINES have

yet to be assessed, but among the most enthusiastic contributors were H. BEDFORD-JONES and Thomas P. Kelley (1905-1982), both natives of Ontario. *Tay John* (**1939**) by Howard O'Hagan (1902-1982) is the earliest Canadian novel of stature that compellingly combines elements of fantasy and realism; it relates the miraculous birth, life and death of a modern HERO in terms of the mindset of the Tsimshian Indians of the West Coast. O'Hagan's writing is seen as "mythopoeic" (a word associated with the criticism of Northrop FRYE); so also is the writing of Gwendolyn MacEwen (1941-1987). The poetry in *The Armies of the Moon* (coll **1972**) and the fables in *Noman* (coll **1972**), to name but two titles, attest to MacEwen's high Celtic imagination, mesmerizing lyricism and fascination with MAGIC.

Fantastic elements appear in the fiction of many Canadian mainstream writers: Margaret Atwood (1939-   ), Robertson DAVIES, Timothy FINDLEY, Pauline GEDGE, Thomas King (1943-   ), Brian MOORE, Jane Urquhart (1949-   ) and others; for instance, Findley's *Not Wanted on the Voyage* (**1984**) is a vigorous, outlandish, inventive reworking of the story of Noah. Without question the most familiar fantasy written by a Canadian is W.P. KINSELLA's *Shoeless Joe* (**1982**), a BASEBALL novel filmed as FIELD OF DREAMS (**1989**).

The highpoint of Canadian HIGH FANTASY is **The Fionavar Tapestry** (**1984-6**) by Guy Gavriel KAY. Kay's prose is by turns eloquent, magical, imaginative, intelligent and romantic. In the trilogy, five University of Toronto students "cross" to the "first world", Fionavar, where they join in an epic struggle against radical EVIL.

In Toronto in the 1980s an AFFINITY GROUP of fantasy/adventure/sf/horror writers dubbed themselves the Bunch of Seven. The group grew to include nine members: Tanya HUFF, Marnie Hughes, Louise Hypher, Shirley Meier (1960-   ), Terri Neal, Fiona Pattan, S.M. Stirling (1954-   ), Mike Wallis and Karen Wehrstein (1961-   ). Three members – Stirling, Meier and Wehrstein – have collaborated on SHARED WORLDS, notably the **Fifth Millennium** series. Huff has written with verve about Toronto VAMPIRES in *Blood Price* (**1991**), *Blood Trail* (**1992**) and *Blood Lines* (**1993**).

In Ottawa, also in the 1980s, another affinity group formed around the House of Speculative Fiction bookstore. The writers include Gordon Derevanchuk, Galad Elflandsson (1951-   ), Charles DE LINT, Charles R. Saunders (1946-   ) and bibliographer and anthologist John Bell (1952-   ). The Ottawa group hosted the 10th World Fantasy Convention in October 1984. The group's most prolific member is de Lint; some of his novels are Ottawa-set URBAN FANTASIES while others follow the adventures of the MINSTREL Cerin Songweaver. *Moonheart: A Romance* (**1984**) combines Celtic motifs and Native American worlds, both real and imaginary.

Other notable Canadian works include: Wayland DREW's *The Wabeno Feast* (**1973**), Ruth Nichols's *Song of the Pearl* (**1976**), Garfield REEVES-STEVENS's *Bloodshift* (**1981**), *A Hidden Place* (**1986**) and others by Robert Charles Wilson (1953-   ), Terence M. GREEN's *The Woman who is the Midnight Wind* (**1987**), Dave DUNCAN's *A Rose Red City* (**1987**), *Fang, the Gnome* (**1988**) in the **Greataway** sequence by Michael G. Coney (1932-   ), Tim Wynne-Jones's *Fastyngange* (**1988**), Antony SWITHIN's *Princes of Sandastre* (**1990**), Ven Begamudré's *A Planet of Eccentrics* (**1990**), Sean Russell's *The Initiate Brother* (**1991**), William Gough's *Chips*

and Gravey (**1991**), Michelle Sagara's *Into the Dark Lands* (**1991**) and Sean STEWART's *Nobody's Son* (**1993**).

The Canadian Science Fiction and Fantasy Association sponsors the Aurora AWARD, a "people's choice" award presented at the annual national convention. If Canadian sf came into its own in the 1980s and early 1990s, it seems that the 1990s and early 2000s are earmarked for the country's fantasy writers.   [JRC]

Further reading: *Canadian Science Fiction and Fantasy* (**1992**) by David Ketterer.

2. French French-language fantasy in Canada is rooted in the old oral tradition of the settlers, in the literary tradition of France itself, and in similar currents of world literature. As part and parcel of the oral tradition, many cautionary tales incorporated, in a Canadian context, the supernatural figures of the Catholic religion – the DEVIL especially, but also damned SOULS metamorphosing into WEREWOLVES or coming back as GHOSTS – as well as references to the mysterious powers of Native American SHAMANS. Such motifs are used in the first novel of French-Canadian letters, *L'Influence d'un livre* (**1837**; trans as *The Influence of a Book* **1993**) by Philippe Aubert de Gaspé Jr. Other 19th-century writers like Honoré Beaugrand, Guillaume Lévesque and Paul Stevens mined the same tradition, though a growing scepticism becomes evident in the later stories of authors like Pamphile Lemay and especially Louis Fréchette, whose ostensibly fantastical tales often subvert the traditional beliefs by remaking them into stories of false PERCEPTION.

In the first half of the 20th century another element of the oral tradition, reflecting the legacy of European FAIRYTALE, was systematically unearthed by ethnographers such as Marius Barbeau. One of the most common HEROES is a typical JACK, who often carries the day through a mixture of guile, luck and virtue rewarded by higher powers. Called Tit-Jean, or "Little John", his name connects him to St John the Baptist, patron saint of French-Canadians, often represented in the Québec iconography of the early 20th century as a young boy carrying a sheep.

The recognition won by MAGIC REALISM in world literature played a role in the resurgence of French-language fantastical literature in Canada after 1960. Similarly, the tremendous popularity in English-speaking countries of GENRE FANTASY and HORROR clearly sparked the writing of works in the same tradition. Finally, the various cults of the New Age have generated literary echoes on the margins of fantasy proper.

In French-speaking Canada, the transition from traditional to modern fantasy may be dated to Yves Thériault's *Contes pour un homme seul* ["Tales for a Man Alone"] (**1944**). His rural tales are crafted sparingly, proceeding directly to a character's inexplicable doom. His daughter, Marie José Thériault, on the other hand, wrote consciously archaic stories, as in *La Cérémonie* (coll **1978**; trans as *The Ceremony* **1984**), all distinguished by a lush prose style.

The fantastical strain has been illustrated by intelligent collections like Claude Mathieu's *La Mort Exquise* ["Death Dreadful"] (coll **1965**) and Jacques Brossard's *Le Métamorfaux* ["The Metamorfalsis"] (coll **1974**), both wonderfully Borgesian in places, and in such poignant collections as Roch Carriers's *Jolis Deuils* ["Beautiful Bereavements"] (coll **1964**) and Claudette Charbonneau-Tissot's *Contes pour hydrocéphales adultes* ["Tales for Adult Hydrocephalics"] (coll **1974**). It has also been marked by novels by Jacques Ferron such as *La Charrette* (**1968**; trans as *The Cart* **1980**) and *L'Amélanchier* (**1972**; trans as *The Juneberry Tree* **1975**), which are even more reminiscent of South American magical realism. Other notable books include André Carpentier's *L'Aigle volera a travers le soleil* ["The Eagle will Fly through the Sun"] (**1978**), Anne Hébert's novel of Parisian VAMPIRES *Héloïse* (**1980** France; trans as *Héloïse* **1982**) and Négovan Rajic's intriguing allegory of political oppression, *Les Hommes-taupes* (**1978**; trans as *The Mole Men* **1980**).

Daniel Sernine probably boasts the most sustained project of Québec fantasy. His deliberately nostalgic **Grandverger** – *Légendes du vieux manoir* ["Tales from the Old Manor House"] (coll **1979**), *Le Trésor du "Scorpion"* (**1980**; trans as *The "Scorpion" Treasure* **1990**), *L'Épée Arhapal* (**1981**; trans as *The Sword Arhapal* **1990**) and *Le Cercle Violet* ["The Purple Circle"] (**1984**) – conflates GOTHIC FANTASY, archetypal Québec settings and the history of Canada to create a true Québec Gothic. In recent times, the most famous example of Québec Gothic is probably found in Anne Hébert's *Les Enfants du Sabbat* (**1975** Paris; trans as *Children of the Black Sabbath* **1977**). The latest Sernine novel, *Manuscrit trouvé dans un secrétaire* ["Manuscript Found in a Secretary Desk"] (**1994**), is a new variation on an intriguing motif in French-Canadian fantasy – the influence of a BOOK within the book. The theme emerges also in the short fiction of 20th-century authors like Michel Bélil, André Carpentier, Claude Mathieu and Marie José Thériault.

GENRE FANTASY, by comparison, is hard to find. A dozen books, by a medley of authors including Sernine and Élisabeth Vonarburg (1947-   ), may qualify as HEROIC FANTASY; tellingly, all were published for a juvenile readership. No author has yet found genre fantasy appropriate for handling and developing major adult themes and stories. However, in supernatural HORROR Joël Champetier has written *La Mémoire du lac* ["The Lake's Memory"] (**1994**), an efficient and suspenseful tale set in northern Québec. In a similar vein, Stanley Péan wrote a Haitian reworking of Edgar Allan POE in *Le tumulte de mon sang* ["The Tumult of my Blood"] (**1991**).

A number of books inhabit the uneasy fringes between fantasy and speculative fiction. Esther Rochon (1948-   ), best known for *Coquillage* (**1986**; trans as *The Shell* **1990**), dwells in this water margin. While clearly closer to fantasy, Annick Perrot-Bishop, author of *Les Maisons de cristal* ["The Crystal-Houses"] (**1990**), may also belong here.

In French Canada short fiction is the natural home of fantasy, and many magazines, notably *Solaris* (ot *Requiem* 1974-current) and *Imagine* (1979), publish such stories. Merging easily with the mainstream of literature, the fantastical also surfaces in collection after collection. [JLT]

CANDYMAN A sequence of (so far) two movies.

1. *Candyman* US movie (**1992**). Polygram/Propaganda/Candyman Films. **Pr** Steve Golin, Alan Poul, Sigurjon Sighvatsson. **Exec pr** Clive BARKER. **Dir** Bernard Rose. **Mufx** Bob Keen. **Vfx** Cruse & Company. **Spfx** Martin Bresin. **Screenplay** Rose. **Based on** "The Forbidden" (1985) by Barker. **Starring** Xander Berkeley (Trevor Lyle), Dejuan Guy (Jake), Kasi Lemmons (Bernadette Walsh), Carolyn Lowery (Stacey), Virginia Madsen (Helen Lyle), Tony Todd (Candyman), Vanessa Williams (Anne-Marie McCoy). 93 mins. Colour.

URBAN-LEGEND researcher Lyle investigates the tale of Candyman, a slave's son who impregnated a landowner's daughter and was punished by having a hand sawn off and

being stung to death by bees: if you say his name five times into a MIRROR, he appears and guts you with the hook he has in place of that hand. This figure is currently linked with a SERIAL KILLER's activities in the slums. The GHOST Candyman appears to Lyle and, as her private life collapses, becomes her sole point of stability. At last she makes a CONTRACT with him: he may take her life if he spares that of an abducted baby. This contract he fulfils, but not before explaining he is merely her psychological projection, and that she has been the killer all along.

This excellent HORROR MOVIE/PSYCHOLOGICAL THRILLER is also a complex fantasy. Of especial note is the lure with which Candyman draws Lyle to her death: the state of being an urban legend is preferable to mere life, for as the former one will be revered through the terror of countless anonymous humans, a name on every lip, without the tedium of actually *being*. [JG]

2. *Candyman 2: Farewell to the Flesh* US movie (**1995**). Propaganda/PolyGram. **Pr** Gregg D. Feinberg, Sigurjon Sighvatsson. **Exec pr** Clive BARKER. **Dir** Bill Condon. **Spfx** John Hartigan, Gary L. King. **Vfx** Paul Gray, John P. Mesa, Andrew Naud, Daniel Radford. **Screenplay** Barker, Mark Kruger, Rand Ravich. **Starring** Timothy Carhart (Paul McKeever), Veronica Cartwright (Octavia Tarrant), Michael Culkin (Phillip Purcell), William O'Leary (Ethan Tarrant), Kelly Rowan (Annie Tarrant), Tony Todd (Candyman). **Voice actor** Russell Buchanan (Kingfish). 95 mins. Colour.

1 told a complete tale, so this sequel is gratuitous; its purpose seems to be, as Kim NEWMAN has observed, to be a launchpad for a series like the NIGHTMARE ON ELM STREET movies, with Candyman elevated to the status of an admired MONSTER somewhere between that series's Freddy Krueger and Pinhead, from the HELLRAISER sequence. Such plot as there is involves much slaughter of lesser characters until plucky teacher Annie Tarrant banishes Candyman, her ancestor, by smashing a MIRROR. What was interesting about the original *Candyman* – its exploitation of the notion of URBAN LEGEND – is here lost. [JG]

CANNELL, CHARLES HENRY Original name and later pseudonym of E.C. VIVIAN.

CANTERVILLE GHOST, THE There have been at least four movies based on the novella "The Canterville Ghost" (1887) by Oscar WILDE.

1. US movie (**1944**). MGM. **Pr** Arthur L. Field. **Dir** Jules Dassin. **Screenplay** Edwin Harvey Blum. **Starring** Charles Laughton (Sir Simon de Canterville), Margaret O'Brien (Lady Jessica de Canterville), Reginald Owen (Lord Canterville), "Rags" Ragland (Big Harry), Robert Young (Cuffy Williams). 95 mins. B/w.

300 years ago Sir Simon was immured by his father for running away from a duel, and cursed ($\lozenge$ CURSES) to haunt Canterville Castle until such time as a Canterville would, wearing Sir Simon's signet RING on his behalf, perform a brave deed. Ever since, Cantervilles have been craven. Now, in 1943, a platoon of US soldiers is billeted in the castle, owner of which is the six-year-old Lady Jessica; one of the soldiers, Cuffy, makes friends with Jessica and the GHOST, proves to be a remote Canterville descendant, and indeed performs a brave deed, thereby ending the HAUNTING. Though enlivened by the performance of O'Brien, this comedy grates: with the odd racist crack against Native Americans, it preaches a form of US cultural imperialism that involves the desecration of other societal values. But the

spfx (uncredited) are good.   [JG]

2. UK/US movie (*1986* tvm). HTV/Poundridge/ Interhemisphere/Columbia. **Pr** Peter Graham Scott. **Exec pr** Patrick Dromgoole, Irwin Meyer, Rodney Sheldon. **Dir** Paul Bogart. **Screenplay** George Zateslo. **Starring** John Gielgud (Sir Simon de Canterville), Harold Innocent (Hummle Umney), Lila Kaye (Mrs Umney), Andrea Marcovicci (Lucy Canterville), Alyssa Milano (Jennifer Canterville), Ted Wass (Harry Canterville), Bill Wallis (Professor Fenton Cook). 96 mins. Colour.

Americans Harry, Lucy and Jennifer inherit the Canterville title, and arrive from Cleveland to occupy the ancestral pile, little realizing it is haunted by the GHOST of Sir Simon, whose cupidity, 300 years ago, caused the death of his daughter and then, through accident, his wife; the ghost has driven away every Canterville ever to try to occupy the castle. Child Jennifer soon makes friends with the ghost, and begs him to drive off Lucy, her new STEP-MOTHER. When it emerges that Harry must sell the castle to a multinational hotel chain if the family is financially to survive, all priorities are changed. Harry takes up Sir Simon's challenge to somehow keep the castle in Canterville hands; Jennifer pleads directly to the Angel of Death that Sir Simon be allowed to die, and herself accepts Lucy into the family.

There is a lot of loose plotting and the feel that this is BEETLEJUICE (*1988*) without the budget – which reflects interestingly on the origins of *Beetlejuice*, which was the later movie – while the nutty parapsychological researcher Cook might have come from POLTERGEIST (*1982*), but nonetheless this has more of the spirit of Wilde's tale than does **1**. The minor parts – notably the butler-and-housekeeper couple the Umneys – are especially well played, but Milano steals the show, even from under Gielgud's nose.
[JG]

3. UK ANIMATED MOVIE (*1990* tvm). Emerald City/ Taffner. **Pr** Al Guest, Jean Mathieson. **Dir** Guest, Mathieson. **Screenplay** Guest, Mathieson. 48 mins. Colour.

A simplified retelling. The Otises buy Canterville Castle and commonsensically confound the GHOST. Variable animation (some good, most not), lousy lip-sync and voicing, an adequate script and much unabashed repeat looping.   [JG]

4. UK/US movie (*1995* tvm). Anasazi/Signboard Hill/Hallmark. **Pr** Robert Benedetti. **Exec pr** Richard Welsh. **Dir** Syd Macartney. **Spfx** Peter Hutchinson. **Screenplay** Benedetti. **Starring** Daniel Betts (Francis), Neve Campbell (Virginia Otis), Cherie Lunghi (Lucille Otis), Leslie Phillips (Lord Canterville), Joan Sims (Mrs Umney), Donald Sinden (Mr Umney), Patrick Stewart (Sir Simon de Canterville, the Ghost), Edward Wiley (Hiram Otis). *c*90 mins. Colour.

In some ways a fairly straightforward retelling, although with interesting overlays concerning the role PERCEPTION plays in the ability to experience a HAUNTING, and observations on the difficulties fathers may have in relating to adolescent daughters, and *vice versa*. Modest yet really rather affecting, this features – in addition to all else – some glorious moments as Stewart (a Shakespearean stage actor before joining *Star Trek – The New Generation*) utters cod-Shakespeare superbly.   [JG]

ČAPEK, KAREL (1890-1938) Czech writer active in many genres, much of whose occasional work remains untranslated or hard to obtain in English or other tongues. His most important plays and novels, however, have long been influential throughout the Western world, where his fertile, feverish wit, his cosmopolitanism, and his dark questioning of the nature of 20th-century progress may have helped disguise the fact that his best-known works are either sf or fantasy. Famous sf by CP includes the play *R.U.R.* (**1920**; trans Paul Selver with Nigel Playfair as *R.U.R.* (*Rossum's Universal Robots): A Fantastic Melodrama* 1923 UK; rev trans Selver alone 1923 US), and *Válka s Mloky* (**1936**; trans M. and R. Weatherall as *War With the Newts* 1937 UK; new trans Ewald Osers 1985 UK). His fantasy is less familiar.

KP began to publish short stories, mostly with his brother Josef Čapek (1887-1945), none translated but several seemingly fantasy. More widely available tales are collected in: *Trapné povídky* (coll **1921**; trans Francis P. Marchant, Dora Round, F. P. Casey and O. Vocadloas as *Money and Other Stories* 1929 UK), in *Tales from Two Pockets* (coll; cut trans **1932** UK; full trans Norma Comrada 1994 US), which assembles *Povídky z jedné kapsy* ["Tales from One Pocket"] (coll **1929**) and *Povídky z druhé kapsy* ["Tales from the Other Pocket"] (coll **1929**), only the uncut translation containing fantasy material; *Devatero Pohádek* (coll **1932**; trans as *Fairy Tales* 1933 UK; new trans Dagmar Herrmann vt *Nine Fairy Tales* 1990 US), FAIRYTALES for older children; and *Kniha apokryfu* (coll **1945**; trans Dora Round as *Apocryphal Stories* 1949 UK). Flickers of ALLEGORY-like significance sharpen the texture of many of these stories, though the ultimate effect of these and other "estrangements" from normal REALITY is of dis-ease and premonition. The supernatural in KČ is a signal of instability – mental, cultural, political – and it is both easy and almost certainly accurate to read much of his fiction as consciously predictive of the catastrophe of WORLD WAR II.

Perhaps the most brilliant of his plays, and the only unquestionable fantasy drama among them, is *Ze života hmyzu* (**1921**; trans Paul Selver as *And So Ad Infinitum (The World of the Insects)* 1923 UK; cut trans Owen Davis vt as *The World We Live In* 1933 US; commonly known as *The Insect Play*; best and first unexpurgated trans [Act 2 only] Tatiana Firkusny and Robert T. Jones as "From the Life of the Insects" in *Toward the Radical Center: A Karel Čapek Reader* coll trans **1990** US) with JosefČapek, which scarifyingly mocks human life and pretensions through a series of BEAST-FABLE episodes in which various insects perform vaudeville versions of human behaviour at the edge of extinction.

With the exception of the weak *Adam strvořitel* (**1927**; trans Dora Round as *Adam the Creator* 1927 UK) with Josef Čapek, the rest of KC's plays and novels are best understood in sf terms. However extravagant their plots, KČ thought that he was describing possible nightmares of history. He was correct.   [JC]

CAPES, BERNARD (1854-1918) UK writer. His career began in 1898 with *The Lake of Wine*, about an accursed jewel. His reputation fell after his death, but he was rediscovered in 1978 by Hugh Lamb (1947-   ). Among more traditional HAUNTED-DWELLING fare, he showed special skill in depicting trapped souls (◊ BONDAGE) – as in "An Eddy on the Floor" (1899), "The Green Bottle" (1902), "The Jade Button" (1906), "The Glass Bell" (1915) and "Poor Lucy Rivers" (1906), the latter being one of the earliest examples of the haunted-typewriter motif. His weird tales are scattered through *At a Winter's Fire* (coll **1899**), *Plots* (coll **1902**), *Loaves and Fishes* (coll **1906**), *Bag and Baggage* (coll **1913**) and *The Fabulists* (coll **1915**). Lamb assembled a selection as *The Black Reaper* (coll **1989**). [MA]

CAPITANO, IL ◊ COMMEDIA DELL'ARTE.

CAPOTE, TRUMAN (1924-1984) US writer. ◊ ANSWERED PRAYERS.

CAPRA, FRANK (R.) (1897-1991) Italian-born US movie director and producer (occasionally screenwriter) who has become identified with a particular strand of feelgood movies, but whose range was in fact far wider than this. His first movie was a short, *Fultah Fisher's Boarding House* (**1922**). Of his 40-plus features, the two of greatest fantasy interest are LOST HORIZON (**1937**) and IT'S A WONDERFUL LIFE (**1946**). Of associational note is *It Happened One Night* (**1934**), which instigated the Hollywood screwball-movie tradition and cheerfully veers into something near fantasy. FC's charm and whimsy were viewed, by the time of his last movie, *Pocketful of Miracles* (**1961**), as mere syrup; but more recently his undoubted virtues have become again appreciated. [JG]

CAPTAIN MARVEL Several COMIC-book SUPERHEROES.

1. The most important and abidingly popular is the full-faced, wavy-haired US comic-book character who wears a bright red costume with a yellow lightning flash on the chest, plus yellow boots, cuffs and waistband and a short, yellow-edged white cloak. He first appeared in Fawcett's *Whiz Comics #1* (1940). CM was created by editor/writer Bill Parker and artist C(harles) C(larence) Beck (1910-1989), and his cleft-chinned, youthful good looks were reputedly based on those of movie actor Fred MacMurray (1907-1991). This first story told of orphaned newsboy Billy Batson's encounter with a mysterious trenchcoated figure who led him to an underground cavern to meet the 3000-year-old WIZARD Shazam. "I am old now," says Shazam: "You shall be my successor. Merely by speaking my name, you can become the strongest and mightiest man in the world – Captain Marvel!" Henceforth, each time young Billy said "Shazam" – an acronym signifying Solomon (wisdom), Hercules (strength), Atlas (stamina), Zeus (power), Achilles (courage) and Mercury (speed) – a clap of thunder and an impressive zag of lightning accompanied his META-MORPHOSIS into "the world's mightiest mortal". The great success of the first few issues was significantly boosted when Otto Binder (1911-1975) replaced Parker as scripter, and a new regular comic book, *Captain Marvel Adventures* (#1-#150 1941-53), is said to have achieved sales in excess of one million every fortnight. CM gained a whole family, including Mary Marvel (Mary Batson, Billy's sister), Captain Marvel Junior (crippled newsboy Freddy Freeman, who became a younger, blue-garbed superhero on speaking CM's name), three Lieutenants Marvel, Uncle Marvel and even Hoppy, the Marvel bunny, "cloned" from CM in *Fawcett's Funny Animals #1* (1942). The CM stories themselves had an appealing whimsy and gentle SATIRE then unique in comic books.

As CM's popularity grew, Binder and Beck assembled a team of writers (Ron Reed, Bill Woolfolk, Bob Kaniger, etc.) and artists (Pete Costanza, Chic Stone, Kurt Schaffenberger, Al Falagy, Mac Raboy, etc.) whom they encouraged to maintain a consistent image of the character. CM later featured in numerous other comic books, including *America's Greatest Comics* (1941-3), *All Hero Comic* (1943) and *Marvel Family* (1945-54).

This was the version of CM who appeared in the serial movie *Adventures of Captain Marvel* (**1941**; dir John English, William Witney; 12 episodes; b/w). In this kitsch outing CM

is in Thailand, where he destroys a baddie called The Scorpion.

2. In 1941 National Periodical Publications (later DC COMICS) brought a lawsuit against Fawcett, claiming CM infringed their SUPERMAN copyright. This dragged on throughout the 1940s until, for financial reasons, Fawcett capitulated in 1953. The UK reprints by L. Miller & Co had, however, been enormously popular, and this company decided to continue independent publication, with artwork by the Mick Anglo studio and giving the character a new name (Marvelman), a smart crewcut hairstyle and a new magic word, "Kimota" ("Atomik" backwards). Artists on this feature included Ron EMBLETON, Don Lawrence (1928-) and George Stokes.

3. 2 was resurrected with an adult storyline by Alan MOORE for Quality Comics's *Warrior*, and became the subject of legal dispute when MARVEL COMICS objected to the use of the word "Marvel" in the title. Marvelman, renamed "Miracleman", was subsequently published unchallenged in the USA by Eclipse from 1981.

4. Two publishing projects (◊◊ **5**) in the 1960s served to apprise the two leading comics publishers in the USA, DC and Marvel, of an odd dilemma. A new superhero with the same name was introduced by MF Enterprises in *Captain Marvel* (#1-#4 1966) and subsequently *Captain Marvel Presents the Terrible Five* (#1-#4 1966-7). These publications contained a number of scurrilous rip-offs of established comics characters, and DC took legal action.

5. Meanwhile, the Milson Publishing Company's comics arm, Lightning Comics, attempted to revive the CM concept in 1967 by commissioning Binder and Beck to create *Fatman, The Human Flying Saucer* (#1-#3 1967) in which a magic word turns a boy into a UFO. This did not last long. DC National's dilemma was that they had assumed ownership of a character whose name included the word "Marvel", so could not sue.

6. MARVEL COMICS were the first to respond to **5**, creating a completely different CM, a warrior-scientist from a distant planet, in *Marvel Superheroes #12* (1968), on the assumption that DC could not publish the original CM without causing undesirable confusion.

7. DC COMICS could not prevent this because they had never actually published a CM but, not wishing to relinquish their entitlement to do so, they created a new comic book, *Shazam* (1972-8), which featured the original CM in new stories; this spawned a live-action tv series, *Shazam* (which ran 1974-7 and starred Jackson Bostwick and, later, John Davey as CM), and a brief animated tv series, *Shazam* (1981).

8. DC began publishing the original Fawcett material in *The Shazam Archives* (graph coll **1992**). To date (1996) only one volume has appeared. A prestigious hardcover publication, *The Power of Shazam* (graph **1994**) by Gerry Ordway (1957-), which retold the origin story, showed considerable respect for Beck's original and foreshadowed a new monthly series. [RT]

CAPTAIN NICE US tv series (1967). NBC. **Pr** Jay Sandrich. **Exec Pr** Buck Henry. **Dir** Richard Kinon, Gary Nelson, Gene Reynolds, Charles Rondeau, Jay Sandrich, Jud Taylor. **Writers** Stan Burns, Buck Henry and many others. **Created by** Henry. **Novelization** *Captain Nice* * (**1967**) by William Johnston. **Starring** William Daniels (Carter Nash/Captain Nice), Liam Dunn (Mayor Finny), Byron Foulger (Mr Nash), Alice Ghostly (Mrs Nash), Ann

Prentice (Sergeant Candy Kane), William Zuckert (Chief Segal). 15 30min episodes. Colour.

This sitcom attempted to do for SUPERHEROES what *Get Smart!* had done for secret agents – not surprising, as Buck Henry was involved in the birth of both series. *CN* featured police scientist Carter Nash, who created Super Juice, a strange concoction that turned him into a superhero. The use of Super Juice inevitably would result in an explosion that destroyed his clothes, so Nash was forced to fight crime in a shabby costume created from his underwear. The series was soon cancelled, as was a similar series from the same season, MR TERRIFIC. [BC]

CAPTAIN SINDBAD (*1963*) ◊ SINBAD MOVIES.

CAR, THE US movie (*1977*). Universal. **Pr** Marvin Birot, Elliot Silverstein. **Dir** Silverstein. **Vfx** Albert Whitlock. **Screenplay** Michael Butler, Dennis Shryack, Lane Slate. **Starring** James Brolin (Wade Parent), Ronny Cox (Luke), Kathleen Lloyd (Lauren). 98 mins. Colour.

Almost certainly the purest example of TECHNOFANTASY the CINEMA has ever produced, this sees WRONGNESS and EVIL, presaged each time by an unnatural WIND, invade a small US town in the form of an anonymous, driverless CAR that mindlessly runs people over. The plot is little more than that. The car is depicted as a MONSTER, and one that cannot enter hallowed ground; interestingly, we sometimes are allowed a glimpse of its PERCEPTION of the world through its yellowed windscreen. When the car is finally "killed" the funnel of flame and smoke that erupts skyward takes the form of a DRAGON, or a DEMON, or even SATAN himself. There is no explanation, although the music soundtrack possibly gives an answer in its various parodies of the *Dies Irae*. [JG]

CARD, ORSON SCOTT (1951-   ) US writer, mainly known as a central figure in sf since the late 1970s; his fantasy is secondary, although his main fantasy sequence is important and a "feel" of fantasy pervades much of his sf work.

OSC's first fantasy novel, *Hart's Hope* (**1983**), is interesting mainly for its protagonist's anti-gift, as he is a nullifier of MAGIC. It was with the **Alvin Maker** sequence – *Seventh Son* (**1987**), *Red Prophet* (**1988**) and *Prentice Alvin* (**1989**), all three assembled as *Hatrack River: The Tales of Alvin Maker* (omni **1989**), plus *Alvin Journeyman* (**1995**) and at least one further volume, «Master Alvin», projected – that OSC established himself as an innovative contributor to US fantasy. The establishing premise could have generated an sf novel: an ALTERNATE WORLD in which there has been no American Revolution, and thus no unified USA. Rather like the romantically Balkanized statelets that so often feature in post-HOLOCAUST sf, a profusion of dominions occupies eastern North America. What qualifies the sequence as fantasy from the beginning is not this LANDSCAPE but the associated premise that MAGIC exists, and that the THINNING of Europe – occasioned by the Scientific Revolution there, and in England in particular by the survival of Oliver Cromwell's Protectorate – has motivated those with TALENTS to emigrate to America, a land irradiated, therefore, by magic. Alvin, the seventh son of a seventh son (◊ SEVEN), seems destined to become (as "Maker") a kind of MESSIAH. He grows up in a late-18th-/early-19th-century environment portrayed by OSC as an almost supernaturally whole world, a LAND-OF-FABLE America whose plenitude is Edenic. But Alvin's growth to maturity, and America's coming to terms with the Native American nations which dominate the

western half of the continent, are darkened by the Unmaker, a figure of desiccating EVIL, one of whose minions is Alvin's brother and SHADOW Calvin. Ancillary figures include William BLAKE, Tecumseh, Honoré de BALZAC and Daniel Webster. The sequence, as it nears its conclusion, appears to be drawing westward (rather as the Mormon leader, Joseph Smith, whom Alvin sometimes resembles, drew his people westward), and possibly a new JERUSALEM will be founded.

Several of the stories assembled in *Maps in a Mirror: The Short Fiction of Orson Scott Card* (coll **1990**; vt in 4 vols as *The Changed Man* 1992, *Flux* 1992, *Cruel Miracles* 1993 and *Monkey Sonatas* 1993) are fantasy; and the intensely mythopoeic, salvationist bent of OSC's sf constantly generates a fantasy sense that the good LAND may be redeemed and that the true STORY (OSC's Mormon faith, itself charged with Story, is central to his work) of the world may be recoverable. For an author like OSC, fantasy subverts the mundane, secular world through that promise of recovery. [JC]

Other works: *Lost Boys* (**1992**), a GHOST STORY.

CARDS It is not known whether playing-cards first arose as a means of gambling or as a method of divination, but they have long fulfilled both functions. Users of the divinatory TAROT pack, of which the conventional deck is a stripped-down version, have constructed many elaborate pseudohistories while attempting to justify their interpretations of the 22 enigmatic picture-cards which form the "Major Arcana". Prosaic explanations link the pack with the game of Tarocchi, but fantasy writers have offered far more fanciful accounts, the most extravagant being found in *The Greater Trumps* (**1932**) by Charles WILLIAMS and the trilogy begun with *God of Tarot* (**1979**) by Piers ANTHONY. Other Tarot novels include Louise COOPER's *The Book of Paradox* (**1973**), Carl SHERRELL's *Arcane* (**1978**), *Coriolanus, the Chariot* (**1978**) by Alan G. Yates (1923-1985) and *The Labyrinth Gate* (**1988**) by Alis A. RASMUSSEN. Italo CALVINO's *The Castle of Crossed Destinies* (**1973**) uses a Tarot deck as a linking device for a series of stories; the deck inspired *Tarot Tales* (anth **1989**) ed Rachel POLLACK and Caitlín Matthews (?1952-   ). Magical cards with Tarot-like "trumps" are a key motif in Roger ZELAZNY's long-running **Amber** series.

The more familiar pack features in many ironic tales of cruel FATE, often involving SATAN or some other diabolical figure; examples are Edgar Allan POE's "The Duc de l'Omelette" (1832), "The Queen of Spades" (1834) by Alexander Pushkin (1799-1837), John MASEFIELD's "A Deal of Cards" (1913), "The Devil Deals" (1938; vt "The King and the Knave") by Carl Jacobi (1908-   ) and the pop song "Spanish Train" (1975) by Chris de Burgh. A borderline fantasy in which a card game exemplifies fate is *The Music of Chance* (**1990**) by Paul Auster (1947-   ). By far the most original and complex fantasy involving playing cards is Tim POWERS's *Last Call* (**1992**). [BS]

CAREY, PETER (1943-   ) Australian writer whose work has always pressed at the edges of mundane worlds, describing them in terms of MAGIC REALISM or through the medium of unreliable narrators. His first tales of fantasy interest appear in *The Fat Man in History* (coll **1974**) and *War Crimes: Short Stories* (coll **1979**), selections from both volumes being published as *The Fat Man in History* (coll **1980** UK; vt *Exotic Pleasures* 1981 UK). Here sf and fantasy themes generally appear not for their own sake but as emblems of alienation. A tale like "Do You Love Me?" describes attempts to generate a REALITY-sustaining MAP of a country which is losing its grip on itself; but attempting this task simply widens the gap between the fallen world and any REALITY from which we may have slipped.

PC's first novel, *Bliss* (**1981**), is a POSTHUMOUS FANTASY if its protagonist can be believed, but under any reading it is about trying to create a HEAVEN, either on Earth or beyond; it was filmed as *Bliss* (**1986**), with a treatment by PC and Ray Lawrence published as *Bliss: The Film* * (**1986**). *Illywhacker* (**1984**) constitutes a kind of Australian CREATION MYTH – if the story of white Australians can be plausibly treated as an instigating one, and if the protagonist's claims can be believed – a huge lifespan, vast and magical experiences, and TALENTS (he professes, during the period he works as a stage MAGICIAN, to be capable of genuine INVISIBILITY). *Oscar and Lucinda* (**1988**), which won the Booker Prize, is technically a mundane novel, but the RIVER at its heart is as saturated and fantastic a SYMBOL of the Australian experience as anything found in a nonmimetic text. *The Tax Inspector* (**1991**) sticks closer to ascertainable reality; but *The Unusual Life of Tristan Smith* (**1994**) is set in an ALTERNATE-WORLD Earth incorporating two IMAGINARY LANDS: the domineering Voorstand and the ARCHIPELAGO called Efica. The story itself, in which a surreal kind of theatrical life emblematizes the difference between this world and ours, stops abruptly; other resemblances to *The Life and Adventures of Tristram Shandy* (**1759-67**) by Laurence Sterne (1713-1768), which likewise stops before finishing, include a protracted pre-birth narrative on the part of the protagonist. *The Big Bazoohley* (**1995**) is a YA tale involving a WITCH. [JC]

CARL, LILLIAN STEWART (1949-   ) US writer who began publishing fantasy with the **Sabazel** HEROIC-FANTASY sequence – *Sabazel* (**1985**), *The Winter King* (**1986**) and *Shadow Dancer* (**1987**) – featuring AMAZONS and the barbarians who surround them. *Wings of Power* (**1989**) has elements of ARABIAN FANTASY: *Ashes to Ashes* (**1989**) and *Dust to Dusk* (**1991**) are SUPERNATURAL FICTION. [JC]
Other works: *Garden of Thorns* (**1992**).

CARLSEN, CHRIS Pseudonym of Robert P. HOLDSTOCK.

CARLYLE, THOMAS (1795-1881) Scottish polemicist, philosopher, historian and editor, of very broad cultural significance in 19th-century thought. He is of fantasy interest through editing and translating a pioneer anthology of German Romantic fiction of the early 19th century, *German Romance: Specimens of its Chief Authors, with Biographical and Critical Notices* (trans anth **1827** 4 vols). Authors represented, often with their best work, include E.T.A. HOFFMANN, J.K. MUSÄEUS and J.L. TIECK. TC's fascination with ROMANTICISM inspired his curiously idiosyncratic SATIRE *Sartor Resartus* (**1833-4; 1836** US), an extended exploration of PERCEPTION and REALITY. [JC]

CARNELL, E(DWARD) J(OHN) (1912-1972) UK editor and literary agent, predominantly in the field of sf. As editor of the magazine SCIENCE FANTASY, EJC encouraged the OTHERWORLD fantasies of John BRUNNER and Kenneth BULMER, and was instrumental in Michael MOORCOCK writing the **Elric** stories and Thomas Burnett SWANN producing his mythological fantasies when there was no market for them in the USA. Among many anthologies EJC edited *Jinn and Jitters* (anth **1946**) and *Weird Shadows from Beyond* (anth **1965**). [MA]

CARNE PER FRANKENSTEIN ot of *Flesh for Frankenstein* (**1973**). ◊ FRANKENSTEIN MOVIES.

CARNIVAL 1. The feast which, in Roman Catholic countries, precedes the austerity of Lent; it is most commonly celebrated on Shrove Tuesday, alias Mardi Gras, and the celebrations often involve parades and masked balls (sometimes involving a tacit dissolution or TOPSY-TURVY of the social hierarchy). Fantasies set on such occasions include "The Masked Ball" (1947) by Seabury QUINN and "Masquerade of a Dead Sword" (1986) and "The Greater Festival of Masks" (1989) by Thomas LIGOTTI. The movie and stage versions of Gaston LEROUX's *The Phantom of the Opera* (**1911**) make much of the MASQUE scene (◊ PHANTOM OF THE OPERA), exploiting a motif pioneered by Edgar Allan POE's "Masque of the Red Death" (1842), which is allegedly outdone by a key scene in the mythical play *The King in Yellow* (◊ Robert W. CHAMBERS).

2. The critic Mikhail Bakhtin drew a distinction between the "dialogical" or "polyphonic" novel and the "monological" novel, tracing the origins of the former – in which the author's narrative voice is ambiguous or confused by opposition – to satire and carnival. "Carnival" and "carnivalesque" have in consequence become technical terms sometimes applied to non-naturalistic fictions in order to characterize their various rhetorical perversities; they are especially pertinent to those involving dramatic transformations of the social order.

3. The more normal sense of the word "carnival" refers, especially in the USA, to a travelling show comprising stalls, sideshows and displays of freaks and animals; it falls somewhere between the European circus and funfair. All three venues have been used extensively by fantasy writers, probably because the overt sham and hoaxery involved in such entertainments evokes ideas of MASKS: the surface is very different from what underlies it, as in the notion of the miserable clown. Noteworthy examples of fantasies set in or around carnivals and fairs are *The* CABINET OF DR CALIGARI (*1919*), Charles BEAUMONT's *The Circus of Dr Lao* (**1935**) – filmed as 7 FACES OF DR LAO (*1964*) – Ray BRADBURY's *Something Wicked This Way Comes* (**1962**) – filmed as SOMETHING WICKED THIS WAY COMES (*1983*) (it is no coincidence that Bradbury's first collection, *Dark Carnival* [coll 1947] was thus titled) – and Arthur Calder-Marshall's *The Fair to Middling* (**1959**). The movie DANTE'S INFERNO (*1935*) centres on a sort of super-carnival. *The Funhouse* * (**1980**; rev 1992) by Dean R. KOONTZ novelizes the movie *The Funhouse* (*1981*) dir Tobe Hooper; a mixture of SUPERNATURAL FICTION and HORROR, it describes what happens to a group of teenagers who are insufficiently credulous about the hauntedness of a carnival's "Haunted House". A large part of Peter S. BEAGLE's *The Last Unicorn* (**1968**) – filmed as *The* LAST UNICORN (*1982*) – memorably focuses on a carnival. The tendency of US carnivals to include "freak shows" has inspired numerous DARK FANTASIES of note, including Tom ROBBINS's *The Unholy Three* (*1917*) and "Spurs" (1926; the basis of the movie *Freaks* [*1932*] dir Tod Browning). Helen CRESSWELL's *The Watchers: A Mystery at Alton Towers* (**1993**) is set in the permanent UK funfair Alton Towers. Perhaps the paradigmatic fantasy set in a circus, as opposed to a carnival, is DUMBO (*1941*). Also of note are: the circus sequences in *The Bride* (**1985**; ◊ FRANKENSTEIN MOVIES), although these are not in themselves fantastic; Bradbury's "The Illustrated Man" (1951) and the linking device of his *The Illustrated Man* (coll **1951**); Michael KURLAND's *The Unicorn Girl* (**1969**), which features a planet-roving circus, complete with UNICORN; Tom REAMY's *Blind Voices* (**1978**); and Angela CARTER's *Nights at the Circus* (**1984**). Fabulous travelling shows are featured also in William KOTZWINKLE's *Fata Morgana* (**1977**) and James BLAYLOCK's *Land of Dreams* (**1988**). In HIGH FANTASY, such as Robert JORDAN's **Wheel of Time** sequence, the tendency for fugitive characters to join a carnival or traveller fair as DISGUISE has become a cliché. [BS/JC/JG]

CARPENTER, LEONARD (1948-   ). US writer. ◊ Robert E. HOWARD.

CARR, JOHN DICKSON (1906-1977) US writer of detective stories, long resident in the UK, who wrote also as Carter Dickson. His fondness for locked-room mysteries and impossible crimes produced eerily atmospheric VAMPIRE novels like *The Three Coffins* (**1935**; vt *The Hollow Man* UK) and *He Who Whispers* (**1946**), which ultimately prove to be RATIONALIZED FANTASIES. Of the Carter Dickson titles, *The Plague Court Murders* (**1934**) plays similarly with GHOSTS and SPIRITUALISM, and *The Curse of the Bronze Lamp* (**1945**; vt *Lord of the Sorcerers* 1946 UK) with apparent dematerialization resulting from a CURSE. *The Burning Court* (**1937**) is a *tour de force* in which dark hints of WITCHCRAFT and REINCARNATION are dispelled by ingenious detective rationalization, only to be later confirmed.

JDC also wrote heavily researched TIMESLIP fantasies confronting modern investigators with crimes in historical LONDON. Of these, the 17th-century *The Devil in Velvet* (**1951**) – featuring also POSSESSION, a PACT WITH THE DEVIL and ensuing QUIBBLES – is the best. *Fear is the Same* (**1956** as by Dickson) visits the 18th century and *Fire, Burn!* (**1957**) the 19th. Some routine SUPERNATURAL FICTION appears amid detections in *The Department of Queer Complaints* (coll **1940**) and *The Door to Doom* (coll **1980**) ed Douglas G. Greene; the latter includes a bibliography. [DRL]

See also: DETECTIVE/THRILLER FANTASY.

CARR, TERRY (GENE) (1937-1987) US writer and editor, best known for his editorial work in sf, although he was equally capable in the field of FANTASY. He was not a prolific writer. Most of his stories are fantasies, including his first, "Who Sups with the Devil" (1962 *F&SF*). His sf shines with the inner glow of fantasy. Stories like "The Dance of the Changer and the Three" (in *The Farthest Reaches* anth **1968** ed Joseph Elder), which considers an alien culture through its own MYTHOLOGY, and *Cirque* (**1977**), which explores the DECADENCE of a FAR-FUTURE city, are sf by subject but fantasy by treatment. His most overt sf pales by comparison, such as *Warlord of Kor* (**1963** dos), a hack space opera.

TC will be best remembered as an editor (in which role he won two Hugo AWARDS). He was a house editor at Ace Books 1964-71 under Donald A. WOLLHEIM, and helped develop writers including Ursula K. LE GUIN, R.A. LAFFERTY and Samuel A. DELANY. He also edited many ANTHOLOGIES (◊ *SFE for full listing*), including several of import to the fields of fantasy and SUPERNATURAL FICTION. His short-lived **New Worlds of Fantasy** series – *New Worlds of Fantasy* (anth **1967**; vt *Step Outside Your Mind* 1969 UK), *#2* (anth **1970**), *#3* (anth **1971**) – was the first to recognize the resurgence in fantastic fiction and to explore and define its redevelopment. He revived the series as **Year's Finest Fantasy** with *Year's Finest Fantasy* (anth **1978**) and *Volume 2* (anth **1979**), retitled **Fantasy Annual** with *Fantasy Annual III* (anth **1981**), *IV* (anth **1981**) and *V* (anth **1981**). Although always critically acclaimed, this impressive series,

despite being much superior to **The Year's Best Fantasy Stories** ed Lin CARTER, sold poorly. Other retrospective anthologies include *The Others* (anth **1969**) and *A Treasury of Modern Fantasy* (anth **1981**; vt omitting 2 stories *Masters of Fantasy* 1992) with Martin H. GREENBERG, restricted predominantly to MAGAZINE supernatural fiction. TC did most of the editorial work on *100 Great Fantasy Short Short Stories* (anth **1984**) with Isaac ASIMOV and Greenberg. Just before his death he became fiction editor on the subsequently abortive *L. Ron Hubbard's To the Stars* sf magazine. [MA]

CARRIE US movie (*1976*). United Artists/Red Bank. **Pr** Paul Monash. **Dir** Brian DePalma. **Spfx** Greg Auer, Ken Pepiot. **Screenplay** Lawrence D. Cohen. **Based on** *Carrie* (**1974**) by Stephen KING. **Starring** Nancy Allen (Chris), Betty Buckley (Miss Collins), Amy Irving (Sue), William Katt (Tommy), Piper Laurie (Margaret White), Sissy Spacek (Carrie White), John Travolta (Billy). 98 mins. Colour.

Bullied by fundamentalist mother and schoolmates alike, Carrie develops psychokinetic TALENTS, and at her MENARCHE these become almost uncontrollable. Taunted beyond endurance at the school prom, she lets rip with pyrotechnic destruction and indiscriminate slaughter. Back home, mother attacks her with a knife for the sin of having gone to the prom; Carrie kills her psychokinetically, then destroys herself. DePalma is rarely a subtle director, and King rarely a subtle writer; however, *C* is among the cinema's best portrayals of the VENGEANCE of the psychically gifted PARIAH ELITE. It is also moving: whatever horrors Carrie commits seem considerably less vile than those perpetrated against her, and her time of joy is deeply poignant because we know how brief it must necessarily be. [JG]

CARROLL, JONATHAN (1949-   ) US-born screenwriter and author, long resident in Vienna, where he has set some of his fiction, though LOS ANGELES, where he was raised and where he does his movie work, remains a central venue. As an author, JC walks alone, in wary solitude: some of his work verges on sf; much of it makes guarded and ironized use of fantasy MOTIFS and OTHERWORLD venues; and he is frequently thought of as primarily a horror writer. His best novels, however, touch on all three genres, illuminating them. Though he received a 1988 WORLD FANTASY AWARD for "Friend's Best Man" (1987), he is less known for his short stories, which have been assembled as *Die Panische Hand* (coll **1989** Germany; exp as *The Panic Hand* **1995** UK).

JC first novel, *The Land of Laughs* (**1980** US), is his most unequivocal fantasy. Two young potential biographers of the dead Marshall France, the most famous of whose books for children is *The Land of Laughs* (◊ BOOKS), anticipate collaborating in their work. Together they visit Galen, Missouri, where France lived all his life; and gradually discover the town is a claustrophobic POLDER whose inhabitants have been possessed by France's STORY of their past and future lives, as continued in his Journal, which carries their tale down the centuries. Hints of GODGAME in the novel are initially provocative but – as is typical of JC's work – the controlling artist/MAGUS fails to grant his players free will or the capacity to learn from their trials. France's self-absorption makes the conclusion of *The Land of Laughs* more like horror than fantasy, where stories tend to transform protagonists more often than to imprison them. JC's seeming distrust of happy endings, a distrust which "lets down" the strong fantasy stories he is entirely capable of beginning, may help explain the dis-ease he causes in some readers.

After the more tentative *Voice of our Shadow* (**1983** US) JC embarked upon his major project to date, an untitled sextet of linked novels: *Bones of the Moon* (**1987** UK; rev 1987), *Sleeping in Flame* (**1988** UK), *A Child Across the Sky* (**1989** UK), *Outside the Dog Museum* (**1991** UK), *After Silence* (**1992** UK) and *From the Teeth of Angels* (**1994** UK), with *Black Cocktail* (**1990** UK) as a loose pendant. The overall title of the series could easily be "Answered Prayers" (◊ ANSWERED PRAYERS); certainly, in volume after volume, artist protagonists both charge their intimates harshly for the privilege of being close to the making of works of art and themselves pay heavily for having their prayers for success answered in the affirmative.

Of the six, *Bones of the Moon* most resembles a fantasy novel. Its female protagonist, post-abortion, is drawn into an OTHERWORLD called Rondua whose REALITY is a barbed manifestation of the internal dramas (◊ JUNGIAN PSYCHOLOGY) of her own psyche, and where she must attempt to save both her kingdom (i.e., herself) and her unborn son from a maniacal killer (and profoundly defective MAGUS) called Jack Chili. Soon Rondua begins to impose itself CROSSHATCH-fashion upon the real world, and the fight for life climaxes in both inner and outer realms. She seems to win, and the story becomes a book, which she writes, called *Bones of the Moon*. *Sleeping in Flame* more problematically faces its artist protagonist in Vienna with a warped magus who is a TWICE-TOLD Rumpelstiltskin, and who seemingly fails in his attempts to disrupt a growing family. In *A Child Across the Sky* a dead man's video (like France's book in *The Land of Laughs*) inspires a film director to afflict those around him with the successful re-making of the dead man's last, failed movie. In *Outside the Dog Museum* an architect must fumblingly attempt to justify his life in terms of the construction of the eponymous EDIFICE for an Eastern potentate. *After Silence* (which arguably contains no supernatural element at all) carries a successful cartoonist into a hell he has (perhaps) constructed for himself, as it may be he has (like a deranged magus) created the sins which he "discovers" in his wife's past. And *From the Teeth of Angels* (which may or may not terminate the sequence) contains strong hints – through the grotesque and haunting presence of a personification of DEATH – that all the previous protagonists may have indeed been victims of a parodic GODGAME, in which the dice are loaded. The sense that everyone in the sequence is spiritually parched may come (it is hinted) from the fact that their lives have been savoured by a GOD who likes the sweet smell of success.

In JC's work, happy endings are truces, not consummations. A huge amount of artistic endeavour is described throughout his work, but few triumphant RECOGNITIONS. There seems little doubt that this is deliberate. JC is a profound, ironic, scathing poet of the terror of the prison of mortality. [JC]

CARROLL, LEWIS Working name of UK writer Charles Lutwidge Dodgson (1832-1898), also known under his own name as a mathematician, lecturer and writer of mathematical texts. LC is best known as author of *Alice's Adventures in Wonderland* (**1865**; vt *Alice in Wonderland* 1944) and its sequel *Through the Looking-glass, and What Alice Found There* (**1871**; vt *Alice Through the Looking-Glass* 1930 US) – the two being assembled first as *Alice's Adventures in Wonderland, and Through the Looking-Glass* (omni **1911**). These books

liberated CHILDREN'S FANTASY from its chains of moral didacticism and presented a world of NONSENSE which lampooned adults at a level that a child could appreciate. The first book had originally been narrated as a series of sketches by LC to the children of the Reverend Liddell (1811-1898), in particular his daughter Alice (1852-1934), during a boating trip in 1862, and LC had first written them down as *Alice's Adventures Under Ground* (written 1862-3; published in manuscript facsimile **1886**), a version that varies in many ways from the final. Although both novels are DREAM fantasies, their creation is so vivid that both WONDERLAND and Looking-glass Land feel absurdly real. This illogical REALITY forces one to accept them as CROSSHATCH fantasies, as access is as instantaneous through the imagination as via the THRESHOLDS of the rabbit-hole or the MIRROR. *Alice Through the Looking-glass* has the added discipline of being based on a CHESS-game, an ingenious contrivance that brings structure to the SURREALISM and makes the sequel a more accomplished and successful work. The second **Alice** novel is quoted more often than the first (though most people believe they are quoting from the first), and features the poems "Jabberwocky" and "The Walrus and the Carpenter". John TENNIEL, who illustrated the **Alice** books, refused to illustrate one chapter in *Through the Looking-Glass* called "The Wasp in a Wig" because he thought it was ridiculous. The chapter was dropped but the proofs survived and it was eventually published as *The Wasp in a Wig: The "Suppressed" Episode of Through the Looking Glass* (**1877** chap). A highly fictionalized account of LC's production of the stories and of his relationship with Alice is DREAMCHILD (**1985**). A number of movies have been made of *Alice in Wonderland*, of which the best-known – although not the best – is DISNEY's ALICE IN WONDERLAND (*1951*); of very considerable interest is Jan SVANKMAJER's ALICE (*1988*). The star-studded ALICE THROUGH THE LOOKING GLASS (*1960* tvm) is really a PANTOMIME.

Although the **Alice** books were always intended for children, there is little doubt that LC entertained himself by satirizing leading members of society, most of these characters being cleverly disguised and only now being rediscovered. Also, true to his mathematical training, he utilized the settings of Wonderland and Looking-glass Land to challenge the norms of society in a way that has caused many critics to regard the books as subversive.

Few authors have ever equalled LC's skills at creating logical nonsense, but the broad parameters of the **Alice** books spawned hosts of imitations and SEQUELS BY OTHER HANDS. Some of the PLOT DEVICES, like the change in size, were picked up by Jean INGELOW in *Mopsa the Fairy* (**1869**), but more overt imitations include: *Davy and the Goblin* (**1884** US) by Charles E. Carryl, set in a Land of Nonsense; *Down the Snow Stairs* (**1887**) by Alice Corkran (? -1916); *Wanted: A King* (**1890**) by Maggie Browne (real name Margaret Andrewes), set in a dreamland full of nursery-rhyme characters; *The Wallypug of Why* (**1895**) by G.E. Farrow, which comes close in imitating the rationalized nonsense; *Dot and the Kangeroo* (**1899** UK) by the Australian Ethel C. Pedley; *David Blaize and the Blue Door* (**1919**) by E.F. BENSON; and, more recently, *The Phantom Tollbooth* (**1961**) by Norton JUSTER. More direct sequels and parodies include (from a much longer list): *A New Alice in the Old Wonderland* (**1895** US) by Anna M. Richards Jr; *Clara in Blunderland* (**1902**) and *Lost in Blunderland* (**1903**) by Caroline Lewis (joint pseudonym of Harold BEGBIE, J.

Stafford Ransome [1860-1931] and M.H. Temple); *Alice in Blunderland* (**1907**) by John Kendrick BANGS; *More 'Alice'* (**1959**) by Yates Wilson; the inevitable erotic version, *Blue Alice* (**1972**) by Jackson Short (real name Peter Hochstein; 1939-   ); *Alice in Blunderland* (**1983** US) by Jack Anderson (1922-   ) and John Kidner (1923-   ); and *Alice Through the Needle's Eye* (**1984**) by Gilbert ADAIR. More recently the world of Alice has been revisited in *Fantastic Alice* (anth **1995**) ed Margaret WEIS and Martin H. GREENBERG.

Between the **Alice** books LC wrote a short FAIRYTALE, *Bruno's Revenge* (1867 *Aunt Judy's Magazine*; **1924** chap), about a bad-tempered FAIRY persuaded to help his sister. LC later developed this into the novel *Sylvie and Bruno* (**1889**), with a sequel *Sylvie and Bruno Concluded* (**1893**) – assembled as *The Story of Sylvie and Bruno* (cut **1904**) ed Edwin Dodgson, LC's brother. Although these have moments of the **Alice** sparkle, the storyline is weighed down by philosophizing. LC considered putting *The Hunting of the Snark* (**1876** chap), his last great nonsense poem, in the former book, and would thereby much have enhanced it. His own edited version of *The Nursery "Alice"* (**1889**), intended for the under-fives, was devoid of the original's virtues. LC had captured the "golden afternoon" of childhood not once, but twice, but even he could not return a third time. [MA]

Other works: *Phantasmagoria* (coll **1869**), poetry, assembled with *The Hunting of the Snark* as *Rhyme? and Reason?* (omni **1883**). Later collections include: *For the Train: Five Poems and a Tale* (coll **1932**) ed Hugh J. Schonfield, drawn from items LC contributed to *The Train* magazine in 1856-7; *The Rectory Umbrella and Mischmasch* (coll **1932**) ed Florence Milner, selected from his private family magazine produced 1850-53, including the ABSURDIST FANTASY "The Walking Stick of Destiny"; *The Complete Works of Lewis Carroll* (omni **1939**; vt *The Penguin Complete Lewis Carroll* 1982); *The Jabberwocky and Other Frabjous Nonsense* (**1964** US); *The Poems of Lewis Carroll* (**1973** US).

Further reading: *The Annotated Alice* (**1960** US), *The Annotated Snark* (**1962** US) and *More Annotated Alice* (**1990** US) all ed Martin Gardner (1914-   ); *Aspects of Alice* (**1971**) ed Robert Phillips; *The Alice Concordance* (**1986** Australia) by Daryl Colquhoun. There are many biographies and studies including: *Lewis Carroll* (**1930**) by Walter DE LA MARE; *The Story of Lewis Carroll* (**1949**) by Roger Lancelyn GREEN; *Lewis Carroll* (**1954**; rev 1976) by Derek Hudson; *The Philosopher's Alice* (**1974**) ed Peter Heath; *Lewis Carroll and His World* (**1976**) by John Pudney; *Beyond the Looking-Glass* (**1982**) by Colin Gordon; *Lewis Carroll* (**1990** US) by Richard Michael Kelly; *The Red King's Dream* (**1995**) by Jo Elwyn Jones and J. Francis Gladstone.

CARRYL, CHARLES E. (1841-1920) US writer. ◊ Lewis CARROLL; CHILDREN'S FANTASY.

CARRY ON SCREAMING (vt *Carry on Vampire*) UK movie (*1966*). ◊ *The* FEARLESS VAMPIRE KILLERS, OR PARDON ME, YOUR TEETH ARE IN MY NECK (*1967*).

CARS Cars appear in some fantasy contexts, especially in fantasies set at the cusp of the present, or actually in the future. Cars can be found in DYING-EARTH tales – typically buried strata deep in the packed soil – where they may serve as emblems of TIME ABYSS; and they can be found in DARK FANTASY and TECHNOFANTASY, where they may well embody or be possessed by DEMONS, manifesting themselves as the distilled malice of a technology become malignant – the purest example of this is the movie *The* CAR (*1977*), but Stephen KING's *Christine* (**1983**), filmed as CHRISTINE (*1983*), can be interpreted in like manner. [JC]

CARTER, ANGELA (OLIVE) (1940-1992) UK writer of fantasy and sf, as well as much other work. She very early gained from the UK literary establishment a fully deserved esteem for her imaginative daring and explorative, Postmodernist cunning. Throughout her career she utilized the language and characteristic motifs of the FANTASTIC to dramatize her sense that the old orders of the Western world were breaking down.

AC's early novels were neither sf nor fantasy, though a tale like *The Magic Toyshop* (**1967**) carries its female protagonist through a RITE OF PASSAGE into adulthood via imagery – especially the use of PUPPETS – that constantly jostles the limits of the mundane. *Heroes and Villains* (**1969**), set in a post-HOLOCAUST world, is perhaps best thought of as sf. Her first fantasy proper, *The Infernal Desire Machines of Doctor Hoffman* (**1972**; vt *The War of Dreams* 1974 US), is a work of powerful and genuine SURREALISM, set initially in an unnamed CITY in an unnamed Latin American country where TIME and REALITY have been distorted by the mysterious Hoffman. Narrated by Desiderio, a bureaucrat in love with Hoffman's cross-dressing daughter Albertina, the tale unfolds as a QUEST for the Doctor through realms of surreality; Desiderio ultimately destroys him (and what he represents) but not before he (and the reader) have passed through various nightmarishly PICARESQUE, CHAOS-threatening LANDSCAPES of the creative imagination. At one point, for example, Desiderio peers through the giant mockup of a woman's vagina (◊ PORTAL) into a kind of interior LITTLE BIG country, at the heart of which lies the ornate EDIFICE of the Doctor; later, he finds the castle and kills the Doctor – who has been using Desiderio's sexual longings (◊ SEX) to fuel his Desire Machines – and the novel ends in a heavily ironized triumph of dreamless REALITY.

The Passion of New Eve (**1977**), though less famous, is perhaps even more dangerous in its acid, cartoonish demands on the reader. It once again presents a dance of realities, conveyed on this occasion through a series of TECHNOFANTASY sex-changes. Evelyn, who is English, comes to a ravaged NEW YORK to find his promised job extinguished, undergoes deranging adventures and is captured in the desert by a cold-blooded female scientist who calls herself Mother and has assembled in her person various attributes of the GODDESS. A sense that she is cognate with the Miss Brunner of Michael MOORCOCK's *The Final Programme* (**1968**) is strengthened when it proves she intends to rape Evelyn, change his sex (and his name to Eve), and impregnate him with his own seed, so that he may give birth to an ambivalent new MESSIAH. The plot continues to thrust GENDER dilemmas at the reader from an extremely tough-minded feminist standpoint; in the end, Eve, having transcended the various impersonations s/he has passed through (◊ METAMORPHOSIS), takes ship westward, en route (maybe) to EDEN.

In AC's final novel of fantasy interest, *Nights at the Circus* (**1984**), another role violator sails through a world which attempts to bind her; this time, she escapes with the greatest of ease. Fevvers is a tall Cockney woman with WINGS and without navel. Though there is no travel through TIME or MULTIVERSE in the book, she resembles a TEMPORAL ADVENTURESS. and has the same capacity to slide through the constraints of the mundane world. The story begins in a

GASLIGHT-ROMANCE version of LONDON, moves for a period to Siberia, and returns home. The book is an exuberant and fluent tirade of images, arguments, and lessons taught and learned.

AC's short fiction was assembled in *Fireworks: Nine Profane Pieces* (coll **1974**; rev 1987), *The Bloody Chamber and Other Stories* (coll **1979**); *Black Venus* (coll **1985**; rev vt *Saints and Strangers* 1986 US) – which incorporates *Black Venus's Tale* (**1980** chap) – and *American Ghosts and Old World Wonders* (coll **1993**). *The Collected Angela Carter: Burning Your Boats* (coll **1995**) assembles all her previous short work. *The Bloody Chamber* is a central collection of REVISIONIST-FANTASY versions of FAIRYTALES and other material, couched in an archaizing DICTION reminiscent of the work of Isak DINESEN – an influence acknowledged by AC – and directed in the main to a rewriting of the traditionally cloistral view of women expressed in the literature. Models revised include BEAUTY AND THE BEAST in "The Courtship of Mr Lyon", and Little Red Riding-Hood, the latter in AC's most famous single story, "The Company of Wolves" (1979), on which was based *The* COMPANY OF WOLVES (*1985*); her screenplay (with Neil Jordan) for this appeared, with other transformations of fairytales, in *Come Unto These Yellow Sands* (coll **1985**). Her shaping interest in fairytales also resulted in several volumes: *The Fairy Tales of Charles Perrault* (coll trans AC **1977**); *Sleeping Beauty and Other Favourite Fairy Tales* (anth trans and ed AC **1982**); *The Virago Book of Fairy Tales* (anth **1980**; vt *The Old Wives' Fairy Tale Book* 1990 US) and *The Second Virago Book of Fairy Tales* (anth **1992**).   [JC]

Other works: *Miss Z, the Dark Young Lady* (**1970** chap), *The Donkey Prince* (**1970** US) and *Moonshadow* (**1982** chap), all for younger children; *Comic and Curious Cats* (**1979**); *The Sadeian Woman: An Exercise in Cultural History* (**1979**; vt *The Sadeian Woman and the Ideology of Pornography* 1979 US); *Nothing Sacred: Selected Writings* (coll **1982**); *Expletives Deleted: Selected Writings* (coll **1992**).

Nonfantasy novels: *Shadow Dance* (**1966**; vt *Honeybuzzard* 1967 US); *Several Perceptions* (**1968**); *Love* (**1971**; rev 1987); *Wise Children* (**1991**).

CARTER, LIN Working name of US writer Linwood Vrooman Carter (1930-1988), moderately well known for his action-oriented sf novels but of much greater importance as an editor of fantasy ANTHOLOGIES, as a pioneering critic of the fantasy genre, and as a fantasy novelist. Most of his fiction deliberately pastiched the work of earlier fantasy writers like Edgar Rice BURROUGHS and Robert E. HOWARD, but did so both openly and with the clear intent of popularizing modes (SWORD AND SORCERY and PLANETARY ROMANCE were his personal favourites) less well understood in the 1960s than they are now. Much of the success of GENRE FANTASY in general – as well as the current renown of writers like Lord DUNSANY and Clark Ashton SMITH – is due to LC's advocacy.

LC began publishing work of genre interest with "Masters of the Metropolis", with Randall GARRETT, for *F&SF* in 1957. His first fantasy novel, *The Wizard of Lemuria* (**1965**; vt *Thongor and the Wizard of Lemuria* 1969), began the **Thongor** sequence of CONAN pastiches, set in a **Barsoom**-like land; the sequence continued with *Thongor Against the Gods* (**1967**), *Thongor in the City of Magicians* (**1968**), *Thongor at the End of Time* (**1968**) and *Thongor Fights the Pirates of Tarakus* (**1970**). The tales are set 500,000 years ago in LEMURIA.

More specifically indebted to Burroughs are three series: the **Green Star Rises** sequence, being *Under the Green Star* (**1972**), *When the Green Star Calls* (**1973**), *By the Light of the Green Star* (**1974**), *As the Green Star Rises* (**1975**), *In the Green Star's Glow* (**1976**) and *As the Green Star Rises* (**1983**); the **Callisto** sequence, being *Jandar of Callisto* (**1972**), *Black Legion of Callisto* (**1972**), *Sky Pirates of Callisto* (**1973**), *Mad Empress of Callisto* (**1975**), *Mind Wizards of Callisto* (**1975**), *Lankar of Callisto* (**1975**), *Ylana of Callisto* (**1977**) and *Renegade of Callisto* (**1978**); and the **Zanthodon** sequence, being *Journey to the Underground World* (**1979**), *Zanthodon* (**1980**), *Hurok of the Stone Age* (**1981**), *Darya of the Stone Age* (**1981**) and *Eric of Zanthodon* (**1982**). In the first two sequences the HERO is more or less magically transported – as in Burroughs – to a planetary-romance venue, identified as a far-off planet; in each case, as in the **Thongor** books, the hero battles his way to power in an S&S environment. The third – clearly intended as HUMOUR – is intensely RECURSIVE, and into its UNDERGROUND setting LC introduces large numbers of characters from various milieux and authors. None of these series could be described as transcending its roots in pulp and other authors' inspirations; but LC's love of his chosen material is evident throughout, and his work helped establish default patterns against which the work of more original explorers of the conventions of S&S, like Michael MOORCOCK, shone brightly.

Some of LC's later work, though remaining hasty, showed increasing ambition; and his last series of note – the **Terra Magica** sequence, being *Kesrick* (**1982**), *Dragonrouge* (**1984**), *Mandricardo* (**1986**) and *Callipygia* (**1988**) – is of considerable interest. Terra Magica itself is a LAND of vast proportions, an *omnium gatherum* territory which contains all the fantasies and legends told on Earth and theoretically in all the IMAGINARY LANDS as well, giving LC an attractive venue for his games of recursion. His singletons, and the various completions of **Conan** material in collaboration with L. Sprague DE CAMP (◊ Robert E. HOWARD *for details*), were less significant.

LC should perhaps be most lauded for his work as the innovative editor who conceived and oversaw the BALLANTINE ADULT FANTASY series, beginning with *Dragons, Elves, and Heroes* (anth **1969**) and continuing with several other anthologies, including the important *Discoveries in Fantasy* (anth **1972**); along with these he edited several selections from important fantasy writers who had slipped out of print in the USA, including *At the Edge of the World* (coll **1970**) by Lord DUNSANY, *Zothique* (coll **1970**) by Clark Aston SMITH and *The Doom that Came to Sarnath* (coll **1971**) by H.P. LOVECRAFT. His advocacy of these writers was intense and persuasive; and their publication in this widely marketed series was one of the more telling signs that fantasy as a commercial genre was beginning to come to the consciousness of publishers, retailers and readers alike.

LC's critical work was similarly couched in terms of advocacy. It cannot be claimed that *Imaginary Worlds: The Art of Fantasy* (**1973**) is very substantial as an argument, but his choice of William MORRIS as the author of the first HIGH FANTASY has weathered scholarly investigation. Few of LC's novels – most are now obscure – may survive; but for his work as a pioneer, a popularizer and an enthusiast his memory remains secure.   [JC]

Other works (fantasy): *The Flame of Iridar* (**1967** chap dos); *Tower at the Edge of Time* (**1968**); *Beyond the Gates of Dream* (coll **1969**); *Lost World of Time* (**1969**); the

Chronicles of Kylix, being *The Quest of Kadji* (**1971**) and *The Wizard of Zao* (**1978**); the **Gondwana Epic**, being *Giant of World's End* (**1969**), *The Warrior of World's End* (**1974**), *The Enchantress of World's End* (**1975**), *The Immortal of World's End* (**1976**), *The Barbarian of World's End* (**1977**) and *The Pirate of World's End* (**1978**); *Dreams from R'lyeh* (coll **1975** chap), poetry; *Tara of the Twilight* (**1979**); *Lost Worlds* (coll **1980**); *Kellory the Warlock* (**1984**); *Found Wanting* (**1985**).

Other works (sf): The **Great Imperium** sequence, being *The Star Magicians* (**1966** dos), *The Man without a Planet* (**1966** dos), *Tower of the Medusa* (**1969**), *Star Rogue* (**1970**) and *Outworlder* (**1971**); *Destination Saturn* (**1967**) with David Grinnell (Donald A. WOLLHEIM); the **Thoth** sequence, being *The Thief of Thoth* (**1968** chap) and *The Purloined Planet* (**1969** chap dos); *Outworlder* (**1971**); *The Black Star* (**1973**); the **Mars** series, being *The Man who Loved Mars* (**1973**), *The Valley where Time Stood Still* (**1974**), *The City Outside the World* (**1977**) and *Down to a Sunless Sea* (**1984**); *Time War* (**1974**); the **Zarkon** sequence of **Doc Savage** pastiches, being *Zarkon, Lord of the Unknown, in The Nemesis of Evil* (**1975**; vt *The Nemesis of Evil* 1978), *Zarkon, Lord of the Unknown, in Invisible Death* (**1975**; vt *Zarkon, Lord of the Unknown and his Omega Crew: Invisible Death* 1978), *Zarkon, Lord of the Unknown, in The Volcano Ogre* (**1976**; vt *Zarkon, Lord of the Unknown and his Omega Crew: The Volcano Ogre* 1978), *Zarkon, Lord of the Unknown, in The Earth-Shaker* (**1982**) and *Horror Wears Blue* (**1987**).

As editor: *The Young Magicians* (anth **1969**); *The Magic of Atlantis* (anth **1970**); *Golden Cities, Far* (anth **1970**); *The Spawn of Cthulhu* (anth **1971**); *New Worlds for Old* (anth **1971**); *Great Short Novels of Adult Fantasy* (anth **1972**) and *Great Short Novels of Adult Fantasy II* (anth **1973**); the **Flashing Swords** series, being *Flashing Swords 1* (anth **1973**), *#2* (anth **1973**), *#3: Warriors and Wizards* (anth **1976**), *#4: Barbarians and Black Magicians* (anth **1977**) and *#5: Demons and Daggers* (anth **1981**); the **Year's Best Fantasy** series, being *The Year's Best Fantasy Stories 1* (anth **1975**), *#2* (anth **1976**), *#3* (anth **1977**), *#4* (anth **1978**), *#5* (anth **1980**) and *#6* (anth **1980**); *Kingdoms of Sorcery* (anth **1976**); *Realms of Wizardry* (anth **1976**); the **Weird Tales** series, being *Weird Tales 1* (anth **1980**), *#2* (anth **1980**), *#3* (anth **1981**) and *#4* (anth **1983**) (◊◊ WEIRD TALES).

Nonfiction: *Tolkien: A Look Behind "The Lord of the Rings"* (**1969**); *Lovecraft: A Look Behind the "Cthulhu Mythos"* (**1972**); *Royal Armies of the Hyborean Age: A Wargamer's Guide to the Age of Conan* (**1975** chap) with Scott Bizar; *Middle-Earth: The World of Tolkien* (graph **1977**) illus David Wenzel (1950-   ), pictures with captions.

Further reading: *Lin Carter: A Look Behind his Imaginary Worlds* (**1991**) by Robert M. Price.

CARTER, MARILYN Pseudonym of Marilyn ROSS.

CASE, JUSTIN [s] ◊ Hugh B. CAVE.

CASPER, THE FRIENDLY GHOST A cute little GHOST whose eager attempts to make friends almost inevitably result in him terrifying the object of his endeavours. The character was created in 1945 by Joseph Oriolo and Seymour Wright for Paramount/Famous Films; in 1950-59 he featured in a series of short ANIMATED MOVIES (**1**), and in 1963-7 and 1979-80 in two tv series (**2** and **3**). There was also a COMIC-book series (**4**) based on the character, and latterly a movie (**5**). Among Casper's accomplices were the Galloping Ghost, the Ghostly Trio and Wendy, the Good

Little Witch. For his part in the creation of Casper Oriolo received a grand total of $175.

1. Casper, The Friendly Ghost US series of 55 short movies, prefaced by three one-off shorts. Paramount/Famous. **Dir** Seymour Kneitel, Isidore Sparber, Bill Tytla. **Voice actors** Gwen Davies, Norma McMillan, May Questel, Cecil Roy (all Casper). Each 6½ mins. Colour; *#21: Boo Moon* (*1954*) was initially released in 3D. The series was syndicated on tv from 1953. [JG]

2. *The New Casper Cartoon Show* (vt *The New Adventures of Casper*) US animated tv series (5 October 1963-2 September 1967). ABC/Famous/Harvey. **Dir** Seymour Kneitel. **Voice actor** Ginny Tyler (Casper). 26 30min episodes, each normally containing three cartoons. Colour. [JG]

3. *Casper and the Angels* US animated tv series (22 September 1979-3 May 1980). HANNA-BARBERA. **Voice actors** Dian McCannon (Maxi), Julie McWhirter (Casper), Laurel Page (Mini), John Stephenson (Hairy Scary). 13 30min episodes, each containing two stories. Colour.

This was set in AD2179. Casper is assigned to look after two space-patrol officers, Mini and Maxi. [RT]

4. US COMIC book. Casper's first comic-book appearance was in *Treasury of Comics* (1948), published by St John Publishing, who continued with three issues of *Casper the Friendly Ghost* (1949-50). Relaunched in 1950 under the same title, the series was taken over after five issues by Alfred Harvey Publications, first as *Harvey Comics Hits #62* (1952) before returning to the original title with the seventh issue and continuing to a total of 70 issues, ending in 1958. It was immediately relaunched as *The Friendly Ghost, Casper* and ran, with breaks (and briefly reverting to its old title), for 260 issues, ending in 1991.

The Harvey series introduced numerous characters for Casper to befriend or fear, some based on animated characters from the Hanna-Barbera series (**3**), including Spooky, the Tuff Little Ghost (1953), Nightmare, the Ghost Horse (1954), Wendy, the Good Little Witch (1954) and The Ghostly Trio – three evil ghosts, one fat and two thin. Various other Harvey comic spin-off titles have appeared over the years, including *Casper's Ghostland* (97 issues 1958-77), *Nightmare and Casper* (later vt *Casper and Nightmare*; 46 issues 1963-74) and *Casper Spaceship* (later vt *Casper in Space*; 8 issues 1972-3), plus many team-ups – *Casper and Spooky*, *Casper and Wendy*, *Richie Rich and Casper*, etc. Casper's popularity appears to have been at its highest in the 1950s and 1960s, with brief periods of revival in 1972-3, 1980, 1986 and, thanks to **5**, now. Harvey have made abundant use of reprints to keep as many as five **Casper** titles running at once.

Since the comic book was aimed at a very young readership, few creators have ever been credited. Some of the earliest episodes were prepared in close collaboration with Famous Studios. Dom (Dominik) Silio, who later worked as an assistant to Will EISNER, is perhaps the best-known Casper artist: he worked on the character in the 1950s and 1960s. [RT]

5. *Casper* US live-action/ANIMATED MOVIE (*1994*). Amblin/Harvey. **Pr** Colin Wilson. **Exec pr** Gerald R. Molen, Jeffrey A. Montgomery, Steven SPIELBERG. **Dir** Brad Silberling. **Vfx** Stefan Fangmeier, Dennis Muren, Industrial Light and Magic. **Anim dir** Eric Armstrong, Phil Nibbelink. **Screenplay** Deanna Oliver, Sherri Stoner. **Starring** Eric Idle (Dibs), Cathy Moriarty (Carrigan Crittenden), Bill Pullman (Dr James Harvey), Christina Ricci (Kat Harvey). **Voice actors** Joe Alaskey (Stinky), Brad

Garrat (Fatso), Joe Nipote (Stretch), Malachi Pearson (Casper). 100 mins. Colour.

Casper is the ghost of 12-year-old Casper McFadden, who died of pneumonia and remained in residence at Whipstaff Manor with three mischievous, domineering uncles, Stretchy, Fatso and Stinky ("The Ghostly Trio"); they are prevented from "crossing over" because of some unfinished business. Carrigan, a cold, greedy vamp, inherits the decaying Gothic mansion from her father and, to rid it of the spirits, she and sycophantic sidekick Dibs hire psychologist James Harvey to psychoanalyse the spirits and thus help them "cross over". Casper befriends Harvey's daughter Kat. It is discovered that Casper's father invented a machine and POTION able to bring a ghost back to life. There is enough potion for one transformation. Casper wants the potion to reincarnate himself, but must sacrifice this opportunity when Harvey meets an untimely end, leaving Kat parentless.

C has charm. There is sparkling humour, plus some excellent characterization. The visual impact is considerable: during the two years of *C*'s making the vfx team made full use of digital imagery, giving the animated characters an unusually long 40 mins of screen-time, and creating the first digital performers to have a full range of facial expressions.   [JT]

CASTLE OF DOOM vt of VAMPYR (*1932*).

CASTLE OF FU MANCHU (*1968*) ◊ FU MANCHU MOVIES.

CAT GIRL UK movie (*1957*). Insignia/Nat Cohen/Stuart Levy. **Pr** Herbert Smith. **Exec pr** Peter Rogers. **Dir** Alfred Shaughnessy. **Mufx** Philip Leakey. **Screenplay** Lou Rusoff. **Starring** Robert Ayres, Martin Boddey, Kay Callard, Edward Harvey, Lilly Kann, John Lee, Jack May, Ernest Milton, Barbara Shelley (Leonora), John Watson, Patricia Webster (movie lacks proper credits). 69 mins. B/w.

A reworking of CAT PEOPLE (*1942*). Leonora is newly wed to already philandering Richard; they travel with friends, including his mistress, to Leonora's old family estate, where they meet also her ex-boyfriend (whom she still loves), Brian, now a happily married psychiatrist. Leonora's batty uncle explains that the family suffers a CURSE: by day they are humans but at night their SOULS become CATS; as his last surviving relative she will inherit the curse when he dies – which she promptly does, inciting his pet leopard to kill him. She becomes identified with the leopard, which kills (or she kills) Richard. After a period in a lunatic asylum under Brian's care she (or the leopard) tries to kill Brian's wife Dorothy. Arriving in his car, Brian inadvertently runs over the leopard; Leonora is discovered dead from the same injuries as the leopard's.   [JG]

CAT PEOPLE US movie (*1942*). RKO. **Pr** Val Lewton. **Dir** Jacques Tourneur. **Screenplay** DeWitt Bodeen. **Starring** Tom Conway (Dr Louis Judd), Jane Randolph (Alice Moore), Simone Simon (Irena Dubrovna), Kent Smith (Oliver Reed). 73 mins. B/w.

Man-about-town Oliver meets Serbian immigrée fashion illustrator Irena at the zoo, courts her and marries her. But she refuses consummation because convinced by old-country legends that physical passion may make her shapeshift (◊ SHAPESHIFTER) into a big CAT and destroy her lover. Reed sends her to disbelieving psychiatrist Judd while himself, in enforced abstinence, swiftly shifting his affections to doting colleague Alice. Yet Irena's fears are justified: unknowing of what occurs when she is in cat form, she threatens Alice and finally, when Judd kisses her to disprove her fears, kills him. She herself dies trying to make common cause with a panther at the zoo.

This is normally billed as a HORROR MOVIE – and was so treated by sensationalist advertising on release – yet, beautifully scripted and photographed, it is in fact a classic example of cinematic DARK FANTASY: nothing is explicit, all merely alluded to, as it takes its traditional premise and, with superb understatement and not a little poignancy, follows it through to a conclusion that seems inevitable.

CP was followed by CURSE OF THE CAT PEOPLE (*1944*), not thematically a sequel; likewise CAT PEOPLE (*1982*) was not a true remake, although CAT GIRL (*1957*) came closer. A TECHNOFANTASY treatment of a similar theme was *Dangerous Desire* (*1992*), in which a man injected with the DNA of a cat develops all sorts of feline characteristics, such as flamboyant promiscuity, hyperacute hearing and loathing of water; he makes a career as a ballet dancer before his amorality brings his downfall.   [JG]

CAT PEOPLE US movie (*1982*). RKO/Universal. **Pr** Charles Fries. **Exec pr** Jerry Bruckheimer. **Dir** Paul Schrader. **Spfx** Tom Del Genio, Pat Domenico, Karl Miller. **Mufx** Tom Burman. **Vfx** Albert Whitlock. **"Catvision" optical fx** Robert Blalack. **Visual consultant** Ferdinando Scarfiotti. **Screenplay** Alan Ormsby. **Novelization** *Cat People* * (*1982*) by Gary Brandner. **Starring** John Heard (Oliver Yates), Nastassia Kinski (Irena), Malcolm McDowell (Paul). 118 mins. Colour.

Although De Witt Bodeen's script for CAT PEOPLE (*1942*) is credited as the basis for *CP*, the connection is slight.

Millennia ago (it emerges) a people habitually made HUMAN SACRIFICE of children to black leopards (panthers; ◊ CATS); the SOULS of the children grew within the panthers until the panthers have by evolution become of human form – except during the height of sexual arousal, when they revert to the feline state, thereafter being able to effect the METAMORPHOSIS back to human form only through the act of killing a human, usually their sex partner. The sole way this vicious circle can be avoided is through incestuous unions of "cat people".

Paul and Irena Gallier, products of such a union, were separated in childhood when their parents suicided; now Paul has tracked down Irena, and she comes to NEW ORLEANS to be reunited with him. He is, in effect, a SERIAL KILLER, periodically having SEX with hookers and then, as a panther, killing them; Irena, still virgin (◊ VIRGINITY), is ignorant of her true SHAPESHIFTER nature. Sexually charged by Irena's presence, Paul again seeks out a prostitute, but fails in his attempt to kill her. Captured by zoo curator Yates and his team, the panther is caged and displayed. Irena discovers the exhibit, and is fascinated by the creature – and in due course by Yates. Panther-Paul kills a zoo attendant and escapes; he confronts Irena, who refuses to believe his explanation; the police discover many human remains in the cellar of Paul's house; Paul attempts to seduce Irena, but she stabs him; reverting to panther form, he tries to kill Yates but himself dies. Irena, now a believer, flees New Orleans, but after a vision returns and sleeps with Yates, hoping she might, after all, not be Paul's sister and thus be free of the CURSE. Not so. After one more killing, she persuades Yates to make love with her a final time so she may dwell as a panther in his zoo.

This very stylish movie is packed with good moments and fledgling ideas, but overall is somehow mediocre – one feels the chief purpose of the exercise was to show lots of film of Kinski nude. This is a shame, because *CP*'s best images – as when the human Paul first leaps catlike onto the bedstead of the sleeping Irena to "devour" her with panther eyes, or the brief sequence when we perceive (◊ PERCEPTION) the world as panther-Irena does – are very good indeed.

A nice nuance occurs when Irena is greeted in a bar by a woman she does not know, who flees in embarrassment . . . but *we* know this woman, for she is Irena Dubrovna (although not played by Simone Simon) from CAT PEOPLE (*1942*). [JG]

CATS These are traditionally WITCHES' familiars and/or agents of the DEVIL, thanks to their strange eyes, aloof grace and lapses into (in human eyes) diabolical sadism when toying with prey. Black cats bring luck, good or bad; being lightfooted, they have nine lives; in ancient EGYPT they were sacred to the cat-headed GODDESS Bast. A cat jumping over your coffin may make you a VAMPIRE, like Florimel in James Branch CABELL's *Jurgen* (**1919**). The aloofness was captured in Rudyard KIPLING's MYTH OF ORIGIN for the cat's near-domestication, "The Cat that Walked by Himself" in *Just So Stories* (coll **1902**). The ruby-hearted glass cat Bungles of L. Frank BAUM's OZ is so determined not to show emotion that when implored to bring help she sets off very slowly and runs only when out of sight. In an exaggerated metaphor of elusiveness, the Cheshire Cat of Lewis CARROLL's *Alice's Adventures in Wonderland* (**1865**) literally fades from view. The cat as TRICKSTER is epitomized as PUSS IN BOOTS. Sexual connotations of furry litheness give an erotic charge to most SHAPESHIFTING into cat form – e.g., "Ancient Sorceries" in Algernon BLACKWOOD's *John Silence* (coll **1908**) and the various CAT PEOPLE movies. There are also innumerable catlike aliens in sf by Fritz LEIBER, Anne MCCAFFREY, Larry NIVEN and many others. Notable cat familiars include Seraphin in Robert A. HEINLEIN's *Magic, Inc.* (**1940**) and Svartalf in Poul ANDERSON's *Operation Chaos* (**1971**). Among memorable nonfamiliars are: SAKI's unpleasantly epigrammatic TALKING ANIMAL Tobermory in *The Chronicles of Clovis* (coll **1911**); the good and bad Nibbins and Blackmalkin in John MASEFIELD's *The Midnight Folk* (**1927**), which are also talking animals; the heroic kitten Gummitch in Leiber's "Space-Time for Springers" (1958), who in his own terms (◊ PERCEPTION) works MAGIC to remove a POSSESSION; and such amiably murderous cats as Throgmorten in Diana Wynne JONES's *The Lives of Christopher Chant* (**1988**) and Greebo in Terry Pratchett's **Discworld**. [DRL]

CATWEAZLE UK tv series (1970-71). LWT. **Pr** Quentin Lawrence, Carl Mannin. **Exec pr** Joy Whitby. **Dir** David Lane, Lawrence, David Reed. **Writer** Richard Carpenter. **Created by** Carpenter. **Starring** Geoffrey Bayldon (Catweazle), Peter Butterworth (Groome), Robin Davies (Carrot Bennett), Elspet Gray (Lady Collingford), Gary Warren (Cedric Collingford), Moray Watson (Lord Collingford). 26 25min episodes. Colour.

A WIZARD from England's early Norman era accidentally falls through TIME into the 20th century and is trapped there. Local farmer's lad Carrot helps him adapt to the wonders of the technological age. Catweazle is returned to his own era at the end of the first series. The start of the second sees the same accident; this time he is befriended by Cedric, son of the local gentry. *C* was much helped by Bayldon's fine, quirky performance in the title role. [JG]

CAULDRON OF PLENTY ◊ ARTHUR; GRAIL.

CAULDRON OF STORY In his essay "High Fantasy and Heroic Romance" (1971 *The Horn Book Magazine*) Lloyd ALEXANDER speaks of incorporating huge amounts of traditional material into his **Chronicles of Prydain** sequence (**1964-8**); he calls this material – which includes names, places, LANDSCAPES, echoes of historical MYTHS and LEGENDS, and plot structures basic to HIGH FANTASY – a Cauldron of Story, taking the term from the Second Branch of the MABINOGION. The phrase was also utilized in the description of FANTASY by J.R.R. TOLKIEN, in "On Fairy-Tales", a talk delivered in 1939 and published in *Essays Presented to Charles Williams* (anth **1947**) ed C.S. LEWIS; Tolkien spoke of "the Pot of Soup, the Cauldron of Story", as always boiling. [JC]

See also: OCEAN OF STORY; STORY; TWICE-TOLD.

CAVALIER, THE ◊ MAGAZINES.

CAVE, HUGH B(ARNETT) (1910-) UK-born US writer, since 1929 author of over 1000 stories and articles for the pulp and slick MAGAZINES. His stories have covered all fields, though most are adventure and mystery, with a high proportion of HORROR and SUPERNATURAL FICTION, and many written under pseudonyms, including H.C. Barnett, Justin Case, Jack D'Arcy, Rupert Knowles and Geoffrey Vace. HBC's work reflects most sensational themes, especially VAMPIRES – in which subgenre "Stragella" (1932 *Strange Tales*) has become a minor classic. HBC's early fiction was collected by Karl Edward WAGNER in *Murgunstruum and Others* (coll **1977**), which won the WORLD FANTASY AWARD. A companion volume of pulpish terrors, *Death Stalks the Night* (coll **1995**) remained unpublished for 18 years. Some of his more grotesque menace stories were collected as *The Corpse Maker* (coll **1988**) ed Sheldon Jaffrey (1934-).

After WWII HBC settled for a while in Haiti, and came into direct contact with VOODOO, about which he wrote several articles. When he returned to writing supernatural fiction in 1977 he used this background in *Legion of the Dead* (**1979**), *The Evil* (**1981**) and *Shades of Evil* (**1982**), all featuring ZOMBIES.

HBC's work is colourful and unashamedly commercial. He received the Life-Achievement BRAM STOKER AWARD in 1991. Cave's reminiscences of his career viewed through his correspondence with Carl Jacobi (1908-) was published as *Magazines I Remember* (**1994**). [MA]

Other works: *The Nebulon Horror* (**1980**); *Disciples of Dread* (**1988**); *The Lower Deep* (**1990**); *Lucifer's Eye* (**1991**).

Further reading: "Hugh B. Cave: Master of Vintage Horror" in *Fantasy Voices* (**1982**) by Jeffrey M. Elliot, detailed interview; *Pulp Man's Odyssey* (**1988**) by Audrey Parente.

CAVE DWELLERS, THE vt of ONE MILLION BC (*1940*).

CAWTHORN, JAMES (1929-) UK critic, writer and illustrator, associated through much of his career with Michael MOORCOCK, who edited what may be the first JC COMICS illustration work of interest, some late issues of *Tarzan Adventures* (1951-9). *The Distant Suns* (1969 *The Illustrated Weekly of India*; exp **1975**), written with Moorcock under the joint pseudonym Philip James, is sf. He wrote occasional reviews for *New Worlds*, his copy sometimes thought to be pseudonymous because his initials (JC) were those of Jerry **Cornelius**. He coscripted with Moorcock *The Land that Time Forgot* (*1975*), and was almost entirely responsible for *Fantasy: The 100 Best Books* (**1988**), jointly credited to Moorcock; the book is sharp and knowledgeable, and shows nostalgia for a time before fantasy became a highly marketable quantity. JC remains best known for his GRAPHIC-NOVEL adaptations of Moorcock

HEROIC-FANTASY texts, including *Stormbringer* (graph **1975**), *The Jewel in the Skull* (graph **1978**), based on the novels with those titles, and *The Crystal and the Amulet* (graph **1987**), based on *Sorcerer's Amulet*. His style is sometimes seemingly primitive, but can convey a sense of almost uncanny decency; through the foursquare contours of his art, a great deal of emotion wells. [JC]

CAZOTTE, JACQUES (1719-1792) French writer in whose *Le Diable Amoureux* (**1772**; trans anon as *The Devil in Love* **1793** UK; new trans anon as *The Enamoured Spirit* 1798 UK; best trans Judith Landry as *The Devil in Love* 1991 UK) a young man, given a SPELL with which to raise the DEVIL, is confronted by the head of a camel who asks him what he wants, then by a mysterious dog, then by a young woman, who is an incarnation of the Devil. Echoes of ARABIAN FANTASY permeate the tale, whose darker moments seem to promise an ARABIAN NIGHTMARE of ever deeper DREAMS; but all turns out well. With Dom Chavis, JC produced a translation of *Arabian Nights* manuscripts as *Suite des milles et une nuits* (as vols 38-41 [**1788-1790**] of the fairytale anthology series **Cabinet des fées**; trans Robert Heron as *The Arabian Tales, or A Continuation of the Arabian Nights* **1792** UK). [JC]

CELIA Australian movie (*1988*). BCB/Seon. **Pr** Gordon Glenn, Timothy White. **Exec pr** Bryce Menzies. **Dir** Ann Turner. **Screenplay** Turner. **Starring** Claire Couttie (Heather), Nicholas Eadie (Ray Carmichael), Mary-Anne Fahey (Pat Carmichael), Amelia Frid (Stephanie), Victoria Langley (Alice Tanner), Margaret Ricketts (Granny), Rebecca Smart (Celia), William Zappa (Sgt John Burke). 103 mins. Colour.

In order to counter the vicissitudes of life in 1957 Australia, ignore her father's adulterous tendencies, and cope with the petty persecutions of local police Sergeant Burke, 9-year-old Celia indulges in fantasies, forming a VOODOO culture with her likewise persecuted friends the Tanners – whose sin is that their parents were once communists – and speaking with the GHOST of her recently deceased grandmother. But her fantasy world is shaded by the lurking figures of the Hobyahs, slithery MONSTERS drawn from one of her school story-books. At last the Hobyahs begin intruding into real life: when Celia finds Sergeant Burke is one she kills him with her father's shotgun. *C* can be seen as a quieter (and arguably more effective) treatment of some of the themes of William GOLDING's *The Lord of the Flies* (**1954**) (another, quite different, resonance is with *The* CURSE OF THE CAT PEOPLE [*1944*]), but it goes further, probing also the origins of FANTASY as not merely ESCAPISM but defence, a deliberate or unconscious switching of PERCEPTIONS in order to make sense of an unpredictable world: the description of the Hobyahs as randomly vicious monsters who persecute without reason or warning could equally apply to most of the adults in Celia's world. [JG]

CELTIC FANTASY The Celtic MATTERS of Ireland, Scotland and Wales – and indeed of Brittany, Cornwall and the Isle of Man – have all boiled together in the CAULDRON OF STORY. From this, many authors distil a FANTASYLAND region peopled with what Diana Wynne JONES's «The Tough Guide to Fantasyland» (1996) calls "Pancelts". The coinage is flippantly intended; but a generalized Celticism can indeed drive powerful fantasies like Paul HAZEL's **Finnbranch** cycle. Katharine KERR's **Kingdom of Deverry** series is likewise Panceltic, to the extent of including an invented "Celtic" language; the term can be applied also to

the Celtic echoes of Charles DE LINT and others.

The Welsh strain of CF has been particularly influential, owing to its links with ARTHUR and its single, accessible and temptingly unpolished TAPROOT TEXT, the MABINOGION. This was first successfully mined by Kenneth MORRIS for his fine **Pwyll** diptych (**1914, 1930**). Evangeline WALTON efficiently novelized the Mabinogion's Fourth Branch as *The Virgin and the Swine* (**1936**; vt *The Island of the Mighty* 1970); her versions of the remaining Branches followed in the 1970s. Meanwhile, Lloyd ALEXANDER had freely rearranged *Mabinogion* elements and Welsh geography for younger readers in his successful **Prydain** quintet (**1964-8**), while Alan GARNER's *The Owl Service* (**1967**) imagined the mythic glare of Llew's and Blodeuwedd's (Fourth Branch) tragedy refracted through modern adolescent relationships. Katherine KURTZ's popular **Deryni** CF series is set in a remapped ALTERNATE-WORLD Wales incorporating traces of Ireland. Susan COOPER drew on CF in her **The Dark is Rising** series, especially *The Grey King* (**1975**), with its Welsh mountain setting and glimpse of Arthur beneath; the Mari Llwyd, a skeleton horse (and literal nightmare) is effectively menacing in *Silver on the Tree* (**1977**). Deep echoes from Welsh/Arthurian CF resound in Robert P. HOLDSTOCK's **Mythago** books; but Arthur's presence tends to absorb CF into the broader Matter of Britain, as also happens with the Cornish segments of Fay SAMPSON's fine REVISIONIST FANTASY **Daughter of Tintagel**.

Irish CF has roots which include the initially exuberant and later darkening MYTHS of the warrior hero CUCHULAIN, and tales of FINN MAC COOL the legendary GIANT. But it is also cursed with leprechauns, frequently encountered in HUMOUR and SLICK FANTASY. James STEPHENS's *The Crock of Gold* (**1912**) is a classic seriocomic treatment of the Little Folk and other figures of Irish myth interacting with a modern world of philosophers and policemen. Lord DUNSANY's *The Curse of the Wise Woman* (**1933**) powerfully evokes the Irish bog country, with a final element of ambiguous fantasy: for a whole stormy night the eponymous WITCH conjures the bog which, as she dies, engulfs the machinery of its exploiters. Flann O'BRIEN wove further Irish myths into his joyously experimental *At Swim-Two-Birds* (**1939**), including Finn Mac Cool and a pooka. L. Sprague DE CAMP and Fletcher PRATT sent their **Incomplete Enchanter** Harold Shea to the LAND-OF-FABLE Ireland in "The Green Magician" (**1954**), which features Druids, geases (◊ CURSES) and a manic-depressive Cuchulain. Mildred Downey BROXON's *Too Long a Sacrifice* (**1981**) interestingly thrusts Irish mythic archetypes into the contemporary Troubles. Peter TREMAYNE's *Raven of Destiny* (**1984**) unusually involves its Irish hero in the Celtic invasion of Greece in 279BC; Pat O'SHEA's *The Hounds of the Mórrigan* (**1985**) pleasingly CROSSHATCHES modern and legendary Ireland in a fast-moving CHILDREN'S FANTASY (◊◊ MORRIGAN). The Cuchulain stories are effectively retold in Gregory FROST's **Tain** series.

Scots CF examples seem less numerous, perhaps owing to the spuriousness of the **Ossian** poems which James Macpherson (1736-1796) claimed to have translated rather than fabricated; these annexed Irish figures for a Scots pseudo-mythos, with Finn Mac Cool becoming Fingal, etc. C.J. CHERRYH's *Faery in Shadow* (**1993** UK) mourns THINNING in a land-of-fable Scotland. *The Lord of Middle Air* (**1994**) by Michael Scott ROHAN builds on 13th-century legends surrounding the scholar Michael Scot, the "Border

Wizard", here a real WIZARD who can open PORTALS to FAERIE; he had already been mythologized in *The Lay of the Last Minstrel* (**1805**) by Sir Walter SCOTT. [DRL]
See also: Kenneth C. FLINT; Paul HAZEL; Morgan LLYWELYN.

CEMETERY DANCE US large-format small-press MAGAZINE, quarterly, December 1988-current, published and ed Richard Chizmar (1965-current).

One of the better of the flush of semiprozines which grew up in the wake of *The* HORROR SHOW and were facilitated by the spread of DTP, *CD* concentrates more on supernatural HORROR than on fantasy. Many stories emphasize physical horror and sexual excess, though only occasionally do they venture into the extremes of Splatterpunk. The supernatural focus is on such traditional themes as VAMPIRES, ZOMBIES and the occult. Early issues had an author theme, with David B. Silva (1950-   ) in *#1*, Richard Christian MATHESON in *#3*, Rick Hautala (1949-   ) in *#5*, J.N. WILLIAMSON in *#6* and Joe R. Lansdale (1951-   ) in *#7*, but rapidly improving production standards led to national newsstand distribution and the attention of many leading horror writers, including Ramsey CAMPBELL, Stephen KING and William F. NOLAN. *CD* also provided a channel for developing writers, like Poppy Z. BRITE, Bentley LITTLE and Norman Partridge (1958-   ). *CD* won the WORLD FANTASY AWARD Special Non-Professional category in 1991. [MA]

C'ERA UNA VOLTA Italian movie (*1967*). ◊ CINDERELLA (*1950*).

CEREBUS THE AARDVARK Stubby-legged, short-tempered star of an epic series of very popular satirical COMIC books independently published by his creator, US writer/artist Dave Sim (1958-   ) under the Aardvark-Vanaheim imprint and elsewhere. Originally intended as a pastiche of MARVEL COMICS's CONAN, as drawn by Barry WINDSOR-SMITH, the series began publication in 1977 as **Cerebus the Barbarian** (1977-9), later retitled **Swords of Cerebus** (1981-4), and is projected to run to 6000 pages.

The numerous stories have been published in *Cerebus* (1988-current) and in EPIC ILLUSTRATED (various issues 1984 and 1985), *Cerebus Jam* (1985), *Teenage Mutant Ninja Turtles* (1986) and *Spawn* (1993). The series has lampooned many established characters from the worlds of HIGH FANTASY and elsewhere, with characters like Elrod the Albino, Red Sophia, Jaka the Cockroach, Oscar the Poet, Pud the Tavern-Owner, Squinteye the Sailor and Margaret Thatcher.

The stories are available in collections entitled *Cerebus* (graph coll **1987**), *High Society* (graph coll **1986**), *Church and State Vol I* (graph coll **1987**) and *Vol II* (graph coll **1988**), *Jaka's Story* (graph coll **1990**), *Melmoth* (graph coll **1991**), *Flight* (graph coll **1993**), *Women* (graph coll **1994**) and *Reads* (graph coll **1995**). [RT]

CERVANTES, MIGUEL DE SAAVEDRA (1547-1616) Spanish writer whose *Don Quixote* (**1605**; Part Two added **1615**) is a TAPROOT TEXT for fantasy, though the book may fairly be read as a sustained assault upon the falsities of ROMANCE. But Don Quixote himself, indissolubly linked with his SHADOW Sancho Panza, is a figure of irresistible potency and an emblem of the melancholy of this world. MDC may have intended his tale as exemplary; but it is also a tale of BONDAGE. Don Quixote dies to the OTHERWORLD and expires in this one. He is the central UNDERLIER figure for the aesthete who baulks at METAMORPHOSIS or transcendence, for the KNIGHT OF THE DOLEFUL COUNTENANCE. [JC]

CHACAL US slick-format small-press MAGAZINE, 2 issues (Winter 1976, Spring 1977), published and ed Arnie Fenner (1955-   ) and Pat Cadigan (1953-   ) – with Byron Roark for *#1*. (The title is French for "jackal".)

Among the first quality SMALL-PRESS magazines, *C* grew from *REH: Lone Star Fictioneer*, a fanzine devoted to Robert E. HOWARD. The focus remained mostly on SWORD AND SORCERY, with stories by David C. SMITH and Karl Edward WAGNER. However, further contributions by Tom REAMY, Howard Waldrop (1946-   ) and Cadigan herself indicated a wider platform for the magazine beyond HEROIC FANTASY. *C* revived as SHAYOL. [MA]

CHALICE ◊ GRAIL.

CHALKER, JACK L(AURENCE) (1944-   ) US writer and editor who has written much fantasy, though best-known for his early sf. He began writing with several books on writers central to US fantasy and HORROR, including H.P.LOVECRAFT – *The New H.P. Lovecraft Bibliography* (**1962** chap; rev vt *The Revised H. P. Lovecraft Bibliography* 1974 chap with Mark Owings (1945-   ), *The Necronomicon: A Study* (**1967** chap), with Owings, and *Mirage on Lovecraft* (**1965** chap) – and Clark Ashton SMITH, with *In Memoriam: Clark Ashton Smith* (anth **1963** chap). *An Informal Biography of $crooge McDuck* (1971 *Markings*; **1974** chap) is enjoyable, and *The Science-Fantasy Publishers: A Critical and Bibliographical History* (**1991**; rev 1992, with subsequent undated revs), along with *The Science-Fantasy Publishers: Supplement One, July 1991-June 1992* (**1992**), remains a vital tool for the study of specialized US small-press publishers.

After successful sf novels like *A Jungle of Stars* (**1976**) and series like the **Well World** sequence, JLC shifted gradually towards SCIENCE FANTASY. His sequences include: **Soul Rider** – *Spirits of Flux and Anchor* (**1984**), *Empires of Flux and Anchor* (**1984**), *Masters of Flux and Anchor* (**1985**), *The Birth of Flux and Anchor* (**1985**), which is a prequel, and *Children of Flux and Anchor* (**1986**) – **Dancing Gods** – *The River of Dancing Gods* (**1984**), *Demons of the Dancing Gods* (**1984**), *Vengeance of the Dancing Gods* (**1985**) and *Songs of the Dancing Gods* (**1990**) – **Rings of the Master** – *Lords of the Middle Dark* (**1986**), *Pirates of the Thunder* (**1987**), *Warriors of the Storm* (**1987**) and *Masks of the Martyrs* (**1988**) – and **Changewinds**, which makes up a single long novel – *When the Changewinds Blow* (**1987**), *Riders of the Winds* (**1988**) and *War of the Maelstrom* (**1988**). Not every volume or sequence follows the same schema; but the underlying pattern is that protagonists are dislocated from their normal environment or world and cast into a WONDERLAND environment whose rules (always arbitrary, and humiliating to women in particular) seem generated as part of a GODGAME by semi-divine manipulators. Again and again, JLC protagonists suffer grotesque METAMORPHOSES in their attempts to escape their fates.

Singletons of fantasy interest include *And the Devil Will Drag You Under* (**1979**), about a QUEST through ALTERNATE WORLDS for five magic jewels. The title story of *Dance Band on the Titanic* (coll **1988**) features a ferry at the heart of a kind of CROSSHATCH vortex. JLC's work always threatens to be more interesting than it is. Unfortunately, prolixity and an inclination to escape conclusions by extending his tales into interminable HEROIC-FANTASY sequences have tended to vitiate the intensity of his basic premises. JLC's reputation has faded somewhat: with one tale told at the pitch he managed earlier, this could change in a moment. [JC]

Other works: *Hotel Andromeda* (anth **1994**).

CHAMBERS, AIDAN (1934-) UK writer, editor and educationist, best-known as a writer for children; he received the Children's Literature Association Award for Criticism in 1978. In his study of children's attitudes to fiction, *The Reluctant Reader* (**1969**), AC suggested that up to 60% of children may be interested in reading but are not sufficiently stimulated to sustain the habit; he advocated simpler but more exciting fiction to encourage young readers, and thus developed the **Topliners** series. AC has a special fascination for GHOST STORIES and has compiled a number of collections and ANTHOLOGIES (◊◊ GHOST STORIES FOR CHILDREN), including the **Ghosts** series: *Ghosts* (anth **1969**) with his wife Nancy Chambers (1936-), *Ghosts 2* (anth **1972**), in which seven of the nine stories are by AC, *Ghost Carnival* (coll **1977**), where AC sought to humanize the ghosts, and *Ghosts Four* (anth **1978**) ed as Malcolm Blacklin, which included a high quota of true hauntings; *Ghosts 3* (coll **1974**) in this series was written by Sam Holroyd. Others were *Ghosts That Haunt You* (anth **1980**), *Ghost After Ghost* (anth **1982**), *Shades of Dark* (anth **1984**), *A Quiver of Ghosts* (anth **1987**) and *A Haunt of Ghosts* (anth **1987**), in which latter five of the 10 stories are by AC.

AC's ghost stories are surprisingly unatmospheric, being presented almost as factual and linking the child protagonist directly with the events. They inspired a short-lived tv series, *Ghosts* (**1980**). His interest in true HAUNTINGS resulted in five further books: *Haunted Houses* (**1971**), *A Book of Ghosts and Hauntings* (**1973**), *More Haunted Houses* (**1973**), *Great British Ghosts* (**1974**) and *Great Ghosts of the World* (**1974**). AC also edited for adults *The Tenth Ghost Book* (anth **1974**) and *The Eleventh Ghost Book* (anth **1975**), both assembled as *The Bumper Book of Ghost Stories* (omni **1976**) (◊ GHOST BOOK). [MA]

Other works as editor: The sf anthologies *World Minus Zero* (anth **1971**) and *In Time to Come* (anth **1973**), both with Nancy Chambers; *Out of Time* (anth **1984**).

CHAMBERS, ROBERT W(ILLIAM) (1865-1933) US artist who in 1894 turned to a highly successful writing career, producing over 70 books, notably *The King in Yellow* (coll **1895**; cut vt *The Mask* 1929), whose first four stories are linked SUPERNATURAL FICTIONS. This highly influential work features an eponymous fictional BOOK, a verse-play which drives its readers into madness and even suicide. Behind this book RWC creates a mythology linked to a shadowy personification of DEATH – drawn in part from the writings of Ambrose BIERCE – which suggests that readers must pass through an INITIATION to enter the land of Carcosa, which is either a SECONDARY WORLD or somewhere perceivable (◊ PERCEPTION) through higher consciousness. This novel likely suggested to Cleveland Moffett (1863-1926) his enigmatic *The Mysterious Card* (**1896** *Black Cat*; fixup **1912**), and certainly influenced many writers of RWC's and the next generation, especially H.P. LOVECRAFT. James BLISH ingeniously reconstructed the fictional play in "More Light" (in *Alchemy and Academe* ed Anne MCCAFFREY **1970**). A fragment of another play, believed to be inspired by RWC's work, was found among the papers of Charles Vaughan (1902-1966) and was published as *The King in Yellow* (**1975**).

Nothing RWC wrote thereafter came close to equalling *The King in Yellow* in either mystery or power. His supernatural writings remain his strongest. He was intrigued by the power of humankind over death, usually through the deployment of BLACK MAGIC; this notion recurs in "The Maker of Moons" (1896 *English Illustrated Magazine*; in *The Maker of Moons* coll **1896**; **1954** chap) and *The Slayer of Souls* (**1920**), both of which involve Oriental sorcery, and "The Messenger" (in *The Mystery of Choice* coll **1897**). "The Key to Grief" (also in *The Mystery of Choice*) is a POSTHUMOUS FANTASY modelled on Bierce's "An Occurrence at Owl Creek Bridge" (1891). RWC also became fascinated with the potential survival of extinct or MYTHICAL CREATURES, and used this as a plot device for a series of light romantic adventures assembled as *In Search of the Unknown* (coll **1904**) and *Police!!!* (coll **1915**), which collections together constitute the best of his work; from these books E.F. BLEILER made an informed selection, *The King in Yellow and Other Horror Stories* (coll **1970**). Some later society novels contain elements of the fantastic; e.g., *The Gay Rebellion* (**1913**), depicting an alternate feminist present, and *The Green Mouse* (1908-9 *Saturday Evening Post*; **1910**) which, with the later *The Talkers* (**1923**), is an uninspired derivative of the Svengali motif. RCW's only other intriguing work is *The Tracer of Lost Persons* (coll of linked stories **1906**), in which a detective's methods sometimes verge on the preternatural (◊ OCCULT DETECTIVES). [MA]

Other works: *The Tree of Heaven* (coll **1907**); *Some Ladies in Haste* (**1908**); *Quick Action* (**1914**) and its sequel *Athalie* (**1915**).

CHAMISSO, ADELBERT VON (1781-1838) Adoptive name of French-born Prussian poet, soldier and botanist Louis Charles Adélaide, Vicomte de Chamisso (1781-1838), sometimes identified as Adelbert von Chamisso de Boncourt. Although on the fringes of German ROMANTICISM, Chamisso was influential through his friendship with E.T.A. HOFFMANN, Friedrich de la Motte FOUQUÉ and Clemens Brentano (1778-1842), and made his mark with his novel *Peter Schlemihls wundersame Geschichte* (**1814**; trans Sir John Bowring [1792-1872] as *Peter Schlemihl* 1824 UK, mistakenly attributed to Fouqué; vt *The Shadowless Man* 1843 UK; new trans as *The Wonderful History of Peter Schlemihl* by Falck-Lebahn 1851 UK; vt *The Marvellous History of the Shadowless Man* 1914 dos UK; new trans as *Peter Schlemihl* by L. von L. Wertheim **1970** UK). The eponymous hero sells his SHADOW to the DEVIL in exchange for a bottomless purse (◊ PACT WITH THE DEVIL), but riches do not replace identity and he becomes an ACCURSED WANDERER. Schlemihl finally disposes of the purse and rejects the devil and thereafter, with the aid of a pair of seven-league boots, dedicates himself to scientific research. The story, apparently written for amusement, has its parallels in FAUST. [MA]

CHAMPIONS, THE UK tv series (1968-69). ATV Midlands. **Pr** Monty Berman. **Dir** Roy Ward Baker, Cyril Frankel, Sam Wanamaker and others. **Writers** Brian Clemens, Terry Nation, Dennis Spooner and others. **Starring** Alexandra Bastedo (Sharron Macready), Stuart Damon (Craig Stirling), William Gaunt (Richard Barrett), Anthony Nicholls (Tremayne). 30 60min episodes. Colour.

Some episodes of this neatly executed series are reminiscent of the earlier *Man from U.N.C.L.E.*, and the three Champions, who work for the organization Nemesis, pursue the same goal: maintaining a balance of power among power-blocs. But the Champions are SUPERHEROES with TALENTS, given to them by a LOST RACE of SECRET MASTERS from Tibet (◊ THEOSOPHY). [JC]

CHANGELINGS Much FOLKLORE relating to FAIRIES alleges they are prone to steal human infants, leaving DOUBLES in place. Folkloristic changelings usually reveal themselves, eventually, to be unhuman entities (or perhaps just blocks of wood), but literary ones are children who seem unnaturally malevolent, angst-ridden or imaginative. Notable literary versions include "Changeling" (1946) by Dorothy K. Haynes (1918-   ) and "The One and the Other", the opening story in Sylvia Townsend WARNER's *Kingdoms of Elfin* (coll **1977**).

The Soul of Kol Nikon (**1923**) by Eleanor FARJEON is a bleak study of a child rejected by family and society. *In Brief Authority* (**1915**) by F. ANSTEY and *The Land of Unreason* (**1942**) by L. Sprague DE CAMP and Fletcher PRATT are comedies in which mistaken identities confuse the traffic between the contemporary world and FAERIE. A changeling and his counterpart function as morally opposed DOPPELGÄNGERS in *The Broken Sword* (**1954**) by Poul ANDERSON. A SHAPESHIFTING changeling is featured in *The Woman who Loved Reindeer* (**1985**) by Meredith Ann PIERCE. [BS]

CHANT, JOY Pseudonym of UK writer Eileen Joyce Rutter (1945-   ). Her first novel was *Red Moon and Black Mountain* (**1970**). Though clearly a CHILDREN'S FANTASY, this is practically the paradigmatic GENRE FANTASY, featuring three COMPANION children recruited by a LIMINAL BEING through a PORTAL to an OTHERWORLD threatened by WRONGNESS; each of the three is the right tool for one part of the plan to oppose the Dark God, and each undergoes a NIGHT JOURNEY. The protagonists do not, ultimately, pay a heavy price for their learning experience; we get instead a near "was it a dream?" ending. The tale owes much to J.R.R. TOLKIEN and a bit to C.S. LEWIS's **Narnia** sequence, and its use of MAGIC is rather conventional; but the intense sense of place and the portrait of the horse-riding Harani (of which the oldest of the children is the HIDDEN MONARCH), as well as a fine high DICTION, set it apart from the slightly later and much more slavish imitators of Tolkien.

The Grey Mane of Morning (**1977**), a stronger and entirely more original work, returns to the first novel's half-Native American and half-Cossack steppe culture; it portrays the folkways of the Harani with deep love and unswerving conviction.

As if to answer feminist critiques of her work, JC's *When Voiha Wakes* (**1983**) presents a female-dominant culture; but it is the male artist who gets to escape the strictures of his society. This is a slighter work than its predecessors. [DK]

Other works: *The High Kings* (**1983**), retelling various tales from the MATTER of Britain.

CHAOS Almost every CREATION MYTH begins with Chaos, a time before TIME when shape had yet to take shape. In Greek MYTH there are at least two stories of primal creation, each beginning in Chaos, who (or which) may be an egg that contains time or an embodiment of the sense of primal beginnings who simply brings forth Nyx (Night), GAIA (or Gaea; Earth) and Erebos (the negative blackness beneath Gaia), whereafter sexual coupling and all our woes almost immediately ensue. At the end of time, too, Chaos takes in all the STORIES that have been told through time. In fantasy, the most fruitful use of the concept can probably be found in the works of Michael MOORCOCK, whose MULTIVERSE balances complexly between the rigours of Law and the dark loosening of the stays of Chaos. It is a vision of BALANCE central to the structure of many fantasy novels, especially those – e.g.,

Judy Allen's *The Lord of the Dance* (**1976**) – in which a cosmic AGON is replicated (◊ AS ABOVE, SO BELOW) in the RITE OF PASSAGE of a young protagonist into a mature person who can accept the complexities of being; a bastardized reification, usually incompletely realized, of this notion of balance between Order (good) and Chaos (bad) can be found in many works of GENRE FANTASY, like the novels by Brian Craig (Brian STABLEFORD), Jack Yeovil (Kim NEWMAN) and others based on the **Warhammer** role-playing GAME. In John GRANT's *The World* (**1992**), Chaos is realized as a macroscopic version, the Mistdom, of the quantum probability field (or virtual-particle sea), dissolution into which implies not only universal destruction but a cleansing, fusion and subsequent rebirth. Generally, however, the TIME ABYSSES of fantasy stop short of contemplating the nothingnesses at the beginning and end.

Chaos is sometimes seen more literally and mundanely as the world system to which FAERIE or DEMONS own allegiance, as the context in which their arbitrary MAGIC – as opposed to the legalistic rational magic of humanity – can operate. This version of the legend is intrinsically linked to the concept of the POLDER and to BELATEDNESS, when the chaotic is seen as a desirable state compared with the THINNING that will follow its demise. [JC]

See also: ELDER GODS; END OF THE WORLD.

CHAPMAN, VERA (MAY) (1898-1996) UK writer, civil servant and founder (1969) of the UK TOLKIEN SOCIETY. VC turned to writing professionally only in her seventies, when the markets for her favourite literature became accessible. The wife of a country vicar, she was for a while a practising Druid; and her interest in pagan MYTHOLOGY drew her towards the LEGEND of ARTHUR, resulting in her first book, *The Green Knight* (**1975**), followed by *The King's Damosel* (**1976**) and *King Arthur's Daughter* (**1976**), all three assembled as *The Three Damosels* (omni **1978**). A decade earlier than Marion Zimmer BRADLEY, VC brought a feminist (◊ FEMINISM) perception to the Arthurian myth, studying the events through the eyes of incidental or invented characters whose importance has (according to the tale) been deliberately masked to disguise their purpose as agents of MERLIN to secure the Arthurian bloodline through his female descendants. VC brings the traditional Arthurian world keenly alive and places a new interpretation upon the standard events. VC returned to the MATTER of Britain in *Blaedud the Birdman* (**1978**), about an early British king who learned to fly; again she invested a scant legend with character and conviction.

Her interests were not confined to ancient Britain, however. Two books were developed from CHAUCER: *The Wife of Bath* (**1979**), which looks at *The Canterbury Tales* in a new light, and *The Notorious Abbess* (coll of linked stories **1993** US; possibly a ghost title), about a nun at the time of the Crusades cursed/blessed with TALENTS. VC has also written two books for children, *Judy and Julia* (**1977**) and *Miranty and the Alchemist* (**1983** US). [MA]

CHARLEMAGNE ◊ ROMANCE.

CHARLES DICKENS' A CHRISTMAS CAROL UK ANIMATED MOVIE (*1971* tvm). ◊ *A* CHRISTMAS CAROL.

CHARME DISCRET DE LA BOURGEOISIE, LE ot of *The* DISCREET CHARM OF THE BOURGEOISIE (*1972*).

CHARMINGS, THE US tv series (1987-8). Embassy Communications/ABC. **Pr** Mark Fink, Danny Kallis, Al Lowenstein. **Exec pr** Roxie Wenk Evans, Prudence Fraser, Robert Sternin. **Dir** Mark Cullingham and others. **Writers**

Douglas Bernstein and others. **Created by** Fraser, Sternin. **Starring** Brandon Call (Thomas Charming), Carol Huston (Snow White Charming in 1988), Caitlin O'Heaney (Snow White Charming in 1987), Garette Ratliffe (Corey Charming), Christopher Rich (Prince Eric Charming), Judy Parfitt (Queen Lillian White), Paul Winfield (The Mirror). 21 30min episodes. Colour.

A sitcom version of SNOW WHITE. Snow White and Prince Charming attempt to dispose of her evil STEPMOTHER Queen Lillian by tossing her into a bottomless pit, but it proves not bottomless: she climbs out and casts a powerful SPELL: the whole family, herself included, falls asleep for centuries. All eventually wake to find themselves living in modern suburbia. Prince Charming becomes a children's author, Snow White a dress designer. Lillian, who lives upstairs, spends much of her time throwing spells and arguing with her Magic MIRROR, which has developed a very sarcastic streak.

Many episodes used elements from other fairytales, such as magic beans and CINDERELLA. But the series lasted barely one season.   [BC]

CHARMS Properly speaking the sung, spoken or written words of a SPELL (◊ MAGIC WORDS); the word comes from the Latin *carmen*, "song". In more modern usage a charm is usually the words of a spell imbued into an object, as for example a TALISMAN or an AMULET, worn by its possessor in order to achieve some magical effect: protection from other MAGIC, INVULNERABILITY. INVISIBILITY. etc. Charms may be given to or bought by a character, and have general application; or they may be provided specifically for the character by a magic-user with the power to create them. Possessing a charm does not imply the owner is able to perform magic; indeed, it usually indicates one cannot. [CB]

CHARNAS, SUZY MCKEE (1939-   ) US writer and former teacher who earned initial fame for her sf novel *Walk to the End of the World* (**1974**) which, with its sequels *Motherlines* (**1979**) – both assembled as *Walk to the End of the World and Motherlines* (omni **1989** UK) – and *The Furies* (**1994**), portrayed a dystopian post-HOLOCAUST future where women seek to recover their status and escape the blame placed upon them for the world's ills (◊ GENDER). Starting with "The Ancient Mind at Work" (1979 *Omni*) SMC began a sequence of stories that depicts a scientifically rationalized VAMPIRE, superior to humans. The series developed into the highly regarded *The Vampire Tapestry* (coll of linked stories **1980**), of which "Unicorn Tapestry" (1980 *New Dimension 11*) won the Nebula AWARD. SMC's portrayal of Dr Weyland, the vampire anthropologist, is one of the most convincing in modern SUPERNATURAL FICTION.

Her remaining longer works include a good TIMESLIP, *Dorothea Dreams* (**1986**), and a traditional YA fantasy adventure, the **Sorcery Hill** trilogy – *The Bronze King* (**1985**), *The Silver Glove* (**1988**) and *The Golden Thread* (**1989**) – in which teenagers seek to fight off marauders from an OTHERWORLD. She returned to a more creative mood with "Boobs" (1989 *Asimov's*), a first-person WEREWOLF story, which received the Hugo AWARD, and the YA *The Kingdom of Kevin Malone* (**1993**), where the eponymous petty-thief-turned-worldmaker becomes a prince in his own created land of FAERIE.   [MA]

Other works: *Listening to Brahms* (1986 *Omni*; **1991** chap); *Moonstone and Tiger Eye* (coll **1992** chap).

CHARRETTE, ROBERT N. (?   –   ) US writer and role-playing GAME designer who initially concentrated on ties for two series owned by FASA: SHADOWRUN (which see

for titles), and **Battletech**, to which sequence he contributed his first novel, *Wolves on the Border* * (**1988**), plus *Heir to the Dragon* * (**1989**) and *Wolf Pack* * (**1992**). The premises governing RNC's own series, **Artos** – *A Prince Among Men* (**1994**) and *A King Beneath the Mountains* (**1995**) – derive from those governing **Shadowrun**, as both are TECHNOFANTASIES set in the middle of a 21st century invaded by a leakage (◊ CROSSHATCH) of MAGIC creatures. But RNC's own world is a dystopia under the sway of giant corporations, and the reborn ARTHUR, who takes to the streets as Artor, is a creation of some vigour.   [JC]

CHARYN, JEROME (1937-   ) US writer who from the start has used NEW YORK, where he was born, as a venue for increasingly mythopoeic FABULATIONS in novels like *Once Upon a Droshky* (**1964**), *The Catfish Man: A Conjured Life* (**1980**) – which transmutes autobiography into fantasy – and the much more panoramic *Panna Maria* (**1982**). Though these novels rarely extend into explicit impossibilities, and their protagonists' frequent PERCEPTION of supernatural agencies rarely requires a literal reading, it can be argued that the New York JC creates, rather like Isaac Bashevis SINGER's Poland – the tales assembled in JC's *The Man who Grew Younger* (coll **1967**) show Singer's deep influence – has been so transformed and heightened that it constitutes a stage for genuine URBAN FANTASIES. The series of hallucinatory detective novels for which JC has become most widely recognized, the **Secret Isaac** sequence – *Blue Eyes* (**1975**), *Marilyn the Wild* (**1976**), *The Education of Patrick Silver* (**1976**) and *Secret Isaac* (**1978**), all assembled as *The Isaac Quartet* (omni **1984** UK), plus *The Good Policeman* (**1990**), *Maria's Girls* (**1992**) and *Montezuma's Man* (**1993**) – is also set in New York, where Isaac Sidel, a wounded MAGUS figure, serves as Commissioner of Police. *Elsinore* (**1991**) shares the same venue and the same sense of implausibility constantly teetering on the edge of impossibility, but never quite falling in.

Eisenhower, My Eisenhower (**1971**) slips into hallucinatory futurity, and *Darlin' Bill* (**1980**) is a hallucinated Western (◊◊ HALLUCINATION). *Pinocchio's Nose* (**1983**) (◊ REAL BOY) similarly transfigures the 21st century. Also of fantasy interest are several GRAPHIC NOVELS, all first published in French: *The Magician's Wife* (graph **1986** Belgium; first English-language version **1987** US) with artist François Boucq; *Billy Budd, K.G.B.* (graph trans Elizabeth Bell **1991**); and *Margot in Bad Town* (graph trans **1992**).   [JC]

CHASE, MARY (1887-1973) UK writer. ◊ CHILDREN'S FANTASY.

CHAUCER, GEOFFREY (?1340-1400) UK poet whose importance to English literature is perhaps second only to that of William SHAKESPEARE, and author of *The Canterbury Tales* (written c1387-1400), a central TAPROOT TEXT. His first work of interest is *The House of Fame* (written c1379-80), a PARODY of DANTE; *The Parliament of Fowls* (written written c1385) is an ESTATES SATIRE done as a BEAST FABLE. But the *Tales* themselves rightly dominate later generations' sense of GC. For fantasy, they share a central shaping role with BOCCACCIO's *Decameron* (written c1350), both powerfully (and lovingly) ironizing the nature of STORY while also telling tales with a narrative grace and intensity rarely found in earlier literature. If GC remains the more influential figure, it is probably because – along with the revolutionary, secular flexibility of his storytelling craft, a fluency most destructive to ALLEGORY – he was also a first-rate portrayer of character. His storytellers, and the people who figure in the tales told, display a psychologically coherent inwardness which later generations have come to expect of characters in fiction but which was radically new in the 14th century. The tales themselves – especially the Nun's Priest's BEAST FABLE about Chanticleer, the Canon's Yeoman's assault on ALCHEMY, the Pardoner's Tale, whose young protagonists fight a personified DEATH, the mocked chivalric ROMANCE recounted by the Squire, and the Wife of Bath's Tale, in which an Arthurian knight rapes a maiden and must marry a hag (◊ GAWAIN) – are memorable. But the heart of GC is the great STORY CYCLE of the *Tales* entire.   [JC]

CHAYKIN, HOWARD (VICTOR) (1950-   ) US COMIC-book illustrator whose b/w work has a raw, bold energetic quality. In much of his painted work, in rich opaque colour, he treats each two-page spread as a single, informal design, in which the sequence of the constituent images is clarified by judicious placing of text and speech balloons.

HC's first published work was for Fritz LEIBER's **Fafhrd and the Gray Mouser** (in *Sword of Sorcery #1* 1973). He both wrote and drew **Cody Starbuck** (in *Star*Reach #1* 1974 and *Cody Starbuck* graph coll **1978**), **The Scorpion** (*#1* 1975) and **Iron Wolf** (1986), and produced painted artwork for three groundbreaking GRAPHIC NOVELS: *Empire* (graph **1978**) by Samuel R. DELANY, *The Swords of Heaven, The Flowers of Hell* (graph **1979**) by Michael MOORCOCK and *The Stars my Destination* (graph *Vol 1* **1979**; complete story graph **1992**), an adaptation of Alfred Bester's *Tiger! Tiger!* (**1956**). Another painted story was *Gideon Faust* (in *Heavy Metal* 1982) written by Len Wein (1948-   ), in which his experiments with page layout were particularly interesting.

Further milestone projects have been **American Flagg** (*#1-#26* 1983-6), **Time²** (in *American Flagg Special* 1986 and *Time²: The Epiphany* 1987), a special four-issue series of **The Shadow** (*#1-#4* 1986; graph coll **1986**) and the erotic **Black Kiss** (*#1-#12* 1988-9), for all of which HC provided both story and art.   [RT]

CHEEVER, JOHN (1912-1982) US writer who occasionally used fantasy elements. In his best-known fantasy, "The Enormous Radio" (1947), a radio begins magically broadcasting the goings-on behind doors in an apartment building. Intrusions of the fantastic into the everyday world are usually rendered more ambiguously, so that it is impossible, for example, to tell whether or not the marriage that falls apart in "The Seaside Houses" (1961) does so as the result of a psychic residue of despair, or how much the increasingly hostile environment encountered by the title character of "The Swimmer" (1964) represents the externalization of his own psychological turmoil. Fantasy is often bound up with ideals of femininity and female sexuality: the female COMPANION conjured by the protagonist of "The Chimera" (1961) is a projection of his need for a sympathetic listener, while the title character of "The Music Teacher" (1959) is cast as a WITCH who knows how to fulfil the longings of all misunderstood husbands. In "Torch Song" (1947), a woman who endures the abuse of a succession of lovers may be a VAMPIRE feeding on their misery.

JC's novels are lighter in tone, usually portraying suburbia as a fallen Paradise built from the thwarted hopes and dreams of its denizens. *Bullet Park* (**1969**) is an ALLEGORY of the struggle between GOOD AND EVIL set in an imaginary suburb of New York City.   [SD]

Other works: *The Wapshot Chronicle* (**1958**) and *The Wapshot Scandal* (**1964**); *The Stories of John Cheever* (coll **1978**); *Falconer* (**1977**); *Oh What a Paradise it Seems* (**1982**);

Thirteen Uncollected Stories by John Cheever (coll **1994**) ed Dennis Franklin.

CHERRYH, C.J. Working name of US writer Carolyn Janice Cherry (1942-   ), who has established a very high reputation as an sf author since publication of her first novel, *Gate of Ivrel* (**1976**), a tale which, though its narrative strategies are fantasy-based, is tied by its underlying premise to sf. This novel begins the **Morgaine** sequence, which continues with *Well of Shiuan* (**1978**) and *Fires of Azeroth* (**1979**), all three being assembled as *The Book of Morgaine* (omni **1979**; vt *The Chronicles of Morgaine* 1985 UK), plus *Exile's Gate* (**1988**). Morgaine is a member of a PARIAH ELITE whose task it is to travel through PORTALS that connect various worlds (each transit involves a shift in TIME as well as space) and to destroy each portal (or gate) they travel through in order to protect the worlds from an alien race, which has used the gates to manipulate time. Any sense that Morgaine and her kin are a kind of Time Patrol is submerged in the tone, specifically in the abiding sense of TIME ABYSS which controls that tone, for Morgaine – to those she encounters – is a figure of STORY, a fearsome manifestation from a legendary past.

CJC's **Arafel** sequence – *Ealdwood* (**1981**; rev vt *The Dreamstone* 1983) and *The Tree of Swords and Jewels* (**1983**), assembled as *Arafel's Saga* (omni **1983**; vt *Ealdwood* 1991 UK) – is more orthodox, being a tale of traditional THINNING. In Ealdwood, a magical FOREST which could also be described as a shrinking POLDER, Arafel, the only surviving ELF, becomes involved with humans and half-humans in various conflicts based on Celtic material (◊ CELTIC FANTASY). Similarly set in a LAND-OF-FABLE Britain, though this time in Scotland, and also crosshatching (◊ CROSSHATCH) FAERIE with an impinging world, *Faery in Shadow* (**1993** UK) works as an elegy for loss of richness, and has been criticized for its bleakness.

The **Rusalka** sequence – *Rusalka* (**1989**), *Chernevog* (**1990**) and *Yvgenie* (**1991**) – also focuses on a crosshatched world, a Land-of-Fable medieval Russia, in which Rusalka, a GHOST, haunts and enamours a mortal; adventures ensue, presented in CJC's usual complex, intense, seemingly rushed but in fact highly crafted pellmell idiom. *Fortress in the Eye of Time* (**1995**) concentrates on a protagonist who is the AVATAR of some UNDERLIER figure of world-shaking importance, though for much of the novel he cannot find out just who; in the end, like a time-bomb, the action explodes. It might be suggested that CJC's idiom is perhaps better suited to the daunting scope and complications of her main sf work, the **Union-Alliance** sequence, which encompasses most of her oeuvre; but her fantasy remains vividly in the mind, and its lack of compromise or nostalgic warmth bodes well for its survival.   [JC]

Other works: *The Paladin* (**1988**); the **Heroes in Hell** SHARED-WORLD sequence, co-generated with Janet E. MORRIS, being *Heroes in Hell* * (anth **1985**), *The Gates of Hell* * (**1986**) and *Kings in Hell* * (**1987**), both with Morris, and *Legions of Hell* * (fixup **1987**); the **Sword of Knowledge** shared-world sequence, being *A Dirge for Sabis* (**1989**) with Leslie Fish, *Wizard Spawn* (**1989**) with Nancy Asire (1945-   ) and *Reap the Whirlwind* (**1989**) with Mercedes LACKEY; *The Goblin Mirror* (**1992**).

CHESBRO, GEORGE C(LARK) (1940-   ) US writer who began publishing material of genre interest with tales like "Strange Prey" for *Alfred Hitchcock's Mystery Magazine* in 1970, and is of fantasy interest almost exclusively for two

OCCULT-DETECTIVE sequences. The first is about **Mongo**, a dwarf criminologist with TALENTS: *Shadow of a Broken Man* (**1977**), *City of Whispering Stone* (**1978**), *An Affair of Sorcerers* (**1979**), *The Beasts of Valhalla* (**1985**), *Two Songs the Archangel Sings* (**1986**), *The Cold Smell of Sacred Stone* (**1988**), *The Second Horseman Out of Eden* (**1989**), *In the House of Secret Enemies* (coll **1990**), *The Language of Cannibals* (**1990**), *The Fear in Yesterday's Rings* (**1991**), *Dark Chant in a Crimson Key* (**1992**), *An Incident at Bloodtide* (**1993**) and *Bleeding in the Eye of a Brainstorm* (**1995**). The earlier volumes are the more interesting: in the later tales it seems a TEMPLATE is being clung to. GC's second sequence – some scenes crossover with **Mongo** – features **Veil Kendry**, a psychic painter: *Veil* (**1986**) and *Jungle of Steel & Stone* (**1988**). *The Golden Child* * (**1986**) is a movie novelization. [JC]

CHERUBIM ◊ ANGELS.

CHESS This ancient boardgame of stylized warfare emerged in recognizable form in India and China by about the 6th century, spreading westward to Europe by the 11th century. It appears early in Celtic MYTH: in the MABINOGION, "Peredur Son of Efrawg" includes a magical chess-set whose pieces move unaided, prefiguring many later stories of anthropomorphized chessmen. One MERLIN legend also features a self-playing chess set, which can checkmate any opponent. Irish and Scots FAIRIES were famous chess hustlers, challenging mortals to THREE games and allowing them to win the first two and choose large prizes – whereupon the Sidhe would win the final game and require some near-impossible task or price. The hustlers in Lord DUNSANY's "The Three Sailors' Gambit" (in *Tales of Wonder* **1916**) do not even know chess, but read their unbeatable moves in a crystal TALISMAN acquired from the DEVIL. But generally the purely intellectual nature of chess seems to give humans a chance against otherwise irresistible foes: hence the tradition of the symbolic chess-game with DEATH, most famously represented in Ingmar BERGMAN's *The* SEVENTH SEAL (*1956*). This theme has often been adapted – e.g., in Roger ZELAZNY's "Unicorn Variation" (1982), where the implacable opponent is a UNICORN.

Living chess pieces appear in many fantasies, notably Lewis CARROLL's *Through the Looking-glass* (**1872**) – whose eccentric chess-moves are usefully traced in *The Annotated Alice* (**1960**) by Martin Gardner (1914-   ), and which exploits the game's built-in UGLY DUCKLING transformation when Alice, a pawn, becomes a queen (◊◊ WONDERLAND). Poul ANDERSON's florid SCIENCE FANTASY "The Immortal Game" (1954) dramatizes the moves of a genuine game between masters, as does John BRUNNER's subtler sf *The Squares of the City* (**1965**). Susan COOPER's *Seaward* (**1983**) enlists its protagonists as players in a chess match which is a GODGAME. *Queenmagic, Kingmagic* (**1986**) by Ian WATSON exuberantly develops the notion that the pieces' long-range abilities of movement and killing are TALENTS, and extends the conceit far beyond the usual, terminal RECOGNITION that this is a game which must end. *The Chess Garden, or The Twilight Letters of Gustav Uyterhoeven* (**1995**) by Brooks Hansen (1965-   ) has an IMAGINARY LAND inhabited by chess figures.

In 1763 Sir William Jones (1746-1794) whimsically invented the MUSE of chess, Caissa, who also presides over the many variations (with changed rules, pieces and/or boards) generically known as FAIRY chess. One such is *jetan*,

played – and fully described – in Edgar Rice BURROUGHS's *The Chessmen of Mars* (**1922**).   [DRL]
Further reading: *Pawn to Infinity* (anth **1982**) ed Fred SABERHAGEN with Joan Saberhagen, sf/fantasy chess stories.

CHESTERTON, G(ILBERT) K(EITH) (1874-1936) UK author, voluminous essayist, poet, illustrator, biographer, editor and Catholic apologist. Though several of his novels are sociological sf (◊ SFE), his best fictions all exude a fantastic glamour, enhanced by a gift for visually evocative prose, perhaps harking back to his Slade School artistic training: luridly symbolic dawns and sunsets glow through all the books.

One basic GKC assertion is that nothing is truly dull or mundane when properly regarded: in his first novel, *The Napoleon of Notting Hill* (**1904**), the grey boroughs of LONDON come to life with revived heraldic pageantry, and Notting Hill is seen in a new light as a place worth dying for. Its defender, simultaneously insane and right, speaks of a fairy wand that can transform grimy streets – and indicates his SWORD. (The implied nostalgia for medieval simplicities is characteristic.) This image of London as a CITY of hidden FAIRYTALE glitter was, as GKC acknowledged, inspired by Charles DICKENS and by Robert Louis STEVENSON's *New Arabian Nights* (coll **1882**).

A second recurring trope is the Happy Surprise or EUCATASTROPHE. The popular **Father Brown** detective stories continually threaten some disquieting supernatural intrusion – BLACK MAGIC, CURSES, INVISIBILITY, psi powers (◊ TALENTS) – as a crime's one possible explanation, only for terror to be dispelled by reason. GKC wrote many fantastic "crime" stories whose eucatastrophe is the revelation that no real crime occurred: examples appear in *The Club of Queer Trades* (coll **1905**) and *Four Faultless Felons* (coll **1930**), which includes an Absurdist RURITANIAN tale of revolution.

Both these themes meet in GKC's masterpiece, *The Man who Was Thursday: A Nightmare* (**1908**), a theological HORROR-comedy of MASKS set in the same fantasticated London, with an excursion to an equally gaudy toy-theatre France. Its narrative exuberance persuasively presents Fin-de-Siècle DECADENCE fused with bomb-throwing anarchy as a vast conspiracy against civilization. The protagonist, Syme, a poet turned policeman, recruited like others by a huge unseen man for "The Last Crusade"; penetrating a secret anarchist conclave, he bluffs his way by sheer rhetoric into the post of Thursday on the conspiracy's General Council, whose members are named for days of the week. It is a gallery of creepy grotesques, Tuesday's bestial hirsuteness being the least alarming example. The aged Friday's decrepitude verges on literal decay, yet he contrives to keep up with Syme's panicky flight through a LABYRINTH of London streets. Saturday, a man of science, has unnervingly hidden eyes: those who climb endless stairs to confront him feel the chill vertigo of mathematical and astronomical infinities. Wednesday seems more obviously diabolic as Syme wounds him four times in a forced duel without drawing blood. Monday has the twisted half-smile of the pure fanatic, and at the book's zenith of terror and paranoia appears to suborn an entire French province to the anarchist cause, merely to crush Syme and his few sane companions.

All these masks are stripped away in a sequence of extravagantly joyful surprises; but there remains Sunday, the President, whose face always borders on being too large to be humanly possible (◊ GREAT AND SMALL; LIMINAL BEING). Prefacing a final antic pursuit, Sunday informs his accusers:

"I am the man in the dark room, who made you all police-men." Contrapuntal discussions of metaphysics continue through the remaining action, with all participants finding Sunday comparable only to PAN or the Universe at large; meanwhile he has pelted them with NONSENSE messages, Nature's reply when the wrong question is asked. When the chase ends, Sunday entertains his pursuers royally at a MASQUE where their own costumes echo the Biblical days of CREATION . . . and finally Sunday's own gigantic visage swells to unbearability (◊ FACE OF GLORY) and dissolves with the words "Can ye drink of the cup that I drink of?" GKC himself tentatively interpreted this vision as tearing the mask off Nature to find GOD behind. The whole is a remarkable narrative confection, compulsively readable yet often profound. A stage adaptation is *The Man who Was Thursday* (**1926**) by Mrs Cecil Chesterton and Ralph Neale.

The Ball and the Cross (**1909** US) is more overtly theo-logical; a Catholic and an atheist passionately wish to duel to the death over RELIGION and the latter's blasphemy in an England which GKC presents as insane since no one else cares enough to fight. Their escapes from police and other interference make for knockabout, sometimes slapstick adventure, punctuated with brilliant debate. The final refuge is a trap: an asylum run by SATAN, who wishes to erase every memory of religious enthusiasm and has there-fore incarcerated everyone involved in the would-be duellists' escapades. Satan tempts the pair with suitably flawed UTOPIAS, which they reject. The prison's architec-tural WRONGNESS (◊◊ EDIFICE) anticipates the N.I.C.E. Objective Room in C.S. LEWIS's *That Hideous Strength* (**1945**), as does the final HEALING conflagration.

Lesser works tend to laborious HUMOUR, as in the mildly surreal *Manalive* (**1911**), where the themes of eucatastrophe and seeing the world with new, childlike eyes are enacted by one "Innocent Smith". *Tales of the Long Bow* (coll of linked stories **1925**) actualizes various metaphors as RATIONALIZED FANTASY: pigs fly, castles float in the air, the Thames is lit-erally set on fire, etc.

Two GKC plays use fantastic themes. *Magic* (**1913**) fea-tures a stage MAGICIAN who performs an impossible trick aided by DEMONS, and to save an observer's sanity must devise a false explanation. *The Surprise* (written *c*1930; **1952**) presents a life-size PUPPET-play which goes awry when its DOLLS are magically granted free will – leading to an ALLEGORY of incarnation as the outraged Author announces his intention of "coming down" to deal with this. *The Coloured Lands* (coll **1938**) ed Maisie Ward assembles a variety of GKC's fantastic writing and drawings, from precocious juvenilia – like an allegorical story of taming the nightmare, written *c*1892 – to mature unpublished work. Some of this material reappears, with uncollected stories, fables and allegories from his copious journalism, in *Daylight and Nightmare* (coll **1986**) ed Marie Smith.

GKC's own ability to regard REALITY with enduring amazement, to carry off melodrama with panache, and to dazzle with unexpectedly apposite similes and metaphors (also alliteration, puns, paradox and ideas that sparked ideas in a kind of controlled free-association) make his fiction persistently readable. His influence is acknowledged by many writers including Neil GAIMAN, Mary GENTLE, R.A. LAFFERTY and Gene WOLFE. Surprisingly many of his more than 100 books are genuinely distinguished, though his rep-utation declined with the decline of the literary essay and persistent accusations of antisemitism (which, though not wholly unfair, were magnified by hindsight). In fantasy, he would be of major importance solely for *The Man who Was Thursday*, an indisputable classic which has remained con-tinuously in print for nearly 90 years. [DRL]

Other works include: *Greybeards at Play* (**1900**), NONSENSE verse; *Charles Dickens* (**1906**) and *Appreciations and Criticisms of the Works of Charles Dickens* (**1911**); *The Ballad of the White Horse* (**1911**), verse retelling of the King Alfred LEGEND, with striking religous VISIONS; *The Flying Inn* (**1914**); *Wine, Water and Song* (**1915** chap), verses from *The Flying Inn*; *The Collected Poems of G.K. Chesterton* (**1927**); *The Return of Don Quixote* (**1927**); *Robert Louis Stevenson* (**1927**); *The Sword of Wood* (**1928** chap); *A Handful of Authors: Essays on Books and Writers* (**1953**) ed Dorothy Collins; *Collected Nonsense and Light Verse* (**1987**) ed Marie Smith.

Further reading: *Autobiography* (**1936**); *Gilbert Keith Chesterton* (**1944**) by Maisie Ward; *G.K. Chesterton: A Centenary Appraisal* (anth **1974**) ed John Sullivan; *G.K. Chesterton* (**1986**) by Michael Ffinch; *G.K. Chesterton: A Half Century of Views* (anth **1987**) ed D.J. Conlon; *The Chesterton Review* (1974-current), journal of the G.K. Chesterton Society.

CHETWIN, GRACE (? -) UK-born writer, in the USA from 1964, though she has also spent much time in New Zealand. She began publishing work of fantasy inter-est with *On All Hallow's Eve* (**1984**), a tale (like most of her work) for the YA market; a Welsh girl in the USA is plunged, at HALLOWE'EN, into an OTHERWORLD where she must fight for her SOUL. The sequel, *Out of the Dark World* (**1985**), verges into sf. The series for which GC remains best known, the **Gom** books – *Gom on Windy Mountain* (**1986**), which serves as a prelude to the **Tales of Gom** tril-ogy, comprising *The Riddle and the Rune* (**1987**), *The Crystal Stair* (**1988**) and *The Starstone* (**1989**) – is much smoother, and traces the growth to maturity and power of Gom, a young WIZARD who comes to adulthood in a LAND OF FABLE that closely resembles the FOREST-haunted Germany of the Brothers GRIMM. A third series, initiated with *The Chimes of Alyafaleyn* (**1993**), combines sf and fantasy, taxing its female protagonist to come to maturity in a world where TALENTS are controlled by technological means. GC's work, though increasingly well couched, tends to be unchallenging. [JC]

Other works: *The Atheling* (**1988**; rev 1991), sf, beginning the unfinished **Last Legacy** sequence; *Mr Meredith and the Truly Remarkable Stone* (**1989**), for younger children; *Box and Cox* (**1990**), for younger children; *Collidescope* (**1990**), sf; *Child of the Air* (**1991**); *Friends in Time* (**1992**).

CHETWYND-HAYES, R(ONALD) (1919-) UK writer, predominantly of GHOST STORIES, though his first novel, *The Man from the Bomb* (**1959**), was sf. His next novel, *The Dark Man* (**1964**; vt *And Love Survived* 1979 US), was about spirit POSSESSION. Although he had sold one story to *The Lady* in 1954, RC-H did not begin to sell short fiction regularly until "The Thing" (in *The Seventh Pan Book of Horror Stories* anth **1966** ed Herbert van Thal), about a GHOST that feeds on the emotions it stimulates. After a few more sales to anthologies, RC-H began to produce his own collections: *The Unbidden* (coll **1971**), *Cold Terror* (coll **1973**), *Terror by Night* (coll **1974**), *The Elemental* (coll **1974**) and *The Night Ghouls* (coll **1975**). Four stories from these were later adapted into the anthology-format movie *From Beyond the Grave* (*1973* UK; vt *The Creatures from Beyond the Grave*).

RC-H's SUPERNATURAL FICTIONS are inventive in applying

new twists to traditional supernatural themes. His ghosts may be malignant or sympathetic, and frequently haunt everyday people in commonplace settings. To heighten the effect, he frequently contrasts HORROR with HUMOUR. "Looking for Something to Suck" (1969) is typical of his titles, introducing an insubstantial entity that seeks humans in order to suck their lifeforce. "The Gatecrasher" (1971) is an effective story of a SÉANCE that summons the spirit of JACK THE RIPPER. "The Ghost Who Limped" (1975; vt "The Limping Ghost") is a fine example of RC-H's ability to see the world sympathetically through the eyes of a ghost, without removing the effect of the horror upon the human characters. One of his best ghost stories is "Which One?" (1981), where six trapped firewatchers have to determine which of them is a ghost. It was adapted effectively for radio in 1983.

RC-H delights in creating a variety of MONSTERS, usually with his tongue firmly in his cheek. Stories like "The Jumpity-Jim" (1974) and "The Catomodo" (1974) led to the collection *The Monster Club* (coll 1976), which included not just VAMPIRES and WEREWOLVES but mocks, shadmocks and humgoos! This book was also made into an anthology movie, *The Monster Club* (*1980* UK), and it resulted in RC-H ghost-editing the sole issue of the ill-fated UK horror MAGAZINE *Ghoul* (1976). RC-H continued the monster theme, creating a whole family of vampires in *Dracula's Children* (coll 1987) and its sequel *The House of Dracula* (coll 1987). He further explored a SECONDARY WORLD of vampires in "Kamtellar" (1980).

RC-H has had two long-running series. The first features **Clavering Grange**, the most HAUNTED DWELLING in the UK, which he introduced in *The Dark Man* and in "The Door" (1973). Three volumes focus exclusively on the house: *The King's Ghost* (1985; vt *The Grange* 1985 US), *Tales from the Hidden World* (coll 1988) and *The Haunted Grange* (1988). RC-H has also told the adventures of the OCCULT DETECTIVE **Francis St Clare**, who first appeared in "Someone is Dead" (1974) and who likewise returns in most later collections; *The Psychic Detective* (1993) details his early career.

RC-H's other collections are: *Tales of Fear and Fantasy* (coll 1977); *The Cradle Demon* (coll 1978); *The Fantastic World of Kamtellar* coll 1980); *Tales of Darkness* (coll 1981); *Tales from Beyond* (coll 1982); *Tales from the Other Side* (coll 1983; vt *The Other Side* 1988 US); *A Quiver of Ghosts* (coll 1984); *Tales from the Dark Lands* (coll 1984); *Ghosts from the Mist of Time* (coll 1985); *Tales from the Shadows* (coll 1986); *Tales from the Haunted House* (coll 1986); and *Shivers and Shudders* (coll 1995). His novels include *The Brats* (1979), *The Partaker* (1980), *The Curse of the Snake God* (1989) and *Kepple* (1992), an interconnected series of stories *Hell is What You Make It* (coll 1994), plus two movie novelizations, *Dominique* * (1979) and *The Awakening* * (1980). Of these, only *The Brats*, a post-nuclear dystopia, and *Kepple*, a supernatural thriller in the style of James HERBERT, are outside his usual fare.

RC-H has also been a prolific anthologist. *Cornish Tales of Terror* (anth 1970) was followed by *Scottish Tales of Terror* (anth 1972) – ed as Angus Campbell – *Welsh Tales of Terror* (anth 1973), *Tales of Terror from Outer Space* (anth 1975), *Gaslight Tales of Terror* (anth 1976) and *Doomed to the Night* (anth 1978). He took over the editorship of the FONTANA BOOK OF GREAT GHOST STORIES from Robert AICKMAN after #9 (anth 1973), producing one volume annually until #20

(anth 1984). Whereas Aickman had focused on period ghost stories, RC-H's selections were mostly modern. He also inaugurated the **Armada Monster Book** series for younger readers – *The First Armada Monster Book* (anth 1975), *Second* (anth 1976), *Third* (anth 1977), *Fourth* (anth 1978), *Fifth* (anth 1979) and *Sixth* (anth 1980).

In 1989 RC-H received both the Life Achievement AWARD from the Horror Writers of America and the British Fantasy Society Special Award for his contributions to the genre. [MA]

Further reading: "A Writer in the Dark Lands" by Stephen JONES and Jo Fletcher in *Skeleton Crew* September 1990.

CHEW, RUTH (1920-   ) US writer. ◊ CHILDREN'S FANTASY.

CHIKARO TO ONNA NO YONONAKA Japanese ANIMATED MOVIE (*1932*). ◊ ANIME.

CHIKYU SAIDAI NO KESSAN vt of *Ghidora, The Three-Headed Monster* (*1965*). ◊ GODZILLA MOVIES.

CHILDE Traditionally a youth of gentle birth, usually as featured in a ballad and usually awaiting knighthood. When capitalized, as Childe, the word serves as a title. The term appears frequently in the 14th and 15th centuries, and was picked up – already with an effect of the archaic – by Edmund SPENSER, who used it in *The Faerie Queene* (1590-96), and by William SHAKESPEARE in a famous line given to the FOOL in *King Lear* (1607): "Childe Rowland to the darke Tower came." Lord BYRON made significant use of the term in *Childe Harold's Pilgrimage* (1812-18), a book-length poem written in Spenserian stanzas; Childe Harold himself, cruising across the chasms and pitfalls of a "haunted, holy" Europe, became a model of the self-exiled, melancholy, romantic aristocrat (◊ KNIGHT OF THE DOLEFUL COUNTENANCE) who gazes to the past and future from various riven aeries. But Robert BROWNING's use of the term has more directly contributed to the iconography of fantasy. "Childe Roland to the Dark Tower Came" (1855) superbly and unforgettably describes a knightly QUEST which leads, almost certainly, deathwards. In Alan GARNER's *Elidor* (1965) young Roland's loyal solitary quest for hidden TALISMANS necessary to heal the LAND is clearly based as much on "Childe Roland" as on the Scots ballad "Childe Roland and Burd Helen"; and Tom, the central male figure in Garner's *Red Shift* (1973), conveys some sense of the derangement often associated with the Childe figure, who can in this context be seen as a Western equivalent of the SHAMAN. Stephen KING's **Dark Tower** FAR-FUTURE fantasy sequence is based directly on Browning; and David A. GEMMELL's Jon Shannow – whose solitary haunted quest for a long-lost JERUSALEM dominates *The Complete Chronicles of the Jerusalem Man* (omni 1995) – is a Childe nonpareil, down to the near-lunatic obsessiveness of his quest.

In the several sf/fantasy novels which valorize BILLY THE KID, the image of the hero as a doomed solitary Childe in a WIND-torn landscape is pervasive. Somewhat less fortunately, the figure takes on a Childe Rimbaud coloration in the work of Samuel R. DELANY, whose Kid protagonists – streetwise, druggy, artistically precocious and sexually alluring – represent the incursion of what might seem an insufficiently examined sentimentality into otherwise hard-edged texts. The protagonist of Bruce Sterling's *The Artificial Kid* (1980) is one of several successor Kids.

As an ICON, the Childe is potentially of great use, for he combines aspects of the HERO, the FISHER KING and the

FOOL, conveys a sense of inbuilt TIME ABYSS, and travels onwards to the DARK TOWER where the answers lie. The object of his quest is likely to be some sort of GRAIL, whose attainment or recovery will have a HEALING function for the world. [JC]

CHILDREN During the 19th century the presumption that FAIRYTALES and FANTASY stories were forms of literature meant for the nursery led to the natural use of children as protagonists in such tales; moreover, whether or not their protagonists were children, authors of 19th-century texts normally addressed their tales downwards to an imagined audience of young folk. This presumption, and the literary strategies dependent upon it, genrated huge amounts of now-unreadable literature.

It is now an axiom in CHILDREN'S FANTASY – as in any writing for the young – that children should not be patronized, although the protagonists of tales designed to appeal to children are usually children. In fantasy written for adults, by contrast, protagonists are often first encountered as children but normally reach adulthood during their adventures: most fantasy novels are shaped around QUESTS which involve their HEROES AND HEROINES undergoing trials, being subjected to METAMORPHOSIS and gaining maturity (◊ MONOMYTH), and it is a natural strategy to dramatize the RITE OF PASSAGE into full empowerment as the literal growth of a child or adolescent into adulthood.

Non-protagonist children in fantasy are very much less common; they are often killed early, or prove to be COMPANIONS whose true nature has been disguised. In SUPERNATURAL FICTION, non-protagonist children tend to be victims of POSSESSION, or may be literal GHOSTS. In HORROR, non-protagonist children are often EVIL. [JC]

CHILDREN OF THE CORN Three movies have derived their inspiration from "Children of the Corn" (1977 *Penthouse*) by Stephen KING, the first being based directly on the story.

1. *Children of the Corn* US movie (**1984**). Angeles/Inverness/Hal Roach Studios/New World/Cinema Group Venture/Gatlin. **Pr** Donald P. Borchers, Terence Kirby. **Exec pr** Earl Glick, Charles J. Weber. **Dir** Fritz Kiersch. **Spfx** Max W. Anderson, Eric Rumsey, SPFX Inc. **Screenplay** George Goldsmith. **Starring** John Franklin (Isaac), Courtney Gains (Malachai [*sic*]), Linda Hamilton (Vicky), Peter Horton (Burt). 93 mins. Colour.

Three years ago in rural, godfearing Gatlin the manifest corn-GOD He Who Walks Behind the Rows instructed his boy-PROPHET Isaac that no one over age 18 should live; the town's older CHILDREN obediently slaughtered all the adults. Now Isaac and his entourage rule, meting out executions and HUMAN SACRIFICES at the frequent whim of the cruel god, a parody of Jahweh. Into this mess stumble young couple Burt and Vicky, who at last succeed in destroying the god – or so they think. *COTC* succeeds, just, in spinning out its short-story plot – similar to but less rich than Thomas TRYON's *Harvest Home* (**1973**) – largely through its sense of place: only rarely is the corn, in some guise, absent from the screen. The movie's subtext is that, while fundamentalists may rail at HORROR MOVIES like this one, mindless fundamentalism creates its own, far greater horrors. [JG]

2. *Children of the Corn II: The Final Sacrifice* US movie (**1992**). Fifth Avenue/Stone Stanley. **Pr** David G. Stanley, Scott A. Stone. **Exec pr** Lawrence Mortorff. **Dir** David F. Price. **Mufx** Image Animation, Bob Keen. **Spfx** Calico Ltd. **Screenplay** Gilbert Adler, A.L. Katz. **Starring** Rosalind

Allen (Angela), Ryan Bollman (Micah), Christie Clark (Lacey), Terence Knox (John Garrett), Paul Scherrer (Danny Garrett). 92 mins. Colour.

A very sequellish sequel. He Who Walks Behind the Rows is not dead, but through a new boy-PROPHET, Micah, inspires CHILDREN and supernatural forces alike to go on a campaign of carnage. Our heroes – journalist John, son Danny and their womenfolk Angela and Lacey – solve the problem by butchering the cult children and setting the corn ablaze. A further sequel was *Children of the Corn III: Urban Harvest* (**1994**). [JG]

CHILDREN'S FANTASY A genre that is clearly separate from ADULT FANTASY at the extremes but which also overlaps. This entry explores first the origins of CF and its discrete emergence, and then considers the main themes in modern CF.

1. Origins For centuries before the early Victorian period, fantasies for children were regarded as immoral and subversive. No fantasies were written specially for children, and those few which became children's books by adoption were morally improving works – best-known was *Pilgrim's Progress* (**1678**) by John BUNYAN. Similarly, the *Fables* of Aesop (6th century BC; **1484**), among the first books printed by William Caxton (1420-1491), were regarded as suitable for children because of their moral and spiritual lessons. FOLKTALES endured in the oral tradition but were not written down. Perhaps the only subversive fantasy to receive any continued support was the anonymous BEAST FABLE *Reynard the Fox* (12th century; **1481** UK), though editions specifically adapted for children did not appear until the 1840s. *Reynard's* history epitomizes the fact that stories in which animals feature were not readily perceived as subversive, and were also believed to appeal especially to the child mentality; to this day they remain among the most popular of CFs.

Also suitable were heroic tales drawn from old SAGAS and ROMANCES which demonstrated valiant and chivalrous attributes. Many were in written form from the 11th century and, though not intended for children, were doubtless retold to them. Most popular in England must have been tales featuring ARTHUR; likewise on the European mainland, stories from the Charlemagne Cycle – particularly those featuring Roland – must have prevailed.

It was only by degrees that FAIRYTALES appeared in print and became accepted as reading matter for children, although they were still generally decried by moralists throughout the 17th and 18th centuries. It may be no surprise that the earliest known in print linked themselves to the Arthurian Cycle, namely *The History of Tom Thumbe, the Little* (**1621** chap) attributed to Richard Johnson (1573-?1659), which, while starting in a ploughman's cottage, soon moves to Arthur's court, and *Jack and the Gyants* (?**1708**; vt *The History of Jack and the Giants* ?1711), which is almost certainly a simplification of Celtic legends which had contributed to the Arthurian Cycle (◊◊ ESTATES SATIRE; JACK).

The development of the literary FAIRYTALE in France in the 1690s also resulted in adaptations of the stories for all ages. Charles PERRAULT's tales were translated as *Histories, or Tales of Past Times* **1729** UK), but by 1768 (probably earlier) the title had been superseded by *Mother Goose's Tales* (coll **1768**). Likewise Madame D'AULNOY's stories, translated as *Tales of the Fairys* (coll **1699** UK) – thereby giving the name "fairytale" to the genre – devolved into the children's volume *Mother Bunch's Fairy Tales* (coll **1773** UK), and Jonathan SWIFT's *Gulliver's Travels* (**1726**) was reduced to a child's book by the 1780s. The primary publisher of children's books in the 18th century, and himself an occasional author, was John Newbery, who established a family firm in 1740 (it was for him that the Newbery Award for distinguished works of children's literature was named in 1922). He popularized the moral tale for children, epitomized by his most successful volume, *The History of Little Goody Two-Shoes* (**1765** chap), which many believe was by Oliver Goldsmith (?1730-1774).

The general antipathy to any fantasies or fairytales for children likely to corrupt their morals continued unabated into the 19th century. The UK was still under the heady influence of Thomas Bowdler (1754-1825), whose *The Family Shakespeare* (coll **1818**) had sought to "purify" SHAKESPEARE even more than had the adaptations by Mary and Charles LAMB in *Tales from Shakespear* (coll **1807**). Even though the publisher Benjamin Tabart had commenced a series of **Popular Stories** for children in 1804, including the seminal *Popular Fairy Tales* (coll **1818**), it was not until the translation of German fairytales, starting with the GRIMM BROTHERS' *German Popular Stories* (coll **1823**) – where the stories were presented as serious academic research – that the folktale became partway respectable. This allowed CF, as a separate subgenre, to begin to detach itself from the fairytale. [MA]

2. Establishment of CF It did not happen overnight. *Phantasmion* (**1837**) by Sara COLERIDGE may be regarded as an extended fairytale, and was the first original CF in novel form in English. The real split began with *Holiday House* (**1839**) by Catherine Sinclair, which broke with the didactic tradition of the moral tale and presented a story for children as children. This allowed an adult to enter into the world of make-believe and, in one chapter, Uncle David tells the children a NONSENSE story. Although regarded with severe caution by many parents, this novel – not a fantasy overall – became extremely popular, and was influential on later CFs. *The Hope of the Katzekopfs* (**1844**) by Francis Edward Paget is a blend of the moral tale and fairytale, wherein a spoilt prince in FAERIE learns self-discipline. Other stories from this period, notably *King of the Golden River* (written 1841; **1850**) by John RUSKIN and *The Rose and the Ring* (**1855**) by William Makepeace THACKERAY, still betrayed the influence of the German fairytale. It was not until the 1860s that two books radically changed the field: *The Water-Babies* (**1863**) by Charles KINGSLEY and *Alice's Adventures in Wonderland* (**1865**) by Lewis CARROLL. The first highlighted the transition: it called itself a fairytale and was intensely moralistic, but also portrayed children in a "real" world and recognized their problems. Carroll's **Alice** stories flouted all past tradition and presented clever SATIRES dressed up as NONSENSE in exactly the form that appealed to children. These two authors changed the world of children's literature and allowed the growth of CF.

During the next few decades writers turned increasingly to specialism in children's books, and many used the worlds and images of the fantastic for their settings. Jean INGELOW visited fairyland in *Mopsa the Fairy* (**1869**), which is far more than a simple fairytale, but the author who cornered the children's market was George MACDONALD, who also produced the first ADULT FANTASY – thus demarcating the two genres. There is no way that *Phantastes* (**1858**) can seriously be regarded as CF, yet it was included alongside "The Light

Princess" and other fairytales in the 10-volume set of Macdonald's *Works of Fancy and Imagination* (coll **1871**). Some of those stories had earlier appeared in *Dealings with the Fairies* (coll **1867**), but it was with *At the Back of the North Wind* (**1871**) that Macdonald made the switch from fairytale to fantasy, and this was further strengthened by *The Princess and the Goblin* (**1871**) and *The Princess and Curdie* (**1883**).

During the period 1870-1900 the CF still struggled to separate itself from the fairytale. It was necessary for the CF to remove itself from a discrete OTHERWORLD for it to claim an identity and avoid classification as a fairytale. Moreover, even after a child is able to distinguish fancy from the mundane, s/he will regard a fantasy set in our REALITY as being as much a story of an imaginary world as one set entirely or partly in a SECONDARY WORLD. Thus many adults recall *The Secret Garden* (**1911**) by Frances Hodgson BURNETT as a fantasy even though it is devoid of overt supernatural trappings (◊ SECRET GARDEN). By this token other such books are regarded as CFs for the purposes of this entry. TRANSFORMATION is a key to CFs: the ability to experience either a transfer of self from place to place or through TIME, or a change in being (from poverty to riches or from beast to beauty). The latter process is particularly important as it allows the child to come to terms with its own change from child to adult.

The emergence of CF was encouraged by Mary Molesworth in *The Cuckoo Clock* (**1877**) and *The Tapestry Room* (**1879**), both of which transport children to dream of fantasy worlds, Carlo COLLODI in *Pinocchio* (**1883**), F. ANSTEY in *Vice Versa, or A Lesson to Fathers* (**1882**; rev 1883) (◊◊ IDENTITY EXCHANGE) and *The Brass Bottle* (**1900**) and Selma LAGERLÖF in *The Wonderful Adventures of Nils* (**1906**) – all stories of transformation. The next major writer of CF was E. NESBIT. Like Anstey's work, Nesbit's was aimed at older children and the stories were more sophisticated. Her first full-length CF was *Five Children and It* (**1902**); then came *The Phoenix and the Carpet* (**1904**), *The Story of the Amulet* (**1906**), which took the children to various periods of ancient history, *The Enchanted Castle* (**1907**) and others through to *Wet Magic* (**1913**). Nesbit allowed the storyline to be dictated by the children's actions, not the parents', and gave them some influence over the supernatural events (but not enough to stifle adventure). This became the model for the many CFs that followed, in particular the form where children discover something strange which takes them into a world of magical adventure (◊ PORTALS).

In the USA, CFs took longer to extricate themselves from the image of the fairytale. Even earlier than Kingsley or Carroll, C.P. Cranch (1813–1892) had written *The Last of the Huggermuggers* (**1855**) and *Kobboltzo* (**1856**), stories inspired more by the voyages of Gulliver than by the fairytale tradition. Popular though they were, they did not have a lasting influence. The work of Julian HAWTHORNE and Howard PYLE tended to be derivative of folklore or the fairytale tradition, although Pyle's *The Garden Behind the Moon* (**1895**) is a moving ALLEGORY of an AFTERLIFE, while that of Charles E. Carryl (1841–1920) – in particular *Davy and the Goblin* (**1884**), inspired by Carroll – is little more than harmless fun. Frank R. STOCKTON showed more originality in his stories, especially those included in *The Bee-Man of Orn and Other Fanciful Tales* (coll **1887**), but it was left to L. Frank BAUM to establish CF in the USA.

Probably the best known CF in the world is not a book but

a play, J.M. BARRIE's *Peter Pan, or The Boy who Would Not Grow Up* (play 1904; **1928**); the novel based on it was *Peter and Wendy* (**1911**). This was based on the story-within-a-story in *The Little White Bird* (**1902**) that had been published separately as *Peter Pan in Kensington Gardens* (**1906**). With this work the CF could claim to have dominated the fairytale for, in a complete role-reversal, the story became regarded as a modern MYTH, with PETER PAN as the icon for eternal childhood, thereby looping back to fairytale. It encouraged at least two other fairy plays: *L'Oiseau bleu* (play **1908**; trans as *The Blue Bird* **1909** UK) by Maurice MAETERLINCK (1862-1949) (◊◊ *The* BLUE BIRD) and *The Starlight Express* (performed 1915) adapted by Violet Pearn (1880-1947) from Algernon BLACKWOOD's *A Prisoner in Fairyland* (**1913**) – plus many fairy-plays by Netta SYRETT.

Also belonging to this period are *Puck of Pook's Hill* (**1906**) and *Rewards and Fairies* (coll **1910**) by Rudyard KIPLING, both of which relied on interaction in TIME, a theme that would later dominate much CF. Kipling's **Jungle Books** were unique for their day and established the theme of children forging a bond with animals (◊◊ TALKING ANIMALS; WAINSCOTS). The enduring popularity of the ANIMAL FANTASY *The Wind in the Willows* (**1908**) by Kenneth GRAHAME led to many SEQUELS BY OTHER HANDS. The ability to converse with animals was the PLOT DEVICE behind the **Doctor Dolittle** books by Hugh LOFTING, and a similar ploy was used by A.A. MILNE in the **Christopher Robin** stories.

CFs dominated children's fiction during the 1920s and, to a lesser extent, the 1930s. Works include: *The Boy Apprenticed to an Enchanter* (**1920**) by Padraic Colum, *The Marvellous Land of Snergs* (**1927**) by E.A. WYKE-SMITH, whose hidden POLDER-land influenced J.R.R. TOLKIEN; *Bambi* (**1923**; trans **1928** UK) by Felix SALTEN; *The Midnight Folk* (**1927**) and *The Box of Delights* (**1935**) by John MASEFIELD; *The Cat who Went to Heaven* (**1930**) by Elizabeth COATSWORTH; *Mary Poppins* (**1934**) and its sequels by P.L. TRAVERS; and the first of the **Worzel Gummidge** stories by Barbara Euphan TODD, *Worzel Gummidge, or The Scarecrow of Scatterbrook Farm* (**1936**). There were also the reprinting in 1921 of Walter DE LA MARE's *The Three Mulla-Mulgars* (**1910**; vt *The Three Royal Monkeys* 1927) and the publication of *The Hobbit* (**1937**) by Tolkien. These last two books saw a return to a genuine SECONDARY WORLD involving QUESTS, in true fairytale tradition, but by now the acceptance of CFs had ensured that they were not classified as fairytale but recognized in a category of their own. The period also saw such excellent collections as *Broomsticks and Other Tales* (coll **1925**) by de la Mare and *My Friend Mr Leakey* (coll 1937) by J.B.S. Haldane (1892-1964), plus the first appearance of RUPERT THE BEAR in COMIC strip in the *Daily Express* in 1920.

After WWII fantasy as a whole was generally ostracized, but it was sustained better for younger readers than for adults; CF thus maintained a continuity that ADULT FANTASY lost. As CF organized itself after 1945 it began to establish a thematic approach that would later influence adult fantasy.

CF can be broken down (as below) into the following primary subgenres: worlds in miniature, secret gardens, time fantasies, otherworlds, wish fulfilment, and animal stories.
[MA]

3. Worlds in miniature The extension of the little-people motif from the earliest days of CF, this may include stories of WAINSCOTS or POLDERS, though there can also be SECRET GARDENS. The particular charm of these books is in the creation of a small world ideal for children to share. In books inspired by this theme the Lilliputians are either real people (or their fantasy equivalents) or animated DOLLS. Soon after WWII came *Mistress Masham's Repose* (**1946**) by T.H. WHITE, wherein an orphaned young girl finds a family of Lilliputians on an old ancestral estate. This story was not especially popular, but its successor firmly established itself as a classic: *The Borrowers* (**1952**) by Mary NORTON (◊◊ *The* BORROWERS [**1973**]). Norton had previously provided the only significant fantasy books of the WWII years, *The Magic Bed-Knob* (**1943** US) and *Bonfires and Broomsticks* (**1947**) – assembled as *Bedknob and Broomstick* (omni **1957**) and filmed much later by DISNEY as *Bedknobs and Broomsticks* (**1971**) (◊◊ ANIMATED MOVIES); both were written in the style of Nesbit. With *The Borrowers* and its sequels Norton developed the definitive WAINSCOT fantasy. Other stories of little creatures and wainscot worlds are in *Stuart Little* (**1945**) by E.B. WHITE, about a TOM-THUMB-size boy, the **Moomin** series by Tove JANSSON, *The Doll's House* (**1947**), *Impunity Jane* (**1954**) and other doll stories by Rumer Godden (1907-   ), the **Noddy** stories by Enid Blyton (1897-1968), *Loretta Mason Potts* (**1958**) by Mary Chase, where wicked DOLLS try to dominate a little girl, *James and the Giant Peach* (**1961**) by Roald DAHL, where a boy travels the world in a huge peach inhabited by giant insects, *The Secret World of Og* (**1962**) by Pierre Berton (1920-   ), where little green men exist under a children's playhouse, *The Twelve and the Genii* (**1962**; vt *The Return of the Twelves* 1963 US) by Pauline CLARKE, which brings back to life the toy soldiers which were once owned by and inspired the Brontës, *The Little People* (**1967**) by John CHRISTOPHER, though the origin of his midgets is given a scientific basis, the eco-friendly **Wombles** created by Elisabeth BERESFORD, *Mindy's Mysterious Miniature* (**1971**) by Jane CURRY, where a dolls' house is the home of real people shrunk years before by a mad professor, *The Carpet People* (**1971**) by Terry PRATCHETT, *Fungus the Bogeyman* (graph **1977**) by Raymond BRIGGS and the **Omri** series by Lynne REID BANKS that began with *The Indian in the Cupboard* (**1980**). Another series about little people, the **Minnipins**, although set entirely in their world, began with *The Gammage Cup* (**1959**; vt *The Minnipins* 1960 UK) by Carol KENDALL; Kendall repeated her success with the similar *The Firelings* (**1981** UK). [MA]

4. Secret gardens This category (◊◊ SECRET GARDEN) overlaps with the WAINSCOT and POLDER themes, particularly in the sense of a hidden world. The seminal modern fantasy in this class is *Tom's Midnight Garden* (**1958**) by Philippa PEARCE, about a garden one can access only when the clock strikes thirteen. There was the very personal world created for herself by the heroine of *Pippi Langstrump* (**1945**; trans as *Pippi Longstocking* 1950 US) by Astrid LINDGREN. Other examples of escape within our own world include: *The Enchanted Wood* (**1939**) by Enid Blyton, the first in a series where children escape to a world of FAERIE (◊◊ INTO THE WOODS); *The Valley of Song* (**1951**) by Elizabeth GOUDGE, where a valley is a manifestation of the Earthly Paradise; *The Phantom Tollbooth* (**1961**) by Norton JUSTER, where a child enters a NONSENSE world; *Beneath the Hill* (**1967**) by Jane CURRY, with its long-deserted fairy CITY; the transcendental *A Walk Out of the World* (**1969**) by Ruth NICHOLS; *The War for the Lot* (**1969**) by Sterling E. Lanier (1927-   ), where a boy helps various TALKING ANIMALS save their remaining woodland from the city rats; *A Castle of

Bone (**1972**) by Penelope FARMER which, along with the moving *Bridge to Terabithia* (**1978**) by Katherine Paterson, about a secret world in the woods from which the children must build a bridge back to REALITY, acknowledges a debt to the **Narnia** books; *The Perilous Gard* (**1974**) by Elizabeth Marie POPE, where a FAERIE world is discovered in the north of England; *The Beginning Place* (**1980**; vt *Threshold* 1980 UK) by Ursula K. LE GUIN; and the evocative *Midnight Blue* (**1990**) by Pauline FISK, where a girl escapes our world via balloon to come to terms with her problems in a MIRROR-world. Also in this category are: *The Neverending Story* (**1979**) by Michael ENDE, where a boy escapes to a BOOK-inspired world; *The Magic Spectacles* (**1991** UK) by James P. BLAYLOCK, where the glasses of the title open up a window to a faerie world; and arguably *Charlie and the Chocolate Factory* (**1964**) by Dahl, where a child wins a chance to indulge in his personal dreams. [MA]

5. Time fantasies A natural step from the SECRET GARDEN is to a secret world in TIME. The past 50 years have seen an increasing growth in stories where children either travel directly to another era – not by scientific means but in DREAM by TIMESLIP – or people from other times enter our century. Escape to a world of the past had previously been used by Alison UTTLEY in *A Traveller in Time* (**1939**). Similar examples include: the various **Magic** books by Andre NORTON, starting with *Steel Magic* (**1965**), where children find some magical artefact which transports them to different eras; *Jessamy* (1967) by Barbara SLEIGH, with a timeslip to 1914; *The Green Hill of Nendrum* (**1969**) by J.S. Andrews (1934-   ), where a child travels back to 10th-century Ireland; *Charlotte Sometimes* (**1969**) by Penelope FARMER, where the contemporary Charlotte swaps places with a girl in 1918; *Over the Sea's Edge* (**1971**) by Jane CURRY, with its IDENTITY EXCHANGE between two boys, one in the 11th century and the other in the present day; *Red Shift* (**1973**) by Alan GARNER, where a stone axe creates a link between its owners spread across 2000 years; the **Claudia and Evan** stories by Betty LEVIN, where children are transported back to different stages of Celtic history; *The House in Norham Gardens* (**1974**) by Penelope LIVELY, where the house is a vortex of all time periods; *The Night Rider* (**1975**) by Tom Ingram (1924-   ); and two books by Helen CRESSWELL, *The Secret World of Polly Flint* (**1982**), where a girl makes contact with the inhabitants of a long-lost village, and *Moondial* (**1987**), where a moondial opens a door to the past.

Stories where the travel is from the past to the present include: *Stig of the Dump* (**1963**) by Clive King (1924-   ), where a Stone Age survivor is found in a rubbish tip; *Over the Hills and Faraway* (**1968**; vt *The Hill Road* 1969 US) by William MAYNE, where a girl from the Dark Ages exchanges places briefly with a girl from today; *The Vision of Stephen: An Elegy* (**1972**) by Lolah BURFORD, where a boy escapes from the brutality of the Dark Ages to the England of 1822; and *The Philosopher's Stone* (**1971**) by Jane Little, where a 12th-century wizard seeks a treasure in the 20th century, a story somewhat reminiscent of the **Catweazle** stories by Richard Carpenter (◊ CATWEAZLE). Associated with this theme is that of the SLEEPER UNDER THE HILL. Some timeslip stories are also treated as ghost stories (◊ GHOST STORIES FOR CHILDREN).

There are also stories set entirely in the past which may include some supernatural elements by association (usually linked with myth or legend) and are, in fact, often recre-

ations of legends or hero tales. The primary writer in this field is Rosemary SUTCLIFF, though it is also worth mentioning Roger Lancelyn GREEN and Rosemary HARRIS.

Finally there are stories of ALTERNATE WORLDS. These verge close to SCIENCE FICTION, though in CF the scientific element is normally absent and the alternate world is created like some OTHERWORLD. Best-known in this subgenre are the stories by Joan AIKEN, starting with *The Wolves of Willoughby Chase* (**1962**). Other alternate worlds are found in the **Changes** trilogy by Peter DICKINSON and the **Chrestomanci** sequence by Diana Wynne JONES.   [MA]

6. Otherworlds The first and one of the most famous otherworld CFs to appear after WWII was *The Lion, the Witch and the Wardrobe* (**1950**) by C.S. LEWIS. This and its six annual sequels not only established a fantasy world complete with its own history and MYTHOLOGY but debatably implanted more effectively than any other work (including J.R.R. TOLKIEN's *The Lord of the Rings* [**1954-5**]) the SECONDARY WORLD as a primary motif in CF. It had its imitators, of which the best was Alan GARNER's *Elidor* (**1965**), although the influence may be seen also in Garner's *The Weirdstone of Brisingamen* (**1960**), *The Moon of Gomrath* (**1963**) and *The Owl Service* (**1967**). The Arthurian theme evident in the **Weirdstone** sequence also influenced Susan COOPER in her **The Dark is Rising** sequence.

During the re-emergence of adult fantasy in the mid-1960s, especially following the paperback reprinting of *LOTR*, many books were marketed as Adult Fantasy which might otherwise have been classified as CF. The classic example is Ursula K. LE GUIN's **Earthsea** sequence, whose first volume, *The Wizard of Earthsea* (**1968**), was published as an adult fantasy in the USA and as a CF in the UK. Richard ADAMS's *Watership Down* (**1972**) was published in UK paperback by Penguin in both its main adult imprint and in the Puffin children's imprint. The following includes many such borderline examples: the **Prydain** sequence by Lloyd ALEXANDER; *The Search for Delicious* (**1969**) by Natalie BABBITT, set in a NONSENSE world; *Mrs Discombobulous* (**1969**) by Margaret MAHY, where the PORTAL to another world is via a washing machine; *Red Moon and Black Mountain* (**1970**) by Joy CHANT; *Garranane* (**1971**) by Tom Ingram (1924-   ); *The Dragon Hoard* (**1971**) and *East of Midnight* (**1977**) by Tanith LEE; *The Throme of the Erril of Sherill* (**1973**) and *The Forgotten Beasts of Eld* (**1974**) by Patricia MCKILLIP; the **Dalemark** books by Diana Wynne JONES; the **Wirrun** trilogy, based on Australian aboriginal myth, by Patricia WRIGHTSON; *The Last Days of the Edge of the World* (**1978**) by Brian STABLEFORD; the **Ash Staff** trilogy by Paul R. FISHER; the **Discworld** series by Terry PRATCHETT; the **Damar** series by Robin MCKINLEY; *Talking to Dragons* (**1985**) and other books by Patricia C. WREDE; and any number of books by Jane YOLEN. It is in this category that many fantasy GAMES books belong, including the **Fighting Fantasy** series by Steve JACKSON and Ian LIVINGSTONE, the **Advanced Dungeons & Dragons** titles and spinoff books by Gary GYGAX, and the **Lone Wolf** books by Joe DEVER (and their novelizations by John GRANT).   [MA]

7. Wish fulfilment In such books, often in the style of E. NESBIT, children discover a magical item, the unleashing of whose properties results in a series of adventures – usually, but not necessarily, in our world and time. A variant concerns stories where children find things happen to them without any necessary explanation, such as in *The Shrinking of Treehorn* (**1971**) by Florence Parry Heide, where a boy finds himself shrinking, but no one notices. The **Treehorn** series continued with two novels much closer to the Nesbit formula: *Treehorn's Treasure* (**1981**), where money really does grow on trees, and *Treehorn's Wish* (**1984**) with a more traditional GENIE in a bottle. Probably the best-known Nesbit imitations are those by Edward EAGER, which began with *Half Magic* (**1954**) and included the particularly clever *Knight's Castle* (**1956**). Other imitations include: the irregular but long-running **Eleanor and Eddy** series by Jane Langton – *The Diamond in the Window* (**1962**), *The Swing in the Summerhouse* (**1967**), *The Astonishing Stereoscope* (**1971**), *The Fledgling* (**1980**) and *The Fragile Flag* (**1984**) – where the children undergo magical adventures with the help of the enigmatic Prince Krishna, though the fantastic element weakens in the later books; *Carbonel* (**1955**; vt *Carbonel, the King of the Cats* 1957 US) and its sequel, by Barbara SLEIGH, in which a girl acquires a WITCH's broom and familiar; *Awkward Magic* (**1964**; vt *The Magic World* 1965 US) and later books by Elisabeth BERESFORD; *The Apple Stone* (**1965**) by Nicholas Stuart GRAY, where a stone inside an apple allows children to talk to inanimate objects; *The Sea Egg* (**1967**) by Lucy M. BOSTON, where two boys find an egg on a beach which hatches into a merman (◊ MERMAIDS); *Magic in the Alley* (**1970**) by Mary Calhoun, where a box found in a junkshop contains various magical objects; *What the Witch Left* (**1973**) by Ruth Chew, where children find a number of magical objects in an old trunk; *William and Mary* (**1974**) by Penelope FARMER, where children discover a shell that allows them access to any pictures or poems of the sea; *The Talking Parcel* (**1975**) by Gerald Durrell (1925-1995), where children find a parcel containing creatures from the land of Mythologia; *Stoneflight* (**1975**) by Georgess McHargue, where a girl befriends a stone griffin, which comes to life; and most of the books by John BELLAIRS, especially *The Figure in the Shadows* (**1975**) and the **Johnny Dixon** series. Arguably *Chitty Chitty Bang Bang* (**1964**) by Ian Fleming is Nesbitesque, though here the magic is invented rather than discovered (◊◊ CHITTY CHITTY BANG BANG [*1968*]). Helen CRESSWELL paid tribute to Nesbit in her continuation of the **Psammead** stories in *The Return of the Psammead* * (**1992**), based on a BBC children's tv series.

Related to this category are those stories where children possess TALENTS rather than find them. This is a motif used more in adult fiction to show the frightening potential of CHILDREN with uncontrolled power. The theme is used sparingly in CF, perhaps because it is thought too corruptive; consequently some stories depict children who want to be rid of their powers and be like other people, such as in *Little Witch* (**1955**) by Anna Elizabeth Bennett. Others seek to make light of occult powers, as in *The Worst Witch* (**1974**) and sequels by Jill Murphy (1949-   ). Yet others find children uniting with those possessed of talents to work on the side of good or ultimate redemption, as in *The Witch Family* (**1960**) by Eleanor Estes, *Wilkin's Tooth* (**1973**; vt *Witch's Business* 1974 US) by Diana Wynne JONES, and *The Changeover* (**1984**) by Margaret MAHY.   [MA]

8. Animal stories This is one of the biggest categories of CF. It is no surprise that children would want to have adventures with animals. In a wide range of fantasies, the animals either interact with humans (usually TALKING ANIMALS in a manner derived from Rudyard KIPLING) or the stories are wholly about anthropomorphically endowed animals, in either BEAST FABLES or ANIMAL FANTASIES. Most of the early stories of this sort were for younger children, the

best-known being the **Peter Rabbit** stories by Beatrix POTTER; their modern equivalents include the **Paddington Bear** stories by Michael Bond (1926-   ), which began with *A Bear Called Paddington* (**1958**). Of more relevance to older readers in the 1950s were *The Hundred and One Dalmatians* (**1956**) by Dodie SMITH, and *The Rescuers* (**1959**) and its sequels by Margery SHARP. Mice are popular in children's stories (and in the movies – MICKEY MOUSE is the prominent example) and reappear in *The Mouse and His Child* (**1967**) by Russell HOBAN, a book for older children (if not for adults) which considers the gradual maturity and possible transformation of a pair of TOY mice; the altercation between the mice and rats is masterfully handled. A similar but less transcendental story is *Manxmouse* (**1968** UK) by Paul GALLICO, in which a toy mouse comes to life, while *Mrs Frisby and the Rats of NIMH* (**1972**) is a SCIENCE FANTASY by Robert C. O'Brien (real name Robert L. Conly; 1918-1973). More recently mice have dominated the world of the beast fable in two popular series: **Redwall** by Brian JACQUES and the **Deptford Mice** stories by Robin JARVIS. The most significant "mouse movie" is probably Don BLUTH's *An AMERICAN TAIL* (*1986*), which again is difficult to classify as a children's rather than an adult fantasy.

There were other animal stories in the 1950s. Gallico produced a number of CAT books, including *The Abandoned* (**1950**; vt *Jennie* 1950 UK), in which a young boy is turned into a cat (◊◊ TRANSFORMATION), and *Thomasina: The Cat who Thought She was God* (**1957**). Still popular is E.B. WHITE's *Charlotte's Web* (**1952**), a rather idiosyncratic fantasy in which a spider saves a pig's bacon. It may be partly derivative from *Animal Farm* (**1945**) by George ORWELL. Pigs, like mice, have ever been popular in children's fiction, at least since the start of the **Freddy** picture-books by Walter R. BROOKS; possibly destined to be one of the most popular of all pig stories is *The Sheep-Pig* (**1983**; vt *Babe*) by Dick KING-SMITH, filmed as *Babe* (*1995*), about a pig with aspirations to be a sheepdog. The book's sequels include *Saddlebottom* (**1985**) and *Ace: The Very Important Pig* (**1990**).

The increase in the number of animal CFs can be counted from the success of *Watership Down* (**1973**) by Richard ADAMS; it inspired a host of imitations including: *The Animals of Farthing Wood* (**1979**) and sequels by Colin DANN; *Duncton Wood* (**1980**) and sequels by William HORWOOD; *Run With the Wind* (**1983**) and sequels by Tom McCaughren; *The Cold Moons* (**1987**) by Aeron CLEMENT; *Hunter's Moon* (**1989**) and *Midnight's Sun* (**1990**) by Garry KILWORTH; plus Adams's own variation on the theme, *The Plague Dogs* (**1977**).   [MA]

9. Movies The topic of CF in the CINEMA is huge, not least because, outside movies concerning DARK FANTASY and HORROR, most filmed fantasies – and almost all ANIMATED MOVIES – are regarded as primarily children's fare. This is a ridiculous view, since some of the best, most innovative and most sophisticated fantasy appears on screen rather than in books; but it is one that we are, for the moment, stuck with. Movie fantasies aimed purely at children are comparatively few and far between: most aim for the "family audience". Movie SUPERNATURAL FICTIONS, by contrast, are generally aimed at adults, although there are plenty of exceptions – HOCUS POCUS (*1993*) is an obvious example.

The major purveyors of CF in the movies have of course been DISNEY. Leaving aside the studio's animation production, Disney has released rafts of fantasies – usually TECHNOFANTASIES – whose standard has not always been high, though often they have been commercially successful; the HERBIE MOVIES are a prime example. Animated shorts from Disney, Warner Bros., MGM and others are often filled with fantasy notions, and are for the most part intended for children. The same goes for Saturday-morning TELEVISION animation, most of which is dire; only Disney and HANNA-BARBERA can regularly be relied upon to produce acceptable fare. For a fuller coverage of CF in the movies ◊ CINEMA. [JG]

10. Conclusion The above categories do not exhaust CFs. There are, for instance, a number of MONSTER stories, such as *The Giant Under the Snow* (**1968**) by John GORDON, and the remarkable *The Iron Man: A Story in Five Nights* (**1968**; vt *The Iron Giant* 1968 US) and *The Iron Woman: A Sequel to the Iron Man* (**1993** chap) by Ted HUGHES. There will always be different and challenging stories for children, for CFs capture and broaden the imagination of a child. Whether they be subversive or instructional, it is by reflecting our world through the imagination that a child can prepare for reality. [MA]

Further reading: *Tellers of Tales* (**1946**; rev 1953) by Roger Lancelyn GREEN; *Written for Children* (**1965**; 3rd rev 1987) by John Rowe Townsend (1922-   ); *Down the Rabbit Hole: Adventures and Misadventures in the Realm of Children's Literature* (**1971**) by Selma G. Lanes; *Pipers at the Gates of Dawn: The Wisdom of Children's Literature* (**1983**) by Jonathan Cott; *The Oxford Companion to Children's Literature* (**1984**; rev 1985) by Humphrey Carpenter and Mari Prichard; *Secret Gardens: A Study of the Golden Age of Children's Literature* (**1985**) by Carpenter; "Twentieth-Century Children's Fantasy" by Mary E. Shaner in *Masterworks of Children's Literature, Volume 8: The Twentieth Century* (**1986**) ed William T. Moynihan and Shaner; *Not in Front of the Grown-Ups: Subversive Children's Literature* (**1990**) by Alison Lurie (1926-   ).

CHIN, M. LUCIE (1947-   ) US writer who began publishing fantasy with "The Best is Yet to Be" for *Galileo* in 1978. Her only novel, *The Fairy of Ku-She* (as "The Snow Fairy" in *Faery!* [anth **1985**] ed Terri WINDLING; exp **1988**) – despite the surface quiet of its setting in a LAND-OF-FABLE Ming dynasty China (◊◊ CHINOISERIE) – is a violent and ruthless tale. The FAIRY, having lost a magic vase necessary to her duties, is exiled to the mortal world, where she gives birth to a son who is immortal (◊ IMMORTALITY) and who, raised to the heavenly OTHERWORLD that is the fairy's own home, begins to ravage both FAERIE and mundane REALITY. [JC]

CHINOISERIE A term used to imply the deployment of Oriental (primarily Chinese) motifs and images in fiction and art. It is a distinct subgenre of ORIENTAL FANTASY, in that it relies predominantly on the mythic, recreating the iconography of an Oriental past that never was, but which is firmly implanted in occidental minds (◊ LAND OF FABLE). The willow pattern design in porcelain is a common example. In fiction, the influence lingered on in fantasy, where the imagery of an ersatz East was an ideal vehicle for creating an atmosphere of distant worlds and times. But it was also used to intrude the mythic into the modern world. Chinoiserie thus works in both fantasy and SUPERNATURAL FICTION; much of it today is DARK FANTASY. In fantasy its initial use derived from the popularity of the stories of the *Arabian Nights* (◊ ARABIAN FANTASY), which heavily influenced much fiction in the 18th century; Horace WALPOLE's "Mi Li: A Chinese Fairy Tale" (1785) is one of the earliest examples of Chinoiserie. Its use faded during the 19th cen-

tury, though one small example is "The Dragon Fang" (*Harper's* 1856) by Fitz-James O'BRIEN. Its popularity re-emerged strongly at the end of the Victorian era, influenced to some degree by the rediscovery of Oriental FAIRYTALES. Its impact manifested itself in all artforms, including OPERA, with *The Mikado* (1885) by W.S. GILBERT and Sir Arthur Sullivan and *Madame Butterfly* (1904) by Giacomo Puccini, and heavily influenced the writings of Lafcadio HEARN and Laurence HOUSMAN and the artwork of Aubrey BEARDSLEY and others. It continued into the 20th century through the work of Kenneth MORRIS and Lord DUNSANY, but generally faded in literature after WWI. Its surviving champion in the UK was Ernest BRAMAH, with his **Kai Lung** stories and, to some extent, Sax ROHMER and Thomas Burke (1886-1945). In the USA, Frank OWEN sustained the motif in his many stories for WEIRD TALES, as did E. Hoffmann PRICE, who returned to the subgenre in the 1970s with *The Devil Wives of Li Fong* (**1979**). Charles G. FINNEY used chinoiserie to significant effect in "The Magician Out of Manchuria" (1968) and, in a way, in *The Circus of Dr Lao* (**1935**), filmed as 7 FACES OF DR LAO (*1964*). By the 1980s, however, the effects of chinoiserie had been rediscovered by a new generation of fantasists, and the field came alive in the creative imaginations of M. Lucie CHIN, Jessica Amanda SALMONSON, Stephen MARLEY and in particular Barry HUGHART. [MA]

CHI O SUU MI ot of *Lake of Dracula* (*1971*). ◊ DRACULA MOVIES.

CHITTY CHITTY BANG BANG UK movie (*1968*). United Artists/Warfield. **Pr** Albert R. Broccoli. **Dir** Ken Hughes. **Spfx** John Stears. **Matte fx** Clive Culley. **Screenplay** Roald DAHL, Ken Hughes. **Based on** *Chitty Chitty Bang Bang, the Magical Car* (**1964**) by Ian Fleming (1908-1964). **Starring** Gert Frobe (Baron Bomburst), Adrian Hall (Jeremy), Robert Helpmann (Child Catcher), Benny Hill (Toy Maker), Sally Ann Howes (Truly Scrumptious), Lionel Jeffries (Grandpa Potts), James Robertson Justice (Lord Scrumptious), Heather Ripley (Jemima), Professor Stanley Unwin (Chancellor), Dick Van Dyke (Caractacus Potts), Max Wall (1st inventor). 145 mins. Colour.

England, about 1910. Eccentric inventor Caractacus converts an old crock into Chitty Chitty Bang Bang, the car of one's dreams. One day he tells children Jemima and Jeremy, plus to-be-girlfriend Truly Scrumptious, the STORY that forms the basis of the movie. In fact, the car is MAGIC: it can go on land, over water and through the air. Wicked Baron Bomburst of Vulgaria covets it, and abducts old Grandpa Potts. The quartet fly the car in pursuit, finding Vulgaria, a land where CHILDREN are banned, a vile Child Catcher rounding up any that might be free. However, a WAINSCOT culture of feral children exists, and with its help our heroes overthrow the tyranny.

CCBB is derivative – influences include MARY POPPINS (*1964*), in which Van Dyke also starred, and *Those Magnificent Men in Their Flying Machines* (*1965*) – and overlong, and the overabundant songs are ill placed. Dahl's influence shows most notably in the depiction of the delightfully nasty Child Catcher, although even that figure seems derivative – of the Coachman in PINOCCHIO (*1940*). The movie is normally dismissed as a stinker, but is better than that. [JG]

CHIVALRY The code of honour supposedly observed by Christian KNIGHTS, whose formal adoption was derived by adaptation of a Germanic RITE OF PASSAGE. The Church employed the device to provide an ideological basis for the Crusades and other military adventures, and it became a major self-justificatory myth of feudalism. Chivalric ideals were central to the *chansons de gestes* and other baronial amusements, and thus to the tradition of imaginative ROMANCE which ultimately produced modern HIGH FANTASY. The most important works of chivalric romance include: *The Song of Roland* and other tales of Charlemagne's knights; the LEGENDS of the GRAIL, popularized by CHRÉTIEN DE TROYES and others; tales of ARTHUR and his knights; and such proto-novels as the 14th-century *Amadis de Gaul*. There was always an element of self-parody in chivalric romance, exhibited by such as *Fergus of Galloway* (c1200; trans 1989) by "Guillaume le Clerc", though its ideals were pilloried far more comprehensively by Miguel de CERVANTES in *Don Quixote* (1605-15).

Neochivalric romance was introduced by Friedrich de la Motte FOUQUÉ, most notably in *The Magic Ring* (1813). Some slight English prose fiction in a similar vein had been written by Nathan Drake, but it was Robert SOUTHEY's rewritten translations of *Amadis of Gaul* (1803) and *Palmerin of England* (1804) that paved the way for William MORRIS's adventures in neochivalric romance, which in turn inspired imitations by G.P. BAKER, E. Hamilton Moore and Henry NEWBOLT.

A reverent but corrosively sceptical account of chivalric ideals is elaborated in the work of James Branch CABELL, beginning with *Chivalry* (coll 1909) and *The Soul of Melicent* (1913; rev vt *Domnei* 1920); other US works in the same ironic vein include John ERSKINE's *Galahad* (1926) and William Faulkner's *Mayday* (written 1926; 1977). European works of similar nature include "Sir Perseus and the Fair Andromeda" (1923) by Robert NICHOLS and "The Non-Existent Knight" (1959) by Italo CALVINO.

The long tradition connecting chivalric romance to modern fantasy is mapped out by Lin CARTER in his *Dragons, Elves and Heroes* (anth 1969) and *Golden Cities, Far* (anth 1970). Modern versions of classic chivalric romances include *Huon of the Horn* (1951) by Andre NORTON, *The Green Knight* (1975) by Vera CHAPMAN and *Parsival* (1977) by Richard MONACO. [BS]

CHOJIN DENSETSU UROTSUKIDOJI Japanese video series (1987). ◊ ANIME.

CHOPIN, KATE (O'FLAHERTY) (1850-1904) US writer and critic. Her fascination for the Creole and Cajun peoples of Louisiana caused her to draw upon their FOLKTALES for her stories in *Bayou Folk* (coll 1894) and *A Night in Acadie* (coll 1897), several of which used local superstition with considerable atmosphere. Much of her work concerned deeply psychological explorations of sexual awakening. Despite critical acclaim KC incurred public indignation over the "immoral" and anti-racist views deployed in *The Awakening* (1899), and found it difficult to sell her work thereafter. [MA]

CHRÉTIEN DE TROYES (?1130-?1190) The father of Arthurian ROMANCE (◊ ARTHUR). Our only knowledge of him comes from his own writings. He was a poet at the court of Count Philip of Champagne, though his real patroness was the Count's wife, Marie, a daughter of Eleanor of Aquitaine. Responding to Marie's wishes, CDT developed the work of the troubadours to produce a new form of narrative poetry which dealt with courtly romance and CHIVALRY. He drew heavily from the Celtic legends prevalent in Brittany and thus is regarded as the founder of

the *romans bretons*. His early work is lost, though he is acknowledged to have worked on translations and adaptations of OVID's *Metamorphoses*. The date of composition of his Arthurian romances is uncertain, and he almost certainly developed and embellished them over the period when he was most active, the 1150s-80s. The most common dates attributed to his works are as follows: *Erec et Enide* (?1170), *Cligés* (?1176), *Lancelot, ou le Chevalier de la charette* (?1177), *Yvain, ou le Chevalier au Lion* (?1177) and *Perceval, ou Le Conte del Graal* (begun ?1182). He refers to having written a sixth romance, about King Mark and Iseult, possibly in the 1170s, but that is lost. The sequence appears to follow a pattern starting with the youthful innocence, young love and thrill of adventure of *Erec et Enide*, to the more deeply romantic adventure and tragedy of *Cligés*, which seems to be modelled on the Tristan legend. *Lancelot* rounds out that tragedy by introducing adulterous love; *Yvain* emphasizes the power and honour of chivalrous love; while *Perceval* takes us into the darker mysteries of Christian love.

These stories share some commonality with stories later collected in the MABINOGION, in particular the relationship between *Erec et Enide* and "Gereint Son of Erbin" (?1250), between *Yvain* and "The Lady of the Fountain" (?1250) and between *Perceval* and "Peredur" (?1250). CDT nevertheless added his own colour. It was he who established courtly romance and chivalry as central to the Arthurian legend, based on the world of the Norman-French court, and thus developed the image of the Arthurian world as we perceive it today. He introduced the character of LANCELOT and developed the romance between him and GUINEVERE. It was also he who established Arthur's court at CAMELOT, clearly a reflection of the glories of the court at Champagne. CDT created the mysterious GRAIL, though he did not depict it as the chalice of Christ, focusing instead on the mysteries of the Grail procession. Through the power of his imagery, the breadth of his vision and the passion of his characters, he developed the romance and gave birth to the fantastic adventure that has developed into the HEROIC FANTASY of today.

Perceval, CDT's last work, was left unfinished. As his most mystical work, it captured the imagination of fellow romancers, who produced various continuations and SEQUELS BY OTHER HANDS. Most are anonymous, but some are signed Manessier and Gerbert de Montreuil; all were written within 50 years of CDT's death. They have been collected and edited by William Roach (1907-) in *The Continuations of the Old French "Perceval" of Chrétien de Troyes* (1949-83 5 vols). [MA]

Further reading: *The Craft of Chrétien de Troyes* (1980) by Norris J. Lacy; *Chrétien de Troyes: A Study of the Arthurian Romances* (1981) by L.T. Topsfield. There have been many translations of CDT's works but the most complete currently available is *Chrétien de Troyes: Arthurian Romances* (coll ed and trans D.D.R. Owen 1987).

CHRIST Known also as Jesus, the MESSIAH of Christianity. Reverence precludes his being widely featured in CHRISTIAN FANTASY, although his promised return is featured in some literary accounts of the APOCALYPSE; there are also several movie and tv versions of his life and ambiguous death, including some that are challenging, such as MONTY PYTHON'S LIFE OF BRIAN (*1979*) and *The* LAST TEMPTATION OF CHRIST (*1988*).

In some plaintive ironic fantasies, including *A Second*

Coming (**1900**) by Richard Marsh, Christ's return is unrecognized and unwelcome, but his viewpoint is more often tacit than incarnate, as in *If Christ Came to Chicago* (**1894**) by W.T. Stead (1849-1912). Victims of DELUSION occasionally think they are he, as in *They Call me Carpenter* (**1922**) by Upton SINCLAIR. Sceptical writers sometimes offer uncharitable accounts of his career and its climax; examples are *My First Two Thousand Years* (**1928**) by George Sylvester VIERECK and Paul Eldridge and *Live from Golgotha* (**1992**) by Gore Vidal. He is reconstructed by Stanley ELKIN in *The Living End* (**1979**). Philip José FARMER's *Jesus on Mars* (**1979**) discovers him in exotic surroundings, while James MORROW's *Only Begotten Daughter* (**1990**) finds him pursuing a mission of mercy in HELL. [BS]

CHRISTIAN FANTASY The Christian faith rapidly accumulated a rich FOLKLORE which thrived in oral culture until written down in such documents as *The Golden Legend*, a 13th-century collection of the lives of saints made by Jacques de Voragine. Many such tales served an important inspirational purpose, often absorbing pre-existent folklore so that their weight could be added to the Christian cause; medieval miracle plays, performed alongside the mystery plays which popularized the scriptures, helped to preserve this heritage. The confusion of British Arthurian romance (◊ ARTHUR) with GRAIL romances imported by the Normans provides a good example of this omnivorous reprocessing. The Reformation sent a sceptical wind gusting through this assemblage, but could not prevent its further growth; S. BARING-GOULD's *Curious Myths of the Middle Ages* (coll **1866**) testifies to its continued elaboration. Those religious writers who were fully conscious of the fact that they were writing fantasies excused their work as ALLEGORY or as propaganda.

The most notable landmark in the early history of CF is DANTE's *Divine Comedy* (written *c*1307-21); the most important precedents in English literature were set by *Paradise Lost* (**1667**; rev **1674**) by John MILTON and *The Pilgrim's Progress* (**1678**; **1684**) by John BUNYAN (◊◊ PILGRIM'S PROGRESS). William BLAKE commented that Milton was "of the Devil's party without knowing it", thus ushering in a new era in which fantastic literature was used as a vehicle for calculated heresies attacking the worldview of the Church. Blake gave new names to GOD and his quasi-Satanic redeemer in his "prophetic books", while Percy Bysshe SHELLEY, who amplified Blake's comment on Milton in his "Defence of Poetry", employed Classical substitutes in such works as "Prometheus Unbound" (**1820**). A tradition of explicit Literary Satanism (◊ SATAN) was eventually founded by Anatole FRANCE, extrapolating from precedents set by Gustave FLAUBERT. Christian writers, not unnaturally, regarded this trend with alarm and horror, and many took it upon themselves to write compensatory fantasies, often involving ANGELS as MIRACLE-working agents of divine intervention. Writers like Laurence HOUSMAN, however, found some difficulty in compiling orthodox CFs; his *All-Fellows* (coll **1896**) and *The Cloak of Friendship* (coll **1905**) are among the best examples of their kind, but their composition led him to the conclusion that "miracles make good fairytales but bad theology". It is noticeable that stern dogmatists like Guy Thorne (1876-1923) usually produced literary works of negligible value, while earnestly uneasy writers like George MACDONALD and T.F. POWYS – who used fantasy as a means of exploring their own painful doubts – sometimes produced masterpieces. The most

effective propaganda was probably that provided by devout writers clever enough and ambitious enough to venture into previously unexplored allegorical territory, especially G.K. CHESTERTON and C.S. LEWIS. Lewis also deserves credit for his cunning in turning the nascent tradition of Literary Satanism back on itself in *The Screwtape Letters* (**1942**), while his friend Charles WILLIAMS brilliantly – if somewhat esoterically – broke new ground in reformatting his allegories as suspenseful thrillers.

For reasons of reverence, CHRIST rarely makes personal appearances in devout CF. Perhaps ironically, such opposed characters as Herod (◊◊ SALOME), Judas and the ANTICHRIST are more often featured, by virtue of their innate melodramatic potential; none, however, is as conspicuous in the literary heritage as the deeply ambiguous figure of the WANDERING JEW. Where agents of Christian virtue are called upon to do more than sadly observe the failures of modern humanity – as Mary memorably does in Upton SINCLAIR's *Our Lady* (**1938**) – they are usually humbler in stature.

Although most of its source materials have been dissolved in the syncretic sea of imagery deployed by US genre fantasists, and many of its motifs are most blatantly paraded in calculatedly blasphemous satires like *Live From Golgotha* (**1992**) by Gore Vidal (1925-), CF has not been entirely relegated to the purist backwaters of inspirational literature. It remains a significant component of popular HORROR, exemplified by such works as *The Case Against Satan* (**1962**) by Ray Russell (1934-) and William Peter BLATTY's *The Exorcist* (**1971**), allegedly written as a gesture of thanks to the Jesuits. The allegorical tradition continues in such works as Walter WANGERIN's *The Book of the Dun Cow* (**1978**), while the Dantean tradition is thoughtfully carried forward in such works as *Chariot of Fire* (**1977**) by E.E.Y. Hales (1908-). [BS].

Further reading: *Christian Fantasy, Twelve Hundred to the Present* (**1992**) by Colin Manlove.

See also: WHISTLE DOWN THE WIND (*1961*).

CHRISTINE US movie (*1983*). Columbia/Polar Film. **Pr** Richard Kobritz. **Exec pr** Kirby McCauley, Mark Tarlov. **Dir** John Carpenter. **Spfx sv** Roy Arbogast. **Screenplay** Bill Phillips. **Based on** *Christine* (**1983**) by Stephen KING. **Starring** Roberts Blossom (LeBay), Keith Gordon (Arnie), William Ostrander (Buddy), Alexandra Paul (Leigh), Robert Prosky (Darnell), Harry Dean Stanton (Jenkins), John Stockwell (Dennis). 110 mins. Colour.

Christine is a CAR possessed (◊ POSSESSION) by an evil SPIRIT, and since manufacture in 1957 has been causing misery and death. In 1978 highschool wimp Arnie – bullied, like the eponymous CARRIE, by peer group and mother – falls in love with the hulk and rebuilds it. Her attainment of new splendour is matched by his own: he becomes the boy all the girls want to date (although he is not pleasant: he too is possessed). One girl, Leigh, is almost killed by a jealous Christine. A gang led by bully Buddy trash the car, but it swiftly repairs itself and kills them. At last Leigh and nice boy Dennis succeed in destroying Christine in a crusher, although Arnie is killed too. *C* is an interesting TECHNO-FANTASY about the double possession; also, as is made explicit in Dennis's final confrontation with the MONSTER, it is about a HERO battling a DRAGON for the hand of a fair princess. [JG]

CHRISTMAS The winter solstice – December 21/22 in the Northern Hemisphere – has long been a time of celebration, welcoming the New Year. The Romans fixed on December

25 as the *Sol Invictus* ("unconquered Sun"), representative of their own unconquerability. It was also the time of their Saturnalia (◊ REVEL), the festival of Saturn, a precursor of SANTA CLAUS. The Nordic tribes celebrated this same period as a 12-day festival called Yule: from this came the concept of Twelfth Night, which marked the end of the celebrations. Twelfth Night was also the time in Italian tradition when the FAIRY Befana rewarded good children with gifts, a practice later grafted onto the Santa Claus myth.

The various features of the Christmas story – the virgin birth (◊ VIRGINITY), the Star of Bethlehem, the ANGELS heralding the birth, the Three MAGI – all feature in the imagery of the fantastic, and have frequently been remoulded in fantasy fiction. In the Northern Hemisphere the solstice is the turning of the year: the landscape is frequently swept clear of all apparent life, before life is reborn. Christmas thus has connections to legends of the GREEN MAN and the RESURRECTION. It is no accident that the BEHEADING MATCH in which GAWAIN partook with the Green Knight took place at Christmas.

The Santa Claus myth is one of the earliest examples of make-believe discovered by many children. This relation between Christmas and fantasy was underscored by the links between Christmas and FAIRYTALES (adapted as PANTOMIMES from the start of the 19th century) and GHOST STORIES. The latter connection was largely established by Charles DICKENS, whose *A Christmas Carol* (**1843**) fixed the Christmas fantasy in Victorian England soon after Victoria's marriage (1840) to Albert, who brought with him to the UK the Germanic beliefs in the Christmas celebration. The first Christmas card was issued in 1843. Within a century the commercialization of Christmas had long suffocated the Christian message, so that the festival was once more largely pagan, a fact recognized in Stella BENSON's biting SATIRE *Christmas Formula* (**1932** chap).

Movie fantasy is big business at Christmas (◊ CINEMA), with several movies and tv features – e.g., *The Snowman* (*1982* tvm) (◊ Raymond BRIGGS), IT'S A WONDERFUL LIFE (*1946*), MIRACLE ON 34TH STREET (*1947*) and various versions of *A CHRISTMAS CAROL* – returning perennially. [MA]

See also: ANNUALS; ANTHOLOGIES; CHRISTMAS BOOKS.

CHRISTMAS BOOKS There have been Christmas ANNUALS in the UK since the turn of the 18th century, and they continue to be produced. Most take the form of ANTHOLOGIES, in which visual material, fiction and nonfiction are mixed; some of the illustrations and fiction are of course fantasy (◊ CHRISTMAS). But the CB proper, as invented by Charles DICKENS with *A Christmas Carol in Prose: Being a Ghost Story of Christmas* (**1843**), is an independent story, bound as a book, published at the Christmas season, and usually of novella length (whether or not they are actually read aloud, CBs have almost invariably been designed to look as though they could be).

The few CBs of lasting importance are FANTASY or SUPERNATURAL FICTION, and are central to any understanding of the shaping of the form through the 19th century; they include tales such as Dickens's own *The Chimes* (dated 1845 but **1844**) and *The Haunted Man and the Ghost's Bargain* (**1848**), John RUSKIN's *The King of the Golden River* (**1851** chap) and William Makepeace THACKERAY's *The Rose and the Ring* (**1855**) as by M.A. Titmarsh. In texts of this sort, the CB variously homaged a SEASON and a consolatory rite, both of which served to surround and to put into a transcending, fantasy-like context the supernatural-fiction

elements so commonly found in early-19th-century fiction. *A Christmas Carol*, the most important of all, demonstrates this clearly. It is technically a GHOST STORY, but – because its GHOSTS have been introduced into the tale to conduct Ebenezer Scrooge through a harrowing NIGHT JOURNEY into the dawn of transformed new life, and because the scenes that follow his TRANSFORMATION are narrated in a manner both heightened and exemplary – the tale, in retrospect, feels as though it is defined by its comedic ending. (It is also left open to us to interpret the text as a fantasy of PERCEPTION.)

When the CB introduces an element of REVEL – dangerous at any period, and particularly unlikely in Victorian times to be "permitted" to turn the world genuinely TOPSY-TURVY – it does so as a prelude to restoration: this too is typical of the movement of the fantasy text. It is moreover the case that the plot of the classic CB almost invariably serves to restore to the world a just and ordered dispensation (◊ THEODICY).

This just and ordered life and landscape – because even in 1840 it was ineradicably nostalgic – helped the development of the mature fantasy vision through the 19th century: during that period of convulsive change, and ever since, fantasy has increasingly treated the ideal society and landscape as something that has been lost (◊ BELATEDNESS).

The CB itself did not flourish as a form, because huge commercial pressures insisted that authors should generate a succession of sentimentalized visions of a world in which – in order to be a good person – it was simply necessary to *wish* to be one; but, whenever transformations are cheap, fantasy becomes rote, and so the CB declined. After the death of Prince Albert in 1861, which effectively terminated the unison song of Christmas as a theodicy of Empire, special Christmas issues of continuing MAGAZINES generally took over from the CB, though occasional examples of the form – several novels by Tom Gallon (1866-1914), including *The Charity Ghost* (**1902**), *The Man who Knew Better* (**1902**) and *Christmas at Poverty Castle* (**1907**), and Marie CORELLI's *The Strange Visitation of Josiah McNason* **1904** chap) – continued to appear, albeit without much success.

Towards the end of his greatest success as a producer of illustrated Christmas gift books – a series which capitalized on the spirit of the Christmas Book, though it significantly used only old stories – Arthur RACKHAM issued a version of Dickens's *A Christmas Carol* (**1915**). The strains of WORLD WAR I – which proved terminal to so much theodicy – occasioned in Rackham a new, nostalgic sentimentality. It was not a good omen for Rackham – whose versions of *The Chimes* (**1931** US) and Ruskin's *The King of the Golden River* (**1932**) were feeble – and an extremely bad one for the CB.

In recent years, however, it has become a reasonably frequent practice for publishers to issue what might be called special Christmas editions of existing seasonally relevant works. *A Christmas Carol* has been thus treated several times within the past decade, most recently in 1995; another example is *A Child's Christmas in Wales* by Dylan THOMAS. Moreover, works like Raymond BRIGGS's *Father Christmas* (graph **1973**) and *The Snowman* (graph **1978**) are genuinely CBs. However, the role of the CB has now almost entirely been taken over by the movies (◊ CINEMA), and these days it is a rare Christmas that does not see the release of at least one high-budget, explicitly seasonally oriented movie, usually fantasy and quite often very cynical, as with *Santa Claus: The Movie* (*1984*). A recent example of the "Christmas Movie" is the remake MIRACLE ON 34TH STREET (*1994*). [JC]

CHRISTMAS CAROL, A At least 14 movies have been based on Dickens's tale.

   1. *Scrooge* UK movie (*1935*). Twickenham. **Pr** John Brahm, Julius Hagen. **Dir** Henry Edwards. **Screenplay** Seymour Hicks, H. Fowler Mear. **Starring** Oscar Asche, Donald Calthrop, Barbara Everest, Maurice Evans, Hicks (Scrooge), Athene Seyler. 79 mins. B/w.

   The preferred version until the release of **3**. Hicks was an impresario as well as a comic actor, and was apparently – we have been unable to trace a viewing copy of this movie – an excellent Scrooge. [JG]

   2. *A Christmas Carol* US movie (*1938*). MGM. **Pr** Joseph L. Mankiewicz. **Dir** Edwin L. Marin. **Screenplay** Hugo Butler. **Starring** Lionel Braham (Christmas Present), Leo G. Carroll (Marley's Ghost), Lynne Carver (Bess), D'Arcy Corrigan (Christmas Future), Terry Kilburn (Tiny Tim), Gene Lockhart (Bob Crachit [*sic*]), Kathleen Lockhart (Mrs Crachit), Barry Mackay (Fred), Reginald Owen (Scrooge), Ann Rutherford (Christmas Past), Ronald Sinclair (young Scrooge). 69 mins. Colour.

   A reasonably faithful version although, seemingly in the interests of economy, some incidents are cut and there is an interminable cheap-to-make set-piece at the Cratchit Christmas dinner. The Cratchit children vie with their English accents and lose, while Owen as Scrooge seems more eager to be a crusty but rather likeable curmudgeon than a genuine VILLAIN. The role of nephew Fred is played up at the expense of Scrooge's SPIRIT-guided NIGHT JOURNEY, which thus becomes almost a subplot; indeed, there is a strong implication that the ghostly encounters are merely ILLUSION or DREAM. The overall effect is rather like a rushed amateur-dramatics production. [JG]

   3. *Scrooge* UK movie (*1951*; vt *A Christmas Carol*). George Minter/Renown. **Pr** Brian Desmond-Hurst. **Dir** Desmond-Hurst. **Phot dir** C. Pennington-Richards. **Screenplay** Noel Langley. **Starring** George Cole (Young Scrooge), Michael Hordern (Marley's Ghost), Patrick Macnee (Young Marley), Miles Malleson (Joe the ragpicker), Alistair Sim (Scrooge), Jack Warner. 86 mins. B/w (a tinted version has been released on video).

   Seemingly made to a restricted budget, this often scantly decorated version plays upon its limitations to convert them into strengths: the austerity of effect and visual starkness are matched by exquisite direction and shot-construction (with superb exploitation of light and shadow) to underline the fundamental harshness and potential terror of Dickens's story. The sets are spare and, among the actors, there is barely a wasted extra; thus full focus is given to the performances of the central characters, among whom Sim and Hordern, in two very different ways, are quite outstanding: there is real *power* in the chilling encounter between Scrooge and Marley's GHOST. Sim, reminiscent here in both performance and mannerism of the sorcerer-scientist Rotwang in Fritz Lang's classic *Metropolis* (*1926*), provides the performance from which all else hangs: juggling pathos and, oddly, charisma in a sophisticated yet traditionally theatrical fashion, he portrays an intriguingly Scottish Scrooge. Aside from a subplot concerning Scrooge's and Marley's early nefarious business dealings, the adaptation is superficially rather faithful to the original, yet draws out a psychoanalytic interpretation of Scrooge's character; also stressed is the malleability of the as-yet-unwritten future REALITY. Short and pacy, this may be the finest *Christmas Carol* of all. [JG]

   4. *Scrooge* UK movie (*1970*). Cinema Center/ Waterbury. **Pr** Robert H. Solo. **Exec pr** Leslie Bricusse. **Dir** Ronald Neame. **Spfx** Wally Veevers. **Screenplay** Bricusse. **Based on** the stage musical with music and lyrics by Bricusse. **Starring** Richard Beaumont (Tiny Tim), David Collings (Bob Cratchit), Frances Cuka (Mrs Cratchit), Edith Evans (Christmas Past), Albert Finney (Scrooge), Alec Guinness (Marley), Michael Medwin (Scrooge's nephew Harry), Kenneth More (Christmas Present), Suzanne Neve (Isobel), Anton Rodgers (Tom Jenkins), Paddy Stone (Christmas Yet to Come). 113 mins. Colour.

   A fairly faithful rendition, distinguished by an exceptional performance from Finney that conjures genuine sentiment and sparks sympathy for even the unreformed Scrooge. Guinness's camp – perhaps too camp – Marley entertains, but Evans and More are surprisingly flat; More's woodenness, in particular, contrasts oddly with his physical presentation as something akin to a PAN-like FERTILITY god. Indeed, the "message" of the movie – spelt out explicitly in the later stages, as a formal choir is persuaded to forsake religious carols in favour of a knockabout roister – is that it is the pagan vision of CHRISTMAS that should be celebrated. The songs are generally weak. [JG]

   5. *A Christmas Carol* US ANIMATED MOVIE (*1970* tvm). **Voice actors** C. Duncan, Ron Haddrick, John Llewellyn, T. Mangan, Bruce Montague, Brenda Senders. *c*60 mins. Colour.

   By all accounts this production, a special episode of the **Famous Classic Tales** series, was modest. [JG]

   6. *Charles Dickens' A Christmas Carol, Being a Ghost Story of Christmas* UK ANIMATED MOVIE (*1971* tvm). A Richard WILLIAMS Production. **Pr** Williams. **Exec pr** Chuck JONES. **Dir** Williams. **Master animator** Ken Harris. **Screenplay** uncredited. **Voice actors** Felix Felton (Christmas Present), Melvin Hayes (Cratchit), Michael Hordern (Marley's ghost), Diana Quick (Christmas Past), Michael Redgrave (Narrator), Alistair Sim (Scrooge), Joan Sims (Mrs Cratchit), Paul Whitsun-Jones (Fezziwig/Joe), Alexander Williams (Tiny Tim). *c*25 mins. Colour.

   This short movie is as beautiful to watch as one might expect from Williams. Because of its small compass, however, the tale is stripped back to the bones, almost as if it were a trailer for a full-length feature; this occasionally works to the telling's advantage – some of the transitions involving the SPIRITS gain considerable effect through their abruptness – but more often, notably in the closing minutes, it gives matters an air of perfunctoriness. Sim and Hordern recreate their roles in **3** as Scrooge and Marley's ghost respectively; the verbal and visual similarity is maintained in the case of Marley, but Sim provides an excellent reinterpretation of his part, and visually Scrooge more resembles Reginald Owen's depiction in **2**. [JG]

   7. *A Christmas Carol* UK movie (*1977* tvm). BBC/Time Life. **Pr** Jonathan Powell. **Dir** Moira Armstrong. **Screenplay** Elaine Morgan. **Starring** Timothy Chasin (Tiny Tim), Paul Copley (Fred), Michael Hordern (Scrooge), Bernard Lee (Christmas Present), John Le Mesurier (Marley), Carol MacReady (Mrs Cratchit), Clive Merrison (Bob Cratchit), Patricia Quinn (Christmas Past). 58 mins. Colour.

   Having played the part of Marley's GHOST twice – in **3** and **6** – Hordern finally took over the principal role in this fairly faithful adaptation, lushly if stagily presented. Budget restrictions led to some odd manoeuvres, notably the representation of several scenes by drawings/paintings, with actors commenting in front of them; nevertheless, the modest spfx are good. Le Mesurier plays Marley's ghost well and Hordern is a delightfully fidgety Scrooge. MacReady excels, making Mrs Cratchit's role surprisingly central; but Merrison's Cratchit – resembling a nob talking common – and Copley's Fred – oddly blokish – grate. For its casting, this is a surprisingly lightweight version. [JG]

8. *A Christmas Carol* Australian ANIMATED MOVIE (*1979*). 72 mins. Colour.

By all accounts this version is modest. [JG]

9. *Mickey's Christmas Carol* US ANIMATED MOVIE (*1983*). Disney. **Dir** Burny Mattinson. **Screenplay** Mattinson. **Voice actors** Wayne Allwine (Bob Cratchit/Mickey Mouse), Eddie Carroll (Christmas Past/Jiminy Cricket), Clarence Nash (Fred/Donald Duck), Pat Parris (Isabel/Daisy Duck), Will Ryan (Christmas Present/Willie the Giant, Christmas Yet to Come/Pete), Susan Sheridan (Tiny Tim/Morty Mouse), Hal Smith (Marley/Goofy), Alan Young (Scrooge/Scrooge McDuck). 26 mins. Colour.

The favourite DISNEY animated characters feature in a very loosely interpreted version, with SCROOGE MCDUCK as Scrooge, MICKEY MOUSE as Cratchit, plus Daisy and DONALD DUCK, etc. The GHOSTS – Jiminy Cricket as Christmas Past, Willie the Giant as Christmas Present, and Pegleg Pete as Christmas Yet to Come – are excellent (notably Pete), but Goofy as Marley's Ghost upstages all. This revival of Mickey's career reportedly cost over $3 million. [JG]

10. *A Christmas Carol* US tvm, also released theatrically (*1984*). Entertainment Partners. **Pr** Alfred R. Kelman, William F. Storke. **Exec pr** Robert E. Fuisz. **Dir** Clive Donner. **Spfx** Martin Gutteridge, Graham Longhurst. **Screenplay** Roger O. Hirson. **Starring** Nigel Davenport (Silas Scrooge), Frank Finlay (Marley), Lucy Gutteridge (Belle), Angela Pleasence (Christmas Past), Roger Rees (Fred Holywell), George C. Scott (Scrooge), Mark Strickson (Young Scrooge), Anthony Walters (Tiny Tim), David Warner (Cratchit), Edward Woodward (Christmas Present), Susannah York (Mrs Cratchit). 100 mins. Colour.

With its all-star cast and its exquisite production values, this version is in constant danger of collapsing into self-conscious lavishness at the expense of involvement and raw sentiment, a danger it does not always avoid: there is a lack of the necessary squalor. The adaptation is faithful to the point of over-deliberation, with some small and not over-subtle additions made for the sake of "contemporary relevance". Overall this version is too conscious of the "classic" status of Dickens's tale. Yet it has considerable virtues, and electrifies during the sequences in which the ghosts of Christmases Past, Present and Future participate: Pleasence admirably conjures combined youth and maturity, and Woodward's hamminess so effectively matches his role that this is probably the definitive portrayal of Christmas Present. There is one very interesting contrast between this and other versions: where they derive much strength from the use of SHADOWS, this relies – to great effect – on *light*, even during the sequences with Christmas Yet to Come, here portrayed most often as merely an ominously cast and brilliantly contrasted shadow. [JG]

11. *A Christmas Carol* US ANIMATED MOVIE (*1984* tvm). Burbank Films. **Voice actors** Barbara Frawley, Ron Haddrick, Philip Hinton, Sean Hinton, Robin Stewart. *c*80 mins. Colour.

By all accounts this is unpretentious but rather good. [JG]

12. *Scrooged* US movie (*1988*). Paramount. **Pr** Richard Donner, Art Linson. **Dir** Donner. **Spfx** Steven Foster, Gary I. Karas, Joe Montenegro. **Mufx** Thomas R. Burman, Bari Dreiband-Burman. **Vfx** Eric Brevig, Dream Quest Images. **Screenplay** Mitch Glazer, Michael O'Donoghue. **Starring** Karen Allen (Claire), John Forsythe (Hayward), David Johansen (Ghost of Christmas Past), Carol Kane (Ghost of Christmas Present), Bill Murray (Frank), Alfre Woodard (Grace). 101 mins. Colour.

The tale recast into the late 20th century. Frank Cross, ruthless tv executive, is masterminding his company's live all-star CHRISTMAS production, *Scrooge*. Sacking a deputy who dares argue with him invokes the familiar procession of GHOSTS: his ex-boss, Lew Hayward, in the Marley role; Christmas Past, a cab-driver (and virtually a double for Christmas Yet to Come in **9**); Christmas Present, a brilliantly, unpredictably violent FAIRY GODMOTHER; and Christmas Yet to Come, a dark, hooded, giant figure. The show is almost stolen by Alfre Woodard as a female Cratchit – Cross's secretary – with a tiny disabled son. The spfx are excellent, the imaginative content surprisingly high (Christmas Past and Present are as good as any ghosts to come out of Hollywood, and *S* has many fine touches of TECHNOFANTASY) and the entertainment rich and witty, despite an overlong finale as Frank invades the *Scrooge* set and renounces, worldwide, his yuppie lifestyle. [JG]

13. *The Muppet Christmas Carol* US puppet/live-action/ANIMATED MOVIE (*1992*). DISNEY/Jim HENSON Productions/Buena Vista. **Pr** Martin G. Baker, Brian Henson. **Exec pr** Frank Oz. **Dir** Brian Henson. **Spfx** David Harris. **Vfx** Paul Gentry. **Miniatures sv** David Sharp. **Screenplay** Jerry Juhl. **Starring** Meredith Braun (Belle), Michael Caine (Scrooge), Steven Mackintosh (Fred), Robin Weaver (Clara). **Voice actors** Donald Austen (Christmas Present/Christmas Yet to Come), Jessica Fox (Christmas Past), Dave Goelz (Gonzo/Robert Marley/Bunsen Honeydew/Betina [*sic*] Cratchit), William Todd Jones (Christmas Past), Jerry Nelson (Tiny Tim/Jacob Marley/Ma Bear/Christmas Present), Frank Oz (Miss Piggy/Fozziwig/Sam Eagle/Animal), Karen Prell (Christmas Past), David Rudman (Peter Cratchit/Old Joe/Swedish Chef), Rod Tygner (Christmas Past/Christmas Yet to Come), Steve Whitmire (Kermit/Rizzo the Rat/Bean Bunny/Belinda Cratchit). 86 mins. Colour.

An exceptionally charming version, almost like a PANTOMIME, this sticks, despite many Muppetish embellishments, surprisingly closely to Dickens's original, the sole major difference being that Scrooge is haunted by the ghosts of two Marleys (brothers Robert and Jacob), who come to gloat rather than warn. The spfx are seamless, Caine is surprisingly good and, among the GHOSTS, that of the childlike Christmas Past is outstanding. All the usual MUPPETS feature, with Kermit as Bob Cratchit, Miss Piggy as Emily Cratchit and the Great Gonzo as Dickens, narrating the story to Rizzo. *Rizzo*: "Whew, that's scary stuff! Hey, should we be worried about the kids in the audience?" *Dickens*: "Nah, it's all right – this is *culture*." [JG]

14. *Brer Rabbit's Christmas Carol* US ANIMATED MOVIE (*1992* tvm). INI Entertainment/Magic Shadows. **Pr** Al Guest, Jean Mathieson. **Dir** Guest, Mathieson. **Screenplay** Guest, Mathieson, William Mathieson. 58 mins. Colour.

Remotely related to Joel Chandler HARRIS's Brer Rabbit, although the character's TRICKSTER elements are clumsily deployed. At CHRISTMAS the mean Brer Fox (*à la* Scrooge; Brer Bear is his Cratchit) refuses to give to the poor. [JG]

CHRISTOPHER, JOHN Working name of UK writer Christopher Samuel Youd (1922-) whose first novel, *The Winter Swan* (**1949**), one of his rare fantasy tales, was published as by Christopher Youd; told in retrograde steps, it employs TIMESLIP techniques to convey the complexity of several intertwined lives. Most of JC's work has been sf or nonfantastic, the latter books being written as by Hilary Ford, Peter Graaf and Samuel Youd. Later fantasy tales include *The Little People* (**1966**), in which an old castle turns out to be a BAD PLACE through which echoes of WORLD WAR II percolate, specifically evoking "medical" experiments in a concentration camp. The **Prince in Waiting** sequence – *The Prince in Waiting* (**1970**), *Beyond the Burning Lands* (**1971**) and *The Sword of the Spirits* (**1972**) – presents what seems to be a fantasy version of post-HOLOCAUST England, though the supernatural "Spirits" who rule this world turn out to have an sf explanation. *A Dusk of Demons* (**1993**), despite its title, is an sf tale set in a similarly devastated UK. [JC]

CHRONICLES OF NARNIA, THE UK tv serial (**1988-90**). BBC. **Pr** Paul Stone. **Dir** Marilyn Fox (first sequence), Alex Kirby (remainder). **Writer** Alan Seymour. **Based on** C.S. LEWIS's **The Chronicles of Narnia** (**1950-56**). **Starring** Sophie Cook (Susan), Richard Dempsey (Peter), Barbara Kellermann (White Witch), Jonathan Scott (Edmund), Sophie Wilcox (Lucy). *The Lion, the Witch and the Wardrobe* 6 30min episodes; *Prince Caspian* 2 30min episodes; *Voyage of the Dawn Treader* 4 30min episodes; *The Silver Chair* 6 30min episodes. Colour.

Presented in three six-episode series, this was one of BBC Television's major excursions into fantasy, with very high production values and a loving and painstaking – in places *too* painstaking – attention to quality throughout. The adaptation was reasonably faithful to the books. Somewhat misjudging its target audience, BBC broadcast the serial during the early evening, in the children's prime-time slot; the video set has been more sensibly marketed. [JG]

See also: *The* LION, THE WITCH AND THE WARDROBE (1967).

CIBOLA ◊ LOST LANDS AND CONTINENTS.

CIENCIN, (MALCOLM) SCOTT (1962-) US writer who works under his own name, under the **Forgotten Realms** housename Richard Awlinson, and as Nick Baron. He began publishing, as Awlinson, with the first two of the **Avatar Trilogy**, *Shadowdale* * (**1989**), and *Tantras* * (**1989**); the third, *Waterdeep* * (**1989**), was by Troy DENNING as Awlinson. As Nick Baron, SC wrote two ties for **Robert Silverberg's Time Tours** – *Glory's End* * (**1991**) and *The Pirate Paradox* * (**1992**), the latter with Greg COX – and four YA horror novels for the **Nightmare Club** sequence: *The Initiation* * (**1993**) and *The Mask* * (**1993**), about a mask (◊ MASKS) which transforms a young girl into a seductress, *Spring Break* * (**1994**) and *Virtual Destruction* * (**1995**). Under his own name, SC published a **Forgotten Realms** tie, *The Night Parade* * (**1992**), being volume 4 of the **Harpers** sequence, the **Wolves of Autumn** sequence – *Wolves of Autumn* (**1992**) and *The Lotus and the Rose* (**1993**) – and the **Vampire Odyssey** sequence – *The Vampire Odyssey* (**1992**), *The Wildlings* (**1992**) and *Parliament of Blood* (**1992**) – in which a young half-VAMPIRE woman and her mother band together to fight off the undead, and (it seems) do some good. [JC]

CINDERELLA Andrew LANG once suggested that the story of Cinderella could be understood only by races whose members wore shoes. It was a 19th-century remark. The appeal of her tale (as we know) is not tied to attire, no matter how resonantly symbolic it may be (in some versions she is identified by a RING). Over the 1000 or more years it has been told, the heart of the story has remained unmodified: a young girl loses her mother and is stripped of her name and position; in her new debased role (◊ DEBASEMENT) – though never during the time she is dressed in a deceitful GLAMOUR of fabulous clothing – she is recognized for who she truly is, and is redeemed.

The first written version (*c*850-60) comes from China, and was best recorded by Arthur WALEY in *Folk-Lore*, in 1947. Young Yeh-hsien's mother dies; her father takes (or promotes) another to fill her place; the new mother treats her badly, arranging for her own daughters to supplant her. In her degradation, Yeh-hsien is consoled by the SPIRIT of her mother, which has entered a fish, which the wicked stepmother soon finds and kills; but Yeh-hsien finds its discarded bones, which now protect her. At the next festival she appears in glittering array and infatuates the local warlord, though she loses a shoe in escaping her vile sisters; the king of a nearby island is sold the shoe, and in his longing to find its owner has all the women of the land try the shoe on, but it fits only Yeh-hsien, who becomes his chief wife. According to Iona and Peter Opie, in *The Classic Fairy Tales* (**1974**), flying stones kill the stepmother and her children; according to Marina WARNER in *From the Beast to the Blonde: On Fairytales and their Tellers* (**1994**), they are stoned to death. Either way, Cinderella comes into her own.

Several hundred versions of the story exist in Europe alone, as demonstrated by Marian Roalfe Cox's *Cinderella: Three Hundred and Forty-Five Variants* (**1893**). In the first to appear in print, "La Gatta Cenerentola" ["The Cat Cinderella"] in the *Pentamerone* (**1634-36**) of Giambattista Basile (1575-1632), young Zezolla murders her stepmother (who has supplanted her mother) but is further humiliated by yet another usurper, whose six daughters lord it over her; in the end, however, the result is the same. The most famous version of the story is that by Charles PERRAULT in the *Histories or Tales of Past Times* (coll **1697**), in which the mother is long dead from the first, and a FAIRY GODMOTHER is introduced, along with the glass slipper and the pumpkin that turns into a coach. It is a highly literary version, and its conceits proved irresistible, not least to Walt DISNEY (◊ CINDERELLA [*1950*]). The underlying story remains the same: a young girl is cast into the dust, where she will remain forever, despite moments of false glory, until a genuine RECOGNITION of her true worth and STORY, which can happen only when she is dressed in ashes and rags.

Because of the universal application of her tale, Cinderella is one of the more effective UNDERLIER figures in modern fantasy; and the felt antiquity of that tale – when properly told – elegantly opens a sense of TIME ABYSS that may deepen and haunt otherwise uninspiring fare. Over and above the numerous TWICE-TOLD and REVISIONIST-FANTASY versions of the story, there are many fantasy novels whose recognition scenes in particular evoke echoes of the dramatic revelation of Cinderella's real place in a renewed world. Novels which echo the plot in more specific ways include Ru EMERSON's *Spell Bound* (**1990**), Nicholas Stuart GRAY's *The Other Cinderella, With Due Acknowledgements to All the Earlier Versions* (**1958**) and David Henry WILSON's *The Coachman Rat* (**1987**). [JC]

CINDERELLA US ANIMATED MOVIE (*1950*). Disney. **Pr sv** Ben Sharpsteen. **Dir** Clyde Geronimi, Wilfred Jackson,

Hamilton Luske. **Special processes** Ub IWERKS. **Based on** the version by Charles PERRAULT. **Voice actors** Eleanor Audley (Lady Tremaine), Lucille Bliss (Anastasia), Verna Felton (Fairy Godmother), June Foray (Lucifer), Betty Lou Gerson (narrator), James Macdonald (Gus, Jaq, Bruno), William Phipps (Prince Charming), Luis Van Rooten (King, Grand Duke), Rhoda Williams (Drizella), Ilene Woods (Cinderella). 74 mins. Colour.

This reasonably faithful comic adaptation – with an added subplot involving a cat and cute mice – is one of the great classic ANIMATED MOVIES, partly through its own merits but perhaps more because of its historical status. In the years since *Bambi* (*1942*), DISNEY had released no full-length animated features, instead concentrating its animation efforts on a rather drab set of compilation movies, which were cheap to make. *C* was thus an emphatic return to form, and its release had almost the same impact as had that of SNOW WHITE AND THE SEVEN DWARFS, back in 1937.

Other movies using the essentials of the tale are a mixed batch. In *Cinderella Jones* (*1946*), a comedy, a girl seeks to wed an intelligent man so that she may claim her inheritance. The musical *Cinderfella* (*1960*) has a male Cinders and was designed as a showcase for Jerry Lewis. *Cinderella* (*1964*) is a Rodgers and Hammerstein musical. *C'era una Volta* (vt *Cinderella – Italian Style*; *1967*) is a surreal comic version. *Cinderella Liberty* (*1974*) sees a sailor falling in love with a prostitute; the reference is to US naval slang for shore-leave, which ends at midnight. *The Slipper and the Rose* (*1976*) is a long, ambitious, sumptuous but wooden musical version (songs by Robert and Richard Sherman). *The Other Cinderella* (vt *Cinderella*; *1977*) is a soft-porn musical version in which a prince seeks the "perfect fit". *Cinderella 2000* (vt *Sex 2000*; *1977*), set in the far future, is a sex-oriented sf adaptation. There are also countless "moral-tale" Hollywood films of the 1940s and 1950s which use the theme without acknowledgement.   [JG]

CINDERELLA vt of *The Other Cinderella* (*1977*). ◊ CINDERELLA (*1950*).

CINDERELLA – ITALIAN STYLE vt of *C'era una Volta* (*1967*). ◊ CINDERELLA (*1950*).

CINDERELLA JONES US movie (*1946*). ◊ CINDERELLA (*1950*).

CINDERELLA LIBERTY US movie (*1974*). ◊ CINDERELLA (*1950*).

CINDERELLA 2000 (vt *Sex 2000*) US movie (*1977*). ◊ CINDERELLA (*1950*).

CINDERFELLA US movie (*1960*). ◊ CINDERELLA (*1950*).

CINEMA The best fantasy movies are just as sophisticated as the best written fantasy, and may sometimes be more so; indeed, there are some fantasy notions (e.g., TOONS) that for reasons obvious or not so obvious made their debut in the movies before being imported into the written corpus. All in all, it is reasonable to argue that, as in no other area except possibly HORROR, fantasy movies and written fantasy represent an integrated genre: they partake of and, more significantly, contribute to the shared CAULDRON OF STORY – the two forms share the same PLAYGROUNDS contemporaneously.

In examining the history of the fantastic cinema it is important to realize that special effects (in this book abbreviated as spfx) are at centre stage. In the very earliest days of the movies – once the original and short-lived wonder of seeing "moving pictures" had worn off (after all, audiences were already used to flickerbooks, shadowgraphs and zoetropes) – moviemakers were keen to show on screen

things that could not be seen in real life. (The very first cinematographic display before an audience occurred on 28 December 1895 in the *salon Indien* of the Grand Café at no. 14 Boulevard des Capucines, Paris. It was by definition an exhibition of spfx – that was its *raison d'être*.) In this respect the early moviemakers were rather like conjurers: they were performance artists intent on dumbfounding their audience – often they personally presented their works, which were usually only a few minutes long – and like conjurers they were intent on keeping the secrets of their tricks; the mechanisms of some of the early spfx have yet to be uncovered.

Spfx, then, drove moviemakers into fantasy. One route was *via* the medium of animation (◊ ANIMATED MOVIES) – which we can briefly regard as being nonstop spfx – while the other was manipulation, in some way or other, of the photographic image to produce the impossible. In the latter context the most important early moviemaker was probably Georges Méliès (1861-1938; ◊ SFE). At a time when most movies – especially "trick" movies – lasted a few minutes at most, he produced *Le Voyage dans la Lune* (*1902*) and *Le Voyage à travers l'Impossible* (*1904*) at then-epic lengths of 21 mins and 30 mins respectively; both involve voyages into outer space. Their appeal was that of the stage illusion: their stories were at best rudimentary, and entirely subservient to the trickery.

This phase of movie history was again short-lived. Audiences became blasé about spfx and began to require them to become a part of rather than the whole of the movies they watched. They wanted movies to be enhanced stage plays; they wanted movies to tell a *story*. And, as technology progressively permitted moviemakers greater scope – notably in terms of length – moviemakers were able to oblige.

Because of the impossibility of giving firm definition to the term FANTASY one cannot sensibly pinpoint the first fantasy movie – that movie when people realized what they were watching was fantasy rather than spfx – but a reasonable candidate must be the nonsupernatural *The Prisoner of Zenda* (*1913*), the ancestor of a string of remakes of the Anthony Hope adventure (◊ RURITANIA). The following year saw an adaptation of Edgar Allan POE's "The Tell-Tale Heart" (1843), *The Avenging Conscience* (*1914*), and the fantasy movie as a distinct subgenre was on its way. *Civilization* (*1916*) was already enough at ease with fantasy to use it in the service of what was intended as a work of pacifist polemic – a mythical country makes war on a neighbour but ceases doing so when a VISION of CHRIST appears on the battlefield – and at about the same time Cecil B. DeMille (1881-1959) waded in with a truly epic (125 mins) version of the JOAN of ARC tale, *Joan the Woman* (*1916*). Two years later came the first of the JUNGLE MOVIES – that subgenre of cinema fantasy in which the African jungle is treated as, in effect, a SECONDARY WORLD – *Tarzan of the Apes* (*1918*); the TARZAN-MOVIE sequence now contains over 90 entrants. *The* CABINET OF DR CALIGARI (*1919*) demonstrated that the interest in fantasy movies extended well beyond the USA, and, with *Leaves from Satan's Book* (*1919*; ◊ Carl DREYER), marked the onset of a peculiarly European school of Expressionist macabre cinema, characterized by angled camera-work and eloquent use of shadow – qualities that were eventually to make their way to the USA in the form of Hollywood *film noir*. In Europe the tradition was swiftly continued in movies like *The* GOLEM (*1920*), *L'Atlantide* (*1921*), *Nosferatu* (*1921*; ◊ DRACULA MOVIES) and, by Fritz Lang (1890-1976), *Doctor Mabuse the Gambler* (*1922*). Most forbidding of all was probably the Swedish moral lesson *Thy Soul Shall Bear Witness* (*1920*), a precursor of much later Hollywood movies like IT'S A WONDERFUL LIFE (*1946*); here a wastrel is "killed" but, on reliving the nadirs of his career, repents even as the sound of DEATH's chariot fills his ears, and is returned to life for a second chance. Fritz Lang and Thea von Harbau (1888-1954) weighed in again with *Destiny* (*1921*), in which Death converses with a young woman.

Meanwhile Hollywood fantasy was exploring two main strands: the macabre – as epitomized by *Dr Jekyll and Mr Hyde* (*1920*; ◊ JEKYLL AND HYDE MOVIES) – and the feel-good: *Sunnyside* (*1919*), made by Charlie Chaplin (1899-1977), centred on a DREAM of a PASTORAL idyll; *Alf's Button* (*1920*) presaged a short series about a soldier who could rub one of his buttons to summon a GENIE (◊ W.A. DARLINGTON); the central characters of *Dream Street* (*1921*) by D.W. Griffith (1875-1948) choose between GOOD AND EVIL through enacting their dreams; and a second version appeared of *The Prisoner of Zenda* (*1922*). Something of this lightheartedness fed back to Europe, where there appeared movies like René CLAIR's *Paris Qui Dort* (*1923*).

In general, however, it is difficult to identify trends in the fantastic cinema during the 1920s (or, for that matter, in any subsequent decade – except perhaps the 1950s – until the mid-1980s to mid-1990s, when a main strand of fantastic cinema devoted itself to exploiting earlier material from various media, including cinema itself): in a sense the moviemakers, on both sides of the Atlantic, saw no firm demarcation between fantasy movies and any other, so fantasy elements could pop up virtually anywhere. During this period virtually any nondocumentary movie was regarded as being an imaginative work first and a depiction of realism only second – this was especially notable in the field of comedy – so moviemakers felt no need to restrain the wilder frontiers of their inventiveness. Traditional legends were exploited to create movies like *The Niebelungen* (*1924*) and FAUST (*1926*). The Bible got the treatment in movies like *Salome* (*1923*), *The Ten Commandments* (*1923*) and *King of Kings* (*1927*). *The* THIEF OF BAGDAD (*1924*) initiated Hollywood's still continuing love affair with ARABIAN FANTASY. PHANTOM OF THE OPERA (*1925*) started another long-lived sequence. Fritz Lang's *Metropolis* (*1926*), although actually a TECHNOFANTASY, gave the SCIENCE-FICTION movie a new gravitas. Soon came *The* HOUSE OF USHER (*1928*) and then *The Mysterious Dr Fu Manchu* (*1929*; ◊ FU MANCHU MOVIES). Also at the end of the 1920s animators like Walt DISNEY were beginning to realize – although they did not yet have the funds or the backing from distributors to do anything about it – that there might be more to ANIMATED MOVIES than mere five-minute knockabout farces.

With the start of the 1930s the output of fantasy movies took an upturn. Sound encouraged remakes, and accordingly *Alf's Button* (*1930*), KISMET (*1930*) – a less ambitious rehash of *The* THIEF OF BAGDAD (*1924*) – and the brilliant Fredric March version of *Doctor Jekyll and Mister Hyde* (*1931*; ◊ JEKYLL AND HYDE MOVIES) were produced. But it would be wrong to regard this period as one of mere consolidation: Luis BUÑUEL, building on the tradition of SURREALISM that had emerged in the European cinema with movies like *Ballet Mécanique* (*1924*), directed *L'Age d'Or* (*1930*); and Hollywood began to tap two rich veins with, on the one

hand, *A* CONNECTICUT YANKEE (*1931*), a fantasy comedy, and on the other *Frankenstein* (*1931*; ◊ FRANKENSTEIN MOVIES), *Dracula* (*1931*; ◊ DRACULA MOVIES) and SVENGALI (*1931*) – not to mention, of course, that **Jekyll and Hyde** remake. Other fantasy-related movies of note during the first half of the 1930s included *The Mummy* (*1932*), *The Old Dark House* (*1932*), *Tarzan the Ape Man* (*1932*) – the first of the celebrated Johnny Weissmuller sequence – VAMPYR (*1932*), *Alice in Wonderland* (*1933*; ◊ ALICE IN WONDERLAND [*1951*]), GABRIEL OVER THE WHITE HOUSE (*1933*), *The* INVISIBLE MAN (*1933*), KING KONG (*1933*), *Tarzan and His Mate* (*1934*) – arguably the best TARZAN MOVIE of all – *The Wandering Jew* (*1933*), *The Bride of Frankenstein* (*1935*; ◊ FRANKENSTEIN MOVIES), DANTE'S INFERNO (*1935*), *The* GHOST GOES WEST (*1935*), MAD LOVE (*1935*), *A* MIDSUMMER NIGHT'S DREAM (*1935*), *The* PASSING OF THE THIRD FLOOR BACK (*1935*), SHE (*1935*) and others too many to list. Indeed virtually all the themes that remain today at the heart of "mainstream" fantasy cinema were established by the end of this industrious half-decade.

The years leading up to the start of WWII saw a great deal of scavenging forays over already well explored territory: the **Tarzan** movies were going rapidly downhill, and the **Frankenstein** and **Dracula** movies matched this, but always a couple of years behind. There were remakes such as *The* GOLEM (*1936*) and, yet again, *The Prisoner of Zenda* (*1937*). SHE (*1935*) had proven that there was mileage to be had out of H. Rider HAGGARD, so the release of *King Solomon's Mines* (*1937*) came as no surprise. But among all this there were some classics: *The* MAN WHO COULD WORK MIRACLES (*1936*), based on H.G. WELLS's story, Frank CAPRA's LOST HORIZON (*1937*), TOPPER (*1937*), based on the Thorne SMITH novel, and *The* WIZARD OF OZ (*1939*), based on L. Frank BAUM's OZ stories – the last of these put paid to any chances for success of the Shirley Temple vehicle *The* BLUE BIRD (*1940*).

But probably the single most important event in fantasy cinema in the late 1930s was the release by DISNEY of SNOW WHITE AND THE SEVEN DWARFS (*1937*). This should not be regarded just as *the* great breakthrough in the history of ANIMATED MOVIES but also as the opening of a door for fantasy movies in general. Through the medium of animation, the wildest fantastications could be presented on the screen; it was incumbent upon the makers of live-action movies to try to follow suit. Once again spfx became of paramount importance: they had to be convincing, and ideally there should be lots of them. Maestros like Willis J. O'Brien and John P. Fulton had shown the way with movies like KING KONG (*1933*) and *The* INVISIBLE MAN (*1933*); now there was a new technological fervour exploring such techniques as stop-motion and double exposure. This did not necessarily make for good movies – *The* THIEF OF BAGHDAD (*1940*), to take an obvious example, is far from a good movie, but it *is* a spectacular one, and some of its spfx are as good as any that would be achieved until the 1980s.

Almost immediately, however, WWII intervened. Europe was in conflict from 1939; this affected Europe's movie industry directly and had almost as great an effect on the US movie industry, whose lucrative European markets abruptly disappeared. Since sophisticated spfx and quality animation generally cost a lot of money, they had largely to be put to one side until the cessation of hostilities. Yet audiences demanded – for reasons of ESCAPISM, if nothing else – that fantasy movies retain the imaginative range spfx had brought. The animators – in this period, so far as features were concerned, the term really refers only to DISNEY – were not able to sustain this: it would be a long time before anyone could afford another PINOCCHIO (*1940*). But live-action movies like *The* DEVIL AND DANIEL WEBSTER (*1941*), HERE COMES MR JORDAN (*1941*), CAT PEOPLE (*1942*), I MARRIED A WITCH (*1942*), BEWITCHED (*1945*), BLITHE SPIRIT (*1945*) and *The* PICTURE OF DORIAN GRAY (*1945*) used different and less expensive means to convey the full frisson of fantasy. That said, there was as always plenty of dross: to pick on one, the comedy HORROR MOVIE *The Vampire's Ghost* (*1945*), while its title tried to touch more than one base, succeeded in being neither comic nor horrifying.

The end of WWII not only meant that more funds (slowly) became available for new cinematic output: it brought an indefinable yet detectable sea-change in the fantasy movie. The prewar noirish tradition was continued with RATIONALIZED FANTASIES like *The* BEAST WITH FIVE FINGERS (*1946*), but these were done better – and in this instance, notably, had better spfx – and more of them at least pretended to a seriousness of purpose; the techniques that Alfred HITCHCOCK had popularized before WWII in the fields of the thriller and PSYCHOLOGICAL THRILLER were absorbed. But other movies appeared which it is hard to conceive being made before WWII: some, like *A* MATTER OF LIFE AND DEATH (*1946*), obviously could not have been, since they represented the introduction of WORLD WAR II itself as a fantasy motif, but in others the difference is one of "feel". In Europe there was Jean COCTEAU's astonishing *La* BELLE ET LA BÊTE (*1946*) while in the USA Frank CAPRA was virtually inventing a new subgenre with IT'S A WONDERFUL LIFE (*1946*). Although initially Capra's achievement was underestimated, MIRACLE ON 34TH STREET (*1947*) continued that particular trend – the reaffirmation of conventional standards through the use of fantasy – which still thrives in the Hollywood of today; others of that period in similar style were *The* GHOST AND MRS MUIR (*1947*), *Miranda* (*1947*) – about a man who establishes a relationship with a MERMAID – and PORTRAIT OF JENNIE (*1948*). Meanwhile, at the other end of the spectrum, JUNGLE MOVIES were undergoing something of a renaissance: not only was the long cycle of TARZAN MOVIES still in full swing, however dreadful they often were, but the BOMBA MOVIES began with *Bomba on Panther Island* (*1946*). It seemed that audiences wanted fantasy either to reassure them or, if it presented violence and adventure, at least to do so in some far distant, imaginary territory – it is hardly surprising that the first of the SINBAD MOVIES, *Sinbad the Sailor* (*1947*), appeared at this time. Yet there was the occasional US movie that went against the escapist or reinforcing tendency; one was *The Boy with Green Hair* (*1948*), which used fantasy to address a very serious subject, the futility of war – a rather courageous movie to make in those (naturally) triumphalist times.

It can sometimes seem the case that an era of movie history rather neatly coincides with a decade. That this is true of the 1950s is unfortunately, in terms of the fantasy movie, no good news, for the decade was marked by a paucity of innovation (the sf movie had a much better time of it). The 1950s certainly started well, with DISNEY making a glorious return to the animated classic tale in the form of CINDERELLA (*1950*), while HARVEY (*1950*), *Pandora and the Flying Dutchman* (*1950*) and ANGELS IN THE OUTFIELD (*1952*) carried on the strand of nostalgic, schmaltzy semi-comedic fantasy that had developed from Capra's *It's a*

Wonderful Life. Disney, still supreme in animation, carried on strongly through much of the decade, although the rot of complacency was beginning to set in by its end; but the average standard of fantasy movies was poor, with series continuations and remakes often seeming desperate in their search for the lowest common denominator through the reduction of potent cinematic myths to crass mundaneness – as evidenced just before the decade's start by *Abbott and Costello Meet Frankenstein* (*1948*; ◊ FRANKENSTEIN MOVIES) and continued well into it by that movie's successors, *Abbott and Costello Meet the Invisible Man* (*1951*; ◊ INVISIBLE MAN), *Abbott and Costello Meet Dr Jekyll and Mr Hyde* (*1953*; ◊ JEKYLL AND HYDE MOVIES) and *Abbott and Costello Meet The Mummy* (*1955*). The FRANCIS MOVIES, about a talking mule (◊ TALKING ANIMALS) explored the lower regions of dimwittedness. And so on. Yet there were a few highlights, although they were more thinly scattered than in the 1940s. *The Man in the White Suit* (*1951*) was a shrewd piece of TECHNOFANTASY that, under a veneer of amiability, warned of the dangers of unplanned progress. *Scrooge* (*1951*; *A* CHRISTMAS CAROL), though its tale had been filmed several times before, is an excellent example of the quality that popular commercial movies can attain. René CLAIR, although he entered his decline fairly early in the decade, was still able to produce *Les* BELLES DE NUIT (*1952*). MONKEY BUSINESS (*1952*) was a good technofantasy, though its makers were clearly more interested in it as a star vehicle. BRIGADOON (*1954*) showed that fantasy musicals had, potentially, a lot to offer – certainly more than *The* SECRET LIFE OF WALTER MITTY (*1947*) had indicated. *Invasion of the Body Snatchers* (*1956*), with brilliant effectiveness, translated a paranoid fantasy theme – POSSESSION – into sf terms. *The* SEVENTH SEAL (*1957*) by Ingmar BERGMAN was one of the finest expressions of what the fantastic cinema could do. DRACULA (*1958*) was a powerful restatement of a modern myth that had been diluted to an almost homoeopathic extreme by the crassness of Hollywood's mindless exploitation; the movie forcefully put HAMMER on the map. *The* FLY (*1958*) demonstrated how good B-movies could be, while TOM THUMB (*1958*) showed how good spfx could be – as did DARBY O'GILL AND THE LITTLE PEOPLE (*1959*) which, despite a slight surfeit of DISNEY syrup, was nevertheless a better movie. Disney also produced one of their finest animated features: SLEEPING BEAUTY (*1959*). But one's overwhelming feeling on scanning any listing of fantasy movies from the 1950s is that the decade was packed out with inferior remakes, second-rate series continuations, piles of unambitious sf B-movies and . . . well, a heck of a lot of FRANCIS MOVIES.

In the movies, as in Western society at large, the 1960s took a while to start swinging: the early years of the decade were indistinguishable from the late 1950s. Throughout the decade Hammer kept plugging away with its horror and other movies, many of which were fantasies and by no means all of which were bad; SHE (*1965*) was a highpoint. DISNEY released ANIMATED MOVIES all through the 1960s but seemed often to forget that it was important to retain the integrity of the STORY upon which the movie was based; the nadir was reached with *The Sword in the Stone* (*1963*), a travesty of T.H. WHITE's vision. They did better with live-action movies, producing lots of adequate fodder – like the first of the HERBIE MOVIES – as well as the hugely successful (and partly animated) MARY POPPINS (*1964*). In general, though, the early years of the 1960s offered little to excite.

Some exceptions were Roger CORMAN's *The* HOUSE OF USHER (*1960*) and others in his Edgar Allan POE cycle, most especially *The* MASQUE OF THE RED DEATH (*1964*). BURN WITCH BURN! (*1961*) was an unusually good B-movie, and *The* INNOCENTS (*1961*) was a chilling reinvention of Henry JAMES's "The Turn of the Screw" (1898); a decade later *The* NIGHTCOMERS (*1971*) would revisit and steamily defantasticate this material. WHISTLE DOWN THE WIND (*1961*) focused on the genesis of fantasy – in this instance CHRISTIAN FANTASY in the minds of children, a theme that would be taken up by some very distinguished movies in this and succeeding decades, like *The* LORD OF THE FLIES (*1963*), *The* SPIRIT OF THE BEEHIVE (*1973*) and CELIA (*1978*). Orson Welles's *The* TRIAL (*1962*) was a significant event. *The* BIRDS (*1963*) saw Hitchcock at his strangest and strongest. 7 FACES OF DR LAO (*1964*) may not have been great art, but it was certainly fun. But perhaps the most interesting development was the growing appreciation in the West of Japanese movies. These had first come to Western notice in the 1950s, with such movies as *The Seven Samurai* (*1954*), but now all sorts not only hit the art-house circuit but were given general release. Some were SUPERNATURAL FICTIONS of a very high order, notably KWAIDAN (*1964*) – based on four tales by Lafcadio HEARN – and *Onibaba* (*1964*), a bizarre medieval HORROR piece that finally devolves into a fantasy of PERCEPTION in which one is left to puzzle the nature of its "ghosts".

The successes of these movies – despite the fact that they were subtitled and often somewhat intimidating because of the alienness of their construction and viewpoint – alerted Western moviemakers, although it was a while before the message sank in, to the fact that at least some audiences were seeking material that was a bit more challenging than, say, the **Dr Goldfoot** movies – two sexist **Bond**/FU MANCHU parodies, *Dr Goldfoot and the Bikini Machine* (*1965*) and *Dr Goldfoot and the Girl Bombs* (*1966*). Similar attractions were probably responsible for the huge success of HAMMER's PREHISTORIC FANTASY, rapidly imitated, ONE MILLION YEARS BC (*1966*). But the youth culture of the later 1960s, open to foreign influences (not merely Japanese but also European) as well as to the SURREALISM that, at least in the UK, was transforming TELEVISION and other comedy (◊ HUMOUR), eventually forced the commercial moviemakers to start releasing some more interesting material. 1967 alone saw ASTERIX THE GAUL (*1967*), BARBARELLA (*1967*), BEDAZZLED (*1967*), *Dr Faustus* (*1967*; ◊ FAUST), *The* FEARLESS VAMPIRE KILLERS, OR PARDON ME, YOUR TEETH ARE IN MY NECK (*1967*), *Playtime* (*1967*) – possibly the most imaginative movie by Jacques Tati (1908-1982) – and QUATERMASS AND THE PIT (*1967*). The year's big flop was, perhaps significantly, an old-style spectacular family musical, DOCTOR DOLITTLE (*1967*); and the following year's similarly targeted CHITTY CHITTY BANG BANG (*1968*) and FINIAN'S RAINBOW (*1968*) fared little better. Ingmar BERGMAN's *The Hour of the Wolf* (*1968*) and Lindsay Anderson's *If . . .* (*1968*) were much more in keeping with the spirit of the times. *The Magus* (*1968*), although it failed to capture the complexity of John FOWLES's original, showed that the effort to bring even the most demanding texts to the screen was commercially worthwhile. Roman POLANSKI had a colossal success with ROSEMARY'S BABY (*1968*), based on the Ira LEVIN novel. The BEATLES-based exercise in psychedelia, YELLOW SUBMARINE (*1968*), drew big audiences. Meanwhile, on the HORROR front, George A. Romero

(1940-   ) was transforming one subgenre with his ZOMBIE MOVIE *Night of the Living Dead* (*1968*). Old-style movies – the remakes and sequels that for so long, in their unambitious way, had propped up the profits of the movie industry, were tending to sink without trace – an example was HAMMER's *The Vengeance of She* (*1968*) – and from about this point the B-movie started its inexorable demise.

The next few years offered thinner pickings, with the accent once more on HORROR MOVIES – some of which in due course influenced fantasy proper. A further influence that should not be ignored was that of the UK director Ken RUSSELL, with such movies as *The* DEVILS (*1970*); his movies are almost all flawed, but his vision opened the eyes of fantasy moviemakers (and fantasy writers) to the notion that there were PLAYGROUNDS yet to be explored. Another equally flawed but equally significant movie of the early 1970s was *Performance* (*1970*), an acid mixture of DECADENCE – for the ideals of the "love generation" had rapidly soured – and violence that, while not in itself fantasy, had much to teach the creators of fantasy. Aside from these, the fantasy cinema of the early 1970s offered fairly standard fare, as if the lessons of the late 1960s had been forgotten. There were a few attempts to capitalize on the success of *Rosemary's Baby*, as might be expected: *The* MEPHISTO WALTZ (*1971*), *The Possession of Joel Delaney* (*1971*), *The Other* (*1972*) – based on Thomas TRYON's much subtler chiller – and finally one that surpassed the original: DON'T LOOK NOW (*1973*). That same year saw the emergence of another movie that spawned imitators: *The* EXORCIST (*1973*), whose thematic offspring included *The* LEGEND OF HELL HOUSE (*1973*), *The Reincarnation of Peter Proud* (*1974*), CARRIE (*1976*) – which started the Stephen KING movie industry – *The* OMEN (*1976*), AUDREY ROSE (*1977*) and EYES OF LAURA MARS (*1978*). One or two were very interesting movies in their own right, and their average standard was reasonable. The melodrama GHOST STORY (*1974*) was an oddball UK entrant to the lists.

But the early 1970s were not all a matter of occult PSYCHOLOGICAL THRILLER and POSSESSION. *Fritz the Cat* (*1971*) was a big hit for the animator Ralph BAKSHI, although it would be a long while yet before he would begin to make his major contributions to movie fantasy. The anarchic musical WILLY WONKA AND THE CHOCOLATE FACTORY (*1971*) belatedly brought Roald DAHL's CHILDREN'S FANTASIES the worldwide recognition they deserved. One of the most foolhardy remakes in movie history was LOST HORIZON (*1972*), recast as a semi-musical. *Slaughterhouse Five* (*1972*) mixed fantasy, sf, social comment, metafiction, TECHNOFANTASY and the surreal in much the same way as had the Kurt Vonnegut (1922-   ) novel on which it was based; *The* FINAL PROGRAMME (*1973*), based on Michael MOORCOCK's novel, was a similar melange. *O Lucky Man* (*1973*) showed Lindsay Anderson at both his best and his most irritating. *The* WICKER MAN (*1973*) offered a fascinating contest between Christianity and paganism, set on a remote ISLAND in a small-scale UTOPIA and challenging us to redefine exactly what we mean by GOOD AND EVIL. *The* SPIRIT OF THE BEEHIVE (*1973*) powerfully recreated the myth of FRANKENSTEIN's MONSTER. *The Phantom of the Paradise* (*1974*; ◊ PHANTOM OF THE OPERA) was unpopular at the time but can now be seen as a pleasing modern version of the FAUST legend. *The* STEPFORD WIVES (*1974*) was a cold TECHNOFANTASY about the negation of the SOUL, and *Young Frankenstein* (*1974*; ◊ FRANKENSTEIN MOVIES) was a fine

PARODY. *Doc Savage, Man of Bronze* (*1975*) was, rightly, a disaster, as was *Jungle Burger* (*1975*), a dire animated PARODY of the TARZAN-MOVIE canon. On the other hand, MONTY PYTHON AND THE HOLY GRAIL (*1975*), the first major cinematic by-product of the successful UK surreal comedy tv series MONTY PYTHON'S FLYING CIRCUS (1969-74) dir Terry GILLIAM, was a significant contribution to the movie-fantasy genre and commented acerbically on the whole MATTER-of-Britain cult. PICNIC AT HANGING ROCK (*1975*) dir Peter WEIR, one of the eeriest and most beautiful fantasy movies, showed how fantasy could be derived from (purportedly) historical events (◊ Joan LINDSAY). *The* ROCKY HORROR PICTURE SHOW (*1975*) echoed the stage event and is still, 20 years later, often screened; audiences treat it less as a movie than as the focus of a costume party – such screenings represent, perhaps, the quintessence of 1990s DECADENCE.

All through this period there was interesting work going on in the field of the TELEVISION movie (tvm). In 1971 there was a very good version of Charles DICKENS's *A* CHRISTMAS CAROL (*1971* tvm). *The* NIGHT STALKER (*1971* tvm) heralded the KOLCHAK: THE NIGHT STALKER tv series (1974-5); KUNG FU (*1972* tvm) was analogous. *The* BORROWERS (*1973* tvm) brought Mary NORTON's WAINSCOT society to the screen. *The Invisible Man* (*1975* tvm; ◊ INVISIBLE MAN MOVIES) was a respectable attempt. *Spiderman* (*1977* tvm) was a successful pilot (but a less successful movie) based on the COMICS character, and the same might be said of *The Incredible Hulk* (*1970* tvm), *The Man from Atlantis* (*1977* tvm) and *Captain America* (*1979* tvm). *The* LION, THE WITCH AND THE WARDROBE (*1978* tvm) was a bad US ANIMATED MOVIE.

And what of the big screen? *Eraserhead* (*1976*) was the first feature from David LYNCH, who would assume fantasy importance in the following decade. KING KONG (*1976*) was regarded as a shoddy imitation, although it is not poor. *Through the Looking Glass* (*1976*) announced that PORNOGRAPHIC FANTASY MOVIES need not be entirely awful, although almost all of its successors have been. ALLEGRO NON TROPPO (*1977*) was a brilliant PARODY of DISNEY's FANTASIA (*1940*) and a significant contribution to the art of the ANIMATED MOVIE. *The* CAR (*1977*), about a mindlessly demonic automobile, represented an austere extreme of TECHNOFANTASY. FREAKY FRIDAY (*1977*) was an early example of the craze for IDENTITY-EXCHANGE movies that would reach its peak 10 years or so later. JABBERWOCKY (*1977*) saw Terry GILLIAM break out of the straitjacket of his animations for MONTY PYTHON'S FLYING CIRCUS. *The Mouse and His Child* (*1977*) – based on the tale by Russell HOBAN – was pleasing. PETE'S DRAGON (*1977*), from DISNEY, was an early TOON movie; *The Rescuers* (*1977*), again from DISNEY, brought Margery SHARP's WAINSCOT society into the cinema; it was sequelled, much later, by *The Rescuers Down Under* (*1990*). *Raggedy Ann and Andy* (*1977*), from Richard WILLIAMS, was a more noteworthy ANIMATED MOVIE. *Star Wars* (*1977*) put the traditional FAIRYTALE into a space-opera milieu; its successors, *The Empire Strikes Back* (*1980*) and *Return of the Jedi* (*1983*), explored the ARTHUR, LANCELOT and GUINEVERE legend – and many other fantasy tropes – in the same context. With WIZARDS (*1977*) Ralph BAKSHI's patchy fantasy career came to a first flowering; soon after, his LORD OF THE RINGS (*1978*) represented one of the fantasy cinema's great disappointments. *The* FURY (*1978*) was an enjoyable TECHNOFANTASY, much later succeeded by the less enjoyable FIRESTARTER (*1984*). MAGIC

(*1978*) was an intriguing fantasy of PERCEPTION. *The* MEDUSA TOUCH (*1978*), based on the novel by Peter Van GREENAWAY, was a grimly enigmatic piece of fantasy that queried whether precognition was a mere passive foretelling of the future or an active bringing of that future about. PATRICK (*1978*) was an unusual technofantasy HORROR MOVIE. WATERSHIP DOWN (*1978*), based on the Richard ADAMS bestseller, proved a rather dull ANIMATED MOVIE once its brilliant first few minutes were over. Mixing animation and live-action to no great effect, *The* WATER BABIES (*1978*) did little justice to Charles KINGSLEY's original. HEAVEN CAN WAIT (*1978*) reprised HERE COMES MR JORDAN (*1941*) and *Invasion of the Body Snatchers* (*1978*) reprised *Invasion of the Body Snatchers* (*1956*).

And then along came a blockbuster, *Superman* (*1978*), a TECHNOFANTASY that altered the course of popular cinema for almost two decades. Its enormous success alerted moviemakers to the fact that the COMICS held great and exploitable riches. Much earlier *Batman – The Movie* (*1966*; ◊ BATMAN MOVIES) had enjoyed modest attention, but it had been based less on the comics than on the tv series BATMAN (1966-8). Aside from the SUPERMAN MOVIES there have been, over the years, *Captain America* (*1979* tvm), FLASH GORDON (*1980*), POPEYE (*1980*), *Swamp Thing* (*1981*) and its sequel *The Return of Swamp Thing* (*1989*), *The Toxic Avenger* (*1985*) and its sequels, HOWARD THE DUCK (*1986*), *Batman* (*1989*) and its sequels (◊ BATMAN MOVIES), *Captain America* (*1989*), *Darkman* (*1990*), DICK TRACY (*1990*), *The Addams Family* (*1991*) and its sequel *Addams Family Values* (*1993*) (◊ ADDAMS FAMILY MOVIES), *The Rocketeer* (*1991*), *The* CROW (*1994*), *Judge Dredd* (*1995*), *Casper* (*1995*) (◊ CASPER, THE FRIENDLY GHOST) and various others. A similar notion was the genesis of *Condorman* (*1981*). Characters from children's literature have not been immune: for example, Captain W.E. Johns's Biggles was transplanted into a TIMESLIP fantasy, BIGGLES (*1986*).

Returning to the late 1970s, there were horror fantasies like *The* FOG (*1979*) and Werner Herzog's exquisite remake, *Nosferatu the Vampyre* (*1979*) (◊ DRACULA MOVIES), as well as an irreverent – although not, as often accused, blasphemous – look at the origins of RELIGION (here Christianity) in MONTY PYTHON'S LIFE OF BRIAN (*1979*). 1980 saw the first in what would come to seem an interminable series, *The* HOWLING (*1980*), and the Stephen KING industry made progress with Stanley Kubrick's *The* SHINING (*1980*). *An* AMERICAN WEREWOLF IN LONDON (*1981*) picked up on the HUMOUR that had been present in HORROR MOVIES like *The Howling* to produce a result that was more humour than horror. CLASH OF THE TITANS (*1981*) represented an overdue swansong for Ray HARRYHAUSEN's spfx, while *Conan the Barbarian* (*1981*; ◊ CONAN MOVIES) was the first of two movies in which Arnold Schwarzenegger – soon to become one of the biggest box-office draws of all – flexed his muscles as the severe, monosyllabic CONAN; the other was *Conan the Destroyer* (*1984*). Also released in 1981 but virtually ignored at the time was one of the best of all SWORD AND SORCERY movies, DRAGONSLAYER (*1981*). EXCALIBUR (*1981*) was John Boorman's excellent but much misunderstood synthesis of the MATTER of Britain: the critics, unfamiliar with the ARTHUR canon, said it was muddled, and they delivered a similar verdict on GHOST STORY (*1981*), based on Peter STRAUB's famous novel. WOLFEN (*1981*) trod the line between fantasy, SUPERNATURAL FICTION and SCIENCE FICTION with initial uncertainty but eventual skill. Steven SPIELBERG – who through movies like *Jaws* (*1975*) and *Close Encounters of the Third Kind* (*1977*) had become the world's most bankable director – took the world by storm with the first of his INDIANA JONES movies, *Raiders of the Lost Ark* (*1981*). These attracted imitators like *King Solomon's Mines* (*1985*), based very loosely on the H. Rider HAGGARD novel. Meanwhile Spielberg carried on to dominate fantasy cinema the way DISNEY had dominated ANIMATED MOVIES decades before: he directed or was otherwise involved in POLTERGEIST (*1982*) and its sequels, *E.T. – The Extraterrestrial* (*1982*), GREMLINS (*1984*) and its sequel, *Back to the Future* (*1985*) and its sequels, *The* GOONIES (*1985*), *Young Sherlock Holmes* (*1985*; ◊ SHERLOCK HOLMES), Don BLUTH's *An* AMERICAN TAIL (*1986*) and its (non-Bluth) sequel, Bluth's *The Land Before Time* (*1988*), WHO FRAMED ROGER RABBIT (*1988*) – done in conjunction with DISNEY – BIG (*1988*), ALWAYS (*1989*), *Arachnophobia* (*1990*), HOOK (*1991*) and *Jurassic Park* (*1993*). It is almost impossible to calculate the number of movies released during this decade and a half that either consciously or automatically incorporated a strong Spielbergian influence – two that spring easily to mind are *The* LOST BOYS (1987) and LADY IN WHITE (*1988*).

In the same year as *Raiders of the Lost Ark* came the first full-fledged fantasy by another director of enormous importance in the field, Terry GILLIAM: TIME BANDITS (*1981*), almost without a doubt the most successful transposition of the INSTAURATION FANTASY to the screen. Gilliam's is too individual a voice to dominate in the way that Spielberg's has done, but his contribution to the fantasy genre – with BRAZIL (*1985*), *The* ADVENTURES OF BARON MUNCHAUSEN (*1989*), *The* FISHER KING (*1991*) and *Twelve Monkeys* (*1996*), as well as, earlier, the **Monty Python** movies – has arguably been even more important.

Jim HENSON released a minor milestone of high fantasy with *The* DARK CRYSTAL (*1982*) and that same year saw Peter GREENAWAY's borderline fantasy *The* DRAUGHTSMAN'S CONTRACT (*1982*). *Jekyll and Hyde: Together Again* (*1982*) hilariously lampooned the JEKYLL AND HYDE MOVIES and much else besides. *The* LAST UNICORN (*1982*), an ANIMATED MOVIE based on the Peter S. BEAGLE novel and scripted by Beagle himself, disappointed; *The* SWORD AND THE SORCERER (*1982*) did likewise. DISNEY entered the TECHNO-FANTASY arena with TRON (*1982*), a movie much disliked on its release; it now seems much better, and likely would have been more successful if released a few years later.

It must already be evident that the 1980s was a decade in which it seemed that every second movie released was fantasy, sf, horror or some blending of the three. Moreover, a considerable number of them had the name of Stephen KING attached, moviemakers having apparently decided this was a guarantee of success. 1983 alone saw *Cujo* (*1983*), *The* DEAD ZONE (*1983*) and CHRISTINE (*1983*) all given major treatment, and there has been little let-up since. Some of these movies have been good but many have been fairly mediocre; certainly there have been far too many of them. Even King himself eventually rebelled, taking various measures to control the flood and ensure at least a modicum of quality.

Some movies came and went almost unnoticed, among them HALLOWEEN 3: SEASON OF THE WITCH (*1983*) – which was unrelated to the rest of the **Halloween** slasher series – *The* KEEP (*1983*) – which was virtually lost for a decade – and *The* HUNGER (*1983*), which was regarded as a triumph of style over content. SVENGALI (*1983*) was a brave experiment. DISNEY offered the live-action SOMETHING WICKED THIS WAY COMES (*1983*), which wasn't much liked, and the animated featurette *Mickey's Christmas Carol* (*1983*; ◊ CHRISTMAS CAROL), which was – it went some way to fill a hiatus in Disney's production of animated features, between *The Fox and the Hound* (*1981*) and the poorly received *The* BLACK CAULDRON (*1985*), the studio's attempt at animated HIGH FANTASY.

1984 had a fair number of minor attractions, like ALL OF ME (*1984*), a brace of further Stephen King-based movies in CHILDREN OF THE CORN (*1984*) and FIRESTARTER (*1984*), the interesting HORROR MOVIE *Child's Play* (*1984*), a surprisingly good version of *A* CHRISTMAS CAROL (*1984* tvm), the second – and weakest – INDIANA JONES movie, *Indiana Jones and the Temple of Doom* (*1984*), *The* NEVERENDING STORY (*1984*), OH GOD! YOU DEVIL! – the last and best in the series begun with OH GOD! (*1977*) – SHEENA (*1984*), which was an attempt at a feminist JUNGLE MOVIE, SPLASH! (*1984*) and *Supergirl* (*1984*; ◊ SUPERMAN MOVIES). But the year saw some major work as well. *The* COMPANY OF WOLVES (*1984*), based on stories by Angela CARTER and co-scripted by her, mixed legends of WOLVES, WEREWOLVES and Little Red Riding-Hood into a sensual cocktail; it is one of the central movies of the modern fantasy cinema. *The* PURPLE ROSE OF CAIRO (*1984*), a rather sombre, wistful tragicomedy, was Woody Allen's most noteworthy contribution to the genre. GHOSTBUSTERS (*1984*) and GREMLINS (*1984*) were big box-office successes, as – on a different scale – was *A* NIGHTMARE ON ELM STREET (*1984*), which brought a new level of imaginativeness to the HORROR MOVIE and spawned an extensive series of sequels, some good in their own right. And *Greystoke: The Legend of Tarzan, Lord of the Apes* (*1984*) showed that, after all these decades which had seemed to prove the contrary, it was possible to make a good – indeed, excellent – TARZAN MOVIE.

The following year did not enjoy such riches, with perhaps only four movies in the front rank. KISS OF THE SPIDER WOMAN (*1985*) was the most powerful of these: a movie *about* fantasy rather than straightforwardly a fantasy movie, it packed considerable punch. BRAZIL (*1985*) saw Terry GILLIAM tackling an ALTERNATE REALITY head-on in a fantasy that assumed its viewers were as conceptually nimble as readers of full-fantasy novels. CLAN OF THE CAVE BEAR (*1985*), though poorly received, was a refreshing attempt to reclaim the PREHISTORIC FANTASY from the realm of bimbos 'n' dinosaurs in which it had been trapped since ONE MILLION YEARS BC (*1966*). And LADYHAWKE (*1985*) was easily the best of the GENRE-FANTASY movies released in a year rich in that subgenre; others included LEGEND (*1985*) and RED SONJA (*1985*). Also of interest were *Return to Oz* (*1985*) – a brave attempt to continue *The* WIZARD OF OZ (*1939*) – the teen comedies TEEN WOLF (*1985*) and the rather sharper and funnier FRIGHT NIGHT (*1985*), *The* GOONIES (*1985*) and RE-ANIMATOR (*1985*), the latter a skewed variant of the **Frankenstein** motif. A couple of bad movies are also worth noting: *The Bride* (*1985*), which sequelled *The Bride of Frankenstein* (*1935*) (◊ FRANKENSTEIN MOVIES), and GHOULIES (*1985*), a dismal horror comedy that imitated *Gremlins* and was itself to be imitated by *Critters* (*1986*). Much more interesting was the Clint Eastwood fantasy/Western CROSSHATCH *Pale Rider* (*1985*) – he had much earlier attempted the same combination with *High Plains Drifter* (*1972*).

The high points of 1986 were David Cronenberg's *The* FLY (*1986*), Don BLUTH's *An* AMERICAN TAIL (*1986*), HOUSE (*1976*), HIGHLANDER (*1986*), *Peggy Sue Got Married* (*1986*), VAMP (*1986*) and, most especially, the obscure tv movie BORN OF FIRE (*1976* tvm); since all the others could be assigned to one or other stock subgenre, this latter, an astonishing piece, was a timely reminder that full fantasy was still alive and well in the movies. *Jumpin' Jack Flash* (*1986*) was a merry movie that can be regarded as a rationalized TECHNOFANTASY; it was marked by one of Whoopi Goldberg's best performances. *Gothic* (*1986*; ◊ FRANKENSTEIN MOVIES) saw Ken RUSSELL made a fool of by Mary SHELLEY. LABYRINTH (*1986*) was a leaden attempt to reprise the high-fantasy successes of the previous years; it is astonishing that Jim HENSON could have made such a mess of this. LITTLE SHOP OF HORRORS (*1986*) was an amiable and perennially popular humorous musical rehash of Roger CORMAN's original *The* LITTLE SHOP OF HORRORS (*1961*). BIGGLES (*1986*) and HOWARD THE DUCK (*1986*) were expensive flops, although the former has probably recouped its investment through repeated tv screenings. BIG TROUBLE IN LITTLE CHINA (*1986*) was a big-budget exercise that failed through flaccidity.

1987 might be characterized as the year in which moviemakers demonstrated their contempt for their public. Around this time Hollywood was bemoaning the fact that cinema audiences were falling off so drastically, and blamed the reduction on various extraneous factors. 1987 saw the exploitative *Masters of the Universe* (*1987*, *Pinocchio and the Emperor of the Night* (*1987*) – an excruciating sequel to PINOCCHIO (*1940*) – *Teen Wolf Too* (*1987*), which was dire, *The Stepford Children* (*1987* tvm; ◊ *The* STEPFORD WIVES [*1974*]), a cynical attempt to draw blood from a long-dead corpse, and numerous movies whose titles began with the word "Return": *The Return of the Living Dead Part II* (*1987*), *The Return of the Shaggy Dog* (*1987*; ◊ DISNEY), *A Return to 'Salem's Lot* (*1987*) . . . There was *Bigfoot and the Hendersons* (*1987*; vt *Harry and the Hendersons*), progenitor of the tv sitcom *Bigfoot and the Hendersons*; feelgood fantasy struggled towards yet another low. Nonetheless, the HORROR MOVIE was producing some good material, with *The Gate* (*1987*), *The* LOST BOYS (*1987*), *Predator* (*1987*) – notable because of the way its viciously carnivorous and invisible alien was treated cinematographically – and especially HELLRAISER (*1987*), a revolting yet conceptually exciting HORROR MOVIE created by Clive BARKER. There was some good commercial animation going on as well; the continuing gap left by DISNEY was filled by *Cat City* (*1987*) and *The* BRAVE LITTLE TOASTER (*1987*), while Japanese ANIME made a big breakthrough with AKIRA (*1987*). *The* WITCHES OF EASTWICK (*1987*) gave a witty account of the impact of the DEVIL on a small US town. The fantasy movie of the year was undoubtedly Wim Wenders' WINGS OF DESIRE (*1987*).

1988 saw some fine and moderately fine movies: BEETLEJUICE (*1988*), BIG (*1988*), *Bill and Ted's Excellent Adventure* (*1988*), *Dream Demon* (*1988*), *Edge of Sanity* (*1988*; ◊ JEKYLL AND HYDE MOVIES), *High Spirits* (*1988*), ELVIRA, MISTRESS OF THE DARK (*1988*), LADY IN WHITE (*1988*), *Scrooged* (*1988*; ◊ CHRISTMAS CAROL), SHADOW DANCING (*1988*), *Sundown* (*1988*), *Vampires in Venice* (*1988*; ◊ DRACULA MOVIES), VICE VERSA (*1988*) and *The* WIND IN THE WILLOWS (*1988* tvm). WILLOW (*1988*) and Ken RUSSELL's *The Lair of the White Worm* (*1988*) gave fantasy a bad name, but the balance was more than redressed by such movies as *The* DEAD

CAN'T LIE (*1988* tvm), *The* NAVIGATOR: A MEDIEVAL ODYSSEY (*1988*) and – each in their way truly exceptional – Jan SVANKMAJER's ALICE (*1988*), CELIA (*1988*), WHO FRAMED ROGER RABBIT (*1988*) and *The* LAST TEMPTATION OF CHRIST (*1988*).

The following year was, by contrast, disappointing: Tim BURTON's BATMAN (*1989*) was a great commercial success but, viewed now, seems pompous; *Captain America* (*1989*), another attempt to revive a comics hero, was bland; *All Dogs Go to Heaven* (*1989*) showed that Don BLUTH's studios were still capable of Disney-quality animation but also that they remained *in*capable of mastering the art of plot-construction; *Sinbad of the Seven Seas* (*1989*) plumbed new depths of badness among the frequently undistinguished series of SINBAD MOVIES; and ALWAYS (*1989*) showed Steven SPIELBERG at his most saccharine. Yet there were some bright spots. Terry GILLIAM's *The* ADVENTURES OF BARON MUNCHAUSEN (*1989*), although a financial disaster – so much so that a book on the subject has been published – has come to be regarded as a central movie of late-1980s fantasy. ERIK THE VIKING (*1989*), again from a **Monty Python** team-member, Terry JONES, captivates through audacity. GHOSTBUSTERS II (*1989*), though disliked by the teenies and the critics alike, added new dimensions of fantasy to what could have been a mere reprise of the basic riff of GHOSTBUSTERS (*1984*). *Honey, I Shrunk the Kids* (*1989*) was one of the first DISNEY live-action movies in the modern era to be a TECHNOFANTASY that did not nauseate; in animation, Disney released *The* LITTLE MERMAID (*1989*), showing that in at least this area of activity the studio was back on song. *The* ICICLE THIEF (*1989*), dir Maurizio NICHETTI, was in the English-speaking world regarded as an Art House movie – a bit "difficult" – but is a classic. *The* WITCHES (*1989*) was an extremely effective and entertaining version of the Roald DAHL tale. *Indiana Jones and the Last Crusade* (*1989*) was the best of the INDIANA JONES series, which had looked in danger of trailing off into mediocrity, as series often do. Most of all, perhaps, FIELD OF DREAMS (*1989*) showed that, through fantasy, even an activity as divorced from the world outside the USA as BASEBALL could be used as the basis for a fantasy movie that was universally affecting.

It is always reassuring, when conducting a historical survey, to regard a new decade as in some way heralding a new dawn, but, as noted, fantasy cinema declines to cooperate. The 1990 crop of fantasy movies was largely undistinguished and unambitious, for the main part representing cashings-in on the successes of the late 1980s. ALMOST AN ANGEL (*1990*), although a nice little movie, could hardly have come into existence had it not been for Paul Hogan's hugely popular **Crocodile Dundee** series; GHOST (*1990*) was a big-budget rehash of earlier ideas; *Frankenstein Unbound* (*1990*) added little to the FRANKENSTEIN-MOVIE *corpus*; GREMLINS 2: THE NEW BATCH (*1990*) was a sequel with a lot of razzmatazz but no real content; LORD OF THE FLIES (*1990*) was a pallid remake; *Neverending Story II: The Next Chapter* (*1990*) showed the story should have ended earlier; TEENAGE MUTANT NINJA TURTLES (*1990*) was better; and so on right down to *Child's Play 2* (*1990*), *The Exorcist III* (*1990*) (◊ *The* EXORCIST), which was actually rather good, *Highlander II: The Quickening* (*1990*) (◊ HIGHLANDER), which wasn't, and *Howling VI* (*1990*) (◊ *The* HOWLING. DICK TRACY (*1990*), DUCKTALES: THE MOVIE – TREASURE OF THE LOST LAMP (*1990*) – both of these based on COMICS (and in the latter case also on a tv series) – EDWARD

SCISSORHANDS (*1990*), the gory NIGHTBREED (*1990*) and the deliberately understated TRULY MADLY DEEPLY (*1990*) – originally a tvm – did something to redress the balance, but overall 1990 was a bad year for cinematic fantasy.

1991 gave us the exact opposite. A mere listing of the titles shows the riches on offer: *The Addams Family* (*1991*) (◊ ADDAMS FAMILY MOVIES), BARTON FINK (1991), BILL & TED'S BOGUS JOURNEY (*1991*), *The Butcher's Wife* (*1991*), DEAD AGAIN (*1991*), DEFENDING YOUR LIFE (*1991*), *The* DOUBLE LIFE OF VÉRONIQUE (*1991*), DROP DEAD FRED (*1991*), FERN-GULLY: THE LAST RAINFOREST (*1991*), *An* AMERICAN TAIL: FIEVEL GOES WEST (*1991*), *The* FISHER KING (*1991*), HOOK (*1991*), NAKED LUNCH (*1991*), PROSPERO'S BOOKS (*1991*), TWIN PEAKS: FIRE WALK WITH ME (*1991*) and VOLERE VOLARE (*1991*); to this list could be added *Terminator 2: Judgment Day* (*1991*), which while an sf movie contains sufficient elements of TECHNOFANTASY – notably the fact that the "bad" Terminator is a SHAPESHIFTER – to be of considerable fantasy interest. Alongside this distinguished list, Blake Edwards's weak attempt at a cross-sexual IDENTITY-EXCHANGE movie, *Switch* (*1991*), looked like a fossil.

The next year almost, but not quite, kept up the standard, as DISNEY weighed in with one of their best ANIMATED MOVIES for a long time, ALADDIN (*1992*). Good fantasy comedies – some of them more than just good – included BUFFY THE VAMPIRE SLAYER (*1992*), DEATH BECOMES HER (*1992*), *Memoirs of an Invisible Man* (*1992*; ◊ INVISIBLE MAN MOVIES) and STAY TUNED (*1992*). *Batman Returns* (*1992*) (◊ BATMAN MOVIES) was a leaden sequel. Two tvms were of note: *Frankenstein: The Real Story* (*1992* tvm; ◊ FRANKEN-STEIN MOVIES) and the bizarre, enigmatic GOLEM: THE WANDERING SOUL (*1992*). TOYS (*1992*) was much underrated by the critics – perhaps perplexed because it seemed to have the premise of a Disney family fantasy yet was quite unDisneyesque – but the most impressive fantasy of the year was ORLANDO (*1992*), an adaptation of Virginia WOOLF's *Orlando: A Biography* (**1928**): it stood far above a by no means unimpressive throng through the brilliance of both Sally Potter's direction and Tilda Swinton's performance in the title role.

As with 1992, 1993 offered a mixed bag. The French comedy *Les* VISITEURS (*1993*) outgrossed in its native land the contemporaneously released *Jurassic Park* (*1993*), Steven SPIELBERG's record-breaking blockbuster, but oddly was not given general release in the English-speaking world. CANDYMAN (*1993*) was an impressively imaginative fantasy-HORROR MOVIE based on URBAN LEGEND, while *Body Snatchers* (*1993*) was the most edge-of-seat version yet of Jack FINNEY's fantasy-themed sf tale – as with *Les Visiteurs*, its appeal was seriously misjudged by UK distributors, and it suffered a direct-to-video release. GROUNDHOG DAY (*1993*) was an enormously popular lightweight feelgood fantasy, while at the time everybody hated the Schwarzenegger vehicle LAST ACTION HERO (*1993*) – an adverse judgement which today, interestingly, critics are generally reversing in a shameless act of revisionist history. Francis Ford Coppola's *Bram Stoker's Dracula* (*1993*; ◊ DRACULA MOVIES) was similarly disliked – and similarly, albeit to a lesser extent, is now more kindly regarded. *The* TRIAL (*1993* tvm) was a brave attempt, but could never match up to the shadows and surreal mystery of Orson Welles's version, *The* TRIAL (*1962*). The outstanding fantasy movie of the year, both commercially and aesthetically, was Tim BURTON's *The* NIGHTMARE BEFORE CHRISTMAS (*1993*),

a rare Hollywood venture into stop-motion animation.

It is difficult, at the time of writing, to judge the merits of fantasy movies released after 1993: as with such works as *Toys* and *Last Action Hero* it can be, as noted above, that initial critical reactions soon seem hopelessly off the mark. To take examples of this, the remake MIRACLE ON 34TH STREET (*1994*) was regarded as solid sugar on release but can now, the publicity hype having mercifully faded from memory, be seen as the rather affecting minor movie it is. Conversely, the same sort of hype made *The* CROW (*1994*) seem like a stylish work, which on repeated viewing one can see that it is not. *Mary Shelley's Frankenstein* (*1994*; ◊ FRANKENSTEIN MOVIES) was disingenuously publicized as very frightening, thereby disappointing moviegoers who expected to have the pants scared off them and deterring others from going to see what was a very lovingly crafted homage to both Mary SHELLEY and the **Frankenstein** movie tradition. INTERVIEW WITH THE VAMPIRE: THE VAMPIRE CHRONICLES (*1994*), eagerly awaited for years by avid Anne RICE fans, was another victim of over-hyping: a respectable piece of horror froth, it fell significantly short of the masterpiece status the publicity machine had claimed for it. *Casper* (*1995*) (◊ CASPER, THE FRIENDLY GHOST) was another to be over-hyped, and the same went for *Batman Forever* (*1995*) (◊ BATMAN MOVIES) – neither were *bad* movies, although the former suffered from cuteness (something almost impossible to achieve when starring the estimable Christina Ricci!) and the latter from plotting incoherence. Exactly this quality was one of the strengths of *Tank Girl* (*1995*), a curious crosshatch of sf, SCIENCE FANTASY and TECHNOFANTASY that was much loathed by the critics (excluding the current writer) but much enjoyed by average moviegoers; it seems a reasonably safe bet that, in a few years' time, the judgement of the latter will come to be the received wisdom. *Congo* (*1995*) was Frank Marshall's partially successful attempt, based on Michael Crichton's *Congo* (**1980**) to recast for a new generation the LOST-RACE movie, this particular race being, for good measure, APES. Hugely over-expensive, widely derided but ultimately commercially respectable was the FAR-FUTURE epic *Waterworld* (*1995*). *The Indian in the Cupboard* (*1995*) (◊ Lynne REID BANKS) was a delightfully inventive children's fantasy that had significant appeal to adults, and *Dr Jekyll and Ms Hyde* (*1995*) carried on a grand tradition (◊ JEKYLL AND HYDE MOVIES) . . . while ensuring that audiences continued to salivate at the prospect of the oft-postponed – and, as it proved, disastrous – *Mary Reilly* (*1996*), based on the Valerie MARTIN novel. *Babe* (*1995*) was a surprisingly charming TALKING-ANIMALS live-action fantasy about a piglet who wanted to be a sheepdog (◊ Dick KING-SMITH). *The Santa Clause* (*1995*) rehashed the idea of a mortal having to take over from SANTA CLAUS. One post-1993 fantasy movie that one can say with certainty will continue to be regarded highly in future decades is Jan SVANKMAJER's FAUST (*1994*).

At the time of writing (early 1996) it is, at least in the UK and USA, difficult at any particular time to find a multi-screen cinema that is not offering at least one fantasy movie – and, if one regards sf movies, ANIMATED MOVIES and HOR-ROR MOVIES as subgenres of fantasy proper – it is probably impossible. Any listing of movies currently on general release shows a multiplicity of works of fantasy interest, and many feature in the box-office Top Ten charts. To be fair, in part this is because Hollywood is currently continuing an orgy of rehashing of earlier material: remakes, developments from the COMICS, resuscitation of decades-ago movie characters, etc. But what should not be forgotten is that a comparatively few directors have in recent years determinedly put fantasy cinema back at the centre of attention: although one can find much to criticize in the output of all of them, moviemakers like Don BLUTH, Tim BURTON, Terry GILLIAM and Steven SPIELBERG – not to mention the various executives of the ever-expanding DISNEY/ Touchstone/Hollywood Pictures organization – have succeeded in making us reassess what going to the cinema is all about. It may be coincidence that moviegoing, which had slumped by 1984-7 to an extent that threatened the future of the industry, is now as popular an activity as it has ever been . . . but one suspects that it is no coincidence at all. [JG]

See also: MONSTER MOVIES; SHAKESPEAREAN FANTASY IN THE MOVIES; VAMPIRE MOVIES; WEREWOLF MOVIES.

CIRCE The most celebrated sorceress of Greek MYTHOLOGY. She turned ODYSSEUS's crew into swine but was ultimately outwitted by him. When Glaucus preferred Scylla she turned that rival into a hideous MONSTER. The most elaborate account of her in modern fantasy fiction is in Eden PHILLPOTTS's *Circe's Island* (**1925**); more ironic depictions can be found in Maurice BARING's "The Island" (1909), Henry KUTTNER's *The Mask of Circe* (1948; **1971**) and Margaret St Clair's "Mrs Hawk" (1950). Interesting de-supernaturalized characterizations are in John ERSKINE's *Penelope's Man* (**1927**) and Storm CONSTANTINE's "Poisoning the Sea" (1992). [BS]

CIRCLE OF FEAR ◊ GHOST STORY (1972-3).

CIRCUS Circuses are quite frequent settings for fantasies (especially DARK FANTASIES), SUPERNATURAL FICTIONS and HORROR stories. Although the details of the backdrop they offer differ from those of the carnival, in essence the two venues are interchangeable, so the main discussion is under CARNIVAL. [JG]

CITY The presence of a city does not necessarily make a text an URBAN FANTASY, where settings tend to be circumambient and where the surrounding city generally exists in the real world. Almost any kind of fantasy may include cities, and in almost any context: they may be seen only from a distance, or they may constitute the primary setting of the tale; they may be come to or left, built or destroyed, forgotten or remembered, demolished or yet to come.

At the same time, not every urban conglomeration mentioned in a CONTEMPORARY FANTASY or in a tale set within a LAND OF FABLE or RURITANIA warrants being called a city – certainly not as the term is used in this encyclopedia. To be so designated, a place must itself be an ingredient in the CAULDRON OF STORY; it must, in other words, embody a set of STORIES – stories which must somehow contain an element of the fantastic. Indeed, it might be argued that a true city could almost be defined *as* a Cauldron of Story, a melting pot where different kinds of world meet (◊◊ INSTAURATION FANTASY). Moreover, the Cauldron must be *known*, the kinds of stories it contains must have been told and TWICE-TOLD; mundane cities whose potential stories have not been conspicuously embodied in texts by more than one writer – an example would be the Ottawa created by Charles DE LINT – may hover at the edge of availability, but have not yet been used sufficiently to be *recognized* as ICONS. Cities which are recognizable as homes of story include BAGHDAD, Byzantium, CAIRO, JERUSALEM, LONDON, LOS ANGELES, NEW ORLEANS, NEW YORK, PARIS, PRAGUE,

ROME, Samarkand, SAN FRANCISCO and VENICE. When visited by characters, they open like caverns into storied outcomes, and they often contain PORTALS; when inhabited, they surround the actions of characters caught in an urban fantasy.

Fantasy cities may be found almost anywhere in SECONDARY WORLDS, in similar OTHERWORLD contexts and within LANDS OF FABLE. (SUPERNATURAL FICTION – which is often set in real cities – comes close to being assimilated into fantasy whenever it takes as its setting a City of the Dead.) These cities are, of course, inherently creations of fiction. They do not – in the obvious sense that Paris or Los Angeles do – contain a mix of the real and the imagined; they may therefore be presented as either chaotic or totally planned (◊ UTOPIA). The cities of pure fantasy might almost seem to subvert our restrictive definition of the term "city", but in fact they do not: if the creator of the fantasy city did not regard it as of iconic importance and theatrical implication, with twice-told stories inherent in its fabric, then almost certainly – outside the ersatz secondary worlds of GENRE FANTASY – the city would not be there at all, unless as a mere way station on the route traversed across the MAP.

Cities in fantasy may be of almost any size, from extended EDIFICES to the megalopolis which encompasses the world. They may exist at any point in time (◊ TIME ABYSS) in relation to the main story, in the deep past or FAR FUTURE, or be visible only when their time or place and our REALITY's intersect; or they may exist in the fantasy only *as* a fantasy. They may be at the actual or symbolic centre of the world or, more frequently, of a lost world (◊ LOST RACES); or they may guard the marches of the known LAND. They may serve primarily as the headquarters of an empire or a DARK LORD; or as a capitol; or as a LAST REDOUBT. They may give space to the living and the dead; they may no longer be inhabited. They may be surrounded by a permeable (or impermeable) THRESHOLD; when this is the case, they are likely to contain or to embody a PORTAL. They may be unapproachable; or they may be impossible to escape. They may serve as a POLDER or a prison from which the protagonist must escape in order to accomplish or continue the QUEST. As in the real world, a city in fantasy tends to be a place where the action converges. A silent city is a sign of death, hardly ever of repose.

Cities abound in fantasy, and no extensive list is necessary. Notable examples include: the capitals of various authors' ATLANTIS; Brian W. ALDISS's Malacia; L. Frank BAUM's Emerald City; the eponymous setting of the **Liavek** SHARED-WORLD anthology sequence ed Emma BULL and Will SHETTERLY; Opar, which appears in various **Tarzan** books by Edgar Rice BURROUGHS; the 55 cities featured in Italo CALVINO's *Le città invisibili* (**1972**); the unnamed city in *Underworld* (**1992**) by Peter Conrad (1948-   ); the eponymous *City of the Iron Fish* (**1994**) by Simon INGS; Paradys, in Tanith LEE's **Books of Paradys**; Fritz LEIBER's Lankhmar; the ruined city of Shalba beneath the pre-deluge Secret Lake in Hugh LOFTING's **Dr Dolittle** sequence; Kadath and Thran, in H.P. LOVECRAFT's *The Dream-Quest of Unknown Kadath* (**1955**), and Innsmouth in his "The Shadow Over Innsmouth" (1939); Terry PRATCHETT's Ankh-Morpork, capital of the **Discworld**; the underground Suicide City in Robert Louis STEVENSON's *New Arabian Nights* (coll **1882**); the eponymous city of death in *The City of Dreadful Night* (**1874**) by James Thomson (1834-1882); Minas Tirith in J.R.R. TOLKIEN's *The Lord of the Rings*

(**1954-5**); and Nessus in Gene WOLFE's *The Book of the New Sun* (**1980-83**). [JC]

ČIURLIONIS, MYKALOJUS KONSTANTINAS (1875-1911) Lithuanian artist and composer whose paintings blended original fantasy and poetic imagery and have had considerable influence on Lithuanian culture. His best works, done 1907-9, include cycles of FAIRYTALES and grand compositions in which he strove to penetrate the secrets of life and the Universe; these are his "sonatas" of *The Sun*, *The Serpent*, *The Pyramids* and *The Stars*. Other works set out to transcend mundane existence – e.g., *Rex*, *Demon*, *Prelude* and *The Fairytale of the Castle* – while in paintings like *Fantasy: Triptych*, *Angel: Prelude* and *The City: Prelude* he depicts fantasy landscapes and worlds. [GeB/RoB]

CLACK BOOK, THE ◊ MAGAZINES.

CLAIR, RENÉ (1898-1981) French moviemaker primarily remembered as the director who adapted French comedy for the talkies. His first movie was *Paris Qui Dort* (*1923*; vt *Le Rayon Invisible*; vt *The Crazy Ray*), which he directed and wrote; in it a mad scientist uses an invisible ray to paralyze Paris. Other early fantasies were *Le Voyage Imaginaire* (*1925*) and *Le Fantôme du Moulin-Rouge* (*1926*); in the latter an unhappy lover learns how to leave his body and, invisible, plays practical jokes all over Paris. RC moved to the UK to make *The* GHOST GOES WEST (*1935*), a comic GHOST STORY. After the non-fantasy *Break the News* (*1938*) RC went to Hollywood, where his movies included I MARRIED A WITCH (*1942*) – based on *The Passionate Witch* (**1941**) by Thorne SMITH and Norman MATSON – in which the ghosts of a WITCH and a sorcerer haunt the descendant of one of their persecutors, and *It Happened Tomorrow* (*1944*), in which an old man habitually shows a youthful reporter what tomorrow's headlines will be, so that he can scoop the competition, which is all fine until he sees his own death reported. Returning after WWII to France RC made, among others: *Le Silence est d'Or* (*1947*; vt *Man About Town*), a tribute to silent movies; the unsuccessful *La Beauté du Diable* (*1949*), a free adaptation of the FAUST legend; and his last movie of significance, *Les* BELLES DE NUIT (*1952*), a comedy TIME FANTASY in which a young composer's DREAMS create alternate REALITIES. Because his late movies were generally disliked, RC fell out of favour for some while, but since his death much of his work has been "rediscovered". [JG]

CLAIRVOYANCE ◊ TALENTS.

CLANGERS, THE UK tv series (1969-73). BBC/Smallfilms. **Pr** Oliver Postgate. **Sets and puppets** Peter Firmin. **Writer** Postgate. **Created by** Postgate. 27 9min episodes. Colour.

Several generations still whistle and coo in memory of this series, whose episodes were repeated over and over again in the early evening, at a time when both adults and children – not necessarily together – might be watching. The Clangers (knitted puppets) were the curious little inhabitants of a tiny blue moon which was under frequent meteoritic bombardment; the burrows in which they sheltered from the meteors were capped by dustbin-lids, the noise of whose hurried replacement gave the species its name. Among the Clangers' various fantasticated associates was the Soup DRAGON.

The Postgate/Firmin team – trading as Smallfilms – for some years more or less cornered the market with this sort of whimsical fare, first made popular in the UK by *The* MAGIC ROUNDABOUT (1963-71 France; screened 1965-77 in the UK). Other Postgate/Firmin collaborations included *The*

Saga of Noggin the Nog (1959-65), *Ivor the Engine* (1962-4, 1976-7), *The Pogles/Pogles' Wood* (1966-7) and *Bagpuss* (1974), but *TC* was almost certainly the greatest success. [JG]

CLAN OF THE CAVE BEAR, THE US movie (*1985*). Producers Sales Organization/Sidney Kimmel/ Guber-Peters/Jozak-Decade/Jonesfilm. **Pr** Gerald I. Isenberg. **Exec pr** Mark Damon, Peter Guber, John Hyde, Jon Peters. **Dir** Michael Chapman. **Spfx** Michael Clifford, Gene Grigg. **Screenplay** John Sayles. **Based on** *The Clan of the Cave Bear* (**1980**) by Jean M. AUEL. **Starring** Curtis Armstrong (Goov), John Doolittle (Brun), Nicole Eggert (teenage Ayla), Daryl Hannah (Ayla), Lycia Naff (Uba), Pamela Reed (Iza), James Remar (Creb), Thomas G. Waites (Broud). 98 mins. Colour.

This little-liked PREHISTORIC FANTASY is subtitled throughout, the performed dialogue being mainly in the form of grunts, gestures and animal calls. Orphaned Cro-Magnon child Ayla is adopted by the Neanderthal Clan of the Cave Bear, which slowly accepts the UGLY DUCKLING despite her blondeness, tallness, independent nature and intelligence – and her impertinence in being an elect of the Cave Lion, a masculine TOTEM. At a clan gathering she and SHAMAN Creb see a VISION of Bear and Lion walking together, but then the Lion walking on beside her newborn Cro-Magnon/Neanderthal son Durc. In due course she leaves the child with the tribe to seek her own destiny. It is hard to see why *TCOTCB* has drawn such critical contempt, unless for its tacit FEMINISM: although the narration is over-expository and the equation of mental versatility with leggy blonde Cro-Magnons, as opposed to shabby Neanderthals, is a cliché, the movie is beautifully shot, well scripted and finely acted. [JG]

CLARE, HELEN Pseudonym of Pauline CLARKE.

CLARENCE US/Canadian/New Zealand movie (*1990* tvm). ◊ IT'S A WONDERFUL LIFE (*1946*).

CLARK, DOUGLAS W. (? -   ) US writer whose **Alchemy** sequence of humorous fantasy novels – *Alchemy Unlimited* (**1990**), *Rehearsal for a Renaissance* (**1992**) and *Whirlwind Academy* (**1993**) – spoofs without challenging the kind of fantasy tale concerning clumsy SORCERERS' APPRENTICES and the minutiae of how they eventually learn to do good MAGIC. In this case, the LAND-OF-FABLE France in which the inept apprentice operates is pestered by bottled DEMONS exported from Spain by the Inquisition. [JC]

CLARKE, A.V. (1922-   ) UK writer. ◊ Kenneth BULMER.

CLARKE, HARRY Working name of Irish illustrator Henry Patrick Clarke (1889-1931) whose theatrical, decadent pen-and-ink ILLUSTRATION owes much to that of Aubrey BEARDSLEY. His drawings have a dark intensity, consisting of strongly delineated figures richly and meticulously embellished with pattern.

HC was the son of a church decorator and stained-glass artist. After brief study in London he attended Dublin Metropolitan School of Art, specializing in stained glass, which he subsequently took up professionally. He also worked as a textile designer and mural painter.

His editions include *Hans Andersen's Fairy Tales* (1916 UK), GOETHE's *Faust* (1925 UK), Edgar Allan POE's *Tales of Mystery and Imagination* (1919 UK), Algernon Swinburne's *Selected Poems* (1928 UK) and *The Fairy Tales of Charles Perrault* (1922). [RT]

Further reading: *Harry Clarke* (**1979**) by Nicola Gordon Bowe; *Harry Clarke: His Graphic Art* (**1984**) by Bowe.

CLARKE, (VICTOR) LINDSAY (1939-   ) UK writer

whose second novel, *The Chymical Wedding* (**1989**), features a complex TIMESLIP between 19th-century and contemporary figures, plus an elaborate presentation of an alchemical QUEST for true understanding. The female MAGUS of *Alice's Masque* (**1994**) seems to be imposing an arduous GODGAME upon the frustrated middle-aged male who lands on her doorstep; but once again a narrative remoteness casts a chill upon proceedings, especially at moments which obscurely replay a cruel Year-King SACRIFICE. The male's slow earning of the right to enter into what seems a RITUAL enactment of the worship of the GODDESS raises the temperature somewhat. [JC]

CLARKE, (ANNE) PAULINE (1921-) UK writer, active under her own name since 1948 as an author of children's stories and since 1953 as Helen Clare, under which name she has published mostly for younger children, mainly the **Five Dolls** sequence about live DOLLS. As HC she wrote *Merlin's Magic* (**1953**), an ARTHUR tale for somewhat older children. Her main accomplishment as an author of CHILDREN'S FANTASY is almost certainly *The Twelve and the Genii* (**1962**; vt *The Return of the Twelves* 1964 US), which again focuses on the animation of dolls, in this case the 12 toy soldiers which inspired the early fantasies of Charlotte BRONTË and her siblings; these are brought to life by a present-day boy and followed through a final journey. *The Two Faces of Silenus* (**1972**) is a RITE-OF-PASSAGE YA fantasy. [JC]

CLASH OF THE TITANS UK movie (**1981**). MGM. **Pr** Charles H. Schneer, Ray HARRYHAUSEN. **Dir** Desmond Davis. **Spfx** Brian Smithies. **Vfx** Steven Archer, Jim Danforth, Harryhausen. **Screenplay** Beverley Cross. **Novelization** *Clash of the Titans* * (**1981**) by Alan Dean FOSTER. **Starring** Ursula Andress (Aphrodite), Claire Bloom (Hera), Judy Bowker (Andromeda), Susan Fleetwood (Athena), Jack Gwillim (Poseidon), Harry Hamlin (Perseus), Donald Houston (Acrisius), Freda Jackson (Stygian witch), Neil McCarthy (Calibos), Anna Manahan (Stygian witch), Burgess Meredith (Ammon), Laurence Olivier (Zeus), Tim Pigott-Smith (Thallo), Siân Phillips (Cassiopeia), Pat Roach (Hephaestus), Flora Robson (Stygian witch), Maggie Smith (Thetis), Vida Taylor (Danaë). 118 mins. Colour.

The LEGEND of Perseus (◊ HEROES) retold as a coupon fantasy (◊ PLOT COUPONS). The most astonishing thing about this movie is its release date: 1981. Aside from a rebarbative mechanical owl – a virtual clone of C3PO from *Star Wars* (**1977**) – this movie reads as if it had been made 20 years earlier, from jerky monsters to creaking plot, from poor pacing to dingy lighting. *COTT* was the end of the line for Harryhausen's brand of spfx; his career might be better remembered had he desisted from making this final movie. [JG]

See also: JASON AND THE ARGONAUTS (*1963*).

CLAYTON, (PATRICIA) JO (1939-) US writer and former teacher, best known for her **Diadem** series of PLANETARY ROMANCES where a young girl draws power from the eponymous headband, a SCIENCE-FANTASY equivalent of a magic RING. The series comprises *Diadem from the Stars* (**1977**), *Lamarchos* (**1978**), *Irsud* (1978), *Maeve* (**1979**), *Star Hunters* (**1980**), *The Nowhere Hunt* (**1981**), *Ghosthunt* (**1983**), *The Snares of Ibex* (**1984**) and *Quester's Endgame* (**1986**), and continues with the **Shadith's Quest** subseries – *Shadowplay* (**1990**), *Shadowspeer* (**1990**) and *Shadowkill* (**1991**) – and the **Shadowsong** trilogy – *Fire in the Sky* (**1995**), *The Burning Ground* (**1995**) and *Crystal Heat* (**1996**).

The series is inventive and colourful, much in the tradition of Leigh BRACKETT and Marion Zimmer BRADLEY. In the same vein is the **Skeen** trilogy – *Skeen's Leap* (**1986**), *Skeen's Return* (**1987**) and *Skeen's Search* (**1987**) – about a female smuggler who seeks refuge through a dimensional PORTAL and is trapped in an OTHERWORLD.

Though the later series are written with vigour and pace they are essentially formulaic. Of more interest are two other series. The first, **Duel of Sorcery** – *Moongather* (**1982**), *Moonscatter* (**1983**) and *Changer's Moon* (**1985**) – is set on an alien magical world where a girl raised and trained by a master WIZARD must now pit her skills against him to free herself. JC later produced an addendum, the **Dancer** trilogy – *Dancer's Rise* (**1993**), *Serpent Waltz* (**1994**) and *Dance Down the Stars* (**1994**).

Drinker of Souls – *Drinker of Souls* (**1986**), *Blue Magic* (**1988**) and *A Gathering of Stones* (**1989**), all three assembled as *The Soul Drinker* (omni **1989**) – contains some of JC's best work, reminiscent of Tanith LEE, with a convincingly delineated Oriental world where GODS and GHOSTS are part of the fabric of the landscape and where a young girl, the souldrinker of the title, goes on a QUEST to rescue her family from an evil king. This series spawned **Wild Magic** – *Wild Magic* (**1991**), *Wildfire* (**1992**) and *The Magic Wars* (**1993**) – which builds inexorably to a battle between the GODS.

Other works include *Shadow of the Warmaster* (**1988**), an sf thriller, and *A Bait of Dreams* (fixup **1985**), drawn from the **Gleia** series (which ran in *Asimov's*) about crystalline drugs which draw one into magical realms but destroy one's SOUL. JC is a highly capable and vivid writer who can create convincing heroines and whose talents are at their best when she moves away from traditional plots into her own worlds. [MA]

CLEMENT, AERON (1936-1989) Welsh businessman and dog-breeder, author of the ANIMAL FANTASY *The Cold Moons* (**1987**). Like Richard ADAMS's *Watership Down* (**1972**) it tells of a QUEST – by a colony of badgers – to escape the brutality of man and find a new home in the land of Elysia. [MA]

CLICHÉS ◊ PLOT DEVICES.

CLIFFHANGERS Overall title for an NBC tv series that featured ongoing stories in serialized fashion. The segments *The* CURSE OF DRACULA and *The* SECRET EMPIRE are of fantasy interest. [BC]

CLIFFORD, LUCY LANE (1853-1929) UK novelist and writer of children's stories; born Sophia Lucy Lane, she wrote as Mrs W.K. Clifford after the death of her husband, mathematics professor William Kingdon Clifford (1845-1879), whom she had married in 1875. Her struggle to raise her two children and survive on her own became a key theme in her fiction, and she caused early notoriety with *Mrs Keith's Crime* (**1885**), which advocated euthanasia for a dying child. She brought this same harsh reality to her children's stories, which are frequently sad and psychologically cruel. She drew her beliefs from the concepts of her husband, who opined that a person consisted of two personas: the physical mind, which was REALITY, and the morality engendered by that individual, which was PERCEPTION. The significance and value of her children's stories are only now being rediscovered (◊ CHILDREN'S FANTASY). *Anyhow Stories for Children* (coll **1882**; exp vt *Anyhow Stories, Moral and Otherwise* 1899) contains the powerful "The New Mother", where a mother punishes her errant children by leaving them and sending a monstrous replacement, fear of meeting whom sends the children fleeing INTO THE WOODS. It has been speculated that this story may have had an influence on Henry JAMES, who was a friend of LLC's, in writing "The Turn of the Screw" (1898). *The Last Touches and Other Stories* (coll **1892**) includes "Wooden Tony" (1890 *English Illustrated Magazine*), about a lazy boy whose idleness causes him to undergo a gradual TRANSFORMATION until he becomes a STATUE, in a bizarre reversal of the PINOCCHIO theme. LLC's well-earned success from her novels caused her to become a literary hostess and a friend of Rudyard KIPLING, George Eliot (real name Mary Ann Evans; 1819-1880) and Mary E. BRADDON, who had encouraged her writing. [MA]

Other works: *Under Mother's Wings* (coll **1885**), *Very Short Stories and Verses for Children* (coll **1886**), *Dear Mr Ghost* (**1895** chap).

Further reading: Chapter 6 of *Not in Front of the Grown-Ups* (**1990**) by Alison Lurie.

CLIMAX, THE US movie (*1944*). ◊ PHANTOM OF THE OPERA.

CLINE, C(HARLES) TERRY Jr. (1935-   ) US writer whose work has tended to cross genres, mixing horror, fantasy and thriller modes, as in his first novel, *Damon* (**1975**), in which a geographically scattered group of children attain full adult physical growth and sexuality by the age of four, then develop TALENTS; the theme is not especially fresh, but is nicely handled. *Death Knell* (**1977**) – sometimes wrongly described as a YA novel – is a grim tale of quasi-POSSESSION, in which the teenaged daughter of a German officer seems taken over by the vengeful spirit of a victim of the Holocaust (◊ WORLD WAR II). *Cross Currents* (**1979**) centres on the latest of a series of REINCARNATIONS of CHRIST, each of whom has died violently before his 33rd birthday. CTC is a writer who seldom fails to be interesting, though his earlier work has a fire missing from his later. If his themes may often be familiar, he intriguingly works them fully out: unlike many writers, he declines to take the easy plotting option. And there is always the scent of the FANTASTIC in his work – even in a straightforward thriller like *Missing Persons* (**1982**). [JC/JG]

Other works: *Mindreader* (**1981**), whose protagonist boasts the TALENT described in the title; *The Attorney Conspiracy* (**1983**); *Prey* (**1985**); *Quarry* (**1987**); *Reaper* (**1989**).

CLOUGH, B(RENDA) W(ANG) (1955-   ) US writer whose work after 1996 will be published as by Brenda Clough. BWC is the author of the **Averidan** series: *The Crystal Crown* (**1984**), *The Dragon of Mishbil* (**1985**), *The Realm Beneath* (**1986**), and *The Name of the Sun* (**1988**). *An Impossumble Summer* (**1992**), a children's novel, deals wittily with TALKING ANIMALS. [GF]

CLUB STORY A CS is made up of two elements: (a) a FRAME STORY, which describes a place (normally a club) and the people (in the 19th century almost always gentlemen) who foregather there, and which serves as a template venue for the telling of as many tales as the author wishes to tell; (b) the tale itself, which is told by a person who probably claims to have participated in the events being narrated, and which has subsequently been written down by one of the club members in attendance.

When it is understood as sf the CS gains some of its appeal from the contrast between the rational club world and the tale itself, which is often highly improbable. When it is understood as fantasy – a genre where impossible events are assumed to occur – the CS loses some of the contrast

between frame and tale but benefits from the greater leeway authors have to tell serious stories. For when the CS structure can be used to generate a sense of worldly verisimilitude – as in Henry JAMES's "The Turn of the Screw" (1898) or Joseph CONRAD's "Heart of Darkness" (1899) – the story itself can range far without seeming to be a TALL TALE, though a sense that one may be hearing a whopper is inherent in the CS structure: stories told aloud by fictional characters are inherently unreliable.

Some TAPROOT TEXTS – like BOCCACCIO's *Decameron* (written *c*1350) and CHAUCER's *Canterbury Tales* (written before **1400**) – clearly prefigure the form; and Count POTOCKI's *Manuscript Found at Saragossa* (written before 1815; full text **1989**) arguably amalgamates those models with the example of the Arabian Nights, and might be considered the first true fantasy CS sequence. The Christmas ANNUALS generated by UK publishers from about the middle of the 19th century had a seemingly similar structure to that of the pure CS, but were in fact ANTHOLOGIES whose contributors wrote tales to fit a pre-existing frame; some contemporary CS titles – like Darrell SCHWEITZER and George Scither's *Tales from the Spaceport Bar* (anth **1987**) – are closer in feel to the annuals than they are to single-author collections. The first single-author book to present the CS ambience to a wide audience was Robert Louis STEVENSON's *New Arabian Nights* (coll **1882** in 2 vols; 1st vol only vt *The Suicide Club, and The Rajah's Diamond* 1894) and its successor, *More New Arabian Nights: The Dynamiter* (coll **1885**), the second being with his wife Fanny Van de Grift Stevenson; and it was his model that late-19th- and early-20th-century writers followed. Titles include: M.Y. HALIDOM's *Tales of the Wonder Club* (coll **1899-1900** 3 vols as by Dryasdust; rev as *Tales of the Wonder Club: First Series* 1903, *Second* 1904 and *Third* 1905, all as by Halidom); Jerome K. JEROME's *After Supper Ghost Stories* (coll **1891**); Andrew LANG's *The Disentanglers* (coll of linked stories **1902**); G.K. CHESTERTON's *The Club of Queer Trades* (coll **1905**); William Dean HOWELLS's *Questionable Shapes* (coll **1903**) and *Between the Dark and the Daylight* (coll **1907**); Robert Hugh BENSON's *A Mirror of Shalott, Composed of Tales Told at a Symposium* (coll **1907**); SAKI's *The Chronicles of Clovis* (coll **1907**); and Alfred NOYES's verse *Tales of the Mermaid Tavern* (coll **1914**).

CSs from after 1918 begin to exude a sense of nostalgia, understandable when clubs themselves began to seem bastions of another age; but by this point the form itself had acquired considerable momentum, and sequences like Lord DUNSANY's **Jorkens** books were clearly successful, though a tendency to revert to the tall tale may have marked the increasing artifice of the form. Other CS texts include: John BUCHAN's *The Runagates Club* (coll **1928**), with various of his novels, such as *The Dancing Floor* (**1926**), deploying the same structure; *The Salzburg Tales* (coll **1934**) by Christina Stead (1902-1983); T.H. WHITE's *Gone to Ground* (coll of linked stories **1935**); Isaac ASIMOV's *Azazel* (coll of linked stories **1988**) and some stories assembled in «Magic: The Final Fantasy Collection» (coll 1996); and John Gregory BETANCOURT's *Slab's Tavern and Other Uncanny Places* (coll **1991** chap). [JC]

CLUTE, JOHN (FREDERICK) (1940-   ) Canadian-born critic and writer, in the UK from 1969. Though most of his work has been sf criticism (◊ SFE), he has also published considerable criticism devoted to fantasy. Collections of nonfiction pieces are *Strokes: Essays and Reviews 1966-*

1986 (coll **1988** US) and «Look at the Evidence: Essays and Reviews» (coll dated 1995 but 1996). He served as Reviews Editor of *Foundation: The Journal of Science Fiction* 1980-90, and was a founder of *Interzone* in 1982; he remains Advisory Editor of that magazine and since 1986 has, most months, contributed a review column. He was Associate Editor of the first edition of *The Encyclopedia of Science Fiction* (**1979**) ed Peter Nicholls (1939-), which won a Hugo AWARD; JC was co-editor with Nicholls of the substantially expanded second edition (1993; rev 1995 US; exp vt *Grolier Science Fiction: The Multimedia Encyclopedia of Science Fiction* 1995 CD-ROM US), which is the text referred to in this encyclopedia as the *SFE*; it received several AWARDS, including the Hugo (again) and a BSFA Special. JC began publishing fiction of genre interest with "A Man Must Die" for *New Worlds* in 1966; a novel, *The Disinheriting Party* (1973 *New Worlds*; much exp **1977**) embeds a FANTASY OF HISTORY within a delusional frame (◊ DELUSION). In 1994 he received the Pilgrim Award.   [JC]
Other works: *Science Fiction: The Illustrated Encyclopedia* (**1995**).
As editor: *The Aspen Poetry Handbill* (portfolio **1965** US), associational; *Interzone: The 1st Anthology* (anth **1985**) with Colin GREENLAND and David PRINGLE; *Interzone: The 2nd Anthology* (anth **1987**) with Greenland and Pringle; *Interzone: The 3rd Anthology* (anth **1988**) with Pringle and Simon Ounsley; *Interzone: The 4th Anthology* (anth **1989**) with Pringle and Ounsley; *Interzone V* (anth **1991**) with Lee Montgomerie and Pringle.

COATLICUE ◊ GODDESS.

COATSWORTH, ELIZABETH (JANE) (1893-1986) US poet and writer, whose 90 or more CHILDREN'S FANTASIES are of interest, though many were written for younger children, including the first, *The Cat and the Captain* (**1927** chap), and the most famous, *The Cat Who Went to Heaven* (**1930** chap) – in which, by painting it into a picture, a Japanese artist sends his cat to HEAVEN. EC remains best-known for the **Incredible Tale** sequence of thematically linked stories – *The Enchanted: An Incredible Tale* (**1951**), *Silky: An Incredible Tale* (**1953**), *Mountain Bride: An Incredible Tale* and *The White Room* (**1958**) – all transfiguring their New England settings into venues subtly irradiated by a sense of the immanence of FAERIE. In the first, a man comes to farm near a mysterious plantation called "The Enchanted", which proves to be a POLDER, and the bride he weds proves to have undergone METAMORPHOSIS from a BIRD. A similar emotional congruence between this and another world governs the much later *Pure Magic* (**1973**; vt *The Werefox* 1975; vt *The Fox Boy* 1975 UK), whose protagonist can metamorphose into a fox, and *Marra's World* (**1975** chap), whose eponymous protagonist proves to be a HALFLING: half-human, half-SELKIE.   [JC]
Other works for younger children (selected): *Knock at the Door* (**1931** chap); *Cricket and the Emperor's Son* (**1932** chap); *The Princess and the Lion* (**1963** chap); *Troll Weather* (**1967** chap); *The Snow Parlor and Other Bedtime Stories* (coll **1971** chap); *All-of-a-Sudden Susan* (**1974** chap).

COBALT, MARTIN Pseudonym of William MAYNE.

COBB, IRVIN S. (1876-1944) US writer, mostly of SUPERNATURAL FICTION, though he wrote some SLICK FANTASY; most of the best of his work was assembled in seven collections: *The Escape of Mr Prim: His Plight, and Other Plights* (coll **1913**); *From Place to Place* (coll **1920**); *Sundry Accounts* (coll **1922**); *The Snake Doctor and Other Stories* (coll

1923); *On an Island That Cost $24.00* (coll **1926**); *This Man's World* (coll **1929**) and *Faith, Hope and Charity* (coll **1934**), the 1930 title story of which , along with "Fishhead" (1913), being his most popular tales.   [JC]

COCHRAN, MOLLY (1949-) US writer who has always worked in collaboration, usually with her husband, Warren MURPHY, largely for the **Destroyer** sequence. Her first non-**Destroyer** collaboration, *Grandmaster* (**1984**) with Murphy, closely reflects the structure of the **Destroyer** series as it features a contemporary HERO with TALENTS who is the REINCARNATION of a GOD (in this case, Brahma); in keeping with *Grandmaster*'s focus on CHESS, the hero is known as the Black King, though his powers are rather more Queen-like, as he is almost invincible. Of greater interest is *The Forever King* (**1992**), again with Murphy, a CONTEMPORARY FANTASY involving the AVATARS of ARTHUR, MERLIN and GALAHAD. The tale is exuberant, gritty and unruly. It ends with a sequel in view.   [JC]
Other works: *High Priest* (**1987**) and *The Hand of Lazarus* (**1988**), both with Murphy.

COCTEAU, JEAN(-MAURICE-EUGÈNE-CLÉMENT) (1889-1963) French artist, poet, writer, journalist and moviemaker who established a reputation before WWI as a dandy and enfant terrible, with volumes of poetry like *La Lampe d'Aladin* (coll **1909**); but who came into his full fame – a fame he treated as a commodity to sustain his lifestyle, but which he mocked – with his involvement in *Parade* (**1917**), a ballet produced by Serge de Diaghilev (1872-1929) with sets by Pablo Picasso (1881-1973), music by Erik Satie (1860-1925), and written and clamorously supervised by a MAGICIAN-like JC (◊◊ COMMEDIA DELL'ARTE). With its camp but sometimes searing iconoclasm, and its dismantling of any secure sense of REALITY, this enterprise shared with the various artistic movements of the time – from Futurism and Dada to SURREALISM – a defiant rejection of versions of the world that pre-dated the apocalypse of WORLD WAR I, while also remaining bound to the MYTHS and ICONS of that old world. For the rest of his long and prolific career (he wrote over 100 books in various genres), JC continued to espouse (and to be trapped by) this aftermath aesthetic.

His first fiction, *Le Potomak* ["The Potomak"] (graph **1919**; rev 1924), a cartoon/text/cartoon presentation verging on Absurdism, centres on the eponymous aquatic monster, whose presence in Paris evokes responses of the utmost frivolity in those aware of its formless menace; a sequel, *Le Fin du Potomak* ["The End of the Potomak"] (**1940**), takes a similar attitude to a civilization lunging into WORLD WAR II. Many of his plays are conspicuously full of elements of the FANTASTIC. The best selection is *The Infernal Machine and Other Plays* (coll trans **1963** UK), which contains: *Les Mariés de la tour Eiffel*, a ballet-script here trans as "The Eiffel Tower Wedding Party"; *Bacchus* (**1952**), in which the central figure in a German CARNIVAL induces his own death; and *La Machine infernale* (**1934**; first trans as *The Infernal Machine* 1936 UK), which retells Sophocles's *Oedipus Rex*. His other works for the theatre include: *Oedipe Roi* (**1927**; trans C. Wildman as *Oedipus-Rex* 1962 US), an earlier adaptation of Sophocles's play which was itself transformed into an opera-oratorio with music by Igor Stravinsky (1882-1971) (◊ OPERA); and *Les Chevaliers de la Table Ronde* (**1937**; trans as *The Knights of the Round Table* 1963 UK), in which a TRICKSTER-like MERLIN subjects "Artus" and his court to a malign ENCHANTMENT (◊ ARTHUR).

Today, however, JC remains most honoured for his work in CINEMA, most importantly for those movies he both scripted and directed. (The nonfantasy *Les Enfants Terribles* [*1950*] was based on his *Les Enfants Terribles* [*1929*].) *Three Screenplays* (coll trans Carol Martin-Sperry **1972** US) contains his scripts for *L'Eternel Retour* (**1943**), which retells the Tristan story (◊ ARTHUR) but which he did not direct himself, *La* BELLE ET LA BÊTE (*1946*) and ORPHÉE (*1949*), which is based on the play *Orphée* (**1926**; trans as *Orpheus* **1933** UK). The sequel to the latter, *Le* TESTAMENT D'ORPHÉE (*1959*), here trans as "The Testament of Orpheus", appears in *Two Screenplays* (coll trans **1968**). These movies combine a sophisticatedly dreamlike intensity with genuine fantasy narrative structures, though the two devoted to a remarkably difficult-to-parse TWICE-TOLD version of ORPHEUS-as-JC are not for the lazy viewer.

JC was not a significant narrative artist, but despite his uneasy contempt for the world which had died around 1914, he had a magpie love for the fragments of that world – for bits of STORY, protrusions of MYTH into mundanity – that did survive. In the end, he was a guardian of these fragments. [JC]

COLD IRON The inherent MAGIC of SMITHS and their forges requires that the resulting iron should have special properties: traditionally it resists or dissolves enchantment. Thus FAIRIES cannot pass a threshold guarded with CI, usually in horseshoe form; the touch of CI terminates well meaning fairy BONDAGE in Rudyard KIPLING's "Cold Iron" in *Rewards and Fairies* (coll **1910**), and the hero of Poul ANDERSON's *Three Hearts and Three Lions* (**1953**) may not take CI into FAERIE – where weapons are of lighter metal, perhaps aluminium (Mercedes LACKEY later imagined Californian ELVES building customized racing cars containing no iron at all). Only the CI goods remain when ELEMENTALS pillage the hero's SHOP in Robert A. HEINLEIN's *Magic, Inc.* (**1940**); only CI chains can bind the goddesses in A.E. VAN VOGT's *The Book of Ptath* (**1947**); magically created GOLD reverts to sand on contact with CI in *The Castle of Iron* (**1950**) by L. Sprague DE CAMP and Fletcher PRATT; a CI bridle tames a Faerie horse in R.A. MACAVOY's *The Grey Horse* (**1987**). *Lords and Ladies* (**1992**) by Terry PRATCHETT suggests that elves rely on magnetic senses and are blinded by proximity to CI. More generally, increasing human use of CI tends to be held responsible for THINNING and the departure of FAIRY folk, and its anti-magical properties are a recurring PLOT DEVICE. [DRL]

COLE, ADRIAN (CHRISTOPHER SYNNOT) (1949-   ) UK writer who began publishing with "Wired Tales" for *Dark Horizons* in 1973, and who has tended to create worlds – generally with a PLANETARY-ROMANCE structure – in which a fantasy tone and fantasy trappings cohabit with sf underpinnings, sometimes rather uneasily. His first sequence of fantasy stories, the tales about an ACCURSED WANDERER named **The Voidal**, peaked with *The Coming of the Voidal* (**1977** chap); they were not serious. AC's first full-blown sequence, the **Dream Lords** series – *A Plague of Nightmares* (**1975** US), *Lord of the Nightmares* (**1976** US) and *Bane of Nightmares* (**1976** US) – was both ambitious and ill-hewn, placing planetary-romance adventures, some extremely dark, in a renamed Solar System. The effect was confused. The **Omaran Saga** – *A Place Among the Fallen* (**1986**), *Throne of Fools* (**1987**), *The King of Light and Shadows* (**1988**) and *The Gods in Anger* (**1988**) – is more put-together, and has been compared to Stephen R.

DONALDSON's **Covenant** sequence because the planetary-romance venue is almost irrecoverably poisonous and because the obsessive planet-wide refusal of belief in MAGIC or TALENTS is conveyed, through the actions of a vigilante group which kills off believers or the talented, with an edge of bleak hysteria almost worthy of Donaldson's harsher insights into the stressful side of inhabiting a world dense with meanings its protagonists may wish to deny. As the sequence progresses, a Celtic tone (◊ CELTIC FANTASY) begins to prevail, and the prevailing darkness of tone begins to acquire TWICE-TOLD resonances, a sense that ancient truths of STORY are being re-enacted with force and clear vision.

One further singleton is of fantasy interest, *Blood Red Angel* (**1993**), in which a DYING-EARTH-like social hierarchy turns out to represent stages in the METAMORPHOSIS of a complex race.

AC's work is not remarkable for concision or stylistic strength; but he is deeply knowledgeable about the sources of his craft and the mythic bent of the stories he wishes to tell. He could easily explode into the first rank of fantasy writers. [JC]

Other works: *Madness Emerging* (**1976**), sf/horror; *Bodoman of Sor* (**1977** chap) as Norma N. Johns, a spoof of John NORMAN's **Gor** novels; *Paths of Darkness* (**1977**), sf/horror; *Longborn the Inexhaustible* (**1978** chap); *The LUCIFER Experiment* (**1981**), sf; *Wargods of Ludorbis* (**1981**), sf; *Moorstones* (**1982**) and *The Sleep of Giants* (**1983**), YA fantasies; the **Star Requiem** sequence, being *Mother of Storms* (**1989**), *Thief of Dreams* (**1989**), *Warlord of Heaven* (**1990**) and *Labyrinth of Worlds* (**1990**), with fantasy PORTALS but essentially sf.

COLE, ALLAN (1943-   ) US tv scriptwriter (e.g., *The Rockford Files*) and author, all of whose work has so far been in collaboration with Chris BUNCH, and who concentrated intially on sf. Their only fantasy work is contained in the **Antero** sequence – *The Far Kingdoms* (**1993**), *The Warrior's Tale* (**1994**) and *Kingdoms of the Night* (**1995**) – a set of HEROIC FANTASIES starring, in Amalric Antero himself, a strapping but melancholic yearner after action and ultimate repose, hopefully the latter to be achieved only after much zestful effort and, with an air of mild transcendence, in some sort of SHANGRI-LA. The venues bear a LAND-OF-FABLE resemblance to the medieval Middle East. [JC]

COLEMAN, CLARE ◊ Clare BELL.

COLERIDGE, SAMUEL TAYLOR (1772-1834) English poet, critic and philosopher whose work, and association with William Wordsworth (1770-1850), sowed the seeds of English ROMANTICISM. An impressionable and precocious child, STC was heavily influenced by his reading of FAIRY-TALES and the *Arabian Nights* (◊ ARABIAN FANTASY), so that by the time he entered his teens he was already "habituated to the Vast", as he later reported. STC conceived with Robert SOUTHEY – who became a close friend and in-law (their wives were sisters) – the concept of a community where all were equal, a Pantisocracy, which they were going to establish on the banks of the Susquehanna River in New England, but the project collapsed through lack of funds. By 1795 he had allied himself to the German Romantics, and his discovery of metaphysics widened his appreciation of the supernatural and his belief in a world beyond his PERCEPTION of REALITY. His increasing reliance on opium further shifted his imagination. All of this cosmic perception was channelled into a series of poems written from the summer of 1797 to the spring of 1798, including "The Rime of the Ancient Mariner" (1798 in *Lyrical Ballads* with Wordsworth; rev 1800), and first drafts of the incomplete "Christabel" (Part 1 1797; Part 2 1800) and "Kubla Khan" (1816), both in *Christabel and Other Poems* (coll **1816**). All make extensive use of supernatural imagery. "The Rime of the Ancient Mariner" (which was plotted with Wordsworth, who contributed a few lines), as well as evoking the pandimensional mysteries of the sea, describes a RITE OF PASSAGE and supernatural (or divine) retribution. The episode of the ghost ship is suggestive of BELATEDNESS and hints at an association between the Ancient Mariner and other ACCURSED WANDERERS like the WANDERING JEW and the FLYING DUTCHMAN. The poem is refreshingly devoid of Gothic trappings, emphasizing STC's importance in the transition from GOTHIC FANTASY to Romanticism. "Christabel" made stronger use of Gothic imagery – indeed, it induced nightmare visions in Percy Bysshe SHELLEY – and in STC's depiction here of the EVIL within humankind, especially the vampirical LAMIA, he sought to distinguish between the *super*natural of "The Rime of the Ancient Mariner" and the *preter*natural, or the abnormal within the natural. STC fell short of completing a trilogy of the fantastic when his anonymous caller (the famed Person from Porlock) interrupted the composition of "Kubla Khan", which was developing into a depiction of HEAVEN of XANADU to set against the HELL of "Christabel" and the PURGATORY of the Ancient Mariner, thus representing a romantic version of DANTE's *Divina Commedia*; some modern critics believe that the "Person from Porlock" tale is a myth, and that "Kubla Khan" is complete as it is. Despite writing for a further 30 years, STC never returned to the peak of supernaturalism marked by this brief outpouring, but it changed the world of SUPERNATURAL FICTION for ever. [MA]

COLERIDGE, SARA (1802-1852) UK poet, translator and writer for children, the only daughter of Samuel Taylor COLERIDGE, whose works she subsequently edited. She produced the remarkable *Phantasmion* (**1837**), modelled on Edmund SPENSER's *The Faerie Queen* (**1589-96**), telling of the adventures of the eponymous Prince in the land of FAERIE. It was the first FAIRYTALE novel written in English. [MA]

COLL, JOSEPH CLEMENT (1881-1921) Influential US illustrator whose work in pen and ink acheived a range and depth comparable to painting. He has been quoted as an inspirational source by many modern masters of fantasy ILLUSTRATION. His line quality ranges from the finest subtle light to bold slashes of black shadow with a masterful use of white areas in a tightly controlled composition. His work shows a strong imaginative quality which is displayed at its best in his illustrations of exotic subjects.

JCC's early work was as a newspaper sketch artist, and he later illustrated several books and contributed work for magazines such as *Collier's*, *Everybody's* and the *Associated Sunday Magazine*, embellishing the work of leading authors including Arthur Conan DOYLE and Sax ROHMER. [RT]
Further reading: *The Magic Pen of Joseph Clement Coll* (1978) by Walt Reed.

COLLIER, JOHN (HENRY NOYES) (1901-1980) UK writer, screenwriter and poet best known for his highly polished, often bitterly flippant magazine stories, many of which are SLICK FANTASY. The novel *His Monkey Wife, or Married to a Chimp* (**1930**) is a barbed comic fantasy whose chimpanzee heroine Emily ($\Diamond$ APES; BEAUTY AND THE

BEAST), though not a TALKING ANIMAL, has like FRANKENSTEIN's MONSTER learned English (and typing); when her beloved human owner is to marry an unsympathetic girl, Emily substitutes herself at the altar and the book ends – wickedly deadpan – on the brink of discreetly blissful consummation.

JC's short work has been rearranged in several collections. *The Devil and All* (coll **1934**) is fantasy. *Green Thoughts* (1931 *Harper's*; **1932** chap), a satirical POSTHUMOUS FANTASY in which victims of a carnivorous plant become pseudo-flowers, and *Variation on a Theme* (**1935** chap) feature in both *Green Thoughts and Other Strange Tales* (coll **1943** US) and the more substantial *Presenting Moonshine* (coll **1941** US). *The Touch of Nutmeg, and More Unlikely Stories* (coll **1943** US) draws on *The Devil and All* and *Presenting Moonshine*; selections from all these appear with new material as *Fancies and Goodnights* (coll **1951** US; cut 1965 US; differently cut vt *Of Demons and Darkness* 1965 UK), which won JC the INTERNATIONAL FANTASY AWARD. A definitive edition is *The John Collier Reader* (coll **1972** US; cut vt *The Best of John Collier* 1975 US), introduced by Anthony BURGESS.

Fantastic elements regularly appear in these finely written TALL TALES and CONTES CRUELS – often featuring murderous wives and/or husbands and often with a sting in the tail. "Bottle Party" complicates the bottled-GENIE legend; "Evening Primrose" depicts (with some pathos) a WAINSCOT society posing as mannequins in a department store ($\Diamond\Diamond$ MANNEQUIN [*1987*]; SHOP); "Fallen Star" plays comically with an ANGEL and DEVIL subjected to psychoanalysis; in "Halfway to Hell" a suicide avoids HELL by trickery and LEARNS BETTER; "The Lady on the Grey" is a quiet HORROR story of a modern CIRCE; "The Devil, George and Rosie" features SATAN, a demented cosmological view of the Universe as a pint of beer in which the nebulae are rising bubbles ($\Diamond$ GREAT AND SMALL), and a rewrite of the ORPHEUS myth; in "Thus I Refute Beelzy" harsh parental scepticism about an INVISIBLE COMPANION proves fatal; "Sleeping Beauty" offers a modern SLEEPING BEAUTY who is a grave disappointment when woken; a dummy used for VENTRILOQUISM comes unhelpfully alive in "Spring Fever" ($\Diamond$ PYGMALION). The famous "The Chaser" is set in a magic shop whose proprietor offers a cheap POTION compelling undying, clinging LOVE, while talking constantly of his expensive and untraceable poison; as a besotted young man departs with the love-philtre, the farewell is *"Au revoir"* – encapsulating JC's smiling misanthropy.

JC's best stories are touched with poetry and real wit, sometimes reminiscent of SAKI's. There are moments of outrageous GRAND GUIGNOL; the occasional sexual naughtiness is far beyond Thorne SMITH in sophistication. JC remains eminently readable. [DRL]
Other works: *No Traveller Returns* (**1931** chap) and *Tom's A-Cold* (**1933**; vt *Full Circle* 1933 US) are sf; *Defy the Foul Fiend* (**1934**), associational; *Witch's Money* (**1940** chap US); *Pictures in the Fire* (coll **1958**); *Milton's "Paradise Lost": Screenplay for Cinema of the Mind* * (**1973**).
See also: HUMOUR; PACTS WITH THE DEVIL.

COLLINS, NANCY A(VERILL) (1959-) US writer whose first novel, *Sunglasses After Dark* (**1989**), introduced **Sonja Blue**, a punk female VAMPIRE with human sensibilities. The book quickly gained a cult following, with the result that the author was soon absorbed into the short-lived Splatterpunk movement and went on to win the BRAM

STOKER AWARD for First Novel and the British Fantasy Society's Icarus Award. NC has continued the **Sonja Blue** cycle with *In the Blood* (**1992**), *Paint it Black* (**1995**) and *A Dozen Black Roses* (**1996**); the first three volumes were assembled as *Midnight Blue: The Sonja Blue Collection* (omni **1995**). She has also written extensively for COMICS, notably for DC/Vertigo's **Swamp Thing** (1991-3) and Verotik Publications' **Verotika** (1995-6), and was founder of the International Horror Critics Guild. NC is married to underground moviemaker and anti-artiste Joe Christ. [SJ]
Other works: *Tempter* (**1990**); *The Tortuga Hill Gang's Last Ride: The True Story* (**1991** chap); *Cold Turkey* (**1992** chap); *Wild Blood* (**1993**); *Nameless Sins* (coll **1994**); *Walking Wolf* (**1995**); *The Fantastic Four: To Free Atlantis* * (**1995**); «Angels on Fire» (**1996**).
As editor: *Forbidden Acts* (anth **1995**) and *Dark Love* (anth **1995**), both with Edward E. Kramer and Martin H. GREENBERG; «Gahan Wilson's Ultimate Haunted House» (**1996**), ghost-edited.

COLLINS, (WILLIAM) WILKIE (1824-1889) UK writer, renowned as the master manipulator of "sensation" fiction of the 1850s and 1860s. He is more important to the development of detective fiction than of SUPERNATURAL FICTION through *The Woman in White* (1859-60 *All the Year Round*; **1860**), with its Gothic residue, and *The Moonstone: A Romance* (**1868**), one of the earliest police procedurals, which also established a subgenre of accursed-jewel stories. Nevertheless, influenced by Edward BULWER-LYTTON and Charles DICKENS, WC delighted in supernatural and terror fiction. In his day he was immensely popular although his private life (he lived with one mistress but sired three children by another) shocked many; only now is he being fully rediscovered.

WC was fascinated with the workings of FATE, which he exploited in early stories "The Ostler" (1855 *Household Words*; rev as "The Dream Woman" 1874; vt "Alice Warlock") and "The Monkstons of Wincot Abbey" (1855 *Fraser's*; vt "Mad Monkton"), both of which featured DREAM premonitions. He used the theme with considerable effect in *Armadale* (1864-6 *Cornhill*; **1866**), which links fate, dreams and MULTIPLE PERSONALITY. He remixed these themes in the less successful romantic novel *The Two Destinies: A Romance* (**1876**), though this makes early use of telepathy (◊ TALENTS), and again in *The Legacy of Cain* (**1888**), where destiny takes on a pseudoscientific veneer through genetics. *The Haunted Hotel: A Mystery of Modern Venice* (1877 *Illustrated London News*; **1878**) is a residual Gothic and over-atmospheric ghost story.

Other shorter GHOST STORIES, including themes of pre-destination, dreams and shadow personalities, appear in the collections *After Dark* (linked coll **1856**), *The Queen of Hearts* (linked coll **1859**), *The Frozen Deep, and Other Stories* (coll **1874**), *The Ghost's Touch* (**1885** US; rev vt *I Say No* 1886 US) and *Little Novels* (coll **1887**). Recent retrospectives include *Tales of Terror and the Supernatural* (coll **1972**) ed Herbert van Thal, *The Illustrated Wilkie Collins* (coll **1989**) ed Peter HAINING and *The Best Supernatural Stories of Wilkie Collins* (coll **1990**) ed Haining.

Most of WC's works can be considered either as early PSYCHOLOGICAL THRILLERS or as RATIONALIZED FANTASIES.
[MA]
Further reading: *Wilkie Collins: A Critical and Biographical Study* (**1977**) by Dorothy L. Sayers; *Wilkie Collins: An Annotated Bibliography 1889-1976* (**1978**) by Kirk H. Beetz.

Other works: *Antonina, or The Fall of Rome: A Romance of the Fifth Century* (**1850**); *Mr Wray's Cash Box, or The Mask and the Mystery* (**1852**); *Basil: A Story of Modern Life* (**1852**); *Hide and Seek* (**1854**); *The Dead Secret* (**1857**); *No Name* (**1862**); *Man and Wife* (**1870**); *Poor Miss Finch* (**1872**); *The New Magdalen* (**1873**); *The Law and the Lady* (**1875**); *A Rogue's Life: From His Birth to His Marriage* (**1879**); *Jezebel's Daughter* * (**1880**), novelization of *The Red Vial* (play 1858); *The Black Robe* (**1881**); *Heart and Science: A Story of the Present Time* (**1883**); *I Say No* (**1884**); *The Evil Genius: A Domestic Story* (**1886**); *The Guilty River* (**1886**); *Blind Love* (**1890**), completed by Walter BESANT.

COLLODI, CARLO Writing pseudonym of Italian writer, editor and freedom-fighter Carlo Lorenzini (1826-1890), best-known as author of *Le Avventure di Pinocchio* (1881 *Giornale dei bambini*; **1883**; trans M.A. Murray as *The Story of a Puppet, or The Adventures of Pinocchio* 1892 UK). CC helped fight for Italian independence during the 1860s and after 1870 endeavoured to settle down as a theatrical censor and magazine editor, but soon turned to CHILDREN'S FANTASY, first translating the FAIRYTALES of Charles PERRAULT in 1875 and then starting a series of didactic fictions. *Pinocchio*, one such moral tale, was by the turn of the century established as a children's classic. It is less read today, having been supplanted in popular consciousness by the DISNEY movie PINOCCHIO (*1940*), but its theme of a naughty child who LEARNS BETTER as he tries to be a REAL BOY is perennial.
[MA]

COLOMBO, JOHN ROBERT (1936-   ) Canadian writer, poet and anthologist whose 125 books, almost all dealing with CANADA, include several of genre interest. *The Great Wall of China: An Entertainment* (coll **1966** chap) and *Mostly Monsters* (coll **1977**) are volumes of found poetry whose intersections are fantastic; ANTHOLOGIES of genre fiction (often comprising the first volumes of such work to be published in Canada) include *Other Canadas: An Anthology of Science Fiction and Fantasy* (anth **1979**), *Friendly Aliens: Thirteen Stories of the Fantastic Set in Canada by Foreign Authors* (anth **1981**), *Not to be Taken at Night: Thirteen Classic Canadian Tales of Mystery and the Supernatural* (anth **1981**), *Windigo: An Anthology of Fact and Fantastic Fiction* (anth **1982**) and *Worlds in Small: An Anthology of Miniature Literary Compositions* (anth **1992** chap).

Years of Light: A Celebration of Leslie A. Croutch (**1982**) examines the life of a Canadian sf fan and writer. JRC's bibliographies include *CDN SF&F: A Bibliography of Canadian Science Fiction and Fantasy* (**1979** chap) with Alexandre L. Amprimoz, John Bell and Michael Richardson; and *Blackwood's Books: A Bibliography* (**1981**), devoted to Algernon BLACKWOOD, now superseded by the work of Mike ASHLEY. JRC's command of his territory is magisterial. [JC]
Nonfiction: *Colombo's Book of Marvels* (**1979**; exp vt *Mysterious Canada: Strange Sights, Extraordinary Events, and Peculiar Places* 1988); *Extraordinary Experiences: Personal Accounts of the Paranormal in Canada* (anth **1989**); *Mysterious Encounters: Personal Accounts of the Supernatural in Canada* (anth **1990**); *Mackenzie King's Ghost; And Other Personal Accounts of Canadian Hauntings* (anth **1991**); *Dark Visions* (**1992**); *The Mystery of the Shaking Tent* (**1993**).

COLOUR-CODING Fantasy novels, especially those designed for readers seeking the comforts of familiarity, present their worlds in black and white. HEROES AND HEROINES are easy to distinguish from DARK LORDS; good

COMPANIONS (though there may be a ringer) do not resemble minions; a well governed country is a different colour from one in the grip of a usurper. These CC distinctions have frequently been derived, perhaps unknowingly, from the works of J.R.R. TOLKIEN, himself influenced by earlier models. According to the unwritten tenets of this form of CC, the mesomorphic body-type and "healthy" light complexion supposedly common to human beings of Northern European ethnic background are generally valued highly; while other body-types and complexions are deemed expressive of villainy, or of lower status, or of "Eastern" origin (whether or not the FANTASYLAND depicted is modelled on our own planet). Blue eyes, unless unnaturally pale, are valued; black eyes, especially in sexually active women, are signs of turpitude. Clothing and other gear follow the same principle: moderately bright but sensible is diagnostic of virtue; black connotes evil. Green fields are good (and hint at good governance), while swamps makes it clear the head of state is rotten.

More interestingly, CC is sometimes built into the explicit nature of the fantasyland, so that parts of the world may literally be limned in different colours. In tales set in WONDERLANDS, whose REALITY is subject to the constant operation of arbitrary rules, this colour-coding comes as no surprise; a modern example is Nancy KRESS's *The Prince of Morning Bells* (**1981**), which describes the land of the Quirks. L. Frank BAUM's famous **Oz** sequence has an underlying wonderland timbre; OZ is famously divided into four colour-coded provinces whose inhabitants share the colour of the land, the capital, Emerald City, being named after its own hue. Among the various homages to **Oz** in *Amnesia Moon* (**1995**) by Jonathan Lethem (1964-) is the colour-coding of its post-HOLOCAUST world. BEAST FABLES (though not ANIMAL FANTASIES) tend to exhibit CC; an example is Brian JACQUES's **Redwall** sequence. FANTASTIC VOYAGES – like François de la Mothe FÉNELON's *The Adventures of Telemachus* (**1699**) – are often colour-coded, though tales set in this world tend to slide from explicit CC as an aesthetic device downwards into anthropological typecasting. Full SECONDARY-WORLD fantasies may have been constructed with CC in mind, but are normally too complexly conceived for this to be more than a nuance, except perhaps in those tales set in ARCHIPELAGO venues, where natural differentiation (as in the Galapagos Islands) may lead to moral distinctions. A recent example is Felicity SAVAGE's *Humility Garden* (**1995**). [JC]

COLUM, PADRAIC (1881-1972) Irish writer. ◊ James Branch CABELL; CHILDREN'S FANTASY.

COLUMBINE ◊ COMMEDIA DELL'ARTE.

COLVIN, JAMES [hn] ◊ Michael MOORCOCK.

COMEDY ◊ HUMOUR.

COMFORT, MONTGOMERY [s] ◊ Ramsey CAMPBELL.

COMICS Before discussing FANTASY and HORROR comics, it is important first to recognize what comics are. Various definitions have been offered over the years. Maurice Horn in *The World Encyclopedia of Comics* (**1976**) offered: "A narrative form containing text and pictures arranged in sequential order (usually chronological)." Scott McCloud in *Understanding Comics* (**1994**) offered "juxtaposed pictorial and other images in deliberate sequence, intended to convey information and/or to produce an aesthetic response in the viewer" – which he subsequently shortened to "sequential art". In its simplest definition, any series of pictures that can convey information or a story if followed in a preset order

can be considered a comic. A number of comic-strip historians have considered the origins of sequential art, back through the Bayeux Tapestry, the exploits of Hercules as shown on Greek artefacts, and the earliest cave drawings. The Egyptians provided their dead with a detailed scroll depicting what dangers they might face in the AFTERLIFE. One could not sensibly describe any of these as comics.

Although comic strips in the USA and UK have become increasingly more sophisticated, sprawling fantasy LANDSCAPES and SECONDARY WORLDS are normally avoided, since a full fantasy environment would require too much explanation and exposition in a medium that relies on weekly or monthly bursts of action. HEROES are defined by their deeds, and characterization is kept to a minimum to allow new readers easy access to a series that might already have been in publication for years. Comics are intended, in general, for a juvenile or adolescent readership; on average serious "comic collectors" in the USA retain their interest in the medium for a mere four years. In mainland Europe, by contrast, comic strips are more often aimed at an intelligent adult reader. Sequences of comics albums – called *bandes dessinées* ["strip drawings"] in France – containing sometimes self-contained successive episodes of long and often complex stories are kept in print for many years, a luxury not conceived in the USA until a few years ago.

While the comic strip draws on several different traditions, these began to coalesce in the mid-19th century. One could take as the starting-point Gustave DORÉ's *Histoire Pittoresque, Dramatique et Caricaturale de la Sainte Russie* (graph **1854**; trans as *The Rare and Extraordinary History of Holy Russia* **1972** UK), a humorous pictorial history with narrative text under each frame. Wilhelm Busch (1832-1908) produced his *Max und Moritz* (graph **1865**; trans as *Max and Moritz – A Story in Seven Tricks* **1874** UK), the illustrated misadventures of a pair of young boys, told by means of a sequence of captioned panels which differ from those in modern comics only by the absence of speech balloons.

The speech balloon was, however, already a well established convention. One of its first incidences was in the Protat Woodcut (1347), in which virtuous words, black-lettered on a scroll, are shown issuing from the mouth of a nobleman. In the 16th century Queen Elizabeth I decreed that all churches should display a copy of Fox's *Book of Martyrs* illustrated by woodcuts showing gory death scenes in which the last utterances of the saints were lettered in balloons. Thomas Rowlandson (1756-1827) and James Gillray (1757-1815) brought the convention into common use in cartoons in the late 18th and early 19th centuries, and periodicals like *McLean's Monthly* (published in the 1830s), but in comic strips it was not until the very end of the century that speech balloons were used.

In 1867, in the UK, Charles Henry Ross drew the first adventures of **Alley Sloper** – Alexander Sloper, F.O.M. (Friend of Man) – in the weekly humour magazine *Judy*. Although owing more to the illustration tradition of *Punch* (weekly from 1841), these stories were republished as *Alley Sloper: A Moral Lesson* (graph **1872**), a precursor of the modern comic book, and **Alley Sloper** was extensively published well into the 20th century in the UK. Nevertheless, the real origins of the comic strip lie in the USA.

By the mid-1890s US newspapers had begun to experiment with gag panels and illustrations. The 5 May 1895 edition of the *New York World* featured two such gag panels by R.F. Outcoult (1863-1928) called **Hogan's Alley**, depicting a group of slum kids; the star was a bald, jug-eared child wearing a nightshirt – the Yellow Kid – and he quickly became enormously popular. In 1896 Outcoult moved to the *New York Journal*, whose proprietor, William Randolph Hearst (1863-1951), suggested adopting a sequential approach to the feature. The artist agreed and **The Yellow Kid** (as it was renamed) effectively became a comic strip a year later. Again at Hearst's instigation, Rudolph Dirks (1863-1928) created **The Katzenjammer Kids**, inspired by *Max und Moritz*, for the *New York Journal*, and the modern form of the popular newspaper comic strip was born.

Others quickly followed, and by the turn of the century there were several strips; most featured slapstick humour in a contemporary setting. There were, however, some changes on the horizon. **The Upside-Downs of Little Lady Lovekins and Old Man Muffaroo** (1903-4; graph coll under same title **1905**) by Gustave Verbeek (1867-1937) can lay some claim to being the world's first fantasy comic strip: its two protagonists adventured in a world of bizarre creatures and strange landscapes. (Verbeek here also innovated the idea of creating a strip that could also be read upside-down – rotating the page gave the conclusion of the *same* six-panel story, each title character becoming the other when inverted – but no other artist has yet had the courage to pick up this particular gauntlet.) Verbeek followed this with the rather more nightmarish **Terrors of the Tiny Tads** (1905-16), which appeared along with another new strip, Winsor MCCAY's LITTLE NEMO IN SLUMBERLAND (1905-27). McCay's strips' mixture of architectural detail, bizarre characters and almost Freudian situations spawned a number of imitators; among the more important were **Nimsby the Newsboy** (1905-6) by George McManus (1884-1954), **Danny Dreamer** (1907-12) by Clare Briggs (1875-1930), **The Explorigator** (1908) by Harry Grant Dart (1869-1938) and **Bobby Make Believe** by Frank King (1883-1969). King later found enormous success as the creator of **Gasoline Alley** which, though very much based in reality – to the extent that its characters actually aged (a first for comics) – still had moments of SURREALISM on its Sunday pages. Similarly Cliff Sterrett (1883-1964) transformed the domestic humour strip **Polly and her Pals** (1912-58) into something quite different. By the 1920s Sterrett had largely abandoned conventional narrative and graphics in favour of a wildly imaginative abstract approach that blended Cubism, Dada and Surrealism to create a quite disorienting effect. Although they were short-lived, the Lionel Feininger (1871-1956) strips **The Kin-der Kids** (1906) and **Wee Willie Winkie's World** (1906) also eschewed a literal approach, opting instead for an abstract, lyrical direction.

Like Feininger, George HERRIMAN had less interest in narrative than in the formal possibilities of the comic medium. His masterpiece, **Krazy Kat** (1916-44), uses its funny-animal milieu as a device with which to explore his interests in composition, abstraction and landscape. Typical of the early strip artists, Herriman was aware that he was drawing for a largely adult audience who, hopefully, would appreciate the more playful poetic elements of his strip rather than simply the eternal triangle of its main characters. There were also surreal elements – e.g., characters cutting through the cartoon frame and hiding outside, and a shifting Arizona landscape where hand-like

rock formations could clap and cause winds.

As the 1920s came to an end the playful Surrealism of the early strips was gradually replaced by humour and soap opera. Then a new genre emerged that appealed to a younger audience: the adventure strip. 1929 saw the initial episodes of both Edgar Rice BURROUGHS's TARZAN, drawn by Harold FOSTER, and the first sf strip, **Buck Rogers** (1929-67), created by Phil Nowlan (1888-1940) and Dick Calkins (1895-1962). The new adventure strips drew on the more figurative tradition of MAGAZINE illustrators, and were much more realistically drawn.

These inspired other fantasy-based adventure strips: **Brick Bradford** (1933-current) by Clarence Grey (1902-1957) and William Ritt (1902-1972); MANDRAKE THE MAGICIAN (1934-current) by Lee Falk and Phil Davis, and *The* PHANTOM (1936-current) by Falk and artist Ray Moore. Foremost was FLASH GORDON (1934-current) by Alex RAYMOND who, along with Foster and Milton Caniff (1907-   ), inspired several generations of artists with the technical brilliance of his art. One last throwback to the earlier tradition of cartooning was **Alley Oop** (1933-current) by V.T. Hamilin (1900-1993), which started as a PREHISTORIC FANTASY but later, with the addition of a time machine, travelled into other fantastic territory.

Whereas the US strip had taken root in newspapers, in the UK the medium flourished in weekly comic papers. The publication in 1890 of Alfred Harmsworth's *Comic Cuts* and *Chips* set the pattern of humour comics for several decades. By the 1920s UK titles had emerged from the Victorian single-panel tradition to the point where they had become comic strips as we would understand that term today (although they were still largely without speech balloons, favouring captions). Their audience was mostly juvenile, but there was enough variety that they frequently included fantasy elements.

For many years UK fantasy comic strips were confined to mischievous TALKING ANIMALS, and any features that might veer towards HORROR – like **Captain Phantom** (*Knockout* 1952) – were in fact more likely to involve adventurous aircraft pilots. The longest-running of the former were the adventures of **Tiger Tim** and his chums, **The Bruin Boys** and **Mrs Hippo's Kindergarten**, which ran for over 80 years in titles such as *Playbox* and *Rainbow*. Other popular talking-animal strips included **Pip, Squeak and Wilfred** (*Daily Mirror* 1919-40, 1947-55) drawn by Austin B. Payne (1876-1959), which also appeared in various annuals (1923-39 and 1953-5) and a series of short ANIMATED MOVIES (*1922*); and **Teddy Tail**, a mouse with a knot in his tail, visited various fantasy lands (*Daily Mail* 1915-40, 1946-60) – he too appeared in a series of books plus annuals and reprints. **Teddy**, a creation of Charles Folkard (1878-1963), was later drawn by Folkard's brother Harry and various others over the years. Further characters were stars of their own comics. The talking-animal tradition continues to this day.

While a succession of gnomes, pixies and ELVES have been stars in UK nursery comics, other fantasy aspects were relegated to the humorous panels of children's strips. Examples are **Sham Poo and His Magic Wand** (*The Butterfly* 1908) by Ernest Webb and **Uncle Dan the Magic Man** (*The Magic Comic* 1939). More substantial elements could be found in **Pansy Potter the Strongman's Daughter** (*Beano* 1947-53), who found a set of steps in a wishing well which led her to a wonderful land full of nursery-rhyme characters; this was drawn by James Clark

(1895-1977). **Little Nemo**-style daydreams were the basis for the feature **Our Ernie** (1939-60), most notably drawn by Hugh McNeil (1910-1979), and **Buster's Dreamworld** (*Buster* 1968-74) by Angel Nadal.

FAIRYTALES and traditional tales were perhaps the area most plundered by UK nursery comics, dating back at least as early as the 1920s: Aladdin was a favourite theme ($\lozenge$ ARABIAN FANTASY), appearing at various times, although as regular were the stories of Hans Christian ANDERSEN. The most important strip of the period was the more original newspaper strip RUPERT THE BEAR, created for the *Daily Express* in 1920 by Mary Tourtel (1874-1940). **Rupert** was in fact originally a single panel with verse running beneath (though the number of panels later varied, often being four); it became effectively a comic strip when the panels were collected into books or annuals. Under Tourtel – and, more particularly, her successor from 1935, Alfred Bestall (1892-1986) – **Rupert**'s adventures took on a fairytale air, with **Rupert** being transported to distant fantastic lands. The strip became populated by DRAGONS, imps and sorcerers.

The first Disney comic strip, syndicated by King Features, appeared in 1930, with MICKEY MOUSE starring in daily strips. The Disney comics are, however, a bit of a law unto themselves, having little effect on and being little affected by the mainstream; they form no further part of this discussion ($\lozenge$ DISNEY; Carl BARKS; SCROOGE MCDUCK).

The comic book established itself early in the UK, but the USA – though Denis Gifford lists about 300 separate comic-book publications before 1933 in his *The American Comic Book Catalogue: The Evolutionary Era 1884-1939* (**1990**) – really had to wait until 1933 for its first true comic book: *Funnies on Parade*. The early US comic books were dominated by newspaper-strip reprints and it was only in 1935 with *New Fun Comics* that much new material was specially produced for the format. Three years later the company responsible for this, National (later DC COMICS), transformed the industry with *Action Comics #1*, starring SUPERMAN by Jerry Siegel (1914-1996) and Joe Shuster (1914-1992), comics's first SUPERHERO; this strip inspired an army of imitators. Throughout WWII superheroes dominated the comic-book scene, and a host sprang up to cash in on Superman's success. DC hit paydirt again with BATMAN, **The Flash**, WONDER WOMAN and many others, while Timely Comics (later MARVEL COMICS) had **Captain America**, **The Human Torch** and **The Submariner**.

Among the throngs of heroes, several had particularly fantasy-based backgrounds. Foremost were two DC heroes: the avenging GHOST **The** SPECTRE (1940-45) by Bernard Baily (1920-   ), and the Egyptian-flavoured sorcerer **Dr Fate** (1940-44) by Howard Shetman and Gardner FOX; both ran in *More Fun Comics*. Magicians of all sorts were commonplace, from ZATARA (*Action Comics #1*) by Fred Guardineer (1913-   ) to **Supermagician** (Street & Smith 1941-7) and **Ibis the Invincible** (Fawcett's *Whiz Comics* 1940-50 and his own title 1943-8). Ibis's stablemate CAPTAIN MARVEL, created by Bill Parker and Charles Clarence ("C.C.") Beck (1910-1989), rivalled if not surpassed **Superman** in popularity at the time. Where **Superman**'s powers were (supposedly) science-based, **Captain Marvel** gained his abilities from a sorcerer, and the strip in general was more whimsical than its rivals. One of **Captain Marvel**'s main scripters was the PULP author Otto Binder (1911-1975); other writers included Alfred Bester (1913-1987), Edmond Hamilton (1904-1977), Henry

KUTTNER and Manly Wade WELLMAN. Further characters of note were: **Kid Eternity** (in Quality's *Hit Comics* 1942-9) by Sheldon Moldoff (1920-   ), a ghost who could summon any hero in antiquity for the cause of good; **The Heap** (in Hillman's *Air Fighters* and *Airboy* 1942-53) by Harry Stein and Mort Leav (1916-   ), an early example of the swamp-creature subgenre; and the humorous **Frankenstein** (published by Prize 1945-54) by Dick Briefer (1915-   ) ($\lozenge$ FRANKENSTEIN).

As the war years waned, so too did the superheroes (temporarily), being supplanted by genres like romance, crime and funny animals; works in this last field contained the occasional fantasy element. Of particular interest were the creations of Walt Kelly (1911-1973) in the comic book *Fairy Tale Parade* (Dell 1942-6), Otto Messmer (1892-1983) in **Felix the Cat** (Dell 1943-61) and George Carlson (1887-1962) in *Jingle Jangle Comics* (Eastern 1942-9).

By the end of the decade a new genre appeared, and it would dominate the 1950s: HORROR. In fact the first horror comic had appeared as early as 1947 (Avon's EERIE), but it was not until late 1948 that a regularly published horror monthly hit the stands, ACG's *Adventures into the Unknown*. Throughout the early 1950s most publishers produced horror comics of some description – even Timely entered the field with the superhero/horror hybrid *Captain America's Weird Tales* (1949). But one name dominated the arena: EC COMICS, which during 1950-55 published three titles – *The Vault of Fear*, *The Vault of Horror* and *Tales from the Crypt* – that set the standard for intelligently written, superbly illustrated mystery tales, invariably with a horror twist. Borrowing the idea of regular hosts from "The Witch's Tale" segment of the radio show *Now Lights Out*, writer/editor Al Feldstein (1925-   ) brought the comic-book field to new heights of sophistication with the aid of an impressive list of artists. Foremost were Johnny Craig (1926-   ), Wallace WOOD, Graham Ingels (1915-?   ), Jack Davis (1926-   ), Reed CRANDALL, George Evans (1920-   ) and Bernard Krigstein (1919-?   ). An early EC artist was Harry Harrison (1925-   ). EC's tales may well have been gory and bloodthirsty but there was invariably a moral element. Other companies were less fastidious: Harvey's *Black Cat Mystery* (1951-4) and *Tomb of Terror* (1952-4), for example, featured some of the most frightening comics material ever published. By 1955 horror was the biggest single genre, commanding 16% of the US market and boasting a diverse number of themes and approaches.

Timely (now called Atlas) was the most prolific publisher: by 1956 it was bringing out 16 regular mystery or fantasy titles, usually featuring sensationalist covers and rather tamer interiors. The company boasted an artist roster every bit as impressive as EC's, with Russ Heath (1926-   ), Joe Maneely (1926-1958), Basil Wolverton (1909-   ) and Bill Everett (1917-1973); the latter's finest hour was *Venus* (1948-52), which started life as a romance book but later embraced sf and finally horror – albeit with a softer edge than most.

Similarly restrained mystery tales were found in DC's products, initially in the ex-superhero titles *Sensation* and *Star Spangled* (both switched genres in 1952) and then in the new *House of Mystery* comic (1952-82). Other comics followed – *My Greatest Adventure* (1955-63), *House of Secrets* (1956-76) and *Tales of the Unexpected* (1956-82) – all drawing on the same talent pool: Mort Meskin (1917-   ), Ruben Moreira (1922-   ), Billy Eby and others. One exception to the rather anonymous stories favoured by DC was the short-lived **Phantom Stranger** (6 issues 1952-3), which starred the eponymous mysterious host, a device that reappeared a number of years later in Harvey's **The Man in Black** (1957-8) and Charlton's **Tales of the Mysterious Traveler** (1956-9) (◊ The MYSTERIOUS TRAVELER), the latter featuring artwork by Steve DITKO.

Ditko's work had also appeared in an earlier Charlton comic, *The Tying* (1952-4), which carried precisely the sort of material Frederick Wertham (1895-1981) had protested about in his *Seduction of the Innocent* (**1954**). In response to the mounting public outcry, in 1955 the companies set up the Comics Code, which ended the horror boom and sent the industry into recession. It certainly put paid to EC's line: while their *Mad Magazine* took off, *Shock Illustrated* (3 issues 1955-6) and *Tales of Terror Illustrated* (2 issues 1955-6) emphatically did not, although these magazines sowed the seeds for the genre's rejuvenation a decade later.

The 1950s spawned other genres, notably PREHISTORIC FANTASY. M.E's *Thun'da* (1952-3) led the pack, principally because of the fine Frank FRAZETTA artwork in #1. More impressive thematically was *Tor* (1953-4) by Joe Kubert (1926-   ), who later went on to the PRINCE VALIANT-flavoured **Viking Prince** (DC's *Brave and Bold* 1955-9), written by Robert Kaniger (1915-   ); but it was Dell's **Turok, Son of Stone** (◊ Alberto GIOLITTI), pitting Native Americans against DINOSAURS, that was to prove the most enduring (130 issues 1954-82), created primarily by Paul S. Newman (writer) and Giolitti (artist).

Truly horrific tales were not an important feature of UK comics until the late 1940s, when some independent companies tried to emulate the style developing in the USA. The most notable artist working with horror was the imaginative William A. Ward, a one-time animator whose wartime strips for the Gerald G. Swan MAGAZINES included the grisly "angel of death" **Krakos the Egyptian** (*New Funnies* 1941 and *Thrill Comics* 1941-4) and the vengeful **The Bat** (*Thrill Comics* 1940 and *Extra Fun* 1940). Also working for Swan at the time was William McCail (1902-1974), whose *Back from the Dead* (*Topical Funnies* 1941) signed "Ron", was the story of Robert Lovett, who died in 1827 and returned in 1949 to wreak VENGEANCE on criminals. McCail's brother John (*c*1896-   ) created **Dane Vernon, The Ghost Investigator** (*Thrill Comics* 1940-46).

These comics caused a similar public outcry in the UK as there had been in the USA. The Horror Comics Bill (1955) effectively banned their publication. More influential was the home-grown and altogether more clean-cut boys' comic *Eagle* (1950-69), published initially by Hulton Press. With its tabloid format and photogravure colour printing, *Eagle* set the standard for a decade. *Eagle*'s star attraction was the sf strip **Dan Dare**, created by Frank Hampson (1918-1985). Following a dispute in 1960, artist Frank BELLAMY proved a worthy successor to Hampson, but worked on **Dare** for only one year before moving on to other features, including the more fantasy-oriented HEROS THE SPARTAN. Artwork on **Dare** was continued by Keith Watson (1935-1994) until 1967, when *Eagle* decided to run only reprints of old stories; the comic died soon after.

Among *Eagle*'s imitators, fantasy strips featured more frequently. *Express Weekly* had **Wulf the Briton** (1957-60) by Jenny Butterworth and Ron EMBLETON. An even more impressive Embleton strip was the first of the **Wrath of the Gods** stories (*Boys' World* 1963-4, briefly thereafter in *Eagle*), a Greek MYTHOLOGY-based series of tales written

by Michael MOORCOCK. Subsequent stories in this series were well drawn by John M. Burns (1939-   ), who also drew Ken Mennelli's charming **Kelpie, The Boy Wizard**, a SORCERER'S-APPRENTICE-style yarn set in Arthurian Britain (*Wham* 1964-5). Another excellent artist in the same painterly tradition, Don Lawrence (1928-   ) worked on a succession of fantasy strips from the pseudohistorical **Orlac the Gladiator** (*Tiger* 1959-64), written by Brian Leigh, and **Carl the Viking** (*Lion* 1960-64), written by Kenneth BULMER, to the more overtly fantastic **Maroc the Mighty** (*Lion* 1964-6), written by Moorcock, and the dynastic SCIENCE FANTASY **The Trigan Empire** (*Ranger* and *Look and Learn* 1965-82), written by Michael Butterworth (1947-   ). This last strip proved immensely popular in both the UK and Europe but, following a dispute with publishers IPC, Lawrence decamped in 1977 to the Dutch comic *Eppo*. His new creation for them was the even more widely acclaimed **Storm** (1977-current), written mostly by Martin Lodewijk, which has been collected in a series of albums and published throughout Europe – although only three volumes have appeared in the UK: *Storm* (graph **1985**), *Storm: The Last Fighter* (graph **1988**) and *Storm: The Pirates of Pandarve* (graph **1988**).

In 1965 City Publishing started adapting Gerry Anderson's tv puppet shows into strip form in the weekly *TV Century 21*, and over the succeeding five years utilized the talents of most of the top UK artists, including Embleton, Lawrence, Bellamy, Burns and Mike Noble. This magazine's high-quality colour presentation attracted a large readership and spawned spinoff comics like *Joe 90* (with which it merged in 1969) and *Lady Penelope* (1966-9). By the end of the 1960s, however, the popularity of Anderson's material had begun to wane, and City (now called Polystyle) introduced a new weekly comic, *Countdown* (later *TV Action*) in 1971, which featured a brightly coloured strip, also named **Countdown**, by Dennis Hooper and John Burns (using spaceship designs from Kubrick's *2001 – A Space Odyssey* [*1969*]); alongside were strip versions of Anderson's later live-action productions and other tv shows like *Dr Who* and *Mission Impossible*. More generally, the penchant for strip versions of tv shows was a feature of many UK comics in the 1960s and 1970s.

Much of the 1950s, 1960s and 1970s UK comics scene was dominated by the vast Amalgamated Press (subsequently renamed Fleetway, then IPC), and their stable of children's weeklies produced many interesting fantasy strips. *Lion* (1952-74) featured the light-hearted adventures of **Robot Archie** by Ted Kearon and **Adam Eterno** by F. Solano-Lopez. This latter, a TIME-TRAVEL strip, had first appeared in *Thunder* (1970-71); it continued in *Lion* until 1974 and then *Valiant* (1974-6) – which also ran **Kelly's Eye** (1963-74) (another Lopez strip), the monstrous **Mytek the Mighty** (1964-75) and, probably, *Valiant*'s most memorable series, the stylish fantasy adventure **The Steel Claw** (1962-70, 1971-6) by Kenneth BULMER and Jesus Blasco. Also worthy of mention from this period are **The Legend Testers** by Graham Balser and José Bernet and the sinister **Cursiter Doom** by Ken Menell and Ray BRADBURY, both of which ran in *Smash* (1966-71).

The late 1970s saw something of a horror boom in UK comics, with Dez Skinn's (1951-   ) *House of Hammer* (1976-84), which featured adaptations of HAMMER movies, and the notorious *Action* (February-October 1976). *Action* – with strips like the bloodthirsty **Hookjaw** (by Ken Armstrong and Ramon Sola), about a maneating shark, and

Kids Rule (by Chris Lowder [1945-   ], Ron TINER and Mike White), which took its basic premise of unruly kids stranded on a desert island from William GOLDING's *The Lord of the Flies* (**1954**) and wallowed salaciously in the resulting vicious behaviour – managed to be both enormously popular among its child readers and universally condemned in the adult media; the brevity of its life was unsurprising. Even the girls' comics were not unaffected by the horror craze, and certainly all had their fair share of fantasies. Alliteratively styled MERMAIDS – e.g., **Myrtle the Mermaid** (*Bunty* 1958) and **Milly the Mermaid** (*Princess Tina* 1967) – appeared alongside stories of inanimate objects coming to life – e.g., **Lucy's Living Doll** (*School Friend* 1963-4 and *June* 1964-74), drawn by Robert McGillivray. TIMESLIP stories abounded beside adaptations (from Charles DICKENS novels to tv shows like BEWITCHED) and a particular vein of Gothic horror not present in boys' comics: tales of cruel step-parents, blindness – in series like **Blind Ballerina** (*Judy* 1963) and **Becky Never Saw the Ball** (*Tammy* 1974), the latter about a blind tennis player – and slavery. Far more significant as fantasy were DC Thompson's *Spellbound* (1976-7) and Fleetway's *Misty* (1978-80), which offered quite potent, often beautifully drawn, schoolgirl-oriented Gothic horror in the style of the early US WARREN PUBLISHING books.

Generally, however, the horror genre was served mostly by the humour titles, which regularly used ghosts and spooks as gag generators. Fleetway humour titles discovered and tapped this rich vein in titles such as *Cor!!* (1970-74) – which featured strips like **Hire a Horror**, **Freddy Fang** and **Fiends and Neighbours** – *Shiver and Shake* (1973-4), which was hosted by two ghosts, and *Monster Fun* (1975-6). The most successful artist in the horror/humour genre was Ken Reid, especially with **Frankie Stein** (*Wham!* 1964-8, *Shiver and Shake* 1973-4, *Whoopee* 1974-5, *Monster Fun* 1975-6, *Buster* 1976-1986), latterly drawn by Robert Nixon.

The late-1950s USA had little time for horror – Gothic or otherwise – but most publishers had at least one mystery title. Harvey's most important offering was the short-lived *Alarming Tales* (6 issues 1957-8), which featured Jack Kirby's first fantasy art since leaving *Black Magic Comic* (created by Kirby with Joe Simon for Crestwood in 1950). The same year, 1957, also saw Kirby at DC creating **Challengers of the Unknown** for their try-out book *Showcase*. The **Challengers** subsequently went on to their own title, *Challengers of the Unknown* (sporadically 1970-78, another short series 1991), and other *Showcase* strips did likewise: **Space Ranger** (1958), **Adam Strange** (1958), **Rip Hunter, Time Master** (1959) and **The Flash** (1956), whose revival in *Showcase #4* is widely credited as having started the second wave of SUPERHEROES, a wave that dominated the market in the 1960s.

Marvel (as Atlas had become) spent the immediate post-Comics Code years churning out a succession of bizarrely titled monster strips by editor Stan LEE and artists Kirby, Don Heck (1929-   ) and Steve DITKO. Ditko was also working for Charlton on adaptations of MONSTER MOVIES – **Konga** (1960-65) and **Gorgo** (1961-5) – though he was unable to draw their third title, **Reptilius** (1961-3). When DC brought out the team book **Justice League of America** (1960-87), Marvel responded with **The Fantastic Four** (1961-current) and followed this with a plethora of highly successful superhero tales, mostly written by Lee and drawn by Kirby. One of the earliest of these characters was **Thor**, who debuted in the previously monster-dominated *Journey into Mystery* (*#83* 1963). That sense of the immensely fantastic reached its apotheosis with **The Silver Surfer** – introduced in *Fantastic Four* (*#48* 1966) and gaining his own book, *Silver Surfer* (1968-72, with later incarnations) – which strip, under Lee and artist John Buscema (1927-   ), had pronounced religious and philosophical undercurrents to match its superheroics, and became something of a counterculture ICON, as did DOCTOR STRANGE by Lee and Ditko (initiated in *Strange Tales #110* 1963). Having already created the immensely popular, archetypal "urban hero with a problem", **Spiderman**, the previous year, Lee and Ditko immersed Strange, the "sorcerer supreme", in a milieu of alien dimensions and incredible creatures; the character is still current (◊◊ SPIDERMAN [*1977* tvm]).

A somewhat different Marvel character – **Kazar** – is testament to the enduring appeal of the lone savage popularized by Edgar Rice BURROUGHS with TARZAN. The 1960s brought other savages including Sam Glanzman's **Kona, Monarch of Monster Isle** (Charlton 1967-9) and Otto Binder's and Frank Thorne's post-HOLOCAUST **Mighty Sampson** (Gold Key 1964-9). Gold Key anticipated a horror revival with its tv tie-ins *Twilight Zone* (1961-2) (◊ *The* TWILIGHT ZONE) and *Boris Karloff's Tales of Mystery* (1963-80).

Previously a publisher of *Famous Monsters of Filmland*, Warren effectively revived the EC style of horror tales in CREEPY (1964-83) and EERIE (1965-83). Under editor/writer Archie Goodwin (1937-   ), the WARREN PUBLISHING line featured almost all the original EC artists along with others including Ditko, Gene Colan (1926-   ) and Neal ADAMS. In 1969 Warren added a third title, VAMPIRELLA (1969-83; revived in the 1990s); Ron GOULART produced some tied novels. Vampirella, a sexy female VAMPIRE from a distant planet, was modelled on **Barbarella**, created in 1962 by French artist Jean-Claude Forest (1930-   ) for *V. Magazine*. **Barbarella**'s moral and sexual ambiguities were indicative of the more aware adult readership that European comics had cultivated. Another Forest strip, **Les Naufrages du Temps** (premiered in *Chou Chou* 1964), drawn by Paul Gillon, proved to be even longer-lasting, being still (1995) current. Written from its fifth volume by Gillon himself, its mixture of hard sf, wild fantasy and frank sexuality has proven extremely successful. Gillon has gone on to write and draw another even more graphically erotic post-apocalyptic strip, *La Survivante* ["The Survivor"].

The European fantasy-comics tradition stretches back to the work of the French artist Rene Pellos (1900-   ) in **Futuropolis** (1937-8) and **Electropolis** (1939), and to the Italian **Saturno Contro la Terra** ["Saturn Against Earth"] (*I Tre Pocellani* and *Il Topolino* 1937-46) by F. Pedrocchi and G. Scottari. Another key early European fantasy strip was the long-running **Les Pionniers de L'Espérance** ["The Pioneers of *The Hope*"] by artist Raymond Poivet (1910-   ) and writer Roger Lecureux, which followed the crew of the spaceship *Espérance*. Similarly hard-sf-based was the excellent French **Valerian** (1967-current) by writer Pierre Christin (1938-   ) and artist Jean-Claude Mezieres (1938-   ). **Mort Cinder** (1962-4) by the South Americans Alberto BRECCIA and Hector Oesterheld, also published widely in Europe, was a masterful blend of TIME TRAVEL, DREAMS and HORROR, first appearing in the Argentinian comic *Misterix*. Breccia went on (in 1968) to apply himself to Oesterheld's remarkable **El Eternauta** ["The Eternaut"], a long-running series about an extra-terrestrial 21st-century philosopher who travels through time and "navigates eternity"; earlier this strip had been, since 1957, more conventionally drawn by Francisco Solano Lopez in the periodical *Hora Cera*, culminating in the ANIMATED MOVIE *El Eternauta* (*1968*). Breccia has had a profound influence on the development of the comic strip in the West.

By the late 1960s, horror had once again begun to take hold in the USA. DC revived The SPECTRE (1967-9) and **The Phantom Stranger** (in *Showcase #80* 1969), but it was the transformation by new editor Joe Orlando (1927-   ) of *House of Mystery* into an EC-style horror anthology that opened the floodgates. Orlando, an EC alumnus himself, created the horror host **Cain** for the previously superhero-oriented comic book and started running horror shorts with a shock ending. With this and the already existing *Unexpected* and *House of Secrets*, plus the new title *The Witching Hour*, Orlando and fellow editors Murray Boltinoff (1911-1994) and Dick Giordano (1932-   ) built a highly influential line of horror books, drawing upon the talents of veterans like writers Carl Wessler and Jack Oleck and artists Alex Toth (1928-   ) and Wally Wood (1927-1981) and Neal ADAMS, plus new US and Filippino creators. By 1974 DC were publishing 16 regular titles – matching Atlas's output two decades earlier – and spanning a wide range of genre hybrids. Alongside *Weird War* and *Weird Western* one could find weird romance (*Dark Mansion of Forbidden Love* 1971-2), weird HUMOUR (*Plop* 1973-6) and the supposedly true tales featured in *Ghosts* (1971-82).

The horror explosion was so pervasive it transformed DC's oldest title, *Adventure Comics*, into *Weird Adventure*, which in its short life featured both Sheldon Mayer's and Tony de Zuniga's mysterious heroine **Black Orchid** (1973; ◊◊ Neil GAIMAN) and Michael Fleischer's and Jim Aparo's seminal revival of **The Spectre** (1974-5). Perhaps DC's most important title of the period was **Swamp Thing** (1972-76), with writers Len Wein (1948-   ) and David Michelinie and artists Bernie WRIGHTSON and Nestor Redondo (1928-1995). An early development of Marvel's similarly burgeoning horror line, the likewise swamp-oriented **Man-Thing** (1974-5, 1979-81), by coincidence hit the stands at the same time. Written primarily by Steve Gerber (1947-   ) and drawn by Mike PLOOG and Jim Mooney among others, this enjoyed a lengthy run (*Fear* 1972-4, *Man-Thing* 1974-5 and 1979-81). A spinoff from the book, **Howard the Duck** (1976-9 and 1986), again by Gerber and artists Frank Brunner and Gene Colan (1926-   ), proved popular enough to spawn the disastrous HOWARD THE DUCK (*1986*).

Marvel's horror line was largely character-based, as opposed to the portmanteau approach favoured by DC. Following the successful *Tomb of Dracula* (1972-6) by Colan and Marv Wolfman, Marvel brought out *Werewolf by Night* (1972-7), GHOST RIDER (1972-83), *Frankenstein* (1973-5), *Morbius* (1974-5) and *Son of Satan* (1977). Running in parallel, the same characters could frequently be found in Marvel's b/w magazine line, which also featured a lengthy series of adaptations of the popular **Planet of the Apes** movies (1974-7), done by Doug Moench, Mike Ploog and Alfredo Alcala, and of the adventures of possibly the decade's most influential character, CONAN. Robert E. HOWARD's archetypal barbarian first appeared in comic form in 1970, done by Roy THOMAS and artist Barry

WINDSOR-SMITH, and went on to win a number of awards and spawn a legion of spinoffs and imitators. Other Howard characters received the Marvel treatment: **Solomon Kane** and **Bran Mak Morn** could be found in the monthly *Savage Sword of Conan* magazine (1974-current), while **King Kull** starred in *Kull the Conqueror* (1971-8; vt *Kull the Destroyer*), initially by the team of Thomas with Marie and John Severin. By far the most popular **Conan** spinoff was **Red Sonja** (tagged "the she-devil with a sword"), previously only a very minor Howard character; the movie RED SONJA (*1985*) was the eventual consequence. **Red Sonja**'s artist, Frank Thorne (1930-   ), went on to draw the very similar but sexually uninhibited **Ghita of Alizarr** for the Warren b/w sf comics magazine *1984/1994* (1978-82). Ghita's Rabelaisian adventures (in which she was every bit as likely to be depicted cavorting in various states of undress as fighting trolls or monsters) may have been shocking to a US audience, but not in Europe, where the SWORD-AND-SOR-CERY strip had gained a more substantial foothold and sexually uninhibited strips were commonplace. Some of the best of these, including the Italian **Il Gioco** ["The Plaything"] (graph colls **1983**, **1990**; trans US as *Click!* **1987**, **1993**) by Milo Manara (1945-   ), the DRUUNA series of sexy sf/horror, and **Morbus Gravis** (graph coll **1985-95** 5 vols; trans *Heavy Metal* 1986-95) by Paolo Eleuteri Serpieri (1944-   ) have begun to see publication in the USA. **Zetari** (1985-7) by Martin Lodewijk and John M. Burns featured much the same sort of material as **Ghita**, as did **Lorna y su Robot** ["Lorna and Her Robot"] (graph colls **1985**, **1992**) by Alfonso Azpiri (1947-   ). The French **Pelisse** (graph **1983-7** 4 vols; trans as *Roxanna* **1987-9** US) by Serge LeTendre and Regis Loisel, featuring the adventures of a plump-breasted nymphet in a charming and inspired fantasy world, was equally uninhibited but owed more to J.R.R. TOLKIEN, as did *Le Grand Pouvoir du Chninkel* ["The Great Power of the Chninkel"] (graph **1988**; trans in *Cheval Noir #13-#18* 1990-91) by Jean Van Hamme (1939-   ) and Grzegorz Rosinski (1941-   ). By contrast, the French Rosinski and Van Hamme's **Thorgal** (1974-8 12 vols) and the Spanish Vincente SEG-RELLES's beautifully painted **Mercenary** series (1980-current; trans sporadically in *Heavy Metal* during the 1980s) had more in common with **Conan**. The prolific Spanish artist Victor de la Fuente (1927-   ) produced several intelligent and highly original S&S epics including the thoughtful **Haxtur** (1973; trans in *Eerie* 1979), the epic **Haggarth** (1975-6; trans in *Eerie* 1980-82) and **Mathai-Dor** (1979).

DC's early entries into the barbarian field also included adaptations: Fritz LEIBER's **Fafhrd and the Gray Mouser** in *Sword of Sorcery* (1973) and Burroughs's **John Carter** and **Pellucidar** in *Weird Worlds* (1972-4). Marvel later picked up the **John Carter** series and also adapted John JAKES's **Brak the Barbarian** in *Savage Tales* (*#5-#8* 1974), Edwin Lester ARNOLD's **Lt Gullivar Jones** (*Creatures on the Loose* 1972-4) and Lin CARTER's **Thongor** (*Creatures on the Loose* 1972-4).

The enormous popularity achieved by the work of Tolkien in the 1970s led inevitably to a rash of comics featuring the adventures of "Little People" – ELVES, etc. Strip adaptations of the original novels have been surprisingly meagre, however, with, aside from a photonovel version of Ralph BAKSHI's *The* LORD OF THE RINGS (*1978*), only Charles Dixon's version of *The Hobbit* (graph **1989** 3 vols US), drawn by David Wenzel (1950-   ), having been published in the English language. A creditable Spanish version of *LOTR*, adapted by Nicola Cuti and drawn by Luis BERMEJO – *El Senor del Anelo* (graph **1979** 6 vols) – retained the spirit of the original while avoiding the pitfall into which all the rest of the comics in this genre stumbled: that of making the characters too cute to be taken seriously by an adult reader. The Dixon-Wenzel series fell into this trap, as did Mike PLOOG and John Buscema in creating Marvel's **Weirdworld** (seen at its best in *Warriors of the Shadow Realm* [graph **1979** 3 vols]). The most successful series in this genre has been ELFQUEST (1978-current) by Wendy and Richard PINI.

The end of the 1970s saw a number of adaptations in book form of the work of leading fantasy and sf writers, usually under the editorship of Byron PREISS: *The Illustrated Roger Zelazny* (graph coll **1978**) by Gray Morrow, *The Illustrated Harlan Ellison* (graph coll **1978**) and a trio of titles drawn by Howard CHAYKIN: Samuel R. DELANEY's *Empire* (graph **1978**), Alfred Bester's *The Stars My Destination* (graph **1979**) and Michael MOORCOCK's *The Swords of Heaven, The Flowers of Hell* (graph **1979**).

These adaptations sold to a more discerning readership, which owed much to the acceptance of the strip medium during the boundary-breaking US underground movement of a decade earlier. Freed from the censorship of the mainstream publishing houses, the late-1960s underground comics dealt initially with the sex'n'drugs lifestyle presumed of its readers, but soon horizons were broadened. Probably the first horror underground was *Bogeyman* (1969-70) by Rory Hayes. It was followed by the more overtly EC-influenced *Insect Fear* (1970-73), *Death Rattle* (1972-3) and Richard CORBEN's *Fantagor* (1970-72) and *Rowlf* (1971). Corben also featured prominently in two outstanding anthology titles: the horror book *Skull* (1970-72) and *Slow Death* (1970-79). Other prominent *Slow Death* creators were the noted historical artist Jack Jackson (1941-) and the writer-artist team Tom Veitch (1941-) and Greg Irons (1947-1984), whose wonderfully titled *Legion of Charlies* (1971) and *Deviant Slice* (1972-3) set new standards in stomach-churning gore – beautifully rendered! Another notable underground artist, Vaughn Bodé (1941-1975) peopled his strips with bizarre lizards and talking hats, but retained the movement's trademark cynicism, albeit tinged with an almost poetic Romanticism.

Mainstream creators generally ignored the underground scene, the main exception being Wallace Wood, whose *Witzend* (1966-81) featured work by Bodé and future Pulitzer Prize-winner Art Spiegelman (1948-) alongside that of more established fantasy artists like Reed CRANDALL and Frank FRAZETTA. Wood's own contribution was the sprawling **Wizard King** epic, typical of the TOLKIEN-inspired strips he would frequently create for Marvel and Warren; this one spawned a sequel, **The King of the World** (1978). The alternative anthology magazines *Hot Stuff* (1974-8) and *Star * Reach* (1974-9) built on *Witzend*'s success, but probably the most significant non-mainstream publication of the period was HEAVY METAL, which was initially conceived as merely an English-language version of the French *Métal Hurlant*. Since 1977 *Heavy Metal* has presented the work of such important artists as MOEBIUS, Philippe Druillet (◊ MOEBIUS), François Schuiten (1956-) and Philippe Caza, all of whose work has been substantially collected and published in album form. A shorter-lived rival, EPIC (1980-86), published some fine

fantasy strips, notably the S&S saga **Marada the She Wolf** by Chris Claremont (1950-) and John BOLTON, which subsequently met with much success in Europe, and a stylish adaptation of Michael MOORCOCK's **Elric** by Roy THOMAS and P. Craig Russell.

A rash of independent companies (companies whose publications sell solely through the specialist market the undergrounds inspired) has widened the range of US comics genres throughout the 1980s and 1990s, starting at the same time as many of the major companies were retreating to a core line of superhero titles. Both Pacific and First continued *Epic's* Moorcock adaptations: **Elric** (1984-9), **Hawkmoon** (1986-9) and **Corum** (1987-9). Anne RICE's **Lestat** novels were successfully adapted by Innovation in *The Vampire Lestat* (1990-91). By far the greatest success was enjoyed by Dark Horse with their licensed titles **Aliens** (1988-current), **Terminator** (1988-current) and **Predator** (1989-current). The quirkier side of Dark Horse has been represented by the charming *Concrete* (1987-current), the kind of eccentric project that inspired the early self-publishers.

One of these was Dave Sim (1958-). His CEREBUS THE AARDVARK started life as a funny-animal PARODY of Barry WINDSOR-SMITH's **Conan**, but matured into a fascinating discourse on religion, society and power. ELFQUEST was another successful self-publishing project. The later **Bone** (1992-current) by Jeff Smith (1960-) presents a surprisingly uncloying alternative to the usual Tolkienesque QUEST epic.

Groo (1982-current) by Mark Evanier (1952-) and Sergio Aragones (1937-) has been published by several companies (most recently Image Comics), and its longevity is proof of the enduring appeal of its S&S spoofing. Darker fare was offered by Pacific's *Twisted Tales* (1982-4) – yet another EC-inspired project – SpiderBaby's TABOO (1988-95) and James O'Barr's **Crow** (1989-90), the basis of *The* CROW (**1994**). OCCULT DETECTIVES were the protagonists of Michael T. Gilbert's **Mister Monster** (1985-91) and Mike MIGNOLA's intriguingly drawn **Hellboy** (1994-current).

Among the many less easily categorizable strips, one of the most praised has been Mark Schultz's *Xenozoic Tales* (1987-current), set in a post-HOLOCAUST world where DINOSAURS and humans coexist – just. Schultz's art shows the influence of Frank FRAZETTA and Alex RAYMOND; the strip has inspired a cartoon and an arcade game. Equally popular was the ridiculously apocalyptic *Hard Boiled* (1990-92) by Frank MILLER and Geoff Darrow, whose later *Big Guy and Rusty the Robot* (1995), a homage to Japanese MONSTER MOVIES, made equally fine use of Darrow's astonishingly detailed artwork.

To get a share of the independent publishers' success the big companies set up their own "mature" imprints: Marvel with Epic and DC with Vertigo. Epic was largely sf-orientated, with *Starstruck* (1985-6) by Elaine Lee and Mike KALUTA and *Timespirits* (1985-6) by Steve Percy and Tom Yeates being perhaps their best titles. Vertigo, by contrast, was primarily a horror or mystery imprint, growing out of the success of Alan MOORE's **Swamp Thing** revival. Popular Vertigo titles have been the **Swamp Thing** spinoff *Hellblazer* (1988), again in the occult-detective vein, **Animal Man** (1988) and **Shade The Changing Man** (1990), both of which were radical revamps of existing characters. The bestselling Vertigo title has been Neil GAIMAN's **Sandman** (1989).

Gaiman is one of the few UK émigrés not to have worked for *2,000 A.D.* From 1977 this sf-oriented comic has used a wide range of influential creators and characters. Initially something of an *Action*-inspired attention grabber – with the two gorefests **Flesh** (1977) and **Shako** (1977) very much in the **Hookjaw** mode – it gradually matured into a broader-based vehicle for more interesting projects. Alongside routine if well crafted future-war stories one found the humorous **Robo Hunter** (1978-current) and the fantasy epic **Nemesis the Warlock** (1980-current); such stories played a part in developing a readership of potential MANGA fans, ready for the first importation and translation of Japanese comics in the early 1990s. *2,000 A.D.*'s first issue was dominated by a competently drawn (by Dave Gibbons) revival of **Dan Dare**, but it was to be **Judge Dredd**, debuting in *#2*, that caught the public's imagination. **Dredd**'s main rival for popularity in the comic is SLAINE, Pat MILLS's Celtic barbarian; this has benefited from a number of talented artists, notably Mike McMahon (also perhaps the definitive **Dredd** artist), Simon BISLEY and Glenn Fabry. *2,000 A.D.*'s success inspired spinoff titles: *Starlord* (1978), *Tornado* (1979) and *Crisis* (1987-91), plus a solo **Dredd** vehicle *Judge Dredd – The Megazine* (1990-current) and a solo monthly reprinting material from *2,000 A.D.*

Of more fantasy significance was a rival publication, Quality Comics's *Warrior* (1982-5). This mixed supermen, sf and fantasy strips to great critical acclaim, and several series were sold to European and US markets. Just as Alan MOORE's **Ballad of Halo Jones** (1984-6) was a high point of *2,000 A.D.*, so were his **Marvelman** and **V for Vendetta** strips at *Warrior*. Of Moore's three ongoing multi-volume projects, only one is of fantasy interest: *Lost Girls*, an exploratory experiment with erotica. Its vehicle magazine, *Taboo* (#5-#7 1990-94), is now defunct and the project has yet (1995) to find another home.

The reluctance of publishers to involve themselves in experimental comics-related projects is a peculiar feature of the 1990s. Although the medium has attracted a great number of original creative minds, most have found difficulty in getting their work into print unless it is related to concepts and characters that are already well established. UK publishers Victor Gollancz released a series of GRAPHIC NOVELS in 1991-2 which broke new ground in graphic narrative with creators such as Alan MOORE, Neil GAIMAN and Dave MCKEAN, Ian MILLER and others, but it was not a commercial success and others have been reluctant to dip a toe into these virtually uncharted waters. Experimentation occurs infrequently, and is likely to virtually cease with the demise (1995) of the 95-year old UK Net Book Agreement – though perhaps not all is gloom: HarperCollins has since released a graphic novel by Doris Lessing (1919-   ), *Playing the Game* (graph **1995**) illustrated by Charlie Adlard.

It is the visual component of the comic strip that is at once its most profoundly expressive aspect and its greatest disadvantage. While no subject matter is denied the text-only book, comic books are easily misrepresented as children's reading matter, and as such can be made the object of the artificial outrage of the tabloid press. Comics publishers fear such diatribes and consequently adopt a play-safe policy. Moore's impressive (albeit clumsily drawn) epic graphic novel about the JACK THE RIPPER crimes, *From Hell* (16 vols **1991**-current), has suffered considerably, as did Peter Milligan's and Brendan McCarthy's *Skin* (1992), about a violent Thalidomide victim, and *The Tale of One Bad Rat* (**1994**) by Bryan Talbot (1952-

), which combined a tale of a lost Beatrix POTTER manuscript with a story about a sexually abused child. These were serious, intelligent attempts to present important social and philosophical issues, but almost foundered because some found them uncomfortable.   [DR/SH/RT]

COMMEDIA DELL'ARTE A term that came into use in the mid-16th century and which, translated literally, means "comedy of professional artists"; it was coined to describe those wandering troupes of actors who presented quasi-improvised repertory comedies with a fixed cast of characters, first in Italy and subsequently throughout Europe, in performances which generally incorporated mime, buffoonery and music. The plots were traditional – the basic stories being traceable back to the Roman theatre – and REVEL-like, generally featuring young lovers who, with the aid of unscrupulous servants, invariably escaped the restraints of their elders, making fools of them in the process; and they incorporated a sense that CYCLES paraded constantly beneath the surface of the farce, a sense that the same STORY – and the same UNDERLIER cast – was infinitely retellable.

Significantly, the CDA very rarely attacked any elder of real power, for the aristocracy was normally (and the monarchy always) exempt. The CDA revel sagaciously fails, therefore, to threaten any genuine change; the form as a whole is of interest to fantasy mainly (a) as a repository of lessons in the creation of plots which reveal themselves to be in reality games, infinitely repeatable, in the manner of the MASQUE, (b) for the dramatis personae who enact these plots, (c) in the development of the sense that true drama lies in the masquerade, or in the family romance (perhaps generations old), that risks exposure on the stage of the world, and (d) in the intuition that the various CDA stories were one perpetually re-enacted Story.

The harlequinade of the 17th century did not differ in essentials from the CDA, though tending more to mime than its direct parent, which was defended as an artform in the 18th century by Carlo GOZZI, who reinvented it in the form of the *fiable*, combining fantasy and CDA conventions, selecting specifically the rigid cast of protagonists and the estranging MASKS. The most famous of these plays are *Fiable dell'amore delle tre melarance* (**1761**; normally trans as *The Love of Three Oranges*) and *Turandot* (**1762**), each of which inspired at least one OPERA of fantasy interest. In the UK, after the 18th century, the harlequinade gradually evolved into the PANTOMIME.

Not all CDAs or harlequinades are toothless. Pierre-Augustin de Beaumarchais (1732-1799) transformed the harlequinade into the subversive and highly literary **Figaro** plays – *The Barber of Seville* (**1775**) and *The Marriage of Figaro* (**1778**). These came too close to attacking the genuine rulers of the 18th-century world, and both were frequently banned. It is hard, moreover, to think that the complex uses made of automata-like Commedia figures, by E.T.A. HOFFMANN are unthreatening. And, insofar as GRAND GUIGNOL admits to its origin in Commedia figures, it too can be thought of as threatening, though the threat is meant to thrill, not to change minds.

In 1917, Guillaume APOLLINAIRE wrote a preface to the premiere of the CDA-derived ballet *Parade* (music Erik Satie, text Jean COCTEAU, sets Pablo Picasso); it was in this preface, as he attempted to explain the frisson generated by CDA techniques at the height of WORLD WAR I, that he coined the term SURREALISM.

The CDA has had a more direct influence on 20th-century fantasy, by providing (as noted) a fixed familial cast (see listing below) who knowingly enact and re-enact a basic tale. This has given the genre a traditional model for the use of UNDERLIER figures as models and AVATARS; has provided lessons in how to treat these in a gaming fashion; and has proved a fruitful group metaphor to use whenever – as in the masquerade-like *Belle Epoque* of MacDonald HARRIS's *Glowstone* (**1987**) – a period of history, or a particular venue (like VENICE), is conceived in terms of theatre. It is, perhaps, in more than one of these senses that Vladimir NABOKOV called his last novel *Look at the Harlequins!* (**1974**).

THE CAST, AND THEIR MODERN ROLES

Harlequin The most important of all, a cunning peasant who has become a servant. He is faithful to his love. He has clear affinities with the god Mercury (◊ HERMES), his "mercurial" nature being reflected in his patchwork garb, and in the invariable MASK which announces his double nature. He is a TRICKSTER, a jester, who surfaces in the form of mutable characters like Michael MOORCOCK's Jerry Cornelius; and he is a LIMINAL BEING, a creature whose presence in a text is a harbinger of METAMORPHOSIS. He is, however, double: he can announce change, or he can – in situations of BONDAGE, in PASTORALS or in REVISIONIST FANTASY – corrode change through PARODY. The climax of Nancy KRESS's *The Prince of Morning Bells* (**1981**) neatly demonstrates this: we are led to believe that the enchanted dog, Chessie, who accompanies the princess on her QUEST for the Heart of the World is a King under a SPELL; but when he finally undergoes metamorphosis "an old, old man" is revealed, "dressed in the pointed shoes and Harlequin tights of a jester . . .". Harlequin is an actor, and cannot be expected to show his real face. *History of Harlequin* (**1926**) by Cyril Beaumont (1891-1976) is a nonfiction study couched as a biography of Harlequin over the centuries.

Brighella Much less known. A kind of parody of Harlequin, he is a masked thief who lacks any doubleness of motive or any decency regarding women.

Il Dottore The pedant, he is allied with the forces of stasis. He tries to prevent marriages (and Spring). He is analogous to the KNIGHT OF THE DOLEFUL COUNTENANCE.

Pantalone He is (in English) a pantaloon, an old dupe and butt – a failed property developer. Perhaps because it is no longer thought the height of wit to depict Jews as grasping usurers, he is not an underlier.

Il Capitano (Miles Gloriosus) The braggadocio who boasts about his military life but is a coward, and has bad luck with women – Falstaff is a version of the figure. Like most CDA figures, Il Capitano can be found underlying HEROIC-FANTASY tales, SWORD AND SORCERY in particular; but here he generally turns out to be in fact brave beneath the veneer of cowardice. In the 20th century he is, in the end, epitomized by the Cowardly Lion (◊ OZ).

Scaramouche The flamboyant, tough-living soldier hoicked up from the ranks. He is commonly found in historical romances, and sometimes in heroic fantasy; he is an S&S staple.

Pulcinella A deformed, sullen, brutal freak. Unadulterated, he becomes the Hunchback of Notre Dame; as the PUPPET Polichinelle, however, he is also capable of sardonic humour, and under the name Guignol was a central character in the late-18th-century French cabarets (◊ GRAND GUIGNOL). Transformed, and transplanted to England, he became Punch.

Columbine A servant girl loved by Harlequin. Like most female characters until recent decades, she lacks character; but in Michael MOORCOCK's **Jerry Cornelius** novels she becomes a TEMPORAL ADVENTURESS.

Pedrolino (Pagliacco; Pierrot) A young lover, graceful and trustworthy. He is easily deceived, both by lovers and by Harlequin. He is often blamed for what he has not done, but always takes on the burden of guilt, tearfully. He does not normally wear a MASK, for he has no duplicity. In 17th-century France – inspired by Molière's play *Dom Juan, ou le festin de Pierre* (**1665**) – he begins to be called Pierrot and to wear a flounced white garment and large floppy hat. He is the KNIGHT OF THE DOLEFUL COUNTENANCE when young. He is also – to use Moorcock's evocative term – a dancer at the end of time. Like Cyril Beaumont in his treatment of Harlequin (see above) Kay DICK shaped her nonfiction study *Pierrot* (**1960**) as the biography of an immortal creature. Throughout the first half of the 20th century, visions of Pierrot are common in music, dance, painting and literature. Edith Sitwell (1887-1964) consciously emphasized her resemblance to a Gothicized version of Pierrot. In her later years, Isak DINESEN actually took to dressing as Pierrot. [JC]

COMMERCIALS Many advertisements have recruited characters from fantasy to sell their products, including: the cookie-baking Keebler ELVES and Rice Krispies' elfish Snap, Crackle and Pop; Lucky the Leprechaun with his Lucky Charms cereal; a version of DRACULA, Count Chocula, selling his eponymous cereal; a charging White Knight with a lance who personified the power of Ajax Cleanser; a recently revived JACK, now depicted as a man with a Jack-in-the-Box head, promoting Jack in the Box restaurants; the dancing scarecrow who sold Country Corn Flakes; and the cherubs of Nice'n'Soft bathroom tissue. Miniature humans (◊ DWARFS) frequently appear, as do GIANTS (e.g., the Jolly Green Giant), and there are images of FLYING people. An inordinate number of animated TALKING ANIMALS are employed in cereal commercials: Sugar Bear (Golden Crisp), Tony the Tiger (Frosted Flakes), a toucan (Froot Loops), a rabbit (Trix), a monkey (Cocoa Krispies), a frog (Smacks), a honeybee (Honey Nut Cheerios), etc. Animated humans appear in many others (e.g., the Tetley's Tea Bag men). More bizarrely, the products themselves, suitably anthropomorphized, may serve as their own spokespersons: Speedy Alka-Seltzer, the living headache tablet; the personified cleanser bottle Mr Clean; walking and talking M&M candies; and – surely the most memorable image from 1950s commercials – dancing cigarette packs. McDonalds commercials regularly feature living food items, like talking hamburgers, fish fillets and chicken nuggets. [GW]

Further reading: *The Best Thing on TV: Commercials* (**1978**) by Jonathan Price.

COMPANIONS HEROES AND HEROINES who embark upon QUESTS, as is normal in GENRE FANTASY, almost always either set off with COMPANIONS or acquire them along the way. There are exceptions, but the extended narrative sweep of most fantasy novels written since J.R.R. TOLKIEN offers ample scope within which secondary characters may act out their destinies. Tolkien's own *The Hobbit* (**1937**) and *The Lord of the Rings* (**1954-55**) possess skilful demonstrations of the usefulness (and appeal) of the band of companions, who fill the scene, bolster the hero, perform feats he or she cannot, depart upon ancillary quests whose accomplishment (◊◊ PLOT COUPONS) will help trigger the climax, and die if necessary.

Two subcategories are used in this encyclopedia to help describe the various groupings characteristic of HEROIC FANTASY as it has developed over the past half-century. They are the DIRTY DOZEN, a group brought together by force, and most often found in MILITARY FANTASY; and the SEVEN SAMURAI, a voluntary association (like those in Tolkien), commonly found throughout adventure fantasy.

Companions may be of the same stock as the hero and heroine, but more usually represent as wide as possible a selection of character types, TALENTS, species and REALITIES of origin. The reasons for this are obvious: variety, pleasure, a reservoir of possible responses when action is required (◊ PLOT DEVICES) and a pool of characters to develop further when required. Companions may therefore be taken from the ranks of available apprentices, BARDS, gifted CHILDREN, DEMONS, DRAGONS and/or other creatures from the BESTIARY, DWARFS and/or ELVES and other visitors from FAERIE, GHOSTS, the GODDESS in disguise, GODS in disguise, JACKS, KNIGHTS, MAGI and/or WIZARDS, TALKING ANIMALS, TEMPORAL ADVENTURESSES, TRICKSTERS, WITCHES, etc. [JC]

COMPANY OF WOLVES, THE UK movie (*1984*). ITC/Palace. **Pr** Chris Brown, Stephen Woolley. **Dir** Neil Jordan. **Mufx** Christopher Tucker. **Animatronic wolf by** Rodger Shaw. **Screenplay** Angela CARTER, Jordan. **Based on** "The Company of Wolves" (1979) by Carter; her own screenplay appears in *Come Unto These Yellow Sands* (coll **1985**). **Starring** Micha Bergese (Huntsman), Graham Crowden (Old Priest), Angela Lansbury (Granny), Sarah Patterson (Rosaleen), Tusse Silberg (Mother), David Warner (Father). 95 mins. Colour.

Early in this elaborate conflation of Little Red Riding-Hood with various WEREWOLF legends, set in a LAND-OF-FABLE Dark or Middle Ages, young Rosaleen's Granny sets out the ground-rules: "Never stray from the path, never eat a windfall apple, and never trust a man whose eyebrows meet." Granny is full of advice for the adolescent girl, and full of WOLF stories, too, three of which are nested into the overall tale. The FRAME STORY is a distorted version of the Red Riding-Hood tale, in which wolves become the symbol of the adult sexuality dawning in the girl-child Rosaleen: the Huntsman is also the wolf (killing Granny), and as such is, too, the cavalier seducer who finally lures Rosaleen from the world of humans to the world of wolves. Yet even this tale is set within a frame, being dreamt (or possibly not) by a modern child in an English country house . . .

TCOW's generally *staged* air – backgrounds tend to resemble backdrops, and some of the acting is theatrical rather than cinematic (bringing a faint reminder of PANTOMIME) – enhances the sense of FAIRYTALE, as does the profusion of sharply observed forest wildlife, with toads and snakes abounding. In a joyously fantasticated ANACHRONISM, the Devil's brightly glowing eyes, emerging from the medieval mists, prove to be the headlights of his silver Rolls Royce. Patterson's performance, mixing ignorance and knowingness in constantly fluid proportions, is outstanding. This movie lies at the heart of fantasy, and beats with its pulse: it is both fantasy and *about* fantasy. [JG]

COMYNS, BARBARA Working name of UK writer Barbara Comyns-Carr (1909-1992), most of whose work deals in heightened but nonfantastic terms with claustrophobic family-romance scenarios. In her best-known work, *The Vet's Daughter* (**1959**), this claustrophobia is literally transcended – though only briefly – through its young protagonist's ability to levitate (◊ TALENTS); in the end, forced to perform for money in a CARNIVAL setting, she plummets fatally to earth. *The Juniper Tree: Adapted from a Children's Fairy Story of the Same Name by the Brothers Grimm, Which is far too Macabre for Adult Reading* (**1985**) is a contemporary TWICE-TOLD version of the FOLKTALE (◊◊ GRIMM BROTHERS). [JC]

CONAN Robert E. HOWARD did not live long enough to get much pleasure out of Conan the Barbarian, whom he created in 1932 for *Weird Tales*, and about whom he completed 16 stories and one novel before his suicide in 1936. It was not until Gnome Press began to publish the Conan tales in book form – beginning with *Conan the Conqueror* (1935-36 *WT* as "The Hour of the Wolf"; rev **1950**; rev [following magazine text] vt *The Hour of the Dragon* 1977) – that he became a central UNDERLIER figure for the sword-bearing, brawling, upwardly mobile barbarian HERO of HEROIC FANTASY.

SEQUELS BY OTHER HANDS came from (in rough chronological order) L. Sprague DE CAMP, Bjorn Nyberg, Lin CARTER, Karl Edward WAGNER, Poul ANDERSON, Andrew J. OFFUTT, Robert JORDAN, John Robert MADDOX, Steve PERRY, Roland GREEN and Leonard Carpenter. Sequences using Conan as underlier include the **Thongor** books by Lin Carter, the **Brak** books by John JAKES and the **Kyrik** books by Gardner F. FOX. [JC]

Further reading: *The Common Reader* (coll **1968**) by L. Sprague DE CAMP, plus THE COMMON SWORDBOOK (anth **1969**) and *The Conan Grimoire* (anth **1972**), both ed de Camp and George H. Scithers, all 3 rev and updated as *The Blade of Conan* (anth **1979**) and *The Spell of Conan* (anth **1980**) ed de Camp alone.

COMICS

Conan first appeared in comic-book form in *Conan the Barbarian* (1970-current), elegantly drawn by Barry WINDSOR-SMITH and scripted by MARVEL COMICS's premiere writer, Roy THOMAS; new tales and Robert E. HOWARD adaptations were interspersed. During the period in which he worked on Conan (1970-73), Windsor-Smith's art developed a sophistication and narrative skill which has rarely been surpassed (reaching a very special peak in *Conan #24* "The Song of Red Sonja" 1973), and Thomas's writing gave the stories tremendous energy, although it was sometimes so enthusiastic that readers could be uncertain whether Conan's adversaries were in less danger from his SWORD than from being *talked* to death. Windsor-Smith was replaced by John Buscema (1927-   ) and Ernie Chua (real name Ernie Chan; 1940-   ), and in their hands Conan became a more conventional Marvel HERO.

The continued popularity of the character has led to the publication of several other long-running comic books, including the b/w title *Savage Sword of Conan* (1974-current), which has featured artwork by many leading artists in the field and some impressive full-colour covers; the standard has remained high through more than 200 issues. Among other series featuring Conan are **King Conan** (1980-89; later vt **Conan the King**), **Conan the Destroyer** (1985) and **The Conan Saga** (1987-current). [RT]

See also: CONAN MOVIES.

CONAN MOVIES Two movies have been based on the creation of Robert E. HOWARD.

1. *Conan the Barbarian* US movie (*1981*). Dino De Laurentiis/Edward R. Pressman. **Pr** Buzz Feitshans, Raffaella De Laurentiis. **Exec pr** D. Constantine Conte, Pressman. **Dir** John Milius. **Spfx** Nick Allder. **Vfx** Frank Van der Veer. **Anim vfx** Peter Kuran, Visual Concepts

Engineering. **Screenplay** Milius, Oliver Stone. **Novelization** *Conan the Barbarian* * (**1982**) by Lin CARTER and L. Sprague DE CAMP. **Starring** Sandahl Bergman (Valeria), Ben Davidson (Rexor), Cassandra Gaviola (Witch), James Earl Jones (Thulsa Doom), Gerry Lopez (Subotai), Mako (Akiro), Arnold Schwarzenegger (Conan), Max Von Sydow (Osric). 129 mins. Colour.

Knowing that he can trust nothing save his own SWORD, Howard's SWORD-AND-SORCERY hero goes through many rigours, magics and women on the way to VENGEANCE against Thulsa Doom, the cult leader who slaughtered Conan's family when he was a child. *En passant*, he undergoes a period as a gladiator; discovers mid-copulation that the glamorous WITCH who has told him, by way of foreplay, that he is fore-ordained, is in fact a VAMPIRE; allies with an eccentric WIZARD, Akiro; battles with GHOSTS; and suffers crucifixion, although he recovers swiftly. Any summary of *CTB* makes the movie seem like a self-PARODY; in fact, though self-consciously ponderous, it is a very respectable attempt to invest Howard's creation with the full burden of epic MYTH. The well drawn character of Valeria (admirably played by Bergman) is one of the earlier prominent examples of the powerful, independent heroines (◊ HEROES AND HEROINES) who would become more prevalent in this subgenre as the 1980s progressed. [JG]

2. *Conan the Destroyer* US movie (**1984**). Dino De Laurentiis/Edward R. Pressman. **Pr** Raffaella De Laurentiis. **Exec pr** Stephen F. Kesten. **Dir** Richard Fleischer. **Technical advisor** L. Sprague DE CAMP. **Spfx** Barry Nolan, John Stirber. **Screenplay** Stanley Mann. **Novelization** *Conan the Destroyer* * (**1984**) by Robert JORDAN. **Starring** Olivia D'Abo (Princess Jehnna), Wilt Chamberlain (Bombaata), Sarah Douglas (Queen Taramis), Grace Jones (Zula), Mako (Akiro), Arnold Schwarzenegger (Conan), Tracey Walter (Malek). 103 mins. Colour.

Conan, as Prince of Thieves, and sneakish sidekick Malek are commissioned by evil Queen Taramis to escort her cutely innocent niece Jehnna on a pre-ordained QUEST to collect a key (in the form of a gem) and use it to retrieve the missing horn of the local GOD Dagoth, thereby bringing him to life; at all costs Jehnna's VIRGINITY must be preserved – although Taramis does not tell Conan that this is because, for the PROPHECY entirely to be fulfilled, Jehnna must be made a virginal HUMAN SACRIFICE. Conan recovers the gem from an island castle of ILLUSION after slaying its SCRYING guardian WIZARD – who, in *CTD*'s most interesting fantastication, appears to him in monstrous form in a hall of MIRRORS, and can be harmed only by attacking his reflections. Treachery abounds before the requisite PLOT COUPONS are collected, the witch-queen and animate god slaughtered, and the maiden saved. Although just as gory as its predecessor, *CTD* lacks the gratuitous sex and cod theology, being clearly designed as a straightforward S&S romp; as such, it succeeds. [JG]

See also: RED SONJA (**1985**).

CONDITIONS These often mark the limits of MAGIC (sometimes a time-limit: CINDERELLA must leave the ball by midnight) or of protection. In the MABINOGION it is a condition of the WIZARD Math's existence that his feet must rest in a maiden's lap except in time of war, while Lleu Llaw Gyffes may be killed only while meeting the bizarre condition that he stand with one foot on a goat and the other on a bathtub. VIRGINITY is a frequent condition of WITCHES' and, less often, WIZARDS' magic, and is invariably required by

catchers of UNICORNS. A GENIE's powers are restricted by the condition of servitude to its bottle's owner. Visitors to FAERIE or HELL should observe the conditions of safety: that they accept no gifts, eat no pomegranates and (in the case of ORPHEUS) not look back. The bewildered GOD Ptath in A.E. VAN VOGT's *The Book of Ptath* (**1947**) is protected by SEVEN conditions which must be fulfilled before a usurping GODDESS can seize his power. In Terry PRATCHETT's *Sourcery* (**1988**) a difficult-to-meet condition (that a born wizard should throw his staff away) is explained as providing the loophole required by FATE in every PROPHECY. In Irish MYTH, binding conditions placed on characters are known as geases or *geasa* (◊ CURSES); one of CUCHULAIN's required him to taste the food at any hearth he passed – he was thus forced to eat dog-flesh before his final battle, so diminishing his powers. [DRL]

See also: PROHIBITIONS.

CONFIDENCE MAN In the world of fantasy, anyone described as a CM is likely to be a TRICKSTER figure, a SHAPESHIFTER whose origins may be divine (and are certainly mysterious); the divine CM in fantasy probably owes much of his nature to the UNDERLIER figure of HERMES. The CM can intrude upon and flim-flam his victims in order to shake them into moral shape (◊ GODGAME); but more often than not is amoral. The most famous example in literature is the shadowy protagonist of Herman MELVILLE's *The Confidence-Man: His Masquerade* (**1857**); the nature of Melville's CM – mundane or supernatural – is never fixed, any more than his motives can be determined or the face he may happen to wear anticipated. Ultimately, he transforms a riverboat into a SHIP OF FOOLS whose occupants lose all trust in REALITY; and at the end of the day, he and the cast embark upon a DANCE OF DEATH.

US examples of the CM are numerous, but his inherent unreliability often works – as with Melville – to make the world of the text itself inherently unreliable. Mocking, ghost-like figures – like Burlingame in John BARTH's *The Sot-Weed Factor* (**1960**), Ben Free in Gene WOLFE's *Free Live Free* (**1984**) and the eponymous SHAMAN in Terry BISSON's *Talking Man* (**1986**) – proliferate in this kind of tale (◊◊ PICARESQUE). In fictions from colonized countries lacking a US-style frontier CMs appear as protagonists of CREATION MYTH: examples are numerous in Latin American literature, and a fine Australian example is Peter CAREY's *Illywhacker* (**1984**). The CM can also be seen as a LIMINAL BEING along the borderline between frontier and settled world. European examples tend – like K. in Franz KAFKA's *The Castle* (**1926**) and the DOUBLE protagonists of Christopher PRIEST's *The Prestige* (**1995**) – to display a more profound entrapment in the worlds whose sense of reality they jostle. Thomas MANN's *Confessions of Felix Krull, Confidence Man: The Early Years* (**1954**) was interrupted by its author's death before it could become clear where and how Krull would reach his apotheosis. [JC]

CONFORD, ELLEN (1942-) US writer whose *And This is Laura* (**1976**) comfortingly conveys to its YA audience the lesson that TALENTS do not necessarily make for happiness, though they may guarantee good highschool grades. Her second novel of fantasy interest, *Genie with the Light Blue Hair* (**1989**), shows that ANSWERED PRAYERS are risky, though to comic effect. The protagonist, Jeannie, acquires a GENIE who resembles Groucho Marx, though without the brains. Lots of WISHES go awry before wisdom is achieved. [JC]

CONJURERS ◊ MAGICIANS.

CONNELL, EVAN E. (1924-) US writer best-known for work outside the fantasy genre; he has won acclaim for his novels, poetry, and nonfiction. *The Alchymist's Journal* (**1991**) comprises seven perspectives on PARACELSUS. Although the novel contains no fantastic elements concerning ALCHEMY (such wonders as Paracelsus's associates claim to have seen may be rationalized), its vivid dramatization of the quest for TRANSMUTATION and transcendence shows its thematic concern with conceptual breakthrough. EEC has also published *A Long Desire* (**1979**) and *White Lantern* (**1980**), essays on antiquities and figures from intellectual history – Paracelsus, PRESTER JOHN – that convey the sense of intellectual QUEST driving *The Alchymist's Journal*. [GF]

CONNECTICUT YANKEE, A At least seven movies have been based on Mark TWAIN's novel *A Connecticut Yankee in King Arthur's Court* (**1889**). In addition, there have been a couple of animated tv half-hour features: *A Connecticut Yankee in King Arthur's Court* (**1970** tvm) and *A Connecticut Rabbit in King Arthur's Court* (**1978** tvm; vt *Bugs Bunny in King Arthur's Court*), the latter being a vehicle for BUGS BUNNY.

1. *A Connecticut Yankee at King Arthur's Court* US movie (**1920**). Fox. **Starring** Harry Myers (Hank). B/w. Silent.

Very little is known about this version, and we have been unable to track down a print. Apparently the subtitles were full of contemporary references, as to the Volstead Act. [JG]

2. *A Connecticut Yankee* US movie (**1931**). Fox. **Dir** Butler. **Screenplay** William Conselman. **Starring** Frank Albertson (Clarence), William Farnum (Arthur), Mitchell Harris (Merlin), Brandon Hurst (Sagramor), Myrna Loy (Morgan Le Fay), Maureen O'Sullivan (Alisande), Will Rogers (Hank). 96 mins. B/w.

Radio-store owner Hank goes to a mysterious old house to install a new battery. The mysterious old householder believes he can communicate with King ARTHUR *via* his radio. A suit of armour topples over, stunning Hank. He DREAMS he has undergone a TIMESLIP back to CAMELOT, where he is almost immediately accused of being a warlock. To protect himself he plays upon his supposed abilities as a seer (◊ PROPHECY) – although much of his prediction is mocked as ludicrous. In the midst of visual gags and the righting of wrongs, he plays a part in the romance between Alisande and Clarence. At the end of it all, he awakens.

Essentially this is a vehicle for the popular comedian Rogers, and has dated badly – not least in its topical references to the Depression and its earnest proselytizing on behalf of democracy. [JG]

3. *A Connecticut Yankee in King Arthur's Court* (vt *A Yankee in King Arthur's Court* UK) US movie (**1949**). Paramount. **Pr** Robert Fellows. **Dir** Tay Garnett. **Spfx** Farciot Edouard. **Screenplay** Edmund Beloin. **Starring** William Bendix (Sagramor), Bing Crosby (Hank Martin), Virginia Field (Morgan Le Fay), Rhonda Fleming (Alisande), Cedric Hardwicke (Arthur), Murvyn Vye (Merlin), Henry Wilcoxon (Lancelot). 106 mins. Colour.

This musical is not so much a remake of **2** as a return to the original book: Hank is this time a blacksmith knocked unconscious in the midst of a fierce storm. The production is lavish, the pacing leisurely, and all the sting which **2** – and Twain's original – possessed has been eliminated: this is a Bing Crosby musical first and a fantasy only second. [JG]

4. *A Connecticut Yankee in King Arthur's Court* US ANIMATED MOVIE (**1970** tvm). Air Programs Intrnational. **Pr** Walter J. Hucker. **Dir** Zoran Janjic. **Screenplay** Michael Robinson. 79 mins. Colour.

This reasonably faithful version of Twain's original mixes a lot of very bad animation with some occasional flashes of brilliance, and the same is true of the voice acting: only in the case of the villainous MERLIN, animated in a quite different style from all the rest, does everything come together. The script is often very witty. [JG]

5. *Unidentified Flying Oddball* (vt *The Spaceman and King Arthur*) US movie (**1979**). DISNEY. **Pr** Ron Miller. **Dir** Russ Mayberry. **Spfx** Cliff Culley. **Screenplay** Don Tait. **Novelization** *Unidentified Flying Oddball* * (**1979**; vt *The Spaceman and King Arthur*) by Heather Simon. **Starring** Rodney Bewes (Clarence), Jim Dale (Mordred), Dennis Dugan (Tom Trimble/Hermes), John le Mesurier (Gawain), Ron Moody (Merlin), Kenneth More (Arthur), Sheila White (Alisande). 93 mins. Colour.

A TECHNOFANTASY version. Brilliant but nerdish NASA technician Trimble is accidentally launched on an interstellar mission alongside a robot, Hermes, that is his exact DOUBLE. But the craft is cast back to the 6th century, where Trimble uncovers and defeats the treachery of Mordred and Merlin at Camelot. There are jolly jokes about being burnt at the stake and stretched on the rack. A battalion of veteran UK comedy actors struggles with appalling tosh. A *ghastly* movie. [JG]

6. *A Connecticut Yankee in King Arthur's Court* US movie (**1989** tvm). Consolidated/Schaefer-Karpf. **Pr** Graham Ford. **Exec pr** Merrill H. Karpf. **Dir** Mel Damski. **Spfx** Gabor Budahazi. **Screenplay** Paul Zindel. **Starring** Rene Auberjonois (Merlin), Hugo E. Blick (Mordred), Michael Gross (Arthur), Bryce Hamnet (Clarence), Whip Hubley (Lancelot), Jean Marsh (Morgana), Keshia Knight Pulliam (Karen Jones), Emma Samms (Guinevere). *c*100 mins. Colour.

The Yankee this time is a 12-year-old black girl, Karen, stunned in a riding accident. The pictures in her schoolbooks provide a sudden technological stimulus to 6th-century England, thanks to adolescent SMITH Clarence; her feminist ideas spread thanks to an enthusiastic Guinevere; artefacts like her Polaroid camera establish her as a powerful WIZARD, and she is knighted as Sir Boss. She and Arthur go in disguise among the peasants and discover how loathed he is because of Mordred's actions. The notion of THINNING is explored: when Karen's "MAGIC" ceases to work, Arthur believes the failure of magic arises from his own failure of governance – he is the LAND. Karen's departure from Camelot homages Dorothy's from Oz in *The* WIZARD OF OZ (**1939**).

It sounds grim, but this, though no major piece, is a delight. The script has wit. Pulliam is a personable young star, and she and the rest of the cast are clearly in love with the project. Budgetary cracks show through the wallpaper, but it hardly matters. This is among the better ARTHUR movies. [JG]

7. *A Young Connecticut Yankee in King Arthur's Court* Canadian/Czech/French movie (**1995**). Filmline/ Images/ Screen Partners/Astral/World International Network. **Pr** Nicolas Clermont. **Dir** R.L. Thomas. **Vfx** François Aubry. **Screenplay** Frank Encarnacao, Thomas. **Starring** Ian Falconer (Lancelot), Paul Hopkins (Galahad), Jack Langedijk (Ulrich), Nick Mancuso (Arthur), Philippe Ross (Hank Morgan/Sir Dude), Theresa Russell (Morgan Le

Fay), David Schaeffer (Clarence), Polly Shannon (Sir Alisande/Alexandra), Michael York (Merlin). 89 mins. Colour.

Young Hank, a high-school wannabee rock-star, hits his head on an amplifier and travels in time. His life is saved by the sole female KNIGHT of the ROUND TABLE, Sir Alisande, a DOUBLE of modern high-school babe Alexandra. He uncovers that Morgan and Ulrich are plotting to usurp the king, and thwarts their plans with the aid of Lancelot and Alisande. Returned to today, he plucks up the courage to ask Alexandra for a date. This movie is not as bad as it sounds; what grates is the deliberate cheapening of the MATTER of Britain through teenage "cleverness". [JG]

CONRAD, JOSEPH Working name of Polish-born writer Jozef Teodor Konrad Nallecz Korzeniowski (1857-1924), a merchant seaman 1874-93 before settling in the UK and becoming one of that country's finest writers. Apart from *The Inheritors* (1901), written with and mostly by Ford Madox FORD, about a race of "Dimensionists" who have superior powers and seek to replace humankind, JC wrote no overt works of fantasy or SUPERNATURAL FICTION. Yet many of his stories have inspired the fantastic in others.

JC's early peak as a writer was with "Heart of Darkness" (1899 *Blackwood's*; in *Youth: A Narrative, and Two Other Stories* coll **1902**), which is also the work of his that comes closest to the overtly supernatural. The story can be read on many levels. The voyage by Marlow, the narrator, along the Congo into darkest Africa is a QUEST, both physically and psychologically, but more potent is the effect of Africa upon Kurtz, the object of Marlow's journey, who has succumbed to the power of the continent and taken on the aspect of a GOD. Native MAGIC plays a minor role in the story, yet becomes a convenient route to making this a RATIONALIZED FANTASY rather than a confrontation with the horrors of REALITY. This balance between fantasy and reality is a feature of many of JC's stories, and identifies those closest to the fantasy medium.

Guilt that creates ILLUSIONS and phantoms is evident in "The Lagoon" (1897 *Cornhill Magazine*) and "Karain" (1897 *Blackwood's Magazine*). Illusions take other forms in "The Secret Sharer" (1910 *Harper's Magazine*), one of JC's best stories, and in *The Shadow-Line* (**1917**), where the fear and tension of a young captain's first command create supernatural manifestations in his mind. In "The Brute" (1906 *Daily Chronicle*), the crew of a doomed ship believe it is diabolic and is seeking to kill them. All these are paradigms of the fantasy of PERCEPTION. JC also wrote several tales of pure HORROR.

The stories mentioned above can be found in *Tales of Unrest* (coll **1898**), *'Twixt Land and Sea* (coll **1912**) and *Within the Tides* (coll **1915**). Also of interest is *Heart of Darkness and Other Tales* (coll **1990**) ed Cedric Watts.

"Heart of Darkness" features at the core of the TECHNO-FANTASY novel *Flicker* (**1991**) by Theodore ROSZAK. According to *Flicker*, a movie version of the novella was to have been co-directed by Orson Welles (1915-1985); the project foundered not merely through lack of funds but because of the quasi-occult activities of Welles's directing colleague. [MA]

CONRAD, PETER (1948-   ) Australian-born writer and critic, in the UK since the late 1960s. His work as a literary critic and as a speculative writer on OPERA is of indirect fantasy interest. More central is *To Be Continued: Four Stories & Their Survival* (**1995**), which treats the corpus of Western literature as a CAULDRON OF STORY, arguing that STORY itself centrally shapes our sense of community and individual identity – an argument familiar to students of FANTASY. PC's one novel, *Underworld* (**1992**), is an uneasy URBAN FANTASY sans fantasy. [JC]

CONSPIRACIES ◊ FANTASIES OF HISTORY.

CONSTANTINE, MURRAY Pseudonym of Katharine BURDEKIN.

CONSTANTINE, STORM (1956-   ) Innovative UK author who has garnered a cult following; many of her novels lie in the borderland between fantasy and sf. The **Wraeththu** trilogy – *The Enchantments of Flesh and Spirit* (**1987**), *The Bewitchments of Love and Hate* (**1988**) and *The Fulfilments of Fate and Desire* (**1989**) – explores sexuality, violence and the nature of GENDER in a post-HOLOCAUST world. A similar background underpins *Hermetech* (**1991**), which combines themes of ecology, sexual healing and MAGIC. The concern with gender is further examined in *The Monstrous Regiment* (**1990**) and *Aleph* (**1991**), a not wholly successful exploration of the potential pitfalls of the extremes of feminist action. SC's more recent novels – *Burying the Shadow* (**1992**), *Sign for the Sacred* (**1993**), *Calenture* (**1994**) and *Stalking Tender Prey* (**1995**) – have moved her work towards DARK FANTASY. Her writing is distinguished by a fine attention to psychological character detail, charting progression and regression through complex inner as well as outer landscapes. A fascination with sexual power, with sexual ambivalence, with REDEMPTION and transcendence, pervades her writing. ANGEL legends provide resonance for social order in both *Burying the Shadow* and *Stalking Tender Prey*, with imagery and texture echoing the Old Testament, the Apocrypha and John MILTON, and forming the core of a complicated narrative web. RELIGION as Mystery and source of action is explored in *Sign for the Sacred*. Her re-inventions of VAMPIRES and angels lie firmly in the realms of danger and delirium. SC's novels are not always comfortable reading, being challenging both stylistically and thematically. *Stalking Tender Prey* marks a new departure, being the first of her novels to have a contemporary UK setting; the initial volume of the projected **Grigori** trilogy, it is concerned with the legend of the nephilim, with salvation and with the responsibilities of psychological control.

A provocative and prolific writer, SC must be considered to occupy a key position in the emergence during the mid-1990s of dark fantasy as a separate subgenre. [KLM]

CONTE CRUEL A type of story named for VILLIERS DE L'ISLE ADAM's classic collection *Contes cruels* (coll **1883** France; trans Hamish Miles as *Sardonic Tales* **1927**), although the author acknowledged the crucial influence of Edgar Allan POE, whose stories – as translated by Charles BAUDELAIRE – were extraordinarily influential in 19th-century France. Some critics use the label to refer only to nonsupernatural stories, especially those which have nasty climactic twists, but Villiers' collection mixed fantasies and nonfantasies, and the kinds of fantasy story that draw uniquely sharp attention to the relentless cruelties of FATE may be conveniently discussed under this rubric. Writers heavily influenced by Villiers who became prolific producers of such tales include Octave Mirbeau (1848-1917) and Maurice Level (1875-1926), but many other writers associated with DECADENCE produced similar items. Frequently these writers also produced fabular *contes* which sarcastically mocked the moralizing tendency of *contes* written for children by calculated perversions of conventional moral order, and the tradition of fantastic CCs owes something to these as well as to Poe's tales of the grotesque and arabesque. The CC was re-imported into the USA by courtesy of Ambrose BIERCE and other "Bohemian" writers; a significant exemplary collection is W.C. Morrow's *The Ape, the Idiot and Other People* (coll **1897**). Many short stories on the borderline between fantasy and HORROR – like those of SAKI – invite consideration under this heading. [BS]

CONTEMPORARY FANTASY By definition, a CF sets the mundanity of the present day in clear opposition to the fantasy premise. A CF is thus a CROSSHATCH in which radically different realms co-exist, or a PORTAL fantasy in which transition between two realms occurs regularly, or a GNOSTIC FANTASY in which the mundane is merely an appearance laid over a higher spiritual truth, or a FANTASY OF HISTORY in which the workings of the mundane world are other than they appear and its real history is not that recorded, or an INSTAURATION FANTASY in which the mundanity of the contemporary is changing into something other and wonderful – or indeed any combination of these. It may well further include revelation of hidden truths about the working of the apparently mundane world such as WAINSCOT societies, PARIAH ELITES or SECRET GUARDIANS. Many texts can be described simultaneously as CF and as URBAN FANTASY.

The term is best used to describe fantasies written from about the end of the 1960s – Peter S. BEAGLE's *Lila the Werewolf* (1969; **1974** chap) is a *locus classicus* – which represent a conscious *return* to the contemporary world (and usually to its CITIES) in a gesture of the imagination which might be called a colonizing move: a colonization from the past, from the realms of the AMERICAN GOTHIC, from the exotic. Earlier examples – like the occult thrillers of Charles WILLIAMS – share a similar movement of opposition to generally accepted mundane values, which is why they enjoyed revival in the late 1960s, but these were for the most part isolated sports.

CFs typically do not rest easy in the contemporary world which they have colonized. In Emma BULL's *War for the Oaks* (**1987**), for example, the mundane city is experienced as both home and as constraint, and the incursion of FAERIE is both an estrangement from home and a liberation from constraint. CF always sets up dichotomies of values and tries to reconcile them; at its best, it does so with subtlety and complexity. Whether the outcome is choice between values or their reconciliation, the dominant mood of closure is almost always in some sense return.

Perhaps because the CF represents an action of colonizing, of making home exotic or making the exotic into a place where home can be refound, it tends to be full of the paraphernalia of modern life. Though the novels of Stephen KING are more readily described in HORROR terms, their cultish obsession with minutiae like brand-names and song titles is typical of this process; making particular, through examples, the consensual mundane world is also a process whereby the fantastic can be authenticated.

CF is perhaps also the branch of fantasy that has learnt most from MAGIC REALISM; the effect of this authenticating process is to make the incursions of the wild more immediate and more involving. When, as often, CF deals with contemporary issues, it does so directly, not merely by analogy. The effect of the fantasy elements in Elizabeth HAND's *Waking the Moon* (**1994**), in the short fiction of Lucius SHEPARD or in the graphic novels scripted by Neil GAIMAN is to heighten our response to the actually existing discourses in politics, perhaps especially sexual politics, to which these works specifically refer. [JC/RK]

CONTRACTS Contracts appear fairly frequently in fantasy. At the most mundane level are the contracts, either written or tacit, between groups of travellers in FANTASY-LAND (who seek protection) and local mercenaries, WIZARDS, etc. More interesting from our point of view are PACTS WITH THE DEVIL, which epitomize the rule that, in fantasy, anyone signing a contract should READ THE SMALL PRINT. [JG]

See also: PACTS; PLOT DEVICES; QUIBBLES.

CONWAY, HUGH Pseudonym of UK writer Frederick John Fargus (1847-1885), who wrote poems and music-hall songs before he became established as a writer of ingenious melodramas with *Called Back* (1883 *Arrowsmith's Christmas Annual*; **1884**), a murder mystery in which a blind man, Vaughan, is witness to the crime. After his sight is restored Vaughan marries a girl who was also a witness to the crime but has lost all memory. Vaughan is able to psychically communicate ($\lozenge$ TALENTS) with his wife by touching hands, and he can see the murder through her eyes, thus solving the case. The book was a bestseller. In a tragically short writing career before his death from tuberculosis, HC wrote 10 books. *Bound Together* (coll **1884**; cut vt *The Secret of the Stradivarius* 1924 chap) contains a number of atmospheric GHOST STORIES. [MA]

COOK, GLEN (CHARLES) (1944-   ) US writer. His first published story was "Song from a Forgotten Hill" in *Clarion* (anth **1971**). The **Dread Empire** sequence – *A Shadow of All Night Falling* (**1979**), *October's Baby* (**1980**) and *All Darkness Met* (**1980**), plus "Soldier of an Empire Unacquainted with Defeat" (1980) – highly coloured and closely plotted, pits a magical empire resembling China ($\lozenge$ CHINOISERIE) against a disorganized group of quasi-European and Indian states which, largely through the military genius of Bragi, leader of a group of mercenaries, fights the empire off. GC has a sense of the emotional cost and degradation involved in both war and MAGIC. The books are also characterized by a sense of FANTASIES OF HISTORY in which all events are being manipulated by a largely off-stage VILLAIN for an implied and unimaginable audience ($\lozenge$ GODGAME); this villain's references to past interventions exploit our sense of a TIME ABYSS to memorable effect.

Closely linked is a quasi-prequel sequence – *The Fire in his Hands* (**1984**) and *With Mercy Toward None* (**1985**) – describing the early career of Bragi and his involvement with El Murid, a prophet loosely based on Mohammed; the prophet is seen in profane terms, but not without sympathy. The sequence comprising *Reap the East Wind* (**1987**) and *An Ill Fate Marshalling* (**1988**) directly sequels the **Dread Empire** books and has some splendidly dark betrayals and magics. Past empires mentioned as throwaways in the earlier books are briefly evoked to memorable effect. This sequence remains incomplete, with Bragi in the hands of his enemies.

The time abyss is omnipresent in GC's second major fantasy sequence, **Chronicles of the Black Company**, of which there are two series: *The Black Company* (**1984**), *Shadows Linger* (**1984**) and *The White Rose* (**1985**) – assembled as *Annals of the Black Company* (omni **1986**) – and the **Books of the South**, being *Shadow Games* (**1989**) and *Dreams of Steel* (**1990**); further volumes are projected. *The Silver Spike* (**1989**) is not directly part of **Books of the**

South, dealing as it does with simultaneous and interacting events in the locale of the first sequence. Inasmuch as the overall title of the sequence pays homage to Arthur Conan DOYLE's *The White Company* (**1891**), it is a double-edged homage. GC's mercenaries are entirely ruthless in their use of magic and massacre and are for much of the first sequence employees of the almost entirely evil ENCHANTRESS Lady. Their eventual allegiance to the cause of what passes for GOOD is a mixture of personal loyalty and expediency, as is the eventual alliance of *The White Rose*'s rebels and the Lady against a MALIGN SLEEPER – her resurrected and even more evil husband, the Dominator. The books are characterized by a hard-boiled cynicism; the sequence is notable for its bracing refusal of the usual moralizing tropes of the genre.

This genial yet sinister cynicism is the besetting tone of GC's rather different **Garrett** series: *Sweet Silver Blues* (**1987**), *Bitter Gold Hearts* (**1988**), *Cold Copper Tears* (**1988**) – these three assembled as *The Garrett Files* (omni **1989**) – *Old Tin Sorrows* (**1989**), *Dread Brass Shadows* (**1990**), *Red Iron Nights* (**1991**), *Deadly Quicksilver Lies* (**1994**) and *Petty Pewter Gods* (**1995**). Private eye Garrett is an OCCULT DETECTIVE, aided by an undead nonhuman Mycroft Holmes figure. The books pay constant homage to Hammett, Chandler and Spillane; Garrett's neurotic sexism is best viewed as ironic homage to GC's sources.

GC has written two noteworthy singletons. *The Swordbearer* (**1982**) is an interesting REVISIONIST-FANTASY commentary on the eternal-champion motif in general and on Michael MOORCOCK in particular, with a young hero transformed into an AVATAR and an ACCURSED WANDERER by his SWORD and finding himself used and abused in a GODGAME by gods, MAGI, the SOULS his sword has drunk and a dwarf COMPANION who is more than he seems. *The Tower of Fear* (**1989**) combines thriller elements with high politics in a complex tale involving IDENTITY EXCHANGE.

GC has often been rebuked for spatchcocked plots and laziness of DICTION; his high rate of production and less than consistent level of achievement are certainly regrettable, but the overall competence and unique flavour of his best work cannot be doubted.   [RK]

Other works: *The Swap* (**1970**) as Greg Stevens, erotica; *The Heirs of Babylon* (**1972**), post-HOLOCAUST sf; the **Starfishers** sequence, comprising *Shadowline* (**1982**), *Starfishers* (**1982**) and *Stars' End* (**1982**), space opera owing debts to Richard WAGNER; the **Darkwar** trilogy, comprising *Doomstalker* (**1985**), *Warlock* (**1985**) and *Ceremony* (**1986**), SCIENCE FANTASY with psi and technology masquerading as magic; *A Matter of Time* (**1985**), time-travel thriller; *Passage at Arms* (**1985**), space opera; *The Dragon Never Sleeps* (**1988**), space opera; *Sung in Blood* (**1990** NESFA limited edition), fantasy/detection.

COOK, HUGH (WALTER GILBERT) (1956-   ) UK-born New Zealand writer. His first novel was the associational *Plague Summer* (**1980** UK), but he is known almost exclusively for a long fantasy sequence originally called the **Wizard War** series, eventually **The Chronicles of an Age of Darkness**: *The Wizards and the Warriors* (**1986** UK; vt *Wizard War* 1987 US), *The Wordsmiths and the Warguild* (**1987** UK; vt in 2 vols as *The Questing Hero* 1988 US and *The Hero's Return* 1988 US), *The Women and the Warlods* (**1987** UK; vt *The Oracle* 1989 US), *The Walrus and the Warwolf* (**1988** UK; cut vt *Lords of the Sword* 1991 US), *The Wicked and the Witless* (**1989**), *The Wishstone and the*

Wonderworkers (**1990**), *The Wazir and the Witch* (**1990**), *The Werewolf and the Wormlord* (**1991**), *The Worshippers and the Way* (**1992**) and *The Witchlord and the Weaponmaster* (**1992**). The various interwoven stories of the sequence – not chronologically told, and not necessarily interdependent – take place initially in a FANTASYLAND venue, with hints that it is in fact a post-HOLOCAUST Earth; but as the series continues signals are given that this venue is in fact a MULTIVERSE. Much of the action concerns attempts at survival and triumph on the part of various HEROES AND HEROINES, WITCHES and WIZARDS and other characters from the armamentarium of GENRE FANTASY; the ever-present background is one of strife, with petty kings and various warlords constantly at odds. Some volumes are devoted to HEROIC-FANTASY routines; in others the world is threatened with change. There are conceits of some considerable interest, such as, in *The Wordsmiths and the Warguild*, the Odex, which transforms creatures of the imagination into real menaces. But the sequence is considerably more incoherent than most readers will accept.   [JC]

Other works: *The Shift* (**1986** UK), comic sf.

COOK, RICK Working name of US writer James Richard Cook (1944-   ), who began publishing work of genre interest with "Mortality" for *Analog* in 1987. He first came to fantasy notice with his **Wizard** series – *Wizard's Bane* (**1989**), *The Wizardry Compiled* (**1989**), *The Wizardry Cursed* (**1991**) and *The Wizardry Consulted* (**1995**) – which takes as UNDERLIER figure the Boss in Mark TWAIN's *A Connecticut Yankee in King Arthur's Court* (**1889**), transporting a computer nerd into an OTHERWORLD where he uses his programming knowledge to work a TECHNOFANTASY transformation on the art of MAGIC, defeats a DARK LORD or two, and wins his battles. *Limbo System* (**1989**) and *Mall Purchase Night* (**1993**) are perhaps less sustained, though the latter, set in a mall built along a line dividing this world from FAERIE, engenders various comical CROSSHATCH encounters.   [JC]

COOKE, CATHERINE Working name of US writer Catherine Marie Cooke Montrose (1963-   ), who began publishing with her first fantasy series, **Eleven Kingdoms** – *Mask of the Wizard* (**1985**), *Veil of Shadow* (**1987**) and *The Hidden Temple* (**1988**). Set in a CELTIC-FANTASY environment, the sequence, whose plot is complicated, may be seen as deriving from a conservative rendering of the Triple GODDESS as a figure masked (and perhaps contorted) by patriarchy; for instance, her priestesses must remain virgin (unlike acolytes of the Goddess in her uncloaked prime), and much of the plot hinges upon the shame and exile experienced by one female protagonist who loses her VIRGINITY, and who must be championed by more than one HERO. A SHAPESHIFTER villain provides much of the interest of the sequence, which ends in a restoration of the proper monarchy. The **Winged Assassin** sequence – *The Winged Assassin* (**1987**), *Realm of the Gods* (**1988**) and *The Crimson Goddess* (**1989**) – similarly hinges on a conservatively rendered Goddess figure.   [JC]

COOL WORLD US live-action/ANIMATED MOVIE (*1992*). Paramount. **Pr** Frank Mancuso Jr. **Dir** Ralph BAKSHI. **Screenplay** Michael Grais, Mark Victor. **Starring** Michele Abrams (Jennifer), Kim Basinger (Holli Would), Gabriel Byrne (Jack Deebs), Brad Pitt (Frank Harris). **Voice actors** Basinger, Byrne and Pitt plus Charles Adler (Nails), Joey Camen (Slash), Michael David Lally (Sparks), Maurice LaMarche (Doc Whiskers, Mash), Candi Milo (Bob, Lonette), Gregory Snegoff (Bash). 99 mins. Colour.

COMICS-artist Deebs has created the Cool World, a place populated by grotesque Doodles – at least, he believes he has created it, but in fact the Cool World is an ALTERNATE REALITY which must forever remain separate from ours if the cosmic balance is to be upheld. One Noid (as humans are known there), Harris, has made the crossing to the Cool World, swapping existences with Doc Whiskers, the Doodle who invented the Spike of Power that made this exchange possible. Now Holli, a nymphomaniac Doodle whom Deebs thinks he has created as the epitome of his own lusts, desires to cross from the Cool World into our REALITY so that she may experience sex in the flesh. Once here she seizes the Spike of Power, opening a PANDORA'S BOX from which the Cool World's grotesques spring nightmarishly. Deebs succeeds in replacing the Spike and negating the threat of universal chaos, but at the cost of himself becoming a comic-strip SUPERHERO pent forever with a vengeful Holli.

CW stands as one of the fantastic CINEMA's most significant achievements, an INSTAURATION FANTASY that reveals greater depths with each viewing. It is also something of a landmark in ANIMATED MOVIES: while the live-action/animation junction is less surely handled than in the comparable WHO FRAMED ROGER RABBIT (*1988*), the manic creativity of *CW*'s animation probably has no peer, with visual grotesqueries and inventiveness – often totally irrelevant to the plot – seemingly threatening to spill out of the screen and into our laps: the climactic scenes of the Doodles pouring into our world are merely confirmation of what has been happening throughout the movie. Many of these nonce-creations are RECURSIVE, with characters differing from their originals by only the width of a breach-of-copyright suit, plus PARODIES: of "That's All Folks" in the movie's finale, of GHOSTBUSTERS (*1984*) as the Doodles stream out from the Spike of Power, and, perhaps rather spitefully, of the opening sequence of DISNEY's *The Rescuers Down Under* (*1990*), culminating in a seedy jail rather than in the romantic shade of Ayer's Rock. The depiction of the Cool World itself is also of interest: from its bizarre architectures and constant thrum of frenetic activity arise a *different* world that is completely realized – almost claustrophobically so. [JG]

COOPER, LOUISE (1952-) UK writer whose first novel was the TAROT-based *The Book of Paradox* (1973). A later novel, *Lord of No Time* (1977), was drastically reworked to form **Time Master**, the first of three linked trilogies, and the first of the two series on which her reputation is principally based.

The original **Time Master** trilogy – *The Initiate* (1985), *The Outcast* (1986) and *The Master* (1987) – involves a REVISIONIST-FANTASY development of the ideas about the perpetual conflict of CHAOS and Order contained in the work of Michael MOORCOCK. Tarod, a young acolyte of Order, is an AVATAR of one of the lords of Chaos and escapes execution by manipulating time (◊ TIME FANTASY); he and his handful of supporters become a persecuted PARIAH ELITE. LC's personified lords of Chaos and Order need the allegiance of human servants to dominate; the long-standing rule of Order has become sterile and tyrannical and the eventual shift of the balance towards domination by Chaos is seen as liberatory, for the moment. The **Chaos Gate** trilogy – *The Deceiver* (1991), *The Pretender* (1991) and *The Avenger* (1992) – is set a few years later and focuses on interventions in human affairs by DEMONS, who owe allegiance to neither Chaos nor Order and whom neither Chaos nor Order can counter directly. The **Star Shadow** trilogy – *Star Ascendant* (1994 UK), *Eclipse* (1994) and *Moonset* (1995) – prequels the other books; it is set in the distant past at a point when a previous ascendancy of Chaos is decaying and a transition to the ascendancy of Order seems imminent.

Most of these books centre on an EDIFICE, the vast complex of temple buildings on the Star Peninsula, or on its secular complement, the High Margrave's castle. LC derives much emotional power from a fairly standard Gothic fascination with the saturnine Tarod and the strong women who are attracted to him.

A strong female character dominates the other series, **Indigo** – *Nemesis* (**1988**; rev 1989 USA), *Inferno* (**1988**; rev 1989 USA), *Infanta* (**1989**; rev 1990 USA), *Nocturne* (**1990**; rev 1990 USA), *Troika* (**1991**), *Avatar* (**1991**), *Revenant* (**1992**) and *Aisling* (**1993**). The heroine, Pandora-like, releases demons from a box, and these kill her entire people; she is obliged, and magically enabled, to wander a variety of locales with an animal COMPANION – a WOLF whose intelligence is specifically nonsupernatural in origin – destroying one demon after another. There is a solid gloomy worth to the series, which endlessly and inventively revises critically its own premises.

LC has evolved into a reliable writer of TEMPLATE fare, who often surprises by her merits without ever transcending the limits she has set for herself. [RK]
Other works: *Blood Summer* (**1976**), a VAMPIRE novel, and its sequel *In Memory of Sarah Bailey* (**1977**); *Crown of Horn* (**1981**); *The Blacksmith* (**1981**); *Mirage* (**1987**); *The Thorn Key* (**1988** USA), YA fantasy; *The Sleep of Stone* (**1991**), *Firespell* (**1996**), YA supernatural; *The Hounds of Winter* (**1996**), YA supernatural; *Daughter of Storms* (**1996**), YA fantasy set in the **Time Master** world; *Blood Dance* (**1996**), YA supernatural; *The Shrouded Mirror* (**1996**), YA supernatural.

COOPER, SUSAN (MARY) (1935-) UK-born writer, in the USA from 1963, who began publishing with *Mandrake* (**1964** UK), an sf novel. Soon after, SC began to publish her most important work to date, the **Dark is Rising** sequence: *Over Sea, Under Stone* (**1965** UK), *The Dark is Rising* (**1973**), *Greenwitch* (**1974**), *The Grey King* (**1975**) and *Silver on the Tree* (**1977**), all five assembled as *The Dark Is Rising Sequence* (omni **1984** UK). Only slowly is the reader brought into the full compass of the tale, in which UNDERLIER figures from the Arthurian saga (◊ ARTHUR) gradually unveil themselves: kindly old Great-Uncle Merriman Lyon turns into MERLIN, one of the Old Ones or SECRET MASTERS, a group of quasi-human beings which exist in a flexible relationship to TIME and have long acted as guardians and mentors in the great conflict between Light (whose knights they are, and which represents the civilizing force of the Arthurian *idea*) and the Dark (which is a kind of personalized entropy, a sickening impulse towards the THINNING of the world into CHAOS). Much of the Arthurian wisdom is transmitted to the main protagonist – young Will Stanton, seventh son of a seventh son (◊ SEVEN), who turns out to be the last of the Old Ones – through precepts conveyed in a fictional BOOK, the *Book of Gramarye*. The sequence, which takes place respectively in summer, winter, spring, autumn and midsummer (◊ SEASONS), is full of characters and icons out of the CAULDRON OF STORY; for example, the GODDESS, known here as The Lady – and owing her nature at least in part to Robert Graves's *The*

White Goddess (**1947**) – is an important figure, and is attended by considerable additional Celtic material (◊ CELTIC FANTASY). The overall tale evolves – not without occasional narrative confusion when time paradoxes and puzzles must be confronted – towards a guardedly affirmative climax in which it seems that the various young protagonists plus Bran Davies (King Arthur's son) may succeed in staving off entropy and totalitarianism.

SC's later work has been more succinct, though not inconsequential. In the remarkably complex *Seaward* (**1983**) two young protagonists find themselves in an OTHERWORLD complexly run on WONDERLAND lines and proving a testing ground for the newly dead (◊ POSTHUMOUS FANTASY). The protagonists, who remain alive, become caught up in a vast CHESS match between a Goddess figure who (ambivalently) represents Death, and her father/brother/son/consort, who (less interestingly) represents Life; the conflict threatens at points to turn the tale into a two-sided GODGAME. Ultimately the novel unfolds through complex RECOGNITIONS of choice and passage into a tale in which QUEST and RITE OF PASSAGE are the same thing; and the two youngsters choose mortal lives, back in the real world. The eponymous spirit who haunts some young protagonists in *The Boggart* (**1993**) longs only for his exile in Toronto to end, and to be allowed back to the enfolding mists of Scotland; eventually he is granted his wish.

Even in her lightest tales, SC conveys an underlying sense of the vital importance of the issues she raises, and she complexly modulates, from moment to moment, the rite-of-passage experiences undergone by all her young protagonists. She gives a sense that the world is important; and that fantasy is a central way of understanding the meaning of the world. [JC]

Other works: *J.B. Priestley: Portrait of an Author* (**1970** UK), study of J.B. PRIESTLEY; *Dawn of Fear* (**1970**), associational; *Jethro and the Jumble* (**1979**), for younger children; several volumes, generally for younger children, which tend to present TWICE-TOLD versions of old tales, including *The Silver Cow: A Welsh Tale* (**1983**), *The Selkie Girl* (**1986**), *Matthew's Dragon* (**1991**), *Tam Lin* (**1991**) and *Danny and the Kings* (**1993**).

COOPERSTOWN US movie (*1992*). ◊ BASEBALL.

COOVER, ROBERT (LOWELL) (1932-   ) US writer whose novels have taken aboard an assortment of non-supernatural fantastic devices while also dealing with contemporary political figures and cultural ICONS, thus establishing him as a key exemplar of Postmodernism. *The Origin of the Brunists* (**1965**) is a study of the psychology and dynamics of a Millennarian cult. *The Universal Baseball Association, Inc., J. Henry Waugh, Prop.* (**1968**) examines an accountant's obsession with a fantasy BASEBALL league ruled by dice, and the manner in which he copes with a crisis in the imaginary world's affairs. *Spanking the Maid* (**1982**) is a vivid erotic fantasy, less intense than the brief *tour de force* "The Babysitter" – which can be found in *Pricksongs and Descants* (coll **1969**) alongside the conjuring extravaganza "The Hat Act". *Pinocchio in Venice* (**1991**) is a heartfelt sequel to COLLODI's original, also echoing Thomas MANN, in which the ageing protagonist reverts to wood. [BS]

Other works: *The Public Burning* (**1977**); *Aesop's Forest* (**1986** chap dos); *A Night at the Movies, or You Must Remember This* (**1987**); *Whatever Happened to Gloomy Gus of the Chicago Bears?* (coll **1988**).

COPPARD, A(LFRED) E(DGAR) (1878-1957) UK writer, almost exclusively of short stories, though *Pink Furniture: A Tale for Lovely Children with Noble Natures* (**1930**) is a CHILDREN'S FANTASY of moderate interest. His work is usually based on a sharply and sensitively described rural England, more often than not without any supernatural elements; many of his more than 100 tales are, however, SUPERNATURAL FICTION (mostly GHOST STORIES), and there are some full-blown fantasies. He began publishing collections with *Adam and Eve and Pinch Me* (coll **1921**), in whose title story a man visits the future as a kind of astral self-projection, finding there his own third child at play. Further collections – all incorporating tales in various modes – include *Clorinda Walks in Heaven* (coll **1922**), *The Black Dog* (coll **1923**), *Fishmonger's Fiddle* (coll **1925**), *Silver Circus* (coll **1928**), *The Gollan* (**1929** chap), *Nixey's Harlequin* (coll **1931**), *Cheefoo* (**1932** chap), *Crotty Shinkwin: A Tale of the Strange Adventure that Befell a Butcher of County Clare; The Beauty Spot: A Tale Concerning the Chilterns* (coll **1932** chap), *Dunky Fitlow* (coll **1933**), *Polly Oliver* (coll **1935**), *Ninepenny Flute* (coll **1937**), *Tapster's Tapestry* (**1938** chap), *You Never Know, Do You?* (coll **1939**), *Ugly Anna* (coll **1944**), *Dark-Eyed Lady* (coll **1947**) and *Lucy in her Pink Jacket* (coll **1954**). *Fares Please!* (omni **1931**) assembles the stories published to that date, and *The Collected Tales of A.E. Coppard* (coll **1947**) represents the author's sense of his best work.

Of more specific interest is *Fearful Pleasures* (coll **1946** US) ed August DERLETH, which assembles most of AEC's supernatural fiction and fantasy to that date. *Fearful Pleasures* stands as a title of considerable significance for 20th-century fantasy. A tale like *The Gollan*, for instance, is a complex, wry, revisionist FAIRYTALE which translates what seems to be rural England into a LAND OF FABLE peopled by leprechauns, one of whom gives the compliant Gollan his WISH – not to be at everyone's beck – by granting him INVISIBILITY when he is awake; this soon costs him dear, as other humans are also invisible to him (◊ ANSWERED PRAYERS). Attempting to rejoin society, he trades eyes, ears and nose with various TALKING ANIMALS; but through these organs can perceive only what they can. Eventually he comes under the command of the king, who requires magical feats of him, the last of which ends in a general catastrophe. Gollan – invisible again – lives on for many years "in great privation". This tale, a mere 10 pages long, contains in germ much of the best of late-20th-century fantasy.

Other fantasies of note include "Cheese" and "Crotty Shinkwin", whose protagonist overturns an ISLAND, revealing to the light a tiny town and a bag of air from EDEN. AEC did not publish in genre MAGAZINES, and so his pioneering work has not been widely recognized. [JC]

COPPELIA ◊ E.T.A. HOFFMANN.

CORBEN, RICHARD (VANCE) (1940-   ) US COMIC-strip artist renowned particularly for his painted fantasy strips featuring bulge-muscled men and enormous-breasted women. He worked in animation with Calvin Productions 1963-72. In 1968 he published his first comic strip in the fanzine *Voice of Comicdom*, and he went on to publish a number of b/w fantasy strips in the underground titles *Slow Death*, *Skull*, *Anomaly*, *Grimwit*, *Death Rattle* and *Fantagor*, signing his work "Rich Corben" or "Gore". His most important creation during this period was *Rowlf* (1971), a story about a dog who is turned into a half-canine, half-human creature by a bungling WIZARD. In 1970 he began drawing for the WARREN PUBLISHING magazines EERIE, CREEPY and VAMPIRELLA.

RC's GRAPHIC NOVELS, many of which first appeared in HEAVY METAL, have been published in many languages in collected and serialized form. They include *Bloodstar* (graph **1976**), *New Tales of the Arabian Nights* (graph **1979**), *Jeremy Brood* (graph **1982**) and an adaptation of Harlan ELLISON's post-HOLOCAUST tale *Vic and Blood* (graph **1989**). His most lasting creation is **Den**, which he began in 1973 with the ANIMATED MOVIE *Neverwhere* (*1973*) and subsequently drew as the full-colour comic strip **Den** (*Heavy Metal* 1977-8; vt *Den: Neverwhere* graph **1978**), plus *Den: Muvovum, Den: Children of Fire, Den: Dreams* and *Den: Elements*. Other works include *Rip in Time* (**1986** 4 vols), *Richard Corben's Art Book, The Bodyssey, Children of Fire* (*#1-#3* 1988), his graphic novel *Mutant World* (1979-80; graph **1981**) and, more recently, its sequel, *Son of Mutant World* (*#1-#5* 1990-91). [RT/SH]
Further reading: *Richard Corben: Flights into Fantasy* (**1982**) by Fershid Barucha; *How to Draw Art for Comic Books* by James Van Hise (**1989**); series of three interviews in *Heavy Metal* (**1981**).

CORBETT, W(ILLIAM) J(ESSE) (1938-   ) UK writer who remains best known for the sequence with which he began his career, the **Pentecost** series of CHILDREN'S FANTASIES: *The Song of Pentecost* (**1982**), *Pentecost and the Chosen One* (**1984**) and *Pentecost of Lickey Top* (**1987**). The eponymous chieftain fieldmouse leads his people out of their threatened homeland into a happier territory adjacent to the pond of a disinherited snake (◊ SERPENTS); a TRICKSTER bug adds complexity. All WJC's other works are likewise ANIMAL FANTASIES. [JC]
Other works: *The End of the Tale and Other Stories* (coll **1985**); *The Bear who Stood on his Head* (**1988**); *Dear Grumble* (**1989**); *Toby's Iceberg* (**1990**); *Little Elephant* (**1991**); *Duck Soup Farm* (**1992**); *The Granson Boy* (**1993**).

CORELLI, MARIE (1855-1923) UK writer and occultist, born Mary Mills (later rendered as Mackay – the circumstances of her possibly illegitimate birth were shrouded in secrecy), who became a bestselling Victorian novelist with moralistic OCCULT FANTASIES despite critical opposition and lack of writing ability. Her first book, *A Romance of Two Worlds* (**1886**), introduced the magician **Heliobas**, who later formed the basis for a thematically linked trilogy. It seeks to bond fading religious belief with a new scientific (or electrical) interpretation of faith, and propounds a hotchpotch of mystical and occult theories with minimal logic; but it had a passion that attracted public interest. Despite her publisher's wishes, MC later returned to the theme in *Ardath, The Story of a Dead Self* (**1889**) and *The Soul of Lilith* (**1892**). *Ardath* is an occult ROMANCE involving spiritual TIME TRAVEL to attain self-fulfilment. All three novels feature astral travel, ANGELS, TALENTS and REINCARNATION, reminiscent of the theosophical novels of H.P. BLAVATSKY, though considerably less cogent. The success of the novels, which hit a public increasingly fascinated with SPIRITUALISM, pandered to MC's egotism, and her tales became even more uncontrolled. *Barabbas, A Dream of the World's Tragedy* (**1893**) is an inept account of CHRIST's final days. *The Sorrows of Satan* (**1895**), a fervently Christian PACT-WITH-THE-DEVIL story, became her biggest seller, benefiting from the public backlash against the DECADENCE of the 1890s. MC saw herself as something of an evangelist, a mood which sustained her work through the 1890s but faltered in the new century, although she wrote for another 20 years. [MA]
Other works: *Cameos* (coll **1896**); *The Mighty Atom* (**1896**);

Ziska, The Problem of a Wicked Soul (**1897**); *The Master-Christian* (**1900**); *The Strange Visitation of Josiah McNason* (**1904**; vt *The Strange Visitation* 1912), a Christmas GHOST STORY based on Charles DICKENS's *A Christmas Carol* (**1843**); *The Devil's Motor* (**1910**); *The Life Everlasting* (**1911**); *The Young Diana: An Experiment of the Future* (**1918**); *The Love of Long Ago* (coll **1920**); *The Secret Power: A Romance of the Time* (**1921**).

CORLETT, WILLIAM (ALBERT) (1938-   ) UK actor, dramatist and writer, initially known for the **Gate** sf sequence and latterly for the more ambitious **Magician's House** sequence – *The Steps up the Chimney* (**1990**), *The Door in the Tree* (**1991**), *The Tunnel Behind the Waterfall* (**1991**) and *The Bridge in the Clouds* (**1992**) – in which a 16th-century MAGUS, involved in his own time in ALCHEMY, comes into the contemporary world in aid of some beleaguered children, with whom he has TIMESLIP and OTHERWORLD adventures. *The Summer of the Haunting* (**1993**) is a SUPERNATURAL FICTION. [JC]

CORLEY, DONALD (1886-1955) US artist, and author of two books of exotic fantasy: *The House of Lost Identity* (coll **1927**) and *The Haunted Jester* (coll **1931**). All his stories affect a flamboyant style, especially those with Arabian or oriental settings (◊ ARABIAN FANTASY; ORIENTAL FANTASY). His tales are much influenced by Lord DUNSANY and James Branch CABELL (who provided an introduction to the first volume) in their fabrication of imaginary worlds. DC's stories are often self-indulgent, but have a captivating flair. [MA]

CORMAN, ROGER (WILLIAM) (1926-   ) US movie director, producer, writer and entrepreneur, famed for his ability to create movies to very low budgets and with exceptional speed. One could mock his *oeuvre* as comprising solely exploitation cheapies, but several of his movies are now regarded as cult classics and numerous of cinema's most hallowed names were given their first break by him: Peter Bogdanovich, Ellen Burstyn, Francis Ford Coppola, Robert De Niro, Dennis Hopper, Jack Nicholson and Martin Scorsese, to name just a few. Of RC's huge output – much too huge to list here – a fair deal is of fantasy interest, almost always on the margins of either sf or horror (◊ HORROR MOVIES). At the same time, his company New World Pictures – which he created after splitting from American International Pictures – was responsible for bringing to the USA various foreign movies (by Ingmar BERGMAN, Federico Fellini, Akira Kurosawa, François Truffaut and others) which more orthodox distributors regarded as "too difficult". In recent years his companies Full Moon and New Horizons have been offering young acting and directorial talents opportunities to work alongside mature – often over-mature – cinema notables; the results can be excruciating (and are more often released direct-to-video than theatrically), but the educationist motive must be extolled and sometimes – as with *Dracula Rising* (*1992*; ◊ DRACULA MOVIES) – the results can be pleasing.

RC's entrance to movies was through co-scripting *Highway Dragnet* (*1954*); the mutilation of his screenplay inspired him to take personal control of subsequent movies, and so he produced *Monster from the Ocean Floor* (*1954*; vt *Monster Maker*) and *The Fast and the Furious* (*1954*) and directed *Five Guns West* (*1954*), the latter a Western. Further directions of fantasy interest include *The Day the World Ended* (*1955*), *It Conquered the World* (*1956*), *Swamp Woman* (*1956*), *The Undead* (*1956*), *Attack of the Crab Monsters* (*1957*), *Not of this Earth* (*1957*), *She-Gods of Shark*

Reef (*1957*; vt *Shark Reef*), *The Viking Women and the Sea Serpent* (*1957*; vt *The Saga of the Viking Women and the Sea Serpent*; vt *Viking Women*), *The Last Woman on Earth* (*1958*), *Out of the Darkness* (*1958*; vt *Teenage Caveman*), *War of the Satellites* (*1958*), *A Bucket of Blood* (*1959*), *The Wasp Woman* (*1959*), *Creature from the Haunted Sea* (*1960*), *The Intruder* (*1961*; vt *I Hate Your Guts*), *The* LITTLE SHOP OF HORRORS (*1961*), *Tower of London* (*1962*), *X – The Man with the X-Ray Eyes* (*1963*), *Gas-s-s-s, or It Became Necessary to Destroy the World in Order to Save It* (*1970*) and *Frankenstein Unbound* (*1990*; vt *Roger Corman's Frankenstein Unbound*; ◊ FRANKENSTEIN MOVIES). Of more significance was his sequence of movies based loosely on works by Edgar Allan POE: these were *The* HOUSE OF USHER (*1960*), *The* PIT AND THE PENDULUM (*1961*), *The Premature Burial* (*1961*), *Tales of Terror* (*1961*), *The Raven* (*1963*), *The Haunted Palace* (*1963*) – which owed more perhaps to H.P. LOVECRAFT's "The Case of Charles Dexter Ward" (1941) than to Poe – *The* MASQUE OF THE RED DEATH (*1964*) and *The Tomb of Ligeia* (*1964*; vt *The House at the End of the Road*; vt *Ligeia*; vt *Last Tomb of Ligeia*). More recently, through Full Moon, he has revisited Poe with *The* MASQUE OF THE RED DEATH (*1989*) and *The* PIT AND THE PENDULUM (*1990*).

The list of movies RC has produced or executive produced is yet more lengthy, but few are of great interest outside *Night of the Blood Beat* (*1958*), *The Brain Eaters* (*1958*), *The Dunwich Horror* (*1969*), *Death Race 2000* (*1975*), *Cannonball* (*1976*; vt *Carquake*), *God Told Me To* (*1976*; vt *Demon*), *The Bees* (*1978*), *Piranha* (*1978*) and *Saturday the 14th Strikes Back* (*1988*); more typical are titles like *Naughty Nurses* (*1973*; vt *Tender Loving Care*). RC has also played bit parts in numerous of his own and some other directors' movies, as in Joe Dante's *The* HOWLING (*1980*). [JG]

CORN DOLLS ◊ DOLLS.

CORMIER, ROBERT (EDMUND) (1925-) US writer of children's books, many being borderline sf. *Fade* (*1988* UK) features a family with inherited powers of INVISIBILITY. [GF]

CORRAN, MARY (1953-) UK writer whose first novel, *Imperial Light* (*1994*), somewhat overelaborately follows through a FANTASYLAND the mission of its female protagonist to combat a PARIAH ELITE of evil priests and to deliver some "lightstones" to places where they will help defeat the Dark and (with the help of the GODS) bring the Light. Her second, *Fate* (*1995*), more interestingly depicts a world in which luck is a measurable quantity; the female protagonist, a surviving TWIN, has twice the normal allocation of it, and successfully defies a mechanical patriarchal hierarchy that, enslaved by superstition, has become a tyranny. [JC]

CORTÁZAR, JULIO (1914-1984) Belgian-born Argentine writer, in Argentina 1918-51, thereafter moving to France, where he lived the remainder of his life. He began to publish early, with a volume of poems, *Presencia* ["Presence"] (coll *1938*) as by Julio Denís. His first work of fantasy interest was *Los Reyes* ["The Kings"] (*1949*), a long narrative poem constituting a meditation on the role and fate of the MINOTAUR in his LABYRINTH. Labyrinths, as with his older contemporary, Jorge Luis BORGES, constantly appear in JC's work, either literally or transformed into linguistic puzzles or games – like that which governs his most famous novel, *Rayuela* (*1963*; trans Gregory Rabassa as *Hopscotch* 1966 US), a tale which engages with the FANTASTIC rather than with fantasy as such. A typical JC story begins in the mundane world, then introduces fantastic elements with an

effect both deadpan and UNCANNY: for JC, the fantastic represents a saving violation of the rules of the real world, and allows through that violation some loosening of the BONDAGE of mere REALITY.

JC's novels of fantasy interest include his first, *Los premios* (*1960*; trans Elaine Kerrigan as *The Winners* 1965 US), in which an ocean liner stranded in port becomes a SHIP-OF-FOOLS microcosm of the world-order, and *62: Modelo para armar* (*1968*; trans Gregory Rabassa as *62: A Model Kit* 1972 US), which carries on from *Hopscotch* the use of linguistic games to shape the lives of the protagonists who – unwittingly or unwillingly – play them. In this case, the game rules; various characters shuttle from a mysterious Zone and the CITY according to GODGAME-like instructions they cannot understand or disobey; VAMPIRES and other supernatural intrusions appear.

It may be that JC's short stories will remain his central legacy. Tales from *Bestiaro* ["The Bestiary"] (coll *1951*), *Final del juego* ["End of the Game"] and *Las armas secretas* ["Secret Weapons"] made considerable impact on their appearance as *End of the Game and Other Stories* (coll trans Paul Blackburn 1963 US; vt *Blow-Up and Other Stories* 1967 – the title change occurring when Michelangelo Antonioni turned one of the tales into his movie *Blow-Up* 1966). Further collections, some overlapping, include: *Historia de cronopios y de famas* (coll *1962*; trans Paul Blackburn as *Cronopios and Famas* 1969 US); *Todos los fuegos el fuego* (coll *1966*; cut trans Suzanne Jill levine as *All Fires the Fire and Other Stories* 1973 US); *Alguien qu anda por ahí* (coll *1977*; trans Gregory Rabassa as *A Change of Light and Other Stories* 1980 US); and *Queremos tanto a Glenda* (coll *1981*; cut trans Gregory Rabassa as *We Love Glenda So Much and Other Tales* 1983 US). Every volume includes some fantasy.

Some of the essays and fantasias assembled in *La vuelta al día en ochenta mundos* (coll *1967*; cut trans Thomas Christensen as *Around the Day in Eighty Worlds* 1987 US), which includes excerpts from *Ultima Round* ["Last Round"] (coll *1969*), are of interest; *Fantomas contra los vampiros multinacionales: Una utopí realizable* ["Fantomas Versus the Multinational Vampires: An Attainable Utopia"] (*1975*) is a nonfiction study. [JC]

COSMIC TREE ◊ WORLD-TREE.

COSMOS SCIENCE FICTION AND FANTASY MAGAZINE ◊ MAGAZINES.

COSTIKYAN, GREG(ORY JOHN) (1959-) US writer and GAME designer who began publishing genre-related work with the role-playing game *The Creature that Ate Sheboygan* (*1979*); he went on to design over two dozen published games, five times winning the Origins Award. His first fiction was "They Want Our Women!" (*Abo* 1988). *Another Day, Another Dungeon* (*1990*) is a humorous ARABIAN FANTASY. *By the Sword* (*1993*) is a HEROIC FANTASY. Sequels to both volumes have been announced. The latter novel is unusual in having first (1991) been serialized, in shorter form, on the computer service Prodigy. [GF]

COULSON, JUANITA (RUTH) (1933-) US writer who began publishing work of genre interest with "Another Rib" for *F&SF* in 1963 with Marion Zimmer BRADLEY, writing together as John Jay Wells. She remains better-known for her sf than her fantasy, but the **Krantin** sequence – *The Web of Wizardry* (*1978*) and *The Death God's Citadel* (*1980*) – is of interest for its overall PLANETARY ROMANCE. The sequence does not otherwise stand out; and JC is clearly more comfortable – as in *Star Sister* (*1990*) – with novels which, though similar in feel, ride upon sf rationales.   [JC]

COUNSELMAN, MARY ELIZABETH (1911-1995) US writer and poet, long resident in the South, which region figures significantly in her work, a fraction of which has been collected in *Half in Shadow* (coll **1978**). MEC's first professional sale was "The Devil Himself" in *My Self* (1931); her most popular was "The Three Marked Pennies" (1934 *WT*), an ALLEGORY in which the titular coins – conveying wealth, travel and death – go to those who will benefit least from them. The folklore and beliefs of Mountain Whites serve as background to "A Death Crown for Mr Hapworthy" (1948 *WT*), where a lump of feathers appears in the death pillow of a good person, "The Unwanted" (1950 *WT*), where a childless woman creates dream children, and "The Tree's Wife" (1950 *WT*), where the spirit of a young man appears in a TREE. An undisguised longing for the antebellum South is visible in such tales as "The Shot-Tower Ghost" (1949 *WT*) and "The Green Window" (1949 *WT*). At her best, MEC was a capable regionalist whose use of the fantastic enhanced her familiar themes.   [RB]

Other works: *African Yesterdays* (coll **1977** chap), a collection of jungle stories from the pulps; *New Lamps for Old* (coll **1978** chap), a vol of mostly new stories; *The Face of Fear and Other Poems* (coll **1984** chap) ed Steve Eng (1940-   ), with intro by Joseph Payne BRENNAN.

COUNT DRACULA Spanish/West German/Italian/Liechtenstein movie (*1970*; vt *El Conde Dracula*; vt *Bram Stoker's Count Dracula*). ◊ DRACULA MOVIES.

COUNT DRACULA'S GREAT LOVE Spanish movie (*1972*; ot *Gran amor del conde Dracula*). ◊ DRACULA MOVIES.

COUNTERATTACK OF THE MONSTER vt of *Gigantis* (*1955*). ◊ GODZILLA MOVIES.

COUNTESS DOLINGEN OF GRATZ French movie (*1981*; vt *Les Jeux de la Comtesse Dolingen de Gratz*). ◊ DRACULA MOVIES.

COUNTESS DRACULA UK movie (*1970*). ◊ DRACULA MOVIES.

COUNT YORGA VAMPIRE US movie (*1970*). ◊ VAMPIRE MOVIES.

COUPON FANTASY ◊ PLOT COUPONS.

COVEN 13/WITCHCRAFT & SORCERY *C13* was a US digest MAGAZINE, 4 issues, bimonthly, September 1969-March 1970, published by Camelot Publishing, Los Angeles; ed Arthur H. Landis (1917-1986). The subscription list was then sold to William L. Crawford (1911-1984), who continued the magazine as a large-format SMALL-PRESS neo-pulp retitled *W&S*, 5 issues, bimonthly (but irregular, January/February, May 1971, [January] 1972, [July] 1972, [July] 1973, [September] 1974), published by Fantasy Publishing Co., Alhambra, California; ed Gerald W. Page (1939-   ).

C13 endeavoured to be a quality magazine, attractively illustrated throughout by William STOUT, but its low budget meant Landis bought fiction from acquaintants, mostly in academia and tv around Los Angeles; best-known contributors were Harlan ELLISON, Ron GOULART, Robert E. HOWARD and Bill Pronzini (1943-   ). *C13* sought to emulate both WEIRD TALES and UNKNOWN, and succeeded moderately. Its fiction focused mostly on tales of WITCHCRAFT, VAMPIRES and the occult with early examples of AMERICAN GOTHIC. It also serialized, through all four issues, the SCIENCE-FANTASY "Let There Be Magick" (exp vt *A World Called Camelot* **1976**) by Landis as James R.

Keaveney, a tale in the style of Christopher STASHEFF.

Revamped as *W&S*, the magazine sought to have stronger, darker fiction and the emphasis gradually shifted from the supernatural to FANTASY, especially HEROIC FANTASY. In this vein it came closer to *WT*, although the quality began to suffer as the budget failed. Well illustrated, and with fiction from Gary Brandner, Glen COOK, August W. DERLETH, R.A. LAFFERTY, Brian LUMLEY, Emil Petaja and E. Hoffmann PRICE, the magazine is now primarily recalled for Price's column of reminiscences about the pulp days, *Jade Pagoda*.   [MA]

COVILLE, BRUCE (1950-   ) US writer of sf and fantasy, almost exclusively for children. Many of his early works were commercial ties and SHARED-WORLD ventures; most are unmemorable, although including the very popular CHILDREN'S FANTASY *My Teacher is an Alien* * (**1989**), which has enjoyed several sequels: *My Teacher Fried My Brains* * (**1990**), *My Teacher Glows in the Dark* * (**1991**), the first three books being assembled as *My Teacher is an Alien: 3 Books in One* * (omni **1995**), and *My Teacher Flunked the Planet* * (**1992**). In recent years he has been engaged in revising and republishing many of his earlier books; this has produced a rather complex bibliography.

BC's best work has probably been his 1990s YA fantasies. These include his **Magic Shop** books – *Jeremy Thatcher, Dragon Hatcher* (**1991**) and *Jennifer Murdley's Toad* (**1992**) – and *Goblins in the Castle* (**1992**) and *Oddly Enough* (coll **1994**). The **Unicorn Chronicles** sequence to date comprises only *Into the Land of Unicorns* (**1994**).   [GF]

Other works: *The Dragonslayers* (**1994** chap); *Fortune's Journey* (**1995**), historical; the **Rod Allbright** series, comprising *Aliens Ate My Homework* (**1993**), *I Left My Sneakers in Dimension X* (**1994**) and *The Search for Snout* (**1995**); the **A.I. Gang** trilogy, comprising *Operation Sherlock* (**1986**; rev 1995), *Robot Trouble* (**1986**; rev 1995) and *Forever Begins Tomorrow* (**1986**; rev 1995); the **Nina Tanleven** series, comprising *The Ghost in the Third Row* (**1987**), *The Ghost Wore Grey* (**1988**) and *The Ghost in the Big Brass Bed* (**1991**). BC has written book and lyrics for four stage musicals, three being fantasies for children: *The Dragonslayers* (produced 1981; professional premiere 1996), *Out of the Blue* (produced 1982) and *It's Midnight! (Do You Know Where Your Toys Are?)* (produced 1983).

As editor: *Bruce Coville's Book of Monsters: Tales to Give you the Creeps* (anth **1993**), *Bruce Coville's Book of Aliens: Tales to Warp Your Mind* (anth **1994**) and *Bruce Coville's Book of Nightmares: Stories to Make You Scream* (anth **1995**).

COX, GREG Working name of US editor and writer William Gregory Cox (1959-   ), who has written some sf ties. His anthologies include *Tomorrow Sucks* (anth **1994**) and *Tomorrow Bites* (anth **1995**), both with T.K.F. Weiskopf; the first is about VAMPIRES, the second about WEREWOLVES. GC is of greatest interest for *The Transylvanian Library: A Consumer's Guide to Vampire Fiction* (**1993**), which is compendious and witty.   [JC]

COX, PALMER US writer (1840-1924). ◊ ELVES.

COYOTE TRICKSTER figure who frequently appears in Native American FOLKLORE. He is to be found in a very wide range of contexts, running from the **Road Runner and Wile E. Coyote** animated shorts (◊ Chuck JONES) to exercises in literary fantasy like Richard Thornley's *Coyote: Comprising a Final Bid from the Maggot, Sundry Tricks, and the Passing of the Soul through Death* (**1994**).   [JC]

CRACE, JIM Working name of UK journalist and writer

James Crace (1946-   ), who began publishing fantasy with "Annie, California Plates" in *The New Review* in 1974. His first novel, *Continent* (fixup **1986**), is a FABULATION set on an imaginary southern continent in an otherwise present-day world; it won the Whitbread First Novel Prize, the Guardian Fiction Prize and the David Higham Prize. JC's spare narrative – the novel comprises six independent sections, none more than short-story length – stands in marked contrast to the expansive narratives that characterize most such works from Tappan WRIGHT's *Islandia* (**1942**) on. *The Gift of Stones* (**1988**) is set at the end of the Stone Age and tells, much like William GOLDING's *The Inheritors* (**1955**), of the end of a community of stoneworkers with the advent of metallurgy. *Arcadia* (**1992**) is again set in an (apparently) imaginary venue: an unnamed CITY, overlooked from his private high-rise by an ageing tycoon who dreams of razing the squalid neighbourhood of his youth to erect a gleaming arcade, a monument to commerce and himself. All of JC's work – including *Signals of Distress* (**1995**), set on England's western coast in 1836 and his only novel to specify time and place – dramatizes the upheavals of social change, with tradition and progress in locked conflict. The doomed forces of tradition are portrayed, unusually for fantasy, without reaction or nostalgia. [GF]

Other works: *The Slow Digestion of the Night* (chap **1995**), five brief stories.

CRAIK, MRS Married name of UK writer Dinah Maria Mulock (1826-1887), most of whose books were originally issued anonymously. Her fantasies were generally for children, but several intended for adults appeared in *Avillion and Other Tales* (coll **1853**) and were reprinted in *Romantic Tales* (coll **1859**). These include two novellas: "Avillion", a VISIONARY FANTASY about the Isles of the Blest, and "The Self-Seer", in which two men trade viewpoints with their spirit DOUBLES. The author's careful piety does not work to the advantage of these tales, although it did wonders for her bestselling nonfantasy *John Halifax – Gentleman* (**1856**). Some of her CHILDREN'S FANTASIES are painfully moralistic. The best is *Alice Learmont: A Fairy Tale* (**1852**), affectionately based on Scottish folklore; *The Little Lame Prince* (**1875**; rev vt *The Little Lame Prince and his Travelling Cloak* 1949 UK) is also notable. [BS]

Other works: *Little Sunshine's Holiday* (**1871**); *The Adventures of a Brownie* (**1872**).

As editor: *The Fairy Book: The Best Popular Fairy Stories Selected and Rendered Anew* (anth **1863**).

CRAM, RALPH ADAMS (1863-1942) US architect, prominently involved in the revival of Gothic architecture. His solitary book of GHOST STORIES, *Black Spirits and White* (coll **1895**), is among the better of its kind. Most of the stories are of HAUNTED DWELLINGS and are highly atmospheric, drawn with an architect's perception for light and shadow. One nonfantasy tale, *The Dead Valley* (1895; **1984** chap), was especially admired by H.P. LOVECRAFT. [MA]

Other works: *Excalibur* (**1895**), a long Arthurian poem.

CRANDALL, REED (1917-1981) Highly skilled and widely respected US COMIC-book artist, with a fine, controlled pen-line style. He drew SUPERHEROES and sf, fantasy and HORROR subjects with equal success. His early influences were Howard PYLE, N.C. WYETH and James Montgomery Flagg (1877-1960).

RC was an editorial cartoonist for the NEA Syndicate before working for the Eisner-Iger partnership (◊ Will EISNER) and then for Quality Comics from 1941. Features he

drew during this period include **Hercules** (*Hit Comics #11-#17* 1941-2), **Firebrand** (*Police Comics #1-#8* 1941-2), **Dollman** (*Feature Comics #44-#63* 1942-3), **The Ray** (*Smash Comics #23-#29* and *#35-#38* 1941-3) and **Blackhawk** (*Military Comics #12-#22* 1942-4). After military service, he continued as the main artist on **Blackhawk** (an almost continuous run in *Modern Comics #46-#83* and subsequently in *Blackhawk #10-#67* **1946-53**).

RC then began working for EC COMICS: his work appeared in *Shock Suspense Stories* (*#9-#18* 1953-5) and *Crime Suspense Stories* (*#18-#26* 1952-5) and several other titles. On the demise of EC he turned more and more towards b/w magazines, and in the 1960s he produced some very highly acclaimed HORROR stories for WARREN PUBLISHING's EERIE and CREEPY. He went on to produce some masterful illustrations for the Canaveral Press series of Edgar Rice BURROUGHS books, augmenting his distinctive pen-and-ink style with a dry-brush and wash technique to achieve some classic renditions of Burroughs themes. RC spent his final years, broken by ill-health and personal problems, working in a fast-food restaurant. [RT]

CRANE, WALTER (1845-1915) UK artist, a senior member of the celebrated triumvirate of 1870s-80s classic nursery-rhyme illustrators. During 1865-76 he designed and illustrated over 40 sixpenny toybooks of nursery-rhymes in colour. Many of these pictures showed the infuence of Japanese art and colour prints. Two of Crane's most ambitious and popular titles were the two songbooks *The Baby's Own Opera* (**1877**) and *The Baby's Bouquet* (**1879**), where the music was perfectly knitted into the decorative scheme. Besides designing the volumes, he calligraphed the text, humorously incorporating the hieroglyph of a crane that acted as his signature to every picture. The two titles were assembled with *The Baby's Own Aesop* (**1887**) as *Triplets* (omni **1889**). *The First of May: A Fairy Masque* (**1881**) by John R.Wise, with text and decorations reproduced by photogravure, was one of Crane's most beautiful and expensive productions. He illustrated the translation *Household Stories from Grimm* (**1882**) by Lucy Crane, his sister. The "Goose Girl" picture from this volume was reproduced in tapestry by William MORRIS, and is now in the Victoria & Albert Museum, London.

WC illustrated many children's books, including 16 by Mrs Molesworth (◊ CHILDREN'S FANTASY). The finest of his adult book ILLUSTRATIONS appear in editions of Edmund SPENSER's *The Fairy Queene* (1894-6) and *The Shepheardes Calendar* (1898). [RD]

CRAVEN, WES (1949-   ) US movie director, writer and producer. He showed an early interest in drawing, 8mm films and music. After earning a BA in English, he was offered a full scholarship at the Johns Hopkins Graduate Writing Seminars, where he obtained a Masters in Writing and Philosophy. That same year WC married, started a family, and began teaching Humanities. He soon bought a 16mm camera and began making amateur movies. Within a year he quit teaching and headed for New York, determined to break into the movies. He started out as a runner and editor for a company making trailers, then raised $70,000 to make his début feature, *Last House on the Left* (*1972*), a disturbing slice of contemporary rape/revenge exploitation loosely based on Ingmar BERGMAN's *The Virgin Spring* (*1959*). Together with Tobe Hooper's *The Texas Chain Saw Massacre* (*1974*), which it predated, the film established a new genre in realistic and graphic HORROR MOVIES. This and

The Hills Have Eyes (*1976*), the latter about a family of mutant psycho-cannibals preying on desert travellers, gained him a cult following, but his first big commercial success was *A* NIGHTMARE ON ELM STREET (*1984*). Inspired by a newspaper article about the inexplicable deaths of young men in their sleep, the movie was budgeted at just $1.5 million by New Line Pictures, who opened it with very little advance publicity. Combining all the elements of the traditional horror movie with impressive spfx, *Nightmare* tells of a group of suburban teenagers who discover they all share a common NIGHTMARE. One by one, as they sleep, they are dispatched by Freddy Krueger, a disfigured killer with a steel-taloned claw, who waits for them in their darkest DREAMS. The movie went on to gross an impressive $30 million at the box office, spawned six sequels, a book and TV series, and turned the character of Krueger into a wisecracking cult success. After a much-publicized disagreement with New Line, Craven was involved only intermittently with the sequels, making a welcome return as co-writer and co-executive producer of *A Nightmare on Elm Street Part 3 Dream Warriors* (*1987*), and writing, directing and playing himself in the film-within-a-film, *Wes Craven's New Nightmare* (*1994*). *The Serpent and the Rainbow* (*1987*) and *The People Under the Stairs* (*1991*) are interesting attempts to do something different with, respectively, the VOODOO and psycho themes. [SJ]

Other works: *Stranger in Our House* (*1978* tvm; vt *Summer of Fear* UK); *Deadly Blessing* (*1980*); *Swamp Thing* (*1981*); *Invitation to Hell* (*1982* tvm); *Chiller* (*1983* tvm); *The Hills Have Eyes Part II* (*1984*); *Casebusters* (*1985* tvm); *Deadly Friend* (*1986*); *Shocker* (*1989*); *Night Visions* (*1990* tvm); *Laurel Canyon* (*1993* tvm), co-creator/exec pr; *Mind Ripper* (*1995*), exec pr; *Vampire in Brooklyn* (*1995*).

CRAWFORD, BETTY ANNE (?   -   ) US writer whose early work appeared under pseudonyms. Three were used on the **Psi Patrol** sequence: *Psi Patrol #1: Sal's Book* (**1985**) as by Sal Liquori; *#2: Hendra's Book* (**1985**) as by Hendra Benoit, and *#3: Max's Book* * (**1985**) as by Maxwell Hurley; the named author of each text is one of the series' three teenaged protagonists, all of whom are learning to use their TALENTS to discover themselves and to fight EVIL. *Two Queens of Lochrin* (**1990**) as by Lee Creighton is set partly in NEW YORK and partly, through TIMESLIP, in a world constructed according to the tenets of CELTIC FANTASY, but whose REALITY is subject to challenge. *The Bushido Incident* (**1992**), as BAC, is sf. [JC]

CRAWFORD, F(RANCIS) MARION (1854-1909) US author who was born and lived mostly in Italy. A capable writer of continental society romances, he is best remembered for "The Upper Berth" (1886 *The Broken Shaft*; in *The Upper Berth* coll **1894**), a powerful tale of a ship's cabin haunted by a suicide, and one of the most reprinted of all GHOST STORIES. Another effective HORROR story of the sea is *Man Overboard* (**1903** chap). His seven SUPERNATURAL FICTIONS are all collected as *Uncanny Tales* (coll **1911** UK; vt *Wandering Ghosts* 1911 US), which includes the VAMPIRE story "For the Blood is the Life".

FMC's longer works tend toward mysticism. His first novel, *Mr Isaacs* (**1882**), uses some of the teachings of THEOSOPHY to enlighten an otherwise mundane romance. The more effective *The Witch of Prague* (**1891**), with Byronic overtones, uses the motif of the FEMME FATALE to explore powers of the mind. *With the Immortals* (**1888**) is a turgid Bangsian Fantasy (and TECHNOFANTASY): an inventor dabbles with electricity in order to summon the GHOSTS of the famous for the purpose of conversation. Much more successful is *Khaled* (**1891**), an ARABIAN FANTASY. *Cecilia: A Story of Modern Rome* (**1902**) is a light romance involving DREAM fulfilment and possible REINCARNATION. [MA]

Further reading: *An F. Marion Crawford Companion* (**1981**) by John C. Moran.

CRAWFORD AWARD ◊ WILLIAM L. CRAWFORD MEMORIAL AWARD.

CRAWSHAY-WILLIAMS, ELIOT (1879-1962) UK writer who had a varied political and military career, retiring with the rank of Lieutenant-Colonel. His initial literary works were plays, collected in *Five Grand Guignol Plays* (coll **1924**) and *More Grand Guignol Plays* (coll **1927** chap). His early novels were realistic, but he made copious use of the fantastic later. *Night in No Time* (**1946**) is a study of mores in which the mysterious Mr Cloxeter allows the protagonist to study the width of the generation gap in 1894 and 1943, thus providing a better context for the grievances he holds against his father in 1920. *The Wolf from the West: Tracing the Glorious Tragedy of Glyndwr* (**1947**) also employs a TIMESLIP to facilitate its purpose. The title story of *The Man who Met Himself* (coll **1947**) is yet another timeslip romance, while "Nofrit", about the revivification of an ancient Egyptian princess, is the best of three VISIONARY FANTASIES therein. *Heaven Takes a Hand* (**1949**) is one of many fantasies inspired by Hiroshima to wonder whether humankind is now worth saving; here, SATAN and Socrates both take an active interest in a HEAVEN-appointed inquiry. *Unusual Eugene* (**1952**) ends with world revolution. [BS]

Other works: *Borderline* (coll **1946**).

CRAY, ROBERTA ◊ Ru EMERSON.

CRAYON, GEOFFREY Pseudonym of Washington IRVING.

CRAZY RAY, THE (*1923*) ◊ René CLAIR.

CREATION MYTHS These are not intended as symbolic answers to scientific questions about the physical origins of the Universe and life. They embody the ideas of a people (or its ruling class) about the relationship between humans and the spiritual and physical world they inhabit now. Writers of fantasy who work in this area are doing the same thing: they employ sf as *social fantasy*, and their alternative creations are a commentary on their contemporary world. CMs show different assumptions about male/female, cyclic/progressive, immanent/ transcendent, cooperation/strife. Yet certain symbols recur across the world, and cannot simply be explained by cultural diffusion: often enough the same image is used to reinforce quite different values.

Among the oldest symbols is that of the Earthmother GODDESS who engenders all life. An Indian hymn praises "the Cause and Mother of the World, the one Primordial Being, Mother of innumerable creatures, Creatrix of the very gods: even of Brahma the Creator, Vishnu the Preserver and Shiva the Destroyer". Often the goddess-mother mates with her son, who may be sacrificed to maintain the cycle of creation. Welsh Madron, the Mother, mourns for Mabon the Son, imprisoned since time beyond memory.

The universal mother may be the primeval deep. In Sumer, she was Nammu, giving birth to An-Ki (Heaven-Earth), joined as a cosmic mountain over the watery abyss. Their son Enlil, the air-god, tore them apart, and then called out the multitude of deities Ki had given birth to, to fill the created world. When the gods tired of doing their own work, they appealed to Mother Nammu. She woke her son, Enki, the wise water-god. At his prompting she pinched off clay from under Ki and shaped a body. Ki went into labour and birth-goddesses delivered the first human, whose descendants would work for the gods.

Here is the worldwide symbol of conjoined parents, separated by their offspring. Space and air are needed for the younger generation to grow and fill the earth with created things. In EGYPT we find another version. Nut is the Sky-Mother arching over Geb the Earth-Father. It was Ra their father who separated them, not their son. This sense of the original androgyny of the Universe is depicted by ARISTO-PHANES, who saw humanity as a two-headed, four-limbed being which had to be split. The two halves desired so much to embrace again the gods had to keep them apart to prevent them starving. With the Dogon of Mali the same image appears. Amma made the Universe, like a potter, from clay. He wanted to copulate with Earth, but was frustrated by an anthill. Instead of the twins she should have borne, she gave birth to a single Jackal. A pair of male/female twins followed. They presented their mother with a life-giving skirt which gave her the power of creative utterance. Jackal raped her to steal the Word. Amma decided to create more beings without her, but the twins were afraid there would be no more pairs like themselves. They drew a male and female on the ground, embracing. Therefore all humaans have twin souls and are bisexual until puberty.

Symbols of the one-becoming-two merge into the universally found ICON of the cosmic egg, or other ovoid. In Oceania, the Naruau story says there was only water, with Old-Spider hovering above. She saw a giant hollow clam and squeezed into its darkness. She found a snail and slept with it under her arm for three days. It opened the shell a little, becoming the MOON. A worm expanded the opening. Its sweat was the salt sea. A larger snail became the SUN. The fully opened halves of the shell are the sky and the Earth.

In the Finno-Ugric myth, the source of ovulation becomes ambiguous. Luonnotar, daughter of the air-god, is a lonely virgin. She lets herself fall into the sea, which makes her fertile. When she has floated for seven centuries a bird appears. He builds a nest on her knees, lays eggs and sits on them. Feeling a scorching heat, she bends her knees and the eggs roll into the "abyss". The lower egg becomes Earth, mother of all, the upper, sky. Yolks make the sun, whites the moon, other fragments are stars and clouds. She shapes the land.

Patriarchal mythologies transfer creative power to a male deity. ZEUS bears Athene from his head. In India, the World Lotus is said to grow from Vishnu's navel, reversing the myth of the Lotus Goddess Padma, whose body was the Universe. When the Babylonians took over Mesopotamia, the creation myth of Sumerian Nammu also changed. The primeval goddess Tiamat gave birth to a son, with Apsu – the Deep – as father. Pairs of gods and goddesses followed. Their disturbance pained the first parents. Apsu was for destroying them; Tiamat advocated tolerance. Wise water-god Ea cast a spell over Apsu, then killed him, and Tiamat produced MONSTERS in revenge. Only the younger-generation god Marduk was brave enough to fight them. He killed Tiamat, splitting her like a shellfish so that her two halves separated the waters above and below the Universe. He made humans to serve the gods from the blood and bones of her evil offspring. The fecund Deep now represents CHAOS, which must be vanquished.

The theme of taking many parts of one body to make a differerentiated world is widespread. In Nordic Europe, the giant Ymir was formed where the glacial streams of Niflheim met the warm air of Muspellheim. As he slept, he sweated. From under his left arm came a man and woman, frost-giants. A primeval cow licked the ice to reveal a man-figure. He fathered the gods ODIN, Vili and Ve. These killed Ymir and made land from his flesh, waters from his blood; his bones became crags, his teeth gravel, his hair trees. The heavens were his skull set on four pillars, lit by sparks from fiery Muspellheim. Such dismemberment carries a memory that death is required for the creation of life.

The North American Apache reverse this symbolism with the creation of the first human by Black Hactcin. He had the birds and animals bring pollen to trace an outline on the ground, and laid minerals inside. The veins were turquoise, blood red ochre, the skin coral, bones of white rock, nails and teeth opal, eyes jet and abalone, marrow white clay. A dark cloud became hair, which would turn to white.

In other stories, creation takes place from a tiny fragment of primeval clay. Often it is necessary to dive for it. The first who attempt this fail. It is usually an animal-hero who succeeds. The North American Iroquois and Hurons tell of the celestial woman Ataentsic who married the Chief-who-owns-the-earth and bore a daughter, Breath-of-Wind. Her jealous husband uprooted the TREE of Life (◊ WORLD-TREE) and threw wife and daughter into the abyss, watched by the water-creatures. First the otter, then the turtle dived. Lastly the musk-rat brought up some soil under his nails. He placed it on the turtle's shell, which swelled enormously to become Earth. BIRDS carried Ataentsic on their wings and set her on it. Breath-of-Wind died when her twin sons fought in her womb, her body becoming the sun and moon. One twin was unjustly accused by the other. Ataentsic threw him out, but he created good things, while his rival twin could produce only monsters. Sometimes this Earth-Diver becomes a TRICKSTER, subverting the creator's work.

The need to explain the more uncomfortable aspects of creation gave rise to stories of cosmic struggle. The Greeks told of a primeval CHAOS and of GAIA, the deep-breasted Earth. She gave birth to Uranus, the starry sky, who covered her. Their children were Titans, Cyclopes and hundred-handed GIANTS. Horrified, Uranus shut them in the depths of the Earth. Gaia first mourned, then raged. Drawing steel from her bosom, she made a sickle. Only Cronus, her Titan son, would volunteer. He castrated his father and threw his genitals into the sea. Blood fell on the Earth and gave rise to the Furies and monsters. From the debris on the sea came the white foam which gave birth to the goddess APHRODITE. Cronus, in turn, had to be defeated by his son ZEUS, continuing the creation and the struggle.

Such myths of antagonism reached their height in Persia. From the beginning, Ahura Mazda, Lord of Light, Wisdom and Light, and Angra Mainyu (Ahriman), Demon of the Lie, competed to create good and bad. Ahura Mazda made spiritual beings without tangible bodies. When Angra Mainyu saw their glory, he rushed to annihilate them. Failing, he created fiends. When Ahura Mazda's creation took physical form, Angra Mainyu flung planets to disrupt the heavenly order, attacked the Earth with drought, covered the ground with poisonous vermin, and pierced to the centre of the Earth, making the road to HELL. Angels restored plant-life and made the Tree-of-All-Seeds in the world-ocean, near the Haoma Tree of IMMORTALITY. Angra Mainyu created a lizard to attack it; 10 fish circled the base to protect it. Angra Mainyu brought EVIL to kill the Sole-

Created Ox. From its seed sprang wholesome things. To vindicate the Ox, Ahura Mazda took the spiritually existing first man and made him sweat briefly to form a physical youth. When this first man died, gold from his body produced a plant which grew into a human couple, so close you could not tell which was which.

Side-by-side with the earthy physicality of many myths is the story of creation by the spoken Word. The two can coexist. A Hindu Upanishad says that the Universe was originally only the Self in the form of a man. His first shout was, "It is I!" He was as large as a man and a woman embracing. He divided himself into male and female to make the human race and, in a SHAPESHIFTING chase of the female, all other animals. His exultant shout was, "I am creation." In Egypt Ptah was the Demiurge who created the god Atum by thought and word, when the divine ibis hatched the world-egg. Atum spat out twins or created them by masturbation. In the earlier, more physical Hebrew creation myth, Yahweh makes Adam out of clay and divides this body by making Eve from a rib. In the later biblical cosmogony, Elohim speaks a word over the watery chaos, where his breath broods like a bird. The expected image of the egg is hidden, as the Word speaks forth light and darkness, and the rest of creation. Man and woman appear finally, both images of Elohim. In the same way, in Polynesia, Kio "mused all potential things and caused his thought to be evoked".

In many, especially Oriental, myths the creator is integrally part of creation and the creature can recognize that divinity in itself. This myth is cyclical: life/death/life. The monotheistic cultures of Judaism, Christianity and Islam move to a creator who transcends creation, existing before, beyond and after it. This is the myth of the world as a once-only, progressive story. The CHRIST-myth is of the God-human reconciling transcendent creator and the created.

Myth-making goes on, not only in fiction. Modern cosmologists now use language which sounds startlingly mythic, even religious. To communicate the results of mathematical calculations conducted in symbolic language, they resort to visual images and poetic metaphors: the big bang, black holes, cosmic worms. Scientific popularizations like Paul Davies's *The Mind of God* (**1992**) hit the bestseller lists. They speak to us through symbols, pictures and metaphorical storylines in a manner as old as those earliest cuneiform tablets of the primeval Sea-Mother. [FS]

CREATURE FROM THE BLACK LAGOON US movie (**1954**). Universal International. **Pr** William Alland. **Dir** Jack Arnold. **Underwater sequences dir** James C. Havens. **Spfx** Charles S. Welbourne. **Screenplay** Harry Essex, Arthur Ross. **Novelization** *Creature From the Black Lagoon* * (**1954**) by Vargo Statten (John Russell Fearn; 1908-1960). **Starring** Julia Adams (Kay Lorris), Ricou Browning (Creature), Richard Carlson (Dr David Reed), Richard Denning (Mark Williams), Antonio Moreno (Carl Maia), Nestor Paiva (Lucas). 79 mins. B/w.

Scientist Maia, working in upper Amazonia, discovers a fossil hand that does not fit into the evolutionary tree. He rounds up a team comprising Lorris, Reed and Williams to investigate further. They discover that an intelligent amphibian from the dawn of time still survives in the Black Lagoon, which lies at the end of one of the Amazon's tributaries. Very little else happens, except that the MONSTER threatens them and kills off a few minor characters, yet this is one of the quintessential MONSTER MOVIES, probably because of the deliberate eroticism of the underwater sequences in which the Creature tracks Lorris, mimicking her swimming motions. [JG]

CRÉBILLON, CLAUDE-PROSPER JOLYOT DE (1707-1777) French writer, known as Crébillon fils because his father, Prosper Jolyot de Crébillon (1674-1762), was also a writer. Crébillon fils is of fantasy interest primarily for *Le Sopha* (**1740**; trans anon as *The Sopha: A Moral Tale* 1742 UK; new trans Bonamee Dobrée 1927; vt *The Divan (Le Sofa): A Morality Story* 1927 US), a volume whose FRAME STORY playfully evokes ARABIAN FANTASY. A sultan has one of his courtiers tell him a series of tales. The courtier, who has had several REINCARNATIONS, describes a time when he was a succession of sofas, a sequential BONDAGE from which he could not be released until two virgin lovers had SEX on him. Before this long-deferred event, he was forced to witness various depravities, which he recounts. [JC]

CREEPSHOW Two US anthology movies – *Creepshow* (*1982*) and *Creepshow 2* (*1987*). ◊ Stephen KING.

CREEPS LIBRARY UK book series published by Philip Allan (1884-1973) and edited by Charles BIRKIN, who remained uncredited throughout. The series began in 1932, though its roots can be traced to a privately printed annual, *The Green Book* (4 issues **1919-22**), issued by a disbanded group of Military Intelligence propaganda veterans, including Allan himself and Edward Heron-Allen (◊ Christopher BLAYRE). Several stories from *The Green Book* later appeared in the CL, which consisted primarily of a thrice-yearly ANTHOLOGY and a series of single-author collections. The anthology series ran to 13 titles: *Creeps* (anth **1932**), *Shivers* (anth **1932**), *Shudders* (anth **1932**), *Nightmares* (anth **1933**), *Quakes* (anth **1933**), *Terrors* (anth **1933**), *Monsters* (anth **1934**), *Panics* (anth **1934**), *Powers of Darkness* (anth **1934**), *Thrills* (anth **1935**), *Tales of Fear* (anth **1935**), *Tales of Death* (anth **1936**) and *Tales of Dread* (anth **1936**); the first three volumes were assembled as *The Creeps Omnibus* (omni **1935**). Although initial volumes printed a number of GHOST STORIES – by Elliott O'Donnell (1872-1965), Tod ROBBINS and H. Russell Wakefield, among others – the later ones focused increasingly on the CONTE CRUEL, and ran mostly original material rather than reprints. CL was, though, the weakest of the notable series of this period, which included NOT AT NIGHT and *The* GHOST BOOK.

Before inaugurating CL Philip Allan had published volumes of WEIRD FICTION, including *Who Wants a Green Bottle?* (**1926**) by Robbins and *They Return at Evening* (**1928**) by Wakefield; it was the publication of the next Wakefield volume, *Imagine a Man in a Box* (**1931**), that sowed the idea for the series in Allan's mind. Among these earlier volumes were: *The Strange Papers of Dr Blayre* (**1932**) by Blayre; *Devil Drums* (**1933**), *Veils of Fear* (**1934**) and *The Curse of Red Shiva* (**1936**) by Vivian Meik (1895-?   ); *The Three Freaks* (**1934**) by Robbins; *Mysteries of Asia* (**1935**) by Achmed ABDULLAH; *Horror on the Asteroid* (**1936**) by Edmond Hamilton (1904-1977); *Tales of the Grotesque* (**1936**) by L.A. Lewis (1899-1961); and *Devil's Spawn* (**1936**) by Birkin. The Blayre and Lewis volumes are the most interesting in terms of original ideas. The series has served as a source for several later anthologies, especially those assembled by Richard DALBY and Herbert van Thal. [MA]

CREEPY Bedsheet-size b/w US HORROR anthology COMIC book (1964-85). In early issues each individual story was introduced by a lean cadaverous ancient, Uncle Creepy. *C* was the first all-strip magazine by WARREN PUBLISHING, and from the start both cover and internal artwork were of the highest standard – with some very impressive contributions by Frank FRAZETTA and most other leading US artists of the genre. The effectiveness of the stories was limited by their brevity: most were routine horror tales with a somewhat clichéd "surprise" ending. In later issues this restrictive format was abandoned.

Throughout the life of *C*, Warren continued to seek out the best artwork available, using artists from Spain, the UK, France, South America and the Philippines. Some of the covers for *C* and its sister magazine EERIE, particularly those by Frazetta and Manuel SANJULIAN, have iconic status.

[RT/DR]

CREPAX, GUIDO (1933-   ) Italian artist with a fine black line style, famous for his erotic COMIC strips and a series of GRAPHIC-NOVEL adaptations of erotic and sadomasochist literature. In 1959 GC began a long association with the medical journal *Tempo Medico*, producing covers, cartoons and illustrations.

His first comic strip, for *Linus* in 1965, was *Neutron*, a somewhat idiosyncratic creation concerning Philip Rembrandt, an artist with supernatural powers, and his clothes-fetishist photographer girlfriend, Valentina, after whom the strip was renamed in 1967. The late 1960s saw GC producing a wide variety of strips, including the sadomasochistic *La Casa Matta* ["The Madhouse"] (1969). He was to become famous for creating sexy victimized heroines (though this victimization is more in their own minds than in REALITY), like *Anita* (1973), *Bianca* (1972) and *Belinda* (1972). **Valentina**, too, falls into this category; she remains the character with whom GC is most closely identified.

GC's stories are told in stream-of-consciousness sequences of events, dreams and imaginings, often self-indulgently erotic, with elements of the surreal and Kafkaesque. The narrative takes place on several levels at once, consecutive frames being often fragmented and interspersed with clusters of images of lips, eyes, nipples, navels and pudenda, along with lettered representations of sounds, resulting in a completely subjective narrative. Although a number of these pieces have been published in English they have not until recently been widely available. The exceptions are two **Valentina** stories serialized in HEAVY METAL in the early 1980s: *Valentina – Reflections* (intermittently December 1980-June 1981) and *Valentina the Pirate* (November 1983-April 1984). In the 1990s NBM have published *Valentina #1* (graph **1994**), *#2: Magic Lantern* (graph **1994**) and an expanding list of GC's further work.

GC's graphic-novel adaptations of classics include *The Story of O* (graph **1975**; trans **1978** US) by Pauline Reage, *Emmanuelle* (graph **1978**; trans **1980** US) by Emmanuelle Arsan, *Justine* (graph **1980**; trans **1981** US) and *Venus in Furs* (graph **1984**; trans **1993** US) both by the Marquis de SADE, *Dr Jekyll and Mr Hyde* (graph **1984**; trans **1994** US) by Robert Louis STEVENSON and *Dracula* (1988) by Bram STOKER – this last was published in the Italian magazine *Corto Maltese*. GC created the character **Becky Lee** (1988) for the Italian magazine *Comic Art* and the adolescent **Francesca** (1992) for *Lupo Alberto Magazine*. He drew two volumes in the Editoriale Cepim series **Un Uomo un'Avventura** ["A Man, An Adventure"]: *L'Uomo di Pskov* ["The Man of Pskov"] (graph **1977**) and *L'Uomo di Harlem* (graph **1979**; trans as *The Man of Harlem* in *Heavy Metal* January-June 1983; graph **1989** US). [RT]

CRESSWELL, HELEN Working name of UK writer Helen Cresswell Rowe (1934-   ), a prolific author of children's stories. Her first novel, *Sonya-by-the-Shore* (**1960**), like several of its immediate successors, contains no overt fantasy elements; and much of her later work, like the **Lizzie Dripping** sequence of tales for younger children – beginning with *Lizzie Dripping* (**1974**) – skirts the supernatural but ultimately avoids it. With *The Night-Watchmen* (**1969**), however, HC began to publish vivid fantasies; in this a young boy encounters two tramps who hail from "There" and who maintain a precarious WAINSCOT existence in our world by pretending to be Night-Watchmen: the crisis – the threat of the supernatural Greeneyes – is averted, and the young boy is given an epiphanic ride on the Night TRAIN. Further fantasy tales include: *Up the Pier* (**1971**), in which a young girl's skewed desires force members of a family to TIMESLIP from 1921; *The Bongleweed* (**1973**), in which the strange eponymous weed causes a young girl to develop MAGIC powers; *The Secret World of Polly Flint* (**1982**), in which a maypole focuses young Polly's attention upon the "Time Gypsies", who are able to slip into an invisible village that occupies a time-POLDER; *Moondial* (**1987**), which conflates timeslip adventures with a girl's adolescent traumas; *The Watchers: A Mystery at Alton Towers* (**1993**), in which the real-life funfair (◊ CARNIVAL) Alton Towers houses ancient creatures, a LOST LAND, a harp which, when played, causes the worlds to happen, and other phenomena; and *Stonestruck* (**1995**), set in WORLD WAR II, which exiles a young Londoner to a castle occupied by GHOSTS and others. [JC]

Other works: *The White Sea Horse* (**1964** chap), for younger children; *The Piemakers* (**1967**); *The Signposters* (**1968**); *A Game of Catch* (**1969** chap), for younger children; *The Wilkses* (**1970** chap; rev vt *Time Out* 1987 chap); *The Beachcombers* (**1972**); *The Winter of the Birds* (**1975**); *The Return of the Psammead* * (**1992**), picking up on E. NESBIT (◊◊ SEQUELS BY OTHER HANDS).

CRESWICK, PAUL (1866-1947) UK author. *The Beaten Path: A Fantasy* (**1924**) is a SUPERNATURAL FICTION featuring GHOSTS. *The Turning Wheel* (**1928**) is a fantasy about ATLANTIS which sees the UK drown, the lost continent return to the surface, and various protagonists learn they are AVATARS of ancient priests and the like. [JC]

CRICHTON, MICHAEL (1942-   ) US writer and movie director. ◊ ANTHROPOLOGY.

CRIES AND SHADOWS vt of *L'Esorcista N.2* (*1975*). ◊ *The* EXORCIST (*1973*).

CROKER, THOMAS CROFTON Irish writer and folklorist (1798-1854). ◊ ANNUALS; FAIRYTALE; FOLKLORE; ILLUSTRATION; W.B. YEATS.

CROMPTON, RICHMAL (1890-1969) UK writer. ◊ John LAMBOURNE.

CROSS, JOHN KEIR (1914-1967) Scottish editor, scriptwriter for radio and author. Most of his work of interest was for children, either under his own name or as Stephen Macfarlane. Those as by JKC are mostly for YA readers, and often sf. Some titles that evoke fantasy – *The Man in Moonlight* (**1947**), *The White Magic* (**1947**) and *The Dancing Tree* (**1955**) in particular – are actually historical tales, except for *The Owl and the Pussycat* (**1946**; vt *The Other Side of Green Hills* 1947 US). As Macfarlane JKC wrote several fantasies for younger children: *The Blue Egg* (**1944** chap), *Lucy Maroon, the Car that Loved a Policeman* (**1944** chap), *Mr Bosanko and Other Stories* (coll **1944**) and *The Strange Tale of Sally and Arnold* (**1944** chap).

JKC remains best known for his one volume of adult tales, *The Other Passenger: 19 Strange Stories* (coll **1944**; cut vt *Stories from The Other Passenger* 1961 US), a volume also memorable for its surreal full-colour illustrations by Bruce Angrave (1912-1983), who also illustrated *Lucy Maroon*, more mildly. Most of the stories here assembled are horror or DARK FANTASY; "The Other Passenger" is an excellent DOPPELGÄNGER tale, and "The Glass Eye" one of the better VENTRILOQUISM tales, though only "Clair de Lune" and "Esmeralda", both GHOST stories, are unequivocally supernatural. JKC's anthologies include *Best Horror Stories* (anth **1957**), *Best Horror Stories 2* (anth **1965**) and *Best Black Magic Stories* (anth **1960**). [JC]

CROSSHATCH Many fantasy tales are set in more than one world; normally one of these worlds is our own and the other (or others) some form of SECONDARY WORLD. TIME may move at different rates in these different worlds (◊ TIME IN FAERIE), especially in tales where THRESHOLDS are sharply demarcated; in tales of this sort, though contiguities may exist, there will be little intermixing of REALITIES between worlds.

However, in many fantasy tales the demarcation line is anything but clearcut, and two or more worlds may simultaneously inhabit the same territory. In tales set in BORDERLAND regions – a well known example is the **Borderland** sequence created by Terri WINDLING – this crosshatching of worlds may be more or less restricted to a ribbon-like region, or a POLDER within which realities can churn together. But in a novel like M. John HARRISON's *A Storm of Wings* (**1980** US) the entire landscape is a crosshatch, quandaries of PERCEPTION are rife, and anything at all may be a TROMPE L'OEIL. In other words, when borderland conventions are absent, there is an inherent and threatening instability (◊ WRONGNESS) to regions of crosshatch; a sense of imminent METAMORPHOSIS. Crosshatches invite journeys: QUESTS lead through them; EDIFICES found at their heart (as is often the case) may have PORTALS leading to various realities or worlds; WAINSCOT societies flourish there; protagonists may find in crosshatch regions echoes and adumbrations of their true nature, and meet their DOUBLES or their SHADOWS; RECOGNITION scenes may be expected, and resolutions of STORY.

The earliest full crosshatch narrative may be William SHAKESPEARE's *A Midsummer Night's Dream* (performed *c*1595; **1600**). Early crosshatch tales of interest include E.T.A. HOFFMANN's *The Golden Pot: A Modern Fairy Tale* (**1815**), George MACDONALD's *At the Back of the North Wind* (**1871**), Kenneth MORRIS's *The Fates of the Princes of Dyfed* (**1914**) and Lord DUNSANY's *The King of Elfland's Daughter* (**1924**); examples have become increasingly common. Stories in which crosshatch regions are central include Manuel MUJICA LAINEZ's *The Wandering Unicorn* (**1965**), John GARDNER's *In the Suicide Mountains* (**1977**), John CROWLEY's *Little, Big* (**1981**), Michael MOORCOCK's *The War Hound and the World's Pain* (**1981**) – the Mittelmark borderland featured here boasting a succession of intersections – Diana Wynne JONES's *Fire and Hemlock* (**1984**), Emma BULL's *The War for the Oaks* (**1987**) and many other URBAN FANTASIES in which FAERIE and the CITY jostle for *Lebensraum*, Raymond E. FEIST's *Faerie Tale* (**1988**), Tanith LEE's **Secret Books of Paradys**, Mercedes LACKEY's **Bedlam's Bard**, Gene WOLFE's *Castleview* (**1990**), A.A. ATTANASIO's *The Moon's Wife: A Hystery* (**1993**), C.J. CHERRYH's *Faery in Shadow* (**1993** UK), Lisa GOLDSTEIN's

Strange Devices of the Sun and Moon (**1993**), Midori SNYDER's *The Flight of Michael McBride* (**1994**) and Sean STEWART's *Resurrection Man* (**1995**). Movies featuring TOONS – like WHO FRAMED ROGER RABBIT (***1988***) and COOL WORLD (***1992***) – also shuttlecock through crosshatch venues. [JC]

CROW, LEVI Pseudonym of Manly Wade WELLMAN.

CROW, THE US movie (***1994***). Miramax/ Entertainment Media Investment. **Pr** Jeff Most, Edward R. Pressman. **Exec pr** Sherman L. Baldwin, Robert L. Rosen. **Dir** Alex Proyas. **Vfx** Andrew Mason. **Screenplay** David J. Schow (1955-), John Shirley (1954-). **Based on** the COMIC by James O'Barr. **Starring** Rochelle Davis (Sarah), Ernie Hudson (Albrecht), Brandon Lee (Eric Draven), Bai Ling (Myca), Michael Wincott (Top Dollar). 101 mins. Colour.

According to legend, when someone dies a crow comes to take away their SOUL; if their death marks something sufficiently EVIL, the crow may return the soul so it may right the wrong. Eric and Shelly are hideously murdered on the eve of HALLOWE'EN by a gang of thugs sent by vile boss Top Dollar. A year later the crow returns, and Eric rises from the grave; invulnerable (◊ INVULNERABILITY), his face painted in a clown's MASK, he gorily slaughters the gang members before, assisted by cop Albrecht, doing likewise to Top Dollar and his incestuous half-sister Myca. His task over, he returns to his grave, where Shelly welcomes him. *TC* is a notably nasty movie; it would perhaps have some style were it not for muddy cinematography and an equally muddy soundtrack. Lee died in the final stages of filming, and audiences flocked in hopes of spotting the moment of genuine, rather than simulated, anguish: such sadism was a matter of life imitating rather poor art. [JG]

CROWLEY, ALEISTER (1875-1947) UK occultist, born Edward Alexander Crowley. He rebelled against the stern religion of his wealthy parents, members of the Plymouth Brethren, and spent his inheritance on travel and high living. He moved by degrees to an opposite extreme that established him as the most flamboyant exponent of LIFESTYLE FANTASY. In 1898 he joined the Order of the GOLDEN DAWN and attempted to take it over, eventually abandoning the resultant splinters in 1908 to form his own Argenteum Astrum, whose creed and ritual he was able to organize from scratch (with the aid of his tutelary spirit Aiwass), incorporating various kinds of sex into its ceremonies. His copious associated writings are foundation stones of modern OCCULTISM, although many became very hard to obtain after rock star Jimmy Page decided to collect them. Like all dedicated lifestyle fantasists AC drew considerable inspiration from literary sources, many of which he listed with tongue-in-cheek annotations in the Curriculum for would-be initiates of the Argenteum Astrum in *Magick in Theory and Practice* (**1929**) as by The Master Therion. He also borrowed inspiration from François RABELAIS in establishing his own Thelema in a Sicilian villa, although he was ultimately expelled by the Italian authorities.

AC's early work includes many volumes of poetry, almost all with mystical and mythological themes. The longest items, both issued by his "Society for the Propagation of Religious Truth", are the verse drama *The Argonauts* (**1904**) and *Orpheus; A Lyrical Legend* (**1905** 2 vols); other substantial volumes were *Songs of the Spirit* (coll **1898**), *Tannhäuser; A Story of All Time* (**1902**) and *Ambergris* (coll **1910**). Pseudonyms used in this period include Khaled Khan,

Frater Perdurabo, H.D. Carr and "A Gentleman of the University of Cambridge". His prose fantasies were the novel *Moonchild* (**1929**), a *roman à clef* (there are characters based on W.B. YEATS and Arthur MACHEN) in which two societies of rival magicians quarrel over an experiment to incarnate a supernatural being, and the items assembled in *The Stratagem and Other Stories* (coll **1930**; with 1 story added 1990), most impressively the horrific POSTHUMOUS FANTASY "The Testament of Magdalen Blair".

AC left a more expansive legacy in the work of others, providing the primary model for 20th-century images of the black magician (◊ BLACK MAGIC). He can be found, lightly disguised, in W. Somerset MAUGHAM's vitriolic *The Magician* (**1908**), in various novels by his one-time acolyte Dion FORTUNE, in *Adrift in Soho* (**1961**) by Colin WILSON, and in several novels by Dennis WHEATLEY. The image was, however, already second-hand, having been borrowed from Eliphas Levi (Alphonse Louis Constant; 1810-1875), whose REINCARNATION AC claimed to be, and one of AC's most notorious projects – a conjuration of PAN employing his Oscar WILDE-style "Hymn to Pan" – was lifted from Edgar JEPSON's thriller *No. 19* (**1910**). These literary connections have been considerably broadened in recent times through Kenneth Grant's insistence that AC's magical theories correspond very closely with the schema of H.P. LOVECRAFT's CTHULHU MYTHOS. [BS]

Further reading: *The Confessions of Aleister Crowley* (**1929-30** 2 vols), autobiography; *The Great Beast* (**1951**) by John Symonds, including a bibliography by Gerald Yorke.

CROWLEY, JOHN (1942-   ) US writer who began publishing work of genre interest with his first sf novel, *The Deep* (**1975**). His reputation as an sf writer was soon eclipsed by the considerable fame he achieved with his first fantasy novel, *Little, Big* (**1981**). *The Deep*, *Beasts* (**1976**) and in particular *Engine Summer* (**1979**) are important sf texts of the 1970s, but it is for *Little, Big* that JC is now recognized as one of the shaping minds of the late-20th-century literature of the fantastic.

The title itself has been appropriated in this encyclopedia (◊ LITTLE BIG) as a shorthand to describe a relationship between an outer world and an inner world in which the latter is larger than the former. It is a phenomenon found very frequently in fantasy tales, sometimes as a simple literal description; but in *Little, Big* the term applies not only literally but to the whole complex metaphysics which structures the text. The further one penetrates that text, the larger it becomes, the huger the vistas on the other side of CROSSHATCH. An early member of the central Bramble family gives a lecture to the Theosophical Society (◊ THEOSOPHY) called "Smaller Worlds Within the Large" in which he proposes that "the world [i.e., FAERIE] inhabited by these beings [i.e., FAIRIES and their kin] is not the world we inhabit. It is another world entirely, and it is enclosed within this one; it is in a sense a universal retreating MIRROR image of this one, which as one penetrates deeper . . . [grows] larger. The further in you go, the bigger it gets."

As with Faerie, so with *Little, Big*, which can be read as a kind of *summae theologica* of modern FANTASY, not solely because of its encircling complexities but because it arguably represents an attempt at marrying UK/European and US ways of apprehending the relationship between the OTHERWORLD and mundane REALITY. The UK mode may be described as centripetal, in that UK fantasies tend to find the otherworld inside the mundane, even though that otherworld may well turn out to be bigger than that which "contains" it. The US mode is, perhaps, more geographical: the fantastic tends to be contiguous with the mundane, rather than hidden within a SECRET GARDEN. The Faerie at the heart of the inner world of *Little, Big* is discovered by travelling further inwards, an INTO THE WOODS journey that typically, in European FAIRYTALES, leads to FAERIE; the near-future world into which the narrative eventually moves, and the reawakened Frederick Barbarossa (◊ SLEEPER UNDER THE HILL), who becomes US President, are reached by travel across the fields we know. The novel closes in a joint manoeuvre, a complexity of RECOGNITIONS unfolding into a HEALING terminus: the central characters accomplish what, most ambivalently, can be understood as a transcendent retreat into Faerie, while the world they leave continues to experience a profound but similarly ambiguous renewal (◊ INSTAURATION FANTASY).

Little, Big is seminal also because its basic structure replicates a basic structure of fantasy itself: Smoky Barnable, and all of the Bramble clan, know or learn that the intertwining lives they are leading constitute parts of a deep STORY which – when told, or understood, or discovered – will tell them who they have been and who they are. They have been actors, or characters; and there is a script.

The initial setting is one familiar to CONTEMPORARY FANTASY. Smoky Barnable has spent his early years in NEW YORK, and falls in love with Daily Alice Drinkwater, the relative of a friend of his; they agree to marry. He leaves the CITY and passes through a complex THRESHOLD into rural New England to arrive at Edgewood, where she and her extensive family live. Edgewood, a classic EDIFICE – larger inside than out, with façades facing seemingly into different REALITIES, as if PORTALS, with the ORRERY at its heart being turned by the Universe and, by analogy, turning the story – is the main venue of the tale. Years pass, during which a large number of fantasy motifs are invoked, quoted and examined. It seems that the kinfolk around Edgewood were shaped by paterfamilias John Storm Drinkwater's stories; this is just as likely as to assume that he simply based his stories on them. Woven into the texture of the tale are references to, and parodies of, the work of Lewis CARROLL, both as teller of tales and in his real person. There are CHANGELINGS and norns (◊ FATES); and there is the figure of Mrs Underhill, who may be a fairy – indeed, the Fairie Queene – or perhaps simply MOTHER GOOSE, or who may combine all these aspects of the ongoing story.

Out beyond Edgewood life goes on in New York, where Auberon (◊ OBERON), a son of Smoky and Daily Alice, holes up in another relative's urban POLDER, a circle of tenements called Old Law Farm in the heart of the city (Old Law tenements actually exist in New York, and are legally protected, polder-like, from landlords), and suffers an excruciating love-affair with a girl whose nickname is Titania. Some of his time is spent in a mysterious park which has resemblances to Edgewood, and (consequently) to the Story of the world; there are also direct similarities to the garden in the later phases of Carroll's *Alice's Adventures in Wonderland* (**1865**). In the meantime, the SECRET MASTERS who run the world – the Noisy Bridge Rod and Gun Club – have discovered that Barbarossa has been reawoken to deal with the crisis of the millennium. But the instauration Barbarossa imposes turns out to be profoundly skewed, and has an obscure (but deeply felt) effect on the crisis of radical THINNING which has been desiccating Fairie as well as our REALITY; and the story ends in the ambivalently redemptive moves towards closure mentioned earlier. Given the orrery at the heart of the home, it may be that where they go (where they are already) is a world in which SEASONS return and tales can begin again (and already have). But this is left unfully spoken by JC.

Many of the themes and motifs of *Little, Big* receive extended treatment in JC's next work of fantasy interest, the ongoing **Aegypt Quartet**, of which so far *Aegypt* (**1987**) and *Love & Sleep* (**1994**) have been published. Though we cannot yet (1996) assess the final shape of this vast story, we can note that **Aegypt** comprises a huge expansion of hints in *Little, Big*, and that the secret history of the world it tells – **Aegypt** being a massively intricate FANTASY OF HISTORY – is a story similar to that inside the world of (and bigger than the outside world of) the earlier book. Rather like Smoky Barnable, the protagonist of the new tale, Pierce Moffett, is an exile full of longing for the real world, a world which has been locked away from him, but which surely abides, for "There is more than one history of the world" – a phrase used many times over the course of the first two volumes.

As the tale progresses, Hermetic motifs begin to coalesce around Moffett's search for happiness and tranquillity in the small country town of Blackbury Jambs. Here he becomes involved (as did Barnable) in an interlinked family, and here (again similarly) his search is governed in part by children's books written by a member of that family, Fellowes Kraft. But many of Kraft's stories concern themselves with far more visibly grave issues. A number of these fictional BOOKS are referred to by name in the text; it turns out that Kraft was supremely interested in two historical figures, John DEE and Giordano Bruno (1548-1600), both important 16th-century philosophers who understood REALITY as perceivable through a visible mundane story which overlays the true, Hermetic Story, which rings like a clarion through the realms of sleep where we mortals live in BONDAGE and awakens those (like Moffett) who know themselves to be exiles in this mundane world of matter. He is a historian, and he has had an intuition that the secret history of the world hinges somehow upon an alternate Egypt, which he calls Aegypt; the plot of the first two volumes carries him into a love affair and implicates him deeply in lives which may (or may not) themselves connect directly to the real Story. The relationship between Aegypt and Kraft's fictional books about Dee and Bruno has not yet been made clear; but the progress of the quartet has a momentum to it which hints that we will eventually be told.

JC's short fiction has been assembled in *Novelty* (coll **1989**) and *Antiquities: Seven Stories* (coll **1993**). The only fantasy in the first collection, "The Nightingale Sings at Night", retells the story of ADAM AND EVE in REVISIONIST-FANTASY terms. *Antiquities* includes: "The Green Child", a fairly straightforward rendering of the GREEN CHILD tale; "Missolonghi 1824", in which Lord BYRON recounts an encounter with a SATYR; "Her Bounty to the Dead", in which it is argued that mortals get not the HEAVEN or the IMMORTALITY they deserve but the one they believe in; and "The Reason for the Visit", whose narrator engages in a kind of TIMESLIP interview with Virginia WOOLF. [JC]
Other works: *Great Work of Time* (**1991**), an sf novella from *Novelty*; *Beasts/Engine Summer/Little, Big* (omni **1991**; vt *Three Novels: The Deep, Beasts, Engine Summer* 1994).

CROWTHER, PETER (1949-   ) UK writer and editor, author of several fantasy and horror stories, but best-known for his ANTHOLOGIES, of which the **Narrow Houses** sequence – *Narrow Houses* (anth **1992**), *Touch Wood* (anth **1993**) and *Blue Motel* (anth **1994**) – is of interest for its thematic concentration on various superstitions. Other anthologies include *Heaven Sent: 18 Glorious Tales of the Angels* (anth **1995** US; vt *Heaven Sent: An Anthology of Angel Stories* 1995 UK) with Martin H. GREENBERG and *Tombs* (anth **1995** US) with Edward Kramer. [JC]

CRUIKSHANK, GEORGE (1792-1878) Influential UK political cartoonist and illustrator, some of whose many satirical prints of life in the reign of George IV were directly responsible for political reforms. He had a cutting wit and a great skill in caricaturing leading figures in political and court life. He continued the tradition of William Hogarth (1697-1764), Thomas Rowlandson (1756-1827) and James Gillray (1757-1815), but was less crude.

GC did much work in partnership with W. Harrison AINSWORTH, his illustrations effectively capturing the fantasy-of-history that Ainsworth sought to achieve. GC split with Ainsworth in 1842 after claiming that Ainsworth was using GC's plot suggestions without credit.

GC illustrated the first UK edition of the GRIMM BROTHERS' *German Popular Stories* (coll **1823** UK), and thereafter his work became closely associated with FAIRYTALES. After his conversion to teetotalism in the late 1840s GC used the fairytale to promote his own ideas of moral rectitude and thus severely edited traditional tales into the form we know now. His series of **Fairy Stories** (1853-64) was later collected as *George Cruikshank's Fairy Library* (coll **1870**). These heavily rewritten versions annoyed traditionalists like Charles DICKENS but found great favour among the pious. [RT/MA]

CRY IN THE DARK, A vt of *L'Esorcista N.2* (**1975**). ◊ *The* EXORCIST.

CRYPT OF CTHULHU US SMALL-PRESS digest MAGAZINE, Hallowmas 1981-current (*#90*, Lammas 1995), published originally by Cryptic Publications, New Jersey, but since 1990 (*#76*) by Necronomicon Press, Rhode Island; ed Robert M. Price (1954-   ). Its issue dates relate to feast days rather than months.

COC is an eclectic, nonacademic magazine dedicated to the study and analysis of the life and career of H.P. LOVECRAFT and associated writers. Most articles – often informative and never staid – are written in a lighthearted style. Of special merit are the articles by S.T. JOSHI and Will Murray (1953-   ). The magazine also contains fiction; of particular interest are the issues which can be regarded as ANTHOLOGIES/collections in their own right. Such issues include: *#10 Ashes and Others* (coll **1982** chap), presenting more recently identified revisions and ghost-writing by Lovecraft; *#16 Tales from the Crypt of Cthulhu* (anth **1983** chap); *#21 Saturnalia and Other Poems* (coll **1984** chap) ed Joshi, a collection of Lovecraft's poems; *#27 Untold Tales* (coll **1984** chap) by Clark Ashton SMITH; *#43 The Tomb Herd and Others* (coll **1986** chap), early tales by Ramsey CAMPBELL; *#44 Medusa and Other Poems* (coll **1986** chap) ed Joshi, a further selection of Lovecraft's poems; *#50 Ghostly Tales* (coll **1987** chap), juvenilia by Campbell; *#54 The Fishers from Outside* (coll **1988** chap) by Lin CARTER; *#59 Forgotten Tales of the Cthulhu Mythos* (anth **1988** chap); *#67 The Plains of Nightmare* (coll **1989** chap) by John Glasby (1928-   ); *#70 The Necronomicon* (coll **1990** chap) by Lin Carter; and *#71 The Brooding City* (coll **1990** chap) by

Glasby. Special issues have studied the works of Robert H. Barlow (1918-1951) *#60*, Robert BLOCH *#40*, Peter Cannon (1951-   ) *#90*, Lin Carter *#36* and *#69*, Robert E. HOWARD *#3*, Henry KUTTNER *#41*, Thomas LIGOTTI *#68*, Frank Belknap LONG *#42*, Brian LUMLEY *#19*, Duane Rimel (1915-   ) *#79*, and Richard L. Tierney (1936-   ) *#24* and *#86*. An index to the first 50 issues was published in *#55*. Price has also produced an anthology derived from the magazine, *Black Forbidden Things* (anth **1992**).   [MA]

CRYSTAL BALL ◊ SCRYING.

CTHULHU MYTHOS A HORROR background or SHARED WORLD initially elaborated by H.P. LOVECRAFT and his circle, including Robert BLOCH, August DERLETH, Robert E. HOWARD, Frank Belknap LONG and Clark Ashton SMITH – hence such recursive (◊ RECURSIVE FANTASY) Lovecraftian jokes as the ascription of a grimoire to the Comte d'Erlette (Derleth) and the driving to insanity of "Robert Blake" (Bloch) in "The Haunter of the Dark" – the latter in mock retaliation for Bloch's Lovecraft pastiche "The Shambler from the Stars" (1935 *WT*).

The CM centres on a pantheon of monstrous "Great Old Ones" whose mere appearance may cause insanity; these, as systematized by Derleth from Lovecraft's shuddering hints, include the MALIGN SLEEPER Cthulhu, the GUARDIAN OF THE THRESHOLD Yog-Sothoth, the obscenely parodic FERTILITY deity Shub-Niggurath, and their HERMES-messenger Nyarlathotep. The essence of Lovecraft's "cosmic terror" is these entities' unconcern for such small fry as humanity in a vast Universe: foolish enquirers are withered or destroyed almost incidentally. Ancillary features are: a recurring set of accursed grimoires such as the *Necronomicon* (◊ BOOKS); various hideous (even when semi-benign) ELDER RACES, not to be confused with the ELDER GODS; lesser horrors like shoggoths, night-gaunts and Hounds of Tindalos; New England's haunted town Arkham (home of Miskatonic University) and decayed seaport Innsmouth; and other mystically charged names – Carcosa, Hali, Hastur – borrowed from Ambrose BIERCE via Robert W. CHAMBERS's *The King in Yellow* (coll **1895**).

Later authors of CM-derived fiction include Ramsey CAMPBELL, Brian LUMLEY and Colin WILSON.

Lin CARTER's *Lovecraft: A Look Behind the Cthulhu Mythos* (**1972**) examines the CM in exhaustive detail and includes a bibliography, as does *The Reader's Guide to the Cthulhu Mythos* (**1969**; rev 1973) by Robert E. WEINBERG and Edward P. Berglund. The retrospective *Tales of the Cthulhu Mythos* (anth **1969**; 1975 2 vols UK) ed Derleth assembles early stories; new contributions appear in *The Disciples of Cthulhu* (anth **1976**) ed Berglund, *New Tales of the Cthulhu Mythos* (anth **1980**) ed Campbell, *Shadows Over Innsmouth* (anth **1994**) ed Stephen JONES – a theme anthology inspired by Lovecraft's "The Shadow Over Innsmouth" (included in the volume) and its sea-dwelling Deep Ones with which human cultists miscegenate – and *Cthulhu 2000: A Lovecraftian Anthology* (anth **1995**) ed Jim Turner.

The GAME *Call of Cthulhu* unwisely quantifies the powers and traits of CM unknowables, leading to such mildly silly spinoffs as *Petersen's Field Guide to Cthulhu Monsters* * (graph **1988**) by Sandy Petersen, Tom Sullivan and Lynn Willis.   [DRL]

CUCHULAIN Also known as Cú Chulainn or Cuchillin, a leading HERO of Irish MYTH. He dominates the "Ulster Cycle" of legends as the archetypal defender of the tribe, an epitome of Celtic nobility whose invincible battle-frenzy is festooned with TALL-TALE exaggeration: fire spurts from his mouth and a fountain of black blood from the top of his head, three vats of freezing water are required to calm him down, etc. Cuchulain's steadily darkening life was hedged with *geasa* (◊ CONDITIONS; CURSES) which finally caused his death. He appears in L. Sprague DE CAMP's and Fletcher PRATT's "The Green Magician" (1954), Pat O'SHEA's *The Hounds of the Mórrígan* (**1985**) and, recurringly, the poetry of W.B. YEATS. His Ulster Cycle stories are retold in Gregory FROST's 1980s **Tain** series and in Kenneth C. FLINT's *A Storm Upon Ulster* (**1981**; vt *The Hound of Culain* 1986 UK) and its sequel.   [DRL]

Further reading: *The Cuchullin Saga in Irish Literature* (**1898**) by Eleanor Hull.

See also: CELTIC FANTASY.

CUDDON, J(OHN) A(ANTHONY) (1928-   ) UK critic, writer and anthologist. *A Dictionary of Literary Terms* (**1977**; rev 1979; rev vt *A Dictionary of Literary Terms and Literary Theory* 1991) is notable among books of this sort in managing to deal with the literature of the fantastic – there are entries on HORROR and SCIENCE FICTION, though none on fantasy as such – without recourse to the pejorative. *The Penguin Book of Ghost Stories* (anth **1984**) and *The Penguin Book of Horror Stories* (anth **1984**) are useful.   [JC]

CUISCARD, HENRI [s] ◊ Charles DE LINT.

CULLEN, SEAMUS (1927-   ) Pseudonym of a US advertising executive and writer, in Ireland from the 1970s, who does not wish his true name divulged. His first novel, *Astra and Flondrix* (**1976** UK), is a relatively rare presentation of eroticism in fantasy. The king of a FANTASYLAND kingdom on a post-HOLOCAUST Earth impregnates his ELF bride; their son Flondrix matures into a SHAPESHIFTER and sexual athlete, and inadvertently awakens a MALIGN SLEEPER. The short sequence *A Noose of Light* (**1986** UK) and *The Sultan's Turret* (**1986** UK) is ARABIAN FANTASY: it less successfully and far less cheerfully subjects its two female protagonists to violent sexual exploitation: one becomes a whore; the latter must disguise herself as a man (◊ GENDER DISGUISE). The telling tends to be both florid and flimsy.   [JC]

CUMMINS, HARLE OREN (1859–1931) US writer whose life remains obscure. His only book, *Welsh Rarebit Tales* (coll **1902**), contains sf and SUPERNATURAL FICTION, mostly involving GHOSTS and other psychic phenomena. Of greatest interest, perhaps, is "The End of the Road", whose protagonist has a terminal encounter with a personalized DEATH.   [JC]

CUNNINGHAM, ALLAN (1784-1842) UK folklorist. ◊ FOLKLORE.

CUPID The Roman god of love, also known as Amor, the equivalent of the Greek Eros. He was often in the company of his mother, Venus (◊ APHRODITE), and was conventionally represented as a cherubic youth firing arrows of desire from his bow.

Cupid's ill-fated marriage to Psyche (◊ CUPID AND PSYCHE) was the literary creation of APULEIUS; modern reconfigurations include *Psiché* (**1671**) by Molière (1622-1673) – which was the basis of *Psyche* (**1675**) by Thomas Shadwell (1642-1692) – *Psyche* (**1898**) by Louis Couperus (1863-1923) and *Till we Have Faces* (**1956**) by C.S. LEWIS. Cupid is also featured in the satirical *Venus & Cupid, or A Trip from Mount Olympus related by the Personal Conductor of the Party* (**1896**) by "The Author of *The Fight at Dame Europa's School*" (Henry W. Pullen) and the comical

Olympian Nights (**1902**) by John Kendrick BANGS. One of his magical darts figures in the sentimental *The Arrow* (**1927**) by Christopher MORLEY. An artistic representation by G.F. Watts (1817-1904) is reinterpreted by Dan SIMMONS in "The Great Lover" (1993). [BS]

CUPID AND PSYCHE CUPID (or Eros) takes the beautiful Psyche as wife, but visits her only by night and insists she must never see his face. When her jealous sisters claim he must be a MONSTER (◊◊ BEAUTY AND THE BEAST, of which CAP is a distant UNDERLIER), Psyche breaks the PROHIBITION by lighting a lamp to examine her sleeping husband, who – woken by a drip of hot oil – vanishes, together with his palace. ZEUS permits a reunion after Psyche has suffered an extended NIGHT JOURNEY. C.S. LEWIS's retelling in *Till We Have Faces* (**1956**) is narrated by Psyche's sister, here loving but dangerously possessive, and focuses on the sister's anguish and redemption. [DRL]

CURRY, JANE LOUISE (1932-   ) US writer, mostly of YA fantasies, who began publishing with the **Beneath the Hill** sequence – *Beneath the Hill* (**1967**), *The Change-Child* (**1969**), *The Daybreakers* (**1970**), *Over the Sea's Edge* (**1971**) and *The Birdstones* (**1977**) – set partly in Wales and partly in the USA, where a community of Welsh ELVES has established a WAINSCOT existence, generally hiding in a huge CITY beneath a hill. The first volume establishes this premise; the second returns to 16th-century Wales, where the Elves are preparing to emigrate westwards; the third TIMESLIPS various protagonists into the deep American past, where the Elves who probably founded the city are in conflict with Mound-Builder Indians. In the third and fourth two protagonists undergo IDENTITY EXCHANGE, switching lives and times, and the contemporary lad becomes involved in the founding of the elf city, which proves to be called AVALON. In the final volume, less engagingly, an INVISIBLE COMPANION turns out to be a real elf.

In the **Mindy** sequence – *Mindy's Mysterious Miniature* (**1970**; vt *The Housenapper* 1971 UK) and *The Lost Farm* (**1974**) – a DOLL house turns out to contain real people who have been shrunk by a professor; the shrunk are eventually unshrunk. In the **Rosemary** sequence – *Parsley Sage, Rosemary, and Time* (**1975**) and *The Magical Cupboard* (**1976**) – Rosemary (a girl) and Parsley Sage (a CAT) go to the 18th century, where among their adventures they save a woman from being burned as a WITCH. The **Tiddi** sequence – *The Wolves of Aam* (**1981**) and *The Shadow Dancers* (**1983**) – returns to the motifs that govern **Beneath the Hill**, though the FANTASYLAND setting of the tales is not innovative: her young protagonist, with animal COMPANIONS befitting the GENRE-FANTASY mode, liberate a long-lost city and find a race of forgotten elves.

JLC's singletons include *The Sleepers* (**1968**), in which the Arthurian cast is come across by some contemporary children – the SLEEPERS UNDER THE HILL including ARTHUR and MERLIN and the enemies including MORGAN LE FAY. Throughout her career, JLC has threatened to publish a classic tale, but has always stopped short of the intensity worthy of her often highly arresting ideas. [JC]

Other works: *Poor Tom's Ghost* (**1977**); *The Bassumtyte Treasure* (**1978**); *Me, Myself and I: A Tale of Time Travel* (**1987**).

CURSE OF DRACULA, THE US tv serial (1979). Universal/NBC. **Pr** Richard Milton, Paul Samuelson. **Exec pr** Kenneth Johnson. **Created by** Johnson. **Writer** Johnson. **Dir** Jeffrey Hayden, Johnson, Milton. **Starring** Carol Baxter (Mary Gibbons), Stephen Johnson (Kurt von [*sic*] Helsing), Michael Nouri (Count Dracula), Louise Sorel (Amanda Gibbons). 10 20min episodes. Colour.

The DRACULA legend removed to modern-day SAN FRANCISCO. Kurt von Helsing, grandson of famous VAMPIRE-hunter Professor von Helsing, believes Dracula is still alive and posing as a college professor. Kurt is helped by Mary Gibbons, whose mother has been vampirized by Dracula. In the final episode, von Helsing kills Dracula and Mary her mother, ending her cursed existence. This serialized story was aired as part of the series CLIFFHANGERS. [BC]

CURSE OF FRANKENSTEIN, THE (*1957*) ◊ FRANKENSTEIN MOVIES.

CURSE OF THE CAT PEOPLE, THE US movie (*1944*). RKO. **Pr** Val Lewton. **Dir** Gunther F. Fritsch, Robert Wise. **Screenplay** DeWitt Bodeen. **Starring** Julia Dean (Mrs Farren), Jane Randolph (Alice Reed, *née* Moore), Elizabeth Russell (Barbara Farren), Simone Simon (Irena Dubrovna), Kent Smith (Oliver Reed). 70 mins. B/w.

A sequel to CAT PEOPLE (*1942*). Oliver and Alice – now married, with a 6-year-old, Amy – live in Tarrytown, near Sleepy Hollow (◊ Washington IRVING), trying to forget his first marriage, to Irena. Amy is an incurable dreamer and fantasist, and hence both the despair of her parents and ostracized by the other kids. She is given a RING by aged actress Mrs Farren, who lives in a reputedly HAUNTED DWELLING with her daughter, Barbara, whom she believes is a CHANGELING. On the ring Amy makes a WISH – for a friend – and thereby conjures up an INVISIBLE COMPANION, who is Irena, clad as a FAIRY QUEEN; she can make the four SEASONS cycle by in as many minutes. Amy, thrashed by her father at CHRISTMAS for stubbornly maintaining that Irena exists, chases into a blizzard after her friend – but the GHOST, rather than wishing her ill, saves her from murder at the hands of a distraught Barbara.

TCOTCP is a GHOST STORY, a FAIRYTALE, a PSYCHOLOGICAL THRILLER and a fantasy of PERCEPTION. In the latter context we can note how Oliver, joyous on recovering Amy from the blizzard, *discovers* how to see Irena; at the same time it is open to us to perceive the movie as a RATIONALIZED FANTASY, with everything we see of Irena being as through Amy's eyes; or we can take matters further and view, instead, the chimera of Irena as nothing to do with Amy but as a personification of Alice's guilty sexual jealousy concerning Oliver's first wife. There are RECURSIVE references to Lewis CARROLL's *Alice's Adventures in Wonderland* (**1865**) and others. This multi-layered movie rewards repeated viewing. [JG]

See also: CELIA (*1988*).

CURSE OF THE FLY (*1965*) ◊ *The* FLY (*1958*).

CURSES The ability to visit misfortune upon the person or property of another by means of a formal recitation of ill-wishes is one of the traditional attributes of WITCHES and other magicians, FAIRIES and DEMONS. Hereditary curses visited on families are staples of legendary lore and OCCULT FANTASY, often featuring in tales of REINCARNATION.

The folkloristic curses most extensively re-examined in literary fantasy include the punishments inflicted on such ACCURSED WANDERERS as the WANDERING JEW and the FLYING DUTCHMAN. Comedies parodying HORROR fiction often employ absurd curses, after the fashion of "The Curse of the Catafalques" (1882) by F. ANSTEY. By the same token, comic fantasy (◊ HUMOUR) often features peculiar curses: "The Affliction of Baron Humpfelhimmel" (1901) by John

Kendick BANGS requires the baron to express all the laughter his immediate ancestors were cursed to avoid, while *The Fakir's Curse* (**1931**) by Kennedy Bruce involves a betwitched camera. In HEROIC FANTASY curses sometimes turn out to be blessings in disguise, like the isolating curse of *The Curse of the Wise Woman* (**1933**) by Lord DUNSANY or the curse of eternal wakefulness in *Slaves of Sleep* (**1948**) by L. Ron HUBBARD. More earnest treatments can be found in *Frost* (**1983**) by Robin W. BAILEY and *The Prince of Ill-Luck* (**1994**) by Susan DEXTER, while "The Curse of the Smalls and the Stars" (1983) by Fritz LEIBER is delicately ambivalent.   [BS]

CYBELE ◊ GODDESS.

CYCLES 1. A term which in literary criticism normally describes a group of works which address a single theme. The theme itself might be trivial; but most cycles accrete around significant subjects, like the stories that make up the BIBLE, the MATTER of Greece (in HOMER's *Iliad* [*c*800BC]), the SAGAS that underlie NORDIC FANTASY, the MATTER of Britain and other constellations of material about the creation of worlds (◊ CREATION MYTHS) and the founding of nations. The term has also been used to describe sequences of stories and poems, from BOCCACCIO's *Decameron* (*c*1350) and Geoffrey CHAUCER's *Canterbury Tales* (*c*1387-1400) down to multi-volume DYNASTIC FANTASIES.   [JC]

2. Philosophies of history can argue that significant patterns recur, that history repeats itself in a non-trivial sense; these are frequently referred to as cyclical. The cyclical theories of history presented in *The Decline of the West* (**1918**) by Oswald Spengler (1880-1936) and *A Study of History* (**1934-61** 12 vols) by Arnold Toynbee (1889-1975) have been influential in sf, less so in fantasy. Students of archaic religions (the most prominent 20th-century scholar in the field is perhaps Mircea ELIADE) normally contrast the cyclical, SEASONS-dominated nature of such faiths with the linear, end-oriented structure of Judaism, Christianity and Islam. A sense that our actions eternally return to a central STORY, and that our lives are renewed through RECOGNITION of our role in that continuance, is deeply congenial to much fantasy.   [JC]

CZECH REPUBLIC The present CR encompasses two historical lands, Bohemia in the west and Moravia in the east. In the 17th century these were forcibly annexed to Austria, as a result of which a strong German – and especially Austrian – influence prevailed until the early 20th century. This influence remained strong after the creation of the new state of Czechoslovakia in 1918, and many of the most important fantasists who have lived and worked in what is now the CR – such as Franz KAFKA and Gustav MEYRINK – actually wrote in German. This entry deals only with writers from the historical Czech lands who wrote in Czech: German-language writers are dealt with either in the entry on AUSTRIA or in their own separate entries, or both.

Czech literature has never been short of writers producing work containing elements of the fantastic. However, Czech fantasy has always tended in the direction of OCCULTISM and what might be termed literary fantasy, rather than HEROIC FANTASY or MYTHOLOGY. In part this may be because the Czech folk tradition lacks convincingly heroic LEGENDS, but more significant historical influences are, first, the romantic tradition of Prague at the time of Rudolf II (1552-1612) and, second, the city's strong Jewish tradition, most famously expressed in the legend of the GOLEM.

The first Czech to produce works that are recognizably fantastic in the modern sense was Václav Rodomil Kramerius (1792-1861). He wrote at least 97 popular chapbooks, including FOLKTALES, fantasies and "true-crime" stories. Most of these were retellings of stories from other writers or oral FOLKTALES; only a few are original. Among them are a retelling of the FAUST legend – *Život, činy a uvržení do pekelné* ["The Life, Deeds and Casting into the Abyss of Hell of Dr Jan Faust"] (**1862**) – and a version of the adventures of Baron MUNCHHAUSEN – *Znamenité a podivné příhody pana Prášílka* ["The Remarkable and Strange Adventures of Little Mr Munchhausen"] (*c*1855).

The first full-length fantastic novel by a Czech was probably *Pekla splozenci* ["The Brood of Hell"] (**1862**) by Josef Jiří Kolár (1812-1896), a well known actor and playwright, also responsible for translations of GOETHE and SHAKESPEARE. Set in the era of Rudolf II, the plot revolves around ALCHEMY and the attempt to create an immortal being (◊ IMMORTALITY).

Fantastic elements can be found also in the works of Jakub Arbes (1840-1914), especially in his short novels – to describe which a new critical term was invented: *romaneto*, a term still occasionally used of some modern Czech literature. (One of Arbes's *romanetos* was translated into English by Jiří Král as "Newton's Brain" in *Clever Tales* [anth **1897** US] ed Charlotte E. Porter and Helen A. Clarke.) Although the fantasy elements are often marginal, the mysterious atmosphere these works evoke and the characteristic way they describe Prague have been very influential, and resurface among many later writers – including, it has been argued, Kafka. Another influential figure of the 19th century was Julius Zeyer (1841-1901), whose long romantic stories are frequently set in distant countries and have something of the feel of myth or legend – many are in fact based on real legends borrowed from a wide variety of sources. His use of pseudo-medieval legend has sometimes led to comparisons with William MORRIS and the PRERAPHAELITES. A useful book on Zeyer is *Julius Zeyer: The Path to Decadence* (**1973**) by Robert B. Pynsent. This earliest period of Czech literary fantasy is encapsulated in *Tajemné příběhy v české krásné próze 19. století* ["Mysterious Stories in Czech 19th-Century Literature"] (anth **1976**) ed Ivan Slavík, which contains representative stories plus an excellent historical introduction.

Two of the most important Czech fantasists of the early part of the 20th century were Jiří Karásek ze Lvovic (1871-1951) and Emanuel Lešehrad (1877-1957) – the latter also wrote as Emanuel z Lešehradu. Karásek, the foremost representative of Czech DECADENCE, wrote a fantasy trilogy, *Romány tří mágů* ["Novels of the Three Mages"] (**1907-25**), whose fragmented narrative describes the tormented wanderings of fey young men in the decaying cityscapes of Prague, Venice and Vienna, and their encounters with the occult (◊ OCCULT FANTASY). Although often discussed in the same context as his friend Karásek, Lešehrad was a far more sober author. His major contributions to the genre were his numerous short stories of fantasy and the irrational, a good reprentative collection of which is *Záhadné životy* ["Mysterious Lives"] (coll **1919**). The contents range from gentle stories on the inevitability of FATE to almost Lovecraftian tales of supernatural terror (◊◊ H.P. LOVECRAFT).

Two further writers whose works may be regarded as classics of Czech fantastic literature are Karel ČAPEK and Jan Weiss (1892-1972). Although today best-known for their contributions to the development of Czech sf, both produced works where the fantastic elements are left unrationalized. Examples might include, from Čapek's *oeuvre*, the mass-production of GOD in *The Absolute at Large* (**1922**), the ELIXIR of IMMORTALITY in *The Macropulos Secret* (**1922**), the BEAST FABLE of *The Insect Play* (**1921**) with Josef Čapek, and the metaphysical drama *Adam the Creator* (**1927**), also with Josef. Weiss was the author of a number of highly literary works which hover on the cusp between sf and fantasy; a good example is *Spáč ve zvěrokruhu* ["The Sleeper in the Zodiac"] (**1937**), a novel describing the life of a schoolteacher whose metabolism changes with the SEASONS, as if he were a plant.

Two other authors of the interbellum made significant contributions to Czech fantasy: the extremely prolific Jan Havlasa (1883-1964) and Josef Šimánek (1883-?). Havlasa was a foremost Czech specialist on Asian culture; he wrote over 20 books containing fantastic elements, most set in exotic locations around the Pacific. Examples include *Zahrada splněné touhy* ["The Garden of Fulfilled Desire"] (coll **1918**), *Souostroví krásy* ["The Archipelago of Beauty"] (**1919**) and *Podivní milenci* ["Strange Lovers"] (coll **1928**). His work often concerns lost worlds (◊ LOST RACES) or is SUPERNATURAL FICTION, but his most important contribution to Czech literature was probably that he introduced Far Eastern elements. An English-language selection of his stories is *Four Japanese Tales* (coll trans by the author **1919** Prague).

Among the works of Josef Šimánek was the effective novella "Háj satyrů" ["Glade of the Satyrs"] (1915), in which a young archaeologist discovers an ancient underground temple to the god PAN and, for his curiosity, is killed by the SATYRS guarding it. Šimánek also wrote a lost-world adventure in the style of H. Rider HAGGARD, *Bratrstvo smutného zálivu* ["The Brotherhood of the Bay of Sadness"] (**1918**), the heroes of which travel across India searching for the seat of a sect that is trying to create a worldwide centre for the SPIRITUALISM movement.

Other, more commercial, fantasy writers from these years were Felix de la Cámara (1897-1945), Karel Piskoř (1879-1945) and Bernard Kurka (1894-1944).

The Czech fantasy tradition was interrupted by the Communist coup of 1948: socialist Realism was the only form of fiction tolerated. Not until the 1960s did elements of the fantastic begin to creep back, usually as a tool of social or political SATIRE. Representative examples are the politician's STATUE that comes to life in *Joachym* (**1967**) by Bohuslav Březovský (1912-1976) and the supernatural folktale figures who encounter modern Czech society in *Bubáci pro všední* ["Bogey Men for Every Day"] (coll **1961**) by Karel Michal (1932-1984).

The cultural liberalization following the fall of Communism in 1989 had far-reaching consequences for the Czech publishing industry: in the succeeding five years more fantasy was published than in the preceding 40 years. Also, fantasy is beginning to divide into subgenres. Some authors are continuing in the GHOST-STORY tradition, like Vladimír Medek in *Krev na Maltézském náměstí* ["Blood on Maltese Square"] (coll **1992**). Others are beginning to work with forms new to Czech literature, like purely mystical, spiritually oriented stories – e.g., the fiction of Eduard Tomáš – and DARK FANTASY with HORROR elements – represented by the stories in Jaroslav Šoupal's debut *Satanova kobka* ["The Cell of Satan"] (coll **1992**).

But at present the most popular form of fantasy is GENRE FANTASY, with strong adventurous plotlines; however, with this is often combined elements of other genres. Thus, when Jaroslav Jiran (1955-) won the Ikaros Award (presented annually by the readers of *Ikarie* magazine) for *Živé meče Ooragu* ["The Living Swords of Oorag"], it was for a novel that stands on the borderline between sf and fantasy. A similar mixture has been created by Vilma Kadlečková (1971-) in her series of novels begun with the Karel Čapek Award winner *Na pomezí Eternaalu* ["On the Borders of the Eternal"] (**1990**) and continued with *Meče Lorgan* ["The Swords of Lorgan"] (**1993**) and *Stavitele věží* ["The Tower-Builders"] (**1994**) – another Karel Čapek Award winner. A further representative of this trend is George P. Walker (Jiří Procházka; 1959-), whose extremely popular *Ken Wood a meč krále D'Sala* ["Ken Wood and the Sword of King D'Sal"] (**1992**) has a US stuntman battling the forces of EVIL on a planet where MAGIC really works. The feminist sf author Carola Biedermannová (1947-) combines fantasy and STEAMPUNK in *Ti, kteří létají* ["Those Who Fly"] (**1992**), while Richard D. Evans – a former house-name associated with the editors of *Ikarie*, now used exclusively by Vlado Ríša (1949-) – has written several adventure-fantasy books. The most popular fantasy of 1993, however, was undoubtedly *Wetemaa* (**1993**) by the previously unknown writer Adam Andres (Veronika Válková; 1970-); her promising debut, somewhat overindebted to J.R.R. TOLKIEN, won the Ikaros Award that same year.

At the other end of the literary spectrum, one of the most interesting figures in contemporary Czech literature is Michal Ajvaz (1949-), who has written two exceptional books of literary fantasy, *Návrat starého varana* ["The Return of the Old Monitor Lizard"] (coll **1991**) and *Druhé město* ["The Other City"] (**1993**); the latter is an ALTERNATE-REALITY tale in which a mysterious "other" CITY begins to intertwine itself with the historic centre of the "real" city of Prague.

The 1990s have seen an enormous growth in the popularity of fantasy among Czech readers, in particular the young. This is in part due to the fact that fantasy, especially genre fantasy, was unavailable in Czech for so many years. Other contributory factors are undoubtedly the publication in 1990-92 of the first Czech translations of the works of Robert E. HOWARD and J.R.R. TOLKIEN and the release, during the same period, of the first Czech role-playing GAMES. 1994 saw the appearance of *Dech draka* ["Dragon's Breath"], the first Czech semi-professional MAGAZINE devoted to role-playing games and fantasy fiction, as well as numerous fanzines centred on role-playing games, plus the formation of Tolkien fan clubs.

Movies. Unlike the case in written fantasy, the use of fantasy motifs was fairly common in Czech movies during the 1960s and 1970s. For the most part these were comedies which made great play of the contrast between FAIRYTALE characters and contemporary society. Among the best were *Dívka na koštěti* ["Girl on a Broomstick"] (*1971*), about a young WITCH who finds friendship in the world of ordinary people; and *Jak utopit doktora Mráčka aneb Konec vodníkův Čechách* ["How to Drown Dr Mráček, or The End of Water-Sprites in the Czech Lands"] (*1974*). Both were dir Václav Vorlíček (1930-). Another successful movie – with an extremely unusual, surreal atmosphere – was *Valerie a týden divů* (*1970*; vt *Valerie and her Week of Wonders* UK) dir Jaromil Jireš (1935-), based on a novel of the same title by the Surrealist poet Vítěslav Nezval (1900-1958), which deals with VAMPIRES. Although rarely screened in its native country, it has enjoyed a cult reputation on the art-house circuit in the UK since its first release, and Angela CARTER claimed to have been inspired by it when drafting her screenplay for *The* COMPANY OF WOLVES (*1984*). The vampire turns up again in TECHNOFANTASY form in *Upír z Feratu* ["The Ferat Vampire"] (*1982*) dir Juraj Herz (1934-), a dark comedy about a bloodsucking CAR, adapted from a story by Josef Nesvadba (1926-).

However, several fantasy movies date back to the very beginnings of Czech cinema. The most important was *Příchozí z temnot* ["He Came from Darkness"] (*1921*; vt *Redivivus*): a 17th-century man drinks an ELIXIR and is preserved in a state of suspended animation until awoken in the present day. [IA/CS]

DADD, RICHARD (1817-1886) UK painter who murdered his father in 1843 and was committed to Bedlam (and later Broadmoor), eventually dying insane. His considerable posthumous fame is particularly for the distressed, nightmarish, hothouse Gothic ROMANTICISM of works like "The Fairy Feller's Master Stroke" (1855-64), an enormously complicated, tapestry-like portrayal of an engorged LANDSCAPE within which a cross-section of the denizens of FAERIE seem caught in BONDAGE. Contradictions of PERCEPTION pervade this work; the whole canvas seems to hover at the edge of an explosive TROMPE L'OEIL; but there is none in view. Instead, centrally located, the "Fairy Feller" seems about to engage in some paroxysm of THINNING – for certainly he is being viewed with deep apprehension by those who notice him at all, and his axe is ready to prune, to demolish, the elaborate sward. Most remarkably, the painting seems to represent a model of a central initiating narrative movement of FANTASY – an overpowering sense of dislocation, of a LABYRINTH of WRONGNESS pointing the viewer towards a resolution, some way through the maze into an OTHERWORLD, some freeing stroke. But ultimately the Faerie RD depicts is entrapped, for there is no RECOGNITION here, and any blow of the axe will bring not transcendence but death; "The Fairy Feller's Master Stroke" can thus be seen as a consummate visual pun in which fantasy wears HORROR's frozen face.

A suite of sketches assembled as *To Illustrate the Passion* (**1853-4**) is also of interest, especially the portrait of Cain and Abel. [JC]

DAEDALUS In Greek MYTHOLOGY, the ARCHETYPE of the Cunning Artificer: he built the LABYRINTH of the MINOTAUR for King Minos, who later imprisoned him in a tower – whereupon Daedalus devised WINGS which saved him but not, famously, his son ICARUS. Dactylos in Terry PRATCHETT's *The Colour of Magic* (**1983**) sympathetically echoes Daedalus; Elof in Michael Scott ROHAN's *The Hammer of the Sun* (**1988**) also constructs wings but harks back to Daedalus's artificer cognate in the KALEVALA, the SMITH Ilmarinen. [DRL]

DAHL, ROALD (1916-1990) UK writer, long famous for a succession of CHILDREN'S FANTASIES running from his first publication, *The Gremlins* (1942 *Colliers*; **1943** chap US) to the end of his life, when tales like *Esio Trot* (**1990** chap) and *The Minipins* (**1991** chap) continued to demonstrate his uncanny ability to think as a child. It is perhaps the huge perceived gap between the clarity of his telling and the seemingly amoral ruthlessness of his writing that has caused many to so distrust him as an author to whom children should be exposed.

For many years RD showed little inclination towards children's literature. *The Gremlins* came about solely through the interest of Walt DISNEY, who flirted over making an ANIMATED MOVIE of the tale (◊ GREMLINS). After the WWII stories assembled in *Over to You* (coll **1946** US) and the arch sf/fantasy novel, *Some Time Never: A Fable for Supermen* (**1948** US), RD began to publish the tales for adults which made him famous, and for which he was long primarily known. They were assembled over the years as: *Someone Like You* (coll **1953** US; exp 1961 UK); *Kiss Kiss* (coll **1960** US); *Twenty-Nine Kisses from Roald Dahl* (coll **1969**), a compilation; *Switch Bitch* (coll **1974** US); *The Best of Roald Dahl* (coll **1978** US); *Tales of the Unexpected* (coll **1979**) and *More Roald Dahl Tales of the Unexpected* (coll **1980**; vt *More Tales of the Unexpected* 1980; vt *Further Tales of the Unexpected* 1981), both assembled as *Roald Dahl's Completely Unexpected Tales* (omni **1986**); *Two Fables* (coll **1986** chap); *Ah, Sweet Mystery of Life* (coll **1989**); the posthumous *The Collected Short Stories* (coll **1991**), which includes some new work; and *Lamb to the Slaughter* (coll **1995** chap UK), which includes nothing new. There is occasionally a tone of SLICK FANTASY in these stories – in this RD resembles authors like SAKI and John COLLIER. Several are SUPERNATURAL FICTIONS whose twists seem, at times, compulsively cruel. Many are mundane studies of characters arraigned by plots which expose their awfulness. The tv series *Roald Dahl's Tales of the Unexpected* (1979) was based on various of his twist-in-the-tale macabre stories, each introduced by him. Further series, sans Dahl and with stories by other writers, were simply called *Tales of the Unexpected* (1980-88). This tv exposure gave his adult stories a new lease of life, so that by the time he died he was a hugely bestselling author in both adult and juvenile markets.

The tales for children began with *James and the Giant*

Peach (**1961** US), filmed in stop-motion animation in 1996. Young James travels across the Atlantic inside a giant peach, befriended by the insects he discovers within. RD's next tale is his most famous. _Charlie and the Chocolate Factory_ (**1964** US), filmed as WILLY WONKA AND THE CHOCOLATE FACTORY (*1971*) – assembled with its sequel, _Charlie and the Great Glass Elevator_ (**1972** US) as _The Complete Adventures of Charlie and Mr Willy Wonka_ (omni **1987**) – is both hated (by some adults) and loved (by many adults and most children) for the exorbitant punishments meted out to gluttons; but the events themselves, though fantastic in the direction of TECHNOFANTASY, do not quite cohere as fantasy, for there is a mildly absurdist, WONDERLAND feel about Charlie's world. Further childen's stories include _The Magic Finger_ (**1966** chap US), _Fantastic Mr Fox_ (**1970** chap), _Danny, the Champion of the World_ (**1975**) – filmed as _Danny, the Champion of the World_ (*1989*) – _The Wonderful Story of Henry Sugar and Six More_ (coll **1977**; vt _The Wonderful World of Henry Sugar_ 1977 US), _The Enormous Crocodile_ (**1978**), _The Twits_ (**1980** chap), _George's Marvellous Medicine_ (**1981**); _The BFG_ (**1982**), _The Witches_ (**1983**) – filmed as _The_ WITCHES (*1989*) – _The Giraffe and the Pelly and Me_ (**1985**) and _Matilda_ (**1988**). Some are fantasy, some verge on other realms. They have given considerable joy, and it is in terms of that – in terms of the exuberance of story they display – that they will, perhaps, finally be judged. [JC]

Other works: _My Uncle Oswald_ (**1979**), erotic TECHNO-FANTASY; _Roald Dahl's Book of Ghost Stories_ (anth **1983**); _Boy: Tales of Childhood_ (**1984**) and _Going Solo_ (**1986**), memoirs.

Further reading: _Roald Dahl_ (**1983**) by Chris Dowling; _Roald Dahl_ (**1988**; exp vt _Roald Dahl: From the Gremlins to the Chocolate Factory_ 1994) by Alan Warren.

DALBY, RICHARD (1949-   ) UK editor, anthologist, writer, bibliographer, literary consultant, antiquarian bookseller and collector of fantasy and SUPERNATURAL FICTION. His first anthology was a collection of GHOST STORIES, _The Sorceress in Stained Glass_ (anth **1971**), for which he unearthed a previously unreprinted story by M.R. JAMES. He compiled three volumes of ghost stories by women writers – _The Virago Book of Ghost Stories_ (anth **1987**), _The Virago Book of Ghost Stories Volume II_ (anth **1991**; vt _Modern Ghost Stories by Eminent Women Writers_ 1992 US) and _The Virago Book of Victorian Ghost Stories_ (anth **1988**; vt _Victorian Ghost Stories by Eminent Women Writers_ 1989 US) – and a series of six Christmas ANTHOLOGIES in which the majority of the stories are ghost, fantasy or supernatural: _Ghosts for Christmas_ (anth **1988**), _Chillers for Christmas_ (anth **1989**), _Mystery for Christmas_ (anth **1990**), _Crime for Christmas_ (anth **1991**), _Horror for Christmas_ (anth **1992**; vt _Mistletoe Mayhem_ 1993 US) and _Shivers for Christmas_ (anth **1995**). Other anthologies include _Dracula's Brood_ (**1987**), being stories written by friends and contemporaries of Bram STOKER, _Ghosts and Scholars_ (anth **1987**), with Rosemary Pardoe (featuring stories by the JAMES GANG), _Tales of Witchcraft_ (**1991**) and _Vampire Stories_ (anth **1992**), with a foreword by Peter Cushing. Returning to his first love, RD edited three anthologies in the **Mammoth** series: _The Mammoth Book of Ghost Stories_ (anth **1990**; cut vt _The Giant Book of Ghost Stories 2_ 1994), _The Mammoth Book of Ghost Stories 2_ (anth **1991**; vt _The Giant Book of Ghost Stories_ 1993), with a foreword by Christopher Lee, and _The Mammoth Book of Victorian and Edwardian Ghost Stories_ (anth **1995**). As a writer and bibliographer RD produced _Bram Stoker: A Bibliography of First Editions_ (**1983**), _The Golden Age of_

Children's Book Illustration (**1991**), featuring biographies and illustrations from over 50 artists who worked between 1860 and 1930, and _The Dervish of Windsor Castle: A Biography of Arminius Vambery_ (**1979**) with Lory Alder, a biography of the famous Hungarian linguist, traveller and explorer Arminius Vambery (a friend of the Prince of Wales [Edward VII] and Bram Stoker; it was probably Vambery who told Stoker about Dracula, Vlad the Impaler and VAMPIRE folklore). RD has also used his immense knowledge and love of the genre to compile a number of collections of long out-of-print ghost-story writers, including _The Best Ghost Stories of H. Russell Wakefield_ (coll **1978**) and _The Collected Ghost Stories of E.F. Benson_ (coll **1992**) (◊ E.F. BENSON). In 1993 he co-founded, with Kat and David Tibet, the Ghost Story Press, to bring back into print forgotten ghost-story classics. For this enterprise he has compiled and introduced _Flaxman Low, Psychic Detective_ (coll **1899**; 1993) by Kate and Hesketh Prichard (◊ E. and H. HERON), _Fear Walks the Night_ (omni **1993**), assembling the three volumes containing the complete ghost stories of Frederick Cowles (1900–1949), _Tales of the Grotesque_ (coll **1934**; 1994) by L.A. Lewis (1899–1961), _Tales of the Supernatural_ (coll **1894**; 1994) by James Platt (1861–1910) and _The Death-Mask and Other Ghosts_ coll (**1920**; 1995) by Mrs H.D. Everett (◊ Theo DOUGLAS). He has edited and introduced numerous SMALL-PRESS publications, including the third in the British Fantasy Society's **Masters of Fantasy** series, _M.R. James_ (**1987**), and has contributed innumerable articles and entries for various encyclopedias (including this one), books and magazines. [JF]

DALÍ, SALVADOR (DOMENECH FELIPE JACINTO) (1904-1989) Spanish painter who became the best-known advocate of SURREALISM, though his actual work is generally thought inferior to that of Max ERNST or René MAGRITTE. Some of his best paintings, most of which date from the 1930s, have become famous for their TROMPE-L'OEIL literalization of DREAM-imagery. His most famous single painting is probably "The Persistence of Memory" (1931), whose oozing soft watches almost seem to tell a profound STORY about TIME. SD's images have become ICONS of the FANTASTIC, signposts (not maps) that point the way inward to that realm. [JC]

DALKEY, KARA (MIA) (1953-   ) US writer, a member of the SCRIBBLIES from its founding in early 1980; she began publishing work of genre interest with "The Hands of the Artist" for _Liavek_ (anth **1985**) ed Emma BULL and Will SHETTERLY, and wrote some further stories for this SHARED-WORLD series. She began publishing novels with the **Sagamore** sequence – _The Curse of Sagamore_ (**1986**) and _The Sword of Sagamore_ (**1989**) – an initially light-hearted spoof of the DYNASTIC FANTASY, set in a nicely realized FANTASYLAND. The CURSE dooms its bearer to become monarch of the Kingdom of Euthymia; the HUMOUR of the first volume turns on the current heir's reluctance to don the mantle. The second features a somewhat more serious QUEST for the now-missing sign of the curse, whose long-dead original generator, Sagamore himself, makes an appearance as REVENANT and COMPANION.

Of more interest are KD's singletons. _The Nightingale_ (**1988**) was published in the **Fairy Tales** series ed Terri WINDLING; it is based on "The Nightingale" by Hans Christian ANDERSEN. Like its fellows, DK's rewritten version is both a TWICE-TOLD tale and a REVISIONIST FANTASY, recasting the original and also examining it. The original nightingale is "played" here by a young girl whose singing entrances the Emperor of 9th-century Heian Japan, but whose actions are governed by a malign GHOST. She exorcises the ghost and saves the Emperor from fatal illness through her song. DK's second singleton, *Euryale* (**1988**), is set in Rome in the 2nd century BC. Euryale is a GORGON who has turned her lover to stone and is searching for a cure; also, as a minor AVATAR of the GODDESS, she is immured in the conflicts which eventually see patriarchal GODS – here represented by the male-oriented Athena – take charge. [JC]

DALTON, ANNIE (1948-) UK writer of YA fantasy and horror who began publishing work of genre interest with *Out of the Ordinary* (**1988**), a CONTEMPORARY FANTASY in which a young protagonist finds herself involved in an OTHERWORLD whose CROSSHATCHES with her suburban environment only she can detect. *Night Maze* (**1989**), powerfully told, is about the WRONGNESS of ALCHEMY when it becomes a life-denying search for "higher" ELEMENTS. *The Alpha Box* (**1991**) invokes a 1980s fantasy cliché – the rock band (◊ MUSIC) which is linked to dark forces – but does so with wit and sympathy. In *Naming the Dark* (**1992**) a drab modern town turns out to coat ATLANTIS (◊◊ TIME ABYSS), and the lads who realize this come face-to-face with the MATTER of Britain. *Swan Sister* (**1992**) evokes FOLKTALES as a young child is drawn further and further into an otherworldly rapport with the swans whose habitat has been destroyed, half-unwittingly, by her father. Eventually she goes with the swans.

Each of AD's tales is challenging, suspenseful, wise, and – in the best tradition of fantasy literature when written seriously – subversive. [JC]

Other works: *The Witch Rose* (**1990**), for younger children; *The Afterdark Princess* (**1990**), for younger children; *Demon-Spawn* (**1991**); the **Tilly Beany** sequence for younger children, being *The Real Tilly Beany* (**1993**) and *Tilly Beany and the Best Friend Machine* (**1993**); *Ugly Mug* (**1994**).

DALTON, JAMES (? -?) UK author active in the 1830s, all of whose work was published anonymously. *The Gentleman in Black* (**1831**) is a comedy whose hero makes a PACT WITH THE DEVIL obliging him to double the amount of time he devotes to sin each year. *The Invisible Gentleman* (**1833**) was the first three-decker fantasy novel; its hero acquires the trick of making himself invisible (◊ INVISIBILITY) from an enigmatic stranger, but his practical jokes soon generate a web of deceit which blights his life, transforming the "gift" into a CURSE. James Forbes Dalton, sometimes differentiated from JD, was active in the same period, and "The Beauty Draught" (1840) is a moralistic fantasy in exactly the same vein as the novels described above; the two are probably the same writer. The *Bentley's Miscellany* contributor who signed himself "Dalton" was, however, the son of Richard Harris BARHAM. [BS]

Other works: *Chartley the Fatalist* (**1831**); *The Robber* (**1832**); *The Old Maid's Talisman and Other Strange Tales* (coll **1834**); *The Rival Demons: A Poem* (**1836** chap).

DALY, HAMLIN [s] ◊ E. Hoffmann PRICE.

DAMIEN: OMEN TWO US movie (**1978**). ◊ *The* OMEN.

DAMN YANKEES US movie (**1958**). ◊ BASEBALL.

DANA, ROSE Pseudonym of Marilyn ROSS (itself a pseudonym).

DANCE OF DEATH The medieval millennial notion – also known as the *danse macabre* – of skeletal DEATH leading the doomed in a processional dance to the grave, as depicted in French and German paintings and woodcuts since about 1424. Hans Holbein (?1497-1543) drew a famous DOD sequence in 1523-6. This image informs Edgar Allan POE's "The Masque of the Red Death" (1842) – filmed at least twice as *The* MASQUE OF THE RED DEATH – and Ingmar BERGMAN's *The* SEVENTH SEAL (*1956*). It is often adapted, as in Peter S. BEAGLE's "Come, Lady Death" (1963), whose Death is a sought-after female dancing partner, and Terry PRATCHETT's *Reaper Man* (**1991**), where Death sweeps the aged Miss Flitworth into a wild rustic folkdance. [JH/DRL]

DANCE OF THE VAMPIRES vt of *The* FEARLESS VAMPIRE KILLERS, OR PARDON ME, YOUR TEETH ARE IN MY NECK (*1967*).

DANE, CLEMENCE Pseudonym of UK dramatist, novelist and editor Winifred Ashton (1888-1965), who edited an sf line late in life for the publisher Michael Joseph but whose own writing is best treated as fantasy. *Legend* (**1919**) is a SUPERNATURAL FICTION in which a dead writer haunts her biographer. In *The Babyons* (**1927**) a family CURSE extends over four generations, the story of each being told in a mode appropriate to the period, beginning in the late 18th century in a tone of sensationalist GOTHIC FANTASY and ending in an early-20th-century venue whose protagonist (who is in a sense the "same" woman) is constricted by the undead past; in the USA, the novel was first published in four small separate volumes. Some of the stories assembled in *Fate Cries Out* (coll **1935**) – like "Frau Holde", a FAIRYTALE based on GRIMM that features a fantastically proportioned EDIFICE, and "Godfather Death", which features a PACT WITH THE DEVIL – are of interest. In *The Arrogant History of White Ben* (**1939**), an animate scarecrow takes on the role of rescuing a near-future UK from political disaster. The sequence of radio plays assembled as *The Saviours* (coll of linked plays **1942**) is narrated by MERLIN, who tells how ARTHUR, in his role as SLEEPER UNDER THE HILL, successively returns to save the UK in the guise of (among others) ROBIN HOOD. CD is a central example of the pervasiveness of the use of fantasy and supernatural-fiction devices by UK mainstream writers. [JC]

Other works: *The Moon is Feminine* (**1938**).

DANGEROUS DESIRE US movie (*1992*). ◊ CAT PEOPLE (*1942*).

DANIEL AND THE DEVIL vt of *The* DEVIL AND DANIEL WEBSTER (*1941*).

DANIELLE Blonde female star of the UK daily newspaper strip *Danielle* (*London Evening News* 1973-4; graph coll **1974** US), drawn by John M. BURNS and scripted by R. O'Neill; she spent most of her time/space travel adventures in a state of bedraggled near-nudity. Scripted by Dennis Hooper, she made a brief (54-day) reappearance in 1978. The basic premise of the seven stories concerned an AMULET that "wrenched" D and her lover, Zabal, "through the unknown dimensions of time and space" to alien planets, Earth at various times in its history and, in the last story, alien spaceships adrift in space. D was notable for the expertness of Burns's drawing. [RT]

DANN, COLIN (MICHAEL) (1943-) UK writer best-known for his **Farthing Wood** sequence of YA ANIMAL FANTASIES – *The Animals of Farthing Wood* (**1979**), *The Ram of Sweetriver* (**1986**) and *In the Path of the Storm* (**1989**), assembled as *The Animals of Farthing Wood Omnibus* (omni **1995**) – which incorporate BEAST-FABLE elements through the unlikely alliance of Fox and Toad, who cooperate to lead their own and other species out of danger to a safe POLDER. Most of CD's other novels – the **Vagabonds**

series, being *King of the Vagabonds* (**1987**) and *The City Cats* (**1991**), plus the singletons *The Beach Dogs* (**1988**) and *A Great Escape* (**1990**) – similarly shift between animal fantasy and beast fable. *The Legacy of Ghosts* (**1991**) is a GHOST STORY. [JC]

DANN, JACK (MAYO) (1945-   ) US writer and anthologist, partly resident in Australia from the early 1990s. Most of the anthologies he has edited with Gardner Dozois (1947-   ) are fantasy. JD began publishing work of genre interest with "Dark, Dark the Dead Star" and "Traps" for *Worlds of If* in 1970, both with George Zebrowski (1945-   ). His novels are all sf except *The Memory Cathedral: A Secret History of Leonardo da Vinci* (**1995**), which is fantasy. At the end of his days, LEONARDO DA VINCI uses the eponymous EDIFICE to re-enter his life and relive it. JD allows some ambiguity as to how literally this re-entry may be taken; but the life revealed itself contributes to a FANTASY OF HISTORY, in which Leonardo invents working airplanes (travelling in one to the Middle East) and variously affects history. [JC]

Other works (all ed with Gardner Dozois): *Aliens!* (anth **1980**); *Unicorns!* (anth **1982**); *Magicats!* (anth **1984**) and *Magicats II* (anth **1991**); *Bestiary!* (anth **1985**); *Mermaids!* (anth **1985**); *Sorcerers!* (anth **1986**); *Demons!* (anth **1987**); *Dogtales!* (anth **1988**); *Seaserpents!* (anth **1989**); *Little People!* (anth **1991**); *Invaders!* (anth **1993**); *Horses!* (anth **1994**) and *Angels!* (anth **1995**).

DANSE MACABRE ◊ DANCE OF DEATH.

DANTE ALIGHIERI (1265-1321) Italian poet whose *La Divina Commedia* (ot *Commedia*; composed 1307-21; **1472**) is regarded as the culminating masterpiece of medieval literature, and is TAPROOT TEXT for fantasy. DA, whose family claimed descent from ancient Italian nobility, lived at the time of the political wars between the Guelphs (who supported the Pope) and the Ghibellines (who supported the Holy Roman Emperor). DA married into a powerful Guelph family and fought at the Battle of Campaldino (1289), where the Ghibellines were defeated. During the next decade, DA led an increasingly active political life, becoming one of the Chief Magistrates of the City and an ambassador to the Pope. His career foundered when the Guelphs split into two factions, the Bianchi ("White") and Neri ("Black"): DA favoured the Bianchi, initially the more dominant, but they were overthrown by the Neri in 1301. DA was banished and sentenced to death if he returned to Florence. He became a political exile wandering throughout Italy and beyond, and began writing in earnest. He had already earned a reputation as a poet after the publication of *Vita Nuova* ["New Life"] (written 1292-4), which poured out his lifelong love for Beatrice Portinari (1265-1290), but he now turned his attention to works on politics, theology, language and culture. *De Monarchia* ["On Monarchy"] (written 1309-13) espoused his views about the relationship between temporal and spiritual government. DA was of the belief that the Holy Roman Empire should function on the same basis as the old Roman Empire, providing a world government, but spiritually and morally responsible to the Pope. It was this same passion that inspired *La Divina Commedia*, which is a dual ALLEGORY: on the progress of the soul toward HEAVEN; and on the anguish of humankind in seeking peace on Earth. Its intention would later be echoed by John BUNYAN in PILGRIM'S PROGRESS (**1678-84**).

La Divina Commedia is divided into three parts. The first and best known is *Inferno* (written ?1307-?1315). It begins in the year 1300. DA has become lost in a wood (◊ INTO THE WOODS), which represents the political turmoil of the time. There he encounters the spirit of VIRGIL, who conducts him through the nine levels of HELL (◊◊ INFERNO). The descent through Hell is in order to free the body of the temptation to sin. DA regarded Virgil as the epitome of human aspiration: only he was of sufficient authority to guide the SOUL. DA also drew upon Virgil's *Aeneid* (**19**BC) for some of the imagery in Hell. At the centre Dante encounters LUCIFER, with three heads gnawing at the ultimate betrayers – Brutus, Cassius and Judas. This powerful scene demonstrates DA's dual allegiance to the Roman Empire and to the Kingdom of Christ. DA believed the Papacy had turned against the true teachings of Christ, and he depicted seven former Popes in Hell. When *Inferno* was completed and copied, DA was proclaimed a heretic and his death sentence was renewed.

Twisting around the waist of Lucifer, Virgil and Dante enter a tunnel which leads to an ocean in which is the mountainous island of PURGATORY. *Purgatorio* was probably written around 1312-17. The ascent of the mountain of Purgatory is to cleanse (or purge) the soul in readiness for Paradise. A terraced path spirals up the mountain, and the less grievous the sin the easier the ascent. At the summit Virgil, who can go no further, leaves DA at the gates of the Earthly Paradise where he passes through EDEN to meet Beatrice, the personification of pure LOVE. She conducts him through the fires of purification, then through the eight concentric spheres of HEAVEN, represented by the celestial bodies, until DA at last is able to gaze upon the supreme radiance of GOD. This final work, *Paradiso*, was written in DA's last years and the manuscript was found only after his death.

Along with the *Summa Theologica* of Thomas Aquinas (1226-1274), on which DA based his theology and cosmology, *La Divina Commedia* became one of the most influential and visionary works of the Middle Ages. Its most popular translation was that by Henry Cary (1772-1844), who issued *Inferno* (trans **1805** UK) first, and later the complete *Divina Commedia* (trans **1814** UK), still in print. A separate translation of *The Inferno* (trans **1961** UK) by Warwick Chipman is considered closer to the style and approach of DA. [MA]

DANTE'S INFERNO US movie (*1935*). Fox. **Pr** Sol M. Wurtzel. **Dir** Harry Lachman. **Spfx** Ralph Hammeras, Fred F. Sersen, Louis J. Witte. **Screenplay** Philip Klein, Robert M. Yost. **Sets** Willy POGANY, inspired by Gustave DORÉ. **Starring** Spencer Tracy (Jim Carter), Claire Trevor (Betty), Henry B. Walthall (Pop McWade). 89 mins. B/w.

A CARNIVAL entrepreneur, Carter, makes his millions through shady dealings and on the strength of his extravagant sideshow "Dante's Inferno"; at last, through heroism, he loses his fortune but regains those he loves. This B-movie would be forgotten were it not for an extraordinary 10min sequence in which lovable Pop shows Carter an edition of *Inferno* illustrated by Gustav DORÉ, and we are suddenly plunged into it. Totally out of place, plopped like a solitary large plum into a stodgy pudding, this astonishing piece of work shows the grim, overpowering landscape of HELL, where tormented SOULS (naked, in defiance of all Hollywood conventions of the day) meet their fiery, inevitable dooms. Why this sequence – which must have been hugely expensive in time, money and ingenuity – was wasted on this movie is a matter for conjecture. [JG]

DARBY, LYNDAN Joint pseudonym of UK writers Ann

Grimsley (? -) and Lynne Kinnerley (? -), whose **Eye of Time** sequence – *Crystal and Steel* (**1988**), *Bloodseed* (**1988**) and *Phoenix Fire* (**1989**) – conveys its protagonist and COMPANIONS through adventures in FANTASYLAND which lead to the eventual restoration of the proper king to his throne.

[JC]

DARBY O'GILL AND THE LITTLE PEOPLE US movie (*1959*). DISNEY. **Pr** Walt DISNEY. **Dir** Robert Stevenson. **Spfx** Peter Ellenshaw, Eustace Lycett. **Anim fx** Joshua Meador. **Screenplay** Lawrence Edward Watkin. **Based on** the **Darby O'Gill** stories by H.T. Kavanagh. **Novelization** *Darby O'Gill and the Little People* * (**1959**) by Watkin (1901–1981). **Starring** Sean Connery (Michael McBride), Janet Munro (Katie O'Gill), Jimmy O'Dea (King Brian), Albert Sharpe (Darby O'Gill). 93 mins. Colour.

Once upon a time, or so old wastrel Darby O'Gill tells his cronies in the pub in rural Irish Rathcullen, he captured King Brian of the leprechauns, who granted him THREE WISHES. But Brian persuaded him to wish a fourth time, thus cancelling out the previous three – which is why Darby has no crock of gold.

One night Darby's recalcitrant horse leads him to the top of the nearby haunted hill, and Darby falls down an old well into the land of the leprechauns – brought there, King Brian announces, as a favour, because life outside holds nothing more for him. But Darby escapes, and uses TRICKSTER wiles to keep Brian in the mortal world all night until the cock crows, when the King's MAGIC ceases to work. Brian promises another three wishes. Darby's first is that Brian stay with him a fortnight until he decides his other two. Brian tricks him out of his second, so Darby tells Brian he will not make his third wish until daughter Katie and newcomer McBride are betrothed, a circumstance Brian tries to engineer. But Katie has a fall on the haunted hill and the howl of the banshee signals her death is imminent. Darby wishes he be taken in her place, and boards the Costa Bower (the Coach of Death). Brian appears one last time and genially tricks him into a fourth wish, thus cancelling the earlier ones. Darby is restored to the living; Katie has recovered; the lovers are united; all ends happily.

This poorly paced movie contains a surprising amount of good material. Despite the final plot twist (by all logic Katie should die), the script is entertaining, showing few signs of staleness even though begun over 20 years earlier. Most impressive of all, despite one or two dreadful lapses, are the spfx, some animated. *DOATLP* was poorly received, but its reputation has been recovering.

[JG]

D'ARCY, JACK [s] ◊ Hugh B. CAVE.

DARK CRYSTAL, THE US puppet movie (*1982*). Universal/ITC. **Pr** Jim HENSON, Gary Kurtz. **Exec pr** David Lazer. **Dir** Henson, Frank Oz. **Spfx** Roy Field, Brian Smithies. **Miniature fx** Brian Smithies. **Screenplay** David Odell. **Novelizations** *The Dark Crystal* * (**1982**) by A.C.H. Smith and *The Tale of the Dark Crystal* * (**1982**) by Donna Bass. **Voice actors** Barry Dennen (Chamberlain), Percy Edwards (Fizzgig), Stephen Garlick (Jen), Michael Kilgarriff (General), Lisa Maxwell (Kira), Brian Muehl (Dying Master), Jerry Nelson (High Priest, Dying Emperor), Joseph O'Conor (Narrator), Billie Whitelaw (Aughra). **Extra credit** "Special Thanks to Dennis Lee and Alan GARNER." 94 mins. Colour.

For 1000 years the cruel Skeksis have ruled the world. Now there are only ten left – soon to be nine, for their Emperor is dying. Far off in the forest the benign and shambling Mystics are likewise waiting for their Master to die. As he does so he charges the (seemingly) last of the Gelfling race, Jen, with finding and returning to its place a lost shard of the once-shattered Dark Crystal; otherwise, according to PROPHECY, the Skeksis will rule forever. Aided by the girl-Gelfling Kira, Jen at last succeeds in his QUEST. The surviving Mystics and Skeksis fuse, one-to-one, and become transcendent creatures; reunited, they leave the world and the Dark Crystal – now the Crystal of Truth – to Jen, Kira and their descendants.

TDC, though not an unalloyed triumph, is in its unabashed (though often derivative) mythopoeia – along with DRAGONSLAYER (*1981*) – among the most successful attempts to bring HIGH FANTASY to the screen. [JG] **Further reading:** *World of the Dark Crystal* * (**1982**) by Brian FROUD, J.J. Llewelyn and Rupert Brown; *The Making of The Dark Crystal* (**1983** chap) by Christopher Finch.

DARK FANTASY In his *Critical Terms for Science Fiction and Fantasy* (**1986**), Gary K. Wolfe indicates that DF is a term sometimes used interchangeably with GOTHIC FANTASY. Though the term has indeed been so used, in recent years many works have been called DF in order – for reasons of perceived prestige – *not* to call them HORROR. Other critics and editors – like Chris MORGAN, in his anthology *Dark Fantasies* (anth **1989**) – have defined the term as more than just a marketing tag; rather they perceive it as describing an affect, rather like "horror" itself; and deploy the term to describe the emotional effect certain stories may have on readers.

For the term DF to be used in conjunction with others it requires a more restricted definition. In this encyclopedia we define a DF as a tale which incorporates a sense of HORROR, but which is clearly FANTASY rather than SUPERNATURAL FICTION. Thus DF does not normally embrace tales of VAMPIRES, WEREWOLVES, SATANISM, GHOSTS or the occult, almost all of which are supernatural fictions (although such tales may contain DF elements, while some DFs contain vampires, ghosts, etc. – an example is Stephen MARLEY's *Mortal Mask* [**1991**]). The term can sensibly be used also to describe tales in which the EUCATASTROPHE normal to most fantasy is reversed – tales in which the DARK LORD is victorious, tales in which the LAND, normally an object of desire, and an arena for the working out of a desired STORY, is itself an object of horror (Stephen DONALDSON's **Covenant** sequence is the prime and definitive example). Clark Ashton SMITH's **Zothique** stories describe a land, Zothique, of this sort, though it can also be understood, in minimally rationalized terms, as a DYING EARTH. And DF can be used to describe certain CROSSHATCHES in which the intersections between this world and an upwelling OTHERWORLD are at least partly described in images and themes out of the worlds of HORROR; an example is Sean STEWART's *Resurrection Man* (**1995**), a tale full of REVENANTS, SPIDERS, open graves, dissections and DOUBLES. [JC]

DARK HALF, THE US movie (*1991*). Orion. **Pr** Declan Baldwin. **Exec pr** George A. Romero. **Dir** Romero. **Spfx** Ed Fountain, Carl Horner Jr. **Vfx** VCE/Peter Kuran, Video Image. **Animatronics** Tom Culnan, Ken Walker. **Mufx** Everett Burrell, John Vulich. **Screenplay** Romero. **Based on** *The Dark Half* (**1989** UK) by Stephen KING. **Starring** Julie Harris (Reggie Delesseps), Timothy Hutton (Thaddeus Beaumont/ George Stark), Robert Joy (Fred Clawson), Amy Madigan (Liz Beaumont), Michael Rooker

(Sheriff Alan Pangborn). 122 mins. Colour.

1968, and young Thad Beaumont wants to be a writer but suffers a brain tumour; when his skull is opened the surgeons find rudimentary organs developed from a TWIN fetus. As the excision is completed, the hospital is attacked by a vast flock of sparrows. 1991, and Beaumont is a respected but obscure novelist, also secretly writing successful thrillers as "George Stark". A blackmailer, Clawson, threatens to reveal all, so Beaumont confesses to the press, gaining enormous publicity; "Stark", he announces over a mock grave, is dead. But Stark, who is almost Beaumont's DOUBLE, has become a tulpa; he hauls himself from the grave and hideously murders Clawson and most of Beaumont's professional colleagues, with the aim of forcing Beaumont to write one more "Stark" book, believing this will allow him continued life in the writer's place. An IDENTITY EXCHANGE begins. But a host of sparrows attacks the house, breaking it apart in scenes reminiscent of *The* BIRDS (*1963*), and devours Stark's flesh before bearing his SOUL away to HELL.

The Dark Half is one of King's more interesting novels, a flawed CROSSHATCH of notions from various fields. The movie adaptation is very faithful, and so suffers the same flaws. Yet it has much to excite the imagination, and its direction is expert. [JG]

DARK LORD In GENRE FANTASIES – particularly those modelled on J.R.R. TOLKIEN's work – the role of principal antagonist is often taken by a DL. DLs are, or aspire to be, the Prince of this world, and as such are a malignant PARODY of the GODS. Alberich in Richard WAGNER's **Ring** cycle (**1850-76**) is a crucial prototype.

The DL may not be explicitly a DEVIL or an ANTICHRIST, though names like Sauron, Ba'alzamon and Lord Foul indicate that he most usually is. Perimal Darkling, in P.C. HODGELL's *Chronicles of the Kencyrath* (from **1982**), is only marginally even a personality; he is, or has become, an almost abstract force of THINNING, DEBASEMENT and entropy. Tad WILLIAMS's Storm King in **Memory, Sorrow and Thorn**, by contrast, has fairly good reason for his machinations – the wrongs done by earlier generations of humans to him and his people.

The DL is usually male, save in PARODY. He operates under constraints or PROHIBITIONS, and is thus often vulnerable to QUIBBLES; he often is, or has been, a MALIGN SLEEPER who has been woken by the investigations of the unwary curious, as in Glen COOK's **Black Company** sequence (exceptional in possessing not one but at least two DLs, one of them female and, in later books, semi-retired and reformed). He is often, or has been, the former servant of some even greater DL, and has often been already defeated but not destroyed eons before.

DLs are frequently rooted in whatever political or religious figures happened to be worrying readers at the time. Tolkien's Sauron, for example, combines features of the dictators of the 1930s, while Terry GOODKIND's Darken Rahl in **The Sword of Truth** sequence (**1994** onwards) combines totalitarian rule, the self-righteous superficial reasonableness of cult leaders like Jim Jones, and the seduction and physical torture of children. Darken Rahl is closer than most DLs to an Antichrist. In *The Lord of the Rings* (**1954-5**) the wounding of the land specifically parallels the effect of the Industrial Revolution – slagheaps, polluted air and water – and of WORLD WAR I. Much genre fantasy simply copies this. Tad Williams manages some interesting variations in **Memory, Sorrow and Thorn**, many of them paralleling

the sense of bad faith coming home to roost that permeates the trilogy. The wounding of the Land is something of a specialty of Stephen R. DONALDSON's Lord Foul, the variety of whose operations interestingly parallels those of Jehovah against Egypt in *Exodus*.

The DL's wounding of the land is usually only part of a broader RITUAL OF DESECRATION. Lord Foul wishes to escape the world of flesh and return to the pure eternity from which he has been expelled; the Storm King wishes to overturn the natural order to facilitate turning back time to when he was alive and not undead; Darken Rahl proves to be the agent of a deathgod who wishes to break into the world of the living. In all these cases, the WRONGNESS originally experienced by the protagonists is only a mundane extension of a far greater spiritual malaise or blasphemy to come.

DLs are often surprisingly stupid; the plots of fantasies which contain them tend to be based on the assumption that EVIL understands good less well than good understands evil. Thus Sauron is supposedly incapable of guessing that the Alliance will try to destroy the Ring rather than use it against him. Darken Rahl, admittedly in a book which takes its title, *Wizard's First Rule* (**1994**), from the perception that people are stupid, allows himself to believe he has trapped his enemy into a position where he cannot possibly lie about the workings of a dangerous ritual (◊ MAGIC). Donaldson's Lord Foul, returned from a first defeat by Covenant, tries to trick, corrupt and gloat over him yet again, with inevitable results. [RK]

DARK MESSENGER READER, THE ◊ MAGAZINES.

DARK SHADOWS Two US tv series.

1. (1966-71). Dan Curtis Productions/ABC. **Pr** Robert E. Costello, George DiCenzo, Peter Miner, Lela Swift. **Exec pr** Dan Curtis. **Dir** Curtis and many others. **Spfx** Dick Smith. **Writers** Joe Caldwell and many others. **Created by** Curtis. **Novelizations** 34 novels by Marilyn ROSS, plus 35 COMIC books (1969-76) released by Gold Key Comics. **Starring** Nancy Barrett (Carolyn Stoddard/ Charity/Millicent Collins/Letitia Faye), Joan Bennett (Elizabeth Collins Stoddard/Naomi Collins/Flora Collins), Betsy Durkin (Victoria Winters 1968-71), Louis Edmonds (Roger Collins/Joshua Collins), David Hennesy (David Collins), Jonathan Frid (Barnabas Collins), David Hennessy (David Collins/Tad Collins), Alexandra Moltke (Victoria Winters, 1966-8), David Selby (Quentin Collins), Diana Walker (Carolyn Stoddard). 1225 30min episodes. Colour.

Undoubtedly the most unusual soap opera ever aired, this dark and moody daily series was set in the small town of Collinsport, Maine. It was there that Victoria Winters was hired as the governess for 10-year-old David Collins, heir to the Collins family fortune. Unknown to Victoria and the townspeople was the fact that David was also heir to the CURSES placed on Collinwood, the eerie family mansion that stood above town on Widow's Hills.

After a beginning that mostly featured a rather standard soap-opera story about plots on Virginia's life, things took an unusual turn. With the ratings falling, executive producer Curtis gambled on the introduction of a 175-year-old VAMPIRE, Barnabas Collins, who quickly became the most popular character in the series. Seemingly overnight, *DS* became a hit. In the second season matters became even more complicated when *DS* jumped back in time, with the same actors recast, to 1796 and the ancestors of those in modern Collinsport. In due course the show also made use of alternate timelines and leaps to the future, with plots winding intricately through the various eras and settings. Many of the stories were loosely based on classics like Robert Louis STEVENSON's *Strange Case of Dr Jekyll and Mr Hyde* (**1886**), Mary SHELLEY's *Frankenstein* (**1818**) and Oscar WILDE's *The Picture of Dorian Gray* (**1891**).

Much like *Star Trek*, *DS* spawned a long series of novels and conventions that continued after its cancellation. 20 years later, **2** brought Barnabas and his friends back to life – or what passed for it.   [BC]

2. (1991-2) MGM-UA/NBC. **Exec pr** Dan Curtis. **Dir** Rob Bowman, Curtis, Paul Lynch, Armand Mastroianni, Mark Sobel. **Spfx** Greg Curtis. **Writers** Jon Boorstin and many others. **Graphic novelization** Nine comic books (1992-3) from Innovation Publishing. **Starring** Barbara Blackburn (Carolyn Stoddard/Millicent Collins), Ben Cross (Barnabas Collins), Stefan Gierasch (Professor Woodard/Joshua), Joanna Going (Victoria Winters/Josette Dupré), Jean Simmons (Elizabeth Collins Stoddard/Naomi Collins), Barbara Steele (Dr Julia Hoffman/Natalie). 2 2hr and 9 1hr episodes. Colour.

This remake of **1** begins with a groundskeeper accidentally opened the Collins family cemetery vault. Out comes Barnabas, who promptly begins feasting on the local women. The hapless VAMPIRE yearns to be cured, and finds help from Hoffman, who labours to create a suitable POTION. Both are endangered by Woodard, who determines to slay Barnabas at any cost – the scenario owes something to Bram STOKER's *Dracula* (**1897**). Like **1**, this short-lived series took place across different timelines.   [BC]

DARK TOWER A classic BAD PLACE in tales of CHIVALRY: the term is now best-known through Robert BROWNING's poem "Childe Roland to the Dark Tower Came" (in *Men and Women* coll **1855**), whose title is a line sung by Edgar during feigned madness in SHAKESPEARE's *King Lear* (performed *c*1604; **1608**). The poem describes, with great intensity, the harried CHILDE's QUEST through a WASTE LAND to the DT, but we never see inside: the symbol is of faceless EVIL. The same is true of Barad-dûr and Minas Morgul in J.R.R. TOLKIEN's *The Lord of the Rings* (**1954-5**) and the Loathly Tower in Gordon R. DICKSON's *The Dragon and the George* (**1957**). Stephen KING's FAR-FUTURE fantasy sequence **The Dark Tower** (**1982-91**) takes its inspiration directly from Browning. C.S. LEWIS's unfinished *The Dark Tower* (written *c*1938; **1977**) updates the concept as a modern EDIFICE, here closely resembling the then-new Cambridge University LIBRARY. This was perhaps the cue for the typical skyscraper DT of later URBAN FANTASY, often found in LONDON; e.g., the Lateinos and Romiith building overshadowing Brentford in Robert RANKIN's *East of Ealing* (**1984**), the sorcerer's DT on Cheapside in Tom HOLT's *Who's Afraid of Beowulf* (**1988**), and the evil media magnate's black-glass Docklands pyramid in Kim NEWMAN's *The Quorum* (**1994**). Diane DUANE's *So You Want to be a Wizard?* (**1983**) features a 90-storey DT, also faced in black glass, in its alternate NEW YORK. An impressive cinematic DT is the Fortress of Ultimate Darkness in TIME BANDITS (**1981**).   [DRL]

DARK VOICES ◊ PAN BOOK OF HORROR STORIES.

DARLING, I AM GROWING YOUNGER ot of MONKEY BUSINESS (*1952*).

DARLINGTON, W(ILLIAM) A(UBREY) (CECIL) (1890-1979) UK theatre critic and writer of F. ANSTEY-style fantasies. A series of morale-building sketches written during WWI for the *Passing Show* was recast as the bestselling *Alf's Button* (**1919**), the first of the **Alf** sequence, in which a working-class soldier fails to make sensible use of a button on his uniform which has been derived from Aladdin's lamp and which conjures up a djinn (◊ GENIES), whom he calls Eustace. The book was the basis of a successful stage play, *Alf's Button, An Extravaganza in Three Acts* (**1925**), and then the movie *Alf* (**1920**), followed by *Alf's Button* (**1930**) and *Alf's Button Afloat* (**1938**). *Alf's Carpet* (**1928**), a sequel to the first movie, had a markedly different ending occasioned by the intervention of Alf's better half Liz. This too became a play, *Carpet Slippers: A Play in Three Acts* (**1937** chap). WWII provided an opportunity for a further sequel, *Alf's New Button* (**1940**), in which Eustace grants Alf six wishes which he and Liz duly waste.

The two comic fantasies which Darlington wrote between *Alf's Button* and its first sequel are somewhat better, by virtue of being set in the middle-class *milieu* which Anstey had employed so successfully. In *Wishes Limited* (**1922**) a would-be novelist finds a FAIRY's attempts to grant his wishes unfortunately restricted by new Trade Union rules. *Egbert* (**1924**) is the story of a barrister turned into a rhinoceros by an offended WIZARD.   [BS]

DARMANCOUR, P. Pseudonym of Charles PERRAULT's son Pierre Perrault (1678-1700).

DATLOW, ELLEN (SUE) (1949-   )US editor, fiction editor of *Omni* from October 1981; she is primarily known for her work in sf, having also edited seven spinoff anthologies of *Omni* material plus five further books of *Omni*-associated work, some of it original. She is secondarily known for her work in HORROR, in particular for the **Year's Best Fantasy and Horror** sequence (anth **1988**-current) co-edited with Terri WINDLING, with Windling handling the fantasy and ED the horror. Also with Windling, ED has edited three volumes of tales – some TWICE-TOLD, some REVISIONIST FANTASY, some combining both – *Snow White, Blood Red* (anth **1993**), *Black Thorn, White Rose* (anth **1994**) and *Ruby Slipper and Golden Tears* (anth **1995**). ED has also edited the erotic VAMPIRE anthologies *Blood is Not Enough* (anth **1989**) and *A Whisper of Blood* (anth **1991**).   [JC]

DAUGHTER OF DR JEKYLL US movie (*1957*). ◊ JEKYLL AND HYDE MOVIES.

DAVEY, (HENRY) NORMAN (1888-?   ) UK writer. *The Pilgrim of a Smile* (coll of linked stories **1921**) and *The Penultimate Adventure* (**1924** chap), both assembled as *The Pilgrim of a Smile* (omni **1933**), are adventures in a LAND-OF-FABLE Europe which has echoes of the old GODS: the protagonist seeks the secret of the Sphinx's smile. In *Judgment Day* (**1928**), a novel mildly reminiscent of T.W. POWYS, the inhabitants of a West Country village are summoned to judgement. In *Pagan Parable: An Allegory in Four Acts* (**1936**) members of the Greek PANTHEON confront modern civilization.   [JC]

Other work: *Yesterday: A Tory Fairy-Tale* (**1924**), sf.

DAVIDSON, AVRAM (JAMES) (1923-1993) US writer and editor. His first nationally published story – some earlier fiction had appeared in Yiddish publications – was "My Boy Friend's Name is Jello" (1954) in *The* MAGAZINE OF FANTASY AND SCIENCE FICTION, which magazine he edited 1962-4. His work was plainly fantasy more than sf, as exemplified by *Or All the Seas with Oysters* (coll **1962**), collecting his early stories. AD's fiction was immediately popular; he was frequently anthologized in the late 1950s, and "Or All the Seas with Oysters" (1958) won a Hugo AWARD.

The mannered, witty, rather baroque style of AD's prose recalls the arch and whimsical fiction of Lord DUNSANY and Lafcadio HEARN, but his early tone, in addition to being contemporary in cast and distinctly US, lacked the affectation of the lapidary Edwardian fantasists. In later work he explored other modes, including SWORD AND SORCERY, ALTERNATE WORLDS and HORROR.

AD's first novel, *Joyleg* (**1962**) with Ward Moore (1903-1978), displays an uncertainty with longer forms, but the series of solo novels that he began publishing in 1964, although all at least nominally sf, show him employing his skill at evoking exotic locales with commercial competence and more panache than his markets required. Of the half dozen novels he published over the next few years, *Rogue Dragon* (**1965**) and *Clash of Star-Kings* (**1966** dos), which use sf rationales to dramatize DRAGONS and Mayan GODS respectively, come closest to fantasy.

The Phoenix and the Mirror (**1969**), which AD worked on for most of a decade, is plainly his magnum opus. His dense and sombre tale of how the sorcerer Vergil (◊ VIRGIL) makes a *speculum majorum* (a virgin MIRROR, in whose face the first to gaze into it may see his heart's desire) possesses the imaginative force of only the most powerful fantasies: we believe in the *cosmos* of the novel, that its existence continues beyond the edge of the page. *The Phoenix and the Mirror* was envisioned as inaugurating a sequence of nine novels concerning Vergil Magus, but AD published only one more, *Vergil in Averno* (**1987**), a darker and more focused work set during Vergil's early manhood and involving a commission to the "very rich city" of Averno, a pre-industrial inferno of hellish manufactories. A third **Vergil** novel exists in draft, but AD's plans for "a trinity of trilogies" may have been foredoomed, given his preference for short forms.

Despite this, in the immediately succeeding years AD began several other series with *The Island under the Earth* (**1969**) and *Peregrine: Primus* (**1971**) both being announced as the first volumes of trilogies. *The Island under the Earth* seems to take its inspiration from the *metope* of the Parthenon, depicting the war between the centaurs and the Lapiths, although it is set not in the mountains of Thessaly but in a fantastic cosmology whose nature has not been made clear by novel's end. *Peregrine: Primus*, a PICARESQUE set during the disintegrating Roman Empire, is broader in its effects and more casual in its structure; a sequel, *Peregrine: Secundus* (**1981**), was eventually published.

AD's other sequences of note include *The Enquiries of Doctor Eszterhazy* (coll of linked stories **1975**; exp vt *The Adventures of Doctor Eszterhazy* 1990), set in a tiny Ruritanian country (◊ RURITANIA) soon to be swept away in WORLD WAR I. Engelbert Eszterhazy, the emperor's WIZARD, who drives about the cobbled streets of the capital in a steam runabout, is an obvious Vergil figure, although the tales show the high spirits of AD's short fiction rather than the grave wonder of the **Vergil** novels. In the mid-1980s AD returned to this venue with a series of eight longer tales about Eszterhazy in his youth (rather as he had lately done with Vergil). By now AD's style had changed: his prose became less tight, more discursive and sometimes prolix. The five **Cornet Eszterhazy** stories are longer than the eight earlier ones, and should be considered as a separate sequence.

AD wrote a second sequence of novelettes concerning young **Jack Limekiller** and his strange encounters in the former colony of British Hidalgo. This series, begun in 1976 and continued the rest of AD's life (with "A Far Countrie" published posthumously), shares the exotic sense of place and love of strange incident that characterize the **Vergil** and **Eszterhazy** stories.

AD never ceased writing short fiction; although as yet uncollected, the stories of his last decade retain the learned sportiveness of his early work, and some (e.g., "The Slovo Stove" 1985) can be counted among his best. Although he had difficulty finding commercial publishers for his books during the last years of his life, he continued to publish in magazines and anthologies; his last two books were issued by a SMALL PRESS, where (like R.A. LAFFERTY) he retained an enthusiastic audience. Nevertheless, at the time of his death almost all of his large body of work was out of print. Accordingly, during 1995-6 his widow, Grania DAVIS, organized the Avram Davidson Award, to be given approximately annually to the "best beloved" work by any author that was, at the time of voting (by professional writers, fans and others), out of print. [GF]

Other works: *Crimes and Chaos* (coll **1962**), nonfiction; *The Best from Fantasy and Science Fiction, 12th Series* (anth **1963**), *13th Series* (anth **1964**), *14th Series* (anth **1965**); *What Strange Stars and Skies* (coll **1965**); *Strange Seas and Shores* (coll **1971**); *And on the Eighth Day* (**1964**) and *The Fourth Side of the Triangle* (**1965**), detections as by Ellery Queen; *Ursus of Ultima Thule* (fixup **1973**); *The Redward Edward Papers* (coll **1978**); *The Best of Avram Davidson* (**1979**) ed Michael KURLAND; *Collected Fantasies* (coll **1982**) ed John Silbersack; *Magic for Sale* (anth **1983**); *Marco Polo and the Sleeping Beauty* (**1988**) with Grania Davis; *Adventures in Unhistory* (1993), spoof essays.

DAVIES, (SIR) PETER MAXWELL (1934-   ) UK composer. ◊ George Mackay BROWN.

DAVIES, (WILLIAM) ROBERTSON (1913-1995) Canadian writer whose work is suffused with a delighted awareness of the FANTASTIC and the grotesque. In almost diametrical opposition to John CROWLEY's use of the techniques of mimetic fiction to insinuate the fantastic into everyday life, RD uses fantasy tropes – e.g., the QUEST for the HERO's true origins or the decoding of gnostic lore which results in a TRANSFORMATION of the worldview – to illumine the beauty and magical potential in the mundane.

In *Tempest-Tost* (**1951**), the first volume of the **Salterton Trilogy** – the others being *Leaven of Malice* (**1954**) and *A Mixture of Frailties* (**1958**), all being assembled as *The Salterton Trilogy* (omni **1986** UK) – the transformation comes about through the attempt to replicate William SHAKESPEARE's *The Tempest* (performed *c*1611; **1623**) in the superficially stolid town of Salterton. In the **Deptford Trilogy** – *Fifth Business* (**1970**), *The Manticore* (**1972** US) and *World of Wonders* (**1975**), assembled as *The Deptford Trilogy* (omni **1983** UK) – a strange murder opens the way for an exploration of hidden longings and fantastical discoveries. Using the realms of MAGIC and MYTH as haloed through the ARCHETYPES of JUNGIAN PSYCHOLOGY, the characters walk a glittering LABYRINTH of plot and subplot which leads to self-knowledge and expiation. In the **Cornish Trilogy** – *The Rebel Angels* (**1981**), *What's Bred in the Bone* (**1985**) and *The Lyre of Orpheus* (**1988**) – the supernatural element is more directly present. In *The Lyre of Orpheus*, for instance, the spirit of E.T.A. HOFFMANN, dwelling in uneasy LIMBO, watches over the completion and presentation of his unfinished OPERA by the Cornish Foundation and its Trustees – the same Trustees who, together with the enigmatic Francis Cornish, were the protagonists of the two earlier volumes. *High Spirits: A Collection of Ghost Stories* (coll **1982**; cut vt *A Gathering of Ghosts* 1995 chap UK) assembles GHOST STORIES couched within the frame of CHRISTMAS; this won a WORLD FANTASY AWARD. *Murther & Walking Spirits* (**1991**) is a POSTHUMOUS FANTASY whose protagonist views his own and his ancestors' lives as a sequence of movies. *The Cunning Man* (**1995**) explores the nature of religious belief and the relationship of mind to body through the story of a doctor – the cunning man of the title – whose life was saved by the magical intervention of a female SHAMAN in his childhood. The interweaving of preternatural event and scientific training leads to a synthesis of the two which might stand as a paradigm of RD's outlook.

RD's largeness of spirit and richly comic view of humanity – as a species of creation absurd yet always capable of the sublime – give his books their distinct and irresistible flavour. [JR/JC]

Other works: *A Masque of Mr Punch* (**1963**), a play.

DAVIS, GRANIA (EVE) (1943-   ) US writer, married to Avram DAVIDSON. She began publishing work of genre interest with "My Head's in a Different Place, Now" for *Universe 2* (anth **1972**) ed Terry CARR, but has not been prolific in short forms. *The King and the Mangoes* (**1975** chap) and *The Proud Peacock and the Mallard* (**1976** chap) are CHILDREN'S FANTASIES about incarnations of the Buddha in various animal forms. *The Rainbow Annals* (**1980**), for adults, again explores the myth-rich intricacies of Buddhism, this time in a tale concerning the MONKEY God and his various incarnations; as the epochs pass in a series of cyclical encounters, there is a sense of the THINNING of the world into history. *Moonbird* (**1986**), set in Malaysia, deals with the dilemmas of a native SHAMAN – whose travels with his GOD make up a significant portion of the novel – when a group of North Americans invade his home. [JC]

Other works: *Dr Grass* (**1978**) and *The Great Perpendicular Path* (**1980**), both associational; *Marco Polo and the Sleeping Beauty* (**1988**) with Davidson.

DAWN OF THE DEAD US movie (*1979*). ◊ ZOMBIES.

DAWSON, CONINGSBY W(ILLIAM) (1883-1959) UK writer, in the USA from 1904, though returning temporarily to fight with a Canadian regiment in WORLD WAR I; most of his early work reflects his war experiences. *The Road to Avalon* (**1911** is a curious ALLEGORY in which Arthurian legend provides the bulk of the apparatus of a modern PILGRIM'S PROGRESS; the hero sets forth to find AVALON in order that ARTHUR may return but is continually seduced from his true path by the enchantress LILITH and her associates. *The Unknown Country* (**1915** chap) is a POSTHUMOUS FANTASY. *When Father Christmas was Late* (coll **1919**) includes among other sentimental CHRISTMAS tales *The Seventh Christmas* (**1917** chap) and *A Christmas Legend of Hamelin Town* (**1965** chap). CD was the son of the Reverend William James Dawson (1854-1958), author of the CHRISTIAN FANTASIES *The House of Dreams* (**1897**) and *A Soldier of the Future* (**1908**). [BS]

Other works: *The Unknown Soldier* (**1929** chap).

DAWSON, WILLIAM J(AMES) (1854-1928) UK cleric and fantasy writer. ◊ Coningsby W. DAWSON.

DAY, J(AMES) WENTWORTH (1899-1983) UK writer and folklorist. ◊ Michael AYRTON.

DAY OF THE DEAD US movie (*1985*). ◊ ZOMBIE MOVIES.

DC COMICS One of the two major US COMIC-book publishing companies, the other being MARVEL COMICS.

Major Malcolm Wheeler-Nicholson (1890-1968) set up his National Comics publishing company in 1935 with the publication of *New Fun*, which ran for five bimonthly issues and was then relaunched the following year as *More Fun* (1936-47). *New Comics* (12 issues 1935-6), another early venture, ran for nearly 50 years under the later titles *New Adventure Comics* (20 issues 1937-8) and *Adventure Comics* (1938-83); *Detective Comics* (1937-current) has been even longer-lived. These two titles were the first US comic books to feature regular characters in a series of adventures, but Wheeler-Nicholson himself could not make a success of them; in 1938 he paid off his printers, Harry Donenfield and Jack Leibowitz, by giving them the company.

Their first new venture was *Action Comics* (1938-current), *#1* of which (June 1938) featured the first appearance of SUPERMAN, created by Jerry Siegel (1914-1996) and Joe Shuster (1914-1992). Then came BATMAN (*Detective Comics #27* May 1939). The successful format of *Detective Comics* – the first regularly scheduled comic book to present all-original material – motivated the company's use of "DC" as a trademark and eventually, in the early 1980s, to change its name to DC Comics Inc.

In 1945 DC absorbed Max Gaines's company All American Comics and thereafter built up an exceptional team of creative talents: Alfred Bester (1913-1987; ◊ *SFE*), Otto Binder (1911-1975; ◊ *SFE*), Gardner Fox (1911-1986; ◊ *SFE*), Edmund Hamilton (1904-1977; ◊ *SFE*), Mort Weisinger (1915-1978) and others. This team created a long list of memorable SUPERHERO characters including **Aquaman**, **The Flash**, **Green Lantern** and WONDER WOMAN.

The 1950s saw an increase in popularity of GHOST STORIES and HORROR, and DC added to its list **The Phantom Stranger**, an enigmatic figure who acted as a supernatural troubleshooter, and who would occasionally re-emerge in various of the company's titles over the next four decades. In 1952 the company, now called National Periodical Publications, launched a regular monthly anthology title, *House of Mystery*, which became a vehicle for SUPERNATURAL FICTION, somewhat in the EC COMICS vein although considerably less gory. The format was succesful enough for DC to create other, similar titles: *Tales of the Unexpected* (1955-67; vt *The Unexpected* 1968-82) and *House of Secrets* (1956-78); further comics of the same ilk were introduced over the next 20 years, including *The Witching Hour* (1967-78), *Secrets of the Haunted House* (1975-82), *Weird Worlds* (1972-4) and brief experiments with books featuring GOTHIC FANTASY (*Dark Mansion of Forbidden Love* [1971-2]) and war stories with an sf edge (*Weird War* [1971-83]). DC's brief flirtations with SWORD AND SORCERY – e.g., **Nightmaster** (*DC Preview Showcase #83-#84* 1969), *Sword of Sorcery* (*#1-#5* 1973) which featured Fritz LEIBER's **Fafhrd and the Gray Mouser**, *Sword of the Atom* (*#1-#4* 1983) and *Ironwolf* (1986) – failed to take off.

In the early 1980s DC responded to the increase in new US retail outlets offered by specialist comics stores by increasing its output of tenuously interlinked superhero titles, until, in 1985, on the company's 50th anniversary, DC felt it necessary to rationalize the "DC Universe" with the publication of *Crisis on Infinite Earths* (*#1-#12* 1985-6). ALTERNATE WORLDS were utilized to explain many evident anomalies which had been brought about by DC's multi-level exploitation of some characters; the series also featured the deaths of several characters DC had decided to discard.

A very substantial factor in DC's phenomenal success in the later 1980s has been its willingness to experiment with some of its more popular characters: SUPERMAN, BATMAN and to a lesser extent **Swamp Thing** – plus offshoot characters and associated series – proved, anew, major moneyspinners. Open-minded editorial attitudes helped nurture substantial new talents (most notably Neil GAIMAN and Frank MILLER) in order to increase DC's range of characters and the number of ways they could be exploited.

But this revealed the need to rationalize another aspect of DC's product: the age-range of its readers. In 1991 DC inaugurated the **Impact** line to cater for younger readers; this revived a number of golden-age heroes, including **The Fly**, **Jaguar**, **The Shield** and **Black Hood**. In 1993 DC's increasing range of fantasy and horror mystery titles was unified under the **Vertigo** imprint and tagged "for mature readers", affording creators an opportunity to present complex and subtle adult storylines.

The long-held shared monopoly of DC and Marvel ended in the 1980s with the founding of a number of new, more creator-influenced companies. The dissipation of Marvel's properties in the mid-1990s placed DC in an unassailable position as the USA's leading comics publisher. [RT]

DEAD AGAIN US movie (*1991*). Paramount. **Pr** Lindsay Doran, Charles H. Maguire. **Exec pr** Sydney Pollack. **Dir** Kenneth Branagh. **Spfx** Steve Foster, Gregg Hendrickson, Frank Toro. **Screenplay** Scott Frank. **Starring** Branagh (Roman Strauss/Mike Church), Andy Garcia (Gray Baker), Gregor Hesse (Frankie), Derek Jacobi (Franklyn Madson), Wayne Knight ("Piccolo" Pete Dougan), Hanna Schygulla (Inga), Emma Thompson (Margaret Strauss/Grace/ Amanda Sharp). 108 mins. B/w and colour.

Decades ago Margaret Strauss was murdered and her husband Roman executed for the crime. Now there comes convincing evidence that an amnesiac woman is a REINCARNATION of Margaret. Calling her "Grace", private eye Church, mysteriously attracted to her, investigates, and eventually believes he is a reincarnation of Roman; in fact regression hypnosis reveals the genders have been swapped – he was Margaret and Grace was Roman. After much tortuous plot, the true murderer is unmasked and the couple look forward to a happy future together. *DA* was released among the wave of attempted emulators of the hugely popular GHOST (*1990*), although it is a very different movie in style, complexity, ambition, psychological insight and theme. The complex plot was little liked by the critics, although it is well worked-out and clearly enough told, and is largely responsible for *DA* being of such interest to those literate in the fantasy genre. [JG]

DEAD CAN'T LIE, THE (vt *Gotham*) US tvm (*1988*). Cannon/Showtime/Phoenix/Keith Addis. **Pr** David Latt. **Exec pr** Addis, Gerald I. Isenberg. **Dir** Lloyd Fonvielle. **Spfx** Michael Kavenagh. **Screenplay** Fonvielle. **Starring** Tommy Lee Jones (Eddie Mallard), Virginia Madsen (Rachel Carlisle-Rand), Denise Stephenson (Debbie), J.B. White (Jimbo). 100 mins. Colour.

Seedy PI Mallard is hired by Charlie Rand to persuade the GHOST of Rand's wife Rachel to stop dogging him. Finding Rachel apparently flesh-and-blood and ignorant of Rand, Mallard becomes sexually obsessed with her. A pattern of scams and counter-scams develops, but of more interest is that it emerges that "Rachel" is indeed a supernatural creature – whether a LAMIA or a SUCCUBUS is unclear – and has

deliberately ensnared Mallard. She begins to manipulate REALITY, so that Mallard becomes justifiably sceptical of the status of all his PERCEPTIONS, but finally he outwits her – and Rand – and escapes. Originally made for cable tv, *TDCL* is a sophisticated piece of fantasy dressed in often parodic Chandleresque guise. It can be seen as a precursor of GHOST (*1990*), and is certainly a better movie. [JG]

DEADLY BEES, THE UK movie (*1966*). ◊ Gerald HEARD.

DEADMAN COMICS character, the GHOST of a murdered circus performer who is granted (a) the right to return to Earth and (b) the ability to enter any living human body at will and animate it for his own purposes. D was created by writer Arnold Drake and artist Carmine Infantino (1925-) for *Strange Adventures #205* (1967), and skilfully developed and experimented with in subsequent issues (#206-#216 1967-8) by Jack Miller and Neal ADAMS.

The series is most significant for Adams's consummate draughtsmanship, his portrayal of Deadman's outrage and frustration (as Deadman gradually uncovers the trivial reason for his death) and his playful experiments with page layout and visual effects. These latter reached a peak in *Strange Adventures #216*, where a cryptographic message is contained within the flames of hellfire and where a 6-panel page has each panel drawn so as to form a full-page portrait of Deadman.

Deadman also appeared during this period in a number of crossover titles in which he became involved with other DC COMICS characters, the sequences featuring him almost always being drawn by Adams. The character was resurrected in *Deadman* (#1-#4 1986), expertly drawn by García López, and again in **Deadman: Love After Death** (1989-90) and **Deadman: Exorcism** (1992), which featured the delightfully contortionistic grotesqueries of artist Kelly Jones (1962-). [RT]

DEAD ZONE, THE US movie (*1983*). Dino De Laurentiis/Lorimar. **Pr** Debra Hill. **Dir** David Cronenberg. **Spfx** Jon Belyeu. **Screenplay** Jeffrey Boam. **Based on** *The Dead Zone* (*1979*) by Stephen KING. **Starring** Brooke Adams (Sarah Bracknell), Herbert Lom (Dr Sam Weizack), Martin Sheen (Greg Stillson), Tom Skerritt (Sheriff George Bannerman), Sean Sullivan (Herb Smith), Christopher Walken (Johnny Smith), Anthony Zerbe (Roger Stuart). 103 mins. Colour.

Johnny Smith is almost killed in a highway accident. Five years later he awakens from coma to discover himself possessed of precognitive/clairvoyant TALENTS. These he uses to help others and in due course to help identify a SERIAL KILLER. He comes into contact with Stillson, a corrupt demagogue who has his sights on the White House. Johnny "flashes" a future scenario in which a megalomaniac President Stillson unleashes WWIII. All his predictions have a "dead zone", comprising those bits that can yet be altered; sensing a dead zone in this one, he attempts to assassinate Stillson. Although he fails and is himself shot, a final "flash" tells him photographs of Stillson using a toddler as a shield against the bullets will be widely published and lead to Stillson's suicide.

TDZ represents an interesting middle ground between Cronenberg's cheaply produced exploitation movies and his more commercial work. In the later parts of the movie there is a definite sense of budgetary restraint; this clashes oddly with some richly – and presumably expensively – conceived scenes earlier. There is also an episodic feel to the movie, the section concerning the serial killer being unconnected with later elements. *TDZ* is enjoyable, if occasionally irritating in that it pretends to a profundity it does not possess.　　　[JG]

DEAN, PAMELA (1953-　　) Working name of US writer Pamela Dyer-Bennett, a founding member of the SCRIB-BLIES. Her CHILDREN'S-FANTASY **Hidden Land** trilogy – *The Secret Country* (**1985**), *The Hidden Lands* (**1986**) and *The Whim of the Dragon* (**1989**) – is a DYNASTIC FANTASY that takes the LANDSCAPE of FANTASYLAND and uses it subversively. A group of young COMPANIONS find the roleplaying GAME they have played for many years becoming real. *Tam Lin* (**1991**) is a nearly quintessential college novel, treating the college experience as a kind of transcendent OTHER-WORLD, realizing the lonely intellectual's fantasy of finding companions exactly like oneself and upping the ante by gradually introducing the supernatural element of the Elvish court drawn from the ballad TAM LIN.

Despite its more modest compass, *The Dubious Hills* (**1994**) is more ambitious. It takes place in a POLDER in the same landscape as the **Hidden Land** trilogy, and describes a peculiar experiment in PERCEPTION: a group of WIZARDS, intending to eliminate war, cast a SPELL over the inhabitants of the polder so that they are innocent of much human knowledge; for example, Arry, the protagonist, is the only one who can feel pain (and she feels everybody's). It is a difficult concept to convey, particularly within the villagers' context, but PD manages with subtlety and clarity. When SHAPESHIFTERS invade and try to wake the villagers from their SPELL there is a philosophical dilemma: are the inhabitants better off with the spell or without it?　　　[DGK]

DEAN, ROGER (1944-　　) UK illustrator whose work as a commercial designer shows in his sf and fantasy commissions, where he is known for paintings whose meticulous technical proficiency gives substance to images that are often extravagant, sometimes learnedly surreal. In the 1960s and 1970s he specialized in album covers, combining SUR-REALISM with glib competence; in fantasy ILLUSTRATIONS – where some narrative content can be assumed to substantiate the leaps of visual imagination – this smoothness was transformed into dreamlike visions in which deeply strange things seem imminent. Much of RD's early work appears in *Views* (graph **1975**). He won a WORLD FANTASY AWARD in 1977 as Best Artist.

Later work appears in *Magnetic Storm* (graph **1984**) with his brother, Martyn Dean, *The Flights of Icarus* (graph **1987**) with Donald Lehmkuhl, and in cover illustrations like that for Colin GREENLAND's *Other Voices* (**1988**).　　　[JC]

DEATH Death, as a character, has its origins in FOLKLORE and RELIGION; he is the King of Terrors, the skeleton man dressed in a black shroud and carrying an hourglass and a scythe; he is, when mounted, the rider of a Pale Horse and as such one of the FOUR HORSEMEN of the APOCALYPSE. In Romantic poetry he is the seducer of young maidens, the rocker of doomed cradles and Field Marshal Death reviewing his troops. He is the Grandfather Death of Russian legend, who may take a fancy to individuals and give them prolonged life and the capacity to cheat him; he is at once the last best friend in German poetry and the ender of all delights in *The Arabian Nights* (◊ ARABIAN FANTASY). He is a pre-eminent version of the LIMINAL BEING.

This traditional version of Death crops up in the movies, notably in Ingmar BERGMAN's *The* SEVENTH SEAL (**1956**) – parodied in Woody Allen's *Love and Death* (**1975**), BILL & TED'S BOGUS JOURNEY (**1991**) and LAST ACTION HERO (**1993**). A rather different image appears in Jean COCTEAU's

ORPHÉE (**1949**) and *Le* TESTAMENT D'ORPHÉE (**1959**).

Death as a character in fiction is usually subjected to some sort of REVISIONIST FANTASY. A comparatively unalloyed version crops up regularly in the stories of Fritz LEIBER, particularly in the **Fafhrd and Gray Mouser** tales, where he is an adversary whom the pair regularly evade, deceive or rob and who periodically attempts supernaturally to kill them. Stories like "Gonna Roll the Bones" (1967) and "The Winter Flies" (1967) present the traditional Skeleton Man in glory; the etiolated protagonist of the sf novel *A Specter is Haunting Texas* (**1969**) masquerades as Death for political purposes; forces such as oil and electricity take on the role of the Dark Gentleman Caller in stories like "The Black Gondolier" (1964), "The Man who Made Friends with Electricity" (1962) and "A Piece of the Dark World" (1962).

In the homoerotic fantasies of Victorian, Edwardian and Georgian writers like Oscar WILDE, SAKI and E.M. FORSTER, Death is predictably a young, good-looking man of foreign or lower-class appearance. He is a myth of consummation, a way of permitting the act that was forbidden by criminal law. More often than not, he is not so much Death the Executioner as Death the Messenger, the psychopomp who conducts the SOUL away from the body's death to some other place. The image of Death as messenger is more recently associated with the idea of Death as a young girl or adolescent woman. Peter S. BEAGLE's "Come, Lady Death" (**1963**) is an early source for this trope. Death is also often referred to as a female in Russian legend – Shostakovitch's Symphony No. 14, a setting of poems about Death, was at one point to be entitled "She". Erwin Schulhoff's OPERA *Flammen* ["The Flames"] (**1932**) to a libretto by Max Brod (◊ Franz KAFKA) has a female Death as the one woman a DON JUAN cannot attain. The image of Death as a young girl was taken up by Neil GAIMAN in the **Sandman** graphic novels (graph **1990-6**) and ongoing associative graphic novels like *The Books of Magic* (graph **1992**) and *Death: The High Cost of Living* (graph **1994**).

Terry PRATCHETT uses a male Death in the **Discworld** series in what is superficially a less revisionist way than Gaiman's, but one which is actually at least as subversive. Pratchett's Death rides a pale horse, but it is called Binkie and likes sugar; he talks in a basso profundo, represented in capital letters; he carries a scythe and hourglass and dresses in black, but would much rather deputize to an apprentice – as in *Mort* (**1987**) – and take time off.　　　[RK]

See also: DANCE OF DEATH.

DEATH BECOMES HER US movie (**1992**). Universal. **Pr** Steve Starkey, Robert ZEMECKIS. **Dir** Zemeckis. **Spfx** Michael Lantieri. **Vfx** Ken Ralston, Industrial Light & Magic. **Special body fx** Alec Gillis, Tom Woodruff Jr. **Makeup design** Dick Smith. **Prosthetic makeup sv** Kevin Haney. **Screenplay** Martin Donovan, David Koepp. **Starring** Goldie Hawn (Helen Stark), Isabella Rossellini (Lisle Von Rhuman), Meryl Streep (Madeline Ashton), Bruce Willis (Ernest Menville). 104 mins. Colour.

A comedy in which various rich socialites are sold by the Wagnerian Von Rhuman the ELIXIR OF LIFE, part of their CONTRACT being that, after they have enjoyed a decade of youth in the public eye, they must live out the remainder of their near-IMMORTALITY in seclusion in case the secret should leak out – which is why, for example, Elvis Presley is still occasionally sighted. Ashton and Stark have always loathed each other, a loathing enhanced when Ashton steals and weds Stark's fiancé Menville. Their attempts to murder each other achieve only mutilation, and at last, grudgingly, they join forces. The moral comes in AD2022, at Menville's funeral: he has achieved immortality through his children, adopted children and charitable works, while Ashton and Stark are reduced to squabbling over a last can of flesh-coloured touch-up paint. Although slight, *DBH* handles its themes with flair and pace, and the spfx, achieved in large part through sophisticated computer graphics, are sensational. [JG]

DEATHREALM US small-press MAGAZINE, Spring 1987-current, digest format until *#14* (1991) and thereafter large-format, currently published by Malicious Press, Greensboro, North Carolina; ed Stephen Mark Rainey (1959-   ).

Initially subtitled *The Gate Where Horror Begins*, later *The Land Where Horror Dwells*, *D* publishes mostly HORROR fiction, both psychological and supernatural, and occasionally physical, though not to the extremes of Splatterpunk. Initially *D* made some attempts to bridge the gap between the followers of H.P. LOVECRAFT's fiction and more recent free-form horror. In early issues its fannish origins were still evident, although with glorious exceptions like the bibliophobic "The Encyclopedia for Boys" by Jeffrey Osier (1954-   ) in *#1*. By *#8* (1989) *D* was placing greater emphasis on nightmares and paranoia, the stories exploring the tensions of the mind with the world around, remaining true to the Lovecraftian roots while exposing itself to modern horrors. The change to larger format allowed better presentation of artwork, especially that by Alan Clark. The switch to a new publisher, Tal Publications, with issue *#18* (Summer 1993), saw imposed a regime of more graphic horror, and it has only been since the move to a third publisher, Lawrence WATT-EVANS's Malicious Press, from *#23* (Spring 1995), that *D* has been able to shift to the occasional subtle horror, as it attracted more work by professional writers including Ramsey CAMPBELL, William F. NOLAN, Karl Edward WAGNER and Manly Wade WELLMAN, as well as material by Douglas Clegg (1958-   ), Ian McDowell (1958-   ), Robert M. Price (1954-   ) and Jessica Amanda SALMONSON. [MA]

DEBASEMENT In SUPERNATURAL FICTION, debasement tends to occur – and is normally visible – as a consequence of a degrading bargain made by a human with the forces of EVIL or of supernatural allure; it is something monstrous, and may well – as in Robert Louis STEVENSON's *Strange Case of Dr Jekyll and Mr Hyde* (**1886**) – manifest itself in terms of deformity. In HORROR, debasement tends to represent a defeat, a point when characters can no longer fight off the forces violating their integrity of being. For protagonists of a FANTASY tale, debasement is forgetting who you are (◊ AMNESIA), and what your STORY is.

Debasement occurs in innumerable ways, some particular to the fantasy genre: flouting of a PROHIBITION, refusing to recognize the WRONGNESS of a new STEPMOTHER or ruler or MAGUS, succumbing to SPELLS or METAMORPHOSIS or other forms of self-alienating BONDAGE.

Examples are easy to find. In Hope MIRRLEES's *Lud-in-the-Mist* (**1926**) Fairy Fruit debases schoolgirls, who forget who they are and are subsequently sold as slaves. In J.R.R. TOLKIEN's *The Lord of the Rings* (**1954-5**) the RING of power debases those who hold it, sucking their true nature dry, so that Frodo almost forgets who he is at the moment of climax; and the ring is destroyed only through the involuntary greed of Gollum, himself a debased relic of his former self.

In Mercedes LACKEY's *By the Sword* (**1991**), honourable warriors are magically transformed into killing machines; the BARD Eric, in Charles DE LINT's *The Little Country* (**1991**), having been drugged out of his true self, becomes a sexual slave.

When both the true protagonist and a PARODY or debased version of that protagonist are simultaneously present in a tale, that tale will normally represent a conflict for the SOUL. In supernatural fiction, the DOPPELGÄNGERS, DOUBLES, GHOSTS and SHADOWS who dramatize that conflict normally represent the debased half of the whole person. In fantasy, they often stand for that which has been lost, remind the protagonists they are debased, and call upon them to become whole again.

In the RECOGNITION scenes through which many fantasy tales climax, the shackles of debasement may dissolve and the metamorphosis be escaped (or finally take place). In Peter S. BEAGLE's *The Last Unicorn* (**1968**), that climax occurs on the beach near King Haggard's castle, where Molly Grue, viewing the scene, thinks the estranged protagonists resemble nothing more than painted playthings. But the moment passes, the cloak of self-alienation lifts, the characters become themselves again, and a world which has suffered THINNING and debasement shows immediate signs of Spring. [JC]

DE BEAUMONT, MARIE LEPRINCE ◊ Marie LE-PRINCE DE BEAUMONT.

DE BEAUVOIR, SIMONE (LUCIIE ERNESTINE MARIE BERTRAND) (1908-1986) French writer, famous for a wide variety of work. Her only novel of genre interest, *Tous les hommes son mortels* (**1946**; trans L. Friedman as *All Men Are Mortal* **1955** US), hovers between sf and fantasy in its scrutiny of the motives and opinions of a 13th-century character, Fosca of Carmona, who drinks the ELIXIR OF LIFE to make himself immortal (◊ IMMORTALITY), and regrets the decision for centuries. [JC]

DE BERNIERES, LOUIS (1954-   ) UK author of two novels of fantasy interest. *The War of Don Emmanuel's Nether Parts* (**1990**) is set in a LAND-OF-FABLE Latin American country riddled with corruption, SHAMANS and CATS who walk through walls. His second novel, *Señor Vivo and the Coca Lord* (**1991**), also a MAGIC-REALISM tale set in an imaginary Latin American country, even more extensively animates the world with TALKING ANIMALS, sorcerers and other forms of MAGIC. Underlying both tales are arguments about the political crisis of the continent, with specific reference to the astonishingly profitable drug trade. [JC]

DE BOSSCHÈRE, JEAN (1878-1953) Belgian poet, novelist, critic, painter, etcher and book illustrator. He is of most fantasy interest for his ILLUSTRATIONS, executed in a decorative line-and-black style influenced by Aubrey BEARDSLEY and the Decadents (◊ DECADENCE), but with a firmer, more fluid line and a feeling for the grotesque. JDB moved among the avant garde both in Brussels and, after 1915, in London: T.S. Eliot (1888-1965) and Ezra Pound (1885-1972) were friends. In addition to his often semi-abstract illustrations for his own works, he illustrated – in a more accessible manner – works by Honoré de BALZAC, BOCCAC-CIO, Miguel de CERVANTES and Gustave FLAUBERT. [RT]

DECADENCE Term borrowed from historians of the Roman Empire to refer to literary works selfconsciously symptomatic of a supposedly parallel phase in 19th-century European culture. Such works characteristically deal with, or are themselves aspects of, Neronic and neurotic quests for eccentric and extreme sensations capable of combating the dire effects of *ennui* and *spleen*. Charles BAUDELAIRE was the archetypal Decadent writer, although he took inspiration from the works of Edgar Allan POE (which he translated into French). Decadent writers were interested in all things abnormal, artificial, morbid, perverse and exotic, and were much given to symbolism; they were inevitably drawn to fantastic themes and bizarre stylistic embellishments, and their best work dramatically expanded the range, the bizarrerie and the grandiloquence of fantasy.

Key figures in the French Decadent movement of the FIN DE SIÈCLE include Paul Verlaine (1844-1896), Stéphane Mallarmé (1842-1898), Octave Mirbeau (1848-1917), Jean Lorrain (1855-1906), Joséphin Péladan (1859-1916), Rachilde (1860-1953), Rémy de GOURMONT, Joris-Karl HUYSMANS and Pierre LOUŸS. The short-lived UK Decadent Movement was nipped in the bud by the catastrophic fall from grace of its tacit leader, Oscar WILDE, but notable fantasy writers influenced by it include Arthur MACHEN, M.P. SHIEL, Vernon LEE, Count Stenbock (1859-1895) and R. Murray Gilchrist (1868-1917). German writers similarly influenced include Thomas MANN (briefly) and Hanns Heinz EWERS. US "Bohemian" writers like James Huneker (1860-1921), Lafcadio HEARN and Ambrose BIERCE had some affinity with the spirit of European Decadence; other US writers who produced Decadent fantasies include George S. VIERECK, Ben HECHT and Emma Frances Dawson (1851-1926). Clark Ashton SMITH, who was heavily influenced by the French Decadent poets as well as their US champion George Sterling (1869-1926), produced the ultimate examples of Decadent exotica, communicating something of that influence to other members of the H.P. LOVECRAFT circle and later writers in a similar vein, including such contemporary figures as Thomas LIGOTTI.

The central document of Decadent prose fiction is Huysmans' sarcastic comedy *À rebours* (**1884**), which was the "yellow book" that provided a guiding light for the hero of Wilde's ironic masterpiece *The Picture of Dorian Gray* (**1891**) and inspired the famous periodical *The Yellow Book* (1894-7), published by John Lane (1854-1925); Lane's **Keynotes** series included Machen's *The Great God Pan and the Inmost Light* (coll **1894**) and Shiel's *Shapes in the Fire* (coll **1896**). Further classics of French Decadent fantasy include Mirbeau's *Le jardin des supplices* (**1899**; trans Alvah C. Bessie as *Torture Garden* **1931** US; new trans Michael Richardson **1995** UK) and Lorrain's *Monsieur de Phocas* (**1901** France; trans Francis Amery **1994**). The most striking example of German Decadent fantasy is Ewers's *Alraune* (**1911**). US Decadence achieved its earliest twin peaks in Hecht's *Fantazius Mallare* (**1922**) and Smith's poem *The Hashish-Eater, or The Apocalypse of Evil* (**1922**; **1989** chap). Anthologies replete with Decadent fantasies include *The Dedalus Book of Decadence (Moral Ruins)* (anth **1990**) and *The Second Dedalus Book of Decadence: The Black Feast* (anth **1992**) both ed Brian STABLEFORD and *The Dedalus Book of German Decadence: Voices of the Abyss* (anth **1994**) ed Ray Furness. [BS]

DECAMERON ◊ BOCCACCIO; CLUB STORY; TRAVELLERS' TALES.

De CAMP, L(YON) SPRAGUE (1907-   ) US writer, married from 1939 to Catherine Adelaide Crook De Camp (CACDC); it has been increasingly been recognized that the two have been a creative team for almost all of LSDC's career. LSDC has been known – from the beginning of that career in 1937, when he published "The Isolinguals" in

Astounding Science Fiction – as an author of both sf and fantasy, much of the latter being SWORD AND SORCERY set in PLANETARY-ROMANCE venues and hard at times to distinguish from his sf adventures set on floridly conceived space-opera planets.

As a writer of fantasy, LSDC was associated with UNKNOWN from its start in 1939, and the RATIONALIZED FANTASIES he wrote for the journal, often in collaboration with Fletcher PRATT, precisely accorded with editor John W. Campbell Jr's penchant for strategies of fantasy in which the irrational or the exorbitant could be controlled, explained, rendered amusingly harmless. Such fantasy – in which METAMORPHOSIS tends to be comic, and in which strange belief-systems are subject to Yankee scrutiny – can at times remarkably resemble lazy sf; and not all of LSDC's work escapes a slapstick flatness, especially in later years when the flattened affect of early heroes like Harold Shea (see below) could no longer be papered over by double-takes.

The best of LSDC's fantasy is contained in the **Incomplete Enchanter** or **Harold Shea** sequence (all but the 1990s titles with Pratt): *The Incomplete Enchanter* (1940 *Unknown*; fixup **1941**; vt *The Incompleat Enchanter* 1979 UK) with Pratt; *The Castle of Iron* (1941 *Unknown*; **1950**) with Pratt; *Wall of Serpents* (fixup **1960**; vt *The Enchanter Compleated* 1980 UK) with Pratt; *Sir Harold and the Gnome King* (**1991** chap); *The Enchanter Reborn* (anth **1992**) ed with Christopher STASHEFF and *The Exotic Enchanter* (anth **1995**) ed with Stasheff. *The Incomplete Enchanter* and *The Castle of Iron* were assembled as *The Compleat Enchanter: The Magical Misadventures of Harold Shea* (omni **1975**), and with *Wall of Serpents* were eventually assembled as *The Intrepid Enchanter: The Complete Magical Misadventures of Harold Shea* (omni **1988** UK; vt *The Complete Compleat Enchanter* 1989 US). This is a playful plethora of titles for the reader to cope with; but it boils down to relatively few individual stories, in each of which Shea is catapulted via PORTAL into a different ALTERNATE WORLD, each being either a world of MYTH or a non-mimetic environment created by an individual author. LSDC makes no distinction between the REALITY levels that might seem to differentiate the relics of a belief system and the world of a writer.

The Incomplete Enchanter assembles two stories. In "The Roaring Trumpet" Shea visits the world of the Norse myths (◊ NORDIC FANTASY), and in "The Mathematics of Magic" (a title delicately attuned to Campbell's sensibilities, and perfectly expressive of the terms under which MAGIC found easy mention in *Unknown*) he enters, through another skewed portal, the world of Edmund SPENSER's *The Faerie Queene* (**1590**). In *The Castle of Iron* Shea moves deeper into the mysteries of being, entering Lodovigo ARIOSTO's *Orlando Furioso* (**1516**), a text which is a kind of ARCHETYPE for Spenser; and indeed he here finds himself facing a sort of problem in practical ontology, as the Belphebe with whom he fell in love in Spenser's world turns out to be an AVATAR (of sorts) of the original Belphegor from Ariosto. *Wall of Serpents* assembles two further tales: "The Wall of Serpents" (1953 *Fantasy Fiction*), in which Shea arrives in the world of the KALEVALA, and "The Green Magician" (1954 *Beyond*), in which he visits the world of CELTIC FANTASY and meets CUCHULAIN.

Other LSDC/Pratt collaborations include *The Land of Unreason* (1941 *Unknown*; **1942**), whose hero is taken by denizens of FAERIE to a land ruled by OBERON where, after suffering TRANSFORMATIONS, he emerges as a HIDDEN

MONARCH, Frederick Barbarossa, and *The Carnelian Cube* (**1948**), whose hero is himself – by virtue of having discovered a particularly proactive PHILOSOPHERS' STONE – responsible for various transits into a series of WONDERLANDS, satirically conceived as places where various kinds of logic run amok. The other main LSDC/Pratt collaborations – the CLUB STORIES assembled as *Tales from Gavagan's Bar* (coll **1953**; exp 1978) – are fresh, inventive and unforcedly comic.

Solo, LSDC wrote several further tales for *Unknown*, some – like "The Wheels of If" (1940 *Unknown*) – being sf. Fantasy stories include "The Undesired Princess" (1942 *Unknown*) – republished in *The Undesired Princess and The Enchanted Bunny* (anth **1990**) with a second story by David A. DRAKE – the hero of which enters yet another SECONDARY WORLD, and *Solomon's Stone* (1942 *Unknown*; **1957**), whose protagonist – possessed by a DEMON – travels to a reality whose inhabitants are dream versions of their earthly selves. After *Unknown*'s demise, and after WWII, LSDC concentrated for some time on sf, although the title story of *The Tritonian Ring and Other Pusadian Tales* (coll **1953**) is a fantasy of novel length. In the late 1950s he began once again to focus on fantasy, concentrating initially on completing works by Robert E. HOWARD (*whom see for details*). Fantasy remains his main field.

Later non-CONAN fantasies include the **Novaria** sequence: *The Goblin Tower* (**1968**), *The Clocks of Iraz* (**1971**) and *The Unbeheaded King* (**1980**), all three assembled as *The Reluctant King* (omni **1985**), plus *The Honorable Barbarian* (**1989**) and a pendant set in the same world, *The Fallible Fiend* (**1973**). That world, a LAND-OF-FABLE Hellenistic Mediterranean, comprises a friendly venue for further rationalized-fantasy adventures. The **Incorporated Knight** sequence – *The Incorporated Knight* (**1987**) and *The Pixillated Princess* (**1991**), both crediting CACDC – carries on the tradition, with a hero who LEARNS BETTER from his misadventures with DRAGONS and the like.

Throughout his extremely long career, a combination of acute intelligence and a flattening indifference to his created worlds has made LSDC, frustratingly, a writer impossible either to ignore or fully to celebrate. [JC]

Other works (fantasy): *The Reluctant Shaman and Other Fantastic Tales* (coll **1970**); *The Best of L. Sprague de Camp* (coll **1977**), some sf; *The Purple Pterodactyls: The Adventures of W. Wilson Newbury, Ensorcelled Financier* (coll of linked stories **1979**); *Footprints on Sand* (coll **1981**) with CACDC; *Heroes and Hobgoblins* (coll **1981**).

Other works (fantasy anthologies): *Swords and Sorcery* (anth **1963**); *The Spell of Seven* (anth **1965**); *The Fantastic Swordsmen* (anth **1967**); *Warlocks and Warriors* (anth **1970**).

Other works (historical fiction): *An Elephant for Aristotle* (**1958**); *The Bronze God of Rhodes* (**1960**); *The Dragon of the Ishtar Gate* (**1961**); *The Arrows of Hercules* (**1965**); *The Golden Wind* (**1969**).

Other works (sf): *Lest Darkness Fall* (1939 *Unknown*; exp **1941**), whose protagonist arrives in ancient Rome via a fantasy TIMESLIP; *Divide and Rule* (coll **1948**), whose title story has been published separately as *Divide and Rule* (1939 *Astounding*; **1990** chap dos); *The Wheels of If* (coll **1948**); *Genus Homo* (1941 *Super Science Stories*; **1950**) with P. Schuyler Miller (◊ SFE); *The Glory that Was* (1952 *Startling Stories*; **1960**); *A Gun for Dinosaur* (coll **1963**); *The Great Fetish* (**1978**), fantasy-like adventures in an sf frame; *The Venom Trees of Sunga* (**1992**); *Rivers of Time* (coll **1993**).

The Viagens Interplanetarias sequence: (nominally sf, but sometimes very close to fantasy), being *Rogue Queen* (**1951**), *The Continent Makers and Other Tales of the Viagens* (coll **1953**), some of the stories in *Sprague de Camp's New Anthology of Science Fiction* (coll **1953** UK), "The Virgin of Zesh" (**1953**) – which appears, with *The Wheels of If* (1940 *Astounding* **1990** chap dos), in *The Virgin and the Wheels* (coll **1976**) – *Cosmic Manhunt* (1949 *Astounding* as "The Queen of Zamba"; **1954** dos; vt *A Planet Called Krishna* 1966 UK; with restored text and with "Perpetual Motion" added rev vt as coll *The Queen of Zamba* 1977 US), *The Search for Zei* (1950 *Astounding* as the first half of "The Hand of Zei"; **1962**; vt *The Floating Continent* 1966 UK), plus *The Hand of Zei* (1950 *Astounding* as the second half of "The Hand of Zei"; **1963**; cut 1963) – both these latter titles superseded by publication of the full original novel, *The Hand of Zei* (1950 *Astounding Science Fiction*; **1982**) – *The Tower of Zanid* (1958 *Science Fiction Stories*; cut **1958**; with "The Virgin of Zesh" added vt as coll *The Virgin of Zesh/The Tower of Zanid* 1983), *The Hostage of Zir* (**1977**), *The Prisoner of Zhamanak* (**1982**), *The Bones of Zora* (**1983**) with CACDC and *The Swords of Zinjaban* (**1991**) with CACDC.

Nonfiction: *Lands Beyond* (**1952**) with Willy Ley, recipient of an INTERNATIONAL FANTASY AWARD; *The Science Fiction Handbook: The Writing of Imaginative Fiction* (**1953**; rev 1973 with CACDC); *Lost Continents: The Atlantis Theme in History, Science, and Literature* (**1954**) (◊ ATLANTIS); *The Conan Reader* (coll **1968**), *Blond Barbarians and Noble Savages* (coll **1975** chap) and *Literary Swordsmen and Sorcerers: The Makers of Heroic Fantasy* (coll **1976**), essays on fantasy; *Spirits, Stars, and Spells: The Profits and Perils of Magic* (**1966**) with CACDC; *The Day of the Dinosaur* (**1968**) with CACDC; *Scribblings* (coll **1972** chap); *Great Cities of the Ancient World* (**1972**); *Lovecraft: A Biography* (**1976**) (◊ H.P. LOVECRAFT); *The Miscast Barbarian: A Biography of Robert E. Howard* (**1975**), much augmented as *Dark Valley Destiny: The Life of Robert E. Howard* (**1983**) with CACDC and Jane Whittington Griffin; *The Ragged Edge of Science* (**1980**); *The Fringe of the Unknown* (coll **1983**).

DeCHANCIE, JOHN (1946-   ) US writer who worked in tv before beginning to publish printed fantasy and sf, and who has since done both. Of fantasy interest is the **Castle Perilous** sequence – *Castle Perilous* (**1988**), *Castle for Rent* (**1989**), *Castle Kidnapped* (**1989**), *Castle War!* (**1990**), *Castle Murders* (**1991**), *Castle Dreams* (**1992**), *Castle Spellbound* (**1992**) and *Bride of the Castle* (**1994**) – all episodes of which are centred on the eponymous EDIFICE. This occupies a POLDER of heightened LITTLE BIG reality at the heart of the worlds, somewhere outside normal space and time, and boasts 144,000 PORTALS to a similar number of worlds, all of which are, **Amber**-like (◊ Roger ZELAZNY), ontologically bound to the centre. Many of the conflicts in the sequence revolve around the castle's ruler, Lord Incarnadine, and his evil TWIN, who runs his own castle on lines which PARODY the good Incarnadine's methods. *MagicNet* (**1993**), a TECHNOFANTASY, features one protagonist whose personality is magically trapped inside a computer program and another who invades a surreal LOS ANGELES in order to defeat the enemy and justify his friend. [JC]

Other works (sf): The **Skyway** sequence, comprising *Starrigger* (**1983**), *Red Limit Freeway* (**1984**) and *Paradox Alley* (**1986**); *Crooked House* (**1987**) with Thomas F. MONTELEONE, horror; *The Kruton Interface* (**1993**); the **Dr Dimenson** sequence with David Bischoff, comprising *Dr*

Dimension (**1993**) and *Dr Dimension: Masters of Spacetime* (**1994**); *Living with Aliens* (**1995**).

DE COMEAU, ALEXANDER (? -) UK writer whose *Fires of Isis* (**1927**) made no impression but whose *Monk's Magic* (**1931**) is a QUEST tale that hovers interestingly between fantasy and SUPERNATURAL FICTION. Ordered to wander the world in search of the ELIXIR OF LIFE, the monk protagonist travels across a LAND-OF-FABLE medieval Europe with a young woman – whom he thinks to be a boy, and with whom he guiltily falls in love (◊ GENDER DISGUISE) – en route encountering GHOSTS, MAGIC and WITCHES, and visiting the land of the dead. [JC]

DEE, [Dr] JOHN (1527-1608) UK mathematician, astrologer, cartographer and Hermetic philosopher, who gained an early reputation through his invention in 1546 of a mechanical flying beetle for a production of ARISTOPHANES' *Peace*. His travels through Europe began as early as 1547, when he met (and was influenced by) Gerardus Mercator (1512-1594); later (*c*1580) he created for the Crown deeply influential charts of newly discovered countries. He drew horoscopes for Mary Tudor after she became queen in 1553, and gave Elizabeth I astrological advice as to the best date for her coronation ceremony in 1559 (◊ ASTROLOGY). His interest in Hermetic philosophy grew throughout his life, and he published prolifically, beginning with *Monas Hieroglyphica* ["The Hieroglyphic Monad"] (**1564** Antwerp), an occult dissertation (◊ OCCULTISM) on the nature of mathematical form as understood through the CABBALA; he proposed that a single hieroglyph (which represented all the planets in an astrally significant shape) reflected the "monas" (or Oneness) of the world, and that through this hieroglyph the mind might gain some sight (as through a MIRROR) of a PORTAL into that Oneness, which is HEAVEN. The text was an important influence on ROSICRUCIANISM.

In 1581 he became associated with Edward Kelley, who undertook to scry for him (◊ SCRYING). JD had for years been profoundly convinced it was possible to converse with ANGELS, and relievedly claimed in 1582 that an angel had visited him (through Kelley's intercession), presenting him with a black stone which served as a MIRROR through which other spheres could be accessed (JD's Mirror or "shew stone" remains in the British Museum). His "experiences" under Kelley's influence are recorded (unreliably) in *A True and Faithful Relation of What Passed for Many Years Between Dr John Dee and Some Spirits* (**1659**) ed Meric Casaubon; a modern text is *John Dee's Actions with Spirits: 22 December 1581 to 23 May 1583* (2 vols **1988**) ed with commentary by Christopher Whitby.

In 1583 JD was warned of a plot against his life, and fled to Poland and later PRAGUE. In 1589, JD and Kelley separated; JD returned to England, where he died old.

JD is an UNDERLIER figure for fantasy writers creating MAGI, a LIMINAL BEING whose appearance in any text signals the nearness of a THRESHOLD between this world and the Real World. Like so many speculative thinkers of the 16th and 17th centuries, he was what Arthur Koestler (1905-1983) called a Sleepwalker – one of the visionaries who created the modern measurement-governed world through their attempts better to explore the prescriptive contours of the old. Because JD stands on the cusp of worlds, and because of the passion with which he attempted to arrive at the truth behind the sleep of matter, he was treated as a precursor figure by THEOSOPHY, and has attracted the attention of several 20th-century authors. There is no evidence that JD ever met Giordano Bruno (1548-1600), though Bruno certainly knew JD's work; nor is there evidence that he met William SHAKESPEARE, though he tutored Fulke Greville (1554-1628), who declared himself "master to William Shakespeare", and it has more than once been suggested that the character of PROSPERO is a portrait of Dee. In *Shakespeare and the Goddess of Complete Being* (**1992**), Ted HUGHES argues that Shakespeare must have been familiar with the work of both Dee and Bruno.

Of the 20th-century responses to JD, Marjorie BOWEN's *I Dwelt in High Places* (**1933**) is a mundane tale of little interest, but Gustav MEYRINK's *Der Engel Vom Westlichen Fenster* (**1927**; trans Mike Mitchell as *The Angel of the West Window* **1991** UK) makes effective use of JD's Prague years in a complex tale involving a contemporary man haunted by JD. JD makes an appearance in Michael MOORCOCK's *Gloriana, or The Unfulfill'd Queen* (**1978**; rev 1993), and his Mirror haunts a contemporary investigator in Simon Rees's *The Devil's Looking-Glass* (**1985**). The most sustained fictional investigation of JD's actual life and purported findings can be found in John CROWLEY's **Aegypt** sequence, where he is the subject of at least two fictional BOOKS written by Fellowes Kraft, each of which is quoted from at very considerable length, and in which JD and Bruno do meet. JD is important to Patrick HARPUR's *Mercurius, or The Marriage of Heaven & Earth* (**1990**), and the contemporary protagonist of Peter ACKROYD's *The House of Doctor Dee* (**1993**) comes to believe that he embodies the spirit of a HOMUNCULUS created by the MAGUS. [JC]

DEEPING, (GEORGE) WARWICK (1877-1950) UK writer of much popular fiction. His genre works include *Uther & Igraine* (**1903**), which approaches the Arthurian Cycle (◊ ARTHUR) through a tale describing the relationship between his parents. Other Arthurian tales include *The Man on the White Horse* (**1934**); *The Man who Went Back* (**1940**), a TIMESLIP story in which an Englishman fighting in WORLD WAR II finds himself in the mind and body of a Roman Briton named Pellias, and rallies his people against the Germanic invaders until he is knocked unconscious and returns to the present battle; and *The Sword and the Cross* (**1957**), which incorporates appearances by a fabulated Artorius. *I Live Again* (**1942**), WD's only fantasy not to treat the MATTER of Britain through Arthur, has a plotline involving multiple REINCARNATIONS. [JC]

DEFENDING YOUR LIFE US movie (*1991*). Geffen/Warner. **Pr** Michael Grillo. **Exec pr** Herb Nanas. **Dir** Albert Brooks. **Screenplay** Brooks. **Starring** Brooks (Dan), Lee Grant (Lena Foster), Meryl Streep (Julia), Rip Torn (Bob Diamond). 111 mins. Colour.

Ad executive Dan, killed in a car crash, finds himself in the AFTERLIFE: Judgment City is like a resort comprised solely of luxury hotels. Here SOULS are judged in a form of court hearing to find if they have conquered their fears and can move on to a higher plane; if not, they face a further REINCARNATION on Earth. Dan eventually admits his life was one of pusillanimity and evasion, and looks set to graduate; but on his last night in Judgment City he declines to bed fellow-soul Julia for fear of blowing his chances, and the prosecutor pounces on this new proof of his frailty . . .

This POSTHUMOUS FANTASY, vacillating between comedy and SATIRE, has its longueurs, but its view of the afterlife as a place of Holiday Inns and Californian tourist attractions, among which adversarial psychoanalysts construct rival interpretations of a soul's prior actions, has refreshing plausibility. [JG]

DEFINITIONS OF FANTASY ◊ FANTASY.

DEFOE, DANIEL (1660-1731) English writer, pamphleteer, merchant and occasional secret agent, born Daniel Foe, best-known for *Robinson Crusoe* (**1719** 2 vols; ot *The Life and Strange Surprizing Adventures of Robinson Crusoe, of York* and *The Further Adventures of Robinson Crusoe*), regarded as the first true English novel. DD's prolific output, begun *c*1691, consisted primarily of political commentary, which earned him both keen supporters and dangerous enemies. He was imprisoned in 1702-3 for his views on government religious reform, and it was upon his release that he changed his name to Defoe and turned to writing for his living. Although he continued with political SATIRE and social observation, he drifted towards fiction, first with *The Consolidator, or Memoirs of Sundry Transactions from the World of the Moon* (**1705**), a form of proto SCIENCE FICTION, and then with *A True Relation of the Apparition of One Mrs Veal, the Next Day After her Death, to One Mrs Bargrave, at Canterbury, the 8th of September 1705* (**1706** chap), one of the earliest fictional GHOST STORIES – though apparently based on a true incident. He wrote a similar piece of reportage about a butler's experiences with some mischievous GOBLINS in *The Friendly Daemon, or The Generous Apparition* (**1726** chap; vt "The Devil Frolics with a Butler"). Anecdotes on hauntings and other manifestations are recounted in *The Political History of the Devil* (**1726-7** 2 vols), *A System of Magick, or A History of the Black Art* (**1727**) and *An Essay on the History and Reality of Apparitions* (**1727**; vt *The Secrets of the Invisible World Disclos'd, or An Universal History of Apparitions* 1729). A compendium was compiled by Carl Withers as *Tales of Piracy, Crimes and Ghosts* (coll **1945** US). [MA]

Other works: *Life and Adventures of Mr Duncan Campbell* (**1720**), concerning a mute conjurer (◊ MAGICIANS); *Captain Singleton* (**1720**); *Moll Flanders* (**1722**); *A Journal of the Plague Year* (**1722**), which arguably influenced Albert Camus (1913-1960); *The History of Peter the Great* (**1722**); *Colonel Jack* (**1722**); *Roxana* (**1724**); *Memoirs of a Cavalier* (**1724**); *A New Voyage Round the World* (**1724**); many others.

Further reading: *Realism, Myth, and History in Defoe's Fiction* (**1983**) by M.E. Novak; *Daniel Defoe: A Life* (**1989**) by Paula R. Backscheider.

DE HAAN, TOM Pseudonym of an unidentified UK writer (?1963-   ), whose *A Mirror for Princes* (**1987**) and *A Child of Good Fortune* (**1989**) are not inherently fantastic but are worth notice for – above their sustained, translucent beauty of telling – the acutely detailed and observed LAND-OF-FABLE kingdom of Brychmachrye in which they are set. [JC]

DE HAVEN, TOM (1949-   ) US writer whose work has been mostly sf, beginning with his first novel, *Freaks' Amour* (**1979**). *Funny Papers* (**1985**), however, creates a kind of COMMEDIA DELL'ARTE vision of NEW YORK, and the **Chronicles of the King's Tramp** sequence – *Walker of Worlds* (**1990**), *The End-of-Everything Man* (**1991**) and *The Last Human* (**1992**) – is a full and uninhibited fantasy. The first volume rather hectically conveys the eponymous walker, fittingly named JACK, and his SEVEN-SAMURAI group of COMPANIONS through a PORTAL into what seems to be a SECONDARY WORLD (called Lostwithal), where much conflict is necessary to save King Sad Agel's throne and for the moment to prevent a collapse of the walls between worlds (Lostwithal and all the other worlds in the series relate to one another as aspects of an underlying REALITY), a collapse which would generate universal CHAOS. The second volume sees a buildup of violence, and the destruction of the Epicene reality introduced earlier, and further exposition of a rigorous school of MAGIC. The third volume is set in an underlying meta-reality called the Undermoment which, acting as entrepot and focus for peoples displaying various TALENTS and for LANDSCAPES out of varying universes, can be perceived as a vast portal-riddled EDIFICE. A further volume would not come as a surprise, and would almost certainly increase the pleasure. [JC]

Other works (sf): *U.S.S.A. Book 1* * (**1987**); *Joe Gosh* (**1988**); *Sunburn Lake* (coll **1988**); *Neuromancer: The Graphic Novel: Volume 1* * (graph **1989**) illus Bruce Jensen; *Pixie Meat* (coll **1990**).

DEITZ, TOM Working name of US writer Thomas Franklin Deitz (1952-   ), whose **David Sullivan** sequence – *Windmaster's Bane* (**1986**), *Fireshaper's Doom* (**1987**), *Darkthunder's Way* (**1989**), *Sunshaker's War* (**1990**), *Stoneskin's Revenge* (**1991**) and *Dreamseeker's Road* (**1995**) – is, like almost all his work, CONTEMPORARY FANTASY set in Georgia. The plots are complex, often involving the teenaged Sullivan's intervention in the CROSSHATCH activities of the denizens of FAERIE; there is also, as in the not dissimilar work of writers like Catherine COOKE and Midori SNYDER, a tendency rather ruthlessly to intermingle elements of CELTIC FANTASY and Native American MYTH. TD's second sequence, the **Soulsmith** series – *Soulsmith* (**1991**), *Dreambuilder* (**1992**) and *Wordwright* (**1993**) – likewise features a teenaged protagonist, this time a TWIN with TALENTS who must battle the head of his family, which has a SECRET-MASTERS relationship to the surrounding mundane world; a sidebar character of some interest is a TRICKSTER figure known under various names.

The Gryphon King (**1989**), in a university setting, features a mummers play whose enactment raises SATAN. *Above the Lower Sky* (**1994**) is a near-future fantasy whose main characters – such as a SELKIE – CROSSHATCH with the real world. At points the complexity works, and gives hints that TD may significantly expand his remit. [JC]

DE LA FONTAINE, JEAN (1621-1695) French fabulist. ◊ FABLE; Charles PERRAULT.

DE LA MARE, WALTER (JOHN) (1873-1956) UK poet and writer who published actively from about 1895 to the end of his life. During the 1920s, when his reputation was at its height, he was deemed in the first rank of English writers; he is today seriously undervalued. From the beginning of his career he wrote in an elusive, nostalgia-drenched style which quite effectively concealed the true, cold, modern worlds at the heart of his work. He wrote voluminously – both as a poet (many hundreds of poems) and an author of short stories (about 100) – while also habitually publishing and republishing this large corpus, under different titles and in variously revised states.

Given the bibliographical tangles WDLM wove, there is little hope of presenting his work in anything like chronological order. His poetry has been assembled as *The Complete Poems of Walter de la Mare* (coll **1969**) ed Leonard Clark *et al.*, with helpful notes, and is perhaps best read thus – though WDLM's inveterate revisions, and the high quality of illustrations accompanying some original editions and reprints, mean that no individual volume can safely be ignored.

Of greatest fantasy interest are the numerous verses written ostensibly for children (◊ CHILDREN'S FANTASY). The subtlety of some of this work is remarkable. Individual volumes include *Songs of Childhood* (coll **1902**) as by Walter Ramal, *The Listeners* (coll **1912**), *A Child's Day: A Book of Rhymes* (coll **1912**), *Peacock Pie: A Book of Rhymes* (coll **1913**), *Down-Adown-Derry: A Book of Fairy Poems* (coll **1922**), *Stuff and Nonsense* (coll **1927**), *Poems for Children* (coll **1930**), *Bell and Grass: A Book of Rhymes* (coll **1941**) and *Collected Rhymes and Verses* (coll **1943**). Certain themes/terms which recur throughout his poetry (dream, snow, sleep, listener, journey, traveller, veil) almost invariably work as markers of WDLM's central territory, which might be described as the complex, shifting, ethereal and essentially boundless THRESHOLD between the BONDAGE of the material world and the world of DREAMS – which may be FAERIE in those poems written most clearly with children in mind, or more commonly perhaps a state, or geography, very close to death. A darkened rendering of the illimitable ins and outs of threshold appears in much of the adult poetry, most specifically – though least interestingly – in some of the more didactic long poems of his later years, like *The Traveller* (**1946** chap) and *Wingèd Chariot* (**1951**); and a sense that our mortal condition is one of BELATEDNESS – a sense common to many writers concerned with childhood – is given its most moving articulation in the last stanza of WDLM's greatest poem of desiderium, "All That's Past": "Very old are we men;/Our dreams are tales/Told in dim Eden/By Eve's nightingales;/We wake and whisper awhile,/But, the day gone by,/Silence and sleep like fields/Of amaranth lie."

WDLM began publishing stories of genre interest (as Walter Ramal) with "Kismet" for *The Sketch* in 1895; these early stories, posthumously assembled as *Eight Tales* (coll **1971** US), are entirely consistent with later stories in their basic attitude towards the world. Putting the numerous nonfantastic tales to one side, that attitude can be described as a sense that the mortal world is a kind of LABYRINTH. WDLM's SUPERNATURAL FICTIONS represent failures to penetrate that labyrinth – which is in this context a synonym for threshold – into a SECONDARY WORLD; his fantasies, almost invariably written as for children, are generally set entirely in a secondary world. The stories in *Eight Tales* run the gamut from "The Village of Old Age", whose protagonist is caught in an unchanging senescent BONDAGE, to "The Moon's Miracle", in which mortals on Earth are able, for a night, to observe a vast supranatural battle on the Moon.

Variously revised and recast in later editions, WDLM's mature supernatural fictions first appear in *Story and Rhyme* (coll **1921**), *The Riddle and Other Stories* (coll **1923**), *Two Tales: I. The Green Room; II. The Connoisseur* (coll **1925** chap), *The Connoisseur and Other Stories* (coll **1926**; cut vt *The Nap* 1936) – the title story is much revised from the version that appeared in the previous year's collection, but both versions are concerned with a mystical shedding of the lures of the material world – *On the Edge* (coll **1930**), *The Wind Blows Over* (coll **1936**) and *A Beginning and Other Stories* (coll **1955**). Further collections of adult stories – including *Seven Short Stories* (coll **1931**) and *Ghost Stories* (coll **1956**) – contain no new material.

"The Creatures", from *The Riddle and Other Stories*, is a tale whose narrator discovers in a valley an EDEN where children are sinless, but he cannot, of course, remain there; "Seaton's Aunt" (written 1909; 1922 *London Mercury*), from the same volume, is a tale of psychic (and possibly literal) vampirism

(◊ VAMPIRES). In "All Hallows", from *The Connoisseur and Other Stories*, a foul TRANSFORMATION is enacted upon a haunted cathedral by supernatural invaders who PARODY – EVIL is here seen as a parody of GOOD – the actual structure of the EDIFICE. There are several GHOST STORIES, including "A Recluse" (1926), "Crewe" (1929) and "The Green Room" (1925), from *On the Edge*, and "A Revenant", from *The Wind Blows Over*, features the shade of Edgar Allan POE. "The House", another from the latter volume, is a POSTHUMOUS FANTASY.

By contrast, in WDLM's fantasies there is no threshold over which to pass. Except for "The Riddle" (1903 *Monthly Review*; in *Story and Rhyme* and *The Riddle and Other Stories*), all WDLM's fantasies appear in two volumes, *Broomsticks and Other Tales* (coll **1925**) and *The Lord Fish* (coll **1933**). Various stories also appear in separate editions: *Miss Jemima* (in *Number One Joy Street* anth **1923**; **1925** chap; vt *The Story of Miss Jemima* 1940 chap US), *Lucy* (in *Number Two Joy Street* anth **1924**; **1927** chap), *Old Joe* (in *Number Three Joy Street* anth **1925**; **1927** chap) and *Mr Bumps and his Monkey* (in *The Lord Fish* as "The Old Lion"; **1942** chap US). All of WDLM's short stories for children subsequently appear, sometimes deleteriously revised, in four further collections: *The Old Lion and Other Stories* (coll **1942**), *The Magic Jacket and Other Stories* (coll **1943**), *The Scarecrow and Other Stories* (coll **1945**) and *The Dutch Cheese and Other Stories* (coll **1946**); these volumes serve as copy texts for *Collected Stories for Children* (coll **1947**) and for all subsequent editions, including *Selected Stories and Verses* (coll **1952**), *A Penny a Day* (coll **1960** US) and *The Magic Jacket* (coll **1962** US).

Notable stories from *Broomsticks and Other Tales* include: "The Three Sleeping Boys of Warwickshire", in which the boys undergo a decades-long magical sleep; "The Lovely Myfanwy"; and "Alice's Grandmother", in which IMMORTALITY is offered and – without WDLM's usual melancholy – turned down. *The Lord Fish* includes several remarkable tales. In the title story – first published in different form as "John Cobbler" (in *Number Four Joy Street* anth **1926**) – a young man finds a MERMAID kept in bondage by the Lord Fish in his strange castle, and frees her; "Dick and the Beanstalk" is a comic sequel to the FOLKTALE. In *Old Joe* variously retitled "Hodmadod" and "The Scarecrow", a scarecrow teaches a lesson about the passage of TIME.

WDLM's first novel, *Henry Brocken: His Travels and Adventures in the Rich, Strange, Scarce-Imaginable Regions of Romance* (**1904**; rev 1924), hovers strangely between the abstract and the particular. Brocken sets out on an allegorical journey (◊ ALLEGORY) to various booklands, where he meets characters like Jane Eyre, Nick Bottom, SLEEPING BEAUTY and Criseyde; but the mild educational benignity of the trip is destroyed, in the end, by Brocken's refusal of Criseyde's erotic needs, and he is cast back, defeated, into a dry mortality. *The Return* (**1910**; rev 1922; further rev 1945) is a supernatural fiction of some intensity, and with the same message: the protagonist is possessed (◊ POSSESSION) by the GHOST of a passionate man, becomes a more vital figure (◊ SHADOW) in his new guise, fails to consummate the love that has been held for centuries in anticipation of this moment, and returns to "life". Only in *The Three Mulla-Mulgars* (**1910**; vt *The Three Royal Monkeys* 1927) – arguably WDLM's single greatest fiction, and certainly one of the central ANIMAL FANTASIES of the 20th century – does he allow the world beyond the threshold full, unlimited, sensuous play. The land through which the three mulla-mulgars conduct their QUEST has some aspects of a LAND OF FABLE (a kind of indeterminate Southeast Asia) and some aspects of a full and mature SECONDARY WORLD (the talking monkeys inhabit an internally consistent environment, have a PANTHEON and MAGIC, and never leave their LAND, whose geography has no direct correspondence with the mundane world's). They are questing for their father, a mulgar of the Royal Blood who has returned to the Valleys of Tishnar, where his brother is king. The sights they see, and the adventures they undergo, are described with a remarkably delicate eye for the transfiguring detail. At novel's end, on the verge of entering Tishnar, they hear a call; their father, "a basket of honeycombs over his shoulder", approaches them from within the longed-for land.

Little of WDLM's work avoids a sense that what humans attempt to describe is either indescribable or lost. More important, WDLM does not depict any secondary land – or its shadow – in a manner that points to any religious backing or sanction. Unlike the stalwarts of CHRISTIAN FANTASY – such as George MACDONALD, who influenced him deeply – WDLM shows absolutely no evidence in his fiction that any form of secondary world can represent a higher, more real form of GOD's handiwork. In this sense, he is far more modern than J.R.R. TOLKIEN, or any of the fantasy writers who make secular use of derivatives of Tolkien's world (◊ FANTASYLAND). The transcendental call of WDLM's secondary worlds issues solely from within.   [JC]

Other works: *Memoirs of a Midget* (**1921**), not fantasy; *Ding Dong Bell* (coll **1924**; exp **1936**), fictionalized ruminations on graveyard inscriptions; *Told Again: Traditional Tales* (coll **1927**), *Stories from the Bible* (coll **1929**) and *Animal Stories: Chosen, Arranged and in Some Part Rewritten* (coll **1939**), three compilations of TWICE-TOLD tales; *A Froward Child* (**1934** chap).

DELANO, JAMIE (195?-   ) UK COMIC-strip writer of horror fantasy. His first work in comic books was a series of text stories featuring **Nightraven**, a pulp-style crimefighter who branded his victims; this appeared in *Daredevils* (1983), *The Mighty World of Marvel* (1983-4) and *Savage Sword of Conan* (1984-5). JD's first comics work was **Captain Britain** (1985; graph coll **1986**).

After a variety of strips and text stories for UK papers and annuals, he was invited by DC COMICS to write *Hellblazer* (1988-current), a then-new comic book starring Alan MOORE's creation **John Constantine**, a chain-smoking, streetwise Londoner well versed in the art of black MAGIC and a magnet for HORROR. Through a series of linked short stories which pitted him against a VOODOO boss, yuppie DEMONS, child-killers and football hooligans, Constantine has been shown by JD as haunted by his own demons, his past exploits having destroyed every relationship he has ever had. Following this highly successful first sequence of tales, JD took the character into more personal areas, journeying with groups of travellers, exploring New Age philosophy, paganism and ecological themes through a series of adventures which involved a psychic horror machine eventually defeated by mystical earth-power. JD continued the saga through SERIAL KILLERS, some lighter moments of literary fantasy, and exploration of Constantine's childhood, writing 37 stories in all (*Hellblazer #1-#24, #28-#31, #33-#40* 1988-92).

Other works include *World Without End* (6 issues 1990-91), an exploration of a FAR-FUTURE Earth made of living flesh, and *House of Cards* (graph **1991**) painted by David

Lloyd, a GRAPHIC NOVEL featuring **Nightraven**. More substantial has been his work on **Animal Man** (*#51-#79* 1992-4), in which JD returned to the theme of ecological horror. JD has shifted the emphasis of the series from crimefighting to studies in the power of Nature, the growing powers within Animal Man's daughter making her a useful, innocent voice to speak out against our inhumanity to those other creatures with whom we share our world.

[RT/SH]

DELANY, SAMUEL R(AY) (1942-   ) US writer and critic, from 1988 Professor of Comparative Literature at the University of Massachusetts, Amherst. His significance as an sf writer is very great; such a reputation seemed assured from the moment his first publication – *The Jewels of Aptor* (**1962** dos; original text 1968; rev 1971 UK) – appeared, even though its initial release had been savagely cut. *Jewels* may be comparatively crude, but in its linguistic exuberance, its Baroque turns of plot, and the mythopoeisis that consistently underlies (◊ UNDERLIERS) both character and plot, it clearly prefigured the great sf novels of SRD's early maturity. In these SRD's increasing use of CHILDE-like protagonists haunted by QUESTS gave a further fantasy tonality to novels whose arguments remained firmly sf-based.

Between those books and the late 1970s, when he began to publish fantasy, SRD became deeply involved in attempts to assimilate late-20th-century literary theories to sf, in particular applying various forms of semiotic analysis to its language and themes. Sf novels like *Triton* (**1976**), and the fantasy tales which soon followed, work (almost, at times, it seems *only* work) as articulate explorations of theory; indeed, most incorporate critical appendices by virtue of which the fictional narratives explode outwards into dialogic encounters with the discourse that subtends, explains, confronts and sometimes eviscerates them. SRD's separately published work in literary theory appears in *The Jewel-Hinged Jaw: Notes on the Language of Science Fiction* (coll **1977**), *The American Shore: Meditations on a Tale by Thomas M. Disch – "Angouleme"* (**1978**), *Starboard Wine: More Notes on the Language of Science Fiction* (coll **1984**), *The Motion of Light on Water: Sex and Science Fiction Writing in the East Village 1957-1965* (**1988**; exp vt *The Motion of Light on Water: East Village Sex and Science Fiction Writing 1960-1965, with The Column at the Market's Edge* 1990 UK), *Wagner/Artaud: A Play of 19th and 20th Century Critical Fictions* (**1988** chap), *The Straits of Messina* (coll **1989**) – which mostly assembles comments on his own work, originally published as by K. Leslie Steiner – and *Silent Interviews: On Language, Race, Sex, Science Fiction, and Some Comics* (coll **1994**).

With the exception of two pornographic novels – *The Tides of Lust* (**1973**; rev vt *Equinox* 1994) and *Hogg* (**1995**) – SRD's fantasy is restricted to the **Nevèrÿon** sequence: *Tales of Nevèrÿon* (coll of linked stories **1979**), *Neveryóna, or The Tale of Signs and Cities* (**1983**), *Flight From Nevèrÿon* (coll of linked stories **1985**; rev 1989 UK) and *The Bridge of Lost Desire* (coll of linked stories **1987**; rev vt *Return to Nevèrÿon* 1989 UK). Set in a LAND-OF-FABLE prehistoric Mediterranean and depicting a time when barbarian cultures are beginning complexly to evolve the signs and physical accomplishments of civilization, the sequence both presents the kind of STORY which tends to adhere to such venues – Gorgik the barbarian, the main protagonist, undergoes various HEROIC-FANTASY ordeals – and subjects those stories to the intense irradiating glare and exposé of

SRD's critical armamentarium. In the same way that the Nevèrÿon world undergoes radical transformation, so the stories that conventionally tell us of that world must also become, as it were, *urbanized*: they must unpack their unconscious burdens, must become self-conscious and alienated, must become *civil* even though inflammatory.

In the first volume, "The Tale of Gorgik" follows Gorgik's slow hegira from ignorant barbarism through a complexly ambivalent and fetishized period of slavery into an understanding of the power (of words, of deeds, of change). A further story, "The Tale of Old Venn", incorporates a teller of stories that incorporate various lessons about change and the meaning of change. "The Tale of Small Sarg" introduces DRAGONS and uses the protagonist's experience of being Gorgik's slave and lover to examine SEX and power in all its aspects (eventually, Gorgik will become the Liberator who ends slavery in the land). In *Neverÿóna*, a young woman undergoes a RITE OF PASSAGE analogous to Gorgik's, and achieves emancipation through art. The last volume is darkened by an extraordinarily effective interplay between the imagined world, which is suffering from a plague, and the consequences of the AIDS epidemic in SRD's own NEW YORK. In the end, however, the last volume dives backwards into the heart of the sequence, without terminating any Story within it. The circling self-reflexivities of the text have no logical terminus.

One additional fantasy tale, *They Fly at Çiron* (**1993**), contains SRD's mutation and expansion of a story written in 1962 and published in *F&SF* in 1971, in collaboration with James Sallis; although its various narrative strands are more than competent, possibly – in describing three ideal ways of life, and locating them in three separate communites, after the fashion of mid-career Ursula K. LE GUIN – he may have rendered the novel as a whole fatally abstract.

Whether or not SRD's major fictions of the past 20 years, the **Nevèrÿon** books, are fantasy or a cauterizing simulacrum of fantasy matters little. The fascination of SRD's work, here and elsewhere, lies in the sense (and the heat) generated when Story mates with discourse. [JC]

Other works: *Heavenly Breakfast: An Essay on the Winter of Love* (**1979**); *The Mad Man* (**1994**), associational; *Atlantis: Three Tales* (coll **1995**), autobiographical fictions.

Further reading: *The Delany Intersection* (**1978**) by George Edgar Slusser; *Worlds Out of Words: The SF Novels of Samuel R. Delany* (**1979**) by Douglas Barbour; *Samuel R. Delany: A Primary and Secondary Bibliography* (**1980**) by M.W. Peplov and R.S. Bravard; *Samuel R. Delany* (**1982**) by Jane Branahan Weedman; *Samuel R. Delany* (**1985**) by Seth McEvoy.

DE LARRABEITI, MICHAEL (1937-   ) UK writer whose **Borribles** sequence – *The Borribles* (**1976**), *The Borribles Go for Broke* (**1981**) and *The Borribles: Across the Dark Metropolis* (**1986**) – describes a WAINSCOT society (the Borribles themselves, ELVES who resemble streetwise human children) in a LONDON whose populous, gargoyle-infested subterranean grimness has fittingly been likened to that of Charles DICKENS (◊◊ URBAN FANTASY); some QUESTS are undertaken, COMPANIONS gained (and cruelly lost), and battles fought. The sequence is remarkable for its threatening contemporaneity. *The Provençal Tales* (coll **1988**), nonfantasy, assembles traditional material with an air of the TWICE-TOLD. [JC]

DE LINT, CHARLES (HENRI DIEDERICK HOEFS-MIT) (1951-   ) Canadian musician and writer – born in the Netherlands to parents who emigrated to CANADA when he was four months old – who began to publish work of genre interest with "The Fane of the Gray Rose" in *Swords Against Darkness IV* (anth **1979**) ed Andrew J. OFFUTT, and who has become the most significant, and almost certainly the most prolific, Canadian fantasy author. Especially in early years, he published stories as by Tanuki Aki, Henri Cuiscard, Jan Penalurick, Cerin Songweaver and Wendelessen; as Samuel M. Key he has published three HORROR novels, *Angel of Darkness* (**1990**), *From a Whisper to a Scream* (**1992**) and *I'll Be Watching You* (**1994**).

CDL's first book, released like several other short texts through his own Triskell Press, was *The Oak King's Daughter* (**1979** chap), which began the loosely connected **Tales of Cerin Songweaver** sequence; further volumes, which with increasing proficiency articulated relationships between STORY and the (Celtic) MUSIC which evokes it, include *A Pattern of Silver Strings: A Tale of Cerin Songweaver* (**1981** chap), *Glass Eyes and Cotton Strings* (**1982** chap), *In Mask and Motley* (**1983** chap), *Laughter in the Leaves* (**1984** chap), *The Badger in the Bag* (**1985** chap), *The Harp of the Grey Rose* (**1979** as "The Fane of The Gray Rose"; exp **1985** US), *And the Rafters Were Ringing* (**1986** chap) and *The Lark in the Morning* (**1987** chap). After *The Calendar of Trees* (**1984** chap) illus Donna Gordon, a long poem about TREES cast in an Anglo-Saxon RIDDLE mode, CDL published his first full-length book, *The Riddle of the Wren* (**1984** US), a fantasy set mostly in a SECONDARY WORLD though accessible from this world through the conventional PORTAL, and derivative of J.R.R. TOLKIEN. CDL has since then generally eschewed full and untrammelled secondary-world settings. From *Moonheart: A Romance* (**1984** US) onwards, most of CDL's novels have been CONTEMPORARY FANTASIES, usually set in Ottawa (◊ CITY; URBAN FANTASY) and other Ontario locations, into which mundanity are woven (◊ CROSSHATCH) Celtic-tinged OTHERWORLDS (◊ CELTIC FANTASY). *Moonheart*, along with some associated tales – *Ascian in Rose* (**1987** chap US), *Westlin Wind* (**1989** chap US), *Ghostwood* (**1990** US) and *Merlin Dreams in the Moondream Wood* (**1990** *Pulphouse*: **1992** chap), all four assembled as *Spiritwalk* (omni **1992** US) – is set in an Ottawa crosshatched into an otherworld itself intricately interpenetrated by fleshly manifestations of various mythological traditions (◊ MYTHAGO). The novel, which involves complex interactions between its female protagonist and various Native American and Celtic mythagoes, is dominated by Tamson House, a large EDIFICE which surrounds a GARDEN at whose heart grows a version of the WORLD-TREE; Tamson House serves as a PORTAL to and from a variety of otherworlds, and features also in all the tales assembled in *Spiritwalk*.

In *Yarrow: An Autumn Tale* (**1986** US) the work-engendering DREAMS of a fantasy writer from Ottawa are stolen by a vampiric dream-thief; she must consequently undertake a QUEST, with some help from her friends, into the otherworld to gain back her creative juices. *Jack, the Giant-Killer* (**1987** US) makes up part of Terri WINDLING's series of TWICE-TOLD fairytales, and exposes Jacky Rowan (a female JACK from Ottawa) to adventures in FAERIE, in all of which she enjoys a TRICKSTER's expected success. Its sequel, *Drink Down the Moon* (**1990**), is not part of Windling's series. Both volumes are assembled as *Jack of Kinrowan* (omni **1995**). *Greenmantle* (**1988**), the last important book from CDL's early career, once again takes place in contemporary

Ontario, this time crosshatched with a wild FOREST that houses a "stagman" intermittently drawn into the mundane world by MUSIC.

Prolific, and suffering a tendency to put nearly indistinguishable protagonists through storylines themselves often very similar, CDL has written weak books; but two novels of the early 1990s represent a growing sureness with his material. *The Little People* (**1991**) complexly interweaves storylines set in a mundane Cornwall and in a turn-of-the-century LAND-OF-FABLE version of the English county; these two Cornwalls, and the intricate plot which connects them, are comprehended (i.e., narrated) through the act of reading various texts (◊ BOOKS) composed by the late William Dunthorn, author of *The Hidden People* (about a WAINSCOT race known as the Smalls), *The Lost Music* (in which music is seen to have engendered the STORY of the world) and *The Little Country* (which tells the intersecting tale of Cornwall as a land of fable).

Memory and Dream (**1994** US) is set in the imaginary CITY of Newford, which resembles Ottawa, and serves as the setting for several URBAN-FANTASY tales. *Uncle Dobbin's Parrot Fair* (1987 *IASFM*; **1991** chap US), *The Stone Drum* (**1989** chap), *Ghosts of Wind and Shadow* (**1990** chap), *Paperjack* (**1991** chap US) and *Our Lady of the Harbour* (**1991** chap US) are all assembled with other work as *Dreams Underfoot: The Newford Collection* (omni **1993** US); and *Mr Truepenny's Book Emporium and Gallery* (**1992** chap US), *The Bone Woman* (**1992** chap), *The Wishing Well* (**1993** chap US) and *Coyote Stories* (**1993** chap), are all assembled with other work as *The Ivory and the Horn: A Newford Collection* (omni **1995** US). In the original novel a successful painter – whose personality and work CDL expertly expounds – turns out to be capable through her art of creating mythagoes, who have a cruelly mixed effect upon the world.

In the war that fantasy authors with fluent tongues almost invariably must wage to avoid the lures of easy consolations, CDL has lost and won some battles. But he is a potent teller of tales, and each fresh book brings a promise of newness. [JC]

Other works: *The Moon is a Meadow: A Tale of Tam Tinkern* (**1980** chap); *De Grijze Roose* ["The Grey Rose"] (coll trans Johan Vanhecke et al **1983** Netherlands); *World Fantasy Convention 1984: Fantasy, An International Genre Celebration* (anth **1984**); *The Three Plusketeers and the Garden Slugs* (**1985** chap); *Mulengro: A Romany Tale* (**1985** US), a DARK FANTASY; *The Drowned Man's Reel* (**1988** chap); *Wolf Moon* (**1988** US), SUPERNATURAL FICTION in which a harper hounds a WEREWOLF; *Svaha* (**1989** US), sf; *Berlin* * (**1989** chap US), set in the SHARED-WORLD **Borderland** sequence ed Terri WINDLING; *Philip José Farmer's The Dungeon, #3: The Valley of Thunder* * (**1989** US) and *#5: The Hidden City* * (**1990** US); *The Fair in Emain Macha* (1985 *Space & Time #68*; exp **1990** dos US); *The Dreaming Place* * (**1990**), a Byron PREISS project; *Hedgework and Guessery* (coll **1991** US); *Café Purgatorium* (coll **1991** US) with stories, separately, by Dana Anderson and Ray Garton; *Into the Green* (**1993**); *The Wild Wood* * (**1994**) in the **Brian Froud's Faerielands** series, illus Brian FROUD.

DELUGE ◊ FLOOD.

DELUSION A delusion leads a person to believe the world is other than it is. Delusions may arise spontaneously or may be induced; the imposition of delusions is a standard trick of literary and stage hypnotism (◊ MESMERISM). Delusory spells used by FEMMES FATALES to attract victims

may be subsumed under the heading of GLAMOUR. Delusions are akin to DREAMS, but delusional fantasies and VISIONARY FANTASIES have quite different effects. Whereas visionary fantasies frequently leave the realm of the real far behind, delusional fantasies remain anchored to reality. The juxtaposition of mundanity and delusion is used to dramatize the limitations of the former and the vaulting optimism or misfortunate treachery of the latter; for this reason delusional tales resist classification as works of fantasy.

The archetypal delusional fantasy is Miguel de CERVANTES's *Don Quixote* (**1605-15**), which assaults the ideals of CHIVALRY. Notable modern delusional fantasies include *The Man with the Black Feather* (**1904**) by Gaston LEROUX, *The Return of Don Quixote* (**1927**) by G.K. CHESTERTON, *Mr Blettsworthy on Rampole Island* (**1928**) by H.G. WELLS and *Nile Gold* (**1929**) by John Knittel (1891-1970). A rich tradition of cinematic delusional fantasies extends from HARVEY (*1950*) through *The Ruling Class* (*1972*) and *The* FISHER KING (*1991*) to *Don Juan DeMarco* (*1995*).

Psychoanalytic literature is less rich in accounts of delusion than anecdotal wisdom supposes, but its concerns are reflected in numerous fictitious case-studies. Robert Lindner's "The Jet-Propelled Couch" (1955) and Rona Jaffe's *Mazes and Monsters* (**1981**) are examples in which imaginative fiction is charged with generating delusions. [BS]

DE MAUPASSANT, (HENRI RENÉ ALBERT) GUY (1850-1893) French writer, mainly of short stories, about a quarter of which may be classified as HORROR stories or CONTES CRUELS; of these about half are SUPERNATURAL FICTION. GDM was influenced by many writers, especially Edgar Allan POE. In his teens he was shown, by the poet Algernon Swinburne (1837-1909), a mummified hand. This image haunted GDM and he used it in "La Main Ecorchée" (1875; trans as "The Flayed Hand"; vt "The Withered Hand"), in which the hand of a WITCH strangles its owner. This was later revised as "La Main" (1883; trans as "The Englishman"). Although he wrote a few genuine GHOST STORIES, including "L'Apparition" (1883; trans as "The Apparition"; vt "The Spectre"; vt "The Ghost"; vt "The Story of a Law Suit"), most of his tales in this vein are psychological, questioning the narrator's sanity. These include "La Morte" (trans as "Was It a Dream?"), "Lui?" (1883; trans as "He?"), "Un Fou" (1884 *Le Figaro*; trans as "Was He Mad?"), and "Qui Sait?" (1890; trans as "Who Knows?"), his last story. His masterpiece was "Le Horla" (1886 *Gil Blas*; trans as "The Horla"), where the narrator becomes overwhelmed by some invisible being that drains his sanity like a psychic VAMPIRE. GDM's bibliography is extremely complicated. No single volume published during his lifetime contained solely supernatural fiction, and early translations included stories misattributed to GDM. The first thorough compilation was *The Complete Short Stories of Guy de Maupassant* (coll **1955** US) ed Artine Artinian. Arnold Kellett compiled a volume in French as *Contes du Surnaturel* (coll **1969** UK; trans rev vt *Tales of Supernatural Terror* coll 1972 UK and *The Diary of a Madman* coll 1976 UK; reassembled as *The Dark Side* coll 1989 UK). Kellett also compiled *A Night on the River and Other Strange Tales* (coll **1976** UK). [MA]

DEMETER ◊ GODDESS.

DEMONS Spiritual beings, usually of an EVIL disposition. The Greek *daimon* and the Latin *daemon* are morally neutral words, but the English equivalent was tainted by Christian belief, which insisted all MAGIC was evil and all WITCHCRAFT involved the instruction of demons. Similar beings are to be found in the FOLKLORE and RELIGIONS of virtually all societies, but those which are benign are more likely to be referred to in English translation by other names, including SPIRITS, FAIRIES and ELEMENTALS. Those retaining the "demon" label are often animistically associated with storms, deserts and other inimical aspects of Nature.

The pattern whereby new religions demonized the deities of those they replaced was set by Zoroastrianism, whose evil "devas" were the GODS of an earlier Indo-Iranian tradition. The Zoroastrian demonic hierarchy, headed by Ahriman, was subsequently replicated within the Christian Mythos (◊ DEVILS). Although Judaism was purged of its demons, the Old Testament contains several residual references to such probable demons as LILITH, Belial and Azazel, many of whom were recovered and revitalized as icons of evil and servants of SATAN by Christianity. Islam similarly recovered a demonic hierarchy headed by Iblis and staffed by various orders of GENIES derived from Arabic folklore.

Demons and humans who worship them, or evoke them by means of BLACK MAGIC, are staples of OCCULT FANTASY and HEROIC FANTASY, although it is rare for demons to become fully manifest and rarer still for them to figure as protagonists; L. Sprague DE CAMP's *The Fallible Fiend* (1973) and Miranda SEYMOUR's *The Reluctant Devil* (1990) are notable exceptions. As modern fantasy writers have become more interested in ANTHROPOLOGY, the range of demons employed in fantasies has widened considerably, with Native American mythology generating particular interest in the USA. The demons of Oriental mythology, only occasionally featured in Western fiction, are more prominent in the cinema of Japan and Hong Kong.

A belated attempt to recover the moral neutrality of the Latin spelling was made by Johann Kepler (1571-1630), who insisted that the "daemon" employed in his VISIONARY FANTASY *Somnium* (1634) derived from a word meaning "knowledge"; he was anxious to avoid charges of SATANISM. [BS]

DE MORGAN, WILLIAM (FREND) (1839-1917) UK writer; also a designer of pottery and stained glass, long involved with William MORRIS. His six novels came late in his life. Two are SUPERNATURAL FICTIONS of some interest. In *Alice-for-Short: A Dichronism* (1907) an orphan is rescued from poverty through proper interpretation of the behaviour of two GHOSTS who re-enact a murder, REVENANT-fashion. In *A Likely Story* a fairly unusual painter-and-model scenario is enacted. His sister Mary de Morgan (1850-1907) was a talented writer of FAIRYTALES; her best were collected as *The Necklace of Princess Fiorimonde* (coll 1880) and *The Windfairies* (coll 1900). [JC]

DE NEUVILLE, ALPHONSE MARIE (1835-1885) French painter celebrated for his battle scenes and pictures of military life. A pupil of Eugène Delacroix (1798-1863), he was awarded the Légion d'Honneur in 1881 for "The Company of Saint-Privat" and "The Despatch-Bearer". He is best remembered for his numerous ILLUSTRATIONS for three of Jules Verne's most popular **Voyages Extraordinaire**, notably the "Nautilus" drawing in *Vingt mille lieues sous la mer* (1870). He contributed 45 illustrations to *Autour de la Lune* (1872) and 80 illustrations for *Le tour de monde en quatre-vingt jours* (1874). [RD]

DENNING, TROY (1958-   ) US author of fantasy ties, including several volumes fitted to the **Forgotten Realms** fantasy-GAME sequence: one of the **Avatar Trilogy**, *Waterdeep* * (1989), as Richard Awlinson; *Dragonwall* * (1990); *The Parched Sea* * (1991) in the **Harpers** sequence; and the **Twilight Giants** sequence – *The Ogre's Pact* * (1994), *The Giant Among Us* * (1995) and *The Titan of Twilight* * (1995). For TSR's **Dark Sun** he contributed to the **Prism Pentad** sequence *The Verdant Passage* * (1991) – where a drastic THINNING is caused by an overuse of MAGIC – *The Crimson Legion* * (1992), *The Amber Enchantment* * (1992) and *The Cerulean Storm* * (1993). [JC]

DENNIS, IAN (1952-   ) Canadian writer. His ARABIAN FANTASY *Bagdad* (1985) and its direct continuation *The Prince of Stars* (1987) make great use of Eastern folklore, retold in an ornate oral storyteller's style; the many individual moral tales related by the characters are ultimately much more satisfying than the FRAME STORY, which is an adventure tale of the overthrow of the caliphate by a political/religious leader, and of a chase across Arabia. [DVB]

DERLETH, AUGUST W(ILLIAM) (1909-1971) US writer, editor and publisher who from his first publication of genre interest – "Bat's Belfry" (*WT* 1926) – was primarily involved as (a) an original creator of, (b) an editor of important early ANTHOLOGIES and collections which publicized the attractions of, and (c) a publisher of the premier SMALL PRESS involved in the release of HORROR fiction. He is as well known for his work in creating a posthumous career for H.P. LOVECRAFT as for any of his own publications, though he was remarkably prolific. In 1939 he co-founded ARKHAM HOUSE with Donald WANDREI, initially in order to publish Lovecraft; it is now the longest-surviving sf/fantasy/horror specialist press in the world. Little of AWD's own work is fantasy, though some of his short stories pastiche Lovecraft's CTHULHU-MYTHOS cycle. Short fiction of interest is assembled in *Someone in the Dark* (coll 1941), *Something Near* (coll 1945), *Not Long for this World* (coll 1948), *The Mask of Cthulhu* (coll 1958), *The Trail of Cthulhu* (coll 1962) and *The Watchers out of Time and Others* (coll 1974) – all Lovecraftian or constituting reworkings of Lovecraft material – *Lonesome Places* (coll 1962), *Mr George and Other Odd Persons* (coll 1963), *Colonel Markesan and Less Pleasant People* (coll 1966) – with Mark Schorer (1908-1977) – and *Dwellers in Darkness* (coll 1976). *The Lurker at the Threshold* (1945) is again Lovecraft-oriented. [JC]
Further reading: *August Derleth: A Bibliography* (1983) by Alison M. Wilson; *SFE*.

DE SADE, MARQUIS Generally used form of the name of Donatien-Alphonse-François, Comte de Sade (1740-1814), French writer and pornographer commemorated in the word "sadism". Among his fantasies of SEX, torture and self-destruction are *Justine, ou les malheurs de la vertu* (1791; trans Helen Weaver as *Justine, or The Misfortunes of Virtue* 1966) – adapted as a graphic novel by Guido CREPAX – and the unfinished *Les 120 journées de Sodome, ou l'école du libertinage* (written 1784; 1931-5 3 vols; trans Austryn Wainhouse and Richard Seaver in *The 120 Days of Sodom, and Other Writings* coll 1966), with its maniacal determination to catalogue and classify – and describe in relentless detail – exactly 600 perversions. MDS is occasionally invoked in fantasy as an incarnation of EVIL or emissary of SATAN; the HORROR MOVIE *The Skull* (1965), based on a story by Robert BLOCH, takes it for granted that MDS's skull will exert a corrupting influence on its owners, while Colin WILSON's *The Mind Parasites* (1967) hints at his POSSESSION by CTHULHU-MYTHOS entities. His repeated literary attempts to justify his desires are (fictionally) continued during TIMESLIP visits to this century in Jeremy REED's *When the Whip Comes Down: A Novel about de Sade* (**1992**). [DRL]

Further reading: *The Marquis de Sade: A Short Account of His Life and Work* (**1934**; exp vt *The Life and Ideas of the Marquis de Sade* 1953) by Geoffrey Gorer; essays on MDS in *The Bit Between My Teeth* (coll **1965**) by Edmund Wilson (1895-1972).

DESTINY ◊ FATE.

DESTROY ALL MONSTERS Japanese movie (*1968*). ◊ GODZILLA MOVIES.

DETECTIVE TALES ◊ WEIRD TALES.

DETECTIVE/THRILLER FANTASY The fantasy analogue of detective/thriller genre fiction: detections in which criminal puzzles are solved by reason, thrillers where miscreants are pursued *via* melodramatic revelation and confrontation, and mixtures of the two – but using fantasy motifs and, normally, a fantasy setting. OCCULT DETECTIVES, who tend to operate in the mundane world, are separately discussed.

An early detective example is Ernest BRAMAH's *The Moon of Much Gladness* (**1932**), which spoofs contemporary crime writers with its Absurdist hunt for the supposed stealer of a Mandarin's pigtail. John Dickson CARR used TIMESLIP fantasy to give his historical mysteries a modern viewpoint character. The **Lord Darcy** ALTERNATE-WORLD series by Randall GARRETT offers classic fair-play detection, with restrained use of fantasy elements: forensic MAGIC reveals roughly as much as science might have, and crimes are committed by mundane means despite magical red herrings – thus the "preservator", the equivalent of a deep-freeze ("The Muddle of the Woad" 1965 *Analog*), in which one ingenious accomplice keeps the body fresh, is fairly introduced in advance. Garrett also amusingly suggests that Great Detectives' typical author-guided intuition is itself a psychic TALENT. Barry HUGHART's **Master Li** CHINOISERIE series features Holmesian deductions with the authentic meretricious dazzle, turning on esoteric scholarship which readers cannot hope to match. Barbara HAMBLY's *The Witches of Wenshar* (**1987**) echoes Agatha Christie's formula of discovering the lone murderer in an isolated and dwindling group of suspects, which here is an informal school of magic.

Thrillers include G.K. CHESTERTON's *The Man who was Thursday* (**1908**), which grounds its metaphysical nightmare in a bizarre cadre of Scotland Yard detectives identified as "The Last Crusade" and briefed to sniff out WRONGNESS in society. More usually, thrillers are LOW FANTASY, like Anthony BOUCHER's comic "The Compleat Werewolf" (**1942** *Unknown*), whose inadvertent WEREWOLF hero tangles with a spy ring. Hapless private investigators stray behind the stage-set of REALITY in Robert A. HEINLEIN's "The Unpleasant Profession of Jonathan Hoag" (**1942** *Unknown*). One unusual short thriller, ghosted by Theodore STURGEON for Leslie Charteris (1907-1993), is "Dawn" (1949; vt "The Darker Drink" 1963), taking Charteris's PICARESQUE hero The Saint into another man's DREAM-world. Chandleresque private-eye fantasies include Glen COOK's lighthearted **Mr Garrett** stories and Simon HAWKE's unlikely exploits of a feline detective in *The Nine Lives of Catseye Gomez* (**1992**), part of his **Wizard of 4th Street** sequence.

A recent hybrid is the police-procedural fantasy thriller,

such as Terry PRATCHETT's **Discworld** farces featuring the harassed men (and others) of the Ankh-Morpork Night Watch: *Guards! Guards!* (**1989**), *Men at Arms* (**1993**) and «Feet of Clay» (1996), the last including an ingenious means of administering poison. Simon R. GREEN's **Hawk & Fisher** series, beginning with *No Haven for the Guilty* (**1990**; vt *Hawk & Fisher* US), stars a tough City Guard DUO whose swordplay is bad news for criminals and innocent bystanders alike. [DRL]

DEVER, JOE (1956-) UK writer of multiple-choice fantasy gamebooks and latterly of interactive computer GAMES; he won the US Advanced Dungeons & Dragons Championships in 1982. He is primarily known for the **Lone Wolf** series of SWORD AND SORCERY gamebooks, set in a FANTASYLAND called Magnamund, of which *#1-#8* were originally published as co-authorships with Gary Chalk (1952-), their illustrator: *#1 Flight from the Dark* (**1984**), *#2 Fire on the Water* (**1984**), *#3 The Caverns of Kalte* (**1984**), *#4 The Chasm of Doom* (**1985**), *#5 Shadow on the Sand* (**1985**), *#6 The Kingdoms of Terror* (**1985**), *#7 Castle Death* (**1986**), *#8 The Jungle of Horrors* (**1986**), *#9 The Cauldron of Fear* (**1987**), *#10 The Dungeons of Torgar* (**1987**), *#11 The Prisoners of Time* (**1987**), *#12 The Masters of Darkness* (**1988**), *#13 Plague-Lords of Ruel* (**1990**), *#14 The Captives of Kaag* (**1990**), *#15 The Darke Crusade* (**1991**), *#16 The Legacy of Vashna* (**1991**), *#17 The Deathlord of Ixia* (**1992**), *#18 Dawn of the Dragons* (**1992**), *#19 Wolf's Bane* (**1993**), *#20 The Curse of Naar* (**1993**), *#21 Voyage of the Moonstone* (**1994**), *#22 Buccaneers of Shadaki* (**1994**), *#23 Mydnight's Hero* (**1995**), *#24 Rune War* (**1995**), with «#25 Trail of the Wolf» (1996), «#26 Fall of Blood Mountain» (1996), «#27 Vampirium» (1996) and «#28 The Hunger of Sejanoz» projected. A tied series of novels was written by John GRANT; a graphic novelization was *The Skull of Agarash* * (graph **1994**) written by JD and drawn by Brian Williams; an appendage to the series was *The Magnamund Companion* * (**1986**) by JD and Chalk. Also set in Magnamund was the **World of Lone Wolf** series of gamebooks by JD and Ian Page (1960-), seemingly largely by Page: *#1 Grey Star the Wizard* * (**1985**), *#2 The Forbidden City* * (**1986**), *#3 Beyond Nightmare Gate* * (**1986**) and *#4 War of the Wizards* * (**1986**). JD also wrote the **Freeway Warrior** series of gamebooks, based in a post-HOLOCAUST USA: *#1 Highway Holocaust* (**1988**), *#2 Slaughter Mountain Run* (**1988**), *#3 The Omega Zone* (**1988**) and *#4 California Countdown* (**1989**). [JG]

DEVEREAUX, ROBERT (1947-) US writer whose short fiction, beginning with "Fructus in Eden" for *Pulphouse #9* (anth **1990**) ed Kristine Kathryn RUSCH, has tended to stylistically adventurous HORROR. His first novel, *Deadweight* (**1994**), also ostensibly horror, perhaps suffers through its GRAND-GUIGNOL depiction of a REVENANT's ornately unpleasant sexual practices; but the burden of the tale is a moderately explicit retelling of the Persephone myth. Having lost her fructifying capacity after her father rapes her, the protagonist undergoes a long NIGHT JOURNEY which leads eventually to both self-redemption and the salving of the world. [JC]

DEVIL AND DANIEL WEBSTER, THE (vt *All That Money Can Buy* UK; vt *Daniel and the Devil*; vt *Here is a Man*) US movie (*1941*). RKO. **Pr** William Dieterle. **Dir** Dieterle. **Spfx** Vernon L. Walker. **Screenplay** Dan Totheroh. **Based on** "The Devil and Daniel Webster" (1936) by Stephen Vincent BENÉT. **Starring** Edward Arnold (Daniel Webster), James Craig (Jabez Stone), Jane Darwell

(Ma Stone), Walter Huston (Mr Scratch), Anne Shirley (Mary Stone), Simone Simon (Belle), Lindy Wade (Daniel Stone). 106 mins. B/w.

Broke New England farmer Jabez Stone idly wishes he could sell his SOUL to the DEVIL for a bit of good luck, and the Devil, in the guise of Mr Scratch, appears to him offering a CONTRACT whereby he can have seven years of it. Initially sceptical, Stone changes his mind as he becomes rich, at the same time transforming from a kindly farmer into a cut-throat businessman. After the Devil's beautiful temptress servant Belle joins the household, Stone casts his own family out. Stone holds a ball in his mansion; it soon becomes more like *The* MASQUE OF THE RED DEATH (*1964*), and he realizes his bizarre, cackling guests are others who have sold their souls to the Devil. Fleeing to his wife Mary he begs forgiveness; she consults the famous lawyer Daniel Webster, who argues Stone's case against Mr Scratch before a jury selected from notorious US villains and traitors, winning the "case" through appealing to their patriotism. The Devil in fury curses Webster to political futility, a CURSE born out by history.

Arnold and Huston have plenty of fun with this enjoyable moral fable – as does Dieterle, whose highpoint this probably was. The photography – by Joseph August – adds to the movie's profound sense of atmosphere: it is one of those movies that exemplifies the power of b/w. The music (by Bernard Herrmann) has much the same effect, and was awarded an Oscar. [JG]

DEVIL AND MAX DEVLIN, THE US movie (*1981*). DISNEY. **Pr** Jerome Courtland. **Exec pr** Ron Miller. **Dir** Steven Hilliard Stern. **Screenplay** Mary Rodgers. **Novelization** *The Devil and Max Devlin* * (**1980**) by Robert Grossbach. **Starring** Susan Anspach (Penny Hart), Julie Budd (Stella Summers), Bill Cosby (Barney Satin), Elliott Gould (Max Devlin), Adam Rich (Toby Hart), Charles Schamata (Nelson). 95 mins. Colour.

Rotten landlord Devlin is run over by a bus and descends to HELL. There he is offered a deal by Barney, a suave MEPHISTOPHELES figure: Devlin will be returned to life for a probationary two months, during which time he must induce three youngsters to sell their SOULS to the DEVIL for reasons of greed. Stella wants to be a rock star; Nelson wants to be a motocross champ; Toby wants a new father to marry his widowed mother. As Barney explains, Devlin can confer on each of his three "targets" magical properties so long as he remains within line-of-sight of them; through his conferred TALENT of teleportation, Devlin somehow manages to become a Svengali to Stella, a successful manager to Nelson and a putative new spouse to Toby's mother Penny. Although he succeeds in getting the CONTRACTS signed, he has in the process discovered decency, and so burns them. Hell rejects him on the basis that his new-found *niceness* might corrupt the rest of the establishment, and all ends happily.

TDAMD has much more bite than the average DISNEY comedy of its period, thanks to a witty script and the seeming ad libs of Gould and Cosby. This is a far better movie than generally reported – even the songs are good. [JG]

DEVIL-DOLL, THE US movie (*1936*). MGM. **Pr** Edward J. Mannix. **Dir** Tod Browning. **Screenplay** Guy Endore, Garrett Fort, Eric von Stroheim. **Based on** *Burn, Witch, Burn!* (1932 *Argosy*; exp **1933**) by A. MERRITT. **Starring** Lionel Barrymore (Paul Lavond), Lucy Beaumont (Mme Lavond), Claire du Brey (Mme Coulvet), Pedro de Cordoba

(Charles Matin), Grace Ford (Lachna), Robert Greig (Emil Coulvet), Arthur Hohl (Radin), Frank Lawton (Toto), Maureen O'Sullivan (Lorraine Lavond), Rafaela Ottiano (Malita), Henry B. Walthall (Marcel). 79 mins. B/w.

Banker Lavond, swindled by his partners, has spent 17 years on Devil's Island. He escapes with archetypal mad scientist Marcel, who has developed a way of miniaturizing animals and even people; those miniatures, however, can function only under the telepathic guiding will of their animator. Marcel dies of a heart attack; Lavond goes with Marcel's equally mad assistant Malita to Paris, where he disguises himself as a toymaker, Mme Mandilip (◊ GENDER DISGUISE), and uses miniaturized people to exact his VENGEANCE on the swindlers. In Merritt's original story the miniaturization was effected by ALCHEMY; Browning chose to adopt a TECHNOFANTASY plot device instead, but the precise mechanism used is soon forgotten in what is, despite its HORROR-MOVIE pretensions, a charming, whimsical fantasy. The excellent spfx, uncredited, were reportedly achieved by a team headed by A. Arnold Gillespie. [JG]

DEVILMAN Japanese tv series (1972 onwards). ◊ ANIME.

DEVIL RIDES OUT, THE (vt *The Devil's Bride* US) UK movie (*1968*). HAMMER/Warner-Pathé/20th Century-Fox. **Pr** Anthony Nelson-Keys. **Dir** Terence Fisher. **Spfx** Michael Stainer-Hutchins. **Screenplay** Richard MATHESON. **Based on** *The Devil Rides Out* (**1935**) by Dennis WHEATLEY. **Starring** Niké Arrighi (Tanith), Paul Eddington (Richard), Gwen ffrangcon-Davies (Countess d'Urfe), Charles Gray (Mocata), Leon Greene (Rex Van Ryn), Rosalyn Landor (Peggy), Christopher Lee (Duc de Richleau), Patrick Mower (Simon Aron). 95 mins. Colour.

A significant cult horror movie, widely regarded as the best of the Hammer crop – and as a great improvement on Wheatley's novel. Aron, a young friend of the white occultist the Duc de Richleau, has become involved in a satanist coven run by the occultist Mocata (based by Wheatley on Aleister CROWLEY). The duc rescues Aron, but Mocato grabs him back using a SPIRIT emissary and prepares to baptize the youth alongside the girl Tanith on Mayday Eve. The duc and his assistant, Van Ryn, arrive in time to drive off the DEVIL and save the two innocents. Mocato then sends a giant SPIDER and the ANGEL of Death, and Tanith is killed. De Richleau retaliates using an incantation that can reverse TIME: the members of Mocata's coven are devoured by flames as time runs back far enough for Tanith once more to be alive.

This project was Christopher Lee's baby: he had been impressed by the novel and pressed Hammer to film it. The key scenes in the movie are the abduction of Aron and Tanith by Mocata, the Mayday Eve conjuration and banishment of the Devil, and the final confrontation between GOOD AND EVIL, represented respectively by our heroes and the Angel of Death, as Mocata sends every form of necromancy he can command against the Good COMPANIONS, who are protected merely by a pentacle.

Some years later, Hammer followed up with TO THE DEVIL A DAUGHTER (*1976*). [JG]

DEVILS Although notionally monotheistic, the Christian mythos not only has an anti-god, variously called SATAN, LUCIFER or simply the Devil, but an entire hierarchy of evil DEMONS mirroring the heavenly hierarchy of ANGELS and saints. Early CHRISTIAN FANTASY made much of the notion that there had been a war in HEAVEN which resulted in the

Devil being cast into HELL with a legion of followers. Some names applied in the Bible to the Devil – notably Beelzebub – were given by later writers to his henchmen, and the same fate overcame the Talmudic demon-king Asmodeus, who was converted into a relatively amiable figure by Alain René Le Sage (1668-1747) in *Le Diable boiteux* (**1707**). The register of minor devils was swelled by the names of the fallen angels listed in the apocryphal *Book of Enoch* and those of the deities – including Moloch, Baal and Astaroth – worshipped by the Hebrews' neighbours. Extensive lists can be found in Richard Bovet's *Pandaemonium* (**1684**) and Lauran Paine's *The Hierarchy of Hell* (**1972**).

Lesser devils often stand in for their master in tales of PACTS WITH THE DEVIL; some – like FAUST's MEPHISTOPHE-LES – have enjoyed long and notable literary careers. The most famous modern literary devil is the writer of *The Screwtape Letters* (**1942**) by C.S. LEWIS. Devils are conventionally depicted by illustrators as horned and bearded manikins with goat's legs and forked tails. [BS]

DEVILS, THE UK movie (*1971*). Warner/Russo. **Pr** Ken Russell, Robert H. Solo. **Dir** Russell. **Spfx** John Richardson. **Screenplay** Russell. **Based on** *The Devils* (play premiered 1961) by John Whiting (1917-1963) and *The Devils of Loudun* (**1952**) by Aldous Huxley (1894-1963). **Starring** Michael Gothard (Barré), Gemma Jones (Madeline), Christopher Logue (Richelieu), Murray Melvin (Mignon), Vanessa Redgrave (Sister Jeanne), Oliver Reed (Grandier), Dudley Sutton (Laubardemont). 111 mins. Colour.

Spared by a whim of Louis XIII, Loudon alone stands for religious tolerance in the France of the early 1630s; its priest, Grandier, is its symbol. To destroy this haven of liberality the State and Church must destroy Grandier, and he has made enemies enough that a charge of heresy would seem supportable: he conducts a dissolute lifestyle, and Mother Jeanne of the contemplative Order of the Angels hates him because she lusts for him. The State's catspaw, the Baron de Laubardemont, orchestrates the campaign: under the persuasion of witchfinder Barré, Grandier's assistant Mignon confesses (sincerely) that he believes Grandier possessed (◊ POSSESSION) by the DEVIL, while Jeanne's erotic fantasies are translated into visitations by an INCUBUS and, in terrified hysteria, her nuns cavort obscenely for lust of Barré himself (who resembles a rock star fighting off groupies). Grandier at last discovers his God through his love for a pure woman, but is burnt at the stake; only Mignon realizes at the last that in reality it is the CITY – not its bricks and mortar but its *self* – that is being consumed. This exemplary CHRISTIAN FANTASY explores how rigid-mindedness – in this instance worship-inspired – causes us to create GODS and DEMONS in our own image. [JG]

DEXTER, SUSAN (ELIZABETH) (1955-   ) US writer who began publishing fantasy with the **Winter King's War** sequence – *The Ring of Allaire* (**1981**), *The Sword of Calandra* (**1985**) and *The Mountains of Channadran* (**1986**) – set in a medievalized FANTASYLAND and initially featuring the SORCERER'S-APPRENTICE mishaps of a young WIZARD in training. Soon enough, though, a comic QUEST (with COMPANIONS) is underway, involving a variety of PLOT COUPONS until eventually the protagonist is proven a HIDDEN MONARCH. Later volumes continue in similar fashion, though the threat of a MAGIC-induced permanent winter adds some gravity. *The Wizard's Shadow* (**1993**), set in the same venue, avoids dynastic problems and works as a

modest HEROIC FANTASY. A second series is the **Warhorse of Esdragon** sequence – *The Prince of Ill Luck* (**1994**) and *The Wind-Witch* (**1995**) with further volumes projected.   [JC]

DIAMOND, GRAHAM (R.) (1945-   ) US writer whose **Haven** series – *The Haven* (**1978**), *Lady of the Haven* (**1978**), *Dungeons of Kuba* (**1979**), *The Falcon of Eden* (**1980**), *The Beasts of Hades* (**1981**) and *Forest Wars* (**1994**) – begins with a LAST-REDOUBT tale in which the Redoubt of All Humanity must preserve itself from the onslaughts of talking WOLVES whose TALENTS are explained in SCIENCE-FANTASY terms as a product of bio-engineering. Eventually the focus of the series shifts, concentrating on the female wolf-leader and her adventures in an increasingly PLANETARY-ROMANCE setting. The **Samarkand** sequence – *Samarkand* (**1980**) and *Samarkand Dawn* (**1981**) – is placed in a LAND-OF-FABLE East, where Samarkand itself serves as an entrepot for Asia's trade and, aided only by a certain amount of MAGIC, must defend itself from barbarian hordes. The **Marrakesh** sequence – *Marrakesh* (**1981**) and *Marrakesh Nights* (**1984**) – is also set in a land of fable, here a somewhat more pixillated ARABIAN-FANTASY venue. Of more interest are GD's first singletons, *The Thief of Kalimar* (**1979**), located in a Earth-like FANTASYLAND, and *Captain Sinbad* (**1980**), which demystifies its hero.   [JC]

Other works: *Cinnabar* (**1985**).

DIANA ◊ GODDESS.

DIBELL, ANSEN Working name of US writer Nancy Ann Dibble (1942-   ), whose work in the sf/fantasy field has been limited to the sequence known as **The Strange and Fantastic History of the King of Kantmorie**: *Pursuit of the Screamer* (**1978**), *Circle, Crescent, Star* (**1981**) and *Summerfair* (**1982**). The stories reveal a clear sf underpinning but the ornately described PLANETARY-ROMANCE setting is strongly fantasy-like.   [JC]

Other works: *Plot* (**1988**), nonfiction.

DICK, KAY (1915-   ) UK writer and editor, author of *They: A Sequence of Unease* (**1977**) and editor of three anthologies of fantasy and SUPERNATURAL FICTION: *The Mandrake Root* (anth **1946**) and *At Close of Eve* (anth **1947**), both ed as by Jeremy Scott, and *The Uncertain Element* (anth **1950**). Her nonfiction *Pierrot* (**1960**) is told as though the biography of an immortal creature (◊ COMMEDIA DELL'ARTE).   [JC]

DICK, R.A. Pseudonym of Josephine LESLIE.

DICKENS, CHARLES (JOHN HUFFHAM) (1812-1870) UK writer. He has a threefold importance for fantasy: (a) he created the CHRISTMAS BOOK; (b) he transmogrified LONDON into a profoundly evocative stage (◊ URBAN FANTASY) for the heightened acting out of human dramas; and (c) he wrote several significant GHOST STORIES.

Christmas Books Of CD's five Christmas Books, four are SUPERNATURAL FICTIONS, though the first and the last of these further transform traditional supernatural motifs into fantasy shape: they are *A Christmas Carol in Prose: Being a Ghost Story of Christmas* (**1843**), *The Chimes: A Goblin Story of Some Bells That Rang an Old Year Out and a New Year In* (dated 1845 but **1844**), *The Cricket on the Hearth: A Fairy Tale of Home* (dated 1846 but **1845**), *The Battle of Life* (**1846**) – the one that is not fantastic – and *The Haunted Man and the Ghost's Bargain: A Fancy for Christmas-Time* (**1848**). All five tales were assembled as *Christmas Books* (omni **1852**; vt *Christmas Stories* 1868 US; best version ed Michael Slater vt *The Christmas Books* 1971 2 vols UK); *A Christmas Carol* has been published in very many forms, cut, variously adapted, dramatized, and often filmed (◊ *A* CHRISTMAS CAROL).

Although technically a GHOST STORY (it can be read also as a fantasy of PERCEPTION), *A Christmas Carol* powerfully conveys a sense of redemptive fantasy through the pacing of the tale, and the rhetorical weighting given to the Edenic festivities which follow upon the moment that Ebenezer Scrooge has earned his way (via a gruelling NIGHT JOURNEY) into the happy land of CHRISTMAS. For the reader of fantasy, the significance of this novella lies not only in its potency and near-perfect calculation of effects but also in its use of supernatural material (the GHOSTS) as agents of fantasy TRANSFORMATION: Scrooge is a central exemplar of the protagonist who gains (or regains) his true nature by wrestling his way through a RECOGNITION of passage, prefigured by moments of indeterminacy and horror – as in the transformation of the doorknob into Marley's ghost countenance. And, in their breadth of release, the final moments of *A Christmas Carol* display the true easement of the fantasy EUCATASTROPHE, though without any hint that the mundane world has been abandoned.

The Chimes and *The Cricket on the Hearth* are of less interest. *The Haunted Man and the Ghost's Bargain* – though also technically constructed as a supernatural fiction – is once again a tale of transformation whose effects range far beyond those ascertainable in more straightforward uses of supernatural props, in this case primarily a DOPPELGÄNGER. The doppelgänger grants the melancholy protagonist his WISH – to lose all memories of the "sorrow, wrong, and trouble" which have plagued him for years. The opening passages of the novella, which lead directly to the granting of this wish, constitute a classic representation of an urban landscape as a theatre in which WRONGNESS can be sensed brooding; and the descent of the protagonist into the ANSWERED PRAYER of AMNESIA similarly presents the THINNING of the fabric of REALITY central to the movement of fantasy. Caught in this thinned world, the protagonist becomes a hollow creature, an artificial, dehumanized caricature of a man; and it is now – for his Answered Prayer also curses those he meets with the same "gift" of amnesia – that one of his stiffening victims asks, "What is it that is going from me again? What *is* this that is going away." Metaphorically, at this point, as CD makes clear, the cast is turning into stone. The tale then progresses into a scene of sickened, surreal REVEL, during which an ancient man, gabbling fiercely to himself, gobbles a holly wreath; it is a savage presentation of the festival of Christmas in PARODY form. Only then does the protagonist begin to wrestle his way clear of the nightmare, to abjure his CURSE, and to enter – frailly – the eucatastrophe of a restored Christmas.

London as Urban-Fantasy Venue The 19th century saw a process of transformation whereby tales set in CITIES evolved into full-blown URBAN FANTASY; CD was a figure of central importance in this process. Early in his career, in novels like *Oliver Twist* (**1839**), he tended to conceive of London in Manichean terms as a prison of EVIL, and though a late novel like *Our Mutual Friend* (**1865**) immensely complexifies that vision, it remains true that, throughout his career, CD treated the city as an almost animate, labyrinthine, serpent-like monster, the coils of which constituted a kind of *theatre* upon which the drama (and melodramas) of life could best be articulated.

Like Eugene SUE, whose *Mysteries of Paris* (**1844**) helped shaped his mature concept of the city as a forum for the enactment of mysteries, CD structured his work around particular institutions – all seen as exemplary platforms or stages where humans exposed their behaviour to view – and peopled these stages with hierarchical societies, high and low, legal and illicit, which mirrored and parodied each other. Softened and romanticized, CD's vision of London has been central to the very numerous GASLIGHT ROMANCES created from *c*1880 onwards by writers like Robert Louis STEVENSON, Arthur Conan DOYLE, H.G. WELLS, Bram STOKER and G.K. CHESTERTON, and in RECURSIVE-FANTASY mode by many more recent writers of sf and fantasy – Joan AIKEN, James P. BLAYLOCK, Tim POWERS and countless others.

Ghost Stories Woven into *The Posthumous Papers of the Pickwick Club* (**1836-7**) are three GHOST STORIES: "The Bagman's Story", "The Story of the Bagman's Uncle" and "The Story of the Goblins who Stole a Sexton". The first two are unremarkable, but the third – on Christmas Eve, GOBLINS kidnap a miser and present him with a vision of what poor people must endure – clearly prefigures *A Christmas Carol*.

Moments of supernatural HORROR percolate, sometimes in the form of interpolations or ultimately rationalized experiences (◊ RATIONALIZED FANTASY), through much of CD's fiction, including, it has been suggested, *The Mystery of Edwin Drood* (**1870**) – CD died before revealing whether this would contain a supernatural resolution. As an author of ghost stories, CD is remembered almost exclusively for the brilliant "No. 1 Branch Line. The Signalman" (in *Mugby Junction: The Extra Christmas Number of All the Year Round* anth **1866** chap). The story is remarkable enough for its atmosphere and suspense, but is also notable for the context within which it first appears, because *Mugby Junction* is a SHARED-WORLD anthology – likely the first of genre interest – comprising tales by CD (four in all) and others, all set at and involving Mugby Junction, a relatively new venue but one which proved highly fruitful (◊ TRAINS).

"The Magic Fishbone" (in *A Holiday Romance* anth **1868** chap) is a FAIRYTALE for children, and is not successful.

CD was not primarily a writer of supernatural fiction or fantasy; but the urgency and plenitude of his imagination, and the subversive anger of his polemical mind, tinged everything he wrote with the dangerousness that early fantasy always threatened to convey. [JC]

Other works (selective): Material pertaining to Gog and Magog in *Master Humphrey's Clock* (coll **1840-41** 3 vols); various extra Christmas numbers of CD's journals, *Household Words* (1850-59) and its retitled reincarnation, *All the Year Round* (1859-67) are shared-world anthologies, beginning with *A Christmas Tree* (anth **1850** chap) and including *The Haunted House* (anth **1859** chap), plus *Mugby Junction*; *To Be Read at Dusk* (1852 *The Keepsake*; dated 1852 but *c*1890 chap); *The Uncommercial Traveller* (coll **1860**); *The Lamplighter's Story; Hunted Down; The Detective Police; and Other Nouvellettes* (coll **1861** US), containing also "Blow Up With the Brig!" by Wilkie COLLINS; *The Uncommercial Traveller and Additional Christmas Stories* (coll **1868** US), which contains the first book publication after *Mugby Junction* of "No. 1 Branch Line. The Signalman"; *The Complete Ghost Stories of Charles Dickens* (coll **1982**) ed Peter HAINING; *A Christmas Carol and Other Christmas Stories* (coll **1984** US), which does not include the other Christmas Books; *The Haunted Man and the Haunted House* (coll **1985**); *The Signalman and Other Ghost Stories* (coll **1988** US); *Charles Dickens's Christmas Ghost Stories* (coll **1992**) ed Haining.

DICKINSON, PETER (MALCOLM DE BRISSAC)
(1927-   ) UK writer, born in what is now Zambia; he is

married to Robin MCKINLEY. PD is well known for his detective fictions, four of which – *Sleep and his Brother* (**1971**), *The Green Gene* (**1973**) and the ALTERNATE-WORLD sequence *King and Joker* (**1976**) and *Skeleton-in-Waiting* (**1989**) – are of sf/fantasy interest. He is equally well known for his YA fiction, including the **Changes** sequence – *The Weathermonger* (**1968**; rev 1969 US), *Heartsease* (**1970**) and *The Devil's Children* (**1971**), assembled as *The Changes* (omni **1975**; vt *The Changes Trilogy* 1985; vt *The Changes: A Trilogy* 1991 US). In this SLEEPER UNDER THE HILL fantasy the sleeper who awakes, and who has been transforming the UK into a kind of LAND-OF-FABLE Dark Ages, is revealed to be MERLIN; his effect on the land and the MATTER of Britain is by no means good. Superstition and WITCH-hunting are rife; and it is fortunate that the young protagonists are able to persuade him to go back to sleep. A later work, *Merlin Dreams* (coll of linked stories **1988**), offers a different Merlin, one who longs to escape the burden of his SHAMAN-like powers in slumber, awakening only at intervals to pass on a tale invoking MOTIFS and UNDERLIERS from the Arthurian cycle (◊ ARTHUR).

There is little sameness in PD's choice of subject matter, though an unfailing civility of utterance sometimes gives a deceptive sense of equipoise to tales which are – especially for a YA audience – clearly meant to challenge preconceptions. *Emma Tupper's Diary* (**1971**) features an ambivalent encounter with the Loch Ness Monster (◊ ANIMALS UNKNOWN TO SCIENCE). In *The Gift* (**1974**) a young man with the TALENT of clairvoyance must defend his family from a destructive lunatic, the family itself is significantly dysfunctional, and the tale leaves no secure base. In *The Blue Hawk* (**1976**) a young initiate into an archaic priesthood in a land-of-fable Egypt causes the death of his pharaoh, and becomes involved in a grinding internecine war between priests and men of war as MAGIC begins to disappear from a world the GODS are leaving. *Healer* (**1983**; vt *The Healer* 1985 US) portrays a young woman whose TALENT is healing, and who is co-opted by a foundation whose motives are both ominous and plausible. *Giant Cold* (**1984** chap) is an ironized but moving FAIRYTALE. *A Box of Nothing* (**1985**) and *Time and the Clockmice, Etcetera* (**1993**) are both FABLES, one about a boy who wants nothing and gets a box of it, the other about a theft of time, and attempts at its recapture.

Some of PD's tales wear an air of bemused civility; but never does his decency of mien ultimately hide the operations of a cool, disillusioned, wise mind. Very quietly, PD has written a canon of central texts. [JC]

Other works: *Sleep and His Brother* (**1971**); *The Iron Lion* (**1972** chap US; rev 1983 chap UK), a fairytale; *Mandog* (**1972**) with Lois Lamplugh (1921-   ); *The Dancing Bear* (**1972**), for children; *Chance, Luck and Destiny* (coll **1975**), which contains some fiction; *Annerton Pit* (**1977**), YA sf; *Walking Dead* (**1977**), which almost rationalizes ZOMBIES; *Hepzibah* (**1978** chap); *The Flight of Dragons* (**1979**), nonfiction; *Tulku* (**1979**), featuring magic in Tibet; *City of Gold* (**1980**); *The Seventh Raven* (**1981**); *Hundreds and Hundreds* (anth **1984**), edited for charity; *Eva* (**1988**), sf; *A Bone from a Dry Sea* (**1992**), associational; *Shadow of a Hero* (**1994**), associational.

DICKINSON, W(ILLIAM) CROFT (1897-1963) UK author, historian and archaeologist, who wrote three magical fantasy novels, all set in Scotland and describing the adventures of two children, Donald and Jean. In *Borrobil* (**1944**) they go INTO THE WOODS and meet a friendly magician, Borrobil, who transports them back to the legendary past – a world of Celtic mythology and Beltane Fires – and encounter the White King of Summer, the Black King of Winter and a poison-breathing DRAGON. Donald and Jean travel through TIME to less fantastic but equally exciting medieval settings in the two sequels: the Eildon Hill above Montrose in *The Eildon Tree* (**1947**) and Flodden Field in *The Flag from the Isles* (**1951**). In the latter novel, they devise their own "magic circle" after consulting an old book of SPELLS. All the characters in this story, especially the alchemist Damien, were based on real people researched by WCD in old documents and "historical warrants".

WCD's penchant for fantasy and magic, combined with his wide knowledge of Scottish history, naturally led him to try his hand at GHOST STORIES. His first short SUPERNATURAL FICTION for adults, "The Sweet Singers" (1947 *Blackwood's Magazine*), which had as its backround the imprisonment of the Covenanters on the Bass Rock, was the basis of his first collection *The Sweet Singers* (coll **1953** chap). Together with nine more ghost stories, it was assembled as *Dark Encounters* (coll **1963**). Ancient clan feuds provide the source of some of the most effective malevolent HAUNTINGS in the collection, notably by the Witch of Morar in "The Return of the Native" and Black Dougal in "Return at Dusk". His last story was "His Own Number". Also memorable is "The House of Balfother", in which the narrator, a university student, seeks shelter at a strange tower-house not recorded on any MAP and meets a naked, incredibly old man secreted away like the Monster of Glamis. [RD]

DICKSON, CARTER Pseudonym of John Dickson CARR.

DICKSON, GORDON R(UPERT) (1923-   ) Canadian-born US author who bulks large in the sf world but has written relatively little fantasy, although his ambitious **Childe Cycle** sf sequence utilizes ARCHETYPES, TALENTS and hints of REINCARNATION. The **Dragon and the George** fantasies comprise *The Dragon and the George* (as "The Dragon and the George" *F&SF* 1957; exp **1976**), *The Dragon Knight* (**1990**), *The Dragon on the Border* (**1992**), *The Dragon at War* (**1993**) and *The Dragon, the Earl, and the Troll* (**1994**). In the first book, ASTRAL-BODY experiments project Jim Eckert and fiancée into an ALTERNATE WORLD medieval-Europe, she as herself and he occupying the body of a DRAGON. Repairing the upset BALANCE requires COMPANIONS – a local WIZARD, KNIGHT, WOLF and others – to help assault a DARK TOWER. There are interesting RATIONALIZED-FANTASY asides. After victory and recovery of his body, Eckert chooses to remain in the magical world. The sequels are sadly ponderous.

GD's **Jamie the Red** character, devised for the SHARED-WORLD series *Thieves' World* (◊ Robert ASPRIN) and used by others (although GD failed to contribute), features in *Jamie the Red* (**1984**) with Roland J. GREEN and in the title story of *Beyond the Dar al-Harb* (coll **1985**). [DRL]

Other work: *The Last Dream* (coll **1987**), including fantasy and SUPERNATURAL FICTION.

As editor: *Rod Serling's Triple W: Witches, Warlocks and Werewolves* (anth **1963**); *Rod Serling's Devils and Demons* (anth **1967**).

DICK TRACY US live-action/ANIMATED MOVIE (*1990*). Touchstone/Silver Screen Partners IV. **Pr** Warren Beatty. **Exec pr** Art Linson, Barrie M. Osborne, Floyd Mutrux. **Dir** Beatty. **Spfx** Bruce Steinheimer. **Vfx** Harrison Ellenshaw, Michael Lloyd. **Mufx** John Caglione Jr, Doug

Drexler. **Art dir** Harold Michelson. **Screenplay** Jim Cash, Jack Epps Jr. **Based on** the **Dick Tracy** comic strip by Chester Gould (1900-1985). **Novelization** *Dick Tracy* * (**1990**) by Max Allan Collins. **Starring** Beatty (Dick Tracy), Glenne Headly (Tess Trueheart), Charlie Korsmo (Kid), Madonna (Breathless Mahoney), Al Pacino (Big Boy Caprice). 103 mins. Colour.

The basic plot of *DT* is not fantasticated, merely exceptionally implausible: it is a gangbusting caper in which Tracy, aided by orphan lad Kid, counters the plans of megalomaniac hood Big Boy to take over the city. The fantastication comes about through the movie's devoted portrayal of the COMIC strip on screen, rejecting – unlike *Batman* (*1989*) and *Superman* (*1978*), for example (◊ BATMAN MOVIES; SUPERMAN MOVIES) – any urge to transform fantastic material into "realism": backgrounds are overtly painted (often part-animated) rather than solid; sets are stripped down to their basics; the colours, done in matt slabs, tend towards the primary; thanks to superb mufx, crooks like Pruneface and Littleface are portrayed in full comic-strip grotesquery; distances (especially in chases) are distorted to conform with the plot; and the dialogue, morality, characterization, motivation and depiction of violence are all of the comic-strip rather than the cinematic genre. The result induces a curious shift of PERCEPTION: after the initial shock, we experience *DT* as if ourselves living for a time in a gaudy comic-strip world. [JG]

DICTION With the exception of J.R.R. TOLKIEN's **The Lord of the Rings** (**1954-5**), which appears to have had part of its origin in his attempt to create an entire imaginary language, it is hard to think that any actual invented languages can be discovered in a fantasy text, though they are very frequently *referred* to, and though there are a good number of invented words and epithets and archaisms which are intended to impart a flavour of the otherness and romance of an invented tongue – this latter technique is particularly well employed in the post-HOLOCAUST movie *Mad Max Beyond Thunderdome* (*1985*). We use the term "diction" in three main senses:

1. Style As such, this can be examined in ways common to the examination of any prose text.

2. Decorum In much fantasy writing we must also address the question of decorum – or, more properly, decorousness. It is a question to be addressed with some interest: whether or not the chaste, clogged language of much HIGH FANTASY derives solely from marketing decisions extrinsic to its true nature, it is suggestive that DARK FANTASY can almost immediately be recognizable through its violation of that chastity of language.

Decorum also provides an explanation for the ways in which the language of HEROIC FANTASY radically differs from one author or book to another. Because this is an impure form, consciously drawing on other genres, it is not felt as an estrangement or an incongruity when a TEMPLATE fantasy like Glen COOK's **Garrett** series (from **1987**) echoes in its language the private-eye thrillers to which it pays homage, in spite of the fact that the novels are set in a medievalized city full of ELVES and trolls. Appropriate decorum here derives from the emotional feel and our reaction to it in terms of other texts, not from the superficial aspects of the FANTASYLAND setting and magical rationale.

3. Archaism The "forsoothery" of much HIGH FANTASY may be nothing more than a reflection of the tendency – derived from William MORRIS and J.R.R. TOLKIEN – to

conduct heroic matters in a Germanic tone of voice, and thereby to convey a sense of the dawn of the world. Attempts at the Saga Voice afflict almost all high fantasy, from Morris to Poul ANDERSON and to most recent writers of GENRE FANTASY. Texts in Saga Voice are full of undigested morsels of "language", but often in no way constitute a genuine attempt at conveying otherness. Where they do, as in the best of Anderson's early work, it is because of a conscious attempt to go back to the sagas and make the imitation of their language personal and new.

Other sources for the effect of archaism in fantasy include the cod Orientalism of Lord DUNSANY and others of the turn of the century, and even the BIBLE. Sometimes this is a deliberate piece of what Bertolt Brecht (1898-1956) called the alienation effect; Dunsany, James Branch CABELL, Ernest BRAMAH and others are announcing, not entirely truthfully, that what they are telling us in this particular and peculiar way is a STORY, and to be taken on Story's terms. What can make later imitations of their language so silly is the absence of a sense of irony, and of the sense that language can have several meanings at once. [JC/RK]

DIETERLE, WILLIAM (1893-1972) German-US movie director/producer. ◊ *The* DEVIL AND DANIEL WEBSTER (*1941*); FAUST: EINE DEUTSCHE FOLKSSAGE (*1926*); KISMET (*1944*); *A* MIDSUMMER NIGHT'S DREAM (*1935*); OMAR KHAYYAM (*1956*); PORTRAIT OF JENNIE (*1948*).

DIETTERLIN, WENDEL (1550-1599) ◊ EDIFICE; Giovanni Battista PIRANESI.

DI FATE, VINCENT (1945-   ) Widely respected US painter and illustrator of sf subjects; he has a well controlled, loose, richly coloured painting style. He began his professional career as a tv animator. His first sf illustration was for *Analog* in 1969, since when his work has appeared in magazines and on book covers. He was one of the NASA artists for the Apollo/Soyuz programme, and received a Hugo AWARD in 1979. He also writes on sf art in his long-running column, **Sketches**, in *Algol* and *SF Chronicle*. [RT]

Further reading: *Di Fate's Catalog of Science Fiction Hardware* (graph **1980**) by VDF and Ian Summers.

DIKTY, THADDEUS E. (1920-1991) US bibliographer. ◊ E.F. BLEILER.

DIME MYSTERY ◊ MAGAZINES.

DINESEN, ISAK Pseudonym of Danish writer Karen Blixen (1885-1962). Her transfiguration of supernatural or mundane material alike into tales profoundly infused with a sense of the shaping importance of STORY marks her as a deeply significant 20th-century figure, a writer whose works have been widely understood as a sustained attempt to redeem a Storied world from the secularizing dehydrations of history. Her work is intricately artificial, with FRAME STORIES and embedded multi-voice narrations presenting material in a fashion so ornate and dancelike that her works seem to manifest themselves most vividly as dreamish COMMEDIA DELL'ARTE. Though her early work – much of it translated with other material as *Carnival: Entertainments and Posthumous Tales* (coll **1977** US) – was first written and published in Danish (beginning 1907), all her mature stories were first published in English, and always in her own words. Collections include *Seven Gothic Tales* (coll **1934** UK/US), *Winter's Tales* (coll **1942** US; cut vt *The Dreaming Child and Other Stories* 1995 chap UK), *Last Tales* (coll **1957** UK/US), *Anecdotes of Destiny* (coll **1958** US) and *Ehrengard* (**1963** chap UK/US). Her one novel, *Gengaeldelsens Veje* (**1944**; trans Clara Selborn as *The Angelic Avengers* **1946**

UK) as by Pierre Andrézel, was not initially known to be by Blixen; it is a nonfantastic melodrama.

Tales of interest from *Seven Gothic Tales* include "The Monkey", in which a mother superior, changed into a MONKEY, gives corrupt advice to a young man, and the long novella "The Dreamers", in which a REVENANT singer fascinates several men who are unaware that it is her passionate art that allows her to appear among them. In "The Sailor-Boy's Tale", from *Winter's Tales*, a young lad saves a BIRD from strangling, an act which later saves his life, as the bird was a SHAPESHIFTER and owes him. In "The Dreaming Child", from the same volume, a young child so powerfully imagines an enriched life for himself that the adults who surround him live his fantasy. In "Echoes", from *Last Tales*, a revenant OPERA singer finds that her voice has come to life again in a choirboy; but puberty intervenes. "The Diver", from *Anecdotes of Destiny*, allegorizes a discussion about love in an ARABIAN-FANTASY frame; "Tempests", from the same volume, fabulates ornately upon the play by William SHAKESPEARE. *Ehrengard*, set in a 19th-century RURITANIA, superficially avoids any fantasy element, while at the same time, at the level of Story, constantly invoking a sense that an archaic TWICE-TOLD tale is "telling" its protagonists who they truly are, and how they must behave.   [JC]

DINOSAURS These reptiles are favourite accessories for LOST-RACE stories – especially those on the margins of sf, like Victor Rousseau's "The Eye of Balamok" (1920). The possibility of their survival into modern times is cherished by such legends as the Loch Ness Monster (◊ ANIMALS UNKNOWN TO SCIENCE). Some SWORD-AND-SORCERY tales set in remote antediluvian eras make room for the occasional surviving dinosaur – one is briefly featured in Robert E. HOWARD's "Red Nails" (1936) – but they are more easily accommodated in various kinds of SECONDARY WORLDS, where they may be drafted to serve as DRAGONS. Ray BRADBURY's *Dinosaur Tales* (coll **1983**) mixes fantasy and sf stories. A fantasy world inhabited by dinosaurs is featured in *Dinotopia* (graph **1992**) by James GURNEY. Since Arthur Conan DOYLE's *The Lost World* (**1912**) was first filmed in 1925, dinosaurs have been favourite subjects of stop-motion animation. They are featured in KING KONG (*1933*), several wildly anachronistic prehistoric romances in which they are juxtaposed with cavemen – such as ONE MILLION BC – and the long series of GODZILLA MOVIES. Reflecting the curious sentimentality with which dinosaurs are often regarded, Godzilla was transformed by degrees into a quixotic HERO defending the world against more malevolent MONSTERS.   [BS]

DIO CHIAMATO DORIAN, IL Italian movie (*1969*). ◊ *The* PICTURE OF DORIAN GRAY.

DIONNET, JEAN-PIERRE (1947-   ) French writer. ◊ MOEBIUS.

DIONYSUS Vine-wreathed Greek GOD of wine and REVEL, Latinized as Bacchus; often contrasted with APOLLO. ARISTOPHANES makes Dionysus (in his subsidiary role as god of drama) the semi-comic hero of *The Frogs* (**405**BC). This god's usual retinue includes PAN, SATYRS and his aged, drunken MENTOR Silenus – who dispenses uncomfortable wisdom in James Branch CABELL's *Jurgen* (**1919**). Introducing Bacchus and Silenus into his CHILDREN'S FANTASY *Prince Caspian* (**1951**), C.S. LEWIS cautiously substitutes fresh grapes for wine but still hints at the dangerous side of bacchanalia; comic fantasies like Thorne SMITH's *The Night Life of the Gods* (**1931**) tend to present the god as a mere genial tippler.   [DRL]

DIRKS, RUDOLPH (1863-1928) US comics artist. ◊ COMICS.

DIRTY DOZEN A group of COMPANIONS gathered together by force, and which normally functions as a some kind of military unit (◊ MILITARY FANTASY). Hierarchy is enforced, the top of the triangle normally being occupied by the original recruiter, who may be the agent of an invisible MAGUS or monarch or DARK LORD. DDs are normally recruited to gain a specific goal, the attainment of which almost certainly involves physical conflict and which exploits the combat skills of the team; fantasy writers, however, tend to equip individual members of the team with a variety of additional TALENTS. Once its goal has been attained, the team may evolve through mutual appreciation into the SEVEN SAMURAI kind of grouping very much more commonly found in fantasy.

The term comes from a nonfantasy novel, *The Dirty Dozen* (**1965**) by E.M. Nathanson (1928-   ), filmed as *The Dirty Dozen* (*1967* US/Spain).   [JC]

DIRTY PAIR Japanese tv series (1985 onwards). ◊ ANIME.

DIS ◊ HADES; HELL.

DISCH, THOMAS M(ICHAEL) (1940-   ) US writer who began his career with sf stories, the first being "The Double-Timer" for *Fantastic* in 1962, and who perhaps remains best-known for his sf novels. His work of fantasy interest is limited to some short stories, two books for children and some SUPERNATURAL-FICTION novels. The children's stories – *The Brave Little Toaster* (in *Fantasy Annual IV* anth **1981** ed Terry CARR; **1986** chap), filmed as *The* BRAVE LITTLE TOASTER (*1987*), and *The Brave Little Toaster Goes to Mars* (**1988** chap) – are told with a straight face, and can be read for their touching ingenuity. In the first volume, the Toaster leads its COMPANIONS on a trek to rescue its long-departed master from danger; in the second the Toaster has adventures in space which culminate in a CHRISTMAS story (and a happy return to the USA). Both volumes can be understood as what one might call counterfactual fables, for they work also as ALLEGORIES of a USA blessed by a technological THEODICY, of a human race succoured by its inventions – which know their place.

None of TMD's other work suggests any sense that he subscribes to so hopeful a view of civilization and its discontents. *On Wings of Song* (**1979** UK) depicts a self-consciously decadent (◊ DECADENCE) 21st-century NEW YORK on the brink of entropic implosion. *The Businessman: A Tale of Terror* (**1984**) is a SUPERNATURAL FICTION set in a contemporary MINNEAPOLIS haunted by (among others) the GHOST of the poet John Berryman (1914-1972), who (in real life) committed suicide in that city; in the central narrative, a murdered wife takes revenge on her husband, the eponymous businessman, though she is raped by him in the process, giving birth to an astonishingly destructive HALFLING. In *The M.D.: A Horror Story* (**1991**), the god Mercury (◊ HERMES) gives to a lad a caduceus whose powers turn out to cost dear; the novel ends, as an AIDS-like plague decimates the USA, in utter bleakness. At its heart, *The Priest: A Gothic Romance* (**1994** UK) is a SATIRE on the moral failings – which in TMD's view ravage the whole – of the Roman Catholic Church; the book features also a grotesque trans-temporal IDENTITY EXCHANGE, in which a medieval priest is liberated from the Inquisition to terrorize MINNEAPOLIS.   [JC]

Other works: *The Tale of Dan de Lion* (**1986** chap), a narrative poem; *The Silver Pillow: A Tale of Witchcraft* (dated 1987 but **1988** chap).

DISCREET CHARM OF THE BOURGEOISIE, THE

(ot *Le Charme Discret de la Bourgeoisie*). French/Italian/Spanish movie (*1972*). **Pr** Serge Silberman. **Dir** Luis BUÑUEL. **Screenplay** Buñuel, Jean-Claude Carrière. **Starring** Stephane Audran (Alice Sénéchal), Julien Bertheau (Monseigneur Dufor), Jean-Pierre Cassel (Henri Sénéchal), Paul Frankeur (François Thévenot), Muni (assassin), Bulle Ogier (Florence), Claude Piéplu (Colonel), Fernando Rey (Raphaël Acosta), Delphine Seyrig (Simone Thévenot). 105 mins. Colour.

Six rich middle-class people endeavour variously to have a meal together, but every time something stops them. It becomes clear that some, at least, of these misadventures are DREAMS of one or other cast-member, and the episodes grow more fantastic; there are even dreams within dreams, giving an ARABIAN-NIGHTMARE tinge to some episodes. Interwoven with the surreal (◊ SURREALISM) events are several GHOST STORIES, told in flashback . . . And, every now and then, we cut to the six walking along a country road that seems to lead them nowhere, though it can certainly be understood as a vision of the Road of Life.

There are elements of social SATIRE in *TDCOTB*, but mostly it is playfully illogical, a random, anarchic collection of plot elements that, together, do not create a meaning. In a way, the movie is a prank: it is the instinct of an audience to search for and theorize about meanings even where there is none: the real butt of Buñuel's satire is thus us and our own aesthetic pretensions. The prank worked perhaps too well: *TDCOTB* won an Oscar. [JG]

DISGUISE

With the exception of GENDER DISGUISE, disguise is less significant in fantasy than the more interesting possibilities of TRANSFORMATION, METAMORPHOSIS or SPELLS of ILLUSION and GLAMOUR. FAIRYTALE characters often go in disguise as pedlars or mendicants, like the Queen in SNOW WHITE, the MAGICIAN in "Aladdin and the Wonderful Lamp" (◊ ARABIAN FANTASY) and the Caliph Haroun al Raschid. The versatile Master of Disguise character is a borrowing from the detective/thriller genre; e.g., Eugène François Vidocq (1775-1857) in his own highly fictionalized *Memoirs* (**1828**) and von Gerolstein in Eugene SUE's *The Mysteries of Paris* (**1844**); the original 1973 version of the COMICS superheroine "Black Orchid" (◊◊ Neil GAIMAN) was a modern Mistress of Disguise. The shrunken but still human-shaped Gray Mouser in Fritz LEIBER's *The Swords of Lankhmar* (**1968**) is sufficiently resourceful with disguise to impersonate a specific rat; Tanith LEE's SWORD-AND-SORCERY detective *Cyrion* (**1982**) uses constant disguises as a way of confusing people into revealing truth. When Puzzle the donkey in C.S. LEWIS's *The Last Battle* (**1956**) is persuaded to wear a lion-skin as Aslan, the success of this disguise, despite its unconvincing crudity, is a measure of Narnia's THINNING; Severian in Gene WOLFE's *The Sword of the Lictor* (**1982**) attends a costume party in his normal cloak and mask as torturer, which is thus assumed to be a disguise (◊◊ TROMPE L'OEIL); the pallid, undead rival heroines of DEATH BECOMES HER (*1992*) must disguise themselves as their living selves. Magical disguises include: Gwydion's placing of "another semblance" on himself and Llew to deceive Arianrhod in the MABINOGION; MERLIN's similar disguise of Uther Pendragon as the lady Igraine's already dead husband, leading to SEX and the birth of ARTHUR; Shea's monstrously tusked disguise as a Djann (◊ GENIES) in *The Castle of Iron* (**1950**) by L. Sprague DE CAMP and Fletcher PRATT; two useful cloaks in Diana Wynne

JONES's *Howl's Moving Castle* (**1986**), whose wearers respectively seem a red-bearded man and a horse; and the WITCH Ursula's disguise of herself as a rival to the heroine of *The LITTLE MERMAID* (*1989*). [DRL]

See also: INVISIBILITY; MASKS; PLOT DEVICES.

DISKI, JENNY

(1947-) UK writer. All her novels, beginning with her first, *Nothing Natural* (**1986**), tend to concentrate on protagonists whose alienation from the course of the world may be expressed in non-naturalistic terms. The assault on normality in *Nothing Natural* is mundane, but *Rainforest* (**1987**) invokes MAGIC-REALISM procedures. *Like Mother* (**1988**) is narrated by Nony, an infant child born without a brain, whose mother's life is similarly empty. *Then Again* (**1990**) is a TIMESLIP tale in which an estranged contemporary woman, whose daughter has disappeared, experiences the much earlier life of a Jewish namesake trapped in a pogrom. *Monkey's Uncle* (**1994**) is an APE-as-human fable. Of the stories assembled in *The Vanishing Princess* (coll **1995**), "The Vanishing Princess or The Origin of Cubism", "Shit and Gold" and "The Old Princess" are of fantasy interest. [JC]

DISNEY

Founded by Walt DISNEY in 1923 in Los Angeles as a partnership with his brother Roy, the Walt Disney Company is a foremost creator of ANIMATED MOVIES, a significant maker of live-action movies, the designer and operator of the world's most successful (generally) theme parks, and a notable producer of COMICS and TELEVISION material. In its meagre beginnings Disney was involved in the production of silent cartoons – the **Alice Comedies** (◊◊ ALICE IN WONDERLAND [*1951*]) and **Oswald the Lucky Rabbit** series. When sound came to the movie industry, Walt was an early exploiter of the new technology: after having been unable to sell two silent cartoons starring a new character, MICKEY MOUSE (devised in conjunction with Ub IWERKS), Walt made a third, *Steamboat Willie* (*1928*), as the first cartoon with synchronized sound. It was an immediate sensation, and enabled Disney to expand production. The **Mickey Mouse** series was soon joined by the **Silly Symphonies**, based on musical themes, the first of which was *The Skeleton Dance* (**1929**), with music by Carl Stalling (not Saint-Saëns, as is often written); in the decade following 1932, the first year the Academy of Motion Picture Arts and Sciences had a Best Cartoon category, a **Silly Symphony** won the award each year.

Significant profits could not be made from short cartoons. Disney thus embarked on the production of animated features, beginning with SNOW WHITE AND THE SEVEN DWARFS (*1937*) – with the profits from which Walt was able to build a new studio in Burbank, California – and continuing with PINOCCHIO (*1940*), FANTASIA (*1940*), DUMBO (*1941*) and *Bambi* (*1942*). WWII caused setbacks, although the studio produced training and propaganda movies for the US Government and created some of its first educational movies. In the late 1940s Disney turned to nature with the **True-Life Adventures**, a live-action natural-history series. Tv beckoned in the mid-1950s and Walt was eager, recognizing that he could use the medium to help market his movies and to earn the money he wanted for building Disneyland, a project that had been in the back of his mind for some years. Disneyland, built in southern California around themes from the Disney movies, was one of the first of what became known as theme parks.

Along with the move into tv, Walt began producing live-action movies, realizing that a studio could get a live-action movie into the theatres much faster than an animated feature, which often took upwards of three years to produce. Movies like *20,000 Leagues Under the Sea* (**1954**), *Swiss Family Robinson* (**1960**) and (with animated sections) MARY POPPINS (**1964**) were remarkably successful, although Disney continued to produce a new animated movie about every three years, thus adding to a library that was ideal for periodical theatre release and lent itself to profitable merchandise licensing.

Walt died in 1966 but his organization continued his policies. In 1971 it opened Walt Disney World on 28,000 acres near Orlando, Florida. EPCOT (Experimental Prototype Community of Tomorrow) Center followed in 1982, Tokyo Disneyland in 1983, the Disney/MGM Studios Theme Park in 1989 and – at least initially unsuccessfully – Euro Disney in 1992.

In 1984 a management shake-up brought on by corporate raiders had seen Michael Eisner become the company's head, and he directed a returned emphasis on moviemaking: movies like SPLASH! (**1984**), *Good Morning, Vietnam* (**1987**), WHO FRAMED ROGER RABBIT (**1988**) and *Dead Poets Society* (**1989**) followed, and new movie labels – Touchstone Films and Hollywood Pictures – were created. Eisner was also responsible for the company's expanded programme of video-cassette sales, the opening of Disney Stores throughout the world, the creation of a new book-publishing company (Hyperion – not to be confused with the non-Disney Hyperion Press, active during the 1970s) and even the purchase of a professional hockey team. The company continues to devote itself to family, rather than exclusively adult, entertainment. [DRS]

LIVE-ACTION MOVIES

Disney has almost always – there are some notable exceptions – displayed a sure touch in the production of fantasy in the form of ANIMATED MOVIES, but until recent years, and still displaying some uncertainty, it has been much less easy with live-action fantasy movies, almost certainly because of the company's exclusive emphasis until the early 1980s on *family* entertainment: this led to a lowest-common-denominator attitude that was surprisingly absent from most of Disney's animated output.

The first predominanyly live-action Disney feature was *Victory Through Air Power* (**1943**), a propagandistic movie projecting the military theories of Major Alexander de Servesky; it contained some animated sequences, just as animated compilation features like *Saludos Amigos* (**1943**) contained some live sequences. But the first feature that can be sensibly described as a live-action movie (although still with animated sequences) was *Song of the South* (**1946**), based on the **Uncle Remus** tales by Joel Chandler HARRIS. The success, both critical and commercial, of this movie brought a return to the same territory – even with the same child stars – in the form of *So Dear to My Heart* (**1949**), based on *Midnight and Jeremiah* (**1943**) by Sterling North; it has little fantasy interest save a few brief animated sequences in which the character Wise Old Owl gives the protagonist moral lessons. This movie made some anti-racist hackles rise because of its (naïve rather than conscious) Uncle Tom-ish representation of blacks; reissues were revised.

Walt DISNEY finally conquered his own reluctance to forgo animation entirely with *Treasure Island* (**1950**), based on the Robert Louis STEVENSON novel; it was well received and proved the first of several live-action movies made by

Disney in the UK using company funds that had been frozen there in the wake of WWII: *The Story of Robin Hood* (**1952**) (◊◊ ROBIN HOOD); *The Sword and the Rose* (**1953**; vt *When Knighthood was in Flower*), a romantic swashbuckler set at Henry VIII's court; and *Rob Roy – The Highland Rogue* (**1954**), an original screenplay rather than an adaptation of Sir Walter SCOTT's novel. The last, generally agreed to be the weakest, ended this run of UK productions.

Disney's next live-action movie was full-blown sf/fantasy: *20,000 Leagues under the Sea* (**1954**), based on the novel by Jules Verne. Though not for the purist, it was exciting and spectacular, and still views well today. Its director was Richard Fleischer, whose father, the animator and animation director Max FLEISCHER, had been one of Disney's main rivals early on. *Davy Crockett, King of the Wild Frontier* (**1955**) was in fact a fixup of sequences culled from the **Disneyland** tv series *Frontierland*; it was sequelled by *Davy Crockett and the River Pirates* (**1956**), both movies fleshing a neo-legendary HERO onto the bare bones of historical fact.

By now Disney – although it was almost always the animated features that drew the critical attention – was a full-fledged live-action producer, with adventure movies running alongside natural-history blockbusters. But it took surprisingly long for the studio to marry the fantasy of its animations to the live-action output. The first live-action fantasy movie proper was *The* SHAGGY DOG (**1959**), wherein a magic RING transforms a young boy into an Old English Sheepdog; its sequels were *The Shaggy D.A.* (**1976**) and *The Return of the Shaggy Dog* (**1987** tvm). This started a train of amiable Disney fantasy comedies, targeted largely at children: the **Flubber** series – *The Absent Minded Professor* (**1961**) and *Son of Flubber* (**1963**), being TECHNOFANTASY about strange substances like flubber, which enables people and objects to levitate, both loosely sequelled by *The Computer Wore Tennis Shoes* (**1970**), which was remade by Disney as a tvm in 1994 and was itself sequelled by *Now You See Him, Now You Don't* (**1972**), about INVISIBILITY, and *The Strongest Man in the World* (**1975**) – and the HERBIE MOVIES. *The Misadventures of Merlin Jones* (**1964**), a comedy technofantasy about a mind-reading machine and hypnotism (◊ MESMERISM) is also in this category of "safe" fantasy; it was sequelled by *The Monkey's Uncle* (**1965**).

Later in the same year as *The Shaggy Dog* came DARBY O'GILL AND THE LITTLE PEOPLE (**1959**), a more significant breakthrough in that it abandoned the reins of "realism" to tell of a mortal straying into the land of the leprechauns; although not unflawed, it is much more watchable today than the stream of "safe" fantasies begun with *The Shaggy Dog*. Moreover, it makes full use of Disney's growing expertise in spfx, born out of the studio's technical sophistication in animation. But it was not commercially successful, and it would be a while before Disney tried anything as conceptually ambitious again – meanwhile contenting itself with stuff like *Swiss Family Robinson* (**1960**), *The Sign of Zorro* (**1960**), *Greyfriars Bobby* (**1961**) and others of vague associational interest, alongside mundane adventures and *The Absent Minded Professor*. *Babes in Toyland* (**1961**) adapted the operetta *Babes in Toyland* by Victor Herbert (1859-1924) for the screen; although set in the world of MOTHER GOOSE (with the introduction of a raygun), it is more a musical than any attempt at fantasy.

Then at last came the whole-hearted return of Disney to fantasy proper: the musical MARY POPPINS (**1964**), based on the tales by P.L. TRAVERS. It is easy to find fault with this movie, which intersperses some brief live-action/animated sequences among a preponderance of live-action, but that would be to ignore its achievement, which was immense – as was its box-office success. A much later attempt to repeat the formula, *Bedknobs and Broomsticks* (*1971*) – based on *The Magic Bedknob* (**1945**) and *Bonfires and Broomsticks* (**1947**) by Mary NORTON – foundered, despite having Angela Lansbury in the lead role as an amateur WITCH who, aided by a group of children, repels a German invasion of the UK.

Those expecting that *Mary Poppins* might herald a renaissance in Disney live-action fantasy were disappointed. There was no real progress until *The Gnome-Mobile* (*1967*) – based on Upton SINCLAIR's *The Gnomobile: A Gnice Gnew Gnarrative with Gnonsense, but Gnothing Gnaughty* (**1936**) – a tale in which a family saves a tribe of gnomes whose forest is at threat from logging.

After Walt's death in December 1966 it might have been expected that the studio would take a new tack. In fact, it continued as before for many years. *Blackbeard's Ghost* (*1968*) tells how a varsity track coach inadvertently conjures up the PIRATE's SPIRIT, invisible to all but himself; Blackbeard assists the coach's dud athletics team to victory. *The Love Bug* (*1969*) was the first of the HERBIE MOVIES. *The Million Dollar Duck* (*1971*) is a TECHNOFANTASY retake on the FABLE of the Goose that Laid the Golden Eggs, the phenomenon coming about through exposure to radiation.

Charley and the Angel (*1973*) was a little more ambitious: somewhat in the spirit of Kapra's IT'S A WONDERFUL LIFE (*1946*), it relates how an ANGEL teaches a hard-nosed businessman the error of his ways, turning his attention back towards those he loves. Also more ambitious, but this time more in the vein of Disney's own *20,000 Leagues Under the Sea*, was *The Island at the Top of the World* (*1974*), based on *The Lost Ones* (**1961**; vt *The Island at the Top of the World* 1974 US) by Ian Cameron (1924-   ), a lost-worlds tale (◊ LOST RACES) in which explorers aboard an airship discover, deep within the Arctic, an island populated by Norsemen still adhering to the old ways.

Escape to Witch Mountain (*1975*), though seemingly produced on a shoestring, was far better: a UFO fantasy based on *Escape to Witch Mountain* (**1968**) by Alexander Key (1904-1979), it tells of how two orphans, possessed of TALENTS, come to realize they are, despite appearances, not human but aliens. They rendezvous with a spacecraft bearing their MENTOR at the eponymous mountain. The sequel, *Return from Witch Mountain* (*1978*), novelized by Key as *Return from Witch Mountain* * (**1978**), more of a TECHNOFANTASY, is conceptually flimsier but also darker in theme and stronger in characterization.

FREAKY FRIDAY (*1977*) was one of the better IDENTITY-EXCHANGE movies. PETE'S DRAGON (*1977*), a live-action/animated outing, was poorly received at first, largely because of a flabby script, but the re-edited reissue has much to recommend it. *The Cat from Outer Space* (*1978*) was a reversion to the bad old days: it featured an extraterrestrial cat in a fantasy/sf adventure suitable for children. The downturn continued with the astonishingly bad *Unidentified Flying Oddball* (*1979*) (◊ A CONNECTICUT YANKEE). From the same year, *The Black Hole* (*1979*) was an almost equally incompetent piece of sf, with elements directly – almost actionably – derived from the *Star Wars* saga. This was so badly received that Disney appears to have had the corporate rethink that would soon lead to the formation of Touchstone Films.

The DEVIL AND MAX DEVLIN (*1981*) was a pleasing fantasy on the theme of a PACT WITH THE DEVIL. *Condorman* (*1981*) is not strictly a fantasy and anyway a mess: a COMICS artist is press-ganged by the CIA and uses ideas drawn from comic strips to persuade a lovely Soviet agent to defect. *The Watcher in the Woods* (*1981*), based on *A Watcher in the Woods* (**1976**) by Florence Engel Randall (1917-   ), is a seeming GHOST STORY in which the rationalization deployed to obviate the supernatural (◊ RATIONALIZED FANTASY) involves the denizens of an ALTERNATE REALITY. The movie had much promise but the studio reportedly lost faith partway through and the result is another mess. TRON (*1982*), although badly received at the time (and commercially unsuccessful), is one of the CINEMA's better conceived pieces of TECHNOFANTASY, and should by no means be consigned to history's dustbin. SOMETHING WICKED THIS WAY COMES (*1983*), was disappointing in that one expected more from a screenplay that was an adaptation by Ray BRADBURY of his own novel.

Then came the first Touchstone movie, SPLASH! (*1984*), and it was immediately evident why so much of what had gone before had been so bland: the self-imposed dictum that all Disney movies must be suitable for the youngest and oldest members of the family had effectively emasculated much of the studio's output. That rule was eased for movies released under the Touchstone label. In fact, there is little in *Splash!* that could offend beyond the admission that unmarried adults do enjoy SEX, plus the (quite innocent) exposure of a fair acreage of Daryl Hannah: this is still a "family" movie, but for the 1980s rather than the 1950s. (*Splash!* was sequelled rather poorly by *Splash, Too* [*1988* tvm].) *Baby . . . Secret of the Lost Legend* (*1985*), although again from Touchstone, was a retrogression to an earlier age but with tougher dialogue and more violence: a community of DINOSAURS is discovered in the African jungle. *My Science Project* (*1985*), another Touchstone issue, is again retrogressive: a TIME-TRAVEL tale for children, it shows the same patronizing assumption that 12-year-olds are too stupid to understand the fundamental of science that had damned earlier efforts like *Unidentified Flying Oddball* and *The Black Hole*.

Return to Oz (*1985*) (◊ *The* WIZARD OF OZ), not Touchstone, was a respectable attempt to rediscover L. Frank BAUM's OZ for a new generation. *One Magic Christmas* (*1985*) was a derivative but well mounted tale in which an ANGEL restores to a widow her hope and her kidnapped children; it incorporates a spectacular sequence in SANTA CLAUS's toy workshop at the North Pole. *Flight of the Navigator* (*1986*) is an interesting but occasionally oversickly piece of sf/UFO fantasy.

From here on, in the discussion of the Disney live-action output, it will be assumed that a movie is from Touchstone unless otherwise stated.

Hello Again (*1987*) was a comedy about the spirit of a dead housewife being recalled by her sister, a medium. *Ernest Saves Christmas* (*1988*) sees SANTA CLAUS appointing a dimwit to take his place; it was one of a series about Ernest, of which *Ernest Scared Stupid* (*1991*) is the only other of fantasy interest. *New York Stories* (*1989*) is an anthology movie by three directors. Woody Allen's contribution, "Oedipus Wrecks", is a Jewish comedy in which a mother-dominated man wishes his nagging mother would disappear – as indeed she does, being "lost" in a conjurer's trick. When she returns, though, it is as an image that fills the NEW YORK skies. *Honey, I Shrunk the Kids* (**1989**) is a comic TECHNO-FANTASY released by Disney proper: a wacky inventor's new device miniaturizes a group of children, who must make their parlous way the full length of the family garden to safety. The movie re-explores scenes made famous in such movies as *The Incredible Shrinking Man* (**1957**) and Lindsay Gutteridge's **Matthew Dilke** series of novels starting with *Cold War in a Country Garden* (**1971**). It was sequelled by *Honey, I Blew Up the Kid* (**1992**), also a Disney release. *Spaced Invaders* (**1990**) was an asinine sf spoof. The same year, though, saw the excellent DICK TRACY (**1990**) as well as the first release under Disney's new Hollywood Pictures label: *Arachnophobia* (**1990**), a splendidly enjoyable ecological fantasy. *Mr Destiny* (**1990**) – a comedy about a middle-aged failure being given a second chance, *via* an elixir, to live his life and discovering that he is just as unhappy being a success – was a disappointment. *White Fang* (**1991**), a Disney release based on the Jack LONDON novel, was unfortunately juvenilized; the sequel was *White Fang 2: The Myth of the White Wolf* (**1994**). *The Rocketeer* (**1991**; vt *The Adventures of the Rocketeer*), surprisingly a Disney release, is a RECURSIVE, INDIANA-JONES-style slice of TECHNOFANTASY based on a 1981 GRAPHIC NOVEL by Dave Stevens. The gadget at the heart of this 1938 adventure is a portable rocket pack that hurtles our hero through various adventures. *Encino Man* (**1992**; vt *California Man* UK), from Hollywood Pictures, is a teen comedy fantasy in which two youths bring to life a prehistoric equivalent of themselves and successfully introduce him into their own partying sphere; it offers few moments of pleasure in its determination to woo the mindless end of the teen market.

The Adventures of Huck Finn (**1993**), a Disney release, was at least the fifth movie version of Mark TWAIN's novel. That same year, SUPER MARIO BROS. (**1993**), from Hollywood Pictures, represented something of a nadir, being both derivative and clumsy; HOCUS POCUS (**1993**), a comedy about WITCHES and GHOSTS released by Disney a few months later, though slight, was a refreshing signal that all was not lost. *Angels in the Outfield* (**1994**) was a remake of the BASEBALL classic ANGELS IN THE OUTFIELD (**1952**). *Ed Wood* (**1994**), not fantasy at all but of associational interest, was Tim BURTON's biopic of the famously awful B-movie director. *Robert A. Heinlein's The Puppet Masters* (**1994**), from Hollywood Pictures, gave one of sf's classics a big-screen treatment. With a similarly structured title – this was the era when other studios were producing movies like *Bram Stoker's Dracula* (**1992**; ◊ DRACULA MOVIES) and *Mary Shelley's Frankenstein* (**1994**; ◊ FRANKENSTEIN MOVIES) – *Rudyard Kipling's The Jungle Book* (**1994**), from Disney, was a remake claiming to be based faithfully on KIPLING's stories. It was very well received, but probably not nearly the commercial success that Disney had anticipated. But, so far as Disney's live-action fantasy is concerned, 1995 may be remembered as the year in which Hollywood Pictures released the disappointing *Judge Dredd* (**1995**).

From the above discussion it will seem clear that, for several decades up to about 1984 and the first Touchstone release, *Splash!*, it was possible to talk of a "Disney live-action fantasy" and to be immediately understood as referring to a particular kind of movie. Since then the company's output has either become chaotic or has successfully diversified, depending on whether one wishes to be negative or positive. What is certainly true is that few Disney fantasies since 1984 have been guilty of that stultifying blandness that affected such a high proportion of the movies that went before: there have been one or two appalling effusions and a few mediocre ones, but most have possessed a refreshing sense of ambition: they have been either *flops d'estimes* or, more frequently, deservedly commercially successful movies.   [JG]

ANIMATED MOVIES

The vast majority of the Disney animated shorts are of course fantasies, if only in that they feature TALKING ANIMALS like MICKEY MOUSE, Minnie Mouse, DONALD DUCK, Goofy, Pluto . . . However, to particularize would be a futile exercise: readers interested in a more extensive coverage should turn to *The Disney Studio Story* (**1988**) by Richard Holliss and Brian Sibley, which contains brief synopses of all shorts except the very few that have appeared since the early part of 1987. It is worth noting here, however, the WONDER TALES, nursery stories and classic fantasies. Some of the very earliest – early enough that their release dates are in doubt – preceded the **Alice Comedies** (◊ ALICE IN WONDERLAND [**1951**]); all dated roughly 1922-3, these were *Bremen Town Musicians*, *Cinderella*, *Goldilocks and the Three Bears*, *Jack and the Beanstalk*, *Little Red Riding Hood* and *Puss in Boots*. It was some time before Walt DISNEY returned to this fertile soil. Later shorts of note in context include *Mother Goose Melodies* (**1931**), *The Spider and the Fly* (**1931**), *The Ugly Duckling* (**1931**), *Babes in the Woods* (**1932**), *Three Little Pigs* (**1933**) – plus its sequels *The Big Bad Wolf* (**1934**), *Three Little Wolves* (**1936**) and *The Practical Pig* (**1939**) – *Old King Cole* (**1933**), *The Pied Piper* (**1933**), *The Grasshopper and the Ants* (**1934**), *Gulliver Mickey* (**1934**), *The Wise Little Hen* (**1934**), *The Tortoise and the Hare* (**1935**), *Mickey's Man Friday* (**1935**), *The Golden Touch* (**1935**), *Who Killed Cock Robin?* (**1935**), *Thru the Mirror* (**1936**), *Little Hiawatha* (**1937**), *Wynken, Blynken and Nod* (**1938**), *The Brave Little Tailor* (**1938**), *Mother Goose Goes Hollywood* (**1938**), the remake of *The Ugly Duckling* (**1939**), *Chicken Little* (**1943**), *The Truth About Mother Goose* (**1957**), *Paul Bunyan* (**1958**), *The Saga of Windwagon Smith* (**1961**), *Winnie the Pooh and the Honey Tree* (**1966**) – and its sequels *Winnie the Pooh and the Blustery Day* (**1968**), *Winnie the Pooh and Tigger Too* (**1974**), *Winnie the Pooh and a Day for Eeyore* (**1983**), plus the educational short *Winnie the Pooh Discovers the Seasons* (**1981**) – *Mickey's Christmas Carol* (**1983**; ◊ *A* CHRISTMAS CAROL) and *The Prince and the Pauper* (**1990**).

In terms of fantasy, however, Disney animation will be remembered more for the features produced by the studio. A number of these have been accorded separate entries in this encyclopedia: SNOW WHITE AND THE SEVEN DWARFS **1937**), PINOCCHIO (**1940**), FANTASIA (**1940**), DUMBO (**1941**), CINDERELLA (**1950**), ALICE IN WONDERLAND (**1951**), PETER PAN (**1953**), SLEEPING BEAUTY (**1959**), *The* BLACK CAULDRON (**1985**), WHO FRAMED ROGER RABBIT (**1988**) (which mixes live-action and animation; other live-action/animated features of note are treated in the previous section of this composite article), *The* LITTLE MERMAID, DUCKTALES: THE MOVIE – TREASURE OF THE LOST LAMP (**1990**), BEAUTY AND THE BEAST (**1991**), ALADDIN (**1992**) and the latter's sequel, *The* RETURN OF JAFAR (**1994**).

Of the other Disney animated features, some belong to the period 1941-9 when Walt DISNEY believed the public would turn up in equal numbers for compilations of shorts as they would for full-length features. This policy almost destroyed the studio's credibility as the cutting edge of animated movies – and would have, had not *Cinderella* been as good as it is. It is painful to watch most of these "compilation features" today, but the individual shorts (many of which have been reissued individually) are sometimes good and of fantasy interest. *The Reluctant Dragon* (**1941**), which includes live-action sequences of Robert Benchley being shown around the Disney studios, contains "Baby Weems", an amusing fantasy, partly SATIRE, about an infant genius, and "The Reluctant Dragon", based on the Kenneth GRAHAME tale from *Dream Days* (**1898**). *Saludos Amigos* (**1943**), again incorporating some live-action, was a WWII propaganda movie aimed at keeping South America neutral; it has little fantasy interest. *The Three Caballeros* (**1945**) was a strategic return to the South American market, suddenly important since the studio's European revenues were frozen; again there is little fantasy interest. *Make Mine Music* (**1946**) contains a treatment of *Peter and the Wolf* (1936) by Sergei Prokofiev (1891-1953) that is one of the classic Disney shorts, "Johnnie Fedora and Alice Bluebonnet", about a romance between two hats, and "The Whale who Wanted to Sing at the Met", an excellent fantasy about a cetacean tenor with ambition (voiced by Nelson Eddy). *Fun and Fancy Free* (**1947**) contains "Bongo", about a talented bear, based on a story by Sinclair Lewis (1885-1951); and "Mickey and the Beanstalk", based on the traditional story "Jack and the Beanstalk" (◊◊ JACK), another classic short. *Melody Time* (**1948**) has little fantasy interest. *The Adventures of Ichabod and Mr Toad* (**1949**) compiles two long shorts (or featurettes): "Mr Toad", based on Kenneth Grahame's *The Wind in the Willows* (**1908**) (◊ *The* WIND IN THE WILLOWS), and "The Legend of Sleepy Hollow", based on the Washington IRVING tale.

Further Disney animated features – some of them classics – are fantasies only insofar as they are set among communities of TALKING ANIMALS. They include: *Bambi* (**1942**), based on *Bambi* (**1929**) by Felix SALTEN; *Lady and the Tramp* (**1955**); *One Hundred and One Dalmatians* (**1961**), based on *The Hundred and One Dalmatians* (**1956**) by Dodie SMITH; *The Jungle Book* (**1967**), an incredibly popular travesty of the Rudyard KIPLING stories; *The Aristocats* (**1970**); *Robin Hood* (**1973**), based on the legend of ROBIN HOOD but acted out by animals; *The Rescuers* (**1977**), a rather good movie based on *The Rescuers* (**1959**) and *Miss Bianca* (**1972**) by Margery SHARP – followed much later by *The Rescuers Down Under* (**1990**), based on Sharp's characters and the only Disney animated feature to be a sequel; *The Fox and the Hound* (**1981**), another good movie, based on *The Fox and the Hound* (**1967**) by Daniel P. Mannix (1911-1984), *The Great Mouse Detective* (**1986**), an excellently buoyant movie, based on *Basil of Baker Street* (**1974**) by Eve Titus; *Oliver & Company* (**1988**), an extremely distorted version of Charles DICKENS's *Oliver Twist* (**1837-8**) set among a pack of stray dogs, Oliver himself being a kitten; and *The Lion King* (**1993**), a hugely popular but eventually rather vacuous ANIMAL FANTASY about competition to be ruler of the plains world. Other animated features include *The Sword in the Stone* (**1963**), based on T.H. WHITE's novel of the early life of ARTHUR, and *Pocahontas* (**1994**), a romanticized and very beautifully animated account of the life of Matoaka (aka Pocahontas; 1595-1617), the Indian princess who twice saved the life of English adventurer John Smith (1580-1631).

As a quite different strand, all on its own, Touchstone released that masterpiece of stop-motion animation, *The* NIGHTMARE BEFORE CHRISTMAS (**1993**), followed by the computer-animated *Toy Story* (**1995**). Although not strictly animated – it has some animation but most of the work is done with PUPPETS (plus Michael Caine) – *The* Muppet Christmas Carol (**1992**; ◊ *A* CHRISTMAS CAROL), deploying the incredible skills of Jim HENSON's Creature Workshop, is another exceptional movie.

It can be seen that, while the Disney output of animated fantasy has been overall a distinguished one, there have been two dreadful hiatuses: between 1941 and 1949, the era of the compilation features, and between 1951 and 1985, when Disney turned most of its attention away from animated features, which were released at long intervals and were often poor. Since 1985, however, when the studio realized what it had been throwing away in the pursuit of illusory short-term gains, production has been much more frequent – on average over one per year – and quality has increased markedly. This has paid commercial dividends: almost all of the Disney animated features released since 1988 have been among the top ten biggest grossers of their year. [JG]

TELEVISION

Since Disney has enjoyed great success with its fantasy movies, it comes as little surprise that the studio has also produced many tv programmes of a similar nature. In fact, Disney's first two experiments in tv, *One Hour in Wonderland* (1950) and *The Walt Disney Christmas Show* (1951), both used the "Slave in the Magic Mirror" from SNOW WHITE AND THE SEVEN DWARFS (**1937**) to promote two upcoming feature films, ALICE IN WONDERLAND (**1951**) and PETER PAN (**1953**).

When Walt DISNEY moved into tv on a regular basis in 1954 with *Disneyland*, his weekly anthology programme, many of the episodes featured his theatrical cartoons and clips from his full-length features. The anthology series provided many examples of fantasy elements over the next 40 years. As an example, *From Aesop to Hans Christian Andersen* (1955) offered several classic FAIRYTALES retold in Disney's optimistic style, including *The Tortoise and the Hare* and *The Ugly Duckling*. That same season also brought *Monsters of the Deep* (1955), featuring Kirk Douglas and Peter Lorre. Like many Disney programmes of the time, this was a barely disguised commercial promotion for one of the company's feature movies, in this case *20,000 Leagues Under the Sea* (**1954**).

The world of fantasy and its obvious artistic possibilities were the subjects of *Adventures in Fantasy* (1957), where various inanimate objects came to life and provided introductions to several animated sequences. A more interesting programme was *I Captured the King of the Leprechauns* (1959), which found Walt Disney looking for leprechauns to star in DARBY O'GILL AND THE LITTLE PEOPLE (**1959**). Several years later, *Fantasy on Skis* (1962) featured a young girl who dreamed of Peter Pan and Captain Hook battling once again, but this time on skis. The same year brought *The Prince and the Pauper* (**1962** tvm): Guy Williams, of *Zorro* fame, co-starred in Mark TWAIN's story.

More than a decade passed until *The Golden Dog* (1977), a tale about a ghostly miner who tried to help his former partners resolve a feud. Another GHOST featured prominently in *Child of Glass* (1978), where the SPIRIT of a murdered girl helped solve both a mystery about her death and a modern crime. In *Shadow of Fear* (1979) Ike Eisenmann's ability to travel outside his body through astral projection (◊ ASTRAL BODY) allowed him to prove that WEREWOLVES were not responsible for the mysterious deaths of local animals. Spirits returned in *The Ghosts of Buxley Hall* (1980), where the ghosts were determined to keep female cadets from entering an all-male military academy.

Disney turned next to the studio's theatrical movies as a source of inspiration. *Beyond Witch Mountain* (**1982** tvm), a sequel to *Escape to Witch Mountain* (**1975**) and *Return From Witch Mountain* (**1978**), was an unsold pilot for a new series; it co-starred Eddie Albert as a man who tried to help the alien children find their missing uncle. *Herbie, The Love Bug* (1982), a short-lived series, reunited Dean Jones with the lively Volkswagen Herbie. Another short-lived series, *Small & Frye* (1983), used a popular TECHNOFANTASY theme: a failed scientific experiment miniaturized a private detective.

After relatively few fantasy episodes over the first 30 years, Disney suddenly began to turn out a wealth of genre programming. Among these were two tv movies for the Disney Channel cable service, *Black Arrow* (**1985** tvm), a ROBIN HOOD-style story based on the Robert Louis STEVENSON novel, and *The Blue Yonder* (**1985** tvm), a TIME-TRAVEL yarn. A change in direction came with *Disney's Wuzzles* (1985) and *Disney's Adventures of the Gummi Bears* (1985), the company's first forays into Saturday-morning cartoon programming. The Wuzzles were strange creatures each composed of two different animals (e.g., Butterbear was part butterfly, part bear). The Gummi Bears were ancient creatures with advanced scientific and magical knowledge who came to the aid of humans in a medieval FANTASYLAND. The latter concept was popular enough to convince Disney there was a market for such shows. Another animated outing was *Disney's Fluppy Dogs* (**1986** tvm), about alien dogs trapped on Earth. The protagonist of *The Last Electric Knight* (**1986** tvm) was a teenager who could summon mystic energies and channel them into martial arts; it was popular enough to return as a weekly series, *SideKicks* (1986). The eponym of *The Richest Cat in the World* (**1986** tvm) was not only wealthy but could talk. IMMORTALITY was explored in *I-Man* (**1986** tvm), where an accident turned a cabdriver into a virtually indestructible government agent. *Mr Boogedy* (**1986** tvm) saw a mild-mannered family battling a ghost for the soul of a young boy; a sequel, *Bride of Boogedy* (**1987** tvm), found the ghost trying to possess (◊ POSSESSION) the family so he could live again.

YOUNG AGAIN (**1986** tvm) was the first of several new IDENTITY-EXCHANGE programmes; it featured Robert Urich and Lindsay Wagner in a tale of a man mysteriously granted his wish to be a teenager again. Another identity-exchange story was *Hero in the Family* (**1986** tvm), where an alien force caused an astronaut to trade minds with a chimp. Strange creatures invisible (◊ INVISIBILITY) to most people were featured in *Fuzzbucket* (**1986** tvm), where a young boy learned that his unusual friend was part of a WAINSCOT society. The next year provided a new identity-exchange story: in *The Return of the Shaggy Dog* (**1987** tvm), a sequel to the movie *The SHAGGY DOG* (**1959**), a cursed RING causes a boy and dog to "trade places". Other genre outings of the period included *Bigfoot* (**1987** tvm), where a Sasquatch kidnaps a young girl, and *Young Harry Houdini* (**1987** tvm), which offers the possibility that the MAGIC of Harry Houdini (1874-1926) was real.

Disney moved into syndicated tv with *DuckTales* (1987), an animated series featuring DONALD DUCK's nephews and their uncle, SCROOGE MCDUCK (◊◊ Carl BARKS). For the next several seasons they would encounter ghostly MUMMIES, a crazed sorceress, evil GENIES, Norse GODS and other menaces, and also make several trips back in TIME. This series was so successful that Disney moved heavily into tv animation; among the new series were *Chip 'n Dale's Rescue Rangers* (1989), *TaleSpin* (1990) – incorporating characters from Disney's *The Jungle Book* (**1967**) – and *Darkwing Duck* (1991).

A teenager who wanted to impress another student by seeming an adult was the theme of *14 Going on 30* (**1988** tvm), in which a machine aged the youngster in a matter of moments – then exploded, leaving him trapped in his new body. A ghostly detective helped solve his own murder in *Justin Case* (**1988** tvm), an unsold pilot starring comedian George Carlin. In a further unsold pilot, *Splash, Too* (**1988** tvm), a sequel to the hit movie SPLASH! (**1984**), the MERMAID and her new husband moved to suburbia and had to free a captive dolphin.

Another sequel was *The Absent Minded Professor* (**1988** tvm), based on the feature movie *The Absent Minded Professor* (**1961**). This TECHNOFANTASY offered a new explanation for the invention of "Flubber", Disney's famous "flying rubber", and was in turn followed by *The Absent Minded Professor: Trading Places* (**1989** tvm), in which the hapless inventor stumbled upon an evil industrialist and a deadly weapon. Both episodes starred comedian and magician Harry Anderson.

Singer/actress Olivia Newton-John was the star of *A Mom for Christmas* (**1990** tvm), where a lonely girl got her wish for a mother when a department store mannequin mysteriously came to life. A weekly series, *Disney Presents The 100 Lives of Black Jack Savage* (1991), followed the adventures of a crooked financier and the ghost of a ancient PIRATE, who unless they could save the lives of 100 people would be doomed to a terrible AFTERLIFE for their past crimes. A lighter theme was found in *Dinosaurs* (1991). Created in partnership with Jim HENSON, the series used elaborate puppet costumes to bring an ancient dinosaur family to life; the twist was that the DINOSAURS lived just as we do today, complete with cars and tv. An INVISIBLE COMPANION was featured in *Dayo* (**1992** tvm), which starred Delta Burke as a woman who was shocked to see her long-forgotten childhood friend return to her life.

The demise of Disney's weekly series has resulted in far fewer fantasy-themed programmes in recent years, but the studio still occasionally ventures into the genre; recent offerings have been remakes of the movies *The Computer Wore Tennis Shoes* (**1970**) and FREAKY FRIDAY (**1976**). With the purchase of the ABC network in 1995, Disney has announced plans for a new weekly anthology series, and thus all but assured there will be further fantasy programmes from the studio in the future. [BC]

COMICS

The first Disney character to appear in comics was **Mickey Mouse**, whose daily strip was distributed to US newspapers by King Features from January 13, 1930. Originally written by Walt and drawn by Ub IWERKS, it was soon entrusted to Floyd Gottfredson, who drew it until his retirement in 1975. Aided by scriptwriters like Ted Osborne, Merrill DeMaris and Bill Walsh, Gottfredson turned Mickey into a full-fledged adventurer, creating memorable continuities until 1955, when a gag-a-day format was adopted. A Sunday page – drawn by Gottfredson until 1938 and then by Manuel Gonzales until 1981 – was added in January 1932. The **Donald Duck** gag-a-day daily strip began February 7, 1938; it was written by Bob Karp until

1975 and drawn by Al Taliaferro until 1969; a Sunday page started December 10, 1939. **Brer Rabbit** (from the movie *Song of the South* [*1946*]) appeared in the **Uncle Remus** Sunday page 1945-72. **Scamp**, son of the stars of *Lady and the Tramp* (*1955*), had his own daily and Sunday strips 1955-88. The **Silly Symphony** (1932-45) and **Treasury of Classic Tales** (1952-87) Sunday series featured adaptations of or characters from most of the Disney short and feature animations. Only **Mickey Mouse** and **Donald Duck**, produced by King Features under Disney's supervision, survive among the newspaper strips.

Disney comic books started featuring original stories in 1941. Jack Bradbury, Carl BARKS and Tony Strobl were the best **Donald Duck** artists until the early 1970s. Paul Murry drew most characters, but is best remembered for the **Mickey Mouse** serial featured in *Walt Disney's Comics and Stories* 1953-73. Other excellent comic-book artists included Al Hubbard (on **Scamp**, **Chip'n'Dale** and feature-animation adaptations), Carl Buettner and Gil Turner (on **The Li'l Bad Wolf**), Harvey Eisenberg (on **Little Hiawatha**) and Ralph Heimdahl (on **Bucky Bug**). More recently, Don Rosa has created great **Donald Duck** stories that emulate the adventurous atmospheres created by Barks.

Original UK Disney comics first appeared in December 1936 with **The De(f)tective Agency**, starring Goofy and Toby Tortoise, started in *Mickey Mouse Weekly*; created by Wilfred Haughton, the series lasted about a year. Original **Donald Duck** adventures appeared in the same magazine 1937-40, written and drawn by William A. Ward, who paired Donald first with a prototype of Daisy Duck, Donna (from the 1937 animated short *Don Donald*), and then with a Scottish sailor named Mac. During the late 1940s and 1950s artist Ronald Neilson contributed a number of beautiful "painted" stories including adaptations of feature animations like CINDERELLA (1950) and (1951-3) episodes starring Mickey Mouse with Eega Beeva.

In Italy Federico Pedrocchi wrote and drew original **Donald Duck** stories from 1937; since then over 100 different artists have worked on Italian Disney comics. Guido Martina was a guiding light: he wrote most stories from 1948 through the 1980s. However, Romano Scarpa and Giovan Battista Carpi, both of whom started in 1953, better absorbed the US Disney comics tradition and mixed it with original Italian influences. Luciano Bottaro adapted **Mickey Mouse** and **Donald Duck** to his ingenious graphic style. In the 1960s and 1970s Giorgio Cavazzano and Massimo De Vita contributed new stylistic approaches to the Disney characters. The artist-writer Marco Rota has rendered the Ducks in a style inspired by Barks.

In France **Mickey Mouse** made his time-travelling appearance in 1952 with **Mickey à travers les siècles**, begun in *Le journal de Mickey* by writer Pierre Fallot and artist Ténas (Louis Saintels); the series was later drawn by Pierre Nicolas, who continued it until 1978. Today France's best Disney artists are Claude Marin and Gen-Clo (Claude Chebille).

In Holland **Donald Duck** first appeared 1953, drawn by Hungarian refugee Ed Lukacs, but only in the 1970s did a "Dutch School" develop, its foremost exponents being Jules Coenen, Daan Jippes, Dick Matena and Ben Verhagen. In Sweden original Disney comics art appeared as early as 1937-8, drawn by Lars Bylund and Birger Allernas. In Denmark the Gutenberghus/Egmont Publishing Group, which controls Disney publications in Northern and Eastern Europe, started its massive production of Disney comics in 1968. Scripts have come mostly from the UK, with most of the artists either being Spanish or working through Spanish studios. Among them, Vicar (Victor Arriagada Rios), from Chile, and Daniel Branca, from Argentina, are best at drawing the Ducks in the classic Barks style. The Spanish artist Miquel Pujol produces excellent work for all the characters, and since the late 1970s has been doing so also for other countries, including France, Holland and Italy.

South America has produced Disney comics since 1945, when **Donald Duck** serials, written and drawn by Luis Destuet, started in the Argentinian magazine *El Pato Donald*. Since 1975 the Argentinian studio headed by Jaime Díaz has been producing art on all Disney characters for both the US comics and the international market. In Brazil, Abril Publishing has released original Disney strips since 1961; among the most popular characters there are **José Carioca**, Donald's screwball cousin **Fethry Duck** and the white-bearded hillbilly **Hard Haid Moe** (the latter two created 1963 by writer Dick Kinney and artist Al Hubbard). Among the best Brazilian artists have been Carlos Herrero, Luiz Podavin and the brothers Moacir and Irineu Soares Rodrigues. [AB]

Further reading: *The Art of Walt Disney* (**1942**) by Robert D. Feild; *The Art of Animation: The Story of the Disney Studio Contribution to a New Art* (**1958**) by Bob Thomas; *The Disney Version* (**1968**; vt *Walt Disney 1968* UK; rev under original title 1985 US) by Richard Schickel; *The Art of Walt Disney* (**1973**; cut 1988) by Christopher Finch; *Disney Animation: The Illusion of Life* (**1981**) by Frank Thomas and Ollie Johnston; *Walt Disney's World of Fantasy* (**1982**) by Adrian Bailey; *The Disney Films* (**1973**; rev 1984) by Leonard Maltin; *Encyclopedia of Walt Disney's Animated Characters* (**1987**; 2nd edn 1993; corrected 1993) by John GRANT; *The Disney Studio Story* (**1988**) by Richard Holliss and Brian Sibley; *Disney's Art of Animation – from Mickey Mouse to Beauty and the Beast* (**1991**) by Bob Thomas.

DISNEY, WALT(ER ELIAS) (1901-1966) US producer and director of ANIMATED MOVIES, and co-creator with Ub IWERKS of MICKEY MOUSE. WD began his career producing short pieces of animation for commercials, then – with Iwerks – set up a small animation studio in Kansas City producing shorts for use as fillers at a local cinema. These **Newman Laugh-O-grams**, or **Lafflets**, typically lasted less than a minute and focused on local issues. In 1922 WD created six more substantial **Laugh-O-grams**: *Cinderella*, *The Four Musicians of Bremen*, *Goldie Locks and the Three Bears*, *Jack and the Beanstalk*, *Little Red Riding Hood* and *Puss in Boots*; the notion of animating and modernizing classic tales guided him throughout his life. These shorts did not pay the bills, and in late 1922 WD accepted a commission to produce a health-education short called *Tommy Tucker's Tooth*. The money from this enabled him to try a new venture: the **Alice Comedies** series, of which there were 57 in all during 1924-7 (including the first, *Alice's Wonderland*, made in 1923 but never properly released). In each, a young live-action girl, Alice – three girls took the role during the series' run – for one reason or another finds herself plunged into a TOON world and interacting with the characters there. The surviving **Alice Comedies** are crude but charming. When they had run their course WD's studio, now in Hollywood, embarked on the **Oswald the Lucky Rabbit** series of 26 shorts (*1927-8*), which enjoyed some success.

Oswald can be seen today as Mickey Mouse with floppy ears and a different tail; there is no real distinction between the later **Oswald** shorts and the early **Mickey Mouse** ones, and the **Oswald** series might have been continued indefinitely had not WD been ripped off by his distributor, Charles Mintz, so that further **Oswald** adventures came from Walter LANTZ. Desperate for a new series character, WD had Iwerks draw up the schema for what was initially Mortimer, then MICKEY MOUSE; WD's contribution to this creation was the "invention of Mickey's character" – WD did indeed voice the Mouse for a considerable period, and his animators based the character's mannerisms on WD's own. Mickey might have come to nothing had it not been for WD's recognition that *sound* was where the future of animation lay. The **Mickey Mouse** cartoon *Steamboat Willie* (*1928*) is recorded as the first animated "talkie", although in fact its soundtrack consists of little more than shrieks and crashes (the short was made as a silent, then the soundtrack was cobbled on) and earlier sound cartoons had been made by Max FLEISCHER and Paul Terry. Yet it was a breakthrough in that WD and his collaborators had ensured the sound was properly synchronized with the on-screen action, so that visuals and sound effects married to create the illusion that they were all of a piece. (Speech came later: Mickey Mouse first spoke in *Karnival Kid* [*1929*].)

WD's other great breakthrough – by this time he recognized that he could employ far better animators than he was himself – was the realization that audiences could well be persuaded to sit through a full-length animated feature. This flew directly in the face of received Hollywood wisdom, which had it that animation was purely for kids, whose attention-spans would not extend above a few minutes. WD launched his studio into what many of its members regarded at the time as "Disney's Folly": SNOW WHITE AND THE SEVEN DWARFS (*1937*), a project that WD originated and then supervised through every stage of its creation. By the time of its release this movie had cost a then-staggering $1,480,000, much of this bill being due to WD's perfectionism – there can be no accurate estimate of the cost of script, animation, sound, etc., rejected and destroyed because WD did not feel it was "quite right". It seemed impossible that this sum could ever be recouped; but of course it was.

It has become popular to downgrade WD's integral role in the further creations of his studio, but in fact he maintained strict control over its output for the rest of his life and personally developed many of its best productions. It has also become popular to assail his private life and ethics. At this remove it is difficult to establish the truth of such attacks, but it seems that indeed he smoked and drank to excess and swore filthily; on the other hand, it seems he was not consciously racist. That he was an ogre in some ways is not questioned, and certainly he was not quite the "favourite uncle" that he publicly portrayed. But WD made fantasy *happen* on screen. [JG]

Further reading: *Walt Disney: An American Original* (**1976**; vt *The Walt Disney Biography* 1977 UK) by Bob Thomas (a hagiography) and *Disney's World* (**1985**; vt *The Real Walt Disney* 1986 UK) by Leonard Mosley (a muckrake).

DISRAELI, BENJAMIN (1804-1881) UK statesman and writer, created the first Earl of Beaconsfield in 1876 during his second term in office as Prime Minister. Most of his fiction, beginning with *Vivian Grey* (**1826-7**), takes a satirical or reformist attitude to UK society; and his stories of the

fantastic tend similarly to be understandable primarily as commentaries on contemporary events. *The Voyage of Captain Popanilla* (**1828**) is a FANTASTIC VOYAGE in the mode of Jonathan SWIFT, though in the opposite direction: the eponymous captain travels to England and finds it remarkably surreal; the existing version constitutes BD's rewriting of his first novel, «The Adventures of Mr Aylmer Papillon in a Terra Incognita», which he had sent in 1824 to the publisher John Murray, who burned it (Murray also burnt BYRON's journal). *The Wondrous Tale of Alroy, and The Rise of Iskander* (**1833**; vt *Alroy: A Romance* 1846) is a FANTASY OF HISTORY in which Alroy, a Prince of the Captivity (i.e., a descendant of King David), founds a world-dominating kingdom centred in 12th-century BAGHDAD. *Ixion in Heaven* (1832-3 *Colburn's New Monthly*; **1925** chap) and *The Infernal Marriage* (1833-4 *Colburn's New Monthly*; **1929** chap) both appeared in various collections, including *Alroy; Ixion in Heaven; The Infernal Marriage; Popanilla* (coll **1845**) and *Ixion; The Infernal Marriage; Popanilla; Count Alarcos* (coll **1853**). The two tales are SATIRES of contemporary England set in the world of the ancient Greek MYTHS. The "infernal marriage" of the second is that of Proserpine to the King of Hell; the grim fate of Ixion – a treacherous monarch who is bound to a wheel of fire in Hell because he boasted of having slept with Hera – is treated by BD, fairly lightly, as an analogue of the treatment meted out to himself as a Jew attempting a political career in 19th-century Europe. [JC]

Other works: *The Tragedy of Count Alarcos* (**1839**), a narrative poem; *Tales and Sketches by the Right Hon. Benjamin Disraeli, Earl of Beaconsfield* (coll **1891**), including several fantasies; *Alroy; Popanilla; Count Alarcos; Ixion in Heaven* (coll **1906**); *Popanilla, and Other Tales* (coll **1926**); *The Dunciad of To-Day: A Satire; and The Modern Aesop* (1826 *The Star Chamber*; **1928** chap) (◊ AESOPIAN FANTASY).

DITKO, STEVE (1927-   ) US COMIC-book artist with a strong, simple line style, noted for his pioneering creative work on several very distinctive comics characters. He did his first comics work for *Black Magic Magazine* (vol 4 #3 1953), then went on to draw further HORROR strips for *Black Magic* and both strips and covers for *Strange Suspense Stories* (#18-#20, #22, #31-#53 1954-7) and *Fantastic Fears* (#5 1954).

SD began to develop his very personal style while drawing *Tales of the Mysterious Traveler* (#2-#11 1956-9) for Charlton (◊ *The* MYSTERIOUS TRAVELER), and soon thereafter started drawing **Captain Atom** in *Space Adventures* (#33-#40 1960). He began a brief but fruitful association with Stan LEE, drawing **Spiderman** (beginning in MARVEL COMICS's *Amazing Fantasy #15* 1962), in which his idiosyncratic portrayals of the paranoid, guilt-ridden SUPERHERO brought the character phenomenal success and its own comic book, *The Amazing Spiderman* (1963-current). Then, in collaboration with Lee and Jack Kirby, SD created DOCTOR STRANGE (in *Strange Adventures #110* 1963). In 1966 he abandoned both strips when a dispute arose with Lee.

SD was then at the height of his creative powers, and produced some outstanding material for WARREN PUBLISHING's EERIE (#3-#10 1966-7) and CREEPY (#9-#16 1966-7 and #25-#27 1969), plus work for other major companies; he revamped two moribund Charlton superhero characters – the two-toned, gadget-minded *Blue Beetle* (vol 2 #1-#5 1967-8) and *Captain Atom* (vol 2 #78-#89 1965-7). For the latter company he also created another superhero, **The Question** (*Blue Beetle #1* 1967). SD's original and provocative handling of this character reached an imaginative highpoint in *Mysterious Suspense #1* (1968), widely considered the finest commercial comic book of the period.

SD decamped to National Periodical Publications (◊ DC COMICS), where he created **The Hawk and the Dove** (*Showcase #75* 1968) – who continued their career in *The Hawk and the Dove #1-#2* (1968) – and **The Creeper** (*Showcase #73* 1967). SD's illness was the reason for the short-livedness of these publications.

SD returned to Charlton in late 1969 to draw for their horror titles. He has, however, continued to address philosophical concepts with **Mr A**, which he drew for a number of limited-edition magazines including *Witzend*, and has published collections of his work including *Mr A* (graph coll **1973**) and *Avenging World* (graph coll **1974**). [RT]

DJINNS ◊ GENIES.

DOC SAVAGE (magazine) ◊ MAGAZINES.

DOCTOR DEATH ◊ MAGAZINES.

DR DOLITTLE US movie (*1967*). Apjac/20th Century-Fox. **Pr** Arthur P. Jacobs. **Dir** Richard Fleischer. **Spfx** L.B. Abbott, Art Cruickshank, Emil Kosa Jr, Howard Lydecker. **Screenplay and songs** Leslie Bricusse. **Based on** the **Doctor Dolittle** series by Hugh LOFTING, mainly *The Voyages of Doctor Dolittle* (**1922**). **Starring** Richard Attenborough (Albert Blossom), William Dix (Tommy Stubbins), Samantha Eggar (Emma Fairfax), Rex Harrison (Dr John Dolittle), Geoffrey Holder (William Shakespeare X), Anthony Newley (Matthew Mugg). 152 mins. Colour.

Puddleby-on-the-Marsh, England, 1845. Young Tommy Stubbins meets Dr Dolittle who, unhappy dealing with people, has for years consorted largely with animals, whose languages he speaks. Escaping committal to Bedlam, Dolittle goes with Stubbins and friends Mugg and Fairfax on a FANTASTIC VOYAGE in QUEST of the Great Pink Sea Snail, which they discover has been living beneath the Floating ISLAND. Dolittle stays behind as the other three sail home aboard the gastropod; he follows astride the Giant Lunar Moth on hearing that all England's animals have gone on strike to have his sentence quashed. This hugely expensive musical was a disaster in every way, but is of some interest in its philistine portrayal of Dolittle as a social inadequate rather than as a brave zoological/ linguistic pioneer. [JG]

DR FAUSTUS UK movie (*1967*). ◊ FAUST.

DOCTOR HACKENSTEIN US movie (*1989*). ◊ FRANKENSTEIN MOVIES.

DR HECKYL AND MR HYPE US movie (*1980*). ◊ JEKYLL AND HYDE MOVIES.

DR JEKYLL AND MR HYDE Various movies (*1920/1931/1941/1968/1973/1980/1981*). ◊ JEKYLL AND HYDE MOVIES.

DR JEKYLL AND SISTER HYDE UK movie (*1971*). ◊ JEKYLL AND HYDE MOVIES.

DOCTOROW, E(DGAR) L(AURENCE) (1931-   ) US writer best-known for *Ragtime* (**1975**), one of a series of novels which attempts to recreate the US past in terms which float – at times uneasily – between ALTERNATE-WORLD devices and MAGIC REALISM. Most of ELD's fiction remains, despite hints of an upwelling otherness, within the mundane world. Exceptions include *Big as Life* (**1966**), which depicts the effects of a sudden unexplained irruption of GIANTS onto the streets of NEW YORK. *The Waterworks* (**1994**), set like *Ragtime* in the myth-engendering US past, is a GASLIGHT ROMANCE in which a mad scientist has learnt how to confer IMMORTALITY, but at terrible cost. [JC]

DOCTOR STRANGE Two very different COMIC-book characters.

1. Crimefighting SUPERHERO scientist star of *Thrilling Comics* 1940-48. He gained limitless power through taking draughts of Alosun, a distillate of sun atoms. At first he fought crime in a business suit, but in *Thrilling #7* (1940) he formulated a new, improved Alosun which endowed him with even greater powers, including the ability to "soar through the air as if winged", at which time he began to affect a red teeshirt, blue riding breeches and black boots. In 1942 (*Thrilling #24*) he gained a boy companion, Mike, and together they fought crime and the Nazi menace. [RT]

2. Arrogant, moustachioed "Master of the Mystic Arts", whose costume consists of an enchanted gold-edged red cloak with a flamboyant high collar. Created by Stan LEE and Steve DITKO in *Strange Tales #110* (1963), Dr Stephen Strange, who lost his amazing surgical skill through nerve damage caused by a car accident, seeks out in the Himalayas the mystical Ancient One, from whom he learns the secrets of white MAGIC. DS goes on to save the Earth from diverse occult aggressors. The beautiful Clea became DS's love interest in *Strange Tales #126* (1964).

Strange Tales was renamed *Doctor Strange* in 1968, and since then DS has featured in a number of comic books, including *The Defenders*, *Marvel Premiere* and *Marvel Fanfare*, as well as a few one-issue titles and short series bearing his name. A new **DS** series introduced in 1988, *Doctor Strange, Sorcerer Supreme*, continues at time of writing (1995). Many of the top names in comics have lent their talents to DS since Lee and Ditko left the character in the 1960s. [RT]

DOLL HOUSE ◊ DOLLS.

DOLLS Dolls have been a common feature of SUPERNATURAL FICTION for over 200 years. There are two main features. On the passive side is doll MAGIC, a branch of sympathetic magic; many tales featuring VOODOO, WITCHCRAFT, obeah and African magic include doll magic; examples are "The Hag Seleen" (1942) by Theodore STURGEON, "Death in Peru" (1954) by Joseph Payne BRENNAN, "Miss Esperson" (1962) by August DERLETH and "Dolls" (1976) by Ramsey CAMPBELL.

The more active use of the motif is where the doll comes alive. Quite commonly in CHILDREN'S FANTASY the doll is friendly; this particular subgenre may trace its origins back to the staid and moralistic *The Adventures of a Pincushion* (?1780) by Mary Ann Kilner; more enterprising were *Memoirs of a London Doll* (**1846**) by Richard Henry Horne (1803-1884) and *The Enchanted Doll* (**1849**) by Mark Lemon (1809-1870). Perhaps the best-known doll – or PUPPET – story from the Victorian age is *Pinocchio* (**1883**) by Carlo COLLODI. TOYS, dolls, puppets and teddy-bears are standard characters in stories for very young children – Enid Blyton's **Noddy** is perhaps the most famous example. It is in books for older children that the theme is best used in fantasy. The most complete writer of doll stories was Rumer Godden (1907-   ), who produced six children's books starting with *The Doll's House* (**1947**), *Impunity Jane* (**1954**) being the most entertaining. Pauline CLARKE has also contributed extensively, firstly with her **Five Dolls** series, written as Helen Clare, and then with *The Twelve and the Genii* (**1962**; vt *The Return of the Twelves* 1963 US), which brings back to life the original toy soldiers owned by Charlotte BRONTË and her siblings.

The exploration of the relationship between dolls and

adult humans is usually done with more sinister intent. This concept in fiction goes back at least as far as *La Poupée* (**1744**; trans *The Fairy Doll* 1925 UK; vt *Amorous Philandre* 1948 US) by Jean Galli de Bibiena (?1710-?1780), where a doll becomes animated by an air ELEMENTAL and agrees to help a young priest in his amorous adventures. The sexual connotations of animated dolls as well as their physical perfection has not been overlooked by generations of writers (◊ SEX). It was the same thinking that inspired E.T.A. HOFFMANN to include in "Der Sandmann" ["The Sandman"] (1816) Olimpia, a beautiful girl who is in fact a clockwork AUTOMATON. Hoffmann returned to the theme in his *Kunstmärchen* "Nußknacker und Mausekönig" ["The Nutcracker and the Mouseking"] (1819), where the nutcracker is a young man under a spell, and leads a band of toy soldiers into battle. Hans Christian ANDERSEN translated this into his own fairytales in "The Steadfast Tin Soldier" (1838). The Tin Woodman, created by L. Frank BAUM in *The Wonderful Wizard of Oz* (**1900**), is a continuation of this same image.

Dolls can be as equally used for EVIL, which is the route that Fitz-James O'BRIEN took in "The Wondersmith" (1859), where mannikins armed with poisoned swords are animated by entrapped SOULS. This is a far more powerful and lasting vision, and has been used by many writers since, including: A. MERRITT, whose *Burn, Witch, Burn!* (1932 *Argosy*; **1933**), filmed as *The* DEVIL-DOLL (**1936**), has a very similar plotline to "The Wondersmith", featuring a witch who captures souls to animate murderous dolls; Fredric BROWN, whose dolls in "The Geezenstacks" (1943) somehow preordain events; Algernon BLACKWOOD, who uses tribal magic to animate a child's doll for revenge in "The Doll" (1946); SARBAN, in the title story of *The Doll Maker, and Other Tales of the Uncanny* (coll **1953**), where a landowner has created his own miniature world through magic and peoples it with dolls who are in fact the dead or ensorcelled re-animated; Thomas LIGOTTI in "Dr Voke and Mr Veech" (1983); and Robert WESTALL in "The Doll" (1989), about an antique doll which retains a residuum of the spirit of its original owner.

Haunted dolls and doll houses have tempted several writers, not least M.R. JAMES, who tells of a doll house modelled on a real one which replicates the scene of a murder in "The Haunted Dolls' House" (1923). Jack Snow has the spirits of two recently deceased children live on in their favourite doll house in "'Let's Play House'" (1947), while in "The Doll's Ghost" (1911) F. Marion CRAWFORD tells of a doll doctor who is visited by the GHOST of a doll to warn him his daughter is in danger.

A favourite doll theme is that of the ventriloquist's dummy (◊ VENTRILOQUISM), where the relationship between the ventriloquist and his doll can lead to schizophrenia and IDENTITY EXCHANGE. Also linked to the doll theme, often using similar aspects of POSSESSION, is that of the UK's bonfire-night guy. The guy, though, is less a doll than something more pagan, to which we may link corn dolls and the wicker man (◊ GREEN MAN).

An anthology of doll stories is *The Haunted Dolls* (anth **1980**) ed Seon Manley (1921-   ) and Gogo Lewis. [MA]

DOLOROUS STROKE In some of the legends of ARTHUR, the stroke or blow delivered by Sir Balin that led to the WASTE LAND and the need for the QUEST for the Holy GRAIL (in other versions the blame is attached to PERCEVAL for failing to ask the FISHER KING the meaning of the Grail). It is first referred to in the *Suite de Merlin* (13th century) and in Book II of Sir Thomas MALORY's *Le Morte Darthur* (**1485**). [MA]

DOMECQ, H. BUSTOS Joint pseudonym of Adolfo BIOY CASARES and Jorge Luis BORGES.

DONALD DUCK Probably the most popular DISNEY animated creation, and the one who has made the most appearances, although, for studio-determined reasons, he has not achieved the same iconic status (◊ ICONS) as MICKEY MOUSE. He first appeared as a supporting character in the short *The Wise Little Hen* (**1934**), and essentially stole the show. In at least 173 movie appearances since then his character has developed from that of an incomprehensibly spluttering crosspatch whose sole joke was his pyrotechnical loss of temper, through being a TRICKSTER whose tricks always backfired, to become a figure of affection, although still as always volatile. His beloved Daisy has, surprisingly, made only 15 screen appearances, the first – where she was called Donna Duck – in *Don Donald* (**1937**). His nephews Huey, Dewey and Louie made their debut in *Donald's Nephews* (**1938**), but their greater significance came when Carl BARKS teamed them with SCROOGE MCDUCK in his comics, whence they returned – with Scrooge – to the screen in the form of the tv series *DuckTales* (from 1987) and an ANIMATED MOVIE, DUCKTALES: THE MOVIE – TREASURE OF THE LOST LAMP (**1990**). DD can be seen as an UNDERLIER in many fantasy texts where short-fused characters prove to have hearts of gold; because of the age-group of modern fantasy writers, one is tempted to regard the DD shorts as a sort of TAPROOT TEXT, to be drawn upon consciously or unconsciously. [JG]

DONALDSON, STEPHEN R(EEDER) (1947-   ) US writer of central significance as an author of demanding and exploratory fantasy novels, beginning with the **Chronicles of Thomas Covenant the Unbeliever**, which appeared as two linked but inherently different sequences; he won the John W. Campbell AWARD for most promising writer in 1979 on the basis of the initial **Covenant** volume. The first (and more impressive) of the two sequences, now entitled **The First Chronicles of Thomas Covenant the Unbeliever**, comprises *Lord Foul's Bane* (**1977**), *The Illearth War* (**1977**) and *The Power that Preserves* (**1977**), assembled as *The Chronicles of Thomas Covenant the Unbeliever* (omni **1993** UK); the **Second Chronicles of Thomas Covenant the Unbeliever** is *The Wounded Land* (**1980**), *The One Tree* (**1982**) and *White Gold Wielder* (**1983**), assembled as *The Second Chronicles of Thomas Covenant the Unbeliever* (omni **1994** UK); a slim pendant to the sequences, *Gilden-Fire* (**1981** chap), is an out-take from *The Illearth War*.

The first sequence concentrates on Covenant's slowly waning refusal to believe in the REALITY of the SECONDARY WORLD into which he has been catapulted. This unbelief is perhaps SRD's most original single invention, for it radically transfigures every moment of the first sequence and profoundly contradicts the reader's normal expectations about the relationships between the HERO and the LAND, the QUEST and his COMPANIONS, plus the overall relationship to the decorum and moral requirements that define the condition of being a Hero. It thoroughly exposes the artifact of the normal fantasy SECONDARY WORLD as a stage-set for the deeds of protagonists whose every act is deeply patriotic, deeply land- and folk-affirming.

In the real world – the 1970s USA – Thomas Covenant is a leper, and he maintains a precarious grip on reality through a constant monitoring of his condition, an act which constitutes a necessary refusal to pretend that his condition does not exist. Every time he contacts something beyond himself, something rubs off from him, and he dies a little more. When he is transported into the LAND, its seductions – as in most secondary worlds, this land is a kind of EDEN – are anathema to him for this reason, and so even here he must continue with his obsessive VSE, the Visual Surveillance of Extremities which lepers, who cannot *feel* their extremities, perform in order to check on whether they remain intact. As far as Covenant is concerned, to surrender, to believe in the new reality, would be to commit suicide: he is a CHILDE who refuses to act as a childe.

The land itself is under threat from Lord Foul, a DARK LORD whose only escape from the BONDAGE of being intrinsic to the land is to dissolve it. He is in a way himself a leper, longing to dissolve his body, and in this sense is Covenant's true dark SHADOW. Only slowly does Covenant understand that to preserve the land is to preserve himself (between each of the three volumes of the **First Chronicles** Covenant returns to the world for a short period, while years pass in the land; and each time he returns the land is significantly more diseased). The surface events resemble those of most HIGH-FANTASY novels since J.R.R. TOLKIEN's *The Lord of the Rings* (**1954-5**), and superficial resemblances between that book and **Covenant** are very numerous. The basic heroics of the surface plot are, therefore, familiar . . . up to a point. For different readers that point will perhaps differ: for some it may come when Covenant rapes a young woman (because his extremities have been reawoken, and because he thinks he remains in a weird fever dream); for others it may come at the conclusion of the second volume, when the land's chief defenders Hile Troy and Elena both fail – one as soldier and hero, the other as wielder of white magic – to do anything but worsen the situation; Elena dies offstage, in a manner presumably horrid, and Troy undergoes TRANSFORMATION to tree form.

As the third volume opens, Covenant returns to find the land desperately diseased, but finally he triumphs over Lord Foul by dint of walking a kind of tightrope of belief/unbelief, by refusing on the one hand completely to surrender to the lures of the land (which would give Foul the chance to dissolve both) and refusing on the other to negate the land utterly (which would have the same result). Defeated in part by the cleansing laughter of a dying COMPANION of Covenant's, Foul subsides until the next sequence. In the **Second Chronicles**, the intimate relationship between Covenant and land is allowed to fade, and the land becomes a more conventional secondary world. At the same time the "superficial" High Fantasy texture is even further subverted. There are no sustained battles to rescue the WASTE LAND from desiccation, no climactic scenes in which right defeats might (◊ MAGGOTS), no survival, in the end, for Covenant himself. He spends much of the trilogy comatose (rather like Jerry Cornelius in the latter parts of Michael MOORCOCK's **Cornelius Chronicles** [1968-77]), and when he is awake is long bemired in guilt for what he has done and for what he is – proof of which he can see all around him, in the horrific destitution of the land. The **First Chronicles** have become a central STORY, thousands of years later, for the inhabitants of the land in the **Second Chronicles**, but that Story has itself become distorted and diseased. Only when Covenant finally comes to understand himself, and the fact that he must die in order that the Land become whole again, can the

sequence end in a kind of subdued HEALING.

SRD's second series, **Mordant's Need** – *The Mirror of Her Dreams* (**1986**) and *A Man Rides Through* (**1987**) – is, compared to the **Covenant** books, a kind of scherzo, and is told in a style much less burdened with image and import. The female protagonist, having been translated through a MIRROR into a secondary world in crisis, gradually comes to grips with herself, with the huge cast of characters, and with her sexual awakening. Much of the tale takes place inside a vast, intricate EDIFICE. Outside, an equally complex war builds in intensity; but in the end all turns out well. SRD is not a lighthearted writer, but **Mordant's Need** develops a very considerable comic momentum before justice and true LOVE triumph.

Each of SRD's three main sequences – the sf **Gap** sequence included – represents an almost obsessional working out of motifs and actions. **Mordant's Need** and **Gap** share a further characteristic: each is a crescendo. SRD's works move, in other words, towards their endings, and it is unsafe to attempt to understand him until he has had the last word. [JC]

Other works: *Daughter of Regals and Other Tales* (coll **1984**; cut vt *Daughter of Regals* 1984); *Epic Fantasy in the Modern World: A Few Observations* (**1986** chap), nonfiction.

Other works (sf): The **Gap** sequence, being *The Gap into Conflict: The Real Story* (**1990** UK), *The Gap into Vision: Forbidden Knowledge* (**1991**), *The Gap into Power: A Dark and Hungry God Arises* (**1992**), *The Gap into Madness: Chaos and Order* (**1994**) and «The Gap into Ruin: This Day All Gods Die» (1996).

As Reed Stephens: An associational detective sequence comprising *The Man who Killed his Brother* (**1980**), *The Man who Risked his Partner* (**1984**) and *The Man who Tried to Get Away* (**1990**).

Further reading: *Stephen R. Donaldson's Chronicles of Thomas Covenant: Variations on the Fantasy Tradition* (**1995**) by W.A. Senior (1953-   ).

DON GIOVANNI ◊ DON JUAN; OPERA.

DONIZETTI, GAETANO (1797-1848) Italian composer. ◊ W.S. GILBERT; OPERA.

DON JUAN The classic Don Juan/Don Giovanni situation comprises two elements: disrespect to the dead (flouting TABOOS) and the exploits of a ruthless sexual athlete. The first motif is ancient. The somewhat intricate combination and equation of impiety and sexuality, however, first became important in the Renaissance, with the drama *El burlador de Sevilla y convidado de piedra* (written 1612-16?) by the Spanish playwright Tirso de Molina (real name Fr. Gabriel Téllez; *c*1571-1648); it survives in two texts. (*Burlador* is variously translatable as "deceiver", "seducer", "playboy", "prankster", "trickster", etc.) Tirso established the main features of the conte: Don Juan Tenorio shabbily seduces women in Italy and Spain, commits general roguery, rejects redemption, insults the dead, and is dragged off to HELL by the STATUE of a man whom he wronged and murdered. Nevertheless, he is a hidalgo of Spain and (apart from with women) is a man of honour. The *Burlador* is episodic and bare by comparison with the richness of contemporaneous English drama, but rapidly became an international work, eventually giving birth to hundreds of imitations, adaptations and comparable plays, novels and poems in most of the languages of Europe. *Dom Juan, ou le festin de Pierre* (**1665**) by Molière (◊ COMMEDIA DELL'ARTE) was accused of being a subtle attack on religion and morals and was withdrawn from the stage. In England Thomas Shadwell (c1642-1692) in his *The Libertine* (**1676**) offered a brutal study of an antinomian sensualist who invoked nature and reason to justify his crimes.

The outstanding exemplification of Don Juan/Don Giovanni has been the OPERA *Il dissoluto punito o Don Giovanni* (**1787**; usually known as *Don Giovanni*) by Lorenzo da Ponte and Wolfgang Amadeus Mozart, where an excellent, witty libretto with clear characterizations served as the base for incomparable music. But the story was operatically popular before *Don Giovanni*: *Don Giovanni o sia il convitato di pietro* (**1787**) by librettist Giovane Bertati and composer Giuseppe Gazzaniga held the stage in Central Europe a few months before the da Ponte/Mozart work. There have been many other uses of the story in music, outstanding among which are Gluck's ballet music for Angiolini's ballet *Le festin de Pierre* ["The Feast of Pierre"] (1761), *Kamennyi Gost* ["The Stone Guest"] (1872) by Pushkin/Dargomyzhsky/Rimsky Korsakov, and Richard Strauss's symphonic poem *Don Juan* (1888), based on Nikolaus Lenau's unfinished *Don Juan*.

As has been observed by Oscar Mandel in *The Theatre of Don Juan* (**1963**), Don Juan has no real, solid identity, but typifies succeeding cultural periods. In the Renaissance work of Tirso he is a believer, though a driven, remorseless, wicked man; in the proto-Enlightenment of Molière and Shadwell he is a rationalist and cynic; in the early 19th century he is a passionate, idealized figure, as in C.D. Grabbe's *Don Juan und Faust* (**1829**), where FAUST, as Intellect, and Juan, as Passion, both seek Donna Anna, the Ideal; in the early 20th century *Man and Superman* (**1905**) by George Bernard Shaw (1856-1950) has Juan is entangled with the lifeforce and paradoxes of sexual pursuit. Later in the 20th century he is likely to be a man obsessed with problems of existence, with prying behind appearances.

Don Juan's fate, too, has been recalculated. For Tirso, Don Giovanni sat down with the statue to face a horrible meal of snakes and scorpions and went to a traditional HELL, while for da Ponte/Mozart the more amiable Don Giovanni's fate is – if George Bernard SHAW is to be believed – the HADES of Persephone and Pluto. In the play *Don Juan Tenorio* (**1844**) by José Zorrilla y Moral (1817-1893), still enormously popular in the Spanish-speaking world, Don Juan, inspired by the GHOST of Doña Ines (the equivalent of Donna Anna), repents and is saved, while in Vernon LEE's sentimental "The Virgin of the Seven Daggers" (1889) Don Juan, though a scoundrel, has such a powerful devotion to the Virgin that she redeems his soul after his death. In *La dernière nuit de Don Juan* ["The Last Night of Don Juan"] (**1921**) by Edmond Rostand (1868-1918) Juan, stripped of illusions by the DEVIL, demands hellfire; the Devil simply places him in a PUPPET show where he will act out his crimes indefinitely. And in *Don Juan, oder Die Liebe zur Geometrie* (**1953**) by Max Frisch (1911-1991) Juan, working in collusion with a cardinal, fakes his descent into Hell in order to escape the women who importune him.

The legend's supernatural aspect, today, has little power. Don Juan is remembered only as the effortless seducer.

[EFB]

Further reading: *The Story of Don Juan* (**1939**) by John Austin, a general, somewhat erratic account; *The Theatre of Don Juan. A Collection of Plays and Views, 1630-1963* (**1963**) ed Oscar Mandel, critical commentary plus translated texts of Tirso, Molière, Shadwell, Grabbe, Frisch and others;

Don Giovanni by Wolfgang Amadeus Mozart (**1964**) by Ellen Bleiler, a study of Mozart's opera, with translated text; *1003 Variationen des Don-Juan-Stoffes von 1630 bis 1934* (**1990**) by Franz Rauhut, good critical summaries of about 150 texts.

DONKEY ◊ ASS.

DONNELLY, IGNATIUS (1831-1901) US writer and politician, author of *Atlantis: The Antediluvian World* (**1882**). ◊ ANTHROPOLOGY.

DON QUIXOTE ◊ Miguel de CERVANTES; KNIGHT OF THE DOLEFUL COUNTENANCE.

DON'T LOOK NOW UK/Italian movie (*1973*). Casey/Eldorado. **Pr** Peter Katz. **Exec pr** Anthony B. Unger. **Dir** Nicolas Roeg. **Screenplay** Chris Bryant, Alan Scott. **Based on** "Don't Look Now" (1971) by Daphne DU MAURIER. **Starring** Julie Christie (Laura Baxter), Hilary Mason (Heather), Clelia Matania (Wendy), Nicholas Salter (Johnny Baxter), Renato Scarpa (Inspector Longhi), Massimo Serato (Bishop Barbarrigo), Donald Sutherland (John Baxter), Leopoldo Trieste (hotel manager), Sharon Williams (Christine Baxter). 110 mins. Colour.

John and Laura Baxter's daughter Christine drowns, an event of which he has some precognitive knowledge, although he is not aware of his latent TALENTS. The couple go to Venice, where he works restoring a church; currently a SERIAL KILLER haunts the city. The couple encounters a DUO of weird sisters: Wendy and blind, psychic Heather. The latter warns that John's life is in danger should he remain in Venice; a near-fatal fall persuades him the PROPHECY has been fulfilled and he stays. Laura goes home briefly because of son Johnny's illness; in her absence John "sees" her aboard a funeral cortège with the two sisters, and, believing her abducted, brings in the police. But his FATE cannot be avoided. John pursues through the night-time streets a small, scuttling figure whom he believes is the GHOST of his dead daughter. When he catches her he has one last, too-late flash of precognition: a hideous dwarf, she is the serial killer, and almost ruefully she cuts his throat.

This GOTHIC FANTASY is one of the most frightening – and most beautiful – movies ever made, its climax even more shattering because of the leisurely way in which the menace and the dislocating effect of John's jumbled future and present visions are built up. Its plot is, one realizes afterwards, highly contrived, yet this hindsight observation does nothing to diminish *DLN*'s unarguable power on further viewings. [JG]

DOOLITTLE, HILDA ◊ H.D.

DOPPELGÄNGER Literally, double-walker. While a protagonist and his/her DOUBLE may have no "blood" relationship, a doppelgänger is always intimately connected to the person in whose footprints he walks. A doppelgänger may therefore be a GHOST double, or an ASTRAL BODY tied to its flesh mirror; and – again unlike the double – may be a projection of the original person whose likeness it takes or mocks (◊ PARODY). The doppelgänger is more common in SUPERNATURAL FICTION than in FANTASY. Robert Louis STEVENSON, in *Strange Case of Dr Jekyll and Mr Hyde* (**1866**), H.G. WELLS, in *The Island of Doctor Moreau* (**1896**), and Bram STOKER, in *Dracula* (**1897**), vividly (and differently) dramatize the polarities between these two aspects. Almost invariably, the doppelgänger represents that which has been repressed.

In 20th-century fiction, and in fantasy as a whole, many of the functions of the doppelgänger have been taken over by the SHADOW. [JC]

DORÉ, GUSTAVE (1832-1883) French illustrator and painter, active in London from 1868. He founded the Doré Gallery in 1869 (in quarters currently occupied by Sotheby's); here his grandiloquent, moody Romanticism attracted a large public – and the obloquy of John RUSKIN. GD's first illustrated text was an edition of François RABELAIS (1854); later illustrated texts included DANTE Alighieri's *Inferno* (1861), Charles PERRAULT's *Fairy Tales* (1862), Miguel de CERVANTES's *Don Quixote* (1863), *Fairy Realm* (1865) by Thomas Hood (1835-1874), *Fables* (**1867**) by Jean de La Fontaine (1621-1695), John MILTON's *Paradise Lost* (1866) and Samuel Taylor COLERIDGE's *The Rime of the Ancient Mariner* (1875). The impassioned, "sublime" chiaroscuros of these illustrations conveyed a sense of simultaneous congestion and vastness, and they have been influential on illustrators of the FANTASTIC ever since, beginning with Aubrey BEARDSLEY and continuing into the time of artists like Virgil FINLAY. In *London: A Pilgrimage* (**1872**), with text by Blanchard Jerrold (1826-1884), GD created in visual terms an analogue of the version of LONDON – as a mephitic Babylon – which dominated the later work of Charles DICKENS. Although not specifically supernatural, these drawings – perhaps mainly through their overpowering theatricality – are central to fantasy's vision of URBAN FANTASY. [JC]

See also: COMICS; DANTE'S INFERNO (**1935**).

DORIAN GRAY: A MUSICAL Hungarian stage presentation (?1994). ◊ *The* PICTURE OF DORIAN GRAY (**1945**).

DOROTHY AND THE SCARECROW OF OZ (**1910**) ◊ *The* WIZARD OF OZ (**1939**).

DORSET, RUTH Pseudonym of Marilyn ROSS (itself a pseudonym).

DOT MOVIES A series of Australian children's live-action/ANIMATED MOVIES of fantasy interest less because of their plots than because the main characters – notably Dot, a young girl – are TOONS operating in a live-action world. The series includes *Dot and the Kangaroo* (**1976**), *Dot and Santa Claus* (**1979**; vt *Around the World with Dot*; vt *Dot Around the World*), *Dot and the Bunny* (**1982**), *Dot and the Whale* (**1985**) *Dot and the Koala* (**1986**), *Dot and Keeto* (**1986**) and *Dot and the Smugglers* (**1987**). In general, the earlier in the series the better the movie, although none are dire. [JG]

DOUBLE LIFE OF VÉRONIQUE, THE (ot *La Double Vie de Véronique*) Polish/French/Norwegian movie (**1991**). Sideral/Le Studio Canal/Tor/Varsovie/Norsk Film-Norvege. **Pr** Leonardo de la Fuente. **Dir** Krzysztof Kieślowski. **Puppet-master** Bruce Schwartz. **Screenplay** Kieślowski, Krzysztof Piesiewicz. **Starring** Sandrine Dumas (Catherine), Halina Gryglaszewska (Auntie), Jerry Gudejko (Antek), Irène Jacob (Veronika/Véronique), Wladyskaw Kowalski (Father), Philippe Volter (Alexandre Fabbri). 98 mins. Colour.

The converse of the IDENTITY-EXCHANGE movie: sexually naive Polish Veronika and sexually sophisticated French Véronique, knowing nothing of each other, share identities and the duplicated boyfriend Antek. In the past Véronique damaged her finger and gave up a career playing piano; Veronika, not knowing why, did likewise. Both become singers, but Veronika dies on stage so Véronique abandons singing to become a music teacher. Always feeling that there has been another of her, she eventually falls for PUPPET-master Fabbri, who understands her entirely – perhaps more than herself, for it is he who accepts more easily than she that, years before when she was on holiday in Krakow, she took a photograph of her DOUBLE. This beautifully made movie is as much about SEX as about fantasy: in their two REALITIES Veronika chooses one course and Véronique another. [JG]

DOUBLES The sinister double, originally as the DOPPELGÄNGER, was one of the central motifs of Gothic fiction (◊ GOTHIC FANTASY) and remained an important theme in weird fiction throughout the 19th century. The notion is connected to various superstitions regarding SHADOWS, MIRROR images and TWINS, but derives much of its psychological power from the fact that we all construct civilized "social selves".

In Edgar Allan POE's classic "William Wilson" (1840) the repressed secret self "escapes" to indulge his darker impulses, whose moral debits are naturally laid at the door of the narrator. *The Devil's Elixirs* (1816) by E.T.A. HOFFMANN and *The Private Memoirs and Confessions of a Justified Sinner* (**1824**) by James HOGG are far more convoluted, as is certainly appropriate to the extrapolation of a relationship far more intimate and confused than any between two different people, however closely they might be bound together by blood ties or erotic obsession. Charles DICKENS adapted the notion to moralistic fantasy in *The Haunted Man and the Ghost's Bargain* (**1848**); Mrs CRAIK followed suit in "The Self-Seer" (1853), but Robert Louis STEVENSON's *Strange Case of Dr Jekyll and Mr Hyde* (**1886**) mercifully failed to reduce the moral problem to mere elements of GOOD AND EVIL, the absurd brutality of such a separation eventually being made starkly clear in Italo CALVINO's *The Cloven Viscount* (**1952**). Oscar WILDE's *The Picture of Dorian Gray* (**1891**) belied the author's introductory claim that there is no such thing as a moral or an immoral book, but also exemplified the difficulties involved in deciding which is which. Henry JAMES was equally deceptive in calling his philosophically ponderous examination of the theme "The Jolly Corner" (1908).

Early-20th-century writers fascinated by doubles include Stefan GRABINSKI, who surrealized the theme in keeping with the literary fashion of his day, and Robert HICHENS, who doggedly explored its links with the theme of IDENTITY EXCHANGE. Less problematic doubles are often featured in TIMESLIP fantasies and tales of REINCARNATION, but the confrontation between a social self formed by one century and its equivalent product of a very different time can raise interesting questions relevant to the philosophical problem of identity, which surface even in such a light-hearted extrapolation of the theme as Edwin Lester ARNOLD's *Lepidus the Centurion* (**1901**). The comic potential was further displayed in *Two's Two* (**1916**) by J. Storer Clouston (1870-1944), and was later extrapolated more thoughtfully and delicately by James Branch CABELL in *There were Two Pirates* (**1946**). In Gerald BULLETT's *Mr Godly Beside Himself* (**1924**) and Vernon KNOWLES's "The Shop in the Off-Street" (1935) men who trade places with their FAIRY doubles regain access to the world of the imagination from which they had somehow become alienated; the farcical ramifications of a similar notion are explored by L. Sprague DE CAMP in *Solomon's Stone* (1942; **1956**).

An important subcategory is that dealing specifically with SHADOWS, where the focus is usually on the loss of an entity which seems superfluous but is in some mysterious sense vital. The precedent was set by Adalbert von CHAMISSO's *Peter Schlemihl* (**1813**), whose theme was reworked by Hans Christian ANDERSEN in "The Shadow" (1847), in turn reworked (in harness with another Andersen tale) by Oscar Wilde in "The Fisherman and his Soul" (1891). The most notable 20th-century shadow-fantasy is *The Charwoman's Shadow* (**1926**) by Lord DUNSANY; another is "The Danger of Shadows" (1941) by Stephen Vincent BENÉT.   [BS]

DOUGLAS, CAROLE NELSON (1944-   ) US writer known for her sf and for a considerable amount of fantasy. She began her publishing career with a historical romance, *Amberleigh* (**1980**); her first work of genre interest is the **Kendri and Irissi** or **Sword and Circlet** sequence: *Six of Swords* (**1982**), *Exiles of the Rynth* (**1984**), *Keepers of Edanvant* (**1987**), *Heir of Rengarth* (**1988**) and *Seven of Swords* (**1989**). The series begins in a thinned (◊ THINNING) Earth from which Irissi, the last survivor of a PARIAH-ELITE race with TALENTS which has vanished along with all MAGIC, must attempt to escape, with her COMPANIONS (including a wise CAT); the second volume follows the cast through a PORTAL into another land, but the sequence begins to assert itself only with the final three volumes, which work as a trilogy, and which introduce, in the evil sorcerer Geronfrey, a formidable enemy. The remainder of the long tale depicts the strife between Irissi and the malicious WIZARD. There are discussions and manifestations of power, suffering and BONDAGE, but the general sense is that ambitious goals have not fully been realized.

The **Taliswoman** sequence – *Cup of Clay* (**1991**) and *Seed Upon the Wind* (**1992**) – is more formidably constructed, and subjects its female protagonist to Veil, an OTHERWORLD run according to a strictly enforced misogyny; moreover, the Littlelost – children abused and abandoned because they are "imperfect" – are exiled from society. The protagonist (in a plot-move typical of CND) begins to argue with and eventually to share her life and adventures with a man who is, in a sense, her DOUBLE, and from whose QUEST she supplants him. In the second volume, the two redeem Veil, which had become desiccated through the loss of its cup. There are moments throughout when substantive issues are faced directly.

CND has also written two series whose fantasy content, though subdued, is central. The **Irene Adler** sequence – *Good Night, Mr Holmes* (**1990**), *Good Morning, Irene* (**1991**), *Irene at Large* (**1992**) and *Irene's Last Waltz* (**1994**) – is GASLIGHT ROMANCE. concerning the one woman who fascinated SHERLOCK HOLMES. The **Midnight Louie** sequence – *Catnap: A Midnight Louie Mystery* (**1992**), *Pussyfoot: A Midnight Louie Mystery* (**1993**), *Cat on a Blue Monday* (**1994**) and *Cat in a Crimson Haze* (**1995**) – is told from the point of view of Midnight Louie, a sentient CAT.   [JC]

DOUGLAS, NORMAN Working name of expatriate UK writer George Norman Douglass (1868-1952). A refugee from Anglo-Saxon puritanism long resident in Capri, he was a notable travel writer and produced two books on aphrodisiacs. His first fantasy, written in 1899, was *Nerinda* (in *Unprofessional Tales* coll **1901** as by Normyx; **1929**), set in Pompeii, a delusional fantasy (◊ DELUSION) about an enigmatic FEMME FATALE. *Unprofessional Tales* also includes "Elfwater" (vt "An Unnatural Feud" 1908 US), while *Experiments* (coll **1925**) includes the Edgar Allan POE-esque "Nocturne" and "Queer!".

ND's tongue-in-cheek novels all celebrate Decadent mores (◊ DECADENCE) in a blithely cavalier fashion. *South Wind* (**1917**) describes the healthy paganism of the inhabitants of Nepenthe (a lightly disguised Capri). The Epicurean parable *They Went* (**1920**) employs muted supernaturalism in its deft reworking of the legend of Lyonesse (◊ IMAGINARY LANDS). *In the Beginning* (**1927**) is a full-blooded fantasy featuring the allegorical adventures of a lusty demigod in the days before the All-Father inflicted morals upon mankind.   [BS]

DOUGLAS, THEO Working name of UK author Mrs H(enrietta) D(orothy) Everett, née Huskisson (1851-1923). At least half her 22 novels had fantasy and supernatural content. In *Iras: A Mystery* (**1896**) an egyptologist unwraps an ancient MUMMY, the beautiful Iras, who awakens from suspended animation. They fall in love and marry, but Iras is gradually transformed back into a mummy as seven magic pendants are removed individually from her necklace by various means. The SOUL of the heroine of *Nemo* (**1900**) possesses (◊ POSSESSION) and animates an AUTOMATON against her will. *One or Two* (**1907**) is a grotesque account of the success of a fat woman in making herself thin by spiritualist means. *Malevola* (**1914**) is a psychic-VAMPIRE story: the mysterious Madame Thérèse Despard is able to draw into herself the beauty and vitality of another during the process of massage. TD reverted to her own name for her final book, a collection of traditional GHOST STORIES, *The Death-Mask, and Other Ghosts* (**1920**; exp 1995).   [RD]

Other works: *Three Mysteries* (**1904**); *A White Witch* (**1908**); *Cousin Hugh* (**1910**); *White Webs* **1912**); *Hadow of Shaws* (**1913**); *The Grey Countess* (**1913**).

DOYLE, [Sir] ARTHUR (IGNATIUS) CONAN (1859-1930). Scottish-born UK writer, nephew of Richard DOYLE. ACD is best-known as the creator of SHERLOCK HOLMES, but he was also a prolific writer of SUPERNATURAL FICTION. His early interest in the paranormal led him to psychic research – he attended SÉANCES from 1879 and joined the Society for Psychical Research in 1893. He later converted to SPIRITUALISM.

ACD's first published story is a RATIONALIZED FANTASY, "The Mystery of Sasassa Valley" (1879 *Chambers's Journal*). This and other early stories written after his studies at Edinburgh University seem imitative of Wilkie COLLINS and R.M. Ballantyne (1825-1894). *The Mystery of Cloomber* (written ?1883; **1888**), an early novel published only once he had become well known, is an OCCULT FANTASY, full of immature sensationalism, about a retired general from the Indian army who finds himself under assault by Indian MAGIC.

Two years as a ship's doctor turned ACD from an adventurous youth into a courageous man with a spiritual and physical zeal. He returned to writing in order to supplement his poor income as a general practitioner, and soon found it the more profitable career. His naval adventures provided plots for two significant stories: "The Captain of the *Polestar*" (1883 *Temple Bar*), where the captain is lured to his death by an Arctic wraith, and "J. Habakuk Jephson's Statement" (1884 *Cornhill*), where ACD provided such a convincing explanation for the mystery of the *Mary Celeste* that many believed it. The best of these early stories were collected as *The Captain of the Polestar and Other Tales* (coll **1890**; rev vt *The Great Keinplatz Experiment* 1894 US). ACD also showed a fascination with the FEMME FATALE, which he linked to his interest in MESMERISM in two stories: "John Barrington Cowles" (1886 *Cassell's Saturday Journal*) and *The Parasite* (1894 *Harper's Weekly*; **1894**).

The success of the **Sherlock Holmes** stories, and ACD's own desire to focus on historical novels, meant that he forsook supernatural fiction for some years, although his psychic researches continued and inspired a few stories – "The Brown Hand" (1899 *Strand*), "Playing with Fire" (1900 *Strand*) and "The Leather Funnel" (1902 *McClure's*) – while a passing interest in egyptology resulted in "The Ring of Thoth" (1890 *Cornhill*) and "Lot No 249" (1892 *Harper's*). He rarely allowed Sherlock Holmes to become involved in cases verging on the supernatural – Holmes was the ultimate sceptic – but, when he did, these were invariably RATIONALIZED FANTASIES, as in *The Hound of the Baskervilles* (1901-2 *Strand*; **1902**) and "The Adventure of the Sussex Vampire" (1924 *Strand*).

During his lifetime ACD's supernatural fictions tended to be scattered through his collections, with the greatest concentration being in *Round the Fire Stories* (coll **1908**) and *The Last Galley* (coll **1911**). Of later collections the most relevant are *Tales of Terror and Mystery* (coll **1922**) and *Tales of Twilight and the Unseen* (coll **1922**), both of which have long since been subsumed into *The Conan Doyle Stories* (omni **1929**). Lesser-known stories, of which a few are supernatural, were included in **The Unknown Conan Doyle** volume *Uncollected Stories* (coll **1982**) ed John M. Gibson and Roger Lancelyn GREEN. Two volumes dedicated to ACD's supernatural fiction are *The Best Supernatural Stories of Arthur Conan Doyle* (coll **1979** US) ed E.F. BLEILER and *The Supernatural Tales of Sir Arthur Conan Doyle* (coll **1987**) ed Peter HAINING.

ACD was able to find an outlet for his fascination with OCCULTISM in his **Professor Challenger** stories: *The Lost World* (**1912**) and its immediate sequel, *The Poison Belt* (**1913**), are early examples of SCIENCE FICTION. The later books, though, veered more toward the supernatural, especially *The Land of Mist* (1925-6 *Strand*; **1926**), which sees an older Challenger won over to Spiritualism, and "When the World Screamed" (1928 *Liberty*), which explores the concept of a living Earth. "The Maracot Deep" (1927-8 *Strand*) begins by describing a scientific exploration of ATLANTIS; in its sequel, "The Lord of the Dark Face" (1929 *Strand*), Maracot invokes an ancient Atlantean power of EVIL. This story was ACD's last, and he seems in it to have returned to the sensationalism of his early work. The stories were assembled in *The Maracot Deep and Other Stories* (coll **1929**); this collection overlaps slightly with *The Professor Challenger Stories* (omni **1952**).

ACD's conversion to Spiritualism led him to open the Psychic Bookshop in London in 1925, and he produced a number of nonfiction books on the subject, including *The History of Spiritualism* (**1926** 2 vols). He often championed *cause célèbres*, of which the most notorious concerned the Cottingley Fairies, a set of faked photographs of FAIRIES created in 1920 by two young girls; ACD believed the pictures genuine, and wrote about the case in *The Coming of the Fairies* (**1922**). Many of his own psychic experiences are recounted in *The Edge of the Unknown* (**1930**), his last book. [MA]
Further reading: *Memories and Adventures* (**1924**), autobiography; *The Life of Sir Arthur Conan Doyle* (**1949**) by John Dickson CARR; *Conan Doyle: His Life and Art* (**1961**) by Hesketh Pearson.
See also: Mark FROST.

DOYLE, DEBRA (1952-   ) US writer who began publishing work of genre interest with "Bad Blood" in *Werewolves* (anth **1988**) ed Jane YOLEN and Martin H. GREENBERG, which she later expanded into a novel (see below). All her books have been written in what seems an equal partnership with James D(ouglas) Macdonald (1954-

). They first came to notice with the **Circle of Magic** sequence – *School of Wizardry* (**1990**), *Tournament and Tower* (**1990**), *City by the Sea* (**1990**), *The Prince's Players* (**1990**), *The Prisoners of Bell Castle* (**1990**) and *The High King's Daughter* (**1990**) – which tells of a young WIZARD in easy language. *Knight's Wyrd* (**1992**) presents the adventures of a young KNIGHT doomed (he fears) to an early death; he meets various supernatural creatures. *Bad Blood* (**1993**) and *Hunter's Moon* (**1994**) make up a children's series about WEREWOLVES and VAMPIRES. [JC]
Other works: Ties include two titles in the **Planet Builders** sequence, *Night of Ghosts and Lightning* * (**1989**) and *Zero-Sum Games* * (**1989**), both as Robyn Tallis; *Horror High: Pep Rally* * (**1991**) as Nicholas Adams; two titles as by Victor Appleton in the 4th **Tom Swift** sequence, being *Monster Machine* * (**1991**) and *Aquatech Warriors* * (**1991**); *Robert Silverberg's Time Tours #3: Timecrime, Inc* * (**1991**); *Daniel M. Pinkwater's Melvinge of the Megaverse #2: Night of the Living Rat* * (**1992**); the **Mageworlds** series – *The Price of the Stars* (**1992**), *Starpilot's Grave* (**1993**), *By Honor Betray'd* (**1994**) and *The Gathering Flame* (**1995**) – the last being a prequel.

DOYLE, RICHARD (1824-1883) UK artist, son of political cartoonist John Doyle (1797-1868) and uncle of Arthur Conan DOYLE. During his seven years at *Punch* (1843-50), for which he designed the logo, he collaborated with John LEECH and other artists on the ILLUSTRATIONS for three of Charles DICKENS's CHRISTMAS BOOKS: *The Chimes* (**1844**), *The Cricket on the Hearth* (**1845**) and *The Battle of Life* (**1846**). RD's fantastic drawings for a new translation of the GRIMM BROTHERS' tales, *The Fairy Ring* (1846), were followed by the tremendously popular *Fairy Tales from All Nations* (**1849**) compiled by Anthony R. Montalba (Anthony Whitehill), in which RD indulged his devotion to FAIRYTALES with many drawings and vignettes of elves, pixies and mythical creatures. His next book illustrations appeared in Mark Lemon's *The Enchanted Doll* (**1849**) and John RUSKIN's *The King of the Golden River* (dated 1851 but **1850**), one of the earliest UK CHILDREN'S FANTASIES. RD's famous picture of the South West Wind with the bugle-like nose was redrawn (with bulbous-shaped nose) for the third edition in late 1851.

His masterpiece was *In Fairyland: A Series of Pictures from the Elf World* (dated 1870 but **1869**), done to a poem by William Allingham (1824-1889). RD was given a free hand to produce his most imaginative pictures (16 colour plates with 36 illustrations and a pictorial title-page) for this glorious folio in decorated green cloth. The volume, which shows RD's secret fairy world at its most enchanting, is one of the finest examples of Victorian book production. The illustrations for *In Fairyland* were adapted for use with a specially written story, *The Princess Nobody: A Tale of Fairy Land* (**1884**) by Andrew LANG.

Among RD's other well known fairy works were *Jack the Giant Killer* (**1888**) and *The Great Sea Serpent*, which toured the UK as a large lantern-slide exhibition. "The Triumphant Entry: A Fairy-tale Pageant", comprising several hundred figures, was acquired by the National Gallery of Ireland; and several other Doyle paintings, notably "The Witch's House" and "Wood Elves Watching a Lady", are in the Victoria & Albert Museum. [RD]
Further reading: *Richard Doyle* (**1948**) by Daria Hambourg; *Richard Doyle* (**1983**) by Rodney K. Engen.

DRACULA Count Dracula appeared for the first time in

Bram STOKER's *Dracula* (**1897**), possibly the most famous SUPERNATURAL FICTION ever written, and the text which gives definitive shape to the figure of the VAMPIRE. The historical Prince Vlad IV (1431-1476) – Vlad the Impaler – is never mentioned in *Dracula*, but the Count is a member of the Szekely family, lives in Transylvania and is 466 years old, and like the warlord Vlad is a military man. He is saturnine, lean and hauntingly demonic in appearance; in this and in his seductive effect on young women he climaxes the vampire tradition John POLIDORI created out of his obsessive relationship with BYRON. But Stoker's Dracula also has hair on his palms, bad breath, unnatural strength and very sharp fingernails, and is cold to the touch. A SHAPESHIFTER, he can survive daylight but much prefers the night. His diet is restricted to human blood, which he sucks from his victims' necks through his fangs. Those he vampirizes eventually die, only to be reborn as undead vampires themselves. Dracula can be made helpless by garlic, a crucifix or a wild rose. A bullet fired into his body, a stake through his heart or decapitation – preferably a combination – will "kill" him. At the end of *Dracula* he was destroyed; Stoker had nothing to do with any of the countless resuscitations.

This version of the vampire has been so dominant over the past century that recent tales depicting vampires otherwise read as REVISIONIST FANTASIES. They include Anne RICE's **Vampire Chronicles** and its imitators, Chelsea Quinn YARBRO's **St Germain** sequence, Suzi McKee CHARNAS's *The Vampire Tapestry* (**1980**), Nancy A. COLLINS's *Sunglasses After Dark* (**1989**) and many YA CONTEMPORARY FANTASIES – Annette Curtis Klause's *The Silver Kiss* (**1990**) is an excellent example – in which teenage vampires try to limit the damage of, or even to kick, their habit. Tales in which Dracula himself appears run a gamut from Fred SABERHAGEN's *The Dracula Tape* (**1975**) and its sequels – in which Dracula is the hero – to Kim NEWMAN's *Anno Dracula* (**1992**) and its sequel, in which Dracula gains horrific sway over the Western world.

In *Dracula* Stoker occasionally uses the term "nosferatu", a Romanian word meaning (approximately) "plague-carrier"; he picked it up from Emily Gerard's travelogue *The Land beyond the Forest* (**1885**), where it is defined as "vampire". [JC]

COMICS

D's first appearance in COMICS was probably in *New Fun Comics #6* (1935), when Jerry Siegel (1914-1996) and Joe Shuster (1914-1992) introduced the mad scientist The Vampire Master as the nemesis of their Dr Occult. Stoker's novel was not among the 169 titles in the long-running and highly successful **Classics Illustrated** series, but there have been a number of attempts to adapt it into comic-strip form. The earliest was published by Avon Periodicals in EERIE *#12* (1952), but a more substantial version was *Dracula* (graph **1966**; vt *The Illustrated Dracula* 1975) drawn by Alden McWilliams (1916-1993) to a script by Otto Binder and Craig Tennis. By far the best to date was the beautifully painted GRAPHIC NOVEL by Fernando FERNANDEZ, *Dracula* (1982-3: CREEPY; graph **1984**). Then came the pretentious *Dracula – A Symphony in Moonlight and Nightmares* (graph **1986**) by John J. Muth (1960-   ).

A number of comics adaptations from DRACULA MOVIES have included *Movie Classics: Dracula* (graph **1962**) and the remarkable adaptation by Mike MIGNOLA of *Bram Stoker's Dracula* (**1992**; ◊ DRACULA MOVIES), *Bram Stoker's Dracula* (graph **1993**); also there have been, in the UK magazine *House of Hammer*, *Dracula* (1976), *Dracula, Prince of*

Darkness (1976) and *Bride of Dracula* (1983).

D has starred in a number of comic-book series, including: **Dracula Lives** (1973-5) from MARVEL COMICS, which featured some interesting D-related stories with artwork by Gene Colan (1926-   ), Neal ADAMS, Tony Dezuniga and others; the multi-award winning **Tomb of Dracula** (1972-9), written by Marv Wolfman and again drawn by Colan; and many more. The most recent is **Dracula – Vlad the Impaler** (1994-5), with script by Roy THOMAS and art by Esteban MAROTO.

A Spanish comics magazine, *Dracula* (graph trans **1972-3** UK; different trans WARREN PUBLISHING graph **1979** US), featured some interesting art by Maroto, Enric Sio (1942-   ) and other Spanish stalwarts, but was marred by puerile scripting: Dracula himself featured in only one frame!   [RT/SH]

See also: *The* CURSE OF DRACULA (1979); DRACULA MOVIES; DRACULA – THE SERIES (1990).

DRACULA MOVIES Of the 40+ **Dracula** movies, some are of considerably more interest than others, as reflected in their treatment below. A complete listing is probably impossible.

1. *Nosferatu* (ot *Nosferatu – Eine Symphonie des Grauens*; reconstructed with sound vt *Die Zwölfte Stunde – Eine Nacht des Grauens* 1930; other vts include *Dracula*; *Nosferatu, A Symphony of Terror*; *Nosferatu, The Vampire*; *Nosferatur, Eine Symphonie des Grauens*; *Eine Symphonie des Grauens*) German movie (*1922*). Prana. **Dir** F.W. MURNAU. **Screenplay** Henrik Galeen. **Plagiarized from** *Dracula* (**1897**) by Bram STOKER. **Starring** John Gottowt (Professor Bulwer), Alexander Granach (Knock), Max Schreck (Graf Orlok), Greta Schröder (Ellen), Gustav von Wangenheim (Hutter). *c*63-*c*80 mins (varies with projection speed), with a 1984 retinted restoration running to 105 mins. B/w plus tints. Silent.

Bremen, 1838. Crazed estate agent Knock sends employee Hutter upriver into darkest Germany to the castle of Graf Orlok on the pretext of selling him a house in Bremen. Hutter finds both castle and count spooky but tolerable, until he cuts his thumb on a table-knife and the count insists on licking the blood. Next morning Hutter awakes assuming he has had bad dreams and that the bites on his neck have been made by spiders or mosquitoes. But that day he reads *The Book of Vampires*, and at sunset he sees Orlok in full vampiric splendour. He swoons and is about to be bled by Orlok when, back home in Bremen, his wife Ellen calls a warning to him *via* clairvoyant DREAM. Next day he discovers Orlok's coffin, with Orlok senseless within it, and that night he witnesses Orlok – now infatuated by a miniature of Ellen – departing with a batch of coffins. Determined to warn Bremen, Hutter makes his slow way home on horseback; Orlok travels by river on the *Demeter*, whose crew he murders. Orlok can manifest himself as hordes of rats and, wherever the ship calls, plague spreads. Knock, now in an insane asylum, talks of his Master's coming; he slays a keeper and briefly escapes. Ellen, reunited with Hutter, despite his warnings reads *The Book of the Vampires* and discovers that the evil of the *Nosferatu* can be assuaged only if a chaste woman freely gives her blood to it and stays with it until dawn. She lures Orlok; he feeds eagerly until cockcrow, when the dawn sunlight causes him to vanish in a puff of smoke.

N, the first VAMPIRE movie, was obviously stolen from Stoker; Stoker's widow Florence sued, and in 1925 gained an order forcing Murnau to destroy all copies. However, pirated prints survived. Moreover, in 1930 Murnau – by now in Hollywood – produced with Waldemar Ronger a reconstructed version, *Die Zwölfte Stunde* ["The Twelfth Hour"], including some scenes excised from the original.

Seen today, *N* is a curious mixture of the risible and the visually impressive, the latter almost exclusively when Orlok (or, importantly, his SHADOW) is on screen. *N* was emulated in **34**; its cine-historical importance can hardly be exaggerated.   [JG]

2. *Dracula* (*1931*) US movie. Universal. **Pr** Carl Laemmle Jr. **Dir** Tod Browning. **Screenplay** Garrett Fort. **Based on** the play by Hamilton Deane and John L. Balderston (although Stoker's novel is given main credit). **Starring** Herbert Bunston (Dr Seward), Helen Chandler (Mina Seward), Frances Dade (Lucy Weston), Dwight Frye (Renfield), Bela Lugosi (Dracula), David Manners (John Harker), Edward Van Sloan (Van Helsing). 84 mins. B/w.

Priggish young English estate agent Renfield travels in Transylvania to pass over to Count Dracula the title deeds to Carfax Abbey, near Whitby; although the occupants of a local inn try to stop him, he continues to Borgo Pass, where the count's weird coachman picks him up to transport him the rest of the way to Castle Dracula. Dracula welcomes his guest, displaying animation only when Renfield cuts his finger on a paperclip; Renfield collapses, his wine having been drugged; Dracula waves off his three "wives" as they advance on the unconscious form and himself vampirizes the man. Soon Dracula and the enslaved Renfield are aboard the *Vesta*, taking with them three crates of Transylvanian soil to England; on arrival Renfield, the vessel's sole survivor, is raving, and is incarcerated as a lunatic in the sanitarium of Dr Seward, near Whitby. Dracula, in London, gains an introduction to Seward, his daughter Mina, her fiancé John Harker and her friend Lucy Weston – who is much taken with the exotic foreigner. Entering Lucy's bedroom as a bat, Dracula vampirizes her. Her mysterious death is investigated by Van Helsing, who immediately realizes a VAMPIRE is at work. Next Dracula goes for Mina, mixing his blood with hers in what she at first thinks is a DREAM. Van Helsing, seeing Dracula has no reflection in a MIRROR, identifies him as the villain, and moves to protect Mina. Dracula hypnotizes Mina's nurse in order to gain access to the sleeping woman, and carries her to ruined Carfax Abbey. Van Helsing and Harker find Dracula's sleeping form, and Van Helsing stakes him; Mina is tormented by the pain, but "cured"; and she and Harker are reunited.

This first US movie version of the tale is interestingly skewed from most others, notably in that Renfield is a central character, with Harker more or less sidelined. Peasantish comedy scenes were introduced to entertain the masses; overall, it is not only the coffin-lids that creak.   [JG]

3. *Dracula's Daughter* US movie (*1936*). Universal. **Pr** E.M. Asher. **Dir** Lambert Hillyer. **Spfx** John P. Fulton. **Screenplay** Garrett Fort. **Starring** Marguerite Churchill (Janet Blake), Gilbert Emery (Sir Basil Humphrey), Nan Gray (Lili), Gloria Holden (Countess Marya Zaleska), Otto Kruger (Jeffrey Garth), Irving Pichel (Sandor), Edward Van Sloan (Van Helsing). 70 mins. B/w.

The story picks up immediately after **2**'s end. Van Helsing is arrested for the murders of Renfield and Dracula. As he languishes, Countess Zaleska, Dracula's daughter, steals with the aid of her manservant Sandor the body of her father, which she cremates in an act of EXORCISM, hoping

thereby to free herself from the curse of vampirism. But there is no such "cure", and soon she preys on a stranger in the London fog. Called in by his old tutor Van Helsing to help the defence, Scotland Yard psychiatrist Garth meets and is attracted by Zaleska. She explains her problem in vague part, and Garth advises that, like an alcoholic, she should put the temptation in front of herself and fight the craving face-on; accordingly, she has Sandor bring a girl, Lili, as an "artist's model" – but is unable to resist the craving . . .

DD is, like **2**, much marred by buffoonish "light relief". Yet the performances of Holden and (in her small part) Gray carry the whole thing off: there are some moments of genuine power, particularly in the final shot of Zaleska's lifeless, still beautiful face. *DD* is widely preferred to its predecessor and often regarded as a classic of the VAMPIRE genre.   [JG]

4. *Son of Dracula* US movie (*1943*). Universal. **Pr** Ford Beebe. **Dir** Robert Siodmak. **Spfx** George Robinson. **Screenplay** Eric Taylor. **Starring** Lon Chaney Jr (Alucard). 80 mins. B/w.

Alucard – "Dracula" backwards – arrives on a Southern US plantation and does some vampirizing. A B-movie that didn't try too hard.   [JG]

5. *House of Dracula* US movie (*1945*). Universal. **Pr** Paul Malvern. **Exec pr** Joe Gershenson. **Dir** Erle C. Kenton. **Spfx** John P. Fulton. **Screenplay** Edward T. Lowe. **Starring** John Carradine (Dracula), Lon Chaney (Larry Talbot, the Wolf Man), Glenn Strange (Frankenstein Monster). 67 mins. B/w.

A MONSTER team-up of little interest. Universal seemed determined to squander the ICONS they had created in FRANKENSTEIN (*1931*), **2**, **3** and *The Wolf Man* (*1940*). *House of Frankenstein* (*1944*; ◊ FRANKENSTEIN MOVIES) was every bit as bad.   [JG]

6. *Blood of Dracula* US movie (*1957*). American International/James H. Nicholson & Samuel Z. Arkoff. **Pr** Herman Cohen. **Dir** Herbert L. Strock. **Mufx** Philip Scheer. **Screenplay** Ralph Thornton. **Starring** Mary Adams (Mrs Thorndyke), Heather Ames (Nola), Jerry Blaine (Tab), Don Devlin (Eddie), Gail Ganley (Myra), Sandra Harrison (Nancy Perkins), Louise Lewis (Miss Branding). 68 mins. B/w.

A crude matching of the PSYCHOLOGICAL THRILLER with an attempt to give vampirism a TECHNOFANTASY rationale, all bundled up in a contemporary teen movie that has little to do with Dracula. Teenage fireball Nancy, dumped in exclusive Sherwood School for Girls, is lured by dotty chemistry teacher Branding, who sees vampirism as the feminist counter to the A-bomb. Nancy accordingly vampirizes people.   [JG]

7. *Dracula* (vt *The Horror of Dracula* US) UK movie (*1958*). HAMMER/Universal. **Pr** Anthony Hinds. **Exec pr** Michael Carreras. **Assoc pr** Anthony Nelson-Keys. **Dir** Terence Fisher. **Mufx** Phil Leakey. **Screenplay** Jimmy Sangster. **Based on** Stoker's novel and **2**. **Starring** Peter Cushing (Van Helsing), Michael Gough (Arthur Holmwood), Christopher Lee (Dracula), Carol Marsh (Lucy Holmwood), Charles Lloyd Pack (Dr Seward), Melissa Stribling (Mina Holmwood), John Van Eyssen (Jonathan Harker). 82 mins. Colour.

The first of the Hammer **Dracula** movies. Englishman Harker arrives at Castle Dracula to work as a librarian. He soon realizes his employer is a VAMPIRE, a fact that much interests his friend Van Helsing, who comes to the local village, Klausenburg, with the express purpose of destroying Dracula. A beautiful vampire woman tries to bite Harker's neck; he is saved from this by the Count, but Dracula then vampirizes him himself. As Dracula goes to London to continue his depredations – starting with Lucy Holmwood, to whom Harker was engaged – Van Helsing arrives at Castle Dracula to discover Harker "sleeping" in a coffin; reluctantly, he stakes him. Mina, the wife of Lucy's brother Arthur, asks Van Helsing for help; but by now Lucy has been fully vampirized, and she rejects his stratagems. Dracula visits her room one last time to kill her. She is put in the family vault but is later seen "alive" and must be staked by Van Helsing, aided by Holmwood. Next Dracula works on Mina, who has to be given a blood transfusion from Holmwood to avoid succumbing to vampirism. Dracula kidnaps her, but this time Van Helsing and Holmwood follow, and eventually succeed in forcing the vampire into the sunlight, where he perishes.

Gory and unabashedly sensual, this short movie is one of the definitive CINEMA versions of the tale. Its genesis was fraught. Universal had drawn up an exclusive contract with Stoker's estate for the rights in the character (much later it was discovered that, through an oversight of Stoker's, the character was in fact out of copyright), and initially resisted attempts by Hammer to make the movie; the eventual agreement was that Universal should have distribution rights. Almost overnight Christopher Lee became the iconic image of Dracula in the popular mind: all subsequent Draculas have been matched against him.   [JG]

8. *Return of Dracula, The* (vt *The Fantastic Disappearing Man* UK) US movie (**1958**). Gramercy/ United Artists. **Pr** Arthur Gardner, Jules V. Levy. **Dir** Paul Landres. **Mufx** Stanley Smith. **Screenplay** Pat Fielder. **Starring** Norma Eberhardt (Rachel), Francis Lederer (Dracula). 77 mins. B/w.

Fleeing the "European Police Authority", Dracula comes to Carleton, California, adopting the identity of an emigrant he has despatched *en route*. The Mayburg family welcome the weird "long-lost cousin Belac", and tolerate his unusual ways, until . . . *TROD* is a not-bad movie. We are shown the shapeshifted Dracula not as a bat but as a cloud of smoke, out of which the human form materializes; this was presumably for the sake of cheapness, but the effect, especially in b/w, adds considerably to the atmosphere of this neatly made trifle.   [JG]

9. *Brides of Dracula, The* UK movie (**1960**). HAMMER-Hotspur/Universal. **Pr** Anthony Hinds. **Exec pr** Michael Carreras. **Assoc pr** Anthony Nelson-Keys. **Dir** Terence Fisher. **Spfx** Sydney Pearson. **Mufx** Roy Ashton. **Screenplay** Peter Bryan, Edward Percy, Jimmy Sangster. **Novelization** *The Brides of Dracula* * (**1960**) by Dean Owen (Dudley Dean McGaughy). **Starring** Victor Brooks (Hans), Peter Cushing (Van Helsing), Marianne Danielle (Yvonne Monlaur), Martita Hunt (Baroness Meinster), Freda Jackson (Greta), Fred Johnson (Curé), Miles Malleson (Dr Tobler), Andree Melly (Gina), Yvonne Monlaur (Marianne Danielle), Henry Oscar (Herr Lang), David Peel (Baron Meinster), Mona Washbourne (Frau Lang). 85 mins. Colour.

Despite its title, not really a **Dracula** movie – and certainly not Hammer's expected sequel to **7**. Young Marianne Danielle is travelling to be a teacher at the Lang Academy in Badstein. *En route* she stops off at the castle of the Meinster family, where she discovers to her horror that the grotesque old Baroness is keeping the young Baron locked up.

Marianne sympathetically releases him, little realizing that in fact the Baroness has confined him because she knows he is a VAMPIRE.   [JG]

10. *Billy the Kid versus Dracula* (vt *Billy the Kid vs Dracula*) US movie (**1965**). Circle/Embassy. **Pr** Carroll Case. **Dir** William Beaudine. **Photographic fx** Cinema Research. **Screenplay** Carl Hittleman. **Starring** John Carradine (Dracula), Virginia Christine (Eva Auster), Chuck Courtney (Billy the Kid), Melinda Plowman (Betty Bentley). 89 mins. Colour.

Pretending to be her uncle, Dracula inveigles his way into the Wild West household of Betty, whose fiancé is the reformed BILLY THE KID. Like everyone else, Billy scoffs at the idea this weird stranger could be a VAMPIRE, but he wakes up to the truth in time to stake Dracula through the heart. This was Beaudine's last movie; earlier the same year he had made the more entertaining *Jesse James Meets Frankenstein's Daughter* (**1965**), which tried the same HORROR/Western mix (◊ FRANKENSTEIN MOVIES).   [JG]

11. *Dracula, Prince of Darkness* UK movie (**1966**). Warner/HAMMER/Seven Arts. **Pr** Anthony Nelson-Keys. **Dir** Terence Fisher. **Spfx** Bowie Films Ltd. **Mufx** Roy Ashton. **Screenplay** John Sansom. **Novelization** *Dracula, Prince of Darkness* * (**1967**) by John Burke. **Starring** Suzan Farmer (Diana Kent), Andrew Keir (Father Sandor), Philip Latham (Klove), Christopher Lee (Dracula), Francis Matthews (Charles Kent), Barbara Shelley (Helen Kent), Charles Tingwell (Alan Kent), Thorley Walters (Ludwig). 90 mins. Colour.

Although Hammer had released **9** in 1960, this was the first proper sequel to the enormously popular **7**. Curiously, given that Lee had already become the archetypal image of Dracula, he is on-screen only briefly, Latham being the primary figure of EVIL in the movie. As finally released, this movie had been heftily axed by the censor.

The Kents – Helen and husband Alan, Diana and husband Charles – are on holiday in Transylvania. The Abbot of Kleinberg, Father Sandor, warns them of the danger of their plans, but inevitably they end up in Castle Dracula, where the sinister butler Klove welcomes them – only to murder Alan that night and sprinkle his blood on Dracula's ashes, reviving the VAMPIRE. Dracula promptly vampirizes Helen, and the two pursue Diana, the Count's real target. Sandor soon stakes Helen. Charles and Dracula prepare to fight it out on the castle's frozen-over moat. Quick-thinking Sandor fires his gun to crack the ice; Dracula falls into the running water and is destroyed.   [JG]

12. *Dracula Has Risen from the Grave* UK movie (**1968**). HAMMER/Warner-Seven Arts. **Pr** Aida Young. **Dir** Freddie Francis. **Spfx** Frank George. **Mufx** Rosemarie McDonald-Peattie, Heather Nurse. **Screenplay** John Elder (Anthony Hinds). **Starring** Barry Andrews (Paul), Norman Bacon (Mute Boy), Veronica Carlson (Maria Müller), George A. Cooper (Landlord), Rupert Davies (Monsignor Ernst Müller), Barbara Ewing (Zena), Ewan Hooper (Priest), Christopher Lee (Dracula), Marion Mathie (Anna Müller), Michael Ripper (Max). 92 mins. Colour.

This has one of the very best Hammer openings. The church of a remote village in the shadow of Castle Dracula is desecrated when one of the VAMPIRE's victims is discovered dead in its bell; thereafter Dracula is cast by the villagers into a frozen stream and presumed dead. But no. *DHRFHG* is, despite its great start, not so much bad as somehow unnecessary.   [JG]

13. *Taste the Blood of Dracula* UK movie (**1969**).

HAMMER/Warner. **Pr** Aida Young. **Dir** Peter Sasdy. **Spfx** Brian Johncock. **Mufx** Gerry Fletcher. **Screenplay** John Elder (Anthony Hinds). **Starring** Ralph Bates (Courtley), Isla Blair (Lucy Paxton), John Carson (Jonathan Secker), Anthony Corlan (Paul Paxton), Linda Hayden (Alice Hargood), Russell Hunter (Felix), Martin Jarvis (Jeremy Secker), Geoffrey Keen (William Hargood), Roy Kinnear (Weller), Christopher Lee (Dracula), Michael Ripper (Cobb), Peter Sallis (Samuel Paxton), Gwen Watford (Martha Hargood). 95 mins. Colour.

Three "virtuous" Victorian gentlemen (Hargood, Samuel Paxton and Jonathan Secker) secretly conduct a life of DECA-DENCE. They are persuaded by disinherited aristocrat Courtley (who has sold his SOUL to the DEVIL) to buy the effects of Dracula from curio-dealer Weller, who witnessed the vampire dying with a stake through his heart in Transylvania. At a form of BLACK MASS with the trio, Courtley ingests his own blood tainted with that of Dracula; he begins a METAMORPHOSIS, and, horrified, they kill him. From his corpse arises Dracula in person, who swears VENGEANCE; he vampirizes Hargood's daughter Alice and Paxton's daughter Lucy, who kill their fathers; Lucy vampirizes young Jeremy Secker, who kills *his* father. Paul Paxton, who loves Alice, finally kills Dracula, saving her from vampirism.

In *TTBOD* HAMMER made the equation between SEX and vampirism clearer than ever before; there are, too, hints of incest in the relationship between Hargood and Alice, so the movie is often seen in cut form. Certainly it is more ambitious than customary for Hammer, as indicated by the long list of principals. Corlan, after changing his name to Higgins, became internationally celebrated for movies like *The DRAUGHTSMAN'S CONTRACT (1982)*. [JG]

14. *Countess Dracula* UK movie (*1970*). Rank/HAMMER. **Pr** Alexander Paal. **Dir** Peter Sasdy. **Spfx** Bert Luxford. **Mufx** Tom Smith. **Screenplay** Jeremy Paul. **Novelization** *Countess Dracula* * (*1970*) by Michel Parry. **Starring** Patience Collier (Julie), Maurice Denham (Fabio), Lesley-Anne Down (Ilona), Sandor Elès (Imre Toth), Nigel Green (Captain Dobi), Peter Jeffery (Captain Balogh), Peter May (Janco), Ingrid Pitt (Countess Elisabeth Nadarosdy). 93 mins. Colour.

Recently widowed, the elderly Countess Elisabeth discovers fortuitously that contact with a young woman's blood restores her youth. Based not on Vlad the Impaler but on the Polish Countess Elizabeth de Báthory (1560-1614), found guilty in 1611 of ordering the murders of countless young women so she might maintain youth by bathing in their blood, *CD* is effective. [JG]

15. *Scars of Dracula* UK movie (*1970*). HAMMER/EMI. **Pr** Aida Young. **Dir** Roy Ward Baker. **Spfx** Roger Dicken. **Mufx** Wally Schneiderman. **Screenplay** John Elder (Anthony Hinds). **Novelization** *The Scars of Dracula* * (*1971*) by Angus Hall. **Starring** Jenny Hanley (Sarah), Anouska Hempel (Tania), Christopher Lee (Dracula), Christopher Matthews (Paul Carlson), Patrick Troughton (Klove), Dennis Waterman (Simon). 96 mins. Colour.

After a peasant girl is slain by Dracula, the local villagers fire the castle. Their glee is short-lived: Dracula retaliates with a plague of biting bats, and the villagers are returned to their oppressed state. A while later, Paul comes to Castle Dracula and is welcomed by the weird woman Tania, whom he later discovers – as she creeps into his bed that night – is a vampire. She tries to kill him; a furious Dracula kills her,

but, threatened by the dawn, abruptly departs. Paul's brother Simon and lover Sarah come to the castle in search of him and are seized by Dracula; Dracula's butler Klove (a relic from **11**) sets them free. After this lacklustre effort, Hammer **Dracula** movies were shifted into the contemporary world (◊ **20** and **25**). [JG]

16. *Dracula versus Frankenstein* (vt *Blood of Frankenstein*; vt *Revenge of Dracula*; vt *They're Coming to Get You*) US movie (*1971*). **Dir** Al Adamson. **Starring** Lon Chaney Jr (Groton the Mad Zombie), Zandor Vorkov (Dracula). 90 mins. Colour.

This began life as *The Blood Seekers* – a black comic TECH-NOFANTASY about drug-crazed hippies and a mad scientist – and was largely made before Dracula and the FRANKEN-STEIN monster were grafted on. The two HORROR-MOVIE icons come to an agreement to guarantee their supplies of fresh blood. This was Chaney's last movie. Not to be confused with the Spanish/German/Italian movie *Dracula vs. Frankenstein* (*1971*; ot *El Hombre que Vino del Ummo*, dir Tulio Demichelli, 87 mins, colour), released the same year, which sees aliens attempting to conquer Earth by reviving various cinematic monsters, including the title two. [JG]

17. *Guess What Happened to Count Dracula?* US movie (*1971*). **Dir** Laurence Merrick. **Starring** Claudia Barron, Des Roberts. 80 mins. Colour.

A comedy in which Dracula terrorizes Hollywood. By all accounts this is rather good. [JG]

18. *Blacula* US movie (*1972*). AIP. **Pr** Joseph T. Naar. **Dir** William Crain. **Screenplay** Raymond Koenig, Joan Torres. **Starring** Elisha Cook Jr, Vonetta McGee, William Marshall, Denise Nicholas. 93 mins. Colour.

In the wake of *Shaft* (*1971*) Hollywood was keen to produce "black" movies, and this was one result. A Caribbean freedom-fighter is vampirized and, centuries later, tries to find his reincarnated wife in Los Angeles so that he can vampirize her and "live" with her happily ever after. *B* was a critical and commercial disaster. [JG]

19. *Count Dracula's Great Love* (ot *El Gran Amor del Conde Dracula*; vt *Cemetery Girls*; *Dracula's Great Love*; vt *Dracula's Love*; vt *Dracula's Virgin Lovers*) Spanish movie (*1972*). **Dir** Javier Aguirre. **Starring** Paul Naschy (Dracula). 91 mins. Colour.

A group of travellers are vampirized by Dracula – all except a young woman with whom he has fallen in love. Rather than submit herself to voluntary vampirization, she suicides. [JG]

20. *Dracula A.D. 1972* UK movie (*1972*). HAMMER/Warner. **Pr** Josephine Douglas. **Dir** Alan Gibson. **Spfx** Les Bowie. **Mufx** Jil (*sic*) Carpenter. **Screenplay** Don Houghton. **Starring** Stephanie Beacham (Jessica Van Helsing), Peter Cushing (Lawrence Van Helsing/Lorimer Van Helsing), William Ellis (Joe Mitcham), Marsha Hunt (Gaynor Keating), Janet Key (Anna Bryant), Michael Kitchen (Greg Pullar), Christopher Lee (Dracula), Philip Miller (Bob Tarrant), Caroline Munro (Laura Bellowes), Christopher Neame (Johnny Alucard). 95 mins. Colour.

In 1872 Lawrence Van Helsing died in London's Chelsea staking Dracula, who crumbled to ash; an acolyte of the vampire collected Dracula's RING, a vial of ash and the stake and reburied the latter two items near St Bartolph's Church. Now, in 1972, Johnny Alucard, a descendant of that acolyte, infiltrates the swingin' crowd surrounding Jessica, granddaughter of Professor Lorimer Van Helsing, Lawrence's grandson. Alucard persuades the others to partake in what he claims is a mock BLACK MASS; in fact, he performs a blood SACRIFICE to raise Dracula from the dead. The vampire seeks to avenge himself on the Van Helsing line by wedding Jessica.

Despite the title, this is a very 1960s movie, and its cool, trendy dialogue makes the rest hard to take seriously. [JG]

21. *Dracula, Prisoner of Frankenstein* (ot *Drácula contra Frankenstein*; vt *Dracula, Prisonnier de Frankenstein*; vt *The Screaming Dead*) Spanish/French movie (*1972*). Interfilme/Fenix/Profit ETS. **Dir** Jesús Franco. **Starring** Dennis Price (Frankenstein), Howard Vernon (Dracula). 85 mins. Colour.

A bizarre, almost dialogue-free movie in which Frankenstein resuscitates Dracula in order to vampirize and thus conquer the world. Some critics regard it all as a mess, others as a triumph of SURREALISM. [JG]

22. *Dracula Saga, The* (ot *La Saga de los Dracula*; vt *Dracula: The Bloodline Continues*; vt *Dracula's Saga*; vt *The Saga of Dracula*; vt *The Saga of the Draculas*) Spanish movie (*1972*). **Dir** Leon Klimovsky. **Starring** Tony Isbert (Dracula). 92 mins. Colour.

Dracula's son is a shambling misfit, unworthy to carry on the line, so the Count invites his pregnant granddaughter to the castle in hopes that her baby will prove a better successor. [JG]

23. *Dracula Sucks* US movie (*1972*). **Dir** Philip Marshak. **Starring** Jamie Gillis, John C. Holmes. 91 mins. Colour.

A porn version, with much gore interspersed among the sex. Not to be confused with the US movie *Dracula: The Dirty Old Man* (*1969*; dir William Edwards, 87 mins, colour), which contains a similar but less explicit mixture. [JG]

24. *Dracula* (vt *Bram Stoker's Dracula*) US/UK movie (*1973* tvm). Universal. **Pr** Dan Curtis. **Dir** Curtis. **Spfx** Kit West. **Screenplay** Richard Matheson. **Starring** Murray Brown (Jonathan), Nigel Davenport (Van Helsing), Penelope Horner (Mina), Fiona Lewis (Lucy), Jack Palance (Dracula). 97 mins. Colour.

A reasonably faithful adaptation of Stoker's novel, this suffers for that very reason: the movie is imbued with a sense of overrespectfulness, so that both script and direction tend to default to pomposity. Worse still, Palance proves totally incapable of portraying a Central European in either accent or mannerism. Keanu Reeves had an analogous difficulty in **40**, but he was playing merely the part of Harker so it did not matter so much.

Not to be confused with the Spanish/Italian/West German/Liechtenstein movie *Bram Stoker's Count Dracula* (*1970*; ot *El Conde Dracula*; vt *Dracula '71*; dir Jesús Franco, 98 mins, colour), which has Christopher Lee as the Count, plus Herbert Lom and Klaus Kinski; despite the cast, this unambitious movie reveals its low budget at every turn. One much-loved moment sees a "rock" bounce off a horse's head on its way to completely crushing a bystander. [JG]

25. *Satanic Rites of Dracula, The* (vt *Count Dracula and his Vampire Bride* US) UK movie (*1973*). HAMMER. **Pr** Roy Skeggs. **Assoc pr** Don Houghton. **Dir** Alan Gibson. **Spfx** Les Bowie. **Mufx** George Blackler. **Screenplay** Houghton. **Starring** Patrick Barr (Lord Carradine), Michael Coles (Inspector Murray), Peter Cushing (Lorimer Van Helsing), William Franklyn (Torrence), Freddie Jones (Julian Keeley), Christopher Lee (Dracula/D.D. Denham), Barbara Yu Ling (Chin Yang), Joanna Lumley (Jessica Van Helsing), Richard Vernon (Matthews). 88 mins. Colour.

A reprise for present-day Lorimer Van Helsing and his granddaughter Jessica (see **20**), the latter played by a fresh actress, this is more of a dark TECHNOFANTASY than the usual Hammer fare. Dracula (in the guise of Howard Hughes-style recluse D.D. Denham) intends to relaunch bubonic plague into the world. He desires Jessica and seizes her; Lorimer saves her, destroys the bacillus, and impales "Denham" on thorns before staking him with a handy fencepost.

This was the last of the Hammer **Dracula** movies, for the simple reason that Lee refused to play the part any longer. From the very first he had been uneasy about Hammer's attitude towards the VAMPIRE – notably the way in which he could be revived at will despite having been comprehensively destroyed in each previous movie. Lee must also have observed that, during the 15 years in which he had been playing the part, audiences had become more sophisticated and wanted something new: the close-up views of his eyes becoming bloodshot now provoked giggles rather than frissons. [JG]

26. *Blood for Dracula* (vt *Andy Warhol's Dracula*; vt *Dracula Cerca Sangue di Vergine e . . . Mori di Sete*; vt *Dracula Vuole Vivere: Cerca Sangue di Vergine!*) French/Italian movie (*1974*). CC Champion & 1/Bryanston. **Pr** Andrew Braunsberg, Carlo Ponti, Jean-Pierre Rassam, Jean Yanne. **Dir** Paul Morrissey, Antonio Margheriti. **Spfx** Roberto Arcangeli, Carlo Rambaldi. **Screenplay** Morrissey. **Starring** Stefania Casini (Rubinia), Joe Dallessandro (Mario), Dominique Darel (Saphiria), Silvia Dionisio (Perla), Arno Juerging (Anton), Udo Kier (Dracula), Maxime McKendry (Marquisa), Roman POLANSKI (a villager), Vittorio De Sica (Marquis), Milena Vukotic (Esmeralda). 103 mins. Colour.

A comedy variant. Dracula can survive only on the blood of virgins (◊ VIRGINITY), and there is a scarcity in Transylvania; he thus goes with valet Anton to Italy – a Catholic country should be *littered* with virgins. He is welcomed by a seedy Marquis eager to marry off his four daughters; alas, as Dracula discovers to his nausea on biting, the lusty gardener, Mario, has in each case got there first. [JG]

27. *Vampira* (vt *Old Dracula*) UK movie (*1974*). Columbia/World Film. **Pr** Jack H. Wiener. **Dir** Clive Donner. **Screenplay** Jeremy Lloyd. **Starring** David Niven (Dracula). 88 mins. Colour.

A silly comedy in which Dracula takes samples of blood from beauty-contest finalists in order to resuscitate his dead wife. On revival, however, she proves to be black rather than white – oh, horrors! – and he tries to sort this out. [JG]

28. *The Flintstones Meet Rockula and Frankenstone* US ANIMATED MOVIE (*1977* tvm). HANNA-BARBERA. **Voice actors** Gay Auterson (Betty Rubble), Mel Blanc (Barney Rubble), Ted Cassidy (Frankenstone), Henry Corden (Fred Flintstone), John Stephenson (Count Rockula), Jean VanderPyl (Wilma Flintstone). 50 mins. Colour.

Above-average Saturday-morning fare in which the Flintstones and the Rubbles win a game-show and have a free holiday in a certain castle in Rocksylvania. [JG]

29. *Count Dracula* UK movie (*1977* tvm). BBC. **Pr** Morris Barry. **Dir** Philip Saville. **Screenplay** Gerald Savory. **Starring** Mark Burns (Jonathan), Frank Finlay (Van Helsing), Louis Jourdan (Dracula), Susan Penhaligon (Lucy), Jack Shepherd (Renfield). 155 mins. Colour.

As far as one can tell, the adaptation of Stoker's original is reasonably faithful; the reason for the doubt is that this is one of the most unusual renditions of the tale, rendered often more in the form of an LSD trip than a straight narrative. Visually very exciting, making adventurous use of film and video combined, this version is not one for chill-seekers. Jourdan is unexpectedly good in the title role. [JG]

30. *Zoltan, Hound of Dracula* (vt *Dracula's Dog*; vt *Zoltan, Hound of Hell*) US movie (*1977*). Albert Band/Frank Ray Perilli/Vic Productions. **Pr** Albert Band, Frank Ray Perilli. **Dir** Band. **Spfx** Sam Shaw. **Screenplay** Perilli. **Novelization** *Zoltan Hound of Dracula* * (1977; vt *Hounds of Dracula* US; vt *Dracula's Dog* US) by Kenneth R. Johnson. **Starring** José Ferrer (Inspector Branco), Arlene Martell (Major Hessle), Reggie Nalder (Veidt Smit), Michael Pataki (Michael Drake/Count Igor Dracula), Jan Shutan (Marla Drake). 88 mins. Colour.

A Romanian Army troop, blasting, uncovers a Dracula family tomb, and a recruit is left as overnight guard. In the darkness, the earth shakes and two coffins eject from their slots. Foolishly the guard pulls the stake from the heart of one, and is slain by the giant dog that erupts, the luminous-eyed Zoltan. The dog pulls the stake from the heart of its master, Smit – a servant of the Dracula slain by the mob in 1927. The part-vampirized Smit can endure daylight and has no craving for blood, but cannot survive without a VAMPIRE master. The revived pair set off in search of Mikhail Dracula, the last of the Dracula line, smuggled out of the country in infancy and now, ignorant of his past, living in California as Michael Drake . . .

Bearing in mind that the fear of large, savage dogs, if nothing else, should make *Z,HOD* a nailbiter, the movie is surprisingly dull. Most of the human cast are wooden: the dogs are good. [JG]

31. *Dracula* US movie (*1979*). Universal/Mirisch. **Pr** Walter Mirisch. **Dir** John Badham. **Spfx** Roy Arbogast. **Screenplay** W. Richter. **Based on** the play by Hamilton Deane and John L. Balderston. **Starring** Trevor Eve (Jonathan), Jan Francis (Mina Van Helsing), Tony Hagarth (Renfield), Frank Langella (Dracula), Kate Nelligan (Lucy Seward), Laurence Olivier (Abraham Van Helsing), Donald Pleasence (Jack Seward). 112 mins. Colour.

A sumptuous production with dazzling spfx and some superb performances (Langella, who had also played the part in a Broadway revival, has been described as the best of all screen Draculas; alas, Olivier chose to ham), this reasonably faithful interpretation of the play has become a "lost" movie. As a HORROR MOVIE it fails; as a romance, with a just-hidden layer of eroticism, it succeeds splendidly. [JG]

32. *Dracula's Last Rites* (vt *Last Rites*) US movie (*1979*). **Dir** Dominic Paris. **Starring** Patricia Lee Hammond, Gerald Fielding, Mimi Weddell. 88 mins. Colour.

A vampiristic undertaker vampirizes a mother and daughter. This has little to do with the **Dracula** *oeuvre* aside from its title, and is by all accounts very bad. [JG]

33. *Love at First Bite* US movie (*1979*). American International. **Pr** Joel Freeman. **Dir** Stan Dragoti. **Mufx** William Tuttle. **Screenplay** Robert Kaufman. **Starring** Richard Benjamin (Jeff Rosenberg), George Hamilton (Dracula), Arte Johnson (Renfield), Susan Saint James (Cindy Sondheim), Dick Shawn (Lt Ferguson). 96 mins. Colour.

An often very funny comedy. Dispossessed of his Romanian castle by the communists, Dracula goes to NEW YORK, home of fashion model Cindy Sondheim, over whose pictures he has for years drooled, recognizing her as his one true, ageless love. With dwarfish, cackling, insect-eating servant Renfield, he soon tracks Sondheim down

and, undeterred by her hardbitten promiscuity, makes romantic – and in due course carnal – love to her, to the distress of her psychiatrist and would-be bedmate Rosenberg, a descendant of Fritz von (*sic*) Helsing. The centuries-outmoded Dracula fails to adapt to the mores of modern NY – a failing that naturally increases his appeal to Sondheim, who incredulously discovers she has fallen in love.

The spfx are appalling – particularly the rubber bat into which Dracula periodically shapeshifts, always just off-camera – but, curiously, this adds to *LAFB*'s attraction. A memorable feature is Hamilton's portrayal of Dracula, assisted by subtly unsubtle make-up done by Tuttle, responsible decades earlier for Lugosi's make-up in **2**. [JG]

34. *Nosferatu the Vampyre* (ot *Nosferatu: Phantom der Nacht*) German/French movie (*1979*). Werner Herzog Filmproduktion/Gaumont/20th Century-Fox. **Pr** Werner Herzog. **Exec pr** Walter Saxer. **Dir** Herzog. **Spfx** Cornelius Siegel. **Screenplay** Herzog. **Based on** Stoker's *Dracula* and **1**. **Novelization** *Nosferatu* * (1979) by Paul Monette. **Starring** Isabelle Adjani (Lucy Harker), Bruno Ganz (Jonathan Harker), Klaus Kinski (Dracula), Walter Ladengast (Van Helsing), Roland Topor (Renfield). 107 mins. Colour.

For the first two-thirds this is a remake of **1**, although with the characters' names altered to approximate those in Stoker's novel and with the omission of most of the scenes aboard the *Demeter* and the subplot concerning Renfield's incarceration and temporary escape; the emulation extends to the physical appearances of the major characters, notably Dracula himself – although Kinski, largely responsible for his own image, claimed that he had not seen **1** before devising it. In this earlier section, however, there is a dimension missing from **1**: the possibility is explicitly stated that Dracula and his castle have no physical REALITY but are, rather, creations of the human mind.

The story diverges later in the movie. On arrival home Harker is insane, and it soon becomes clear he is now himself a VAMPIRE. Dracula comes to Lucy and sadly confesses that he wishes he could die, but is instead condemned to IMMORTALITY, as is Harker; he would sacrifice everything to feed on the LOVE that Lucy and Harker share. But Lucy, a much more dominant personality than in **1**, brusquely rejects him, and equates him with the plague infesting the city, which plague she in turn identifies, on witnessing the DANCE OF DEATH in the town square, as one not of disease but of EVIL (Herzog himself drew the analogy with the plague of Nazism). Crumbling consecrated wafers to form a ring of crumbs around Harker, thereby imprisoning him, she lures Dracula as in **1**; the eroticism of his feeding on her is here made profound, although not explicit. The dawn light brings not immediate death but protracted agony and collapse to Dracula; Lucy herself dies, her task fulfilled. Van Helsing, arriving on the scene, stakes Dracula and is arrested for the murder. Harker persuades a maid to sweep away the imprisoning crumbs, reveals himself as a vampire, and sets forth to spread the plague across the world.

This is a meticulously crafted, exquisitely conjured, finely paced and visually rich piece of moviemaking, and almost certainly the most *beautiful* version of *Dracula*. [JG]

35. *Dracula Exotica* US movie (*1981*). **Dir** Warren Evans. **Starring** Jamie Gillis (Dracula). 78 mins. Colour.

A porn variant set in New York, this starts off with a plot but soon loses it amid turmoiling buttocks. [JG]

36. *Mamma Dracula* Belgian-French movie (*1988*). **Dir**

Boris Szulzinger. **Starring** Louise Fletcher, Maria Schneider. 80 mins. Colour.

Based, like **14**, on the crimes of Countess Elizabeth de Báthory rather than Stoker's creation, this occasionally amusing comedy, set in the present, sees a vampiress employing a scientist to make artificial virgins' blood for her to bathe in. [JG]

37. *Vampire of Venice* (ot *Nosferatu a Venezia*; vt *Vampires in Venice*; vt *The Vampires of Venice*) Italian movie (*1988*). Scena/Reteitalia. **Pr** Augusto Caminito. **Exec pr** Carlo Alberto Alfieri. **Dir** Caminito. **Screenplay** Caminito. **Starring** Elvire Audrey (Uta), Ciunga (Queen of the Gypsies), Maria C. Cumani (Matilde Camins), Barbara De Rossi (Countess Helietta Canins/Letitia), Klaus Kinski (Nosferatu), Donald Pleasence (Don Alvise), Christopher Plummer (Paris Catalano). 106 mins. Colour.

Set in the present day, with Plummer as a Van Helsing figure, *VOV* is frequently listed as a sequel to **34**, which it is not, despite having Kinski and (a much changed) Nosferatu in common. The disjointed tale devolves too easily into a series of PLOT COUPONS interlaced with graphic sex scenes. There are strong echoes of The EXORCIST (*1973*) in the confrontations with Nosferatu and of DON'T LOOK NOW (*1973*) in the sensuous evocation of Venice, while the direction and the (beautiful) cinematography clearly owe much to **34**. [JG]

38. *Dracula's Widow* US movie (*1989*). **Dir** Christopher Coppola. **Starring** Sylvia Kristal. 86 mins. Colour.

Just three years before his father entered the **Dracula** arena (◊ **40**), Christopher Coppola tried his directorial hand for the first time with this very minor, would-be erotic version. Hiding out in a Hollywood wax museum, Mrs Dracula vampirizes people. [JG]

39. *Rockula* US movie (*1990*). **Dir** Luca Bercovici. **Starring** Toni Basil, Dean Cameron. 91 mins. Colour.

A vampire becomes a rock star in order to regain his reincarnated lover. [JG]

40. *Bram Stoker's Dracula* US movie (*1992*). Columbia/American Zoetrope-Osiris. **Pr** Francis Ford Coppola, Fred Fuchs, Charles Mulvehill. **Exec pr** Michael Apted, Robert O'Connor. **Dir** Francis Ford Coppola. **Vfx** Roman Coppola, Fantasy II Film Effects, 4-Ward Productions, Gary Gutierrez, Alison Savitch. **Mufx** Greg Cannom. **Screenplay** James V. Hart. **Novelization** *Bram Stoker's Dracula* * (*1992*) by Fred SABERHAGEN and James V. Hart. **Starring** Bill Campbell (Quincey P. Morris), Cary Elwes (Arthur Holmwood), Sadie Frost (Lucy Wessenra), Richard E. Grant (Jack Seward), Anthony Hopkins (Abraham Van Helsing), Gary Oldman (Dracula), Keanu Reeves (Jonathan Harker), Winona Ryder (Elisabeta/Wilhelmina [Mina] Murray), Tom Waits (Renfield). 123 mins. Colour.

The stated aim of *BSD* was to return to Stoker's original story, ignoring the countless fresh layers that had been added to the myth by earlier HORROR MOVIES. The plot is therefore both familiar and unfamiliar at the same time. The 15th-century Prince Vlad fought doughtily, impaling thousands, for the sake of the Roman Catholic Church; yet, when his beloved wife Elisabeta, believing him dead, suicided, the Church said her SOUL was damned. Vlad, furious, took a blood oath to return from the dead and, in league with the powers of darkness, avenge this verdict. Now, in 1897, young Jonathan Harker is sent by his company from London to Transylvania. The centuries-old count behaves

towards Harker with a mixture of aggression and unpleasant affection, realizing Harker's fiancée Mina Murray is a REINCARNATION of Elisabeta. Arranging for himself and a couple of dozen trunks of earth to be sent to Carfax Abbey, he enslaves Harker among a trio of SUCCUBI, the Brides, and sets off to seduce Mina, currently staying at Hillingham with the rich family of her friend Lucy Wessenra. En route to London, Dracula transmutes into a vicious beast, destroying the crew and then, on land, savagely debauching Lucy, recently affianced to Lord Arthur Holmwood. Dr Jack Seward, superintendent of the asylum and infatuated with Lucy, unable to treat her consequent medical condition, sends for his old mentor, blood-disease expert Professor Abraham Van Helsing (a richly comic performance by Hopkins), who deduces the presence of a VAMPIRE. Dracula returns to Lucy's invalid bed and, as a wolf, in a climactic scene kills her. Van Helsing, Holmwood, Seward and another swain of Lucy's, Quincey P. Morris, confront her undead form and finally stake and behead it. Harker, meantime, has escaped the Brides and sends from Transylvania for Mina; they wed and return to London. But before her departure she was infatuated by the rejuvenated Prince Vlad, and the pleasure of her guilty love lingers. When he comes to her and partakes of her blood she (knowing through her DREAMS that she is indeed his beloved Elisabeta, and thus loving him) seduces him further, drinking his blood and thereby infecting herself with vampirism. Led by Van Helsing and Jonathan (now grey-haired, despite his youth) and using a divided-self Mina (◊ MULTIPLE PERSONALITY) as first guide and then decoy, the friends track the fleeing Dracula back to his castle, where they wound him grievously. However, at the last Van Helsing stays their hands: it is the duty of Elisabeta/Mina, with her ancient husband's consent, to drive the blade through his heart and cleave his head from his shoulders, thereby releasing him and lifting the vampiric CURSE from herself.

BSD is a hard movie to assess. Coppola seems to revel in old-fashioned narrative techniques. Few superstitions concerning vampires are left untouched: MIRRORS shatter rather than bear Dracula's reflection, and flowers wilt at his passing; he can defy gravity, alter the weather and project his semblance over distance, so that Mina and Lucy see him in VISIONS; crucifixes, the ritual of EXORCISM, holy water and garlic hamper him, but little more than that. There is also much play with SHADOWS, which need not move in parallel with their casters; Dracula often pantomimes his unrealized desires. All told, though, this *BSD* generates a powerful portrayal. [JG]

41. *Dracula Rising* US/Bulgarian movie (*1992*). New Horizons. **Pr** Roger Corman. **Assoc pr** Steven Rabiner. **Co-pr** Mary Ann Fisher. **Dir** Fred Gallo. **Spfx dir** John Eppolito. **Mufx** Everett Burrell, John Vulich. **Screenplay** Rodman Flender, Daniella Purcell. **Starring** Christopher Atkins (Vlad), Stacey Travis (Theresa), Zahari Vatahov (Vlad the Impaler), Doug Wert (Alec). 87 mins. Colour.

Picture restorer Theresa is captivated by mysterious foreign visitor Vlad. Subsequently she is summoned to a creepy Transylvanian monastery to restore a portrait of Vlad the Impaler, and finds Vlad there, as well as vile abbot Alec. Subsequently she undergoes a series of TIMESLIPS, whereby she learns she is the REINCARNATION of a 15th-century peasant girl, burnt as a WITCH at Alec's instigation because of her carnal affair with the monk Vlad, the non-vampiric son of the Impaler. On her death, Vlad renounced the Church; his father persuaded him that through becoming a vampire he could achieve IMMORTALITY and wait for Theresa to be incarnated once more. Director Gallo does wonders with his budget. [JG]

Further reading: *House of Horror: The Complete Hammer Films Story* (**1973**; rev with Jack Hunter 1994; rev 1995) by Robert Adkinson, Allen Eyles and Nicholas Fry gives good coverage of all the Hammer **Dracula** movies and a reasonable survey of the others; *The Annotated Dracula* (**1975**) by Leonard Wolf is helpful; *Hollywood Gothic: The Tangled Web of "Dracula" from Novel to Stage to Screen* (**1990**) by David J. Skal; *Bram Stoker's Dracula: The Film and the Legend* (**1992**) by Francis Ford Coppola and James V. Hart is good on **40**; *The Monster Show* (**1993**) by David J. Skal contains much of relevant interest.

DRACULA – THE SERIES US syndicated tv series (1990). RHI. **Exec pr** David J. Patterson, Robert Halmi Jr. **Pr** Glenn Davis, William Laurin. **Dir** René Bonniere and many others. **Spfx** John Gadjecki. **Writers** Phil Bedard and many others. **Starring** Bernard Behrens (Gustav Helsing), Geordie Johnson (Alexander Lucard/DRACULA), Mia Kirshner (Sophia Metternich), Joe Roncetti (Christopher Townsend), Jacob Tierney (Maximilian Townsend), Geraint Wyn Davies (Klaus Helsing). 21 30min episodes. Colour.

Alexander Lucard, an enigmatic billionaire dwelling in a European castle, is the target of Gustav Helsing, a fourth-generation VAMPIRE-killer. Several episodes deal with Helsing's efforts to rescue his son, Klaus, who has been vampirized by Lucard. Also involved are Sophia, a student boarding with Helsing, and the Townsend brothers, who discover Lucard's deadly secret and join Gustav's quest. The series, filmed in Luxembourg, centred on their suspicions and subsequent efforts to draw Lucard out into the open, and on the surprisingly large number of other vampires who crossed their paths. [BC]

DRAGONBANE ◊ MAGAZINES.

DRAGONFIELDS ◊ MAGAZINES.

DRAGONLANCE Created and marketed by TSR as part of their "Advanced Dungeons & Dragons" enterprise, **DragonLance** is a role-playing GAME that has given its name to a large set of series of tied novels. The central creators of these books have been Tracy HICKMAN and Margaret WEIS, who work as a team. Other contributors include Nancy Varian BERBERICK, Mary KIRCHOFF, Richard A. KNAAK and Douglas NILES. The **DragonLance** GAME-WORLD is a FANTASYLAND in which various fairly violent HEROIC-FANTASY tales take place. [JC]

DRAGONS In most Mediterranean and European MYTHOLOGIES, SERPENTS are associated with evil, and dragons, a sort of super-serpent, are more evil still. Centuries before St George slew his dragon, APOLLO and Hercules were disposing of giant reptiles of various kinds and being celebrated for it. The dragons of Chinese mythology, by contrast, are usually benevolent. This tradition has facilitated REVISIONIST FANTASY about dragons of the Western sort.

In the Christian tradition, dragons occur in religious texts as a metaphor for the DEVIL and as a result in secular works they are the ultimate antagonist which a hero faces at the culmination of his career: BEOWULF, for example, in old age dies killing one. To kill a dragon is often to become a king – in J.R.R. TOLKIEN's *The Lord of the Rings* (**1954-5**) the

dragon Smaug's archer slayer founds a line of kings at Esgaroth. In Tad WILLIAMS's **Memory, Sorrow and Thorn** (**1988-93**) it is PRESTER JOHN's supposed slaying of the dragon of the Hayholt which gives him the right to rule there, from a throne made of its bones; when the HIDDEN MONARCH Simon almost accidentally tackles a dragon during his journeys, it is one of the first things that signals to us that he is more than a kitchen boy.

In Melanie RAWN's DYNASTIC FANTASIES and in Michael MOORCOCK's **Elric** stories dragons are essentially no more than useful semi-sapient fighting beasts; in Samuel R. DELANY's **Nèverÿon** books (**1979-87**) they are little more than totemic animals rarely viewed. In the **Pern** PLANETARY ROMANCES by Anne MCCAFFREY (from **1968**) they are again semi-sapient, capable of an emotional bonding to their selected riders that is quasi-sexual. Terry PRATCHETT's *Guards! Guards!* (**1989**) contrasts a large traditional dragon of great malevolence with a bunch of little pet dragons prone to auto-destructive bouts of hiccups. Another comic dragon is the fish-eating specimen in Avram DAVIDSON's *Peregrine: Secundus* (**1981**).

As often as not, whether intelligent or bestial, dragons are the hunter, not the hunted. Standing as they do as a gate between life and death and as flesh-and-blood beings that are nonetheless magical in their essence, they are LIMINAL BEINGS often connected with the getting of wisdom rather than merely enemies to be confronted. A conversation with a dragon is always a kind of duel, a struggle to refuse hypnotism or mastery. Even a moribund and largely petrified dragon like the one in Lucius SHEPARD's **Dragon Griaule** sequence of novellas (1984-8) has the power to influence events, thinking slow dragonish thoughts that compel human action to the same pattern. The blood of a dead dragon makes Richard WAGNER's Siegfried in the **Ring** sequence of OPERAS (**1853-74**) a liminal being, at once able to understand the language of birds yet still unable to comprehend the trickery of human beings, invulnerable and yet susceptible to a draught that brings forgetfulness and a spear in the part of his back that the blood never reached – a variant of the ACHILLES' HEEL motif.

Dragons are emblems of covetousness – when, in *The Voyage of the Dawn Treader* (1952) C.S. LEWIS's Eustace is turned into one, it is by thinking covetous thoughts about the hoard he has come across. Wagner's Fafner has similarly opted to change into a dragon in order better to guard the CURSE-ridden hoard for which he has already sacrificed his brother. Though dragons like Tolkien's Smaug are typologically related to the Satanic dragon of Christianity, their endless pursuit of anyone who has stolen from their hoard derives from the Norse version of dragonishness. This is at once one of their defining characteristics and their Achilles' Heel; it is because he has suffered a theft from his hoard that Smaug emerges, and is thus killed.

Dragons' legendary habit of devouring maidens (◊ VIRGINITY) is something many fantasists have tried to rationalize (◊ RATIONALIZED FANTASY). Because dragons are seen as solitary, they have to have some sort of sexuality, and eating virgins fits the bill. The best movie portrayal of this theme is in DRAGONSLAYER (*1981*).

Ursula K. LE GUIN's dragons, in *A Wizard of Earthsea* (**1968**), *The Farthest Shore* (**1963**) and *Tehanu* (**1992**), are perhaps the most high-minded and superior, and maybe the most beautiful, embodying as they do physical perfection, wisdom and a sense of danger that has little to do with the fear they might eat you. She establishes them so totally that when, in *The Farthest Shore*, they prove vulnerable to the malignity of an evil WIZARD, the effect is one of the most tragic moments in modern fantasy.

What Le Guin's dragons never for a moment become is cosy. There is a long tradition in children's fantasy of misunderstood and cosy dragons, starting perhaps with Kenneth GRAHAME's *The Reluctant Dragon* (**1938** chap). The oddest combination of this cuteness with a traditional morality stance comes perhaps in Gordon R. DICKSON's **Dragon and the George** sequence (from **1976**) where dragons and human beings (georges) live together in intermittent disharmony but as allies against the forces of real evil.

In many respects the most terrifying dragon of recent years – aside from, perhaps, DISNEY's metamorphosed (◊ METAMORPHOSIS) Maleficent in SLEEPING BEAUTY (**1959**) and the MONSTER in Terry GILLIAM's JABBERWOCKY (**1977**) – is the one in Michael SWANWICK's *The Iron Dragon's Daughter* (**1993** UK) – a dragon that is also a machine (◊ TECHNOFANTASY) and a TRICKSTER. Swanwick shows us a Universe that is a trap from which his heroine may awaken but never escape; the dragon is terrifying for both his malice and his manipulations, and because, in this Universe, he is ultimately a minor pest, powerless and flicked away. [RK]

See also: WORM.

DRAGONSLAYER US movie (**1981**). Paramount/ Disney. **Pr** Hal Barwood. **Exec pr** Howard W. Koch. **Dir** Matthew Robbins. **Spfx sv** Brian Johnson. **Vfx** Dennis Muren, Industrial Light & Magic. **Chief sculptor** Derek Howarth. **Screenplay** Barwood, Robbins. **Novelization** *Dragonslayer* * (**1981**) by Wayland DREW. **Starring** Sydney Bromley (Hodge), Caitlin Clarke (Valerian), Peter Eyre (King Casiodorus), John Hallam (Tyrian), Emrys James (Blacksmith), Peter MacNicol (Galen Bradwarden), Ralph Richardson (Ulrich), Chloe Salaman (Princess Elspeth). 110 mins. Colour.

SORCERER'S APPRENTICE Galen, on the apparent suicide of his master Ulrich, joins forces with youthful warrior Valerian in an attempt to destroy a DRAGON, *Vermithrax pejorative*, which is held from devastating the nation of Urland only through a CONTRACT whereby it is fed a virgin (◊ VIRGINITY) annually. Working against them is Tyrian, right-hand man of King Casiodorus. The king explains the wisdom of the contract: better a single lottery-selected virgin dies each year than that thousands die. (It is to *D*'s great credit that this argument is presented persuasively.) Casiodorus's daughter, Elspeth, discovers the lottery has been rigged to spare the aristocracy, and rigs it so that she will herself be next victim. Aided by Valerian (who proves a girl reared as a boy to avoid the lottery, and whom Galen gleefully disqualifies from further lotteries; ◊ GENDER DISGUISE) and his own clumsy attempts at MAGIC, Galen tries to save Elspeth and maims, but cannot kill, the dragon. At last he recalls instructions transmitted to him from Ulrich, and raises the sorcerer – who kills the dragon but himself dies.

The dragon owes much at times to the metamorphosed Maleficent in SLEEPING BEAUTY (**1959**); it is "real" enough that, on occasion, it can draw pathos. *D* was a box-office failure, largely because the Disney name made potential audiences believe it was a children's movie, yet is one of the best HIGH-FANTASY movies made, blending, often with wit, many archetypal motifs – notably, aside from those mentioned above, the idea that the death of the dragon symbolizes the THINNING of the GOLDEN AGE, with crudescent Christianity lurking in the wings. [JG]

DRAKE, BURGESS ◊ H.B. DRAKE.

DRAKE, DAVID A(LLEN) (1945-   ) US writer and attorney who has become known primarily for his military sf but who has also written a wide range of fantastic fiction, seeming equally at home in the genres of SWORD AND SORCERY, SCIENCE FICTION, SUPERNATURAL FICTION and historical fantasy (◊ HISTORY IN FANTASY). His first professional publication, "Denkirch" (in *Travellers by Night* anth **1967** ed August DERLETH), was an H.P. LOVECRAFT pastiche about the cosmic origins of humankind. He has produced several new stories in the CTHULHU MYTHOS, and has compiled a volume of Robert E. HOWARD's **Cthulhu** stories, *Cthulhu: The Mythos and Kindred Horrors* (coll **1987**). DAD was for a while assistant editor of WHISPERS, to which he contributed several stories including the first of his **Old Nathan** series, about a grumpy ghost-hunter in the 19th-century US backwoods (◊ OCCULT DETECTIVES). These stories, collected with new material as *Old Nathan* (coll of linked stories **1991**), were an affectionate tribute to the work of Manly Wade WELLMAN.

Some of DAD's fiction draws from his experiences in military service, giving a hard edge to his treatment of classic fantasy themes, plus a strong sense of alienation, suggestive of a darker power that we may occasionally unleash at our peril but which normally we perceive only tangentially. This mood pervades *From the Heart of Darkness* (coll **1983**); both its title story and "Out of Africa" deliberately evoke Joseph CONRAD's classic tale of the overwhelming otherness of the dark continent.

DAD has a fascination for ancient Rome and the collapse of civilization. He captured an authentic feel in his realistic ARTHUR novel, *The Dragon Lord* (**1979**), and his collection *Vettius and His Friends* (coll **1989**). He blends several genres in *Birds of Prey* (**1984**), a thriller about a TIME traveller ensnared in intrigue in ancient Rome, and *Ranks of Bronze* (**1986**), where aliens land there. *The Eternal City* (anth **1990**) with Martin H. GREENBERG and Charles G. Waugh is a volume of sf stories exploring the Roman Empire throughout time.

DAD's only traditional HEROIC FANTASY, *The Sea Hag* (**1988**), was the first in the **World of Crystal Walls** sequence; further volumes are awaited. [MA]

DRAKE, H(ENRY) B(URGESS) (1894-1963) UK writer and teacher who wrote also as Burgess Drake. He spent several years teaching English at universities in Japan, Korea and China, the latter being the setting of his most successful novel, *Chinese White* (**1950**). His early *The Remedy* (**1925**; vt *The Shadowy Thing* 1928 US) deals with malevolent SPIRITUALISM, POSSESSION and the release of a tortured SOUL. *Hush-A-Bye Baby* (**1952**) is a GHOST STORY about a woman haunted by the POLTERGEIST spirits of her unborn TWINS, who died during a miscarriage; this novel was dedicated to his friend Mervyn PEAKE, who illustrated Drake's fantasy *The Book of Lyonne* (**1952**) and his short story about a Korean ghostly monster, "Yak Mool San" (1949). [RD]

DRAUGHTSMAN'S CONTRACT, THE UK movie (1982). British Film Institute/Channel 4. **Pr** David Payne. **Dir** Peter GREENAWAY. **Screenplay** Greenaway. **Starring** Neil Cunningham (Thomas Noyes), Michael Feast (Statue), Hugh Fraser (Louis Talmann), Anthony Higgins (R. Neville), Anne Louise Lambert (Sarah Talmann), Janet Suzman (Mrs Virginia Herbert). 108 mins. Colour.

Perhaps Greenaway's best-known example of genre-crossing. It is 1694. The draughtsman Neville is hired by Mrs Herbert to produce 12 drawings of the family estate, Anstey, as a gift to her husband, who departs for a fortnight's womanizing in Southampton; Neville's fee, formalized in a CONTRACT, is £8 per drawing, full board and lodging, and the daily sexual favours of Mrs Herbert. Neville delights in his domination, but underestimates the ingenuity of both her and especially her daughter, Sarah Talmann, the frustrated wife of an impotent boor; both women, unknown to him, seek an heir to Anstey. Mrs Talmann points out to Neville that extraneous details in his drawings seem to point to a plot injurious to the still-absent Herbert, and persuades him to enter a contract whereby he has daily SEX with her, too. Herbert's corpse turns up in the pond, and accusations fly. Yet all seems to have blown over when Neville returns to Anstey to make a 13th drawing. He is seduced yet again by Mrs Herbert, and she and Mrs Talmann explain how they have manipulated him; they then manipulate the men of the household into murdering him.

Aside from the MAGIC REALISM infused into this chilly movie by direction and camerawork – and not least by the brittle dialogue – *TDC*'s primary fantasy component concerns an anonymous male nude who, his body blacked appropriately, poses habitually as different parts of the estate's statuary; only the servants, a child and Neville actually *see* this man, the other characters being too lost in their own artificial PERCEPTIONS to observe him. In turn, the "statue" is an observer of all that proceeds; yet, like an ELEMENTAL, he plays no part in the mortals' games beyond passive observation.   [JG]

DREADSTONE, CARL Pseudonym of Ramsey CAMPBELL.

DREAM A LITTLE DREAM US movie (*1989*). Vestron/Lightning. **Pr** D.E. Eisenberg, Marc Rocco. **Exec pr** Lawrence Kasanoff, Ellen Steloff. **Dir** Rocco. **Screenplay** Eisenberg, Daniel J. Franklin, Rocco. **Starring** Corey Feldman (Bobby Keller), Piper Laurie (Gina Ettinger), Jason Robards (Coleman Ettinger), Meredith Salenger (Leni). 110 mins. Colour.

Eccentric Coleman is researching the DREAM state, believing that if he and wife Gina can enter the dream existence they might achieve IMMORTALITY. As they experiment one night, highschool bombshell Leni and almost-dropout Bobby bump into each other nearby and suffer mild concussion. On awakening, Bobby's body now contains Coleman's mind. Something of a similar IDENTITY EXCHANGE appears to have occurred between Leni and Gina: Leni is still Leni, but is troubled by some of Gina's memories. Coleman and Bobby meet several times in a dream REALITY (much like ours, but shot through a blue filter and with what looks like toilet paper draped everywhere): Coleman is keen to regain his true identity, Bobby less so. There are various adventures of potential interest to ethnologists studying courtship displays among US highschool students before the *status quo ante* is restored by means of an unusually feeble plot device. *DALD* is jerkily directed, poorly plotted and has nothing new to add beyond the (excellent) forays into a reified dreamland; there is a sense that an originally intelligent screenplay may have been, during filming, compromised into mediocrity and beyond.   [JG]

DREAMCHILD UK movie (*1985*). Thorn EMI/Pennies From Heaven. **Pr** Rick McCallum, Kenith Trodd. **Exec pr**

Verity Lambert, Dennis Potter. **Dir** Gavin Millar. **Creature fx** Jim HENSON's Creature Shop. **Screenplay** Potter, with lines from the works of Lewis CARROLL. **Starring** Jane Asher (Mrs Liddell), Coral Browne (Alice Hargreaves), Nicola Cowper (Lucy), Peter Gallagher (Jack Dolan), Ian Holm (Charles Dodgson), Amelia Shankley (Alice Liddell). **Voice actors** Alan Bennett (Mock Turtle), Tony Haygarth (Mad Hatter), Fulton Mackay (Gryphon), Julie Walters (Dormouse), Ken Campbell (March Hare), Frank Middlemass (Caterpillar). 94 mins. Colour.

Alice Liddell, now the widowed Alice Hargreaves and nearly 80, comes to New York in 1932 to receive an honorary doctorate at Columbia to celebrate Carroll's centennial. With her she brings downtrodden personal companion Lucy; Alice has become a mean-spirited, avaricious bitch. What the Americans want from her, in the midst of the Depression, is something of the hope and dreams they assume Carroll gave her; what she soon comes to want from them is money. But the experience changes her – or perhaps her memories change ($\lozenge$ PERCEPTION), for through flashbacks (representing her thoughts), some into her real childhood and some into WONDERLAND, we discover first her growing suspicion that the relationship between her and Carroll had been pedophile and then her realization (true or false) that all the stuttering, uncertain, middle-aged Carroll had wanted was to be loved by the enchanting coquette she then was. Through this realization she at last discovers within herself the sweet child whom Carroll thought he saw.

This – Potter's second venture into this territory after *Alice* (*1965* tvm; $\lozenge$ ALICE IN WONDERLAND [*1951*]) – is a deeply rewarding piece of fantasy at many levels. Script, direction, music (by Stanley Myers) and cast are flawless: all contribute to an artwork that is extremely affecting.   [JG]

DREAMS The most elementary form of fantasy is conscious daydreaming, but the hallucinatory dreams we experience in sleep are what reveal the true range and strangeness of our fantasy life. Dreams are often bizarre but rarely so detached from the everyday as to offer a reliable haven of escape; what we most often find in Dreamland are distorted reflections of our anxieties.

When literary fantasies masquerade as dreams ($\lozenge$ VISIONARY FANTASY) they cannot help but import the kinds of meaning for which we search our actual dreams in vain. The device of justifying a narrative by representing it as a dream is the most elementary in fantastic literature, fundamental to such medieval subgenres as the dream ALLEGORY, although climactic awakenings came to be seen as the ultimate narrative cop-out once the notion of literary fantasy had become familiar and such apologetic moves no longer seemed necessary.

The suspicion that life itself may be a kind of dream, and we mere figments of it, is endorsed by several Far Eastern religious systems and incorporated into the fascinating worldview of the Australian Aborigines ($\lozenge$ DREAMTIME). It remains unsettling even when it is broached as lightly as in Lewis CARROLL's *Through the Looking-glass* (**1872**). The notion that we might become lost in a dream so mazy that all our apparent awakenings are merely renewals is so uniquely ominous as to give Robert IRWIN's *The Arabian Nightmare* (**1983**) its title.

Many notable literary fantasies grew from seeds planted by dreams, the most famous examples being Mary SHELLEY's *Frankenstein* (**1818**), Robert Louis STEVENSON's *Strange*

Case of Dr Jekyll and Mr Hyde (**1886**) and Bram STOKER's *Dracula* (**1897**). This was to be expected in the 19th century, when laudanum was the only available painkiller; its influence on the literary work of Samuel Taylor COLERIDGE, Thomas de Quincey (1785-1859) and others is notorious. Several French writers, including Théophile GAUTIER and Charles BAUDELAIRE, forged literary fantasies out of dreams induced by opium and hashish, while Jean Lorrain (◊ DECADENCE) reconfigured dreams induced by drinking ether. Later writers were able to medicate themselves more carefully, but several of the most significant 20th-century writers of fantasy have made assiduous use of their dreams as a resource. Lord DUNSANY's penchant for so doing, most clearly evidenced in *A Dreamer's Tales* (coll **1910**), was echoed and elaborated by H.P. LOVECRAFT, whose exploits in this vein are described and exemplified in *Dreams and Fancies* (coll **1962**).

Fantasies involving the submission of dreams to some kind of technical investigation or strategic control are common. Notable fantasies are "The Wonderful Glass" by ERCKMAN-CHATRIAN and *Sphinx* (**1923**) by David LINDSAY. Dreamland is represented in these stories – tacitly, at least – as a kind of metaphysical ALTERNATE WORLD, a status which it is also afforded in *The Wonderful Visit* (**1895**) by H.G. WELLS, *Smirt* (**1934**) and its sequels by James Branch CABELL, *Marianne Dreams* (**1958**) by Catherine Storr (1913-), *A NIGHTMARE ON ELM STREET* (**1984**) and its sequels, *Dreamwatcher* (**1985**) by Theodor ROSZAK, *The Bridge* (**1986**) by Iain Banks (1954-), *Bones of the Moon* (**1987**) by Jonathan CARROLL, PAPERHOUSE (**1988**) and *Only Forward* (**1994**) by Michael Marshall Smith (1965-).

Collections of stories allegedly based on dreams, or which simulate dreams, include *Dreams* (coll **1891**) by Olive Schreiner (1855-1920) and *Strange Dreams* (anth **1993**) ed Stephen R. DONALDSON. Nightmares, as featured in *Nightmares of Eminent Persons* (coll **1954**) by Bertrand Russell (1872-1970) and *A Book of Contemporary Nightmares* (anth **1977**) ed Giles Gordon (1940-), have the advantage of innate melodrama. [BS]

Further reading: *Nightmare* (**1979**) by Sandra Shulman; *Dreamers* (**1984**) by John GRANT.

DREAMTIME An Australian Aboriginal term referring to a time when the spirit ancestors of humans, plants and animals invested themselves into the territory of the world (i.e., what we know as Australia), shaping its geography and giving it life in various ways, animal and vegetable. The LANDSCAPE today is a reminiscence of the true landscape of the Dreamtime and is isomorphic with the Dreamtime, which can be perceived (◊ PERCEPTION) by real people (not the Whites, who represent a fatal THINNING of the world, and who destroy anything they touch). The D exists, therefore, in what may be called a TROMPE L'OEIL relationship to the mundane world; in this respect it is an ALTERNATE REALITY. Several novels – e.g., *The Second Bridegroom* (**1991**) by Rodney Hall – take as their subject the destructive relationship between White civilization and the "CHAOS" of the Dreamtime world. *The Songlines* (**1987**) by Bruce Chatwin responsibly depicts a world in which the Dreamtime is manifest for those who are able to look. Some of B. WONGAR's tales make reference as well; and an exploration of the Dreamtime occupies the heart of Michaela ROESSNER's *Walkabout Woman* (**1988**). In Frank Questing's *Nobody Dick in Dreamtime* (**1989**) the PICARESQUE adventures of a surreal contemporary DUO are couched as a sequence of episodes in the Dreamtime. A theme anthology is *Dream Time* (anth **1989**) ed Toss Gascoigne, Jo Goodman and Margot Tyrrell. [JC]

DREAM WORLD (magazine) ◊ FANTASTIC; MAGAZINES.

DREHER, SARAH (ANNE) (1937-) US writer. ◊ BILLY THE KID; CHILDE.

DREW, WAYLAND (1932-) Canadian teacher and writer whose first novel, *The Wabeno Feast* (**1973**), can be read as sf, but which benefits as well from a fantasy understanding. The Wabeno himself is a SHAMAN who represents a heart of darkness for the 18th-century white man who encounters him beyond the circle of a civilization which is already fatally THINNING the world; in the other two narrative strands – one set in the present, the other in the near future – the wilderness gradually comes to represent the last breath of the ruined planet. WD's next novel, *Dragonslayer* * (**1981**) – novelizing DRAGONSLAYER (*1981*) – is an efficient expansion upon the original. After an sf trilogy, WD published a second movie tie, *Willow* * (**1988**), which competently translates WILLOW (*1988*).

"The Old Soul" (in *Once Upon a Time: A Treasury of Modern Fairy Tales* anth **1985** ed Lester del Rey and Risa Kessler) – in which the FOREST, which is "Life", and the CITY, which is THINNING the world, come into conflict – is typical of WD's longer work in its ease of access but also its proneness, perhaps augmented by the FAIRYTALE format, to sentimentality. WD has not recently been active as an author of commercial fiction. [JC]

DREYER, CARL TH(EODOR) (1889-1968) Danish movie director and screenwriter. The first movie he directed was *Praesidenten* ["The President"] (*1919*). With his next, *Blade af Satans Bog* (*1919*; vt *Leaves from Satan's Book*), whose screenplay he based on Marie CORELLI's *The Sorrows of Satan* (**1895**), he moved into the territory of fantasy, showing SATAN's activities in various epochs from the time of CHRIST onwards; the movie was attacked on several fronts, notably for its portrayal of Christ. After several other movies and much lauded in France, CTD made in that country *La Passion de Jeanne d'Arc* (*1927*; vt *The Passion of Joan of Arc*), a movie made all the more powerful by the fact that its screenplay (by CTD) was based on the extant trial records. Almost all further cinematic representations of JOAN OF ARC have been affected by the imagery of this silent movie; its script appears in CTD's *Four Screenplays* (trans **1970**). Despite the fact that this was swiftly recognized as a classic of cinema – a status it still enjoys – CTD was on his uppers when he accepted private financing from Baron Nicholas de Gunzberg to make VAMPYR (*1932*), one of the oddest of all fantasy movies. His next movie of interest was *Vredens Dag* (*1943*; *Day of Wrath*), in which an innocent 17th-century woman is burnt as a WITCH; she pronounces a CURSE on the local cleric, who soon dies; his wife, who loves another, is accused of witchcraft . . . and so the vicious round of human stupidity and vindictiveness continues. *Ordet* (*1954*; vt *The Word*) – by now CTD was much less prolific – was another complex tale, and probably a considerable influence on Ingmar BERGMAN. It is just about viewable as a realistic drama in which revelatory religious experiences are merely a matter of PERCEPTION, but it plays with the *viewer*'s perception to make one realize that the supernatural is truly at work. *Gertrud* (*1964*) uses Surrealist techniques born purely of the cinema, rather than derived from painted SURREALISM, to convey emotional anguish in a carefully understated tale of emotional deprivation.

It was CDT's last movie: he died while planning *Jesus*, for which his extensive research included touring Israel at length and learning Hebrew. Not all of CDT's movies were well received at the time, and many have become hard to find; but all share a sense of the mysterious that is very close to the heart of fantasy – while yet being entirely cinematic, owing little or nothing to the written word. [JG]

DROP DEAD FRED US movie (*1991*). Working Title/PolyGram. **Pr** Paul Webster. **Exec pr** Tim Bevan, Carlos Davis, Anthony Fingleton. **Dir** Ate De Jong. **Vfx** Pete Kuran. **Screenplay** Davis, Fingleton, Elizabeth Livingston. **Starring** Phoebe Cates (Elizabeth), Ron Eldard (Mickey Bunce), Marsha Mason (Polly Cronin), Tim Matheson (Charles), Rik Mayall (Fred), Ashley Peldon (Young Elizabeth). 99 mins. Colour.

Insultingly dumped by vile, faithless husband Charles, Elizabeth finds her life invaded once more by the almost TOON-like INVISIBLE COMPANION of her childhood, Drop Dead Fred. Under his anarchic influence she learns how to break free of the stultifying dominance of her mother, Polly, and the insecurity that drives her to try to make up with Charles. At last she learns, too, how to release herself from Drop Dead Fred – only to discover that the divorced childhood friend, Mickey, with whom she is interested in starting a new life, has a daughter who also knows Drop Dead Fred . . .

Cates turns in a performance of skin-deep charm; Mayall is too constrained by considerations of taste to be fully effective, and is hampered by the fact that he works better with Peldon (the child Elizabeth, seen in copious flashbacks) than with Cates. *DDF* is memorably forgettable, despite some fine fantasy moments. [JG]

See also: RITE OF PASSAGE; SECRET SHARER.

DROWNED WORLDS ◊ LOST LANDS AND CONTINENTS.

DRUILLET, PHILIPPE (1944-　　) French writer. ◊ MOEBIUS.

DRUMMOND, HAMILTON (1857-1935) UK writer, mostly of historical romances, some of whose uncollected stories – like "A Secret of the South Pole" (1902 *Windsor Magazine*) – were sf. He is of fantasy interest for *Gobelin Grange* (coll **1896**), whose FRAME STORY features a storytelling DEVIL; some of the tales are fantasies. The protagonist of *The Chain of Seven Lives* (**1906**) is given a POTION by the MAGUS Albert Magnus, and undergoes seven REINCARNATIONS. The protagonist of *The Three Envelopes* (**1912**) is sent on missions involving the supernatural by the mysterious Society for Promoting Queer Results (whose initials, SPQR, hint that SECRET MASTERS run it). [JC]

DRUMS OF FU MANCHU US movie (*1940*). ◊ FU MANCHU.

DRUUNA Beautiful, dark-haired, full-breasted star of the **Druuna** series of morbid sf/horror/fantasy GRAPHIC NOVELS created and drawn by Italian artist Paolo Eleuteri-Serpieri (1944-　). The first story, *Morbus Gravis* (*Comic Art* 1985; trans HEAVY METAL 1986 US) introduces a rundown rubble-strewn CITY in which an epidemic disease causes the inhabitants to mutate into ghastly, tentacled cannibalistic monsters. After some horrifying adventures and gruesome sexual experiences, Druuna discovers the city is a gigantic derelict spaceship, aimlessly adrift. Further volumes have appeared at regular intervals: they include *Morbus Gravis II* (graph **1988**; trans **1992** US), *Creatura* (graph **1991**; trans **1993** US) and *Carnivora* (graph **1993**; trans **1994** US).

The series' success owes much to the sexually explicit nature of many of Druuna's experiences and to Eleuteri-Serpieri's lovingly rendered art. His earlier work had consisted almost solely of strips about the US West, including contributions to the Larousse *Histoire du Far West* ["History of the Far West"] (graph **1975** France) and *I Grandi Miti del West* ["Great Legends of the West"] (graph coll **1978-80** France; trans as *Storie del West* **1980** Italy), although his later work in this genre included fantasy elements – e.g., *Sciamano* ["Shaman"] (graph **1983**) and *L'Indiana Bianca* ["The White Indian"] (graph **1984**). He also collaborated with Victor de la Fuente (1927-　　), Raymond Poivet (1910-　　) and others on a very highly regarded COMIC-strip version of *The Bible* (graph **1984** 8 vols France; trans **1995-6** 24 vols UK). An impressive volume of his **Druuna**-related studies and sketches is *À la Recherche de Druuna: Obsession* (graph **1990**; trans as *Druuna X* **1995** US). [RT]

DRYADS The tree-nymphs of Greek mythology, originally associated with oaks, although the term has been broadened to encompass all wood-spirits. Those inhabiting specific TREES perished when the trees died; those associated with groves could endure much longer, like the ones in *The Dryad* (**1905**) by Justin McCARTHY (1860-1936) and *Forbidden Marches* (**1929**) by E.V. de Fontmell; both feature love affairs between men and dryads and are determined ALLEGORIES of THINNING. Other human/dryad love stories include "The Hamadryad" (1909) by Bernard Capes (1870-1918), "The Woman of the Wood" (1926) by A. MERRITT and "Forsaking All Others" (1939) by Lester del Rey; J.K. MUSÄUS's "Libussa" (c1780) is about the offspring of such a union. Kindly dryads are found in "Old Pipes and the Dryad" (1887) by Frank R. STOCKTON and "The Dear Dryad" (1924) by Oliver ONIONS, which features a matchmaking oak. Dryads are assimilated to the population of FAERIE in the title-story of Frank White's *The Dryads and Other Tales* (coll **1936**); they play important roles in several fantasies by Thomas Burnett SWANN, most significantly "Where is the Bird of Fire?" (1962; exp **1976** as *Lady of the Bees*) and its sequel *Green Phoenix* (**1972**). Moral tales include *Utinam* (**1917**) by William Arkwright (1857-1925) and "When Pan was Dead" (1905) by Laurence HOUSMAN, a heartrending tale of a "woodling" who fails to adapt to the advent of Christianity. "The Hardwood Pile" (1940) by L. Sprague DE CAMP is a comedy. Celtic tree-spirits feature in "The Annirchoille" (1896) by Fiona MACLEOD. [BS]

DRYASDUST ◊ M.Y. HALIDOM.

DUANE, DIANE E(LIZABETH) (1952-　　) US writer, married to Peter MORWOOD, with whom she lives in Ireland. She has published considerable sf, much of it for the **Star Trek** enterprise. She began publishing work of interest with her first fantasy series, the **Tale of the Five** – *The Door into Fire* (**1979**), *The Door into Shadow* (**1984**) and *The Door into Sunset* (**1992**), with «The Door into Starlight», projected – the title of which refers to five COMPANIONS whose fates interweave in SWORD-AND-SORCERY fashion against a FANTASYLAND backdrop. The central MAGIC is a kind of TALENT, a "Fire" of enormous power usually reserved for women (the CREATION MYTH underlying the series is based on the GODDESS); but one of the five finds himself also able to generate the Fire, though he cannot at first control it. After opening a PORTAL which allows a fire ELEMENTAL (named Sunspark) access to his world, he does eventually gain control and tames Sunspark. As the series progresses the five realize their own self-potential, gain power, defend their land, and undertake to rule it.

The **Wizardry** sequence – *So You Want to be a Wizard?* (**1983**), *Deep Wizardry* (**1985**) and *High Wizardry* (**1990**), all three assembled as *Support Your Local Wizard* (omni **1990**), plus *A Wizard Abroad* (**1993**) – more lightly, for a YA audience and in a TECHNOFANTASY vein, follows the adventures of several apprentice WIZARDS who, in their fight against the "Lone Power" who generates deadly entropy, enter (for instance) an ALTERNATE-WORLD Manhattan (◊ NEW YORK) in which machines are inimical. The element of RITE OF PASSAGE in the sequence, as the various youths mature, is handled with aplomb.

With Morwood she has written *Keeper of the City* * (**1989**), a contribution to the **Guardians of the Three** SHARED-WORLD created by Bill Fawcett, set in a world in which human-like creatures are in conflict with sentient reptilians. [JC]

DUBLIN UNIVERSITY MAGAZINE ◊ J. Sheridan LE FANU; MAGAZINES.

DUCKTALES: THE MOVIE – TREASURE OF THE LOST LAMP US ANIMATED MOVIE (*1990*). Disney Movietoons. **Pr** Bob Hathcock. **Dir** Hathcock. **Spfx** Alan Howarth, Mel Neiman. **Screenplay** Alan Burnett. **Based on Disney's Ducktales** comics, in part reprinting work by Carl BARKS, and the **DuckTales** tv series. **Voice actors** Joan Gerber (Mrs Beakley), Richard Libertini (Dijon), Christopher Lloyd (Merlock), Chuck McCann (Duckworth), Terence McGovern (Launchpad McQuack), Rip Taylor (Genie), Russi Taylor (Huey, Dewey, Louie, Webbigail), Alan Young (Scrooge McDuck). 74 mins. Colour.

SCROOGE MCDUCK and his associates discover in the Middle East the treasure of Collie Baba, despite the attempts of evil sorcerer Merlock and sidekick Dijon to get there first. Part of the treasure is a battered old lamp, given to Webbigail as a dolls' teapot but in fact containing a GENIE. Back in Duckburg, the VILLAINS use foul means to try to gain the lamp; in an astonishingly dramatic finale Merlock is destroyed. DISNEY seemed embarrassed by this movie, releasing it under a new label, Disney Movietoons, to distance it from other animated features; US critics vilified it, some because they saw it as a travesty of Barks. Yet, despite obvious flaws, it is verbally hilarious, its climax is as fine a piece of animation as anything in ALADDIN (*1992*) – of which it is a precursor – and its fond PARODY of INDIANA JONES-style adventures, and the traditions that gave them birth, is assured. [JG]

DUFAYS, BAUDELAIRE Pseudonym of Charles BAUDELAIRE.

DULAC, EDMUND (1882-1953) French-born illustrator, in the UK from about 1903; from 1923 until his death he lived with Helen BEAUCLERK. With Arthur RACKHAM, ED was one of the central illustrators of the Edwardian period, a time when fantasy ILLUSTRATION reached a peak of sophistication. As in Rackham's case, his prime stretched from about 1905 through WWI. From the first, though ED shared Rackham's exuberant capacity to create intricately peopled fantasy LANDSCAPES, his work was more painterly, and he had an unmatched knowledge of how to use book-production technology to gain expressive colour effects. The influence of Persian and Indian art – especially miniatures – was always evident; later, Chinese influences also showed in the increasingly jewelled precision of his effects.

ED's first illustrations appeared in France in 1897; work of fantasy interest began to appear after his move to England,

the first being illustrations to *Fairies I Have Met* (**1907**; exp vt *My Days with the Fairies* 1913) by Mrs Rodolph Stawell. In that year he began to produce an annual CHRISTMAS BOOK for Hodder & Stoughton, the firm which had published Rackham's first Christmas gift-book. The sequence, for which ED did much of his best fantasy work, comprises *Stories from the Arabian Nights* (coll **1907**), text by Laurence HOUSMAN, *Shakespeare's Comedy of the Tempest* (1908), *The Rubáiyát of Omar Khayyám* (**1909**) by Edward Fitzgerald (1809-1883), *The Sleeping Beauty and Other Fairy Tales* (coll **1910**), adapted from the Brothers GRIMM and Charles PERRAULT by Arthur QUILLER-COUCH, *Stories from Hans Christian Andersen* (coll **1911**), *The Bells and Other Stories* (coll **1912**) by Edgar Allan POE, *Princess Badoura* (**1913**), text by Housman (again from *The Arabian Nights*), and *Sinbad the Sailor & Other Stories from the Arabian Nights* (coll **1914**). WWI broke the sequence, the next title being a compilation for charity, *Edmund Dulac's Picture-Book for the French Red Cross* (graph **1915**; cut vt *Edmund Dulac's Picture-Book* 1919). The sequence proper then continued with *The Dreamer of Dreams* (**1915**) and *The Stealers of Light* (**1916**), both by Queen Marie of Roumania (1875-1938), *Edmund Dulac's Fairy-Book – Fairy Tales of the Allied Nations* (coll **1916**) and Nathaniel HAWTHORNE's *Tanglewood Tales* (coll 1918; cut 1938).

After illustrating William Butler YEATS's *Four Plays for Dancers* (coll **1921**) and Beauclerk's first two novels, ED released *A Fairy Garland, Being Fairy Tales from the Old French* (coll **1928**). Some of his late work included illustrations for *Gods and Mortals in Love* (coll **1936**), with text by Hugh Ross Williamson, *The Daughters of the Stars* (coll **1939**), with text by Mary C. Crary, and John MILTON's *Comus* (1954 US). [JC]

DUMAS, ALEXANDRE père (1802-1870) French dramatist and writer who composed – on his own, and with the aid of anonymous collaborators – hundreds of works, many enormously long and enormously popular. The central corpus makes up a romantic fictional history of France; within this corpus, which eschews supernatural elements, best-known is the **Musqueteers** sequence, beginning with *Les Trois mousquetaires* (**1844**; trans anon as *The Three Musqueteers* **1846** UK) with Auguste Maquet (anon). His most famous singleton, *Le Comte de Monte-Cristo* (1844-45 *Journal des Débats*; **1846**; trans anon as *The Count of Monte-Cristo* **1846** UK) with Maquet (anon), is likewise not a supernatural fiction, but the Château d'If, where Edmond Dantès is unjustly imprisoned, has many characteristics of the fantasy EDIFICE; the escaped Dantès's transformation into the infinitely rich, revengeful but justice-seeking count underlies dozens of SUPERHEROES; his exploits are no less remarkable than those of BATMAN, and the several long-delayed RECOGNITION scenes which end the text transform his story into a revelatory legend, while he himself goes to continue his attempts at HEALING a troubled world.

CHILDREN'S FANTASIES aside, AD began writing SUPERNATURAL FICTION with tales like "Un Bal Masqué" (1833 *Le Centaur*; trans as "The Masked Ball" in *Peter Cushing's Tales of a Monster Hunter* anth **1977** ed Peter HAINING). His first novel of fantastic interest is probably *Le Château d'Eppstein* (**1844**; trans Alfred Allinson as *The Castle of Eppstein* 1903 UK; new trans Norma Lorre Goodrich 1989 US), a GHOST STORY. In *Les Frères Corse* (**1844**; trans anon as *The Corsican Brothers* 1845 US; new trans in *The She-Wolves of Machecoul and The Corsican Brothers* coll **1895** UK) Siamese TWINS, separated at birth, maintain a psychic knowledge of each other's dire fates. *Isaac Laquedem* (**1853**; trans as *Isaak Lakadam* **1853** UK; cut trans vt *"Tarry Till I Come!"* 1897) is a WANDERING-JEW tale. AD's most famous supernatural fiction, *Le Meneur de Loups* (**1857**; trans Alfred Allinson as *The Wolf-Leader* **1904** UK), follows the complex life of a young man who agrees a PACT WITH THE DEVIL, which backfires; but, just when his damnation is nigh, he is given the chance to become a WEREWOLF, though the dice are loaded against him. He wishes for death in order that his beloved may live again, and the novel closes with the Devil frustrated.

Much of AD's shorter fiction of interest was assembled as *Les Mille et un Fantomes* (coll **1848-51**; part trans anon 3 vols as *Tales of the Supernatural* **1907** UK – new trans of this vol Alan Hugh Walton as *Horror at Fontenay* **1975** – *Tales of Strange Adventure* **1907** UK and *Tales of Terror* **1909** UK), a title which reflects the ARABIAN-FANTASY structure – usually superficial – of its frame, and the vague Orientalizing impulse which shapes some of its contents. Some of his literary FAIRYTALES were assembled as *The Phantom White Hare and Other Stories* (coll trans Douglas Munro **1989** UK). Plays of interest include *Le Vampire* ["The Vampire"] (**1851**), which sums up the tradition of the Byronic VAMPIRE instituted by John POLIDORI in 1819. Children's stories include: *Histoire d'un casse-noisette* (**1845**; trans anon as *The Story of a Nutcracker* **1846** chap UK; new trans, vt *The Nutcracker of Nuremberg* 1930 UK), adapted from "Nussknacker unde Mausekönig" ["Nutcracker and Mouseking"] (1816) by E.T.A. HOFFMANN; *La Bouillie de la comtesse Berthe* (**1845**; trans anon as *Good Lady Bertha's Honey Broth* **1846** chap UK; new trans *The Honey Feast* 1980 chap UK); *La Jeunesse de Pierrot* (**1854**; trans anon as *When Pierrot was Young* **1975** chap US); and *The Dumas Fairy Tale Book* (coll trans **1924** UK). Much remains unplumbed. [JC]

DU MAURIER, DAPHNE (1907-1989) UK writer best-known for nonsupernatural Gothic novels like *Rebecca* (**1938**). The protagonist of her first novel, *The Loving Spirit* (**1931**), is as in much of DDM's fiction haunted by a REVENANT, though in this case the HAUNTING is literal. *Castle Dor* (**1962**), a completion of A.C. QUILLER-COUCH's last novel, is a TWICE-TOLD tale in which the story of Tristan and Iseult is replicated in 19th-century Cornwall. *The House on the Strand* (**1969**) is her most remarkable and sustained fantasy traversal of the darkness that underlies almost all her work. Although the FRAME STORY provides an exiguous sf rationale – the protagonist's experiences are caused by an experimental drug – the core is an extremely bleak romance of TIMESLIP. Cast back into 14th-century Cornwall, but invisible to those alive then, the protagonist increasingly finds life there more vivid and more real than 20th-century existence, and falls in love with a young girl (significantly named Isolde); at novel's close, when he plunges backwards again, he finds her dead, and his SHADOW – through whom he has come to feel alive – dying of plague.

DDM's short SUPERNATURAL FICTION has become well known through movie versions of two stories: "The Birds", from *The Apple Tree: A Short Novel and Some Stories* (coll **1952**; vt *Kiss Me Again, Stranger* 1953 US), was made into *The* BIRDS (*1963*) dir Alfred HITCHCOCK; and "Don't Look Now", from *Not After Midnight* (coll **1971**; vt *Don't Look Now* 1971 US), was made into DON'T LOOK NOW (*1973*) dir Nicholas Roeg. Here, as in her novels, the overriding sensation is of ineradicable sadness; her protagonists are not so

much threatened as brought to an awareness of their own haunted natures. [JC]

Other works: *The Progress of Julius* (**1933**); *My Cousin Rachel* (**1951**); *The Breaking Point* (coll **1959**); *Rule Britannia* (**1972**); *Echoes from the Macabre: Selected Stories* (coll **1976**); *Rendezvous and Other Stories* (coll **1980**); *Classics of the Macabre* (coll **1987**; vt *Daphne Du Maurier's Classics of the Macabre* 1987 US) illus Michael FOREMAN.

See also: *The* BREAKTHROUGH (*1993*).

DUMBO US ANIMATED MOVIE (*1941*). Disney. **Pr** Walt Disney. **Sv dir** Ben Sharpsteen. **Story** Joe Grant, Dick Huemer. **Based on** *Dumbo, the Flying Elephant* (   ) by Helen Aberson and Harold Pearl. **Novelization** *Dumbo* * (**1975**) by Derry Moffatt. **Voice actors** Ed Brophy (Timothy Mouse), Sterling Holloway (Stork), Malcolm Hutton (Skinny), Billy Sheets (Joe/clown), Margaret Wright (Casey Jr). 63½ mins. Colour.

Dumbo, an archetypal UGLY-DUCKLING figure, is mocked by the other elephants at the CIRCUS because of his huge ears. Yet, inspired by pal Timothy Mouse and four hep crows, Dumbo discovers he can use those ears as wings, and soon he makes the circus the most famous in the land. Simple and short – DISNEY was strikebound through much of its production – D is among the most enduringly popular movies of all time. [JG]

DUMMIES ◊ DOLLS; VENTRILOQUISM.

DUNCAN, DAVE Working name of Scottish-born Canadian writer David John Duncan (1933-   ), whose first novel was the fantasy *A Rose-Red City* (**1987**). His work in the fantasy field is dominated by three sequences. **The Seventh Sword** sequence – *The Reluctant Swordsman* (**1988**), *The Coming of Wisdom* (**1988**) and *The Destiny of the Sword* (**1988**) – seems at first a fairly standard PORTAL fantasy in which a dying man from the mundane world comes to life in the body of a HERO who has already failed, and is given a mission by the GODDESS; but it turns out to be a much more thoughtful work than its high-energy action plot might suggest.

Two closely interlinked sequences share location, characters and themes – **A Man of His Word**, comprising *Magic Casement* (**1990**), *Faery Lands Forlorn* (**1991**), *Perilous Seas* (**1991**) and *Emperor and Clown* (**1991**), and **A Handful of Men**, comprising *The Cutting Edge* (**1992**), *Upland Outlaws* (**1993**), *The Stricken Field* (**1993**) and *The Living God* (**1994**). The world of these two series seems at first to have elements of a GAMEWORLD; as the series progress, however, this comes to seem a conscientious decision to stylize and make use, often subversive, of stock conventions as a shorthand. Another element in the books which resembles but escapes cliché is the operation of MAGIC; magical power is conveyed in varying degrees by the possession of up to four Words of Power (◊ MAGIC WORDS) – possession of more than four is apparently fatal and the level of power at each level depends on how many others each word is shared with. These words turn out to be the TRUE NAMES of a group of aboriginals farmed so that they may give up their names at death; this is one of the most obvious examples of the use and abuse of power which proves to be what DD is interested in investigating.

A singleton of interest is *The Reaver Road* (**1992**), a HEROIC FANTASY whose travelling storyteller narrator/protagonist Omar may be the AVATAR of a TRICKSTER god, suffering from AMNESIA. He finds himself caught up in the complicated dealings of potential avatars of another god. DD's

hallmarks – perpetual good humour, plus moral seriousness and competent story-management – are attractively on display here. *The Hunter's Haunt* (1995) is not so much a sequel as an independent singleton featuring the same protagonist in its FRAME STORY. *Past Imperative* (1995) – the first volume of the projected **Great Game** sequence – is a FANTASY OF HISTORY involving a WAINSCOT society of transdimensional intelligence operatives functioning in the Europe of WORLD WAR I and a FANTASYLAND where deities, who may be less than they seem, are involved in a GODGAME. [RK]
Other works: *Shadow* (1987), science fantasy; *West of January* (1989), sf; *Strings* (1990), sf; *Hero!* (1991), sf.

DUNDERKLUMPEN! Swedish live-action/ANIMATED MOVIE (*1974*). GK-Film Visar/Tecknad/AB Europa Film Ljudstudio. **Pr** Gunnar Karlsson. **Dir** Per Åhlin. **Screenplay** Beppe Wolgers. **Starring** Beppe Wolgers (Father), Jens Wolgers (Jens). **Voice actors** Halvar Björk (Dunderklumpen), Gósla Ekman (En Dum En), Håkan Serner (Lejonel), Lotten Strömstedt (Dockan), Toots Thielemans (Pellegnillot). 80 mins. Colour.

The TROLL-like (and animated) Dunderklumpen, starved of companionship, emerges from the forest – his OTHERWORLD – to kidnap four TOYS (also animated), from the live-action Wolgers household. Father and son (Jens) Wolgers, assisted by a "mean streets" bumblebee, pursue through a landscape partly animated and partly real. *D* contains some excellent animation among the dross: often very beautiful, this movie is perhaps best watched with the sound turned off. Its primary interest is as an early, and in animation terms rather good, example of the TOON movie. [JG]

DUNGEONS AND DRAGONS US animated series (1983-7). CBS. **Pr** Gary GYGAX, Bob Richardson. **Exec pr** David DePatie, Lee Gunther. **Dir** Gerry Chiniquy, John Gibbs, Milt Gray, Tom Ray, Nelson Shin. **Voice actors** Willie Aames (Hank/Ranger), Peter Cullen (Venger), Ted Field III (Bobby/Barbarian), Toni Gayle Smith (Diana/Acrobat), Katie Leigh (Sheila/Thief), Sidney Miller (Dungeon Master), Donny Most (Eric/Cavalier), Adam Rich (Presto/Magician), Frank Welker (Uni). 27 30min episodes. Colour.

Inspired by the famous GAME, this begins with six teenagers on a carnival ride that transports them into the past and a very different world. A being known as the Dungeon Master informs them they have been summoned to help defeat the evil Venger, a WIZARD intent on conquering the world; only after his downfall will they be allowed home. Luckily the teenagers are given new NAMES and powers to help in their QUEST; Sheila, for example, becomes Thief, with the power of INVISIBILITY. They are aided by a UNICORN that can teleport them (◊ TALENTS) – but only once per day. During their battles with Venger, they face dangers including DRAGONS, an unexpected TRANSFORMATION caused by a seemingly innocent flower, magic RINGS and Venger's efforts to master the power of time travel. [BC]

DUNN, [Mrs] GERTRUDE (1884-?) UK writer whose three fantasy novels – *Unholy Depths* (1926), *The Mark of the Bat* (1928) and *So Forever* (1929) – deal respectively with GHOSTS, VAMPIRES and the ELIXIR OF LIFE. She has frequently been confused with a quite different Mrs Gertrude Dunn (1855-1926), *née* Renton, who wrote as Mrs Baillie-Weaver, Gertrude Renton Weaver and G. Colmore, under the latter name writing the OCCULT FANTASY *A Brother of the Shadow* (1926). [RD]

DUNNE, J(OHN) W(ILLIAM) (1875-1949) UK writer responsible for the notion of Serial Time. ◊ John BUCHAN.

DUNSANY, LORD Working name of Irish writer Edward John Moreton Drax Plunkett (1878-1957), 18th Lord Dunsany, who wrote widely and much. Through his stories, plays and novels he was instrumental in creating the essential autonomous venues within which modern FANTASY could be told. Not only does any tale which CROSSHATCHES between this world and FAERIE (or any self-contained OTHERWORLD) owe a Founder's Debt to LD, but the SECONDARY WORLD created by J.R.R. TOLKIEN – from which almost all FANTASYLANDS have devolved – also took shape and flavour from LD's example. The style in which he told his tales – an ultimately *sui generis* mix of rhythms and vocabulary from the King James Bible and Celtic revival poets and tale-tellers (◊ CELTIC FANTASY) like William Butler YEATS, plus the FIN-DE-SIÈCLE ambience of *The Yellow Book* (◊ Aubrey BEARDSLEY; DECADENCE; MAGAZINES; Oscar WILDE), plus the specific example of previous writers of fantasy like George MACDONALD and William MORRIS – also proved highly influential, to less happy effect. The steely delicacy of LD's descriptions of the ineffable turned into the pulp Orientalism of writers from H.P. LOVECRAFT, A. MERRITT and Robert E. HOWARD onwards: choked with "hushed" references to high-sounding godlings and topographies. This style continues to plague GENRE FANTASY.

LD's work as a whole cannot be reduced to simple patterns of attack and withdrawal, but the model is sufficiently valid to permit its use in an attempt to sort the hundreds of short stories, the dozens of plays and the numerous novels. It can be argued that, in each of the modes he successively engaged upon, he began with work set in fantasy venues and written with considerable intensity, and then progressively tended to move his settings towards mundanity while, in parallel, his style tended to lose (or divest itself of) its more elaborate attributes. Moreover, he began each new mode with work less deeply involved in the fantastic than had been the initial works composed in the previous mode; and the work in each new mode progressively moved closer to the present time. LD's oeuvre, as a whole, can be seen as a recognition – and representation – of THINNING. Though there are chronological overlaps, it will be convenient to follow each mode separately.

Stories LD began to publish work of genre interest with the tales assembled as *The Gods of Pegana* (coll **1905**) under the immediate inspiration of seeing *The Darling of the Gods* (**1902**), a florid melodrama about Japan by David BELASCO and John Luther Long. Although the play is not itself supernatural, it clearly provided LD with a model from which to create a wholly autonomous OTHERWORLD: Pegana, land of the GODS. The tales are poetic fragments, invocatory and scented, and do not deal with mortals. As with most of his early books, *The Gods of Pegana* was illustrated by S.H. SIME; elsewhere the normal author-artist relationship was sometimes reversed, a picture by Sime inspiring LD to compose the matching story. The tales assembled in *Time and the Gods* (coll **1906**) illustrated by Sime are narratives featuring both gods and humans, often intertwined to ironic effect, as in "The Relenting of Sarnidac", in which a lame dwarf – lamenting the departure of the gods from the world – is mistaken for the one god who has relented and decided to remain. The various descriptions of the nature of the Universe offered to King Ebalon in "The Journey of the King" add up to a moving compendium of fantasy HEAVENS, PANTHEONS and MYTHS OF ORIGIN, all presented through imagery deeply evocative of the TIME ABYSS.

The Sword of Welleran and Other Stories (coll **1908**) illustrated by Sime includes LD's most complex and perhaps most sustained short stories. *The Fortress Unvanquishable, Save for Sacnoth* (**1910** chap), first published in this collection, almost singlehandedly created the SWORD AND SORCERY genre, though without any excesses of plot: a LAND which has been cursed needs HEALING; a HERO wrests the eponymous SWORD from the body of the DRAGON which has protected it, advances across a vertiginous THRESHOLD and through guarded PORTALS into the labyrinthine EDIFICE where the evil MAGUS dwells, and defeats him; and thus the land is healed. All is told with absolute assurance; other writers, for almost a century, have expanded upon this inspiration.

Further volumes of stories show a slightening of effect, though *A Dreamer's Tales* (coll **1910**) illustrated by Sime, *The Book of Wonder* (coll **1912**) illustrated by Sime, *Fifty-One Tales* (coll **1915**; vt *The Food of Death* 1974 US), *Tales of Wonder* (coll **1916**; vt *The Last Book of Wonder* 1916 US) illustrated by Sime, *Tales of War* (coll **1918**) and *Tales of Three Hemispheres* (coll **1919** UK) contain many fine works, though almost always with an ET IN ARCADIA EGO sense that old-seeming tales could no longer be articulated without the ironies of FABULATION.

LD's next stories were entirely different. The long sequence of **Jorkens** CLUB STORIES – *The Travel Tales of Mr Joseph Jorkens* (coll **1931** UK), *Mr Jorkens Remembers Africa* (coll **1934** UK; vt *Jorkens Remembers Africa* 1934 US), *Jorkens Has a Large Whiskey* (coll **1940** UK), *The Fourth Book of Jorkens* (coll **1948** UK) and *Jorkens Borrows Another Whiskey* (coll **1954** UK) – are mostly TALL TALES, fantasy episodes set almost always in this world, and provide whimsy and GRAND GUIGNOL in a generally alluring mix. Further volumes of stories rehearsed, sometimes without much energy, the inspiration of earlier years; they include *The Man who Ate the Phoenix* (coll **1948** UK) and *The Little Tales of Smethers* (coll **1952** UK); some previously uncollected work appears in *The Ghosts of the Heavyside Layer and Other Fantasms* (coll **1980** US) ed Darrell Schweitzer.

Plays LD wrote over 40 plays, and some earlier examples are of strong interest. *Five Plays* (coll **1914**), the most impressive collection, includes *King Argimenes and the Unknown Warrior* (produced 1910), in which the discovery of a SWORD revolutionizes an imaginary Eastern kingdom, and *The Gods of the Mountain* (produced 1912), in which impostors pretending to be GODS are – after the real gods have turned them to stone (◊ BONDAGE) – finally believed to be gods. *Alexander* (written 1912), which appears in *Alexander and Three Small Plays* (coll **1926** UK), traces the inexorable THINNING of the emperor's life after he dismisses APOLLO from his service. *Plays of Gods and Men* (coll **1917**) includes *The Laughter of the Gods* (produced 1919), in which a monarch calls the bluff of gods he no longer respects, forcing them into terrible acts, and *A Night at an Inn* (produced 1916), in which thieves who steal an idol are confronted by the god himself. *Plays of Near and Far* (coll **1922** UK) includes *Fame and the Poet* (produced 1919), in which an immortal being visits an insufficiently grateful human. *If* (**1921** UK) features TIME TRAVEL into the past and the creation of an ALTERNATE WORLD. Plays of small fantasy interest appeared in *Seven Modern Comedies* (coll **1928** UK) and *Plays for Earth and Air* (coll **1937** UK). *Lord Adrian* (**1933** UK) is sf.

Novels Again the overall pattern recurs. LD's first novels unequivocally set out to occupy what might be called the domain of fantasy; later volumes gradually reduced the temperature. That said, his first was relatively modest. *The Chronicles of Rodriguez* (**1922** UK; vt *Don Rodriguez: Chronicles of Shadow Valley* 1922 US) conveys its young disinherited protagonist through a fantasized Spain, gifting him with a Sancho Panza companion (◊ DUOS), good luck with magicians, and a castle. His second novel, however, is one of the seminal fantasies of the century. *The King of Elfland's Daughter* (**1924** UK) is a CROSSHATCH in which – almost for the first time, and certainly for the first time with any conviction – the two poles (our world and FAERIE) are of equal weight. Responding to the communal WISH of his people to "be ruled by a magic lord", the crown prince of Erl travels to Elfland in search of its King's daughter, and wins her through MAGIC; but when they return to Erl they discover that a decade has passed (◊ TIME IN FAERIE). Erl finds the imposition of magic rule a mixed blessing (◊ ANSWERED PRAYERS). The prince wanders through a bereft BORDERLAND in search of his princess (who has returned to Elfland); and only when the King invokes a final RUNE, which encompasses the TRANSFORMATION of Erl into an aspect of Faerie, does the spiralling tale come to an ambiguous resolution. *The King of Elfland's Daughter* is thus the first significant INSTAURATION FANTASY.

There are further impressive novels. *The Charwoman's Shadow* (**1926** UK) is an exemplary FAIRYTALE whose HERO wins a bride by recapturing her SHADOW from a magician, all the while thinking she is an old charwoman. *The Blessing of Pan* (**1927** UK) portrays English rural life under a sign of paganism, after the fashion of writers like T.F. POWYS. *The Curse of the Wise Woman* (**1933** UK) is set in an Ireland haunted by TIR-NAN-OG, but which remains out of sight. *Rory and Bran* (**1936** UK) is narrated by the dog Bran (◊ TALKING ANIMALS), *My Talks with Dean Spanley* (**1936** UK) is a set of interviews with a spaniel reincarnated as a clergyman (◊ REINCARNATION), and *The Strange Journeys of Colonel Polders* (**1950** UK) surreally recounts various adventures of the eponymous soldier, who suffers transmigration into various animals as a punishment for disbelief (◊ LEARNS BETTER).

LD was too copious a writer for his oeuvre to be grasped whole with any ease, and he had a liquid ability to generate insightful passages or entire works almost at will over a career which lasted more than half a century. He remains a father of GENRE FANTASY, an imp ancestor of much that is good and much that is bad. He is much less read than he warrants. [JC]

Other works: *Selections from the Writings of Lord Dunsany* (coll **1912**) ed William Butler YEATS; *Nowadays* (**1918** chap US), speech on poetry; *Unhappy Far-Off Things* (coll **1919** UK), associational; *The Compromise of the King of the Golden Isles* (**1924** chap US), a play; *The Old Folk of the Centuries* (**1930** UK), a play; *If I Were Dictator: The Pronouncement of the Fraud Macaroni* (**1934** UK); *Mr Faithful* (**1935** UK), a play; *The Story of Mona Sheehy* (**1939** UK); *Up in the Hills* (**1936** UK), associational; *Guerrilla* (**1944** UK), associational; *The Donnelly Lectures 1943* (**1945** chap UK), literary talks; *The Last Revolution* (**1951** UK), sf; *His Fellow Men* (**1952** UK), associational; *The Sword of Welleran and Other Tales of Enchantment* (coll **1954** UK), a compilation which differs from the 1908 volume; three important compilations ed Lin CARTER, being *At the Edge of the World* (coll **1970**

US), *Beyond the Fields We Know* (coll **1972** US) and *Over the Hills and Far Away* (coll **1974** US); *Gods, Men and Ghosts* (coll **1972** US) ed E.F. BLEILER; various volumes of poetry.

Further reading: *Patches of Sunlight* (**1938**) and *While the Sirens Slept* (**1944**), autobiographies; *Lord Dunsany: A Biography* (**1972**) by Mark Amory; *Pathways to Elfland: The Writings of Lord Dunsany* (**1989**) by Darrell SCHWEITZER; *Lord Dunsany: A Bibliography* (**1993**) by S.T. JOSHI and Schweitzer.

DUOS Partners in adventure have existed in fantasy and its TAPROOT TEXTS since the time of GILGAMESH. A distinction needs to be drawn between (a) duos whose contrasting partners may provide readers anything from light relief to revealing insights about human interactions, but who in the end remain together in order to continue adventuring, and (b) partners whose main function is to dramatize profound oppositions, or to represent (◊ JUNGIAN PSYCHOLOGY), in terms of STORY, the great interior battles of the SOUL on its QUEST or NIGHT JOURNEY towards self-RECOGNITION. In this encyclopedia we use the term "duos" to refer to the former category: duos are equal partners in adventure.

Don Quixote and Sancho Panza, the indissolubly married opposites in Miguel de CERVANTES's *Don Quixote* (**1605-16**), might logically be so described, but the burden of their interaction – the profundity of the vision of human life they represent – is inadequately conveyed by the thought that they might form a duo, like SHERLOCK HOLMES and Dr Watson. Arthur Conan DOYLE's famous pair are a paradigm duo, and centrally affected the modelling of duos in popular literature from the end of the 19th century.

The distinction between duo and SHADOW continues to be valid throughout the 20th century. Frodo and Samwise, in J.R.R. TOLKIEN's *The Lord of the Rings* (**1954-5**) interact profoundly, but not as a duo; Fafhrd and the Gray Mouser, in Fritz LEIBER's lifelong **Fafhrd and the Gray Mouser** sequence, may also interact on a profound level, but they serve more saliently as warrants that their mutual tale will continue. The FANTASYLAND they inhabit, and the HEROIC-FANTASY adventures they enjoy, are both conducive to the creation of duos and strengthened when duos appear within them. Within template structures, duos have proved to be an essential building block: split plots, trilogy-management, rescue missions, fights, reconciliations, hidden missions, infodumps of backstory, and any number of PLOT DEVICES – all are easier to fabricate when pairs are involved, and when, as often, one partner needs to be told what the other partner has learned. They may also shape a group of COMPANIONS, or a SEVEN-SAMURAI gathering.

Holmes and Watson have provided a model for one particular category, in which a genius of some sort is paired with a SENSIBLE MAN, who may tell the tale. There are many other categories. There are Defiant Ones, pairs locked together (sometimes in a DIRTY DOZEN squad) who, while escaping a mutual fate, and while fighting like cat and dog, eventually become intimate. There are Dick and Myrna duos – from the protagonists of *The Thin Man* (**1934**) by Dashiell Hammett (1894-1961) – who advance the plot (and their relationship) through cross-talk, a banter of solidarity against a conspiring world. There are Odd Couples, and Robin Hood and Maid Marian duos, and Batmans and Robins, and reverse duos (whose partners do not know they are partnered), and dozens more. They are numerous because they are engines of Story. [JC]

DÜRER, ALBRECHT (1471-1528) German artist. ◊ ILLUSTRATION.

DURRELL, GERALD (1925-1995) UK zoologist and writer. ◊ CHILDREN'S FANTASY.

DWARFS The two most popular beings to be included in HEROIC FANTASY as either COMPANIONS to or enemies of humans are dwarfs and ELVES, yet the origins of these two groups of beings are confusing. In Nordic mythology the Alfar (elves) comprise one of the four main groups of dwarfs, but in Celtic mythology the elves are part of the land of FAERIE, distinct from the dwarfs, who are creatures of the Earth. In most GENRE FANTASY the dwarfs are craftsmen and inventors. They are often associated with MINES, and may be portrayed as greedy and covetous, particularly of gold. They are small, but solidly built and strong, almost always bearing beards and wielding axes. Sometimes they are described as drawing their power from the Earth. In this sense they may be synonymous with gnomes, and to a lesser extent with kobolds (◊ GOBLINS) and leprechauns. All these strands emphasize the diminutive and mischievous aspects, but dwarfs are also warlike. It is possible the notion originated in Nordic perceptions of the Asian races that invaded Europe in the 4th and 5th centuries AD, particularly the Huns – Attila, for example, was apparently only just over 4ft tall.

The traditional dwarf-figure is drawn from NORDIC FANTASY, particularly the *Volsunga Saga* and *Nibelungenlied* (◊ SAGAS). Here the dwarfs were closely associated with the AESIR. The most famous was Alberich (or Andvari), who guarded the treasure of the Nibelungs. Dwarfs are also depicted as being among the powers of EVIL who lurk beneath the roots of the WORLD-TREE Yggdrasil. As the sagas devolved into FOLKTALES dwarfs were regularly depicted as scheming and cunning, and in this form they found their way into FAIRYTALES, of which "The Yellow Dwarf" (in *Les contes de fées* coll **1698**) by Madame d'AULNOY is one of the earliest. Here the dwarf becomes betrothed to the princess All-Fair against her wishes, but is determined to marry her at all costs; the tragic ending is in keeping with current attitudes towards dwarf villainy. This image was utilized by the German Romantics (◊ ROMANTICISM), especially J.K. MUSÄUS and Ludwig TIECK, and translated into popular fairytales by the GRIMM BROTHERS, who depicted evil dwarfs in "Snow White and Rose Red" and the helpful but cunning gnome in "Rumpelstiltskin". Interestingly, the dwarfs in "Snow White and the Seven Dwarfs" are friendly and intensely loyal to the young princess (◊ SNOW WHITE). The goblins in *The Princess and the Goblin* (**1872**) by George MACDONALD are akin to dwarfs, and that image pervaded most Victorian CHILDREN'S FANTASY. A lighter-hearted image was given to the dwarflike beings, the Munchkins, in L. Frank BAUM's *The Wonderful Wizard of Oz* (**1900**).

Modern treatments of dwarfs can be traced to J.R.R. TOLKIEN, who drew upon both Nordic myth and some of the mischievous aspects in the works of E.A. WYKE-SMITH to depict his dwarves (as he spelled it) in *The Hobbit* (**1937**); these have all the aspects of traditional dwarfs, including squabbling belligerence, but are essentially good. The influence of Tolkien's works, especially *The Lord of the Rings* (**1954-5**), on modern GENRE FANTASY has meant that dwarfs have now become stock characters in most HEROIC FANTASY (notably the works of Terry BROOKS, Niel HANCOCK, Guy Gavriel KAY and, tongue-in-cheek, Terry PRATCHETT). Something of this same image is perpetuated in Terry GILLIAM's TIME BANDITS (**1981**). The evil nature of diminutive folk (not necessarily of supernatural origin), however, continues to be a trope, as in Daphne DU MAURIER's "Don't

Look Now" (1971) and William HJORTSBERG's *Alp* (**1969**): in both cases dwarfs are the source of violent death. [MA]

DWYER, JAMES FRANCIS (1874-1952) Australian writer and traveller, active from the early years of the century as an author of adventure stories, some of them TALL-TALE fantasy; a selection assembled as *"Breath of the Jungle"* (coll **1915** US) includes a FLYING-DUTCHMAN tale of some interest, "The Phantom Ship of Kirk van Tromp". His novels tend to be set in various unexplored regions of the world, and to hint at LOST-RACES scenarios; *The White Waterfall* (**1912** US), for instance, places its surface love story on a South Pacific ISLAND, where a lost world is duly uncovered. *The Spotted Panther* (**1913** US) plunges into a LAND-OF-FABLE early-20th-century Orient, engaging its tough rapscallion protagonists in a QUEST for a lost artifact whose recovery is necessary if civilization is not to be threatened by the forces of the East. Other work in book form (much remains uncollected) includes the **Spillane** series – *The Lady with Feet of Gold* (**1937** UK) and *The City of Cobras* (**1938** UK) – which reinvokes lost worlds. [JC]

Other works: *Evelyn: Something More than a Story* (**1929** US); *Hespamora* (**1935** UK).

DYALHIS, NICTZIN (1880-1942) US writer whose slender output appeared primarily in WEIRD TALES. He first earned popular notice for his SCIENCE FANTASY "When the Green Star Waned" (1925) and its sequel, "The Oath of Hul Jok" (1928), but is best-known for a handful of romantic fantasies that embody his beliefs in OCCULTISM. In "The Eternal Conflict" (1925) and "The Dark Lore" (1927) adventurers in the ASTRAL PLANE encounter the traditional Judeo-Christian HELL. "The Red Witch" (1932), "The Sapphire Goddess" (1934; vt "The Sapphire Siren"), "The Sea-Witch" (1937) and "The Heart of Atlantan" (1940) all feature SOULS that transmigrate across TIME or undergo RE-INCARNATION to enact conflicts from the past. [SD]

Further reading: "Nictzin Dyalhis: Mysterious Master of Fantasy" by Sam Moskowitz in *Echoes of Valor III* (1991) ed Karl Edward WAGNER.

DYING EARTH Most literary images of the FAR FUTURE are conveniently discussed as SCIENCE FICTION, but a significant group of fantasies are based on the premise that MAGIC entities and forces hypothetically removed from the Earth's past by THINNING might enjoy a spectacular resurgence in the senile world's "second childhood". Clark Ashton SMITH switched his attention from the imaginary past of HYPERBOREA to the historical terminus of Zothique because the latter was better adapted to his fantasies of extreme DECADENCE. Earlier precedents had been set by William Hope HODGSON in *The Night Land* (**1912**) and other languorous accounts of the END OF THE WORLD. Jack VANCE's *The Dying*

Earth (coll **1950**) provided a milieu which he was to use extensively; it was borrowed by Michael SHEA for *A Quest for Simbilis* (**1974**), and weakly imitated by Lin CARTER in his **World's End** series. Other milieux of a similar stripe have been used by Robert Silverberg (◊ *SFE*) in the allegorical *Son of Man* (**1971**), by Michael MOORCOCK in the **Dancers at the End of Time** sequence and by Gene WOLFE in his **New Sun** and **Long Sun** sequences. [BS]

DYNASTIC FANTASY Many fantasy series begin as an EPIC FANTASY in which a LAND is founded or defended or taken over by a HERO (or heroine). For diverse reasons, this tale is often extended into a series, during the course of which the hero dies or becomes a SLEEPER UNDER THE HILL, the Land turns into a template FANTASYLAND – usually conceived on medieval lines – and many years pass. The hero's descendants now rule, and begin to squabble; or are suborned by the DARK LORD who was seemingly defeated by the founding father. Fantasy novels focusing on these conflicts can fairly be called DFs. With their necessity for continuity, DFs tend to avoid METAMORPHOSIS, either of characters or in the fabric of REALITY, which remains a stable backdrop for the adventures taking place stage-front.

David EDDINGS's long **Belgariad** sequence clearly demonstrates the typical shift from epic fantasy into DF. [JC]

DYRENFORTH, JAMES (? -?) ◊ WONDERLAND.

DYSTOPIAS ◊ UTOPIAS.

DZIEMIANOWICZ, STEFAN R(ICHARD) (1957-) US editor and critic who has concentrated on HORROR and SUPERNATURAL FICTION, co-editing *Necrofile: The Review of Horror Fiction* from its inception in 1991; it won a 1995 BRITISH FANTASY AWARD. His anthologies include *Weird Tales: 32 Unearthed Terrors* (anth **1988**), *The Rivals of Weird Tales* (anth **1990**) with Martin H. GREENBERG and Robert WEINBERG, *Famous Fantastic Mysteries* (anth **1991**) with Greenberg and Weinberg, *The Mists from Beyond* (anth **1993**) with Greenberg and Weinberg, *To Sleep, Perchance to Dream . . . Nightmare* (anth **1993**) with Greenberg and Weinberg, the **100** series, all with Greenberg and Weinberg – *100 Ghastly Little Ghost Stories* (anth **1993**), *100 Creepy Little Creature Stories* (anth **1994**), *100 Wicked Little Witch Stories* (anth **1995**) and *100 Vicious Little Vampire Stories* (anth **1995**) – *Sea-Cursed* (anth **1994**) with Greenberg and T. Liam McDonald, and *A Taste for Blood* (anth **1995**) with Greenberg and Weinberg.

SD's *The Annotated Guide to Unknown and Unknown Worlds* (**1991**) is a valuable issue-by-issue commentary (◊ UNKNOWN). [JC]

Other work: *The Core of Ramsey Campbell: A Bibliography & Reader's Guide* (**1995** chap) with Ramsey CAMPBELL and Greenberg.

EAGER, EDWARD (1911-1964) US playwright and lyricist who began writing children's books to please his son; the first published was *Red Head* (**1951**). His debt to E. NESBIT in plot, style and atmosphere is obvious. Typically, EE wrote of a group of book-loving children who access magical powers, imperfectly understood, and struggle to grasp the rules of MAGIC and achieve their desires within a limited time. In *Half Magic* (**1954**) the magic resides in a coin with the power of granting half a WISH; in *The Time Garden* (**1958**) the Natterjack, a Cockney toad, enables them to travel in TIME (by sniffing thyme). *Knight's Castle* (**1956**) builds on and improves the basic premise behind Nesbit's *The Magic City* (**1910**) and brings the bane of modern culture to Sir Walter SCOTT's *Ivanhoe* (**1819**). EE's penchant for RECURSIVE FANTASY reaches a logical extreme in his last book, *Seven Day Magic* (**1962**), in which the enabling TALISMAN is a library book that must be returned after a week of literary wishes. At their best, the books sparkle with wit and have a lasting charm. [LT]

Other works: *Mouse Manor* (**1952**); *Magic or Not?* (**1959**); *The Well-Wishers* (**1960**).

See also: CHILDREN'S FANTASY.

EARTHDAWN ◊ Christopher KUBASIK.

EARTH MOTHER ◊ GODDESS.

EASTON, M(ALCOLM) COLEMAN (1942-) US writer who began to write work of genre interest with "Superflare" for *F&SF* in 1980 as by Coleman Brax; he lives with Clare BELL, and most of his sf (◊ SFE) is with her, jointly as Clare Coleman. He began writing fantasy with the **Kyala** sequence – *Masters of Glass* (**1985**) and *The Fisherman's Curse* (**1987**) – set in a medieval-style FANTASYLAND and featuring a PARIAH ELITE of WIZARDS who manufacture glass beads that are the exact hue of various animal species' eyes, allowing the animals to be communicated with and controlled. *Iskiir* (**1986**) is an ARABIAN FANTASY; the eponymous hero develops a TALENT to save the unspecified venue from an incursion of vast stone monoliths, which threaten to thin (◊ THINNING) the world into a desert.

Of more interest is *Spirits of Cavern and Hearth* (**1988**), set in a land occupied by two peoples, the nomadic Chirudak and the urbanized Hakhan. The former have an edgy but vital relationship, through trances, with the ELDER GODS, who maintain a remote suzerainty; the Hakhan, by contrast, treat any contact with the "mythical" gods as a sign that the victim has been "god-striken" and must be exiled. The protagonist, a Hakhan physician, is afflicted by a disease which makes him young again but also makes visible to him the WAINSCOT spirits. [JC]

EBIRAH, TERROR OF THE DEEP (ot *Nankai No Daiketto*; vt *Big Duel in the North Sea*; vt *Godzilla Versus the Sea Monster*) Japanese movie (**1966**). ◊ GODZILLA MOVIES.

EC COMICS Publishers of a line of COMIC books in the early 1950s which proved influential in the development of comic-strip fantasy. The Educational Comics Company was formed in the mid-1940s by Max C. Gaines, and passed into the hands of his son, William M. Gaines (1922-), after Max's death in 1947. EC was transformed in late 1949 with the introduction, by Gaines and Albert Feldstein (1925-), of HORROR strips into one of their **Police** titles, *Crime Patrol* (#15 1949). Enthusiastic reader response inspired them radically to revamp their entire list: *Crime Patrol* became *Crypt of Terror* (from #17 1950; vt *Tales from the Crypt* from #20), *War Against Crime* became *Vault of Horror* (from #12 1950), and *Gunfighter* became *Haunt of Fear* (from #15 1950). Borrowing from popular radio shows the gimmick of a host to introduce each story, Gaines and Feldstein fashioned a succession of intelligently written, twist-in-the-tail strips that were to prove immensely popular.

Simultaneously, they launched two sf titles (*Weird Fantasy* and *Weird Science*), and later the same year two more "new trend" comics emerged: *Crime Suspense Stories* ed Feldstein and the pacifist war title *Two-Fisted Tales* ed Harvey Kurtzman (1924-). In mid-1951 another war title, *Frontline Combat*, emerged, followed in early 1952 by *Shock Suspense Stories*.

EC's output was characterized by mature writing and art by the top people in the field, including Jack Davis (1926-), Wally Wood (1927-1981), Graham INGELS, Johnny Craig (1926-), Al WILLIAMSON, Reed CRANDALL, Bernie Krigstein (1919-), George Evans (1920-) and the team of John Severin (1921-) and Will Elder (1922-). After "borrowing" a plot from Ray BRADBURY,

Gaines and Feldstein were flattered when Bradbury suggested they officially adapt his stories; this led to a series of memorable strips.

Another characteristic of EC was the unflinching depiction of gore and violence; in the McCarthyite period they inevitably came to the notice of conservative pressure groups – indeed, Gaines was called in to testify before the Senate subcommittee in 1954. EC, like many other publishers, saw its sales plummet, and by early 1955 Gaines had cancelled almost all his titles, replacing them with a slew of Comics Code Approved titles. Although these were not without merit, the "new direction" titles like *Piracy*, *Valor*, *Impact* and *Psychoanalysis* were emasculated, and by 1956 all had folded. As a last stab, EC entered the magazine field, but the "picto-fiction" mixture of text and illustrations found few friends, and with the May issue of *Shock Illustrated* EC closed the door as a publisher of fantasy. Only one title survived the change to magazine format, *Mad Magazine*, which went on to enjoy sales countable in millions. Initiated by Kurtzman in 1952, *Mad*'s irreverent, anti-authoritarian stance swiftly caught on, and in many ways sowed the seeds of the underground comics of the 1960s – many of whose artists saw Kurtzman as their mentor. EC's horror comics inspired James Warren to start publishing his HORROR-magazine line (◊ WARREN PUBLISHING) and DC COMICS to enter the "mystery" field in the late 1960s.

The original EC comics have been reprinted a number of times, most notably in the hardback series from Russ Cochran (**1985-6** 12 vols) and, in the 1990s, in comic-book format from Gladstone and Gemstone. [DR/RT]

ECO, UMBERTO (1932-   ) Italian literary academic and bestselling writer, most famed for *Il nom della rosa* (**1980**; trans William Weaver as *The Name of the Rose* **1983** US/UK). At the heart of this medieval detective story (◊ DETECTIVE/THRILLER FANTASY), nonfantastic yet with deep fantasy resonances including telling DREAMS and VISIONS, is a clash between grimly fanatical RELIGION and reason allied with HUMOUR. Reason solves the strange crimes patterned on the LAST JUDGEMENT in *Revelations*, and traces a MCGUFFIN (the lost second volume of Aristotle's *Poetics*); but irrationality has the last word as this BOOK – with the whole marvellous monastery LIBRARY, which is a LABYRINTH – is destroyed by fire. *Il pendolo di Foucault* (**1988**; trans William Weaver as *Foucault's Pendulum* **1989** US) examines FANTASIES OF HISTORY and specifically the notion of SECRET MASTERS. Three weary Italian editors decide to spoof the dismal occult books they publish, using a computer to recombine such randomly selected notions as "The Templars have something to do with everything"; this is developed with immense, sometimes comic, erudition and discursiveness. Lacking any "real" secret, the hoaxers fall victim to cultists who desperately wish there to be a secret – exemplifying John CROWLEY's remark in *Aegypt* (**1987**) that, although secret societies have not had power in history, the *idea* of their power has. UE deploys insights of semiotics (the gap between the sign and the signified, the MAP and the territory) with knowing expertise. [DRL]

Other works: *Postille a Il nome della rosa* (**1983** chap; trans William Weaver as *Reflections on The Name of the Rose* **1984** US), nonfiction; *Travels in Hyperreality* (coll **1986** US trans Weaver and others; vt *Faith in Fakes* 1986 UK), essays; *Il bomba e il generale* (**1989** chap trans Weaver as *The Bomb and the General* **1989** chap UK/US) and *Tre Cosmonauti* (trans Weaver as *Three Astronauts* **1989** chap US/UK), children's

books; *L'Isola del giorno primo* (**1994**; trans Weaver as *The Island of the Day Before* **1995**).

See also: FABULATION; ITALY.

EDDINGS, DAVID (CARROLL) (1931-   ) US writer whose first novel was the associational *High Hunt* (**1973**), and who began in 1995 to credit his wife, Leigh Eddings (LE), as co-author, indicating she had been an active partner throughout. In the ensuing discussion, therefore, it will be tacit that works listed as by "DE" solo are by both.

DE's first fantasy series was the **Belgariad** sequence, comprising – by internal chronology – a prequel, *Belgarath the Sorcerer* (**1995**) with LE, *Pawn of Prophecy* (**1982**), *Queen of Sorcery* (**1982**) and *Magician's Gambit* (**1983**), all three of these assembled as *The Belgariad: Part One* (omni **1985**; vt *The Belgariad 1* 1985 UK), plus *Castle of Wizardry* (**1984**) and *Enchanters' End Game* (**1984**), both assembled as *The Belgariad: Part Two* (omni **1985**; vt *The Belgariad 2* 1985 UK). A second series, the **Malloreon** books, with much the same cast, followed immediately: *Guardians of the West* (**1987**), *King of the Murgos* (**1988** UK), *Demon Lord of Karanda* (**1988**), *Sorceress of Darshiva* (**1989**) and *The Seeress of Kell* (**1991**).

The central character of the first sequence is young Garion, a farmboy UGLY DUCKLING (of whose HIDDEN-MONARCH role the reader is soon aware), who gradually traverses the vast FANTASYLAND in which the entire sequence is set, accumulating COMPANIONS, learning about his sorcerous TALENTS, and settling into a QUEST to achieve his destiny. That destiny is climactically intertwined with 7000 years of history, giving an intermittent sense of a J.R.R. TOLKIEN-like depth of back-story; Garion's growth and quest also echo ARTHUR's and PERCEVAL's. But, because so much attention is paid to the interactions between Garion and his companions, because the tone of the narrative is frequently lighthearted, and because digresssions are not infrequent (though so well plotted as not to distract), the **Belgariad** generates an elated and rather comic sense of HEROIC FANTASY while moving always towards the HIGH-FANTASY conclusion that awaits the cast. By the end of the third volume, Garion has gained an ISLAND throne, and eventually he becomes Belgarion, Overlord of the West and central ruler in the complicated and divisive nest of dynasties that cover the land, has long returned the MAGIC orb (a TALISMAN whose loss had allowed CHAOS to threaten) to the pommel of the SWORD in which it belongs, and has defeated the DARK LORD, the failed GOD Torak. All seems well.

The second sequence focuses on DYNASTIC-FANTASY issues, with a new quest soon under way to recover the kidnapped son of King Garion. The SEVEN SAMURAI of the first sequence, with occasional new participants, regain their camaraderie, and are eventually successful. The humour, the narrative ingenuity and the picturesque variety of cultures fade almost not at all. At the end there are scenes of somewhat perfunctory TRANSFORMATION, and a new GOD gazes upon the world. The overall sense is that all is now very well indeed.

DE's next sequence also came in two separate series: the **Elenium** books, being *The Diamond Throne* (**1989**), *The Ruby Knight* (**1990** UK) and *The Sapphire Rose* (**1991** UK), all three assembled as *The Elenium* (omni **1993**), and the **Tamuli** books, being *Domes of Fire* (**1992**), *The Shining Ones* (**1993**) and *The Hidden City* (**1994** UK), with further volumes projected. Though there are superficial distinctions – the protagonist is a veteran KNIGHT, and the quest is to find a magic stone which will awaken his young queen from an enchanted sleep – the mixture is in fact rather as before. There is a complex fantasyland, back and forth over which a band of companions ranges; there is a Dark-Lord-cum-failed-god in the background, hoping to gain the magic stone for himself, and henceforth rule the world; there is humour, intricate plotting, violent action, and a sense that all will be well in the end.

It is easy to make DE sound negligible through a description of his PLOT DEVICES, and any analysis of the use to which he puts them would generate a sense that his worlds were sophomoric. But his considerable popularity – though not perhaps hurt by the safeness of his overall world (◊ GENRE FANTASY) – does in the end derive from his compulsive storytelling skills. Although he is in no sense a theorist of STORY, he is a telling example of the centrality of Story in the sustained fantasy epic.   [JC]

Other works: *Two Complete Novels* (omni **1994**) assembling *High Hunt* and *The Losers* (**1992**), crime.

EDDISON, E(RIC) R(UCKER) (1882-1945) UK civil servant, writer and scholar of Old Norse who concealed the intensity of his imaginative life by creating an external existence and career of superficial calm. This tranquil surface no more describes ERE than it does his near contemporary, J.R.R. TOLKIEN; and there may be some point in emphasizing the fact that both writers were profoundly alienated by the destruction of the LAND of Britain, that they both clearly *refused* the 20th century. In common with others of their generation, both were also psychically wounded and estranged by the apocalypse of WORLD WAR I; ERE's first published work, *Poems, Letters, and Memories of Philip Sidney Nairn* (coll **1916**), which he edited and to which he supplied an eloquent 100-page memoir, reflects this sense of loss and destruction. (Nairn, a minor poet, died during WWI.) ERE's secret immersion in the creation of a SECONDARY WORLD can be seen – like Tolkien's – to represent an attempt to claw back a lost ARCADIA, a world before the War.

But ERE differs radically from Tolkien through his refusal to accord any supporting or defining role to RELIGION, either in *The Worm Ouroboros* or the **Zimiamvia** sequence; and for this reason he never found full favour with either Tolkien or C.S. LEWIS, both of whom knew him in his last years. In any case, ERE owes nothing to these advocates of CHRISTIAN FANTASY; and his work therefore stands to one side of fantasy's main line of development, which (perhaps unconsciously) owes much to Lewis's and Tolkien's sense that fantasy requires both a conflict between GOOD AND EVIL and a moral ending (◊ EUCATASTROPHE). ERE seems mainly to have been influenced by the Jacobean revenge tragedy (for plot) and by the works and example of William MORRIS (for LANDSCAPE and DICTION); ERE's style is accordingly ornate, dense and heavily cadenced. Like Morris, he also became deeply intimate with the Norse SAGAS, and published, in *Egil's Saga: Done into English Out of the Icelandic with an Introduction, Notes, and an Essay on Some Principles of Translation* (**1930**), a highly competent translation of one of them.

ERE's first fiction – and his most popular work – is *The Worm Ouroboros* (**1922**), a tale which sounds in synopsis like elaborate SWORD AND SORCERY but which reads as *sui generis*. As the title indicates, the STORY constitutes a kind of CYCLE, but one without any "redeeming" quality, a story with no point beyond the pleasure its protagonists take in

living through it. *The Worm Ouroboros* is a masterpiece of hedonism, a tale which ultimately justifies itself in aesthetic terms, and can still shock an audience used to colour-coded conflicts between Good and Evil (◊ COLOUR-CODING). Accusations that the ending is arbitrary and abrupt seem to originate in misreadings of the text.

There is a FRAME STORY, or seems to be (though the frame is not closed at the end). A human protagonist, Lessingham, is transported via a PORTAL from his home on Earth to a fantasy Mercury, which is a SECONDARY WORLD in all but name. Here Lessingham begins to observe the unfolding of a vast AGON-like war between factions whose reasons for fighting are much less important than the fact of the conflict. The names ERE gives to the warring nations – the lands of Witches, Demons, Ghouls and Goblins – have no relationship to normal usage. Lessingham soon fades out of the tale, and the war continues until Gorice, the sorcerer king of Witchland, is definitively beaten. At this point the assembled victors, feeling a staleness in the air, ask the GODS for TIME to bite its tale like the Worm Ouroboros and turn backwards, so that the conflict can recur. This boon is granted and, in a brilliant SLINGSHOT ENDING, the Agon dawns.

After a historical novel, *Styrbiorn the Strong* (**1926**), whose closing chapter is set in VALHALLA, ERE began to publish his major work, the **Zimiamvia** sequence: *Mistress of Mistresses: A Vision of Zimiamvia* (**1935**), *A Fish Dinner in Memison* (**1941** US) and *The Mezentian Gate* (**1958**), the latter uncompleted; all three were assembled as *Zimiamvia: A Trilogy* (omni **1992** US), which includes some previously unpublished material relevant to the third volume. The internal chronology reverses that of publication, but the three can be read in any order. Beyond the presence of Lessingham, their main connection with *The Worm Ouroboros* is that Zimiamvia is a kind of afterworld of the earlier novel; Zimiamvia is also connected to Earth, which is a pocket universe created by King Mezentius (in *A Fish Dinner in Memison*) to provide a sphere within which to experiment with existences bound by unalterable law and lived on the wrack of an arrow of time which moves only forward. Once Mezentius creates the planet, he and his son Barganax incorporate themselves on Earth in the identity of Lessingham, who returns to Zimiamvia (at the beginning of *Mistress of Mistresses*) only after he has died, where he encounters (or re-encounters) the true base of being, who is the GODDESS in the form of Aphrodite. Once he is safely there the experiment of Earth, which is only a SHADOW of the *real* world (◊ REALITY), is ended.

The four main protagonists – Lessingham (who had been Mezentius) and his mistress Antiope, Barganax and his mistress Fiorinda – engage in two levels of activity. All four are involved in an intricate Agon revolving around a dynastic squabble, in which everyone participates with passionate dispassion; some passages come close to creating a LAND-OF-FABLE version of the Renaissance while others are translucent with play. More profoundly, all four protagonists engage in a sustained mutual exploration and discovery of "inscape", which leads to the revelation that they are AVATARS of the two god-principles which govern the Universe (ERE names these universals ZEUS and APHRODITE). As the sequence deepens they become more than avatars: godhood shines through their human raiments. There is much lovemaking; the **Zimiamvia** books are suffused with a joyful erotic glow which, along with their "amoral" refusal to lament a lost Arcadia, must have doubly alienated Lewis and Tolkien.

Because of their high-handed disdain about everything except glory and beauty, ERE's four novels are a highly dangerous example for fantasy; had they been more widely read they could perhaps be blamed for much of the failure of imagination of the GENRE FANTASY. But ERE's language has proven too knotted for pillagers to unravel, and his work remains unechoed, except by accident. Every re-reading of ERE is a rediscovery. [JC]

Further reading: "Superman in a Bowler: E.R. Eddison" in *Literary Swordsmen and Sorcerers: The Makers of Heroic Fantasy* (**1976**) by L. Sprague DE CAMP; "The Zimiamvian Trilogy" by Brian ATTEBERY in *Survey of Modern Fantasy Literature* (anth **1983**) ed Frank Magill; "E.R. Eddison" by Attebery in *Supernatural Fiction Writers* (anth **1985**) ed E.F. BLEILER.

EDEN The Garden of Eden is a significant mythical motif; it is closely related to other images of primeval rural innocence like ARCADIA. The possibility – or tragic impossibility – of rediscovering such a state of being is a significant element of ecological mysticism, as reflected in such novels as *Green Mansions* (**1904**) by W.H. HUDSON and the title story of *Beneath the Surface and Other Stories* (coll **1918**) by Gerald Warre Cornish (1875-1916). Eden maintains a presence in such post-Adamian Biblical fantasies as *Blow, Blow Your Trumpets* (**1945**) by Shamus FRAZER and influences such images of the AFTERLIFE as Neil GUNN's *The Green Isle of the Great Deep* (**1944**). Fruit from the two TREES of knowledge occasionally appear in contemporary fantasies, ironically in "The Apple" (**1897**) by H.G. WELLS and earnestly in David LINDSAY's *The Violet Apple* (**1978**). The notion promulgated by some 19th-century occultists that Eden might be located at the then-inaccessible North Pole is echoed in M.P. SHIEL's *The Purple Cloud* (**1901**), and is thought by some scholars to have been in Edgar Allan POE's mind when he penned the even-more-enigmatic conclusion of *The Narrative of Arthur Gordon Pym* (**1837**). [BS]

See also: ADAM AND EVE.

EDGERTON, TERESA (ANN) (1949-   ) US writer whose works are mostly CELTIC FANTASY, gathered into two linked sequences: the **Green Lion** series – *Child of Saturn* (**1989**), *The Moon in Hiding* (**1989**) and *The Work of the Sun* (**1990**) – and the **Celydonn** series – *The Castle of the Silver Wheel* (**1993**), *The Grail and the Ring* (**1995**) and *The Moon and the Thorn* (**1995**). The setting is a FANTASYLAND with Celtic colouring, and the first trilogy in particular is recognizably GENRE FANTASY. There are, therefore, few surprises as an adolescent girl matures (◊ RITE OF PASSAGE), goes on a QUEST, grows in power as a female WIZARD despite the enmity of an evil princess, and saves the kingdom. But much of the detail is fresh, alertly recounted; the SEX is modestly convincing; the CHRISTIAN-FANTASY elements are, as is appropriate in a Celtic Fantasy, subdued; the male co-protagonist, a SHAPESHIFTER, interestingly teaches himself – via an ANIMAL-FANTASY sequence with a wolf-pack – how to cope; and the tale grows in intensity. In the **Celydonn** books, a similar sprightliness enlivens a plot whose larger contours are unsurprising.

In the **Goblin** sequence – *Goblin Moon* (**1991**) and *The Gnome's Engine* (**1991**) – TE creates a somewhat more original environment, an ALTERNATE-WORLD medieval Europe in which a variety of species (except for the dire GOBLINS) live harmoniously together. A MALIGN SLEEPER is aroused by an inept student of ALCHEMY; there is derring-do. In the second volume, a somewhat pixillated TECHNOFANTASY engine

is created, through which various foes are defeated. TE remains a commercial writer, but one who could at any time abandon ship. [JC]

EDIFICE An edifice is more than a house and less than a CITY, though it may resemble a house from the outside and a city from within. From without, an edifice may seem self-contained and finite; from within, it may well extend beyond lines of vision, both spatially and temporally. In almost every possible way, edifices manifest a principle central to the description of most physical structures in fantasy: there is always more to them than meets the eye (◊ LITTLE BIG).

Because of the centripetal force they exert on the mind's eye, edifices tend to dominate their LANDSCAPE. In a physical sense, this is obvious; but it is also the case that those who rule edifices (whether they be DARK LORDS, KNIGHTS OF THE DOLEFUL COUNTENANCE, mages or monarchs – or simply fathers) tend also to *be ruled by them*, in the sense that they become tied to – and identified with – their abodes; that those who work and live in edifices tend to be described in the language of ESTATES SATIRE; that protagonists raised within edifices must escape (◊ RITE OF PASSAGE); and that protagonists who come to edifices do so in the furtherance of QUESTS. Edifices, therefore, occupy the very heart of the geographies of fantasy.

Any fantasy edifice conforms to at least some of the following range of descriptions: it is named or significantly "nameless"; it was built by a person mentioned in the text, or has always existed; it is larger inside than out; it contains or is a LABYRINTH, one which quite possibly reveals (though not perhaps immediately) the bilateral symmetry of the WORM OUROBOROS; its various façades each provide a different "reading" of the nature of the structure as a whole; it is the omphalos or navel of the CITY or the world, and is a capitol or cathedral or both; it is a MICROCOSM of the world; it is a POLDER or a PORTAL (or even a set of portals); it is coextensive with the UNDERWORLD and/or the HEAVENS above (◊ AS ABOVE, SO BELOW; MIRROR); it is the WORLD TREE; it is coextensive with the mind of its builder or ruler (◊◊ MALIGN SLEEPER); it is alive; it occupies simultaneously the past, the present, and the future; it is a LIBRARY and a MAP; it tends to undergo METAMORPHOSIS at the turning-point (or RECOGNITION) of the Story, a metamorphosis which may have been hinted at throughout (◊◊ TROMPE L'OEIL); it is a three-dimensional representation of STORY; it represents, in the end, a consort of manifest or discoverable (rather than repressed) REALITIES, this final attribute distinguishing the fantasy edifice from its HORROR sibling.

Edifices are found frequently throughout world literature, probably, in the West, beginning with the Tower of BABEL (treated, most memorably, by Jorgé Luis BORGES as a LIBRARY) or the Labyrinth at Knossos (◊ DAEDALUS; MINOTAUR). CAMELOT is possibly too vacuous to be deemed an edifice, but the sorcerer's illusory castle in ARIOSTO's *Orlando Furioso* (**1516**) is a labyrinth whose coils entrap the minds of its victims. The astonishing phantasmagoria on the five orders of architecture composed by Wendel Dietterlin (1550-1599) for his *Architectura* (graph **1593-4**; exp 1598) is a comprehensive analysis of the façade as an interface out of dream; it is also an uncanny prefiguration of the work of Giovanni PIRANESI.

The first edifice to be described in a narrative context directly relevant to fantasy (and HORROR) is almost certainly that in Horace WALPOLE's *The Castle of Otranto* (**1765**), described as being essentially animate and containing honeycombs of labyrinths beneath the ground that mirror the heaven-mimicking corridors of the higher regions. Lovel Castle, in Clara REEVE's *The Champion of Virtue* (**1777**; vt *The Old English Baron* 1778), introduces the forbidden inner chamber to the Gothic mix. The eponymous opposing keeps in Ann RADCLIFFE's *The Castles of Athlin and Dunbayne: A Highland Story* (**1789**) seem part of an animate circumambient landscape. Edifices are of course extremely common in GOTHIC FANTASY – and perhaps the key to that subgenre. Edgar Allan POE's "The Fall of the House of Usher" (1839) underlines the relevance of the edifice to horror and DARK FANTASY, making adroit claustrophobic use of the isomorphic relationship between SOUL and surrounding bodily shell. A recent horror example (one of many) is the underground labyrinth-without-exit featured in *Daemonic* (**1995**) by Stephen Laws (1952-   ).

As the 19th century advanced, fantasy examples began to proliferate: the Snow Queen's Castle in Hans Christian ANDERSEN's *Snedrondingen* (**1844**); the castle in George MACDONALD's *The Princess and the Goblin* (**1871**), which features a FAIRY GODMOTHER in the tower and GOBLINS in the roots of the mountains below; the valley of The Masters in H.P. BLAVATSKY's *The Secret Doctrine* (**1888**), underneath which vast caverns house a library containing the secret history known only to THEOSOPHY; the Nome King's palace in L. Frank BAUM's *Ozma of Oz* (**1907**); the eponymous structure in *Fortress Unvanquishable, Save for Sacnoth* (1908; **1910** chap) by Lord DUNSANY; and the seeming tenement in Gustav MEYRINK's *The Green Face* (**1916**), whose frontage resembles a skull and which contains a magic SHOP, the WANDERING JEW and mysterious passageways which debouch upon a theatre whose performances constitute a REVEL.

20th-century examples are numerous. Among them are Badger's underground home in Kenneth GRAHAME's *The Wind in the Willows* (**1908**), which is built over the ruins of an ancient city; Locus Solus, the surreal villa in Raymond Roussel's *Locus Solus* (**1914**), which surrounds an enormous garden; the impregnable fortress of Hefeydd Hen, which it takes Pwyll Pen Annwn 50 subjective years to penetrate in Kenneth MORRIS's *The Fates of the Princes of Dyfed* (**1914**); Storisende in James Branch CABELL's **Biography of Manuel**; the Nature Theatre of Oklahoma in Franz KAFKA's *Amerika* (**1927**) and the eponymous castle in his *Das Schloss* (**1926**); the palace of Namirrha, larger inside than out, in Clark Ashton SMITH's "The Dark Eidolon" (1935); the huge, ancient, many-chambered mansion in Clemence DANE's "Frau Holde" (1935), which heals its lodgers; Memison, the summer palace in E.R. EDDISON's *A Fish Dinner in Memison* (**1941**); the department store in John COLLIER's "Evening Primrose" (1941); Mervyn PEAKE's Gormenghast; the Schloss of Count Johann von Hackelnberg, Reich Master Forester in SARBAN's *The Sound of his Horn* (**1952**), which features large trees "actually knitted into the fabric of the building"; Aslan's How in the **Narnia** books by C.S. LEWIS; The Last Homely House East of the Sea in the heart of Rivendell in J.R.R. TOLKIEN's *The Lord of the Rings* (**1954-5**), along with many other fortresses and refuges to be found throughout the trilogy's immense back-story, the most formidable of which is perhaps Utumno, built 25,000 years earlier by Sauron the DARK LORD's predecessor and master; the Hall of Justice in Orson Welles's movie of Kafka's *The Trial* (**1962**); the Spiral

Castle in Lloyd ALEXANDER's *The Book of Three* (**1964**); King Haggard's Castle in Peter S. BEAGLE's *The Last Unicorn* (**1968**); the School of Roke in Ursula K. LE GUIN's *A Wizard of Earthsea* (**1968**); Penelope FARMER's eponymous *The Castle of Bone* (**1972**); the palace in Michael MOORCOCK's *Gloriana* (**1978**); at least three examples in Michael ENDE's *The Neverending Story* (**1979**) – the Ivory Tower "as big as a whole city" at the heart of Fantastica, the Temple of a Thousand Doors, each door being a PORTAL, in the primordial desert that precedes Fantastica, and the huge structure that suddenly exfoliates around the Worm Ouroboros which enfolds the Water of Life; the apartment blocks in Dario Argento's movie INFERNO (*1980*); the House Absolute in Gene WOLFE's *The Book of the New Sun* (**1980-83**); the eponymous maze-like bureaucratic palace, featuring a huge inner palace within the outer structure, in *The Palace of Dreams* (**1981**) by Ismail Kadare; Edgewood in John CROWLEY's *Little, Big* (**1981**); the Quirkian Hold in Nancy KRESS's *The Prince of Morning Bells* (**1981**), which extends within a cliff-face downwards to a great library; the Fortress of Ultimate Darkness in TIME BANDITS (*1981*), in which the Supreme Being has incarcerated the Universe's Evil Genius and his cohorts; the London Pantheon in John M. FORD's *The Dragon Waiting: A Masque of History* (**1983**); the "new palace" of the PUPPET emperor in Alasdair Gray's *Five Letters from an Eastern Empire* (**1995** chap); the WORLD-TREE first joked about and then encountered, full of Americans in the top branches, in Steve ERICKSON's *Rubicon Beach* (**1986**); Tamson House in *Moonheart* (**1984**) and *Spiritwalk* (omni **1992**) by Charles DE LINT; the hugely intricate palace where most of Stephen DONALDSON's **Mordant's Need** sequence (**1986-8**) is set; the eponymous **Castle Perilous** with 144,000 portals in John DECHANCIE's series starting with *Castle Perilous* (**1988**); the palace in Mark HELPRIN's *Swan Lake* (**1989**), with its 17,500 rooms, in the largest of which "the Duke of Tookiaheim used to fly his glider"; the vast, many-roomed Undermoment that underpins and interweaves the worlds in Tom DE HAVEN's **Chronicles of the King's Tramp** (**1990-92**); the intensely animate Castle Banat in Lucius SHEPARD's *The Golden* (**1993**); the Winchester Mystery House (a real-world edifice examined in fantasy terms) in Michaela ROESSNER's *Vanishing Point* (**1993**); the building in Diana Wynne JONES's *Howl's Moving Castle* (**1986**) and *Castle in the Air* (**1990**); the Gallery of Bone (where Old Father Time lives) in Simon R. GREEN's *Shadows Fall* (**1994**); the Shrine of the Archangels and St John the Divine in Elizabeth HAND's *Waking the Moon* (**1994**); the St Petersburg Arms in William Browning SPENCER's *Zod Wallop* (**1995**), being a hotel which is transformed into Grimfast, a DARK-FANTASY locus of threat, horror and METAMORPHOSIS; the mountaintop castle of Torra Alta, in the labyrinth beneath which a DRAGONS' hoard contains weapons of magic power, in *The Runes of War* (**1995**) by Jane Welch (1964-   ); and the eponym of Jack DANN's *The Memory Cathedral: A Secret History of Leonardo da Vinci* (**1995**), a structure containing "hundreds of thousands" of images of Leonardo's life under its domes and in its labyrinths, and, at the knotted heart of it all, "as if it were a trompe l'oeil", himself. There are many more. [JC]

EDWARDS, AMELIA B. (1831-1892) UK writer. ◊ GHOST STORIES.

EDWARDS, CLAUDIA J(ANE) (1943-   ) US writer whose debut novel, *Taming the Forest King* (**1986**), opened her loosely linked **Forest King** series of lightly feminist fantasies. It was followed by *A Horsewoman in Godsland* (**1987**) and *Bright and Shining Tiger* (**1988**). A fourth book, *Eldrie the Healer* (**1989**), is not part of the series but is similar in subject. [DP]

EDWARDS, LES (1949-   ) UK illustrator of fantasy and horror, whose work has been used for book covers, movie posters and game packaging. LE in the early 1970s joined the agency Young Artists, with whom he created many distinctive images for paperback fantasy and horror books. He produced posters for the movies NIGHTBREED (*1990*) and *The Thing* (*1982*). He has also done two GRAPHIC NOVELS based on works by Clive BARKER: *Son of Celluloid* (graph **1991** US) and *Rawhead Rex* (graph **1994** US). [RT]

EDWARD SCISSORHANDS US movie (*1990*). 20th Century-Fox. **Pr** Tim BURTON, Denise Di Novi. **Exec pr** Richard Hashimoto. **Dir** Burton. **Spfx/mufx** Stan Winston Studio, Michael Wood. **Vfx** VCE, Peter Kuran. **Screenplay** Caroline Thompson. **Starring** Johnny Depp (Edward Scissorhands), Anthony Michael Hall (Jim), Robert Oliveri (Kevin Boggs), Vincent Price (Inventor), Winona Ryder (Kim Boggs), Dianne Wiest (Peg Boggs). 98 mins. Colour.

On a hill above a pastel-coloured suburb is a castle. Here dwelt a FRANKENSTEIN-style inventor, who died just before completing his MONSTER. Peg, an Avon Lady desperate for clients, calls and finds a timid youth with two masses of blades in place of hands. She takes him into her family and names him Edward; soon he is lionized for his useful dexterity with the blades, and Peg's daughter Kim comes to love him. But public opinion turns: the UGLY DUCKLING is made a scapegoat and must flee back to the sanctuary of the castle. A vastly aged Kim, who has been telling this story to her granddaughter (◊ FRAME STORY), adds the explanation that he must still be there, for the snowflakes falling outside the house are in reality splinters from the ice-statues he carves with those ever busy scissorhands.

ESH is an attempt to construct a modern-day FAIRYTALE and so, as in many fairytales, the ambience is stripped down to mere basics: the suburb is never related to any greater world, and the passage of time is fitful (one moment it is summer, the next Christmas). Again traditionally, *ESH* incorporates many RECURSIVE elements. The net effect of such ploys, designed presumably to make the tale universal and immediate, is unfortunately to distance us from it. [JG]

EERIE Two US COMIC books and one US comic-strip magazine.

1. Pre-Comics Code HORROR comic book published by Avon Periodicals, memorable for little except the (almost certainly) first adaptation of *Dracula* (**1897**) by Bram STOKER (in *#12* 1952) and a few bondage strips by artist Wallace Wood (1927-1981). A single issue in 1947 was followed by a continuous run 1951-4, when the name changed to *Strange Worlds*. [RT]

2. Short-lived comic book published by IW Enterprises in 1964. Some fantasy stories were reprinted from elsewhere. [RT]

3. Large-format anthology-style horror-fantasy comic book published by WARREN PUBLISHING 1965-83; in early issues each story was introduced and rounded off by a facetious homily from a warty-faced grotesque called Cousin Eerie. A companion to CREEPY, *E* adopted an identical format, and had the same high standards in cover artwork and content. In the mid-1970s the storylines became more substantial, with some occupying almost a complete issue while others formed a series concerning one character. There were also novel-length translated reprints in serial form of GRAPHIC NOVELS from Europe and elsewhere; these included *Dax* (from *#59* 1974) by Esteban MAROTO, *El Cid* (from *#66* 1975) by Bill DuBay, Budd Lewis and Gonzalo Mayo, *Spacewrecked* (*#129-#136* 1982) – trans from the first two volumes of *Les Naufrages du Temps* by Jean-Claud FOREST – by Paul Gillon and **Haxtur** and **Haggarth** (*#111-#136* 1980-82) by Victor de la Fuente, two SWORD-AND-SORCERY strips translated from Spanish.

A full list of contributors reads like a compendium of the finest artists in comics: they included Neal ADAMS, Frank FRAZETTA, Jeffrey JONES, Will EISNER, Roy G. KRENKEL, Reed CRANDALL, Mike PLOOG, Alex NINO, Alex Toth (1928-   ), Al WILLIAMSON, Wallace Wood and Bernie WRIGHTSON, plus leading artists from Spain and the Philippines. Cover artists make an equally impressive list: Frazetta, James Steranko (1938-   ), Richard CORBEN, Gray Morrow (real name Dwight Graydon Morrow; 1934-   ) and SANJULIAN. [RT/DR]

EERIE COUNTRY ◊ WEIRDBOOK.

EGAN, DORIS (1955-   ) US writer whose first sale – "Timerider" (1986), about a group of time-travelling art thieves – appeared in *Amazing Stories*. *The Gate of Ivory* (**1989**), *Two-Bit Heroes* (**1992**) and *Guilt-Edged Ivory* (**1992**) are SCIENCE FANTASY; the series' protagonists include a sorcerer from a planet where MAGIC works. Unusually for the subgenre, the three novels, although linked by place and characters, are not a trilogy but a series of linked singletons. As Jane Emerson DE has written the epic space opera «City of Diamond» (1996), the first in a projected trilogy. [JF]

EGBERT, FLOYD C. [s] ◊ Kenneth MORRIS.

EGGLETON, BOB (1960-   ) US artist. ◊ ILLUSTRATION; REALMS OF FANTASY; WEIRD TALES.

EGYPT, ANCIENT It could be claimed – unprovably – that the ancient Egyptians were the first to write fantasy. Certainly Egypt offers the oldest known examples of the FAIRYTALE, the prose short story and several narrative devices. Both fanciful and realistic stories were written as early as the 20th century BC; the writers of the former may well have seen a difference and so known themselves as fantasists. There is no such possibility in any other of the world's literatures until centuries later.

Native Egyptian rule ended in 525BC, and the country thereafter produced Greek, Jewish, Christian and Muslim works of the first importance. Here we are concerned solely with the two Egyptian literatures prior to Egypt's conversion to Christianity, which created significant literary discontinuities. The first strand produced narratives *c*1950-*c*1050BC; the narratives of the second were written 500BC-AD300. Neither is at all well known, nor is it clear how much connection there is between them. Most Egyptian stories have been found in only one papyrus or inscription, anonymous and datable only by that copy's date, not the (possibly much earlier) date of composition. Many are fragmentary or incomplete – including every story yet found from the later period. A full collection in translation would run to only a few hundred pages. Many of the most important works can be found in Miriam Lichtheim's excellent *Ancient Egyptian Literature* (anth **1973-80** 3 vols).

Quite realistic, quasi-religious autobiographies inscribed in tombs or on stone are the oldest narratives, beginning about 2300BC. In the 20th century BC an older genre of wisdom literature diversified and acquired narrative frames or fully narrative form. In contrast, the story of Sinuhe transfigured the autobiographic genre. A powerful realistic tale of exile and forgiveness, it includes examples of several other literary genres in an artful structure which well disguises its likely propagandistic purpose. It was often copied until the 13th century BC. Other fantasies of that period seem artless by comparison, but show from the beginning a grasp of the basic technique of story-within-story (◊ FRAME STORY); they do not obviously derive from any older genre. In "The Shipwrecked Sailor" (20th century BC) the title character appealingly and didactically relates his rescue and enrichment by a regal, kindly DRAGON (who in turn has a story to teach). Papyrus Westcar (*c*1600BC) gives the oldest examples of stories about magicians.

In the subsequent centuries of empire, many fantastic stories appeared. Those known include a charming GHOST STORY, an ALLEGORY with fairytale elements; and the wildly bombastic Kadesh inscriptions (1270BC), in which Ramesses II claims to have slain hundreds of thousands in a day's battle, single-handed. Meanwhile "The Doomed Prince" (*c*1300BC) is the oldest known fairytale. It remoulds "Sinuhe", breaking from autobiography into the third person, to depict an exile fleeing fates which prove to be TALKING ANIMALS. "The Two Brothers" (*c*1200BC), perhaps by Ennana, is among the greatest works of these centuries. From a mundane opening it becomes a rich fairytale, then moves to transformation and apotheosis. A number of MYTHS also survive from this era. "The Contendings of Horus and Seth" (*c*1145BC), a farcical tale of the last phase of the great **Osiris** cycle, crude in both style and content, is the earliest known example of myth-PARODY. Both it and "The Two Brothers" contain several motifs later found in the BIBLE.

Very little narrative at all survives from 1050-550BC. The only relevant example is a fragment from a student's slate (before 730BC) which tells of a lawsuit between the belly and the head for primacy over the body. This is arguably a FABLE, the oldest written one known, though Egyptian art and Mesopotamian literature allude to fables much earlier.

In the later narratives, mostly written in the newer Demotic script, wild fairytales and sober realism both vanished in favour of ROMANCES, sometimes fantastic, sometimes not. The change is already apparent in the UNDERWORLD story of the magician Merire (*c*500BC; ed and trans into French by Georges Posener as *Le Papyrus Vandier* **1986**) and the fragmentary texts from Saqqara (4th century BC; ed and trans H.S. Smith and W.J. Tait as *Saqqara Demotic papyri 1* **1983**), which involve the GODS more than most later romances. *Saqqara Text 1* seems very like a secular romance of separated lovers' adventures such as later became popular in Greek (◊ GREEK AND LATIN CLASSICS).

The best-preserved Demotic romances (as well as many of the fragments) fall into two cycles. There are martial ones, generally without fantastic elements, about the heirs of Inaros; and there are two tales of a magician called Setna. *Setna 2* (2nd century AD) includes an underworld tour and a magician's duel across millennia. *Setna 1* (1st century BC) concentratedly and with some eloquence contrasts Setna's lust and wrongdoing with ancient mages' true love and suffering; several copies are known.

None of the later works, save presumably the BEAST FABLES now found in numbers, were necessarily written as fantasy; Egypt was then, after all, a centre of magical study.

Contemporaneous myths in Egyptian often seem like sacred dramas, as such constituting the only genre with any resemblance to the last masterpiece known to us: the "Kufi text" (2nd century AD). This tells of the GODDESS Kufi's angry departure from Egypt and Thoth's efforts to persuade her to return. In this cause Thoth mixes rank flattery, drawing on the language of hymnody, with moralizing built from both wisdom traditions and fables; the goddess forces him to tell many of the latter, of which three are intact. The whole melds multiple genres at the end of ancient Egyptian literature as skilfully as did "Sinuhe" at its start, with the addition of HUMOUR, and was often copied and even translated into Greek before Egypt's Christian conversion in the 3rd and 4th centuries submerged the native tradition. [JB]
Further reading: Relevant essays in *Civilizations of the Ancient Near East* (**1955**) ed Jack M. Sasson; "Egyptian Fiction in Demotic and Greek" by John Tait in *Greek Fiction* (**1994**) ed J.R. Morgan and Richard Stoneman.

EISENSTEIN, PHYLLIS (LEAH KLEINSTEIN) (1946-) US writer. Her first story, "The Trouble with the Past" with Alex Eisenstein for *New Dimensions 1* (anth **1971**) ed Robert SILVERBERG, was sf. Her first fantasy series, the **Alaric the Minstrel** sequence – *Born to Exile* (1971-4 *F&SF*; fixup **1978**) and *In the Red Lord's Reach* (**1989**) – depicts the slow coming of age of Alaric, an orphan with a TALENT (teleportation) whose experiences verge on cliché but escape by dint of PE's emphasis on the complexities of the process of growing up, and on a sense that the worlds he comes across – like the nomadic WATER-MARGIN world of the Red Lord in the second volume – have in a way been *chosen* by him because they MIRROR his own condition. (The second volume also interestingly portrays an alternative science of strange but consistent quasi-magnetic effects.) Her second series, the **Cray Ormeru** sequence – *Sorcerer's Son* (**1979**) and *The Crystal Palace* (**1988**) – is again devoted to a young man's coming of age, though in this case more complexly, as Cray is the son of a WITCH and a SUCCUBUS-cum-INCUBUS (actually a DEMON disguised as a KNIGHT) bearing the sperm of a paranoid sorcerer. His search for his father, who is a destructive KNIGHT OF THE DOLEFUL COUNTENANCE, is fraught with risks, even in a world of rationalized MAGIC. When he discovers the truth, Cray responds by granting freedom to the demon, a freedom which serves to emancipate his mother and to demonstrate his maturity. In the second volume the adult Cray embarks on a QUEST to save a maiden he has scried in a magic MIRROR; their relationship is engagingly spiky.

Her short fiction, which is uncollected, includes the fantasies "Subworld" (1983) – an intriguing URBAN FANTASY set in a NEW YORK whose subways support a WAINSCOT society – "The Demon Queen" (1984) and "The Amethyst Phial" (1984). Each of PE's stories is singular, densely but deftly composed. The infrequency of her publications has often been regretted. [JC]

EISNER, WILL (1917-) US COMICS artist and writer who created the MASKED AVENGER **The Spirit**. WE's roots were in Brooklyn, the inhabitants of whose crowded tenement blocks have formed the subject matter of much of his graphic storytelling. His first comics work appeared in *Wow* (1936). Forming a partnership with former Hearst cartoonist Jerry Iger (1910-), he set up a comic-book production shop, mass-producing comics for Quality, Fiction House and Fox; artists working with him at this time included Bob KANE, Mort Meskin (1916-), Bob

Powell (1917-1967) and Lou Fine (1915-1971). Among features by the Eisner-Iger studios were **Hawks of the Sea**, written and drawn by WE himself, and **Sheena, Queen of the Jungle**, conceived by WE and Iger and drawn by Meskin. Other strips created by the studios were **Espionage** (1938-9) by WE and **Dollman** (1939-50), conceived by WE and drawn by Fine. WE went on to work for Quality Comics, editing *National* and *Military Comics*. In 1940 he created his masterwork, **The Spirit**: this ran as an 8-page supplement in Sunday newspapers until 1952.

WE went on to produce comic strips for military, advertising and educational purposes. Then he published four **Gleeful Guides**: humorous illustrated instruction books: *Communicating with Plants* (**1974**), *Occult Cookery* (**1974**), *Facts, Statistics and Trivia* (**1974**) and *Living with Astrology* (**1974**). Later came *The Spirit's Casebook of True Haunted Houses and Ghosts* (**1976**).

In 1978 WE published the first of his remarkable GRAPHIC NOVELS. *A Contract with God* (graph **1978**) tells of a Russian Jew, Frimme Hersh, who makes a CONTRACT with GOD. When his adoptive daughter dies he considers that God has broken the contract, and furiously devotes his life to meanness and dishonesty. When he is very rich he goes to the religious elders to draw up a better contract, but as he reads it and plans a changed life he dies of a heart attack. WE went on to write and draw other stories about tenement life, mostly focused on an imaginary tenement block at 55 Dropsie Avenue. His multi-layered narratives are often profoundly moving and thought-provoking, revealing a touching insight into the lives of ordinary people.

In 1983 Kitchen Sink Press began publication of *Will Eisner's Quarterly* (#1-#8 1983-6) as a companion periodical to the **Spirit** reprints they were then publishing. These comics contained new work, reprints and articles and interviews by WE.

WE remains a significant figure on the US comics field and has continued to produce graphic novels, including *Life on Another Planet* (*Spirit Magazine #19-#26* 1979-80; exp vt *Signal from Space* graph **1983**), *A Life Force* (*Will Eisner's Quarterly #1-#7* 1983-5; graph **1988**), *The Dreamer* (graph **1986**), *New York, The Big City* (graph **1986**), *The Building* (graph **1987**), *Heart of the Storm* (graph **1991**), *Invisible People* (graph **1993**) and *Dropsie Avenue* (graph **1995**). [RT]
Further reading: *Comics and Sequential Art* (**1985**) and *City People Notebook* (**1985**), both by WE.

ELDER GODS Various PANTHEONS have elder and younger GODS – as with the AESIR and Vanir of Nordic mythology – but the phrase EG is most closely linked to the CTHULHU MYTHOS, where the EGs are portrayed as near immortal beings from a distant star, accessible via an alternate dimension. They are neither good nor evil, but *different*. Some were banished from their original home because of a rebellion, and these Great Old Ones came to Earth and established the earliest known civilizations (◊ ATLANTIS; LEMURIA; MU), whose ruins still survive. The Great Old Ones (of whom Cthulhu was the chief deity) are now in slumber awaiting the time to reawaken (◊ MALIGN SLEEPER). The identities of the EGs are not always clear, though Azathoth and Yog-Sothoth are usually counted among them, as is Nodens, though he is also described as being of an opposing faction. Lovecraft's scenario was ideally suited to other works of fantasy set during the early days of Earth, and the concept of EGs was readily adopted by Lin CARTER and, in particular, Brian LUMLEY, who developed the PANTHEON to a considerable degree. [MA]

ELDER RACES Races which have been in the world – or the SECONDARY WORLD – for so much longer than contemporary folk that their still-living representatives, in their very persons, constitute a TIME ABYSS. ERs usually date back to the period when the basic underlying STORY which secretly or openly governs the present world was first told; and they have normally played a significant role in that basic Story (unlike some of the more frivolous denizens of FAERIE). An ER is not, therefore, simply a race older than other races; it is normally one that has helped shape the races which follow. Contemporary folk may be able to trace their descent from ERs; on the other hand, ERs may be of different stock entirely, like the ELVES and dwarves (◊ DWARFS) who feature in J.R.R. TOLKIEN's *The Lord of the Rings* (**1954-55**). The continued existence of an ER may be unknown or known. It may take the shape of a PARIAH ELITE or inhabit a secret POLDER; it may dwell openly in the new world or in WAINSCOT fashion, or inhabit an OTHERWORLD, visiting the current one only on occasion. An individual member may be recognized or unrecognized, and may be introduced into the text as a FOOL, a strangely wise COMPANION, a GOD, a MAGUS, a TRICKSTER, or any other figure whose gifts and/or presence is not fully explained by his or her ostensible role or TALENTS. But whatever the contemporary role of an ER, there will be an underlying tendency for that role to knit together the present and the past, so that present events ultimately affirm the true nature of the underlying Story that shapes the contemporary world.

ERs are also commonly found in tales of cosmic HORROR by writers like H.P. LOVECRAFT, whose CTHULHU MYTHOS elaboratedly describes members of an ER who awaken from MALIGN-SLEEPER repose and terrify unto death contemporary humans. In Lovecraft, and in his imitators and admirers, ERs are normally presented in terms of a time abyss whose effect is infinitely to disparage contemporary humans, in direct contrast to the effect in fantasy. In SUPERNATURAL FICTION, ERs may be nonhuman (VAMPIRES and WEREWOLVES are often relics of ERs) or may have some ancestral relationship to humans (as in most tales which make use of THEOSOPHY or other confabulations). ERs may appear in FANTASIES OF HISTORY, but in a fashion which affirms a sense that their Story unveils the MATTER of the world. [JC]

ELDORADO A legend that obsessed the Spanish *conquistadores* in South America: *El hombre dorado*, the Golden Man, was said to be chief of a rich tribe whose RITUAL was to coat him daily with GOLD that he washed off in a lake – which thanks to this and other gold offerings became a repository of fabulous wealth. El Dorado was generalized as a sought-after IMAGINARY LAND of treasure. [DRL]

ELDRIDGE, PAUL (1888-1982) US writer. ◊ George VIERECK.

ELDRITCH TALES ◊ MAGAZINES.

ELEMENTALS Magical personifications of natural forces associated with the ELEMENTS of ALCHEMY: earthquakes and landslides for Earth, gales for Air, floods and tidal waves for Water (also, in collaboration with Air, storms), and Fire in any aspect which feature heavily in the stories by Algernon BLACKWOOD. Fritz LEIBER's "Smoke Ghost" (1941) suggests a CITY elemental; John BRUNNER's *The Traveller in Black* (coll of linked stories **1971**) features elementals of qualities like sharpness and oblivion; Fred SABERHAGEN's *Changeling Earth* (**1973**) adds the desert elemental of sandstorms, and the subtler prairie elemental that expands distance and lengthens travellers' paths; Alan MOORE reconstructed Swamp Thing from a GOLEM-like creature of earth and mud to a plant elemental attuned to ecology and the "Green". Generally a true elemental is non-anthropomorphic, as opposed to the personifications which are GODS; thus Robert A. HEINLEIN's *Magic, Inc.* (**1940**) has a simple ball of flame as its fire elemental (or salamander), and a watery blob reeking of low tide as the water elemental (here called an UNDINE); but, yielding to anthropomorphism, Heinlein's earth elementals are gnomes. By this token, those TROLLS which are animated stone may be earth elementals; Terry PRATCHETT's *The Colour of Magic* (**1983**) offers a complementary water elemental with its liquid "sea troll" Tethis. The most famous of air/wind elementals is the WENDIGO; each wind has its own controlling elemental in Elizabeth A. LYNN's *The Red Hawk* (**1983** chap). Notable fire elementals appear in: Poul ANDERSON's "Operation Salamander" (**1956**), as a lizard-shape hidden in flame; Roger ZELAZNY's *Lord of Light* (**1967**), as static lightning-bolts; Diane DUANE's *The Door into Fire* (**1979**) and its sequels, in horse form; and Diana Wynne JONES's *Howl's Moving Castle* (**1986**), as a fiery face in the hearth which is a literal fallen STAR. [DRL]

ELEMENTS In ancient cosmology (e.g., Aristotle's), ALCHEMY and much MAGIC, these are earth, water, air and fire – regarded as mystic fundamentals of the Universe rather than as mere states or phases corresponding to the solid, liquid, gas and plasma of physics. A fifth element, SPIRIT, is sometimes added. The ancient elements are still symbolically used in ASTROLOGY – e.g., Aries is a "fire sign". Often they are personified as ELEMENTALS: in the magical system of Tanya HUFF's *Sing the Four Quarters* (**1994**), power over each element is gained by commanding its countless elemental inhabitants through TALENT and SONG. In ancient Chinese science there were the five elements earth, fire, metal, water and wood. [DRL]

ELEUTERI-SERPIERI, PAOLO (1944-   ) Italian COMICS artist. ◊ DRUUNA.

ELFQUEST Epic cycle of COMIC-strip adult FAIRYTALES concerning a race of immortal, SHAPESHIFTING, telepathic ELVES, created by husband-and-wife team Wendy and Richard PINI. The MYTH-OF-ORIGIN story told of a group of immortal beings (◊ IMMORTALITY) forced to leave their dying home planet to search for a new world. One group finds a planet which they call The World of Two Moons, inhabited by a human-like race with a medieval society. This race has a MYTHOLOGY involving elves, and the immortals alter their appearance to resemble these and their spaceship to resemble a huge EDIFICE. However, they are inadvertently thrown back in TIME to a prehistoric age, before the mythology was established, and find themselves feared and attacked by cavemen; those not killed are driven away from their ship.

The story then shifts in time to an early medieval period (before that in which the spaceships landed) when there is a war between the elven descendants and the natives. The elves have been scattered into several different tribes by now, and the tales focus on one of these, The Wolfriders, who live in a forest environment and have formed very close links with WOLVES. Their leader is Cutter, whose adviser and confidant is Skywise.

The elves can communicate telepathically – "sending" (◊ TALENTS) – and are impelled to reproduce with a member of the opposite sex who has exactly compatible strengths and

talents, whom they instantly discern upon first meeting – a system which sometimes falls out well and sometimes not.

The saga goes on to relate how the tribe is driven from the forest and encounters others of its kind in a desert, where Cutter recognizes his predestined mate, Leetah. The series draws in many other plot threads and continues to expand in scope and complexity.

E began life in *Fantasy Quarterly #1* (**1978**), but when its publishing company, Independent Publishers Syndicate, went out of business the Pinis went on to produce their saga independently as WaRP Graphics, launching **Elfquest** (#1-#21 1978-85) with #2 (1979), retrospectively reissuing the first story as #1 alongside #4 (1979). After MARVEL COMICS picked up the title and reprinted the entire first run in their **Epic** line (1985-8), further stories were released by WaRP, following several plot threads in different subtitles simultaneously. These include **Elfquest: Gatherum** (#1-#2 1985), **Elfquest: Siege at Blue Mountain** (#1-#8 1987-8), **Elfquest: Kings of the Broken Wheel** (#1-#9 1990-92), **Elfquest: The Hidden Years** (1992-current), **Elfquest: New Blood** (1992-current), **Elfquest: Blood of Ten Chiefs** (1993-current), **Elfquest: Wavedancers** (1993-current) and **Elfquest: Shards** (1994-current). The creators brought all the threads of the series to a close early in 1996, intending to take it in a new direction. [RT]

ELIADE, MIRCEA (1907-1986) Romanian writer, also a philosopher of comparative religion. He spent most of his life in exile elsewhere in Europe and, latterly, in the USA, where he was chairman of the Department of History of Religions at the University of Chicago.

Beginning his career in the early 1920s, he was an intensely prolific author of fiction and nonfiction alike, publishing over 1300 pieces over 60 years. Due to the wide range of his scholarship and to his peripatetic life, his bibliography is intricate. His doctoral dissertation, for instance, was written in English in 1931 for Calcutta University, translated into French as *Yoga: Essai sur les origines de la mystique indienne* (**1936** France), and eventually recast as *Le Yoga: Immortalité et Liberté* (**1954** France; trans Willard R. Trask as *Yoga: Immortality and Freedom* **1958** US; rev 1969 US); and in that final form stands, along with *Le Chamanisme et les techniques archaïques de l'extase* (**1951** France; exp version 1300 trans Trask as *Shamanism: Archaic Techniques of Ecstasy* **1964** US), as a standard and definitive examination of materials of deep interest to many readers of fantasy (◊ SHAMANISM). Other works of interest include *Le Mythe de l'éternal retour: archétypes et répétition* (**1949** France; trans Trask as *Cosmos and History: The Myth of the Eternal Return* **1954** US), *Das Heilige und das Profane* (**1957** Germany; trans Trask as *The Sacred and the Profane: The Nature of Religion* **1959** US) and *Mythes, Rêves et Mystères* (**1957** France; trans Philip Mairet as *Myths, Dreams and Mysteries* **1960** US). Running through ME's work is the elaborately articulated conviction that the cyclic patterns of archaic RELIGIONS sacralized the world in a fashion no longer available, and that through the understanding of the relationship between the twin poles of the sacred and the profane it is possible to begin to understand the intimacy between archaic humans and their sense of ultimate Being.

ME wrote fiction only in Romanian. His first works of fantasy interest were *Lumina ce se stinge* ["The Light that Fails"] (**1931**; exp 1934), involving MAGIC and an encounter with a demonic figure, and *Domnisoara Christina* ["Miss Christina"] (**1936**), about a VAMPIRE which possesses the eponymous woman, whose unsuccessful subsequent attempt to seduce the protagonist is soon understood by him to represent an opening into the unknown that he has failed to accept. Some of ME's later work is available in English: *Secretul Doctorului Honigsberger* (coll **1940**; trans William Ames Coates as *Two Tales of the Occult* **1970** US: vt *Two Strange Tales* **1986** US) contains "Doctor Honigsberger's Secret" and "Nopti la Serampore" (**1939**), trans as "Nights at Serampore"; *Fantastic Tales* (coll trans E. Tappe **1969** UK) contains "Un om mare" (**1948**), trans as "A Great Man", and "Douasprezece mii de capete de vite" (**1952**), trans as "Twelve Thousand Head of Cattle", both originally assembled with other material in *Nuvele* ["Stories"] (coll **1963** Spain); *Forêt Interdite* (**1955** France; ot *Noaptea de Sânziene* **1971** France; trans Mac Linscott Ricketts and Mary Park Stevenson as *The Forbidden Forest* **1978** US); *Tales of the Sacred and the Supernatural* (coll trans Ricketts and William Ames Coates **1981** US) contains "La Tigancu" (**1962**), trans as "With the Gypsy Girls", and "Les Trois Grâces" (**1976**); *Pe strada Mântuleasa* (**1968** France; trans as *The Old Man and the Bureaucrats* **1979** US); *Youth Without Youth and Other Novellas* (coll trans Ricketts **1988** US) contains *Die Pelerine* (**1976** chap Germany), trans as "The Cape", "Tinerete far de tinerete" (**1978**), trans as "Youth Without Youth", and *Nouasprezece trandafiri* (**1980** France), trans as "Nineteen Roses".

Much of this work might be described as SUPERNATURAL FICTION; but ME utilized a conventional repertory of motifs and figures – demons, SERPENTS, GHOSTS, searches for IMMORTALITY, TIMESLIPS – to dramatize tightly argued metaphysical concerns. Much of his fiction hovers at the edge of the FANTASTIC and revolves around a sense that to penetrate into the fantastic and to unwrap the cloak of Being are similar activities. In this context, TIME is simply another cloak of ILLUSION, another "camouflage", and tales like "Nights at Serampore" and "Doctor Honigsberger's Secret" utilize timeslip devices to engross their protagonists in transcendent realities. "With the Gypsy Girls" – the protagonist of which spends a night with three dancers and awakens 12 years later – makes play with the TIME-IN-FAERIE motif. "The Cape" ironically opposes a totalitarian regime and a temporally unfixable message, which when deciphered calls upon the dreamers of the world – i.e., those who understand that the world is a veil – to unite. *The Forbidden Forest*, ME's most significant novel, parallels an individual search for IMMORTALITY and a CREATION MYTH for a Romania of dreams.

Some of the translations of ME's work are unreliable, and much remains unavailable to the English-language reader; his considerable importance to the 20th-century literature of the fantastic is not yet fully appreciated. [JC]

ELIOT, T(HOMAS) S(TEARNS) (1888-1965) US-born UK poet. ◊ ARTHUR; James FRAZER; WASTE LAND.

ELIXIR OF LIFE One of the two traditional goals of ALCHEMY; the term has broadened out to encompass any draught which extends life or protracts youth. The quest to obtain it is often frustrated, as in Edward BULWER-LYTTON's *A Strange Story* (**1861**), Nathaniel HAWTHORNE's unfinished *Septimius* (**1872**) and Alexander DE COMEAU's *Monk's Magic* (**1931**).

Even when successful, it rarely satisfies the optimistic expectations of the seekers. The many negative wish-fulfilment fantasies involving the EOL include William GODWIN's *St Leon* (**1799**), Honoré de BALZAC's "The Elixir

of Life" (**1830**), Mary SHELLEY's "The Mortal Immortal" (**1834**), Nathaniel Hawthorne's "Dr Heidegger's Experiment" (**1837**), W. Harrison AINSWORTH's unfinished *Auriol* (**1850**), Richard GARNETT's "The Elixir of Life" (**1881**), S.B. Alexander's "The Living Dead" (**1887**), E. NESBIT's *Dormant* (**1911**), George Allan England's "The Elixir of Hate" (**1911**), Claude Farrère's *The House of the Secret* (**1923**), Frederick Carter's "The Skeleton" (**1935**) and Neil BELL's "Mr Albert Finkleman" (**1953**). [BS]

ELKIN, STANLEY (LAWRENCE) (1930-1995) US novelist; he won the National Book Critics Circle Award and (posthumously) the National Book Award. Best-known as a writer of unsentimental and extremely funny contemporary novels, some of them FABULATIONS, SE is not generally associated with genre fiction. Nevertheless, from "Perlmutter at the East Pole" (**1963**), with its IMAGINARY LANDS, to "Town Crier Exclusive, Confessions of a Princess Manque: `How Royals Found Me "Unsuitable" to Marry Their Larry'" (**1993**), which comprises the tell-all memoirs of the Prince of England's spurned fiancée in a faintly ALTERNATE REALITY, SE has consistently told stories that are too outlandish to be considered mimetic. *The Making of Ashenden* (**1973** chap) tells of a man who finds himself in an erotic entanglement with a bear. *The Living End* (**1979**) is a black – or bleak – AFTERLIFE comedy; its cosmology ironically revises DANTE's. *George Mills* (**1982**) tells the story of a man whose forebears have for 1000 years been cursed (◊ CURSES) with haplessness. The protagonist of *The MacGuffin* (**1991**), a characteristically beleaguered man who adopts "the MacGuffin" as a private metaphor for his troubles, finds the MacGuffin, once named, occasionally speaking to him. [GF]

Other works: *The First George Mills* (**1980** chap); *The Magic Kingdom* (**1985**); *The Six-Year-Old Man* (**1987**).

ELLIOTT, JANICE (1931-1995) UK writer. Most of her novels – from *Cave with Echoes* (**1962**) on – lack any overt element of the fantastic, but almost invariably they give off an ambience of the inexplicable and the menacing. Her sole sf novel, *The Summer People* (**1980**), places a cast in a holiday resort, where they remain as the polar icecaps begin to melt. Her SUPERNATURAL FICTION includes: *Magic* (**1983**), in which some characters can leave their bodies and commune with the Virgin Mary (◊ GODDESS); *Dr Gruber's Daughter* (**1986**), which incorporates hints of SHAPESHIFTING; *The Sadness of Witches* (**1987**), whose WITCHES are presented with such subtlety that the points of congress between supernatural and mundane worlds are impossible to fix; and *City of Gates* (**1992**), set in a JERUSALEM guesthouse which works as a kind of PORTAL to a variety of epochs. The Arthurian **Sword and the Dream** YA sequence – *The King Awakes* (**1987**) and *The Empty Throne* (**1988**) – arouses the ONCE AND FUTURE KING into a grim post-HOLOCAUST UK. JE also wrote some tales for younger children, including *The Birthday Unicorn* (**1970** chap), illustrated by Michael FOREMAN. [JC]

Other works: *The Country of her Dreams* (**1982**), set within an ambience of the threat of nuclear war.

ELLIOTT, KATE Pseudonym of Alis A. RASMUSSEN.

ELLISON, HARLAN (JAY) (1934-) US writer and editor, known mostly for his sf. Some significant collections and stories are, however, clearly fantasy.

In *Strange Wine* (coll **1978**), "Croatoan" (**1975**) tells of a lost colony and the need for absolute personal responsibility (◊◊ URBAN LEGENDS). *Shatterday* (coll **1980**) contains

"Jeffty is Five" (1977), in which a child never ages, a story shot through with a wrenching sense of romance and innocence (◊ CHILDREN; IMMORTALITY). "Shatterday" (1975) itself, a harrowing piece about compassion and identity, was later adapted for *The* NEW TWILIGHT ZONE, for which HE served as creative consultant.

The Essential Ellison (coll **1987**) contains many of the stories which made HE's reputation. "One Life, Furnished in Early Poverty" (1970), in which a man encounters his younger self, reveals a necessary harshness. The book won a 1988 BRAM STOKER AWARD. *Angry Candy* (coll **1988**) is the most recent of HE's full-length collections. Of particular note are "Paladin of the Lost Hour" (1985 *TZ*), featuring a timepiece counting towards the END OF THE WORLD, "Laugh Track" (1984 *WT*), a TECHNOFANTASY in which a GHOST is heard amid canned sitcom laughter, and "The Avenger of Death" (1988 *Omni*), whose protagonist sets out to kill DEATH.

Of HE's works in the 1990s, two are of significance. *Mefisto in Onyx* (**1993**) is a short novel about a man able to "jaunt" into other people's mental landscapes. *Mind Fields: The Art of Jaček Yerka/The Fiction of Harlan Ellison* (graph coll **1994**), comprises 33 stories, each inspired by a painting – or, in one case, paintings – by Jaček Yerka, with the corresponding art reprinted opposite the text. The stories are extremely short, but display craftsmanship. Additionally, the 1990s has seen the release of a number of issues of *Harlan Ellison's Dream Corridor*, a COMIC which contains graphic adaptations of HE's work and features original fiction.

Because of HE's promotional nature, it has become fashionable to keep track of those titles announced but not yet seen. The most famous of these is the anthology «The Last Dangerous Visions». Others include «Ellison Under Glass», a collection of stories written in public, and «Slippage», listed among HE's publications as early as 1991. The last title seems the most likely to see print. [WKS]

ELVES The word "elf" comes from the Saxon *ælf*, derived from the Nordic "Alfar", one of several groupings of DWARFS. However, in Celtic myth elves are more closely related to the world of FAERIE, which makes them creatures of light and air, whereas dwarfs are creatures of darkness and earth. In early FOLKTALES there is considerable overlap between the two "species", and the mischievous attributes of dwarfs may be equally possessed by elves – especially brownies and pixies. GOBLINS and leprechauns are more dwarflike beings, but have many of the attributes of faerie folk. Elves, certainly as depicted by J.R.R. TOLKIEN but also as portrayed in some early FAIRYTALES, tend to be more graceful than dwarfs, are seemingly ageless, and though mischievous are not warlike. The elf race splits between good and bad, just like the faerie realm. In Nordic myth the good elves live in Alfheim while the black (bad) elves live in Svartheim. The Svarts are shown as dwarfs or goblins; this manifestation was used by Alan GARNER in *The Weirdstone of Brisingamen* (**1960**) and *The Moon of Gomrath* (**1963**).

In *Die Elfen* (**1811**) Ludwig TIECK depicted an Elfland indistinguishable from Faerie and with the same time displacement (◊ TIME IN FAERIE). Throughout the Victorian period, elves and fairies are interchangeable. A subset of fictional elves – brownies – were considered to appeal particularly to children. Brownies derive from Scottish FOLKLORE, where they are depicted as helpful faerie folk who attach themselves to a household and assist in running it; if they are offended, though, their mischievous side surfaces

and they become hobgoblins (◊ GOBLINS). They are portrayed in this way as elves by the GRIMM BROTHERS in "The Elves and the Shoemaker" (1812); James HOGG described them in "The Brownie of Black Haggs" (1828); memories of these tales inspired Juliana Ewing (1841-1885), who depicted them in "The Brownies" (1865 *Monthly Packet*). This suggested the idea of calling helpful children "brownies", which is how the name was eventually adopted into the Girl Guides in 1919. Brownies were also popularized in the USA by Palmer Cox (1840-1924) with his illustrated brownie poems in *St Nicholas Magazine*, which later appeared in the first of several books, *The Brownies: Their Book* (coll **1887**). The popularity of these books meant that the brownie was firmly entrapped in the realm of CHILDREN'S FANTASY.

It was not until the 20th century that authors sought to establish elves as a distinct part of Faerie, especially as it gave an opportunity to use faerie folk but to distinguish the stories from FAIRYTALES and brownie tales. Elves thus became acceptable adult "packaging" for fairies, and in that sense brought certain adult attributes with them. 20th-century elves ceased to be playful and mischievous: they became secret guardians of Faerie, aristocratic and full of the wisdom of the ancient world. Lord DUNSANY took readers some way there in *The King of Elfland's Daughter* (**1924**), and Poul ANDERSON went much further in *The Broken Sword* (**1954**), but the transformation was completed by J.R.R. TOLKIEN, particularly in *The Lord of the Rings* (**1954-5**). Tolkien's image of elves now dominates most GENRE FANTASY. Typical examples are: the parallel elfin world in *Kingdoms of Elfin* (coll **1977**) by Sylvia Townsend WARNER; the haughty Sidhe in Greg BEAR's **Songs of Earth and Power** sequence; the **Alfar** novels by Elizabeth BOYER; the **Elvish** sequence by Nancy BERBERICK; the **Arafel** tales by C.J. CHERRYH, about the last surviving elf; *Lords and Ladies* by Terry PRATCHETT; and the ELFQUEST stories by Wendy and Richard PINI.

Other authors have portrayed WAINSCOTS of elves surviving in our world; Jane CURRY's **Beneath the Hill** sequence is a finely structured example. In his creation of the **Borribles** in *The Borribles* (**1976**) and its sequels, Michael DE LARRABEITI returned to the original mischievous image of elves.

Different aspects of elves and their kin are covered in the anthologies *Little People!* (anth **1991**) ed Jack DANN and Gardner Dozois (1947-   ) and *Smart Dragons, Foolish Elves* (anth **1991**) ed Alan Dean FOSTER and Martin H. GREENBERG. [MA]

ELVIRA, MISTRESS OF THE DARK US movie (*1988*). New World/NBC/Queen "B" Productions. **Pr** Eric Gardner, Mark Pierson. **Exec pr** Michael Rachmil. **Dir** James Signorelli. **Spfx** Dennis Dion, Rich Ratliff. **Screenplay** Sam Egan, John Paragon, Cassandra Peterson (aka Elvira). **Starring** Elvira (Elvira), Daniel Greene (Bob), Susan Kellermann (Patty), Edie McClurg (Chastity Pariah), W. Morgan Sheppard (Vincent Talbot). 96 mins. Colour.

Neo-Goth tv presenter of late-night B-movies Elvira inherits her aunt's house in Fallwell, New England. Arriving, she finds herself reviled by the adults in the puritanical small town, but loved by its adolescents and lusty swain Bob. She discovers, too, that her aunt was a powerful white WITCH: Elvira has inherited also her powers and her grimoire. Necromantic Uncle Vincent yearns for the BOOK, and watches happily as his niece faces burning for witchcraft. But she is saved in the nick of time to receive the Las Vegas debut of her lifelong dreams . . .

EMOTD is a romp, often pleasurably tasteless, and a fine, knowing PARODY of the HORROR genre, notably of the alienation of its settings and circumstances from anything remotely resembling real life (except, perhaps, religious fundamentalism). [JG]

EMBLETON, RON(ALD SYDNEY) (1930-1988) Prolific UK artist and illustrator of COMIC strips, FAIRYTALES and historical and mythical subjects, with a meticulous, smooth, tightly controlled style. He approached Apex Publishing (later Scion Ltd) in 1947, where he began producing a wide range of strips for such titles as *Big Hit Comic*, *Big Noise Comic* and *Big Pirate Comic*, signing his work "by Ron" or "Byron". Over the next 10 years he produced many strips for all the major UK publishers.

His most remarkable early achievement was **Wulf the Briton** (*Express Weekly* 1957-60), a story set in Roman-occupied Britain, with mythical undertones. This was followed by **Wrath of the Gods** (1963), written by Michael MOORCOCK, a story set in a LAND-OF-FABLE Ancient Greece. It was now that RE began to develop his sleek painting style, unique for comic strips of that time. His interest in military history led him to work on educational publications, painting meticulously researched historical pictures as well as comic-strip versions of the Norse myths, **Legends of the Rhineland**, **Tristan and Isolde**, **Tannhauser** and others (*Look and Learn* 1972-4), plus a single story in the long-running sf fantasy *The Trigan Empire* (*Look and Learn* 1969). He drew a number of sf strips based on characters created by Gerry Anderson.

RE provided charming ILLUSTRATIONS for many children's books. His historical illustrations were published in various collections for both children and adults, of which *Hadrian's Wall in the Days of the Romans* (graph **1984**) is a prime example.

RE's adult work included a long running sexy-satirical saga in *Penthouse* called **Oh Wicked Wanda** (intermittently 1973-8; graph coll **1976**) written by Frederic Mullally (1920-), which was followed by **Sweet Chastity** (1981-8) written by Bob Guccione, also for *Penthouse*. [RT/SH]

EMERALD CITY ◊ OZ; *The* WIZARD OF OZ.

EMERSON, JANE Pseudonym of Doris EGAN.

EMERSON, RU (1944-) US writer, who also writes as Roberta Cray. She began publishing fantasy with her first novel, *The Princess of Flames* (**1986**), which like most of her work combines elements of the family romance and of DYNASTIC FANTASY. In this tale (as in several of her other novels), young members of a royal family are deposed or otherwise forced into exile, where they accumulate COMPANIONS and/or PLOT COUPONS in a FANTASYLAND and return in triumph. Here (as later) the influence of William SHAKESPEARE can be felt; the title refers to the use of the TAROT throughout.

RE's first sequence, the **Tales of Nedao** series – *To the Haunted Mountains* (**1987**), *In the Caves of Exile* (**1988**) and *On the Seas of Destiny* (**1989**) – again features a princess in exile who soon becomes proficient with weapons, finds a POLDER where she and her folk can become strong again and leads her people to eventual triumph, but this time the story is narrated by a CAT. The **Night-Threads** series – *The Calling of the Three* (**1990**), *The Two in Hiding* (**1991**), *One Land, One Duke* (**1992**), *The Craft of Light* (**1993**) and *The Art of the Sword* (**1994**) – marketed for a YA audience, introduces three young people from Earth via a PORTAL into a fantasyland whose rightful rulers have, once again, been deposed. "Night-Threads" is a kind of usable MAGIC. Various characters find they have TALENTS. All set about to restore the LAND to its proper status.

Of RE's singletons, two are of interest. *Spell Bound* (**1990**) is set in a LAND-OF-FABLE 16th-century Germany, in whose numerous principalities magic continues to work; in a complex plot, a WITCH avenges her mother's death through a potentially malign re-creation of the CINDERELLA story, but the victims who act out this TWICE-TOLD experience live happily ever after. *The Sword and the Lion* (**1993**) as by Roberta Cray is set in a historical-seeming deep past: it deals with the interaction of a GODDESS figure and the folk on whose side, in a war against a patriarchal conquering horde, she proves intermittently helpful. [JC]

Other works: *Beauty and the Beast: Masques* * (**1990**), tied to the tv series BEAUTY AND THE BEAST; *The Bard's Tale: Fortress of Frost and Fire* * (**1993**) with Mercedes LACKEY, tied to the SHARED-WORLD **Bard's Tale** sequence; *Midwife's Nightcap* (**1994** chap dos).

EMERY, CLAYTON (1953-) US writer whose *Tales of Robin Hood* (**1988**) is set in a Britain rooted in MAGIC, and features a ROBIN HOOD intimate with those roots. The **Rune Sword** sequence, begun with *Outcasts* * (**1990**), is routine; but CE's contributions to the **Magic: The Gathering** sequence – *Whispering Woods* * (**1994**), *Shattered Chains* * (**1995**), *The Final Gathering* * (**1995**) and «*Final Sacrifice*» * (**1996**) – are competent novelizations of the Wizards of the Coast card GAME. [JC]

EMETT, ROWLAND (1906-1990) UK artist and inventor. A fine cartoonist, he was also a draughtsman and engineer. He became known for his succession of large, incredibly intricate "Gothic-Kinetic" inventions. Unlike William Heath ROBINSON, who merely drew his eccentric contraptions, RE regularly created three-dimensional working models.

The amazing success of his Far Tottering and Oyster Creek Railway at the Festival of Britain in 1951 led to many more commissions, including permanent constructions like "The Rhythmical Time Fountain" at Nottingham, UK, and models built for CHITTY CHITTY BANG BANG (*1968*). His "Honeywell-Emett Forget-Me-Not" resides at the Ontario Science Museum in Toronto.

RE's imaginative fantasy about a china pig and a railway engine, *Anthony and Antimacassar* (**1943**), with text by his wife Mary Emett, was followed by a series of books collecting his bizarre drawings of railways (and on related themes) from *Punch*: *Engines, Aunties & Others* (**1943**), *Sidings, and Suchlike* (**1946**), *Home Rails Preferred* (**1947**), *Saturday Slow* (**1948**), *Buffers End* (**1949**), *Far Twittering* (**1949**), *High Tea* (**1950**) and *The Forgotten Tramcar* (**1952**). A US omnibus was *Emett's Domain – Trains, Trams and Englishmen* (**1953**). RE's most popular engine became the subject of a "special" story he wrote and illustrated, *Nellie Come Home* (**1952**; vt *New World for Nellie* US). [RD]

Other works: *The Early Morning Milk Train* (**1976**); *Alarms and Excursions* (**1977**); *Emett's Ministry of Transport* (**1981**).

EMSHWILLER, CAROL (FRIES) (1921-) US writer who began to publish work of genre interest with "This Thing Called Love" for *Future* in 1955. Her work can be divided into three periods: (a) her literary apprenticeship, during which she published primarily sf; (b) from 1960 to perhaps 1985, during which she developed her reputation for witty and highly inventive FABULATIONS, which often partook of sufficient fantastic coloration to appear in such genre venues as Damon Knight's **Orbit** and Harlan ELLISON's *Dangerous Visions* (anth **1967**); and (c) her later work, most containing much sharper (though not more conventional) fantasy elements while developing stronger – and longer – narratives than before. Such distinctions, however, belie the unity of her mature work, which is marked by a dogged but sunny FEMINISM, considerable verbal playfulness, and – what marks her work as fantastic more essentially than the occasional offworld or supernatural elements – a fluid sense of identity, in which the narrative voice may dissolve and reform between the first persons singular, plural and archetypal.

The titles of CE's most characteristic stories – e.g., "Maybe Another Long March Across China 80,000 Strong", "Biography of an Uncircumcised Man (Including Interview)" and "Expecting Sunshine and Getting It" – suggest both her playfulness and her longtime indifference to conventional narrative. Her two collections of the 1990s, *Verging on the Pertinent* (coll **1989**) and *The Start of the End of It All* (coll **1990** UK; vt with 4 stories cut and 4 added *The Start of the End of It All and Other Stories* 1991 US) are nominally divided into nongenre and genre work respectively, but the distinction is largely arbitrary.

Carmen Dog (**1988** UK), a feminist fantasy, draws on both OVID and David GARNETT in its witty depiction of women turning into animals and animals into women, with bemusing result to the male establishment. *Venus Rising* (**1992** chap), avowedly based on the anthropological speculations of Elaine Morgan, dramatizes recent hypotheses concerning the development of *Homo sapiens* (◊ APES). *Ledoyt* (**1995**), set in the US West in the first decade of this century, is not fantasy. CE's most recent short fiction, also set in the US West, suggests she has, at least for the moment, turned away from fantasy.   [GF]

ENCHANTMENT The influencing, through SPELLS, GLAMOUR or other MAGIC, of a person's behaviour or, especially, PERCEPTION. As with MESMERISM, the effects are essentially subjective. The victim believes that an ugly woman is a beautiful princess (or *vice versa*, as in *The Wedding of Sir Gawen and Dame Ragnell* [?**1450**]); or that a hovel is a palace, as in Jack VANCE's *The Eyes of the Overworld* (coll **1966**); or that he or she has been turned into a frog, as in Terry PRATCHETT's *Witches Abroad* (**1991**) – without any actual TRANSFORMATION occurring.   [CB/DRL]

ENCHANTRESS Although there is no universally agreed hierarchy of MAGIC users in fantasy, an enchantress or sorceress tends to differ from a WITCH in being considerably more powerful, socially elevated, physically attractive and well dressed – though the last two attributes may be only magical GLAMOUR. If EVIL, she may heartlessly manipulate men through desire and SEX as well as ENCHANTMENT. Examples include CIRCE, MORGAN LE FAY, the LADY OF THE LAKE, Achren in Lloyd ALEXANDER's **Prydain** sequence (and perhaps also her distant MABINOGION original, Arianrhod), and the sorceress Iris – mistress of ILLUSION – in Piers ANTHONY's *A Spell for Chameleon* (**1977**).   [DRL]
See also: LAMIA.

ENDE, MICHAEL (ANDREAS HELMUTH) (1929-1995) German theatrical director and writer, son of the Surrealist painter Edgar Ende (1901-1965), by whom he was deeply influenced. After some years writing songs and sketches for literary cabarets, he began to publish work of interest with the **Jim Knopf** CHILDREN'S FANTASY sequence – *Jim Knopf und Lukas der Lokomotivführer* (**1960**; trans

Renata Symonds as *Jim Button and Luke the Engine Driver* 1963 US; new trans Anthea Bell 1990 US) and *Jim Knopf und die Wilde 13* ["Jim Knopf and the Wild 13"] (**1962**). Jim, a small black child, arrives by post in the tiny ISLAND of Morrowland, which has no room for him; so he and Luke and the locomotive Emma go travelling. They visit a LAND-OF-FABLE China, free a maiden from a DRAGON, and return home bearing an annexe large enough to hold everyone. In the sequel, Jim searches for his origins and defeats various enemies, whom he converts into useful citizens; the end of tale – with a GIANT working as a lighthouse, among other TRANSFORMATIONS – combines TECHNOFANTASY and PASTORAL. Indeed, throughout his career, ME powerfully advocated a utopian search for "multidimensional" solutions to problems in the world, as well as generating a series of convincing happy endings in his fiction. His thoughts on how to rework both world and fiction appear in *Phantasie/Kultur/Politik: Protokolleines Gresprächs* ["Fantasy/Culture/Politics: The Protocol of a Conversation"] (**1982**).

ME's next novel, *Momo* (**1973**; trans Frances Lobb as *The Grey Gentlemen* 1974 UK; new trans J. Maxwell Brownjohn as *Momo* 1985 US), eloquently addresses these issues. The eponymous girl lives in an amphitheatre in an unnamed CITY threatened by the Grey Gentlemen, the "time thieves" who transform humans into "adults" by persuading them to quantify the timing of their lives. After everyone but Momo has been transformed into an alienated urban monster, she undertakes a QUEST to get TIME back; on her successful return there is a general HEALING.

Die unendliche Geschichte (**1979**; trans Ralph Manhein as *The Neverending Story* 1983 US) remains ME's most successful attempt to mix ALLEGORY and fantasy – in this instance a compelling flow of STORY carries children through the long narrative, while at the same time the author advances fully adult arguments about personal growth, the good society, and the nature of story itself. Young Bastian Balthasar Bux steals a BOOK called *The Neverending Story*, in which he begins to read what turns out to be both his own Story as well as a PORTAL into a SECONDARY WORLD, the threatened land of Fantastica. As soon as he opens the book into the tale, a whiff of WRONGNESS assails him: four messengers from different races are hurrying to the Ivory Tower (◊ EDIFICE) to warn the Empress that a terrible "Nothing" is eating the LAND away (the THINNING depicted here is frighteningly literal); but the Empress, who does not govern but "is the centre of all life" in Fantastica, is herself wasting away, and unless a HERO undertakes a QUEST to cure her she and the land will die. In the end, Bastian himself turns out to be the one human child capable of giving her what is necessary, a new NAME, which he bestows upon her halfway through the book. He calls her Moon Child, a name which evokes the GODDESS, and the two set about creating the world anew from Nothing. The story then darkens as – despite his initial disbelief – Bastian passes INTO THE WOODS of Perilin, the Night Forest (which he has himself created), and becomes intrinsic to the land. Only when Bastian completes his RITE OF PASSAGE into genuine adulthood will the Story be properly told, the knot untied (in a scene at the heart of Fantastica where he must enter within the WORM OUROBOROS and come to the Waters of Life and recognize himself again), the dark MIRROR escaped, the relationship between fantasy and REALITY properly negotiated. The book was filmed, rather poorly, as *The* NEVERENDING STORY (*1984*), with two weak sequels.

Later novels move away from their ostensible grounding in CHILDREN'S FANTASY. *Der Spiegel im Spiegel: Ein Labyrinth* (coll of linked stories **1984**; trans J. Maxwell Brownjohn as *Mirror in the Mirror* **1986** US) is a kind of ARABIAN NIGHTMARE with hope at the end. Constructed as a series of FABLES which take their cue from a sequence of lithographs and etchings by Edgar Ende, the book moves from MINOTAUR to ICARUS, from JANUS-figure to ORPHEUS. The effect is like that generated by Franz KAFKA in some of his parables, or by Luis BUÑUEL's *Phantom of Liberty* (*1974*). In the end, though, as the host of ARCHETYPES increasingly mirror one another, there is a sense of integration. The YA *Der satanarchäolügenialkohöllische Wunschpunsch* (**1989**; trans Heike Schwarzbauer and Rick Takvorian as *The Night of Wishes, or The Satanarchaeolidealcohellish Notion Potion* **1992** US) is something of a romp, the eponymous POTION being used by a comic sorcerer to corrupt the world: two TALKING ANIMALS save the day.

ME's fantasies are ultimately about FANTASY. They are parables of integration, and they argue that the stories they tell are in themselves forms of guidance. His tales are also expressive and entertaining. He was a major figure of 20th-century fantasy. [JC]

Other works: *Das Gauklermärchen* ["The Juggler's Fairy Tale"] (**1982**), a fantasy play; *Der Goggolori* (**1984**), a libretto featuring the eponymous GOBLIN; *Lirum Larum* (**1995**).

END OF THE WORLD Most stories of the EOTW are reckoned SCIENCE FICTION; the main group of exceptions deal with the Christian APOCALYPSE. There remains, however, a significant subcategory of rather dreamlike accounts of the world's end (or cataclysmic interruption) – some satirical, others selfconsciously exotic or grotesque – which comfortably fit neither of these categories. Examples include BYRON's poem "Darkness" (1816), *The Night Land* (**1912**) by William Hope HODGSON, *The End of all Men* (**1922**) by C.F. RAMUZ, "The End of the World" (**1922**) by Dino Buzzati (1906-1972), "Up and Out" (1957) and "Cataclysm" (written 1960; 1985) by John Cowper POWYS, "Wednesday, Noon" (1968) by Ted White (1938-) and *God's Grace* (**1982**) by Bernard Malamud (1914-1986). The notion that Earth's senility might be attended by a return of the MAGIC eroded by THINNING provides an underlying logic for many fantasies of the DYING EARTH. [BS]

See also: ESCHATOLOGY.

ENDORE, (SAMUEL) GUY (1900-1970) US novelist, screenwriter and biographer, best-known for his historical novels, which often verged on the sensational. The only one to be overtly supernatural is *The Werewolf of Paris* (**1933**), which drew upon the historical case of Sergeant Bertrand, a notorious 19th-century murderer and cannibal. The book was one of the first complete studies of lycanthropy (◊ WEREWOLVES) in modern mainstream fiction. It was part of a sequence of books he wrote about unusual or obsessed historical characters, beginning with *Casanova* (**1930**) and including the novels *Babouk: The Story of a Slave* (**1935**), which contains some references to VOODOO, *Satan's Saint* (**1966**) about the Marquis DE SADE, and *King of Paris* (**1967**) about Alexandre DUMAS. *The Man from Limbo* (**1931**) and *Methinks the Lady* (**1945**; vt *The Furies In her Body* 1951; vt *Nightmare* 1957) are further studies of the dark side of the human psyche.

Much of GE's shorter fiction has not been collected. It includes several warning stories about the perils of science dabbling with the unknown. Most reprinted are "The Day of the Dragon" (1934), where a scientist reintroduces DINOSAURS, and "Men of Iron" (1940), where machines become sentient. GE also adapted *Les Mains d'Orlac* (**1920**) by Maurice Renard (1875-1940) for the screen as MAD LOVE (*1935*) and assisted in the adaptation of Abraham MERRITT's *Burn, Witch, Burn!* (**1933**) as *The* DEVIL DOLL (*1936*). He also translated *Alraune* (**1911**; trans **1929** US) by Hanns Heinz EWERS. [MA]

EPIC FANTASY An epic is a long narrative poem which tells large tales, often incorporating a mixture of LEGEND, MYTH and folk history, and featuring HEROES whose acts have a significance transcending their own individual happiness or woe. The classic epic tells the story of the founding or triumph of a folk or nation. Examples include GILGAMESH (*c*3000BC), HOMER's *Iliad* and *Odyssey* (both *c*1000BC), VIRGIL's *Aeneid* (*c*30-19BC) and BEOWULF (8th century). Later epics include many TAPROOT TEXTS, like ARIOSTO's *Orlando Furioso* (**1516**) and Edmund SPENSER's *The Fairie Queene* (**1589-96**). More recent examples of fantasy interest include *The Idylls of the King* (**1842-85**) by Alfred Lord Tennyson (1809-1892) (◊ ARTHUR), the KALEVALA (**1834**; exp 1849) by Elias Lonnrot (1802-1884) and *Hiawatha* (**1855**) by Henry Wadsworth Longfellow (1807-1882); and – if one admits prose fictions to the canon – Herman MELVILLE's *Moby-Dick* (**1851**) and James JOYCE's *Finnegans Wake* (**1939**). But EF in the 20th century – certainly if one insists the tale be told in verse – is uncommon. Examples include *The Dynasts* (**1903-08**) by Thomas Hardy (1840-1928), Nikos KAZANTZAKIS's *The Odyssey* (**1938**) and Martyn SKINNER's *Merlin* (**1951**) and *The Return of Arthur* (**1955**). Prose fictions which might be called EF include several of the central SECONDARY-WORLD tales central to the development of fantasy over the past 100 years – e.g., much of the work of Kenneth MORRIS, E.R. EDDISON, J.R.R. TOLKIEN and Stephen R. DONALDSON. Any fantasy tale written to a large scale which deals with the founding or definitive and lasting defence of a LAND may fairly be called an EF. Unfortunately, the term has been increasingly used by publishers to describe HEROIC FANTASIES that extend over several volumes, and has thus lost its usefulness. [JC]

EPIC ILLUSTRATED Bimonthly bedsheet-size 98-page, COMIC-strip and art magazine (1980-86), ed Archie Goodwin. Subtitled *The Marvel Magazine of Fantasy and Science-Fiction*, *EI* was created in response to the popularity of HEAVY METAL. Its aim was to offer US comics writers and artists a wider scope for their creations than was possible in MARVEL COMICS's regular line, and in this sense it was successful, being one factor that led Marvel to set up its **Epic** series of comic books. Nevertheless *EI*'s content was uneven, although many issues had covers by some of the best US comics and other artists, including Frank FRAZETTA, Neal ADAMS, Barry WINDSOR-SMITH, Jeffrey JONES, Mike KALUTA and John BOLTON.

The content was mostly fantasy, with some sf; frequently several pages were devoted to FANTASY-ART portfolios. Contributors of strips included Adams, Vaughn Bode (1941-1975), Howard CHAYKIN, Frank Brunner (1949-), Richard CORBEN, Kaluta, Bernie WRIGHTSON and Charles VESS. [RT]

EQUINOX, THE ◊ MAGAZINES.

ERASERHEAD US movie (*1976*). ◊ David LYNCH.

ERCKMANN-CHATRIAN Working name of collaborating French writers Emile Erckmann (1822-1899) and

Alexandre Chatrian (1826-1890). Their first book was
Contes Fantastique (**1847**). Their best-known work in translation is *Le Juif Polonais* (**1871**; trans as *The Polish Jew*),
describing the psychological decline of a murderer; in its
long-running stage version, *The Bells*, this became Henry
Irving's most celebrated role.

They wrote many supernatural and fantasy short stories
(including "L'Oeil Invisible", "L'Araignee Crabe" and "Le
Blanc et le Noir") and novellas including "La Maison
Forestière" and the WEREWOLF classic "Hugues le Loup".
These were much admired – in their original French texts –
by M.R. JAMES and other connoisseurs of French romantic
Gothic fiction.

Most of these stories were translated into English in var-
ious collections: *The Forest House* (coll **1871** UK), *Popular
Tales and Romances* (coll **1872** UK), *Confessions of a Clarinet
Player* (coll **1874** UK), *Stories of the Rhine* (coll **1875** UK),
The Man-Wolf(coll **1876** UK) and *The Wild Huntsman* (coll
1877 UK) and *Strange Stories* (coll **1880** US). The best
modern collection is *The Best Tales of Terror of Erckmann-
Chatrian* (**1981** UK) ed Hugh Lamb.

After over 40 years of collaboration, the two men quarrelled violently in 1889 and, following an acrimonious
lawsuit, became bitter enemies.   [RD]

ERICKSON, STEVE Working name of Stephen Michael
Erickson (1950-   ), US journalist and novelist. He won
the Samuel Goldwyn award for fiction while a student at the
University of California, Los Angeles, in 1972. His fiction
makes use of typical Postmodernist tropes and techniques,
which often acquire a fantastic allure. Barriers between
worlds are broken down, TWINS and HAUNTING devices proliferate, the very shape of the world is corrupted, and the
distinction between fiction and reality becomes hazy. Most
of his fiction takes the form of a moral quest to discover the
soul of either the USA or the 20th century.

In *Days Between Stations* (**1985**), which takes the entire
20th century as its stage, a dysfunctional couple pursue an
affair in a distorted landscape in which Los Angeles is
flooded by sand and the canals dry up in Venice. Embedded
within this is the story of a leading moviemaker of the early
years of the century. The two stories are linked by a pair of
blue eyes in a bottle. There is a similar topographical distortion in SE's next novel, *Rubicon Beach* (**1986**), in which
three stories seem to be set in different REALITIES but are
nevertheless linked: a released prisoner in a changed Los
Angeles plays out his sense of guilt with visions of a girl cutting off the head of a man he only belatedly comes to
recognize as himself; a girl with near-magical powers travels from the remote Amazon to contemporary Hollywood,
where she embroils a failed scriptwriter in an insoluble
moral quandary; the prisoner and the girl come together in
the final tale of a man who has discovered a new whole
number between 9 and 10.

SE's themes find their best statement in *Tours of the Black
Clock* (**1989**), in which his characters QUEST for the blueprint
of the 20th century, on which is revealed a secret room
where the soul of the century is hidden. At the centre of the
story is Jainlight, an American who becomes Hitler's private
pornographer and in the sexual fantasies he spins creates an
ALTERNATE WORLD: Hitler did not embark on his disastrous
invasion of Russia, so that by the 1960s WORLD WAR II is still
being fought. Jainlight sets out to smuggle Hitler into the
USA and in so doing discovers the girl from a different
branch of the timestream who inspired his pornography.

This notion – that there are separate realities whose
boundaries can be pierced in the course of a moral quest –
also informs SE's one book-length work of nonfiction and
the novel that sprang from it. In 1988, SE followed the
course of the US presidential election from the primaries
onwards. *Leap Year: A Political Journey* (**1989**) was, however, far from a traditional work of journalism. His journey
is haunted by the GHOST of Sally Hemings, the slave who
was also the lover of Thomas Jefferson; in SE's version of
events, her decision to remain with Jefferson when he
returned from Paris, accepting slavery over freedom, was the
moral turning point which defines the US soul, for Jefferson
did not then abolish slavery as he might have done.

This theme is sustained in *Arc d'X* (**1993**), which again
takes Sally Hemings as a central character, here in a variety
of guises across different realities which reflect the moral
taint that slavery has spread across US history. The fault
lines spreading from this central moral failure have created
a number of ALTERNATE REALITIES, most notably a severe
theocracy where liberties are, literally, hidden away UNDER-
GROUND and one main character has the significant,
Orwellian task of rewriting HISTORY.   [PK]

ERICSON, ERIC (1925-   ) UK author. ◊ INCUBUS.

ERIK THE VIKING UK movie (*1989*). John Goldstone/
Prominent Features/Erik the Viking Productions. **Pr** John
Goldstone. **Exec pr** Terry Glinwood. **Dir** Terry JONES.
Spfx Richard Conway. **Screenplay** Jones. **Starring** Gary
Cady (Keitel Blacksmith), John Cleese (Halfdan the Black),
Simon Evans (Odin), Freddie Jones (Harald the
Missionary), Terry Jones (King Arnulf), Eartha Kitt
(Freya), Tim McInnerny (Sven the Berserk), Richard
Ridings (Thorfinn Skullsplitter), Tim Robbins (Erik),
Danny Schiller (Snorri the Miserable), Antony Sher (Loki),
John Gordon Sinclair (Ivar the Boneless), Imogen Stubbs
(Aud). 108 mins. Colour.

An urbane Viking, Erik, depressed by slaughter, is told by
Freya the Seer that the wolf Fenrir has swallowed the Sun,
that the world is in FIMBULWINTER, and that the GODS are
asleep, but that he can wake them if he QUESTS across the sea
to High Brasil for the Horn Resounding, which he must
blow three times: first to be taken to Asgard, second to wake
the gods, and third to be brought home. The expedition he
raises through the Gates of the World, battles the
Jormungand and attains High Brasil, which soon sinks
beneath the waves. At the first blowing of the Horn the
ship topples over the edge of the world and floats down to
Asgard; the second blowing gathers the stars from the skies
to form Bifrost. But in VALHALLA the gods prove to be (literally) malicious children: they banish Erik and his
COMPANIONS to Hel (◊ HELL) – which turns out to be their
own village of Ravensfjord.

ETV mixes **Monty Python** HUMOUR with much from
Norse MYTHOLOGY, not always easily; even so, its imaginative scope is impressive, as is its erudition. Particularly
interesting are the interplays between REALITIES, as when
Harald the Missionary, being a Christian, is unable to perceive (◊ PERCEPTION) Valhalla, walking unknowing through
walls that are solid rock to the Vikings.   [JG]

ERNST, MAX (1891-1976) German artist and critic, resident in Paris for most of his career, active and influential
from about 1910. During and after WORLD WAR I, which
had a profound effect upon him, he was involved in the
Dada movement, and was influential in the artistic ferment
that finally resulted, c1924, in full-blown SURREALISM; he

contributed to the first Surrealist exhibition in 1925, and illustrated a number of texts by Surrealist writers like Paul Éluard (1895-1952) and André Breton (1896-1966). As fully as René MAGRITTE, he incorporated a narrative syntax into his visual work: an urgent "and then . . ." haunts his depictions of the LANDSCAPES of DREAM. It is probably for this reason that his work has been used so effectively for pictorial covers and illustrations for tales of the fantastic, especially by novelists intimate with surrealism like J.G. BALLARD, whose *The Crystal World* (**1966**) has an Ernst cover for its UK edition, or Alan Burns (1929-   ), whose *Europe After the Rain* (**1967**) is illustrated by, and named after, an Ernst painting. In paintings like this, ANACHRONISM and TIME ABYSS meet in a landscape which seems to incorporate – and to be about to move into – any tense of storytelling; they seem to enfold past, present and future in one sight. But they are not UNCANNY: however estranged ME's worlds may seem, they are fully embodied. ME is important to fantasy in particular because his landscapes, which are impossible, seem real. Stories can wind through them.

ME's GRAPHIC NOVELS include: *La femme 100 têtes* ["The 100-Headless Woman"] (graph **1929**), a novel in 147 pictures, with captions; *Rêve d'une petite fille qui voulut entrer au Carmel* ["Dream of a Little Girl who Wanted to be a Carmelite Nun"] (graph **1930**), 79 pictures with captions; *Une semaine de bonté, ou Les sept éléments capitaus* ["A Week of Goodness, or The Seven Capital Elements"] (graph **1934**; rev 1963 Germany), 182 pictures with captions; and *Les chiens ont soif* ["The Dogs are Thirsty"] (graph **1964**), text by Jacques Prévert, 33 pictures. These works are funny, gravid with implications, sexy, and full of a mocking element of the *non sequitur*. Their style is imitated, modestly, by Jocelyn Brooke (1908-1966) in *The Crisis in Bulgaria; Or Ibsen to the Rescue!* (graph **1956**).

Many books have been written about ME. *Max Ernst* (**1984**) by Edward Quinn, copiously illustrated, was begun in collaboration with the artist.   [JC]

ERSKINE, BARBARA (1944-   ) UK writer, most of whose fantasy involves contemporary women in TIMESLIPS to earlier periods. *Lady of Hay* (**1986**) accomplishes this through MESMERISM. In *Kingdom of Shadows* (**1988**), rather more interestingly, two women (one the mistress of Robert the Bruce) find themselves interchanging lives and passions. Some of the 44 stories in *Encounters* (coll **1990**) are fantasy. *Child of the Phoenix* (**1992**), set in 13th-century Wales, features a girl whose TALENTS bring her into danger. *Midnight is a Lonely Place* (**1994**) is a GHOST STORY.   [JC]

ERSKINE, JOHN (1879-1951) US writer. His early essay, "Magic and Wonder in Literature", is reprinted in *The Moral Obligation to be Intelligent and Other Essays* (coll **1915**). Many of his works recycle well known MYTHS and LEGENDS (their supernatural components eliminated or drastically reduced) such that the actions of their protagonists are satirically related to contemporary social etiquette. *The Private Life of Helen of Troy* (**1925**) describes Helen's readjustment to family life following the fall of Troy; the story is told almost entirely in dialogues in which Helen runs rings around her critics. *Galahad, Enough of His Life to Explain his Reputation* (**1926**) reveals the perfect GALAHAD to be a perfect prig. *Adam and Eve, Though He Knew Better* (**1927**) is really the story of LILITH and Eve (◊ ADAM AND EVE); the weak-willed Adam eventually forsakes the free-spirited Lilith for the nagging emotional blackmailer Eve.

Penelope's Man: The Homing Instinct (**1927**), perhaps the most telling of JE's SATIRES, shows a hypocritical ODYSSEUS disappointing a whole series of philosophical coquettes. *Uncle Sam, in the Eyes of his Family* (**1930**) is concerned with myths of a more modern stripe. The stories in *Cinderella's Daughter and Other Sequels and Consequences* (coll **1930**) rework the motifs of classic FAIRYTALES to comment on various social issues; the sharpest – "The Patience of Griselda", "Sleeping Beauty" and "Beauty and the Beast" – deal with sexual politics, as does the most fantastic of JE's novels, *Venus, the Lonely Goddess* (**1949**), which describes rather melancholily the hapless goddess's attempts to figure out exactly what she stands for.   [BS]
Other works: *Solomon, My Son!* (**1935**).

ESCAPE OF MEGAGODZILLA vt of *Mekagojira No Gyakushu* (*1975*). ◊ GODZILLA MOVIES.
ESCAPE TO WITCH MOUNTAIN US movie (*1974*). ◊ DISNEY; UFOS.
ESCAPISM A derogatory term much used in descriptions of genre literatures in general and fantasy in particular. It is a term, however, which more accurately describes the motives of the reader than the nature of what is read. The works of Jane Austen (1775-1817) are notoriously read for escapist motives. Within the structure of fantasy, the term Escape has a more particular meaning – not "the flight of the deserter, but the escape of the prisoner", as J.R.R. Tolkien puts it.   [JC]
ESCHATOLOGY The aspect of MYTHOLOGY pertaining to "last things", including descriptions of the END OF THE WORLD and accounts of the fate of the SOUL after death. The Christian APOCALYPSE and the Norse *Götterdämmerung* (◊ NORDIC FANTASY) offer the most prolific world-end motifs for modern fantasy. Personal eschatology is much more commonly featured, in mock-Dantean CHRISTIAN FANTASIES (◊ HEAVEN; HELL), like those by Harry BLAMIRES and R.H. Mottram (1883-1971), and in a host of POSTHUMOUS FANTASIES. Many of the latter – such as those by Mrs OLIPHANT and Elizabeth Stuart PHELPS – are associated with SPIRITUALISM, but the best are more adventurous. The elaborate personal eschatologies of the Greek and Egyptian mythologies are also significant contributors of imagery to modern fantasy, the former often in connection with Orphean journeys into the UNDERWORLD (◊ ORPHEUS) and the latter in connection with MUMMIES. An interesting commentary on eschatological images of the future and their social significance can be found in *The Image of the Future* (**1973**) by Fred Polak.   [BS]
ESHBACH, LLOYD ARTHUR (1910-   ) US editor and writer who has been a central figure in sf publishing for over half a century, and most of whose early fiction – including his first, "The Man with the Silver Disc" for *Scientific Detective* in 1930 – was sf. Some of his early work was, however, fantasy, including *The Elfin Lights* (**1934** chap); this material tended to reflect the influence of major writers of fantasy like A. MERRITT, and is not now much read, though some examples were included in *Tyrants of Time* (coll **1955**). His major fantasies came in the 1980s, when he returned to writing with the **Gates of Lucifer** sequence: *The Land Beyond the Gate* (**1984**), *The Armlet of the Gods* (**1986**), *The Sorceress of Scath* (**1988**) and *The Scroll of Lucifer* (**1990**). A PORTAL sends the series' protagonist to four different UNDERWORLDS – the first and second amalgamate CELTIC FANTASY and NORDIC FANTASY, with Celts and Vikings and others tangling together; the third is

vaguely Bablyonian; the fourth faces the protagonist with LUCIFER, who wants to reconquer the world. [JC]

Further reading: *Over my Shoulder: Reflections on a Science Fiction Era* (**1983**) by LAE.

ESCHER, M(AURITS) C(ORNELIS) (1898-1971) Dutch graphic artist, famous for his distinctive, usually monochrome, engravings and lithographs. Fantastic elements appear in early prints like "Dream" (1935), featuring a gigantic locust (◊ GREAT AND SMALL); but it was MCE's lifelong experimentation with quasi-mathematical patterns and extremes of perspective that frequently compelled fantasy in his work. Thus "Circle Limit IV (Heaven and Hell)" (1960) combines the ideas of tiling the plane with repeated figures and representing infinity in a finite area, giving a circular universe entirely filled with receding, interlocking ANGELS and DEMONS. IMAGINARY ANIMALS emerge from flat tessellation into 3-D reality and sink back again (◊ CYCLE) in "Magic Mirror" (1946) – except that, in a tension with which MCE often teasingly plays, the "reality" is of course also in the 2-D picture plane (◊◊ TROMPE-L'OEIL); this work, in that it forms a progression from left to right, exemplifies the notion that almost all central FANTASY ART is a narrative form. Several MCE works actualize the impossible geometries described by H.P. LOVECRAFT and others as a signal of WRONGNESS: there are eye-hurting contradictions of architectural perspective in "Convex and Concave" (1955), "Belvedere" (1958) and "Waterfall" (1961) – whose aqueduct runs visibly downhill to a point two storeys above its beginning – and the more subtly disconcerting Penrose stairs in "Ascending and Descending" (1960), where cowled figures plod around a quadrilateral loop of staircase going forever up, or down. *The Graphic Work of M.C. Escher* (graph coll **1961**; exp 1967; trans John E. Brigham **1967** UK) assembles 76 plates, each with brief comments by MCE. His art has appeared on postage stamps and as murals in Dutch public buildings; it is well known in fantasy, an early UK example being the use of "Stars" (1948) on the jacket of *Best Fantasy Stories* (anth **1962**) ed Brian W. ALDISS. MCE is one of the rare artists to have created his own universally recognizable iconography of wonder. [DRL]

Other works: *The Pop-Up Book of M.C. Escher* (graph coll **1991**).

Further reading: *The Magic Mirror of M.C. Escher* (**1976**; trans John E. Brigham 1976 US) by Bruno Ernst.

ESORCISTA N.2 Italian movie (*1975*). ◊ *The* EXORCIST (*1973*).

ESPÍRITU DE LA COLMENA, EL ot of *The* SPIRIT OF THE BEEHIVE (*1973*).

ESPRIT DE L'EXIL, L' ot of *Golem: The Wandering Soul* (*1992*). ◊ *The* GOLEM.

ESTATES SATIRE Any anatomy of the social world which describes it in terms of "estates" is an ES. The form flourished during the Middle Ages. As far as medieval Britain was concerned, society could be divided into three traditional estates: Lords Spiritual, Lords Temporal and the Commons, with women of all categories at the bottom. The Press was dubbed the Fourth Estate by Edmund Burke (1729-1797). At least in medieval times, the ES was written as a challenge not to the concept of a hierarchical society (◊ THEODICY), or with any intention of unravelling in a social sense the Great Chain of Being, but to reprobate individuals who failed to act according to the requirements laid down for them by their position within the estates structure; for example, the mode provides the controlling context for

the morality play *Ane Pleasant Satire of the Thrie Estaitis* (**1540**) by Sir David Lindsay (1486-1555), which specifically targets the clergy. [JC]

ESTES, ELEANOR (1906-1988) UK writer. ◊ CHILDREN'S FANTASY.

ESTES, ROSE (? -) US writer who began her career as an employee of TSR, a significant generator of GAMES and texts, for whom she created the **Endless Quest** sequence of multiple-choice gamebooks; her similar **Find Your Fate** sequence was done for a different firm. More original were two series of PREHISTORIC FANTASIES: **Saga of the Lost Lands** – *Blood of the Tiger* (**1987**), *Brother to the Lion* (**1988**) and *Spirit of the Hawk* (**1988**) – involves MAGIC in an Ice Age setting; in the **Hunter** sequence – *The Hunter* (**1990**), *The Hunter on Arena* (**1991**) and *The Hunter Victorious* (**1992**) – aliens capture an Ice Age hunter for use as a gladiator, but he eventually wins through. The CONTEMPORARY-FANTASY **Troll** sequence – *Troll-Taken* (**1993**) and *Troll-Quest* (**1995**) – places a society of TROLLS under Chicago, from which they are prone to kidnap children. [JC]

Other works: The **Endless Quest** gamebooks, being *Dungeon of Dread* * (**1982**), *Mountain of Mirrors* * (**1982**), *Pillars of Pentegarn* * (**1982**), *Return to Brookmere* * (**1982**), *Circus of Fear* * (**1983**), *Dragon of Doom* * (**1983**), *Hero of Washington Square* * (**1983**), *Revenge of the Rainbow Dragons* * (**1983**) and *Revolt of the Dwarves* * (**1983**); the **Find Your Fate** gamebooks, being *Indiana Jones and the Lost Treasure of Sheba* * (**1984**), which is also tied to the INDIANA JONES movies, *The Children of the Dragon* * (**1985**), *The Trail of Death* * (**1985**) and *The Mystery of the Turkish Tattoo* * (**1986**); *Case of the Dancing Dinosaur* (**1985**); the **Greyhawk** sequence of game ties, being *Master Wolf* * (**1987**), *The Price of Power* * (**1987**), *The Demon Hand* * (**1988**), *The Name of the Game* * (**1988**) and *The Eyes Have It* * (**1989**); *Skryling's Blade* * (**1990**) with Tom Wham and *The Stone of Time* * (**1992**), both contributions to the **Runesword** SHARED-WORLD sequence; *Elfwood* (**1992**); *Iron Dragons: Mountains and Madness* * (**1993**), game tie.

ETCHISON, DENNIS (1943-) US writer born and resident in Southern California; his fiction transcends the HORROR genre to explore the dark underside of the West Coast lifestyle. His first professional sale was an sf story, "Odd Boy Out" (1961), in *Escapade*. Since then he has been a multiple winner of both the WORLD FANTASY AWARD and the BRITISH FANTASY AWARD for writing and editing, and his short fiction has been widely published. Many of his finest stories – such as "The Dead Line" (1979), "Deathtracks" (1981), "The Dark Country" (1981), "The Woman in Black" (1984) and "The Olympic Runner" (1986) – are collected in *The Dark Country* (coll **1982**), *Red Dreams* (coll **1984**) and *The Blood Kiss* (coll **1988**). He novelized *The* FOG (*1979*) as *The Fog* * (**1979**) and, as Jack Martin (the name of a recurring character in his fiction), he wrote *Halloween II* * (**1981**), *Halloween III: Season of the Witch* * (**1982**) – based on HALLOWEEN III: SEASON OF THE WITCH (*1983*) – and *Videodrome* * (**1983**). His own novels, such as *Darkside* (**1986**), *Shadowman* (**1993**), *California Gothic* (**1995**) and «*Double Edge*» (1996), are notable for their sharp, sparse prose but have never quite captured the power of his short fiction. He has contributed (uncredited) to several movie scripts and was a staff writer for the tv series *The Hitch Hiker* (1985). His ZOMBIE story "The Late Shift" (**1980**) was adapted by Damian Harris for the short movie *Killing Time* (*1984*). [SJ]

As editor: *Cutting Edge* (anth **1986**); *Masters of Darkness* (anth **1986**); *Masters of Darkness II* (anth **1988**); *Lord John Ten* (anth **1988**); *The Complete Masters of Darkness* (anth **1991**); *Masters of Darkness III* (anth **1991**); *MetaHorror* (anth **1992**).

ETERNAL CHAMPION ◊ Michael MOORCOCK.

ET IN ARCADIA EGO The phrase translates as "And I too in Arcadia". Its source is unknown, but it first came into currency in 17th-century Italy, where it served as shorthand for a sentiment which could be rendered as "Even in Arcadia I, Death, can be found". In contemporaneous paintings set in ARCADIA the phrase is sometimes shown ornamentally inscribed on tombs set in sylvan glades, and can be understood as a straightforward *memento mori*. Soon, however – in particular through paintings by Nicholas Poussin (1594-1665) – the meaning began to shift, and the phrase started to refer not to Death but to the occupant of the tomb, who was thought fortunate. "Et in Arcadia ego" was by now a deliciously melancholy boast that can be translated as "I too lived in Arcadia". This meaning is today primary and has some relevance to fantasy, serving as a catch-phrase for a nostalgia for a GOLDEN AGE that has – perhaps irrecoverably – passed. Much sentimental fantasy exploits this nostalgia, though relatively few 20th-century texts refer specifically to Arcadia itself; but the impulse to cast one's longing gaze backwards to a place of safety and tranquillity and (it may be) eternal life remains powerful. Because "Et in Arcadia ego" points back in time, and focuses upon an imaginable (though perhaps entirely imaginary) place, its meaning is almost precisely the opposite of SEHNSUCHT. [JC]

EUCATASTROPHE Term coined by J.R.R. TOLKIEN in his essay "On Fairy-Stories" (1947) as an opposite of "Tragedy" to argue that the uplifting effect of FAIRYTALES – and thus of FANTASY in general – is the highest of its three functions (Recovery, Escape and Consolation). It refers to the final "turn" of a plot which gives rise to "a piercing glimpse of joy, and heart's desire, that for a moment passes outside the frame, rends indeed the very web of story". Insofar as the essay remains central to theoretical accounts of fantasy, the term has become an important element of the genre's critical discourse. [BS]

EUHEMERISM In his *Sacred History*, Euhemerus (*fl*300BC), describes an ISLAND, Panchaea, where kings and HEROES are treated to posthumous deification by those they had ruled over or saved. He then advances the theory that it is in this manner that all the GODS have been created. This theory is not given much credence in modern anthropology or comparative religion. Some FANTASIES OF HISTORY may suggest that the rulers of ATLANTIS or of some other prehistoric POLDER have provided models for our later gods. Some SCIENCE FANTASY tales – like Roger ZELAZNY's *Lord of Light* (**1967**) – play fruitfully with the concept. [JC]

EURYDICE ◊ ORPHEUS.

EURIPIDES (*c*485BC-406BC) Greek dramatist. Little is known of his life: most of what passes for information about it derives from contemporary comedians' slanders (as in ARISTOPHANES' *The Acharnians*, *Thesmophoriazusae* and *The Frogs*). Of perhaps 88 plays he wrote, 14 survive essentially complete; two – *Electra* (**413**BC) and *The Bacchae* (written 407BC; **405**BC) are somewhat damaged – and two – *The Phoenician Women* (*c*412BC) and *Iphigenia in Aulis* (written 407BC; **405**BC) – have suffered great changes. Substantial fragments of several other plays have been found. *Cyclops* (undated) is a farce; the rest, technically, are tragedies; all are based on MYTHS. Each of Euripides' surviving plays includes elements we today would see as fantastic, but the extent to which he so saw them is debatable.

Ancient attacks on him depicted him as a misogynist and criticized him for depicting low subjects. In fact he concentrated on women more than any other known ancient writer, and repeatedly condemned class prejudice. His *Medea* (**431**BC) is a powerful early image of a WITCH, but heroic and ordinary women are more common in his work, as in *Alcestis* (**438**BC). His other sexual and racial prejudices are, however, disconcertingly prominent.

He was called a Sophist – a spokesman for the often amoral new ideas of rhetoricians and logicians. In fact, logic assumed a new prominence in his plays. Most contain scenes strongly reminiscent of the new rhetoric. But he also forced his dramas into considerably stricter narrative logic than his predecessors had used, freely reshaping the myths to this end, but putting this logic in turn in the service of probing moral studies. This aspect of his work was immensely influential on later Classical writers and, through them, on Western literature generally.

Other influential (and controversial) innovations concerned plot outlines. *Orestes* and *Electra* are the oldest known melodramas, while *Alcestis*, *Iphigenia in Tauris* (*c*414BC), *Helen* (**412**BC) and *Ion* (*c*414BC) are more or less fanciful romances with happy endings. *Helen* is particularly fantastic in content and style.

In his comedy Euripides appears atheist. Although *The Bacchae*, in particular, troubles this claim, and many of the plays express a passionate religious searching, it is true that he never forgave the myths their inconsistencies nor the gods their cruelties. This is notable relatively early, in *Hippolytus* (**428**BC), with its scheming goddesses and misfiring divine gift, but becomes pronounced in the later romances and melodramas. *Ion* amounts to an attack on ORACLES. In *The Trojan Women* (**415**BC) the strain of disbelief is stronger, and in *Heracles* (*c*415BC) the protagonist's doubts about his father, ZEUS, and his heroic mission become the stuff of one of the greatest Greek tragedies.

Soon after his death Euripides became the most read and performed of the Greek dramatists, and remained so throughout antiquity and the Middle Ages. [JB]

EVE ◊ ADAM AND EVE.

EVERETT, H.D. Real name and then pseudonym of Theo DOUGLAS.

EVERYMAN The title of the most popular of all medieval British morality plays (*c*1509-19), named after its central character. Called by DEATH, he learns that various of his friends – figures from ALLEGORY with names like Fellowship and Kindred – refused to accompany him to HEAVEN, though Good Deeds remains loyal. In fantasy terms, Everyman is a JACK figure. Hugo von HOFMANNSTHAL's *Jedermann* (**1911**) reworks the story; as does, in a non-supernatural mode, *Morality Play* (**1995**) by Barry Unsworth (1930-). [JC]

EVIL In fantasy Evil is most often present in a state of uneasy BALANCE with Good (◊ GOOD AND EVIL). Outside CHILDREN'S FANTASY it is quite rare for Good to be present without a counterbalancing force of Evil, but there are many instances where Evil is counterbalanced not by Good but by the morally unreliable, especially although not exclusively in HORROR and SWORD AND SORCERY: it would be hard to describe Michael MOORCOCK's Elric, Robert E. HOWARD's Conan or Stephen DONALDSON's Thomas

Covenant as virtuous, yet certainly they are the HEROES. Many interesting fantasies, SUPERNATURAL FICTIONS and PSYCHOLOGICAL THRILLERS derive much of their interest from the reader's slow realization that a protagonist, although possibly morally superior to the Evil adversary, is by no means a saint: *The Vanishment* (**1993**) by Jonathan Aycliffe (1949-   ) is a supernatural fiction of this type, although there are countless others, while most of the psychological thrillers of Ruth Rendell (1930-   ), under that name or as Barbara Vine, pivot on the same theme, a notable example being *The Bridesmaid* (**1989**), as Rendell. Tim BURTON's two BATMAN MOVIES are among many movies that have played in the same area of moral dubiety. [JG]

See also: PARODY.

EVIL OF FRANKENSTEIN, THE UK movie (*1964*). ◊ FRANKENSTEIN MOVIES.

EWERS, HANNS HEINZ (1871-1943) German novelist and essayist. Although a member of the Nazi Party, HHE was later accused of being a Jewish sympathizer and his work was banned. His writing career began in 1901 with a set of satirical rhymes, but not until the appearance of *Der Zauberlehring* (**1907**; trans Ludwig Lewisohn as *The Sorcerer's Apprentice* **1927** US) – the first of the **Frank Braun** trilogy – did he became established. Braun becomes involved with a religious cult, and hypnotizes his mistress into believing she is a saint. The mistress willingly becomes a martyr, with Braun delivering the final blow. By the second volume, *Alraune* (**1911**; trans Guy ENDORE **1929** US), Braun has become a dissolute and debtor. He encourages his uncle to experiment in the creation of a degenerate form of humanity. The result is a woman who has a fatal effect on any whom she contacts (◊ FEMME FATALE). Strong in erotic (◊ SEX) and VAMPIRE imagery and intensely decadent (◊ DECADENCE), the novel was immensely popular and was five times filmed (in 1918 twice, 1928, 1930, 1952). The final volume, *Vampir* (**1921**; trans as *Vampire* 1934 US), depicts Braun as a nonsupernatural vampire; the novel is an excuse for further erotic and sadistic episodes. All three volumes propound the inferiority of non-Teutonic races. HHE's short fiction is often sadistic with a preference for the CONTE CRUEL. His work was influenced by Edgar Allan POE, though he also shows the effect of Théophile GAUTIER, Charles BAUDELAIRE and J.-K. HUYSMANS. Although HHE's stories were collected in three volumes in Germany – *Dar Grauen* ["The Gruesome"] (coll **1908**), *Die Besessenen* ["Obsessions"] (coll **1909**) and *Nachtmahr* ["Nightmares"] (coll **1922**) – few have been translated into English; a limited selection has been issued as *Blood* (trans coll **1977** chap). His best-known story, "The Spider" (1915 *International Magazine*), is a genuinely atmospheric tale of WITCHCRAFT and MAGIC.

HHE appears as a character in «The Bloody Red Baron» (1996) by Kim NEWMAN, where he gets a comeuppance many will cheer. However, HHE's works were a product of their time, and in due course he came to his senses, leaving Germany before WWII. He died in the USA. [MA]

EWING, JULIANA H (ORATIA) (1841–1885) UK writer ◊ CHILDREN'S FANTASY; ELVES; FAIRYTALES

EXCALIBUR King ARTHUR's SWORD, sometimes confusingly equated with the SWORD IN THE STONE but in fact the sword given to Arthur by the LADY OF THE LAKE. It symbolized Arthur's strength and INVULNERABILITY, though the scabbard was more important than the sword: after the scabbard was stolen by MORGAN LE FAY, Arthur's star began to fade. The sword (called Caliburnus by GEOFFREY OF MONMOUTH) also stood for Arthur's sovereignty, and with his passing it was returned to the Lady of the Lake by Sir Bedivere. Excalibur is the focus of *King Arthur's Sword* (**1968**) by Errol Le Cain (1941-1989). The name has lent itself to the title of another children's novel, *Excalibur* (**1973**) by Sanders Anne LAUBENTHAL, where it turns up in the USA, the movie EXCALIBUR (*1981*) and an anthology, *Excalibur* (anth **1995**) ed Richard Gilliam (1950-   ), Martin H. GREENBERG and Edward Kramer. [MA]

EXCALIBUR US movie (*1981*). Orion/Warner. **Pr** John Boorman. **Exec pr** Robert A. Eisenstein, Edgar F. Gross. **Dir** Boorman. **Spfx** Peter Hutchinson, Alan Whibley. **Screenplay** Boorman, Rospo Pallenberg. **Based on** *Le Morte Darthur* (**1485**) by Sir Thomas MALORY. **Novelization** *Excalibur!* * (**1980**) by Gil Kane and John JAKES. **Starring** Robert Addie (Mordred), Katrine Boorman (Ygrayne), Gabriel Byrne (Uther), Nicholas Clay (Lancelot), Paul Geoffrey (Perceval), Cherie Lunghi (Guenevere), Helen Mirren (Morgana), Liam Neeson (Gawain), Niall O'Brien (Kay), Clive Swift (Ector), Nigel Terry (Arthur), Nicol Williamson (Merlin). 140 mins. Colour.

E astonishingly succeeds in conflating almost the entire ARTHUR mythos (◊◊ MATTER), largely as recounted by Malory, into a single small compass; much of its incidental music comprises appropriate passages by Richard WAGNER. With Arthur as the embodiment of the LAND, its sickness or health bound up with his, the tale falls into four parts: (a) the prelude to Arthur's drawing of the SWORD IN THE STONE; (b) the events leading up to the mortal blow dealt to the king/land by Lancelot's and Guenevere's adultery and by Morgana's seduction of the unwitting Arthur; (c) the QUEST for the GRAIL, which Arthur institutes as a cure for these ills, conducted in his winter (◊ FISHER KING) and the land's (◊ WASTE LAND), and bedevilled by Morgana and the sadistic Mordred; and (d) Arthur's final attempt to drive the EVIL of Mordred from the land and thereby cure it (◊ HEALING). Arthur himself is rarely the focus: MERLIN is the pivot of the first two parts and Perceval that of the third. The Merlin presented here is quirky to the point, sometimes, of buffoonery, thereby conveying a nice difference between him (and his priorities) and the mortals of whose world he has grown weary. This sense of the seepage of MAGIC from the world (◊ THINNING) is carried on through the portrayal of Morgana (conflating Malory's MORGAN LE FAY, Morgause and Nimue), Merlin's pupil and eventual nemesis. Arthur and Guenevere, by contrast, are depicted as naive country folk of – even in Arthur's case – limited vision; in one striking shot Arthur declares his intention to build the Round Table and bring everlasting peace while behind the small, raggledy band of his supporters we see an infinitely vast, infinitely starry night sky. The scene could stand as *E*'s epitome: whatever the grandeur of their ideas, none of the protagonists – save Merlin – can ever stir very far from the grime of the real world. [JG]

EXORCISM A rite by which unwanted or evil SPIRITS are expelled. Usually they have been in POSSESSION of a human SOUL. In fantasy novels, rites of exorcism may be elaborated upon with the utmost freedom. In SUPERNATURAL FICTIONS, where the rite may well constitute part of a genuine belief system being imparted to the reader, exorcisms are less imaginative. In HORROR, where novels like William Peter

BLATTY's *The Exorcist* (**1971**) – filmed as *The* EXORCIST – have been popular, the rite of exorcism is rarely very successful, and the evil spirit is as likely to incorporate the exorcist as to be driven out into the cold.   [JC]

EXORCIST, THE A series of three movies (plus one sport) derives from *The Exorcist* (**1971**) by William Peter BLATTY.

1. *The Exorcist* US movie (*1973*). Hoya/Warner Bros. **Pr** Blatty. **Dir** William Friedkin. **Mufx** Dick Smith. **Screenplay** Blatty. **Starring** Linda Blair (Regan MacNeil), Ellen Burstyn (Chris MacNeil), Jason Miller (Father Damien Karras), Max von Sydow (Father Merrin), Kitty Winn (Sharon). 121 mins. Colour.

Probably the most famous HORROR MOVIE of all time, this shocked many because of its sensationalism; in fact, it is based on a recorded case of POSSESSION, that of 14-year-old Washington youth Douglass Deen in 1949. Young Regan suffers fits and other weird manifestations, and medical science cannot help. Matters worsen: she is a focus of wild POLTERGEIST effects, and also is revoltingly physically transformed ($\Diamond$ TRANSFORMATION). When mother Chris in desperation brings in young Father Karras things get worse: the child blasphemes in a hoarse, unnatural voice, can swivel her head through 360°, and produces green projectile vomit. Karras, dubious in his faith but certain that Regan has been possessed by the DEVIL, calls on senior priest Merrin to perform an EXORCISM. This at last succeeds, but at the cost of Merrin's life.

TE is a disturbing movie on several levels. Ignoring the grue of the impressive spfx, it invokes several of our primal fears, not least the notion that this might happen to *our* child; in this respect, it can be seen as a vastly exaggerated version of the RITE OF PASSAGE every adolescent makes when moving from childhood into a rebellious form of pseudo-adulthood. The movie's structure (like the novel's) creaks, though; while we can empathize with Regan and her mother, the other characters seem always to be only temporarily on the stage, so that even Merrin's violent death is an emotionally arid affair.

Video release of *TE* was disallowed by the UK censors on the grounds that it could damage viewers psychologically, although the fiction was maintained that it was the movie industry which had voluntarily withheld the release. Appeals in the 1990s against the decision were met with obfuscation.

Further reading: *The Story Behind the Exorcist* * (*1974*) by Peter Travers and Stephanie Reiff.

2. *Un Urlo dalle Tenebra* (vt *L'Esorcista N.2*; vt *A Cry in the Dark*; vt *Cries and Shadows*; vt *Naked Exorcism*; vt *The Exorcist 3: Cries and Shadows*) Italian movie (*1975*). **Dir** Angelo Pannacciò. **Starring** Richard Conté, Françoise Prévost, Jean-Claude Verné. 88 mins. Colour.

Unconnected with the rest except through its ripoff vts and included here merely for clarification. A nun attempts to exorcize her brother.   [JG]

3. *Exorcist II: The Heretic* (vt *The Heretic*) US movie (*1977*). Warner Bros. **Pr** John Boorman, Richard Lederer. **Dir** Boorman. **Spfx** Jim Blount, Wayne Edgar, Chuck Gaspar, Jeff Jarvis, Roy Kelly. **Vfx** Van der Veer Photo, Albert J. Whitlock. **Mufx** Dick Smith. **Screenplay** William Goodhart. **Starring** Belinha Beatty (Liz), Ned Beatty (Edwards), Linda Blair (Regan MacNeil), Richard Burton (Father Philip Lamont), Louise Fletcher (Dr Gene Tuskin), Joey Green (Young Kokumo), Paul Henreid (Cardinal), James Earl Jones (Adult Kokumo), Max von Sydow (Father Lancaster Merrin), Kitty Winn (Sharon Spencer). 117 mins. Colour.

Some years have passed since the events of **1**, and Regan, in New York, is receiving psychiatric help from Tuskin, though she seems completely adjusted; in the temporary absence of her mother she is being cared for somewhat shakily by mother's secretary Sharon, more affected by the memories than Regan herself. Lamont, a pupil of Merrin's (from **1**), having failed in a South American EXORCISM, is detailed by his Cardinal to investigate the mystery surrounding Merrin's death. Lamont at last recognizes he is dealing with a DEMON. After many plot twists Lamont, with Regan in tow, reaches the Washington house where the events of **1** occurred and defeats the demon, but at the cost of his own life.

EII received an appalling press on first release, and was much recut by Boorman for release in the UK. In fact, it is at least as imaginative and interesting as **1** (the above synopsis does not do its complicated plot justice) although, lacking gore and shock obscenity, it does not have the same jolt-you-out-of-your-seat power. The TECHNOFANTASY aspects are contrived, but intercut flashback African scenes are broodingly impressive, and the movie's flow of ideas never flags. The direction is exciting, as is much of the photography (some by Oxford Scientific Films). This is not a landmark fantasy movie but it is a good one.   [JG]

4. *The Exorcist III* US movie (*1990*). 20th Century-Fox/Morgan Creek. **Pr** Carter De Haven. **Exec pr** James G. Robinson, Joe Roth. **Dir** Blatty. **Spfx** Bill Purcell. **Mufx** Greg Cannom. **Screenplay** Blatty. **Based on** *Legion* (**1983**) by Blatty. **Starring** George DiCenzo (Stedman), Brad Dourif (James Venamun), Nancy Fish (Nurse Julie Allerton), Ed Flanders (Father Joseph Kevin Dyer), Don Gordon (Ryan), Mary Jackson (Mrs Clelia), Jason Miller (Father Damien Karras/Patient in Cell 11), George C. Scott (Lt Bill Kinderman), Nicol Williamson (Father Paul Morning), Scott Wilson (Dr Freeman Temple). 110 mins. Colour.

15 years ago, around the time of the events in **1**, a SERIAL KILLER called James Venamun, nicknamed the Gemini Killer, was caught and executed. Now, impossibly, he has renewed his crimes, still confining his murders to people with a name starting "K" but choosing his victims among those connected with the EXORCISM of Regan. Lt Kinderman, investigating, finds as his first mystery that the fingerprints around the mangled corpses differ each time; also, to prolong their agony, they have been injected before mutilation with a medically correct dose of a paralysing drug. Kinderman's old friend Joseph Kevin Dyer is taken into Georgetown General Hospital for a check-up and becomes the next victim. During the investigation at the hospital, a staff doctor, Temple, guides Kinderman to Cell 11 in the secure block, where a catatonic patient, brought in 15 years ago, has been confined these past few weeks since "waking up" and is becoming violent. Kinderman recognizes Father Karras, who died at the end of **1**; but as he interviews the patient the face shapeshifts ($\Diamond$ SHAPESHIFTER) into that of Venamun, who in due course explains that his SOUL after electrocution was – as revenge for the exorcism – inserted by "The Master" into Karras's dead body.

This, although it leaves many unanswered questions, is a powerful movie and an improvement on Blatty's routine novel, which it much simplifies; its construction of atmosphere is immaculate – notably aided by attention to incidental sound effects. Blatty billed it as the first true sequel to **1**, but the word "sequel" seems misapplied to what is a related but separate tale. [JG]

EX-PRIVATE X Pseudonym of A.M. BURRAGE.

EYES OF LAURA MARS US movie (*1978*). Columbia. **Pr** Jon Peters. **Exec pr** Jack H. Harris. **Dir** Irvin Kershner. **Spfx** James Liles. **Stills** Rebecca Blake, Helmut Newton. **Screenplay** John Carpenter, David Zelag Goodman. **Novelization** *Eyes of Laura Mars* * (**1978**) by H(arriet) B. Gilmour (1939-   ). **Starring** Rene Auberjonois (Donald Phelps), Brad Dourif (Tommy Ludlow), Faye Dunaway (Laura Mars), Tommy Lee Jones (Lt John Neville), Raul Julia (Michael Reisler). 104 mins. Colour.

Photographer Mars's sexually violent images are the latest chic in New York. She starts having clairvoyant DREAMS and VISIONS of watching from a SERIAL KILLER's vantage as he murders her associates by stabbing through the eyes. Incredulous cop Neville points out further that some of her previous photographs uncannily reproduce real scenes from unsolved murders. The murders continue; cop and photographer predictably become lovers. It proves Neville is a MULTIPLE PERSONALITY whose *alter ego* – whom he believes is the spirit of his murdered mother, possessing him (◊ POSSESSION) – is the killer.

Highly suspenseful, full of style and brio, *EOLM* is at a loss to supply a resolving conclusion; no explanation is offered for the telepathic link between Mars and Neville. This aside, *EOLM* has a lot to offer; it is of especial interest in that, unlike most others of its ilk, it focuses on clairvoyance (◊ TALENTS) rather than precognition. These visions are exceptionally convincingly rendered, seemingly by use of hand-held video. [JG]

FABIAN, STEPHEN (1930-   ) US illustrator. ◊ Virgil
FINLAY.

FABLE A short narrative which is often a commentary upon
society or the human condition, presented as an ALLEGORY
or parable, almost always with a hidden (though not
obscure) message. It may share common idioms with the
FOLKTALE or FAIRYTALE, particularly in its moral counsel,
but is not TWICE-TOLD. It often uses animals to symbolize
aspects of humankind, so that, although the original fables
were SATIRES for an adult audience appreciative of their
more subtle imagery (◊ BEAST FABLE), their anthropomorphism also lent an appeal to younger audiences – hence the
appeal of the **Uncle Remus** stories by Joel Chandler HARRIS.

The earliest known fables are those of Aesop (6th century
BC), a Greek slave, although most of them were probably by
later Greek writers, including Gaius Julius Phaedrus
(?15BC-AD50), Claudis Aelianus (?170-235), Aulus Gellius
(117-180) – who wrote the well known fable "Androcles
and the Lion" – Babrius (2nd century AD) and Flavius
Avianus (4th century AD). Frequently these fables were seen
as subversive (◊ AESOPIAN FANTASY): Phaedrus was severely
punished during the tyranny of Sejanus. Other early fables
are found in the Indian collection the *Panchatantra* (6th
century AD) and a similar collection by Bidpai or Pilpay (8th
century AD), now lost. Many of these tales found their way
into other medieval texts, especially the *Gesta Romanorum*
["Deeds of the Romans"] (13th century).

By the time William Caxton (*c*1422-*c*1491) produced the
first published text of Aesop's *Fables* (coll **1484**), it was difficult to attribute source with any accuracy, and this was
further blurred by Jean de la Fontaine, whose own collection, *Fables choisies mises en vers* (coll **1668**; exp 1678; exp
1693), drew heavily on earlier sources, not always with due
credit; they needed to be severely rewritten when William
GODWIN produced his *Fables Ancient and Modern* (anth
1805), the first modern compilation aimed specifically at
children. Other noted writers include John Gay (1685-
1732) with *Fables* (coll **1727** first series), Gotthold Lessing
(1729-1781) with *Abhandlungen über die Fabel* ["Treatise
about Fables"] (**1759**), Jean-Pierre Claris de Florian (1755-
1794) with *Fables* (coll **1792**), Ivan Krilof (1768-1844) with

Fables (coll **1809**), Robert Louis STEVENSON with *Fables*
(coll **1896** US) and Ambrose BIERCE in the rather more sardonic *Fantastic Fables* (coll **1899**). The fable continues to be
a popular medium, though its use today focuses more on
political commentary or satire; its appearance in fantastic
fiction is less visible. Authors who have used fables to good
effect include Marcel AYMÉ, Jorges Luis BORGES, Italo
CALVINO, Lafcadio HEARN, John MORRESSY, Jessica Amanda
SALMONSON and James THURBER.

Two comprehensive compilations of fables are *An
Anthology of Fables* (anth **1913**) ed Ernest Rhys (1859-1946)
and *Ride the East Wind* (anth **1973**) ed Edmund C. Berkeley
(1912-   ).   [MA]

FABLED BEASTS ◊ MYTHICAL CREATURES.

FABLED LAND ◊ LAND OF FABLE.

FABULATION In fantasy any fabulation which significantly
undermines the felt reality of the world it depicts will be
responded to by most readers as aberrant. At the same time,
fabulation foregrounds STORY: not only does it makes readers aware that they are reading one, but – most fittingly in
fantasy texts – it articulates the sense that the RECOGNI-
TIONS central to the satisfactory resolution of many
full-fantasy texts are in fact recognitions of the nature of the
Story being told, whether the underlying tale is a CREATION
MYTH or BEAUTY AND THE BEAST or the SEVEN SAMURAI's
Brilliant Adventure. Fabulation techniques foreground this
fundamental sense that fantasy stories take place in a Story-
shaped world, and that they may reach apogee at the point
this becomes evident.

Full-blown fabulations are rarely perceived as occupying
the centre ground of fantasy (or sf). And it comes as no sur-
prise that fabulators who write fantasy are often the same
people as fabulators who write sf. Fabulators of fantasy
include Reinaldo ARENAS, John BARTH, Donald BARTHELME,
Adolfo BIOY CASARES, Jorge Luis BORGES, Scott BRADFIELD,
Richard BRAUTIGAN, Ed BRYANT, Mikhail BULGAKOV,
Anthony BURGESS, William BURROUGHS, Italo CALVINO,
Angela CARTER, Jerome CHARYN, Barbara COMYNS, Robert
COOVER, Jim CRACE, Tom DE HAVEN, Thomas M. DISCH,
E.L. DOCTOROW, Katherine DUNN, Umberto ECO, Carol
EMSHWILLER, Steve ERICKSON, Karen Joy FOWLER, Carlos

FUENTES, MacDonald HARRIS, M. John HARRISON, William HJORTSBERG, Russell HOBAN, Rachel INGALLS, Franz KAFKA, William KOTZWINKLE, Michael MOORCOCK, Vladimir NABOKOV, Flann O'BRIEN, John Cowper POWYS, Christopher PRIEST, Thomas PYNCHON, Peter REDGROVE, Leon ROOKE, Salman RUSHDIE, Josephine SAXTON, Lucius SHEPARD, Clive SINCLAIR, Stefan THEMERSON, Rex WARNER, William WHARTON and Gene WOLFE. [JC]

FABULOUS ADVENTURES OF BARON MUN-CHAUSEN, THE (ot *Les Fabuleuses Aventures du Baron de Munchausen*) French animated movie (*1979*). ◊ MUNCHHAUSEN.

FABYAN, ROBERT (? -1513) English chronicler. ◊ EDIFICE.

FACE OF FU MANCHU, THE UK movie (*1965*). ◊ FU MANCHU MOVIES.

FACE OF GLORY The term is a translation of a Sanskrit name, *Kirttimukha*, which is given to a particular JANUS-faced statue of the elephant god Ganesa found in Java. The front face of this statue shows the kindly side of the pot-bellied, large-trunked god. But the back face has bulbous eyes, savage canines like those in Balinese MASKS, a gaping hole where a lower jaw should be, and overall a sense of lion-like ferocity – like the Greek Gorgoneion, "the fearful face".

The backward-grimacing face of Ganesa is an aspect of the god Siva, a kind of DEMON generated by Siva in order to attack an insolent invader, who immediately surrenders when threatened. Thwarted of its feast, the demon asks Siva for something to eat, and is instructed to devour its own body, which it does, until only the head – the FOG – remains. As a reward, Siva gives the head a name, Kirttimukha; and a role, that of "door-keeper", or the Guardian of the Threshold (◊ LIMINAL BEING), a threshold visage which (or who) warns PILGRIMS, HEROES and others engaged upon life-transforming QUESTS of the risks of passage into regions where they may recognize the nature of their search, the point of the STORY they have been living, and which/who wards off EVIL. The Janus-faced Ganesa embodies that role, a role which, in mutated form, underlies Tanith LEE's rendering of the two-faced god Chuz in her **Tales from the Flat Earth** sequence (**1978-87**). But it is in the face of the Greek GORGON – specifically in the mortal gorgon Medusa – that the iconographic density of the FOG is most evident, and most retentive.

The Gorgon herself seems to be traceable back to various Eastern images of the GODDESS (prior to her transformation into the victim of Perseus), the Goddess figure frequently portrayed, in her incarnation as Cybele, in the *potnia theron* or Mistress of the Beasts pose, with lions in attendance. The Gorgon whom Perseus beheads, turning her into a FOG whose apotropaic functions Athena gladly utilizes, is in this light an inverted Cybele, the Goddess as Kirttimukha, whose bulging eyes, protruding tongue, foliated skin (◊ FOLIATE HEAD), tusks and beard all proclaim her BONDAGE.

At the same time, the central image of Medusa has never entirely lost an ambivalent allure, and in 20th-century literature she has clearly come to represent a great deal more than HORROR, petrification, justified bondage and guard-dog servitude at the gates through which the hero passes. Russell HOBAN's *The Medusa Frequency* (**1987**) frequently conflates imagery of Medusa with associated imagery of ORPHEUS after *his* head is severed; but as often it transforms the image into a genuine FOG. "The Gorgon's head, the

face of Medusa, shimmered luminous in a silence that crackled with its brilliance. Her mouth was moving." Or as Hans BEMMANN puts it in *The Broken Goddess* (**1990**): "I realized how much a stranger you still are to me, even more of a stranger than the lioness on the steppes beneath the tower. Your face was a wild, trackless forest in which I would probably get lost, but that wasn't going to stop me. It seemed that only now was I really beginning my quest . . ." Or, according to A.S. BYATT in *The Djinn in the Nightingale's Eye* (coll **1994**): "She was the earth and the lions."

More complex (as far as fantasy is concerned) is the use to which Elizabeth HAND puts the imagery in *Waking the Moon* (**1994** UK), at the point at which the Goddess is (perhaps) about to be reawakened. The protagonists are in a tower, in which an ORRERY is spinning, more and more rapidly, until it becomes a bodiless Mask of the MOON (where the Goddess sleeps, or which is in fact the Goddess in a state of bondage). More and more rapidly it whirls, until it seems capable of turning the world into a new dispensation. The orrery pulses "like a swimming medusa" with "a flaming eye; like a demonic moon". It is here that one thing becomes two, that an INSTAURATION FANTASY begins, though later the novel decoys instauration into SUPERNATURAL FICTION. In passages of this sort, the FOG takes on a complex polyvalency (a confluence of MAP and territory which authors may themselves not entirely control), for it is at one and the same time a Guardian, a MIRROR, a LABYRINTH, a mandala, a warning and a pointer inwards, a rictus of bondage and the loosening smile of SEX. It is a face we may wear in DREAMS. It is a face we will encounter there, if we wish to earn passage. [JC]

FAERIE Faerie, or Faërie, is the land of the FAIRIES, or fairyland. It is therefore an OTHERWORLD perhaps linked to ours, though access is seldom physical in the normal sense. The most common methods of access are by going INTO THE WOODS, by RIVER, by being transported by the WINDS (usually the North Wind) or by DREAMS (though Faerie is seldom portrayed as a dreamland; rather, memory of being there is as of a dream).

Since fairies were once believed part of our own world, the concept of Faerie was a late development in mythography. Faerie was more closely associated with the UNDERWORLD where one passed after death, whether a HEAVEN or a HELL. The worlds were ruled by spirits equated with fairies, such as Gwynn ap Nudd in Celtic mythology, or Arawn the King of Annwn. AVALON, ruled by MORGAN LE FAY, was another Faerie-realm linked to the AFTERLIFE, as was TIR-NAN-OG. In *Huon of Bordeaux* (15th century; trans Lord Berners ?**1534** UK), OBERON is depicted as the King of the Fairies whom Huon meets in a FOREST, while in Edmund SPENSER's *The Faerie Queen* (**1596**) fairyland is ruled by Queen Gloriana.

One of the main features of Faerie is its timelessness (◊ TIME IN FAERIE). To cite just one example, in *De Nugis Curialium* (12th century; ed M.R. JAMES 1915), Walter Map (?1137-1209) tells the story of "Herla's Ride", about a king who is invited to a wedding in fairyland; on his return several centuries have passed.

The very elusiveness of Faerie has been compared to that of the pot of gold at the end of the rainbow, and thus fairyland may be perceived as over the rainbow, the equivalent of L. Frank BAUM's land of OZ. The borders of Faerie shift according to the power of the land's magic and its balance with the encroachment of humankind's science and civilization (◊ THINNING). The concept thus held an appeal for the PRERAPHAELITES, who yearned for a GOLDEN AGE and utilized the idea of Faerie to depict a timeless paradise attainable only by halting humanity's material progress. This idea was used by William MORRIS in his fantasies from as early as "The Hollow Land" (1856), and particularly in his novels *The Wood Beyond the World* (1894) and *The Well at the World's End* (1896), where his fantasy worlds (almost akin to Faerie) are on unknown shores reached only by sea. George MACDONALD explored fairyland in *Phantastes* (1858), this time making it more of a dreamland but also requiring, for access, an encounter in the woods (◊ INTO THE WOODS).

In modern fantasy there have been several novels and stories exploring the relationship between our world and Faerie. These include *The King of Elfland's Daughter* (1924) by Lord DUNSANY, *Lud-in-the-Mist* (1926) by Hope MIRRLEES, *Land of Unreason* (1942) by L. Sprague DE CAMP and Fletcher PRATT, *The Broken Sword* (1954) by Poul ANDERSON, *The Kingdoms of Elfin* (coll 1977) by Sylvia Townsend WARNER, *The Elfin Ship* (1982) and sequels by James P. BLAYLOCK, *War for the Oaks* (1987) by Emma BULL, *Faerie Tale* (1988) by Raymond E. FEIST, *Thomas the Rhymer* (1990) by Ellen KUSHNER and *Faery in Shadow* (1993) by C.J. CHERRYH. The concept of Faerie has grown more popular since the 1980s because of its association with New Age beliefs and the increase in interest in CELTIC FANTASY. It provides a ready-made semi-believable world, in contradistinction to the fabricated FANTASYLANDS generated by too many writers influenced by J.R.R. TOLKIEN.   [MA]

FAERIE QUEENE ◊ FAIRY QUEEN.

FAGUNWA, D(ANIEL) O(LORUNFEMI) (1900/1903-1963) Nigerian teacher and writer, the most popular Nigerian writer of the 1940s-60s. Although DOF's fantasy novels, written in Yoruba, look like somewhat disorganized FAIRYTALES, they are not juveniles. Most of his fantastic adventures are – unlike those of other African fantasists like Amos TUTUOLA – based not on Yoruba folklore but on fresh inventions.

Fagunwa's major work, *Ogboju ode ninu igbo irunmale* (1938; trans Wole Soyinka as *The Forest of a Thousand Daemons* 1968), tells of a brave hunter who undertakes, on the orders of a king, a dangerous expedition to a deep FOREST full of supernatural beings. There he fights various SPIRITS and GHOSTS, and even falls in love with a beautiful WITCH. Loosely connected volumes, similar in style and content, are *Igbó Olódumare* ["The Jungle of the Almighty"] (1949) and *Irinkerindo ninu igbó Elégbeje* ["Wanderings in the Forest of Elégbeje"] (1954).

Less popular fantastic works are *Ireké oníbudó* ["The Cane of the Guardian"] (1949) and *Adiitu Olódumare* ["The Secret of the Almighty"] (1961); in the latter the realistic elements take over from the fantastic. A few short stories were published in *Asayan itan* ["Selected Stories"] (coll 1959).   [JO]

See also: BLACK AFRICAN FANTASY.

FAIRIES "Fairy" is an anglicization of the French *faerie* (enchantment), which absorbed and largely displaced the Anglo-Saxon "elf" (◊ ELVES) after the Norman Conquest. The term first became common in the 13th century, although many relevant FOLKTALES involving such beings had already been recorded by chroniclers such as Walter Map (?1137-1209), Giraldus Cambrensis (c1146-c1223) and Gervase of Tilbury (c1150-c1220). The Celtic mythology of the mound-dwelling Sidhe was, of course, accommodated within the framework, which was a confused multicultural

ragbag long before literateurs got to work on it.

In the UK a good deal of fairy poetry was produced in the Elizabethan era, when Edmund SPENSER borrowed extravagantly from ARIOSTO's *Orlando Furioso* (1516; final rev 1532) in producing *The Faerie Queene* (1596), and SHAKESPEARE provided the definitive imagery of *A Midsummer Night's Dream* (performed c1595; 1600). A rich tradition of British fairy poetry – enlivened by input from Wales, Scotland and Ireland, where oral culture was not so rapidly superseded as in England – extended from these beginnings to the early 20th century, eventually petering out in the work of such writers as William Butler YEATS and Alfred NOYES. New literary versions of traditional tales began to be produced on a prolific scale in France and Britain in the wake of Charles PERRAULT's *Histoires ou contes du temps passé* (1697; trans 1729) and Madame d'AULNOY's *Contes des fées* (coll 1696-8). Some familiar tales deriving from collections such as these do not actually feature fairies (◊ FAIRYTALES; WONDER TALES).

In the wake of Perrault, fairies became commonplace in French 18th-century fiction, sometimes even intruding into fashionable tales of contemporary high society – although their most frequent milieu was ARABIAN FANTASY, whose demonic peris were transformed into a species of fairy. This tradition extended into the 19th century in such works as Charles NODIER's *Trilby* (1822) and *La Fée aux miettes* (1832). The German Romantic movement revived interest in Teutonic versions of fairy mythology; J.K. MUSÄEUS published five volumes of *Volksmärchen der Deutschen* (coll 1782-7) and Ludwig TIECK produced his own *Volksmärchen* (omni 1797). Several writers associated with the movement, including Novalis (real name Friedrich von Hardenberg; 1772-1801) and GOETHE dabbled in the production of *Kunstmärchen* ("art fairytales"). Folklorists (◊ FOLKLORE) inspired by the GRIMM BROTHERS' *Kinder- und Hausmärchen* (vol 1 1812; vol 2 1814) attempted – or at least pretended – to recapture the authentic *Volksgeist* of the tales they recorded, partly in the hope of helping to clarify a notion of "national identity" appropriate to a nation-state still in the process of formation. The most significant attempt to codify and explicate British folklore was *Fairy Mythology* (1828; rev 1850) by Thomas Keightley (1789-1872), whose *Tales and Popular Fictions* (1834) had analysed similarities between different folkloristic traditions in diffusionist terms. By now imitation folktales were being produced in great profusion by Hans Christian ANDERSEN and others.

Edwin Hartland's *The Science of Fairy Tales* identified several major categories, among the most important being tales of "swan-maidens" (◊ BIRDS) and other FAIRY BRIDES, tales of timeless sojourns in fairyland (◊ TIME IN FAERIE), and tales of CHANGELINGS. More recent attempts to produce grand theories decoding the hidden meanings of fairytales and accounting for the remarkable endurance of the most familiar motifs include *The Erotic World of Faery* (1972) by Maureen Duffy, *The Uses of Enchantment* (1976) by Bruno Bettelheim and *Fairy Tales and the Art of Subversion* (1983) by Jack ZIPES, but it is not clear that any kind of all-encompassing theory is necessary; K.M. BRIGGS's *The Fairies in Tradition in Literature* (1967) carefully steers clear of any such commitment.

Fairies became popular in 19th-century UK painting, partly because nude female fairies were more acceptable to the Victorians than nude women, although the only material difference between the two was the addition of WINGS.

Artists like Richard DADD, however, brought into focus a more sinister side of fairy life and helped retain the darker aspects of the traditional lore for use by writers for adults. Sara COLERIDGE's *Phastasmion* (**1837**), John RUSKIN's *The King of the Golden River* (**1851**) and George MACDONALD's *Phantastes* (**1858**) set important precedents in making the kind of fairy mythology which had been re-popularized by painters available for use in adult fantasy, the latter being quickly followed by Christina ROSSETTI's erotic verse fantasy "Goblin Market" (1862). Many writers of CHILDREN'S FAN-TASIES followed Macdonald in producing tales of considerable sophistication (◊ FAIRYTALE). Once the fad for fairy painting had passed, the insectile wings became less common, but they have never quite become extinct in fairy images designed for children.

The retreat of Britain's fairies to some far-off exile was evoked nostalgically in "Titania's Farewell" (1882) by Walter BESANT and James Rice and satirically in Andrew LANG's and May Kendall's *That Very Mab* (**1885**), but the loss became much more keenly felt in the decade after WWI, which produced several fine works lamenting the loss of a kind of illumination for which fairyland could easily stand as a symbol. Significant examples include Gerald BULLETT's *Mr Godly Beside Himself* (**1924**), Lord DUNSANY's *The King of Elfland's Daughter* (**1924**) and Hope MIRRLEES's *Lud-in-the-Mist* (**1926**), all of which contrast strongly with the underlying ideology of F. ANSTEY's *In Brief Authority* (**1915**), in which the rejection of the values of *Märchenland* is slyly endorsed. Although the acute sense of danger these works encapsulate soon evaporated, the march of progress merely paved the way for its return in the wake of WWII, when a new generation of adult fantasies – spearheaded by J.R.R. TOLKIEN's *The Lord of the Rings* (**1954-5**) but also encompassing Poul ANDERSON's *The Broken Sword* (**1954**) – reclaimed and renewed the Old English heritage of Elfland in the service of a new and spectacularly successful bid to ensure that the magic of Faerie could not be extinguished.

The post-Tolkienian elves which have displaced Frenchified fairies are of many subtly different species, important variants being introduced by *The Kingdoms of Elfin* (coll **1977**) by Sylvia Townsend WARNER – the most comprehensive account of the manners and politics of FAERIE since John Hunter Duvar's highly eccentric *Annals of the Court of Oberon* (**1895**) – and such Sidhe-starring CELTIC FANTASIES as *The Dreamstone* (**1983**) by C.J. CHERRYH and *King of Morning, Queen of Day* (**1991**) by Ian McDonald (1960-   ), but they are almost invariably finer folk than humans. The 19th-century image of the fairy is sarcastically "updated" in *The Good Fairies of New York* (**1992**) by Martin Millar, but more earnest versions are much rarer in adult fantasy than in children's fiction; one magnificent exception is *Little, Big* (**1981**) by John CROWLEY.   [BS]

FAIRY BRIDE The human male who marries a supernatural bride, promising he will respect some TABOO whose eventual breakage results in her reversion to her "true nature", is a common figure in FOLKLORE. The reversion is usually to animal form, as in tales of swan-maidens (◊ BIRDS) or the classic story of the fairy LAMIA Melusine, or to an aquatic way of life, as in la Motte FOUQUÉ's *Undine* (**1811**) and some tales of MERMAIDS and seal-maidens (◊ SELKIES). A subtle but powerful modern deployment of the motif is in *The Owl Service* (**1967**) by Alan GARNER.   [BS]

FAIRY GODMOTHER The archetypical FG appears in "Cinderella" (1697) by Charles PERRAULT, where she provides the magical means by which CINDERELLA is able to go to the ball. Perrault used the feature of fairies attending christenings in his first FAIRYTALE, "Sleeping Beauty" (1696), where it was the oversight in neglecting to invite one of the FAIRIES that caused the CURSE to be laid on the baby. The FG has its origin in the Greek legend of the FATES: SLEEPING BEAUTY falls into her deep sleep after her finger is pricked by the needle held by an old lady who is spinning. However, the FG as developed by Perrault has aspects of the guardian ANGEL and patron saint (◊ PATRONS) – a supernatural being who has dominion and protection over our fate.
     [MA]

FAIRYLAND ◊ FAERIE.

FAIRY QUEEN Several names have been attributed to the Queen of the Fairies. In Greek MYTHOLOGY Diana the huntress, with her attendant nymphs, was their queen, and this name was corrupted into Titania, the wife of OBERON. MORGAN LE FAY was queen in AVALON. The original phrase Faerie Queene may well have been invented by Edmund SPENSER for *The Faerie Queen* (**1596**). This tells of an Arthurian QUEST for Gloriana, or Belphoebe, the eponymous FQ, who is symbolic of Elizabeth I (1533-1603). The main storyline is obfuscated by many subtexts and adventures, which characteristic in itself demonstrates the complications and machinations of the Elizabethan court. However, the tale also captures the love the nation has for the FQ, offsetting it against her retention of distance and an extensive power base. This image of a powerful but distant FQ has remained part of the iconography of FAERIE ever since.

The development and maturity of the FQ is explored with considerable flair in Jean INGELOW's *Mopsa the Fairy* (**1869**), an early feminist FAIRYTALE. *That Very Mab* (**1885**) by Andrew LANG and May Kendall (real name Emma Goldworth; 1861-?1931) tells of an FQ who left England after Puritanism took hold and who returned to find a much sadder state. Mab is a name often attributed to the FQ, though the origin of the title "queen" (from the Saxon *cwén*) really meant just "woman", so the name might be equally expressed as "Lady" Mab. "Mab" may have come from Medb or Maeve, the Warrior Queen of Connacht in Irish mythology. Her supernatural adventures are recast by James STEPHENS in *In the Land of Youth* (coll of linked stories **1924**).   [MA]

FAIRYTALE The term "fairytale" has come to cover almost anything fanciful and, since the Victorian period, has been largely identified with children. This has limited an appreciation of the literary artform of the fairytale at its best and misinterpreted its purpose. Here we distinguish the fairytale from the FOLKTALE and from its nonfantastic counterparts. We also distinguish it from the FABLE, where the emphasis is on message not story, but we do recognize the fairytale's contribution toward CHILDREN'S FANTASY. We would prefer to differentiate between fairytales (tales involving FAIRIES) and WONDER TALES, a more embracing term; unfortunately the distinction has yet to enter popular usage.

To define our terms, then: The fairytale is a written story that relies upon the FANTASTIC, although it need not involve fairies or FAERIE. It is usually set somewhere distant in place or time or both – in "once upon a time" or, as J.R.R. TOLKIEN called it, the "Perilous Realm". It almost always involves a TRANSFORMATION, either physically or through self-discovery, so that by its end people or circumstances have changed, generally allowing a "happy ever after" conclusion. The emphasis is on STORY, particularly the TWICE-TOLD, both highlighting the timelessness and the continuity of the tale and allowing education from generation to generation. In this respect it has borrowed from the oral tradition but codified it within a format that encourages the instructive suspension of disbelief. The literary fairytale is thus the core of the WONDER TALE (in that it relies totally on the marvellous and the creation of awe) and can be distinguished from the HORROR tale, where the supernatural tends to a destructive end. While never intended to be wholly believable, it contains sufficient sense to ensure the audience understands its application in the real world. Thus stories like "Little Red Riding-Hood" and "Hansel and Gretel", while containing the potential to horrify, seek to frighten within a world that is initially welcoming and familiar and (usually) becomes so again. The first transformation is usually linked to an INTO-THE-WOODS device, marking the transcendence from this world to one more fantastic. Fairytales are thus often CROSSHATCH fantasies, if not set wholly in the SECONDARY WORLD of FAERIE.

The oldest known fairytale "The Doomed Prince" (c1300BC) appeared in ANCIENT EGYPT. Many Greek and Celtic LEGENDS reappear as fairytales, which thus have much in common with other early literature, such as the *chansons de geste* and the medieval ROMANCE. The *Gesta Romanorum* ["Deeds of the Romans"] (13th century) became an accumulation of ALLEGORIES, fables and stories, totalling over 200, originally composed in Latin and drawing from Roman and Oriental sources, which served as lessons given by preachers. It was first printed in 1473 and was a source for many medieval writers, including SHAKESPEARE, who used elements from it in *King Lear* (performed 1606; **1608**) and *The Merchant of Venice* (**1600**). It is also the primary source for the story of "Androcles and the Lion" (attributed to Aulus Gellius [117-180]).

The fairytale form had been used by earlier writers, the earliest surviving example being CUPID AND PSYCHE in *The Golden Ass* (cAD155) by APULEIUS, but it was not a consciously composed idiom. The credit for this goes to the Italian writer Giovanni Straparola (?1480-1558), whose *Le Piacevoli Notti* ["The Delectable Nights"] **1550-53**) brought together over 70 stories and anecdotes – several presented in fairytale form – told in the vernacular, including "The Pig Prince", a seminal story of transformation. Straparola used the framework-story device of BOCCACCIO's *Decameron* (**1358**). This was also the inspiration for *Lo Cunto de li Cunti* ["The Story of Stories"] (**1634**; vt *The Pentameron* 1674) by Giambattista Basile (1575-1632), whose volume of over 50 fairytales became an inspiration across Europe.

The fairytale form was now taken up by French society, in particular Madame d'AULNOY and Charles PERRAULT in their literary *salons*. They used the mode for contemporary satire. D'Aulnoy's "The Blue Bird" first introduced the character Prince Charming. The English translation of her collection *Les contes de fées* (coll vols 1-3 **1696-7**, vol 4 **1698**; trans as *Tales of the Fairys* **1699** UK) gave the name "fairytale" to the form. Perrault's *Histoires ou Contes du temps passé; Avec des Moralitez* (coll **1697**; trans Guy Miège as *Histories, or Tales of Past Times* **1729** UK), on the other hand, through its simplicity and directness, contained far more memorable stories, and provided the seminal texts for many of our best-known UNDERLIERS, including SLEEPING BEAUTY, "Little Red Riding-Hood", PUSS IN BOOTS and CINDERELLA. Although not particularly aimed at children, Perrault's tales immediately attracted a younger audience. This was partly because each of Perrault's stories had a moral; though this form had been used by earlier writers, Perrault's morals were more direct and applicable to children. It was thus that the children's fairytale and the literary fairytale began to divide, a division made even more obvious by the nursery ghettoization of the medium in the UK, though fairytales for children became frowned upon with the growth of moral rectitude at the start of the 19th century. The fairytale continued to develop throughout Europe, further inflated by the Oriental stories which had become so popular in the wake of *The Arabian Nights* (◊ ARABIAN FANTASY). A series of anthologies of fairytales published in France and the Netherlands as **Le Cabinet des Fées** ["The Fairy Library"] started out with three volumes in 1731 and had expanded to 41 volumes by 1785, when the definitive set was published by Charles Mayer. This period of the French fairytale has been recreated in *Beauties, Beasts and Enchantments* (anth **1989**) ed Jack ZIPES, which includes the first full-length translation into English, as "The Story of Beauty and the Beast", of "La Belle et la Bête", of "La Belle et la Bête" (in *Les contes marins ou la jeune Américaine* coll **1740**) by Gabrielle-Suzanne de Villeneuve (1695-1755) – whose original version has been occluded by the shorter and more accessible "Beauty and the Beast" (1757) by Jeanne-Marie LEPRINCE DE BEAUMONT – which Zipes suggests may be the best-known fairytale in the world (◊ BEAUTY AND THE BEAST).

As the fairytale form faded in France it became adopted by the German Romantics (◊ ROMANTICISM). Their use of the word *Märchen* describes fairytales in the widest sense, featuring the supernatural, but not necessarily fairies. This was further distinguished by *Volksmärchen* (folktales) and *Wundermärchen* (wonder tales, or literary fairytales), which latter could be further identified as *Kindermärchen* (children's fairytales). The earliest collectors of such folktales were Johann MUSÄUS, with *Volksmärchen der Deutschen* (**1782-7** 5 vols; cut trans as *Popular Tales of the Germans* **1791** UK; vt *Popular Tales* 1826 UK) – which included "Richilde", an early version of SNOW WHITE – and Ludwig TIECK, with *Volksmärchen* (omni **1797**), which included his own versions of "Bluebeard" and PUSS IN BOOTS. The retrieval of folklore roots was crucial to the Romantic movement and, encouraged by Achim von Arnim (1781-1831) and Clemens Brentano (1778-1842), who collaborated on collections of folksongs and tales, the GRIMM BROTHERS began a systematic accumulation of oral tales, eventually collected as *Kinder- und Hausmärchen* (coll **1812**; vol 2 **1814**; rev 1819; rev 1822). What distinguished the Grimms from earlier raconteurs is that they took a redactive approach, transcribing the tales literally and ascribing their sources. Many of their stories are brief and not really fairytales, although others, including "Rapunzel", "Hansel and Gretel" and "Snow White and the Seven Dwarfs", immediately entered the fairytale canon. The first English translation (1823) was immensely popular, revolutionizing attitudes toward the fairytale. Thereafter the fairytale was reinstated in the UK nursery, although controversy continued over the allowable degree of violence. Bowdlerization of fairytales continued in the UK for the rest of the century.

The efforts of the Grimms were replicated throughout Europe by Thomas Crofton Croker (1798-1854) in Ireland, Robert Chambers (1802-1871) in Scotland, Peter Asbjörnsen (1812-1885), Jörgen Moe (1813-1882) in

Norway and Ludwig Bechstein (1801-1860) in Germany, but all merely consolidated the collection of folktales. One of the better writers of the period to create his own derivative but distinctive fairytales was Wilhelm HAUFF, in *Märchen für Söhne und Töchter Gebildeter Stände* ["Tales for the Sons and Daughters of Gentlefolk"] (coll **1825**). The next major European influence was Hans Christian ANDERSEN. Although he began in *Eventyr fortalte for Børn* (coll **1835**) by retelling folkloristic material, he rapidly began to create stories of his own, and the contents of his later annual **Eventyr og Historier** were almost all new. Although the fairytale as satire remained prevalent throughout Europe, especially in Germany, it was not until the 1970s that the adult appeal of the fairytale re-emerged in the UK and USA.

The translation of Andersen's fairytales into English in 1846 met with a rapturous response. The UK was experiencing a fairytale revolution, and many leading writers were using the form, such as Sara COLERIDGE with *Phantasmion* (**1837**), the first original fairytale written in English, Robert SOUTHEY with "The Three Bears" (1837), Catherine Sinclair (1800-1864) with *Holiday House* (**1839**), which flouted convention and treated children as children, Robert BROWNING with his poem "The Pied Piper of Hamelin" (1842), and John RUSKIN with *King of the Golden River* (written 1841; **1850**), a story in direct imitation of the Grimms.

The period 1840-75 was rich for the development of the UK fairytale, seeing it evolve from the traditional folktale through the nursery tale to become the CHILDREN'S FANTASY. All legends and myths were plundered. In many cases the stories were laden with moral and Christian messages, as in *Crowquill's Fairy Book* (coll **1840**) by Alfred Crowquill (real name Alfred Henry Forrester; 1804-1872), or were edited to the point of banality, a fate that befell many gathered in Mrs CRAIK's *The Fairy Book* (coll **1863**). It was at this stage that the Christmas PANTOMIME began to develop its distinct UK form, utilizing the fairytale for storylines. The pantomime in turn began to influence the composition of later tales, especially *The Rose and the Ring* (**1855**) by William Makepeace THACKERAY.

The UK fairytale became more ambitious. Significant works include: *The Hope of the Katzekopfs* (**1844**) by Francis Edward Paget (1806-1882), set entirely in Faerie, where a spoiled child learns the error of his ways; *The Enchanted Doll* (**1849**) by Mark Lemon (1809-1870), the first editor of *Punch*; and *Granny's Wonderful Chair* (**1856**) by Frances Browne (1816-1879), where a chair tells fairytales and takes its young occupant to far lands. Thereafter transitional forces came to work. *The Water-Babies* (**1863**) by Charles KINGSLEY calls itself "A Fairytale for a Land-Baby" but is not a fairytale. *Phantastes* (**1858**) by George MACDONALD is subtitled "A Faerie Romance for Men and Women" and uses all of the motifs of the fairytale, though heavily allegorized, to create a CHRISTIAN FANTASY which is arguably the first serious ADULT FANTASY. Macdonald went on to produce his own genuine fairytales, many collected as *Dealing with the Fairies* (coll **1867**), which includes what is often regarded as the best of all Victorian fairytales, "The Golden Key". He also wrote some transitional children's fantasies (regarded by him as fairytales) – *At the Back of the North Wind* (**1871**), *The Princess and the Goblin* (**1872**) and *The Princess and Curdie* (**1882**) – and encouraged Lewis CARROLL to publish *Alice's Adventures in Wonderland* (**1865**), which further developed the CHILDREN'S FANTASY away from the fairytale.

The fairytale, as distinct from the embellished folktale, became further relegated to the nursery. To their credit, many writers continued to bring significant and creative ideas to the fairytale, but their work was pigeonholed and dismissed by later generations until their rediscovery in recent years. They include: Anne Isabella Ritchie (1837-1919), the daughter of Thackeray, with *Five Old Friends and a Young Prince* (coll **1868**); Jean INGELOW, with *Mopsa the Fairy* (**1869**); Juliana H. Ewing (1841-1885), particularly with *The Brownies and Other Tales* (coll **1870**) and *Old-Fashioned Fairy Tales* (coll **1882**); Mary de Morgan (1850-1907), with *On a Pincushion* (coll **1877**), *The Necklace of Princess Fiorimonde* (coll **1888**) and *The Windfairies* (coll **1900**); Harriet Childe-Pemberton, in her heavily didactic and revisionist *Fairy Tales of Every Day* (coll **1882**); the long series by E.H. KNATCHBULL-HUGESSEN; and Lucy CLIFFORD with the remarkably atmospheric *Anyhow Stories* (coll **1882**) and *The Last Touches* (coll **1892**). The fairytale reached its final Victorian fling in works by Oscar WILDE and Andrew LANG. In addition to his own highly readable stories – *The Princess Nobody* (**1884** chap), *The Gold of Fairnilee* (**1889** chap) and the **Prince Prigio** series – Lang edited the long-running series of coloured **Fairy Books**, starting with *The Blue Fairy Book* (anth **1889**). These volumes, comprising a collection of folktales from around the world, served to confirm how interlinked the folktale and fairytale had become in the public mind. Their popularity begat imitations. *The Strand* regularly ran fairytales and children's stories, and these were eventually collected in four anthologies: *The Golden Fairy Book* (anth **1894**), *The Silver Fairy Book* (anth **1895**), *The Diamond Fairy Book* (anth **1897**) and *The Ruby Fairy Book* (anth **1900**), all ed anon.

By the turn of the century, with children's fantasy established as a separate medium, only Laurence HOUSMAN, Lord DUNSANY, Kenneth MORRIS – and later Walter DE LA MARE – sustained the fairytale form, aimed sometimes at children, sometimes at adults. But by WWI the relegation of the fairytale to the nursery was at last complete.

This was also true to some extent in the USA, but there the fairytale had a less rigid pedigree and evolved by degrees rather than natural selection. By mid-century the children's fantasy was emerging in the USA with *The Last of the Huggermuggers* (**1855**) and its sequel *Kobboltzo* (**1856**) by C.P. Cranch. The fairytale survived on very shallow soil, children's stories generally dealing with either frontier adventure or more homely family circumstances. Only with the emergence of the *St Nicholas Magazine* in 1873 did the fairytale begin to appear on a more regular basis, thanks mostly to the writings of Julian HAWTHORNE, Frank R. STOCKTON and Charles E. Carryl (1841-1920). Stockton's stories, especially in *Ting-a-Ling* (coll **1870**) and *The Floating Prince* (coll **1881**), have endured better than those of his contemporaries, though Hawthorne's *The Yellow Cap and Other Fairy Tales* (coll **1880**) is undeservedly neglected. Carryl, in *Davy and the Goblin* (1884) and *The Admiral's Caravan* (**1891**), admirably blended fairytale motifs into children's fantasies. L. Frank BAUM developed two books of fairytales: *American Fairy Tales* (coll **1901**), an extremely idiosyncratic volume which does all it can to break the mould while just staying within it, and *The Life and Adventures of Santa Claus* (**1902**), a delightful book that creates a life for SANTA CLAUS as one long fairytale.

The new century and the emergence of SCIENCE FICTION further relegated the fairytale. It may be argued that the emergence of HEROIC FANTASY during this period, especially in the US pulp MAGAZINES, constituted a development of the fairytale, and certainly some of these stories could be regarded as modern fairytales, but by now the model had become confusing, the original fairytale having long been supplanted by new forms with differing purposes. The genuine fairytale was rare. Dunsany's work remained the closest to the true form, and mostly US imitations of this – by Donald CORLEY, Vernon KNOWLES, H.P. LOVECRAFT, Clark Ashton SMITH, etc. – continued the thread of the fairytale. An attempt at revisionist fairytales as satires on modern society failed in *The Fairies Return* (anth **1934**) ed anon Peter Davies, despite the presence of A.E. COPPARD, Clemence DANE and Dunsany.

In the UK the next development came with J.R.R. TOLKIEN's *The Hobbit* (**1937**), although its impact was not evident at the time. This is a sustained fairytale in the purest sense; its sequel, *The Lord of the Rings* (**1954-5**), more resembles a medieval ROMANCE. This same dichotomy affected Tolkien's shorter fantasies: "Leaf By Niggle" (1945) and *Farmer Giles of Ham* (**1949** chap) are both ALLEGORIES of life, but the former assumes the fairytale form and the latter the style of romance.

Post-WWII adult fantasy used the fairytale form on occasion, particularly in FANTASTIC, where stories by Ursula K. LE GUIN were finely crafted fairytales. As modern fantasy established itself, so writers turned again to the fairytale to explore and reutilize the themes and motifs. Nicholas Stuart GRAY was a pioneer, his stories of the 1950s-60s superbly recapturing the purity of the fairytale. The leading new writers of original fairytales from the 1970s onward include Joan AIKEN, Steven BRUST, Louise COOPER, Charles DE LINT, Dianna Wynne JONES, Nancy KRESS, Tanith LEE, Patricia MCKILLIP, Robin MCKINLEY, Vivian VANDE VELDE, Patricia C. WREDE, Jane YOLEN and Mary Frances Zambreno (1954-   ). Many of these writers have revisited the original tales and sought to make them contemporary (◊ REVISIONIST FANTASY), some retaining the fairytale form, though most have forsaken it. The most popular tale is BEAUTY AND THE BEAST. Robin McKinley recast it in *Beauty* (**1978**). The success of the tv series BEAUTY AND THE BEAST (1987-90) led to several spinoff novels. SLEEPING BEAUTY was replayed in *Briar Rose* (**1992**) by Jane Yolen, in *Beauty* (**1991**) by Sheri S. TEPPER, and (pornographically) in the **Sleeping Beauty** series by A.N. Roquelaure (Anne RICE). The story of CINDERELLA was merged with that of the PIED PIPER in *Ashmadi* (**1985** Germany; trans as *The Coachman Rat* **1987** UK) by David Henry WILSON. The tale of SNOW WHITE was recast into an Elizabethan setting by Patricia C. Wrede in *Snow White and Rose Red* (**1989**), while the FROG PRINCE and "The Twelve Dancing Princesses" are retold in Robin McKinley's *The Door in the Hedge* (coll **1981**). Hans Christian Andersen's story of the Seven Swans formed the basis of Nicholas Stuart Gray's *The Seventh Swan* (**1962**) and of *Swan's Wing* (**1984**) by Ursula Synge (1930-   ). The tale of JACK the Giant-Killer formed the basis of Charles de Lint's *Jack, the Giant-Killer* (**1987**) and *Drink Down the Moon* (**1990**).

Other modern renditions of fairy- and folktales are found in *The King's Indian* by John GARDNER (**1974**), *The Iron Wolf and Other Stories* by Richard ADAMS (coll **1980**), *Fire and Hemlock* (**1984**) by Diana Wynne Jones, *The Sun, the Moon, and the Stars* (**1987**) by Steven Brust, *Provençal Tales* (coll **1988**) by Michael DE LARRABEITI, *Rusalka* (**1989**) and its

sequels by C.J. CHERRYH and *Thomas the Rhymer* (**1990**) by Ellen KUSHNER. Sylvia Townsend WARNER recrafted the fairytale in *Kingdoms of Elfin* (coll **1977**) and Angela CARTER not only used the fairytale to explore the darker side of human nature in *The Bloody Chamber* (coll **1979**) and *Fireworks* (coll **1987**) but also brought its brighter side back into the daylight in a couple of fairytale anthologies. A.S. BYATT brought a cosmopolitan breadth to the traditional fairytale in *The Djinn in the Nightingale's Eye* (coll **1994**).

The rediscovery of the values of the fairytale in the past 20 years owes much to Ursula K. LeGuin, Jane Yolen, Terri WINDLING, who brought her knowledge of the field to developing markets for new writers, and Jack ZIPES, who re-established the pedigree of the fairytale tradition. At its best the modern fairytale contains some of the most beautiful and memorable imagery in all fantasy, and remains true to its spirit.

Important recent anthologies include: *Twelve Dancing Princesses and Other Fairy Tales* (anth **1964**) ed Alfred and Mary David; *Tales of Kings and Queens* (anth **1965**) ed Herbert van Thal (1904-1983); *The Classic Fairy Tales* (anth **1974**) ed Iona Opie (1923-   ) and Peter Opie (1918-1982); *Elsewhere* (anth **1981**), *Elsewhere Vol. II* (anth **1982**) and *Elsewhere Vol. III* (anth **1984**) ed Terri Windling and Mark Alan Arnold; *A Christmas Carol and Other Victorian Fairy Tales* (anth **1983**) ed U.C. Knoepflmacher (1931-   ); *Faery!* (anth **1985**) ed Windling; *The Victorian Fairy Tale Book* (anth **1988**) ed Michael Patrick Hearn (1950-   ); *Victorian Fairy Tales* (anth **1987**), *Spells of Enchantment* (anth **1991**; vt *The Penguin Book of Western Fairy Tales* 1993 UK) and *The Outspoken Princess and the Gentle Giant* (anth **1994**) ed Jack ZIPES; *Isaac Asimov's Magical Worlds of Fantasy: Faeries* (anth **1991**) ed Isaac ASIMOV, Martin H. GREENBERG and Charles G. Waugh; *Once Upon a Time* (anth **1991**) ed Lester del Rey (1915-1993) and Risa Kessler, all original stories; *Forbidden Journeys* (anth **1992**) ed Nina Auerbach (1943-   ) and Knoepflmacher; *Caught in a Story* (anth **1992**) ed Christine Park and Caroline Heaton; *The Oxford Book of Modern Fairy Tales* (anth **1993**) ed Alison Lurie (1926-   ); *A Fairy Tale Reader* (anth **1993**) ed John and Caitlín Matthews; *Wonder Tales* (anth **1994**) ed Marina WARNER; and «The Mammoth Book of Fairy Tales» (anth 1997) ed Mike ASHLEY.

In addition, many volumes of feminist fairytales have recently been compiled, including *Don't Bet on the Prince* (anth **1986**) ed Jack Zipes plus the revisionist series **Fairytales for Feminists**, ed anon, though steered for the most part by a collective including Anne Claffey, Linda Kavanagh and Sue Russell: *Rapunzel's Revenge* (anth **1985**), *Ms Muffet and Others* (anth **1986**), *Mad and Bad Fairies* (anth **1987**), *Sweeping Beauties* (anth **1989**), *Cinderella on the Ball* (anth **1991**), all with original stories, plus the retrospective *Ride on Rapunzel* (anth **1992**). Further REVISIONIST FANTASIES based on fairytales with a high feminist consciousness include the series ed Ellen DATLOW and Terri Windling: *Snow White, Blood Red* (anth **1993**), *Black Thorn, White Rose* (anth **1994**) and *Ruby Slippers, Golden Tears* (anth **1995**). [MA]

FAIRY TIME ◊ TIME IN FAERIE.

FAKINOU, EUGENIA (1945-   ) Greek author of four novels, of which only *To Evdhomo Roucho* (**1983**; trans as *The Seventh Garment* **1991** UK) has appeared in English. An imaginative recapitulation of the sorrows of modern Greek history, this employs numerous elements of MAGIC REALISM, including supernatural events, elements from MYTHOLOGY and the unifying device of "the Tree", a (sapient) figure of Greek identity and endurance. [GF]

FALCONER, LANOE Working name of UK author Mary Elizabeth Hawker (1848-1918). In a brief literary career blighted by illness she achieved great success with two much-reprinted short novels, *Mademoiselle Ixe* (**1890**) and the psychological GHOST STORY *Cecilia de Noël* (**1891**). In the latter a resident GHOST affects individual visitors to an English country house. [RD]

FALKNER, J(OHN) MEADE (1858-1932) UK armaments manufacturer and writer whose first novel, *The Lost Stradivarius* (**1895**), is a compellingly told SUPERNATURAL FICTION. The 19th-century protagonist, after playing an anonymous 18th-century piece of music, finds himself haunted by the GHOST of a musician and profligate from that period. He traces the ghost to a sealed-over cupboard, which when opened reveals the eponymous violin. The protagonist's life goes downhill from this point. JMF imparts a message similar to that conveyed by Vernon LEE: that art – in particular the art of 18th-century Italy – is deeply alluring but deadly. The terror underlying the work of both authors is not in fact of POSSESSION but of transcendence. [JC]

FALL ◊ SEASONS.

FALL FROM GRACE ◊ ADAM AND EVE; EDEN.

FALL OF THE HOUSE OF USHER, THE vt of *The House of Usher* (**1960**). ◊ *The* HOUSE OF USHER.

FAMOUS FANTASTIC MYSTERIES US pulp MAGAZINE, 81 issues, September/October 1939-June 1953; published originally by Frank A. Munsey (1854-1925), then from March 1943 by Popular Publications ed Mary Gnaedinger (1898-1976).

Primarily a reprint magazine, *FFM* presented stories from the early Munsey pulps (especially *All-Story Weekly*, *Argosy* and *Cavalier*). Although *FFM* was aimed at the growing sf readership, it also reprinted stories drawing on the occult and "borderland science". It placed special emphasis on the lost-world stories (◊ LOST RACES) and PLANETARY ROMANCES of J.U. GIESY, Abraham MERRITT and Charles B. Stilson (1880-1932) and the scientific romances of Ray Cummings (1887-1957), George Allan England (1877-1936), Homer Eon Flint (1888-1924) and Austin Hall (1880-1933). When *FFM* was taken over by Popular Publications the policy switched to reprinting stories that had previously been published only in book form. This resulted in the appearance of many UK authors, especially E.F. BENSON, Algernon BLACKWOOD, G.K. CHESTERTON, Lord DUNSANY, H. Rider HAGGARD, C.J. Cutcliffe Hyne (1866-1944), Arthur MACHEN, Sax ROHMER, E.C. VIVIAN and, through arrangements with August W. DERLETH, William Hope HODGSON, which was the first US popularization of his work. The new policy also resulted in the appearance of names normally unknown to fantasy magazines, like Warwick Deeping (1877-1950), C.S. Forester (1899-1966), Franz KAFKA, Ayn Rand (1905-1982) and Edward Shanks (1892-1953). Under Popular's regime *FFM* also included the occasional original story by Ray BRADBURY, Henry KUTTNER, C.L. MOORE and others. The magazine's artwork, especially that by Virgil FINLAY and Lawrence Sterne Stevens (1886-1960), was notable. Despite a brief flirtation with a large-digest format in 1951, *FFM* remained a pulp to the end. A representative anthology is *Famous Fantastic Mysteries* (anth **1991**) ed Stefan DZIEMIANOWICZ, Robert WEINBERG and Martin H. GREENBERG.

FFM had two similar companions, *Fantastic Novels* (2 series: 5 issues July 1940-April 1941; 20 issues March 1948-June 1951) and *A. Merritt's Fantasy Magazine* (5 issues December 1949-October 1950). [MA]

FANCIFUL TALES ◊ MAGAZINES.

FANTASIA US ANIMATED MOVIE (**1940**). Disney. **Production** sv Ben Sharpsteen. **Dir** James Algar ("Sorcerer's Apprentice"), Samuel Armstrong ("Toccata and Fugue in D Minor", "Nutcracker Suite"), Ford Beebe, Jim Handley and Hamilton Luske ("Pastoral Symphony"), Norm Ferguson and T. Hee (Walt Disney) ("Dance of the Hours"), Wilfred Jackson ("Night on Bald Mountain" and "Ave Maria"), Bill Roberts and Paul Satterfield ("Rite of Spring"). **Voice actor** Deems Taylor (narrator). 120 mins. Colour.

F grew from a meeting between Walt DISNEY and Leopold Stokowski (1887-1977). Initially Stokowski's collaboration was to be on a DISNEY **Silly Symphony** short based on Dukas's *The Sorcerer's Apprentice* (1897), but both men had a grander idea: a full-length feature-movie anthology of such extracts. The Dukas interpretation became perhaps *F*'s most famous, with MICKEY MOUSE cast as the SORCERER'S APPRENTICE who disobeys his master Yen Sid ("Disney" backwards); originally Dopey (from SNOW WHITE AND THE SEVEN DWARFS [**1937**]) had been proposed for the role. Almost equally famous are sections based on: Igor Stravinsky's *The Rite of Spring* (1913), with lumbering DINOSAURS; Modeste Moussorgsky's *Night on Bald Mountain* (composed *c*1866), with Chernabog waking from the hillside to receive the worship of assorted GHOSTS, ghouls, imps, etc.; and, for all the wrong reasons, Ludwig van Beethoven's 6th Symphony (1808), which has capering centaurs, over-cute CUPIDS and a rollicking Bacchus. In all instances the music was played by the Philadelphia Orchestra, conducted by Stokowski.

F polarizes opinions, with some seeing it as crass and others as a masterpiece. Initially regarded as impossibly long by its distributors, RKO, *F* was cut for first release to 81 min and was not very successful. Among other bitternesses associated with the movie was Stravinsky's horror over unauthorized "improvements" made to his score. But during the 1960s a re-release brought *F* sudden popularity, its psychedelic SURREALISM appealing to hippy culture. [JG]

Further reading: *Walt Disney's "Fantasia"* (**1983**) by John Culhane.

See also: ALLEGRO NON TROPPO (**1976**).

FANTASIES OF HISTORY Tales which uncover a Secret History of the World, doing so with the aid of fantasy devices. These fantasy elements include invocations of ELDER GODS, plumbings of fictional BOOKS, and discoveries that the rulers of the world are immortal (◊ IMMORTALITY), comprise PARIAH ELITES, are SECRET MASTERS and/or inhabit societies in a WAINSCOT relationship to the world we commonly know. An FOH is set in the mundane world, either now or at some point in the past, no matter how fundamentally our understanding of the mundane world may be rewritten through the revelations of the tale. FOHs cannot occur in SECONDARY WORLDS, FANTASYLANDS or WONDERLANDS.

The FOH can resemble sf; and in some STEAMPUNK and GASLIGHT ROMANCE texts, notably by Tim POWERS and James BLAYLOCK, certainly does. That connection with history and historical process is one, essentially, of control: the FOH is interested only in a Secret History of the World which invokes a controlling ruler or faction, or one seriously aspiring to control. Indeed, the conspiracy subtext of many FOHs *necessitates* successful conspirators. Text, rune and labyrinth are central to the texture of FOHs, enforcing a sense that in these texts the history of the world can be *deciphered*. FOHs tend to feel as though they must be unlocked, or subjected to *unpacking*, rather than merely read.

In "The Secret Masters of the World", Michael Dirda describes Lawrence NORFOLK's *Lemprière's Dictionary* (1991) as an "antiquarian fantasy", which he defines as a text which tends to juxtapose the present and the past, disclose awesome, frequently gamelike conspiracies at work in history, draw heavily on some branch of arcane learning (CHESS, Renaissance hermeticism), and provide a trail of scholarly "documents". It is this element of play and the feeling of an intelligent mind in control that prevents such fantasies being only a variant of SUPERNATURAL FICTION, even when the protagonists can do no more than discover the extent of the conspiracy and realize their own powerlessness against it. A central examination and embodiment of the form is Umberto ECO's *Foucault's Pendulum* (1989), which meditates on the ways in which humanity's constant generation of theories about the origins of power is itself a principal source of that power.

The principal overlap between FOH and SUPERNATURAL FICTION is in the case of ELDER GODS. H.P. LOVECRAFT and his disciples elaborated the CTHULHU MYTHOS in which the Earth is surrounded in space and other dimensions by powerful beings who dislike humanity; perhaps inevitably, the mythos came to serve for Lovecraft as a symbolic parallel with various paranoid dreads about the world, including degeneration and multi-racial society. FOHs have enough in common with genuine conspiracy theories that they can generate real conspiracy theories – it is believed by some occultists, for example, that Lovecraft only *thought* he was making it all up.

The issue is further confused by the fact that various writers whose work includes FOHs – e.g., Sax ROHMER, Arthur MACHEN – were occultists; when Machen writes of malevolent FAERIE as a LOST RACE of Picts with supernatural powers, it is possible he believed in such things. An active religious belief also figures in some CHRISTIAN FANTASIES that, like those of Charles WILLIAMS, resemble FOHs.

Generally, when FOHs do not end in the destruction, cooption or mere survival of their protagonists, they tend towards INSTAURATION FANTASY. The assertion of the agency of the protagonist in the struggle with principalities and powers is almost inherently a statement of renewal, save when that agency, as in Tim POWERS's *The Stress of her Regard* (1989), is merely a clearing of the board; the alliance between the silicon-based LAMIAS and the Austro-Hungarian empire is broken, but there is not enough energy left in Powers's drained protagonist to affect the subsequent direction of history. In Elizabeth HAND's *Waking the Moon* (1994 UK), Sweeney is the human observer who perceives a battle between aspects of the GODDESS which have possessed her former friends – her perception and sympathy may be held to be agency enough and all that humanity is capable of in the face of power. She is explicitly contrasted in her acceptance of powerlessness with the patriarchal Benandanti, whose ruthless conspiratorial efforts produce results contrary to their desires. SECRET MASTERS are impotent in the face of the Goddess. John CROWLEY's *Little, Big* (1981) is clearly an FOH as well

as an instauration fantasy, given that it includes conspiracies – the Rod and Gun club, Secret Masters of the USA – Elder Gods (in the shape of FAERIE) and a variety of esoteric texts, notably the Least Trumps themselves and the various recensions of Drinkwater's *Architecture of Country Houses*.

There are often elements of FOH present in genre thrillers, whether those that deal with surviving Nazi conspirators – e.g., Ira LEVIN's *The Boys from Brazil* (1976) – or those which deal with strange and murderous cults – e.g., Michael Dibdin's *Dark Spectre* (1995). The protagonists or villains of such thrillers act as if they were living in an FOH, whatever the true state of affairs. [JC/RK/GW]

FANTASTIC Since both words derive from the same root, "fantastic" might be logically regarded as the adjective form of "fantasy"; in practice the word is rarely used in that way. When first applied as a critical term in the sf community of the 1930s-40s, "fantastic" functioned as a blanket description of both sf and fantasy works, as seen in the titles of E.F. BLEILER's *The Checklist of Fantastic Literature* (1948) and the magazine FAMOUS FANTASTIC MYSTERIES (1939-53). In a similar but more expansive spirit, "the fantastic" has been recently adopted by critics as a general term for all forms of human expression that are not realistic, including fantasy and sf, MAGIC REALISM, FABULATION, SURREALISM, etc. Thus there is the annual International Conference on the Fantastic in the Arts and its associated publication, *The Journal of the Fantastic in the Arts*.

In the 1970s Tzvetan TODOROV defined "the fantastic" in a more restrictive manner to describe stories in which unusual events might have either a natural or a supernatural explanation; Henry JAMES's *The Turn of the Screw* (1898) is the favourite example. In Todorov's scheme "the fantastic" occupies an intermediate, and privileged, position between "the uncanny" – stories where unusual events are clearly assigned a natural explanation – and "the marvellous" – stories where unusual events are clearly assigned a supernatural explanation. Although Todorov thus seems to marginalize or even trivialize traditional fantasy and sf, his ideas have been frequently employed by modern academic critics, with occasionally fruitful results. [GW]

FANTASTIC US digest MAGAZINE, companion to *Amazing Stories* (◊ SFE), 208 issues Summer 1952-October 1980, published by Ziff-Davis, New York, until June 1965, thereafter by Ultimate Publishing, New York; ed Howard Browne (1908-) Summer 1952-August 1956, Paul W. Fairman (1916-1977) October 1956-November 1958, Cele Goldsmith (1933-) December 1958-June 1965 (as Cele G. Lalli from July 1964), Joseph Ross (real name Joseph Wrzos; 1929-) September 1965-November 1967, Harry Harrison (1925-) (January-October 1968), Barry N. Malzberg (1939-) (December 1968-April 1969), Ted White (1938-) (June 1969-January 1979), Elinor Mavor (1936-) April 1979-October 1980 (initially under pseudonym Omar Gohagen). *F* was largely bimonthly but monthly February 1957-June 1965 and quarterly after 1976. It also underwent minor title changes, including *Fantastic Science Fiction* (April 1955-September 1960) and *Fantastic Stories* during Ultimate's regime, though it was always registered as *F*. It should not be confused with the large format *Fantastic Science Fiction* (2 issues August-December 1952) ed Walter B. Gibson.

F was launched to cash in on the wider interest fantastic fiction had gained by the early 1950s. Browne was more interested in fantasy than sf, and believed he could produce a magazine to rival the slicks, especially *The New Yorker* and *The Saturday Evening Post*; unsurprisingly, some of the contents may be classified as SLICK FANTASY, especially contributions by Isaac ASIMOV, Ray BRADBURY, Horace L. GOLD and Kris Neville (1925-1980). Browne presented new and reprint material by Truman Capote (1924-1984), Raymond Chandler (1888-1959), Shirley JACKSON, B. Traven (real name Otto Feige; 1882-1969), Cornell Woolrich (1903-1968) and Evelyn Waugh (1903-1966), aiming to capture a wider market, but the core was SUPERNATURAL FICTION by, e.g., Anthony BOUCHER, Fritz LEIBER, Richard MATHESON and particularly Theodore STURGEON. There was a UK edition during this period from Strato Publications (8 undated issues December 1953-February 1955).

By 1955 the budget had been cut and *F*'s quality plummeted. Browne remained editor but left the task largely to Paul Fairman, who was also one of the major pseudonymous contributors. Although most of the stories were hackwork, with the emphasis on sf and humorous fantasy, the fertile minds of a new generation of writers, especially Harlan ELLISON, Randall GARRETT, Milton Lesser (1928-) and Robert SILVERBERG, allowed for some ingenuity of concept if not of treatment. Several issues covered DREAM fantasies and wish-fulfilment, resulting in a short-lived companion *Dream World* (3 issues February-August 1957).

F began to regain some of its former glory under Cele Goldsmith, especially with the November 1959 issue, which was dedicated to Fritz LEIBER and saw the revival of **Fafhrd and the Gray Mouser**. Their popularity made *F* a focus for stories of SWORD AND SORCERY, including imitative material by John JAKES and Michael MOORCOCK, although the strength of *F* in the early 1960s lay in the diversity of its fiction. It was this freedom that encouraged the emergence of Piers ANTHONY, Thomas M. DISCH, Ursula K. LE GUIN, Norman Spinrad (1940-) and Roger ZELAZNY (whose **Dilvish** stories first appeared here).

In the period 1960-65 *F* was the premier fantasy magazine; apart from the UK SCIENCE FANTASY it was almost a single voice. It published idiosyncratic stories by J.G. BALLARD, Neal Barrett Jr (1929-), Rosel George Brown (1926-1967), Philip K. Dick (1928-1982), Daniel F. Galouye (1920-1976), Randall Garrett, Ron GOULART, Arthur Porges (1915-) and Jack Sharkey (1931-1992), few of which would have appeared elsewhere. Some stories were affected by the paranoia of the Cold War, and explored the realities hidden behind the façade of life, but others were just effervescent romps enjoying the chance to broaden the boundaries of fantasy and supernatural fiction.

During 1965-70, however, under Ultimate's regime, *F* was a reprint magazine, which included some good stories from its early days and from FANTASTIC ADVENTURES, but also much weak material. Gradually these stories were siphoned off into other all-reprint magazines, of which *Science Fantasy* (4 issues [Summer] 1970-Spring 1971; *#1* published as *Science Fantasy Yearbook*), *Strange Fantasy* (6 issues Spring 1969-Fall 1970, though numbered *#8-#13* as it continued from another reprint title, *Science Fiction Classics*), *The Strangest Stories Ever Told* (1 issue Summer 1970) and *Weird Mystery* (4 issues Fall 1970-Summer 1971) drew their contents from *Fantastic Adventures* and *F*. This allowed *F* to return to publishing new material, and under Ted White the magazine again blossomed. Although it slanted itself strongly towards S&S, especially with new

Conan stories by L. Sprague DE CAMP and Lin CARTER, and Carter's own **Thongor** series, *F*'s poor distribution meant it never benefited from the upsurge of interest in this general subgenre during the 1970s. *F* also featured much fantasy and supernatural fiction in the style of UNKNOWN, especially in the works of Juanita COULSON, Avram DAVIDSON, de Camp, Marvin KAYE, Keith Laumer (1925-1993) and Thomas Burnett SWANN, and was also strong in off-trail stories, notably those by Grant Carrington (1938-), Geo Alec Effinger (1947-), Gordon Eklund (1945-) and Richard Lupoff (1935-). This period gave rise to another brief companion publication, *Sword & Sorcery Annual* (1 issue 1975). During this period *F* was at an artistic height, especially with its covers, by Stephen Fabian (1930-), Jeff JONES and Esteban MAROTO.

After White's resignation *F* returned to reprints, and though for a few issues there was a sign of recovery, with good fantasies by Marvin Kaye, Darrell SCHWEITZER and Wayne Wightman (1946-), finances forced *F*'s merger with *Amazing Stories*. The title remained on the masthead for some years, and *Amazing* began to publish a high quota of fantasy, but the aura and image of the old *F* were gone. *The Best from Fantastic* (anth **1973**) ed Ted White and *Fantastic Stories: Tales of the Weird and Wondrous* (anth **1987**) ed Martin H. GREENBERG and Patrick Lucien Price are good, but not representative. [MA]

FANTASTIC ADVENTURES US pulp MAGAZINE, 128 issues May 1939-March 1953, published by Ziff-Davis, New York, initially bimonthly, but monthly January-June 1940 and again from May 1941; ed nominally Raymond A. Palmer (1910-1977) May 1939-December 1949, and Howard Browne (1908-) January 1950-March 1953, though William L. Hamling (1921-) was managing editor November 1947-February 1951.

Like FANTASTIC, *FA* was launched as a fantasy companion to *Amazing Stories* (◊ SFE), although early issues contained much immature sf, but with a wider remit to include LOST-RACE stories. Palmer was a fanatical devotee of Edgar Rice BURROUGHS, whom he was able to lure to *FA* from *#2*. Nevertheless, *FA* almost folded in 1940; it was only the reader reaction to **Jongor** in "Jongor of Lost Land" (October 1940) by Robert Moore Williams (1907-1977) – imitation TARZAN – that saved it. During WWII *FA* provided a diet of lighthearted whimsical fantasies adequately produced by Nelson S. BOND, Robert BLOCH, William P. McGivern (1924-1982), David Wright O'Brien (1918-1944) and Leroy Yerxa (1915-1946). Series included Bloch's **Lefty Feep** and McGivern's **Enchanted Bookshelf**, in which the Three Musketeers return to life. *FA* was visually attractive, especially the covers by Harold McCauley (1913-1983).

After WWII, steered by Hamling, *FA* turned to more darkly SUPERNATURAL FICTIONS, still mostly written by Ziff-Davis's Chicago stable but also including stories by Ray BRADBURY, August W. DERLETH, Geoff St Reynard (real name Robert W. Krepps; 1919-) and Theodore STURGEON. A brief success came with the **Toffee** stories, about a DREAM girl, by Charles F. Myers (1922-), beginning with "I'll Dream of You" (1947). In its final years, *FA* published some strong material by Lester del Rey, Fritz LEIBER, Sturgeon and William Tenn (1920-). Its liberal policy allowed the introduction of new, more experimental writers, and it was in *FA* that John JAKES and Mack Reynolds (1917-1983) first appeared. But, with the success of Browne's slick-styled FANTASTIC, *FA*'s days were numbered, and the two magazines were merged in May/June 1953.

Unsold issues of *FA* were rebound and published as *Fantastic Adventures Quarterly* (2 series: 8 issues Winter 1941-Fall 1943; 11 issues Summer 1948-Spring 1951). Two separate UK editions were published in cut form, 2 issues (undated July, September 1946) and 24 issues (undated June 1950-February 1954). No anthology has been based solely on *FA*, although *Fantastic* reprinted heavily from *FA* in the period 1965-70, as did the companion reprint magazines *Fantastic Adventures Yearbook* (1 issue 1970), *Strange Fantasy* (6 issues Spring 1969-Fall 1970, though numbered *#8-#13* as it continued from another reprint title, *Science Fiction Classics*), *The Strangest Stories Ever Told* (1 issue Summer 1970) and *Weird Mystery* (4 issues Fall 1970-Summer 1971). Hamling continued his formula from *FA* briefly in *Imagination* (63 issues October 1950-October 1958) and *Imaginative Tales* (26 issues September 1954-November 1958; *#24-#26* retitled *Space Travel*), but both of these became solely sf from 1955. [MA]

Further reading: *The Annotated Guide to Fantastic Adventures* (1985) by Edward J. Gallagher (1940-   ).

FANTASTIC ADVENTURES QUARTERLY ◊ FANTASTIC ADVENTURES.

FANTASTIC ADVENTURES YEARBOOK ◊ FANTASTIC ADVENTURES.

FANTASTIC DISAPPEARING MAN, THE UK vt of *The Return of Dracula* (*1958*). ◊ DRACULA MOVIES.

FANTASTIC JOURNEY, THE US tv series (1977). Bruce Lansbury Productions/Columbia Pictures/NBC. **Pr** Leonard Katzman. **Exec pr** Bruce Lansbury. **Dir** Barry Crane and many others. **Writers** D.C. Fontana (◊ SFE), Katzman and many others. **Starring** Ike Eisenmann (Scott Jordan), Carl Franklin (Dr Fred Walters), Jared Martin (Varian), Roddy McDowall (Dr Jonathan Willoway), Katie Saylor (Liana). 1 90min and 9 1hr episodes. Colour.

A research ship in the Bermuda Triangle is drawn into a strange cloud and wrecked. The survivors find themselves on a strange, uncharted ISLAND where different TIME-lines exist together, complete with beings from other years and races. To complicate matters, fierce timestorms sweep the area, tossing the travellers between eras and leaving chronal chaos in their wake. The travellers meet Varian, a man from the 23rd century, who tells them of Evoland, a rumoured way home, and together search for a way out, only to find themselves endangered by relic people from ATLANTIS. With the assistance of Liana, an Atlantean who can communicate with animals, they escape, but in the second episode most are killed. Jordan, Varian and Liana, with a rebellious scientist, Willoway, battle across the island towards the elusive Evoland . . . but the series ended before they got there. [BC]

FANTASTIC NOVELS (magazine) ◊ FAMOUS FANTASTIC MYSTERIES.

FANTASTIC SCIENCE FICTION (magazine) ◊ FANTASTIC.

FANTASTIC STORIES (magazine) ◊ FANTASTIC.

FANTASTIC UNIVERSE ◊ FEAR; MAGAZINES.

FANTASTIC VOYAGES TRAVELLER'S TALES are among the oldest forms of narrative discourse. Journeys of discovery play a crucial part in FOLKLORE, MYTHOLOGY, epic poetry and medieval ROMANCE, and similar ALLEGORIES of maturation remain central to modern fantasy. It can be argued that the principal value of CHILDREN'S FANTASY is its ability to provide a sentimental education appropriate to those who do not yet possess an adult consciousness of the world; in this view, the FVs of children's fantasy mirror and provide counsel for the child's problematic progress to adulthood.

Two basic patterns of the FV, the former most familiar in the stories of the Argonauts (◊ GOLDEN FLEECE) and Sinbad (◊ ARABIAN FANTASY) and the latter in the tale of ODYSSEUS, are the expeditionary QUEST and the peregrinations of an ACCURSED WANDERER unable to reach his destination. FAIRYTALES often use a variant in which a son leaves home to "seek his fortune". The imaginary voyage became an important instrument of SATIRE following the example set by *Utopia* (**1551**) by Thomas More (1478-1535), and still remains a useful satirical strategy. Many VISIONARY FANTASIES also take the form of FVs, often incorporating elements of satire and allegory.

Classics of fantasy cast in this mould include DANTE's *Divine Comedy* (written 1307-21), *The Travels of Sir John Mandeville* (*c***1360**) ◊MANDEVILLE, *Amadis of Gaul* (*c***1450**), ARIOSTO's *Orlando Furioso* (**1532**), *The Pilgrim's Progress* (**1678-84**) by John BUNYAN, *Gulliver's Travels* (**1726**) by Jonathan SWIFT, *Rasselas* (**1759**) by Samuel Jonson (1709-1784), *The Princess of Babylon* (**1768**) by Voltaire (1694-1778), *Vathek* (**1786**) by William BECKFORD, *She* (**1886**) by H. Rider HAGGARD, *The Well at the World's End* (**1896**) by William MORRIS and *The Wonderful Wizard of Oz* (**1900**) by L. Frank BAUM. Outstanding later works in a similar vein include *A Voyage to Arcturus* (**1920**) by David LINDSAY, *The Phantom Tollbooth* (**1961**) by Norton JUSTER and *The Infernal Desire Machines of Doctor Hoffman* (**1972**) by Angela CARTER. [BS]

Further reading: *The Imaginary Voyage in Prose Fiction* (**1941**) by Philip Babcock Gove.

FANTASY Much world literature has been described, at one time or another, as fantasy. "Fantasy" – certainly when conceived as being in contrast to Realism – is a most extraordinarily porous term, and has been used to mop up vast deposits of story which this culture or that – and this era or that – deems unrealistic. In the late 20th century, however, the term FANTASTIC has more and more frequently been substituted for "fantasy" when modes are being discussed. As a term of definition, "fantasy", though a term which continues to lack the specificity of SCIENCE FICTION, does designate a *structure*. Fantasy is not a form – like HORROR – named solely after the affect it is intended to produce.

This encyclopedia's central focus is on fantasy, although many entries (like AFTERLIFE, ALLEGORY, DARK FANTASY, FABULATION, FAIRYTALE, FOLKLORE, FOLKTALES, HORROR, SCIENCE FANTASY, SCIENCE FICTION, SUPERNATURAL FICTION, SURREALISM, TAPROOT TEXTS and WONDERLANDS) deal at some length with material within the broader realm of the fantastic. But fantasy's specific location in the spectrum of the fantastic is a matter of constant critical speculation; there is no rigorous critical consensus over the precise definition and "reach" and interrelation of any of the terms referred to above. As Brian ATTEBERY has indicated through his description of fantasy as a "fuzzy set", it may be that fantasy is *inherently* best described and defined through prescriptive and exploratory example. That is why this encyclopedia includes entries on material which many critics and readers might not consider pure fantasy, and which the definition of fantasy suggested below makes no attempt to encompass.

DEFINITION OF FANTASY

A fantasy text is a self-coherent narrative. When set in this world, it tells a story which is impossible in the world as we perceive it (◊ PERCEPTION); when set in an otherworld, that otherworld will be impossible, though stories set there may be possible in its terms.

Some of the terms used here warrant explanation.

Text Any format in which a fantasy story can be told: the written word; COMICS and GRAPHIC NOVELS; ILLUSTRATION and FANTASY ART; CINEMA and TELEVISION; MUSIC (notably OPERA and SONG).

Self-coherent Here a contrast between fantasy and other forms of the fantastic can perhaps be suggested. Certain kinds of narrative presentation of the unreal – like DREAM tales, SURREALISM and Postmodernism – manifestly decline to take on the nature of STORY, though episodes of full stories may be part of their complex, challenging textures, their dismantling of the reader's sense that a coherent world is being presented through the text. Modernist and Postmodernist texts *use* elements of fantasy, but are not designed to be lived within in the way a fantasy text clearly invites its readers to co-inhabit the tale. It is not just that Modernism or Postmodernism question the nature of Story (much sophisticated fantasy does that); it is that they are profoundly subversive of the "naive" connective tissue that permits narrative consequences to *follow on* from narrative beginnings.

Because of its numerous fantasticated sequences, James JOYCE's *Ulysses* (**1922**) might otherwise loosely qualify as a fantasy, as might Thomas MANN's *The Magic Mountain* (**1924**) or Franz KAFKA's *The Castle*, almost any work by Samuel BECKETT, Eugene IONESCO or Gertrude Stein (**1874-1946**) (◊ ABSURDIST FANTASY) and almost any FABULATION by post-WWII writers like John BARTH, Donald BARTHELME, Robert COOVER, Georges Perec (**1936-1982**), Thomas PYNCHON, etc. – or the entire works of Jorge Luis BORGES, or indeed almost any MAGIC-REALISM tale by Julio CORTAZAR or Carlo FUENTES or Gabriel GARCIA MARQUEZ or a hundred others. Indeed, almost any 20th-century novel which stands aside from – or which puts to the question – the presumptions of the mimetic novel will almost certainly contain elements of the fantastic (and many 20th-century authors of this kind of work are given entries here). But clearly to call so much of 20th-century literature fantasy is radically to misunderstand the enterprises of Modernism and Postmodernism, and thereby to strip the term "fantasy" of any specific meaning.

Story Much that is said about "self-coherence" could be said here as well, because a STORY is by definition a self-coherent narrative. Stories are traditionally transparent: they do not conceal the fact that something is being told, *and then* something else, *and then* we reach the end. This transparency of Story, which is typical of fantasy, creates what Brian Wicker describes in the title of his *A Story-Shaped World* (**1975**): "We may say that the characters in fairytales [to which it is possible to add characters in Fantasy] are 'good to think with' . . . [and that] the job of the fairytale is to show that Why? questions cannot be answered except in one way: by telling stories. The story does not contain the answer, it is the answer."

Fantasy is a way to tell stories about the fantastic.

Perceive as impossible Before the beginning of the scientific revolution in Western Europe in the 16th century, most Western literature contained huge amounts of material 20th-century readers would think of as fantastical. It is, however, no simple matter to determine the degree to which various early writers distinguished, before the rise of science, between what we would call fantastical and what we would call realistic. Nor is it possible with any certainty to determine how much various early writers perceived stories which adhered to possible events and stories which did not as being different. There is no easy division between realism and the fantastical in writers before 1600 or so, and no genre of written literature, before about the early 19th century, seems to have been constituted so as deliberately to confront or contradict the "real". Though fantasy certainly existed for many centuries before, whenever stories were told which were understood by their authors (and readers) as being impossible, it is quite something else to suggest that the perceived impossibility of these stories *was their point* – that they stood as a counter-statement to a dominant worldview.

SCIENCE FICTION can be distinguished from fantasy on several grounds; but in our terms the most significant difference is that sf tales are written and read on the presumption that they are *possible* – if perhaps not yet.

Otherworld In "The Fantastic Imagination" – in *A Dish of Orts* (coll **1893**) – George MACDONALD comes close to creating a full definition of the OTHERWORLD or SECONDARY WORLD: "The natural world has its laws, [which] themselves may suggest laws of other kinds, and man may, if he pleases, invent a little world of his own, with its own laws." In "Boiling Roses: Thoughts on Science Fantasy" – in *Intersections: Fantasy and Science Fiction* (anth **1987**) ed George E. Slusser and Eric S. Rabkin – Robert Scholes suggests that Macdonald's "invented world, with laws of other kinds" is the "key" to modern fantasy. If it is not the key, it certainly points towards the natural venue for the self-coherent impossible tale; i.e., an internally coherent impossible world in which that tale *is* possible. Almost all post-TOLKIEN fantasy inhabits this region.

STRUCTURE OF FANTASY

The working definition of fantasy described above has shaped this encyclopedia, and is constructed so as to include within the remit of the book many texts which we call fantasy almost solely because the OTHERWORLD in which they are set is itself, by definition, impossible. Many GENRE FANTASIES (a term which encompasses almost all DYNASTIC FANTASY and HEROIC FANTASY) boast storylines which could – with almost no alteration – be transferred from FANTASYLAND to a mundane venue. We do not claim these texts are not fantasy, nor that they are inherently inferior to more ambitious attempts to exploit the freedoms and obligations of the genre. We do, however, suggest that the greatest fantasy writers – George MACDONALD, William MORRIS, L. Frank BAUM, E. NESBIT, Lord DUNSANY, H.P. LOVECRAFT, Kenneth MORRIS, E.R. EDDISON, Clark Ashton SMITH, J.R.R. TOLKIEN, L. Sprague DE CAMP, Fritz LEIBER, C.S. LEWIS, Mervyn PEAKE, Ray BRADBURY, Alan GARNER, Peter S. BEAGLE, Ursula K. LE GUIN, Stephen R. DONALDSON, John CROWLEY, Mark HELPRIN and others – almost invariably engage deeply with the transformative potentials of fantasy.

A fantasy text may be described as the story of an earned passage from BONDAGE – via a central RECOGNITION of what has been revealed and of what is about to happen, and which may involve a profound METAMORPHOSIS of protagonist or world (or both) – into the EUCATASTROPHE, where marriages may occur, just governance fertilize the barren LAND, and there is a HEALING.

The initial state of bondage, of REALITY-distorting constriction, is normally signalled in fantasy by WRONGNESS, by a sense that the world as a whole has gone askew, that the story of things has been occluded. The Hobbits' first sight of the Nazgûl in J.R.R. TOLKIEN's *The Lord of the Rings* **(1954-5)** shockingly opens their eyes to darkness, almost tangibly informs them that any return to the world that has been fogged over may be profoundly taxing; that the world (and the stories that tell it) is about to undergo a dangerous and painful THINNING of texture, a fading away of beingness. This thinning may manifest itself through a loss of MAGIC, or the slow death of the GODS, or a transformation of the LAND into desert, or a blockage of METAMORPHOSIS (so that nothing can change or grow), or an AMNESIA (the protagonist's, or the world's) about the true nature of the self or history or the SECONDARY WORLD, or of any of the consequences of the rule of a DARK LORD, whose diktats almost inevitably represent an estranging PARODY of just governance.

We use the term RECOGNITION frequently to describe the moment at which – after penetrating the LABYRINTHS of story-gone-astray, the protagonist finally gazes upon the shrivelled heart of the thinned world and sees what to do. After this moment of transformative recognition comes the transition into what Tolkien calls "consolation" but which we (more secularly) call HEALING, a transition often accomplished (though not in Tolkien's work) through literal METAMORPHOSES.

STORY is central throughout. Fantasy can almost be defined as a genre whose protagonists reflect and embody the tale being told, and who lead the way through travails and reversals towards the completion of a happy ending. (Tragic fantasy exists, but is uncommon.) GENRE FANTASY, which dominates the marketplace, is normally structured so as to defer completion indefinitely, to lead readers into sequel after sequel; and it is for this reason, too, that our working definition of fantasy must give *lebensraum* to texts which have so little fantasy in them. (At the same time it needs to be recognized that a great SWORD-AND-SORCERY author like Fritz LEIBER gains many of his finest effects through a kind of parodic flirtation with "full" narratives, dodging their moves to closure.) This Story-driven urge to comedic completion also distinguishes full fantasy from its siblings, SUPERNATURAL FICTIONS and HORROR, whose plots often terminate – shockingly – before any resolution can be achieved. This is deliberate, but the *feel* and the reality are different.

As the terms are used in this encyclopedia, SUPERNATURAL FICTIONS tend to focus on the experience of WRONGNESS in the world and HORROR stories tend to focus on the experience of THINNING, when the body and the world are progressively violated, lessened, brought to despair. A supernatural fiction which passes through its natural habitat into the full rigours of thinning tends to be thought of as horror; supernatural fictions and horror stories which pass through their natural habitats into the transformed world of healing tend to be thought of as fantasies (or DARK FANTASIES). When supernatural fictions or horror stories become fantasies, they become stories which can be completed. [JC]

FANTASY (two magazines) ◊ MAGAZINES.

FANTASY ART FANTASY is unusual among the genres in that there is no clearcut distinction between its four primary forms: the written word, COMICS, CINEMA and art. Certainly the modes of expression differ, but the underpinning is the same. Notably, FA is, like the other modes, essentially a *narrative* form: although the image is solitary and, of course, static, in the best of "realistic" FA that image is understood as a moment taken from a STORY that began some time before and will continue for some time afterwards; more surrealist (◊ SURREALISM) or abstract FA can perform the trick too, or it may convey to the viewer an encapsulation of a story, a collection of images that the viewer can unpick.

The earliest examples of FA date back to long before there was such a thing as "fantasy": the cave paintings at sites like Altamira and Lascaux show what appear to be MYTHICAL CREATURES, and anyway are fantasy in that they (one assumes) depict hunting and other scenes on the sympathetic-MAGIC basis that imagining something can make it happen. Early religious FA includes a depiction of the Sumerian half-man, half-fish quasi-GOD Oannes as well as numerous South American artworks which Erich von Däniken (1935-   ) and his disciples have assumed to portray ancient astronauts.

Religious art has played a large part in the history of FA, to the extent that the genre as a whole can really be traced to religious paintings by 15th- and 16th-century artists such as Pieter BRUEGEL, Matthias GRÜNEWALD and, by far the most notably, Hieronymus BOSCH, whose *The Garden of Earthly Delights* has served as a source of inspiration to fantasy artists to the present day. Even so, for some centuries thereafter there was a reluctance on the part of Western artists to indulge in the full flood of fantasy: there were "realistic" depictions of religious and Classical scenes, but there was little by way of true FA. The next major landmark was probably Henry FUSELI's painting *The Nightmare* (1781; various other versions), which transports the viewer clear out of mundane reality to create a sense that all is truly not as we perceive it. A few decades later J.M.W. Turner (1775-1851), some of whose paintings can be considered fantasticated, was at work, and John MARTIN was creating huge and spectacular canvases on subjects like ARMAGEDDON. The age of FA had begun.

Martin worked also as an illustrator – he did an edition of John MILTON's *Paradise Lost* – and by about this time ILLUSTRATION was beginning to have a mutually profitable relationship with FA. Later in the 19th century was seen the rise of imaginative fiction – with writers like George MACDONALD and Lewis CARROLL coming to the fore – and illustrators like Sir John TENNIEL (for Carroll) pointed the way for fantasy artists. Some had already followed that way: Richard DADD is a prime example, as is Sir Lawrence Alma-Tadema (1836-1912), who created a sort of fantasticated quasi-historical eroticism than is hard to describe. The PRE-RAPHAELITES, such as Dante Gabriel ROSSETTI, introduced their own very potent brand of eroticism.

In the USA around the same time Howard PYLE was enormously influential on the history of the genre, giving to it the preoccupation with the human figure as focus that is still visible today in the works of artists like Frank FRAZETTA. Maxfield PARRISH, working a little later than Pyle, created an Arcadian tradition that, likewise, still has an overt influence on the many fantasy artists who habitually portray a tiny figure dwarfed by a vast and detailed landscape. Grant Wood (1892-1942) and Andrew Wyeth (1917-   ; ◊◊ N.C. WYETH) generated images that, while they seemed to portray reality, were nevertheless highly fantasticated.

As US artists like Wood and Wyeth were working, so too were Europeans, who created the school of SURREALISM, which can be regarded as the apex, to date, of FA. Notable among these artists were Max ERNST, Salvador DALÍ and René MAGRITTE; to the list could be added the very much later artist H.R. GIGER, but in truth the school still continues in the work of countless currently practising artists. Another major influence on late-20th-century FA has been M.C. ESCHER; he was arguably a graphic designer rather than a pure artist, creating impossible REALITIES often through the use of pattern, but his line technique is clearly evident in the work of current artists like Ian MILLER.

FA today takes so many forms that any synopsis would be a travesty. There are the elaborate TECHNOFANTASY and SCIENCE-FICTION structures of artists like Jim BURNS and David A. Hardy (1936-   ); there are the more straightforward depictions by people like the multiple-award-winning Michael WHELAN; artists such as Frazetta and Boris VALLEJO, both influencing and influenced by the COMICS, portray posed scenes that relate to CONAN and that barbarian's many clones; there is the ultra-realism of artists like Harvey G. Parker; the superb moodiness of artists like Brian FROUD . . . the list could go on forever. What is certain is that the FA genre, even ignoring ILLUSTRATION, is part of the mainstream of today's art, and is likely to remain so.

[JG/RT]

See also: Edward BURNE-JONES; George CRUIKSHANK; Gustav DORÉ; Michael FOREMAN; Norman LINDSAY; John Everett MILLAIS; Bruce PENNINGTON; Gerald SCARFE; Ronald SEARLE; Ralph STEADMAN; Tim WHITE; Gahan WILSON; Patrick WOODROFFE.

FANTASY ASSOCIATION Founded 1973 by Mythopoeic Society members disgruntled by that society's perceived overemphasis on the INKLINGS (◊ MYTHOPOEIC AWARDS). The major activity was the publication of the monthly newsletter *Fantasiae* ed Ian Slater (102 issues April 1973-October 1981), containing essays, reviews and news. A companion quarterly, *The Eildon Tree* ed Donald G. Keller, produced only two issues (both 1974). The FA died in 1981.

[DGK]

FANTASY BOOK Two unrelated US small-press MAGAZINES.

1. 8 issues, irregular, undated July 1947-January 1951, published and ed William L. Crawford (1911-1984), Los Angeles.

This concentrated on sf, though with some borderline fantasies by Andre NORTON and A.E. VAN VOGT.   [MA]

2. 23 issues, quarterly, October 1981-March 1987, published and ed Dennis Mallonee and Nick Smith, Pasadena.

This focused solely on fantasy, in all its forms, hoping to fill a void left by the passing of FANTASTIC. The first issues suggested close links with UNKNOWN, not just because of stories by H.L. GOLD and L. Ron HUBBARD but also in the style of its fiction, which slanted toward lighthearted SUPERNATURAL FICTION, with bizarre but nonthreatening invasions of reality. There was an emphasis on MYTHICAL CREATURES, especially DRAGONS and UNICORNS, which featured strongly in the cover art, particularly by Corey Wolfe (1953-   ) and Janny WURTS. HIGH FANTASY appeared less regularly, though was represented by works by Raul Garcia Capella (1933-   ), Katherine KURTZ and Darrell SCHWEITZER; most authors preferred historical (◊ FANTASIES OF HISTORY; HISTORY IN FANTASY) and ORIENTAL FANTASIES. Later issues took a slightly darker turn, with

supernatural HORROR stories from Hugh B. CAVE, Brian LUMLEY (continuing H.P. LOVECRAFT's CTHULHU MYTHOS stories), William F. NOLAN and Ian WATSON, but humorous fantasy (◊ HUMOUR) remained dominant. *FB* published some of the first professional stories by Esther FRIESNER, Mercedes LACKEY and Josepha SHERMAN. Other regular contributors included Alan Dean FOSTER, Janet Fox (1940-   ), Stephen Goldin (1947-   ), Jessica Amanda SALMONSON, Al Sarrantonio (1952-   ) and Terri Pinckard (1930-   ). *FB* was attractively illustrated throughout, but this cost contributed towards the suspension of publication in 1987. Two attempts to revive *FB* have been abortive. The magazine's approach has been closely followed by MARION ZIMMER BRADLEY'S FANTASY MAGAZINE.

[MA]

FANTASY FICTION (magazine) ◊ FANTASY MAGAZINE.

FANTASY GAMES ◊ GAMES.

FANTASY ISLAND US tv series (1978-84). Columbia/ABC. **Pr** Don Chaffey, Michael Fisher, Don Ingalls, Arthur Rowe, Skip Webster, Don Weis. **Exec pr** Leonard Goldberg, Aaron Spelling. **Dir** Earl Bellamy and many others. **Writers** Ingalls and many others. **Starring** Samantha Eggar (Helena Marsh), Christopher Hewett (Lawrence 1983-4), Roddy McDowall (Devil), Ricardo Montalban (Mr Roarke), Michelle Phillips (Mermaid), Wendy Schaal (Julie 1981-4), Herve Villechaize (Tattoo 1978-83). 152 60min episodes. Colour.

After the two successful FANTASY ISLAND MOVIES, this weekly series enjoyed a six-year run. Each week, Mr Roarke and Tattoo (famous for his greeting arriving guests by excitedly yelling out "The plane! The plane!") would welcome their guests at the dock. Nattily dressed in a white suit, Roarke was the epitome of charm and grace as he greeted his guests. Usually two different stories were interwoven in each episode; story titles included "The World's Most Desirable Woman", "The World's Greatest Escape Artist" and "Charlie's Cherubs", the latter tale focusing on women who fantasized about being glamorous private detectives. Each fantasy was made more exciting by the fact that the guests were totally immersed in the experience, and that the dangers presented were apparently real. For example, a story set in WORLD WAR II Germany featured Nazis bearing machine guns and other deadly weapons; the implication was always that the guests could indeed perish if their fantasy went awry.

Exactly how Mr Roarke managed to stage these elaborate fantasies was never explained, although in later years there were increasingly strong hints about TALENTS – indeed, in the final year Roarke himself became involved in a battle with the DEVIL for his SOUL. Most stories ended with the guests realizing their real lives were far better than they had known, and that their fantasies could safely be left behind.

[BC]

FANTASY ISLAND MOVIES Two tvms preceded the long-running series FANTASY ISLAND (1978-84).

1. *Fantasy Island* US movie (*1977* tvm). Columbia/ABC. **Pr** Leonard Goldberg, Aaron Spelling. **Screenplay** Gene Levitt. **Dir** Richard Lang. **Starring** Bill Bixby (Arnold Greenwood), Sandra Dee (Francesca), Peter Lawford (Grant Baines), Carol Lynley (Liz Hollander), John McKinney (Hunter), Ricardo Montalban (Mr Roarke), Hugh O'Brian (Paul Henley), Eleanor Parker (Eunice Baines), Victoria Principal (Michelle), Dick Sargent (Charles Hollander), Christine Sinatra (Connie Raymond),

Herve Villechaize (Tattoo). 120 mins. Colour.

Guests pay $50,000 each for a three-day stay on a mysterious ISLAND where their WISHES just might be granted. Somehow the staff, led by Mr Roarke, the island's enigmatic owner, and his dwarf assistant, Tattoo, manage to make even the wildest dream come true. Featured here are a man who wants to rekindle a World War II romance, a hunter who would like to be the prey, and a woman who fakes her death so she can see what life will be like when she is gone.   [BC]

2. *Return to Fantasy Island* US movie (*1978* tvm). Columbia/ABC. **Pr** Michael Fisher. **Exec pr** Leonard Goldberg, Aaron Spelling. **Screenplay** Marc Brandel. **Dir** George McCowan. **Starring** Horst Buchholz (Charles Fleming), Joseph Campanella (Brian Faber), Patricia Crowley (Louise Faber), George Maharis (Lyle Benson), Ricardo Montalban (Mr Roarke), Karen Valentine (Janet Fleming), Herve Villechaize (Tattoo). 120 mins. Colour.

This sequel to **1** features three new sets of visitors to the island. Benson has arrived in hopes of seducing his boss, a beautiful workaholic. Brian and Louise Faber desperately want to see the child they were forced to give up to adoption 12 years, and Fleming, who suffers AMNESIA, hopes to discover her true identity. Once again Mr Roarke fulfils their wishes. As with **1**, this enjoyed high ratings; hence the series.   [BC]

FANTASYLAND The basic venue in which much GENRE FANTASY is set. This location has been given many names: it may be called the World (as John GRANT has done) or the GAMEWORLD (as John CLUTE has done), or any of a dozen other designations, but it is as recognizable a locale as RURITANIA. A typical Fantasyland will display – often initially by means of a prefatory MAP – a selection, sometimes very full, from a more or less fixed list of landscape ingredients, which includes the following features: a continent (or two), one or more inland seas and an ocean (or two), ARCHIPELAGOS, mountains, isolated ISLANDS, fjords, steppes, pastures, deserts, FORESTS (but rarely jungles, for they are too far south – though Fantasyland authors taking elements from the SWORD AND SORCERY tales of Robert E. HOWARD do incorporate them) and realms of ice, EDIFICES and CITIES, usually ancient, sometimes abandoned. POLDERS and sites evoking a sense of TIME ABYSS are rare in Fantasyland.

These features persevere wherever the author claims her or his Fantasyland is actually located: it may ostensbly be set in an ALTERNATE REALITY (as in Andrew M. Greeley's *God Game* [**1986**], or Gael BAUDINO's **Dragonsword** sequence, or Greg BEAR's *Songs of Earth and Power* [omni **1992** UK]); it may be set in prehistory (as in Julian MAY's **Saga of the Pliocene Exiles**); it may be set on Mars or Venus (as in Edgar Rice BURROUGHS's numerous PLANETARY ROMANCES), or on a distant world (like many of Jack VANCE's novels, or in Terry PRATCHETT's **Discworld** sequence); it may be set in the future (as in John CHRISTOPHER's **Prince in Waiting** trilogy) or the LAND OF FABLE in the deep past (like Howard's **Conan** books, and consequently much of the S&S which imitates Howard), or in some other SCIENCE-FANTASY venue; it may be set inside the HOLLOW EARTH (once again Burroughs provides examples); it may ostensibly be set in land-of-fable venues like Arthurian Britain (almost the only author *not* to set his or her ARTHUR saga in Fantasyland is T.H. WHITE, whose *The Once and Future King* [**1958**] argues with the venue) or any Nordic Twilight location; or it may simply default to the

venue which underlies the surface appearance of all late-20th-century Fantasylands: this venue (or condition of being) is the SECONDARY WORLD.

The source of most late-20th-century Fantasylands is, naturally enough, the central secondary world of modern fantasy: J.R.R. TOLKIEN's Arda, in the centre of which Middle-Earth abides. *The Lord of the Rings* (**1954-5**) is set in Middle-Earth itself; Tolkien's other fantasies, all describing events prior to the climactic tale told in *LOTR*, cover all of Arda, its oceans and isles and transfigurations, over the 30,000 years of the full story. Two aspects of Tolkien's subcreation are relevant to any understanding of Arda, though only the first may be relevant to the creators of Fantasyland, all of whom reflect Arda in their own subcreations: (a) the central Middle-Earth LANDSCAPE depicted by Tolkien, though not original to him in any particular, is imagined with such detail and solidity that its geography has become a template, as described above; and (b) Arda is a world constantly – as befits the world of a full fantasy – undergoing METAMORPHOSIS.

Full-fantasy narratives are stories of profound, all-transforming change. *LOTR* is such a narrative; and is subject, as noted, to the constant metamorphic meaning-drenched interplay between setting and tale which is essential to any definition of full fantasy. In *LOTR* world (or LANDSCAPE) and STORY are inherently intertwined: one cannot exist without the other, and each modifies the other. But Fantasyland's relationship to the secondary world parallels the relationship of genre fantasy to full fantasy, or of S&S to the Monomyth. Fantasyland, it follows, is a secondary world which is fixed in place; it is inherently *immobile*; it is backdrop, not actor; and, because it has already been "solved", it cannot be transformed. Though fine stories can readily be set in a landscape so fixed – a good example being *Assassin's Apprentice* (**1995** UK) by Megan LINDHOLM (writing as Robin Hobb) – it is still the case that Fantasyland is a natural home for unambitious tales.

The Fantasyland which serves as backdrop or arena for the various genre fantasies that have flourished in the late 20th century is profoundly dissociated from the actions played out upon it, frequently in the form of indefinitely repeated AGONS (◊ GAMEWORLDS; PARODY; THINNING); and the genre fantasies set there are often fantasies in name only. For Fantasyland can never, in any genuine sense, be the true subject of a story set in it. As soon as a plot begins to intersect with the world in which it is set – an example might be E.R. EDDISON's *The Worm Ouroboros* (**1922**) – then it becomes evident that a genuine fantasy may be unfolding, and Fantasyland is transformed into Story.

Fantasyland is a particularly useful thought-free setting for authors of SHARED-WORLD enterprises and extended series.   [JC/JG]

FANTASY MAGAZINE US digest MAGAZINE, 4 issues, bimonthly (but dated February/March, June, August, November 1953), published by Future Publications, New York; ed Lester del Rey (last issue under housename Cameron Hall). Although issues after *#1* were titled *Fantasy Fiction*, the magazine is always remembered as *FM*, partly so as not to confuse it with an earlier *Fantasy Fiction* (2 issues May-November 1950, *#2* retitled *Fantasy Stories*).

FM was like BEYOND though, with the inclusion of HEROIC FANTASY, less sophisticated. The presence of L. Sprague DE CAMP (with two **Conan** revisions and, with Fletcher PRATT, a **Harold Shea** novella) also forged links with WEIRD TALES and UNKNOWN. Contributors included Algis Budrys (1931-   ), Philip K. Dick (1928-1982), Randall GARRETT, Harry Harrison (1925-   ), Robert Sheckley (1928-   ), Clark Ashton SMITH and John Wyndham (1903-1969), with fiction that at times resembled SLICK FANTASY. All covers were by Hannes BOK, adding to the value of the magazine. It folded due to the fickleness of its publisher. Del Rey sought to recapture the essence of the magazine in WORLDS OF FANTASY, but with less success.   [MA]

FANTASY OF MANNERS Term coined in 1991 to describe the shared sensibility of a group of fantasy writers emerging in the 1980s, including Steven BRUST, Emma BULL, Ellen KUSHNER, Delia SHERMAN, Caroline STEVERMER and Terri WINDLING. Born in the 1950s, these writers were the first generation to be influenced by tv; movies like The BEATLES' *A Hard Day's Night* (**1964**) also had a great deal of impact. Children's literature remained an interest for them (both as readers and as writers) long after childhood. The Regency romances of Georgette Heyer (1902-1974) – and her source, Jane Austen (1775-1817) – and the flamboyantly complex historical melodrama of the **Lymond Chronicles** (**1960-75**) by Dorothy Dunnett (1923-   ) were further influences.

Their main genre predecessor is Fritz LEIBER, particularly as begetter of URBAN FANTASY and CONTEMPORARY FANTASY, and for his DICTION – simpler and slangier than the previous generation of fantasists but nonetheless not overtly contemporary. Both Michael MOORCOCK – especially for *Gloriana* (**1978**) – and M. John HARRISON – with the **Viriconium** stories – had something to say to these writers.

Among the characteristics of these works are: the negotiability of social structures (a peculiarly US view of European institutions); the importance of DISGUISE, particularly GENDER DISGUISE, as a method of that negotiation, of changing one's life; the importance of childhood in the formation of the adult; the necessity to find one's place in the world (particularly as an artist, whether musician or swordsman) by being true to one's own nature; and the cruciality of manners not only in fashion and behaviour but also in language – because control of words and of tone is power. What their characters say is even more important than what they do, and the range of diction, from lowest to highest, that these writers make use of is large.   [DGK]

FANTASY STORIES (magazine) ◊ FANTASY MAGAZINE.

FANTASY TALES UK small-press MAGAZINE, 17 issues, irregular (approx twice yearly) Summer 1977-Summer 1987, then relaunched in book format as an anthology series, 7 issues, Autumn 1988-Winter 1991. All issues ed Stephen JONES and David A. Sutton (1947-   ), who published the magazine issues; book issues were published by Robinson Books, London, and from Spring 1990 jointly in the USA by Carroll & Graf, New York.

FT was launched as a deliberate latterday tribute to WEIRD TALES and, though limited by finance, was moderately successful in this; it focused primarily on either LOW FANTASY or Lovecraftian fiction (◊ H.P. LOVECRAFT). The magazine was avidly supported by the editors' close colleagues, including Ramsey CAMPBELL and Karl Edward WAGNER, whose contributions elevated the magazine above the norm, resulting in it being an almost annual winner of the BRITISH FANTASY AWARD from 1979, and winning the WORLD FANTASY AWARD in 1984. Later issues evoked a strong *WT* atmosphere, with new stories by Hugh B. CAVE, H. Warner MUNN and Manly Wade WELLMAN, plus reprints from

Robert BLOCH and Fritz LEIBER. The new generation of writers were amply represented by Clive BARKER, Dennis ETCHISON, Joel Lane (1963-   ), Thomas LIGOTTI, Brian LUMLEY and Richard Christian MATHESON. The anthology series continued in much the same vein, with a slight shift towards visceral HORROR, satisfying its small core of readers but never becoming financially viable. A representative selection from past issues was published as *The Best Horror from Fantasy Tales* (anth **1988**) ed Jones and Sutton, while some of the material acquired for but unused in the magazine and series was issued as *The Anthology of Fantasy & the Supernatural* (anth **1994**) ed Jones and Sutton.   [MA]

FANTÔME DU MOULIN-ROUGE, LE French movie (*1926*). ◊ René CLAIR.

FAR FUTURE A period more often found in sf and SCIENCE FANTASY than in fantasy proper. Brian STABLEFORD has distinguished two categories of FF imagery: the "future of prophecy", which depicts the consequences of warnings or predictions, is clearly an sf concern; the "future of destiny", on the other hand, which depicts "images of the ultimate future", can just as clearly be a concern of fantasy. Examples (most of which hover between fantasy and science-fantasy structures of verisimilitude) include: Clark Ashton SMITH's **Zothique** sequence; Henry KUTTNER and C.L. MOORE's *Earth's Last Citadel* (1943; **1964**); Jack VANCE's **Dying Earth** sequence; and Gene WOLFE's *The Book of the New Sun* (**1980-1983**).   [JC]

See also: DYING EARTH.

FARJEON, B(ENJAMIN) L(EOPOLD) (1833-1903) UK writer, father of Eleanor FARJEON, of the literary critic and dramatist Herbert Farjeon (1887-1945) and of J. Jefferson Farjeon (◊ *SFE*). He was the author of at least two CHRISTMAS BOOKS, *Shadows on the Snow: A Christmas Story* (**1866** chap New Zealand) and *Christmas Angel* (**1884**), and other CHRISTMAS stories including *Christmas Stories* (coll **1874**) and *The Shield of Love* (in *Arrowsmith's Christmas Annual* anth **1884**; **1891**); and he contributed to a round-robin Christmas novel, *Seven Xmas Eves* (**1894**), whose other contributors included Mrs Campbell PRAED. He was best-known for detective thrillers; most of his work of fantasy interest is SUPERNATURAL FICTION like *Devlin the Barber* (**1888**) and *A Strange Enchantment* (**1889**), in which a rather didactic "Mystic" uses strange TALENTS to solve a mystery. Other fiction includes *The Last Tenant* (**1893**), about the GHOST of a CAT, *Something Occurred* (**1893**) and *The Clairvoyante* (**1905**).   [JC]

FARJEON, ELEANOR (1881-1965) UK writer known mostly for her poems and stories for children, but also a formidable author of works for adults, several of which are fantasies. She began publishing with *Floretta* (**1899** chap), a libretto for music by her brother Herbert Farjeon (1887-1945). Her first independent book was *Pan-Worship and Other Poems* (coll **1908** chap), the title poem of which is a typical Edwardian effusion about PAN caught in stone BONDAGE in a suburbanized ARCADIA. Her first novel, *The Soul of Kol Nikon* (c**1914**; slightly rev 1923), is a fantasy account, told in a mode remotely evocative of NORDIC FANTASY, of the eponymous CHANGELING's doomed attempts to find a SOUL through the power of his MUSIC, which allows him to go through a forced IDENTITY EXCHANGE with the fiancé of a young woman, whom he seduces; he is eventually stoned to death. Other adult fantasies include *Gypsy and Ginger* (**1920**), a THEODICY-tinged *jeu d'esprit* about LONDON, where the eponymous newlyweds set themselves up in a house in Trafalgar Square, from which they listen as the animals and mythical folk of the city lament the THINNING of their world. In *The Fair of St James: A Fantasia* (**1932**) the protagonists enter an OTHERWORLD in which a glorious fair is taking place, and which serves as a kind of FRAME STORY for various tales. *Ariadne and the Bull* (**1945**) sees the classic story TWICE-TOLD in a modern US setting.

EF's reputation now rests on her work for children, beginning with the **Martin Pippin** sequence: *Martin Pippin in the Apple-Orchard* (coll of linked stories **1921**) – some individual tales being released as *The King's Barn, or Joan's Tale* (**1927** chap), *The Mill of Dreams, or Jennifer's Tale* (**1927** chap) and *Young Gerard, or Joyce's Tale* (**1927** chap) – and *Martin Pippin in the Daisy Field* (coll of linked stories **1937**). In both volumes the eponymous PUCK figure solves RIDDLES and tells fantasy tales – including the famous "Elsie Piddock Skips in her Sleep" – to either the six original girls involved in complex folkloristic games with Martin or (in the second volume) their six children. As his name implies, Martin Pippin is not only a TRICKSTER but an analogue of the GREEN MAN, and in that guise is a spirit of vegetation; despite superficial resemblances, therefore, he is most unlike PETER PAN. Other volumes similarly – though never with quite the intensity – present various tales through frame stories narrated by LIMINAL BEINGS; they include *Kaleidoscope* (coll **1928**), *The Old Nurse's Stocking-Basket* (coll **1931**) and *Jim at the Corner* (coll **1934**). Fantasy novels for children include *The Silver Curlew: A Fairy Tale* (**1953**) and *The Glass Slipper* * (**1955**); the first is a twice-told version of RUMPELSTILTSKIN and the second – based on the play *The Glass Slipper* (**1946**) by EF and Herbert Farjeon – is a version of CINDERELLA. [JC]

Other works for adults: *Arthur Rackham: The Wizard at Home* (**1914** US), about Arthur RACKHAM; *Faithful Jenny Dove and Other Tales* (coll **1925**); *Humming Bird: A Novel* (**1936**); *A Room at the Inn: A Christmas Masque* (**1956** chap).

Other works for children (selected): *Nursery Rhymes of London Town* (coll **1916** chap) and *More Nursery Rhymes of London Town* (coll **1917** chap); *All the Year Round* (coll **1923**; vt *Around the Seasons* 1969), poems; contributions to the **Basil Blackwell "Continuous Stories" Series** for younger children, beginning with *Tom Cobble* (**1925** chap) and ending with *Jim and the Pirates* (**1936** chap); *Joan's Door* (coll **1926**), poems; *Italian Peepshow and Other Tales* (coll **1926** US; vt *Italian Peepshow and Other Stories* 1934 UK; rev vt *Italian Peepshow* 1960 UK); *One Foot in Fairyland: Sixteen Tales* (coll **1938**); *Come, Christmas* (coll **1927**), poems; *An Alphabet of Magic* (**1928**), poem; *The Tale of Tom Tiddler* (**1929**); *Westwoods* (**1930** chap); *Sing For Your Supper* (coll **1938**), poems; *The New Book of Days* (coll **1941**); *The Starry Floor* (coll **1949** chap); *Silver-Sand & Snow* (coll **1951**), selected poems; *The Little Book-Room* (coll **1955**) and *Eleanor Farjeon's Book: Stories, Verses, Plays* (coll **1960**), both compilations, both illus Edward ARDIZZONE.

FARMER, PENELOPE (JANE) (1939-   ) UK writer who began publishing fantasy, initially for a YA audience, with the **Charlotte and Emma** sequence: *The Summer Birds* (**1962**), *Emma in Winter* (**1966**) and *Charlotte Sometimes* (**1969**). In the first volume, the sisters Charlotte and Emma Makepeace meet a boy who teaches them the art of flying, but who cannot persuade them to enter his OTHERWORLD country at summer's end in order to save his folk; this task is left to another girl. In the second tale Emma lives a life in a DREAM world, but returns. In the third Charlotte TIMESLIPS

repeatedly into the life of a girl in 1918, but finds, in the end, her way home again.

The realism of these stories, which is bracing but melancholy, pales in comparison to the matured hardness of vision expressed in *A Castle of Bone* (**1972**), in which what initially seems a fantasy adventure – a young boy finds a cupboard in a magic SHOP which not only works as a PORTAL but subjects those who pass through it to METAMORPHOSIS – proves a grave and transformative tale of maturation. The castle of bone is both an EDIFICE in the OTHERWORLD and the protagonist's own body (◊ RECOGNITION); and the cast must, in a cruel RITE OF PASSAGE within that castle, somehow transcend the eternal demand of the GODDESS for a sacrificial consort. Some of the thematic material in this book, which remains PF's most complex and sustained single fantasy, reappears in *Year King* (**1977**).

Some of PF's later work, like *Eve: Her Story* (**1985**), which retells the ADAM AND EVE story from a feminist perspective, more explicitly enters adult territory, though with no gain in complexity over her remarkable YA tales. Other stories of interest include *Glasshouses* (**1988**) and *Thicker than Water* (**1989**), in which a GHOST from the Industrial Revolution seeks a resolution of old wrongs. [JC]

Other works: *The Magic Stone* (**1964** US); a set of retellings of well known MYTHS, being *Daedalus and Icarus* (**1971**), *The Serpent's Teeth: The Story of Cadmus* (**1971**), *The Story of Persephone* (**1972**) and *Heracles* (**1975**); *William and Mary* (**1974**); *Stone Croc* (**1991**).

FARMER, PHILIP JOSÉ (1918-   ) US writer known primarily for his sf, but whose outright fantasy novels are of strong interest, and whose PLANETARY ROMANCES have long been enjoyed for their flamboyance and their forthrightness about matters of SEX and RELIGION. Examples of this sort of tale run throughout his career, from his first novel, *The Green Odyssey* (**1956**), on through the loose **Wold Newton Family** sequence, based on the sf premise that a vast extended family of supermen (it includes TARZAN and Doc Savage) was begun when an 18th-century meteorite irradiated a number of pregnant women in England. But PJF's intention in these stories is to present fantasy-like material within an sf frame, even though he sometimes stresses that frame to the point of (deliberate) absurdity.

As a fantasy author, PJF is most known for two other sequences. The first, the **World of Tiers** series, comprises *The Maker of Universes* (**1965**; rev 1980), *The Gates of Creation* (**1966**; rev 1981), *A Private Cosmos* (**1968**; rev 1981) and *Behind the Walls of Terra* (**1970**; rev 1982), all four assembled as *The World of Tiers* (omni 2 vols **1981**; vt *World of Tiers #1* 1986 UK and #2 1986 UK), plus *Red Orc's Rage* (**1991**) – an associational sidebar to the series – and *More Than Fire: A World of Tiers Novel* (**1993**). The basic venue is an intricately (and at times arbitrarily) interlinked set of pocket universes, which PJF uses as a PLAYGROUND, manipulating his REALITIES according to the central "law" of the MULTIVERSE (as created by Michael MOORCOCK), which is that Anything Goes.

The varying realities are under the irresponsible governance of a number of "Lords"; their names are taken from the works of William BLAKE, and they occasionally attempt to create GODGAMES within their creations, though they do not normally have the attention-span to follow any such programmes through to a satisfactory conclusion. The first protagonist of the sequence is Robert Wolff, a classic HERO who enters the Tier Worlds through a PORTAL, and whose

AMNESIA cloaks the fact that he himself is one of the Lords; eventually he recognizes (◊ RECOGNITION) his TRUE NAME and the nature of the STORY he is living, and fades slowly from view. The second main protagonist is again a man from Earth who turns out to be a god, Kickaha (a name clearly meant to remind the reader of Native-American TRICKSTER myths); romping from world to world, he confounds friend and foe, beds Anana (perhaps a reference to the Sumerian Inanna; ◊ GODDESS), and in the final volume confronts his dark SHADOW, the evil trickster god Red Orc (having to do so more than once, because, in a TECHNO-FANTASY spoof, Red Orc is cloned). *Red Orc's Rage* is an odd sidebar tale, readable as neither sf nor fantasy, in which the **World of Tiers** is used therapeutically for extremely vivid – but imaginary – role-playing activities on the part of the psychologically deformed young protagonist.

More ambitious is the **Riverworld** sequence: *To Your Scattered Bodies Go* (1965-6 *Worlds of Tomorrow*; fixup **1971**), *The Fabulous Riverboat* (1967-71 *If*; fixup **1971**), *The Dark Design* (**1977**), *Riverworld and Other Stories* (coll **1979**), *The Magic Labyrinth* (**1980**), *Riverworld War: The Suppressed Fiction of Philip José Farmer* (coll **1980**), *The Gods of Riverworld* (**1983**) and *River of Eternity* (**1983**), the last being a rediscovered version of «I Owe for the Flesh» (1952), a novel that in the mid-1950s disappeared before publication, plus two SHARED-WORLD anthologies, *Tales of Riverworld* * (anth **1992**) and *Quest to Riverworld* * (anth **1993**), both ed PJF with Edward Kramer, Richard Gilliam and Martin H. GREENBERG. Along the banks of a vast RIVER, somewhere on another planet, all of humanity is resurrected. Though there are POSTHUMOUS-FANTASY moments, when a character attempts to sort out his or her mortal life in terms of the new existence, the sequence is essentially an AFTERLIFE fantasy, depicting ongoing lives and conversations far more closely resembling those in a SECONDARY WORLD than the allegorized configurations of a SOUL in crisis. PJF's direct model for **Riverworld** is the **Houseboat on the Styx** sequence (**1895-9**) by John Kendrix BANGS. In accordance with that tradition, PJF's protagonists tend to be historically recognizable figures: the cast includes Sir Richard Burton (who successfully woos Lewis CARROLL's Alice), Cyrano de Bergerac, Alfred Jarry, King John, Jack LONDON, the movie cowboy Tom Mix, Mark TWAIN and Yeshua (CHRIST). The sequence is mainly devoted to the interacting QUESTS of Burton and Twain/Clemens, both of whom long for varying reasons to find the DARK TOWER in which the secret of their afterlife existence may be found. In the end, Clemens is unsuccessful but Burton undergoes a dark journey to the Tower, where the TECHNOFANTASY nature of the sequence is foregrounded through the revelation that a computer has been maintaining the world, but is now, at the end of *The Magic Labyrinth*, running down. With the help of Alice, Burton manages to reset the machine, and "life" continues; a complex explanation, involving half-sane godlike entities, is perhaps somewhat estranging. Subsequent volumes lack the narrative drive of the first four. The anthologies tend to feature RECURSIVE jokes.

PJF is not a fully comfortable author of fantasy, but joy comes when his urgent testing of the barriers of genre and TABOO leads tales in unexpected and sometimes dangerous directions. [JC]

FARRINGTON, GEOFFREY Pseudonym of UK writer Geoffrey Smith (1955-   ). *The Revenants* (**1983**) is an

exercise in VAMPIRE existentialism similar in spirit to Anne RICE's *Interview with the Vampire* (**1976**). *The Acts of the Apostates* (**1990**), is a vivid historical fantasy set in Rome and Judea in the time of Nero: the hero is unwillingly involved with the arcane PROPHECIES of Nabaim. GF edited *The Dedalus Book of Roman Decadence: Emperors of Debauchery* (anth **1993**). [BS]

FAR SEEING ◊ SCRYING; TALENTS.

FATA MORGANA ◊ MORGAN LE FAY.

FATE In SUPERNATURAL FICTION and SLICK FANTASY texts, Fate often serves as a comeuppance. It punishes those who fail to READ THE SMALL PRINT, or hope to trick the GODS, or who enter into any form of AGON with destiny. CHESS matches in particular provide an ideal level playing field upon which Fate (which or who often appears as a personification of DEATH) may demonstrate its (or his, or her) ineluctable writ.

In Algernon BLACKWOOD's "By Water" (1914), a man told to fear the effects of water dies in the deserts of Egypt. In the "Appointment in Samarra" FABLE, a man runs away from Death to Samarra – where Death has a prior appointment with him. In Ingmar BERGMAN's *The* SEVENTH SEAL (***1956***), a knight, foreknowing the outcome, plays chess with Death/Fate. [JC]

FATES THREE women of Greek MYTH who exist outside both the ordinary world and the PANTHEON. Ultimately, they control everybody's FATE or weird, including that of the GODS themselves. Clotho spins threads representing each individual life; Lachesis weaves each thread into a pattern, representing the course of the life; finally, Atropos cuts off the thread with shears. The Fates (or Parcae) are omniscient, and may be consulted as ORACLES; they offer information unwillingly, usually under threat, and traditionally should not be approached except as a last resort. Their Nordic cognates, the Norns, are Urthr, Verthandi and Skuld, respectively ruling the past, present and future. [CB/DRL]

FATHER CHRISTMAS ◊ SANTA CLAUS.

FAULCON, ROBERT Pseudonym of Robert P. HOLDSTOCK.

FAUNS ◊ SATYRS.

FAUST The original of the Faust legend was Georgius Faustus (or Georgius Sabellicus, Faustus Junior) (*c*1480–1540/1), recorded in official documents, memoirs and letters from the first third of the 16th century. A practising supernaturalist who wandered about much of southern Germany, he claimed to be a university Master of Arts, an astrologer, an expert in various types of medieval and Renaissance MAGIC, and a skilled alchemist. There are records of his being consulted for horoscopes and divinations of the future. Working the markets and fairs, he was an aggressive self-advertiser and promoter, passing out literature (none of which survives) that detailed his accomplishments. One of his claims was that, if the books of Plato and Aristotle were lost, he could recreate them from memory. He had a few highly placed patrons (who later rejected him) but a bad reputation; he was banned from several cities and was generally regarded as a rogue, swindler and mountebank.

Beyond this little is known except that his contemporaries associated him with the area or university of Heidelberg. His true identity is not known. He used to be identified with a Johannes Faust who matriculated at Heidelberg University in 1505, but the chronology does not support this. More recently he has been identified by some scholars with a

Georgius Helmstetter, who attended Heidelberg somewhat earlier. Other authorities have speculated that there were two Faustuses, perhaps father and son, and that the claims to an MA were false. Part of the problem lies in the name Faustus itself. It is not known whether it is from Latin *faustus* ("fortunate", "happy"), perhaps with reference to earlier men named Faustus, or the fairly common German surname Faust (fist). In any case, Georgius Faustus seems to have died *c*1540.

The historical Faustus exemplified a not unfamiliar type of the day: a man who brought university classical training into the world of popular literature and folklore. NOSTRADAMUS also emerged from such a mixture of Humanism and folkways, as did François RABELAIS.

After Faustus's death his legend grew, perhaps because he became posthumously involved in the polemics of the Reformation, which coincided with heavy witchcraft persecutions. Faustus was taken by the Lutherans as the prime example of a DEMON-sponsored VILLAIN. Anecdotes, some of Classical origin, some folkloristic, began to accrete around his name, which now became Johannes Faustus. According to such later stories Faustus sold his soul to the Devil (◊ PACTS WITH THE DEVIL), travelled with two familiars, a horse and a dog, and was strangled by the Devil when his term was up.

As a result of this legend-development, the Faust that became a powerful force in literature and social thought has little to do with the historical Faustus. The literary Faust came into being with *Historia von D. Johan[nes] Fausten dem weitbeschreyten Zauberer unnd Schwartzkuenstler* (**1587** chap) attributed to Johann Spies (d. 1607), published in Frankfurt-am-Main; this has formed the substructure for almost all later writings about Faust. A primitive episodic novel, it described Faustus as a university professor who struck a bargain with Mephostophiles [*sic*] for 24 years of service in exchange for his soul. With Mephostophiles' guidance Faustus travelled widely and instantaneously; he visited the court of the Emperor Charles V, where he gave exhibitions of his magic (◊ *The* GOLEM); he invaded the seraglio of the Grand Turk and had intercourse with his wives; he conjured up Helen of Troy, with whom he cohabited; he performed various magical pranks; he rejected repentance and prayer; and at the expiration of his contract he was torn to shreds by his servitor/master. Capitalizing on both the subterranean reputation of the historical Faustus and the general interest in magic and the occult, the book contained other popular elements: the intense interest in travel, the japes of a TRICKSTER figure, and colorful erotica. It became a contemporary bestseller, was translated almost immediately into foreign languages, and appeared (with some modifications) in English as *History of the Damnable Life and Deserved Death of Doctor John Faustus* (*c*1588). This work, usually called the *English Faustbuch*, formed the basis for *Doctor Faustus* (**1604**) by Christopher MARLOWE and his collaborators.

Marlowe's play (written *c*1589-92), dropped the trivialities, misdirections and padding from the *Faustbuch*. It offered: a psychological study of the breakdown of a learned egotist who is torn between GOOD AND EVIL in his soul; a morality play in the long English tradition; a second statement of hubris and overreaching in the mode of *Tamburlaine* (**1590**); a glorification of the forbidden; and (probably most important of all historically) an occult thriller (◊ OCCULT FANTASY).

English strolling players carried this play back to Germany, where it gave rise to a host of adaptations, imitations, parodies and rechannellings, even to popular PUPPET shows (a tradition embodied in Jan ŠVANKMAJER's movie FAUST [*1994*]). Thus Faustus and his fate remained very much alive in the German public consciousness for the next two centuries. By the 1750s, when Gotthold Lessing (1729-1781) wrote his lost **Faust** play, the magician's name had shifted from Georgius or Johannes Faustus MA to Dr Johannes Faust, its present form.

The second great embodiment of Faustus/Faust in literature is the play *Faust, Eine Tragoedie* ["Faust: A Tragedy"] (**1808-32**), which GOETHE worked on intermittently from the middle 1770s until his death in 1832. The final two-part *Faust* has been interpreted variously, but it is generally taken to be a consciously metaphoric statement of aspects of (then) modern Man. Faust, who is ever striving, is not just a man who has run out of scholarly resources, like Marlowe's Faustus, but a man who recognizes the emptiness of his knowledge, desires to savour all human experience and passes through tragic moments, but grows with his failures and trivial pursuits. His associate, MEPHISTOPHELES, is no Christian DEVIL but, in a way, a statement of flux, an equivalent of Siva. He destroys so that new and better things can emerge. He is the spirit who denies, who tries to work EVIL but always works good.

The bond between Faust and Mephistopheles is not a matter of years. As originally posited by Faust, it will hold until "I say to the passing moment, stay a while, you are so beautiful" – or satiation. This sensualism soon disappears. The hidden point of the drama, however, is Mephistopheles's growing, unwitting predicament: Faust must live so that the Devil can receive his due, yet, the longer Faust lives, the surer the man is of salvation – which turns out to be the case. His eternal striving has outweighed his sins.

In addition to totally redirecting the **Faust** theme, Goethe added much: the tragico-sentimental romance of Margarete and her salvation; Faust's visit to the Brocken on Walpurgisnacht; the winning of Helen and the birth of poetry (Euphorion) from wisdom and beauty; Faust's would-be UTOPIA; and his final salvation. Faust's redemption is a result partly of his growth and partly of Goethe's stated principle that he could not send anyone to damnation.

Goethe's *Faust* created a new persona for the DEVIL. He was no longer a crude, bellowing teratological monstrosity but a suave, insinuating, sarcastic, sardonic, fascinating, debonair gentleman who could be at home in the parlours of polite society. The **Faust** lithographs of Eugène Delacroix (1798-1863) placed this image, which we still largely retain, into pictorial form. The adjective "mephistophelian" has a quite different meaning from "satanic".

Faust has also had a brilliant career in music, in most cases derived from Goethe's drama. Hector Berlioz's *La damnation de Faust* (1845-6), Franz Liszt's *Eine Faust-Symphonie* (1854-7), Richard WAGNER's unfinished *Eine Faust-Ouvertuere* (1840; rev 1856), Charles Gounod's OPERA *Faust* (1859), Arrigo Boito's opera *Mefistofele* (1868) and Ferrucio Busoni's unfinished opera *Doktor Faust* (1925) are examples.

In addition to the direct line of Faust and his fate, the general theme of the Faustian Pact has become important in SUPERNATURAL FICTION (◊ PACTS WITH THE DEVIL).

Today the **Faust** story is still very much alive in literature, music, dance, popular art and other areas. Both Marlowe's play and Goethe's symbolic work are still vital on the stage. Gounod's *Faust* remains one of the most frequently performed operas. [EFB]

More recent versions As one of the four great literary ARCHETYPES identified by the critic George Steiner as central to the European consciousness – the others being DON JUAN, Hamlet (◊ William SHAKESPEARE) and Don Quixote (◊ Miguel de CERVANTES) – Faust might seem to have a permanent place in the imaginations of Western writers. Over the past past century or so, however, two fundamental shifts have diminished his importance: the growth of the scientific ethic, through which the search for knowledge is understood as an essential intellectual task; and the loss of a moral universe whose precepts might control (or shackle) that imperative post-Faustian drive. Without the moral universe, without a threat of genuine damnation, Faust becomes a mad scientist, his story defaults from myth to HORROR, and Mephistopheles becomes a monster foiled by QUIBBLES.

Nevertheless, much modified, the drama continues. The best early-20th-century fantasia upon Faust is probably Eugene LEE-HAMILTON's *The Lord of the Dark Red Star* (1903); less impressive, and presented from a more orthodox CHRISTIAN-FANTASY viewpoint, is *The Devil to Pay* (1939), a play by Dorothy L. Sayers (1893-1957). In the greatest 20th-century version, Thomas MANN's *Doctor Faustus* (1947), Faust himself becomes the composer Adrian Leverkühn, and Mephistopheles manifests as a genius-inducing disease (probably syphilis); significantly, parts of the novel are told in an archaic form of German, and other distancing devices are employed. Later attempts to take Faust seriously include *John Faust* (1958 chap) by Anthony Borrow and *The Devil's Own Work* (1991) by Alan Judd (1947-   ). William HJORTSBERG's *Falling Angel* (1978), Robert NYE's *Faust* (1980) and Emma TENNANT's *Faustine* (1992) also use Faust as an UNDERLIER for contemporary fantasy, where the bargain, the debate, and the consequences are presented indirectly; and Kim NEWMAN's *The Quorum* (1994), a horror novel, searingly dramatizes the late-20th-century lack of a moral universe by presenting a pact and a Mephistopheles, but rendering the Faust figure as a conglomerate of doomed conspirators.

More commonly, the metaphysical drama between Faust and Mephistopheles devolves into SLICK-FANTASY contests, some of them, like Stephen Vincent BENÉT's *The Devil and Daniel Webster* (1937 chap), extremely well conceived. Novels like *The Missing Angel* (1947) by Erle Cox (1873-1950) reveal the poverty of the Faust myth when treated lightly; some late comedies, like *If at Faust You Don't Succeed* (1993(by Robert Sheckley and Roger ZELAZNY and Tom HOLT's *Faust Among Equals* (1994), more or less deftly defuse the drama into surreal farce. [JC]

Movies It is widely reported that there have been over 40 CINEMA versions of the **Faust** legend, but in fact most of these are, rather, tales of PACTS WITH THE DEVIL. It could of course be argued that *all* such tales owe their inspiration ultimately to the **Faust** story, but, while the basis of such an argument is sound, the conclusion is false; for example, *The* DEVIL AND DANIEL WEBSTER (*1941*) and the Stephen Vincent Benét's story obviously show consciousness of their origins, but cannot be regarded as retellings – rather, they are adaptations of a theme. Much the same could be said of Oscar WILDE's *The Picture of Dorian Gray* (1891), which few would instinctively class as a **Faust** tale.

The **Faust** story itself has been filmed remarkably few times, although there were several silent versions, most now lost – although a masterpiece, FAUST: EINE DEUTSCHE VOLKSSAGE (*1926*), dir F.W. MURNAU, survives. *La Beauté du Diable* (*1949*) by René CLAIR was a comic variation on the theme. More recently there has been Jan ŠVANKMAJER's FAUST (*1994*) which, while surrealistically treated and uprooted into the 20th century, can be classed as a retelling. An earlier movie consciously based on **Faust** is the comedy BEDAZZLED (*1967*), in which everything is thrown TOPSY-TURVY: the protagonist is no bearded sage but a short-order cook, and his Faustian voyages serve merely to frustrate him (and are designed by the Mephistopheles figure so to do). In the same year there appeared the filmed stage adaptation *Dr Faustus* (*1967*), dir Richard Burton and Nevill Coghill, written by Coghill and based more closely on Marlowe, with Burton as Faust, Andreas Teuber as Mephistopheles and Elizabeth Taylor as a scantily clad Helen of Troy; it was excoriated by the critics. OH GOD! YOU DEVIL (*1984*) is another Faustian comedy. Both *Phantom of the Opera* (*1983*) and *Phantom of the Opera* (*1989*) make much of the fact that the opera currently in production is *Faust*; the latter version and *Phantom of the Paradise* (*1974*) are in their very different ways more deeply concerned with Faustian pacts (◊ PHANTOM OF THE OPERA). [JG]

Further reading: *Faust in Literature* (1975) by J.W. Smeed; *Doctor Faustus from History to Legend* (1978) by Frank Baron; *Faust through Four Centuries/Vierhundert Jahre Faust* (1989) ed Peter Boerner and Sidney Johnson; *Doctor Faustus: A- and B-texts (1604-1616)* (1993) by Christopher Marlowe, ed David Bevington and Eric Rasmussen; *The English Faust Book* (1994) by John Henry Jones.

FAUST French/Czech live-action/puppet/stop-motion ANIMATED MOVIE (*1994*). Athanor/Heart of Europe Prague K/Lumen/BBC Bristol/Koninck/ Pandora. **Pr** Jaromir Kallista. **Exec pr** Karl Baumgartner, Keith Griffiths, Michael Havas, Hengameh Panahi, Colin Rose. **Dir** Jan ŠVANKMAJER. **Anim dir** Bedrich Glaser. **Screenplay** Švankmajer. **Based on** libretto by Jules Barbier and Michel Carré for the OPERA *Faust* (1859) by Charles Gounod, and on the FAUST recountings by GOETHE, Christian Dietrich Grabbe, Christopher MARLOWE and the Czech Folk Puppeteers. **Starring** Peter Cepek (Faust/Mephistopheles) plus Jan Kraus, Vladimir Kudla, Jiri Suchy, Antonin Zacpal. **Voice actor** Andrew Sachs (English-language version). 92 mins. Colour.

To synopsize this exceptionally complicated, highly Surrealist (◊ SURREALISM) and often very funny movie would not be sensible. The central character – just another citizen of a modern city, the scruffy KAFKA-esque figure – is, we discover, the next in an endless cycle of Fausts, lured by two emissaries of LUCIFER to a derelict EDIFICE which he penetrates to discover is both a theatre (in fact, *two* theatres: PUPPET and operatic) and the alchemical laboratory (◊ ALCHEMY) of Dr Faustus, where he witnesses and aids the creation of a HOMUNCULUS. In this cloven locale – and in part in the real world – he is driven, in what has become a GODGAME, both to *play* Faust for an audience and to *be* Faust, ever following the script which sometimes he finds lying around (◊ STORY). MEPHISTOPHELES, when summoned, proves to be "Faust" – a MIRROR of the man. The damnation is predetermined, as "Faust" comes to understand too late to deviate from the script: in his instance it comes when, fleeing the theatre, he is run over by an empty car.

The tale is told using a stimulatingly bewildering variety of media, with "Faust" himself being at one moment a puppet and the next a human being (though even then it is made clear that he is in a different sense still a puppet). Other main characters are likewise always ambiguous: aside from the emissaries Cornelius and Valdez, who remain human, they seem to be puppets yet their scale is ever uncertain and they are sometimes capable of independent action far from the strings of the never-seen puppeteer. The setting shares this uncertainty: as if the stage were a PORTAL, sometimes it briefly blossoms to become a real-world exterior, which in turn can sometimes be seen to be merely a puppets' stage.

There is much of nightmare in this movie as it locks one in its paranoid grip, just as "Faust" is locked in Lucifer's. Apart from one or two longueurs, it is so dense with fantasy notions, icons and imageries that repeated viewing is necessary before one has the feeling one has watched it completely. *F* has the completely unsettling effect that is unique to the best of FANTASY.

It should be noted that the excellent dubbing of the English-language version, with all voices played by Andrew Sachs, is in itself no mean feat. [JG]

See also: CONTRACT; PACT WITH THE DEVIL.

FAUST: EINE DEUTSCHE VOLKSSAGE (vt *Faust: A German Popular Legend*) German movie (***1926***). **Dir** F.W. MURNAU. **Spfx** Carl Hoffmann. **Screenplay** Hans Kyser. **Based on** Part 1 of *Faust* (**1808**) by GOETHE. **Novelization** *Faust* * (*c***1927**) by Hayter Preston and Henry Savage. **Starring** Wilhelm Dieterle (Valentin), Gösta Ekmann (Faust), Werner Fuetterer (Archangel), Camilla Horn (Margarethe/Gretchen), Emil Jannings (Mephisto). 100 min. B/w. Silent.

An archangel bets the DEVIL, with the Earth as stake, that the virtuous alchemist (◊ ALCHEMY) FAUST cannot be subverted to EVIL. First the Devil casts plague over Faust's town; then he sends Mephisto (◊ MEPHISTOPHELES) to tempt the man with promises that he may cure the plague if he signs PACT WITH THE DEVIL. Faust agrees to a 24-hour trial period. At first his cures work; then the citizens, seeing he is powerless to touch a crucifix, realize he is allied with SATAN, and stone him. Retreating to his chambers, he again bargains with Mephisto, now being given youth and the abducted Herzogin of Parma, reputedly the most beautiful woman in Italy. Just as he is about to enjoy her, the 24 hours is up, and he agrees to the contract becoming permanent.

Some while later, tired of hedonism, he commands Mephisto to take him back to his hometown, just as it was. He is immediately besotted with the virtuous young woman Gretchen/Margarethe, and Mephisto uses MAGIC so that Faust may debauch her, but also frames Faust for the murder of Gretchen's brother Valentin. Ten months later, at Christmas, Gretchen is lost in the snow with her infant; delirious, she imagines a snowdrift is a cot, and puts the baby in it. Condemned to the stake for the baby's murder, she screams for Faust, who at the last returns with love on his lips to join her on the pyre. Because of his repentance and the purity of his LOVE, his SOUL reverts to the Lord, and the Devil has lost his bet.

The tale is told with a tremendous sense of melodrama and not a little clumsy HUMOUR. Where the movie is genuinely powerful is in the strength of Murnau's vision. Frame after frame is beautifully contrived, as if he were creating individual pieces of artwork. The spfx are grimly impressive – technically as good as most that would appear during the next several decades, but, more importantly, so boldly conceived as to be beyond any evaluation but awe. [JG]

See also: BEDAZZLED (***1967***); FAUST (***1994***).

FEAR Two horror MAGAZINES.

1. Distinguished by the exclamatory: *Fear!*, US digest magazine, bimonthly, 2 issues May-July 1960, published by Great American Publications; ed Joseph L. Marx.

A companion to the sf magazine *Fantastic Universe*; some sf appears, though the contributors came mostly from the publisher's mystery magazines. The stories are routine, with the supernatural element being usually BLACK MAGIC. [MA]

2. UK slick-format magazine, 33 issues July/August 1988-September 1991, initially bimonthly, then monthly from July 1989, published by Newsfield; ed John Gilbert.

A heavily illustrated glossy magazine, *F* concentrated on HORROR MOVIES and physical HORROR in all of its forms. It was strong on author and personality profiles, and was briefly the UK's only horror-fiction magazine. It gave wide coverage to horror fiction and encouraged new writers, though little of quality emerged; *F*'s best fiction, only marginally supernatural, came from more accomplished writers including John BRUNNER, Ramsey CAMPBELL, Jonathan CARROLL, Christopher FOWLER, Thomas LIGOTTI, Darrell SCHWEITZER, Brian STABLEFORD and Ian WATSON. Although *F* was profitable, its publisher suffered financial losses on other magazines, which included *Frighteners* (3 issues July-September 1991), whose *#1* had to be withdrawn from sale because of the story "Eric the Pie" by Graham Masterton (1946-   ). Newsfield's collapse also saw the demise of *GMI*, a companion GAME magazine. [MA]

FEARLESS VAMPIRE KILLERS, THE; OR PARDON ME, YOUR TEETH ARE IN MY NECK (ot *Dance of the Vampires*) UK movie (***1967***). MGM/Cadre/ Filmways. **Pr** Gene Gutowski. **Exec pr** Martin Ransohoff. **Dir** Roman Polanski. **Mufx** Tom Smith. **Screenplay** Gerard Brach, Polanski. **Starring** Alfie Bass (Shagal), Fiona Lewis (Maid), Jack MacGowran (Professor Abronsius), Ferdy Mayne (Count von Krolock), Polanski (Alfred), Jessie Robins (Rebecca), Sharon Tate (Sarah). 107 mins (initial US release cut to 91 mins). Colour.

A sporadically amusing PARODY of the DRACULA-MOVIE canon. Abronsius (cf Van Helsing) and servant Alfred come to Transylvania in search of VAMPIRES. Both men fall for the charms of their innkeeper's lovely daughter Sarah, but she is snatched away by von Krolock (cf Dracula) to his castle. After much clowning they rescue her and leave Transylvania, not realizing that she too is now a vampire and will spread the bloodsucking "plague" worldwide.

Although it has impressive moments (as when our heroes are recognized as mortals at a ghostly ball because they are the only ones whose reflections appear in a MIRROR), this movie defaults too easily into buxom-bathing-belles humour, and is elsewhere leaden; it is a step up from the previous year's *Carry On Screaming* (vt *Carry On Vampire*), but only a step. The movie was initially released in Europe as *Dance of the Vampires*, but in the USA Ransohoff, concerned there was too much HORROR for a comedy, cut 16 mins, dubbed out the Yiddish/East European accents (Polanski's included) and instituted the title by which the movie is now universally known. The movie flopped in Ransohoff's version but was successful in Polanski's, under both titles. [JG]

FEDER, ROBERT ARTHUR (1909-1986) US movie producer. ◊ Robert ARTHUR.

FEIST, RAYMOND E(LIAS) (1945-   ) US writer who has largely concentrated on three sequences set in the same complicated FANTASYLAND. The first of these – the **Riftwar Saga** sequence, comprising *Magician* (**1982**; vt in 2 vols as *Magician: Apprentice* 1985 and *Magician: Master* 1986; the whole rev in 1 vol 1992), *Silverthorn* (**1985**), *A Darkness at Sethanon* (**1986**), *Prince of the Blood* (**1989**) and *The King's Buccaneer* (**1992**) – shows the influence of REF's earlier work as a designer of role-playing GAMES, for both good and ill. The fantasyland is made up of two neatly contrasted worlds: medieval Midkemia, beset by DYNASTIC-FANTASY disputes, and Oriental militaristic Kelewan. The "rifts" between them serve as PORTALS constructed to allow convenient shifts of personnel when the underlying rules of the GAMEWORLD-like narrative so demand. The UGLY-DUCKLING protagonists are also conveniently dual, and follow similar patterns in their individual QUESTS. Pug becomes a great WIZARD and Tomas becomes a mighty HERO, with the aid of a friendly DRAGON. Both worlds offer COMPANIONS galore – from ELVES to ROBIN HOOD clones, and in the second volume a TRICKSTER thief – for the protagonists to choose from.

The underlying structure of the sequence gradually reveals itself to be a continuation of an age-old CHAOS War between the Valheru, who are themselves intrinsic with Being, and the GODS, who have created domains upon the world. Tomas is by essence (it turns out) a Valheru, and his search for a master sorcerer who will answer certain questions conflates with more secular conflicts, engaging a rightful king against the forces of the Dark. The conclusion of these quests and battles, which comes at the end of *A Darkness at Sethanon*, occasions a shift in focus in following volumes, which are essentially HEROIC FANTASIES set in the pre-existing Fantasyland.

The **Empire** sequence, with Janny WURTS – *Daughter of the Empire* (**1987**), *Servant of the Empire* (**1990**) and *Mistress of the Empire* (**1992**) – is likewise adventurer fantasy. It involves the female protagonist in an intricate gamelike set of conflicts (military and financial) with various families who oppose her own betrayed House of Acoma. Eventually she wins through, gaining wealth, husband and heirs, and even copes with a loose cadre of magicians, who refuse to obey the laws of the game but fortunately disagree among themselves.

REF's third series, the ongoing **Serpent Wars Saga** – *Shadow of a Dark Queen* (**1994**) and *Rise of a Merchant Prince* (**1995**), with further volumes projected – is again set in the same basic universe, some time before the climactic events that close *A Darkness at Sethanon*. Once more it combines heroic-fantasy routines with potentially world-changing plot events. On that larger stage, a SERPENT-like race threatens to invade Midkemia, but is opposed by the young protagonist.

REF's only singleton, *Faerie Tale* (**1988**), is a DARK FANTASY whose protagonist and family move into a house, Old Kessler Place, which turns out to be a BAD PLACE: denizens of FAERIE are warring here, and they oppress and assault the entrapped humans in the course of their quarrel. The action eventually shifts to one of the protagonists' children, whose TWIN has been abducted into Faerie and substituted by an unpleasant CHANGELING, and who must undertake a quest to the Erl King's court in order to retrieve him. He is

there subjected to the usual QUIBBLES, but successfully transacts the RITE OF PASSAGE and regains his brother. REF is an adept manipulator of standard material.   [JC]

FELITTA, FRANK (PAUL) DE (1921-   ) US writer of novels that tend to inhabit the borderlands between fantasy, HORROR and SUPERNATURAL FICTION. His first novel, *Audrey Rose* (**1975**), remains his best-known; it was filmed as AUDREY ROSE (*1977*) – which Felitta coproduced and wrote – and sequelled by *For Love of Audrey Rose* (**1982**). Felitta also wrote the screenplay for the movie version of his next novel, *The Entity* (**1978**) – *The Entity* (*1981*). His other novels have been *Golgotha Falls: An Assault on the Fourth Dimension* (**1984**), a fantasy, and *Funeral March of the Marionettes* (**1990** UK), which is horror.   [JG]

FELL, ANDREW [s] ◊ Robert ARTHUR.

FELL, H(ERBERT) GRANVILLE (1872-1951) UK illustrator, active from about 1895 in a style which derives more from the curvilinear Celtic of William MORRIS than from the stiffly hieratic Walter CRANE, though he did imitate the latter whenever noble subjects were depicted. Books illustrated by HGF include versions of *Ali Baba and the Forty Thieves* (1895) and *Cinderella and Jack and the Beanstalk* (coll 1895), *The Book of Dragons* (**1900**) by E. NESBIT, *Wonder Stories from Herodotus* (coll 1900) and the anonymous *Sir Thomas Thumb, or The Wonderful Adventures of a Fairy Knight* (1907). In this last volume HGF showed a sharp, fluent gracefulness even in the depiction of horrors – an example being a giant SPIDER's attack on TOM THUMB.   [JC]

FELLOWSHIPS ◊ COMPANIONS.

FEMINISM Providing GENRE-FANTASY women with adequate roles as protagonists, mages or rulers has not been – given the medieval fixity of FANTASYLAND as a venue – a task very plausibly accomplished. But it is a task that needs constant doing. Therefore, it is welcome to see female warriors; AMAZONS who do not commit suicide when their men die; young girls triumphantly going into menarche and coming out the other side with TALENTS and lovers and power; older women who like SEX (as older men have always been allowed to); and so forth. But even serious fantasy novels may be overburdened by the weight of traditional STORY, and may at times struggle too visibly to make REVISIONIST-FANTASY hay out of intractable material. Again, it is a task that needs constant doing.   [JC]

See also: GENDER.

FEMME FATALE A woman whose powers of sexual attraction are so great (though not necessarily supernatural) that her paramours become utterly careless of their own well-being, usually perishing. Some of the many Classical models, like the SIRENS, took no active part in destroying the men who fell under their spell; others, such as the blood-drinking LAMIAS, exercised their lures in order to devour their victims; SUCCUBI drained men's SOULS by night; while CIRCE had the symbolically appropriate habit of turning her victims into swine. The archetypal *femme fatales* in the Judeo-Christian tradition are LILITH and SALOME. Literary *femmes fatales* were prolifically produced by writers associated with ROMANTICISM during the late 18th and early 19th centuries; a comprehensive analysis of their significance in this context can be found in *The Romantic Agony* (**1933**) by Mario Praz. Some such images are figures of menace, but most are regarded with decided ambivalence; those who are sadistically contemptuous of their easy prey are frequently provided with victims who accept that the intensity of the erotic experience more than adequately compensates for its brevity. A relevant anthology is *The Dedalus Book of Femmes Fatales* (anth **1992**) ed Brian STABLEFORD. [BS]

FÉNELON, FRANÇOIS DE SALIGNAC DE LA MOTHE (1651-1715) French cleric and writer, whose *Les aventures de Télémaque* (**1699**; trans I. Littlebury as *The Adventures of Telemachus* **1699** UK; preferred trans John Hawkesworth **1768** UK) carries the son of ODYSSEUS on a journey through a Greece (conceived on PASTORAL lines as a GOLDEN AGE) in a search for his lost father. The goddess Minerva accompanies him, cross-dressed as a wise, male Mentor (◊ GENDER DISGUISE). At least two UTOPIAS are encountered: one, the Kingdom of Salente, features a colour-coded paternalism (◊ COLOUR-CODING); the second, La Bétique, is pure pastoral. [JC]

FENN, LIONEL Pseudonym of Charles L. GRANT.

FENRIR/FENRIS ◊ NORDIC FANTASY.

FERAL CHILDREN Human children require education, and thought-experiments asking what might become of a child denied "civilization" are common in fantasy. Most such stories, recalling the Greek legend of Romulus and Remus, propose that lost infants might be found and reared by animals; there are claims this has happened in reality. The early literary history of the theme is discussed in an essay by Rudolph Altrocchi in *Sleuthing the Stacks* (coll **1944**). The most famous literary examples are Rudyard KIPLING's MOWGLI, raised by WOLVES in a series begun in *The Jungle Book* (coll **1894**) and collected in *All the Mowgli Stories* (coll **1933**), and Edgar Rice BURROUGHS's TARZAN, raised by APES in the long series begun with *Tarzan of the Apes* (**1912**; **1914**); a significant earlier work is *The Child of Ocean* (**1889**) by Sir Ronald Ross (1857-1932). Tarzan spawned many imitations, most of which substitute some other foster-parent species: MONKEYS in *Jungle Boy* (**1925**) by John Eyton; lions in *The Lion's Way* (**1931**) by C.T. Stoneham; leopards in *Lord of the Leopards* (**1935**) by F.A.M. Webster; bears in *Hawk of the Wilderness* (**1936**) by William L. Chester; and dolphins in *Dolphin Boy* (**1966**) by Roy Meyers. Imitations of Kipling include *Shasta of the Wolves* (**1928**) by Olaf Baker and *Hathoo of the Elephants* (**1943**) by Post Wheeler. Female feral children are raised by lions in *Wild Cat* (**1935**) by H.M.E. Clamp, jackals in *The Jungle Goddess* (c**1935**) by Orme Sackville and hyenas in *Kala* (**1990**) by Nicholas LUARD. A useful theme anthology is *Mother Was a Lovely Beast* (anth **1974**) ed Philip José FARMER; Farmer has written numerous tales "updating" and revising the myth of Tarzan, none of them as sweeping in its revisions as *The Death of Tarzana Clayton* (**1985** chap) by Neville Farki. A rare satirical variant is "The Death of an Apeman" (trans 1970) by Josef Nesvadba. [BS]

FERGUSSON, BRUCE (CHANDLER) (1951-) US writer whose fantasy work consists of two titles in his **Six Kingdoms** sequence, which seems open-ended. The general venue is a medieval FANTASYLAND, somewhat grimmer than the norm; but – most unusually – the main setting within this venue is urban (◊ URBAN FANTASY) and extremely bleak. The first tale, *The Shadow of his Wings* (**1987**), features a HIDDEN MONARCH who is inclined not to accept the throne and a QUEST which he can succeed in terminating only at heavy cost to others. The protagonist of *The Mace of Souls: A Novel of the Six Kingdoms* (**1989**) is a criminal whose initial reasons for becoming involved in his quest are to acquire sufficient funds to get to a nicer climate; the general tenor of the story is mean-street and weary. Both novels, after much duelling with SHADOWS and episodes of personal growth,

end positively; but it is clear that BF was not entirely comfortable with the conventions of HEROIC FANTASY. It is hoped his silence will end. [JC]

FERNANDEZ, FERNANDO (1940-) Self-taught Catalan artist and writer whose art has appeared as book covers and illustrations and whose fantasy and sf COMIC strips have received worldwide acclaim. His writing includes comics criticism as well as comic strips and educational books. FF often uses several art styles in successive frames on the same page of comic strip.

He had several menial jobs before being taken on as an art assistant for **Sparky: The Phantom Horseman's Son** in 1955. He then began to produce strips, and gained professional experience inking artwork for the studio/agency Selecciones Illustradas (SI), then beginning to provide artwork for the UK comics market. By 1957, FF was doing strips for UK romance titles like *Romeo* and *Valentine*.

In 1958 his family migrated to Argentina. Here he met Hugo Pratt (1927-1995), Arturo del Castillo (1925-1992) and Alberto BRECCIA. He came back to Spain the following year and continued producing strips for the UK market including some war stories for Fleetway, and then in 1965, along with a few colleagues at SI (including José Maria Miralles, Felix Mas and Lopez Espi), he gave up comics to produce painted covers for books and magazines. But by 1970 he returned to comics, with a satirical humour strip, **Flies** (1970-72), and began work on a long series of short pieces with a fantasy or sf bias which were published very widely in many languages, including English (by WARREN PUBLISHING).

In 1975-7 FF wrote and illustrated four educational books for the publisher Afha in the series **Ciencia y Aventura** ["Science and Adventure"]. These took the form of comic-strip adventures featuring a group of pioneering cosmonauts embarking on voyages of discovery: FF's titles were *Conocimientos del Cuerpo Humano* ["Facts about the Human Body"], *Viaje al Mundo Secreto de los Insectos* ["Journey into the Secret World of Insects"] – awarded the Premio Nacional for Best Illustrated Book in 1977 – *Viaje a la Preistoria* ["Voyage into Prehistory"] and *Viaje a las Estrellas* ["Journey to the Stars"]. FF then embarked on the impressive *L'Uomo di Cuba* ["The Man of Cuba"] (graph **1979**), one of Cepim's **Un Uomo Un'Avventura** ["A Man, An Adventure"] series of comic-strip books which presented stories about individuals caught up in the great events of history.

FF's masterpiece, the GRAPHIC NOVEL *Zora y los Hibernautas* (1980; trans as *Zora and the Hibernauts* in HEAVY METAL 1982-3; graph **1984**) was published in the Spanish magazine *1984*. This luxuriantly coloured SCIENCE FANTASY was followed by the richly painted *Dracula* (1984; trans graph **1984**), based on Bram STOKER's *Dracula* (**1897**).

FF has continued to write for the Spanish comics magazine *Cimoc* and to produce impressively drawn comic strips in collaboration with his wife, Rosa Lleida. [RT]

FERNGULLY: THE LAST RAINFOREST Australian ANIMATED MOVIE (*1992*). FAI/Youngheart. **Pr** Peter Faiman, Wayne Young. **Exec pr** Robert W. Cort, Ted Field. **Dir** Bill Kroyer. **Screenplay** Jim Cox. **Based on** the **FernGully** stories by Diana Young. **Voice actors** Tim Curry (Hexxus), Samantha Mathis (Crysta), Robert Pastorelli (Tony), Christian Slater (Pips), Jonathan Ward (Zak Young), Robin Williams (Batty Koda), Grace Zabriskie (Magi Lune). 76 mins. Colour.

Long ago the volcano Mount Warning erupted, driving humans from the rainforest, whose community has come to regard humans as MYTHICAL CREATURES; at the same time the rainforest's good WITCH, Magi Lune (seemingly based on Angela Lansbury), imprisoned the local SPIRIT of EVIL, Hexxus, in a gnarled TREE. But now she is old, and wishes to pass on her MAGIC to her reluctant apprentice, the FAIRY Crysta. Now, too, humans – in the shape of a voracious logging company – are threatening the rainforest. Disbelieving this, Crysta goes in company with an escaped laboratory fruitbat, Koda, to see for herself. They encounter Zak, a logger who has just marked Hexxus's gnarled prison for felling; Crysta casts a SPELL clumsily, thereby accidentally reducing him to her own size (◊ GREAT AND SMALL). The gnarled tree is felled, and Hexxus – a resinous SHAPESHIFTING mass that thrives on oil, smoke and fumes – is released. The rainforest denizens, plus Zak, at last defeat the MONSTER and save the rainforest. As a piece of animated-feature making, *F:TLR* competes with DISNEY on Disney's home territory, and wins. As a fantasy with an ecological moral it works well; it is also an unusually well constructed animated feature, right down to its underpinning: much thought has gone into the "magical ecosystem". [JG]

FERTILITY A central concern of most archaic RELIGIONS is the RITUAL control of fertility, which is almost invariably understood as being tied to the SEASONS. Fertility – of humans, beasts and vegetation alike – is understood as part of a great CYCLE, and its regular return is understood as being in the gift of the GODS. Fertility is not, therefore, thought of as following upon natural or unthinking behaviour; it is something mysterious, triumphant, earned, and given.

Historical fantasies abound in images of fertility, though rarely examined to much effect. In SUPERNATURAL FICTION – as befits its 19th-century period of literary triumph – fertility is generally thought of as obscenely procured, and as a signal of WRONGNESS. Fantasies and supernatural fictions which focus on GODDESS figures tend to be suspicious of fertility – and the rituals associated with it – until well into the 20th century. In HORROR novels, any attention to questions of the seasons or their fructification is likely to be understood in terms of blasphemous PARODY. In full fantasy, questions of fertility tend to be aetherialized into mythic patterns, in which a period of THINNING of the LAND is succeeded by a HEALING, and the WASTE LAND becomes fertile once again.

Fertility is one of the three aspects of the GODDESS (◊◊ APHRODITE), the other two being VIRGINITY and agedness. [JC]

FESSIER, MICHAEL (1907-1988) US writer. His first novel, *Fully Dressed and in his Right Mind* (**1935**), is a curious tale whose luckless hero is persecuted by a murderous midget with a direly intimidating stare, but finds succour by virtue of his involvement with a supernatural female. *Clovis* (**1948**) is a comedy about a smart-talking parrot who finds it frustratingly difficult to build a career as an educator of mankind. MF's uncollected short stories include the SHAPESHIFTING fantasies "Bewitched" (1950) and "The Blue-Eyed Horse" (1955), the first about "the cat in every woman" and the second about a gambler's comeuppance. [BS]

FICTION (magazine) ◊ FRANCE; *The* MAGAZINE OF FANTASY AND SCIENCE FICTION.

FIELD, EUGENE (1850-1895) US journalist and poet.

Most of his fantasy consists of whimsical CHILDREN'S FANTASY. His most famous fantasy for adults was the title story of *The Holy Cross and Other Tales* (coll **1893**; exp as vol V of *The Writings in Prose and Verse of Eugene Field* **1896** 10 vols), a CHRISTIAN FANTASY about the WANDERING JEW; others of note therein are "The Pagan Seal-Wife" and (from the expanded version) the comedy "The Platonic Bassoon". Volumes II and X of the posthumous *Writings*, *A Little Book of Profitable Tales* (coll **1889**) and *Second Book of Tales* (coll **1896**), contain his other children's stories, the latter also including "The Werewolf". One tale which escaped the collected works was "Story of the Two Friars" (1900; vt *How One Friar Met the Devil and Two Pursued Him* **1900** chap; vt *The Temptation of Friar Gonsol*; in *Eugene Field: An Autoanalysis*; *How One Friar Met the Devil and Two Pursued Him* coll 1901). [BS]

Other works: *The Symbol and the Saint: A Christmas Tale* (**1886** chap); *Christmas Tales and Christmas Verse* (coll **1912**); *The Divell's Chrystmass* (**1913** chap); *The Mouse and the Moonbeam* (**1919** chap).

FIELD, GANS T. Pseudonym of Manly Wade WELLMAN.

FIELD, RACHEL (1894-1942) US writer primarily noted for realistic adult and children's novels, though she made one noteworthy contribution to fantasy – *Hitty, Her First Hundred Years* (**1929**), the memoirs of a sentient (though immobile) DOLL – for which she became the first woman to win the Newbery Medal for Children's Literature. Happy with her existence, Hitty is a singular example of a doll who does *not* want to be human (◊ REAL BOY), and her story provides interesting commentary on the joys and liabilities of IMMORTALITY. [LL]

FIELD OF DREAMS US movie (*1989*). Universal/Gordon. **Pr** Charles Gordon, Lawrence Gordon. **Exec pr** Brian Frankish. **Dir** Phil Alden Robinson. **Screenplay** Robinson. **Based on** *Shoeless Joe* (**1982**) by W.P. KINSELLA. **Starring** Dwier Brown (John Kinsella), Timothy Busfield (Mark), Kevin Costner (Ray Kinsella), Gaby Hoffman (Karin Kinsella), James Earl Jones (Terence Mann), Burt Lancaster (Dr Archibald "Moonlight" Graham), Ray Liotta (Shoeless Joe Jackson), Amy Madigan (Annie Kinsella), Frank Whaley (Archie Graham). 106 mins. Colour.

Iowa farmer Ray Kinsella fell out of love with father John years before the man died. One night a Voice says to Ray: "If you build it, he will come." The incident recurs, and he believes he is losing his mind; he is furious on discovering daughter Karin watching HARVEY (*1950*) on tv. At last come VISIONS of a BASEBALL ground on his land and of Shoeless Joe, his father's baseball hero. Ray ploughs under much of his corn to carve a pitch. Months later, the GHOST of Shoeless Joe indeed appears, bringing with him others; they use the pitch as a practice-ground, though cannot cross its perimeter (◊ HAUNTING). The Voice speaks to Ray again, saying "Ease his pain", and Ray concludes the sufferer is 1960s guru writer and pacifist Terence Mann, now an embittered recluse; Ray and Mann go to Chisholm, Minnesota, to contact minor baseball player "Moonlight" Graham, but discover Graham is dead. Ray enters a TIMESLIP, returning to 1972, where the aged Graham, a much-loved doctor and benefactor, explains he thanks God his baseball career was cut so short. Driving back to Iowa, they pick up a baseball-mad youth called Archie Graham; once they have returned to the "field of dreams" Archie is invited to play with the past stars, who regard the field as HEAVEN. But Karin has a choking fit and Archie exits the field to aid her, becoming the elderly Doc Graham and eventually joining the baseball players in the AFTERLIFE; Mann, too, is invited there, and goes. The last shade left on the field is John Kinsella, and father and son rediscover each other through baseball practice.

FOD shows every sign of having been crafted out of love. Its major theme is the power of belief: belief, it says, can alter PERCEPTION; and perception has the power to alter REALITY. [JG]

FIENDISH PLOT OF DR FU MANCHU, THE US movie (*1980*). ◊ FU MANCHU MOVIES.

FIEVEL GOES WEST vt of *An* AMERICAN TAIL: FIEVEL GOES WEST (*1991*).

FILM ◊ CINEMA.

FIMBULWINTER In Norse MYTH and NORDIC FANTASY, the deep, unnatural winter afflicting Earth for several years before the LAST BATTLE (◊◊ RAGNAROK). L. Sprague DE CAMP's and Fletcher PRATT's "The Roaring Trumpet" (1940) is set in Fimbulwinter. Other fantasies borrow the term for less apocalyptic blizzards and cold generated by MAGIC, as in Alan GARNER's *The Weirdstone of Brisingamen* (1960). [DRL]

FINAL CONFLICT ◊ LAST BATTLE.

FINAL CONFLICT, THE: OMEN 3 (*1981*) ◊ *The* OMEN.

FINAL PROGRAMME, THE (vt *The Last Days of Man on Earth* US) UK movie (*1973*). Goodtimes/Gladioli. **Pr** John Goldstone, Sandy Lieberson. **Exec pr** Roy Baird, David Puttnam. **Dir** Robert Fuest. **Screenplay** Fuest. **Based on** *The Final Programme* (1968) by Michael MOORCOCK. **Starring** Jon Finch (Jerry Cornelius), Jenny Runacre (Miss Brunner). 89 mins. Colour.

A high-camp TECHNOFANTASY caper set in the time leading up to the END OF THE WORLD. After much **James Bond**-parody chasing, Jerry Cornelius and Miss Brunner (an awesomely efficient, awesomely strong near-superhuman who initially at least resembles Cruella De Vil in *One Hundred and One Dalmatians* [*1961*] and who, like a SPIDER, devours her sexual partners) are fused by means of the Final Programme to form the shambling, immortal (◊ IMMORTALITY) Neanderthalish MONSTER who will inherit the new world. Moorcock's **Jerry Cornelius** novels are not noted for their adherence to narrative conventions: *TFP*, its plot complicated by what may be judged either gallant SURREALISM or irrelevant trendy psychedelia, perpetuates the tradition. [JG]

FIN DE SIÈCLE The specific meaning of *siècle* is "century", but it has a more general sense; thus, the phrase FDS carries a double meaning when used to refer to the art of the 1890s, because one of the significant movements in French letters at that time was DECADENCE, which compared the modern metropolis to Rome immediately before its ruined empire fell to the barbarians. It is sometimes argued that the 1990s are similarly haunted by a sense of termination, often likened to the Millennarian panic alleged to have swept Europe as AD1000 approached; the phrase is occasionally used in that context to apply to contemporary fantasies of moral dereliction and the APOCALYPSE. [BS]

FINDLEY, NIGEL (1959-1995) US writer, reportedly of at least 100 titles tied to various GAMES; among those which have been traced are three titles for the SHADOWRUN series of TECHNOFANTASIES (◊◊ SHARED-WORLDS); plus *Spellhammer: Into the Void* * (*1991*) and *Torg: Out of Nippon* * (*1992*). [JC]

FINDLEY, TIMOTHY (IRVING) (1930-   ) Canadian

actor, dramatist and writer who began publishing fiction in 1956 with the nongenre "About Effie". His first work of fantasy interest came much later. *Famous Last Words* (**1981**) is like many RECURSIVE FANTASIES set in an ALTERNATE-WORLD version of the 20th century in that it mixes fictional characters (like Ezra Pound's Hugh Selwyn Mauberley) with real figures (like Charles Lindbergh) in its analysis of the rise of fascism in Europe (a typical focus of the modern recursive fantasy), but does not explicitly fantasticate the world itself. *Not Wanted on the Voyage* (**1984**) is a flamboyant TWICE-TOLD contemporary rendering of the story of Noah (◊ FLOOD), which moves into REVISIONIST-FANTASY territory in its assaults upon the patriarchal order enforced within the Ark. *Headhunter* (**1993**) again introduces a fictional character into the real world, but this time within a fantasy frame; a librarian unleashes Mr Kurtz – from Joseph CONRAD's "Heart of Darkness" (1902) – into contemporary Toronto, which he begins to dominate. [JC]

FINGER, BILL (1914-1974) US COMICS writer. ◊ BATMAN.

FINLAY, VIRGIL (WARDEN) (1914-1971) US artist and one of the most popular illustrators for the pulp MAGAZINES, especially WEIRD TALES and FAMOUS FANTASTIC MYSTERIES. He was a master of stippling; his results, often reminiscent of the work of Gustav DORÉ, were highly atmospheric. VF succeeded in capturing a mood of the outré in his artwork unlike any other magazine illustrator at the time.

VF's illustrations first appeared in the December 1935 *WT*, and he graduated to covers with the February 1937 issue. His pictures were immediately popular with readers and authors alike, many of the latter – including H.P. LOVECRAFT – clamouring to have their work illustrated by him. A. MERRITT hired him as staff artist on *The American Weekly* in late 1937. During 1940-69 VF's work appeared in almost all fantasy and sf magazines. *Famous Fantastic Mysteries*, realizing his sales potential, issued three portfolios of his work as premia to subscribers in 1941, 1942 and 1948. He also illustrated the dustjacket of the first book issued by ARKHAM HOUSE, *The Outsider and Others* (coll **1939**) by Lovecraft.

VF illustrated few books. Most of his best fantasy illustrations appeared in *WT*, *Famous Fantastic Mysteries*, FANTASTIC, FANTASTIC ADVENTURES, *If* and *Startling Stories*; in the 1960s he worked also for a number of astrological magazines. Since his death his style has been imitated by George Barr and Stephen Fabian (1930-   ).

VF's work has been reissued in a number of compilations. The first were *Virgil Finlay* (graph coll **1971**) and *Virgil Finlay: An Astrology Sketchbook* (graph coll **1975**) both ed Donald M. Grant (1927-   ). From Gerry de la Ree (1924-1993) came a pair of books, *Virgil Finlay: A Portfolio of His Unpublished Illustrations* (graph coll **1971** chap) and *Finlay's Lost Drawings* (graph coll **1975** chap), and then a more thematic series – *The Book of Virgil Finlay* (graph coll **1975**), *The Second Book of Virgil Finlay* (graph coll **1978**), *Third* (graph coll **1979**), *Fourth* (graph coll **1979**), *Fifth* (graph coll **1979**) and *Sixth: The Astrology Years* (graph coll **1979**) – plus *Virgil Finlay Remembered: The Seventh Book of Virgil Finlay: His Art and Poetry* (graph coll **1981**). These 7 vols were indexed by Ian Bell in *Virgil Finlay Indexed* (**1986** chap). Other compilations include *Finlay's Femmes* (graph coll **1977** chap), *Virgil Finlay's Women of the Ages* (graph coll **1992**), *Virgil Finlay's Strange Science* (graph coll **1993**) and *Virgil Finlay's Phantasms* (graph coll **1993**). [MA]

FINIAN'S RAINBOW US movie (*1968*). Warner/Seven Arts. **Pr** Joseph Landon. **Dir** Francis Ford Coppola.

Screenplay E.Y. Harburg, Fred Saidy. **Based on** the musical *Finian's Rainbow* (staged 1947) by Harburg and Saidy, based in turn on James STEPHENS's *The Crock of Gold* (**1912**). **Starring** Fred Astaire (Finian McLonergan), Petula Clark (Sharon McLonergan), Don Francks (Woody Mahoney), Al Freeman Jr (Howard), Barbara Hancock (Susan [Mahoney] the Silent), Tommy Steele (Og), Keenan Wynn (Senator Billboard Rawkins). 145 mins. Colour.

Finian and daughter Sharon come from Ireland to small US community Rainbow Valley bearing a crock of gold (which is also a crock of WISHES) that Finian has "borrowed" from a leprechaun, Og; Finian believes that in the USA, because of its riches, gold planted in the ground will naturally grow – especially here, near Fort Knox. Og pursues, for the lack of his crock is turning him mortal. Between them the three cast light into hearts that were before filled only with darkness, bring a measure of prosperity to Rainbow Valley, and in their different ways find LOVE.

Coppola has referred to this musical as a "disaster" and Astaire called it a "disappointment", yet time has favoured it: for example, the lack of formal choreography, regarded on *FR*'s release as a failing, seems today a strength. *FR*'s antiracism may seem trite today, but it is partly because of movies like *FR* that this has become so. Throughout *FR* there are themes of blarneyesque mythopoeia (the cod Oirish is tolerable, just) and of the MAGIC to be discovered in the everyday.   [JG]

FINNEY, CHARLES G(RANDISON) (1905-1984) US newspaperman and writer remembered mainly for his first novel, *The Circus of Dr Lao* (**1935**), famously illustrated by Boris Artzybasheff (1899-   ) in a hard-edged surreal mode which perfectly captured the hard impersonal clarity of the tale itself. In writing the story, CGF was clearly influenced by his military service (he had served 1927-30 in China). To Abalone, Arizona, sometime early in the 20th century, comes Dr Lao's CIRCUS, announced beforehand by a poster claiming that visitors will be able to see a chimera, Medusa, MERLIN, a MERMAID, PAN, the Loch Ness MONSTER, a sphinx, a UNICORN and a WEREWOLF. The three caravans arrive, seemingly from nowhere, and disgorge far more than they seem capable of containing. The circus begins; discords soon mount between the small-town Abalonians and the creatures on display. Dr Lao himself is a TRICKSTER, and seems immortal, though the denizens of his circus show the signs of THINNING natural to creatures of MYTH locked into a secular world. Soon the show turns into a black REVEL, and before the night is out 11 humans are dead. The climax – the HUMAN SACRIFICE of a virgin (◊ VIRGINITY) to the ancient GOD of a LOST LAND – passes quickly, the promise that Lao's libidinously evocative creatures will liberate Arizona fades, and the circus leaves. The movie version – 7 FACES OF DR LAO (*1964*) – introduces new protagonists and a plot with a more "congenial" ending, and allows the inference that the denizens of the circus may be projections of Dr Lao's own being.

The book has been deeply influential on writers like Ray BRADBURY, whose small towns owe much to *Dr Lao*, and Peter S. BEAGLE, whose circus in *The Last Unicorn* (**1968**) reads as a direct homage. CGF was perhaps the first US writer to depict the MARVELLOUS and the UNCANNY in terms so mercilessly and hauntingly deadpan.

CGF's later work is of less interest. In *The Unholy City* (**1937**) a resident of Abalone crashlands in Asia, and is guided through mysterious Heilar-Wey where surreal ways of living expose themselves. Some of the stories in *The Ghosts of Manacle* (coll **1964**) are fantasy, including in particular "The Life and Death of a Western Gladiator", in which a SERPENT narrates its own demise. *The Magician Out of Manchuria* (in *The Unholy City*, coll **1968**; **1976** UK) is an intermittently graceful exercise in CHINOISERIE.   [JC]

FINNEY, JACK Working name of US writer Walter Braden Finney (1911-1995), whose fiction, beginning with "Such Interesting Neighbors" for *Collier's Weekly* in 1951, occupied several genres, including sf and fantasy. From the beginning, his style had a slick clear polish, which may be why he was sometimes treated as an impersonal creator of classic works. He is best known for *The Body Snatchers* (**1955**; vt *The Invasion of the Body Snatchers* 1973; rev 1978), which has been filmed as *Invasion of the Body Snatchers* (*1956*), *Invasion of the Body Snatchers* (*1978*) and *Body Snatchers* (*1993*); the novel and the movies can be read as fantasy. In straight fantasy, JF is remembered most for the best of his TIMESLIP tales, *Time and Again* (**1970**) which – with its competent sequel, *From Time to Time* (**1995**) – is one of the most important, and most moving, timeslip texts yet composed.

Though sometimes garnished with perfunctory sf explanations, timeslip (almost always pastwards) usually constitutes an act of entry into something like EDEN, a movement inherent to fantasy. A number of short stories also deal with timeslips; with other tales, these are assembled in *The Third Level* (coll **1957**; vt *The Clock of Time* 1958 UK) and *I Love Galesburg in the Springtime: Fantasy and Time Stories* (coll **1963**), while *About Time* (coll **1986**) assembles time stories from the previous two volumes. The most famous of these tales is "The Third Level" (1952), whose narrator finds the eponymous lowest platform at Grand Central Station (◊ NEW YORK) is a PORTAL to 1894, into which year he departs with a whole heart. Similar routes into the past are found in "Quit Zoomin' Those Hands Through the Air" (1952), "Second Chance" (1957) and "The Love Letter" (1959). *The Woodrow Wilson Dime* (**1968**), unusually, moves sideways into an ALTERNATE WORLD; and *Marion's Wall* (**1973**), filmed as MAXIE (*1985*), clearly homages Thorne SMITH in telling the tale of a 1920s flapper and silent-film starlet who possesses (◊ POSSESSION) a contemporary woman.

The concept of the timeslip receives its fullest explanation in *Time and Again*, where it is argued (by the director of an ominous government TIME-TRAVEL project) that our BONDAGE to any one time is a consequence of our overwhelming sensory experience of being when we are, and that it simply requires a similar intensity of awareness of another time in order to slip the shackles of *now* – to think oneself into the longed-for past – to depart thence. The protagonist of the novel, once he succeeds in operating this technologized nostalgia, begins to find the New York of 1882 much preferable to the polluted, congested, deafening world of 1970. In *From Time to Time* the same protagonist, now in 1912, attempts to prevent WORLD WAR I.

Because JF rendered his vision of timeslip with such skill and devotion, he is an important figure in modern fantasy.   [JC]

Other works: *Five Against the House* (**1954**); *The House of Numbers* (**1957**); *Assault on a Queen* (**1959**); *Good Neighbor Sam* (**1963**); *The Night People* (**1977**); *Forgotten News: The Crime of the Century and Other Lost Stories* (coll **1983**); *3 by Finney* (omni **1987**), assembling *The Woodrow Wilson*

Dime, *Marion's Wall* and *The Night People*.

FINN MAC COOL Also Finn or Fionn Mac Cumhal (or Cumhaill); a warrior-poet hero of Irish MYTH and last leader of the Fianna, an order of CHIVALRY comparable to the ROUND TABLE. Finn's ancestors included FAIRIES and GODS, but though unnaturally long-lived he was mortal; he had powers of PROPHECY and is often described as a GIANT. His "Fianna Cycle" exploits have been much retold and echoed as CELTIC FANTASY – e.g., in Paul HAZEL's **Finnbranch** and Kenneth C. FLINT's **Finn McCumhail** sequences – and he appears as a storyteller in Flann O'BRIEN's *At Swim-Two-Birds* (**1939**). [DRL]

FIRBANK, (ARTHUR ANNESLEY) RONALD (1886-1926) UK writer whose stories are intensely and elaborately designed as artifices of DECADENCE. He wrote nine novels, all short, all at a high pitch, and mostly detailing LIFESTYLE-FANTASY careers. *The Artificial Princess* (written *c*1910; **1934**) presents a TWICE-TOLD version of the story of Salome, along with an episode in which the DEVIL tempts a secondary character into sin. *Prancing Nigger* (**1924** US; vt *Sorrow in Sunlight* 1925 UK) is of modest fantasy interest for its setting, the Caribbean ISLAND of Tacarigua ($\lozenge$ IMAGINARY LANDS). *Concerning the Eccentricities of Cardinal Pirelli* (**1926**) follows the eponymous cleric, reported to the Vatican for baptizing a dog, in a search for transcendence that ends (successfully) in death. [JC]

FIRCHOW, STEVE (1966-) New Guinea-born US artist and illustrator with a rich imagination and a rich, painterly style. He works in acrylics and inks, often augmented with coloured pencils. His work has appeared on book and comic-book covers and game packaging. [RT]

FIRE AND ICE US ANIMATED MOVIE (**1982**). 20th Century-Fox/Producers Sales Organization. **Pr** Ralph BAKSHI, Frank FRAZETTA. **Dir** Bakshi. **Screenplay** Gerry Conway, Roy Thomas. **Voice actors** Leo Gordon (King Jarol), Sean Hannon (Nekron), Cynthia Leake (Teegra), Randy Norton (Larn), William Ostrander (Taro), Steve Sandor (Darkwolf). 81 mins. Colour.

A fairly standard SWORD-AND-SORCERY tale in which wicked WIZARD Nekron's beautiful (i.e., good) daughter Teegra is kidnapped by dusky (i.e., bad) subhumans and all looks lost until the blond (i.e., good) HERO Darkwolf saves the day. The movie's COLOUR-CODING has been much objected to, though really that comes with the S&S territory. In general this is a much less interesting and imaginative excursion than Bakshi's WIZARDS (**1977**) or even his murky LORD OF THE RINGS (**1978**). [JG]

FIRE MONSTER, THE vt of *Gigantis* (**1955**). $\lozenge$ GODZILLA MOVIES.

FIRESIDE GHOST STORIES $\lozenge$ MASTER THRILLER SERIES.

FIRESTARTER US movie (**1984**). Universal/Dino de Laurentiis. **Pr** Frank Capra Jr. **Dir** Mark L. Lester. **Spfx** Jeff Jarvis, Mike Wood. **Screenplay** Stanley Mann. **Based on** *Firestarter* (**1980**) by Stephen KING. **Starring** Drew Barrymore (Charlie McGee), Art Carney (Irv Manders), Antonio Fargas (Taxi Driver), Louise Fletcher (Norma Manders), Moses Gunn (Dr Herman Pynchot), Freddie Jones (Dr Joseph Wanless), David Keith (Andy McGee), Heather Locklear (Vicky McGee), George C. Scott (John Rainbird), Martin Sheen (Captain Hollister). 114 mins. Colour.

Vicky and Andy met when paid subjects of an experiment on a supposed hallucinogen done by The Shop, a covert US Government agency; as a result they gained moderate psi TALENTS of telepathy. Their daughter Charlie inherits these talents in extreme form, and is capable of pyrokinesis – indeed, she has difficulty controlling it. Agents of The Shop try to seize her, killing Vicky; Andy and Charlie go on the run but are eventually captured by The Shop's psychopathic assassin Rainbird. Charlie's last act before escaping is to ignite The Shop's HQ and large numbers of its employees.

This all sounds very familiar if one knows *The Fury* (**1976**) by John Farris or the movie based on it, *The* FURY (**1978**); it has been reported that King wrote *his* version because dissatisfied with that movie's treatment of the theme. It is ironic, then, that *F* is certainly the lesser of the two movies – and lesser, too, than CARRIE (**1976**). Although the climactic scene of the small girl advancing slowly across The Shop's estate spreading devastation on all sides is curiously impressive, and although Scott is good as the genial psychopath, *F* lacks original ideas. [JG]

FIRTH, VIOLET M. $\lozenge$ Dion FORTUNE.

FISCHER, HARRY (OTTO) (1910-1986) US writer who originated the **Fafhrd and Gray Mouser** characters. $\lozenge$ DUOS; Fritz LEIBER.

FISCHER, MARJORIE (1903-1961) UK anthologist. $\lozenge$ ANTHOLOGIES.

FISHER, PAUL R. (1960-) US writer whose **Ash Staff** sequence – *The Ash Staff* (**1979**), *The Hawks of Fellheath* (**1980**) and *The Princess and the Thorn* (**1980**) – provides its YA audience with an initially unremarkable tale, whose young protagonists come to maturity ($\lozenge$ RITE OF PASSAGE) in a FANTASYLAND through testing conflicts with a DARK LORD. The later volumes, by concentrating on character development, are surprisingly subtle within their limiting frame. *Mont Cant Gold* (**1981**) is a singleton. [JC]

FISHER KING The Guardian of the GRAIL, so-called because, having been wounded by a lance, he is unable to hunt and must fish. He is first referred to in *Perceval, ou Le Conte de Graal* by CHRÉTIEN DE TROYES, although he clearly symbolizes a greater protector. His wounds and title suggest a CHRIST figure, while his role as protector of the LAND is suggestive of the Celtic legend of Bran the Blessed. He is sometimes also called the Wounded King or Maimed King, and he bears the ills of the land; his many NAMES indicate the many sources from which his legends are taken. In Sir Thomas MALORY's *Le Morte Darthur* (**1485**) the FK is called Pelles, and is the father of Elaine, who bore GALAHAD to LANCELOT. When PERCEVAL encounters him at the Grail Castle and witnesses the Procession of the Grail he fails to ask the meaning of the Bleeding Lance or of the Grail, and his failure to do so means that the land will be laid to waste ($\lozenge$ WASTE LAND); it is this inaction that prompts the QUEST for the Grail.

In some legends the FK was Joseph of Arimathea's brother-in-law, Bron (suggestive of Bran), who accompanied Joseph to Britain with the Grail. The "fish" association is thus linked to the earlier emblem of Christianity, and the FK becomes the symbol for the sanctity of Britain. It is in this shadowy role as guardian and protector of the spirit of Britain with the ability of HEALING that the FK is most represented in fiction – see particularly *That Hideous Strength* (**1945**) by C.S. LEWIS, and *The Drawing of the Dark* (**1979**) and *Last Call* (**1992**) by Tim POWERS. In *The Paper Grail* (**1991**) by James P. BLAYLOCK the FK forms part of a complex quest, while Anthony Powell (1905-) uses the concept symbolically in *The Fisher King* (**1986**). The most complete fantasy novel about the FK is *The Grail of Hearts* (**1992**) by Susan SHWARTZ. The movie *The* FISHER KING (**1991**) uses some of the legend in a contemporary setting. [MA]

FISHER KING, THE US movie (**1991**). Columbia TriStar. **Pr** Debra Hill, Lynda Obst. **Dir** Terry GILLIAM. **Spfx** Dennis Dion. **Screenplay** Richard LaGravenese. **Novelization** *The Fisher King* * (**1991**) by Lenore Fleischer. **Starring** Jeff Bridges (Jack Lucas), Amanda Plummer (Lydia), Mercedes Ruehl (Anne), Robin Williams (Parry). 137 mins. Colour.

Manhattan radio DJ Lucas stupidly spurs an insane phone-in caller to mass murder, and his life goes on the skids. Virtually down and out, he meets tramp Parry (for Parsifal), who is convinced Lucas has been sent to him as "The One" who will help retrieve the GRAIL (disguised as a presentation goblet) from an East Side plutocrat's castle-like home. But the questers have enemies, personified in the flame-belching figure of the Red Knight (much like the mysterious horned rider in Gilliam's earlier TIME BANDITS [*1981*]), who pursues Parry (or is pursued by him) through the Manhattan streets, invisible to others; the Red Knight is also equated with the slaughter Lucas inspired three years earlier, in which Parry saw his wife's brains blown out. All seems lost for the two men when Parry, chased again by the Red Knight, is beaten into a coma by thugs; but Lucas, driven by shame over his own selfishness, steals the Grail, thereby curing Parry and reuniting both men with their loved ones.

It is possible to read *TFK* at face value as not a fantasy at all, or at most as a fantasy of PERCEPTION, its fantastications being not real but Parry's crazed DELUSIONS. However, a different reading shows *TFK* as INSTAURATION FANTASY of a very high order: when we see the Red Knight we are experiencing a perceptual shift that allows us access to a perfectly valid alternative REALITY; and the kitsch goblet does indeed restore the Fisher King to life. [JG]

FISK, PAULINE (1948-   ) UK writer whose fantasy is restricted to two YA novels. In *Midnight Blue* (**1990**) a distressed young woman steals a balloon which takes her to "the land beyond the sky", where, in a hallucinated MIRROR world, she meets her near-DOUBLE, negotiates with strange GODS and goddesses about her evil grandmother, and enacts a RITE OF PASSAGE which takes her back to the ambivalences of her own home, where she can continue the HEALING. *Tyger Pool* (**1994**) similarly pitches a young girl against an older family member, a conflict that escalates into a battle which shakes the fabric of the worlds. PF is a writer of considerable power: in her hands fantasy is genuinely threatening and offers a sense of genuine REDEMPTION. [JC]

FISKE, TARLETON [s] ◊ Robert BLOCH.

FITZGERALD, F(RANCIS) SCOTT (KEY) (1896-1940) US writer, famed for *The Great Gatsby* (**1926**). Two of the stories in *Tales of the Jazz Age* (coll **1922**) are sf, and *The Fantasy and Mystery Stories of F. Scott Fitzgerald* (coll **1991**) ed Peter HAINING includes some fantasy of interest. [JC]

FITZGERALD, JOHN ANSTER (1832-1906) UK artist ◊ ILLUSTRATION.

FITZPATRICK, JIM (1948-   ) Self-taught Irish artist and illustrator, mainly of Celtic mythology. He works in a strong outline with some linear modelling, coloured with transparent and opaque watercolour; his pages are richly embellished in the manner of medieval Celtic manuscript decoration.

FP worked in advertising before producing and publishing a series of drawings entitled *Celtia* (graph coll **1974**); one of these, "Morrigan", was purchased by the Victoria & Albert Museum. This success led to the publication of these drawings in book form along with further pieces under the title *Jim Fitzpatrick: Celtia* (graph coll **1975**), which in turn led to the creation of his three-volume illustrated retelling of the folktales, myths and battle sagas of the Tuatha Dé Danann: *The Book of Conquests* (graph **1978** UK), *The Silver Arm* (graph **1981** UK) and *Son of the Sun* (graph **1987**). [RT] **Other work:** *Erinsaga: The Mythological Paintings of Jim Fitzpatrick* (graph **1985**).

FIVE MILLION YEARS TO EARTH US vt of QUATER-MASS AND THE PIT (**1967**).

FLASH, THE US tv series (1990-91). Pet Fly Productions, Warner Bros./CBS. **Pr** Don Kurt, Michael Lacoe, Steven Long Mitchell, Craig W. Van Sickle. **Exec pr** Danny Bilson, Paul DeMeo. **Dir** Mario Azzopardi and many others. **Spfx** Robert D. Bailey, Philip Barberio, Bill Schirmer. **Writers** Bilson, Howard CHAYKIN, Michael REAVES and many others. **Starring** Richard Belzer (Joe Kline), David Cassidy (The Mirror Master), Vito D'Ambrosio (Officer Bellows), Alex Desert (Julio Mendez), Mike Genovese (Lt Warren Garfield), Mark Hamill (The Trickster), Biff Manard (Officer Murphy), Dick Miller (Fosgate), Amanda Pays (Dr Tina McGee), John Wesley Shipp (Barry Allen/The Flash). 120min pilot and 20 60min episodes. Colour.

Another COMIC-book hero came to life with the arrival of Flash, known also as "The Scarlet Speedster" in the comics. This version, reasonably close to the original, finds police scientist Allen doused with a strange brew of chemicals that somehow gives him super-speed. Literally faster than a speeding bullet, he becomes The Flash, protector of Central City. Aided by McGee, a brilliant scientist, he gets involved with a mutating drug that creates superhumans (but has horrific side-effects), a brain-controlling implant, suspended animation, an invisible thief, an evil clone and, in the most popular episodes, a deranged criminal known as The Trickster. Another popular episode placed The Flash several years in the future, where his absence has caused civilization to break down (shades of IT'S A WONDERFUL LIFE [*1946*]). Despite some great spfx and costumes, the series lasted only one season. [BC]

FLASH GORDON Hero of the still extant SCIENCE-FAN-TASY comic strip created by artist Alex RAYMOND and writer Don Moore in 1934 for King Features Syndicate for weekly publication in Sunday newspapers. In 1944, another illustrator with a strong line style, Austin Briggs (1909-1973), took over the art, followed rather less ably in 1948 by Emanuel "Mac" Raboy (1916-1967). A daily FG strip began in 1940, also written by Moore and illustrated by Briggs, and then in 1951 a rather different daily was created with a more modern sf flavour, drawn by Dan Barry (1923-   ); in 1967, Barry also took on the Sunday page. Since then Barry has often farmed out the work to other creators.

The first FG story told of a "strange new planet" which was rushing towards Earth, whose inhabitants prepare for doom; Dr Hans Zarkov is working night and day to perfect "a device with which he hopes to save the world". Coincidentally, an airliner is struck by a meteor while passing directly over Zarkov's house, and among the passengers parachuting to safety are FG (world-renowned polo player and a Yale graduate) and the beautiful Dale Arden. "A dishevelled, wild-eyed figure [Zarkov]" forces them at gunpoint into a rocketship which takes off "with a deafening roar" to land on the wayward planet, Mongo. Here the three encounter weird MONSTERS, strange humanoid races and the dictator Ming the Merciless. [RT]

See also: FLASH GORDON MOVIES.

FLASH GORDON MOVIES A number of movies have been based on the COMIC-strip characters and scenarios created by Alex RAYMOND.

1. *Flash Gordon* US serial movie (*1936*). Universal. **Pr** Henry MacRae. **Dir** Frederick Stephani. **Spfx** Norman Drewes. **Screenplay** Basil Dickey, Ella O'Neill, George Plympton, Stephani. **Starring** Richard Alexander (Prince Barin), Buster Crabbe (Flash Gordon), Priscilla Lawson (Princess Aura), John Lipson (King Vultan), Charles Middleton (Ming the Merciless), James Pierce (King Thun), Jean Rogers (Dale Arden), Frank Shannon (Dr Alexis Zarkov), Duke York Jr (Kala). 13 *c*20min chapters. B/w.

Earth lies in the path of the wandering planet Mongo. Flash, Dale and Zarkov fly by spaceship to Mongo, hoping to divert it from its course. There, however, they are seized and imprisoned under the tyrannical rule of Ming the Merciless, who hopes to become emperor of the Universe. But their lives are spared because Ming lusts for Dale while his daughter Aura lusts for Flash. Flash deposes the tyrant, and the Earthlings return home in triumph.

The creaking spfx are familiar from this era of serial movies, but *FG* had a comparatively large budget; even so, the music (by Franz Waxman) was drawn largely from *The Bride of Frankenstein* (*1935*; ◊ FRANKENSTEIN MOVIES), as were many of the sets. Middleton is an exceptional Ming and Rogers a very lovely Dale, although her acting abilities are not overtaxed. Crabbe, hair bleached to match the COMIC-strip prototype, became the definitive image of Flash.

There were two serial sequels, *Flash Gordon's Trip to Mars* (*1938*), edited as the feature *The Deadly Ray from Mars* (*1938*; vt *Flash Gordon: Mars Attacks the World*), and *Flash Gordon Conquers the Universe* (*1940*), edited as the feature *Purple Death From Outer Space* (*1940*), as well as a spectacular remake, **2**; a soft-porn spoof was the rather tedious *Flesh Gordon* (*1974*), inevitably sequelled. *FG* itself was re-released in various edited forms as *Rocket Ship*, *Spaceship to the Unknown* (*1936*), *Perils from the Planet Mongo* (*1936*), *Space Soldiers* and *Atomic Rocketship*. [JG]

2. *Flash Gordon: The Greatest Adventure* US ANIMATED MOVIE (*1979* tvm). Filmation. **Voice actors** Bob Holt (Ming), David Opatoshu (Zarkov), Diane Pershing (Dale Arden), Robert Ridgely (Flash). *c*100 mins. Colour.

Very little is known about this movie; we have been unable to obtain a reference copy. This was the pilot of the series described under **4**. [JG]

3. *Flash Gordon* US movie (*1980*). EMI/Starling/Famous Films. **Pr** Dino De Laurentiis. **Exec pr** Bernard Williams. **Dir** Mike Hodges. **Spfx sv** George Gibbs. **Screenplay** Lorenzo Semple Jr, adapted Michael Allin. **Based largely on 1**. **Starring** Melody Anderson (Dale), Brian Blessed (Vultan), Timothy Dalton (Barin), Sam J. Jones (Flash), Mariangela Melato (Kala), Ornella Muti (Aura), John Osborne (High Priest), Topol (Hans Zarkov), Max Von Sydow (Ming), Peter Wyngarde (Klytus). 115 mins. Colour.

A SCIENCE FANTASY that resites many motifs from heroic FOLKTALE. Read this way, the story is that through sheer decency, HERO Flash wins the hand of the beautiful princess

(Dale), cures the EVIL princess Aura of her wickedness, unites all Mongo's feuding lordlings to defeat the evil DARK LORD and technofantastic WIZARD Ming, and saves two worlds. Rather heavy-handed in its attempts at PARODY and using stark garishness to compensate for appalling spfx, *FG* is a gaudy cliché whose charm should not be under-estimated. [JG]

4. A series of six *FG* ANIMATED MOVIES was released, in colour, during 1989-90 by Filmation/King Features, direct to video; these were based on a tv series broadcast 1979-80. Uniformly badly animated, plotted and voiced, they are: *Flash Gordon* US movie (?*1989*; 56 mins), *Flash Gordon in A Planet in Peril* (*1989*; 60 mins), *Flash Gordon in To Save Earth* (*1989*; 60 mins), *Flash Gordon in The Frozen World* (*1989*; 60 mins), *Flash Gordon in Blue Magic* (*1990*; 60 mins) and *Flash Gordon in Castaways in Tropica* (*1990*; 60 mins). The credits offered are incomplete: **Pr** Don Christensen. **Exec pr** Norm Prescott, Lou Scheimer. **Anim dir** Gwen Wetzler. **Screenplay** Ted Pedersen, Sam Peeples. **Voice actors** Melendy Britt (Princess Aura), Allen Melvin, Allen Oppenheimer, Diane Pershing (Dale Arden), Bob Ridgely (Flash Gordon). [JG]

FLAUBERT, GUSTAVE (1821-1880) French writer. Obsessive perfectionism severely restricted his output in later years. Several of the works he produced in more slap-dash fashion during his teens are extravagant fantasies, including "Rêve d'enfer" (written 1837; first reprinted in the 1922 edition of the *Oeuvres complètes*), in which an alchemist (◊ ALCHEMY) encounters SATAN. Satan appears briefly also in the rhapsodic "La danse des morts" (written 1838; reprinted in the 1885 *Oeuvres complètes*; trans as "The Dance of Death" in the *Complete Works* of 1904) and gives a very elaborate account of himself in the drama "Smarh" (written 1839; reprinted in the 1885 *Oeuvres complètes*). *Salammbô* (**1863**; trans M. French Sheldon **1886** UK) is a brilliant romance of antiquity; its eponymous heroine is a self-sacrificing priestess of Tanit who saves Carthage from disaster when the GODDESS's sacred veil is stolen by Mathô in the hope of ensuring that the city will fall to his siege. More overtly fantastic is *La tentation de Saint Antoine* (**1874**; trans D.F. Hannigan as *The Temptation of St Anthony* **1895** UK), a landmark CHRISTIAN FANTASY which redeployed the material of GF's earlier exercises in Literary Satanism to great effect, although some critics prefer an even more extravagant and exotic version written 1849-56 and published as *La première tentation de Saint Antoine* (**1908**; trans René Francis as *The First Temptation of St Anthony* **1910** UK). *Trois contes: Un coeur simple, La légende de Saint Julien l'Hospitalier, Hérodias*" (coll **1877**; trans George Bernard Ives as *Gustave Flaubert* **1903** in the **Little French Masterpieces** series ed A. Jessup; new trans Arthur McDowell vt *Three Tales* 1923 UK) includes a much gentler fantasy of the same species (the second item) as well as the melodramatic concluding tale. [BS]

FLECKER, (HERMAN) JAMES ELROY (1884-1915) UK poet, dramatist and writer of fantasy interest primarily for *The King of Alsander* (**1914**), whose mild-mannered English protagonist is directed by a LIMINAL BEING to travel to Alsander, a land described in terms which combine RURITANIA and ARABIAN FANTASY. There he soon becomes king, after a pixillated revolution. Some of the stories assembled in *Collected Prose* (coll **1920**) are fantasy, and *Hassan* (**1922**) – a play – again evokes the world of Arabian fantasy, including GHOSTS. [JC]

FLEISCHER, MAX (1883-1972) Viennese-born US animator, the inventor of rotoscoping (whereby animated movies could seem more realistic because drawings could be based on the movements of live actors) and in his heyday regarded as Walt DISNEY's primary rival; he collaborated closely with various other family members, notably his brother Dave Fleischer (1894-1979) – who was the original live-action model for Koko the Clown, the character featured in MF's first successful series, the **Out of the Inkwell** shorts (*1915-27*), which mixed animation with live action, MF himself being part of the live action. Either MF or Dave also invented the "bouncing ball" to accompany the lyrics of movies to which the audience is expected to sing along. MF created, with Lee De Forest (1873-1961), what was probably the first animated movie to have a soundtrack – *My Old Kentucky Home* (*1924*) – predating Disney's *Steamboat Willie* (*1928*) by several years. Not until 1929, though, did MF enter the field of the talkies seriously, with his **Talkartoon** series. His greatest moment should have come in 1939 when, prompted by the success of Disney's SNOW WHITE AND THE SEVEN DWARFS (*1937*), he released his riposte, *Gulliver's Travels* (*1939*; ◊ GULLIVER MOVIES). Making considerable use of the rotoscope, this possesses much fine material yet fails, ultimately, to cohere. It did, however, earn some money; the same was not true of MF's second and last animated feature, *Mr Bug Goes to Town* (*1941*; vt *Hoppity Goes to Town*), an ecologically oriented tale of a community of insects threatened by humans. Its failure signalled the end of MF's and Dave's running of the studio: Paramount, the Fleischer distributors, were no longer willing to pour money into it under its current management. The brothers – by now on very bad terms – worked for other studios thereafter, but without much distinction.

Long before this, however, the Fleischer Studio had launched two of animation's most significant characters. **Betty Boop** initially appeared in *Dizzy Dishes* (*1930*), portrayed as half-girl, half-dog; she soon became all-girl as MF realized the potential of the character. **Betty Boop** shorts were produced prolifically until 1939, and she became and still is an ICON; Betty has a bit part in WHO FRAMED ROGER RABBIT (*1988*). Even more prolific was **Popeye**, devised in 1919 for the COMICS by Elzie Segar and bought for the screen by MF in 1932: the **Popeye** series continued until 1957, and a live-action feature movie appeared much later as POPEYE (*1980*). Also of note was the **Superman** series of shorts (◊ SUPERMAN MOVIES). [JG]

FLEMING, IAN (1908-1964) UK writer. ◊ CHILDREN'S FANTASY; CHITTY CHITTY BANG BANG (*1968*).

FLESH FOR FRANKENSTEIN (*1973*; ot *Carne per Frankenstein*) ◊ FRANKENSTEIN MOVIES.

FLETCHER, GEORGE U. Pseudonym of Fletcher PRATT.

FLINT, [Sir] W(ILLIAM) RUSSELL (1880-1969) UK artist, one of the very few book and magazine illustrators to blossom into a painter of the first rank. After producing 32 illustrations for H. Rider HAGGARD's *King Solomon's Mines* (**1905**), WRF was commissioned to illustrate a long series of ancient and historical classics from *The Song of Songs* (1909) to *The Odyssey* (1924). His most ambitious project was the ten-guinea set of Sir Thomas MALORY's *Le Morte Darthur* (1910-11 4 vols) with 48 mounted colour plates. Medievalism was then much in vogue, and Flint's painting of "The Passing of King Arthur" won him a silver medal at the Paris Salon. His editions of Charles KINGSLEY's *The Heroes* (1912) and Geoffrey CHAUCER's *The Canterbury Tales*

(1913 3 vols) were equally poplular and beautifully produced. During the last half of his life, WRF became world-famous for his paintings of landscapes, gipsies and nudes. [RD]

Further reading: *In Pursuit, An Autobiography* (**1970**).

FLINT, KENNETH C(OVEY) (? -) US writer. All of his work, including two novels as by Casey Flynn, have dealt in one way or another with Ireland, except for one late tale, *Otherworld* (**1992**), a contemporary DARK-FANTASY thriller. KCF began publishing CELTIC FANTASY with *A Storm upon Ulster* (**1981**; vt *The Hound of Culain* 1986 UK), which tells the story of CUCHULAIN; Cuchulain defends Ulster from Meave (i.e., Meadhbh), queen of Connacht. A later novel, *Isle of Destiny: A Novel of Ancient Ireland* (**1988**), prequels Cuchulain's life before the main conflict.

KCF's first series, the **Sidhe** sequence – *Riders of the Sidhe* (**1984**), *Champions of the Sidhe* (**1984**) and *Master of the Sidhe* (**1985**) – is based on a central narrative cluster of Irish myth, the story of the divine Tuatha Dé Danann (◊ GODS), who become rulers of Ireland with the aid of the Irish hero Lugh Lamfada (i.e., Lamhfhada), originally a TRICKSTER god associated with SEASON rites. In KCF's hands this backdrop of divinity serves (as in most of KCF's work) mainly as an UNDERLIER bolstering of characters whose actions he tends to explain in terms fairly close to RATIONALIZED FANTASY.

The **Finn MacCumhal** sequence – *Challenge of the Clans* (**1986**), *Storm Shield* (**1986**) and *The Dark Druid* (**1987**) – is based on a third Irish complex of tales, the Fianna Cycle, whose central figure, Finn (or Fionn) Mac Cumhaill (◊ FINN MAC COOL), here takes on a fairly straightforward role as a PARIAH-ELITE hero, distrusted for his FAIRY blood, who makes his appearance out of the WATER MARGINS, gathers COMPANIONS about him, and defends Ireland from invaders; the sequence ends rather inconclusively with Finn – after a SHAPESHIFTING dispute with an evil Druid – safely in the arms of his beloved.

Later novels – like *Cromm* (**1990**), *Legends Reborn* (**1992**) and *The Darkening Flood* (**1995**) – tend to confront the contemporary world with figures out of Celtic fantasy. The **Gods of Ireland** series as by Casey Flynn is *Most Ancient Song* (**1991**) and *The Enchanted Isles* (**1991**).

KCF is a bland writer whose subject matter is savage. Only rarely does that underlying content make its mark. [JC]

FLOOD Among the myths of the Old Testament only ADAM AND EVE's exploits in EDEN have inspired more echoes in modern fantasy than the Flood. Noah and his family are featured in such works as *Seola* (1878) by Mrs J. Gregory Smith (1818-1905), *Blow, Blow Your Trumpets* (1945) by Shamus FRAZER, *Two by Two* (1963) by David GARNETT, *The Moon in the Cloud* (1968) by Rosemary HARRIS, *Not Wanted on the Voyage* (1985) by Timothy FINDLEY, *Boating for Beginners* (1986) by Jeanette WINTERSON and "Bible Stories for Adults, No. 17: The Deluge" (1988) by James MORROW. A new deluge is featured in *All Aboard for Ararat* (**1940**) by H.G. WELLS, but merely threatens in *The Elephant and the Kangaroo* (**1947**) by T.H. WHITE and "Alfred's Ark" (1965) by Jack VANCE. The Ark is a powerful symbol in its own right, commonly evoked in sf and in such surreal fantasies as *The Ark Sakura* (**1984**) by Kobo ABÉ.

Catastrophic floods are naturally at the heart of many ATLANTIS stories, sometimes linked to the flood-references in Greek MYTHOLOGY and the tale of GILGAMESH. *The Lost Continent* (1900) by C.J. Cutcliffe Hyne is a notable example of this syncretizing tendency. Other mythical lands supposedly lost by drowning include the kingdoms of Ys and Lyonesse. The folktale of the inundation of Ys following the seduction of the king's daughter was repopularized in France by the successful OPERA *Le roi d'Ys* (**1888**) by Edouard Blau (1836-1906) and Edouard Lalo (1823-1892), from which Robert W. CHAMBERS borrowed the setting for "The Demoiselle d'Ys" (1895); the tale had previously been retold (in a Welsh setting) in *The Misfortunes of Elphin* (**1829**) by Thomas Love Peacock (1785-1866) and was ironically recast by Norman DOUGLAS in *They Went* (**1920**).

[BS]

FLY, THE Two series of TECHNOFANTASIES (**1-3** and **4-5**) drawing inspiration from "The Fly" (1957) by George Langelaan (1908-   ).

1. *The Fly* US movie (**1958**). 20th Century-Fox. **Pr** Kurt Neumann. **Dir** Neumann. **Spfx** L.B. Abbott. **Screenplay** James Clavell (1925-   ). **Starring** Al Hedison (André Delambre), Charles Herbert (Philippe Delambre), Patricia Owens (Hélène Delambre), Vincent Price (François Delambre). 94 mins. Colour.

Scientist André develops the "Disintegrator-Integrator", a matter transmitter, but in experimenting on himself overlooks a fly in the laboratory; his and the fly's atoms mix, and he becomes a MONSTER within which human and fly instincts war. He convinces wife Hélène that knowledge of matter transmission is too dangerous for humankind, and destroys all evidence of the experiment, persuading her to kill his monstrous body in a hydraulic press. Police and brother François are sceptical until François discovers the man-headed fly in the garden.

TF leaves much to our imagination until its final scenes: oddly, while the fly-headed human holds few terrors, the trapped fly with its horribly human head strikes a deep chord. An early moment smacks of purest fantasy: André has failed to transmit the family cat, which vanishes who knows where; but for hours there remains a spectral, echoing, fading squawl. [JG]

2. *Return of the Fly* US movie (**1959**). 20th Century-Fox. **Pr** Bernard Glasser. **Dir** Edward L. Bernds. **Screenplay** Bernds. **Starring** David Frankham, Brett Halsey, Vincent Price (François Delambre), Dan Seymour, John Sutton (movie lacks proper credits). 80 mins. B/w.

André's son (Halsey) unearth's André's matter-transmission equipment and determines to reactivate it. By astonishing coincidence, his assistant (Frankham), a crook, thinks to put a fly in the pod with Halsey, and the inevitable happens. This time Uncle François, a sort of one-man Greek chorus for much of the movie, is able to sort things out, reconstituting Halsey and destroying the MONSTER fly. A movie that seems much longer than its 80 mins. [JG]

3. *Curse of the Fly* US/UK movie (**1965**). **Pr** Robert L. Lippert, Jack Parsons. **Dir** Don Sharp. **Spfx** Harold Fletcher. **Screenplay** Harry Spalding. **Starring** George Baker, Brian Donlevy, Carole Gray. 86 mins. B/w.

A bleakly, depressingly poor movie that takes the TECHNOFANTASY of its predecessors and turns it into schlock-HORROR/sf. [JG]

4. *The Fly* US movie (**1986**). 20th Century-Fox/Brooksfilms. **Pr** Stuart Cornfeld. **Dir** David Cronenberg. **Spfx** Louis Craig, Ted Ross. **Screenplay** Cronenberg, Charles Edward Pogue. **Starring** Joy Boushel (Tawny), Geena Davis (Veronica Quaife), John Getz (Stathis Borans), Jeff Goldblum (Seth Brundle). 100 mins. Colour.

This is not a remake of **1**. Idiosyncratic scientist Brundle has been experimenting with matter transmission, but his attempts to transmit flesh are messy failures. Through his sexual relationship with journalist Quaife, especially through a chance remark of hers, he realizes his error: he has never taught the transmitter's computer about the *poetry* of flesh. That done (it is not clear how), he proceeds, in due course inadvertently combining himself with a fly. Almost immediately he gains prodigious physical, notably sexual, vigour; slower to come are physiological changes. His obsessive self-preoccupation intensifies as the TRANSFORMATION accelerates: he becomes able to walk on the ceiling, and must predigest his food by vomiting on it. Quaife, who discovers herself pregnant, he now regards as a betrayer; he seizes her from an abortion clinic and plans one last mad experiment. But, when Quaife's ex-boyfriend appears on the scene, Brundle – or now, more accurately, The Fly – begs piteously to be killed.

TF's subtext, that MONSTERS lurk within us all (Brundle's mental changes are more repugnant than his physical ones), is a cliché, but the movie's sparse direction and overall moodiness make it seem fresh; the electricity between Davis and Goldblum contributes much. The movie, with its emphasis on the "poetry of flesh", bears all the trappings of philosophical profundity without in fact being profound, yet it bears them with some integrity. [JG]

5. *The Fly II* US movie (**1989**). 20th Century-Fox/Brooksfilms. **Pr** Steven-Charles Jaffe. **Exec pr** Stuart Cornfeld. **Dir** Chris Walas. **Spfx** Chris Walas Inc. **Screenplay** Frank Darabont, Mick Garris, Jim & Ken Wheat. **Starring** Gary Chalk (Scorby), Harley Cross (young Martin), John Getz (Stathis Borans), Jeff Goldblum (Seth Brundle), Ann Marie Lee (Dr Jainway), Lee Richardson (Mr Bartok), Eric Stoltz (adult Martin Brundle), Frank C. Turner (Dr Shepard), Daphne Zuniga (Beth Logan). 104 mins. Colour.

This conventional HORROR MOVIE tells of Seth Brundle's (◊ **4**) UGLY-DUCKLING son Martin, brought up as an experimental subject in Bartok Industries. He displays astonishing physical and intellectual growth, being "adult" by age 5. Trusted surrogate father Bartok asks him to take over Seth's matter-transmission researches. Martin makes great strides, helped by colleague and lover Logan, until he discovers that Bartok has had their lovemaking videotaped. His wrath triggers superhuman abilities and accelerates his transformation from human to insectile form. [JG]

FLYING DUTCHMAN The FD is a particular subset of the ACCURSED WANDERER. As the LEGEND is now commonly rendered, he is the captain of a ship doomed eternally to sail the seas in search of landfall, due to an ill-advised ultimatum he once delivered to the GODS – that he would round a cape (or gain a harbour) in spite of wind and weather, "though I should beat about here until the Day of Judgement". On the rare occasions he comes within hailing distance of another ship, the captain tends to try to send letters home to his long-dead family. The legend – which puts into a maritime venue the story of the WANDERING JEW – ends (sometimes) in spiritual redemption, and release from the BONDAGE of IMMORTALITY.

There seems no clear literary provenance for the specific FD story before the first years of the 19th century, although early references to the legend seem to assume it is familiar. Sir Walter SCOTT, in the notes to his poem *Rokeby* (**1813**), indicates that the original ship – which seems to have no name of its own, being usually called, like its captain, *The Flying Dutchman* – had been carrying a cargo of gold (which caused a murder) and a plague, and a subsequent ban on its ever reaching port.

In an introduction to Charles MATURIN's *Melmoth the Wanderer* (**1820**), Alethea Hayter suggests that the FD legend was one of the author's sources. The anonymous "Vanderdecken's Message Home, or The Tenacity of Natural Affection" (1821 *Blackwood's Magazine*) emphasizes the BELATEDNESS of the tale – the sense that Vanderdecken cannot register the passing of TIME – and indicates that the captain uttered his fateful ultimatum sometime in the middle of the 17th century in the region of the Cape of Good Hope. Edward Fitzball's *The Flying Dutchman, or The Phantom Ship: A Nautical Drama* (produced 1826) repeats this material.

The first literary sign of the possibility of redemption seems to come in "The Memoirs of Herr Von Schnabelewopski" (1834 *Salon*) by Heinrich Heine (1797-1856), whose fictional narrator tells of a possibly genuine theatrical performance in Amsterdam, during which the DEVIL takes Vanderdecken's oath as a pact, and tells him he'll have to sail the seas until the LAST JUDGEMENT, unless redeemed by a woman's love. Frederick MARRYAT, in *The Phantom Ship* (**1839**), also offers redemption to the Dutchman, in the form of a fragment of the True Cross. Richard WAGNER – who acknowledges Heine – soon transformed this material into the definitive rendering of the legend in his OPERA *Der Fliegende Hollander* (**1843**).

Later uses of the legend tend to reflect Wagner. W. Clark RUSSELL's *The Death Ship: A Strange Story* (**1888**; vt *The Flying Dutchman* 1888 US) involves a tedious shipboard romance. In Edgar Turner's *The Submarine Girl* (**1909**) the eponymous lass in her super-sub awakens the FD, marries him, and they live together in South Africa. He settles down in E.A. WYKE-SMITH's *The Marvellous Land of Snergs* (**1927**). He features in *Master Davy's Locker: A Story of Adventure in the Undersea* (**1935**) by Ernest Wells (1902-   ). In the novel-length "The Shadow" (in *Crimes, Creeps and Thrills* anth **1936** ed anon John GAWSWORTH) by E.H. VISIAK his appearance lacks much serious point. The narrator of *The Drunken Sailor* (**1947** chap), a poem by Joyce Cary (1888-1957), is a version of the Dutchman. Richard MATHESON's "Death Ship" (1953) re-enacts the story on another planet. The **Master Mariner** sequence – *Running Proud* (**1978**) and *Darken Ship: The Unfinished Novel* (**1980**) – by Nicholas Monsarrat (1910-1979) is based on the legend, and was meant to carry its protagonist from the moment he is cursed by a WITCH in the 16th century up to a redeemed death in 1978, but the author died too soon to bring the narrative into focus. The FD makes a cameo appearance in Diana Wynne JONES's *The Homeward Bounders* (**1981**), and Tom HOLT's *Flying Dutch* (**1991**) plays a humorous riff on the legend as a whole.

In general, any accursed wanderer in any fantasy who is associated with ships and who is caught in time and who cannot get home is almost certainly intended to echo the FD.
[JC]

FLYNN, CASEY ◊ Kenneth C. FLINT.

FLYNN, DANNY (ANTHONY) (1958-   ) UK painter and illustrator of fantasy, sf and horror subjects, with an incisive line and rich colour style. His work has appeared mainly on book covers. [RT]

Further reading: *Only Visiting This Planet: The Art of Danny Flynn* (graph **1994**), with foreword by Arthur C. Clarke.

FOG, THE US movie (*1979*). Avco-Embassy/EDI. **Pr** Debra Hill. **Exec pr** Charles B. Bloch. **Dir** John Carpenter. **Spfx** A&A Special Effects, Richard Albain Jr, Rob Bottin, Dean Cundey. **Mufx** Bottin. **Vfx** James F. Liles. **Screenplay** Carpenter, Hill. **Novelization** *The Fog* * (**1979**) by Dennis ETCHISON. **Starring** Tom Atkins (Nick Castle), Adrienne Barbeau (Stevie Wayne), Bottin (Blake), Jamie Lee Curtis (Elizabeth Solley), Charles Cyphers (Dan O'Bannon), Hal Holbrook (Father Malone), John Houseman (Mr Machen), Janet Leigh (Kathy Williams), Nancy Loomis (Sandy Fadel), Ty Mitchell (Andy Wayne). 91 mins. Colour.

The small Californian town of Antonio Bay was founded in 1880 using gold stolen from a shipful of lepers, lured onto the rocks by six conspirators. In 1980, as the citizens, ignorant of this, prepare for the centenary celebrations, the GHOSTS/ZOMBIES of the lepers seek vengeance in the form of six sacrificed lives and the return of their gold. For two nights the town is invaded by a mysterious Fog, from which death strikes in a series of chilling vignettes. Thanks to Carpenter's direction, notably his cross-cutting, the movie is substantially better than its plot and screenplay, which tend to oscillate between HORROR-MOVIE and *Airport*-style-disaster mode. Included are some RECURSIVE jokes: reference to nearby Bodega Bay (◊ *The* BIRDS [*1963*]), and a bit-part for a doctor called Phibes. There is one moment of real power: trapped in the lighthouse from which she operates her local radio station, focal character Stevie Wayne sees – as do we – the Fog spilling over a hillcrest before washing inexorably down towards her. It is a memorable image. [JG]

FOLIATE HEAD The carved image of a head, from whose mouth thrust two tusk-like stems and whose ears, nostrils and eyes also typically sprout foliage. This relatively straightforward version of the FH can be found in many English churches, most often in those from the 14th century. The significance of this form of the FH to FOLKLORE lies primarily in its association with the JACK in the Green, the version of the GREEN MAN whose wickerwork image is paraded in traditional Mayday processions, and who marks the beginning of Spring. Occasionally the FH is composed entirely of carved leaves and other foliage, a rendering which generates a TROMPE L'OEIL effect very much like that verbally generated in many fantasy texts to depict TRANSFORMATIONS. This version of the FH, which is known as the *Tête de Feuilles*, is clearly similar to the many heads portrayed, in various series, by Giuseppe ARCIMBOLDO, but comes closest to his portrait of the Emperor Rudolph II or Vertumnus, the Roman vegetation god, who heralds the SEASONS.

It seems very clear that the FH has a more than accidental resemblance to the FACE OF GLORY. [JC]

FOLKLORE The tried and trusted accumulated wisdom of generations. Its teachings may not always be *right*, but they are *believed*. They consist of anecdotes, LEGENDS, sayings, proverbs and much of what is often referred to as Old Wives' Tales. The way the lore was remembered was often in the form of simple sayings or rhymes, many relating to the weather, health or moral sense. Little of folklore is itself either fantasy or supernatural, although much may relate to SUPERSTITION, but as it formed the basis for FOLKTALES and LEGENDS, so these became more fantastic. Those most steeped in folklore were often regarded as "wise", a word that has the same source as the word WIZARD (the old English *wis*) and a similar association with WITCH, so that folklore was linked with a belief in MAGIC and ENCHANTMENT.

Folklore's links with superstition and witchcraft meant that in the UK knowledge of folklore was anathema during the Puritan domination of the 17th century. A few serious antiquarians did assemble collections of folk traditions, especially William Camden (1551-1623) and John Aubrey (1626-1697), but this did not become a serious scientific pursuit until the 19th century. The word "folklore" itself was coined in 1846 by W.J. Thomas (1803-1885), the founder of *Notes and Queries*. The Folk-Lore Society was founded in 1878, its leading lights including Joseph Jacobs (1854-1916) and Andrew LANG.

The serious study of folklore began to emerge in Germany with the work of Clemens Brentano and Achim von Arnim, particularly with their *Des Knaben Wunderhorn* ["The Boy's Wonderhorn"] (coll 1805; exp 1808), which collected over 700 German folksongs and rhymes; it was they who encouraged the GRIMM BROTHERS in their scholarly pursuit of folktales and tradition. The interest aroused by the publication of the Grimms' *Kinder- und Hausmärchen* (coll 1812) lit a flame that soon spread across Europe (◊ FOLKTALE).

Folklore is always being added to, as we learn from each generation. Much modern folklore is known as URBAN LEGEND, but we each learn through experience and may convey this learning to others by way of advice. Sometimes this may be misinformed, and such advice will enter into folklore; a current example is the folklore developing about AIDS. Frequently the power of what we want to believe usurps the scientific fact.   [MA]

Further reading: *The British Folklorists: A History* (1968) by Richard M. Dorson; *Larousse Dictionary of Folklore* (1995) by Alison Jones.

FOLKTALE A narrative derived from oral tradition which relates a well understood and recognizable story. Usually dependent upon a nation's FOLKLORE, the folktale will reflect a cultural identity and be part of that nation's heritage. Folktales may be about national heroes – like ROBIN HOOD, ARTHUR, William Tell or Siegfried, whose deeds have become part of LEGEND – but they can also be about everyday people and focus on moral rather than adventure. Many of these stories are TAPROOT TEXTS for FAIRYTALES. The story may once have had a basis in fact, but has been subject to generations of embellishment, resulting in a familiar and often many-layered text (◊ TWICE-TOLD). Folktales are almost always set in this world, usually in the past, and while the supernatural may dictate the events its presence is often accepted rather than sensationalized. Folktales are thus precursors of SUPERNATURAL FICTION rather than OTHERWORLD fantasy. Not all folktales involve the supernatural or fantastic but many rely upon a feeling of WRONGNESS drawn from SUPERSTITION.

Many early folktales spread with tribes across Europe and Asia and are known in different versions in divers lands; this spread of a tale into many cultures is a key element in defining a folktale. The tale upon which CINDERELLA is based has been identified in almost 700 versions spread over 1000 years. Folktales are always being created and passing into national currency.

There is a distinction to be drawn between the literary fairytale and the folktale. The fairytale is a literary artform that brings structure and style to the folktale, which is otherwise an unencumbered transcription of a simple tale. While folktales still have the essence of being twice-told, some lack the depth of STORY and appear anecdotal. A folktale is not as brief or pointed as a FABLE, and as a story may be even shallower – though some can be immensely imaginative.

Folktales, like LEGENDS, have been collected since earliest times. The Papyrus Westcar (*c*1600BC) from ancient EGYPT may be the oldest known such compilation. HOMER's *Iliad* and *Odyssey* are other early examples.

Most of the well known traditional fairytales began as folktales but were sculpted into literature. The best-known English folktales concern Jack and the Beanstalk, Jack the Giant Killer (◊ JACK), TOM THUMB, Dick Whittington, Lady Godiva and ROBIN HOOD. A selection is found in *English Fairy Tales* (coll 1890) and *More English Fairy Tales* (coll 1894), both assembled as *English Fairy Tales* (omni 1979) by Joseph Jacobs (1854-1916) and more recently in *Alan Garner's Book of British Fairy Tales* (coll 1984) (◊ Alan GARNER). English language and literature has been widely influenced by Celtic, Nordic, Greek and Arabian tales and LEGENDS (◊◊ ARABIAN FANTASY; CELTIC FANTASY; GREEK AND LATIN CLASSICS; MABINOGION; NORDIC FANTASY).

The original *Arabian Nights Entertainment* is essentially folktale. The development of these stories by Antoine Galland into the texts we know today shows the conversion of folktales into fairytales. Galland translated and freely adapted the stories, and seven volumes of *Les mille et une nuit* appeared 1704-8; five more volumes were added over the remainder of Galland's life, with the final volume appearing in 1717.

This same process affected the stories now generally known as Grimms' Fairytales. The GRIMM BROTHERS were assiduous in collecting local folktales and determined not to embellish them but to capture the oral tradition. Their first volume of *Kinder- und Hausmärchen* (coll 1812) comprises literal transcriptions of local tales. It was only later, as friction developed between the brothers about how pure the stories should remain, that Wilhelm began to embellish, and thereby develop some of the best-known FAIRYTALES.

The scholarly work of the Grimms encouraged others to research and compile national tales and legends. Allan Cunningham (1784-1842) produced *Traditional Tales of the English and Scottish Peasantry* (coll 1822); Thomas Crofton Croker (1798-1854) assembled *Fairy Legends and Traditions of the South of Ireland* (coll 1825); Thomas Keightley (1789-1872) produced *The Fairy Mythology* (1828; rev 1850) and *Tales and Popular Fiction* (coll 1834); and James Orchard Halliwell-Phillips (1820-1889) compiled the seminal *Nursery Rhymes of England* (coll 1842) and *Popular Rhymes and Nursery Tales* (coll 1849). Perhaps the best-known example of folktales being developed into a structured form to produce a national literature is the Finnish KALEVALA.

It is quite common for "fairytale" and "folktale" to be indistinguishable in book titles, even though each is a distinct form. *Fairy Tales of All Nations* (anth 1849; vt *The Doyle Fairy Book* 1890) ed Anthony R. Montalba and illustrated by Richard DOYLE is an early example. The muddle was perpetuated in the 12-volume series of coloured **Fairy Books** by Andrew LANG, starting with *The Blue Fairy Book* (anth 1889), which is notable for its diversity in collecting folktales from cultures worldwide. Lang's research (and the indefatigable translations of his acquaintances) brought together one of the most accessible and well read series of folktales yet published. The confusion of terms continues today. For this reason many critics prefer the term WONDER TALE to "fairytale".

Writers are more likely to draw upon folktales when seeking to establish some form of universal truth, whereas the fairytale is reserved for exploring the MARVELLOUS. The folktale form thus lends itself to religious and cultural stories and parables. An example is the way that Rudyard KIPLING used the folktale form in his *Just-So Stories* (coll **1902**). Folktales are particularly relevant to the wider heritage of diverse cultures seeking their roots, as when Joel Chandler HARRIS drew upon the Negro folktales of the Georgian plantations for his **Uncle Remus** stories. Other examples include: the stories in *The Celtic Twilight* (coll **1893**) and similar volumes by W.B. YEATS; the Native American stories in *Mystic Women: Their Ancient Tales and Legends* (coll **1991**), *The Mysterious Doom and Other Ghost Stories of the Pacific Northwest* (coll **1993**) and *Phantom Waters* (coll **1995**), all by Jessica Amanda SALMONSON, and the anthology *Tales from the Great Turtle* (anth **1994**) ed Piers ANTHONY and Richard Gilliam (1950-   ); and the African folktales retold by Geraldine Elliot (1899-   ) in *The Long Grass Whispers* (**1939**), *Where the Leopard Passes* (coll **1949**), *The Hunter's Cave* (coll **1951**) and *The Singing Chameleon* (coll **1957**). In similar vein are the writings of Lafcadio HEARN, who during his days as a journalist often retold local tales from New Orleans, later collected as *Fantastics* (coll **1914**) ed Charles Woodward Hutson. In the same way Kate CHOPIN used local Creole and Cajun beliefs in *Bayou Folk* (coll **1894**) and *A Night in Acadie* (coll **1897**) and John Bennett collected the rather more grotesque folktales and legends of Old Charleston in South Carolina in *The Doctor to the Dead* (coll **1946**). Hearn went on to salvage folktales from China and Japan in *Shadowings* (**1900** US) *Kotto* (**1902** US) and *Kwaidan* (**1904** US) (◊◊ KWAIDAN [*1964*]).

There are too many collections of folktales to detail here. A comprehensive bibliography is contained in the series **Index to Fairy Tales**, commenced by Mary Huse Eastman in 1915. There have been many series of folktales and legends, often based on national or cultural themes, of which the best-known are the 12-volume **Myths and Legends** series published by George Harrap (◊ LEGENDS) and the more recent **Myths and Legends** series published by Oxford University Press. All-round compendia are *Best-Loved Folktales of the World* (anth **1982**) ed Joanna Cole (1944-   ) and the **Folktales of the World** series ed Richard M. Dorson. [MA]

Oxford Myths and Legends: *African Myths and Legends* (anth **1962**) ed Kathleen Arnott; *Armenian Folk-tales and Fables* (anth **1972**) trans Charles Downing; *Chinese Myths and Fantasies* (coll **1961**) by Cyril Birch; *English Fables and Fairy Stories* (coll **1954**) by James Reeves; *French Legends, Tales and Fairy Stories* (coll **1955**) and *German Hero-Sagas and Folk Tales* (coll **1958**) both by Barbara Leonie Picard; *Hungarian Folk-Tales* (coll **1960**) by Val Bíro; *Indian Tales and Legends* (coll **1961**) by J.E.B. Gray; *Japanese Tales and Legends* (coll **1958**) by Helen and William McAlpine; *Tales of Ancient Persia* (coll **1972**) by Picard; *Russian Tales and Legends* (coll **1956**) by Downing; *Scandinavian Legends and Folk-tales* (coll **1956**) by Gwyn Jones; *Scottish Folk-tales and Legends* (coll **1954**) by Barbara Ker Wilson; *Welsh*

Legends and Folk-tales (coll **1955**) by Jones; *West Indian Folktales* (coll **1966**) by Philip Sherlock.

Other titles of interest: *West African Folktales* (coll **1917**) by William H. Barker; *The Rain-God's Daughter and Other African Fairy Tales* (coll **1977**) by Amabel Williams-Ellis; *Australian Legendary Tales* (coll **1896**) and *More Australian Legendary Tales* (coll **1898**) by Kate Langloh Parker; *Canadian Wonder Tales* (coll **1918**) by Cyrus Macmillan; *Celtic Fairy Tales* (coll **1892**) and *More Celtic Fairy Tales* (coll **1894**) both assembled as *Celtic Fairy Tales* (omni **1994**) ed Joseph Jacobs; *Celtic Wonder Tales* (coll **1910**) by Ella Young; *Myths and Legends of China* (coll **1922**; vt *Ancient Tales and Folklore of China* 1995) by Edward T.C. Werner; *Chinese Fairy Tales and Fantasies* (coll **1979**) by Moss Roberts; *Danish Fairy and Folk Tales* (coll **1899**) by J. Christian Bay; *Egyptian and Sudanese Folk-tales* (coll **1978**) by Helen Mitchnik; *Legends and Folk Tales of Holland* (coll **1963**) by Adele de Leeuw; *Régie Magyar Mondák* ["Hungarian Folk Tales"] (coll **1972**) by Dénes Lengyel; *Icelandic Folk-tales and Legends* (coll **1972**) by Jacqueline Simpson; *Old Deccan Days, or Hindoo Fairy Legends* (coll **1868**) by Mary Frere; *Indian Fairy Tales* (coll **1892**) by Joseph Jacobs; *Indian Fairy Tales* (coll **1946**) by Mulk Raj Anand; *Tales and Legends of India* (coll **1982**) by Ruskin Bond; *Indonesian Folk Tales* (coll **1970**) by A. Koutsokis; *The Japanese Fairy Book* (coll **1903**) by Yei Theodore Ozaki; *Ancient Tales and Folklore of Japan* (coll **1918**) by Richard Gordon Smith; *Japanese Fairy Tales* (coll **1962**) by Juliet Piggott; the Nigerian *Ikolo the Wrestler and Other Tales* (coll **1947**) by Cyprian Ekwenski; *Russian Popular Legends* (coll **1871**; trans and adapted W.R.S. Ralston as *Russian Folk Tales* **1873**) by Alexander Afanasief; *Russian Fairy Tales* (coll **1893**) and *Cossack Fairy Tales and Folk Tales* (coll **1894**) by R. Nisbet Bain; *Fairy Tales from Turkey* (coll **1946**) by Margery Kent.

Critical/historical works: *Morphology of the Folktale* (**1928**; trans Laurence Scott 1958 US; rev 1968 US) by Vladimir Propp; *Standard Dictionary of Folklore, Mythology and Legend* (**1949**; rev 1972) ed Maria Leach; *The Motif-Index of Folk-Literature* (**1955-8** 6 vols) by Stith Thompson.

FONTANA BOOK OF GREAT GHOST STORIES Annual ANTHOLOGY series: 20 vols, **1964-84**. The first eight vols were ed Robert AICKMAN 1964-72 (missing 1965) and the remainder were ed R. CHETWYND-HAYES. Aickman's volumes relied wholly on reprinted material (apart from his own stories), including some Victorian, but selected primarily from the golden age of GHOST STORIES in the years immediately before and after WWI. His erudite introductions established a pedigree for the ghost story, identifying it as an artform. The examples he chose – including works by E.F. BENSON, Algernon BLACKWOOD, L.P. HARTLEY, M.R. JAMES and Vernon LEE – were presented as superior; these volumes remain among the best selections of ghost stories ever published. Fontana, however, felt sales suffered from the inclusion of old material and brought in Chetwynd-Hayes to update the series. Although he retained some reprints, by *#12* (anth **1976**) the contents were predominantly new, though the quality was seldom as high. In addition to Chetwynd-Hayes's own stories, the series presented new material by Sydney J. Bounds (1920-   ), Ramsey CAMPBELL, Steve Rasnic Tem (1950-   ), Rosemary Timperley (1920-1988) and Elizabeth Walter.

A companion annual (missing 1974 and 1976) series was **The Fontana Book of Great Horror Stories** (anth 17 vols **1966-84**), *#1-#4* ed Christine Bernard (1926-   ) and the rest by Mary Danby (1941-   ). This series also ran ghost and other SUPERNATURAL FICTIONS, but relied mostly on nonfantastic fiction. The same editors also produced a junior version, **The Armada Ghost Book** (anth 15 vols **1967-83**), with *#1-#2* ed Bernard and the rest ed Danby.   [MA]

Bibliography: *The Fontana Book of Great Ghost Stories* (anth **1964**), *Second* (anth **1966**), *Third* (anth **1966**), *Fourth* (anth **1967**), *Fifth* (anth **1969**), *Sixth* (anth **1970**), *Seventh* (anth **1971**), *Eighth* (anth **1972**), these 8 ed Robert AICKMAN; *Ninth* (anth **1973**), *Tenth* (anth **1974**), *Eleventh* (anth **1975**), *Twelfth* (anth **1976**), *Thirteenth* (anth **1977**), *Fourteenth* (anth **1978**), *Fifteenth* (anth **1979**), *Sixteenth* (anth **1980**), *Seventeenth* (anth **1981**), *Eighteenth* (anth **1982**), *Nineteenth* (anth **1983**) and *Twentieth* (anth **1984**), these 12 ed R. CHETWYND-HAYES.

The Fontana Book of Great Horror Stories (anth **1966**), *Second* (anth **1967**), *Third* (anth **1968**), *Fourth* (anth **1969**), these 4 ed Christine Bernard; *Fifth* (anth **1970**), *Sixth* (anth **1971**), *Seventh* (anth **1972**) *Eighth* (anth **1973**), *Ninth* (anth **1975**), *Tenth* (anth **1977**), *Eleventh* (anth **1978**), *Twelfth* (anth **1979**), *Thirteenth* (anth **1980**), *Fourteenth* (anth **1981**), *Fifteenth* (anth **1982**), *Sixteenth* (anth **1983**), *Seventeenth* (anth **1984**).

The First Armada Ghost Book (anth **1967**), *Second* (anth **1968**), both ed Christine Bernard; *Third* (anth **1970**), *Fourth* (anth **1972**), *Fifth* (anth **1973**), *Sixth* (anth **1974**), *Seventh* (anth **1975**) *Eighth* (anth **1976**), *Ninth* (anth **1977**), *Tenth* (anth **1978**), *Eleventh* (anth **1979**), *Twelfth* (anth **1980**), *Thirteenth* (anth **1981**), *Fourteenth* (anth **1982**), *Fifteenth* (anth **1983**).

FOOD OF THE GODS, THE US movie (*1976*). ◊ H.G. WELLS.

FOOL Four types of fools are important in fantasy.

1. The court JESTER who traditionally, as in the plays of William SHAKESPEARE, utters wisdom.

2. The usage associated with the PERCEVAL motif: a naive innocent fails to ask the questions which would resolve a situation – in Perceval's instance the wounds of the FISHER KING – and who subsequently LEARNS BETTER and redeems himself and the LAND. By extension, fools in this category may fail to answer RIDDLES, or may ask or answer questions in a wrong spirit or context.

3. The figure whose preparedness to diverge from the QUEST to assist (e.g.) wounded animals is often seen by the rest as a fool; Androcles is an UNDERLIER. This is a frequent CHRISTIAN-FANTASY motif in which worldly folly becomes seen as heavenly wisdom.

4. The fool of the TAROT pack.   [RK]

FORD, FORD MADOX (1873-1939) UK writer and editor, born Joseph Leonard Ford Hermann Madox Hueffer; he signed his work Ford Madox Hueffer until 1919, changing then to FMF as a protest against German behaviour in WWI. He was a man of letters, founder and editor of the great *English Review* 1908-9 and the *Transatlantic Review* after WWI; of his 75 books, the best-known remain *The Good Soldier* (**1915**) and the four **Tietjens** novels assembled as *Parade's End* (omni **1950** US). He had a pronounced bent towards the fantastic, beginning with his first story, *The Brown Owl* (**1892** chap), a FAIRYTALE for younger children.

After *The Inheritors: An Extravagant Story* (**1901**) with Joseph CONRAD, which is sf, FMF's first novel of fantasy interest – aside from the marginal *Mr Apollo* (**1908**) – was

The "Half Moon": A Romance of the Old World and the New (**1909**), a SUPERNATURAL FICTION featuring a WITCH who, sexually obsessed by the man who rejects her, bedevils him with various forms of MAGIC. *Ladies Whose Bright Eyes* (**1911**), a TIMESLIP tale, is similar in some ways to Mark TWAIN's *A Connecticut Yankee in King Arthur's Court* (**1889**). The contemporary protagonist, a businessman, finds himself in 1326, and tries at first to modernize medieval life; but far more interestingly becomes involved in a kind of erotic tournament with two ladies who are AVATARS of the White GODDESS. A similar progression occurs in *The Young Lovell: A Romance* (**1913**), whose protagonist timeslips backwards from medieval times into a GOLDEN AGE, drawn there by the goddess APHRODITE in the guise of the White Lady.   [JC]

Other works: *The Simple Life Limited* (**1991**) and *The New Humpty-Dumpty* (**1912**), both as by Daniel Chaucer; *Vive le Roy* (**1936** US).

For children: *The Feather* (**1892** chap); *The Queen who Flew* (**1894** chap); *Christina's Fairy Book* (coll **1906**).

FORD, H(ENRY) J(USTICE) (1860-1941) UK artist, sometimes credited as Henry Ford, best-known as the illustrator of Andrew LANG's 12 coloured **Fairy Books**. HJF's drawings contributed greatly to the success of this hugely popular series. He collaborated with other artists on the first two titles, *The Blue Fairy Book* (anth **1889**) with G.P. Jacomb-Hood and *The Red Fairy Book* (anth **1890**) with Lancelot Speed, but all later volumes were illustrated wholly by HJF. Several related volumes by Lang and his wife were given richly decorated covers by HJF: *The Animal Story Book* (**1896**), *The Arabian Nights Entertainments* (**1898**), *The Red Book of Animal Stories* (**1899**), *The Book of Romance* (**1902**), *The Red Book of Romance* (**1905**; vt *The Red Romance Book*) and *Tales of Greece and Troy* (**1909**).

Ford's widely imaginative graphic drawings, combining realism and fantasy, and full of action, perfectly complemented Lang's selections of FAIRYTALES, myths and legends. In later volumes Ford included colour plates (in addition to numerous b/w drawings) which strongly recalled the PRE-RAPHAELITES in their attention to detail and brilliant rosy colours. The dreamlike air of fantasy which pervades much of his work resulted from the strong influence of his friend Sir Edward BURNE-JONES.

After Lang died in 1912, Ford illustrated, among others, M.R. JAMES's *Old Testament Legends* (coll **1913**) and E.F. BENSON's *David Blaize and the Blue Door* (**1918**).   [RD]

FORD, JOHN M. (1957-   ) US writer who has worked in various genres, beginning to publish sf with "This, Too, We Reconcile" for *Analog* in 1976; he remains best-known for his sf. As a fantasy author he is known almost exclusively for *The Dragon Waiting: A Masque of History* (**1983**), winner of the 1984 WORLD FANTASY AWARD for Best Novel; it is set in an ALTERNATE-WORLD medieval Europe dominated by the threat of an expanding Byzantine Empire (the dragon of the title). With Christianity being only one of many competing sects, there is little of the THINNING normally found in fantasies set in this period: vampirism – in the form of a communicable disease – exists (◊ VAMPIRES), as does MAGIC; WIZARDS congregate; echoes of the MATTER of Britain (◊ ARTHUR) swell under the surface of events; and at least one of the four protagonists – Hywell Peredur – bears with cause a name of mythic significance (◊ PERCEVAL). The story is almost unduly complex, involving Peredur and three others from various parts of Europe in a long, intrigue-filled campaign against the advance of Byzantium.

JMF's other fantasy work is of less interest. Two of the tales assembled in *Casting Fortune* * (coll **1989**) were contributed to **Liavek** SHARED-WORLD anthologies (ed Emma BULL and Will SHETTERLY), and "The Illusionist", original to the volume, is another **Liavek** tale.   [JC]

FORD, RICHARD (1948-   ) UK writer whose **Faradawn** sequence – *Quest for the Faradawn* (**1982**), *Melvaig's Vision* (**1984**) and *Children of Ashgaroth* (**1986**) – begins as an ANIMAL FANTASY in which the protagonists undertake a great QUEST upon whose success rests the fate of the world, which has been contaminated and thinned (◊ THINNING) by humans. A human HERO is raised from infancy in order to help with the quest for the three "Faradawn", currently held by ELVES in their kingdom. Unfortunately, World War III intervenes. After the HOLOCAUST young Melvaig survives partly by mystic recourse to *Quest for the Faradawn* itself, in this context a fictional BOOK, but not a maker of continuing STORY. In the third volume, the animals return, tweely. The ultimate effect is of manipulation.   [JC]

FOREMAN, MICHAEL (1938-   ) UK painter, illustrator and author of several stories which he himself has illustrated. Much of his ILLUSTRATION has been on books for children with texts by other writers; these books are not treated in detail here. Texts both written and illustrated by MF include *The Two Giants* (graph **1967**), *The Great Sleigh Robbery* (graph **1968**), *Moose* (graph **1971**), *War and Peas* (graph **1974**) and *Panda's Puzzle, and his Voyage of Discovery* (graph **1977**) and its sequels – *Trick a Tracker* (graph **1981**), *Land of Dreams* (graph **1982**), *Ben's Baby* (graph **1987**) and *Angel and the Wild Animal* (graph **1988**). His best work is unmistakable: over a delicately worked watercolour wash background, precisely (but wildly) imagined figures move with grace or venom or both. He has long served as (unpaid) art editor of the small journal *Ambit*, with which J.G. BALLARD has also long been associated; and has been a central figure in UK fantasy illustration from about 1970.

MF began illustration work with *The General* (graph **1961**), with text by his first wife, Janet Charters. Other illustrated books include: *The Birthday Unicorn* (**1970** chap) by Janice ELLIOTT; *Mr Noah and the Second Flood* (**1973** chap) by Sheila Burnford (1918-1984); *Rainbow Rider* (**1974** chap) by Jane YOLEN; *Hans Andersen: His Classic Fairy Tales* (coll **1976**) ed Eric Haugaard; *The Stone Book* (**1976** chap), *Granny Reardun* (**1977** chap), *Tom Fobble's Day* (**1977** chap) and *The Aimer Gate* (**1978** chap), all by Alan GARNER, along with several of Garner's works for younger children; *Popular Folk Tales* (coll 1978) by the GRIMM BROTHERS; *The Pig Plantagenet* (**1980**) by Allen ANDREWS; *The City of Gold* (coll **1980**) by Peter DICKINSON; *Terry Jones' Fairy Tales* (coll **1981**) and other titles by Terry JONES; *The Sleeping Beauty* (coll **1982**) by Angela CARTER; *A Christmas Carol* (1983) by Charles DICKENS; *Letters from Hollywood* (**1986**) by Michael MOORCOCK, nonfiction; *The Jungle Book* (1987 edition) and *Just So Stories* (1987) by Rudyard KIPLING; *Classics of the Macabre* (coll **1987**) by Daphne DU MAURIER; and *Arthur, High King of Britain* (**1994**) by Michael Morpurgo.   [JC]

FORESTS Forests represent a barrier, especially one that is dark, mysterious and impenetrable. Woods, by contrast, convey a connotation of encounter and TRANSFORMATION. If you do make it INTO THE WOODS you may not return or, if you do, things will have changed. Forests have a grander and more ancient connotation. Not only are they bigger, and thus more of a barrier, but they have existed almost forever,

and are thus more likely to contain creatures or spirits from the dawn of time, surviving while THINNING continues, as in *The Not-World* (**1975**) by Thomas Burnett SWANN and in the **Arafel** sequence by C.J. CHERRYH, particularly *Ealdwood* (**1981**; rev vt *The Dreamstone* 1983). The association of forests with the land of FAERIE is strong – as in *A Midsummer Night's Dream* (performed *c*1595; **1600**) by William SHAKESPEARE – and occurs in many FOLKTALES; other examples are *Die Elfen* (**1811**) by Johann Ludwig TIECK and "The Golden Key" (1867) by George MACDONALD. In Arthurian legend (◊ ARTHUR) the enchanted Forest of Broceliande is where MERLIN was imprisoned by Viviane; this forest was also the scene of a mighty battle between the forces of HEAVEN and HELL. This sense of the ancient nature of forests is crucial to the mythography of J.R.R. TOLKIEN, who used forests to considerable effect; in *The Lord of the Rings* (**1954-5**) there is both the sinister enchantment of Mirkwood and the ancient world of the Old Forest, the home of Tom Bombadil. The power, governance and secrets of the forest are also utilized by Robert HOLDSTOCK in his **Mythago Wood** sequence, by Charles DE LINT in various novels – especially *Greenmantle* (**1988**) – and by Diana Wynne JONES in *Hexwood* (**1993**).

There is another sense in which the forest, because of its vastness, takes on an innate sentience. This may derive from the ancient spirits in the wood, or from the very TREES themselves. This is particularly evident in the wild and uncivilized areas of the world where the powers of the Earth still prevail – the vast coniferous forests of northern Canada, as explored in stories by Algernon BLACKWOOD, the depths of the African (or Indian) jungle (personified by Joseph CONRAD as the Heart of Darkness and explored by writers like H. Rider HAGGARD, Rudyard KIPLING and Edgar Rice BURROUGHS), or the tropical rainforests, as depicted in FERNGULLY: THE LAST RAINFOREST (*1992*).

Forests do not have to be alien. They can be places of refuge, and if this also leads to an affinity with the spirits of the forest, the results can be powerful. The obvious example is the legend of ROBIN HOOD and Sherwood Forest, which is closely associated with that of the GREEN MAN. A thematic anthology is *Enchanted Forests* (anth **1995**) ed Katharine KERR and Martin H. GREENBERG.   [MA]

FORGOTTEN FANTASY ◊ MAGAZINES.

FORSTER, E(DWARD) M(ORGAN) (1879-1970) UK writer whose best-known works are novels like *Howards End* (**1910**) and *A Passage to India* (**1924**). In *The Longest Journey* (**1907**), not explicitly a SUPERNATURAL FICTION, the numerous sudden deaths and the constant (though imagined) incursions of a sexually ambivalent PAN generate a sense of achieved transcendence of the trammelled, mundane world. EMF's most famous single short story, "The Machine Stops" (1909), is sf; but a considerable proportion of the remainder – assembled in *The Celestial Omnibus and Other Stories* (coll **1911**) and *The Eternal Moment* (coll **1928**) – are supernatural fictions, several of them invoking Pan and HERMES, and most of them concerned to capture the UNCANNY moment when a THRESHOLD can be passed and transcendence gained. Hermes is a dominant figure throughout; in his introduction to *The Collected Stories* (coll **1947**) EMF suggests that the volume could be dedicated to Hermes Psychopompus, "messenger, thief, and conductor of souls to a not too terrible hereafter". In "The Story of a Panic" (1904), an imperceptive narrator almost fails to register the transfigurative presence of the god; Pan also appears in "Other Kingdom" (1909). The guide who brings epiphanic fulfilment to a boy and death to a philistine in "The Celestial Omnibus" (1908) is Hermes. "The Point of It" (1911) is a POSTHUMOUS FANTASY whose protagonist is redeemed when he is allowed return to a single moment whose missed significance had shaped his life. [JC]

FORTUNE ◊ FATE.

FORTUNE, DION Working name of UK writer and occultist Violet Mary Firth Evans (1890-1946), who also published as Violet M. Firth and as V.M. Steele; the pseudonym is derived from her family motto, "Deo, Non Fortuna". She became a member of the Order of the GOLDEN DAWN in 1919, and from 1923 was a central figure in the Theosophical Society (◊ THEOSOPHY), from which she spun off her own independent order, the Fraternity of the Inner Light, in 1927. Her main aim in these endeavours was to Westernize the societies' cod Orientalism, specializing in her nonfiction on relating ALCHEMY and hermeticism to modern concerns. Much of her fiction can be understood as the sort of SUPERNATURAL FICTION designed to convey concealed truths to the mundane world, and to convert that world.

The Secrets of Dr Taverner (coll **1926**; exp 1979), is a set of OCCULT-DETECTIVE tales facing the eponymous doctor with various cases involving the supernatural. The Watson narrator – a fellow doctor – has himself been affected by the estranging trauma of WORLD WAR I, and several of the stories are charged by a sense of profound dislocations stemming from that war; in "Blood-Lust" (1922 *Royal*), which is typical, an ex-soldier is haunted by the GHOST of a German sorcerer whose demands for energy turn his victim into a VAMPIRE. In the end, the Watson figure has a VISION of the essential aliveness of every aspect of the world, and wakes from the dead post-war world into a new life in a manner evocative of hermetic doctrine. In *The Demon Lover* (**1927**) a young woman is manoeuvred into the ASTRAL PLANE, where she spies on a cadre of hopeful SECRET MASTERS and becomes involved with a psychic VAMPIRE who is also a WEREWOLF. *The Winged Bull* (**1935**) again focuses on a protagonist whose life has been derailed by WWI, but is redeemed through proper understanding of the Rite of the Winged Bull. Similar arguments infuse *The Goat-Foot God* (**1936**), whose title refers directly to PAN.

DF's last two novels, the **Le Fay Morgan** sequence – *The Sea Priestess* (**1938**) and *Moon Magic* (**1956**) – are an early attempt in popular fiction to depict a GODDESS figure who is autonomous, not a consort or nagging SHADOW of a patriarchal GOD. A 20th-century worshipper, Miss Le Fay Morgan, a REINCARNATION of MORGAN LE FAY, engages a half-willing male in a spiritual analogue of the Year-King ritual celebrated in ATLANTIS by followers of Ishtar.

The nonfiction *Avalon of the Heart* (**1934** as by Violet M. Firth; exp by other hands vt *Glastonbury: Avalon of the Heart* 1986 as DF) presents arguments for the spiritual centrality of Glastonbury. [JC]

Other works (nonfiction): *The Esoteric Philosophy of Love and Marriage* (**1923**); *Esoteric Orders and their Work* (**1928**); *Sane Occultism* (**1929**); *The Training and Work of an Initiate* (**1930**); *Spiritualism in the Light of Occult Science* (**1931**); *Psychic Self-Defence* (**1931**), about psychic vampirism; *Through the Gates of Death* (**1932**); *The Mystical Qabalah* (**1935**); *Practical Occultism in Daily Life* (**1935**); *The Cosmic Doctrine* (**1949**); *Applied Magic* (**1962**); *Aspects of Occultism* (**1962**); *The Magical Battle of Britain* (coll **1984**), articles

written during WWII calling upon occultists to attempt to evoke the ancient guardians of Britain in order to fight off the psychic assaults of the Nazis.

As Violet M. Firth: *Violets* (coll **1904** chap) and *More Violets* (coll **1906** chap), poetry; *Machinery of the Mind* (**1922**).

As V.M. Steele: *The Scarred Wrists* (**1935**); *Hunters of Humans* (**1936**); *Beloved of Ishmael* (**1937**).

FORTUNE-TELLING ◊ PROPHECY.

FOSTER, ALAN DEAN (1946-   ) US writer who began publishing with "Some Notes Concerning a Green Box" for *The Arkham Collector* in 1971, and who has become well known as an author of sf and of many tv and movie novelizations, most of the latter also being sf. He has written some fantasy, including occasional novelizations of fantasy movies, like *Clash of the Titans* * (**1981**) and *Krull* * (**1983**). Almost all of his own fantasy appears in the **Spellsinger** sequence – *Spellsinger at the Gate* (**1983**; vt in 2 vols as *Spellsinger* 1983 and *The Hour of the Gate* 1984) and *The Day of the Dissonance* (**1984**), both assembled as *Season of the Spellsong* (omni **1985**), plus *The Moment of the Magician* (**1984**), *The Path of the Perambulator* (**1985**) and *The Time of the Transference* (**1986**), all three assembled as *Spellsinger's Scherzo* (omni **1987**), plus *Son of Spellsinger* (**1993**) and *Chorus Skating* (**1994**) – during the course of which the human protagonist is yanked from Earth to a SECONDARY WORLD with a FANTASYLAND backdrop, where he becomes a HERO, acquires various COMPANIONS, engages in various QUEST routines and collects various PLOT COUPONS. Grotesqueries and doubletakes are common. [JC]

Other works: *The Horror on the Beach* (**1978** chap), a CTHULHU-MYTHOS tale; *Shadowkeep* * (**1984**), a game tie; *Into the Out Of* (**1986**), horror; *Metrognome and Other Stories* (coll **1990**), containing some fantasy.

As editor: *Smart Dragons, Foolish Elves* (anth **1991**) with Martin H. GREENBERG; *Betcha Can't Read Just One* (anth **1993**) with Greenberg.

FOSTER, HAROLD R(UDOLPH) (1892-1981) US artist, often referred to as the father of the US COMIC strip because of the pioneering work he did on two newspaper strips during the latter 50 or so years of his life. After working as a prizefighter, gold prospector and guide, and attending art classes in Chicago, HRF began working at an advertising agency, gaining a reputation for ILLUSTRATION and poster design. In 1928 he was approached to provide the illustrations for a daily TARZAN newspaper strip. He produced a comic-strip adaptation of Edgar Rice BURROUGHS's first **Tarzan** novel, and then went on to draw new stories for the Sunday newspaper feature before quitting this to create his own Sunday feature, PRINCE VALIANT.

HRF's great strength was his realistic figure drawing, which he was able to imbue with a sense of nobility uniquely suitable to the subjects he portrayed. Later comics artists have almost certainly more often listed HRF as an influence than any other artist. [RT]

FOUNTAIN OF YOUTH Legendary youth-restoring spring (◊◊ ELIXIR OF LIFE). The 16th-century Spanish conquistador Juan Ponce de León (1460-1521) supposedly disappeared in the Americas while seeking the FOY and ELDORADO. Peter Lambourn Wilson's TIME FANTASY "Fountain of Time" (1986 *Interzone*) sees de León washed up by the tides of time in a modern Florida Everglades bar; his cognate "Ponce da Quirm" in Terry PRATCHETT's *Eric* (**1990**) actually drinks at the FOY, fatally, because – an unwise tourist – he neglects to boil it first. Tim POWERS's PIRATE novel *On Stranger Tides* (**1987**) includes a FOY QUEST episode in 18th-century Florida. In some Breton traditions, MERLIN is finally reduced to a babbling infant by the FOY in the FOREST of Broceliande, entrapped by the local "lady of the fountain" Vivian or Nimuë. Such FOY-induced infancy reappears as a PLOT DEVICE in Thorne SMITH's *The Glorious Pool* (**1934**), Piers ANTHONY's *Dragon on a Pedestal* (**1983**) and Tim POWERS' *On Stranger Tides* (1987).

[JH/DRL]

FOUQUÉ, FRIEDRICH (HEINRICH KARL), BARON DE LA MOTTE (1777-1843) German soldier and writer, best-known for his FANTASY *Undine* (**1811**; trans George Soane as *Undine: A Romance* **1818** UK), which has become one of the classics of German literature. It tells of a husband and wife whose daughter is lost, believed drowned. They take in a CHANGELING – in fact, an ELEMENTAL – whom they call Undine. She falls in love with a knight, Huldbrand, and thereby acquires a SOUL. Huldbrand subsequently loves another and Undine is banished, but she returns on his wedding night to exact her VENGEANCE. FF adapted his story as the libretto for the OPERA *Undine* (1816) by E.T.A. HOFFMANN. (◊◊ UNDINE.)

Undine drew upon the Melusine legend (◊ LAMIA) so beloved by the German Romantics (◊ ROMANTICISM), of whom FF was a leading light. His works in general engendered national pride during the Napoleonic Wars, and for a period he was Germany's most popular writer. He was the first to adapt the legend of Siegfried in his play *Sigurd* (**1808**), to which he later added *Sigurd's Rache* (**1810**) and *Aslaugas Ritter* (**1810**; trans as "Aslauga's Knights" in *German Romance* ed Thomas CARLYLE anth **1827** UK), the three forming the trilogy **Der Held des Nordens** ["The Hero of the North"], which hugely influenced Richard WAGNER.

Der Zauberring (**1813**; trans as *The Magic Ring* **1825** UK), about a knight who acquires a magical RING and uses it in his many QUESTS against evil forces, is the archetypal novel of HEROIC FANTASY. *Sintram* (**1815**; trans as *Sintram and His Companions* **1820** UK) – inspired by the engraving "The Knight, Death and the Devil" (1513) by Albrecht Dürer (1471-1528) – is an ALLEGORY on human life. FF wrote a number of shorter stories, most of which also drew on legend, although his best-known, "Das Galgenmännlein" (**1814**; trans as "The Bottle Imp" in *Popular Tales and Romances of the Northern Nations* anth **1823** UK), is a further example of the THREE-WISHES motif; Robert Louis STEVENSON later wrote his own version.

FF assembled his collected works as *Ausgabe letzter Hand* ["Collected Works"] (coll **1841** 12 vols). The English-language equivalent was the 6-vol *Fouqué's Works* (**1845** UK) ed James Burn, whose second volume, *Romantic Fiction*, includes most of FF's shorter fantasies. FF's main works are more readily available as *Undine, and Sintram and his Companions* (coll **1845** US; vt *Sintram and his Companions, and Undine* 1896 UK) and *Sintram and his Companions; Aslauga's Knight* (coll **1887** UK). [MA]

FOUR HORSEMEN The harbingers of Christian APOCALYPSE: in the BIBLE's *Revelation vi* they are Conquest, War, Famine and DEATH (though only the last is named), their horses being respectively white, red, black and – most famously – "pale". Popular iconography substitutes Pestilence for Conquest, as does fantasy in general. The Horsemen are noted for their doom-laden symbolic

appearances in *The Four Horsemen of the Apocalypse* (**1921**), the silent WORLD WAR I movie starring Rudolph Valentino. REVISIONIST FANTASY updates include Nancy SPRINGER's *Apocalypse* (**1989**), which features Four Horsewomen, and Neil GAIMAN's and Terry PRATCHETT's *Good Omens* (**1990**), where Pestilence has yielded to Pollution and the riders are now bikers. [DRL]

FOWLER, CHRISTOPHER (1953-) UK writer who also runs a movie marketing company, The Creative Partnership. Most of his stories are based in and around LONDON, which he first explored in his bestselling début novel, *Roofworld* (**1988**), about a secret WAINSCOT society of misfits who live high on the London rooftops. He has continued to use the cityscape as an evocative backdrop for such SUPERNATURAL FICTIONS as *Rune* (**1989**), *Red Bride* (**1992**) and *Darkest Day* (**1993**); all four novels loosely comprise a **London Quartet**, in which characters cross over into each other's tales. *Spanky* (**1994**) is about a PACT WITH THE DEVIL and a young man's demonic DOPPELGÄNGER; it and CF's subsequent novels, *Psychoville* (**1995**) and «Disturbia» (1996), are again set in the author's distinctive milieu. His short stories, which like his books often deal with themes of urban paranoia, broken relationships and loss of identity, have been collected in *City Jitters* (coll **1987**), *City Jitters Two* (coll **1988**), *The Bureau of Lost Souls* (coll **1989**), *Sharper Knives* (coll **1992**), *Flesh Wounds* (coll **1995**) and *Dracula's Library* (coll **1997**). More recently, he has experimented with "Tales of Britannica Castle I: Ginansia's Ravishment" (1995) and "II: Leperdandy's Revenge" (1995), a pair of fantasy stories inspired by the exotic grotesqueries of Mervyn PEAKE. "The Master Builder" (1989) was filmed as *Through the Eyes of a Killer* (**1992** tvm US), and his first short story, "Left Hand Drive" (1987), was produced as an award-winning short in 1993. [SJ]

FOWLES, JOHN (ROBERT) (1926-) UK writer whose first novel, *The Collector* (**1963**), and his third, *The French Lieutenant's Woman* (**1969**), became well known movies. His second novel, *The Magus* (**1965** US: rev 1977 UK), also filmed – as *The Magus* (**1968**) – is of strong fantasy interest, though in its final outcome proves to be a nonfantastic GODGAME. The venue and the broad plot are based on William SHAKESPEARE's *The Tempest* (performed *c*1611; **1623**): to an ISLAND dominated by a mysterious PROSPERO figure is called a callow young man, who falls in love with a Miranda figure and finds himself increasingly lost in an intricate psychic LABYRINTH, a tangle of PERCEPTIONS that prove illusory, a knot he can untie only through self-knowledge, through a knowledge of which STORY is dictating his being. The term "godgame" seems to have been invented by JF, whose draft title for *The Magus* was "The Godgame".

Later novels and tales tend intermittently to convey a similar doubleness of texture. *Mantissa* (**1982**), for instance, evokes the GODDESS in modernized MUSE aspects through veils of narrative dissembling. *A Maggot* (**1986**) conveys some of the same pressure of revelation. JF is an author who tells fantasy writers and readers what fantasy can aspire to. [JC]

FOX, GARDNER F(RANCIS) (1911-1986) US author and COMICS writer, active in the latter capacity from 1937 for DC COMICS and later MARVEL COMICS, beginning with *Detective Comics*, scripting stories for BATMAN, **The Flash**, **Green Lantern**, **The Atom**, ZATARA, **Hawkman**, The SPECTRE, **The Justice Society** (subsequently **Justice League) of America** and **Crom the Barbarian**, which he created as a rake-off of Robert E. HOWARD's CONAN. GFF also created **Adam Strange** in the 1950s. He is estimated to have written over 4000 scripts. He published his first story of fantasy interest, "The Weirds of the Woodcarver" for WEIRD TALES in 1944, and from the end of WWII was active in many genres, writing at least 160 books under a variety of names. Of those novels not marketed as sf or fantasy some – like *Woman of Kali* (**1953**), which plays with the Hindu GODS, and *One Sword for Love* (**1956**), which is about PRESTER JOHN – are of interest; as is *The Druid Stone* (**1965**) as by Simon Majors, a singleton featuring a contemporary man who is called to defend an OTHERWORLD.

GFF's fantasy is mostly restricted to three sequences. The first, the **Llarn** sequence – *Warrior of Llarn* (**1964**) and *Thief of Llarn* (**1966**) – is a PLANETARY ROMANCE. The **Kothar** sequence – *Kothar – Barbarian Swordsman* (coll of linked stories **1969**), *Kothar of the Magic Sword!* (**1969**), *Kothar and the Demon Queen* (**1969**), *Kothar and the Conjuror's Curse* (**1970**) and *Kothar and the Wizard Slayer* (**1970**) – is derived from **Conan** in most particulars. The **Kyrik** sequence – *Kyrik: Warlock Warrior* (**1975**), *Kyrik Fights the Demon World* (**1975**), *Kyrik and the Wizard's Sword* (**1976**) and *Kyrik and the Lost Queen* (**1976**) – is similar, though darker. Both barbarian-warrior sequences are competent, occasionally amusing, and entirely professional. [JC]

FRAGGLE ROCK US tv series (1983-8). Jim Henson Productions/HBO. **Pr** Duncan Kenworthy, Lawrence S. Mirkin. **Exec pr** Jim HENSON. **Dir** Henson and many others. **Spfx** Faz Fazakas, Tim McElcheran. **Writers** Jerry Juhl & Susan Juhl and many others. **Created by** Henson. **Comics adaptation** *Fraggle Rock* * (14 issues 1985-8) from MARVEL COMICS. **Starring** Gerry Parkes (Doc). 96 30min episodes.

The Fraggles are a band of small, gaily coloured creatures who live underneath the workshop of Doc, a handyman, who is unaware of their existence (his dog, Sprocket, has discovered the Fraggles but can never catch one to show Doc). In fact, there are three WAINSCOT communities UNDERGROUND: the Fraggles themselves, the Doozers, and the fearful Gorgs. The Fraggles are generally friendly, mostly content to live in their subterranean world. The Doozers, quite different, have as their sole purpose in life the building of huge, elaborate structures across the cavern, structures which serve the Fraggles as snacks. The Gorgs think themselves the rulers of a kingdom, but in actuality there are just the three of them: a father, mother and son, Junior. They try to catch the Fraggles to make them their servants, but always their efforts are thwarted. The Fraggles would prefer to stay away from the Gorgs, but have to pass through Gorg territory to reach their oracle, the all-knowing Trash Heap.

Like other efforts from Henson, each story contains a moral of some sort. Overseas versions featured local hosts in place of Doc and Sprocket; in the UK, for example, the Fraggles lived underneath a lighthouse. The original version aired on the HBO cable service; and a cartoon version followed on NBC in 1987-8. Numerous children's books were based on the series. [BC]

FRAME STORY A story which contains another story or series of stories. FSs are usually constructed to both precede and follow the story or stories they frame, which may or may not have been written by the author responsible for the frame – they may be traditional or anonymous works assembled by the frame author, or they may make up an anthology which the frame author has edited. In STORY

CYCLES like BOCCACCIO's *Decameron* (*c***1350**), Geoffrey CHAUCER's *The Canterbury Tales* (*c***1380-1400**), the *Arabian Nights* (◊ ARABIAN FANTASY) or in the CLUB-STORY convention, it is common for FS passages also to appear between individual tales. APULEIUS, Boccaccio and Chaucer used the FS both as a convenient forum for the assembly of tales and as a focusing device through which the stories could be made to seem relevant to a particular occasion or to some overarching theme or moral thesis – a declaration of relevance that seemed only natural, as it was only in the 19th century that short stories began to be perceived as autonomous works of art.

Authors like Count POTOCKI, Ludwig TIECK, E.T.A. HOFFMANN, Charles DICKENS (in *Master Humphrey's Clock* [coll **1841-42**]) and Robert Louis STEVENSON also used FSs in order to shape the perceptions of readers; and traditional material presented in such assemblies – Potocki's *Manuscript Found at Saragosse* (**1804-15**), for instance – will have been transformed through the new perspective. In recent decades this process of transformation has become axiomatic. FSs are now normally used as devices to ironize or distance the material encompassed. *Arabesques: More Tales of the Arabian Nights* (anth **1988**) ed Susan SHWARTZ, like most anthologies based on ARABIAN FANTASY, emphasizes the TALL-TALE comicality of material which cannot be taken neat. Many of Isak DINESEN's stories, and all the stories Sylvia Townsend WARNER assembled as *The Cat's Cradle Book* (coll **1940** US), are irradiated with ironic indecisions and narrative slow curves.

FSs can also be used as narrative portals into a larger tale which, once commenced, carries the main weight of meaning. PLANETARY ROMANCES, from the time of Edgar Rice BURROUGHS on, typically frame the adventures of their heroes through an introductory frame which soon becomes irrelevant. A more sophisticated author, like E.R. EDDISON in *The Worm Ouroboros* (**1922**), uses an FS to launch his tale, and never bothers to return to it at all. [JC]

FRANCE Strangely, while France has been a wellspring of fantasy, the history of fantasy there is one of long and steady attrition. During the 12th century, BARDS helped spread the oral tradition of Ireland to Wales, and then to the kingdom of the Plantagenets, where the political, religious and poetic context proved fertile. Thus was the *roman* (i.e., novel) born: the famous MATTER of Britain should better be called the Matter of Brittany. This was not the first epic cycle devised about some charismatic sovereign; it was in fact a collaborative movement, as the poets working on it considered themselves caretakers of the material, rather than originators. The first, Marie de France, about whom little is known, not even whether she was a woman or a single writer, turned the material into verse in *Les lais de Marie de France* (*c***1170**).

Then CHRÉTIEN DE TROYES started *c***1180** what would be the first cycle of Arthurian novels. Through a wealth of fantastic episodes and courtly intrigues, they expressed the ideals of the Plantagenet regime, a feudal GOLDEN AGE as opposed to the Capetian centralism. Chrétien's *Perceval* (begun ?1182) was left unfinished, as was his *Lancelot* (?**1177**), prompting others to exercise their imagination, all through the 12th and 13th centuries, both inside and beyond the borders of the kingdom of France, right up to Wolfram von Eschenbach's *Parzival* (**1225**) and Thomas MALORY's *Le Morte Darthur* (**1469**).

By the end of the 13th century, the Crusades had brought

in Europe a consciousness of the wider world, and this led to the popularity of TRAVELLERS' TALES. *The Voiage and Travayle of Syr John Maundeville, Knight* (*c*1366) (◊ MANDEVILLE), variously attributed to Jean d'Outremeuse or a real Jean de Mandeuille, is a plagiarization of sources such as Pliny the Elder's *Historia Naturalis* (AD77). By the 14th century, these literary hoaxes were ready to be satirized. François RABELAIS obliged. In *Pantagruel* (**1532**) and *Gargantua* (**1534**) he used a family of GIANTS as the focus of a series of tales combining fantasy, political farce, philosophical musings and humour.

QUESTS were Arthurian legacies. In novels they had to some extent degenerated into *romans galants* (love novels) and *romans pastorals* (country idylls). These quests, often in imaginary countries, proved an enduring genre, from the *Roman de la Rose* (started by Guillaume de Lorris in the early 13th century and finished by Jean de Meung in the 14th) to such examples as *L'Astrée* (**1607-28**) by Honoré d'Urfé (1568-1625). The end of this form more or less coincided with the vogue for FAIRYTALES, which blossomed in the 18th century. Among the best-known practitioners were Madame D'AULNOY and Madame LEPRINCE DE BEAUMONT. At about the same time, Charles PERRAULT chose to collect "naive fairytales" transmitted by oral tradition, setting the course for future endeavours, such as by the German GRIMM BROTHERS.

The gathering of traditional FOLKTALES by the Grimms hit a chord with German nationalism and was among the roots of the German Romantic movement. French ROMANTICISM was heavily affected by the German movement, and the writing of fantasy continued through the 19th century, from Charles NODIER, George Sand (1804-1876), Victor Hugo (1802-1885), Gustave FLAUBERT and Alexandre DUMAS to Théophile GAUTIER and Honoré de BALZAC.

After this bright period, French fantasy has almost completely disappeared. The popularity of Jules Verne's and H.G. WELLS's books and a proclivity to rational thinking have led French authors, both popular and "literary", towards scientific romance and sf. The Arthurian tradition still shows signs of vitality, with such writers as Xavier de Langlais – *Le Roman du Roi Arthur* (**1965**) – René Barjavel – *L'Enchanteur* (**1984**) – Hersart de la Ville-Marqué, Jacques Boulenger, Gilles Servat – *Les Chroniques d'Arcturus* (**1986**) – and Florence Trystam – *Lancelot* (**1987**). Otherwise, while there is a strong sf school in France, there is no comparable movement in fantasy. True fantasy from recent years is restricted to five works of importance: *Khanaor* (**1983**) by Francis Berthelot, *Succubes* (**1983**) by Jean-Marc Ligny, *Célubée* (**1986**) by Isabelle Hausser, *Les Flammes de la Nuit* (**1986-7**) by Michel Pagel and *La Grande Encyclopédie des Lutins* (**1992**) by Pierre Dubois and Roland Sabatier. [PM/AFR]

FRANCE, ANATOLE Pseudonym of French writer Anatole-François Thibault (1844-1924), winner of the Nobel Literature Prize. His first long fantasy was a moralistic novella in the sentimental tradition of Charles NODIER's *Trilby*, "The Honey-Bee" (in *Balthazar* coll **1889**; trans Mrs John Lane as *Balthasar* **1909**; vt *Honey-Bee* **1911**; vt *Bee* and – in Lin CARTER's *Great Short Novels of Adult Fantasy* anth **1972** – "The Kingdom of the Dwarfs"). The eponymous princess is abducted by DWARFS while her beloved fosterbrother is imprisoned by nixies (water spirits); the king of the dwarfs becomes enamoured of her, but gradually discovers an altruistic determination to reunite the lovers.

AF's classic CHRISTIAN FANTASY *Thaïs* (**1890**; trans Charles Carrington **1901** UK) is closely related to Gustave FLAUBERT's *The Temptation of St Anthony* (**1874**), following the QUEST of Anthony's disciple Paphnuce to save the soul of the eponymous actress; he persuades her to enter a nunnery, but is subsequently forced to realize that his desire to "save" her was really a perverted sublimation of lust. More attacks on the life-denying asceticism of orthodox Christianity are featured in *L'étui de nacre* (coll **1892**; trans Henry Pene du Bois as *Tales from a Mother-of-Pearl Casket* **1896**; vt *Mother of Pearl* **1908** UK), including "Amycus and Celestine", in which the friendship between a hermit and a faun contrives an eclectic fusion of Epicurean and Christian ideals, and "Leslie Wood", in which a similar modification is achieved in the relationship between a man and his wife's GHOST. The argument was further extrapolated in *Le puits de Sainte Clare* (coll **1895**; trans Charles Carrington as *The Well of Santa Clara* **1903**; vt *The Well of St Clare* **1909** UK). "San Satiro" (trans as "Saint Satyr") explains how the tomb of a mistakenly beatified SATYR became a refuge for the last remnants of a Classical glory eroded by THINNING, while the brilliant novella "L'humaine tragédie" (new trans Alfred R. Allinson as *The Human Tragedy* **1917** UK) explains how a saintly medieval monk imprisoned by corrupt churchmen discovers that SATAN is his only friend.

The Rabelaisian *La rôtisserie de la reine Pédauque* (**1893**; trans Mrs Wilfrid Jackson as *At the Sign of the Reine Pédauque* **1912** UK; vt *At the Sign of the Queen Pedauque* US) offers a satirically sceptical account of the follies of 18th-century occultism, while the VISIONARY FANTASY *Sur le pierre blanche* (**1905**; trans Charles E. Roche as *The White Stone* **1909** UK) is a philosophical discourse on the difficulties plaguing attempts to foresee the future. Another extended *conte philosophique* is *L'île des pingouins* (**1908**; trans A.W. Evans as *Penguin Island* **1909** UK), which traces in savage SATIRE the history of an ISLAND race of penguins mistakenly baptised by a myopic saint, paying particular attention to their version of the Dreyfus affair. *Les sept femmes de le Barbe-Bleu et autres contes merveilleux* (coll **1909**; trans Mrs D.B. Stewart as *The Seven Wives of Bluebeard and Other Marvellous Tales* **1920** UK; vt *Golden Tales of Anatole France* **1926** US) features stories in a lighter vein, concluding with the fine novella "The Shirt", in which emissaries of an unhappy king search unsuccessfully for the shirt of a happy man with which to redeem his melancholy spirit.

The various strands of France's fantastic fiction culminated in his second adventure in Literary Satanism, written on the eve of WWI, which became the archetype and masterpiece of that subgenre: *La révolte des anges* (**1914**; trans Mrs Wilfrid Jackson as *The Revolt of the Angels* **1914** UK). The story tells how a guardian ANGEL, Arcade, is converted to free thought by Lucretius' summary of Epicurean philosophy *De rerum natura* and organizes a new revolution of the fallen angels, most of whom are teachers and artists – but when he offers Satan (who is now a humble gardener) a commanding role his offer is politely declined, on the grounds that the fight must be won in the hearts and minds of mortals, not on the field of battle. [BS]

Other work: *Crainquebille, Putois, Riquet et plusieurs autres récits profitables* (coll **1904**; trans Winifred Stephens as *Crainquebille, Putois, Riquet and Other Profitable Tales* **1924** UK).

FRANCIS MOVIES Series of US movies, based on the **Francis the Talking Mule** series of books by David J.

Stern (1909-1971), in which a dimwit undergoes farcical adventures with a talking mule called Francis (◊ ASS). The voice of Francis was done throughout by Chill Wills. Mickey Rooney was offered the lead role but turned it down; it was taken by Donald O'Connor for the first six movies, Rooney taking it for the last. Almost unwatchable today, the FM were highly popular in their day. The series ran: *Francis* (**1950**), *Francis Goes to the Races* (**1951**), *Francis Goes to West Point* (**1952**), *Francis Covers Big Town* (**1953**), *Francis Joins the WACS* (**1954**), *Francis Joins the Navy* (**1955**) and *Francis in the Haunted House* (**1956**). A later incarnation was the tv series MR ED, in which FM director Arthur Lubin had the original idea of making Mr Ed a horse rather than a mule. [JG]

See also: CINEMA; TALKING ANIMALS.

FRANK, FREDERICK S(TILSON) (1935-) US academic whose investigations – both critical and bibliographical – into Gothic literature (◊ EDIFICE; GOTHIC FANTASY) have been of seminal importance. *Guide to the Gothic: An Annotated Bibliography of Criticism* (**1984**; cut vt *Gothic Fiction: A Master List of Twentieth Century Criticism and Research* 1988) is useful; but *The First Gothics: A Critical Guide to the English Gothic Novel* (**1987**) is an essential, searchingly annotated, cunningly selected bibliography of the field, focusing on 500 of an estimated 5000 eligible titles. *Montague Summers: A Bibliographical Portrait* (**1988**) presents a portrait of that self-taught scholar in the Gothic and supernatural. *Through the Pale Door: A Guide to and Through the American Gothic* (**1990**) repeats for US literature (1798-1990) the task *The First Gothics* performs for the parent tradition. [JC]

FRANKENHOOKER (*1990*) ◊ FRANKENSTEIN MOVIES.

FRANKENSTEIN Character created by Mary SHELLEY in *Frankenstein, or The Modern Prometheus* (**1818**; rev 1831; vt *Frankenstein* 1897): he epitomizes the scientist who experiments first and thinks about the consequences afterwards. The MONSTER he creates is not monstrous because of physical appearance (although Shelley indicates this is grim) but – at least in later versions of what has become a LEGEND – because he lacks a SOUL; he proves intelligent and (later) highly articulate, but becomes embittered and vengeful as his ugliness leads to repeated rejections. The novel can best be read as either a GOTHIC FANTASY or a TECHNOFANTASY, or preferably as both. This has not stopped Brian W. ALDISS and many others from treating it as the first SCIENCE-FICTION novel, which it manifestly is not: scientific rationalization is entirely missing from the first version, and only a token is provided in the revision. The FRANKENSTEIN MOVIES – of which there have been many – generally rely on cod technology to explain how Frankenstein could have made the Creature from dead material, and often portray the Creature as *intrinsically* evil. A version of the Creature appeared as Herman in the popular tv series *The* MUNSTERS (1964–6), and the Creature has had various incarnations in COMICS; the many SEQUELS BY OTHER HANDS include the juveniles *Frankensteins faster* (**1978**; trans Joan Tate as *Frankenstein's Aunt* 1980 UK) and *Frankensteins faster – igen!* (**1989**; trans Tate as *Frankenstein's Aunt Returns* 1990 UK) by Allan Rune Pettersson (1936-), plus *Frankenstein's Bride* (**1995**) by Hilary Bailey (1936-). The Creature, an UNDERLIER figure himself, is in turn underlain by the GOLEM. [JG]

FRANKENSTEIN MOVIES There have been innumerable movies based on the premise of *Frankenstein, or*

The Modern Prometheus (**1818**) by Mary SHELLEY or on the play *Frankenstein* (**1927**) by Peggy Webling, itself based on Shelley's original. These movies are of varying interest, as indicated by the discussion below. Some, like *Frankenstein's Great-Aunt Tillie* (**1983** Mexico/USA), have been omitted either deliberately or through ignorance; the pool of Frankenstein movies is virtually bottomless.

1. *Frankenstein* US movie (**1931**). Universal/Carl Laemmle. **Pr** Carl Laemmle Jr. **Dir** James WHALE. **Spfx** Kenneth Strickfaden. **Mufx** Jack P. Pierce. **Screenplay** John L. Balderston, Francis Edwards Faragoh, Robert Florey, Garrett Fort. **Based on** Webling's play. **Starring** John Boles (Victor Moritz), Mae Clarke (Elizabeth), Colin Clive (Henry Frankenstein), Dwight Frye (Fritz), Marilyn Harris (Maria), Boris Karloff (Monster), Edward Van Sloan (Waldman). 71 mins. B/w.

Aside from a 1910 silent short by Thomas Edison with Charles Ogle as the MONSTER, this seems to have been the tale's first screen outing, and is a fine example of both GOTHIC FANTASY and early TECHNOFANTASY. Henry Frankenstein and dwarfish servant Fritz rob graves and gallows because Henry wants to create life from dead material. Fritz steals a brain from Goldstadt Medical College, not realizing it is that of a psychotic. Henry's fiancée Elizabeth, friend Victor and Dr Waldman arrive at his windmill laboratory in time to see the Monster vivified using electricity drawn from a storm; they become Henry's co-conspirators. A few days later Fritz inadvertently terrifies the Monster; the Monster is thrust into a cell and taunted by and kills Fritz. Waldman convinces Henry his creation must be put down, but instead prepares to vivisect the Monster, who kills him, too. Roaming wild, the Monster encounters child Maria, whom through misunderstanding he drowns. Henry confronts and is vanquished by his creation, who drags him to the windmill, on which a mob converges. The Monster tosses Henry to them, as if in supplication, but they torch the building and, presumably, the Monster.

Parts of the movie grate, but it is full of poignant and abiding images: Henry's boyish exhilaration as the Monster's hand twitches into life, the charm of shared innocence as the Monster and Maria play together with flowers, the advancing tide of torches as the mob ascend a nighttime hillside – all have become ICONS, as has the sorrowing visage of the shambling, much-wronged Creature. [JG]

2. *The Bride of Frankenstein* US movie (**1935**). Universal/Carl Laemmle. **Pr** Carl Laemmle Jr. **Dir** James WHALE. **Spfx** John P. Fulton. **Screenplay** John Balderston, William Hurlbut. **Novelization** *The Bride of Frankenstein* * (**1936**) by Michael Egremont (Michael HARRISON). **Starring** Colin Clive (Henry Frankenstein), Dwight Frye (Karl), Gavin Gordon (BYRON), O.P. Heggie (Hermit), Valerie Hobson (Elizabeth), Boris Karloff (Monster), Elsa Lanchester (Bride/Mary SHELLEY), Una O'Connor (Minnie), Ernest Thesiger (Dr Pretorius), Douglas Walton (Percy Bysshe SHELLEY). 80 mins. B/w.

Widely regarded as a masterpiece, this sequels **1**, although the dovetailing is imperfect. The Monster, surviving, kills the father (here called Hans, rather than Ludwig) and mother of the drowned child Maria, and sets off into the countryside. Henry, recuperated at home and now Baron, dreams of repeating his feat, but is dissuaded by Elizabeth. Sinister alchemist Dr Pretorius arrives, and forces himself as collaborator on the unwilling Henry; Pretorius has succeeded in growing life "from seed" to make charming living miniatures. He insists they must make a mate for the Monster. That individual, meanwhile, is caught and imprisoned; he bursts his chains and makes his slaughterous escape, finally reaching a blind hermit's cottage; Monster and hermit find friendship and a release from loneliness until two hunters arrive. Fleeing again, the Monster hides in a mausoleum, where he encounters Pretorius selecting a body to revive as the Monster's mate. They seize Elizabeth and use her as hostage to force Henry to complete the revivification. But the Bride is a more sophisticated creation than her intended mate: she rejects immediately the advances of the Monster. Mournfully, the Monster tells Henry and Elizabeth to flee, then short-circuits the equipment to destroy Pretorius, Bride and (presumably) himself.

The story is prefaced by a scene in which Byron and Shelley persuade Mary to continue her story. Elsa Lanchester's dual portrayal of Mary and the Bride is noteworthy. The movie is sequelled by **31**. [JG]

3. *Son of Frankenstein* US movie (**1939**). Universal. **Pr** Rowland V. Lee. **Dir** Lee. **Spfx** John P. Fulton. **Mufx** Jack P. Pierce. **Screenplay** Willis Cooper. **Starring** Lionel Atwill (Inspector Krogh), Donnie Dunagan (Peter von Frankenstein), Josephine Hutchinson (Elsa von Frankenstein), Boris Karloff (Monster), Bela Lugosi (Ygor), Basil Rathbone (Baron Wolf von Frankenstein). 96 mins. B/w.

The hunchback Ygor was hanged for bodysnatching but, although his neck was broken, he has survived, and has been making the Monster (still alive) kill the jurors who condemned him. Henry's son Wolf Frankenstein comes to the village to inherit, and gradually discovers the situation. At last Wolf kills Ygor in self-defence; the Monster vengefully prepares to cast Wolf's son Peter into a boiling pool of sulphur beneath the laboratory, but is hurled into the sulphur himself. Wolf and his family hand over the Frankenstein estates to the villagers, and depart. The movie is stolen by Lugosi's eagerly cackling performance as Ygor. (The character of Krogh was hilariously parodied in **28**.) With the debatable exception of **4**, this was the last of Universal's **Frankenstein** movies to display integrity. [JG]

4. *The Ghost of Frankenstein* US movie (**1942**). Universal. **Pr** George Waggner. **Dir** Erle C. Kenton. **Mufx** Jack P. Pierce. **Screenplay** W. Scott Darling. **Starring** Evelyn Ankers (Elsa Frankenstein), Lionel Atwill (Dr Theodor Bohmer), Ralph Bellamy (Erik Ernst), Lon Chaney Jr (Monster), Janet Ann Gallow (Cloestine), Sir Cedric Hardwicke (Dr Ludwig Frankenstein), Bela Lugosi (Ygor), Barton Yarborough (Dr Kettering). 67 mins. B/w.

Ygor and the Monster have survived being killed in **3**. The pair reach the village of Vasaria and the clinic of a fresh Frankenstein – Ludwig, Wolf's younger brother – and his surgical colleagues Kettering and Bohmer. Young Kettering is killed by the Monster. Ludwig seizes on the notion that transferring Kettering's brain into the Monster will render it a force for good rather than EVIL. The Monster would prefer the brain of Cloestine, a village girl, and Ygor bribes Bohmer to substitute *his* – i.e., Ygor's – brain in place of Kettering's. Ygor's plan succeeds, but his implanted brain is rejected by the Monster's tissues, and most of the main cast perish in his final, maddened fury.

Chaney took over from Karloff as the Monster for this movie; Karloff never played the part again (although he did appear in **6** and, as Frankenstein, in **11**). [JG]

5. *Frankenstein Meets the Wolf Man* US movie (**1943**). Universal. **Pr** George Waggner. **Dir** Roy William Neill.

Spfx John P. Fulton. **Mufx** Jack P. Pierce. **Screenplay** Curt Siodmak. **Starring** Lon Chaney Jr (Larry Talbot/Wolf Man), Patric Knowles (Dr Frank Mannering), Bela Lugosi (Monster), Ilona Massey (Elsa Frankenstein), Maria Ouspenskaya (Maleva). 72 mins. B/w.

Killed by his father at the end of *The Wolf Man* (**1941**), the WEREWOLF Larry Talbot is dug up and inadvertently revived by a pair of graverobbers. He flees to Europe, taking up with Maleva, mother of the werewolf who infected him and whom he killed. She reckons Dr Ludwig Frankenstein could cure him, but discovers he has been destroyed by the events of **4**. In an icy cavern under the Frankenstein castle, however, Talbot finds the Monster and, in time, tracks down Ludwig's daughter Elsa. She guides the pair, plus Mannering, to her father's hidden cache of secrets and equipment, and Mannering restores the Monster to full capability. Wolf Man and Monster are battling to the death when a cunning villager breaks the dam over the estate, thus drowning them both. [JG]

6. *House of Frankenstein* US movie (**1944**). Universal. **Pr** Paul Malvern. **Exec pr** Joseph Gershenson. **Dir** Erle C. Kenton. **Spfx** John P. Fulton. **Mufx** Jack P. Pierce. **Screenplay** Edward T. Lowe. **Starring** John Carradine (Dracula), Lon Chaney Jr (Larry Talbot/Wolf Man), Boris Karloff (Dr Gustav Niemann), J. Carrol Naish (Daniel), Glenn Strange (Monster). 70 mins. B/w.

Lunatic surgeon Niemann and equally lunatic hunchback Daniel – who hopes Niemann will transplant his brain into a new body – succeed in reviving DRACULA, whom they harness to Niemann's vengeful aims. Later, in the cellars of the Frankenstein castle, they discover the frozen bodies of the Wolf Man and the Monster, both of whom they revive. All ends in tears. [JG]

7. *House of Dracula* US movie (**1945**). Like **6**, a Dracula-Frankenstein-Wolf Man team-up. ◊ DRACULA MOVIES.

8. *Abbott and Costello Meet Frankenstein* US movie (**1948**). Universal. **Pr** Robert Arthur. **Dir** Charles T. Barton. **Spfx** Jerome H. Ash, David S. Horsley. **Mufx** Bud Westmore. **Screenplay** John Grant, Robert Lees, Frederic I. Rinaldo. **Starring** Bud Abbott (Chick Young), Lon Chaney Jr (Larry Talbot/Wolf Man), Lou Costello (Wilbur Grey), Bela Lugosi (Dracula), Glenn Strange (Monster). **Voice actor** Vincent Price (Invisible Man). 82 mins. B/w.

In fact, the comedians have to cope with not only the Monster but Dracula; aided by the Wolf Man, they thwart a plot by Dracula to have Costello's brain transplanted into the Monster's body. [JG]

9. *The Curse of Frankenstein* UK movie (**1957**). Hammer/Warner. **Pr** Anthony Hinds. **Exec pr** Michael Carreras. **Dir** Terence Fisher. **Mufx** Phil Leakey. **Screenplay** Jimmy Sangster. **Starring** Hazel Court (Elizabeth), Peter Cushing (Victor Frankenstein), Valerie Gaunt (Justine), Paul Hardtmuth (Bernstein), Melvyn Hayes (Young Victor), Noel Hood (Aunt Sophie), Christopher Lee (Monster), Robert Urquhart (Paul Krempe). 83 mins. Colour.

Victor Frankenstein, imprisoned, tells his tale to a priest. He and tutor Krempe probed the mysteries of life until Victor conceived the notion of creating it. As he gathered bodily pieces, the revolted Krempe stayed solely for the love of Victor's cousin Elizabeth – although she loved Victor, whose fancy was, in turn, Justine the housemaid. Victor murdered Professor Bernstein for his brain; Krempe, fighting with him, damaged it. The vivified

Monster, accordingly homicidal, escaped, killed a blind man, and was shot dead by Krempe, who now left, satisfied Elizabeth was safe. But Victor exhumed and revived the Monster, and had it kill a pregnant and vengeful Justine. Krempe, returning, discovered all and ran for the police; the Monster escaped onto the roof; Elizabeth went to investigate; Victor set fire to the Monster, who fell into a bath of acid and was dissolved entirely. Believing all this the ravings of a madman, the authorities guillotine Victor for Justine's murder.

TCOF oozes B-movie, but its huge success spurred the long HAMMER **Frankenstein** series. Added interest lies in its introducing a Baron who is murderous rather than merely unwise. [JG]

10. *I Was a Teenage Frankenstein* (vt *Teenage Frankenstein* UK) US movie (*1957*). American International/James H. Nicholson & Samuel Z. Arkoff. **Pr** Herman Cohen. **Dir** Herbert L. Strock. **Mufx** Philip Scheer. **Screenplay** Kenneth Langtry. **Starring** Whit Bissell (Frankenstein), Robert Burton (Karlton), Phyllis Coates (Margaret), Gary Conway (Monster). 72 mins. B/w, with short colour finale.

Orthodox science knows organ transplants impossible – but not according to Professor Bill Frankenstein. He and assistant Karlton assemble teenage body parts in the traditional way, feeding the leftovers to a tame alligator. The Monster escapes and kills a woman. Returning, he is persuaded by Frankenstein to kill the latter's fiancée Margaret, who has learnt too much, and a young man, so the Monster may have his face. Just before being dismembered for export to England, the Monster breaks loose, feeds Frankenstein to his own alligator and electrocutes himself.

This was the follow-up to *I Was a Teenage Werewolf* (*1957*). [JG]

11. *Frankenstein '70* (vt *Frankenstein 1970*) US movie (*1958*). Allied Artists. **Pr** Aubrey Schenck. **Dir** Howard W. Koch. **Screenplay** Richard Landau, G. Worthing Yates. **Starring** Boris Karloff (Frankenstein), Mike Lane (Monster). 83 mins. Colour.

A tv company descends on Castle Frankenstein, eager to make a movie; the current Baron accepts their money because he wishes to buy an atomic reactor in order to create a new Monster – to whose body, in due course, the slaughtered tv crew "contribute". This was a US response to the success of the UK-produced **9**, but flopped. [JG]

12. *The Revenge of Frankenstein* UK movie (*1958*). Columbia/HAMMER. **Pr** Anthony Hinds. **Exec pr** Michael Carreras. **Assoc pr** Anthony Nelson Keys. **Dir** Terence Fisher. **Mufx** Phil Leakey. **Screenplay** Hurford Janes, Jimmy Sangster. **Starring** Peter Cushing (Baron Victor Frankenstein), Eunice Gayson (Margaret Conrad), Michael Gwynn (New Karl), Francis Matthews (Dr Hans Kleve), Oscar Quitack (Hunchback Karl). 89 mins. Colour.

The direct sequel to **9**. Frankenstein, aided by hunchbacked Karl – to whom he has promised a new body – has the attendant priest guillotined instead. The two flee to Carlsbrück, where Frankenstein practises as Dr Victor Stein, doing much work at the Poor Hospital, from whose patients he collects organs and limbs. Local Dr Kleve, recognizing him, demands to be taken on as assistant. The operation is a success, but trauma to the new Karl's transplanted brain transforms him into a psychopathic cannibal; also, it leads to the return of his deformity and, soon, death. As he dies he divulges Frankenstein's true identity, and soon all Carlsbrück knows. Frankenstein is beaten to death by his

own hospital patients; before he dies, his brain is rescued by Kleve and transplanted into a new cobbled-together body. [JG]

13. *The Evil of Frankenstein* UK movie (*1964*). HAMMER/Universal. **Pr** Anthony Hinds. **Dir** Freddie Francis. **Spfx** Les Bowie. **Mufx** Roy Ashton. **Screenplay** John Elder (Hinds). **Starring** Peter Cushing (Frankenstein), Sandor Elès (Hans), David Hutcheson (Burgomaster), Kiwi Kingston (Monster), Duncan Lamont (Police Chief), Peter Woodthorpe (Zoltan), Katy Wild (Mute). 84 mins. Colour.

A decade ago in Karlstaad, Frankenstein created a living Monster, but it was shot by the authorities and he was exiled. Now, expelled from another town, he returns with assistant Hans, but finds his château has been looted by the Burgomaster. He and Hans run to the hills, where in the cave of a mute beggar girl they find the Monster perfectly preserved in glacier ice. Aided by drunken hypnotist Zoltan, they revive the Monster. But Zoltan sends the Monster out to terrorize the town – *à la The* CABINET OF DR CALIGARI (*1919*) – then orders it to kill Frankenstein. The Monster kills Zoltan instead and drunkenly destroys the laboratory.

This was the first HAMMER movie to use a semblance of Pierce's makeup for the Monster (◊ **1**), which Universal had copyrighted. There is an excellent performance from Wild as the mute. [JG]

14. *Furankenshutain Tai Baragon* (vt *Frankenstein Conquers the World*; vt *Frankenstein and the Giant Lizard*; vt *Frankenstein Versus the Giant Devil Fish*) Japanese/US movie (*1965*). Toho/American International. **Pr** Reuben Bercovitch, Henry Saperstein, Tomoyuki Tanaka. **Dir** Inoshiro Honda. **Spfx** Eiji Tsuburaya. **Screenplay** Kaoru Mabuchi. **Starring** Tadao Takashima (Frankenstein). 95 mins. Colour.

The first of two Japanese assays at the legend (the other was **16**). In 1945 the Nazis bring the heart of the Monster to Hiroshima, where it is exposed to the nuclear flash. A boy finds the heart and eats it, and is transformed into a 10m-tall nipponized replica – Furankenshutain – of the Monster. Although effectively created from the twin evils of Nazism and nuclear warfare, the creature is good; he saves Japan from the depredations of the giant lizard Baragon. This is less a **Frankenstein** movie than one in the tradition of the GODZILLA MOVIES. [JG]

15. *Frankenstein Meets the Space Monster* (vt *Duel of the Space Monsters*) US movie (*1965*). Vernon-Seneca. **Pr** Robert McCarty. **Dir** Robert Gaffney. **Screenplay** George Garret. 78 mins. B/w.

A bad sf movie involving neither Frankenstein nor his monster. [JG]

16. *Furankenshutain No Kaija – Sanda Tai Gailah* (vt *Duel of the Gargantuas*; vt *The War of the Gargantuas*) Japanese/US movie (*1966*). Toho/American International. **Pr** Henry Saperstein, Tomoyuki Tanaka. **Dir** Inoshiro Honda. **Spfx** Eiji Tsuburaya. **Screenplay** Honda, Kaoru Mabuchi. **Starring** Russ Tamblyn (Dr Stewart), Kenji Sahara (Sanda), Jun Tazaki (Gailah). 88 mins. Colour.

The sequel to **14**. The original Furankenshutain, here called Sanda, lost a hand during the fighting; he has regenerated his hand, but the separated hand has generated a new body, Gailah, a green and nasty duplicate of Sanda. The two MONSTERS fight it out, incidentally destroying Tokyo. [JG]

17. *Jesse James Meets Frankenstein's Daughter* US movie (*1966*). Circle/Embassy. **Pr** Carroll Case. **Dir** William

Beaudine. **Spfx** Cinema Research Corp. **Screenplay** Carl Hittleman. **Starring** Rayford Barnes, Cal Bolder, Jim Davis, Estelita, William Fawcett, Steven Geray, John Lupton, Narda Onyx, Nestor Paiva (movie lacks proper credits). 82 mins. Colour.

Frankenstein's granddaughter (not daughter) Maria has with her brother Rudolf set up a laboratory in New Mexico. After a bungled stagecoach raid, James and wounded side-kick Hank Tracy are led by local girl Juanita to the Frankensteins; Maria transplants an artificial brain into the ailing Hank. At last the Monster kills its maker. This is an impressively bad movie; in the same year Beaudine directed the equally excruciating *Billy the Kid Versus Dracula* (*1965*; ◊ DRACULA MOVIES). [JG]

18. *Frankenstein Created Woman* UK movie (*1967*). HAMMER/Seven Arts/Associated British Pathé/20th Century-Fox. **Pr** Anthony Nelson-Keys. **Dir** Terence Fisher. **Spfx** Les Bowie. **Mufx** George Partleton. **Screenplay** John Elder (Anthony Hinds). **Starring** Peter Blythe (Anton), Peter Cushing (Frankenstein), Susan Denberg (Christina), Derek Fowlds (Johann), Alan MacNaughton (Kleve), Robert Morris (Hans), Thorley Walters (Dr Hertz), Barry Warren (Karl). 92 mins. Colour.

A rather good low-budget movie with a modified IDENTITY-EXCHANGE theme. Frankenstein experiments with the transmigration of SOULS. Junior assistant Hans loves disabled, facially scarred Christina. Three young aristocrats (Anton, Johann, Carl), whose sport is humiliating her, murder her father and let Hans be guillotined for the crime. Christina suicides in grief. Frankenstein and senior assistant Hertz isolate Hans's soul and transfer it into Christina's body, which they also repair and beautify. The amalgam has no specific memories, but enough survives that s/he gorily enacts VENGEANCE on the killers. The title is a somewhat blasphemous pun on that of the hugely successful Bardot vehicle *And God Created Woman* (*1957*). [JG]

19. *Frankenstein Must Be Destroyed* UK movie (*1969*). Warner-7 Arts/HAMMER. **Pr** Anthony Nelson-Keys. **Dir** Terence Fisher. **Spfx** Studio Locations Ltd. **Mufx** Eddie Knight. **Screenplay** Bert Batt, Nelson-Keys. **Starring** Veronica Carlson (Anna Spengler), Peter Cushing (Victor Frankenstein), Freddie Jones (Professor Richter), George Pravda (Dr Frederick Brandt), Simon Ward (Dr Karl Holst). 96 mins. Colour.

Brandt and Frankenstein have corresponded about their researches into brain transplantation. Frankenstein takes lodgings at the boarding-house run by Spengler. Soon he discovers both that Spengler's fiancé Holst is dealing in drugs and that Brandt is in the local madhouse, one of his doctors being Holst. Frankenstein blackmails Holst and Spengler into helping him abduct Brandt, murder the madhouse's surgeon Richter, and transfer Brandt's brain (rendered newly sane by use of a bradawl) into Richter's body. Richter/Brandt awakes after the operation, is almost murdered by Spengler, who is wilfully murdered by Frankenstein. Richter/Brandt destroys Frankenstein (but only until **20**), after the Baron has been grievously battered by a vengeful Holst.

Fisher said that he was particularly proud of this movie, but it is hard to understand why. An exceptionally crude rape scene is wisely omitted from most prints. [JG]

20. *The Horror of Frankenstein* UK movie (*1970*). EMI/HAMMER. **Pr** Jimmy Sangster. **Dir** Sangster. **Mufx** Tom Smith. **Screenplay** Jeremy Burnham, Sangster.

Starring Bernard Archard (Professor Heiss), Ralph Bates (Victor Frankenstein), Veronica Carlson (Elisabeth Heiss), Jon Finch (Henry Becker), Graham James (Wilhelm), Kate O'Mara (Alys), Dave Prowse (Monster). 95 mins. Colour.

Psychopathic Victor murders his father so that he can go to university in Vienna. There he impregnates the dean's daughter, and flees for home with friend Wilhelm; *en route* they save Elisabeth and her father, the Professor, from highwaymen, one of whom Victor kills and beheads, taking the head for experimentation. Their first successful revivification is of the Professor's pet tortoise Gustave. Victor has a graverobber steal human parts; when Wilhelm threatens exposure, Victor murders him. Professor Heiss is murdered for his brain, but his death bankrupts Elisabeth; she comes to Schloss Frankenstein – which maddens housekeeper/bedmate Alys, who knows much of the truth. Victor next murders the graverobber, but not before the latter has accidentally damaged the professor's brain; the animated MONSTER, immensely strong, is thus psychopathic. The body-count rises until at last the Monster is inadvertently destroyed by a meddlesome child.

This seems intended as black comedy, but lacks wit. Prowse, who plays the Monster here and in **25**, later played Darth Vader in the **Star Wars** movies. [JG]

21. *Dracula versus Frankenstein* (*1971*) ◊ DRACULA MOVIES.

22. *Dracula, Prisoner of Frankenstein* (*1972*) ◊ DRACULA MOVIES.

23. *Andy Warhol's Frankenstein* (vt *Carne per Frankenstein*; vt *The Devil and Dr Frankenstein*; vt *Flesh for Frankenstein*; vt *The Frankenstein Experiment*; vt *Up Frankenstein*; vt *Warhol's Frankenstein*) Italian/French movie (*1973*). CC Champion & 1/Bryanston. **Pr** Andrew Braunsberg, Carlo Ponti, Jean-Pierre Rassam, Jean Yanne. **Dir** Paul Morrissey, Antonio Margheriti. **Screenplay** Morrissey. **Starring** Joe Dallesandro, Arno Jürging, Udo Kier (Frankenstein), Monique Van Vooren. 100 mins. Colour.

A sick-joke version, the companion to *Blood for Dracula* (◊ DRACULA MOVIES). Incestuously married Frankenstein tries to create the progenitors of a new super-race, but fails amid a bloodbath. The movie was filmed in 3D, but the gore level was such that distributors stuck to a 2D format. [JG]

24. *Blackenstein* (vt *Black Frankenstein*) US movie (*1973*). **Dir** William A. Levey. **Starring** John Hart, Andrea King, Liz Renay, Ivory Stone. 92 mins. Colour.

Like *Blacula* (*1972*; ◊ DRACULA MOVIES), this was an attempt to cobble together a horror motif with the "black-movie" subgenre created by the popular *Shaft* (*1971*); it flopped. A disabled Vietnam veteran is reconstructed by the doctor who loves his girlfriend. The result is a murderous Monster. [JG]

25. *Frankenstein and the Monster from Hell* UK movie (*1973*). HAMMER. **Pr** Roy Skeggs. **Dir** Terence Fisher. **Mufx** Eddie Knight. **Screenplay** John Elder (Anthony Hinds). **Starring** Shane Briant (Dr Simon Helder), Peter Cushing (Dr Victor Frankenstein), Dave Prowse (Monster), Madeline Smith (Sarah Klauss), John Stratton (Director Adolf Klauss). 99 mins. Colour.

The last of Hammer's **Frankenstein** outings sees young Dr Helder arrested for trying to repeat the Baron's experiments and sent to the lunatic asylum where Frankenstein was committed years ago. Although officially dead, Frankenstein has become the asylum's physician, and is also – aided by the beautiful Sarah (nicknamed The Angel), mute since her father, the asylum's Director, years ago tried to rape her – building a new creature from deceased inmates. Discovering that Frankenstein's hands are wounded so that he can no longer perform surgery himself, Helder pitches in. The resultant Monster in due course goes berserk and is ripped to gobbets by the other inmates. Cushing seems bored. [JG]

26. *Frankenstein's Castle of Freaks* (vt *Dr Frankenstein's Castle of Freaks*; vt *House of Freaks*; vt *Monsters of Dr Frankenstein*; vt *Terror Castle*) Italian movie (*1973*). **Dir** Robert H. Oliver. **Starring** Rossano Brazzi, Sylvia Koscina, Edmund Purdom, Christiane Royce. 81 mins. Colour.

A cheaply produced rehash of the basic theme. [JG]

27. *Frankenstein: The True Story* UK tv miniseries/movie (*1973*). Universal/NBC. **Pr** Hunt Stromberg Jr. **Dir** Jack Smight. **Spfx** Roy Whybrow. **Screenplay** Don Bachardy, Christopher Isherwood, published as *Frankenstein: The True Story* * (1973). **Starring** Tom Baker (Sea Captain), John Gielgud (Chief Constable), Clarissa Kaye (Lady Fanschawe), Margaret Leighton (Françoise DuVal), David McCallum (Dr Henri Clerval), James Mason (Dr Polidori), Agnes Moorehead (Mrs Blair), Nicola Pagett (Elizabeth Fanschawe), Ralph Richardson (Mr Lacey), Michael Sarrazin (The Creature), Jane Seymour (Agatha/Prima), Leonard Whiting (Dr Victor Frankenstein), Michael Wilding (Lord Fanschawe). *c*200 mins (cut for cinematic release to 123 mins). Colour.

A substantive revision. Young Victor comes to London from his adoptive home with the Fanschawes near York. Returning to his medical studies he encounters the eccentric surgeon Clerval, who has resurrected dead animals and plans the creation of a man from dead parts, using solar energy. Victor assists and, when Clerval dies on the eve of the experiment, not only completes it but incorporates Clerval's brain into the Creature. That Creature is at first beautiful, but soon degenerates. Victor, initially a fond friend, emotionally rejects him, and he tries to suicide by throwing himself from a cliff. Surviving, he is befriended by the blind Lacey, with whose granddaughter Agatha the Creature becomes besotted. But she is horrified and, fleeing, is run over by a coach. The Creature takes her corpse to Victor's laboratory, there finding instead the handless hypnotist Polidori. Polidori blackmails Victor, on the night of his wedding to Elizabeth, into sewing Agatha's head onto an unmutilated body, which Polidori reanimates (by chemical means) as the beautiful Prima. On returning from honeymoon, the Frankensteins discover Prima installed in the Fanschawe household; she is a far more perfect creation than Frankenstein's Creature, yet is EVIL. Polidori, it emerges, has FU MANCHU-like ideas of world domination (and Asiatic attendants to match) through manipulation of Prima . . . The plot thickens until all the principals die, Creature and Victor entering what is almost a suicide pact.

F:TTS is largely black comedy (◊ HUMOUR), with McCallum especially revealing a brilliant comedic streak. There is intriguing ambiguity as to the identity of the Creature: is he Victor's *alter ego*, his physique decaying like the PICTURE OF DORIAN GRAY while his creator's moral integrity does likewise; or is he a manifestation of Clerval? This reworking is thoughtful, handsome and often, amid the humour or horror, moving. [JG]

28. *Young Frankenstein* US movie (*1974*). Gruskoff/ Venture/Crossbow/Jouer/20th Century-Fox. **Pr** Michael Gruskoff. **Dir** Mel Brooks. **Screenplay** Brooks, Gene Wilder. **Novelization** *Young Frankenstein* * (1974) by Gilbert Pearlman. **Starring** Peter Boyle (Monster), Marty Feldman (Igor), Teri Garr (Inga), Gene Hackman (Blind Man), Madeline Kahn (Elizabeth), Cloris Leachman (Frau Blücher), Kenneth Mars (Inspector Kemp), Wilder (Dr Frederick Frankenstein). 108 mins. B/w.

Though eager to forget his family history, young Frankenstein travels to Transylvania on inheriting grandfather Victor's estate. Discovering his grandfather's laboratory and equipment, he succumbs to the family obsession and constructs a Monster. After a merry melange of parodied clichés, he saves the Monster from the mob by transferring part of his own genius into it. The Monster marries Frankenstein's frightful fiancée Elizabeth (who comes more and more to resemble Elsa Lanchester's portrayal of the Bride in 2), while Frankenstein finds joy with lovely assistant Inga. The various set-pieces – notably the scenes with the little girl and the blind man – represent highpoints of movie PARODY. The parody of 3's Krogh (here called Kemp) is blissfully cruel. [JG]

29. *Victor Frankenstein* Irish/Swedish movie (*1975*). **Dir** Calvin Floyd. **Starring** Per Oscarsson (Creature), Leon Vitali (Frankenstein). 92 mins. Colour.

A painstakingly careful recreation of Shelley's original. Although beautifully made, this tends to plod because of its fidelity – in particular, the original's concern with the miseries of the Creature. [JG]

30. *Frankenstein's Island* US movie (*1982*). **Dir** Jerry Warren. **Starring** Tain Bodkin, Steve Brodie, John Carradine, Robert Clarke, Andrew Duggan, Laurel Johnson, Cameron Mitchell. 91 mins. Colour.

A balloon party is blown adrift and arrives on an island ruled by a bikini-clad descendant of the Baron. As bad as it sounds. [JG]

31. *The Bride* US/UK movie (*1985*). Columbia/ Victor Drai. **Pr** Victor Drai. **Exec pr** Keith Addis. **Dir** Franc Roddam. **Spfx** Peter Hutchinson. **Screenplay** Lloyd Fonvielle. **Novelization** *The Bride* * (1985) by Vonda N. McIntyre. **Starring** Jennifer Beals (Bride [Eva]), Clancy Brown (Monster [Victor]), Quentin Crisp (Zahlus), Phil Daniels (Béla), Cary Elwes (Josef), Geraldine Page (Mrs Baumann), David Rappaport (Rinaldo), Alexei Sayle (Magar), Sting (Charles Frankenstein). 118 mins. Colour.

This sequels 2, starting from the laboratory scene as Frankenstein (here Charles rather than Henry) and Dr Zahlus (rather than Pretorius, although Crisp almost perfectly mimics Thesiger) awaken their female creation. After her rejection of him, the Monster destroys the tower; he, she and Frankenstein escape. Fleeing, the Monster befriends dwarf Rinaldo; while the Baron names (Eva) and educates the Bride – with a view to the making of a New Woman, as intelligent as any man – Rinaldo names (Victor) and educates the Monster on the road (◊◊ TRUE NAMES). About here the movies loses its way. Sporadically linked telepathically to Eva, Victor goes with Rinaldo to Budapest to join a CIRCUS. Returning, he finds both Frankenstein and cavalry officer Josef wooing Eva; in face of such competition he despairs, and is imprisoned for a murder. Meanwhile Frankenstein tells Eva the truth of her origin and attempts rape. Her distress, communicated telepathically to Victor, makes him break his chains and rush to her rescue. Monster and creator battle over Eva, Victor winning and claiming his Bride.

Although a mess, and enfeebled by Sting's woodenness, *TB*

has some merits beyond its visual opulence, notably the identification of Eva not only with Mary SHELLEY but also, less obviously, with the Monster – in that, to the intellectually emasculated Frankenstein, she becomes another MONSTER of his own creation. *TB*'s most effective sequences, aside from its spectacular opening, come in the long (but irrelevant) subplot involving Rinaldo and Victor. [JG]

32. *Gothic* UK movie (**1986**). Virgin. **Pr** Penny Corke. **Exec pr** Al Clark, Robert Devereux. **Dir** Ken RUSSELL. **Screenplay** Stephen Volk. **Novelization** *Gothic* (**1987**) by Stephen Volk. **Starring** Gabriel Byrne (BYRON), Myriam Cyr (Claire Clairmont), Natasha Richardson (Mary SHELLEY), Julian Sands (Percy Bysshe SHELLEY), Timothy Spall (Dr John POLIDORI). 90 mins. Colour.

The quintet at the Villa Diodati, after consuming much laudanum and scaring themselves with GHOST STORIES, hold a SEANCE; in truth the visitors are allowing themselves to be pawns in Byron's psychological games. The night becomes a nightmare, as HALLUCINATIONS take on physical status. The five convince themselves that, through their impertinence to the God in whom they claim not to believe, and their arrogant attempt to annex his prerogatives, they have created a supernatural MONSTER that cannot now be nullified. With the morning there is a banishment of fears, yet Mary retains the germ of her novel. [JG]

33. *Doctor Hackenstein* US movie (**1989**). **Dir** Richard Clark. **Starring** David Muir, Anne Ramsey, Logan Ramsey. 90 mins. Colour.

Weak black comedy, set in the present day. The doctor's wife is dead but he has kept her head alive; he rebuilds her body using bits of passing hitch-hikers. [JG]

34. *Frankenhooker* US movie (**1990**). Shapiro/Glickenhaus. **Pr** Edgar Levens. **Dir** Frank Henenlotter. **Screenplay** Henenlotter, Robert Martin. **Starring** James Lorintz, Patty Mullen. 90 mins. Colour.

The girlfriend of a mad scientist has a dismembering encounter with a lawnmower, so he picks up prostitutes, kills them, and recreates her body. Derivative comedy. [JG]

35. *Frankenstein Unbound* (vt *Roger Corman's Frankenstein Unbound*) US movie (**1990**). Warner/Mount. **Pr** Roger CORMAN, Kobi Jaeger, Thom Mount. **Dir** Corman. **Vfx** Syd Dutton, Bill Taylor. **Mufx** Nick Dudman. **Spfx** Renato Agostini. **Screenplay** Corman, F.X. Feeney. **Based on** *Frankenstein Unbound* (**1973**) by Brian W. ALDISS. **Starring** Nick Brimble (Monster), Catherine Corman (Justine Moritz), Bridget Fonda (Mary SHELLEY), John Hurt (Joseph Buchanan), Michael Hutchence (Percy Bysshe SHELLEY), Raul Julia (Victor Frankenstein), Jason Patric (BYRON), Catherine Rabett (Elizabeth). 85 mins. Colour.

Buchanan, working in AD2031 on a particle-beam weapon, falls with his robotic CAR through TIME and finds himself in 1817 near Geneva, where almost the first person he meets is Frankenstein. Soon he discovers the MONSTER is alive and causing mayhem. Rather futilely, Buchanan attempts to enlist the aid of the occupants of the Villa Diodati to stop a young girl, Justine, being hanged as a WITCH guilty of one of the Monster's crimes; however, he succeeds in bedding Mary. The Monster insists that Frankenstein create a bride for it; denied, it butchers its maker's mistress Elizabeth. Frankenstein resurrects her; as he does so, Buchanan gimmicks his car to jerk them all into a glacial far future. Carnage starts when the Monster realizes that the revived Elizabeth is intended for Frankenstein, not itself; the climax

is a duel between it and Buchanan inside the electronic brain of a ruined city.

By taking Aldiss's *jeu d'esprit* RECURSIVE FANTASY at face value, *FU* transforms a TECHNOFANTASY into schlock. This undermines the movie's moralizing: the analogy between Frankenstein's creation and the "monster" Buchanan was creating is hammered home mercilessly, as is the presumed belief of scientists that they are outside all ethics. [JG]

36. *Frankenstein: The College Years* US movie (**1991** tvm) 20th Century-Fox/Spirit/FNM. **Pr** Bob Engelman. **Exec pr** Richard E. Johnson, Scott D. Goldstein. **Dir** Tom Shadyac. **Spfx** Charlie Belardinelli, Tom Bellissimo. **Screenplay** Bryant Christ, John Trevor Wolff. **Starring** Christopher Daniel Barnes (Jay Butterman), Andrea Elson (Andi Richmond), Vincent Hammond (Monster), Larry Miller (Albert Loman), William Ragsdale (Mark Chrisman), Patrick Richwood (Blaine Muller). *c*90 mins. Colour.

Medical professor Lippzigger leaves his effects, including Frankenstein's logbook and frozen Monster, to star student Mark. Mark and buddy Jay reanimate the Monster, Frank. The rest of the movie concentrates on their efforts, helped by their girlfriends, to conceal Frank's identity in the face of ambitious Professor Loman's eagerness, aided by moronic Muller, to steal the fruits of Lippzigger's research. After a poor start, this becomes enjoyable. [JG]

37. *Frankenstein: The Real Story* US movie (**1992** tvm). Turner. **Exec pr** David Wickes. **Dir** Wickes. **Spfx** Graham Longhurst. **Mufx** Mark Coulier. **Screenplay** Wickes. **Starring** Patrick Bergin (Victor Frankenstein), Fiona Gillies (Elizabeth), Ronald Leigh Hunt (Alphonse, Baron Frankenstein), John Mills (De Lacey), Jacinta Mulcahy (Justine), Randy Quaid (Monster), Timothy Stark (William Frankenstein), Lambert Wilson (Clerval). 112 mins. Colour.

A reasonably faithful adaptation, told appropriately in flashback from the Arctic, where the Monster and Frankenstein have their final confrontation. One significant change from convention is that, rather than build the Monster from fragments of corpses, Frankenstein "grows" him in a nutrient soup – in effect cloning his creation from himself, and thereafter being always empathetically linked with him (there is a JEKYLL AND HYDE subtext). This is a classy production enhanced by some good performances, notably Quaid's. [JG]

38. *Mary Shelley's Frankenstein* US movie (**1994**). Columbia TriStar/American Zeotrope/Japan Satellite Broadcasting/IndieProd. **Pr** Francis Ford Coppola, James V. Hart, John Veitch. **Exec pr** Fred Fuchs. **Dir** Kenneth Branagh. **Spfx** Lulu Morgan. **Vfx** Richard Conway. **Mufx** Mark Coulier, Daniel Parker, David White. **Screenplay** Frank Darabont, Stephen Lady. **Novelization** *Mary Shelley's Frankenstein* * (**1994**) by Leonore Fleischer. **Starring** Helena Bonham Carter (Elizabeth), Branagh (Victor), John Cleese (Professor Waldman), Robert De Niro (Monster), Robert Hardy (Professor Krempe), Ian Holm (Victor's Father), Tom Hulce (Henry Clerval), Cherie Lunghi (Victor's Mother), Trevyn McDowell (Justine), Aidan Quinn (Walton), Ryan Smith (William). 123 mins. Colour.

A fairly faithful translation. The structure of the movie is roughly as in **37**, with the tale told as flashback after Walton's encounter in the Arctic with Frankenstein. The concentration is largely on spectacle, and certainly some of the images are memorable: the monstrous Elizabeth, cobbled together by Frankenstein from her own corpse and Justine's, rushing in flames through the castle's corridors; the lynching of Justine, which contrasts cinematic beauty with the sordidness of the act; the final set-piece of the Monster igniting Frankenstein's pyre on a broken ice-floe. There are some excellent performances, especially that by De Niro. Somehow, nevertheless, the overall effect is flat. Branagh's portrayal of Frankenstein as a sort of LAST ACTION HERO cum brilliant scientist is untenable. Bonham Carter, hair coiffed in a style reminiscent of Lanchester's in 2, makes a rather tedious Elizabeth – although an excellent female Monster. About Quinn's performance the less said the better. The inability to suspend our disbelief is compounded by a compulsive staginess: even the best of the sets are *sets*. The omnipresent music has a rich soupiness that cloys. This is a good **Frankenstein** movie, but no landmark. [JG]

Further reading: *The Annotated Frankenstein* (1977) by Leonard Wolf; *It's Alive!: The Classic Cinema Saga of Frankenstein* (1981) by Gregory William Mank covers the eight Universal features in considerable detail and summarizes much of the remainder of the Monster's career; *Hideous Progenies: Dramatizations of Frankenstein from Mary Shelley to the Present* (1990) by Steven Earl Forry; *The Illustrated Frankenstein Movie Guide* (1994) by Stephen JONES.

See also: RE-ANIMATOR (*1985*); *The* ROCKY HORROR PICTURE SHOW (*1975*); *The* SPIRIT OF THE BEEHIVE (*1973*).

FRANKLIN, CHERYL J(EAN) (1955-   ) US writer who has published both sf and fantasy, though her sf tends to be constructed so as to allow tales of a fantasy hue to exist within an sf explanation. This is clearest in the **Network/Consortium** sequence – *The Light in Exile* (1990) and *The Inquisitor* (1992) – which is set in a PLANETARY-ROMANCE venue, and concentrates on the plus side of psychic vampirism (◊ VAMPIRES), depicting the parasitic Mirlai from outer space as spiritually enhancing for those who have been possessed (◊ POSSESSION). The **Fire Lord** series – *Fire Get* (1987), *Fire Lord* (1989) and *Fire Crossing* (1991) – is more purely fantasy, and is unremarkable except for a sense, confirmed in the third volume, that this and the sf sequence are linked. In *Sable, Shadow, and Ice* (1994), a singleton, oppositions between good and bad mages, similar to those which fuel CJF's series, cause strife in a post-HOLOCAUST world where MERLIN-worshipping polytheists find themselves opposed by a revived monotheism. [JC]

FRASER, CLAUD LOVAT (1890-1921) UK illustrator, involved for much of his short life with the Curwen Press; he died of heart failure after an operation. He worked in woodcut and in pen and ink, with wash effects and sudden touches of colour; the effect was of a small exuberance. Books illustrated include *Peacock Pie* (1924) by Walter DE LA MARE and *The Luck of the Bean-Rows* (trans **1921**) and *The Woodcutter's Dog* (trans **1921**), both by Charles NODIER. [JC]

FRASER, ERIC GEORGE (1902-1983) UK illustrator, significantly active from the 1920s; his dense black-ink or scraperboard drawings, almost always in sharp contrast to a white overlay, became familiar in advertisements for British Rail and Shell. Books illustrated include *The Book of the Thousand Nights and One Night* (1958, 2 vols only), *The Art of Love* (1971) by OVID, *Tales of the British People* (**1961**) by

Barbara Léonie Picard (1917-   ), *The Hobbit* (1977) and *The Lord of the Rings* (1977) by J.R.R. TOLKIEN, and *The Ring* (1976) by Richard WAGNER. [JC]

FRASER, [Sir] RONALD (ARTHUR) (1888-1974) UK writer and civil servant. His fantasies express the hopeful conviction that some spiritually blessed human beings are capable of achieving a bountiful transcendence of the ordinary human condition, while dwelling painfully on the recognition that such gifts would inevitably alienate them from their less-favoured loved ones. In early works like the mildly satirical fantasy of levitation *The Flying Draper* (**1924**), the rather effete ORIENTAL FANTASY *Landscape with Figures* (**1925**) and the magnificently bizarre botanical fantasia *Flower Phantoms* (**1926**) the theme is treated lightheartedly, but he was in earnest in *Miss Lucifer* (**1939**), whose heroine is able to perceive a much greater and finer REALITY than those around her and whose visions inform her about former incarnations of her SOUL and the metaphysical context which defines the highest human morality. *The Fiery Gate* (**1943**) is a marginal fantasy of muted superhumanity. RF's later works borrow sf imagery in the service of mercurial metaphysical speculation. In the TECHNOFANTASY *Beetle's Career* (**1951**), experiments in nuclear physics lead to the development of an ultimate weapon and a device for photographing the SOUL. In a series begun with *A Visit from Venus* (**1958**) the conflict between the mystical worldview favoured by RF and the positivism of science is developed in dialogues and discussions provoked by the revelations of a mechanical Eye which exploits "hyperphysical light". Benevolent "planetary spirits" add their voices to the chorus, calling an interplanetary conference in *Jupiter in the Chair* (**1958**). *Trout's Testament* (**1960**) and *The City of the Sun* (**1961**) complete the series in an eccentric spirit. His final fantasy, *A Work of Imagination: The Pen, the Brush, The Well* (**1974**), again embraces a more earnest occultism. [BS]

FRAYLING, CHRISTOPHER (1946-   ) UK writer and academic, Head of the School of Humanities at the Royal College of Art in London since 1979. *Vampyres* (anth **1979**; exp vt *Vampyres: Lord Byron to Count Dracula* 1991) is a useful and wide-ranging presentation of VAMPIRE matters. *Strange Landscape: A Journey Through the Middle Ages* (**1995**) provides insights into fantasy LANDSCAPE, particularly with reference to SWORD AND SORCERY tales, in which medieval backdrops tend to be utilized, sometimes mercilessly. Hints about the broader landscapes of FANTASYLAND in general – particularly when GENRE FANTASIES are located there – can also be gleaned. [JC]

FRAYN, MICHAEL (1933-   ) UK novelist, satirical essayist and playwright. His chief fantasy excursion is the genially barbed AFTERLIFE novel *Sweet Dreams* (**1973**), whose aspiring architect hero drives through a red traffic-light, and unknowingly through DEATH onto the ten-lane expressway approaching a modernized City of HEAVEN. Here the saved are rewarded according to their capacity: for the protagonist, a middle-class social circuit identical to Earth's, with a wife, a pleasant lover, a rewarding career, etc. MIRACLES are possible but regarded as socially gauche. Since Heaven is outside TIME, jobs include the design of the Alps and even of humankind; there are opportunities for social climbing towards business partnership with GOD. It is an amusing, deceptively light parable of human longings and limitations.

Some of MF's shorter squibs likewise satirically address problems of RELIGION and theology. [DRL]

Other works: *At Bay in Gear Street* (coll **1967**) and *Listen to This* (coll **1990**) contain relevant pieces.

FRAZER, [Sir] JAMES (GEORGE) (1854-1941) UK anthropologist, a Fellow of Trinity College, Cambridge 1879-1941, far more influential in the literary world than in the field of anthropology, where his conclusions – all based on vast sifts through secondary scholarship (he never made a field trip) – have received a decreasing amount of attention. Though he was a prolific writer, much of what he published is ancillary to his central work, the various editions of *The Golden Bough: A Study in Comparative Religion* (**1890** in 2 vols; exp 1900 in 3 vols; much exp 1911-15 in 12 vols; cut vt *The Golden Bough: A Study in Magic and Religon: Abridged Edition* 1922), which was followed by *Aftermath: a Supplement* (**1936**). Parts of the third edition were released as a series of separate volumes, whose titles include *The Magic Art* (**1911**), *Taboo and the Perils of the Soul* (**1911**), *The Dying God* (**1911**), *Spirits of the Corn and of the Wild* (**1912**), *The Scapegoat* (**1913**), *Balder the Beautiful* (**1914**) and *Adonis, Attis, Osiris* (**1914**). Given the powerful literary resonance of such titles, it is not surprising that *The Golden Bough* has proved enormously rewarding for 20th-century writers in search of mythic provenances for their work. Modernist writers like T.S. Eliot (1888-1965), D.H. LAWRENCE and Ezra Pound (1885-1972) all used JF's sonorously articulated speculations in their own search for order; for example, the FISHER KING and WASTE-LAND motifs that shape Eliot's *The Waste Land* (**1921** US) were inspired in part by JF.

The GOLDEN BOUGH itself is the branch which a newly elected priest of Diana must pluck from a TREE in the grove sacred to her; afterwards, in his guise of King of the Wood, he must slay his predecessor. What is going on, according to JF, is the sacrificial killing of the Year King by his successor in a rite that invokes the SEASONS through a concept of FERTILITY and renewal tied to vegetation RITUALS. The king must die so that – in fantasy terms – the LAND can begin to experience HEALING.

JF's literary pieces – including a solitary fiction, "The Quest of the Gorgon's Head: A Fantasia" (1920) – were assembled in *Sir Roger de Coverley and Other Literary Pieces* (coll **1920**). [JC]

Other works (selected): *Psyche's Task* (**1909**; rev vt *The Devil's Advocate* 1928); *Totemism and Exogamy* (**1910** 4 vols); *The Belief in Immortality and the Worship of the Dead* (**1913-24** 3 vols); *Folk-Lore in the Old Testament* (**1918** 3 vols); *The Worship of Nature* (**1926**); *The Gorgon's Head* (**1927**); *The Fear of the Dead in Primitive Religion* (**1933-6** 3 vols).

See also: Jessie L. WESTON.

FRAZER, SHAMUS Working name of UK writer James Ian Arbuthnot Frazer (1912-1966). His titles include *Acorned Hog* (**1933**) and *A Shroud as Well as a Shirt* (**1935**). *Blow, Blow Your Trumpets* (**1945**) is a SATIRE in which a WONDERLAND-style pre-FLOOD world is described at considerable length. ANGELS, who are not very clever, interfere constantly, passing on all sorts of undue technology to humans. Noah is a faithful husband in a culture full of pleasure palaces. A DEMON-fomented war begins. GOD steps in, and determines that after his Flood there will be a general THINNING: MAGIC will no longer be permitted. [JC]

FRAZETTA, FRANK (1928-   ) Influential US artist of HEROIC-FANTASY subjects with a classic painterly style and subtle colour sense. His distinctive depictions of strong-muscled men and voluptuous, softly rounded women have brought him worldwide recognition. FF works in oils on canvas in the great tradition of US ILLUSTRATION; the influence of Howard PYLE, Frank Schoonover and N.C. WYETH can be detected in his mature work, along with that of the Czech artist Zdeneč Burian.

FF's first professional job was as assistant to sf illustrator John Giunta, and his first COMIC strip, **Snowman**, was published in Bailey Publishing's *Tally Ho Comics #1* (1944). In 1947 FF began to draw funny-animal strips for Standard Publishing – including **Hucky Duck** and **Bruno Bear** – and a year after drew a complete episode of **Judy of the Jungle** in *Exciting Comics #59*. Later that year he began working for Magazine Enterprises (ME), signing his work "Fritz"; and it was for this company that he produced one of his longest-running strips, **Dan Brand and Tipi**. During this period he also drew strips for Standard, Hillman and National (later DC COMICS), and in 1951 drew regularly for the Famous Funnies titles *Heroic Comics* and *Personal Love*; a short-lived newspaper strip, **Johnny Comet**, came in 1952.

That year ME published FF's four classic **Thun-da, King of the Congo** strips in *Thun-da #1* (1952), about a TARZAN-style character pitted against prehistoric beasts. There followed the equally memorable **White Indian** (*#11, #12, #13* 1953). FF then began to work for EC COMICS, but drew only one story by himself, *Squeeze Play* (in *Shock Suspense Stories* March 1954) – although he collaborated on several others with artists like Al WILLIAMSON, Angelo Torres and Roy G. KRENKEL. In 1953-62 FF worked as an assistant to Al Capp on **Lil' Abner**, and then went on to men's magazines (*Gent, Dude* and *Cavalcade*) and a short stint on *Playboy*'s **Little Annie Fanny** with Harvey Kurtzman, Jack Davis and Will Elder.

In the early 1960s Ace Books began reprinting the books of Edgar Rice BURROUGHS and FF undertook (largely at the instigation of Krenkel) to paint the cover illustrations. These were enormously successful (one, the cover for *Back to the Stone Age*, won the New York Society of Illustrators' Award of Excellence). He did cover paintings for Lancer's series of CONAN reprints and for WARREN PUBLISHING's *EERIE, CREEPY* and *Blazing Combat*.

FF still produces sf and fantasy paintings for a long list of publications, as well as posters and portfolios. His character **Death Dealer** has been the basis of a series of novels by James R. SILKE and the comic book *Deathdealer* (*#1* 1995), drawn by Simon BISLEY. FF won a Hugo AWARD in 1966. [RT]

Other works: Many compilations of FF's work, from all sources, have been put together. The widest range of styles is represented in *The Fantastic Art of Frank Frazetta* (graph coll **1975**) with intro by Betty Ballantine, *Frank Frazetta: Book Two* (graph coll **1977**), *Book Three* (graph coll **1978**), *Book Four* (graph coll **1980**), *Book Five* (graph coll **1985**), these latter four ed Ballantine. FF's funny-animal art is featured in *Small Wonders: The Funny Animal Art of Frank Frazetta* (graph coll **1992**). Reprinted strips include *Thun'da, King of the Congo* (graph coll **1973**). *Frank Frazetta, The Living Legend* (graph coll **1981**) is by FF himself, while *The Frazetta Pillow Book* (graph coll **1994**) features a collection of some of his erotica. *Frank Frazetta* (graph coll **1994**), produced as an auction catalogue, provides one of the best showcases of his work. His most recent work is featured in *Illustrations Arcanum* (graph coll **1994**).

Further reading: Interview with FF in *The Comics Journal #174* (1995).

FREAKY FRIDAY US movie (*1976*). Disney. **Pr** Ron Miller. **Dir** Gary Nelson. **Spfx** Art Cruickshank, Danny Lee, Eustace Lycett. **Screenplay** Mary Rodgers. **Based on** *Freaky Friday* (**1972**) by Rodgers. **Starring** John Astin (Bill Andrews), Jodie Foster (Annabel Andrews), Barbara Harris (Ellen Andrews), Sparky Marcus (Ben Andrews), Brooke Mills (Lucille Gibbons). 95 mins. Colour.

It is Friday 13. 13-year-old Annabel and her mother Ellen simultaneously make the mock-WISH that they could each spend just one day enjoying the lifestyle of the other. Their personalities swap; Ellen proceeds to make a mess of a schoolgirl's day, while Annabel is a disastrous housewife – although suddenly a dream mother to Annabel's younger brother Ben. When finally the two wish simultaneously to have their own bodies back, the wish is granted but each is dumped into a position hazardous for them but, in effect, the process has been a HEALING. Then father Bill and young Ben simultaneously wish they could be each other . . .

Harris is completely convincing and very funny, without being patronizing, as the 13-year-old in an adult body; and Foster's complementary performance is only marginally less adept. *FF* is among the best and funniest IDENTITY-EXCHANGE movies.

In 1994 Disney remade *FF* as a tvm. [JG]

FREAS, (FRANK) KELLY (1922-   ) US illustrator best-known for his sf artwork, though his ability to portray fantastic creatures has endeared him to devotees of fantasy. His first professional appearance was on the cover of WEIRD TALES for November 1950, and he was the featured artist in the Fall 1990 issue. His portfolios include *Frank Kelly Freas* (graph **1957**), *Frank Kelly Freas: The Art of Science Fiction* (graph **1977**) and *Frank Kelly Freas: A Separate Star* (graph **1984**). [MA]

FREDDY'S DEAD: THE FINAL NIGHTMARE (*1991*) ◊ *A* NIGHTMARE ON ELM STREET.

FREEDOM FOR US (*1931*) ◊ René CLAIR.

FREEMAN, MARY E(LEANOR) WILKINS (1852-1930) US writer. Her first stories were for children. Of her 200 or so stories for adults, most are realist tales set in New England; only about a dozen are SUPERNATURAL FICTIONS. Some are scattered in otherwise realist collections like *A Humble Romance and Other Stories* (coll **1891**), *A New England Nun and Other Stories* (coll **1891**) and *Silence and Other Stories* (coll **1898**); some MEWF assembled herself as *The Wind in the Rose-Bush and Other Stories of the Supernatural* (coll **1903**), while *Collected Ghost Stories* (coll **1974**) assembles 11 tales. A texture of BELATEDNESS infuses most of her supernatural fictions. "Luella Miller" is a tale of psychic vampirism (◊ VAMPIRES). Only one tale, "The Hall Bedroom" (1905), is fantasy: a PICTURE in a boarding-house is a PORTAL to a benign ALTERNATE WORLD, into which the main character of the tale disappears. [JC]

FREYA/FREYJA ◊ GODDESS.

FRIEDMAN, MICHAEL JAN (1955-   ) US writer, mostly associated with sf, though he began his career with the **Vidar** sequence – *The Hammer and the Horn* (**1985**), *The Seekers and the Sword* (**1985**) and *The Fortress and the Fire* (**1988**) – which mixes NORDIC FANTASY and CONTEMPORARY FANTASY. Vidar, a man in this world, is also an AVATAR of Vidar, the son of ODIN; and must pass back and forth between world and OTHERWORLD. *The Glove of Maiden's Hair* (**1987**), set in contemporary NEW YORK, features ELVES there. [JC]

FRIESNER, ESTHER M. Working name of US writer

Esther Mona Friesner-Stutzman (1951-   ), who began publishing stories of genre interest with "The Stuff of Heroes" for *IASFM* in 1982, and almost all of whose work has been fantasy, the main exception being a **Star Trek: Deep Space Nine** tie, *Warchild* * (**1994**). She has written relatively little short fiction, and has published only two collections, *Ecce Hominid* (coll **1991**) and *It's Been Fun* (coll **1991**); her career therefore really begins with her first series, the **Chronicles of the Twelve Kingdoms**, set in an ARABIAN-FANTASY venue: *Mustapha and His Wise Dog* (**1985**), *Spells of Mortal Weaving* (**1986**) – the first written – *The Witchwood Cradle* (**1987**) and *The Water King's Laughter* (**1989**). Each volume features a QUEST plot, a sense of the serpentine immensities of power and lore (specifically that attaching to MAGIC), and an intensely invoked LANDSCAPE. Though there is a tendency for her plots to sink into GENRE-FANTASY cliché, the flowing graveness of the **Twelve Kingdoms** books redeems the haste of some of the plotting, though the last – featuring a comic UGLY DUCKLING destined to reveal himself as having been a HIDDEN MONARCH – rather breaks the storytelling mood. The sequence was clearly intended to extend to 12 volumes, one per kingdom.

Most of EMF's work has been in the realm of comedy (◊ HUMOUR). The **New York** series of CONTEMPORARY FANTASIES – *New York by Knight* (**1986**), *Elf Defence* (**1988**) and *Sphynxes Wild* (**1989**) – uses its CITY settings as an effective backdrop for plots in which creatures of FAERIE intersect clashingly with our world, though rarely with sufficient intensity to generate a sense that full-scale URBAN FANTASY is being attempted. The first, in which a DRAGON and a KNIGHT war against the skyline, is the most effective; the others are funnier. The **Demons** sequence – *Here Be Demons* (**1988**), *Demon Blues* (**1989**) and *Hooray for Hellywood* (**1990**) – is mostly humorous SUPERNATURAL FICTION, though the RECURSIVE second volume, in which members of the SOCIETY FOR CREATIVE ANACHRONISM become COMPANIONS of an IMMORTAL Richard the Lionheart, is of broad fantasy interest. The **Gnome** sequence – *Gnome Man's Land* (**1991**), *Harpy High* (**1991**) and *Unicorn U* (**1992**) – engages its protagonist with a welter of CROSSHATCH creatures, all of whom interfere with his education and who engage him on tours of the UNDERWORLD. The **Majyk** series – *Majyk by Accident* (**1993**), *Majyk by Hook or Crook* (**1994**) and *Majyk by Design* (**1994**) – involves a SORCERER'S-APPRENTICE protagonist and various other figures in dealings with majyk, a "substance" necessary to any workings in magic. *The Psalms of Herod* (**1995**) initiates a new series, set in a fundamentalist-Christian post-HOLOCAUST world in which women are treated as pariahs, and all life is deemed sacred until the moment of birth, when the "defective" are killed off.

Her singletons are various. *Harlot's Ruse* (**1986**) plays rather persistent jokes on its sexual content; but *Druid's Blood* (**1988**) is a sharply and amusingly told GASLIGHT ROMANCE set in an ALTERNATE-WORLD Victorian England, with RECURSIVE references galore, and a plot centring on investigations by analogues of SHERLOCK HOLMES and Watson. *Yesterday We Saw Mermaids* (**1991**) is a tale of THINNING set in 1492; a ship – separate from those under Columbus, and full of figures from the backstory of Western civilization – travels west, finds PRESTER JOHN and witnesses the Church's frustration of the birth of a second MESSIAH and the departure of the magic folk from our ken.

There is a sense of excessive and hasty fertility to much of

EMF's work, but again and again a sharp poetic density of image alerts the reader to the fact that a high intelligence waits in the wings. [JC]

Other works: *The Silver Mountain* (**1986**); *Ecce Hominid* (**1991** chap); *Split Heirs* (**1993**) with Lawrence WATT-EVANS; *Wishing Season* (**1993**), YA ARABIAN FANTASY; *Alien Pregnant by Elvis* (anth **1994**) with Martin H. GREENBERG, comprising original stories; *The Sherwood Game* (**1995**); *Chicks in Chain Mail* (anth **1995**).

FRIGHTENERS (magazine) ◊ FEAR; MAGAZINES.

FRIGHT NIGHT There have been two movies in this series.

1. *Fright Night* US movie (**1985**). Columbia-Delphi IV/Vistar. **Pr** Herb Jaffe. **Dir** Tom Holland. **Spfx** Richard Edlund, Entertainment Effects Group, Albert Lannutti, Michael Lantieri, Darrell Pritchett, Clayton Pinney. **Mufx** Ken Diaz. **Screenplay** Holland. **Novelization** *Fright Night* * (**1985**) by John M. Skipp, Craig Spector. **Starring** Amanda Bearse (Amy Peterson), Stephen Geoffreys ("Evil" Ed Thompson), Roddy McDowall (Peter Vincent), William Ragsdale (Charley Brewster), Chris Sarandon (Jerry Dandridge), Jonathan Stark (Billy Cole). 105 mins. Colour.

Chubby teenaged horror fan Charley discovers that new neighbour Dandridge is a VAMPIRE and SERIAL KILLER. The police laugh at his information, as initially do girlfriend Amy, smart-aleck college friend "Evil" Ed, and camp Peter Vincent, phony vampire-killer and just-fired host of tv horror show *Fright Night*, whose aid Charley eventually enlists. After much adventuring, Charley and Vincent succeed in slaying Dandridge. *FN* is rare among HORROR comedies in being effective in both genres: it is genuinely funny and genuinely scaring, with brilliant spfx. The performances, especially by Sarandon, Stark and McDowall (whose finest latter hour this may be), are superb: a memorable URBAN-FANTASY sequence sees Dandridge, with seeming inexorability, pursuing Charley and Amy through a hellish nightlife of discos and clubs where people can no longer perceive each other as human beings. The movie's homo-erotic content is brave and perfectly handled. One can read *FN* as a recasting of Bram STOKER's *Dracula* (**1897**), with Dandridge as DRACULA, Amy as Mina, Charley as Harker and Vincent as Van Helsing. [JG]

2. *Fright Night, Part 2* US movie (**1988**). Vista. **Pr** Mort Engelberg, Herb Jaffe. **Dir** Tommy Lee Wallace. **Spfx** Gene Warren Jr. **Screenplay** Tim Metcalfe, Miguel Tejada-Flores, Wallace. **Starring** Merritt Butrick (Richie), Julie Carmen (Regine), Russell Clark (Belle), Jonathan Gries (Louie), Traci Lin (Alex), Roddy McDowall (Peter Vincent), William Ragsdale (Charley Brewster), Ernie Sabella (Dr Harrison), Brian Thompson (Bosworth). 104 mins. Colour.

This direct sequel is tricksier, scarier, funnier and better acted . . . yet is somehow the lesser movie. Extensive psychotherapy has cured Charley of his "delusions" about VAMPIRES. However, a quartet of vampires, led by the exquisite Regine, vengeful sister of Dandridge, has taken up residence in the apartment above Vincent's. Soon their murderous activities are terrifying the neighbourhood. In the end Charley and Vincent succeed, largely through the love between Charley and new girlfriend Alex, in destroying the four. [JG]

FRISWELL, (JAMES) HAIN (1825-1878) UK author of *Ghost Stories and Phantom Fancies* (coll **1958**), a collection of SUPERNATURAL FICTIONS which includes a CHRISTMAS tale,

"All Alone on Christmas Day", and a number of GHOST STORIES. A further tale, "The King of the Gnomes", is the first translation into English of "Viy" (1835), by Nikolai GOGOL. [JC]

FRITH, NIGEL (ANDREW SILVER) (1941-   ) UK writer whose first book, *The Lover's Annual* (coll **1965**), consisted of poems written in a troubador idiom, and who began publishing material of genre interest with *The Legend of Krishna* (**1975**; vt *Krishna* 1985) which, like *The Spear of Mistletoe: An Epic* (**1977**; vt *Asgard* 1982), attempts to recuperate a sense of the original Being of the GODS whose biographies are here recounted with TWICE-TOLD faithfulness. His **Pangaia** sequence – *Jormundgand* (**1986**), *Dragon* (**1987**) and *Olympiad* (**1988**) – retells other MYTHS and LEGENDS. A play, *Commedia* (performed **1987**), is a re-creation of the COMMEDIA DELL'ARTE style. *Snow* (**1993**) is a GHOST STORY set at a contemporary university; most of its characters are plagued by UNDERLIER echoes they cannot cope with. [JC]

"FRITZ" Pseudonym of Frank FRAZETTA.

FRITZ THE CAT US ANIMATED MOVIE (*1971*). ◊ Ralph BAKSHI.

FROG PRINCE Classic FAIRYTALE collected by the GRIMM BROTHERS. The prince has suffered TRANSFORMATION by a WITCH into frog form, and this BONDAGE will end only upon fulfilment of a CONDITION: that a princess takes the frog into her bed or, in other versions, gives it a kiss. This story has been repeatedly retold and subjected to PARODY – e.g., by Anthony ARMSTRONG. [DRL]

FROST, ARTHUR BURDETTE (1851-1928) US artist. ◊ ILLUSTRATION.

FROST, GREGORY (DEE) (1951-   ) US writer whose first work of genre interest – "Rubbish" for *F&SF* in 1984 – is sf, and whose most remarkable single novel, *The Pure Cold Light* (**1993**), is also sf. Most of his other work has been fantasy. His first novel, *Lyrec* (**1984**), transforms an apparently straightforward SWORD-AND-SORCERY tale into a complex PLANETARY ROMANCE involving ALTERNATE WORLDS, conflicted GODS and a variety of TALENTS. The **Tain** series – *Tain* (**1986**) and *Remscela* (**1988**) – retells with considerable energy myths concerning CUCHULAIN. [JC]

FROST, MARK (1953-   ) US screenwriter, director and novelist. He was writer and associate producer of the movie *The Believers* (*1987*), based on Nicholas Conde's novel *The Religion* (**1982**; vt *The Believers* 1988) about a weird Catholic ritual taking over NEW YORK's occult underworld. Thereafter MF co-created TWIN PEAKS with David LYNCH, writing or co-writing several episodes, one of which he directed; he also co-produced with Lynch TWIN PEAKS: FIRE WALK WITH ME (*1992*). MF wrote and directed the movie *Storyville* (*1992*). His novel, *The List of 7* (**1993**), is a helter-skelter RECURSIVE FANTASY involving the young Arthur Conan DOYLE – as well as other historical characters including H.P. BLAVATSKY and Bram STOKER – in a GASLIGHT-ROMANCE adventure trying to prevent an entity much like the GREAT BEAST from incarnating in a mortal child; more fantasy motifs than can sensibly be listed are blended into the stew – including the TECHNOFANTASY creation of ZOMBIES. The driving premise is that the young Doyle has completed and is trying to sell a novel (◊ BOOKS) called *The Dark Brotherhood*, based on the writings of Blavatsky, and that his tale turns out to come too close to describing a SECRET-MASTERS confederation called, indeed, The Dark Brotherhood. In MF's second novel, *The 6*

Messiahs (**1995**), the conflict continues in the US West. Doyle is helped throughout by the man on whom he will base SHERLOCK HOLMES. [JC/JG]

FROUD, BRIAN (1949-   ) UK illustrator of fantasy and fairytale subjects, much of whose work is in the line and sub-dued colour-wash style of traditional English book ILLUSTRATION, with many elements showing the distinct influence of Arthur RACKHAM. His creatures, however, are uniquely his own, having a charm and originality that gives his work a distinctive personal quality. He also works in a fully rendered, meticulously detailed style reminiscent of natural-history watercolourists like Beatrix POTTER, for some of his fairy paintings; in these he often depicts blue- or green-skinned children with gossamer WINGS among wild flowers and woodland undergrowth. He is also a model-maker, and has worked as designer on fantasy movies.

BF illustrated several children's books, including Margaret MAHY's *The Wind Between the Stars* (**1975** chap) and *Are All the Giants Dead?* (**1977** chap), before collabo-rating with Alan LEE on *Faeries* (graph **1978**). He was a conceptual designer on the movies *The DARK CRYSTAL* (*1982*) and LABYRINTH (*1986*). He is illustrator of the series **Brian Froud's Faerielands**, which to date includes *The Wild Wood* (graph **1994**) with text by Charles DE LINT and *Something Rich and Strange* (graph **1994**) with text by Patricia A. MCKILLIP.

Other works: *The World of the Dark Crystal* * (**1982**) with J.J. Llewellyn and Rupert Brown; *Goblins* (graph **1983** chap); *The Goblins of the Labyrinth* (graph **1986**) and «The Goblin Companion» (graph 1996 chap), both with Terry JONES; *Lady Cottington's Pressed Fairy Book* (graph **1994**) with Jones, which won a Hugo AWARD. [RT]

FRYE, (HERMAN) NORTHROP (1912-1991) Canadian literary critic. NF's greatest achievement, and a book deeply relevant to fantasy, is his *Anatomy of Criticism: Four Essays* (**1957**), where he attempts to establish a structure to incor-porate and analyse all works of literature. In the third essay, "Theory of Myths", he posits as a general background an upper, benign world of "Innocence" and a lower, inimical world of "Experience", and sees four basic narrative pat-terns, or *mythoi*, operating against this background: ROMANCE, which moves circularly within the realm of Innocence; irony/SATIRE, which moves circularly within the realm of Experience; comedy (◊ FANTASY; HUMOUR), which proceeds linearly from Experience to Innocence; and tragedy, which proceeds linearly from Innocence to Experience. These four *mythoi* can be further united into one grand, circular narrative analogous to the four SEASONS: comedy = spring, romance = summer, tragedy = autumn and irony/satire = winter. In the first essay, "Theory of History", NF posits a different sort of cycle moving against this backdrop as a culture proceeds through its literary his-tory, characteristically beginning with MYTH (stories focused on GODS) and moving downward via Romance (demigods and HEROES) and High Mimetic (noble people) to the Low Mimetic (common people) and finally to Irony (commonplace or ignoble people); then beginning again with a new age of myth. [GW]

Other works: *T.S. Eliot* (**1963**); *The Well-Tempered Critic* (**1963**); *Fables of Identity: Studies of Poetic Mythology* (**1963**); *The Educated Imagination* (**1964**); *A Natural Perspective* (**1965**) and *Fools of Time* (**1967**), both on William SHAKE-SPEARE; *The Return of Eden* (**1965**), on John MILTON; *The Modern Century* (**1967**); *The Critical Path* (**1971**); *The Bush*

Garden (**1971**); *The Secular Scripture* (**1976**).

FRYER, DONALD S(IDNEY) (1934-   ) US poet and editor whose *Songs and Sonnets Atlantean* (coll **1971**) is pre-sented as a translation of surviving literary works from ATLANTIS. His nonfiction has been mostly involved with the oeuvre of Clark Ashton SMITH, including the unfortu-nately unindexed *Emperor of Dreams: A Clark Asthon Smith Bibliography* (**1978**) and several editions of work by Smith: *Poems in Prose* (coll **1960**), *Other Dimensions* (coll **1970**) and *Strange Shadows: The Uncollected Fiction and Essays of Clark Ashton Smith* (coll **1989**), the last with Steve Behrends and Rah Hoffman. [JC]

FUENTES, CARLOS Working name of Mexican writer and diplomat Carlos Manuel Fuentes Macías (1928-   ), active in both careers from about 1950, when he became secretary to the Mexican representative at the International Law Commission of the United Nations, in Geneva. From his first book – *Los días enmaskarados* (coll **1954**; part trans Margaret Sayers Peden as *Burnt Water* **1980** US) – CF's fiction plays elaborate, MYTH-saturated, MAGIC-REALISM games around the problem of the MYTH OF ORIGIN of the land of Mexico, both ancient and modern. His first novel, *La región más transparente* (**1958**; trans Sam Hileman as *Where the Air is Clear* **1960** US), is narrated by an Indian who is also an AVATAR of the Aztec GOD of war, but who behaves in the modern world more like a TRICKSTER than an entity it would be death for mortals to gaze upon. A play, *Todos los gato son pardos* ["All Cats Are Grey"] (**1970**), features a conflict of gods, each wearing the MASK of a man.

CF's most famous single novel, *La muerte de Artemio Cruz* (**1962**; trans Sam Hileman as *The Death of Artemio Cruz* **1964** US), hints at POSTHUMOUS FANTASY in the re-conciliation of various DOUBLES of the eponymous protagonist. Doubles – changes of skin (◊ SKINNED) – appear throughout CF's work: gods double men, and vice versa; men and women double each other; avatars haunt creatures of the future. A typical example is *Aura* (**1962** chap; trans Lysander Kemp **1965** US), a SUPERNATURAL FICTION in which a man is entrapped by the eponymous GHOST so that his identity may be taken over (◊ POSSES-SION) by the long-dead husband of the WITCH who has controlled the action: in this case, echoing many 19th-century DOPPELGÄNGER tales, the double eats him. The complexities of doubling in *La cabeza de la hidra* (**1978** Spain; trans Peden as *The Hydra Head* **1978** US), which slip in and out of supernatural realms, defy synopsis.

Terra Nostra (**1975** Spain; trans Peden **1976** US) is an enormously complex, deeply ambitious attempt to incar-nate the myth of Mexico within the scope of a single – albeit vast – novel. ALTERNATE REALITIES – an early exploration of Mexico is confronted by an indigenous God of the land – interpolate and are interpolated by sf-like perspectives, con-veyed mainly through a FRAME STORY set in 1999, at the close of which, in a kind of LICENZA, various characters dis-robe themselves of some of the identities they have worn, and a wedding ensues. *Gringo viejo* (**1985**; trans Peden as *The Old Gringo* **1986** US) presents a version of the last months of Ambrose BIERCE which hedges into the supernatural. Nearly as complex as *Terra Nostra*, *Cristóbal nonato* (**1987**; trans Alfred MacAdam and CF as *Christopher Unborn* **1989** US) presents in nine sections the prenatal previsions of a namesake of the famous Columbus, who contemplates a near-future redemptive revolution in Mexico. Even at his most dazzlingly experimental, CF hews close to his central concern, which is to continue constructing a STORY to shape his land. [JC]

Other works: *Cambio de piel* (**1967**; trans Sam Hileman as *A Change of Skin* **1968** US); *Un familia lejana* (**1980**; trans Peden as *Distant Relations* **1982** US); *Constancia y otras novelas para virgenes* (coll **1989**; trans Thomas Christensen as *Costancia, and Other Stories for Virgins* **1990** US); *The Buried Mirror: Reflections on Spain and the New World* (**1992** US), a nonfiction study written in English.

FULLER, JOHN (LEOPOLD) (1937-) UK poet and novelist, son of the poet Roy Fuller (1912-1991), active as an author of poetry from his first book, *Fairground Music* (coll **1961**); *The Illusionists* (**1980**), a book-length poem, hovers at the edge of the fantastic. In his first fiction of interest, *Flying to Nowhere: A Tale* (**1983** chap), set in a LAND-OF-FABLE Middle Ages, a religious emissary comes to an ISLAND where a possibly heretic miracle has long before taken place, and finds – at the heart of an underground LABYRINTH – the abbot of the order busy dissecting bodies in a search for the SOUL. Once this RECOGNITION has been passed, the abbot urges, then we will be free. In the end, though ambivalently, the harness of this world seems to hold.

The Adventures of Speedfall (coll **1985**) features an eccentric philosophy professor in various adventures, some of them supernatural. The various protagonists of *Look Twice: An Entertainment* (**1991**) leave a RURITANIA called Gomsza in search of safety, but are ensnared in an ARABIAN NIGHTMARE of STORY by the fellow-traveller they encounter on the last train to leave the Duchy. *The Worm and the Star* (coll **1993**) is a sequence of FABLES and short fantasies, many heavy with message. Though not widely known as a genre writer, JF is a conscious and entertaining manipulator of genre material, and warrants notice. [JC]

FU MANCHU More properly, Doctor Fu Manchu. A Chinese master criminal, corrupted sage, head of the dread "Si-Fan", a secret society dedicated to EVIL, and an emblem of the Yellow Peril. He was created by Sax ROHMER in 1912, and appeared in many novels, beginning with *The Mystery of Dr Fu Manchu* (fixup **1913**; vt *The Insidious Dr Fu Manchu* 1913 US). The early **Fu Manchu** novels have a GASLIGHT-ROMANCE flavour, as does the recent SEQUEL BY OTHER HANDS, *Ten Years Beyond Baker Street* (**1984**) by Cay Van Ash (1918-1994), in which FM fights SHERLOCK HOLMES. FM's normal foe is Nayland Smith, a white man with enormous pluck. There were radio and COMICS adaptations, a US tv series – *The Adventures of Fu Manchu* (1955-6) – and the FU MANCHU MOVIES. [JC]

FU MANCHU MOVIES Two series of movies based on the character FU MANCHU. The first comprises *The Mysterious Dr Fu Manchu* (**1929**), *The Return of Fu Manchu* (**1930**), *Daughter of the Dragon* (**1931**), *The Mask of Fu Manchu* (**1932**) and *Drums of Fu Manchu* (**1940**). Warner Oland played the doctor in the first three of these UK b/w movies; Boris Karloff took over for the fourth and Henry Brandon for the last. A second series, starring Christopher Lee, came much later: *The Face of Fu Manchu* (**1965**), *The Brides of Fu Manchu* (**1966**), *The Vengeance of Fu Manchu* (**1967**), *Blood of Fu Manchu* (**1968**) and *Castle of Fu Manchu* (**1968**). *The Fiendish Plot of Fu Manchu* (**1980**), starring Peter Sellers, is a spoof of the *oeuvre*. [JG]

FUREY, MAGGIE (1955-) UK writer whose fantasy sequence, the **Artefacts of Power** – to date *Aurian* (**1994**), *Harp of Winds* (**1994**) and *The Sword of Flame* (**1995**) – is a heavily plotted, fast-moving QUEST tale, perhaps rather

overloaded with PLOT COUPONS. The much-harried female protagonist survives bereavements, pregnancy and the kidnapping/death of more than one lover, but looks likely to prevail over the DARK LORD who has been manipulating her life and her world, which is a fairly unremarkable FANTASYLAND. [JC]

FURLONG, MONICA (TEAVIS) (1930-) UK writer whose two YA fantasies comprise a loose sequence, the second preceding the first. In *A Year and a Day* (**1990**; vt *Juniper* 1991 US) a young princess must spend the eponymous period with her WITCH godmother in an indeterminate CELTIC-FANTASY venue, where there are invocations of Arthurian MATTER. In *Wise Child* (**1987**) another child is raised by the same witch – opposed by the THINNING onslaught of a Christian priest – and trained to use old gifts to maintain justice. [JC]

FURNISS, HARRY (1854-1925) Irish caricaturist and illustrator, in the UK from 1873. His only fantasy book, *How Smilestown Became Glumstown: A Simple Story Told in Sketches* (graph *c*1920), describes the effects on England when alcohol is prohibited and a fake ELIXIR takes its place. Books illustrated by HF include: F.M. ALLEN's *Brayhard* (**1890**); Lewis CARROLL's *Sylvie and Bruno* (**1889**) and *Sylvie and Bruno Concluded* (**1893**); *Travels in the Interior* (**1887**) by L.T. Courtenay; G.E. Farrow's *The Wallypug of Why* (**1895**) and *The Wallypug Book* (**1905**); and Edward Abbott PARRY's *Gamble Gold* (**1907**). Much of HF's work appeared in *Punch* during 1873-94; his witty, fluent hyperbolic drawings soon became famous. [JC]

FURY, THE US movie (*1978*). 20th Century-Fox. **Pr** Frank Yablans. **Exec pr** Ron Preissman. **Dir** Brian DePalma. **Spfx** A.D. Flowers. **Mufx** Rick Baker. **Screenplay** John Farris. **Based on** *The Fury* (**1976**) by Farris. **Starring** John Cassavetes (Childress), Kirk Douglas (Peter), Charles Durning (Dr Jim McKeever), Amy Irving (Gillian Bellaver), Fiona Lewis (Dr Susan Charles), Carol Rossen (Dr Ellen Lindstrom), Carrie Snodgress (Hester), Andrew Stevens (Robin). 117 mins. Colour.

A covert US agency, headed by the vile Childress, attempts to assassinate one of its own agents, Peter, so his psychokinetically gifted son Robin may be experimented upon. Peter strives to track Robin down through the Paragon Institute, where Robin was once tested. Equally gifted – although her TALENTS, notably psychokinesis and psychometry, are as yet uncontrollable – is a newcomer to the Institute, Gillian. The girl eventually leads Peter to the secret establishment where Robin's talents are being honed. But the agency's efforts have turned the lad into an unstable, paranoid MONSTER: he kills his mentor and lover, Dr Charles, then attempts likewise with the arriving Peter; but himself dies, transferring his powers to – or taking POSSESSION of (it is unclear which) – Gillian. After Peter's suicide, Gillian (or Gillian/Robin) blows Childress to pieces using raw mind-power.

DePalma had two years earlier scored a hit with another movie about teenage psi powers, CARRIE (*1976*), in which Irving had also starred, and *TF* is generally dismissed as a return to the same trough. In fact, the two movies are quite different, *TF* being much more of a thriller with HORROR-MOVIE elements. Irving's focal performance, while it cannot quite camouflage serious plot flaws, is mesmerizing. [JG]

See also: FIRESTARTER (*1984*).

FUSELI, HENRY (1741-1825) Swiss painter, born Johann Heinrich Füssli. He lived in Italy 1770-78 and subsequently in the UK, where he enjoyed considerable success, becoming Professor at the Royal Academy in 1799; the Italian form of his surname is normally used. Very early in his career, he translated *Gedanken über die Nachahmung der griechischen Werke* (**1755**) by Johann Winckelmann (1717-1768) as *On the Imitation of Greek Works* (trans **1765**), and never fully repudiated Winckelmann's Neoclassical approach to the canons of art, retaining to the end of his own career a sense of proportion and distancing that gives his most violently Romantic paintings an impersonal, narrative clarity. Though his most famous works – like *The Nightmare* (1781; many later versions exist) – are not illustrations as such, they are not, at the same time, subjective visions. What is depicted in *The Nightmare*, for instance, is *not* envisioned by the sleeper, who sprawls with her arms hanging down and her head thrown back, her sleeping face turned away from the ape-like GOBLIN squatting on – almost between – her thighs, and from the horse (i.e., the "night mare") gazing through the curtains: none of the "actors" in the scene are looking at each other. The nightmare is what those who view the painting catch sight of; it is, in other words, a frozen moment out of STORY. More than one fantasy writer subsequently attempted to put that icon into words: Charles NODIER's *Smarra* (**1821**) is an example, as is Fitz-James O'Brien's "What Was It?" (1859).

HF was deeply attracted to the works of William SHAKESPEARE, and his versions of scenes from *A Midsummer Night's Dream* (performed *c*1595; **1600**) – most remarkably perhaps, *Titania and Bottom* (*c*1790) – once again alarmingly freeze moments in canvases that seem almost to burst with held Story. The erotic intensity of his portrayals of women, which is clear in his Shakespeare illustrations, becomes engagingly explicit in his numerous portraits of courtesans, many using his wife as model. She is directly portrayed in *Mrs Fuseli Seated by a Fireplace (The Rosy-Cheeked Medusa)* (1799), in which Medusa herself can be seen in a MIRROR: the fact that Mrs Fuseli is gazing at the *real* Medusa, but has not turned to stone, complicatedly relates this drawing to the Romantic obsession with the FEMME FATALE, which was about to burgeon. [JC]

FUTURE FANTASY AND SCIENCE FICTION ◊ MAGAZINES.

GAARDER, JOSTEIN (1952-   ) Norwegian philosopher and writer, only two of whose books have to date appeared in English. *Kabalmysteriet* (**1990**; trans Sarah Jane Hails as *The Solitaire Mystery* **1996** UK) traces the late 20th-century QUEST of a young boy and his father for the mother and wife who had abandoned them; *en route*, a mysterious DWARF gives the boy a magnifying glass by which he is able to read the tiny print in a BOOK given him by a baker, who turns out to be his long-lost grandfather. This book tells the tale of a sailor shipwrecked in 1842 on an ISLAND inhabited by animate playing cards, subject of a GODGAME created by a previous castaway; but it also tells the story of the boy's quest in terms dictated by patterns foretold 150 years earlier by the animate cards, leading to the philosophical inference that human beings are themselves actors in some inescapable, greater STORY or godgame. The dwarf turns out to be the immortal Joker (◊ TRICKSTER) from the original deck.

Sofies verden (**1991**; trans Paulette Møller as *Sophie's World* **1994** US), which is more complexly sustained, starts with 14-year-old Sophie receiving periodic letters on the cyclical (◊ CYCLES) history of philosophy from an anonymous philosopher; while much of the text continues as an episodic popularization of this subject, the FRAME STORY deploys fantasy tropes extensively and, for the most part, knowledgeably. Sophie and the philosopher, Alberto Knox, prove to be characters in a book written as a birthday present by a man for his daughter Hilde, who is the same age as Sophie; *Sophie's World* is thus not only a book but a book within a book. Sophie and Knox, discovering the state of their own REALITY, plot to find a way of continuing their lives beyond the end of the story of which (as in the earlier book) they are the focus and by which they are driven – and at last do so, although in an ALTERNATE REALITY, mapped onto our own, inhabited by fictional, legendary, mythological and FAIRYTALE characters; Hilde conspires to help them attain this state. Discovery of the core of philosophy is symbolized by a journey INTO THE WOODS by Sophie and Knox to reach a house that exists in their world but is also the home of Hilde and her parents. Mundane reality is depicted as a sort of FAERIE, and the different timescale – because Hilde reads the book far quicker than Sophie and Knox live it – as an equivalent of TIME IN FAERIE; in this context, the relationship between the two realities eventually reverses. A MIRROR is a quasi-PORTAL between the two realities; DREAMS allow a degree of interaction between them; Hilde's father is playing a godgame with Sophie's reality, while Knox is a MAGUS within it. Countless other fantasy themes are invoked.

The didactic bulk of the book is lightly – probably too lightly – rendered, and there are some curious omissions from the history, particularly bearing in mind the nature of the fantasy: Kurt Gödel (1906-1978) and Werner Karl Heisenberg (1901-1976) are among the notable absentees. The name Sophie derives from Sophia, the name given by early Judeo-Christians to the feminine aspect of GOD (◊◊ GODDESS) and also the Greek word for "wisdom".   [JC]

GABALDON, DIANA Working name of US writer Diana Jean Gabaldon Watkins (1952-FRANKENSTEIN MOVIES), whose **Outlander** series – *Outland* (**1991**; vt *Cross Stitch* 1991 UK), *Dragonfly in Amber* (**1992**) and *Voyager* (**1994**), with further vols projected – is an extended TIMESLIP romance carrying a nurse from the end of WORLD WAR II back to 1743, where she becomes involved with Jamie Fraser, a man who profoundly reminds her of her husband and who is involved with Bonnie Prince Charlie and the doomed Jacobite Rebellion. In the second volume the nurse returns to Jamie, who dies – she thinks – at the Battle of Culloden (1746). In the third volume, though, he turns out to have survived, but now both are 20 years older (she is a widow with Jamie's child); once this is sorted out between them the sequence shifts rousingly to the Caribbean.   [JC]

GABRIEL OVER THE WHITE HOUSE US movie (*1933*). MGM/Cosmopolitan. **Pr** Walter Wanger. **Dir** Gregory LaCava. **Screenplay** Bertram Bloch, Carey Wilson. **Based on** *Rinehard* (1933) by Thomas F. TWEED. **Starring** Walter Huston (President Judson Hammond), Karen Morley (Pendola Molloy), Franchot Tone (Hartley Beekman). 87 mins. B/w.

Elected for unkeepable promises, new US President Hammond settles in for a term of *laissez-faire*: huge unemployment, widespread starvation, rampant organized crime, etc., are none of his concern, being merely local issues; people must stand on their own two feet (in short, as much later a UK Prime Minister would say, "there is no such thing as society"). But a car crash puts him in a coma, and he emerges as an altered character, seemingly possessed (◊ POSSESSION) by the ANGEL Gabriel. Assisted by secretary Beekman and ex-mistress (soon to be Beekman's fiancée) Molloy, he takes direct action to end unemployment and poverty, to eradicate gangsters (by firing squad) and to attain world disarmament and thereby inaugurate universal peace. At this achievement, he dies. This bizarre political fantasy is surprisingly socialist in tone (albeit veering towards National Socialism). [JG]

GAIA ◊ GODDESS.

GAIMAN, NEIL (RICHARD) (1960-FRANKENSTEIN MOVIES) UK writer of GRAPHIC NOVELS, US fantasy COMIC books and fiction; his fertile imagination has given new life to several moribund comic-book characters and created a number of new ones. NG has also published one text novel, *Good Omens, the Nice and Accurate Prophecies of Agnes Nutter, Witch* (**1990**) with Terry PRATCHETT, a collection of short stories, *Angels and Visitations: A Miscellany* (coll **1993**), three nonfiction books – *Duran Duran* (**1985**), *Ghastly Beyond Belief* (**1985**) with Kim NEWMAN and *Don't Panic: The Official Hitch Hiker's Guide to the Galaxy Companion* (**1988**) – and several anthologies, including *Now We are Sick: A Sampler* (anth **1986** chap; exp vt *Now We Are Sick: An Anthology of Nasty Verse* **1991**) with Stephen JONES; plus five titles (which he "devised" or "co-devised" rather than edited): *Temps* (anth **1991**) with Alex Stewart, *The Weerde: Book I* * (anth **1992**) with Mary GENTLE and Roz KAVENEY, *Eurotemps* (anth **1992**) with Stewart, *The Weerde II* * (**1993**) with Gentle, Kaveney and Stewart, and *Villains!* * (**1992**) with Gentle. He has also published short stories and verse in a number of periodicals and anthologies.

NG worked as a journalist, interviewer and reviewer before becoming a freelance writer. He came to prominence in the comics field on publication of *Violent Cases* (graph **1987**; original colour restored 1991), the first of four collaborations with artist Dave MCKEAN, and the almost simultaneous appearance of the anthology comic book *Outrageous Tales of the Old Testament* (graph anth **1987**), for which he produced many of the scripts. The former brought him fame and the latter notoriety.

NG went on to establish himself in US comic books with *Black Orchid* (#1-#3 1988-9; graph **1989**), in which he breathed new life into the minor DC COMICS superheroine created by Sheldon Mayor (1917-1991). His greatest success has been with **The Sandman** sequence (◊ SANDMAN) (1989-1996) – for one episode of which, *A Midsummer Night's Dream* (#19 1990), he was awarded the 1991 WORLD FANTASY AWARD for Best Short Story. His other major works include *The Books of Magic* (1990-91; graph **1993**) and *Death: The High Cost of Living* (graph **1993**).

NG has a wide-ranging romantic imagination and has created a uniquely personal gallery of fantasy characters: DEATH, the dark maiden, is a streetwise punk teenager (besides Death, **Sandman**'s brooding lead character DREAM has five more "Endless" siblings: Destiny, Destruction, Despair, Desire and Delirium); Mad Hettie hid her heart and forgot where it was; Calliope, the captured MUSE, is imprisoned in the attic of a writer to whose advances she is compelled to submit in order to enable him to write prizewinning novels; and so on.

NG's **Sandman** graphic novels are *The Doll's House* (graph 1990), *Preludes and Nocturnes* (graph 1991) – which comes first by internal chronology – *Dream Country* (graph 1991), *Season of Mists* (graph **1992**), *A Game of You* (graph 1993), *Fables and Reflections* (graph 1993), *Brief Lives* (graph 1994), *Worlds' End* (graph **1994**), *The Kindly Ones* (graph 1995) and «The Wake» (graph 1996). His graphic novels with McKean, aside from *Violent Cases* and *Black Orchid*, are *Signal to Noise* (in *The Face* #10-#17 1989-90; graph 1992), about a dying movie producer planning an apocalyptic film he knows will never be made, and *The Tragical Comedy or Comical Tragedy of Mr Punch* (graph **1994**), another enigmatic account of a childhood experience. A BBC tv series, *Neverwhere* (**1996**), is about an UNDERGROUND London WAINSCOT. [RT/NG]

Other works: *Miracleman: The Golden Age* (1992; graph 1993); *Snow, Glass, Apples* (**1994** chap).

GAINES, WILLIAM M. (1922-1992) US publisher and editor. ◊ COMICS.

GALAHAD The most perfect KNIGHT of the ROUND TABLE. The son of LANCELOT, he alone was sufficiently pure of body and spirit to achieve the QUEST for the Holy GRAIL and to cure the WASTE LAND. Lancelot's baptismal name was also Galahad, suggesting that the two characters may originally have been the same, though there is also the inference that Lancelot, through his adultery with GUINEVERE, had become impure, and thus had to make way for his son (born in deceit, but guiltless of blame). Galahad not only symbolizes perfection and chivalry but also someone who will champion a righteous cause. The first complete telling of his story was in the *Queste del Saint Graal*, part of the **Vulgate Cycle** (◊ ARTHUR) set down by anonymous writers during the period 1215-35, where he is identified as the grandson of Pelles, the FISHER KING. He features in many Arthurian novels and stories (◊ ARTHUR), though only *Galahad* (**1926**) by John ERSKINE, *Brother to Galahad* (**1963**) by Gwendolyn Bowers and the irreverent *Too Bad Galahad* (**1972**) by Matt Cohen (1942-FRANKENSTEIN MOVIES) focus on him solely. [MA]

GALLAND, ANTOINE (1646-1715) French orientalist. ◊ ANTHOLOGIES; ARABIAN FANTASY; FOLKTALES.

GALLICO, PAUL (WILLIAM) (1897-1976) US writer, whose extremely popular writings on sport were assembled in *Farewell to Sport* (coll **1938**), and who spent most of his remaining career writing fiction, beginning with *The Adventures of Hiram Holliday* (coll **1939**), a set of TALL TALES that come close to fantasy, as does his WORLD WAR II fable, *The Snow Goose* (**1941** chap). His first fantasy proper – and the first of several tales featuring CATS – is *The Abandoned* (**1950**; vt *Jennie* 1950 UK), in which a young boy suffers TRANSFORMATION into a cat, is befriended by another cat (Jennie), has adventures in her tutelary company, is mortally wounded fighting over her, and is reborn in human form, wiser. Other stories with cats include *Thomasina: The Cat who Thought She was God* (**1957**), whose heroine feline thinks she has been reborn as an Egyptian goddess (◊ REINCARNATION), *The Silent Miau: A Manual for Kittens, Strays, and Homeless Cats* (**1964**), fictionalized musings, and *Manxmouse* (**1968** UK), in which a TOY mouse is animated (◊ REAL BOY) and proves himself worthy of the companionship of a rough-hewn Manx cat.

Of some interest as SUPERNATURAL FICTION are the **Alexander Hero** books – *Too Many Ghosts* (**1959**) and *The Hand of Mary Constable* (**1964**) – featuring the experiences of an OCCULT DETECTIVE. *The Man who was Magic: A Tale of Innocence* (**1967**), in which an individual capable of genuine MAGIC threatens a town full of charlatans, fails through PG's decreasing ability to control his sentimentality. In *The House that Wouldn't Go Away* (**1979** UK) a house which is itself a GHOST gives visions of the Victorians who once lived in it. [JC]

Other works: *The Small Miracle* (coll **1951** UK); *Snowflake* (**1952** chap UK): *The Foolish Immortals* (**1953**), sf; *Ludmila: A Story of Leichtenstein* (**1954** chap Liechtenstein), which appears also in *Three Stories* (omni **1964** UK; vt *Three Legends: The Snow Goose; The Small Miracle; Ludmila* 1966 US) and in *Ludmila and The Lonely* (coll **1967** UK); *The Poseidon Adventure* (**1969**) and *Beyond the Poseidon Adventure* (**1978**); *The Best of Paul Gallico* (coll **1988** UK).

GALLISTER, MICHAEL Pseudonym of H. BEDFORD-JONES.

GALLON, TOM (1866-1914) UK author. ◊ CHRISTMAS BOOKS.

GALT, JOHN (1779-1839) Scottish writer. ◊ ASTROLOGY.

GAMES, FANTASY Today there are hundreds of fantasy games of many different types: board and card games, wargames, live and table-top role-playing games (RPGs), multiple-choice gamebooks, computer games, and many others. These are not developed in isolation: there is considerable feedback between designers and authors in different companies and different areas of the hobby. Writers are frequently freelances, and may be employed by several companies, and products are often the work of several authors and designers, whose individual contributions it is frequently impossible to assess. One additional cause of confusion is the existence of two influential author/designers named Steve JACKSON.

The first important fantasy games appeared in the late 1960s. While there had been some attempts at original games on sf themes, little had been done with fantasy concepts aside from games were aimed exclusively at children and based on popular juvenile characters. Essentially these were simple variants of conventional boardgames.

The first signs of adult interest in the fantasy genre were various amateur-run postal games. *Slobbovia* (1969), a humorous *Diplomacy* variant with storytelling elements, still runs as an amateur press association (APA); players write in with game moves and related articles and fiction, which are periodically published by one or another member. Set in a world based loosely on a COMIC strip, Al Capp's **Lil' Abner**, its background includes a complex feudal society (whose basic unit of currency is the serf), plus various RELIGIONS and nonhuman races. *Armageddon* (1970), a German fantasy wargame played partly by post and partly face-to-face, was demonstrated at the 1970 World Science Fiction Convention in Heidelberg. A UK version, *Midgard*, soon appeared, and was subsequently run in the USA and Australia (where it is still commercially run). Many other postal games followed.

Many commercial fantasy games trace their origins to the 1970s, when attempts to depict characters, events or the atmosphere of popular fiction – in particular, the works of J.R.R. TOLKIEN – led to a rapid expansion of the field. The earliest – e.g., *Battle of Helm's Deep* (Fact & Fantasy Games 1974) and *Siege of Minas Tirith* (FFG 1975) – were played with counters and dice; both these examples were essentially conventional wargame systems with added rules for MAGIC and MONSTERS. *Sorcerer* (SPI 1975), a more abstract

boardgame, treated magic as the basic mechanism of the game rather than as an afterthought. Such developments eventually led to the production of wargames which tried to portray important characters; for example, *War of the Ring* (SPI 1977) used special cards to represent the abilities of Gandalf, Sauron, etc. (This game should not be confused with *War of the Ring* [FGU 1976], a *Diplomacy* variant.)

Meanwhile miniatures wargamers were attempting to develop their own rules for fantasy battles played with models. One system was *Chainmail* (TSR 1972-3), which originated some rules mechanisms later used in the first fantasy RPG, *Dungeons and Dragons* (*D&D*; TSR 1974) by Gary GYGAX and Dave Arenson (who had previously had contact with UK *Midgard* players).

Role-playing games RPGs revolve around individuals, not armies. Each player controls a "character", defined by various parameters – lists of equipment, spells, TALENTS, degree of physical strength, etc. One player is the referee (in *D&D* called the Dungeon-master) and acts as storyteller, describing the setting and whatever else the characters may encounter: monsters, traps, peasants, etc. Often the referee also designs additional elements of the game's world. The outcome of encounters is determined partly by the intentions of the players and referee and partly by a randomizing process, usually dice. Miniature figures are often used to represent the characters and whatever they meet. Players generally act out a personality for each character, and often look on this personality as an *alter ego*.

D&D was a generic fantasy system with elements derived from the works of Tolkien, Jack VANCE, Robert E. HOWARD and Fritz LEIBER. It was immediately successful, developing a loyal and proliferating following among students and sf/fantasy fandom in the USA; the craze soon spread to the rest of the English-speaking world. It was followed by a succession of rules-supplements, extra adventures and revisions. Its successor, *Advanced Dungeons and Dragons* (*AD&D*; TSR 1978-9; rev 1989), is still by far the most popular RPG.

While early RPGs used generic and poorly described fantasy worlds, emphasizing combat and the acquisition of treasure at the expense of character development, most modern systems devote a good deal of attention to backgrounds and personalities. An early example is *Empire of the Petal Throne* (TSR 1975) by M.A. Barker. Based on *D&D*'s rules, it features a complex society, Tekumel, derived from Mayan and Arabic sources, with an emphasis on linguistics and class structures. Another important successor was *Runequest* (Chaosium 1978), set in a world first described in Greg Stafford's fantasy wargame *White Bear, Red Moon* (1976). Again the background is a detailed fantasy world, in this case Glorantha, in which GODS are real and often manifest. From the 1980s onward *AD&D* also acquired more detailed backgrounds, published as supplements to the game; the DRAGONLANCE setting is the best-known, through the highly successful novelizations.

Since many RPGs set out to simulate the world of fantasy fiction, there has naturally been considerable interest in products based directly on fictional sources. An early pioneer in this field was Chaosium, with *Call of Cthulhu* (1981), based on H.P. LOVECRAFT's stories, *Stormbringer* (1981), based on Michael MOORCOCK's **Elric** novels, and *Elfquest* (1984), based on ELFQUEST. *The Middle Earth Role-Playing System* (Iron Crown 1982) is based on the works of Tolkien, while *Amber* (Phage 1991) derives from Roger ZELAZNY's **Amber** series.

The effort and expense needed to develop commercial RPGs, often several man-years per game, can rarely be justified by the profits of selling the game alone. RPG rulebooks are mainly sold to create a market for additional material, which can be developed at less expense, and without these subsequent sales a game can easily bankrupt its publishers. TSR are by far the most prolific in this respect; their current catalogue for *AD&D* alone includes over 100 adventures, rule books and character packs, plus special dice, posters, CDs, magazines and dozens of other products. There is also a lucrative market in spinoff fiction, such as the **Dragonlance** and **Dark Sun** novels. Other companies have followed TSR's formula for success, or use a single rules system for a wide variety of genres; this makes games easier to develop and easier for players to learn, thus encouraging continued use of the company's products. For instance, the *Runequest* system was later adapted for *Call of Cthulhu*, *Stormbringer*, *Superworld* (1983), *Ringworld* (1984) and *Elfquest*. The rules and much of the background of *Vampire: The Masquerade* (White Wolf 1991) were incorporated into later games – *Werewolf: The Apocalypse* (1991), *Mage: The Ascension* (1993) and *Wraith: The Oblivion* (1994). GURPS (Generic Universal Role-Playing System; Steve Jackson Games 1978) was designed from the outset to make such conversions easy, and uses a set of "universal" rules with supplements covering sf, fantasy, horror, espionage, Westerns, historical fiction, Arthurian myth and military operations. This system also licences many popular fictional backgrounds, most notably sf but also fantasies like *Robert E. Howard's Conan* (1987) and *Andre Norton's Witch World* (1988).

Outside the USA, *Warhammer Fantasy Role Playing* (Games Workshop 1986, Hogshead Publishing 1995) is the only major UK fantasy system to emerge (there have been several sf RPGs). This was a spinoff from a wargame, *Warhammer Fantasy Battles* (1983), which is supported by a wide range of figures and rules supplements, and also has an sf spinoff, *Warhammer 40,000* (1991). Australia has produced *Lace and Steel* (BTRC 1992), a fantasy game in which FAERIE coexists with black-powder firearms and swashbuckling swordsmen. Elsewhere, there have been games from Germany, France, Italy, Japan, Poland and Norway; *In Nomine* (1991), a French horror game now translated into English, is probably the best-known.

Today's fantasy RPG scene is polarized into several main genres: SWORD AND SORCERY, typified by *AD&D* and *Runequest*; hybrid SCIENCE FICTION/fantasy, such as SHADOWRUN (FASA 1992) and *Magitech* (TSR 1993); contemporary HORROR, as in *Call of Cthulhu*, *Vampire* and *Dark Conspiracy* (GDW 1992); Arthurian, as in *Pendragon* (Chaosium 1985) and *GURPS Camelot* (1993); and swashbuckling/STEAMPUNK fantasy, as in *Lace and Steel* and *Castle Falkenstein* (R. Talsorian 1994). It seems likely that the future will see increasing specialization – as well as the extinction of some unpopular systems, since the total market for RPGs is slowly shrinking.

Numerous magazines cater to the hobby, but most are short-lived; the most regular are TSR's *Dragon* and *Dungeon*, Steve Jackson Games's *Pyramid* and Games Workshop's *White Dwarf*, although the latter now deals only with Games Workshop's own products. *Dragon* publishes fiction in most issues. *Valkyrie* (Partizan Press) and *Arcane* (Future Publishing) are independent UK RPG magazines; *Valkyrie* also publishes fiction. *Interactive Fantasy* (formerly *Inter*Action*) is the only scholarly journal covering the field and related areas, most notably storytelling and freeform games (see below), and the recreational, therapeutic, industrial and educational uses of these systems. Regular commercially organized RPG conventions include Origins and Gencon in the USA and Euro-Gencon in the UK. There are also many fanzines and fan-organized conventions.

While most RPGs are played as table-top games, there are also live games, in which players act out adventures using costumes and padded weapons. Another variant is the freeform game: players are assigned roles in a plot (or several interconnected plots), given several goals and secrets, and left to interact among themselves. The referee acts as adjudicator rather than controlling all aspects of the game. Usually combat is banned. These games have been run with hundreds of players and multiple referees, although groups of 10-30 players are more usual. A nonfantasy commercial equivalent is the "murder weekend" run at some hotels.

Gamebooks The RPG hobby has had many spinoffs, one of them being role-playing gamebooks, usually aimed at children. At their simplest – for example, the *D&D* **Endless Quest** series (**1983-5**) – they set the scene in an opening chapter, then present various options that each lead to a different numbered paragraph. A branching tree of decisions takes the reader through to a conclusion, which may be victory or death. *The Warlock of Firetop Mountain* (**1982**) by (the UK) Steve Jackson and Ian LIVINGSTONE and its sequels in the **Fighting Fantasy** series use this mechanism, but allow players to map the adventure and retrace their steps. Dice are used for combat and other activities, with the results determining the paragraph selected. *Sorcery* (**1984**), by Jackson alone, was aimed more at adults, and added a magic system which required a separate "spell book"; it flopped. Probably the most successful series, and the closest to an RPG, is the ongoing **Lone Wolf** series (**1984**-current) by Joe DEVER, which uses most of these mechanisms plus record-keeping, allowing weapons and magical items to be used from one book to another.

Other forms Storytelling games comprise a separate but related genre. One early example was *Tales of the Arabian Nights* (West End Games 1985), a boardgame in which each move involved a reference to a paragraph in a booklet describing various situations, such as abduction by a GENIE. Today's games are mostly played with cards, marked with words or phrases which must be used in a continuing story. Sometimes the object of the game is simply to produce a coherent and enjoyable narrative; for example, *Into the Dragon's Cave* (The Magellanica Company 1993) and *Once Upon a Time* (Atlas Games 1993) have no scoring mechanism whatever, although the latter allows the use of multiple cards and is won by the player finishing first. Others are played for points based on the type of card used; for instance, an action sequence might be worth more than a descriptive passage. A typical example is *Dark Cults* (Kenneth Rahman 1983), a horror game for two players.

Fantasy wargames are less popular than RPGs. The most widely played is *Warhammer Fantasy Battles* (Games Workshop 1983), based on the earlier *Reaper* (Table Top Games early 1970s), but its dominance is largely the result of Games Workshop's immense marketing strength. Surprisingly, TSR's *AD&D Battle System* (1985; rev 1989) was never a success. The field is otherwise the province of smaller publishers, including many semi-professional and amateur presses. Typical are *Hordes of the Things* (WRG 1991), a generic fantasy battle system, and *Tusk* (Irregular Miniatures 1994), a game pitting cavemen against mammoths and dinosaurs. *Ragnarok*, published bimonthly by the Society of Science Fiction and Fantasy Wargamers, is the leading fanzine.

Play-by-mail games largely fall into one or other of the classes described above. Players fill in "turn sheets" describing the moves they wish to make, and return them to a referee. It is also possible to communicate with other players, via the referee or directly, and make alliances. Many games are run on an amateur basis, often as APAs, news and fiction, but there is also a thriving commercial sector. Typical commercial games include: *Saturnalia*, a fantasy RPG; an official *Middle Earth* wargame (Allsorts Games), based on the *Iron Crown* RPG; and *Delenda Est Carthago* (Inferno Games), in which players control noble families in a medieval ALTERNATE WORLD, with play involving intrigue and diplomacy. E-mail games of these types are also common.

There are many other fantasy board and card games. *Talisman* (Games Workshop 1983) was a traditional boardgame, with movement around a fixed course; players used special cards to speed their own movement or impede others'. In *Lost Worlds* (Mayfair 1983) each player had an illustrated book showing a warrior or monster in different combat positions; players selected a move, rolled dice to determine the outcome, and turned to a numbered page for an illustration of the result. *Heroquest* (Milton Bradley 1989, designed by Games Workshop) uses RPG-like mechanics, with figures representing characters, monsters and obstacles set out on a map board. *Credo* (Chaosium 1994) is a card game about the evolution of religions; players try to ensure that their own doctrine succeeds while those of the other players fail. These are a few examples among hundreds.

Collectable card games are a recent innovation. Normal card games are played with packs of fixed composition, but in collectable games each player builds a pack from a vast range of cards sold by the manufacturer; packs must obey the rules of the game, but otherwise players are free to accumulate cards indefinitely, and often own hundreds. The rarest cards are often the most powerful. Cards are sold in opaque packets; players have no way of knowing if they are buying rarities or common cards. Rich players are thus most likely to win; they can afford to buy more packs until they have everything they need. Naturally there is a thriving unofficial market in individual rare cards, usually at extortionate prices; recently (1995) one of the rarest was sold for £150 (c$225). Since the equipment needed to make the cards is very expensive, the manufacturers must ensure that the games continue to sell, so there is always a steady flow of new card releases, special editions, etc. It is thus almost impossible for a single player to own every possible card, although many try. The first of these games was *Magic: The Gathering* (Wizards of the Coast 1993); other examples are *Spellfire* (TSR 1994), *Jyhad* (WOTC 1994) and *Illuminati: New World Order* (Steve Jackson 1994).

A variant is *Dragon Dice* (TSR 1995), a collectable dice game in which players build up "armies" made of dice marked with special colours and symbols.

Computer games The period of development of the games described above, from the late 1960s onward, is also the period in which computers began to enter widespread use. Since there was considerable overlap between the users of computers, sf/fantasy fandom and fantasy games players, it was natural that computerized fantasy games would eventually emerge. The first computer games were simple logic puzzles and sf-related games, such as *Spacewar* (1968), which could be played via the teletype terminals then used. *Adventure* (vt *Colossal Cave* c1972) by Crowther and Wood was the first fantasy game; players used two-word commands to navigate around a complex of caves and solve various logic puzzles. There was little progress until home computers became available; *Adventure* was soon converted to run on these machines, and was quickly followed by more games of the same general type, usually improving on *Adventure*'s parsing (understanding of instructions) and presentation. An early one was *Zork* (Infocom 1982).

Naturally improvements in computers led to improvements in games. Adventures added graphics, sound and eventually speech, until now text-only games are virtually extinct; they survive mainly in games played via computer networks with a large number of participants (a typical example is *MUD* [Multi-User Dungeon; Exeter University 1982]). Today's adventures often come on CD-ROM, with several hundred megabytes of sound and graphics. A recent example is *Discworld* (Psygnosis 1995), based on Terry PRATCHETT's novels. An on-screen character (Rincewind) interacts with various objects and personalities as directed by the player through mouse movements. All characters speak, and there are sound effects in every scene.

Naturally there are many other types of fantasy computer game, some derived from arcade systems, others from tabletop games and RPGs. A common variety is the platform game, in which the player is represented by an on-screen character which must jump, climb and fight. A classic in this genre is *Prince of Persia* (Brøderbund 1990), an ARABIAN-FANTASY adventure in which the hero has an hour to escape from a dungeon and save his princess from an evil vizier. Several recent games use a first-person viewpoint and perspective graphics to present a world which must be explored, with monsters to kill and treasure and weapons as rewards. By far the best known is *Doom* (ID Software 1993), a hybrid sf/fantasy in which a Martian colony is mysteriously transported to HELL, and the viewpoint character must destroy hundreds of DEMONS to escape. (It was designed by Sandy Peterson, a noted RPG writer and principal author of *Call of Cthulhu*.) *Doom* has several imitators, and its "engine" (programming) has been adapted to four other games: *Doom II* (ID 1994), *Heretic* (Raven 1994) and *Hexen* (Raven 1995), both pure fantasy games, and *Dark Forces* (LucasArts 1995), an sf variant based on the **Star Wars** movies. All but the last let up to four players participate via a network, so they can cooperate or fight each other. *Hexen* additionally allows players to choose different areas of specialization – e.g., magician, fighter – as in many RPGs. *Doom* and *Heretic* are unusual in allowing players to design their own worlds, which can easily be added to the game. There are hundreds in the public domain, and numerous design programs.

Naturally there are also computer fantasy wargames and RPGs. The former have never been very popular, mainly because it is difficult to show all the pieces on screen. Nonwar examples include a computer version of *Heroquest* and several games based on *AD&D*; *Pool of Radiance* (SSI 1989) is typical.

The future of fantasy computer games undoubtedly involves more realistic high-resolution graphics, improved sound and eventually some form of virtual reality. Better artificial intelligence should eventually be added to computer adventures and RPGs, giving on-screen characters a more lifelike personality. *Doom* has shown that it is possible to write networked games with convincing real-time graphics; while *Doom* and its successors do not allow players to interact, except by fighting monsters or each other, the technology could easily be extended to increase the capabilities – for example, objects might be passed from one character to another. Eventually this should lead to *MUD*-style networked RPGs with first-person viewpoint graphics and sound. Unfortunately the current limitations of computer hardware, especially CD-ROMS, are beginning to cause problems; 550 megabytes once seemed an inexhaustible amount of disk space, but already some games need more, and many computers are lagging badly behind the memory and speed requirements of the latest games. This problem has occurred throughout the history of computing, and there is little chance that it will ever change.

[MLR]

Further reading: *SFE* includes an equivalent summary of sf-related games. *Inter*Action #1* (1994) includes historical and "state of the art" pieces by several leading authors.

GAMEWORLDS Several terms in this encyclopedia are used to describe fantasy worlds which operate according to rules. A gameworld is a fantasy world created for an actual GAME. It may be represented literally by a board, or more symbolically through the models and other paraphernalia which help articulate role-playing games. Given the nature of any game, gameworlds are relatively simple TEMPLATES which do not permit breaking of the rules.

Other rule-operated fantasy worlds include: the WONDERLAND (worlds built on CHESS matches are best thought of as wonderland venues); the FANTASYLAND (home of GENRE FANTASY, and designed not to crack under the stress of repeated storylines); and whatever world a GODGAME might be set in. [JC]

GANDALF AWARD An award created by Lin CARTER to honour lifetime fantasy achievement, and given within the framework of the Hugo Awards (◊ *SFE*) 1974-81. The award was presented the year following the year designated. Winners were:

1973: J.R.R. TOLKIEN (posthumously)
1974: Fritz LEIBER
1975: L. Sprague DE CAMP
1976: Andre NORTON
1977: Poul ANDERSON
1978: Ursula K. LE GUIN
1979: Ray BRADBURY
1980: C.L. MOORE. [JC]

GANPAT Pseudonym of Anglo-Indian writer Martin Louis Alan Gompertz (1886-1951), most of whose tales of interest are lost-race novels (◊ LOST LANDS), normally set somewhere in Asia. In his first novel, *Stella Nash* (**1924**), the protagonists seek treasure from a lost city in the jungles of India. Like many lost-race novelists, G slides with some lack of discrimination from sf to fantasy. For instance, his **Sakaland** sequence – *Harilek: A Tale of Modern Central Asia* (**1923**) and *Wrexham's Romance: Being a Continuation of "Harilek"* (**1935**), assembled as *Adventures in Sakaland* (omni **1978** US) – offers a cod racial explanation for the persistence of a Classic Greek culture in the middle of the Gobi Desert: Indo-Europeans who had been visited by Greeks in the 5th century maintain an uneasy hegemony over the "native" clan of Mongol-like SHAMANS. An element of fantasy enters, back-door, through the existence of TALENTS like telepathy. The second volume is dominated by love matters.

Other lost-race novels include *Snow Rubies* (**1925**), *The Voice of Dashin: A Romance of Wild Mountains* (**1926**), *Mirror of Dreams: A Tale of Oriental Mystery* (**1928**), *The Speakers in Silence* (**1929**) and *Fairy Silver: A Traveller's Tale* (**1932**). In *The War Breakers* (**1939**) a WORLD WAR II is conjured up by occult forces. [JC]

Other works: *High Snow* (**1927**); *Dainra* (**1929**); *The Three R's* (**1930**), an sf novel in which an international conspiracy tries to build an atomic bomb; *Walls Have Eyes* (**1930**); *The Second Tigress* (**1933**); *Seven Times Proven* (**1934**); *The War Breakers* (**1939**).

GARCÍA MÁRQUEZ, GABRIEL (1928-) Colombian writer, generally considered the premier Latin American literary figure of the last half of the 20th century. He is the central contemporary creator of texts in the mythopoeic mode known loosely as MAGIC REALISM, and won the Nobel Literature Prize in 1982. GGM began publishing fiction with "La tercera resignación" ["The Third Resignation"] for *El Espectador* (Bogotá) in 1947, and has remained prolific for half a century. Like many later tales, this first effort – a POSTHUMOUS FANTASY set in an exorbitantly conceived afterworld – translates "classic" 20th-century influences like Franz KAFKA into a densely tropical *mise en scène*, a world where the relationship between REALITY and MAGIC is no longer simply a question of clear PERCEPTION, but where the two intermingle incessantly and arbitrarily. It is this sense of the epistemological *permeability* of the nature of REALITY that, above all, makes magic realism in general, and GGM's work in particular, difficult to assess as fantasy. If FANTASY narratives can (in part) be defined as self-coherent STORIES that readers know are fantasy, while at the same time giving allegiance to the telling, then GGM is a writer of pure fantasy only at rare moments.

Further GGM stories – some realistic, some fantastic – are assembled in *Los funerales de la Mamá Grande* ["Big Mama's Funeral"] (coll **1962**), which appears with *El coronel no tiene quien le escriba* (**1961**) as *No One Writes to the Colonel and Other Stories* (coll trans J.S. Bernstein **1968** US); in *Leaf Storm and Other Stories* (coll trans Gregory Rabassa **1972** US), the title story being a translation of GGM's first novel, *La hojarasca* (**1955**), which has supernatural elements, and adding four stories from *La increíble y triste historia de la cándida Eréndira y de abuela desalmada* (coll **1972**); the three remaining stories from this coll appear in *Innocent Eréndira and Other Stories* (coll trans Gregory Rabassa **1978** US); all three volumes of translations are reassembled as *Collected Stories* (omni **1984** US). A further volumes is *Doce cuentos peregrinos* (coll **1992**; trans Edith Grossman as *Strange Pilgrims: Twelve Stories* **1993** US). Of these stories, several are of particular fantasy interest. "Isabel viendo llover en Macondo" (trans as "Monologue of Isabel Watching it Rain in Macondo" in *Leaf Storm*) is another posthumous fantasy, and represents GGM's first use of the town of Macondo, a fantasticated MYTH-irradiated POLDER which appears again and again in his most famous works. "Big Mama's Funeral" depicts a congested, particoloured REVEL. In "El mar de tiempo perdido" (trans as "The Sea of Lost Time"), from *Innocent Eréndira*, two men descend undersea to a transfigured world, where the live and the dead commingle. In "El ahogado más hermosos del mundo" (trans as "The Handsomest Drowned Man in the World"), from the same collection, a dead Gulliver-like GIANT is washed up on a beach, where he occasions an efflorescence of mythopoeisis.

But it is with his novels that GGM comes most fully into his own. The greatest of them tend to work through structures that commingle images of the eternal return of the CYCLE of the SEASONS and linear narratives through which timeless worlds are exposed to the secular desiccations of history (◊◊ THINNING). There is a sense in which *Cien anos de soledad* (**1967**; trans Gregory Rabassa as *One Hundred Years of Solitude* **1970** US/UK) transfigures the polder of Macondo almost literally into a CAULDRON OF STORY. Within its compass, the family chronicle featuring the founders and residents of Macondo becomes a CREATION MYTH of Colombia. The passage from prehistory to linear TIME is conveyed through a description of a plague of AMNESIA, by virtue of which the inhabitants of Macondo must learn to recognize who they are by memorizing their Story. That Story becomes secular, and a notion of the increasing *congestion* of the world is conveyed through GMM's increasing use of images involving LABYRINTHS and MIRRORS. This sense of the recursive complexity of the world is strengthened by the revelation – as the vast spiralling cycle of Story reaches its climax – that a Sanskrit manuscript now being deciphered is a fictional BOOK whose contents are in fact identical to the real book it contains. *One Hundred Years of Solitude* is, in other words, all STORY.

Although varied in texture and substance, later novels have a similar redemptive take on the story and history of the LAND; they include *El otono del patriarca* (**1975**; trans Gregory Rabassa as *The Autumn of the Patriarch* **1976** US), *El amor en los tiempos del colera* (**1985**; trans Edith Grossman as *Love in the Time of Cholera* **1988** US) and *El general en Su Laberinto* (**1989**; trans Edith Grossman as *The General in the Labyrinth* **1990** US). The first again treats a funeral in terms of revel: it contains a long and comically savage portrayal of the eponymous dictator, probably of a Caribbean ISLAND, whose stultifying rule and egregious life are conveyed through TIMESLIPS and satirical absurdities. Various cycles lead always back to ruin and death (DEATH presides over the Patriarch's final demise, when he is aged somewhere between 107 and 232 years). GMM has been for many years a writer dangerous to politicians (many of his works have been banned) and vital to the imagination of the world.

[JC]

GARCIA Y ROBERTSON, R(ODRIGO) (1949-) US historian and writer who began publishing fantasy with "The Flying Mountain" for *Amazing Stories* in 1987, an ALTERNATE-WORLDS tale set in a LAND-OF-FABLE medieval Turkey threatened by Mongols in balloons. Most of his work similarly explores the effects of threats upon recognizable historical venues, and can be read as SUPERNATURAL FICTION: WITCHES are common in the WATER MARGINS of his stories, their existence being perceived by normal people as both inimical and unnatural. His first novel, *The Spiral Dance* (**1991**), is set in Elizabethan England, and initially follows the attempts of Princess Anne to regain her throne through witchcraft, aided by a WEREWOLF from Scotland.

[JC]

GARD, JOYCE Pseudonym of UK teacher, potter and writer Joyce Reeves (1911-), sister of James REEVES. She began publishing her YA novels with *Woorroo* (**1961**), in which a child is taught by the Bat People the art of flying. She has set several stories – including *The Dragon of the Hill* (**1963**), *Talargain, the Seal's Whelp* (**1964**) and *The*

Mermaid's Daughter (**1969**) – in LAND-OF-FABLE versions of Roman or Dark-Age Britain. In the second of these a young orphan boy is brought up by SELKIES and TIMESLIPS into modern England to speak of this and of King Aldfrith's (i.e., King Alfred's) attempts to unite the land. The protagonist of the last of them – perhaps his finest tale – is a covert agent of the GODDESS in her Great Mother aspect, and attempts to convey to the Romans at Caerleon something of the fresh-minted sacredness of life in the Scilly Isles. [JC]

Other works: *The Hagwaste Donkeys* (**1976**).

As translator: *Journey to the Centre of the Earth* (trans **1961**) by Jules Verne (1828-1905).

GARDEN There can be no gardens in this world, or in any OTHERWORLD, unless there is something beyond their boundaries which is not a garden. Gardens are enclosures; they are shaped to express a meaning, a vision of the ideal world, a dream of escape or seclusion. In fantasy, they may be physically distinct from the surrounding world, or magically distinct, as POLDERS are; they may be unreachable from the surrounding world (◊ SECRET GARDEN), or reachable only if those who search for them trace a LABYRINTH into their heart; they may transform those who enter them (◊ INTO THE WOODS); they may constitute a PORTAL into an otherworld.

At one time it was thought UK fantasy homed in on the garden while US fantasy moved outwards into the wilderness or the world. There is an element of historical truth in the thought; but English-language fantasy writers, on both sides of the Atlantic, have tended to become less easily distinguishable in the years since J.R.R. TOLKIEN.

In Biblical typology, the garden (which represents EDEN) stands at the opposite pole to the CITY (which is JERUSALEM); but does so as part of a complex theological drama in which the first becomes last, Eden becomes Jerusalem, time past becomes time future. An UNDERLIER resonance of this drama can be assumed to operate, however obscurely, whenever gardens are highlighted in the literatures of the West. [JC]

GARDNER, CRAIG SHAW (1949-) US writer who began writing stories of fantasy interest with "A Malady of Magicks" for *Fantastic Magazine* in 1978, a tale which inspired his first novel, *A Malady of Magicks* (fixup **1986**), which with its two sequels – *A Multitude of Monsters* (**1986**) and *A Night in the Netherhells* (**1987**) – makes up the **Ebenezum** sequence of comic GENRE FANTASIES set in a crudely depicted FANTASYLAND. The **Ballad of Wuntvor** sequence – *A Difficulty with Dwarves* (**1987**), *An Excess of Enchantments* (**1988**) and *A Disagreement with Death* (**1989**), assembled as *The Wanderings of Wuntvor* (omni **1989**) – continues the adventures, which are filled with slapstick and bad puns.

More interestingly, the **Cineverse** sequence – *Slaves of the Volcano God* (**1989** UK), *Bride of the Slime Monster* (**1990**) and *Revenge of the Fluffy Bunnies* (**1990**), assembled as *The Cineverse Cycle* (omni **1991**; vt *The Cineverse Cycle Omnibus* 1992 UK) – features a young man from Earth who discovers that various movie genres (◊ STORY; WONDERLAND) reflect the natures of the various ALTERNATE WORLDS which make up the Cineverse. With his Captain Crusader Decoder Ring he traverses these worlds, fighting off the evil plans of Dr Dread, almost succumbing to cliffhanger perils, etc. The **Arabian Nights** sequence – *The Other Sinbad* (**1991** UK), *A Bad Day for Ali Baba* (**1991** UK) and *Scheherazade's Night Out* (**1992** UK; vt *The Last Arabian*

Night 1993 US) – is a PARODY of the *Arabian Nights* (◊ ARABIAN FANTASY). In the first volume, a GENIE confuses the real Sinbad and a street porter with the same name, precipitating various adventures taken from the original; in the second, Ali Baba and Aladdin become involved, along with SCHEHERAZADE; with pleasing intricacy, the first two volumes turn out to be tales told by Sinbad and Ali Baba in a storytelling contest, which Scheherazade tops by herself narrating the actual third volume, in which she describes herself narrating it. All ends well.

About CSG's fourth sequence, the nonhumorous **Dragon Circle** series – *Raven Walking* (**1994** UK; vt *Dragon Sleeping* 1994 US) and *Dragon Waking* (**1995** UK), with further volumes projected – opinions remain mixed. There is a welcome straightforwardness about the telling and the setting of this tale, during the course of which a number of humans from Earth begin to work out their family traumas as they make good in FANTASYLAND; but the plot is unwieldy and the cast overlarge. From the beginning of his career, CSG has also published NOVELIZATIONS, including: *The Lost Boys* * (**1987**), which novelizes *The* LOST BOYS (*1987*); *Wishbringer* * (**1988**), a game tie; two BATMAN ties: *Batman* * (**1989**) and *Batman Returns* * (**1992**) (◊ BATMAN MOVIES); and two **Back to the Future** ties, *Back to the Future II* * (**1989**) and *Back to the Future III* * (**1990**). *The Batman Murders* * (**1990**), is based on the COMIC. [JC]

GARDNER, JOHN (CHAMPLIN Jr) (1933-1982) US academic and writer. The relationship between his work and GENRE FANTASY – or any other genre – is conspicuously distant, given the indifference to genre made evident in *On Moral Fiction* (**1978**), a tract against the "nihilism" of Postmodern literature: for JG, writing in genre formats might well have seemed tantamount to surrendering to inauthenticity. He is best-known for *Grendel* (**1971**), his first fantasy novel, which constitutes a kind of TWICE-TOLD version of BEOWULF from the point of view of the MONSTER, and although JG's attempts to re-create in contemporary prose the rhythms of the *Beowulf* poet are stylistically congested – he has, for instance, nothing of the control over diction and rhythm of a genuinely fine re-creator like Kenneth MORRIS – the monster's embittered (and prophetic) assessment of the meaningless of his existence, and of the fatal pretentiousness underlying the human rage for order (◊ THINNING), are powerfully conveyed. *Jason and Medea* (**1973**) also begins to present its material in a twice-told manner, down to the verse format in which it is narrated; but times and venues shift, the tale turns into a NEW YORK legend for a space, and continues to the END OF THE WORLD.

Most of the stories assembled in *The King's Indian: Stories and Tales* (coll **1974**) go together to make up a set of FABLES modelled on recognizably US ways of rendering STORY. The title novella is told in a manner evocative of the TALL TALE, while invoking Edgar Allan POE's *The Narrative of Arthur Gordon Pym* (**1838**) and Herman MELVILLE's *Moby-Dick* (**1851**) in a complexly self-referential artifact of storytelling, which takes its unreliable narrator on a ship into southern waters in search of his DOUBLE. *In the Suicide Mountains* (**1977**), which takes elements from various Russian FOLKTALES, features the coming together of three suicidal protagonists who LEARN BETTER through observing and listening to each other; there is MAGIC involved, and an abbot plays the role of LIMINAL BEING.

Some of JG's later novels also incorporate, loosely, fantasy elements. The interior QUEST saga told by the eponymous GIANT narrator of *Freddy's Book* (**1980**) seems ultimately delusional, if the text is meant literally as fantasy; but the book is an engrossing examination of its protagonist's deep need for a fantasy structure to explain – and to in effect allegorize – his life. *Mickelsson's Ghosts* (**1982**) plays some of the same games in terms of SUPERNATURAL FICTION. "Shadows" – an uncompleted novel assembled in *Stillness and Shadows* (coll **1986**) – features a detective haunted by a presence his AMNESIA prohibits him from identifying.

JG was a scholar of medieval literature, and much of his work in this field – typical examples being *The Complete Works of the Gawain-Poet in a Modern English Version* (trans **1965**) and *The Gawain-Poet* (**1967**), a study – are of importance.

JG is not to be confused with the UK thriller writer John Gardner (1926-). [JC]
Other works: Three volumes of tales for younger children, being *Dragon, Dragon, and Other Timeless Tales* (coll **1975**), *Gudgekin the Thistle Girl and Other Tales* (coll **1976**) and *The King of the Hummingbirds and Other Tales* (coll **1977**); three librettos for OPERAS by Joseph Baber, being *Rumpelstiltskin* (**1978** chap), *Frankenstein* (**1979** chap) and *William Wilson* (**1979** chap), all assembled as *Three Libretti* (omni **1979**); *Vlemk the Box-Painter* (**1980**); *The Art of Living and Other Stories* (coll **1981**).
Other works (nonfiction): *The Forms of Fiction* (anth **1962**) with Lennis Dunlap; *Le Morte Arthure* (**1967**); *The Alliterative Morte Arthure, The Owl and the Nightingale, and Five Other Middle English Poems, in a Modernized Version* (trans **1971**); *The Construction of the Wakefield Cycle* (**1975**); *The Poetry of Chaucer* (**1977**); *The Life & Times of Chaucer* (**1977**); *Gilgamesh* (trans **1984**) with John Maier.

GARFIELD, LEON (1921-1966) UK writer, mainly for children. Most of his numerous books are set in history, with only the occasional rumour of fantasy or SUPERNATURAL FICTION. Some stories in *Mr Corbett's Ghost and Other Stories* (coll **1969**) – the title story being republished as *Mr Corbett's Ghost* (**1982** chap) – are of fantasy interest (◊ GHOST STORIES FOR CHILDREN), and *The Ghost Downstairs* (**1972**) describes a PACT WITH THE DEVIL. Of more interest is *The God Beneath the Sea* (**1970**) with Edward Blishen (1920-), illustrated by Charles KEEPING, in which the baby GOD Hephaestus – hurled into the sea from OLYMPUS because of his ugliness – is rescued by the goddesses Thetis and Eurynome, who tell him the long STORY CYCLE of the Greek PANTHEON (◊ MYTHS) into which he has been born; the novel closes with his second fall from Olympus, which he also survives, ultimately in the company of HERMES. *The Golden Shadow* (coll **1973**), also with Blishen, again presents Greek myths, this time in a more usual TWICE-TOLD fashion. LG also produced a "completion" of Charles DICKENS's unfinished *The Mystery of Edwin Drood* (**1980**). [JC]
Other works: *The Pleasure Garden* (**1976**) (◊ REVEL); *Empty Sleeve* (**1988**).

GARNER, ALAN (1934-) UK writer. Like William MAYNE he has a reputation, not wholly unearned, for difficulty, for both address the full range of human concerns, and any simplicity in their strategies for doing so is deceptive. AG published some adolescent stories – beginning with "The Obelisk" for *Ulula* in 1952 – but began his career proper with the **Alderley Edge** sequence: *The Weirdstone of Brisingamen: A Tale of Alderley* (**1960**; vt *The Weirdstone: A Tale of Alderley* 1961 US: rev 1963 UK) and *The Moon of*

Gomrath (**1963**). These two novels, along with his third, *Elidor* (**1965**), are his most explicit fantasies; and his whole career can be viewed as a progression (or departure) from fantasy into works which treat the "mundane" world with mythic intensity, burrowing more and more deeply into the numinousness of that world. The CROSSHATCH structure of his fantasy work becomes, in his later tales, a pattern intricately tied to things of the Earth. As an author of original fiction, AG has fallen silent since about 1980, almost as though he himself had become so bound into the world that no new story could escape. At the same time that world does, clearly, for AG, constitute the kind of sacred drama propounded by Mircea ELIADE, who has influenced his work; and his silence may not be a result of philosophy.

But the **Alderley Edge** books are unfettered by any premonitions of this fate. In the first, young Colin and Susan arrive for the summer at Alderley Edge, a region of Cheshire where AG has lived much of his life and where much of his subsequent nonfantasy work has been set. Inspired by a local legend, they search for and find the "Iron Gates" to an UNDERWORLD region where 140 KNIGHTS (◊ SLEEPER UNDER THE HILL) must be prevented from awakening prematurely. Echoes of NORDIC FANTASY, Welsh-tinged CELTIC FANTASY and the Arthurian cycle (◊ ARTHUR) in particular are omnipresent. There is also a sense – which comes to the fore in later novels – that TIME is a CYCLE, that all times either co-exist or that an eternal return confronts epochs and DOUBLES, UNDERLIERS and AVATARS with one another, constantly.

In the first novel, Colin and Susan, under the management of the MERLIN-like WIZARD whose task it has been to act as GUARDIAN over the sleepers, lose the eponymous TALISMAN to Grimnir, the wizard's dark double who, unable to master the talisman, takes it to a WITCH – who also fails; from the wreckage of her spells the children recover the jewel, and the sleepers continue to sleep. In the second they are taxed far more severely; each undergoes NIGHT JOURNEYS that double one another, and their mutual RITE OF PASSAGE into adulthood carries them also into a condition of estrangement with the world. First Susan is possessed (◊ POSSESSION) by a Brollachan, an evil spirit associated with sexual matters, and then Colin, in the process of exorcising her, permits the WILD HUNT access to the world. By the end – as though these invaders had been in fact MYTHAGOES representing their needs and natures – they cannot return to normality.

The eponymous SECONDARY WORLD featured in AG's next novel, *Elidor*, is depicted with a clarity that deceives, and the numerous analogies with the Arthurian cycle are ultimately ironic. Four siblings wander into an abandoned church in a slum area of Manchester, drawn there by a TRICKSTER fiddler, a LIMINAL BEING whose initial function is to translate them across the THRESHOLD into Elidor, and who is soon revealed to be Malebron, King of Elidor, a lame FISHER KING ruling what proves to be a WASTE LAND. The three older children are caught in BONDAGE, like flies in amber, in a TIME trap: they have been in Elidor for a huge span, and can only be freed by Roland (◊ CHILDE), the youngest of the siblings, who must enter into a QUEST for the DARK TOWER (a burial mound) which contains both them and the four treasures seemingly necessary to bring a HEALING to the LAND. The four children escape back to Manchester with the four treasures (a GRAIL, a spear, a SWORD and a stone), which turn or appear to turn into mundane objects. But Elidor is not yet secure. Findhorn the UNICORN must sing:

the scenes in which he appears in this world, succumbs to Roland's virgin sister, is mortally wounded by the enemies of Elidor, and sings his death, are deeply troubling. In the end, Elidor is saved; but the children have been used ruthlessly by Maleborn. *Elidor* is an astonishing book; its effect is both poisonous and healing.

The Owl Service (**1967**), AG's last novel which can be unequivocally deemed a fantasy, is a central text of the field. The action is contemporary, and takes place mainly in this world, in the Welsh valley where the story of Llew Llaw Gyffes and Blodeuwedd and Gronw Pebyr, from the fourth branch of the MABINOGION, is reckoned to have taken place. In that story, because his son Llew Llaw cannot have a human wife, the WIZARD Gwydion has created Blodeuwedd for him out of meadowsweet, broom and the flowers of the oak. But she has no love for the husband inflicted upon her, and becomes enamoured of Gronw, who kills Llew Llaw, who suffers METAMORPHOSIS into a bird. His father then arranges for him to kill Gronw in the same manner that Gronw killed him, and punishes Blodeuwedd by changing her into an owl. The violent energies created by this tragedy have warped the valley into a WASTE LAND, and periodically require discharging. In the contemporary story these energies enter three children at the edge of young adulthood. Alison and her stepbrother Peter are English; Gwyn is Welsh. The children discover the eponymous plates, which bear the design of a dismembered owl made out of flowers. A portrait of a woman made out of flowers (◊ FOLIATE HEAD; PERCEPTION) appears on an inner wall of the Welsh house where the action is centred. Both plates and painting represent earlier generations' attempt to "lock" Blodeuwedd into BONDAGE, so that she cannot continue to rage as an owl through the community in an eternal return of her tragedy. In the end, the haunting is purged through forgiveness, and she is allowed to be flowers. AG adapted the tale for tv in 1969.

In *Red Shift* (**1973**), AG moved away from anything like straightforward fantasy, though the three interlocked tales making up the book – each carrying similar and similarly named characters through similar anguish in the same Alderley region in three different epochs (Roman Britain, the Interregnum, and now) – impinge upon one another in ways which can readily be understood as fantastic. The three male protagonists are, in a sense, DOUBLES of one another; and Macey, the Roman version of Cromwellian Thomas and contemporary Tom, suffers visions of the future. Plot elements from the three periods constantly intersplice in a fashion not remotely amenable to a mimetic reading; they can be understood as a rendering of J.W. Dunne's vision of TIME as an accessible network comprising past, present, and future. AG's tv version of the novel was shown in 1978.

The **Stone Quartet** – *The Stone Book* (**1976** chap), *Tom Fobble's Day* (**1977** chap), *Granny Reardon* (**1977** chap) and *The Aimer Gate* (**1978** chap), all four assembled as *The Stone Quartet* (omni **1983**), and all illustrated by Michael FOREMAN – is not fantasy, though its concerns and sense of place make it of interest to readers of the earlier works. Since its composition, AG has restricted himself to a series of collections of TWICE-TOLD tales extracted from various FOLKLORES and other sources. These collections are difficult to describe – they could as easily be designated anthologies – because they variously contain straight reprints of other material, modified reprints, and almost totally new recraftings. Here listed as collections, they include: *The Hamish Hamilton Book of Goblins* (coll **1972**; vt *A Cavalcade of Goblins* 1969 US; vt *A Book of Goblins* 1972 UK); *The Guizer: A Book of Fools* (coll **1975**); *The Girl of the Golden Gate* (**1979** chap), *The Golden Brothers* (**1979** chap), *The Princess and the Golden Mane* (**1979** chap) and *The Three Golden Heads of the Well* (**1979** chap), all four assembled as *Alan Garner's Fairy Tales of Gold* (omni **1980**; rev vt *Fairytales of Gold* 1989), all illustrated by Foreman; *The Lad of the Gad* (coll **1980**); *Alan Garner's Book of British Fairy Tales* (coll **1984**); *A Bag of Moonshine* (coll **1986**).

There is a feeling that AG has not so much departed the field of fantasy as ended a quest in search of his own identity and his own place. In 1996 there are no rumours of his return. [JC]

Other works: *Holly from the Bongs: A Nativity Play* (**1966** chap) and *Holly from the Bongs: A Nativity Opera Text* (**1974** chap), differing texts; *The Old Man of Mow* (graph **1967**) photo-illus Roger Hill; *Potter Thompson* (**1975** chap), libretto; *The Breadhorse* (graph **1975**) illus Albin Trowski, for younger children.

Further reading: *A Fine Anger: A Critical Introduction* (**1981**) by Neil Philip.

GARNETT, CONSTANCE (1862-1946) Highly regarded UK translator from the Russian, mother of David GARNETT and wife of Edward GARNETT. [JC]

GARNETT, DAVID (1892-1981) UK writer, member of a literary family whose members include his grandfather, Richard GARNETT, and his parents, Constance GARNETT and Edward GARNETT. He is best-known for his first novel under his own name, *Lady into Fox* (**1922** chap), which depicts the TRANSFORMATION of a newly wed Victorian into a vixen. Her BONDAGE, which can end only with her violent death, unavoidably generates a sense of powerful WRONG-NESS. This wrongness can be understood as an ALLEGORY of the position of women in a patriarchal world, for whom indentured marriage is a kind of immurement. After giving birth to a litter of pups, Silvia is eventually killed by hounds: as hounds are to foxes – one is bound to conclude, despite the distancing "wit"-laden 18th-century tone of the tale – so patriarchal society is to women.

Lady into Fox, famously parodied by Christopher WARD as *Gentleman into Goose . . .* (**1924** chap), remains a central 20th-century text; but it is unfair to DG to think of him as a one-book author. He published about 35 further books, many of them fiction. *A Man in the Zoo* (**1924**) and *The Grasshoppers Come* (**1931**), while fantastical, are not actually fantasies. Some of DG's late novels, however, are of direct genre interest. *Two by Two: A Story of Survival* (**1963**) tells of the FLOOD and of the part played by Noah in terms distinctly unfriendly to the Old Testament GOD who drowns his people; Noah himself is almost certainly based on T.H. WHITE; a later volume of letters between DG and White, *The White/Garnett Letters* (**1968**) ed DG, illuminates both figures. *Ulterior Motives* (**1966**) flirts in a gingerly manner with matter transmission, treating the phenomenon, uninterestingly, in terms reducible to sf or fantasy, or both.

DG's last novel of interest is a BEAST FABLE of some power. The first half of *The Master Cat: The True and Unexpurgated Story of Puss in Boots* (in *A Chatto & Windus Almanack* anth **1926** as "From 'Puss in Boots'"; much exp **1974**) is a REVISIONIST FANTASY version of the Charles PERRAULT fable of 1697, set in a LAND-OF-FABLE Northumberland about 1000 years ago and treating the miller's

son with some reserve. In the second half, the son – whom PUSS-IN-BOOTS has given high birth, wealth and a king's daughter – denounces Puss to a visiting delegation of Christians from Ireland. Puss eventually escapes, after torture and an attempted *auto-da-fé*, and resolves never to speak again to a human being. He travels the world as the King of the Cats, spreading his message of distrust. He is heeded. By virtue of this sequel, DG transforms the old FAIRYTALE into an extremely pointed lesson in THINNING, for the triumph of the Christians (DG's estimate of Christianity was always scathing), who loathe and fear beings such as Puss, leaches the richness of the world. *The Master Cat* is another example of the subversive impulses tappable in fantasy. [JC]

Other works: *A Terrible Day* (**1932** chap); *Purl and Plain, and Other Stories* (coll **1973**).

GARNETT, EDWARD (1868-1936) UK editor and writer, son of Richard GARNETT, married to Constance GARNETT; their son was David GARNETT. As a publisher's reader he was enormously influential from the 1890s on, stimulating the early careers of writers like Joseph CONRAD, E.M. FORSTER, W.H. HUDSON and D.H. LAWRENCE. He was a minor playwright but a significant critic. His only fiction of genre interest is *Papa's War* (coll dated 1918 but probably **1919**), a set of 22 sf and fantasy fables, usually containing some SATIRE addressed towards correcting the failures of civilization that led to WORLD WAR I. [JC]

GARNETT, RICHARD (1835-1906) UK writer and scholar who spent much of his career working for the British Museum library. Active as an editor, biographer and poet, he also produced the deft and delicate fantasies in *The Twilight of the Gods* (coll **1888**; exp from 16 items to 28 1903; illus with intro by T.E. Lawrence [1888-1935] 1924). The tales gather motifs eclectically from various mythologies, and display them in whimsical fables whose moral is that Epicurean humanism is always to be preferred to Puritan intolerance. The title story tracks the career of PROMETHEUS after his liberation, when he lives with the last priestess of APOLLO – who takes a more active role in several other stories, including "The Poet of Panopolis". "The Demon Pope", "The Bell of St Euschemon" and "Alexander the Ratcatcher" (1897) are fine examples of English Literary Satanism (◊ SATAN). [BS]

GAROU, LOUIS P. Pseudonym of Stephen BOWKETT.

GARRETT, (GORDON) RANDALL (PHILLIP DAVID) (1927-1987) US writer of much competent but routine sf, chiefly for magazines. His finest achievement was the **Lord Darcy** series of DETECTIVE/THRILLER FANTASIES, set in an ALTERNATE-WORLD England whose history diverged at the time of Richard I (1157-1199) and where MAGIC rather than science has been harnessed and codified, subject to laws based on those of Sir James FRAZER. The **Darcy** series comprises *Too Many Magicians* (**1967**), *Murder and Magic* (coll **1979**), *Lord Darcy Investigates* (coll **1981**), and "The Spell of War" in *Thor's Hammer* (anth **1979**) ed Reginald Bretnor (1911-1992). The first three volumes were assembled as *Lord Darcy* (omni **1983**). A typical **Darcy** story has a twofold interest as period detection, with his lordship in the role of SHERLOCK HOLMES, and as an exposition of RATIONALIZED FANTASY, with his assistant Master Sean o Lochlainn clarifying but never quite solving the crime using forensic sorcery. Sean's processes and lectures are narrative highlights, cannily echoing scientific method while developing plausible sympathetic-magical laws such as those of Relevancy and Synecdoche. Like Dr Watson with his revolver, Master Sean is also a useful wielder of defensive spells; other entertainingly RECURSIVE detective elements abound. For example, not only does the title *Too Many Magicians* echo several thriller titles by Rex Stout (1886-1975), but its obese Marquis of London and his legman Lord Bontriomphe are obvious cognates of Stout's Nero Wolfe and Archie Goodwin; the strictly prosaic though psychically aided murder method gestures ironically towards the similar improbable stroke in *The Judas Window* (**1938**) by Carter Dickson (John Dickson CARR). The series' sympathetic treatment of RELIGION and priestly magic – shown to complement secular magic with TALENTS of PERCEPTION and healing, including psychiatry – reminds us that RG spent several years in Holy Orders.

The 1981-6 **Gandalara** HEROIC-FANTASY series, though partly plotted and nominally co-authored by RG, was actually written by his third wife Vicki Ann HEYDRON after viral meningitis had left him hospitalized and amnesiac. After RG's death the **Darcy** sequence was extended by Michael Kurland (1938-   ) with the somewhat inferior *Ten Little Wizards* * (**1988**) and *A Study in Sorcery* * (**1989**). The original series remains a fantasy landmark.   [DRL]

Other works: *Pagan Passions* (**1959**) with Laurence M. Janifer (1933-   ), a comic fantasy featuring the gods of classic MYTHOLOGY in a futuristic setting.

GARTH Fair-haired, square-jawed, large-muscled TIME-travelling fantasy troubleshooter in UK newspaper COMIC strips. He first appeared in the *Daily Mirror* in 1943. Although an occasional story has had pure fantasy elements involving G's eternal love, the goddess Astra (to whom he is remarkably unfaithful: most stories involve some kind of sexual encounter), most belong more properly to sf. A number of writers have worked on the strip: the best has been Peter O'Donnell (1920-   ), who provided some very tightly plotted and imaginative tales in 1953-66 including – in *The Last Goddess* (1965; graph **1966**) – the introduction of Astra. Artists have included Steve Dowling, John Allard, Frank BELLAMY and Martin Asbury.   [RT]

GASCOIGNE, BAMBER (1935-   ) UK broadcaster and writer. ◊ AESOPIAN FANTASY.

GASCOIGNE, MARC (?   -   ) US writer with Carl Sargent of ties in the SHARED-WORLD series of TECHNOFAN-TASIES, SHADOWRUN, and of a volume in the much looser **Fighting Fantasy Gamebook** series, with Ian LIVING-STONE (*whom see for details*).   [JC]

GASKELL, JANE Working name of UK writer Jane Gaskell Lynch (1941-   ), who began writing at a very early age: her first published novel, *Strange Evil* (**1957**), was written when she was 14. Printed almost without alteration from her original – and thus showing some overwriting and naïveties – it is an astonishingly imaginative piece of fantasy by any standards. Its young female protagonist discovers that her hitherto-unknown cousin and the cousin's spouse are among a number of FAIRIES currently present in our world for diverse purposes. They take her to a FAERIE quite unlike either the Victorian conception or the Celtic MYTH: this is a full-blown ALTERNATE REALITY attainable by invisible PORTALS, transport within it being by a moving silver road that arches across the sky. Here she finds the fairies engaged in a war that is partly racist, partly economic and partly religious (◊ RELIGION); central to the latter aspect is the GOD Baby, a vastly overgrown infant whose spoilt, tantrum-born outpourings are regarded as divine commands. This

wildly original Faerie might have offered a new PLAY-GROUND for others to explore – the novel has the feel of being the first of a new subgenre – but none, not even JG, has chosen to enter it.

King's Daughter (**1958**) is more professionally written but less interesting and vivid. Set 200,000 years ago, after the departure of a previous Moon but before the arrival of the one we know, it tells of the daughter of an exiled Atlantean princess struggling to reach her deceased mother's land (◊ ATLANTIS). This novel prefigured the series of fantasies for which JG is best known, the **Atlan** series: *The Serpent* (**1966**; vt in 2 vols as *The Serpent* and *The Dragon* 1977 US), *Atlan* (**1966**), *The City* (**1966**) and, much later, *Some Summer Lands* (**1977** US). Set some while after *King's Daughter*, the first trilogy follows the progress across South America of the Princess Cija – revolted by but at the same time passionately attracted to a half-man, half-reptile conqueror, General Zerd – to warn the citizens of remote, vacuum-shielded off-shore Atlan, home of an ancient, pacifistic culture living in harmony with nature, of Zerd's tyrannical intentions. She arrives to find the conquest a *fait accompli*. She marries Zerd, but is soon banished in favour of his first wife to have adventures including a weird PARODY of the FRANKENSTEIN theme. Yet the invaders, Cija included, are rejected by the LAND itself – all except her elder child, born of an incestuous union with her half-brother, who is independently descended from divinity. Ousted, most of the protagonists return to the theocratic dictatorship of Cija's mother, where Cija undergoes standard HIDDEN-MONARCH adventures before settling for the love of an APE. Though finally "rescued" from her new life, she is bearing the ape's child. A sequel beckoned, but never came; instead there was the addendum of *Some Summer Lands*, an often incoherent, genitalia-obsessed novel wherein Cija and her daughter (by Zerd) are taken to and now welcomed by Atlan and its people. Some of the characters from *King's Daughter* anachronistically reappear.

The saga as a whole is characterized by a girlish breathlessness and an increasing preoccupation with ever-odder SEX – the curiosities of *Some Summer Lands* represent merely the extreme conclusion, more surprising because of the 11-year hiatus, of a process that was already well under way in the initial trilogy. The vividness and conceptual playfulness that marked *Strange Evil* had largely vanished by the time of *The Serpent* and almost entirely by *Atlan*. Nevertheless, JG's evident enthusiasm for her central character keeps at least the first two books ticking along; by the third, her interest was flagging, as does ours. Despite its flaws, the initial trilogy was an ambitious and mainly successful essay at world-creation: JG's was no standard FANTASYLAND.

Between *King's Daughter* and the **Atlan** series JG published various comedies of sexual manners, all non-genre save *The Shiny Narrow Grin* (**1964**), a VAMPIRE novel that can be seen as a Swinging Sixties precursor of modern novels by authors like Nancy A. COLLINS. Most of JG's subsequent novels, increasingly sloppily written, their protagonists ageing alongside JG herself, are of no fantasy interest, although sometimes with slight paranormal content – she is a professional astrologer – as in *Sun Bubble* (**1990**), concerned with the death of a loved one.   [JG]

GASLIGHT ROMANCE There is a growing habit whereby almost every fantasy which deals with the Gaslight Period is labelled STEAMPUNK. It is useful, though, to limit that term to what are in effect historical TECHNOFANTASIES – to books which fit directly into the form developed by Tim POWERS, K.W. Jeter and James P. BLAYLOCK from models derived by Michael MOORCOCK, Christopher PRIEST and others – books whose principal plot-driver is technological ANACHRONISM. GR is accordingly our default term for URBAN FANTASIES (and other generic fictions) set in the high Victorian or Edwardian period, usually but not invariably in LONDON; their tone is often melancholic and there is often an underlying sense of the transitoriness of imperial glory. Many GRs directly refer to the losses of innocence at the beginning of the period – through the AMERICAN CIVIL WAR and the Indian Mutiny – or to the bloodletting of WORLD WAR I at its end. Greyness, twilight and fog are more than local colour in GRs; they are its pervading metaphor.

Both Steampunk and the GR can easily be set in ALTER-NATE WORLDS; both can represent FANTASIES OF HISTORY. RECURSIVE FANTASIES featuring iconic figures from the gaslight era – SHERLOCK HOLMES, JEKYLL AND HYDE, DRAC-ULA, H.G. WELLS's Time Traveller, G.K. CHESTERTON's Man who was Thursday, FU MANCHU, etc. – are GRs. There are also a number of books – like Kim NEWMAN's *Anno Dracula* (**1993**), E.L. DOCTOROW's *The Waterworks* (**1994**) and F. Gwynplaine MacIntyre's *The Woman between the Worlds* (**1994**) – which examine the period and its tropes with a modern or Postmodernist sensibility, rather than merely recycling the old tropes. [RK/JC]

See also: JACK THE RIPPER; RURITANIA.

GATES ◊ PORTALS.

GAUTIER, JUDITH (1845-1917) French writer, daughter of Théophile GAUTIER. ◊ ANTHROPOLOGY.

GAUTIER, THÉOPHILE (1811-1872) French writer, a leading figure in the Romantic movement. Many of his works, including the lushly erotic *Mademoiselle de Maupin* (**1835**) – which is prefaced by his manifesto for "l'art pour l'art" – are so selfconsciously exotic as to qualify as border-line fantasies even though without supernatural content. This applies particularly to baroque historical *nouvelles* like *Une nuit de Cléopâtre* (**1838**; **1894** chap; trans Lafcadio HEARN as *One of Cleopatra's Nights* **1882** US) and *Le roi Candaule* (**1844**; **1893** chap; trans Hearn as *King Candaules* **1893** US; vt *The Wife of King Candaules* **1942**). *One of Cleopatra's Nights* also includes a classic FEMME-FATALE story about a female VAMPIRE, "La morte amoureuse" (**1836**; here trans as *Clarimonde* **1899** chap; new trans by Andrew LANG and Paul Sylvester as *The Dead Leman* **1889**; **1903** chap; further vts "The Vampire", "The Beautiful Vampire", "The Priest"). Hearn's collection included also the TIMESLIP romance "Arria Marcella; souvenir de Pompeii" (**1852**; trans here as "Arria Marcella"; vt "The Tourist") and two *contes* in a lighter vein: "Omphale" (**1834**; trans here as "Omphale: A Rococo Story"; vt "The Adolescent") and "Le pied de la Momie" (**1840**; here trans as "The Mummy's Foot"; vt "The Foot of the Princess Hermonthis").

TG's works were collected as *Oeuvres* (coll **1855-74** 22 vols; trans F.C. Sumichrast as *The Works of Théophile Gautier* **1900-1903** 24 vols or 12 vols). Other fantasies include three notable *nouvelles*: *Avatar* (**1856**) tells the story of an ill fated IDENTITY EXCHANGE; *Jettatura* (**1857**; trans M. de L. **1888**; new trans vt "The Evil Eye") examines the plight of a man who unwittingly falls prey to an unfortunate CURSE; *Spirite; nouvelle fantastique* (**1866**; trans anon as *Spirite: A Fantasy* **1877**; vts *Spirit Love*, *Stronger than Death*) is a sen-timental tale of a love affair between a young man and a

female GHOST. TG wrote the last-named item for the dancer Carlotta Grisi (1819-1899), for whom he also wrote the libretto of the ballet *Giselle*, but it was Carlotta's sister Ernesta who actually lived with him and bore his daughter Judith (1845-1917). Judith, who married Catulle Mendès (1841-1909), herself wrote several extravagant Oriental tales. Many collections culled from the collected works were issued subsequently, including *The Works of Gautier* (coll **1928** US).

TG's other long prose works with slight fantasy elements are *La belle Jenny* (**1865**; trans F.C. Sumichrast as "The Quartette" 1901; vt "The Four-in-Hand"), *Le roman de la Momie* (**1858**; trans Anne T. Wood as *The Romance of the Mummy* 1863; vts *The Romance of a Mummy*, *The Mummy's Romance*), *Fortunio* (**1837**; trans Alexina Loranger *c*1890) and *Le capitaine Fracasse* (**1863**; trans Ellen Murray Beam as *Captain Fracasse* 1880). A recent collection of Gautier's short fantasies is *My Fantoms* (coll trans Richard Holmes **1976**). [BS]

Other works: *La mille et deuxième nuit* (**1894** chap; trans F.C. Sumichrast as "The Thousand and Second Night" 1902).

GAWAIN One of the leading KNIGHTS of the ROUND TABLE, Gawain was the son of King Lot of Lothian and Orkney, and his origins may be more closely associated with early Scottish history under his Celtic name Gwalchmai. He first features strongly in the Arthurian story as related by GEOF-FREY OF MONMOUTH in *Historia Regum Britanniae* (**1136**), where he is one of ARTHUR's most valiant knights. In the later romances, especially those by the French and German writers, Gawain is less of a successful HERO, being instead rather a hot-head who blunders into adventures regardless of success. However, in the anonymous medieval poem *Sir Gawain and the Green Knight* (14th century) he is portrayed as near perfection, a courageous knight prepared to sacrifice his life in order to face a challenge and meet his obliga-tions. In this story CAMELOT is visited on New Year's Day by a giant of a man dressed all in green and riding a green horse. He challenges Arthur's men to a beheading match. Gawain takes up the challenge and strikes off the knight's head. The knight, though, recovers the head and departs, reminding Gawain of his promise. The following year Gawain sets out to find the knight at the Green Chapel. He is at first unsuccessful, and spends Christmas at a castle with its lord, his wife and an old lady. He learns that the Green Chapel is nearby and, after three days, after being subject to various temptations, is led there, where he meets the knight. The knight strikes three blows: the first two are feints and the third slightly cuts Gawain's neck. Gawain now learns that the Green Knight is his host from the Castle, Sir Bertilak, who reveals that Gawain has passed the test of the temptations, the slight cut being inflicted because he almost fell for one of them. Gawain also learns that the whole episode was contrived by MORGAN LE FAY to test the ROUND TABLE. The story is evidently descended from earlier Nature tales, personifying the CYCLE of the year (◊ GREEN MAN), but has been developed to a high degree.

A similar portrayal of an honest and virtuous Gawain appears in the anonymous medieval verse romance *The Wedding of Sir Gawen and Dame Ragnell* (?**1450**), where Gawain rescues King Arthur but to do so must promise to marry a hideous old crone. On their wedding night, how-ever, the dame becomes young and beautiful. This latter image of Gawain most appealed to Thomas BERGER, who incorporated it in *Arthur Rex* (**1978**).

Gawain's character interests many writers, and so he often appears as an ancillary character in Arthurian novels and stories (◊ ARTHUR). He does, however, take centre stage in some. *The Green Knight* (**1975**) by Vera CHAPMAN retells the medieval tale with a modern interpretation. The **Gwalchmai** sequence by Gillian BRADSHAW is in effect Gawain's biography. In "King of the World's Edge" (1939 *WT*; **1966**), *The Ship from Atlantis* (**1967**; vt as fixup with "King of the World's Edge" *Merlin's Godson* 1976) and *Merlin's Ring* (**1974**) by H. Warner MUNN Gawain (as Gwalchmai) is portrayed as the son of a Roman centurion who, after Arthur's death, accompanies his father and Merlin to the New World and beyond. [MA]

See also: GAWAIN AND THE GREEN KNIGHT.

GAWAIN AND THE GREEN KNIGHT At least two movies have been based on the anonymous 14th-century poem *Sir Gawain and the Green Knight* (◊ GAWAIN).

1. UK movie (*1973*). Scancrest/United Artists. **Pr** Philip Breen. **Dir** Stephen Weeks. **Spfx** Les Millman. **Graphics and titles** Richard WILLIAMS. **Screenplay** Breen, Weeks. **Script consultant** Rosemary SUTCLIFF. **Starring** Nigel Green (Green Knight), Robert Hardy (Sir Bertilak), Murray Head (Gawain), David Leland (Humphrey), Ciaran Madden (Linet), Anthony Sharp (Arthur). 93 mins. Colour.

The court at CAMELOT is becoming corrupt; ARTHUR rails to little effect. One evening the GREEN KNIGHT – a phantasmic figure – enters the great hall and challenges the knights to a beheading match. The only taker is the squire GAWAIN, who hacks off the Green Knight's head with a single blow; the Green Knight, replacing his head, reminds Gawain that in a year's time he must present his neck for the Green Knight's blow. That year is spent by Gawain (now Sir Gawain) in traditional QUEST, during which he loves and loses and rediscovers the fair ENCHANTRESS Linet. At the end the Green Knight cannot strike Gawain but instead rots away into the earth, a symbol of the THINNING of the world and of the inevitable corruption that time brings.

Not for the purist, this is an adventure movie of some vigour and integrity, although badly marred by an over-repetitive music track. As an entertainment it largely succeeds; as a rendition of the poem – and despite Sutcliff's contribution – it compromises. [JG]

2. UK movie (*1991* tvm). Thames. **Pr** John Michael Phillips. **Exec pr** Ian Martin. **Dir** Phillips. **Screenplay** David Rudkin. **Starring** Jason Durr (Gawain), Valerie Gogan (Linet/The Lady), Malcolm Storry (Green Knight/Red Lord). *c*90 mins. Colour.

Although this is evidently a shoestring production, it is of interest in that it is remarkably faithful to the original poem, with much of its dialogue and narration being in the form of direct translation. The Green Knight is portrayed as a FERTILITY figure, and his wife Linet, in the GODGAME both are playing with Gawain, is a sort of skewed version of the Madonna (◊ GODDESS) whom Gawain worships – and whose image is painted on the inside of his shield. Gawain finds himself a more full person, because he LEARNS BETTER – as with other men, his purity is not inviolable. [JG]

GAWSWORTH, JOHN Pseudonym of UK poet, writer, bibliographer and anthologist (Terence Ian) Fytton Armstrong (1912-1970), who signed his real name for some work, but most of whose literary publications of genre interest were as JG. An avid collector, JG became close friends with many UK writers and, like August DERLETH,

became something of an author's champion through his ANTHOLOGIES, MAGAZINES and SMALL-PRESS ventures; he also compiled two bibliographical studies, *Ten Contemporaries* (**1932**) and *Ten Contemporaries (Second Series)* (**1933**), which concentrated on writers, like John COLLIER, Oliver ONIONS and M.P. SHIEL, with whom he had become associated. With the exception of *Strange Assembly: New Stories* (anth **1932**) as Fytton Armstrong and *Full Score: Twenty Five Stories* (anth **1933**), his anthologies were anonymous. They include *New Tales of Horror by Eminent Authors* (anth **1934**), *Thrills, Crimes and Mysteries: A Specially Selected Collection of Sixty-Three Stories by Well-Known Writers* (anth **1935**), *Thrills: Twenty Specially Selected New Stories of Crime, Mystery and Horror* (anth **1936**), *Crimes, Creeps and Thrills* (anth **1936**), *Masterpiece of Thrills* (anth **1936**) and *Twenty Tales of Terror* (anth **1945** India), the last compiled from earlier volumes. Over and above their general high quality, these large anthologies of HORROR and SUPERNATURAL FICTION – ranging from quality GHOST STORIES to mediocre CONTES CRUELS – are notable for the amount of original material they contain, much of this coming from authors of considerable interest, including Oswell Blakeston (?1907-?1985), Thomas Burke (1886-1945), Frederick Carter, Louis GOLDING, Edgar JEPSON, Arthur MACHEN, Richard MIDDLETON, Eimar O'Duffy (1893-1935), M.P. Shiel and E.H. VISIAK – as well as JG himself, none of whose short stories have been collected separately. His early stories, like *Above the River* (**1931**) and "Scylla and Charybdis" (1934) are wistful nature fantasies in the style of Algernon BLACKWOOD, but his later "collaborations" with Shiel, Visiak and particularly Jepson ("The Shifting Growth" 1936) are more sinister and portentous, shadowing death.

JG was close to Shiel, editing *The Best Short Stories of M.P. Shiel* (coll **1948**). In 1947, Shiel proclaimed JG the heir to the genuine Caribbean island of Redondo. After Shiel's death, JG proclaimed himself Juan I, and distributed dukedoms to his colleagues with enthusiasm. [MA/JC]

Other works: *Collected Poems* (coll **1948**).

Other works as editor: *The Pantomime Man* (coll **1933**) by Richard MIDDLETON; *The Best Stories of Thomas Burke* (coll **1950**).

Further reading: "John Gawsworth" by Steve Eng in *Night Cry* Spring 1987; "John Gawsworth Unforgotten: A Tribute Anthology" compiled by Steve Eng in *The Romantist #6-#8* (1986), a series of reminiscences; *SFE*.

GEARY, PATRICIA (CAROL) (1951-) US writer and academic whose first novel, *Living in Ether* (**1982**; cut 1987), wittily portrays Southern California as a land all but coextensive with the AFTERLIFE (the narrator, a professional medium, receives tantalizing communications from her dead brother), whose residents employ yoga, psychics and martial-arts training to deal with otherworldly complications. *Strange Toys* (**1987**), which won the Philip K. Dick AWARD, is more overtly a fantasy; the complex plot involves VOODOO, incantations and malevolent sorcery. [GF]

GEASES (Irish *geis*, plural *geasa*) ◊ CONDITIONS; CURSES; TABOOS.

GEDGE, PAULINE (ALICE) (1945-) New Zealand-born writer, in Canada from the mid-1950s. Most of her novels have been historical fiction, including *Child of the Morning* (**1977**) and *The Twelfth Transforming* (**1984**), both set in ancient EGYPT, and *The Eagle and the Raven* (**1978**), set in Celtic Britain. A later novel, *Scroll of Saqqara* (**1990**; vt *Mirage* 1991 US), also set in ancient Egypt, is a fantasy involving REINCARNATION and a scroll with mystical powers. PG is of note as a fantasy writer primarily for *Stargate* (**1982**), a tale of vast scope. The creator of the Universe has become its Unmaker, through rage at the Law-Giver who controls his acts of creation; and systematically destroys world after world, each of them ruled by a sun-god and inhabited by mortals previously innocent of sin. The only way to stop the Unmaker, who as DARK LORD is a kind of PARODY of himself, is to close the eponymous PORTALS between the worlds. Although a single world survives at the end of the complex tale, a sense of irremediable loss remains. This is unconnected with the movie *Stargate* (*1994*). [JC]

GEMMELL, DAVID A(NDREW) (1948-   ) UK writer whose fantasy novels have become extremely popular. His first sequence, the GENRE-FANTASY **Drenai Saga**, comprises: *Legend* (**1984**; vt *Against the Horde* 1988 US), *The King Beyond the Gate* (**1985**) and *Waylander* (**1986**), all three assembled (with a novella, "Druss the Legend") as *Drenai Tales* (omni **1991**); *Quest for Lost Heroes* (**1990**), *Waylander II: In the Realm of the Wolf* (**1992**), *The First Chronicles of Druss the Legend* (coll **1993**) and «The Legend of Deathwalker» (**1996**); plus a GRAPHIC NOVEL, *Legend: A Graphic Novel* (graph **1993**), text by Stan Nicholls, illustrated by Fangorn. The first volume of the sequence, *Legend*, remains DAG's best-known book, in both original and graphic-novel versions. His strengths as a fantasy author are immediately evident: a technical ability to press ahead with revelation and action at a pace whose intensity can seem at times almost surreal; an interest in depicting HEROES who are weathered, seemingly past their prime, reluctantly but profoundly charismatic; and a focus on bands of COMPANIONS – usually SEVEN-SAMURAI volunteers attracted by the allure of the hero – engaged in seemingly impossible guerrilla warfare against usurper DARK LORDS and/or invading hordes.

Drenai itself is a decadent medieval empire which is in near-terminal decline; succeeding instalments give a sense, at times rather insistent, that the battle against foes is unending, and perhaps unwinnable. In the first volume Druss, a legendary warrior near retirement, gathers a band of companions around him and defends a pass to the death. He is given the ambivalent aid of a PARIAH ELITE of martial priests with TALENTS. The empire is saved from the northern barbarians. Further volumes tend to rehearse, with very considerable skill and exhilarating momentum, the same basic tale, though *Quest for Lost Heroes* varies the mix by featuring a young protagonist who undergoes the normal RITE OF PASSAGE into manhood in the company of a much older band of companions, and *Waylander II: In the Realm of the Wolf* returns to an earlier figure in the sequence who, now in retirement, must defend himself, with the aid of companions, from a passel of assassins. There are debates *inter alia* about pacifism, which does not win the argument.

DAG's second sequence is a more complicated affair. The **Sipstrassi** books, though sharing the same underlying premise and a FANTASY-OF-HISTORY relationship to our own world, are in fact two separate series. The **Jerusalem Man** series comprises *Wolf in Shadow* (**1987**; vt *The Jerusalem Man* 1988 US), *The Last Guardian* (**1989**) and *Bloodstone* (**1994**), all three assembled as *The Complete Chronicles of the Jerusalem Man* (omni **1995**), plus a graphic novel, *Wolfe in Shadow: The Graphic Novel* (graph **1994**), text by Stan Nicholls, illustrated by Fangorn. The **Stones of Power** sequence comprises *Ghost King* (**1988**) and *Last Sword of Power* (**1988**).

Chronologically the second of these two sequences, being set in Dark-Age Britain (◊ ARTHUR), comes first. The High King of Britain is murdered while the Enchanter Maedhlyn (◊ MERLIN) is away, leaving young Thuro, heir to the throne, in danger. But Maedhlyn is not simply a MAGICIAN; he is an immortal (◊ IMMORTALITY), a member of a group of SECRET MASTERS which owes its power to the "Sipstrassi", fragments of a meteor which originally impacted upon ATLANTIS, and which serve as a focus for TALENTS. Maedhlyn's nurturing of young Thuro – who grows up to become Uther Pendragon – is part of the Secret Masters' long campaign against the powers of darkness, all of whom long to obtain sole use of the Sipstrassi and rule the world. The Dark Lord in this instance is a combined Moloch and Wotan (◊ ODIN).

The **Jerusalem Man** tales, by contrast, are post-HOLOCAUST, set some 300 years after the Earth's axis has tilted, and feature the obsessed Jon Shannow, a Bible-reading CHILDE figure locked into a QUEST for JERUSALEM, where he hopes for surcease from the ravaged world. There are sf elements in the sequence – 20th-century artifacts, including both guns and the *Titanic*, tend to reappear whenever needed – but the underlying involvement of the SECRET MASTERS from Atlantis gives a TECHNOFANTASY gloss to that which is not treated, fairly straightforwardly, in terms of RATIONALIZED FANTASY. Shannow is identified, by one of the Secret Masters, as a Rolynd (i.e., Childe Roland) from Atlantis. In their long war to maintain BALANCE over CHAOS, he comes to play a significant role, first destroying the armies of DEVIL-worshipping victims of wrongly used Sipstrassi stones and finally tricking the Dark Lord into a TIME-TRAVEL quest to 1945, where he is incinerated at Alamogordo.

DAG's next series, the **Macedon** sequence – *Lion of Macedon* (**1990**) and *Dark Prince* (**1991**) – invokes the Sipstrassi stones briefly in a story set in Greece in the 4th century BC. Alexander the Great is supernaturally linked to the forces of CHAOS, and the second volume carries him and his general – the series' main protagonist – into an ALTERNATE WORLD dominated by the Chaos Lord. Good triumphs. The **Pallides Saga** – *Ironhand's Daughter* (**1995**) and *The Hawk Eternal* (**1995**) with a further volume projected – features a female protagonist, and so far has a HEROIC-FANTASY air. DAG's only singleton of interest, *Knights of Dark Renown* (**1989**), again features an ageing warrior, a band of companions and a threat to the land.

It cannot be denied that mechanical effects sometimes drown out the original elements in DAG's work; at the same time, the grit and speed of his best tales override almost any reader's initial reluctance to be borne away. He is one of the central entertainers of the genre. [JC]

Other works: *Morningstar* (**1992**), a YA fantasy; *White Knight, Black Swan* (**1993**) as by Ross Harding, a gangster thriller.

GENDER Fantasy as a genre is generally perceived as more hospitable to women than SCIENCE FICTION and HORROR, and more flexible in the choices it offers than historical fiction or romance, yet the standard patriarchal bias imposes limitations which are seldom subverted or even questioned. Whereas sf has the potential to question gender roles and try to envision new ways of living, fantasy looks to the past, seeking out patterns and archetypes.

Writers in search of different, more empowering, roles for women, rather than defining a woman in terms of her relationship to a man, have looked to matriarchal cultures, imagined or real, and particularly to the realm of the GODDESS. CELTIC FANTASY has been particularly attractive to women writers because, according to Nickianne Moody (in *Where No Man Has Gone Before* [**1991**] ed Lucie Armitt), "The images found in Celtic myth . . . provide a women's fantasy of freedom and equality which has historical support." Although the ARTHUR saga centres on a Warrior King, the presence in the legends of a number of interesting female characters, and the possibility of writing about women who are active and important not only in the domestic sphere, in ways not possible in the standard romance, gives Celtic Fantasy its strong contemporary appeal.

One obviously heroic role for women is that of the AMAZON, and the 1980s saw the development of a subgenre of Amazonian SWORD AND SORCERY. In a critical overview in *The New York Review of Science Fiction #20* (1990) Jessica Amanda SALMONSON declared that this was a reactionary development, that the depiction of the Amazon by most contemporary writers confirmed rather than debunked or expanded feminine stereotypes. She pointed out that the historical image of the Amazon was of a *society* of women for whom valour and physical prowess were the norm; thus the Amazon was not, as most writers in the 1980s presented her, an anomaly, the exception to the "rule" of passive, unarmed women, and thereby alienated from the society in which she lived. Even the feminist (⚥ FEMINISM) heroines created by Joanna Russ (*Alyx* [**1976**]) and Mary Mackey (*The Last Warrior Queen* [**1983**]) were seen as isolated in a patriarchal culture. The only exception she cited was the character Raven from Samuel R. Delany's *Tales of Nevèrÿon* (coll of linked stories **1979**) along with a handful of short stories collected in her own WORLD FANTASY AWARD-winning anthology *Amazons!* (anth **1979**).

The SWORD is an equalizer, allowing individual women free passage in male-dominated worlds; MAGIC is another. In the past, female magic users were seen in archetypal terms, according to their effect on the hero of the tale. Either they were evil or, as "good FAIRIES", they were effectively de-sexed by being presented as extremely small, old, ugly or immaterial. In modern fantasy the female magic user may well be the protagonist, often from a PARIAH ELITE, persecuted for her powers or her RELIGION. In Margaret BALL's *Flameweaver* (**1991**) and *Changeweaver* (**1993**) the women of Gandhara rule their country with a benevolent use of magic. Only women have access to magic, and their power must be "grounded" to the service of society by marriage and children. Women who cannot have children are not controllable, so are exiled. In the **Inland** series by Ann Halam (Gwyneth JONES), women are trained to use their innate magical abilities for the good of society; men's magic also exists, but is considered a useless matter of "tricks" – much as women's magic is perceived in the more traditional HIGH FANTASY, where men alone are allowed to be mages and sages, as exemplified in Ursula K. LE GUIN's *A Wizard of Earthsea* (**1968**), where traditional sayings include "weak as woman's magic" and "wicked as woman's magic".

Even when women's magic is not perceived as fundamentally different from that of men's ("natural" or earth-magic used for household affairs, as contrasted with the training, SACRIFICE and book-learning aimed at controlling universal powers), women's relationship to this, as to other powers, is coded differently. Unless she is perceived as evil, a woman

uses her powers for the good of others, either to help her community or to provide back-up strength to the HERO (⚥⚥ COMPANIONS). Hence the many fantasies in which a young woman becomes aware of a TALENT within her and subsequently finds her purpose in helping some young man to reach a destiny she would never seek for herself. In *The Women who Walk Through Fire: Women's Fantasy & Science Fiction* Vol 2 (**1990**) ed Susanna J. Sturgis, Sturgis cites Kathleen Cioffi to the effect that feminist fantasy is concerned with the female hero's attempt to "reintegrate" with her community after the successful completion of her QUEST. Knowing too much, or knowing something different, makes a woman an outsider; while a man, accustomed to defining himself in oppositional terms of power and freedom, will consider the achievement of a quest to be the end of the story – his own story – women, who customarily define themselves within a web of mutually reciprocal relationships and responsibilities, cannot rest until the impact of their actions has been accepted by others. *Lavondyss* (**1988**) by Robert HOLDSTOCK is exceptional in presenting a believable female figure who goes on a quest for much the same reasons and in the same way as the male characters in his earlier *Mythago Wood* (**1984**).

In the early 1970s feminists were highly critical of FAIRYTALES for their culturally normative function. Yet "household" or "old wives'" tales were most often told by women, and the range of female roles in the originals was far greater than the extremes of wicked old witch/beautiful passive princess known to modern consumers of the DISNEY versions. The virtues most often rewarded in fairytales tend to be the same in men and women – values like patience, kindness and cleverness – and, although women today might perceive the inevitable marriage of the happy ending as conservative, Angela CARTER has commented – in *The Virago Book of Fairy Tales* (anth **1990**) – that, given the context of the societies from which most of these tales come, the emphasis on FERTILITY and marital happiness should be seen rather as optimistic, even utopian.

Claims are often made for the universal psychological truths to be found in fairytales, but such readings should be applied with caution. It can be difficult sometimes to recognize the originally intended message. Jack ZIPES, in *The Trials and Tribulations of Little Red Riding Hood* (**1983**), demonstrated how an oral tale about a girl who manages to outwit and escape a WEREWOLF was transformed almost out of recognition by Charles PERRAULT in 1697 into a fable about male sexuality and the need to control women and their desires. Today the sexuality and violence of Perrault's tale has been so watered down that the little girl has become stupid rather than licentious. Similar caveats should apply to attempts to read texts as if they were of universal human relevance. Ungendered souls are usually male; it is the female who's marked out, signified as *other*.

Heroes in our culture have been gendered male, so "the hero-tale has concerned the establishment or validation of manhood", as Ursula K. LE GUIN put it in *Earthsea Revisioned* (**1993**). After writing three **Earthsea** novels in the tradition of HEROIC FANTASY, Le Guin abandoned the "pseudo-genderless male viewpoint of the heroic tradition" and returned to the world she had created to re-imagine it through the eyes of an ageing woman. This novel, *Tehanu* (**1990**), disappointed many readers, who felt denied the traditional pleasures of fantasy; at least one critic felt Le Guin was supporting the oppression of women by having her character choose marriage and obscurity rather than continuing her studies in magic. Le Guin herself has said that her aim was to set her characters free: "The deepest foundation of the order of oppression is gendering . . . To begin to imagine freedom, the myths of gender, like the myths of race, have to be exploded and discarded. My fiction does that by these troubling and ugly embodiments."

Androgyny, once the goal of many feminists, is now unfashionable; feminist belief in the irrelevance of gender is unusual. Yet there have been important changes in the old, patriarchal, bipolar way of thinking, if only because gender roles have been so expanded and redefined in the past several decades. Women are offered much more latitude in what is considered acceptable in a female. The old-fashioned passive princess is passé; even in retellings of older tales, in which her only role was as the hero's reward, the heroine today is spunkier, depicted in more egalitarian terms, and given more to do – even if the plot isn't actually changed by it, as in Disney's ALADDIN (*1992*).

Woman as spunky COMPANION is an easy option, a way of seeming up-to-date and flattering readers without diminishing the role or values of the male hero. Woman as hero is more difficult to depict and to define: is there another way to approach her beyond the self-sacrificing, supportive heroine or the man-manqué?

The hero of Geoff RYMAN's *The Warrior who Carried Life* (1985) is a woman who magically shapeshifts (◊ SHAPESHIFTERS) into the form of a powerful male warrior because "a man could take revenge, where she could not". At an important moment in the plot she is recognized as being neither man nor woman – hence able to accomplish what no man or woman could ever do – and subsequently as being both man *and* woman, hence the first fully human being since the original sin and fall from grace. Ryman is rare in his treatment of gender as something which can be put on or taken off like a suit of magic armour – although Peter S. BEAGLE has a male-to-female shapeshifter in *The Innkeeper's Song* (1993), and both were preceded by L. Frank BAUM's *The Marvelous Land of Oz* (1904) and Virginia WOOLF's *Orlando* (1928). While a new subgenre of magical sex-change stories seems unlikely, Ryman's protagonist may point the way towards the possibility of a new type of hero, celebrating the best of both genders without devaluing either.   [LT]

Further reading: There are many other works of fiction that pertain to the subject of gender. A very selective list of additional references is: the **Atlan** saga (**1966-77**) by Jane GASKELL; *Les Guerillères* (trans David Le Vay **1971** UK) by Monique Wittig; *The Wanderground* (**1979**) by Sally Miller Gearheart; *Don't Bet on the Prince: Contemporary Feminist Fairy Tales in North America and England* (anth **1986**) ed Jack ZIPES; *Carmen Dog* (**1988**) by Carol EMSHWILLER; *Unquenchable Fire* (**1988**) by Rachel POLLACK; *Sarah Canary* (**1991**) by Karen Joy Fowler; *The Iron Dragon's Daughter* (**1993**) by Michael SWANWICK.

GENDER DISGUISE In most genres of fiction women dressed as men (and vice versa, though much less common) are frequently found. In the 18th and 19th centuries, such disguises flourish in innumerable Gothic tales, and in the "Mysteries" tales which urbanized the Gothic and were made famous by writers like Eugène SUE, Alexandre DUMAS and Charles DICKENS. In the 20th century, they show up in detective novels – quite often, for instance, in the **Brother Cadfael** mysteries by Ellis Peters (Edith PARGETER), which

are set in what is almost a LAND-OF-FABLE 12th-century England – and in various kinds of thriller. The example of Sue and further fabulators of the Mysteries of PARIS and other CITIES points to the usefulness of GD as a plot device in the costume drama of urban life; but the example of the **Brother Cadfael** stories demonstrates more universal points: the threatening allure of the impostor; the subversive consequences of the need of women to act as protagonists in their own stories; the arousal of discovery.

In fantasy, straightforward GD is less common than it could be, for the very good reason that the conventions of fantasy permit much more profound means of concealment. The travelling warrior in Peter S. BEAGLE's *The Innkeeeper's Song* (**1993**) does not *disguise* himself as a woman: he shapeshifts (◊ SHAPESHIFTERS) into one. Cara, in Geoff RYMAN's *The Warrior who Carried Life* (**1985**) magically transforms herself into a man to become strong and nasty enough to avenge the torment and mutilation of her family and herself. The threatening TROMPE L'OEIL indeterminacy generated by this figure derives not from clothing but from essence: what it loses as commentary on role typing it gains through its obedience to the fundamental metamorphosing structure of fantasy texts. Nor is Ozma of Oz disguised as Tip in *The Marvelous Land of Oz* (**1904**): she *is* Tip until the time comes for him to be transformed back into a princess (◊ HIDDEN MONARCH). In DEAD AGAIN (*1991*) the two main protagonists are gender-reversed REINCARNATIONS of their earlier selves.

But GDs do appear, especially in SWORD-AND-SORCERY tales, where the disguise helps create narrative suspense. There is, of course, precedent in MYTHOLOGY and LEGEND: Achilles famously "hid himself among women", and Thor had to disguise himself as Freya as a ruse to recover his stolen hammer. In the joyful melodramatic excesses of Joan AIKEN's **James III** sequence, set in an ALTERNATE-WORLD England, the discovery that Cris – in *The Cuckoo Tree* (**1971**) – is actually a girl seems no more than part of the general texture, but there are more profound examples. In J.R.R. TOLKIEN's *The Lord of the Rings* (**1954-5**) Éowyn joins the army of Rohan in disguise as Dernhelm. In John Myers MYERS's *Silverlock* (**1949**) Rosalette goes incognito as the male Nicolind partly for safety in the wild and partly to check out her boyfriend Aucando. The young protagonist of Rhoda LERMAN's *The Book of the Night* (**1984**) arrives on Iona disguised as a boy. In the movie DRAGONSLAYER (**1981**), a girl dresses as a male in order to escape the annual lottery among virgins (◊ VIRGINITY) to determine the year's sacrifice to a DRAGON; and in *The Book of Joanna: A Fantasy Based on Historical Legend* (**1947**) by George Borodin (real name George Milkomane; 1903-   ), a woman disguised as a man makes her way in the world, eventually becoming Pope, which is upsetting to the ANGELS in HEAVEN. Any fantasy involving POPE JOAN is obviously rooted in GD. Patience, the protagonist of Orson Scott CARD's SCIENCE FANTASY *Wyrms* (**1987**), spends much of her time disguised as a male: she is, after all, the rightful king of her land. The protagonist of Lucy Cullyford BABBITT's **Melde** sequence likewise deems it sensible to go dressed as a boy. In François de Salignac de la Mothe FÉNELON's *Les aventures de Télémaque* (**1699**; trans as *The Adventures of Telemachus* **1699** UK) the goddess Minerva disguises herself as a male to accompany and advise the protagonist. The character Alyss, who appears in many of John GRANT's novels, is frequently mistaken for a boy, and does not always disabuse people of their error. The occasional trope of a male hero nurturing strange lusts for a pretty young male COMPANION – before discovering that this love may dare speak its name – can be exemplified by Alexander DE COMEAU's *Monk's Magic* (**1931**).

Men disguised as women are less common, and are often treated humorously, as with Toad-as-Washerwoman in Kenneth GRAHAME's *The Wind in the Willows* (**1908**), the ghostly King Smoit of James Branch CABELL's *Jurgen: A Comedy of Justice* (**1919**) – who recalls "climbing out of gaol windows figged out as a lady abbess" – and Madmartigan, in WILLOW (**1988**), who seems habitually to disguise himself as female in order to dodge the wrathful husbands of his adulterous lovers. In *The* DEVIL-DOLL (**1936**), however, there is no humorous intent in Lavond's disguise as Mme Mandelip. Although not fantasies, both *Tootsie* (**1982**) and *Mrs Doubtfire* (**1993**) are fantastications whose plot centres on transvestite men. In ARABIAN FANTASY, of course, there are various examples of men adopting female guise in order to inveigle themselves into harems. And in the UK version of Elizabeth HAND's *Waking the Moon* (**1994** UK), one young man, sexually victimized by the GODDESS, lives subsequently for many years as a woman.

Women Who Wear the Breeches: Delicious & Dangerous Tales (anth **1995**) ed Shahrukh Husain assembles FOLKLORE and FAIRYTALE examples of gender disguise. [JC/DRL/JG]

GENIES MAGIC supernatural beings from Islamic FOLKLORE, some good, some bad, most adept at SHAPESHIFTING. They help or hinder, sometimes frivolously. A well known genie is the female version played by Barbara Eden in the US tv series, I DREAM OF JEANNIE (1965-70).

Genies are frequently confused or identified with djinns, who are supernatural beings specifically (though not necessarily accurately) identified with Arabia by Western users of traditional material. These also manifest themselves as good or evil, and shapeshift. They may be INVISIBLE; they can be trapped, and can be forced to obey those who release them from bottles or lamps, etc. They are lower than ANGELS, but not necessarily much. They appear in many ARABIAN FANTASIES. The djinn in ALADDIN (**1992**) is described as a genie. [JC]

GENRE FANTASY This is an exceptionally difficult term to define – and even more so to define without pejorative implications. GF is almost always HIGH FANTASY, HEROIC FANTASY or SWORD AND SORCERY, and its main distinguishing characteristic is that, on being confronted by an unread GF book, one *recognizes* it; one has been here before, and the territory into which the book takes one is familiar – it is FANTASYLAND. The characters, too, are likely to be familiar: HIDDEN MONARCHS, UGLY DUCKLINGS, DWARFS, ELVES, DRAGONS . . . In short, GF is not at heart fantasy at all, but a comforting revisitation of cosy venues, creating an effect that is almost anti-fantasy. An allied point is that GFs cater in large part for unimaginative readers who, through the reading of a GF, can feel themselves to be, as it were, vicariously imaginative. This goes exactly counter to the purpose of the full fantasy, which is to release or even to catapult the reader into new areas of the imagination.

All this might seem to suggest that GF is universally imitative dross, but this is not the case: while much is indeed formulaic stuff emitted by publishers to fulfil their monthly quotas and bought by readers who seek reassuring works with which they can in effect hold a phatic discourse (or just while away a long train journey) – here there be no tygers –

various authors have used the mode knowledgeably to produce works that are of undoubted interest.

The hallmark of GF is that it is set in a SECONDARY WORLD (in the broadest sense of the term). In less imaginative works this is just a granted – a Fantasyland derivative of J.R.R. TOLKIEN's Middle-Earth. But it can more interestingly be an ALTERNATE REALITY or ALTERNATE WORLD, accessible from this one via some kind of PORTAL or shift in PERCEPTION. Alternatively, it can be in the distant past (◊ PREHISTORIC FANTASY); in an outstandingly interesting piece of GF Julian MAY creates in her **Saga of the Exiles** series a prehistoric Fantasyland that is accessible only through the use of a future technology, that future being not recognizably our own. The future itself is a useful place to put one's secondary world, providing a locale where, typically, MAGIC works or misunderstood science is so sophisticated that it doubles for magic; an example is John CHRISTOPHER's **Prince in Waiting** trilogy. Millennia further into the future, the secondary world may be another planet – which is the province of SCIENCE FANTASY – or it may be our own, but now vastly changed (◊ DYING EARTH). Other possible locales include Arthurian Britain, LANDS OF FABLE, GAMEWORLDS and the interior of the HOLLOW EARTH.

The constraints of GF can bring out the best in some authors. Tad WILLIAMS, in his **Memory, Sorrow and Thorn** series, mixes traditional themes, many drawn from FOLKTALES, to create a story that, while set in a Fantasyland, is able nevertheless to present a philosophical argument. The early novels of Colin GREENLAND – notably *Other Voices* (**1988**) – *used* the conventions of GF and Fantasyland to make ironic points about our mundane existence. Terry PRATCHETT's **Discworld** series plunders GF and its conventions for purposes of PARODY and often something more: *Small Gods* (**1992**), for example, is a blistering SATIRE on RELIGION. Richard MONACO's *Parsival, or A Knight's Tale* (**1977**) exuberantly exploits the Arthurian Fantasyland to produce a vicious REVISIONIST FANTASY.

The depressing truth, however, is that GF is by and large poor and that a very great deal of it is published – to the detriment of full fantasy, which is often presented in an indistinguishable format. Quite how much commercial damage publishers are doing to the fantasy genre as a whole through this short-termism is hard to establish – and likely will be for some years – but there is considerable anecdotal evidence to suggest the wound is deep. [JG]

GENTLE, MARY (ROSALYN) (1956-) UK writer well known for her sf, but who has spent most of her career writing fantasy, including her first novel, the YA *A Hawk in Silver* (**1977**; rev 1985 US). Within a POLDER in southern England lives in exile a remnant of the folk of ATLANTIS; to humans they manifest themselves as denizens of FAERIE. A group of adolescents interacts with them. There is nothing in the tale to indicate that MG would not continue with further books of the same sort.

After the **Orthe** sf sequence – *Golden Witchbreed* (**1983**) and *Ancient Light* (**1987**) – she moved on to a very much more challenging kind of fantasy, the **White Crow** sequence: *Rats and Gargoyles* (**1990**), *The Architecture of Desire* (**1991**) and the main story in *Left to His Own Devices* (coll **1994**), plus two of the stories assembled in *Scholars and Soldiers* (coll **1989**). Most of *Rats and Gargoyles* takes place in the "city called the heart of the world" in an ALTERNATE REALITY where giant rats (◊ MICE AND RATS) are dominant, and which is governed by a committee-like PAN-THEON of GODS, 36 in number. The main characters are human, including the female MAGUS and soldier White Crow, and they become involved in a subterranean human plot to overturn the rule of the rats; but the central interest of the novel lies in its use of hermetic principles – mainly the thesis of correspondence (◊ AS ABOVE, SO BELOW; MICRO-COSM/MACROCOSM) according to which each part of the Universe corresponds with every other part, when properly ordered so as to reveal the correspondence. In the novel, Lord Architect Casaubon's GARDEN and LABYRINTHS work both to facilitate understanding of the Universe and to manifest the harmony of that Universe. "Beggars in Satin" and "The Knot Garden", from *Scholars and Soldiers*, address the same problems of attaining conjunction.

The Architecture of Desire is set in an ALTERNATE-WORLD version of Elizabethan England, where White Crow (◊ TEMPORAL ADVENTURESS) operates as a physician. While Casaubon (here and now her husband) attempts to understand the metaphysical corruption which prevents the construction of a temple in LONDON, White Crow rapes a young woman who is in her medical care. The woman suicides, and the fantasy turns into a tragedy whose implications savagely question the HEROIC-FANTASY *sang-froid* of the heroine. The title novella of *Left to His Own Devices* carries White Crow and Casaubon into a TECHNOFANTASY near-future London where CHAOS is increasing. Underlying the tale, and much of the sequence as a whole, are quotes and echoes from, and imitations of, Jacobean drama. The sense of exorbitant unreality thus generated works as a counter to the densely felt texture of MG's scene-setting; and may well consciously operate to free both readers and protagonists from any undue adherence to any one world. The **White Crow** sequence may have ended; but because its various worlds very much resemble a MULTIVERSE, no narrative terminus can be seen as final.

MG's only other sustained fantasy, *Grunts!* (**1992**), is an extended spoof of various conventions, within and without the field, treating the LAST BATTLE typical of GENRE FANTASY from the viewpoint of the orcs who serve on the "wrong" side. With Roz KAVENEY, and with the help of Neil GAIMAN, MG edited and contributed to three SHARED WORLD anthologies: **The Weerde** – *The Weerde: Book 1* * (anth **1992**) and *The Weerde II* * (anth **1993**), the latter also with Alex Stewart – and *Villains!* * (anth **1992**). The first follows a PARIAH ELITE of alien SHAPESHIFTERS in various adventures through history; the second takes the side of the VILLAINS in genre fantasy.

MG is a writer of violence: her style is muscular and thrusting; her conclusions tend to be dangerous; and her stories are never comfortable. [JC]

GEOFFREY OF MONMOUTH (?1100-?1154) Welsh chronicler and churchman, probably of Breton stock, commissioned by the newish Norman masters of England to write a history. Claiming his work was a translation of old Welsh manuscripts which no one else had seen – comparisons with Joseph Smith (1805-1844) and Colonel James Churchward (◊ MU) are inescapable – with borrowings from the MABINOGION, he wrote *Historia Regum Britanniae* (**1136**), generally known today as *The History of the Kings of Britain*, an almost completely fictitious history of British rulers, whom GOM claimed to have traced back to the Trojans. As a sidebar he produced *Prophetiae Merlini* ["Prophecies of Merlin"] (**1134**), and much later he wrote

Vita Merlini ["Life of Merlin"] (?**1155**). The importance of these books is that they brought the LEGENDS of ARTHUR and MERLIN into the forefront of European literature; moreover that they blended those two legends, even though GOM seems to have realized that the Welsh MAGUS Merlin (Myrddin) lived a century or so after Arthur. In terms of influence, *Historia Regum Britanniae* is one of the most important medieval texts: TAPROOT-TEXT writers affected by it include William SHAKESPEARE, François RABELAIS and particularly Edmund SPENSER, who essentially plagiarized large parts of it for *The Faerie Queene* (**1589**), and Sir Thomas MALORY. GOM's influence continues today. [JG]

See also: ANTHOLOGIES; AVALON; CAMELOT; EXCALIBUR; GAWAIN; MATTER; MORGAN LE FAY.

GERBER, STEVE US COMICS artist (1947-). ◊ HOWARD THE DUCK.

GERMANY The best start for a survey of German fantasy is perhaps the late 18th century, when ghostly novels of knights and robbers – widely translated into French and English – were a seminal influence on the development of the early Gothic novel, and were in turn influenced by it (◊ GOTHIC FANTASY). Nearly all the hundreds of such novels are forgotten, although they enjoyed a tremendous popularity in their day. One of the most successful authors was Christian Heinrich Spiess (1755-1799), an actor and playwright. Typical of his work is *Das Petermännchen* (**1792**) which influenced Matthew Lewis's *The Monk* and Ann RADCLIFFE. It is the story of an evil GHOST who assumes the form of a DWARF and entices a knight to rape six innocent girls, to live with his own daughter in incestuous matrimony, and to kill more than 70 people, until the DEVIL takes him and tears him apart. *Der alte Überall und Nirgends* ["The Old Everywhere and Nowhere"] (**1792-3**) tells of a knight of Charlemagne who is sentenced to death for his assumptions and begins a ghostly existence, always led astray by his passions. These novels abound in perversions. The author ended in madness, after having eloquently warned against the causes of madness in his *Biographien der Wahnsinnigen* ["Biographies of the Mad"] (**1795-6**). Even more successful were the prolific Karl Gottlob Cramer (1758-1817) and the Austrian Joseph Alois Gleich (1772-1841), who anonymously and pseudonymously authored more than 100 novels. A host of imitators included Georg Karl Ludwig Schöpfer (1811-?), Karl Friedrich Kahlert (1758-1813), Ignaz Ferdinand Arnold (1774-1812) and Carl Grosse (1774-1847), whose *Der Genius* (**1774-94**; trans P. Will as *Horrid Mysteries* **1796** UK), is a labyrinthine novel of ubiquitous secret societies, was a favourite of BYRON's generation and is still reprinted.

The next and superior wave of German fantasy arrived with the Romantic interest in the nightside of human nature. It was wittily transformed in the FOLKTALES of J.K. MUSÄUS, in the stories of the great Romanticists Ludwig TIECK, Wilhelm HAUFF and Achim von Arnim (1781-1831). Friedrich de la Motte FOUQUÉ's historical romances appear stilted, but his unforgettable tale of the unfortunate water spirit UNDINE is a classic, as is Adelbert von CHAMISSO "Peter Schlemihls wundersame Geschichte" (**1814**; trans as *Peter Schlemihl* **1824** UK). But the greatest fantasist of them all was E.T.A. HOFFMANN, whose work combines elements of HORROR (especially in "Der Sandmann" (1817) with the whimsical and the poetically romantic. "The Sandman" was subject of a famous interpretation by Sigmund Freud (1856-1939) on the UNCANNY (the plucking out of eyes as symbolic castration). The touching fairytale "Der goldene Topf" ["The Golden Flower-Pot"] (1813-14) contrasts a philistine everyday world with the fantasy world of ATLANTIS, the goal of artistic aspirations and the yearnings for a purer state of being, symbolized in the love of the naive dreamer Anselmus for the etheric Serpentina. *Die Elixiere des Teufels* (**1815-16**; trans as *The Devil's Elixir* **1824** UK) is an impressive psychological study, in which the DOPPELGÄNGER dramatizes the inner conflict in the protagonist's character; one side of it aspires to become a better human being while the other wants to live out in full his sensual nature. Wilhelm Meinhold (1797-1851) is remembered as the author of two important WITCH novels, *Die Bernsteinhexe* (**1843**; trans as *The Amber Witch* **1844** UK; vt *Mary Schweidler*) and *Sidonia von Bork, die Klosterhexe* (**1847-8**; trans as *Sidonia the Sorceress* **1847** UK), written in a documentary vein. Ludwig Bechstein (1801-1860), whose collections of fairytales were more popular than those of the GRIMM BROTHERS, wrote also a volume of *Hexengeschichten* ["Witch Tales"] (coll **1854**). Of lesser stature were the ghost stories written by Johann August APEL and Friedrich Laun, issued in four volumes of *Gespensterbuch* ["Ghost Book"] (colls **1810-12**), followed by three volumes of *Wunderbuch* ["Miracle Book"] (**1815-17**). Apel's retelling of the folktale "Der Freischütz" formed the basis for Friedrich Kind's libretto for Weber's OPERA of the same name about the charmed hunter who uses magic bullets.

The most important 19th-century German writer of fantastic stories was probably Peter Alexander, Freiherr von Ungern-Sternberg (1806-1868), a Baltic author who wrote numerous stories that reflect the division of the self between progress and reaction, the Classic and the Romantic, man and woman. His best collection is *Die Nachtlampe* ["The Night Lamp"] (coll **1854**), but many stories remain uncollected.

Around the turn of the century there began a strong German fantasy wave, a movement of black Romanticism. The Austrian Karl Hans Strobl (1877-1946) published his first book, *Aus Gründen und Abgründen* ["From Ground and Abysses" or "Of Reasons and Unreasons"] (**1901**) and credited himself as the renovator of German fantasy, but there were others before him – among them Karl von Schlözer (1854-1916) with *Aus Dur und Moll* (**1885**), Bodo Wildberg (1862-1942) with *Tötliche Triebe* ["Deathly Instincts"] (coll **1894**), and especially Oskar Panizza (1853-1921) who spent a year in prison for blasphemy for his fantastic play "Das Liebeskonzil" ["The Love Council"] (**1894**) before being put in an asylum (where he had earlier been a psychiatrist), which he never left. Two of his early collections were *Dämmerungsstücke* ["Pieces for Twilight"] (coll **1890**) and *Visionen* ["Visions"] (coll **1893**). Panizza's stories are bitingly anti-religious but also sometimes marred by a violent antisemitism. Also abundant at the time were occult writings; the subgroup of the didactic occult novel is represented by *Das Kreuz am Ferner* ["The Cross at the Ferner"] (**1897**) by Karl du Prel (1839-1899).

The rise of fantasy, then usually called "strange stories", sprang from the spirit of the FIN DE SIÈCLE and French DECADENCE. J.K. HUYSMANS, Théophile GAUTIER, Barbey D'Aurevilly (1808-1889), Gérard de NERVAL and VILLIERS DE L'ISLE-ADAM exerted their influence. Edgar Allan POE excited interest, and there was a re-evaluation of E.T.A. HOFFMANN following his French fame. It is interesting to note that, while French writers were widely translated into German, at this stage UK fantasists, aside from Oscar

WILDE, played no role. *Dracula* (**1897**) by Bram STOKER was translated in 1908 but went largely unnoticed, and apart from the adventure novels of H. Rider HAGGARD there was only Richard Marsh's *The Beetle: A Mystery* (**1897**), but no Algernon BLACKWOOD, no Arthur MACHEN and no William Hope HODGSON. Their discovery was left to the tide of translated fantasy in the 1970s.

Many Austrian writers – e.g., Gustav MEYRINK and Karl Hans Strobl (1879-1946) – played an important role in this fantasy revival; it is anyway worthless to distinguish according to nationality, for the German-speaking countries have always formed a common book market, and many Austrian writers published in Germany. Meyrink first captured his audience with satirical and uncanny stories published in the Munich satirical paper *Simplicissimus* – later collected in several volumes – and went on with *The Golem* (**1915**), one of the big German bestsellers of WWI, and continued with a series of increasingly didactic novels advocating occult doctrines. The leading role among the German nationals was undoubtedly played by Hanns Heinz EWERS. Like the Polish-German writer Stanislaw Przybyszewski (1868-1927), Ewers was a decadent *par excellence*. Ewers's flirtation with the Nazis did not save him from falling out of favour with them, and his books were soon forbidden. He embodied his dreams of a Nietzschean hero (who is nevertheless easily seduced by LILITH figures) in **Frank Braun**, the protagonist of a trio of fantastic novels.

Other fantasy writers of the time included the prolific Georg von der Gabelentz (1868-1940), Paul Ernst 1866-1933), Isolde Kurz (1853-1944), Kurt Münzer (1879-1944), the Swiss Gustav Renker (1889-1967), Willy Seidel (1887-1934), Oscar A.H. Schmitz (1873-1931), Wilhelm von Scholz (1874-1969), Hermann Esswein (1877-?   ), Roland Betsch (1888-1945) and Werner Bergengruen (1892-1964).

Paul Scheerbart (1863-1915) wrote a large number of whimsically cosmic, Oriental and other fantasies, often with mini-stories included in a FRAME STORY; his works are unique, creating a private cosmos. Scheerbart starved to death in protest, it is claimed, against WWI: he had always written about the martial follies of disgusting beings that inhabited the Earth's crust.

The numerous romances of Eufemia Adlersfeld-Ballestrem (1854-1941) contain often supernatural motifs and episodes (visions, prophetic dreams, ghostly appearances) and the paraphernalia of GOTHIC FANTASY's (secret rooms, oubliettes, the poisoned rings of Italian Renaissance FEMME FATALES). Two of her novels, *Ca'Spada* (**1904**) and *Die Dame in Gelb* ["The Lady in Yellow"] (**1904**) are outright GHOST STORIES in the UK manner.

A recent anthology of German "strange tales" from the beginning of the century is Robert N. Bloch's (1950-   ) *Jenseits der Träume* ["Beyond Dreams"] (coll **1990**).

The flourishing of German fantasy increased even after WWI, when it was felt that stronger thrills were needed following the mass slaughter, and that new sensibilities had developed. The major figures of German fantasy edited their own series: Ewers with the 8-volume **Galerie der Phantasten** ["Gallery of Fantasists"], containing material by Hoffmann, Poe, Balzac, Ewers himself, Strobl, Panizza, Alfred Kubin (1877-1959) and the Spaniard Gustavo Adolfo Becquer (1836-1870); Strobl with the 6-volume **Geschichten um Mitternacht** ["Tales for Midnight"], containing material by Strobl himself, Ewers, Poe, Villiers de l'Isle-Adam, Gogol and Hoffmann; Meyrink with the 5-volume **Romane und Bücher der Magie** ["Romance and Books of Magic"]. Strobl also edited the beautifully illustrated magazine *Der Orchideengarten* ["The Garden of Orchids"] (1919-1921), which was likely the first specialized fantasy magazine in the world. Publishers especially active in the fantasy field were Georg Müller, Albert Langen, Drei-Masken-Verlag (with their **Sindbad Books**) and Rikola. Many others published some fantastic fiction at all levels of literary quality, from the heights of Franz KAFKA to forgotten amateur writers. There existed also an even wider field of occult writings, including those by the notorious Aryan racist occultist Lanz von Liebenfels, the eponym of Wilfried Daim's *The Man who Gave Hitler his Ideas* (1958). While most writers of fantasy were rather conservative or even reactionary, an exception was Alexander Moritz Frey (1881-1957). Although he had served with Hitler in the trenches during WWI and was offered by him a high position at the Nazi paper *Der völkische Beobachter*, Frey, who was a strict antimilitarist, declined and emigrated in 1933 to Austria and then in 1938 to Switzerland, where he lived in poverty, a forgotten writer whose books were banned by the Reich. Besides a satirical novel, *Solneman der Unsichtbare* ["Nameless the Invisible"] (1914), he wrote many quiet, controlled and mysterious short stories, sometimes with an Expressionist touch, and collected in many volumes, including *Der Mörder ohne die Tat* ["The Murderer without the Murder"] (coll 1918), *Spuk des Alltags* ["Spookery of Everyday Life"] (coll 1920) and *Aussenseiter* ["Outsiders"] (coll 1928).

Also important during that period was the fantastic cinema, which, according to Siegfried Kracauer's *From Caligari to Hitler* (1947), provided the imaginative prehistory of the Nazi system. Although fantastic movies formed only a minority of cinematic productions, a number were important, from Ewers's DOPPELGÄNGER study *Der Student von Prag* ["The Student of Prague"] (*1913*) via the GOLEM movies by Paul Wegener to F.W. MURNAU's adaptation of *Dracula*, called for copyright reasons *Nosferatu* (◊ DRACULA MOVIES). 1919 saw R. Wiene's famous *The* CABINET OF DR CALIGARI.

The rise of the Nazis ended this rich flowering of the arts. Unlike sf, which continued to be published right up to the end of the Third Reich, fantastic literature of the uncanny, the morbid and the outré virtually ceased to be published, with only a few exceptions, mostly by mainstream writers, such as the humorous John COLLIER-like tales of Kurt Kuseberg (1904-1983) – in *La Botella* (coll 1940) and *Der blaue Traum* ["The Blue Dream"] (coll 1942) – and the work of Heinrich Schirmbeck (1913-).

After WWII fantasy in Germany was mostly a matter of translation, with all the important English-language writers appearing. Book series of interest included **Die Bibliothek des Hauses Usher** ["Library of the House of Usher"] (1969-75); these 26 volumes, ed Kalju Kirde for Insel Verlag, introduced classic authors of weird fiction – H.P. LOVECRAFT, Arthur MACHEN, J.P. HODGSON, M.R. JAMES, Algernon BLACKWOOD and Clark Ashton SMITH – but also the Belgians Jean Ray (1887-1964) and Thomas Owen (1910-) and the Polish fantasist Stefan GRABINSKI. The 14 volumes of **Bibliotheca Dracula** (1967-74) published by Hanser Verlag focused on Gothic novels, but presented also four anthologies on WEREWOLVES, artificial beings, VAMPIRES and the BLACK MASS.

Fantasy by German authors was mostly restricted to juveniles and *Romanhefte* – the saddle-stiched publications distributed through newsstands – but there was a strong streak of fantasy in mainstream fiction, from Günter GRASS to Wolfgang Hildesheimer (1916-) and Michael Schneider (1943-), with *Das Spiegelkabinett* ["The Room of Mirrors"] (**1980**). Especially important was the fantasy by writers of the former German Democratic Republic: Rolf Schneider (1932-), Irmtraud Morgner (1933-), Günter Kunert (1929-) and Anna Seghers (1900-1983). The first German novel after WWII to be internationally acclaimed and to be translated into several languages was *Die Gesellschaft vom Dachboden* (**1946**; trans Robert Kee as *The Attic Pretenders* 1948 UK) by Ernst Kreuder (1930-1972), which tells of a group of young people who have found refuge in an attic from the mundanity of normal life. Some remarkable ghost stories were written by Marie Luise Kaschnitz (1901-1974), especially in the collection *Lange Schatten* ["Long Shadows"] (coll **1960**). There is a strong element of fantasy and the grotesque in the work of Herbert ROSENDORFER, who burst onto the literary scene with *Der Ruinenbaumeister* (**1969**; trans Mike Mitchell as *The Architect of Ruins* 1993 UK), an involved novel built from stories-within-stories. *Stephanie und das vorige Leben* (**1977**; trans Mike Mitchell as *Stephanie, or A Previous Existence* 1996 UK) tranfers a personality into the past.

The most successful German fantasy author has been Michael ENDE. He consistently used fantastic themes in his CHILDREN'S FANTASIES. *Jim Knopt und Lukas der Locomitvefürher* (**1960**; trans Maurice S. Dodd as *Jim Button and Luke the Engine Driver* 1990 US) was well received, but *Die unendliche Geschichte* (**1979**; trans Ralph Manheim as *The Neverending Story* 1983 US) was a major hit.

Also eminently successful, but at a much lower literary level, is the work of the prolific Wolfgang E. Hohlbein (1953-). He writes all kind of fantasy, from SWORD AND SORCERY to HORROR. Hohlbein started with stories in fan magazines, went on to write *Hefte* – the occult series **Professor Zamorra** and **Damona King** – and won the Uebereuter fantasy competition with *Märchenmond* ["Fairy Moon"] (**1993**). This juvenile novel sold more than 100,000 copies, and Hohlbein has written one fantasy juvenile every year since. His horror novel *Das Druidentor* ["Gate of Druids"] (**1993**) was the huge Bertelsmann bookclub's bestselling book in its category.

On a more literary level is Hans BEMMANN, whose *Stein und Flöte* (**1983**; trans Anthea Bell as *The Stone and the Flute* 1986 UK) was an international success.

Unlike sf, which in Germany is strictly restricted to paperbacks, genre fantasy regularly appears in hardback, following the success of Tolkien, Ende and Marion Zimmer BRADLEY, and there have also been fantasy series in hardback and quality paperback (Klett-Cotta's "Hobbit-Presse", with Lord DUNSANY, Mervyn PEAKE, Peter S. BEAGLE and similar writers). A paperback fantasy series has been **Terra Fantasy** (94 releases 1974-82). All of the big paperback publishers issue plenty of fantasy, though this is almost exclusively translated material. A German *Hefte* fantasy series was **Dagon** (1973-5). More successful have been horror-fantasy; among long-running series has been **Dämonenkiller**, which was both a *Hefte* and a paperback series: it appeared first as a subseries within **Vampir-Horrorroman** (451 issues 1972-81), then as **Dämonkiller-Gruselroman** (143 issues 1974-7) and **Dämonenkiller-Horror-Serie** (175 issues 1983-6), with paperbacks appearing as **Dämonenkiller-Taschenbuch** (63 releases **1975-80**). The *Hefte* were by German authors, the paperbacks mostly translations. The most prolific and successful German horror writer of the *Hefte* is Helmut Rellergerd (1945-   ), who writes as Jason Dark about a ghosthunter named **John Sinclair**, whose adventures appeared first in a series of **Gespenster-Krimis** and since 1978 in its own series. Since 1981 there has also been a monthly **John Sinclair** paperback.

A theoretical discussion of fantastic literature was the five **Phaïcon** ["Image of the Fantastic"] almanacs (anths **1974-82**) ed Rein A. Zondergeld. These combined theoretical essays with fiction. A notable survey, including a bibliography, of German-language fantasy between Decadence and Fascism by Jens Malte Fischer appeared in #3 (anth **1980**). The critical anthology *Phantastik in Literatur und Kunst* ["The Fantastic in Literature and Art"] (anth **1980**) ed Christian W. Thomsen and Jens Malte Fischer combines theoretical works with author studies. Winfried Freund's *Literarische Phantastik* ["The Literary Fantastic"] (**1990**) traces the development of the fantastic novella. Peter Cersowsky, in *Phantastische Literatur im ersten Viertel des 20 Jahrhunderts* ["Literary Fantasy in the first Quarter of the 20th Century"] (**1983**) concentrates on Kafka, Kubin and Meyrink, while Marianne Wünsch in *Die Fantastische Literatur der frühen Moderne* ["The Fantastic Literature of Early Modern Times"] (**1991**) discusses the occult context of fantasy literature. Stephen Berg in *Schlimme Seiten, Böse Räume* ["Bad Times, Evil Spaces"] (**1991**) conducts a discourse on the structures of time and space in Meyrink, Kubin, Leo PERUTZ and Alexander LERNET-HOLENIA, among others. Clemens Ruthner discusses in *Unheimliche Wiederkehr* ["Uncanny Return"] (**1993**) the ghostly characters in the novels of Strobl, Franz Spunda (1890-1963), Otto Soyka (1882-1955), Meyrink and Ewers. Gero von Wilpert wrote on the motifs, narrative form and development of the German ghost story in *Die deutsche Gespenstergeschichte* ["The German Ghost Story"] (**1994**).

Most useful is *Bibliographie der utopischen und phantastischen Literatur 1750-1900* ["Bibliography of Utopia and the Fantastic in Literature 1750-1900"] (**1984**) compiled by the antiquarian Robert N. Bloch. Excellent is the brief and personal *Lexikon der phantastischen Literatur* ["Lexicon of Fantastic Literature"] (**1983**) by Rein A. Zondergeld. Profiles and bibliographies of many important German authors are included in *Bibliographisches Lexikon der utopischen Literatur* (1984-current) ed Joachim Körber. Especially valuable in it are the contributions of Robert N. Bloch, who has also written many descriptions of old texts in *Werkführer durch die utopischen phantastische Literatur* (1989-current) ed Michael Koseler and Franz Rottensteiner. Fantasy of the Tolkien kind is primarily covered in publications of the Erste Deutsche Fantasy Club ["The First German Fantasy Club"]. The Inklings Gesellschaft für Literatur und Aethetik publishes a printed yearbook with some interesting material on Charles WILLIAMS, George MACDONALD, C.S. LEWIS and J.R.R. TOLKIEN.   [FR]

GHIDORAH SANDAI KAIJU CHIKYU SAIDAI NO KESSAN ot of *Ghidora, The Three-Headed Monster* (*1965*). ◊ GODZILLA MOVIES.

GHIDORA, THE THREE-HEADED MONSTER (ot *Ghidorah Sandai Kaiju Chikyu Saidai No Kessan*; vt *Chikyu Saidai No Kessan*; vt *The Biggest Battle on Earth*; vt *The Biggest Fight on Earth*; vt *Ghidrah*; vt *Monster of Monsters*) Japanese

movie (*1965*). ◊ GODZILLA MOVIES.

GHIDRAH vt of *Ghidora, The Three-Headed Monster* (*1965*). ◊ GODZILLA MOVIES.

GHOST US movie (*1990*). Paramount/Howard W. Koch. **Pr** Lisa Weinstein. **Exec pr** Steven-Charles Jaffe. **Dir** Jerry Zucker. **Vfx** Laura Buff, Richard Edlund, Terry Frazee, Industrial Light & Magic, Kathy Kean, Bruce Nicholson, John Van Vliet. **Screenplay** Bruce Joel Rubin. **Starring** Rick Aviles (Willie Lopez), Whoopi Goldberg (Oda Mae Brown), Tony Goldwyn (Carl Bruner), Demi Moore (Molly Jensen), Vincent Schiavelli (Subway Ghost), Patrick Swayze (Sam Wheat). 127 mins. Colour.

Not long after discovering a financial fraud, yuppie Sam is killed by a "mugger", Lopez. As a GHOST he haunts the apartment he and lover Molly shared. He sees the "mugger" searching there for something and planning to kill Molly. He enlists fake medium Oda Mae (◊ SPIRITUALISM) – who is aghast to discover she *can* actually hear a spirit voice – to warn Molly of the danger. Molly, reluctantly persuaded, tells all to Sam's quondam best friend Carl, in fact the criminal behind it all. Thereafter the tale is a fairly orthodox thriller, its fantasy interest lying primarily in: the lessons, given to Sam by a paranoid ghost haunting the subway, on how to manipulate matter; the appearance of DEMONS to take the SOULS of the dead to HELL; Oda Mae's suddenly increased clientele among the spirits with whom she is now able to converse freely; and the discovery by Sam and Oda Mae alike of the phenomenon of POSSESSION, whereby Sam can briefly adopt her body but at great penalty in terms of his own exhaustion.

The spfx are generally splendid and *G* was hugely successful; it sparked off a small subgenre of dead-lover movies, including TRULY MADLY DEEPLY (*1990* tvm), GHOSTS CAN'T DO IT (*1990*) and DEAD AGAIN (*1991*). Notable among precursors to them all were *The* GHOST AND MRS MUIR (*1947*) and *The* DEAD CAN'T LIE (*1988*).   [JG]

GHOST, THE (magazine) ◊ *The* RECLUSE.

GHOST AND MRS MUIR, THE US movie (*1947*). 20th Century-Fox. **Pr** Fred Kohlmar. **Dir** Joseph L. Mankiewicz. **Spfx** Fred Sersen. **Screenplay** Philip Dunne. **Based on** *The Ghost and Mrs Muir* (**1947**) by R.A. Dick (Josephine LESLIE). **Starring** Edna Best (Martha Huggins), Vanessa Brown (adult Anna), Rex Harrison (Captain Daniel Gregg), George Sanders (Miles Fairley), Gene Tierney (Lucy Muir), Natalie Wood (Anna). 104 mins. B/w.

An exquisite FIN DE SIÈCLE romantic fantasy. Widow Lucy – with daughter Anna and maid Martha – rents seaside Gull Cottage, which proves a HAUNTED DWELLING. The GHOST, that of a sea captain, Gregg, tries at first to scare her away as he has done earlier tenants, but she defies him. In due course they fall in love. He dictates to her his autobiography which (*TGAMM* is fantasy, after all) a London publisher immediately buys on excellent terms. But she also meets suave lothario Fairley (a children's author, "Uncle Neddy"), and falls for him. Gregg does the decent thing and ceases to haunt her, deliberately twisting her PERCEPTION of events such that she believes their experiences together were all a DREAM. But Fairley proves a deceiver, so Lucy settles down to a life of decent widowhood. Long after, however, the now grown-up Anna confesses that she, too, used to speak with Gregg; Lucy spends her final years expecting her lover to be there to meet her when she dies, which sure enough he does.

TGAMM was one of a spate of filmed romantic GHOST

STORIES – including BLITHE SPIRIT (*1945*) and PORTRAIT OF JENNIE (*1948*) – and is probably the best, largely because of the screenplay, which has greater depth than most, and includes some interesting speculation as to what ghosts actually *are*. [JG]

GHOST BOOK, THE UK ANTHOLOGY series. The first volume, *The Ghost Book* (anth **1926**) ed Cynthia ASQUITH was intended as a one-off. It contained mostly literary GHOST STORIES, either original or reprinted from select sources; authors included Algernon BLACKWOOD, Clemence DANE, Walter DE LA MARE, L.P. HARTLEY, D.H. LAWRENCE, Arthur MACHEN, Oliver ONIONS, May Sinclair (1863-1946) and Hugh Walpole (1884-1941). The stories were not traditional, instead exploring the more psychological aspects of POSSESSION. The volume was a critical success. Asquith did not return to TGB for over 25 years, then producing *The Second Ghost Book* (anth **1952**; vt *A Book of Modern Ghosts* 1953 US), which further updated the ghost story and set a firm basis for later volumes. *The Third Ghost Book* (anth **1955**), with its critically important introduction by L.P. HARTLEY, was Asquith's last. The series was revived by James Turner (1909-1975) for *Fourth* (anth **1965**), then continued by Rosemary Timperley (1920-1988) for *Fifth* (anth **1969**), *Sixth* (anth **1970**; vt in 2 vols as *The Blood Goes Round and Other Stories* 1972 and *The Judas Joke and Other Stories* 1972), *Seventh* (anth **1971**), *Eighth* (anth **1972**) and *Ninth* (anth **1973**). Further editors were: Aidan CHAMBERS for *Tenth* (anth **1974**) and *Eleventh* (anth **1975**), these two vols being assembled as *The Bumper Book of Ghost Stories* (omni **1976**); Polly Parkin (uncredited) for *Twelfth* (anth **1976**); and James Hale with *Thirteenth* (anth 1977), assembled with *Twelfth* as *The Second Bumper Book of Ghost Stories* (omni **1978**). Under Hale's editorship the series was retitled, starting with *The Midnight Ghost Book* (anth **1978**; vt *The Third Bumper Book of Ghost Stories* 1979). It changed publisher from Barrie & Jenkins (by now part of the same conglomerate) for *The After Midnight Ghost Book* (anth **1980**; vt *The Fourth Bumper Book of Ghost Stories* 1981), technically the last of the series, although Hale's *The Twilight Book* (anth **1981**), from Gollancz, can be considered a spiritual continuation. All are faithful to Asquith's original series, presenting literary ghost stories that explore psychological aspects in modern-day settings; most stories are original to the series. Regular contributors included Joan AIKEN, George Mackay BROWN, John F. Burke (1922-   ), Elizabeth Fancett, Giles Gordon (1941-   ), Dorothy K. Haynes (1918-1987), William Sansom (1912-1976), Rosemary Timperley and William Trevor (1928-   ). [MA]

GHOSTBUSTERS There have been two movies in this series.

1. *Ghostbusters* US movie (*1984*). Columbia. **Pr** Ivan Reitman. **Exec pr** Bernie Brillstein. **Dir** Reitman. **Spfx** John Bruno, Richard Edlund, Chuck Gaspar, Mark Vargo. **Screenplay** Dan Aykroyd, Harold Ramis. **Novelization** *Ghostbusters* * (**1985**) by Richard Mueller. **Starring** William Atherton (Walter Peck), Aykroyd (Ray Stantz), Ernie Hudson (Winston Zeddemore), Rick Moranis (Louis Tully), Bill Murray (Peter Venkman), Annie Potts (Janine Melnitz), Ramis (Egon Spengler), Sigourney Weaver (Dana Barrett). 107 mins. Colour.

Three parapsychology researchers – Spengler, Stantz and Venkman – are fired from their cozy academic sinecure, and in desperation found Ghostbusters, a commercial company offering the spectral equivalent of vermin clearance: using an invention of Spengler's they can capture GHOSTS electronically and imprison them in a containment system. As it happens, the city is suffering a plague of GHOSTS; new employee Zeddemore believes this to be a sign of the imminence of the LAST JUDGEMENT, and is proved approximately right: the building in which dwells their first customer, Barrett, proves an EDIFICE deliberately constructed as a sort of psychic nexus to facilitate the apocalyptic return of the Babylonian GOD Gozer. A mad bureacrat, Peck, releases the contents of the Ghostbusters' containment system, and NEW YORK is ravaged. However, the Ghostbusters' technology proves in the end superior to the god, and all ends happily.

G was originally intended as an Aykroyd-John Belushi vehicle called *Ghostsmashers*, but Belushi died before shooting started; Murray was drafted in for the hastily rewritten movie, and stole it. Even so, the often witty dialogue (and excellent spfx) fail to disguise a flaccid plot: only in the final sequence, when Gozer presents him/herself, is there any true frisson of otherness. [JG]

2. *Ghostbusters II* US movie (*1989*). Columbia. **Pr** Ivan Reitman. **Exec pr** Bernie Brillstein, Michael C. Gross, Joe Medjuck. **Dir** Reitman. **Vfx** Dennis Muren. **Screenplay** Dan Aykroyd, Harold Maris. **Novelizations** *Ghostbusters II* * (**1989**) by Ed Naha and *Ghostbusters II* * (**1989** chap) by B.B. Hiller. **Starring** Aykroyd, Hudson, Moranis, Murray, Potts, Ramis and Weaver as in **1**, plus Kurt Fuller (Jack Cartermeyer), Peter MacNicol (Janosz/Vigo). 102 mins. Colour.

Five years after the events of **1**, Barrett has married and divorced, and has an eight-month-old baby, Oscar. A psychic force is clearly assailing the child, and she calls in the Ghostbusters team – now on the verge of bankruptcy – to investigate. Soon they discover a river of pink slime infesting the sewers beneath New York: clearly a major spectral event is imminent – and in due course they realize the slime is a manifestation of the negative emotions of Manhattan's citizens. At the Manhattan Museum of Art (where Barrett is working temporarily) the PICTURE of sorcerer tyrant Vigo the Carpathian in fact contains Vigo's SOUL. It seeks to re-enter and rule the world, but must find a child to possess (◊ POSSESSION); it also inspires the slime. The Ghostbusters, to counter all this, energize the Statue of Liberty and march it to save Barrett and Oscar and banish Vigo's spirit back to Hell.

The spfx are better and the screenplay funnier and much more inventive, yet this failed to achieve anything like the popularity of **1**. Although the way was left open for a *GIII*, and despite the continued success of the animated tv series, such a movie has failed to materialize. [JG]

GHOST DANCE In the history of the Native American nations, the phenomenon of the GD represents a final and almost wholly desperate attempt to resist the cultural genocide being visited upon them by the advance of white civilization. It was invented by a SHAMAN named Wovoka (1856-1932), a Paiute from what is now Nevada, who went through a typical shaman's trip to HEAVEN in 1886, returning from his visit to GOD and the host of the dead with the movements of the GD, and a doctrine: after a great catastrophe (◊ HOLOCAUST AND AFTER), the whites would disappear from the face of the Earth; the bison would return, greening the WASTE LAND the Whites had created; and all the dead ancestors would return bearing HEALING gifts. A new GOLDEN AGE would begin. Within a few years, various Rocky Mountain and Plains nations were performing the GD, a circular slow shuffle which followed the course of the SUN. One of the Sioux leaders who converted to the GD religion was Sitting Bull, who was killed in 1890 by the US military. Within a few months, his followers – who believed their "ghost shirts" would protect them from soldiers – were massacred at Wounded Knee.

In fantasy, where MAGIC exists, and where GODS may intervene to help the worthy at the last moment, versions of the GD may underlie particularly moving moments when the weak and the honest humble their innumerable foes, or perhaps escape into an OTHERWORLD through a PORTAL opened by the pattern of the dance. In the mundane world, the GD terminated in dead silence. [JC]

GHOST GOES WEST, THE UK movie (*1935*). London/United Artists. **Pr** Alexander Korda. **Dir** René CLAIR. **Spfx** Ned Mann. **Screenplay** Geoffrey Kerr (uncredited), Robert Sherwood. **Based on** "Sir Tristram Goes West" by Eric Keown. **Starring** Ralph Bunker (Ed Bigelow), Robert Donat (Murdoch Glourie/Donald Glourie), Everley Gregg (Gladys Martin), Elsa Lanchester (Miss Shepperton), Elliot Mason (Miss McNiff), Eugene Pallette (Joe Martin), Jean Parker (Peggy Martin), Hay Petrie (The McLaggan), Morton Selten (The Glourie). 85 mins. B/w.

Scotland, 200 years ago. Womanizing Murdoch Glourie died ignominiously and, on reaching LIMBO, was condemned by his dour father to return to haunt the family castle until he could make a representative of the hated Clan McLaggan humiliate himself.

Today. The last of the Glouries, Donald, sells the castle to US millionaire Joe Martin, who dismantles it to take it to Sunnymede, Florida, where Donald will supervise its rebuilding. But Murdoch is bound to the stones; on discovering the Awful Truth about the USA, the GHOST is mercifully allowed by his father to adopt INVISIBILITY. Martin, embarrassed the ghost no longer appears, holds a launch party where Donald will fake the HAUNTING; but Martin's business rival Bigelow proudly announces he is of Scots lineage – he is the last of the McLaggans. Murdoch manifests, forces the requisite concession from Bigelow, and is allowed to leave this plane.

This lightweight screwball GHOST STORY, Clair's first movie outside France, was a hit but represents an imaginative decline for Clair. Some of *TGGW*'s themes reappeared decades later in *High Spirits* (*1988*), where an impoverished Irish minor aristocrat tries to increase takings from US tourists by faking a castle ghost, only to spark off a real HAUNTING.

Lanchester received star billing in *TGGW* but had only a bit part; that same year, 1935, she had a far more significant (dual) role as Mary SHELLEY and the Bride in *The Bride of Frankenstein* (◊ FRANKENSTEIN MOVIES). [JG]

GHOST MOVIES ◊ HORROR MOVIES.

GHOST OF FRANKENSTEIN, THE (*1942*) ◊ FRANKENSTEIN MOVIES.

GHOST RIDER Five fantasy COMIC-book characters.

1. White-garbed cowboy vigilante with a white cloak and featureless white MASK, inspired by Vaugn Monroe's recording *Ghost Riders in the Sky* (1949). GR is the secret identity of US Marshal Rex Fury (aka The Calico Kid), who uses phosphorescent paint on both himself and his horse, Spectre, to enhance his ghostly image. He first appeared in

A1 Comics #27 (1950). Drawn by Dick Ayers (1924-   ), this GR featured in 14 issues of *A1* 1950-54, with further appearances in *Red Mask* (1954-5), *Best of the West* (1951), *Bobby Benson's B-Bar-B Riders* (1952) and *Tim Holt Comics* (1950-54). In the early stories GR was pitted against bandits and badmen, but later tales introduced supernatural and HORROR elements. Many covers featuring GR were drawn by Frank FRAZETTA. [RT/SH]

2. In 1967 Ayers revived the character, with scripter Gary Friedrich, for MARVEL COMICS as a costumed Western crimefighter (*Ghost Rider #1-#7* 1967-8), but this time GR was the secret identity of schoolteacher Carter Slade. These stories were reprinted in 1974, but by then Marvel had introduced **3**, so **2** was renamed Night Rider in the reprints. [RT/SH]

3. Motorcycle-riding avenger with a fiery skull head, created by Gary Friedrich and artist Mike PLOOG, possibly inspired by an earlier comic book character, The Blazing Skull. This GR first made his appearance in *Marvel Spotlight #5* (1972) as the bizarre ghostly *alter ego* of stuntman Johnny Blaze, forced to make a PACT WITH THE DEVIL in order to save the life of his friend Crash Simpson. By the CONTRACT Blaze loses his SOUL and gains his flaming skull head. Friedrich made full use of GR's bizarre appearance in some interesting SUPERNATURAL FICTIONS, including conflicts with Damien Hellstrom, son of SATAN; succeeding writers made GR more of a routine SUPERHERO. A regular comic book devoted entirely to GR (*Ghost Rider #1-#81* 1973-83) developed the concept, and the details of Blaze's deal with Satan (later changed to Mephisto) were altered such that GR was now defined as Zarathos, a DEMON inhabiting Blaze's body. In the final issue Zarathos was banished to make a happy ending. [RT/SH]

4. *Alter ego* of Danny Kvetch, retaining the flaming skull. This version first appeared in *Ghost Rider #1* (1990), with art by Javier Saltares; the title is still current (1995). A slicker, more modern look and a high-tech bike were matched by a grittier, more modern style of story. Marketing ploys have played a part in the success of this incarnation – e.g., a glow-in-the-dark cover on *#15*, a popup section in *#25*, reprints of the earlier series with gold ink on the covers, and a host of crossover stories featuring other Marvel characters. [RT/SH]

5. A 21st-century GR appeared in *Guardians of the Galaxy #13-#14* (1992), tagged as The Spirit of Vengeance. [RT/SH]

GHOSTS Ghosts are spirits of (generally) the dead who remain earthbound and may be perceived by humans. Ghosts are normally in SPIRIT form, and may be visible only to those who are psychic. Spirits may communicate via mediums (◊ SPIRITUALISM) or may manifest themselves in physical form, drawing either upon the spirit substance, ectoplasm, or on the energies of others present (which usually results in a sense of coldness). Ghosts may attach themselves to surroundings known to them in life, mostly commonly HAUNTED DWELLINGS; but they may equally attach themslves to other objects or people. This attachment may become sufficiently strong that the spirit takes the body over (◊ POSSESSION).

Ghosts may be manifestations of other spirits, especially DEMONS. Ghosts may prey more on the mind and become a psychological manifestation which causes ILLUSION, if the HAUNTING is real, or DELUSION if wholly psychological.

Ghosts may also be spirits of the living, perceived as a

result of astral projection (◊ ASTRAL BODY). They may even be a projection of one's own spirit, or the spirit of one's DOUBLE (◊ DOPPELGÄNGERS), who often come as a warning of impending doom. Animal ghosts are common in FOLK-LORE and particularly in tribal MYTHOLOGY (◊◊ TOTEMS).

[MA]

See also: GHOST STORIES; GHOST STORIES FOR CHILDREN; POLTERGEISTS; REVENANTS.

GHOSTS AND GOBLINS ◊ MASTER THRILLER SERIES.

GHOSTS CAN'T DO IT US movie (*1990*). Epic/Sarlui-Diamant/Taprobane/Crackajack. **Pr** John Derek. **Dir** Derek. **Vfx** Cinema Research Corp. **Screenplay** Derek. **Starring** Leo Damian (Fausto Garibaldi), Bo Derek (Katie Scott), Don Murray (Winston Hill), Anthony Quinn (M.B. "Great" Scott). 83 mins. Colour.

Elderly magnate Scott commits suicide. His GHOST becomes an INVISIBLE COMPANION to his young widow Katie. Vile beach bum Garibaldi seeks her body, her mind clearly being *terra incognita*. Scott and Katie plan to kill Garibaldi so Scott may possess (◊ POSSESSION) the youthful body at the moment of death, but she baulks at murder. However, Garibaldi drowns while attempting to rob Katie, so Scott effects the possession and all ends with a bang. Like most John/Bo Derek outings (e.g., *Tarzan the Apeman* [*1981*]; ◊ TARZAN), this weak comedy is largely an excuse for its star to make savings in the costume budget. Similar economies are evident elsewhere: Quinn's scenes consist almost entirely of solo chunks of dialogue declaimed from a cut-price AFTERLIFE, separately and presumably quickly shot.

[JG]

GHOST STORIES The most common form of SUPERNAT-URAL FICTION, usually featuring an encounter with a disembodied SPIRIT or personality. The spirit is most often of someone dead, but can be of someone living (◊ DOPPEL-GÄNGER) or of someone unborn. The GHOST may take a physical form or be solely in the mind of the protagonist – the latter form of GS is usually called the psychological ghost story; its leading example is "The Turn of the Screw" (1898) by Henry JAMES. Stories involving HAUNTING by other spirits or supernatural agencies such as DEMONS and nature spirits may also be composed as GSs.

The literary GS, as either entertainment or warning, had a long gestation. GSs as discrete elements of legends and hero tales are found in all cultures. In the BIBLE's *Samuel I*, Saul instructs a medium to raise the spirit of Samuel. In HOMER's *Odyssey* (9th century BC), Odysseus visits the Underworld to consult the dead seer Tiresias. Chinese GSs abound; over 300 were captured by P'u Sung-Ling (1640-1715) in *Liao-Chai chih i* (coll 1679; selection trans Herbert A. Giles as *Strange Stories from a Chinese Studio* 1908 UK; cut vt *Strange Stories from the Lodge of Leisures* 1913 UK; while Lafcadio HEARN retold many Oriental GSs.

GSs based on either folklore or apparent true hauntings are not fiction in the strictest sense, though in the rise of the GS it is often difficult to distinguish fiction from apocrypha. One of the earliest GSs purportedly based on fact was related by Pliny the Younger (62-114) in his "Letter to Sura". This ghost, of an emaciated bearded old man, complete with rattling chains and manacles, had all the trappings of the Gothic ghost that would haunt fiction during the 18th and 19th centuries. Likewise, *A True Relation of the Apparition of One Mrs Veal, the Next Day After her Death, to One Mrs Bargrave, at Canterbury, the 8th of September 1705* (**1706**) by Daniel DEFOE was essentially reportage.

During the Elizabethan and Jacobean periods, the ghost was seen as a melodramatic warning, a harbinger of doom, or an outward manifestation of an individual's guilt. This was how William SHAKESPEARE depicted the ghosts in *Julius Caesar* (performed *c*1599; **1623**) and *Hamlet* (performed *c*1601; **1603**), Ben Jonson (1572-1637) in *Catiline, His Conspiracy* (**1611**), and John Webster (?1580-?1634) in *The White Devil* (**1614**), each using the ghost more as a stage device than as a central character.

When Horace WALPOLE used all the apparatus of GOTHIC FANTASY in *The Castle of Otranto* (**1765**) he created a literary vehicle for ghosts without establishing the GS: the spectres were merely a stage effect. But they now became the stock-in-trade of the Gothic school of writing. In these novels a haunted EDIFICE was an inevitable centrepiece. In most cases the hauntings were contrived by human agency or the ghost had a walk-on part in order to provoke the action; in much fewer cases were the ghosts genuinely supernatural, with a more significant role. Most of these books were formulaic and repetitive, but a few are worthy of note; these include *The Haunted Castle* (**1794**) by George Walker (1772-1847), *The Castle of Ollada* (**1795**) by Francis Lathom (1777-1832), *The Castle of Hardayne* (**1795**) by John Bird (1768-1829) and *The Spirit of the Castle* (**1800**) by W.C. Proby. Although the better Gothics also utilized ghosts, these were usually rationalized (◊ RATIONALIZED FANTASY). The use of sheeted spectres, wailing phantoms, rattling chains and gibbering skeletons reduced the supernatural element to burlesque, and the Gothic GS was soon parodied in such works as *Nightmare Abbey* (**1818**) by Thomas Love Peacock (1785-1866) and *Northanger Abbey* (**1818**) by Jane Austen (1775-1817). Rudolph Ackermann (1764-1823) published *Ghost Stories Collected with a Particular View to Counteract the Vulgar Belief in Ghosts and Apparitions* (anth **1823**) in an effort to counter the excesses of the Gothic story.

It required the emergence of the short story to allow a more artistic development of the GS. Initially these works bore all the clutter of the Gothic tale, but some of the more masterful German writers developed stories that were powerful and effective. Paramount among them were J.K. MUSÄUS, Johann August APEL and E.T.A. HOFFMANN, and their works had a considerable influence upon US writers like Washington IRVING, Nathaniel HAWTHORNE and Edgar Allan POE and, via the anthology *Tales of the Dead* (anth **1813** UK), drawn from Apel's work, influenced Mary SHEL-LEY, Percy Bysshe SHELLEY, Lord BYRON and John POLIDORI. Hoffmann in particular, in such stories as "Das Majorat" (1817) and "Ein Fragment aus dem Leben dreier Freunde" (1819), utilized supernatural retribution to more personal effect, hinting at psychological suggestion.

In the UK the GS was as much influenced by folktales and legend as by the Gothic romance. A number of GSs were collected in the same manner as the GRIMM BROTHERS compiled their FOLKTALES, and presented in narrative form. The main folklorists reporting these tales included James HOGG in *Winter Evening Tales* (**1820**) and *The Shepherd's Calendar* (**1828**), the brothers John (1798-1842) and Michael Banim (1796-1874) in *Revelations of the Dead-Alive* (**1824**) and T. Crofton Croker (1798-1854) in *Fairy Legends and Traditions of the South of Ireland* (**1825**). Their stories take the form of reportage of folk legends, but became increasingly humanized. The best-remembered collector of GSs was Catherine Crowe (1790-1876), with her highly influential *The Night Side of Nature* (coll **1848**), *Light and Darkness* (coll **1850**) and *Ghosts and Family Legends* (coll **1858**). These set a trend for "real" GSs related in fictional narrative form, an approach that became very popular in the 1890s and was subsequently used extensively by Jessie Adelade Middleton, Elliott O'Donnell (1872-1965), Shane Leslie (1885-1971) and Lord Halifax (1839-1934).

The most important writer to draw upon folktales, and who bridged the gap between Gothic fiction and the modern GS, was J. Sheridan LE FANU. He was not significantly influenced by Gothic fiction, but instead used local legend in a series of GSs for the *Dublin University Magazine* starting in 1838, later collected as *The Purcell Papers* (coll **1880**). However, in "A Strange Event in the Life of Schalken the Painter" (1839) he began to fuse elements of the Gothic and the folktale to produce a more sinister humanized story of supernatural VENGEANCE. With "The Watcher" (1851; rev vt "The Familiar") he took the step of describing a purely personal haunting with the ghost audible but not visible (though occasionally *felt*) to only one man. He perfected this in "Green Tea" (1869), the seminal psychological GS.

The Gothic GS had run its course by the 1820s, though vestiges of it may be seen in *Rookwood* (**1834**) by W. Harrison AINSWORTH, *The Phantom Ship* (**1839**) by Frederick MARRYAT (◊◊ FLYING DUTCHMAN), and *The Castle of Ehrenstein, Its Lords Spiritual and Temporal, Its Inhabitants Earthly and Unearthly* (**1847**) by G.P.R. James (1799-1860). It was soon to be recycled in more popular format by Charles DICKENS. He repackaged the standard techniques in more homely surroundings in a number of his Christmas stories, especially *A Christmas Carol* (**1843**). Dickens was adept at using modern, local settings for his hauntings, and produced one of the most popular of all GSs, "The Signalman" (1866), with the setting of a railway tunnel. Such industrial, manmade non-Gothic features were rapidly adopted by Dickens's successors, particularly Amelia B. Edwards (1831-1892), although the supreme artist of industrial hauntings was L.T.C. Rolt (1910-1974), with *Sleep No More* (**1948**).

The transformation of the haunted Gothic castle into the haunted house was given a significant boost by BULWER-LYTTON in "The Haunted and the Haunters, or The House and the Brain" (1859) – again based on a true haunting – which used occult practices as the premise for the ghost. The acceptance of the occult (especially with the growth in ROSICRUCIANISM) allowed a pseudoscientific rationalization for the GS without removing the supernatural *frisson*. One can almost see the Gothic past dropping away and the new rationalism emerging as the GS developed during the 1850s-70s. On the one hand there were writers like Wilkie COLLINS who, in "Mad Monkton" (1855 as "The Monkstons of Wincot Abbey"), "John Jago's Ghost" (1873) and *The Haunted Hotel* (**1878**), liked to retain the paraphernalia of the Gothic ghost; on the other was a new wave of writers, epitomized by Mrs J.H. RIDDELL in such books as *Fairy Water* (**1873**), *The Uninhabited House* (**1875**) and *Weird Stories* (coll **1882**), who were able to bring the haunting into the family and allow us to perceive the supernatural as pervading the world about us.

Many of this new wave were women, and the period 1870-1910 marked the first Golden Age of the GS. In addition to Elizabeth Gaskell (1810-1865), Amelia B. Edwards, George Eliot (real name Mary Ann Evans; 1819-1880) and Florence Marryat (1838-1899), all of whom produced the occasional

GS, were many women writers now primarily remembered for their GSs. They include: Mrs Henry [Ellen] Wood (1814-1887), whose *The Shadow of Ashlydyat* (**1863**) and various **Johnny Ludlow** stories mark a clear transition from the Gothic to Victorian form; Mary E. BRADDON, in *Ralph the Bailiff* (coll **1867**); Rhoda Broughton (1840-1920) in *Tales for Christmas Eve* (coll **1873**); Mrs OLIPHANT in *A Beleaguered City* (**1879**) and *Stories of the Seen and Unseen* (coll **1889**); Mary Molesworth (1839-1921) in *Four Ghost Stories* (coll **1888**) and *Uncanny Tales* (coll **1896**); Vernon LEE in *A Phantom Lover* (**1886**) and *Hauntings* (coll **1890**); Rosa Mulholland (1841-1921) in *The Haunted Organist of Hurly Burly* (coll **1891**); and E. NESBIT in *Grim Tales* (coll **1893**) and *Fear* (coll **1910**). Representative selections of GSs by Victorian women writers include *The Gentlewomen of Evil* (anth **1967**) ed Peter HAINING and *The Virago Book of Victorian Ghost Stories* (anth **1988**) ed Richard DALBY.

Many US women, too, became noted for their GSs. Prime was Mary Wilkins FREEMAN with *The Wind in the Rose-Bush* (coll **1903**), but others of note include Harriet Prescott Spofford (1835-1921) with *Sir Rohan's Ghost* (**1860**) and *The Amber Gods* (coll **1863**), Elizabeth Stuart PHELPS in *Men, Women and Ghosts* (coll **1869**), Sarah Orne Jewett (1849-1909) with *Old Friends and New* (coll **1879**), Emma Frances Dawson (1851-1926) with *An Itinerant House* (coll **1896**), Elia Wilkinson Peattie (1862-1935) with *The Shape of Fear* (coll **1898**), and Edith WHARTON, whose *Tales of Men and Ghosts* (coll **1910**) was the first of several powerful collections to appear. A representative selection of their fiction is *Haunted Women* (anth **1985**) ed Alfred Bendixen (1952-).

These writers established a pedigree within the genre that has continued to this day, bringing a more homely, human and often more incisive perspective to the GS. Their ghosts are seldom as malicious as those created by male writers (◊ GENDER) – and can often be, indeed, benevolent, especially where children are involved – but are no less frightening when intended.

By the 1890s male writers, not just in the UK and USA but throughout Europe and Australia, were developing further strands of the GS. UK writers needed to break out of the mould they had created of earthbound spirits of the dead, a form that had not entrapped writers elsewhere. Guy DE MAUPASSANT, for instance, succeeded in creating a number of psychological GSs showing a mounting insanity along the lines of Le Fanu, something that would not be picked up in the UK until Henry James and Oliver ONIONS turned to the field. The UK seemed tired of the GS, as evidenced by H.G. WELLS's lampoon of the genre in "The Red Room" (1896), which is nevertheless genuinely frightening, while Oscar WILDE wrote an affectionate but scolding parody in *The Canterville Ghost* (1887; **1906** chap). It was at this period that other subgenres of supernatural fiction became more distinct, especially stories of VAMPIRES or of the occult. One subgenre blossoming at this stage concerned the OCCULT DETECTIVE, spawned by the twin interests in the detective story and psychic research, although Arthur Conan DOYLE, intimately involved with both, avoided the theme. Doyle did produce some original GSs, like "The Captain of the 'Polestar'" (1883), and depicted the effects of SPIRITUALISM in "Playing with Fire" (1900). The period 1890-1910 can be seen as one of divergence for the GS, with the DECADENCE of M.P. SHIEL, the psychological GS of Henry James, the mysticism of Arthur MACHEN and the pantheism of Algernon BLACKWOOD.

The traditional version received a vital boost of energy in the form of the antiquarian GS created by M.R. JAMES with *Ghost-Stories of an Antiquary* (coll **1904**). Influenced by Le Fanu, James used ancient churches and books to establish a familiar setting in which traditional ghosts might be expected, but James's ghosts are anything but traditional. His technique has been often copied but seldom matched (◊ JAMES GANG).

Good representative selections of Gothic and Victorian GSs are: *The Oxford Book of Gothic Tales* (anth **1992**) ed Chris Baldick (1954-   ); *Five Victorian Ghost Novels* (anth **1971**) and *Three Supernatural Novels of the Victorian Period* (anth **1975**) both ed E.F. BLEILER; *Victorian Ghost Stories* (anth **1991**) ed Michael Cox (1948-   ) and R.A. Gilbert (1942-   ); *The Mammoth Book of Victorian and Edwardian Ghost Stories* (anth **1995**) ed Richard DALBY; *Gothic Tales of Terror* (anth **1972**; vt in 2 vols as *Great British Tales of Terror* 1973 and *Great Tales of Terror from Europe and America* 1973) ed Peter HAINING; *Tales from a Gas-Lit Graveyard* (anth **1979**), *Victorian Nightmares* (anth **1977**) and other period anthologies ed Hugh Lamb (1946-   ); and *Victorian Ghost Stories* (anth **1933**) ed Montague Summers (1880-1948).

After WWI the public clamour for a belief in the AFTERLIFE increased interest in Spiritualism, and the GS reflected this – particularly in the USA, where the emergence of specialist fiction magazines saw a number dedicated to spiritualist-based fiction, especially the magazine GHOST STORIES. In the UK the GS entered a second Golden Age in the 1920s, partly due to the editorial work of Cynthia ASQUITH and the emergence of *The* GHOST BOOK. Thanks to the writings of L.P. HARTLEY, Hugh Walpole (1884-1941), Walter DE LA MARE, William F. HARVEY, Oliver ONIONS, E.F. BENSON, Elizabeth Bowen (1899-1973), Marjorie BOWEN and Margaret IRWIN the GS became established as a literary artform.

The GS has advanced little in the UK since the 1930s. The only significant GS of the 1940s was *Uneasy Freehold* (**1941**; vt *The Uninvited* 1942 US) by Dorothy Macardle (1889-1958), a convincing HAUNTED-DWELLING novel, filmed as *The Uninvited* (*1943*). Interest in supernatural fiction waned after WWII, and the re-establishment of the genre since the 1970s has hinged primarily on HORROR. Those modern writers who have sought to avoid the horror angle have tended to model themselves upon M.R. James. This is especially true of the works of Robert AICKMAN, regarded by many as the UK's leading writer of GSs in the second half of the 20th century, Ramsey CAMPBELL and Jonathan Aycliffe (1949-   ). Other works of note are Kingsley AMIS's *The Green Man* (**1969**) and to a lesser extent Norah LOFTS's **Gad's Hall** novels and the later works of James HERBERT. Otherwise, with the possible single exception of R. CHETWYND-HAYES, almost all other recent UK GS-writers have looked backwards to the two Golden Ages, either portraying traditional ghosts, but in modern (often regionalistic) settings, or emphasizing the psychological aspect. Works include *Snowfall* (coll **1965**) and other volumes by Elizabeth Walter; *The Dark Land* (coll **1975**) and other collections by Mary Williams; *Shadow at Midnight* (coll **1979**) by Michael Sims (1952-   ) and Len Maynard; *Stories of the Strange and Sinister* (coll **1983**) by Frank BAKER; *Seven Ghosts in Search* (coll **1983**) by Fred Urquhart (1912-1995); *Ghost Train* (**1985**) and *Spectre* (**1986**) by Stephen Laws (1952-   ); *Shades of Darkness* (**1986**) by Richard

Cowper (real name John Middleton Murry; 1926-   ); *Winterwood, and Other Hauntings* (coll **1989**) by Keith ROBERTS; and *Element of Doubt* (coll **1992**) by A.L. Barker (1918-   ). Basil Copper (1924-   ) was successful in blending influences from GOTHIC FANTASY, H.P. LOVECRAFT and the thriller genre into effective supernatural fictions, especially those in *Not After Nightfall* (coll **1967**), *From Evil's Pillow* (coll **1973** US) and *And Afterward, the Dark* (coll **1977** US), the last two from ARKHAM HOUSE.

Some of the best advances have been made in writing for children (◊ GHOST STORIES FOR CHILDREN), in particular the works of Leon GARFIELD, Philippa PEARCE and Robert WESTALL.

In the USA the mainstream GS became a vehicle for HUMOUR, as in the **Topper** series by Thorne SMITH. Genre fiction in the 1920s-1940s was dominated by the pulp MAGAZINES. WEIRD TALES and UNKNOWN featured many traditional GSs; not until the work of Fritz LEIBER and to some extent Manly Wade WELLMAN and August DERLETH, in the 1940s, did the GS take any significant step forward. Leiber successfully urbanized the ghost, starting with "Smoke Ghost" (1941), translating the imagery of M.R. James into inner-city USA. Nevertheless the GS field remained relatively fallow, though ploughed occasionally by Joseph Payne BRENNAN and Russell Kirk. A lone example of what was to come was *The Haunting of Hill House* (**1959**) by Shirley JACKSON, a PSYCHOLOGICAL THRILLER which explores the effects of fear on a group of individuals who stay in a HAUNTED DWELLING; it was filmed with considerable atmosphere as *The Haunting* (*1963*). This shock-horror approach emerged in strength in the 1970s with the success of *The* EXORCIST (*1973*), which saw an explosion of interest in HORROR and coincided with the emergence of Stephen KING; *The* LEGEND OF HELL HOUSE (*1973*), echoing the title of Jackson's book and based on a Richard MATHESON original, was one offshoot, and there have been countless others. This allowed publishers to consider the GS an acceptable medium, though in the USA it still is more closely linked to terror and violence. The prime example is *Ghost Story* (**1979**) by Peter STRAUB – filmed as GHOST STORY (*1981*) – with others including: *Ghost House* (**1979**) and its sequel *Ghost House Revenge* (**1981**) by Clare McNally; *The Night Boat* (**1980**) by Robert McCammon (1952-   ); *Moon Lake* (**1982**) and *Midnight Boy* (**1987**) by Stephen Gresham (1947-   ); the **Blackwater** sequence, being *The Flood* (**1983**), *The Levee* (**1983**), *The House* (**1983**), *The War* (**1983**), *The Fortune* (**1983**) and *Rain* (**1983**), by Michael McDowell (1950-   ); various works by J.N. WILLIAMSON; *A Manhattan Ghost Story* (**1984**) and its sequel *The Waiting Room* (**1986**) by T.M. Wright (1947-   ); *Soulstorm* (**1986**) by Chet WILLIAMSON; *House Haunted* (**1991**) by Al Sarrantonio (1952-   ); and *Lost Boys* (**1992**) by Orson Scott CARD.

Rarer are more subtle GSs. Examples include: *Cast a Cold Eye* (**1984**) by Alan Ryan (1943-   ); *Time Out of Mind* (**1986**) by John R. Maxim (1937-   ), which like many others overlaps with the TIMESLIP tale; *Summer's End* (coll **1987**) by Ann LAWRENCE; *A Truce With Time* (**1988**) and "The Fire When it Comes" by Parke GODWIN; *Night Relics* (**1994**) by James P. BLAYLOCK; *Women and Ghosts* (coll **1994**) by Alison Lurie (1926-   ); and *Expiration Date* (**1995**) by Tim POWERS.

Anthologies seeking to explore the contemporary GS, some with new stories, include: *The Literary Ghost* (anth

1991) ed Larry Dark; *Frights* (anth **1976**; vt in 2 vols *Frights 1* 1979 UK and *Frights 2* 1979 UK) ed Kirby McCauley; *Post Mortem: New Tales of Ghastly Horror* (anth **1989**) ed Paul F. Olson and David B. Silva (1950-); and *Triumphs of the Night* (anth **1989**; vt *The Omnibus of 20th Century Ghost Stories* 1989 UK) ed Robert S. Phillips (1938-).

Other anthologies include: *The* FONTANA BOOK OF GREAT GHOST STORIES; *The Penguin Book of Indian Ghost Stories* (anth **1993** India) ed Ruskin Bond; *The Oxford Book of English Ghost Stories* (anth **1986**) ed Michael Cox and R.A. Gilbert; *The Penguin Book of Ghost Stories* (anth **1984**) ed J.A. CUDDON; *The Mammoth Book of Ghost Stories* (anth **1990**; cut vt *The Anthology of Ghost Stories* 1994), *The Mammoth Book of Ghost Stories 2* (anth **1991**), *The Virago Book of Ghost Stories* (anth **1987**), *The Virago Book of Ghost Stories: The Twentieth Century, Volume Two* (anth **1991**), all ed Richard DALBY; *Ghosts* (anth **1981**; cut vt *A Classic Collection of Haunting Ghost Stories* 1993 UK) ed Marvin KAYE; *The Haunted Omnibus* (anth **1937**; cut vt *Great Ghost Stories of the World* 1941) ed Alexander Laing (1903-1976); *The Oxford Book of Canadian Ghost Stories* (anth **1990**) ed Alberto MANGUEL; *Great American Ghost Stories* (anth **1991**) ed Frank D. McSherry Jr (1927-), Charles G. Waugh (1943-) and Martin H. GREENBERG; *What Did Miss Darrington See?: An Anthology of Feminist Supernatural Fiction* (anth **1989**) ed Jessica Amanda SALMONSON; *Lost Souls* (anth **1983**) ed Jack Sullivan (1946-); *The Supernatural Omnibus* (anth **1931**; 9 stories cut and 6 added 1932 US; cut in 2 vols 1967 UK; cut differently 2 vols 1976 UK) ed Montague Summers; *The Fireside Book of Ghost Stories* (anth **1947**) ed Edward WAGENKNECHT; and *Great Tales of Terror & the Supernatural* (anth **1944**) ed Herbert A. Wise (1890-1961) and Phyllis Fraser (1915-).

The GS retains an enduring popularity. Almost anyone who has written a substantial amount of fiction has tackled a GS at some time. While the traditional ghost may be relegated to an artistic backwater, ghosts continue to feature strongly in much modern horror. [MA]

Further reading: *The Supernatural in Modern English Fiction* (**1917**) by Dorothy Scarborough (1878-1935); *The Haunted Castle: A Study of the Elements of English Romanticism* (**1927**) by Eino Railo; *The Supernatural in Fiction* (**1952**) by Peter Penzoldt; *Night Visitors: The Rise and Fall of the English Ghost Story* (**1977**) by Julia Briggs (1943-); *Elegant Nightmares: The English Ghost Story from Le Fanu to Blackwood* (**1978**) by Jack Sullivan (1946-); *The Guide to Supernatural Fiction* (**1983**) by E.F. BLEILER; *The Haunted Dusk: American Supernatural Fiction, 1820-1920* (anth **1983**) ed Howard Kerr (1931-), John W. Crowley and Charles W. Crow (1940-); *Specter or Delusion? The Supernatural in Gothic Fiction* (**1987**) by Margaret L. Carter (1948-); *Haunted Presence: The Numinous in Gothic Fiction* (**1987**) by S.L. Varnado (1929-); *The Supernatural and English Fiction* (**1995**) by Glen Cavaliero (1927-).

See also: GHOST STORIES FOR CHILDREN; HORROR MOVIES.

GHOST STORIES US MAGAZINE, 64 issues, monthly (but last 3 issues bimonthly), July 1926-December 1931/January 1932, published by Bernarr Macfadden (1868-1955), New York, under different imprints. Details of editorial responsibility are hard to unravel, but overall editorial director throughout was Fulton Oursler (1893-1952), assisted in sequence by Harry A. Keller, W. Adolphe Roberts (1886-1962), George Bond, Daniel Wheeler and Arthur B.

Howland, each serving for about a year, before Harold Hersey (1893-1956), one-time editor of *The* THRILL BOOK, took over as publisher from April 1930, assisted by Stuart Palmer (1905-1968). The magazine was initially a large-format slick, switching to standard PULP format in July 1928, with a brief period (April-December 1929) as a large-format pulp.

GS focused almost entirely on GHOST STORIES, pandering chiefly to the vogue for SPIRITUALISM; many tales sported posed photographs. Tales were written in the first person in a "true confessions" style. The stories were usually by staff writers (most often Harold Standish Corbin [1887-1947]) and attributed to a bogus narrator. Only rarely did other pulp writers appear, chief among them being Victor Rousseau (real name Victor Rousseau Emmanuel 1879-1960), but occasional contributors included Hugh B. CAVE, Ray Cummings (1887-1957), Paul Ernst (1899-1985), Robert E. HOWARD, Carl Jacobi (1908-) and Frank Belknap LONG. Macfadden had an arrangement with UK publishers, including Walter Hutchinson (1887-1950), and stories were exchanged with *The Sovereign Magazine* and *Mystery-Story Magazine* (◊ HUTCHINSON'S MAGAZINES). *GS* also regularly reprinted many classic ghost stories. The magazine's demise came about almost certainly because of reader boredom with the predictability of the stories.

GS had a short-lived weird-fiction companion, *True Strange Stories* (8 issues March-November 1929) ed Walter B. Gibson (1897-1985), which had a similar format though a wider remit than *GS*. [MA]

Further reading: *GS* is indexed in *Stories of Ghosts* (**1970**) by James R. Sieger; this includes a short history by Sam Moskowitz, who has also written about the magazine in *Fantasy Commentator #41* (Fall 1990). No specific anthologies exist, although *Prize Ghost Stories* (1 issue 1963) and *True Twilight Tales* (2 issues Fall 1963, Spring 1964) drew their contents entirely from *GS*. Two representative stories were included in *When Spirits Talk* (anth **1990** chap) ed Mike ASHLEY.

GHOST STORIES FOR CHILDREN Stories written deliberately to frighten CHILDREN, as distinct from instructing them (as was common in FAIRYTALES), were frowned on in Victorian society. Thus children's GHOST STORIES are predominantly a 20th-century phenomenon, although there are antecedents. Hans Christian ANDERSEN, for instance, wrote a very effective ghost story in "Auntie Toothache" (1872). Mary Molesworth (1839-1921) was a prolific and popular writer for children, and though her ghost stories were intended for adults it is quite likely that some were shared with children, such as "The Shadow in the Moonlight" (1896 *Uncanny Tales*), where children are the main protagonists. The same was probably true of E. NESBIT's ghost stories, especially those in *Grim Tales* (coll **1893**), many of which are today frequently reprinted in children's ghost-story ANTHOLOGIES, and possibly of Francis Hodgson BURNETT, whose ghost novella *The White People* (**1917**) would have been acceptable to children.

It was at this same time that M.R. JAMES began to tell his ghost stories to his students (firstly at Cambridge and later at Eton). His *The Five Jars* (**1922**) is a CHILDREN'S FANTASY, but James never produced a genuine children's ghost story, although "An Evening's Entertainment" (1926) and *Wailing Well* (**1928**) – written for the Eton College Boy Scouts – come close.

The first author seriously to produce GSFC was Walter

DE LA MARE, starting with *Broomsticks and Other Tales* (coll **1925**) and *The Lord Fish* (coll **1933**). These and later revised compilations – *The Old Lion and Other Stories* (coll **1942**), *The Magic Jacket and Other Stories* (coll **1943**), *The Scarecrow and Other Stories* (coll **1945**) and *The Dutch Cheese and Other Stories* (coll **1946**), plus the omnibus *Collected Stories for Children* (coll **1947**) – contain a mixture of fairytales and ghost stories. The growth in children's fiction generally during the 1920s meant that many writers produced some SUPERNATURAL FICTION for children.

It was not until the 1950s that the children's ghost story emerged as a distinct genre. The earliest recognized collection was *A Pad in the Straw* (coll **1952**) by Christopher Woodforde (1907-1962), the Chaplain of New College Oxford (and a member of the JAMES GANG), who would tell ghost stories to the children in the Choir School – the "eight boys" of his dedication. The next example was the first of the **Green Knowe** series by Lucy M. BOSTON, *The Children of Green Knowe* (**1954**). A boy's great-grandmother tells him stories of their ancestors and soon he becomes aware of these ghostly children. The **Green Knowe** stories have a similar atmosphere to certain TIMESLIP stories, and bear comparison with, for example, *A Traveller in Time* (**1939**) by Alison UTTLEY, *The Sherwood Ring* (**1958**) by Elizabeth Marie POPE, and *The Ghosts* (**1969**; vt *The Amazing Mr Blunden* 1972) by Antonia Barber (real name Barbara Anthony; 1932-   ). In a similar vein is *The Spring Rider* (**1968**) by the US regionalist writer John Lawson, about the ghosts of Civil War soldiers whose memories become intermeshed with those of a present-day child living near the site of a battlefield, and *The House on Parchment Street* (**1973**) by Patricia MCKILLIP, also featuring ghosts from the Civil War. In all of these stories the ghosts are friendly. Other such cosy ghost stories include *The Ghost of Thomas Kempe* (**1973**) by Penelope LIVELY and *The Court of the Stone Children* (**1973**) by Eleanor CAMERON.

Leon GARFIELD preferred the evil ghost. *Mr Corbett's Ghost* (**1968** chap US; exp as *Mr Corbett's Ghost and Other Stories* coll 1969) is a Dickensian story about a wicked employer who is killed and seeks his revenge on his apprentice. Garfield's work opened up the field to a wider range of ghost stories. Among the more effective sinister GSFCs since then have been: *The Wild Hunt of Hagworthy* (**1971**; vt *The Wild Hunt of the Ghost Hounds* 1972 US) by Penelope Lively, where a young boy and girl find themselves the victims of ghostly hounds summoned up by a local vicar; *Castaways on Long Ago* (**1973**) by Edward Ormondroyd (1925-   ), in which children become involved with the menacing spirit of a drowned girl; *The Shadow Guests* (**1980**) by Joan AIKEN, where a boy must come to terms with death in the family; *The Haunting of Cassie Palmer* (**1980**) by Vivien Alcock (1924-   ), where a girl inadvertently summons up the spirit of an evil man; *The Clock-Tower Ghost* (**1981**) by Gene Kemp (1926-   ); the highly atmospheric *The Haunting* (**1982**) by Margaret MAHY; *Emer's Ghost* (**1981**) by Catherine Sefton (real name Martin Waddell; 1941-), where the ghost takes the form of a DOLL, and the same author's *The Ghost Girl* (**1985**), where a young girl pieces together the story of a dead girl; and *The Ghost Messengers* (**1985**) by Robert Swindells (1939-   ), where a girl's dead grandfather tries to communicate with her through her sleep.

Like Garfield, Robert WESTALL brings an immediacy to his children's stories that is at once both surprising and effective. His first full-length GSFC was *The Watch House* (**1977**), and he piles effect upon effect in *The Scarecrows* (**1981**), *Ghost Abbey* (**1988**) and *The Promise* (**1990**), as well as many excellent short stories. Westall is arguably the best modern writer of supernatural fiction for children. Not as powerful, but equally imaginative, are the ghostly novels of John GORDON, especially *The House on the Brink* (**1970**), *The Ghost on the Hill* (**1976**) and *Gilray's Ghost* (**1995**).

As for shorter ghost stories, memorable assemblies include: *A Whisper in the Night* (coll **1982**), *Past Eight O'clock: Goodnight Stories* (coll **1986**), *A Fit of Shivers* (coll **1990**) and others by Joan AIKEN; *Ghostly Companions* (coll **1984**) by Vivien Alcock (1924-   ); *Ghost Carnival* (coll **1977**) and others by Aidan CHAMBERS; *The Spitfire Grave and Other Stories* (coll **1979**), *Catch your Death and Other Ghost Stories* (coll **1984**) and *The Burning Baby and Other Ghosts* (coll **1992**) by John GORDON; *Seven Strange and Ghostly Tales* (coll **1991**) by Brian JACQUES; *The Shadow-Cage* (coll **1977**) and *Who's Afraid?* (coll **1986**) by Philippa PEARCE; *A Nasty Piece of Work* (coll **1983**), *The Darkness Under the Stairs* (coll **1988**) and *Beware, This House is Haunted!* (coll **1989**) by Lance Salway (1940-   ); and *Break of Dark* (coll **1982**), *The Haunting of Chas McGill* (coll **1983**), *Ghosts and Journeys* (coll **1988**), *The Call, and Other Stories* (coll **1989**), *A Walk on the Wildside* (coll **1989**) and *The Stones of Muncaster Cathedral* (coll **1991**) by Robert WESTALL.

Acceptance of ghost and horror stories for children changed radically in the 1980s, spurred particularly by the **Point Horror** and **Archway Books** series. The premier contributors to these were R.L. Stine (1943-   ) and Christopher Pike (real name Kevin McFadden). Stine's **Fear Street** and **Goosebumps** sequences made him one of the biggest-selling authors in the world. The success of these books, and of movies like the GHOSTBUSTERS series, has caused other publishers to develop more horror and supernatural books for children, such as the **Hippo Ghost** series.

There have been many ghost-story ANTHOLOGIES for children, but as yet only one has sought to trace the development of the children's ghost story: *Dread & Delight: A Century of Children's Ghost Stories* (anth **1995**) ed Philippa Pearce. [MA]

GHOST STORY UK movie (*1974*). Stephen Weeks Co. **Pr** Stephen Weeks. **Dir** Weeks. **Screenplay** Philip Norman, Rosemary SUTCLIFF, Weeks. **Starring** Anthony Bate (Dr Frederick Borden), Larry Dann (Gregory Talbot), Marianne Faithfull (Sophy Crickworth), Penelope Keith (Miss Rennie), Leigh Lawson (Robert Crickworth), Vivian Mackerell (Duller), Murray Melvin (McFayden), Barbara Shelley (Matron). 89 mins. Colour.

England, the early 1930s. Three men come together for a shooting holiday organized by one of them, McFayden, in a remote mansion he has just inherited. McFayden and Duller are bullying snobs, and not very intelligent; working-class Talbot, the butt of their japes, is much brighter, and it is he who from the very outset is made terrifyingly aware that the house is filled with WRONGNESS (◊ HAUNTED DWELLINGS), his movements being dogged uncannily by an old blue-dressed DOLL. The doll pulls him (in one especially striking sequence it literally drags him) into a sequence of GRAND-GUIGNOL events that occurred in and around the house some 50 years ago (◊ TIMESLIP). He discovers, piecemeal, a horrific tale of past incest and mass murder. The STORY told/re-enacted, the doll kills Talbot; and at movie's end is preparing to do likewise to McFayden.

This is the supernatural HORROR MOVIE remade for the genteel, and moments of stunning effectiveness alternate with long periods of stilted, stagy selfconsciousness; the plot has too many holes to list (surprising, with Sutcliff among the writing credits). Yet when this movie is good, it is very good. It has no connection with GHOST STORY (*1981*). [JG]

GHOST STORY US movie (*1981*). Universal. **Pr** Burt Weissbourd. **Dir** John Irvin. **Spfx** Henry Millar Jr. **Vfx** Albert Whitlock. **Mufx** Dick Smith. **Screenplay** Lawrence D. Cohen. **Based on** *Ghost Story* (1979) by Peter STRAUB. **Starring** Fred Astaire (Ricky Hawthorne), Jacqueline Brookes (Milly), Melvyn Douglas (John Jaffrey), Douglas Fairbanks Jr (Edward Wanderley), Miguel Fernandes (Gregory Bate), Lance Holcomb (Fenny Bate), John Houseman (Sears James), Alice Krige (Alma Mobley/Eva Galli), Patricia Neal (Stella), Craig Wasson (Don Wanderley/David Wanderley). 110 mins. Colour.

50 years ago Hawthorne, Sears, Wanderley and James were infatuated by sophisticated Eva Galli, an Englishwoman come to live in the small New England town of Milburn. Finally Wanderley bedded her, but proved impotent; when she was about to blurt this out to the others he hit her and believed he'd killed her. The four bundled her body into a car and sank it at the bottom of the lake; at the last moment, too late for rescue, they realized she was alive. Now the four old men – the Chowder Society – tormented by nightmares, meet regularly, never to talk about their guilt but instead to tease it by telling each other GHOST STORIES. Unknown to them, Wanderley's son Don has, in a plethora of SEX, thrown his career away for a mysterious, seductive Englishwoman, Alma. At last, bemused by her WRONGNESS, he has thrown her over; almost immediately she has moved on to his brother David. When she reveals her true horror to David he plunges from a high window, and this tragedy brings Don to Milburn to comfort his father.

Eva's GHOST – sometimes acting through her agents, escaped lunatics Gregory and Fenny Bate – toys with the four old men, leading Wanderley and then Jaffrey to their deaths. Hawthorne, James and Don, aware that "Alma" is Eva's ghost, go to Eva's now-derelict HAUNTED DWELLING for a confrontation; but before the ghost can finally be laid, through the recovery of the drowned car and the release from it of the corpse, James has lost his life, Don has nearly lost his wits, and Hawthorne has had to confront his old guilt.

Straub's impressive novel is a convoluted analysis of the power of STORY, and requires careful unpicking; the movie version, forced to abridge while yet attempting to retain the feel of the original's complexities, almost inevitably lapses into something close to incoherence and displays an irritating form of shallowness – in that one is aware that beneath the surface are great depths of which one is shown barely a glimmer. Yet *GS* is in many ways an excellent movie, especially in its sense of setting. It may not attain the art/HORROR MOVIE heights it clearly aimed for, but very few movies do.

GS has no connection with GHOST STORY (*1974*). [JG]

GHOST STORY US tv series (1972-3). Columbia Pictures/NBC. **Pr** Joel Rogosin. **Exec pr** William Castle. **Dir** Robert Day, Richard Donner and many others. **Starring** Sebastian Cabot (Winston Essex). 22 1hr episodes. Colour.

This anthology series was anchored by Cabot as the mysterious head of a gloomy hotel. Each week he introduced stories about GHOSTS, VAMPIRES, WITCHES and other supernatural entities. Plots included: a young woman battles a ghost that wants to drive her out of her new house; a ghost wants her TWIN sister to join her in death; a VOODOO practitioner conspires against his former son-in-law; and a college professor is, outside the classroom, a vampire. After 13 episodes *GS* changed format, dropping Cabot and placing less emphasis on the supernatural. Re-titled *Circle of Fear*, the new version lasted a further nine episodes. [BC]

GHOUL (magazine) ◊ R. CHETWYND-HAYES.

GHOULIES US movie (*1985*). Empire/Charles Band. **Pr** Jefery Levy. **Exec pr** Charles Band. **Dir** Luca Bercovici. **Mufx** John Carl Buechler, Mechanical & Makeup Imageries Inc. **Screenplay** Bercovici, Levy. **Starring** Michael Des Barres (Malcolm), Peter Liapis (Jonathan), Jack Nance (Wolfgang), Lisa Pelikan (Rebecca), Peter Risch (Grizzel), Tamara de Treaux (Greediguts). 84 mins. Colour.

In infancy Jonathan was saved by his medallion and at the cost of his mother's life from HUMAN SACRIFICE at the hands of his father Malcolm in a satanic ritual. Raised by grizzled family retainer Wolfgang, he inherits the decrepit ancestral pile on his father's ghastly demise, and moves into it with girlfriend Rebecca. Soon, dabbling in his father's grimoires (◊ BOOKS), he develops luminous green eyes, calls up sundry DEMONS (a powerful pair, Grizzel and Greediguts, being dwarfish humanoids, the rest being cackling little MONSTERS, the Ghoulies) and resurrects Malcolm. Mayhem follows.

Band and Buechler made two attempts in 1985 to hitch onto the bandwagon of the Steven SPIELBERG/Joe Dante GREMLINS (*1984*); *G* was the better, the other being *Troll* (*1985*), but is still poor. However, it attained enough popularity to spawn two sequels (both released direct to video), *Ghoulies 2* (*1987*), in which the hairy little imps infest a carnival show, and *Ghoulies 3 – Ghoulies Go to College* (*1991*), which mixes soft-porn college-movie clichés into the usual formula. [JG]

GIANTS Outsize humanoids are rooted in ancient MYTH. Earth's first ELDER-RACE inhabitants are often identified as giants: e.g., the Titans, the giant-father Ymir in Nordic MYTHOLOGY, the BIBLE's "There were giants in the earth in those days" (*Genesis vi*), GOG AND MAGOG, and the root races of THEOSOPHY. Early giants have affinities with earth ELEMENTALS (◊ TROLLS): Atlas supports the Earth, Antaeus drew strength from contact with it. Traditionally, giants are somewhat brutish and hostile, often with a taste for human flesh: Polyphemus the Cyclops; Blunderbore, killed by JACK the Giant Killer and tricked by the BRAVE LITTLE TAILOR; Galapas, killed by ARTHUR; Giant Despair of Doubting Castle in John BUNYAN's PILGRIM'S PROGRESS; George MACDONALD's unnamed giant in "The Giant's Heart" (1863); the Nordic frost and fire giants in L. Sprague DE CAMP's and Fletcher PRATT's "The Roaring Trumpet" (1940); the giants of Harfang in C.S. LEWIS's *The Silver Chair* (**1953**); the ogre Throop of the Three Heads in Jack VANCE's *Lyonesse III: Madouc* (**1989**); and many more. (Some magical taxonomies differentiate between giants and ogres.) But there is a separate tradition of the giant whose size reflects heroic stature: Brân the Blessed who wades across the Irish Sea in the Second Branch of the MABINOGION, FINN MAC COOL and Paul Bunyan. Jonathan SWIFT's giants of Brobdingnag in *Gulliver's Travels* (**1726**) are more

neutral and humanlike, thus sharpening the SATIRE; Gargantua in François RABELAIS's *Gargantua and Pantagruel* (**1532-52**) takes a different satirical line of wild exaggeration. REVISIONIST FANTASY treatments, challenging the cliché of the giant as mindlessly wicked and merely there to be killed, appear in: Oscar WILDE's "The Selfish Giant" (1888), whose lapse into kindness is rewarded by CHRIST; A.A. MILNE's "A Matter-of-Fact Fairy Tale" (1912); Norton JUSTER's *The Phantom Tollbooth* (**1961**), whose Gelatinous Giant represents indecision and fear of new ideas; J.G. BALLARD's "The Drowned Giant" (1964), where the move from wonder to exploitation of the huge carcase savagely illustrates THINNING; John GORDON's *The Giant Under the Snow* (**1968**), whose hidden giant is a GREEN MAN; Raymond BRIGGS's *Jim and the Beanstalk* (**1970**); Roald DAHL's *The BFG* (**1972**); Piers ANTHONY's *A Spell for Chameleon* (**1977**), whose unseen giant (◊ INVISIBILITY) is ultimately heroic; and Gene WOLFE's *The Book of the New Sun* (**1980-83**), where the practical problem of (nonmagical) giants' size-weight ratio is met by having them live underwater. [DRL]
Further reading: *The Hamish Hamilton Book of Giants* (anth **1968**) ed William MAYNE; *Giants* (graph anth **1979**) ed David Larkin; *Isaac Asimov's Magical Worlds of Fantasy #5: Giants* (anth **1985**) ed Isaac ASIMOV, Martin H. GREENBERG and Charles G. Waugh.

GIESY, JOHN ULRICH (1877-1948) US pulp writer best-known for sf but who also wrote, in collaboration with lawyer and fellow astrologer Junius B. Smith (1883-1945), a number of fantasies featuring the OCCULT DETECTIVE Semi Dual, the name under which Prince Abduel Omar of Tehran, Persia, solves crimes, rescues women, thwarts criminals and captures spies who would menace the world. The physically impressive Semi Dual initially appeared as a series character in *Cavalier* (1912-13), moved to *All-Story Weekly* (1914-20), *People's Magazine* (1917-18) and *Top Notch* (1918), ending his career in *Argosy* (1920-24). His adventures are vivid and increasingly improbable, but the ASTROLOGY is presented as a rigorous, scientifically valid method. None of the **Semi Dual** stories has been collected. [RB]

GIGANTIS (ot *Gojira No Gyakushu*; vt *Counterattack of the Monster*; vt *The Fire Monster*; vt *Godzilla Raids Again*; vt *Godzilla's Counterattack*; vt *The Return of Godzilla*; vt *The Volcano Monster*) Japanese movie (*1955*). ◊ GODZILLA MOVIES.

GIGER, H.R. Working name of Swiss artist and designer Hasruedi Giger (1940-   ), whose weirdly fascinating and disturbing paintings and structures gained worldwide acclaim with the release of Ridley Scott's *Alien* (*1979*). Usually executed in a range of blue-greys, HRG's works, evidently influenced by the Surrealists Max ERNST and Antonio Gaudí (1852-1926), plumb the most disturbing regions of the subconcious, combining organic and mechanoid forms in a morbidly obsessive manner.

HRG provided designs for, aside from *Alien*, *Poltergeist II* (*1986*; ◊ POLTERGEIST) and *Species* (*1995*); his work has been imitated in many other movies. [RT]
Other works: *A Rh+* (graph **1971**); *H.R. Giger* (graph **1976**); *H.R. Giger's Necronomicon* (graph **1977**); *H.R. Giger's Alien* (graph **1979**); *H.R. Giger: New York City* (graph **1981**); *H.R. Giger: Retrospektive, 1964-1984* (graph **1984**); *H.R. Giger's Necronomicon II* (graph **1986**); *H.R. Giger's Biomechanics* (graph **1988**; trans **1990** US); *H.R. Giger's Species Design* (graph **1995**).

GILBERT, WILLIAM (1804-1890) UK writer, father of W.S. GILBERT; also one of the pioneers of psychiatric medicine. He drew upon his work in that field in two notable collections of hypothetical case-studies, *Shirley Hall Asylum, or The Memoirs of a Monomaniac* (coll of linked stories **1863**; issued anon) and *Doctor Austin's Guests* (coll of linked stories **1866**). Although not fantasy as such, the tales offer interesting early descriptions of the various kinds of DELUSIONS to which people may fall victim: the first volume includes an imaginary HAUNTING and an account of religious mania; the second features a man who believes he is getting younger and a man who believes himself possessed (◊ POSSESSION) by a DEMON. Authentic fantasies can be found in *The Magic Mirror: A Round of Tales for Young and Old* (coll of linked stories **1866**), a compendium of cautionary moral tales set in the 15th century; the eponymous MIRROR grants WISHES, with either farcical or horrific consequences. *The Wizard of the Mountain* (coll of linked stories **1867**) likewise mixes elements of HORROR and comedy in its moralistic fantasies, in which various characters who seek the aid of its gifted protagonist are eventually served according to their deserts. The guilty perish but the innocent receive only modest assistance – an imbalance dutifully explained in "The Innominato's Confession", which proposes that MAGIC is essentially diabolical, tempting mortals to rebel against divine providence. [BS]
Other works: *The Washerwoman's Foundling* (**1867**); *The Seven-Leagued Boots* (**1869** chap).

GILBERT, (SIR) W(ILLIAM) S(CHWENCK) (1836-1911) UK dramatist, librettist, poet and author of some short fiction. Almost all of his writings are ostensibly comic, much being PARODY of other authors' works. His **Bab Ballads** sequence of narrative poems – including *The "Bab" Ballads: Much Sound and Little Sense* (coll **1867**) and *More "Bab" Ballads* (coll **1873**), both assembled with other work in *The Bab Ballads: With Which are Included Songs of a Savoyard* (coll **1898**), plus *Lost Bab Ballads* (coll **1932**) – is a thorough conspectus of logical absurdities and fantastic rigmaroles of plot and circumstance (◊ TOPSY-TURVY) amounting to a sometimes nightmarish vision of a world subject to fantasy mechanisms gone haywire. These ballads are far more cruel than any of the famous Savoy Operas (◊ OPERA) he wrote, with music by Sir Arthur Sullivan (1842-1900), some of which are of fantasy interest – notably the pre-Savoy Theatre *Thespis, or The Gods Grown Old* (**1871** chap), which mocks the THINNING of the ancient world; *Iolanthe, or The Peer and the Peri* (performed 1882; **1885** chap), a SATIRE in which FAERIE is invaded by the House of Lords and vice versa; and *Ruddigore, or The Witch's Curse!* (**1887** chap), about a haunted castle and a CURSE. The first third of Marvin KAYE's *The Incredible Umbrella* (**1979**) and part of *The Amorous Umbrella* (**1981**) are set in a topsy-turvy world where characters from the **Savoy Operas** jostle, as the tale continues, with other figures from a GASLIGHT-ROMANCE version of Victorian England.

WSG's other work is less known, but is of a piece with the **Bab Ballads** and generally more scathing than the **Savoy Operas** about human nature, which he deemed irretrievably selfish. His favourite dramatic turn, which became known as the "lozenge plot", was the utilization of some sort of MAGIC device which seems to change people utterly but which (in his early work at least) reveals them to retain, no matter how radically they seem to have been transformed, their essential selfish nature; in the grip of their essential being, humans are

– for WSG – like puppets. In *Dulcamara, or The Little Duck and the Great Quack* (**1866** chap) – his first play, based on the OPERA *L'elisir d'amore* (**1832**) by Gaetano Donizetti (1797-1848) – the "lozenge" is an ELIXIR. The eponymous magic building in the verse play *The Palace of Truth: A Fairy Comedy* (**1871** chap) – based on the fairy plays of James Robinson PLANCHÉ – makes it possible for the protagonists to hear what sycophants really wish to say. The characters in *A Sensation Novel in Three Volumes* (**1871** chap), driven by demonic compulsions, search for a plot which will satisfy them. *Happy Arcadia* (**1872** chap) features an IDENTITY EXCHANGE. *The Wicked World: An Entirely Original Fairy Comedy* (**1873** chap) and *The Happy Land: A Burlesque Version of "The Wicked World"* (**1873** chap), the latter by WSG writing as F. Tomline with Gilbert Arthur àBECK-ETT, are political SATIRES. *Eyes and No Eyes* (**1875** chap) is a harlequinade (◊ COMMEDIA DELL'ARTE) based on Hans Christian ANDERSEN's "The Emperor's New Clothes".

WSG wrote little prose fiction of interest, though *Foggerty's Fairy and Other Tales* (coll **1890**) attempts to deflate various FAIRYTALES through exorbitance and absurdity; as *Foggerty's Fairy* (**1881** chap) the title story was presented on the stage. Some uncollected stories were assembled as *The Lost Stories of W.S. Gilbert* (coll **1982**) ed Peter HAINING. [JC]

Other works: *Harlequin, Cock-Robin and Jenny Wren, or Fortunatus and the Water of Life, the Three Bears, the Three Gifts, the Three Wishes, and the Little Man Who Woo'd the Little Maid: Grand Christmas Pantomime* (**1867** chap), a musical; *Ages Ago: A Musical Legend* (performed 1869; **1895** chap); *The Gentleman in Black* (**1870** chap), a musical; *Pygmalion and Galatea* (**1871** chap); *Topsyturvydom* (performed 1874; **1931** chap); *Broken Hearts: An Entirely Original Fairy Play* (**1876** chap), set on a magic ISLAND; *The Mountebanks* (**1892** chap), a "lozenge" musical; *The Fairy's Dilemma* (**1904** chap); *Fallen Fairies* (**1909** chap), a musical; *Gilbert Before Sullivan* (coll **1969**).

GILGAMESH The hero of a famous Sumerian epic, one of the oldest surviving HEROIC FANTASIES (◊◊ TAPROOT TEXTS). Gilgamesh was a part-divine, part-human king of Erech, or Uruk (current Warka in Iraq), in the years after the FLOOD. He deposed Tammuz to recapture the city of his father, Lugulbanda, but his reign was initially cruel. He was befriended by the beast-man Enkidu. While collecting timber for new building projects Gilgamesh defeated Humbaba, the God of the Forest, who breathed fire and had an eye that could petrify. Their actions incurred the wrath of the GODS, particularly En-lil and Ishtar (whom Gilgamesh spurned), resulting in a seven-year drought (◊ WASTE LAND). Gilgamesh had to kill the Bull of Heaven to lift the drought. His success led to tragedy, however, as the gods decided that either Gilgamesh or Enkidu must die for their many offences. The two drew lots and Enkidu lost. Griefstricken, Gilgamesh embarked on a QUEST for wisdom and IMMORTALITY. He underwent many adventures, some similar to later episodes in Greek and Jewish LEGENDS, before returning home an older and wiser man.

It is almost certain that Gilgamesh was a historical Sumerian hero-king who lived sometime before 2500BC and to whom legends clung. His adventures were recounted throughout the empires of the Hittites, Assyrians and Babylonians. The *Epic*, in oral tradition from the mid-25th century BC, was probably first recorded on clay tablets during the reign of King Ur-Nammu of Ur around 2100BC.

Other stories were added, until brought together possibly by the priest Sin-leqe-unnini, credited as the scribe of the stone tablets found in the library of the Assyrian king Assurbanipal (668-626BC). These were translated by Henry Rawlinson (1810-1895) and George Smith (1840-1876); the first full English translation was *The Epic of Gilgamesh* (**1917** US) by Stephen Langdon.

Gilgamesh is occasionally referred to in fantastic fiction as a heroic UNDERLIER, but rarely figures as a character, though he is the subject of *Gilgamesh the King* (**1984**) by Robert SILVERBERG. Silverberg's contributions to the **Heroes in Hell** SHARED-WORLD series (◊ Janet E. MORRIS), novelized as *To the Land of the Living* (fixup **1989**), describe the adventures of Gilgamesh in the AFTERLIFE. Andrew SINCLAIR wrote about Enkidu in *Inkydoo, the Wild Boy* (**1976** chap). [MA]

GILLIAM, TERRY (VANCE) (1940-   ) US movie director and scriptwriter, earlier an animator. TG first came to wide notice with his animations for MONTY PYTHON'S FLYING CIRCUS; he perhaps continued his connection with the **Python** crew too long. He partook in the **Python** compilation movie *And Now For Something Completely Different* (*1971*), co-directed (with Terry JONES) MONTY PYTHON AND THE HOLY GRAIL (*1975*) and did various work on MONTY PYTHON'S LIFE OF BRIAN (*1979*) and *Monty Python's The Meaning of Life* (*1983*). His first solo directorial venture was JABBERWOCKY (*1977*); here, though he still clung to the **Python** team in the form of the actor Michael Palin and others, he began to speak with his own voice. This voice became resoundingly clear with TIME BANDITS (*1981*), one of the core works of fantasy CINEMA. As if this were not enough, TG went on to create what is arguably both the best TECHNOFANTASY and the best ALTERNATE-WORLD fantasy in cinema – BRAZIL (*1985*) – and then an outstanding piece of TWICE-TOLD storytelling in the form of The ADVENTURES OF BARON MUNCHAUSEN (*1988*), which was a commercial flop due to budgetary mis-estimation. The FISHER KING (*1991*) is an excellent stab at CONTEMPORARY FANTASY, bringing the legends of PERCEVAL and the GRAIL into the modern USA. *Twelve Monkeys* (*1996*) is a brilliant TIME FANTASY that largely ignores its sf trappings while echoing in part the technofantasy of *Brazil*. TG is one of the most significant late-20th-century creators of fantasy. [JG]

GILLILAND, ALEXIS A(RNALDUS) (1931-   ) US writer and cartoonist, better known for sf than for fantasy, but whose **Wizenbeak** sequence – *Wizenbeak* (**1986**), *The Shadow Shaia* (**1990**) and *The Lord of the Troll-Bats* (**1992**) – is of interest, partly because the eponymous WIZARD made his first appearances in books of cartoons like *The Iron Law of Bureaucracy* (graph coll **1979**), *Who Says Paranoia Isn't "In" Anymore?* (graph coll **1985**) and *The Waltzing Wizard* (graph coll **1989**), although the Wizenbeak of the acidly satirical cartoons has little in common with the genial book character. Beginning as a comic romp, featuring the moderately incompetent Wizenbeak in a world where rationalized MAGIC rules, it gradually darkens into a political commentary on power, politicians and just rule. The trollbats who actually do most of Wizenbeak's work turn out, absorbingly, to be members of a neighbouring species in whose homeland humans are slaves. [JC]

GILLRAY, JAMES (1757-1815) English satirical artist. ◊ COMICS.

GILLULY, SHEILA (?   -   ) US writer whose fantasy comprises two series set in the same FANTASYLAND composed out of CELTIC-FANTASY elements as filtered through

J.R.R. TOLKIEN. The **Greenbriar Queen** sequence – *Greenbriar Queen* (**1988**), *The Crystal Keep* (**1988**) and *Ritnym's Daughter* (**1989**) – follows the life of young Ariadne, an UGLY DUCKLING soon exposed as a HIDDEN MONARCH whose throne is held by a usurper king and his evil WIZARD sidekick. Aided by a SEVEN-SAMURAI group of COMPANIONS, she gains the throne, survives the loss of her husband in the second volume, and brings her headstrong son to maturity in the third. The second sequence, **Book of the Painter** – *The Boy from the Burren* (**1990**), *The Giant of Inishkerry* (**1992**) and *The Emperor of Earth-Above* (**1993**), with further volumes projected – is far more resolutely, although no more interestingly, Celtic.   [JC]

GILMAN, GREER ILENE (1951-   ) US writer whose *Moonwise* (**1991**) remarkably attempts to compose an entire long fantasy at a pitch and density of language reminiscent of Gerard Manley Hopkins (1844-1899). It is a work of quite mind-clouding complexity, whose narrative is literally its most superficial layer. Deeper, it describes a mythology of the cycle of the SEASONS, with the light and dark of the Moon characterized as warring sister-goddesses; they are also a thorn-tree and a standing stone (representing life and death). The dark sister has sought to turn a circle of world-dancers from time's beginning into stone in order to freeze TIME in winter and give her eternal victory. On another level the entire novel is a web of wordplay and symbols, of doublings and mirrorings and correspondences that spell the pattern of the text. GIG's deep knowledge of English etymology (including dialectal variations) charges every word with all its possible meanings.

Protagonists Ariane and Sylvie form a DUO whose sobriquet "silly sisters" connects them directly to the folk album *Silly Sisters* by Maddy Prior and June Tabor; Ariane and Sylvie are thus explicitly marked as singers, artists. (Traditional folk music is another significant layer of the text.) They are also themselves aspects of the WITCHES Malykorne and Annis. In one explicit doubling, Ariane and Sylvie go INTO THE WOODS and once return to Sylvie's house, then go back to the woods to enact a solstice RITUAL calling forth the OTHERWORLD of Cloud which they had invented as childhood friends. Ariane's much-commented-on clumsiness causes her to be left behind when Sylvie crosses the THRESHOLD.

There follows a nearly fatal stasis of narrative with Ariane sleeping, dreaming, waking, sleeping again; she dreams of finding in the woods a LIMINAL BEING she names Craobh, whom, when she wakes, she indeed finds; Craobh is the last dancer not turned to stone, and she gives Ariane her soul-stone to carry. At great length they set forth on a QUEST (accompanied by a tinker, Cloudwood) to find Sylvie in Cloud and prevent Annis's victory. The final wedding is a LICENZA which firmly roots the mythological back into the everyday.

Moonwise is a work of inexhaustible richness.   [DGK]

GILMER, ANN Pseudonym of Marilyn ROSS (itself a pseudonym).

GIOLITTI, ALBERTO (1923-1993) Italian COMIC-strip artist, most of whose (copious) works, characterized by a strong, clear, realistic drawing style, have been published in the USA. He is best remembered for **Turok, Son of Stone** (1954-82). AG's first comics work appeared in *Vittorioso* in 1939. After WWII he met an American who told him to seek work in the USA, and he set out in January 1946. His first attempts to gain admission as an immigrant were unsuccessful, and he went to Buenos Aires to work for Editorial Lainez before returning to the USA in 1950.

AG's first US work was drawing Westerns for Dell's *Indian Chief*, followed by *Tonto* and *Sergeant Preston*. He went on to draw comics adaptations of several tv series, and worked with other long-established comics characters including *Zorro*, *Jungle Jim*, FLASH GORDON and TARZAN. AG drew a great number of adaptations from classical literature, including *Alexander the Great*, *The Christmas Story*, *Aladdin and the Wonderful Lamp* and *Gulliver's Travels*; for this last he received the 1955 Thomas Alva Edison Award for Children's Literature.

In 1960 he went back to Italy and set up the Giolitti Studio to provide comics artwork for US and UK publishers. For the UK he drew the fantasy series **Flame of the Forest** (*Lion* 1965-6), **The Fiery Furnaces** (*Lion* 1965-6) and **Enchanted Isles** (*Tammy* 1965-6); he also drew the true-life Western **Blood on the Prairie** (*Ranger* 1965-6). For the US market he drew **Turok, Son of Stone**, a series of stories concerning two Native Americans lost in a mysterious valley of DINOSAURS. He drew also an extensive list of movie- and tv-related titles, including *Star Trek*, *Voyage to the Bottom of the Sea* and *Beneath the Planet of the Apes*, plus a GRAPHIC-NOVEL adaptation (graph **1970**) of KING KONG (*1933*).

For the Italian market he provided artwork for *The* PHANTOM and MANDRAKE, the pornographic titles *Jacula* and *Cosmine*, and the HORROR titles *Oltretomba* and *Terror*. [RT]

GIRAUD, JEAN ◊ MOEBIUS.

GIRL WITH SOMETHING EXTRA, THE US tv series (1973-4). Columbia Pictures/NBC. **Pr** Mel Swope. **Exec pr** Bob Claver. **Created by** Bernard Slade. **Starring** John Davidson (John Burton), Sally Field (Sally Burton). 22 30min episodes. Colour.

John, a conservative attorney, is surprised on his wedding night when his bride Sally delivers the news that she can sometimes read people's minds (◊ TALENTS). Much to his dismay, she admits she can read *his* mind. Most episodes of this sitcom find Sally using her ESP in attempts to advance Jack's career, or trying to undo the damage he has caused by misreading other people's minds. The show was a little-disguised imitation of BEWITCHED (1964-72).   [BC]

GLAMOUR This word, now used only metaphorically, originally meant a kind of ENCHANTMENT employed to confuse the sense of sight, often to make a person or a place supernaturally attractive in spite of actual decrepitude. SPELLS of this kind are often cast by FEMMES FATALES, after the fashion of those featured in *Maker of Shadows* (**1938**) by Jack Mann (E. Charles VIVIAN) and "Glamour" (1939) by Seabury QUINN.   [BS]

GLASTONBURY ◊ AVALON.

GNOMEMOBILE, THE US movie (*1967*). ◊ Upton SINCLAIR.

GNOSTIC FANTASY The Gnostic sects (◊ RELIGION) which flourished in the 2nd century AD laid claim to esoteric spiritual knowledge (Greek *gnosis*) resulting from special revelation. Thus ROSICRUCIANISM and THEOSOPHY are in a sense gnostic, and GF may in fact loosely describe any FANTASY OF HISTORY whose emphasis is on genuinely spiritual rather than political SECRET MASTERS. Another frequent tenet of Gnosticism, that the physical world is an imperfect and even EVIL creation (thus leading to MANICHEISM), is complemented by the central notion of an ideal REALITY – the Pleroma or celestial totality – of which our world is only a SHADOW. Roger ZELAZNY's **Amber** sequence revolves around this trope – for Earth is only one of many shadows of Amber – and E.R. EDDISON's **Zimiamvia** books share it. Even C.S. LEWIS introduces this GF element when the Narnia of *The Last Battle* (**1956**) proves to be a mere shadow of its "more real" original in HEAVEN; there is also something of GF in the special revelations, mediated by ANGELS, vouchsafed to Lewis's Ransom in his CHRISTIAN FANTASY *That Hideous Strength* (**1945**) – reminding us that, to modern eyes, Gnosticism seems very close to the early Christianity which denounced it as heresy. John CROWLEY piercingly evokes the GF yearning for that true and ultimate REALITY in *Aegypt* (**1987**); Franz KAFKA and Jorge Luis BORGES can be interpreted as gloomy modern Gnostics for whom this fallen world is an endless LABYRINTH of texts and sophistries, without any exit to transcendence.   [DRL]

GOBLE, WARWICK (WATERMAN) (1862-1943) UK artist, the foremost illustrator of sf in the pages of *Pearson's Magazine* 1896-1903. He contributed 66 illustrations to the original serialization of *The War of the Worlds* (1897; **1898**) by H.G. WELLS in *Pearson's*. WG also illustrated several sf and fantasy stories by George Griffith (1857-1906) and Fred M. White (1859-?   ), as well as F. Provand Webster's LOST-RACE novel *The Oracle of Baal: A Narrative of Some Curious Events in the Life of Professor Horatio Carmichael, M.D.* (**1896**).

WG abandoned sf illustration in 1903 to concentrate on Edmund DULAC-style colour-plate gift books.   [RD]

GOBLIN Any form of mischievous or evil spirit. The name entered FOLKLORE in the 14th century; the name of the Ghibellines who, a century earlier, had supported the Holy Roman Emperor against the Papacy (◊ DANTE ALIGHIERI) may have encouraged the usage by association, but the word comes from the Greek *kobalos*, through which root the Germans also derived the word "kobold". Early appearances of goblins in FOLKTALES present them merely as tormenters, often of children. They appear as boggarts, bogles or bogey-beasts. Charles DICKENS used goblins thus in the story within *Pickwick Papers* (**1836-7**) entitled "The Goblins Who Stole a Sexton" (1836), and they are the influential spirits who encourage Toby Veck to his good deeds in *The Chimes* (**1844** chap). Goblins became more distinct in the poem *Goblin Market* (**1862** chap) by Christina ROSSETTI, where they try to tempt children to eat poisonous fairy fruit. They are more closely associated with kobolds and DWARFS in George MACDONALD's *The Princess and the Goblin* (**1872**); here they were once human, now degenerate through millennia of living UNDERGROUND. The illustrations by Arthur Hughes (1823-1915) to this volume and by Arthur RACKHAM to later editions of *Goblin Market* have coloured our perceptions of goblins.

Those with a better nature are called hobgoblins, retaining the mischievousness without the maliciousness. This version is usually represented by PUCK or ROBIN GOODFELLOW, and was given form in SHAKESPEARE's *A Midsummer Night's Dream* (performed 1596; **1600**). It is this variety that appears in *Davy and the Goblin* (**1884**) by Charles E. Carryl (1841-1920). In the **Hob** series by William MAYNE, Hob is a form of brownie (◊ ELVES). The main image today derives from J.R.R. TOLKIEN's *The Hobbit* (**1937**) and *The Lord of the Rings* (**1954-5**), by which time he had renamed them orcs. Tolkien's goblins/orcs were similar to the Nordic svarts, hideously evil creatures created by magic, who appear in Alan GARNER's *The Weirdstone of Brisingamen*

(**1960**). Brian FROUD produced his own book of *Goblins* (graph **1983** chap) and teamed up with Terry JONES to produce *The Goblins of Labyrinth* (graph **1986**) and «The Goblin Companion» (graph **1996** chap). A representative anthology is *The Hamish Hamilton Book of Goblins* (anth **1972**; vt *A Cavalcade of Goblins* 1969 US; vt *A Book of Goblins* 1972 UK) ed Alan Garner.   [MA]

GOD The supreme being of monotheistic religions, signified in the Judeo-Christian tradition by the tetragrammaton variously expanded as Yahweh and Jehovah, and called Allah by the followers of Islam. Reverence demands He maintain a dignified absence from most CHRISTIAN FANTASIES, which often employ ANGELS as His agents, but He is much more frequently glimpsed – usually in unflattering guise – in antireligious fantasies and exercises in Literary Satanism (◊ SATAN), especially those offering irreverent accounts of the LAST JUDGEMENT. Stories which work towards a climactic confrontation with God are often calculatedly bathetic; stories in which He puts in a long-anticipated but less-than-satisfactory appearance include *Jurgen* (**1919**) by James Branch CABELL, *The Adventures of the Black Girl in Her Search for God* (**1932**) by George Bernard SHAW and *The Jehovah Contract* (**1987**) by Victor Koman (1944-   ). In Lester del Rey's heretical fantasy "For I Am a Jealous People" (1954) He decides that humans are not His chosen people after all, and who can blame Him? *The Man who was Thursday* (**1908**) by G.K. CHESTERTON is supposed to provide a more inspiring climactic confrontation, but opinions as to its propriety vary. His appearances in the work of T.F. POWYS, often as "Jar" but sometimes as "Mr Weston", are far more extensive but distinctly ambivalent; other works in which He makes substantial and moderately impressive appearances include *The Green Isle of the Great Deep* (**1944**) by Neil M. GUNN, "The Adventures of God in His Search for the Black Girl" (1973) by Brigid BROPHY, *The Living End* (**1979**) by Stanley ELKIN and *Towing Jehovah* (**1993**) by James MORROW. *God: The Ultimate Autobiography* (**1988**) by Jeremy Pascall is unrepentantly silly.   [BS]

GODDESS A thin partition divides fantasy writers and readers from those – they include some feminists (◊ FEMINISM), New Agers, devotees of Wicca and speculative theorists – who argue that the UNDERLIER figure we may name the Goddess is as real as the GOD – who, according to the most accepted story of Her fate, supplanted Her *c*2000BC – and who treat her subjugation and/or disappearance as due to a conspiracy on the part of an oppressive world order.

Unfortunately, God – i.e., the Judeo-Christian-Moslem figure whom it is usual for Westerners to designate as God – is a far better-documented figure of belief than the Goddess. This may be simply because history is told by winners, but there is little evidence for anything like a pre-Indo-European belief in a Great Mother – and even less that any Goddess-dominated religious life was necessarily (or even usually) matriarchal, on lines promulgated by the Goddess's less cautious late-20th-century advocates. We know, in fact, very little indeed about the relation between female deities and political power in the prehistoric world. Nor is it safe for us to assume that GAIA (one of the Goddess's better-known early names) promoted a now-lost rapport with the Earth, a rapport forcibly violated (in ARCHETYPE-generating scenes of rapine) by Indo-European blonds.

However, given the basic understanding that writers of fantasy know they are writing *fiction*, it does not matter that the Goddess, as a historically worshipped figure, is a concept severely questioned by modern anthropologists, archaeologists and scholars of religion, nor that the rapport of her people with the Earth is no more plausible than that supposed (by sentimentalists) to have been enjoyed by Native Americans. What matters for writers and readers of fantasy is the fact that the Goddess is an ample source of STORY.

That Story begins in the Palaeolithic, perhaps 50,000 years ago. Most of the carvings extant from this period are of women, usually with exaggerated sexual features. Images that seem to associate the vulva with openings into the Earth are common. These images depict – it is useful for fantasy writers to think – a primal Earth Goddess: Gaia (according to early Greek myths) or Nut (according to early Egyptian versions). As time passes, and we come into near prehistory, the Earth Goddess takes on greater individuality, and in becoming the Great Mother – a term used to describe the Sumerian Inanna, or the Canaanites' Astarte, or the Egyptian Maat, or the Aztec Coatlicue, or the Pueblo Spider Woman, whose story shapes the narrative structure of Sheri S. TEPPER's *Shadow's End* (**1994**) – she becomes the creator of the world, the fertile source of all new life, the guardian of all that lives and grows. But the Goddess is not wholly benign. She contains in her nature the darkness and violence that designate the end of things, and she continues the CYCLE. She has not yet been divided into two, into the good Goddess aboveground, and her SHADOW in the UNDERWORLD; she has certainly not yet become the Triple Goddess: maiden (symbol of VIRGINITY), mother (symbol of FERTILITY) and crone. Men are not yet important; neither virginity nor marital fidelity (in the PANTHEON or, by hopeful extension earthwards, in life) have yet become, therefore, issues of much interest.

As soon as a consort enters the picture, however, it becomes necessary (for those who believe in the Goddess) to begin to decipher the "true" story out of the interstices of the victors' version of the nature of the world, their rewriting of the stories of creation and birth and guardianship. Inanna becomes Ishtar. Figures like the Anatolian Cybele (who according to some scholarship represents a different tradition, deriving from the Iranian Anahita, but one which ultimately combines characteristics with other Goddess figures), or the Indian Parvati, the Egyptian Isis, the Greek Demeter, the Norse Freyja/Freya and the Celtic Rhiannon, are all participants in a much more complex Story; they share essential characteristics, and fantasy literature tends not to be particular as to which version of the Goddess most legitimately represents which part of the Story. These versions of the Goddess tend to have consorts (who may, as in the tale of Isis and Osiris, be siblings, or, as in the tale of Cybele and Attis, sons) and (like Inanna/Ishtar's Dumuzi/Tammuz) they tend to make SEASON-renewing journeys to the Underworld (◊◊ GOLDEN BOUGH), where their shadow-sisters dwell, in order to bring their consort back to the light, if only for a while. They continue to lie at the centre of things, and to generate all life from their wombs – but they can be dismembered, they can suffer anguish, and their berserker side increasingly comes to be seen as an opposing principle rather than as merely an aspect. It is this version of the Goddess that underlies, according to *Shakespeare and the Goddess of Complete Being* (**1992**) by Ted Hughes (1930-), the Venus-and-Adonis myth that was a central engine of the genius of William SHAKESPEARE.

But time does not stand still for the Goddess. Grossly to simplify almost unimaginably complex histories, the slow irruption of Indo-European cultures into older worlds, from India to the Atantic Ocean, arguably marks a transition from Goddess-dominated to male-dominated worldviews. It is hard to imagine a testosterone-choked god like ZEUS, with his constant philandering, coming into existence in a world much earlier than that of the Greeks, who created him in all his rampageous glory in the first millennium BC. In the world of Zeus, the Goddess is Hera, whose shrewishness and obsession with the married state neatly – as David Leeming and Jake Page suggest in *Goddess: Myths of the Female Divine* (**1994**), and as Hughes insists in his book on Shakespeare – demonstrate how completely patriarchy has won the war against the Goddess whom it (in Hughes's language) "abhors", for she lends "female approbation to the very institution by which the old matrilineal rights were most clearly usurped by the patriarchy that saw wives as belonging, like other valuable objects, to their husbands". Hera's concerns are the concerns of her master: control of female sexuality is necessary to maintain paternal lines of descent. Female virginity is (according to the advocates of the Goddess) of innate interest to men for the same reason.

When the Babylonian Tiamat, descended from early Goddess figures, causes chaos in HEAVEN, young King Marduk kills her outright. When GILGAMESH is tempted by Ishtar, he decides not to become a Year King and declines her "favours". When LILITH turns out to be an unsuitable spouse for Adam, because too vividly reminiscent of the Goddess of whom she is a late version, Adam takes a second wife, and Lilith turns into the SERPENT (and also is identified as a LAMIA). The all-replenishing vulva of the earlier Goddess becomes, in the very late story of Pandora, a "box" which, when opened, brings to the light-irradiated world of men all the moist infections of the mortal world. The Goddess becomes Chimera; becomes Medusa; becomes CIRCE.

Once tamed – but always dangerous – the Goddess enters the literate Mediterranean world as Athena/Minerva, Artemis/Diana and Aphrodite/Venus. Athena is sexless; Artemis destroys Actaeon for looking at her naked; and Aphrodite, whose morals are loose, is treated by men with a gingerly respect neatly characterized by Thorne SMITH in *The Night Life of the Gods* (**1931**). She becomes Maeve, who in turn becomes the FAIRY QUEEN. She evolves (it may be) from the MORRIGAN to MORGAN LE FAY; the protagonist of two novels by Dion FORTUNE – *The Sea Priestess* (**1938**) and *Moon Magic* (**1956**) – is Miss Le Fay Morgan, a priestess and avatar of the Goddess, who is (perhaps for the first time in popular fiction) seen here as autonomous, rather than the consort of a patriarchal god. The Goddess also becomes the Virgin Mary/Madonna, who is intended to be not worshipped but (less dangerously) "revered". She is an UNDERLIER figure helping to shape the FEMME FATALE who dominates much of the erotic or erotized fantasy of the 19th century, and becomes the "She" who dominates so many LOST-RACE novels. She is, of course, many beings. In the imaginations of many 20th-century fantasy writers, however, the Goddess is one, however cloaked by the machinations of patriarchy. The storylines of much modern fantasy move therefore – as on a plumbline into the depths of dream – towards a declaration of Her underlying unity.

Tales in which the Goddess variously figures are very numerous. They include: Lynn ABBEY's **Rifkind** series (**1979-80**) and her **Unicorn and Dragon** series (**1987-8**);

Lloyd ALEXANDER's **Chronicles of Prydain** (**1964-8**); Constance ASH's **Glennys** series (**1988-92**); A.A. ATTANASIO's *The Moon's Wife: A Hystery* (**1993**); Peter S. BEAGLE's *The Folk of the Air* (**1986**); Hans BEMMANN's *The Broken Goddess* (**1990**); Marion Zimmer BRADLEY's *The Forest House* (**1993**); *The Goddess of Atvatabar* (**1892**) by William R. Bradshaw (1851-1927), one of many lost-world novels that could be instanced; Anthony BURGESS's *The Eve of Saint Venus* (**1964**); Moyra CALDECOTT's *The Lily and the Bull* (**1979**); Catherine COOKE's **Eleven Kingdoms** series (**1985-8**); Susan COOPER's **The Dark is Rising** series (**1965-77**) and *Seaward* (**1983**), whose JANUS-faced Goddess figure engages in a GODGAME-like CHESS match with her father/brother/son; Kara DALKEY's *Euryale* (**1988**), which pits the eponymous Goddess/Gorgon figure against a patriarchy-defending Athena, and her «Blood of the Goddess» sequence beginning with «Goa» (**1996**); E.R. EDDISON's **Zimiamvia** sequence; Ford Madox FORD's *Ladies Whose Bright Eyes* (**1911**) and *The Young Lovell: A Romance* (**1913**); Robert Graves's *The White Goddess* (**1947**), ostensibly nonfiction, intensely mythopoeic; Joyce Ballou GREGORIAN's **Tredana Trilogy** (**1975-87**); H. Rider HAGGARD's **Ayesha** books, primarily *She* (**1886**) and *Ayesha* (**1905**); Elizabeth HAND's *Waking the Moon* (**1994**); Ursule Molinaro's *The New Moon with the Old Moon in her Arms* (**1990**); Alis A. RASMUSSEN's *The Labyrinth Gate* (**1988**); Fay SAMPSON's **Daughter of Tintagel** series (**1989-92**) and *Star Dancer* (**1993**), the former about Morgan and the latter about Inanna; Linda Lay SHULER's **Time Circle** series (**1987-92**); various novels by Thomas Burnett SWANN; "Dream on Monkey Island" (**1967**), in *Dream on Monkey Island and Other Plays* (coll **1970**) by Derek Walcott (1930-), in which the White Goddess is executed as part of a black man's emancipation from the white hegemony which has thinned (◊ THINNING) his people's culture into near extinction; and Gene WOLFE's **Latro** series (**1986-9**) and *There Are Doors* (**1988**). [JC]
See also: GODS.

GODGAME There is a sense in which almost any written story can be described as a godgame – one being played by its author. That aside, a considerable proportion of FANTASY narratives can be described as tales whose protagonists search – almost always successfully – for the underlying STORY which explains their nature and their world, a Story whose natural outcome is a structurally complete – and therefore all-encompassing – ending. A Story can thus be defined as a godgame whose rules govern its protagonists.

The most extensive analysis of the godgame appears in R. Rawdon Wilson's *In Palamedes' Shadow: Explorations in Play, Game, & Narrative Theory* (**1990**), where he credits John FOWLES with inventing the term to describe the action of *The Magus* (**1965** US; rev **1977** UK), whose protagonist, Conchis, ensnares the young teacher who narrates the tale in a LABYRINTH of ILLUSIONS which can be escaped only at Conchis's behest, after he has judged the teacher to have become a competent human being; Fowles's original, unused title for the novel was in fact «The Godgame». "In a godgame," Wilson says, "one character (or several) is made a victim by another character's superior knowledge and power. Caught in a cunningly constructed web of appearances, the victim, who finds the illusion to be impenetrable, is observed and his behaviour is judged." It is, Wilson continues, "a narrative category that has existed since the tales of ancient mythology", in which a figure like

Semele might be "played" by Hera into forcing APOLLO to do the one thing – reveal himself to her – that will guarantee her death, guarantee in the most final sense that she has been found wanting. The essence of the godgame, for Wilson, is threefold: (a) there must be a victim; (b) there must be a plot through which the victim must struggle, his/her every action in truth a reaction; and (c) there must be an *owner* of the game (a MAGUS, a *magister ludi*, a GOD) who is in some sense present while the game is being played, *and who stands in judgement*.

To use the term "godgame" is not, therefore, to describe in a general sense the relationship between the creator and the created, or to emphasize the central power of Story in most full fantasy narratives; a godgame is a tale in which an actual game (which may incorporate broader implications) is being played without the participants' informed consent, and which (in some sense) is being scored by its maker. As such, the term may be used to describe a large number of works.

An encompassing example comes very early. *The Tempest* (performed *c*1611; **1623**) by William SHAKESPEARE is set on an ISLAND (a natural venue for godgames, because controllable: *The Magus*, too, is set on an island) governed by PROSPERO, who controls the actions of the entire cast, manipulates their responses, and judges them all. It is almost certain that any novel which featuring a character named Prospero, or which evokes him, will either be a godgame tale or will flirt with the possibility – and this aspect is brought even more to the foreground in Peter GREENAWAY's movie PROSPERO'S BOOKS (*1991*). Other Shakespeare plays with godgame elements include *A Midsummer Night's Dream* (performed *c*1595; **1600**) and *Macbeth* (performed *c*1606; **1623**).

The literature of the fantastic is full of godgames. The works of authors like Jorge Luis BORGES, Hermann HESSE, Franz KAFKA and Thomas MANN (and many others) manipulate the concept in very various ways; Wilson's central example of the importance of the godgame in recent literature is Thomas PYNCHON.

Almost any POSTHUMOUS FANTASY – any tale in which a human SOUL must work out the rules that had governed its mortal life – implies a godgame, though in most cases that implication is undeveloped. An exception is Susan COOPER's *Seaward* (**1983**), whose live protagonists find themselves in a land which can be defined, at least in part, as a CHESS board; in the long game between the GODDESS figure who represents Death and her father/brother/son who represents Life the two protagonists can be seen as acting out the moves of two conflicting godgames, which become reconciled only when both youngsters pass through the morally complex RECOGNITION of choice and passage which concludes the book.

Most FANTASIES OF HISTORY hint at or directly incorporate godgames. The novels of Philip José FARMER – the **Tierworlds** sequence in particular – are often godgames. The **Star Trek** tv series (◊ *SFE*) comes close to fantasy in the very numerous episodes involving godgames only Kirk and Spock (and later lead characters) can hope to solve. The UK tv series **The Prisoner** (1967-8; ◊ *SFE*) is centred on its godgame elements, as is Ursula K. LE GUIN's sf novel *City of Illusions* (**1966**). Godgames in HORROR – except the separate category of novels about SATAN – are not of much interest, a possible exception being Kim NEWMAN's *The Quorum* (**1994**). Recent fantasy novels incorporating godgame elements include William KOTZWINKLE's *Fata Morgana* (**1977**), Nancy KRESS's *The Prince of Morning Bells* (**1981**), Hans BEMMANN's *The Broken Goddess* (**1990**), Graham JOYCE's *House of Lost Dreams* (**1993**) – a novel strongly evocative of *The Magus* – John GRANT's TECHNOFANTASY *The Hundredfold Problem* * (**1994**) and Lindsay CLARKE's *Alice's Masque* (**1994**). [JC]

GODS Most religions have a PANTHEON of gods and, even though the Christian religion has the one GOD, He is often associated with the Trinity, suggesting a mini-pantheon. Only the Jewish faith has a true single god.

Most gods known to Western cultures derive from the Greek/Roman, Scandinavian or Celtic mythologies (◊ MYTHS). Prior to the emergence of GENRE FANTASY, the Greek gods were the most used in fiction. The Greek pantheon is quite complicated, and though most will recognize ZEUS as King of the Gods, he became that only after deposing his father Kronos, king of the Titans. The central Greek pantheon (with their Roman identities where applicable) were Zeus (Jove), Hera (Juno), APHRODITE (Venus), APOLLO, Ares (Mars), Artemis (Diana), Athena (Minerva), Demeter (Ceres), DIONYSUS (Bacchus), Hephaestus (Vulcan), HERMES (Mercurius) and Poseidon (Neptune). There were lesser gods, of whom Eros (CUPID), HADES (Pluto), Helios (Sol) and PAN (Faunus) are best-known. These gods may occur in fiction singly – as in the revival of Aphrodite in *The Tinted Venus* (**1885**) by F. ANSTEY and Pan's magical appearance in *The Wind in the Willows* (**1908**) by Kenneth GRAHAME – or in groupings, as in Richard GARNETT's *The Twilight of the Gods* (coll **1888**; exp 1903). They also quite commonly appear as a pantheon, especially when the author is lampooning modern society (◊ SATIRE): examples are *Olympian Nights* (coll of linked stories **1902**) by John Kendrick BANGS and *The Night Life of the Gods* (**1931**) by Thorne SMITH. Eden PHILLPOTTS was kinder to the memory of the Greek gods in bucolic sketches including *Pan and the Twins* (**1922**), *The Miniature* (**1926**) and *Arachne* (**1927**). Today they are less utilized, with writers generally favouring the Scandinavian or Celtic gods – except, of course, in the many novels devoted to famous Greek LEGENDS. The most complete series about the Greek gods is the **Titans** trilogy by Patrick H. ADKINS.

The Scandinavian Gods are also well known to Western cultures. The main pantheon is the AESIR, under the leadership of ODIN (Woden to the Teutons who introduced these gods to England via the Saxons). Their adventures are best known through the *Volsunga Saga* and the *Nibelungenlied* (◊ NORDIC FANTASY). J.R.R. TOLKIEN drew heavily upon this pantheon. (◊◊ LOKI.)

The Celtic Gods have become more prominent with the rise of CELTIC FANTASY since the 1970s. While their names, which include Lir, Lugh, Cernunnos and Mabon, are less well known, their adventures have frequently been adapted from the MABINOGION. The Tuatha Dé Danann invaded Ireland and defeated the previous occupants, the Fir Bholg. The Tuatha had magical skills and rapidly established themselves, with their capital at Tara. Their leader, Nuadha, lost his arm in battle with the Fir Bholg and was fitted with a silver one, becoming called Nuadha Airgedlámh (of the Silver Hand). He later abdicated in favour of Lugh. The Tuatha were eventually defeated by the Milesians, the first true Gaelic Irish, who later worshipped the Tuatha as gods. The Tuatha goddess Ériu gave her name to Ireland, as Eire, because she proclaimed the Milesians would inherit the land. (◊ Morgan LLYWELYN, Paul HAZEL, Keith TAYLOR and particularly Kenneth C. FLINT for relevant works.)

Other gods who sometimes appear in fiction include: the Egyptian pantheon – Re (or Ra), Isis, Osiris, Horus and Thoth; the Hindu Pantheon – Brahma, Siva and Vishnu; Quetzalcoatl, from the Aztec pantheon; Enlil, the Sumerian god of the air; Marduk, the chief Babylonian deity; Ashtoreth, the Philisitine FERTILITY goddess based on the Sumerian Ishtar (◊ GODDESS); and the Phoenician Baal and Dagon. As understandings of past cultures develop, so their gods are borrowed more meaningfully into fiction. Roger ZELAZNY adapted the Hindu pantheon in *Lord of Light* (**1967**) and deployed the Egyptian gods in *Creatures of Light and Darkness* (**1969**). Carlos FUENTES has drawn upon the Aztec gods, while Chinese and Japanese gods frequent much ORIENTAL FANTASY.

As the older gods gave way to Christianity, so the old-world MAGIC began to fade (◊ THINNING). This passing is explored best in the works of Thomas Burnett SWANN. That the old gods might still be re-awakened is a theme increasingly popular in stories of white magic, particularly in Nature myths (◊◊ Algernon BLACKWOOD; Charles DE LINT; Robert HOLDSTOCK). The old gods may also be summoned for BLACK MAGIC, and many stories of the occult rely on worship of the ELDER GODS (◊◊ Aleister CROWLEY; Arthur MACHEN; Dennis WHEATLEY).

Any genuine SECONDARY WORLD or autonomous OTHERWORLD requires its own RELIGION and thus its own gods. Lord DUNSANY realized this from the outset with *The Gods of Pegana* (coll **1905**). E.R. EDDISON created a complex theogony in his **Zimiamvia** series, with gods and men indistinguishable other than through their world-making abilities. J.R.R. TOLKIEN went furthest in the creation of a complex genealogy of gods for Middle-Earth. Other writers who have treated gods seriously are L. Sprague DE CAMP, Philip José FARMER, Michael MOORCOCK, Terry PRATCHETT, Michael SHEA, Jack VANCE and Roger ZELAZNY. H.P. LOVECRAFT is unusual in creating a pantheon within our own world (◊ CTHULHU MYTHOS).

In anthropomorphic fiction, animals have their own gods, which may be based on the human world. Richard ADAMS created a rabbit mythology in *Watership Down* (**1972**).

The interaction between gods and humans can at times become intense. In the finale of Stephen Donaldson's *White Gold Wielder* (**1983**) Covenant realizes he must transcend into a godlike state in order to defeat Lord Foul. Conversely, gods may become human to understand the human condition. A.E. VAN VOGT explores this on a vast timescale in *The Book of Ptath* (**1947**). [MA]

Further reading: *The Encyclopedia of Gods* (**1992**) by Michael Jordan.

GODWIN, PARKE (1929-   ) US writer. His first work was a detective novel, *Darker Places* (**1971**). He began publishing work of genre interest with "Unsigned Original" for *Brother Theodore's Chamber of Horrors* (anth **1975**) ed Brother Theodore and Marvin KAYE, with whom he also wrote some sf novels. After stories like "The Lady of Finnegan's Hearth" (1977), which is about Isolde, PG published his first fantasy sequence, the **Firelord** series – *Firelord* (**1980**) and *Beloved Exile* (**1984**), plus an associated volume, *The Last Rainbow* (**1985**) – which treats the story of ARTHUR in a realistic mode, though with an overlay of CELTIC FANTASY. This element is strongest in the final volume, in which a figure modelled on Saint Patrick encounters FAERIE in Ireland; but the series, like the fine later **Robin Hood** sequence – *Sherwood* (**1991**) and *Robin and the King* (**1993**) – takes an essentially historical view of its protagonists, and intrusions of fantasy seem intended mainly to signal a THINNING of the traditional world. That world is also depicted in *A Memory of Lions* (**1976**), in which no vestige of fantasy persists. The vision of Arthur and of ROBIN HOOD as figures with mystically binding relationships to the MATTER of Britain is, in PG's hands, elegiac.

The Tower of Beowulf (**1995**) fully integrates the story of the MONSTER Grendel into the grim life-history of BEOWULF, who becomes an alienated CHILDE figure, literally obsessed by a DARK TOWER which (at least in his DREAMS) represents his own lonely state. PG's portrayal of Grendel, who is killed halfway through, rather resembles John GARDNER's in *Grendel* (**1971**), for both are hugely embittered, estranged and doomed. Other novels include *A Cold Blue Light* (**1983**) with Marvin KAYE (who produced a solo sequel), and *A Truce with Time: (A Love Story with Occasional Ghosts)* (**1988**), a GHOST STORY whose GHOSTS help two contemporary lovers gain self-knowledge and wary happiness. "The Fire When it Comes" (1981) – reprinted in *The Fire When It Comes* (coll **1984**), containing mostly SUPERNATURAL FICTION – won the 1982 WORLD FANTASY award.

PG is, by choice, not a central figure in the field of fantasy; this is a placement which could change. [JC]

GODWIN, WILLIAM (1756-1836) UK philosopher, political journalist and writer, father of Mary SHELLEY. His first book was *The Life of Chatham* (**1783**). The polemic which made him famous was *An Enquiry Concerning Political Justice* (**1793**; rev 1796), in which an almost Hermetic belief in the perfectability of Man is argued in an entirely rational manner, reaching a devastatingly anarchic conclusion: the enemy of truth turns out to be any political institution. WG's reputation as a novelist rests primarily upon *Things as They Are, or The Adventures of Caleb Williams* (**1794**), though he wrote considerable later adult fiction and children's books, some as by Edward Baldwin. These latter include *Fables Ancient and Modern* (coll **1805**) and the anonymous *Dramas for Children* (coll **1809**), both consisting of TWICE-TOLD tales.

Caleb Williams is not a SUPERNATURAL FICTION, but the political system it depicts, entirely in keeping with WG's expressed views, is so nightmarish that the world the oppressed Caleb descends into – in his attempts to avoid astonishingly relentless pursuit – is very close to a genuine UNDERWORLD. A later novel, *St Leon: A Tale of the Sixteenth Century* (**1799**), is a genuine supernatural fiction: the cold-hearted Reginald St Leon meets a wandering Jewish alchemist who gives him the PHILOSOPHERS' STONE, by virtue of which he can make gold, and the ELIXIR OF LIFE, by virtue of which he gains IMMORTALITY. But St Leon now finds himself perpetually at odds with those he encounters and exploits, and must wander Europe bereft of contentment, meeting at one point a GIANT who imprisons him. In the end (the alchemist died once his "gift" had been passed on), St Leon's resemblance to the WANDERING JEW increases, and he disappears. [JC]

Other works: *Lives of the Necromancers* (**1834**).

See also: FABLES.

GODZILLA (ot *Gojira*; vt *Godzilla, King of the Monsters*) Japanese movie (**1954**). ◊ GODZILLA MOVIES.

GODZILLA FIGHTS THE GIANT MOTH vt of *Godzilla Versus Mothra* (**1964**). ◊ GODZILLA MOVIES.

GODZILLA, KING OF THE MONSTERS vt of *Godzilla* (**1954**). ◊ GODZILLA MOVIES.

GODZILLA MOVIES "Godzilla" is an anglicization of the original Japanese "Gojira"; in keeping with our general practice, however, we treat these movies according to the titles under which they will most likely be encountered by Western viewers. Nowhere is this a more practical step than with this series of TECHNOFANTASIES, most of which are graced with a bewildering array of vts. Variant versions proliferate almost as much as do the vts: aside from cuts and changes, English-language versions of most were produced with not only dubbing but the splicing in of extra scenes featuring US actors. All the originals were produced by the studio Toho, latterly as Toho-Eizo.

A vast, fire-breathing *Tyrannosaurus rex* stirred from hugely long slumber by an atomic test, Godzilla was capable of trampling city buildings underfoot, and generally did so – although he later switched from VILLAIN to HERO, defending Earth against creatures even more destructive in what became almost a MONSTER-MOVIE soap opera. His first appearance was in *Godzilla* (**1954**), which showed his genesis, his devastation of much of Tokyo, and the first of his several deaths. *Gigantis* (**1955**) saw him defeat another monster, Angurus, then devastate Osaka before dying in an avalanche. *King Kong Versus Godzilla* (**1963**) has much the plot one might expect, although it has an attempted political/ethical subtext; KING KONG, although of course a cultural import from the USA, is regarded as Good, ranged against the Evil of US imperialist aggression, as represented by Godzilla. *Godzilla Versus Mothra* (**1964**) has a quite bizarre plot – it is the most fantasticated of the series – that sees Godzilla again attacking Japan; a giant moth is persuaded to try to stop him and fails, but its newly hatched larvae succeed. *Kaiju Daisenso* (**1965**) – none of whose English-language titles has ever really caught on – sees an abrupt shift in attitude towards Godzilla and the monster pterodactyl Rodan (which had made its debut in *Rodan* [**1956**]): far from being destroyers they are now pitted as Earth's defenders against a three-headed interstellar DRAGON, Ghidora. They thwart it (after a very entangled plot), but it was back later the same year in *Ghidora, The Three-Headed Monster* (**1965**), this time being defeated by a team comprising Godzilla, Rodan and Mothra. *Ebirah, Terror of the Deep* (**1966**) has the giant crab Ebirah defeated by Godzilla and Mothra. *Son of Godzilla* (**1967**) is blatantly aimed at the children's market; father and newly hatched son team up to defeat a big SPIDER, Spigon. *Destroy All Monsters* (**1968**) paraded not only Godzilla but 10 other monsters from the Toho stable in what was intended as a spectacular, but proved forgettable. *Godzilla's Revenge* (**1969**) is a half-hearted affair, consisting largely of rerun footage from *Ebirah, Terror of the Deep* and *Son of Godzilla*. *Godzilla Versus the Smog Monster* (**1971**) took a new theme, pollution: the monster Hedora is born from an oceanic sludge of industrial waste, and must be destroyed by Godzilla. This is generally regarded as the worst of the series. *Gojira Tai Gaigan* (**1972**) continues the ecological thrust, with Godzilla and Angurus (from *Gigantis*) fighting off an invasion of Earth by the monsters Ghidora (again) and Gaigan, which are the weapons of a race whose planet is dying through pollution. *Godzilla Versus Megalon* (**1973**) has Godzilla fighting a giant cockroach (Megalon) and a giant hen (Borodan); it is better than it sounds. *Godzilla Versus the Bionic Monster* (**1974**) mixes SWORD AND SORCERY with sf as Godzilla battles a cyborg replica of himself which is under alien control; it was sequelled by *Mekagojira No Gyakushu* (*1975*), which was more of the same, and ended the **Godzilla** series proper. There was, however, an addendum: *Godzilla 1985* (*1985*) is a homage to the original *Godzilla* (*1954*) and a thematic remake of it.

It cannot be said that any of these movies are good – and some are diabolical – yet the canon as a whole represents something of a feat of consistent vision that has not been matched by any series of MONSTER MOVIES in the West. [JG] **Further reading:** A fuller treatment of the **Godzilla** movies – and of the **Gamera** movies, for which there is no space in this book – is given by Peter Nicholls in *SFE*. Good coverage of all these movies is offered by *The Aurum Film Encyclopedia: Science Fiction* (**1984**; vt *Science Fiction: The Complete Film Sourcebook* 1985 US; rev 1991) ed Phil Hardy.

GODZILLA 1985 (vt *Gojira 1985*) Japanese movie (*1985*). ◊ GODZILLA MOVIES.

GODZILLA RAIDS AGAIN vt of *Gigantis* (*1955*). ◊ GODZILLA MOVIES.

GODZILLA'S COUNTERATTACK vt of *Gigantis* (*1955*). ◊ GODZILLA MOVIES.

GODZILLA'S REVENGE (ot *Oru Kaiju Daishingeki*) Japanese movie (*1969*). ◊ GODZILLA MOVIES.

GODZILLA VERSUS GIGAN Japanese movie (*1972*). ◊ GODZILLA MOVIES.

GODZILLA VERSUS HEDORA vt of *Godzilla Versus the Smog Monster* (*1971*). ◊ GODZILLA MOVIES.

GODZILLA VERSUS MEGALON (ot *Gojira Tai Megalon*) Japanese movie (*1973*). ◊ GODZILLA MOVIES.

GODZILLA VERSUS MOTHRA (ot *Mosura Tai Gojira*; vt *Gojira Tai Mosura*; vt *Godzilla Fights the Giant Moth*; vt *Godzilla Versus the Giant Moth*; vt *Godzilla Versus the Thing*; vt *Mothra Versus Godzilla*) Japanese movie (*1964*). ◊ GODZILLA MOVIES.

GODZILLA VERSUS THE BIONIC MONSTER (ot *Gojira Tai Mekagojira*) Japanese movie (*1974*). ◊ GODZILLA MOVIES.

GODZILLA VERSUS THE GIANT MOTH vt of *Godzilla Versus Mothra* (*1964*). ◊ GODZILLA MOVIES.

GODZILLA VERSUS THE SEA MONSTER vt of *Ebirah, Terror of the Deep* (*1966*). ◊ GODZILLA MOVIES.

GODZILLA VERSUS THE SMOG MONSTER (ot *Gojira Tai Hedora*; vt *Godzilla Versus Hedora*) Japanese movie (*1971*). ◊ GODZILLA MOVIES.

GODZILLA VERSUS THE THING vt of *Godzilla Versus Mothra* (*1964*). ◊ GODZILLA MOVIES.

GOETHE, JOHANN W(OLFGANG) (1749-1832) German poet, scientist and philosopher, the single greatest influence on the development of literature in GERMANY (and to a large extent throughout Western Europe) at that time, particularly affecting ROMANTICISM; he was raised to the nobility in 1782 as von Goethe. Although his name is immutably linked with his verse tragedy based on the FAUST legend, this was but one element of a complex of compositions that JWG experimented with and developed throughout his life.

His first productive period came during the 1770s, at the height of the Sturm und Drang movement of literary pyrotechnics. In 1775 he had been invited to the court of Karl August, Duke of Weimar, which court saw the flowering of literature in Germany. Heavily influenced by the Classics and the works of Jean-Jacques Rousseau (1712-1778), JWG established himself with the tragic novel *Die Leiden des jungen Werthers* (*1774*; rev 1787; trans Richard Graves as *The Sorrows of Young Werther* 1779 UK), and completed several plays based on the Classics; these included *Götter, Helden und Wieland* ["Gods, Heroes and Wieland"] (*1774*), which lampoons *Alceste* (*1773*) by Christoph Wieland (1733-1813) and involves a FRAME STORY where Wieland, magically transported to witness his ancient Greek characters (◊ GREEK AND LATIN CLASSICS), is astonished at the vibrancy of the old world. The combination of these interests, plus JWG's own thirst for knowledge, led to his initial thoughts on the character of Faust, and he drafted an early version of the play during 1773-6. The text of this version was discovered in 1887 and published as *Goethes Faust in ursprünglicher Gestalt* ["Goethe's Original Faust"] (**1887**) ed Erich Schmidt (1853-1913), a text usually known as the *Urfaust*. The play continued to evolve; a later draft was published as "Faust, ein Fragment" in *Schriften* ["Writings"] (coll **1790**), but it was only through the persuasion and encouragement of Friedrich Schiller (1759-1805) that JWG drew sufficient inspiration to complete the first part of *Faust – Der Tragödie erster Teil* (**1808**) – and eventually the second part, *Der Tragödie zweiter Teil* (**1832**). The English-language translation first appeared as *Faust: A Tragedy* (trans Warburton Davies **1838** 2 vols UK) in a limited edition of 50 copies.

The composition of the play stretched over 60 years, and incorporates all the moods and development of JWG's writings. The basic plot is about the testing of the human spirit. The DEVIL, in the shape of MEPHISTOPHELES, obtains permission from GOD to tempt Faust. Faust is dissatisfied with the state of human knowledge and understanding and desires to know more, and stands as the central UNDERLIER figure expressive of this paradigmatically "Western" theme. He summons the spirit of the Earth (◊ GODDESS), which proves too powerful to control. Faust almost commits suicide but reconsiders, and soon after meets Mephistopheles (or Mephisto), with whom he makes his pact (◊ PACTS WITH THE DEVIL). Mephisto agrees to support Faust in his pursuit of scientific knowledge provided Faust continues to develop morally and spiritually; any slip on Faust's part and Mephisto will claim him. Mephisto, of course, endeavours to trap Faust by every means of temptation, and there are graphic descriptions of witches' SABBATS and all manner of supernatural creatures and beasts from LEGEND. Although Faust frequently falters as his sexual desires take over, his thirst for scientific advancement never fails. For all its supernatural trappings the play may be considered a form of proto-SCIENCE FICTION, and certainly reflects the spirit of the age in its shift from SUPERSTITION to science, a change in which JWG played a significant part. It includes the creation of a synthetic man (◊ HOMUNCULUS) which cannot exist outside its medium but which has a powerful intellect, sufficient to guide Faust through his trials and dilemmas.

Faust is a towering achievement, and has overshadowed most of JWG's other works, especially those in the field of SUPERNATURAL FICTION. Although he was not a direct member of the Romantic movement (◊ ROMANTICISM), his works were a significant influence, including his version of Reynard the Fox, "Reineke Fuchs" (1794 in *Goethes Neue Schriften*; trans as *History of Reynard the Fox* **1840** UK), and his *Bildungsroman*, *Wilhelm Meisters Lehrjahre* (**1795-6** 4 vols; trans Thomas CARLYLE as *Wilhelm Meister's Apprenticeship* **1824** UK) and its sequel *Wilhelm Meisters Wanderjahre* (**1821**; trans as *Wilhelm Meister's Travels* **1824**

UK). The latter includes a number of stories which JWG had written earlier; e.g., "Die Neue Melusine" ["The New Melusina"] (written 1807; first published in *Taschenbuch für Damen* anth **1817**), which offers a twist on the Melusine legend (◊ LAMIA). A barber falls in love with a beautiful lady who is in reality a DWARF. The barber agrees to be transformed into a dwarf but soon realizes his mistake. As in *Faust*, regardless of the sexual urge, life will win out over LOVE.

During the turbulent literary period of his lifetime – the modern novel and short story were taking form – JWG was instrumental in contributing key works that either influenced these developments or became acclaimed examples of the new forms. These included *Unterhaltungen deutscher Ausgewanderten* ["Conversations with the German Emigrants"] (**1795**), a set of seven tales told within a FRAME STORY about a group of refugees from the French Revolution seeking shelter in a castle. In addition to the GHOST STORIES "Die Sängerin Antonelli" ["Antonelli the Singer"] and "Der Klopfgeist" ["The Knocking Ghost"], the volume includes "Das Märchen" (trans as "The Tale" in *Popular Tales and Romances of the Northern Nations* 1823 UK; as *The Tale* chap **1877** US; vt *The Parable* chap 1963 US; most recently as "The Fairy Tale" in *Spells of Enchantment* anth **1991** ed Jack ZIPES), an extremely complex form of FAIRYTALE, essentially an ALLEGORY for the survival of life and the attainment of happiness. Another example is "Novelle" (in *Ausgabe letzter Hand* ["Collected Works"] **1828**; trans as *Goethe's Novel* chap **1837** UK), which typifies the *nouvelle*, or story with a surprise ending. Although essentially nonfantastic it has the form of a FAIRYTALE.

JWG's work and influence were key to the development of fantastic fiction in Germany beyond its Gothic roots, especially through the Romantic movement and the works of Ludwig TIECK, Friedrich de la Motte FOUQUÉ and E.T.A. HOFFMANN, which in turn influenced the course of fantastic fiction throughout the West.   [MA]

Other editions: Translations and editions of JWG's works have appeared in profusion. The main collected volumes are *Novels and Tales* (coll **1854** UK) and *Goethe's Works* (coll **1885** 5 vols US).

Further reading: JWG's autobiography is *Aus meinem Leben: Dichtung und Wahrheit* ["My Life and Family: Poetry and Truth"] (**1811-14** 3 vols; trans as *Mémoires de Goethe* **1823** France; trans from the French as *Memoirs of Goethe* **1824** UK; with 4th vol added **1833**; the whole trans by John Oxenford [1812-1877] as *Autobiography* **1848-9** UK); this includes the fairytale "Der Neue Paris" ["The New Paris"] written as early as 1763. Also revealing of JWG's views in his later years is *Gespräche mit Goethe* (**1837** 2 vols; **1838** vol 3; trans Margaret Fuller as *Conversations with Goethe* **1839** UK) by Johann Eckermann (1792-1854). Other books of note are *The Life and Work of Goethe* (**1932**) by J.G. Robertson, *Goethe: A Collection of Critical Essays* (coll **1967**) ed Victor Lange, *Goethe and the Novel* (**1976**) by Eric Blackall and *Figures of Identity: Goethe's Novels and the Enigmatic Self* (**1984**) by Clark S. Muenzer.

GOG AND MAGOG Two GIANTS, or perhaps GODS, who may be brothers of the Irish god/giant Daghda (◊ CELTIC FANTASY), or who may have some linguistic connection to the Biblical nations, Gog and Magog, which are scheduled to be in cahoots with SATAN when ARMAGEDDON arrives. As far as legend is concerned, however, they are the giants captured by the Trojan, Brute, and brought to LONDON, which

he founded, calling it Troy-novant. The Goemagog of Edmund SPENSER's *The Fairie Queene* (**1590**) comes from an alternate tradition, in which there is only one giant, known also as Gogmagog. The famous statues at London's Guildhall show the two giants in the role of LIMINAL BEINGS, who act as GUARDIANS of the town, a role given prominence in F.W. Fairholt's *Gog and Magog: The Giants in Guildhall: Their Real and Legendary History* (**1859**). It is in this light, though comically, that Charles DICKENS introduces them into the FRAME STORY of *Master Humphrey's Clock* (coll **1841-2**). As part of the URBAN FANTASY texture of London they are of some importance. Occasionally – as in Andrew SINCLAIR's **Gog** sequence (**1967-88**) – a broader significance is accorded them: they warn and ward, and bespeak the THINNING of the old STORY of Britain.   [JC]

GOGOL, NIKOLAI (1809-1852) Russian writer, one of the central 19th-century figures of Russian literature; he has been described as the founder of Russian realism, though his version of realism is profoundly fantasticated. *Vechera na khutore blix Dikan'ki* (coll **1831-2**; trans Constance GARNETT as *Evenings on a Farm near Dikanka* 1926 UK), is, however, a collection of TWICE-TOLD tales of the Ukraine, generally fantastic. *Mirgorod* (coll **1835**; trans Garnett **1928** UK), contains "Viy", a famous tale about a female VAMPIRE, and "Taras Bulba", an historical epic into which Cossack LEGENDS are interwoven. *Arabeski* ["Arabesques"] (coll **1835**), contains various stories later translated, including "The Portrait" and "The Nose". It is tempting to understand *Mertvyye Dushi* (**1842**; various trans as *Dead Souls* or *Tchitchikoff's Journeys* from **1887**) primarily in terms of its surreal dislocations, its animate metonymies (coats are more alive than their forgotten wearers), and the constant juxta-positioning of life and death which leads to a climactic DANCE OF DEATH when HELL is let loose. As a tour of Hell, *Dead Souls* can be understood as an important text in the development of 19th-century fantasy.   [JC]

GOJIRA ot of *Godzilla* (**1954**). ◊ GODZILLA MOVIES.

GOJIRA 1985 ot of *Godzilla 1985* (**1985**). ◊ GODZILLA MOVIES.

GOJIRA MOVIES ◊ GODZILLA MOVIES.

GOJIRA NO GYAKUSHU ot of *Gigantis* (**1955**). ◊ GODZILLA MOVIES.

GOJIRA NO MUSUKO ot of *Son of Godzilla* (**1967**). ◊ GODZILLA MOVIES.

GOJIRA TAI GAIGAN (vt *Gojira Tai Gigan*; vt *War of the Monsters*) Japanese movie (**1972**). ◊ GODZILLA MOVIES.

GOJIRA TAI GIGAN vt of *Gojira Tai Gaigan* (**1972**). ◊ GODZILLA MOVIES.

GOJIRA TAI HEDORA ot of *Godzilla Versus the Smog Monster* (**1971**). ◊ GODZILLA MOVIES.

GOJIRA TAI MEGALON ot of *Godzilla Versus Megalon* (**1973**). ◊ GODZILLA MOVIES.

GOJIRA TAI MEKAGOJIRA ot of *Godzilla Versus the Bionic Monster* (**1974**). ◊ GODZILLA MOVIES.

GOJIRA TAI MOSURA vt of *Godzilla Versus Mothra* (**1964**). ◊ GODZILLA MOVIES.

GOLD, HORACE L(EONARD) (1914-1996) Canadian-born writer and editor, long resident in the USA and holding dual nationality. HLG is best-known for SCIENCE FICTION, in particular for his editorship of *Galaxy Magazine*, which he founded in 1950 and edited until 1961. HLG edited a short-lived companion, BEYOND FANTASY FICTION, which sought to emulate UNKNOWN in presenting adult treatments of fantasy.

HLG's first story, "Inflexure" for *Astounding Science Fiction* in 1934, was one of five published under the pseudonym Clyde Crane Campbell (he also used the names Dudley Dell, Harold C. Fosse, Christopher Grimm, Leigh Keith and Richard Storey); it considers all time as simultaneous. His first story under his own name, "A Matter of Form" (1938 *Astounding*), is about IDENTITY EXCHANGE between a man and a dog. His work was ideally suited to *Unknown*; it is a pity that only four of his stories appeared there. All of HLG's best fantasies are in a similar format: an unpretentious person either acquires or discovers he has some special TALENT. The stories always focus on the "little man". This has some special significance in HLG's case as he became increasingly agoraphobic and by the 1950s was conducting all of his business from his home.

"Trouble With Water" (1939) is a classic. With just the right balance of HUMOUR and logic, it shows a man cursed (◊ CURSES) by a gnome so that all water will avoid him. "Day Off" (1939), where the protagonist's imagination creates people, was less successful, but "Warm, Dark Places" (1940), where a man is again cursed, this time to be confronted with all the filth and detritus he creates, was on form. HLG's only novel-length work was "None But Lucifer" (1939), which was revised by L. Sprague DE CAMP for publication (to HLG's annoyance). It tells of a man who believes Earth is HELL and who not only confronts LUCIFER but begins to outdo him in creating misery in the world. Among HLG's later fantasies are: "And Three to Get Ready" (1952) about a man who can kill anyone merely by speaking their name THREE times (◊ MAGIC WORDS; TRUE NAMES); "Don't Take it to Heart" (1953) where a man realizes that SUPERSTITIONS are there for a purpose; and "What Price Wings?" (1962) about a man who is too angelic for his own good. HLG's fantasies have not been collected as a single volume; only two appear in *The Old Die Rich* (coll **1955**).   [MA]

GOLDBERG, RUBE Working name of US cartoonist Reuben Lucius Goldberg (1883-1970), who gave his name to the National Cartoonists Society's award, the Reuben, which he designed. He also added his name to the language (as in "Rube Goldberg contraption") for the many fanciful, gravity-defying, spoof mechanical contraptions he invented. His drawing had a beguiling simplicity, but was firmly constructed and executed in a strong, flexible line.   [RT]

GOLDEN AGE An idyllic past age from which the present represents a sad fall or THINNING. The term's first recorded use is in the ancient Greek Hesiod's *Works and Days* (8th century BC; ◊◊ GREEK AND LATIN CLASSICS); the BIBLE's garden of EDEN represents another lost GA. The GA is conceptually distinct from magic OTHERWORLDS, but the two are often mixed in fantasy's legendary pasts. Thus the popularity of Arthurian romance (◊ ARTHUR) in the female-ruled courts of Eleanor of Acquitaine and Marie de Champagne has been attributed – e.g. in Jean Markale's *Women of the Celts* (**1975**) – to nostalgia for the pivotal roles played in Celtic legend by magical "ladies of the foundation" and "sovereignty" givers: MORGAN LE FAY, Vivian, or the LADY OF THE LAKE. The success of modern works like Marion Zimmer BRADLEY's *The Mists of Avalon* (**1983**) suggests the attractiveness, for female writers and readers, of rewriting legendary histories into an ALTERNATE-WORLD past GA where women's roles are more prominent (◊◊ GODDESS). Indeed, Edmund SPENSER's verse epic *The Faerie Queen* (**1590-96**) shows a "Gloriana" flatteringly based on Elizabeth Tudor, reigning in a nostalgic quasi-Arthurian

GA. Temporal CYCLES may promise a new GA – as in P.B. SHELLEY's *Hellas* (**1822**), where "The golden years return . . ." – but one which will again give way to iron and leaden years.   [JH/DRL]

See also: ARCADIA; DREAMTIME; ET IN ARCADIA EGO.

GOLDEN ASS ◊ APULEIUS; ASS.

GOLDEN BOUGH In Classic MYTH, the branch which a newly elected priest of Diana must pluck from a TREE in the grove sacred to her; then – as King of the Wood – he must slay his predecessor. Aeneas (◊ VIRGIL) breaks a branch from the same tree before his descent into the UNDERWORLD. What is going on, according to *The Golden Bough* (**1890**) by Sir James FRAZER, is the ritual SACRIFICE of the king (◊ CHRISTIAN FANTASY) by his successor so that the LAND may be healed (◊ HEALING).   [JC]

See also: CHRISTIAN FANTASY; GODDESS; SEASONS; TREES.

GOLDEN DAWN The Hermetic Order of the Golden Dawn was founded 1888 by William Wynn Westcott (1848-1925), Samuel Liddell Mathers (1854-1918) – better known as MacGregor Mathers, and the main controlling force in the GD – and William R. Woodman. The purpose of the order was the study of occult science and MAGIC, especially as laid down by Hebrew doctrine; it drew heavily from ROSICRUCIANISM. Many of its members were also either active or lapsed Theosophists, and likely the GD was established to provide a body for the study of the Western alchemical tradition (◊ ALCHEMY) in the same way that THEOSOPHY drew upon Eastern wisdom. By a process of study and examination it was possible for initiates to rise through a series of Grades to the level of Philosophus in the First Order. Only a few were allowed through to the Second Order, whose members were known as Adepts. The first temple opened in London in March 1888, and was followed by temples in Bradford, Weston-super-Mare, Edinburgh and Paris. The GD later spread to the USA.

The GD proper lasted only until 1902. Dr Robert Felkin (1858-1922), by then the Imperator of the London Temple, renamed the GD the Order of the Stella Matutina, but another leading member, A.E. WAITE, established his own Holy Order of the Golden Dawn, which lasted until 1914. The original GD continued to function through the Stella Matutina and other temples until 1972, although various offshoots still exist.

The GD attracted a number of members with literary interests. Most notorious was Aleister CROWLEY, who joined in 1898 and contributed to its initial collapse. Probably the best-known member was W.B. YEATS, who was not only an Adept but held the post of Imperator 1900-1902. Little of Yeats's writings directly reflect his experiences in the GD, although most of his poetry and tales produced during the 1890s reflect his joint interests in the GD and Theosophy, particularly some of the stories in *The Secret Rose* (coll **1897**). John William Brodie-Innes (1848-1923), a Scottish lawyer, was an active member from the time he founded the Edinburgh Temple in 1894 until his death. He was also a student of FOLKLORE and witchcraft; his four occult novels focus more on those subjects than on the GD, although there is some interesting background in *Old as the World* (**1909**) and *For the Soul of a Witch* (**1910**).

The writer who best expressed the GD beliefs in his fiction was Algernon BLACKWOOD, who joined in 1900 and later shifted to Waite's Order. Several of Blackwood's early stories utilize GD teachings, such as "With Intent to Steal" (1906), but the two most detailed expositions arising from his studies in the GD are the stories in *John Silence* (coll **1908**) and especially *The Human Chord* (**1910**), which explores in detail the effects of harnessing the power of the Word (◊ MAGIC WORDS). It is quite likely that both M.L.W., the original of John Silence, and Richard Skale, the clergyman in *The Human Chord*, are based on unidentified members of the GD. It is also probable that some of the episodes described in *Julius LeVallon* (**1916**) are drawn from Blackwood's GD experiences.

Arthur MACHEN joined the GD in 1899, but probably soon lost interest. Like Blackwood, he was more concerned with the mystical aspects of the supernatural, in particular the GRAIL legend, than in hermeticism. Most of his more creative SUPERNATURAL FICTION had been produced before he joined the GD, though it is possible that "The White People" (1904) utilized some of the impressions arising from his GD studies, while *The Hill of Dreams* (**1907**) draws some autobiographical parallels with his researches.

A.E. Waite had already produced his first fiction, *Prince Starbeam* (**1889**), before he joined the GD in 1891 and neither this nor *The Golden Stairs* (coll **1893**) draw especially upon GD teachings, although his subsequent mystical beliefs did figure in his revision of these books as *The Quest of the Golden Stairs* (fixup **1927**). The fiction of Evelyn UNDERHILL, a member of Waite's later Order, is likewise too mystical and obscure to throw much light on the GD. The last major writer to emerge from the GD was Dion FORTUNE, who joined the revived Order in 1919 and then established her own Fraternity of the Inner Light in 1922. All of her fiction centres on her occult studies, and her OCCULT DETECTIVE, **Dr Taverner**, is based on an amalgam of characters including GD Adepts.

Other members of interest include Violet Tweedale (1862-1936), who produced a few occult novels plus some saccharine romances and a number of books about GHOSTS and psychic research, Maude Gonne, the actress Florence Farr and, perhaps surprisingly, the wife of Oscar WILDE, Constance Lloyd, who was one of the earliest members. Authors associated but not in fact members were Bram STOKER, Sax ROHMER and Arthur Conan DOYLE. Both William Sharp (◊ Fiona MACLEOD) and E. NESBIT had connections with GD initiates. Later writers who drew upon their knowledge of the GD (often secondhand) were Dennis WHEATLEY and Charles WILLIAMS. [MA]

Further reading: *The Magicians of the Golden Dawn* (**1972**; rev 1985) by Ellic Howe; *Yeats's Golden Dawn* (**1974**) by G.M. Harper; *Sword of Wisdom, MacGregor Mathers and The Golden Dawn* (**1975**) by Ithell Colquhoun; *The Golden Dawn: Twilight of the Magicians* (**1983**) by R.A. Gilbert.

GOLDEN FLEECE In Greek MYTHOLOGY, the fleece of a winged ram given by HERMES to Nephele, the wife of Athamas and mother of Phrixus and Helle. Athamas had abandoned Nephele and married Ino, who sought to kill her stepchildren. Phrixus and Helle escaped on the ram, although Helle became scared and fell into the sea now called the Hellespont. Phrixus arrived in Colchis and, in gratitude to the gods, sacrificed the ram to ZEUS. He gave the fleece to Aeëtes, King of Colchis, who set it, guarded by a DRAGON, in the grove of Ares. The fleece became legendary and Pelias, king of Iolcus, wishing to be rid of Jason, who had come to claim his birthright, sent him to fetch it. Thus begins the famous story of the Argonauts, one of the most famous QUESTS in legend. Jason, during his journey, met and married the sorceress Medea, whose

powers enabled him to complete tasks set by Aeëtes and claim the fleece. Jason returned to Iolcus, killed Pelias and gained his kingdom.

The story, in the oral tradition from *c*800BC, was first written down in the epic poem *Argonautica* (?250BC) by Apollonius of Rhodes (?295-215BC), which remains our primary source. It is recounted in *The Heroes* (coll **1856**) by Charles KINGSLEY, *Tales of the Greek Heroes* (**1958**) by Roger Lancelyn GREEN, and many others, and is explored in greater depth by Robert Graves (1895-1985) in *The Golden Fleece* (**1944**) and Henry TREECE in *Jason* (**1961**). In *The Mask of Circe* (**1971**) Henry KUTTNER has a modern descendent of Jason travel through a PORTAL to a world of Greek myths where he finds the fleece to be an alien artifact. The legend itself was filmed as JASON AND THE ARGONAUTS (*1963*).

The fleece is representative of any object whose attainment brings the keys to the kingdom or other great prize. It is not cognate with the GRAIL, whose quest is for spiritual perfection – indeed, it might be seen as an opposite, since the quest is for material gain. Although success may bring an immediate reward, it usually ends in tragedy – Jason was betrayed by Medea, his children were killed, and he died when part of the *Argo* fell on him. The fleece may thus be seen as a THING BOUGHT AT TOO HIGH A COST. [MA]

GOLDEN VOYAGE OF SINBAD, THE (*1973*) ◊ SINBAD MOVIES.

GOLDING, LOUIS (1895-1958) UK writer whose tales of genre note are mostly SUPERNATURAL FICTION. In *The Miracle Boy* (**1927**), the dead are brought to life again. *The Pursuer* (**1936**) is sf. *Honey for the Ghost* (**1949**) is a tale of POSSESSION. His collections include *The Doomington Wanderer* (coll **1934**; vt *This Wanderer* 1935 US; cut vt in 2 vols as *The Call of the Hand and Other Stories* 1944 chap UK and *The Vicar of Dunkerly Briggs* 1944 chap UK) and *Pale Blue Nightgown: A Book of Tales* (coll **1944**), both including GHOST STORIES. [JC]

Other works: *Bareknuckle Lover and Other Stories* (coll *c*1947 chap); *The Frightening Talent* (**1973**).

GOLDING, [Sir] WILLIAM (GERALD) (1911-1993) UK writer, winner of the Nobel Literature Prize in 1983, best-known for his first novel, *Lord of the Flies* (**1954**), filmed twice as LORD OF THE FLIES. Like almost all his work, this tale generates a sense that the underlying STORY being told is both opaque and crystalline; his work seems – more perhaps than that of any other contemporary UK mainstream novelist – to be telling FABLES from deep waters. Though told anew, each WM novel seems to unfold a story which pre-exists its telling: his work *feels*, therefore, very much like fantasy. Only one of his novels, however, comes close to the form. *Pincher Martin* (**1956**; vt *The Two Deaths of Christopher Martin* 1957 US) is almost certainly a POSTHUMOUS FANTASY, though it is possible to read the story as occupying the very short period between the sinking of the protagonist's ship in WWII, and his death by drowning. A more resonant reading of the tale understands it to represent Martin's ranting attempts to sort out the shambles of the life he has unwittingly departed. In this light, when he discovers that the tiny ISLAND to which he has been clinging is identical in shape to a diseased tooth he has been feeling with his tongue, he may be thought to be closing in posthumously on wisdom. [JC]

GOLDMAN, WILLIAM (W.) (1931-) US screenwriter and novelist, well known for his scripts for movies like

Harper (**1966**), *Butch Cassidy and the Sundance Kid* (**1969**) and *The* STEPFORD WIVES (**1974**). He has also adapted several of his own novels, most notably *Marathon Man* (**1974** book; **1976** movie), whose unfilmed sequel, *Brothers* (**1987**), is sf. Beginning with *The Temple of Gold* (**1957**), he has written novels in various genres, with the **Morgenstern Fables** sequence – *The Princess Bride: S. Morgenstern's Classic Tale of True Love and High Adventure, The "Good Parts" Version, Abridged by William Goldman* (**1973**) and *The Silent Gondoliers: A Fable by S. Morgenstern* (**1983**; 1985 as by WG) – being of most fantasy note. The former – recreated as the movie *The* PRINCESS BRIDE (**1987**) – is an interesting example of a book which is itself a fictional BOOK, since it purports to be WG's recension of a BOOK called *The Princess Bride* by one S. Morgenstern, which had been read to him in full when he was a sick child. In its exaggeration of motifs it can be understood as REVISIONIST FANTASY. Princess Buttercup, unwillingly agrees to marriage with the evil Prince Humperdinck, but is rescued by young Westley, who combines all the characteristics of the HERO – notably, he is her long-lost childhood sweetheart who has made good or, as it were, bad, because he has been a renownedly swashbuckling PIRATE. Much action ensues, and a metafictional ending – in the course of which WG attacks the fantasy STORY because it inherently strives for EUCATASTROPHE – closes proceedings. *The Silent Gondoliers* is a tale in the same mode, but without the bite.

Other novels of interest include *Magic* (**1976**), filmed as MAGIC (**1978**) with a WG script, in which a ventriloquist (◊ VENTRILOQUISM) suffering a schizophrenic disorder comes to believe that his dummy is his DOUBLE; the text is interestingly ambiguous at points (◊ PERCEPTION), though everything can ultimately be explained as delusional (◊ DELUSION; RATIONALIZED FANTASY). *Control* (**1982**) is a complex SUPERNATURAL FICTION.   [JC]

GOLDSTEIN, LISA (1953-) US writer of fantasy and sf. Her first work of genre interest was fantasy. *The Red Magician* (**1982**) is set in a LAND-OF-FABLE Eastern Europe, in a Jewish village whose location might be described as occupying a shifting BORDERLAND between Hungary, Czechoslovakia and Russia; the time is some point before WORLD WAR II, and the story concerns a conflict between two WIZARDS, one of whom is intent on maintaining his village as a POLDER, the other haunted by previsions of what is to come. It won the American Book Award for 1982. In her second novel, *The Dream Years* (**1985**), a 1920s student in Paris, deeply involved in SURREALISM, becomes obsessed with a young woman, whom he follows via TIMESLIP into 1960s PARIS, at the verge of its failed flower-child "revolution". The two Parises soon begin to CROSSHATCH together, and the novel closes in a fairly lighthearted excursion, on the part of most of the cast, into the 21st century, where hope dawns – maybe.

A Mask for the General (**1987**), set in an oppressed 21st-century USA, though centred on SAN FRANCISCO, is LG's most sf-like fiction, though even here there can be discerned the pattern of doubling (◊ AS ABOVE, SO BELOW; DOUBLES) which runs throughout her work. Here this doubling pits the authoritarian general who rules the USA against the West Coast tribal MASK-makers and artists who, by tapping tribal depths, are able to create TOTEMS to transfigure their friends (and enemies). Their attempts to get the general to wear his crow mask – and thus to open his being to a more fitting and profound REALITY – lie at the

heart of the plot. *Tourists* (**1989**; rev 1994) – which is unconnected to an early short story, "Tourists" (1984) – is an ARABIAN-NIGHTMARE tale, set in a land of fable called Amaz, which in this case is located in the Middle East (though the location shifts in other tales, like "Tourists", which are set in the same Amaz); here a family representative of a standard-kit USA searches for a 1000-year-old document, which may guide them to the reality-shifting SWORD of the Jewel King. This becomes a nightmarish exploration of two conflicting versions of Amaz, which crosshatch. The unnerving topology is doubled by the internal collapse of the family.

LG's relatively few short stories are assembled in *Daily Voices* (coll **1989**) and *Travelers in Magic* (coll **1994**). Many of them are fantasy, and the best include reworked material similar to that given full play in the novels; an example is "Breadcrumbs and Stones", a TWICE-TOLD version of Hansel and Gretel set after the HOLOCAUST.

In the 1990s, LG has published two further fantasy tales, each of strong interest. *Strange Devices of the Sun and Moon* (**1993**) is set in a LONDON that much resembles the London of 1590 (though the date of Christopher MARLOWE's death has been rejigged to fit the story), but which is crosshatched by FAERIE. As in most FANTASIES OF HISTORY, the elder folk are in the process of departing the mundane world and an acceleration of THINNING is in the offing. The story is concerned with the attempts of the FAIRY QUEEN to find her son – whose name is ARTHUR, and who was exchanged with a human baby at birth (◊ CHANGELINGS) to safeguard him from the Red King – and, with Arthur's help, to win a final fairy battle. But Arthur is not easy to discover; one human character, trailing him into a series of fairylands, finds "that the Land itself was a sort of door" for the vagrant, icy-souled elf. In the end all is sorted, and the thinning proceeds.

Summer King, Winter Fool (**1994**) is LG's only novel to be set in a full SECONDARY WORLD. The land is dominated by god-doubles: the god of summer and the god of winter. We meet neither god until late in the text, which is mainly devoted to the intersecting adventures of two young men (they themselves double one another) whose fortunes rise and fall with great speed. The novel reads a little like a compressed DYNASTIC-FANTASY trilogy, an intensification which – along with the constant theatricals involved, the habit protagonists have of wearing MASKS and changing roles, and a sense that the tale being told may be as twice-told and indefinitely retellable as any fable of the SEASONS – generates a powerful COMMEDIA DELL'ARTE atmosphere. LG is a deceptively quiet writer.   [JC]

GOLEM In Hebrew legend, the golem is a man made of clay (replicating God's creation of Adam) brought to life through words of power called a *shem* (◊ MAGIC WORDS) written on paper and inserted into the golem's mouth. The golem is thereby controlled by its master. The physical creation of a golem, though supposedly linked to Jewish and rabbinical lore, is most closely associated with Judah Loëw or Lowe (1512-1609), who is believed to have created a golem to protect the Jews of the Prague ghetto against a pogrom imposed by the Hapsburg Emperor, Rudolf II (1552-1612). The golem is only semi-human: it has no understanding of GOOD AND EVIL, it cannot speak, and it cannot reproduce itself. The tale of the golem has similarities to that of FRANKENSTEIN, particularly in a tale associated with Elijah of Chelm, in Poland, in the mid-16th century, who created a golem to protect the town, but the golem became too powerful and Elijah had to destroy it by recovering the *shem*.

While the legend forms the background to Gustav MEYRINK's *Der Golem* (1913-14 *Die wessen Blatter*; **1915**; cut trans Madge Pemberton as *The Golem* 1928 UK; full trans 1977 US), it is not the golem *per se* that haunts 19th-century Prague but the power behind it. Golem imagery is ideally suited to works of GASLIGHT ROMANCE, such as *Dan Leno and the Limehouse Golem* (**1994**) by Peter ACKROYD and *Irene's Last Waltz* (**1994**) by Carole Nelson DOUGLAS, one of her **Irene Adler** SHERLOCK HOLMES pastiches. The golem did not feature much as a story MONSTER in the intervening years. A modern treatment is *The Master of Miracle: A New Novel of the Golem* (**1971**) by Shulamith Ish-Kishor (1896-1977); a more traditional one is Isaac Bashevis SINGER's *The Golem* (trans **1982**). Sean STEWART used the golem as a potent image of magic returning to the world in *Resurrection Man* (**1995**), when golems start to appear in WORLD WAR II deathcamps. The modern version of the golem may be more appropriately linked with the cyborg or android, a link made by Marge Piercy (1936-   ) in *He, She and It* (**1991**; vt *Body of Glass* 1992 UK).   [MA]

See also: *The* GOLEM.

GOLEM, THE Several movies have been based on the legend of the GOLEM, including a German version in 1913 (sequelled 1917) and a Czech version in 1951. The three discussed here are probably the most important.

1. *The Golem* (ot *Der Golem, Wie Er in die Welt Kam*) German movie (**1920**). Projections/A.G.Union. **Dir** Carl Boese, Paul Wegener. **Spfx** Boese. **Screenplay** Wegener. **Starring** Lother Müthel (Knight Florian), Lyda Salmonova (Miriam Lowe), Albert Steinrück (Rabbi Lowe), Wegener (Golem). 85 mins. B/w, silent.

One of the most startlingly impressive Expressionist movies, concerned almost less with its plot than with its image – in this it resembles *The* CABINET OF DR CALIGARI (**1919**). The starting point is the traditional tale of Rabbi Lowe, who creates a man of clay to defend his people. From here the tale diverges. In one strand the Rabbi's daughter Miriam is seduced by the Emperor's messenger, Knight Florian; the end of this strand (and the end of the movie) is that Lowe's assistant, loving Miriam, has the Golem kill Florian, after which it goes on a murderous rampage until being destroyed by an unwitting child (in scenes forerunning the encounter between Maria and the Monster in *Frankenstein* [**1931**]; ◊ FRANKENSTEIN MOVIES). In the other strand Lowe travels to the court of the Emperor, gives a powerful demonstration of MAGIC (cf FAUST), and saves the lives of the Emperor and his throng – for which service the Emperor cancels his edict against the Jews. As will be noted, the Golem itself is somewhat sidelined; yet the movie's use of visual imagery to create both dread and a sense of the TWICE-TOLD is stunning.   [JG]

2. *The Golem* (vt *The Legend of Prague*) Czechoslovakian/French movie (**1936**). **Dir** Julien Duvivier. **Screenplay** André-Paul Antoine, Duvivier. **Starring** Harry Baur, Charles Dorat, Ferdinand Hart, Roger Karl. 95 mins. B/w.

Apparently (we have been unable to obtain a viewing copy of this movie) a fairly straightforward remake of **1**, but lacking much of the skill.

3. *Golem: The Wandering Soul* (ot *L'esprit de l'exil: Golem*; vt *Golem: The Spirit of Exile*) German/French/Dutch/Italian/UK movie (**1992**). Agav/Friedlander Film Produktion/Allarts/Nova/Rai 2/Groupe TSF/ Channel 4/Centre Nationale de la Cinématographie. **Pr** Laurent Truchot. **Dir** Amos Gitai. **Screenplay** Gitai. **Starring** Fabienne Babe, Bernardo Bertolucci, Antonio Carallo, Bernard Eisenschitz, Samuel Fuller, Philippe Garrel, Sotigui Kouyate, Bernard Levy, Marceline Loridan, Alain Maratrat, Vittorio Mezzogiorno, Mireille Perrier, Bakary Sangare, Ophra Shemesh, Hanna Schygulla (movie lacks proper credits). *c*100 mins. Colour.

This conflates the legend of the GOLEM with the events of the *Book of Ruth*; though it is set in modern times, almost all the dialogue comes directly from various books of the Old Testament. A man uses the letters of the alphabet to raise a (female) golem from riverside mud, and charges her with seeking out and aiding the persecuted of the world. She comes to Paris (the land of Moab) and settles on a Jewish family comprising Elimelech, his wife Naomi, their sons Mahlon and Chilion, and the sons' spouses – one of whom is Ruth. The golem appears to them as a VISION which they never *consciously* see or hear; certainly she seems to bring them little aid. First Elimelech dies; then the two sons are murdered by antisemites. Naomi tells the women to return to their homes, but Ruth refuses: ". . . for whither thou goest, I will go . . . thy people shall be my people . . ." The two women, golem in tow, journey to a far land, where the golem – her voice for once obeyed – instructs Ruth to marry Boaz and bear Naomi a grandchild. Beautifully directed and photographed, this is by any standards a striking movie. Yet it seems to rejoice in self-imposed obscurity. Viewers ignorant of the Judeo-Christian tradition must find it incomprehensible (ironic, in that it purports to proclaim that we should all embrace "outsiders").   [JG]

GOOD AND EVIL Neither Good, nor EVIL, *per se*, is dealt with in this entry. The phrase good and evil describes a *dynamic opposition* which drives much fantasy, SUPERNATURAL FICTION and horror; the Good and the Evil themselves are seldom examined. In a manner which reflects his deep understanding of medieval ROMANCE and ALLEGORY, J.R.R. TOLKIEN dramatizes the linkage – through vividly presented contrasts of character and STORY and LAND – in order to manifest his strong, conservative, Roman Catholic understanding of Good and Evil (a PARODY of Good); but he also does so in order to tell a fine story. Many who have followed him, on the other hand, and who have transformed his SECONDARY WORLD into FANTASYLAND, use the GAE contrast as a convenient PLOT DEVICE, which is generally colour-coded (◊ COLOUR-CODING) for ease of understanding. HEROES (until recently) tend to be "Aryan" in appearance, and DARK LORDS tend to the swarthy.

Though the phrase implies an equal responsibility for turning the wheel of Story, in normal practice it is the Evil characters who erupt from their dark coverts or lands, avidly hungry to devour and waste the plenitude of the Good. In fantasy stories, Good reacts to Evil.   [JH/JC]

GOOD HEAVENS US tv series (1976). ABC. **Pr** Austin Kalish, Irma Kalish, Mel Swope. **Exec pr** Carl Reiner. **Dir** Peter Bonerz, John Erman, Reiner, James Sheldon, Swope. **Writers** Jay Folb and others. **Created by** Bernard Slade. **Starring** Reiner (Mr Angel). **Guests** Don Ameche, Clu Gulager, Huntz Hall, Florence Henderson, George Maharis, Julie Newmar, Loretta Swit, Paul Williams and many others. 13 30min episodes. Colour.

This sitcom featured Carl Reiner as an ANGEL who gave out MIRACLES each week, without the recipients knowing it.

The WISHES were granted as rewards for acts of kindness or generosity. [BC]

GOODKIND, TERRY (1948-) US author of the **Sword of Truth** sequence, to date *Wizard's First Rule* (**1994**) and *Stone of Tears* (**1995**). These are redeemed from the stock level of commercial HIGH FANTASY by a rather bleak moral complexity and by some moments of (usually unpleasant) inventiveness. In the first volume, the UGLY-DUCKLING Richard is chivvied into his QUEST by the murder of his supposed father and finds himself pursued by agents, human and otherwise, of the DARK LORD Darken Rahl. Richard spends the second volume setting to rights the magical consequences of his victory over Darken Rahl, which proves, potentially, a THING BOUGHT AT TOO HIGH A COST. [RK]

GOONIES, THE US movie (*1985*). Warner/Steven Spielberg. **Pr** Harvey Bernhard, Richard Donner. **Exec pr** Kathleen Kennedy, Frank Marshall, Steven SPIELBERG. **Dir** Donner. **Spfx** Matt Sweeney. **Vfx** Micheal McAlister, David Carson, Industrial Light & Magic. **Screenplay** Chris Columbus. **Novelization** *The Goonies* * (**1985**) by James Kahn (1947-). **Starring** Sean Astin (Mikey Walsh), Josh Brolin (Brand Walsh), Jeff Cohen (Chunk), Robert Davi (Jake Fratelli), Corey Feldman (Mouth), Kerri Green (Andy), John Matuszak (Sloth), Joe Pantoliano (Francis Fratelli), Martha Plimpton (Stef), Ke Huy Quan (Data), Anne Ramsey (Mama Fratelli). 111 mins. Colour.

Based on a story by Spielberg, this is an attempt to produce an INDIANA JONES adventure for children. A bunch of little boys from the wrong side of town (Chunk, Data, Mikey, Mouth) plus Mikey's elder brother (Brand) find a map to PIRATE One-Eyed Willie's treasure, and seek it, *en route* roping in two girls (Andy, Stef). But the QUEST entangles the COMPANIONS with the clumsily murderous Fratelli gang and countless lethal booby traps. Aided by the monstrously deformed Fratelli brother Sloth, until now kept always chained (there are conscious echoes of FRANKENSTEIN's MONSTER), they recover the loot.

There is nothing supernatural in this TALL TALE, but many traditional elements of the FANTASTIC VOYAGE play a part – not least that there are SEVEN companions. There is some play on the need for FANTASY and on the power of reified dreams, but most is drowned out by the incessant cacophony of raised childish voices: Donner seems to have encouraged his young actors to improvise, with the result that many important moments are lost. [JG]

GOPALEEN, MYLES NA Pseudonym of Flann O'BRIEN.

GOR Italian movie (*1987*). ◊ John NORMAN.

GORDON, FRANCES Pseudonym of Bridget WOOD.

GORDON, JOHN (WILLIAM) (1925-) UK writer, mostly for YA audiences and mostly of SUPERNATURAL FICTION. He is of fantasy interest for his first novel, *The Giant Under the Snow* (**1968**), and its sequel, *Ride the Wind* (**1989**). The GIANT of the first novel is in fact a GREEN MAN, a figure evoked out of the chthonic mounds and hollows that shape an otherwise innocent English countryside; his appearance, at the behest of a supernatural "Warlord" trying to regain the Golden Treasure which will give him dark sovereignty once again, is countered by a good WITCH and some children. The sequel repeats the threat, but without the Green Man. Other fantasies include *The Edge of the World* (**1983**), whose protagonists are sent by a LIMINAL BEING into an OTHERWORLD where they must perform feats to rescue one of their siblings.

JG's frequent supernatural-fiction tales, many of which involve GHOSTS and other invaders of the normal world, include *The House on the Brink* (**1970**), which effectively evokes a sense of WRONGNESS centring on an old black log that is in fact a MUMMY. *The Ghost on the Hill* (**1976**), *The Waterfall Box* (**1978**), which invokes ALCHEMY, *The Spitfire Grave and Other Stories* (coll **1979**), the title story of which effectively evokes a sense of the surreality of WORLD WAR II, *Catch Your Death and Other Ghost Stories* (coll **1984**), *The Quelling Eye* (**1986**), *The Burning Baby and Other Ghosts* (coll **1992**) and *Gilray's Ghost* (**1995**). [JC]

"GORE" Pseudonym of Richard CORBEN.

GOREY, EDWARD (ST JOHN) (1925-) US artist, illustrator and writer, most of whose over 70 chapbooks are collected in the omnibus volumes *Amphigorey* (graph coll **1972**), *Amphigorey Too* (graph coll **1975**) and *Amphigorey Also* (graph coll **1983**). Typically EG produces engraving-like picture stories in black ink, often with baroque Victorian/Edwardian backgrounds and costumes. Characteristics include effective composition, somewhat sketchy characters' faces, precise hand-lettered captions – frequently in doggerel – and a generally sombre yet surreal mood (as in the work of Charles ADDAMS, though EG's style is rarely cartoon-like). EG often injects macabre elements into CHILDREN'S FANTASY templates; a regular if unexplained prop is the ominous Black Doll. Joke pseudonyms recur, including Eduard Blutig and anagrams like Ogdred Weary.

Of special fantasy note are: *The Doubtful Guest* (graph **1957**), where an inexplicable beast resembling a furry penguin joins a perturbed Victorian family, never to leave; *The Object-Lesson* (graph **1958**), an exercise in darkly humorous SURREALISM; *The Insect God* (graph **1963**), whose narrative is light verse but which culminates with a kidnapped child's sacrifice to the eponymous deity; *The Sinking Spell* (graph **1964**), where the creature that sinks slowly and alarmingly from roof to cellar is never depicted; *The Evil Garden* (graph **1966**), subverting the children's-story format of a merry outing (caption: "A hissing swarm of hairy bugs/Has got the baby and its rugs"), an approach repeated in EG's popup book *The Dwindling Party* (graph **1982**); *The Inanimate Tragedy* (graph **1966**), a *reductio ad absurdum* of anthropomorphism with "characters" like the Two-Holed Button and Glass Marble plotting darkly; *The Utter Zoo* (graph **1967**), one of several EG exercises in the NONSENSE-alphabet vein of Edward LEAR – here a bizarre BESTIARY; the very odd *The Epiplectic Bicycle* (graph **1969**), featuring TALKING ANIMALS, a talking bicycle and apparent TIME TRAVEL; *The Disrespectful Summons* (graph **1971**), showing the PACT WITH THE DEVIL and orthodox downfall of middle-class WITCH Miss Squill; *[The Untitled Book]* (graph **1971**), whose cheerful nonsense captions belie the disturbing uncuddliness of the creatures a child sees dancing in the garden; and *The Raging Tide, or The Black Doll's Imbroglio* (graph **1987**) reducing the multiple-choice "gamebook" format (◊ GAMES) to absurdity – "If you loathe prunes more than you do turnips, turn to 22."

EG drew jacket and interior illustrations for many other works, notably Lear's "The Jumblies" and "The Dong with a Luminous Nose" as collected in *Gorey x 3* (graph coll **1976**), and T.S. Eliot's *Old Possum's Book of Practical Cats* (1982 US). His sets and costumes for the 1977 Broadway production of DRACULA won a Tony award and are mimicked in his *Dracula: A Toy Theatre* (graph **1979**). He received the WORLD FANTASY AWARD as Best Artist in 1985 and 1989.

Though aficionados relish his characteristic mix of whimsy and deadpan menace, EG himself insists: "I write about everyday life." [DRL]

Other works, excluding chapbooks in *Amphigorey* collections: *The Other Statue* (graph **1968**); *The Black Doll: A Silent Film* (graph **1973**); *Category: Fifty Drawings* (graph **1973**); *Gorey Posters* (graph **1979**); *Dancing Cats and Neglected Murderesses* (graph coll **1980**); *Le Mélange Funeste* (graph **1981**); *The Water Flowers* (graph **1982**); *E.D. Ward A Mercurial Bear* (graph **1983**); *The Tunnel Calamity* (graph **1984** accordion-fold novelty); *The Prune People II* (graph **1985**); *The Improvable Landscape* (graph **1986**); *The Dripping Faucet: Fourteen Hundred and Fifty Eight Tiny, Tedious and Terrible Tales* (graph **1989**); *The Helpless Doorknob: A Shuffled Story* (graph **1989**); *Q.R.V.* (graph **1989** vt *The Universal Solvent* 1990); *The Fraught Settee* (graph **1990**); *La Balade Troublante* (**1991**); *The Betrayed Confidence: Seven Series of Dogear Wryde Postcards* (graph coll **1992**); *Figbash Acrobate* (graph **1994**); *The Retrieved Locket* (graph **1994**); *The Unknown Vegetable* (graph **1995**); various others.

As editor: *The Haunted Looking Glass: Ghost Stories Chosen and Illustrated by EG* (anth **1959**; vt *Edward Gorey's Haunted Looking Glass* 1984).

Further reading: "The Albums of Edward Gorey" (**1959**) in *The Bit Between My Teeth* (coll **1965**) by Edmund Wilson (1895-1972); self-interview in *Gorey Posters*.

GORGONS In Greek MYTH, the THREE Gorgons – mortal Medusa and her immortal sisters Stheno and Euryale – were MONSTER figures: women with snakes for hair, one glimpse of whose faces converted onlookers to STATUES. A marble Gorgon MASK of the 6th century BC is an early exemplar of what this encyclopedia terms the FACE OF GLORY. The HERO Perseus escaped by looking only at Medusa's reflection in his shield, and beheaded her; his legend is retold in CLASH OF THE TITANS (*1981*). The petrifying gaze and MIRROR defence are recurring fantasy PLOT DEVICES. C.L. MOORE's SCIENCE-FANTASY "Shambleau" (1933 *WT*) revises the Gorgon as a tentacle-haired alien psychic VAMPIRE.
[JH/DRL]

GOSCINNY AND UDERZO Writer Réné Goscinny (1928-1977) and Albert Uderzo (1926-   ), creators of ASTERIX THE GAUL and other COMICS characters.

Goscinny was born in Paris but lived in Argentina until 19, when he decided to become a cartoonist for Walt DISNEY, and went to the USA. Failing to realize this ambition, he worked in various jobs, including for *Mad Magazine* in the early 1950s; here he met the cartoonist Morris (real name Maurice de Bevère; 1923-   ), with whom he went to France to begin providing scripts for **Lucky Luke**. He became a prolific scriptwriter for the magazines *Tintin* and *Spirou*, and in 1959 cofounded the comic weekly *Pilote*, in the first issue of which, with Uderzo, he created **Asterix**. He was for many years the highest-paid comics writer in France, receiving several awards and honours including Chevalier of Arts and Letters in 1967.

Uderzo was born in Italy but spent most of his life in France. He showed an early talent for drawing. In 1945 he moved to Paris to become one of the first cartoonists on the comic weekly *OK*, for which he created, among other features, **Aris Buck**, a tale about an invincible Gaul – clearly a forerunner of **Asterix**. He became an advertising artist in the early 1950s, but meanwhile created many short-lived comics features. In the early issues of *Pilote* he drew two series: **Asterix** and the Air Force strip **Michel Tanguy**, the

latter written by the creator of **Blueberry**, Jean-Michel Charlier (1924-1989). After the death of Goscinny, Uderzo also wrote some scripts. [RT]

GOTHAM ◊ BATMAN; *The* DEAD CAN'T LIE (1988); NEW YORK.

GOTHAM vt of *The* DEAD CAN'T LIE (*1988*).

GOTHIC UK movie (*1986*). ◊ FRANKENSTEIN MOVIES.

GOTHIC FANTASY The starting point of Gothic literature is usually given as *The Castle of Otranto* (**1765**) by Horace WALPOLE, but its antecedents are evident in much earlier work, especially the plays of John Webster (?1580-?1634) and William SHAKESPEARE. Although all Gothic fiction is tragedy, its key component is the EDIFICE; Walpole's novel introduced all the main PLOT DEVICES. Gothic fiction usually takes place in an ancient castle or abbey whose owner discovers his noble line is doomed, usually because some past misdemeanour has caused the family to be cursed (◊ CURSES). All FATE seems against him as he strives to overcome massive odds (frequently of either genuine or fabricated supernatural origin), and he usually fails. The story may be told from the viewpoint of another character, one dispossessed of his inheritance, who may regain it. Walpole's *Otranto* abounds in supernatural manifestations, and is thus a true GF; but not all Gothic fiction is genuinely supernatural. The devices may be rationalized (◊ RATIONALIZED FANTASY). Pure or High Gothic aims to terrify; the supernatural may appear in Low Gothic without the need to terrify – indeed, towards the end of the Gothic Age the novels tended to self-PARODY. Later writers have used the setting for atmosphere and as a device to simplify explanations to readers well versed in the material.

The usual pedigree of the GF passes from Walpole to Clara REEVE's *The Champion of Virtue* (**1777**; vt *The Old English Baron* 1778), to *The Recess: A Tale of Other Times* (**1783-85**) by Sophia Lee (1750-1824), which established the Historical Gothic, through William BECKFORD's *Vathek* (**1786**), which merged ORIENTAL FANTASY with the Gothic, to Ann RADCLIFFE's *The Mysteries of Udolpho* (**1794**), to Matthew LEWIS's *The Monk* (**1796**), the peak of the Gothic HORROR novel, to *The Midnight Bell* (**1798**) by the prolific Francis Lathom (1777-1832), culminating in *Melmoth the Wanderer* (**1820**) by Charles MATURIN, which mingled all the elements to produce the ultimate Gothic romance. By this time the Gothic trappings were being supplanted by a deeper psychological exploration of the doomed hero (◊ ACCURSED WANDERER), a furrow already ploughed by Mary SHELLEY in *Frankenstein* (**1818**). The tragic elements of the Gothic remained but, with the removal of the EDIFICE, the Gothic flame dimmed. Its influence continued, however: many Victorian HAUNTED-DWELLING stories and melodramas are Gothic in treatment, especially the sensation novels of the late Georgian and early Victorian periods – by writers like W. Harrison AINSWORTH, Wilkie COLLINS and Lord BULWER-LYTTON – and the penny dreadfuls of the mid-19th century, typified by the works of G.W.M. Reynolds (1814-1879), James Malcolm RYMER, Thomas Pecket Prest (1810-1859) and G.P.R. James (1799-1860). During its ascendancy the Gothic influence was parodied by Jane Austen (1775-1817) in *Northanger Abbey* (**1818**); it was desensationalized and adopted into the mainstream by Charlotte BRONTË in *Jane Eyre* (**1847**) and Emily Brontë in *Wuthering Heights* (**1847**).

The same brooding atmosphere occurred in many works of German ROMANTICISM, especially by Johann Friedrich von Schiller (1759-1805) with the first part of *Der Geisterseher* (**1786**; trans as *The Ghost-Seer* **1795** UK) and Lawrence Flammenburg (real name Karl Friedrich Kahlert; 1765-1813) with *Der Geisterbanner* (**1792**; trans Peter Teuthold as *The Necromancer* **1794** UK), and in the short stories of Johann Karl MUSÄUS, Johann APEL and Freidrich Schulze (1770-1849). Their work in turn influenced Washington IRVING, Charles Brockden BROWN, Nathaniel HAWTHORNE and Edgar Allan POE in the USA, while the Gothic movement in France, picked up partly from the work of Beckford, continued through the Marquis DE SADE, Honoré de BALZAC and Eugène SUE.

The Gothic influence continued with the growth of SUPERNATURAL FICTION at the turn of the 19th century, especially that fuelled by DECADENCE – evident in the work of Robert Louis STEVENSON, M.P. SHIEL and Bram STOKER – but the day of the doom-laden GF had long passed. The Gothic mode shifted toward romantic fiction, and was revived strongly in the work of Daphne DU MAURIER, who built on the work of the Brontës with *Rebecca* (**1938**) to lay the foundation for the modern Gothic romance. Occasionally supernatural elements may intrude, especially in the works of Edwina Noone (Michael Avallone), Lyda Long (Frank Belknap LONG) and Marilyn ROSS. Even Anne MCCAFFREY has written such novels, though mostly non-supernatural. The return of the Gothic to HORROR can probably be traced to Alfred HITCHCOCK and his adaptation of Robert BLOCH's *Psycho* (*1960*). In the 1970s and 1980s there was an increase of Gothic elements in SUPERNATURAL FICTION, mostly evident in the work of writers like Mary STEWART, Tanith LEE and Angela CARTER, and in the unique work of Thomas LIGOTTI. A fine example of modern GF was the title story of *Fengriffen* (coll **1971**) by David Case (1937-   ); it was filmed as *And Now the Screaming Starts* (*1972*). Another writer well able to blend the supernatural and Gothic is Basil Copper (1924-   ), especially in *Necropolis* (**1980** US), set in a vast cemetery. In the USA Joyce Carol OATES's contribution may be allied to the genre of AMERICAN GOTHIC, which applies the GF mood to crumbling colonial mansions of the South. Patrick McGrath (1950-   ) – who has produced his own neo-Gothic fiction in *Blood and Water* (coll **1988**) and *The Grotesque* (**1989**) – highlighted the change in his anthology *The New Gothic* (anth **1991**) with Bradford Morrow (1951-   ). The cityscape has replaced the old castle and URBAN FANTASY is the new Gothic.

Representative anthologies include: *Seven Masterpieces of Gothic Horror* (anth **1963**) ed Robert Donald Spector (1922-   ); *Three Gothic Novels* (omni **1966**) ed E.F. BLEILER; *Three Gothic Novels* (omni **1968**) ed Peter Fairclough; *Gothic Tales of Terror* (anth **1972**) ed Peter HAINING; *The Evil Image* (anth **1981**) ed Patricia L. Skarda (1946-   ) and Nora Crow Jaffe (1944-   ); *The Oxford Book of Gothic Tales* (anth **1992**) ed Chris Baldick; *Four Gothic Novels* (anth **1994**) ed anon. [MA]

GOUDGE, ELIZABETH (DE BEAUCHAMP) (1900-1984) UK popular novelist and children's writer. She firmly believed in her mother's psychic TALENTS and her own ability to sense and occasionally see GHOSTS, both of which she discusses in her autobiography, *The Joy of the Snow* (**1974**). Of over 40 titles – including novels, collections and religious nonfiction – her best-known fantasy is the whimsical children's book *The Little White Horse* (**1946**), which won the Carnegie Medal in 1947. Set in the 19th century, the tale of

Maria Merryweather, who comes to Moonacre, a mysterious manor house in Cornwall and discovers her inherited QUEST to right ancient wrongs and bring happiness back to the family, the valley and the village of Silverydew, mixes fantasy and realism. EG's other CHILDREN'S FANTASIES are *Smoky-House* (**1940**), a story of smuggling in an idyllic Cornish village in which FAIRIES save the day, *Henrietta's House* (**1945**), which is part of her **Cathedral Trilogy** – the other two are the adult novels *A City of Bells* (**1936**) and *The Dean's Watch* (**1960**) – *The Valley of Song* (**1951**), a fable of 18th-century boat-builders, and *Linnets and Valerians* (**1964**), about a family sundered then drawn together by WITCHCRAFT. Of her other adult books, only the historical novel *The White Witch* (**1958**), set in 1642 at the start of the English Civil War, features overt fantasy, in the shape of the heroine, who uses white witchcraft to heal (and who is based on the ghost of a former occupant of EG's Henley cottage). Many others, however, use LEGEND and SUPERSTITION either as the basis for the story, as in *Gentian Hill* (**1949**), or as interludes. Prime examples of the latter include her first novel, *Island Magic* (**1934**), *Towers in the Mist* (**1938**) and EG's most famous and most successful novel, *Green Dolphin Country* (**1944**; vt *Green Dolphin Street* US), which won a Literary Guild Award and was filmed as *Green Dolphin Street* (*1947*). [JF]

Other works: *The Middle Window* (**1935**); *Sister of the Angels* (**1939**); *The Castle on the Hill* (**1942**); *Make-Believe* (**1949**); *The Bird in the Tree* (**1940**), *The Herb of Grace* (**1948**; vt *Pilgrim's Inn* US) and *The Heart of the Country* (**1953**), which make up the series **The Eliots of Damerosehay**; *God So Loved the World* (**1951**); *The Heart of the Matter* (**1953**); *The Rosemary Tree* (**1956**); *St Francis of Assisi* (**1959**); *The Scent of Water* (**1963**); *The Chapel of the Blessed Virgin* (**1966**); *I Saw Three Ships* (**1969**); *The Child from the Sea* (**1970**).

Collections: *The Fairies' Baby and Other Stories* (coll **1919**); *The Pedlar's Pack and Other Stories* (coll **1937**); *Three Plays* (coll **1939**); *The Golden Skylark and Other Stories* (coll **1941**); *The Ikon on the Wall* (coll **1943**); *Songs and Verses* (coll **1947**); *At the Sign of the Dolphin* (coll **1947**); *The Reward of Faith* (coll **1950**); *White Wings: Collected Short Stories* (coll **1952**); *A Christmas Book* (coll **1967**); *The Ten Gifts* (coll **1969**); *The Lost Angel* (coll **1971**).

As editor: *A Book of Comfort* (anth **1964**); *A Diary of Prayer* (anth **1966**); *A Book of Peace* (anth **1968**); *A Book of Faith* (anth **1976**).

GOULART, RON(ALD JOSEPH) (1933-   ) US writer of, principally, sf in planetary settings with a uniformly Californian flavour, full of deadpan HUMOUR and one-line gags. His **Chameleon Corps** sf stories star SHAPESHIFTING intelligence agents in *The Sword Swallower* (**1968**), *The Chameleon Corps and Other Shape Changers* (coll **1972**) – including several fantasies with the same theme – and *Flux* (**1974**). *The Fire-Eater* (**1970**), though notionally sf, involves comic capers on a world where MAGIC works, WEREWOLVES and SORCERY abound, etc. *Ghost Breaker* (coll **1971** dos) collects RG's fantasy tales of the OCCULT DETECTIVE **Max Kearney**, whose clients' problems include TRANSFORMATION into an elephant on US public holidays, GHOSTS haunting tv sets, a water ELEMENTAL in a Californian swimming-pool, a harassing employer who inflicts INVISIBILITY on those seeking another job, and other modern magical outbreaks. RG has also written several books of nonfiction on pulp MAGAZINES and COMICS; he edited, and largely wrote, the quirky *The Encyclopedia of American Comics* (**1990**). [DRL]

See also: VAMPIRELLA.

GOULD, F(RANCIS) CARRUTHERS (1844-1925) UK illustrator and writer, much of whose work – like his copious illustrations for six *Truth Christmas* ANNUALS (**1886-1891**) – was sf SATIRE. He was active 1879-1914 as a political cartoonist for journals like the *Westminster Gazette*. *"Who Killed Cock Robin?"*, *and Other Stories for Children Young and Old* (coll **1896**) gathers various satirical BEAST FABLES, both written and illustrated by FCG. *The Agnostic Island* (**1891**) satirizes UTOPIAS; *Explorations in the Sit-tee Desert, Being a Comic Account of the Supposed Discovery of the Ruins of the London Stock Exchange some 2000 Years Hence* (**1899** chap) is another sf satire. [JC]

GOURMONT, RÉMY DE (1858-1915) French writer, a major figure of DECADENCE and a leading literary critic of his day; writing became his only real channel of communication when he was forced to become a virtual recluse following facial disfigurement by discoid lupus erythematosus. Several fantasies are included in his early *Histoires magiques* (coll **1894**; trans Francis Amery as "Studies in Fascination" in *The Angels of Perversity* coll **1992** UK), which begins with the extraordinarily vivid erotic fantasy "Péhor". *Proses moroses* (coll **1894**), *Le pèlerin du silence* (coll **1896**) and *D'un pays lointain: miracles; visages de femmes; anecdotes* (coll **1898**) likewise contain some fantasy items, the last – written in the shadow of his illness – including his darkest and most powerful exercises in that vein. His longer fantasies are the dramatic excercise in Literary Satanism *Lilith* (**1892**; trans John Heard as *Lilith: A Play* **1946** chap US), the last of his novels – partly rendered in dramatic form – *Une nuit au Luxembourg* (**1906**; trans Arthur RANSOME as *A Night in the Luxembourg* **1912** UK), and the whimsical *Lettres d'un Satyr* (**1913**; trans John Howard [Jacob Howard Lewis] as *Mr Antiphilos, Satyr* **1922** US). All this work develops theories of sexuality minutely explored in RDG's realistic novels and laid out in a quaintly rhapsodic nonfictional form in *Physique de l'amour: Essai sur l'instinct sexuel* (**1903**; trans Ezra Pound [1885-1972] as *The Natural Philosophy of Love* **1922** US). [BS]

GOZZI, CARLO (1720-1806) Italian dramatist, author of numerous *"fiable"*, plays which married fantasy and the COMMEDIA DELL'ARTE. The most famous (though the least fantastical) of these is *Turandot* (**1765**) (◊ OPERA). He was subject of a study by Vernon LEE, "Carlo Gozzi and the Venetian Fairy Comedy". His *Fiable dell' amore delle tre melarance* (**1761**) was adapted by Sergei Prokoviev (1891-1953) as the opera whose whole title is generally rendered in English as *The Love for Three Oranges* (produced 1921), and his *Il re cervo* ["King Stag"] (**1762**) was made into the opera *König Hirsh* (produced 1956) by Hans Werner Henze (1926-   ). [JC]

GRABIEN, DEBORAH (1954-   ) US writer, partly resident in the UK, where most of her work is set. She began publishing work of interest with *Woman of Fire* (**1988**; vt *Eyes in the Fire* **1989** US), a tale whose two female protagonists – one in contemporary England, one from pre-Roman times – begin to have TIMESLIP experiences of one another's worlds, though neither makes a transition. The action is muted, but under the surface of events pulses a vision of the LAND.

Her second novel, *Plainsong* (**1990**), a MILLENNIAL FANTASY, much intensifies the sense of the Land of England

as being a central theatre upon which the drama of the world may be played. Though it is again quietly told, the story itself is a deeply blasphemous – and, with regard to the usual action of the FANTASY text, structurally intriguing – assault on normal expectations. For 2000 years, the patriarchal rule of GOD (the Father) and CHRIST has caused a disastrous THINNING of REALITY. The novel opens some time after a plague has eliminated almost all adults; the world is beginning to recover (◊ HEALING), the land to burgeon, and animals to reclaim the gift of nonvocal speech. The wheel has come full circle, and it is time for Christ to be evicted from the patriarchal solitude of his reign. To accomplish this, the WANDERING JEW – whose SHADOW Jesus is – sleeps with a woman, who becomes pregnant with the new (female) MESSIAH, who is due to be born in rural Wiltshire. In his attempts to have the new holy family murdered, Jesus behaves almost like a DARK LORD, though a BEAST-FABLE episode in a zoo, when the animals refuse to kill for him, chastens him; and the SECRET GUARDIANS of the CYCLE of the world – Gad the CAT, Grandfather Trout and Simon the sheep, all three being immortal LIMINAL BEINGS – also ensure his frustration. In the end, everyone is reconciled; and Jesus and the Wandering Jew, revelling in their dawn-fresh mortality, go off arm-in-arm.

And Then Put Out the Light (**1993**), describes a romance in terms evocative of SUPERNATURAL FICTION. She is not a powerful writer, but can, once the essentially nondramatic flow of her stories is accepted, convey a considerable charge. [JC]

Other works: *Fire Queen* (**1990**).

GRABIŃSKI, STEFAN (1887-1936) Polish writer. *The Dark Domain* (coll trans Miroslaw Lipinski **1993** UK), which marks his first appearance in English, assembles work published throughout his active career, which ran only 1918-c1930. Collections include *Na wzgóru ro* ["On the Hill of Roses"] (coll **1918**), *Demon ruchu* ["The Demon of Motion"] (coll **1919**), *Szalony pątnik* ["The Frenzied Pilgrim"] (coll **1920**), *Kşelga ognia* ["The Book of Fire"] (coll **1922**) and *Niesamowita opowieść* ["An Uncanny Story"] (coll **1922**); his work was advocated by Stanislaw Lem (1920-   ). Most of SG's stories are SUPERNATURAL FICTION, and very frequently use DOPPELGÄNGER figures to dramatize the plights of fractured consciousnesses. A sense of the permeating ubiquity of anguished supernatural voices and presences, and of the unquenchable REALITY of such voices and presences, is reminiscent of the work of Gustav MEYRINK. He was personally obsessed by trains – at least one tale in *The Dark Domain* unmistakably sexualizes their thrust and motion, and references to them appear throughout his work: trains conveying passengers into OTHERWORLDS; trains haunted by GHOSTS; insane locomotives (◊ ANIMATE/INANIMATE); trains which evoke erotic frenzy. [JC]

GRAHAME, KENNETH (1859-1932) UK banker and writer who began composing light nonfiction pieces as a pastime. Some were assembled as *Pagan Papers* (coll **1893**), a title which implies a rather more iconoclastic set of contents than KG was willing to deliver. His life is notorious for its conflicts and dis-ease. Unhappy (though eminently successful) as a man of business, and uneasy (though ultimately world-famous) as an author of fictions, KG was a classic Edwardian writer. He was deeply conservative, and was instinctively alarmed at the loss of the THEODICY-irradiated rural England that he, and so many other

Edwardians of similar mind, conceived to have once existed. And, like some other writers at the cusp of the new century – J.M. BARRIE is an example – he conflated the loss of childhood with the loss of that England, that EDEN-like past. So he did not wish to grow up. His marriage was disastrous, his career was a straitjacket, the world was heating up to some sort of nightmare apocalypse; and he put all his conflicted passions into three books, ostensibly written for children, plus one short story published in FIN DE SIÈCLE gear as a Bodley Booklet. When WWI came he stopped writing.

The Headswoman (**1898** chap), though not fantasy, occupies a clearly impossible TOPSY-TURVY medieval world, where a female executioner's clients quip and flirt on their way to death because she is beautiful; the tone lies somewhere between W.S. GILBERT's **Savoy Operas (1876-93)** and the weekend essays of A.A. MILNE. KG's first two full-size books – *The Golden Age* (coll **1895**) and *Dream Days* (coll **1898**; rev 1899), one of the tales from the latter being published separately as *The Reluctant Dragon* (**1938** chap) – comprise stories told to a family of children by an unnamed narrator. They are very various. In the distressing "The Argonauts", which appears in the first volume, some children inadvertently CROSSHATCH from a suburban brook into a confrontation with a mad Medea at the edge of a rough and mythopoeic sea; on the other hand, "A Saga of the Seas", in the second volume, is a gay tale of swashbuckling in the South Seas.

The Wind in the Willows (**1908**) is one of the most popular and famous CHILDREN'S FANTASIES of the 20th century. Debates are frequent as to whether this almost excessively complex text should be regarded as a children's book, but huge numbers of children have been intoxicated by most parts of the book, though they often skip the chapter called "Piper at the Gates of Dawn" (◊◊ MUSIC), which is a typical Edwardian paean to PAN; while at the same time it is potent for adult readers because of its nostalgia, because it is so alluring as a depiction of ARCADIA, and (perhaps) because it is so sophisticated in its creation of a non-moral SECONDARY WORLD – hence (perhaps) J.R.R. TOLKIEN's high-minded disregard for the book.

Despite its overt parallels to HOMER's *Odyssey*, the text is not seemingly well constructed, as the main tale it contains – how Toad's obsession with motorcars leads him into imprisonment, from which he escapes into the Wild Woods and with the help of his COMPANIONS regains Toad Hall from *Untermenschen* stoats – begins only partway through, is interrupted more than once, and stars, in Toad, a TALKING ANIMAL who could easily have been the VILLAIN. But these failures of construction matter little in the event; and can be seen as contributing to the strange mix of the arbitrary and the deeply reassuring that so marks the book. *The Wind in the Willows* is an ANIMAL FANTASY, but many of its animals are merely dumb, and some are eaten for breakfast by Mole, Rat or Badger. Time and space fluctuate in a manner almost akin to the shifts of perspective endemic to WONDERLANDS, where absurdist rules govern "REALITY": the prison Toad enters is medieval, as are his jailors, while the roads he travels are 20th-century; the protagonists are variously the size of a rat or a mole, or of human bulk; the region in which they live either has no boundaries, or is a suburb of London, but is certainly never a POLDER because it fluctuates too severely; almost every character in the book is fixed into role, but nameless.

In the end the texture of this complex and undulant world seems sustained by its creator's longing to believe that something like an upstream prelapsarian Thames Valley, where *The Wind in the Willows* is clearly set, is attainable, if only by a stretch of imagination. What remains altogether remarkable about the tale is how, while it turns its back on the outside and the future, it prefigures both throughout. [JC]

Other works: *First Whisper of The Wind in the Willows* (**1944**) presented by Elspeth Grahame; *Paths to the River Bank* (coll **1983**) ed Peter HAINING.

Further reading: *Kenneth Grahame: A Biography* (**1959**; cut rev vt *Beyond the Wild Wood: The World of Kenneth Grahame* 1982) by Peter Green; *Kenneth Grahame: An Innocent in the Wild Wood* (**1994**) by Alison Prince; *The Wind in the Willows: A Fragmented Arcadia* (**1994**) by Peter HUNT.

GRAIL The original meaning of "grail" (or *graal* in Old French) referred to a serving platter at a dinner. The term was introduced in this sense to the Arthurian canon by CHRÉTIEN DE TROYES in his unfinished *Perceval, ou Le Conte del Graal* (begun ?1182), where it appears as part of some unexplained ceremony performed to PERCEVAL in the castle of the FISHER KING. The basis of the Grail is not explained – indeed, it is because of Perceval's failure to explore the mysteries of the Grail procession that ARTHUR's world is now blighted. Chrétien does not link the Grail with CHRIST – in fact, he was probably adapting some ancient Celtic FERTILITY procession. In an early Celtic Arthurian story, *Preiddeu Annwfyn* ["The Spoils of Annwn"] (?**900**), Arthur and his men steal a cauldron of plenty from the Irish UNDERWORLD. Such stories may reflect a folk memory from a time in the early years of Arthur's reign of plenty and good harvest, but which was followed in the later years by the plague and pestilence that seemingly scythed through Europe in the mid-6th century. The Cauldron of Plenty, sometimes called the Cauldron of Rebirth, was one of four objects of power in Celtic mythology, along with the SWORD Fragarach the Defender, the Stone of Destiny and the Spear of Lugh.

All this symbolism changed (or was distorted) when Robert de Boron (? -1212) linked the Grail to the chalice from which Christ dined at the Last Supper, so that it became the Holy Grail (or Sangraal, Sangrail or Sangreal). In *Joseph d'Arimathie* (?**1200**; vt *Le Roman de l'Estoire dou Graal*) Robert specifically imbued the Grail with Christian symbolism, making it the vessel which held Christ's blood; it thus came to symbolize life. Writing at the time of the Crusades, when Christianity was perceived as under threat, he made the Grail a powerful image in depicting Christ's sacrifice to give humankind eternal life. Within a generation of Robert de Boron's account, the anonymous author of the *Perlesvaus* (early 13th century) gave the Grail the power of rejuvenation, and the anonymous authors of the Prose *Lancelot* in the **Vulgate Cycle** of Arthurian stories (composed 1215-35) converted it from a dish to a chalice (◊ ARTHUR).

All the main elements were there by the time Sir Thomas MALORY completed *Le Morte Darthur* (**1485**), which established the Grail story we now know. Arthur's kingdom has entered a period of decline and pestilence. Some say this was caused by the DOLOROUS STROKE delivered by Sir Balin when he killed King Pellam with the Lance of Longinus; since this lance, a development of the Bleeding Lance (a phallic symbol for the giving of life), has been used to take a life, a blight settles over the land (◊ WASTE LAND). This blight is personified by the Maimed King, in some legends equated to the FISHER KING, who lives in the Grail Castle.

Arthur prays for a sign, and at Camelot he and his KNIGHTS have a vision of the Holy Grail. One by one his knights set out on their QUESTS to seek the Grail. It is not their purpose physically to hold the Grail, merely to understand its significance. All the leading knights seek the Grail. GAWAIN and LANCELOT are unsuccessful because they are impure. Some legends have Perceval and Sir Bors successful, but the true Grail Knight is GALAHAD, who solves the riddle of the Grail, restores the Maimed King and thus brings prosperity back to the land.

The Grail legend has had a significant impact upon Arthurian fiction in general, and within fantastic fiction. Authors have utilized it either directly, in reworking the Grail story for modern readers, or have used the symbolism to make the Grail represent something attainable only by perfection.

The basic Grail story turns up in most retellings of the Arthurian legend, notably in *Perronik the Fool* (**1926**) by George Moore (1852-1933), *Kinsmen of the Grail* (**1963**) by Dorothy James Roberts, *The King's Damosel* (**1976**) by Vera CHAPMAN, the initial **Parsival** trilogy (**1977-80**) by Richard MONACO, *Perceval and the Presence of God* (**1978**) by Jim Hunter (1939-   ); *The Light Beyond the Forest* (**1979**) by Rosemary SUTCLIFF (a more traditional version for younger readers), *The Magic Cup* (**1979**) by Andrew M. Greeley (1928-   ), which presents the Celtic version of the quest, and *The Grail of Hearts* (**1994**) by Susan SHWARTZ, which links the Fisher King to the WANDERING JEW. A collection of mostly new stories is «The Chronicles of the Holy Grail» (1996) ed Mike ASHLEY.

The metaphysical aspects of the Grail and its significance today have attracted a different range of writers. Arthur MACHEN, a mystic who wrote extensive Grail studies, used Grail imagery in several of his stories, and it was the focal point of *The Great Return* (1915), in which the reappearance of the Grail results in a series of miracles. The Grail likewise reappears in *War in Heaven* (1930) by Charles WILLIAMS, where it becomes a focus for the battle between GOOD AND EVIL. A similar background is used in Susan COOPER's **The Dark is Rising** sequence, particularly in the first book, *Over Sea, Under Stone* (1965), where a group of children recover the Grail. The search for the Grail is the theme of the **Grail Quest** series of role-playing gamebooks (◊ GAMES) by J.H. BRENNAN. Aspects of the Grail, both directly and indirectly, are found in Michael MOORCOCK's **Eternal Champion** series, especially *The War Hound and the World's Pain* (1981), Jack VANCE's **Lyonesse** series, especially *Madouc* (1989), and Diane DUANE's *A Wizard Abroad* (1993).

The Grail has become symbolic for the attainment of perfection which can become a life's quest rather than a physical journey. The Grail motif has thus adapted itself into both HIGH FANTASY and metaphysics as a quest for the ultimate in life. It can also be mirrored in many of those quest stories where the protagonist is seeking a sacred object. An anthology on this theme is *Grails: Quests, Visitations and Other Occurrences* (anth 1992; exp in 2 vols as *Grails: Quests of the Dawn* 1994 and *Grails: Visitations of the Night* 1994) ed Richard Gilliam (1950-   ), Martin H. GREENBERG and Edward E. Kramer.

The Grail has had three recent cinematic treatments which between them have illustrated much of its appeal: *Indiana Jones and the Last Crusade* (*1989*) portrays the continuing lure and desire for the Grail (◊ INDIANA JONES); *The FISHER KING* (*1991*) shows how the power of belief in the Grail can still heal; and MONTY PYTHON AND THE HOLY GRAIL (*1975*) is sufficiently irreverent to spoof the entire Arthurian genre while retaining an intelligent understanding of the motifs. [MA]

GRAINVILLE, JEAN (BAPTISTE FRANÇOIS XAVIER) COUSIN DE (1746-1805) French writer. ◊ APOCALYPSE; LAST MAN.

GRAND GUIGNOL A French term, now often used to describe plays which feature cruel quick-fire mayhem and melodrama, frequently with a supernatural element; GG plays are normally one act in length. The name originated in late-18th-century France, where the COMMEDIA DELL'ARTE character Pulcinella had for some time been known as Polichinelle, in which guise his quick wits had rendered his grotesque shape and actions both more stomachable and more ingenious, and turned into a kind of witticism that erasure of the boundary between animal and human which marks the grotesque as a mode. Around the turn of the century, it is thought, a PUPPET-master known as Mourquet (1744-1844) transformed Polichinelle into a humorously violent puppet named Guignol. Soon cabarets were starring Guignol, and these became known as Guignol theatres. In 1888 the most famous of these, the Théâtre du Grand Guignol, was founded. Most GG plays are HORROR; a high proportion feature mad doctors and other berserk scientists; and many are set in institutions.

The speed and the far-fetchedness of the typical GG plot tend to generate a sense of farce, and at the end of the 20th century GG tales or presentations are likely to elicit simultaneously a frisson and a giggle. The term is thus time-bound: applied to a story by (say) H.P. LOVECRAFT, it presumes no element of camp; applied to a modern tale, it implies knowingness. [JC]

GRANDVILLE, ISIDORE Working name of French engraver Jean-Ignace Isidore Gérard (1803-1847), credited as the world's first sf illustrator for *Un Autre Monde* ["Another World"] (graph **1844**) as by Taxile Delord, which contained illustrations depicting weird alien creatures, monstrous beings and fantastic feats of interplanetary engineering. His graphic style is very much that of the early-19th-century steel engravers, but his imagination is individual.

IG was born in Nancy and illustrated editions of several famous works, including de la Fontaine's *Fables* (1838) and Jonathan SWIFT's *Gulliver's Travels* (1838), which latter contained his most enduringly infamous image, of a leather boot stamping on a crucifix. His other works of fantasy interest were *Scènes de la Vie Privée et Publique des Animaux* (graph **1842**; trans as *Public and Private Lives of Animals* **1877** UK), which features some very original early images of anthropomorphosed animals, and the now rather sentimental-seeming *Les Fleurs Animées* ["Animated Flowers"] (graph **1847**). [RT]

GRANT, CHARLES L(EWIS) (1942-   ) US writer and editor, married to Kathryn GRANT. He began writing sf in the late 1960s, publishing several novels in that genre in the 1970s. Since then CLG has concentrated on HORROR, crafting subtle, atmospheric works that eschew overt violence in favour of the powerful terrors of the imagination. He has acquired a deserved reputation for creating strong, credible female protagonists, a quality especially apparent in his **Oxrun Station** series, which comprises 11 books and several stories. Oxrun Station is an imaginary town in Connecticut beset by a variety of evils, including an ancient Welsh religious cult in *The Sound of Midnight* (**1978**), a satanic cult in *The Hour of the Oxrun Dead* (**1977**), and a fierce elemental wind that assumes the form of a large beast in *The Bloodwind* (**1982**). Other titles are *The Last Call of Mourning* (**1979**), *The Grave* (**1981**), a 19th-century trilogy internal to the **Oxrun** sequence – *The Soft Whisper of the Dead* (**1982**), *The Dark Cry of the Moon* (**1986**) and *The Long Night of the Grave* (**1986**) – *Nightmare Seasons* (fixup **1982**) – a 1983 WORLD FANTASY AWARD-winner as Best Collection – *The Orchard* (**1986**), *Dialing the Wind* (**1989**) and *The Black Carousel* (**1994**).

CLG's other horror and DARK-FANTASY novels typically contain a meaningful subtext and explore a variety of themes, such as the Caucasian-Native American conflict at the heart of *The Nestling* (**1982**), in which a flying creature terrorizes a small Wyoming community. In *The Pet* (**1986**), the teenage protagonist's angst and frustration take the physical form of a supernatural black stallion that acts out the boy's aggressions.

Although CLG is a talented novelist, his careful utilization of language and manipulation of mood are perhaps best showcased at shorter lengths. Prime fantasy examples include: "When All the Children Call My Name" 1977 *Year's Best Horror Stories*, wherein a group of young children acquire MAGIC abilities that allow them VENGEANCE against older tormentors; "Come Dance with Me on My Pony's Grave" 1973 *F+SF*, in which magic is again used by a child for vengeance; and "The Three of Tens" 1975 *F+SF*, involving a magical artifact which, thrown into a river, proceeds to give life to a river-bottom corpse (◊ REVENANT).

As Lionel Fenn, CLG has written, among others, the four-volume **Quest for the White Duck** comic-fantasy series (◊ HUMOUR): *Blood River Down* (**1986**), *Web of Defeat* (**1987**), *Agnes Day* (**1987**) and *The Seven Spears of the W'dch'ck* (**1988**). As Geoffrey Marsh CLG wrote the four-volume **Lincoln Blackthorne** series, which features an INDIANA JONES-style protagonist whose adventures often involve the fantastic and the supernatural: *The King of Satan's Eyes* (**1984**), *The Tale of the Arabian Knight* (**1986**), *The Patch of the Odin Soldier* (**1987**) and *The Fangs of the Hooded Demon* (**1988**).

As editor, CLG is responsible for over 20 anthologies, most notably the **Shadows** series, comprising 11 original volumes and a "best of" collection; one volume won a 1979 World Fantasy Award. The series specializes in "quiet horror" and includes some excellent fantasy, including work by John CROWLEY, Jack DANN, Tanith LEE and Lisa TUTTLE. [BM]

Other works: *The Curse* (**1977**); *A Quiet Night of Fear* (**1981**); *Tales from the Nightside* (coll **1981**); *A Glow of Candles and Other Stories* (coll **1981**); *Night Songs* (**1984**); *The Teaparty* (**1985**); *Black Wine* (coll **1986**) with Ramsey CAMPBELL; *For Fear of the Night* (**1988**); *In a Dark Dream* (**1989**); *Stunts* (**1990**); *Fire Mask* (**1991**); *Something Stirs* (**1991**); *Raven* (**1993**); *Jackals* (**1994**); *The X Files: Goblins* * (**1994**) and *The X Files: Whirlwind* * (**1995**), tied to the tv series.

As Felicia Andrews: *River Witch* (**1979**); *Moon Witch* (**1980**); *Mountain Witch* (**1980**).

As Steven Charles: The **Private Academy** sequence of YA sf/horror novels: *Nightmare Session* (**1986**), *Academy of Terror* (**1986**), *Witch's Eye* (**1986**), *Skeleton Key* (**1986**), *The Enemy Within* (**1987**) and *The Last Alien* (**1987**).

As Lionel Fenn: The **Kent Montana** series of comic sf tales, which invoke Hollywood ICONS through the adventures of a failed actor – *Kent Montana and the Really Ugly Thing from Mars* (**1990**), *Kent Montana and the Reasonably Invisible Man* (**1991**), *Kent Montana and the Once and Future Thing* (**1991**), *The Mark of the Moderately Vicious Vampire* (**1992**) and *668: The Neighbor of the Beast* (**1992**) – and the **Diego** series, featuring a gunslinger who travels through time: *Once Upon a Time in the East* (**1993**), *By the Time I Get to Nashville* (**1994**) and *The Semi-Final Frontier* (**1994**).

As Simon Lake: The **Midnight Place** sequence: *Midnight Place: Daughter of Darkness* (**1992**), *#2: Something's Watching* (**1992**), *#3: Death Cycle* (**1993**) and *#4: He Told Me To* (**1993**).

As Deborah Lewis: *The Eve of the Hound* (**1977**); *Kirkwood Fires* (**1977**); *Voices out of Time* (**1977**); *The Wind at Winter's End* (**1979**).

Sf: The **Parric Family** series: *Shadow of Alpha* (**1976**), *Ascension* (**1977**), *Legion* (**1979**); *The Ravens of the Moon* (**1979**).

Nonfiction: *Writing and Selling Science Fiction* (anth **1976**).

As editor (series): The **Shadows** anthologies, comprising *Shadows* (anth **1978**; vt *Shadows II* 1987 UK), *#2* (anth **1979**), *#3* (anth **1980**), *#4* (anth **1981**; vt *Shadows* 1987 UK), *#5* (anth **1982**), *#6* (anth **1983**), *#7* (anth **1984**), *#8* (anth **1985**), *#9* (anth **1986**) and *#10* (anth **1987**) plus *Final Shadows* (anth **1991**) and *The Best of Shadows* (anth **1988**); two **Night Visions** anthologies, *Night Visions 2* (anth **1985**; vt *Night Visions: Dead Image* 1987; vt *Night Terrors* 1989 UK) and *Night Visions 4* (anth **1987**; vt *Night Fears* 1989 UK), uncredited for the UK versions; the **Greystone Bay** anthologies, comprising *The First Chronicles of Greystone Bay* * (anth **1985**), *Doom City* * (anth **1987**), *The Sea Harp Hotel* * (anth **1990**) and *In the Fog: the Final Chronicle of Greystone Bay* (anth **1993**).

As editor (singletons): *Nightmares* (anth **1979**); *Horrors* (anth **1981**); *Terrors* (anth **1983**); *The Dodd, Mead Gallery of Horror* (anth **1983**; vt *Gallery of Horror* 1983 UK); *Midnight* (anth **1985**); *After Midnight* (anth **1986**).

GRANT, JOAN Pseudonym of UK writer and occult theorist Joan Marshall Kelsey (1907-1989), whose books give off a powerful sense of otherness, whether or not they actually incorporate elements of fantasy or (more likely) SUPERNATURAL FICTION. Her first novel, *Winged Pharaoh* (**1937**), is fantasy only if lectures on THEOSOPHY – delivered to the protagonist as part of her ongoing education in being a priestess and a pharaoh at the same time – render any text fantastical; EGYPT is viewed as a successor to ATLANTIS. Further novels by JG intensified a sense that she herself, under whatever name or REINCARNATION, was their main character, usually through the exercise of "far memory", whether the setting was Renaissance Italy, as in *Life as Carola* (**1939**), or in a (less ancient) Egypt in the **Ra-Ab Hotep** sequence – *Eyes of Horus* (**1942**) and *Lord of the Horizon* (**1943**) – or slightly forward again in the 2nd-century Greece depicted in *Return to Elysium* (**1947**). [JC]

Other works: *Redskin Morning and Other Stories* (coll **1944**); *Scarlet Feather* (**1945**); *The Laird and the Lady* (**1949**; vt *Castle Cloud* 1971 US); *So Moses Was Born* (**1952**); much nonfiction concerning "far memory".

GRANT, JOHN Usual working name of Scottish editor and writer Paul (le Page) Barnett (1949-   ); he has published nonfantasy under his own name and other pseudonyms, notably Eve Devereux. He entered the field with "When All Else Fails" in *Lands of Never* (anth **1983** ed Maxim

Jakubowski) but has not been a prolific short-story writer, instead concentrating on books, of which he has written over 40. His first two fantasy novels were *#1* and *#2* in the **Legends of Lone Wolf** SWORD-AND-SORCERY series, loosely based on gamebooks (◊ GAMES) by Joe DEVER and published as co-authorships: *#1: Eclipse of the Kai* * (**1989**) and *#2: The Dark Door Opens* * (**1989**), assembled as *Legends of Lone Wolf Omnibus* (omni **1992**). These introduced the characters Alyss and Qinefer, both of whom became fundamental to JG's *corpus*. The series continued with *#3: The Sword of the Sun* * (**1989**; vt in 2 vols *The Tides of Treachery* 1991 US and *The Sword of the Sun* 1991 US), *#4: Hunting Wolf* * (**1990**), *#5: The Claws of Helgedad* * (**1991**), which introduced his humorous character Thog the Mighty, *#6: The Sacrifice of Ruanon* * (**1991**), *#7: The Birthplace* * (**1992**), which is central to an understanding of JG's preoccupations, *#8: The Book of the Magnakai* * (**1992**), *#9: The Tellings* * (coll **1993**), *#10: The Lorestone of Varetta* * (**1993**), *#11: The Secret of Kazan-Oud* * (**1994**) and *#12: The Rotting Land* * (**1994**). *History Book – A Thog the Mighty Text* (**1994** chap), a limited-edition addendum, contains revised material from *The Rotting Land*.

Elsewhere JG was publishing the two novels on which his reputation as a fantasy writer rests: *Albion* (**1991**) and *The World* (**1992**). The first is of only moderate interest; set in a LAND whose common people have only short-term memory except for rote activities, it depicts the tyranny imposed by those equipped with full memories, its eventual defeat, and the breaching of the walls of the POLDER. *The World*, closely related, is significantly more ambitious. On the surface, it describes the collision of our Universe with the created SECONDARY WORLD in which *Albion* was set; in fact it is an exploration of the relationship between physical and created REALITIES, both of which JG depicts as parts of the "polycosmos", a MULTIVERSE whose almost infinite coexistent universes can be travelled between by Alyss (who is both integral and extraneous to the polycosmos), consciously by rare mortals like Qinefer (who through her encounter with transcendence in *The Birthplace* discovers the *tao* that leads to the overarching reality of the polycosmos), and by all mortals during DREAMS and after death, viewed as the transition from one physical or created universe to REINCARNATION in the next. Moments of RECOGNITION abound. This INSTAURATION FANTASY ends with the two universes fused, so that reality is arbitrary, memory treacherous and indeterminacy paramount.

JG's fiction since has been sparse. *The Hundredfold Problem* * (**1994**), featuring the COMICS character **Judge Dredd**, is a joky SCIENCE FANTASY set in a Dyson sphere. *Dr Jekyll and Mr Hyde* * (**1995**) is a children's retelling of the Robert Louis STEVENSON tale.

In nonfiction JG is noted as author of *Encyclopedia of Walt Disney's Animated Characters* (**1987**; rev 1993) (◊ DISNEY), and as Technical Editor of the 2nd edition of *The Encyclopedia of Science Fiction* (1993) ed John CLUTE and Peter Nicholls (◊ SFE), for which work he shared the 1994 BSFA Special Award. He is joint editor of *The Encyclopedia of Fantasy*.

He is not to be confused with the John Grant who writes mysteries as Jonathan Gash. [JG]

Other works (fiction): *Aries 1* (anth **1979**), sf; *The Truth About the Flaming Ghoulies* (**1983**), humorous epistolary fringe-sf novel; *Sex Secrets of Ancient Atlantis* (**1984**), pseudoscience spoof; *Earthdoom!* (**1987**) with David LANGFORD, spoof disaster novel.

Other works (nonfiction): *The Book of Time* (**1980**) ed with Colin WILSON; *A Directory of Discarded Ideas* (**1981**); *The Directory of Possibilities* (**1981**) ed with Wilson; *Dreamers: A Geography of Dreamland* (**1983**); *Great Mysteries* (**1988**); *An Introduction to Viking Mythology* (**1989**); *Monsters* (**1992**; vt *Monster Mysteries* US); *The Encyclopedia of Fantasy and Science Fiction Art Techniques* (**1996**) with Ron TINER; many others.

See also: AFTERLIFE; ALTERNATE REALITIES.

GRANT, KATHRYN (1952-   ) US writer and journalist, married to Charles L. GRANT. KG started writing historical romances as Kathryn Atwood, Kathleen Maxwell and Anne Mayfield before turning to fantasy and HORROR. **The Land of Ten Thousand Willows** is a HIGH-FANTASY trilogy set in a LAND-OF-FABLE pre-Manchu China: *The Phoenix Bells* (**1987**), *The Black Jade Road* (**1989**) and *The Willow Garden* (**1989**) combine Chinese, English and Russian folklore, fantasy and real history. Under her maiden name, Kathryn Ptacek, she has written the horror novels *Shadoweyes* (**1984**), which won the Porgie Award (Silver Medal), *Blood Autumn* (**1985**), which won the Porgie (Gold Award), *Kachina* (**1986**), *In Silence Sealed* (**1988**), *Ghost Dance* (**1990**) and *The Hunted* (**1993**). [JF]

As editor: *Women of Darkness* (anth **1988**); *Women of the West* (anth **1990**); *Women of Darkness II* (anth **1990**); *The Gila Queen's Guide to Markets*, a monthly market report and newsletter for writers and artists.

GRANVILLE-BARKER, HARLEY (1877-1946) UK actor, playwright, author and producer. ◊ Dion Clayton CALTHROP.

GRAPHIC NOVEL Term coined, probably in the early 1970s, to distinguish between the COMIC book and the attempts then being made to create works of literary merit using a similar amalgam of words and pictures. The term "comic book" had other disadvantages in this context (it is not a book but a magazine) and "adult comic" had been commandeered by the pornography market. The increasing sophistication of the narrative techniques employed by comics artists and writers throughout the 1970s made evident a growing need to attract an intelligent adult reader who could appreciate the nuances of a multi-layered verbal/visual narrative. The difficulty in attracting such readers did not exist in France, where the term *bandes dessinées* usefully avoided both the dichotomy and the intimations of juvenilia. Indeed, it was in France that the most interesting experiments with the medium were being published, and there the burgeoning market for picture-story albums for both adults and children was very different from that of the US comic book, whose readership was generally considered to be limited to a very narrow adolescent ageband.

The notion of a visual narrative is not a modern one. The most remarkable early example is Trajan's column, a 2nd-century sculpted pillar covered with a sequence of relief sculptures depicting the emperor Trajan's war against the Dacians. Almost 1000 years later, the seamstresses who created the Bayeux tapestry adopted a similar approach in pictorial terms – a left-to-right movement with consecutive incidents shown in a continuously unfolding tale – but with the addition of a verbal explanation of each event. Many medieval paintings show consecutive scenes from BIBLE stories. Two series of prints by William Hogarth (1697-1764), *The Harlot's Progress* (c1731) and *The Rake's Progress* (1735), depicted individuals brought low through dissipated living. But books in which pictures carry the burden of the narrative are uncommon before the 20th century, a singular example being Gustave DORÉ's *Histoire Pittoresque,*

Dramatique et Caricaturale de la Sainte Russie (graph **1854**). Mention should also be made of the work of Heinrich Zille (1858-1929), whose semi-autobiographical *Hurengespäche Gehört* ["Stories Told by Prostitutes"] (graph **1913**), *Kinder der Strasse* ["Street Children"] (graph **1923**) and *Zwanglosen Gerschichten* ["Stories without Restraint"] (graph **1925**) consisted of lively drawings of German proletariat life linked by a sparse handwritten text. In the 1920s the Swiss Franz Masereel (1889-1972) produced a sequence of wordless "novels" in the form of woodcuts in which he cast himself as "everyman"; these include *Le Soleil* (graph **1920**; trans as *The Sun* **1989** UK), *Histoire sans Parole* (graph **1920**) and *L'Idée* (graph **1920**) – these two assembled as *Story without Words and The Idea* (trans graph omni **1990** UK). The last of the sequence was *La Ville/Die Staadt* (graph **1926** France/Germany; trans as *The City* **1988** UK). In these, however, the pictures show only consecutive individual scenes, and pictorial narrative is minimal, as in the cycle of autobiographical woodcuts in similar style, *Childhood* (trans graph **1931** UK), by Helena Borcharâková-Dittrichová.

In the USA, Lynd Ward (1905-1985) produced a woodcut "novel" in a far more sophisticated and selfconcious style: *God's Man* (graph **1930**). But it was the development of the comic book and an awareness of CINEMA that led to a realization that subtle nuances of meaning could be conveyed by judicious use of varied picture composition and viewpoint, along with dramatic lighting, fragmented images, etc. In this way pictorial narrative techniques developed which could convey increasingly complex story ideas.

The publication of *A Contract with God and Other Stories of Tenement Life* (graph **1978**), the first of a series of "sequential art" books by Will EISNER, served to give the term GN definition and focus. It was quickly taken up as a useful marketing term when, in the early 1980s, specialist comics shops began to attract more discerning readers and publishers perceived a market for reprinted comic-book material in glossy covers; they labelled such collections GNs in order to lend them credibility. The useful adult/juvenile distinguishing element would have been lost had not the mid-1980s seen the publication of two remarkable works, both of which first appeared as short comic-book series: *Batman: The Dark Knight Returns* (1986; graph **1986**) by Frank MILLER (◊ BATMAN) and *Watchmen* (1986-7; graph **1987**) by Alan MOORE and Dave Gibbons. Both had the structure of the conventional novel, as did a third milestone book publication of the period, *Maus* (1980-85 *Raw*; graph **1987**) by Art Spiegelman (1948-). These three works served to re-establish the GN as distinct from a comic book. Other publications of the time include *Violent Cases* (graph **1987**) by Neil GAIMAN and Dave MCKEAN, several books in the series **Love and Rockets** by the brothers Hernandez, and a number of *bandes dessinées* translated from the French, among them *The Magician's Wife* (graph **1986**; trans **1987** US) by François Boucq and Jerome CHARYN, *Pelisse* (graph **1983-7** 4 vols; trans as *Roxanna* **1987-9** US) by Letendre and Loisel, and *Bell's Theorem* (**1986-9** 3 vols; trans **1987-9** US) by Matthias Schulheiss. Further excellent examples are regularly published in HEAVY METAL. Other significant GNs have been translated from the Japanese (◊ MANGA), early examples being *Hadashi no Gen* (graph **1972-3**; trans as *Barefoot Gen* **1987** US) by Keiji Nakazawa, about the bombing of Hiroshima, and *Akira* (graph **1984-8**; trans **1988-90** 34 vols US) by Katsuhiro OTOMO.

Comics writers and artists, inspired by the variety and expressiveness of these, began to test the boundaries of their medium and experiment with narrative techniques, many derived from the cinema. The UK publisher Victor Gollancz issued a list of GNs of considerable merit, including *A Small Killing* (graph **1991**) by Alan MOORE and Oscar Zarate, *The City* (graph **1994**) by James HERBERT and Ian MILLER and *The Minotaur's Tale* (graph **1992**) by Al Davison. But these admirable experiments were commercially disappointing, and the series was discontinued. GNs are now usually produced and distributed by the companies who deal with general comic books: they may be published as multi-volume partworks, in episodes/ long features in magazines like HEAVY METAL, or as complete volumes in sturdy covers.

The dividing line between a comic book and a GN remains indistinct, signifying a process of "growing up" for the comic-strip medium. It is significant that many of the titles published by DC COMICS in their "adult" Vertigo line are clearly GNs (a notable example being Gaiman's and McKean's *Mr Punch* [graph **1994** US]) while others equally clearly are comic books. The difference is in the maturity of the concepts, the complexity and subtlety of the narrative, and the perceptiveness of the intended reader. [RT]

GRASS, GÜNTER (WILHELM) (1927-) Danzigborn German sculptor, poet, dramatist and writer, very much better known for his activities as a novelist than for his work in other fields, most of which preceded the publication of *Die Blechtrommel* (**1959**; trans Ralph Manheim as *The Tin Drum* **1962** US), the novel which made him famous. Through the eyes of one Oskar Matzerath – who has decided as a very small child to respond to the cruelties of German history by refusing to grow – the novel traverses much of the 20th century in terms so surrealistic and intense that fantasy and REALITY seem to fuse in its pages; Oskar communicates solely through the eponymous drum, which makes a tocsin sound throughout the depicted world. GG's second full-length novel, *Hundejahre* (**1963**; trans Ralph Manheim as *Dog Years* **1965** US), covers the same period, concentrating on the sculptor Amsel, who makes first scarecrows then AUTOMATA in the shape of SS-men; and also focusing on several generations of dogs who, like Oskar, turn their backs on the vileness of history. Later novels – like *Der Butt* (**1977**; trans Ralph Manheim as *The Flounder* **1978** US) and *Die Rattin* (**1986**; trans Ralph Manheim as *The Rat* **1987** US) – continue to scarify history through epic narratives in which humans and animals interchange roles, in a deliberately grotesque hypertrophication of the BEAST FABLE. *The Flounder* is narrated by a logorrhoeic fish; *The Rat*, narrated from space after a dream-like APOCALYPSE by a she-rat, is almost incoherently misanthropic. GG's fury at the human condition, and at the institutions (the Church, the military, the state, etc.) which worsen our lot, is not always controlled; but it is never his intention to construct a fantasy story whose resolution will seem redemptive. [JC]

GRAUSTARK ◊ RURITANIA.

GRAY, NICHOLAS STUART (1922-1981) Scottish actor and writer, much of whose written work was for the theatre; the first of his many plays (several of them fantasy), *The Haunted* (produced 1948), has not been published. His others include *Beauty and the Beast* (**1951**) (◊ BEAUTY AND THE BEAST), *The Tinder-Box* (**1951**) and *The Imperial Nightingale* (**1957**) and *New Clothes for the Emperor* (**1957**) – all based on

Hans Christian ANDERSEN tales – *The Princess and the Swineherd* (**1952**), *The Hunters and the Henwife* (**1954**), *The Marvellous Story of Puss in Boots* (**1955**) (◊ PUSS-IN-BOOTS), *The Other Cinderella, With Due Acknowledgements to All the Earlier Versions* (**1958**) (◊ CINDERELLA), *The Seventh Swan* (**1962**), *The Wrong Side of the Moon* (**1968**) – based on a tale by the GRIMM BROTHERS – *New Lamps for Old* (**1968**) (◊ ARABIAN FANTASY) and *Gawain and the Green Knight* (**1969**) (◊ GAWAIN). In book form, these plays come with comprehensive stage directions and can be read with pleasure for their sensitive and playable transformations of FOLKLORE and FAIRYTALE.

NSG's prose fiction, beginning with *Over the Hills to Fabylon* (**1954**), also reveals a certain contrast between an amiable surface and a sense of the graver issues evoked by traditional material. In *The Seventh Swan: An Adventure Story* (**1962**), a prose version of his play *The Seventh Swan* (**1962**), the poignancy of a man's BONDAGE as a swan is told with pleasing wit, but the drama underlying the process by which he LEARNS BETTER how to be a man is strongly conveyed. *Grimbold's Other World* (**1963**) and *The Stone Cage* (**1963**) both feature CATS who act as COMPANIONS; in the latter (and more interesting) tale Tomlyn first tells the story of Rapunzel, then narrates his own NIGHT JOURNEY to exile on the far side of the MOON, where he heals a WITCH of her misogyny. Other novels, like *Down in the Cellar* (**1961**) and *The Apple-Stone* (**1965**), carry groups of adventurous children into growth-inducing experiences: in the latter a MAGIC stone has the power of bringing life to dead things, with very mixed consequences. Collections of interest include *Mainly in Moonlight: Ten Stories of Sorcery and the Supernatural* (coll **1965**), *The Edge of Evening* (coll **1976**) and *A Wind from Nowhere* (coll **1978**). [JC]

Other works: *The Sorcerer's Apprentices* (**1965**); *The Boys* (**1968**); *The Further Adventures of Puss in Boots* (**1971**); *The Wardens of the Weir* (**1978**); *The Garland of Filigree* (**1979**).

GREAT AND SMALL Alterations of scale are a stock device of FOLKLORE and MYTHOLOGY, where giants and "little people" are commonplace; FAIRIES are conventionally miniaturized in art and fiction. The adventures in Lilliput and Brobdingnag remain the most enduring images of *Gulliver's Tavels* (**1726**) by Jonathan SWIFT, while the stories of TOM THUMB, JACK the Giant-Killer and Jack and the Beanstalk are traditional tales. The relative ease with which changes of scale can be simulated by spfx has led to their frequent use in CINEMA and tv. Giants and little people are also standard motifs of FANTASY ART, the latter being of particular importance; idiosyncratically conceived little people have been a principal theme of several artists, including Richard DADD and Richard DOYLE.

Notable fantasies about giants include *Pantagruel* (**1532**) and *Gargantua* (**1534**) by François RABELAIS, "The Last of the Giants" (1828) by John STERLING, *A Spell for Old Bones* (**1949**) by Eric LINKLATER, *All or Nothing* (**1960**) by John Cowper POWYS, *The BFG* (**1982**) by Roald DAHL, "Caves" (1984) by Jane GASKELL, *The Little People* (**1985**) by MacDonald HARRIS and *Towing Jehovah* (**1994**) by James MORROW. Notable fantasies about little people include *Phantasmion* (**1837**) by Sara COLERIDGE, "The Diamond Lens" (1858) by Fitz-James O'Brien (1828-1862), *The Water-Babies* (**1863**) by Charles KINGSLEY, *Mistress Masham's Repose* (**1946**) by T.H. WHITE, *The Borrowers* (**1952**) and its sequels by Mary NORTON, and *Truckers* (**1989**) and its sequels by Terry PRATCHETT; a notable

theme anthology is *Little People!* (anth **1991**) ed Jack DANN and Gardner Dozois.

The idea that the microcosm and the macrocosm mirror one another in some way is a significant element of occult lore (◊ AS ABOVE, SO BELOW; MICROCOSM/MACROCOSM). Occult lore also produced the notion of the HOMUNCULUS. A fascinating reinterpretation of the complicity of the macrocosm and the microcosm, deftly fused with various FAIRYTALE motifs, can be found in John CROWLEY's *Little, Big* (**1981**). Another metaphysical fantasy involving changes of scale is James P. BLAYLOCK's *Land of Dreams* (**1987**). [BS]

GREAT BEAST In *Revelation xiii*, APOCALYPSE is heralded by various signs, including this SEVEN-headed 10-horned Beast with elements of leopard, bear, lion and DRAGON. Its prophet (described as a second beast) is the ANTICHRIST, and its number is famously 666 (◊◊ NUMEROLOGY); occasionally it is identified with SATAN. The associated PROPHECY that "no man might buy or sell" without the mark or number of the beast has generated URBAN LEGENDS about 666 being concealed in bar codes, etc., as featured in Robert RANKIN's *East of Ealing* (**1984**). The Beast is cheekily depicted as part of HEAVEN's furniture in James Branch CABELL's *The Silver Stallion* (**1926**). Aleister CROWLEY exploited the shock value of giving himself the title Great Beast. [DRL]

See also: *The* OMEN.

GREATEST AMERICAN HERO, THE US tv series (1981-3). Stephen J. Cannell Productions/ABC. **Pr** Alex Beaton, Babs Greyhosky, Frank Lupo, Christopher Nelson. **Exec pr** Juanita Bartlett, Stephen J. Cannell. **Dir** Robert Culp and many others. **Writers** Bartlett, Cannell, Culp, Greyhosky, Lupo and many others. **Created by** Cannell. **Starring** William Bogert (Carlyle), Don Cervantes (Paco Rodriquez), Culp (Bill Maxwell), Jesse D. Goins (Cyler Johnson), Faye Grant (Rhonda Harris), William Katt (Ralph Hinkley), Michael Paré (Tony Villacona), Connie Sellecca (Pam Davidson). 2hr pilot plus 43 1hr episodes. Colour.

Ralph Hinkley, a mild-mannered schoolteacher, encounters a UFO in the desert. The alien visitors give him a bright red suit, complete with tights and cape, and claim it will give him superpowers; they tell him he should use these for the good of mankind, then pair him with Bill Maxwell, a down-on-his-luck FBI agent. Unfortunately, Ralph loses the suit's instruction manual, so the putative SUPERHERO and his sidekick undergo several chapters of accidents.

When the show finished, the producers tried to launch a new version by passing the suit on to a woman, but *The Greatest American Heroine* failed to fly (though the pilot has been included in the *Greatest American Hero* syndication package). A proposed animated version for Saturday mornings fared equally poorly.

Largely because of a lawsuit from DC COMICS, *TGAH* failed to generate any merchandise or novelizations, but the theme song, written by Mike Post and Stephen Geyer and sung by Joey Scarbury, was a major hit in 1981. [BC]

GREAT GABBO, THE US movie (*1929*). ◊ MAGIC (*1978*).

GREAT MOTHER ◊ GODDESS.

GREEK AND LATIN CLASSICS The Classics are among the major sources of modern fantasy's imagery and ideas. Classical MYTHOLOGY lacks such a basic item in the genre's repertoire as the WIZARD, while its GODS have none of the tragic grandeur of their northern European counterparts. Nevertheless, besides the influence the Classics have had on

Western literature as a whole, they have an important role in the story of fantasy.

The Greeks gave us fantasy. They did not invent fantasy, which was probably consciously written in EGYPT centuries before Greek literature began. But in the Graeco-Roman world it was first identified consciously – indeed, *phantasia* was a technical term in the study of poetic techniques for representing these stories, and ancient literary criticism for the first time drew a clear distinction between the possible and the "mythical" or "fabulous". For centuries, Classical literature was dominated by a set of stories widely considered to have impossible aspects.

The literature of Greece and Rome extended beyond the mythological in both time and genre. Classical literature began with HOMER, whose identity remains a problem; but from this point forward the Greeks manifested a concern, unique in the ancient Mediterranean world, for preserving authors' true names, a concern probably related to the highly competitive Greek spirit.

With respect to fantasy, Classical literature may be divided into three eras, which differ from the standard periodization of Classical literature. The first era extends from Homer to 400BC, ending with the apparent victory of philosophers' scepticism about the MYTHS. Homer can be read either as a believer, a fantasist, or as both; our ignorance of his antecedents forbids any definite classification. The oldest surviving Greek works after him generally share uncertainties over the morality of fiction. Narratives from these centuries are few, but both Hesiod (c675BC) and Pindar (518-c445) speak directly of their desire to avoid poetic "lies". Hesiod wrote an epic *Theogony*, a rather livelier poem about farming containing several myths, and some other mythological works known only in fragments, all trans by Richmond Lattimore as *Hesiod* (**1959**). Bacchylides (c510-c450) and Pindar both wrote choral poetry, of which some victory odes with embedded mythological narratives survive. Pindar's (trans by Frank J. Nisetich as *Pindar's Victory Songs* **1980**) are difficult, allusive and grand, while Bacchylides' (trans Robert Fagles **1961**) are plainer and more narrative and seductive.

After Homer, the greatest and most famed Greek art of myth is the Athenian tragedy of the 5th century BC. 33 plays by three tragedians survive. Aeschylus (c525-456) and Sophocles (c496-c405) did not use many fantastical elements, beyond gods, ORACLES and the like. *Prometheus Bound* is a strikingly fantastical exception, whose attribution to Aeschylus is questioned. In the tragedies of EURIPIDES, newer attitudes toward myth begin to prevail; and as the second period began, in the 4th century BC, the myths were widely acknowledged as implausible in the writings of Plato (c428-c348), Thucydides (c455-c399) and Aristotle (384-322); these and later writers believed the old stories, but considered their fantastical elements either ALLEGORY or invention. Similarly, writers held widely varying views of the GODS. But they usually acknowledged that poets spoke of the gods' deeds as poets, not historians. Such authors' views were certainly not universal, in a largely illiterate population. All this is ably presented by D.C. Feeney in *The Gods in Epic* (**1991**), a work invaluable for any understanding of fantasy in older Classical literature.

The second period is otherwise hard to characterize as a unity, since its painfully limited remains consist mostly of either Greek literature from before about 200BC or Latin literature from after 50BC. This was the great age of the

epigram, the allusion and scholarship, in which epics lost primacy; it was also a time when realistic fictions assumed new prominence.

The most important shaper of the Hellenistic aesthetic was probably a man whose own work in the fantastic genres no longer survives. Callimachus of Cyrene (c310-c240) has left only six difficult and allusive hymns, some epigrams and some fragments; but the fragments attest to his founding role in several areas. His *Aitia* ["Origins"] began a tradition of poetry about origin myths; *Hecale* typified the epyllion, or miniature epic, which focused on a side-episode in mythology and then digressed at length from even that story; and he apparently wrote the first paradoxography, or collection of implausible information. What remains is all trans by Stanley Lombardo and Diane Rayor in *Callimachus* (**1988**). The surviving epyllia of other writers make a truly jewelled collection; the technique of digression became a powerful tool for mood-setting and foreshadowing.

Contemporaneously, Apollonius of Rhodes (?295-215 BC) wrote an *Argonautica*, the first surviving epic after Hesiod. Other than in scale, it exemplifies the Callimachean aesthetic, which has been read particularly for its vivid and moving portrait of the young Medea in love. The epic is trans Barbara Hughes Fowler in *Hellenistic Poetry* (**1990**), along with the later Sicilian poet Moschus's "Europa", a sophisticated and sensitive epyllion, sharply bringing home the impact of divine lust on an innocent girl.

Callimachus is known to have praised a quite different sort of epic, too: the *Phenomena* (trans G.R. Mair **1921**) of Aratus of Soloe (315-240 BC) in Turkey. This is a poetic handbook of astronomy, whose first part tells some of the myths of the constellations. Later the *Catasterismi* ["Constellations"] of Eratosthenes of Cyrene (c275-c195) compiled these myths more fully in prose. The oldest surviving paradoxography, the *Collection of Marvellous Stories* of Antigonus of Carystus (c240BC) also comes from this century. Another nonfiction writer important to the future of fantasy was Euhemerus, who considered the gods merely ancient human rulers (◊ EUHEMERISM).

In the first century BC, Latin literature attained the degree of perfection – as seen by its own practitioners – that Greek had reached in the days of Sophocles and Plato. Rome was by now ruler of the Mediterranean, and almost alone in the luxury of imaginative literature. Gaius Valerius Catullus (c85-c55) left, amid much love poetry of a new kind, a studiedly brilliant love-epyllion. Many of his contemporaries also wrote epyllia, but the form's day was ending. Publius Vergilius Maro (VIRGIL), Sextus Propertius (c48-c15) and Publius Ovidius Naso (OVID) all used short mythic narratives, but left no traditional epyllion. Propertius' tales are mostly rather mundane, but his lover's GHOST does appear in two of his best poems. Virgil is better known for his epic the *Aeneid* (**19**BC) and Ovid's *Metamorphoses* (cAD8) defies all true classification. *Metamorphoses* is also, as Feeney shows, explicitly written as fantasy fiction, the first of these works to attract this description.

Later, Ovid's supple Latin style profoundly affected that of Lucius Annaeus SENECA (c4BC-AD65), whose tragedies are perhaps the first body of supernatural HORROR; Seneca in turn influenced his nephew Lucan (Marcus Annaeus Lucanus; 39-65) in his epic poem *Bellum Civile* ["Civil War"] (vt *Pharsalia*). Lucan set out to disavow fantasy, and stuck reasonably closely to historical sources for his story. Lucan's work, politically impassioned, highly charged and sometimes hysterical, has had varying fortunes. Brian W. ALDISS credits it with helping inspire Mary SHELLEY's *Frankenstein* (**1818**), but after two centuries of neglect it is only now returning to favour. The *Thebaid* of Publius Papinius Statius (*c*50-*c*95), who learned much from Lucan and left a poem praising him, is if anything still darker, a powerful intentional fantasy with something to say, the first such in our literature. Other epics and epyllia, though of much less interest, survive from the first century.

The third period in the history of Classical fantasy had already begun, with Seneca's pupil, the emperor Nero (ruled 54-69). Nero is of interest because his suicide was widely disbelieved, and expectations of his return (a horrific variant on the ONCE AND FUTURE KING) contributed greatly to the idea of the ANTICHRIST; but he was also patron of literary trends which led, over the following two centuries, to the most distinctive works of fantasy in Classical literature.

Two aspects of the third period's literature are particularly relevant. First, much more of it survives; one can rarely speak with certainty of an innovation of this era, as opposed to an ongoing phenomenon whose earlier stages are lost. Allowing for such cautions, a second characteristic still stands out: the confusion of categories. The literary audience widened in both geography and class. New genres arose (prose dominated for the first time), and older ones miscegenated. Storytelling burst all generic bounds to enter nonfiction works as various as grammars and geographies. And, most germane, religious unity vanished; new faiths proliferated and, to the delight of satirists and the dismay of preachers, one faith's central beliefs routinely reappeared as another's fairytales. In this context, fantasy could be found everywhere.

No less an authority than the apostle Paul conceded as much in *Corinthians I* (*c*56) at the period's start; modern critics have also discussed the *New Testament* in relation to fantasy. Its books vary considerably in this regard, from *Mark*'s spare account of signs and wonders to *Matthew*'s mythic resonances and the rather fuller, more elegant tale of *Luke* and *Acts*. Perhaps the most influential *New Testament* book for fantasy has been *Revelation*, whose mysterious story and vivid images defined the Christian cosmology.

The *New Testament* books have been the most lastingly influential examples of a much wider literature of the time. In an outpouring of apocalypses and testaments, gospels, acts and theological treatises, such fundamental elements of the medieval world view as salvation, HEAVEN and HELL, DEMONS and ANGELS, martyrs and saints, miracle-workers and magicians neared the mainstream of Classical literature. A variety of religions contributed to the flow, but the non-Christian survivals are mostly dry treatises – APULEIUS's *Metamorphoses* and Philostratus' *Life of Apollonius of Tyana* (*c*220; trans F.C. Conybeare 2 vols **1912**) are the chief exceptions.

Most of the Christian works can be found in *The Old Testament Pseudepigrapha* and in *The Apocryphal New Testament* (**1993**) ed J.K. Elliott; of these, the most interesting include a portrait of Jesus as a troubled juvenile delinquent in the *Infancy Gospel of Thomas* and an early precursor to DANTE's *Inferno* (*c*1320) in the Apocalypse of Peter. But the texts best understood as fantasy are apocryphal Acts of Apostles from the 3rd century. *Acts of Peter* conveys a sense of grim fervour, *Acts of Xanthippe and Polyxena* (trans W.A. Craigie in *Ante-Nicene Christian Library*, additional volume IX ed Allan Menzies **1897**)

includes an entertaining adventure tale, and *Acts of John* careers exuberantly from MIRACLE to miracle, creating a world whose every death is reversible.

Most of these works have been described as romances, and Apuleius's *Metamorphoses* is always so treated. Other works from the main stream of the Romance genre range from the strictly mundane to the wildly fanciful. Antonius Diogenes' *The Wonders Beyond Thule*, another 1st-century work, though known to us only from fragments and a much later summary, is as fantastic a FANTASTIC VOYAGE as could be sought, with proto-VAMPIRES prominent among its characters. *The Ass*, attributed to LUCIAN, is a simpler tale of MAGIC.

The five Greek love and adventure stories considered central to the Romance genre include three with fantasy elements. Those in Achilles Tatius' *Leucippe and Clitophon* (*c*175) are peripheral. Magic doings are somewhat more essential to Heliodorus' *Ethiopiaca* (date uncertain), otherwise a moderately exciting account clearly ancestral to the modern novel. But the triumph of fantasy in Greek Romance is Longus's *Daphnis and Chloe* (*c*200). Written with art, elegance and languor, and set (unlike the others) entirely in or near its characters' home, this story of a couple's upbringing and awakening to LOVE is mostly realistic. Yet the protagonists are always under the care of local deities, a protection which explodes into a vivid scene featuring PAN on a rampage, and which subtly reshapes the rest of the story into a satisfying fantasy PASTORAL.

Romance spilled over into history as well as religion, producing one of the Middle Ages' favourite works, the Alexander romance (3rd century), falsely attributed to Callisthenes. While much of this is a straightforward story of royal conniving and conquest, it opens with a tale of magic and offers, in the middle, a magnificent though confusedly told fantasy of Alexander's attempt to reach the heavens which bears comparison with *Gilgamesh*. Nor would any discussion of Classical fantasy be complete without reference to Lucian's *True History*, an eloquent parody of the whole genre.

Perhaps the extreme example of the mixture of genres is the orator T. Flavius Cocceianus Dio (Dio Chrysostom) of Turkey (*c*45-*c*120), who turned from the respected pursuit of oratory to popular philosophy, and whose lively and flowing speeches (trans J.W. Cohoon *et al.* 5 vols **1922-51**) are in turn pervaded by stories long and short, including a "Trojan oration" which claimed Egyptian history as its source. Dio's fellow popularizer Mestrius Plutarchus of Greece (*c*46-*c*120) is generally known for his realistic *Lives*, but his *Moralia* (trans Frank Cole Babbitt *et al.* 16 vols **1927-69**) include numerous variously entertaining works which play inventively with MYTH, including a tale of the inhabitants of the Moon, a dialogue between Odysseus and a contented pig, and the poignant "On the Passing of Oracles", which includes the famous story "Great Pan is Dead".

Plutarch illustrates a decline in rationalism among the literary elite. Another often cited landmark in this process is the compendious *Natural History* (*c*76) by Gaius Plinius Secundus (23-79), whose encyclopedic contents indeed include some marvels. Of more consistent interest are two later works. *De Mirabilibus Auscultationibus* (*c*130; trans Launcelot D. Dowdell **1909**) is a sort of ancient *Ripley's Believe It or Not*, several hundred short and mostly untrue items. Caius Julius Solinus's *Collectanea Rerum Memorabilium* (3rd century; trans Arthur Golding **1587**), while a genuine geography, is thoroughly riddled with details of DRAGONS, manticores and other fantastic phenomena. [JB]

GREEN, ROGER (GILBERT) LANCELYN (1918-1987) UK scholar and writer, particularly of children's books. As a scholar he produced a number of studies useful to the fantasy enthusiast, in particular *Tellers of Tales* (**1946**, rev 1953), a series of essays on children's writers, *Andrew Lang: A Critical Biography* (**1946**) and *Andrew Lang* (**1962** chap), about a writer whose career had much in common with RLG's (◊ Andrew LANG), *The Story of Lewis Carroll* (**1949**) and *Lewis Carroll* (**1960** chap; rev in *Three Bodley Head Monographs* omni **1968**), *Fifty Years of Peter Pan* (**1954**) and *J.M. Barrie* (**1960** chap), *Authors & Places: A Literary Pilgrimage* (**1963**), and two biographies of C.S. LEWIS, whom RLG knew at Oxford: *C.S. Lewis* (**1963** chap; rev in *Three Bodley Head Monographs* omni **1969**) and the longer *C.S. Lewis: A Biography* (**1974**) with Walter Hooper (1931-   ). He also produced an early study of spaceflight in fiction, *Into Other Worlds* (**1957**). His work as an anthologist and editor focused mostly on children's books, though his scholarly approach makes these equally appealing to older readers. They include *Modern Fairy Stories* (anth **1955**), *Thirteen Uncanny Tales* (anth **1970**), *The Hamish Hamilton Book of Dragons* (anth **1970**), *A Book of Magicians* (anth **1973**; vt *A Cavalcade of Magicians* US), *Strange Adventures in Time* (anth **1974**) and two collections of works by Victorian writers, *Fairy Stories* (coll **1958**), containing tales by Mary Molesworth (1839-1921), and *The Complete Fairy Tales of George Macdonald* (coll **1977**).

Of his own fiction, *From the World's End* (**1948**) is the most mystical. Two children explore a remote empty house which is so steeped in the past that it evokes visions of the ancient world (◊ TIME ABYSS). *The Land Beyond the North* (**1958**) follows the Argonauts to Britain. He is best known for his retelling of legends for children, which are numerous; they include *King Arthur and his Knights of the Round Table* (**1953**), *The Adventures of Robin Hood* (**1956**), *Tales of the Greek Heroes* (**1958**), *Heroes of Greece and Troy* (**1960**) and *The Saga of Asgard* (**1960**; vt *Myths of the Norsemen* 1970). These works offer an excellent introduction to MYTHS and LEGENDS. [MA]

See also: CHILDREN'S FANTASY.

GREEN, ROLAND J(AMES) (1944-   ) US writer who has in recent years concentrated on sf, but whose career began with the first volume in the **Wandor** sequence of SWORD-AND-SORCERY adventures: *Wandor's Ride* (**1973**), *Wandor's Journey* (**1975**), *Wandor's Voyage* (**1979**) and *Wandor's Flight* (**1981**). There is an overall structure to the sequence which distinguishes Wandor's exploits from HEROIC FANTASY; he is an UGLY DUCKLING and a HIDDEN MONARCH whose various feats and QUESTS are shaped to demonstrate his growing fitness for his ultimate role. Unfortunately, the series was terminated before any fitting climax. While working on the **Wandor** books, RJG also wrote most of the **Richard Blade** sequence of SCIENCE-FANTASY adventures under the house name Jeffrey Lord. Special agent Blade travels by PORTAL to a variety of ALTERNATE REALITIES – some are other dimensions of an sf sort, some are TECHNOFANTASY venues, some are pure FANTASYLAND – where he commits various exploits by virtue of his martial and other skills. RJG's contributions to the sequence include all volumes from #9 on, with the exception of #30 *Dimension of Horror* (**1979**), by Ray Nelson. The first

volumes, all by Manning Lee Stokes, are #1 *The Bronze Axe* (**1969**), #2 *The Jade Warrior* (**1969**), #3 *Jewel of Tharn* (**1969**), #4 *Slave of Sarma* (**1970**), #5 *Liberator of Jedd* (**1971**), #6 *Monster of the Maze* (**1972**), #7 *Pearl of Patmos* (**1973**) and #8 *Undying World* (**1973**). RJG's contributions are #9 *Kingdom of Royth* (**1974**), #10 *Ice Dragon* (**1974**), #11 *Dimension of Dreams* (**1974**), #12 *King of Zunga* (**1975**), #13 *The Golden Steed* (**1975**), #14 *The Temples of Ayocan* (**1975**), #15 *The Towers of Melnon* (**1975**), #16 *The Crystal Seas* (**1975**), #17 *The Mountains of Brega* (**1976**), #18 *Warlords of Gaikon* (**1976**), #19 *Looters of Tharn* (**1976**), #20 *Guardians of the Coral Throne* (**1976**), #21 *Champion of the Gods* (**1976**), #22 *The Forests of Gleor* (**1977**), #23 *Empire of Blood* (**1977**), #24 *The Dragons of Englor* (**1977**), #25 *The Torian Pearls* (**1977**), #26 *City of the Living Dead* (**1978**), #27 *Master of the Hashomi* (**1978**), #28 *Wizard of Rentoro* (**1978**), #29 *Treasure of the Stars* (**1978**), #31 *Gladiators of Hapanu* (**1979**), #32 *Pirates of Gohar* (**1979**), #33 *Killer Plants of Binaark* (**1980**), #34 *The Ruins of Kaldac* (**1981**), #35 *The Lords of the Crimson River* (**1981**), #36 *Return to Kaldac* (**1983**) and #37 *Warriors of Latan* (**1984**).

Other fantasy novels by RJG include *Jamie the Red* (**1984**) with Gordon R. DICKSON, set in a LAND-OF-FABLE medieval Europe; and *Throne of Sherran: The Book of Kantela* (**1985**) with Frieda Murray (RJG's wife), in which young Queen Kantela learns how to keep her throne. He also contributed several novels to the **Conan** enterprise (◊ Robert E. HOWARD). His recent work – with the exception of a tie, *Dragonlance: Knights of the Crown* * (**1995**) – has been sf. [JC]

GREEN, SHARON (1942-   ) US writer whose early work comprised two PLANETARY-ROMANCE series: the **Terrilian** sequence, featuring the sadomasochistic plight of a bound, stripped, beaten, ultimately obedient woman on another planet, and the **Amazon Warrior** sequence, whose protagonist fights back harder, but has the same things done to her. The obvious parallel with the work of John NORMAN was underlined by the publishing company, which released the works of both authors. Of more fantasy interest is the **Far Side of Forever** sequence – *The Far Side of Forever* (**1987**) and *Hellhound Magic* (**1989**) – in which a flirtatious heroine and a SEVEN-SAMURAI group, one of them a SHAPESHIFTER, use PORTALS to search various worlds for their goal. In *Dawn Song* (**1990**), the King of the Sun and the Daughter of the Moon collaborate in a QUEST. *Silver Princess, Golden Knight* (**1993**) and its sequel, *Wind Whispers, Shadow Shouts* (**1995**), carry a female shapeshifter via portals into various worlds in search of a man who will master her. [JC]

Other works: *Lady Blade, Lord Fighter* (**1987**); *The Hidden Realms* (**1993**); *Enchanting* (**1994**).

GREEN, SIMON R. (1955-   ) UK writer. SG first came to attention with the **Hawk & Fisher** series of DETECTIVE/THRILLER FANTASIES – *Hawk & Fisher* (**1990** US; vt *No Haven for the Guilty* 1990 UK), *Winner Takes All* (**1991** US; vt *Devil Take the Hindmost* 1991 UK), *The God Killer* (**1991** US), *Wolf in the Fold* (**1991** US; vt *Vengeance for a Lonely Man* 1992 UK), *Guards Against Dishonor* (**1991** US) and *The Bones of Haven* (**1992** US; vt *Two Kings in Haven* 1992 UK). Set in an interestingly depraved CITY, Haven, these involve a DUO of lovers in which the woman is the tougher and more ruthless.

Blue Moon Rising (**1991**) deals expansively with the QUEST of a young prince for the means to defeat the DEMON hordes that threaten his father's kingdom. The lightheartedness of some of the tone – particularly of the banter between him, a princess and a sarcastic UNICORN – does not very well fit the extreme emotional bleakness of a war in which his father's castle rapidly becomes a LAST REDOUBT and in which betrayal is the order of the day. This is a prevailing fault of SG: too often he flips between attemptedly smart-aleck schoolboy humour and lasciviously described, quite astonishingly sadistic personal violence, as if inflicting and witnessing torturous death were somehow jolly good fun. *Blood and Honour* (**1992**; vt *Blood and Honor* 1993 US) is more unified in tone; its BRAVE LITTLE TAILOR hero is pre-vailed upon to impersonate a lost prince and becomes worthy of rule in the process. *Down Among the Dead Men* (1993) is set in an abandoned castle where a group of soldiers is picked off in circumstances that combine detection and HORROR tropes.

Shadows Fall (**1994**) is SG's most ambitious fantasy to date. Set in a POLDER to which legends, including some TOONS, go to retire and ultimately fade away, it deals with a SERIAL KILLER, with an incursion of mundane religious cru-saders and, ultimately, with the manipulations of the DEVIL. It combines, with mixed success, tropes from INSTAURA-TION FANTASY and horror.

SG is often an interesting writer, but his overt self-assurance is not always justified. [RK]

Other works: *Mistworld* (1992), *Ghostworld* (1993 US) and *Hellworld* (**1993** US) all space opera; their sequel *Deathstalker* (**1995**), first of a tetralogy using the same back-ground and some of the same characters, followed by «*Deathstalker Rebellion*» (1996); *Robin Hood, Prince of Thieves* * (**1991**), movie tie; *Dark Mirror, Dark Dreams* (**1994**), fantasy.

GREEN, TERENCE M(ICHAEL) (1947-   ) Canadian writer whose novels have been sf or associational but many of whose stories – including his first, "Of Children in the Foliage" for *Aurora: New Canadian Writing 1979* (anth **1979**) ed Morris Wolfe, and "Ashland, Kentucky" (1985 *IASFM*) – are fantasy. Most were collected in *The Woman who is the Midnight Wind* (coll **1987**). [JC]

GREENAWAY, KATE Working name of UK illustrator Catherine Greenaway (1846-1901). Her quaint costume designs and pictures of mob-capped little girls and frilly-trousered short-jacketed boys so captivated the public that it was claimed KG "dressed the children of two continents". After illustrating many books by other writers, she achieved international success with her own collection of verses and drawings, *Under the Window* (coll **1878**). John RUSKIN gave her much encouragement. She was also acclaimed for her paintings of fruit and flowers, seen to best effect in her *Language of Flowers* (**1884**) and *Marigold Garden* (**1885**). The very popular *Kate Greenaway Almanacks* appeared every year 1883-95, concluding with a final *Almanack and Diary for 1897* (**1896**). One of KG's best-known works was her edi-tion of Robert BROWNING's *The Pied Piper of Hamelin* (**1888**). The (UK) Library Association Kate Greenaway Award, presented annually in recognition of distinguished works of children's book illustration, is named in her honour. [RD]

Other works illustrated (selective): *Puck and Blossom* (**1874**) by Rosa Mulholland; *The Fairy Spinner* (**1875**) by Miranda Hill; *Seven Birthdays; A Fairy Chronicle* (**1876**) by Kathleen Knox; *Mother Goose or The Old Nursery Rhymes* (**1881**) by Bret Harte; *The*

Royal Progress of King Pepito (**1889**) by Beatrice Cresswell.
Further reading: *Kate Greenaway* (**1905**) by M.H. Spielmann and G.S. Layard; *The Kate Greenaway Book* (**1976**) by Bryan Holme.

GREENAWAY, PETER (1942-   ) UK movie director and screenwriter, whose cold eye can make his movies almost impossible to watch yet who works at the heart of fantasy. After various respected shorts he came to interna-tional attention with *The* DRAUGHTSMAN'S CONTRACT (*1982* tvm), which one can regard as either fantasy or a mimicry of fantasy. *A Zed and Two Noughts* (*1985*) falls into roughly the same category: husbands whose wives die in a car crash dis-cover new relationships as they jostle over the survivor of the crash, an amputee. *The Belly of an Architect* (*1987*) is a sort of psychofantasy, and the same could be said of *Drowning by Numbers* (*1988*), about a coroner obsessed with three gen-erations of women who drown their husbands, and *The Cook, The Thief, His Wife and Her Lover* (*1989*), which depicts an almost unbearably brutal gangster (played unfor-gettably by Michael Gambon) and the murder of the stranger his wife picks up as a lover, were it not for the fact that PG deliberately employs dissociational techniques to remind us all the while that this is fantasy, rather than real-ity. *A TV Dante: Cantos I-VIII* (*1989* tvm) showed PG, in this adaptation of DANTE, experimenting with some boldness with video techniques; this was carried to an extreme in PROSPERO'S BOOKS (*1991*), probably PG's most successful yet most "difficult" movie to date, reinterpreting the GODGAME played by PROSPERO in William SHAKESPEARE's *The Tempest* (performed *c*1611; *1623*). [JG]

GREENBERG, MARTIN H(ARRY) (1941-   ) US aca-demic and editor, immensely prolific in the latter capacity for over two decades; most of this work has been in the field of sf. MHG's contributions to fantasy, though he would rank as prolific by most standards, are comparatively uncommon. ANTHOLOGIES of interest include: *Hollywood Unreel* (anth **1982**); the **Barbarians** sequence, *Barbarians* (anth **1985**) and *Barbarians II* (anth **1988**) with Robert ADAMS and Charles G. Waugh (1943-   ); *Fantastic Stories: Tales of the Weird and Wondrous* (anth **1987**) with Patrick Lucien Price; *Vamps* (anth **1987**) with Waugh; *Cinemonsters* (anth **1987**), comprised of stories upon which various MONSTER MOVIES were based; *House Shudders* (anth **1987**) with Waugh; *Red Jack* (anth **1988**) (◊ JACK THE RIP-PER); *Hunger for Horror* (anth **1988**) with Pamela Crippen Adams (1961-   ) and Robert Adams; *Werewolves* (anth **1988**) with Debra DOYLE and Jane YOLEN; *Phantoms* (anth **1989**) with Rosalind M. Greenberg; *Mummy Stories* (anth **1990**); *Phantom Regiments* (anth **1990**) with Pamela Crippen Adams and Robert Adams; *The Rivals of Weird Tales* (anth **1990**) with Stefan R. DZIEMIANOWIC (*whom see for further* HORROR-*based anthologies with this collaborator*) and Robert WEINBERG ; *Cults of Horror* (anth **1990**) with Waugh; *Devil Worshipers* (anth **1990**) with Waugh; *Civil War Ghosts* (anth **1991**) with Waugh and Frank D. McSherry; *Back from the Dead* (anth **1991**) with Waugh; *The Fantastic Adventures of Robin Hood* (anth **1991**) (◊ ROBIN HOOD); *Fantastic Chicago* (anth **1991**) (◊ URBAN FANTASY); *Nightmares on Elm Street: Freddy Krueger's Seven Sweetest Dreams* (anth **1991**); *Horse Fantastic* (anth **1991**) with Rosalind M. Greenberg; *New Stories from the Twilight Zone* (anth **1991**) (◊ TWILIGHT ZONE); *Smart Dragons, Foolish Elves* (anth **1991**) with Alan Dean FOSTER; *A Taste for Blood* (anth **1992**) with Dziemianowicz and Weinberg; *Dracula:*

Prince of Darkness (anth **1992**) (◊ DRACULA); *After the King: Stories in Honor of J.R.R. Tolkien* (anth **1992**), a series of homages honouring the memory, but not the style or very often the themes, of J.R.R. TOLKIEN; *Tales of Riverworld* * (anth **1992**) and *Quest to Riverworld* * (anth **1993**), with Philip José FARMER, Edward Kramer and Richard Gilliam (1950-   ); *Frankenstein: The Monster Wakes* (anth **1993**) (◊ FRANKENSTEIN); *Betcha Can't Read Just One* (anth **1993**) with Foster; *A Newbery Halloween* (anth **1993**) with Waugh; *Confederacy of the Dead* (anth **1993**) with Gilliam and Edward E. Kramer; *Alien Pregnant by Elvis* (anth **1994**) with Esther FRIESNER; *Peter S. Beagle's The Immortal Unicorn* (anth **1995**) with Peter S. BEAGLE and Janet Berliner; *Sisters in Fantasy* (anth **1995**) with Susan SHWARTZ; *Heaven Sent: 18 Glorious Tales of the Angels* (anth **1995**) with Peter CROWTHER; *Tombs* (anth **1995**) with Crowther Edward E. Kramer, MHG anon.

In the 1990s, an increasing proportion of MHG's titles appeared with his name missing from the title page; usually, but not always, he would be listed on the copyright page. The reason for this modesty may have been a market response to readers' sense that MHG had saturated the ANTHOLOGY market. The listing above should be treated as a selection only. MHG also edited numerous fantasy anthologies with Isaac ASIMOV.   [JC]

GREEN BOOK, THE ◊ CREEPS LIBRARY.

GREEN CHILD The legend of the Green CHILDREN, recounted by the 12th-century monk William of Newburgh, originated in Woolpit, Suffolk. Two strange children, brother and sister, emerged at harvest-time from the ancient pits surrounding the village. Their skins were entirely green, they spoke no known language and the only food they would eat was beans. Gradually they were weaned onto bread; their skins whitened and they began to learn English. The boy died but his sister survived and married, living into old age. All she would say about their origins was that they were inhabitants of the Land of St Martin. In the ALLEGORY *The Green Child* (**1935**) by Herbert Read (1863-1968) the hero finds a fragile, green-skinned woman whom he rescues from her sadistic husband. She leads him back to her OTHERWORLD via the millstream from which she emerged. Sylvia Townsend WARNER also uses the motif in "Elphenor and Weasel" (in *Kingdoms of Elfin* coll **1977**). *The Green Gene* (**1973**) by Peter DICKINSON uses the notion to PARODY racism; *The Boy with Green Hair* (**1948**) uses a part of the idea with similar intent, and also to demonstrate the stupidity of war.   [JR]

GREEN KNIGHT Sir Bertilak, who challenges GAWAIN to the beheading match in *Sir Gawain and the Green Knight* (14th century).   [MA]

See also: GREEN MAN.

GREENLAND, COLIN (1954-   ) UK critic and writer whose PhD thesis in sf at Oxford University became in revised form his first book, *The Entropy Exhibition: Michael Moorcock and the UK "New Wave"* (**1983**). Michael MOORCOCK is given considerable attention in the text, which focuses also on the work of J.G. BALLARD and Brian W. ALDISS. Further critical work includes *Storm Warnings: Science Fiction Confronts the Future* (anth **1987** US) with Eric S. Rabkin and George E. Slusser, and a second take on Moorcock in the shape of a book-length interview, *Death is No Obstacle* (**1992**).

CG began to publish fiction of genre interest with "Miss Otis Regrets" for *Fiction Magazine* in 1982, and soon published his first three fantasy novels – *Daybreak on a Different Mountain* (**1984**), *The Hour of the Thin Ox* (**1987**) and *Other Voices* (**1988**) – which make up a very loose series, mainly through being set in the same austerely depicted FANTASYLAND. Because so much of his work borrows moods and idioms from earlier work, interrogating that earlier work in the process, almost all of CG's earlier fiction can be thought of as REVISIONIST FANTASY, even where models may be hard to pin down. Fittingly, the first volume of the loose sequence not only is in this sense a PARODY of fantasy but actually describes what might be called a parody of QUEST: two aristocrats are sent to find the prophet of the GOD Gomath, both of them needing to deny that either is in fact that prophet, to no avail. One turns out to be the prophet, and the other, having developed TALENTS, the god.

The second and third tales also involve impostures, inconclusive gestures, emotional states that peter out rather than – as in the models being examined – prefiguring and justifying the TRANSFORMATION of the world into EUCATASTROPHE. CG then moved to sf, with the **Tabitha Jute** sf sequence – *Take Back Plenty* (**1990**), which won the 1991 ARTHUR C. CLARKE AWARD, *In the Garden: The Secret Origin of the Zodiac Twins* (**1991** chap) and *Seasons of Plenty* (**1995**) – though in the middle of composing this ambitious (and incomplete) narrative he produced in *Harm's Way* (**1993**) his most successful and exuberant single novel. As with so much sophisticated work of the 1990s, it is not easy to make useful generic distinctions: it could be treated as an sf exercise in alternate cosmology or as a fantasy excursion into an ALTERNATE-WORLD 19th-century England making use of STEAMPUNK and GASLIGHT ROMANCE conventions in the depiction of the consequences of living in a Universe filled with aether, so that sailing ships can travel from planet to planet. The style intermittently pastiches Charles DICKENS; the plot takes the daughter of a murdered London whore on a hegira to confront (and eventually to kill) her father, the head of the guild of aether pilots. The whole enterprise of *Harm's Way* is irradiated by nostalgia of a highly conscious sort. CG is a novelist whose touchstones – entropy and nostalgia – are belated. But he uses these touchstones, with increasing freshness, as a language. [JC]

Other works: *Magnetic Storm: The Work of Roger and Martyn Dean* (**1984**); *Interzone: The First Anthology* (anth **1985**) ed with John CLUTE and David PRINGLE; *The Freelance Writer's Handbook* (**1986**) with Paul Kerton.

GREEN MAN A FOLKLORE figure symbolizing FERTILITY, often carved in old churches as a FOLIATE HEAD. His mythic provenance is suggested by *Sir Gawain and the Green Knight* (14th century), wherein a literally green GIANT stalks into ARTHUR's court challenging all comers to behead him; when GAWAIN does so, the GM replaces his severed head and requires Gawain to accept a return blow a year later. Echoes of the severed, regrowing head appear in the folksong "John Barleycorn" and the maize legend in *The Song of Hiawatha* (**1858**) by Henry Wadsworth Longfellow (1807-1882). As with the benevolent "Person with Horns" in Elizabeth GOUDGE's PASTORAL *The Herb of Grace* (**1948**), the GM seems sometimes conflated with the Horned God of animal fertility.

In J.R.R. TOLKIEN's *The Lord of the Rings* (**1954-5**), the Ents provide a different vision of personified vegetation, partly based on Arthur RACKHAM's drawings of gnarled, twisted TREES. The literal GM of Kingsley AMIS's *The Green Man* (**1969**), is a woodland ELEMENTAL of animated branches (◊ ANIMATE/INANIMATE).

Julius Caesar's black-propaganda report that British druids set fire to wicker giants containing live HUMAN SACRIFICES inspired *The* WICKER MAN (*1973*). Celtic LEGENDS do not support this tale, but abound with HEROES who go mad and run INTO THE WOODS to live as beasts: examples are Suibhne Geilt (featured as Sweeney in Flann O'BRIEN's *At Swim-Two-Birds* [**1939**]), Bladud the Birdman, LANCELOT, MERLIN and Tristan. The GM, Wild Man and JACK in the Green have boiled together in the CAULDRON OF STORY, as implied in Robert HOLDSTOCK's **Ryhope Wood** sequence.

[JH/DRL]

GREENWOOD, JAMES (1832-1929) UK writer and journalist; during his long career he was best-known as a crusading reporter investigating UK low-life in the footsteps of Henry Mayhew (1812-1887), producing several important studies on this subject, notably *The Seven Curses of London* (**1869**) and *The Wilds of London* (**1876**). He had a parallel career as author of several books for "young readers" which mainly concentrated on savage cruelty, often involving animals and Native Americans, including *Curiosities of Savage Life* (**1867**) and *The Bear King* (**1868**), accompanied by graphically bloodthirsty illustrations; this last title was an acknowledged influence on Rudyard KIPLING. Among JG's most fantastic works is *The Adventures of Seven Four-Footed Foresters, Narrated by Themselves* (**1865**), in which a character can converse fluently with savage animals and finds himself turning into a WOLF. [RD]

Other works: *Reminiscences of a Raven* (**1865**); *Silas the Conjurer: His Travels and Perils* (**1866**); *The Adventures of R. Davidger* (**1869**); *A Queer Showman, and Other Stories* (**1885**); *Jaleberd's Bumps: A Phrenological Experiment* (**1891**).

GREGORIAN, JOYCE BALLOU (1946-1991) US writer whose father was an Armenian immigré from Iran. Her **Tredana Trilogy** – *The Broken Citadel* (**1975**), *Castledown* (**1977**) and *The Great Wheel* (**1987**) – is set in a LAND OF FABLE that bears some resemblance to Persia, and is attainable by young Sib from Massachusetts when she stumbles through one PORTAL or another as each volume begins. Each novel reflects a new stage in Sib's maturation, and on re-entering the Tredana Empire afresh she finds it has been evolving rapidly. In a sometimes overcomplex storyline – shot through with passages of Tredanian folklore, verse and history – Sib interacts meaningfully with Leron, young heir to the throne, and Semiramis, who is perhaps an AVATAR of the GODDESS; one of the themes of Tredanaian MYTH is the perhaps impermanent defeat of the Great Mother (◊ GODDESS) and victory of the First Father. In *Castledown* – the title is also the name of a game, and Player gods are playing this game with Tredana as board – Sib's growing up (◊ RITE OF PASSAGE) is reflected in hints that the land is developing technology; the final volume confirms a sense of irrevocable change. [JC]

GREMLINS Mischievous little creatures, usually invisible, who interfere with the functioning of airplanes. The myth seems first to have emerged during WORLD WAR II among the pilots and mechanics of the Royal Air Force. Roald DAHL, himself a fighter pilot, wrote a short fantasy about them, *The Gremlins* (**1943** chap), but otherwise they have appeared little in fantasy – the GREMLINS movies use the name but are unrelated. One episode of *The* TWILIGHT ZONE, "Nightmare at 20,000 Feet", scripted by Richard MATHESON, was based on the idea. [JH/JG]

GREMLINS There have been two movies in the **Gremlins** series.

1. *Gremlins* US movie (*1984*). Warner/Amblin. **Pr** Michael Finnell. **Exec pr** Kathleen Kennedy, Frank Marshall, Steven SPIELBERG. **Dir** Joe Dante. **Spfx sv** Bob MacDonald Sr. **Gremlins created by** Chris Walas. **Screenplay** Chris Columbus. **Novelization** *Gremlins* * (**1984**) by George Gipe. **Starring** Hoyt Axton (Rand Peltzer), Phoebe Cates (Kate Beringer), Corey Feldman (Pete), Zach Galligan (Billy Peltzer), Polly Holliday (Ruby Deagle), Keye Luke (Chinese Shopkeeper), Frances Lee McCain (Lynn Peltzer), Glynn Turman (Roy Hanson). 106 mins. Colour.

Crank inventor Rand Peltzer buys in a Chinese junkstore a cute (and highly intelligent) little animal, a Mogwai, as a Christmas present for son Billy; he is instructed not to expose the creature to bright light, give it water or feed it after midnight. Billy's schoolboy friend Pete spills water on the creature (christened Gizmo) and it buds, producing five offspring just like itself but nasty. By accident, Billy feeds them after midnight, and they pupate, metamorphosing into much larger scaly versions with voracious appetites and destructive humours. Further accidents lead to hundreds of monsters rampaging all over town. Billy and girlfriend Kate destroy the mob; as they relax afterwards, the old shopkeeper comes to reclaim Gizmo: US culture, he declaims, is too immature to be permitted Mogwais. In fact, the monstrous but childlike Mogwais are collectively an unsubtle and quite savage SATIRE of one facet of that culture: intelligent enough to meddle with things they do not understand, they thereby cause immense destruction; their greed must be satisfied instantly without thought for consequences.

G is laden with RECURSIVE references. Among the several films briefly seen on tv are IT'S A WONDERFUL LIFE (*1946*) and *Invasion of the Body Snatchers* (*1956*), the latter in particular commenting on *G*'s plot, and there are visual allusions also to SPIELBERG's POLTERGEIST (*1982*) and *E.T. – The Extra-Terrestrial* (*1982*); local skinflint Ruby Deagle first appears like Miss Gulch in *The* WIZARD OF OZ (*1939*) and threatens Billy's dog just as Miss Gulch threatened Toto, and her death, later on, is depicted much like the Wicked Witch of the West's; Deagle has a pre-Christmas exchange with a debtor that seems drawn almost word for word from any of the versions of *A* CHRISTMAS CAROL/*Scrooge*; in their pre-metamorphosis state the Mogwais owe considerable debts visually to E.T. and conceptually to the furry, cute but threateningly fecund creatures in the **Star Trek** episode "The Trouble with Tribbles" (1967) by David Gerrold (the Gremlins were themselves "homaged" in the two movie series beginning with GHOULIES [**1985**] and *Critters* [**1986**]); and, of course, their name is taken from the WORLD WAR II myth of the GREMLIN. Their monstrous form seems consciously derived from the Hopkinsville Goblins, UFO creatures reported as having terrorized a Kentucky farmhouse in 1955; indeed, the rampaging Mogwais trash a farmhouse using an appropriated snowplough of the Kentucky Harvester marque. These and other allusions are perhaps of greater interest than *G*'s story, which leaves loose ends dangling and seems designed merely as an excuse for all concerned to have pyrotechnic fun. [JG]

2. *Gremlins 2 – The New Batch* US movie (*1990*). Warner/Amblin. **Pr** Michael Finnell. **Exec pr** Kathleen Kennedy, Frank Marshall, Steven SPIELBERG. **Dir** Joe Dante. **Spfx** Ken Pepiot. **Vfx** Dennis Michelson. **Gremlin and Mogwai fx** Rick Baker. **Title anim written/dir** Chuck JONES. **Screenplay** Charlie Haas. **Novelization** *Gremlins 2: The New Batch* * (**1990**) by David F. Bischoff. **Starring**

Phoebe Cates (Kate), Zach Galligan (Billy), John Glover (Daniel Clamp), Christopher Lee (Dr Catheter), Dick Miller (Murray Futterman), Haviland Morris (Marla Bloodstone), Keye Luke (Mr Wing), Robert Picardo (Forster), Robert Prosky (Grandpa Fred), Gedde Watanabe (Katsuji). **Voice actors** Jeff Bergman (Bugs Bunny/Daffy Duck), Howie Mandel (Gizmo), Tony Randall ("Brain"). 105 mins. Colour.

A direct sequel to **1**. Millionaire Daniel Clamp is buying up all New York's Chinatown for development. Gizmo, the original little Mogwai from *Gremlins*, is seized by staffers from Splice of Life Inc, a genetic-research organization based in Clamp's premier block. Working for Clamp are **1**'s Billy and Kate. Billy saves Gizmo from the knife of Splice of Life's soulless boss Dr Catheter, but inevitably the Mogwai is accidentally wetted and the building becomes infested with Gremlins – multiplying to such numbers, and mutating thanks to Splice of Life's stocks, that they will threaten New York as a whole if freed. Our heroes destroy the threat.

More is not necessarily better, and this movie collapses under the weight of its own in-jokes. There are clips from and allusions to PHANTOM OF THE OPERA, the **Rambo** cycle (starting with *First Blood* [*1982*]), E.T. – *The Extra-Terrestrial* (*1982*), *The* FLY (*1986*), BATMAN (*1989*) and others, while Prosky, as late-night horror presenter Grandpa Fred, duplicates his role from tv's *The* MUNSTERS; the animated prologue and epilogue feature BUGS BUNNY and Daffy Duck. At one point the Gremlins even interrupt the course of the movie itself, invading other movies and tv shows before being banished (by John Wayne) back into this one. Overall the constant deployment of excess in the pursuit of HUMOUR voids *G2* of all power to affect. [JG]

GRENDEL ◊ BEOWULF.
GREYSTOKE: THE LEGEND OF TARZAN, LORD OF THE APES UK movie (*1984*). ◊ TARZAN MOVIES.
GRIFFITHS, ALAN (? -?) UK author of three satirical fantasies. *Strange News from Heaven* (**1934**) offers a farcical account of a rebellion in a HEAVEN in which only those who die young retain their youth. *Spirits Under Proof* (**1935**; vt *Authors in Paradise* US) tracks the career of a medium who passes off the posthumous literary endeavours of the late and great as his own. *The Passionate Astrologer* (**1936**) describes what follows when a lovestruck ANGEL allows pages from the book of destiny to fall into earthly hands. [BS]
GRILE, DOD Pseudonym of Ambrose BIERCE.
GRIMM BROTHERS Jacob Ludwig Carl Grimm (1785-1863) and Wilhelm Carl Grimm (1786-1859), German philologists, folklorists and writers, generally and correctly treated as a team, though Jacob concentrated on linguistic studies (he devised the principle of consonantal shifts in pronunciation known as Grimm's Law) and Wilhelm was primarily a literary scholar. They were both highly productive in various associated fields, but remain best-known for the various versions of *Die Kinder-und Häusmarchen* ["Children's and Household Tales"] (coll **1812-15** 3 vols; trans Edgar Taylor as *German Popular Stories* **1823-6** 2 vols UK) illustrated by George CRUIKSHANK, a gathering of tales from FOLKLORE sources which has been published in many editions, usually as *Grimm's Fairy Tales*. *The Complete Fairy Tales of the Brothers Grimm* (trans coll Jack ZIPES **1987** US) provides a useful analysis of the seven German editions; the 241 tales Zipes assembles include 32 dropped by the Grimms from various of these editions.

As young men, both brothers were affected by German ROMANTICISM; their scholarly endeavours were shaped by a desire to provide an intellectual justification for Romantic assumptions about the unique and intertwined relationship between the "original" German tongue and the FOLKTALES whose origins were – they felt – coeval with the origins of German culture. Jacob's first book, *Über den altedutschen Meistergesang* ["On the Ancient German Master-song"] (**1811**) began to argue the case; as did the *Fairytales* (◊ FAIRYTALES), which they had begun to assemble several years before the first edition appeared, their associated collection of LEGENDS, *Deutsche Sagen* (coll **1816-18**; trans as *The German Legends of the Brothers Grimm* **1981** 2 vols US) and Jacob's *Deutsche Mythologie* ["German Mythology"] (**1835**), in which archaic pre-medieval Germany was seen as a GOLDEN AGE.

Both brothers argued that FOLKLORE should be recorded and presented in print in a form as close as possible to the original mode, but in practice modified those originals in varying ways, a habit in which Wilhelm indulged increasingly as the years passed. In later editions of the *Fairytales* – which were primarily his responsibility – he introduced more and more "literary" values into the stories, and succumbed to the bowdlerizing instinct. But the *Fairytales* remain, as a whole, a huge and progressive achievement.

The GB star in George Pal's Cinerama biopic, *The Wonderful World of the Brothers Grimm* (*1962*). [JC]
GRIMOIRES ◊ BOOKS.
GRIM REAPER ◊ DEATH.
GRIMSHAW, BEATRICE (ETHEL) (1870-1953) Irish writer and journalist, resident in Papua from 1907 and in Australia from 1936, popular in her day for novels and stories set in the South Seas. BG allied herself to the feminist movement of the 1890s, a mood reflected in her first novel, *Broken Away* (**1897**). She was the first white woman to ascend the Sepik river in New Guinea; her photographic studies of the area were published as *In the Strange South Seas* (**1907**). Her **South-Sea** series started with *Vaiti of the Islands* (1906-7 *Pearson's*; **1907**), which told of the entrepreneurial activities of the daughter of a white sea captain and a Polynesian princess. Her most popular book was *Conn of the Coral Seas* (**1922**), a love story set among cannibals. Only two novels come close to SUPERNATURAL FICTION: *The Sorcerer's Stone* (**1914**), featuring local MAGIC, and *The Terrible Island* (**1920**), a RATIONALIZED FANTASY about an accursed treasure. *Victorian Family Robinson* (**1934**) contrasts the lives of a Victorian family shipwrecked on a desert island with those of descendants of survivors shipwrecked a century earlier. The supernatural surfaces more frequently in her short stories. Her most rewarding collections are *The Valley of Never-Come-Back* (coll **1923**), *The Beach of Terror* (coll **1931**), *The Long Beaches* (coll **1933**) and *Pieces of Gold* (coll **1935**). BG tells of her own experiences in *Isles of Adventure* (**1930**). [MA]
Other works: *When the Red Gods Call* (**1911**; cut 1921); *Nobody's Island* (**1917**); *The Coral Palace and Other Tales* (coll **1920**); *My Lady Far-Away* (**1929**); *The Mystery of Tumbling Reef* (**1932**).
GRIMSHAW, NIGEL (GILROY) (1925-) UK writer. His YA **Wildkeepers** sequence – *Bluntstone and the Wildkeepers* (**1974**) and *The Wildkeepers' Guest* (**1976**) – confronts the WAINSCOT society of child-sized Wildkeepers with an incursion of modern reality in the form of Bluntstone, a building contractor whose machines threaten to destroy their domain. They evoke the powers of MAGIC,

but – unusually – Bluntstone is by no means a VILLAIN, and the two sides come to terms. [JC]

GRIMWOOD, KEN(NETH) (?1945-   ) US radio journalist and writer who wrote three nongenre novels before publishing *Replay* (**1986**), a highly successful TIME FANTASY which won the WORLD FANTASY AWARD. In 1988 a 43-year-old radio journalist dies suddenly, reawakening in 1963 in his 18-year-old body with all his memories intact. He relives his life, but this time not making his earlier mistakes, and becomes immensely rich. So far, this is classic wish-fulfilment. But in 1988, at the age of 43, he dies suddenly again, and reawakens again, although somewhat later than before. This time he knows when he will die. The plot gradually and intricately thickens, with his discovery of others forced to replay their lives; but the heart of the tale is the swift sophistication of its unpacking of the great dream. Particularly interesting is the fact that each of the "replays" effectively generates an ALTERNATE REALITY or ALTERNATE WORLD, yet these are sequential rather than, as is almost always the case in alternate-world fictions, "parallel". KG's second novel of genre interest, *Into the Deep* (**1995**), is metaphysical sf, not dissimilar in some of its implications to Greg BEAR's *Blood Music* (**1985**). [JC]

GRIPE, MARIA KRISTINA (1923-   ). Swedish author. Her first fantasy, *Glasblåsarnas barn* (**1964**; trans Sheila La Farge as *The Glassblower's Children* 1973 US), was, like *I klockornas tid* (**1965**; trans La Farge as *In the Time of the Bells* **1976** US) and *Landet utanför* (**1967**; trans La Farge as *The Land Beyond* 1974 US), a medieval fantasy, in part inspired by MKG's fascination with the Edda. In particular, *Landet utanför* is a complex, innovative novel, telling the same story twice in different ways and with changed emphasis.

MKG returned to fantasy of a new, more enigmatic kind with *Tordyveln flyger i skymningen* (**1978**), based on a radio play, *Agnes Cecilia - en sällsam historia* (**1981**) and the so-called **Shadows** series: *Skuggan över stenbänken* (**1982**), *Och de vita skuggorna i skogen* (**1984**), *Skuggornas barn* (**1986**) and *Skugg-gömman* (**1988**). In these novels, ethereal and minimalist suggestions of mysticism and a restatement of MKG's fundamental theme of the basic importance of imagination to life and sanity are combined in a highly literate and intellectually challenging manner. The **Shadows** tetralogy is a towering achievement; beginning in 1911, it tells of young Berta and the new maid, Caroline, whose identity remains the series' central mystery, but it is also a story of shifting REALITIES, of sexual and psychological identities, and of RITES OF PASSAGE. [JHH]

GRISET, ERNEST (1844-1907) French illustrator, for some years in the UK – during which time he worked for *Punch* – who had a special flair for creating anthropomorphic animals. Greatly influenced by the work of Isidore GRANDVILLE, he worked in a traditional pen-line style, though many of his drawings were published in the UK as engraved by the Dalziel brothers. His later work was often coloured, which increased its charm considerably. He contributed drawings for books of nursery rhymes and illustrated editions of *Aesop's Fables* (1869), *Robinson Crusoe* (1869) and *Reynard the Fox* (1872). A collection of his early work was *Griset's Grotesques* (graph coll **1866**). [RT]

GROUNDHOG DAY US movie (*1993*). Columbia. **Pr** Trevor Albert, Harold Ramis. **Exec pr** C.O. Erickson. **Dir** Ramis. **Spfx** Tom Ryba. **Screenplay** Ramis, Danny Rubin. **Starring** Andie MacDowell (Rita), Bill Murray (Phil Connors). 101 mins. Colour.

Second-rate Pittsburgh newscaster Phil is sent to cover the Punxsutawney Groundhog Day, February 2 (on which day a groundhog is supposed capable of predicting the coming weather). A blizzard stops his crew returning to Pittsburgh that night. Next morning, Phil wakens to find it is once again Groundhog Day, and the same happens again, and again . . . At first, after getting over his panic, he uses the time loop for selfish purposes – robbing banks, bedding pretty women, making repeatedly unsuccessful attempts to seduce Rita, his producer. To win her he must, he realizes, make himself seem a better person than he is, and he employs his repeated Groundhog Days in educating himself to that end. In so doing, of course, he becomes that better person. Presumably because LOVE conquers all, the cycle is broken and the pair enter February 3 together.

GD is a slight fantasy; it is of interest primarily because, largely through an adroit screenplay, it avoids the tedium its theme of repetition might all too easily have incurred. [JG]

GRUBB, DAVIS (ALEXANDER) (1919-1980) US author, best-known for *The Night of the Hunter* (**1953**), filmed as an effective PSYCHOLOGICAL THRILLER by Charles Laughton as *The Night of the Hunter* (*1954*). Some of the imagery from this novel, particularly the psychological battle between GOOD AND EVIL, reappears in DG's last novel, *Ancient Lights* (**1982**), which has parallels with Stephen KING's *The Stand* (**1978**) in its portrayal of supernatural forces preparing for a spiritual ARMAGEDDON. DG often depicts his mounting horror through the eyes of CHILDREN; this is particularly effective in his short stories, collected as *Twelve Tales of Suspense and the Supernatural* (coll **1964**; vt *One Foot in the Grave* 1966 UK), which includes "Where the Woodbine Twineth" (1964), in which a child is abducted by her INVISIBLE COMPANIONS. "One Foot in the Grave" (1948) is almost a rewrite of "The Beast with Five Fingers" (1928) by William F. HARVEY, but with a severed foot rather than hand. DG's later supernatural horror stories are slightly more mystical and esoteric; they are found in *The Siege of 318: Thirteen Mystical Stories* (coll **1978**) and *You Never Believe Me* (coll **1989**). [MA]

GRUE US SMALL-PRESS digest MAGAZINE, irregular, 1985-current), published by Hell's Kitchen's Productions; ed Peggy Nadramia.

One of the computer-published HORROR magazines that responded to the heightened interest in horror fiction in the 1980s. It carries no fantasy; the SUPERNATURAL FICTIONS it does publish follow traditional themes, especially BLACK MAGIC and VAMPIRES, but with considerable emphasis on sex and violence. Nadramia received the WORLD FANTASY AWARD in 1990. [MA]

GRUNDY, STEPHAN (1967-   ) US writer whose only work of fantasy, *Rhinegold* (**1994**), has met with considerable critical acclaim for the relentlessness of its TWICE-TOLD rendering of the story of the Walsungs, the tragedy of the Ring of the Nibelungen (◊ NORDIC FANTASY). The overall tale is familiar from Richard WAGNER's **Ring** cycle (**1851-76**), and SG is initially very faithful to the sources he and Wagner share. The early parts of the tale, when the AESIR mingle with humans, hew closely to accepted versions. But gradually the SG version, as the story more and more concentrates on its human carriers, shifts into a depiction of something half-recognizable as history (◊ THINNING), sometime around the end of the Roman Empire. The moment when Attila the Hun enters the picture is not without humour; but the overall accomplishment is substantial. [JC]

GRÜNEWALD, MATTHIAS (c1470-1528) German painter who seems never to have been known as Grünewald during his lifetime; his real surname may have been Gothardt, though he also used the surname Neithardt, which may have been his wife's name. His given name is sometimes rendered as Mathis. MG's obscurity is due partially to the times in which he lived – he was a Protestant, and may have been involved in the Peasants' War of 1524-5 – and partially to the remarkable anguish of the work he produced. A late-medieval Gothic sensibility is agonizingly dramatized in his best work, through an expert knowledge of Renaissance perspective and other innovations. Not much survives; the only piece of direct fantasy interest is the right-hand panel of the third stage of the *Isenheim Altarpiece* (c1513-15), which depicts "The Temptation of Saint Anthony" as a CARNIVAL in which every touch of talon or paw to flesh seems carnivorous. In Hieronymus BOSCH's comparable "Temptation of Saint Anthony" there seems some chance the protagonist still retains an element of choice; in MG's vison, the saint seems on the verge of being ingested. [JC]

GUARDIAN OF THE THRESHOLD This term could as well be rendered Keeper of the Gate. In the MONOMYTH evolved by Joseph CAMPBELL in *The Hero With a Thousand Faces* (1949), the HERO leaves the home zone and confronts a Guardian of the THRESHOLD "at the entrance to the zone of magnified power", within which begin the great adventures pertaining to the phase of Initiation. In this encyclopedia the term is generally cross-referred to LIMINAL BEINGS. [JC]

GUESS WHAT HAPPENED TO COUNT DRACULA (*1971*) ◊ DRACULA MOVIES.

GUEST, LADY CHARLOTTE (1812-1895) UK writer. ◊ ARTHUR; MABINOGION.

GUINEVERE The wife of King ARTHUR; her LOVE for Sir LANCELOT led to the downfall of the ROUND TABLE and the eventual decline of Arthur's kingdom. Guinevere (Gwenhwyfar) was daughter of the King of Cornwall, and was the most beautiful woman in Britain; a more historical application suggests she may have been of Pictish origin (see *Guinevere* [*1991*] by Norma Lorre Goodrich). Her role is central to the Arthurian romances. It was CHRÉTIEN DE TROYES who first introduced the relationship between Guinevere and Lancelot in his court romance *Lancelot* or *Le Chevalier de la Charrete* ["The Knight of the Cart"] (?*1177*) – there was no earlier basis for this in the Celtic tales – and thereby developed a love triangle that swept the Anglo-Norman world. As the stories developed to their final consolidation in Sir Thomas MALORY's *Le Morte Darthur* (*1485*) Guinevere was portrayed first as a seductress and then as a victim, married to a king she could not love but unable to declare her love for Lancelot and, ultimately, raped by Mordred (Modred) when he usurped Arthur's throne.

Guinevere features in most Arthurian fictions, and is the focus of attention in some. In *Launcelot* (**1926**) by Ernest Hamilton and *The Little Wench* (**1935**) by Philip Lindsay (1906-1958) she is seen as scheming and divisive. Both the **Guinevere** sequence by Sharan NEWMAN and the **Guinevere** trilogy by Persia WOOLLEY retell the Arthurian saga from Guinevere's viewpoint, depicting her as a strong-willed individual who nevertheless becomes a victim of both

love and power. In *Firelord* (**1980**) and *Beloved Exile* (**1984**) Parke GODWIN contrasts Arthur and Guinevere. *The Idylls of the Queen* (**1982**) by Phyllis Ann KARR is an Arthurian murder mystery with Guinevere as the prime suspect.

The nature of Guinevere's role means her presence in fiction is usually to develop a romantic or tragic theme rather than to contribute any plot elements – she is a catalyst rather than a doer: the desire of Lancelot and others to be her champion often results in her being the reason for heroic QUESTS. [MA]

GULLIVER MOVIES Various movies have been based on *Gulliver's Travels* (**1726**) by Jonathan SWIFT. Some are of more interest than others, as reflected below. Most concentrate solely on the first or first two voyages.

1. *The New Gulliver* USSR ANIMATED MOVIE (*1933*). **Dir** Alexandr Ptoushko, A. Vanitchkin. **Screenplay** Ptoushko, B. Roshal. 85 mins. B/w.

An obscure propaganda movie, made in stop-motion animation with puppets, in which Swift's SATIRE is bent to become an attack on capitalism. By all accounts, the movie is – despite such dire augurs – both charming and amusing. [JG]

2. *Gulliver's Travels* US ANIMATED MOVIE (*1939*). Paramount. **Pr** Max FLEISCHER. **Dir** Dave Fleischer. **Screenplay** Dan Gordon, Cal Howard, Ted Pierce, Edmond Seward, I. Sparber. **Voice actors** Jessica Dragonette (Princess Glory singing), Lanny Ross (Prince David singing) – no other voices credited. **Character model** Sam Parker (Gulliver). 74 mins. Colour.

This was the Fleischer response to DISNEY's hugely successful SNOW WHITE AND THE SEVEN DWARFS (*1937*), and did moderately well. Today *GT* is largely forgotten, although many of its images – especially that of the giant Gulliver surrounded by minuscule Lilliputians, and the sequence in which he tows the Blefuscan fleet – are surprisingly familiar, through their frequent appearance as stills; while the song "It's a Hap-Hap-Happy Day" has, thanks to being reprised frequently in Paramount animated shorts, the same quasi-traditional status as "Hi Ho". Although it enjoys a welter of often brilliant "business", it is singularly lacking in *event*, with perhaps its first three-quarters seeming to be Prelude.

What there is of the plot is drawn approximately from the first part of Swift's original, covering Gulliver's time in Lilliput. [JG]

3. *The Three Worlds of Gulliver* US/Spanish movie (*1959*, dated 1960). Columbia/Morningside. **Pr** Charles H. Schneer. **Dir** Jack Sher. **Spfx** Ray HARRYHAUSEN. **Screenplay** Arthur Ross, Sher. **Starring** Sherri Alberoni (Glumdalclitch), Grégoire Aslan (King Brob), Mary Ellis (Queen of Brobdingnag), Kerwin Mathews (Gulliver), Jo Morrow (Elizabeth), Marian Spencer (Empress of Lilliput), Basil Sydney (Emperor of Lilliput), June Thorburn (Gwendolen). 97 mins. Colour.

The voyages to Lilliput and Brobdingnag (the "three worlds" are those two plus the mundane world), are in large part faithful to Swift's original although with the SATIRE homoeopathically diluted, and with the addition of romantic interest – Gulliver's fiancée Elizabeth stows away with him and, although missing Lilliput, is joined by and married to him in Brobdingnag. After escaping the wrath of the Brobdingnagian court – who, because confusing his science with MAGIC, wish to burn him as a WITCH (cf *A CONNECTICUT YANKEE*) – Gulliver wakens with Elizabeth on an

English seashore: they have experienced a shared DREAM.

Harryhausen's spfx are generally good, even by today's standards, although the MONSTERS are jerky and unreal, and suffer from not being DINOSAURS: gargantuan squirrels lack *frisson*. Alberoni's Glumdalclitch, done with solemn charm, upstages everyone. [JG]

4. *Gulliver's Travels Beyond the Moon* (ot *Garibah No Uchu Ryoko*) Japanese ANIMATED MOVIE (*1966*). Toei. **Pr** Hiroshi Okawa. **Dir** Yoshio Kuroda. **Screenplay** Shinichi Sekizawa. 78 mins. Colour.

A cheerful piece of work, having little to do with Swift's original. [JG]

5. *Gulliver's Travels* Belgian/UK live-action/ANIMATED MOVIE (*1976*). EMI/Valeness-Belvision. **Pr** Derek Horne, Raymond Leblanc. **Exec pr** Josef Shaftel. **Dir** Peter Hunt. **Screenplay** Don Black. **Starring** Richard Harris (Gulliver). 81 mins. Colour.

Tedious musical version of Gulliver's first voyage, with the Lilliputians animated and their country and architecture crudely modelled. All teeth are pulled from the SATIRE, and all possible excitement and sense of fantasy are eliminated by the unambitiousness of the animation (some limited, and with some use of repeated sequencing) and screenplay. [JG]

6. *Gulliver in Lilliput* Austrian/UK/US movie (*1981* tvm). **Dir** Barry Letts. **Starring** Andrew Burt, Jonathan Cecil, George Little, Linda Polan, Elisabeth Sladen. 105 mins. Colour.

Derived from a tv serial, and apparently good (we have been unable to obtain a viewing copy). Only the first voyage is treated. [JG]

7. *Gulliver's Travels Part 2* (vt *Land of the Giants: Gulliver's Travels Part 2*) Spanish ANIMATED MOVIE (*1983*). Estudios Cruz Delgado/Art Animation. **Pr** Druk Delgado. **Dir** Cruz Delgado. **Screenplay** Gustavo Alcalde, adapted into English by Karen Morgan. **Voice actors** (English version) Alexa Bates (Glundalitch), Nelson Modlin (Gulliver). 90 mins. Colour.

A sequel to 2. Found on the Brobdingnag shore by the fisherman father of Glundalitch (*sic*), the childlike Samuel (*sic*) Gulliver is sold to a CARNIVAL and thence to the vain Prince Felina, whose fool, Bufo, he supplants. Glundalitch arrives to save him from Bufo's vengeance, and is adopted by Felina as Gulliver's companion. But Bufo sets Sylvester, the castle gorilla, on the little man and – after a PARODY of KING KONG atop a palace tower – Gulliver and Glundalitch flee back to her home. She sends him to safety with a pigeon as his carrier. The animation, though not top-flight, is appealingly vigorous. Interestingly, it is made explicit that Brobdingnag is in an ALTERNATE WORLD. [JG]

8. *Gulliver's Travels* US/UK movie (*1995* tvm). Hallmark/Henson. **Pr** Duncan Kenworthy. **Dir** Charles Sturridge. **Spfx** Matthew Cope, Fiona Wallinshaw. **Screenplay** Simon Moore. **Starring** Ted Danson (Lemuel Gulliver), Warwick Davis (Grildrig the Dwarf), James Fox (Dr Bates), Kate Maberly (Glumdalclitch), Peter O'Toole (Emperor of Lilliput), Omar Sharif (Sorcerer), Mary Steenbergen (Mary Gulliver), Thomas Sturridge (Tom Gulliver), Alfre Woodard (Queen of Brobdingnag). *c*180 mins. Colour.

This is the definitive version to date, and has the unique distinction of covering all five voyages, with brilliant intercutting between those adventures and the experiences of Gulliver on his return to England, an added subplot: Dr

Bates, lusting for Gulliver's wife Mary, has Gulliver committed to an asylum, where his further descriptions of his travels succeed only in convincing the other doctors that he is indeed insane. Everything about this production is splendid, and Danson, best-known before as a comic actor, turns in an astonishingly good performance at the head of a highly distinguished, largely UK cast (even Sir John Gielgud has a bit part). The spfx are superb. [JG]

GUNN, NEIL M(ILLER) (1891-1973) Scottish writer noted for realistic accounts of Highland life. Several stories in *Hidden Doors* (coll **1929**) involve visions of some sort; the title story and "Such Stuff as Dreams" are marginal fantasies. The historical novel *Sun Circle* (**1933**) borders on fantasy by virtue of its depiction of Druidic ritual and RELIGION. *The Silver Darlings* (**1941**) was the first of several novels in which the endeavours of some characters echo the exploits of legendary figures. Another is *Young Art and Old Hector* (**1942**), which includes some exemplary items of FOLKLORE and whose sequel, *The Green Isle of the Great Deep* (**1944**), was NMG's only full-blown fantasy. Its two protagonists enter the land of the dead to find its condition mirrors the predicament of the actual Highlands, taken over by alien administrators who have rendered the fruit of the TREE of knowledge inedible and forbidden its consumption. All NMG's subsequent works involved themselves with the quest to pin down the essence of the freedom which Art and Hector have to reclaim; in his last book, a spiritual autobiography, he called it eponymously *The Atom of Delight* (**1956**). *Second Sight* (**1940**), the most fantastic of his later novels, features a symbolic hunt for a stag which fulfils a precognitive vision. *The Silver Bough* (**1948**) and *The Well at the World's End* (**1951**) take their titles from legendary motifs which are of symbolic significance in the QUESTS which the protagonists undertake in search of a magical illumination. *The Other Landscape* (**1954**), a similar quest story, forsakes the imagery of Celtic folklore for the ideologies of Eastern mysticism which fascinated NMG in his latter years. [BS]

GUNNARSSON, THORARINN (1957-) US writer whose **Starwolves** series – *The Starwolves* (**1988**), *Battle of the Ring* (**1989**), *Tactical Error* (**1991**) and *Dreadnought* (**1992**) – is sf, but whose other work is fantasy, always in series. The **Song** series – *Song of the Dwarves* (**1988**) and *Revenge of the Valkyrie* (**1989**) – is NORDIC FANTASY, incorporating elements of the Nibelungen saga. The **Dragons** sequence – *Make Way for Dragons!* (**1990**), *Human, Beware!* (**1990**) and *Dragons on the Town* (**1992**) – makes lighthearted play of the plight of a young DRAGON stranded in the OTHERWORLD of California. TG has contributed *Dragonlord of Mystara* * (**1994**) and *Dragonking of Mystara* * (**1995**) to the **Mystara** segment of the **Dragonlord Chronicles**.[JC]

GUON, ELLEN (SUE) (1964-) US GAMES-designer and writer, most of whose work of interest has been in collaboration with Mercedes LACKEY. Her prequel to the **Bedlam's Bard** sequence with Lackey, *Bedlam Boyz* (**1993**), is an URBAN FANTASY set in a LOS ANGELES subject to conflict between the police and a CROSSHATCH gang of ELVES called the Unseleigh Court. [JC]

GURNEY, JAMES (1958-) UK illustrator and painter, mainly of archaeological subjects, with a strongly coloured realistic style. JG's first professional job was as a background artist on the animated movie FIRE AND ICE (*1983*). A deep interest in archaeology drew commissions from the National Geographic Society to paint reconstructions of Etruscan, Moshe and Kushite cultures. A 1990 print entitled

"Dinosaur Parade" led to his writing and illustrating the fantasy *Dinotopia: A Land Apart from Time* (graph **1992**), followed by *Dinotopia: The World Beneath* (graph **1995**). These tell of a Darwinian voyage of exploration by a professor and his family, from which they never return, having become immersed in the life of a fantastical world in which DINOSAURS and other prehistoric beasts feature.   [RT]

Other work: *Dinotopia Pop-up Book* * (graph **1994**).

GUY NAMED JOE, A US movie (*1944*). ◊ ALWAYS (*1989*); POLTERGEIST.

GUYS Effigies burnt in the UK each November 5 to celebrate the foiling of a plot to blow up the Houses of Parliament on November 5, 1605. One of the conspirators was Guy Fawkes (1570-1606) – hence the name. ◊ DOLLS.
[JG]

GYGAX, (ERNEST) GARY (1938-   ) US entrepreneur and writer, co-creator of the role-playing GAME *Dungeons & Dragons*, and co-founder in 1974 of TSR, which publishes the game, and writer of associated books. As an author of fantasy novels, GG has been active from the mid-1980s, though his first book came much earlier. *Victorious German Arms: An Alternate Military History of World War II* (**1973** chap) with Terry Stafford describes what might have happened had Germany won the Battle of Stalingrad.

The **Sagard the Barbarian** sequence, all with Flint Dille, comprises *The Ice Dragon* * (**1985**), *The Green Hydra* * (**1985**), *The Crimson Sea* * (**1985**) and *The Fire Demon* * (**1986**); the **Greyhawk** sequence comprises *Saga of Old City* * (**1985**) and *Artifact of Evil* * (**1986**); the **Gord the Rogue** sequence comprises *Sea of Death* * (**1987**), *Night Arrant* * (coll **1987**), *City of Hawks* * (**1987**), *Come Endless Darkness* * (**1988**) and *Dance of Demons* * (**1988**); and the **Dangerous Journeys** sequence comprises *The Anubis Murders* * (**1992**), *The Samarkand Solution* * (**1993**) and *Death in Delhi* * (**1993**). *The Temple of Elemental Evil* * (**1985**) with Frank Mentzer is a singleton.   [JC]

Other works (nonfiction): *Advanced Dungeons & Dragons: Monster Manual* (**1977**); *AD&D: Players Handbook* (**1978**); *AD&D: Dungeon Masters Guide* (**1979**; rev 1989); *AD&D: Monster Manual II* (**1983**); *Official AD&D Unearthed Arcana* (**1985**); *Oriental Adventures* (**1985**) with David Cook and François Marcela-Froideval; *Role-Playing Mastery* (**1987**); *Master of the Game* (**1989**).

HADES The Greek god of the UNDERWORLD, known also as Pluto and Dis, the latter names being primarily used by the Romans. Although he supervised the punishments to which such individuals as Sisyphus had been condemned, he was not a malevolent figure like the Christian SATAN. The name Hades is also applied to the Underworld itself; it was popularly employed as a euphemistic substitute for HELL in the days when that word was considered indecent, and is used thus in some infernal fantasies, including those of Frederick Arnold KUMMER. John Kendrick BANGS's works in the same vein retain some of the better-known geographical features of the Classical Hades, notably the river Styx, which formed its boundary; a slightly fuller account is included in *O Men of Athens* (**1947**) by A.C. Malcolm. Another of the RIVERS of Hades was Lethe, whose waters induced AMNESIA. The precise relationships between Hades and the paradisal Elysian Fields and between Hades and the lightless Tartarus are unclear. The most significant myths featuring the god Hades are the story of Persephone and of ORPHEUS's descent into the Underworld; the best literary accounts of Hades are to be found in recapitulations and transfigurations of the latter MYTH. [BS]

HAGGARD, [Sir] H(ENRY) RIDER (1856-1925) UK civil servant, barrister, politician and writer, knighted in 1912, who spent 1875-81 in the Colonial Service in Africa, an experience which provided him with background material for his best fiction. His basic attitude to issues of imperialism and race can be understood as conservative, but hints of FIN-DE-SIÈCLE cultural pessimism, and the stresses of his private life (married to another, he lived for years close to the woman he had always loved), frequently undermine any sense that he was a simple advocate of the still-expanding British Empire, or that his attitude to women and to "other" races was straightforward. The uneasy timbre of his work as a whole became evident in the mid-1880s with his third and fourth novels, *King Solomon's Mines* (**1885**) and *She: A History of Adventure* (**1886** chap). Lost-world venues (◊ LOST RACES) are often central to his best tales, and an obsessive conflation of IMMORTALITY, REINCARNATION, LAMIA and GODDESS motifs reoccurs throughout his oeuvre.

The urgent forward thrust of his storytelling genius can give, in other words, the mistaken impression that he is at heart a propagandist for the Empire – an impression no more (or less) accurate that given by the works of Rudyard KIPLING, who was a friend of HRH's and shared with him the intuition that the White Man's pomp was an imposture. Generally speaking, HRH's storylines are most evocative when they turn their back on the bluster and regalia of empire and move into the unknown, the lost, the Heart of Darkness, the eternally recurring. The gaze of the best HRH novels is towards the past. In view of this essential escapist orientation, it is not surprising that when Edgar Rice BURROUGHS, the 20th-century writer HRH most visibly influenced, created **Tarzan** he was able to penetrate to the wish-fulfilment heart of the HRH protagonist; Tarzan is not, in essence, a "timeless" occupant of the Heart of Darkness but a great lord, with TALENTS, on permanent holiday.

Much of HRH's best work is contained in the **Quatermain** sequence, which appeared over a 40-year span. Titles are given here in order of internal chronology (in each case, the date of the action precedes the title): 1835-8 *Marie* (**1912**); 1842-69 *Allan's Wife* (**1887** US), which was incorporated into *Allan's Wife and Other Tales* (coll **1889**); 1854-6 *Child of Storm* (**1913**); 1857 *A Tale of Three Lions* (**1887** chap US) – assembled with "Hunter Quatermain's Story" from *Allan's Wife and Other Tales* as *Allan the Hunter: A Tale of Three Lions* (coll **1898** US); 1859 *Maiwa's Revenge* (**1888** US); 1870 *The Holy Flower* (**1915**; vt *Allan and the Holy Flower* 1915 US); 1871 *Heu-Heu, or The Monster* (**1924**); 1872 *She and Allan* (**1921** US), also linked to the **Ayesha** sequence; 1873 *Treasure of the Lake* (**1926** US); 1874 *The Ivory Child* (**1916**); 1879 *Finished* (**1916**); 1879 "Magepa the Buck" in *Smith and the Pharaohs and Other Tales* (coll **1920**); 1880 *King Solomon's Mines* (**1885**); 1882 *The Ancient Allan* (**1920**); 1883 *Allan and the Ice Gods: A Tale of Beginnings* (**1927**); and 1884-5 *Allan Quatermain: Being an Account of his Further Adventures and Discoveries in Company with Sir Henry Curtis, Bart., Commander John Good, and One Umslopogaas* (**1887**; cut by other hands vt *Allan Quatermain and the Lost City of Gold* * 1986 as a movie novelization). *Nada the Lily* (**1892** US), not directly connected to the sequence, deals with the early life of the Zulu hero Umslopogaas, who in later years becomes one of Allan Quatermain's faithful COMPANIONS.

There are inconsistencies in the series. Quatermain died at the end of *Allan Quatermain* (**1887**), and HRH responded to the popularity of the character by retrofitting new instalments into sometimes implausible gaps in Quatermain's previous life. The results are various, and for fantasy readers several of the instalments are of little interest. Some – including *Maiwa's Revenge*, *Marie*, *Child of Storm* and *Finished* – deal almost exclusively with the Zulu nation's doomed resistance to the advance of the white empires; others are straightforward adventure tales. The central text remains the first, *King Solomon's Mines*, which establishes Quatermain as an UNDERLIER for the intrepid HERO who is called into the kind of PLANETARY-ROMANCE world which replaced the lost world as a usable venue. Burroughs's **John Carter** of Mars, in the **Barsoom** series, is the first and most important example. *King Solomon's Mines* is also important to the history of fantasy because the impossible lost world it depicts underlies later visions of the longed-for land from the deep past, because it offers a vivid model for the QUEST tale, and because the subplot dealing with Umbopa, the HIDDEN MONARCH of the kingdom of the Kukuanas, is shaped (though lacking circumstances) as a fantasy STORY. All of these qualities were ditched from the movie *King Solomon's Mines* (**1985**), which was an adventure yarn in pale emulation of the INDIANA JONES sequence. Earlier versions were *King Solomon's Mines* (**1937**), which is ponderous and worthy, and *King Solomon's Mines* (**1950**), which is drastically underplotted but pretty.

Some of the later volumes in the sequence move into remoter regions of the imagination. These include *The Ancient Allan* and *Allan and the Ice-Gods*, in which Quatermain TIMESLIPS (in a DREAM) to, in the first book, ancient EGYPT and, in the second, to prehistory (◊ PREHISTORIC FANTASY); *Heu-Heu, or The Monster*, in which a GOD is exposed as fraudulent but the native sorcerer has genuine powers. In *Treasure of the Lake* Quatermain is tricked into helping instal a new Year King in a plot clearly derived from the work of Sir James FRAZER.

HRH's **Ayesha** sequence – *She: A History of Adventure* (1886-7 *The Graphic*; cut **1886** chap US; text restored 1887 UK; cut 1896 UK), also published as *The Annotated She: A Critical Edition of H. Rider Haggard's Victorian Romance* (**1991** US) ed Norman Etherington, with unreliable notes but a variorum text; *Ayesha: The Return of She* (**1905**; vt *The Return of She: Ayesha* 1967 US); *She and Allan* (**1921** US), which provides a link with the **Quatermain** series; and *Wisdom's Daughter: The Life and Love Story of She-Who-Must-Be-Obeyed* (**1923**) – is of strong fantasy interest. The first novel was rewritten as a movie tie by Don Ward (1911-) as *She: The Story Retold* * (**1949** US), though the movie itself seems not to have been made. Other movie versions of *She* have been released, however (◊ SHE).

Ayesha is a minor incarnation of the GODDESS, partaking in her nature of both Isis and Aphrodite, which causes unending internal warfare as Isis calls her spirit into realms described by HRH in occult terms and Aphrodite requires her to act like a LAMIA. But HRH's portrayal of her descends only rarely to unction or caricature, and Ayesha remains – along with Umslopogaas – his most successful character creation. In *She* the immortal Ayesha rules the Lost World of Kôr in the heart of Africa, to which are drawn young Leo Vincey and his companions. He is either a descendant or a REINCARNATION of Kallikrates, the ancient Egyptian priest whom Ayesha loved and killed. Leo is both revolted and tempted by her extraordinary beauty, her violent imperious person, and by the promise of IMMORTALITY. In the end, She seems to perish – ageing instantaneously into a monkey-like creature – in the Pillar of Life that Leo has balked at entering. In the direct sequel, *Ayesha*, Leo once again is drawn to seek out the goddess, taking 18 years this time to accomplish his QUEST, finding her at last in Turkestan, where she rules a lost world, reincarnated this time as an ancient woman. In passages reminiscent of ordeals described in many FOLKTALES (◊◊ GAWAIN), Leo is asked to choose the crone over a younger woman, and completes his NIGHT JOURNEY into her arms by choosing correctly. Unfortunately She kills him with a kiss, and as the novel closes she has departed for the Land of the Dead to find her lover. It is in this novel that She becomes an important UNDERLIER. The further volumes in the sequence are less central.

Other HRH novels of interest are singletons. They include: *The World's Desire* (**1890**) with Andrew LANG, in which Helen of Troy plays a role similar to that of Ayesha; *Eric Brighteyes* (**1891**; vt *The Saga of Eric Brighteyes* 1974 US), a NORDIC FANTASY told in a sustained prose imitation of SAGA style, and as cruel as most Nordic fantasy; *The People of the Mist* (**1894**), a lost-world tale set in Africa; *Heart of the World* (**1895** US), a lost-world tale set in Mexico; *Stella Fregelius: A Tale of Three Destinies* (**1903** US), sf with occult touches; *Benita: An African Romance* (**1906**; vt *The Spirit of Bambatse: A Romance* 1906 US), a SUPERNATURAL FICTION; *The Yellow God: An Idol of Africa* (**1908** US), whose evil, multiply incarnated priestess is a parody of Ayesha; *The Mahatma and the Hare: A Dream Story* (**1911**), a kind of POSTHUMOUS FANTASY in which the eponymous mahatma converses with a hare on the Great White Road the dead take, and is horrified by the hare's description of his death at the hands of English hunters; and *Red Eve* (**1911**), in which Murgh, a personification of DEATH, traverses a plague-afflicted Europe. There are fantasticated or supernatural elements in some of the other novels, such as *Cleopatra: Being an Account of the Fall and Vengeance of Harmachis, the Royal Egyptian, as Set Forth by his Own Hand* (**1889** US).

HRH could shift from subtlety to coarseness, from original insight to tendentious cliché within a single paragraph. His tales remain powerfully in the memory, but tend to disappoint on being reread. His influence comes from his capacity to create images of adventure and unattainable romance, from the mythopoeic vividness of the underliers he created. But there is almost always a lingering sense that an HRH tale could have been told better. Over the past century, many writers have tried to do this. [JC]

Other works: *Beatrice* (**1890**); *Montezuma's Daughter* (**1893**); *The Wizard* (**1896**), also incorporated into *The Wizard, and Black Heart and White Heart* (coll **1907**; vt *Black Heart and White Heart, and The Wizard* 1924); *Swallow: A Tale of the Great Trek* (**1899** US); *Elissa, the Doom of Zimbabwe: Black Heart & White Heart* (coll **1900** US; rev vt *Black Heart and White Heart, and Elissa* 1900 Germany; "Elissa" only vt *Elissa, or The Doom of Zimbabwe* 1917 UK); *Lysbeth: A Tale of the Dutch* (**1901** US); *Pearl-Maiden: A Tale of the Fall of Jerusalem* (**1903**); *The Brethren* (**1903**); *The Ghost Kings* (**1908**; vt *The Lady of the Heavens* 1908 US); *The*

Lady of Blossholme (**1909**); *Morning Star* (**1910**); *Queen Sheba's Ring* (**1910**); *The Wanderer's Necklace* (**1914**); *Moon of Israel: A Tale of the Exodus* (**1918**); *When the World Shook: Being an Account of the Great Adventure of Bastin, Bickley, and Arbuthnot* (**1919**), sf; *The Missionary and the Witch-Doctor* (**1920** chap US); *The Virgin of the Sun* (**1922**); *Queen of the Dawn: A Love Tale of Old Egypt* (**1925**); *Mary of Marion Isle* (**1929**; vt *Marion Isle* 1929 US); *Belshazzar* (**1930**); *The Best Short Stories of H. Rider Haggard* (coll **1981**) ed Peter HAINING. There are also various omnibuses.

Further reading: *H. Rider Haggard: A Bibliography* (**1987**) by D.E. Whatmore; *Rider Haggard and the Fiction of Empire: A Critical Study of British Imperial Fiction* (**1987**) by Wendy R. Katz; *Rider Haggard and the Lost Empire* (**1993**) by Tom Pocock.

HAHN, MARY DOWNING (1937-) US writer who specializes in YA tales, mainly SUPERNATURAL FICTION. In *Time of the Witch* (**1982**) a WITCH dangerously tempts the young female protagonist from a broken home with the fulfilment of her deepest WISHES. *Wait Till Helen Comes* (**1986**) is a GHOST STORY. *The Doll in the Garden* (**1989**) is a TIME-TRAVEL tale whose protagonists gain access to the 19th century through a hedge which opens into a mysterious GARDEN. In *Time for Andrew: A Ghost Story* (**1994**) a young man TIMESLIPS into 1910, exchanging bodies with his sick DOUBLE, who threatens not to return from the future. *Look for Me by Moonlight* (**1995**) features VAMPIRES. [JC]

HAINING, PETER (ALEXANDER) (1940-) UK writer and – more significantly – anthologist, under his own name and as William Pattrick, Richard Peyton and Sean Richards; he has also ghost-edited as Alfred HITCHCOCK. A high proportion of his books deal with SUPERNATURAL FICTION and HORROR; but many of his ANTHOLOGIES, and several of his nonfiction studies, concern FANTASY. His first book, *Devil Worship in Britain* (**1964**) with A.V. Sellwood, was an examination of BLACK MAGIC; his first anthology, *The Hell of Mirrors* (anth **1965**; rev 1974; vt *Everyman's Book of Classic Horror Stories* 1976) was horror. Several nonfiction titles are of fantasy interest, including: *Ghosts: The Illustrated History* (**1974**) and *A Dictionary of Ghosts* (**1981**); *Ancient Mysteries* (**1977**); *The Legend and Bizarre Crimes of Spring-Heeled Jack* (**1977**) (◊ SPRING-HEELED JACK); *The Mystery and Terrible Murders of Sweeney Todd* (**1977**); *The Sherlock Holmes Compendium* (**1980**; rev 1994); *The Leprechaun's Kingdom* (**1980**); and *The Scarecrow: Fact and Fable* (**1988**).

Anthologies of fantasy interest include *The Gentlewomen of Evil: An Anthology of Rare Supernatural Stories from the Pens of Victorian Ladies* (anth **1967**), which includes work by Margaret OLIPHANT and others; *The Clans of Darkness* (anth **1971**; vt *Scottish Stories of Fantasy and Horror* 1988 US); *Gothic Tales of Terror* (anth **1972** US: vt in 2 vols *Great British Tales of Terror* 1972 UK and *Great Tales of Terror from Europe and America* 1972 UK); *The Magicians: Occult Stories* (anth **1972**); *The Dream Machines* (anth **1972**), about balloons, with some fiction; *The Monster Makers: Creators and Creations of Fantasy and Horror* (anth **1974**); *The Magic Valley Travellers: Welsh Stories of Fantasy and Horror* (anth **1974**); *The Ancient Mysteries Reader* (anth **1975** US); *The Fantastic Pulps* (anth **1975**); *The Hashish Club #1: Founding the Modern Tradition* (anth **1975**) and *#2: The Psychedelic Era* (anth **1975**); *The Ghost's Companion: Stories of Personal Encounters with the Supernatural* (anth **1975**); *Greasepaint and Ghosts: An Anthology of Strange and Supernatural Stories from the World of Theatre* (anth **1982**); *Christmas Spirits: Ghost Stories of the*

Festive Season (anth **1983**); *Ghost Tour: An Armchair Journey through the Supernatural* (anth **1984**); *The Ghost Ship* (anth **1985**); *Supernatural Sleuths: Stories of Occult Investigators* (anth **1986**); *Tales of Dungeons and Dragons* (anth **1986**); *The Ghost Now Standing on Platform One* (anth **1990**; vt *Journey into Fear* 1991 US) as by Richard Peyton; *Great Irish Stories of the Supernatural* (anth **1992**) and *Great Irish Tales of the Unimaginable: Stories of Fantasy and Myth* (anth **1994**); and *The Vampire Omnibus* (anth **1995**).

Edited editions of individual authors include: *The Gaston Leroux Bedside Companion* (coll **1980**); *The Final Adventures of Sherlock Holmes* (coll **1981**), *Sherlock Holmes and the Sussex Vampire* (coll **1981**), *Sherlock Holmes and the Devil's Foot* (coll **1986**) and *The Supernatural Tales of Arthur Conan Doyle* (coll **1987**) (◊ Arthur Conan DOYLE); *The Best Short Stories of Rider Haggard* (coll **1981**) (◊ H. Rider HAGGARD); *Shades of Dracula* (coll **1982**) and *Midnight Tales* (coll **1990**) (◊ Bram STOKER); *The Complete Ghost Stories of Charles Dickens* (coll **1982**) and *Charles Dickens's Christmas Ghost Stories* (coll **1992**) (◊ CHRISTMAS BOOKS; Charles DICKENS); *Paths to the River Bank* (coll **1983**) (◊ Kenneth GRAHAME); *The Complete Supernatural Stories of Rudyard Kipling* (coll **1987**) (◊ Rudyard KIPLING); *A Book of Learned Nonsense* (coll **1987**) (◊ Edward LEAR); *The Supernatural Tales of Thomas Hardy* (coll **1988**); *The Best Supernatural Tales of Wilkie Collins* (coll **1990**) (◊ Wilkie COLLINS); *The Fantasy and Mystery Stories of F. Scott Fitzgerald* (coll **1991**) and *The Best Supernatural Stories of John Buchan* (coll **1991**) (◊ John BUCHAN).

PH is sometimes superficial, but some books – *The Dream Machines* and *The Scarecrow* are two excellent examples – are both fascinating and extremely useful. And an anthology like *The Vampire Omnibus*, because of its thoroughness and its ample annotations, can shine new light on old territory. PH is one of the essential popularizers of the field. [JC]

Other works

Nonfiction studies: *Witchcraft and Black Magic* (**1971**); *The Anatomy of Witchcraft* (**1972**); *The Warlock's Book* (**1972**); *The Witchcraft Papers* (**1974**); *An Illustrated History of Witchcraft* (**1975**); *Terror!* (**1976**; vt *The Art of Horror Stories* 1986 US); the **True Mysteries** sequence, comprising *The Monster Trap* (**1976**), *Restless Bones* (**1978**), *The Screaming Skulls* (**1979**), *The Hell Hound* (**1980**) and *The Vampire Terror* (**1981**); *Fun to Know About Ghosts* (**1979**) as by Sean Richards; *Superstition: An Illustrated History* (**1979**); *The Man who was Frankenstein* (**1979**); various **Doctor Who** studies, including *Doctor Who: A Celebration* (**1983**), *Doctor Who: The Key to Time* (**1984**), *Doctor Who: The Time-Travellers' Guide* (**1987**) and *Doctor Who: 25 Glorious Years* (**1988**); *Eyewitness to the Galaxy* (**1985**); *The Race for Mars* (**1986**); *The Television Sherlock Holmes* (**1986**); *The Dracula Centenary Book* (**1987**; vt *The Dracula Scrapbook* 1992, not to be confused with the 1976 anth); *James Bond: A Celebration* (**1987**); *Meals on Wheels: W. Heath Robinson* (**1989**); *The Supernatural Coast* (**1992**).

Anthologies

As PH: *Where Nightmares Are* (anth **1966**); *The Craft of Terror: Extracts from the Rare and Famous Gothic "Horror" Novels* (anth **1966**); *Beyond the Curtain of Dark* (anth **1966**); *Summoned from the Tomb* (anth **1966**) and *Legends for the Dark* (anth **1968**), both assembled as *Summoned from the Tomb* (omni **1973**); *Dr Caligari's Black Book: An Excursion into the Macabre, in Thirteen Acts* (anth **1968**; rev 1969); *The Evil People, Being Thirteen Strange and Terrible Accounts of Witchcraft, Black Magic, and Voodoo* (anth **1968**); *The*

Midnight People (anth **1968**; vt *Vampires at Midnight* 1970 US); *The Unspeakable People, Being Twenty of the World's Most Horrible Horror Stories* (anth **1969**); *The Satanists* (anth **1969**); *The Witchcraft Reader* (anth **1969**); *The Wild Night Company: Irish Tales of Terror* (anth **1970**; vt *Irish Tales of Terror* 1988 US); *The Freak Show* (anth **1970**); *The Hollywood Nightmare* (anth **1970**; rev 1971 US); *A Circle of Witches: An Anthology of Victorian Witchcraft Stories* (anth **1971**); *The Ghouls* (anth **1971**; vt in 2 vols as *The Ghouls 1* 1974 and *The Ghouls 2* 1974); *Nightfrights: An Anthology of Macabre Tales that Have Frightened Three Generations* (anth **1972**); *The Lucifer Society* (anth **1972**; vt *Detours into the Macabre* 1974; vt *Masters of the Macabre: The Best of the 20th Century* 1993); *The Sherlock Holmes Scrapbook* (anth **1973**); *The Nightmare Reader* (anth **1973**); *Christopher Lee's New Chamber of Horrors* (anth **1974**; vt in 2 vols *Christopher Lee's New Chamber of Horrors* 1976 and *More of Christopher Lee's New Chamber of Horrors* 1976); *The Penny Dreadful, or Strange, Horrid & Sensational Tales* (anth **1975**); *The Dracula Scrapbook* (anth **1976**), nonfiction; *The Black Magic Omnibus* (anth **1976**; vt in 2 vols *Black Magic 1* 1977 and *Black Magic 2* 1977); the **Unknown** sequence comprising *Unknown Tales of Horror* (anth **1976**; vt *The First Book of Unknown Tales of Horror* 1976), *The Second Book of Unknown Tales of Horror* (anth **1978**; vt *Tales of Unknown Horror* 1978) and *More Tales of Unknown Horror* (anth **1979**; vt *The Third Book of Unknown Tales of Horror* 1980); *The Edgar Allan Poe Scrapbook* (anth **1977**), nonfiction; *The Frankenstein File* (anth **1977**); *Deadly Nightshade: Strange Stories of the Dark* (anth **1977**); *The Jules Verne Companion* (anth **1978**), nonfiction; *The Ghost Finders: Tales of Some Famous Phantoms* (anth **1978**); *The Shilling Shockers: Stories of Terror from the Gothic Bluebooks* (anth **1978**); *The H.G. Wells Scrapbook* (anth **1978**), nonfiction; *Classic Horror Omnibus, Volume 1: Five Classic Novels of Terror* (anth **1979**), no vol 2 published; *Dead of Night: Horror Stories from Radio, Television and Films* (anth **1981**); *Nightcaps and Nightmares: Ghosts with a Touch of Humour* (anth **1983**); *Hallowe'en Hauntings: Stories About the Most Ghostly Night of the Year* (anth **1984**); *Tune in for Fear* (anth **1985**); *Zombie* (anth **1985**; vt *Stories of the Walking Dead* 1986); *Vampire: Chilling Tales of the Undead* (anth **1985**); *Werewolf: Horror Stories of the Man-Beast* (anth **1987**); *Poltergeist: Tales of Deadly Ghosts* (anth **1987**); *Movie Monsters* (anth **1988**); *The Mummy: Stories of the Living Corpse* (anth **1988**); *Murder on the Menu* (anth **1991**); *The Television Detectives Omnibus* (anth **1992**; vt *Great Tales of Crime and Detection* 1993 US; vt *The Armchair Detectives* 1993 UK), associational; *The Television Late Night Horror Omnibus* (anth **1993**; vt *The Armchair Horror Collection: Great Tales from TV Anthology Series* 1994); *Frankenstein: The Monster Wakes* (anth **1994**); *Tales from the Rogues' Gallery: A Guided Tour* (anth **1994**).

As Alfred Hitchcock: *This Day's Evil* (anth **1967**); *Behind the Locked Door and Other Strange Tales* (anth **1967**); *Meet Death at Night* (anth **1967**); *Guaranteed Rest in Peace* (anth **1967**); *The Graveyard Man* (anth **1968**).

As William Pattrick: *Mysterious Railway Stories* (anth **1984**); *Mysterious Sea Stories* (anth **1985**); *Mysterious Air Stories* (anth **1986**); *Mysterious Motoring Stories* (anth **1987**; vt *Duel, and Other Horror Stories of the Road* 1987).

As Richard Peyton: *Deadly Odds* (anth **1986**; vt *At the Track* 1988 US; vt in 2 vols as *Deadly Odds* 1988 UK and *Deadlier Odds* 1988 UK; vt in 1 vol as *Great Racing Stories* 1993 US); *Sinister Gambits* (anth **1991**).

As Sean Richards: *The Elephant Man and Other Freaks* (anth **1980**); *The Barbarian Swordsmen* (anth **1981**), containing SWORD-AND-SORCERY tales.

HAKUJADEN Japanese ANIMATED MOVIE (*1958*). ◊ ANIME.

HALDANE, CHARLOTTE (FRANKEN) (1894-1969) UK writer best-known for her sf novel, *Man's World* (**1926**), but also the author of a fantasy of considerable interest: *Melusine, or Devil Take Her!* (**1936**) is a FANTASY OF HISTORY set in 12th-century France, where the destinies of the potentially significant rulers of the earldom of Poitiers are controlled by SECRET MASTERS whose goal is to oppose the continued rise of Christianity (◊ THINNING), advocating in its stead a devotion to the values of the old Earth. *The Shadow of a Dream* (**1952**) is SUPERNATURAL FICTION. [JC]

HALDEMAN, LINDA (WILSON) (1935-1988) US writer whose first fantasy novel, *The Lastborn of Elvinwood* (**1978**), imports MERLIN into a CONTEMPORARY-FANTASY setting where a proposed exchange between a FAIRY and a human child must be dealt with. *Esbae: A Winter's Tale* (**1981**) features a conflict between the ANGEL-like eponymous spirit and the DEMON Asmodeus, who has been conjured up from HELL by an incompetent university student. LH's work, not excessively ambitious, was told with wit. [JC]

Other works: *Star of the Sea* (**1978**).

HALE, JAMES (? -) UK anthologist. ◊ *The* GHOST BOOK.

HALFLINGS Beings whose parentage is half-human and half-other, with no presumption that one half is inferior, wicked or cursed. For this reason, the term should probably not be used to describe hybrid beings in SUPERNATURAL FICTION or HORROR, unless – as in Thomas M. DISCH's *The Businessman* (**1984**) – some irony or moral comment is implied. Most often, the parents of a halfling are an ELF and a human girl, and halflings are often found in CROSSHATCH venues. J.R.R. TOLKIEN also used the term for his (no-crossbred) hobbits. [JC]

HALIDOM, M.Y. Pseudonym of an unidentified UK writer who also wrote as Dryasdust, and who flourished around the end of the 19th century, though his archaic style has led to the suspicion that much of his work had been composed earlier; Mike ASHLEY has suggested that more than one writer may have been responsible for the Dryasdust/Halidom oeuvre. The major work is the *Tales of the Wonder Club* (coll **1899-1900** 3 vols as by Dryasdust; rev in 3 vols as *Tales of the Wonder Club: First Series* 1903, *Second Series* 1904 and *Third Series* 1905, all as by MYH), a set of CLUB STORIES told in a haunted INN, with fantasy only in the first two volumes. Typically of the genre, most of the stories are light in tone, and some are TALL TALES – like "The Mermaid", in which a jolly tar must agree to an amputation and a prosthetic tail before he can marry his beloved MERMAID. *The Gipsy Queen* (**1903**) is a nonfantasy play from vol 3. *The Wizard's Mantle* (**1902** as by Dryasdust; rev 1903 as by MYH) is a tale involving INVISIBILITY. Almost all of MYH's later work is HORROR; it includes *The Spirit Lovers and Other Tales* (coll **1903**), *A Weird Transformation* (**1904**), *The Woman in Black* (**1906**), *Zoe's Revenge* (**1908**), about an animate DOLL, *The Poet's Curse* (**1911**), in which the desecration of SHAKESPEARE's tomb wreaks disaster, and *The Poison Ring* (**1912**). [JC]

HALL, FRANCES (TEBBETTS) (1914-) US writer. ◊ Piers ANTHONY.

HALL, RODNEY (1935-) Australian writer. ◊ DREAMTIME.

HALLOWE'EN In the pagan CALENDAR, the date on which the SPIRITS of those who have died during the previous year finally depart this world for the AFTERLIFE. It is thus the day on which PORTALS to the afterlife are opened; movement from there to here is likewise possible. Much SUPERNATURAL FICTION makes play with this; the title of Charles WILLIAMS's *All Hallows' Eve* (**1945**) signals its complex interactions between living and dead. [CB]

HALLOWEEN III: SEASON OF THE WITCH US movie (*1983*). Universal/Dino De Laurentiis/ Moustapha Akkad. **Pr** John Carpenter, Debra Hill. **Exec pr** Joseph Wolf, Irwin Yablans. **Dir** Tommy Lee Wallace. **Spfx** Jon G. Belyeu. **Mufx** Tom Burman. **Screenplay** Wallace (and Nigel Kneale, uncredited). **Novelization** *Halloween III: Season of the Witch* by Jack Martin (Dennis ETCHISON) * (**1982**). **Starring** Tom Atkins (Dan Challis), Stacey Nelkin (Ellie Grimbridge), Dan O'Herlihy (Conal Cochran), Garn Stephens (Marge Gutman), Ralph Strait (Buddy Kupfer), Wendy Wessberg (Teddy). 98 mins. Colour.

Halloween (*1978*) and *Halloween II* (*1981*) were HORROR MOVIES about a SERIAL KILLER. *HIII* is an unconnected TECHNOFANTASY.

California, a few days before HALLOWE'EN. The Silver Shamrock Novelties TOY factory is running a massive promotion of Halloween MASKS. It proves that Shamrock's mad Celtic owner, Cochran, has loaded the masks' brand decals with microcircuitry based on fragments of a sarsen stolen from Stonehenge; these fragments are loaded with occult energy so that, in response to light strobing at the correct frequency, the device blasts the brains of the mask's wearer, conjuring up in her/his place scuttling insects and poisonous SERPENTS. Cochran's motive is rooted in WITCHCRAFT: at a time of correct planetary alignment it is necessary to perform HUMAN SACRIFICE, preferably of children, on a vast scale in order to . . . well, *HIII* follows his motives no further than that.

HIII is gory and confused, but is well enough directed and contains enough interesting ideas to hold the attention. Although Kneale objected to the gore sufficiently to have his name withdrawn from the credits, *HIII* continues the same line of horror/ technofantasy displayed in his QUATERMASS AND THE PIT (1958-9 tv; *1968*) and *Quatermass* (1979 tv). [JG]

HALLOWEEN WITH THE NEW ADDAMS FAMILY ot of *The Addams Family* (*1977* tvm). ◊ ADDAMS FAMILY MOVIES.

HALLUCINATION The PERCEPTION (in fantasy usually visual, though other senses may be invoked) of an object which does not exist or the experience of a false sensation. In SUPERNATURAL FICTION and HORROR, plots commonly turn on whether a hallucination is not in fact a hallucination at all but has been created by some intruder into the real world. The structure of fantasy stories does not normally depend on the correct interpretation of hallucination, ILLUSION being much more common. [JC]

HAMBLY, BARBARA (1951-   ) US writer. All of BH's fantasies take place in worlds that are explicitly part of a MULTIVERSE, as is her Earth; protagonists from Earth have discovered full power and agency in other worlds where magic is more operational, and at least one of her mages has found his full strength here. These worlds have in common that those gifted with magic are a PARIAH ELITE, often pursued by Church and State; only occasionally, as in *Stranger at the Wedding* (**1994**; vt *Sorcerer's Ward* UK 1994), has BH

provided a salutary sense of the power of magic to work petty personal evil.

Such considerations are crucial to her early **Darwath** trilogy – *The Time of the Dark* (**1982**), *The Walls of Air* (**1983**) and *The Armies of Daylight* (**1983**) – where two standard California misfits, a woman scholar and a boy biker, find themselves shoulder to shoulder with WIZARDS and warriors against a resurgence of powerful beings that mingle H.P. LOVECRAFT with the **Alien** movies, and against backstabbing intrigues by courtiers and inquisitors. Both Gil and Rudy are standard BRAVE-LITTLE-TAILOR figures discovering new strengths as swordswoman and magus.

The **Sun Wolf** books – *The Ladies of Mandrigyn* (**1984**), *The Witches of Wenshar* (**1987**), these two assembled as *The Unschooled Wizard* (omni **1987**), and *The Dark Hand of Magic* (**1990**) – are TEMPLATE fantasies in that Sun Wolf, one of the few surviving WIZARDS in a world where an evil wizard has extirpated his own kind, seeks instruction in powers he can only partially control. His laconic lover is Star Hawk; in *The Witches of Wenshar* the DUO find themselves solving that most unlikely of things, an S&S country-house murder mystery.

BH's most explicitly REVISIONIST FANTASY is *Dragonsbane* (**1986**), whose middle-aged warrior and witch protagonists killed a DRAGON once, found it a messy job, and resent being required to do it again. This is at once her funniest book and her saddest, as her heroine faces the choice of mediocre human happiness with husband and children or full glorious power and the loss of humanity. The GASLIGHT ROMANCE *Those who Hunt the Night* (**1988**; vt *Immortal Blood* 1988 UK) forces a UK scholar-agent and his lover to investigate the serial killing (◊ SERIAL KILLERS) of London's VAMPIRES (whose status is partly rationalized; ◊ RATIONALIZED FANTASY). In a sequel, *Travelling with the Dead* (**1995** UK), the same duo and their vampire MENTOR/adversary travel severally to Istanbul in an attempt to prevent a vampiric dimension in the looming WWI.

One of BH's great strengths is that she both gives her readers what they want and then explores the ethics. When in *The Dark Hand of Magic* Sun Wolf and Star Hawk meet up with their old company, they find themselves caught up with quandaries about the mercenary calling. In *Bride of the Rat God* (**1994**), a supernatural fantasy set in 1920s Hollywood (◊ LOS ANGELES), we long for the overbearing silent-movie star to whom the heroine acts as companion to get a comeuppance, but find ourselves concerned for her when that comeuppance takes the form of the supernatural menace of a DEMON.

The two remaining series rehearse similar themes, both dealing with persecuted mages and interactions with Earth. **Sun-Cross** – *The Rainbow Abyss* (**1991** UK) and *The Magicians of Night* (**1992**), assembled as *Sun-Cross* (omni **1992**) – takes the protagonist Rhion on a guided tour of oppression at home and then enables him to escape to what he thinks is a safe haven, but which turns out to be the Third Reich (◊ WORLD WAR II). In the **Antryg Windrose** series – *The Silent Tower* (**1986**) and *The Silicon Mage* (**1988**), both assembled as *Darkmage* (omni **1988**), plus *Dog Wizard* (**1993**) and *Stranger at the Wedding* – local magic blends interestingly with science stolen from Earth in a complicated tale in which IDENTITY EXCHANGE and elements of TECHNOFANTASY are added to the usual mix.

BH uses TEMPLATES and numerous standard PLOT DEVICES to tell stories which constantly examine their own premises in humane terms. When she writes, as she often does, of UGLY DUCKLINGS, it is with a real and unannealed pain of frustrated aspiration that she imbues them, not the easy consolatory self-identification of GENRE FANTASY. But the popularity of HIGH-FANTASY epics in the 1980s left even as prolific a writer as BH rather less popular than she merited; the refusal to skew her work in a more conventional and perhaps remunerative direction echoes the lonely integrity of her protagonists. [RK]

Other works: *The Quirinal Hill Affair* (**1983**; vt *Search the Seven Hills* 1987), a historical whodunnit; *Ishmael* * (**1985**), *Ghost-Walker* * (**1991**) and *Crossroad* * (**1994**), all **Star Trek** ties; *Beauty and the Beast* * (**1989**) and *Beauty and the Beast: Song of Orpheus* * (**1990**), novelizing the tv series BEAUTY AND THE BEAST (1987-90); *Star Wars: Children of the Jedi* * (**1995**); *Sisters of the Night* (auth **1995**) with Martin H. GREENBERG.

HAMILTON, CLIVE Pseudonym of C.S. LEWIS.

HAMILTON, JOHN Pseudonym of Achmed ABDULLAH.

HAMILTON, LAURELL K. (? -) US author of the entertaining **Anita Blake, Vampire Hunter** series of DETECTIVE/THRILLER FANTASIES. Blake is a tough female PI working in an ALTERNATE-WORLD USA where VAMPIRES and WEREWOLVES have gained civil rights as oppressed minorities. The series comprises *Guilty Pleasures* (**1993**), *The Laughing Corpse* (**1994**), *Circus of the Damned* (**1995**) and «The Lunatic Café» (1996). [JR/DRL]

Other works: *Nightseer* (**1992**); *Ravenloft: Death of a Darklord* * (**1995**), GAME tie.

HAMMER Successful UK movie production company (more properly Hammer Films) whose name became synonymous with horror, fantasy and action-adventure movies for over two decades. Hammer Productions was created in 1934 by William Hinds (pseudonym Will Hammer), who had formed the Exclusive movie-distribution company with cinema-owner Enrique Carreras. The first Hammer production was a comedy, *The Public Life of Henry the Ninth* (**1935**). It was followed by *The Mystery of the Marie Celeste* (**1936**; vt *Phantom Ship* US), which featured imported US star Bela Lugosi as a crazed killer, and two more movies. Hammer was no longer listed as an active company by the outbreak of WWII. In 1947, however, when the ABC cinema circuit was looking for a company to supply supporting features, Hammer was reformed. Under the guidance of Enrique's son Sir James Carreras (and later his son Michael and William's son Anthony Hinds) the company turned out numerous low-budget pictures, usually released through Exclusive and often adapted from established radio and tv plays. These included three features based on the BBC radio series *Dick Barton* (*Dick Barton Special Agent* [**1948**; vt *Dick Barton Detective* US], *Dick Barton Strikes Back* [**1949**] and *Dick Barton at Bay* [**1950**]), each starring Don Stannard as the investigator thwarting the plans of mad scientists. Valentine Dyall recreated his radio role in *The Man in Black* (**1950**) and played a sinister lodger suspected of being JACK THE RIPPER in *Room to Let* (**1950**). With *Stolen Face* (**1952**), director Terence Fisher foreshadowed his own FRANKEN-STEIN-MOVIE series and the medical horrors of the European cinema when a plastic surgeon (Paul Henreid) discovered beauty was only skin-deep. Fisher's *Four-Sided Triangle* (**1953**) and *Spaceways* (**1953**) were both firmly rooted in the burgeoning 1950s SCIENCE-FICTION boom. With *The Quatermass Xperiment* (**1955**; vt *The Creeping Unknown* US) Hammer finally made the move towards the type of full-blown HORROR MOVIE with which it would always be identified. Hammer quickly followed up with *X The Unknown* (**1956**) and *Quatermass 2* (**1956**; vt *Enemy from Space* US). *The Abominable Snowman* (**1957**; vt *The Abominable Snowman of the Himalayas* US) starred Peter Cushing. The same year, Hammer decided audiences were interested in more human monsters and that the time was right to produce the first colour version of Mary SHELLEY's classic novel. Directed by Terence Fisher and scripted by Jimmy Sangster, *The Curse of Frankenstein* (**1957**) starred Cushing as the ruthless Baron Victor Frankenstein and Christopher Lee as his horribly scarred creation. This went on to become the highest-grossing movie produced by a UK studio that year and ushered in the Hammer Age of Horror. Hammer quickly had Fisher and Sangster reteam its star duo in *Dracula* (**1958**), with Cushing playing Van Helsing and Lee playing Dracula. The result proved an even bigger box-office hit, and the quartet was reunited for colour versions of *The Hound of the Baskervilles* (**1959**) and *The Mummy* (**1959**). Meanwhile Hammer had already embarked on a series of sequels to its two hit horror remakes: because Lee's Monster had been dissolved in a bath of acid, the studio ingeniously decided to make Cushing's Victor Frankenstein the recurring character; the Baron continued his experiments in several further Hammer FRANKENSTEIN MOVIES. Cushing was also back as Van Helsing to confront the Count's disciples in *The Brides of Dracula* (**1960**); Lee donned cape and fangs again to return as the eponymous VAMPIRE for *Dracula Prince of Darkness* (**1965**) and further DRACULA MOVIES. In an attempt to inject some freshness into the series, Hammer updated the Count's exploits to contemporary London for *Dracula A.D. 1972* (**1972**) and *The Satanic Rites of Dracula* (**1973**), once again adding Cushing to the mix as Van Helsing's grandson. The actor portrayed the character one final time, battling Kung Fu zombies and John Forbes-Robertson's Dracula in *The Legend of the 7 Golden Vampires* (**1974**).

Until 1967, Hammer was based at Bray Studios, a converted country house, where it continued to remake classic horror stories with *The Two Faces of Dr Jekyll* (**1960**; vt *House of Fright* US; ◊ JEKYLL AND HYDE MOVIES), *The Curse of the Werewolf* (**1960**), *The* PHANTOM OF THE OPERA (**1962**) and *The Plague of the Zombies* (**1966**), and created new monsters in *The Gorgon* (**1964**) and *The Reptile* (**1966**). The company was joined in its revival of the horror movie by rivals Amicus and Tigon in the UK and American International in the USA. As the decade continued, Hammer's output came to depend not only on sequels and revisions of earlier successes, such as *The Kiss of the Vampire* (**1962**; vt *Kiss of Evil* US), *The Curse of the Mummy's Tomb* (**1964**), *The Mummy's Shroud* (**1967**) and QUATERMASS AND THE PIT (**1968**; vt *Five Million Years to Earth* US), but diversified with numerous war movies, comedies, PSYCHO-LOGICAL THRILLERS, historical dramas, sf and such big-budget fantasies as remakes of H. Rider HAGGARD's SHE and of ONE MILLION BC. In the early 1970s, as horror movies became less popular, Hammer responded by adding more nudity and violence to its pictures – as in the **Karnstein** trilogy, being *The Vampire Lovers* (**1970**), *Lust for a Vampire* (**1970**) and *Twins of Evil* (**1971**) – and experimented with new variations on established themes, such as *The Horror of Frankenstein* (**1970**), *Countess Dracula* (**1970**), *Hands of the Ripper* (**1971**), *Blood from the Mummy's Tomb* (**1971**), *Dr Jekyll & Sister Hyde* (**1971**), *Vampire Circus* (**1971**) and *Captain Kronos Vampire Hunter* (**1972**).

Hammer returned to its roots with a series of low-brow comedies based on popular tv shows. Although it had almost 170 films and numerous shorts to its credit, Hammer found it harder to compete in the international marketplace, and the financial crash of 1974 was the final nail in the coffin. A German co-production of TO THE DEVIL A DAUGHTER (*1976*), based on the novel by Dennis WHEATLEY, was the end. Although there have been various attempts by current owner Roy Skeggs to keep the name alive – with the short-lived tv series *Hammer House of Horror* (1980), a package of lacklustre tvms broadcast as *Hammer House of Mystery and Suspense* (1984), and an often-announced feature production slate during the 1990s – it seems unlikely Hammer will ever again recapture the Gothic elegance and box-office success it enjoyed during its heyday.   [SJ]

Further reading: *The House of Horror: The Complete Story of Hammer Films* (**1973**; rev 1981; rev vt *House of Horror: The Complete Hammer Film Story* 1994) ed Allen Eyles, Robert Adkinson and Nicholas Fry; *A Heritage of Horror* (**1973**) by David Pirie; *Hammer and Beyond: The British Horror Film* (**1993**) by Peter Hutchings; *Hammer, House of Horror* (**1996**) by Howard Maxford; *Hammer Films: An Exhaustive Filmography* (**1996**) by Tom Johnson and Debrah Del Vecchio.

See also: JOURNEY TO THE UNKNOWN (1968-9).

HAMMOND, KEITH Pseudonym of Henry KUTTNER.

HANCOCK, NEIL (ANDERSON) (1941-   ) US writer, almost all of whose work has been in his three **Atlantean Earth** sequences; any connection to ATLANTIS is remote, though the overall tale does not yet seem fully to have been told. The first sequence – the **Circle of Light** series, comprising *Greyfax Grimwald* (**1977**), *Faragon Fairingay* (**1977**), *Calix Stay* (**1977**) and *Squaring the Circle* (**1977**) – is the one most visibly influenced by J.R.R. TOLKIEN's *The Lord of the Rings* (**1954-5**), though the next two series, being all backstory, likewise follow the Tolkien pattern. In **Circle of Light** three COMPANIONS – a Bear, an Otter (◊ ANIMAL FANTASY) and a DWARF – return to a Middle-Earth-like Earth to guard the Arkenchest whose retention is vital if the good GODDESS in NH's extremely complex cosmology is to prevail in the Upper Worlds (and elsewhere) over the bad goddess. The **Wilderness of Four** series – *Across the Far Mountain* (**1982**), *The Plains of the Sea* (**1982**), *On the Boundaries of Darkness* (**1982**) and *The Road to the Middle Islands* (**1982**) – describes the early years of the **Atlantean Earth** and details the lives of the creatures who prefigure the heroes in the **Circle of Light** books. The **Windameir Circle** series – *The Fires of Windameir* (**1985**), *The Sea of Silence* (**1987**), *A Wanderer's Return* (**1988**) and *The Bridge of Dawn* (**1991**) – continues diffusely to present backstory.

A singleton, *Dragon Winter* (**1978**), repeats the basic story: animal companions are threatened by an evil principle opposed to the health of the LAND.   [JC]

HAND, ELIZABETH (1957-   ) US writer who began publishing work of genre interest with "Prince of Flowers" for *Twilight Zone Magazine* in 1988, and whose **Winterlong** sequence – *Winterlong* (**1990**), *Aestival Tide* (**1992**) and *Icarus Descending* (**1993**) – is sf in its underlying structure, though its timbre is fantasy-like. The DYING-EARTH cadences of her story and her narrative rhythm generate a sense that her tale – which in fact takes place in the eastern USA, after a plethora of cataclysms – occupies a PLANE-TARY-ROMANCE venue.

So it was not surprising that EH's first singleton, *Waking the Moon* (**1994** UK; cut and preferred version 1995 US), would be fantasy. The long tale begins as a CONTEMPORARY FANTASY set in Washington, DC, around 1970, with the protagonist entering The University of the Archangels and St John the Divine, which is run by *benandanti* – an historical society – who turn out to be SECRET MASTERS of the world whose task is to guard against a reawakening of the GODDESS and the reimposition of her matriarchal rule. Like most serious secret-master tales, the basic subject of *Waking the Moon* is the MATTER of the world itself; halfway through, the tale turns (or threatens to turn) into a full-scale INSTAURA-TION FANTASY, when it seems inevitable that the goddess, immanent in the MOON, is about to awaken out of BONDAGE (◊◊ FACE OF GLORY). If she awakes it will be a new world, but she does not, and the novel turns into a SUPERNATURAL FIC-TION, with the protagonist (and the folk of Earth) threatened by the invasive seductions of the still-thwarted goddess as she plots men's downfall through her acolytes and AVATAR. The book closes, almost dutifully, with a scene – typical of supernatural fictions – in which the temple of evil (in Washington) is brought down. However, though the narrative points in more directions than it can follow, EH's glad energy is evident throughout. She won a 1995 WORLD FANTASY AWARD for her novella "Last Summer at Mars Hill" (1994 *F&SF*). Her further work is awaited eagerly.   [JC]

Other works: *12 Monkeys* * (**1995**), novelizing the movie.

HANDS OF A STRANGLER vt of *The* HANDS OF ORLAC (*1960*).

HANDS OF ORLAC, THE (ot *Orlacs Hände*) Austrian movie (*1924*). ◊ MAD LOVE (*1935*).

HANDS OF ORLAC, THE (vt *Les Mains d'Orlac*; vt *Hands of a Strangler*) (*1960*). UK/French movie. Riviera/Pendennis/Britannia/Continental. **Pr** Steven Pallos, Donald Taylor. **Dir** Edmond T. Gréville (French version Jacques Lemare). **Screenplay** John Baines, Gréville, Taylor. **Novelization** *The Hands of Orlac* * (**1961**) by Robert Bateman. **Starring** Felix Aylmer (Dr Francis Cochrane), Dany Carrel (Li-Lang), Mel Ferrer (Stephen/Steven Orlac), Christopher Lee (Néron), Lucile Saint Simon (Louise Cochrane), Yanilou (Emilie). 105 mins, cut to 95 mins. B/w.

Generally listed as a remake of MAD LOVE (*1935*), this has a quite different story. US pianist Orlac crashes while flying to Paris to visit fiancée Louise. Brilliant surgeon Dr Volcheff either (it is left ambiguous) miraculously repairs Orlac's badly damaged hands or replaces them using those of guillotined strangler Louis Vasseur; whichever, Orlac has an HALLUCINATION that newspaper headlines show the latter to be the case. Later, lovemaking with Louise, he almost strangles her; he flees with his guilty hands. Much blood and tears are shed before the police discover Vasseur was innocent – so Orlac's hands, whatever their origins, are not those of a strangler.

THOO was shot simultaneously in French and English, with some clumsy dubbing. It is raised above humdrum through its ambiguity as to the source of Orlac's "new" hands, which gives it interest as both a PSYCHOLOGICAL THRILLER and fantasy of PERCEPTION.   [JG]

HANK IN MAGGS [s] ◊ Kenneth MORRIS.

HANNA-BARBERA Animation company set up in 1957 by William Hanna (1910-   ) and Joseph Barbera (1911-   ), who together created, at MGM, the characters Tom & Jerry (initially Jasper and an unnamed mouse) in *Puss*

Gets the Boot (**1940**). Oddly, MGM producer Fred Quimby was dubious about the possibilities of this cartoon, and released it only reluctantly. For the next 17 years the Hanna-Barbera team produced a string of **Tom & Jerry** hit cartoons; when they departed MGM, leaving the characters behind them, no other director – even Chuck JONES – proved capable of picking up the baton; a nadir was reached with the full-length ANIMATED MOVIE *Tom and Jerry: The Movie* (**1992**), dir Phil Roman.

HB became most famous for their limited-animation tv work, which included the series **The Flintstones** (1960-66), **Huckleberry Hound** (1958-61), **Top Cat** (1961-2) and **Yogi Bear** (1961-2, 1973-5, 1988-9); all were characterized by fast one-liners that more than made up for the economy of the animation. Other series have been **The ABC Saturday Superstar Movie** (1972-3; vt **The New Saturday Superstar Movie** 1973-4), **Abbott and Costello, The** ADDAMS FAMILY (1973-5), **The Adventures of Gulliver** (1968-79), **The Adventures of Jonny Quest** (1964-5), **The All-New Popeye Hour** (1978-81), **The All-New Scooby and Scrappy-Doo Show** (1983-4), **The All-New Super Friends Hour** (1977-8) – which featured SUPERHEROES from DC COMICS – **The Amazing Chan and the Chan Clan** (1972-4), **Amigo and Friends** (1980-82), **The Atom Ant Show** (1968) and **The Secret Squirrel Show** (1967) and **The Atom Ant/Secret Squirrel Show** (1965-7), **The Banana Splits Adventure Hour** (1968-70), **The Best of Scooby-Doo** (1983-4), **Birdman and the Galaxy Trio** (1967-8), **The Biskitts** (1983-4), **Buford and the Galloping Ghost** (1979), **Butch Cassidy and the Sun Dance Kids** (1973-4), **Captain Caveman and the Teen Angels** (1980), **Captain Inventory** (1973), **Casper and the Angels** (1979-80) (◊ CASPER, THE FRIENDLY GHOST), **The Cattanooga Cats** (1969-70), **C.B. Bears** (1977-8), **Challenge of the Gobots** (1984), **Clue Club** (1976-7), **The Completely Mental Misadventures of Ed Grimley** (1988-9), **Dastardly and Muttley and their Flying Machines** (1969-71), **Drak Pack** (1980-82) – featuring youthful descendants of the FRANKENSTEIN monster, DRACULA and the Wolf Man – **The Dukes** (1983), **Dynomutt, Dog Wonder** (1978), **Fantastic Four** (1967-70), **Fantastic Max** (1988), **The Flintstone Kids** (1986-8), **Fonz and the Happy Days Gang** (1980-82), **Foofur** (1986-8), **Frankenstein Jr and The Impossibles** (1966-8), **The Funky Phantom** (1971-2), **The Galaxy Goof-Ups** (1978-9), **Galtar and the Golden Lance** (1985), **The Godzilla Show** (1979) and **The Godzilla Power Hour** (1978; vt **Godzilla and the Super 90**), **Goober and the Ghost Chasers** (1973-5), **The Harlem Globetrotters** (1970-73) and **The Super Globetrotters** (1979), **Help! It's the Hair Bear Bunch** (1971-2), **The Herculoids** (1967-9), **Hong Kong Phooey** (1974-6), **Inch-High Private Eye** (1973-4), **Jabberjaw** (1976-8), **Jeannie** (1973-5) – a spinoff from I DREAM OF JEANNIE (1965-70) – **The Jetsons** (1962-4), **Jokebook** (1982), **Josie and the Pussycats** (1970-72; vt **Josie and the Pussycats in Outer Space** 1972-4), **The Kwicky Koala Show** (1981-2), **Laurel and Hardy** (1966), **Laverne and Shirley with The Fonz** (1982-3), **Lippy the Lion** (1962), **The Little Rascals/Richie Rich Show** (1982-4), **The Magilla Gorilla Show** (1964), **Moby Dick & the Mighty Mightor** (1967-9), **Monchichis** (1983-4), **Mork and Mindy** (1982-3), **Motormouse and Autocat** (1970-71), **The New Adventures of Huck Finn** (1968-9), **The New Scooby-Doo Comedy Movies** (1972-4), **The**

New Scooby-Doo Mysteries (1984-5), **The New Shmoo** (1979), **The Pac-Man Show** (1982-3), **Paddington Bear** (1989), **Partridge Family: 2200 AD** (1974-5), **The Paw Paws** (1985), **Peter Potamus and his Magic Flying Balloon** (1964), **Pound Puppies** (1986-7) and **All-New Pound Puppies** (1987-8), **A Pup Named Scooby-Doo** (1988-90), **Quick Draw McGraw** (1959-62), **Roman Holidays** (1972-3), **The Ruff and Reddy Show** (1957-64), **Sealab 2020** (1972-3), **Shazzan!** (1967-9; ◊◊ CAPTAIN MARVEL), **The Shirt Tales** (1982-4), **Sinbad Jr** (1965), **The Skatebirds** (1977-8), **The Snorks** (1984-8), **Space Ghost and Dino Boy** (1966-8), **Space Kiddettes** (1966-7), **Space Stars** (1981-2), **Speedy Buggy** (1973-4), **Super Friends** (1973-5) and **Challenge of the Super Friends** (1978-9) and **The World's Greatest Super Friends** (1979-80), **Super Powers Team: Galactic Guardians** (1985-6), **These Are the Days** (1974-6), **The Three Robinic Stooges** (1978-81), **Touché Turtle** (1962), **The Trollkins** (1981-2), **Valley of the Dinosaurs** (1974-6) and **Wildfire** (1986-7). Most of these are journeyman stuff, suitable only for a Saturday morning, but almost without exception they contain material of fantasy interest. [JG]

HARD FANTASY A useful term for stories where MAGIC is regarded as an almost scientific force of Nature, and subject to the same sorts of rules and principles. This was the type of fantasy championed by John W. Campbell Jr in UNKNOWN – represented by stories like Robert A. HEINLEIN's "Magic, Inc." (1941) and H.L. GOLD's "Trouble with Water" (1939). By analogy with SCIENCE FICTION, the term HF might refer to fantasy stories equivalent to the form of hard sf known as the "scientific problem" story, where the hero must logically solve a problematic magical situation (◊ QUIBBLES). One example is Laurence YEP's *Dragon Cauldron* (**1991**), whose hero is trapped on a magic ISLAND where all objects that attempt to float away are driven back to shore: he reasons that the SPELL must make an exception for dirt (otherwise, silt from the surrounding river would eventually build a landbridge from the island to the shore), so builds a raft made of clay pots (clay being a form of dirt) and thereby escapes the trap. [GW]

HARDING, ROSS Pseudonym of David A. GEMMELL.

HARDING, SIMON (? -) UK writer whose **Streamskelter** sequence – *Streamskelter* (**1994**) and *Changeling Hearts* (**1995**) – concerns the CROSSHATCH relationship between a contemporary small town in the west of England and a FANTASYLAND called Trafarionath. Much of the plot is generated by matters typical of DYNASTIC FANTASY. The sequence is marked by a jarring flippancy of tone, though some genuine storytelling energy does break through. [JC]

HARDY, LYNDON (MAURICE) (1941-) US physicist and writer whose fantasy has been restricted to the **Arcadia** sequence: *Master of the Five Magics* (**1980**), *Secret of the Sixth Magic* (**1984**) and *Riddle of the Seven Realms* (**1988**). Rigorously applying the precepts of RATIONALIZED FANTASY to the laws of MAGIC, LH has created an attractive, internally coherent, impossible science; although his novels avoid the WONDERLAND logic-chopping and attenuated sense of REALITY typical of fantasy tales whose worlds are built and operated according to rules, he never strays far from his argument. In the first volume magic is divided into five categories – thaumaturgy, alchemy, magic, sorcery and wizardry – all meticulously distinguished. In the second volume a corrosive alternate form of magic invades Arcadia, and must be countered. In the third a TRICKSTER protagonist careenes through a variety of REALITIES contiguous to Arcadia's. [JC]

HARGREAVES, [Sir] GERALD (DELAPRYME) (1881-1972) UK writer in whose long play, *Atalanta: A Story of Atlantis: A Fantasy with Music* (**1949**), ACHILLES is sent in disguise to ATLANTIS to woo and otherwise convince Princess Atalanta not to wage war on the Greeks, who have just destroyed Troy, an Atlantean outpost. [JC]

HARLEQUIN ◊ COMMEDIA DELL'ARTE.

HARLEQUINADE ◊ COMMEDIA DELL'ARTE.

HARMAN, ANDREW (1964-   ) UK writer of comic fantasies, the first five comprising the **Firkin** sequence – *The Sorcerer's Appendix* (**1993**), *The Frogs of War* (**1994**), *The Tome Tunnel* (**1994**) and *Fahrenheit 666* (**1995**) – set in the opposing medieval FANTASYLAND kingdoms of Rhyngill and Chranachan. There is a good deal of MAGIC, usually botched, and some TWICE-TOLD fairytales. *101 Damnations* (**1995**) shifts venue to our world's present. AH's humour tends to slapstick. He is one of several UK writers to generate work of this sort on the model – but without the wit – of Terry PRATCHETT. [JC]

HARPER'S MONTHLY ◊ MAGAZINES.

HARPUR, PATRICK (1950-   ) UK writer in whose first novel, *The Serpent's Circle* (**1985**), a secret monastic order called The Little Brothers of the Apostles unleashes its old RELIGION against the Roman Catholic Church. In *Mercurius, of the Marriage of Heaven and Earth* (**1990**) a country clergyman involved in ALCHEMY and a search for the PHILOSOPHERS' STONE finds his QUEST interacting with that of another in another time, under the auspices of Mercurius, or HERMES. The sense in which this text is meant as fantasy – i.e., as a tale of the impossible – is made unclear, if only retroactively, by *Daimonic Reality: A Field Guide to the Underworld* (**1994**), a nonfiction study in which experiences of this sort are granted, at the very least, an agnostic understanding. [JC]

HARRIS, CHRISTIE (LUCY IRWIN) (1907-   ) Canadian writer whose **Mouse Woman** sequence – *Mouse Woman and the Vanished Princesses* (coll of linked stories **1976**), *Mouse Woman and the Mischief Makers* (coll of linked stories **1977**) and *Mouse Woman and the Muddleheads* (coll of linked stories **1979**) – retells Native American LEGENDS about the eponymous mouse (◊ MICE AND RATS), who is also a woman, and who busies herself in aid of errant humans. Other collections in which legends are transformed into fantasy include *Once Upon a Totem* (coll **1963**) and *Once More Upon a Totem* (**1973**). *Secret in the Stlalakum Wild* (**1972**) is a CHILDREN'S FANTASY set in British Columbia; spirits of the Salish nation send a young girl INTO THE WOODS to seek treasure. [JC]

HARRIS, DEBORAH TURNER (1951-   ) US writer whose **Mages of Garillon** sequence – *The Burning Stone* (**1986**), *The Gauntlet of Malice* (**1987**) and *Spiral of Fire* (**1989**) – is a DYNASTIC FANTASY set in a medievalized Scotland-like LAND OF FABLE. The impetuous MAGUS of the first volume is a figure of some interest, as he gradually becomes a person who LEARNS BETTER; but the vividness is not consistently sustained. *Caledon of the Mists* (**1994**) shares a similar venue. DTH collaborated with Katherine KURTZ on the **Adam Sinclair** sequence: *The Adept* (**1991**), *The Lodge of the Lynx* (**1992**), *The Templar Treasure* (**1993**) and *Dagger Magic* (**1995**). [JC]

HARRIS, GERALDINE (RACHEL) (1951-   ) UK

writer whose **Seven Citadels** sequence – *Prince of the Godborn* (**1982**), *The Children of the Wind* (**1982**), *The Dead Kingdom* (**1983**) and *The Seventh Gate* (**1983**) – follows the QUEST of the young son of the ruler of the Empire of Galkis (which lies in a SECONDARY WORLD) in his attempt to find SEVEN sorcerers who hold seven keys to open seven gates (◊ PORTALS) that bar him from the potential saviour of the land. The PLOT-COUPON nature of the tale tends at times to detract from GH's quiet presentation of the slow maturation of the protagonist; but this understory does, eventually, prevail, and the protagonist's knot of passage into full adulthood (◊ RECOGNITION) neatly makes possible the HEALING of the land. [JC]

HARRIS, JOEL CHANDLER (1848-1908) US newspaperman and writer who wrote several novels for adults and the long series of BEAST-FABLES told to a small boy by **Uncle Remus**. The first, "Negro Folklore. The Story of Mr Rabbit and Mr Fox, as told by Uncle Remus" (1879 the *Constitution*) was followed immediately by the most famous, "Brer Rabbit, Brer Fox, and the Tar Baby" (1879 the *Constitution*), which memorably retells (within the **Uncle Remus** FRAME STORY) a FOLKTALE known worldwide, and immortalizes Brer Rabbit and Brer Fox. These stories, and 32 others, were assembled as *Uncle Remus: His Songs and His Sayings: The Folklore of the Old Plantation* (coll **1881**; vt *Uncle Remus and His Legends of the Old Plantation* 1881 UK; vt *Uncle Remus, or Mr Fox, Mr Rabbit, and Mr Terrapin* 1881 UK; rev under original title 1895). This first volume was followed by *Nights with Uncle Remus: Myths and Legends of the Old Plantation* (coll **1883**), *Uncle Remus and his Friends: Old Plantation Stories, Songs and Ballads* (coll **1892**), *Told by Uncle Remus: New Stories of the Old Plantation* (coll **1905**), *Uncle Remus and Brer Rabbit* (coll **1907**), *Uncle Remus and the Little Boy* (coll **1910**), *Uncle Remus Returns* (coll **1918**), *The Witch Wolf: An Uncle Remus Story* (**1921** chap), *Stories from Uncle Remus* (coll **1934**) and *Seven Tales of Uncle Remus* (coll **1948**). These volumes are conveniently assembled as *The Complete Tales of Uncle Remus* (omni **1955**); some of the associated poems appear in *The Tar-Baby and Other Rhymes of Uncle Remus* (coll **1904**). The basic frame was incorporated into the DISNEY live-action/ANIMATED MOVIE *Song of the South* (**1947**).

Over and above their acuteness, the tales are marked by an acute sensitivity to Black dialects, and by a remarkable lack – compared with the late-19th-century norm – of condescension. [JC]

Other works: *Daddy Jack the Runaway and Short Stories Told After Dark* (coll **1889**); *Little Mr Thimblefinger and His Queer Country: What the Children Saw and Heard There* (**1895**), the "country" being an underground WONDERLAND; *Wally Walderon and His Story-Telling Machine* (**1901**).

HARRIS, MABEL ROLLINS (?   -   ) US artist. ◊ OLIVIA.

HARRIS, MacDONALD Working name of US writer and academic Donald William Heiney (1921-1993) for all his fiction, much of which hovers on the cusp of the FANTASTIC. In *Bull Fire* (**1973**) mythopoeic resonances gradually dominate as the narrator of the tale is revealed as a MINOTAUR figure. *Pandora's Galley* (**1979**), ostensibly an historical novel, slowly transforms the VENICE of 1797 into a surreal DREAM of cityscape, with emanations of SHADOW and a "male animal" odour in the wet air.

Herma (**1981**), MH's first full fantasy (◊ HERMES), tells the story of Herma, a female OPERA singer able to change her

SEX by staring into a MIRROR and willing her METAMORPHO-SIS. As a woman, she plays all three female roles in *The Tales of Hoffmann* (**1880**) by Jacques Offenbach (1819-1880) (◊ E.T.A. HOFFMANN); as a man, she becomes a fighter pilot in WORLD WAR I. Only Marcel Proust (1871-1922) rumbles the dual nature of this LIMINAL BEING. *Screenplay* (**1982**) is a TIMESLIP fantasy whose protagonist enters a 1920s Hollywood, a black-and-white POLDER armed against time and decay. In the RECURSIVE FANTASY *Tenth* (**1984**) – the title refers to Beethoven's hypothetical 10th symphony – a contemporary scholar becomes involved in the works of Adrian Leverkühn, the composer who, in Thomas MANN's *Doctor Faustus* (**1947**), gains genius through a PACT WITH THE DEVIL. In *The Little People* (**1986**) hints of CROSSHATCH between this world and FAERIE are ingeniously presented, but prove delusional. *Glowstone* (**1987**), much of which, like *Herma*, is set in PARIS, treats Belle Époque France as a COMMEDIA DELL'ARTE masquerade. [JC]

HARRIS, ROSEMARY (JEANNE) (1923-   ) UK writer who early concentrated on nongenre novels for adults, beginning with *The Summer-House* (**1956**), but much of whose work for a YA audience is fantasy, beginning with the **Reuben** sequence: *The Moon in the Cloud* (**1968**), *The Shadow on the Sun* (**1970**) and *The Bright and Morning Star* (**1972**). In this compelling tale, Reuben is tricked by Ham into voyaging into EGYPT to acquire animals wanted for Noah's Ark. Luckily, because Ham and his wife have meantime died, Reuben and his own wife are allowed on the Ark and survive the FLOOD; in subsequent volumes they undergo increasingly complex adventures in the service of the Egyptian Pharaoh, the great Merenkere. Much of the overall tale is commented upon by a talking CAT.

Later work includes *The Seal-Singing* (**1971**), in which the ambiguous influence of an ancestor WITCH causes some contemporary children to save seals at risk, and *A Quest for Orion* (**1978**) and its sequel, *Tower of the Stars* (**1980**), both sf. [JC]

Other works (mostly for younger children): *The Child in the Bamboo Grove* (**1981**) and *The Little Dog of Fo* (**1976**), TWICE-TOLD Chinese legends; *The King's White Elephant* (**1973**); *The Lotus and the Grail: Legends from East to West* (coll **1974**; cut vt *Sea Magic and Other Stories of Enchantment* 1974 US); *The Flying Ship* (**1975**); *I Want to be a Fish* (**1977**); *Beauty and the Beast* (**1979**); *Green Finger House* (**1980**); *Tower of the Stars* (**1980**); *The Enchanted Horse* (**1981**); *Janni's Stork* (**1982**); *Zed* (**1982**).

HARRIS, (THEODORE) WILSON (1921-   ) Guyana-born writer, in the UK from 1959, who applies an intense magic-realist (◊ MAGIC REALISM) colouring and vigour to his stories, set in Guyana – some of which concern themselves with the creation of a MYTH OF ORIGIN for the nation. His first series, the **Guyana Quartet** – *Palace of the Peacock* (**1960**), *The Far Journey of Oudin* (**1961**), *The Whole Armour* (**1962**) and *The Secret Ladder* (**1964**), with *Heartland* (**1964**) as a pendant – treats Guyana as a kind of tympanum on which the eternal return of past peoples and unending conquests makes a kind of dance: WH's prose is highly rhythmical, heavily foregrounded, and REALITY and its PERCEPTION constantly shiver to the beat. A second, loosely connected sequence comprises *The Eye of the Scarecrow* (**1965**), *The Waiting Room* (**1967**), *Tumatumari* (**1968**) and *Ascent to Omai* (**1970**). Other novels, like *Black Marsden* (**1972**), *Companions of the Day and Night* (**1975**) and *The Angel at the Gate* (**1982**), are also loosely connected;

the **Carnival** trilogy – *Carnival* (**1985**), *The Infinite Rehearsal* (**1987**) and *The Four Banks of the River of Space* (**1990**) – is more tightly constructed, treating modern life as a CYCLE patterned on DANTE's *Divine Comedy* (written *c*1320); the third volume also introduces a number of characters whose lives are splintered presentations of the UNDERLIER figure of ODYSSEUS. *Resurrection at Sorrow Hill* (**1993**), like many of WH's novels, uses the image of a voyage up-RIVER to give order and flow to inveterate conflations of history and myth. [JC]

Other works: *The Sleepers of Roraima* (coll **1970**) and *The Age of the Rainmakers* (coll **1971**), presenting Native American material in a TWICE-TOLD fashion; *Da Silva da Silva's Cultivated Wilderness, and Genesis of the Clowns* (coll **1977**), two novellas; *The Tree of the Sun* (**1978**).

HARRISON, JANE E. (1850-1928) UK writer. ◊ Hope MIRRLEES.

HARRISON, MICHAEL Name adopted by UK writer Maurice Desmond Rohan (1907-1991), who contributed to various genres. Short stories of interest are assembled in *Transit of Venus* (coll **1936**). *Higher Things* (**1945**) is told in the manner H.G. WELLS used mostly in the 1920s and 1930s when he wished to sugarcoat harsh lessons about history and the World State, but also owes much to books like Ronald FRASER's *The Flying Draper* (**1924**). In MH's version of the mode, a young man finds he can levitate (◊ TALENTS), lifts himself out of the muddle of the UK, and tells Hitler (disguised as the "Dictator") a few home truths. *The Darkened Room: An Arabesque* (**1951**), set like *Higher Things* in the imaginary city of Rowcester, introduces an sf element. *The Brain* (**1953**) may be understood as feeble sf; but the eponymous sentient mushroom cloud, which offers the protagonist the FAUST-style bargain of unlimited knowledge and power, is perhaps best thought of as fantasy. *The Exploits of the Chevalier Dupin* (coll **1968** US; exp vt *Murder in the Rue Royale* 1972 UK) continues the detective exploits of Edgar Allan POE's sleuth. MH was also a noted SHERLOCK HOLMES scholar, publishing several books in the mode that assumes Holmes was a historical character. [JC]

HARRISON, M(ICHAEL) JOHN (1945-   ) UK rock-climber and writer who began to publish work of genre interest with "Baa Baa Blocksheep" in 1968 for *New Worlds*, a journal with which he soon became closely identified. He wrote stories for the **Jerry Cornelius** SHARED WORLD created by Michael MOORCOCK, and – usually as Joyce Churchill – contributed book reviews to the journal; for a time he also served as its literary editor. Some of his work has been sf, but he has written relatively little in that genre since the 1970s, though «Signs of Life» (**1996**) has an sf premise.

The central argument of MJH's fantasy can be reduced to some fairly simple propositions: that the worlds of FANTASY are a distortion and denial of REALITY; and that those who inhabit or imagine those worlds (MJH is increasingly disinclined to differentiate between fictional inhabitants of fantasy worlds and readers of fantasy texts) are themselves creatures whose grasp on reality is dreadfully frail. If they suffer some form of AMNESIA – many behave as though they do – their affliction may be defined not as a failure to remember who they were, but as a failure to remember – or to *see* – where they *are*. Most of MJH's characters are, consequently, locked to the useless past; and are incapable of making any move into the world. ESCAPISM is, for MJH, BONDAGE.

In retrospect, these propositions seem built into the **Viriconium** sequence from its very beginning, though MJH has revised some of the tales, and the inevitability of the movement of the sequence, from SECONDARY WORLD to the salutary mundanity of this world, is in part an artifact of hindsight. The sequence comprises *The Pastel City* (**1971**), *A Storm of Wings* (**1980** US) and *In Viriconium* (**1982**; vt *The Floating Gods* 1983 US), plus *Viriconium Nights* (coll **1984** US; very much rev 1985 UK). The latter's revision is radical; one story from the new version later appeared as *The Luck in the Head* (1983 *Interzone*; graph with rev text **1991** illus Ian MILLER).

The Pastel City is set in a fairly conventional DYING-EARTH venue. Viriconium, the CITY of the title, lies at the heart of a world littered midden-like with the detritus of aeons; it is ruled by a queen, surrounded by enemies, and defended by Lord tegeus-Cromis, who is (like most of MJH's protagonists) a KNIGHT OF THE DOLEFUL COUNTENANCE: his weapons fail to be MAGIC and he ultimately turns his back on the war. *A Storm of Wings* is a kind of visual pun on the first book, repeating the basic story but this time in CROSSHATCH terms, with the giant insectile invaders' PERCEPTION of the world literally vying with humans' perception of the same terrain. *In Viriconium* completes the process; through a PARODY of Arthurian motifs like that of the FISHER KING and the WASTE LAND, it conveys a sense of the uselessness of STORY or UNDERLIER when the LAND itself is a phantasm, otiose and paralysed with entropy. But here – unlike the ultimate refusal of Despite which redeems the land in Stephen R. DONALDSON's first **Chronicles of Thomas Covenant the Unbeliever** (**1977**) – the proper response is, precisely, a kind of Despite. Unless Viriconium is refused, the souls who manufacture it out of their own refusal to see the true world have no hope. At the end, marking some kind of exit from dolour, the painter Audsley King abandons his Fisher-King-like pose and begins to paint that real world, which is ours. The tales in the revised version of *Viriconium Nights* continue the movement away from fantasy; in the last of them, "A Young Man's Journey to Viriconium", nothing remains of Viriconium in this world but accidents of perception.

MJH's only other fantasy novel – *Climbers* (**1989**) is set entirely in the contemporary UK – is *The Course of the Heart* (**1992**), which may be his finest single title. The relationship between this world and an imagined (or real) other one here becomes more complex. Between this world and the "Pleroma" – which the protagonists have gained sight of during a disastrous college RITUAL under the guidance of a fake MAGUS – lies an IMAGINARY LAND created by two of those protagonists, and through which they construct a precariously meaningful shape to their lives. The third protagonist – another Knight of the Doleful Countenance – fails to grasp that meaning, and his life turns into a profoundly depressing cul-de-sac. The stories assembled in *The Ice Monkey and Other Stories* (coll **1983**) variously gnaw at similar failures (and very occasional, very partial successes) in the task of living real lives.

In his 1990s work MJH shows some signs of not punishing his protagonists for their attempts to make sense of things. If this pattern evolves, both MJH and his protagonists will have earned any dream to which they can hold. [JC]

Other works (sf): *The Committed Men* (**1971**); *The Centauri Device* (**1974** US); *The Machine in Shaft Ten and Other Stories* (coll **1975**).

HARRYHAUSEN, RAY (1920-   ) US moviemaker, particularly renowned for his skill in creating screen MONSTERS; he was also an animator. He served an apprenticeship with Willis H. O'Brien (1886-1982) before taking charge of spfx for *Mighty Joe Young* (**1949**), followed by *The Beast from 20,000 Fathoms* (**1953**), *It Came from Beneath the Sea* (**1955**) – his first movie done in conjunction with producer Charles H. Schneer, with whom his partnership was virtually lifelong – *Earth Versus the Flying Saucers* (**1956**) and *Twenty Million Miles to Earth* (**1957**), for which he wrote the story. *The Seventh Voyage of Sinbad* (**1958**; ◊ SINBAD MOVIES), *The Three Worlds of Gulliver* (**1959**; GULLIVER MOVIES), JASON AND THE ARGONAUTS (**1963**), *One Million Years BC* (**1966**; ◊ ONE MILLION BC), *The Golden Voyage of Sinbad* (**1973**; ◊ SINBAD MOVIES) and *Sinbad and the Eye of the Tiger* (**1977**; ◊ SINBAD MOVIES) followed. By this time RH's spfx were looking increasingly creaky; CLASH OF THE TITANS (**1981**), yet again done with Schneer, was an overdue swansong. [JG]

HARVEY US movie (**1950**). Universal. **Pr** John Beck. **Dir** Henry Koster. **Screenplay** Mary Chase with Oscar Brodney. **Based on** the play *Harvey* by Chase. **Starring** Victoria Horne (Myrtle Mae Simmons), Josephine Hull (Veta Louise Simmons), James Stewart (Elwood P. Dowd). 104 mins. B/w.

Rich, eccentric bachelor Elwood has an INVISIBLE COMPANION – a 6ft 3½ in invisible white rabbit called Harvey. Assuming him crazy, his sister Veta Louise and her daughter Myrtle Mae attempt to have him committed, but in due course it becomes evident that, just as Elwood has been maintaining, Harvey is not an ILLUSION but a pooka. At last, to please Veta Louise, Elwood agrees to submit himself to an injection that will "banish the delusion"; but, when she realizes it will also banish Elwood's kindness and generosity of spirit, she retracts her request and agrees that, if living with Harvey is the price she must pay, pay it she shall.

Essentially *H* has two plots, one fantastic and the other a fairly routine comedy of errors. The case for Harvey's being a pooka is well expounded, and we see physical evidence (swinging doors, the altered text of a book, etc., though never the rabbit himself). In one scene Elwood spells out a few of Harvey's abilities, notably that of stopping the clock for a while so that with him one can, as it were, take a break from REALITY. There are further interesting asides on the natures of objective and subjective reality: "I've wrestled with reality for 35 years," says Elwood, "and I'm happy, doctor: I finally won out over it." [JG]

HARTLEY, L(ESLIE) P(OLES) (1895-1972) UK writer, mostly known for nonfantastic works like *The Go-Between* (**1953**). Throughout the **Eustace and Hilda** trilogy – *The Shrimp and the Anemone* (**1944**; vt *The West Window* 1945 US), *The Sixth Heaven* (**1946**) and *Eustace and Hilda* (**1947**) – hints of supernatural events almost penetrate into REALITY, but always only along the border. Much of his short work is SUPERNATURAL FICTION, with GHOST STORIES appearing in collections from the first, *Night Fears and Other Stories* (coll **1924**), to the last, *Miss Carteret Receives and Other Stories* (coll **1971**). A US collection, *The Travelling Grave and Other Stories* (coll **1948**; vt *Night Fears and Other Supernatural Tales* 1994 UK), is devoted exclusively to supernatural fiction; most of his work of interest also appears in *The Complete Short Stories of L.P. Hartley* (coll **1973** 2 vols). [JC]

Other works: *The Killing Bottle* (coll **1932**); *The White*

Wand and Other Stories (coll **1954**); *Facial Justice* (**1960**); *Two for the River* (coll **1961**).

HARTWELL, DAVID G(EDDES) (1941-   ) US editor, critic and anthologist who has for most of his career been primarily associated with sf; in recent years he has also produced several anthologies of importance to fantasy. The most important is almost certainly *The Dark Descent* (anth **1987**; vt in 3 vols as *The Dark Descent #1: The Colour of Evil* 1990 UK, *#2: The Medusa in the Shield* 1991 UK and *#3: A Fabulous, Formless Darkness* 1991 UK), which concentrates on HORROR. In his introduction, DGH divides horror into three categories: (a) the moral/allegorical, usually involving an intrusion of EVIL into consensus reality; (b) "stories of aberrant human psychology embodied metaphorically", usually in the form of a MONSTER; and (c) the fantastic, in which REALITY is treated as essentially ambiguous, potentially absurd. The third category is clearly the most relevant to the concerns of this encyclopedia, for which DGH is a Consultant Editor.

Other titles of fantasy interest include a sequence of CHRISTMAS anthologies: *Christmas Ghosts* (anth **1987**) with Kathryn Cramer (1962-   ), *Spirits of Christmas* (anth **1989**), *Christmas Stars* (anth **1992**), *Christmas Forever: All New Tales of Yuletide Wonder* (anth **1993**) and *Christmas Magic* (anth **1994**). The two **Masterpieces** titles – *Masterpieces of Fantasy and Enchantment* (anth **1988**) and *Masterpieces of Fantasy and Wonder* (anth **1989**), the latter again with Cramer – are enjoyable and learned. *The Screaming Skull and Othe Great American Ghost Stories* (anth dated 1994 but **1995**) is also of interest.   [JC]

HARVEY, WILLIAM F(RYER) (1885-1937) UK writer and journalist who severely damaged his already weak health during WWI saving a comrade's life; thereafter he was a semi-invalid. He started writing before WWI, his first stories being collected in *Midnight House and Other Tales* (coll **1910**); his two later collections of strange stories were *The Beast with Five Fingers and Other Tales* (coll **1928**) and *Moods and Tenses* (coll **1933**). WFH was a skilled writer of the psychological GHOST STORY, of which "The Ankardyne Pew", arguably the best, shows the influence of M.R. JAMES. The supernatural seldom enters his fiction overtly but ripples around the edges, suggesting madness. One of his earliest stories, "August Heat" tells of a man who sees his own gravestone and ends on the verge of madness awaiting his pre-ordained death. WFH's fascination with death, FATE and coincidence occurs again in "Across the Moors" and "Peter Levisham". His best-known story is "The Beast with Five Fingers" (1919 in *The New Decameron*; much rev 1928), filmed as *The* BEAST WITH FIVE FINGERS (*1946*), another PSYCHOLOGICAL THRILLER in which a man is haunted by a disembodied hand. His best stories were collected as *Midnight Tales* (coll **1946**; vt *The Beast with Five Fingers* 1947 US) ed Maurice RICHARDSON, who provided a useful biographical introduction. A final collection of previously unpublished material was *The Arm of Mrs Egan & Other Stories* (coll **1952**). WFH also wrote a children's book, *Caprimulgus* (**1936**), and an autobiographical account of his Quaker childhood, *We Were Seven* (**1936**).   [MA]

HASFORD, (JERRY) GUSTAV (1947-1993) US writer who began publishing fantasy stories with "Heartland" for *Orbit 16* (anth **1975**) ed Damon Knight, but most of whose work of interest was generated by his time as a combat correspondent in Vietnam, where he observed the 1968 Tet offensive. Like Norman Mailer (1923-   ), he had gone to war in his teens with the express purpose of writing about it, and his first novel, *The Short-Timers* (**1979**), successfully deploys HORROR and fantasy elements to describe a world as crazy as the one that allowed the war to happen. GHOST tanks roll across blasted landscapes, M-14s whisper instructions to their adolescent familiars, and DEMON snipers pick off entire patrols. For the movie version – *Full Metal Jacket* (*1987*) dir Stanley Kubrick – GH received co-credit for the screen adaptation. In the book's sort-of-sequel, *The Phantom Blooper* (**1990**), the voice of a turncoat Marine haunts a ravaged Vietnam LANDSCAPE.

GH was a maverick of enormous energy, part of it spent in stealing a reported 800 library books from around the world (for which he eventually served time in prison). His death was hastened by untreated diabetes.   [SB]

HAUFF, WILHELM (1802-1827) German writer whose several novels include *Liechtenstein* (**1826** Liechtenstein; trans F. Woodley and W. Lander 1846 UK), an historical romance in the style of Sir Walter SCOTT, into which supernatural elements penetrate. He wrote three cycles of literary FAIRYTALES – "Die Karavane" ["The Caravan"], "Der Scheik von Allesandria und seine Sklaven" ["The Sheikh of Alexandria and his Slaves"] and "Das Wirthshaus im Spessart" ["The Inn in the Spessart"] – published in three *Märchenalmanache* (anth **1826**; anth **1827**; anth **1828**), selections from which have been variously translated as *The Caravan and Other Tales* (coll trans **1840s** UK), *Arabian Days' Entertainments* (coll trans **1869** Germany), *Tales of the Caravan, Inn, and Palace* (coll trans **1881** US), *Tales* (coll trans J. Mendel **1886** UK), which is relatively complete, *The Little Glass Man* (coll trans **1893** UK), *Fairy Tales* (coll trans **1896** US) and *Hauff's Tales* (coll trans **1905** UK). A good recent translation is *Fairy Tales* (coll trans J.R. Edwards **1961** UK). Many of the tales are in an *Arabian Nights* mode; TRANSFORMATIONS proliferate, sometimes very frighteningly. Hans Werner Henze (1926-   ) adapted one tale from the second cycle – "Der Affe als Mensch" ["The Monkey as a Man"] – into an OPERA, *Der Junge Lord* ["The Young Lord"] (**1965**).

Novels of interest include *Memoiren des Satan* ["The Memoirs of Satan"] (**1825-6**), a SATIRE in the mode of E.T.A. HOFFMANN, in which the DEVIL tours contemporary Germany making observations, at one point meeting the WANDERING JEW, and in another episode acting as the DOPPELGÄNGER of an elderly man; and *Phantasien im Bremer Ratskeller* ["Fantasies in the Bremer Tavern"] (**1827**), which is also lightheartedly gruesome.   [JC]

HAUNTED DWELLINGS Haunted houses and other buildings which lack the special architectural or cosmic significance of an EDIFICE. The WRONGNESS which makes the HD a BAD PLACE generally lies in its contents, usually a GHOST or REVENANT (as in numberless GHOST STORIES), or a psychic disturbance of the inhabitants, as in Rudyard KIPLING's "The House Surgeon" (1909) and most tales of POLTERGEIST phenomena, or the impression of some powerful emotion which has been absorbed by the fabric of the HD, as in H.G. WELLS's "The Red Room" (1896), whose HAUNTING consists of pure fear. In GOTHIC FANTASY and DARK FANTASY the HD may physically mirror the wrongness within: Edgar Allan POE's "The Fall of the House of Usher" (1839) flamboyantly reflects the insane morbidity of Roderick Usher with a mansion façade which is literally cracked, while the house in *The Haunting of Hill House* (**1959**) by Shirley Jackson (1919-1965) – filmed as *The*

Haunting (*1963*) and reprised as *The* LEGEND OF HELL HOUSE (*1973*) – has many windowless rooms and no precise right angles. In terms of the CINEMA, the most sustained sequence concerning HDs has been the **Amityville** sequence, so far comprising *The Amityville Horror* (*1979*) – based on a supposedly real case, unlike its successors – *Amityville II: The Possession* (*1982*), *Amityville 3-D* (*1984*), *Amityville: The Evil Escapes* (*1989*), *The Amityville Curse* (*1990*) and *Amityville: A New Generation* (*1993*). The first of the HOUSE sequence is also of interest, though really it is in the province of DARK FANTASY rather than haunting. GHOST STORY (*1974*) and *The* PAGEMASTER (*1994*) are further movies involving, in very different ways, HDs. [DRL/JG]

HAUNTED PALACE, THE US movie (*1963*). ◊ *The* HOUSE OF USHER.

HAUNTING Haunting has links with memory, the power of place and ultimately POSSESSION. It is very focal: haunts are locations which have the power to lure, attract and entrap – such as the haunts of the WENDIGO or other nature spirits. The music and presence of PAN in "Piper at the Gates of Dawn" in Kenneth GRAHAME's *The Wind in the Willows* (1908) exercise a powerful haunting over Mole, Ratty and the baby Portly. The term has much wider connotations than the obvious links with GHOSTS, REVENANTS and HAUNTED DWELLINGS. [MA]

HAUNT OF HORROR, THE ◊ MAGAZINES.

HAUNTS ◊ MAGAZINES.

HAUPTMANN, GERHART (1862-1946) German dramatist and prose-writer, winner of the 1912 Nobel Prize for Literature. His first plays, beginning with *Vor Sonnenaufgang* (**1889** trans L. Bloomfield **1909** US), were uneasily naturalistic, but by the time of *Hanneles Himmelfahrt* (**1893**; trans William Archer as *Hannele: A Dream Poem* **1894** UK) he had begun to move into Expressionist realms of fantasy interest. In the second act of this play, though the child Hannele may not be literally dead, a POSTHUMOUS-FANTASY atmosphere powerfully imbues her description of HEAVEN, from where she reports. *Die Versunkene Glocke* (**1896**; trans Charles Henry Meltzer as *The Sunken Bell* **1898** UK), *Michael Kramer* (**1900**) and *Und Pippa tanzt* (**1906**; trans as *And Pippa Dances* **1907** UK) continue in this mode; and to contemporary audiences can seem vacuous. But *Der Bogen des Odysseus* (**1914**; trans as *The Bow of Odysseus* **1917** US) offers a less dreamy portrait of ODYSSEUS.

His novels move closer to explicit fantasy. In *Der Narr in Christo Emanuel Quint* (**1910**; trans Thomas Seltzer as *The Fool in Christ, Emanuel Quint* **1911** US) the protagonist re-enacts CHRIST's life and woes in a narrative with an unmistakable aura of the TWICE-TOLD. *Atlantis* (**1912**; trans Adele and Thomas Seltzer **1912** US) transforms an Atlantic liner into a SHIP OF FOOLS. *Phantom* (**1923**; trans Bayard Quincey Morgan **1923** US) treats a criminal's confession as manifesting supernatural BONDAGE, for the fantasies which drive him come from elsewhere. *Die Insel der grossen Mutter, oder Das Wunder von Ile des Dames: Eine Geschichte aus dem Utopischen Archipelagus* (**1924**; trans Willa and Edwin Muir as *The Island of the Great Mother: The Miracle of the Île des Dames: A Tale from the Utopian Archipelago* **1925** US) is sf, even though the shipwrecked worshippers of the GODDESS attribute a continual flow of children to Her, rather than to the young men with whom they make love. *Till Eulenspiegel* (**1928**), in epic-verse form, takes its TRICKSTER hero, in this incarnation disillusioned by WORLD WAR I, on a kind of

Grand Tour of European myth and fantasy, granting him conversations with figures like Amfortas and allowing him centuries of SEX and discourse in a version of ancient Greece. A further untranslated epic, the **Atriden-Tetralogie** ["The Atrides Tetralogy"] play cycle – *Iphigenie in Delphi* (**1941**), *Iphigenie in Aulis* (**1944**), *Agamemnons Tod* (**1948**) and *Elektra* (**1948**) – takes a grim view of the gods. [JC]

HAVILAND, VIRGINIA (1911-1988) UK writer. ◊ Raymond BRIGGS.

HAVLASA, JAN (1883-1964) Czech writer. ◊ CZECH REPUBLIC.

HAWKE, SIMON (1951-   ) US writer, born Nicholas Valentin Yermakov, but whose legal name is now SH. His career began with some sf adventure novels and the **Boomerang** series – all as by Nicholas Yermakov, and all in a baroque idiom. He wrote several unpretentious series as SH from about 1984, and military sf as J.D. Masters and S.L. Hunter.

Not all his series are sf. The **Wizard** sequence – *The Wizard of 4th Street* **1987**), *The Wizard of Whitechapel* (**1988**), *The Wizard of Sunset Strip* (**1989**), *The Wizard of Rue Morgue* (**1990**), *The Samurai Wizard* (**1991**), *The Wizard of Santa Fe* (**1991**), *The Wizard of Camelot* (**1993**) and *The Wizard of Lovecraft's Castle* (**1993**) – is initially set in an sf-like 23rd-century post-HOLOCAUST USA, though one which is actually an ALTERNATE WORLD in which rationalized MAGIC rules, where the protagonist must gain the aid of MERLIN. Several of the tales are URBAN FANTASY, focusing (in order) on NEW YORK, LONDON, LOS ANGELES, Tokyo and Santa Fe. *The 9 Lives of Catseye Gomez* (**1991** chap; exp vt *The Nine Lives of Catseye Gomez* 1992), a pendant, focuses on an intelligent CAT from the *Santa Fe* volume. [JC]

Other works: Three **Friday the 13th** ties, *Friday the 13th #1 * (1987), #2 * (1988)* and *#3 * (1988)*; a **Batman** tie, *To Stalk a Specter * (1991)*; a comic-fantasy sequence, being *The Reluctant Sorcerer* (**1992**) and *The Inadequate Adept* (**1993**); the **Chronicles of Athas** series, contributed to the **Dark Sun** world, being *The Outcast * (1993)*, *The Seeker * (1994)*, *The Nomad * (1994)* and *The Broken Blade * (1995)*; *Birthright: The Iron Throne * (1995)*, another tie; and *The Whims of Creation* (**1995**), where a generation starship is invaded by denizens of FAERIE.

HAWKWOOD, ALLAN [s] ◊ H. BEDFORD-JONES.

HAWTHORNE, HILDEGARDE (1871-1952) US writer. ◊ Julian HAWTHORNE.

HAWTHORNE, JULIAN (1846-1934) US writer, the son of Nathaniel HAWTHORNE; he lived in London for some years. He frequently rearranged, retitled and sometimes rewrote his works for their UK and US publications; his bibliography is inordinately complex. He began publishing short fiction in 1870, much of his early work – including "Otto of Roses" (1871; rev vt "The Rose of Death" 1876) – being fantastic. His first novel, *Bressant* (**1873**), briefly involves a GHOST but is not nearly as strange as the marginally supernatural Gothic tale *Idolatry* (**1874**), whose plot involves two fateful RINGS. His most notable full-length fantasies are the dual-personality story *Archibald Malmaison* (**1879**) and the metaphysical fantasy *The Professor's Sister* (**1888**; vt *The Spectre of the Camera* UK). An interesting collection of fantasy novellas, ostensibly for children, *Yellow-Cap* (coll **1880**), includes the peculiar ALLEGORY "Calladon".

Many of JH's novellas and short stories are melodramatic

weird tales of CURSES and apparitions, some drawing inspiration from his Swedenborgian faith. Those of most fantasy interest are the cosmic-vision story "The New Endymion" (1879) and the timeslip VAMPIRE tale "Ken's Mystery" (1883; in *David Poindexter's Disappearance and Other Stories* coll **1888**). *Kildhurm's Oak: A Strange Friend* (**1889**) had earlier appeared in *Ellice Quentin and Other Stories* (coll **1880**), which also includes "The New Endymion". The title novella of *The Laughing Mill and Other Stories* (coll **1879**) was reprinted as the second item in *Prince Saroni's Wife and The Pearl Shell Necklace* (coll **1884**); another item from the former collection became the second element in *Constance and Calbot's Rival* (coll **1889**). *Sinfire* (**1888**) has no supernatural content, but is sometimes listed as fantasy. Of interest is *The Golden Fleece* (**1892**). Several SUPERNATURAL FICTIONS are included in the CLUB-STORY collection *Six Cent Sam's* (coll **1893**; vt *Mr Dunton's Invention and Other Stories*). JH's progressive ideas regarding social reform were summarized in the TIMESLIP story "June 1993" (1893 *Cosmopolitan*).

JH's career was interrupted by a jail term for his (probably unwitting) involvement in a land fraud. He moved to California, where he wrote for newspapers but found it difficult to publish books. The best of his subsequent fantasies is "The Delusion of Ralph Penwyn" (1909); most of the remainder appeared in the pulp MAGAZINE *All-Story Weekly*. These include the futuristic fantasy novel "The Cosmic Courtship" (1917) and the **Martha Klemm** series of tales of SPIRITUALISM, which concluded with the novel "Sara was Judith?" (1920). He tried unsuccessfuly to sell this and other projects to the movies.

JH edited a notable series of anthologies which included many translations of supernatural stories from various European languages: **Library of the World's Best Mystery and Detective Stories** (1908 6 vols; exp vt **The Lock and Key Library: Classic Mystery and Detective Stories** 1909 10 vols). He was probably the uncredited editor of *One of Those Coincidences and Ten Other Stories by Julian Hawthorne and Others* (anth **1899**), in which JH's story and one other are fantastic. He edited a novel cobbled together from documents left behind by his father, *Doctor Grimshawe's Secret: A Romance* (**1884**), one of several versions of a projected novel about a quest for the ELIXIR OF LIFE; others had earlier been published as *Septimius* (**1872**) and *The Dolliver Romance* (**1876**).

JH's daughter Hildegarde (1871-1952) wrote some fantasies, including two sentimental stories featuring ghostly children, "Perdita" (1897) and "Unawares" (1908); these can be found in *Faded Garden: The Collected Ghost Stories of Hildegarde Hawthorne* (coll **1985** chap) ed Jessica Amanda SALMONSON. [BS]

HAWTHORNE, NATHANIEL (1804-1864) US writer, a central literary figure of the 19th century; father of Julian HAWTHORNE. The darkness of his vision of the human psyche gives to almost everything he wrote, even works which were not SUPERNATURAL FICTION or fantasy, a sense that its protagonists are acting in obedience to the Gothic manipulations of the dead but shaping past, that they can never simply flourish in the here and now. It is in something like this sense that so much of his work seems to have been treated as ALLEGORY: his characters are so in BONDAGE to the STORIES they have been appointed to undergo that they seem to "stand" in an allegorical relationship to symbolic events, rather than to live them.

After an early novel, *Fanshawe* (**1828**), which he repudiated, NH began his publishing career with stories like "The Hollow of the Three Hills" for the *Salem Gazette* in 1830, his first. It is typical of his mature work in both style and subject matter: an adulteress falls dead after a WITCH has revealed to her VISIONS of the life she has left behind. This and many similar tales – including "The Prophetic Pictures" and "Edward Randolph's Portrait", both of which express a sense that a PICTURE can imprison and/or prefigure REALITY – were assembled in *Twice-Told Tales* (coll **1837**; exp in 2 vols 1842), a collection whose title refers not to the nature of STORY in FANTASY but to the sense that contemporary lives are bound to repeat patterns laid down for them (◊◊ TWICE-TOLD). Further collections include *Mosses from an Old Manse* (coll **1846**; exp 1854), which contains NH's most famous tales of the supernatural – "Young Goodman Brown" (1835), whose protagonist journeys INTO THE WOODS where a vision turns his life off at the root, "Rappaccini's Daughter" (1844), an ironic allegory in which the poisonousness of the demonic is hardly worse than a blighting rationalism, and "Feathertop: A Moralized Legend" (1852), about a scarecrow who persuades the world he is human as long as he does not look into a MIRROR – and *The Snow-Image and Other Twice-Told Tales* (coll **1852**), whose title story plays sorrowfully on the ANIMATE/INANIMATE girl of snow constructed by two children and killed by their parents' rational disbelief. *The Snow Image and Uncollected Tales* (coll **1974**), which is volume 11 of the centenary edition of NH's works, contains several uncollected stories, including a tale of MERLIN, "The Antique Ring" (1842). *A Wonder-Book for Girls and Boys* (coll **1852**) (◊ WONDER TALE) and *Tanglewood Tales for Girls and Boys* (coll **1853**) retell the Greek MYTHS in a simple, chaste, literary style.

A sense of supernatural ordination pervades NH's most famous novel, *The Scarlet Letter* (**1850**), though the VISIONS its characters suffer are perhaps best understood in thisworldly terms. *The House of the Seven Gables* (**1851**) similarly deploys the apparatus of GOTHIC FANTASY without – quite – allowing the supernatural into the story. Early in his career, NH had written an ELIXIR-OF-LIFE tale, *Dr Heidegger's Experiment* (1837 *Salem Gazette*; **1883** chap); towards the end of his life, NH composed at least three fragmentary drafts of stories again dealing with this subject: published posthumously, they were *Septimius: A Romance* (**1872** UK; vt *Septimius Felton, or The Elixir of Life* 1872 US), the title story of *The Dolliver Romance and Other Pieces* (coll **1876**), and "Septimius Norton" (1890), all three assembled as *The Elixir of Life Manuscripts* (omni **1977**), which is volume 13 of the Works. The first of these fragments, the most complete, is another Gothic fantasy that does not quite venture over the edge – the elixir, when found, turns out to be missing an essential ingredient.

In the end, the NH ROMANCE of predetermination casts a long shadow over the American Dream, telling us that we must both dream very hard and surrender absolutely. [JC]

Other works: *The Ghost of Doctor Harris* (**1900** chap), a GHOST STORY; many posthumous re-sortings of NH's short work, including *The Complete Short Stories of Nathaniel Hawthorne* (coll **1959**), containing 72 tales, *The Celestial Railroad and Other Stories* (coll **1963**) and *Young Goodman Brown and Other Short Stories* (coll **1992**).

HAY, TIMOTHY Pseudonym of US writer Margaret Wise BROWN.

HAZEL, PAUL (1944-   ) US writer and teacher whose **Finnbranch** sequence – *Yearwood* (**1980**), *Undersea* (**1982**) and *Winterking* (**1985**), all three assembled as *The Finnbranch* (omni **1986** UK) – presents again and again, in cyclical form, versions of the birth, heroic adulthood and death of a hero named Finn. Though much of the imagery accompanying and intensifying the reiterations of Finn's life-course derives from CELTIC FANTASY, and inevitably relates to the FINN MAC COOL cycle, the sequence is in no clear sense obedient to any single underlying Northern STORY, Celtic or Arthurian (◊ ARTHUR) or Teutonic (◊ NORDIC FANTASY), though elements from these and other cycles appear throughout. Over the course of the first two volumes, which are both set in OTHERWORLD-tinged Celtic LANDSCAPES, Finn undergoes several of the METAMORPHOSES typical of Celtic fantasy at its most unrelenting. In the third volume, told in the guise of an ALTERNATE-WORLD tale set in a displaced 20th-century USA, Finn surfaces into a more socialized world but remains tied to the wheel of the central story that arguably generates heroic cycles: the THRESHOLD-crossing wheel of life that turns birth into triumph into death and (in some traditions) into birth again. The language throughout is strikingly intense, allusive, dislocatingly dreamlike, vivid.

The Wealdwife's Tale (**1993**) is once again understandable as a meditation on the wheel of life, here in the guise of a fantasia very loosely based on the Christmas carol "Good King Wenceslas". Waldo Wenceslas is not a king but a duke, a scholarly, vague, haunted KNIGHT OF THE DOLEFUL COUNTENANCE; and his tiny fiefdom is impinged upon both by a mundane landscape which closely resembles Colonial New England and by a deathly MIRROR fiefdom which seems reachable only by members of the Wenceslas family, each of whom acts in strict accordance with a pattern or story laid down generations earlier. Obsessed by the death of his wife Elva decades earlier, Wenceslas goes literally INTO THE WOODS (the novel is a classic example of this TOPOS) in search of her, but does not recognize the old woman he meets as Elva. She brings him to a young girl, Alve, and accompanies them back to the fiefdom, where they prepare to marry. Wenceslas's three sons then go into the woods, there dying and being reborn as a "brollachan" – a tripartite hedge, a cruel embodiment of the greenwood in the guise of a grotesque PARODY of the GREEN MAN. His daughter, too, vanishes into the woods, and is rescued by young Odlaw (Waldo reversed) from the other side, who weds her. Tragedies ensue.

Both **Finnbranch** and *The Wealdwife's Tale* operate at a level of originality rarely found in fantasy.   [JC]

H.D. Pseudonym of US poet and novelist Hilda Doolittle (1886-1961), who lived in the UK or Switzerland from 1911, and who became a significant Imagist poet from the publication of *Sea Garden* (coll **1916**). From early on she was absorbed by Greece and Classic Greek literature (◊ GREEK AND LATIN CLASSICS), which shaped much of her work. An early novel, *Hedylus* (**1928**), makes complex play with Hellenistic motifs. Meditations on GODDESS themes appeared frequently. She wrote little fantasy as such, but *Helen in Egypt* (**1961**) – a book-length narrative poem told as by the Sicilian poet Stesichorus (*fl* early 6th century BC) – intricately amalgamates the MYTHS and LEGENDS of Troy and of the GODS who intervened there; her Helen is both an immortal being and an UNDERLIER for HD's vision of modern woman, whose sexual nature must be vivid and sustaining.   [JC]

HEALING 1. Straightforwardly, the GENRE-FANTASY analogue of medical treatment. The most usual form is a laying-on of hands, sometimes accompanied by chanting or a spoken SPELL. Herbs, POTIONS or poultices can speed the healing process. More dramatic magical healing may be available – e.g., the touch of a UNICORN's horn.   [CB]
2. More interestingly, one of several central terms which help describe the narrative course of the fully structured fantasy tale. Of these terms – the others are WRONGNESS, THINNING, RECOGNITION and STORY (◊ FANTASY *for full discussion*) – "healing" is the most straightforward. It is what occurs after the worst has been experienced and defeated. It is the greening of the WASTE LAND or the recovery from AMNESIA on the part of the HERO or the escape from BONDAGE and the METAMORPHOSIS into the desired shape and fullness of those who have been wounded/imprisoned by the DARK LORD. In the language of J.R.R. TOLKIEN, it is the EUCATASTROPHE. In *The Lord of the Rings* (**1954-5**) the harrowing of the Shire is a form of healing; in Peter S. BEAGLE's *The Last Unicorn* (**1968**) the land immediately blossoms after the death of King Haggard, the KNIGHT OF THE DOLEFUL COUNTENANCE who cannot pass through the necessary purging self-recognition into the new world. It is what Covenant, in Stephen R. DONALDSON's **Thomas Covenant** sequence, finally permits himself to grant the LAND which has been his disease. It is what the story of fantasy "wishes" to tell.   [JC]

HEARD, GERALD Working name of UK author and journalist Henry Fitzgerald Heard (1889-1971), used by him for all his UK work; in the USA, where he lived from 1937, he wrote as H.F. Heard. Much of his work vacillates between metaphysical sf, hopeful studies in mysticism in association with his friend Aldous Huxley, and credulous examinations of topics like UFOS. In the SHERLOCK HOLMES pastiche *A Taste for Honey* (**1941**; vt *A Taste for Murder* 1955), filmed as *The Deadly Bees* (**1966**), there is a touch of ambiguously supernatural HORROR; but GH is mostly of genre interest for some of the fantasy tales and SUPERNATURAL FICTIONS assembled in *The Great Fog and Other Weird Tales* (coll **1944**; vt *Weird Tales of Terror and Detection* 1946; rev with 2 stories added and 1 dropped under original title 1947 UK) and *The Lost Cavern and Other Tales of the Fantastic* (coll **1948**). In the first volume, "Dromenon" combines GHOST STORY with theological speculation, interestingly suggesting that mystical truths are attainable by acts of PERCEPTION which unravel the patterns within actual cathedrals, and "The Cat, 'I Am'" describes a supernatural CAT. In the second volume, "The Cup" pits a GRAIL chalice against metaphysical EVIL. *The Black Fox: A Novel of the Seventies* (**1950** UK) faces a Victorian canon with profound moral problems; his sister, dying in his stead, saves him from the grasp of an affronted ANUBIS.   [JC]
Other works: *Gabriel and the Creatures* (**1952**).

HEARD, H.F. Pseudonym of Gerald HEARD.

HEARN, (PATRICIO) LAFCADIO (TESSIMA CARLOS) (1850-1904) Greek-born US translator and writer who lived in Japan from 1891, and whose work – almost all SUPERNATURAL FICTION – can generate a feeling of fantasy through the sense it conveys that GHOSTS in Japan and China are inherently part of the literal LANDSCAPE. LH's early work, which tends to melancholic prose poetry, appears in *Fantastics* (coll **1914** US), a posthumous assemblage, and in *Some Chinese Ghosts* (coll **1887**). His most mature work, composed with a delicate sinewy spareness, was assembled in *Kwaidan: Stories and Studies of Strange Things* (coll **1904** US); some of the tales included, like "The Dream of Akinosuk" (1904), in which a man enters a kind of TIME IN FAERIE after drinking some wine, is fantasy of very strong interest. Much of LH's work constitutes TWICE-TOLD versions of Japanese FOLKTALES; four of these GHOST STORIES were filmed as KWAIDAN (*1964*). [JC]

Other works: *In Ghostly Japan* (coll **1899** US); *Shadowings* (coll **1900** US); *A Japanese Miscellany* (coll **1901** US); *The Romance of the Milky Way and Other Studies and Stories* (coll **1905** US); *Karma and Other Stories* (coll **1921** UK); *The Selected Writings of Lafcadio Hearn* (coll **1949** US), which is very thorough.

HEAVEN The abode of the Old Testament GOD, which also became – in the JEWISH RELIGIOUS LITERATURE of the 2nd and 1st centuries BC, and then in the Christian mythos – the place to which the souls of the virtuous were elevated after the LAST JUDGEMENT. Various apocryphal documents, including *Enoch*, offer alternative accounts of a War in Heaven in which God expelled a host of rebel ANGELS, whose leader became the Christian SATAN. Analogues of Heaven in other religious traditions include the Chinese T'ien and the Buddhist Nirvana.

Having less melodramatic potential, Heaven is more rarely featured in literary fantasy than HELL; many imitations of DANTE's *Divine Comedy* become distinctly awkward when they progress to the *Paradiso* phase, and the most reassuring fantasies of SPIRITUALISM tend to be irritatingly vague. Many allegorical journeys heavenward, including those in Harry BLAMIRES's *Blessing Unbounded* (**1955**) and Theodore Zeldin's *Happiness* (**1988**), pay far more attention to the stops *en route* than to the ultimate destination.

Fantasy's heavens often turn out dismal, as in Robert NATHAN's *There is Another Heaven* (**1929**), or perversely inconvenient, as in Mark TWAIN's *Extracts from Captain Stormfield's Visit to Heaven* (**1909**), A.E. COPPARD's "Clorinda Walks in Heaven" (1922), Alan GRIFFITHS's *Strange News from Heaven* (**1934**) and Lester Del Rey's "Hereafter, Inc." (1939). Works of this stripe reproduce the "sour grapes" philosophy of many tales of mundane IMMORTALITY. The protagonist of Nelson BOND's "Union in Gehenna" (1942) is kicked out of Hell for making trouble, but finds Heaven far less congenial. Paradisal variants of the rich tradition of infernal comedies sometimes use the Elysian Fields in much the same way that writers like John Kendrick BANGS and Frederick Arnold KUMMER use HADES; Eric LINKLATER's plays *The Cornerstones* (**1942**) and *Crisis in Heaven* (**1944**) are notable examples.

The heavens of CHRISTIAN FANTASY are frequently modelled on the same kind of rural paradise as EDEN; they are rather effete by comparison with the lushly sensual paradise of Islam. The embarrassingly condescending image of a US Negro Heaven in *The Green Pastures* (**1929**) by Marc Connelly (1890-1980) is also markedly less refined, and an uptight Anglican Heaven is sharply contrasted with other cultures' ideas of the good life in Andrew LANG's "In the Wrong Paradise" (1886). A determinedly earnest attempt to sophisticate the quasi-Edenic Christian image can be found in *The Great Divorce* (**1945**) by C.S. LEWIS. The main rival to the quasi-Edenic image of Heaven is rather more in the vein of Eastern mysticism, involving a transcendental realm of light in which disembodied spirits dwell in unimaginable ecstasy. By their nature, such realms can be only briefly and distantly glimpsed in literary fantasy, as

in *Time Must Have a Stop* (**1944**) by Aldous Huxley.

An unusually elaborate heavenly realm is mapped out in Steven BRUST's *To Reign in Hell* (**1984**), but this is essentially a conventional Earth-clone of the kind favoured in GENRE FANTASY. A Heaven *exactly* like Earth, except that everything works out for the best, is satirically displayed in *Sweet Dreams* (**1973**) by Michael FRAYN. Sceptics sometimes argue that *any* imaginable Heaven would be so boring as to qualify as Hell – a case put with particular clarity in "The Choice: An Allegory of Blood and Tears" (1929) by S. Fowler Wright (1874-1965). [BS]

See also: AFTERLIFE; POSTHUMOUS FANTASY.

HEAVEN CAN WAIT US movie (*1943*). 20th Century-Fox. **Pr** Ernst Lubitsch. **Dir** Lubitsch. **Spfx** Fred Sersen. **Screenplay** Samson Raphaelson. **Based on** the play *Birthday* by Lazlo Bus-Fekete. **Starring** Don Ameche (Henry Van Cleve), Charles Coburn (Hugo Van Cleve), Laird Cregar (The Devil), Gene Tierney (Martha Strable Van Cleve). 112 mins. Colour.

Henry Van Cleve, at the end of a long life, arrives in HELL to be interviewed by the DEVIL. He recounts his autobiography, and it proves that – though he regards himself as a sinner – his sins have been small ones; he is thus sent onwards to HEAVEN. Almost all of the movie is concerned with the lifelong romance between Henry and his wife Martha. [JG]

HEAVEN CAN WAIT US movie (*1978*). Paramount. **Pr** Warren Beatty. **Exec pr** Howard W. Koch Jr, Charles H. Maguire. **Dir** Beatty, Buck Henry. **Spfx** Robert MacDonald. **Screenplay** Beatty, Elaine May. **Based on** HERE COMES MR JORDAN (*1941*), itself based on the play *Heaven Can Wait* by Harry Segall. **Novelization** *Heaven Can Wait* * (**1978**) by Leonore Fleischer. **Starring** Warren Beatty (Joe Pendleton/Leo Farnsworth/Tom Jarratt), Dyan Cannon (Julia Farnsworth), Julie Christie (Betty Logan), James Mason (Mr Jordan), Jack Warden (Max Corkle). 100 mins. Colour.

A remake of HERE COMES MR JORDAN (*1941*), but with the central character, Pendleton, now an LA Rams quarterback rather than a boxer: his dream is to play, and win, at the Superbowl. This dream is brought true when, finally, his substitute, Jarratt, dies on the field and Pendleton takes over his body. In the interim, there are a few changes, mainly PC ones: Logan is now an English schoolteacher come to the USA to persuade Farnsworth not to build a polluting power station, and the turnabout in his company that Pendleton/Farnsworth effects is not merely moral but environmental. Played as a lighter comedy than its predecessor, this works very effectively. Beatty handles the IDENTITY EXCHANGES well, while Warden, as his eventually convinced trainer, excels. [JG]

HEAVY METAL Glossy full-colour fantasy and sf COMIC-strip magazine inspired by the French magazine of similar format, *Métal Hurlant* ["Screaming Metal"], and containing translations of material from this and other similar French, Spanish and Italian magazines alongside US material by Richard CORBEN, Bernie WRIGHTSON and others; *HM* makes a special feature of the work of the French artist MOEBIUS. *HM* was published monthly April 1977-December 1985, quarterly January (Winter) 1986-Winter 1989, then bimonthly March 1989-current. In the monthly issues material was in serialized form, with stories of sometimes 50+ pages being uncomfortably segmented in long and short episodes. The change to quarterly publication signalled a change of policy; since then *HM* has featured only complete stories and full-length GRAPHIC NOVELS.

HM has maintained a consistently high standard of production and content, and has claimed a readership in excess of two million. It has been responsible for introducing to English-language readers many of the top comics creators from the Latin countries, and has thereby been a significant force in the internationalization of the medium. Examples of *HM* material of fantasy interest are noted in the entries on Alberto BRECCIA, Richard CORBEN, Guido CREPAX, MOEBIUS, Alex NINO and Bernie WRIGHTSON; others include *Ulysses* (1978) by Georges Pichard, *Conquering Armies* (1978) by Gal, *Cody Starbuck* (1981) and *Gideon Faust, Warlock at Large* (1982) by Howard CHAYKIN (1981 and 1982), *Urm the Mad* (1978), *Lone Sloane* (1980) and *Salammbo* (1984) by Phillip Druillet, *Zora* (1982; graph **1984**) by Fernando FERNANDEZ, *Polonius* (1977) by Jaques Tardi, DRUUNA (1986, 1988, 1994; graph coll **1993-5** 4 vols) by Eleuteri Serpieri, *The Waters of Deadmoon* (1990, 1991) by Adamov and Cothias, *Dieter Lumpen* (1989) by Pellejero and Zentner, *The Towers of Bois Maury* (1990) by Herman, SLAINE by Mills and Simon BISLEY.   [RT]

HECHT, BEN (1893-1964) US journalist and writer, bestknown as a playwright (often in collaboration with Charles MacArthur) and screenwriter. The fervent radicalism which had to be held in check in his commercial work gained expression in other, scaldingly sarcastic fictions, of which the best and most extreme is the Joris-Karl HUYSMANS-inspired *Fantazius Mallare: A Mysterious Oath* (**1922**), which offers a lurid account of the eponymous artist's descent into madness. The illustrations by Wallace Smith captured the book's selfconscious DECADENCE perfectly, and got the book banned. In the phantasmagoric sequel, *The Kingdom of Evil: A Continuation of the Journal of Fantazius Mallare* (**1924**), the disparate fragments of Mallare's shattered personality prepare a temple for the GOD Synthemus. BH's interest in MULTIPLE PERSONALITIES was further reflected in his baroque murder mystery *The Florentine Dagger* (**1923**). The stories in *A Book of Miracles* (coll **1939**), mostly reprinted in *The Collected Short Stories of Ben Hecht* (coll **1945**), include several skittish comedies – e.g., "The Heavenly Choir", in which the spirits of the dead interrupt radio broadcasts – and two moralistic novellas based in religious folklore: "Death of Eleazer", about the supernaturally diverted trial of a US Nazi, and a notable exercise in literary Satanism (◊ SATAN), "Remember Thy Creator", in which Archangel Michael is sent to Earth to bring mortals closer to GOD but realizes they have little to gain from such closeness. The sentimental *Miracle in the Rain* (**1943** chap) presumably began life as a treatment for the 1954 movie it eventually became; BH's other work for the movies included scripts for several PSYCHOLOGICAL THRILLERS – including *Spellbound* (**1945**) dir Alfred HITCHCOCK, *The Specter of the Rose* (**1946**) dir BH, and a quirky fantasy of redemption, *The Scoundrel* (**1935**), written with Noël Coward (1899-1973).   [BS]

Other works: *The Wonder Hat: A Harlequinade in One Act* (**1920** chap) with Kenneth Sawyer Goodman; *The Bewitched Tailor* (**1941** chap); *The Cat that Jumped out of the Story* (**1947** chap).

HEIDE, FLORENCE PARRY (1919-   ) US writer. ◊ CHILDREN'S FANTASY.

HEINLEIN, ROBERT A(NSON) (1907-1988) US writer of major importance in the development of modern

SCIENCE FICTION. RAH's methodical approach to scientific extrapolation was equally effective when he bent to FANTASY and SUPERNATURAL FICTION, which he wrote especially for UNKNOWN. The fantasy is so rationalized in "Magic, Inc." (1940 *Unknown* as "The Devil Makes the Law!"; in *Waldo and Magic, Inc.* coll **1950**; vt *Waldo: Genius in Orbit* 1958) – set in an ALTERNATE WORLD where MAGIC operates by strict laws and codes – that the story reads like sf (◊ RATIONALIZED FANTASY). "Waldo" (1942 *Astounding*) is a mixture of hard sf and GNOSTIC FANTASY: a phsyical weakling finds a way of manipulating the spirit world in order to benefit our world. RAH also produced a couple of paranoia fantasies: "They" (1941 *Unknown*), perhaps the ultimate solipsist fantasy (a man is convinced the world is a puppet show), and "The Unpleasant Profession of Jonathan Hoag" (1942 *Unknown* as by John Riverside), where a man seeks help in solving what he does during the daytime, discovering he is a sort of art critic of the flawed cosmos. These were assembled with others in book form as *The Unpleasant Profession of Jonathan Hoag* (coll **1959**; vt *6 x 6* 1961), which also included "'. . . And He Built a Crooked House'" (1941 *Astounding*), where a house has topologically shifted to become a PORTAL to other worlds (◊◊ EDIFICE). RAH produced some definitive TIME FANTASIES in "By His Bootstraps" (1941 *Astounding* as by Anson MacDonald) and "'All You Zombies'" (1959 *F&SF*) – a theme he referred to in his later sf novels about **Lazarus Long**, especially *Time Enough for Love* (**1973**), where RAH uses time to sustain IMMORTALITY.

RAH concentrated on sf for the next 20 years, and his first return to fantasy was critically disliked. *Glory Road* (**1963**), dismissed as a weak attempt at SWORD AND SORCERY, is actually a rousing adventure novel that would have been at home in *Unknown* except for the liberal doses of SEX. It concerns a soldier who becomes infatuated with a beautiful girl who transports him to an OTHERWORLD in the MULTIVERSE to aid her in her QUEST to steal the Egg of the Phoenix, guarded by the Eater of Souls. The work was in the forefront of the revival of HEROIC FANTASY that gathered pace two years later. There are echoes of *Glory Road* in *Job: A Comedy of Justice* (**1984**), where a man and a woman find themselves thrust into the AFTERLIFE – here seen as an ALTERNATE WORLD – and have to rely on their wits and on divine intervention to survive.

RAH was one of the first writers successfully to meld the substance of sf and fantasy into an integral whole without compromising either genre. His experiments were later developed by others, in particular Roger ZELAZNY and Gene WOLFE.   [MA]

Other works: Much sf (◊ SFE).

HELL The place to which the SOULS of the damned are consigned after death or the LAST JUDGEMENT. Proto-Hells, like the Greek HADES and the Hebrew Sheol, tend to be dark and dismal but not particularly nasty; the Norse Niflheim (also known as Hel) is similarly bleak. The idea that all SOULS would share the same dreary fate was first put about by Zoroastrians and later by Jewish apocalyptic writers (◊ JEWISH RELIGIOUS LITERATURE), Christians and followers of Islam. The Zoroastrian arena of posthumous punishment was cold, but the Jewish Gehenna was fiery, and the idea of hellfire was adopted with enthusiasm by Christians. Hell then became the place to which SATAN and his rebel ANGELS were consigned after the war in HEAVEN, some becoming DEMONS charged with subjecting the damned to endless torture.

As the ultimate weapon of Christian spiritual terrorism Hell is much more fascinating to the literary imagination than HEAVEN; the *Inferno* is by far the best-known part of DANTE's *Divine Comedy* (written *c*1320) and Book I of John MILTON's *Paradise Lost* (**1667**), in which the expelled Satan raises the palace of Pandemonium, is very striking. Hellfire lurks in the wings of most Renaissance fantasy – notably Christopher MARLOWE's *Dr Faustus* (**1604**) – and much GOTHIC FANTASY, but is rarely described in any detail; William BECKFORD's account of the Islamic Hell in *Vathek* (**1786**) is an exception. Much fuller descriptions are featured in works which treat the idea less seriously, especially many comedies and SATIRES; examples include *A Houseboat on the Styx* (**1895**) and its sequels by John Kendrick BANGS, *Satan's Realm* (**1899**) by Edgar C. Blum, *Efficiency in Hades* (**1923**) by Robert B. Vale, *Ladies in Hades* (**1928**) and its sequel by Frederick Arnold KUMMER, several of the stories in *The Devil and All* (coll **1934**) by John COLLIER, *Hell's Bells* (**1936**) by Marmaduke Dixey, *Cold War in Hell* (**1955**) by Harry BLAMIRES and *Chariot of Fire* (**1977**) by E.E.Y. Hales (1908-   ). The later and more heavily ironic novels follow the example of many other writers of POSTHUMOUS FANTASY in taking it for granted that the world in which we live has much about it that is quite Hellish enough – a line of thought extrapolated in *Morwyn* (**1937**) by John Cowper POWYS, in "None but Lucifer" (1939) by H.L. GOLD and L. Sprague DE CAMP, in *All Hallows' Eve* (**1945**) by Charles WILLIAMS, in C.S. LEWIS's *The Great Divorce* (**1945**), in *To Hell, with Crabb Robinson* (**1962**) by R.H. Mottram (1883-1971) and in Stanley ELKIN's *The Living End* (**1979**).

The Hells of Spiritualist fantasy (◊ SPIRITUALISM) are often rather anaemic, sometimes recalling the gloomy pre-Christian lands of the dead; examples include *Letters from Hell* (1887) by "L.W.J.S." (V. Thisted) and "On the Dark Mountains" (1888) by Mrs OLIPHANT. Elements of the classic Dantean image are preserved in such earnestly moralistic works as *After* (**1930**) by Lady Saltoun (1863-1940) and melodramas like C.L. MOORE's "Black God's Kiss" (1934). Dante's Hell is recapitulated in all its horrific glory in *Inferno* (**1976**) by Larry NIVEN and Jerry Pournelle (1933-   ) and *Only Begotten Daughter* (**1990**) by James MORROW.

In the long series of SHARED-WORLD adventures begun with *Heroes in Hell* * (anth **1986**; ◊ Janet MORRIS), Hell becomes an arena in which all the interesting people in history can come together to continue the relentless pursuit of their various ends. Various versions of Hell are used as settings for HEROIC FANTASY in Lloyd Arthur ESHBACH's series begun with *The Land Beyond the Gate* (**1984**).

A notable nonfiction work tracking the evolution of the idea is *The History of Hell* (**1993**) by Alice K. Turner. [BS]

See also: AFTERLIFE; INFERNO; POSTHUMOUS FANTASY.

HELLBLAZER ◊ COMICS; Jamie DELANO.

HELLBOUND: HELLRAISER II ◊ HELLRAISER.

HELLRAISER Sequence of three HORROR MOVIES with unusually rich fantasy content.

1. *Hellraiser* UK movie (**1987**). New World/Cinemarque/Film Futures. **Pr** Christopher Figg. **Exec pr** Mark Armstrong, David Saunders, Christopher Webster. **Dir** Clive BARKER. **Mufx** Bob Keen. **Screenplay** Barker. **Based on** *The Hell-Bound Heart* (1986 *Night Visions 3* ed George R.R. MARTIN; **1991**) by Barker. **Starring** Doug Bradley (Pinhead), Sean Chapman (Frank), Clare Higgins (Julia), Ashley Laurence (Kirsty), Andrew Robinson (Larry), Oliver

Smith (Frank as Monster). 93 mins. Colour.

Unpleasant adventurer Frank buys a curious puzzle-box which, on manipulation, opens a PORTAL to the other-dimensional world of the Cenobites, seekers after the utmost in pleasure and pain. Some time after he has been snatched away to their tortures, his brother Larry and Larry's second wife Julia (once Frank's lover) move into the family home. After a minor accident, some of Larry's blood spills onto the attic floorboards; it is enough for Frank to begin to regenerate himself. But more blood is needed, so Julia lures men back to the house and murders them for Frank to feast on. Last killed is Larry, whose disguising skin (◊ SKINNED) Frank wears in an attempt to deceive Kirsty, Larry's daughter by his first wife. Kirsty has accidentally manipulated the box, but bargained with the Cenobites: her life or Frank's return. In a gory ending, Frank kills Julia but is reclaimed by the Cenobites.

Hugely successful, *H* mixes much horror cliché with some genuinely adventurous DARK FANTASY (there is on occasion a fine sense of being on the borders of the transcendent). The leading Cenobite – known colloquially as Pinhead – became a minor fantasy ICON.   [JG]

2. *Hellbound: Hellraiser II* UK movie (**1988**). New World/Cinemarque/Film Futures. **Pr** Christopher Figg. **Exec pr** Clive BARKER, Christopher Webster. **Dir** Tony Randel. **Mufx** Image Animation. **Screenplay** Peter Atkins (1955-   ). **Starring** Doug Bradley (Pinhead/Elliott), Imogen Boorman (Tiffany), Kenneth Cranham (Dr Channard), Clare Higgins (Julia), Ashley Laurence (Kirsty). 93 mins. Colour.

Although again rich in cliché, the first sequel to **1** is a significant work of the fantastic. Weird psychiatrist Dr Channard uses the flesh and blood of a patient to revive Julia from the gory mattress on which she died, and – solving the puzzle-box through the medium of mute-savant child Tiffany – follows Julia to a DANTE-style HELL, ruled by Leviathan, Lord of the LABYRINTH. In pursuit come Kirsty – eager to rescue her father – and Tiffany. Julia betrays Channard, but she becomes the most powerful Cenobite of all, destroying the earlier ruling quartet (here more clearly identified as DEMONS, though much is made of their human origins). At last, Julia having been SKINNED, he himself is destroyed by the two girls and by Leviathan. There are conscious resonances with the works of Hieronymus BOSCH, Salvador DALI and M.C. ESCHER and with FAIRYTALE – "I'm no longer just the wicked STEPMOTHER," says Julia to Kirsty: "Now I'm the evil queen."   [JG]

3. *Hellraiser III: Hell on Earth* UK/US movie (**1992**). Fifth Avenue/Lawrence Mortorff. **Pr** Lawrence Mortorff. **Exec pr** Clive BARKER. **Dir** Anthony Hickox. **Spfx** Bob Keen. **Screenplay** Peter Atkins. **Starring** Kevin Bernhardt (J.P. Monroe), Doug Bradley (Pinhead/Elliott), Ken Carpenter (Doc), Terry Farrell (Joey Summerskill), Ashley Laurence (Kirsty), Paula Marshall (Terry). 93 mins. Colour.

Essentially a video nasty. The movie's sole interesting piece of fantasy comes when tv journalist Joey steps through LIMBO's layers of REALITY in company with the GHOST of WWI veteran Captain Elliott Spencer, who wishes to stop the sadistic activities of the entity he himself later became, Pinhead; this can be done only by Joey with the puzzle-box. At last Joey is able to despatch Pinhead/Elliott to HELL.   [JG]

Further reading: *The Hellraiser Chronicles* (photo coll 1992) by Clive BARKER with Peter Atkins and Stephen JONES.

HELPRIN, MARK (1947-   ) US writer whose first book, *A Dove of the East and Other Stories* (coll **1975**), contains work – almost all nonfantastic – from as early as 1969. One tale from this volume, "A Jew from Persia", manifests the shuttlecock shifts between realism and MAGIC REALISM typical of MH's later work, his narrative urgency, and his clarity of texture.

MH is very much a writer whose imagination is at home in the 20th century, as his finest single work attests. *Winter's Tale* (**1983**) is an URBAN FANTASY set in NEW YORK; but unlike Jerome CHARYN's *Panna Maria* (**1982**) or *Pinocchio's Nose* (**1983**), for instance, it attempts to engage with the CITY as a STORY to be celebrated rather than as a vast prison within which fables are trapped. It is an INSTAURATION FANTASY, not a SUPERNATURAL FICTION, and as such closely resembles John CROWLEY's *Little, Big* (**1981**), especially when MH describes the city, embraced in unending winter snow, in the language of THEODICY. The first part of the tale introduces the book's main protagonist, parentless Peter Lake, who is both a TRICKSTER and a FISHER KING, and whose coracle disgorges him (like Moses) into the welcoming golden LABYRINTH of late-19th-century New York, where he is like a fish in water. With his closest COMPANION, a magic flying horse, he traverses the flourishing, insanely energetic, glowing city; and profits from his adventures, even though constantly on the run from a gangland boss. In the second part, as the millennium approaches, Peter and his COMPANIONS become more and more intimately involved in attempting to tune the great world change, during which New York either will or will not spearhead a HEALING of the world. The final parts of the book move into a fantasticated near future. *Winter's Tale* is, justifiably, often claimed as one of the seminal works of late-20th-century fantasy.

Swan Lake (**1989** chap) is a tale with TWICE-TOLD resonances set in a RURITANIA, and features an enchanting EDIFICE.   [JC]

Other works: *Refiner's Fire: The Life and Adventures of Marshall Pearl, a Foundling* (**1977**); *Ellis Island and Other Stories* (coll **1981**), some supernatural; *A Soldier of the Great War* (**1991**).

HENIGHAN, TOM Working name of US-born academic and writer Thomas Joseph Henighan (1934-   ), in Canada from 1965. His nonfiction includes "Tarzan and Rima: The Myth and the Message" (1969 *Riverside Quarterly*) and *Natural Space in Literature* (**1980**); he also edited *Brave New Universe* (anth **1980**), being essays on science and art. *Tourists from Algol: Stories of the Unexpected* (coll **1983**) assembles sf and fantasy stories. In his impressive *The Well of Time* (**1988** UK) a Viking settlement in North America is threatened by a ZOMBIE-like race of risen dead; in her QUEST for a saving ELIXIR, the heroine must undergo METAMORPHOSIS into a WENDIGO, and answer tough RIDDLES posed her by a fickle ODIN. TH's collection of DARK FANTASIES, *Strange Attractors* (coll **1991**) similarly finds mythic echoes in the natural spaces of the great Canadian wilderness. *Home Planet* (coll **1994**) is poetry.   [JC]

HENRY, ALAN [s] ◊ John W. JAKES.

HENSON, JIM Working name of US puppeteer, stop-motion animator and movie/tv director/producer James Muary Henson (1936-1990), known primarily for his creation of the Muppets. These anarchic characters first came to worldwide attention with the children's educational programme *Sesame Street* (from 1969). They eventually starred in their own tv series, *The* MUPPET SHOW (1976-81), and in

several movies: *The Muppet Movie* (**1979**), *The Great Muppet Caper* (**1981**), *The Muppets Take Manhattan* (**1984**) and, after JH's death, *The Muppet Christmas Carol* (**1992**; ◊ *A* CHRISTMAS CAROL). His other major tv programme was FRAGGLE ROCK (1983-8), while his notable fantasy movie work included *An* AMERICAN WEREWOLF IN LONDON (**1981**), in which he played a part, and, much more significantly, *The* DARK CRYSTAL (**1982**) and LABYRINTH (**1986**), both of which he directed from his own stories. Further movies directed by JH were *Hey, You're As Funny As Fozzie Bear!* (**1988**), *Mother Goose Stories* (**1988**), *Neat Stuff to Know and to Do* (**1988**), *Peek a Boo* (**1988**) and *Wow, You're a Cartoonist* (**1988**). Because of the medium in which he worked – primarily tv – and the exuberance of his imagination and creativity, it is by no means foolish to suggest that JH was one of the most important fantasists of the later 20th century. His company is now merged with the DISNEY organization.   [JG]

HERA ◊ GODDESS.

HERBERT, JAMES (1943-   ) UK writer generally associated with HORROR, but most of whose works are also SUPERNATURAL FICTION and/or DARK FANTASY. At the same time, JH often offers cod sf explanations for supernatural events, which reduces his impact on the fantasy genre. He began with what has remained his most popular work, the **Rats** sequence – *The Rats* (**1974**; vt *Deadly Eyes* 1983 US), *Lair* (**1979**), *Domain* (**1984**) and *The City* (graph **1994** illus Ian MILLER); the first instalment was filmed as *The Rats* (**1982**). In a sense an URBAN FANTASY set in a devastated LONDON, the sequence traces an escalating war between humans and the huge mutant rats (◊ MICE AND RATS) who attack us from their UNDERGROUND redoubts. JH's second series, the **David Ash** books – *Haunted* (**1988**) and *The Ghosts of Sleath* (**1994**) – features a latter-day OCCULT DETECTIVE.

Singletons began with *The Fog* (**1975**), in which a malign yellow fog – its existence inspired by REVENANTS – transforms all those who breathe it into homicidal maniacs. In *Fluke* (**1977**) a man is reincarnated as a dog (◊ REINCARNATION). In *The Spear* (**1978**) neo-Nazis use the spear that pierced CHRIST's side to resurrect Heinrich Himmler (◊◊ HITLER WINS). *The Dark* (**1980**) traces the effect of an ancient malignity on modern folk. The police officer referred to in the title of *The Jonah* (**1981**) discovers he is a bad-luck carrier. In *Shrine* (**1983**) a miraculously cured girl turns out, perhaps, to be a victim of the DEVIL. *Moon* (**1985**) features a man of TALENT linked telepathically to a SERIAL KILLER. *The Magic Cottage* (**1986**) is a GHOST STORY and, although (or perhaps because) modest, possibly JH's best book. In *Sepulchre* (**1987**) an ELDER GOD hopes to gain POSSESSION of contemporary folk. *Creed* (**1990**) is supernatural horror. *Portent* (**1992**), set *c*1999, pits various troubled Britishers against a giant black woman from abroad (◊ COLOUR-CODING) who wishes to invoke the GODDESS as rightful inheritor of the polluted world.   [JC]

Further reading: *James Herbert: By Horror Haunted* (anth **1992**) ed Stephen JONES.

HERBERT, MARY H. (?   -   ) US writer whose fantasy has been restricted to the **Gabria** sequence – *Dark Horse* (**1990**), *Lightning's Daughter* (**1991**), *Valorian* (**1993**) and *City of Sorcerers* (**1994**) – set in a high-plains FANTASYLAND in which a young girl upsets the patriarchal applecart, with the aid of a MAGIC black horse, and herself becomes a magic-user.   [JC]

HERBIE MOVIES The intelligent Volkswagen who starred

in four DISNEY movies and a brief tv series (HERBIE, THE LOVE BUG [1982]), earned himself a huge audience. Essentially a TECHNOFANTASY character, Herbie was the first CAR to have its tyre-marks recorded outside Grauman's Chinese Theater, and the first to be issued with a US passport. There were minor overlaps between this series and the **Shaggy Dog** and **Flubber** series (◊ DISNEY). The movies were *The Love Bug* (**1969**) – novelized as *The Love Bug* * (**1969**) by Mel Cebulash – *Herbie Rides Again* (**1974**) – novelized as *Herbie Rides Again* * (**1974**) by Mel Cebulash – *Herbie Goes to Monte Carlo* (**1977**) novelized as *Herbie Goes to Monte Carlo* * (**1977**) by Vic Crume – and *Herbie Goes Bananas* (**1980**) – novelized as *Herbie Goes Bananas* * (**1980**) by Joe Claro. [JG]

HERBIE, THE LOVE BUG US tv series (1982). DISNEY/CBS. **Pr** Kevin Corcoran. **Exec pr** William Robert Yates. **Dir** Bill Bixby, Charles S. Dubin, Vincent McEveety. **Spfx** Michael Edmundson. **Writers** Arthur Alsberg, Don Nelson, Don Tait. **Created by** Alsberg, Nelson. **Based on** characters created by Gordon Buford. **Starring** Douglas Emerson (Robbie MacLane), Patricia Hardy (Susan MacLane), Dean Jones (Jim Douglas), Nicky Katt (Matthew MacLane), Richard Paul (Bo Phillips), Claudia Wells (Julie MacLane). 5 1hr episodes. Colour.

Disney's *The Love Bug* (**1969**; ◊ HERBIE MOVIES) was the highest grossing movie of its year, a fact that led to several sequels as well as this series. Here Dean Jones, star of the original movie, was reunited with the CAR for a mid-season replacement series. The premise is that Jones's character, Jim, has given up racing to open a driving school. Barely able to keep ahead of the bankers, Jim's only asset is Herbie, who is not above helping hapless students pass their tests just to aid Jim. Herbie also helps Jim meet an attractive widow, and soon she and her children are involved with Herbie and his misadventures. The series was no great success, but Disney have announced that another small-screen **Herbie** series will be launched soon. [BC]

HERE COMES MR JORDAN US movie (*1941*). Columbia. **Pr** Everett Riskin. **Dir** Alexander Hall. **Screenplay** Sidney Buchman, Seton I. Miller. **Based on** the play *Heaven Can Wait* by Harry Segall. **Starring** John Emery (Tony Abbott), James Gleason (Maxie Corkle), Edward Everett Horton (Messenger #7013), Rita Johnson (Julia Farnsworth), Evelyn Keyes (Bette Logan), Donald MacBride (Inspector Williams), Robert Montgomery (Joe Pendleton/Bruce Farnsworth/Ralph K.O. Murdock), Claude Rains (Mr Jordan). 93 mins. B/w.

Prizefighter Pendleton is snatched prematurely from his plummeting aircraft by an overanxious ANGEL, Messenger #7013; his body is swiftly cremated by his grieving manager, Corkle. The angel's boss, the saturnine Mr Jordan, leads Pendleton back to Earth to find a replacement body/life. As a GHOST Pendleton falls in love with Logan, whose father has been defrauded by playboy banker Farnsworth. To kill two birds with one stone, Pendleton possesses (◊ POSSESSION) Farnsworth's body when the banker is drowned by his wife and her lover, Abbott. Yet his destiny, according to Jordan, is to become world boxing champion and live until 1991 . . .

Couched as a TALL TALE told by Corkle, this is not so much a POSTHUMOUS FANTASY as an IDENTITY-EXCHANGE story. Its somewhat confused moral is that everything is preordained and that no one is cheated of the destiny they deserve. LIMBO is presented as a cloud-strewn featureless

plain from which aircraft transport SOULS to their final destination – an image picked up later in *A* MATTER OF LIFE AND DEATH (*1946*).

In a musical semi-sequel to *HCMJ*, *Down to Earth* (*1947*), an attempt was made to recreate Mr Jordan, now played by Roland Culver: Jordan facilitates Terpsichore's efforts to help a Broadway producer mount a stage show featuring her persona. The title of Segall's original play was used for the unrelated movie HEAVEN CAN WAIT (*1943*). *HCMJ* was remade as HEAVEN CAN WAIT (*1978*), with Pendleton as an LA Rams quarterback rather than a prizefighter. [JG]

HERE IS A MAN vt of *The* DEVIL AND DANIEL WEBSTER (*1941*).

HERETIC, THE vt of *Exorcist II: The Heretic* (*1977*). ◊ *The* EXORCIST.

HERM ◊ HERMES.

HERMAN, WILLIAM Pseudonym of Ambrose BIERCE.

HERMES Greek name for the GOD known in Roman MYTHOLOGY as Mercury. He is the messenger of the gods, patron of those who go on journeys and of those who simply wander, and patron also of thieves. He is a marker of THRESHOLDS, a LIMINAL BEING, the original psychopomp in charge of guiding dead souls through HADES, a TRICKSTER – "the thief-god, the god of roadways and night journeys [◊ NIGHT JOURNEY], the god of here-and-gone, the easer through the shadows, the finder in the dark", according to Russell HOBAN's *The Medusa Frequency* (**1987**). As a figure who signals by craft and DREAM the possibility of fundamental change – and one who (by inspired confusion with his son Hermaphroditus) bears a heavy charge of sexual ambivalence – Hermes often appears in texts where the longed-for change must be accomplished against the way of the world. In stories written around the FIN DE SIÈCLE, therefore, he can be seen guiding trapped homosexual men into the epiphany of an OTHERWORLD, possibly indistinguishable from death – as in Thomas MANN's *Death in Venice* (**1913**). Several stories by E.M. FORSTER reflect Hermes's complex, blasphemous appeal.

In his guise as Harlequin (◊ COMMEDIA DELL'ARTE) he is a mocker, a scamp, and a dissolver of worlds. [JC]

HERNE THE HUNTER In Celtic mythology, the Lord of the Forest and of the Forest Animals, usually represented with stag's horns. He is probably derived from an older deity, Cernunnos, the Horned God, likewise Lord of the Animals. The symbol of Cernunnos/Herne was so important to the Celts that the Christians adopted the horned image as that of the DEVIL and the epitome of EVIL. Herne has been linked to the WILD HUNT, although this confuses him with Gwynn ap Nudd, Lord of the UNDERWORLD. Herne is also mistakenly associated with ROBIN HOOD as the Wild Man of the Woods, another confusion of ARCHETYPES. Herne was a LIMINAL BEING who guarded the sanctity of the FORESTS and the PORTAL that led to the OTHERWORLD. Today Herne is closely linked to the forest in Windsor Great Park, where he is still believed to exist as the spirit of the Great Oak; he is depicted in this form in *Windsor Castle* (**1843**; vt *Herne the Hunter* 1920) by W. Harrison AINSWORTH. He also figures as a representative of ancient powers in *The Moon of Gomrath* (**1963**) by Alan GARNER and *Too Long a Sacrifice* (**1981**) by Mildred Downey BROXON; in both novels his spirit is reawakened (◊ SLEEPER UNDER THE HILL). He is recognized as part of British heritage in *The Box of Delights* (**1935**) by John MASEFIELD, but is most richly depicted in *A Wizard Abroad* (**1993**) by Diane DUANE. In modern fantasy that embraces New Age perspectives with Celtic imagery (◊ CELTIC FANTASY), Herne is a useful ICON of the old world's worship of Nature. He is always likely to appear, or at least be referred to, in works drawing upon those beliefs, such as those of Charles DE LINT. [MA]

Further reading: *Herne the Hunter, A Berkshire Legend* (**1972**) by Michael John Petry.

HEROES AND HEROINES The most common protagonists of fantasy, which draws much of its mythopoeic strength from TAPROOT TEXTS in which the values of a heroic age are taken more or less for granted, and which as an escapist genre (◊ ESCAPISM) derives much of its emotional appeal from the sense of identification with a strong individual. Heroes and heroines are fairly essential to HIGH FANTASY and HEROIC FANTASY. They may be, singly or severally, AVATARS of GODS, BRAVE LITTLE TAILORS, CHANGELINGS, CHILDE figures, DUOS, HIDDEN MONARCHS, KNIGHTS OF THE DOLEFUL COUNTENANCE, persons who LEARN BETTER, TEMPORAL ADVENTURESSES or UGLY DUCKLINGS. This list is not exhaustive. [RK]

HEROIC FANTASY There may be a useful distinction between HF and SWORD AND SORCERY, but no one has yet made it. The term itself – like DARK FANTASY, often used as an "up-market" term for HORROR – was almost certainly coined in an attempt to avoid the garishness of the S&S tag. This seems an inadequate reason to adopt a term which might seem synonymous with EPIC FANTASY (itself a term little used in this encyclopedia, and for similar reasons). A term like Hero Fantasy might be more arguable, because a central thread in any analysis of S&S is the understanding that it is a kind of GENRE FANTASY which features a HERO. In this encyclopedia, the term Sword and Sorcery, being familiar and self-evident, appears often; the type of fantasy of which S&S is a facet is neologistically referred to as ADVENTURER FANTASY. [JC]

HERON, E. & H. Joint working name of Hesketh Prichard (1876-1922), UK author, sportsman, explorer and naturalist, and his mother Kate Prichard (née Ryall; 1851-1935). They also collaborated as "K. and Hesketh Prichard", and created three very popular series characters: the bandit **Don Q**, who featured in many adventures – Douglas Fairbanks starred in the movie *Don "Q", Son of Zorro* (**1925**) – **November Joe**, detective of the Canadian backwoods, and **Flaxman Low**, the first professional full-time OCCULT DETECTIVE (he preceded **John Silence** and **Carnacki** by a decade). Clearly influenced by SHERLOCK HOLMES, Low was an Oxford-trained psychologist. E&HH's 12 *Real Ghost Stories* first appeared in *Pearson's Magazine* (January-June 1898, January-June 1899). [RD]

HERON-ALLEN, EDWARD Real name, used on some books, of Christopher BLAYRE.

HEROS THE SPARTAN Series of eight long and three short COMIC-strip myth-style historical fantasies set in Caesarian Rome and featuring a centurion of the Roman army. Written by Tom Tulley, the long stories were published in weekly episodes as colour centre-spreads in the UK comic *Eagle* 1962-6 and provided an impressive showcase for artists Frank BELLAMY and Luis BERMEJO. The general pattern of the stories was of a QUEST, in course of which HTS encountered strange peoples, evil MAGIC and opportunities to display his swordsmanship.

Plans to reprint the series as a collection have been hindered by the double-page format – only one of the three, less memorable, short stories (which were published in annuals) could be easily reprinted in book form – and the fact that much of the original artwork was stolen from the IPC archive in the late 1970s. [RT]

HERRIMAN, GEORGE (JOSEPH) (1880-1944) US COMICS artist and writer who began to publish visual work as early as 1897. By the first years of the new century GH had actively begun to create and/or write and illustrate comic strips. His first weekday strip, *Home Sweet Home*, appeared in 1904, the year he began his lifelong relationship with William Randolph Hearst (1863-1951), who gave him unprecedented artistic freedom for his work; after 1910 GH worked as a comics artist for no one but Hearst. That year he began to draw the cat and mouse who became the main figures in **Krazy Kat**, which ran from 1913 until his death. It is one of the 20th century's greatest fantasy sequences, as well as almost certainly the longest COMICS sequence of any fantasy importance to have been solely the creation of one person.

In the first half of the century, most comic strips were under the ultimate control of the newspaper or syndicate which ran them, and tended to continue, drawn by others, after their initial creator's retirement or death. It may be that **Krazy Kat**'s lack of mass appeal was the reason for its termination when GH died (in 1944 it was circulating to only 35 newspapers), but the sequence was, in fact, inimitable.

Its basic BEAST-FABLE premise, reiterated literally thousands of times, is a simple love triangle. Krazy Kat loves Ignatz Mouse, who does not love her in return, and torments her by throwing bricks at her. For Krazy Kat, the brickbats are signs of devotion, but Offissa B. Pupp, who loves her, "defends" her from Ignatz Mouse by sticking him in the jailhouse hundreds (perhaps thousands) of times. Krazy Kat does not seem to know that Pupp loves her. From this shifting triangle, GH spins a surreal (◊ SURREALISM) fantasy world of great intensity, set from the mid-1920s in a dream-like rendering of Coconino County, Arizona; it is a LANDSCAPE which borrows some mesas and high-desert chiaroscuro from nearby Monument Valley, but which in the end amalgamates geography and vision into a unique and immediately recognizable inscape. The DICTION of the strip is half *Katzenjammer Kids* and half James JOYCE, and the individual episodes are recounted with a loving savour that has occasioned comparisons to Miguel de CERVANTES, Charlie Chaplin (1889-1977), Charles DICKENS and others. GH's visual style occasioned a similar range of comparisons to 20th-century artists. In later years – especially after the large Sunday pages went to colour in 1935 – the increasing concentration of GH's work tended to transform individual episodes into autonomous works of art, not easily decipherable but haunting.

In 1916, the world's first ANIMATED MOVIE to feature a CAT, *Krazy Kat* (**1916**), was made by International Film Service. Many more **Krazy Kat** animated shorts followed. A ballet suite by John Alden Carpenter was produced in 1922, with scenario and libretto by GH; the score was later published as *Krazy Kat* * (**1923**). *Krazy Kat and Ignatz Mouse in Koko Land* (graph **1934**) was a Little Big Book. The first **Krazy Kat** collection proper – *Krazy Kat* (graph coll **1946**), with an introduction by e e cummings (1894-1962) – gave a scattered selection of strips, poorly reproduced. More recently, attempts have begun to make more of the strip available. Though its overall title is a misnomer, as it covers only Sunday pages, the **Komplete Kat Komics** sequence –

The Komplete Kat Komics #1: 1916: Krazy Ignatz (graph coll **1988**), *#2: 1917: The Other-Side to the Shore of Here* (graph coll **1989**), *#3: 1918* (graph coll **1989**), *#4: 1919: Howling Among the Halls of Night* (graph coll **1989**), *#5: 1920: Pilgrims on the Road to Nowhere* (graph coll **1920**), *#6: 1921: Sure as Moons is Cheeses* (graph coll **1990**), *#7: 1922: A Katnip Kantata in the Key of K* (graph coll **1991**) and *#8: 1923: Inna Yott on the Muddy Geranium* (graph coll **1991**), with further vols projected – is invaluable. Of the **Komplete Kolor Krazy Kat** so far only *The Komplete Kolor Krazy Kat #1: 1935-1936* (graph coll **1990**) has appeared. The best study of the strip is *Krazy Kat: The Comic Art of George Herriman* (**1986**) by Patrick McDonnell, Karen O'Connell and Georgia Riley de Havenon.

Although he rarely worked outside his chosen format, GH did famously illustrate the last two of Don MARQUIS's **archy and mehitabel** books, *archy's life of mehitabel* (coll **1933**) and *archy does his part* (coll **1935**). [JC]

HERZMANOVSKY-ORLANDO, FRITZ VON (1877-1954) Austrian writer. ◊ AUSTRIA.

HESCOX, RICHARD (1949-   ) US fantasy illustrator, with a traditional painterly style inspired by Howard PYLE, N.C. WYETH and Frank FRAZETTA. His first job was as a portrait painter in Disneyland. He went on to paint several covers for MARVEL COMICS' *Savage Sword of Conan* and *Vampire Tales*, and then covers for the publishers DAW. He has also provided artwork for the movies. [RT]

HESPERIDES ◊ LOST LANDS AND CONTINENTS.

HESSE, HERMANN (1877-1962) German-born writer, a Swiss citizen from 1923, winner of the 1946 Nobel Prize for Literature. He began his career with *Romantische Lieder* ["Romantic Poems"] (coll **1898**), a book that combined a certain fastidious austerity with a palpable taste for DECADENCE. Most of his early works of fantasy interest are literary FAIRYTALES, the first, "Lulu", appearing as an inset story in *Hinterlassene Schriften und Gedichte von Hermann Lauscher* ["The Posthumous Writings and Poems of Hermann Lauscher"] (**1900**); along with other tales, "Lulu" is in *Pictor's Metamorphoses and Other Fantasies* (coll trans Rika Lesser **1982** US). *The Fairy Tales of Hermann Hesse* (coll trans Jack ZIPES **1995** US), a fuller volume, assembles tales from 1904 through 1919, including five from *Märchen* (coll **1919**; trans as *Strange News from Another Star and Other Tales* **1972** US), and excludes only five late tales (1922-1933). Many of these stories are SUPERNATURAL FICTION, though some are of direct fantasy interest. In "Faldum" (1916), for instance, an ODIN-like GOD gives everyone in the eponymous city WISHES, some of which turn out disastrous, some beneficial: one artist becomes a mountain which transforms the city over the aeons he remains in this form; only when Faldum has become a myth is the artist allowed a final transcendence into death.

A similar longing for transcendence, couched as a search for wholeness, also infuses *Siddhartha* (**1922**; trans Hilda Rosner **1954** UK). For the eponymous worker of MIRACLES, in the slightly earlier *Demian* (**1919**; trans **1965** UK), any search for wholeness is truncated by WORLD WAR I, which kills him. By the time of *Der Steppenwolf* (**1927**; trans Basil Creighton as *Steppenwolf* **1929** UK), HH had come to dramatize the search for self-awareness through a tale whose protagonist, the nocturnal loner who calls the dark DOUBLE within him "Steppenwolf", is profoundly riven. The THRESHOLD dividing him from this inner self (or vice versa) is a kind of metaphor of the sort of threshold normally

found in a fantasy text, as *Steppenwolf* can be understood doubly: as a series of self-redeeming phantasies, and as narrative of the protagonist's journey through Expressionist otherworlds with the same goal in view, a whole self. So the fictional BOOK he uses as a vade mecum, the "Treatise on the Steppenwolf", is either a fiction, or a book, or perhaps both. In the end, in a MAGIC theatre whose MIRRORS give forth innumerable versions of his condition, the protagonist finally begins to heal.

As a UTOPIA, *Das Glasperlenspiel* (**1943**; trans M. Savill as *Magister Ludi* **1949** US; preferred trans Richard and Clara Winston as *The Glass Bead Game* **1969** US) is best addressed as sf. [JC]

HETEROTOPIA ◊ UTOPIAS.

HEYDRON, VICKI ANN (1945-   ) US writer, from 1978 married to Randall GARRETT, with whom she had planned the **Gandalara** sequence of PLANETARY ROMANCES before 1979, when he contracted meningitis and was subsequently unable to produce work with any consistency. The sequence – *The Steel of Raithskar* (**1981**), *The Glass of Dyskornis* (**1982**) and *The Bronze of Eddarta* (**1983**), all three assembled as *The Gandalara Cycle 1* (omni **1986**), plus *The Well of Darkness* (**1983**), *The Search for Kä* (**1984**) and *Return to Eddarta* (**1985**), all three assembled as *The Gadalara Cycle 2* (omni **1986**), plus *The River Wall* (**1986**) – was written by VAH, though both writers were credited throughout. The sequence, clearly showing the influence of Edgar Rice BURROUGHS, features a terminally ill university professor who is translated from Earth to Gandalara, where in desert climes a variety of cultures flourish and clash; there he enjoys a HEROIC-FANTASY life in a new body. MAGIC works in the RATIONALIZED-FANTASY mode; TALENTS like telepathy are the norm; and giant CATS serve as both mounts and COMPANIONS. [JC]

HICHENS, ROBERT (SMYTHE) (1864-1950) UK writer, prolific for half a century but now almost forgotten except for *The Green Carnation* (**1894**) and *The Garden of Allah* (**1904**), neither fantasy though both written in a style so heated, and so confused about the loathsome allure of SEX, that they seem infused with the supernatural. Most of his work of interest is SUPERNATURAL FICTION, with explicit reference to OCCULTISM, and most of it concentrates on various forms of psychic BONDAGE – through HAUNTING, or transfers of malign influence between DOUBLES, or other transactions of the SPIRIT. Relevant fiction appears in various collections, including: *The Folly of Eustace* (coll **1896**); *Bye-Ways* (coll **1897**), which includes "A Tribute of Souls" about a PACT WITH THE DEVIL; *Tongues of Conscience* (coll **1900**), containing stories in which intolerable guilt tends to accompany imprecisely described sins, as well as RH's best single tale, "How Love Came to Professor Guildea", about the fatal attraction of a dimwitted lovesick spirit for the eponymous scientist; *The Black Spaniel* (coll **1905**), the title story being a typical turn-of-the-century vivisection tale, in which a guilty doctor's spirit descends into a dog, which is duly tortured; *Snake-Bite* (coll **1919**); *The Last Time* (coll **1923**); *The Gardenia* (coll **1934**); and *The Man in the Mirror* (coll **1950**).

Throughout his work RH tends to punish any love relationship between man and woman, but abhors any hint of homosexual love. Both his supernatural novels – *Flames: A London Phantasy* (**1897**) and *The Dweller on the Threshold* (**1911**) – depict the malign consequences of intensifying male relationships through spiritual means. The DOUBLES so created – in the latter book, so intimate is the intermingling that the two victims almost seem to become involved in a full IDENTITY EXCHANGE – are against nature, and the sadism of the resulting interactions seems unmistakably sexual, though deeply coded (♀♀ DEBASEMENT). Most vividly, perhaps, in the "Parable of the Footprints" – which constitutes the narrative heart of *The Dweller on the Threshold* – RH's work is haunted by a horror of exposure. [JC]

Other works: *The Daughters of Babylon* (**1899**) with Wilson Barrett, historical novel with fantasy elements; *The Prophet of Berkeley Square* (**1901**), romance dealing mundanely with ASTROLOGY.

HICKMAN, TRACY (RAYE) (1955-   ) US author and role-playing GAMES designer who writes in collaboration with Margaret WEIS. [JC]

HIDDEN MASTERS ♀ SECRET MASTERS; THEOSOPHY.

HIDDEN MONARCH When King Richard returns incognito from exile and unveils himself to ROBIN HOOD, a monarch who had been hidden is recognized, and the LAND is healed (♀ HEALING). But Richard already *is* a monarch, and knows he is. As a PLOT DEVICE or TOPOS in fantasy, the HM motif commonly follows a somewhat different pattern: the HM is a youngster who does not know his (less frequently her) identity or destiny. The most famous HM in English literature is the young ARTHUR; and the definitive tale of the RITE OF PASSAGE into adulthood, responsibility and healing power is almost certainly T.H. WHITE's *The Sword in the Stone* (**1938**).

Most HMs are UGLY DUCKLINGS when first met; Wart in *The Sword in the Stone* is certainly one. They are children occupying a lowly position in the world; they are badly dressed; they are mocked; they have unappeasable longings for some other state, but their BONDAGE – their immolation in the wrong identity – seems unloosenable, often having been imposed by MAGIC at the behest of a jealous DARK LORD. Tip, the young lad who becomes Princess Ozma of OZ in L. Frank BAUM's **Oz** sequence, fulfils all these conditions, as do Taran the Assistant Pig-Keeper in Lloyd ALEXANDER's **Chronicles of Prydain**, Garion the farmboy in David EDDINGS's **Belgariad**, and Simon the scullion in Tad WILLIAMS's **Memory, Sorrow and Thorn**.

The wish-fulfilment function of the HM device is obvious. An HM is a figure easy for young readers to identify with. But his function is wider than that in some texts, where his ascension to the throne carries a promise of profound transformation, a confirmation of HEALING – a healing function which may in fact be literal: Covenant, in Stephen R. DONALDSON's **Thomas Covenant** sequence, has miraculous curative powers, as does Severian in Gene WOLFE's *The Book of the New Sun* (**1980-83**). When Aragorn is crowned as Elessar in J.R.R. TOLKIEN's *The Lord of the Rings* (**1954-5**), after the unjust have been cast down, he properly takes on a demigod role. [JC/RK]

HIGGINS, GRAHAM (1953-   ) UK illustrator and cartoonist. ♀ Terry PRATCHETT.

HIGH FANTASY Fantasies set in OTHERWORLDS, specifically SECONDARY WORLDS, and which deal with matters affecting the destiny of those worlds. [JC]

See also: ADVENTURER FANTASY; FANTASY; GENRE FANTASY; HEROIC FANTASY.

HIGHLANDER There have been three movies (so far) in the **Highlander** sequence, plus a tv series (♀ *The* HIGHLANDER).

1. *Highlander* UK/US movie (**1986**). EMI/Highlander.

Pr Peter S. Davis, William N. Panzer. **Exec pr** E.C. Monell. **Dir** Russell Mulcahy. **Spfx** Martin Gutteridge. **Mufx** Bob Keen. **Vfx** Optical Film Effects Ltd. **Screenplay** Peter Bellwood, Larry Ferguson, Gregory Widen. **Novelization** *Highlander* * (**1986**) by Garry Douglas (Garry KILWORTH). **Starring** Clancy Brown (The Kurgan), Sean Connery (Juan Sanchez Villa-Lobos Ramírez), Beatie Edney (Heather), Sheila Gish (Rachel Ellenstein), Roxanne Hart (Brenda J. Wyatt), Christopher Lambert (Connor MacLeod), Alan North (Lt Frank Moran), Hugh Quarshie (Sunda Kastagir). 111 mins. Colour.

For much of *H* two stories – one set in the 16th-century Scottish Highlands and the other in 1985 NEW YORK – are interleaved. In 1536 Connor MacLeod was killed by a mysterious outsider during a clan battle against the Frasers, yet rose within the night; assumed a WITCH, he was driven off. Far away in the Highlands he married Heather. Into their almost eremitic life stormed a 2437-year-old Egyptian, Ramírez, who explained he and McLeod were both immortal (♀ IMMORTALITY) unless decapitated, and told him there were others like themselves: strongest of all was The Kurgan, a sadistic "perfect warrior" from the Steppes; centuries in the future instinct would bring the surviving immortals together for the Gathering, a LAST BATTLE that only one could win. The Kurgan arrived in MacLeod's temporary absence, killed Ramírez and raped Heather.

1985 is the time of the Gathering. Attacked by another immortal beneath Madison Square Gardens, MacLeod beheads him. Wyatt, a police forensic metallurgist, discovers his sword – inherited from Ramírez – to be an historical paradox; she investigates, finds the truth, and becomes MacLeod's lover. The Kurgan kills the third-last immortal, Kastagir, and seizes Wyatt. In a pyrotechnic flurry of swords, MacLeod kills The Kurgan, saves Wyatt and – surrounded by swirling id-monsters – attains the ability to be at one with all living things.

H, weirdly crosshatching CONTEMPORARY FANTASY, HIGH FANTASY, URBAN FANTASY and martial-arts movies, is widely castigated as incoherent, illogical and unresolved – all true, but unimportant in the context of this cult movie's undoubted power and glamour; its frenetic glitz complements and enhances the sense of MAGIC. [JG]

2. *Highlander II: The Quickening* US movie (**1990**). Ziad El-Khoury & Jean Luc Defait/Lamb Bear. **Pr** Peter S. Davis, William N. Panzer. **Exec pr** Guy Collins, Mario Sotila. **Dir** Russell Mulcahy. **Spfx** John Richardson. **Vfx** Sam Nicholson. **Screenplay** Peter Bellwood. **Starring** Sean Connery (Ramírez), Michael Ironside (Katana), Christopher Lambert (MacLeod), John C. McGinley (David Blake), Virginia Madsen (Louise Marcus). 100 mins. Colour.

A mess that tries but fails to provide a coherent TECHNOFANTASY rationale for **1**, the immortals being now, apparently, aliens (♀ SFE).

3. *Highlander III: The Sorcerer* US movie (**1994**). Entertainment International/Transfilm/Lumiere/Fallingcloud/Peter S. Davis & William Panzer. **Pr** Claude Léger. **Exec pr** Guy Collins, Charles L. Smiley. **Co-pr** Eric Altmayer, Jean Cazes, James Daly. **Dir** Andy Morahan. **Screenplay** Paul Ohl. **Starring** Christopher Lambert (Connor MacLeod), Deborah Unger (Alex Johnson/Sarah), Mario Van Peebles (Kane). 96 mins. Colour.

HIII wisely forgets about **2** and sequels **1**. 400 years ago MacLeod left Europe for a time to train in Japan with a fellow-immortal, a Japanese sorcerer and martial-arts expert. In his wake came the most EVIL of all the immortals, the sadistic Kane, who beheaded the sorcerer and thereby inherited his powers of ILLUSION, notably becoming a proficient SHAPESHIFTER; yet the sorcerer's last action was to collapse his cavernous lair, trapping Kane. Now, in 1994, commercial excavations open up the cave, releasing Kane to hunt down MacLeod – who only *thought* he had become "The One" when killing The Kurgan (in **1**), and who now lives with his adopted son John in Marrakesh. Alerted to Kane's re-emergence, he goes to New York. Archaeologist Johnson, who was present at the Japanese site, just happens to be the DOUBLE (aside from hair-colour) of Sarah, a woman MacLeod loved at the time of the French Revolution. Kane makes oddly futile attempts to behead MacLeod – his powers of MAGIC are such that it is hard to see how he could fail – and MacLeod and Johnson become lovers in Scotland (to the authentic strains of Canadian Loreena McKennitt singing the Irish song "Bonny Portmore"). At last Kane uses his imitative shapeshifting powers to seize John; MacLeod wins the ensuing swordfight and this time truly achieves transcendence as "The One"; he and Johnson go to live with John back in that remote Scottish glen.

In terms of plot, *HIII* is a nonsense, but the plotting chaos becomes almost a strength: interestingly interwoven are countless traditional motifs, of which Kane's shapeshifting (e.g., to become a crow, or MacLeod's DOPPELGÄNGER) is just one – a significant other is the association of MacLeod with the figure of the WANDERING JEW. [JG]

HIGHLANDER, THE US tv series (1992-current). Rysher Entertainment & Gaumont International/ Syndicated. **Pr** Ken Gord. **Exec pr** Peter Davis and Bill Panzer, Christian Charret and Marla Ginzberg. **Dir** Ray Austin and many others. **Writers** David Abramowitz and many others. **Based on** the HIGHLANDER movies. **Created by** Gregory Widen. **Novelizations** *Highlander: The Element of Fire* (**1995**) by Jason Henderson; *Highlander: Scimitar* (**1996**) by Ashley McConnell. **Starring** Philip Akin (Charlie DeSalvo), Jim Byrnes (Joe Dawson), Elizabeth Gracen (Amanda), Lisa Howard (Dr Anne Lindsey), Stan Kirsch (Richie Ryan), Christopher Lambert (Connor MacLeod – pilot only), Adrian Paul (Duncan MacLeod), Alexandra Vandernoot (Tessa Noël), Peter Wingfield (Adam Pierson/Methos). Color. 78 60-minute episodes to date.

Despite the movies, in this world there are many Immortals, the Gathering has just begun, and Connor MacLeod did not win the Prize back in 1985 or 1994. The series centres on Duncan MacLeod, clansman to Connor, born 400 years ago in Scotland and now battling EVIL. Duncan discovers that he and the other Immortals have been under the scrutiny of a group known as The Watchers. While most Watchers are benign, a splinter group, The Hunters, is dedicated to killing Immortals. Not all stories involve the Watchers or Hunters; many concern people from Duncan's past who have resurfaced, often with a grudge against him or the Immortals in general. There is much swordplay. [BC]

HIGH SPIRITS UK/US movie (**1988**). ◊ *The* GHOST GOES WEST (**1935**).

HIGHWAY TO HEAVEN US tv series (1984-9). Michael Landon Productions/NBC. **Pr** Kent McCray. **Exec pr** Michael Landon. **Dir** Victor French, Landon. **Spfx** Ray Robinson. **Writers** Paul W. Cooper, Dan Gordon, Vince R.

Guitierrez, Landon, David Thoreau. **Starring** French (Mark Gordon), Landon (Jonathan Smith). 1hr episodes. Colour.

Veteran actor Landon created this popular series about Jonathan Smith, a trainee ANGEL assigned to aid humans while proving himself ready for duties in HEAVEN (rather like Clarence in IT'S A WONDERFUL LIFE [**1946**]). Instead of MAGIC he generally uses persuasion to convince people to help themselves. He is joined in his adventures by Mark Gordon, an ex-policeman who has had his life turned around by Jonathan and has vowed to help others. Together, they travel across the country as itinerant workers, a PLOT DEVICE that allows them to meet a steady stream of needy possibilities.

The plots were quite varied: in one Edward Asner appeared as an angel on probation for his past screw-ups; in another Bob Hope was cast as an angel handing out assignments; in a further episode Jonathan tried to help a man who had made a PACT WITH THE DEVIL to save a boy hit by a car. Most were more mundane. *HTH* was a long-running hit. [BC]

HILDEBRAND BROTHERS Twins Tim and Greg Hildebrand (1939-), whose successful collaboration from childhood on a number of fantasy projects ended *c*1982, when they began to seek work as individuals. They both work in acrylics in bright, strong colours.

The brothers studied briefly at Meinzinger's Art School in Detroit before taking a job with the animation company Jam Handy for four years, after which they relocated in 1962 to New York to make documentaries for a religious organization. They then began illustrating children's educational books for Holt-Reinhart-Winston, also doing advertising work and album covers for RCA Victor. In 1975 they were commissioned to produce a cover for J.R.R. TOLKIEN's *Smith of Wooten Major & Farmer Giles of Ham* (1949), which led to commissions to paint a series of Tolkien Calendars; these brought them wide acclaim, and their reputation was enhanced by their poster for the movie *Star Wars* (*1977*). Their next project, *Urshurak* (**1979**), was a copiously illustrated fantasy novel. They ceased to work in concert for financial reasons, but have continued to follow parallel careers in fantasy illustration. [RT]
Further reading: *The Fantasy Art Techniques of Tim Hildebrandt* (graph **1991**) by Jack E. Norton; GH is featured in *Great Masters of Fantasy Art* (graph **1986**) by Eckart Sackmann.

HILGARTNER, BETH (1957-) US writer in whose first novel, *A Necklace of Fallen Stars* (**1979**), a young princess baulks at the wrong marriage. Of rather more interest is the **Dreamweaver** YA sequence – *Colors in the Dreamweaver's Loom* (**1989**) and *The Feast of the Trickster* (**1991**) – in which an adolescent girl crosses the THRESHOLD into a SECONDARY WORLD, where she is known as the Wanderer, and, with the aid of a SEVEN-SAMURAI group of COMPANIONS possessing various TALENTS, works to keep things from falling apart. In the second volume, back on Earth and afflicted with AMNESIA concerning her sojourn, she must find the TRICKSTER god who is putting the secondary world out of joint. [JC]
Other works: *A Murder for Her Majesty* (**1986**), associational.

HILL, SUSAN (ELIZABETH) (1942-) UK novelist. *The Woman in Black* (**1983**), adapted by Nigel KNEALE in 1989 for tv, is a SUPERNATURAL FICTION in which a GHOST kills children to avenge the death of her own; the narrative is of relentless bleakness, and ironizes any elements of pastiche in the tale. Other titles with supernatural elements include *The Mist in the Mirror* (**1992**), which is a rather ineffectual GHOST STORY, and *Mrs De Winter* * (**1993**), a sequel to Daphne DU MAURIER's *Rebecca* (**1938**) (◊◊ SEQUELS BY OTHER HANDS). SH also edited *The Walker Book of Ghost Stories* (anth **1990**; vt *The Random House Book of Ghost Stories* **1991** US). [JC]

HILMAN, F. MCHUGH [s] ◊ Kenneth MORRIS.

HILTON, JAMES (1900-1954) UK writer, in the USA from 1935, where he worked mainly as a screenwriter. He became famous with his 13th novel, *Lost Horizon* (**1933**), which Frank CAPRA made into an equally famous movie, LOST HORIZON (*1937*). The movie changes some aspects but remains remarkably faithful in spirit to the book's marriage of cultural anxiety and escapist longing. Conway, the protagonist, is a vicarious victim of WORLD WAR I, a spiritually deadened Walking Wounded of a sort found quite frequently in novels of the 1920s; *Lost Horizon* itself is one of the last "aftermath" tales written, and perhaps the most resonant of all.

The tale told within the frame is not particularly complex. Escaping from troubles in northern India. Conway and his companions find themselves passengers on a hijacked plane that takes them deep into Himalayan Tibet, where it crashes. The party is rescued by a group led by an aged man, and they come to SHANGRI-LA, a lamasery set high above a POLDER valley which Conway soon realizes has been hidden from the outside world (◊ LOST RACES) for centuries. (Though in the novel Shangri-La is the name of the lamasery only, not of the Valley of the Blue Moon which contains it, the term has come to represent the entire polder.) Here Conway enjoys the air of studious, timeless serenity rather similar to that which, in Madame BLAVATSKY's *The Secret Doctrine* (**1888**), attends the SECRET MASTERS who expound THEOSOPHY from their Tibetan fastness. This atmosphere profoundly answers Conway's need for surcease from a world whose centre no longer holds; and on meeting Father Perrault, the centuries-old High Lama, he soon discovers that Perrault considers him his natural heir, which is why the plane was hijacked in the first place. [JC]

See also: MAGIC MOUNTAIN.

HIMMEL ÜBER BERLIN, DER ot of WINGS OF DESIRE (*1987*).

HIS MAJESTY, THE SCARECROW OF OZ (*1914*) ◊ The WIZARD OF OZ (*1939*).

HISTORY IN FANTASY Fantasy as a genre is almost inextricably bound up with history and ideas of history, reflected and reworked more or less thoroughly according to the needs, ambitions and intentions of individual authors. Its roots can be regarded as lying in the swashbuckling historical adventure stories of Alexandre DUMAS, Robert Louis STEVENSON, H. Rider HAGGARD and others; in the backward-looking romanticism of the PRERAPHAELITES; and in academic Classical and medieval scholarship. To many writers and readers, a fantasy novel should be set against a quasi-historical (very often quasi-medieval) background, and the boundaries between historical novels and fantasy can be thin. The influence of historical writing upon fantasy is dynamic, and indeed a number of modern fantasy authors have been trained as historians (including Judith TARR and Katherine KURTZ), and a number, like Tarr and Pauline

GEDGE, write successfully within both genres. In addition, characters and archetypes drawn from literature (e.g., Miguel CERVANTES's Don Quixote), quasi-historical figures (e.g., ARTHUR), genuine historical figures (e.g., Alexander the Great) and legendary heroes (e.g., CUCHULAIN) continue to provide material and inspiration for writers in certain subgenres of fantasy.

HIF, while predominantly associated with medieval imagery, is not restricted to a single historical era, nor to the Western European tradition. Fantasy novels have been set against backgrounds drawn from or reflecting the European Renaissance, the Enlightenment, the French Revolution, 19th-century Britain, and WORLD WAR II. Cultural backgrounds include the history of Korea (Judith Tarr), Japan (Jessica Amanda SALMONSON; Peter MORWOOD; C.J. CHERRYH), China (Barry HUGHART; Kathryn GRANT; Stephen MARLEY), Russia (Morwood), India (Margaret BALL), and 19th-century North America (Karen Joy Fowler [1950-]). Other authors have used fantasy to explore themes drawn from historical experience – including colonialism (Colin GREENLAND), emigration and exploitation (Karen Joy Fowler; Guy Gavriel KAY) and cultural ANTHROPOLOGY (Rosemary KIRSTEIN).

It is possible to identify four fluid and overlapping categories of the use of HIF. The first comprises novels set largely in a genuine historical context and often drawing upon actual historical events, but blending these with fantasy tropes. Thus Barbara ERSKINE and Anya Seton (?1904-1990) have used the idea of REINCARNATION to explore themes of personal commitment and political responsibility in parallel modern and historical contexts. Judith Merkle RILEY has blended psychic healing and precognition with authentic 14th- and 17th-century backgrounds. The standard of research in such novels is usually high and the books are often marketed as historical.

The second category is in many ways closely linked with the first. It consists of "what if?" ALTERNATE-WORLD novels, adding a major new element to a historical period. Tarr, in her **The Hound and the Falcon** trilogy, uses the complex politics of the late 12th century as a background, but postulates the existence of an additional player in the shape of an elven kingdom, whose existence further complicates the ecclesiastical problems of the time. In *The Dragon Waiting*, John M. FORD postulates a late-15th-century England in which Earl Rivers survives the accession of Richard II, in which VAMPIRES exist and are politically highly active, and in which Henry Tudor never becomes king. The **Lord Darcy** series by Randall GARRETT has a background based on 19th-century England, but is predicated upon the survival of the Angevin empire and upon the development of a strong tradition of magical, rather than scientific, research.

Novels which seek to create their own internal, coherent, invented history for an imaginary world or kingdom form the third category. The works of J.R.R. TOLKIEN are pre-eminent here. Tolkien, a medieval scholar, borrowed largely from early literature, but transformed it to create his own unique mythology, history and culture on a scale unrepresented elsewhere: indeed, in many respects the creation of the background took precedence over the narrative; other writers upon whom he has had a profound influence have largely worked the other way round (◊ FANTASYLAND). Imaginary histories have been created to make political points or to explore social and gender roles, but more usually the "historical" material is there to provide an exciting problem for the characters to solve. Some authors have attempted to generate histories over a series of novels, often by writing "back" events: the results in terms of the creation of a cohesive sense of history are highly variable, and the impression created is often more one of a sequence of episodes than of a genuine history.

Many fantasy authors borrow historical events and ideologies, and transform them to provide the basis of a new world. This is the largest of the four categories, and ranges widely in style and depth of borrowing, from the adoption of a simple generic Fantasyland, usually although not necessarily quasi-medieval (often highly inaccurate), to the appropriation of a specific historical practice, event or detail to lend resonance to an imaginary world, as with Kurtz's painstaking borrowing of medieval liturgy.

History is a rich vein for fantasy authors. Inevitably, degrees of accuracy vary, with consequent results for the conviction and success with which a given background is presented. The use of historical imagery has never been free from romanticism and idealism. The forms of borrowing have tended to vary over time as writers reflect their own contemporary political and social concerns. Current trends to depict early cultures as feminist havens ($ GODDESS) or to explore the idea of child abuse as a major issue for communities living at subsistence level sit uncomfortably against a medieval background. Some modern fantasy writers tend to see the past as a better as well as a different country.

[KLM]

See also: ANACHRONISM; CELTIC FANTASY; CHIVALRY; EGYPT; GASLIGHT ROMANCE; GOLDEN AGE; GOTHIC FANTASY; FANTASIES OF HISTORY; LAND OF FABLE; NORDIC FANTASY; ORIENTAL FANTASY; ROMANTICISM.

HISTORY OF FANTASY ◊ FANTASY.

HITCHCOCK, [Sir] ALFRED (JOSEPH) (1899-1980) UK movie director who worked for most of his career in Hollywood. His only movie to be overtly fantasy is *The* BIRDS (*1963*) – one of several based on tales by Daphne DU MAURIER, others being *Jamaica Inn* (*1939*) and *Rebecca* (*1940*) – but very few do not have a *frisson* of fantastication, even if in the event these PSYCHOLOGICAL THRILLERS prove to be RATIONALIZED FANTASIES. Among many examples are: *The 39 Steps* (*1935*), based on the John BUCHAN novel, which in AH's hands becomes a semi-comedic yet nightmarish NIGHT JOURNEY; *The Lady Vanishes* (*1938*), an exemplary exploration of paranoid (though ultimately rather badly rationalized) fantasy; *Spellbound* (1945), another astonishingly effective psychological thriller; *Rear Window* (*1955*), playing upon urban paranoia; *Vertigo* (*1958*), in many ways the oddest of all AH's movies, concentrating obsessively on obsession; and, based on the Robert BLOCH novel, *Psycho* (*1960*), a flawed but awe-inspiring work in which AH, apart from anything else, puts the notion of the EDIFICE to devastating use. The list could be extended.

[JG]

HITLER WINS Most novels in which Adolf Hitler wins WORLD WAR II are set in sf ALTERNATE WORLDS; *SS-GB* (*1978*) by Len Deighton (1929-) is one example and *Fatherland* (*1992*) by Robert Harris (1957-) another. In some SUPERNATURAL FICTIONS, occult procedures are used either to defend the Nazis in their own time, as in *Werewolves* (*1990*) by Jerry Ahern and Sharon Ahern, or to defeat them, as in Dennis WHEATLEY's *Strange Conflict* (*1941*), or to resurrect the Reich, as in James HERBERT's *The Spear* (*1978*). In *The Bargain* (*1990*) by Jon Ruddy

DRACULA vampirizes Eva Braun, who – in a SLINGSHOT ENDING – in turn vampirizes Hitler, thus giving him IMMORTALITY and the likelihood of eventual victory. Some novels – like Katherine BURDEKIN's *Swastika Night* (**1937**) as by Murray Constantine, which is a future history, and SARBAN's *The Sound of his Horn* (**1952**) – portray Hitler's victory in terms so remote that it is a legend for the characters in the foreground. Some of the stories assembled in *Hitler Victorious* (anth **1987**) ed Gregory Benford and Martin H. GREENBERG are supernatural fiction. [JC]

HJORTSBERG, WILLIAM (REINHOLD) (1941-) US writer whose first novel, *Alp* (**1969**), is set in an Alpine POLDER inhabited by stir-crazy AVATARS of figures from FOLKLORE like Gretel, by US tourists, and by a cannibal DWARF. It is not quite fantasy, and neither is the rather similar *Toro! Toro! Toro!* (**1974**). *Gray Matters* (**1971**) is sf. *Falling Angel* (**1978**) is a virtuoso DARK FANTASY set in an hallucinated 1950s NEW YORK; it was filmed as *Angel Heart* (**1987**), with the venue changed to NEW ORLEANS. The protagonist – detective Harry Angel – is hired by a man who turns out to be SATAN and who is searching for Johnny Favorite, who defaulted on a PACT WITH THE DEVIL; but Angel finds, as he falls closer to his fate, that he is victim of a devastating SECRET-SHARER relationship with Favorite, who long before underwent a VOODOO RITUAL in order to take on his new identity as a kind of ANGEL. *Nevermore* (**1994**) features Arthur Conan DOYLE and Harry Houdini (1874-1926) as an Odd Couple (◊ DUOS) in the 1920s USA, quarrelling over the nature of the supernatural until (to Doyle) the GHOST of Edgar Allan POE apears in conjunction with a number of murders based on his stories. [JC]

Other works: *Symbiography* (**1973** chap), sf; *Tales & Fables* (coll **1985** chap).

HOBAN, RUSSELL (CONWELL) (1925-) US-born illustrator and writer, resident from 1969 in the UK. He began his professional career as an illustrator in 1951, moving into copywriting, and only later publishing his first book, the nonfiction *What Does It Do and How Does it Work?* (**1959**). For the first decades of his writing career, RH wrote exclusively for children, the most successful of these early tales being the **Frances** series, about a badger child: *Bedtime for Frances* (**1960** chap), *A Baby Sister for Frances* (**1964** chap), *Bread and Jam for Frances* (**1964** chap), *A Birthday for Frances* (**1968** chap), *Best Friends for Frances* (**1969** chap), *A Bargain for Frances* (**1970** chap) and *Egg Thoughts and Other Frances Songs* (coll **1972** chap), the last being poetry. The emotional and linguistic intensity of RH's children's books prefigures his most ambitious efforts.

In *The Mole Family's Christmas* (**1969** chap), illustrated by RH's first wife, Lillian Hoban, a mole becomes obsessed with the stars he cannot see, and writes a letter to SANTA CLAUS, who grants his wish for a telescope. A tiger-hunting Rajah, who blasphemes the way of the jungle by playing "light classics" on his cassette while out hunting in *The Dancing Tigers* (**1979** chap), illustrated by David Gentleman, is tricked into watching his prey dance: by dawn, after they have "danced the moon down low and pale into the morning", the Rajah and his beaters are dead from the sight, for the tigers are in effect the tygers of William BLAKE. In *The Marzipan Pig* (**1986** chap), illustrated by Quentin Blake, a sweet in the shape of a pig falls behind a sofa, suffers, is eaten by a mouse who becomes transfixed by romantic longings and is eaten by an owl, who is likewise infected and falls in love with a taxi meter; eventually, another mouse in a hibiscus-petal frock dances on the Albert Embankment, and is not eaten. In 30 brief pages, a profound fantasy of TRANS-FORMATION is enacted. (See listing below for a full list of RH's work for younger children.)

RH's most famous single work for children, *The Mouse and His Child* (**1967**), is for an older audience. The mouse and his child, who are wind-up clockwork creatures, are bound together back to back. When they are not wound up, they are helplessly immobile; when wound, they cannot stop. They are, in other words, caught in a profound BONDAGE, and seem incapable of transcending that state. After years as CHRISTMAS toys – for RH, Christmas is a frequently evoked, highly dangerous passage of the SEASONS – the mice are broken and abandoned. Fixed by a tramp so they can progress in a straight line, they find themselves in the thrall of a Rat King – Manny Rat – who uses wind-up toys as indentured labour; after many adventures, during which they continue to wear out, they find themselves in the precarious miniature POLDER of an abandoned DOLL house, where a reformed Manny Rat jiggers their workings so they can wind each other up – but not forever: the toll of friction is explicitly acknowledged. The book was adapted as an ANIMATED MOVIE, *The Mouse and His Child* (**1977**), which proved to be more worthy than entertaining.

Another tale of complex fascination is *The Sea-Thing Child* (**1972**), in which an animal, washed up on a beach, has intense conversations with other creatures, including a crab whose haunting "face" prefigures *The Medusa Frequency* (see below).

RH's adult novels are of various fantasy interest. The first, *The Lion of Boaz-Jachin and Jachin-Boaz* (**1973**), is dominated (as its title hints) by a sense of a MIRROR-like identity between the father, long-absent on a QUEST for the ultimate MAP which he had been promised, and the son who goes on his own quest, for the older man; both are haunted by the eponymous lion, who is in one sense long-dead but in another a manifestation of the Being of the world who transcends the map searched for. The eponymous protagonist of *Kleinzeit* (**1974**) inhabits an internal world broken in upon constantly by fantasy representations (DEATH lurks outside his door) that monitor his increasing incapacity to find himself in the world. *Riddley Walker* (**1981**) is sf; it and *The Mouse and His Child* perhaps represent the two peaks of RH's career.

In *Pilgermann* (**1983**), the presence – he is more a device of fabulation than a GHOST – of an 11th-century European Jew, likely called Pilgermann (it is never stated), tells his story; one of the characters he encounters appears also in *The Flight of Bembel Rudzuk* (**1982** chap), a book for younger children. *The Medusa Frequency* (**1987**), which may be both the most enigmatic and the funniest of all RH's adult books, is a quite savage portrayal of a writer, Orff, who attempts to cope with his creative block – and his erotic nostalgia for a woman who left him nine years earlier – by undertaking a long dialogue with the severed head of ORPHEUS. At the same time, Orff ploughs deeper and deeper into dark waters in his search for various women, all of whom he thinks are Eurydice, and who most profoundly manifest their UNDER-LIER Eurydice aspect (◊ FACE OF GLORY) as a dread figure of the THRESHOLD between the protagonist's mortal state and a further place he cannot reach. In the end, *The Medusa Frequency* is a tale of defeat, a tale whose protagonist – in this it is typical of RH's work – is caught in a STORY which cannot find a proper conclusion and so lames him, though some ambivalence remains.

Though often hilarious, RH is not a writer of comedy (◊ HUMOUR). His importance for the genre, beyond the visible merits of his large oeuvre, lies in his presentation of the anguished complexities attending his heroes' attempts to penetrate the unpassable thresholds that mark our mortality. [JC]

Other works (for children): *Herman the Loser* (**1961** chap); *The Song in my Drum* (**1962** chap); *London Men and English Men* (**1962** chap); *Some Snow Said Hello* (**1962** chap); *The Sorely Trying Day* (**1964** chap); *Nothing to Do* (**1964** chap); *Tom and the Two Handles* (**1964** chap); *The Story of Hester Mouse who Became a Writer and Saved Most of her Sisters and Brothers and Some of her Aunts and Uncles fom the Owl* (**1965** chap); *What Happened when Jack and Daisy Tried to Fool the Tooth Fairies* (**1966** chap); *Henry and the Monstrous Din* (**1966** chap); *The Little Brute Family* (**1966** chap) and its sequel, *The Stone Doll of Sister Brute* (**1968** chap); *Save my Place* (**1967** chap); *Charlie the Tramp* (**1967** chap); *Harvey's Hideout* **1969** (chap); *Ugly Bird* (**1969** chap); *Emmet Otter's Jug-Band Christmas* (**1971** chap); *Letitia Rabbit's String Song* (**1973** chap); *How Tom Beat Captain Najork and his Hired Sportsmen* (**1974** chap) and its sequel, *A Near Thing for Captain Najork* (**1975** chap); *Ten What?* (**1974** chap); *Crocodile & Pierrot* (**1975** chap); *Dinner at Alberta's* (**1975** chap); *Arthur's New Power* (**1978** chap); *The Twenty-Elephant Restaurant* (**1978** chap); *La Corona and the Tin Frog* (in *Puffin Annual* anth **1974**; **1979** chap); *Flat Cat* (**1980** chap); *Ace Dragon Ltd* (**1980** chap); *They Came from Aargh!* (**1981** chap); *The Great Fruit Gum Robbery* (**1981** chap; vt *The Great Gumdrop Robbery* 1982 US); *The Battle of Zormla* (**1982** chap); *The Serpent Tower* (**1983** chap); the **Ponders** sequence, comprising *Big John Turkle* (**1983** chap), *Jim Frog* (**1983** chap), *Charlie Meadows* (**1984** chap) and *Lavinia Rat* (**1984** chap), all 4 assembled as *Ponders* (omni **1988**); *The Rain Door* (**1986** chap); *Moustery* (**1989** chap); *Jim Hedgehog and the Lonesome Tower* (**1990** chap) and its sequel, *Jim Hedgehog's Supernatural Christmas* (**1992** chap); *The Court of the Winged Serpent* (**1994** chap).

Other works (for adults): *Turtle Diary* (**1975**); *The Moment Under the Moment* (coll **1992**).

HOBB, ROBIN Pseudonym of Megan LINDHOLM.

HOBSON, [Sir] HAROLD (1904-1992) UK drama critic, active in that capacity from 1932 with *Christian Science Monitor* and then 1947-76 for the London *Sunday Times*. His only fiction, *The Devil in Woodford Wells: A Fantastic Novel* (**1946**), is told in a manner deliberately reminiscent of Sir Max BEERBOHM, even including a reference to Beerbohm's famous character Enoch Soames. Focusing on an early-19th-century cricket match, the tale speculates on the role of the DEVIL in ruining a young man's career. [JC]

HOCUS POCUS ◊ MAGIC WORDS.

HOCUS POCUS US movie (**1993**). DISNEY. **Pr** Steven Haft, David Kirschner. **Exec pr** Ralph Winter. **Dir** Kenny Ortega. **Spfx** Terry Frazee. **Vfx** Peter Montgomery. **Screenplay** Neil Cuthbert, Mick Garris. **Voice actor** Jason Marsden (Thackery Binx). **Starring** Thora Birch (Dani Denison), Doug Jones (Billy Butcherson), Omri Katz (Max Denison), Karyn Malchus (headless Billy Butcherson), Bette Midler (Winifred Sanderson), Sean Murray (Thackery Binx), Kathy Najimy (Mary Sanderson), Sarah Jessica Parker (Sarah Sanderson), Vinessa Shaw (Alison). 96 mins. Colour.

Salem, 1693, and Thackery Binx is cursed (◊ CURSES) by Winifred Sanderson with IMMORTALITY as a CAT for trying to save his sister Emily from being leached of youth by the three WITCH Sanderson sisters. The trio are soon hanged, promising to return should ever a VIRGIN light their black-flamed candle at HALLOWE'EN.

Halloween 1993, and young Max goes with moppet sister Dani and highschool belle Alison to the creepy Sanderson House Museum, where he cockily lights the candle – despite protests from lurking cat-Binx. Grabbing the witches' sentient grimoire (◊ BOOKS), the three children plus the TALKING ANIMAL flee the reborn Sandersons through a town full of Halloween partiers; even the cemetery's consecrated ground offers no protection, for Winifred raises there the ZOMBIE of her quondam lover Billy. The witches conceive the aim of leaching youth from all Salem's children and nearly succeed before dawn, but our plucky heroes thwart them and release Thackery's SOUL into the AFTERLIFE, to join Emily's.

There is some pleasing TECHNOFANTASY, the witches initially regarding 20th-century technology as MAGIC. Midler, Najimy and Parker succeed in being both riotously funny (much dialogue seems ad-libbed) and, on occasion, quite frightening, with even a sort of skew-witted grandeur. [JG]

HODGELL, P(ATRICIA) C(HRISTINE) (1951-   ) US writer, first notable for her essay "The Night Journey Motif in 20th Century English Fantasy" (*Riverside Quarterly* 1977) (◊ NIGHT JOURNEY). Her subsequent career has been taken up with the **Kencyrath** sequence: *God Stalk* (1982), *Dark of the Moon* (1985) – these two assembled as *Chronicles of the Kencyrath* (omni 1988 UK) – *Bones* (1993 chap), a previously unpublished sequence from *God Stalk*, *Child of Darkness* (1993 chap) and *Seeker's Mask* (1994). Further novels are projected, as is a volume of linked short fiction.

The protagonist is one of the more interesting in implicitly feminist REVISIONIST FANTASY, a classic ACCURSED WANDERER and OBSESSED SEEKER whose search for her own identity and whose growth towards status as champion of her people, the Kencyrath, against a DARK LORD, are constantly trammelled by sexism (◊ GENDER). For much of the first book, she is active in two roles, as apprentice (in GENDER DISGUISE) in the Thieves' Guild and as dancer in an INN. Her subsequent journeys regularly involve her in conflict with the assumption universally made by her own people and others that she belongs in purdah, learning womanly arts. Her people – a subrace of magically gifted aristocrats, another of stolid yeomanry and a largely lost group of wise giant CATS – have fled from one ALTERNATE WORLD to another, creating POLDERS there. Their hereditary foe, Perimal Darkling, is at once a Dark Lord and a personified process of THINNING and DEBASEMENT.

PCH is among the most aware fantasy writers of the tropes and topoi of the genre and the extent to which the combination of story modules can of itself generate STORY. [RK]

HODGSON, WILLIAM HOPE (1877-1918) UK writer who spent nine years (1891-9) as a merchant seaman. He turned to writing fiction with "The Goddess of Death" (1904 *Royal Magazine*), a rather weak mystery story, but soon began to draw on his maritime experiences. Not all his sea stories are supernatural, but all atmospherically evoke both the remoteness of a ship at sea and the strangeness of what lies beneath the waves. His most effective sea stories are about TRANSMUTATION, as in "The Voice in the Night" (1907 *Blue Book*), where castaways are overcome by a fungus, and "The Derelict" (1912 *Red Magazine*), where an ancient

ship mutates into a living organism. Some of WHH's best sea stories were collected as *Men of the Deep Waters* (coll 1914; exp vt *Deep Waters* ed August DERLETH coll 1967).

In *The Boats of the "Glen Carrig"* (1907) survivors of a shipwreck encounter strange lifeforms in a seaweed-engulfed environment – this novel inspired Dennis WHEATLEY's *Uncharted Seas* (1938) – and in *The Ghost Pirates* (1909), WHH's most successful example of sustained HORROR, a fated ship becomes haunted by an infradimensional craft.

All of WHH's fiction seeks a natural, often scientific explanation, which makes his work closer to sf than fantasy, though the explanation is not always a solution, so that the supernatural is left as an option (◊ PERCEPTION; RATIONALIZED FANTASY). This was most evident in his OCCULT-DETECTIVE series about **Carnacki**, collected as *Carnacki the Ghost-Finder* (coll 1913). Although commissioned by the publisher to cash in on the success of Algernon BLACKWOOD's *John Silence* (1908), WHH produced an investigator more scientist than psychic, though **Carnacki** relies on ancient manuscripts as much as modern equipment.

WHH's wrote two great works. *The House on the Borderland* (1908), inspired by H.G. WELLS's *The Time Machine* (1895), involves a house on a transdimensional THRESHOLD, whose occupant is first haunted by strange hog-like creatures and then drawn into a vision of the far future. Iain Sinclair's *Radon Daughters: A Voyage, Between Art and Terror, from the Mound of Whitechapel to the Limestone Pavements of the Burren* (1994) began life as an intended sequel to WHH's novel; many echoes and reflections of that project are retained in the texture of this URBAN FANTASY about LONDON. WHH's *The Night Land: A Love Tale* (1912) is a highly individualistic and idiosyncratic novel of the DYING EARTH: the world is in perpetual darkness and the land has been overcome by monstrosities. Although written as a love story, it is essentially a QUEST.

In his last decade WHH turned to writing full-time, producing much hackwork in order to make a living. Novels gave way to short stories, and fewer of these were supernatural. Later works were collected as *The Luck of the Strong* (coll 1916) and the nonfantastic *Captain Gault* (coll 1917). WHH was also adept at protecting his copyright in the USA, and issued abridged, very limited editions of some of his books; these volumes are scarce. They include *The Ghost Pirates, A Chaunty and Another Story* (1909 US chap), *Carnacki the Ghost-Finder and a Poem* (1910 chap), *The Captain of the Onion Boat* (1910 *Nash's Magazine*; 1911 chap US), *Poems and the Dream of X* (1912 US chap) – the latter part being an abridgement of *The Night Land*, later reissued in its own right as *The Dream of X* (1977 US) – *Impressionistic Sketches* (1913 chap US) and *Cargunka and Poems and Anecdotes* (1914 US chap), being an abridgement of the **Captain Gault** stories.

Several stories appeared posthumously, as did two volumes of poetry, *The Calling of the Sea* (1920) and *The Voice of the Ocean* (1921). Although cheap editions of his books remained in print in the UK for some years, his work was gradually forgotten until the inclusion of stories by him in two ANTHOLOGIES – *They Walk Again* (anth 1931) ed Colin de la Mare and *A Century of Horror Stories* (anth 1935) ed Dennis WHEATLEY – led to his rediscovery, especially in the USA. After August W. DERLETH established ARKHAM HOUSE he issued an impressive omnibus of WHH's novels, *The*

House on the Borderland and Other Novels (omni **1946** US), and an expanded version of *Carnacki the Ghost Finder* (coll **1947** US). Sam Moskowitz (1920-   ) has produced two collections, each with an extensive biographical introduction: *Out of the Storm* (coll **1975** US) and *The Haunted "Pampero"* (coll **1991** US), with a third planned «Terrors of the Sea» (1996). R. Alain Everts (?   -   ) has produced a short biography, *Some Facts in the Case of William Hope Hodgson: Master of Phantasy* (1973 *Shadow* UK; as vol 2 of *William Hope Hodgson: Night Pirate* **1987** chap Canada) and has issued 15 chapbooks of individual stories (listed below) which were also collected as *The Room of Fear* (coll **1988** US). Other retrospective volumes are: *Masters of Terror, Volume 1: William Hope Hodgson* (coll **1977**) and *William Hope Hodgson: A Centenary Tribute* (coll **1977** chap), both ed Peter TREMAYNE; *Poems of the Sea* (coll **1977**); *The Haunted "Pampero"* (1918 *Short Stories*; **1980** chap) and the nonfantasy *Tales of Land and Sea* (coll **1984**), all three ed George Locke (1936-   ); *Spectral Manifestations* (**1984** chap) ed Ian Bell (1959-   ); and *Demons of the Sea* (coll **1992** US chap) and *William Hope Hodgson At Sea* (coll **1993** US chap), both ed Sam Gafford, who is also compiling a WHH bibliography.

WHH's work bridges the gap between the supernatural horrors of the 19th century and the scientific wonders of the 20th, and demonstrates that both can produce horrors of equal bewilderment. [MA]
Other works: A series of single-story chapbooks, issued in envelopes, based on the original manuscripts by WHH: *The Baumoff Explosive* (1919 *Nash's Weekly*; **1988** US chap), also known as "Eloi, Eloi Lama Sabachthani"; *Fifty Dead Chinamen All in a Row* (**1988** US chap); *From the Tideless Sea* (1906 *Monthly Story Magazine*; **1988** US chap); *The Goddess of Death* (1904 *Royal*; **1988** US chap), *The Heaving of the Log* (**1988** US chap), *Homeward Bound* (1908 *Putnam's* as "The 'Shamraken' Homeward-Bounder"; **1988** US chap); *The Mystery of the Ship in the Night* (1914 *Red Magazine* as "The Stone Ship"; **1988** US chap); *Old Golly* (1919 *Short Stories*; **1988** US chap); *The Phantom Ship* (1973 *Shadow*; vt "The Silent Ship"; **1988** US chap), a variant epilogue to *The Ghost Pirates*; "The Riven Night" (1973 *Shadow*; **1988** US chap); *The Room of Fear* (1983 *Etchings & Odysseys*; **1988** US chap); *Sea-Horses* (1913 *London Magazine*; **1988** US chap); *The Terrible Derelict* (1907 *The Story-Teller* as "The Mystery of the Derelict"; **1988** US chap); *The Valley of Lost Children* (1906 *Cornhill Magazine*; **1988** US chap); *The Ways of the Heathens* (**1988** US chap).
Further reading: *William Hope Hodgson: Night Pirate, Volume One: An Annotated Bibliography of Published Works 1902-1987* (**1987** chap Canada) by Joseph Bell; *William Hope Hodgson: Voyages and Visions* (**1988** chap) by Ian Bell.

HOFFMANN, E(RNST) T(HEODOR) A(MADEUS) (1776-1822) German composer, lawyer and writer; he changed his third given name, originally Wilhelm, to Amadeus in 1813 in homage to Wolfgang Amadeus Mozart (1756-1791). As a composer he was most active in the decade before 1809, when he began to publish fiction with "Ritter Gluck", a tale in which the composer Christoph Gluck (1714-1787) appears as a GHOST. Much of ETAH's own music is of merit, and some of it – like his last opera, *Undine* (**1816**) – is relevant to fantasy. Also, a number of later composers have based ballets and OPERAS on ETAH's fiction; best-known is probably *The Tales of Hoffmann* (**1881**) by Jacques Offenbach (1819-1880).

It is for his many short stories, novellas and two full-length novels that ETAH is of greatest fantasy interest. Almost all the tales which have been translated, sometimes several times (often badly), appear in various ETAH collections, including: *Fantasiestücke in Callots Manier* ["Fantasy Pieces in the Style of Callot"] (coll **1814** 4 vols; rev 1819), among the contents of which are "Ritter Gluck" and (occupying the whole of vol 3) *Der Goldne Topf: Ein Märchen aus der neuen Zeit* (**1814**; trans Thomas CARLYLE as "The Golden Pot" in *German Romance: Specimens of its Chief Authors* coll **1827** UK); *Nachtstücke* ["Night Pieces"] (coll **1816-17** 2 vols), which contains "Der Sandmann" and others; the tales assembled within a FRAME STORY as *Die Serapions-Brüder* (coll **1819-22** 4 vols; trans Alexander Ewing as *The Serapion Brethren* **1886-92** 2 vols UK), which contains "Nussknacker unde Mausekönig" (1816; trans in *Nutcracker and Mouse King, and The Educated Cat* coll trans Ascott R. Hope **1892** US; separately vt *Nutcracker* **1985** UK), "Die Automate" (1817), "Klein Zaches genannt Zinnober" (1819), "Die Königsbraut" (1821; trans as *The King's Bride* **1959** chap UK), and many more; and *Die Letzten Erzählungen* ["The Last Tales"] (coll **1825** 2 vols).

The tales of most fantasy interest are of two kinds. The LITERARY FAIRYTALES are perhaps the more popular, though not always the most deeply reflective of ETAH's profound sensitivity to stress and abnormal psychology. This category does, however, include what may be his finest single piece of fiction, "The Golden Pot: A Modern Fairy Tale", whose protagonist, Anselmus, lives on two levels of REALITY – mundane and fairytale – which CROSSHATCH with dreamlike liquidity. Dresden and ATLANTIS intersect constantly, an interaction ETAH conveys through a constant flicker of TROMPE L'OEIL moments. Gradually Anselmus's bourgeois life is swallowed by Atlantis, though he must undergo a NIGHT JOURNEY in preparation for his bliss, after a moment of doubt causes his BONDAGE in a glass bottle, which seals him from Paradise until he is forgiven by the presiding MAGUS – who is a salamander from the morning of the world, and with whose daughter (who flickeringly appears as a SERPENT) Anselmus falls in love.

Other tales of this sort include: "The Nutcracker", whose young heroine enters a miniature OTHERWORLD where she defends the eponymous prince; "Little Zaches Called Cinnabar", whose protagonist lives under a fairy SPELL until, fatally, it is shattered; and *Princess Brambilla: Ein Capriccio nach Jakob Callot* ("Princess Brambilla: A Caprice in the Manner of Jakob Callot"] (dated 1821 but **1820**), set at CARNIVAL time in ROME, where the story is told of a prince and princess so blind to their true nature that they must see themselves in a magic MIRROR, where each is confronted by the denied DOUBLE aspect of selfhood necessary for a whole life.

Not fully distinguished from this, but generally more problematical about REALITY as seen (or projected) by their protagonists, are tales like "The Sandman", whose protagonist's love for two women who occupy differing realities has a grimmer outcome than Anselmus's, for this time the other is an automaton, and has possibly been foisted on him by the DEVIL. The eponymous monster who psychically tortures an innocent man in "Ignaz Denner" (1817) has gained his supernatural venom through drinking an ELIXIR OF LIFE derived from the blood of children. And "Die Bergwerke zu Falun" (from *The Serapion Brethren*; normally trans as "The Mines of Falun") once again features a personality split between two worlds, and follows him downwards into the mine where his artistic (and perhaps insane) PERCEPTION gives him what turns out in the end to be a fatal VISION. Rarely in ETAH's serious work does the artist manage to live whole: either the world takes him or he dies into something other.

ETAH wrote two long novels. *Die Elixiere Des Teufels* (**1815-16**; trans R.P. Gillies as *The Devil's Elixir* **1824** UK; preferred trans Ronald Taylor as *The Devil's Elixirs* **1963** UK) explores with savagery and depth the theme of the double or DOPPELGÄNGER through the hyperbolically conflicted life of Medardus, a monk and sensualist whose dance of identity with his double – whom he kills, who returns, who kills his bride-to-be (whose brother and stepmother Medardus has already murdered) – is a remarkably sophisticated and haunted vision of the gnawed Romantic self. *Lebens-Ansichten des Katers Murr nebst fragmentarischer Biographie des Kapellmeisters Johannes Kriesler in zufälligen Makulaturblättern* (**1820-21**; trans Leonard J. Kent and Elizabeth C. Knight as *The Life and Opinions of Kater Murr, Including a Fragmentary Biography of the Kapellmeister Johannes Kriesler on Miscellaneous Pieces of Scrap Paper*, vol 2 of their *Selected Writings of E.T.A. Hoffmann* coll **1969** US) is rather milder, though its mockery of the confessional mode exemplified by GOETHE's *The Apprenticeship of Wilhelm Meister* (**1796**) is pointed (◊ PARODY): Kater Murr, after all, is a CAT. The structure of the book is markedly complex. Murr (as "editor", ETAH, explains) has written his own rather philistine confession upon the sole existing copy of a biography of Kriesler; the two texts jostle together contradictingly in a manner that forecasts 20th-century plays on textuality – the wording of the title makes reference to *The Life and Opinions of Tristram Shandy, Gentleman* (**1759**) by Laurence Sterne (1713-1768).

For a century or so after his death, ETAH tended to be thought of as an unwholesome eccentric whose fiction was uncontrolled and whose use of the supernatural detracted from his serious intentions. He is now recognized as a skilled and conscious creator, some of whose visions prefigure the darker dreams of this century. [JC]

Other translations (selective): *Hoffmann's Strange Stories* (coll trans anon **1855** US), unreliable; *Hoffmann's Fairy Tales* (coll trans L. Burnham **1857** US), from French translations; *Weird Tales* (coll trans T.J. Bealy **1885** UK), unidiomatic; *Stories* (coll trans **1908**) ed Arthur RANSOME, whose intro reappears in *The Tales of Hoffmann* (coll with old trans **1943** US); *Tales of Hoffmann: Retold from Offenbach's Opera* (coll trans Cyril Falls **1913** UK); *Tales of Hoffmann* (coll trans James KIRKUP **1966** UK); *The Best Tales of Hoffmann* (coll with some new trans **1967**) ed E.F. BLEILER; *Three Märchen* (coll trans **1971** US).

HOFFMAN, NINA KIRIKI (1955-   ) US writer who began publishing with "Drawing on the Kitchen Table" for *Snapdragon* in 1983. *Legacy of Fire* (coll **1990**) contains nonfantasy work (like that story) as well as fantasy, SUPERNATURAL FICTION and sf. Her short work – which is increasingly various, and which conveys through contemporary settings a sense of driven, underlying STORY – has also appeared in *Courting Disasters and Other Strange Affinities* (coll **1991**). Her first longer work was *Child of an Ancient City* (**1992** chap) with Tad WILLIAMS, an ARABIAN FANTASY. In *The Unmasking* (**1992**) – which like much of NKH's work contains radically reworked echoes of writers like Charles G. FINNEY and Ray BRADBURY – the residents of

a small town are forced to reinhabit their own worst moments. *The Thread that Binds the Bones* (**1993**), which begins a sequence of loosely linked novels, rather more ambitiously describes a somewhat Bradburyesque Pacific Rim family with TALENTS which has been occupying a position in the world – half WAINSCOT, half SECRET MASTERS – for centuries, but is now threatened by a loss of FERTILITY. *The Silent Strength of Stones* (**1995**) focuses on an adolescent boy who becomes fascinated by members of this family, one of them a WEREWOLF, one with the power to control or "own" other people; his erotically charged resistance to the latter's talent is sensitively told. [JC]

HOFMANNSTAHL, HUGO VON (1874-1929) Austrian poet and dramatist, best-known for his libretto for the OPERA *Der Rosenkavalier* (**1911**) by Richard Strauss (1864-1949). In the fantasy vein he wrote other librettos for Strauss, including (see also below) *Ariadne auf Naxos* ["Ariadne on Naxos"] (**1912**), *Die Frau ohne Schatten* ["The Woman without a Shadow"] (**1919**) and *Die Ägyptische Helena* ["The Egyptian Helen"] (**1928**).

HVH's early writings are full of the FIN-DE-SIÈCLE melancholy that pervaded much of the literature of DECADENCE in the 1890s, with a particular emphasis on death. His play *Der Tor und der Tod* (**1894**; trans Max Blatt as *Death and the Fool* **1913** US) explores the tension between artistic aestheticism and moral obligation as an elaborate DANCE OF DEATH. There followed a sequence of experimental stories later collected as *Das Märchen der 672. Nacht und andere Erzählungen* (coll **1905**; trans as *Four Stories* **1968** UK), all of which explored different encounters with DEATH. The title story, "Das Märchen der 672. Nacht" (1895 *Die Zeit*; cut trans Alan Trethewey as "The Tale of the Merchant's Son and His Servants" in *The Lion Rampant* anth **1969** UK; trans as "The Tale of the 672nd Night" in *German Literary Tales* anth **1983** US ed Frank G. Ryder and Robert M. Browning), is about a merchant's son who withdraws from the world; but a letter accusing his servant of a crime draws him back to experience the nightmare of the CITY and be killed. Its companion piece, *Reitergeschichte* (**1899** chap; trans Basil Creighten as "Cavalry Patrol" in *Tellers of Tales* anth **1939** ed W. Somerset MAUGHAM; new trans Mike Mitchell as "Sergeant Anton Lerch" in *The Dedalus/Ariadne Book of Austrian Fantasy* anth **1993** UK ed Mitchell), tells of a soldier who sees his own DOPPELGÄNGER and, affected by the incident, hesitates to carry out his superior's command so is promptly shot. Both stories emphasize that death is sudden and inescapable, but their deeper meanings (◊ ALLEGORY) have been the subject of discussion for a century.

Richard Strauss became attracted to HVH's work and the two began to collaborate in 1906 when Strauss chose HVH's tragedy *Elektra* (**1904**), the first of HVH's **Oedipus** trilogy, as the libretto for *Elektra* (**1909**); the partnership lasted 20 years. HVH was cofounder of the Salzburg Festival in 1917, and it was here that his play *Jedermann* (**1911**) was first performed in 1920. A religious parable, it explores a further confrontation with DEATH. HVH returned to his enigmatic study of life and death in *Der Turm* ["The Tower"] (**1925**; rev 1927), a symbolic historical play about the heir to the throne who is, for many years, imprisoned in a tower (◊ EDIFICE), wherein he passes through madness to worldly wisdom.

HVH's one other major fantasy, generally known as *Andreas*, was never completed. A first part appeared as *Fragmente eines Romans* (**1931** chap Germany) and an expanded version as *Andreas oder die Vereinigten* (**1932** Germany; trans Marie D. Hottinger as *Andreas, or The United* 1936 UK). Andreas's personality is split between the spiritual and the physical, but a series of encounters with mysterious individuals begins to rebuild him.

A literary maestro, HVH early discovered that language was inadequate to express the poetic experience, and channelled his creative energies into drama. His few attempts at the fantastic challenge the reader to explore with HVH the mystery and purpose of life.   [MA]

Other works: *Der Kaiser und die Hexe* ["The Emperor and the Spell"] (play **1900**); *Das Bergwerk von Falun* ["The Mines of Falun"] (play **1933**) based on the story by E.T.A. HOFFMANN; *Selected Writings* (coll **1963** 3 vols US) trans/ed Marie D. Hottinger, Tania Stern and James Stern.

HOGARTH, BURNE (1911-1996) US artist, writer and teacher, renowned for his work on the TARZAN Sunday newspaper strip. He became an assistant cartoonist, aged 15, for Associated Editors Syndicate, and drew his own feature, **Famous Churches of the World**, in 1927. In 1929 he created his first COMIC strip, **Ivy Hemmenshaw**, for Bonnet Brown Company; this was followed by **Odd Occupations and Strange Accidents** for Leeds Features in 1930.

In 1933 he taught art history for the Works Progress Administration. A year later he travelled to New York to work as an assistant at King Features. In 1935 he took over the drawing of a PIRATE feature, **Pieces of Eight**, written by Charles Driscoll, for McNaught Syndicate. 1936 brought the opportunity to draw the **Tarzan** strip, which he did continuously until 1945.

BH then created (for Robert Hall Syndicate) **Drago**, an action-adventure strip set in an exoticized Argentina; it ran for less than two years, and BH returned to **Tarzan**, this time also writing the scripts. He simultaneously created the short-lived, Walter Mitty-style humour strip **Miracle Jones**. He left **Tarzan** in 1950 to devote himself to teaching at the School of Visual Arts (which he had founded with Silas Rhodes in 1947). BH also applied himself to painting and etching.

His contribution to the development of the US comic strip has been considerable. Despite the high polish of his later artwork on **Tarzan**, it retains a tremendous vitality and expressiveness, and remains, along with the work of Hal FOSTER, among the major influences still evident in the US comic strip. He returned to the **Tarzan** theme with two graphic novels: *Tarzan of the Apes* (graph **1972**) and *Jungle Tales of Tarzan* (graph **1976**).   [RT]

Other works: *Dynamic Anatomy* (**1958**), *Drawing the Human Head* (**1965**), *Dynamic Figure Drawing* (**1970**), *Drawing Dynamic Hands* (**1974**), *Dynamic Light and Shade* (**1981**) and *Dynamic Wrinkles and Drapery* (**1993**).

Further reading: *Tarzan, Jungle Lord* (**1968**); interview in *Comics Journal #166-#167* (1994).

HOGG, JAMES (1770-1835) Scottish poet and writer, known as the "Ettrick Shepherd" because of his early, illiterate life as a shepherd in Ettrick parish, south of Edinburgh. He began publishing poetry around the turn of the century, and in *The Pilgrims of the Sun* (**1815**) put into verse form a FANTASTIC VOYAGE, describing a trip to various planets by a young woman and her mysterious companion; it was dedicated to BYRON. Of greater interest are the tales based on Scottish FOLKLORE assembled in: *The Brownie of Bodsbeck and Other Tales* (coll **1818**), which comprises three full-length novels; *Winter Evening Tales Collected Among the Cottagers in the South of Scotland* (coll **1820**); *The Shepherd's Calendar: Tales Illustrative of Pastoral Occupations, Country Life, and Superstitions* (coll **1828**); *Altrive Tales, Collected Among the Peasantry of Scotland, and from Foreign Adventures* (coll **1832**); and *Tales and Sketches of the Ettrick Shepherd* (coll **1837** 6 vols).

Though many of these tales are technically SUPERNATURAL FICTIONS, the abiding sense of JH's tales is of so pervasive an interaction between the mundane and the supernatural that little tension – or distinction – between the two can normally be detected. In the novel which gives its title to the first collection, for instance, the interaction between the WAINSCOT sprite and the young woman he protects is intricately woven into a tale about 17th-century Scotland; and "The Hunt of Eildon", from the same volume, features two sisters who, metamorphosed into white hounds (◊ METAMORPHOSIS), protect the king from various dangers, both mundane and supernatural, and pass over, after their work has been done, into FAERIE.

Outside Scotland JH remains best remembered for *The Private Memoirs and Confessions of a Justified Sinner, Written by Himself, With a Detail of Curious Traditionary Facts, and Other Evidence, by the Editor* (**1824**; vt *The Suicide's Grave, or Memoirs and Confessions of a Sinner* 1828; vt *The Confessions of a Justified Sinner* 1898), the complexly told and ambivalently confessional narrative of a man who believes his salvation to have been predestined (hence his self-description as a "justified sinner"), and who is seduced into sin by the DEVIL.

Condescended to by his contemporaries, JH has only in the last half-century been recognized as a writer of significance.   [JC]

HOLDSTOCK, ROBERT P(AUL) (1948-   ) UK writer who gained an MSc in medical zoology from the London School of Hygiene and Tropical Medicine in 1971 and spent the next three years in medical research. During this period he published his first story, "Pauper's Plot", for *New Worlds* in 1968, and was moderately prolific in the form before 1976, when he turned to full-time writing; most of these early shorts are assembled in *In the Valley of the Statues* (coll **1982**).

For the next decade, his writing career was twofold. Under his own name he wrote three sf novels of some merit: *Eye Among the Blind* (**1976**), *Earthwind* (**1977**) and *Where Time Winds Blow* (**1981**). Rather more interestingly, in fantasy terms, RH wrote a number of novels under various names, most hurried commercial efforts but almost all so exuberantly full of action that they seem almost libidinous with release. Those written as by Ken Blake (a housename) have no fantasy interest. As Robert Black he wrote two film ties, *Legend of the Werewolf* * (**1976**) and *The Satanists* * (**1977**). As Richard Kirk (another housename) he contributed to the **Raven** sequence of SWORD AND SORCERY tales, writing *Swordsmister of Chaos* (**1978**) with Angus WELLS, *A Time of Ghosts* (**1978**) and *Lords of the Shadows* (**1979**). As Chris Carlsen he wrote the **Berserker** series – *Shadow of the Wolf* (**1977**), *The Bull Chief* (**1977**) and *The Horned Warrior* (**1979**) – in the second volume of which his constantly reborn Eternal Champion hero (◊ Michael MOORCOCK) comes to life in 5th-century Ireland, where he assists ARTHUR so vigorously that Arthur is impelled to kill him.

The most important of these pseudonymous works is the DARK-FANTASY **Night Hunter** sequence, written as Robert

Faulcon: *The Stalking* (**1983**) and *The Talisman* (**1983**), both assembled as *The Stalking* (omni **1987**), *The Ghost Dance* (**1984**) and *The Shrine* (**1984**), both assembled as *The Ghost Dance* (omni **1987**), and *The Hexing* (**1984**) and *The Labyrinth* (**1987**), both assembled as *The Hexing* (omni **1988**). It is here that RH first made significant use of plots in which sons or fathers, having been sundered from one another, must attempt to reunite. In this case the protagonist's wife and son have been abducted by Satanists and he must become a deadly master of the arts of MAGIC in order to recapture them; it takes him all six novels to complete this, the process coming to a climax in the eponymous LABYRINTH of the final volume.

During this period RH wrote one similar tale, *Necromancer* (**1978**), which appeared under his own name; but the two strands of his career did not properly converge until he began to publish the **Ryhope Wood** sequence: *Mythago Wood* (**1984**), *Lavondyss: Journey to an Unknown Region* (**1988**), *The Hollowing* (**1993**) and *Merlin's Wood* (coll of linked stories **1994**), plus the title story of *The Bone Forest* (coll **1991**). The sequence as a whole is a central contribution to late-20th-century fantasy, and is almost embarrassingly dense with fantasy tropes. The narrative energy is throughout very considerable, and there is ever a sense that RH has fixedly not allowed the underlying rigour of his conception to slide into the easy solutions of GENRE FANTASY. **Ryhope Wood** as a whole makes play with the motifs of CELTIC FANTASY and never shrinks from what may be that subgenre's central element: the unflinching *hardness* of the ordeal-choked, METAMORPHOSIS-ridden SECONDARY WORLD it depicts. The world depicted in **Ryhope Wood** is similarly hard; the NIGHT JOURNEYS into metamorphosis that its various protagonists undergo are cruel, are devastating to those protagonists and to those who have lost them, and do not lead back to daylight.

The basic premise unfolds rapidly enough in *Mythago Wood*, which won the 1986 WORLD FANTASY AWARD. Ryhope Wood, in the English county of Herefordshire, is from the outside a 3-mile-square fenced-in rural woodland; but any person who passes into its shade travels INTO THE WOODS and finds themselves within a primeval, impossibly intricate LABYRINTH of TREES – for Ryhope Wood is larger inside than out (◊ LITTLE BIG); as one penetrates further, TIME itself opens downwards (◊ TIME ABYSS). Moreover, the further one penetrates the more inevitable is one's encounter with a MYTHAGO – one of the metamorphic figures that tend to attack those who attempt to invade the POLDER and that take their essence from the collective unconscious of the British people. The mythagoes wear the MASKS of various HEROES and other darker persons whose lives, real or imagined, have been central to the MATTER of Britain. Some are relatively human – the ROBIN HOOD and ARTHUR mythagoes, for instance, recognizably represent the acceptable face of STORY – but the GREEN MAN is more distressingly chthonic and the even more primeval wodewose (a primal creature from the Stone Age called Urscumug, various versions of whom appear furtively throughout the sequence) is even less amenable to anthropological surmise.

The storyline of *Mythago Wood*, as with its successors, works as a kind of inward spiral. The narrator – a SENSIBLE MAN whose dead father had already sacrificed his life and family to the wood – returns to England after the trauma of WORLD WAR II to find his brother obsessively attempting to recapture his mythago wife, who has disappeared back inside the wood. Her name is Guiwenneth (◊ GUINEVERE), and she is a kind of GODDESS figure. She is dead, but the narrator's brother still plunges deeper into the woods under the conviction that she will return. Beyond Guiwenneth is the heartwood, which cannot be reached, and which may be as infinite in size as a whole SECONDARY WORLD. A wind of SEHNSUCHT blows from the heartwood and ultimately governs the actions of most of the sequence's protagonists, who obsessively try to gain access to that "unknown region" where the human story begins to germinate and where the meaning of life is born.

In the second volume, *Lavondyss*, a new character, Tallis, grows up haunted by LIMINAL BEINGS who call to her from the edge of the wood, the circumference of which she obsessively traverses; there she encounters Ralph Vaughan Williams (1874-1958). Eventually, thanks to her menarche, she penetrates Ryhope's deep interior, searching inwards, through the "hollowings" she is able to create, for the heartwood; but the knot at the core of her inward search for REDEMPTION is not so much a further PORTAL but a profound metamorphic MAP of her own life, which she must recognize and enter as a kind of DRYAD. Ultimately she transforms utterly into something partly like a mythago in GREEN-MAN guise, partly like a map (◊◊ FACE OF GLORY) and partly like Ryhope itself. In a fashion perhaps best elucidated through an ambitious and through-composed fantasy novel like *Lavondyss*, she becomes something rich and strange.

The Hollowing backs slightly away from the harsh heart of *Lavondyss* but continues to take ICONS from the CAULDRON OF STORY. BABEL and GAWAIN and Jason, as well as various GHOSTS and TRICKSTERS, make flickering appearances; there are continuing hints of the WILD HUNT; and the basic sundered family is once again central. The eponymous wood in *Merlin's Wood* is a smaller version of Ryhope Wood, and is sited in Brittany; here the Matter of Britain continues to hold sway (◊ MERLIN). *The Fetch* (**1991**) and «Ancient Echoes» (**1996**), RPH's only recent novels not directly associated with **Ryhope Wood**, again feature families in disarray. A sense remains that RPH, having created in **Ryhope** a notably original structure of story through which the most taxing fantasies can be told, will continue to tell those stories. [JC]

Other works: *Elite: The Dark Wheel* * (**1984** chap), sf novella tie to a computer game; *Bulman* * (**1984**) and *One of Our Pigeons is Missing* * (**1984**), tv novelizations; *The Emerald Forest* * (**1985**), novelizing the film by John Boorman.

As Ken Blake (housename): *Cry Wolf* * (**1981**), *The Untouchables* * (**1982**), *Operation Susie* * (**1982**) and *You'll be All Right* * (**1982**), novelizing the tv series **The Professionals**.

As Steven Eisler: Linking texts for 2 vols of reprinted illustrations, being *Space Wars Worlds and Weapons* (**1979**) and *The Alien World* (**1980**).

Nonfiction: *Encyclopedia of Science Fiction* (**1978**), Consultant Editor; *Alien Landscapes* (**1979**), *Tour of the Universe: The Journey of a Lifetime – The Recorded Diaries of Leio Scott and Caroline Luranski* (**1980**), *Magician: The Lost Journals of the Magus Geoffrey Carlyle* (**1982**), *Realms of Fantasy* (**1983**) and *Lost Realms* (**1985**), all with Malcolm Edwards (1949-   ).

As editor: *Stars of Albion* (anth **1979**) with Christopher

PRIEST; the **Other Edens** series of original anthologies, all with Christopher Evans (1951-   ), being *Other Edens* (anth **1987**), *Other Edens II* (anth **1988**) and *Other Edens III* (anth **1989**).

HOLLAND, CLIVE Pseudonym of UK writer Charles James Hankinson (1866-1959). His LOST-RACE novel *Raymi, or The Children of the Sun* (**1889**) is of mild interest. In *An Egyptian Coquette* (**1898**; rev vt *The Spell of Isis* 1923) a young woman falls into suspended animation after committing a crime; the tale takes us to ancient EGYPT. *The Hidden Submarine, or The Plot that Failed* (**1917**) verges on sf in its description of an unsuccessful German invasion plot. [JC]

HOLLOW EARTH There has always been an UNDER-WORLD in Western literature (◊ HADES). As late as DANTE's *The Divine Comedy* (written *c*1307-1321), it remained possible to put into serious imaginative form the traditional Christian notion that HELL lies below the surface of the world, just as HEAVEN is above. But understanding of the fact that the Earth is a globe became widespread, so that a writer like Giacomo Casanova (1725-1798), in *Icosaméron* (**1788**), allegorized the notion to depict an EDEN-like *interior* world without much attempt at verisimilitude. Meanwhile, exponents of SCHOLARLY FANTASY transformed the old flat-Earth concept into a vision of the surface of the globe as simply the outermost of a series of nested spheres, at the heart of which a small central sun shone.

So described, the HE may not be a plausible notion, but it was certainly *arguable*, and most tales set within our globe provide some form of explanation for the hollows within. HE stories are, therefore, almost invariably sf, like *Journey to the Centre of the Earth* (**1863**) by Jules Verne (1828-1905), though an author like Edgar Rice BURROUGHS, in his **Pellucidar** sequence, might well place tales there that have as little to do with the justification of scientific argument as any of his PLANETARY ROMANCES set in the skies. *The Hollow Earth* (**1990**) by Rudy Rucker (1946-   ) is a RECURSIVE FANTASY set largely in the HE and featuring Edgar Allan POE.

James P. BLAYLOCK's *The Digging Leviathan* (**1984**) invokes Pellucidar, but one of its jokes is that the action never in fact goes UNDERGROUND. [JC]

HOLLYWOOD ◊ LOS ANGELES.

HOLME, CONSTANCE (1880-1955) UK writer best-known for *The Lonely Plough* (**1914**), a novel set like most of her work in rural Westmorland, northern England. In *The Old Road from Spain* (**1915**; vt *The Homecoming* 1916 US) a CURSE laid down by a survivor of the Spanish Armada entails the heads of an English farming family knowing their death is nigh when sheep come into the local park. *He-Who-Comes?* (**1930**) likewise embraces the circumambient LAND, giving to its storyline – a WITCH hovers between good and bad practices – a sense of autonomy and rightness that shifts it from the normal timbre of SUPERNATURAL FICTION. Some of the tales assembled in *The Wisdom of the Simple* (coll **1937**) are supernatural. [JC]

HOLOCAUST AND AFTER In SCIENCE FICTION, SUPER-NATURAL FICTION, HORROR and FANTASY alike, the future holocaust is a world-changing disaster in whose survivors must begin civilization again from scratch. Sf tales normally offer arguments from history about the holocaust's cause. The other genres may show the end of the world in non-secular terms which might reflect a belief in damnation, but which do not, for most readers today, constitute arguments from history; for example, the holocaust, as remembered in

David GEMMELL's **Jerusalem Man** sequence, is tied not to history as we know it but to an argument that unholy VENGEANCE has somehow been cast forward from the time of ATLANTIS.

Fantasy stories – which do not claim to offer arguments from history about the events they depict – are not often set in any particular year, and very rarely in a specific near future, so post-holocaust fantasies are rarely concerned with the event itself (details of which have likely been lost to the protagonists' known history). Fantasy tales set in something approximating the near future include: Robert ADAMS's **Horseclans** sequence, depicting a romanticized post-holocaust LANDSCAPE; Alan BRENNER's **Dance of the Gods** tales; John CHRISTOPHER's **The Prince in Waiting** sequence; Storm CONSTANTINE's **Wraeththu** books; and Fred SABERHAGEN's **Empire of the East** SWORD-AND-SORCERY sequence, in which atomic fireballs have been turned into DEMONS. Fantasy set in DYING-EARTH venues (◊◊ SCIENCE FANTASY) may or may not incorporate references to the holocaust in terms reminiscent of the MYTH OF ORIGIN; M. John HARRISON's **Viriconium** books, for instance, contain coded references of this sort. The series started with *Black Trillium* (**1990**) by Marion Zimmer BRADLEY, Julian MAY and Andre NORTON is set in a post-holocaust SECONDARY WORLD, which is possibly another planet, colonized from Earth – which might make this long tale a RATIONALIZED FANTASY in two senses, in that in this world MAGIC works.

In full FANTASY, from J.R.R. TOLKIEN's *The Lord of the Rings* (**1954-5**) onwards, the opposition between GOOD AND EVIL is often made literal by contrasting the primal healthy LAND with the WASTE LAND created subsequently by a sterile DARK LORD envious of the plenitude of the Good. That waste land may be described in terms analogous to those used to describe post-holocaust landscapes, and some critics have, indeed, understood Tolkien's novel to present an ALLEGORY of the devastations of WORLD WAR I. Such suggestions are only moderately enticing. [JC]

HOLT, TOM Working name of UK writer Thomas Charles Louis Holt (1961-   ), whose first book, *Poems by Tom Holt* (coll **1973**), assembled poetry written while he was still a child. His writing career proper began with *Lucia in Wartime* * (**1985**) and *Lucia Triumphant* * (**1986**), sequels to E.F. BENSON's famous **Mapp and Lucia** sequence, plus *Goatsong* (**1989**) and its sequel, *The Walled Orchard* (**1990**), the latter two set in ancient Greece. TH began publishing the comic fantasies which have made him famous with *Expecting Someone Taller* (**1987**), which established in broad terms the pattern of all its successors. Most are set in a contemporary England invaded by UNDERLIER figures or ensembles from Earth's mythic past. In the first, feckless but nice young Malcolm runs over a badger, which turns out to be a Frost GIANT who had been "expecting someone taller", but who dutifully hands over the Tarnhelm and the RING of the Nibelungen (◊ NORDIC FANTASY; OPERA; Richard WAGNER), making Malcolm the ruler of the world. This irritates Wotan (◊ ODIN), and there is some apocalyptic action; but all ends moderately well.

Though the jokes tend to improve with the years, the novels themselves are only variously successful. *Who's Afraid of Beowulf?* (**1988**) brings a SLEEPER UNDER THE HILL and his men back to life, where they come into conflict with the DARK LORD who has been running England for centuries. In *Flying Dutch* (**1991**) the intruder is the FLYING DUTCHMAN, whose ELIXIR OF LIFE unfortunately gave him appalling body odour, hence his inconspicuousness over the centuries; but he does have a centuries-old bank account, with compound interest. *Ye Gods!* (**1992**) introduces a baby Hercules into a suburban home. *Overtime* (**1993**), despite some fine jokes about TIME TRAVEL, generates a sense of congestion as Blondel ricochets through a modern world run – typically of TH – by a cackhanded conspiracy. *Here Comes the Sun* (**1993**) stands aside from its stablemates by virtue of its depiction of a fantasy cosmology in which the Universe is literally worked by the heavenly hierarchy, which is not doing the job very well.

Many of these novels, despite the humour, are laced with what seems to be a genuine pessimism about the human condition. Later novels continue, with a gradually increasing sense of *joie de vivre*, in the same mode. *Grailblazers* (**1994**) imports another sleeper into modern England. *Faust Among Equals* (**1994**) features a sorcerer FAUST on the lam from HELL. *Odds and Gods* (**1995**) puts various PANTHEONS in an old folks' home. *Djinn Rummy* (**1995**) plays with ARABIAN FANTASY. [JC]

HOLY GRAIL ◊ GRAIL.

HOMER The name given to the author(s) of the *Iliad* and the *Odyssey*, which, with the books of Moses, are the founding works of Western literature.

The *Iliad* tells of a few weeks in the siege of Troy, during which the hero Achilleus (◊ ACHILLES), offended by the siege's leader Agamemnon, refuses to fight, at great cost to the siege and ultimately to himself. The action is overwhelmingly grim – one-third of the epic consists of a single day's battle – and, in the depiction of Achilles' TRANSFORMATION through suffering, sowed the seeds of later Greek Tragedy. What relief there is often lies in the sometimes comic depiction of the GODS. Lin CARTER, in *Tolkien: A Look Behind "The Lord of the Rings"* (**1969**), called the *Iliad* "probably the first psychological novel in literature".

The *Odyssey* contrasts sharply. The overall story offers conflict and meaning, but in much less concentrated form. It concerns the hero ODYSSEUS's return home two decades after his departure for Troy, and his family's search for him; it has been called the first Greek ROMANCE. A sixth of the epic consists of Odysseus' tales of his wanderings: in these pages we find the West's first FANTASTIC VOYAGE, filled with cyclopes, SIRENS and MAGIC, with greater and lesser GODS acting in rather arbitrary ways, and with a desperate struggle for survival. There are ample signs that the singer intended these as fantasy, and later Greeks certainly saw them as such. These epics' surpassing skill and beauty made them the Greeks' national MYTHOLOGY, establishing for literature a set of themes beyond merely local interest. Criticism of Homer's portrayal of the gods soon arose, and within a few centuries the poems were generally understood as histories embellished with either ALLEGORY or sheer fantasy. Their authority, however, remained, as did their popularity. Homer was the first scholarly concern of the Museum of Alexandria (3rd and 2nd centuries BC), whose critical editions shaped the versions we read today. [JB]

See also: GREEK AND LATIN CLASSICS.

HOMUNCULUS A miniature man produced by artificial means; the term, a diminutive of the Latin *homo*, was also employed in the protobiological theory of spermaticism to describe the formative entity supposedly carried by the sperm.

A recipe for the manufacture of homunculi was recorded by PARACELSUS; its successful application is imagined in *The Artificial Man* (**1931**) by John Hargrave (1894-1982), *The Homunculus* (**1949**) by David H. KELLER and *Schimmelhorn's Gold* (**1986**) by Reginald Bretnor (1911-1992), but in the first two cases the result is a full-sized individual. The "homunculi" in *To the Devil – A Daughter* (**1953**) by Dennis WHEATLEY are also full-sized. Magically generated homunculi of smaller dimension are featured in *The Magician* (**1908**) by W. Somerset MAUGHAM and the movie *The Bride of Frankenstein* (**1936**; ◊ FRANKENSTEIN MOVIES); the soul-animated DOLLS of "The Wondersmith" (1859) by Fitz-James O'BRIEN and *Burn, Witch, Burn!* (**1933**) by A. MERRITT might also qualify as homunculi. The androgynous *übermensch* which Salome is trying to create in *Salome, the Wandering Jewess* (**1930**) by George VIERECK and Paul Eldridge (1888-1982) is called a homunculus by virtue of an evolutionist extrapolation of the spermaticist notion first proposed by GOETHE, while the eponymous individual in *Homunculus* (**1986**) by James BLAYLOCK is apparently an alien. [BS]

HOMUNCULUS Pseudonym of William Makepeace THACKERAY.

HONTE DE LA JUNGLE, LA vt of *Jungle Burger* (**1975**). ◊ TARZAN MOVIES.

HOOK US movie (**1991**). Columbia TriStar/Amblin. **Pr** Kathleen Kennedy, Frank Marshall, Gerald R. Molen. **Exec pr** Dodi Fayed, Jim V. Hart. **Dir** Steven SPIELBERG. **Spfx** Michael Lantieri. **Vfx** Eric Brevig, Industrial Light & Magic. **Screenplay** Hart, Malia Scotch Marmo. **Based loosely on** the books by J.M. BARRIE and on PETER PAN (**1953**). **Novelization** *Hook* * (**1991**) by Terry BROOKS. **Starring** Dante Basco (Rufio), Caroline Goodall (Moira), Dustin Hoffman (Hook), Bob Hoskins (Smee), Julia Roberts (Tinkerbell), Maggie Smith (Wendy), Robin Williams (Peter). 144 mins. Colour.

Years ago PETER PAN fell in love with Wendy's granddaughter Moira and determined to stay in our world and grow up. Now he has forgotten his former existence and is a predatory US corporate lawyer. Hook reaches from the ALTERNATE REALITY of Never Land to steal Peter's children. Wendy knows what has happened, but Peter refuses to believe such "nonsense"; he is still disbelieving when Tinkerbell comes to fetch him away to the rescue. Hook taunts him with the children, then sentences him to death; but Tinkerbell reminds Hook of the glory he would have if he defeated Peter in war. Given three days to train her man, she takes him to the Lost Boys, now led by punkish swaggerer Rufio, who resents the intruder. Peter's training goes badly, because he has lost his imagination. He gets it back in perhaps the most interesting of many good fantasy sequences in *H*: a feast is served to the Lost Boys, who devour joyously, but all Peter can see are empty plates; Rufio starts to twit him, and this develops into a full-scale flyting, which Peter, after a slow start, wins through the imaginativeness of his insults; with that, his PERCEPTION shifts so that he can at last see the gaudy feast and, amid a wild food-fight, he discovers also several of his old Pannish qualities, notably use of the SWORD. There is further progress – he remembers the art of flying through locating a Happy Thought, the birth of his son Jack – but now Hook, egged by Smee (played by Hoskins in direct imitation of the character in PETER PAN [*1953*], so that he becomes almost a three-dimensional TOON), is in process of seducing Jack away from his real father, presenting himself as surrogate; Jack begins to forget who he is . . . but all ends well.

H was much disliked by the critics, who regarded it as a cynical exercise. In fact *H* is probably overextended at its end, but elsewhere for the most part this CROSSHATCH fantasy bubbles along imaginatively and involvingly. Although there are some moments of quiet sensitivity, as a whole *H* refreshingly lacks subtlety or delicacy – which is as it should be, for Barrie did not design his original tale for sissies of either sex. [JG]

See also: NEVER-NEVER LAND.

HORIZON OF EXPECTATIONS A term devised by critic Hans Robert Jauss to describe the context within which a given generation of readers will understand a work. Within that context – that horizon – the particular generation may come to view certain kinds of material (e.g., FAERIE, the supernatural and HELL) as being clustered together into a genre, where previously such material may have borne no genre import and not have been perceptually associated. Any attempt to define FANTASY must suggest a point in literary history at which readers (and writers) began to define previously scattered topics as being clustered together within an HOE. [JC]

See also: GENRE FANTASY.

Further reading: *Literary History as a Challenge to Literary Theory* (1967) by Hans Robert Jauss; *A Dictionary of Literary Terms* (1977; latest rev vt *A Dictionary of Literary Terms and Literary Theory* 1991) by J.A. CUDDON.

HORLICK'S MAGAZINE ◊ MAGAZINES.

HORNED GOD ◊ DEVILS.

HORROR Unlike FANTASY, SUPERNATURAL FICTION and SCIENCE FICTION – terms which describe generic structures – horror is a term which describes an *affect*. A horror story makes its readers *feel* horror. There are two important distinctions to make, however.

Distinctions. Horror stories can be set in entirely mundane worlds and be simple exercises in sadism; this category of horror does not concern us. They can also be set in any of the regions of the FANTASTIC, though usually in the one we think of as the supernatural.

Second, what we are calling "pure" horror can be distinguished from stories which convey a sense of horror while continuing to fulfil other genre requirements. Fantasies which convey a sense of horror are better called DARK FANTASY, and supernatural fictions with a horror "feel" are better called WEIRD FICTION. Sf horror stories, which are relatively rare, boast no label in particular.

The shape of horror. A "pure" horror tale may occupy the same region as a supernatural fiction – this world is being encroached upon by another – but is shaped *primarily* to convey the affect of horror. Thus the "pure" horror story is normally structured so that its protagonist and its readers share the same reactions. This shared horror is evoked in the text through the joining of two simultaneous elements: the recognition of a threat to one's body and/or culture and/or world; and a sense that there is something inherently monstrous and wrong (◊ WRONGNESS) in the invasive presence.

Both are necessary for the affect of horror. It is not enough for the mundane world to be invaded, asssaulted, seduced, taught or inveigled from another sphere, as generally happens in supernatural fiction; nor is it enough for MONSTERS to exist – as they often do in fantasy without threatening the fabric of the OTHERWORLD. What generates the frisson of horror is an overwhelming sense that the invaders are obscenely, transgressively impure. The monsters of horror are befoulers of the boundaries that mark us off from the Other. Like FRANKENSTEIN's Monster, they may be neither one thing or another, so that they violate the decorum of species, of role, of fitness to place. They may be either suffocatingly too full of rotten being, like a fallen ANGEL or Dorian Gray; or they may represent a haemorrhage of joining, like JEKYLL AND HYDE, two distinct beings who are, as though by virtue of TROMPE L'OEIL, obscenely one; or they may be manifestly incomplete, like the great majority of monsters in horror tales and HORROR MOVIES. Especially when incomplete, they are ravenous, and will eat anything: the body and the SOUL; the CITY and the state; the FERTILITY of the LAND.

Many authors who are mainly known for horror also write fantasy, dark fantasy, supernatural fiction or weird fiction; those given entries in this encyclopedia include Scott BAKER, Clive BARKER, Charles BEAUMONT, E.F. BENSON, Ambrose BIERCE, William Peter BLATTY, Robert BLOCH, Poppy Z. BRITE, Ramsey CAMPBELL, Nancy A. COLLINS, August DERLETH, Dennis ETCHISON, Charles L. GRANT, James HERBERT, Stephen KING, T.E.D. KLEIN, Dean R. KOONTZ, Sheridan LE FANU, Matthew LEWIS, Thomas LIGOTTI, Bentley LITTLE, Frank Belknap LONG, H.P. LOVECRAFT, Brian LUMLEY, Arthur MACHEN, Richard MATHESON, Charles Robert MATURIN, Edgar Allan POE, Anne RICE, James Malcolm RYMER, SARBAN, Dan SIMMONS, Peter STRAUB, Bram STOKER, Karl Edward WAGNER, Donald WANDREI, Chet WILLIAMSON, J.N. WILLIAMSON and F. Paul WILSON. [JC]

See also: PSYCHOLOGICAL THRILLER.

HORROR MOVIES It is difficult, when discussing the written word, to differentiate neatly between the genres of FANTASY, SUPERNATURAL FICTION, the PSYCHOLOGICAL THRILLER and HORROR – not to mention such subgenres as DARK FANTASY and WEIRD FICTION. The task becomes yet more difficult when dealing with the CINEMA, because very few fantasy movies do not stray into one of the other genres (including SCIENCE FICTION, although usually in the form of TECHNOFANTASY), while many HMs contain a considerable deal of straightforward – and often very good – fantasy: most VAMPIRE MOVIES, for example, concern themselves with what is in effect a WAINSCOT society even if, as with most of the DRACULA MOVIES, that society has a population of only one.

HMs that have no element of the FANTASTIC are clearly outwith our remit. Nevertheless, a considerable number of movies that would often be considered HMs are discussed within these pages. Aside from those in extended series like the **Dracula** movies and the technofantastic FRANKENSTEIN MOVIES and JEKYLL AND HYDE MOVIES, these include *An AMERICAN WEREWOLF IN LONDON* (*1981*), AUDREY ROSE (*1977*), *The BIRDS* (*1963*), the CANDYMAN series, CAT PEOPLE (*1942*), *The DEVIL RIDES OUT* (*1968*; vt *The Devil's Bride*), the EXORCIST series, the FLY series, GHOST STORY (*1974*), GHOST STORY (*1981*), the HELLRAISER series, the HOUSE series – although only the first, *House* (*1986*) is of much interest – the various HOUSE OF USHER movies, the HOWLING series, *The HUNGER* (*1983*), INTERVIEW WITH THE VAMPIRE: THE VAMPIRE CHRONICLES (*1994*), *The INVISIBLE MAN* (*1933*), *The KEEP* (*1983*), the two LITTLE SHOP OF HORRORS movies, the two MASQUE OF THE RED DEATH movies, the NIGHTMARE ON ELM STREET series, the OMEN series, PAPER-

HOUSE (*1988*), the various PHANTOM OF THE OPERA movies, *The* PICTURE OF DORIAN GRAY (*1945*), the two PIT AND THE PENDULUM movies, *The* SHINING (*1980*), TO THE DEVIL A DAUGHTER (*1976*), VAMP (*1986*) and *The* VAMPYR: A SOAP OPERA (*1992* tvm), not to mention numerous movies based on the works of Stephen KING. It is notable how many of these are VAMPIRE MOVIES; for some reason WEREWOLF MOVIES – like WOLF (*1994*) and WOLFEN (*1981*) – seem easier to accept as supernatural fictions or even techno-fantasies. Differently, although some of the HMs noted above do involve GHOSTS, the figure of the ghost seems more palatable for general family consumption: at one end of the scale are the various CANTERVILLE GHOST movies, which can chill but do not seek to horrify, and at the other is the POLTERGEIST series, which can be very frightening but which lacks horror's sense that the WRONGNESS is likely to remain forever unrighted.   [JG]

See also: HAMMER; ZOMBIE MOVIES.

HORROR SHOW, THE US small-press MAGAZINE, 28 issues, quarterly, November 1982-Spring 1990, published and ed David B. Silva (1950-   ), Phantasm Press, Oak Run, California.

One of the more important SMALL-PRESS publications of the 1980s, *THS* acquired national distribution and encouraged a new generation of writers. Inspired more by Dean R. KOONTZ than Stephen KING, Silva strove to publish a magazine that ran the whole gamut of HORROR fiction. The magazine indirectly developed the Splatterpunk writers. The early issues used basic computer production, but from Summer 1986 the quality improved. *THS* released several special author and artist issues, including: Dean R. Koontz (Summer 1986), Steve Rasnic Tem (1950-   ) (Fall 1986), J.K. Potter (1956-   ) (January 1987), Robert R. McCammon (1952-   ) (Spring 1987), Dennis ETCHISON (Winter 1987), William F. NOLAN (Summer 1988) and Harry O. Morris (Winter 1988). It became the primary magazine for new writers Poppy Z. BRITE, Nancy A. COLLINS and Bentley LITTLE.

Silva won the BALROG AWARD in 1985 and the WORLD FANTASY AWARD in 1988 for *THS*. He produced a representative anthology, *Best of The Horror Show* (anth **1987**; exp vt *The Definitive Best of the Horror Show* 1992). *THS* has a spiritual successor in CEMETERY DANCE.   [MA]

HORWOOD, WILLIAM (1944-   ) UK writer best-known for the **Duncton Chronicles**, a sequence of ANIMAL FANTASIES: *Duncton Wood* (**1980**), *Duncton Quest* (**1988**), *Duncton Found* (**1989**), *Duncton Tales* (**1991**), *Duncton Rising* (**1992**) and *Duncton Stone* (**1993**). Set in rural England near Oxford, and telling the saga through many generations of moles, the sequence has some resemblance to Richard ADAMS's *Watership Down* (**1972**): the two are similar in their versimilitudinous settings, the sense of the precariousness of the lives depicted, and their underlying QUEST structure, but differ profoundly in that WH's sequence is also a BEAST FABLE. His moles are mole-like only to a point, beyond which their intelligence, the complexity of their RELIGION, the ardour of their quests, the violence of their internecine disputes and the consuming intensity of their sexual activities all make them clearly allegorical ($\Diamond$ ALLEGORY) of human beings. The final effect although not particularly subtle, gives the sequence as a whole a sense of brooding estrangement: its protagonists, neither mole nor human, echo both conditions of being.

Most of WH's other tales deal with animals. In *The Stonor Eagles* (**1982**) the last sea-eagle in Scotland flies to Norway, where she has mystical experiences and dreams of returning home. *Callanish* (**1984**), not a sequel, deals with a golden eagle who masterminds an escape of his kin from London Zoo. *The Willows in Winter* * (**1993**) sequels Kenneth GRAHAME's *The Wind in the Willows* (**1908**) and has moments of surprising power, especially when the Wild Wood is seen as part of a threatening larger world ($\Diamond$ SEQUELS BY OTHER HANDS). **The Wolves of Time**, begun with *Journeys to the Heartland* (**1995**), focuses on the wolves of Europe, who are mystically called upon to undergo a cleansing quest eastwards for the Heartland, which is dominated by an evil figure. Only one WH tale, *Skallagrigg* (**1987**), focuses primarily on humans; it deals with the eponymous MESSIAH figure in terms that allow and disallow his literal reality.

Writers of animal fantasies do not normally get very seriously treated by critics; WH deserves rather more attention than he has been accorded to date.   [JC]

HOTEL Hotels are essentially urban constructions, and are not found with any great frequency in fantasy texts; even in URBAN FANTASIES they do not appear often and, when they do, they generally exhibit most of the characteristics that define an EDIFICE; their staffs may well be described in terms of ESTATES SATIRE. Rare examples of working fantasy hotels feature in Gene WOLFE's *There Are Doors* (**1988**) and William Browning SPENCER's *Zod Wallop* (**1995**); both are edifice-like. The remote, snowbound Outlook Hotel in Stephen KING's *The Shining* (**1977**) is another example; vast and eerily empty, it is as much a character in the story as any of the human players. This point was stressed even more in the movie version, *The* SHINING (*1980*). By contrast, the hotel in *The* PASSING OF THE THIRD FLOOR BACK (*1935*) is claustrophobically cramped.

The GENRE-FANTASY form of the hotel is the INN.   [JC/JG]

HOUGHTON, CLAUDE Working name of UK writer Claude Houghton Oldfield (1889-1961). Much of his work has no genre content; some is sf. *Neighbours* (**1926**) hovers at the edge of SUPERNATURAL FICTION, though the DOUBLE on whom the protagonist spies is in fact a projection. *Julian Grant Loses his Way* (**1933**) is a POSTHUMOUS FANTASY whose protagonist's unpleasant life, of which he is forced to relive bits, mirrors CH's sense of the decadence of England between the World Wars. *Three Fantastic Tales* (coll **1934**) contains ALLEGORIES whose carriers are profoundly estranged from the shabby world. *This was Ivor Trent* (**1935**) and *The Man who Could Still Laugh* (c1935 in unknown Czech magazine; **1943** chap) are sf.   [JC]

HOUSE There have been three movies in the **House** sequence, of which only the first is of interest.

1. *House* US movie (*1986*). New World. **Pr** Sean S. Cunningham. **Dir** Steve Miner. **Creature spfx** James Cummins, Backwood Film. **Mechanical spfx** Tassilo Baur. **Vfx** Dream Quest Images. **Screenplay** Ethan Wiley. **Starring** William Katt (Roger Cobb), Richard Moll (Big Ben), Mary Stavin (Tanya), George Wendt (Harold Gordon). 93 mins. Colour.

Divorced HORROR novelist Roger moves into the house where his Aunt Elizabeth recently suicided and where, longer ago, his son Jimmy mysteriously vanished. Elizabeth's GHOST soon appears to warn him the house is EVIL. She is wrong: it is possessed ($\Diamond$ POSSESSION) by the vengeful spirit of Roger's psychopathic army buddy Big Ben, tortured to death by the Viet Cong. While learning this

Roger encounters POLTERGEIST effects, closet-dwelling MONSTERS, SHAPESHIFTERS and his neighbours to either side, the alarmingly helpful Harold and the bedroom-eyed single mother Tanya. A terrifying surreal painting reveals that Jimmy is trapped behind a bathroom MIRROR; this Roger smashes to find a PORTAL into a pitch-dark otherness, at length rescuing Jimmy from the midst of a Vietnam jungle. Passing back through a different portal, the garden swimming pool, they reach what seems safety until the undead Big Ben appears for a final confrontation.

The overriding quality of *H* is its energetic inventiveness, its confident rapid-fire exploitation of rafts of fantasy notions, some of which are unusually sophisticated for low-budget HORROR MOVIES, especially HAUNTED-DWELLING movies – its blending of past, present and otherly REALITIES becomes almost like the shuffling of a deck of cards.

2. The two follow-ups have been *House II: The Second Story* (**1987**) and *House III: The Horror Show* (**1989**). Neither bear relation to 1. [JG]

HOUSE AT THE END OF THE ROAD, THE vt of *The Tomb of Ligeia* (**1964**). ◊ Roger CORMAN; *The* HOUSE OF USHER.

HOUSEHOLD, GEOFFREY (EDWARD WEST) (1900-1988) UK writer whose best-known novel remains *Rogue Male* (**1938**); he wrote several sf novels. His fantasies, which tend to a certain carelessness of construction, included *The Cats to Come* (**1975**), about a future world run by CATS, *The Sending* (**1980**), a SUPERNATURAL FICTION involving murders, curses, WITCHCRAFT and a protagonist who baulks at his SHAMAN status, and *Summon the Bright Water* (**1981**), which invokes ATLANTIS and its WAINSCOT descendants in the contemporary world. [JC]

HOUSEHOLD WORDS ◊ MAGAZINES.

HOUSE OF DARK SHADOWS US movie (**1970**). ◊ Marilyn ROSS; VAMPIRE MOVIES.

HOUSE OF USHER, THE At least five movies have been based on "The Fall of the House of Usher" (1839) by Edgar Allan POE. All have appeared titled both *The House of Usher* and *The Fall of the House of Usher*. They have been *The House of Usher* (**1928**), *The House of Usher* (**1949**), *The Fall of the House of Usher* (**1960**), *The House of Usher* (**1982** tvm) and *The House of Usher* (**1989**). Two are interesting.

1. *The Fall of the House of Usher* US movie (**1960**). American-International. **Pr** Roger CORMAN. **Exec pr** James H. Nicholson. **Dir** Corman. **Spfx** Pat Dinga. **Screenplay** Richard MATHESON. **Starring** Mark Damon (Philip Winthrop), Harry Ellerbe (Bristol), Myrna Fahey (Madeline Usher), Vincent Price (Roderick Usher). 85 mins. Colour.

The first of Corman's **Poe** movies sees Philip come to the Usher home – rocked by tremors as it subsides into the swamp – to claim his fiancée Madeline. Her brother Roderick is cursed by hyperacute senses and lurching towards insanity. Madeline, obsessed by death, seemingly dies at the height of an argument with her brother, and, though Roderick knows she is merely in a cataleptic fit, is put in the crypt alongside her mad and dangerous ancestors. Now totally insane, she escapes her coffin, tries to kill Philip and does kill Roderick. All ends in flames as Philip flees the accursed house.

Shot in almost comic-book colours, this is the default version of the Poe tale, its gloomy power being perhaps partly born from its equivocation over GOOD AND EVIL and from the slow growth of our awareness that Roderick's doomed

love for his sister is nigh incestuous. A DREAM sequence in which Philip battles the shades of past Ushers in pursuit of Madeline is an impressive highpoint. Much of this, notably the PREMATURE BURIAL, was to be thematically reprised in CORMAN's next **Poe** movie, *The* PIT AND THE PENDULUM. [JG]

2. *The House of Usher* UK movie (**1989**). 21st Century/Breton Film. **Pr** Harry Alan Towers. **Exec pr** Avi Lerner. **Dir** Alan Birkinshaw. **Spfx** Scott Wheeler. **Screenplay** Michael J. Murray. **Starring** Norman Coombes (Clive Derrick), Carole Farquhar (Gwendolyn), Philip Godewa (Dr Bailey), Donald Pleasence (Walter Usher), Oliver Reed (Roderick Usher), Anne Stradi (Mrs Derrick), Rufus Swart (Ryan Usher), Romy Windsor (Molly McNulty). 87 mins. Colour.

Molly comes with fiancé Ryan Usher to the dismal Usher mansion at the behest of his Uncle Roderick. Before they can reach the house, two spectral CHILDREN appear in the road ahead of their car, causing a crash in which Ryan apparently dies. What follows is a mishmash of GHOSTS, PREMATURE BURIAL (of Ryan), CURSES, HALLUCINATION, secret passages, psychopathy and more: the script leaves no cliché of HORROR unexploited, and the cast give performances to match. Much proves to have been a plot engineered by Roderick, who wishes to sire on Molly an heir to the Usher line. At last she escapes the blazing house . . . and we are back with her and Ryan in the car before the crash: it has all been a DREAM. [JG]

HOUSMAN, CLEMENCE (1861-1955) UK artist, poet and writer. She illustrated some works by her brother Laurence HOUSMAN. Her extended prose-poem *The Werewolf* (**1896**) is a symbolic erotic fantasy in which two brothers disagree about a FEMME FATALE; the one who thinks her EVIL pursues her to a climactic meeting where their fates are decided. It is among the most notable products of the short-lived English Decadent Movement (◊ DECADENCE). CH's novel *The Unknown Sea* (**1898**) also features an enigmatic *femme fatale*; its fantastic element is muted and its style conventionally verbose. [BS]

HOUSMAN, LAURENCE (1865-1959) UK poet, dramatist and writer; his elder brother was the poet A.E. Housman (1859-1936) and his sister was Clemence HOUSMAN. LH was a prolific writer from the early 1890s until the 1950s; his FEMINISM, along with his ironical treatment of RELIGION and the English royal family, ensured that many of his plays were banned by the Lord Chamberlain. His later work became less energetic; it has been speculated that his brother's slowly increasing fame sapped LH's confidence in his own work, which was never more than highly competent.

Many of his early stories are FAIRYTALES, ostensibly for children, in the mode of Oscar WILDE. Early titles include *A Farm in Fairyland* (coll **1894**), *The House of Joy (Fairy Tales)* (coll **1895**), *The Field of Clover* (coll **1898**) illustrated by Clemence Housman and *The Blue Moon* (coll **1904**); these four volumes were resorted as *Moonshine and Clover* (coll **1922**) and *A Doorway in Fairyland* (coll **1922**), and later examples were assembled as *What-O'Clock Tales* (coll **1932**). After LH's death a further sorting produced *The Rat-Catcher's Daughter: A Collection of Stories* (coll **1974** US). Compared to much that was published at the end of the 19th century, these tales are notable for their lack of cant, the occasionally prominent role given to female protagonists and an underlying sense of rather laid-back decency.

LH's short fiction for adults is various. *All-Fellows: Seven*

tion to *The Soft Side*, collections containing supernatural fictions include *A Passionate Pilgrim and Other Tales* (coll **1875**), *The Lesson of the Master. The Marriages. The Pupil. Brooksmith. The Solution. Sir Edmund Orme* (coll **1892**), *The Private Life. The Wheel of Time. Lord Beaupre. The Visit. Collaboration. Owen Wingrave* (coll **1893**), *The Real Thing and Other Tales* (coll **1893**), *Embarrassments: The Figure in the Carpet/Glasses/The Next Time/The Way it Came* (coll **1896**), and *The Travelling Companions* (coll **1919**). All HJ's supernatural fictions are assembled as *The Ghostly Tales of Henry James* (coll dated 1948 but **1949** US; cut vt *Ghostly Tales of Henry James* 1963; full text vt *Stories of the Supernatural* 1970) ed Leon Edel, using wherever possible HJ's revised texts; the tales also appear in *The Complete Tales of Henry James* (coll **1962-4** 12 vols) ed Edel.

HJ's closest approach to full fantasy is *The Sense of the Past* (**1917**), an uncompleted novel. The protagonist of this 20th-century TIMESLIP tale, obsessed with his past, enters the early 19th century – through a PORTAL in a door – where he occupies the body of a counterpart, who himself, by way of BALANCE, has come forward in time. Just as the complexities begin to mount, the story ends. Partially drafted not long after he completed *The Sacred Fount* (**1901**), and added to in early 1914, HJ's fragment arguably represents a reactive and prophetic response to a world sliding into the moral abyss of WORLD WAR I.

After the outbreak of WWI, HJ wrote no more fiction. [JC]

JAMES, JOHN (?1924-?1993) UK writer of historical novels, several of which are fantasy. *Votan* (**1966**) and *Not For All the Gold in Ireland* (**1968**) are humorous retellings of Norse and Celtic MYTHOLOGY respectively, and are notable for their acceptance of the everyday importance of the supernatural to people of the time, and for their solid historical grounding. *The Bridge of Sand* (**1976**), more of a straight Romano-British historical novel narrated by the soldier-poet Juvenal, also includes magic occurrences as a normal part of life. [DVB]

JAMES, M(ONTAGUE) R(HODES) (1862-1936) UK scholar and GHOST-STORY writer. MRJ was raised in Suffolk, a setting which featured in many of his stories. He had a distinguished academic career as a linguist and biblical scholar, becoming Provost of King's College, Cambridge, then of Eton. His first published story, "Canon Alberic's Scrapbook" appeared in 1895. As his popularity as a writer of ghost stories grew, many of his later tales were written for magazines like *Atlantic Monthly* and *Cambridge Review*: "The Haunted Dolls' House" was originally written for the library of Her Majesty the Queen's Dolls' House. Nevertheless, most of his stories were originally written to be read to friends, usually around Christmas. In his stories the ordered world of cathedral close or college library would be disrupted by an incursion of the irrational, in the face of which all the learning of his protagonists was usually ineffectual. In "Oh Whistle, and I'll Come to You, My Lad" an academic unearths an ancient whistle on the Suffolk coast which summons a dreadful apparition given substance by the bedclothes in his room. In "The Stalls of Barchester Cathedral" an ambitious cleric schemes his way to a senior position in the cathedral, only to have his triumph destroyed by a ghostly animal. In "The Mezzotint" the purchaser of the PICTURE of an old house can only watch as a shadowy figure appears in the scene and progresses towards the house where an invisible but doubtless terrible act is

perpetrated. MRJ's first collection, *Ghost Stories of an Antiquary* (coll **1904**), was followed by *More Ghost Stories of an Antiquary* (coll **1911**), *A Thin Ghost and Others* (coll **1919**) and *A Warning to the Curious* (coll **1925**). *Collected Ghost Stories* (coll **1931**) contains a few extra stories, and one or two others have been turned up subsequently by researchers like Michael Cox, Richard DALBY and Rosemary Pardoe. MRJ's influence on the English ghost story has been immeasurable. He also contributed to a revival of interest in earlier examples of the genre with an edition of forgotten stories by J. Sheridan LE FANU, *Madam Crowl's Ghost and Other Tales of Mystery* (coll **1923**). He collected ghost stories from Latin manuscripts as *Twelve Medieval Ghost Stories* (coll **1922**), presented in English for the first time in *The Man-Wolf* (coll **1978**) ed Hugh Lamb.

MRJ's only novel was *The Five Jars* (**1922**), an attractive CHILDREN'S FANTASY in which the young protagonist discovers a series of jars containing ointments which open up his senses to the world of animals and FAIRIES. A projected sequel was never written, though some of the characters reappear in one of MRJ's ghost stories, "After Dark in the Playing Fields" (1924). [PK]

Further reading: *M.R. James: An Informal Portrait* (1983) by Michael Cox; *Montague Rhodes James* (1980) by R.W. Pfaff.

JAMES, PHILIP Joint pseudonym of James CAWTHORN and Michael MOORCOCK.

JAMES GANG An AFFINITY GROUP of GHOST-STORY writers working in the tradition of M.R. JAMES. The term was first used in print by Mike ASHLEY (in "The James Gang: The Disciples of M.R. James" 1988 *Horrorstruck 10*), though it had been in use among Jamesian scholars for years. Membership of the JG varied from those who were friends, such as A.C. BENSON and E.F. BENSON – though their ghost stories showed only occasional Jamesian influence – to others who wrote extensively in the Jamesian manner, such as Arthur Gray (1852-1940), Master of Jesus College, Cambridge, whose *Tedious Brief Tales of Granta and Gramarye* (coll **1919**) shows a clear Jamesian influence. Other noted members of the group include: R.H. Malden (1879-1951), Dean of Wells and a friend of James for 30 years, whose *Nine Ghosts* (coll **1943**) is explicit in its Jamesian influence; A.N.L. Munby (1913-1974), librarian of King's College, Cambridge, and author of *The Alabaster Hand* (coll **1949**); L.T.C. Rolt (1910-1974), author of *Sleep No More* (coll **1948**) and many others; E.G. Swain (1861-1938), Chaplain of King's College, Cambridge, and a close friend of James, author of *The Stoneground Ghost Tales* (coll **1912**); and H. Russell Wakefield (1888-1964), a prolific writer of ghost stories whose work includes *Ghost Stories* (coll **1932**) and *A Ghostly Company* (coll **1935**). Among contemporary writers members of the JG are John GORDON, notably for *The House on the Brink* (**1970**), and Ramsey CAMPBELL, some of whose ghost stories show an evident Jamesian influence, particularly "The Guide" (1989), an out-and-out tribute to James. [PK]

Further reading: *The James Gang: A Bibliography of Writers in the M.R. James Tradition* (**1991** chap) by Rosemary Pardoe.

JAMES WARREN PUBLISHING ◊ WARREN PUBLISHING.

JANSSON, TOVE (MARIKA) (1914-   ) Finnish writer and artist, belonging to the Swedish-speaking minority in Finland. In eight novels, a short-story collection, three large-format full-colour verse books and a daily COMIC strip (from 1954), TJ has chronicled the Moomin Valley and its charming, chilling and totally unique characters and creatures. The Moomin family's warm, flowering house is set in a world inhabited also by the incomprehensible, the alien and the frightening; overlying the texts is TJ's zany, absurd whimsy, which is comparable to that of Lewis CARROLL.

TJ's novels are deceptively mild: it is easy to misperceive them as bland or harmless. Nothing could be further from the truth. In the Moomin world there are darkness and fear, loneliness and exposure. There are the totally alient hattifatteners, truly alive only when electrically charged; there is the Groke, a totally evil being; there are Snufkin, the unpredictable anarchist who refuses ever to be tied down, and a host of others.

The first Moomin level, *Småtrollen och den stora översvämningen* (**1945**), is very much a children's book and is not considered by the author as part of the **Moomin** series proper.

Several of the books have gone through numerous rewrites: their bibliography is complex. These are: *Kometjakten* (**1946**; trans Elizabeth Portch as *Comet in Moominland* **1951** UK; rev vt *Mumintrollet på kometjakt* 1956; rev vt *Kometen kommer* 1968); *Trollkarlens hatt* **1948**, this version trans Ernest Benn as *Finn Family Moomintroll* **1950** UK vt *The Happy Moomins* 1952 US; Finnish text rev 1968); *Muminpappans bravader* (**1950**; rev 1952, this version trans Thomas Warburton as *The Exploits of Moominpappa* **1952** UK; rev vt *Muminpappans memoarer* 1968); and *Farlig midsommar* (**1954**, this version trans Warburton as *Moominsummer Madness* **1961** UK; Finnish rev 1969). The last four books in the series have not undergone rewriting: *Trollvinter* (**1957**; trans Warburton as *Moominland Midwinter* **1958** UK), *Det osynliga barnet* (coll **1962**; trans Warburton as *Tales from Moomin Valley* **1963** UK), *Pappan och havet* (**1965**; trans Warburton *Moominpappa at Sea* **1966** UK), and *Sent i november* (**1970**; trans Kingsley Hart as *Moominvalley in November* **1971** UK). The last novel is a deliberate farewell to the Moomin family: the family itself is absent, and the other creatures of Moomin valley must solve their own problems rather than turn to the family for help.

Two of the three book-long **Moomin** poems have not been translated: *Hur gick det sen?* ["What Happened Then?"] (**1952**), *Vem ska trösta knyttet?* (**1960**; trans Warburton as *Who Comforts Toffle* 1960) and *Den farliga resan* ["The Dangerous Journey"] (**1974**).

The **Moomin** series comprises an imaginative feat of depth and complexity. The quiet animism, the lyrical portrayals of nature and of natural phenomena, the subtle shifts in REALITY – as in the scene in *Moominpappa at Sea* where the Moominmomma steps into her beloved garden, which she has recreated in paint on the walls of the lighthouse to which the family has moved – and the constant presence of strange and incomprehensible mysteries are all effortlessly integrated into a world that is superficially simple and comfortable.

TJ drew the **Moomin** comic strip in its early years (*London Evening News* 1954-9); she was succeeded as artist by her brother Lars Jansson and then by others. She has written several stage plays set in the Moomin Valley. The **Moomin** characters are now licensed, and TJ has had little to do with the recent avalanche of comic books, animated movies and other **Moomin** products.

TJ has illustrated fantasy works other than her own, including translated editions of Carroll's *Alice's Adventures in Wonderland* (**1865**) and *The Hunting of the Snark* (**1876** chap) and J.R.R. TOLKIEN's *The Hobbit* (**1937**).   [JHH]
Further reading: *Bildhuggarens dotter* (**1968**), autobiography.

JANUS A Roman deity who guards gates and doorways. He has two faces, one looking backwards (at the year which has just ended; the month of January is named after him), and one gazing forward. One face frowns, the other smiles; normally, the frowning face looks backwards. He is usually depicted with a clutch of keys (hence the word "janitor"). He is a GUARDIAN OF THE THRESHOLD, a LIMINAL BEING who signals the existence of a PORTAL from this world to an OTHERWORLD, an UNDERLIER, a MASK, a FACE OF GLORY. He evokes CARNIVAL, CHRISTMAS and the SEASONS.   [JC]

JAPAN Japanese literature's long tradition begins with the country's oldest book, *Kojiki* ["Record of Ancient Matters"] (anth **712**), which includes many mythical episodes. But in our context the most notable work of antiquity is *Konjaku Monogatari* ("Tales of Ancient Times"] (anth **1120**), which comprises over 1000 FOLKTALES in which various MONSTERS and SPIRITS appear. Since the late 17th century, influenced by Chinese GHOST STORIES, many Japanese authors have written SUPERNATURAL FICTION; best known is *Ugetsu Monogatari* ["Tales of Moonlight and Rain"] (coll **1776**) by Ueda Akinari (1734-1804).

Realism came to dominate Japanese literature at the end of the 19th century, but even so some masterly fantasy was produced by well respected writers. Kyôka Izumi (1873-1939) deployed a rich vocabulary in his many supernatural short stories; "Kôya Hijiri" ["A Priest of Kôya"], his most important, deals with the METAMORPHOSIS of man into animal through the charm of a FEMME FATALE. Rohan Kôda (1867-1947) ranks alongside Izumi. *Yume Jûya* ["The Dreams in Ten Nights"] (coll **1908**) by Sôseki Natsume (1867-1916), among Japan's foremost writers, is imbued with Surrealist images. The naturalized Englishman Lafcadio HEARN (1850-1904) – known in Japan as Yakumo Koizumi – collected and retold Japanese supernatural folktales in his *Kwaidan* (coll **1904** US) (◊◊ KWAIDAN [**1964**]).

During the interwar years several great fantasy writers made impressive debuts. The most unique talent was that of Taruho Inagaki (1900-1977), whose first book, *Issen-ichibyô Monogatari* ["Tales of A Thousand and One Seconds"] (coll **1923**), contained NONSENSE tales featuring intelligent planets, comets, stars and MOON. Kyûsaku Yumeno (1887-1936) made his debut with "Ayakashi No Tuzumi" ["The Evil Drum"] (1926) in the magazine *Shinseinen* ("New Youth"), which was basically a mystery magazine but published some fantasy. His best-known work, *Dogura-Magura* ["Abracadabra"] (**1935**), is psychological HORROR in which a man involuntarily re-enacts a dreadful crime committed by an ancestor, then afterwards suffers AMNESIA over what he has done. Another noteworthy contributor to *Shinseinen* was Mushitarô Oguri (1901-1946); his *Kokushikan Satsujinjiken* ["The Murder at Black Death Mansion"] (**1935**) is an occult mystery written in an ornate style. *Shinseinen's* most famous contributor, Edogawa Rampo (real name Tarô Hirai; 1894-1965) – the pseudonym is a punning homage to Edgar Allen POE – included some fantasies among his many mysteries.

After WWII the mystery magazine *Hoseki* ["Jewel"] was launched; as with *Shinseinen*, several authors of both mystery and fantasy debuted here. Sigeru Kayama (1909-1975) used his knowledge of palaeontology as a basis for many fantasies; his best-known work – *Gojira* (**1954**), filmed as

GODZILLA (*1954*) – is not his best. The most gifted writer to start in *Hoseki* was Shin-ichi Hoshi (1926-   ), who has produced over 1000 short-short stories; like Ray BRADBURY, he is often termed an sf writer though most of his work is fantasy. In fact, most Japanese sf writers produce fantasy as well. Yasutaka Tsutsui (1934-   ), another to debut in *Hoseki*, is noted for his black humour and slapstick; his recent works have veered towards metafiction and slipstream.

Hayakawa Shobo launched *SF Magazine* in 1960. This began as a Japanese edition of *The* MAGAZINE OF FANTASY AND SCIENCE FICTION, but gradually the ratio of Japanese material increased, and this has become a good vehicle for fantasy as well as sf. But there was no specialist fantasy MAGAZINE until *Gensô To Kaiki* ["Roman and Fantastique"] appeared in 1973. Mainly translations but with a few original stories, *Gensô To Kaiki* lasted only 12 issues, but was remarkable for the quality of its selection. At about the same time book publishers began the systematic translation of the best of Western fantasy.

Domestic authors were busy creating an indigenous form of fantasy. In his most important novel, *Musubinoyama Hiroku* ["The Secret History of the Holy Mountain"] (**1973**), Ryô Hanmura (1933-   ) reinterpreted Japanese history from the perspective of a LOST RACE possessed of strange TALENTS. Kazumasa Hirai (1938-   ) began a successful series about a WEREWOLF hero with *Ôkamiotoko Dayo* ["I Am a Werewolf"] (coll **1969**).

In the 1970s the reading public began to notice fantastic literature. In mainstream literary circles, writers of visionary fiction like Kobo ABÉ, Yumiko Kurahashi (1935-   ) and Hideo Nakai (1922-1992) gained vast popularity. Tatsuhiko Shibusawa (1928-1987) – a popularizer of Western occultism and a translator of the Marquis DE SADE and J.-K. HUYSMANS – became a cult hero.

In the 1980s fantasy became ever more widespread. Kaoru Kurimoto (1953-   ) began her bestselling HEROIC-FANTASY series **Guin Saga** with *Hyôtô No Kamen* ["A Mask of Panther"] (**1979**). Baku Yumemakura (1951-   ) and Hideyuki Kikuchi (1949-   ) are both very popular writers of violent heroic fantasies. Hiroshi Aramata (1947-   ) – respected also as a critic of fantastic literature and as a translator of Robert E. HOWARD's **Conan** series – wrote the monumental *Teito Monogatari* ["The History of the Imperial City"] (**1985-9** 12 vols), in which he reconstructed the history of Tokyo from an occult perspective. This novel became a bestseller and was adapted for movies and COMICS. In 1982 Japan's second specialist fantasy magazine was launched: *Gensô Bungaku* ["Fantastic Literature"] contains mainly criticism but a few short stories.

Osamu Tezuka (1928-1929), Japan's most notable MANGA artist, produced some fantasy comics: *Banpaia* ["Vampire"] (graph **1966**) features infant WEREWOLVES and *Dororo* (graph **1967**) is a weird story; both were adapted for animated series on TELEVISION. Sigeru Mizuki (1924-   ) has specialized in weird comics; the hero of his *Gegege No Kitaro* ["Kitaro the Witch-Boy"] (from 1966) fights evil supernatural beings. Kazuo Umezu (1936-   ) is another specialist in weird comics; he is noted for his bizarre drawing. Moto Hagio (1949-   ) is among the most popular comics artists/writers; her *Pô No Ichizoku* ["A Clan of Poe"] (from 1972) features beautiful VAMPIRES and has been compared with Anne RICE's **Vampire Chronicles** series. Daijiro Moroboshi (1949-   ) has often dealt with Asian or Oceanian mythologies in his fantasy comics. Go Nagai (1945-   ) produced a famous weird comic, *Debiruman* ["The Devil Man"] (1972-3), which was adapted for a tv animated series. Japanese fantasy animation (◊ ANIME), by contrast with sf animation, is rather a minor genre, although *Tonari No Totoro* (**1988**; vt *My Neighbour Totoro*) by Hayao Miyazaki (1941-   ) was very popular. The Japanese contribution to the MONSTER-MOVIE genre should certainly not go unmentioned; aside from the GODZILLA MOVIES and various other series, Japan has made a couple of unique – if to Western eyes baffling – contributions to the sequence of FRANKENSTEIN MOVIES.

Since the late 1980s several publishers have produced light YA fantasy novels, and many writers have made their debut with these. More important, however, is the Japanese Fantasy Novel Contest, held annually since 1989 by the leading publisher Shinchosha. Ken-ichi Sakemi (1963-   ) was the first prizewinner with his first novel *Kôkyû Shôsetsu* ["A Tale of the Imperial Harem"] (**1989**), which wittily created an alternative history of China. The 1991 prizewinner was Aki Satô (1962-   ) with her first novel *Barutazâru No Henreki* ["Travels of Balthazar"] (**1991**), in which she dealt with the DOPPELGÄNGER theme. Both of these gifted writers are still at work in the field.   [ShM]

JAROŠ, PETER (1940-   ) Slovak writer. ◊ SLOVAKIA.

JARVIS, ROBIN (?   -   ) UK writer of generally YA fantasies, always in series. The **Deptford Mice** sequence – *The Dark Portal* (**1989**), *The Crystal Prison* (**1989**) and *The Final Reckoning* (**1990**) – is an ANIMAL FANTASY set in a subterranean LONDON, with MICE pitted against a vicious supernatural CAT. The **Deptford Histories** – *The Alchymist's Cat* (**1991**), *The Oaken Throne* (**1993**) and *Thomas* (**1995**) – is again set in London, as is *The Woven Path* (**1995**), which begins the **Tales from the Wyrd Museum** sequence, centring on an East End museum in which have been secreted various dark secrets concerning the metropolis. RJ's other series, the **Whitby** sequence – *The Whitby Witches* (**1991**), *A Warlock in Whitby* (**1992**) and *The Whitby Child* (**1994**) – is a somewhat more conventional set of SUPERNATURAL FICTIONS.   [JC]

JASON ◊ GOLDEN FLEECE.

JASON AND THE ARGONAUTS UK movie (*1963*). Columbia. **Pr** Charles H. Schneer. **Assoc pr** Ray HARRYHAUSEN. **Dir** Don Chaffey. **Spfx** Harryhausen. **Screenplay** Beverley Cross, Jan Read. **Starring** Todd Armstrong (Jason), Honor Blackman (Hera), John Cairney (Hylas), Nigel Green (Hercules), Jack Gwillim (Aeetes), Michael Gwynn (Hermes), Nancy Kovack (Medea), Niall MacGinnis (Zeus), Laurence Naismith (Argus), Gary Raymond (Acastus), Douglas Wilmer (Pelias). 104 mins. Colour.

Based on Greek MYTHOLOGY and a sort of thematic sequel to Schneer's and Harryhausen's *The Seventh Voyage of Sinbad* (*1958*; ◊ SINBAD MOVIES), this tells episodically of the quest for the GOLDEN FLEECE, and the adventures not only of Jason and his HEROES but also of the GODS on OLYMPUS, where ZEUS and Hera are playing a board-GAME that prescribes the fates of mortals. Among various adventures, the heroes battle a gigantic bronze Titan, save blind Phineas, sail between the Clashing Rocks (aided by a magically created GIANT), rescue Medea, gain the fleece (in so doing slaying the Hydra) and fight with the army of skeletons that springs up when Medea's wrathful father Aeetes sows the Hydra's teeth.

Harryhausen's spfx are the main reason for viewing *JATA* today; although the control of colour is uncertain, they remain imaginative and tremendously watchable.   [JG]

JEFFERIES, MIKE (1943-   ) UK illustrator and writer. He has illustrated several of his own novels, which begin with his first **Loremasters of Elundium** sequence – *The Road to Underfall* (**1986**), *Palace of Kings* (**1987**) and *Shadowlight* (**1988**). A second sequence was begun with *The Knights of Cawdor* (**1995**). Given its roughness of execution, the series has, perhaps unsurprisingly, gained little critical recognition, though the basic story – in the first volume, an UGLY DUCKLING saves a LAND in peril from dark forces who are represented by the villainous Kruel, and becomes king – may please some. In the second sequence, though the evil Kruelshards have been defeated, a new danger threatens the kingdom. MJ's second series, **Heirs of Gnarlsmyre** – *Glitterspike Hall* (**1989**) and *Hall of Whispers* (**1990**) – features a female heir to the city of Glor. Her difficulties (males from the WATER MARGINS surrounding the city oppose her because of her GENDER and because their leader Mertzork is ambitious) are eventually surmounted.

MJ's singletons tend towards DARK FANTASY. In *Shadows in the Watchgate* (**1991**), set in contemporary East Anglia, a taxidermist named Strewth animates his animals in error and havoc nearly ensues in Norwich. *Stone Angels* (**1993**) also involves the inanimate coming to life in Norwich: this time they are two stone ANGELS, one good and one evil. In *Hidden Echoes* (**1992**) a fantasy writer is abducted into a WONDERLAND where TIME is stored. *Children of the Flame* (**1994**) deals with a centuries-old CURSE.   [JC]

JEFFERIES, RICHARD Working name of UK essayist and writer John Richard Jefferies (1848-1887), still best-known for his sf novel *After London, or Wild England* (**1885**) (◊ SFE). *The Early Fiction of Richard Jefferies* (coll **1896**) contains "A Strange Story", a minor SUPERNATURAL FICTION involving a DOPPELGÄNGER. Of more importance as fantasy is his **Bevis** sequence – *Wood Magic: A Fable* (**1881**; cut vt *Sir Bevis: A Tale of the Fields* 1889) and *Bevis: The Story of a Boy* (**1882**) – in whose first volume RJ dramatizes an animistic sense of Nature through the experiences of young Bevis, who is able to talk with animals.   [JC]

JEKYLL AND HYDE Two people who are in fact one person, reflecting the polarized GOOD AND EVIL in the human psyche. In Robert Louis STEVENSON's *Strange Case of Dr Jekyll and Mr Hyde* (**1886**), Jekyll is a respectable doctor who "becomes" the loathsome Hyde at night. The book – and the nightmare which started it – was probably inspired by the play *Deacon Brodie, or The Double Life: A Melodrama, Founded on Facts* (**1880**; rev 1888), which Stevenson wrote with W.E. Henley (1849-1903); this tells of William Brodie (1741-1788), publicly a respectable Edinburgh councillor and privately a thief and rakehell. *Deacon Brodie* is purely mundane; in *Jekyll and Hyde*, however, Stevenson so intensifies the translation from Jekyll to Hyde that it can be read as a literal and tragic METAMORPHOSIS from one state of BONDAGE to another; though not quite. Because the transition is drug-induced and described in just-possible physiological terms, the J&H figure is not a satisfactory UNDERLIER for fantasy heroes caught in circular hells of metamorphosis. What J&H do represent is the DOUBLE, and with preternatural vividness. Hyde "hides" inside the corruptible external Jekyll; it is a vision of astonishing intimacy, like a very haunting dream. Probably for Stevenson, and certainly for many of his readers, the double represents

an unclean and compulsive vision of transgressive union of the self to the undersoul, a marriage which proved exceedingly nightmarish for Victorian readers (partly from a misreading of Darwin, partly perhaps from the cultural guilt consequent upon the distension of an empire which sought to control much of the non-white world).

Later readers may perceive J&H in terms of the GASLIGHT ROMANCE. This is to miss the driven, horrific salience of the true story.   [JC]

JEKYLL AND HYDE MOVIES At least 22 movies (including *Mary Reilly* [**1996**]) have been based on the notion of *Strange Case of Dr Jekyll and Mr Hyde* (**1886**) by Robert Louis STEVENSON and/or on the play *Dr Jekyll and Mr Hyde* by Thomas R. Sullivan (1849-1916) (◊ JEKYLL AND HYDE).

1. *Dr Jekyll and Mr Hyde* US movie (**1920**). Paramount/ Artcraft/Famous Players/Lasky. **Dir** John S. Robertson. **Screenplay** Clara S. Beranger. **Starring** John Barrymore (Jekyll/Hyde), Brandon Hurst (George Carew), Charles Lane (Dr Richard Lanyon), Martha Mansfield (Millicent Carew), Nita Naldi (Gina). 63 mins. B/w with tints. Silent.

This brings the sexuality of the tale more to the fore than Stevenson dared, introducing not only the chaste Millicent but also the fallen woman Gina; this pattern is repeated not only in most other JAHMs but also in the very Jeckyll-and-Hydish *The* PICTURE OF DORIAN GRAY (**1945**). An intriguing twist is that Millicent's father, Sir George, is the one who initially sets the staid Jekyll on the path of debauchery; concerned for the future of his daughter, he explains that the only way to rid oneself of temptation is initially to give in to it. It is after giving in to a deal of it that Jekyll creates his POTION. This was one of the early movies in which Hollywood adopted the Expressionism that the European moviemakers of the time had so triumphantly deployed.   [JG]

2. *Doctor Jekyll and Mister Hyde* US movie (**1931**). Paramount/Rouben Mamoulian/Adolph Zukor. **Pr** Rouben Mamoulian. **Dir** Mamoulian. **Spfx** Wally Westmore. **Screenplay** Percy Heath, Samuel Hoffenstein (published **1975** ed Richard J. Anobile, with many inaccuracies). **Starring** Holmes Herbert (Dr Lanyon), Rose Hobart (Muriel Carew), Halliwell Hobbes (Brigadier-General Carew), Miriam Hopkins ("Champagne" Ivy Pearson [spelt thus in the movie but "Pierson" in cast list]), Arnold Lucy (Utterson), Fredric March (Dr Henry [Harry] M. Jekyll/Hyde), Edgar Norton (Poole). 98 mins (cut to 90 mins). B/w.

The philanthropic Jekyll seeks to marry Muriel, but her pompous father insists on delay. Out walking with conservative friend Lanyon, Jekyll saves flirty Ivy from an attacker, and is nearly seduced by her. Some days later he first quaffs a POTION he has devised to release the human id, and briefly METAMORPHOSES into a Neanderthalish version of himself. Muriel is taken away by her father to Bath; hearing she is to be away for a full month, Jekyll swallows another dose and, as Hyde, tracks down Ivy. A month later, Hyde is rarely seen; Hyde is virtually living with Ivy, whom he brutalizes. On Muriel's return, Jekyll resolves henceforth to remain Jekyll, but spontaneously metamorphoses, and brutally murders Ivy. At the last he is shot dead as Hyde, but reverts in death to his Jekyll form.

This was lost for nearly three decades as a result of MGM's idiotic decision to suppress all opposition to **3**. The standard abridged version was unearthed in 1967 and the full version only in 1994.

This is an astonishing movie. The link between the suppressed instincts and sexuality is clearly stated; the sadistic Hyde is the inner Jekyll, not Jekyll's counterpart or other self, and takes sexual and sadistic revenge upon Ivy for the celibacy which Muriel (through her father) is imposing on the outer Jekyll. March is brilliant as Hyde and received an Oscar, but the true stars are director Mamoulian and cinematographer Karl Struss. Much use is made of split-screen techniques, superimposed images and point-of-view shots: our first view of Jekyll and later of Hyde comes when they examine themselves in the mirror, while the initial transformation is seen through Jekyll/Hyde's eyes as a baffling whirl of images. Indeed, MIRRORS play a major role: it is hard to forget the shot in which Ivy, believing herself forever rid of Hyde, toasts herself in the mirror as, over her reflected shoulder, the door inches open to reveal her leering oppressor. This movie is a classic of the CINEMA and a milestone in both HORROR and TECHNOFANTASY genres. Its most terrifying feature is that we all know Hydes.   [JG]

3. *Dr Jekyll and Mr Hyde* US movie (*1941*). MGM. **Pr** Victor Saville. **Dir** Victor Fleming. **Spfx** Peter Ballbusch, Warren Newcombe. **Screenplay** John Lee Mahin. **Based on 2.** **Starring** Ingrid Bergman (Ivy Peterson), Donald Crisp (Sir Charles Emery), Peter Godfrey (Poole), Ian Hunter (Dr John Lanyon), Frances Robinson (Marcia), Spencer Tracy (Jekyll/Hyde), Lana Turner (Beatrix Emery). 127 mins. B/w.

The lesser of the two "classic" versions is a reasonably close remake of its predecessor, with some inscrutable name-changes and the occasional overlayering of pseudo-Freudian imagery. The inherent sexuality of the tale is again brought to the surface, though something of the dynamic is lost in that Bergman is so obviously *not* a Cockney music-hall good-time girl. Hyde's incidental brutalities are almost entirely omitted in favour of concentration on the murder of Ivy and an attack on chaste Beatrix. One transformation scene – the first time Jekyll involuntarily becomes Hyde – is especially well handled, the change being initially signalled by Jekyll's inability to stop whistling Hyde's catch-tune. This rather good movie pales beside **2.**   [JG]

4. *Son of Dr Jekyll* US movie (*1951*). Columbia. **Dir** Seymour Friedman. **Screenplay** Mortimer Braus, Jack Pollexfen. **Starring** Louis Hayward, Alexander Knox, Jody Lawrance. 77 mins. B/w.

As with the DRACULA MOVIES and the FRANKENSTEIN MOVIES, Hollywood, having made a couple of fine movies, tried to cash in by cheapening the underlying idea to create HORROR MOVIES. In this outing Dr Jekyll's son falls for the family vice.   [JG]

5. *Daughter of Dr Jekyll* US movie (*1957*). Allied Artists. **Dir** Edgar G. Ulmer. **Screenplay** Jack Pollexfen. **Starring** John Agar, Arthur Shields, Gloria Talbott. 74 mins. B/w.

Another cheapening of the original, this sees a young woman accused of being a WEREWOLF because she is (apparently) Dr Jekyll's daughter.   [JG]

6. *The Ugly Duckling* UK movie (*1959*). HAMMER/Columbia. **Exec pr** Michael Carreras. **Dir** Lance Comfort. **Screenplay** Sid Colin, Jack Davies. **Starring** Reginald Beckwith (Reginald), Bernard Bresslaw (Henry Jekyll/Teddy Hyde), Maudie Edwards (Henrietta Jekyll), John Pertwee (Victor Jekyll). 84 mins. Colour.

A rather amusing update of the tale to the present day – a descendant of Jekyll (Bresslaw) unwillingly transforms into a teddyboy.   [JG]

7. *Two Faces of Dr Jekyll, The* (vt *House of Fright* US; vt *Jekyll's Inferno* US) UK movie (*1960*). Columbia/HAMMER. **Pr** Michael Carreras. **Assoc pr** Anthony Nelson-Keys. **Dir** Terence Fisher. **Mufx** Roy Ashton. **Screenplay** Wolf MANKOWITZ. **Starring** Dawn Addams (Kitty Jekyll), David Kossoff (Professor Ernst Litauer), Christopher Lee (Paul Allen), Norma Marla (Maria), Paul Massie (Henry Jekyll/Edward Hyde). 88 mins. Colour.

London, 1874. Unusually, Jekyll is married; while he spends long hours in the laboratory attempting to create a higher order of man who transcends GOOD AND EVIL, wife Kitty dallies with scrounging lover Allen. Transformed into Hyde (by hypodermic injection) – the sole physical change is the disappearance of his beard – he encounters the pair in a dancehall and, unrecognized, flirts with Kitty and nearly murders a drunk (the young Oliver Reed in a bit part). The two selves argue with each other, and the restored Jekyll vows never to let Hyde live again – a vow soon broken.

There are some interesting moments towards the end when Jekyll confronts his MIRROR reflection, which is Hyde; Hyde explains that *Jekyll himself* is guilty of all the crimes, and it seems as if the movie is suddenly to develop a subtext. But then, swiftly, we return to the by-the-numbers grue.   [JG]

8. *Nutty Professor, The* US movie (*1963*). Paramount. **Pr** Ernest D. Glucksman. **Dir** Jerry Lewis. **Spfx** Paul K. Lerpae. **Mufx** Jack Stone. **Screenplay** Lewis, Bill Richmond. **Starring** Lewis (Professor Julius F. Kelp/Buddy Love), Del Moore (Dr Warfield), Stella Stevens (Stella Purdy). 107 mins. Colour.

Stumbling idiot university botany professor Kelp (who bears more than a passing resemblance to Fredric March's Hyde in **2**), is bullied by his football-playing students and reviled by his dean, Warfield, but liked by wide-eyed blonde student Stella. After a crash course of bodybuilding fails he turns to chemistry and METAMORPHOSES (via an intermediate stage in which he is a lumbering Neanderthal) into the debonair, handsome, '50s-chic but obnoxiously egotistic Buddy Love (who looks not unlike Lewis's long-time colleague Dean Martin). Stella, initially revolted, falls for the overt sexual charge of the man she meets as Love. But the two personalities start periodically intruding into each other's existences at inconvenient moments, and on Senior Prom night Kelp swallows an inadequacy of the formula, so that Love very publicly reverts to Kelp. Naturally, Stella discovers it is the UGLY DUCKLING she loves.

This has its moments, notably some Tati-esque sight gags and sound effects. It is generally regarded as Lewis's best movie.   [JG]

9. *Doctor Jekyll and Mister Hyde* (vt *The Strange Case of Doctor Jekyll and Mister Hyde*) Canadian/US movie (*1968* tvm). **Dir** Charles Jarrot. **Starring** Denholm Elliott, Leo Glenn, Oscar Homolka, Duncan Lamont, Jack Palance, Billie Whitelaw. 136 mins. Colour.

Derived from a mini-series, this tends to worthiness and fidelity.   [JG]

10. *I, Monster* UK movie (*1971*). Amicus/British Lion. **Pr** Max J. Rosenberg, Milton Subotsky. **Dir** Stephen Weeks. **Mufx** Harry Frampton. **Screenplay** Subotsky. **Starring** Peter Cushing (Frederick Utterson), Richard Hurndall (Dr Lanyon), Susan Jameson (Diane Thomas), Marjie Lawrence (Annie), Christopher Lee (Charles Marlowe/Edward Blake), George Merritt (Poole), Mike Raven (Enfield). 75 mins. Colour.

Strangely, the names of Jekyll and Hyde are changed; Lanyon, Enfield, Poole and Utterson all have their accustomed rôles. Taciturn Dr Marlowe has developed a drug that, when injected, destroys one of the id, ego and superego (*which* one depends on the individual), allowing the others to come to the fore. He tests it on his cat, which becomes a wild creature and has to be destroyed; then on prim, repressed Miss Thomas, who displays nymphomania until he hastily administers the antidote. Various others of his patients show different reactions. Finally he experiments on himself and becomes leering, impulsive, amoral moron Edward Blake. From here *IM* is a pretty straightforward version, with some scenes adopting Stevenson's own words. One interesting note is that in the early experiments Blake is physically little different from Marlowe, but as the weeks progress Blake's life of dissipation is reflected on his face, which becomes steadily uglier and more inhuman – a skewed reminder of Oscar WILDE's *The Picture of Dorian Gray* (**1891**). Cheaply but beautifully made, this is among the best versions.   [JG]

11. *Dr Jekyll & Sister Hyde* UK movie (**1971**). EMI/Hammer. **Pr** Brian Clemens, Albert Fennell. **Dir** Roy Ward Baker. **Mufx** Trevor Crole-Rees. **Screenplay** Clemens. **Starring** Dorothy Alison (Mrs Spencer), Ralph Bates (Henry Jekyll), Martine Beswick (Mrs Hyde), Susan Brodrick (Susan Spencer), Lewis Fiander (Howard Spencer), Gerald Sim (Professor Robertson). 97 mins. Colour.

Jekyll seeks a universal panacea. Roué friend Robertson tells him he may die before attaining this. Jekyll realizes he must thus first achieve his own IMMORTALITY, and accordingly develops an ELIXIR OF LIFE. Female hormones are the key to this: he buys corpses of young females from a morgue and hires bodysnatchers to increase the supply. When finally he tries the concoction himself he transforms into a beautiful, amoral woman, whom he explains to the neighbouring Spencer family is his widowed sister, Mrs Hyde. To obtain further fresh corpses Jekyll perforce becomes (initially himself, later effectively in GENDER DISGUISE as Hyde) JACK THE RIPPER, reasoning this is a small EVIL when compared to the immense benefits his researches may bring. Meanwhile Jekyll has fallen for Susan Spencer and Hyde less chastely for her pompous brother Howard . . .

DJ&SH is a rather stodgy HORROR MOVIE, somehow losing sexual tension through making the tale's underlying sexuality overt. The fortuitous facial resemblance between Bates and Beswick, both HAMMER stalwarts, adds a measure of interest.   [JG]

12. *Doctor Jekyll and Mister Hyde* US movie (**1973**). **Dir** David Winters. **Starring** Kirk Douglas, Susan George, Susan Hampshire, Michael Redgrave. 90 mins. Colour.

A musical.   [JG]

13. *Dr Heckyl and Mr Hype* US movie (**1980**). Cannon/Golan-Globus. **Pr** Yoram Globus, Menahem Golan. **Dir** Charles B. Griffith. **Spfx** J.C. Buechler, Tim Doughten. **Mufx** Steve Neill. **Screenplay** Griffith. **Starring** Corinne Calvet (Pizelle Puree), Sharon Compton (Mrs Quivel), Maia Danziger (Miss Finebum), Denise Hayes (Liza Rowne), Sunny Johnson (Coral Careen), Oliver Reed (Dr Henry Heckyl/Mr Hype), Mel Welles (Dr Vince Hinkle), Kedric Wolfe (Dr Lew Hoo). 99 mins. Colour.

Impossibly ugly chiropodist Heckyl swallows a colleague's instant-slimming formulation and turns into the devastatingly handsome Hype – or, at least, Oliver Reed. Narcissistic,

paranoid and psychopathic, he becomes a SERIAL KILLER; he also fails to win lovely Coral, who prefers the hideous Heckyl. One's heart usually sinks at the prospect of a Golan-Globus production, but in the case of this deeply unfunny, badly made movie it doesn't sink far enough.[JG]

14. *Doctor Jekyll and Mister Hyde* UK movie (**1981** tvm). BBC. **Dir** Alastair Reid. **Starring** Ian Bannen, Lisa Harrow, David Hemmings (Jekyll/Hyde). 115 mins. Colour.

A typically lavish BBC "classic" production, with fine performances and a generally inventive script, this unfortunately suffers from stolidity.   [JG]

15. *The Strange Case of Dr Jekyll and Miss Osbourne* (ot *Docteur Jekyll et les femmes*; vt *Bloodbath of Dr Jekyll*; vt *The Blood of Dr Jekyll*; vt *Le Cas Étrange du Dr Jekyll et de Miss Osbourne*; vt *Doctor Jekyll and Miss Osbourne*) French movie (**1981**). **Dir** Walerian Borowczyk. **Starring** Clément Harari, Udo Kier, Patrick Magee, Marina Pierro, Howard Vernon. 92 mins. Colour.

An interesting semi-porn version that concentrates on the sexuality of the tale, distorting it considerably to this end. It starts with a young woman being murdered as she arrives at Jekyll's house; then, as Jekyll becomes a monster, he prowls what is represented as not just a house but an EDIFICE in sexually predatory fashion. As Chris Peachment has written of this, "God knows what the raincoat trade makes of it . . ."
                                          [JG]

16. *Jekyll and Hyde . . . Together Again* US movie (**1982**). Paramount/Titan. **Pr** Lawrence Gordon. **Exec pr** Joel Silver. **Dir** Jerry Belson. **Spfx** Dewey G. Grigg. **Vfx** R/Greenberg Associates. **Mufx** Mark Busson. **Screenplay** Belson, Monica Johnson, Michael Leeson, Harvey Miller. **Starring** Bess Armstrong (Mary Carew), Mark Blankfield (Dr Daniel Jekyll/Mister Hyde), Krista Errickson (Ivy Venus), Michael McGuire (Carew), Tim Thomerson (Dr Knute Lanyon). 87 mins. Colour, some b/w.

Designedly tasteless, over-the-top comedy version, transported to the modern USA. Saintly, naive Jekyll gives up a brilliant career in surgery to find a drug that will bring forth our primitive selves and thus enable us to cure our own ailments – a move opposed by the rich dean of the Our Lady of Pain and Suffering Hospital, Carew, and by Carew's daughter Mary, to whom Jekyll is engaged. Jekyll inadvertently snorts a line of formula #143 and becomes the hippy rock'n'roll Mister Hyde – the transformation is climaxed by the extrusion of gaudy jewellery from his flesh. As Hyde he indulges in unbridled sex with nightclub singer Ivy; as Jekyll he eventually succeeds in bedding Mary. He is persuaded to conduct one last operation – a total organ transplant for the world's richest man – but midway through it Hyde asserts himself. At length, after a rooftop chase in foggy, 19th-century London, Mary decides she loves Hyde and Ivy likewise Jekyll, and the two agree to share him.

The gags are scattershot, but most are funny enough that the duds are forgivable.   [JG]

17. *The Jekyll Experiment* (vt *Dr Jekyll's Dungeon of Darkness*; vt *Dr Jekyll's Dungeon of Death*) US movie (**1982**). **Dir** James Wood. **Starring** John Kearney, Dawn Carver Kelly, James Mathers, Tom Nicholson. 88 mins. Colour.

We have been unable to obtain a viewing copy of this movie, which appears to have been released direct to video. Jekyll lethally experiments on criminals with his new drug.
                                          [JG]

18. *Dr Jekyll and Mr Hyde* Australian ANIMATED MOVIE (**1986** tvm). Burbank. **Exec pr** Tom Stacey. **Anim dir**

Warwick Gilbert. **Screenplay** Marcia Hatfield. **Voice actors** John Ewart (Gabriel Utterson), Max Meldrum (Dr Henry Jekyll), David Nettheim (Edward Hyde). 49 mins. Colour.

A fairly straightforward adaptation, pleasingly rendered. One odd feature is that the METAMORPHOSIS sequences, for which animation would seem to be the ideal medium, occur off-camera, as if this were a live movie economizing on the spfx. A nice piece of comedy sees a mouse swallowing some of the POTION and intimidating the laboratory cat.   [JG]

19. *Edge of Sanity* UK movie (*1988*). Palace/Allied Vision. **Pr** Edward Simons, Harry Alan Towers. **Dir** Gérard Kikoine. **Screenplay** J.P. Felix, Ron Raley. **Starring** Glynis Barber, Ben Cole, Sarah Maur-Thorp, Anthony Perkins (Jekyll/Jack). 90 mins. Colour.

A GRAND-GUIGNOL version in which Jekyll accidentally sniffs a puff of the chemical with which he has been experimenting on a laboratory monkey, and turns into JACK THE RIPPER. The acting is almost universally ham and the script little better, but underwear fetishists will enjoy themselves as the Ripper tours London's brothels.   [JG]

20. *Jekyll and Hyde* US movie (*1990* tvm). **Dir** David Wickes. **Starring** Joss Ackland (Lanyon), Michael Caine (Jekyll/Hyde), Cheryl Ladd, Ronald Pickup. 120 mins. Colour.

A lavishly made spectacular in which Caine is insanely cast – being unconvincing as the gentle Jekyll and hilarious as the bestial Hyde – while much of the GOOD AND EVIL dynamic has been excised through the fact that Jekyll is little less of a lecher than Hyde. But the movie's general ambience is pleasing.   [JG]

21. *Dr Jekyll and Ms Hyde* US movie (*1995*). Rastar/Leider Shapiro/Savoy/Rank. **Pr** Jerry Leider, Robert Shapiro. **Exec pr** John Morrissey. **Dir** David F. Price. **Vfx** Tim Landry, Melissa Taylor. **Screenplay** Oliver Butcher, William Davies, Tim John, William Osborne. **Starring** Lysette Anthony (Sarah Carver), Tim Daly (Dr Richard Jacks), Harvey Fierstein (Yves DuBois), Stephen Tobolowsky (Oliver Mintz), Sean Young (Helen Hyde). 90 mins. Colour.

A revamp of **11** and in a way of **16**, all dressed (or undressed) up as a sort of **Carry On** comedy. The setting is the present-day USA. Jacks, a cosmetics scientist and a descendant of Jekyll, inherits the latter's scientific notes, and believes that if female hormones were inserted into the POTION it would create someone who possessed the best of both sexes. In fact, on trying this, he immediately becomes utterly amoral Helen Hyde. Before order is restored, there are lots of cross-dressing jokes.   [JG]

JENKS, ALMET (1892-1966) US writer whose short work often appeared in *The Saturday Evening Post*. A hint of SLICK FANTASY thus unsurprisingly adheres to *The Huntsman at the Gate* (1952), a POSTHUMOUS FANTASY in which a professional huntsman is killed (as he only subsequently discovers) jumping a mysterious gate; but the central scene bites rather deeper than is normal for slick fantasy. Set in a country house, the tale takes the form of a BEAST-FABLE, for the huntsman's enigmatic hosts are in fact, though in the guise of humans (◊ SHAPESHIFTERS), foxes he has been responsible for killing. In the end, they forgive him.   [JC]

JENKS, JACQUETTA Pseudonym of William BECKFORD.

JENNIE UK vt of PORTRAIT OF JENNIE (*1948*).

JEPSON, EDGAR (1863-1938) UK writer, a close friend of the poet Ernest Dowson (1867-1900) and thus on the fringe of the English Decadent Movement (◊ DECADENCE). This spirit is echoed in *The Horned Shepherd* (1904), a calculatedly heretical fantasy in which an avatar of the Lord of the Forest (i.e., PAN) must meet his appointed fate regardless of the intrusions of an intolerant churchman. This book stands in stark contrast to all EJ's subsequent writings, its manner and private publication by "The Sons of the Vine" suggesting a seriousness of which there is no hint in his autobiography, *Memoirs of a Victorian* (1933); in adapting ideas from Sir James FRAZER's *Golden Bough* to the depiction of clandestine pagan FERTILITY cults surviving within the bounds of Chistendom, it anticipates the central thesis of *The Witch-Cult in Western Europe* (1921) by Margaret Murray (1863-1963). One of the short stories in *Captain Sentimental and Other Stories* (coll 1911), "The Resurgent Mysteries", similarly sets out – in a somewhat less sensational form – the thesis which became central to Murray's *The Divine King in England* (1954).

The Mystery of the Myrtles (1909) is a thriller in which HUMAN SACRIFICES are offered to dangerous ELEMENTALS in a magically protected suburban garden. It was followed by the similar but more striking *No. 19* (1910; vt *The Garden at Number 19* US), in which an ambitious suburban MAGUS succeeds in summoning Pan but cannot obtain his polite cooperation. Like the novel which presumably inspired these two conscientiously unsympathetic accounts of society occultism – W. Somerset MAUGHAM's *The Magician* (1908) – *No. 19* is included in the list of recommended reading in Aleister CROWLEY's *Magick in Theory and Practice* (1929).

EJ spent the greater part of his career producing dull detective stories and frothy comedies; one later exception is *The Moon Gods* (1930), a lively LOST-RACE novel. EJ translated several novels from the French, including *The Man with the Black Feather* (trans 1912) by Gaston LEROUX. His son Selwyn Jepson (1899-1989) wrote numerous crime thrillers.   [BS]

Other work: *The Edge of the Empire* (coll **1899**) with David Eames.

JEROME, JEROME K(LAPKA) (1859-1927) UK writer and editor (notably of *The Idler*), best-known for humorous pieces. *Told After Supper* (coll 1891) contains parodies of Christmas GHOST STORIES. The title story of *The Passing of the Third Floor Back* (coll 1907) – in which an enigmatic man illuminates the lives of the heartsick inhabitants of a boarding-house – was successfully adapted for the stage in 1908 and twice filmed, in 1918 and 1935 (◊ *The* PASSING OF THE THIRD FLOOR BACK [*1935*]); the collection also includes "The Soul of Nicholas Snyders, or the Miser of Zandam", a striking fabular *conte*, and "His Time Over Again", whose hero pusillanimously passes up a second chance at life. Three anecdotal fantasies excerpted from *Novel Notes* (fixup 1893) and two strange tales from *John Ingerfield* (coll 1894) are included with one item from *Told After Supper* and various stories by Barry PAIN and JKJ's fellow *Idler* editor Robert BARR in *Stories in the Dark* (anth 1989) ed Hugh Lamb. JKJ's earnest romance *Paul Kelver* (1903) has a painfully cute fantasy prologue commenting on its FAIRY-TALE-like structure.   [BS]

Other works: *Malvina of Brittany* (coll **1916**).

JERUSALEM The central city of three religions – Judaism, Christianity and Islam – and as such the focus of much religious symbolism and typological thinking. Two patterns are of particular interest: the medieval Christian conception that Jerusalem occupies not only the spiritual but also the literal centre or navel of the world; and the physical, temporal and typological contrast between GARDEN (which is EDEN) at the beginning of TIME and CITY (Jerusalem) at the other end of things. At the heart of Eden is the TREE, and at the heart of Jerusalem is the EDIFICE, the Temple of Solomon. As the type of the ideal city, Jerusalem often comes to represent and to underlie visions of UTOPIA, in many of which the presentation of models for Paradise on Earth has an air of the almost literal cultivation of a New Jerusalem. William Alexander McClung's *The Architecture of Paradise: Survivals of Eden and Jerusalem* (**1983**) is a useful study of the potently polarized structure of idea and dream.

In fantasy texts, this ideal harmony may be sought, and a character like John Shannon in David GEMMELL's **Jerusalem Man** sequence shapes his life around a search for the mythical City; but Jerusalem itself has not been much used as a venue for URBAN FANTASY. Rare exceptions include Edward WHITTEMORE's **Jerusalem Quartet**, Stuart JACKMAN's series begun with *The Davidson Affair* (**1966**), Janice ELLIOTT's *City of Gates* (**1992**) and Graham JOYCE's *Requiem*. More usually, the city serves as a venue for CHRISTIAN FANTASY like *Jerusalem Delivered* (**1591**) by Torquato Tasso (1544-1595), an important TAPROOT TEXT, or for stories of the WANDERING JEW. [JC]

JESTERS A stock motif which fantasy took over from the historical novel and historical romance. In reality, the licensed wit in motley was a phenomenon limited to a few courts in a period that lasted little over a century. The existence of some wise or half-wise court fools in literature from that century, notably the various fools in the plays of SHAKESPEARE, helped set the pattern of the jester as the possibly knowing source of the truth others dare not tell the king.

Victor Hugo's Triboulet in *The King Amuses Himself* (**1844**), best-remembered in the play's adaptation by Piave and Verdi as *Rigoletto* (**1851**), and the protagonist of Edgar Allen POE's "Hopfrog" kill or attempt to kill the ruler who has dishonoured their womenfolk. W.S. GILBERT's Jack Point, in *The Yeoman of the Guard* (**1888**), is a sentimentalized version of the Romantic jester, who dies of a broken heart after losing his beloved to the hero; Point's sarcastic comments on a jester's conditions of employment have found echoes in fantasy, not least in Terry PRATCHETT's *Wyrd Sisters* (**1988**).

Jesters in fantasy come equipped with all this cultural baggage. They may simply be local colour indicating the decadence and depravity of a court; they may be the secret wise counsellor of the ruler or the ally of an intending usurper; they may be the HIDDEN MONARCH enduring the shame of motley as disguise or humiliation (the lost king in Guy Gavriel KAY's *Tigana* [**1990**] has been transformed into an idiot and deformed jester by his magician enemy, and can be redeemed and returned to normality only by death). They may, like Verence in *Wyrd Sisters*, be not so much the rightful king as the only competent person left standing.

The jester is the king's SHADOW, like him set apart by costume and able, up to a point, to do anything he likes. [RK]

JESUS ◊ CHRIST.

JEWISH RELIGIOUS LITERATURE Many of the most basic myths of the West originate, or take their oldest surviving form, in the ancient sacred texts of Judaism or in those later works which, engaging those texts, fundamentally shaped Christianity. These legends and tales, though rarely (if ever) written as fantasy, are inexorably part of fantasy's history.

One of the oldest surviving Jewish writings is the Yahwist, or J, source of the BIBLE's *Genesis*, *Exodus* and *Numbers* (?*c*925BC); Harold BLOOM has argued that this oldest Western story was indeed meant as fantasy. It includes such basic Western myths as EDEN, the FLOOD, the tower of BABEL, the destruction of Sodom and Gomorrah, the burning bush, the plagues on EGYPT, and manna. Bloom, in *The Book of J* (**1990**), claims that the whole book was intended as ironic fantasy, and its GOD as a character only. This position seems unlikely in view of J's extensive national propaganda, but cannot be ruled impossible. Regardless, J's monotheism clearly leaves room for more varied marvels than does later Judaism's. And J's strangest passages – involving half-divine HEROES (*Genesis vi*) and Egyptian MAGIC (*Exodus viii*), for example – provided tropes for millennia of fantasy within monotheistic faiths.

Other old stories, certainly not intended as fantasy, have also profoundly informed the genre's history. The books of *Samuel* combine the grimly realistic *Succession Narrative of David* (?*c*800BC) with the hero-cycles of Samuel, Saul and David (? before *c*900BC), whose few marvels include Goliath and the WITCH of Endor. This combination – probably effected around 600BC – makes David's tale a prototype for later stories of a king's rise and fall, most obviously those about ARTHUR. The Elisha stories of *2 Kings ii-xxiii* (*c*850BC) were probably first written as political propaganda and understood as realistic, but nevertheless present the oldest Western tales of a wonder-working holy man.

Later biblical narratives grew steadily less fantastical. The Elohist, or E, source of the *Torah* (?*c*800BC), replaces J's personal God with one who works through DREAMS and ANGELS. The Deuteronomistic History (*c*600BC) encases the marvels of *Deuteronomy*, *Joshua*, *Judges*, *Samuel* and *Kings* within a moralistic framework; the Priestly, or P, source encases those of *Genesis* through *Numbers* within a ritualistic one. Finally *Chronicles* (*c*460BC) revises *Kings*, sanitizing it (though building up the legend of Solomon), and *Ezra* (*c*460BC) and *Nehemiah* (*c*440BC) are entirely realistic.

Meanwhile a visionary streak, expressed in short passages full of marvels with little or no narrative content, grew in Hebrew literature. From rare early examples such as E's gateway to HEAVEN (*Genesis xxviii*) and *Isaiah*'s vision of God enthroned (*c*700BC), these became almost common after the Babylonian Exile of 587BC.

An era of enormous literary productivity in Hebrew, Aramaic and Greek came after 300BC, and in such works the spare legends of the Bible became the backdrop to new and powerful myths. While the actual origin of each is debated, these works crucially conveyed to the Christian world such basic tropes of romance and fantasy as DEMONS, martyrs, secret knowledge, wicked magicians, and above all HEAVEN and HELL. Similarly, the concepts and of APOCALYPSE are presaged in older works from other cultures, but most fully instantiated in the Jewish works (and their Christian descendants), from which the term was defined. The Ethiopic Apocalypse of Enoch, or *1 Enoch*, for example, is practically an encyclopedia of the new cosmology, collects some of the oldest texts concerned with it, and is still canonical today for the Ethiopian Jews. The apocalypse genre is usually defined as a narrative one, in which a supernatural figure reveals secret knowledge to a mortal, and in which that knowledge includes information about the end of times. It has been carefully compared to sf by Frederick A. Kreuziger in *Apocalypse and Science Fiction* (**1982**), but the interest for fantasy derives largely from the impetus apocalyptic literature gave to elaborations of a specifically moral and eschatological cosmos, basic to mythopoeic fantasy. Since the moral and eschatologic aspects are typically stereotyped, those apocalypses most relevant are those which develop cosmic ideas, typically about Heaven and/or Hell: the Slavonic Apocalypse of Enoch (*2 Enoch*), which includes a remarkable birth-tale at its end; the wildly visionary Apocalypse of *Zephaniah*; and the dazzlingly descriptive Greek Apocalypse of Baruch (*3 Baruch*). The most directly influential apocalypse, of course, and the best-written, is the canonical book of *Daniel* (*c*165BC).

Another encyclopedic work, the Biblical Antiquities attributed to Philo (date uncertain), compiles legends of the Biblical history and works them cohesively into the newer sort of cosmology. The quasi-historical *2 Maccabees* (late 2nd century BC), besides introducing the idea of martyrs to Jewish literature, offered several MIRACLES.

Other works elaborated stories of Biblical characters. Some were entirely realistic; the best with fantastic elements are *Tobit* (2nd century BC) – a charming romance of angelic deliverance – *Daniel*, and the lives of ADAM AND EVE (1st century), which depict the first humans struggling for physical and spiritual survival in a story whose misogyny fails to ruin its power. Others of genuine interest include the *Testament of Job* (*c*1st century), the *Testament of Abraham* (*c*100), whose interplay of human and divine will is typical of later Jewish legend, the romance *Joseph and Aseneth* (date uncertain), and the brief *Paralipomena of Jeremiah* or *4 Baruch* (*c*100), which is reminiscent of Washington IRVING's "Rip Van Winkle".

Judaism, after the destruction of its second temple, retreated steadily from Greek, and abandoned sustained narrative as well as the memory of the post-Biblical literature. Storytelling remained pervasive, but became a fugitive presence for centuries; ironically, this occurred exactly when storytelling was separated from strict sacred status as part of the *haggadah* of rabbinic literature. A superb guide to the extensive and bewildering remains, including the only translations of the main texts, is *Rabbinic Fantasies* (anth **1990**), ed David Stern and Mark Jay Mirsky. More extensive collections of shorter legends from the rabbinic classics include *Gates to the Old City*, (anth **1981**) ed and trans Raphael Patai and the classic condensation of Louis Ginzberg, *The Legends of the Jews* (text **1909-13** 4 vols, notes and index **1925-38** 3 vols). The latter includes only legends concerned with the Bible, and is succeeded by *Jewish Legends of the Second Commonwealth* (**1983**) by Judah Nadich. A collection of later legend with comparable status is *Mimekor Yisrael* (**1938-45** 6 vols; trans I.M. Lask **1976** 3 vols US) by Micha Joseph bin Gorion, ed Emanuel bin Gorion. Other works of interest include: *The Book of Jasher* (*c*1100), a cohesive medievalized retelling of the sacred history; *The Chronicle of Moses* (*c*1100; trans Oliver Shaw Rankin in *Jewish Religious Polemic* anth **1956**); and the varied UTOPIAS reported by Eldad ha-Dani (*c*875), trans in *Jewish Travellers* (**1930**) by Elkan Nathan Adler.

Samaritans honour only the five books of Moses as scripture. In the long period covered by the preceding paragraph, their great poet Marqe (4th century) retold the story of Moses in the first book of *Tibat Marqe* (trans John MacDonald in vol 2 of *Memar Marqah* **1965**), while a much later writer provided a Samaritan version of post-Mosaic history in the *Chronicle of Joshua* (trans Oliver Turnbull Crane as *The Samaritan Chronicle* **1890**).

Although introductions to the Hebrew Bible abound, and books about early rabbinic literature are also fairly common, *Back to the Sources* (anth **1984**) ed Barry W. Holtz is valuable for combining these tasks with a discussion of Hasidic works, critical bibliographies, and attention to the literary values of the narratives. For the rejected literature of post-Biblical times, the standard introduction is *Jewish Literature Between the Bible and the Mishnah* (**1981**) by George W. Nickelsburg. [JB]

JINNS ◊ GENIES.

JOAN OF ARC (?1412-1431) French national heroine, created a saint by the Roman Catholic Church in 1920. The Church required, before her canonization could be authorized, that MIRACLES be authentically ascribed to her; but it is not for this reason that she is of note here. Her cross-dressing valour as a national symbol of resistance, and her ability to "hear" voices, makes her an UNDERLIER figure for the female heroine (◊ GENDER; GENDER DISGUISE), though not for the TEMPORAL ADVENTURESS. *Saint Joan* (**1924**) by George Bernard Shaw (1856-1950) places her briefly in the AFTERLIFE; and movies like Carl Theodor DREYER's *La Passion de Jeanne d'Arc* (*1928*) similarly make metaphysical capital out of her short life. [JC]

JODOROWSKI, ALEJANDRO (1930-) Bolivian moviemaker. ◊ MOEBIUS.

JOHNS, JACOB [s] ◊ John W. JAKES.

JOHNSON, CHARLES (RICHARD) (1948-) US writer for children. *The Sorcerer's Apprentice* (coll **1986**; vt *The Sorcerer's Apprentice: Tales and Conjurations* 1994) presents some surprisingly dark and surreal tales of METAMORPHOSIS. The **Footprints in Time** sequence, begun with *Pieces of Eight* (**1989**), carries its young protagonists by TIMESLIP into PIRATE adventures with Blackbeard. [JC]

JOHNSON, RICHARD (1573-?1659) UK writer. ◊ CHILDREN'S FANTASY; DWARFS; TOM THUMB.

JOHNSTON, GUNNAR (? -) UK writer of whom nothing is known but that he published some thrillers, three of which were fantasy. The **Silber** sequence – *The Claws of the Scorpion* (**1935**) and *The Two Kings* (**1936**) – features Silber, an occult figure who inspires supernatural events in otherwise routine adventure tales. *Soria Moria Castle: The Manuscript of Donald Gayforth Forbest* (**1938**) more interestingly depicts a sinister East-European castle, an EDIFICE paired with a MIRROR version of itself and lying athwart UNDERGROUND passages of great intricacy. *Perilous Discovery* (**1938**) is sf. [JC]

JOHNSTON, MARY (1870-1936) US writer, best-known for historical fiction like *To Have and to Hold* (**1900**). Most of her works of interest are SUPERNATURAL FICTION, starting with *The Witch* (**1914**), set in 17th-century England and dealing with WITCHCRAFT. The estate in *Sweet Rocket* (**1920**) has the capacity to focus psychic messages from various places and times. [JC]

Other works: *The Wanderers* (**1917**); *The Exile* (**1927**).

JONATHAN LIVINGSTON SEAGULL US movie (*1973*). ◊ Richard BACH.

JONES, DIANA WYNNE (1934-) UK writer who began publishing with *Changeover* (**1970**), and whose early fantasy tales – like *Wilkins' Tooth* (**1973**; vt *Witch's Business* 1974 US), *The Ogre Downstairs* (**1974**) and *Eight Days of Luke* (**1975**), Luke proving to be LOKI – were decreasingly tentative efforts at finding her own voice. The **Dalemark**

sequence – *Cart and Cwidder* (**1975**), *Drowned Amnet* (**1977**), *The Spellcoats* (**1979**) and *The Crown of Dalemark* (**1993**) – has elements of traditional fantasy, with QUEST plots; the tales are rousing. The **Chrestomanci** sequence – *Charmed Life* (**1977**), *The Magicians of Caprona* (**1980**), *Witch Week* (**1982**) and *The Lives of Christopher Chant* (**1988** US) – is set in a series of ALTERNATE WORLDS in which MAGIC works, administrated by Chrestomanci, who combines guardianship and MAGUS status; he tends to intervene in the series as a kind but tetchy *deus ex machina*. The first and fourth tales are set in a GASLIGHT-ROMANCE England; the second is set in Italy; and the third is another alternate-world tale – one where WITCHES exist and are persecuted – this time set in a high school.

Her third series comprises the **Howl** tales. In *Howl's Moving Castle* (**1986**) a young woman is locked into the BONDAGE of an old woman's body, and must cope with Wizard Howl, who may or not be cursed himself. The sequel is *Castle in the Air* (**1990**).

Singletons of interest include *The Time of the Ghost* (**1981**), a GHOST STORY written from the point of view of the GHOST, and *Archer's Goon* (**1984** US), perhaps her most successful novel, which mixes elements of fantasy and sf in a SECRET-MASTERS tale. *Hexwood* (**1993**) again mixes sf and fantasy with deceptive ease: TIME paradoxes abound, as a small English wood – which embodies the animate spirit of the ancient British FOREST – brings down a galactic empire. Each book contains a protagonist who is, unwittingly, an AMNESIA-ridden member of the opposition, and eventually its nemesis.

The protagonist of *Fire and Hemlock* (**1984** US) must work her way through a painful NIGHT JOURNEY before she comes to a RECOGNITION that she must recreate the travails of the heroine of the TAM LIN ballad. Also a CONTEMPORARY FANTASY, *Black Maria* (**1991**; vt *Aunt Maria* 1991 US) is set in a small seaside town controlled, through magic, by women, and interrogates issues of FEMINISM. In *A Sudden Wild Magic* (**1992** US), the secret masters of Earth, a High Council of male and female witches, must defend the world from an alternate world called Arth whose male witches have been fomenting disasters here – the 20th century is described as an assemblage of these disasters – in order to steal our solutions, thus avoiding the costs of progress back home; this complexly intriguing idea is executed with some awkwardness.

At her best, DWJ has a suppleness, wit and storytelling ability that make her the equal of any living fantasy writer. [NG/JC]

Other works (selective): *Dogsbody* (**1975**); *The Power of Three* (**1976**); *The Homeward Bounders* (**1981**); *Warlock at the Wheel and Other Stories* (coll **1984**); *A Tale of Time City* (**1987**), sf; *Everard's Ride* (coll **1995**); *The Tough Guide to Fantasyland* (**1996**), a pseudo-encyclopedia parodying GENRE FANTASY; «Wild Robert».

Other works (edited): *Hidden Turnings* (anth **1989**); *Fantasy Stories* (anth **1994**).

JONES, JEFFREY (1944-   ) US painter and illustrator whose work, on many fantasy book covers in the late 1960s and 1970s, showed the strong influence of Frank FRAZETTA – although easily distinguishable by its sensitivity and less formalized composition. Other influences include N.C. WYETH and Gustav Klimt (1862-1918). JJ also produced a number of idiosyncratic COMIC strips. His paintings have an appealing calmness and a gentle romantic eroticism.

JJ produced some drawings for fanzines before travelling to New York in 1967 to seek work as a professional illustrator. He painted over 120 fantasy and sf book covers for Ace, Avon, Belmont, Fawcett, Dell, Tower and others. These were painted in oils with subtle, muted colour, and were very popular. During this period he also drew and painted covers for comic books (for *Wonder Woman* and DC COMICS mystery books) and fantasy and sf magazines like *Amazing Science Fiction*. In association with Vaughn Bodé he did some impressive fantasy comic strips and GRAPHIC NOVELS, for WARREN PUBLISHING's EERIE and VAMPIRELLA, plus others like **Alien** (*Witzend* 1968), FLASH GORDON (*#13* 1969) with Bernie WRIGHTSON, **Nightmaster** (*DC Preview Showcase #83-#84* 1969) with Wrightson, *An Axe to Grind* (graph **1970**), an adaptation of Washington IRVING's *Legend of Sleepy Hollow* (graph **1974**) and a book of fey short pieces entitled *Spasm* (graph coll **1973**). Some interestingly illustrated poems, some by JJ himself, appeared in avant-garde magazines like *Phase* and *Imagination*.

In 1971 JJ began to publish his strip **Idyl** (*National Lampoon* 1971-5; graph coll **1979**). This starred a naked, pregnant young woman who had desultory verbal exchanges about art, philosophy and life with talking flowers, a permanently perplexed chimpanzee, and a menagerie of other intellectual fauna. Equally puzzling was his strip **I'm Age** (HEAVY METAL 1981-3), featuring the same female character, now no longer pregnant. JJ has done little published work since then apart from a few portfolios of drawings.

Books showcasing his work include *The Studio* (graph **1979**), *Yesterday's Lily* (graph **1980**) and *Age of Innocence* (graph **1994**). [RT]

JONES, JENNY (HUWS) (1954-   ) UK writer who began publishing fantasy with the **Flight Over Fire** sequence – *Fly by Night* (**1990**), *The Edge of Vengeance* (**1991**) and *Lies and Flame* (**1992**) – which can be read as one sustained tale. The protagonist, a young woman, has been called in error to the PLANETARY-ROMANCE venue of Chorolon, a world caught in sterile stasis through the actions of a WIZARD who has trapped the "Bird of Time", thus making it impossible for natural processes to continue. Everything now works by MAGIC, which proves deadening, particularly for those literally out of the sun (which no longer shines on the "unjust"). The protagonist finds herself in the shadows, with acolytes of the MOON goddess (called Astarte), whose conflict with the male advocates of SUN worship has touches of the conflict between GODDESS and priest-ridden patriarchy common to much late-20th-century fantasy. But the protagonist soon upsets the apple cart, and by the end of the tale – after much love and violence, and the intervention of several ruthless sorcerers – a dynamic BALANCE has finally been restored.

In *The Blue Manor* (**1995**), a singleton, a man comes to the eponymous BAD PLACE and finds that the novel he is writing begins to tell its STORY and that of its generations of inhabitants; a GODGAME atmosphere permeates the tale, but proves ultimately deceptive. The sophistication of this novel is considerably greater than that of JJ's trilogy, and may point to a considerable and growing talent. [JC]

Other works: *The Webbed Hand* (**1994**) and *Firefly Dreams* (**1995**), both YA.

JONES, J(ULIA) V(ICTORIA) (1963-   ) UK writer, in the US from the mid-1980s, whose **Book of Words** sequence – *The Baker's Boy* (**1995**) and «A Man Betrayed» (**1996**), with a further volume projected – has many of the airs and devices of a conventional HIGH-FANTASY epic set in a FANTASYLAND, complete with the eponymous UGLY DUCKLING of the first volume, whose name is JACK, and a massively intriguing EDIFICE, Castle Harvell, in which the old King lies dying. [JC]

JONES, PETER ANDREW (1951-) UK painter and illustrator of fantasy and sf subjects; he has a rich, subtle colour sense and a soft, smoothly rendered painting style. He has also designed tv title sequences and backdrops. His first published art was for Puffin Books; his work has since appeared on book covers and GAMES packaging. [RT]
Further reading: *Solar Wind* (graph **1979**) with Roger and Martyn Dean.

JONES, STEPHEN (GREGORY) (1953-) UK editor, columnist, illustrator and movie publicist and consultant. He first established himself in UK fantasy fandom with his work on *Dark Horizons* 1974-6, which helped make the magazine a focal point for the still fledgling British Fantasy Society. At this time he was more productive as an artist (often in the style of Virgil FINLAY) and a poet (usually as Steven Gregory). This period gave SJ a grounding in magazine production, which he used to good effect when in 1977, with David A. Sutton (1947-), he launched FANTASY TALES, the UK's first semi-professional magazine of FANTASY and SUPERNATURAL FICTION. It modelled itself on WEIRD TALES and rapidly proved popular, winning its editors the BRITISH FANTASY AWARD seven times and the 1984 WORLD FANTASY AWARD. In 1988 it was converted into a professional magazine in paperback format. Its two spinoff ANTHOLOGIES, *The Best Horror from Fantasy Tales* (anth **1988**) and *The Anthology of Fantasy & the Supernatural* (anth **1994**), were both ed SJ with Sutton.

Through his role as publicist and promoter SJ helped Pan Books repackage their horror and fantasy lines. This included re-energizing the long-running PAN BOOK OF HORROR STORIES, which he relaunched with Clarence Paget (1909-1991) as **Dark Voices**. The first volume was a retrospective of the series, *Dark Voices: The Best From the Pan Book of Horror Stories* (anth **1990**). The series was continued with Sutton with #2 (anth **1990**), #3 (anth **1991**), #4 (anth **1992**), #5 (anth **1993**) and #6 (anth **1994**). It rapidly became a focal point for modern HORROR, with stories (original and reprinted) by many of today's leading writers. #5 won SJ and Sutton a British Fantasy Award in 1994. After #6 the series was dropped by Pan and transferred to Gollancz as *Dark Terrors* (anth **1995**), with «Dark Terrors 2» (anth 1996).

As an anthologist SJ also inaugurated the **Best New Horror** annual selection, the first 5 vols compiled with Ramsey CAMPBELL. It runs *Best New Horror* (anth **1990**), #2 (anth **1991**), #3 (anth **1992**), #4 (anth **1993**), #5 (anth **1994**), #6 (anth **1995**) and «#7» (anth 1996). The first 3 vols were assembled as *The Giant Book of Best New Horror* (cut omni **1993**); selections from #3 and #4 were assembled as *The Giant Book of Terror* (cut omni **1994**). #1 won the editors both World Fantasy Award and British Fantasy Award in 1991. SJ also compiled a popular set of thematic anthologies in the **Mammoth Book** series: *The Mammoth Book of Terror* (anth **1991**; vt *The Anthology of Horror Stories* 1994), *The Mammoth Book of Vampires* (anth **1992**; vt *The Giant Book of Vampires* 1994), *The Mammoth Book of Zombies* (anth **1993**; vt *The Giant Book of Zombies* 1995), *The Mammoth Book of Werewolves* (anth **1994**; vt *The Giant Book of Werewolves* 1995) and *The Mammoth Book of Frankenstein* (anth **1994**; vt

The Giant Book of Frankenstein 1995). His interest in the works of H.P. LOVECRAFT inspired two further anthologies: *H.P. Lovecraft's Book of Horror* (anth **1993**) with Dave Carson, a compilation of Lovecraft's favourite stories, and *Shadows Over Innsmouth* (anth **1994** US), new stories inspired by Lovecraft's classic work. All of this work was for Robinson Books, for whom SJ was editorial director 1993-5, establishing the specialist Raven Books imprint for supernatural fiction and publishing the works of Les Daniels, Dennis ETCHISON, Robert WEINBERG and Nancy Kilpatrick.

SJ has been particularly adept at packaging special-interest books. Some, like *Horror: 100 Best Books* (**1988**) with Kim NEWMAN – which won the BRAM STOKER AWARD in 1990 – and *Now We Are Sick* (anth **1991**) with Neil GAIMAN, a volume of grotesque poetry, are of novelty interest. *James Herbert: By Horror Haunted* (anth **1992**), however, is a more evaluative tribute to James HERBERT. In like vein is «Clive Barker's A-Z of Horror» (1996), a heavily illustrated but more serious overview of the horror genre tied to a tv series. SJ has co-produced with Clive BARKER several illustrated volumes related to Barker's work: *Clive Barker's The Nightbreed Chronicles* (graph **1990**), *Clive Barker's Shadows in Eden* (anth **1991**) – which won the Bram Stoker Award in 1992 – and *The Hellraiser Chronicles* (graph **1992**) (◊ HELLRAISER; NIGHTBREED [*1990*]). He also also compiled «The Vampire Stories of R. Chetwynd-Hayes» (coll 1996) (◊ R. CHETWYND-HAYES).

SJ's work as a movie publicist and consultant (and occasional director and production coordinator on tv programmes and videos) has resulted in a series of highly illustrated guides: *The Illustrated Vampire Movie Guide* (**1993**), *The Illustrated Dinosaur Movie Guide* (**1993**), *The Illustrated Frankenstein Movie Guide* (**1994**; vt *The Frankenstein Scrapbook* 1995 US) and «The Illustrated Werewolf Movie Guide» (1996). [MA]
Other works: *Gaslight & Ghosts* (anth **1988**) ed with Jo Fletcher (1958-) for the 1988 World Fantasy Convention.

JONES, TERRY Working name of UK writer, moviemaker and actor Terence Graham Parry Jones (1942-), who first came to notice as a member of MONTY PYTHON'S FLYING CIRCUS. An Oxford graduate in medieval literature, he rejected academia, though his *Chaucer's Knight: The Portrait of a Medieval Mercenary* (**1980** US) is a clear attempt to justify in academic terms the portrait suggested by the subtitle. Throughout his creative career he has constantly expressed an interest in the medieval world, notably in MONTY PYTHON AND THE HOLY GRAIL (*1975*), which he co-wrote and, with Terry GILLIAM, co-directed, and ERIK THE VIKING (*1989*), which he directed, wrote, took a major role in, and put in book form as *Erik the Viking* * (**1989**; vt *Erik the Viking: The Screenplay* 1989 US); the latter movie's credits specifically state that the screenplay was not based on TJ's *The Saga of Erik the Viking* (**1983**). He directed and acted in MONTY PYTHON'S LIFE OF BRIAN (*1979*) as well as *Monty Python's Meaning of Life* (*1983*), and wrote the screenplay for LABYRINTH (*1986*). Among other movies directed by TJ is *Personal Services* (*1987*), about the career of a brothel madam; *Consuming Passions* (*1988*) is based on a tv play he wrote with Michael Palin, *Secrets* (broadcast 1983), in which a young executive gains glory by discovering that the addition of corpses to the mix makes his company's chocolate sell better.

Of almost as much interest are TJ's books, which include: *The Complete and Utter History of Britain* * (**1969**) with Michael Palin, a tv tie; *Bert Fegg's Nasty Book for Boys and Girls* (**1974**; rev vt *Dr Fegg's Encyclopedia of All World Knowledge* 1985) with Palin; *Their Finest Hours: Underhill's Finest Hour; Buchanan's Finest Hour* (coll **1976**) with Palin, two plays; *Ripping Yarns* * (coll **1978**) and *More Ripping Yarns* * (coll **1978**), both with Palin, from the 1976-9 tv series; *Fairy Tales* (coll **1980** US; vt *Terry Jones' Fairy Tales* 1986 UK) illustrated by Michael FOREMAN; *Nicobobinus* (**1985**); *The Curse of the Vampire Socks* (coll **1988**), poetry for children; and three titles with Brian FROUD: *The Goblins of the Labyrinth* * (graph **1986**), based on LABYRINTH, *The Goblin Companion* (graph 1996) and *Lady Cottington's Pressed Fairy Book* (graph **1994**), which contains images of FAIRIES who have been pressed like leaves, and which won a Hugo AWARD. TJ also contributed to the various **Monty Python** books (◊ MONTY PYTHON'S FLYING CIRCUS).   [JG/JC]

JONSON, BEN(JAMIN) (*c*1572-1637) UK poet and dramatist. ◊ ALCHEMY.

JORDAN, ROBERT Pseudonym of US writer James Oliver Rigney Jr (1948-   ), a decorated soldier (for service in Vietnam) with a degree in physics; he writes dance reviews as Chang Lung and has written as Reagan O'Neal and Jackson O'Reilly. Apart from some CONAN sharecrops set in the world created by Robert E. HOWARD (◊ SEQUELS BY OTHER HANDS), RJ's fiction is restricted to the **Wheel of Time** sequence, only superficially a GENRE FANTASY little out of the routine save for its already considerable length. To date the sequence occupies seven large volumes – *The Eye of the World* (**1990**), *The Great Hunt* (**1990**), *The Dragon Reborn* (**1991**), *The Shadow Rising* (**1992**), *The Fires of Heaven* (**1993**), *Lord of Chaos* (**1994**) and *A Crown of Swords* (**1996**) – with at least three more volumes projected.

The back-story structuring the sequence, which is dominant and coherently presented, works as an effective TIME ABYSS, in terms of which the protagonists act out (and come to understand) their destinies, a process which extends over 1000s of pages. Millennia before the tale begins, humans had won, at great cost, a vast war against a DARK LORD: in the wake of this conflict, all male power-wielders have been cursed or "tainted" so that any man who attempts to use MAGIC afterwards will go insane. Only women, who operate a complementary form of magic, are immune. But both magics are necessary if BALANCE is to be maintained, and this unbalanced, precarious world – as the narrative begins – is clearly vulnerable to the risk that the Dark One, in BONDAGE all this while as a MALIGN SLEEPER, may re-emerge.

The **Wheel of Time** is built from conventional genre fantasy sequences, but these sequences are assembled with notable architectonic skill into an EPIC FANTASY whose momentum (despite longueurs) is very considerable. At the very start, three male and two female UGLY DUCKLING or BRAVE LITTLE TAILOR peasant principals are conscripted by those aware that they will be crucial in the forthcoming struggle with the Dark One, who is apparently soon to break free from his long bondage. The central HERO, Rand, turns out to be the AVATAR of the man who had originally immured the Dark One; he is always therefore at the edge of becoming insane, and much of the suspense of the sequence is generated through his acting out of the combined roles of MESSIAH and ACCURSED WANDERER. He, his village friends and various other COMPANIONS – there are seven in all, and

they constitute a literal as well as operational SEVEN SAMURAI grouping – all acquire additional powers and prowess as they advance through a combination of apprenticeships and NIGHT JOURNEYS; become major players in the politics of contending kingdoms (one of them is a HIDDEN MONARCH); and gradually eliminate one after another of the Forsaken, evil human henchmen of the Dark One, originally imprisoned with him but already free.

The plotting is dominated by a SECONDARY-WORLD equivalent of the FANTASY OF HISTORY. The Aes Sedai order of warrior-WITCH nuns, at once a PARIAH ELITE and SECRET GUARDIANS, have manipulated history for 3000 years and are opposed by crusader-witchfinders, the Children of Light, whose RELIGION regards all magic as evil. Also of note are the warrior people from beyond the mountain, the Aiel, paradoxically determined to expiate in military prowess the shame of abandoning their original pacifism; and the Seanchan, invaders from a colony of those driven from the major landmass in a civil war and seeking revenge. Many of these orders and races are derived from various sources; RJ draws material with particular frequency from CELTIC FANTASY, and from the ARTHUR in particular. When complete, the sequence will almost certainly constitute one of the major epic narratives of modern fantasy.

RJ's faults are obvious. He is not a careful writer ("Egwene's stomach sank into her feet" – *The Fires of Heaven*), and he clearly writes long rather than short. At times, his characters seem to be endlessly in transit from one PLOT COUPON to the next, in constantly (but not germanely) changing combinations of PLOT DEVICES. And his sense of moralized LANDSCAPE rarely transcends the commonplace.

But the **Wheel of Time** conveys, as a whole, a surprising emotional charge. This may stem in part from RJ's repeated plumbing of the relationship between humiliation and hierarchy, a scrutiny which is appropriate when applied to a nostalgic and conservative genre like fantasy, and which seems over-worked only when he deals with the internal discipline of the Aes Sedai, who are oddly fond of corporal punishment. Furthermore, a sense of an intelligent creative enterprise is sustained throughout the sequence, most notably perhaps in the ingenuity with which standard plot devices, backgrounds and characters are subjected to constant and sophisticated modification; and in the overall sense RJ gives that he is in full control of his enormous narrative structure. Later volumes of the sequence, though they remain stylistically awkward at times, increasingly demonstrate the pleasures to be derived from RJ's mastery of long-breathed storytelling.   [RK/JC]

Other works: the **Conan** sequence: *Conan the Invincible* * (**1982**), *Conan the Defender* * (**1982**) and *Conan the Unconquered* * (**1983**), all three assembled as *The Conan Chronicles* * (omni 1995); *Conan the Triumphant* * (**1983**); *Conan the Destroyer* * (**1984**); *Conan the Magnificent* * (**1984**); *Conan the Victorious* * (**1984**).

JOSHI, S(UNAND) T(RYAMBAK) (1958-   ) US critic, scholar and bibliographer who has concentrated for much of his career on the works and life of H.P. LOVECRAFT. Books on Lovecraft by STJ include *H.P. Lovecraft* (**1982** chap), *H.P. Lovecraft and Lovecraft Criticism: An Annotated Bibliography* (**1981**) and the ambitious *H.P. Lovecraft: The Decline of the West* (**1990**), which attempts to argue a significant connection between 20th-century currents of thought in the Western world, Lovecraft's own ideas (he was a materialist, an atheist, and often accused of antisemitism) and the fiction. A book of this sort must almost certainly constitute an advocacy of its subject's opinions; STJ is sometimes overemphatic in this respect. Further titles include *An Index to the Selected Letters of H.P. Lovecraft* (**1991** chap) and *An Index to the Fiction and Poetry of H.P. Lovecraft* (**1992** chap). Books about Lovecraft edited by STJ include *Lovecraft: Four Decades of Criticism* (anth **1980**), *Selected Papers on H.P. Lovecraft* (anth **1989**), *The H.P. Lovecraft Centennial Conference Proceedings* (anth **1991** chap), *An Epicure in the Terrible: A Centennial Anthology of Essays in Honor of H.P. Lovecraft* (anth **1991**) with David E. Schultz, and *Caverns Measureless to Man: 18 Memoirs of H.P. Lovecraft* (anth **1996**). He has in addition edited both the journal *Lovecraft Studies* and the more general *Studies in Weird Fiction* (1986-current), from their inception; both are published by Necronomicon Press, which specializes in Lovecraftiana. Lovecraft collections ed STJ include *H.P. Lovecraft: Autobiographical Writings* (coll **1992**), *H.P. Lovecraft: Letters to Robert Bloch* (coll **1993** chap), *H.P. Lovecraft in the Argosy: Collected Correspondence from the Munsey Magazines* (coll **1994** chap), *The H.P. Lovecraft Dream Book* (**1994** chap) with David E. Schultz, *H.P. Lovecraft: Letters to Samuel Loveman & Vincent Starrett* (coll **1994** chap) with Schultz and *Miscellaneous Writings by H.P. Lovecraft* (coll **1995**).

STJ has more recently focused on two other writers, Lord DUNSANY and Ramsey CAMPBELL. *Lord Dunsany: A Bibliography* (**1993**) with Darrell SCHWEITZER is competent and thorough. *Lord Dunsany: Master of the Anglo-Irish Imagination* (**1995**) represents almost the first serious attempt to put the life and prolific career of Dunsany into a perspective by which it may be possible to begin to understand his early fame and subsequent oblivion. *The Count of Thirty: A Tribute to Ramsey Campbell* (anth **1993** chap) and *The Core of Ramsey Campbell: A Bibliography & Reader's Guide* (**1995** chap) with Campbell and Stefan DZIEMIANOWICZ are extremely useful.

In a theoretical study, *The Weird Tale* (**1990**), STJ attempts to describe the work of several writers – including Ambrose BIERCE, Algernon BLACKWOOD, Dunsany, M.R. JAMES and Arthur MACHEN – as precursors of Lovecraft; the individual studies, though, are acute. [JC]

JOURNEY BACK TO OZ (*1971/74*) ◊ *The* WIZARD OF OZ.
JOURNEY TO THE UNKNOWN UK/US tv series (1968-9). HAMMER/20th Century-Fox/ABC. **Pr** Anthony Hinds. **Exec pr** Joan Harrison. **Created by** Harrison. **Based on** stories by Charles BEAUMONT, Robert BLOCH Richard MATHESON and others. 17 1hr episodes. Colour.

This anthology series featured a number of stories of the supernatural. Unfortunately, many of the plot elements were rather routine; e.g., a girl can predict death through DREAMS, a department store mannequin comes to life (◊ PYGMALION), a man is seen in the crowd just before several disasters, a woman programs a computer to kill her husband. [BC]

JOVE ◊ ZEUS.
JOYCE, GRAHAM (WILLIAM) (1954-   ) UK writer whose fiction has been marketed as HORROR but which mostly constitutes a set of deft DARK FANTASIES. His first novel, *Dreamside* (**1991**), follows an experiment in shared dreaming (◊ DREAMS), with grave consequences over the years for the college students involved. *Dark Sister* (**1992**), which won the 1992 BRITISH FANTASY AWARD, conflates WITCHCRAFT, a BAD PLACE and a DOUBLE in a family romance. *House of Lost Dreams* (**1993**), opening out some-

what, takes its protagonists to Greece where they embark (as it seems) upon a reenactment of the ORPHEUS and Eurydice myth, but are diverted into a testing of life-roles with a MAGUS-figure whose enigmatic behaviour is rather GODGAME-like, though ultimately not quite; comparisons with John FOWLES's *The Magus* (**1966**; rev 1977) are inevitable. *Requiem* (**1995**) – though its protagonist's discovery of a Dead Sea Scroll telling a different version of the life of CHRIST hints at a FANTASY OF HISTORY – turns out to be a psychological study.

GJ is a writer of skill, and gives a sense that he is tracing a course of discovery through the genres he visits. It will be intriguing to see if he finds a home. [JC]

JOYCE, JAMES (AUGUSTINE) (1882-1941) Irish writer, long resident in Paris, whose impact on 20th-century literature, especially with *Ulysses* (**1922** France), is perhaps unsurpassed. The influence of JJ upon fantastic and other literature includes his technical innovations and his extreme brilliance of language. This influence is broad but diffuse: many writers have felt it without having encountered his work other than glancingly. Richard Ellmann's observation that "We are still learning to be James Joyce's contemporaries" points up the unassimilability of his genius: *Ulysses* is still, 75 years later, a radical text.

Finnegans Wake (1924-32 var chap; fixup **1939** UK) is the work of JJ's that can most usefully be considered a type of fantasy: although Modernist critics long sought to identify it as an essentially "realistic" novel, describing in subjective terms the dream-life of a sleeping Irish publican on a single night, more recent critics have seen it (as well as the later sections of *Ulysses*) as radically post-narrative, atomizing both character and event in a manner that has much, in various ways, in common with myth, fable and phantasia, but very little in common with the Victorian "Great Tradition" in narrative literature, which JJ wished to explode.

Much of JJ's influence has been technical and superficial: the dream-frame of Earwicker's "nightmaze" has informed the structure of numerous other fantasies, including Norman MAILER's abortive cycle of dream-novels, just as the Wake's circular form is imitated by such Postmodernist works as Samuel R. DELANY's *Dhalgren* (**1975**). Brian ALDISS's *Barefoot in the Head* (fixup **1969**) makes more sophisticated use of JJ's stylistic innovations, while James BLISH's "Common Time" (1953) attempts to use JJ's combined strategy of interior monologue, ironically transfigured QUEST tale and atomized individuation, all within the compass of a novelette. Richard GRANT's *Views from the Oldest House* (**1989**), like Thomas PYNCHON's *Gravity's Rainbow* (**1973**), presents narrative consciousness unmoored from ego and dissolving through the landscape, a more significant use of JJ's imaginative universe. That all these works are at least debatably sf suggests that modern fantasy writers, though they have made use of some of JJ's more detachable and less disturbing innovations, are not anxious to become JJ's contemporaries. [GF]

Other works: *Dubliners* (coll **1914** UK); *A Portrait of the Artist as a Young Man* (**1916** US).

J.R.R. TOLKIEN'S THE LORD OF THE RINGS vt of *The* LORD OF THE RINGS (*1978*). ◊◊ Ralph BAKSHI.
JUDGEMENT DAY ◊ LAST JUDGEMENT.
JUNGIAN PSYCHOLOGY The theory of "analytic psychology" argued by Carl Gustav Jung (1875-1961) may not be as central to the 20th century as that advocated by his teacher and rival, Sigmund Freud (1856-1939), but Jung's

essentially narrative delineation of the contours of the human psyche – and of the route each human must follow to attain full maturity – has proved enormously suggestive for writers of fantasy.

Jung's career lasted from the beginning of the century until his death, and his terminology and concepts inevitably shifted over this extended period; any of his central texts – e.g., *Wandlungen und Symbole der Libido* (**1912**; trans Beatrice M. Hinkle as *Psychology of the Unconscious* **1916** US; 4th ed much rev vt *Symbole der Wandlung* 1952; this version trans R.F.C. Hull, with author's further revs, as *Symbols of Transformation: An Analysis of the Prelude to a Case of Schizophrenia* 1956 US) – underwent significant alterations. This may have caused some difficulty to his colleagues; but creative writers in search of evocative images have, naturally, paid relatively little attention to any inconsistencies.

Very roughly, in JP the conscious surface structure of the mind or ego coats or (depending on the metaphor) orbits a central inner self, which is allied to the "collective unconscious". In turn, the dark, profound, beckoning, populous "collective unconscious" registers on the self – rather in the way that the MYTHAGOES of Robert P. HOLDSTOCK's **Ryhope Wood** sequence, which owes much to Jung, register on characters who orbit the woods – through the agency of protean but persistent images, whose shape is determined by a kind of imploring *conversation* between the ego/self and the larger inner world.

But, before travelling INTO THE WOODS – into the region where one's ego/self may conduct an integrative conversation with that greater community within – one must learn how to recognize some aspect of one's self capable of guiding the conscious ego inward. As Ursula K. LE GUIN suggests in "The Child and the Shadow" (1975), this guiding aspect of the ego/self might be thought of as a kind of LIMINAL BEING at the THRESHOLD to the interior, a being which takes the shape of that which the surface ego has repressed, but which the self, Antaeus-like, continues to generate – all the more powerfully when that denied shape has been stringently repressed. Jung gives an explicit name to this "being": as Le Guin phrases it, "The first step is often the most important, and Jung says that the first step is to turn around and follow your own shadow."

This SHADOW, Le Guin suggests, may have appeared in literature in many guises: Cain, CALIBAN, Enkidu (the TRICKSTER friend of GILGAMESH), FRANKENSTEIN's MONSTER, Gollum (Frodo's shadow in J.R.R. TOLKIEN's *The Lord of the Rings* [**1954-1955**]), DANTE's VIRGIL, etc. More abstractly, it is the DOPPELGÄNGER who haunts the dreams of the ego; the DOUBLE who stares back through the MIRRORS of the mind's eye. It is the bright-eyed TALKING ANIMAL that does not tell a lie. It attends the WRONGNESS that warns the protagonist the world is becoming untuned; it also attends the THINNING that desiccates the LAND which has forgotten its true nature, the fading out of and retreat from REALITY that wastes the protagonist who has forgotten his full self (◊◊ AMNESIA).

First there is a NIGHT JOURNEY to make. The child must follow the shadow through the gates (◊ ORPHEUS) into the interior, where can be found deep ARCHETYPES. These figures or clusters of imagery include the ANIMA, which is the feminine side of a man's nature (the analogous masculine side of a woman's is the animus), and which may manifest itself as a LAMIA, MUSE, crone or – certainly for many fantasy writers – as the GODDESS entire (Jung was less fertile in

describing aspects of the animus). Other archetypes include the father, who is also the MAGUS, and the mother. It is only here, where the SYMBOLS of TRANSFORMATION can be recognized and sorted, that the integration of self and shadow can be accomplished, perhaps in the end through an experience of METAMORPHOSIS. In fantasy tales this metamorphosis may well be literal.

Later Jungian explorations into ALCHEMY and the works of PARACELSUS may enrich this central narrative, but do not significantly alter it. Jung was much less interested in what he called exoteric alchemy – in attempts at TRANSMUTATION – than in what he referred to as esoteric alchemy: secret studies in the path upwards into wholeness and true being. His ideas on UFOS – very roughly, that they are mental constructs essential to human belief processes – form a PLAYGROUND that fantasy writers have been rather slow to exploit; however, for example, *The Flipside of Dominick Hide* (**1980** tvm) by Alan Gibson and Jeremy Paul, though generally regarded as maverick sf, can also be read as a TECHNOFANTASY commentary on Jungian ideas. [JC]

JUNGLE BURGER (*1975*) ◊ TARZAN MOVIES.

JUNGLE JIM MOVIES ◊ JUNGLE MOVIES.

JUNGLE MOVIES Hollywood's love-affair with the LAND-OF-FABLE jungle began properly with the TARZAN MOVIES, the first of which was *Tarzan of the Apes* (**1918**). The most famous run of these began with *Tarzan, The Ape Man* (**1932**), starring Johnny Weissmuller and Maureen O'Sullivan, but in fact this movie represented a desperate attempt by MGM to get back some of the money they had expended on what seems to have been the first JM talkie, *Trader Horn* (**1930**), an Oscar nominee but a commercial disaster; wildlife scenes from *Trader Horn*, shot expensively on location in Africa, were recycled throughout the MGM sequence of **Tarzan** movies. The first major offshoot from the Weissmuller **Tarzan** movies was the sequence of BOMBA MOVIES, starting with *Bomba the Jungle Boy* (**1948**), starring Johnny Sheffield, who had played Tarzan's son Boy. Almost immediately afterwards Weissmuller, now well into middle age, likewise sidled out of the series to become the star of the **Jungle Jim** movies, where he played not an apeman but a white man adventuring in the jungle. The first of these, *Jungle Jim* (**1948**), was a remake of *Jungle Jim* (**1937**); other titles were *The Lost Tribe* (**1949**), *Captive Girl* (**1950**), *Mark of the Gorilla* (**1950**), *Pygmy Island* (**1950**), *Fury of the Congo* (**1951**), *Jungle Manhunt* (**1951**), *Jungle Jim in the Forbidden Land* (**1952**), *Voodoo Tiger* (**1952**), *Killer Ape* (**1953**), *Savage Mutiny* (**1953**), *Valley of the Head-Hunters* (**1953**), *Cannibal Attack* (**1954**), *Jungle Maneaters* (**1954**), *Devil Goddess* (**1955**) and *Jungle Moon Men* (**1955**). These movies were undistinguished.

But the same is not true of all JMs: two of distinction are *The African Queen* (**1951**) and *Gorillas in the Mist* (**1988**). The difference is that neither is concerned with the anachronistic land-of-fable jungle. The former uses the jungle merely as backdrop for the tale of a spiky relationship; the latter is almost a documentary. [JG]

JUPITER ◊ ZEUS.

JUSTER, NORTON (1929-) US writer and architect whose reputation rests on his first book, *The Phantom Tollbooth* (**1962**), a fine CHILDREN'S FANTASY. It describes the adventures of Milo, a depressed young boy who comes home to find a cardboard tollbooth which is a PORTAL to a WONDERLAND. He visits the warring cities of Dictionopolis (ruled by Azaz the Unabridged) and Digitopolis (ruled by

Azaz' brother the Mathemagician), and sets forth on a QUEST to bring home the royal princesses Sweet Rhyme and Pure Reason, so that the country of Wisdom will be at peace. The wonderland venue, the unflagging verbal invention and the ongoing play with mathematics all recall Lewis CARROLL. The movie version, *The Phantom Tollbooth* (*1969* US), containing live action and – mostly – animation, is disappointing; although it has some superb moments, too often it seems like an extended educational section from *Sesame Street*.

None of NJ's later books have the same stature. *The Dot and the Line: A Romance in Lower Mathematics* (**1963** chap) is a heavily (and cleverly) illustrated picture-book love story; it was made into an animated short. *Alberic the Wise and Other Journeys* (coll **1965** chap) comprises three slight but charming didactic fables. *AS: A Surfeit of Similes* (**1989**) is a Dr SEUSS-like exploration of the figure of speech of the title.

[DK]

JUST OUR LUCK US tv series (1983-4). Lorimar/ABC. **Pr** Bob Comfort, Charles Gordon, Les Gordon, Rick Kellard, Linda Morris, Victor Raueso. **Exec pr** Ronald E. Frazer. **Dir** John Astin, Alan Bergmann, Bruce Bilson, Bob Sweeney. **Writers** James Berg and many others. **Created by** Charles Gordon, Lawrence Gordon. **Starring** T.K. Carter (Shabu), Richard Gilliland (Keith Barrow). 11 30min episodes. Colour.

This sitcom version of the legend of Aladdin (◊ ARABIAN FANTASY) and the magic lamp took an unusual twist by pairing a black GENIE with a white master. To start, tv weatherman Barrow breaks a souvenir bottle and is forced to buy it. Inside is Shabu, a genie who has been trapped there for 196 years. The genie is unhappy with his master, for he feels that Keith is too afraid to have fun. In one episode, for example, Keith refuses to let Shabu use MAGIC to pay for a vacation to Las Vegas, so Shabu splits himself into a quartet to win money on a tv quizgame.

The NAACP felt it was objectionable to show a black man serving a white master, and the stories were changed as a result.

[BC]

KA ◊ SHADOWS.
KAANGA, JUNGLE LORD ◊ TARZAN.
KABBALA ◊ CABBALA.
KABINETT DES DR CALIGARI, DAS ot of *The* CABINET OF DR CALIGARI (**1919**).
KADARE, ISMAIL (1936-   ) Albanian novelist. ◊ EDIFICE.
KADATH Two SMALL-PRESS fantasy MAGAZINES named for the Antarctic city of the CTHULHU MYTHOS.

1. US magazine, 1 issue 1974, published and ed Lin CARTER. Slim (24pp) but expensive, *K* had negligible circulation and is almost unknown. The only item of interest is a new story by Hannes BOK, "Jewel Quest".   [MA]

2. Italian large-format glossy magazine, 6 issues, irregular, October 1979-Fall 1984, published and ed Francesco Cova, Genoa. Although issues *#1-#2* were mostly in Italian, they had small English-language sections; issues *#4-#6* were entirely in English. The magazine ran a mixture of Lovecraftian HORROR (◊ H.P. LOVECRAFT) and LOW FANTASY, mostly produced by Cova's UK and US supporters. From *#3*, issues became thematic: Brian LUMLEY (November 1980), a WEIRD TALES issue (July 1981), OCCULT DETECTIVES (July 1982) and HEROIC FANTASY (Fall 1984). *K* was beautifully produced, but the fiction was of variable standard. The magazine's atmosphere of excitement and rediscovery always prevailed.   [MA]
KAFKA, FRANZ (1883-1924) Czech writer, of German-Jewish background, who wrote in German. Along with Jorge Luis BORGES he is of central importance to the literature of the Western World and of seminal significance to the evolution of fantasy. No more than Borges, however, was FK a creator of SECONDARY WORLDS or of tales with CROSSHATCH structures; in FK's work, anything that has the semblance of a THRESHOLD will likely be revealed as some façade or cul-de-sac in a surrealistic maze of bureaucracy, a LABYRINTH through which there is no passage. If fantasy is a literature which resolves anxiety through the medicament of STORY, through the assurance that – at least in imagination – the REALITY of other worlds can be trusted, then FK is not a writer of fantasy. Some of his many stories are among the most brilliant FABLES and parables ever writ-

ten, but – significantly, for one of the salient features of the fantasy story is that it generally *ends* – the best are fragments, haunted and anxiety-ridden strobes into the darkness. There is a sense that, as with Borges, FK's vision of the world can be understood as gnostic (◊ GNOSTIC FANTASY): the material world may be an error, a falling from true being, but if there ever was a GOD from whom we fell there is none now; the Pleroma (which has some distant affinity to the more *real* REALITY many fantasy writers – e.g., C.S. LEWIS – describe as intensifying their secondary worlds) is beyond reach, or null. The world is recursive, and it refers back only to itself. If FK can be understood as a religious writer, his religion addresses a Universe from which God has been evacuated.

To fantasy writers and readers, the four main FK texts are probably: *Die Verwandlung* (**1915** Germany; trans A.L. Lloyd as *The Metamorphosis* chap **1937** UK), whose protagonist suffers TRANSFORMATION into a cockroach; *Amerika* (first chapter as *Der Heizer* ["The Stoker"] **1913** chap Germany; remainder written 1911-14; **1927** Germany; trans Willa and Edwin Muir as *America* **1938** UK), whose protagonist (Karl) undergoes PICARESQUE adventures in a theatrical LAND-OF-FABLE USA. *Der Prozess* (written 1914-15; **1925** Germany; trans Willa and Edwin Muir as *The Trial* **1937** UK; exp rev with new material trans E.M. Butler 1956), whose protagonist (K.) is arrested one morning and subjected to profound anxiety and guilt as his "case" progresses unfathomably through an arcane legal bureaucracy in the direction of his inevitable death; and *Das Schloss* (written 1921-2; **1926** Germany; trans Willa and Edwin Muir as *The Castle* **1930** UK; exp rev with new material trans Ernst Kaiser and Eithne Wilkins 1953), whose rather more formidable CONFIDENCE-MAN protagonist (again named K.) constantly fails in his attempts to have his existence recognized by the authorities in the eponymous Castle to which he believes he has been summoned. All except *The Metamorphosis* were left incomplete at FK's death; they were published by his executor, the writer Max Brod (1884-1968), who fortunately ignored instructions to destroy the manuscripts. Movie versions have appeared as *The* TRIAL (**1962**), *The Castle* (**1969**) and *The* TRIAL (**1992** tvm).

FK did, however, publish much of his shorter work. *Betrachtung* ["Meditation"] (coll **1913** Germany) contains mostly short sketches. *Das Urteil* ["The Judgement"] (**1913** chap Germany) is a parable in which a father orders a son to kill his friend and DOUBLE, which he does. The title story of *Ein Landarst* ["A Country Doctor"] (coll **1919** Germany) is a brilliant and horrifying parable in which – in the manner of a CONTE CRUEL – nightmarish immensities of misfortune attend a doctor's response to a call, until at the end he has become an ACCURSED WANDERER. *In der Strafkolonie* ["In the Penal Colony"] (**1919** chap Germany) features a machine designed to etch a description of his crime into the flesh of the prisoner, but which merely causes death. *Ein Hungerkünstler* ["A Hunger Artist"] (coll **1924** Germany) contains four exemplary tales, each profoundly ironic. These fictions – plus a new translation of *Metamorphosis*, as "The Transformation" – were assembled as *The Penal Colony: Stories and Short Pieces* (coll trans Willa and Edwin Muir **1948** US; vt with slightly differing contents *In the Penal Settlement: Tales and Short Prose Works* **1949** UK); a similar coverage of FK's stories is presented in *The Transformation and Other Stories: Works Published During Kafka's Lifetime* (coll trans Malcolm Pasley **1992** UK).

An important posthumous collection, *Beim Bau der Chinesischen Mauer* (coll **1931** Germany; trans Willa and Edwin Muir as *The Great Wall of China and Other Pieces* **1933** UK), contains the BEAST FABLE "Forschungen eines Hundes" (here trans as "Investigations of a Dog"), the famous title story, and a number of parables and fables. The book was later issued, with additional material, as *Description of a Struggle, and The Great Wall of China* (coll with new material trans Tania and James Stern **1960** UK). Further posthumous material appears in *Hochzeitsvorbereitungen auf dem Lande* (coll **1953**; trans Ernst Kaiser and Eithne Wilkins as *Wedding Preparations in the Country and Other Posthumous Prose Writings* **1954** UK).

FK's oeuvre seems to trail off, but fragmentation is at the heart of his vision of a duration without meaning or God or stop.   [JC]

KAIDAN vt of KWAIDAN (*1964*).

KAIJU DAISENSO (vt *Battle of the Astros*; vt *Invasion of Planet X*; vt *Invasion of the Astro Monster*; vt *Monster Zero*) Japanese movie (*1965*). ◊ GODZILLA MOVIES.

KAIJU SOSHINGEKI ot of *Destroy All Monsters* (*1968*). ◊ GODZILLA MOVIES.

KALEVALA "Land of Heroes" – the TAPROOT TEXT of Finnish MYTH, an epic SONG-cycle assembled from ancient oral tradition. The definitive version is ed Elias Lönnrott (**1835**; exp 1849; trans W.M. Crawford **1887** UK), running to 22,800 lines. Its background is of magic-laden conflict between Kalevala (Finland) and Pohja or Pohjola (Lapland); the latter is ruled by the ENCHANTRESS Louhi, who commands storms and deep cold, and at one stage imprisons the SUN. Principal heroes include Väinämöinen, who is specially adept in the universal Kalevala MAGIC of MUSIC, SONG and the gaining of power over persons and things by knowing their genealogy or origins (◊ TRUE NAME); his brother the SMITH Ilmarinen, forger of the coveted *sampo* (apparently a magic mill that creates whatever one desires); and the rash seducer and TRICKSTER figure Lemminkäinen. Tuonela, the Kalevala's UNDERWORLD, is a gloomy HADES rather than a place of punishment; the swan of its RIVER Tuoni, which Lemminkäinen fatally attempts to kill for Louhi, is commemorated in the symphonic tone-poem "The Swan of Tuonela" by Jean Sibelius (1865-1957). L. Sprague DE CAMP and Fletcher PRATT take their **Incomplete Enchanter** to the Kalevala world in *Wall of Serpents* (**1960**), whose action echoes Lemminkäinen's expedition against Pohjola in revenge for not being invited to Louhi's daughter's wedding; the need to know the relevant ancestry before working magic is used as a PLOT DEVICE. The *Kalevala* sequence by Emil Petaja (1915-   ) is sf. Michael Scott ROHAN's **The Winter of the World** makes Louhi one of the chief powers behind a threatened ice age; the smith hero opposing the Ice emerges as an AVATAR of Ilmarinen. Ian WATSON's **The Books of Mana** offer a witty, inventive and exuberant SCIENCE-FANTASY reworking of numerous Kalevala strands, with such character names as Lucky (Louhi), van Maanen and Minkie Kennan.   [DRL]

KALUTA, MICHAEL WILLIAM (1947-   ) US fantasy COMIC-book artist and illustrator, with a fine line style; much of his work is inspired by 1930s and 1940s pulp illustration, and in many respects stylistically resembles it. MWK has been influenced by Roy G. KRENKEL, Charles VESS, Al WILLIAMSON, Bernie WRIGHTSON and other leading artists with whom he has worked.

MWK's first comics work was a four-page strip published in *Flash Gordon #18* (1970), although he had assisted Wrightson on the artwork for two strips in *DC Showcase* (*#83* and *#84* 1969). His other important early strip work included a number of Burroughsian pieces in *Korak* (*#46-#56* 1972-3), TARZAN (*#230* 1974), *Weird Worlds* (*#4* 1973) and *Tarzan Family* (*#60-#65* 1975-6), plus a short series on *The Shadow* (*#1-#6* 1973-4) and subsequently *The Shadow Graphic Novel* (graph **1988**), a three-part *Spawn of Frankenstein* tale (*Phantom Stranger #23-#25* 1973), and some short fantasy tales in *Superman* (*#240* 1971), *Witching Hour* (*#7* 1970), *House of Mystery* (*#185*, *#195*, *#200*, *#221* 1971-4), *House of Secrets* (*#87*, *#98* 1970, 1972) and *Doorway into Nightmare* (*#1-#5* 1978).

For the next 10 years most of MWK's work was in the form of comic-book covers for *Batman*, *Detective*, *House of Mystery*, *House of Secrets*, *Time Warp* and *Conan the King*, plus spot illustrations for adaptations of Robert E. HOWARD's *Lost Valley of Iskandar* (graph **1974**) and *Swords of Shaharazar* (graph **1976**), and in comic books.

He became involved in the designing and building of stage sets for the 1980 and 1982 off-Broadway productions of Elaine Lee's **Starstruck**, and subsequently produced several comic-strip pieces based around **Starstruck** themes for HEAVY METAL (November 1982-February 1983) and *Starstruck* (*#1-#6* 1985-6).

For many years MWK has continued his association with the 1930s pulps crimefighting character **The Shadow**, coscripting and producing covers for a number of **Shadow** comic books and drawing the **Shadow** graphic novel *Hitler's Astrologer* (1988). Many portfolios of his work have been published.   [RT]

Other works: *Abyss* (graph **1970**); *Flash Gordon* (graph **1971**); *Edgar R. Burroughs* (graph **1974**); *Phantasmagoria* (graph **1974**); *Wet Dreams* (graph **1975**); *Dante's Inferno* (graph **1975**); *MWK Fantasy Portfolio* (graph **1978**); *Children of the Twilight* (graph **1979**); *Heroes; Heavies and Heroines* (graph **1980**); *Starstruck* (graph **1980**); *Friends of Ol' Gerber* (graph **1982**) and *Bird of Death* (graph **1984**).

Further reading: *The Studio* (**1979**) by MWK; *The Michael Wm Kaluta Treasury* (**1988**) by MWK; *The Mike Kaluta Sketchbook* (**1993**) by MWK.

KANE, BOB (1916-   ) US COMIC-book artist renowned as co-creator, with writer Bill Finger, of BATMAN. BK joined the Eisner-Eiger comic-book workshop in 1937, drawing gag cartoons and a comedy adventure strip called *Peter Pupp*. His first work published by National (later DC COMICS) included the two-page filler "Oscar the Gumshoe" for *Detective Comics* and the Milton Caniff-influenced "Rusty and his Pals", written by Finger, for *Adventure Comics*. The remainder of his professional life, however, has been associated with BATMAN.   [RT]
Further reading: *Batman & Me* (**1989**) by BK and Tom Andrae.

KANGILASKI, JAAN (1936-   ) US author of the **Seeking Sword** sequence – *The Seeking Sword* (**1977**) and *Hands of Glory* (**1981**) – about an ancient Berserker SWORD which is possessed (◊ POSSESSION) by the spirit of its first owner and seeks revenge through the centuries.   [MA]

KARR, PHYLLIS ANN (1944-   ) US writer, noted also for Arthurian scholarship, in which field she has compiled *The King Arthur Companion* (**1983**) and written the Arthurian murder-mystery *The Idylls of the Queen* (**1982**).

PAK's first professional sales were historical mysteries, to *Ellery Queen's Mystery Magazine* in 1974. In fantasy, she created the character **Torin the Toymaker**, a magician turned toymaker, starting with "Toyman's Trade" (1974 *Literary Magazine of Fantasy & Terror*). None of the **Torin** stories has been collected in book form, but *At Amberleaf Fair* (**1986**) sets him on a wider canvas. The **Torin** world is far removed from the heroics of HIGH FANTASY, being generally bucolic and PASTORAL.

PAK's series about **Frostflower and Thorn** are straight-forwardly HEROIC FANTASY, with a strong feminist slant (◊ FEMINISM). Frostflower is a sorceress who in the first book, *Frostflower and Thorn* (**1980**), helps the female warrior Thorn through a rapid and unconventional pregnancy that leads to their banishment. Each lives by her own code, which causes friction between them and the people they encounter in their travels. *Frostflower and Windbourne* (**1982**) explores their relationship further. Other stories have appeared in magazines and anthologies.

Although PAK is most closely associated with fantasy, she has also written several romantic novels, including a completion of Jane Austen's *Lady Susan* (**1980**).   [MA]
Other works: *Wildraith's Last Battle* (**1982**).

KASHCHEI ◊ KOSHCHEI.

KÄSTNER, ERICH (1899-1974) German writer, best-known for his classic children's book *Emil und die Detektive* (**1929**; trans as *Emil and the Detectives* **1931** UK). His other books include *Die Konferenz Der Tiere* (**1949**; trans as *The Animals' Conference* **1953** UK), a biting SATIRE on Man's inability to achieve lasting peace, in the mode of George ORWELL's *Animal Farm* (**1945** chap).   [RD]

KATHA SARIT SAGARA ◊ OCEAN OF STORY.

KATZ, WELWYN (WILTON) (1948-   ) Canadian writer. Her first novel of interest, *The Prophecy of Rau Ridoo* (**1982**), was a YA fantasy set in a SECONDARY WORLD that has long been in the thrall of dark forces; with the help of a WITCH, the young protagonists cure the LAND. More typical of WK's work in general is her second novel, *Witchery Hill* (**1984**), set in this world, and involving an invasive use of WITCHCRAFT. In *Sun God, Moon Witch* (**1986**), witchery and the powers of good are opposed on a more cosmic scale. *False Face* (**1987**) treats a magic Indian MASK in terms of its effect on this world. *The Third Magic* (**1988**) carries its

young contemporary protagonist, Morgan Lefevre (◊◊ MORGAN LE FAY), into a full secondary world.

WK's later novels – except *Time Ghost* (**1995**), which is sf – are SUPERNATURAL FICTIONS. Though acutely told, they depend too frequently on the effects of incursive EVIL on the mundane world.   [JC]
Other works: *Whalesinger* (**1990**); *Come Like Shadows* (**1993** US).

KAUFFER, EDWARD McKNIGHT (1890-1954) US illustrator, in the UK 1914-40, where he became well known for commercial work, notably as a main designer of posters for London Underground. His illustrations for the Earl of Birkenhead's nonfiction *The World in 2030 AD* (**1930**) are typical in their adroit use of Modernist techniques like Cubism to give representational images a sense of sustained kinetic drive. This strategy does somewhat less well for fantasy, though he did work of interest for editions of Miguel de CERVANTES's *Don Quixote* (1930), W.H. HUDSON's *Green Mansions* (1944) and Edgar Allan POE's *Complete Poems and Stories* (coll 1946).   [JC]

KAVAN, ANNA Name used by French-born UK writer Helen Woods (1901-1968), initially for an autobiographical character in two early novels written under her first married name, Helen Ferguson: *Let Me Alone* (**1930**) and *A Stranger Still* (**1935**). Later she took the name AK by deed-poll, and all her work from *Asylum Piece and Other Stories* (coll **1940**) was published thus. In this collection – and in *I Am Lazarus: Short Stories* (coll **1945**), *A Bright Green Field and Other Stories* (coll **1958**), *Julia and the Bazooka* (coll **1970**) and *My Soul in China* (coll **1975**), both posthumous colls ed Rhys Davies – she used the language of SURREALISM, fantasy and sf to articulate worlds which corresponded to the daemonic visions caused by her heroin addiction, and by her violently unhappy early life. That life is also reworked in fantasy terms in *The House of Sleep* (**1947** US; vt *Sleep has his House* 1948 UK), in which a NIGHT JOURNEY into the self generates baroque inscapes. Similarly, *Who Are You?* (**1963**) is set in a tropic environment whose inhabitants behave as though they were in HELL; *Ice* (**1967**) reshapes a future ice age to like effect. *Eagles' Nest* (**1957**), which features an absurdist CHILDE on a desert QUEST for the eponymous DARK TOWER, reads as deeply bleak fantasy, though again it tends to fall back into the pain of this world.   [JC]
Other works: *Mercury* (**1994**).

KAVENEY, ROZ (1949-   ) UK writer, editor and critic, best-known for sf criticism, though much of her work since the early 1980s has been in fantasy. She is a Contributing Editor to this encyclopedia. Her first fiction of genre interest was "A Lonely Impulse" in *Temps: Volume One* * (anth **1991**) "devised by" Neil GAIMAN and Alex Stewart. Works edited include: *Tales from the Forbidden Planet* (anth **1987**) and *More Tales from the Forbidden Planet* (anth **1990**); the **Weerde** sequence of SHARED-WORLD anthologies comprising *The Weerde: Book 1* * (anth **1992**) and *The Weerde: II* * (anth **1993**), both with Mary GENTLE, and focusing on a clutch of SHAPESHIFTERS who live WAINSCOT lives on Earth while infiltrating our legends and history; and *Villains!* * (anth **1992**), also with Gentle.   [JC]

KAY, GUY GAVRIEL (1954-   ) Canadian writer who worked for a time during the 1970s as assistant to Christopher Tolkien in editing J.R.R. TOLKIEN's *The Silmarillion* (**1977**). GGK's own fantasies, however, are original in conception and execution, and he seems to have learned from the experience of working on *The Silmarillion* how to combine worldbuilding with the narrative vigour of mythic plot structure, giving his work a strength and intelligence missing from most sub-Tolkienian epics. In terms of his scene-setting and use of dialogue he probably learned as much by producing and writing the tv series *The Scales of Justice* (1982-9) for CBC Canada. These lessons are most clearly displayed in his first work, **The Fionavar Tapestry**: *The Summer Tree* (**1985**), *The Wandering Fire* (**1986**) and *The Darkest Road* (**1986**).

The story tells of five contemporary US students transported to the world of Fionavar, where they find themselves amid a conflict between a patriarchy and a matriarchy. At first the mythic figures whom the five encounter seem an ill-matched mishmash of characters from a host of legendary sources, notably British and Norse myth. However, Fionavar is the core world from which others have sprung, a sort of Platonic Ideal, and our own myths and legends are debased forms of this original REALITY. This notion bears comparison with that of Robert HOLDSTOCK's **Ryhope Wood** sequence. Just as Holdstock's mythagoes are coarser and more complex than their retold forms, so GGK attempts to create a grittier, harsher world than is common within fantasy: one of his student heroes is killed, another is raped. Even so, the involvement of the students in the Fionavar adventure brings HEALING to the land and, in the end, to the people they were before their arrival.

GGK's subsequent novels display a willingness to seek widely for inspiration. Thus *Tigana* (**1990**), which in many ways feels like a pendant to **The Fionavar Tapestry**, has all the superficial appearances of a standard fantasy novel, but underlying the conflict is the sense of changing systems, with patriarchy and matriarchy in conflict for the soul of their world – and again GGK displays a willingness to kill his more attractive characters. The central story is the familiar one of countries that follow the old way being invaded by a new regime, but by the time of the inevitable LAST BATTLE it is no longer clear which is right and which is wrong. The SACRIFICE which finally settles the matter actually leaves an even greater ambiguity.

The setting of *A Song for Arbonne* (**1992**) is apparently a FANTASYLAND but actually based on the Court of Love of troubadour lyrics. The troubadours who provided one of the central motifs of *Tigana* are here set in their natural environment; there is a slightly uneasy tension between fantasy and historical novel, with the setting vividly presented but the foreground not so carefully worked out.

GGK attempts much the same blending of historical and fantasy novel, with more success, in *The Lions of Al-Rassan*, set in an analogue of medieval Spain, where the struggle for power between Asharites, Jaddites and Kindath clearly mirrors the fate of Moors, Christians and Jews. As ever, leading characters suffer and die, and GGK gives the impression that his world is a real place, as nasty and brutish as our own, rather than an escapist refuge. [PK]

KAYE, MARVIN (NATHAN) (1938-) US writer and anthologist who began his career with detective novels like *A Lively Game of Death* (**1972**), produced a horror anthology, *Brother Theodore's Chamber of Horrors* (anth **1975**) with Brother Theodore, and came into his own as a genre writer with an sf sequence, the **Masters of Solitude** series – *The Masters of Solitude* (**1978**) and *Wintermind* (**1984**) – with Parke GODWIN. His second sequence, the **Fillmore** series – *The Incredible Umbrella* (fixup **1979**) and *The Amorous Umbrella* (**1981**) – is of fantasy interest for its sophistication

of the comic model established by L. Sprague DE CAMP and Fletcher PRATT in their **Incomplete Enchanter** sequence. Fillmore's travels through umbrella-engineered PORTALS to various ALTERNATE REALITIES are conducted at a very considerable rate, are extremely funny at times, and develop a complex relationship between Fillmore and the worlds he enters. In the first, for instance, which is based on the TOPSY-TURVY worlds of the **Savoy Operas** (◊ W.S. GILBERT; OPERA), Fillmore finds that when he actually sings along in order to expedite an ongoing plot, he becomes immured (◊ BONDAGE) in the STORY, which must be told before he can escape to the next world.

The **Aubrey House** sequence – *A Cold Blue Light* (**1983**) with Godwin and *Ghosts of Night and Morning* (**1987**) – introduces a passel of OCCULT DETECTIVES into a HAUNTED DWELLING inhabited by a GHOST capable of generating, in each of them, a unique confrontation with the OTHERWORLD; in the second volume, the ghost is so introjected that the tale can be read as a mundane examination of personalities under stress. *Fantastique* (**1992**) may be MK's most interesting novel. The tale mirrors that of the *Symphonie Fantastique* (**1830**) by Hector Berlioz (1803-1869), its five chapters corresponding in structure and relative length to the five movements of the symphony (◊ MUSIC). Berlioz's *belle dame sans merci* (◊ LAMIA) and his trip to HELL recur here, in a contemporary theatre setting. It is hoped MK can continue with work of similar ambition.

[JC]

Other work: *The Possession of Immanuel Wolf and Other Improbable Tales* (coll **1981**).

As editor: *Fiends and Creatures* (anth **1975**); *Ghosts: A Treasury of Chilling Tales Old and New* (anth **1981**; cut vt *A Classic Collection of Haunting Ghost Stories* 1993 UK) with Saralee Kaye; *Masterpieces of Terror and the Supernatural: A Treasury of Spellbinding Tales Old and New* (anth **1985**) with Saralee Kaye; *Devils & Demons: A Treasury of Fiendish Tales Old and New* (anth **1987**) with Saralee Kaye; *Weird Tales: The Magazine that Never Dies* (anth **1988**) with Saralee Kaye (◊◊ WEIRD TALES); *13 Plays of Ghosts and the Supernatural* (anth **1990**); *Haunted America: Star-Spangled Supernatural Stories* (anth **1991**) with Saralee Kaye; *Lovers and Other Monsters* (anth **1992**); *Masterpieces of Terror and the Unknown* (anth **1993**); *Angels of Darkness* (anth **1995**).

KAZANTZAKIS, NIKOS (1883-1957) Greek novelist, poet and playwright, who also served twice in the Greek government, as Director General of the Ministry of Public Welfare in 1919 and as Minister of National Education in 1946. In his spiritual autobiography, *Anafora Ston Greko* (**1961**; trans P.A. Bien as *Report to Greco* 1965 US), NK describes his life as a QUEST inspired at various times by CHRIST, Marx, Nietzsche, Buddha, ODYSSEUS and Ghiorgios Zorbas.

It was Odysseus who set the themes, preoccupations and mythic underpinnings of NK's work in *O Odyssey* (**1938**; trans Kimon Friar as *The Odyssey: A Modern Sequel* 1958 US), the huge epic poem that was NK's masterpiece. This sequel to HOMER stands comparison with James JOYCE's *Ulysses* (**1922**) for both its formal and linguistic experimentation (it was written in 24 books, one for each letter of the Greek alphabet, and of exactly 33,333 lines) and for the way it revitalizes the ancient LEGEND. Odysseus is unable to settle in Ithaca and sets out on his voyages once more. The fantastic incidents which marked Homer's original are echoed on this journey, which takes Odysseus first to Egypt, then to the source of the Nile, and eventually to the South Pole. Along the way there are encounters with phantasms and strange beings, though most tellingly the journey becomes a spiritual contest with DEATH and GOD.

This notion of life as a spiritual contest is central to all NK's novels, as in his most famous work, *Zorba the Greek* (**1946**; trans Carl Wildman **1952** US), in which his partner in an ill-fated mining project in NK's native Crete becomes the embodiment of the zest for life which is the abiding feature of all NK's work. NK barely escaped excommunication for *Christ Recrucified* (**1949**; trans Jonathan Griffin **1954** UK), set in a remote Greek village under Turkish occupation, where the villagers put on a Passion Play and find themselves taking their roles on into their everyday lives.

Fantasy rarely intrudes directly into NK's novels, though where it does – as in *The Last Temptation* (**1953**; trans P.A. Bien **1961** US; vt *The Last Temptation of Christ*) – it is used to emphasize the primacy of the zest for life. In this novel CHRIST is a Nietzschean superman who achieves spiritual salvation through force of will after being tempted by visions of the life he could have if he descended from the cross. NK presents a vivid portrait of Christ leading a normal, happy family life – the most subtle and telling temptation. Christ must succumb to these temptations in order truly to triumph over them. Both the novel and the movie version, *The* LAST TEMPTATION OF CHRIST (*1988*), though profoundly religious, generated bitter condemnation.

These novels, which infuse their depictions of reality with the shape and tenor of myth, established Kazantzakis as one of the foremost writers of his era. [PK]

Further reading: *Nikos Kazantzakis* (**1968**) by Helen Kazantzakis.

KA ZAR, LORD OF THE HIDDEN JUNGLE ◊ TARZAN.

KAZE NO TANI NO NAUSICAA Japanese ANIMATED MOVIE (*1983*). ◊ ANIME.

KEARNEY, PAUL (1967-) Northern Irish writer, now resident in Denmark. His first novel, *The Way to Babylon* (**1992**), features an ex-soldier turned fantasy author who, following the death of his wife, faces writer's block. He eventually finds himself drawn by a stranger into the ALTERNATE REALITY of his own FANTASYLAND; now, though, the LAND is poisoned. HEALING the land enables him to return to our world and start healing himself. The general feeling is of ALLEGORY.

Both of PK's next two novels follow the same basic pattern. In *A Different Kingdom* (**1993**) the protagonist ventures into an OTHERWORLD to find his long-lost cousin. He never does find her, but in the QUEST he grows to manhood as a warrior. When he returns to our world no time has passed (◊ TIME IN FAERIE). In *Riding the Unicorn* (**1994**) an unpleasant prison officer is drawn into a primitive world and is forced to take part in a damaging war. Injuries inflicted on him in the fantasy world are just as real as in this world; only by pushing himself close to death can he resolve the conflicts he faces in both worlds.

By contrast, *Hawkwood's Voyage* (**1995**) – beginning the **Monarchies of God** sequence – is set entirely within its SECONDARY WORLD, a curious amalgam of elements from our own history. The novel promises an inventive fantasy adventure in succeeding volumes. [PK]

KEATS, JOHN (1795-1821) English poet who made frequent use of Greek MYTHS in his poems. Of his long works,

Endymion: A Poetic Romance (**1818**) is minor, and the more ambitious fragments "Hyperion" (1820) and "The Fall of Hyperion: A Dream" (written 1819; 1848) were abandoned. All these works owe a debt to OVID, although the actual source for *Endymion* is more obscure. While JK set greatest store in his long poems, his shorter narratives are more accomplished and have exerted a greater influence over modern fantasy. "Lamia" (1820), telling the story – from *Anatomy of Melancholy* (**1621**) by Robert Burton (1577-1640) – of a Greek philosopher who unwittingly marries the eponymous serpent woman (◊ LAMIA), and "La Belle Dame Sans Merci" (1820) are tales of erotic entrapment and devastation; the latter poem, with its medieval venue and spare form, is widely echoed in modern fantasy. "The Eve of St Agnes" (1820), also probably derived from Burton, gives a happy ending to a story of erotic enchantment: its lovers, both human, escape an "elfin-storm from faery land", cheating, very unusually, the world of FAERIE.

JK has been dramatized in fantasy on various occasions, as in Tim POWERS's *The Stress of Her Regard* (**1989**) and Dan SIMMONS's **Hyperion** books. [GF]

KEEP, THE US movie (*1983*). Paramount. **Pr** Gene Kirkwood, Howard W. Koch Jr. **Exec pr** Colin M. Brewer. **Dir** Michael Mann. **Spfx** Nick Allder. **Vfx** Wally Veevers, Robin Browne. **Mufx** Nick Maley. **Laser fx** John Carr, Ken Goddard. **Screenplay** Mann. **Based on** *The Keep* (**1981**) by F. Paul WILSON. **Starring** Gabriel Byrne (Kaempffer), Michael Carter (Radu Molasar), Scott Glenn (Glaeken), Ian McKellen (Dr Theodor Cuza), Jürgen Prochnow (Woermann), Robert Prosky (Father Mikhail Fenescu), Alberta Watson (Eva Cuza). 93 mins. Colour.

German soldiers led by Woermann are sent in 1943 to occupy a grim fortress in the Dinu Pass, Romania. The locals – led by Father Fenescu – warn them to leave it alone, but will not say why. Almost at once two sentries discover a vault far vaster than the Keep itself (◊ LITTLE BIG), but are swiftly and gorily killed by an emergent miasma, Molasar, which has been in BONDAGE. The death-toll rises, but at last an enigmatic quasi-mortal called Glaeken uses a TALISMAN to re-incarcerate Molasar, at the cost of incarcerating himself alongside.

This impressive but somewhat impenetrable art/HORROR MOVIE cross (its plot is far more complicated than described above) is widely regarded as a "lost masterpiece". It is in fact no masterpiece, though worth watching for the use of light and colour alone. Yet no amount of visual splendour or cod philosophical underpinning can obscure the fact that the basic story is hokum; seemingly aware of this, Mann has stripped it back until too little is explained for us ever to have any confidence as to what exactly is going on. Even so, *TK*'s visual beauty and ambience – both echoing the work of Andrei Tarkovsky (1932-1987), as does the stolidly measured pace – impress. [JG]

KEEPING, CHARLES (WILLIAM JAMES) (1924-1988) UK artist, illustrator and writer, active from 1953, when he began a four-year stint doing a daily cartoon for the London *Daily Herald*; his first book ILLUSTRATIONS appeared the same year, but not until his work for Rosemary SUTCLIFF's *The Silver Branch* (**1957**) did he begin to become known for his radical approach. Henceforth CK consistently treated illustration as a form of co-creation, generating a pattern of images (in his unmistakable, superbly draughted, powerfully and "primitively" emotional style) that presented its own version of the textual tale. Many of the dozen or so he illustrated dealt with Sutcliff historical subjects, but in her *Beowulf* (**1961**; vt *Dragon Slayer* 1966) he tellingly prefigured his version of Kevin Crossley-Holland's *Beowulf* (**1984**), in which he envisaged Grendel (◊ BEOWULF) as a creature with as much autonomy, and as ruthlessly judged by humans, as CALIBAN. CK illustrated other Crossley-Holland tales like *The Wildman* (**1976** chap). His seven collaborations with Henry TREECE had similar impact – especially in late tales like *The Dream-Time* (**1967**) – and the savage strokes of his representation of Viking life and MYTH have influenced subsequent writers and illustrators of NORDIC FANTASY. His later illustrations for a complete edition of Charles DICKENS (1981-8) made LONDON appear at times as though it, too, loomed out of the mists of time, bearing monsters and prodigies to our view; *The Christmas Books* (omni 1988) is of direct fantasy interest.

Further work of interest includes illustrations for H. Rider HAGGARD's *King Solomon's Mines* (1961), Kenneth GRAHAME's *The Golden Age, and Dream Days* (1961), Mollie HUNTER's *The Kelpie's Pearls* (**1964**) and other titles, Alan GARNER's *Elidor* (**1965**) and other titles, *Celtic Folk and Fairy Tales* (anth **1966**) ed Eric Protter and Nancy Protter, Nicholas Stuart GRAY's *Mainly in Moonlight* (**1967**) and other titles, Robert Louis STEVENSON's *Strange Case of Dr Jekyll and Mr Hyde* (1967) and other titles, Aldous Huxley's *Time Must Have a Stop* (1968), *The God Beneath the Sea* (**1970**) by Edward Blishen (1920-   ) and Leon GARFIELD and other titles by these two, *The Ghost Stories of M.R. James* (coll **1973**) (◊ M.R. JAMES) ed Nigel Kneale, Helen Hoke's *Weirdies* (anth **1973**) and other titles, P.L. TRAVERS's *About the Sleeping Beauty* (**1977**), *The Mermaid's Revenge* (**1979**) by Forbes Stuart, Rudyard KIPLING's *The Beginning of the Armadilloes* (coll **1982**) and other titles, Neil PHILIP's *The Tale of Sir Gawain* (**1987**), Bram STOKER's *Dracula* (1988) and Mary SHELLEY's *Frankenstein* (1988).

Of CK's own picture books, *Charley, Charlotte and the Golden Canary* (graph **1967**) and *Alfie and the Ferry Boat* (graph **1968**; vt *Alfie Finds the Other Side of the World* 1968 US) are vivid fantasies of London for younger children. *Charles Keeping's Book of Classic Ghost Stories* (anth **1986**) and *Charles Keeping's Classic Tales of the Macabre* (anth **1987**) are strong selections. [JC]

KEIGHTLEY, THOMAS (1789-1872) UK folklorist. ◊ FAIRIES.

KELLEHER, VICTOR (MICHAEL KITCHENER) (1939-   ) UK-born academic and writer who lived first in South Africa and New Zealand, then from 1976 in Australia; he writes for a YA audience, and is as well known for sf as for fantasy, beginning with *Forbidden Paths of Thual* (**1979** UK). His most interesting early novel, *Master of the Grove* (**1982** UK), carries its young protagonist through a taxing RITE OF PASSAGE into maturity; it is set in a SECONDARY WORLD of considerable interest. *The Beast of Heaven* (**1984**), which won the 1985 Ditmar AWARD, is a TECHNO-FANTASY concerning a post-HOLOCAUST land whose inhabitants worship UNDERGROUND computers. *The Red King* (**1989**) is set in a land ruled by a wicked MAGICIAN. *Brother Night* (**1990** UK) sets TWINS, one of them an UGLY DUCKLING, not unpredictably into conflict with one another in a bleak world. VK's work never lacks symbol or PORTENT; there is at times, however, a certain absence of fluency. [JC]
Other works: *The Hunting of Shadroth* (**1981** UK); *Ringwood* (**1986**); *Baily's Bones* (**1988**), involving POSSESSION;

Del-Del (**1991**); *To the Dark Tower* (**1992** UK).

KELLER, DAVID H(ENRY) (1880-1966) US writer better known for his sf, starting with "The Revolt of the Pedestrians" (1928 *Amazing Stories*), than for his fantasy and horror, which is in fact superior. From 1928 DHK was a regular contributor to WEIRD TALES, for which his best-known story is "The Thing in the Cellar" (1932 *WT*; **1940**), a PSYCHOLOGICAL THRILLER or fantasy of PERCEPTION in which a boy dies of fright because he believes something exists in the cellar, that belief being sufficient to create the entity. *WT* also began publication of DHK's **Tales from Cornwall** with "The Battle of the Toads" (1929), a series (15 stories) never published in full, though the MAGAZINE OF HORROR published 10 of the stories in internal chronological order, starting with "The Oak Tree" (1969). The series is a form of DYNASTIC FANTASY, set in our world, following the Hubelaire family from 200BC to 1914. The early stories draw heavily upon FOLKLORE and tradition, and were influenced by James Branch CABELL. The Hubelaire or **Hubler** family appears in others of DHK's stories, including *The Devil and the Doctor* (**1940**), DHK's first novel, where the DEVIL, in the shape of ROBIN GOODFELLOW, tells Dr Hubler the real story of the Creation, making Jehovah the villain. The tale is superficially a twist on the PACT-WITH-THE-DEVIL story, but in development is more deeply philosophical on the nature of GOOD AND EVIL.

DHK's other major series featured the OCCULT DETECTIVE **Taine of San Francisco**. Early stories were sf, starting with "The Menace" (1928 *Amazing Stories Quarterly*), but later ones are more weird and supernatural, including *Wolf Hollow Bubbles* (**1933** chap), where cancer cells grow to an enormous size, and "The Tree of Evil" (1934 *Wonder Stories*), where an entire town succumbs to the effects of a powerful psychedelic drug. Drugs were used to similar effect in the non-**Taine** novella "The Abyss" (in *The Solitary Hunters and The Abyss* coll **1948**), where a drug which can separate the conscious from the unconscious is introduced into chewing-gum sold in NEW YORK, whose populace descends into barbarity and finds itself under the power of a new surreal RELIGION.

DHK was a compulsive writer for over 60 years. His first published story, "Aunt Martha" (1895 *Bath Weekly*) appeared in a country newspaper. At college he collaborated in a small literary magazine, *The White Owl*, which published several of his stories as by Henry Cecil; he also used the name Amy Worth. DHK seldom wrote directly for a market (other than the sf magazines), and much of his output remains unpublished. Several works appeared first in French, including *La Guerre du Lierre* ["The Ivy War"] (coll **1936** France) and *The Sign of the Burning Hart: A Tale of Arcadia* (coll of linked stories **1938** France; in English **1948**). DHK was also happy to offer stories free to amateur magazines and presses, and so many of his tales appeared in obscure publications; his bibliography is thus complicated. Apart from *The Devil and the Doctor*, all of DHK's book publications have been as amateur booklets or through specialist SMALL PRESSES. One of these, Prime Press, issued DHK's only other novels to appear in book form: *The Eternal Conflict* (1939 *Les Primaires* France as "Le Duel sans Fin", part only; **1949**), which explores the power of a GODDESS who can create and destroy worlds at will, *The Homunculus* (**1949**), in which a doctor's successful attempts to create a HOMUNCULUS are subverted by the intervention of PAN and LILITH, and *The Lady Decides* (**1950**), concerning an allegorical quest through Spain. All three reflect DHK's increasingly anti-feminist views. Many of his short stories, like "Binding De Luxe" (1934 *Marvel Tales*), "Tiger Cat" (1937 *WT*) and "The Bridle" (1942 *WT*), feature FEMMES FATALES.

The best of DHK's short fiction is collected in *Life Everlasting and Other Tales of Science, Fantasy and Horror* (coll **1947**) ed Sam Moskowitz and Will Sykora, with an informative introduction by Moskowitz, *Tales from Underwood* (coll **1952**), *The Folsom Flint* (coll **1969**), with another fine introduction by Paul Spencer, and *The Last Magician* (coll **1978** chap) ed Patrick H. Adkins; this last volume, the first of two in the incomplete **David H. Keller Memorial Library**, prints DHK's autobiographical "Half a Century of Writing". [MA]

Other works: *Songs of a Spanish Lover* (coll **1924** chap), poetry; *The Thought Projector* (**1929** chap); *Men of Avalon* (**1935** chap dos); *The Waters of Lethe* (**1937** chap); *The Television Detective* (**1938** chap); *The Final War* (**1949** chap); *A Figment of a Dream* (**1962** chap). *Fanomena* March 1948 is entirely dedicated to DHK's work.

KELLEY, EDWARD (1555-1595) Irish scryer. ◊ John DEE.

KELLEY, THOMAS P. (1905-1982) Canadian writer. ◊ CANADA.

KELLY, KEN (1946-   ) US painter and illustrator of epic fantasy. He studied under Frank FRAZETTA, from whom he inherited his tableau-style compositions and general ethos, though the influence of Boris VALLEJO is also strongly in evidence. He works in oils in a generally subdued and subtle colour range. His figures are massively muscled and exaggeratedly heroic. Among his earliest published works were several covers for WARREN PUBLISHING. His paintings now appear mostly on book covers; a series for books by Robert E. HOWARD received particular accolade. [RT]

KELLY, WALT Working name of US cartoonist Walter Crawford Kelly (1913-1973), who began his career with DISNEY in the early 1940s and who contributed to COMICS like *Fairy Tale Parade* (Dell 1942-6); he gained his central importance in US fantasy for his **Pogo** comic strip. The strip – begun October 4, 1948, and written and drawn by WK from its inception – was preceded by more than one comic-book incarnation of Pogo and his COMPANIONS, beginning with the first issue of *Animal Comics* (**1942**). The strip continued until WK's death; members of his family continued it for a period afterwards, but without success.

Set in a richly drawn LAND-OF-FABLE Okeefenokee Swamp in the state of Georgia – which WK transformed into a POLDER not always immune from invasions from the outer world – the **Pogo** sequence presents a BEAST-FABLE commentary whose sharpness about the contemporary USA often caused distribution difficulties: newspapers frequently excised strips satirizing politicians like Senator Joseph McCarthy ("Simple J. Malarkey"), the John Birch Society – these were famously assembled as *The Jack Acid Society Black Book* (graph coll **1962**) as by Pogo – or Lyndon B. Johnson, in strips assembled as *Prehysterical Pogo (in Pandemonia)* (graph coll **1967**), which takes place in a WONDERLAND inhabited by a wide range of fantasy figures. A phrase coined by WK with reference to ecological issues – "We have met the enemy and he is us" – has become so well known that its origin in comics has often been forgotten.

The strip's linguistic inventiveness, the intense cleverness of WK's transformation of Disney-esque visual drolleries into an absolutely unmistakable and individual vision, and

the genial surrealism with which he expressed his love for and anger with the USA – all make **Pogo** one of the central US fantasy sequences. Pogo the possum, Albert the alligator, Howland Owl, Churchy la Femme the turtle and about 600 other named characters occupy one of the most complexly sustained fantasy worlds yet conceived. Along with George HERRIMAN's **Krazy Kat**, **Pogo** demonstrates the range and pathos and depth of feeling attainable in comics form.

Much of the strip was assembled, with the originals revised to form coordinated assemblies. These include: *Pogo* (graph coll **1951**); *I Go Pogo* (graph coll **1952**); *Uncle Pogo So-So Stories* (graph coll **1953**); *The Pogo Papers* (graph coll **1953**); *The Pogo Stepmother Goose* (graph coll **1954**); *The Incompleat Pogo* (graph coll **1954**); *The Pogo Peek-A-Book* (graph coll **1955**); *Potluck Pogo* (graph coll **1955**); *The Pogo Sunday Book* (graph coll **1956**); *The Pogo Party* (graph coll **1956**); *Pogo's Sunday Punch* (graph coll **1957**); *Positively Pogo* (graph coll **1957**); *The Pogo Sunday Parade* (graph coll **1958**); *G.O. Fizzickle Pogo* (graph coll **1958**); *The Pogo Sunday Brunch* (graph coll **1959**); *Ten Ever-Lovin' Blue-Eyed Years with Pogo: 1949-1959* (graph coll **1959**); *Beau Pogo* (graph coll **1960**); *Pogo Extra (Election Special)* (graph coll **1960**); *Pogo à la Sundae* (graph coll **1961**); *Gone Pogo* (graph coll **1961**); *Instant Pogo* (graph coll **1962**); *Pogo Puce Stamp Catalog* (graph coll **1963**); *Deck Us All with Boston Charlie* (graph coll **1963**); *The Return of Pogo* (graph coll **1965**); *The Pogo Poop Book* (graph coll **1966**); *Equal Time for Pogo* (graph coll **1968**); *Pogo: Prisoner of Love* (graph coll **1969**); *Impollutable Pogo* (graph coll **1970**); *Pogo: We Have Met The Enemy and He is Us* (graph coll **1972**). Posthumous compilations include: *The Best of Pogo* (graph coll **1982**); *Pogo Even Better* (graph coll **1984**); *Outrageously Pogo* (graph coll **1985**); *Pluperfect Pogo* (graph coll **1987**); *Phi Beta Pogo* (graph coll **1989**); *Pogo Files for Pogophiles* (graph coll **1992**). A proposed complete republication of the strips in their original order has begun with *Pogo: Volume One* (graph coll **1992**) and *Pogo: Volume Two* (graph coll **1994**). [JC]

KENDALL, CAROL (SEEGER) (1917-   ) US writer whose career began with detective novels like *The Black Seven* (**1946**) and *The Other Side of the Tunnel* (**1956** UK), but who is best-known for her **Minnipins** sequence of fantasy novels: *The Gammage Cup* (**1959**; vt *The Minnipins* 1960 UK) and *The Whisper of Glocken* (**1965**). Clearly influenced by J.R.R. TOLKIEN – her Hobbit-like Minnipins live in the Shire-like Land Between the Mountains – the first volume of the sequence engages a SEVEN-SAMURAI band of exiled heroes, armed with magic SWORDS, in a conflict with their traditional GOBLIN-like enemy. The second again features a group of heroes, this time on a QUEST to find out the reason for the flooding of the local river. Both novels are ostensibly for older children, and are warmly and expertly crafted. In *The Firelings* (**1981** UK) a young boy escapes becoming a HUMAN SACRIFICE to the volcano god worshipped by his fellow Firelings. [JC]

KENDALL, GORDON Joint pseudonym of S.N. Lewitt and Susan M. SHWARTZ.

KENDALL, MAY (real name Emma Goldworth; 1861-?1931) UK writer. ◊ Andrew LANG.

KENNEALY, PATRICIA ◊ Patricia KENNEALY-MORRISON.

KENNEALY-MORRISON, PATRICIA (1946-   ) US writer who until 1994 published her fiction as Patricia Kenneally, but now uses her full name, derived from a "private religious ceremony" of marriage to rock singer and composer Jim Morrison (1943-1971). Her **Keltiad**

sequence, being two trilogies, could be described as HIGH FANTASY set in a SCIENCE-FICTION frame. Its initial premise mixes both genres: persecuted by humans for their MAGIC and strange customs, the alien survivors of ATLANTIS, known as the Danaans, follow their leader Brendan through space in search of a safe haven, which turns out to be Keltia, one of several planets PK-M utilizes as high-fantasy LAND-SCAPES. The first trilogy, the **Aeron** sequence – *The Copper Crown* (**1985**), *The Throne of Scone* (**1986**) and *The Silver Branch* (**1988**), this last prequelling the first two volumes – tells of Queen Aeron of Keltia's fight against invaders from other planets, during which she must invoke the memory of the local ARTHUR (himself a re-enactment of the original fig-ure from Earth) and recover his Thirteen Treasures in order to prevail against the foe. The second trilogy, the **Tales of Arthur** sequence – *The Hawk's Gray Feather* (**1990**), *The Oak Above the Kings* (**1994**) and *The Hedge of Mist* (**1996**) – deals with early events in PKM's PLANETARY-ROMANCE venue through a slow and complex re-enacting of the Arthurian Cycle, during which young Arthur Pernarvon, a HIDDEN-MONARCH figure ignorant of his lin-eage but helped along by a MERLIN character, comes to lead his people against the long oppressive rule of the Danaans' hereditary foes.

PKM employs a slow, archaizing DICTION throughout, especially in the novels dealing directly with the Arthur fig-ure, and her sense of plotting is not innovative; the **Keltiad** remains memorable mainly for its thorough mixing of genres. [JC]

KENNEDY, RICHARD (JEROME) (1932-   ) US writer, most of whose earlier works were fantasies for younger children. Tales like *The Blue Stone* (**1976** chap), in which a MAGIC stone causes various METAMORPHOSES – peo-ple into animals, a sparrow into an ANGEL, etc. – are strongly inventive. Other books for the same audience include: *Come Again in the Spring* (**1977** chap), in which a man bargains with DEATH for a few extra months in which to take care of his birds; *Inside my Feet: The Story of a Giant* (**1979** chap), in which a child's parents have been stolen by a GIANT; *The Mouse God* (**1979** chap), a BEAST FABLE in which a cat dresses up as the GOD of mice and orders them into his "church" so he can eat them; and *The Boxcar at the Center of the Universe* (**1982** chap), in which a hobo on a TRAIN narrates the tale of his search for the centre of the Universe. Volumes for YA audiences include the very com-plex *Amy's Eyes* (**1985**), in which an adolescent girl's love of a TOY sailor causes it to come to life (◊ ANIMATE/INANI-MATE) and her to become a toy (◊ BONDAGE); and *Richard Kennedy: Collected Stories* (coll **1987**), which assembles vari-ous pieces – including *Crazy in Love* (**1980** chap), about a man who thinks his wife is crazy because she has conversa-tions with a donkey (◊ ASS) – that ingeniously manipulate a wide variety of traditional modes of storytelling. [JC]

Other works for younger children (selective): *The Dark Princess* (**1978** chap; *The Lost Kingdom of Karnica* (**1979** chap); *The Leprechaun's Story* (**1979** chap).

KENOSIS In Christian theology, the theory that, when CHRIST took human form, he emptied himself of divine attributes. More generally, the word can describe the process by which the Divine (who may be one of the GODS, or the primordial myth) becomes incarnate, like Severian in Gene WOLFE's *The Book of the New Sun* (**1980-83**); by anal-ogy, it may also be used to describe the process whereby an IMMORTAL or unageing being becomes mortal, like Arwen in

J.R.R. TOLKIEN's *The Lord of the Rings* (**1954-5**) – who accepts the "Doom of Men" to live as Aragorn's queen – Ged in Ursula K. LE GUIN's **Earthsea** sequence, Schmendrick in Peter S. BEAGLE's *The Last Unicorn* (**1968**), and Callabrion, the god of Summer in Lisa GOLDSTEIN's *Summer King, Winter Fool* (**1994**), who becomes enamoured of the world and descends into it, stopping the SEASONS. In Michael MOORCOCK's **Eternal Champion** sequence, keno-sis is constant but revocable; and is, in the end, ironized deliberately into meaninglessness.

As an act of BONDAGE, whether or not voluntary, kenosis tends to mark a THINNING of the relevant world. [JC]

KENT, KELVIN Pseudonym shared by Arthur K. Barnes (1911-1969) and Henry KUTTNER.

KERBY, SUSAN ALICE Pseudonym of UK writer Alice Elizabeth Burton (1908-   ), long resident in Canada. Her gentle fantasy *Miss Carter and the Ifrit* (**1945**) is one of the few novels imitative of F. ANSTEY to use a female protagonist; it cosily sublimates the frustrations of a non-combatant ground down by wartime rationing. In *Mr Kronion* (**1949**) a refugee from OLYMPUS takes up brief res-idence in an Oxfordshire village and helps defend it against the creeping erosions of post-WWII reform. *Gone to Grass* (**1948**; vt *The Roaring Dove* US), describing a similarly mild rebellion, is a satirical RURITANIAN romance. [BS]

KERNAGHAN, EILEEN (SHIRLEY MONK) (1939-   ) Canadian writer whose **Journey to Aprilioth** sequence – *Journey to Aprilioth* (**1980** US), *Songs from the Drowned Lands* (**1983** US) and *The Sarsen Witch* (**1989**) – involves a QUEST through a FANTASYLAND. *Dance of the Snow Dragon* (**1995**) is set in 18th-century Bhutan. [JC]

KERR, KATHARINE (1944-   ) US author whose rela-tively complex **Deverry** sequence (still in progress) began with *Daggerspell* (**1986**; rev 1993). Set in the Kingdom of Deverry – an imaginary Western European post-Roman country – between the 7th and 11th centuries, the **Deverry** novels tell a complicated story of several characters whose destinies entangle through successive REINCARNATIONS. The series is notable less for its topos – which is populated, although sparsely, with fairly conventional dwarves, elves and dragons – than for the intelligence with which KK limns the details of her world, especially its linguistic aspects. The eight novels so far published fall into two quar-tets, with a third projected. The remaining volumes of the **Kingdom of Deverry** sequence are *Darkspell* (**1987**; rev 1994), *The Bristling Wood* (**1989**; vt *Dawnspell: The Bristling Wood* 1989 UK) and *The Dragon Revenant* (**1990**; vt *Dawnspell: The Southern Sea* 1990 UK). The **Westlands Cycle** – *A Time of Exile* (**1991**), *A Time of Omens* (**1992** UK), *A Time of War: Days of Blood and Fire* (**1993** UK; vt *Days of Blood and Fire* 1993 US) and *A Time of Justice: Days of Air and Darkness* (**1994** UK; vt *Days of Blood and Fire* 1994 US) – is the better of the two. The later volumes are more polished and assured than the earlier ones.

Freezeframes (coll of linked stories, dated 1994 but **1995** UK; vt *Freeze Frames* 1995 US), which includes an earlier short novel, *Resurrection* (**1992** chap), combines elements of sf with supernatural elements. It was announced as the first of a longer sequence. Like the individual **Deverry** novels, it comprises a closely knit series of novellas: KK seems to be attracted to multi-volume sequences braided from novellas arranged several per volume. [GF]

Other works: *Polar City Blues* (**1990**), sf; *Weird Tales from Shakespeare* (anth **1994**) and *Enchanted Forests* (anth **1995**), both ed with Martin H. GREENBERG.

KERRUISH, JESSIE DOUGLAS (1884-1949) UK writer who specialized mainly in romantic, historical and Arabian tales, including her first two popular novels, *Miss Haroun-al-Raschid* (**1917**) and *The Girl from Kurdistan* (**1918**), and a collection of ARABIAN FANTASIES, *Babylonian Nights' Entertainments* (**1934**). *The Undying Monster: A Tale of the Fifth Dimension* (**1922**) is the most regularly reprinted UK WEREWOLF novel from the period. It tells of a CURSE that breaks out horribly in successive generations of the Hammand family, and of a female OCCULT DETECTIVE who discovers the truth. It was filmed as *The Undying Monster* (**1943**; vt *The Hammond Mystery* UK).

Among her uncollected short fantasies is "The Seven-Locked Room" (1933), about the discovery of the GRAIL.

[RD]

KERSH, GERALD (1911-1968) UK writer active in several genres from the 1930s, beginning with the family chronicle *Jews Without Jehovah* (**1934**); he was most famous for novels of LONDON like *Night and the City* (**1938**) and *Fowler's End* (**1957** US), and for WWII tales like *They Die with their Boots Clean* (**1941**). His sf was restricted to some short stories and to *The Great Wash* (**1953**; vt *The Secret Masters* 1953 US), though the SECRET-MASTERS plot of this novel – involving a planned inundation of the world – is built on a FANTASIES-OF-HISTORY assumption that the Masters are already in place to benefit from a second FLOOD. Other novels – like *The Weak and the Strong* (**1945**), set in an unnatural-seeming UNDERGROUND venue, and *An Ape, a Dog, and a Serpent: A Fantastic Novel* (**1945**), about the fantasy world of moviemaking – have slim elements of fantasy.

Short fantasies and SUPERNATURAL FICTIONS appeared throughout his career, the later years of which were darkened by ill-health and drink. Individual volumes of note include: *I Got References* (coll **1938**), which contains "Comrade Death"; *Selected Stories* (coll **1943**; cut vt *The Battle of the Singing Men* **1944** chap), both versions of which contain "The Extraordinarily Horrible Dummy", a tale of POSSESSION that appears also in *The Horrible Dummy and Other Stories* (coll **1944**) and was filmed as *Dead of Night* (**1945**); *Clean, Bright and Slightly Oiled* (coll **1946**), mostly associational; *Neither Man Nor Dog* (coll **1947**), which includes the POSTHUMOUS FANTASY "In a Room without Walls"; *Sad Road to the Sea* (coll **1947**), which contains the GHOST STORY "The Scene of the Crime"; *Clock without Hands* (coll **1949**); *The Brazen Bull* (coll **1952**); *The Brighton Monster and Others* (coll **1953**), whose title story sees a Japanese man cast back centuries by nuclear TIMESLIP and thought a MONSTER, and which contains also "Whatever Happened to Corporal Cuckoo", whose protagonist – though made immortal (◊ IMMORTALITY) by a POTION – is incapable of benefiting from his condition; *Guttersnipe: Little Novels* (coll **1954**); *Men Without Bones* (coll **1955**; different selection with the same title 1962 US), which contains "The Oxoxoco Bottle", about a purported Ambrose BIERCE manuscript describing a LOST RACE; *The Ugly Face of Love* (coll **1960**), which contains "River of Riches"; *The Terribly Wild Flowers* (coll **1962**); *More Than Once Upon a Time* (coll **1964**); and *The Hospitality of Miss Tolliver* (coll **1965**). Two US compilations usefully present fantasy material: *On an Odd Note* (coll **1958** US) and *Nightshade and Damnations* (coll **1968** US) ed Harlan ELLISON. *The Best of Gerald Kersh* (coll **1960**) is also useful. [JC]

KESTEVEN, AMBROSIUS [s] ◊ Kenneth MORRIS.

KEY, SAMUEL M. Pseudonym of Charles DE LINT.

KEY, UEL Working name of UK writer Reverend Samuel Whittell Key (1874-1948). After serving as a chaplain in the Royal Armoured Corps throughout WWI, he embarked on a brief literary career, writing five short stories and a novel all featuring a OCCULT DETECTIVE Dr Arnold Rhymer: *The Broken Fang, and Other Experiences of a Specialist in Spooks* (coll **1920**) and *Yellow Death* (**1921**). "Broken Fang" itself features German vampiric ZOMBIES (Key's fiction was virulently anti-German); "A Sprig of Sweet Briar" concerns MESMERISM. [RD]

KEYNE, GORDON ◊ H. BEDFORD-JONES.

KILWORTH, GARRY (DOUGLAS) (1941-   ) UK writer who in the mid-1970s retired as from 18 years' RAF service as a cryptographer. He began publishing work of genre interest with "Let's Go to Golgotha!" for the *Sunday Times Weekly Review* in 1974, having won an associated competition. Most of his subsequent work has been sf, and some recent work – in particular *Angel* (**1993**) and its sequel *Archangel* (**1994**), set in a noirish LOS ANGELES – combine sf and HORROR. He also publishes as Garry Douglas and F.K. Salwood. By 1996 he had published at least 100 stories, many fantasy, assembled in *The Songbirds of Pain: Stories from the Inscape* (coll **1984**), *In the Hollow of the Dark Sea Wave: A Novel and Seven Stories* (coll **1989**), *Dark Hills, Hollow Clocks: Stories from the Otherworld* (coll **1990**) – mostly TWICE-TOLD tales from various FOLKLORES for a YA audience – *In the Country of Tattooed Men* (coll **1993**) and *Hogfoot Right and Bird-Hands* (coll **1993** US). The most famous of his fantasy tales is "The Ragthorn" (1991) with Robert HOLDSTOCK, which won a 1992 WORLD FANTASY AWARD for best novella; it verges on FANTASIES-OF-HISTORY territory in its depiction of a search through various texts for clues to the location of the TREE of life.

GK is very widely known for his ANIMAL FANTASIES, each of which pays rigorous attention to the actual life-patterns of the animals given sentience within the texts. They include *Hunter's Moon* (**1989**; vt *The Foxes of First Dark* 1990 US), about foxes, *Midnight's Sun: A Story of Wolves* (**1990**), *Frost Dancers: A Story of Hares* (**1992**) and *House of Tribes* (**1995**), about mice (◊ MICE AND RATS). They have been likened (perhaps inevitably) to Richard ADAMS's *Watership Down* (**1972**), but are far more strict in their avoidance of BEAST-FABLE analogues to humankind.

Although his venues widely vary – e.g., *Spiral Winds* (**1987**), which contains hints of SUPERNATURAL FICTION, is set inland of Aden – GK's main home seems to be the southeast of England, particularly East Anglia, whose landscape he has richly described. YA novels like *The Drowners* (**1991**), a GHOST STORY set near 19th-century Winchester, and *Billy Pink's Private Detective Agency* (**1993**), set in 19th-century Essex and featuring a detective will o' the wisp, both give off a powerful sense of place. *The Phantom Piper* (**1994**) invokes Scotland in a TIMESLIP tale whose powerfully effective intricacies are perhaps reminiscent of the work of William MAYNE.

Sometimes GK's work may thrust too haphazardly at the edges of genres he seems to be proposing to inhabit, but the end result is a complexly active style and substance. His worlds are intensely alive. [JC]

Other works (fantasy): *The Wizard of Woodworld* (**1987**), for children; *The Voyage of the Vigilance* (**1988**), for children; *Trivial Tales* (coll **1988** chap); *The Rain Ghost* (**1989**), for children; «The Gargoyle» (1996), for children; «A

Midsummer's Nightmare» (1996); the «Navigator Kings» sequence of three fantasies set in a LAND-OF-FABLE Polynesian ARCHIPELAGO, beginning with «The Roof of Voyaging» (1996).

Other works (selected nonfantasy): *In Solitary* (**1977**), GK's first novel (*for other sf and associational titles* ◊ *SFE*); *Highlander* * (**1986**), movie novelization, and *The Street* (**1988**), HORROR, both as by Garry Douglas; the associational **Essex Saga** as by F.K. Salwood, being *The Oystercatcher's Cry* (**1993**), *The Saffron Fields* (**1994**) and *The Ragged School* (**1995**).

KING, BERNARD (JOHN HOWARD) (1946-) UK writer who began publishing with a translation (from French) on MAGIC: *The Grimoire of Pope Honorius III* (**1984** chap). BK's first published novel was *Starkadder* (**1985**), a NORDIC FANTASY about a late-Dark-Age Swedish warrior blessed by ODIN with three lives and cursed by THOR to end each in dishonourable betrayal. The novel initiates the interlinked stories of Hather Lambisson and of the stolen dwarfish SWORD Tyrfing, cursed to kill any mortal bearer: in *Vargr-Moon* (**1986**) Hather's wife tries to become a *vargr* (a WEREWOLF with great magical powers); in *Death-Blinder* (**1988**) Hather destroys a Lappish sorcerer who has taken one of Odin's names and seeks to establish an evil empire. The **Chronicles of the Keeper** – *The Destroying Angel* (**1987**), *Time-Fighters* (**1987**) and *Skyfire* (**1988**) – is an ambitiously constructed FANTASY-OF-HISTORY trilogy whose central notion is that Thule (◊ LOST LANDS) was not a physical place but a prehistoric empire of successor GODS, some of whom sought to destroy humanity; the benign faction planted RELIGION in human minds, projected Thule into "alterjective" REALITY, and appointed the first in a continuing succession of Keepers, who must maintain the separation, moulding human history where necessary. In *Witch-Beast* (**1989**), which shares a minor character with the **Keeper** trilogy, the canine SPIRIT Black Shuck tries to establish itself as the new god and usher in an Age of Dogs. *Blood Circle* (**1989**) is HORROR involving BLACK MAGIC. King's work is fertile in ideas but clumsily written, barring him from a wider audience. [JG]
See also: LIBRARY.

KING, CLIVE (1924-) UK writer. ◊ CHILDREN'S FANTASY.

KING, FRANK (1883-1969) US comics creator. ◊ COMICS.

KING, JESSIE M(ARION) (1875-1949) Scottish illustrator whose early influences included Charles Rennie Mackintosh (1868-1928) and Aubrey BEARDSLEY. Most of her work of fantasy interest was done in the early years of her career, which extended from 1899 until her death. Her best work gave off a spidery, perspectiveless, medieval ambience; later work employed washes without much positive effect. Works of strongest fantasy interest include her illustrations for editions of *The High History of the Holy Graal* (1903), *The Defence of Guinevere* (1904) by William MORRIS and *Goblin Market* (1907 chap) by Christina ROSSETTI. Many of the pamphlets she illustrated were on similar subjects. [JC]

KING, STEPHEN (1947-) US writer, primarily of HORROR – which is almost always DARK FANTASY and sometimes TECHNOFANTASY – as well as of some straightforward fantasy; he is married to the novelist Tabitha King (1949-). He started with ambitions to be an sf writer (rumours that he early published five pseudonymous pornographic novels have been revealed as a hoax) and his first published work, "The Glass Floor" (1967 *Startling Mystery Stories*), is

in this genre. Most of his early sf stories are contained in *Night Shift* (coll **1978**), and it is clear from them that the sf is merely the vehicle used to convey what are really horror stories. Some of his novels deploy sf motifs, they are so shrouded in fantasy and the supernatural that they can hardly be described as SCIENCE FICTION.

Much of his fiction concerns people with TALENTS, and this is the case with his first published novel, *Carrie* (**1974**), about a bitterly unhappy girl approaching MENARCHE whose uncontrolled psychic abilities make her a focus for spectacular POLTERGEIST activity. It was filmed as CARRIE (*1976*), the first of a plethora of movies to be based – sometimes very loosely – on SK's work. Already in this novel SK's strengths and weaknesses are on display: an undoubted narrational power drives the tale through sections of sloppy writing and a tendency to default, when in doubt, to splatter. It is the narrational power that has carried SK's career ever since, for he is rarely a thematic innovator, instead generally concentrating on the reworking of old riffs. This is manifest in his second novel, *'Salem's Lot* (**1975**) – filmed as 'SALEM'S LOT (1979 tv miniseries; *1979* tvm) – which tells of the arrival of a VAMPIRE in a small New England town, and of efforts to stave off a virtual holocaust of vampirism there. SK's ability, through the overlayering of seemingly irrelevant mundane details, to generate a sense of WRONGNESS found its first full flowering in this novel. The telling is all, as in *The Shining* (**1977**) – filmed as *The* SHINING (*1980*) – a long and complicated tale of a HAUNTING.

The Stand (**1978** cut; restored vt *The Stand: The Complete & Uncut Edition* 1990) is a more original work. Its early parts are derived fairly directly from the post-HOLOCAUST subgenre of sf, as a manmade plague virus, accidentally released from a military establishment, destroys civilization; the survivors must regroup and start anew. Then, however, it mutates into fantasy proper, as it becomes clear that the post-holocaust scenario is to be merely the setting for a tale about the LAST BATTLE between GOOD AND EVIL (◊◊ ARMAGEDDON). The book is overlong (a frequent failing of SK's), and the restored version is, as it were, overlonger, but it cannot be denied that the overall effect is impressive. *The Stand* was made into a successful tv miniseries (1994).

The Dead Zone (**1979**) marked a return to more conventional territory, mixing a political thriller with a straightforward tale of PRECOGNITION; it was filmed as *The* DEAD ZONE (*1983*). This was the first of SK's novels to be set in Castle Rock, a new England small town indistinguishable from his other small-town settings except by name. Even so, it is clearly an important piece of territory so far as SK is concerned; his company designed to exploit the movie rights in his work is named Castle Rock. *Firestarter* (**1980**) reprises the central notion of *Carrie*: a little girl has the power of pyrokinesis, and is sought by a faceless, clandestine US Government organization, which wishes to put this talent to military or espionage use. There are anecdotal reports that SK wrote this book because he felt the theme had been badly tackled in The FURY (*1978*), based on John Farris's novel *The Fury* (**1976**). Whatever the truth of this, SK's variation is itself filmed as FIRESTARTER (*1984*). *Cujo* (**1981**) tells of a rabid dog threatening Castle Rock; it was filmed as *Cujo* (*1983*).

The Dark Tower: The Gunslinger (**1982**) was the first of the picaresque **Dark Tower** series – continued in *The Dark Tower II: The Drawing of the Three* (**1987**) and *The Dark Tower III: The Waste Lands* (**1991**) – a transposition of elements of Robert BROWNING's "Childe Roland to the Dark Tower Came" into a bleakly fantasticated, underpopulated landscape of the future (although it is an imprecise, fantasist's future rather than an sf one), where strange events occur among the decaying detritus of a lost technological age. Although these are clearly SK novels – his voice in them is unmistakable – they represent something of a detour from the mainstream of his writing. He had published four earlier such diversions under the pseudonym Richard Bachman: *Rage* (**1977**), *The Long Walk* (**1979**), *Roadwork* (**1981**) and *The Running Man* (**1982**) – filmed as *The Running Man* (**1985**). When "Bachman's" identity was revealed at the time of publication of *Thinner* (**1984**), the four earlier books were reissued as *The Bachman Books: Four Early Novels by Stephen King* (omni **1985**; vt *The Bachman Books: Four Novels by Stephen King* UK). These first four are more like long novellas than novels. Only the fifth, *Thinner*, is a novel proper, and only it is fantasy: a man behaves appallingly to an old gypsy, who puts on him a CURSE such that he gradually, and irreversibly, loses weight.

Although SK was still publishing fairly prolifically, there is a sense of hiatus about his output during the first years of the 1980s, as if he had lost direction. This came to an end when he returned to what one might call Stephen King Territory with *Pet Sematary* (**1983**) and *Christine* (**1983**) – filmed as PET SEMATARY (**1989**) and CHRISTINE (**1983**) respectively. The first is an intriguing DARK FANTASY: the dead, when interred in an ancient Indian burial ground, come back to life, but "soured". *Christine* is a TECHNOFANTASY about a car that is possessed (◊ POSSESSION) and in turn possesses its owners. *The Talisman* (**1984**), with Peter STRAUB, is a modern QUEST tale through ALTERNATE REALITIES; highly ambitious, it is not very successful, Straub's more literary style clashing with King's less formal narration, to the detriment of both.

Again there came something of a hiatus, this time ended by *It* (**1986**), a vast and slightly chaotic novel in which a group of children become aware of and eventually defeat the malignant spirit that dwells in the tunnels and sewers under their small New England town. It was made into a tv miniseries, *It* (1990). *Misery* (**1987**) was a complete contrast: a taut, controlled, comparatively short PSYCHOLOGICAL THRILLER in which a psychopath insists violently that a novelist not kill off, as planned, his series character. Although not fantasy, it reads like a nightmare, and is often rated as SK's best novel. It was filmed as *Misery* (*1990*) and adapted as a stage play. *The Tommyknockers* (**1987**) was another long, rambling novel; a TECHNOFANTASY about the discovery of a long-ago crashed UFO and the evil that emanates from it, it seems to owe much to Nigel Kneale's QUATERMASS AND THE PIT (1958-9 tv serial UK). *The Tommyknockers* (1994) was the miniseries.

The Dark Half (**1989** UK) is a first-rate fantasy and its theme one of the most original to appear in a HORROR novel. A respected but commercially unsuccessful novelist has written under a pseudonym a series of nasty but best-selling thrillers. Now he wants to lay the pseudonym – and the series' psychopathic VILLAIN – to rest (◊ SHADOWS). However, the villain refuses to die, reifying himself and brutally murdering several of the novelist's acquaintances in an attempt to force him either to resume writing the series or to pass along the baton of writing to the villain himself. The movie was *The* DARK HALF (*1992*).

SK's next book was a collection of four novels, *Four Past*

Midnight (coll **1990**). "The Langoliers" is a rather silly sf fantasy involving TIME TRAVEL into the past – which is depopulated because humans and other animals exist only in the "now". "Secret Window, Secret Garden" tackles once more the themes of *Misery* and *The Dark Half*. "The Library Policeman" is one of King's best and most imaginative horror novels, an uncanny fantasy mixing childhood terrors with a fairly sombre take on such topics as child sexual abuse and alcoholism. In "The Sun Dog" a Polaroid camera takes photographs that show a terrifying canine spirit progressively invading our REALITY.

In his preface to "The Sun Dog" SK explains that it is intended as the second of a trilogy that stands as his farewell to Castle Rock: the first was *The Dark Half* and the third was *Needful Things* (**1991**). Clearly SK expected much of this novel – about how a shopkeeper, probably the DEVIL, sets various members of Castle Rock's population against each other in an escalation of violence – but the end-product was an overlong, episodic and eventually tiresome piece. It was filmed as NEEDFUL THINGS (*1993*). *Gerald's Game* (**1992** UK) and *Dolores Claiborne* (**1992**), the latter filmed in 1995, are nonfantastic.

It was time for SK to reassert himself. Although it may have disappointed many of his regular readers in that it contains little horror, *Insomnia* (**1994**) is an excellent – and huge – fantasy novel. Its thesis is that there are four forces that drive the Universe – Life, Death, the Purpose and the Random – and that agents of the Purpose are responsible for ending people's lives at the "correct time" while leaving room for the corresponding agent of the Random, who may snatch them away at any moment. But now the Random has imbalanced the relationship between itself and the Purpose, and the Purpose agents draw in an elderly human couple to help sort things out. Central to the plot is the topic of abortion: SK conscientiously presents both sides of the argument with some fairness, although his "pro-lifers" eventually prove psychopathic. *Insomnia* was followed by *Rose Madder* (**1995**) which mixes a standard SERIAL-KILLER tale (a woman married to a psychopath flees after years of physical abuse, and is pursued by him) with a complex piece of fantastication: the fugitive enters a picture to discover her *alter ego* (which is also her courage) and later induces her husband into the picture where he is murdered by, in effect, himself.

As noted, SK's work has been much filmed. Movies not mentioned above include *Creepshow* (*1982*), *Cat's Eye* (*1984*, CHILDREN OF THE CORN (*1984*), *Silver Bullet* (*1985*), *Maximum Overdrive* (*1986*) – dir by SK himself – *Stand by Me* (*1986*), *Creepshow II* (*1987*), *Graveyard Shift* (*1990*), *The Lawnmower Man* (*1992*) – although here the relationship was so tenuous that SK dissociated himself from the movie – *Sometimes They Come Back* (*1991*), *Sleepwalkers* (*1992*) and – not fantasy at all, but one of the best – *The Shawshank Redemption* (*1995*). Sequels based on his situations and/or characters have been *Return to 'Salem's Lot* (*1987*), *Creepshow II* (*1987*), *Children of the Corn II: The Final Sacrifice* (*1992*) and *Pet Sematary II* (*1992*).

SK is probably the most successful English-language writer of this century, and his total sales in all languages are likely incalculable. Although he cannot be said to have contributed much new to the world of fantasy, as a synthesist and reworker of existing ideas he has played an important part in bringing such ideas to a vaster audience than might otherwise have encountered them: the underlying notion of, say, *The Dark Half* might have appeared in a dozen other fantasies before, but none of those had any noticeable impact on the public consciousness. He has had numerous imitators in the HORROR field – indeed, it could be maintained that he is responsible for the resurgence of horror as a commercial genre since the late 1970s – but no obvious disciples within fantasy proper. [JG]

Other works: *Frankenstein, Dracula, and Dr Jekyll and Mr Hyde* (omni/anth **1978**), ed; "The Mist" (1980 in *Dark Forces* [anth **1980**] ed Kirby McCauley), short horror novel; *The Monkey* (**1980** chap); *Danse Macabre* (coll **1981**), essays on horror fiction; *The Raft* (**1982** chap); *The Plant, Part 1* (**1982** chap), *Part 2* (**1983** chap) and *Part 3* (**1985** chap); *Creepshow* (coll **1982**); *Different Seasons* (coll **1982**), the last story of which was also published separately as *The Breathing Method* (**1984** chap); *Cycle of the Werewolf* (**1983**; vt as coll with movie screenplay *Silver Bullet* 1985); *Skeleton Crew* (coll **1985**; exp by 1 story 1985); *The Eyes of the Dragon* (**1985** limited edn; rev 1987 trade edn); *My Pretty Pony* (**1988** chap), nonfiction; text for *Nightmares in the Sky* (coll graph art **1988**) by F-Stop Fitzgerald; *Dolan's Cadillac* (**1989** chap); *Nightmares and Dreamscapes* (coll **1993**).

Further reading: *The Stephen King Companion* (anth **1989**) ed George W. Beahm; *The Shape Under the Sheet: The Complete Stephen King Encyclopedia* (**1991**; vt *The Complete Stephen King Encyclopedia* 1992) by Stephen J. Spignesi; *The Work of Stephen King: An Annotated Bibliography and Guide* (**1991**) by Michael R. Collings; *The Films of Stephen King* (**1994**) by Ann Lloyd; very many others.

KING, TAPPAN (WRIGHT) (1950-   ) US editor and writer who worked for Bantam Books before becoming editor (1986-9) of ROD SERLING'S THE TWILIGHT ZONE MAGAZINE, where he tended to restrict his fantasy to tales ranging from SLICK FANTASY through DARK FANTASY. His first novel was *Nightshade* * (**1976**) with his wife Beth Meacham (1951-   ), tied to the **Weird Heroes** sequence of fantasies. *Down Town: A Fantasy* (**1985**) with Viido Polikarpus (1946-   ), who also illustrated the book, is an URBAN FANTASY whose young protagonist finds underneath NEW YORK an OTHERWORLD called Down Town, whose evil ruler he defeats. [JC]

KING KONG MOVIES A number of movies have been based on the premise of **1**. In early 1996 it was reported that another remake was on its way, this time from Universal.

1. *King Kong* US movie (**1933**). RKO. **Exec pr** David O. Selznick. **Dir** Merian C. Cooper, Ernest B. Schoedsack. **Chief technician** Willis H. O'Brien. **Screenplay** James Creelman, Ruth Rose. **Novelizations** *King Kong* * (**1932**) by Delos W(heeler) Lovelace (1894-1967), a *King Kong* (graph **1970**) by Alberto GIOLITTI – which was a GRAPHIC-NOVEL version – and an illustrated version, *King Kong* (graph **1994**) by Anthony Browne. **Starring** Robert Armstrong (Carl Denham), Bruce Cabot (Jack Driscoll), Fay Wray (Ann Darrow) and "King Kong (The Eighth Wonder of the World)". 100 mins. B/w.

Ruthless moviemaker Denham takes his team, including starlet Darrow, to an uncharted ISLAND, legendary home of the god Kong. The LOST RACE they find there worships the giant APE-god, and Darrow is seized from the ship's deck as a HUMAN SACRIFICE. Denham's men, led by ship's mate Driscoll, pursue, fighting DINOSAURS in a TARZAN-esque jungle until Driscoll saves her. Kong is captured and shipped back to New York, trained, and put on stage – a scene parodied in *Young Frankenstein* (**1974**; ◊ FRANKENSTEIN MOVIES).

Flashbulbs terrify the chained, seemingly crucified MONSTER; he panics, and charges through the streets of New York, creating mayhem. Seizing Darrow, he climbs to the pinnacle of the Empire State Building – a symbol of civilization which dwarfs even the huge jungle king – from where he is shot down by aeroplanes.

KK is a classic of the CINEMA, the forefather of the MONSTER MOVIE and, arguably, of the JUNGLE MOVIE (although in neither case the first), a significant new variation on the old theme of BEAUTY AND THE BEAST, a triumph (for its day) of spfx . . . It becomes hard to see it for what it is: an extremely good movie. Wray's performance is far better than usually remembered, and the emotional dynamic of the plot far more sophisticated. Kong's touching demonstrations of apparent love for the struggling, shrieking Darrow (or is it only that he's fascinated by this endlessly mobile trinket? or by her blondeness?) are juxtaposed ably with scenes of him thrusting screaming extras into his great maw. It is the triumph of *KK* that our sympathies come to reside with the monster.

KK's credits acknowledge it to be based on a story by Cooper and Edgar Wallace, but it is unclear how much Wallace, who died in 1932, was involved. *KK* was hurriedly sequelled by **2**, and was remade as **3**, in turn sequelled by **4**. *King Kong Versus Godzilla* (**1963**; ◊ GODZILLA MOVIES) and *King Kong Escapes* (**1967**) exploited the name. [JG]

Further reading: *The Making of King Kong: The Story Behind a Film Classic* (**1975**) by Orville Goldner and George E. Turner; *On the Other Hand* (**1989**) by Fay Wray (1907-   ).

2. *Son of Kong* US movie (**1933**). Radio. **Exec pr** Merian C. Cooper. **Assoc pr** Archie Marshek. **Dir** Ernest B. Schoedsack. **Spfx** Willis O'Brien. **Screenplay** Ruth Rose. **Starring** Robert Armstrong (Carl Denham), Helen Mack (Helen), John Marston (Helstrom), Frank Reicher (Captain Engelhorn), Victor Wong (Charlie). 69 mins. B/w.

The sequel to **1**. Rarely has a direct sequel been so poor compared to its original: in desperation it was billed as "a serio-comic phantasy". The entrepreneur Denham, fleeing from his creditors in New York after the devastation caused by King Kong, returns to Kong Island and discovers a smaller version of the original. An earthquake destroys the island, but Kong Jr saves Denham so that he can marry pretty Helen, picked up along the way. [JG]

3. *King Kong* US movie (**1976**). Paramount/Dino de Laurentiis. **Pr** Dino de Laurentiis. **Dir** John Guillermin. **Spfx** Carlo Rambaldi, Glen Robinson, Frank Van Der Meer. **Screenplay** Lorenzo Semple Jr, published 1977. **Based on 1**. **Starring** René Auberjonois (Roy Bagley), Rick Baker (King Kong), Jeff Bridges (Jack Prescott), Charles Grodin (Fred S. Wilson), Jessica Lange (Dwan). 134 mins. Colour.

A loving recrafting of **1**, rewritten in order to update the story and render the circumstantial details more "plausible": notable differences of detail are that the expedition is mounted by an oil company (Petrox) rather than a movie producer, that there are no DINOSAURS on the island (although a giant SERPENT appears), that the actress Dwan – picked up from a sinking pleasure yacht – is somewhat hard-bitten rather than an ingenue, that the chief protagonist, Prescott, is a stowed-away primate-sociology professor, and that the denouement involves the twin towers of the World Trade Centre rather than the Empire State Building. Oddly, the "realistic" framing has the effect of giving the central plot a greater sense of fantastication than in **1**, an effect enhanced by the direction: Guillermin had directed and cowritten *Tarzan's Greatest Adventure* (*1959*) and *Tarzan Goes to India* (*1962*), and the scenes on the island are distinctly reminiscent of TARZAN MOVIES, with Bridges in the role of the Jungle Lord. While this is no cinematic landmark, it is a considerable movie, with a sparky screenplay, several good performances and Oscar-winning spfx.

All copies of the UK/Italian sex comedy *Queen Kong*, released about the same time, were apparently bought by De Laurentiis to avoid perceived competition. [JG]

4. *King Kong Lives* US movie (*1986*). De Laurentiis. **Pr** Martha Schumacher. **Dir** John Guillermin. **Screenplay** Steven Pressfield, Ronald Shusett. **Starring** John Ashton (Nevitt), Peter Elliot (King Kong), Linda Hamilton (Amy C. Franklin), Brian Kerwin (Hank Mitchell), George Yiasomi (Lady Kong). 105 mins. Colour.

Little-liked sequel to KING KONG (*1976*), at whose end Kong was, we find, not dead but comatose: a team led by Dr Franklin maintains him on life-support while waiting for the completion of an artificial heart. INDIANA JONES-style adventurer Mitchell discovers a female, Lady Kong, and brings her to the USA where she is used to give Kong a blood transfusion as the artificial heart is installed. On recovering, Kong whiffs her pheromones, breaks free and liberates her, and the pair head for the hills. The army under Nevitt recaptures Lady Kong, and Kong is presumed dead; but months later he emerges to storm an army base, free the now pregnant Lady Kong, and see the birth of his son before dying in a hail of bullets. She and child are given a reserve in Borneo in which to find their destiny. [JG]

KING KONG VERSUS GODZILLA (ot *King Kong Tai Gojira*) Japanese movie (*1963*). ◊ GODZILLA MOVIES; KING KONG MOVIES.

KINGSLEY, CHARLES (1819-1875) English clergyman, author, academic and Christian Socialist, remembered for his historical novels *Westward Ho!* (*1855*) and *Hereward the Wake* (*1865*) and, especially, for his CHILDREN'S FANTASY *The Water Babies* (1862-3 *Macmillan's Magazine*; *1863*). The story is about a series of TRANSFORMATIONS occurring to Tom, an orphan chimney-boy. The first is when he encounters the girl Ellie in a big mansion and realizes for the first time that he is dirty. This compels him to wash in a river: he drowns and is spiritually transformed into a water baby. The analogy with baptism and the cleansing power of water is self-evident. Not until he learns that he has been selfish and repents is he allowed to see other water babies. A further transformation takes place on an ISLAND, where he learns about GOOD AND EVIL. He is finally joined by Ellie, who drowned in a rock pool, but in order to be with Ellie and share her purity he must go on a QUEST to rescue his old bullying master, and this allows him at last to attain HEAVEN. Despite the strong moral overtones, this was written more for entertainment than instruction and was the first modern English children's fantasy novel which did not draw from FOLKLORE roots (◊ FAIRYTALES). It was because of this book's success that Macmillan published Lewis CARROLL's *Alice's Adventures in Wonderland* (*1865*), urged on by recommendations from George MACDONALD, a friend of Kingsley's and a fellow Christian reformer. The book has been filmed somewhat disappointingly as *The* WATER BABIES (*1978*), with a more saccharine ending.

The Water Babies was something of a one-off for CK. His only other children's fantasy was *The Heroes* (*1856*), a retelling of the Greek LEGENDS of Perseus, the Argonauts

and Theseus, done in a more direct and realistic fashion than the earlier renditions by Nathaniel HAWTHORNE. *The Water Babies* itself was immensely influential – its mark can be seen on the works of such later children's writers as Mary Molesworth and E. NESBIT, and even Joan AIKEN – and ushered in the Golden Age of CHILDREN'S FANTASY.

CK's brother, Henry Kingsley (1830-1876), the black sheep of the family, also turned to writing. He produced the children's fantasy *The Boy in Grey* (*1871*), modelled on *Alice in Wonderland* and essentially an ALLEGORY about a boy's childhood search for the Boy in Grey, who represents CHRIST. Although the QUEST takes the boy three days it has in our REALITY taken 11 years (◊ TIME IN FAERIE).

CK's daughter Mary (1852-1931), as Lucas Malet, wrote mostly social romances, though did produce one children's book, *Little Peter* (*1887*), and two romantic GHOST STORIES – *The Gateless Barrier* (*1900*) and *The Tall Villa* (*1919* US) – in both of which LOVE proves powerful over death. [MA]

Further reading: *Canon Charles Kingsley: A Biography* (*1949*) by Una Pope-Hennessy; *The Beast and the Monk* (*1974*) by Susan Chitty; *The Novels of Charles Kingsley* (*1981*) by Allan J. Hartley.

KINGSLEY, HENRY (1830-1876) UK writer. ◊ Charles KINGSLEY.

KINGSLEY, MARY (1852-1931) UK writer. ◊ Charles KINGSLEY.

KING-SMITH, DICK Working name of UK writer Ronald Gordon King-Smith (1922-   ), most of whose 80 or more tales are fantasies; most are for younger children, and thus not treated here in detail. He began writing with "Alphabeasts" for *Punch* in 1965; this set of comic verses was illustrated by Quentin Blake. His first novel, *The Fox Busters* (*1978*), is like many of its successors a BEAST FABLE whose protagonists live on a farm and sometimes exhibit (DK-S has a sharp eye for genuine animal behaviour) ANIMAL-FANTASY traits. His most interesting books tend to elicit comparisons between animal and human behaviour in a traditional beast-fable manner. He is best-known for *The Sheep-Pig* (*1983*; vt *Babe: The Gallant Pig* 1985 US), filmed as *Babe* (*1995*), about a pig that wants to be a sheepdog. Other titles of interest include *Noah's Brother* (*1986*) and *The Schoolmouse* (*1994*). [JC]

KINSELLA, W(ILLIAM) P(ATRICK) (1935-   ) Canadian writer whose first stories, beginning with *Dance Me Outside* (coll *1977*) deal with Native Canadians, centring on the figure of **Simon Ermineskin**. He remains best-known, however, for his BASEBALL fantasies, the most famous being *Shoeless Joe* (*1982* US), expanded from the title story of *Shoeless Joe Jackson Comes to Iowa* (coll *1980*) and filmed as FIELD OF DREAMS (*1989*). The novel differs from the movie mainly in the presence of a fictionalized J.D. Salinger (1919-   ), whose longing for happiness is rewarded when he accompanies the GHOST baseball players to enter some sort of AFTERLIFE; the analogous figure in the movie is called Terence Mann. *The Iowa Baseball Confederacy* (*1986* US) sends its baseball-haunted protagonists by TIMESLIP to the turn of the century, where they participate in an exhibition match which lasts 40 days and 40 nights while around them the world becomes more and more phantasmagoric. Short stories with fantasy content – usually concerning baseball – are assembled in *The Thrill of the Grass* (coll *1984*) and *The Further Adventures of Slugger McBatt: Baseball Stories* (coll *1988*). *Red Wolf, Red Wolf* (coll *1987*) assembles stories about neither Native Canadians nor baseball.

It is hard to judge how significant it is that WPK, a Canadian, has so intensely embraced and recreated the THEODICY-choked myth of US baseball; perhaps outsiders have an edge in longing. [JC]

KIPLING, (JOSEPH) RUDYARD (1865-1936) UK author and poet, in 1907 the first Briton to receive the Nobel Prize for Literature; he was regarded as unofficial Poet Laureate. His vivid, sometimes harshly over-brilliant short stories cross genre boundaries to explore fantasy, sf and SUPER-NATURAL FICTION; those of fantasy interest are scattered through RK's many collections. Early childhood and adult journalism in India coloured all RK's life: his masterpiece, *Kim* (**1901**), is not fantasy, despite its subplot of religious QUEST for a kind of GRAIL, but lovingly paints the subcontinent as an almost enchanted LAND.

Plain Tales from the Hills (coll **1888** India) contains the RATIONALIZED FANTASIES "In the House of Sudhoo", featuring a lurid fake-magical RITUAL, and "The Gate of the Hundred Sorrows", with its opium dreams. *The Phantom 'Rickshaw, and Other Tales* (coll **1888** India; rev **1890** UK) takes its title from a GHOST STORY and also includes the hallucinated "The Strange Ride of Morrowbie Jukes" (1885), set in a rationalized AFTERLIFE – a nightmarish "Village of the Dead" where officially dead Indians are exiled; this collection was incorporated into *Wee Willie Winkie* (coll **1889**). *Soldiers Three and Other Stories* (coll **1888**) includes the minor ghost story "The Solid Muldoon". *Life's Handicap, Being Stories of Mine Own People* (coll **1891**) contains "At the End of the Passage", whose distanced HAUNTING by a blind, weeping face culminates in the photographing of the now-dead victim's eyes (◊ URBAN LEGENDS) to reveal some fearful but undisclosed image; in "The Mark of the Beast", insulting the monkey GOD Hanuman brings POSSESSION by a WOLF's spirit. *Many Inventions* (**1893**) contains: "The Finest Story in the World" (1891), a teasing tale of REIN-CARNATION and elusive, fading past-life memories; "A Matter of Fact" (1892), an atmospherically plausible TRAV-ELLERS' TALE of a sea-serpent (◊ SEA MONSTERS); "The Lost Legion" (1892), with its ghost army almost unnoticed by the English as it terrifies their Afghan foes (◊ PERCEPTION); and "The Children of the Zodiac", a powerfully symbolic FABULATION personifying the ZODIAC creatures as GODS of life and death.

RK's best-known fantasy comprises the powerfully imagined ANIMAL-FANTASY stories in *The Jungle Book* (coll of linked and other stories **1894**) and *The Second Jungle Book* (coll of linked and other stories **1895**). The growing-up of the boy MOWGLI, reared by wolf foster-parents and with a bear and panther as MENTORS (◊◊ TALKING ANIMALS), subject to the Law of the Jungle which is RK's strongest statement of BALANCE, is a new-created MYTH. Mowgli's RITES OF PASSAGE from "cub" to full human include mastering the "Red Flower" of fire which even the wolves fear, proving immune to the MESMERISM of the SERPENT Kaa, destroying the tiger Shere Khan that laid claim to his life, pronouncing a sentence of bloodless oblivion on a village that maltreated his (adoptive) parents and which is memorably swallowed by the jungle, and organizing wolves and other animals in bloody resistance to an oncoming horde of devouring *dhole*-dogs. The DISNEY animated movie *The Jungle Book* (*1967*) is a feeble travesty of the original stories' sometimes shocking intensity.

The Day's Work (coll **1898**) includes: "The Bridge-Builders", in which a doped and delirious European civil engineer overhears the gods of India debating his blasphemy (or otherwise) in imposing the BONDAGE of a bridge on the sacred RIVER Ganges; "The Ship that Found Herself" (1895) and ".007" (1897), eccentric anthropomorphic tales of a ship's component parts and of railway engines (◊ TRAINS); and *The Brushwood Boy* (1895; **1899** chap), with its vivid and plausibly irrational geographies of recurring, shared DREAMS. *Just So Stories for Little Children* (coll **1902**; cut vt *How The Leopard Got his Spots and Other Just So Stories* 1993) contains TALL TALES meant to be read aloud to children, many being humorously bizarre MYTHS OF ORIGIN for the elephant's trunk, camel's hump, leopard's spots, etc.; RK's own unpolished but meticulous illustrations complement the stories. *Traffics and Discoveries* (coll **1904**) includes: the well known "Wireless" (1902), where experiments in telegraphy seem to facilitate a resonance across time (◊ TIME FANTASIES) between a tubercular chemist's assistant and John KEATS, with the former's stumbling efforts at poetry unknowingly echoing "The Eve of St Agnes" and "Ode to a Nightingale"; and *They* (**1905** chap), which elusively and sentimentally tells of a blind woman whose home has become a refuge for children's GHOSTS.

Puck of Pook's Hill (coll **1906**) and *Rewards and Fairies* (coll **1910**) are fantasy chiefly for their FRAME STORY, in which PUCK introduces children to characters from history whose finely told personal sagas touch on the MATTER of Britain – a teaching which for some reason Puck promptly erases *via* MEMORY WIPE. In the first book, "'Dymchurch Flit'" reports the departure of the FAIRIES; in the second, the eponymous COLD IRON of "Cold Iron" is forged by Thor and its touch ends a boy's adoption into FAERIE, substituting human BONDAGE; the prehistoric hero of "The Knife and the Naked Chalk" trades an eye (◊ ODIN) for the secret of metal knives, and pays the harsh price of being thought a god; and "A Doctor of Medicine" is a *tour de force* of medical reasoning based on ASTROLOGY.

Actions and Reactions (coll **1909**) contains one of RK's subtlest ghost stories, "The House Surgeon", with its twofold haunting as a live person's morbid obsession oppresses the house and even troubles its dead. *A Diversity of Creatures* (coll **1917**) has "In the Same Boat" (1911), another story of psychic horrors which becomes RATIONALIZED FANTASY with the discovery that they reflect prenatal trauma, and "Swept and Garnished" (1915), RK's grimmest supernatural fiction, personifying German WORLD WAR I guilt in the figures of children dead from shelling who bloodily appear to a Berlin widow, but to her alone (◊ PERCEPTION). In *Debits and Credits* (coll **1926**), "The Enemies to Each Other" retells ADAM AND EVE in ARABIAN-FANTASY mode; "A Madonna of the Trenches" (1924) is another interesting ghost story; "The Wish House" (1924) is a sombre tale of what Charles WILLIAMS later called Substitution, with an ageing woman taking on herself (or believing she does so) the malign ulcer that would have afflicted the man who, unrequitedly, she loved; "The Gardener" unobtrusively features CHRIST; "On the Gate: A Tale of '16" presents the machinery of AFTERLIFE as a benign bureaucracy where ANGELS, DEATH, SATAN, the WANDERING JEW and others all work together – like friendly Army quartermasters – to wangle SOULS into HEAVEN even when the rules consign them to HELL (◊ AS ABOVE, SO BELOW); a similar story in this vein is "Uncovenanted Mercies" in *Limits and Renewals* (coll **1932**). *Thy Servant a Dog* (**1930**), told with relentless sentimentality from an adoring dog's viewpoint, is not highly regarded.

Many relevant RK stories are usefully assembled in *The Complete Supernatural Stories of Rudyard Kipling* (coll **1987**) ed Peter HAINING. This overlaps with two volumes ed John BRUNNER: *Kipling's Fantasy* (coll **1992**) and *Kipling's Science Fiction* (coll **1992**; vt *The Science Fiction Stories of Rudyard Kipling* 1994 US).

RK is a major figure of English literature who used the full power and intensity at his command during excursions into fantasy. Sometimes he seems wilfully obscure; a master of "less is more" narration, he recommended a repeated cutting of inessentials from stories, and occasionally cut too much – while his phonetic reproductions of lower-class and Irish accents can be embarrassing. But his works repay close attention. [DRL]

Further reading: *Something of Myself* (**1937**), a taciturn and unreliable fragment of autobiography; *Rudyard Kipling* by Charles Carrington (**1955**); *The Art of Rudyard Kipling* (**1959**) by J.M.S. Tompkins; *Kipling and the Children* (**1965**) by Roger Lancelyn GREEN; *Rudyard Kipling* (**1966**) by J.I.M. Stewart (1906-1994); *The Strange Ride of Rudyard Kipling* (**1977**) by Angus Wilson (1913-1991); *Rudyard Kipling and His World* (**1977**) by Kingsley AMIS; *Rudyard Kipling* (written 1948 but suppressed; **1978**) by Lord Birkenhead (1907-1975).

KIRBY, JOSH (1928-   ) UK artist and illustrator whose teeming, ebullient oil paintings have been used most famously since 1984 for the Gollancz, Corgi and Doubleday covers for the **Discworld** books by Terry PRATCHETT. These illustrations create fantasy worlds peopled by a tumbling mass of clumsy ogres, gesticulating wizards, chunky barbarians and busty nymphettes, all with enormous *joie de vivre*. His rich colour range and monumental compositions show the influence of the mural painter Frank Brangwyn (1867-1956).

JK early gained a reputation as a portraitist, but quickly turned to book-cover ILLUSTRATION, discovering a flair for sf. His early illustrations – for editions of works by Lin CARTER (DAW Books 1979-81), Ron GOULART (DAW Books 1980-82), Robert SILVERBERG (Pan Books 1983) – differ in nature from his later covers, for Craig Shaw GARDNER (Headline 1987-9) and Pratchett. He has also painted movie posters for *Starflight One* (**1981**), *The Beastmaster* (**1982**), KRULL (**1983**), *Morons from Outer Space* (**1985**) and *Return of the Jedi* (**1983**). [RT]

Other works: *The Josh Kirby Poster Book, As Inspired by Terry Pratchett's Discworld Novels* (portfolio **1989**); *Eric* (**1990**), published as a co-authorship with Pratchett; *In the Garden of Unearthly Delights: The Paintings of Josh Kirby* (graph coll **1991**) with text by Nigel Suckling; *The Josh Kirby Discworld Portfolio* (portfolio **1993**).

KIRCHOFF, MARY L(YNN) (1959-   ) US author of various ties for the **DragonLance** enterprise, including two titles in the **DragonLance Preludes** sequence, *Kendermore* * (**1989**) and *Flint, the King* * (**1990**) with Douglas NILES. Titles in the main **Dragonlance** sequence include *Wanderlust* * (**1991**), *The Black Wing* * (**1993**) (tied to the **Villains** game) and the **Defenders of Magic** ancillary sequence comprising *Night of the Eye* * (**1994**), *The Medusa Plague* * (**1994**) and *The Seventh Sentinel* * (**1995**). [JC]

KIRKUP, JAMES (1918-   ) UK poet, novelist and teacher, who has spent many years abroad in the latter capacity. His first book, *A Cosmic Shape: An Interpretation of Myth and Legend with Three Poems and Lyrics* (coll **1946**) with Ross Nichols, contains speculative nonfiction about the relationship between the fabulous and the mundane,

plus several poems intended as illustrative. *The True Mistery of the Passion* (**1961**) is a fantasy about CHRIST's crucifixion. *Queens Have Died Young and Fair: A Fable of the Immediate Future* (**1993**) is a fantasy novel set in AD2001 on Liberalia, a newly discovered Atlantic ISLAND. The plot, hyperbolically satirical of many subjects, is loosely based on William SHAKESPEARE's *The Tempest* (performed *c*1611; **1623**). JK's translations include Erich KÄSTNER's *Der Kleine Mann* (**1963**; trans as *The Little Man* **1966** UK) and *Tales of Hoffmann* (coll trans **1966**) (◊ E.T.A. HOFFMANN). [JC]

KIRSTEIN, ROSEMARY (1953-   ) US author of a series beginning with *The Steerswoman* (**1989**) and *The Outskirter's Secret* (**1992**). RK employs the familiar format of a QUEST to explore epistemology, responsibility and cultural assimilation. The protagonist, the "steerswoman" Rowan, is a purveyor and disseminator of knowledge to the varied cultures of a SECONDARY WORLD (in fact, a post-HOLOCAUST one). The novels chart her need to confront and deal with problems which arise when knowledge is hidden, exploited or perverted, as well as wider issues surrounding the nature of knowledge itself, and the responsibilities incumbent upon those who possess it. RK used the byline R.R. Kirstein for her first story, "Act Naturally" (1982 *Isaac Asimov's Science Fiction Magazine*), but in fact has no middle initial. [KLM]

KISMET Six movies have been based directly or indirectly on the ARABIAN-FANTASY play *Kismet: An "Arabian Night" in Three Acts* (**1911**) by Edward Knoblock (1874-1945). Little is known about the French production (*1919*) – indeed, it is uncertain that prints survive.

1. *Kismet* UK movie (*1914*). Zenith. **Dir** Leedham Bantock. **Screenplay** Bantock. **Starring** Oscar Asche (Hajj), Lily Brayton (Marsinah), Herbert Grimwood (Wazir), Caleb Porter (Sheik Jawan), Frederick Worlock (Caliph Abdallah). B/w, silent.

Hajj, arrested for street-theft, makes a CONTRACT with the Wazir of BAGHDAD: he will regain freedom in exchange for assassinating the new caliph, Abdallah. The assassination attempt fails, and Hajj is thrown into prison – where he finds himself sharing a cell with his old foe Jawan, whom he strangles. Dressed in Jawan's clothing, he escapes to discover his daughter, Marsinah, has been abducted into the Wazir's harem; meanwhile the caliph, who has been wooing her in the guise of a humble gardener, declares himself. All is solved when Hajj murders the Wazir, frees Marsinah, is banished by the caliph for his various crimes, and covertly returns to the Baghdad streets, where he continues his twin careers of petty theft and death.

Asche got the movie part of Hajj because he had premiered the stage role (1911) in London. We have been unable to trace a print of this movie. [JG]

2. *Kismet* US movie (*1920*). Robertson-Cole. **Dir** Louis Gasnier. **Screenplay** Gasnier. **Starring** Leon Barry (Caliph Abdallah), Marguerite Comont (Nargis), Elinor Fair (Marsinah), Herschall Mayall (Jawan), Hamilton Revelle (Wazir Mansur), Cornelia Otis Skinner (Miskah), Otis Skinner (Hajj), Rosemary Theby (Kut al-Kulub). B/w, silent.

This was a fairly faithful remake, but this time the role of Hajj was taken by Skinner, who had premiered the part (1911) in New York. Once more, we have been unable to trace a print. [JG]

3. US movie (*1930*). First National. **Dir** John Francias Dillon (with simultaneous German version **dir** William

Dieterle). **Screenplay** Howard Estabrook. **Starring** Sidney Blackmer (Wazir Mansur), Mary Duncan (Zaleekha), David Manners (Caliph Abdallah), Otis Skinner (Hajj), Loretta Young (Marsinah). 90 mins. B/w.

Skinner reprised his role as Hajj in this sound remake, the last to show fidelity to Knoblock's play: the subsequent remakes altered the plot more considerably. Of interest is the fact that this was made simultaneously in English and German, with minor variations between the two versions. [JG]

4. (vt *Oriental Dream*) US movie (*1944*). MGM. **Pr** Everett Riskin. **Dir** William Dieterle. **Spfx** Warren Newcombe. **Screenplay** John Meehan. **Starring** Edward Arnold (Grand Vizier Mansur), Florence Bates (Karsha), Hobart Cavenaugh (Moolah), Ronald Colman (Hafiz), James Craig (Caliph), Harry Davenport (Agha), Marlene Dietrich (Jamilla), Hugh Herbert (Feisal), Joy Ann Page (Marsinah), Robert Warwick (Alfife). 100 mins. Colour.

Hafiz the TRICKSTER, self-styled King of BAGHDAD's Beggars, spends his evenings disguised as a prince and conducting various clandestine liaisons, notably with Jamilla, enforced and unconsummated wife of the loathed Grand Vizier. Meanwhile the new Caliph enjoys wandering the streets at night in the guise of his gardener's son; as such he woos Marsinah, Hafiz's daughter. After a Gilbertian froth of a plot, the two pairs of lovers become officially recognized items. Owing considerable debts to *The* THIEF OF BAGDAD (*1924*) and to *The Prince and the Pauper* (*1937*; based on the tale by Mark TWAIN), this bubbles along merrily enough for the first half-hour, then dissipates into a meringue of badly synchronized dance routines and astonishingly badly dubbed songs, almost always abjuring fantasy. Dieterle had been here before, directing the German version of **3**. [JG]

5. US movie (*1955*). MGM. **Pr** Arthur Freed. **Dir** Vincente Minnelli. **Spfx** Warren Newcombe. **Screenplay** Luther Davis, Charles Lederer. **Based on** 1953 stage musical (itself based on both Knoblock's play and **4**) by Davis and Lederer, with music and lyrics by George Forrest and Robert Wright based on themes by Alexander Borodin (1833-1887). **Starring** Ann Blyth (Marsinah), Sebastian Cabot (Grand Wazir), Vic Damone (Caliph), Jay C. Flippen (Jawan), Dolores Gray (Lalume), Howard Keel (Hajj), Monty Woolley (Omar). 113 mins. Colour.

A broke Omar Khayyám figure in BAGHDAD discovers begging is more profitable than poetry, assumes the identity of the beggar Hajj, and is at once threatened by the bandit Jawan, who seeks his long-lost son. Jawan's son proves to be the pompous Grand Wazir . . . Various plot complications ensue until various pairs of lovers are united. The hugely popular stage musical did less well as a movie. The settings are lavish, but cannot make up for the paucity of imagination and daring. Newcombe, responsible for **4**'s spfx, repeats the task here. [JG]

KISS OF THE SPIDER WOMAN Brazilian/US movie (*1985*). HB/FilmDallas. **Pr** David Weisman. **Exec pr** Francisco Ramalho Jr. **Screenplay** Leonard Schrader. **Based on** *El beso de la mujer araña* (*1976*; trans Tomas Colchie as *Kiss of the Spider Woman* 1979 US) by Manuel Puig (1932-1990). **Starring** Sonia Braga (Leni/Marta/Spider Woman), William Hurt (Luis Molina), Raul Julia (Valentin Arregui). 121 mins. Colour and leached colour.

Incarcerated in an anonymous Latin American dictatorship are ostentatiously gay Molina and tough, straight journalist Valentin, imprisoned respectively for paedophilia and for aiding a revolutionary. To escape their misery, Molina recounts an old romantic Nazi-propaganda movie melodrama, scenes from which we see in colour leached to emulate sepia. Valentin gradually melts, and the two men become united in ESCAPISM; although Molina, for privileges, has agreed to betray to their masters any information Valentin might let slip. The night before his parole, Molina tells a different tale, that of a beautiful woman held on an ISLAND by the spiderweb she herself produces, and of how one day a man, Valentin, is washed up on the beach and would die were it not for her sweet KISS. That night the two men make love, and in the morning Valentin gives Molina a telephone number and a message. This secret Molina does not betray; freed, he calls the number and agrees to a meeting, not knowing he is under police surveillance. Caught in the crossfire, he refuses, having been given the gift of dignity by Valentin, to reveal the number to the police, and dies. For his part, although still under torture in prison, Valentin has been given by Molina the DREAM of the Spider Woman to sustain him.

Resting almost entirely on the performances of Hurt and Julia, *KOTSW* is an exceptionally moving portrayal of the wellsprings of FANTASY. [JG]

KLEIN, T(THEODORE) E(IBON) D(ONALD) (1947-) US editor and writer, often referred to as Ted Klein; he has also written as Kurt Van Helsing. He was founding editor of ROD SERLING'S THE TWILIGHT ZONE MAGAZINE, editing it 1981-5 and publishing a wide range of horror and fantasy, though generally avoiding full-fantasy tales. *Great Stories from Rod Serling's The Twilight Zone Magazine: 1983 Annual* (anth **1983**) is a representative selection.

The Events at Poroth Farm (1972 *From Beyond the Dark Gateway*; rev **1990** chap), TEDK's first and best-known story, deposits its protagonist, a graduate student interested in SUPERNATURAL FICTION, in the isolated Poroth Farm, a BAD PLACE seemingly the focus for a resurgence of an unknowable EVIL. In the expansion, *The Ceremonies* (**1984**), this unknowable evil is unveiled – partly through an extensive map of references to writers like H.P. LOVECRAFT and Arthur MACHEN – as a resurgence of the ELDER GODS in the shape of a DRAGON, ushered into the contemporary world by a vicious acolyte who arranges the necessary RITUALS. The sense that the elder gods are MONSTERS and that humans are little more than mayflies in a malign CYCLE turns the horrific THINNING imagery of tale and ending into a sustainable vision of the Universe as a whole. Most of TEDK's later fiction is assembled in *Dark Gods* (coll **1985**) – a collection including "Nadelman's God", which won a 1986 WORLD FANTASY AWARD for Best Novella – and focuses on typical DARK FANTASY venues like NEW YORK, as in "Children of the Kingdom" (1980), in which a race of blind monsters surfaces during a power blackout. [JC]

Other works: *Raising Goosebumps for Fun and Profit* (**1988** chap), nonfiction.

KNAAK, RICHARD A(LLEN) (1961-) US writer involved at the beginning of his career in the **DragonLance** enterprise, for which he wrote several ties, including contributions to the **DragonLance Heroes** sequence: *The Legend of Huma* * (**1988**) and *Kaz, the Minotaur* * (**1990**). His own similar series, the **Dragonrealm** sequence, set in a FANTASYLAND governed by colour-coded DRAGONS, comprises *Firedrake* (**1989**), *Icedragon* (**1989**), *Wolfhelm* (**1990**), *Shadow Steed* (**1990**), *The Crystal Dragon* (**1993**) and

The Dragon Crown (**1993**). *The Shrouded Realm* (**1991**), *Children of the Drake* (**1991**) and *Dragon Tome* (**1992**) make up a subsidiary sequence describing the **Origin of Dragonrealm**. The COLOUR-CODING is thorough: the red dragon governs volcanic territories, the green dragon the FOREST, etc. *King of the Grey* (**1993**) is an URBAN FANTASY set initially in Chicago; its protagonist finds himself in a CROSS-HATCH relationship to invading creatures from the "Grey", into which he eventually plunges and where he becomes king. [JC]

Other works: *Frostwing* (**1994**); *The Janus Mask* (**1995**), a DARK FANTASY.

KNATCHBULL-HUGESSEN, E(DWARD) H(UGESSEN) (1829-1893) UK politician, created 1st Lord Brabourne in 1880, and author of 14 books of FAIRYTALES; his surname, Knatchbull, was expanded in 1849. The first volume, *Stories for My Children* (coll **1869**), contained fairly standard Victorian tales, in imitation of Hans Christian ANDERSEN and the GRIMM BROTHERS. The reception accorded to this book encouraged EHKH to produce a volume or two of fairytales annually for the next six years: *Crackers for Christmas* (coll **1870**), *Moonshine: Fairy Stories* (coll **1871**), *Tales at Teatime* (coll **1872**), *Queer Folk: Seven Stories* (coll dated 1874 but **1873**), *River Legends, or Father Thames and Father Rhine* (coll **1874**) illus Gustav DORÉ; *Whispers from Fairy-Land* (coll **1874**; cut as *Some Whispers from Fairyland* 1933) and *Higgledy-Piggledy, or Stories for Everybody and Everybody's Children* (coll **1875**) illus Richard DOYLE. After this period he turned to other writings, in particular a life of Cromwell, returning to fairytales in 1878 with *Uncle Joe's Stories* (coll **1878**), *Other Stories* (coll **1879**), *The Mountain-Sprite's Kingdom and Other Stories* (coll **1880**), *Ferdinand's Adventure* (coll **1882**) and *Friends and Foes from Fairyland* (coll **1885**). EHKH's final stories were *The Magic Oak-Tree and Prince Filderkin* (coll **1894**), comprising volume 12 of **The Children's Library** (1892-4 23 vols). [MA]

KNICKERBOCKER, DIEDRICH ◊ Washington IRVING.

KNIGHT, ERIC (MOWBRAY) (1897-1943) UK writer, novelist and cartoonist; his most famous book was *Lassie Come Home* (**1940** US). Although he lived and worked largely in the USA, most of his stories and novels were set in his native Yorkshire. Before Lassie, his best-known creation was **Sam Small**, who first appeared in the satirical novella *The Flying Yorkshireman* (**1936**), in which Sam discovers that he can fly, thanks to the power of faith. *Sam Small Flies Again* (coll **1942**; vt *The Flying Yorkshireman* 1946 US) contains further humorous fantasies – some of the best of their period. [RD]

KNIGHT OF THE DOLEFUL COUNTENANCE In Chapter 19 of Part One of *Don Quixote* (**1605-15**) by Miguel de CERVANTES, Sancho Panza describes the Don as "El Caballero de la Triste Figura", which can be translated as "Knight of the Doleful Countenance". The phrase has a complex effect on many readers, partly because Don Quixote's response is itself complex. Panza has so dubbed him (he says) because the author of the STORY which is being told about them needed a soubriquet at this point, and so wrote it in. This metafictional sophistication certainly makes it natural to assign to post-Cervantes versions of the KOTDC a high degree of selfconsciousness about being entrapped in a told story.

More immediately, the phrase clearly describes the comical half-crazed gaunt simpleton KNIGHT of Part One, the *quixotic* old gentleman who believes he has been called

upon to redress wrongs and who, in a world far more mundane than his chivalric romances have depicted, finds himself tilting against windmills in the belief that they are GIANTS. In the behaviour of this wasted ectomorph – whom illustrators have been wont to picture as hyperactive and praying-mantis-like – there is no glimmer of consciousness that he occupies an embarrassing position in a world which he *takes* to be TWICE-TOLD but which – in fact, despite his assumption that he is *being written* – is without STORY.

The KOTDC is also the saddened old man who, at the end of Part Two, now understands that the world he has been attempting to defend is nothing but a play of SHADOWS, and that secular reality was far harsher – and immeasurably less interesting – than the world of his dreams, the world he thought was being told. He has come to the end of his book. He turns his face and dies.

But what if he had lived on? As the term is used in this encyclopedia, the KOTDC is a figure of sustained, selfconscious BELATEDNESS, a figure caught in a THINNING world who has not only failed to pass through any knot of RECOGNITION into EUCATASTROPHE (◊◊ HEALING) but who has set his face against attempting to do so. He lives, therefore, in the aftermath of a profound disappointment in the nature of the world, and bears his nature as a kind of wound – FISHER-KING imagery tends to suffuse descriptions of the KOTDC, even though he may be defined, like King Fisher in Michael TIPPETT's OPERA *The Midsummer Marriage* (**1955**) or King Haggard in Peter S. BEAGLE's *The Last Unicorn* (**1968**), as one who in fact refuses the role – which necessitates his attempts to *recline* within an aesthetic posture of world-weariness, or to rule a world he blights through his refusal to permit growth. He is an aristocrat who feigns to abjure a world whose contempt he has not actually earned. He remains quixotic, in the sense that he tends to retain an inveterate (though mournful) interest in the minutiae of science and in contriving aesthetic solutions to the problems of getting through the day; but the moment of truth for him is always something sunk deep into the past (◊ STEMMA), and the formulae which give shape to his day-to-day existence are entered into with a profound sigh.

The KOTDC re-entered literature in the first decades of the 19th century as a kind of TWIN to ACCURSED WANDERERS like the protagonist of Charles MATURIN's *Melmoth the Wanderer* (**1820**); he also very closely resembles the first literary VAMPIRES – figures like Lord Ruthven (based on Lord BYRON) in John POLIDORI's *The Vampyre: A Tale* (**1819** chap). But the KOTDC reached full flower as an ICON in the form of the White Knight in Lewis CARROLL's *Through the Looking-glass, and What Alice Found There* (**1871**), through his BONDAGE to the progress of the CHESS match whose rules shape the text (and cause him constantly to fall off his horse in a parody of the Knight's Move), because of the unavailing fertility of his mind (for none of his inventions work) and through his romantic melancholy. The White Knight, as portrayed by Sir John TENNIEL (who was closely briefed by Carroll), is clearly modelled on Don Quixote.

KOTDC figures surface intermittently in the literature of DECADENCE: in Joris-Karl HUYSMANS's *À rebours* (**1884**; trans as *Against the Grain* 1922), whose hero, Des Esseintes, attempts to seek aesthetic release; or in VILLIERS DE L'ISLE ADAM's *Axel* (**1890**; trans **1925**), whose eponymous hero wards off the secular world from within the POLDER of an impregnable castle, inside which he engages in arcane dilettantism – he is famous for his closing utterance: "Live? Our servants will do that for us." He is parodied – as Axel Heist – in *Victory* (**1915**) by Joseph CONRAD. The elderly aristocrat known as the Penguin in Gustav MEYRINK's *Walpurgisnacht* (**1916**; trans **1993**), dancing like a spastic water spider over the ruins of his past life, is clearly a KOTDC, as are the marquises of Carabas (◊ PUSS-IN-BOOTS) in Sylvia Townsend WARNER's "The Castle of Carabas" (in *The Cat's Cradle Book* coll **1940** US), for the original Marquise's guilt at betraying Puss has locked his descendents into a state of paralysed refusal – whenever one of them sees a cat, for instance, he swoons. A far more comprehensive figure is Lord Sepulchrave, the father of Titus in Mervyn PEAKE's **Gormenghast** sequence (**1946-59**). Sepulchrave's profound bondage to ritual, the enormously complicated pattern of his daily life, his posture, his distance, his haunted weariness: all amounts to a definitive portrait of the type.

Sepulchrave has almost certainly been a central influence – along with the White Knight – in the creation of many fantasy lords and mages at whose hearts despair gnaws, like King Haggard (see above), whose woundedness creates a genuine WASTE LAND; and of numerous DYING-EARTH characters in the works of writers like Michael MOORCOCK, whose **Dancers at the End of Time** sequence is rich in such figures, as are M. John HARRISON's **Viriconium** books and Elizabeth HAND's **Winterlong** sequence. The quixotism of Caesar Grailly, in *The Knight on the Bridge* (**1982**) by William Watson (1931-   ), devolves under the pressure of history into a "sane" refusal of transcendental gesture. The guilt-ridden protagonists of Tanith LEE's "Bite-Me-Not or Fleur de Fur" (1984), of Richard GRANT's *Rumors of Spring* (**1987**), of Alexander Jablokov's sf *Carve the Sky* (**1991**) and of Paul HAZEL's *The Wealdwife's Tale* (**1993**) all show evidence of a continuing (and evolving) image. [JC]

KNIGHTRIDERS US movie (*1981*). United Artists/ Laurel. **Pr** Richard P. Rubinstein. **Exec pr** Salah M. Hassanein. **Dir** George A. Romero. **Spfx** Larry Roberts. **Screenplay** Romero. **Starring** Brother Blue (Merlin), Ken Foree (Little John), Christine Forrest (Angie), Ed Harris (King Billy), John Hostetter (Tuck), Amy Ingersoll (Linet), Gary Lahti (Alan), Tom Savini (Morgan), Patricia Tallman (Julie). 145 mins. Colour.

King Billy leads a band of biker "knights" who make a living by jousting at pseudo-medieval fairs across the USA, with the full accoutrement of lances and other nonlethal but nevertheless fairly heavy weapons. Billy (for whom read ARTHUR) is interested more in the chivalric than the financial aspects, a pose that leads to his downfall, for his lieutenant, Morgan (for whom read Mordred), is keen to take the crown. Bravest of the king's KNIGHTS – although far from ablest – is Alan (for whom read LANCELOT), who in the end claims Billy's woman Linet (for whom read GUINE-VERE). Morgan eventually attains the crown; Billy avenges the honour of his knights on a corrupt police officer and dies in a road accident, dreaming of a knight from the Age of Chivalry.

K was released at the same time as EXCALIBUR (*1981*), and both were excoriated by the critics: now *Excalibur* is highly regarded and *K* is deservedly a popular cult movie. *K* is often muddled (complete with the introduction of some characters from the ROBIN HOOD rather than the Arthur mythos), but has vigour, ambition and verve. [JG]

KNIGHTS The medieval knight of ROMANCE reaches his apogee in the Arthurian cycle (◊ ARTHUR; ◊◊ GALAHAD; GAWAIN; LANCELOT; PERCEVAL); as the wildly overdrawn

hero of innumerable Spanish ROMANCES of chivalry, particularly popular in the 16th century, he reaches his nadir.

It is this ludicrous figure of the knight errant that Miguel de CERVANTES mocks (and cherishes) in *Don Quixote* (**1605-15**). As a warrior and representative of the Lords Temporal (◊ ESTATES SATIRE) the knight was by this time a figure of the past. As one of an organized society of warriors – the Knights Hospitalers, the Knights Templar or the Teutonic Knights – he was feared and hated. And as an ICON of the chivalric submission of the HERO to an honourable code of conduct, he was a figure of fun. All in all, the idea of the knight was too tied to circumstance and history to seem very usable by writers of anything but historical fiction.

In a sense, the situation remains the same in the worlds of fantasy. If the special case of the Arthurian cycle be put aside – along with the MATTER of various other countries – there is little direct carry-over of the armoured knight, or the order of chivalry he espouses, into fantasy. But there is much indirect carry-over. Lewis CARROLL's White Knight is somewhat outwith our current discussion (◊ KNIGHT OF THE DOLEFUL COUNTENANCE). The figure of the chivalrous knight as Lord Temporal, suitably distanced from courtier-hood, becomes a hero, or a COMPANION; many such figures appear in HEROIC FANTASY. As the member of a quasi-religious order he will be found in FANTASIES OF HISTORY, and in DYNASTIC FANTASIES where the presence of SECRET MASTERS or a PARIAH ELITE is required for the sake of continuity over troubled centuries. As a solitary man of honour, he becomes a CHILDE.

If there is a paucity of knightly tales in 20th-century written fantasy, however, the situation is almost the opposite in the cinema, where the figure of the knight remains a powerful ICON, with even popular entertainments like *Indiana Jones and the Last Crusade* (*1989*; ◊ INDIANA JONES) recognizing the emotional force of the image; Terry GILLIAM's TIME BANDITS (*1981*) and *The* FISHER KING (*1991*) both use a riding knight not as a character in any real sense but as a symbol that brings great emotive impact. Comedies like the various CONNECTICUT YANKEE movies and MONTY PYTHON AND THE HOLY GRAIL (*1975*) of course ridicule the chivalric pose; but in the latter, and in JABBERWOCKY (*1977*) – again directed by Gilliam – there is much respect for the *notion*. Arthurian movies – examples are *Lancelot and Guinevere* (*1962*), EXCALIBUR (*1981*) and the GAWAIN AND THE GREEN KNIGHT movies, but there are many others – obviously often have the whole basis of chivalric knighthood as the focus of their concern; the same could be said even of fairly pedestrian movies like PRINCE VALIANT (*1954*). But the true tragedy of the knight as a figure out of his time – which is perhaps today the knight's most potent iconic characteristic – is best expressed in two utterly dissimilar movies: Ingmar BERGMAN's *The* SEVENTH SEAL (*1956*), set in a LAND-OF-FABLE Middle Ages, and George A. Romero's KNIGHTRIDERS (*1981*), which captures a similar sense of loss among modern "knights" who joust on motorbikes. [JC/JG]

See also: DRAGONS.

KNOWLES, RUPERT [s] ◊ Hugh B. CAVE.

KNOWLES, VERNON (1899-1968) Australian writer, later resident in the UK, who produced several collections (often overlapping) of calculatedly quaint and mostly rather effete fantasies. Four of the nine stories in *The Street of Queer Houses and Other Stories* (coll **1924** US) are reprinted, with 11 new items, in *The Street of Queer Houses and Other*

Tales (coll **1925** UK); the title novella features an architecturally eccentric street which attracts appropriate inhabitants. None of the five omitted stories is included in *Here and Otherwhere* (coll **1926**), which does contain the celebratory "The Shop in the Off-Street" and the structurally convoluted novella "A Set of Chinese Boxes", or in *Silver Nutmegs* (coll **1927**). The latter includes the novella *The Ladder* (**1929** chap), in which the population of a village is seduced into climbing a ladder into the clouds; their subsequent fate is frustratingly unspecified. Eight stories from the later collections were reprinted, along with four new ones, in *Two and Two Make Five* (coll **1935**). [BS]

KOLÁR, JOSEF JIŘÍ (1812-1896) Czech writer. ◊ CZECH REPUBLIC.

KOLCHAK MOVIES These two movies gave rise to the tv series KOLCHAK: THE NIGHT STALKER (1974-5).

1. *The Night Stalker* US movie (*1971* tvm). ABC. **Pr** Dan Curtis. **Dir** John Llewellyn Moxey. **Mufx** Jerry Cash. **Screenplay** Richard MATHESON. **Based on** story by Jeff Rice. **Starring** Claude Akins (Sheriff Warren Butcher), Barry Atwater (János Skorzeny), Carol Lynley (Gail Foster), Darren McGavin (Carl Kolchak), Ralph Meeker (Bernie Jenks), Simon Oakland (Tony Vincenzo). 73 mins. Colour.

A SERIAL KILLER is loose in Las Vegas; the young female victims are found drained of blood and with human bitemarks at their throats. Hardbitten reporter Kolchak of the *Daily News* pursues the story, believing the killer a maniac who thinks he is a VAMPIRE; girlfriend Gail convinces him the vampirism might be more than delusional. At last the authorities accept Kolchak's theory, and hammers, stakes and crosses are issued to the police. But Kolchak finds the vampire first and, with FBI friend Jenks, exposes him to sunlight and stakes him. For this he is, to shut him up (vampirism being bad for the tourist trade), threatened with arrest for murder and thrown out of town. [JG]

2. *The Night Strangler* US movie (*1972* tvm). ABC. **Pr** Dan Curtis. **Dir** Curtis. **Spfx** Ira Anderson. **Mufx** William J. Tuttle. **Screenplay** Richard MATHESON. **Starring** Richard Anderson (Dr Richard Malcolm), Scott Brady (Captain Schubert), John Carradine (Llewellyn Crossbinder), Wally Cox (Titus Berry), Margaret Hamilton (Professor Crabwell), Darren McGavin (Carl Kolchak), Simon Oakland (Tony Vincenzo), Jo Ann Pflug (Louise Harper), Nina Wayne (Charisma Beauty). 74 mins. Colour.

The success of **1** gave rise to this sequel, a pilot for the tv series. Kolchak's old editor Vincenzo, now working on the Seattle *Daily Chronicle*, employs him. His first assignment is to report the murder of a belly dancer. Soon there is a second victim, with some blood drawn off from near the brain by hypodermic and a residue of dead human flesh being left around the neck. Kolchak finds out that similar series of murders have been committed in Seattle's Pioneer Square area every 21 years since 1889; each time six young women are murdered over an 18-day period. He next discovers that an ingredient of the ELIXIR OF LIFE sought in ALCHEMY was human blood. And an 1867 reference describes a local physician, Dr Richard Malcolm, telling Mark TWAIN he has found the key to IMMORTALITY. Identifying this man with the modern-day philanthropist Malcolm Richards, Kolchak discovers beneath the Richards Clinic a subterranean CITY inhabited by corpses plus, indeed, Dr Malcolm. Malcolm has killed six times, but has yet to take the sixth dose of elixir; before he can do so, Kolchak destroys it, and Malcolm enters accelerated ageing.

Both **1** and **2** transcend their cheapness and the oversmartass characterization of Kolchak to stand as excellent SUPERNATURAL FICTIONS, the former as a worthy treatment of a standard theme, the latter as a fascinatingly inventive extension of the formula. [JG]

KOLCHAK: THE NIGHT STALKER US tv series (1974-5). Universal/ABC. **Pr** Paul Playdon, Cy Chermak. **Exec pr** Darren McGavin. **Created by** Jeff Rice. **Dir** Allen Baron and many others. **Writers** Bill S. Ballinger and many others. **Starring** Jack Grinnage (Ron Updyke), Ruth McDevitt (Emily Cowles), Darren McGavin (Carl Kolchak), Simon Oakland (Tony Vincenzo). 20 60min episodes. Colour.

After the events of the KOLCHAK MOVIES, reporter Kolchak moves to Chicago, where he works for a small news service. He continues to unearth strange stories that defy all common sense and logic. While others might suspect a missing person is just a runaway or random crime victim, Kolchak always suspects, correctly, that something more sinister is at work. He is continually thwarted in his efforts to prove his theories, either by authorities who wish to avoid a panic or by supernatural forces beyond his control, and all his proofs conveniently disappear at the end of each episode. His foes include the original JACK THE RIPPER, a SUCCUBUS, a WEREWOLF, a VAMPIRE . . . Throughout all, Kolchak remains convinced that someday someone will believe him. [BC]

KOONTZ, DEAN R(AY) (1945-   ) US writer who established an early reputation for sf, beginning his career with "Kittens" for *Readers & Writers* in 1966. He started to publish energetic, brooding adventure tales with "Soft Come the Dragons" for *F&SF* in 1967; it became the title story of *Soft Come the Dragons* (coll **1970**), one of at least 20 sf books of interest. In his later career, DRK has concentrated on HORROR – since the 1970s he has become of central importance in that genre – and has had decreasing recourse to supernatural explanations for his tales. He has also written as David Axton, Brian Coffey, John Hill, Leigh Nichols, Anthony North, Richard Paige, Owen West and Aaron Wolfe; in recent years he has republished much of this pseudonymous output under his own name.

Though DRK is not of particular fantasy importance, and (as in his pure horror tales) often uses TECHNOFANTASY imagery to provide a "rational" explanation for seemingly supernatural events (◊ RATIONALIZED FANTASY), some of his DARK FANTASIES are of interest. They include: *The Crimson Witch* (**1971**), in which a young American is translated by PORTAL into a FANTASYLAND where he encounters a talking DRAGON and other unsurprising figures; *The Haunted Earth* (**1973**), in which aliens cause CROSSHATCH chaos by imposing a mix of the mundane and the supernatural; *Darkness Comes* (**1984** UK; vt *Darkfall* 1984 US), in which supernatural rats (◊ MICE AND RATS) assault the world; *Twilight Eyes* (**1985**; exp 1987 UK), whose protagonist possesses the TALENT of being able to perceive (◊ PERCEPTION) GOBLINS who inhabit human shape, but flicker; *Strangers* (**1986**), whose cast foregathers in a Nevada motel to discover they form a gestalt in communication with good aliens; the children's book *Oddkins* (**1988**), in which the eponymous line of TOYS comes to life (◊ ANIMATE/INANIMATE) when its maker dies; *Dragon Tears* (**1993**), which features a fabulated LOS ANGELES venue, TALKING ANIMALS, a Rat-King which constructs itself into a GOLEM, protagonists and antagonists with TALENTS, and an abiding sense of thoroughly secularized

MILLENNIAL FANTASY; and *Mr Murder* (**1993**), which is really sf/horror, though the relationship between a persecuted writer and the manufactured DOUBLE which wants to take over his life (◊ REAL BOY) gives the tale an air of dark fable – the book can be seen as a TECHNOFANTASY response to Stephen KING's *The Dark Half* (**1989**). Over the years DRK has increasingly complexified his use of various genres, sometimes confusingly, sometimes intriguingly.

[JC]

Other works (selected): *The Funhouse* * (**1980**; rev 1992), first published as by Owen West, movie novelization; *The Mask* (**1981**), first published as by West; *Twilight* (**1984**; vt *The Servants of Twilight* 1985 UK; vt *Twilight* 1988 US), first published as by Leigh Nichols; *The Bad Place* (**1990**); *Hideaway* (**1992**).

KOSHCHEI (or Kashchei) In Russian FOLKLORE, an evil MAGICIAN or DEMON who retains his MAGIC by hiding his heart. He appears in *Mlada* (**1892**) and *Kashchei the Immortal* (**1902**) by Nicolai Rimski-Korsakov (1844-1908) (◊ OPERA), in *The Fire-Bird* (**1910**) by Igor Stravinsky (1882-1971), and in James Branch CABELL's *Jurgen* (**1919**), where, without the demonic aspect, he plays chief GOD. In his traditional guise he serves as an UNDERLIER figure in some more recent works, including: Peter MORWOOD's *Prince Ivan* (**1990**), where he appears as Koshchey the Undying; in «One for the Morning Glory» (1996) by John Barnes (1957-   ), as a DARK LORD who hides his heart to retain his malign power over the LAND; in the short story "The Death of Koschchei the Deathless (A Tale of Old Russia)" (1996) by Gene WOLFE; and in «The Memory Palace» (1996) by Gill Alderman (1941-   ), where (heart safely hidden) he manipulates the BOOK into which its author (the protagonist) has been enveloped.

Curiously, in the CINEMA it is various ARABIAN-FANTASY movies that have picked up the idea: *Captain Sindbad* (*1963*; ◊ SINBAD MOVIES) is one example. [JC]

KOSSAK-SZCCZUCKA, ZOFIA (1890-1968) Polish novelist. ◊ POLAND.

KOSTER, R.M (1934-   ) US writer, resident in Panama since the early 1960s. His first novel, *The Prince* (**1972**), set in the imaginary Central American Republic of Tinieblas, is a FABULATION, evoking a modern world filled with corruption, misery and some MAGIC. *The Dissertation* (**1975**) – a "doctoral dissertation" on Tinieblan political history – and *Mandragon* (**1979**), the most nearly fantastic of the three, complete a triptych. *Carmichael's Dog* (**1992**), an extremely funny fantasy set in an ALTERNATE-WORLD present, is about an author possessed (◊ POSSESSION) by DEMONS. [GF]

KOTZWINKLE, WILLIAM (E.) (1938-   ) US writer who has worked in several genres since starting with tales for children, first being *The Fireman* (**1969**). By 1990 he had published another 18 or so titles, some of direct fantasy interest, like *Elephant Boy: A Story of the Stone Age* (**1970**), which is a PREHISTORIC FANTASY, *The Ants who Took Away Time* (**1978**), in which the Solar System must be searched for scattered parts of the Great Gold Watch that has been taken apart by ants, and *Trouble in Bugland: A Collection of Inspector Mantis Mysteries* (coll **1983**), a set of SHERLOCK HOLMES parodies for a YA audience, about such BEAST FABLES. WK's first adult books gave a sense that mundane material had been transformed into FABULATION, though on examination novels like *Hermes 3000* (**1972**), *The Fan Man* (**1974**) and *Queen of Swords* (**1984**) were secretly rather sober, beneath an air of pixillation. His sf has been more or less

restricted to the movie ties he produced in the early 1980s, including a set of ties to Stephen SPIELBERG's *E.T. – The Extra-Terrestrial* (*1982*) as well as *Superman III* * (**1983**) (◊ SUPERMAN MOVIES).

WK's works that can be defined as fantasy, like his work in other categories, tend to make play with the boundaries of genre. *Doctor Rat* (**1976**) – which won the 1977 WORLD FANTASY AWARD – can probably best be understood as a beast fable set in a TECHNOFANTASY venue. Doctor Rat speaks as apologist for the experiments being inflicted upon his fellows in a huge laboratory which – when he himself escapes – turns into a LABYRINTH of rooms and passages (◊ EDIFICE) where he has some PICARESQUE adventures. But counterpointing this narrative are shorter narratives told by members of various species, during a vast emigration of all the animals to central meeting-places where they expect the gestalt thus formed to convert humanity from its murderous ways. At last, though, all the animals are butchered.

Fata Morgana (**1977**), set initially in Paris around 1860, begins as a vaguely STEAMPUNK evocation of La Belle Époque through the eyes of a large, dour, sexually irresistible police inspector named Picard; soon his investigations into the doings of an apparent charlatan turns into ARABIAN NIGHTMARE, because the charlatan is in fact a MAGUS who sends Picard into an obsessive search through Eastern Europe, each new venue constituting a further descent from REALITY. In the end Picard is snapped back into our present, but has no way of knowing whether or not his journeys downwards have been a trick of PERCEPTION or a warning from the magus that there are illimitable darknesses beyond the THRESHOLD. Similar in feel is *Herr Nightingale and the Satin Woman* (**1978**), with many illustrations by Joe Servello, which follows a detective through Europe, in the aftermath of WORLD WAR I, as he pursues a German officer and his companion; as the book progresses, the cast increasingly enters a surreal world (◊ SURREALISM). *Seduction in Berlin* (**1985**), a book-length poem, occupies the same kind of territory. Later tales tracing the perils of detection tend – like *The Midnight Examiner* (**1989**) – only to flirt with the supernatural; but *The Game of Thirty* (**1994**), though technically not fantasy, involves a game enacted on the streets of NEW YORK that dates from the time of the Pharaohs (◊ EGYPT; FANTASIES OF HISTORY).

WK's third tale of central interest, *The Exile* (**1987**), is a DARK FANTASY of IDENTITY EXCHANGE, carrying a contemporary Hollywood star by TIMESLIP into helpless residence in the body of a German WORLD WAR II gangster who – trying to help a Jewish girl escape the Final Solution – is captured by the Gestapo and tortured. The experience drives the gangster out of his body and into the star's, leaving the latter trapped in the torture chamber.

Collections are: *Jewel of the Moon* (coll **1985**); *Hearts of Wood and Other Timeless Tales* (coll **1985**), which – like *The Oldest Man and Other Timeless Stories* (coll **1971**) for children – includes TWICE-TOLD tales; and *The Hot Jazz Trio* (coll **1989**), which contains "Boxcar Blues", a long tale in which hoboes evade DEATH by riding TRAINS into ALTERNATE REALITIES. WK, because he floats away from committing himself to any one genre, receives far less attention than he should. [JC]

Other works (for children): *The Ship that Came Down the Gutter* (**1970**); *The Day the Gang Got Rich* (**1970**); *Return of Crazy Horse* (**1971**); *The Supreme, Superb, Exalted, and Delightful, One and Only Magic Building* (**1973**); *Up the Alley* with *Jack and Joe* (**1974**); *The Leopard's Tooth* (**1976** chap); *Dream of Dark Harbor* (**1979**); *The Nap Master* (**1979**); *The World is Big and I'm So Small* (**1986**); *The Empty Notebook* (**1990**).

KRAMERIUS, VÁCLAV RODOMIL (1792-1861) Czech writer. ◊ CZECH REPUBLIC.

KRASINSKI, ZYGMUNT (1812-1859) Polish dramatist. ◊ POLAND.

KRENKEL, ROY G(ERALD) (1918-1983) Fantasy artist and illustrator, claimed as an influence by many of today's leading fantasy artists, including Reed CRANDALL, Frank FRAZETTA, Jeff JONES, Mike KALUTA, William STOUT and Al WILLIAMSON. Some of his book ILLUSTRATIONS change hands for considerable sums.

RGK's abundant creativity resulted in comparatively few finished artworks by his own hand. He preferred to work in collaboration with other artists, providing sketches and idea drawings on which they based their paintings. The most notable beneficiary was Frazetta, for whom RGK drew compositional layouts on which Frazetta based some of his early paintings for the covers of a series of reprints of Edgar Rice BURROUGHS's TARZAN and Robert E. HOWARD's CONAN.

His early influences were sf pulps, the writings of H.P. LOVECRAFT and Clark Ashton SMITH and the art of Virgil FINLAY, Hal FOSTER, Norman LINDSAY, J. Allen St John (◊ TARZAN), Joseph Clement Coll and Czech DINOSAUR artist Zdeneč Burian. He studied for a year under the anatomist George Bridgeman. After service in WWII, he attended the Cartoonists' and Illustrators' School of Burne HOGARTH, where he met other young artists who would, like him, subsequently go on to provide artwork for EC COMICS. His only solo strip for EC was *Time to Leave* (*Incredible Science Fantasy* 1955), but he collaborated with Williamson and Frazetta on EC classics like *Food for Thought, I, Rocket, 50 Girls 50* and many others. In 1951-8 RGK collaborated with Williamson on a total of 23 strips for EC, Atlas, ACG and Famous Funnies; he was also a guiding creative force on the early issues of CREEPY and EERIE, producing story concepts, cover design sketches and **Monster Gallery** features.

In addition to COMICS work RGK painted memorable covers and internal illustrations for editions of Edgar Rice BURROUGHS and Robert E. HOWARD, as well as for many pulp magazines and historical novels. Most of his later work appeared in fanzines like *Amra, Erbdom* and *Squa Tront*. Other books he illustrated include *Great Cities of the Ancient World* (**1972**) by L. Sprague DE CAMP. Portfolios include *Roy G. Krenkel: Portfolio #1* (graph **1983**) and *Seven Wonders of the Ancient World* (graph **1975**).   [RT/DR]

Further reading: *Cities and Scenes from the Ancient World* (graph **1974**) contains a detailed description of RGK's working methods; *Swordsmen and Saurians* (graph **1989**) reprints many of his illustrations from fanzines, pulps and the Burroughs books; RGK's composition sketches for Warren covers were reprinted in *Squa Tront #7* (1977); RGK's Burroughs covers were reprinted in the Russ Cochran hardback collections of EC Comics (graph colls **1985-6**).

KRENSKY, STEPHEN (ALAN) (1953-   ) US writer who began publishing fantasy with *A Big Day for Scepter* (**1977**), for children. His **Wynd** sequence – *The Dragon Circle* (**1977**), *The Witching Hour* (**1981**) and *A Ghostly Business* (**1984**) – concerns a WAINSCOT family who live in rural Massachusetts and whose TALENTS are very various. Their balancing of life in the mundane world against the perils of their revanchist kin – who long to re-establish the

hegemony of the old ways – is neatly told. Other books of interest include *The Perils of Putny* (**1978**), featuring a GIANT on a knightly QUEST, *Castles in the Air and Other Tales* (coll **1979**) and *A Troll in Passing* (**1980**).   [JC]

KRESS, NANCY (ANNE) (1948-   ) US writer who began publishing work of genre interest with "The Earth Dwellers" for *Galaxy* in 1976 and whose first three novels were fantasy. Though she is now much better known for her sf (for which she has won the Hugo and Nebula AWARDS), these early novels are of considerable interest.

The first, *The Prince of Morning Bells* (**1981**), is a QUEST fantasy whose protagonist, Princess Kirila, sets out to find the Heart of the World. Tricked by her own obdurate nature, she immolates herself three times in different forms of BONDAGE – (a) a WONDERLAND culture devoted to discovering the rules that make the world work, (b) a village culture subject to a drastic THINNING through its obsession with SEASON rites which evoke ELDER GODS, and (c) a 25-year marriage to a sports-minded prince. Her animal COMPANION, a great purple dog who believes he is a prince under a SPELL, returns for her after her husband's death; and together they find the Heart of the World. She becomes fully herself, and he – in a striking PARODY scene – is transformed (◊ TRANSFORMATION) back into the jester (◊ FOOL) he always was, a truth concealed from him long past by a WIZARD indulging in a particularly cruel GODGAME. The novel is sprightly and moving.

The Golden Grove (**1984**) is far more visibly serious, and in its inversion of the quest structure far more explicitly revisionist in nature (◊ REVISIONIST FANTASY). The Mediterranean setting is again a land of fable: a fantasized Greek culture underlies the tale – which is set on a remote ISLAND that none of the main characters ever leaves, for the book is about maintaining a world from within, not about bringing salvation back from beyond the horizon. At the heart of the island is the eponymous sacred grove (◊ GOLDEN BOUGH), where holy SPIDERS spin webs from which cloth (and the life pattern of the island's inhabitants) is woven. This pattern has been THINNING dangerously. The fate of the protagonist, who is obsessed with the grove and with the nature of the SACRIFICE she must herself make to bring about a HEALING, is perhaps more than adequately announced by her name, which is Arachne.

The White Pipes (**1985**) is revisionist in yet another fashion, because its protagonist Fia is a mother, and the presence of her child is integrated into the tale, which is set in a medieval secondary world. Fia is a "Storygiver", one of those from the Silver Cities who can create visual representations of tales for the pleasure of others, but whose power more dangerously evokes the capacity of STORY to call the tune. The pipes, which assault the SOULS of those who hear them, have a similar entrancing and coercive effect.

Each of these novels constitutes an exploration of GENDER issues, and a working analysis of types of fantasy character and plot; they are fantasies about FANTASY. They are not relaxing tales and are not intended to be. It is to be hoped that NK will challenge fantasy again.   [JC]

Other works (all sf): *Trinity and Other Stories* (coll **1985**); *An Alien Light* (**1988**); *Brain Rose* (**1990**); the **Beggars in Spain** sequence, being *Beggars in Spain* (**1991**; much exp into full novel 1993) and *Beggars & Choosers* (**1994**); *The Price of Oranges* (1989 *IASFM*; **1992** chap); *The Aliens of Earth* (coll **1993**); *Beginnings, Middles & Ends* (**1993**), nonfiction, about writing sf.

KRULL UK movie (*1983*). Columbia. **Pr** Ron Silverman. **Exec pr** Ted Mann. **Dir** Peter Yates. **Spfx** Robin Browne, Derek Meddings, Paul Wilson and many others. **Mufx** Nick Maley. **Modelmaking sv** Terry Reed. **Screenplay** Stanford Sherman. **Novelization** *Krull* * (**1983**) by Alan Dean FOSTER. **Starring** Francesca Annis (Widow of the Web), Lysette Anthony (Princess Lyssa), Bernard Archard (King Eirig), Alun Armstrong (Torquil), David Battley (Ergo the Magnificent), Bernard Bresslaw (Raoul, the Cyclops), Tony Church (King Turold), Freddie Jones (Ynyr), Graham McGrath (Titch), Ken Marshall (Prince Colwyn), John Welsh (The Blind Seer). 121 mins. Colour.

Despite occasional TECHNOFANTASY trappings, this is SWORD AND SORCERY set in a SCIENCE-FANTASY venue – in what is in effect a SECONDARY WORLD, the planet Krull. Led by the Beast, the Slayers arrive on Krull to enslave the people there. The kings Eirig and Turold reluctantly agree to join their kingdoms and abdicate in favour of their respective children, Colwyn and Lyssa, who are in LOVE and wish to marry. But the forces of the Beast (for whom read the DARK LORD) capture Lyssa. Colwyn embarks, with the help of wise oldster Ynyr, on a QUEST to rescue her and save the world, gathering COMPANIONS as he goes. The Black Fortress, a rocky EDIFICE that is not only the lair of the Beast but the Beast's spaceship and, internally, the Beast's body, spontaneously shifts position on Krull's surface with every dawn, making the quest more difficult. At last, after many – often original – fantasy twists, Colwyn slays the Beast and thus destroys the fortress and the Slayers as a whole.

K is HIGH FANTASY at its best. Undoubtedly produced as a UK response to the string of successful US SCIENCE FANTASIES begun with *Star Wars* (*1977*), it transcends that medium. Some moments are derivative – echoes of even *The Prisoner of Zenda* (*1952*) are detectable – but the whole glorious mixture produces a rattling adventure.   [JG]

KUBASIK, CHRISTOPHER (? –) US writer of ties – one for the SHADOWRUN sequence (*Changeling* * [**1992**]), one for TSR's **Battletech** (*Battletech: The Ideal War* * [**1993**]); and three so far for FASA's **Earthdawn** series, also based on a GAME – *The Longing Ring* * (**1993**), *Mother Speaks* * (**1994**) and *Poisoned Memories* * (**1994**) – which graphically unpack the violent family lives and S&S adventures of characters deployed against a FANTASYLAND backdrop.   [JC]

KUBIN, ALFRED (1877-1959) German/Austrian artist, illustrator and writer. ◊ AUSTRIA; GERMANY.

KUBLA(I) KHAN (1214-1294) ◊ Samuel Taylor COLERIDGE; XANADU.

KUMMER, FREDERIC ARNOLD (1873-1943) US writer who concentrated for much of his career on detective fiction, beginning with *The Green God* (**1911**). *The Second Coming: A Vision* (**1916** chap) with Henry P. Janes is a WORLD WAR I fantasy in which CHRIST appeals for peace to the Kaiser. FAK also wrote two AFTERLIFE fantasies, *Ladies in Hades: A Story of Hell's Smart Set* (**1928**) and *Gentlemen in Hades: The Story of a Damned Debutante* (**1930**). In the first, Eve establishes a club in HELL; in the second, a young woman who has just been killed makes good in Hell, flirting with Benjamin Franklin (1706-1790), Noah and eventually SATAN. A mild eroticism permeates both books, in the fashion of Thorne SMITH.

FAK's son, Frederic Arnold Kummer Jr (1913-?), was a frequent contributor to sf magazines around 1939-42, becoming inactive by about 1950.   [JC]

Other works: *The Earth's Story #1: The First Days of Man, as Narrated Quite Simply for Young Readers* (**1922**), a PREHISTORIC FANTASY.

KUNDERA, MILAN (1929-) Czech novelist and playwright, resident in France since 1975, famous for his novels of contemporary Czechoslovakia, which ran afoul of the authorities after 1968. Although these early novels – of which the first, *Zert* (**1967**; cut trans David Hamblyn and Oliver Stallybrass as *The Joke* **1968** UK; restored trans Michael Henry Heim **1982** US; "definitive edition" revised MK **1992** US), is the most famous – are essentially realistic, those published after his expatriation have tended more to the condition of *discours* than narrative: the text braids a narrative thread with ironical auctorial commentary and, often, scenes of FABULATION. *Kniha smichu a zapomneni* (first publication as *Le Livre du rire et de l'oubli* **1979** France; trans Michael Henry Heim as *The Book of Laughter and Forgetting* **1980** US; Czech text **1981** Canada), although not essentially fantastic, contains levitation and the ironic evocation of ANGELS. *Nesmrtelnost* (trans Peter Kussi as *Immortality* **1991** US; Czech text **1993**), however, less ambiguously presents an AFTERLIFE encounter between Ernest Hemingway (1899-1961) and GOETHE; and *La lenteur* (**1995** France; trans Linda Asher as *Slowness* **1996** US) deals more centrally with a TIMESLIP encounter between its protagonist and a figure from an 18th-century novel.

In *L'art du roman* (**1986** France; trans Asher as *The Art of the Novel* **1988** US) and *Les Testaments trahis* (**1993** France; trans Asher as *Testaments Betrayed: An Essay in Nine Parts* **1995** US) MK describes his interest in the novel as a vehicle for poetic expression and philosophical inquiry rather than as a vessel for storytelling. This affinity is made clear in his play *Jakub a jeho pan: pocta Denisi Diderotovi* (**1981** France in translation; trans Michael Henry Heim from the French as *Jacques and His Master: An Homage to Diderot* **1985**; Czech text **1992**).   [GF]

See also: CZECH REPUBLIC.

KUNG FU The **Kung Fu** tv enterprise has spawned two series and two pilots.

1. *Kung Fu* US tv series (1972-5). Lou Step Productions/ Warner Bros./ABC. **Pr** Alex Beaton, John Furia Jr, Herman Miller. **Exec pr** Jerry Thorpe. **Dir** Robert Butler, David Carradine and many others. **Writers** Kathryn Barton, Gene L. Coon, Dorothy C. Fontana and many others. **Created by** Ed Spielman, Herman Miller. **Starring** Philip Ahn (Master Kan), Carradine (Kwai Chang Caine), Season Hubley (Margit McLean 1974-5), Keye Luke (Master Po), Stephen Manley (Caine aged 6), Radames Pera (Young Caine). Warner Bros. 62 60min episodes plus 90min pilot. Colour.

Caine, born to a Chinese mother and a US father, is an orphan drifter in the 1870s USA. He studied to be a Shaolin priest but was forced to flee China after killing a man there, for his victim was a member of the royal family. While Caine has been taught to practice nonviolence, he has also been taught the martial art of kung fu in case he must defend himself. This is lucky, for each week he finds himself having to resort to violence when all attempts to escape trouble fail.

The fight scenes are best remembered for their unusual use of slow-motion photography; additionally, the series featured a large number of flashbacks, in which the young Caine was seen with his learned tutor, Master Po. The unusual formula worked for, though many of the stories themselves were rather uninteresting, the series enjoyed a healthy run.   [BC]

2. *Kung Fu: The Movie* US movie (**1986** tvm). Lou Step Productions/Warner Bros./CBS. **Pr** Skip Ward, David Carradine. **Screenplay** Durrell Royce Crays. **Based on 1**. **Dir** Richard Lang. **Starring** Carradine (Kwai Chang Caine), Keye Luke (Master Po), Brandon Lee (Caine's son). 120 mins. Colour.

Pilot for an unsold series revival. Its failure prompted the production company to try another format (◊ **3**).   [BC]

3. *Kung Fu: The Next Generation* (vt *The Way of the Dragon*) US movie (**1987** tvm). Lou Step Productions/ Warner Bros. **Pr** Danny Bilson, Paul DeMeo, Ralph Riskin. **Dir** Tony Wharmby. **Screenplay** Bilson, DeMeo. **Starring** Brandon Lee (Cain's Grandson). 90 mins. Colour.

Pilot for an unsold series intended to feature the adventures of Caine's offspring. Brandon Lee, son of martial-arts expert Bruce Lee, achieved posthumous fame as star of *The* CROW (**1994**).   [BC]

4. *Kung Fu: The Legend Continues* US syndicated tv series (1993-current). Warner Bros. **Pr** Phil Bedard, Martin Borycki, David Carradine, Larry Lalonde, Reuben Leder, Susan Murdoch. **Exec pr** Michael Sloan. **Dir** Mario Azzopardi, William Shatner and many others. **Writers** Phil Bedard and many others. **Starring** David Carradine (Kwai Chang Caine), Kim Chan (The Ancient), Robert Lansing (Captain Paul Blaisdell), Nathaniel Moreau (Young Peter), Chris Potter (Peter Caine). 60min episodes. Colour.

In a modern-day setting, Carradine plays the grandson of the original Kwai Chang Caine from **1**; he believes he is still carrying the dishonour of his ancestor. Eventually making his way to San Francisco, he finds his own grandson, Peter Caine, a police officer. Together, they use the disciplines of the Shaolin teachings in a series of battles against local criminals and renegades from their own religion. While most episodes feature rather mundane adversaries, the story "Gunfighter" found Caine trading places with his grandfather in the past, where he had to fight to save both of their existences. In later episodes, hints are given that Kwai Chang may be the original Caine and not his grandson; if so, he is over 130 years old.   [BC]

KURLAND, MICHAEL (JOSEPH) (1938-   ) US writer who began publishing work of genre interest with "Elementary" with Laurence M. Janifer for *F&SF* in 1964, and who is better known for sf than for fantasy. His two GASLIGHT ROMANCES on SHERLOCK HOLMES themes, *The Infernal Device* (**1979**) and *Death by Gaslight* (**1982**), are fantasy; both concentrate on Moriarty and the presumption he has made a PACT WITH THE DEVIL. MK has also written two tales set in the universe Randall GARRETT created for his **Lord Darcy** sequence: *Ten Little Wizards* * (**1988**) and *A Study in Sorcery* * (**1989**) – the second again invokes Holmes. *Button Bright* (**1990**) combines sf, fantasy and HORROR.   [JC]

KURTZ, KATHERINE (IRENE) (1944-   ) US writer who worked as an instructional designer for the Los Angeles Police Department 1969-81 and who began to publish fantasy with the initial volume of the **Deryni** sequence, *Deryni Rising* (**1970**), the first original text to be published by Lin CARTER in his BALLANTINE ADULT FANTASY list. This sequence, set in a SECONDARY WORLD somewhat like a LAND-OF-FABLE Britain, has occupied most of her career. Its internal chronology, which is carefully organized, differs markedly from the order of publication of individual subseries: the **Legends of Camber of Culdi** sequence, comprising *Camber of Culdi* (**1976**), *Saint Camber* (**1978**)

and *Camber the Heretic* (**1980**), which won the 1982 BALROG award; the **Heirs of Saint Camber** sequence, comprising *The Harrowing of Gwynedd* (**1989**), *King Javan's Year* (**1992**) and *The Bastard Prince* (**1994**); the **Chronicles of the Deryni** sequence, comprising the original *Deryni Rising*, *Deryni Checkmate* (**1972**) and *High Deryni* (**1973**), all three assembled as *The Chronicles of the Deryni* (omni **1985**); and the **Histories of King Kelson**, comprising *The Bishop's Heir* (**1984**), *The King's Justice* (**1985**) and *The Quest for Saint Camber* (**1986**). Pendants include *The Deryni Archives* (coll **1986**), whose contents range over the whole chronology, and *Deryni Magic: A Grimoire* (**1990**), a text which describes the very extensive (and intricately detailed) RITUALS necessary to operate MAGIC throughout the sequence.

Unsurprisingly, given that the first volume of the sequence appeared only half a decade after the great J.R.R. TOLKIEN boom in the USA created a market for HIGH FANTASY, the **Deryni** sequence resembles *The Lord of the Rings* (**1954-5**) in many ways. Both secondary worlds are clear ET IN ARCADIA EGO homages to a land which never existed and has never suffered the ravages of the 20th century. The TIME ABYSS Tolkien creates, by virtue of which *LOTR* seems to float on top of a STORY fathomlessly deep, is far more profound than the back-story KK creates through prequels for her **Deryni** world; the conflict between GOOD AND EVIL in the older writer has all the melancholy undertow of the best CHRISTIAN FANTASY, while in KK's hands the conflict comes close to becoming a TEMPLATE; and KK's constant presentation of the workings of MAGIC does not generate quite the complexity of imagined texture as was accomplished through Tolkien's creation of a nest of imaginary (but seemingly workable) tongues. KK's own linguistic experiments are restricted.

Despite the Celtic names and the Celtic venue, the **Deryni** sequence in one vital respect adheres much more closely to Tolkien than to Celtic models: her characters, like Tolkien's Frodo or Strider, do not suffer METAMORPHOSIS (like, say, the heroes of the MABINOGION, or the protagonists of novels by Kenneth MORRIS or Paul HAZEL) but instead change by growing into their roles. The **Deryni** books are, therefore, central to the development of GENRE FANTASY. The pattern they exemplify – repeated RITE-OF-PASSAGE plots which are set in FANTASYLAND and which climax in an assumption of the burdens of power (i.e., magic or adulthood) – has become a template for hundreds of successors. But KK was a very early visitor to Fantasyland, whose fixed parameters she after all helped establish, and thus her books continue to seem relatively fresh.

The sequence, though superficially complicated, follows the model of personal growth with some consistency. This basic tale is told within the context of the long conflictual relationship between humans and the Deryni, a nonhuman but human-seeming race with multiple TALENTS (including an adept control over PORTALS). Some of the Deryni are arrogant and tyrannical; some, like Camber MacRorie – who starts the sequence off by establishing humans on the throne of Gwynedd (i.e., Wales) – are compassionate and knowing . . . and after his disappearance Camber becomes Saint Camber. Many of the humans are mere spearcarriers in the drama; others, like most of the religious figures whose Church resembles Christianity in its need to eradicate magic (◊ THINNING), are culpable (bishops in particular). Over the first books of the sequence, the human rule harshens, and the Deryni become a PARIAH ELITE; they come to behave more like SECRET MASTERS, eventually ruling the world through the agency of King Kelson, who is half-human and half-Deryni. Later volumes are full of DYNASTIC-FANTASY conflicts and – like most examples of that form – are difficult to read without extensive knowledge of the family trees whose founders dominated earlier volumes.

KK's second series, the **Adam Sinclair** sequence with Deborah Turner HARRIS – *The Adept* (**1991**), *The Lodge of the Lynx* (**1992**), *The Templar Treasure* (**1993**) and *Dagger Magic* (**1995**) – are SUPERNATURAL FICTIONS starring an occult doctor and protector who defends the UK against various supernatural incursions, including (in *Dagger Magic*) Tibetan villains who control a German submarine from WORLD WAR II. An early singleton, *Lammas Night* (**1983**), likewise deals with occult forces in WWII: Hitler is dabbling in black MAGIC, and the hero, prefiguring Adam Sinclair, uses his knowledge of white magic to defend the sacred Isle of Britain. [JC]

Other works: *The Legacy of Lehr* (**1986**) YA sf; *Tales of the Knights Templar* (anth **1995**), stories about the Order from a FANTASIES-OF-HISTORY perspective. The **Knights of the Blood** sequence – *Knights of the Blood #1: Vampyr-SS* (**1993**) and *#2: At Sword's Point* (**1994**) – by Scott Macmillan, KK's husband, is described as "created by Katherine Kurtz" and is an extravaganza involving opposing sets of VAMPIRES, more fantasies of history, WORLD WAR II, and PARIAH ELITES galore.

KUSHNER, ELLEN (1955-) US writer and editor. She first appeared in print as editor of *Basilisk* (anth **1980**). Its mixture of original stories and reprints, poems and stories, each with a full-page illustration by Terri WINDLING, served as a model for the three **Elsewhere** volumes ed Windling, in the initial volume of which appeared EK's first work of fiction, "The Unicorn Masque". With its (bizarre and slightly clumsy) GENDER DISGUISE, its obsession with MASKS, its reliance on witty dialogue, its focus on the court new-comer making his way to the centre of the society and its demonstration of the dangers of intimacy, this was a harbinger of the FANTASY OF MANNERS.

EK quit editing in 1981 to write full-time. The first result was *Swordspoint* (**1987**), which received considerable praise. The novel is essentially a DYNASTIC FANTASY set in a RURITANIAN 18th-century CITY; its total lack of supernatural paraphernalia might seem to blur its generic identity, but in a sense the structure of the depicted society, which has no exact historical analogue, is itself the fantastic element. *Swordspoint* is flamboyant but controlled; the manipulation of viewpoint is impressive. A pendant, "The Swordsman whose Name was not Death" (**1991**), centres on a complex gender disguise and an exploration of GENDER roles in its society. Despite a climactic sword duel, it is the aftermath verbal duel that generates the devastating conclusion.

Where *Swordspoint* is a display-piece like a jig or a reel, *Thomas the Rhymer* (**1990**) is a lovely slow air of a book. EK sets this retelling of the classic ballad of a harpist taken to FAERIE by its queen in what she calls Ballad-Land. The portion of the book set in Elfland uses the ballad "Famous Flower of Serving Men" as a tightly woven subplot in EK's strongest manner; throughout, its theme of the MINSTREL whose gift is to speak only the truth conveys EK's burning conviction that the purpose of fantasy is to do just that. This won the MYTHOPOEIC AWARD and tied for the WORLD FANTASY AWARD. [DGK]

Other works: *Outlaws of Sherwood Forest* * (**1985**), *The Enchanted Kingdom* * (**1986**), *Statue of Liberty Adventure* * (**1986**), *The Mystery of the Secret Room* * (**1986**), *Knights of the Round Table* * (**1988**), all in the **Choose Your Own Adventure** gamebook series.

KUTTNER, HENRY (1915-1958) US writer, best-known for his sf. He was briefly associated with the H.P. LOVECRAFT circle, and much of his early fiction was in the subgenres favoured by WEIRD TALES, including the SWORD-AND-SORCERY *Elak of Atlantis* (1938-41 *WT*; coll of linked stories **1985**) and *Prince Raynor* (1939 *Strange Stories*; coll of linked stories **1987** chap). He wrote several notable short stories for UNKNOWN, including "The Misguided Halo" (1939), "Design for Dreaming" (1942) and "Compliments of the Author" (1942); his quirky sense of humour was ideally suited to the magazine. The TIME-TRAVEL comedies starring **Pete Manx** which HK and Arthur K. Barnes (1911-1969) wrote – both separately and in collaboration – for *Thrilling Wonder Stories* as Kelvin Kent are in a similar vein; these have never been collected.

Before marrying C.L. MOORE in 1940 HK collaborated with her on the **Jirel of Joiry/Northwest Smith** story "Quest of the Starstone" (1937 *WT*). After their marriage the two collaborated on a number of lushly exotic melodramas, heavily influenced by A. MERRITT, which may be regarded as archetypal examples of SCIENCE FANTASY: *Earth's Last Citadel* (1943 *Argosy* as by HK and CLM; **1964** as by CLM and HK); *The Dark World* (1946 *Startling Stories* as by HK; **1965**); *Valley of the Flame* (1946 *Startling Stories* as by Keith Hammond; **1964** as by HK); "Lands of the Earthquake" (1947 *Startling Stories* as by HK); *The Mask of Circe* (1948 *Startling Stories* as by HK; **1971**); *The Time Axis* (1949 *Startling Stories* as by HK; **1965**); *Beyond Earth's Gates* (1949 *Startling Stories* as "The Portal in the Picture" by HK; **1954** dos as by Lewis Padgett); and *Well of the Worlds* (1952 *Startling Stories*; **1953** as by Lewis Padgett; vt *The Well of the Worlds* 1965 as by HK). Presumably HK and Moore would have written far more outright fantasy had there been ready markets for such work, but the sf disguises worn by these works are by no means merely tokenistic compromises. [BS]

KWAIDAN (vt *Kaidan*) Japanese movie (*1964*). Bungei-Ninjin/Toho. **Pr** Shigeru Watasuke. **Dir** Masaki Kobayashi. **Screenplay** Yoko Mizuki. **Based on** stories by Lafcadio HEARN. **Starring:** "Black Hair": Michiyo Aratama (First Wife), Rentaro Mikuni (Samurai), Misako Watanabe (Second Wife); "Snow Woman": Keiko Kishi (Oyuki/Snow Woman), Mariko Okada (Minokichi's Mother), Tatsuya Nakadai (Minokichi); "Hoichi the Earless": Katsuo Nakamura (Hoichi); "In a Cup of Tea": Kanemon Nakamura (Kannai), Noboru Nakaya (Shikibu Heinai). 164 mins (US release cut to 125 mins by the omission of "Snow Woman"). Colour.

Four GHOST STORIES; in this movie, as in Hearn's tales, GHOSTS have physical substance. "Black Hair" is an ALLEGORY of the impossibility of returning to the past to correct one's errors. An impoverished samurai, to improve his lot, divorces his wife and marries a rich girl, but regrets his folly. Years later he returns home to find his first wife and the house just as he left them, and enjoys a passionate reconciliation. By dawn all has decayed, and where she was are only bones and hair. As he flees in terror, his own unnatural youthfulness deserts him. (This tale can be interpreted also as a fantasy of PERCEPTION.) In "Snow Woman" two woodcutters, the ancient Mosake and the boy Minokichi, are lost in a blizzard. Sheltering, they are visited in the night by the VAMPIRE Snow Woman, who breathes death into Mosake but spares Minokichi on condition he tells nothing. A year later Minokichi marries the mysterious Oyuki, the Snow Woman's DOUBLE. While he ages and they have children, she remains young. One day his memories flood back, and he recounts the tale to her. But "Oyuki" is in fact the Snow Woman . . . "Hoichi the Earless", by far *K*'s longest segment, tells how, long ago, two samurai clans fought to the death. Now, blind musician Hoichi is nightly led by a samurai spirit to a ghostly gathering of one of the clans, singing to them the ballad of their defeat. To protect him, priests cover him with sacred writing so that he will appear as a ghost to the ghosts; but they miss his ears, which the samurai spirit tears off. Thereafter Hoichi's vocation is to sing to the SOULS of the dead, thereby placating them. In "In a Cup of Tea" an aristocrat's guard, Kannai, sees the face of a samurai, Shikibu Heinai, reflected in the tea he is about to drink; but he drinks it anyway, and so becomes haunted by the samurai's spirit. This last, unfinished tale is framed by a narrative in which Hearn discovers the STORY.

K is a stunningly beautiful movie, and its soundtrack is so brilliantly manipulated as to become virtually a part of the spfx. Its first two segments are ponderous but riveting, the latter two more ponderous than riveting; yet the whole has a haunting affect.   [JG]

LABYRINTHS Mazes and labyrinths have a long history: the archetypal maze was a pattern, usually cut in turf, to be traversed in a religious or magical ceremony, while the archetypal – though not of course the first – labyrinth was that built by DAEDALUS to hold the MINOTAUR. Usage has blurred the distinction, but mazes tend to be submitted to voluntarily as a GAME or RITUAL – William SHAKESPEARE's "quaint mazes in the wanton green", and even traditional hedge-mazes like Hampton Court's, do not offer serious physical barriers; thus the narrator of Alasdair Gray's "Five Letters from an Eastern Empire" (1979) can choose not to play, stepping over the inches-high hedge. Further mazes in this sense include the Pattern in Roger ZELAZNY's **Amber**, the shadow-maze of Gene WOLFE's "A Solar Labyrinth" (**1983**), and Destiny's garden in Neil GAIMAN's SANDMAN graphic novels – a knowing concretization of Jorge Luis BORGES's "The Garden of Forking Paths" (**1941**), whose maze or labyrinth ramifies not spatially but through ALTERNATE-WORLD pathways in time.

The labyrinth tends to be (a) quite often three-dimensional, like the spaghetti corridors of the Mile-High Tower housing the "Egg of the Phoenix" in Robert A. HEINLEIN's *Glory Road* (**1963**); (b) roofed or UNDERGROUND or enclosed in an EDIFICE, as with the Stone Lanes region of Gormenghast in Mervyn PEAKE's *Titus Groan* (**1946**); (c) sometimes natural or part-natural, like the labyrinthine caverns threaded in Barry HUGHART's *Bridge of Birds* (**1984**); (d) always full of walls that block the intending solver's view; (e) physically imprisoning until solved. Some further labyrinth examples are: the MINES of Moria in J.R.R. TOLKIEN's *The Lord of the Rings* (**1954-5**); the deeply claustrophobic mineworkings in Alan GARNER's *The Weirdstone of Brisingamen* (**1960**); the infinite labyrinth of Avram DAVIDSON's *Masters of the Maze* (**1965**), whose PORTALS open on all of space and time; the subterranean labyrinth focusing old powers of Earth and darkness in Ursula K. LE GUIN's *The Tombs of Atuan* (**1971**); the Empty Palace mirror-labyrinth in Susan COOPER's *Silver on the Tree* (**1977**); the LIBRARY in Umberto ECO's *The Name of the Rose* (**1980**); the oily, ridged labyrinth which is the MORRIGAN's thumbprint on her magic MAP (◊ GREAT AND SMALL) in Pat O'SHEA's *The Hounds of the*

Mórrígan (**1985**); and the booby-trapped Labyrinth of Ephebe (the **Discworld** analogue of Classical Greece) in Terry PRATCHETT's *Small Gods* (**1992**). There are many more. [DRL]

LACKEY, MERCEDES (RITCHIE) (1950-   ) US writer who has become remarkably prolific since "A Different Kind of Courage" in *Free Amazons of Darkover* (anth **1985**) ed Marion Zimmer BRADLEY, with whom she later collaborated on one of her few sf novels; she is married to artist Larry Dixon, with whom she has collaborated on some novels. Her fantasy work breaks into two very broad areas, which intersect thematically: her solo works, which are of the greater interest; and her collaborations, which are numerous. Central to her solo output is the **World of Valdemar** sequence, organized into individual series: **Heralds of Valdemar**, comprising *Arrows of the Queen* (**1987**) – ML's first novel – *Arrow's Flight* (**1987**) and *Arrow's Fall* (**1988**), all three assembled as *Queen's Own* (omni **1993**); **Vows and Honor**, comprising *The Oathbound* (fixup **1988**) and *Oathbreakers* (**1989**), both assembled as *Vows and Honor* (omni **1994**); **The Last Herald-Mage**, comprising *Magic's Pawn* (**1989**), *Magic's Promise* (**1990**) and *Magic's Price* (**1990**), all three assembled as *The Last Herald-Mage* (omni **1990**); **Mage Winds**, comprising *Winds of Fate* (**1991**), *Winds of Change* (**1992**) and *Winds of Fury* (**1993**); **Mage Storms**, comprising *Storm Warning* (**1994**) and *Storm Rising* (**1995**); and **Mage Wars**, comprising *The Black Gryphon* (**1994**) and *The White Gryphon* (**1995**), both with Larry Dixon, and which, though most recently published, are set some 1500 years before the main stories. *By the Sword* (**1991**) is a singleton set in the same world.

The world constructed in the **Mage Wars** books is a FANTASYLAND based on a genetic-engineering substratum and lays the groundwork – for those not previously familiar with the series – for a wide range of tales, exploiting DYNASTIC-FANTASY and HEROIC-FANTASY models in particular, with an abiding emphasis on MAGIC: how it operates, and how it is learned by young people undertaking their RITE OF PASSAGE into the proper use of the TALENTS ML has, very frequently, bestowed upon them. The governance of the world of Valdemar as a whole depends upon the relationship between those who are chosen as "Heralds of Valdemar" (i.e., those who act as *consiglieri* to the reigning royals, and who are mystically networked so that all know when one is in dire straits or dead) and the horses with talents who act as their COMPANIONS and MENTORS. In the first-published trilogy, for instance, young Talia – whose talent is empathy – must learn how to do magic while at the same time coping with the fractious young female royal to whom, as Queen's Herald, she has been assigned. The female protagonists – one a mercenary and one a magician – of the **Vows and Honor** books share minds and bodies through a series of adventures in which they enjoy ultimate success. The protagonist of the **Last Herald-Mage** books is a brilliant young herald whose talents may be overwhelmingly great, and who falls in love with (and loses) a similarly romantic hero. Much of the overall sequence offers, as is clear, role models for young readers which extend beyond heterosexual conventions. There is no reason for the **Valdemar** books to stop; the TEMPLATE fantasyland against which sometimes intricate tales are played out remains available for further use.

ML's second series, the **Diana Tregarde** DARK-FANTASY sequence – *Burning Waters: A Diana Tregarde Investigation* (**1989**), *Children of the Night* (**1990**) and *Jinx High* (**1991**) – features a female OCCULT DETECTIVE and WITCH who helps police solve RITUAL murders, cope with VAMPIRES, etc. Again the detail-work is sometimes very considerable. The **Bedlam's Bard** sequence – *Knight of Ghosts and Shadows: An Urban Fantasy* (**1990**) and *Summoned to Tourney* (**1992**), both with Ellen GUON, and assembled as *Bedlam's Bard* (omni **1992**) – is set initially in a LOS ANGELES suffering from a CROSSHATCH invasion of bad ELVES. The **SERRAted Edge** sequence – comprising *Born to Run* (**1992**) with Larry Dixon, *Wheels of Fire* (**1992**) with Mark Shepherd, *When the Bough Breaks* (**1993**) with Holly LISLE and *Chrome Circle* (**1994**) with Larry Dixon (*Elvendude* [**1994**] being by Shepherd alone) – likewise involves elves, who have formed the South Eastern Road Racing Association (or SERRA), around which various URBAN-FANTASY tales revolve. ML's contributions to the **Bard's Tale** sequence, which she supervises, are *Castle of Deception* * (**1992**) with Josepha SHERMAN, *Fortress of Frost and Fire* * (**1993**) with Ru EMERSON and *Prison of Souls* * (**1993**) with Mark Shepherd; these are TECHNOFANTASIES tied to a computer game. The **Bardic Voices** sequence – *The Lark and the Wren* (**1992**), *The Robin and the Kestrel* (**1993**) and *The Eagle and the Nightingale* (**1995**), plus *A Cast of Corbies* (**1994**) with Josepha Sherman – is set in a FANTASYLAND composed of humans and elves and subject to censorious THINNING by a Church whose dictates resemble those of Christianity.

Of ML's singletons, *Sacred Ground* (**1994**) stands out for the quality of its research (into Native American MYTHS), for the vigour of its storytelling, and for the original use of a Native American woman as an OCCULT DETECTIVE.

There is no real doubt that ML writes too fast and too much; despite the active strength of her mind, despite the number of issues she effectively addresses (FEMINISM being perhaps paramount), and despite the thrust of story in her best works, her prose fails, time and again, to realize the virtues that spring onto the careless page. [JC]

Other works: *Reap the Whirlwind* * (**1989**) with C.J. CHERRYH, a contribution to the **Sword of Knowledge** SHARED-WORLD sequence created by Cherryh; the **Halfblood Chronicles** with Andre NORTON, being *The*

Elvenbane: An Epic High Fantasy of the Halfblood Chronicles (**1991**) and *Elvenblood: An Epic High Fantasy* (**1995**); *The Fire Rose* (**1995**).

Other works (sf): *Wing Commander: Freedom Flight* (**1992**) with Ellen Guon; *Rediscovery: A Darkover Novel* * (**1993**) with Marion Zimmer Bradley; *If I Pay Thee Not in Gold* (**1993**) with Piers ANTHONY; *The Ship who Searched* (**1993**) with Anne MCCAFFREY.

LADRI DI SAPONETTE ot of *The* ICICLE THIEF (*1989*).

LADYHAWKE US movie (*1985*). 20th Century-Fox/ Warner. **Pr** Richard Donner, Lauren Shuler. **Exec pr** Harvey Bernhard. **Dir** Donner. **Spfx** John Richardson. **Screenplay** Edward Khmara, Tom Mankiewicz, Michael Thomas. **Novelization** *Ladyhawke* * (**1985**) by Joan D. VINGE. **Starring** Matthew Broderick (Phillipe Gaston), Rutger Hauer (Étienne Navarre), Ken Hutchison (Marquet), Leo McKern (Brother Imperius), Alfred Molina (Cezar), Michelle Pfeiffer (Isabeau d'Anjou), John Wood (Bishop of Aquila). 124 mins. Colour.

A LAND-OF-FABLE medieval France. Two years ago, when the Bishop discovered that the woman he desired, Isabeau, and the Captain of his Guards, Navarre, were in love, he made a PACT WITH THE DEVIL to CURSE them: henceforward Navarre would be a wolf by night and a man by day and Isabeau a hawk by day and a woman by night, so the two would have together as humans only the instants of dawn and sunset. One day Navarre saves the sneak-thief lad Phillipe from the Bishop's soldiers, led by Marquet. The soldiers wound the hawk, and Navarre sends Phillipe with it to the care of Imperius, the monk who drunkenly revealed the lovers' secret to the Bishop. But now Imperius knows a solar eclipse is due – a night that will also be day, so that both lovers will be human simultaneously. If they can confront the Bishop in human form together the curse will be broken – which, with the aid of Phillipe and Imperius, they do.

L is an epic of the old-fashioned kind: superbly photographed, often ponderous, and overall impressive. The SHAPESHIFTING scenes are convincing without being specific. There are no profound subtexts here: they would be out of place. [JG]

LADY IN WHITE US movie (*1988*). Goldwyn/New Sky. **Pr** Frank LaLoggia, Andrew G. La Marca. **Exec pr** Charles M. LaLoggia, Cliff Payne. **Dir** Frank LaLoggia. **Vfx** Ernest D. Farino, Gene Warren Jr. **Screenplay** LaLoggia. **Starring** Len Cariou (Phil), Lukas Haas (Frankie), Henry Harris (Willie), Katherine Helmond (Amanda Harper), Joelle Jacobi (Melissa), Jason Presson (Geno), Alex Rocco (Angelo). 113 mins. Colour.

Halloween 1962 in small town Willowpoint Falls. 10-year-old Frankie, locked by pranksters in the school cloakroom overnight, sees the GHOST of a little girl, Melissa, reenact her murder; the murderer returns in real life to try to recover his dropped graduation ring from a grille and attacks Frankie, who has a near-death vision in which Melissa begs help finding her mother. A scapegoat is accused of being the SERIAL KILLER of 11 children – of which Melissa, the first, was found near a cliff site haunted by the spectral Lady in White, the ghost of her mother, who threw herself from the cliff on the child's death. Frankie, regularly visited by Melissa's ghost, regains the ring. Exploring the Lady in White's house, he finds her perfectly preserved bedroom and encounters "the ghost" – actually mad local recluse Miss Harper, Melissa's aunt. Finally Frankie discovers the killer is his foster-uncle Phil, from whose clutches he is saved by first Miss Harper and then the Lady in White herself, who thereby earns reunion with her daughter.

Thanks in good measure to Haas (events are seen largely through Frankie's eyes) and Helmond, and to its Ray BRADBURY/Thomas TRYON-like evocation of small-town childhood, *LIW* survives several dreary selfconscious emulations of Steven SPIELBERG, patchy spfx and occasionally heavy-handed direction as a compelling and genuinely moving GHOST STORY: what makes it exceptional is the sense that there are real griefs and real passions amid the melodrama. [JG]

LADY OF THE LAKE An enchantress (or one of several) who has a manipulative role in the Arthurian romances (◊ ARTHUR); she is called variously Viviane, Niniane, Nimuë, etc. She does not appear in the earliest Arthurian stories, but was probably adopted into the canon from other legends, especially those of lake spirits (◊ UNDINE). Her part emerges in the **Vulgate Cycle** of Arthurian stories (compiled 1215-35) and most completely in Sir Thomas MALORY's *Le Morte Darthur* (**1485**). She is seen as a student of MERLIN's who takes on his function as Arthur's protector. It is she who gives Arthur EXCALIBUR, and who receives it back as Arthur lies dying. She feigns love for Merlin and uses the spells she has learned from him to ensnare and trap him. She becomes the foster-mother of LANCELOT when he is abandoned, and subsequently delivers him to Arthur's court. An almost certainly different Lady of the Lake is killed by Sir Balin, who thereby incurs the wrath of Arthur.

The Lady is often portrayed as a benign spirit or FAIRY, as distinct from the darker MORGAN LE FAY, although both are probably based on the same early beliefs. In most Arthurian fiction that seeks "realism" her character is downplayed, but in the more traditional Arthurian stories her role is allowed to develop. She is best displayed in *The Mists of Avalon* (**1982**) by Marion Zimmer BRADLEY, which explores her part in the rise of Christianity and its impact on the Old Religion. In *King Arthur's Daughter* (**1978**) by Vera CHAPMAN she is the loving wife of Merlin; their descendant marries Arthur's daughter and thus perpetuates the bloodline. In the **Daughter of Tintagel** sequence by Fay SAMPSON – in particular *Black Smith's Telling* (**1990**) and *Taliesin's Telling* (**1991**) – she is depicted as leader of a band of rebel women warriors (◊ AMAZONS) who use Arthur as their champion to protect the Old Religion against Christianity. [MA]

LaFARGE, TOM Working name of US writer Thomas LaFarge (1947-   ); he is married to Wendy WALKER. His first published stories, "The Image Breaker" (1991) and "Amont, Aval" (1991) are fantasies based upon the BESTIARY of Marie de France (*fl*1160-90). These and other formally innovative reworkings of "fables" are collected in «Terror of Earth» (coll **1996**). *The Crimson Bears, Part 1* (**1993**) and its sequel *A Hundred Doors* (**1994**) comprise a mannered and elaborate fantasy set in an imaginary and richly populated world of TALKING ANIMALS. [HW]

LAFFERTY, R(AFAEL) A(LOYSIUS) (1914-   ) US writer who worked in the electrical business until retiring in 1971; he began to write only in his early 40s, his first story of genre interest being "Day of the Glacier" for *Original Science Fiction Stories* in 1960, and retired from writing in 1984, aged 70. His early work – examples of which appear in *The Early Lafferty* (coll **1988** chap Canada) and *The Early Lafferty II* (coll **1990** chap Canada) – are uneccentric in

style and content, but soon RAL's unique voice could be heard. It found its first outlet in the sf magazines of the early 1960s.

To treat RAL as an eccentric, ribald sf writer is to misunderstand his genius. It would be difficult to assign any appropriate genre to his work, though his best novels and stories can probably be read most easily as (broadly) fantasy. However, it is almost certainly more rewarding to treat RAL in terms of specific literary analogues or influences. He is Catholic, and his work is in many respects similar to that of G.K. CHESTERTON, whose flamboyance also can move swiftly into a sometimes thin SURREALISM. But the writer with whom he can best be compared is probably Flann O'BRIEN, whose *The Third Policeman* (**1967**) shifts from a crazy-quilt mundane into hyperbolic fantasy with an ease and off-handed dispatch; part of its fabric of "explanation", as with much of RAL's work, ruthlessly and joyfully ransacks the grammar of science and of sf.

A mature RAL work can almost always be identified as by him within a sentence. His prose revels in easy paradox and is ebullient and baroque; and the line of thought moves with an utterly natural-seeming obliquity, through TALL TALE and, by leaps and bounds which it is almost impossible to analyse, into the sublime. But perhaps the most important touchstone of his language is the demotic base from which everything else is built – or perhaps, more accurately, leaps.

RAL's characters befit this language. Many protagonists tend to be introduced as blue-collar US working men, no matter how much larger than life they turn out to be (◊ AVATAR). They have the innocence and monstrosity of CHILDREN; the actual children in RAL's stories – such as *The Reefs of Earth* (**1968**) or "Ginny Wrapped in the Sun" (1967 *Galaxy*) – are undoubtedly monsters, though in a state of grace. In their adult manifestations, these protagonists may well stand in for, or literally embody, opposing forces (who resemble GODS) in the cosmic duels between the fallen and unfallen that structure *Fourth Mansions* (**1969**), *Not to Mention Camels* (**1972**), and the **Devil is Dead** sequence – *The Devil is Dead* (**1971**), *Archipelago* (**1979**) and *More Than Melchisedech*, the last published in three volumes as *Tales of Chicago* (**1992** Canada), *Tales of Midnight* (**1992** Canada) and *Anamnesis* (**1992** chap Canada).

Past Master (**1968**) features Sir Thomas More (1478-1535), who is transported to another world where his dark joke, *Utopia* (**1515**), has been used as a grotesque societal model. *Space Chantey* (**1968**) is a TWICE-TOLD version of HOMER's *Odyssey* (*c*9th century BC). "Where Have you Been, Sandaliotis?", one of the two book-length stories published in *Apocalypses* (coll **1977**), is a DREAM-like story about the return to the Mediterranean of a Fortean nation which had linked Sardinia to the coast. Later works of interest include the **In a Green Tree** sequence, the first part of which is *My Heart Leaps Up*, a novel published in five segments: *Chapters 1-2* (**1986** chap), *3-4* (**1987** chap), *5-6* (**1987** chap), *7-8* (**1988** chap) and *9-10* (**1990** chap); and the second part of which, *Grasshoppers & Wild Honey* (1928-1942), starts on the same basis with *Chapters 1-2* (**1992** chap).

Many of RAL's short stories remain scattered, but most can be found in various collections. Books include *Nine Hundred Grandmothers* (coll **1970**), *Strange Doings* (coll **1972**), *Does Anyone Else Have Something Further to Add?* (coll **1974**), *Ringing Changes* (coll **1984**; first published in Dutch trans as *Dagan van Gras, Dagan van Stro* ["Days of Grass, Days of Straw"] **1979**), *Golden Gate and Other Stories* (coll

1982), *Through Elegant Eyes: Stories of Austro and the Men who Know Everything* (coll **1983**), *Lafferty in Orbit* (coll **1991**) and *Iron Tears* (coll **1992**). Pamphlets include *Funnyfingers & Cabrito* (coll **1976** chap), *Horns on their Heads* (coll **1976** chap), *Snake in his Bosom and Other Stories* (coll **1983** chap), *Four Stories* (coll **1983** chap), *Heart of Stone, Dear, and Other Stories* (coll **1983** chap), *Laughing Kelly and Other Verses* (coll **1983** chap), *The Man who Made Models and Other Stories* (coll **1984** chap), *Slippery and Other Stories* (coll **1985** chap), *Strange Skies* (coll **1988** chap Canada; verse), *The Back Door of History* (coll **1988** chap Canada), *The Elliptical Grave* (**1989**; 1 story added to limited edition to make coll 1989) and *Mischief Malicious (and Murder Most Strange)* (coll **1991** chap Canada), which contains work from as early as 1961.

If there is one contemporary figure with whom RAL shares a similar incapacity to fit the expectations of readers, it might be Avram DAVIDSON. Unlike most modern writers of fantasy or sf, both were obsessed by, and believers in, traditional religions: Roman Catholicism for RAL and Judaism for Davidson. And there is a sense that one must know the whole of each writer before coming to an understanding of any one part. [NG/JC]

Other works: The **Coscuin Chronicles**, comprising *The Flame is Green* (**1971**) and *Half a Sky: The Coscuin Chronicles 1849-1854* (**1984**); *The Fall of Rome* (**1971**; vt *Alaric: The Day the World Ended* 1993), a historical novel; *Arrive at Easterwine: The Autobiography of a Ktistec Machine* (**1971**), an engrossingly bizarre TECHNOFANTASY; *Okla Hannali* (**1972**); *Aurelia* (**1982**); *Annals of Klepsis* (**1983**); *Serpent's Egg* (**1987** UK; 1 story added to limited edition to make coll 1987); *East of Laughter* (**1988** UK; with 1 story added to limited edition to make coll); *Sindbad: The 13th Voyage* (**1989**).

LA FONTAINE, JEAN DE (1621-1695) French fabulist. ♦ FABLE; Charles PERRAULT.

LAFORGUE, JULES (1860-1887) Innovative French poet. The posthumous *Moralités légendaires* (coll **1887**; trans Frances Newman as *Six Moral Tales by Jules Laforgue* 1928 US; new trans William Jay Smith as *Moral Tales* 1985 US) presents six parodic reinterpretations of traditional tales. Hamlet, Lohengrin, SALOME, Perseus and PAN are featured in five of them; the sixth is a mock CHRISTIAN FANTASY. Two of Smith's translations had earlier appeared in *Selected Writings of Jules Laforgue* (coll **1956**), which he edited. John ERSKINE and many others have produced works similar in spirit to the *Moral Tales*, but no one has ever duplicated JL's effervescent style, whose example helped Alfred Jarry (1873-1907) and Guillaume APOLLINAIRE build the literary bridge between DECADENCE and SURREALISM. [BS]

LAGERLÖF, SELMA (OTTILIANA LOVISA) (1858-1940) Swedish novelist, awarded the Nobel Prize in 1909. Her career was launched with the episodic novel *Gösta Berling's Saga* (**1891**; trans Pauline Bancroft Flach as *The Story of Gösta Berling* 1898 UK), which set the pattern for SL's work by constructing a contest between pietistic ideas of Christian virtue and wilder codes of conduct based in local tradition and supported by local FOLKLORE. *Antikrists Mirakler* (**1897** Denmark; trans Pauline Bancroft Flach as *The Miracles of Antichrist* 1898 UK) and *Jerusalem* (**1901-2** 2 vols; trans Jessie Bröchner 1903 UK) are ALLEGORIES in which the supernatural never becomes explicit, but many of SL's shorter works are sentimental exercises in CHRISTIAN FANTASY. Works of this kind can be found in four collections: *Osynliga länkar* (coll **1894**; 12 of the 20 stories from

this collection plus 2 others make up *Invisible Links* coll trans Flach **1899** UK); *En herrgårdssägen* (coll **1899**; trans Bröchner as *From A Swedish Homestead* 1901 UK; cut vt *Drottningar i Kunghälla; jamte andra berättelser* 1900; trans C. Field as *The Queens of Kunghälla and Other Stories* 1930 UK), which includes a novella about St Olaf, "Queens at Kunghälla", and a curious dialogue between "Our Lord and St Peter"; *Kristuslegender* (coll **1904**; trans Howard as *Christ Legends and Other Stories* coll 1908 US); and *En saga om en saga, och andra sagor* (coll **1904**; trans Howard as *The Girl From the Marsh Croft* coll 1910 UK), which is mostly mundane but includes the most famous of SL's religious fables, *The Legend of the Christmas Rose* (**1934** chap US).

Although *Herr Arnes penningar* (**1904**; trans Arthur G. Chater as *Herr Arne's Hoard* 1923 UK; vt *The Treasure* 1925 US) features supernatural intrusions in its account of the visitation of divine vengeance upon three Scottish soldiers, SL's first wholehearted fantasy novel was the children's story *Nils Holgerssons underbara resa genom Sverige* (**1906-7** 2 vols; trans Howard as *The Wonderful Adventures of Nils* 1907 US and *The Further Adventures of Nils* 1911 US), which was commissioned by the Primary School Board as an aid to the teaching of Swedish geography; it is the story of a farmer's son who is changed into an elf as punishment for his cruelty to animals and joins a flock of migrating geese. The moralistic VISIONARY FANTASY *Körkarlen* (**1912**; trans William Frederick Harvey as *Thy Soul Shall Bear Witness!* 1921) is based on a legend alleging that the last man to die on New Year's Eve must drive DEATH's cart during the coming year. *Liljecronas hem* (**1911**; trans Anna Barwell as *Lilicrona's Home* 1913 UK) is a slightly supernaturalized story about a pastor's daughter whose wicked STEPMOTHER might be a dispossessed water-sprite. *Löwensköldska ringen* (**1925**; trans Francesca Martin as *The General's Ring* 1928 UK; new trans L. Schenck as *The Lowenskold Ring* 1991 US) is a historical fantasy in which all who acquire a RING stolen from a general's corpse are cursed by his GHOST; the other two novels making up the trilogy reprinted in *The Ring of the Lowenskölds* (omni **1930** UK) have no supernatural content. SL's collected works were issued as *Skrifter* (**1933** 12 vols). [BS]

Other works: *The Legend of the Sacred Image* (**1914** chap US); *Harvest* (coll trans Florence and Naboth Hedin **1935** US); *The Changeling* (**1992** chap US).

LAGERKVIST, PÄR (1891-1974) Norwegian poet, dramatist and writer, active from c1914, winner of the 1951 Nobel Literature Prize. His early work, especially his plays, employs a variety of expressionistic techniques to argue issues of high seriousness, as shaped by J.G. FRAZER's *The Golden Bough* (**1890-1915**), though sometimes bald in the utterance. *Det eviga leendet* (**1920**; trans anon as *The Eternal Smile* 1932 chap UK) is a SUPERNATURAL FICTION; it is assembled, along with *Gäst hos verkligheten* ["A Guest in the Actual World"] (**1925**), as *Guest of Reality* (coll trans E. Mesterton and D.W. Harding 1936 UK; rev vt *The Eternal Smile and Other Stories* 1971 UK; vt *The Eternal Smile: Three Stories* 1971 US). The DWARF in *Dvärgen* (**1944**; trans A. Dick as *The Dwarf* 1945 US) becomes no more than a SHADOW – a mocking PARODY – of his master the Prince; but the plot hovers this side of the supernatural. *Barabbas* (**1950**; trans Alan Blair 1952 US) similarly invokes, but does not enter, worlds beyond the mundane. The WANDERING JEW features in *Sibyllan* (**1956**; trans N. Walford as *The Sybil* 1958 UK) and *Ahasverus' död* (**1960**; trans Walford as *The*

Death of Ahasuerus **1962** UK), being allowed to die in the second volume as reward for abandoning his obsessive religious concerns. Other tales of interest include *Pilgrim på havet* (**1962**; trans Walford as *Pilgrim at Sea* **1964** UK), *Det heliga landet* (**1964**; trans Walford as *The Holy Land* **1966** UK) and *Mariamne* (**1967**; trans N. Walford as *Herod and Mariamne* **1968** US; under orig title 1968 UK). [JC]

LAINEZ, MANUEL MUJICA ◊ Manuel MUJICA LAINEZ.

LAIR OF THE WHITE WORM, THE UK movie (*1988*). ◊ Ken RUSSELL.

LAKE OF DRACULA (ot *Chi o Suu Mi*) Japanese movie (*1971*). ◊ DRACULA MOVIES.

LAMARK, DREW Pseudonym of André LAUNAY.

LAMB, CHARLES (1775-1834) UK essayist and writer, best-known for the essays he published under the name Elia; the first book publication was *Elia* (coll dated 1823 but **1822**). His work for children is mostly TWICE-TOLD, and is neatly done: it includes *The Adventures of Ulysses* (**1808**) – which appears with some other short fantasies in *Eliana: Being the Hitherto Uncollected Writings* (coll **1864**) – *Prince Dorus: A Poetical Version of an Ancient Tale* (**1811** chap) and *Beauty and the Beast* (**1811** chap). *Satan in Search of a Wife, and Who Danced at the Wedding* (**1831**) as by An Eyewitness is poetry. *Tales from Shakespear* * (coll **1807**), written with CL's sister Mary (Ann) Lamb (1764-1847), adapted the plays of William SHAKESPEARE as stories for children; this volume was published by William GODWIN in his **Juvenile Library** series and marked CL's first commercial success. [JC]

LAMB, HAROLD (ALBERT) (1892-1962) US writer whose fiction and nonfiction tended to focus on the Near East and Asia. He began publishing fantasy with *Marching Sands* (**1920**), a LOST-RACE tale set in the Gobi Desert, where two groups of Westerners are oppressed by Buddhists who hate the White Crusaders who have secretly inhabited the lost city of Sungan for many centuries. *The House of the Falcon* (**1921**), *The Three Paladins* (1923 *Adventure*; **1977**) and *A Garden to the Eastward* (**1947**) are similar. *Durandal* (1926 *Adventure*; cut as Part 1 only of *Durandal* **1931**; text restored and published separately 1981; text of Part 2 restored and published separately vt *The Sea of the Ravens* 1983; Part 3 remains unrestored), set in the 13th century, pits the Western bearer of a sacred SWORD called Roland against the paynim. HL was read by, and was an influence upon, Robert E. HOWARD. [JC]

LAMBERT, S.H. Pseudonym of Neil BELL.

LAMBOURNE, JOHN Pseudonym of UK writer John Battersby Crompton Lamburn (1893-?) for his fiction; he retained his full name and the alias John Crompton for his nonfictional studies of insects and spiders. JL spent time in South Africa (as a policeman) and China (as a shipping clerk) before settling back in the UK in 1932. His travels provided background colour for his work, which included two LOST-RACE adventures in the style of H. Rider HAGGARD and featuring the character **Professor Ellis**: *The Kingdom that Was* (**1931**) and *The Second Leopard* (**1932**).

His sister, who wrote as Richmal Crompton (1890-1969), was famous for her nonfantasy **William** stories, and also produced several stories about malevolent GHOSTS, including the novel *The House* (**1926**; vt *Dread Dwelling* 1926 US) and the collection *Mist, and Other Stories* (coll **1928**). [MA] **Other works:** *The Unmeasured Place* (**1933**).

LAMIAS A species of empusas (demons sent by Hecate to waylay and devour travellers) whose members had the trick of disguising themselves as young women in order to beguile men before supping their blood. Some accounts suggest their natural form was serpentine; a tale of this kind reproduced from Philostratus in Robert Burton's *Anatomy of Melancholy* (**1621**) inspired John KEATS's poem "Lamia" (1820). Lamias provide one of the basic models of the FEMME FATALE and the female VAMPIRE. [BS] **See also:** LILITH; SUCCUBUS.

LANCE, KATHRYN (1943-) US writer who also writes as by Lynn Beach; she is known for the **Pandora** sf sequence as KL and for a number of fantasy novels as by Beach for YA readers, including the **Phantom Valley** sequence – *Phantom Valley #1: H.O.W.L. High* (**1991**), *#2: The Evil One* (**1991**), *#3: The Dark* (**1991**), *#4: Scream of the Cat* (**1992**), *#5: Stranger in the Mirror* (**1992**) and *#6: The Spell* (**1992**) – the first featuring the adventures of a HALFLING who is half-warlock, the second featuring a demon DOLL, etc. Other titles, tied to various juvenile series, include *Secrets of the Lost Island* * (**1984** chap), *The Attack of the Insecticons* * (**1985** chap), *Conquest of the Time Master* * (**1985**), *The Haunted Castle of Ravencurse* * (**1985**) and *Invaders from Darkland* * (**1986**). [JC]

LANCELOT Most valiant of the KNIGHTS of King ARTHUR, and the lover of Queen GUINEVERE, Lancelot first appears in the 12th-century romance *Lancelot or Le Chevalier de la Charrete* ["The Knight of the Cart"] (?**1177**) by CHRÉTIEN DE TROYES, to whom the idea was suggested by his patroness Marie de Champagne. No earlier reference to Lancelot survives, though it has been conjectured that he was a 6th-century Pictish ruler. In the legend the castle he eventually attains, Joyous Gard(e), is usually placed in the north of England. He is named as the son of King Ban of Benwick (in Brittany); when his father dies of a broken heart, the young Lancelot is reared by the LADY OF THE LAKE, who then sends him to Arthur's court. Lancelot overcomes Arthur's initial suspicion and becomes his most loyal and heroic knight, and the champion of Queen Guinevere.

It was Chrétien's version of Lancelot's adventures – and their subsequent prose adaptation in the **Vulgate Cycle** (compiled 1215-35) – that was used by Sir Thomas MALORY in *Le Morte Darthur* (**1485**) In the German *Lanzelet* (?**1195**) by Ulrich von Zatzikhoven Lancelot is unaware of his origins but learns he must prove his valour in order to regain his inheritance. In Malory's version, Lancelot's love for Guinevere eventually leads to the dissolution of the ROUND TABLE, the war between Arthur and Lancelot, and the usurpation of the throne by Mordred (Modred). Lancelot is of course unable to attain the GRAIL because of his adultery. After Arthur's defeat and Guinevere's death, Lancelot pines away.

Lancelot seldom features strongly in modern Arthurian fiction, although he almost certainly was the model for the PRINCE VALIANT comic strip by Hal FOSTER. He is centre-stage in: *Launcelot* (**1926**) by Ernest Hamilton; *Launcelot and the Ladies* (**1927**) by Will Bradley (1868-1962); *The Ill-Made Knight* (**1940**) by T.H. WHITE; *The Little Wench* (**1935**) by Philip Lindsay (1906-1958); *Launcelot My Brother* (**1954**) by Dorothy James Roberts (1903-1990); *The Queen's Knight* (**1955**) by Marvin Borowsky; and *Lancelot* (**1978**) by Peter Vansittart (1920-). A particularly touching portrayal of Lancelot wandering the Earth searching for Guinevere's grave centuries later is given by Jane YOLEN in "The Quiet Monk" (1988 *IASFM*). Lancelot's love for Guinevere is the focal point of the movies *Lancelot and*

Guinevere (*1963*) and *First Knight* (*1995*), and plays a major part in *Excalibur* (*1981*). [MA]

LANCOUR, GENE Working name of US writer Gene Lancour Fisher (1947-   ), whose last novel to date, *The Globes of Llarum* (**1980**), is a space opera. He began publishing with the **Dirshan the God-Killer** SWORD-AND-SORCERY sequence – *The Lerios Mecca* (**1973**), *The War Machines of Kalinth* (**1977**), *Sword for the Empire* (**1978**) and *The Man-Eaters of Cascalon* (**1979**). [JC]

LAND The term "land", when cross-referred to this entry, has a restricted meaning. A land is a venue located in a SECONDARY WORLD – and in certain kinds of secondary world only. The mere backdrop to the action, as in almost any novel set in FANTASYLAND, is not a land but simply a venue. For our purposes a land may be defined as a secondary-world venue whose nature and fate are central to the plot: a land is not a protagonist, but has an analogous role. Some or all of the following will almost certainly be the case: the land may evince WRONGNESS; it may be subject to THINNING; it may be a WASTE LAND; it may suffer (or be saved by virtue of) a fundamental TRANSFORMATION; it can be healed; and it is almost certainly, in some sense, alive.

The central figures in the history of the development of the secondary world all created some form of land: the various domains travelled through in William MORRIS's *The Well at the World's End* (**1896**) are increasingly land-like; the Mercury where E.R. EDDISON's *The Worm Ouroboros* (**1922**) unfolds, though ostensibly another planet, soon intensifies into a land; J.R.R. TOLKIEN's Middle-Earth is a land in every respect, as is the eponymous WASTE LAND whose condition must be transformed in Alan GARNER's *Elidor* (**1965**). But the most comprehensive example of a land is almost certainly that in Stephen R. DONALDSON's **First Chronicles of Thomas Covenant the Unbeliever** (**1977**), where the dynamic tension between Covenant's Unbelief and land can be read as a nearly articulate debate between a protagonist and the world which both defines him and is defined by him. Other novels in which the land is central include Michael ENDE's *The Neverending Story* (**1979**), where Fantastica is being devoured by the "Nothing", which will not retreat until the ruling Empress, who "is the centre of all life" there, is given a new name and thus cured of her wasting illness. [JC]

LAND OF FABLE LOFs can be found somewhere between the identifiable real world – in which all SUPERNATURAL FICTION and many kinds of fantasy are set – and impossible worlds: an LOF always has its roots in this world, but in at least some aspects is an impossible one. For instance, the Middle East featured in much ARABIAN FANTASY is clearly an impossible realm. What distinguishes the LOF from a full-blown ALTERNATE REALITY or ALTERNATE WORLD is the fact that the reality-status of the LOF, or its relationship to our world, is not examined. The LOF, usually quasi-historical, stands to this world as FANTASYLAND stands to the fully developed SECONDARY WORLD: it is a TEMPLATE of the real world, designed to operate as a backdrop for adventures. If an author creates a territory which appears to be a LOF, but then proceeds to suggest reasons for its having diverged in some way from our real world, s/he has created an alternate world; if s/he suggests that the characters' PERCEPTIONS of their world are somehow intrinsic to its reality, then s/he has created an ALTERNATE REALITY.

HEROIC FANTASIES are often set in LOF versions of the prehistoric world; the same relationship prevails between

fantasies about Native Americans and pre-Columbian America, fantasies derived from SAGA and Scandinavia, CELTIC FANTASIES/ARTHUR tales and Britain, tales of the Greek GODS and ancient Greece, CHINOISERIE and China, etc. The LOF has all the colour of the full-fantasy venue, but not the confrontational, METAMORPHOSIS-prone substance. [JC]

LAND OF FARAWAY, THE (vt *Mio in the Land of Faraway*) Norwegian/Swedish/Russian movie (**1987**). Nordisk Tonefilm/Gorky/Norway Film Development/Swedish Film Institute/Filmhuset/Sovinfilm. **Pr** Ingemar Ejve. **Dir** Vladimir Grammatikov. **Vfx** Derek Meddings. **Screenplay** William Aldridge. **Based on** *Mio, My Son* (**1954**) by Astrid LINDGREN. **Starring** Christian Bale (Benke/Jum-Jum), Timothy Bottoms (The King), Igor Isulovitch (Eno), Christopher Lee (Kato), Sverre Anker Ousdal (The Swordmaker), Nicholas Pickard (Bo/Mio), Geoffrey Staines (The Spirit), Linn Stokke (Mrs Lundin), Susannah York (The Weaver Woman). 95 mins. Colour.

Bullied Stockholm orphan Bo dreams of his long-lost father. Titania-like local shopkeeper Mrs Lundin asks him to mail a postcard for her at a particular mailbox. But the postcard is addressed to the King of the Land of Faraway, and announces that Bo is the one this king has been seeking – his long-lost son. She has given him also a MAGIC apple, which proves the key to the PORTAL leading to the Land of Faraway, the gatekeeper being a GENIE (in effect) who springs from an empty bottle left in the gutter. There, on Green Meadow Island, Bo is welcomed by his father, the king, as Mio – his TRUE NAME. There, too, Mio and Jum-Jum – who in our REALITY is Bo's best friend Benke – undergo various adventures before driving the murderous Kato, Faraway's child-hating DARK LORD, from the land.

The early parts of *TLOF*, where it is an URBAN FANTASY based on the wish-fulfilment PERCEPTIONS of young Bo – are beautifully effective. Later, though, *TLOF* becomes cliché-ridden and turgid. The commercial territory *TLOF* is attempting to explore is that of *The* NEVERENDING STORY (**1984**). Lindgren herself ventured into this territory long before Michael ENDE, yet *TLOF* rewards her pioneering efforts with a movie that seems tired, imitative and cynical. [JG]

LAND OF OZ, THE (**1910**) ◊ *The* WIZARD OF OZ (**1939**).

LAND OF THE GIANTS: GULLIVER'S TRAVELS PART 2 vt of *Gulliver's Travels Part 2* (**1983**). ◊ GULLIVER MOVIES.

LANDOLFI, TOMMASO (1908-1979) Italian writer of sf interest for *Cancroregina* (coll **1950**; trans Jack Murphy as "Cancroregina" 1950 *Botteghe Oscure*; new trans Raymond Rosenthal in *Cancerqueen and Other Stories* 1971 US), in which a mad astronaut is imprisoned in a living starship, and finds – not surprisingly, from an author much influenced by Edgar Allan POE, Nikolai GOGOL (whom he translated into Italian) and Franz KAFKA – that the groteque world in which he is imprisoned is all he will ever get. Most of TL's fantasy was assembled in *Dialogo dei massimi sistemi* ["Dialogue Concerning the Chief World Systems"] (coll **1937**), *Il mar delle blatte* ["The Sea of Roaches"] (coll **1939**) and *Le Labrene* ["The 'Lipzards'"] (coll **1974**). Many of these tales are fantasies in the same sense that Gogol and Kafka wrote fantasies; translated tales are assembled in *Cancerqueen and Other Stories* and *Gogol's Wife and Other Stories* (coll trans Raymond Rosenthal, John Longrigg and Wayland Young 1963 US); the best presentation of his work is probably that posthumously edited by Italo CALVINO, *La più belle* *pagine di Tommasso Landolfi* ["The Very Best Pages of Tommasso Landolfi"] (coll **1982**; cut trans Kathrine Jason as *Words in Commotion and Other Stories* **1986** US). Throughout, a vertigo of real and threatened METAMORPHOSIS, rendered without the safety net of fantasy conventions, constantly reminds readers that TL is a dangerous, enticing Modernist. [JC/CP]

LANDSCAPE In fantasy novels, to leave a CITY is almost invariably to enter a landscape. Outside CONTEMPORARY FANTASY and URBAN FANTASY, there are few suburbs or purlieus in the worlds of fantasy. Landscapes almost invariably convey a sense, though often the effect is subliminal at best, that every nook and cranny, every chasm and crag, every desert and fertile valley is potentially *meaningful*. And how a landscape is described in fantasy is what that landscape *means*. A wasted landscape (◊ WASTE LAND) almost necessarily signals THINNING, or the desiccating attentions of a DARK LORD – Sauron in J.R.R. TOLKIEN's *The Lord of the Rings* (**1954-5**) hates even vegetable life and has virtually sterilized his own land of Mordor. A fertile landscape almost necessarily signals an ARCADIA, or a POLDER, or a wasted landscape which has enjoyed HEALING. A populous landscape thrives; an empty one awaits the return of the folk. Landscapes which waver from one state to the other are likely to represent the outskirts of a region where different versions of REALITY may conflict, where landscapes may be seen as puns, pointing towards two different meanings of the world (◊ TROMPE L'OEIL).

The first landscapes of any significance for writers of fantasy were probably those painted by Hieronymus BOSCH, who came very early in the European move towards understanding that it was possible and legitimate to treat the "backdrop" to human action – i.e., the "natural" world – as a subject worth intense concern. Some of the great TAPROOT TEXTS of this period – e.g., the works of ARIOSTO – are much concerned with landscape. "Pure" landscape painting came later, and has had little influence on fantasy. Painters after Bosch who, like him, join actors and world in meaningful discourse include Pieter BRUEGEL, Matthias GRÜNEWALD, John MARTIN and Richard DADD. These painters were to varying degrees influential upon the great fantasy illustrators who followed them – Gustave DORÉ, Arthur RACKHAM, Edmund DULAC, etc. – all of whom plastically interweave landscape and action. This interweaving was embodied in prose by William MORRIS in *The Well at the World's End* (**1896**), whose unrolling of endless landscape has been hugely influential.

That tradition has continued. It is only when landscape becomes an end in itself that fantasy is not being served: in true fantasy, landscape and action are different aspects of the same STORY. [JC]

LANG, ANDREW (1844-1912) Scottish man of letters who wrote novels, poetry, belles-lettres, influential book reviews, anthropological studies and children's books, and who edited a famous and influential ANTHOLOGY series of traditional FABLES and FAIRYTALES retold for children (under the supervision of his wife, Leonora Lang, who assiduously censored his renderings), with some added hagiographical and historical material, all of which was new to the market. This sequence was *The Blue Fairy Book* (anth **1889**), *The Red Fairy Book* (anth **1890**), *The Green Fairy Book* (anth **1892**), *The Yellow Fairy Book* (anth **1894**), *The Grey Fairy Book* (anth **1900**), *The Violet Fairy Book* (anth **1901**), *The Crimson Fairy Book* (anth **1903**), *The Brown Fairy Book*

(anth **1904**), *The Orange Fairy Book* (anth **1907**), *The Olive Fairy Book* (anth **1907**) and *The Lilac Fairy Book* (anth **1910**). His adult fiction is now forgotten, as are his CHILDREN'S FANTASIES; but these anthologies, which were fairly heavily illustrated, continue to be used. *The Rainbow Fairy Book* (anth **1993**) ed and illus Michael Hague, presents a selection of 31 tales from the originals, with new illustrations. AL's introduction to his edition of Charles PERRAULT, *Perrault's Fairy Tales* (coll **1888**), gives his thoughts on the origins of the kind of material he adapted.

As a thinker in the field of anthropology, AL tended to a somewhat reductive common sense, arguing (for instance) that FOLKTALES simply represent primitive beliefs in dramatic form (an argument which depreciates any sense of the inherent power of STORY), and animadverting against the mythopoeic theorizing of Sir James FRAZER in *The Golden Bough* (**1890**). If AL is now unread in this field, it is perhaps because he was neither right nor interestingly wrong.

Of his over 120 titles, AL's best-remembered original works are his fairytales for children. *The Princess Nobody: A Tale of Fairy Land* (**1884** chap; rev vt *In Fairyland* 1979 chap) is a prose fantasy based on Richard DOYLE's illustrations to *In Fairy Land: A Series of Pictures from the Elf-World* (graph **1869**), a poem by William Allingham (1824-1889), also making use of some illustrations omitted from the 1869 volume. *The Gold of Fairnilee* (**1888** chap), much more substantial, tells of a young man abducted into the GARDEN of FAERIE, where he remains seven years (◊ TIME IN FAERIE) until his true love rescues him; in a central episode, his PERCEPTION of Faerie is altered by a magic liquid, and he sees the immortal land as a profoundly attenuated WASTE LAND; it is a significantly vivid presentation of a version of Faerie which has since become common. The **Pantouflia** sequence – *Prince Prigio* (**1889**) and *Prince Ricardo of Pantouflia: Being the Adventures of Prince Prigio's Son* (**1893**), both assembled as *My Own Fairy Book* (omni **1895**), plus some stories from *Tales of a Fairy Court* (coll **1906**) – rather more frivolously recounts the adventures of Prigio (and later his son, Ricardo) in a Faerie somewhat vitiated by jokes which reduce the venue to a LAND-OF-FABLE Britain. Some of the adventures are amusing. *The Gold of Fairnilee and Other Stories* (coll **1967**) conveniently assembles the best work.

Some of AL's adult fiction is of interest. *"That Very Mab"* (**1885**) with May Kendall (real name Emma Goldworth; 1861-?1931), published anon, tells of the 19th-century adventures of the FAIRY QUEEN, who left England after the Puritan THINNING of the isle and now returns from Samoa to find her home sadly transformed. The title story of *In the Wrong Paradise and Other Stories* (coll **1886**) dramatizes the dictum that one man's HEAVEN is another man's HELL; in "The End of Phaeacia", also in this volume, a missionary shipwrecked on a South Sea ISLAND discovers it to be the Homeric Phaeacia. Some of the pieces collected in *Old Friends: Essays in Epistolary Parody* (coll **1890**) offer a forerunner for the writing of RECURSIVE fiction.

AL was a close friend of H. Rider HAGGARD. He anonymously parodied *She* (**1887**) in *He, by the Author of It* (**1887**; vt *He, A Companion to She* 1887 US) with Walter Herries POLLOCK. After this jape, AL collaborated with Haggard on *The World's Desire* (**1890**), which combines Haggard's crude, sometimes haunting vigour with AL's chastely pastel classicism; despite occasional longueurs, the resulting tale of ODYSSEUS's last journey to find Helen in Egypt is a moving, frequently eloquent romance, the best AL was ever involved in, and effectively comes to a climax with Odysseus's discovery that Helen is the AVATAR of Ayesha and his death at the hands of his son.

Copious, but flawed in general by a disheartening dilettantism, AL's work lies just the wrong side of major ranking in the fantasy field – exactly as in his other areas of concentration. [JC]

Other works: *Much Darker Days* (**1884**; rev 1885) as by A Huge Longway, which parodies *Dark Days* (**1884**) by Hugh CONWAY; *The Mark of Cain* (**1886**), sf; *Pictures at Play, or Dialogues of the Galleries* (coll **1888**) with W.E. Henley (1849-1903), as by Two Art-Critics; *A Monk of Fife: Being the Chronicle Written by Norman Leslie of Pitcullo, Concerning Marvellous Deeds that Befell in the Realm of France, in the Years of our Redemption, MCCCCXXIX-XXXI* (dated 1896 but **1895**), about JOAN OF ARC; *The Disentanglers* (coll of linked stories **1901** chap US; much exp 1902 UK), CLUB STORIES; *The Story of the Golden Fleece* (**1903**); *The Story of Joan of Arc* (**1906**); *When it was Light: A Reply to "When it was Dark"* (**1906**), anon, responding to Guy Thorne's 1903 novel; *Tales of Troy and Greece* (coll **1907**).

Nonfiction: *Custom and Myth* (**1884**; rev 1885); 6 vols of translations and studies of HOMER and associated figures, being *The Odyssey* (trans **1879**) with S.H. Butcher, *The Iliad* (trans **1883**) with Walter Leaf and Ernest Myers, *Homer and the Epic* (**1893**), *The Homeric Hymns* (**1899**), which includes translations, *Homer and his Age* (**1906**) and *The World of Homer* (**1910**); *Myth, Ritual, and Religion* (**1887**; rev 1899) in 2 vols; *Modern Mythology* (**1897**); *Cock-Lane and Common Sense* (coll **1894**); *The Book of Dreams and Ghosts* (**1897**); *Magic and Religion* (**1901**); *The Puzzle of Dickens's Last Plot* (**1905**), about Charles DICKENS's *Edwin Drood* (**1870**); *The Secret of the Totem* (**1905**); *Method in the Study of Totemism* (**1911**).

As editor: *The Dead Leman and Other Tales from the French* (anth trans AL and Paul Sylvester **1889** UK), containing some SUPERNATURAL FICTION; *The Blue Poetry Book* (anth **1891**); *The Animal Story Book* (anth **1896**); *The Nursery Rhyme Book* (anth **1897**); *The Arabian Nights Entertainments* (anth **1898**); *The Red Book of Animal Stories* (anth **1899**); *The Book of Romance* (anth **1902**); *The Red Romance Book* (anth **1905**).

LANGFORD, DAVID (ROWLAND) (1953-   ) UK writer, critic and former scientist, most of whose fiction is sf or sf-like. His contribution to *The Necronomicon* (anth **1978**) ed George Hay (1922-   ) was a spoof account of deciphering the CTHULHU MYTHOS text which the book semi-seriously reconstructs (◊ BOOKS). DL's collections of reviews, *Critical Assembly* (coll **1987** chap; rev 1992) and *Critical Assembly II* (coll **1992** chap) contain much humorous and acerbic discussion of GENRE FANTASY, as do the reprinted essays and sf convention speeches in *Let's Hear It for the Deaf Man* (coll **1992** chap US; exp vt «The Silence of the Langford» 1996 US) ed Ben Yalow. He has written relatively little fantasy, though genre authors are lampooned in his PARODY collection *The Dragonhiker's Guide to Battlefield Covenant at Dune's Edge: Odyssey Two* (coll **1988**). His "The Arts of the Enemy" in *Villains!* * (anth **1992**) ed Mary GENTLE and Roz KAVENEY is a wry reconsideration of BALANCE. In "Waiting for the Iron Age", in *Tales of the Wandering Jew* (anth **1991**) ed Brian STABLEFORD, DL injects TECHNOFANTASY into the WANDERING JEW legend, and he deals similarly with themes from H.P. LOVECRAFT in *Irrational Numbers* (coll **1994** chap US). Several of his HORROR stories

were selected for the *Year's Best Horror* anthologies ed Karl Edward WAGNER. He assisted informally with the *SFE*, and is a Contributing Editor to *The Encyclopedia of Fantasy*. [DRL]

Other works: *Earthdoom!* (**1987**) with John GRANT, a disaster-novel spoof, featuring ARTHUR and an anthropomorphic DEATH; *The Unseen University Challenge** (**1996**), quizbook based on Terry PRATCHETT's **Discworld** novels.

LANGLEY, NOEL (A.) (1911-1980) South African-born writer, long resident in the UK, but more successful in the USA (where he died) as a screenwriter and movie director. He provided the basic script for *The* WIZARD OF OZ (*1939*) and also scripted *Scrooge* (*1951*; ◊ *A* CHRISTMAS CAROL) and SVENGALI (*1954*). *The Tale of the Land of Green Ginger* (**1937**; vt *The Land of Green Ginger* 1947; cut 1966) is an ARABIAN FANTASY which has influenced writers such as Neil GAIMAN and Diana Wynne JONES. *The Rift in the Lute* (**1952**) is also fantasy. *Tales of Mystery and Revenge* (coll **1950**) contains mostly HORROR. [JC]

LANGRAN, MAURICE [s] ◊ Kenneth MORRIS.

LANGTON, JANE (1922-   ) US writer. ◊ CHILDREN'S FANTASY.

LANTZ, WALTER (1900-   ) US animator who joined the Hearst animation studio at the precocious age of 16; there he worked alongside such greats as Burt Gillett, Jack King, I. Klein, Grim Natwick and Ben Sharpsteen; his development was rapid. When the studio closed in 1918, WL moved first to Barré/Bowers and then to John Bray, where he soon became studio manager and produced his **Dinky Doodle** series of live-action/animated shorts. When this studio in turn closed in 1927, WL was employed as a gag writer by Mack Sennett. In due course Universal's Carl Laemmle asked him to set up an animation studio, his first project being to continue the **Oswald the Lucky Rabbit** series created by Walt DISNEY and Ub IWERKS. The commercial details of this are murky, and best left so; WL himself was blameless. In the wake of Disney's SNOW WHITE AND THE SEVEN DWARFS (*1937*) he proposed to create for Universal a rival feature, «Aladdin and his Wonderful Lamp», but this never appeared – largely because WL wisely realized that, at the time, no one could take on DISNEY at their own game and win (Max FLEISCHER was less wise).

Over the years WL had created various TALKING ANIMALS for his animated shorts: the one he is renowned for, **Woody Woodpecker**, seems in fact to have been created by Ben Hardaway (◊ BUGS BUNNY) for the short *Knock Knock* (*1940*); the raucous laugh was done first by Mel Blanc, then by various others briefly until in 1951 Grace Stafford, WL's wife, adopted it in perpetuity. The character had really been launched with *The Barber of Seville* (*1944*), when Shamus Culhane took over the direction.

Over the years WL also, in imitation of Disney's **Silly Symphonies** series, was responsible for **Cartune Classics**, **Swing Symphonies** and **Musical Miniatures**. He eventually gave up animation in 1972, apparently because it no longer made commercial sense to produce shorts. Although he was never a real innovator, his genuine wit and his ability to spot and hire talented animators – any list would fill a column of this book – make him a landmark figure in the history of screened fantasy. [JG]

LANYARD, FORTESCUE [s] ◊ Kenneth MORRIS.

LA PLANTE, RICHARD (1948-   ) US writer, in the UK since 1976. His **Tegné** sequence – *Tegné: Warlord of Zendow* (**1988**) and *Tegné: The Killing Blow* (**1990**) – is

SWORD AND SORCERY set in a vaguely post-catastrophe LAND-OF-FABLE Orient. A second series – *Mantis* (**1992**) and *Leopard* (**1993**) – is HORROR. [JC]

LARSEN, JEANNE (LOUISE) (1950-   ) US academic and writer whose *Silk Road: A Novel of Eighth-Century China* (**1989**), won the 1990 WILLIAM L. CRAWFORD MEMORIAL AWARD for Best First Fantasy Novel. Set in a LAND-OF-FABLE ancient China (◊◊ CHINOISERIE), it follows the PICARESQUE life of a pearl transformed into a human child by an irritated Jade Emperor, whose game of Go has been interrupted (◊ ANIMATE/INANIMATE; REAL BOY); the human Pearl loses her mother to a god UNDER THE SEA, and must QUEST for her, aided by a varying band of COMPANIONS. The picture of China is ornate, detailed and convincing. JL's second novel, *Bronze Mirror* (**1991**), set in a similar China, creates an intricate interplay between a bored passel of GODS and the earthly STORY they create. Eventually the Story acquires a momentum which cannot safely be tampered with, for the lives of mortals do not constitute a proper playground for the gods. [JC]

LARSON, RANDALL D(OUGLAS) (1954-   ) US editor and writer, a contributor to and editor of *Cinemascore*, a journal which deals with movie music. *Musique Fantastique: A Survey of Film Music in the Fantastic Cinema* (**1985**) is a necessary study. His other work of interest concerns Robert BLOCH: *The Complete Robert Bloch: An Illustrated, Comprehensive Bibliography* (**1986**); *Robert Bloch* (**1986**), a broad study; and *The Robert Bloch Companion: Collected Interviews 1969-1989* (coll **1990**). [JC]

LASDUN, JAMES (1958-   ) UK poet, writer and anthologist, resident in the USA. *The Silver Age* (coll **1985**) and *Three Evenings and Other Stories* (coll **1992**) both contain lucid, somewhat surreal explorations into the fantastic. *After Ovid: New Metamorphoses* (trans anth **1994**) with Michael Hofmann is a useful assemblage of TWICE-TOLD adaptations of OVID in verse form. JL himself translates the tale of Erisychthon – who destroys a sacred grove (◊ GOLDEN BOUGH) and is punished by a hunger so unappeasable that he eventually eats himself – into the 20th century, where a New England property developer suffers the same fate. [JC]

LAST ACTION HERO US movie (*1993*). Columbia/Steve Roth/Oak. **Pr** John McTiernan, Steve Roth. **Exec pr** Arnold Schwarzenegger. **Dir** McTiernan. **Spfx** Connie Brink Sr. **Vfx** Richard Greenberg. **Screenplay** David Arnott, Shane Black. **Novelization** *Last Action Hero* * (**1993**) by Robert Tine. **Starring** Charles Dance (Mr Benedict), Ian McKellen (Death), Frank McRae (Dekker), Tom Noonan (himself/Ripper), Austin O'Brien (Danny Madigan), Robert Prosky (Nick), Anthony Quinn (Vivaldi), Schwarzenegger (himself/Jack Slater). 131 mins. Colour.

Young Danny is addicted to movies featuring tough maverick cop Jack Stalker. Aged local projectionist Nick invites Danny to a sneak preview of *Jack Stalker VI* and gives him a MAGIC golden ticket, once given to Nick by Harry Houdini (1874-1926). The ticket plunges Danny into the heart of the movie and a plot involving criminal Godfather Vivaldi and sinister one-eyed assassin Benedict, as well as a host of movie stereotypes. But the ticket works both ways: Benedict and his henchman psychopath The Ripper (killer of Stalker's son) escape into our REALITY, Benedict to conquer the world using reified movie MONSTERS (e.g., Freddy Krueger from the NIGHTMARE ON ELM STREET series) and the Ripper charged with killing Schwarzenegger (and hence

his movie persona, Stalker). Stalker and Danny follow, the former (like Benedict) discovering that our reality is a much nastier, more painful place than the movie one. Meanwhile the ticket interacts with a screening of Bergman's *The* SEVENTH SEAL (*1956*), so that the figure of DEATH stalks Manhattan's streets. Danny and Nick, reassured by Death that Stalker cannot die – having never lived – use the other half of the magic ticket to return the HERO to the movie reality, where the injuries he has sustained in our reality represent merely a minor flesh wound.

LAH, played largely as comedy – Schwarzenegger outrageously parodying Schwarzenegger movies – is a highly self-referential RECURSIVE FANTASY, featuring countless split-second glimpses of characters from other Schwarzenegger and non-Schwarzenegger movies. Laurence Olivier is seen briefly in a school screening of *Hamlet* (*1948*), but is swiftly transformed through a mutation of Danny's PERCEPTION (triggered by the teacher's description of Hamlet as "the first action hero") into a gun-totin' Schwarzenegger in a fine PARODY. *LAH* is, by its very nature, intimately concerned with STORY and ARCHETYPES: within the movie reality conventions govern the behaviour and destinies of characters, and the intruder Danny finds himself made to conform to what are in effect TWICE-TOLD plotlines. If the late 1980s and early 1990s saw a peak in movie TECHNOFANTASY, then *LAH* can be seen as its topmost cairn. [JG]

See also: ALTERNATE REALITIES; TOONS.

LAST BATTLE APOCALYPSE and the END OF THE WORLD are traditionally associated with a final armed struggle, like Christianity's ARMAGEDDON – often conflated with World War III, as in James BLISH's *Black Easter* (**1968**) – or the battle on the thousand-league-square field of Vigrid which is the climax of RAGNAROK. Michael MOORCOCK's *Stormbringer* (**1965**) features a spectacular LB between GODS of Law and CHAOS. It is not vast scale which defines the LB, but that it marks the passing of an old world: the hopeless final defence in J.R.R. TOLKIEN's *The Lord of the Rings* (**1954-5**) and the mere skirmish of C.S. LEWIS's *The Last Battle* (**1956**) signal, respectively, the end of the Third Age and of Narnia. [DRL]

LAST DAYS OF MAN ON EARTH, THE US vt of *The* FINAL PROGRAMME (*1973*).

LAST JUDGEMENT In the Christian mythos, the tribunal set to follow the APOCALYPSE, and in fictional accounts frequently combined with it. Fantasy writers often express doubts as to the quality of the justice that might be handed down; *Judgment Day* (**1928**) by Norman Davey (1888-?   ) is more optimistic than most. "The Judgement Seat" (1929) by W. Somerset MAUGHAM suggests GOD might have more interesting things to do than arbitrate in humanity's petty moral squabbles. "The House of Judgement" (1893) by Oscar WILDE, *Portrait of Gideon Power* (**1944**) by S.H. Lambert (Neil BELL) and *Heaven Takes a Hand* (**1949**) by Eliot CRAWSHAY-WILLIAMS attempt in different ways to illustrate the difficulties inherent in the judgement business, as does "The Irreverence of God" (1933) by Shane Leslie (1885-1971), in which the Last Judgement is also the Last Laugh. [BS]

LAST MAN It is obviously convenient for writers about the END OF THE WORLD to stop short of annihilating the entire human race; the preservation of a single observer conserves a narrative viewpoint whose anguish can be lavishly indulged. The significant prototype is *Le dernier homme*

(**1805**; trans **1805**) by Jean Cousin de Grainville (1746-1805). *The Purple Cloud* (**1901**) by M.P. SHIEL is a transfiguration of the story of Job as well as that of ADAM AND EVE. In "Spikenard" (1930) by C.E. Lawrence (1870-1940) the survivors of world destruction are the ACCURSED WANDERERS of the Christian mythos. In *The Last Man* (**1940**) by Alfred NOYES, GOD obligingly spares a Benedictine monastery to provide spiritual guidance for the eponymous hero and his Eve; but *The Last Adam* (**1952**) by Ronald Duncan (1914-1982) takes a scornful view of all such providence. [BS]

LAST REDOUBT Any BUILDING or EDIFICE can serve as such if assaulted from without, but a true LR was almost certainly constructed long before (perhaps aeons before) the conflict at hand, and may be a POLDER of toughened reality, so that the adversary's forces wash against it – at least initially – in vain. The LR may well house the last remnants of the old world, and be a repository of last MAGIC in particular. The period in which the story is set may well be towards the END OF THE WORLD, in which case an atmosphere of BELATEDNESS is likely to be overpowering. If the tale is set in more normal times, the LR's defenders are likely to be faithful to the true cause and/or to include a HIDDEN MONARCH. The reader may well expect – until disabused – any LR to incorporate features characteristic of an EDIFICE, and will almost certainly hope that it contains in particular a PORTAL at its heart, through which the protagonist is destined to pass in order to redeem himself, or the SECONDARY WORLD, or both.

LRs are numerous. They include the Great Redoubt in William Hope HODGSON's *The Night Land* (**1912**), Aslan's How in C.S. LEWIS's *Prince Caspian* (**1951**), the Citadel in *Earth's Last Citadel* (**1943** *Argosy*; **1964**) by Henry KUTTNER and C.L. MOORE, Hagedorn in Jack VANCE's *The Last Castle* (**1966** chap dos), the City of the Pyramid in Michael MOOR-COCK's *The Queen of the Swords* (**1971**), the Redoubt of All Humanity in Graham DIAMOND's **Haven** sequence and the Keep of Dare in Barbara HAMBLY's **Darwath** sequence. [JC]

LAST TEMPTATION OF CHRIST, THE US movie (**1988**). Universal/Cineplex Odeon. **Pr** Barbara De Fina. **Exec pr** Harry Ufland. **Dir** Martin Scorsese. **Spfx** Dino Galliano. **Screenplay** Paul Schrader. **Based on** *The Last Temptation of Christ* (**1955**) by Nikos KAZANTZAKIS. **Starring** Victor Argo (Peter), Michael Been (John), Verna Bloom (Mary the Mother), David Bowie (Pontius Pilate), Juliette Caton (Angel), Willem Dafoe (Jesus), Randy Danson (Mary, sister of Lazarus), Peggy Gormley (Martha), Andre Gregory (John the Baptist), Barbara Hershey (Mary Magdalene), Harvey Keitel (Judas), Harry Dean Stanton (Saul of Tarsus). 163 mins. Colour.

Tormented by the knowledge that he is the Chosen One of GOD – or sometimes convinced he is possessed by SATAN, who tells him to aggrandize himself – Jesus (◊ CHRIST) the humble carpenter builds crosses for the Romans and collaborates in the crucifixion of Jewish rebels, hoping God will come to hate him and seek elsewhere. His sole friend, Judas, loathes his actions but seeks to redeem him. At last bowing to the truth, Jesus begs forgiveness of the prostitute whom he loves, Mary Magdalene, and goes to seek spiritual guidance among desert hermits. There he succeeds in casting out the SERPENTS of EVIL from within himself; there, too, comes Judas, sent to slay the collaborator but soon convinced by Jesus's new gospel of LOVE to become the first disciple of the MESSIAH. Together, having stopped the stoning of Mary

Magdalene, they gather apostles. Meeting John the Baptist, a desert fanatic, Jesus is told by him that love is not enough: there must also be righteous anger; seeking counsel from God about this in the desert, Jesus is three times tempted by the DEVIL and returns assured John was right. He travels the land performing EXORCISM, HEALING and MIRACLES. Entering Jerusalem, with the possibility of leading the mob to cast out the Romans, he develops the stigmata and realizes the STORY of which he is the focus cannot be thwarted: he commands Judas, his truest friend, to betray him, and is seized to be crucified on Golgotha.

But on the cross, as he begs mercy of God, an ANGEL comes in the guise of a young child and releases him from the burden of godhood. With her as his constantly present guardian, he weds Mary Magdalene and then, when she dies, another Mary, Lazarus's sister, becoming the lover also of her sister Martha. Life is good until, in late middle age, he encounters the freshly converted Saul who, discovering Jesus's identity, tells him he is unimportant beside the fictional Jesus who died and was resurrected: *that* Jesus may be a lie, but one that will save the world. And on his deathbed Jesus is visited by three of the apostles and by Judas, who condemns him for his betrayal of them – "Your place was on the cross" – and reveals the girl-angel to him as Satan in disguise: Jesus has succumbed to the Devil's last temptation, that of being a man rather than God. But God grants his final wish to be returned to the cross, where he expires content in the knowledge that the Story has reached its accomplishment.

Kazantzakis said he wrote this CHRISTIAN FANTASY out of deep love for Jesus; the movie is redolent of that same integrity. The Holy Land is portrayed as an alien world, comprehensible to us only through the humanity we share with the Story's protagonists: jarring at first, the characters' street-US accents soon cement that bond. The acting and photography are exquisite; the music, by Peter Gabriel, is as fine. The fallible, uncertain, imperfect Jesus emerges as a figure of great stature – for he lacks the sin of pride.

TLTOC was widely campaigned against by many Christian fundamentalists, who perhaps saw themselves depicted here not as disciples but as Romans. [JG]

LAST TOMB OF LIGEIA vt of *The Tomb of Ligeia* (**1964**). ◊ *The* HOUSE OF USHER.

LAST UNICORN, THE US ANIMATED MOVIE (**1982**), animated in Japan. ITC. **Pr** Jules Bass, Arthur Rankin Jr. **Exec pr** Martin Starger. **Dir** Bass, Rankin. **Anim dir** Katsukisha Yamada. **Screenplay** Peter S. BEAGLE. **Based on** *The Last Unicorn* (**1968**) by Beagle. **Voice actors** Alan Arkin (Schmendrick), Jeff Bridges (Lir), Mia Farrow (Last Unicorn/Amalthea), Paul Frees (Cat), Tammy Grimes (Molly Grue), Robert Klein (Butterfly), Angela Lansbury (Mommy Fortuna), Christopher Lee (King Haggard), Keenan Wynn (Cully). 88 mins. Colour.

Long ago, the Red Bull drove all the UNICORNS to the edges of the world – all but one, who dwells in a forest rendered enchanted by her presence. She QUESTS, aided by incompetent WIZARD Schmendrick, to find the rest of her kind. Together with softhearted strumpet Molly Grue they reach the Hagsgate, the castle of miserable King Haggard, the Red Bull's master (◊ KNIGHT OF THE DOLEFUL COUN-TENANCE). By the Hagsgate the Red Bull attacks; to save her, Schmendrick transforms (◊ TRANSFORMATION) the unicorn into a mortal woman, the vacuous Amalthea – for this good deed she rebukes him in a characteristic whine; her rough COMPANIONS must supply the nobility she lacks. Haggard confesses that long ago he had the unicorns transformed into breaking waves (or "white horses"), so he might watch them perpetually strive for a shore they cannot reach. At last Amalthea, turned back into a unicorn, finds the courage to defy the Red Bull, driving it into the sea and so releasing the world's unicorns from their BONDAGE.

Beagle's script has sparks of wit but is otherwise forsoothly. The twee animation is often stilted, with rudimentary lip-sync, although occasionally the more abstract use of colour and pattern is enchanting.   [JG]

LATIN AMERICA In LA literature the fantastic did not follow the Neoclassic-Romantic-Gothic sequence that it did in France, England and Germany. Large parts of the Americas were still Spanish colonies in the 18th century, a time when the mother country was undergoing a period of obscurantism. Isolation kept Spain from exposure to the Enlightenment and, consequently, the flowering of ROMANTICISM and of the Gothic novel (◊ GOTHIC FANTASY), a circumstance that was exported to its colonies. But at the beginning of the 19th century, when independence movements took root in South America, the situation changed. The educated liberal classes became interested in MYTHS – Indian (i.e., indigenous), black and creole – and invoked them to emphasize the difference between their lands and Spain. These myths were cast in the realistic and even naturalistic form inherited from the mother country. At the same time, a generation of LA intellectuals and liberals, many educated in France, were open to the elements of fantasy born of Romanticism. According to Spanish critic Rafael Llopis, LA fantasy is a synthesis of these two factors.

Argentina's Juana Manuela Gorriti (1818-1892) was probably the first LA fantasy writer to be published in book form. Her *Sueños y realidades* ["Dreams and Realities"] (coll **1865** 2 vols) feeds on native myths, on incidents of the Spanish conquest – when the existence of underground Indian cities concealing vast treasure was suspected – and on her interest in certain scientific discoveries also exploited by E.T.A. HOFFMAN and Edgar Allan POE. Another Argentine follower of these writers and of Jules Verne (1828-1905) and Camille Flammarion (1842-1925) was Eduardo Ladislao Holmberg (1852-1937), author of *Viaje maravilloso del señor Nic-Nac* ["Mr Nic-Nac's Marvelous Journey"] (**1875**); his interest in phrenology, psychopathy and spiritualism shows in his *Cuentos fantásticos* ["Fantasy Stories"] (coll **1957**). A taste for the exotic, the morbid and the esoteric also appears in *Cuentos malévolos* ["Spiteful Tales"] (coll **1904**) by Clemente Palma (1872-1946), Peru's first writer of fantasy; his novel *XYZ* (**1931**) is seen by some as a forerunner of Adolfo BIOY CASARES' *The Invention of Morel* (**1940**).

A fondness for the occult, parapsychological states and scientific experiments culminates in Leopoldo Lugones (1874-1938), an Argentine Modernist poet and short-story writer who greatly influenced LA literature during the first three decades of this century. His remarkable *Las fuerzas extrañas* ["Strange Forces"] (coll **1906**) includes "An Essay on a Cosmogony in Ten Lessons", in which he takes up some of the ideas in his stories, which foreshadow both sf and MAGIC REALISM. Much in the same vein is the work of Uruguayan-born Horacio Quiroga (1878-1937), a master of the short story, who spent a good part of his life in Argentina, especially in the tropical regions of the northeast. A reader of Poe, Guy DE MAUPASSANT, Anton Chekhov (1860-1904), Rudyard KIPLING, Joseph CONRAD and W.H.

HUDSON, Quiroga gave shape to his literary theories in a famous "Decalogue of the Perfect Storyteller" (1925). His fantasy stories, which often depict a lonely man fighting only to postpone his inevitable death, appear in *Cuentos de la selva* (coll **1918**; trans Arthur Livingston **1922** US), *Cuentos de amor, de locura y de muerte* ["Tales of Love, Madness, and Death"] (coll **1917**), and *Más allá* ["Beyond"] (coll **1935**). *The Decapitated Chicken and Other Stories* (trans Margaret Sayers Peden **1976** US) and *The Exiles and Other Stories* (trans J. David Danielson **1987** US) are selections from various Spanish-language collections.

In the late 1930s LA fantasy, especially that written in the River Plate area, found a voice entirely its own. The publication in *Sur* magazine of the first stories by Jorge Luis BORGES from 1939 and throughout the 1940s ushered in Spanish America's golden age of fantasy. Paradoxically, while Borges was admired by the avant-garde of Europe and the USA, he proclaimed himself decidedly old-fashioned; indeed, the writers who most influenced his fiction were Kipling, G.K. CHESTERTON, Robert Louis STEVENSON and H.G. WELLS. In the decade 1939-49 Borges wrote some of the century's best fantasy stories. These are collected in his two most famous books, *Ficciones* (coll **1944**; exp **1956**; trans ed Anthony Kerrigan **1962** US) and *El Aleph* (coll **1949**; exp 1952); the latter appears in English in two selections: *Labyrinths* (trans ed James E. Irby and Donald Yates **1962** US) and *The Aleph and Other Stories 1933-1969* (trans ed Norman Thomas di Giovanni and author **1970** US). Borges also compiled a fantastic bestiary, *Manual de la zoología fantástica* (**1957**; exp vt *El libro de los seres imaginarios* 1967; the latter trans di Giovanni and author as *The Book of Imaginary Beings* **1969** US). Some later fantasies are included in *El libro de arena* (coll **1975**; trans di Giovanni as *The Book of Sand* **1977** US).

In 1940 Adolfo BIOY CASARES published a now famous novel, *La invención de Morel* (**1940**; trans Ruth L.C. Simms in *The Invention of Morel and Other Stories* coll **1964** US, which includes also *La trama celeste* coll **1948**); and edited with his wife Silvina Ocampo (1903-1994) and Borges *Antología de la literatura fantástica* (anth **1940**; rev 1977; trans as *The Book of Fantasy* **1976** US; rev 1988 with intro by Ursula K. LE GUIN), which included the type of fantasy stories they were reading in other languages along with samples in Spanish, mostly by Argentine writers. Bioy Casares, who collaborated with Borges on five books and three screenplays, became an important fantasy writer in his own right with *Plan de evasión* (**1945**; trans Suzanne Jill Levine as *A Plan for Escape* **1975** US), *El sueño de los héroes* (**1954**; trans Diana Thorold as *The Dream of the Heroes* **1987** UK), *Diario de la guerra del cerdo* (**1969**; trans Gregory Woodruff and Donald Yates as *Diary of the War of the Pigs* **1969** US), *Dormir al sol* (**1973**; trans Suzanne Jill Levine as *Asleep in the Sun* **1978** US). His shorter fiction is found in *Historia prodigiosa* ["A Prodigious Story"] (coll **1956**), *El lado de la sombra* ["The Shady Side"] (coll **1962**), *El gran serafín* ["The Great Seraph"] (coll **1967**), *Historias fantásticas* ["Fantasy Stories"] (coll **1972**), *Historias desaforadas* ["Outrageous Stories"] (coll **1986**) and *Una muñeca rusa* (coll **1991**; trans Suzanne Jill Levine as *A Russian Doll and Other Stories* **1992** US).

During the 1930s-40s, on both sides of the River Plate, two men were writing stories that would become influential only decades later. Uruguayan Felisberto Hernández (1902-1964), a music teacher and pianist who performed in bars and cinemas, composed a handful of odd short pieces that years later would be highly regarded by Julio CORTAZAR and would lead Italo CALVINO to say that Hernández "wrote like no one else". His first two slim books, for whose publication he paid himself, were *Fulano de Tal* ["Mr So-and-So"] (coll **1925**) and *Libro sin tapas* ["Book without Covers"] (coll **1938**). Only his third book, *Nadie encendía las lámparas* ["No One Had Lit a Lamp"] (coll **1946**), managed to arouse interest. A translated selection, drawn mainly from the last, is *Piano Stories* (trans Luis Harss **1993** with intro by Calvino). Only in the 1960s did Hernández begin to get the attention he deserved, which led to translations into French and Italian. About the same time in Buenos Aires, well into his 50s, Macedonio Fernández (1874-1952), an eccentric thinker who used to preside over informal gatherings of writers in bars and was greatly admired by Borges, published his first two books, miscellanies of fantasy and philosophical writing: *No toda es vigilia la de los ojos abiertos* ["We're Not Always Awake when Our Eyes Are Open"] (coll **1928**) and *Papeles de Recienvenido* ["Newcomer's Papers"] (coll **1929**). He was discovered after the posthumous publication of his *Museo de la novela de la eterna* ["Museum of the Novel of the Eternal"] (coll **1967**), a collection of 20 chapters prefaced by 56 different forewords.

In Mexico, Juan José Arreola (1918-   ) published in the 1940s the stories that later would appear in *Varia invención* ["Varied Invention"] (coll **1949**) and *Confabulario* (coll **1952**; trans George D. Schade as *Confabulario and Other Inventions* **1962** US). His Kafkaesque "El guardagujas" ["The Switchman"] is one of the most anthologized stories in Latin-American literature.

Argentine Julio CORTÁZAR wrote in the late 1940s and the 1950s most of the pieces that would make him one of the best-known writers in the Spanish language. Before the publication in 1963 of *Rayuela* (trans Gregory Rabassa as *Hopscotch* **1966** US) – which, along with work by Gabriel GARCÍA MÁRQUEZ, Mario Vargas Llosa (1936-   ), Carlos FUENTES and others, initiated what would be called the "Latin-American literary boom" – Cortázar was already much admired for his fantasy short stories, the first of which, "Casa tomada" ["Taken House"], was published in 1946 by Borges when he was editor of *Los Anales de Buenos Aires*. Cortázar's fantasies can be found in *Bestiario* ["Bestiary"] (coll **1951**), *Final del juego* ["End of the Game"] (coll **1956**; exp 1964) and *Las armas secretas* ["The Secret Weapons"] (coll **1959**). These last three have been assembled in English as *End of the Game and Other Stories* (trans Paul Blackburn **1967** US) and *Blow-up and Other Stories* (trans Blackburn **1967** US). JC's other fantasies are in *Historias de cronopios y de famas* (coll **1962**; trans Blackburn as *Cronopios and Famas* **1969** US), *Todos los fuegos el fuego* (coll **1966**; trans Suzanne Jill Levine as *All Fires the Fire and Other Stories* **1971** US), *Octaedro* ["Octahedron"] (coll **1974**), *Alguien que anda por ahí* (coll **1977**; trans Gregory Rabassa as *A Change of Light and Other Stories* **1980** US), *Queremos tanto a Glenda* (coll **1981**; trans Gregory Rabassa as *We Love Glenda So Much and Other Tales* **1983** US) and *Deshoras* ["Unhours"] (coll **1982**).

The term MAGIC REALISM, first used in 1925 by German critic Franz Roh to describe the work of certain post-Expressionist painters, was picked up by Juan Ramón Jiménez in 1942 to refer to the poetry of Pablo Neruda (1904-1973). In 1947 Venezuelan writer Arturo Uslar Pietri

(1906-   ) applied it for the first time to LA fiction. The term has also been used, differently, by Cuban Alejo Carpentier (1904-1980). "In order to give us an illusion of irreality," wrote Argentine writer and critic Enrique Anderson Imbert, "a magic-realistic narrator pretends to escape from nature and relates an action that, however inexplicable it may be, we find oddly disturbing . . . Between the dissolution of reality (magic) and the copy of reality (realism), magic realism amazes itself as if it were attending the spectacle of a new Creation. Seen with new eyes by the light of a new morning, the world is, if not marvellous, at least disturbing." The expression "magic realism" has been used mostly in relation to the exuberant masterpieces of the Mexican Juan Rulfo (1918-1980), the Cuban Alejo Carpentier and Nobel-Prize winners Miguel Ángel ASTURIAS and Gabriel GARCÍA MÁRQUEZ.

OTHER AUTHORS

Argentina Santiago Dabove (1889-1951), *La muerte y su traje* ["Death and Its Dress"] (coll **1961**; foreword by Borges). Enrique Anderson Imbert (1910-   ), *El grimorio* ["The Conjurer's Book"] (coll **1961**), *El gato de Cheshire* ["The Cheshire Cat"] (coll **1965**). Manuel Mujica Láinez (1910-1984), *Crónicas reales* ["Royal Chronicles"] (coll **1967**), *El brazalete y otros cuentos* ["The Bracelet and Other Stories"] (coll **1978**). Silvina Ocampo (1903-1994), *La furia* ["The Fury"] (coll **1959**), *Las invitadas* ["The Guests"] (coll **1961**), *Los días de la noche* ["The Days of Night"] (coll **1970**); her best stories are in *Leopoldina's Dream* (trans Daniel Balderston **1988** UK, with a preface by Borges). Eduardo Stilman (1938-   ), *Jugar a ciegas* ["Blind Game"] (coll **1984**), *El samovar de plata* ["The Silver Samovar"] (coll **1993**). Angélica Gorodischer (1928-   ), *Las pelucas* ["The Wigs"] (coll **1968**), *Bajo las jubeas en flor* ["Under the Flowering Jubeas"] (coll **1973**), *Kalpa imperial* ["Imperial Kalpa"] (coll **1983-4** 2 vols). Vlady Kociancich (1941-   ), *La octava maravilla* ["The Eighth Wonder"] (**1982**), *Todos los caminos* ["All Roads"] (coll **1991**). Elvio E. Gandolfo (1947-   ), *Dos mujeres* ["Two Women"] (coll **1992**), *Ferrocarriles Argentinos* ["Argentine Railroads"] (coll **1994**). Ana María Shua (1951-   ), *La sueñera* ["Sleepiness"] (coll **1984**), *Viajando se conoce gente* ["Travelling You Get to Know People"] (coll **1988**). Carlos Gardini (1948-   ), *Mi cerebro animal* ["My Animal Brain"] (coll **1983**), *Sinfonía Cero* ["Zero Symphony"] (coll **1984**). Rogelio Ramos Signes (1949-   ), *Las escamas del señor Crisolaras* ["Mr Crisolaras' Scales"] (coll **1983**). Spanish-born Marcial SOUTO, *Para bajar a un pozo de estrellas* ["Climbing Down into a Well of Stars"] (coll **1983**), *Trampas para pesadillas* ["Traps for Nightmares"] (coll **1988**). Cristina Siscar (1947-   ), *Reescrito en la bruma* ["Rewritten in the Mist"] (coll **1987**), *Lugar de todos los nombres* ["The Place of All Names"] (coll **1988**). Eduardo Abel Giménez (1954-   ), *El fondo del pozo* ["The Bottom of the Well"] (**1985**). Sergio Gaut vel Hartman (1947-   ), *Cuerpos descartables* ["Disposable Bodies"] (coll **1985**). Luisa Axpe (1945-   ), *Retoños* ["Sprouts"] (coll **1986**).

In an attempt to explain why Argentina produced so much fantasy, Julio Cortázar suggested: "A cultural polymorphism that arises out of the contributions of many different groups of immigrants; and a geographical immensity that acts as a factor of isolation, monotony and boredom, with a consequent appeal for the unusual, an *anywhere cut out of the world*."

Chile Juan Emar (real name Álvaro Yáñez; 1893-1964),

Diez ["Ten"] (coll **1934**; reprinted 1971 with foreword by Pablo Neruda). María Luisa Bombal (1910-1980), *La última niebla* ["The Last Fog"] (**1934**), *La amortajada* ["The Shrouded One"] (**1941**). Alejandro Jodorowsky (1927-), *Cuentos pánicos* ["Panic Stories"] (coll **1963**), El loro de siete lenguas ["The Parrot with Seven Tongues"] (**1991**).

Cuba Virgilio Piñera (1912-1979), *Cuentos fríos* ["Cold Stories"] (coll **1956**), *El que vino a salvarme* ["The One Who Came to Save Me"] (coll **1970**). Germán Piniella (1935-), *Polífagos* ["Polyphagous"] (coll **1967**).

Mexico Alfonso Reyes (1889-1959), *El plano oblicuo* ["The Slanting Plane"] (coll **1920**). Elena Garro (1920-), *La semana de colores* ["The Colored Week"] (coll **1964**). Carlos FUENTES, *Los días enmascarados* ["The Masked Days"] (coll **1954**; trans Margaret Sayers Peden, together with *Agua quemada* [see below], as *Burnt Water* **1980** US) *Aura* (**1962**; trans Lysander Kemp **1966** US), *Cantar de ciegos* ["Blind Men's Song"] (coll **1964**), *Terra nostra* (**1975**; trans Peden **1976** US), *La cabeza de la hidra* (**1978**; trans Peden as *The Hydra Head* **1978** US), *Agua quemada* ["Burnt Water"] (coll **1981**), *Cristóbal nonato* (**1987**; trans Alfred MacAdam and author as *Christopher Unborn* **1989** US). José Emilio Pachecho (1939-), *El viento distante* ["The Distant Wind"] (coll **1969**). Mauricio-José Schwarz (1955-), *Escenas de la realidad virtual* ["Scenes from Virtual Reality"] (coll **1991**).

Uruguay José Pedro Díaz (1921-), *Tratados y ejercicios* ["Treaties and Exercises"] (coll **1967**). Mario Levrero (1940-), *La ciudad* ["The City"] (**1970**), *La máquina de pensar en Gladys* ["The Machine for Thinking About Gladys"] (coll **1970**), *París* (**1980**), *Todo el tiempo* ["All the Time"] (coll **1982**), *Aguas salobres* ["Salty Waters"] (coll **1983**), *Los muertos* ["The Dead Ones"] (coll **1986**), *Espacios libres* ["Free Spaces"] (coll **1987**), *El lugar* ["The Place"] (**1991**). [MH/MS]

Further reading: *Historia natural de los cuentos de miedo* (**1974**) by Rafael Llopis; *El realismo mágico y otros ensayos* (**1976**) by Enrique Anderson Imbert; *Literatura fantástica de lengua española* (**1987**) by Antonio Risco; *El relato fantástico en España e Hispanoamérica* (**1991**) ed E. Morillas Ventura.

LATVIA Latvian fantasy is based mostly on FOLKLORE and MYTH. It was important in defining Latvia's nationhood during both independence and occupation.

An important early modern HEROIC FANTASY is the epic poem *Lacplesis* ["Bearfighter"] (**1888**) by Andrejs Pumpurs (1841-1902), which weaves in figures from folklore and incidents from Baltic history. Lacplesis is an ancient Latvian legendary HERO, whose enormous strength, held in his ears, is inherited from a she-bear. He fights evil powers who seek help from his greatest enemy, the Dark Knight, a symbolic representation of medieval German crusaders. Janis Rainis (real name Janis Plieksans; 1865-1929) created a drama based on Pumpurs's poem: *Uguns un nakts* ["Fire and Night"] (**1908**) includes DEVILS, WITCHES and a DRAGON. The protagonists are symbols representing the nation through the last millennium, their views and attitudes changing with each succeeding act, depicting the spiritual and intellectual development of hero and nation. The ideology of Rainis's play, a classic of European symbolist drama, differs considerably from that of Pumpurs's original, having been written during the attempted russianization of Latvia by Imperial Russia. Pumpurs's Dark Knight becomes Rainis's Black Knight, who "comes from the Tatars who trample every land under the hooves of their horses".

Lacplesis is now the chosen hero to stand against *all* evil forces, whether from West or East. Rainis's other play of fantasy note is the tragedy *Speleju, dancoju* ["I Played and I Danced"], a symbolic DARK FANTASY.

Rainis's wife, the poet, playwright, feminist and freedom-fighter Aspazija (real name Elza Plieksane; 1868-1943), used Lithuanian and Latvian mythology in her drama *Zalksa ligava* ["The Serpent's Bride"] (**1928**; trans Astrida B. Stahnke in *Aspazija: Her Life and Her Drama* **1984** US), based on a tale about a water SERPENT who marries a human woman. Aspazija's most successful play was *Sidraba skidrauts* ["The Silver Veil"] (**1905**). The heroine possesses a silver veil from the GODS, through which she can see past and future and learn people's secrets. Supporting justice, she faces conflict with a tyrant, a conflict aggravated by her LOVE for a prince. Aspazija's verse plays *Vestal* ["The Priestess"] (**1894**) and *Ragana* ["The Witch"] (**1895**) are also fantasies.

Karlis Skalbe (1879-1945) and Janis Veselis (1896-1962) founded modern Latvian prose fantasy. Skalbe wrote over 60 short stories, most using classic FAIRYTALE form but with a modern ethical content. Veselis based his fantasies on the ancient Baltic PANTHEON. His 29-story cycle covers the period from the creation to modern Latvian civilization, and mixes gods and people as in ancient Greek mythology. In the 20th century stories the gods influence history. The early stories were published in the magazine *Daugava* until 1936, later stories being written after the 1940 occupation. Veselis emigrated to Germany, then the USA. The Chicago-based Latvian magazine *Zintis* published some of his stories in English 1962-5, and some appeared also in *Latvian Literature* (trans anth **1964** Canada) ed Aleksis Rubulis.

As Latvian fantasy was associated with the nationalist movement in earlier years, it was not encouraged during the Soviet occupation. [IB]

LAUBENTHAL, SANDERS ANNE (1943-) US writer, much respected for her one novel, *Excalibur* (**1973**), a powerful fantasy which conflates the MATTER of Britain with an AMERICAN-GOTHIC version of Mobile, Alabama. This posits that the legendary 12th-century Welsh prince MADOC was a descendant of ARTHUR and, when he crossed the Atlantic *c*1170, brought with him to the New World the GRAIL and EXCALIBUR. Excalibur survives unmolested until the 20th century, when the current Pendragon – in the guise of a youthful archaeologist – must once again attempt to seek it out, his efforts being opposed by an attractively conceived MORGAN LE FAY. [JC]

LAUNAY, ANDRÉ (JOSEPH) (1930-) Writer born in the UK of French parents and now resident in Spain; he writes as Droo Launay, Drew Lamark and Andrew Laurance, as well as under his own name. His first book was a collection of cartoons, *I Married a Model* (graph coll **1960** UK) as by Droo Launay; his other Droo Launay titles are all thrillers. *The Snake Orchards* (**1982** UK) and *The Medusa Horror* (**1983** UK), both as by Drew Lamark, are HORROR, as are AL titles like *The Latchkey Children* (**1985** UK), *The Harlequin's Son* (**1986** UK) and *Seance* (**1991** UK). The **Blood of Nostradamus** sequence as by Laurance – *Premonitions of an Inherited Mind* (**1979** UK; vt *The Blood of Nostradamus: The Premonition* 1991 US), *Embryo* (**1980** UK; vt *The Blood of Nostradamus: The Unborn* 1991 US) and *The Link* (**1980** UK; vt *The Blood of Nostradamus: The Link* 1991 US) – concerns a family descended from NOSTRADAMUS who have inherited his memory and prophetic powers.

Other titles as by Laurance include *The Hiss* (**1981** UK; vt *Catacomb* 1991), *Ouija* (**1982** UK) and *The Black Hotel* (**1983** UK), all horror. [JC]

LAUNAY, DROO ◊ André LAUNAY.

LAURANCE, ANDREW ◊ André LAUNAY.

LAUTRÉAMONT, COMTE DE Pseudonym – borrowed from a sensational novel by Eugène SUE – of French writer Isidore Ducasse (1846-1870). Born in Montevideo, he was sent to Paris to be educated but died there after three years, having spent his time writing *Les chants de Maldoror* (**1874**; trans John Rodker as *The Lays of Maldoror* 1924 UK), which is perhaps better regarded as an avant-garde novel than a string of prose-poems, and *Poésies* (**1870** as ID), a series of aphorisms adopting a viewpoint sternly contrary to that expressed by the calculatedly scabrous rantings of the ANTI-HERO Maldoror. The two works are juxtaposed in *Oeuvres complètes* (coll **1927**; trans Paul Knight as *Maldoror and Poems* coll **1978** UK). Maldoror's sadism and wild indulgence in nightmarish fantasies established a paradigm for the writers of DECADENCE. L is the central character of Jeremy REED's novel *Isidore* (**1991**). [BS]

LAWHEAD, STEPHEN R. (1950-   ) US writer, resident in the UK, known mainly for his CELTIC FANTASIES. His **Pendragon** series – *Taliesin* (**1987** US), *Merlin* (**1988** US), *Arthur* (**1989** US), *Pendragon* (**1994** US), «Grail» and «Avalon» – in fact starts in a POLDER that is a surviving fragment of ATLANTIS. His Arthurian Britain is realistic 6th century rather than chivalric 12th century, and he avoids many of the failings of other US writing in this subgenre, but his evangelical Christian stance leads to a too-great polarization between GOOD AND EVIL, while his christianizing of all the major characters (including Taliesin and MERLIN) rewrites the entire mythic subtext of the ARTHUR cycle. His **Song of Albion** series – *The Paradise War* (**1991**), *The Silver Hand* (**1992**) and *The Endless Knot* (**1993**) – explores non-Arthurian Celtic mythology. [DVB]

Other works: The **Dragon King** trilogy, being *In the Hall of the Dragon King* (**1982**), *The Warlords of Nin* (**1983**) and *The Sword and the Flame* (**1984**); *Dream Thief* (**1983**); *The Search for Fierra* (**1985**) and *The Siege of Dome* (**1986**), assembled as *Emphyrion: The Search for Fierra, The Siege of Dome* (omni **1990**); *Howard Had a Spaceship* (**1986**), juvenile sf.

LAWRENCE, ANN (MARGARET) (1942-1987) UK writer whose first genre fiction of interest was the remarkable *Tom Ass, or The Second Gift* (**1972**), a BEAST FABLE set in medieval England. Young Tom is given two gifts by an ELF-lady: whatever he begins to do at dawn he will continue to do all day; and he will "be" whatever his future wife decides he will be. The first gift means that certain activities (like counting money) generate riches; the second, after he has been turned into an ASS, necessitates a NIGHT JOURNEY in the company of his competent future wife to LONDON, where he eventually LEARNS BETTER, demonstrates his human value to her, and becomes her human husband. Nothing else AL wrote had quite the glow of this tale, though *The Half-Brothers* (**1973**), set in FAERIE, tells a FAIRYTALE-like story with happy conviction: once again, it presents a marriage-choice, and a strong young woman makes the correct decision.

In *The Conjuror's Box* (**1974**) two children find themselves the keepers of a LIMINAL BEING who has been imprisoned (◊ BONDAGE) as a china jug; under the direction of this being, and of a mysterious Fiddler, they are involved in a struggle to prevent the subversion of real TOYS and proper childhood,

a THINNING caused by the malign influence of the Green Lady, an evil survival of the ELDER GODS who has retreated through the eponymous box – which is a PORTAL – into an ALTERNATE WORLD. The book, like Alan GARNER's *Elidor* (**1963**) and Susan COOPER's **Dark is Rising** sequence, sophisticatedly draws from the CAULDRON OF STORY, and the text is resonant with echoes and refigured PLOT DEVICES.

The eponymous imp of *The Good Little Devil* (**1978**) is initially tamed by the monks who discover him, but is ultimately given his moral freedom again. *The Hawk of May* (**1980**), which concerns GAWAIN, and *Merlin the Wizard* (coll of linked stories **1986**) are Arthurian tales (◊ ARTHUR). *Summer's End: Stories of Ghostly Lovers* (coll of linked stories **1987**) imparts an autumnal tone to a sequence of SUPER-NATURAL FICTIONS about growing up, told via the FRAME STORY to some feisty adolescent siblings. *Beyond the Firelight* (coll **1983**), collecting ARABIAN FANTASIES, and *Tales from Perrault* (trans coll **1988**), modernizing Charles PERRAULT, are both competent exercises in the TWICE-TOLD. There is a risk that AL's early death may condemn her work to obscurity. [JC]

Other works: The **Oggy the Hedgehog** sequence for younger readers, comprising *The Travels of Oggy* (**1973**), *Oggy at Home* (**1977**) and *Oggy and the Holiday* (**1979**); *The Half-Brothers* (**1973**), associational; *Mr Robertson's Hundred Pounds* (**1976**); *Between the Forest and the Hills* (**1977**); *There and Back Again* (coll **1985**).

LAWRENCE, D(AVID) H(ERBERT) (1885-1930) UK writer, best-known for nonfantastic works. He wrote some SUPERNATURAL FICTION, including the famous "The Rocking Horse Winner" (in *The Ghost Book* anth **1926** ed Cynthia ASQUITH), about a boy who can predict winning horses; the tale was filmed as *The Rocking Horse Winner* (*1949*). In the title story of *The Woman who Rode Away and Other Stories* (coll **1928**) Native Americans demonstrate their consanguinity with the old dark life principles; "The Last Laugh", from the same volume, is a CONTE CRUEL with fantastic elements. *The Man who Died* (1928 *The Forum* as "The Escaped Cock"; rev **1929**) features a resurrected CHRIST, powers of healing intact, who comes into ecstatic oneness with his whole body in an act of sexual union with a priestess of Isis. [JC]

LAWRENCE, LOUISE Working name of UK writer Elizabeth Rhoda Wintle Holden (1943-   ), who has concentrated since her first novel, *Andra* (**1971**), which is sf, on sf and fantasy for young adults. *The Wyndcliffe* (**1974**) depicts the growing love between an unhappy young woman and a GHOST; its sequel, *Sing and Scatter Daisies* (**1977**), set some while later, intensifies the situation through the jealousy of another relative and the slow realization that the central female character is dying. *The Earth Witch* (**1981**) is also about LOVE: a young man enamoured of an older woman is amazed when she grows younger as the SEASONS turn toward Spring. Further titles of fantasy interest include *Cat Call* (**1980**) and *The Dram Road* (**1983**). [JC]

Other works (sf): *The Power of Stars* (**1972**); *Star Lord* (**1978**; rev 1987); *Calling B for Butterfly* (**1982**); *Children of the Dust* (**1985**); *Moonwind* (**1986**); *The Warriors of Taan* (**1986**); *Extinction is Forever and Other Stories* (coll **1990**); *Ben-Harran's Castle* (**1992**; vt *Keeper of the Universe* 1993 US); *The Disinherited* (**1994**; vt *The Patchwork People* 1994 US).

LAWRENCE, MARGERY (1889-1969) UK writer who specialized in GHOST STORIES and who spent much of her time commenting upon the supernatural in nonfiction terms. In many articles and books, like *Fifty Strangest Stories Ever Told* (coll **1937**), she related various "true" ghostly happening from her own experience, and most of her best tales were closely based on reports and observations. *Ferry Over Jordan* (**1944**) is a study in the occult. She took part in a great many "clearings" of HAUNTED DWELLINGS, as discussed in her foreword to *Ghosts Over England* (**1953**) by R. Thurston Hopkins (1884-1958).

Her first collections – the **Round Table** sequence, told within a FRAME STORY and comprising *Nights of the Round Table* (coll **1926**) and *The Terraces of Night* (coll **1932**) – assemble SUPERNATURAL FICTIONS, usually featuring malevolent and horrific visitations. In "The Curse of the Stillborn" (1925) an outraged Egyptian god takes revenge; "The Woozle" sees a ghostly creature in a nursery cupboard seemingly created by the power of suggestion in a child's mind (◊ INVISIBLE COMPANION); "Morag-of-the-Cave" is dominated by a toad-white shape, ghastly and "obscenely awful to see", which communicates in a strange and terrible tongue; and "Mare Amore" (1931) describes the ultimate power and vengeance of "the spirit of the sea". Her second series is the **Miles Pennoyer** OCCULT-DETECTIVE sequence: *Number Seven, Queer Street* (coll **1945**) and *Master of Shadows* (coll **1959**).

Several of ML's later novels also deal with the occult and with the SPIRITUALISM that particularly infuses the **Round Table** tales. They include: *The Bridge of Wonder* (**1939**), which is about mediums; *The Rent in the Veil* (**1951**), a tale of REINCARNATION, in which a 1940s woman falls in love with a long-dead Roman centurion; *The Tomorrow of Yesterday* (**1966**), in which a Martian conveys, via SÉANCE, the story of the founding and destruction of the Martian colony of ATLANTIS; *Bride of Darkness* (**1967**), in which a man realizes his loving wife is a WITCH devoted to SATAN; and *A Residence Afresh* (**1969**), about communications from the "Other Side". [RD]

Other works: *The Floating Cafe and Other Stories* (coll **1936**); *Strange Caravan* (coll **1941**).

LAWS, STEPHEN (1952-   ) UK writer of HORROR novels. ◊ EDIFICE; Giovanni Battista PIRANESI. [JC]

LAWSON, ROBERT (1892-1957) US illustrator and writer who became active in the early 1920s with illustrations for *The Wonderful Adventures of Little Prince Toofat* (**1922**) by George Randolph Chester (1869-1924); other illustrated tales include *The Unicorn with Silver Shoes* (**1932**) by Ella YOUNG, *The Story of Ferdinand* (**1935**) and others by Munro Leaf (1905-1976), *Mr Popper's Penguins* (**1938**) by Florence Atwater (1896-1979) and Richard Atwater (1892-1938), *One Foot in Fairyland* (**1938**) by Eleanor FARJEON and *Poo-Poo and the Dragons* (**1942**) by C.S. Forester (1899-1966). His work is strongly narrative, and he conveys his effects through a tightly etched technique.

RL's own fiction, which he also illustrated, includes four TALKING-ANIMAL fantasies *Ben and Me* (**1939**) – filmed by DISNEY as an animated featurette in 1953 – in which a mouse (◊ MICE AND RATS) describes how he kept Benjamin Franklin from trouble; the similar *I Discover Columbus* (**1941**); *Mr Revere and I* (**1953**); and *Captain Kidd's Cat* (**1956**). The **Rabbit Hill** BEAST-FABLE sequence for younger children runs *Rabbit Hill* (**1944**), *Robbut, a Tale of Tales* (**1948**), *Edward, Hoppy and Joe* (**1952**) and *The Tough Winter* (**1954**). Other titles include: *They Were Strong and Good* (**1940**), associational; *Mr Wilmer* (**1945**); *Mr Twigg's*

Mistake (**1947**), about a bear-sized mole; *McWhinney's Jaunt* (**1947**), a TALL TALE featuring an itinerant peddler; *The Fabulous Flight* (**1949**), sf; *Smeller Martin* (**1950**); and *The Great Wheel* (**1957**). [JC]

LEAR, EDWARD (1812-1888) UK artist and poet, the first significant purveyor of the limerick and a composer of some of the finest NONSENSE verse ever written in English. Although his poetry makes no "literal" or mundane sense, many of his narrative poems have an internal consistency – and indeed pathetic intensity – that gives his work close affinity to later fantasy. EL's poems appeared initially in *The Book of Nonsense* (coll **1846**; exp 1861; exp 1863; exp 1870). Later volumes included: *Nonsense Songs, Stories, Botany and Alphabets* (coll **1871**), which includes "The Owl and the Pussycat" and "The Jumblies", both of which are sustained fantasy narratives in verse; *More Nonsense Pictures, Rhymes, Botany, Etc.* (coll **1872**); *Laughable Lyrics: A Fourth Book of Nonsense, Poems, Song, Botany, Music, Etc* (coll **1876**), which contains, in "The Dong with the Luminous Nose", a verse tale whose pathos – the Dong being clearly EL himself – quite overshadows the "absurdity" of the events depicted; and *Nonsense Songs and Stories* (coll **1894**). *Queery Leary Nonsense* (coll **1911**) assembles old material, with some additions; a convenient later assembly of his work is *The Complete Nonsense of Edward Lear* (coll **1947**) ed Holbrook Jackson (1874-1948).

The Victorians treated EL as an author for children, a dismissive bias which extended well into the 20th century. Today we are fortunate to inhabit a climate of opinion that recognizes his genius. [JC]

LEARNS BETTER Robert A. HEINLEIN provocatively claimed that all STORY plots fall into just three categories: Boy Meets Girl, the BRAVE LITTLE TAILOR, and the Man (or Person) Who Learns Better. Typically this person is an obnoxious, deluded or evilly misled character (◊ ANTIHERO), like C.S. LEWIS's Edmund in *The Lion, the Witch and the Wardrobe* (**1950**), who is shamed into mending his ways, or like Eustace in *The Voyage of the Dawn Treader* (**1952**), who has his nose rubbed in logical extremes of selfishness and greed by TRANSFORMATION into a DRAGON. Pippin in J.R.R. TOLKIEN's *The Lord of the Rings* (**1954-5**) is driven by foolish curiosity to steal and use a SCRYING-stone, and duly learns better; here the initial folly proves fortunate, since it misleads the DARK LORD. A subtler LB example is Cat in Diana Wynne JONES's *Charmed Life* (**1977**), whose sister borrows and misuses his MAGIC power with his unconscious support – withdrawn only after his belated RECOGNITION of what is happening. Many fantasies deal with the LB process in terms of WISHES which, once granted, prove less than welcome (◊ ANSWERED PRAYERS); E. NESBIT uses this theme repeatedly and effectively. Sometimes the author's definition of "better" may be idiosyncratic: in Charles WILLIAMS's *The Place of the Lion* (**1931**), the heroine's LB insight is that her harassment by a vile pterodactyl ARCHETYPE is deserved, because she has studied ancient texts on (e.g.) ANGELS academically rather than with willingness to believe (and, subtextually, has thus trespassed on male intellectual territory). TIME FANTASIES may allow a second chance in which a now wiser protagonist can, like Florian in James Branch CABELL's *The High Place* (**1923**), choose a different path – though here moral didacticism is entirely absent, and Florian repeats all his life's crimes with perfect confidence of success, up to the turning point at age 36; he then pragmatically reforms. [DRL]

LEAVES FROM SATAN'S BOOK vt of *Blade af Satans Bog* (*1919*). ◊ Carl-Theodor DREYER; VAMPYR (*1932*).

LE CLÉZIO, J(EAN)-M(ARIE) G(USTAVE) (1940-) French writer who spent time in Thailand and Central America, both of which locations affected his work. His first novel, *Le Procès-Verbal* (**1963**; trans Daphne Woodward as *The Interrogation* **1964** UK), features a protagonist whose name, Adam Pollo, manages to conflate two ARCHETYPES or UNDERLIER figures; the novel does not, though, violate the mundane world. *Le Déluge* (**1966**; trans Peter Green as *The Flood* **1967** US) replays the tragedy of Oedipus, though without an ORACLE. *Les Géants* (**1973**; trans Simon Watson Taylor as *The Giants* **1975** US) is hyperbolic sf. *Voyages de l'autre côté* ["Voyages to the Other Shore"] (**1975**) carries its surreal protagonist through innumerable METAMORPHOSES into a series of escapes from this Universe; it has not been translated, and nor have any of the author's more recent works. [JC]

LEE, ALAN (1947-) UK fantasy illustrator, with a delicate, subdued, misty watercolour style, in the manner of the traditional English watercolourists Thomas Girtin (1775-1802) and John Sell Cotman (1782-1842). He also produces sensitively rendered drawings in soft pencil. The imaginative quality of his work owes much to Arthur RACKHAM and the influence of AL's associate Brian FROUD, but there is evidence of a truly original – if at times rather gloomy – mind at work. His imaginative figure drawing is strong, with a charming sense of the grotesque; all his best work has a sense of timeless antiquity.

With Froud AL produced the enormously popular and charming *Faeries* (graph **1978**). The financial success of this venture allowed him to embark on an illustrated version of *The* MABINOGION (**1982**). He also painted covers for the 1982 edition of Mervyn PEAKE's **Gormenghast** trilogy. *Castles* (graph **1984**) written by David Day was an attempt to capitalize on *Faeries*, but was less commercially successful. Other projects include *The Mirrorstone: A Ghost Story with Holograms* (graph **1986**) by Michael Palin (1943-) and the covers for *Merlin Dreams* (**1988**) by Peter DICKINSON and the 1991 printing of J.R.R. TOLKIEN's *The Lord of the Rings*. Since the last endeavour he has published affiliated work, notably illustrating *Tolkien's Ring* (**1994**) by David Day. [RT]

LEE, JOHN (ARTHUR WALTER) (1939-) Writer and publicist, born in Iraq of UK parents, resident in the USA since 1963. His **Strand** sequence includes *The Unicorn Quest* (**1986**), *The Unicorn Dilemma* (**1988**), *The Unicorn Solution* (**1991**), *The Unicorn Peace* (**1993**) and *The Unicorn War* (**1995**). Set on the world of Strand, which mixes SCIENCE-FANTASY elements out of PLANETARY ROMANCE with a FANTASYLAND ambience – including an ISLAND at the centre of things where UNICORNS have their home – this is essentially a long DYNASTIC FANTASY, involving two or three royal houses and told with moderate good humour. Eventually, MAGIC sustains itself against the THINNING encroachment of a machine age. The GODDESS (treated unsympathetically) and the Mother who officiates for her on Strand are antipathetic to magic, and favour the advocates of progress. [JC]

LEE, STAN (1922-) US COMIC-book writer and executive, whose ebullient approach to comics editing had a profound effect on the remarkable popularity of Marvel Comics. Born Stanley Lieber, he changed his name and began to establish himself in comics before WWII, joining Timely Comics Inc (eventually renamed Marvel in 1963). He served as editor 1942-72 and publisher and editorial director from 1972. His professional life has been intimately bound up with the fortunes of that company (◊ MARVEL COMICS). [RT]

LEE, TANITH Working name of UK writer Tanith Lee Kaiine (1947-), most of whose work has been fantasy, though she is active also as an author of HORROR and SCIENCE FICTION. She began publishing work of genre interest with *The Betrothed* (**1968** chap), a short story privately printed by a friend, but started her career proper with several CHILDREN'S FANTASIES. Of these, *The Dragon Hoard* (**1971**), her first novel, is a comic fantasy (◊ HUMOUR) in which an affronted ENCHANTRESS compels the QUEST-ridden protagonist to shapeshift humiliatingly into a raven at unpredictable moments (◊ SHAPESHIFTERS); *Princess Hynchatti & Some Other Surprises* (coll of linked stories **1972**) puts its cast through various travails; in *Companions on the Road* (**1975**) the COMPANIONS are the VILLAINS, a trio of hellish REVENANTS who kill through their control of DREAMS as they search for the holders of a magic chalice; and *The Winter Players* (**1976**) – assembled with the previous book as *Companions on the Road and The Winter Players: Two Novellas* (omni **1977** US) – dramatizes the interaction between a young woman and the ACCURSED WANDERER whom she ultimately redeems. Even in these early works, several characteristic motifs dominate: the RITE OF PASSAGE whereby a young protagonist (most of TL's viewpoint characters are young) comes to terms – often via METAMORPHOSIS – with his or her extraordinary nature, and strives for BALANCE in a riven world; vivid, but indeterminate, LANDSCAPES serving as almost interchangeable backdrops for psychic dramas; and a fine indifference to any moralistic settling of scores, her tales tending to close with GOOD AND EVIL characters settling into uneasy equipoise.

TL's adult work began with the first volume of the **Birthgrave Trilogy** – *The Birthgrave* (**1975** US), *Vazkor, Son of Vazkor* (**1978** US; vt *Shadowfire* 1978 UK) and *Quest for the White Witch* (**1978** US) – whose ending rewrites as sf what had seemed a SWORD-AND-SORCERY tale set in a highly stylized PLANETARY-ROMANCE venue. The first volume stars a woman – born adult but amnesiac (◊ AMNESIA) in the heart of an erupting volcano – who must undertake a long QUEST, which includes her death and rebirth, in search of her TRUE NAME, Karrakaz, and a solution to the mystery of her birth. The two sequels, which comprise a single narrative, are told from the viewpoint of Karrakaz's son, who is also a healer (◊ HEALING) and revenant and, with his mother (for whom he searches, and with whom he has incestuous sex), a surviving member of an ancient PARIAH ELITE; together they represent a MYTH-saturated rebirth, through their mutual rewriting of the Oedipus story, of a matriarchal PANTHEON, but they must also learn to exercise their sexuality (from the first, TL's use of SEX is disturbingly exploratory, and unadulterated by any form of political correctness) and their power in a new and complex world.

The dominant motifs in this sequence persist throughout her work, as in her second series, the **Wars of Vis** – *The Storm Lord* (**1976** US) and *Anackire* (**1983** US), assembled as *The Wars of Vis* (omni **1984** US), plus *The White Serpent* (**1988** US) – which, in another planetary-romance setting, follows the trauma-beset rite of passage of the half-breed Raldnor, who in the first volume gains but then (like many a self-questioning TL protagonist) abandons the GOD-role of Storm Lord. His children, in the next volume, continue to dominate the various intersecting races of Vis through an exceedingly complex plot involving incest and internecine strife. The final volume, typically of TL's work, renders complex REALITIES (seen through a plot involving LOVE, sex, miscegenation and empire) as a dance-like AGON, whose central point may be that, to be fully human, one must attempt to occupy as many positions as possible on the gameboard of life.

Tales from the Flat Earth – *Night's Master* (coll of linked stories **1978** US), *Death's Master* (**1979** US) and *Delusion's Master* (**1981** US), all three assembled as *Tales from the Flat Earth: The Lords of Darkness* (omni **1987**), plus *Delirium's Mistress: A Novel of the Flat Earth* (**1986** US) and *Night's Sorceries* (coll **1987** US), both assembled as *Tales from the Flat Earth: Night's Daughter* (omni **1987**) – is set in an ALTERNATE-WORLD version of Earth so deeply sunk in time (◊ TIME ABYSS) that it is still flat and still governed by a self-contemplating Upperearth PANTHEON. On the platform of this Flat Earth, mortals and DEMONS from Underearth – the Lords of Darkness – dance out intricate and deadly ballets. Of the three Lords, Chuz – Delusion's Master – is perhaps the most profoundly terrifying, for the two sides of his face (◊ MASK) present JANUS-like contrasts of youth and age, both unbearably vivid; viewed directly he presents a grotesque, toothy, staring FACE OF GLORY, a sight that drives any mortal – and even other gods – insane. He is a TRICKSTER, a bringer of CHAOS and a tester of the fragile BORDERLANDS of Flat-Earth reality, and the threat of his presence in the sequence – which comprises novel-length stories, linked episodes, and quasi-independent FAIRYTALES – does much to ironize and (at the same time) give bite to the ongoing proceedings.

It is typical of TL, with whom reversal of stock expectations is almost predictable, that the protagonist of the series should be this mythology's SATAN, Azhrarn, and that he should save the world from an all-consuming embodiment of hatred that he himself created by accident. TL's somewhat thorny relationship with the Christian mythos is embodied here in the facts that Azhrarn dies and is reborn to save the world of humanity when the gods (who regard world and mankind as a mistake not to be repeated) stand by and that his motives are mixed: love; a dare by the spectre of a lover he killed; a sheer need for humanity as a punchbag for his whims – all these are present and all are plausible. When Azhrarn exposes himself to the sun and dies, TL's language – as so often at her moments of epiphany – echoes the Oscar WILDE of the fairytales and *Salome* (**1893**).

Azhrarn is also the typical example of a tendency on TL's part to identify the hottest of SEX with consensual male ravishment. This is a tendency not confined to her – Storm CONSTANTINE and Anne RICE have similar obsessions. In a patriarchal society, it would seem, only men, ultimately, are sexual and noble, so that some women writers may come to regard the only unproblematic sexual representation as sex involving men only.

The **Secret Books of Paradys** sequence – *The Book of the Damned* (coll of linked stories **1988**) and *The Book of the Beast* (**1988**), both assembled as *The Secret Books of Paradys* (omni **1991**), plus *The Book of the Dead* (coll **1991** US) and *The Book of the Mad* (coll **1993** US), both assembled as *The Secret Books of Paradys III & IV* (omni **1993**) – centres on Paradys, a CITY which resembles PARIS and which occupies a CROSSHATCH relationship between this world and the supernatural; the stories set within its REALITY-blurring ambience and in various historical periods tend to feature artists and poets and to involve these characters in erotic tangles with REVENANTS and demons (◊ SHAPESHIFTERS). Paradys bears some resemblance to cities like M. John HARRISON's Viriconium which crosshatch between reality and DREAM, between now and the time of the DYING EARTH. The final volume, through the introduction of ALTERNATE-WORLD versions of the city, including one peculiarly savage dystopia, darkens the portrayal of this central concept of urban life as a form of drama. The three narratives contained in *The Book of the Mad*, one of TL's finest texts, dovetail into what registers as a final reconciliation.

The **Blood Opera** sequence – *Dark Dance* (**1992**), *Personal Darkness* (**1993**) and *Darkness, I* (**1994**), with further volumes projected – is cast in a HORROR mode, with an isolated heroine undergoing sexual torment in a mansion on a moor, a process which is incorporated into her RITE OF PASSAGE into an adulthood that seems to involve her becoming a VAMPIRE.

There are several singletons of interest. *Volkhavaar* (**1977** US) is a HIGH FANTASY involving female slavery, SORCERY and GOOD AND EVIL. *Kill the Dead* (**1980** US) features a ghost-killer who concentrates on liberating the dead from BONDAGE as much as on protecting the living – complex reversals of expectation are spun on this material in a short space. *Lycanthia, or The Children of Wolves* (**1981** US) and the much later *Heart-Beast* (**1992**) are WEREWOLF tales; the latter features some of TL's purplest prose – it is a trap into which she is prone to fall. *Sung in Shadow* (**1983** US) tells the Romeo and Juliet story (◊ William SHAKESPEARE) in fantasy terms. *A Heroine of the World* (**1989** US), among TL's most impressive novels, is set in an ALTERNATE-WORLD version of early-19th-century Europe; it tells the life-story of a woman whose heroism is to become fully human, fully bound to her condition – despite the title, she is everything the TEMPORAL ADVENTURESS is not, for her humanity is earned within the limitations of mortality. *The Blood of Roses* (**1990**) reads like a dance, an immense gavotte of VAMPIRES through time and space around the central wound they have suffered, and whose focus lies within a central METAMORPHOSIS-ridden Wild Wood (◊ MYTHAGO): the wound here is the death of the World TREE. *Elephantasm* (**1993**) once again subjects a young heroine to a rite of passage involving sexual abuse: she finds herself exacting the vengeance of the elephant god Ganesha against the English, whose abuse of her mirrors their rape of India.

The contents of TL's several collections generally enact in concentrated form the dream-like spirals of plot characteristic of TL's full-length work, forcing an awareness upon the reader that the "dream" is under artistic control. Collections include: *Unsilent Night* (coll **1981** US); *Cyrion* (coll of linked stories **1982** US); *Red as Blood, or Tales from the Sisters Grimmer* (coll **1983** US) (◊ REVISIONIST FANTASY); *Tamastara, or The Indian Nights* (coll **1984** US), whose concerns prefigure *Elephantasm*; *The Gorgon and Other Beastly Tales* (coll **1985** US), stories assembled to make up a fantasy BESTIARY, the title story winning the WORLD FANTASY AWARD; *Dreams of Dark and Light: The Great Short Fiction of Tanith Lee* (coll **1986** US); *Forests of the Night* (coll **1989**), an impressive retrospective volume; *Women as Demons: The Male Perception of Women Through Space and Time* (coll **1989**); and *Nightshades: Thirteen*

Journeys into Shadow (coll **1993**). Of these, *Red as Blood* recycles in a commercial direction the revisionist insights into FAIRYTALES of Angela CARTER, again often combining SLICK-FANTASY ideas – SNOW WHITE as VAMPIRE – with neo-Christian decadent sentiments of a Wildean kind, for the prince who wakes Snow White is CHRIST. The **Cyrion** stories combine some deeply fetishistic explorations of polymorphous perversity and vague androgyny with interesting HEROIC-FANTASY detective material.

25 years into her career, TL has encompassed every genre of the fantastic (except hard sf) with supple attentiveness and an ongoing exuberance of invention which transcends – or sometimes swamps – genre constraints. The exoticism of her early work has become part of an utterly assured vocabulary, a royal flush of imaginable venues. In the 1990s, new explorations continue to emerge. [JC]

Other works:
YA fantasies: *Animal Castle* (**1972**); *East of Midnight* (**1977**); *The Castle of Dark* (**1978**) and *Prince on a White Horse* (**1982**), both assembled as *Dark Castle, White Horse* (omni **1986** US); *Shon the Taken* (**1979**); *Black Unicorn* (**1991** US).
Adult fantasies: *Day by Night* (**1980** US); *The Beautiful Biting Machine* (**1984** chap US); *Into Gold* (1986 *IASFM*; **1991** chap US); *Madame Two Swords* (**1988** chap US).
Sf and horror: *Don't Bite the Sun* (**1976** US) and its sequel, *Drinking Sapphire Wine* (**1977** US), assembled as *Drinking Sapphire Wine* (omni **1979**); *Electric Forest* (**1979** US); *Sabella, or The Blood Stone* (**1980** US), an interstellar VAMPIRE tale; *The Silver Metal Lover* (**1981** US), sf; *Days of Grass* (**1985** US), sf; *Eva Fairdeath* (**1994**), sf.
Further reading: *Daughter of the Night: A Tanith Lee Bibliography* (**1993** chap) by Jim Pattison and Paul A. Soanes.

LEE, VERNON Working name of Violet Paget (1856-1935), born in France of UK parents, half-sister of Eugene LEE-HAMILTON; she spent much of her life in Italy, where almost all her best fiction is set. Through *Studies of the Eighteenth Century in Italy* (**1880**) she laid claim (though only 24 when it reached print) to a land and a period that constituted, for her, a lost ARCADIA. Two chapters in particular – "The Comedy of Masks" (◊ COMMEDIA DELL'ARTE; MASKS) and "Carlo Gozzi and the Venetian Fairy Comedy" (◊ Carlo GOZZI) – retain a vivid relevance. She was a writer of stubborn intellect and obsessive concern.

VL is of most interest as an author of SUPERNATURAL FICTION. Her first book, *Tuscan Fairy Tales (Taken Down from the Mouths of the People)* (coll **1880**), belies its parenthetical caveat through the somewhat selfconscious literary polish imparted to the FAIRYTALES it assembles. After the negligible *The Prince of the Hundred Soups* (**1883**) came her first important supernatural fiction: *A Phantom Lover: A Fantastic Story* (**1886** chap) – set, oddly enough, in England. Along with the later "The Turn of the Screw" (1898) by her friend Henry JAMES, *A Phantom Lover* fits remarkably well into Tsetvan TODOROV's definition of the FANTASTIC. A painter becomes obsessed with the similarity (almost certainly a matter of his own PERCEPTION) between the woman he is painting and her ancestor, who was involved in a family tragedy. By the end of the tale, he has generated an uncanny further tragedy.

In contrast, the supernatural fictions in VL's finest single volume, *Hauntings: Fantastic Stories* (coll **1890**), do not hover on that Todorovan cusp, and almost invariably treat the urge to transcendence as transgressive; she is a deeply

punitive writer. In "Amour Dure: Passages from the Diary of Spiridion Trepka", the protagonist is haunted by the GHOST of a Renaissance FEMME FATALE; his reward for obeying her commands is death. The eponymous AVATAR of a pagan goddess in "Dionea" burns away the dross of an equivocating mortal marriage merely by allowing the couple sight of her. VL's equivocal sense of the arts – they allure, but in so doing arouse an amoral desperation for bliss – governs "A Wicked Voice", in which heart-rendingly beautiful music is sung by a ghost castrato from the 18th century. The best story in *Pope Jacynth, and Other Fantastic Tales* (coll **1904**) is "Prince Alberic and the Snake Lady", in which young Prince Alberic is haunted by an ancient tapestry which depicts an ancestor embracing a LAMIA in a LANDSCAPE which evokes the *genius loci* of antique Italy. Exiled to a ruined castle, the prince soon finds himself threading the landscape – "as though the tapestry had become the whole world" – until he arrives at a deep well, in the bright water of which an underworld sky glows; in this MIRROR he sees the Lamia. They wed.

The tales assembled in *For Maurice: Five Unlikely Stories* (coll **1927**), written long before, are lesser, though "The Virgin with the Seven Daggers", a POSTHUMOUS FANTASY featuring Don Giovanni (◊ DON JUAN), is of interest. Later work includes *The Ballet of the Nations: A Present-Day Morality* (**1915**; exp vt *Satan the Waster: A Philosophic War Trilogy with Notes & Introduction* 1920; rev 1930), an ALLEGORY excoriating warfare. VL spent her last years on works of aesthetics. [JC]

Other works:
Fiction: *Vanitas: Polite Stories* (coll **1892**), contains one supernatural tale; *Ariadne in Mantua* (**1903** chap), play; *Sister Benvenuta and the Christ Child: An Eighteenth Century Legend* (**1906** chap); *Louis Norbert: A Two-Fold Romance* (**1914**), whose obscure plot contains implications of TIMESLIP.
Posthumous compilations (selective): *The Snake Lady, and Other Stories* (coll **1954** US: rev vt *Supernatural Tales: Excursions into Fantasy* 1955 UK; vt *The Virgin of the Seven Daggers* 1962 UK); *Pope Jacynth, and More Supernatural Tales* (coll **1956**; with 1 story cut vt *Ravenna and her Ghosts* 1962).

LEECH, BEN Pseudonym of Stephen BOWKETT.

LEECH, JOHN (1817-1864) UK cartoonist and illustrator, a pupil of George CRUIKSHANK, working with him for *Bentley's Magazine*, for which he illustrated Richard Harris BARHAM's *The Ingoldsby Legends* from 1837. Soon after the founding of *Punch* in 1841, he became that magazine's most prominent illustrator. His "Mr Briggs" – who appears in *Mr Briggs and his Doings* (graph **1860**) – is a Pickwickian character whose exaggerated adventures are gently spoofed in JL's slightly awkward but entirely accessible manner. Texts illustrated include the first volume of Barham's *Ingoldsby Legends* (coll **1840**), *Jack the Giant Killer* (**1843**) by Leigh Percival (1813-1889), most famously Charles DICKENS's *A Christmas Carol* (**1843**), and *Nursery Ditties from the Lips of Mrs Lullaby* (coll **1844** chap). [JC]

LEE-HAMILTON, EUGENE (1845-1907) Writer of English descent who, like his half-sister Vernon LEE, resided in Europe. He is remembered mainly for *The Lord of the Dark Red Star: Being the Story of the Supernatural Influences in the Life of an Italian Despot of the Thirteenth Century* (**1903**), a DARK FANTASY whose protagonist descends to HELL to strike a bargain there with Iblis, who had fathered him on a WITCH. Once the bargain is broken –

by which token DEATH wins the GAME he has been conducting with Sin – the protagonist is killed. In *The Romance of the Fountain* (**1905**) the discovery in the New World of perpetual youth proves (as usual) to be a barbed gift (◊ FOUNTAIN OF YOUTH). [JC]

LE FANU, J(OSEPH THOMAS) SHERIDAN (1814-1873) Irish writer who spent much of his life as proprietor of various Dublin newspapers and journals. He is of central importance to the history of 19th-century SUPERNATURAL FICTION, though none of his 14 novels does more than hint at supernatural materials. JSLF wrote relatively little FANTASY; his bibliography, which is vast and quite remarkably complex – especially in view of his tendency to incorporate recastings of earlier work into various contexts – is here rendered only in part.

Of JSLF's 40-45 stories, about 28 are supernatural fictions, beginning with his first story, "The Ghost and the Bone-Setter" (1838 *Dublin University Magazine*), a GHOST STORY whose condescending "Irishry" now seems offensive. But JSLF's relationship to his native Ireland was by no means simple. As a conservative Protestant from a prosperous landed family, he was violently opposed to any measures which might loosen the political union between Ireland and the rest of the UK, and resisted any form of local rule; he also fought against the disestablishment of the Protestant Church. It has been suggested that much of his mature fiction can be understood, at one level, as a deeply pessimistic (and indeed guilt-ridden) presentation of his class and religion as locked into a decaying redoubt within Ireland. This may be why his best stories frequently turn on the disastrous consequences of sins committed – years or decades before the price must be paid – against now-decadent families, and against the Protestant religion which serves as a bulwark against dissolution. A sense of chill BELATEDNESS – a sense that those now living are PUPPETS of dead sinners – governs these stories, which often gain their most powerful and most haunting effects when characters act out prior dramas of violation.

Of those tales written before 1853, when he became relatively inactive for some years, the best is probably "Strange Event in the Life of Schalken the Painter" (1839 *Dublin University Magazine*), in which a young painter loses his loved one to a corpse-like DEMON. It appears in JSLF's first collection, *Ghost Stories and Tales of Mystery* (coll **1851**), along with "The Watcher", first published here. (Much transformed, it appeared again as "The Familiar" in 1872. We will not provide further examples of these recastings, because JSLF's reworkings sometimes extend far beyond simple revisions.) Further early work of interest includes "Spalatro, from the Notes of Fra Giacomo" (1843), in which a DOPPELGÄNGER represents the cold *priority* of the past, and "The Mysterious Lodger" (1850 *Dublin University Magazine*), in which apostasy from Protestant values is punished by the eponymous DEMON.

It is for the short stories that JSLF wrote between 1866 and his death – a mere seven-year timespan – that he is now esteemed. They include: "Squire Toby's Will" (1868 *Temple Bar*), in which the death of a house (JSLF's HAUNTED DWELLINGS are exceptionally evocative) is rendered inevitable by a son's refusal to obey the ghost of his father; *Green Tea* (1869 *All the Year Round*; **1943** chap), featuring a malevolent ghost monkey, and arguing the case that bad thoughts *infect* the thinker, allowing evil or repressed supernatural entities to invade the real world;

"The Child that Went with the Fairies" (1870 *All the Year Round*), a rare – and horrific – fantasy about TIME IN FAERIE; *Carmilla* (1871-2 *Dark Blue*; **1971** US), one of the first successful VAMPIRE stories, in which vampirism is linked to lesbianism; "Dickon the Devil" (1872 *London Society*), which features a vengeful ghost willing to kill in order to maintain his long-perished will; and "Mr Justice Harbottle", which appears in *In a Glass Darkly* (coll **1872**; cut vt *Green Tea* 1907), and whose protagonist sends an innocent man to be hanged, beds the widow, and is now trying to shake off the guilt.

Of JSLF's novels, *Uncle Silas: A Tale of Bartram-Haugh* (**1864**) is perhaps the most powerfully atmospheric, and its VILLAIN the closest to supernatural in the demonic intensity of his actions; it was memorably filmed as *Uncle Silas* (**1947**; vt *The Inheritance* US). But it is in his stories – some of them of nearly novel length – that JSLF remains vital, and where his sense that the world was sliding into chaos most vividly continues to chasten. [JC]

Other works: *Chronicles of Golden Friars* (coll **1871**); *The Purcell Papers* (coll **1880**), assembling early work; *The Watcher and Other Weird Stories* (coll **1894**); *Madame Crowl's Ghost* (coll **1923**) ed M.R. JAMES; *Green Tea and Other Ghost Stories* (coll **1945** US); *Sheridan La [sic] Fanu: The Diabolic Genius* (coll **1959** US); *Best Ghost Stories of J.S. Le Fanu* (coll **1964** US) ed E.F. BLEILER; *Carmilla and The Haunted Baronet* (coll **1970** US); *The Best Horror Stories* (coll **1970**); *Vampire Lovers, and Other Stories* (coll **1970**); *Irish Ghost Stories* (coll **1973**); *Ghost Stories and Mysteries* (coll **1975** US); *The Hour After Midnight* (coll **1975**); *The Purcell Papers* (coll **1975** US) ed August DERLETH, differing contents from the 1880 coll; *Borrhomeo the Astrologer* (1862 *Dublin University Magazine*; **1985** chap); *The Illustrated J.S. Le Fanu* (coll **1988**).

Further reading: *Joseph Sheridan Le Fanu* (**1971**) by Michael H. Begnal.

LEGEND US movie (***1985***). 20th Century-Fox/Universal/Legend Productions. **Pr** Arnon Milchan. **Dir** Ridley Scott. **Spfx** Nick Allder. **Mufx** Rob Bottin. **Screenplay** William HJORTSBERG. **Starring** Billy Barty (Screwball), David Bennent (Gump), Tom Cruise (Jack), Tim Curry (Darkness), Cork Hubbert (Brown Tom), Anabelle Lanyon (Oona), Alice Playten (Blix), Mia Sara (Lili). 94 mins. Colour.

Lili, a princess, loves forest boy Jack. He takes her to see the last pair of UNICORNS, and she touches the stallion, apparently corrupting him with her mortality, despite her VIRGINITY; in fact, a party of GOBLINS sent by the Lord of Darkness (i.e., SATAN) – who wishes unicorns exterminated so everlasting night may fall – have shot the creature with a poisoned dart. As the goblins hack the horn from its forehead, Winter falls upon the world. Jack befriends the woodland SPIRIT Gump with his COMPANIONS Screwball and Brown Tom. The FAIRY Oona – usually just a blob of light but capable of SHAPESHIFTING – leads him to an ancient suit of armour and a golden SWORD, for he is to be their Champion. Lili and the last unicorn are snared by the goblins and taken to the Great TREE, the heart of the world's EVIL; our heroes follow and are tipped into a small-scale HELL. Eventually they use sunlight and Heath Robinson gadgetry to destroy Darkness.

The plot of *L* resembles an amalgam of bits from numerous other FAIRYTALES. This same feeling persists visually: the set-pieces sing with familiarity. Yet the concentrated synthesis of derivative components achieves a freshness that is hard to convey. This quality is encapsulated in the exceptionally otherworldly performances of Bennent and Lanyon.

[JG]

LEGEND OF HELL HOUSE, THE UK movie (*1973*). Academy/20th Century-Fox. **Pr** Albert Fennell, Norman T. Herman. **Exec pr** James H. Nicholson. **Dir** John Hough. **Spfx** Tom Howard, Roy Whybrow. **Screenplay** Richard MATHESON. **Based on** *Hell House* (**1971**) by Matheson. **Starring** Pamela Franklin (Florence Tanner), Gayle Hunnicutt (Ann Barrett), Roddy McDowall (Ben Fischer), Clive Revill (Dr Lionel Barrett). 94 mins. Colour.

Physicist Lionel Barrett heads a team of three sent by a mysterious magnate to investigate the Belasco house, renowned for the HAUNTING it has experienced since the 1920s; his wife Ann accompanies. The evil SPIRIT occupying the HAUNTED DWELLING essentially plays a GODGAME with the quartet, constantly deluding them as to its real nature. Tanner is eventually killed by POLTERGEIST activity, having been misled into consenting to SEX with a spirit she has misidentified. Barrett uses a TECHNOFANTASY device to exorcise the house (◊ EXORCISM), but fails to penetrate the lead-shielded room behind the chapel, from which the vile energy continues to emanate. It is left to Fischer – sole sane survivor of a previous investigation – to overcome his terrors and expose the secret of the original Belasco, thus banishing the spirit.

Matheson's original SUPERNATURAL FICTION was a rather directionless mixture of HORROR tropes. The movie tightens this up considerably, but still gives the feeling that an attempt has been to throw everything into the cocktail – including a would-be killer black CAT. Even so, the ride, while seeming to lack any clear destination, is a thoroughly enjoyable one; the pseudo-drama-documentary presentation assists in building up the thrills. [JG]

LEGEND OF PRAGUE, THE vt of *The Golem* (*1936*). ◊ *The* GOLEM (*1913*).

LEGEND OF SLEEPY HOLLOW, THE Section, later released individually, of the ANIMATED MOVIE *The Adventures of Ichabod and Mr. Toad* (*1949*). ◊ *The* WIND IN THE WILLOWS.

LEGENDS Events or stories which have grown to mythic proportions. Legends are closely associated with FOLKTALES, but usually on a heroic scale, as with ROBIN HOOD. A collection of legends related to a culture's foundations and HERO figures becomes a MYTHOLOGY. Legends generally are not born of nothing – there is usually some historical basis – but embellishment transforms them into a form of MYTH. (URBAN LEGENDS may be quite another kettle of fish.) They provide many of the TAPROOT TEXTS for fantastic fiction and frequently offer UNDERLIERS. The most significant usage in GENRE FANTASY concerns those derived from Celtic legend (◊ CELTIC FANTASY) – which include the many legends of ARTHUR – Nordic and Teutonic legends (◊ NORDIC FANTASY; SAGA), and Greek, Roman, Chinese and Japanese legends. Legends found their ways into the medieval sagas and ROMANCES and through them were picked up by William MORRIS and developed into modern ADULT FANTASY. Legends contribute most heavily to HEROIC FANTASY.

Legends have also given us many LOST LANDS AND CONTINENTS. Best-known is ATLANTIS, but others include HYPERBOREA, LEMURIA and Mu, settings used diversely by Robert E. HOWARD, Clark Ashton SMITH and Lin CARTER.

Legends are such a backbone of a culture's belief that they

are also essential to any SECONDARY WORLD. This was recognized by J.R.R. TOLKIEN, who created an entire mythology for Middle-Earth. No other author has repeated this to such an extent, although writers from E.R. EDDISON through Michael MOORCOCK and Terry PRATCHETT to Roger ZELAZNY have used the power of legend to support the worlds they have created. [MA]

Further reading: *The Encyclopedia of Myths and Legends of All Nations* (**1962**) ed Barbara Leonie Picard.

LE GUIN, URSULA K(ROEBER) (1929-) US writer, one of the two or three most important US sf authors of the second half of the 20th century, though the fact that she tends to speculate less in the hard than in the human sciences has led some critics, unwisely, to describe her as mainly a writer of fantasy. She is not. The **Hainish** sequence, which contains her best-known work, is thoroughly sf; even *The Lathe of Heaven* (**1971**) – one of the works she has called "psychomyths" – offers an sf explanation for the reality-changing power of its protagonist's DREAMS.

Nevertheless, UKLG is an important fantasy writer – for her work in short forms, which runs a gamut from SLICK FANTASY through FAIRYTALE and parable to erudite (and sometimes arduous) metafictions; for the **Earthsea** sequence, set in a full SECONDARY WORLD; for the **Orsinia** books, set in a minimally altered LAND-OF-FABLE Europe; and for individual novels like *The Beginning Place* (**1980**). Throughout all this work, an underlying pattern can be traced: the QUEST for one's opposite, with whom one must join together to become whole (◊ BALANCE; JUNGIAN PSYCHOLOGY; YIN AND YANG). In 1979 she was awarded the GANDALF AWARD.

UKLG began publishing with "An die Musik" for *Western Humanities Review* in 1961 – further essentially nongenre work was assembled in *Orsinian Tales* (coll **1976**) – and began releasing work of genre interest in 1962 with "April in Paris" (*Fantastic*), a slick fantasy in which a variety of people from different eras are brought together by TIME TRAVEL into a glowing 15th-century PARIS, where their WISHES come true in a series of marriages with fitting others. In "Darkness Box" (1963), a prince is given the power to free TIME, which allows him properly to confront his brother, who is also his SHADOW. In "The Word of Unbinding" (1964), a MAGUS begins to understand that his enemy is keeping him alive because he fears him dead, and after unbinding his own SOUL defeats his foe in the AFTERLIFE. "The Rule of Names" (1964), a tale which seems to present a normal THEODICY of its medieval FANTASYLAND venue, turns into a subversive reversal of all these easy values. "The Ones Who Walk Away from Omelas" (1973), which won a 1974 Hugo AWARD, puts into a fantastic frame a moral lesson about the funding of UTOPIA. UKLG's earlier short fiction is assembled in *The Wind's Twelve Quarters* (coll **1975**) and *The Compass Rose* (coll **1982**).

For many years her interest in short stories was intermittent, though the ANIMAL FANTASIES and BEAST FABLES assembled as *Buffalo Gals, and Other Animal Presences* (coll **1987**; vt *Buffalo Gals* 1990 UK) – "Buffalo Gals, Won't You Come Out Tonight" itself won both a Hugo and a WORLD FANTASY AWARD – are highly energetic attempts to dramatize a proper relationship (and balance) between the original American LAND and the peoples who have invaded it. But not until the early 1990s did she begin once again to produce classic tales with any consistency. The most notable of these later works may be "The Poacher", from *Xanadu* (anth **1993**) ed Jane YOLEN, whose ET IN ARCADIA EGO conclusion is moving, evocative and ambivalent. A young man explores deeper and deeper INTO THE WOODS until finding a magic POLDER, where SLEEPING BEAUTY and her court continue to slumber. He reads her STORY in a BOOK of fairytales he finds in the castle, decides not to awaken her, and spends the rest of his long life enjoying the sleepers, protected by TIME IN FAERIE, eerie and calm. He realizes that he is not *in* a DREAM but in fact a dream dreamed by Story.

As a fantasy writer, UKLG remains best-known for the **Earthsea** sequence – *A Wizard of Earthsea* (**1968**), *The Tombs of Atuan* (**1971**) and *The Farthest Shore* (**1972**), all three assembled as *Earthsea* (omni **1977** UK; vt *The Earthsea Trilogy* 1979 UK), plus *Tehanu: The Last Book of Earthsea* (**1990**). The first three volumes, which remain central to UKLG's reputation, follow the life of young Ged, who begins as an UGLY DUCKLING on one of the outlying ISLANDS of the great ARCHIPELAGO which constitutes this SECONDARY WORLD, throughout which evocatively conceived DRAGONS fly. Ged soon studies to become a WIZARD, and ends his active life as a MAGUS. During this career he resurrects (and must confront) his SHADOW; he learns how to cooperate with a young woman in the furtherance of his goal, which is to achieve BALANCE between light and dark, male and female, life and death, MAGIC and its cost (every act of magic necessarily puts balance at risk). Finally he travels to the ends of the world and to the UNDERWORLD to combat a balance-destroying religious claim that SOULS are immortal. In the much later fourth volume, a further sorting out of proper balance and a certain "correction" of the male-oriented fantasy world of the earlier books, is attempted; in *Earthsea Revisioned* (1992 lecture as "Children, Men and Dragons"; **1993** chap), UKLG makes the case for this recasting. Throughout the sequence, the intimate clarity of her conception and the pellucid authenticity of her descriptions of her world and its mortal folk make **Earthsea** one of the most deeply influential of all 20th-century fantasy texts.

UKLG's other work of interest includes the **Orsinia** sequence – *Orsinian Tales* (coll **1976**) and *Malafrena* (**1979**) – whose relatively unadventurous texture tends to confirm speculation that the tales were at least drafted long before publication; and *The Beginning Place* (**1980**; vt *Threshold* 1980 UK), in which two adolescents pass through a PORTAL into a subfusc OTHERWORLD, where they undergo a RITE OF PASSAGE, and return home together. As usual, UKLG transforms cliché into wise, unfettered observation.

Several of the essays assembled in *The Language of the Night: Essays on Fantasy and Science Fiction* (coll **1979**; rev 1989 UK) ed Susan Wood (1948-1980) concern fantasy – most notably its republication of *From Elfland to Poughkeepsie* (**1973** chap), which contains informed appreciations of Lord DUNSANY, of E.R. EDDISON and of Kenneth MORRIS, whose reputation this essay re-established. Others are "Why Are Americans Afraid of Dragons?" (1974), in which UKLG argues that one "matures" not by outgrowing fantasy but by growing up into it; and "The Child and the Shadow" (1975), about JUNGIAN PSYCHOLOGY. Some of the essays in *Dancing at the Edge of the World: Thoughts on Words, Women, Places* (coll **1989**) are also of strong interest.   [JC]

Other works: *The Water is Wide* (**1976** chap); *The Visionary: The Life Story of Flicker of the Serpentine* (**1984** chap dos); much sf (◊ *SFE*).

Plays and poetry: Plays include one of fantasy interest,

King Dog (**1985** dos). Prolific poetry begins with *Wild Angels* (coll **1975** chap).

For younger children: *Leese Webster* (**1979** chap); the **Adventures in Kroy** sequence, being *The Adventures of Cobbler's Rune* (**1982** chap) and *Solomon Leviathan's Nine Hundred and Thirty-First Trip Around the World* (**1983** chap); *A Visit from Dr Katz* (**1988** chap; vt *Dr Katz* 1988); the **Catwings** sequence, being *Catwings* (**1988** chap), *Catwings Return* 1989 chap) and *Wonderful Alexander and the Catwings* (**1994** chap); *Fire and Stone* (**1989** chap); *Fish Soup* (**1992** chap); *A Ride on the Red Mare's Back* (**1992** chap).

Nonfiction: *Dreams Must Explain Themselves* (coll **1975** chap); *The Way of the Water's Going* (**1989**), with photographs by Ernest Waugh and Alan Nicholson; *Myth and Archetype in Science Fiction* (**1991** chap); *Talk About Writing* (**1991** chap); *Findings* (**1992** chap); *The Art of Bunditsu* (**1993** chap).

LEIBER, FRITZ (REUTER Jr) (1910-1992) US writer of major importance in both fantasy and sf. His best-loved work is the long-running HEROIC FANTASY series featuring the paradigmatic DUO: **Fafhrd and the Gray Mouser** (F&GM). These characters, respectively a tall northern barbarian of unbarbaric intelligence and a small, mercurial TRICKSTER figure, were suggested by FL's friend Harry Otto Fischer (1910-1986) in 1934; FL acknowledged this when introducing his first collection, *Night's Black Agents* (coll **1947**; 1 story cut vt *Tales from Night's Black Agents* 1961; original exp with 2 stories added 1978), going into greater detail in his essay "Fafhrd and Me" (1963; exp in *The Second Book of Fritz Leiber* coll **1975**; further exp *Fafhrd & Me* coll **1991** chap). F&GM intentionally had something in them of the very tall FL and the small, ingenious Fischer.

The PICARESQUE sequence begins with FL's first published story, "Two Sought Adventure" (1939 *Unknown*; vt "The Jewels in the Forest" 1957), and was eventually organized as what has also been called the **Swords** series. (FL is thought to have coined the term SWORD AND SORCERY.) This comprises: *Swords and Deviltry* (coll **1970**), placed first since it details the heroes' early lives and first meeting – the tale "Ill Met in Lankhmar" (1970 *F&SF*) won Hugo and Nebula AWARDS; *Two Sought Adventure* (coll **1957**; exp rev vt *Swords Against Death* 1970); *Swords in the Mist* (coll **1968**); *Swords Against Wizardry* (coll **1968**) – including "The Lords of Quarmall" (1964 *Fantastic*) with Fischer – *The Swords of Lankhmar* (first part 1961 *Fantastic* as "Scylla's Daughter"; exp **1968**); *Rime Isle* (coll **1977**; exp from 2 to 8 stories vt *Swords & Ice Magic* 1977); and *The Knight and Knave of Swords* (coll **1988**). The first three titles were assembled as *The Three of Swords* (omni **1989**); the next three – the third in the exp vt form – as *Swords' Masters* (omni **1990**); FL's three favourite F&GM shorts make up *Bazaar of the Bizarre* (coll **1978**).

The F&GM SECONDARY WORLD of Nehwon ("No-when") emerges with considerable stylish wit: it has PORTALS to elsewhere and elsewhen, accounting for the ancient Mediterranean setting of the early "Adept's Gambit" (1947 in *Night's Black Agents*), the comic German time-traveller riding a SEA MONSTER in "Scylla's Daughter", and the sinister SHOP in "Bazaar of the Bizarre" (1963 *Fantastic*) which sells rubbish by deceiving PERCEPTION. Other recurring features are the sunken land of Simorgya, all-purpose barbarian hordes cheekily called Mingols, the WIZARDS Ningauble and Sheelba (F&GM'S unreliable MENTORS), a personified DEATH who repeatedly tries to harvest the elusive duo . . . and above all the CITY of Lankhmar with its eccentric Overlords, Silver Eel tavern, Thieves' Guild, exotic pleasure-areas like the Plaza of Dark Delights, countless warring temples in the Street of the GODS (with a piquant distinction between these "Gods *in* Lankhmar" and the feared, unworshipped "Gods *of* Lankhmar") and general sense of sleazy, colourful inexhaustibility. *The Swords of Lankhmar* sees Lankhmar threatened by rats (◊ MICE AND RATS) whose extensive WAINSCOT society, Lankhmar Below, elaborately reflects the city. Major plot strands share a knowing erotic perversity: the Overlord is a risible voyeur whose female servants are compulsorily nude, totally shaven, and subject to arbitary whippings by a dominatrix; Fafhrd becomes entangled with a female "Ghoul" whose invisible but tangible flesh makes her seem a living skeleton; the Mouser is besotted by a tantalizing human/rat hybrid, pursuing her into Lankhmar Below when Sheelba's magic POTION to "put [him] on the right footing" to deal with the rat plague unexpectedly shrinks him to rat-size (◊ GREAT AND SMALL; READ THE SMALL PRINT). The novel is a tasty confection, and one of FL's best.

FL continued to be unpredictable: heroes' success with willing or unwilling women is taken for granted in S&S, but "Under the Thumbs of the Gods" (1975 *Fantastic*) sees F&GM punished for hubris with a series of sexual humiliations. FL moved further from the adventurer TEMPLATE with the darker "Rime Isle" (1977 *Cosmos*): the duo ultimately save the eponymous northern ISLAND from ice-magic, invading Mingols and the interventions of ODIN and LOKI, but at high cost when Fafhrd loses a hand to magical snares intended to provide Odin with a multiple HUMAN SACRIFICE. In "The Mouser Goes Below" (inset novel in *The Knight and Knave of Swords*) the now older heroes have settled on Rime Isle and it is the Mouser's turn to suffer: he experiences a phantasmagoric NIGHT JOURNEY whose sexual episodes and torments are no longer so adequately leavened with wit.

FL also wrote notable SUPERNATURAL FICTION and some HORROR. "Smoke Ghost" (1941 *Unknown*) imagines a city ELEMENTAL of smoke and grime which, as an incipient god, demands worship. *Conjure Wife* (1943 *Unknown*; in *Witches Three* omni **1952** ed Fletcher PRATT; **1953**) is set on a 20th-century US university campus where male career advancement depends on magical conflict between SECRET MASTERS: academic wives who are (like all women) WITCHES. By making his wife abandon this "superstition", the professor-hero destroys his own defences. A cement DRAGON moves (◊ ANIMATE/INANIMATE) to threaten him; the wife's SOUL is stolen, with POSSESSION and IDENTITY EXCHANGE to follow; the protagonist is forced to abandon scepticism and discover the underlying equations of magic, via symbolic logic (◊ RATIONALIZED FANTASY). It is an effectively uneasy exercise in the paranoid. It has been filmed as *Weird Woman* (**1944**) and BURN WITCH BURN (**1961**).

The Sinful Ones (1950 *Fantastic Adventures* as "You're All Alone"; cut vt as title story of *You're All Alone* coll 1972; text restored **1980**) is a paranoid solipsist fantasy in which most people on Earth are automata who never depart from their life-scripts, and the few who have "awakened" jealously – and murderously – guard their privileged status. "A Bit of the Dark World" (1962 *Fantastic*) depicts a disturbingly alien, and non-anthropomorphic supernatural intrusion, differently perceived by different viewers. "Gonna Roll the Bones" (1967) is a fine tale of a gambling addict playing dice with the DEVIL for his SOUL, the whole adventure being an ILLUSION devised by his wife; this won both Hugo and Nebula AWARDS. "Belsen Express" (1975) grimly hints at a TIMESLIP visit to WORLD WAR II death camps; this won the Lovecraft and Derleth awards. *Our Lady of Darkness* (**1977**), a smoothly underplayed URBAN FANTASY with echoes of H.P. LOVECRAFT (including a new addition to the *Necronomicon* library, the prophetic work *Megalopolisomancy*; ◊◊ BOOKS), has elements of autobiography and of ironic ALLEGORY – when the irrational MONSTER finally takes form, it is composed of shredded paper from the books to which the protagonist has given too much of his life, and is dispersed by reciting rationalist names of power including those of Pythagoras, Newton and Einstein.

FL additionally received the 1975 GANDALF AWARD, the 1976 Life Achievement Lovecraft Award, and the 1981 Grand Master Nebula Award – the latter for sf as well as fantasy. The sheer variety of fine work (often subversive of clichés and received ideas) which he produced over such a lengthy career is remarkable. At his best, his stylistic confidence unifies quirky notions, mannered passages sometimes verging on prose poetry, unexpected imagery and a frothy sense of fun, into stories that glitter and sing.   [DRL]

Other works (selective): *Night Monsters* (coll **1969** chap dos; exp 1974 UK); *The Book of Fritz Leiber* (coll **1974**); *The Best of Fritz Leiber* (coll **1974** UK); *The Worlds of Fritz Leiber* (coll **1976**); *Heroes and Horrors* (coll **1978**); *Ervool* (**1980** chap); *Quicks Around the Zodiac: A Farce* (**1983** chap); *The Ghost Light* (coll **1984**) including the specially written "Not Much Disorder and Not So Early Sex: An Autobiographic Essay"; *The Leiber Chronicles: Fifty Years of Fritz Leiber* (coll **1990**) ed Martin H. GREENBERG; *Conjure Wife/Our Lady of Darkness* (omni **1991**); GRAPHIC NOVEL adaptations *Fafhrd and the Gray Mouser Book 1* (graph **1991**) and *Book 2* (graph **1992**) by Howard V. CHAYKIN; much sf.

See also: CATS; CHESS; SWORDS; SFE.

LELAND, CHARLES GODFREY (1824-1903) US writer, best-known during his lifetime by his pseudonym Hans Breitmann, under which, beginning in 1857 with the famous "Hans Breitmann's Barty", he wrote large amounts of comic verse in German-US dialect. His fantasy, all under his own name, included: *Johnnykin and the Goblins* (**1876**), in which Johnnykin gains artistic success through his kindness to the GOBLINS; *Arcadia: The Gospel of the Witches* (**1899**), which presents some of his own poetry as genuinely transmitted from an ancient WITCH cult; and *Flaxius: Leaves from the Life of an Immortal* (coll of linked stories **1902**) – Flaxius is an Etruscan MAGICIAN granted IMMORTALITY by a FAIRY and who tells various anecdotes as he travels up through the centuries, eventually to reach the future.   [JC]

LELLAND, FRANK Pseudonym of A.M. BURRAGE.

LEMURIA Lemuria was originally a scientific hypothesis created to explain the sporadic distribution of lemurs in Africa, Madagascar and India. In the 1870s the German biologist Ernst Haeckel (1834-1919) speculated that a land-bridge or small continent had once connected these regions. The discovery of continental drift rendered the hypothesis unnecessary. However, H.P. BLAVATSKY worked Haeckel's Lemuria into her Theosophical cosmology (◊ THEOSOPHY), where it was a LOST LAND preceding ATLANTIS: the Lemurians were semi-material, semi-solid beings of monstrous shape and size, who had developed enormous magical abilities. They perished in a cataclysm, a few survivors evolving into Atlanteans. In later occult literature

Lemurians are often identified with ELEMENTALS – vague, hostile, malevolent supernatural beings.

Often confused with Lemuria is Mu, a purported lost continent in the Pacific Ocean, whereas Lemuria was properly in the Indian Ocean. Mu was the creation of Colonel James Churchward in *The Lost Continent of Mu* (**1926**; rev 1931), *Cosmic Forces of Mu* (**1932**), and other books. Based partly on Blavatsky's writings and more on his own imagination, Churchward's books described an ancient culture possessed of high occult wisdom that left traces in various parts of the world, notably Mexico. Churchward's forte was the interpretation of geometric designs as solemn messages about Mu. It has never been clear whether Churchward was a crank or a hoaxer or both. In any case, he seems to have been the source for the motif that lost continents were destroyed by the explosion of subterranean gas caverns.

In fantastic fiction, Lemuria/Mu has been far less important than Atlantis. Drawing on occult sources, Charles VIVIAN's *City of Wonder* (1922) describes surviving Lemurian beings, primal forces not wholly intelligent or individualized, who threaten a LOST RACE. In most instances, however, Lemuria is simply a resonant placename. Relevant titles are Lin CARTER's **Wizard of Lemuria** series and Henry S. WHITEHEAD's "Scar Tissue" (1946) and "Bothon" (1946). A. MERRITT's "The Conquest of the Moon Pool" (1919) describes a huge cavern world beneath Micronesia inhabited by descendants of ancient Lemurians. Richard Shaver's paranoid *I Remember Lemuria* (1948) invokes both Atlantis and Mu, but with a private interpretation of both.   [EFB]

L'ENGLE, MADELEINE Writing name of Madelaine (*sic*) L'Engle Camp Franklin (1918-), who, after a short period of acting and writing for the stage in the early 1940s, began producing novels for adults with *The Small Rain* (**1945**) and then for children with *And Both Were Young* (**1949**), neither fantasy. Her later works shifted towards CHILDREN'S FANTASY. Her best-known work, winner of the Newbery Medal in 1963, is *A Wrinkle in Time* (**1962**), the first in a series about **Meg Murry**, the daughter of a scientist. In this novel Meg's father vanishes through the fifth dimension to an alien planet and is captured by an evil brain. Although written as a rebellion against Christian piety, the book is an intensely religious ALLEGORY of GOOD AND EVIL. The series offers a curiously appropriate blend of hard science and mysticism: *A Wind in the Door* (**1973**), *A Swiftly Tilting Planet* (**1978**), *Many Waters* (**1986**) and *An Acceptable Time* (**1989**).

ML first hit success with her highly moralistic children's novel *Meet the Austins* (**1960**), set in small-town New England. This and its immediate sequel, *The Moon By Night* (**1963**), are nonfantastic, but the next, *The Young Unicorns* (**1968**), shows the disintegration of the family under both social and alien pressures. As with *A Wrinkle in Time*, it portrays CHILDREN who rise above their disabilities to combat evil (in the shape of a scientist intent on taking over NEW YORK), and again has a strong religious theme. Both this and *The Arm of the Starfish* (**1965**), linked to the **Meg Murry** series, share some common characters.   [MA]

LEONARDO DA VINCI (1452-1519) Florentine painter, sculptor, inventor and military engineer, celebrated since the Enlightenment as the epitome of Renaissance creative genius. Leonardo not infrequently appeared among 19th-century novels concerned with Cesare Borgia or Lorenzo Visconti – *Leonora d'Orco, or The Times of Caesar Borgia* (**1857**) by George James (1801-1860) is typical – but the first

novel to present Leonardo as protagonist may be Dmitri Merezhkovsky's *Voskresennie Bogi: Leonardo da-Vinchi* (**1901**; trans as *The Forerunner: The Romance of Leonardo da Vinci* **1902** UK; complete trans Bernard Guilbert Guerney as *The Romance of Leonardo da Vinci (The Gods Resurgent)* **1928** US; rev 1964). Although Merezhkovsky's Leonardo inhabits a world of supernatural events, the evocation of the man is essentially rationalist, and what might be called the Vinciad – a subgenre today comprising the majority of all stories written about Leonardo – has always emphasized Leonardo's ability to create marvels through mechanical means.

Manly Wade WELLMAN's *Twice in Time* (1940 *Startling Stories*; cut 1957; rev with text restored and 1 story added coll **1988**) offers a perspective on Leonardo's Florence via TIME TRAVEL, by which – in a plot-twist used again with regard to Leonardo in Robert A. HEINLEIN's *The Door into Summer* (**1957**) – the modern protagonist eventually becomes Leonardo. A variant is the story in which Leonardo is brought forward into the present – "Mister da V" (1962) by Kit Reed (1932-) is an example – but most stories dealing with Leonardo eschew TIME TRAVEL, leaving him in Renaissance Italy where – and this is the defining element of the Vinciad – he actually does build one or more of his fantastical designs.

Gerald Kersh's "The Dancing Doll" (1955) exemplifies the specific appeal of the Vinciad: his Leonardo realizes how to build breathing tubes that will allow soldiers to cross a shallow river while remarking on the probosces of mosquitoes in pestilential swamps (and incidentally almost discovering the role of insects in disease transmission), and explains this to his patron duke while designing a toy doll for the duke's ailing son (whom he also cures).

Vinciads have been popular in mainstream fiction as well as in the sf and fantasy genres, although with the difference that Leonardo's inventions in these cases usually do not surpass 20th-century technology. *The Medici Guns* by Martin Woodhouse and Robert Ross (**1974**) – sequelled by *The Medici Emeralds* (**1976**) and *The Medici Hawks* (**1978** US) – is a typical example, with Leonardo being called upon by Lorenzo de Medici to create weaponry to defend Florence against the invading papal armies. Vinciads have also proven popular outside print fiction, usually with a more fantastic cast. Vertigo/DC COMICS published a GRAPHIC NOVEL, *Chiaroscuro: The Private Lives of Leonardo da Vinci* (May 1995-April 1996 in 10 instalments) created by Charles Truog, David Rawson and Pat McGreal; an earlier **Batman** graphic novel proposed that Leonardo might have donned a Bat costume in order to wreak VENGEANCE in Florence. The movie *Hudson Hawk* (**1991**) dir Michael Lehmann dramatizes a modern-day QUEST to recover an invention by Leonardo (who in fact detested ALCHEMY) to transmute lead into gold. Leonardo has appeared, in variously rationalized sf forms, in episodes of **Star Trek** ("Requiem for Methuselah" by Jerome BIXBY) and **Doctor Who** ("City of Death" by Douglas Adams).

Treatment of the theme by more literary writers has tended toward the elliptical and figurative, and appears almost exclusively in the work of fabulists. Jeanette WINTERSON's *Art and Lies* (**1994**) briefly evokes Leonardo (in the consciousness of a disturbed painter). Leonardo does not in fact appear in Guy Davenport's *Da Vinci's Bicycle* (coll **1979**).

In the past decade the Vinciad has become popular in

genre sf and fantasy, in part perhaps because the clockwork Universe of Renaissance cosmology (Leonardo's drawings included designs for hydraulic screws and gearworks) has affinities with the spirit of STEAMPUNK. *Pasquale's Angel* (**1994**) by Paul J. McAuley (1955-   ) sees an ALTERNATE WORLD in which Leonardo's genius creates an early industrial dystopia in Renaissance Italy. Jack DANN's *The Memory Cathedral* (**1995**) is a convergent alternate history that dramatizes Leonardo's invention of the flying machine and other marvels during an undocumented period of his life. The Vincian fantasy seems to have become a potent trope.

[GF]

Further reading: *The Deluge* (**1954**) by Robert Payne, a disaster novel credited to Leonardo "edited by" Payne, is based on fragmentary material by Leonardo; *The Second Mrs Giaconda* (**1975**) by E.L. Konigsburg, for young readers, is a historical novel that emphasizes the synthetic nature of Leonardo's genius.

LEPRINCE DE BEAUMONT, MADAME JEANNE-MARIE (1711-1780) French writer whose version of BEAUTY AND THE BEAST has become the standard text. Born and raised in France, JMLDB settled in England in 1746 after an unhappy marriage. England remained her home until 1762; she retired to Switzerland in 1768, continuing to write. Her voluminous production included work for both children and adults. Almost all of her FAIRYTALES are in *Le Magasin des Enfants* (coll **1756** UK; trans *The Young Misses Magazine* 1757). The most immediately successful was "La Belle et la Bête", which JMLDB had adapted from a much longer, more intricate story of the same title by Gabrielle-Suzanne Barbot de Gallon de Villeneuve (1695-1755) in *Les contes marins ou la jeune Américaine* (coll **1740**). JMLDB's version emphasizes the moral triumph of love and virtue over beauty and riches. [MA]

LERMAN, RHODA (1936-   ) US writer whose *The Book of the Night* (**1984**) conveys, through an exorbitant plot, some sense of the violence attending the RITE OF PASSAGE through MENARCHE of a young girl, who is brought to the Celtic ISLAND of Iona disguised as a boy (◊ GENDER DISGUISE) and whose passage (◊ RECOGNITION), via self-METAMORPHOSIS, into a white cow awakens the ELDER GODS by shaking the thin partitions between worlds in this LAND-OF-FABLE venue. [JC]

LERNER, ALAN JAY (1918-1986) US dramatist whose most famous adaptation was the musical *My Fair Lady* (play **1956**) – done with his long-time partner, the composer Frederick Loewe (1904-1988), and based on *Pygmalion* (play **1913**) by George Bernard SHAW. AJL is of fantasy interest mainly for *Brigadoon* (play **1947**), filmed as BRIGADOON (**1954**). The Scottish village of Brigadoon is a POLDER protected from the rest of the world by a SPELL according to which it appears in our REALITY only once a century, and then for only a single day. The partners' later fantasy effort, *Camelot: A New Musical* (produced 1960; **1961**), based loosely on T.H. WHITE's *The Once and Future King* (omni **1958**), was saccharine; it was filmed as *Camelot* (**1967**). Also of fantasy import was *On a Clear Day You Can See Forever* with Burton Lane, which was filmed disastrously by Vincente Minnelli in 1970: a modern woman TIMESLIPS into 19th-century London, where – echoing the situation in *My Fair Lady* – she is a Cockney lass trying to act the part of a lady. [JC]

LERNET-HOLENIA, ALEXANDER (1897-1976) Austrian writer of elegant fantasy and other work. *Baron Bagge* (**1936**; trans Jane B. Greene) and *Graf Luna* (**1955**;

trans Richard and Clara Winston as *Count Luna*) were published together in English as *Count Luna: Two Tales of the Real and Unreal* (omni **1956** US). *Baron Bagge* is a POSTHUMOUS FANTASY whose narrator – who is engaged in a suicidal cavalry charge in WORLD WAR I, and who refuses to cross a golden bridge – awakens in the knowledge that everyone has been killed. *Count Luna* is a kind of GHOST STORY in which a concentration-camp victim haunts the man who put him there. *Mars im Widder* ["Mars in Aries"] (**1941**) is a WORLD WAR II fantasy, told in terms reminiscent of Franz KAFKA.

[JC]

LEROUX, GASTON (1868-1927) French writer, known for mysteries like *Le Mystere de la Chambre Jaune* (**1908**; trans as *The Mystery of the Yellow Room* 1908 UK; vt *Murder in the Bedroom* 1945 Belgium), a pioneer Locked Room mystery.

He remains most famous, however, for *Le Fantome de l'Opera* (**1910**; trans Alexander Texeira de Mattos as *The Phantom of the Opera* 1911 UK). Although there is nothing supernatural in his text, the charged GRAND-GUIGNOL ambience GL created *seems* to point to a non-naturalistic explanation of the story; movie versions (◊ PHANTOM OF THE OPERA) have tended to render those hints in literal terms. But GL assured his readers that the story was based on fact. Erik, a Middle Eastern ex-torturer and half-crazed musician, lurks in the LABYRINTH of passages and catacombs that make the gigantic Paris Opera House a genuine EDIFICE. He creates a series of mysterious events to further the career of Christine Daaé, a beautiful young singer; and, when the Opera managers refuse to make her a star, inflicts savage revenge. Certain scenes – the MASQUE which evokes Edgar Allan POE's "The Masque of the Red Death" (1842), and the moment when the "phantom's" own MASK is stripped off – have become HORROR paradigms. The story as a whole is a central URBAN FANTASY.

Most of GL's other work of genre interest is sf, though his first novel, *La Double Vie de Theophraste Lonquet* (**1904**; trans as *The Double Life* 1909 US; rev vt, trans Edgar JEPSON as *The Man with the Black Feather* 1912 UK), is a tale of POSSESSION: a long-dead villain attempts to commit further felonies in the 20th century, and an UNDERGROUND race of mutants, with ears like horns, features. Other titles include: *Le Fauteuil hante* (**1911**; trans as *The Haunted Chair* 1931 US), horror; *Balaoo* (**1912**; trans Alexander Texeira de Mattos 1913), which was filmed as *The Wizard* (**1927**) and as *Dr Renault's Secret* (**1942**), plus its sequel, *Les Fils de Balaoo* ["Balaoo's Sons"] (**1937**), about a half-man, half-ape; *L'Epouse du Soleil* (**1913**; trans as *The Bride of the Sun* 1915 US), a LOST-RACE tale; *The Man who Came Back from the Dead* (**1916** UK; trans, preceding 1st French publication, of *L'Homme qui Revient de Loin* 1917); *Aventures Effroyables de Herbert de Renich* (**1920** in 2 vols: *Le Capitaine Hyx*, trans Hannaford Bennett as *The Amazing Adventures of Carolus Herbert* 1922 UK; and *La Bataille Invisible*, trans as *The Veiled Prisoner* 1923 UK), a saga obviously inspired by Jules Verne and featuring a marvellous submarine equipped with an array of fantastic devices; *Le Coeur Cambriole* (**1922**; trans with 2 CONTES CRUELS in coll *The Burgled Heart* 1925 UK); and *La Poupée Sanglante* (**1924**; trans as *The Kiss that Killed* 1934 US), with its sequel, *La Machine à Assassiner* (**1924**; trans as *The Machine to Kill* 1935 US), sf. A Peter HAINING anthology – *The Gaston Leroux Bedside Companion* (coll **1980**; exp vt *Real Opera Ghost and Other Tales* 1994) – assembles most of GL's short work of interest, much of it originally translated for *WT*. [RD/JC]

LEŠEHRAD, EMANUEL (1877-1957) Czech writer who also wrote as Emanuel z Lešehradu. ◊ CZECH REPUBLIC.

LESLIE, JOSEPHINE (AIMÉE CAMPBELL) (1898-1979) Scottish writer who wrote several novels and two plays as R.A. Dick, scoring considerable success with *The Ghost and Mrs Muir* (**1945**), which tracks the relationship between a widowed mother and the GHOST of an old sea-captain; the story was the basis of *The* GHOST AND MRS MUIR (*1947*) and a tv series, but did not save JL's writing career from slow decline. In a more earnest vein, *The Devil and Mrs Devine* (**1975**) as by JL features another widow who can be redeemed from a reckless PACT WITH THE DEVIL only by the love of a good man – which is, of course, not easy to come by. [BS]

Other works: *Witch Errant: An Improbable Comedy in Three Acts* (**1959**).

LETHE ◊ HADES; RIVERS.

LEVEL, MAURICE (1875-1926) French writer. ◊ CONTE CRUEL.

LEVIATHAN ◊ SEA MONSTERS.

LEVIN, BETTY (1927-   ) US writer, mostly for YA readers, whose first work of interest is the **Claudia and Evan** sequnce – *The Sword of Culann* (**1973**), *A Griffon's Nest* (**1975**) and *The Forespoken* (**1976**) – in which two children TIME TRAVEL into various eras of CELTIC-FANTASY interest; they are, however, more spectators than participants, and the action seems sometimes remote. The contemporary protagonist of *The Keeping-Room* (**1981**) also visits the past, in this case the 19th century. Other singletons include: *A Binding Spell* (**1984**), about a GHOST horse; *The Ice Bear* (**1986**), a SECONDARY-WORLD fantasy; *The Trouble with Gramary* (**1988**), an associational novel whose protagonist's name is Gramary; *Mercy's Mill* (**1992**); and *Starshine and Sunglow* (**1994**). [JC]

LEVIN, IRA (1929-   ) US writer whose first novel of genre interest, *Rosemary's Baby* (**1967**), was filmed as the extremely influential ROSEMARY'S BABY (*1968*). In both book and movie, the innocent young Rosemary conceives, gives birth to, and begins to mother the ANTICHRIST – although in both versions the door is left open for us to read this as a fantasy of PERCEPTION rooted in the immature Rosemary's subconscious fear of SEX, which is *messy*. *The Stepford Wives* (**1972**) is SATIRE dressed as TECHNOFANTASY; it was filmed as *The* STEPFORD WIVES (*1974*). *This Perfect Day* (**1970**) is sf; *The Boys from Brazil* (**1976**) – filmed in 1978 – is marginal sf; *Sliver* (**1991**) – filmed in 1993 – is an erotic PSYCHOLOGICAL THRILLER. [JC/JG]

LEVITATION ◊ TALENTS.

LEWIS, C(LIVE) S(TAPLES) (1898-1963) Belfast-born UK writer, critic, Christian (Anglican) apologist and Oxbridge don; a Fellow of Magdalen College, Oxford, 1925-54, and then Cambridge Pofessor of Mediaeval and Renaissance English until his death. He was the central figure of the INKLINGS, and a close friend of fellow-members J.R.R. TOLKIEN and Charles WILLIAMS. His first full-length fantasy was the complexly allusive *The Pilgrim's Regress: An Allegorical Apology for Christianity Reason and Romanticism* (**1933**; rev 1943), an eccentric ALLEGORY of CSL's personal path to RELIGION, whose title salutes John BUNYAN but which lacks Bunyan's universality of application; nevertheless there are effective touches, like the Freudian GIANT who compels a visually revolting PERCEPTION of people's internal organs.

The **Cosmic Trilogy** or **Ransom Trilogy** comprises CSL's best-known adult fiction, beginning as sf and slip-streaming into CHRISTIAN FANTASY. *Out of the Silent Planet* (**1938**) is a pure PLANETARY ROMANCE, touring an evocatively described Mars; its mystical elements remain within the sf tradition of religion as PLAYGROUND, and the almost invisible tutelary beings called eldils are not yet fully identified as ANGELS. In *Perelandra* (**1943**; vt *Voyage to Venus* 1953), the significantly named series protagonist Ransom is transported between planets not by spaceship, as before, but in a eldil-borne wax coffin. The floating ISLANDS of Perelandra (the planet Venus) are beautifully imagined, but prove to be the laboratory for a religious thought-experiment revisiting the temptation in EDEN. This world's first-created Lady of course plays Eve (◊ ADAM AND EVE; the Adam figure is kept offstage by GOD); the earlier book's amoral scientist Weston – now a puppet of dark eldils – returns as the tempter, coaxing "Eve" to disobey divine PROHIBITIONS. Ransom argues the case for obedience, with waning success; it is unquestioned that the woman, however romanticized and revered, cannot resist the tempter unaided. Ultimately, stripped of intellectual defences by the former Weston's alternating casuistry and childish nastiness (e.g., pointless torture of animals), Ransom disposes of this "SERPENT" by physical violence. Venus remains unfallen; angels join in a celebration.

The tale moves to Earth in *That Hideous Strength: A Modern Fairy Tale for Grown-Ups* (**1945**; cut 1955; cut version vt *The Tortured Planet* 1958 US), whose supernatural action and moral-theological debates are reminiscent of Charles Williams (e.g., the romantic subplot's stress on female "obedience"). CSL's most effective creation is the KAFKA-like nightmare of the N.I.C.E., a state-funded scientific organization whose inner circle plans to take control of human evolution and to annexe BLACK MAGIC as science. One result is the artificially and repulsively sustained human head intended as an experiment in IMMORTALITY but in fact a channel (◊ SPIRITUALISM) for dark eldils: this punning "Head of the Institute" speaks directly for SATAN. Indeed, the N.I.C.E.'s hatefulness goes beyond plausibility; there is a cruel caricature of the dying H.G. WELLS. But CSL effectively shows a lazy-minded academic being drawn in by what he called the temptation of the Inner Ring – the simple desire to become an insider. As Jorge Luis BORGES noted of William BECKFORD's *Vathek* (**1786**), HELL is here both punishment and temptation: the N.I.C.E.'s innermost circle offers nothing but the privilege of entering, and being damned. Its "Objective Room", used in INITIATIONS that destroy all human emotion, recalls the asylum in G.K. CHESTERTON's *The Ball and the Cross* (**1909**). Meanwhile Ransom, incurably wounded in the heel from his Venusian struggle and too appropriately renamed FISHER-KING, opposes the N.I.C.E. in his new role as the Pendragon – SECRET MASTER of an inner Britain called Logres (◊ FANTASIES OF HISTORY). He finds an ally in MERLIN, long entombed as a SLEEPER UNDER THE HILL nearby; Merlin becomes a magical conduit for the planetary eldils of Mercury, Mars, Venus, Jupiter and Saturn. The term "that hideous strength" alludes to the hubristic tower of BABEL, and Merlin inflicts the appropriate CURSE of gibberish on a N.I.C.E. banquet (whose feasters are then killed by freed experimental animals) before loosing heavenly fires and earthquakes which destroy the N.I.C.E. and himself. The finale pairs off surviving humans and animals in a ritual of LOVE. There are some very fine scenes, but this is a curate's egg of a novel. The series was assembled as *The Cosmic Trilogy* (omni **1990**).

The epistolary *The Screwtape Letters* (**1942**; exp vt *The Screwtape Letters and Screwtape Proposes a Toast* 1961) was a particularly successful book of apologetics, thanks to its device (worthy of Jonathan SWIFT) of preaching by ironic implication *via* a senior DEMON's advice to a junior on how best to tempt and ensnare; the "Lowerarchy" of HELL also allows SATIRE on bureaucracy. Another and more vivid work of apologetics is *The Great Divorce: A Dream* (**1946**). This moralizes effectively, its fantastic premise being the theological notion of the *Refrigerium* – cooling or mitigation – supposedly allowed to the damned, who may briefly visit HEAVEN. But here the damned are self-damned, almost all preferring the neverending suburban dreariness of Hell to the spiritual tempering required by Heaven (◊ PURGATORY); in a metaphor for this small-mindedness, Hell's vastness occupies the tiniest of cracks in the ground, which only CHRIST could make himself small enough to enter (◊ GREAT AND SMALL; LITTLE BIG). CSL narrates in his own persona as one of the damned (though only in a DREAM), whose MENTOR in Heaven is George MACDONALD.

The most popular and influential fantasy sequence produced by CSL is the **Chronicles of Narnia**, written for children (◊ CHILDREN'S FANTASY): *The Lion, the Witch and the Wardrobe: A Story for Children* (**1950**), *Prince Caspian: The Return to Narnia* (**1951**), *The Voyage of the "Dawn Treader": A Story for Children* (**1952**), *The Silver Chair: A Story for Children* (**1953**), *The Horse and his Boy* (**1954**), *The Magician's Nephew* (**1955**) and *The Last Battle: A Story for Children* (**1956**) – this last a Carnegie Medal winner. This publication order is CSL's recommended reading order, though by internal chronology *The Magician's Nephew* comes first and *The Horse and His Boy* takes place during rapidly summarized decades towards the first-published book's close.

The Lion, the Witch and the Wardrobe was clearly written in haste, beginning as a romp: its SECONDARY WORLD of Narnia, to which the wardrobe is a PORTAL, has a variegated FANTASYLAND population including Bacchus, centaurs, dryads, dwarfs, fauns, giants, SANTA CLAUS and – its particular trademark – TALKING ANIMALS. The four sibling CHILDREN – Peter, Susan, Edmund and Lucy – who enter Narnia find it frozen in FIMBULWINTER by a White WITCH (descended from LILITH rather than Eve) fond of converting her foes to STATUES (◊ BONDAGE). The children's arrival heralds change and winter's end with the return of the lion Aslan (Turkish for "lion"), who, it is made increasingly clear, is the Narnian aspect of CHRIST. Edmund having betrayed the others, his life is forfeit to the Witch; Aslan offers himself as a replacement sacrifice. Following a grim death scene amid the Witch's gloating hordes, an ancient PROPHECY is fulfilled and the voluntary sacrifice is rewarded by RESURRECTION; the Witch is destroyed in battle, Edmund LEARNS BETTER and there is general HEALING. After subsequent long years of ruling Narnia, the now adult visitors stumble back through the wardrobe, returning to Earth as children after a real-time absence of mere minutes (◊ TIME IN FAERIE).

Prince Caspian sees the same children summoned back after one year – which for Narnia has been a TIME ABYSS of many centuries, featuring outside conquest, the dwindling of the children's GOLDEN-AGE rule from history to LEGEND, and widespread THINNING. Even Aslan is more elusive, initially appearing only to Lucy and requiring the others to have faith in her as an appointed prophet. The eponymous prince wins the usurped Narnian throne; the elder children are warned that they cannot return to Narnia but should seek Aslan's Earthly aspect. Thus only Edmund and Lucy return to join Caspian in *The Voyage of the "Dawn Treader"*, along with an unpleasant child, Eustace, fated to Learn Better after TRANSFORMATION into a DRAGON. The voyage is a QUEST for lost Narnian lords, involving SEA MONSTERS, MERMAIDS and marvellous ISLANDS. Ultimately, the children and an engaging talking mouse reach the world's edge and the shores of "Aslan's Country", where the lion pointedly emphasizes his nature by appearing as a lamb before sending the children home.

Eustace and a schoolmate are summoned for *The Silver Chair*, and are issued by Aslan with precise-seeming instructions which prove hard to follow (◊ READ THE SMALL PRINT); e.g., Eustace must immediately greet an old friend, but fails – since 70 Narnian years have passed – to recognize the now ancient Caspian. They set out with the lugubrious "Marsh-Wiggle" Puddleglum to rescue Caspian's son Rilian from BONDAGE to a SHAPESHIFTING witch who is also a SERPENT, and who keeps Rilian bemused by GLAMOUR when he is not confined to the eponymous chair or imprisoning magic armour. Confronted in her UNDERGROUND realm, she works a similar glamour on the rescuers (a sly religious thrust, for her denial that surface Narnia exists is isomorphic with the argument that HEAVEN is imaginary since always described in earthly imagery), until Puddleglum extinguishes her befuddling incense and shocks himself awake by stamping barefoot on the fire. There is a posthumous appearance of Caspian, restored to youth as a denizen of Heaven.

The Horse and his Boy opens in Calormen, an ARABIAN-FANTASY kingdom south of Narnia. The horse is a Narnian talking horse, the enslaved boy the missing heir (◊ HIDDEN MONARCH) to a Narnian-allied throne; they are joined by a Calormene girl and her horse for what becomes a cross-desert race to warn the north of treacherous Calormene attack. Aslan repeatedly appears incognito to stage-manage the plot, most disconcertingly when his roar impels the horses to (needed) extra speed and his claws slash the girl's back in punishment for thoughtless cruelty.

The final books deal with Narnia's beginning and end. *The Magician's Nephew* looks back to LONDON when "Mr SHERLOCK HOLMES was still living in Baker Street". A petty self-made WIZARD has misused magic passed down since ATLANTIS to create RINGS that carry their wearers between worlds. He pusillanimously tests these on two children (one being his nephew), who find themselves in a numinous Wood Between the Worlds (◊ INTO THE WOODS) full of portal-pools leading everywhere in the MULTIVERSE. They visit a dead world and foolishly wake its destroyer, who will become the White Witch of Narnia. She follows them to London and briefly runs riot until (*via* the rings) the children transport her and others to what proves to be the dark void preceding Narnia's creation. Aslan then movingly sings the world and its creatures into being; this Eden already has its SERPENT, the intruding Witch. Finally, by the time of *The Last Battle*, Narnia is so wasted by THINNING that an ANTICHRIST figure (a duped ASS cloaked in lion-skin DISGUISE by a cunning APE) gains acceptance. A RITUAL OF DESECRATION follows as Narnian trees are felled and their DRYADS thus killed, by order of the puppet Aslan; Calormen invades and its evil god Tash manifests as a Satan figure. In the futile-seeming LAST BATTLE which is Narnia's ARMAGEDDON, its own denizens are divided by bewilderment and unbelief. Aslan no longer appears to the living; for the children who are as usual called to aid, this is POSTHUMOUS FANTASY since they have unknowingly died on Earth. The remaining portal out of Narnia opens on Heaven: there is a LAST JUDGEMENT as all creatures choose to accept Aslan, or not. Narnia is laid waste and Father Time puts out the sun; it is the END OF THE WORLD, comprehensive and unforgettable. The "happy ever after" conclusion shows the "real" Narnia – its Platonic archetype in Heaven, linked to the archetypes of Earth and all other worlds.

Despite slapdash passages and Tolkien's disparagement, the first **Narnia** book succeeded through the genuine power of its death-and-resurrection theme, and CSL's gift for eidetic imagery: the faithful mice nibbling at dead Aslan's bonds, the cracking of the Stone Table of sacrifice, the new life spreading like flame over the Witch's statue-victims. The sequels are less haphazard, with further glowing images and a powerful grasp of STORY more than compensating for the intermittently rough-hewn writing and glossed-over inconsistencies; Pauline Baynes's illustrations also have a unifying effect. Children take the stories at speed, often not detecting the Christian agenda until, perhaps, the last book. Tv versions include *The* LION, THE WITCH AND THE WARDROBE (1967 UK) and *The* CHRONICLES OF NARNIA (1988-90 UK); there has also been an ANIMATED MOVIE, *The* LION, THE WITCH AND THE WARDROBE (*1978* tvm US).

CSL's last fantasy novel, *Till We Have Faces: A Myth Retold* (**1956**), is a serious retelling of the MYTH of CUPID AND PSYCHE, wherein Psyche becomes the god's lover subject to the CONDITION of never seeing his face. Egged on by her sisters, she disobeys. CSL makes one sister his narrator; Psyche's role is overshadowed by this painful study of the sister's too-possessive LOVE and its destructive effect.

An Experiment in Criticism (**1961**) has brief but penetrating chapters on myth and FANTASY. Other relevant critical pieces are assembled, with short stories (one featuring a lunar GORGON) and "After Ten Years" – a fragment of an unfinished romance about the Trojan War and its aftermath – in *Of Other Worlds: Essays and Stories* (coll **1966**) ed Walter Hooper (1931-   ). The fiction was later rearranged, with another short story and the long title fragment (found amid CSL's papers) as *The Dark Tower and Other Stories* (coll **1977**) ed Hooper, while *Of This and Other Worlds* (coll **1982**; vt *On Stories – And Other Essays on Literature* US) ed Hooper expands the selection of essays. The "The Dark Tower" fragment (written *c*1938) represents a road not taken: seven inferior chapters of a sequel to *Out of the Silent Planet*, whose premise of TIME TRAVEL shifts to an ALTERNATE-WORLD/IDENTITY-EXCHANGE plot, involving much academic discussion of J.W. Dunne's TIME theories and the central image of a "Stingingman" whose lubriciously described, forehead-mounted sting erases its victims' individuality and instils mindless cheeriness. The symbolism verges on the embarrassing, as the Inklings seem to have told CSL – who added a kind of disclaimer and then abandoned the work. Its appearance did his reputation little good and eventually triggered an attack, *The C.S. Lewis Hoax* (**1988**) by Kathryn Lindskoog (1934-   ), accusing Hooper of forging *The Dark Tower*; but handwriting analysis authenticated the MS.

CSL's shrewd eye for human folly and self-deception is as effective in fiction as in theology, but a less welcome crossover from the books of apologetics is the occasional bullying rhetoric – a too-triumphant scoring of debating points. Other CSL strengths are ability to marshal and synthesize many Classical sources (and sometimes newer ones): **Narnia** contains several nods to E. NESBIT), power of imagery, gift for the telling metaphor, sense of the numinous, and genuine feel for EVIL. The **Narnia** sequence's popularity among children and the **Ransom Trilogy**'s adult cult following seem likely to continue indefinitely. [DRL]

Other works (selective): *Dymer* (**1926**), verse as by Clive Hamilton; *The Allegory of Love* (**1936**), a seminal discussion of ALLEGORY; *A Preface to "Paradise Lost"* (**1942**), nonfiction; *Surprised by Joy: The Shape of My Early Life* (**1955**), religious autobiography; *Poems* (coll **1964**), including "Narnian Suite" and other fantasy verses; *Boxen: The Imaginary World of the Young C.S. Lewis* (coll **1985**), ANIMAL FANTASY juvenilia ed Walter Hooper.

As editor: *Essays Presented to Charles Williams* (coll **1947**); *Arthurian Torso: Containing the Posthumous Fragment of The Figure of Arthur by Charles Williams and A Commentary on the Arthurian Poems of Charles Williams by C.S. Lewis* (**1948**).

See also: *C.S. Lewis: A Biography* (**1974**) by Roger Lancelyn GREEN and Walter Hooper; *C.S. Lewis: A Biography* (**1990**) by A.N. Wilson (1950-   ) has a bibliography listing some of the many other books about CSL.

LEWIS, HILDA (WINIFRED) (1896-1974) UK writer best-known for her first CHILDREN'S FANTASY, *The Ship that Flew* (**1936**), in which four children discover a TOY ship in a magic SHOP. They soon find the ship can grow to hold them, that it was once ODIN's gift to the god Frey, and that it is prepared to take them to Asgard and other places – such as Sherwood Forest (◊ ROBIN HOOD). HL wrote much else, mostly historical novels for adults; *The Witch and the Priest* (**1956**) is a SUPERNATURAL FICTION. [JC]

LEWIS, MATTHEW GREGORY (1775-1818) UK novelist, playwright and translator, best-known as the author of the archetypally lurid Gothic novel *The Monk: A Romance* (**1796**; exp but bowdlerized 1798; vt *Ambrosio, or The Monk* 1798 US; vt *Rosario, or The Female Monk* US). The bibliography of this work is complicated, many abridged and chapbook editions reprinting sections of the text under various combinations of the titles *The Castle of Lindenburg, The Legend of the Bleeding Nun* and *The History of Raymond and Agnes*; this element of the text was also dramatized as *Raymond and Agnes* (*c*1820 chap). Most 20th-century editions restore the passages censored from the 1798 version.

MGL was a prolific writer of stage melodramas, some supernatural; many were never formally published, although scripts have survived. *The Castle Spectre* (**1798**) is Gothic HORROR, but *One o'Clock!, or The Knight and the Wood Daemon* (**1811** chap; vt *The Wood Daemon, or The Clock Has Struck!* US) is a fantasy. Some of the stories in *Romantic Tales* (coll **1808** 4 vols; cut 1838 1 vol) are also of fantasy interest, although most are adaptations of works by other hands. "Mistrust, or Blanche and Osbright, A Feudal Romance" is a further Gothic piece, but *The Four Facardins* (**1899**) begins with a translation (from the French) of an elaborate but incomplete parodic ORIENTAL FANTASY by Count Anthony Hamilton, to which MGL added a second part longer than the first; the volume includes also an alternative (but very short) second part by one Monsieur de Levis. The only fantasy among the three novellas reprinted in the 1838 edition of *Romantic Tales*, "Amorassan, or The

Spirit of the Frozen Ocean", is a tediously moralistic Oriental tale.

Yet more bibliographical confusion surrounds MGL's ballad anthology *Tales of Wonder* (anth **1801** 2 vols). The contents of the first volume, consisting of works by MGL and several others, were later reprinted in the Morley's Universal Library volume *Tales of Terror and Wonder* (anth **1887**) along with a collection of parodies which had appeared as *Tales of Terror* (coll **1801**), falsely advertised as MGL's work. There are numerous chapbook editions of selections or single items from the MGL anthology, which was an influential work; it contains versions of GOETHE's "The Erl-King", "Leonora" and "The Wild Huntsmen" by Gottfried August Bürger (1747-1794), and the famous ballad TAM LIN, plus significant early work by Walter SCOTT.

[BS]

Other works: *The Isle of Devils: A Historical Tale* [in verse] (**1827** chap).

LEWIS, (PERCY) WYNDHAM (1882-1957) UK writer and painter, founder of Vorticism, a radical programme for modernizing the visual arts which he espoused in the journal *Blast: Review of the Great English Vortex* (**1914-5**) co-ed with Ezra Pound (1885-1972), and in *The Caliph's Design: Architects! Where is Your Vortex?* (**1919** chap). His illustrations for Naomi MITCHISON's *Beyond This Limit* (**1935** chap) are in his full-fledged mature style, which sometimes resembles a moderately tamed Cubism. *Enemy of the Stars* (1914 *Blast*; **1932**) is a play in which figures emblematic of flesh and spirit square off against each other, surreally. Although it has been read as sf, WL's most ambitious fiction, **The Human Age** – *The Childermass* (**1928**; rev 1956) and *Monstre Gai; Malign Fiesta* (coll **1955**; vt in 2 vols *Monstre Gai* 1966 and *Malign Fiesta* 1966) – begins as a POSTHUMOUS FANTASY of very great scope, though it soon evolves into an AFTERLIFE narrative which carries the original SOULS-as-protagonists into several venues. The overall structure refers but ultimately owes little to DANTE's *The Divine Comedy* (written 1307-21). The two protagonists begin their posthumous existence in a vast WASTE LAND, where they must await the decision of the Bailiff as to whether they can pass through the PORTAL to the Magnetic City, but in the end do not follow him. Later they are carried elsewhere: first into an infernal PARODY of the British Welfare State where the Bailiff is re-introduced as a criminal czar who rules this UNDERWORLD; and second into HELL, which is called Matapolis, and is ruled by a dangerously plausible fallen ANGEL. After a discordant REVEL, the sequence stops in mid-flight as the hosts of HEAVEN prepare their assault. A further section was never written.

WL had an unerring capacity to offend socially and politically – he espoused antisemitic theories in the 1930s, which he recanted barely in time in *The Jews: Are They Human?* (**1939**) – and his style seems at times too truculently rebarbative. But he remains a major 20th-century Modernist.

[JC]

LEYENDECKER, J.C. (1874-1951) US artist. ◊ ILLUSTRATION.

LEYTON, E.K. Pseudonym of Ramsey CAMPBELL.

LIBRARY An EDIFICE of respectable size should certainly contain a library, perhaps the most mysterious and magical part of the structure and often carrying a whiff of sadness. This may stem from the author's wistful desire for such a private collection – when young, George MACDONALD catalogued the library of a great house in Scotland, and the image of library-as-THRESHOLD haunts his work. Deeper sadness is associated with the fragility of accumulated knowledge. The burning of the library at Alexandria echoes through history as a shocking crime, even when reworked humorously in Terry PRATCHETT's *Small Gods* (**1992**). It is not surprising that the fiery destruction of his books unhinges Lord Sepulchrave in Mervyn PEAKE's *Titus Groan* (**1946**), and in Umberto ECO's *The Name of the Rose* (**1980**; trans **1983**) the library-LABYRINTH's blaze reeks of apocalypse.

A more modern and topical pathos arises from the awareness that BOOKS already contain far more than any single human mind can assimilate. James Branch CABELL's *The Silver Stallion* (**1926**) features a lifelong search through an infernal collection where amid endless trivia only one ironic truth appears: that time ruins everything. The crushing weight of accreted information is most felt in FAR-FUTURE libraries like those of the Museum of Man in Jack VANCE's *The Dying Earth* (**1950**), which lacks a master index, or of the Citadel in Gene WOLFE's **Book of the New Sun**, which may extend underground far beyond its city. Jorge Luis BORGES provides a definitive perspective with his *gedanken* experiment in "The Library of Babel", which includes all possible book-length permutations of letters and is thus finite (though not containable in our physical universe), completely exhaustive, and useless. David LANGFORD's "The Net of Babel" (*Interzone*, **1995**) demonstrates that computerizing such a library for instant access would merely heighten its futility. The above-cited Eco and Wolfe novels nod to Borges's profession and infirmity by including a blind librarian.

Certain fantastical libraries are mainly of interest for their fictional BOOKS, as in Cabell's *Beyond Life* (**1919**) or Neil GAIMAN's homage to this in *Sandman #22* (**1990**), whose Library of DREAM also contains "every story that has ever been dreamed". WIZARDS' libraries, like PROSPERO's, tend to be small, perilous, and strong on grimoires (◊◊ PROSPERO'S BOOKS [**1991**]); the dangers of handling these are amusingly spoofed in Pratchett's accounts of Unseen University Library on **Discworld**. Other thematic collections are the Library of Cocaigne in Cabell's *Jurgen* (**1919**), devoted wholly to erotica; the library in Bernard KING's *The Destroying Angel* (**1987**), which holds only books on LOST LANDS, notably Ultima Thule; and Death's archive of self-writing biographies in Pratchett's *Mort* (**1987**). Further genre libraries include those of Miskatonic University in Arkham, Massachusetts, famous for its locked collection of CTHULHU MYTHOS texts; of Chrestomanci Castle in *The Lives of Christopher Chant* (**1988**) by Diana Wynne JONES, which is magically ordered and compressed; and of Castle Banat in Lucius SHEPARD's *The Golden* (**1993**), a broad and mile-deep circular stair lined with book-shelves. [DRL]

See also: BABEL.

LICENZA A licenza is technically an insertion into, or an epilogue to, an OPERA or other stage work, and makes reference to a patron's birthday or to a special festive occasion. The melodrama of the preceding plot, which has seemed to be leading inexorably towards a tragic outcome, is both forgiven (everyone is sorted out) and celebrated (as appropriate to the festival); the players are shown to be, in truth, mere players. Operas which incorporate explicit licenzas include Luigi Rossi's *Orfeo* (**1647**), George Friedrich Handel's *Atalanta* (**1736**) and Gioacchino Rossini's *Il Viaggio a Reims* (**1825**). Operas whose plots are controlled by magi, like Handel's *Orlando* (**1732**) and Mozart's *The Magic Flute*

(**1791**), tend to close in scenes resembling explicit licenzas.

In fantasy, any text whose plot is controlled by a MAGUS or other figure of authority behind the scenes – GODGAME texts in particular – may have licenza scenes towards its conclusion. Shakespeare's *The Tempest* (performed *c*1611; **1623**) can be read in this light, as can (perhaps) John FOWLES's *The Magus* (**1966**). C.S. LEWIS's *Perelandra* (**1943**) much more overtly presents the dance which concludes the action as a celebration of the throned and observing GOD.   [JC]

LIFESTYLE FANTASY If an individual has adopted a lifestyle that is noticeably affected by belief in MAGIC, mysticism, divination or any of the other trappings of fantasy fiction, that individual can be said to be living an LF.

The most accomplished lifestyle fantasists of the 19th century were French, although they relied to some extent on the example of such UK exports as "Beau" Brummell (1778-1840), Algernon Swinburne and Oscar WILDE. By the end of the century, however, the fashionable OCCULTISM of Paris was being exported to London on a grand scale, its "traditions" enthusiastically taken up and carried forward by such quasi-Rosicrucian societies (◊ ROSICRUCIANISM) as the GOLDEN DAWN and the imitative organizations founded by Aleister CROWLEY and Dion FORTUNE. Like Madame BLAVATSKY and her Theosophist successors (◊ THEOSOPHY), those involved with various kinds of ritual magic drew heavily on fantasy and occult fiction in customizing their particular brands of "wisdom"; later, writers of fantasy were only too glad to reclaim what had been borrowed – with interest. Today, the UK probably hosts at least as many lifestyle fantasists as France, but far fewer than California.

The LF which has had the most conspicuous influence on modern fantasy fiction is the New Paganism, which has colonized imaginative territory carved out by the SCHOLARLY FANTASIES of Jules Michelet (1798-1874) and Margaret Murray (1863-1963); its followers have co-opted the entire history of WITCH-persecution, often enriched with a substantial slice of FAIRY mythology, and the restyled materials have been copiously fed back into literary fantasy. The prolific resurgence of REINCARNATION fantasies among lifestyle fantasists has not been matched by any dramatic increase in literary interest.

No matter what popular humour may allege, few lifestyle fantasists are found in mental hospitals; even the most devout and the most extraordinary are quite sane.   [BS]

LIGEIA vt of *The Tomb of Ligeia* (**1964**). ◊ Roger CORMAN; *The* HOUSE OF USHER.

LIGOTTI, THOMAS (1953-   ) US writer and publisher who has established a cult following for his special brand of HORROR. TL first came to attention in the SMALL-PRESS magazine *Nyctalops* with "The Chymist" (1981); that magazine's publisher, Harry O. Morris, issued TL's first collection, *Songs of a Dead Dreamer* (coll **1986**; exp 1989 UK). The book remains the best example of TL's fiction. Although his work is Gothic in mood (◊ GOTHIC FANTASY), it is surreal (◊ SURREALISM) in perception, with most events seen through the distorted perspective of its doomed narrator. TL has borrowed many images from the early German CINEMA, especially in "Dr Voke and Mr Veech" (1983 *Grimoire*), about a doctor who brings dummies to life (◊ DOLLS), which is reminiscent of *The* CABINET OF DR CALIGARI (**1919**). His work shows influences from Edgar Allan POE, H.P. LOVECRAFT, Robert AICKMAN, Henry JAMES and Sheridan LE FANU, which he has managed to incorporate into a mixture with with he can paint variegated

portraits of fear. TL's fiction is the closest to dark MAGIC REALISM of any North American writer. His concentration of images does not work at novel length, although his second book, *Grimscribe: His Life and Works* (coll **1991** UK), was presented as a novel, consolidating a series of first-person narratives. TL's other collections are *The Agonizing Resurrection of Victor Frankenstein & Other Gothic Tales* (coll **1994** chap) and *Noctuary* (coll **1994** UK). A retrospective selection, with some new stories, is «The Nightmare Factory» (coll 1996).   [MA]

Further reading: *Dagon #22/#23*, September/ December 1988; *Crypt of Cthulhu #68*, Hallowmass 1989; *WT* Winter 1991/2.

LILITH A DEMON or species of GENIE whose name is translated as "screech-owl" in the Authorised Version of the Bible (*Isaiah* 34:13); in Babylonia the name was associated with WITCHCRAFT and applied to the first wife of Adam (◊ ADAM AND EVE); her existence is recorded in the Talmud and various apocryphal documents, though not in *Genesis*. This Lilith was allegedly expelled from EDEN for refusing to accept Adam's domination, and vowed revenge upon her meeker replacement, becoming a stealer and mutilator of babies. Fantasies re-examining the CREATION MYTH – notably Rémy de GOURMONT's *Lilith* (**1892**), George MACDONALD's *Lilith* (**1895**) and John ERSKINE's *Adam and Eve* (**1927**) – sometimes restore Lilith's crucial role but differ markedly in their judgement as to the extent to which she might have been unjustly maligned by tradition. She is treated with considerable sympathy in David H. KELLER's *The Homunculus* (**1949**), where she is the sterile twin sister of PAN. Lilith is a name frequently worn by FEMMES FATALES, and its mythical significance has recently been renewed and transfigured by her adoption as a symbolic ancestor-figure in many revisionist VAMPIRE stories.   [BS]

LIMBO An AFTERLIFE region identified by DANTE as the First Circle of HELL, to which the SOULS of virtuous pagans and the unbaptized are relegated – not tormented, but excluded from the presence of GOD. Piers ANTHONY's **Incarnations of Immortality** sequence addresses the injustice (to modern eyes) of this situation, and makes Limbo-cum-PURGATORY the HQ of a computerized celestial bureaucracy. Various POSTHUMOUS-FANTASY movies depict Limbo as a metaphorical waiting-room where souls await assignation to HEAVEN or Hell – or, implausibly often, return to Earth (◊ AFTERLIFE).   [DRL]

LIMINAL BEINGS An LB exists at the THRESHOLD of two states; this gives LBs both wisdom and the ability to instruct, while also rendering them dangerous and uncanny. Centaurs are often LBs. In classical MYTHOLOGY, the centaur Chiron was the MENTOR of heroes like Hercules, while his coeval Nessus was a VILLAIN whose intelligence enabled him to enact VENGEANCE on Hercules from beyond the grave. Centaurs are also often seen as in BONDAGE to their animal natures – hence prone to drunkenness and rape. Many other LBs are in bondage, limited by their double nature to an essentially passive or catalytic role and often physically confined to a particular area, or mentally limited to particular kinds of discourse. They are not capable of change or growth – indeed, their double nature may be the product of a specific act of METAMORPHOSIS interrupted at a halfway point; they may as a result be figures of pathos.

An LB can be both dead and alive, like Gandalf in the later stages of J.R.R. TOLKIEN's *The Lord of the Rings* (**1954-5**), or both animal and human, like the centaurs, or both animal and intelligent, like many DRAGONS, or both animal and vegetable, like the Ents in *LOTR*, or both GOOD AND EVIL, or an androgyne, or existing in a perverse relationship to past and future, like T.H. WHITE's MERLIN in *The Once and Future King* (omni **1958**), or existing outside time altogether at the hinge of the year, like the Christmas Ghosts in Charles DICKENS's *A Christmas Carol* (**1843**). GODS become LBs when they dwell in the world of humanity – ODIN is an obvious example (and also often a TRICKSTER). Harlequins (◊ COMMEDIA DELL'ARTE) are often LBs.

LBs may instruct by their very existence rather than as mentors; the GREEN MAN, whether mobile or a sessile FOLIATE HEAD, instructs humanity in its rootedness in Nature and the rhythms of Nature, even if he does not engage in direct interaction of an instructive kind, as in *Sir Gawain and the Green Knight* (*c*1370) (◊ GAWAIN). The FACE OF GLORY is both divine and demonic, both full and empty; again, it stands as a sign of the voracity of intellectual and creative joy.

A distinction needs to be drawn between LBs as plot functions and full-blown LBs. The latter are comparatively rare in HEROIC FANTASIES and TEMPLATE FANTASIES. Ningauble of the Seven Eyes and Sheelba of the Eyeless Face, in Fritz LEIBER's **Fafhrd and the Gray Mouser** series, have most of the aspects and roles of LBs but exist in plots where they are merely part of the TEMPLATE. Similarly, the undead Mycroft who advises Glen COOK's eponymous detective in the **Garrett** series exists at the threshold of life and death, and is uncanny, but his role as Garrett's mentor is to solve his cases rather than change his life. DARK LORDS and their henchpersons are often both alive and dead, as are a variety of lesser VILLAINS – like Suraklin in Barbara HAMBLY's *The Silicon Mage* (**1988**) – who have liminality but are not LBs. The same can be said of most VAMPIRES and WEREWOLVES.

Liminality is often seen as a dangerous state. Angier, in Christopher PRIEST's TECHNOFANTASY *The Prestige* (**1995**), is as a result of Nikola Tesla's teleportation machine simultaneously alive and dead; he is a conjurer (◊ MAGICIANS) whose tricks are more than an appearance – he is at risk for stretching the margins of REALITY. WIZARDS who turn into animals (◊ SHAPESHIFTERS) are at risk of losing their identity and becoming permanently and irrevocably beasts. In Robin Hobb's (Megan LINDHOLM's) **The Farseer Trilogy**, mental communication with animal companions is correctly seen as dangerous because potentially blurring the distinction between animal and human; condemned for this, the character Fitz escapes *via* a NIGHT JOURNEY in which he is doubly liminal, both animal and human and both alive and dead.

Protagonists often acquire a measure of liminality in the course of their adventures but by definition cannot be an LB in their *own* story, though two or more principals may serve as LBs to each other, as with Dracula and Mina in *Bram Stoker's Dracula* (**1992**; ◊ DRACULA MOVIES). Covenant, in Stephen DONALDSON's **The Second Chronicles of Thomas Covenant the Unbeliever**, serves as portal, mentor and source of the uncanny in the adventures of his co-protagonist Linden Avery. Neil GAIMAN's **Sandman** and **Death** GRAPHIC NOVELS, by contrast, deal in large measure with the private lives of characters – Death, Dream, Desire, etc. – who are occasionally LBs in the lives of their liminal siblings, as well as in those of the human and divine characters with whom they interact.

In John CROWLEY's *Little, Big* (**1981**), Alice is instructed by

the LB Grandfather Trout – a shapeshifted victim of punitive metamorphosis – while herself acting as an LB to her husband Smoky Barnable. Paradoxically, it is Smoky's single nature, inability to change and consequent death that make the replacement of the FAERIE in which he was unable quite ever to believe by his wife and family a progress rather than merely a stage of a CYCLE. Where the earlier Faerie was uncanny and irresponsible, its replacements are chastened by his loss.   [RK]

LINDGREN, ASTRID (ANNA EMILIA) (1907-   )
Swedish writer. She wrote her first novel in the late 1930s, but it was rejected by numerous Swedish publishers and she put it aside. Her first published novel was the mundane YA *Britt-Mari lttar sitt hjrta* ["Britt-Mari Unburdens Her Heart"] (**1944**). The following year, however, the rejected first novel was published as *Pippi Lngstrump* (**1945**; widely trans as *Pippi Longstocking*), and AL became controversial in Sweden, accused of undermining parental and school authority and of fostering juvenile delinquency and anarchy. There is no doubt that Pippi was and remains an uncommon heroine. She is immensely wealthy, having access to a chestful of PIRATE's gold inherited from her supposedly deceased seafarer father; she owns her own house and lives there without adults but with a horse and a monkey; she is strong enough to lift her horse with one hand, and has no interest in going to school and even less in adhering to adult authority. Her adventures are further chronicled in *Pippi Lngstrump grombord* (**1946**) and *Pippi Lngstrump i Sderhavet* (**1948**). The tale has been filmed several times, first in Sweden as *Pippi Lngstrump* (**1949**); the best movie version is the Swedish-German *Pippi Lngstrump* (**1969**), dir Olle Hellbom and scripted by AL.

AL's later fantasy is of yet greater stature. *Nils Karlsson Pyssling* (coll **1949**) is a lyrical book of stories detailing the friendship between a lonely boy in a Stockholm apartment and the miniature boy living beneath his floorboards. Conceptually similar to Mary NORTON's **Borrowers** series, AL's book is nevertheless totally individual in tone – sympathetic, low-key, and substituting tenderness for Norton's comedy.

AL's first HIGH FANTASY was *Mio, min Mio* ["Mio, My Son"] (**1954**). Orphaned Bo Vilhelm Olsson is brought by a genie to the Far Country, where his true father is king and where he is preordained to fight the evil Squire Kato in his black fortress. The movie version, *The* LAND OF FARAWAY (*1987*), is sadly lacklustre. *Brderna Lejonhjrta* ["The Brothers Lionheart"] (**1973**) again caused controversy. As the novel begins, Karl lies dying; the house where he lives starts burning and his older brother Jonatan saves Karl but is killed jumping through a window with his brother. Two months later Karl dies and the brothers are reunited in Nangijala, an OTHERWORLD where they manage to free the inhabitants from an evil DRAGON and a DARK LORD. Again killed, the two brothers are again resurrected, this time in Nangilima, a further fantasy realm.

In her third and last high fantasy, *Ronja Rvardotter* ["Ronja, the Robber's Daughter"] (**1981**), AL finally stepped entirely into a SECONDARY WORLD; there are no longer any links with our reality. On a stormy night in a robber's keep, Ronja is born to a world of magical beings and great adventures; her mission is finally to make peace between the two robber clans sharing the same valley and eternally at feud. This is a more cheerful and adventurous story than the earlier fantasies.

AL has recently stated that she now feels too old to write, but her niche in CHILDREN'S FANTASY remains both secure and exalted. Her stories and images can never be forgotten.

[J-HH]

Other works: *Lillebror och Karlsson p taket* (**1955**), *Sunnanng* (coll **1959**), *Karlsson p taket flyger igen* (**1962**), *Karlsson p taket smyger igen* (**1968**).

Further reading: *Barndom i Smland* (**1967**), autobiography.

LINDHOLM, MEGAN Working name of US writer Margaret Astrid Lindholm Ogden (1952-   ), who began her career with the conventional **Windsingers** trilogy: *Harpy's Flight* (**1983**), *The Windsingers* (**1984**) and *The Limbreth Gate* (**1984**), assembled as *The Windsingers* (omni **1986** UK); a belated sequel was *Luck of the Wheels* (**1989**). The URBAN FANTASY *The Wizard of the Pigeons* (**1986**) was the work that brought ML wider attention: the WIZARD of the title is a magically talented Vietnam veteran living among the street people of Seattle. This success was followed by the PREHISTORIC FANTASY *The Reindeer People* (**1988**) and its sequel *Wolf's Brother* (**1988**), assembled as *A Saga of the Reindeer People* (omni **1989**). *Cloven Hooves* (**1991**) is an engaging singleton, and *Gypsy* (**1992**), with Steven Brust, is another venture into a contemporary urban setting. Adopting the pseudonym Robin Hobb, ML launched the **Farseer** trilogy of DYNASTIC FANTASIES with *Assassin's Apprentice* (**1995** UK) and *Royal Assassin* (**1996** UK). A restless and varied fantasist, ML possesses real talents which may be coming into full focus in the latest trilogy.

[DP]

Other works: *Alien Earth* (**1992**), sf; *The Silver Lady and the Fortyish Man* (chap **1994**).

LINDSAY, DAVID (1878-1945) UK writer. Forced by penury to shelve academic ambitions, DL became a successful businessman but, after serving in WWI he settled down with his wife of two years – 20 years his junior and for whom he had jilted a previous fiancée of 14 years' standing – to be a writer. His *A Voyage to Arcturus* (**1920**) is a masterpiece of allegorical fantasy (◊ ALLEGORY) whose hero is taken to Tormance, a planet of Arcturus, where he encounters many strange beings and undergoes a series of painful METAMORPHOSES while struggling to comprehend the creative force of Shaping and its crucial relationship to the symbolic figures of Crystalman and Surtur; the metaphysics thus elaborated applies a harsh metaphorical Darwinism to the processes of personal development. The novel sold badly, and DL spent the rest of his career trying to find more commercially acceptable ways of communicating his esoteric insights.

In *The Haunted Woman* (**1922**) a stairway revealed by TIMESLIPS gives occasional access to the Saxon EDIFICE which once stood on the site of a modern house, offering its owner brief intervals of liberation from the burden of repression and constraint to which centuries of civilization have subjected human consciousness. He and the young heroine can love one another readily enough in the OTHER-WORLD, but find it almost impossible to import their LOVE into the degraded and derelict form of experience that is 20th-century life. *Sphinx* (**1923**) places muted metaphysical images of the same type in a conventional domestic drama: a hapless young man becomes entangled with two women while trying to perfect a machine to record the deep, unremembered DREAMS that contain the hidden truth of human existence. *The Violet Apple* (in *The Violet Apple and The Witch* coll **1975** US; separate publication **1978** UK), attempted

something similar in a tale of engagements painfully severed by virtue of the unexpected revelations experienced after eating a fruit from the Tree of Knowledge which ADAM AND EVE did not sample in EDEN; it, too, did not sell.

Chastened by this failure, DL wrote the historical novel *The Adventures of M. de Mailly* (**1926**; vt *A Blade for Sale* US), returning to fantasy only some years later with *Devil's Tor* (**1932**), in which various mystery-story shenanigans delay the bringing together of the two halves of a powerful TALISMAN, whose union is supposed to precipitate an apocalyptic return of the GODDESS. DL here sketched out an eclectic mythological system, recasting some of the metaphysical notions detailed in *A Voyage to Arcturus*. He died without completing the fuller elaboration of his revised thesis which he intended «The Witch» to be; only a few fragments survived to be printed in the 1975 collection noted above. Although DL's works clearly reflect – and might be regarded as an attempt to justify – his dramatic transformation when he met his wife, they nevertheless constitute a remarkable series of essays in speculative metaphysics. The first two are dazzlingly brilliant, and the rest are not trivial. It is a tragedy that the development of this canon was so direly hampered by exactly the kind of incomprehension DL was trying to subvert.

[BS]

Further reading: *The Strange Genius of David Lindsay* (anth **1970**) by J.B. Pick, E.H. VISIAK and Colin WILSON; *David Lindsay* (**1982** chap) by Gary K. Wolfe; *David Lindsay's Vision* (**1991** chap) by David Power.

LINDSAY, NORMAN (ALFRED WILLIAM) (1879-1969) Australian writer and artist, best-known for his early CHILDREN'S FANTASY *The Magic Pudding: Being the Adventures of Bunyip Bluegum and his Friends Bill Barnacle and Sam Sawnoff* (**1918**), which he wrote and illustrated; verse from this was reprinted as *Puddin' Poems* (coll **1977** chap). The pudding is indeed MAGIC, being able to renew itself constantly, no matter how thoroughly it has been eaten. The protagonists of *The Flyaway Highway* (**1936**) have various slightly racy adventures in the company of PAN, who here calls himself Sylvander Dan. NL's adult work tends to propagandist excesses, where his misogyny and antisemitism make it difficult to countenance his Nietzschean exalting of the male creative spirit. Woman's role is to decorate, allure and titivate – NL's numerous drawings of nude females are sentimental but pleasingly sex-intoxicated – and the closest he came to a female protagonist is in *The Cautious Amorist* (**1934** UK), which spoofs the attempts of some ineffectual males to cope with an alarmingly nubile female. A sample of his graphic work is found in *Siren and Satyr: The Personal Philosophy of Norman Lindsay* (graph **1976**).

[JC]

LINKLATER, ERIC (ROBERT RUSSELL) (1899-1974) Scottish writer whose work is marked by an ironic intelligence and sharp wit. His early play *The Devil's in the News* (**1929**), concerns a SÉANCE whose participants are possessed by the spirits of real people (Cromwell and Napoleon) and fictional characters (many from the cast of John Gay's *The Beggar's Opera* [1728]). The fantasies (with other material) in *God Likes Them Plain* (coll **1935**), *Sealskin Trousers* (coll **1947**) and *A Sociable Plover* (coll **1957**) include mock-FAIRY-TALES, irreverent CHRISTIAN FANTASIES and comedies based in Classical MYTHOLOGY. He wrote a series of philosophical dialogues for the BBC in which historical characters from different eras debate moral and political issues of relevance to WWII – e.g., *The Cornerstones; a Conversation in Elysium*

(1941 chap), a dispute between Lenin, Lincoln and Confucius. The series continued with *The Raft and Socrates Asks Why* (coll **1942** chap), *The Great Ship and Rabelais Replies* (coll **1944** chap) and *Crisis in Heaven: An Elysian Comedy* (**1944**).

Before the outbreak of WWII, EL had written a modern version of ARISTOPHANES' *Lysistrata* as *The Impregnable Women* (**1938**). The political ALLEGORY *A Spell for Old Bones* (**1949**) is about the unfortunate conflicts generated by the clumsy GIANTS of ancient Scotland. *Husband of Delilah* (**1962**) was the second and more fantastic of his two novels based on BIBLE stories, following *Judas* (**1939**). His last marginal fantasy, *A Terrible Freedom* (**1966**), involves the invasion of the protagonist's life by figments of his DREAMS.

EL wrote two CHILDREN'S FANTASIES, *The Wind on the Moon* (**1944**) and *The Pirates in the Deep Green Sea* (**1949**); the former won the Carnegie Medal. He also wrote a non-fiction book about the Icelandic SAGAS, *The Ultimate Viking* (**1955**). [BS]

LINS, OSMAN (1924-1978) Brazilian writer who produced several mainstream works and some complex and well crafted urban fantasies, beginning with *Avalovara* (**1973**; trans Gregory Rabassa **1980** US), a mystical love story whose spacetime structure is based on geometrical figures (a "magical square", a spiral); it was to be compared to Julio CORTÁZAR's *Rayuela* (**1968**; trans as *Hopscotch* US). In the short stories of *Nove, Novena* ["Nine, Novena"] (coll **1966**) OL continued his search for a sort of "spatial time" where all events would be simultaneous and the reader could move through them at will, while in *A Rainha dos Cárceres da Grécia* ["The Queen of the Jails of Greece"] (**1976**) a man reads the manuscript of a novel that the woman he loved wrote before dying; he is gradually absorbed by the universe of her book. OL's work creates a fascinating blend between medieval art, mysticism and current-day cosmology. [BT]

LINTON, LYNN Working name of UK writer Mrs Eliza Lynn Linton (1822-1898), who also published under her maiden name, Eliza Lynn, and some of whose works also appeared under her married name. Her first novel, *Azeth: The Egyptian* (**1846**), anon, is historical romance set about 800BC, with a long fantasy sequence describing a rite of INITIATION. Much of her later work was SUPERNATURAL FICTION, including the stories and articles published as *Witch Stories* (coll **1861**); some of the contents of *The Mad Willoughby's and Other Tales* (coll **1875**) are supernatural, and *With a Silken Thread and Other Stories* (coll **1880** Germany) assembles most of her remaining work of interest, though not "The Fate of Madame Cabanel" (1880), a VAMPIRE tale of interest for its association of imputed vampirism with female sexual allure. *The True History of Joshua Davidson* (**1872**), anon, depicts CHRIST as a Communist. In *The Second Youth of Theodora Desanges* (**1900**) a woman experiences spontaneous rejuvenation. [JC]

LION, THE WITCH & THE WARDROBE, THE 1. UK tv series (1967). ATV. **Pr** Pamela Lonsdale. **Writer** Trevor Preston. **Based on** *The Lion, the Witch and the Wardrobe* (**1950**) by C.S. LEWIS. **Starring** Elizabeth Crowther, Edward McMurray, Preston, Zuleika Robson, Elizabeth Wallace, Paul Waller. 10 episodes. Colour.

This was a modest but admirable attempt to capture the essence of Lewis's novel; it was restricted by the fact that it was assumed to be merely children's fodder, and the budget was calculated accordingly – which was unfair to Preston's lovingly crafted screenplay. This version has been eclipsed

by the BBC's much more ambitious *The* CHRONICLES OF NARNIA (1988-90). [JG]

2. UK/US ANIMATED MOVIE (*1979* tvm). Children's Television Workshop/Bill Melendez Productions/Episcopal Radio-TV Foundation/Pegbar. **Pr** Steven Cuitlahuac Melendez. **Exec pr** David D. Connell. **Dir** Bill Melendez. **Screenplay** Connell, Melendez. **Voice actors** Simon Adams (Edmund), Beth Porter (Jadis, the White Witch), Susan Sokol (Susan), Stephen Thorne (Aslan), Rachel Warren (Lucy), Reg Williams (Peter). 96 mins. Colour.

Lucy emerges from the wardrobe after first discovering Narnia; her epic journey is depicted in flashback. There are various further legitimate distortions and contortions of the plot: other failings are less forgivable. The ubiquitous US accents for this quintessentially English tale are an immediate act of vandalism: the effect is akin to giving Dorothy (of *The* WIZARD OF OZ [*1939*]) a Cockney twang. The animation is of Saturday-morning tv standard. [JG]

LISLE, HOLLY (1960-   ) US writer who began publishing with the **Faia** FANTASYLAND sequence – *Fire in the Mist* (**1992**), *Bones of the Past* (**1993**) and *Mind of the Magic* (**1995**). The protagonist, a young woman with great though undeveloped TALENTS, goes through an elaborate training in MAGIC. The middle volume of the trilogy focuses not on Faia but on another woman whose talent lies in her ability to talk to the dead. With Mercedes LACKEY HL wrote a volume in the **SERRAted Edge** sequence, *When the Bough Breaks* (**1993**). HL's singletons include *Minerva Wakes* (**1994**) and *Mall, Mayhem and Magic* (**1995**) with Chris Guin, both CONTEMPORARY FANTASIES set in US suburbia, with excursions into OTHERWORLDS and (in the second) a dangerous grimoire (◊ BOOK) on the loose. The amusing «Sympathy for the Devil» (1996) is set in HELL. [JC]

LISLE, JANET TAYLOR (1947-   ) US writer, mostly of YA fantasy, beginning with *The Dancing Cats of Applesap* (**1984**), whose young protagonist must save magical dancing CATS from the loss of their home. In *The Great Dimpole Oak* (**1987**) a big OAK draws around it, from near and far, those who are sensitive to its myth-laden resonance. JTL remains best-known for *Afternoon of the Elves* (**1989**), which plays with CROSSHATCH expectations that ELVES, whom only a child can see (◊ PERCEPTION), may genuinely exist. In *The Lampfish of Twill* (**1991**) a maelstrom carries humans into a deathly world UNDER THE SEA. In *Forest* (**1993**) a WAINSCOT relationship exists between upper and lower levels of a great FOREST: squirrels above, humans below. [JC]

Other works: The **Investigators of the Unknown** sequence: *The Gold Dust Letters* (**1994**) and *Looking for Juliette* (**1994**).

LITERARY FAIRYTALES As the FAIRYTALE evolved, emphasis began to be given to it as STORY. Madame d'AULNOY and Charles PERRAULT were the first to write deliberate fairytales, intended as much for adults as for children. These stories were the first LFs and one of the geneses for written FANTASY outside the oral tradition.

LFs almost always draw on the MOTIFS, PLOT DEVICES and images of FOLKTALES in order to contextualize their story. Nearly all REVISIONIST FANTASIES are LFs. In that continuum of the WONDER TALE which runs from the oral to the literary tradition, the LF is the highest form of development. Many FABLES, too, are LFs. [MA]

See also: MÄRCHEN.

LITHUANIA It is only in recent years that Lithuanian

readers have discovered GENRE FANTASY: during the Soviet occupation, while the West was enjoying the works of J.R.R. TOLKIEN and his many imitators, Lithuanians were restricted to a diet of realist fictions, only a few of which broke out of their shackles to become allegorical, surreal or symbolist fictions. In the related field of sf, Lithuanian writers followed the same line as their Russian counterparts: only optimistic UTOPIAS and hard sf were tolerated.

Even before the Soviet occupation, Lithuanian fantasy had not been especially rich. Vydūnas (real name Vilhelmas Storosta; 1868-1953) wrote some symbolic mysteries, like *Amžinoji ugnis* ["Eternal Fire"] (**1913**), *Ragana* ["Witch"] (**1918**) and *Jūros varpai* ["The Bells of the Sea"] (**1920**). Later, during the interbellum, several writers produced novels in which fantasy was a dominant element. Justas Piliponis (1907-1947), famous as an adventure writer, often set his fictions in something close to FANTASYLANDS, although they were described as real but exotic countries. His *Kunigaikštis be praeities* ["Knight Without a Past"] (**1936**) shows the strong influence of Edgar Rice BUR-ROUGHS; *Amžinas žydas Kaune* ["The Eternal Jew in Kaunas"] (**1934**) depicts a Biblical character existing in the real world. The most interesting work of this period was *Meilės prakeiktos sielos* ["Damned Souls in Love"] (**1934**) by the Brothers Tomdykas (joint pseudonym of Alfonsas Buzčikas [1906-?] and Jonas Buzčikas [1912-?]). After the death of his fiancée a doctor studies religion, mysticism and spiritualism and eventually builds a "coffin of the feelings", through whose use people can be killed and then raised from the dead. He tries it himself and reaches a SECONDARY WORLD which proves to be HELL.

The occupation years were not entirely without fantasy interest. Possibly the closest to genre fantasy was the novelette "Kentauro herbo giminė" ["Family of Centaur Symbol"] (1978) by Saulius Tomas Kondrotas (1953-): centuries ago a duke discovers his wife making love with a servant and decides to kill her; she, crying for mercy, suddenly turns into a huge lizard, as does the duke, and his son determines he must kill these MONSTERS. The purpose of the tale was ALLEGORY.

In some of Kondrotas's other stories the incredible happens. "Rūke mano siela" ["My Soul in the Mist"] (1977) concerns a car accident caused by a boy running at the same speed as the car. "Vėjas" ["Wind"] (1977) depicts a girl from whom the wind blows in all directions. The protagonist of "Kolekcionierius" ["Collector"] (1976) collects sunsets. *Ir apsiniauks žvelgiantys pro langa* ["And Would Frown the Ones who Look Through the Window"] (**1985**) describes a secondary world and is largely dictated by a dead man.

"Šaunusis ketvertukas prasmegusioje pilyje" ["The Brave Four in the Vanished Castle"] (1988) by Romualdas Kalonaitis (1941-) is based on the LEGEND of a castle that vanished in Lake Plateliai. Four children swim to an island in the lake and discover there a castle-LABYRINTH from which they can escape only by solving mathematical and logical RIDDLES.

Jurga Ivanauskaitė (1961-) is another whose stories venture well beyond the mundane although, as in Kondrotas's works, she deploys fantasy elements as auxiliaries rather than main themes. She shows no interest in *why* the protagonist of "Apie tai, kaip . . ." ["About This, How . . .?"] (1985) is suddenly able to hear the thoughts of others, or *why* a young nude Japanese woman, lacking a SHADOW and not visible to everyone, appears in "Regėjimai geltonoje

šviesoje" ["Spectacles in Yellow Light"] (1985), *why* a character turns into a photograph in "Labai nemalonus atsitikimas" ["Very Unpleasant Accident"] (1985), *why* a girl grows WINGS in "Diena, kurios nebuvo" ["The Day Which Wasn't"] (1985), and so on. In these stories Ivanauskaitė's interest is, rather, in exploring the real problems of youth through the introduction of surreal elements. Elsewhere she deploys fantasy analogously to explore the hopelessness of the "lost" generation – "Kada ateis Godo?" ["When Will Godo Come?"] (1985) – terror of the future – "Koncertas Nr. 1" ["Concert No 1"] (1985) – or the hippy worldview – "Pakalnučiu metai" ["The Time of Lilies"] (1985).

Several writers have ventured into SCIENCE FANTASY. In "Svarbiausias atradimas" ["The Main Discovery"] (1990) by Mindaugas Peleckis (1977-) some boys fall asleep and waken on the Planet of Neon Robots. In Peleckis's work in general one encounters WITCHES, GODS, ZOMBIES, WERE-WOLVES, MAGIC, DRAGONS, a skull that can kill people telepathically . . . Kazys Paulaskas, in *Laukesos aitvaras* (**1983**) used Lithuanian FOLKTALES about brownies as the basis for a UFO fantasy.

Some largely realist stories were driven by fantasy elements. Loreta Latonaitė (1934-) wrote several. In "Paskutinysis faraono Cheopso paslapties saugotojas" ["The Mystery of the Last Guard of Pharaoh Cheops"] (1983) the narrator has been damned by the gods of Ancient EGYPT. In "Prisiminimu ežeras" ["Reminiscence Lake"] (1983) the characters briefly venture into a secondary world. In "Akmens ašaros" ["Tears of Stone"] (1990) people in the mountains discover either living stone or a creature that lives in stone. These are really fantasies of PERCEPTION: it is left moot as to whether the events really occurred. The same might be said of the allegorical *Vienaragio išdaigos* ["Unicorn's Tricks"] (**1982**) and *Triragio Pinklės* ["Threehorn's Intrigues"] (**1988**) by Anelius Markevičius (1923-). In these SATAN comes to our world and seduces people into making PACTS WITH THE DEVIL, but the Devil is really an ALLEGORY of our own evil natures.

Despite their synopses, this balancing act between realism and the fantastic can also be found in some of the works of Ričardas Gavelis (1950-). In "Žvaigždžiu Kvepėjimas" ["The Smell of Stars"] (1982) the king Belerofontas overcomes a monstrous chimera, but the chimera resurrects again and again and pursues him all his life. In *Vilniaus Pokeris* ["Vilnius Poker"] (**1989**) three dead souls are reincarnated (◊ REINCARNATION) as a dog, tree and pigeon, and observe living people.

The prospects for genre fantasy in Lithuania look good, especially since fantasy GAMES are increasingly popular. Also, since liberation, there has been a growing nostalgia for Lithuania's legendary past, which offers a broad PLAY-GROUND for the next generation of fantasists. [GB/RB]

See also: Mykalojus Konstantinas ČIURLIONIS.

LITTLE, BENTLEY (1960-) US writer who began publishing work of genre interest with "The Backroom" in *Bringing Down the Moon* (anth **1985**) ed Jani Anderson. Most of his work is fantasy/HORROR, like *The Revelation* (**1990**), which won the 1991 BRAM STOKER AWARD for Best First Novel: a small town is invaded by "hellspawn" whose ultimate defeat does little to redeem a sense of irretrievable THINNING. *The Mailman* (**1991**) is a powerful DARK FAN-TASY in which a nonhuman mailman destroys a community; it is tauter and better than Stephen KING's comparable *Needful Things* (**1991**). *The Summoning* (**1993**) is another weird fantasy: a small town is invaded by a host of VAMPIRES posing as Christian fundamentalists. His other novels follow in the same lines, sometimes without supernatural elements; they include *Death Instinct* (**1992** as by Phillip Emmons; vt *Evil Deeds* 1994 UK as BL), *Night School* (**1994** UK; vt *University* 1995 US) and *Dark Dominion* (**1995**).   [JC]

LITTLE BIG John CROWLEY's fantasy novel, *Little, Big* (**1981**), is built around a movement and PERCEPTION which is common to fantasy and uncommon anywhere else. We use the term generically to describe both a REALITY and the movement which reveals the nature of that reality to figures in the story, and to readers. In fantasy almost anything that can be entered – a body, BOOK, rabbit hole or other PORTAL, GARDEN, SHOP, LABYRINTH, UNDERGROUND cavern, EDIFICE, ocean liner, FOREST, ISLAND, POLDER or OTHERWORLD – may well be bigger inside than out. Anything from Dr Who's *Tardis* to the Palace of the Goddess (◊ GODDESS) in Michael SWANWICK's *The Iron Dragon's Daughter* (**1993** UK), an edifice which is a topological model of all space and which encloses the Universe around it, can reveal itself through a little-big transition from the outside into the opening world of the interior.

The eponymous MAGIC object in Jorge Luis BORGES's "The Aleph" (1945), which contains within it the entire Universe, is an example of the movement, one which demonstrates the plastic potential of the little-big trope to open into metaphysical realms as well as physical. Other examples of this relationship between outer and inner worlds include: the LANDSCAPE visible within the mockup of a woman's vagina in Angela CARTER's *The Infernal Desire Machines of Doctor Hoffman* (**1972**); almost every movement towards the heart of the STORY in Crowley's own *Little, Big* (**1981**), where it is stated more than once that "The further in you go, the bigger it gets"; the WIZARD's house inside the oak in Charles DE LINT's *Into the Green* (**1993**); Ryhope Wood in Robert HOLDSTOCK's sequence; the stable in C.S. LEWIS's *The Last Battle* (**1956**); Dr Dolittle's garden in Hugh LOFTING's series; the gourds in Piers Anthony's **Xanth** sequence, which each contain the (same) entire world of DREAM; and the Botanical Gardens in Gene WOLFE's *The Book of the New Sun* (**1980-3**).   [JC]

See also: GREAT AND SMALL.

LITTLE MERMAID, THE US ANIMATED MOVIE (***1989***). Disney. **Pr** Howard Ashman, John Musker. **Dir** Ron Clements, John Musker. **Screenplay** Clements, Musker. **Based loosely on** "The Little Mermaid" (1937) by Hans Christian ANDERSEN. **Voice actors** Christopher Daniel Barnes (Eric), Jodi Benson (Ariel), Pat Carroll (Ursula), Buddy Hackett (Scuttle), Jason Marin (Flounder), Kenneth Mars (Triton), Ben Wright (Grimsby), Samuel E. Wright (Sebastian). 82 mins. Colour.

Apart from its general theme, *TLM* is not an interpretation of Andersen. The youngest of the underwater King Triton's SEVEN daughters, Ariel, is far too interested in matters terrestrial, notably human beings; investigating a shipwreck she spies and falls in love with the insipid Prince Eric. The sea-WITCH Ursula sees this as an opportunity to gain VENGEANCE against Triton: she draws up a CONTRACT with Ariel whereby the MERMAID may be human for three days, remaining so if receiving Eric's true-love KISS before the end of that time; failing that, Ariel's SOUL will be forfeit to Ursula. Eric falls for Ariel, but the kiss remains unkissed. Ursula disguises herself as a woman much like Ariel but of stronger will, Vanessa, and immediately ensnares Eric, who

forgets Ariel. Luckily, his wedding with Vanessa is disrupted at the last moment by Ariel and her friends.

TLM's subtext is more sombre than expected from a DISNEY animated movie. The story can be read as of the sexual awakening of a young woman, her fascination for the world of "alien" men, and the difficult RITE OF PASSAGE she must undergo in order to enter both that world and full social and sexual adulthood. This subtext is reinforced by the figure of Triton, whose resistance to the notion of his daughter meddling with humans can be interpreted in terms of sexual jealousy; in the end, however, he must graciously surrender her to her rightful lover.   [JG]

LITTLE NEMO: ADVENTURES IN SLUMBERLAND US/Japanese ANIMATED MOVIE (***1993***). Hemdale/TMS. **Pr** Yutaka Fujioka. **Dir** Masami Hata, William Hurtz. **Screenplay** Chris Columbus, Fujioka, Jean MOEBIUS Giraud (credited *sic*), Richard Outten. **Based on** LITTLE NEMO IN SLUMBERLAND (1905-14, 1924-7) by Winsor MCCAY. **Concept for screen** Ray BRADBURY. **Story consultants** David Hilberman, Oliver Johnston, Koji Shimizu, Frank Thomas, Robert Towne. **Conceptual design** Moebius. **Visual image development** John Canemaker. **Voice actors** Rene Auberjonois (Professor Genius), Gabriel Damon (Nemo), Bernard Erhard (King Morpheus), Danny Mann (Icarus), William E. Martin (Nightmare King), Laura Mooney (Princess Camille), Mickey Rooney (Flip). 84 mins. Colour.

King Morpheus sends Professor Genius to summon Nemo to be heir to Slumberland and also a playmate for Princess Camille. Genius gives Nemo a key, telling him not to unlock the door that bears the image of the key; so Nemo, spurred on by TRICKSTER Flip, discovers and opens that door, which proves a PORTAL that allows the Land of Nightmare to invade. Morpheus is seized by the Nightmare King, and Nemo, Camille, Genius and Flip – plus Nemo's cute chipmunk pal Icarus – set off on a QUEST to the Land of Nightmare to rescue him.

A recurring device is that Nemo wakes up in what he assumes is his own bedroom – i.e., that everything has been "just" a DREAM – only to discover that dreamland REALITY intersects with ours (◊ ARABIAN NIGHTMARE). On his final "wakening" he discovers his previously remote parents to be loving and caring, so we are left with the suspicion that the STORY of his dreaming is not over.

The credits line-up for this movie is astonishing (to add to the above, Brian FROUD worked on "design development") so it is inexplicable that it was not more widely distributed. Any paranoid suspicion that some kind of DISNEY stranglehold might have been exerted is banished by the fact than Johnston and Thomas – two of Walt DISNEY's Nine Old Men, and still good friends of the company – participated. The animation itself is varied: usually brilliantly inventive, visually stunning and technically exemplary, every now and then it lapses. What really impresses is the sense that everyone working on this movie really *understands* fantasy; it merits repeated viewing, for by no means all of its virtues are on the surface.   [JG]

LITTLE NEMO IN SLUMBERLAND Remarkable early example of the US newspaper COMIC strip, recounting the DREAM adventures of a small boy, created and drawn by Winsor MCCAY. It appeared in the *New York Herald* (1905-11), then under the series title **In the Land of Wonderful Dreams** in Hearst newspapers (1911-14) and, later, the *Herald-Tribune* (1924-7).

The story begins with a messenger from King Morpheus summoning Nemo from his bed to visit the kingdom of Slumberland. Nemo mounts the spotted night horse, Somnus, and rides off into the night sky, where he encounters fantastic animals, who race him until he falls off and lands on the floor beside his bed. Each subsequent episode has a similar format: Nemo experiences incredible adventures in fantasticated locations peopled by fabulous creatures and weird and whimsical characters; in a final small frame we find him sitting up in bed or, more often, falling out of it.

He eventually reached Slumberland on 4 March 1906 and discovered why he had been summoned: to be a playmate for the king's daughter, the Princess of Slumberland. Other COMPANIONS become Flip, a green-faced dwarfish hooligan and a taciturn jungle cannibal, Impy, a refugee from another of McCay's strips.

McCay's elegant Art Nouveau linework and abundant imagination brought the strip great success, and a stage musical based on it travelled the country in 1908. McCay made a pioneering ANIMATED-MOVIE version in 1911. A modern animated version – with MOEBIUS one of the principal creators – is LITTLE NEMO: ADVENTURES IN SLUMBERLAND (*1993*). Many book versions and collected reprints have been published (*see* Winsor MCCAY *for bibliographical details*).

With its luxurious handling and tremendous sense of scale, this was one of the most imaginative and innovative comic strips ever produced, and almost a century after its first appearance its appeal remains undiminished.   [RT]

LITTLE PEOPLE ◊ FAERIE.

LITTLE RED RIDING-HOOD ◊ FAIRYTALES.

LITTLE SHOP ◊ SHOP.

LITTLE SHOP OF HORRORS, THE Two movies, the latter a musical PARODY of the former, which is itself a parody of the HORROR-MOVIE genre.

1. US movie (*1960*). Filmgroup. **Pr** Roger CORMAN. **Dir** Corman. **Screenplay** Charles B. Griffith. **Starring** Jonathan Haze (Seymour Krelboin), Jackie Joseph (Audrey), Mel Welles (Gravis Mushnick). **Voice actor** Griffith (Audrey Jr). 71 mins. B/w.

The plot of this is very much like that of the better-known **2**: the sole major differences are that the role of the dentist, here called Dr Farb and not sexually involved with Audrey, is much played down; the carnivorous plant has not come from outer space (it is, instead, a bizarre Venus Fly-Trap cross); and Seymour does not survive the escapade. There is some highly amusing parody of, in addition to horror, *film noir*; also, a very young Jack Nicholson does a pleasing Jerry Lewis impersonation in the role of a masochistic dental patient.

This is by far the funnier of the two versions. The screenplay was apparently written in a week by Griffith and Corman together (although Griffith is given sole credit) and the movie shot in two days because the main set (the shop), left over from another movie, was due to be torn down. It is probably precisely because of this haste that *TLSOH* is such a masterpiece of black comedy.   [JG]

2. US movie (*1986*). Geffen/Warner. **Pr** David Geffen. **Dir** Frank Oz. **Spfx** Bran Ferren, with Audrey II designed and created by Lyle Conway. **Screenplay** Howard Ashman. **Based on** the stage musical by Ashman (book/lyrics), Alan Menken (music), itself based on **1**. **Starring** Vincent Gardenia (Mr Mushnik), Ellen Greene (Audrey), Steve Martin (Orin Serivello), Rick Moranis (Seymour Krelborn). **Voice actor** Levi Stubbs (Audrey II). 91 mins. Colour.

Seymour, inept assistant in Mushnik's Skid Row flower shop, buys a plant that looks like a hybrid between a phallus and Kermit the Frog and comes from outer space. He names it Audrey II for the flower-arranger colleague he silently loves; he discovers it thrives only on fresh human blood. Mushnik's shop suddenly becomes hugely successful. Eventually, now vast, the plant demands Seymour bring it a corpse. Seymour determines to kill Serivello, Audrey's biker/sadistic-dentist boyfriend, but before he can do so Serivello conveniently overdoses on laughing gas; Seymour chops up the body and feeds it to Audrey II. Seymour's boss Mushnik knows all and tries blackmail, but the plant devours him whole. At last Seymour realizes he must destroy Audrey II and its offspring lest horticulturists propagate the strain worldwide. This proves difficult.

This movie has a lot of charm, generated largely by Greene and Moranis as the two UGLY-DUCKLING lovers; affectingly, their pastel-coloured DREAM life is born from the pages of *Better Homes & Gardens*. Of several RECURSIVE elements the funniest appears in the incidental music: a brief homage-PARODY of *Tubular Bells*, extensively used in *The EXORCIST* (1973). Cameo roles are by James Belushi, John Candy, Miriam Margolyes and Bill Murray.   [JG]

Further reading: *The Little Shop of Horrors Book* * (*1988*) by John McCarty and Mark Thomas McGee.

LIVELY, PENELOPE (MARGARET) (1933-   ) UK writer, born in Cairo, who won the Booker Prize for *Moon Tiger* (*1987*) and was awarded an OBE in 1989. A constant theme in her work is the impact of the past on the present; in her work for adults this theme is developed naturalistically, in terms of memory and heritage, save for the occasional peripheral TIMESLIP – as in "The Pill-Box" (in *Corruption* coll *1984*) – but in her children's books the past routinely intrudes upon the present in the form of supernatural manifestations. *Astercote* (*1970*) features a medieval village depopulated by the Black Death and the QUEST to recover a lost relic, though does not stray into outright fantasy. In *The Wild Hunt of Hagworthy* (*1971*; vt *The Wild Hunt of the Ghost Hounds* US), however, the ancient RITUAL revived by a country vicar to provide his village with "local colour" unleashes powerful supernatural forces (◊◊ WILD HUNT). *Whispering Knights* (*1971*) involves the accidental summoning of the spirit of MORGAN LE FAY.

The Driftway (*1972*) is a much quieter account of history come to life. Its delicate moralistic element was more effectively recapitulated in *The Ghost of Thomas Kempe* (*1973*), in which an Elizabethan WIZARD's influence over a young boy is misdirected because he cannot get to grips with the facts of modern life; the novel won the Carnegie Medal. *The House in Norham Gardens* (*1974*) involves an even greater cultural divide, in that its haunters are New Guinea tribesmen in search of a sacred relic plundered in Victorian times. *A Stitch in Time* (*1976*), which won a Whitbread Award, takes the line of thought a step further, involving the evolutionary past as well as the historical past, redeploying the reflective analysis of LANDSCAPE which was the subject of PL's nonfiction *The Presence of the Past* (*1976*).

Most of PL's subsequent children's books are for a younger audience and much less serious. One notable comic fantasy for older children is *The Voyage of QV66* (*1978*; vt *The Voyage of QV Sixty-Six* US), in which the animals that have inherited the Earth after a new FLOOD try to help one of their number figure out what (or who) he is. *The Revenge of Samuel Stokes* (**1981**) returns to more familiar ground, offering an eerie account of a HAUNTING. [BS]

Other works: *Boy Without a Name* (**1975**); *The Stained Glass Window* (**1976**); *Fanny's Sister* (**1976** chap), *Fanny and the Monsters* (**1978** chap) and *Fanny and the Battle of Potter's Piece* (**1980** chap); *Dragon Trouble* (**1984** chap); *Uninvited Ghosts and Other Stories* (coll **1984**); *Pack of Cards: Stories 1978-86* (coll **1986**); *Debbie and the Little Devil* (**1987**); *A House Inside Out* (**1987**); *Judy and the Martian* (**1992**); *Princess by Mistake* (**1993**); *The Cat, the Crow and the Banyan Tree* (**1994**); *Good Night Sleep Tight* (**1994**).

LIVINGSTON, MARJORIE (PROUT) (1893-?) UK writer. Her first novel of genre interest, *The Future of Mr Purdew* (**1935**), is a POSTHUMOUS FANTASY whose protagonist encounters the AFTERLIFE as a vaguely sf-like fourth dimension. *Moloch* (**1942**) describes a search for the home venue of the eponymous Phoenician SUN-god. ML is best-known for her **Karmic Destiny** sequence of ATLANTIS tales: *Island Sonata* (**1944**), *Muted Strings* (**1946**) and *Delphic Echo* (**1948**). The approaching end of the island kingdom is hinted at primarily through the complicated WRONGNESS of the marriages engaged in, against the structure of custom, by the various protagonists of the first volume; their escape to EGYPT is too late for the female protagonist, who dies as her consort finishes building the first pyramid, within which the Mysteries of Atlantis are stored. In the second volume, the protagonists are now, through REINCARNATION, members of the Egyptian royal family, and suffer savage persecution to balance the bad karma of their earlier marital extravagances. In the third volume, this time reincarnated in Ancient Greece, they suffer further tragedies. [JC]

LIVINGSTONE, IAN (1949-) UK writer and promoter of fantasy GAMES. His *Dicing with Dragons: An Introduction to Role-Playing Games* (**1983**; rev **1984** US) is a useful guide. His other books, many in collaboration with Steve JACKSON, are largely confined to titles in the **Fighting Fantasy Gamebook** sequence: *Fighting Fantasy Gamebook #1: The Warlock of Firetop Mountain* (**1982**) with Steve Jackson, *#3: The Forest of Doom* (**1983**), *#5: City of Thieves* (**1983**), *#6: Deathrap Dungeon* (**1984**), *#7: Island of the Lizard King* (**1984**), *#8: Scorpion Swamp* (**1984**) with Jackson, *#9: Caverns of the Snow Witch* (**1984**), *#13: Freeway Fighter* (**1985**), *#14: Temple of Terror* (**1986**), *#19: Demons of the Deep* (**1986**) with Jackson, *#21: Trial of Champions* (**1986**), *#22: Robot Commando* (**1986**) with Jackson, the unnumbered *Crypt of the Sorcerer* (**1987**) and *Shadowmaster* (**1992**) with Marc GASCOIGNE. Associated titles include *Titan: The Fighting Fantasy World* (**1986**) and *The Fighting Fantasy Poster Book* (graph **1990**), both with Jackson. There is one independent title: *Casket of Souls* (**1987** chap). [JC]

LLEIDA, ROSA Catalan COMICS artist. ◊ Fernando FERNANDEZ.

LLOYD, A(LAN) R(ICHARD) (1927-) UK journalist and writer of popular history whose first novel, not fantasy, was *The Eighteenth Concubine* (**1972**; as Alan Lloyd). He used his knowledge of the natural world to good effect in the **Kine Saga** ANIMAL-FANTASY trilogy for younger readers: *Kine* (**1982**; vt *Marshworld*), *Witchwood* (**1989**) and *Dragonpond* (**1990**). These recount the adventures of the eponymous weasel. [DP]

Other works: *The Last Otter* (**1984**; vt *The Boy and the Otter*), *The Farm Dog* (**1986**) and *Wingfoot* (**1993**), all non-fantastic novels about animals.

LLOYD, CHARLES ◊ Charles BIRKIN.

LLYWELYN, MORGAN (1937-) Irish writer, born in New York. Although she has written historical novels, starting with *The Wind from Hastings* (**1978** US), and romantic fiction – some as by Shannon Lewis – most of her work falls into the category of CELTIC FANTASY. The earlier books tend to be straight historical fiction, with the supernatural (usually drawn from Celtic LEGENDS) evident only as part of the commonplace, as in *Lion of Ireland* (**1979** US), which is the story of Brian Boru (?926-1014), High-King of Ireland, and its sequel, «Pride of Lions» (**1996** US). ML has retold the story for the YA market in *Brian Boru: Emperor of the Irish* (**1990**). It was with *The Horse Goddess* (**1982** US) that ML moved firmly into the mythic, bringing Celtic mythology alive. This novel takes place at the dawn of Celtic history, using as characters the mythic HEROES AND HEROINES of the Celts. The novel helps explain the origins of the druids and the Celtic mastery of horses. ML's work is building into a history, though not chronological, of the Celtic people. The first part of *The Elementals* (coll of linked stories **1993**) considers the FLOOD and the settlement of Ireland; *Bard: The Odyssey of the Irish* (**1984**) retells the life of the semi-legendary Amergin, who comes from Iberia to settle in Ireland and discovers that the earlier settlers have TALENTS, giving rise to the legend of the FAIRIES; *The Isles of the Blest* (**1989**) is the story of Connla, who follows the fairy Blaithine to FAERIE, seen as a paradise; *Finn MacCool* (**1994**) retells the story of FINN MAC COOL as straight fiction; *Red Branch* (**1989**; vt *On Raven's Wing* **1990** UK) concerns CUCHULAIN and his relationship with the MORRIGAN; *Druids* (**1991**) brings us back into the historical past and explores the conflict between Julius Caesar and the Celts in Gaul.

Most of ML's work develops an undercurrent of the supernatural which powers and drives events without necessarily becoming overt. She balances the excesses of the fantastic with the rigour of history. [MA]

Other works: *Grania: She-King of the Irish Seas* (**1986**) and *Xerxes* (**1988**), both historical novels; *Silverhand: The Arcana* (**1995**) with Michael SCOTT, first in a series where a student mage must find a series of magical objects in order to defeat evil twin rulers.

LOCUS AWARDS These were created in 1971 by Charles N. Brown, publisher and editor of *Locus Magazine: The Journal of the Science Fiction Field*. Awarded annually, they are voted for by readers of the magazine. The LA have some claim to present the fullest range of reader and professional response of any sf or fantasy AWARD. Initially they excluded either fantasy or horror as specific categories; the listing below starts in 1978, when the Fantasy Novel category began. Awards are given for work published in the previous calendar year. [JC]

Best Fantasy Novel

1978: *The Silmarillion* by J.R.R. TOLKIEN
1979: no award
1980: *Harpist in the Wind* by Patricia MCKILLIP
1981: *Lord Valentine's Castle* by Robert SILVERBERG
1982: *The Claw of the Conciliator* by Gene WOLFE
1983: *The Sword of the Lictor* by Gene WOLFE
1984: *The Mists of Avalon* by Marion Zimmer BRADLEY
1985: *Job: A Comedy of Justice* by Robert A. HEINLEIN
1986: *Trumps of Doom* by Roger ZELAZNY
1987: *Soldier of the Mist* by Gene WOLFE
1988: *Seventh Son* by Orson Scott CARD
1989: *Red Prophet* by Orson Scott CARD

1990: *Prentice Alvin* by Orson Scott CARD
1991: *Tehanu: The Last Book of Earthsea* by Ursula K. LE GUIN
1992: *Beauty* by Sherri S. TEPPER
1993: *Last Call* by Tim POWERS
1994: *The Innkeeper's Song* by Peter S. BEAGLE
1995: *Brittle Innings* by Michael BISHOP
Best Dark Fantasy/Horror Novel (variously described)
1989: *Those Who Hunt the Night* by Barbara HAMBLY
1990: *Carrion Comfort* by Dan SIMMONS
1991: *The Witching Hour* by Anne RICE
1992: *Summer of Night* by Dan SIMMONS
1993: *Children of the Night* by Dan SIMMONS
1994: *The Golden* by Lucius SHEPARD
1995: *Fires of Eden* by Dan SIMMONS

LOFTING, HUGH (JOHN) (1886-1947) UK writer, resident in the USA from 1919, where most of his books were first published. HL began writing children's stories while serving in WORLD WAR I, creating the first versions of some of the **Dr Dolittle** (in some editions rendered as **Doctor Doolittle**) stories while in the trenches. This sequence of ANIMAL FANTASIES comprises *The Story of Dr Dolittle: Being the History of his Peculiar Life and Astonishing Adventures in Foreign Parts* (**1920**), *The Voyages of Dr Dolittle* (**1922**), *Dr Dolittle's Post Office* (**1923**), *Dr Dolittle's Circus* (**1924**), *Dr Dolittle's Zoo* (**1925**), *Dr Dolittle's Caravan* (**1926**), *Dr Dolittle's Garden* (**1927**), *Dr Dolittle in the Moon* (**1928**), *Dr Dolittle's Return* (**1933**), *Dr Dolittle and the Secret Lake* (**1948**), *Dr Dolittle and the Green Canary* (coll **1950**) and *Dr Dolittle's Puddleby Adventures* (coll **1952**). *Gub Gub's Book: An Encyclopedia of Food in Twenty Volumes* (**1932** – in fact in just 1 vol) features one of the **Dr Dolittle** cast but is otherwise unconnected to the main series. The entire sequence is illustrated by HL, in a style whose simplicity is deceptive; he was in fact an illustrator of genius. His delicate spidery line, his Oriental horizon-effects, and his use of chiaroscuro all come together in images of a profound, prelapsarian openness.

The central sequence of tales, those published 1920-29, confirms this sense. Although the Doctor has some connection to the real world of Puddleby-on-the-Marsh – often, for instance, being strapped for funds – his main life points elsewhere. The garden behind his home – which seems to expand indefinitely (◊ LITTLE BIG) and which serves as an impregnable refuge against the world – is a genuine POLDER, and much of the action takes place within it. It is here that the various animals whose languages the Doctor has learned live in safety, tell their tales, and inspire his wanderlust. Dolittle's travels take him to various LAND-OF-FABLE parts of Africa, most notably in *The Voyages of Dr Dolittle* and the undervalued *Dr Dolittle and the Secret Lake*; the heart of the latter consists of the long story of Mudface the Turtle, whose life began before Noah's and who survived the FLOOD. Elements from the tales were incorporated in a musical-comedy movie adaptation, DOCTOR DOLITTLE (*1967*).

The prime of HL's active career lasted only a decade, and during that period he wrote only one other fantasy, *The Twilight of Magic* (**1930**), a slightly moralistic tale set in a land-of-fable medieval Europe; it is clear that only in the character of Dolittle was he able to refashion trauma into PASTORAL. He was superbly successful; but when the immediate impact of WWI faded HL became increasingly afflicted with a cultural pessimism that made it hard for him to work. *Dr Dolittle and the Secret Lake* is the only escapee from this despair. [JC]

Other works: *The Story of Mrs Tubbs* (**1923**) and *Tommy, Tilly and Mrs Tubbs* (coll **1936**), for younger children; *Porridge Poetry* (coll **1924**), poems.

LOFTS, NORAH (ETHEL ROBINSON) (1904-1983) UK writer, most of whose works were in the crime and historical genres, some of the latter being SUPERNATURAL FICTION with HORROR ingredients: HAUNTED DWELLINGS, family CURSES and EXORCISMS figure in tales like *Afternoon of an Autocrat* (**1956**; vt *The Deadly Gift* 1976 US), *The Devil's Own* (**1960** US as by Peter Curtis; vt *The Witches* 1966 UK as NL; vt *The Little Wax Doll* 1970 US) – whose protagonist finds a WITCH-cult in rural England – *Is There Anybody There?* (**1974**; vt *Hauntings: Is There Anybody There?* 1975 US) and the **Gad's Hall** sequence – *Gad's Hall* (**1977**) and *Haunted House* (**1978**; vt *The Haunting of Gad's Hall* 1979), assembled as *Gad's Hall and The Haunting of Gad's Hall* (omni **1979**) – which concerns the eponymous BAD PLACE. [JC]

Other works: *The Claw* (**1981**); *The Old Priory* (**1981**).

LOGSTON, ANNE (? –) US author of the **Shadow** sequence of fantasy adventures: *Shadow* (**1991**), *Shadow Hunt* (**1992**), *Shadow Dance* (**1992**) and *Greendaughter* (**1993**), a prequel to the main series. Shadow, a female ELF, goes with assorted COMPANIONS on enjoyable QUESTS typical of series set in FANTASYLAND. A subsequent series, **Shadow – The Next Generation** – *Dagger's Edge* (**1994**) and *Dagger's Point* (**1995**) – set in the same world, follows the adventures of Dagger as she prepares to become monarch of her domain and attempts to gain the portion of her SOUL which remains missing. [JC]

LOKI The TRICKSTER god among the Norse AESIR, associated with fire, malice and betrayal. Despite being ODIN's blood brother, Loki is fated to fight on the opposing side in RAGNAROK; he is a recurring ANTIHERO figure in NORDIC FANTASY. His visit with Thor to the GIANT Utgardaloki – who outwits them with ILLUSION – is adapted as an episode of L. Sprague DE CAMP's and Fletcher PRATT's *The Incomplete Enchanter* (**1941**). John JAMES's *Votan* (**1966**) retells some of Loki's exploits as RATIONALIZED FANTASY. Diana Wynne JONES's *Eight Days of Luke* (**1975**) sees him freed from his traditional imprisonment UNDERGROUND (where a SERPENT drips venom in his eyes) to become the grateful, charming, yet still dangerous COMPANION of the modern boy who accidentally releases him; the Loki in Neil GAIMAN's «The Kindly Ones» (graph coll 1996) seems truer to the original, being driven by corrosive malice to destroy his liberator rather than remain under an obligation. [DRL]

LONDON Fantasy writers in English rarely see London whole, from the outside, or as a vista to be described; the CITY serves them, above all, as a venue for works of URBAN FANTASY, with NEW YORK (and debatably NEW ORLEANS) coming in a fairly distant second. For fantasy writers, to evoke London is to conjure a set of ICONS and LEGENDS of unparalleled depth in time, all set within a frame whose complex, theatrical immensity seems inexhaustible.

There is much for fantasy writers to draw on. Lewis SPENCE, in *Legendary London* (**1937**), provides a convenient overview of the place of the city in the national MYTH OF ORIGIN of the British people, though he notes in passing that the Arthurian cycle more or less totally excludes London from its version of the MATTER of Britain. GOG AND MAGOG (he intimates) should properly be thought of as GUARDIANS protecting London from the rest of the world. But London is too large, and fades too slowly into regions whose density for the imagination lessens significantly, to have a THRESH-OLD, or to be defined as a POLDER. Its effect on the imagination of writers is centripetal: once inside its bournes, stories tend to remain there.

A fair number of fantasies have been set in pre-19th-century London. Lisa GOLDSTEIN's *Strange Devices of the Sun and Moon* (**1993**) is set in Elizabethan times, as are various tales featuring John DEE, like Peter ACKROYD's *The House of Doctor Dee* (**1993**) and John CROWLEY's **Aegypt** sequence. Virginia WOOLF's *Orlando* (**1928**) begins about the same time, as does Jeanette WINTERSON's *Sexing the Cherry* (**1989**). But most urban fantasies are set no earlier than the early half of the 19th century, the period when London was growing into the first world city.

The first great wave of stories set between 1850 and the devastating outbreak of WORLD WAR I was unsurprisingly written by contemporaries. Charles DICKENS is of course pre-eminent; but Wilkie COLLINS set work of interest in London. From around 1880, London – or Babylon-on-the-Thames – became a kind of SHARED WORLD for writers like Robert Louis STEVENSON, whose *New Arabian Nights* (coll **1882**) and *More New Arabian Nights: The Dynamiter* (coll **1885**) with Fanny Van de Grift Stevenson were of great importance for importing of ARABIAN-FANTASY elements into the urban-fantasy mix. Stevenson's *Strange Case of Dr Jekyll and Mr Hyde* (**1888**) seems no more conceivable outside its East End frame than the savageries of JACK THE RIPPER. Arthur Conan DOYLE's SHERLOCK HOLMES tales are generally set in London, or in its WATER-MARGIN suburbs. Oscar WILDE's *The Picture of Dorian Gray* (**1891**) reflects a London become overripe. Fergus HUME's *Aladdin in London* (**1892**) – like F. ANSTEY's *The Brass Bottle* (**1900**), Barry PAIN's *The One Before* (**1902**) and many more – continue the interplay between Arabian fantasy and the London venue. Joseph CONRAD's *Heart of Darkness* (**1902**) conflates the interior of Africa with the Thames estuary, where Marlowe tells the tale within the darkness of the purlieus of the world city. G.K. CHESTERTON's *The Napoleon of Notting Hill* (**1904**) and *The Man who Was Thursday: A Nightmare* (**1908**) are both set in London, and could be set nowhere else. Wyndham LEWIS's *Mrs Duke's Millions* (written *c*1908; **1977** Canada) is similarly locked into the world of the city.

STEAMPUNK stories are normally set in the 19th century – and very frequently in London. London steampunk includes novels like James P. BLAYLOCK's *Lord Kelvin's Machine* (**1992**) and William Gibson's and Bruce Sterling's *The Difference Engine* (**1990** UK) – though it would obviously be silly to exclude a novel like Tim POWERS's *The Anubis Gates* (**1983**) from the category. Many of the GASLIGHT ROMANCES so popular in the last two decades of the 20th century – e.g., Mark HARRIS's **Dark Brotherhood** sequence – are set in London.

London itself is less frequently the subject of stories. Many of Michael MOORCOCK's tales – pre-eminently the **Jerry Cornelius** sequence and *Mother London* (**1988**) – could be so described, as could all of Iain SINCLAIR's poetry and fiction, the most comprehensive examples of which are *Downriver, or The Vessels of Wrath* (**1991**) and *Radon Daughters: A Voyage, Between Art and Terror, from the Mound of Whitechapel to the Limestone Pavements of the Burren* (**1994**). But most contemporary fantasies use the city as an all-surrounding given. They include: Martin AMIS's *Other People: A Mystery Story* (**1981**); Angela CARTER's *Nights at the Circus* (**1984**); Michael DE LARRABEITI's **Borribles** sequence;

Eleanor FARJEON's *Gypsy and Ginger* (**1920**); Christopher FOWLER's *Roofworld* (**1988**); James HERBERT's **Rats** sequence; Russell HOBAN's *The Medusa Frequency* (**1987**); Robin JARVIS's **Deptford Mice** and **Deptford Histories** sequences; Tanith LEE's *Reigning Cats and Dogs* (**1995**); C.S. LEWIS's *The Great Divorce: A Dream* (**1946**); Kim NEWMAN's *Anno Dracula* (**1992**); Brian STABLEFORD's **Werewolves of London** sequence; Bram STOKER's *Dracula* (**1897**); and P.L. TRAVERS's **Mary Poppins** books. There are many more.   [JC]

LONDON, JACK Working name of US writer John Griffith London (1876-1916). His imaginative fiction made a considerable contribution to the development of US SCIENCE FICTION, some of his works being cast as VISIONARY FANTASIES. His most important was *The Star Rover* (**1915**; vt *The Jacket* 1915 UK), whose protagonist learns to dissociate his mind from his body while undergoing painful punishment and experiences various past lives (◊ REINCARNATION). *Before Adam* (**1906**) is a PREHISTORIC FANTASY. A more concrete form of ATAVISM is featured in "When the World was Young" (1910), which appeared in *The Night-Born* (coll **1913**) and is reprinted in both *Curious Fragments: Jack London's Tales of Fantasy Fiction* (coll **1975**) and *Selected Science Fiction & Fantasy Stories* (coll **1978**). The second title includes some WEIRD FICTION in which the dead exert supernatural influences upon the living, the story of most fantasy interest being "The Eternity of Forms" (1911) from *The Turtles of Tasman* (coll **1916**). *Hearts of Three* * (**1918**) is a novelization of a unproduced movie script done with Charles William Goddard (1879-1951); it involves a LOST RACE and drug-induced VISIONS.   [BS]

LONG, FRANK BELKNAP (1901-1994) US writer whose reputation is inextricably linked to that of his friend and mentor, H.P. LOVECRAFT. Lovecraft recommended his work to Farnsworth WRIGHT, who published FBL's first professional sale, "The Desert Lich", in WEIRD TALES in 1924. FBL became one of the magazine's regular names, contributing to it into the 1950s. Most of FBL's early weird fantasies are built around exotic settings and adventures typical of the period, although some feature gruesome horrors and sf elements. All have the same streak of romanticism that runs through the verse collected in his first two books, *A Man from Genoa and Other Poems* (coll **1926**) and *The Goblin Tower* (coll **1935**), most of whose contents were reproduced in *In Mayan Splendor* (coll **1977**).

FBL's "The Space-Eaters" (1928) and "The Hounds of Tindalos" (1929) are acknowledged today as the seeds of the CTHULHU MYTHOS. FBL's **Mythos** tales established the pattern of embodying conceptions of the alien in extradimensional monstrosities modelled on Lovecraft's; "The Horror from the Hills" (1931 *WT*) incorporates a lengthy paraphrase of one of Lovecraft's dreams. His most valuable contributions to Lovecraft studies are perhaps his numerous defences of Lovecraft against critics, which form the basis of his anecdotal memoir *H.P. Lovecraft: Dreamer on the Nightside* (1975).

In the prewar years, FBL placed stories in most pulp fantasy magazines, and was one of the few *WT* writers to find a niche in UNKNOWN; here he specialized in lyrical stories where elements of Classical myth worked their way incongruously into modern settings. Stories from these years comprise his best work; a representative sample was collected in his first book of fiction, *The Hounds of Tindalos* (coll **1946**; cut as *The Dark Beasts* 1963; cut material issued as *The Black Druid and Other Stories* 1975 UK).

As markets for fantasy shrank after WWII, FBL increased his output of sf, which he had been writing since the 1930s. The cream of his sf is in *John Carstairs: Space Detective* (coll of linked stories **1949**), *The Rim of the Unknown* (coll **1972**) and *Night Fears* (coll **1979**) ed Roy Torgeson. He also wrote for the COMICS in the 1940s, and served as an editor for several sf and mystery magazines in the 1950s. In the 1960s, he made the transition to the paperback-original market with *It was the Day of the Robot* (**1963**), *This Strange Tomorrow* (**1966**) and other competent sf novels. As Lyda Belknap Long, FBL penned a series of Gothic romances with supernatural overtones, beginning with *To the Dark Tower* (**1969**). His final major work of fantasy as FBL was *The Night of the Wolf* (**1972**). In the last two decades of his life, the merits of his sporadic forays into fantasy were often eclipsed by his renown as a Lovecraft protégé and veteran of the pulp era; his reminiscences in the introduction to *The Early Long* (coll **1976**) and in *Autobiographical Memoir* (**1985**) are informative documents of this era. In 1978 he received a WORLD FANTASY AWARD and in 1988 a BRAM STOKER AWARD, both for Lifetime Achievement. He has been memorialized as the protagonist of Peter Cannon's mystery novel *Pulptime* (**1984**), and recently has become a focus of interest for SMALL-PRESS editor Perry M. Grayson, whose chapbooks *Return to Tomorrow* (coll **1995**), *The Eye Above the Mantel and Other Stories* (coll **1995**) and *The Darkling Tide: Previously Uncollected Poems* (coll **1995**) are the first of a series resurrecting FBL's less familiar writings.   [SD]
Other works: *When Chaugnar Wakes* (**1978** chap), poem; *Rehearsal Night* (**1981** chap); much sf.
As Lyda Belknap Long: *Fire of the Witches* (**1971**); *The Shape of Fear* (**1971**); *The Witch Tree* (**1971**); *The House of Deadly Nightshade* (**1972**); *Legacy of Evil* (**1973**); *Crucible of Evil* (**1974**).

LONG, JOHN LUTHER (1861-1927) US writer. ◊ David BELASCO; Lord DUNSANY.

LOOK WHAT'S HAPPENED TO ROSEMARY'S BABY (vt *Rosemary's Baby 2*) US movie (**1973** tvm). ◊ ROSEMARY'S BABY (**1968**).

LORD, JEFFREY [hn] ◊ Roland J. GREEN.

LORD OF THE FLIES Two movies have been based on *The Lord of the Flies* (**1954**) by William GOLDING.
   1. *Lord of the Flies* UK movie (**1963**). British Lion/Allen-Hodgdon/Two Arts. **Pr** Lewis Allen. **Exec pr** Al Hine. **Assoc pr** Gerald Feil. **Dir** Peter Brook. **Screenplay** Brook. **Starring** James Aubrey (Ralph), Tom Chapin (Jack), Hugh Edwards (Piggy), Tom Gaman (Simon). 91 mins. B/w.
   An aircraft bearing schoolboys evacuated from London at the start of a war (perhaps WWIII) crashes on an ISLAND. The survivors, from two schools, elect Ralph as their chief; the unsuccessful candidate, Jack, appoints himself head of the Hunters. Ralph, assisted by the asthmatic, myopic Piggy – the profoundly irritating voice of sweet reason – attempts to establish a workable community, with much effort directed towards the possibility of rescue; he bases his leadership on a fetish – a conch-shell which is generally agreed to permit whoever bears it to speak to the assembly. But Jack rebels against rationality, eventually setting up his own tribe, which thrives on totemism (◊ TOTEMS), SACRIFICE and RITUAL, and is bound together by Jack's own manipulation of others' fear to create a crude RELIGION and a mythological (◊ MYTHOLOGY) enemy, the SHAPESHIFTING Beast supposed to inhabit the high rocks. Soon all but Ralph and Piggy are in Jack's tribe, and have become little more

than wild animals – parasitic upon the other two, for Piggy's spectacles are their sole means of creating fire. At last the "tribesmen" murder Piggy, and hunt down Jack with the intention of killing him also – for he is the last reminder of their conscience. Rescue arrives just in time to save his life.
   TLOTF is not easy to watch: it offers a deeply depressing view in microcosm of the cultural evolution of humankind towards the bestial. In the course of that evolution the central characters develop into symbols of societal functionaries: they become, in effect, representatives of stock characters used by HIGH FANTASY. Simon, both mystic and scientific inquirer, represents the MAGUS (his given name can be no accident); Jack is the barbarian warrior; Piggy is the stolid chronicler; and Ralph is EVERYMAN. An austere movie, shot in b/w, *TLOTF* is powerfully affecting: the killings of Simon and Piggy are two of the most shocking and poignant moments in CINEMA. The image of Piggy's myopic scowl, one spectacle lens broken by the bullying Jack, is almost a cultural ICON signifying human despair in the face of humanity's own EVIL. The movie is technically not well made, yet watching it remains an acutely unnerving experience.
   TLOTF is one of a clutch of movies concerned with the origins of FANTASY through the evocation of childhood: others include WHISTLE DOWN THE WIND (*1961*), *The* SPIRIT OF THE BEEHIVE (*1973*) and CELIA (*1988*).   [JG]
   2. *The Lord of the Flies* US movie (*1990*). Castle Rock/Nelson/Jack's Camp/Signal Hill. **Pr** Ross Milloy. **Exec pr** Lewis Allen, Peter Newman. **Dir** Harry Hook. **Screenplay** Sara Schiff. **Starring** Badgett Dale (Simon), Chris Furrh (Jack), Balthazar Getty (Ralph), Michael Greene (The Pilot), Daniel Pipoly (Piggy). 90 mins. Colour.
   The story is roughly as in **1**, but modernized and Americanized: the schoolboys are here teenaged US military cadets. The pilot is found not dead but alive, though so evil-minded because of injury that he comes to be regarded as a supernatural creature whose wrath must be assuaged through sacrifice. This is technically better made than **1**, but out with the amateurishness goes any sense of mythopoeia: what we have is an adventure tale, almost an anti-fantasy.   [JG]

LORD OF THE RINGS, THE (vt *J.R.R. Tolkien's The Lord of the Rings*) US ANIMATED MOVIE (*1978*). Fantasy Films/United Artists. **Pr** Saul Zaentz. **Dir** Ralph BAKSHI. **Screenplay** Peter S. BEAGLE, Chris Conkling. **Based on** *The Lord of the Rings* (**1954-5**) by J.R.R. TOLKIEN. **Fotonovel** *J.R.R. Tolkien's The Lord of the Rings* * (**1979**), anon. **Voice actors** Norman Bird (Bilbo), Simon Chandler (Merry), Annette Crosbie (Galadriel), Michael Graham-Cox (Boromir), Christopher Guard (Frodo), Dominic Guard (Pippin Took), John Hurt (Aragorn), Fraser Kerr (Saruman), Andre Morell (Elrond), Michael Scholes (Sam Ganji), William Squire (Gandalf), Peter Woodthorpe (Gollum). 133 mins. Colour.
   The lesson of *TLOTR* is that a book as lengthy, complex and rich as Tolkien's cannot be compressed into a single movie, even one as long as this one. The moviemakers discovered this themselves before the end: *TLOTR* finishes mid-action, with almost everything still left unresolved. Even then, only those familiar with *LOTR* – which the movie follows with reasonable fidelity but much compression – would know what was going on. (The movie also suffers from that very fidelity: much that was original to

Tolkien has since become cliché.) However, if we consider the movie not as an act of storytelling but as an artwork – as, perhaps, a set of moving, talking illustrations to Tolkien's tale – then it has much to commend it. Through its murkiness (later emulated by DISNEY in *The* BLACK CAULDRON [*1985*]), the ambience of a world in decay (◊ THINNING) is admirably conveyed, as on occasion is the notion of variant REALITIES in overlap. Bakshi succeeds admirably in communicating the pathos of Gollum, and many moments in the movie are genuinely scaring; the RING's power to corrupt is skilfully imparted. Most noteworthy is the excellence of the voice acting. Bakshi has been widely decried as the wrong animator for the job; yet it is hard to think of anyone who might have made a better attempt at the impossible.   [JG]

LORELEI ◊ MERMAIDS; SIRENS.

LORRAH, JEAN (*c*1938-   ) US writer and academic; she has written no singletons. The **Savage Empire** sequence – *Savage Empire* (**1981**), *Dragon Lord of the Savage Empire* (**1982**), *Captives of the Savage Empire* (**1984**), *Flight to the Savage Empire* (**1986**) with Winston A. Howlett, *Sorcerers of the Frozen Isles: A Tale of the Savage Empire* (**1986**), *Wulfston's Odyssey: A Tale of the Savage Empire* (**1987**) with Howlett, and *Empress Unborn: A Tale of the Savage Empire* (**1988**) – is set in a LAND-OF-FABLE late-Roman-Empire Europe, distinguished from the mundane world by a certain amount of MAGIC, characters with various TALENTS, and so forth. Action predominates.   [JC]

Other works: Various volumes in Jacqueline Lichtenberg's **Sime/Gen** sequence, those in collaboration with Lichtenberg including *First Channel* * (**1980**), *Channel's Destiny* * (**1982**) and *Zelerod's Doom* * (**1986**), plus the solo *Ambrow Keon* * (**1986**); various **Star Trek** and **Star Trek: The Next Generation** ties.

LORRAIN, JEAN (1855-1906) French writer. ◊ DECADENCE.

LOS ANGELES Nobody knows where LA starts or stops; vast cityscape views of LA – like that at the close of TRON (*1982*) – do not give perspectives on a CITY but only glimpses of the interminability of the interstices of the URBAN FANTASY. The reference to CINEMA is of course significant: from the 1920s, Hollywood (which lies in the megalopolis) has been the movie centre of the English-speaking world. LA itself, ever since 1920s silent movies, has served as a backdrop and agent on innumerable occasions; *film noir* versions of the darker side of the American Dream are urban fantasies in everything but the presence of the literally fantastic. The bleak underlying realities against which WHO FRAMED ROGER RABBIT (*1988*) plays – LA's streetcars *were* in actuality purchased by giant corporations and shut down – are more vividly brought to mind in the cinema by audiences who remember *Chinatown* (*1974*). The story of LA is – as BARTON FINK (*1990*) identifies – inseparable from the movie of LA.

Fantasies in which the mundane world and the movie world intersect can, therefore, be found without difficulty; examples are Macdonald HARRIS's *Screenplay* (**1982**), *Suspects* (**1985**) by David Thomson (1941-   ) and Theodore ROSZAK's *Flicker* (**1991**); authors with stories in *Hollywood Unreel* (anth **1982**) ed Martin H. GREENBERG include Robert BLOCH, Ray BRADBURY, Thomas M. DISCH and Jack FINNEY.

Fantasies in which Hollywood reality does not explicitly intervene include James P. BLAYLOCK's *The Digging Leviathan* (**1984**), Francesca BLOCK's *Witch Baby* (**1991**),

much of the work of Scott BRADFIELD and Jonathan CARROLL, John DECHANCIE's *MagicNet* (**1993**), Steve ERICKSON's *Rubicon Beach* (**1986**) and «Amnesiascope» (1996), Ellen GUON's *Bedlam Boyz* (**1993**), Garry KILWORTH's *Angel* (**1993**) and *Archangel* (**1994**), Mercedes LACKEY's **Bedlam's Bard** sequence with Ellen Guon, *They Thirst* (**1991**) by Robert McCammon (1952-   ) and *Golden Days* (**1986**) by Carolyn See (1934-   ).   [JC]

LOST BOYS, THE US movie (*1987*). Warner. **Pr** Harvey Bernhard. **Exec pr** Richard Donner. **Dir** Joel Schumacher. **Spfx** Greg Cannom, Matt Sweeney. **Screenplay** Jeffrey Boam, Janice Fischer, James Jeremias. **Novelization** *The Lost Boys* * (**1987**) by Craig Shaw GARDNER. **Starring** Corey Feldman (Edgar), Jami Gertz (Star), Edward Herrmann (Max), Barnard Hughes (Grandpa), Brooke McCarter (Paul), Jamison Newlander (Allan), Jason Patric (Michael), Kiefer Sutherland (David), Dianne Wiest (Lucy), Alex Winter (Marco), Billy Wirth (Dwayne). 98 mins. Colour.

Divorcee Lucy comes with elder son Michael and younger son Sam to live with her father (Grandpa) in waterside fun city Santa Carla, the "Murder Capital of the World". The high murder rate is in fact because this BAD PLACE is plagued by a gang of four stylish, punk-dressing motorcycling VAMPIRES (Dwayne, Marco, Paul and leader David). The two lads and Grandpa, plus the Frog brothers Edgar and Allan (who proclaim themselves professional vampire-killers and are like GHOSTBUSTERS in miniature) and the girl Michael falls for, Star, eventually succeed in destroying the vampires and their seemingly respectable master, Max.

Reportedly born from an aborted project to create a vampiristic version of J.M. BARRIE's *Peter Pan* (**1911**), the Spielbergish *TLB* is a triumph of style over content; it is largely responsible for the modern subgenre featuring streetwise VAMPIRES. The hip Lost Boys are ancestors of Sonja Blue, in Nancy A. COLLINS's *Sunglasses After Dark* (**1989**), and descendants of Lestat in Anne RICE's **Vampire Chronicles**. Yet *TLB* has a reactionary, paranoid subtext: punkish youths are a frighteningly different species, sucking blood from established society.   [JG]

LOST HORIZON Two movies have been based on *Lost Horizon* (**1933**) by James HILTON.

1. US movie (*1937*). Columbia. **Pr** Frank CAPRA. **Dir** Capra. **Vfx** Ganahl Carson, E. Roy Davidson. **Screenplay** Robert Riskin. **Starring** Ronald Colman (Robert Conway), Edward Everett Horton (Lovett), John Howard (George Conway), Sam Jaffe (Perrault), Isabel Jewell (Gloria), Margo (Maria), Thomas Mitchell (Barnard), H.B. Warner (Chang), Jane Wyatt (Sondra). 132 mins (released cut to 118 mins). B/w.

1935, China. Diplomat Robert Conway, plus brother George, palaeontologist Lovett, dying floozie Gloria and fraudster Barnard flee a massacre. But their pilot proves an impostor, and they are taken not to "civilization" but deep into the Himalayas to the remote Valley of the Blue Moon, dominated by the lamasery of SHANGRI-LA. Courteous elderly lama Chang tells Conway the lamasery was founded some two centuries earlier by a lost Belgian priest, Father Perrault, who stumbled on the valley and its occupants and discovered the place held the secret to perfect health as well as ample gold for trade. Called to audience with the High Lama, Conway realizes this is Perrault, still alive; Perrault explains his vision of Shangri-La as an oasis of culture designed to survive the imminent destruction, by greed and materialism, of the rest of humanity, and so to carry civilization forward until the end of the dark age. As he expires, Perrault passes the baton of high-lamahood to Conway, in effect offering the lovely valley-dweller Sondra as bait.

By now the other members of the party have changed for the better. But George, abetted by a rebellious resident, the girlish Russian Maria, wants to escape, even though she knows that leaving the valley brings accelerated ageing and death; they persuade Conway that all is a cruel sham, and the trio depart in the company of bribed porters. Only Conway reaches civilization; having told rescuing diplomat Gainsford the truth he flees back to Shangri-La.

LH is faithful to the spirit of Hilton's novel. However, much is lost in the translation into the new medium. In particular, aside from most of the FRAME STORY, omitted is the novel's depiction of the chillness of mind inculcated by near-IMMORTALITY – as epitomized by Hilton's rather cold observation of the Chinese (not Russian) apparent girl with whom Conway (not George) becomes entangled: that she has fallen in love with him because it is the courteous thing to do, a service that she has performed before with other immigrants. Much moral dichotomy is likewise glossed over: there is no trace of the novel's suggestion that Shangri-La is unknown to the outside world because would-be emigrants are quietly murdered. And, of course, almost all of the book's "orientalism" is lost through casting. Such objections should not be taken as detractions from *LH*'s merits in its own terms, however: this is a movie of breathtaking ambition.

LH, previewed at 132 mins, was almost immediately cut to 118 mins. The truncated version, a box-office blockbuster, was further shortened during WWII because of its pacifism. The original nitrate negative had by the mid-1960s, when interest in *LH* resurfaced, become unscreenable and potentially lethal; no complete prints survived. From 1973 to 1986 a team under Robert Gitt located prints worldwide and pieced together an almost complete version – full 132min soundtrack; 125 mins of pictures. It is this, the lost 7 mins infilled using stills, that currently stands as the best version.

Return to Shangri-La * (**1987**) by Leslie Halliwell (1929-1989) is a sequel to Capra's movie, not to the original novel.

                                                  [JG]

2. *Lost Horizon* US movie (**1972**). Columbia. **Pr** Ross Hunter. **Dir** Charles Jarrott. **Spfx** Butler-Glounerjine. **Screenplay** Larry Kramer. **Starring** Charles Boyer (High Lama/Perrault), Peter Finch (Richard Conway), John Gielgud (Chang), Olivia Hussey (Maria), Sally Kellerman (Sally Hughes), George Kennedy (Sam Cornelius), James Shigeta (Brother To-Lenn), Liv Ullmann (Catherine), Bobby Van (Harry Lovett), Michael York (George Conway). 143 mins. Colour.

To remake one of the classics of popular cinema as a Burt Bacharach/Hal David musical might seem a fool's errand, and it was. Although the particulars of the stranded characters have changed, the remake is reasonably faithful – albeit modernized, with added spectacle plus overt inspirational infusions from the movies *Airport* (**1970**) and, incredibly, *The Sound of Music* (**1965**). The all-star cast do their best in the bits between the embarrassing song-breaks, but the cause is lost.

                                                     [JG]

LOST LANDS AND CONTINENTS The idea that certain civilizations and continents have been obliterated by disaster, and their cultural heritage lost, is widespread in

MYTH. The fact that the world map has indeed been transformed by continental drift, and that some human settlements have been wiped out by tectonic upheavals, lends plausibility to such myths as that of ATLANTIS. Those lost lands which people actually search for – in both fiction and reality – tend to be those whose legends are linked to fabulous wealth, like the Biblical Ophir and the ELDORADO of the *conquistadores*. The former provides a basis for H. Rider HAGGARD's *King Solomon's Mines* (**1885**) and for *By Airship to Ophir* (**1910**) by Frank Aubrey (1840-1927), as Fenton Ash, who also wrote *The Devil Tree of El Dorado* (**1896**). Eldorado (or El Dorado) is usually recognized as a tempting illusion, as in Voltaire's *Candide* (**1759**); like the fabled kingdom of PRESTER JOHN, it might be classified by pedants as a land which could not be "found" because never actually "lost". What characters in fantasy usually find in lost lands, if anything, is not gold *per se* but the peace and innocence of the GOLDEN AGE which our ancestors supposedly enjoyed before corrupting themselves with civilization. James HILTON's *Lost Horizon* (**1933**) is a particularly neat encapsulation of this kind of hyperinflated nostalgia.

The lost continent of LEMURIA was invented by zoologists attempting to understand similarities between the ecosystems of Madagascar and India. The notion was taken up by the occultist Helena BLAVATSKY and integrated into an elaborate evolutionary history of the seven "root races" of humankind, along with the sub-Arctic realm of HYPERBOREA, popularized by Pliny the Elder. *The Last Lemurian* (**1898**) by G. Firth Scott is one of the earliest fantasy novels to pick up the notion. A full account of the Theosophical Lemuria can be found in *The Lost Lemuria* (**1904** chap), a SCHOLARLY FANTASY by W. Scott-Elliott. Some later scholarly fantasists moved Lemuria from the Indian Ocean to the Pacific, and a new variant of it was popularized in *The Lost Continent of Mu* (**1926**) by Colonel James Churchward; Mu's fictional spinoffs include *Mukara* (**1930**) by Muriel Bruce, *The Monster of Mu* (**1932**) by Owen Rutter and *Exiles of Time* (**1940**; **1949**) by Nelson S. BOND. Lemuria's mythology was further elaborated in "I Remember Lemuria" (**1945**) by Richard S. Shaver. Other lost continents of scholarly fantasy include John Newbrough's Pan and Lewis SPENCE's Antillia, but these have little relevance to fantasy.

Lost lands are useful repositories of cultures ($\lozenge$ LOST RACES) and species long extinct in the known world. Their primary function in fantasy is to facilitate such re-creations; many involve DINOSAURS. Several writers of the WEIRD TALES school of HEROIC FANTASY made extensive use of lost civilizations as magic-laden settings. A hypothetical geography of the "Hyborian age" emerged by degrees from the work of Robert E. HOWARD, and Clark Ashton SMITH made productive use of Hyperborea. H.P. LOVECRAFT utilized similar devices *en passant*, as did A. MERRITT in the general-fiction pulps. Later imitators included L. Sprague DE CAMP, in the series which includes *The Tritonian Ring* (**1951**; **1968**), and Lin CARTER, as in the series begun with *The Wizard of Lemuria* (**1965**).

Many mythical lands can easily be drawn into the borderlands of geography, as AVALON is in "Avillion" (1853) by Mrs CRAIK, St Brendan's Isle in *The Water Babies* (**1863**) by Charles KINGSLEY, and Cibola (the "seven cities" from which the Aztecs allegedly sprung) in *The City of Frozen Fire* (**1950**) by Vaughan Wilkins (1890-1959). ISLANDS are, of course, particularly easy thus to accommodate, which is why they are scattered with reckless abandon in such SATIRES as LUCIAN's *True History* (*c*150) – which includes the legendary Isles of the Blessed – Jonathan SWIFT's *Gulliver's Travels* (**1726**) and *The Isles of Wisdom* (**1924**) by Alexandr Moszkowski (1851-1934). Other notable accounts of the "rediscovery" of imaginary islands include *The Land that Time Forgot* (**1916**) by Edgar Rice BURROUGHS and *The Island of Captain Sparrow* (**1928**) by S. Fowler Wright (1874-1965). The motif has gradually fallen into disuse by virtue of increasing geographical knowledge; these days lost lands have to be very well hidden indeed or displaced beyond some kind of magical or dimensional boundary. Such displacement – e.g., in Vaughan Wilkins's second **Cibola** story, *Valley Beyond Time* (**1955**) – so transforms their significance that they are better thought of as SECONDARY WORLDS or OTHERWORLDS. [BS]

LOST RACES Lost, forgotten or deliberately hidden civilizations occupying undersea or UNDERGROUND realms or hidden valleys, or some other forbidden enclave on or beneath our Earth. The LR tale was established as a major popular form by H. Rider HAGGARD in *King Solomon's Mines* (**1885**), *She* (**1886** US) and *Allan Quatermain* (**1887**), but of course there were many earlier occurrences of this motif-cluster, which descends from TRAVELLERS' TALES and FANTASTIC VOYAGES of the 18th century and before. There is a sense in which many early fantasy (and almost all Scientific Romance) was about unknown lands or undiscovered societies. Nevertheless, it only becomes meaningful to talk of the LR story as a distinct subgenre in the period after the globe was fully mapped and hence geographically "closed": the period of its emergence in this sense was the last third of the 19th century.

From the 1880s to the 1930s an enormous number of LR novels were published in book form or serialized in pulp MAGAZINES, young people's story-papers, etc. The "lost" MOTIF – extending to lost temples, lost arks, lost grails – still retains some vigour in the late 20th century, as demonstrated by the success of Steven SPIELBERG's INDIANA JONES sequence. Other types of fiction overlapping with LR tales, and sometimes confused with them, include RURITANIAN Romances (set in imaginary principalities, etc.), parallel-worlds stories (◊ ALTERNATE WORLDS) and latter-day ISLAND utopias. Another related form is the PLANETARY ROMANCE – essentially the LR story transplanted to an alien planet.

Although many examples of the form – including some of the earliest, like Jules Verne's *Journey to the Centre of the Earth* (**1864**) and Edward BULWER-LYTTON's *The Coming Race* (**1871**) – may be claimed as sf, the vast bulk belong more properly to fantasy. Not only do many examples, most notably Haggard's *She*, feature supernatural motifs (the Flame of Life, REINCARNATION), but it has been argued (by, e.g., Darko Suvin) that the essence of the form resides exactly in the depiction of primitive and hierarchical societies which have little or no awareness of science and the possibilities of technological progress. Such arguments may somewhat over-strenuously insist that sf is intrinsically contemptuous of nostalgia, hierarchy or characters out of the worlds of adventure fiction; but it is does properly emphasize the inward- and backward-looking nature of the LR.

In the wake of Haggard, LR novels of the more fantastic kind include direct pastiches – e.g., *He* (**1887**) by Andrew LANG and Walter Herries POLLOCK, *He* (**1887**) (the coincidence of title and date can confuse), *"It"* (**1887**) and *King*

Solomon's Treasures (**1887**), all three by John De Morgan, and *King Solomon's Wives* (**1887**) by Hyder Ragged (Henry Chartres Biron) – as well as more original variations on the theme: *The White Man's Foot* (**1888**) by Grant ALLEN, *The Cavern of Fire* (**1888**) by Francis W. Doughty, *Beneath Your Very Boots* (**1889**) by C.J. Cutcliffe Hyne, *The Aztec Treasure House: A Romance of Contemporaneous Antiquity* (**1890**) by Thomas A. Janvier, *The Goddess of Atvatabar* (**1892**) by William R. Bradshaw, *The Lost Valley of the Toltecs* (**1893**) by Charles S. Seeley, *The Wonderful City* (**1894**) by J.S. Fletcher, *The Land of the Changing Sun* (**1894**) by Will N. Harben, *Devil-Tree of El Dorado: A Romance of British Guiana* (**1896**) and *A Queen of Atlantis: A Romance of the Caribbean* (**1898**) by Frank Aubrey, *The Great White Queen: A Tale of Treasure and Treason* (**1896**) and *The Eye of Istar: A Romance of the Land of No Return* (**1897**) by William Le Queux, *The City of Gold* (**1896**) by Edward Markwick, *The White Princess of the Hidden City* (**1898**) by D. Lawson Johnstone, *In Oudemon: Reminiscences of an Unknown People* (**1900**) by Henry S. Drayton, *Thyra: A Romance of the Polar Pit* (**1901**) by Robert Ames Bennet, *The Great White Way* (**1901**) by Albert Bigelow Paine, *The Sunless City* (**1905**) by Joyce E. Preston Muddock, *By the Gods Beloved* (**1905**; vt *The Gates of Kamt*) by Baroness Orczy, *The Smoky God, or A Voyage to the Inner World* (**1908**) by Willis George Emerson, *The White Waterfall* (**1912**) by James Francis Dwye and *The Return of Tarzan* (**1915**) by Edgar Rice BURROUGHS. With the last-named, the second volume in Burroughs's 24-book Tarzan series, we arrive at the work of Haggard's most popular and influential successor. Burroughs was to write dozens of lost-race tales, both within his best-known series (*Tarzan and the Jewels of Opar* [**1918**], *Tarzan and the Ant Men*, [**1924**], etc.), and outside it (*At the Earth's Core* [**1914**; **1922**], *The Land that Time Forgot* [**1918**; **1924**], etc.).

French writers, as well as English-language ones, worked in the subgenre: Gaston LEROUX's *The Bride of the Sun* (**1913**); Pierre Benoit's oft-filmed *L'Atlantide* (**1919**) and J.-H. Rosny's *L'Etonnant voyage de Hareton Ironcastle* (**1919**; adapted rather than trans Philip José FARMER as *Ironcastle* **1976**) are examples. But it was mainly in the US pulp magazines that the tradition thrived in the years during and after WWI. Examples include *Under the Andes* (1914 *All-Story*; **1984**) by Rex Stout, *The Seal of John Solomon* (1915; **1924**) by Allan Hawkwood (H. BEDFORD-JONES), *Polaris of the Snows* (1915 *All-Story*; **1965**) by Charles B. Stilson, *The Bowl of Baal* (1916-17; **1975**) by Robert Ames Bennet, *The Golden City* (**1916**) by A. Hyatt Verrill, *The Citadel of Fear* (1918; **1970**) by Francis STEVENS, *The Moon Pool* (**1919**) by A. MERRITT, *Marching Sands* (**1920**) by Harold LAMB, *The Temple of the Ten* (1921; **1973**) by H. Bedford-Jones and W.C. Robertson, *The Garden of Eden* (**1922**) by Max Brand, *The Pathless Trail* (**1922**) by Arthur O. Friel and *Om: The Secret of Ahbor Valley* (**1924**) by Talbot MUNDY. UK pulps of the interbellum reprinted many of these and added some notable examples by UK writers, like E. Charles Vivian's sequence *City of Wonder* (**1922**), *Fields of Sleep* (**1923**), *People of the Darkness* (**1924**), *The Lady of the Terraces* (**1925**) and *A King There Was* (**1926**). Other works by UK writers include *Wine of Death* (**1925**) by Anthony ARMSTRONG, *The Glory of Egypt: A Romance* (**1926**) by Louis Moresby (L. Adams Beck), *The City in the Sea* (**1926**) by H. De Vere Stacpoole, *The Moon Gods* (**1930**) by Edgar JEPSON, *Beyond the Rim* (**1932**) by S. Fowler Wright (**1932**) and, most famously, *Lost Horizon* (**1933**) by James HILTON. Rider Haggard himself was still pursuing the theme with works as late as *The Treasure of the Lake* (**1926**).

Despite their increasing implausibility, stories of the familiar type continued to appear throughout the 1930s and 1940s: *Golden Blood* (1933 *WT*; rev **1964**; rev 1978) by Jack Williamson, *Lost City of Light* (**1934**) by F.A.M. Webster, *The Fabulous Valley* (**1934**) by Dennis WHEATLEY, *The Secret People* (**1935**) by John Beynon (John Wyndham), *Hidden World* (1935; **1957**) by Stanton A. Coblentz, *The Vampire of N'Gobi* (**1935**) by Ridgwell Cullum, *Queen of the Andes* (**1935**) by Barbara Gilson (Charles Gilson), *In the Sealed Cave: Being a Modern Commentary on a Strange Discovery Made by Captain Lemuel Gulliver* (**1935**) by Louis Herrman, *Dian of the Lost Land* (**1935**) by Edison Marshall, *Land Under England* (**1935**) by Joseph O'Neill, *Hawk of the Wilderness* (**1936**) by William L. Chester, *Inland Deep* (**1936**) by Richard Tooker, *The Smoking Land* (1937; **1980**) by Max Brand, *City of Cobras* (**1938**) by James Francis Dwyer, *Ivory Valley: An Adventure of Captain Kettle* (**1938**) by C.J. Cutcliffe Hyne, *Biggles Flies South* (**1938**) by W.E. Johns, *Jongor of Lost Land* (1940; **1970**) by Robert Moore Williams, *A Yank at Valhalla* (1941; **1950**) by Edmond Hamilton, *The Man who Missed the War* (**1945**) by Dennis Wheatley, *Valley of the Flame* (1946; **1964**) by Henry KUTTNER, *The Valley of Doom* (**1947**) by Mary Richmond, *Stones of Enchantment* (**1948**) by Wyndham Martyn, *The City of Frozen Fire* (**1950**) by Vaughan Wilkins, and many more.

By the 1950s, lost worlds seemed to have been mined to death, although they still made occasional appearances, mainly in juvenile fiction. Examples are *The Perilous Descent into a Strange Lost World* (**1952**) by Bruce Carter, *Forbidden Kingdom* (**1955**) by Elleston Trevor and *The Island at the Top of the World* (**1961**) by Ian Cameron. However, the 1960s and 1970s saw a small revival, principally in the form of nostalgic pastiche, following the posthumous boom in Edgar Rice Burroughs's popularity in paperback reprints. Latter-day examples inspired by Burroughs include *Tarzan and the Valley of Gold* * (**1966**), a movie novelization by Fritz LEIBER, *The Sunbird* (**1972**), a tribute to Haggard rather than Burroughs by Wilbur Smith, and *Journey to the Underground World* (**1979**) by Lin CARTER and its sequels *Zanthodon* (**1980**), *Hurok of the Stone Age* (**1981**), *Darya of the Bronze Age* (**1981**) and *Eric of Zanthodon* (**1982**). Other recent examples include *Congo* (**1980**) by Michael Crichton, *The People Beyond the Wall* (**1980**) by Stephen Tall, *The Undying Land* (**1985**) by William Gilmour, *The Haunted Mesa* (**1987**) by Louis L'Amour, *Kala* (**1990**) by Nicholas LUARD, *The Last Camel Died at Noon: An "Amelia Peabody" Mystery* (**1991**) by Elizabeth Peters and various series pastiches such as *Indiana Jones and the Seven Veils* (**1991**) by Rob MacGregor and *Python Isle: A "Doc Savage" Adventure* (**1991**) by Kenneth Robeson (in this case Will Murray, using an unused 1930s outline by Lester Dent). Still more recently, Edward MYERS has produced an ambitious lost-race trilogy with a Latin American setting, *The Mountain Made of Light* (**1992**), *Fire and Ice* (**1992**) and *The Summit* (**1994**). The remote Himalayas still offer writers the possibility of a Yeti LR, as in *Brother Esau* (**1982**) by John Gribbin and Douglas Orgill and in Luard's *Himalaya* (**1992**). [DP]

LOST WORLDS ◊ LOST LANDS AND CONTINENTS; LOST RACES.

LOTI, PIERRE Working name of French writer Louis Marie Julien Viaud (1850-1923). ◊ ANTHROPOLOGY.

LOTR Unattractive but convenient contraction for *The Lord of the Rings* (**1954-5**) by J.R.R. TOLKIEN, a text of seminal importance that is very frequently referred to in any discussion of modern FANTASY. [JC]

LOUŸS, PIERRE (1870-1925) French writer who shot to fame with *Aphrodite* (**1895**; trans Stanley Reynolds **1900** France), a lush erotic fantasy set in ancient Alexandria. It was followed by *Les Aventures du roi Pausole* (**1901**; trans Mitchell S. Buck as *The Adventures of King Pausole* 1926 US), a Rabelaisian comedy in which the king of Tryphême sets out with an irreverent servant in pursuit of his eloping daughter; a luckless steward tries in vain to keep his harem in order during his absence. Six prose-poems with motifs drawn from Greek MYTHOLOGY, including *Lêda, ou la louange des bienheures ténèbres* (**1893** chap), *Ariane, ou le chemin de la paix eternelle* (**1894** chap) and *Byblis changé en fontaine* (**1898** chap), were collected as *Le crepuscule des nymphs* (coll 1925; trans Phyllis Duveen as *The Twilight of the Nymphs* 1928 UK). Further short stories, issued as *Sanguines* (coll **1903**; trans James Cleugh 1932 US), included several ironic fantasies. *Notes sur "Aphrodite"* (coll **1928**), one of several posthumous pornographic works, provided a complete menu of the services available in the temple garden where Chrysis plied her trade. *Les chansons de Bilitis* (coll **1895**; trans Horace Machester Brown as *The Songs of Bilitis* 1904 US) comprises erotic poems supposedly written by a contemporary of Sappho. [BS]

Other works: *The Collected Works of Pierre Louÿs* (coll **1932** US).

LOVE Love is frequently represented as if a quasi-supernatural force, so irresistible that in the morally ordered world of fiction it readily becomes a progenitor of MIRACLES. Gods often featured in modern fantasy to represent the awkward waywardness of erotic feelings include APHRODITE and CUPID, while love POTIONS are a chief stock-in-trade of WITCHES. The most luridly heartfelt of all fantasy novels are sentimental stories in which love defies death and transcends time (◊ TIMESLIPS; REINCARNATION); examples include *Pharaoh's Daughter* (**1889**) by Edgar Lee and *Love Eternal* (**1918**) by H. Rider HAGGARD, not to mention the latter's *She* (**1886**).

Notable accounts of fantasticated love include *Spirite* (**1865**) by Théophile GAUTIER, *Peter Ibbotson* (**1891**) by George du Maurier (1834-1896), *Nephelé* (**1896**) by F.W. Bourdillon, *Aphrodite* (**1900**) by Pierre LOUŸS, *Going Home* (**1921**) by Barry PAIN, *The King of Elfland's Daughter* (**1924**) by Lord DUNSANY, *Portrait of Jennie* (**1940**) (◊◊ PORTRAIT OF JENNIE [*1948*]), *So Love Returns* (**1958**) by Robert NATHAN, *Bid Time Return* (**1975**) by Richard MATHESON, *The Dream Years* (**1985**) by Lisa GOLDSTEIN and *Replay* (**1987**) by Ken GRIMWOOD. A few of these have been adapted for the cinema, which has also treated the theme in such movies as *The* GHOST AND MRS MUIR (*1947*), ALL OF ME (*1984*), ALWAYS (*1989*), GHOST (*1990*), TRULY MADLY DEEPLY (*1991*), GROUNDHOG DAY (*1993*) and the various versions of SVENGALI – countless others could be cited.

In HIGH FANTASY the importance of love is generally as a plot driver. This has been the case ever since the earliest FOLKTALES and FAIRYTALES – i.e., since before the genre of FANTASY itself emerged. The love between LANCELOT and GUINEVERE, and of ARTHUR for his Queen, is crucial to the version we now accept of the Arthurian cycle; the situation is parallelled in the legend of Tristan and Iseult. Of the tales recorded by the GRIMM BROTHERS, "Rumpelstiltskin" is one of many whose plot is underpinned by the power of love, and Rapunzel let down her long gold hair for reasons of love. Love of a totally non-erotic form is what drives Gerda on her QUEST to rescue Kai in Hans ANDERSEN's "The Snow Queen", and a similar chaste dynamic moves the tale of Irene and Curdie in George MACDONALD's *The Princess and the Goblin* (**1872**). It is impossible to estimate how many GENRE FANTASIES centre on a QUEST by the HERO to regain his true love, either to rescue her from some DARK LORD (and thereby, as an aside, save the world) or to demonstrate his prowess and thus stop her marrying another. Love of a non-sexual kind can bind together DUOS, or sexual love can eventually burgeon between them – indeed, in modern fantasy, wherever love is a primary motif in a tale, SEX is generally not far behind (though it may, as it were, not catch up until the page after the end of the novel), but usually it is love rather than lust that drives the tale. [BS/JG]

LOVE AT FIRST BITE US movie (*1979*). ◊ DRACULA MOVIES.

LOVE BUG, THE US movie (*1969*). ◊ HERBIE MOVIES.

LOVECRAFT, H(OWARD) P(HILLIPS) (1890-1937) US author whose admirers regard him as a kind of 20th-century Edgar Allan POE. His mother made him wear his hair long until the age of six, and treated him like a girl; although possessive, she seems to have been unaffectionate, and often remarked that he was ugly, laying the foundation for a life-long inferiority complex. His father, a travelling salesman, went mad (probably from syphilis) when HPL was two, and died when the child was seven; HPL was brought up in the home of his maternal grandmother, surrounded by women, and thoroughly spoilt.

HPL was intellectually precocious, acquiring a taste for Gothic thrillers (◊ GOTHIC FANTASY) from his grandfather, and reading the *Arabian Nights* (◊ ARABIAN FANTASY) and *The Age of Fable* (**1855**) by Thomas BULFINCH at an early age. Because he was nearsighted and suffered from headaches his mother kept him away from school until he was eight, and after a year withdrew him. But during that year he discovered the work of Poe, and realized that the "ghoul-haunted woodland of Weir" was his natural intellectual climate. He began writing stories with titles like "The Mysterious Ship" and "The Mystery of the Graveyard".

At high school, in his early teens, he developed a love of science that seems to have been an emotional reaction against his naturally morbid ROMANTICISM, and it led him to take a certain pleasure in the notion that men are mere insects whose only defence against a harsh and meaningless Universe is self-delusion.

After two and a half years of high school he had a "nervous collapse", and was again withdrawn. Confined in the stale and bookish environment of home, his reclusiveness increased. He longed for friendship and a wider range of experience, but felt they would be forever denied him. At 27 he was still spending most of his time in his room, like a lonely teenager; he was also writing gloomy stories – with titles like "The Tomb" (1922) – in a wildly melodramatic style, peppered with obscure words. "Dagon" (1923), another of these early stories, describes how a man is shipwrecked on a mud-covered island hurled up from the seabed; here he encounters signs of a civilization of fishmen, and a scaly but obviously intelligent monster. A few of these early stories were published in small magazines; most were not.

His mother died when he was 31, and he continued to live with his two aunts. In 1924 he married Sonia Greene, whom he had met at a writers' convention. She was seven years his senior, and the marriage lasted only until 1926, breaking up largely because HPL disliked sex; the fact that she was Jewish and he was prone to antisemitic rants cannot have helped. He spent those two years in New York, which he hated and where he developed strongly racist views. Thereafter he spent the rest of his life in Providence, RI, where he had been born and raised.

In 1924 HPL achieved a kind of notoriety in New York when he inadvertently saved WEIRD TALES from bankruptcy. He had rewritten a story by one C.M. Eddy; "The Loved Dead" is about a necrophile who becomes a sex murderer. The story created a scandal, and a whole issue of *WT* had to be withdrawn from the newsstands . . . but the next issue sold out within hours. HPL was offered the editorship of *WT*, but the idea of so much involvement in the real world appalled him. Surviving correspondence suggests that he would have made a first-class editor.

By now he had become a regular contributor to *WT* and other pulp MAGAZINES, and had developed his unique fantasy world set around "Arkham" (based on Providence), with its gloomy hills, legends of MAGIC and DEVIL-worship, and its air of decay and DECADENCE. In his most famous story, "The Call of Cthulhu" (1928), he creates his basic myth of the mysterious ELDER RACE that once dominated the Earth but largely destroyed itself through sorcery, and whose members now lie sleeping somewhere UNDER THE SEA (or UNDERGROUND). He called them "the ancient old ones" – typically failing to note the tautology – or more often the Great Old Ones, not to be confused with the Great Race in "The Shadow Out of Time" (see below), nor indeed the Old Ones in Antarctica. HPL's terminology was erratic, but it can be argued that Cthulhu and his gang are unpleasant ELDER GODS rather than an elder race; unfortunately this distinction is also confusing since August DERLETH's systematization of the CTHULHU MYTHOS includes vaguely benevolent Elder Gods. HPL was always inclined to hurl around words like "eldritch", "monstrous", "miasmic" and "gibbous" with the abandon of a tachist artist flinging paint at the canvas.

The best of HPL's stories include "The Rats in the Walls" (1924), *The Colour out of Space* (1927; **1982** chap), "The Thing on the Doorstep" (1937) and "The Whisperer in Darkness" (1931). Unlike Poe, whose imagination was obsessed by death, HPL was fascinated by sliminess, corruption, disintegration; his work seems to exude the smell of rotting vegetation. Most of the horrible "creatures" in his stories – tentacled MONSTERS who drip green slime – have been conjured up by sinister old recluses using a BOOK of ancient magic called *The Necronomicon*, written by the "mad Arab Abdul Alhazred".

With his increasing circle of correspondents and his success in *WT*, HPL began to travel (as much as his limited means would allow), and the morbidity started to evaporate from his work. So far he had been essentially a Romantic in the same tradition as E.T.A. HOFFMANN or the early W.B. YEATS, turning his back on the real world and preferring a world of imagination – with the difference that HPL enjoyed exciting HORROR rather than wonder or dreamy nostalgia. But by the age of 40 he was beginning to outgrow his desire to scare his readers out of their wits. A tale like *At the Mountains of Madness* (cut 1936; 1939; **1990** chap) is closer to traditional sf, with a touch of fantasy in the manner of Lord DUNSANY. The short novel *The Shadow over Innsmouth* (**1936**) is unconvincing because HPL himself had ceased to be convinced by his own horrors; he could now see that the idea of a race of amphibious semi-humans is interesting rather than terrifying. And some stories, like "The Haunter of the Dark" (1936), have a distinctly tongue-in-cheek flavour – this one in particular being partly a literary game, with HPL killing off Robert Blake (i.e., Robert BLOCH) in mock-retaliation for a Bloch pastiche of HPL.

Professionally speaking, HPL's belated attainment of adulthood was something of a disaster. He had made his reputation for his own unique brand of horror fantasy. Now, just as it was beginning to look as if he could make a living from writing, he was starting to find it all rather silly.

The real solution would have been to give up horror in favour of his individual type of sf. In fact, this is what is beginning to happen in the novelette "The Shadow Out of Time" (cut 1936; 1939), his last finished work, written at about the time that he learned he had cancer. The story is about a professor from "Miskatonic University" (based on Brown University, Providence) whose body is taken over by some alien intelligence from the distant past which needs to undertake research in the 20th century. The professor's psyche meanwhile finds itself in the conical, betentacled body of one of the "Great Race". At story's end, back in his own body, he finds the library of the "ancient old ones" in an underground city in Australia and comes upon some notes in his own handwriting – written millions of years ago.

HPL died on March 15, 1937. Two years later his friend Derleth published, under the ARKHAM HOUSE imprint, a collection of HPL's stories, *The Outsider and Others* (coll **1939**); this prevented HPL from being totally forgotten. In fact, within two decades HPL had become a cult figure and Arkham House, largely because of its republication of his works, a successful publishing house. Derleth also worked up various fragments that HPL had left, publishing them as co-authorships; *The Lurker at the Threshold* (**1945**) is a notable example.

The influence of HPL, and in particular of the CTHULHU MYTHOS, on later fantasy writers has been patchy. Of his contemporaries or near contemporaries, the names of Robert BLOCH, Derleth, Robert E. HOWARD, Henry KUTTNER, Frank Belknap LONG, E. Hoffman PRICE and Clark Ashton SMITH stand out. More recently Lin CARTER, Fritz LEIBER, Manly Wade WELLMAN, Colin WILSON and especially Ramsey CAMPBELL and Brian LUMLEY have been indebted to him. [CW/DRL]

Other works (selective): *Fungi from Yuggoth* (coll **1941**), poetry, not to be confused with vt of 1963 collection noted below; *The Lurking Fear* (coll **1947**; vt *Cry Horror!* 1958), not to be confused with either *The Lurking Fear* (coll **1964** UK) or *The Lurking Fear* (coll **1971**), all 3 with differing contents, or with *The Lurking Fear* (1928 *Weird Tales*; **1977** chap), which reprints the story alone; *The Curse of Yig* (coll **1953**); *Dreams and Fantasies* (coll **1962**); *The Dream-Quest of Unknown Kadath* (1943; **1955**), not to be confused with *The Dream-Quest of Unknown Kadath* (coll **1970**) ed Lin Carter; *Something About Cats, and Other Pieces* (coll **1949**), revisions, essays, notes, etc.; *Collected Poems* (coll **1963**; cut vt *Fungi from Yuggoth and Other Poems* 1971); *Selected Letters 1911-1937* (coll **1965-76** 5 vols); *Uncollected Prose and Poetry* (coll **1978**) ed S.T. JOSHI and Marc Michaud.

LOVEHILL, C.B. Pseudonym of Charles BEAUMONT.

LOWDER, JAMES (1963-   ) US writer, exclusively of works for the TSR fantasy GAME group, specializing in **Forgotten Realms Fantasy Adventure** stories. At least one of these, *Tantras* * (**1989**) with Scott CIENCIN, has appeared under the housename Richard Awlinson. Solo titles include *Crusade* * (**1991**), *The Prince of Lies* * (**1993**) and *The Ring of Winter* * (**1993**). JL edited *Realms of Valor* * (anth **1993**) and *Realms of Infamy* * (anth **1994**). *Knight of the Black Rose* * (**1991**) is a DARK FANTASY for the same firm. [JC]

LOW FANTASY The introduction to *The Fantastic Imagination* (anth **1977**) ed Robert H. Boyer and Kenneth J. Zahorski defines HIGH FANTASY and implies LF as an antonymic description of fantasies not set in SECONDARY WORLDS, nor elevated in their literary style. E.g., the **Samella** sequence by John Brosnan (1947-   ) contains determinedly low comedy about SEX, flatulence and lavatories throughout, but its first book, *Damned and Fancy* (**1995**), is not LF since the action occurs in a FANTASYLAND, while the sequel, «Have Demon, Will Travel» (1996), is LF set in LONDON. [DRL]

See also: ADVENTURER FANTASY; HEROIC FANTASY; SWORD AND SORCERY.

LOZENGE PLOT ◊ W.S. GILBERT.

LUARD, NICHOLAS (LAMBERT) (1937-   ) UK soldier, theatrical impresario, publisher, travel writer and novelist, initially best-known for his thrillers (some as by James McVean). His fantasies, all large novels promoted as exotic mainstream bestsellers, are *Gondar* (**1988**), *Kala* (**1990**) and *Himalaya* (**1992**). The first two, set in 19th-century Africa, are reminiscent of H. Rider HAGGARD, involving dynastic struggles in hidden kingdoms; the eponymous heroine of the second is a girl raised by hyenas (◊◊ FERAL CHILDREN). The third utilizes similar subject matter in a Yeti-haunted Asian setting (◊ ANIMALS UNKNOWN TO SCIENCE). The books are colourful and well told, but basically derivative. [DP]

Other work: *Silverback* (**1995**).

LUCIAN (?115-?200) Greek-Syrian writer, essayist and satirist who for part of his life was an advocate and ended his days as a procurator in Egypt. He is usually called Lucian of Samosata, from his birthplace. It is difficult to date his writings accurately; they probably developed over a period of years, being completed sometime between 160 and 185. He was one of the first great fantasists, producing material that was knowingly fiction, satirizing the old GODS but using supernatural PLOT DEVICES. His most important works were his **Dialogues**, a form he derived from Plato (427-347BC); they have been imitated by scores of writers from the 15th century on. In *Deorum dialogi* ["Dialogues of the Gods"], *Marinorum dialogi* ["Dialogues of the Sea Gods"] and *Mortuorum dialogi* ["Dialogues of the Dead"] he supposes a series of discussions with the divine and departed spirits in order to lampoon the old RELIGION and to highlight the shortcomings and vanities of Man and the futility of power. His **Menippus** sequence, *Menippus* (vt *Necyomantia*) and *Icaromenippus*, shows an old philosopher endeavouring to discover the meaning and realities of life, first through discussions with the dead in the UNDERWORLD and then with the gods on OLYMPUS. (It is from this sequence that the term "menippea" was derived.) In *Charon* the ferryman over the Styx leaves the underworld to explore life. It is perhaps Lucian's most expressive form of reverse PERCEPTION, again highlighting the pettiness of humanity. *Kataplous* ["The Voyage to the Underworld"] describes the attitudes of the recently dead, and bears some surprising comparisons with John Kendrick BANGS's satire *A Houseboat on the Styx* (**1895**). Others of Lucian's **Dialogues** are more philosophical than fantastic, although in *Gallo* ["The Cock"; also known as "The Dream"] a cobbler is rendered invisible by a cock's magic tail feathers so he can spy on the rich and discover they are less happy than he. *Philopseudes* ["Lover of Lies"] uses the STORY-CYCLE in an early form of CLUB-STORY collection. A doctor visits an ailing friend, Eucrates; Eucrates and other friends present try to convince him of the truth of the supernatural by each telling a story. The sequence includes GHOST STORIES, walking STATUES, WIZARDS and the earliest known version of the SORCERER'S APPRENTICE, which may be original to Lucian.

Lucian's most famous work is *Verae historiae* ["True History"], which takes its intrepid voyagers to the Moon and past the Sun as well as to many distant ISLANDS; it is a PARODY of the TRAVELLERS' TALES that were already multitudinous in Lucian's day. Lucian is also attributed with writing a version of *Lucius e onos* ["Lucius or the Ass"]; though this is doubtful, he may have drawn upon the same earlier text, now lost, as APULEIUS, who wrote the more famous version at about the same time.

Lucian's works were translated into Latin by Erasmus (1466-1536) and later writers, and were a significant influence on François RABELAIS, Sir Thomas More (1478-1535), Johannes Kepler (1571-1630), Cyrano de Bergerac (1619-1655), Jonathan SWIFT and others. The first English translation was *Certaine Select Dialogues of Lucian, Together with his True Historie* (**1634** UK) by Francis Hickes (1566-1630). One of the best modern renditions is *Works of Lucian of Samosata* (**1905** 4 vols UK) trans H.W. and F.G. Fowler. [MA]

Further reading: *Culture and Society in Lucian* (**1986**) by C.P. Jones.

LUCIFER "Light-bearer" in Latin; used in Classical mythology with reference to the planet Venus as a morning star. The name appears in *Isaiah* 14:12 – "How art thou Fallen from heaven, O Lucifer, son of the morning!" – as a translation of the Hebrew *hillel* ("light-bringer"), and the misinterpretation of this passage resulted in Lucifer being added to the list of names associated with SATAN; it became popular in this sense following John MILTON's use of it in *Paradise Lost* (**1667**). [BS]

LUCK ◊ FATE.

LUKEMAN, TIM (?1954-   ) US writer whose ORIENTAL FANTASY **Khe'chin** series – *Rajan* (**1979**) and *Koren* (**1981**) – was mildly promising. A projected sequence set in a FANTASYLAND called Therrilyn paused at the first volume, *Witchwood* (**1984**), in which, to gain womanhood, a young girl with TALENTS must come to terms with them and with the DARK LORD who wishes to use her. [JC]

LUMLEY, BRIAN (1937-   ) UK writer. BL encountered the work of H.P. LOVECRAFT while a teenager; his early Lovecraftiana was published in *The Arkham Collector*, house organ of ARKHAM HOUSE. These stories formed the core of his first collection, *The Caller of the Black* (coll **1971**), and linked his early reputation to the CTHULHU MYTHOS. BL's first two novels, *Beneath the Moors* (**1974**) and *The Burrowers Beneath* (**1974**), liberally reinterpret the philosophical concepts behind Lovecraft's fiction, anthropomorphizing Lovecraft's monstrous incarnations of universal CHAOS and disorder as evil adversaries in a cosmic Cold War where humanity is forever getting the upper hand. He pursued

this theme in *The Transition of Titus Crow* (**1975**), *The Clock of Dreams* (**1978**), *Spawns of the Winds* (**1978**) and *In the Moons of Borea* (**1979**), a cycle of novels featuring the OCCULT DETECTIVE **Titus Crow**, who penetrates to the extradimensional void inhabited by Lovecraft's MONSTERS and finds it a fantasy landscape populated by fallible GODS and heroic human beings.

BL has a very masculine storytelling style and his protagonists are usually idealized embodiments of virility and strength. Nonetheless, they depend on brains as much as brawn to fight their battles. Titus Crow, whose earthly exploits are assembled as *The Compleat Crow* (omni **1987**), relies on his superior knowledge of occult arcana to triumph over supernatural nemeses. The hero of the **Psychomech** trilogy – *Psychomech* (**1984**), *Psychosphere* (**1984**) and *Psychamok* (**1985**) – harnesses the brain power bestowed on him through sf means to fight a series of villains endowed with advanced psychic powers. In *Demogorgon* (**1987**), the offspring of SATAN labours to use his latent supernatural powers to free himself from his father's clutches and avoid succumbing to their dark side.

A considerable amount of BL's fiction is SWORD AND SORCERY. *Hero of Dreams* (**1986**), *Ship of Dreams* (**1986**), *Mad Moon of Dreams* (**1987**) and *Elysia: The Coming of Cthulhu* (**1989**) are set nominally in the world of Lovecraft's *The Dream Quest of Unknown Kadath* (**1943**) but have more the flavour of the HEROIC FANTASIES of Edgar Rice BURROUGHS and Robert E. HOWARD in their elaborations of the adventures of two ordinary men who live vivid DREAM lives as fantasy warriors. *Iced on Aran and Other Dreamquests* (coll **1990**) collects their shorter exploits. The stories in *The House of Cthulhu and Other Tales of the Primal Land* (coll **1984**; vt *Tales of the Primal Land, Volume 1* 1990), *The Complete Khash, Volume 1: Never a Backward Glance* (**1991**; vt *Hrossak!: Tales of the Primal Land, Volume 2* 1992), and *Sorcery in Shad: Tales from the Primal Land 3* (**1993**) take place at the dawn of human civilization; although abundant with Lovecraftian elements, they are stylistically more indebted to Clark Ashton SMITH.

The **Necroscope** series is BL's best-known and most vividly imagined work. *Necroscope* (**1986**) introduces an entire subculture of psychics and paranormals who have opened a new frontier in Cold War espionage. *Necroscope II: Vamphyri!* (**1988**; vt *Wamphyri!*), *Necroscope III: The Source* (**1989**), *Necroscope IV: Deadspeak* (**1990**) and *Necroscope V: Deadspawn* (**1991**) are laced with elements of traditional Gothic HORROR and pit a UK special agent capable of conversing with the dead against a series of increasingly diabolical enemies in league with VAMPIRES from another dimension and backed by the USSR. The **Vampire World** trilogy, spun off from this series – *Vampire World #1: Blood Brothers* (**1992**), *#2: The Last Aerie* (**1993**) and *#3: Bloodwars* (**1994**) – is an exotic fantasy epic set in the complex and detailed extradimensional vampire world.

BL's non-series short fiction, primarily supernatural horror, has been collected in *Fruiting Bodies and Other Fungi* (coll **1992**), *The Last Rite* (coll **1993**), *Dagon's Bell and Other Discords* (coll **1994**) and *The Second Wish and Other Exhalations* (coll **1995**). These stories show BL's competence at handling a variety of traditional and nontraditional horror themes. [SD]

Other works: *The Horror at Oakdeene and Others* (coll **1977**); *Khai of Ancient Khem* (**1981**); *Ghoul Warning and Other Omens* (coll **1982**), poetry; *Synchronicity, or Something* (chap **1989**), GAME-related novelette; *The House of Doors* (**1990**); *Return of the Deep Ones and Other Mythos Tales* (coll **1994**); *Necroscope: The Lost Years* (**1995**); *Necroscope: Resurgence. The Lost Years 2* (**1996**).

Further reading: *Brian Lumley: A New Bibliography* (**1984**) by Leigh Blackmore.

LUNDWALL, SAM J(ERRIE) (1941-   ) Swedish writer, editor, critic and translator. ◊ WONDERLAND.

LUPERCALIA ◊ PAN.

LUSTBADER, ERIC ◊ Eric VAN LUSTBADER.

LYNCH, DAVID (1946-   ) US movie director and scriptwriter who trained as a painter. After *Six Figures Get Sick* (**1967**), a 30sec ANIMATED MOVIE designed to be run as an endless loop, and a similar live-action/animated loop that has been lost, his first movie proper the 4min live-action/animated *The Alphabet* (**1968**), which tells of a sick child vomiting blood; such is art. *The Grandmother* (**1970**), again live-action/animated, is more ambitious in both length (34 mins) and plot: thrashed by his father for bed-wetting, a little boy plants a seed in his damp sheets and grows from it a Grandmother to be his INVISIBLE COMPANION. For some years DL worked on *Eraserhead* (**1976**), his first feature, in his spare time; a curious filmed nightmare, not easily subject to analysis, it blends fantasy and the HORROR MOVIE to unsettling effect, and quickly acquired cult status.

His first mainstream movie, supported by Mel Brooks, was *The Elephant Man* (**1980**), based on the true story of the hideously deformed Joseph Merrick and the man who, in late-19th-century London, saved him from life as a freak-show exhibit, Frederick Treves. Its considerable success made DL a sought director, but his next movie, the sf epic *Dune* (**1984**), based on Frank Herbert's *Dune* (fixup **1965**), was a disaster – the book being too long and complex for a 140min movie (or even for the 190min version that was later televised, with additional narration and the use of cut footage; DL disavowed this version, and the direction is credited to the pseudonym Allen Smithee). DL recovered reputation with *Blue Velvet* (**1986**), a surreal movie that is part small-town URBAN FANTASY, part a fantasy of PERCEPTION, and part an essay in *film noir* for the 1980s. Both it and his next movie – *Wild at Heart* (**1990**) a slightly fantasticated road movie featuring a surprise visit by the Good Witch

from *The* WIZARD OF OZ (**1939**) – are plentifully supplied with graphic sex, which is probably why these "art" movies were so generally successful.

In the meantime, with Mark FROST, DL had been creating for tv the surrealistic soap opera/detective serial TWIN PEAKS (two series, 1989 and 1991), and it is for this that he is best-known. His only feature movie since then has been TWIN PEAKS: FIRE WALK WITH ME (**1992**), which was very badly received; shot in 1.85:1 format, rather than in wide-screen's customary 2.35:1, it was clearly intended by DL to capture the "feel" of a tv broadcast rather than a theatrical movie.

It is hard to classify DL's output as fantasy unless one is watching one of the relevant movies: then there is no doubt. And certainly his movies are identifiably *his*: even a short clip is usually enough to reveal that this is DL's work. The only moviemaker who has anything like the same "eye" is his daughter, Jennifer Chambers Lynch, director and scriptwriter of BOXING HELENA (**1993**). [JG]

Further reading: *David Lynch* (**1992**; trans Robert Julian **1995**) by Michel Chion; *The Films of David Lynch* (**1993**) by John Alexander.

LYNDELL, CATHERINE Pseudonym of Margaret BALL.

LYNES, (JOSEPH) RUSSELL (Jr) (1910-1991) US writer. ◊ AESOPIAN FANTASY.

LYNN, ELIZA Maiden name, under which she published some work, of Lynn LINTON.

LYNN, ELIZABETH A(NNE) (1946-   ) US writer whose debut novel, *A Different Light* (**1978**), was sf. She swiftly turned to fantasy with the **Chronicles of Tornor** trilogy: *Watchtower* (**1979**), *The Dancers of Arun* (**1979**) and *The Northern Girl* (**1980**). Interweaving narratives on the themes of GENDER politics and pacifism, these books received much praise, and the first volume won a WORLD FANTASY AWARD. Unfortunately, EAL has written little fantasy of substance since. [DP]

Other works: *The Woman who Loved the Moon, and Other Stories* (coll **1981**); *The Red Hawk* (**1983** chap); *The Silver Horse* (**1984**), for children; *Tales from a Vanished Country* (coll **1990**).

LYONESSE ◊ ARTHUR; FLOOD; IMAGINARY LANDS; LOST LANDS AND CONTINENTS; Jack VANCE.

LYTTON, LORD ◊ Edward BULWER-LYTTON.

MAAT ◊ GODDESS.

MABINOGION Collection of Welsh-language tales offering a glimpse into the largely lost MYTHOLOGY of Britain (◊ MATTER). The texts are found in *The White Book of Rhydderch* (early 14th century) and *The Red Book of Hergest*, a century later. The trappings are medieval, the sources ancient, with recurrent Celtic themes. The name was coined by Lady Charlotte Guest (1812-1895) for the first English-language translation and adaptation, *The Mabinogion* (**1838-49**). This included the SHAPESHIFTING tale *Hanes Taliesin*, in which the boy Taliesin takes the inspiration of poetry from the GODDESS Ceridwen. (This, while not regarded as authentic, is of fantasy interest for its chase in which both pursued and pursuer flit through a variety of different shapes.) The definitive English translation by Gwyn Jones and Thomas Jones (**1948**; rev 1974; rev 1982) contains 11 stories.

It begins with the *Four Branches of the Mabinogi*, dating from the 11th century; the term *mabinogi* means "tales". The original hero appears to have been Pryderi, and there are tales of his conception, youth, OTHERWORLD journey and death, but these have been overlaid with other legends. In "Pwyll, Prince of Dyfed" a human king changes places with Arawn, king of the otherworld Annwn. He lies chastely in Arawn's marriage-bed and defeats Arawn's enemy in a ritual battle. On his return, he is seated on an enchanted mound when Rhiannon rides past on a magic horse (this is probably Rigantona, the Great Queen, and a horse-goddess). She outwits an unwelcome suitor to marry Pwyll. When her newborn son, Pryderi, is stolen in the night she is accused of infanticide. Her punishment is to offer to carry visitors to the court like a horse. Teyrnon (once perhaps Tigernonos, the Great King) has lost a newborn colt every May-eve. When he cuts off a claw which comes through the window, the colt is saved and the baby discovered and restored to his parents.

Young Pryderi takes part in "Branwen, Daughter of Llŷr" but is overshadowed by the Children of the sea-god Llŷr. Brân the Blessed (Bendigeidfran) is the gigantic king of Britain. When the British set sail for Ireland to avenge his humiliated sister Branwen, Brân wades across the sea because no ship can hold him. In the ensuing battle, another brother bursts the magic cauldron which restores dead Irish warriors to life (◊ Lloyd ALEXANDER; *The* BLACK CAULDRON [**1985**]). Only seven men escape with Branwen. Brân, mortally wounded, orders them to sever his head. It brings them happiness and protects Britain, until they disobey his injunctions.

"Manawydan, Son of Llŷr" tells how Manawydan marries the widowed Rhiannon. Enchantment creates a WASTE LAND. In Lloegr (England), Manawydan and Pryderi make saddles, shields and shoes, until driven out by jealous craftsmen. Back in Dyfed, Pryderi and Rhiannon are successively snared by enchantment and spirited away. The crops of Manawydan and Pryderi's wife are stolen by MICE. Manawydan catches a fat one and determines to hang her; she proves the pregnant wife of the enchanter, who is avenging a trick Rhiannon once played on a suitor. Pryderi and Rhiannon are released from captivity, in which Rhiannon was made to wear the collar of an ASS.

"Math, Son of Mathonwy" tells of the death of Pryderi in battle, but chiefly concerns the children of Dôn, a Celtic goddess. Her son, abetted by his brother Gwydion, slayer of Pryderi, rapes the virgin footholder of their uncle, Math. They are sentenced to three years' SHAPESHIFTING, alternately male and female. They propose their sister Arianrhod as virgin footholder, but she fails the test: stepping over Math's wand she drops a boy-child – who makes for the sea – and something else, which Gwydion snatches up and hides. It is another boy, whom Arianrhod refuses to acknowledge or name. Disguised as a shoemaker, Gwydion tricks Arianrhod into naming the boy Lleu Llaw Gyffes, the Fair One with the Deft Hand. She swears he will never bear arms unless she arms him. Lleu in turn tricks her. She swears he shall never have a human wife. Math and Gwydion make Blodeuwedd for him from flowers, but Blodeuwedd falls in love with Gronw and lures Lleu into telling her the secret of the RITUAL slaughter by which alone he can be killed. When Gronw commits this murder, Lleu disappears as an eagle. Gwydion finds him in a treetop, his flesh rotting. Healed, Lleu kills Gronw in the same way. Blodeuwedd is turned into an owl. Her legend was the basis for Alan GARNER'S *The Owl Service* (**1967**).

Four independent stories follow. "The Dream of Macsen Wledig" and "Llud and Llefelys" are short tales. Macsen Wledig is Magnus Maximus, a 4th-century Roman commander in Britain, who invaded Gaul and ousted the Emperor Gratian. Llud and Llefelys are brothers who free Britain from three plagues.

The most primitive of all the stories is "Culhwch and Olwen". Culhwch, set on winning Olwen, the GIANT's daughter, seeks help at ARTHUR's court. Arthur is here a wild warrior-chief, and has an enormous retinue with supernatural attributes. His chief companions are Bedwyr and Cei. The giant sets Culhwch a bizarrely long list of impossible tasks: he must get supernatural help to obtain everything for the wedding feast and also the equipment to shave the giant, including the comb and shears from between the ears of the boar Twrch Trwyth. A fantastic expedition scours Britain in pursuit of these items; it involves the discovery of the oldest animal and the release of the god Mabon, the Son, from age-old BONDAGE. (There are signs of missing episodes – tasks not accomplished, helpers not used.) The giant is brutally shaved, then beheaded.

"The Dream of Rhonabwy" begins in a filthy hall. Rhonabwy DREAMS himself back to heroic times (i.e., TIMESLIPS). At Arthur's camp he witnesses a game of *gwyddbwyll* (a boardgame) between Arthur and Owein. Arthur refuses to stop playing, though his men are wounding Owein's ravens. Owein orders his standard raised. His ravens slaughter Arthur's warriors. He in turn plays on, until Arthur crushes the gaming pieces. Thus Arthur's Saxon enemies are defeated.

The last three romances are of later origin. The context is Norman-French; the tales no longer show detailed knowledge of the topography of Britain, and characters and adventures follow stock patterns. Similar stories appear in CHRÉTIEN DE TROYES's 12th-century *Yvain*, *Perceval* and *Erec*. Both probably draw on earthier British originals.

In "The Lady of the Fountain" Owein meets the Lord of the Animals. Well-water thown on a stone summons a champion. Owein pursues him, but is trapped. The maiden Luned rescues him with a magic RING. When the knight dies, Owein marries the Lady. Shamed for failing his promise to return from Arthur's court, he runs mad in the forest. Eventually he rescues Luned from death and is restored to the Lady.

"Peredur, Son of Efrawg" introduces the GRAIL hero Peredur/PERCEVAL. His mother tries to keep him from knowledge of knighthood, but he sees three knights and determines to follow them. He is mocked at Arthur's court by Cei, but hailed as the flower of knights by DWARFS, whom Cei abuses. (Cei, earlier a hero, is now shown as a boor. Gwalchmei [GAWAIN] is, by contrast, the model of a courteous knight.) Peredur leaves in anger, meets a lame king (◊ FISHER KING), and sees a bleeding lance and a severed head on a salver borne through a hall. Often he makes mistakes through following advice literally. A hideous damsel comes to court and upbraids Peredur for not asking the meaning of the bleeding lance, by which question he could have healed the Grail king and his land. He slays the WITCHES who once taught him to use arms, but his story ends with the Grail king still not healed (◊ HEALING).

"Gereint, Son of Erbin" is a Calumniated Wife tale. Gereint marries Enid, whom he has delivered from unjust impoverishment. He overhears her weeping about accusations that he is growing soft for love of her, and orders her to put on her

worst dress and ride out ahead of him, without speaking. Repeatedly she disobeys, warning him of danger. He defeats his attackers, but castigates her. When their host strikes Enid for refusing his advances, Gereint rouses, kills the man and grieves that he has wronged her. In a typically Celtic final exploit, he passes through a hedge spiked with severed heads, surrounded by mist, defeats the guardian and sounds a horn hung on an apple tree to end the ENCHANTMENT.   [FS]

MACABRE Small-press MAGAZINE. ◊ Joseph Payne BRENNAN; MAGAZINES; WEIRD TALES.

MacAULEY, CHARLES RAYMOND (1871-1934) US writer. ◊ John Kendrick BANGS.

MACAVOY, R(OBERTA) A(NN) (1949-   ) US writer who gained notice with her first novel, *Tea with the Black Dragon* (1983), a finely wrought CONTEMPORARY FANTASY about a woman's friendship with a centuries-old DRAGON masquerading in human form. As with most of RAM's novels, this is a witty tale which does not cover familiar ground. The novel mixes mystery, Tao and computers. It was followed by *Twisting the Rope* (1986).

The **Trio For Lute** trilogy – *Damiano* (1983), *Damiano's Lute* (1984) and *Raphael* (1984), assembled as *A Trio for Lute* (omni 1985) – follows Damiano Delstrego, a young Italian "witch", his familiar (a talking dog) and the Archangel Raphael through an ALTERNATE-WORLD Renaissance Italy. By the third volume, this changes from a RITE-OF-PASSAGE story about the at times annoyingly feckless Damiano into a much more substantial work in which Raphael takes on human form.

The **Lens of the World** trilogy – *Lens of the World* (1990), *King of the Dead* (1991) and *Winter of the Wolf* (1993; vt *The Belly of the Wolf* 1994) – begins with more coming-of-age material: Nazhuret has humble beginnings as an orphan, is instructed by Powl – a most unorthodox MENTOR – and becomes an adviser to the king. The quality of RAM's quirky, evocative prose and her willingness to play with conventions raise the trilogy well above standard GENRE FANTASY.

RAM's singletons include *The Book of Kells* (1985), a TIME-TRAVEL story steeped in things Celtic (◊ CELTIC FANTASY), set in both the present and 10th-century Ireland, and *The Grey Horse* (1987), a fine mix of fantasy and romance.   [WKS]

Other works: *The Third Eagle* (1989), sf with Native-American elements.

McBAIN, ED Best-known pseudonym of US writer Evan Hunter (1926-   ), born S.A. Lombino and who also wrote as D.A. Addams, Hunt Collins, Richard Marsten and Ted Taine during the 1950s, when he was a prolific contributor to crime, mystery and sf MAGAZINES. As EM, he writes the popular **87th Precinct** police-procedural novels, most of which are fixed in the real world, although *Ghosts* (1980) concerns apparently genuine psychic manifestations. Most of EM's stories of genre interest appeared under the Hunter name in 1951-4, before he established himself as a mainstream writer with *The Blackboard Jungle* (1954), a seminal novel of juvenile delinquency. Although he concentrated on sf, a few of his early stories verged on the fantastic, especially "Robert" (1953 *Thrilling Wonder*) and "The Miracle of Dan O'Shaughnessy" (1954 *Imagination*) about people with special TALENTS. A few fantasies are found in *The Last Spin* (coll 1960 UK) and *Happy New Year, Herbie* (coll 1963), both as Hunter, the name under which he also wrote the screenplay for Alfred HITCHCOCK's *The* BIRDS (*1963*).   [MA]

McBRIDE, ANGUS (1931-   ) Self-taught London-born Scottish artist and illustrator of historical and fantasy subjects, with a clean, crisp, richly coloured painting style. His art, done in gouache, has a remarkably attractive clarity and vitality. AM worked in advertising studios in London and South Africa before turning to illustration in 1961, specializing in historical reconstruction. He has done many paintings based on the works of J.R.R. TOLKIEN for the role-playing GAMES of Iron Crown Enterprises.

Other work: Angus McBride's Characters of Middle-Earth (graph **1990**).   [RT]

McCAFFREY, ANNE (INEZ) (1926-   ) US writer, now resident in Eire, whose work, since her first story, "Freedom of the Race" for *Science Fiction Plus* in 1953, has been sf (◊ SFE). Her **Pern** sequence, begun with *Dragonflight* (**1968**), utilizes images of fantasy – not just the DRAGONS but the highly detailed feudal society which has developed on the planet Pern. Many of AM's characters in this and also the **Pegasus** series, starting with *To Ride Pegasus* (fixup **1973**), and the **Rowan** books – which start with *The Rowan* (**1990**) but draw on "Lady in the Tower" (1959 *F&SF*) – are telepathic or use psionic powers (◊ TALENTS). All these stories, though, are scientifically rationalized. The **Killashandra** series – *Crystal Singer* (1974-5 in **Continuum** series ed Roger Elwood; fixup **1982** UK), *Killashandra* (**1985**) and *Crystal Line* (**1992**), assembled as *The Crystal Singer Trilogy* (omni **1996**) – is perhaps the closest AM's sf comes to fantasy: the heroine can use the power of MUSIC and voice over matter. AM employs her extensive musical training in the development of this concept. Despite its title, *The Mark of Merlin* (**1971**) is a contemporary murder mystery.

AM has written fantasy for the YA market. *The Girl who Heard Dragons* (**1985** chap) is incorporated into *The Girl Who Heard Dragons* (coll **1994**). *An Exchange of Gifts* (**1995** chap) is about a princess with a TALENT for growing things who needs help from a young boy to help her live her own life. *Alchemy & Academe* (anth **1970**) is an anthology of MAGIC and ALCHEMY.   [MA]

McCARTHY, CORMAC (1933-   ) US writer. ◊ BILLY THE KID.

McCARTHY, JUSTIN (1830-1912) Irish writer. ◊ Justin Huntly MCCARTHY.

McCARTHY, JUSTIN HUNTLY (1860-1936) Irish writer, son of writer, journalist and MP Justin McCarthy (1830-1912), who wrote several nonfantasy novels with Mrs Campbell PRAED. JHM's historical novels tend to be rather more colourful than his father's. His 13th-century romance *The Dryad* (**1905**) is an archetypal account of THINNING: the last immortal favoured by ZEUS falls in love with a young prince but has to go to extraordinary lengths to rescue him from the clutches of a rival sorceress. *Our Sensation Novel* (**1886**) is a blithely unrationalized parody of Wilkie COLLINS's *The Woman in White* (**1860**), one of whose women in white may be a LAMIA. JHM edited *The Thousand and One Days: Persian Tales* (anth **1892** 2 vols), which contains ARABIAN FANTASIES.   [BS]

McCARTY, DENNIS (1951-   ) US writer whose **Thlassa Mey** sequence – *Flight to Thlassa Mey* (**1986**), *Warriors of Thlassa Mey* (**1987**), *Lords of Thlassa Mey* (**1989**), *Across the Thlassa Mey* (**1991**) and *The Birth of the Blade* (**1993**) – sets a series of chivalric adventures in a FANTASYLAND whose cast and place-names have an Arthurian ring.   [JC]

McCAUGHREAN, GERALDINE (1951-   ) UK editor and writer, initially for children, more recently for wider audiences. Her adaptations of TAPROOT TEXTS – including versions of *One Thousand and One Arabian Nights*, Geoffrey CHAUCER's *Canterbury Tales* and *El Cid* and *The Odyssey* (**1993** chap) – have been well received. In *A Pack of Lies: Twelve Stories in One* (coll of linked stories **1988**), a mysterious figure – apparently materialized from a volume of *Wisden's Cricketing Year* but in reality (perhaps) a divine being – comes to work in an antique shop, where he tells prospective customers a STORY which forces them either to buy or (if they are unworthy) not to buy various items on sale. The stories themselves, neatly and compactly told, are likewise mostly fantasy. Two novels for adults – *The Maypole* (**1989**) and *Fires' Astonishment* (**1990**) – work as fantasias based on traditional English ballads (◊ SONG): "Little Musgrove and Lady Barnard" for the first, which is associational, and "The Laily Worm and the Machrel of the Sea" for the latter; in a LAND-OF-FABLE 11th-century England, a stepmother METAMORPHOSES her husband's son into a DRAGON, who spends much of the novel moping over his condition. A REVISIONIST sharpness permeates GM's work in general. She is, in the end, and not entirely to the benefit of her craft, a no-nonsense writer.   [JC]

McCAUGHREN, TOM (1936-   ) Irish writer. ◊ CHILDREN'S FANTASY.

McCAY, WINSOR Working name of US COMICS artist and creator of ANIMATED MOVIES, whose full name is sometimes given as Winsor Zenic (or Zenic Winsor) McCay (1867-1934). He is of seminal importance in both fields. His earliest years were obscure (it is not known where he was born; his year of birth has also been given as 1869 or 1871) but he can be traced back to 1889, when he was employed in Chicago as an engraver in a printing firm. During the 1890s he worked as a freelance poster painter and as an in-house artist at Cincinnati's Vine Street Dime Museum before, in 1898, starting his newspaper career by doing editorial cartoons for the *Cincinnati Commercial Tribune*. By 1900 WM had switched papers and was drawing his first comic strip, **Tales of the Jungle Imps**, signed Felix Fiddle.

From 1902 he was in New York, working as a cartoonist for *Life*, and beginning to work for the two New York papers owned by James Gordon Bennett (1841-1918): the *New York Herald* as WM and the *New York Telegram* as "Silas". Several humorous ALLEGORIES followed, including **A Pilgrim's Progress by Mr Bunion**, **Hungry Henrietta**, **Poor Jake** and **Little Sammy Sneeze**, selections from the last of which appeared as *Little Sammy Sneeze* (graph coll **1906**). 1904 saw the début of WM's **Dreams of the Rarebit Fiend**, which carried its adult characters into a variety of very frightening dyspepsia-generated nightmare experiences; it appeared in book form as *Dreams of a Rarebit Fiend* (graph coll **1905** as by "Silas"; bowdlerized 1973 as by WM), which reprinted 61 strips; a 1913 sequence of strips was reprinted as an appendix to volume 6 of the **Little Nemo** collected edition (*see below*).

The success of this strip inspired his masterpiece, LITTLE NEMO IN SLUMBERLAND, which appeared in the *New York Herald* (1905-11), then for William Randolph Hearst papers under the title *In the Land of Wonderful Dreams* (1911-14), then for the *Herald-Tribune* (1924-7) under the original title. The first sequence was perhaps the most innovative and inspired; selections were reprinted as *Little Nemo in Slumberland* (graph coll **1906**) and *Little Nemo in*

Slumberland (graph coll **1909**). Later titles included an adaptation by Edna Sarah Levine, *Little Nemo in Slumberland* * (**1941**) illus WM; and the **Complete Little Nemo in Slumberland** set of collections: *The Complete Little Nemo in Slumberland #1: 1905-1907* (coll **1989**) ed Richard Marschall, *#2: 1907-1908* (coll **1989**) ed Marschall, *#3: 1908-1910* (coll **1990**) ed Marschall and *#4: 1910-1911* (coll **1990**) ed Marschall, reprinting the original sequence in its original colours; plus *#5: In the Land of Wonderful Dreams: 1911-1912* (coll **1991**) ed Marschall and *#6: In the Land of Wonderful Dreams: 1913-1914* (coll **1993**) ed Bill Blackbeard, which reprints the second sequence in original colours, also including some **Rarebit Fiend** colour strips; further volumes are projected. Many of the first-sequence episodes – drawn in WM's fluent, hallucinatory, meticulously crafted, architectonic, poster-like Art Nouveau style – were straightforward dream fantasies; but later sustained sequences – like those dealing with Shantytown, with Befuddle Hall, and with a 1909 excursion by airship into outer space – were genuine FANTASTIC VOYAGES; as pioneering explorations into the techniques of narrating complex visions through sequential drawings, the strip as a whole was of vital importance.

WM was intensely prolific, and at the same time as writing and drawing **Little Nemo** he also continued with other graphic work, including many individual drawings, those making up the **Spectrophone** series of visions of the future being of particular sf interest. After he moved to Hearst, he began concentrating on political cartoons from the conservative point of view required by the proprietor; but continued to issue enormously detailed prophetic drawings involving vast airships, cityscapes and catastrophes. Some of these have been assembled as *Daydreams & Nightmares: The Fantastic Visions of Winsor McCay* (graph coll **1988**) ed Richard Marschall.

WM also took a central role in the development of the animated cartoon – indeed, some claim that he invented the art of animation. In whatever medium he worked, he drew with incredible speed; this gave rise to the vaudeville act he presented from 1906, during which he executed a series of 40 chalk drawings, one every 30 seconds, showing a man and a woman ageing while the orchestra played a suitable melody. From here it was a logical step to animation. With astonishing industry, he hand-painted each frame of his cartoons; beginning in 1909 he produced 10 short films: *Little Nemo* (1911), which required about 4000 drawings; *The Story of a Mosquito* (**1912**; vt *How a Mosquito Operates*); *Gertie, the Dinosaur* (**1914**), which required *c*10,000 drawings; *The Sinking of the Lusitania* (**1918**), the most ambitious, requiring *c*25,000 drawings done in much more detail than in the earlier films; *The Centaurs*, a fantasy movie, *Flip's Circus* and *Gertie on Tour*, these three being done *c*1918-21 and surviving only as fragments; and three **Dreams of the Rarebit Fiend** shorts, all released in 1921: *The Pet, Bug Vaudeville* and *The Flying House*. In *The Pet*, household animals drink an elixir and swell to huge proportions; a 10-storey cat ravages a city and, King Kong-style, is pestered by airships. *Bug Vaudeville* is a (pre-DISNEY) **Silly Symphonies**-style fantasy. In *The Flying House* a couple, escaping creditors, fit out their house with wings and a propeller and fly off into outer space.

It is not certain why WM gave up animation after these successes, but it was possibly because he thought – wrongly, as was soon proven by **Felix the Cat** and Walt Disney's

Alice and **Oswald the Lucky Rabbit** – that animation, as an artform, was a cul-de-sac to whose end he had come. He continued to produce newspaper strips and illustrations until the end of his life. [JC/JG/SW]

Further reading: "Winsor McCay" by John Canemaker in *The American Animated Cartoon: A Critical Anthology* (anth **1980**) ed Danny Peary and Gerald Peary; *Of Mice and Magic* (**1980**; rev 1987) by Leonard Maltin; *Winsor McCay: His Life and Art* (**1987**) by John Canemaker; *Comic Artists* (**1989**) by Richard Marschall.

McCORMACK, ERIC P. (1938-) Scottish-born writer, resident in Canada, whose first collection, *Inspecting the Vaults* (coll **1987**), assembles stories of various kinds, including some MAGIC REALISM tales. *The Paradise Motel* (**1989**) is a story of dour BELATEDNESS whose protagonist – though he has left Scotland – cannot divest himself of the hauntedness of his homeland. [JC]

McCRUMB, SHARYN (1948-) US writer, mostly of detections, two satirizing sf fandom and fantasy GAME-playing: *Bimbos of the Death Sun: Murder Most Fun at the Ultimate Fantasy Con* (**1987**) and *Zombies of the Gene Pool* (**1992**); neither novel, though amply RECURSIVE, strays into the fantastic. Some of SM's other tales are SUPERNATURAL FICTION: *The Hangman's Beautiful Daughter* (**1992**) – the title is taken from the Incredible String Band album (1968) – is set in a richly envisioned Appalachian Tennessee (◊ LANDSCAPE), where a family tragedy precipitates a complex AMERICAN GOTHIC drama involving funerals, slow deaths, routine MAGIC among women – prefiguring some of the subject matter of Rebecca ORE's *Slow Funeral* (**1994**) – and a general sense that contemporary events replay STORIES out of FOLKLORE. *She Walks These Hills* (**1994**), whose title quotes an Appalachian ballad, generates a similar ambience, with the significant addition of GHOSTS. In her fantasy work, SM proves a skilled evoker of mood and teller of tales. [JC]

MacDONALD, GEORGE (1824-1905) Scots author, poet, minister (briefly), lecturer and Christian apologist whose nonfantasy – including several novels – has generally not survived. In fantasy, though, GM was a remarkable innovator, adept in reproducing the strange logic of DREAMS and in creating early versions of SECONDARY WORLDS whose quality (in his admirer C.S. LEWIS's phrase) "hovers between the allegorical and the mythopoeic". His literary influences included John BUNYAN, E.T.A. HOFFMANN and the German romantic/mystic poet and novelist "Novalis" (Friedrich Leopold Hardenberg; 1772-1801).

GM's first prose work was *Phantastes, A Faerie Romance for Men and Women* (**1858**), whose uneven exuberance suggests an author revelling in the literary freedom of fantasy as he moved away from a strict Calvinist upbringing. The episodic story begins with a memorable dream-transition: the protagonist Anodos (Greek, "upward path") finds his mundane bedroom transforming to the outdoor woods of FAERIE, inhabited by DRYADS. After some twee observations of flower-FAIRIES, Anodos walks INTO THE WOODS by moonlight and sees the shadow of the malign, invisible Ash-spirit's hand. He reasons that he can observe the spirit by lying in this shadow and looking towards the MOON: the resulting vision is horrid, but the episode reflects the sense that Faerie has self-consistent rules which may be explored and analysed. Thus, detecting a woman's shape trapped in a marble block (◊ BONDAGE) but unable to kiss her awake owing to intervening stone, Anodos sings her free (◊ SONG).

Later, wishing to join a magical dance whose dancers freeze into statuary at his mere *thought* of entering their room, he must discipline himself to act on random impulse.

Breaking another rule – that warnings and PROHIBITIONS must be heeded – Anodos meets difficulties with Alder and Ash, and acquires an unwanted SHADOW of his own which dulls the sense of wonder and may be responsible for his optically distorted PERCEPTION of a village's inhabitants. In a strange LIBRARY whose BOOKS allow readers to experience the authors' or characters' feelings at first hand, he visits a tertiary world where women are armless but winged, and lives through an inset supernatural story revolving around the image of a woman inhabiting the reflected room behind a MIRROR. Magic doorways open onto Anodos's revisited childhood and other sad times. He helps fight GIANTS and briefly rides out as a KNIGHT in the classic mould, before his unworthiness is exposed by confrontation with the DOPPEL-GÄNGER who is his Shadow. Rescued by a woman whom he wronged under the Shadow's influence, he serves briefly as esquire to a better knight, and dies heroically exposing a religious RITUAL that conceals the feeding of innocents to a wolf-like MONSTER. Though dead, Anodos relates his own burial and glimpse of AFTERLIFE bliss before awakening on Earth . . . where the easy "it was all a dream" is undermined when he learns he has been missing for weeks.

The rich variety of sometimes confused symbolism in *Phantastes* cannot and should not be tied down to any particular ALLEGORY, though the Shadow is hardly enigmatic. GM concludes with the thoroughly Christian sentiment (◊ CHRISTIAN FANTASY) that apparent EVIL is merely a disguise for the best possible good (◊ PARODY). Without the Shadow, there can be no getting free of the Shadow.

GM's short fantasies begin in *Adela Cathcart* (coll with novel frame **1864**; without inset stories 1882 USA), whose stories were collected with additions (including "The Golden Key") as *Dealings with the Fairies* (coll **1867**; cut as *The Light Princess, and Other Fairy Stories*, 1890). Particularly striking is "The Golden Key", a deceptively lucid ALLEGORY of travel from childhood dreams of the rainbow's end to, eventually, the high "country whence the shadows fall" (or of which things below are but shadows) – with haunting imagery and incidental inventions like rainbow-feathered "air-fish" eager to enter the cooking-pot and be transfigured. This is permeated with the sense of life as a NIGHT JOURNEY full of irrevocable choices: "You must throw yourself in. There is no other way."

Further stories of note include: "The Shadows", depicting the WAINSCOT realm of living shadows acting as guardian angels; "The Giant's Heart" (1863), an enjoyably blood-thirsty romp whose giant-villain likes children ("forked radishes") raw, and believes he has hidden his heart in a safe place (◊ KOSHCHEI); and "The Light Princess", which, though amusingly parodic of FAIRYTALE conventions, indicates that the princess hilariously cursed at her christening with complete loss of gravity – both physical and emotional – is a hollow person until finally enabled to cry a little. This was adapted as *The Light Princess* (**1985** tvm), a 56min live-action/animated movie.

At the Back of the North Wind (**1871**) was the first of GM's three popular CHILDREN'S FANTASIES. The boy Diamond is befriended by the North Wind – manifesting as a beautiful woman of widely variable stature (◊ GREAT AND SMALL), compassionate yet implacable and duty-bound to sink ships, etc. Being alive only when blowing south, she cannot enter the magic country at her back but sits frozen on its North Pole doorstep – a LIMINAL BEING through whom Diamond may pass. GM explains that this country is not the HYPER-BOREA of Herodotus, but allusively identifies it with the "land of love" glimpsed in James HOGG's poem "Kilmeny" and with the Earthly Paradise of DANTE's *Purgatorio*. (These references presumably escaped child readers of *Good Words for the Young*, the magazine – edited for a time by GM himself – which serialized this and *The Princess and the Goblin*.) Diamond's visit to this country lasts seven days, which for him seem many years (◊ TIME IN FAERIE), while to his family he lies dangerously ill at home . . . indicating the nature of the THRESHOLD he did not fully cross. He emerges as a kind of holy FOOL, sentenrically doing good works amid Victorian London's underclasses. Perhaps sensing that the fantasy interest had here begun to flag, GM interposed DREAM sequences (one with an interestingly TOPSY-TURVY journey down an UNDERGROUND staircase whose lower end emerges above the sky, where one can dig for stars) and a fairytale slightly resembling SLEEPING BEAUTY. The book ends edifyingly with the still-young Diamond's unfearing death.

The Princess and the Goblin (**1872**) is set in a fairytale OTHERWORLD with a generally medieval flavour. The GOBLINS are once-humans whose generations underground have caused ugly DEBASEMENT. Comically malicious rather than evil, they plan to kidnap the surface people's young Princess Irene as a spouse for their rebarbative prince. Curdie, a boy worker in the MINES, discovers the goblins' plans and their ACHILLES' HEEL: despite invulnerable skulls they have tender feet, susceptible to being stamped on. Overseeing Irene's welfare is her claimed great-great-grandmother, one of the many ageless wise women in GM's fantasies. She provides a spider-silk thread which improves on Ariadne's by always leading Irene by the best route to the right place – including the heart of the goblin caves in order to rescue the captured Curdie, and then out again. When the goblin attack comes, Curdie's warnings go unheeded by the royal guards (his measured punishment for scepticism about the wise woman and the thread, neither of which he can perceive; ◊ PERCEPTION). But the kidnap plan is frustrated in a welter of stamping on feet. The goblins' fall-back strategy is to destroy the human palace and mines by flood: thanks to Curdie's warning the waters are turned back by a buttress of masonry and it is the goblins who drown *en masse*. This novel was the basis of the ANIMATED MOVIE *The* PRINCESS AND THE GOBLIN (**1992**).

Years later, in *The Princess and Curdie* (**1883**), matters have gone somewhat to the bad. Curdie has grown morally slack and uses his bow to shoot a pigeon, an event momentarily as doom-laden as the Ancient Mariner's crime: the bird belongs to the wise woman, who, after trials of recognition, obedience and pain, forgives Curdie and sends him out as her agent; the last trial requires him to place his hands in a fire which is also roses, burning them clean (even of callouses) and granting the ability to sense others' dangerous spiritual METAMORPHOSES. Accompanied by a doglike MONSTER (which may be a debased human), Curdie travels as instructed to the capital city Gwyntystorm, which (with the decay of the good king's rule) is inhospitable and intolerant. Evil courtiers are weakening the king with poison disguised as medicine: Curdie and Irene soon end this, and clear out the palace with the gleefully described aid of 49 comic-grotesque monsters befriended by the "dog" *en route*.

Invaders from an adjacent kingdom join the unpleasant townsfolk in a battle where Curdie's creatures and handful of human supporters are hopelessly outnumbered: the wise woman intervenes with flocks of pigeons which peck at, blind and eventually panic the opposition.

In *The Princess and the Goblin*, malevolent beings were all conveniently identifiable by ugliness. Contrariwise, it is a continuing theme of the sequel that appearances deceive, though not forever: "Fairest things turn foulest by their deeds." For its intended child audience the book is soured by misanthropy: decent human beings are rare, and the whole kingdom goes rotten again after Irene's and Curdie's brief GOLDEN AGE.

Lilith (**1895**; rev by excision of incidental verse 1924) is GM's second and last adult fantasy novel; more unified than *Phantastes*, it has a curious dreamlike intensity. Many pet tropes are revisited. Led by a raven, the narrator, Mr Vane, passes through a mirror PORTAL into a highly symbolic OTHERWORLD where ADAM AND EVE (the former also being the raven, the latter yet another ageless woman) watch over the sleeping dead. Vane declines their hospitable offer of death and explores the dreamland: the Bad Burrow where grotesque monsters emerge from the soil but are impotent in moonlight, the Evil Wood, and a region beyond inhabited by "Little Ones" – eternal children with a distressing line in baby-talk – and "Bags", the dull adults which Little Ones may become through moral laziness. LILITH herself is glimpsed at intervals, appearing for example at a hallucinatory DANCE OF DEATH. A beautiful woman with a dark rot-spot on her side, whose drought has made this place a partial WASTE LAND, she is a SHAPESHIFTER often seen as a spotted leopardess – preying on children in response to the PROPHECY that a child will be her downfall (her own daughter Lona leads the Little Ones). Finding her temporarily stricken and withered by opposing magic, Vane incautiously nurses her to health and becomes unwitting prey to her VAMPIRE traits. She rejects his advances; he follows her to the cruel, greedy CITY Bulika where she is princess. (Further complications are a white leopardess opposing the spotted one, and the Shadow which is Lilith's original corrupter, SATAN.)

Though Vane learns to distrust Lilith, he forgets Adam's warning about obeying anyone not trusted, and is induced to climb a tree which abruptly becomes the high fountain-jet outside his own house (◊ RECOGNITION): the whole journey must be done again. This time, actively rejecting Adam's advice, he leads the armed Little Ones against Bulika; the princess is captured but first kills Lona. Back at Adam's House of Death, Lilith is painfully redeemed and passes with Lona and the other children into the sleep of death, from which one awakens to eternity. After a ritual task which restores water to the dry land, Vane follows them; but his glimpse of HEAVEN gives way to an awakening in his own library.

Lilith is a powerfully mysterious book. However, its underlying cross-currents of sexual symbolism suggest that GM may not have been wholly in control of his material. The author's influence has been acknowledged by J.R.R. TOLKIEN and C.S. LEWIS, whose allegory *The Great Divorce* (**1945** chap) features GM as Lewis's crusty spiritual mentor. GM was also highly regarded by Lewis CARROLL, who lent him the original MS of *Alice in Wonderland* and asked his opinion on publication; GM duly read it to his children, who were rapturous.

Even in his best work GM was prone to routine Victorian sentimentality, but his potent gift of "dream realism", and his ability to elaborate moral or religious ALLEGORY into a larger subcreation which resists simple allegorical decoding, make him a landmark figure of pre-GENRE FANTASY and especially of CHRISTIAN FANTASY. [DRL]

Other works: *The Portent: A Story of the Inner Vision of the Highlanders, Commonly Called the Second Sight* (**1864**; vt *Lady of the Mansion* 1983 US); *Cross Purposes, and The Shadows: Two Fairy Stories* (coll **1890**); *The Wise Woman: A Parable* (**1875**; vt *A Double Story* 1876 US; vt *Princess Rosamund* US; vt *The Lost Princess, or The Wise Woman* 1895 UK); *The Gifts of the Child Christ* (coll **1882** 2 vols; 1 vol vt *Stephen Archer and Other Tales* **1882/3**); *The Flight of the Shadow* (**1891**); *The Fairy Tales of George MacDonald* (coll **1904** 5 vols) ed Greville Macdonald; *The Portent and Other Stories* (coll **1924**); *The Visionary Novels of George MacDonald: Lilith; Phantastes* (omni **1954** US; vt *Phantastes; and Lilith* 1962 UK); *The Complete Fairy Stories of GM* (coll **1962** US); *The Gifts of the Child Christ: Fairy Tales and Stories for the Childlike* (coll **1973** 2 vols US) ed Glenn Edward Sadler; *At the Back of the North Wind; The Princess and the Goblin; The Princess and Curdie* (omni **1979**); a 4-vol set of the short fantasy as *The Golden Key and Other Fantasy Stories* (coll **1980**), *The Gray Wolf and Other Fantasy Stories* (**1980**), *The Light Princess and Other Fantasy Stories* (coll **1980**), *The Wise Woman and Other Fantasy Stories* (coll **1980**); *The Gold Key and the Green Life* (anth **1986**), mixing stories by GM and Fiona MACLEOD; *The Day Boy and the Night Girl* (**1988** chap); *The Light Princess* * (**1988** chap), "adapted" by Robin MCKINLEY; *Little Daylight* (**1988** chap), inset fairytale from *At the Back of the North Wind*.

Further reading: *George MacDonald and His Wife* (coll **1924**) by Greville Macdonald (GM's son).

MacDONALD, GOLDEN Pseudonym of Margaret Wise BROWN.

MACDONALD, JAMES D(OUGLAS) (1954-) US writer. ◊ Debra DOYLE.

McEVILLEY, THOMAS (1939-) US essayist and novelist, much of whose nonfiction, like *Art & Discontent: Theory at the Millennium* (coll **1991**), addresses questions of Modernism and Postmodernism in the visual arts. *North of Yesterday, or Flowers of Waz* (**1987**) is a DREAM novel featuring a variety of UNDERLIERS like ORPHEUS and DEATH. Central to the text is the 2000-year-old "author's" sequel to HOMER. [JC]

MacEWEN, GWENDOLYN (1941-1987) Canadian writer. ◊ CANADA.

MacFALL, HALDANE (1860-1928) UK writer. ◊ Dion Clayton CALTHROP.

MACFARLANE, STEPHEN Pseudonym of John Keir CROSS.

McGIRT, DAN (1967-) US author whose **Jason Cosmo** sequence is fantasy HUMOUR. *Jason Cosmo* (**1989**) introduces the hero, who is befriended by a mirrorshaded wizard and farcically pursued through a FANTASYLAND studded with anachronistic gags, droll names like "the Incredibly Dark Forest", and invincible VILLAINS, who are duly vanquished. Further and similar escapades are chronicled in *Royal Chaos* (**1990**) and *Dirty Work* (**1993**) – whose ploys include deliberately overdone padding, Cosmo's knowing asides about being a character in fiction, and RECURSIVE references to Terry PRATCHETT. The series remains open-ended. [DRL]

McGOWEN, TOM Working name of US advertising executive and writer Thomas McGowen (1927-), who began publishing work of genre interest with *Sir MacHinery* (**1970**), a TECHNOFANTASY: various COMPANIONS, including a WIZARD and a WITCH, come across a large box labelled "Machinery" in a scientist's residence. It contains a robot, which the companions mistake for a KNIGHT in armour; they enlist it in their fight against evil TROLLS.

Though its first pages seem set in a SECONDARY WORLD, the **Armindor** sequence – *The Magician's Apprentice* (**1987**), *The Magician's Company* (**1988**) and *The Magician's Challenge* (**1989**) – is sf, set on a post-HOLOCAUST Earth where the technologies of the past seem like MAGIC. *The Shadow of Fomor* (**1990**), in which two young boys are transported into a Celtic secondary world, is fantasy. The **Age of Magic** sequence – *The Magical Fellowship* (**1992**), *A Trial of Magic* (**1992**) and *A Question of Magic* (**1993**) – is SCIENCE FANTASY, set in a LAND-OF-FABLE prehistoric world populated by DRAGONS and ELVES and being invaded by aliens.   [JC]

McGUFFIN A variety of PLOT COUPON whose prime purpose is to motivate searches, chases and conflicts. The term was invented by Alfred HITCHCOCK: celebrated movie examples include the Maltese Falcon and, in *Raiders of the Lost Ark* (*1981*; ◊ INDIANA JONES), the Ark of the Covenant. A McGuffin need not be successfully tracked down, need not even exist. The PICARESQUE action of Michael SHEA's *In Yana, the Touch of Undying* (**1985**) is driven by the ultimately unattained McGuffin of IMMORTALITY; Barry HUGHART's *Bridge of Birds* (**1984**) effectively entwines a McGuffin hunt – for a ginseng root that has HEALING properties – with a gradually realized and far more significant QUEST to cure an ancient, cosmic WRONGNESS.   [DRL]

McHARGUE, GEORGESS (1941-) US writer. ◊ CHILDREN'S FANTASY.

MACHEN, ARTHUR (LLEWELYN) (1863-1947) Welsh writer whose reputation has always been high, though he never became widely known. AM began his writing career with hack translations and fantasy pastiche in the 1880s, continuing to produce work into WWII; he was employed first as an actor with the Benson Shakespeare Repertory Company 1901-9 and then as a London journalist until 1921. At the turn of the century he joined the Order of the GOLDEN DAWN, but his participation in the "orders" was always somewhat sceptical.

In his long career, AM embraced and mastered two of the main categories of the fantastic: SUPERNATURAL FICTION and FANTASY. At first under the influence of his friend A.E. WAITE as well as his own moody mysticism, AM wrote GHOST STORIES and other supernatural fiction in tune with the **Yellow Book** DECADENCE of the late Victorians. Most of these early tales of WRONGNESS, of the intervention of the fantastic and the horrible into a firm, everyday reality, combine what his reclusive Ambrose in *The House of Souls* (written in the 1890s; **1906**) speaks of as "Sorcery and sanctity . . . these are the only realities. Each is an ecstasy, a withdrawal from the common life." The best-known and most influential of these supernatural fictions is "The Great God Pan" (1890; exp in *The Great God Pan and The Inmost Light* coll **1894**), in which a young girl is, after a quasiscientific surgical procedure, made to see the hideous reality of PAN, or the DEVIL himself.

Perhaps AM's most successful book in this mode is *The Three Impostors, or the Transmutations* (fixup **1895**), which is structured as a set of fantastic narratives about intervention of Celtic "little people" and other marvels told by very rational late-Victorian London gentlemen – in other words, AM adopted the CLUB-STORY mode in order to heighten the wonder of his tales.

AM never quite gave up supernatural fiction. When he was working for the *London Evening News* during WORLD WAR I his story "The Bowmen" (1914) won him a remarkable notoriety because of its hopeful and positive supernatural intervention of the ghosts of ancient British archers to help at the Battle of Mons. He wrote some other fantastic legends about WWI to go with this; the sequence was assembled as *The Angel of Mons, The Bowmen and Other Legends of the War* (coll **1915** chap), followed by *The Terror* (**1917**), in which the apocalyptic fears aroused by WWI are confirmed by a revolt of the animals against the corrupt rule of humanity.

In his important nonfiction book about literature, *Hieroglyphics* (**1902**), AM speculates on the role of style and writerliness in fiction that he admires from CERVANTES to De Quincey and gives the indication that he most wants to write high fantasy. The word "ecstasy" was a favourite with him in some of his supernatural fictions; here he uses it as the label for the style of the highest literature or the highest fantasy. His best books in the mode of high fantasy are fantasized or spiritual autobiography in which Celtic Wales, ARTHUR and the "ecstasy" writer himself are his real subjects (◊ CELTIC FANTASY). These autobiographical books do contain the urgency of STORY for AM because they must be at least TWICE-TOLD for him. In other words, the supernatural writer who sounds like Robert Louis STEVENSON is effective on the BELATEDNESS of a confident one-level view of reality that hardly needs the "storying" of the lost age of heroes; and AM knew he was part of such belatedness and even viewed it with some humour, as in "The Bowmen". But AM wanted, also, to tell a life with multiple levels of REALITY, and believed in this inner life enough to produce the high fantasy of his spiritual autobiographies. *The Hill of Dreams* (**1907**) is his best book in this mode. The protagonist Lucian Taylor is much like AM, with the same Welsh youth, the same effort to write in London, the same dreaming about ancient Rome and about hills back in Wales and about Arthur. In fact, AM first titled this text "The Garden of Avallaunius" as a coinage from a Roman British name he had uncovered in research, Vallaunius, his version meaning "the man of Avalon". He also says that he wanted to write a "Robinson Crusoe of the soul" in which a man is lonely in the "midst of millions". He came back to this mode of belief with *The Secret Glory* (**1922**). Finally, then, in a strange way, this eager young **Yellow Book** hack of the 1890s grew into a sort of father figure for 20th-century fantasists.   [DMH]

Other works: *Eleusinia* (**1881**), poem; *The Anatomy of Tobacco* (**1884**) and *The Chronicle of Clemendy* (**1888**), medieval pastiche; *The Heptameron of Marguerite of Navarre* (**1886**) and *The Memoirs of Jacques Casanova* (**1894**), translations; *The House of the Hidden Light* (**1904**); *Dr Stiggins* (**1906**), journalism; *The Great Return* (**1915**); *War and the Christian Faith* (**1918**); *Far Off Things* (**1922**) and *Things Near and Far* (**1923**), autobiography; *The Shining Pyramid* (coll **1923**; with differing contents 1924 UK); *Dog and Duck* (**1924**), journalism; *The Green Round* (**1933**); *The Children of the Pool and Other Stories* (coll **1936**); *The Cosy Room and Other Stories* (coll **1936**); *Tales of Horror and the Supernatural* (coll **1948** US).

Further reading: *Arthur Machen* (**1964**) by Wesley D. Sweetser; *Arthur Machen and Montgomery Evans: Letters of a*

Literary Friendship, 1923-1947 (**1994**) ed Sue Strong Hassler and Donald M. Hassler.

McKEAN, DAVE Working name of UK artist David McKean (1963-   ), who has worked on COMIC books, GRAPHIC NOVELS and book and CD covers, and latterly a set of TAROT cards. DM works in a wide range of media: each artwork may include painted and drawn elements with photographic prints and found objects attached; much of his recent work has involved computer-enhanced photography. A recurring feature in many of his comic-book covers has been a printer's font tray, in the receptacles of which he places objects associated with the book's storyline. His contact with US artist Barron Storey (1940-   ), beginning in the late 1980s, has had a profound effect on his work.

DM first came to prominence with the publication of the GRAPHIC NOVEL *Violent Cases* (graph UK **1987** b/w; **1991** colour) written by Neil GAIMAN. DM produced some haunting cover paintings for two series of comic books: *Hellblazer* (for *#27* and *#40* he also provided internal artwork) and SANDMAN, and painted artwork for two GRAPHIC NOVELS featuring mainstream comic-book characters: the three-part *Black Orchid* (graph **1988** US; omni **1991** US) written by Gaiman, and the bestselling *Arkham Asylum* (graph **1989** US) written by Grant MORRISON. His other work includes two further GRAPHIC NOVELS written by Gaiman: *Signal to Noise* (1989 *The Face*; graph **1992** UK) and *Mr Punch* (graph **1994** US), the latter about childhood experiences with seaside entertainers. DM is writing and drawing *Cages* (1991-current US), a projected 500-page comic GRAPHIC NOVEL about creativity. His Tarot set was published in 1995 by DC Vertigo (◊ DC COMICS). [RT]

McKENNA, STEPHEN (1888-1967) UK writer, mostly of lighthearted romances that often make metaphorical use of FAIRYTALE motifs. *The Sixth Sense* (**1915**) features a "Seraph" whose extra sense helps track his kidnapped sweetheart. SM's one fantasy novel of significance is *The Oldest God* (**1926**), in which an earnest debate about the problem of EVIL and the Christian response to it is ironically subverted by PAN. SM's subsequent attempts to produce more serious work include *Beyond Hell* (**1931**), about a near-future experiment in penology, and *Superstition* (**1932**), a story about the power of suggestion and the efficacy which it may lend to CURSES. [BS]

MACKEY, MARY (LOU McGUINESS) (1945-   ) US writer. ◊ GENDER.

McKIERNAN, DENNIS L(ESTER) (1932-   ) US writer of HEROIC FANTASY who spent 31 years in R&D for antiballistic missile defence systems and telephone software before becoming a full-time writer in 1977, while recovering from a car crash. His first book was split by its publisher into **The Iron Tower** trilogy, being *The Dark Tide* (**1984**), *Shadows of Doom* (**1984**) and *The Darkest Day* (**1984**). This unashamed homage to J.R.R. TOLKIEN, set in the magical land of Mithgar, deploys the FIMBULWINTER concept. His second book, a direct sequel, was split into **The Silver Call** sequence, *Trek to Kraggen-cor* (**1986**) and *The Brega Path* (**1986**), using a QUEST story to explore the naivety that war can be romantic. In *Dragondoom* (**1990**) DM looks at issues of racism and intolerance. *The Eye of the Hunter* (**1992**) explores the impact of humanity on the environment as well as IMMORTALITY; its prequel, *Voyage of the Fox Rider* (**1993**), discusses the nature of EVIL. *Caverns of Socrates* (**1995**), the only novel to date not set in Mithgar, is essentially a TECHNOFANTASY: a group is trapped in a virtual-reality game

unless they can beat the computer; DM's philosophical questions here concern the nature of REALITY and consciousness. His short fiction has been collected as *Caverns of Mithgar* (**1994**). He wrote the GRAPHIC NOVEL *Tales from the One-Eyed Crow: The Vulgmaster* (**1991**). Projected are «The Dragonstone» and «Hel's Crucible», both **Mithgar** novels. [JF]

McKILLIP, PATRICIA A(NNE) (1948-   ) US writer whose early works were for children. *The House on Parchment Street* (**1973**) features GHOSTS from the English Civil War and *The Throme of the Erril of Sherill* (**1973** chap; reissued with "The Harrowing of the Dragon of Hairsbreath" [1982] coll 1984) is comic fantasy in the manner of James THURBER. *The Forgotten Beasts of Eld* (**1974**), a stylish moralistic fantasy about the sentimental education of an ENCHANTRESS, was again marketed as a juvenile, but won the first WORLD FANTASY AWARD (1975) as Best Novel. The **Riddle-Master** trilogy – *The Riddle-Master of Hed* (**1976**), *Heir of Sea and Fire* (**1977**) and *Harpist in the Wind* (**1979**), assembled as *Riddle of Stars* (omni **1979**; vt *The Chronicles of Morgon, Prince of Hed* 1981 UK) – a more orthodox HEROIC FANTASY, is similarly well wrought. The intellectual and emotional maturation of its mild-mannered hero and independent-minded heroine are handled with scrupulous delicacy. *Stepping from the Shadows* (**1982**) is an interesting naturalistic novel about the making of a fantasy writer, which presumably draws on PAM's own experiences.

PAM then digressed into sf for some years before returning to fantasy. *The Changeling Sea* (**1988**) was aimed at younger readers, but *The Sorceress and the Cygnet* (**1991**) is a complex and thoughtful QUEST fantasy for adults. It compares and contrasts the intellectual odysseys of a stigmatized boy, an assiduous but reckless female scholar and a pragmatic swordswoman in a strange world whose constellations embody a series of tutelary deities; PAM's customary lyricism is here delicately leavened with HUMOUR. The adventures and philosophical inquiries begun in this volume are continued in *The Cygnet and the Firebird* (**1993**), in which the harmony established at the end of its predecessor is assaulted by surreal distortions of PERCEPTION which must be tracked to their magical source. The fine novella *Something Rich and Strange* (**1994**) was issued as part of **Brian Froud's Faerielands** illustrated by Brian FROUD; a romance in which a female artist and her half-hearted lover are separately seduced by sea-spirits which turn out to be more powerful and more sinister than they had imagined, it won the 1995 MYTHOPOEIC AWARD. The narrative is exquisitely detailed, adding several wry convolutions to the kind of plot traditionally deployed in tales of MERMAIDS.

PAM is one of the most accomplished prose stylists working in the fantasy genre; she always brings a keen and refreshingly idiosyncratic intelligence to her employment of its motifs. [BS]

Other works: *The Night Gift* (**1976**), marginal; *Moon-Flash* (**1984**), sf; *The Moon and the Face* (**1985**), sf; *Fool's Run* (**1987**), sf; *The Book of Atrix Wolfe* (**1995**).

McKINLEY, ROBIN (1952-   ) US writer, resident in the UK; married to Peter DICKINSON. She caused a considerable stir with her first book, *Beauty: The Retelling of the Story of Beauty and the Beast* (**1978**), which grew out of a short-story idea she had for recasting BEAUTY AND THE BEAST (◊◊ REVISIONIST FANTASY), a mode continued in *The Door in the Hedge* (**1981**), a collection of two original fairytales and two more retellings (of "The Twelve Dancing

Princesses" and "The Princess and the Frog"). She moved into SWORD-AND-SORCERY with *The Blue Sword* (**1982**) and its prequel, *The Hero and the Crown* (**1985**), which won the Newbery Award. Both are set in the medieval FANTASYLAND of Damar and utilize RK's knowledge and love of fairytales and legends as backbone for stories of high romance and heroic QUESTS. In *The Outlaws of Sherwood* (**1988**) RK portrays ROBIN HOOD in spirited fashion as a rather diffident hero bullied into setting up a band of outlaws in Sherwood Forest after accidentally killing a forester. *Deerskin* (**1993**) – based loosely on Charles PERRAULT's "Donkeyskin" (in its unbowdlerized form) – is her darkest and most adult fantasy, tackling issues of incest, abuse, self-doubt, madness and obsession. The mythic elements march hand-in-hand with more prosaic TRANSFORMATIONS as a princess escapes from a forced marriage to her father into a wilderness both physical and emotional; her survival depends as much on her own strength as on the gifts of her patron GODDESS. This was followed by a second collection, *A Knot in the Grain and Other Stories* (**1994**), which includes two tales set in Damar and, unusually for RM, one story set in the real world. [JF]
As editor: *Imaginary Lands* (anth **1985**), which won the WORLD FANTASY AWARD.

MacLEISH, ARCHIBALD (1892-1982) US poet and dramatist, very eminent during his lifetime. Much of his work comprises narrative poems with dramatic elements, or plays written in verse. Of the former, *The Pot of Earth* (**1925**) conflates various RITES OF PASSAGE, including a dramatic rendering of the death of the FERTILITY god Artemis, and *The Hamlet of A. MacLeish* (**1928**) reincarnates Hamlet's father. Of AM's plays – several written for radio – *Nobodaddy* (**1926**) is based on ADAM AND EVE; *The Fall of the City* (**1937**), produced for radio by Orson Welles, sees an Aztec-like city warned of a fascist invasion by a Truth-Telling woman (◊ ORACLE) who rises four times from the dead to predict disaster; *J.B.* (**1958**), by far his most famous work (it won for him his third Pulitzer Prize), is set in a CIRCUS and features a supernatural Voice from the Whirlwind; *Herakles* (**1967**) conflates the HERO's life with that of Modern Man; and *Scratch* * (**1971**) dramatizes "The Devil and Daniel Webster" (1936) by Stephen Vincent BENÉT. In *The Great American Fourth of July Parade* (**1975**), Presidents John Adams (1735-1826) and Thomas Jefferson (1743-1826) comment on the contemporary USA. [JC]

MacLEISH, RODERICK (1926-) US broadcaster and writer, author of several books, though only one – *Prince Ombra* (**1982**) – is of fantasy interest. Recounted within a FRAME STORY, it depicts the life of an Eternal Champion who has been reborn (for the 1001st time) in 1978 to fight the eponymous DARK LORD, whose previous incarnation was as Adolf Hitler (◊ WORLD WAR II) and on whose defeat the history of the world turns. Although the novel hovers, at points uneasily, between fantasy and SUPERNATURAL FIC-TION – the Dark Lord's nature is explained in details that evoke occult philosophies – it remains a fluent and vivid demonstration of the essential movement of the typical fantasy narrative from WRONGNESS (Ombra's corrupting infusion of his foul, nightmare-creating self into the New England setting is eloquently described) to HEALING, as the victorious HERO transforms his home town, and the world, back to a prelapsarian GARDEN state. [JC]

MacLEOD, FIONA Pseudonym of William Sharp (1855-1905) for his mystical and religious CELTIC FANTASIES. He was intensely secretive about this alter ego, retaining the separate identity until his death. The works as FM are so different from Sharp's own relatively unmemorable books as to suggest a split personality. Most of the FM stories, novels and poems were concentrated into four years (1893-6), produced almost with the fervour of religious experience. All have a dreamlike spiritual quality, capturing the folk-memory of Celtic myth and legend. Their popularity contributed to the Celtic revival of the 1890s. The best short fiction and sketches appeared as *The Sin-Eater and Other Tales* (coll **1894**) and *The Washer of the Ford and Other Legendary Moralities* (coll **1896**), which are mostly stories of spiritual TRANSFORMATION and second-sight (◊ TALENTS). In both books the title story features mythic characters whose actions consume or purify the sins of others. Later collections, especially *The Dominion of Dreams* (coll **1899**) and *Where the Forest Murmurs* (coll **1906**), contain mystical nature pieces that often lack a storyline but are strong on atmosphere. The early stories were reassembled as *The Shorter Stories of Fiona Macleod, Vol. I Spiritual Tales, Vol. II Barbaric Tales, Vol. III Tragic Romances* (coll **1895**); the complete *Works of Fiona Macleod* (coll **1910-12** 7 vols) was ed Elizabeth Sharp (Sharp's wife). A more recent selection forms the second part of *The Gold Key and the Green Life* (coll **1986**) ed Elizabeth Sutherland.

FM's novels – *Pharais* (**1894**), *The Mountain Lovers* (**1895**) and *Green Fire* (**1896**) – are essentially nonfantastic, though all three may be perceived as allegorical in their use of death (of the adults) and rebirth (through the children) as a transformation from an earthly Hell to some other Heaven. *Pharais* is Gaelic for Paradise; this paradise cannot be earned without trial. The closest to fantasy is *Green Fire*, a nature myth about two sisters, one of whom reflects the darkness of earth while the other (Annaik) has the feyness of spirit. The latter, after an incestuous relationship with her brother, re-unites with the FORESTS. The novel may have influenced Algernon BLACKWOOD in *The Bright Messenger* (**1921**). FM always intended to return to the character of Annaik to tell her story in full, but never did. [MA]
Further reading: *William Sharp (Fiona Macleod): A Memoir* (**1910-12** 2 vols) by Elizabeth A. Sharp; *William Sharp – "Fiona Macleod", 1855-1905* (**1970** US) by Flavia Alaya.

McNUTT, CHARLES Pseudonym of Charles BEAUMONT.

MACPHERSON, JAMES (1736-1796) Scottish poet and literary hoaxer ("Ossian"). ◊ CELTIC FANTASY.

MACROCOSM ◊ MICROCOSM/MACROCOSM.

MAD ABOUT MEN UK movie (*1954*). ◊ SPLASH! (*1984*).

MAD LOVE (vt *The Hands of Orlac*) US Movie (*1935*). MGM. **Pr** John W. Considine Jr. **Dir** Karl Freund. **Screenplay** John Balderston, P.J. Wolfson. **Based on** adaptation by Guy ENDORE of *Les Mains d'Orlac* (**1920**) by Maurice Renard (1875-1940). **Starring** May Beatty (Françoise), Edward Brophy (Rollo), Colin Clive (Stephen Orlac), Frances Drake (Yvonne Orlac), Sara Haden (Marie), Ted Healy (Reagan), Peter Lorre (Dr Gogol). 70 mins. B/w.

Genius surgeon Gogol is besotted with Yvonne Orlac, star of a GRAND-GUIGNOL-style theatre in Montmartre, but she is married to brilliant pianist Stephen. Stephen's hands are mangled in a train crash. Yvonne manipulates Gogol's obsession with her to make him save the hands; he does the next best thing, covertly stitching on instead those of CIRCUS knife-thrower and convicted murderer Rollo. Stephen finds his new hands lousy at piano-playing but uncannily skilled at knife-throwing. Gogol frames Stephen for murder, persuading even Stephen of his guilt. Through fingerprinting, the police come to believes the hands, not Stephen, were guilty. They rush to Gogol's clinic, where they find Gogol strangling Yvonne. Only Orlac's hands, throwing a knife, can save her.

This TECHNOFANTASY, a remake of *The Hands of Orlac* (*1924*), is claimed as a major influence on Orson Welles's *Citizen Kane* (*1941*). An important subplot concerns the waxwork of Yvonne which Gogol buys from a theatre foyer and has his drunken housemaid Françoise tend, in the belief that the myth of Galatea (◊ PYGMALION) might be re-enacted – as indeed it seems to him later to be, with the real Yvonne masquerading as the waxwork. [JG]

See also: *The* HANDS OF ORLAC (*1960*).

MAD MAX BEYOND THUNDERDOME (*1985*). Australian movie. ◊ DICTION.

MADOC 12th-century Welsh prince, sometimes called Madog, believed to have sailed the Atlantic and discovered the New World, landing in Mobile Bay (in today's Alabama) in *c*1170. His historicity is hard to establish, and it is possible that both he and his voyage – recounted in, for example, *Divers Voyages Touching the Discovery of America* (*1582*) by Richard Hakluyt (*c*1552-1616) – are MYTHS. In the 17th century there were presumed TRAVELLERS' TALES of a tribe of White Amerindians, the Mandans, who spoke a language similar to Welsh. The figure of Madoc has been used in various fantasies, notably Sanders Anne LAUBENTHAL's *Excalibur* (*1973*). [JG]

MADOG ◊ MADOC.

MADONNA ◊ GODDESS.

MAETERLINCK, MAURICE (1862-1949) Belgian dramatist and essayist associated with DECADENCE and the Symbolist movement. The bibliography of his dramatic works is complicated. Many, including several fantasies, were used as bases for OPERAS.

MM's first fantasy work was the Edgar Allan POE-like "Onirologie" (1889; *1936* chap). His early plays – *L'intruse* (1890), *Les aveugles* (1890) and *La princesse Maleine* (1889; trans Richard Hovey as *Princess Maleine* 1894 US) – were initially translated in mixed pairs: *Blind & The Intruder* (coll trans Mary Vielé 1891 UK) and *Princess Maleine & The Intruder* (coll trans Gerard Harry 1892 US). These Symbolist plays involve the gradual unfolding of a tragic scheme, apparently at the behest of malign FATE. Three of his shorter plays in this vein were issued as *Alladine et Palomides; Intérieur; La mort de Tintagiles; trois petits drames pour marionettes* (coll *1894*; trans Alfred Sutro and William Archer as *Alladine and Palomides, Interior, and The Death of Tintagiles: Three Little Plays for Marionettes* 1899 US; vt *Three Plays* UK; vt *Three Little Dramas* US). The third of these plays is exceptionally harrowing.

MM's other plays of fantasy interest include two downbeat mock-chivalric romances (◊ CHIVALRY), *Pelléas et Mélisande* 1892; trans Erving Winslow as *Pelleas and Melisanda* 1894 US) and *Aglavaine et Sélysette* (*1896* France; trans Sutro as *Aglavaine and Selysette* 1897 US), and several quasi-folkloristic romances: *Les sept princesses* (*1891* chap; trans Charlotte Porter and Helen A. Clarke as "The Seven Princesses" 1894 US; new trans William Metcalfe as *The Seven Princesses* 1909 chap UK); *Ariane et Barbe-Bleu* (*1901*) and *Soeur Béatrice* (*1901*), assembled as *Sister Beatrice and Ardiane & Barbe Bleue* (trans coll Bernard Miall *1901* UK); *Joyzelle* (*1903* France; trans Alexander Texeira de Mattos *1903* Canada).

MM's most famous play is the allegorical FAIRYTALE *L'oiseau bleu* (*1909* France; trans Texeira de Mattos as *The Blue Bird* *1909* US); his most optimistic work, it concludes with a mystical transcendence reflective of MM's profound interest in SPIRITUALISM. It was so popular in the USA that its sequel, *The Betrothal* (*1918* trans Texeira de Mattos US), was first produced there and was only belatedly published as *Les fiançailles* (*1922* France). A children's version of the former play was prepared by the actress Georgette Leblanc (1869-1941), who was MM's companion 1895-1918 and took the leading role in most of his plays; this was issued in English as *The Children's Blue Bird* (trans Texeira de Mattos *1913* US; vt *The Blue Bird for Children* US). Texeira de Mattos did his own prose version of the sequel as *Tyltyl* (*1920* US; vt *The Bluebird Chooses* US). The tale was filmed twice as *The* BLUE BIRD.

Once MM had parted from Leblanc, the subject-matter of his plays shifted markedly towards naturalistic psychological melodrama, but the farcical *Le Miracle de Saint Antoine* (*1919* France; trans Texeira de Mattos as *The Miracle of Saint Anthony* 1918 US) and the second item in *The Cloud that Lifted, and The Power of the Dead* (trans coll trans F.M. Atkinson *1923* US) are fantasies. MM's curious brand of contemplative fatalism is abundantly expressed in his many essays. [BS]

MAGAZINE OF FANTASY AND SCIENCE FICTION, THE US digest MAGAZINE, Fall 1949-current (534 issues to December 1995), originally quarterly, then bimonthly from February 1951, monthly from August 1952. Since 1991 the October/November issues have been combined. First published by Lawrence Spivak's Mercury Press, then bought out by General Manager Joseph W. Ferman (1906-1974) in 1954; he published it until 1965; his son, Edward L. Ferman (1937-   ) has published it since. Ed Anthony BOUCHER and J. Francis McComas (1911-1978) Fall 1949-August 1954, by Boucher alone until August 1958, Robert P. Mills (1920-1986) September 1958-March 1962, Avram DAVIDSON April 1962-November 1964, Joseph W. Ferman December 1964-December 1965, Edward L. Ferman January 1966-June 1991, Kristine Kathryn RUSCH since July 1991.

F&SF has for nearly 50 years been the primary magazine of fantasy. It began as *The Magazine of Fantasy* (#1 only); "Science Fiction" was introduced to the title to meet the growing post-WWII sf market. Although for a period in the late 1950s *F&SF* became sf-dominated, it has always sustained a varied and healthy diet of fantasy. One level – the one at which *F&SF* is strongest – was the cultivation of SLICK FANTASY. *F&SF* wanted to distance itself from the image of the pulps and be acceptable to readers of *The New Yorker*, *Good Housekeeping* and *The Saturday Evening Post*. Boucher and McComas thus reprinted many examples from these magazines and elsewhere – including fictions by Robert Graves (1895-1985), Gerald HEARD, F. Tennyson Jesse (1889-1958), James STEPHENS, James THURBER and P.G. Wodehouse (1881-1975) – and encouraged other contributors. The shining example from the early years was Shirley JACKSON, and the *F&SF* approach attracted a high proportion of women writers, including Mildred Clingerman (1918-   ), Miriam Allen DeFord (1888-1975), Carol EMSHWILLER, Zenna Henderson (1917-1983), Kit Reed (1932-   ), Joanna Russ (1937-   ), Margaret St Clair (1911-   ) – whose alter ego Idris Seabright was one of Boucher's favourite contributors – and Evelyn E.

Smith (1927-   ). Male writers often seemed to opt for the CLUB STORY or TALL TALE, exemplified by the **Gavagan's Bar** series by L. Sprague DE CAMP and Fletcher PRATT, and including the **Murchison Morks** stories by Robert ARTHUR, the **Papa Schimelhorn** tales by Reginald Bretnor (1911-1992), and later the **Brigadier Ffellowes** adventures by Sterling Lanier (1927-   ).

At another level *F&SF* could be seen as a continuation of WEIRD TALES, primarily in its focus on supernatural HORROR and GHOST STORIES. In this vein it not only inherited some of *WT*'s writers – e.g., Ray BRADBURY, August W. DERLETH and Manly Wade WELLMAN – and continued the resurrection and reprinting of stories by the older generation of supernaturalists – e.g., H. Russell WAKEFIELD and Lord DUNSANY – it also rapidly developed a strong stable of its own, especially Richard MATHESON, Bruce Elliott (1914-1973) and Russell Kirk (1918-1994), while more recently it has drawn upon the talents of Robert AICKMAN, Charles L. GRANT, J. Michael REAVES, Michael SHEA and Karl Edward WAGNER.

Early in its life *F&SF* was able to exemplify the way ancient horrors could be modernized with Donald A. WOLLHEIM's story "The Rag Thing" (1951 as David Grinnell), though the greatest exponents of this development were Fritz LEIBER and Theodore STURGEON. Both these writers, like Boucher, had been contributors not only to *WT* but also to UNKNOWN, and *F&SF* soon emerged as a fusion of these two great magazines; but its content was raised to the next level through the ever-deepening veneer of slick fantasy. Thus within a short time *F&SF*'s content had created a middle ground. The magazine was not alone in doing this – much of the same development had been happening at BEYOND, FANTASTIC and FANTASY MAGAZINE – but it was fortunate in having the financial support and the vision of its publisher and editors in seeing this through. Its maturing was underscored in an ironic fashion when Robert BLOCH's "That Hell-Bound Train" (1958) became the first SUPERNATURAL FICTION to win the Hugo AWARD.

From the outset Boucher banned HEROIC FANTASY, singling out Robert E. HOWARD's work in particular, and this policy was continued by Mills and Davidson. As a result *F&SF*'s WEIRD FICTION throughout the 1950s was entirely in the supernatural vein, though it never stagnated and writers showed a tremendous ability to explore new themes. Among the most original was Manly Wade Wellman, who drew upon local Appalachian LEGENDS for his **John the Balladsinger** stories and turned to Native American beliefs for his tales as Levi Crow. Later editors entertained heroic fantasy if it developed stories from legend and was treated realistically or lightheartedly. Poul ANDERSON made the breakthrough with *Three Hearts and Three Lions* (1953; rev **1961**), which was also *F&SF*'s first serial. Later Gordon R. DICKSON contributed "St Dragon and the George" (1957; exp vt *The Dragon and the George* **1976**).

Boundaries were pushed back further in the 1960s under the more liberal editorship of Edward Ferman, especially with tales by Roger ZELAZNY, whose work imported mythic resonances into the mechanics of sf. Through *F&SF* Zelazny presented *This Immortal* (1965 as ". . . And Call Me Conrad"; exp **1966**), "Dawn" (1967) and "Death and the Executioner" (1967) – the latter two being self-contained extracts from *Lord of Light* (**1967**) – and *Jack of Shadows* (**1967**). It is interesting that, once Zelazny's tales became more rooted in tradition – as with the **Amber** series – he

became a less frequent contributor to *F&SF*; likewise, when Stephen KING established himself in the HORROR field, it was not his horror that *F&SF* pursued but his more mythic fantasy, the **Dark Tower** sequence, which began with "The Gunslinger" (1978). Although SWORD AND SORCERY stories began to appear in the magazine from the mid-1960s – including Jack VANCE's **Cugel the Clever**, Fritz Leiber's **Fafhrd and the Gray Mouser**, Phyllis EISENSTEIN's **Alaric the Minstrel** and John MORRESSY's **Kedrigern** stories – these were tales that tested and expanded the existing subgenre. Throughout its life *F&SF* has opened to writers who explored this mythic quality of fiction, examples of which are prevalent in the works of Thomas M. DISCH, Harlan ELLISON, Philip José FARMER, Robert HOLDSTOCK (whose **Mythago Wood** sequence – launched with "Mythago Wood" [1981 *F&SF*] – is a natural extension of this development), Tanith LEE, Leiber, Richard MCKENNA, Walter M. Miller Jr (1922-1996), Tom REAMY, Lucius SHEPARD and Thomas Burnett SWANN.

This richness and diversity means that *F&SF* tends to have a wide appeal, and consequently its stories have frequently won awards. Those closer to the fantasy milieu include: "That Hell-Bound Train" (1958) by Bloch; "The Doors of His Face, the Lamps of His Mouth" (1965) and ". . . And Call Me Conrad" (1965) by Zelazny; "Ship of Shadows" (1969), "Ill Met in Lankhmar" (1970) and "Catch That Zeppelin" (1975) by Leiber; "The Queen of Air and Darkness" (1971) by Anderson; "Pages from a Young Girl's Journal" (1973) by Robert AICKMAN; "The Deathbird" (1973), "Adrift Just Off the Islets of Langerhans . . ." (1974) and "Jeffty is Five" (1977) by Harlan ELLISON; "Born With the Dead" (1974) by Robert SILVERBERG; "San Diego Lightfoot Sue" (1975) by Tom REAMY; "A Crowd of Shadows" (1976) by Charles L. GRANT; "Stone" (1978) by Edward BRYANT; "The Persistence of Vision" (1978) by John Varley (1947-   ); "Cassandra" (1978) by C.J. CHERRYH; "The Bone Flute" (1981) by Lisa TUTTLE; "Souls" (1982) by Joanna Russ; "Another Orphan" (1982) and "Buffalo" (1991) by John Kessel (1950-   ); "Black Air" (1983) by Kim Stanley Robinson (1952-   ); "Buffalo Gals Won't You Come Out Tonight" (1987) by Ursula K. LE GUIN; "Kirinyaga" (1988) and "Seven Views of Olduvai Gorge" (1994) by Mike Resnick (1942-   ); "Guide Dog" (1991) by Mike Conner (1951-   ); "Ma Qui" (1991) by Alan BRENNERT; "Graves" (1992) by Joe Haldeman (1943-   ); and "The Night We Buried Road Dog" (1993) by Jack Cady (1932-   ). *F&SF* itself won the Hugo for the Best Magazine in 1958, 1959, 1960, 1963, 1969, 1970, 1971 and 1972; when that category was dropped in favour of Best Editor, Edward L. Ferman won in 1981, 1982 and 1983 and Rusch in 1994.

F&SF has published occasional celebratory issues featuring a chosen author. Those honoured have been Theodore STURGEON (September 1962), Ray BRADBURY (May 1963), Isaac ASIMOV (October 1966; who also contributed a science column November 1958-February 1992), Charles BEAUMONT (June 1967), Fritz Leiber (July 1969), Anderson (April 1971), James BLISH (April 1972), Frederik Pohl (September 1973), Silverberg (April 1974), Ellison (July 1977) and Stephen KING (December 1990).

The magazine has been regularly mined for anthologies. The formative years were covered in a tribute volume, *The Eureka Years* (anth **1982**) ed Annette Peltz McComas (1911-1994). Starting in 1952, Boucher and McComas edited *The Best from Fantasy and Science Fiction* (anth **1952**), which continued under that title with *Second Series* (anth **1953**) and *Third Series* (anth **1954**); then, ed Boucher alone, *Fourth Series* (anth **1955**), *Fifth Series* (anth **1956**), *Sixth Series* (anth **1957**), *Seventh Series* (anth **1958**) and *Eighth Series* (anth **1959**); then, ed Robert P. Mills, *Ninth Series* (anth **1960**; cut vt *Flowers for Algernon and Other Stories* 1960), *Tenth Series* (anth **1961**), *Eleventh Series* (anth **1962**), *12th Series* (anth **1963**), *13th Series* (anth **1964**), *14th Series* (anth **1965**); then, ed Edward L. Ferman, *15th Series* (anth **1966**), *16th Series* (anth **1967**), *17th Series* (anth **1968**), *18th Series* (anth **1969**), *19th Series* (anth **1971**), *20th Series* (anth **1973**), *A Special 25th Anniversary Anthology* (anth **1974**) – not numbered, but treated as if so – *22nd Series* (anth **1977**), *23rd Series* (anth **1980**) and *24th Series* (anth **1982**). Other retrospectives are: *A Decade of Fantasy and Science Fiction* (anth **1960**) ed Robert P. Mills; *Once and Future Tales from the Magazine of Fantasy and Science Fiction* (anth **1968**) ed Edward L. Ferman; *Twenty Years of the Magazine of Fantasy and Science Fiction* (anth **1970**) ed Ferman and Mills; *The Magazine of Fantasy and Science Fiction: A Thirty Year Retrospective* (anth **1980**) ed Ferman; and *The Best from Fantasy & Science Fiction: A 40th Anniversary Anthology* (anth **1989**) ed Ferman. *The Magazine of Fantasy and Science Fiction, April 1965* (anth **1981**) ed Ferman and Martin H. GREENBERG is an enhanced reprint of the original contents of that issue, with added articles by the contributors; this issue was selected because it was the first to be edited solo by Ferman. Of special relevance are: *The Best Fantasy Stories from the Magazine of Fantasy and Science Fiction* (anth **1986**) ed Ferman, and *The Best Horror Stories from the Magazine of Fantasy and Science Fiction* (anth **1988**; in 2 vols US 1989; vt *The Best of Modern Horror* 1989 UK) ed Ferman and Anne Jordan.

Selective UK editions ran October 1953-September 1954 (12 issues) and December 1959-June 1964 (55 issues). The UK *Venture Science Fiction* (28 issues September 1963-December 1965) also comprised reprinted material from *F&SF* and *F&SF*'s former companion magazine *Venture SF* (2 series: 10 issues January 1957-July 1958, 6 issues May 1969-August 1970). There was a selective Australian reprint edition (14 undated issues 1954-8), and there have been several foreign-language editions, of which the most important was the French *Fiction* (412 issues, October 1953-February 1990), which soon developed into a separate magazine and became the backbone of the French sf and fantasy scene (◊ FRANCE). [MA]

MAGAZINE OF HORROR, THE US digest MAGAZINE, irregular (usually bimonthly or quarterly), 36 issues August 1963-April 1971, published by Health Knowledge Inc., New York; ed Robert A.W. Lowndes.

Primarily a reprint magazine, *MOH* was launched to cash in on the popularity of the PAN BOOK OF HORROR STORIES. Lowndes had a greater passion for the pulps, especially WEIRD TALES and STRANGE TALES, than for modern HORROR, and so *MOH* became a haven for lesser-known reprints, its contents predominantly reflecting the pulp era of the 1920s and 1930s. Among the scarce stories reprinted were fictions by Arthur J. Burks (1898-1974), Hugh B. CAVE, Robert E. HOWARD, Clark Ashton SMITH and Henry S. WHITEHEAD, To save money Lowndes reprinted mostly out-of-copyright material, including classics by Ambrose BIERCE, Robert W. CHAMBERS, Rudyard KIPLING, Richard Marsh (real name Bernard Heldmann; 1867-1915) and

Mary Wilkins FREEMAN. In this respect *MOH* had much in common with AVON FANTASY READER. Lowndes did the field a great service in bringing back into print "lost" stories, particularly works by Robert E. Howard, David H. KELLER and Wallace West (1900-1980). *MOH* printed very few original stories; exceptions were by Joseph Payne BRENNAN, August W. DERLETH, Emil Petaja (1915-   ) and Donald WOLLHEIM, and by new writers like Janet Fox (1940-   ), Stephen Goldin (1947-   ), Joanna Russ (1937-   ) and Roger ZELAZNY.

MOH had several companion magazines. *Startling Mystery Stories* (18 issues Summer 1966-March 1971) printed non-fantastic fiction alongside some weirder material, concentrating on mystery fiction with a strange slant. It reprinted many of the **Jules de Grandin** stories by Seabury QUINN, and also published the first stories by Stephen KING and F. Paul WILSON. The others were *Weird Terror Tales* (3 issues Winter 1969/70-Fall 1970) and *Bizarre Fantasy Tales* (2 issues Fall 1970-March 1971). [MA]

MAGAZINES Magazines as a recognizable medium first appeared in France in 1665 with *Le Journal des savants*, which discussed art, science and literature (no fiction). Fantasy, in the form of FAIRYTALES, first appeared in *Le Mercure Galant* in February 1696, with "La Belle au Bois Dormant" ("The Beauty Sleeping in the Wood") by Charles PERRAULT.

The fascination for fairytales, particularly Oriental ones, meant that FANTASY remained a fairly regular item in society magazines in both France and the UK. The Oriental SATIRES of John Hawkesworth (1715-1773) appeared regularly in *The Adventurer* (1752-4), and the first all-fiction magazine published in the UK, *New Novelists Magazine* (2 vols 1786-7), featured ORIENTAL FANTASIES as a regular part of its content. This continued into the 19th century, when the Gothic revival (◊ GOTHIC FANTASY) saw the steady emergence of the ghostly HORROR story. Several chapbooks and serial publications carried Gothic stories, though no single magazine was devoted to them; the closest was *The Marvellous Magazine* (1822), published in Ireland, which featured unauthorized versions, heavily edited, of popular Gothic novels and stories like Ann RADCLIFFE's *The Mysteries of Udolpho* (**1794**). A similar compendium magazine was *The Romancist and Novelist's Library* (1841-2) ed William Hazlitt the Younger (1811-1893), although this time the stories were authorized. This attempted to be a complete ANTHOLOGY of all exotic and adventure stories and novels then available, issued in a format similar to today's weekly partworks.

Magazines featuring fiction as a regular part of the content are really measured from BLACKWOOD'S MAGAZINE (1817-1980). It often published GHOST STORIES. Other 19th-century UK magazines to feature fantasy and SUPERNATURAL FICTION were *The New Monthly Magazine* (1814-84), *Bentley's Miscellany* (1837-68), *Dublin University Magazine* (1833-77) and in particular Charles DICKENS's *Household Words* (1850-59) and *All the Year Round* (1859-95), the latter promoting a special CHRISTMAS BOOK, with its festive offering of ghost stories by Wilkie COLLINS, Amelia B. Edwards (1831-1892), Elizabeth Gaskell (1810-1865) and others. *Temple Bar* (1860-1906), *Belgravia* (1866-99) and *London Society* (1862-98) were the main sources for SUPERNATURAL FICTIONS in the second half of the century. Ghost stories and fantasies were a regular component of US periodicals as well, especially in family magazines like *Harper's*

Monthly (1850-current) and *Atlantic Monthly* (1857-current), which in turn inspired the popular turn-of-the-century UK magazines, among them *The Strand Magazine* (1891-1950), *Pearson's Magazine* (1896-1939) and *The Idler* (1892-1911). Of special merit was *The Pall Mall Magazine* (1893-1937, merged with *Nash's Magazine* from 1914), which in the 1890s carried fantasy and supernatural fiction in almost every issue, including works by Grant ALLEN, F. ANSTEY, W.W. Astor (1848-1919), E.F. BENSON, Algernon BLACK-WOOD, Marjorie BOWEN, Bernard CAPES, R. Murray Gilchrist (1868-1917), Laurence HOUSMAN, M.R. JAMES, Fiona MACLEOD, M.P. SHIEL and H.G. WELLS.

Pall Mall was influenced by the FIN-DE-SIÈCLE mood that pervaded much literature of the 1890s. The Aesthetic movement that arose from this resulted in several magazines that relied heavily on the fantastic, primarily in art and poetry but also in fiction. The leading publication was *The Yellow Book* (1894-7) ed Henry Harland (1861-1905), best-remembered for its artwork by Aubrey BEARDSLEY who, after his dismissal from the magazine during the furore of the Oscar WILDE trial, went on to provide even more scandalous artwork and fiction for *The Savoy* (January-December 1896). The occult movement was also well under way in the 1890s and, though most of the occult and mystical magazines concentrated on nonfiction, a few became the home for hermetic fiction. The most notorious was *The Equinox* (10 vols March 1909-September 1913), a biannual hardcover magazine published and ed Aleister CROWLEY, and containing among the mumbo-jumbo a variety of his stories. Another source for OCCULT FANTASY was *Horlick's Magazine* (12 issues January-December 1904) ed Arthur MACHEN and carrying several of his own fantasies.

The Yellow Book had its equivalent in the USA: *The Clack Book* (12 issues April 1896-June 1897), ed Frank G. Wells, specialized in bohemian poetry and fiction, including works by John Kendrick BANGS, Edgar Fawcett (1847-1904) and Elia Wilkinson Peattie (1862-1935). Another in the same vein was *The Black Cat* (316 issues October 1895-April 1923; vt *The Thriller* from October 1919). This is sometimes credited as the first weird-fiction magazine, but it was not. It did specialize in stories of the unusual and inexplicable, but these were not necessarily fantastic – indeed, fantasy was specifically underplayed. It is best-known for having purchased the first fiction from Jack LONDON.

The first magazine to devote itself entirely to fantastic fiction was the German *Der Orchideengarten* ["The Orchid Garden"] (51 issues April 1919-May 1921) published and ed Karl Hans Strobl in Munich. It was heavily illustrated and carried a wide range of new and reprinted material from all over Europe, including the UK. Focusing on the surreal and the macabre, it was part of a growing movement in Germany at that time for satirical and GOTHIC FANTASY.

At the same time in the USA, the pulps were reaching their zenith, and the proliferation of specialist titles had started. The publishing company of Frank A. Munsey (1854-1925), who had issued the first regular pulp magazine, *The Argosy* (December 1882-November 1988 as *The Golden Argosy*; December 1888-September 1917 as *The Argosy*; October 1917-July 1920 as *Argosy Weekly*; thereafter merging with *All-Story Weekly* to become *Argosy All-Story Weekly* July 1920-September 1929), after converting it from a children's magazine in 1896, developed a number of titles which regularly ran weird and fantastic stories as part of their content, especially *The All-Story* (January 1905-March

1914; as *All-Story Weekly* March-May 1914; as *All-Story Cavalier Weekly* [merged with *The Cavalier*] May 1914-May 1915; as *All-Story Weekly* May 1915-July 1920; then merging with *The Argosy*), *The Scrap Book* (March 1906-January 1912) and *The Cavalier* (October 1908-January 1912; as *The Cavalier Weekly* January 1912-May 1914; then merged with *All-Story Weekly*), which began in 1905, 1906 and 1908 respectively. However, the first magazine to specialize in the form was *The* THRILL BOOK (March-October 1919), which picked up the post-WWI mood for the occult. The first magazine wholly to concentrate on fantasy and HORROR was WEIRD TALES (May 1923-September 1954, with later revivals). Under the editorship of Farnsworth WRIGHT in 1924-39, *WT* was the backbone of the fantastic in magazine form (as distinct from sf, which it also published but which established its own genre with *Amazing Stories* in 1926) and the main market for all the leading pulp writers of WEIRD FICTION; it established the careers of many, particularly Robert BLOCH, Ray BRADBURY, Robert E. HOWARD, Henry KUTTNER, H.P. LOVECRAFT and Seabury QUINN. Although never a significant commercial success, it made for itself a niche from which it ruled supreme. This encouraged a few rivals, including: *Tales of Magic and Mystery* (5 issues December 1927-April 1928) ed Walter B. Gibson (1897-1985), which focused more on stage magic and the occult; STRANGE TALES, the only significant rival to *WT* in its first decade; and *Mind Magic* (6 issues June-December 1931; retitled *My Self* from November 1931), which was more into SPIRITUALISM and psychic adventures, though attracting fiction from leading authors like Mary Elizabeth COUNSELMAN, Ralph Milne Farley (real name Roger Sherman Hoar; 1887-1963) and Manly Wade WELLMAN. A further title was GHOST STORIES, which contained purportedly true tales of HAUNTINGS and SÉANCES.

The weird and fantastic influenced many of the hero pulps of the 1930s, although most of these endeavoured to rationalize their often bizarre events as sf. This was true, too, of the various weird-menace pulps of the time, chiefly *Terror Tales*, *Horror Stories* and *Dime Mystery*. The only single-character pulp to be wholly supernatural in content was *Doctor Death* (3 issues February-April 1935), whose arch-villain, Dr Rance Mandarin, used occult powers in attempts to destroy civilization. The author of the three short novels involved was Harold Ward (1879-1950), writing as Zorro.

The sole other magazine of interest in the mid-1930s was *The Witch's Tales* (2 issues November-December 1936), based on the successful US radio series *The Witch's Tale* (1931-8), though bad management and poor distribution saw the magazine's early demise. Apart from original fiction by Alonzo Deen Cole (1897-1971) and Laurence D. Smith, both associated with the radio series, the magazine contained only reprints, with stories selected from the US edition of *Pearson's Magazine*.

During the 1930s the interest in SUPERNATURAL FICTION waned in favour of mystery and sf, but by the end of the decade there was a revival. Three magazines that appeared in early 1939 showed the three faces of fantasy: STRANGE STORIES, an unambitious imitation of *WT* which featured most of the same authors; UNKNOWN ed John W. Campbell Jr (1910-1971), which aimed to achieve for fantasy what *Astounding* had for sf; and FANTASTIC ADVENTURES, a lighter-hearted magazine ed Raymond A. Palmer (1910-1977), which at its rare best probably came closest to the pulp equivalent of SLICK FANTASY. The latter pandered primarily to US servicemen, with stories of LOST RACES and MAGIC powers. *Unknown*, however, genuinely changed the shape of fantasy.

The success of *Unknown* and *Fantastic Adventures* had a minor effect on the sf magazines. From its first issue, *Stirring Science Stories* (4 issues February 1941-March 1942) ed Donald A. WOLLHEIM had featured a self-contained and generally superior section called *Stirring Fantasy Fiction*, with work by Cyril Kornbluth (1923-1958), Robert A.W. Lowndes, David H. KELLER and Clark Ashton SMITH. Similarly, the hitherto sf-based *Future Fiction*, begun in 1939, changed its title to *Future Fantasy and Science Fiction* in October 1942 under the editorial direction of Robert A.W. Lowndes; although this new name lasted only three issues (until April 1943), it heralded a marked increase in the quality of the magazine's fiction, with work by Hannes BOK, Donald A. WOLLHEIM, Ross Rocklynne (1913-1988) and Lowndes himself.

Few of these magazines survived WWII. In the mid-1940s *WT* and *Fantastic Adventures* were the only fantasy magazines aside from the rather anomalous FAMOUS FANTASTIC MYSTERIES, primarily a reprint magazine, initially drawing upon the archives of the Munsey magazines and emphasizing the work of A. MERRITT. But there was then a revival of interest, spearheaded initially by the success of AVON FANTASY READER, really an ANTHOLOGY series but often regarded as a magazine. By now the pulps were starting to give way to digest magazines, and the success of *Avon Fantasy Reader* led to Mercury Publications issuing *The Magazine of Fantasy*, retitled *The* MAGAZINE OF FANTASY AND SCIENCE FICTION from #2. Although sf dominated its content for much of the 1950s, *F&SF* remains the major fantasy magazine of the past 40 years.

Academic interest emerged when August W. DERLETH issued *The Arkham Sampler* (◊ ARKHAM HOUSE), a short-lived but important title, the first to discuss SUPERNATURAL FICTION seriously; it printed both new and reprint fantasy and GHOST STORIES.

During the early 1950s there was a boom in the publication of digest magazines, especially of sf and mystery; fantasy and horror often took refuge within these. There were some specialist titles, however, of which the best were FANTASY MAGAZINE ed Lester del Rey and BEYOND ed Horace L. GOLD. Both emulated *Unknown*, although *Fantasy Magazine* focused more on fantastic adventure and *Beyond* on SLICK FANTASY. Hybrid sf/fantasy magazines included: *Imagination* (63 issues October 1950-October 1958), ed after #2 William L. Hamling (1921-   ), which followed in the style of *Fantastic Adventures*; *Avon Science Fiction and Fantasy Reader* (2 issues January-April 1953) ed Sol Cohen, a magazine continuation of AVON FANTASY READER; *Cosmos Science Fiction and Fantasy Magazine* (4 issues September 1953-July 1954) ghost-ed Laurence M. Janifer (1933-   ); *Mystic* (16 issues November 1953-July 1956; thereafter retitled *Search* and devoting itself to nonfiction) – a companion to Raymond A. Palmer's *Other Worlds* and *Fate*, it concentrated on paranormal and OCCULT FANTASY, and shifted more towards nonfiction after its early issues; and *Fantastic Universe* (69 issues June/July 1953-March 1960), which always projected an sf image but published much fantasy, notably SWORD AND SORCERY, especially during the editorship of Hans Stefan Santesson (1914-1975). *Fantastic Universe* had a brief fantasy companion, FEAR! (2 issues May-July 1960), which featured competent but

routine WEIRD FICTION. A similar magazine was *Shock* (3 issues May-September 1960). Both of these had much in common with three earlier short-lived magazines linked to popular radio series: *Suspense* (4 issues Spring 1951-Winter 1952), *The Mysterious Traveler Magazine* (5 issues November 1951-June 1952; #5 retitled *The Mysterious Traveler Mystery Reader*; ◊ MYSTERIOUS TRAVELER), and *Tales of the Frightened* (2 issues Spring-August 1957). The first two carried some fiction of interest, but only the first sought to go beyond traditional tales of the weird and unusual.

The other US magazine primarily containing fantasy in the 1950s was FANTASTIC. This began in 1952 mainly as a vehicle for slick fantasy, but budgets were cut and by the mid-1950s it was concentrating on sf, with occasional issues on wish-fulfilment; this last theme inspired a short-lived companion magazine, *Dream World* (3 issues February-August 1957). *Fantastic* returned twice, first in 1959-65 under Cele Goldsmith (1933-   ) and again in 1969-79 under Ted White (1938-   ); in the mid- to late 1970s it was a major market for S&S.

Fantasy was almost anathema in the US magazine market in the mid- to late 1950s, but an audience, albeit a small one, remained in the UK, which had had no specialist fantasy magazines during the pulp era. Some early general-fiction pulps featured WEIRD FICTION; e.g., *The Novel Magazine*, which in 1913-22 had a regular "uncanny tale" feature. For a period this magazine was edited by E.C. VIVIAN; he left in 1922 and joined Hutchinson's to start *Adventure-Story* and *Mystery-Story* (◊ HUTCHINSON'S MAGAZINES). Both carried much fantasy, especially *Mystery-Story*, which traded tales with *WT* and later *Ghost Stories*. After these magazines folded in 1929 there was no specialist market in the UK, although in the 1930s the pulp MASTER THRILLER SERIES (1933-9) included among its titles a number of one-off issues like *Tales of the Uncanny* (1934) and *Tales of Terror* (1937). During WWII the enterprising publisher Gerald G. Swan (1902-1980) issued several ephemeral magazines, all thin because of paper rationing. These started with *Weird Story Magazine* (1 issue August 1940), with contents written almost solely by William J. Elliott (1886-   ), but the magazines were soon masquerading as US publications under the general banner **Yankee Magazine**, of which three issues were entitled *Yankee Weird Shorts* (nd but January 1941, July 1941 and May 1942). Swan also issued *Weird Shorts* (1 issue nd but ?1943) and *Occult* (#1 nd but 1941; #2 titled *Occult Shorts* nd but ?1946), and reissued *Weird Story Magazine* (2 issues 1946). Swan's publications were still appearing in 1960, when he put out three issues of *Weird and Occult Library*, using material of WWII vintage.

Outlands (1 issue, Winter 1946) ed Leslie J. Johnson (1914-   ) was the only fantasy magazine proper in the UK until 1950. Despite their title the two separate magazines called *Fantasy* – the first had 3 issues 1938-9, the second 3 issues December 1946-August 1947 – were wholly sf. The latter, ed Walter Gillings (1912-1979), resurfaced in 1950 as SCIENCE FANTASY, initially as an sf magazine; but, under the editorship of John CARNELL, it became the UK's main fantasy market during the 1950s and 1960s. It published much *Unknown*-style fantasy by Kenneth BULMER and John BRUNNER, the historical fantasies of Thomas Burnett SWANN and the **Elric** stories of Michael MOORCOCK. It was the only significant UK market for fantasy in the late 1950s – the only other UK magazines to carry fantasy, mostly GHOST STORIES, were the short-lived PHANTOM and the much longer-lived *Supernatural Stories* (109 issues 1954-67), written almost entirely by R.L. Fanthorpe (1935-   ) and John S. Glasby (1928-   ), both writing at phenomenal pace.

The success of the HAMMER horror movies in the late 1950s saw a gradual resurgence of interest in horror fiction, spearheaded in the UK by the anthology series PAN BOOK OF HORROR STORIES (from 1959). This stimulated the re-emergence of horror ANTHOLOGIES in the UK, but had little impact on the magazine market. In the USA, however, the success was noticed by a small-time publisher, Health Knowledge Inc., who decided to issue *The* MAGAZINE OF HORROR in 1963. Although containing mostly reprints and heavily influenced by *WT*, under the skilful editorship of Robert A.W. Lowndes this magazine – and its later companions, *Startling Mystery Stories*, *Bizarre Fantasy Tales* and *Weird Terror Tales* – sustained and began to rejuvenate interest in WEIRD FICTION. *Startling Mystery Stories* carried the first professional fiction by Stephen KING and F. Paul WILSON. The resurgence of interest in HIGH FANTASY, following the success in paperback of the works of J.R.R. TOLKIEN and Robert E. HOWARD, had little impact on the magazines, although a short-lived magazine called WORLDS OF FANTASY was initially moderately successful.

It is ironic that, at the height of the fantasy paperback boom, the magazines dwindled in circulation – even *Fantastic*, which under Ted White published some of the most innovative fantasy of the 1970s. Of new titles, only COVEN 13 (later retitled *Witchcraft & Sorcery*) made any effort to be original. *The Haunt of Horror* (2 issues June-August 1973) was a highly illustrated digest-sized production that tried to repeat as a fiction magazine the success of MARVEL COMICS' horror COMICS; it carried some good fiction, but *#1* did not make a profit and so the publisher killed off the magazine. Fantasy remained in the hands of the reprint magazines, of which *Strange Fantasy* (6 issues Spring 1969-Fall 1970), *Weird Mystery* (4 issues Fall 1970-Summer 1971) and *The Strangest Stories Ever Told* (1 issue Summer 1970) drew solely from the archives of *Fantastic Adventures* and *Fantastic*. Better fare was offered by *Forgotten Fantasy* (5 issues October 1970-June 1971) ed Douglas Menville (1935-   ), which sought to reprint rarer (and out-of-copyright) material, and was more in the vein of *Famous Fantastic Mysteries*.

What now became noticeable was the emergence of the SMALL-PRESS magazine. These can generally be described as amateur magazines, in the sense that profit is not the sole reason for publication, and they sell mostly by subscription, newsstand distribution being limited. Some of the higher-circulation magazines may be classed as semi-professional; there is no clearcut distinction. The grandfather of these was *The* RECLUSE, published by W. Paul Cook (1880-1948) in 1927. In the 1930s the magazines produced by William L. Crawford (1911-1984) – *Unusual Stories* (3 issues March 1934-Winter 1935) and *Marvel Tales* (5 issues May 1934-Summer 1935) – although aimed primarily at the sf market, included some WEIRD FICTION. *Marvel Tales* published the first story by Robert BLOCH, as well as fiction by August W. DERLETH, H.P. LOVECRAFT, David H. KELLER, John Beynon Harris (1903-1969), Carl Jacobi (1908-   ) and Ralph Milne Farley (real name Roger Sherman Hoar; 1887-1963). *Fanciful Tales* (1 issue Fall 1936) ed Donald A. WOLLHEIM was devoted solely to weird fiction.

Although other amateur magazines appeared throughout

the 1940s, some professionally printed and assiduously edited – the best was *The Acolyte* (14 issues Fall 1942-Spring 1946) – it was only in the late 1950s that the small-press magazine as a vehicle for professional-quality fiction began to appear. Leading the way was *Macabre* (23 issues 1957-76), produced by Joseph Payne BRENNAN, intended to fill the gap left by the demise of *WT* in 1954.

The same motive saw the launch of WEIRDBOOK in 1968 and WHISPERS in 1973. These two became the leading small-press magazines of the 1970s and inspired many imitations. Both emulated *WT* in publishing a mixture of fantasy and supernatural fiction. The most lavishly produced small-press magazine was ARIEL, which received professional distribution; it placed the emphasis on fantasy art and graphics. Other quality titles in the fantasy field in the 1970s and early 1980s were CHACAL (later reissued as SHAYOL) and FANTASY TALES, and in the 1980s and early 1990s there were ARGOSY, FANTASY BOOK, MARION ZIMMER BRADLEY'S FANTASY MAGAZINE and, most recently, REALMS OF FANTASY.

The re-emergence of HORROR fiction as a market genre in the mid-1970s led to new magazines that focused more on this. They were led by ROD SERLING'S THE TWILIGHT ZONE MAGAZINE, which appeared in 1981, and its short-lived companion magazine NIGHT CRY. The success of commercial horror magazines in the professional market has remained limited, however, because of the age-old problems of distribution and shelf-life; since the 1980s, with the emergence of desk-top publishing, the main new fiction magazines have come from the small presses. Many lavishly produced, these have proliferated; notable are *The* HORROR SHOW, CEMETERY DANCE, *2AM* (21 issues Fall 1986-Spring 1993), GRUE and DEATHREALM. Others worthy of remark include: *Iniquities* (3 issues Autumn 1990-Autumn 1991) and *Midnight Graffiti* (7 issues June 1988-Fall 1992), both of which developed around the Splatterpunk movement; *Nyctalops* (19 issues May 1970-April 1991), initially nonfiction centred on Lovecraft but later developing into a fiction magazine of the *WT* school; *Eldritch Tales* (Winter 1975-current; *#1* titled *The Dark Messenger Reader*), which again seeks to emulate *WT* with predominantly Lovecraftian fiction; *The* CRYPT OF CTHULHU, a scholarly magazine concerned with Lovecraftiana; *Haunts* (1984-current), which only occasionally rises above the visceral into more spectral horror; and the latest revival of *WT*, retitled WORLDS OF FANTASY AND HORROR from 1994 (when it lost the licence to the title).

Beyond the small-press field, there is no specialist fantasy magazine, although both *The* MAGAZINE OF FANTASY AND SCIENCE FICTION and ASIMOV'S SCIENCE FICTION MAGAZINE feature fantasy and horror as a regular part of their content, as does *Interzone* (Spring 1982-current) in the UK. For a short period the UK had the very successful FEAR, which relied mostly on features and HORROR-MOVIE reviews but during 1989-90 was the UK's only regular market for horror fiction, although *Skeleton Crew* (6 issues July-December 1990), which grew out of a small-press magazine, and *Frighteners* (3 issues, July-September 1991) briefly joined the fray.

Outside the USA and the UK, few other English-speaking countries have had specialist magazines. Canada had *Uncanny Tales* (21 issues November 1940-September/October 1943), which mixed sf and fantasy and carried some native material, though relying mostly on US reprints.

More recently, Canada has had some good SMALL-PRESS magazines. In the fantasy field there has been the sequence published by Charles DE LINT: *Dragonbane* (1 issue Spring 1978) and *Beyond the Fields We Know* (1 issue Autumn 1978), the two merging as *Dragonfields* (2 issues Summer 1980-Winter 1983). These magazines focused on mythic fantasy in the FOLKTALE tradition. In the realm of DARK FANTASY there has been *Borderland* (4 issues 1984-6) ed Robert S. Hadji, which reprinted rare fiction alongside new material in the traditional vein.

Australia has never produced a straight fantasy magazine, though in 1970 *Sword and Sorcery*, a putative companion to Ronald E. Graham's *Vision of Tomorrow*, reached dummy stage before a poor financial deal killed it. *Void* (5 issues 1975-7), an sf magazine, published occasional fantasy. Not until *The Australian Horror & Fantasy Magazine* (5 issues Summer 1984-Fall 1985) did a specialist publication emerge in the small-press field, though it concentrated mostly on horror, in imitation of *WT*. The same applied to *Terror Australis* (3 issues Fall 1988-Summer 1992), which emphasized graphic visceral horror.

In Italy Francesco Cova produced the largely English-language KADATH (1979-82), but this interesting experiment has never been repeated.

Between the 1920s and the mid-1950s the specialist fiction magazine dominated the short-story market, but for a long time now the paperback ANTHOLOGY (and for younger readers the COMIC book) has been paramount – and increasingly GRAPHIC NOVELS (not to mention home videos) have played their part. Specialist fiction magazines have come to be seen as an anachronism, and survive predominantly by subscription and by distribution through specialist shops. Many devotees of horror and fantasy in book form are unaware of the magazines, although these remain the main testing ground for new writers, most of whom graduate to books only after first selling to magazines. Although prophets proclaimed magazines would become extinct in the 1970s, they continue to appear, and the leading ones remain moderately healthy. [MA]

Further reading: *Monthly Terrors: An Index to the Weird Fantasy Magazines Published in the United States and Great Britain* (**1985**) by Frank H. Parnell and Mike ASHLEY; *Science Fiction, Fantasy, and Weird Fiction Magazines* (**1985**) by Marshall B. Tymn and Ashley.

MAGGOTS We use the word in the sense of "fad or crotchet". FANTASY is often characterized by the persistence of outmoded notions as authors unthinkingly echo earlier texts. Some of these notions are dead science, others outmoded cultural assumptions. Individual authors can display personal maggots – their *idées fixes*.

Racism (of the pseudoscientific variety founded in SCHOLARLY FANTASY) and the closely linked notion of ATAVISM are two such, surviving as concepts in popular fiction, including fantasy, long after they had been discredited as science. Robert E. HOWARD, for example, takes for granted an incoherent set of ideas about the intrinsic superiority of whites. Ideas about innate racial talents and the specifics of prehistoric folk-wanderings inform much of Howard's HEROIC FANTASY, often mixed up with the racial and geographical speculations of THEOSOPHY; Howard's imitators have not always avoided echoing all this. It is noticeable that in the fantastic CINEMA there are remarkably few black HEROES AND HEROINES; this trend is far more pronounced than in mainstream cinema, and would seem to reflect the written

form: the FANTASYLANDS of GENRE FANTASY are almost exclusively populated by whites, except for the occasional VILLAIN. That said, there has in recent years been a backlash in written fantasy, possibly begun by Ursula K. LE GUIN's **Earthsea** series – with its Aryan-style villains and Polynesian-style good guys – so black (or yellow or brown) heroes and heroines have become more prevalent; this may soon permeate into the movies (◊◊ COLOUR-CODING). Even so, James HERBERT's recent HORROR novel *Portent* (**1992**), where sweaty black villains vie with nice white heroes, though excoriated by some critics, raised almost as few ripples among readers as similar racial stereotyping decades earlier in the novels of Dennis WHEATLEY. The Herbert example is interesting in that there is no suggestion at all of him being personally racist (quite the opposite): the maggot was perpetrated unconsciously.

It is specifically and consistently argued in the work of H.P. LOVECRAFT that it is precisely the degenerate state of modern Western humanity – as embodied in modernism and jazz and the criminal physiognomies of slum dwellers – which has left the world vulnerable to incursions by the ELDER GODS. Lovecraft's ideas about class and race, and his fears of SEX, thoroughly imbue the CTHULHU MYTHOS; it is usually possible to tell in his work the nonhuman or partly human infiltrating humanity by their possession of various stigmata of physical degeneracy derived from Cesare Lombroso (1836-1909) or from various advocates of racism. Too many of those who have adopted the mythos as a sort of SHARED WORLD have failed to consider the origins of some of these ideas and their ultimate consequences in the real world. Lombroso's ideas are still actively used by COMICS artists, who generally convey villainy through facial appearance.

In both Howard and Lovecraft, and generally in the work of LOST-RACE writers like A. MERRITT, we find the complex of ideas about primordial civilizations and LOST LANDS popularized by THEOSOPHY and by a variety of other cults. These often reflect the crudest – and largely discredited – form of archaeology's diffusionist hypothesis, whereby inventions are made once and once only and then are transmitted by the enlightened to the lesser breeds. Many lost-race novels have an intrinsically racist subtext, since the lost race usually has as its neighbours and nemeses crude caricatures of actual tribal peoples. The TARZAN MOVIES are very often guilty of this.

Various sexist stereotypes are informed by the discredited scientific orthodoxy of the late 19th century – the idea that women could cultivate intellect or physical prowess only at the expense of sexuality: the lovely heroine always twists an ankle when running away.

Less harmfully, various URBAN LEGENDS and discredited ideas – the idea that a dead person's eyes photograph the last thing they see, for example – were treated in all seriousness in earlier fiction and have to be regarded as maggots when they recur non-ironically. Various contemporary ideas – the complex around survivalism, Social Darwinist libertarianism, and belief in genetically programmed human aggressiveness is a case in point – are already discernible as maggots in the making. [RK]

MAGI ◊ MAGUS.

MAGIC Although notions of magic differ slightly from writer to writer, there is a remarkable consensus among fantasy writers, especially writers of GENRE FANTASY: magic, when present, can do almost anything, but obeys certain rules according to its nature. Generally ideas as to its nature are left undefined. Attempts to write to a system or define the rules – as in Lyndon HARDY's *Master of the Five Magics* (**1980**), E. NESBIT's *Five Children and It* (**1902**), Randall GARRETT's *Too Many Magicians* (**1967**) and Piers ANTHONY's **Xanth** series – can produce shallow and simplistic fantasies (◊ RATIONALIZED FANTASY). In the more generalized field of fantasy there is a huge, tangled complex of ideas concerning magic and magical practices; many varieties of magic are depicted, several of which tend to occur together, and all of which tend to melt into one another. Good examples of this tangling or melting are Geoffrey CHAUCER's "Squire's Tale" (*c*1390), George MACDONALD's *The Princess and Curdie* (**1877**), Anne Lawrence's *The Conjurer's Box* (**1975**), Fritz LEIBER's **Swords** series, Michael Scott ROHAN's *Cloud Castles* (**1993**) and almost any book by Katharine KERR or Terry PRATCHETT. All these make use of magic that involves two or more different assumptions. The most one can do when trying to pin down the use of magic in fantasy is to follow certain lines – certain *sequences* or *vectors* – of ideas. These are of two kinds and concern: what magic is seen to *be*; and what is *done* with magic.

The primary assumption is that magic is possible in the world of the fantasy, and the exact nature of this ambient magic strongly influences the narrative. The way this influence works is most easily seen in one of the best-known worlds of magic: the WONDERLAND of *Alice's Adventures in Wonderland* (**1866**) by Lewis CARROLL. Because Carroll was a mathematician, he started the book with ideas of logic (in the form of logic-chopping) and concomitant wordplay. Before the book has gone very far, we have the bread-and-butterfly and Humpty Dumpty busy with words. *Alice* illustrates also one of the most basic views: magic inheres in a particular SECONDARY WORLD or universe, but not in our own; Alice returns to mundane life at the end. This world or universe can be in a POLDER, as in the Land of OZ, to take another of the best-known examples, where magic is constantly present as a possibility (as also in Pamela DEAN's **Hidden Land** series); or it can be secretly present in our own world, when it is liable to be seen as a dwindling resource – K.M. BRIGGS's *Hobberdy Dick* (**1955**) shows this tendency with great clarity. If the magic is not THINNING in this way, then it tends to be derived from legend or history – as in Pat O'SHEA's *The Hounds of the Mórrígan* (**1985**). In other cases – like Terri WINDLING's **Borderland** series – the magic is probably also based on LEGEND but is brought in from outside. Emma BULL, in *War for the Oaks* (**1987**), took up the same notion of Elven Magic irrupting into the normal world, but the idea occurred much earlier in such books as Lord DUNSANY's *The King of Elfland's Daughter* (**1924**) and Hope MIRRLEES's *Lud-in-the-Mist* (**1926**), where the place with inherent magic is a land bordering on this mundane one, in a sort of reverse of the POLDER situation. In most cases, the magic sort of seeps in.

This segues into the notion of magic as a reservoir of power (as in Jo CLAYTON's **Diadem** series or Robert A. HEINLEIN's "Waldo" [**1942**]), usually in some between-world, which can be tapped, or stored by some individuals (as for instance in Sheri S. TEPPER's **Gameworld** series, where sorcerers act as batteries for magic). In most CELTIC FANTASIES a trance is needed in order to tap this reservoir – an obvious example of this approach is Patricia KENNEALY-MORRISON's **Keltiad** series. The most favoured method, however, is by RITUAL and invocation, as in Robert

HOLDSTOCK's *Thorn* (**1984** chap). Here magic is sometimes taken to inhere in the TRUE NAMES of things, people and GODS, as in Ursula K. LE GUIN's **Earthsea** series, or to reside in RUNES and symbols, though these can also be be used more mechanically as tools or passwords to unlock the magic, as in the preparation for the grand ritual in Barbara HAMBLY's *Dog Wizard* (**1993**).

From here it is a short step to seeing magic as bound up in SPELLS that make a certain pattern – often elegant ones, as in Steven BRUST's Mandarin elves' workings. This takes us on to worlds in which magic itself is seen as a pattern, usually of glowing threads (as in Ru EMERSON's **Nightthreads** series and Roger ZELAZNY's *Madwand* [**1981**]) which the operator must weave or pull. Magic is seen as even more solid where it is present in autonomous banks of mist – as in Patricia C. WREDE's *The Seven Towers* (**1984**) and P.C. HODGELL's *Seeker's Mask* (**1994**) – or in waves of a magic backlash – as in Mercedes LACKEY's *Storm Warning* (**1994**). Even solider is the notion that the magic is present in a person (amoral as in Zelazny's **Amber** or almost any VAMPIRE, evil as in J.R.R. TOLKIEN's Mordor, good as in his Gandalf or in Robert WESTALL's Cuddy) or a God or a sand fairy. Animals and mythical creatures are particular favourites as vehicles of magic – as in Elizabeth GOUDGE's *The Little White Horse* (**1946**), R.A. MACAVOY's *Tea with the Black Dragon* (**1983**), Mary BROWN's *Pigs Don't Fly* (**1994**) and Peter S. BEAGLE's *The Last Unicorn* (**1968**). Most solidly of all, magic can be present in a stone circle or megalith – as in Penelope LIVELY's *The Whispering Knights* (**1971**) – an EDIFICE – as in Charles DE LINT's *Moonheart* (**1984**) – a CITY – as in *The Far Kingdoms* (**1993**) by Allan COLE and Chris BUNCH – or an object – like Tolkien's RING. Quite often, two or more of these views are present at the same time.

Magic can also be viewed as a skill or ability, inborn or not. The most common assumption is that it takes inborn TALENT to work magic, as in Caroline STEVERMER's *A College of Magics* (**1994**), though a normal person can sometimes learn to do it mechanically. This ability tends to need an arduous education to develop and nearly always results in longevity, but it may also have to be trained without help (as in Peter DICKINSON's *The Gift* [**1974**]) or not trained at all, so that the user must proceed by guess (as in Andre NORTON's *Knave of Dreams* [**1975**] and Tanith LEE's *The Birthgrave* [**1975**]). Quite often the exercise of *ability* can be accidental and magic is then performed by shedding tears, or laughing, or forgiving an injury – as in the anonymous *St Erkenwald* (*c*1400), where the tears of a bishop inadvertently baptize a righteous heathen. An extension of this line of notions is the magic of innocence, where a child or childlike person can perform magic by simply doing or asking for what they think is right – this occurs in, for example, Susan Cooper's *Greenwitch* (**1974**). Where the *ability* is not inborn, it can be acquired suddenly by magical means (as in Dave DUNCAN's works) or as the gift of a magical being or object (as in Nesbit's *The Enchanted Castle* [**1907**]); or a person can learn by rule of thumb. Magic acquired by rote in this way normally invites disaster, as for the SORCERER'S APPRENTICE. Lastly, there is the complete lack of magic or ability, as when GAWAIN tries to use a girdle he thinks is a CHARM.

Then, too, magic can be seen as a sequence that runs through white magic, green magic, grey magic, black magic. Here the magic, however present, is seen as a *neutral* force used for differing ends, from extremely good to very evil, with green magic as a benevolent ecological activity. This same line can be viewed from another angle so that, starting again from the idea that magic is neutral, one has high ritual of a benevolent kind at one end, moves on to low or domestic magics in the centre, and thence arrives at dubious witchcraft and finally BLACK MASSES. Both these conceptions tend to occur together in the same books, as in Diana Wynne JONES's *A Sudden Wild Magic* (**1992**) and in *Good Omens* (**1990**) by Neil GAIMAN and Terry PRATCHETT. But all writers concur that nonhumans like DRAGONS and ELVES use magic either not present on these two scales or, as it were, into the infrared and ultraviolet. Here, it can be seen, views of the nature of magic are beginning to melt into notions about its use.

Transitional between the two is the notion of magic as largely ILLUSION. A few writers – like Chaucer in "The Franklin's Tale" and Paula VOLSKY in *Illusion* (**1994**) – claim that magic is entirely an illusion of great persuasiveness, but this position is hard to maintain. Most adopt a sequence that runs from pure illusion (in "Don't notice me" spells) through mixed transformation spells, where the illusion can be broken by the victim or an interested party seeing the truth, as in Robin MCKINLEY's *Beauty* (**1978**), to true transformational magic, where the change seen is actual; Tanya HUFF's *Sing the Four Quarters* (**1994**) is an almost perfect example of this entire sequence . . . which sequence melts into mind magic.

Quite a large body of fantasy takes mind magic along two parallel vectors, both concerned largely with practicality and both often present together. First is purely mental working, where the practitioner simply concentrates her/his potent thoughts to a desired end. At the centre of this sequence are the creation of magelight, SHAPESHIFTING, telepathy/mindspeech, telekinesis, precognition and mindreading (◊ TALENTS). Belgarath, in the **Belgariad** series by David EDDINGS, operates almost entirely on the central section of this sequence. But the sequence also stretches backwards to include memory, where recall of true facts either breaks a spell or brings power – as in Diana Wynne Jones's *Fire and Hemlock* (**1985**) – and forwards to take in knowledge as power. Knowledge of facts in the past, present or future is usually an important feature of mind magic. Innumerable WIZARDS of the Belgarath type are represented as very learned and often capable of some kind of PROPHECY or foreknowledge. The knowledge encompasses spells, weakness of an adversary, or the true names of objects or entities, whose powers may then be used – and so becomes the vector along which we find necromancy and the conjuring of DEMONS. The second parallel sequence takes the path of the spirit, where the practitioner is often in a trance, dealing with the maze, LABYRINTH or spiral of life and death, or dreaming (◊ DREAMS), either to acquire knowledge or to take direct action in a way which will affect the mundane level. Holdstock's *Lavondyss* (**1988**) operates almost entirely along this vector. In the centre of the sequence come magical INITIATIONS (as in Monica FURLONG's *Wise Child* [**1987**] and its sequel), druidism and SHAMANISM, but it also includes dealings with demons on the spiritual level at one end and fetches, mythagos, spirits, souls and deities at the other. Diane DUANE's *The Door into Fire* (**1979**) and its sequels make great use of this vector of mind magic, whereas her **Wizard** series operates almost entirely in the parallel practical kind.

The mind-magic sequence is criss-crossed by several other major ones. Today one of the most important is that

of MUSIC, which in most cases is seen as having the same functions as mind magic – as in Gael BAUDINO's *Gossamer Axe* (**1990**) – with a strong infusion from Celtic sources. Bards are frequent and usually magic-users. But, possibly because of the association of mind magic with Celticism, where the GODS are seen as behind only a thin veil, the powerful vector of cosmic magic crosses this sequence more or less at right angles. Here, always on a vast scale, the Earth, the Sun and often powers immeasurably further distant are brought into play as magical entities. *Good Omens*, A.A. ATTANASIO's *The Dragon and the Unicorn* (**1994**) and C.S. LEWIS's *That Hideous Strength* (**1945**) use this vector, and also share the inevitable consequence of it: that human beings, being so much smaller and weaker, tend to have a miserable time. Entities at the opposite ends of this scale, who can be good/bad or simply opposed to each other, not only clash but also bring into play gods, AVATARS, demons and lesser beings, who use the mundane world as their battleground. Oddly enough, the point where this vector intersects with others is usually over nature magic. Strong feeling for the LAND and its products are usually central. Patricia A. MCKILLIP's *The Riddle-Master of Hed* (**1976**) and its sequels show this admirably, and also show how this sequence tends to entangle itself in another that centres on sympathetic magic.

The sympathetic-magic sequence, which is intersected by mind magic as well, concerns correspondences between the acts of the practitioner and the physical world. Sympathetic magic, where tormenting a wax DOLL will torment the victim – or using hair/fingernails for the same purpose – is the central concept here. Writers who wish to put this more mystically invoke the notion of AS ABOVE, SO BELOW. This, of course, is the main principle of ALCHEMY, but the general idea of doing something in one place to have an effect in another also pulls in protective magic, usually concerning the setting of wards, SEX magic (as in Diana L. PAXSON's *Brisingamen* [**1984**]), MOON magic, planetary influences (◊ ASTROLOGY) and seasonal magic. These segue into more purely physical magics such as dancing, acts of worship, healing, nature magic (such as herblore and treelore) and earth magic. This is where this sequence entangles with cosmic magic, as it does very markedly in Tepper's **Ginian Footseer** series. Also along it we find the land-tie of kings, weather magic, dealings with ELEMENTALS, magic POTIONS and frequent assertions as to the potency of human exudations – tears, sweat, spit, piss and blood – though a sort of civility in most writers tends to confine them simply to tears and blood.

Blood magic forms a potent sequence on its own. Blood is seen by all writers as a great source of power. Dark practitioners will make a sacrifice, animal or human, to obtain this power; but a wizard or WITCH will also take blood, without killing the victim, in order to bond a human servant – as in, say, Hambly's *The Silicon Mage* (**1988**) – or use her/his own blood to power a spell, as when the witch in Hans ANDERSEN's "The Little Mermaid" (1937) remarks that she needs to add blood to the voice the MERMAID has given her. VAMPIRES are at the centre of this vector. Those of the DRACULA type take blood for food and power, while the more recent good vampires – as in Huff's **Blood** series – either drink only from criminals or, in great need, from a willing friend. This is just a step from the voluntary blood-bonding of blood brotherhood, or the voluntary sacrifice of blood to a spirit or god. At its extreme, this becomes the voluntary sacrifice of the *self* – as in Elizabeth MOON's **The Deed of Paksenarian** series. This is the source of the greatest power for good, but also approaches curiously closely to the practices of black magic. Here, what is *done* with the magic is paramount.

The same applies to a slightly lesser extent to the giving or taking of lifeforce or energy. A demon or black practitioner will drain a victim of lifeforce, while a good magician will frequently link hands with willing donors for greater magical power.

Slightly outside the main tangle of notions are the magic of deities, when they are seen as powerful but not omnipotent – such as the godlings in Rudyard KIPLING's "The Sing-Song of Old Man Kangaroo" (1902) – and the action of FATE, as in Kipling's *Rewards and Fairies* (coll **1910**), which acts to bring law to Britain. Manifestations of old and barely understood entities such as the WILD HUNT (as in Susan Cooper's *The Dark is Rising* [**1973**] and QUATERMASS AND THE PIT [**1958-9**]) seem to defy all efforts to involve them along most vectors.

There are also numerous oddball magics: things reduced in size to fit in nuts; cats produced from cauldrons; sailing in sieves and eggshells; flying brooms; spells cast to animate furniture (as in Nicholas Stewart GRAY's *The Apple-Stone* [**1965**]); and a myriad one-offs that defy categorization. Many of these derive from FOLKLORE or MYTHOLOGY and, like the Wild Hunt, seem to find it hard to blend with the magic of most fantasy. But all obey the basic tenet that, as mentioned at the start, magic, if present, can do almost anything. [DWJ]

MAGIC US movie (*1978*). 20th Century-Fox/Joseph E. Levine. **Pr** Joseph E. Levine, Richard P. Levine. **Exec pr** C.O. Erickson. **Dir** Richard Attenborough. **Spfx** Robert MacDonald Jr. **Consultant ventriloquist** Dennis Alwood. **Screenplay** William GOLDMAN. **Based on** *Magic* (**1978**) by Goldman. **Starring** Ann-Margret (Peggy Ann Snow), Anthony Hopkins (Corky Withers), Ed Lauter (Duke), Burgess Meredith (Ben Greene). 107 mins. Colour.

Struggling MAGICIAN Corky is told by his dying MENTOR that he is doomed to failure unless he can get some *charm*. Two years later he has that charm, in the shape of parodically matched ventriloquist dummy Fats, which he comes to believe is a living entity (◊ VENTRILOQUISM). Fame beckons, but fear of failure drives Corky into hiding. There he meets Peggy Ann, the girl he yearned for as a youth. He – or is it Fats? – commits a brace of murders. Peggy Ann, ignorant of the killings, plans to elope with Corky, but then Fats blows the gaffe. Rejected, Corky is urged by Fats to kill her, but instead kills himself . . . and thereby also Fats.

M inevitably loses in subtlety by comparison with Goldman's novel, which relies largely on a constant confusion of the reader's PERCEPTION for its effect; the plot itself is not new, having cinematic precursors at least as early as *The Great Gabbo* (**1929**). But *M* is a fine PSYCHOLOGICAL THRILLER and, although a RATIONALIZED FANTASY, involving enough that we come to share Corky's perception and to disbelieve the rationalization. Yet *M* is honest with us: through facial appearance and dress Fats is labelled as Corky's evil TWIN, his scatology and sexual innuendo being the thoughts Corky dare not utter himself; the equation of the two characters is further stressed by the fact that Hopkins did indeed ventriloquize Fats's lines. At movie's end we are left questioning our own understanding of what is and is not MAGIC. [JG]

MAGICAL MYSTERY TOUR UK movie (*1967* tvm).

Apple. **Pr** The BEATLES. **Dir** The Beatles. **Starring** The Beatles. 50 mins. Colour.

A party of people go off by bus on a magical mystery tour. During the day they see or experience various bizarre events – stage-managed by WIZARDS who live in the sky in a "secret place where the eyes of man have never set foot" – interspersed with several Beatles songs and one from The Bonzo Dog Doo Dah Band. State-of-the-art in its time, *MMT* now makes a pleasant period piece, with several sections retaining visual interest. The SURREALISM makes *MMT* an early ancestor of the modern ROCK VIDEO, the style of humour foreshadowed that in MONTY PYTHON'S FLYING CIRCUS. [JG]

See also: YELLOW SUBMARINE (*1968*).

MAGICIAN A term used in various ways in various places, and often interchanged with other terms, like sorcerer, MAGUS and WIZARD. But when the term is used in this encyclopedia it generally concerns conjuring – i.e., stage ILLUSION. Whether or not they perform real MAGIC, conjurers in works like MAGIC (*1978*), Jack Curtis's *Conjure Me* (**1992**), Christopher PRIEST's *The Prestige* (**1995**) and Lisa GOLDSTEIN's «Walking the Labyrinth» (**1996**) are magicians. [JC]

MAGICK Variant spelling of MAGIC attributed to Aleister CROWLEY; it has been suggested the additional "K" was intended by him as the initial letter of the Greek word for the female genitalia. Use of the term generally implies either that magic is being practised with malign or sexual intent or that the author wishes to "mystify" her/his ideas about magic. [CB]

MAGIC MOUNTAIN The English title of Thomas MANN's greatest novel, *Der Zauberberg* (**1924**; trans H.T. Lowe-Porter as *The Magic Mountain* **1927** US/UK). It designates the main venue of the tale, the Alpine sanatorium which Hans Castorp visits for a short stay, but where he remains for seven years (◊ TIME IN FAERIE), gaining worldly and metaphysical wisdom, seemingly safe from the world – until the onset of WORLD WAR I forces him to leave, in an apocalyptic climax to his world-historical RITE OF PASSAGE. But until then the MM is a POLDER, and has come to represent a secluded arena in which SECRET MASTERS of knowledge can quarrel and teach, though a great war looms. The MM is, in other words, tied to history in the end; and is neatly bracketed, from the other side of WWI, by James HILTON's SHANGRI-LA, which represents an ultimate escape from history. [JC]

MAGIC NUMBERS ◊ NUMEROLOGY.

MAGIC REALISM Term first used in 1920s Germany to describe some contemporary painters, and at first not easily distinguishable from SURREALISM, though distinctions did emerge. In Surrealism, there is no real concern for a default region or rhetoric which may be described as "real", no framing REALITY; surreal paintings and texts may be deemed FANTASTIC but are rarely FANTASY. In MR, by contrast, the regions of the real may be irradiated with dream imagery, dislocations in time and space, haunting juxtapositions, etc., but reality is the frame within which the narration, whether visual or textual, proceeds. The more extreme examples of MR are often fantastic; but many MR texts – in particular MYTH OF ORIGIN tales by the Latin American writers with whom the term is now generally identified – can be treated as fantasy.

By the early 1980s, MR had become a term mostly applied to writers like Jorge Luis BORGES, whose *Universal History of Infamy* (coll **1935**) subjects its various subjects to manipulations that make the fictional seem true, the historical seem imagined: but always within an ultimate frame that acknowledges the ongoing world. In the hands of Borges and those who followed him, MR is a technique of interpretation, and in fictions is almost invariably tied to STORY. MR is a way of telling the story of reality, and its deep popularity in LATIN AMERICA reflects the fact that old European modes had failed to capture the complex, interwoven, fabulous history of that vast region – MR is, in short, a way of approaching the MATTER of Latin America. English-language writers who have been described as magic realists include Brian ALDISS, James P. BLAYLOCK, Peter CAREY, Angela CARTER, E.L. DOCTOROW, John FOWLES, Mark HELPRIN, Salman RUSHDIE and Emma TENNANT.   [JC]

MAGIC ROUNDABOUT, THE French tv series (1963-71), with new soundtrack for BBC (1965-77) and Channel 4 (1992). **Dir** Serge Danot. **Writer** (of English-language version) Eric Thompson. **Created by** Danot. **Voice actor** Thompson (first series), Nigel Planer (second series). 160 5min episodes. Colour.

Although broadcast for children, this puppet/stop-motion animated series was widely watched by adults. The English-language narration by Thompson was what gave the series its great appeal: Danot's images were sufficiently loose that Thompson could supply virtually any story he wished – from surreal to topically satiric. In a brightly coloured garden live the characters Florence (a DOLL), Dougal (a bouncy dog), Dylan (a numbskull rabbit), Zebedee (a sort of jack-in-the-box without the box) and various other characters. "'Time for bed,' said Zebedee" became a catchphrase. *Pollux et le Chat Bleu* (*1970*; vt *Dougal and the Blue Cat* 1972 UK) was a feature-movie version.   [JG]

MAGIC SHOP ◊ SHOP.

MAGIC WORDS Familiar magical catchwords include "abracadabra", a near-palindrome used since the 2nd century as a CHARM written in AMULETS, and "hocus-pocus", possibly from the Welsh for a trick by a *pwca* (pooka) but also interpreted as a gibe at the Latin communion's "hoc est corpus"; both are now associated, like "Hey Presto!", with trickery and stage magic. MWs are often condensed SPELLS protected by pronunciation hazards: the Tetragrammaton conceals the NAME of GOD behind the letters YHWH or JHVH; in L. Frank BAUM's *The Magic of Oz* (**1919**), the MW whose correct pronunciation allows METAMORPHOSIS into any animal is "pyrzqxgl"; SUPERMAN's imp-like magical tormentor Mr Mxyzptlk can be returned to his OTHERWORLD dimension only by inducing him to pronounce his reversed name; the MW "Absarka" which can rehumanize the WEREWOLF hero of Anthony BOUCHER's "The Compleat Werewolf" (**1942**) is unfortunately not sayable by a wolf. The Deplorable Word uttered by Queen Jadis, which destroyed all life on the world Charn in C.S. LEWIS's *The Magician's Nephew* (**1955**), is wisely not printed; the words of power uttered in *Howl's Moving Castle* (**1986**) by Diana Wynne JONES are always lost in rolls of thunder. MWs of SUPERHEROES are personalized: CAPTAIN MARVEL's "Shazam" and Marvelman's "Kimota" work only for their owners.

MWs may also be passwords or magic keys, the most famous being "Open Sesame". Saying "mellon" (Elvish for "friend") opens the secret gate of Moria in J.R.R. TOLKIEN's *The Lord of the Rings* (**1954-5**), being actually written above the doors; the passwords XYZZY and PLUGH from the original fantasy computer GAME "Colossal Cave" are still remembered; an animated doorknocker in Terry

PRATCHETT's *Mort* (**1987**) parries an imperious request for admission with demands for the MW, which here is "Please".   [DRL]

MAGOG ◊ GOG AND MAGOG.

MAGRITTE, RENÉ (-FRANÇOIS-GHISLAIN) (1898-1967) Belgian artist who became one of the best-known painters of SURREALISM. His most famous images are created with painstaking, almost naive realism but show jarringly symbolic discrepancies and juxtapositions. TRANSFORMATIONS are implied in such works as "Découverte" ["Discovery"] (1927), a woman whose skin is partially wood, and "Le chant de la violette" ["The Song of the Violet"] (1951), whose business-suited men are STATUES. "Tentative de l'impossible" ["Attempting the Impossible"] (1928) shows RM himself creating a living woman in paint. Various TROMPE L'OEIL landscapes are obscured yet not obscured by depicted paintings of the same scene, as in "La condition humaine" ["The Human Condition"] (1933). "La durée poignardée" ["Time Transfixed"] (1938) has a railway engine bursting from a fireplace to float in mid-air; "La chambre d'écouté" ["The Listening Room"] (1952) upsets GREAT AND SMALL expectations with an apple that fills a room; French loaves levitate in "La Legende dorée" ["The Golden Legend"] (1958); another apple famously obscures the face of a bowler-hatted man – RM's trademark image – in "Le fils de l'homme" ["The Son of Man"] (1964). RM constantly teases viewers with reminders that his strange subjects are only paint on canvas: hence "La trahison des images" ["The Treachery of Images"] (1929), the celebrated picture of a pipe whose inscription "Ceci n'est pas une pipe" is not a paradox but simple truth. Michael BISHOP's surreal homage "A Spy in the Domain of Arnheim" (in *Pictures at an Exhibition* **1981** ed Ian WATSON) uses a variety of RM's props as scenery.   [DRL]

Further reading: *Magritte* (graph coll **1972**) ed David Larkin assembles 40 paintings; *Magritte* (graph coll **1992**) by Sarah Whitfield reproduces many major works with commentary and bibliography.

MAGUS One of several terms whose meanings often converge. Other terms in this grouping include MAGICIAN, seer, sorcerer and WIZARD.

The Magi were a priestly caste dominant in ancient Persia; Zoroaster (c660BC-c580BC) may have been one. Certainly the religious beliefs of the Magi, as far as they are known, have marked similarities to Zoroastrianism: there is a conflict between GOOD AND EVIL, with Good ultimately triumphant; and there is some anticipation of a MESSIAH. The wise men in the BIBLE who anticipate the coming of CHRIST are referred to as the Three Magi.

Over the centuries since, the term has gradually accreted a constellation of associations, as evidenced by the range of figures, both real and fictional, who have been called magi: Moses, Solomon, VIRGIL, Pythagoras, Simon Magus, Theophilus of Adana, Apollonius of Tyana, MERLIN, Roger Bacon, Cagliostro, PROSPERO, Saint-Germain, FAUST, John DEE, Sarastro, etc. A magus is likely to be male, elderly, wise, powerful and manipulative. He may stand as a LIMINAL BEING between the young man or woman and a goal, which may be maturity (◊ RITE OF PASSAGE; GODGAME), or wisdom or a kingdom. But, because he is dedicated to the gaining of knowledge, to the penetrating of arcana which may be forbidden, he may be dangerous; he may be a SECRET MASTER; he may be the head of an order or caste, either openly or in secret (◊ PARIAH ELITE); he may or may not be capable of MAGIC, but that ability will be secondary to his magus-hood.

Fantasy tales in which magi appear include Peter ACKROYD's *The House of Doctor Dee* (**1993**) and other novels about John DEE, Peter S. BEAGLE's *The Innkeeper's Song* (**1993**), which features a magus who (as Schmendrick in *The Last Unicorn* [**1968**]) was a wizard in earlier life, M. John HARRISON's *The Course of the Heart* (**1992**), E.T.A. HOFFMANN's *The Golden Pot* (**1814**), William KOTZWINKLE's *Fata Morgana* (**1977**), Amanda PRANTERA's *The Cabalist* (**1985**), Herbert ROSENDORFER's *The Architect of Ruins* (**1969**) and of course William SHAKESPEARE's *The Tempest* (written *c*1611; **1623**) and John FOWLES's *The Magus* (**1965**).   [JC]

MAHABHARATA ◊ SANSKRIT LITERATURE.

MAHY, MARGARET (MAY) (1936-   ) New Zealand writer, especially prolific in her earlier career for younger children, beginning with *The Dragon of an Ordinary Family* (**1969** US); there have been at least 30 more, including several collections: *A Lion in the Meadow* (coll **1969** US; exp vt *A Lion in the Meadow and Five Other Favourites* 1976 UK), *Leaf Magic* (coll **1976** UK), *The Great Chewing-Gum Rescue and Other Stories* (coll **1982** UK), *The Downhill Crocodile Whizz and Other Stories* (coll **1986** UK), *Mahy Magic* (coll **1986** UK; vt *The Boy who Bounced and Other Magic Tales* 1988) and a compilation, *The Girl with the Green Ear: Stories About Magic in Nature* (coll **1992** US). Her short work is sometimes reminiscent of Joan AIKEN. Fantasies for younger children include *The Man whose Mother Was a Pirate* (**1972** UK) and *The Boy who Was Followed Home* (**1977**), the followers being an increasing number of hippopotamuses.

MM is of greatest fantasy interest for her YA novels, usually centring on large families under stress. In *The Haunting* (**1982** UK) a young boy is haunted by his WIZARD uncle, who seems to have selected him as successor; but in fact the new wizard turns out to be a girl. It won a Carnegie Medal, as did *The Changeover: A Supernatural Romance* (**1984** UK), in which another family is disrupted by the coming into power of a member of the younger generation; in this case, a young woman becomes a WITCH through encounters with a young male witch to whom she is attracted, and saves her brother from an evil witch who is sucking his lifeforce. In *The Tricksters* (**1986**) a family is threatened by what seem the GHOSTS of previous owners of their summer home. In *Dangerous Spaces* (**1991** UK) two adolescent girls are drawn into a supernatural world.

MM's subject matter is relatively limited, but the swift fluent vivacity of her style, and the engrossing accuracy of her portraits of adolescents, makes her remarkable.   [JC]

Other works: *The Catalogue of the Universe* (**1985** UK), associational, about Tycho Brahe (1546-1601); *Aliens in the Family* (**1986**), sf; *The Greatest Show Off Earth* (**1994** UK), sf.

MAIDEN ◊ VIRGINITY.

MAILER, NORMAN (1923-   ) US writer and public figure who has been at the centre of US intellectual life since the publication of his first novel, *The Naked and the Dead* (**1948**). Although not identified with genre fiction, all of NM's later works contain elements of fantasy; he returns repeatedly to an imaginative world containing spirits, paranormal phenomena and magical correspondences. *Barbary Shore* (**1951**), which largely comprises a psychodrama between a secret policeman and a former revolutionary, takes an unexpected leap into FABULATION when the "little object" whose return the secret policeman seeks proves to be the revolutionary's ideals, and the novel dissolves into

ALLEGORY. *The Deer Park* (**1955**) was first projected as one of seven volumes that would constitute the successive DREAMS of an artist manqué whose compromised existence was reflected, though reshaped by mythopoeia, in his dream-life; "The Man who Studied Yoga" (1957), a long story, was the prologue to this (later abandoned) novel cycle.

Much of NM's career has been devoted to his "long novel": an enormous work that would force "an entrance into the mysteries of murder, suicide, incest, orgy, orgasm and Time". Most of the major works he has published since 1955 have been successive fragmentary attempts to write this book. "Advertisements for Myself on the Way Out" (1959) and "The Time of her Time" (1959) are fragments from one such assault on this, which is evidently to be a POSTHUMOUS FANTASY narrated by a murder victim. *Why Are We in Vietnam?* (**1967**) – a short novel fraught with telepathy and magical thinking (◊ TALENTS) – was written, NM later acknowledged, as the prologue for a long and violent novel about a senseless crime, from the writing of which he finally recoiled. *Ancient Evenings* (**1983**), his first long novel after *The Naked and the Dead*, proved to be another posthumous fantasy: it is the recollection, by one of the seven SOULS of a murdered young man in ancient Egypt, of hearing his great-grandfather relate the wonders of his long life and three earlier lives through a late evening in the company of the Pharoah. *Harlot's Ghost* (**1991**), even longer, presents the memoirs of a CIA agent whose career spans the history of that organization – a FRAME STORY involves the apparent GHOST of his dead MENTOR. Both novels are the opening volumes of uncompleted sequences. *Tough Guys Don't Dance* (**1984**) is a murder mystery charged with spirits, clairvoyance and MAGIC.

NM's career as a novelist has been shaped – perhaps deformed – by his struggle to resolve his own artistic sensibility with the tenets of the great literary Modernists who reigned during his formative years. The place of his fiction in his large and uneven oeuvre – and of the role of the fantastic in his fiction – remains to be assessed.   [GF]

Other works: *An American Dream* (**1965**), a novel filled with magic; *The Executioner's Song* (**1979**), nonfiction, though NM called it a novel; *Of Women and their Elegance* (graph **1980**), text (accompanying photographs) "narrated" by Marilyn Monroe (1926-1962).

MAITZ, DON (1953-   ) Award-winning US fantasy and sf painter and illustrator whose work is distinctive for its strong draughtsmanship and rich colour. His compositions are bold and simple, usually featuring a single figure or small group in an ornate, atmospheric setting, often with a touch of sly humour.

DM's first published work was a b/w advertisement (1975) in MARVEL COMICS's *Kull and the Barbarians* magazine. He painted some unusual covers for WARREN PUBLISHING's HORROR magazines, and created Captain Morgan for the Seagram's Rum advertising campaign, but the main bulk of his work has been for fantasy book covers. Some of his best-known work was for the original editions of Gene WOLFE's *The Book of the New Sun* (**1980-83**). In 1980 he was awarded both the Silver Medal of the American Society of Illustrators and the WORLD FANTASY AWARD, and he won the Hugo AWARD as Best Professional Artist in 1990 and 1993. Collections are *First Maitz: Selected Works of Don Maitz* (graph **1988**) and *Dreamquests: The Art of Don Maitz* (graph **1993**).   [RT]

MAJORS, SIMON ◊ Gardner F. FOX.

MALET, LUCAS Pseudonym of Mary Kingsley. ◊ Charles KINGSLEY.

MALIGN SLEEPER A notion particularly found in H.P. LOVECRAFT-derived HORROR, but frequently present in FANTASY. The forces of EVIL often include chthonic beings who can be awoken by accident – the classic high-fantasy examples are the Watcher in the Lake and the Balrog in J.R.R. TOLKIEN's *The Lord of the Rings* (**1954-5**), who are awoken partly by meddling adventurers in Moria and partly by a general rising tide of evil. Rarely, DARK LORDS may be awoken by meddlers, as with the Dominator in Glen COOK's *The White Rose* (**1985**).

Supernatural forces fought by OCCULT DETECTIVES like William Hope HODGSON's **Carnacki** have often lain asleep until awoken by those prying into That Which Humanity Is Not Meant To Know. In Lovecraft, what is awoken usually has something to do with the CTHULHU MYTHOS gods or with the lost races who worshipped or fought them; the awakening is accordingly a double threat, since the changes invoked in the awakeners' PERCEPTION of humanity's place in the Universe may well cause madness.

A good sf example of an MS is the Blight in Vernor VINGE's *A Fire Upon the Deep* (**1992**). [RK]

MALLARMÉ, STÉPHANE (1842-1898) French writer. ◊ DECADENCE.

MALORY, [Sir] THOMAS (?1416-?1471) English author of *Le Morte Darthur* (written before 1471; coll **1485**, including distinguished Preface by the book's printer, William Caxton [*c*1422-1491]), the title being usually written thus not on the basis of the first edition's title page (it has none) but in accordance with the colophon that ends the text. TM is often identified with the colourful Sir Thomas Malory of Newbold Revel in Warwickshire, although this identification is not certain, and a variety of other candidates have been proposed. *Le Morte Darthur* itself provides very little information about him, other than that he wrote it while in prison. The work occupies a key position in the development of Arthurian literature (◊ ARTHUR), and has widely become regarded as almost the canonical version. He drew mainly upon French sources, in particular the 13th-century prose ROMANCES known as the Vulgate Cycle. He also drew upon the Middle English alliterative poem *Morte Arthure*, the French *Perlesvaus*, and the 15th-century English stanzaic poem *Le Morte Arthur*. But TM's *Le Morte Darthur* is far from a simple translation, and he introduced many elements of characterization and motivation, as well as improving the consistency of the tales. It remains uncertain if any part of the work is his own invention, although a strong case has been put forward for "The Tale of Sir Gareth". The influence of TM on later writers of Arthurian material in English has been profound: many authors have based their account directly upon his. The definitive modern edition is *The Works of Sir Thomas Malory* (coll **1947** 3 vols; rev 1967; further rev by P.J.C. Field 1990) ed Eugène Vinaver (1899-1979). [KLM]

Further reading: *The Life and Times of Sir Thomas Malory* (1993) by P.J.C. Field; *A Companion to Malory*, (1996) ed E. Archibald and A.S.G. Edwards.

MALVERN, JEFFERSON D. [s] ◊ Kenneth MORRIS.

MAN AND HIS MATE vt of ONE MILLION BC (*1940*).

MANARA, MILO (1945-   ) Italian COMIC-strip artist, with a strong, realistic and economical line style. His work includes a series of political allegories, graphic-narrative examinations of and improvisations on historic events, and fantasy and sf stories, all with a strong erotic content. MM's drawing is interesting for its impressive realism and simplicity, in a style strongly influenced by the pen-line work of MOEBIUS.

MM worked as an assistant to the Spanish sculptor Berrocal before studying architecture. Inspired by the politics of such comics as Jean-Claude Forest's **Barbarella** and Pierre Barier's and Guy Peellaert's *Les Aventures de Jodelle* (graph **1966** France), he began drawing soft-porn strips, including the masked secret agent *Genius* (graph **1969**) and the female pirate *Jolanda de Almaviva* (graph **1969**). In partnership with Silverio Pisu he created the satirical monthly *Telerompo*. In 1974 he began producing historical strips for the weekly *Corriere dei Ragazzi* ["Children's Courier"] in the two series **Il Fumetto della Realta** and **La Parola ai Giurati**. In 1977 he created the "easy rider" strip **Chris Lean**, which featured a version of the actor James Dean (1931-1955).

In 1976, for the experimental strip monthly *Alterlinus*, MM drew the Chinese mythical stories of **The Monkey King** (◊ MONKEY) under the title **Lo Scimmiotto** (trans as **The Ape** in HEAVY METAL 1982-3; graph coll **1988** US), based on a script by Pisu, and the following year he adapted Joseph Roth's *Die Fluchte Ohne Ende* (**1927**) as *Alessio, il Borghese Rivoluzionario* ["Alessio, the Revolutionary Bourgeois"] (graph **1977**). He did two historical volumes for Larousse and one for Mondadori, plus a volume in the landmark series **Un Uomo un'Avventura** ["A Man, an Adventure"], *L'Uomo delle Nevi* (graph **1978**; trans as *The Snow Man* **1982** US). There followed *L'Uomo di Carta* (1981 *Pilote*; trans as *The Paper Man* **1986** US) and *Tutto Ricomincio con un'Estate Indiana* (1983 *Corto Maltese Magazine*; trans as *Indian Summer* **1987** US).

MM then began work on a series of complicated, surreal political allegories, **Le Avventure di Giuseppe Bergman**, which examines the predicament of the adventurer-outsider (in the sense that the term is used by Colin WILSON in *The Outsider* [**1956**]). In these sometimes chaotic stories he acknowledges an inspirational debt to Hugo Pratt ("H.P.") (1927-1995). Through the DREAM-like adventures of the ingénue Bergman (MM himself), he explores the role of the artist-adventurer who, in remaining outside the normal routine of capitalist society, acts as a revolutionary against the alienating and dehumanizing influence that society imposes. Four vols have been published in English: *HP and Giuseppe Bergman* (trans graph **1988** US), *An Author in Search of Six Characters* (trans graph **1989** US), *Dies Irae* (trans graph **1992**) and *Perchance to Dream* (trans graph **1992** US).

A similar quality pervades *Gita al Tulum* (trans as *Trip to Tulum* **1994** US), which was scripted by Federico Fellini (1920-1993). MM's SCIENCE-FICTION erotica includes *Il Gioco* ["The Game"] (**1983**; trans as *Click!* **1992** US), *Il Gioco II* (**1992**; trans as *Click 2* **1994** US), *Il Profumo del'Invisibile* ["The Scent of the Invisible"] (**1987**; trans as *Butterscotch: The Flavor of the Invisible* **1993** US), *Hidden Camera* (trans graph **1993** US) and *Nuove Civette* ["New Flirtations"] (**1989**; trans as *Shorts* **1994** US).

Other books in English include *The Art of Spanking* (graph **1994** US) and *The Art of Manara* (graph **1994** US). MM won the Yellow Kid Award at the Lucca Comics Festival in 1978. [RT]

MANDEVILLE, [Sir] JOHN Pseudonym of the unknown author of a collection of TRAVELLERS' TALES, *The Travels of*

Sir John Mandeville (*c*1357), written in French and rapidly translated and reprinted throughout Europe. An English edition appeared in 1360, and it was first printed by Wynkyn de Worde in 1499. There are several hypotheses concerning the authorship, and a Flemish physician, Jehan de Bourgogne (*d*1372), actually claimed it on his deathbed. Also, there was an Irish knight, Sir John Mandeyville, who killed an earl in 1333 and had to flee England; no more is known of him.

The book tells of a knight who left England in 1322 and travelled throughout the near and far East, including India and northern Africa, until forced by health to return home in 1357. The author used other sources extensively; it is not known how many of the place-descriptions were derived from experience – the elements concerning the Holy Land are candidates. What marked JM's *Travels* as different was the territory it covered. Nothing like it had been written since Marco Polo's *Travels* some 60 years earlier. Although it contained many accurate facts, JM was either quite happy to invent or was readily misled. It is from his book that we derive much of the legend of PRESTER JOHN. JM credits the occupants of the island Dundeya as having no heads but faces in their chests, while others there have a single eye in the middle of the forehead. In the land of Natumeran the people have heads like dogs. Near the city of Polumbum JM drank from the well of youth (◊ FOUNTAIN OF YOUTH).

This was the most popular travel book of the Middle Ages – consulted by Columbus prior to his first voyage – and the source of many MAPS by which medieval voyages were plotted. It was immensely influential in creating an image of the wonders of distant lands, and inspired travel and exploration.   [MA]

Further reading: *The Travels of Sir John Mandeville* (**1983**) trans and ed C.W.R.D. Moseley; *Sir John Mandeville: The Man and His Book* (**1949**) by M. Lettes; *The Rediscovery of Sir John Mandeville* (**1954**) by J.W. Bennett.

MAN FROM ATLANTIS, THE US tv series (1977-8). Solow Productions/NBC. **Pr** Herman Miller. **Exec pr** Herbert F. Solow. **Dir** Dick Benedict and many others. **Spfx** Tom Fisher. **Writers** Larry Alexander, Jerry Sohl and many others. **Novelizations** *The Man from Atlantis: Sea Kill* * (**1977**) and *The Man from Atlantis: Death Scouts* * (**1977**) by Richard Woodley. **Comics adaptation** *The Man from Atlantis* * (7 issues 1978) from MARVEL COMICS. **Starring** Victor Buono (Mr Schubert), Patrick Duffy (Mark Harris), Belinda J. Montgomery (Dr Elizabeth Merrill). 12 60min episodes plus 120min pilot. Colour.

ATLANTIS still exists. A violent storm washes a strange man ashore, and he is discovered by members of the Foundation for Oceanic Research. Suffering AMNESIA, he cannot remember where he came from, but his webbed hands and feet and his uncanny ability to breathe underwater through gills all point to his being from Atlantis. Taking the name Mark Harris, he joins the researchers on their various expeditions, hoping to discover a way back to Atlantis. Many of the stories pit the researchers against a master-VILLAIN, Mr Schubert, who is intent on ruling the world. Others feature aliens, an ELF, a MERMAID and, in one of the more unusual episodes, a trip through TIME with a re-telling of *Romeo and Juliet* (◊ William SHAKESPEARE). All these adventures are punctuated by the fact that Mark has to get back in the water every 12 hours or die. Mr Schubert and other villains do their best to dry him out.   [BC]

MANGA Japanese term first coined by the artist and teacher Hokusai (1760-1849) to describe the books of sketches and drawings he produced for the use of his students. Translated as "irresponsible pictures", the term may also be used to refer to cartoons or comic strips, and in this latter meaning has gained currency in the West. In Japan the term *gekiga* ("drama pictures") is often used to refer to the more serious comic books, as is the hybrid word *komikksu*.

Comics first began to flourish in Japan after WWII, since when the popularity of the distinctive, chunky small-format *manga* has been phenomenal, with examples reaching into all areas of Japanese life. Stories deal with a very wide range of subjects – fantasy, sf, horror, humour, romance, business, economics, sociology, sport, crime, cookery and pornography. A great many are adapted into animated movies (◊ ANIME).

The first examples translated into English mostly had an sf bias and included *Akira*, published in 34 vols (1988-92) (◊◊ AKIRA (**1987**), *Mai, the Psychic Girl* (1987-8 US), recounting the adventures of a girl with psychic abilities, and *Appleseed* (1988-94 US), a hard-sf series. *Manga* have gained greater and greater popularity in the 1990s, and this trend looks set to continue, with an increasing number of examples (often featuring gutsy, sexy nymphettes) appearing on newsstands.

As English-language *manga* proliferate in both the UK and the USA, a Japanese influence has begun to appear in indigenous comic books, with interesting examples of a hybrid drawing style becoming evident.   [RT]

Further reading: *Manga! Manga!: The World of Japanese Comics* (**1983** US) by Frederik L. Schodt.

MANGUEL, ALBERTO (1948-) Argentine-Canadian anthologist and fiction-writer. Jorge Luis BORGES left a mark on the young AM, who worked as a writer and editor in Milan and Tahiti before moving in 1982 to Toronto, where he was active as a cultural journalist. In 1993 he moved to Alsace; he currently lives in Paris. With Gianni Guadalupi he wrote *The Dictionary of Imaginary Places* (**1980**; exp 1987), which offers detailed descriptions, as if for the traveller, of realms that exist in writers' imaginations. AM's encyclopedic knowledge of Latin American literature (◊ LATIN AMERICA), his catholic tastes, his concern for new fiction writers from Canada and elsewhere as well as his translations from the Romance languages all inform the large anthologies *Black Water: The Anthology of Fantastic Literature* (anth **1983**) and *Black Water II: Further Fantastic Literature* (anth **1990**). Other theme anthologies are *The Oxford Book of Canadian Ghost Stories* (anth **1991**). His sole novel, *News from a Foreign Country Came* (**1991**), is nonfantasy.   [JRC]

MANICHEISM Syncretic religion founded in Babylonia by Mani (216-*c*276) which fused Christian and Buddhist ideas with the Dualist tradition of Zoroaster to produce alternative acccounts of the War in HEAVEN, the creation of ADAM AND EVE and many other aspects of the Christian mythos. The Christian Church condemned Mani's doctrines, and tended thereafter to consider all forms of Dualism – assertions that the Universe is the battleground of more-or-less equal forces of GOOD AND EVIL – as variants of the Manichean heresy.

Individual writers of considerable importance in the fantasy tradition who knowingly embraced some form of quasi-Manichean philosophy include William BLAKE and John Cowper POWYS, but it can readily be seen that the framing assumptions of modern GENRE FANTASY are

Manichean, typically involving incarnate forces of Good and Evil. The Zoroastrian opposition of Ormazd and Ahriman – symbolizing Light and Darkness – is recalled in *The Cosmic Puppets* (**1957**) by Philip K. Dick (1928-1982) and echoed in such Day/Night oppositions as the one featured in Roger ZELAZNY's *Jack of Shadows* (**1971**), while Thomas Burnett SWANN's Biblical fantasies *How are the Mighty Fallen* (**1974**) and *The Gods Abide* (**1976**) oppose Yahweh to Ashtaroth. Michael MOORCOCK's many SWORD AND SORCERY series set a significant trend by representing the fundamental conflict as Order versus Chaos, but the DARK LORD who symbolizes absolute EVIL remains a staple of HEROIC FANTASY. J.R.R. TOLKIEN's Sauron provided the most influential prototype; other significant AVATARS include Stephen R. DONALDSON's Lord Foul, David EDDINGS's Torak and Guy Gavriel KAY's Rakoth Maugrim. In these cases and almost all others the powers of Light are much less well defined, their hypothetical parent-figure remaining remote and unseen while HEROES AND HEROINES function as redeemers. [BS]

MANKOWITZ, WOLF (1924-   ) UK author and scriptwriter, now resident in Ireland, who began writing with *A Kid for Two Farthings* (**1953**), the tale of a young boy who mistakes his goat for a UNICORN. WM's work of genre interest includes: *The Biggest Pig in Barbados* (**1965**), a SUPERNATURAL FICTION involving VOODOO; *The Day of the Women and the Night of the Men* (coll **1977**), a sequence of FABLES told in a sophisticated version of Aesopian language (◊ AESOPIAN FANTASY); *Raspberry Reich* (**1979**) and *!Abracadabra!* (**1980**), both sf SATIRES; *The Devil in Texas* (**1984**); *Exquisite Cadaver* (**1990**), an exuberant POSTHUMOUS FANTASY; and *A Night with Casanova* (**1991**), consisting mainly of a conversation between the eponymous chevalier and the WANDERING JEW. Movies WM has scripted include *The Day the Earth Caught Fire* (*1963*), *Casino Royale* (*1966*) and *Dr Faustus* (*1967*). [JC]

MANN, JACK Pseudonym of E.C. VIVIAN.

MANN, THOMAS (1875-1955) German writer, one of the central figures of 20th-century literature; he won the Nobel Prize for Literature in 1929. His first stories, some with elements of SUPERNATURAL FICTION, appeared as early as 1894, and were assembled as *Der kleine Herr Friedemann* ["Little Herr Friedmann"] (coll **1898**; trans H.T. Lowe-Porter in *Stories of Three Decades* coll **1936** US/UK); his first novel, the famous *Buddenbrooks* (dated 1901 but **1900**; trans H.T. Lowe-Porter **1924** US/UK) is associational. His first major work of fantasy interest is *Der Tod in Venedig* (**1913**; trans Kenneth Burke in *Death in Venice and Other Stories* coll **1925** US; new trans H.T. Lowe-Porter as *Death in Venice* **1930** US/UK), in which the writer Gustav von Aschenbach engages in a possibly solipsistic *liebestod* in a VENICE haunted by a beckoning HERMES figure who may be DEATH. *Der Zauberberg* (**1924**; trans Lowe-Porter as *The Magic Mountain* **1927** US/UK) is not fantasy, but presents a vision of early-20th-century civilization that is so all-encompassing that Hans Castorp (the protagonist) can serve as an UNDERLIER figure for those who undergo a transformative RITE OF PASSAGE in fantasy narratives, and that the sanatorium or MAGIC MOUNTAIN where he spends seven years (◊ TIME IN FAERIE) can serve as a lesson in the imaginative use of the POLDER as a venue for the intensest learning possible (◊◊ SHANGRI-LA). The novel as a whole may represent a definitive defence of the civilization that WORLD WAR I devoured.

Joseph und Seine Brüder – published in four parts as *Die Geschichten Jaakobs* (**1933**; trans Lowe-Porter as *Joseph and His Brothers* **1934** US; vt *The Tales of Jacob* **1934** UK), *Der junge Joseph* (**1934**; trans Lowe-Porter as *Young Joseph* **1935**; vt *The Young Joseph* UK), *Joseph in Ägypten* (2 vols **1936**; trans Lowe-Porter as *Joseph in Egypt* **1938** US/UK), and *Joseph, der Ernährer* (**1943** Sweden; trans Lowe-Porter as *Joseph the Provider* **1944** US), all assembled as *Joseph and his Brothers* (omni **1960** US) – takes the BIBLE as a kind of PLAYGROUND, an arena of speculation about the nature of MYTH in which the characters themselves – their behaviour clearly dictated by STORY – act with knowing complicity. *Dei vertauschten Köpfe* (**1940** Sweden; trans Lowe-Porter as *The Transposed Heads* **1941** US) is based on an Indian LEGEND in which two men's heads are literally transposed through the agency of the GODDESS Kali, causing the same suicidal confusion that made them behead themselves in the first place; it was made into an OPERA by Peggy Glanville-Hicks (1912-1990).

Of more direct fantasy interest is *Doktor Faustus: Das Leben des deutschen Tonsetzers Adrian Leverkühn, erzählt von einem Freunde* (**1947** Sweden; trans Lowe-Porter as *Doctor Faustus: The Life of the German Composer Adrian Leverkühn as Told by a Friend* **1948** US), within the complex texture of which reposes a TWICE-TOLD version of the original FAUST story, overlaid by the main body of the text, which is set in the 20th century and depicts the PACT WITH THE DEVIL by virtue of which Leverkühn becomes a great composer (◊ MUSIC); his fate and the fate of Germany through two World Wars are seen as intertwined. [JC]

Other works: *Mario und de Zauberer* (**1930** chap; trans H.T. Lowe-Porter as "Mario and the Magician" in *Stories of Three Decades* coll **1930** US/UK), associational ALLEGORY of the psychology of totalitarianism; *Der Erwählte* (**1951**; trans Lowe-Porter as *The Holy Sinner* **1951** US); *Die Betrogene* (**1953**; trans Willard R. Trask as *The Black Swan* **1954** US).

MANNEQUIN US movie (*1987*). Gladden/20th-Century-Fox. **Pr** Art Levinson. **Exec pr** Joseph Farrell, Edward Rugoff. **Dir** Michael Gottlieb. **Spfx** Phil Cory. **Screenplay** Gottlieb, Rugoff. **Starring** G.W. Bailey (Felix), Kim Cattrall (Mannequin/Emmy), Andrew McCarthy (Jonathan Switcher), Meshach Taylor (Hollywood). 89 mins. Colour.

Sculptor *manqué* Switcher, employed menially at a department store, finds in its window the mannequin he made during a brief previous job. She comes to life – though only when they are alone – and tells him she is possessed (◊ POSSESSION) by the spirit of an Ancient Egyptian princess, Emmy, freed by the GODS from her patriarchal society and granted IMMORTALITY in this limited form. After many adventures and some sex, Emmy is granted the gift of normal life, so that she and Switcher may marry.

A routine but good-humoured fantasy (it has the feel of a tvm), *M* owes something to John COLLIER's "Special Delivery" – where the fantasy is more PERCEPTION-based – and a lot to the movie *One Touch of Venus* (*1948*), based on a play by Ogden Nash and S.J. Perelman, in which a windowdresser falls in love with a statue of Venus that obligingly comes to life; the basic theme is, of course, that of PYGMALION. *M* was successful enough to spawn a limp sequel, *Mannequin Two: On the Move* (*1991*), with a different cast. [JG]

MANNING-SANDERS, RUTH ◊ Raymond BRIGGS.

MAN WHO COULD WORK MIRACLES, THE (vt *H.G. Wells' The Man who Could Work Miracles*) UK movie

(*1936*). London/UA. **Pr** Alexander Korda. **Dir** Lothar Mendes. **Spfx** Ned Mann. **Screenplay** H.G. WELLS. **Based on** "The Man who Could Work Miracles" (1898) by Wells. **Starring** Joan Gardner (Ada Price), Joan Hickson (Effie), Ralph Richardson (Colonel Winstanley), Ernest Thesiger (Rev. Silas Maydig), Roland Young (George McWhirter Fotheringay). 82 mins. B/w.

Three GODS observe Earth, and one determines to give humans godlike powers; caution prevails, and he experiments by endowing only meek haberdashery assistant Fotheringay, who is just entering the pub. There the conversation turns to MIRACLES, and Fotheringay tries to demonstrate their impossibility by instructing the oil-lamp to flip over – which, to his astonishment, it does. His playful miracle-working – he finds the one thing his powers will not do is make local belle Ada love him – is next day noted by his employer and a banker, who suggest to him a partnership to exploit this for profit. Contrariwise, local vicar Maydig sees his miracles as opening up a future of peace and plenty for all. But Fotheringay rejects all advice. He conjures up a mighty palace, summons the leaders of the world and instructs them to sort out all inequities and injustices by nightfall. Being informed the Sun is about to set, he orders the Earth to stop spinning. As lone survivor of the resultant disaster, he miraculously restores all and finally ordains he shall no longer be able to work miracles.

For a comedy, this has a surprisingly sombre script, but there are also lightly inventive moments, as when Fotheringay impatiently tells a local constable to "go to Blazes", and we follow the unfortunate indeed to HELL (later commuted to SAN FRANCISCO). Young's performance exploits Wells' script to bring proceedings a not inconsiderable depth of dignity and poignancy. The movie ends with the moral dilemma unresolved: an incompetent god Fotheringay may have been, but was he any worse than the restored world-leaders will be? [JG]

MAPS In a number of allegorical TAPROOT TEXTS, notably DANTE's *The Divine Comedy* (written 1304-21), moralized landscapes are described in ways which involve geographical layout, and maps based on these descriptions were increasingly supplied in later editions. Jonathan SWIFT's *Gulliver's Travels* (**1726**) made a specific point of fitting its imaginary locations into unknown spaces in mundane geography, satirizing human knowledge by the implication that this authenticating device was as reliable, or otherwise, as those in TRAVELLERS' TALES. The adventure stories of the late 19th and early 20th centuries – e.g., those of Robert Louis STEVENSON and H. Rider HAGGARD – regularly came supplied with actual maps both as an authenticating device and in order to facilitate understanding of the text. J.R.R. TOLKIEN, in *The Hobbit* (**1937**) and *The Lord of the Rings* (**1954-5**), provided a map in much the same spirit that he provided endless glossaries and appendices. In imitation, almost all modern GENRE FANTASIES come equipped with a map, to the extent that maps are only much noticed when absent.

It has been remarked by Diana Wynne JONES in her «The Tough Guide to Fantasyland» (1996) that to see an extensive map at the front of a trilogy is to know that "you must not expect to be let off from visiting every damn' place shown on it". [RK]

MARA, BERNARD Pseudonym of Brian MOORE.

MÄRCHEN (plural *Märchen*) The word used in Germanic literature, and adopted into English, to describe the

FAIRYTALE in the widest sense. The word is often linked as *Volksmärchen*, meaning FOLKTALES, stories drawn from oral tradition. One of the earliest collectors of such stories was Johann MUSÄUS, with *Volksmärchen der Deutschen* ["German Folktales"] (**1782-7** 5 vols). The Germans created the word *Wundermärchen* (WONDER TALE) to describe those fairytales involving the supernatural and drawing on the same mythic or folkloristic roots but set in a SECONDARY WORLD or a timeless past. Although not describing them as such, Ludwig TIECK developed the *Wundermärchen* in his stories collected as *Volksmärchen* (omni **1797**). The primary German example of the *Kunstmärchen*, or LITERARY FAIRYTALE, is "Das Märchen" (1795) by GOETHE. When the GRIMM BROTHERS came to collect their FOLKTALES aimed at the public rather than academics, they called their first collection *Kinder- und Hausmärchen* ["Folktales for Children and the Home"] (coll **1812**; trans variously as *German Popular Stories*). The *Märchen* has continued to be a popular form of story in German literature. [MA]

MARCH OF THE MONSTERS, THE vt of *Destroy All Monsters* (**1968**). ◊ GODZILLA MOVIES.

MARGROFF, ROBERT (ERVIEN) (1930-   ) US writer. ◊ Piers ANTHONY.

MARIE DE FRANCE (*fl c*1160-90) French poet. ◊ ROMANCE.

MARION ZIMMER BRADLEY'S FANTASY MAGAZINE US large format MAGAZINE, quarterly, Summer 1988-current (#28 Summer 95), published and ed Marion Zimmer BRADLEY.

MZBFM is something of a spiritual successor to Dennis Mallonee's FANTASY BOOK, except that most of the contents are set in SECONDARY WORLDS. Most of the stories are HIGH FANTASY, drawing upon traditional images from FAIRYTALES and LEGENDS, most often of European origin, but occasionally Oriental (◊ ORIENTAL FANTASY) and Native American. There are a few SUPERNATURAL FICTIONS, usually tales of WITCHCRAFT. Some stories have strong feminist overtones (◊ FEMINISM); many are humorous or light fantasies, and even the macabre ones go easy on violence. The magazine supports new writers, many of whom Bradley is also developing in her companion ANTHOLOGY series SWORD AND SORCERESS, but more prominent names also appear, in particular Jo CLAYTON, Parke GODWIN, Phyllis Ann KARR, Mercedes LACKEY, Diana L. PAXSON, Jennifer ROBERSON, Darrell SCHWEITZER, Laura J. Underwood (1954-   ) and Elisabeth Waters (1952-   ). *MZBFM* is heavily illustrated, making a feature of its covers, and has used high-quality material by Alicia Austin (1942-   ), George Barr (1937-   ), David Cherry (1949-   ), Stephen Fabian (1930-   ), Ron Walotsky (1943-   ) and Janny WURTS; both Barr and Wurts have also contributed fiction. A representative anthology is *The Best of Marion Zimmer Bradley's Fantasy Magazine* (anth **1994**) ed Bradley. [MA]

MARL, DAVID J. (?   –   ) ◊ David ARSCOTT.

MARLEY, STEPHEN (MICHAEL JOSEPH) (1946-   ) UK writer. *The Life of the Virgin Mary* (**1988**), a fictionalized biography of the Madonna (◊ CHRISTIAN FANTASY), was ambiguously published and is thus often listed as nonfiction. It was followed by *Spirit Mirror* (**1988**), the first of SM's **Chia Black Dragon** series about a quasi-immortal Chinese VAMPIRE; the novel is rather over-richly written, but has much interest. *Mortal Mask* (**1991**), continuing the series, is SM's masterpiece to date: densely textured prose gives chilling effect to what is at once a piece of CHINOISERIE and a first-rate GHOST STORY – yet not a SUPERNATURAL FICTION but most decidedly a DARK FANTASY. SM continued the series with *Shadow Sisters* (**1993**), a disappointing romp. Since this novel SM has engaged in ties (see below) and CD-Interactive GAMES; it is to be hoped he will return soon to the rich potential offered by **Chia Black Dragon**. [JG]

Other works: Two **Judge Dredd** ties, *Dreddlocked* * (**1993**) and *Dread Dominion* * (**1994**), the first of merit and the latter of considerable merit; *Dreamweb* * (**1994** chap), game tie; *Managra* * (**1995**), a **Doctor Who: The Missing Adventures** tie.

MARLOWE, CHRISTOPHER (1564-1593) English playwright and, apparently, secret agent. The success of Marlowe's historical tragedy *Tamburlaine the Great* (performed 1587; **1590**) helped establish the Elizabethan theatre and made his name. All his plays are pseudohistorical, few containing any strong fantasy elements. *Dido, Queen of Carthage* (**1594**) with Thomas Nashe (1567-1601) draws upon VIRGIL's *Aeneid*. His most famous play, though, is fantasy – *The Tragicall History of Doctor Faustus* (performed ?1589; **1604**). CM closely followed *Historia von D. Johann Fausten* (**1587** chap) by Johann Spies (*d*1607) (◊ FAUST).

The conjecture over CM's life, death and possible role in writing SHAKESPEARE's plays make him a colourful character in historical novels; he appears also in *Strange Devices of the Sun and Moon* (**1993**) by Lisa GOLDSTEIN – where his secret-agent activities force him into adventures at court – and in various other fantasies. [MA]

MAROTO, ESTEBAN (1942-   ) Spanish COMIC-strip artist, with a fine decorative line style; his sensuous, moody, evocative fantasy artwork has been published widely and received many awards, including (1971) the Academy of Comic Book Arts Award for Best Foreign Artist.

In the early 1960s EM became art assistant to Manuel Lopez Blanco, with whom he produced episodes of *Las Aventuras del FBI* ["The Adventures of the FBI"]. He subsequently formed his own studio with Carlos Giminez (1941-   ), producing book covers, illustrations and comic strips including the popular Western **Buck John** and **El Principe de Rhodes** ["The Prince of Rhodes"], and some highly rated romance strips for the UK. In 1963 he joined the agency Selecciones Illustradas, and in 1967 he created the long and complex sf saga **Cinco por Infinito** (trans as *Zero Patrol #1-#10* 1984-8 US). In 1971 he created the SWORD-AND-SORCERY tale **Wolff: La Reina de los Lobos** ["Wolff: Queen of the Wolves"] for the magazine *Dracula* (trans *#1-#12* 1972-3 UK; graph coll **1973** US); in similar vein came *Manly* (trans as **Dax the Warrior** in EERIE *#39-#50* 1972-3). He became a prolific contributor to the WARREN PUBLISHING's EERIE, CREEPY and VAMPIRELLA, and has worked on the MARVEL COMICS characters **Satanna**, **Conan** and **Red Sonja** and the DC COMICS characters **Amethyst** and **Aquaman**.

EM is best when creating SECONDARY WORLDS, and seems distinctly uncomfortable with gory violence, as in *Dracula: Vlad the Impaler* (#1-#3 1993 US). His most recent US work has been *Savage Sword of Conan* (#218 1994). Meanwhile he continues to produce more delicate and decorative fantasy comics for the Spanish market. [RT]

MARQUIS, DON(ALD ROBERT PERRY) (1878-1937) US journalist, poet and writer who began his career as a columnist for Joel Chandler HARRIS's *Uncle Remus Magazine*, and who by 1912 had established himself as a newspaper columnist in New York. He remains famous for his **archy and mehitabel** sequence of BEAST FABLES in verse form, starring the CAT mehitabel and told by archy, a cockroach possessed by the soul of a dead poet, who types his narrative poems by jumping from key to key of a typewriter (hence the lack of capital letters). The sequence began in the *Herald Tribune* in 1916, and was assembled as *archy and mehitabel* (coll **1927**), *archy's life of mehitabel* (coll **1933**) and *archy does his part* (coll **1935**); the last two volumes were illustrated by George HERRIMAN, and all three were assembled as *The Lives and Times of archy and mehitabel* (omni **1940**). Individual episodes incorporate topical SATIRE, reports from mehitabel's cat-life, jokes, aphorisms and moments of existential melancholy. Most are set in NEW YORK, with an interval in LOS ANGELES. Other poems of fantasy interest appeared in *Noah an' Jonah an' Cap'n Smith: A Book of Humorous Verse* (coll **1921**), the title work being an AFTERLIFE fantasy. [JC]

MARRYAT, (CAPTAIN) FREDERICK (1792-1848) UK naval officer and, after retirement, writer. FM wrote at least 25 books during the last two decades of his life, most famously tales like *Mr Midshipman Easy* (**1836**), *Masterman Ready* (**1841**) – the first of his books for children – and *The Children of the New Forest* (**1847**). His first work of genre interest, *The Pacha of Many Tales* (coll of linked stories **1835**), was a PARODY of the *Arabian Nights* (◊ ARABIAN FANTASY) in which some TALL TALES of fantasy interest – the seven "Voyages of Huckaback" (◊ FANTASTIC VOYAGES), several of them to mysterious ISLANDS – are recounted to the entertainment-seeking pasha. *Snarleyyow, or The Dog Fiend* (**1837**) is a HORROR tale about a dog which cannot be destroyed.

The Phantom Ship (**1839**), FM's most important work of fantasy interest, is based on the FLYING-DUTCHMAN story. Episodic, melodramatic, and surprisingly grim (Vanderdecken's virtuous though pagan daughter-in-law is burned at the stake), the novel is a significant rendering of the TOPOS of the BONDAGE of IMMORTALITY. The protagonist (the Dutchman's son) spends his life – accompanied by the antic and vicious Schriften, a SHADOW figure – attempting to release his father through the agency of the fragment of the True Cross he carries with him. He dies at the moment of success. [JC]

MARSH, GEOFFREY Pseudonym of Charles L. GRANT.

MARSHALL, ALAN (1902-1984) Australian writer. ◊ B. WONGAR.

MARTENS, PAUL Pseudonym of Neil BELL.

MARTIN, GEORGE R(AYMOND) R(ICHARD) (1948-   ) US writer and editor whose first published story of genre interest appeared in 1971 and who quickly rose to prominence. GRRM's background as a journalist is reflected in his fiction, particularly the longer works. A matter-of-fact style of reportage is employed in *Fevre Dream* (**1982**), a historical VAMPIRE novel set in the 19th century on a Mississippi paddleboat. In *The Armageddon Rag* (**1983**) a former underground journalist investigates a murder and finds himself being led back into the shadowy realms of his previous career. Amid much sentimentality and nostalgia, he uncovers a supernatural plot to revive the "revolution" of the 1960s, revolving around the resurrection of Nazgûl, a near-mythical rock group from that era, and ultimately geared towards APOCALYPSE. In "The Ice Dragon" (1980) a little girl's emotional isolation is tied to the appearances of a DRAGON, which ultimately gives its life to save the girl's father, restoring both warmth to the girl and her relationship with her father. Loneliness is the central emotion in "The Lonely Songs of Laren Dorr" (1976), wherein a woman travelling via PORTALS from world to world in search of her lost lover encounters an inhabitant whose sense of desertion dwarfs her own. "Remembering Melody" (1981) is another tale of nostalgia (harking back to 1970), but this time marked by sadness as the characters are literally haunted by their past in the person of Melody, the ghost of a college friend who never achieved success or happiness. "The Monkey Treatment" (1983) is a black comedy about weightloss. An obese character is also at the heart of the offbeat "The Pear-Shaped Man" (1987 *Omni*; **1991** chap); GRRM won a BRAM STOKER AWARD for this tale. He has also won Hugo AWARDS for three of his short stories.

In the late 1980s, GRRM began devoting more time to tv work, writing for *The* NEW TWILIGHT ZONE (1985-8) and then for BEAUTY AND THE BEAST (1987-90). He also began developing (and editing) the SHARED-WORLD anthology series **Wild Cards**, which seeks to take a realistic look at how SUPERHEROES would affect and interact with society at large. The series comprises *Wild Cards: A Mosaic Novel* * (anth **1987**; vt *Wildcards* 1987), *II: Aces High* * (anth **1987**), *III: Jokers Wild* * (anth **1987**), *IV: Aces Abroad* * (anth **1988**), *V: Down and Dirty* * (anth **1988**), *VI: Ace in the Hole* * (anth **1990**), *VII: Dead Man's Hand* * (**1990**) with John M. Miller, *VIII: One-Eyed Jacks* * (anth **1991**), *IX: Jokertown Shuffle* * (anth **1991**), *X: Double Solitaire* * (**1992**) by Melinda SNODGRASS (assistant **Wild Cards** editor since *VI*), *XI: Dealer's Choice* * (anth **1992**), *XII: Marked Cards* * (anth **1994**) and *XIII: Black Trump* (anth **1995**). The associated **Card Sharks** sequence started with *Wild Cards: Card Sharks* * (coll of linked stories **1993**). The companion graphic-story series runs *Wild Cards #1: Heart of the Matter* (graph **1990**), *#2: Diamond in the Rough* (graph **1990**), *#3: Welcome to the Club* (graph **1990**) and *#4: Spadework* (graph **1990**), collected as *Wild Cards* (graph omni **1991**). [BM]
Other works: *A Song for Lya and Other Stories* (coll **1976**); *Songs of Stars and Shadows* (coll **1977**); *Dying of the Light* (**1977**), sf; *Windhaven* (fixup **1981**) with Lisa TUTTLE; *Sandkings* (coll **1981**); *Songs the Dead Men Sing* (coll **1983**; cut 1985 UK); *Nightflyers* (coll **1985**); *Tuf Voyaging* (coll of linked stories **1986**); *Portraits of his Children* (coll **1987**).
As editor: *Night Visions 3* (anth **1986**; vt *Night Visions* 1987 UK); ◊◊ SFE.
Further reading: *George R.R. Martin, The Ace from New Jersey: A Working Bibliography* (last rev **1989** chap) by Phil Stephensen-Payne.

MARTIN, GRAHAM DUNSTAN (1932-   ) UK writer. His first two novels, *Giftwish* (**1980**) and *Catchfire* (**1981**) were CHILDREN'S FANTASIES set in FANTASYLAND; two neighbouring kingdoms separated by an old spell are reunited by teenage protagonists. In the very dark *Soul Master* (**1984**), his first adult novel, an evil ruler uses MAGIC to take over the minds of the population.

GDM's later novels are more complex and more satisfying, with clear if slightly formal writing and an enjoyable Scottishness. *Time-Slip* (**1986**) deals with ALTERNATE WORLDS in a CONTEMPORARY-FANTASY manner, concentrating on the philosophy of a new RELIGION with no guilt, because all actions must happen in one world or another. *Half a Glass of Moonshine* (**1988**), his most interesting work to date, explores the overlaps between psychology, philosophy, parapsychology and metaphysics as it asks whether

appearances of the protagonist's dead husband represent a GHOST or if there is some other explanation; the question is left unresolved. [DVB]
Other works: *The Dream Wall* (**1987**), sf; *Shadows in the Cave* (**1990**), nonfiction about mysticism and quantum physics; several academic works.

MARTIN, JOHN (1789-1854) UK painter and illustrator, with a spectacularly melodramatic imagination, whose vast canvases depicting cataclysmic biblical scenes have informed much modern fantasy ILLUSTRATION. Even his delicate mezzotint illustrations for such editions as John MILTON's *Paradise Lost* (1824) have an epic quality. His paintings typically feature multitudes of tiny figures and fantastic architecture under turbulent skies. Known as "mad John Martin", he was the younger brother of Jonathan Martin (1782-   ), who was imprisoned in a London mental hospital for setting fire to York Minster in 1829.

JM was born of poor parents in Northumberland and apprenticed to a coachbuilder in Newcastle, learning to paint heraldic devices on the side panels. Released from his indentures, he finally settled in London, exhibiting at the Royal Academy from 1811. The paintings that made him famous were doom-laden Old Testament catastrophies such as "Joshua Commanding the Sun to Stand Still" (1816), "The Fall of Babylon" (1819), "The Deluge" (1826) and "The Great Day of His Wrath" (*c*1853). Despite the great popularity of his illustrations for Milton and the Bible, his career faltered, though revived when he painted "The Coronation of Queen Victoria" (1839). Thereafter his reputation declined again, but has recently risen once more. In the intervening years, much of his work has been lost. [RT]

MARTIN, JOHN Pseudonym of Dennis ETCHISON.

MARTIN, J(OHN) P(ERCIVAL) (1880-1966) UK author and Methodist minister whose popular **Uncle** sequence of absurdist CHILDREN'S FANTASIES comprises *Uncle* (**1964**), *Uncle Cleans Up* (**1965**), *Uncle and His Detective* (**1966**), *Uncle and the Treacle Trouble* (**1967**), *Uncle and Claudius the Camel* (**1969**) and *Uncle and the Battle for Badgertown* (**1973**). Uncle is an elephant characterized by immense wealth, lofty rhetoric and exotic dressing-gowns. His residence is a fantastically elaborate EDIFICE called Homeward, set in a post-industrial NEVER-NEVER LAND: its unexplored towers and outbuildings are full of strange machinery, secret doors and memorable eccentrics in the English NONSENSE tradition (many of them TALKING ANIMALS). Nearby, in slum-like Badfort, the unregenerate Beaver Hateman and his gang swill "Black Tom", wear revolting sack outfits, scribble furiously in their "hating books", and continually harass Homeward's inhabitants: many comic clashes ensue. JPM has been accused of excessive violence (for example, Uncle's cartoon-like habit of kicking foes high into the air when provoked) and also of "classism" . . . but this is to miss the subversive elements. Lip-service is paid to Uncle's rank and privilege, yet he is gently revealed as pompous, patronizing and not terribly bright – while the opposition is full of clever schemes and entertainingly wicked energy.

The stories, originally narrated by JPM to his children, tend to be rambling and episodic: *Uncle and His Detective*, with its treasure hunt for "dlog" (a cryptogram which Uncle alone cannot unravel), perhaps comes closest to having a plot. JPM's daughter Helen Estella Currey edited the manuscripts left after his death, producing the three posthumous novels. All six books are quirkily and appropriately illustrated by Quentin Blake. [DRL]

MARTIN, VALERIE (1948-   ) US writer whose first novel of fantasy interest, *A Recent Martyr* (**1987**), sets a tale of sexual and religious obsession against the rich, URBAN-FANTASY background of a NEW ORLEANS gripped by a mysterious blight. Her other relevant novels display a Postmodernist interest in the reinterpretation of existing texts (◊ RECURSIVE FANTASY). *Mary Reilly* (**1990**) is a delightfully matter-of-fact tale in which the JEKYLL AND HYDE story is recounted by Jekyll's maid; its subtext focuses, unexpectedly, not on the sexuality of Jekyll/Hyde but on that of the maid, so that the mundane is rendered more fantastical than the bizarre events going on in Jekyll's anatomy theatre. It was filmed as *Mary Reilly* (*1996*).

The Great Divorce (**1994**) reworks some of the material of Val Lewton's CAT PEOPLE (*1942*) and its remake, CAT PEOPLE (*1982*), into three interconnected narratives in which different kinds of what could be regarded as POSSESSION intertwine: a contemporary young woman in New Orleans, employed as a zoo attendant, develops (in the end fatally) the mindless sexual habits of a feline; a 19th-century woman undergoes, thanks to VOODOO, an IDENTITY EXCHANGE with an escaped jaguar just long enough to kill her brutal husband; a 20th-century researcher, who himself behaves like a strutting tomcat – and whose abandoned wife, the zoo's vet, is the tale's centre of gravity – is obsessed by the history of the murderess. The GRAND-GUIGNOL elements of VM's imagination may occasionally seem overstretched; but her tales more and more impressively make use of these elements to explore SEX, innocence and obsession. Her unobtrusive writing style is a delight. [JC/JG]
Other works: *The Consolation of Nature* (coll **1988**).

MARTINE-BARNES, ADRIENNE (1942-   ) US writer who has generally used her maiden name, Adrienne Zinah Martinez, for her contemporary and nonfiction works. While still in school she had two one-act plays produced, but then had a 20-year hiatus before publishing, as AM-B, her first novel, *Never Speak of Love* (**1982**), a contemporary tale. It was followed by her first fantasy novel, *The Dragon Rises* (**1983**), based on the premise that ARTHUR and DRACULA are aspects of the same soul. (It is the first book in a proposed tetralogy dealing with ARCHETYPES.) The tetralogy *The Fire Sword* (**1984**), *The Crystal Sword* (**1988**), *The Rainbow Sword* (**1989**) and *The Sea Sword* (**1989**) mixes historical fantasy and ALTERNATE WORLDS as the protagonists are drawn from the far future to fight various forces of darkness which are threatening to overthrow Albion. With Diana PAXSON she wrote a CELTIC-FANTASY trilogy about FINN MAC COOL: *Master of Earth and Water* (**1993**), *The Shield Between the Worlds* (**1994**) and *Sword of Fire and Shadow* (**1995**). She collaborated with Marion Zimmer BRADLEY on a *Darkover* novel, «The Shadow Matrix». An Arthurian fantasy about Sir Bors, «The Last Grail Night», is solo. [JF]

MARVEL COMICS US publisher of COMIC books, founded 1939 as Timely Comics by Martin Goodman (1910-1992). In the 1950s it adopted the name of its distribution company, becoming Atlas Comics; in 1963 it took a new name that honoured its first publication, decades earlier – *Marvel Comics #1*, November 1939. That publication had introduced **The Human Torch**, an android who could turn himself into a figure of fire, and **Prince Namor, the Sub Mariner**, a bellicose undersea monarch. The abiding popularity of these two, plus the masked superpatriot **Captain America** (introduced in *Captain America Comics #1*, March

1941), had proved a major factor in the company's consistent success. Another was the considerable communication and PR skills of editor and mainstay writer Stan LEE, who referred to his creative team as "The Bullpen".

During the 1950s most of the company's output consisted of fairly unremarkable anthology comic books, although the work of some artists, such as Gene Colan (1926-   ), Gray Morrow (1934-   ) and Al WILLIAMSON, generally scripted by Lee, was outstanding. In 1958 Lee began to inject more sf and HORROR into the stories, and then, in 1961, in collaboration with Jack Kirby (1917-   ), revitalized the SUPERHERO genre with the creation of **The Fantastic Four** (in *Fantastic Four #1*, November 1961), a group of superheroes constantly beset by personal problems while they fought various criminals. This notion of the superhero with relationship difficulties was further exploited in **Spider-Man** (introduced in *Amazing Fantasy #15*, August 1962), **The Hulk** (introduced in *The Incredible Hulk #1*, May 1962), **The Mighty Thor** (introduced in *Journey into Mystery #83*, August 1962), **The Avengers** (introduced in *The Avengers #1*, September 1963), **The X-Men** (introduced in *X-Men #1*, November 1963) and **Daredevil** (introduced in *Daredevil #1*, April 1964). Lee used his entrepreneurial skills to clarion his revamped comics line, headlining "The Dawn of the Marvel Age" and crediting the creative team with alliterative epithets ("Smiling" Stan Lee, Jack "King" Kirby, etc.). Regular readers ("True Believers") were awarded a "No Prize" for spotting errors; he created a readers' club called the "Merry Marching Marvel Society" with its own fanzine, *Foom* (Friends of Ol' Marvel). In so doing he established a unique rapport with his readers and took MC to the top of the sales charts. Another Lee/Kirby innovation was the "Marvel Universe", a consistent scenario in which their superheroes could encounter one another in crossover stories and team-ups (◊ SUPERHERO TEAMS).

Recent writers have tended to credit Kirby as the sole creative genius during this period, with Lee merely acting as front man: the view has some credibility but, at the very least, Lee embellished these ideas with tremendous flair.

Also among their creations were **The Watcher**, an intergalactic storyteller, **The Inhumans**, a superhuman WAINSCOT race, and **Galactus**, a planet-eating god whose herald, **The Silver Surfer** (introduced in *Fantastic Four #48* March 1966), is another durable MC character – visually and emotionally appealing, he became a kind of CHRIST metaphor for Lee, who used him to expound his own somewhat naive philosophies in *The Silver Surfer* (*#1-#18* 1968-70).

In October 1970 MC introduced their comics version of Robert E. HOWARD's SWORD-AND-SORCERY characters **Conan** (in *Chamber of Darkness #4*, April 1970) and **King Kull** (in *Creatures on the Loose #10*, March 1970). The former met with great success in *Conan the Barbarian* (◊ Barry WINDSOR-SMITH) and spawned many spinoff titles and characters including *Red Sonja* (◊ Frank THORNE).

The mid- to late 1970s were a low period for the company, with many fantasy and horror titles – e.g., *Amazing Adventures* (*#1-#39* 1970-76), *Astonishing Tales* (*#1-#36* 1970-76) – being cancelled, but in 1979 a new boost came with the introduction of Frank MILLER to the creative team, with *Daredevil #158*. He brought a new maturity to the storylines and narrative technique, thus extending the age-range of MC's readers. Other factors were attracting more mature readers to US comic books at this time, notably the rise of small, more creator-oriented publishers. MC responded to this challenge by introducing a De Luxe format, with slick paper, in some titles and the creation of the **Epic** line of quality fantasy. The high production quality was extended to MC's reprint titles in 1983 with their Special Editions, the precursors of what are now termed Trade Paperbacks (reprinted collections more durably bound and often billed as GRAPHIC NOVELS). Further team ups and spinoffs from MC's established lines proliferated, particularly associated with the **X Men**, but some new titles and character ideas, like *The Dazzler* (*#1* 1981), attained high sales. Talented artists like Walt Simonson (1946-   ) and Todd McFarlane (1961-   ) were able to breathe new life into old characters like **Thor** and **The Hulk**. A **New Universe** series introduced concepts unassociated with MC's traditional superheroes, but was not, for the most part, a success.

The 1990s saw "gimmick" covers with metallic finishes or bearing a hologram, some books being available inside a choice of covers, with "completist" collectors encouraged to buy every variant. In 1992 MC launched **2099 Universe** to link new innovations concerning characters like **Doom**, **Spiderman**, **Punisher** and **Ravage**.

MC remained a lively and innovative force in US comics publishing, with a market share often allegedly in excess of 60%, until it was acquired by an asset-stripper in 1994 and floated on the stock market. The number of monthly titles was slashed, and a consequent fall in quality created enormous resentment among both readers and suppliers. Profits fell: many of MC's creative staff were dismissed and leading Marvel characters and titles were subcontracted to rival companies.   [RT]

MARVELLOUS The marvellous represents a transcendence of the norm. The result of the marvellous will inspire and instil awe, not solely by the result but by the awareness of its achievement. To achieve the marvellous requires abilities which might normally be regarded as supernatural or superhuman, and certainly beyond the comprehension of EVERYMAN. The marvellous is thus akin to the miraculous (◊ MIRACLES), but with the added aura of the glorious. It is not the simple realization that MAGIC works, but the awe of appreciating that there is a power that makes it work. The marvellous underpins many of those TRANSFORMATIONS which are key to the WONDER TALE, and is particularly appropriate to CHILDREN'S FANTASY. It is harder to achieve in ADULT FANTASY, where the reader's cynicism combats the sense of wonder. It seldom works in HIGH FANTASY, an exception being in the climax to **Second Chronicles of Thomas Covenant the Unbeliever** by Stephen R. DONALDSON, where Covenant's eventual transcendence is akin to the marvellous. It also works at a more mundane level, as in *The Circus of Dr Lao* (1935) by Charles G. FINNEY, where the allure of the marvellous allows a circus to capture its audience through its wonders.   [MA]

MARVEL TALES ◊ MAGAZINES.

MARVELLOUS MAGAZINE, THE ◊ MAGAZINES.

MARY POPPINS US live-action/ANIMATED MOVIE (*1964*). Disney. **Dir** Robert Stevenson. **Spfx** Peter Ellenshaw, Eustace Lycett, Robert A. Mattey. **Anim dir** Hamilton S. Luske. **Screenplay** Don DaGradi, Bill Walsh. **Based on** the **Mary Poppins** series by P.L. TRAVERS. **Starring** Julie Andrews (Mary Poppins), Karen Dotrice (Jane), Matthew Garber (Michael), David Tomlinson (Mr Banks), Dick Van Dyke (Bert/Chairman of the Board), Ed Wynn (Uncle Albert). 139 mins. Colour.

London, 1910, and the Banks children, Jane and Michael, are so naughty that their nanny walks out. Mr Banks advertises for a replacement; they devise a "better" advertisement, but this he throws on the fire. The scraps of paper, borne skywards, reach the magical Mary Poppins, who routs all other applicants and gets the job. Soon they love her: she uses MAGIC to transform even the dullest chores into games. When they meet her friend Bert, a pavement artist and chimneysweep, she takes them into a LANDSCAPE he has chalked, and all four cavort among TOONS. Another day they visit Poppins's Uncle Albert, whose laughter causes literal levitation. Banks complains all this fun is bad for the children but soon after, on being sacked, realizes he has been an unaffectionate father and mends his ways. Poppins, her work done, departs.

MP gained five Oscars, including that for Special Visual Effects: the two animation/live-action sequences were astonishingly sophisticated for their time. While the image of Poppins floating through the air beneath her umbrella has become iconic (◊ ICONS), some of the rest today seems rebarbative rather than genial, as epitomized by Van Dyke's gratingly inadequate version of a Cockney accent. Reportedly, difficulties of negotiation with Travers, refreshingly unimpressed by DISNEY's reputation, prevented sequels.   [JG]

MARY SHELLEY'S FRANKENSTEIN (*1994*) ◊ FRANKENSTEIN MOVIES.

MASEFIELD, JOHN (EDWARD) (1878-1967) UK poet and writer, popular as a poet of the sea from his first book, *Salt-Water Ballads* (coll **1902**); Poet Laureate 1930-67. Some of his verse narratives – notably *Reynard the Fox, or The Ghost Heath Run* (**1919**) and *Minnie Maylow's Story and Other Tales and Scenes* (coll **1931**), a collection of FAIRYTALES – are of fantasy interest, though little read now, perhaps because JM's poetry as a whole fell profoundly out of favour after about the end of WWI. Some of his many plays – notably *Tristan and Isolt* (produced 1923; **1927**), a CHRISTIAN FANTASY trilogy comprising *The Trial of Jesus* (**1925**), *The Coming of Christ* (**1928**) and *Easter: A Play for Singers* (**1929**), and *A Play of Saint George* (**1948**) – are of some interest. His early adult novels – like *Multitude and Solitude* (**1909**) and *The Street of To-Day* (**1911**) – though nonfantastic, merit attention for an intensity and realism his later work conspicuously eschews.

Two collections – *A Mainsail Haul* (coll **1905**; exp 1913) and *A Tarpaulin Muster* (coll **1907**) – contain some SUPERNATURAL FICTION, and *Oldtaa* (**1926**) is set in a mundane IMAGINARY LAND. JM's main interest for fantasy lies in his work for older children. In *A Book of Discoveries* (**1910**) two young lads travel – in their imaginations – through a range of adventures, some supernatural. But the **Kay Harker** series – *The Midnight Folk* (**1927**) and *The Box of Delights, or When the Wolves Were Running* (**1935**) – is an important English fantasy sequence, told in a voice rather like that of a modernized, speeded-up Walter DE LA MARE. Young Kay Harker, a descendant of the eponymous hero of *Sard Harker* (**1924**), has lost his parents, and finds himself embroiled in a hunt for PIRATE treasure in the grounds of his family home. Aided by several TALKING ANIMALS and a WITCH, he travels into a "midnight" OTHERWORLD, a CROSSHATCH with this world and apprehensible, De La Mare-fashion, through DREAMS; he is threatened by BLACK MAGIC; he finds a new family. In the sequel, several years later, the adolescent Kay is once again pursued by the villains of the first tale, now in search of the eponymous magical box. In the course of his adventures, Kay encounters HERNE THE HUNTER, ARTHUR, SHAPESHIFTER wolves, etc. At points the impact is scattered, and JM's dis-ease with adult or near-adult characters is limiting, but overall the two tales make up a central English fantasy of longing. [JC]

MASK, THE US movie (*1994*). New Line/Dark Horse. **Pr** Bob Engelman. **Exec pr** Michael De Luca, Mike Richardson, Charles Russell. **Dir** Russell. **Spfx** Industrial Light & Magic. **Mufx** Greg Cannom. **Screenplay** Mike Werb. **Based on** the COMIC (1991-current) by Mike Richardson. **Novelizations** *The Mask* * (*1994*) by Steve Perry and *The Mask* * (*1994*) by Madeline Dorr. **Starring** Jim Carrey (Stanley Ipkiss/The Mask), Cameron Diaz (Tina Carlyle), Peter Greene (Dorian), Richard Jeni (Charlie), Peter Riegert (Lt Kelloway), Amy Yasbeck (Peggy Brandt). 93 mins. Colour.

Divers in the river disturb some old chests, and a wooden mask floats to the surface; it is the MASK into which ODIN banished LOKI for his eternal mischief. Lowly bank clerk and ANIMATED-MOVIE fan Ipkiss finds the mask. That night he dons it and is transformed pyrotechnically into The Mask – a TOON-type character capable of supernatural feats of speed and METAMORPHOSIS. Love for improbably beautiful nightclub singer Carlyle drags Ipkiss into a gang feud; coincidentally, The Mask is drawn into the same dispute. After many adventures, Ipkiss and doughty dog Milo save the day and the gal, and deliver the baddies to the law. Ipkiss and Carlyle – who has confessed it is gawky Ipkiss rather The Mask whom she loves – throw the mask away.

Extremely funny, *TM* is a highly sophisticated piece of fantasy open to several different readings aside from its straightforward magical-conversion-to-SUPERHERO veneer. Clues abound to suggest that all that alters when Ipkiss dons the mask is his own self-PERCEPTION; other clues suggest The Mask is a reification of Ipkiss's DREAMS, with one clue pointing to both these interpretations (a fragment of cloth cut from The Mask's tie transmutes into a shred from Ipkiss's pyjamas). It is also of interest to read *TM* as a converse-TOON movie: whereas toons are generally depicted as vulnerable intruders into our REALITY, The Mask is a solid creation, a toonified human being who is *more real* than the mundane world into which he explodes. [JG]

MASKED AVENGER Some HEROES are primarily inspired by their desire to seek VENGEANCE on those who have wronged them or theirs. One common pattern involves the hero donning a MASK and assuming a new identify to pursue evildoers; nonfantasy examples include **The Scarlet Pimpernel**, **The Green Hornet**, **Zorro** and **The Lone Ranger**; an MA with fantasy overtones is the COMIC-strip hero The PHANTOM. But the MA has also veered into pure fantasy, first with **The Shadow** – a crimefighter without supernatural powers in his pulp MAGAZINE, but granted the hypnotic power to "cloud men's minds" in his popular radio series – and then BATMAN.

In one version of the pattern, the hero is literally murdered but returns to life in a new identity: the Sunday-supplement comics hero **The Spirit** (◊ Will EISNER) gains a new identity but no new powers from his RESURRECTION, but the comic-book hero The SPECTRE is returned to Earth by GOD with virtually unlimited magical powers to fight crime; when revived in the 1970s, the character became a truly

grim avenger, usually killing evildoers. Other examples are the 1960s comic-book hero DEADMAN, a murdered circus acrobat who takes over other people's bodies, and the comic-book and movie hero of *The* CROW (*1994*).

Alternatively, the hero is not killed but horribly disfigured, the disfigurement being in effect a mask. The early comic-book figure **The Heap**, created by writer Harry Stein and artist Mort Leav in *Hillman's Air Fighters #3* (1942), was one such. His more famous descendant was the comic-book and movie hero **Swamp Thing**; also relevant is the hero of the movie *Darkman* (*1990*). A prominent avenger who is both deformed and masked is, of course, Gaston LEROUX's PHANTOM OF THE OPERA. In such incarnations the MA resembles the divided JEKYLL AND HYDE figure, as observed most clearly, perhaps, in the comic-book and tv hero The **Incredible Hulk**.

The MA offers rich possibilities for exploration, since he essentially is a good character created by evil circumstances – a point driven home in the climactic confrontation between Batman and The Joker in *Batman* (*1989*; ◊ BATMAN MOVIES). His obsessive drive for vengeance can make such a hero seem more like an ANTIHERO, or even a VILLAIN; the idea arises that a fight against EVIL pursued with excessive zeal might itself become evil. [GW]

MASK OF FU MANCHU US movie (*1932*). ◊ FU MANCHU.

MASKS The mask, in its multifaceted glory and significance, is a central thread in the story of the human species. The first law in any attempt to understand its use in fantasy is that, unless force has been applied, a mask does not remain fixed in place. It may be the covering that conceals a face of flesh (or, as often in fantasy, not-flesh) or it may be the face that the covering hides; but when it represents a covering the movement of the text will be to uncover what is hidden, and when the mask is an empty shell the movement of the text will be to ensure that it soon covers a countenance. At any one point in a tale, a mask *means more than it apparently does*; for, in literature, a mask almost always contains (and threatens to reveal) its opposite. To the reader, the introduction of a mask into a text inevitably arouses an expectation of change.

Very roughly, masks can be thought of as agents of transition, transformation, bondage or aspect-maintenance.

Transition The transitions signalled by masks can be further broken down roughly into three categories.

(a) When a transition is registered from the viewpoint or agency of any character with whom the reader is identifying, that transition can be seen as a passage from *here* to *beyond*. Masks in this sense represent (and in fantasy terms clearly *are*) PORTALS that enable significant THRESHOLDS to be crossed: between human and animal; between human and GODS and DEMONS; between this world and FAERIE or any other ALTERNATE WORLD (a task signalled, along with other forms of metamorphic transition, by the ten masks featured in the first half of Robert HOLDSTOCK's *Lavondyss* [1988], part of the **Ryhope Wood** cycle); between the present and the past (◊ TIME ABYSS); and between the present – often via a NIGHT JOURNEY – and the future EUCATASTROPHE which resolves the tale. The crossing of these various thresholds is fundamental to the QUEST structure of most GENRE FANTASY, and when masks are invoked in this kind of tale they are, therefore, most often seen as enabling devices.

(b) When the reverse is the case – when the passage is through the mask from *beyond* to *here* – masks represent (and, especially in HORROR, clearly *are*) portals between the animal and the human (as in the masks worn by various MYTHAGOS in Holdstock's **Ryhope Wood** books); between GOD or AVATAR or MALIGN SLEEPER and human (as in Arthur QUILLER-COUCH's "Oceanus" [1900], in which all the suffering humans in the great arena washed by God's tears wear masks which display one face only, that face being CHRIST's); between other REALITIES and this world (as in A.E.'s "The Mask of Apollo" [1904] and Roger ZELAZNY's and Thomas T. Thomas's *The Mask of Loki* [**1990**]); between the past and the present; between the primitive and the civilized (as in H. Rider HAGGARD's *The Yellow God* [**1908**] and innumerable other stories in which "native" masks encroach upon the boundaries that define civilized realities); and – unusually – between the future and the present.

(c) When the passage represents a movement of the world itself across a threshold, then the fantasy deployments of the mask merge with traditional patterns of usage throughout human culture. Any use within a fantasy text of masks to mark a passage of SEASONS may therefore generate a sense that the "real" world is being quoted, and that the mask may be in this context of less intrinsic significance in understanding how the particular story intends to proceed. In the end, we should perhaps keep in mind that, in human cultures, masks have at least two functions: they act to mark and to enable the passage, to make it liminal, so that it can be noticed and celebrated; and by their fixity they emphasize a pre-Christian sense that seasons and history (and the patterns of individual lives) return upon themselves, that they comprise CYCLES – that the STORY returns.

Transformation When the threshold is within the self and the sense of literal passage is therefore rendered metaphorical, masks represent a TRANSFORMATION, like that enjoyed by the protagonist of **The Mask**, either in comic-book form or as filmed (◊ *The* MASK [**1994**]) or by Uther Pendragon in the movie EXCALIBUR (**1981**), when Merlin transforms him into his enemy (so he can seduce Ygrayne, his enemy's wife), a magical operation signalled when the mailed helmet covering Uther's face changes suddenly into another; or the transition from illness to health (◊ FISHER KING) or vice versa; or the imposition of that internal change from an outside source, as in any story featuring a Medusa figure (◊ GORGONS) or a TOTEM-derived imposition of "true" being on a character, as in Lisa GOLDSTEIN's *A Mask for the General* (**1987**), Gloria Hatrick's *Masks* (**1992**) or Stephen MARLEY's *Mortal Mask* (**1991**); or the bearing, by human or god, of both poles of transformation simultaneously (◊ FACE OF GLORY), an image conveyed by any mask which can be understood right-side-up or upside-down, or any double mask which laughs or weeps depending on which way it is turned (◊◊ JANUS; TROMPE L'OEIL).

Bondage There is always a danger that the passage will not succeed, that the real – and perhaps abominable – visage will remain locked beneath the mask, until the pressure grows too great; it is a pressure of this sort that gives Edward LEAR's "The Dong with the Luminous Nose" (1877) its intensity of pathos, as the nose is a mask which disguises the poet himself. The pressing of a loathsome reality against a mask is a central motif of HORROR and defines much of the fantasy appeal of Gaston LEROUX's PHANTOM OF THE OPERA.

In fantasy any mask with a frozen grimace showing signs of tears is certain to have been presented consciously as a sign of bondage, as a sign of a being (or succession of beings, as in Henry KUTTNER's *The Mask of Circe* [1948;

1975], where generations of CIRCES are frozen into one identity by a single mask) which has become fixed. MALIGN SLEEPERS may on being awoken emerge from masks whose expressions seem fixated in grief or ire (the CTHULHU MYTHOS features many carven images which represent the ELDER GODS and their ardent lust to re-enter the world). The formal MASQUE, in which faces are covered, often represents a venue which has become stuck in time (masques are commonly found in DYING-EARTH tales), or which oppresses reality with its persistence. In Edgar Allan POE's "The Masque of the Red Death" (1842), masks perform at least two linked functions: the masked Prince Prospero and his courtiers clearly oppress reality with their persistence; but the Red Death itself is a mask, a horrific incursion of transformative truth.

Aspect-maintenance When masks are used as disguises (◊ MASKED AVENGER), they tend to serve simultaneously to obscure the face and to present a desired aspect of the disguised persons, enabling them to seem – indeed, in fantasy terms, to become – what they wish. The protagonist of **The Mask** clearly presents an aspect of his submerged, ideal self, as does BATMAN – and as does the protagonist of Max BEERBOHM's *The Happy Hypocrite: A Fairy Tale for Tired Men* (**1897** chap), who adopts a mask of goodness and finds that it has stuck. In general, the mask as aspect of the whole self serves to *streamline* that self for dangerous operations, adventures, impostures and roles, passages through portals, and rites.

An anthology whose contents reflect various of these uses is Ray Russell's *Masks* (anth **1971**). [JC]

Further reading: *Masks, Transformation, and Paradox* (**1986**) by A. David Napier, supplying an anthropological perspective; *The Mask of Glass* (**1954**) by Holly Roth (1916-1964), nonfiction concentrating on horror.

MASQUE Originally spelled "mask", a term used in the theatre to describe a form of entertainment, popular at the English court in the late 16th and early 17th centuries, which was apparently earlier known as a "disguising", and even earlier as a "mumming". It was often presented to royalty: the dramas and intrigues acted out in this artificial manner were, as it were, *gifts* for the monarch to delectate (and any concluding LICENZA-like scenes of forgiveness might also flatter the monarch). It was initially danced and acted in dumb show, with masks and costumes (as in the contemporaneous COMMEDIA DELL'ARTE) designating the identity and nature of the various participants; the tight predictable plot would normally involve elements of ALLEGORY, and the presence of various GODS. In the end, a dance involved both actors and audience.

The great innovator within the form was Ben Jonson (1572-1637), the first to use the French spelling "masque" – in the prefatory matter to his *The Fountaine of Self-Love; or Cynthia's Revels* (**1601**) – and whose most famous masque is *Oberon the Fairy Prince* (**1611**). In Jonson's hands, the form became very much more elaborate. Hugely intricate scenery and costumes were required, as were professional musicians; and in 1609 Jonson introduced the "anti-masque" (also known as the "antic masque"), in which a violent REVEL would be enacted and danced, creating a state of CHAOS only the full daylight masque might resolve (◊◊ PARODY). In all of this, the masque worked ultimately to flatter and validate the royal hierarchy – though John MILTON's *Comus* (performed 1634; **1637**) depicts the overthrow of the eponymous WIZARD. The heyday of the masque was brief;

after the start of the English Civil War in 1642, few were performed.

For 20th-century readers and spectators, it is probable that the masque is less interesting than its use as an ICON, or as an inserted sequence, in other kinds of performance or text. There are elements in several of William SHAKE-SPEARE's plays, most notably *The Tempest* (performed *c*1611; **1623**), where PROSPERO uses masque-like sequences – like the literal masque in Act IV – to impose GODGAME rituals on the cast. The masques in Edgar Allan POE's "The Masque of the Red Death" (1842) and in Gaston LEROUX's *The Phantom of the Opera* (**1910**) are centrally iconic (◊◊ *The* PHANTOM OF THE OPERA). To the extent that the life of inner cities is perceived as a form of theatre, URBAN FAN-TASIES tend to incorporate the complex artifice of the masque in plots involving disguises, ruses, foregatherings of masked citizens, REVELS, counterplots, revolutions, HIDDEN MONARCHS, crimes or VENGEANCE. G.K. CHESTERTON's *The Man who Was Thursday* (**1908**) closes with a literal masque. DYING-EARTH tales almost always incorporate an element of the masque, in which the wearing of MASKS and DISGUISES can stand for the capacity of those at the end of time to take on any identity they wish. [JC]

MASQUE OF THE RED DEATH, THE Roger CORMAN has made two movies with this title.

1. UK/US movie (**1964**). Anglo Amalgamated. **Pr** George Willoughby. **Exec pr** Nat Cohen, Stuart Levy. **Dir** Corman. **Spfx** George Blackwell. **Screenplay** Charles BEAUMONT, R. Wright Campbell. **Based on** "The Masque of the Red Death" (1842), with a subplot based on "Hop Frog" (1849), both by Edgar Allan POE. **Novelization** *The Masque of the Red Death ** (**1964**) by Elsie Lee. **Starring** Jane Asher (Francesca), Hazel Court (Juliana), Nigel Green (Ludovico), Patrick Magee (Alfredo), Vincent Price (Prospero), John Westbrook (Red Death), David Weston (Gino). 89 mins. Colour.

The land is ridden with plague, the Red Death, sent out to each village in the form of a rose by a mysterious red-robed stranger. Tyrant Prince Prospero quarantines his castle against all comers, and prepares with his guests for an orgiastic MASQUE. But the red-robed man also appears at the masque and, after identifying himself to Prospero as DEATH, or *a* Death – and not, as Prospero had hoped, an emissary of SATAN – spreads the plague among the guests, who lurch into a macabre DANCE OF DEATH; last to die is Prospero himself. In the movie's final sequence the red-robed man plays CARDS with a child – one of the six people spared by the plague – before joining company with four other Deaths of different colours to continue their duties elsewhere.

This is Corman's stab at mixing HORROR MOVIE with art movie, and a fairly good stab. The influences of Ingmar BERGMAN (most overtly, the red-robed man is a cognate of Death in *The* SEVENTH SEAL [**1956**]) and of Luis BUÑUEL are obvious, while the rather impressive VISION sequence experienced by Prospero's mistress Juliana after she has betrothed herself to Satan could be traced to any of a number of directors. The ethical dialectic that occasionally emerges is again Bergmanesque. *TMOTRD*'s appearance is often striking; this is generally attributed to its cinematographer, Nicolas Roeg, later an esteemed director in his own right (of, for example, DON'T LOOK NOW [**1973**]), but it is a simplistic assignment of credit to assume that Corman had no control over his own movie. [JG]

2. US movie (**1989**). Concorde. **Pr** Corman. **Dir** Larry Brand. **Screenplay** Brand, Daryl Haney. **Starring** Clare Hoak (Julietta), Patrick Macnee (Machiavel), Paul Michael (Benito), Jeff Osterhage (Claudio), Adrian Paul (Prospero), Tracy Reiner (Lucrecia). 79 mins. Colour.

A shallower, nastier and mercifully shorter rehash of most of the ingredients of *The* MASQUE OF THE RED DEATH (**1964**), marked by wooden acting (Osterhage and Reiner notably excepted) and an overall lack of ambition. The glorifyingly evil Prospero of the earlier movie is replaced by a Prospero who believes that sadistically wielding the power of life and death over others is part of his job. The eventual moral is that no mortal, only DEATH himself, has the right to command death. [JG]

MASTER, THE US tv series (1984). NBC. **Pr** Joe Boston, Nigel Watts. **Exec pr** Michael Sloan. **Dir** Ray Austin and many others. **Spfx** Phil Cory. **Writers** Tom Sawyer, Michael Sloan, Susan Woollen. **Starring** Sho Kosugi (Okasa), Lee Van Cleef (John Peter McAllister, The Master), Vincent Van Patten (Max Keller). 13 1hr episodes. Colour.

McAllister has lived in Japan and studied ninjutsu for 30 years, becoming a ninja master. He receives a letter from a daughter he never knew he had, and returns to the USA to search for her, thereby earning the wrath of his fellow ninjas. They order him killed for turning his back on their teachings, and send Okasa, another skilled ninja, to find and kill him. The series revolved around McAllister's search and his efforts to avoid Osaka. [BC]

MASTER THRILLER SERIES UK pulp series, with 32 individual titles, published quarterly by World's Work, Surrey; ed H. Norman Evans. Each issue was complete in itself and represented a different fictional category, so might arguably be called a pulp anthology series.

Most issues featured reprints from US MAGAZINES with occasional new UK stories (more prevalent in later issues). The series began with *Tales of the Foreign Legion* (anth July **1933**; all issues were undated). Those that published SUPER-NATURAL FICTION were *#6 Tales of the Uncanny* (anth September **1934**), *#17 Tales of Terror* (anth July **1937**), *#20 Tales of the Uncanny 2* (anth April **1938**), *#23 Tales of the Uncanny 3* (anth January **1939**) and *#32 Tales of Ghosts and Haunted Houses* (anth December **1939**).

Outside the series were two associated single-issue pulps: *Fireside Ghost Stories* (anth **1938**) and *Ghosts and Goblins* (anth **1938**). Although unremarkable, these were the UK's first specialist GHOST-STORY magazines. [MA]

MATHESON, CHRIS(TIAN LOGAN) (1958-) US writer of movie and tv scripts, son of Richard MATHESON, brother of Richard Christian MATHESON. His main credits of interest are for *Bill & Ted's Excellent Adventure* (**1989**), BILL & TED'S BOGUS JOURNEY (**1991**) and *Mom and Dad Save the World* (**1992**), all with Ed Solomon. [JC/JF]

MATHESON, RICHARD (BURTON) (1926-) US author of stories, novels, filmscripts and teleplays, whose work combines elements of sf, fantasy and suspense, but is coloured darkly by a mood of paranoid DARK FANTASY. His first published story, "Born of Man and Woman", appeared in *F&SF* in 1950 and was hailed as a groundbreaking sf mutant tale, even though he had not written it with the sf market in mind. The story of a freak child who schemes to escape its imprisonment by its parents, it introduced the theme that dominates most of Matheson's fiction: the individual alone in a hostile Universe, trying to survive. It served as the title story of his first collection, *Born of Man* *and Woman* (coll **1954**; with 4 stories cut vt *Third from the Sun* 1961), whose 17 stories showed his mastery at achieving a sense of realism through simple descriptions rendered in sleek, economical prose.

RM's first two novels, *Someone is Bleeding* (**1953**) and *Fury on Sunday* (**1953**), are crime tales but, like his later *Ride the Nightmare* (**1959**), they abound with concealed identities, hidden pasts and moments of horror that shock characters out of their hitherto secure worldviews. The notion of the ordinary world turning unexpectedly menacing serves also as a basis for his first two sf novels. *I Am Legend* (**1954**; vt *The Omega Man: I Am Legend* 1971) is the story of the last mortal in a near-future world ravaged by a plague that has turned everyone else into VAMPIRES: its power lies in its ingenious inversion of traditional vampire-story morality, in that the hero is ethically as suspect as the vampires. *The Shrinking Man* (**1956**) is concerned with a character exposed to a toxic cloud who begins shrinking irreversibly in size and finds the ordinary objects of daily life he hitherto took for granted becoming dangerous to his existence (◊ GREAT AND SMALL). Both novels display the free blending of genres that distinguishes much of RM's writing.

A Stir of Echoes (**1958**) introduced a theme that recurs in several of RM's novels: paranormal experience. A domestic drama in which a man discovers that the upsetting visions destroying his life and throwing his sanity into doubt are caused by the psychic residue left in his house by an unsolved neighbourhood murder, it is both a period paranoid fantasy on the dark underbelly of suburban life and a precursor of later works that feature more overt depictions of paranormal phenomena. In *Hell House* (**1971**) a team of occult investigators seeks the source of supernatural power that slaughtered previous teams who investigated a notorious HAUNTED DWELLING. The clumsily written erotic horror novel *Earthbound* (**1982** as by Logan Swanson; text restored as by RM 1989 UK) tells of a SUCCUBUS who siphons energy from victims in order to attain physical form and seduce them. These dark tales are counterbalanced by optimistic and often romantic later works. *Bid Time Return* (**1975**; vt *Somewhere in Time*), which won a WORLD FANTASY AWARD, concerns a LOVE that transcends time and unites the souls of two people separated by a century. *What Dreams May Come* (**1978**) is an AFTERLIFE fantasy supposedly drawn from reports of near-death experiences. RM has outlined the metaphysical underpinnings of his interest in the afterlife and psychic phenomena in a nonfiction book, *The Path: Metaphysics for the 1990s* (**1993**).

RM's best-known work is his short fiction. Most of his stories have been collected in *The Shores of Space* (coll **1957**), *Shock!* (coll **1961**; vt *Shock 1: Thirteen Tales to Thrill and Terrify* 1979), *Shock II* (coll **1964**), *Shock III* (coll **1966**), *Shock Waves* (**1970**), *Shocks IV* (coll **1980** UK) and *Richard Matheson: Collected Stories* (coll **1989**). His fiction output dwindled considerably in the late 1960s and 1970s, during which time he devoted his efforts to writing for tv and CINEMA. RM's earliest screen credits were adaptations of his novels. *I Am Legend* was filmed as *L'Ultimo Uomo Della Terra* (**1964**; vt *The Last Man on Earth*), but RM disavowed its rewritten screenplay; it was filmed again as *The Omega Man* (**1971**), scripted by others and disliked even more by RM. *The Shrinking Man* was filmed as *The Incredible Shrinking Man* (**1957**) from his script and won a Hugo AWARD in 1958. RM also did (among much other tv work) 14 screenplays for *The* TWILIGHT ZONE and scripted the

KOLCHAK MOVIES. His most interesting work has debatably been his adaptations from other writers, particularly BURN WITCH BURN (**1961**), a treatment of Fritz Leiber's *Conjure Wife* (**1943**), and his renderings of Edgar Allan POE's stories for Roger CORMAN, among them *The* HOUSE OF USHER (**1960**), *The* PIT AND THE PENDULUM (**1961**), *Tales of Terror* (**1962**) and *The Raven* (**1963**). But his collaboration with William F. NOLAN on the teleplay for *Trilogy of Terror* (**1975** tvm), based on three of his own stories, and adaptations of *Hell House* (as *The* LEGEND OF HELL HOUSE [**1973**]) and *Bid Time Return* (as *Somewhere in Time* [**1980**]) are also notable.

RM returned to fiction writing in the 1990s with Westerns, among them *Journal of the Gun Years* (**1991**) and *The Gunfight* (**1992**). *Now You See It* (**1995**), a Locked Room mystery in a stage-magic ambience, shows that he has not lost his interest in paranoid scenarios centred on deceptive identities and false realities. RM's awareness of the very qualities by which his work is recognized is most evident in *7 Steps to Midnight* (**1992**), a suspense thriller and self-conscious pastiche in which the protagonist becomes embroiled in a series of interlinked intrigues that he realizes are taken from the plots of the many different sf, fantasy, horror, crime and romance novels he reads.   [SD]

Other works: *The Beardless Warriors* (**1960**); *Through Channels* (**1989** chap); *Shadow on the Sun* (**1994**); *By the Gun* (coll **1994**); *The Memoirs of Wild Bill Hickok* (**1996**).

As editor: *The Twilight Zone: The Original Stories* (anth **1985**), with Martin H. GREENBERG and Charles G. Waugh.

Further reading: *Richard Matheson: He is Legend: An Illustrated Bio-Bibliography* (**1984** chap) by Mark Rathbun and Graeme Flanagan.

MATHESON, RICHARD CHRISTIAN (1953-) US author and screenwriter, the son of Richard MATHESON and brother of Chris MATHESON. Most of his work has been as scriptwriter and story consultant on many tv series, including *The A-Team*, *Quincy*, *Knightrider*, *Magnum*, *The Incredible Hulk* and AMAZING STORIES; he shifted to movies in the mid-1980s. He spent 18 months as a parapsychologist at UCLA in 1975-6, a period which provided some authenticity to his fiction, which started to appear with "Graduation" (1977 *Whispers*), exploring the degradation of a student in a college which starts to possess (◊ POSSESSION) its students. RCM became connected with the Splatterpunk movement, rather by association with fellow writers than by his own fiction, which is more psychological (◊ PSYCHOLOGICAL THRILLER) and shows the influences of not only his father but such others as Ray BRADBURY and Truman Capote (1924-1984). His best fiction blurs humour and the bizarre, as in *Holiday* (1982 *Rod Serling's The Twilight Zone Magazine*; **1988** chap), where a man meets SANTA CLAUS while he is on holiday. Most of his works are exercises in PERCEPTION, as his protagonists attempt to make sense of their increasingly strange lives. A collection is *Scars, and Other Distinguishing Marks* (coll **1987**; exp 1988). *Created By* (**1993**) is a nonfantastic horror novel.   [MA]

MATSON, NORMAN (1893-1965) US writer whose work is best-known today by virtue of its spinoffs into other media; *The Passionate Witch* (**1941**), which he completed from a posthumous fragment by Thorne SMITH, was filmed as I MARRIED A WITCH (**1942**) dir René CLAIR, which years later became a basis for the tv series BEWITCHED (1964-72). NM's first solo fantasy, *Flecker's Magic* (**1926**), is an unadventurous tale of a US art student in Paris who acquires a

magic RING but is too pusillanimous to use it. *Doctor Fogg* (**1929**), in which James Elroy FLECKER appears in a minor role, is the tale of a scientist whose apparatus conjures up a beautiful naked blonde; it too tends towards the moral that the ordinary is to be preferred to the extraordinary. Smith certainly did not believe this, so *The Passionate Witch* runs contrary to its initiator. The compensation partly made by the movie version was supplemented by NM in his solo sequel *Bats in the Belfry* (**1943**), in which the hero begins to regret preferring his secretary to the WITCH as a life-partner.
[BS]

MATTER A Matter – the 12th-century French poet, Jean Bodel, was the first to use the term, describing the Arthurian legends (◊ ARTHUR) as the Matter of Britain – is an accumulated network of MYTHS, LEGENDS and FABLES about a nation or a culture. A Matter is understood by those who write or sing it, or who listen to recountings of it, or who read it, etc., as the *true* STORY of that nation or culture. Whether it is factual is secondary. It is likely to contain or refer to a MYTH OF ORIGIN; it is likely to describe the father of the nation or culture as a HERO.

Because it incorporates either a MYTH OF ORIGIN (which usually involves a GOD or PANTHEON), or a line of descent from the days of that myth, a Matter points to the past, incorporating sanctioning myths about that past. A Matter will normally also incorporate large elements of the supernatural.

Several Matters can be identified, beyond the Matter of Britain. There is the Matter of Troy and Classical Greece, which is as close as can be to a Matter written by the enemy; the Matter of Hellenistic Greece, telling of Alexander the Great and his companions; the Matter of Rome, which focuses on its founding; the Matter of France, which focuses on its defence at the time of Roland (*d*778); the Matter of America, which is in essence a frontier myth; the Matter of each country of LATIN AMERICA, being divers myths of origin, usually composed in MAGIC-REALISM terms by novelists like Miguel ASTURIAS or Gabriel GARCÍA MÁRQUEZ; and there is a fantasy Matter, which may be called the Matter of the World as unfolded in the INSTAURATION FANTASY. There is, oddly, no Matter of Arabia (◊ ARABIAN FANTASY).
[JC]

MATTER OF LIFE AND DEATH, A (vt *Stairway to Heaven* US) UK movie (**1946**). Archers. **Pr** Michael Powell, Emeric Pressburger. **Dir** Powell, Pressburger. **Spfx** Percy Day, Henry Harris, Douglas Woolsey. **Screenplay** Powell, Pressburger. **Novelization** *A Matter of Life and Death* * (**1946**) by Eric Warman. **Starring** Marius Goring (Conductor 71), Kim Hunter (June), Roger Livesey (Dr Frank Reeves), Raymond Massey (Abraham Farlan), David Niven (Squadron Leader Peter David Carter). 104 mins. Colour and duotone b/w.

May 2, 1945. Carter's bomber is ablaze in thick fog over the English Channel; he radios to operator June that he will bale out parachuteless rather than burn alive. In HEAVEN there is consternation: Conductor 71, who should have retrieved Carter's SOUL, lost him in the fog, with the result that Carter survives. Conducter 71 is sent to fetch him. But Carter refuses to come, having now fallen in love with June. Carter describes his experience to June; they and local GP Reeves attribute the affair to brain damage. In a second visit Conductor 71 tells Carter an appeal against his death will be heard in three days' time, with anglophobe Abraham Farlan (the first American killed by an English

bullet in 1775) as prosecutor. To Reeves, mundanely, Carter's report of this is evidence he is approaching a neurological crisis: it is thus psychiatrically important that Carter "win his case". Conductor 71, a TRICKSTER, tries to lure Carter up the great moving stairway to the AFTERLIFE by devious means. Reeves is killed in a motorcycle crash, and is thus able to be Carter's defence lawyer at the trial, held in a galaxy-sized courtroom. The case hinges on the fact that the grief caused by separating Carter from June would be the responsibility of the celestial Records Office, for without the latter's slip-up the pair would never have met and fallen in love – but how deep is that love? The court descends the moving stairway to Earth to interrogate the two; as it is immediately evident they would die for each other, Carter is given a new lease of life.

Owing much to HERE COMES MR JORDAN (*1941*) but produced on a hugely more lavish scale, *AMOLAD* is presented as a RATIONALIZED FANTASY: it is stated at the outset that the "other world" depicted is purely a product of Carter's PERCEPTION – and, by implication, that the supernatural events are indeed an ALLEGORY of brain illness. The use of duotone b/w for heavenly scenes and colour for terrestrial ones works well; on coming to Earth the flamboyantly romantic Conductor 71 observes gratefully, "One is starved for Technicolor up there." The only real flaw in this affecting piece of first-rate schmaltz is that such a weak case is presented for the prosecution.
[JG]

MATTHEWS, RODNEY (1945-) UK fantasy and sf illustrator, with a distinctively bizarre drawing style, often featuring gourd-like and insectile forms liberally embellished with thorny decorations and excrescences. His early interest in nature study is always evident: even his cities and spaceships have shapes reminiscent of insect and plant life. He favours soft "pastel" hues, with violet and turquoise often predominating. He makes liberal use of airbrush, tightened with line, plus coloured pencil and gouache. His early work was drawn mainly with a technical pen laid over with coloured ink washes, or painted in acrylic gouache on canvas. He has also built a considerable reputation as a designer of alphabets and logos.

RM worked in advertising before accepting freelance commissions for record sleeves. Since then his output has included posters, prints, calendars and covers and internal illustrations for fantasy and sf books. He worked on a series of posters featuring themes from the writings of Michael MOORCOCK in the mid-1970s, also designing the author's logo for *Legends from the End of Time* (coll **1976** US); this led to a lavishly illustrated story, *Elric at the End of Time* (graph **1980**). Other RM-illustrated books have included volumes of Greek myths and, somewhat disappointingly, Felicity Brooks's *Tales of King Arthur* (**1994**). He has done much work for the manufacturers of role-playing GAMES, including packaging design and books.

Collections of RM's work include *In Search of Forever* (graph **1985**) with text by Nigel Suckling – from which was extracted *R.M.* (graph **1994** chap) – *Last Ship Home* (graph **1989**) with text by Suckling, *The Rodney Matthews Portfolio* (graph **1991**) and *The Second Rodney Matthews Portfolio* (graph **1993**).
[RT]

MATTINGLY, DAVID (1956-) US sf and fantasy illustrator, with a richly embellished and brightly coloured painting style; he is particularly adept at representing surface textures and intricate incidental detail. DM's first job was as a matte artist for DISNEY. He produced work for *The Black*

Hole (*1979*), TRON (*1982*) and DICK TRACY (*1990*), among others. His work now mostly appears on book covers. [RT]

MATURIN, CHARLES R(OBERT) (1782-1824) Irish novelist, playwright and curate, of Huguenot descent, who came under the spell of Gothic fiction and produced a string of melodramatic and highly atmospheric novels, starting with *The Fatal Revenge, or The Family of Montorio* (**1807**), as Dennis Jasper Murphy. *The Fatal Revenge* includes a spectacle of supernatural experience at the outset, though all is subsequently rationalized (◊ RATIONALIZED FANTASY); CRM seldom took recourse to the horrors and spectres that littered much GOTHIC FANTASY, preferring the grimness and natural evils of humanity. His play *Bertram, or The Castle of Saint Aldobrand* (**1816**) stands as a master Gothic work for the Victorian stage; the character Bertram was the first to blend successfully the tragic Byronic hero (◊ Lord BYRON) with the traditional Gothic villain.

CRM took this one stage further with his most significant achievement, *Melmoth the Wanderer* (**1820**), the definitive novel of the ACCURSED WANDERER. Inspired to some degree by the works of Matthew LEWIS and William BECKFORD, CRM combined all of the Gothic elements. Melmoth enters a PACT WITH THE DEVIL, gaining prolonged life in exchange for his SOUL, but the debt can be passed on if he can find someone prepared to assume it. The novel thus becomes a QUEST as Melmoth searches among increasingly depraved and hopeless individuals, but no one is willing to take over his burden. Honoré de BALZAC continued the story (◊ SEQUELS BY OTHER HANDS) in "Melmoth réconcilié" (1835; trans as "Melmoth Reconciled" 1896). Oscar WILDE took on the name Melmoth during his exile in Paris at the end of his life.

CRM's final novel, *The Albigenses* (**1824**), with little super-natural content, marks the transition from the Gothic historical novel to the more heroic historical romance developed by Walter SCOTT. [MA]

Other works: Of associational Gothic interest are *The Wild Irish Boy* (**1808**), *The Milesian Chief* (**1812**), the plays *Manuel* (**1817**) and *Fredolfo* (**1819**) and the story "Leixlip Castle" (**1825** *The Literary Souvenir*).

Further reading: *Charles Robert Maturin: His Life and Works* (**1923**) by Niilo Idman.

MAUGHAM, W(ILLIAM) SOMERSET (1874-1965) UK writer, best-known for *Of Human Bondage* (**1915**), a rather grim *Bildungsroman*, partially autobiographical. "The Choice of Amyntas", which appears in *Orientations* (coll **1899**), is a long FABLE whose 18th-century protagonist, cast out of England to make his living, finds a magic UNDER-GROUND palace occupied by four allegorical maidens, each representing a life-choice. He selects Love: disappointed, War, Riches and Art return to the surface, where they sig-nificantly shape the 19th century. *The Magician* (**1908**) presents a savage portrait of Aleister CROWLEY, easily rec-ognized as Addo, a student of OCCULTISM who forces a young woman to marry him by casting a GLAMOUR upon her, and who eventually sacrifices her in order to create a HOMUNCULUS. There are some SUPERNATURAL FICTIONS in *Cosmopolitans: Very Short Stories* (coll **1936**) and *The Mixture As Before* (coll **1940**). [JC]

MAUPASSANT, GUY DE ◊ Guy DE MAUPASSANT.

MAXIE US movie (*1985*). Orion/Aurora/Elsboy/ Carter De Haven. **Pr** Carter De Haven. **Exec pr** Rich Irvine, James L. Stewart. **Dir** Paul Aaron. **Vfx** Bill Taylor. **Screenplay** Patricia Resnick. **Based on** *Marion's Wall* (1973) by Jack

FINNEY. **Starring** Glenn Close (Jan/Maxie Malone), Ruth Gordon (Trudy Lavin), Mandy Patinkin (Nick). 90 mins. Colour.

The time is 1985. A young couple (Jan and Nick) discover the apartment they've just moved into was once occupied by 1927 potential movie star Maxie, killed in a car accident before she could get her big break. For the next few weeks Maxie occasionally occupies Jan's body – so that Nick has an odd form of adultery – while once again clawing herself to a starring role. That achieved, Maxie returns to whatever AFTERLIFE she came from.

This is a minor IDENTITY-EXCHANGE tale – readable also as a tale of POSSESSION – in which no one really shines, although Close does well with her dual role. [JG]

MAXON, REX (1892-1973) US COMIC-strip artist. ◊ TARZAN.

MAXWELL DAVIES, [Sir] PETER (1934-   ) UK com-poser. ◊ George Mackay BROWN; OPERA.

MAY, JULIAN Working name of US writer Julian May Dikty (1931-   ), who began publishing work of genre interest with "Dune Roller" for *Astounding Science Fiction* in 1951, but who concentrated for the next 30 years on a wide variety of publishing enterprises. With her husband, T.E. Dikty (1920-1991), she was an important sf SMALL-PRESS publisher; under her own name and several pseudonyms she wrote some 200 books, mostly nonfiction texts for children. Of her work during this period, some titles are of fantasy interest, including a variety of movie ties as by Ian Thorne and *A Gazeteer of the Hyborian World of Conan* (**1977**) as by Lee N. Falconer.

In 1982 JM returned to adult sf and fantasy with the extremely impressive and influential **Saga of Pliocene Exile**: *The Many-Colored Land* (**1981**) and *The Golden Torc* (**1982**), assembled as *The Many-Colored Land & The Golden Torc* (omni **1982**), plus *The Nonborn King* (**1983**) and *The Adversary* (**1984**), assembled as *The Nonborn King & The Adversary* (omni **1984**), and *The Pliocene Companion* (**1984**), a guide to the sequence. Whether or not the sequence is sf or fantasy is debatable; the issue hardly matters, for JM's main narrative is complex, florid and thoroughly engrossing. Subsequently came related works that are certainly sf, pre-quelling the **Saga**: *Intervention* (**1987**; vt in 2 vols as *The Surveillance* 1988 and *The Metaconcert* 1988) and the **Galactic Milieu** sequence, which follows on immediately, and so far constitutes *Jack the Bodiless* (**1992**), *Diamond Mask* (**1994**) and «*Magnificat*» (**1996**).

Intervention recounts the rise in the early 21st century of the psi-powered Remillard family, who bring down upon Earth the Great Intervention, an imposition of alien super-vision of our struggling civilization. This imposition is ultimately challenged, in the **Galactic Milieu** books, by the Remillard family, who lose the fight and are sent into exile through a one-way PORTAL – described in TECHNOFANTASY terms – into the world of the **Saga of Pliocene Exile**. Here, as *The Many-Colored Land* begins, they find that Pliocene Earth has been turned into a vast arena for the combative but playful AGON between two subraces – known as the Tuag and the Firvulag – of a galaxy-spanning civilization. The Remillard family's interactions with these godlike but frivolous beings are told in terms – if the vast FRAME STORY is forgotten – more or less indistinguishable from fantasy. Between them, the two subraces are a CAULDRON OF STORY out of which most human MYTHS and LEGENDS have grown – e.g., UNDERGROUND kingdoms occupied by SHAPESHIFTERS, creatures who prefigure all the denizens of FAERIE, and visions of the WILD HUNT conducted by HEROES on flying horses. The relationship between humans and aliens flickers between GODGAME activities and the psychic depths of JUNGIAN PSYCHOLOGY. Throughout, the play of sf with and against fantasy is of great interest. The **Saga** is a quintessential late-genre accomplishment.

Of rather less import is the TECHNOFANTASY **Trillium** sequence, partly in collaboration with Marion Zimmer BRADLEY and Andre NORTON: *Black Trillium* (**1991**) with Bradley and Norton; *Blood Trillium* (**1992** UK) by JM alone; *Golden Trillium* (**1993**) by Norton alone; *Lady of the Trillium* (**1995**) by Bradley alone. The series follows the interlinked lives and careers of triplet Princesses, each of whom must search a FANTASYLAND (possibly a far-future colonized planet) for one TALISMAN; all three relics, when brought together, will bring back health to the LAND. [JC]

As Ian Thorne: *Frankenstein* * (**1977** chap), movie tie; *Godzilla* (**1977** chap), nonfiction; *Dracula* * (**1977** chap), movie tie; *King Kong* (**1977** chap), nonfiction; *Mad Scientists* (**1977** chap), nonfiction; *The Wolf Man* * (**1977** chap), movie tie; *The Mummy* * (**1981** chap), movie tie; *Frankenstein Meets Wolfman* * (**1981** chap), movie tie; *Creature from the Black Lagoon* * (**1981** chap), movie tie; *The Blob* * (**1982** chap), movie tie; *The Deadly Mantis* * (**1982** chap), movie tie; *It Came from Outer Space* * (**1982** chap), movie tie.

MAYNE, WILLIAM (JAMES CARTER) (1928-   ) UK author of children's/YA novels that are characterized by subtle narrative development and a strong sense of character, landscape and the presence of the past; he has also written as Martin Cobalt, Dynely James and Charles Molin. The Yorkshire moors where WM was brought up provide a frequent background, as do versions of Canterbury Cathedral (whose choir school he attended in youth). His fantastic touches often tend to be oblique or eventually explicable – like the eerie sound of the "marsh-dragon" in *The Member for the Marsh* (**1956**), which dwindles to a pumping engine but ushers in the true secret of the marsh, an Iron Age settlement.

There is real MAGIC in the fine *Earthfasts* (**1966**): the accidental TIME TRAVEL of an 18th-century drummer boy through an underground PORTAL to the present day begins a cascade of WRONGNESS centred on the cold-burning candle he found below and brought into the real world. Upheaval follows, partly literal: earth and standing stones move, creatures of myth and history are abroad, and a strange cataclysm (described with a piercing, mystical clarity that does not explain) snatches a modern boy out of TIME. His friend must return the now impossibly heavy candle to its own subterranean reality, that of ARTHUR as SLEEPER UNDER THE HILL: the path is nightmarish and significant blood is spilled before order is, mostly, restored.

A Game of Dark (**1971**) shuttles its young protagonist Donald between grim family life and the simpler but no more soluble problems of an analogous medieval fantasy world – easily read as entirely a function of Donald's PERCEPTION, which self-defensively creates this game of dark. In the GAMEWORLD, the foul, icy and limbless WORM that must be slain corresponds to Donald's censorious, chilly, crippled father (embodying his dislike for his father, but also his guilt at failing to love). Other real-world figures appear in the fantasy's distorting mirror. When at last fantasy-Donald kills the Worm he need no longer hate, and it is indicated

that the real father very soon dies: the book ends on a worryingly ambiguous note of consolation.

A much sunnier fantasy is *It* (**1977**), whose quirky schoolgirl heroine Alice idly gropes through earth for the buried foot of a stone cross – whereupon a dry hand takes hers. She has awakened a kind of POLTERGEIST, which is bound to her . . . perhaps because, during a "later" TIMESLIP into a period centuries beforehand, she has killed the WITCH whose familiar the creature is. The joy of the book lies in Alice's highly practical handling of this situation.

For younger readers there is the **Hob** series of stories about the eponymous domestic brownie-like spirit who lives beneath the stairs and secretly helps protect a family home. These are *The Blue Book of Hob Stories* (coll **1984** chap), *The Green Book of Hob Stories* (coll **1984** chap), *The Red Book of Hob Stories* (coll **1984** chap), *The Yellow Book of Hob Stories* (coll **1984** chap) – all these collected as *The Book of Hob Stories* (omni **1991**) – and *Hob and the Goblins* (**1993**). The last is a rather darker and more extended tale, in which Hob defends his family's new rural home – which is a Portal to an UNDERWORLD reeking of Wrongness – from GOBLIN invaders. A similar but richer vein is tapped in *The Blemyah Stories* (coll **1987** chap), a strange year-long cycle about WAINSCOT creatures called Blemyahs who carve wooden decorations for a medieval abbey.

Cuddy (**1994**) is particularly complex and challenging, with its barely signalled time-transitions between present-day Durham – whose cathedral holds the tomb of the eponymous St Cuthbert – and Cuthbert's own era, the 7th century. The girl protagonist Ange has suffered from lifelong tinnitus, which is the saint's bell calling for help to return him, in some figurative sense, to the island where he died. His SEVEN scattered possessions must be assembled: not mere gathering of PLOT COUPONS but the weaving of a complex web across time. Ange and her supporting cousins encounter surreally horrific guardians and undergo various animal METAMORPHOSES *en route*. Even her protecting teddy-bear is *also* the real bear which Cuthbert once admonished for part-devouring a nun (who persists within the bear as a querulous SECRET SHARER). Close attention to the narrative is demanded.

WM's longer fictions can occasionally seem a little too elliptical or fey for his ostensible audience, and critics have quarrelled about his stylized, impressionistic yet effective dialogue. But he is a writer of remarkable resources and imaginative power whose work has become more concise and more powerful – though also more elusive – over his long career. He stands in the first rank of authors not just of CHILDREN'S FANTASY but of children's fiction as a whole.

[DRL]

Other works (some associational): *Follow the Footprints* (**1953**); *The World Upside Down* (**1954**); *A Swarm in May* (**1955**); *Choristers' Cake* (**1956**); *The Blue Boat* (**1957**); *A Grass Rope* (**1957**), a Carnegie Medal winner; *The Long Night,* (**1958**); *Underground Alley* (**1958**); *The Gobbling Billy* (**1959**; reissued 1969 as by WM and Caesar), as Dynely James, with Dick Caesar; *The Thumbstick* (**1959**); *Cathedral Wednesday* (**1960**); *The Fishing Party* (**1960**); *Thirteen O'Clock* (**1960**); *The Changeling* (**1961**); *The Glass Ball* (**1961** chap); *Summer Visitors* (**1961**); *The Last Bus* (**1962** chap); *The Twelve Dancers* (**1962**); *The Man from the North Pole* (**1963**); *On the Stepping Stones* (**1963**); *A Parcel of Trees* (**1963**); *Plot Night* (**1963**); *Words and Music* (**1963**); *A Day Without Wind* (**1964** chap); *Sand* (**1964**); *Water Boatman* (**1964**); *Whistling*

Rufus (**1964**); *The Big Wheel and the Little Wheel* (**1965**); *No More School* (**1965**); *Pig in the Middle* (**1965**); *The Old Zion* (**1966** chap); *Rooftops* (**1966**); *The Battlefield* (**1967**); *The Big Egg* (**1967**); *The House on Fairmount* (**1968** chap); *Over the Hills and Far Away* (**1968**; vt *The Hill Road* 1969 US); *The Toffee Join* (**1968** chap); *The Yellow Aeroplane* (**1968**); *Ravensgill* (**1970**); *Royal Harry* (**1971**); *The Incline* (**1972**); *Robin's Real Engine* (**1972** chap); *Skiffy* (**1972**); *The Swallows* (**1972**; vt *Pool of Swallows* 1974 US), as Martin Cobalt; *The Jersey Shore* (**1973**); *Robin's Real Engine and Other Stories* (omni **1975** containing title story, *The Big Egg* and *The Toffee Join*); *A Year and a Day* (coll **1976** chap); *Max's Dream* (**1977** chap); *Party Pants* (**1977** chap); *While the Bells Ring* (**1979** chap); *The Mouse and the Egg* (**1980** chap); *Salt River Times* (coll of linked stories **1980**); *The Patchwork Cat* (**1981** chap); *All the King's Men* (coll **1982**); *Skiffy and the Twin Planets* (**1982**); *Winter Quarters* (**1982**); *The Mouldy* (**1983** chap); *A Small Pudding for Wee Gowrie and Other Stories of Underground Creatures* (**1983** chap); *Drift* (**1985**); the **Animal Library** sequence, comprising *Barnabas Walks* (**1986** chap), *Come, Come to My Corner* (**1986** chap), *Corbie* (**1986** chap), *Tibber* (**1986** chap), *A House in Town* (**1987** chap), *Mousewing* (**1987** chap), *Lamb Shenkin* (**1987** chap) and *Leapfrog* (**1987** chap); *Gideon Ahoy!* (**1987**); *Kelpie* (**1987** chap); *Tiger's Railway* (**1987**); *Antar and the Eagles* (**1989**); *Netta* (**1989** chap); *The Farm that Ran Out of Names* (**1990** chap); *Netta Next* (**1990** chap); *The Second-Hand Horse and Other Stories* (coll **1990**); *Thursday Creature* (**1990** chap); *Rings on Her Fingers* (**1991** chap); *Low Tide* (**1992**), a *Guardian* Children's Fiction Award winner; *And Netta Again* (**1992** chap); *The Egg Timer* (**1993** chap); *Oh Grandmama* (**1993** chap); *Bells on Her Toes* (**1994** chap); *Cradlefasts* (**1995**), and *Earthfasts*; *Fairy Tales of London #1: Upon Paul's Steeple* (**1995**) and «*#2 C-Saw Sacradown*» (**1996**); *Pandora* (**1995** chap); «*Lady Muck*» (**1996**); «*The Fox Gate and Other Stories*» (**1996**).

As Charles Molin: The **Dormouse Tales** sequence, comprising *The Lost Thimble* (**1966**), *The Steam Roller* (**1966**), *The Picnic* (**1966**), *The Football* (**1966**) and *The Tea Party* (**1966**); *Ghosts, Spooks, Spectres* (anth **1967**).

As editor: *Over the Horizon, or Around The World in Fifteen Stories* (anth **1960**); *The Rolling Season* (anth **1960**); *The Hamish Hamilton Book of Kings* (**1964**, vt *A Cavalcade of Kings* 1965 US) and *The Hamish Hamilton Book of Queens* (**1964**; vt *A Cavalcade of Queens* 1965 US), both with Eleanor FARJEON; *The Hamish Hamilton Book of Heroes* (anth **1967**; vt *William Mayne's Book of Heroes* 1968 US; vt *A Book of Heroes* 1970); *The Hamish Hamilton Book of Giants* (anth **1968**; vt *William Mayne's Book of Giants* 1969 US; vt *A Book of Giants* 1972); *Ghosts: An Anthology* (anth **1971**); *Supernatural* (anth **1995**).

As composer: Score for first version of Alan GARNER's *Holly from the Bongs* (**1966**).

Further reading: "Games of Dark: William Mayne" in *Not in Front of the Grown-Ups* (**1990**) by Alison Lurie (1926-).

MAZES ◊ LABYRINTHS.

MEAGHER, MAUDE (?1895-?1977) UK writer. ◊ AMAZONS.

MEDIUMS ◊ SPIRITUALISM.

MEDUSA ◊ GORGONS.

MEDUSA TOUCH, THE UK/French movie (*1978*). Coatesgold/Sir Lew Grade/Elliott Kastner/Arnon Milchan/ITC. **Pr** Anne V. Coates, Jack Gold. **Exec pr** Arnon Milchan. **Dir** Gold. **Spfx** Brian Johnson. **Vfx** Doug Ferris. **Screenplay** John Briley. **Based on** *The Medusa Touch* (**1973**) by Peter Van Greenaway. **Starring** Richard Burton (John Morlar), Michael Byrne (Duff), Lee Remick (Dr Zonfeld), Lino Ventura (Brunel). 109 mins. Colour.

A visitor to the flat of writer Morlar clubs him almost to death. His brain inexplicably survives the awful damage. As he lies in intensive care, seemingly a vegetable despite his EEG traces, Inspector Brunel, investigating, is led to Morlar's psychiatrist, Zonfeld. Largely through her he discovers, at first incredulously, that Morlar has throughout his life been telekinetically (◊ TALENTS) able to create disaster around him, usually unconsciously causing the deaths of those who earn his hatred; to prove his ability to her Morlar caused a plane crash and the failure of a NASA Moon shot. At this second proof she attempted to kill him. But Morlar has forced his mind to live on for one last crime . . .

Van Greenaway's novels, as Morlar's are implied to be, were too original for the comprehension of the literary establishment, who therefore largely ignored them, and this intelligent movie, a CROSSHATCH between fantasy and thriller genres (the latter occasionally parodied), has suffered the same fate. Yet, despite some rebarbative moments, it functions well at both levels, the swirling moral subtexts of the novel (Morlar is not so much a villain as in large part a wish-fulfilment figure, and his actions are a debate between GOOD AND EVIL) being translated well to the screen. *TMT* is a movie not easily forgotten. [JG]

MEIER, SHIRLEY (MARIE) (1960-) Canadian writer whose first work, *The Sharpest Edge* (**1986** US), with S.M. Sterling (1954-), the second volume in Sterling's **Snowbrother** sequence of tales set in a post-HOLOCAUST Earth, is essentially SWORD AND SORCERY. The continuation of this **Fifth Millennium** series – *The Cage* (**1989** US) with Sterling, *Shadow's Daughter* (**1991** US) and *Shadow's Sun* (**1991** US) – follows the adventures of two female adventurers, who become lovers and get involved in complex, treacherous disputes. [JC]

MEKAGOJIRA NO GYAKUSHU (vt *The Escape of Megagodzilla*; vt *Monsters from the Unknown Planet*; vt *Terror of Mechagodzilla*) Japanese movie (*1975*). ◊ GODZILLA MOVIES.

MELVILLE, HERMAN (1819-1891) US writer best-known for *The Whale* (**1851** UK; vt *Moby-Dick* 1851 US), a vast, emblem-ridden allegorical novel. It was preceded by several book-length tales in which heightened travelogue – as in *Typee: A Peep at Polynesian Life* (**1846**) – gradually turned to the metaphysical fantasy of *Mardi and a Voyage Thither* (**1849**), in which realistic narrative, SATIRE and outright fantasy intermingle. Mardi itself is an ARCHIPELAGO which, as the story progresses, becomes indistinguishable from and co-extensive with the world. The story begins as two sailors leave their ship and rescue a white woman from HUMAN SACRIFICE; after she disappears, they start an ornate QUEST for her. The protagonist takes on godly qualities, and the girl increasingly seems to combine prelapsarian virtues with the LAMIA-like allure of a DOUBLE who haunts her seekers. Eventually five comrades become involved in the quest, visiting at least 13 allegorically conceived islands (one clearly the USA) in the great archipelago; but they never reach their goal, and the tale slingshots (◊ SLINGSHOT ENDING) the heroes into further travels.

Moby-Dick contains little fantasy *per se*, though it is imbued with such an intensity of STORY that its every incident glows with mythic doubleness, and each character seems doubled by (or embodied into) mythical roles. This intense doubling of effect, plus the metaphorical immensity of Melville's sea, the preternatural hubris of Captain Ahab's damnation-bound quest and the potent image of the white whale itself have all made the novel into an extremely rich source of imagery and plot for fantasy writers ever since. Philip José FARMER's *The Wind Whales of Ishmael* (**1971**) is an unsuccessful sequel.

The eponymous shyster in *The Confidence-Man: His Masquerade* (**1857**) once again seems doubled by a supernatural ambience; but in this pungent fable the supernatural becomes explicit, for the CONFIDENCE MAN who visits the greedy passengers on a Mississippi riverboat is a godlike TRICKSTER, and he changes his identity not only to confuse and mock his victims but also radically to undermine their – and the reader's – trust in mundane REALITY. At the end of the tale he shuts the day down and leads the cast in a DANCE OF DEATH darkwards.

The stories assembled in *The Piazza Tales* (coll **1856**) tend to hover between HORROR and fantasy. "Bartleby the Scrivener" (1853 *Putnam's Monthly Magazine*) describes the eponymous clerk's refusal to do his duties or leave the office where he is employed; the surreal BELATEDNESS of the tale prefigures much in US literature, including Gene WOLFE's "Forlesen" (1974). "The Encantadas, or The Enchanted Isles" (1854 *Putnam's Monthly Magazine*) as by Salvator R. Tarnmoor comprises 10 connected sketches which fantasticate the Galápagos Islands. But, in most of HM's works, it is more the ambience than the plot itself that gives them their extraordinary generative power. Even when obscurity renders it unintelligible, his work has the chill, electrifying taste of MYTH. [JC]

See also: AMERICAN GOTHIC.

MEMORY WIPE Partial AMNESIA inflicted by some outside agency, via MESMERISM or MAGIC. MW is a recurring device in CHILDREN'S FANTASY – sometimes vaguely justified as a necessary security precaution of SECRET MASTERS, SECRET GUARDIANS, WAINSCOT societies, etc.; sometimes merely implying a sense of auctorial tidiness, as though, once children's magical adventures are over, even the memories must be lost. In either case the violation of a mind conveys unease, as when PUCK in Rudyard KIPLING's *Puck of Pook's Hill* (coll **1906**) gives two children many remarkable TIMESLIP history lessons, but repeatedly withdraws the knowledge. The selective MWs imposed on Miramon Lluagor in James Branch CABELL's *The Silver Stallion* (**1926**) by a hasty WISH and on the heroine of Diana Wynne JONES's *Fire and Hemlock* (**1984**) by the FAIRY QUEEN are properly shown as WRONGNESS, and are overcome; Susan COOPER's *The Dark is Rising* (**1973**) features a MW which is a kind of HEALING, relieving characters of terrible knowledge that could break their minds. But in Cooper's *Silver on the Tree* (**1977**), Pat O'SHEA's *The Hounds of the Mórrígan* (**1985**), and many further examples it seems slightly unfair that children who have helped defeat EVIL should have even the memory of their achievement taken away. It is different when they themselves choose to forget, like Susan in C.S. LEWIS's **Chronicles of Narnia** – who finds the memory of Narnia incompatible with her grown-up worldliness, and rejects it. [DRL]

MENARCHE The first appearance of menstruation. It is not the same as puberty, which applies to both male and female, and which is defined as the capacity to beget or to bear children.

The ritual significance of menarche in human societies varies widely. When the term is used in descriptions of fantasy, describes a momentous event which tends radically to affect the life of the girl undergoing it, perhaps most radically in Robert HOLDSTOCK's *Lavondyss* (**1988**), whose young protagonist's menarche coincides with the book's halfway point, and signals an extraordinary intensifying of the tale. Frequently menarche is associated with the sudden blossoming of a TALENT, as in Stephen KING's *Carrie* (**1974**), filmed as CARRIE (*1976*). More vaguely, it can be associated with the first steps towards empowerment in fantasy novels whose protagonists are young women destined to gain great gravitas in their world. [JC]

MENTORS Younger HEROES AND HEROINES of fantasy are very often prepared for their coming tasks by the teachings of a mentor. MERLIN is the UNDERLIER for most WIZARD mentors, and fills this role for ARTHUR in T.H. WHITE's *The Sword in the Stone* (**1938**). Gandalf in J.R.R. TOLKIEN's *The Lord of the Rings* (**1954-5**) indicates the paths his hobbit charges must follow, and is then separated from them by turns of STORY: heroes cannot remain tied to their mentor's apron-strings. But Ged in Ursula K. LE GUIN's *A Wizard of Earthsea* (**1968**) is over-hasty in leaving the teacher Ogion who gave him his NAME: only much later does he appreciate Ogion's Zen-like wisdom, echoing it when Ged himself becomes a mentor in *The Farthest Shore* (**1973**). The Mastersmith of Michael Scott ROHAN's *The Anvil of Ice* (**1986**) exemplifies the evil mentor who trains the hero in order to exploit him. Suraklin, the disguised Dark Mage in Barbara HAMBLY's **Antryg Windrose** sequence, functions as an evil-wizard mentor. R.A. MACAVOY's Powl in *The Lens of the World* (**1990**) is, interestingly, a natural philosopher. Pupil and mentor tend to be of the same GENDER; Greg BEAR varies this in *The Infinity Concerto* (**1984**), whose hero is trained by THREE female half-ELVES. Nonhuman mentors with a LIMINAL-BEING flavour include PUCK in Rudyard KIPLING's *Puck of Pook's Hill* (coll **1906**) and the unclassifiable Golux of James THURBER's *The Thirteen Clocks* (**1950**). [DRL]

See also: PATRONS.

MEPHISTOPHELES The DEMON who acts as agent for FAUST's PACT WITH THE DEVIL, best-known for the speech about the limitlessness of HELL credited to him by Christopher MARLOWE. GOETHE's icily sarcastic Mephistopheles is less impressive but more widely echoed in modern Faustian fantasies. [BS]

MEPHISTO WALTZ, THE US movie (*1971*). 20th Century-Fox/QM. **Pr** Quinn Martin. **Dir** Paul Wendkos. **Spfx** Howard A. Anderson Co. **Screenplay** Ben Maddow. **Based on** *The Mephisto Waltz* (**1969**) by Fred Mustard Stewart (1936-). **Starring** Alan Alda (Myles Clarkson), Jacqueline Bisset (Paula Clarkson), Pamelyn Ferdin (Abby Clarkson), Curt Jurgens (Duncan Ely), Barbara Parkins (Roxanne DeLancey). 109 mins. Colour.

Music journalist Myles is suddenly taken up by aged, world-famous pianist Ely; wife Paula is perplexed then frightened by the excess of generosity, and repelled by Ely and especially his sinister daughter Roxanne. Indeed, they have a PACT WITH THE DEVIL such that, on Ely's imminent death from leukaemia, his SOUL takes POSSESSION of Myles's body and continues his incestuous relationship with Roxanne. Paula – ignorant of the conspiracy – is both bewildered and sexually excited by Myles's "new" character, but then her child dies . . . Finally, in desperation, she makes her own CONTRACT with SATAN. Killing her own body, she possesses Roxanne's and thus, at least physically, regains her husband.

Stewart's rather superficial novel (it reads like a novelization) cannot really suspend disbelief throughout its somewhat Byzantine proceedings. *TMW* makes a better job of it, exploring macabre sidelights (e.g., a fancy-dress party whose guests are clad as FERTILITY animals) in order to deflect our attention from plot absurdities. Very stylishly directed and sharply scripted, *TMW* builds up a genuine feeling of decadent EVIL. [JG]

MERCURE GALANT, LE ◊ MAGAZINES.

MERCURY ◊ HERMES.

MEREDITH, OWEN Pseudonym of Robert Bulwer-Lytton. ◊ Edward BULWER-LYTTON.

MERLIN Probably the most famous of all WIZARDS; the mage and advisor of ARTHUR and the architect of his kingship. Not linked to Arthur in the original Celtic tales, Merlin seems to have sprung full-blown from the imagination of GEOFFREY OF MONMOUTH in his *Historia Regum Britanniae* ["History of the Kings of Britain"] (**1136**). Geoffrey claimed to be translating an older Welsh text into Latin. While working on this translation or pseudo-translation, Geoffrey brought out a smaller book, *Prophetiae Merlini* ["Prophecies of Merlin"] (**1134**), later incorporated into the *Historia*. In a sequence of now hopelessly obscure statements, Merlin prophesies a Celtic resurgence which will overthrow the conquering Saxons. In the *Historia* Geoffrey provides more detail about Merlin. As a child Merlin is found by the soldiers of Vortigern, the High King, who is trying unsuccessfully to build a fortress in Wales. Merlin explains why Vortigern is not succeeding and also foretells Vortigern's downfall. His prophecies come to be. Now the advisor of Ambrosius Aurelianus, Merlin erects Stonehenge in celebration of Ambrosius's kingship. Later he helps Uther Pendragon deceive Ygraine, wife of the Duke of Cornwall, and Ygraine bears Arthur. Merlin educates Arthur and raises him for the kingship.

The interest shown in Merlin caused Geoffrey to research further and he produced a third volume, the *Vita Merlini* ["Life of Merlin"] (?**1155**). Although Geoffrey had now discovered that the character on whom he had based Merlin, Myrddin Wyllt or Myrddin the Wild, lived over a generation later than Arthur (the latter half of the 6th century), he fudged the references for the sake of continuity, but retained the story that Merlin had been driven wild by the death of his king and roamed the Caledonian Forest like a wild animal. Merlin's story was now taken up by the French poet Robert de Boron (? -1212) who, in *Merlin* (?**1200**), brought in all of the elements with which we are now familiar. Boron makes Merlin the son of a nun and a DEMON and links him to the GRAIL story, introduces the motif of the SWORD IN THE STONE, and makes him the creator of the ROUND TABLE. Much of Robert's original poem is now lost, but the anonymous *Suite de Merlin* (early 13th century) is probably based on it, with some embellishments. This tells of Merlin's enchantment by Niniane (◊ LADY OF THE LAKE), which results in his imprisonment in an OAK (or, in some versions, a cave). Merlin, therefore, like Arthur, is not dead, only trapped (◊ BONDAGE), and this allows his resurrection in later stories.

Merlin was a favoured character in medieval tales because his MAGIC was a convenience to the storyline. Thus he and Arthur figure prominently in *The Faerie Queene* (**1590**; exp

1596) by Edmund SPENSER; he is also the enchanter who arranges the birth of TOM THUMB in *The History of Tom Thumb the Little* (**1621**) published by Richard Johnson (1573-1659).

Although Merlin's role is linked to Arthur's, he is sufficiently independent that a separate strand of fiction has developed to explore his character and exploits, and these tales are generally more closely related to GENRE FANTASY than some other Arthurian fiction. He is the central character in *The Sword in the Stone* (**1938**) by T.H. WHITE, and White's portrayal of this scatter-brained, white-bearded old magician has become the archetypal image of the fantasy wizard; it is this image upon which J.R.R. TOLKIEN established Gandalf, especially in *The Hobbit* (**1937**). Merlin has a lesser role in the other books that make up White's *The Once and Future King* (**1958**), although White did return to him for the less successful *The Book of Merlyn* (written 1940; **1977**), in which Merlin seeks to re-educate Arthur in his final days.

Merlin survives Arthur in "King of the World's Edge" (**1966**) by H. Warner MUNN and its sequels, *The Ship from Atlantis* (**1967**; fixup with "King of the World's Edge" vt *Merlin's Godson* 1976) and *Merlin's Ring* (**1974**). A stronger historical portrayal was made by John Cowper POWYS in *Porius* (**1951**), drawing upon the historical Myrddin Wyllt to depict a sinister though sad and ageing shaman. The historically rationalized version was further explored by Henry TREECE in *The Green Man* (**1966**).

The first major book to tell Merlin's life from his own viewpoint was *The Crystal Cave* (1970) by Mary STEWART; the tale continued in *The Hollow Hills* (**1973**) and *The Last Enchantment* (**1979**). Stewart explored Merlin's character in depth, considering him in the traditional Arthurian world but investing him with human traits in addition to his power of Second Sight (◊ TALENTS). Books (most but not all GENRE FANTASY) in which either Merlin is the lead character or his machinations are widely explored include; *Merlin* (**1978**) by Robert NYE, who presents a bawdy interpretation of his life, a viewpoint supported by Robert HOLDSTOCK in his **Mythago** series, but especially *Merlin's Wood* (**1994**); *The Mists of Avalon* (**1982**) by Marion Zimmer BRADLEY, the first to investigate the conflict between the old and new RELIGIONS and the dilemma between Merlin and Arthur, a subject returned to in *Merlin* (**1988**) by Stephen LAWHEAD; *Child of the Northern Spring* (**1987**) by Persia WOOLLEY, and *Black Smith's Telling* (**1990**) by Fay SAMPSON – all books forming part of longer Arthurian sequences. The eccentricities of Merlin are admirably unfolded in *Merlin's Booke* (coll of linked stories **1986**) by Jane YOLEN, while his own memories are recounted in *Merlin Dreams* (coll of linked stories **1988**) by Peter DICKINSON. The most complete fictional study of Merlin is the sequence by Nikolai TOLSTOY, *The Coming of the King* (**1988**) and «Merlin and Arthur» (**1996**), who looks in depth at the Celtic base for the real Myrddin and seeks to rationalize the link between him and the historic Arthur. A children's book exploring the relationship between Merlin and Vortigern is *Wizard of Wind and Rock* (**1990**) by Pamela F. Service (1945-   ). Merlin is the pivotal character in the movie EXCALIBUR (**1981**), where he is presented as a quirky, eccentric figure – rather like T.H. White's version – but also with a chilling sense of otherness about him: we grow aware that his agenda is quite alien from those of the less magical characters around him.

The survival of Merlin to the modern day is a common motif. David H. KELLER brought Merlin alive with EXCALIBUR to help the Allies in WWI in *Men of Avalon* (**1935** chap dos). Early efforts for the pulp MAGAZINES were rather facile – e.g., "The Enchanted Week End" (1939) by John MacCormac, where Merlin works wonders at a sports' day, and "Wet Magic" (1943) by Henry KUTTNER, where Merlin is released and aids the War effort. More serious use was made of Merlin's resurrection in *That Hideous Strength* (**1945**) by C.S. LEWIS, where Merlin's powers are unleashed in an apocalyptic climax. There is a humorous portrayal of Merlin in *The Elixir* (**1971**) by Robert NATHAN, while a more romantic view is depicted in *The Enchanter* (**1990**) by Christina Hamlett. In *The Return of Merlin* (**1995**) Deepak Chopra interweaves the old world and the new, while Fred SABERHAGEN twines a post-Arthurian world with that of the 21st century in *Merlin's Bones* (**1995**).

Merlin appears as a kindly but secretive old gentleman in *The Weirdstone of Brisingamen* (**1960**) and *The Moon of Gomrath* (**1963**) by Alan GARNER, where he takes the role of Cadellin, and in the **Dark is Rising** sequence by Susan COOPER, where he figures as Uncle Merriman; in both instances Merlin returns to save the world from rising supernatural EVIL. More personalized and introspective narratives of a living Merlin are found in "Merlin's Oak" (1932 *Cornhill*) by C.E. Lawrence, *The Lastborn of Elvinwood* (**1980**) by Linda HALDEMAN and *Merlin Dreams in the Mondream Wood* (**1992** chap) by Charles DE LINT. A collection is *The Merlin Chronicles* (anth **1995**) ed Mike ASHLEY.

[MA]

MERMAIDS Chimerical beings, human above the waist and fish below (although there is a famous painting by René MAGRITTE of a "reversed mermaid"); mermen are of lesser importance (*but see* SELKIES). Mermaids' role in FOLKLORE is that of sea-dwelling equivalents of FAIRIES and UNDINES – they often replace BIRDS in fairy-bride stories about fisherfolk – but their fondness for seductive singing and reputation as OMENS of disaster link them to SIRENS and to the Lorelei. Two fine versions of the mermaid-bride motif are Hans Christian ANDERSEN's "The Little Mermaid" (1837) and Oscar WILDE's "The Fisherman and his Soul" (1891); another effective 19th-century image is contained in "The Forsaken Merman" (1849) by Matthew Arnold (1822-1888). *The Sea Lady* (**1902**) by H.G. WELLS and "The Floating Café" (1936) by Margery LAWRENCE are FEMME-FATALE stories reflecting the more sinister element of the folkloristic tradition, but *Loona, A Strange Tail* (**1931**) by Norman Walker, "Nothing in the Rules" (1939) by L. Sprague DE CAMP, *Peabody's Mermaid* (**1946**) by Guy and Constance Jones and "Mr Margate's Mermaid" (1955) by Robert BLOCH are much lighter; the jokes deployed in the movie version of the Joneses' novel, *Mr Peabody and the Mermaid* (*1948*) set the pattern for subsequent mermaid movies like SPLASH! (*1984*). Theodore STURGEON's "A Touch of Strange" (1958) and Ray BRADBURY's "The Shoreline at Sunset" (1959) are effective sentimental parables, while *The Merman's Children* (**1979**) by Poul ANDERSON is a fine epic novel which restores all the gravity of the folkloristic tradition in a striking account of the inexorable THINNING of the realm of FAERIE; *Mermaid's Song* (**1989**) by Alida Van Gores attempts something similar in its account of conflict between the "merra" and the "mogs". "The Whimpus" (1933) by Tod Robbins (1888-1949) is an amusing account of a strange fish with mermaid-like

attributes, and "The Malaysian Mer" (1982) by Jane YOLEN describes an unusually lively example of the fake mermaids which sailors used to mock up by stitching together the relevant parts of monkeys and fish. A good theme anthology is *Mermaids!* (**1985**) ed Gardner Dozois and Jack DANN. [BS]

See also: *The* LITTLE MERMAID (*1989*).

MERRITT, A(BRAHAM) (1884-1943) US journalist – latterly editor of William Randolph Hearst's *American Weekly* – and writer of lush fantasies which took exotic ESCAPISM to extremes unexplored even by Edgar Rice BURROUGHS. His fiction was among the most popular ever published in the pulp MAGAZINES, and occupied such a significant place in the reprint magazines FAMOUS FANTASTIC MYSTERIES and *Fantastic Novels* as to cause the founding of a companion, *A. Merritt's Fantasy Magazine* (1949-50). A brief adventure, *Thru the Dragon Glass* (1917 as "Through the Dragon Glass"; **1932** chap), and *The People of the Pit* (1918; **1948** chap) were followed by "The Moon Pool" (1918), an intriguing account of the mysterious guardian of a PORTAL beyond which – it is implied – all the treasures of the human imagination might be found. Unfortunately, the novelette lost most of its force in the version rewritten to form a prelude to the rather pedestrian "The Conquest of the Moon Pool" (1919) in *The Moon Pool* (fixup **1919**). Its sequel, *The Metal Monster* (1920; rev 1927 as "The Metal Emperor"; rev **1946**) substitutes an authentic alien lifeform for the seemingly supernatural guiding entity of the earlier novel, but describes its wonders in the same lavishly purple prose; AM was never wholly satisfied with the story. "The Face in the Abyss" (1923), which recapitulates the basic theme of "The Moon Pool", is a much better evocation of the allure of the exotic.

The Ship of Ishtar (1924; **1926**) foreshadows modern HEROIC FANTASY in its use of a SECONDARY WORLD and an ongoing struggle between GOOD AND EVIL, here personified as Ishtar and Nergal. The victory ultimately won by Good is Pyrrhic, and the happy ending seems contrived for propriety's sake. This downbeat tendency in AM's work became increasingly evident, although his editors did their utmost to thwart it. An atypical thriller in the vein of Sax ROHMER, *Seven Footprints to Satan* (**1928**), proved very popular; but AM returned to his own metier in a belated sequel to "The Face in the Abyss", "The Snake-Mother" (1930). This is the gaudiest and most elaborate of all AM's work, but the book combining the two stories, *The Face in the Abyss* (fixup **1931**), is unsatisfactory. AM's love of the exotic is here extrapolated into a heartfelt critique of the awful mundanity of human nature, allegedly ruled by the twin principles of Greed and Folly. An effective ALLEGORY embodying the same message is "The Woman of the Wood" (1926; vt "The Women of the Wood"; rev as *The Woman of the Wood* chap **1948**).

Dwellers in the Mirage (**1932**; rev 1941) is a late LOST-RACE story. Its hero's psychological conflicts are mirrored after the fashion of H. Rider HAGGARD in two women, one a striking FEMME FATALE. Unfortunately, AM was not allowed to resolve this conflict as he wished; the early versions carried a false ending grafted on by an editor and AM's preferred conclusion was not revealed until the novel was reprinted in *Fantastic Novels* in 1941; some subsequent paperback editions still retain the false ending. AM's subsequent work consists of thrillers with supernatural embellishments. *Burn, Witch, Burn!* (**1933**), filmed as *The* DEVIL DOLL (*1936*), pits a doctor and a gangster against murderous DOLLS devised by a WITCH; its sequel *Creep, Shadow!* (**1934**) involves an ancient CURSE relating to the destruction of the legendary city of Ys.

AM left a number of fragments. Two – *The Fox Woman and the Blue Pagoda* (**1946**) and *The Black Wheel* (**1947**) – were filled out after his death by Hannes BOK; the results are not particularly Merrittesque. The first of these fragments and some others were included with the author's most notable short fiction in *The Fox Woman and Other Stories* (coll **1949**), the remainder being reprinted, along with poetry and a biography of the author, in *A. Merritt: Reflections in the Moon Pool* (coll **1985**) ed Sam Moskowitz.   [BS]

Other works: *Three Lines of Old French* (1919; **1937** chap); *The Drone Man* (1934 as "The Drone"; **1948** chap); *Rhythm of the Spheres* (1936; **1948** chap); *Seven Footprints to Satan and Burn Witch Burn!* (omni **1952**); *Dwellers in the Mirage and The Face in the Abyss* (omni **1953**).

MERWIN, W(ILLIAM) S(TANLEY) (1927-   ) US poet and translator, frequently resident in Europe. WSM's first collection of poems, *A Mask for Janus* (coll **1952**; later coll in *The First Four Books of Poems* omni **1975**), shows his attraction to what W.H. Auden (1907-1973) called WSM's "mythological" sensibility: classical figures, animals (especially BIRDS) and MYTHICAL CREATURES populate a formal and meditative but deeply impersonal verse. For *The Carrier of Ladders* (coll **1970**) he won a Pulitzer Prize. But his major interest as a fabulist lies in his imaginative prose, published as *The Miner's Pale Children* (coll **1970**) and *Houses and Travelers* (coll **1977**). These short pieces, mostly 1-5 pages long, occasionally contain elements of sf and seem sometimes to approach Borgesian fantasy ("The Animal who Eats Numbers", "The Camel Moth"), but their carefully modulated evocations of strangeness are more ontological than dramatic. WSM's later volume of three long novellas, *The Lost Upland* (coll **1992**), set in the peasant farmlands of southern France, eschews fantasy.

WSM's fantasies, although notable, belong finally more to the realm of poetry than to that of fiction.   [GF]

MESMERISM Invented by Franz Mesmer (1734-1815) and the forerunner of modern hypnotism, this involved the transmission between individuals of "animal magnetism", a kind of lifeforce which could induce "trance" states. It was quickly adopted as a staple of OCCULT FANTASY by such writers as E.T.A. HOFFMANN and was recruited as a device to add plausibility to VISIONARY FANTASIES like "A Mesmeric Revelation" (1844) by Edgar Allan POE and *The Soul of Lilith* (1892) by Marie CORELLI. Mesmeric theory also provided mechanisms for IDENTITY EXCHANGE, as in *The Doubts of Dives* (1889) by Walter BESANT, and psychic vampirism (◊ VAMPIRES), as in *The Parasite* (**1894**) by Arthur Conan DOYLE.

The literary image of the hypnotist was largely determined by the characterization of Svengali in George du Maurier's *Trilby* (**1894**) (◊ SVENGALI), whose imitations include *The White Witch of Mayfair* (**1902**) by George Griffith (1857-1906) and *The Sorcerer's Apprentice* (**1927**) by Hanns Heinz EWERS. Hypnotic "regression" is employed in some tales of REINCARNATION, including a few by H. Rider HAGGARD, but the use of the motif in fantasy has now declined dramatically – although a form of mesmerism is part of the standard armoury of many WIZARDS, and their ILLUSIONS may have all the trappings of mass hypnosis.   [BS]

MESOPOTAMIAN EPIC No known criterion distinguished epic, myth or any other genre of narrative poem within Mesopotamian culture itself, yet among the surviving works are long hero-poems that are today usually called epics. These works survive because they were written on clay. Perhaps because of this permanence, Mesopotamian writing had extraordinary continuity: tablets of the 6th century BC frequently carry 17th-century BC words – hence it is often impossible to date a work within less than five centuries. It is also possible that entire lost genres were carried on through media other than cuneiform.

The works generally considered oldest are in Sumerian and mostly 18-century copies of works believed to date from the 21st century. The copious Sumerian MYTHS vary widely in interest. One, known as *Lugal-E*, foreshadows the common later myth pattern of a good deity saving the Universe by defeating a much-feared MONSTER. The modern word "epic", however, is usually reserved to short narratives about GILGAMESH, Lugalbanda and Enmerkar, deified kings of earlier centuries whom the kings of Ur considered relatives. These are hero stories, but rarely battle stories; the best relates a contest of wits between Enmerkar and his foe which ends in the invention of trade. The kings are shown as human but in close contact with the GODS. These poems suffer relatively little from Sumeria's characteristically enthusiastic use of the devices of repetition and simile.

They are called "epics" partly because some of the **Gilgamesh** ones went on to be incorporated in the Mesopotamian epic *par excellence*, *Gilgamesh*. According to Jeffrey Tigay, in *The Evolution of the Gilgamesh Epic* (**1982**), the unified epic dates from the early 2nd millennium BC. Although not complete, the extensive parts compose the masterwork of their era, depicting Gilgamesh's TRANSFORMATION in the face of death, in a world of DEMONS and jewels growing on trees.

This and other epics of, perhaps, the 18th-16th centuries were written in Akkadian, the Semitic language of Babylonia and Assyria. The others include: *Anzu*, in the *Lugal-E* tradition, and a number of Mesopotamia's most interesting hero-stories: *Etana*, about a king who flies to heaven; *Atrahasis*, telling of human CREATION and of the FLOOD; and *Sargon, King of Battle*, a life story of another legendary king. In the Cuthµan *Legend of Naram-Sin*, Sargon's son is falsely blamed for that dynasty's fall. Only *Atrahasis* is at all well known from tablets this old. It is strikingly theological. *Etana* is rather like a folktale and *Sargon* like boasting royal inscriptions. All, so far as preserved, occupy a world nearly as fantastical as that of *Gilgamesh*.

In the following centuries (15th-10thBC the mature poetic style which dominated Semitic poetry (including Hebrew) throughout the ancient era seems to have crystallized. This used long lines broken by a chiasmus, in which parallelism between half-lines, lines or couplets was a principal device. Narrative flow suffered accordingly. Moreover, outright repetition became increasingly prominent, and older works such as *Gilgamesh* were revised, replacing parallelisms with repetitions. (Simile did, however, become rather more natural in frequency.) Interest shifted from narrative to wordplay and learning, and Mesopotamian literature grew steadily narrower in its aesthetic outlook.

Very few tablets survive from these centuries aside from fragments of those mentioned above. Others include *Nergal and Ereshkigal*, a myth of love in the UNDERWORLD, and *Adapa*, about the gods tricking the wisest of men. Another genre, the royal epic, offers the first realistic narratives known from the region (other than relatively terse inscriptions). A further novelty of the era was ESCHATOLOGY, placed in the mouth of a god or ancient king.

There is also the culmination of the *Lugal-E* motif, the famous "Epic of Creation" *Enuma Eliš*. We know this work's purpose – at least in the 7th century, whence the surviving tablets come: it was recited by a priest, alone in a closed temple room, once a year at a festival, and celebrates the rise of Marduk to rule the gods. It excels in precisely those ways ME least appeals to the modern mind, being full of recondite religious information and elaborate punning etymologies, and using repetition as its principal narrative device. Yet the central story of Marduk's fight with Tiamat has appealed to some ever since the work's discovery.

The ancient Semitic poetic style could also be used more naturally, as the myths of Ugarit (*c*1360BC) in Syria show. These, the only surviving literature of the Canaanites, include: *Kirta* (vt *Keret*), focusing on the role of the king; the **Baal** cycle, portraying a brash young storm god and his protective sister as they take over the cosmos and defeat Sea and Death; and *Aqhat*, the beautifully written story of a beloved son's death.

In the first millennium BC, cuneiform literature gradually lost contact with the increasingly Aramaic-speaking peoples who supported it. Most extant tablets date from this era (including portions of all the aforementioned Akkadian works except *Adapa*), but there are few original compositions. The only epic among these is the striking *Erra and Ishum* (vt *Irra*), in which the lord of the underworld unleashes destruction on mankind (an obsessive Mesopotamian theme), but relents. Like many earlier myths, it is easiest read not simply as a cosmic story but also as describing the ritual actions of idols. This is the work of one Kabti-ilani-marduk, probably writing in the 8th century, who presented it as a sacred revelation. It is extremely innovative in form and style, though in ways opaque to modern readers' concerns, and became popular. Other works first attested in this era include revelations and THEODICIES as well as realistic comic prose stories.

Mesopotamia originated many ideas central to Western religious tradition, including, in the epic sphere, the FLOOD story. Mesopotamian writings provide essential context for reading the BIBLE; links with Greece, though much speculated on, can rarely be firmly established (◊ GREEK AND LATIN CLASSICS). Some of the epic characters and motifs survived a millennium to reappear in Muslim legend. In modern times the grimness of the rediscovered Mesopotamian worldview, as exemplified in the story of Tiamat and in *Gilgamesh*, has perennially fascinated. [JB]

MESSIAHS In the Old Testament the Messiah is the saviour promised to the Jews by sacred PROPHECY (◊◊ JEWISH RELIGIOUS LITERATURE); the Christian mythos is founded on the proposition that CHRIST was the Messiah and will return someday to precipitate the APOCALYPSE. The term may be applied by analogy to any SLEEPER UNDER THE HILL. Folklore often attaches such promises to heroes like King ARTHUR and Frederick Barbarossa (*c*1123-1190), and HEROIC FANTASY frequently deals with such figures; one of its favourite themes is the displacement of a human into a SECONDARY WORLD where a quasi-messianic destiny is unexpectedly thrust upon her/him, often in an unkindly manner.

A. MERRITT's *Dwellers in the Mirage* (**1932**) was an important prototype of this formula and some of his imitators –

including Henry KUTTNER and C.L. MOORE – adapted it to SCIENCE FANTASY. In C.S. LEWIS's **Narnia** tales the protagonists require the assistance of Aslan, and in Lewis's *That Hideous Stength* (**1945**) MERLIN has to be drafted to help out the hero, but most such HEROES must do the job alone; Stephen R. DONALDSON's Thomas Covenant, who is direly hampered by his own lack of faith, provided a new and conspicuously modern archetype. Other notable quasi-messianic heroic fantasies include *Draught of Eternity* (**1924**) by Victor Rousseau, *Three Hearts and Three Lions* (1953; **1961**) by Poul ANDERSON and – in a more subtle fashion – *The Traveler in Black* (coll of linked stories **1971**) by John BRUNNER.

There is an interesting subspecies of messianic fantasies which deliberately flirts with heresy in featuring unorthodox redeemers whose "return" is frequently unappreciated. These include *The Horned Shepherd* (**1904**) by Edgar JEPSON, *The Man in the Tree* (**1983**) by Damon Knight, *Godbody* (**1986**) by Theodore STURGEON and *Only Begotten Daughter* (**1990**) by James MORROW. [BS]

METAMORPHOSIS The word translates from the Greek as simply "change of shape", but in English it has always implied *radical* change – from one kind of being to another. Western myths in general, and OVID's *Metamorphoses* (written *c*AD1-8) in particular, further suggest that, when something or someone has undergone metamorphosis, the event must have occurred through MAGIC.

Fantasy uses a variety of terms for changes of shape and nature. In this encyclopedia's taxonomy, metamorphosis has the old meaning of magical and radical change experienced by the subject, who may well have initiated as well as lived through the process; such change can be involuntary and/or inherent in the subject's nature. The keys to SHAPESHIFTING are reversibility and repeatability, as seen when WEREWOLVES change with each full MOON. TRANSFORMATION implies and emphasizes an external agent of change: the FROG PRINCE is transformed by a WITCH. TRANSMUTATION concerns changes in the nature of *inanimate* material.

The active/passive distinction between *metamorphosing* and *being transformed* is too easily blurred to be a rigid dividing line. Tallis in Robert HOLDSTOCK's *Lavondyss* (**1988**) painfully metamorphoses into a tree; or perhaps she is transformed by the magic of the heartwoods. The nymph Daphne is transformed into a laurel to save her from Apollo's pursuit; but the change can be read as a chosen metamorphosis. In a world of magic, any transforming action of the gods may be as "natural" a metamorphosis as the (evidently god-mediated) change from caterpillar to butterfly.

Thus, in FANTASY, metamorphosis tends not to be arbitrary. Often it reveals the real nature of the subject, as with Arachne – the prideful weaver whom Athene transformed into a SPIDER – or the protagonist of Franz KAFKA's *Metamorphosis* (**1915**), who awakens one morning in the shape of a cockroach. (In both cases metamorphosis can be understood, in terms of JUNGIAN PSYCHOLOGY, as a triumph of the SHADOW.) It may represent a RITE OF PASSAGE or constitute a release from BONDAGE: being taken into a princess's bed fulfils a CONDITION and frees the Frog Prince. But however a metamorphosis can be understood – whether longed-for or abhorred by the subject – it does not happen by accident: it comes from the nature of the subject. An abhorred metamorphosis is likely to have generated the

STORY; a longed-for metamorphosis is likely to resolve the Story through a RECOGNITION of the true identity of the protagonist. [JC/DRL]

METEMPSYCHOSIS Literally, the changing of SOULS: transmigration of souls to new bodies (human, animal or even vegetable) after DEATH. Some purists call the process "metensomatosis", or changing of bodies; this encyclopedia's preferred term is REINCARNATION. [DRL]

See also: IDENTITY EXCHANGE.

MEYRINK, GUSTAV (1868-1932) Austrian writer, born Gustav Meyer, from 1889 mostly resident in Prague, where he worked as a banker until a financial scandal (he was innocent) and the FIN-DE-SIÈCLE decadence of his lifestyle (which he proclaimed) ended that career. He published his first story, "Die heisse Soldat" ["The Ardent Soldier"], in the satirical weekly magazine *Simplicissimus* in 1901. It led off *Der heisse Soldat und andere Geschichten* ["The Ardent Soldier and Other Stories"] (coll **1903**), which was later assembled – along with *Orchideen: Sonderbare Geschichten* ["Orchids: Strange Stories"] (coll **1904**), *Das Wachsfigurenkabinett: Sonderbare Geschichten* ["The Wax Museum: Strange Stories"] (coll **1907**) and some additional stories – as *Des deutschen Spiessers Wunderhorn* ["The German Philistine's Magic Horn"] (omni **1913**); selections from this omnibus were published as *The Opal (and Other Stories)* (coll trans Maurice Raraty **1994** UK). Further stories were assembled as *Fledermause* ["Bats"] (coll **1916**) and *Goldmachergeschichten* ["Tales of Alchemists"] (coll **1925**). Those stories published before its outbreak seemed to prefigure the watershed disaster of WORLD WAR I through their apocalyptic savagery, their mockeries of the dying era conveyed in an Expressionist idiom reminiscent of that of E.T.A. HOFFMANN.

GM's first novel, *Der Golem* (1913-14 *Die wessen Blatter*; **1915**; cut trans Madge Pemberton as *The Golem* **1928** UK; full trans 1977 US; new trans Mike Mitchell 1995 UK), also prefigures that disastrous climax to a century of growth and change. The phantasmagorical, threatened PRAGUE of this novel – its WAINSCOTS haunted by a GOLEM who is more like a psychic fog than an actual entity, and the false POLDER of its ghetto withering under the baleful light of a new century – makes *The Golem* into a pure URBAN FANTASY: a tale whose setting, like some vast subterranean EDIFICE, seems literally alive, organic, all-encompassing, Gothic. It is, however, like most 20th-century urban fantasies distinguished from its Gothic ancestors (◊ GOTHIC FANTASY) through a sense that the fall of the city will be a bad omen for its inhabitants, not a release. At the same time, the pending destruction of Prague (and of Europe itself) is treated by GM with a deep ambivalence, and his next novels all tend to present the APOCALYPSE as an occult cleansing of the material world so that higher unions of the spirit can be achieved.

This opposition is very clearly evident in *Das grüne Gesicht* (**1916**; trans Mike Mitchell as *The Green Face* **1992** UK), an INSTAURATION FANTASY set after the end of WWI in an Amsterdam whose inhabitants – including the WANDERING JEW – engage in a virtual DANCE OF DEATH as a great wind scours Europe clean of its past. From the protection of a literal polder, those destined to survive into the new world gaze calmly upon the devastation. GM's second WWI novel, *Walpurgisnacht* (**1917**; trans Mike Mitchell **1993** UK), offers its cast no similar sanctuary as the precarious and corrupt world of Prague collapses into CARNIVAL under the influence of a strange, comical figure capable of

becoming a literal MIRROR of the other main characters, taking on their physiques and their natures, and whose skin serves as a human DRUM to call the lower classes into revelry and rebellion.

The post-WWI novels are quieter. *Der weisse Dominikaner* (**1921**; trans Mike Mitchell as *The White Dominican* **1994** UK) is set entirely in a small town which serves as polder and focus for the protagonist's eventual transcendence of the deluding, corporeal world. The contemporary protagonist of *Der Engel Vom Westlichen Fenster* (**1927**; trans Mike Mitchell as *The Angel of the West Window* **1991** UK), perhaps GM's most complex work, discovers he is a partial AVATAR of John DEE; as Dee he becomes profoundly embroiled in the toils of this world (represented, as often in GM's work, through the maleficent allure of a FEMME FATALE, in this case a goddess) and must tussle to achieve the higher one.

A sense of struggle marks GM's work from beginning to end, and the fantastic elements permeating it are most profitably understood as the insignia of that struggle, for the golems and witches and goddesses and apocalyptic scourings are presented with some casualness. The consequent muffling of effects, and elusive narrative sequences, make GM a sidebar figure to modern fantasy; but the cultural resonances of his best novels merit continued attention. [JC]

MICE AND RATS Mice are often found in fantasy, very frequently as TALKING ANIMALS inhabiting WAINSCOTS, like the mice in Margery SHARP's **Miss Bianca** series or the eponymous scholar in Dick KING-SMITH's *The Schoolmouse* (**1994**). BEAST FABLES – like Russell HOBAN's *The Mouse and his Child* (**1967**), the movie *An* AMERICAN TAIL (**1986**), and the GRAPHIC NOVEL *Maus* (graph **1987**) by Art Spiegelman (1948-   ) – often feature EVERYMAN mice. They are popular faithful wee COMPANIONS, like the mice who nibble at the ropes binding Aslan in C.S. LEWIS's *The Lion, the Witch and the Wardrobe* (**1950**). At least two sequences – the **Redwall** books by Brian JACQUES and the **Deptford Mice** books by Robin JARVIS – describe entire mouse societies. They are perhaps the most common animal to be found in ANIMATED MOVIES. MICKEY MOUSE is only one of many cartoon mice.

Perhaps because of their size and potential dangerousness, rats in fantasy contexts tend to occupy more ambiguous roles. The eponymous narrator of William KOTZWINKLE's *Doctor Rat* (**1976**), for instance, is a figure of both pathos and menace; and CINDERELLA's coachman, whose story is told in *The Coachman Rat* (**1985**) by David Henry WILSON, is a similarly ominous personage – much different from the singing mice in DISNEY's CINDERELLA (**1950**) The venal, greedy cellar-rat of John MASEFIELD's *The Midnight Folk* (**1927**), though helpful, has to be bribed. Rats can be the VILLAINS of mouse worlds, as with Warren T. Rat in the animated movies *An American Tail* and Ratigan in *The Great Mouse Detective* (**1986**). A plague of devouring rats is often recognized as unstoppable without supernatural intervention, as in the story of the PIED PIPER, Richard GARNETT's "Alexander the Rat-Catcher" (1897) and Fritz LEIBER's *The Swords of Lankhmar* (**1968**). Robert SOUTHEY's poem "Bishop Hatto" commemorates German LEGENDS of oppressive rulers whose punishment is to be eaten by swarming mice or rats.

Rats also come into their own in SUPERNATURAL FICTION or DARK FANTASY, where they tend to represent invasive EVIL, insatiate appetite – as in H.P. LOVECRAFT's "The Rats in the Walls" (1924) – DECADENCE or obscene submission to the DEVIL. A sequence like James HERBERT's **Rats** would have had rather less effect had its opening title been «The Mice». The connection between DRACULA and plagues of rats is made particularly explicit in *Nosferatu the Vampyre* (*1979*; ◊ DRACULA MOVIES). Only occasionally do hordes of mice appear, as in Roald DAHL's *The Witches* (**1983**), filmed as *The* WITCHES (**1989**). [JC]

MICKEY MOUSE Animated character, created by Walt DISNEY and Ub IWERKS, who has become a worldwide ICON. The same team had earlier created a series star called Oswald the Lucky Rabbit, who in most respects closely resembled MM; had the DISNEY studio not been swindled out of the rights to Oswald, MM would probably never have been brought into existence.

MM first appeared in *Steamboat Willie* (**1928**), a short ANIMATED MOVIE undistinguished except insofar as it featured a soundtrack – albeit an unsophisticated one. (It was not the first animated short with a soundtrack: Max FLEISCHER had performed the trick with *My Old Kentucky Home* [**1924**], and may not have been the first.) The public went "Mickey-mad", and further MM shorts followed with dizzying frequency until about 1940, when the pace slowed. After *The Simple Things* (**1953**) there was a 30-year hiatus before MM's screen career restarted with *Mickey's Christmas Carol* (**1983**; ◊ *A* CHRISTMAS CAROL); during these years, however, MM was omnipresent in the DISNEY comics as well as in the company's theme parks, while MM merchandise was widely sold throughout the world – MM watches being especially prized.

MM has in fact appeared in fewer movies (and probably fewer COMICS) than Disney's other great "property", DONALD DUCK, but the latter has never been promoted to the same extent as a company symbol. This is probably because the character of MM was based in large part on that of Walt Disney himself (who also voice-acted MM for many years): thus the corporate image was not so much MM as Walt.

MM is a jealously guarded copyright character, and therefore for legal reasons rarely appears "in clear" in written fantasy. [JG]

Filmography:
Mickey Mouse shorts: B/w: *Steamboat Willie* (**1928**); *Gallopin' Gaucho* (**1928**); *Plane Crazy* (**1928**); *The Barn Dance* (**1928**); *The Opry House* (**1929**); *When the Cat's Away* (**1929**); *The Barnyard Battle* (**1929**); *The Plow Boy* (**1929**); *The Karnival Kid* (**1929**); *Mickey's Follies* (**1929**); *Mickey's Choo-Choo* (**1929**); *The Jazz Fool* (**1929**); *Jungle Rhythm* (**1929**); *The Haunted House* (**1929**); *Wild Waves* (**1929**); *Just Mickey* (**1930**; ot *Fiddlin' Around*); *The Barnyard Concert* (**1930**); *The Cactus Kid* (**1930**); *The Fire Fighters* (**1930**); *The Shindig* (**1930**); *The Chain Gang* (**1930**); *The Gorilla Mystery* (**1930**); *The Picnic* (**1930**); *Pioneer Days* (**1930**); *The Birthday Party* (**1931**); *Traffic Troubles* (**1931**); *The Castaway* (**1931**); *The Moose Hunt* (**1931**); *The Delivery Boy* (**1931**); *Mickey Steps Out* (**1931**); *Blue Rhythm* (**1931**); *Fishin' Around* (**1931**); *The Barnyard Broadcast* (**1931**); *The Beach Party* (**1931**); *Mickey Cuts Up* (**1931**); *Mickey's Orphans* (**1931**); *The Duck Hunt* (**1932**); *The Grocery Boy* (**1932**); *The Mad Dog* (**1932**); *Barnyard Olympics* (**1932**); *Mickey's Revue* (**1932**); *Musical Farmer* (**1932**); *Mickey in Arabia* (**1932**); *Mickey's Nightmare* (**1932**); *Trader Mickey* (**1932**); *The Whoopee Party* (**1932**); *Touchdown Mickey* (**1932**); *The Wayward Canary* (**1932**); *The Klondike Kid* (**1932**); *Mickey's Good Deed* (**1932**); *Building a Building* (**1933**); *The Mad Doctor* (**1933**); *Mickey's Pal Pluto* (**1933**); *Mickey's Mellerdrammer* (**1933**); *Ye Olden Days* (**1933**); *The Mail Pilot* (**1933**); *Mickey's Mechanical Man* (**1933**); *Mickey's Gala Premiere* (**1933**); *Puppy Love* (**1933**); *The Steeple Chase* (**1933**); *The Pet Store* (**1933**); *Giantland* (**1933**); *Shanghaied* (**1934**); *Camping Out* (**1934**); *Playful Pluto* (**1934**); *Gulliver Mickey* (**1934**); *Mickey's Steam Roller* (**1934**); *Orphan's Benefit* (**1934**); *Mickey Plays Papa* (**1934**); *The Dognapper* (**1934**); *Two-Gun Mickey* (**1934**); *Mickey's Man Friday* (**1935**); *Mickey's Service Station* (**1935**); *Mickey's Kangaroo* (**1935**). Colour: *The Band Concert* (**1935**); *Mickey's Garden* (**1935**); *Mickey's Fire Brigade* (**1935**); *Pluto's Judgment Day* (**1935**); *On Ice* (**1935**); *Mickey's Polo Team* (**1936**); *Orphan's Picnic* (**1936**); *Mickey's Grand Opera* (**1936**); *Thru the Mirror* (**1936**; also released in b/w); *Mickey's Rival* (**1936**); *Moving Day* (**1936**); *Alpine Climbers* (**1936**); *Mickey's Circus* (**1936**); *Mickey's Elephant* (**1936**); *The Worm Turns* (**1937**); *Magician Mickey* (**1937**); *Moose Hunters* (**1937**); *Mickey's Amateurs* (**1937**); *Hawaiian Holiday* (**1937**); *Clock Cleaners* (**1937**); *Lonesome Ghosts* (**1937**); *Boat Builders* (**1938**); *Mickey's Trailer* (**1938**); *The Whalers* (**1938**); *Mickey's Parrot* (**1938**); *Brave Little Tailor* (**1938**); *Society Dog Show* (**1939**); *The Pointer* (**1939**); *Tugboat Mickey* (**1940**); *Pluto's Dream House* (**1940**); *Mr Mouse Takes a Trip* (**1940**); *The Little Whirlwind* (**1941**); *The Nifty Nineties* (**1941**); *Orphans' Benefit* (**1941**; remake of *Orphan's Benefit* [**1934**]); *Mickey's Birthday Party* (**1942**); *Symphony Hour* (**1942**); *Mickey's Delayed Date* (**1947**); *Mickey Down Under* (**1948**); *Mickey and the Seal* (**1948**); *Plutopia* (**1951**); *R'coon Dawg* (**1951**); *Pluto's Party* (**1952**); *Pluto's Christmas Tree* (**1952**); *The Simple Things* (**1953**); *Mickey's Christmas Carol* (**1983**; ◊ *A* CHRISTMAS CAROL); *The Prince and the Pauper* (**1990**).

Other shorts in which MM appeared: *The Fox Hunt* (**1938**); *Mickey's Surprise Party* (**1939** commercial); *The Standard Parade* (**1939** commercial); *A Gentleman's Gentleman* (**1941 Pluto** cartoon); *Canine Caddy* (**1941 Pluto** cartoon); *Lend a Paw* (**1941 Pluto** cartoon, remake of *Mickey's Pal Pluto* [**1933**]); *All Together* (**1942** commercial); *Pluto and the Armadillo* (**1943**); *Squatter's Rights* (**1946**); *Pluto's Purchase* (**1948**); *Pueblo Pluto* (**1949**); *Crazy Over Daisy* (**1950**).

Features in which MM appeared: *The Hollywood Party* (**1934**), a non-Disney "all-star" mess, none of whose eight directors permitted his name to appear in the credits; FANTASIA (**1940**); *Fun and Fancy Free* (**1947**); WHO FRAMED ROGER RABBIT (**1988**).

Further reading: *Mickey Mouse: Fifty Happy Years* (anth **1977**) ed David Bain and Bruce Harris; *Mickey's Golden Jubilee* (**1979**) by Francis Weber; *Walt Disney's Mickey Mouse – His Life and Times* (**1986**) by Richard Holliss and Brian Sibley.

See also: MICE AND RATS.

MICKEY'S CHRISTMAS CAROL US movie (**1983**). ◊ *A* CHRISTMAS CAROL.

MICKIEWICZ, ADAM (1798-1855) Polish poet. ◊ POLAND.

MICROCOSM/MACROCOSM A prevalent notion throughout much of human history has been that the human being (the microcosm) was a miniature replica of the Universe (the macrocosm); the Universe was then seen as geocentric, so the perceived relationship was in fact with the Earth (and its components) and the more visible celestial objects. Thus human eyes equated with the sky's two great lights – the Sun and Moon – and likewise bones with rocks, hair with vegetation, the pulse with the tides, etc.

Even into the 18th century the notion inspired ideas that the Earth itself might be a living organism; "When the World Screamed" by Arthur Conan DOYLE is a very late story with this premise.

A related idea was that there was a "chain of being" that ran in ascending order of biological complexity from fossils and the simplest "animalcules" up to humanity and then beyond, to the ANGELS and GOD. Were any link in this chain to be missing, the set-up of the Universe would be impossible; e.g., were there no such creature as a dog, no creature more biologically complex than a dog could exist. Thus the importance of humanity was obvious: were there no humanity, there could be neither angels nor God.

Fantasy often draws on these notions, but probably often without the awareness of the author. The version most often encountered is the Aristotelian one that, though the great may be observably entirely different from the small, there can nevertheless be a one-to-one relationship between their constituent parts. (This is the idea at the core of the theory of transubstantiation: the bread clearly does not magically turn into a gobbet of flesh, yet it *is* Christ's flesh because the one-to-one relationship makes the two identical.) Thus the king can become the LAND, and his lack of personal HEALING may turn his country into a WASTE LAND. [JG]

See also: AS ABOVE, SO BELOW; GREAT AND SMALL.

MIDDLETON, HAYDN (1955-) UK writer in whose first sequence is the **People** series – *The People in the Picture* (**1987**) and *The Collapsing Castle* (**1990**) – contemporary protagonists are influenced by the world of CELTIC FANTASY; in the second volume, a particular Celtic STORY controls the dire events taking place in a 20th-century English village. HM's second series, the **Mordred Cycle** – *The King's Evil* (**1995**) and «The Queen's Captive» (**1996**), with further volumes projected – tells in DARK-FANTASY terms the story of ARTHUR from the viewpoint of Mordred. Singletons include *Son of Two Worlds: A Retelling of the Timeless Celtic Saga of Pryderi* (**1987**) and *The Lie of the Land* (**1989**), associational. [JC]

MIDDLETON, MARTIN (1954-) Australian author. ◊ AUSTRALIA.

MIDDLETON, RICHARD (BARHAM) (1882-1911) UK poet and writer who claimed descent from Richard Harris BARHAM. Although his work met with much critical acclaim he was financially insecure, and committed suicide when on the verge of potential greatness. A sensitive man, he poured his emotions into his poetry and essays. Most of his short stories are mood sketches. Some are GHOST STORIES, best-known being the humorous "The Ghost Ship" (1912 *Century*), but better still are his DREAM fantasies, where people escape from the horrors of the present into a personal SECRET GARDEN. The best are "The Bird in the Garden" (ot "The Boy in the Garden" 1911 *The Academy*) and "Children of the Moon". "On the Brighton Road", superficially humorous, is in fact a more tragic story of a tramp who is perpetually reincarnated (◊ REINCARNATION). These stories were collected as *The Ghost Ship and Other Stories* (coll 1912 intro Arthur MACHEN). RM left behind many unpublished and uncollected stories; posthumous collections of prose and poetry are *The Day Before Yesterday* (coll **1913**) and *The Pantomime Man* (coll **1933**) ed John GAWSWORTH, who championed RM's work and included other material in his own anthologies, in particular *New Tales of Horror* (anth **1934**). [MA]

MIDNIGHT GRAFFITI ◊ MAGAZINES.

MIDSUMMER NIGHT Midsummer in the Northern Hemisphere falls on June 21, although Midsummer Day is traditionally on the feast day of St John the Baptist, June 24. This period, particularly St John's Eve, is ripe with LEGENDS and FOLKLORE, some no doubt arising from the notion of Midsummer madness, traditionally attributed to the workings of the supernatural. On Midsummer's Eve all the sprites, hobgoblins and FAIRIES become manifest and play their tricks. Similar stories relate to Mayday Eve and Walpurgisnacht. The most famous use of this is in *A Midsummer Night's Dream* (performed 1596; **1600**) by William SHAKESPEARE, where OBERON and PUCK play their games on Theseus's wedding party. Other examples include: "St John's Eve" (1831; vt "The Witch") by Nikolai GOGOL, where the DEVIL plays to humanity's greed and evil on that night; "Midsummer" (in *Ornaments in Jade* coll **1924**) by Arthur MACHEN, where a young man witnesses a ghostly procession; and "How Pan Came to Little Ingleton" (in *Nights of the Round Table* coll **1926**; vt "Mr Minchin's Midsummer") by Margery LAWRENCE, where PAN tempts a rigidly puritan vicar back to more homely ways. J.B. PRIESTLEY wrote the libretto for Sir Arthur Bliss's comic opera *The Olympians* (1949), about the Greek GODS coming alive once a century on Midsummer Day; he also chose Midsummer for his play *Summer Day's Dream* (performed 1949; **1950**), a VISIONARY FANTASY of mankind returning to Nature. [MA]

MIDSUMMER NIGHT'S DREAM, A US movie (*1935*). Warner Bros. **Pr** Max Reinhardt. **Dir** William Dieterle, Reinhardt. **Spfx** Byron Haskin, Fred Jackman, Hans Koenekamp. **Screenplay** William SHAKESPEARE (ed and cut). **Music** adapted from *A Midsummer Night's Dream* (1826/1842) by Felix Mendelssohn (1809-1847). **Starring** Ross Alexander (Demetrius), Joe E. Brown (Flute), James Cagney (Bottom), Hobart Cavanaugh (Philostrate), Otis Harlan (Starveling), Olivia de Haviland (Hermia), Hugh Herbert (Snout), Ian Hunter (Theseus), Victor Jory (OBERON), Anita Louise (Titania), Frank McHugh (Quince), Grant Mitchell (Egeus), Jean Muir (Helena), Dick Powell (Lysander), Dewey Robinson (Snug), Mickey Rooney (PUCK), Verree Teasdale (Hippolyta), Arthur Treacher (Epilogue). 133 mins, often cut. B/w.

This all-singing, all-dancing musical old-style Hollywood spectacle version of Shakespeare's play – with a casting that must on paper have seemed insane – is, remarkably, charming; there is quite a lot of singing, but even the spoken dialogue is matched with Mendelssohn's music in such a way that one has the sense of watching an OPERA. The plot is more or less as in the original, so need not be synopsized. What the movie does very well is to convey a sense of genuine MAGIC; in this Louise – as Titania – and Rooney – as the TRICKSTER Puck – are exceptional, portraying successfully creatures that are quite otherworldly. Cagney is a surprisingly effective Bottom, while Muir and de Haviland play well against rather weak leading men. Titania is identified with the GODDESS rather than merely with the FAIRY QUEEN, while Oberon is the Horned God (◊ DEVILS). This LOVE comedy gave some visual impetus to the *Ave Maria* section of FANTASIA (*1940*) and possibly influenced the concluding visuals of Steven SPIELBERG's *E.T. – The Extra-Terrestrial* (*1982*). The moment when FAIRIES are born out of the ground-hugging mist is one of the high points of fantastic cinema.

There have been at least two other movies based on the

play. The Czech ANIMATED MOVIE *A Midsummer Night's Dream* (**1961**; Animania) dir Jiri Trnka – mixing mainly stop-motion techniques with painted ones, concentrates largely on the *Pyramus and Thisbe* play put on by the artisans. In an interesting shift of PERCEPTION, for a while the play becomes real rather than merely a performance. The English-language version was narrated by Richard Burton, and the cast of voice actors was distinguished. The UK/Spanish *A Midsummer Night's Dream* (**1984**) dir Celestino Coronado is a live-action version based on a stage production dir Lindsay Kemp. Here the lovers not only swap allegiances but also sexual orientation. The Woody Allen movie *A Midsummer Night's Sex Comedy* (**1982**) is unrelated, being based instead on the Ingmar BERGMAN movie *Smiles of a Summer Night* (**1955**; ot *Sommarnattens Leende*), also the basis for Stephen Sondheim's *A Little Night Music* (produced 1973). [JG]

MIFFKIN, PATTON H. [s] ◊ Kenneth MORRIS.

MIGHTY JOE YOUNG US movie (**1949**). ◊ Ray HARRYHAUSEN.

MIGNOLA, MICHAEL J. (1960-   ) US COMIC-book artist whose distinctive evocative drawing – in a sharp, lucid, deceptively minimal line-and-black style – became increasingly sophisticated and atmospheric as it matured in the late 1980s and early 1990s. The beguiling simplicity of his work is achieved through a process of refinement and distillation.

MJM spent a short period inking the work of other artists before drawing a number of fairly routine comic books for MARVEL COMICS, including **Sub-Mariner** and **The Human Torch** (*Marvel Fanfare #43* 1985), **Incredible Hulk** (*#310-#313* 1985), **Solomon Kane** (*#3* 1985) and **Cloak and Dagger** (*#8* 1986). He then drew two stories in the series **Chronicles of Corum** (*#1-#6* 1987-8), based on Michael MOORCOCK's character, the GRAPHIC NOVEL *Triumph and Torment* (graph **1989**) – which featured two long-established Marvel characters – and eight stories in the series **Fafhrd and the Gray Mouser** (*#1-#4* 1990-91), based on Fritz LEIBER's characters. His GRAPHIC-NOVEL adaptation of *Bram Stoker's Dracula* (**1993**; ◊ DRACULA MOVIES) (*#1-#4* 1992-3; graph coll **1993**) showed a remarkable talent for creating gloomy atmospheres, which he has gone on to exploit very effectively in his OCCULT-DETECTIVE series **Hellboy** (*Dark Horse Presents* and *Hellboy: Seeds of Destruction #1-#4* 1994-5; graph coll **1995**). [RT]

MILÁN, VICTOR (WOODWARD) (1954-   ) US writer who has published a number of post-holocaust military-sf novels as by Richard Austin, and may have written other pseudonymous work. As VM he began publishing work of genre interest with the long **War of Powers** SWORD-AND-SORCERY sequence, all with Robert E. VARDEMAN (*whom see for details*), set in a FANTASYLAND and featuring much MAGIC. In his own right, VM is best-known for the sf **Samurai** sequence – *The Cybernetic Samurai* (**1985**) and *The Cybernetic Shogun* (**1990**). In *Runespear* (**1987**) with Melinda SNODGRASS three adventurers become embroiled in 1936 with Nazi Germany, whose leaders' occult obsession about finding the eponymous spear of ODIN leads to perilous scrapes. VM's later work has been largely sf. [JC]

MILES GLORIOSUS ◊ COMMEDIA DELL'ARTE.

MILITARY FANTASY Much GENRE FANTASY concerns itself with warfare. Though J.R. TOLKIEN's *The Lord of the Rings* (**1954-5**) can be seen as largely concerned with warfare, to think of it as MF would be to misread – which does not, of course, rule out *LOTR*'s having been an influence on MF; the anachronistic specifics of its battle scenes and individual trials of arms, which combine the medieval/archaic with the modern, have been reproduced endlessly and faithfully. Mercenaries in (e.g.) the works of Glen COOK and Barbara HAMBLY combine attributes from those of the Hellenistic world, the early Italian Renaissance and the Thirty Years' War. Accordingly, the sense of warfare as a part of the historical process which we find in such books always approaches the GAMEWORLD even when as competently and insightfully done as by these authors. Also, Cook's and Hambly's characters retain moral sense even if compelled by circumstances to perform dreadful acts. Not all MFs are as likeable. Tolkien's MAGGOTS – e.g., his assumption that vice, if not virtue, can be intrinsic to race or species identity – are often reproduced more or less unthinkingly in MF (◊ COLOUR-CODING). Much of this is parodied by Mary GENTLE in *Grunts!* (**1992**).

More attractively, warfare is often seen as a school which teaches hard and necessary lessons, as a moulder of character; it is likely to be a crucial experience for UGLY DUCKLINGS, BRAVE LITTLE TAILORS and people who LEARN BETTER. It is a set of situations in which the SENSIBLE MAN has the opportunity to show his sense and in which competence is a life-or-death matter. In the work of David GEMMELL and many like him it is seen as intrinsically tragic; competent men are both created and destroyed by warfare, usually at the hand of others of their own kind. Richard MONACO's *Parsival* trilogy depicts this dilemma in bitingly satirical terms; Bernard KING's NORDIC-FANTASY sequence begun with *Starkadder* (**1985**) almost wallows in it.

Most MF pays insufficient attention to logistic problems and to, for example, the high-protein diet necessary to sustain a heavily armoured KNIGHT. Hambly's *Dark Hand of Magic* (**1990**) is notable for being one of the few genre fantasies to pay serious attention to the practicalities of quartermastering. *Albion* (**1991**) by John GRANT is rare in recognizing that an army made up in part of women should carry material for sanitary towels.

As noted, MF draws rather too eclectically on real-world military history and to forget that strategy and tactics reflect the class and economic structures, religion, ideology, technology and animal husbandry of the societies from which they spring. For example, Robert JORDAN's cavalry charges in various of the battles in **The Wheel of Time** sequence (**1990** onwards) are essentially drawn from the American Civil War, even though the battles in which they take place involve medieval or early-Renaissance levels of armament. Genre fantasy like this anyway usually adds MAGIC to the range of armaments available without any clear idea of how to represent it in terms of analogy with the real world. Magic is problematic partly because it is used as the equivalent of heavy artillery, chemical/bacteriological weapons or air power. It is hard for virtuous characters to retain sympathy when deploying such lethal weaponry except in the most desperate cases, but if the magic remains unintegrated it is rendered irrelevant to the main issue, throwing the purpose of making the book a fantasy into question – except where, as in Glen Cook's **Dread Empire** sequence (**1979** onwards), the argument is that, in FANTASYLAND as everywhere else, the point of a battle is ultimately that the poor bloody infantry have to gain or hold terrain, and all else is top-dressing. It is because of these difficulties that many fantasies announce arbitrarily the difficulty of using magic for military purposes or subject it to ideological or ethical constraints. Similar considerations often apply to the use of DRAGONS, whether intelligent or otherwise.

Almost all MF deals in land warfare. Naval combats usually take place off-stage except in narratives that deal with PIRATES – Tim POWERS's *On Stranger Tides* (**1987**), for example – and even these are rare. The **Liavek** series of SHARED-WORLD anthologies (**1985-90**) ed Emma BULL and Will Shatterley contains stories on naval themes by Gene WOLFE and Walter John WILLIAMS. Air combat, even when managed with dragons or levitation spells, is even rarer in anything more than the crudest form; two recent exceptions are Kim NEWMAN's *The Bloody Red Baron* (**1995**) and Felicity SAVAGE's *Ever* (**1996**), two of whose principals are fighter aces in a magical equivalent of WORLD WAR I where airplanes are demon-powered. [RK]

MILLAIS, [Sir] JOHN EVERETT (1829-1896) UK artist, a central member – along with his mentors William Holman Hunt (1827-1910) and Dante Gabriel ROSSETTI – of the PRERAPHAELITES. His "Ophelia" (1852) is the first of many florid Preraphaelite portrayals of the drowning of Hamlet's betrothed. In general his subject matters have little to do with the language or movement of fantasy; he is of interest, and his influence is most evident, for the superb moody tangibility of his LANDSCAPES. Those featured in "The Blind Girl" (1856) and "Autumn Leaves" (1856), for instance, seem to evoke LAND-OF-FABLE medieval worlds, whose foregrounds later illustrators glady filled with ICONS of the OTHERWORLD. [JC]

MILLAR, H(AROLD) R(OBERT) (1869-1940) Prolific UK illustrator, with a fine pen-line style. His work, mainly for children's books, has a lively, imaginative charm and a distinctive sense of design. He is particularly remembered for his illustrations to the works of Rudyard KIPLING and E. NESBIT. HRM was a great collector of Eastern works of art and ancient weapons, and these enthusiasms are often reflected in the strong sense of history evident in his work. [RT]

MILLENNIAL FANTASY A term which has sometimes been applied to fantasies in which the END OF THE WORLD is accomplished by a return of CHRIST, who presides over the LAST JUDGEMENT, which may be condemnatory or salvationary. As very few fantasies deal in any literal sense with this ending – though very many are shaped by considerations of ESCHATOLOGY – the term is not frequently used. [JC]

MILLER, FRANK (1957-) US writer/artist with an impressive list of ground-breaking and phenomenally successful COMIC books and several CINEMA scripts to his credit. FM utilizes a distinctive fragmented narrative technique and, by skilful manipulation of the visual and verbal elements of the comic-strip medium, tells tightly plotted multi-level stories that deal with both the springs of violence and its effects. His work treats small tragedies and great events with equal perceptiveness and insight, and displays an expert control of narrative pace and scope. FM's recent artwork has also proved influential: from a somewhat wooden depiction of the human figure in his work of the early 1980s, drawn with a strong but indecisive line, he has developed, in his 1990s **Sin City** series, a very forceful "bleached out" style of b/w drawing in which line is almost entirely absent. These pages pack a powerful emotional punch and have tremendous vitality.

FM's first published work was for MARVEL COMICS's **Daredevil**; this was later re-released in three collections: *Child's Play* (graph coll **1986**), *Marked for Death* (graph coll **1990**) and *Gang War* (graph coll **1992**). He went on to produce two landmark apocalyptic dramas for DC COMICS: **Ronin** (#1-#6 1983-4) and the revolutionary BATMAN story **The Dark Knight Returns** (#1-#4 1985-6; graph coll **1986**). The phenomenal success of this latter led FM to begin writing scripts for several established comic books and short series, including **Daredevil** – *Born Again*, drawn by David Mazzuchelli (1960-) (*Daredevil #228-#233* 1985-6; graph coll **1987**) – *Love and War*, drawn by Bill SIENKIEWICZ (graph **1986**), *Batman – Year One*, drawn by Mazzuchelli (*Batman #404-#407* 1987; graph coll **1988**), *Elektra Assassin* (#1-#8 1986-7; graph coll **1987**), with painted art by Sienkiewitz, *Give Me Liberty* (#1-#5 1991-2; graph coll **1992**), drawn by Dave Gibbons (1949-), *Hard Boiled* (1990-2), drawn by Geoff Darrow, and a collaboration with Walt Simonson on a series tied to the **Terminator** movies.

FM's other work includes *Elektra Lives Again* (graph **1990**) and the hard-hitting **Sin City** series, into which he has injected a *noir* PI with great intensity. The series first appeared in *Dark Horse Presents* (#51-#53 and #55-#59 1991-2) and then in book form as *Sin City* (graph coll **1992**), *Sin City: A Dame to Kill For* (#1-#6 1993; graph **1994**), *Sin City: The Babe Wore Red and Other Stories* (graph coll **1995**), *Sin City: The Big Fat Kill* (#1-#6 1994; graph **1995**) and «*That Yellow Bastard*» (graph 1996).

FM worked on the screenplays for *Robocop 2* (**1992**) and *Robocop 3* (**1994**). [RT]

MILLER, IAN (1946-) UK artist and illustrator, mainly of fantasy and sf; his bizarre, intricate line and colour creations often contain both Surreal and STEAMPUNK elements and motifs. The meticulous execution gives his work an obsessive quality, and certain images frequently recur, among them complex mechanical fish and flies, puppet strings and robotic structures. He draws in inks with a technical pen, often with ruled, almost mechanical cross-hatching, overlaid with light washes of watercolour and some coloured pencil work.

IM studied at Northwich School of Art and at St Martin's School of Art, then produced book and magazine ILLUSTRATIONS, including some work for WARREN PUBLISHING. In 1975 he went to Hollywood to do scene origination and background design for Ralph BAKSHI's WIZARDS (**1977**). His first book, *Green Dog Trumpet and Other Stories* (graph **1979**), was intended to include a text storyline; this was in the event omitted, his art providing sufficient narrative. Other books include *Secret Art* (graph **1980**) and *Ratspike* (graph **1990**), this latter with John Blanche. In 1983 he produced some gloriously amorphous drawings of Mervyn PEAKE's castle of Gormenghast. He has co-created two GRAPHIC NOVELS: *The Luck in the Head* (graph **1991**) with M. John HARRISON and *The City* (graph **1992**). [RT]

Further reading: *The Guide to Fantasy Art Techniques* (graph **1984**) ed Martin Dean.

MILLHAUSER, STEVEN (1943-) US writer whose first novel, *Edwin Mullhouse: The Life and Death of an American Writer, 1943-1954, by Jeffrey Cartwright* (**1972**), remains his best-known; although not fantasy, its intensely hilarious recounting of the life of a child in the form of a hagiographic literary biography constantly urges otherwise mundane material into the fantastic. *Portrait of a Romantic* (**1977**) performs a similar feat with an adolescent. SM's third novel, *From the Realm of Morpheus* (**1986**), is a full-blown fantasy whose narrator descends through a neighbourhood PORTAL into the UNDERWORLD, where he enters an environment which (though substantial to his senses) derives its logic from DREAM and literary analogue. Some episodes – as in a LIBRARY which clearly refers to the work of Jorge Luis BORGES – are without satisfying narrative closure; others – as in the sequence in ATLANTIS – are redolent of fantasy's grounding in STORY. The overall brokenness of the text – which has reminded critics of Herman MELVILLE's *Mardi and a Voyage Thither* (**1849**) – may have been caused by rumoured severe editorial cutting of the original manuscript.

SM's short fiction is more coherent. The stories assembled in *In the Penny Arcade* (coll **1986**) reflect with concision upon the conventions which govern our PERCEPTIONS of what is real and what is fantastic, in life and in art. *The Barnum Museum* (coll **1990**) assembles several stories that retell and rather abstractly revise some central fantasy tales, including Lewis CARROLL's *Alice in Wonderland* (**1864**). *Little Kingdoms: Three Novellas* (coll **1993**) contains, in its longest tale – "The Little Kingdom of J. Franklin Payne" – an analysis in fantasy terms of a cartoonist whose life and art seem clearly to reflect that of Winsor MCCAY, and who passes literally over into the world of his imagination. As a writer of literary fantasies SM may treat such passages with ambivalence, but the dream-like potency of his best work calls most powerfully, as though the Other were indeed true. [JC]

MILLS, PATRICK (1949-   ) Innovative, award-winning UK editor and writer of UK and US COMIC strips; he is credited with creating the highly successful UK weekly *2,000 A.D.*

PM began working for UK comics publisher D.C. Thomson as a subeditor on the romance title *Romeo* before turning freelance and writing, in collaboration with John Wagner, a long series of strips for juvenile comics. In 1975 the two created the popular war comic *Battle Picture Weekly* (673 issues 1975-88) and the violent, controversial *Action* (36 issues 1976). Both were innovative, using contemporary tv and movies as springboards for story ideas; they had remarkably high sales figures. In 1977 PM was invited to create a third title; this became *2,000 A.D.*, whose centrepiece character, **Judge Dredd**, has become one of the most widely recognized comics ICONS in the UK and, with the 1995 movie, worldwide.

It was through his writing in *2,000 A.D.*, where writer and artist credits were introduced (from #36 1977) for the first time in the UK for many years, that PM's name became widely known. He had anonymously written the scripts for one of the UK's most widely respected war stories, **Charlie's War** (1977-85 *Battle Picture Weekly*; graph coll 2 vols **1986**). Its honesty and stark portrayal of the military incompetence of the UK campaign in WORLD WAR I caused some editorial problems, and towards the end of its run both script and artwork were being censored. Concurrently, PM was writing two sf strips, both with a thread of cynical humour and SATIRE, for *2,000 A.D.*: **Ro-Busters** and **ABC Warriors**.

Much of PM's work is set in a self-contained universe, with characters often switching between strips. This constantly broadening base allows him to encompass concepts within an established framework, and brought immediate popularity to such strips as **Nemesis the Warlock** (*2,000*

A.D. from #222 1981). Another meticulously researched creation was SLAINE. Further creations include *Marshal Law* (#1-#6 1990-1), the graphic novel *Metalzoic* (graph **1986** US), the disturbing **Third World War** (*Crisis* intermittently 1988-90), **The Punisher** (*Toxic* 1991), **Sex Warrior** (*Toxic* 1991), **Finn** (*2,000 A.D.* 1992 and 1993), **Punisher 2099** (1992-current US), **Ravage 2099** (1993-current US) and **Terrorists** (1993-current US).

Collected editions of his UK work include *ABC Warriors* (graph coll 1983-8 4 vols), *Nemesis the Warlock* (graph coll 1984-9 9 vols), *Ro-Busters* (graph coll **1983** 2 vols), *Marshall Law* (graph colls **1990**, **1994**) and the **Slaine** series.

[SH/RT]

MILNE, A(LAN) A(LEXANDER) (1882-1956) UK writer, humorist and playwright whose greatest successes were the CHILDREN'S FANTASIES classics *Winnie-the-Pooh* (coll **1926**) and *The House at Pooh Corner* (coll **1928**), and the related verse collections *When We Were Very Young* (coll **1924**) and *Now We Are Six* (coll **1927**), forming the **Christopher Robin** sequence.

AAM developed his skills at *Punch* magazine, of which he was assistant editor 1906-18, and occasional short fantasy appears in his collected *Punch* HUMOUR: *The Day's Play* (coll **1910**), *The Holiday Round* (coll **1912**), *Once A Week* (coll **1914**) and *The Sunny Side* (coll **1921**), assembled as *Those Were the Days* (omni **1929**). Some of this – e.g., thoughts on the geometrical difficulties of attaining a destination via seven-league boots which take literal 21-mile strides – was recycled in his first substantial fantasy, the comic FAIRYTALE pastiche *Once On a Time* (**1917**; vt *Once Upon a Time* 1988 US), based on a lost 1915 play written to entertain WWI troops. Supposedly written for adults but enjoyed by children, this deals amusingly with RURITANIAN squabbles complicated by cloaks of INVISIBILITY, a RING granting WISHES upon certain CONDITIONS, the TRANSFORMATION of an unworthy prince into an IMAGINARY ANIMAL composed of lion, lamb and rabbit, etc. In a gambit prefiguring William GOLDMAN's *The Princess Bride* (**1973**), AAM addresses the reader in "personal" asides which dispute the imagined official history of those times by one "Roger Scurvilegs".

The extraordinarily popular **Christopher Robin** stories are set in Ashdown FOREST, where AAM then lived; they feature his son Christopher Milne (1920-1996) amid various TALKING ANIMALS and animated versions of his TOYS. E.H. SHEPARD's drawings of all these are inseparable from the prose. The lead character, Winnie-the-Pooh, is Christopher's teddy-bear, and several of his small adventures with Piglet, Tigger, the wondrously pessimistic Eeyore and others are chronicles of childhood games or real-life incidents. While confining himself to language which young children can understand (◊◊ DICTION), AAM describes ARCHETYPES of character and situation which make the stories quotable – and applicable as parables – to an extent perhaps surpassed in fantasy only by Lewis CARROLL's **Alice** books. *The Pooh Perplex* by Frederick C. Crews (coll **1963**) illustrates Pooh's protean nature with crushing SATIRE by providing 12 variously plausible critical interpretations: Freudian, Leavisite, Marxist, as CHRISTIAN FANTASY with Eeyore as CHRIST, etc.

Some readers dislike the stories' whimsy and *The House at Pooh Corner*'s sentimental close, which addresses the specific THINNING of a child beginning to put away childish things; but AAM never in fact descends to the baby-talk implied in the notorious dismissal by Dorothy Parker (1893-1967):

"Tonstant Weader Fwowed up." Movie adaptations began with *Winnie the Pooh and the Honey Tree* (1966) from DISNEY; Pooh's US accent and the new animal "Gopher" were poorly received by lovers of this quintessentially English work. The entire **Christopher Robin** sequence has remained in print since first publication, leading to such further spinoffs as *The Hums of Pooh* (with music by H. Fraser-Simson **1929**), the Latin translations *Winnie Ille Pu* (trans Alexander Lenard 1958; **1960**) and *Domus Anguli Puensis* (trans Brian Staples **1980**), and a facsimile of the original manuscript as *Winnie-the-Pooh* (**1971**) – as well as much merchandise, including many Disney rather than AAM ties. Pooh is an industry.

Most AAM plays are nonfantastic. The frothy *Make-Believe* (performed 1918; in *Second Plays* coll **1921**) comprises three loosely linked RECURSIVE-FANTASY playlets for children, spoofing FAIRYTALE tests of suitors, a NEVER-NEVER LAND with PIRATES (◊ J.M. BARRIE), and CHRISTMAS. *The Ivory Door: A Legend in a Prologue and Three Acts* (**1928** US) is a FABLE of PERCEPTION and the power of belief over REALITY. Legend says that a fairytale castle's eponymous door (whose ivory suggests false DREAMS) leads only to death; when the Prince enters and returns unscathed, his own people reject him as "evidently" an impostor or evil SPIRIT. *Toad of Toad Hall* (written 1921; **1929**), a children's play based on Kenneth GRAHAME's *The Wind in the Willows* (**1908**), has, ironically, survived all AAM's original dramatic work.   [DRL]

Other works: *Prince Rabbit and the Princess who Could Not Laugh* (coll **1966**), children's stories; *The World of Christopher Robin* (omni **1991**), comprising *When We Were Very Young* and *Now We Are Six*; *The Complete Winnie-the-Pooh* (omni **1991**), comprising *Winnie-the-Pooh* and *The House at Pooh Corner*; other editions too numerous to list.

Further reading: *It's Too Late Now* (**1939**; vt *Autobiography* US); *The Enchanted Places* (**1974**) by Christopher Milne; *A.A. Milne: His Life* (**1990**) by Ann Thwaite (1932-   ); *The Brilliant Career of Winnie-the-Pooh: The Story of A.A. Milne and His Writing* (**1994**) by Thwaite; *The Pooh Dictionary: The Complete Guide to the Words of Pooh and All the Animals in the Forest* (**1995**) by A.R. Melrose.

MILTON, JOHN (1608-1674) English poet and pamphleteer, government official during the Cromwell Protectorate and the single most influential English poet since William SHAKESPEARE. His renown, in both English literature and genre fantasy, is based primarily on *Paradise Lost* (**1667**; rev 1674), a long verse account of the biblical Fall and consequent expulsion from EDEN. Enormous in scope, the poem describes the entire Christian cosmology, dramatizes conversations between divine beings, and proposes famously to "justify the ways of God to men" – an ambition that JM, who spent his youth preparing to be a great poet, felt fully equal to do. The work's influence on English-language poetry, although enormous, is indirect. It served as the model for many epic poems in the century following its publication, of which John KEATS's fragment "Hyperion" (1820) alone is remembered. JM's influence is more lastingly seen in the adoption of his "grand style", which, while unsuited for any further attempts at epic poetry, proved remarkably congenial to the mock-heroic: *Absalom and Achitophel* (**1681**) by John Dryden (1631-1700) is the first work to employ JM's elevated manner for purposes of satiric contrast; *The Rape of the Lock* (**1714**) and *The Dunciad* (**1729**) by Alexander Pope (1688-1744) both contain explicit echoes of *Paradise Lost*.

The poem's influence upon modern fantasy is at least threefold. First is JM's characterization of SATAN, who, finding it "Better to reign in hell than serve in heav'n", is conceived in terms of heroic defiance and with an imaginative complexity not previously seen in the centuries-old tradition of dramatizing devils in English literature. Satan is a figure not only of defiance but of selfconsciousness, and his portrayal as a reflecting figure capable of wondering about (then acting upon) his relationship to the cosmos is more compelling than any since Hamlet's. A great number of infernal and doomed figures in modern fantasy, from J.R.R. TOLKIEN's Saruman to the satanic "Darkness" in Ridley Scott's LEGEND (**1985**), derive from *Paradise Lost*.

In addition, the remarkable *plottiness* of JM's poem – the revolt against Heaven is dramatized in great detail, with skirmishes and the employment of "devilish Engines" described – has exerted a great influence on EPIC FANTASY. Tolkien certainly wrote *The Lord of the Rings* (**1954-5**) with JM's poem much upon his mind; some critics have speculated that the character of Sauron is kept off-stage in part as a response to the tendencies of then-current JM criticism, which regarded his Satan favourably. It is impossible for the modern reader to encounter *Paradise Lost* without being struck how some of its scenes read like modern fantasy. John COLLIER's *Milton's "Paradise Lost": Screenplay for the Cinema of Mind* (**1973**) dramatizes the poem in cinematic terms; were such a movie to be in fact produced, it would certainly be regarded as a fantasy.

JM's third influence upon fantasy lies in what must be called his ideology. Historians have demonstrated how the vicissitudes of JM's reputation between his death and 1940 were influenced by political feeling; and William BLAKE's famous declaration that Milton was "a true Poet and of the Devil's party without knowing it" was, like Percy Bysshe SHELLEY's assertion that JM's Satan was morally superior to his God, an attempt to claim the force of JM's imaginative power for polemical purposes. In this century, both Charles WILLIAMS and C.S. LEWIS attempted (in essays) to reclaim JM for Christian orthodoxy, while Tolkien took pains to avoid being impressed into the Devil's party. Whether JM's Satan is the "archangel ruined" of the Romantic imagination or the comic and egotistical failure that Lewis finds is an issue that every serious fantasy writer concerned with GOOD AND EVIL must ponder.

Lord BYRON's "The Vision of Judgment" (1822) is explicitly set in the heaven of *Paradise Lost*, to satiric effect. JM's cosmology is a model in Steven BRUST's *To Reign in Hell* (**1984**), and while the cosmos in James BLISH's *The Day After Judgment* (**1971**) is DANTE's, the speech that Satan gives at its climax is cast in Miltonic verse. Milton himself appears in Peter ACKROYD's alternate-history *Milton in America* (**1996**) and in Blake's visionary and highly eccentric *Milton* (c**1815**), which evokes him as a thoroughly Miltonic figure. JM is villainously portrayed in Robert Graves's *Wife to Mr Milton* (**1944**).   [GF]

MINERVA ◊ GODDESS.

MINES Mines, when they appear in fantasy, are usually redolent with meanings beyond the industrial. A mine may be a PORTAL to an ALTERNATE WORLD or to a SECONDARY WORLD; it may be a passage to the UNDERWORLD, or to some other place of trial where the protagonist must undergo a NIGHT JOURNEY; it may contain a MALIGN SLEEPER or the ONCE AND FUTURE KING; it may open the eyes of the protagonists to the true age of the world (◊ TIME

ABYSS), and may well be worked by DWARFS whose history extends backwards into the dawn of time; it may be a POLDER or a LAST REDOUBT, or underlie an EDIFICE; it may have countless other roles. Mines crop up in all kinds of fantasy, and are part of the standard geography of FANTASYLAND.

There are mines in George MACDONALD's *The Princess and the Goblin* (**1872**). Moria, in J.R.R. TOLKIEN's *The Lord of the Rings* (**1954-5**), and Vincula, in Gene WOLFE's *The Book of the New Sun* (**1980-83**), are further examples. The most famous mine in all fantasy is almost certainly the one in which the Seven DWARFS labour (◊ SNOW WHITE; SNOW WHITE AND THE SEVEN DWARFS [*1937*]).   [JC]

MINNEAPOLIS A CITY which has become a popular fantasy venue. ◊ CONTEMPORARY FANTASY; SCRIBBLIES; URBAN FANTASY.

MINOTAUR Properly speaking, the name of the son of Pasiphaë (wife of King Minos of Crete) who had mated with a bull. The word "minotaur" – which has become a generic term for any such beast – translates as "bull son of King Minos", which is not biologically valid.

It is a sad story. King Minos asks Poseidon for a token to demonstrate his legitimacy as king in a dispute with his brothers, and Poseidon sends him a beautiful white bull, demanding that Minos sacrifice it to him immediately. But Minos values the animal too highly, and substitutes another. Poseidon, enraged, intoxicates Pasiphaë with lust for the bull. She has DAEDALUS construct a hollow cow, into which she positions herself for intercourse. The bull mounts her, and she gives birth to the Minotaur, a MONSTER with a human body and the head and tail of a bull. Minos then has Daedalus construct the LABYRINTH beneath Knossos, and hides the Minotaur at its heart, where as a LIMINAL BEING the creature opens the PORTAL only to death. And here – as a profound metaphor of BONDAGE, of stalled METAMORPHOSIS – it remains eternally.

The rest of the story is central to Greek MYTHOLOGY. It turns out that the Minotaur can eat only human flesh, and that the anguish of its cry of hunger terrifies all of Crete, broadcasting the bondage of its state into the social world. So Minos feeds it the human tribute he receives yearly from Athens. The story now becomes that of the HERO Theseus, who threads his way into the maze and kills the Minotaur.

The Minotaur is perhaps too complete an image of tragic bondage to be of much direct use in fantasy narratives, unless the figure is extricated from its defining situation, as Thomas Burnett SWANN did in his **Minotaur** sequence (**1966-77**), which sentimentalizes the central figure into an emblem of the ARCADIA on Earth that precedes the THINNING of things as history bites. In the **Saga/Heroes** track of the **DragonLance** GAME enterprise, bull-headed warrior Minotaurs – in tales like Richard A. KNAAK's *Kaz, the Minotaur* * (**1990**) – are capable of gaining sagacity points. John Farris's *Minotaur* (**1985**) is HORROR. The figure of the Minotaur has occasioned some literary fantasies, most notably perhaps MacDonald HARRIS's *Bull Fire* (**1973**), though Robert Sheckley's *Minotaur Maze* (**1990**) transforms the figure into a victim – sometimes disguised as a Paris taxicab, "its tires whispering of atrocious pain and meaningless retribution" – in a labyrinth that encompasses the Universe. Most often, though, the Minotaur is seen as the adversary and victim of the HERO, as in "The Tale of the Student and his Son", embedded into Gene WOLFE's *The Book of the New Sun* (**1980-83**), or in Dave DUNCAN's *A*

Rose-Red City (**1987**), where the hero must fight a Minotaur in a FAERIE full of UNDERLIER figures reawakened (as it seems) for AGON.

In the graphic arts of the 20th century, the Minotaur came to be identified with the work (and the self-projection as artist and male force) of Pablo Picasso (1881-1973); Michael AYRTON also incorporated the Minotaur into his overarching obsession with DAEDALUS.   [JC]

MINSTRELS The figure of the minstrel, carried over from historical fiction into fantasy, draws on and often illegitimately combines a number of historically discrete sources. The BARD was a recognized figure of Celtic society, ranked in a system that overlapped with that of the druid priesthood so that he possessed a measure of the druids' magic. In Scandinavian society, the skald was an entertainer and maker of flattering or satirical verse which enhanced or diminished the lord's status; the skald himself had status only if otherwise respected as a fighter or adviser. The jongleur was a professional entertainer in medieval Western European society; he or she might well also juggle, ropewalk or sell sexual favours. The troubadour was a noble who made music and verses as part of courtly love; he might well also be a grizzled military commander like Bertrand de Born. Similar figures in later medieval and early Renaissance society were either members of theatrical troupes or itinerant professional musicians; women musicians were likely to be either members of religious orders or courtesans – they were unlikely to travel except as part of a group.

The figure of the minstrel in GENRE FANTASY draws on these types eclectically, though usually with an emphasis that depends on the particular part of history from which the author has chosen to construct the FANTASYLAND with which s/he is working. There is also a default setting which draws on all of these without much thought, treating minstrelsy as a TEMPLATE attribute which moves a protagonist around from location to location. Bards are found most often in CELTIC FANTASY, but may occur elsewhere. Fritz LEIBER's Fafhrd in **Fafhrd and the Gray Mouser** is primarily a warrior, but also explicitly capable of earning his living as a skald. Poul ANDERSON's Cappen Vara starts as a jongleur with pretensions to troubadour status who has wandered into a Northern land where he is treated as a skald. Guy Gavriel KAY's *A Song for Arbonne* (**1992**) is unusual in treating the cult of courtly love and its troubadours sympathetically and at some length.

Minstrels are not often protagonists but more usually the COMPANIONS of heroes, though minstrelsy is often an attribute of the hero, or a disguise adopted by him; ROBIN HOOD has Alan'A Dale and himself passes as a minstrel on occasion. Minstrelsy is occasionally a useful GENDER DISGUISE, explaining why someone of boyish appearance with a high voice is on the road. ELVES are often minstrels, and were extensively presented as such by J.R.R. TOLKIEN.   [RK]
See also: MUSIC.

MINTON, JOHN (1917-1957) UK artist. ◊ Michael AYRTON.

MIRACLE IN THE RAIN US movie (*1954*). ◊ Ben HECHT.

MIRACLE ON 34TH STREET There have been two movies called *Miracle on 34th Street*, **2** being based on **1**.

1. (vt *The Big Heart*) US movie (*1947*). 20th Century-Fox. **Pr** William Perlberg. **Dir** George Seaton. **Spfx** Fred

Sersen. **Screenplay** Valentine Davies, Seaton. **Novelization** *Miracle on 34th Street* * (**1947**) by Davies. **Starring** Edmund Gwenn (Kris Kringle/Santa Claus), Porter Hall (Granville M. Sawyer), Natalie Wood (Susan Walker), Maureen O'Hara (Doris Walker), John Payne (Fred Gailey). 96 mins. B/w.

In emergency, Doris Walker, Macy's Assistant Toys Manager, hires Kringle, resident of Brooks Memorial Home for the Aged, as this year's SANTA CLAUS. An ultra-rationalist because of her messy divorce, she is horrified when Kringle claims to be the real Santa; Macy's management is likewise horrified when Kringle starts directing parents to other stores for Christmas toys Macy's does not stock, until this is proved a brilliant commercial strategy. Kringle earns the enmity of store psychologist Sawyer, who has him committed to Bellevue. Gailey, an entrepreneurial lawyer in love with Walker (and her 9-year-old daughter Susan), takes the case to court: Kringle is "proved" to be Santa when – to get rid of them – the US Mail delivers all this year's "Santa Claus" letters from children to Kringle at the court. Kringle afterwards delivers a final proof: Susan identifies a house as the one she drew when asking Santa for it as a Christmas present, and Walker and Gailey determine to marry and live there; then they see Kringle's cane propped against the fireplace. But this is not the movie's most important MIRACLE: Walker – whose arch-rationalism had almost driven the imagination out of Susan and turned her into an offensive little prig – has become a believer.

This is *the* US Christmas movie, and Gwenn (1875-1959) is *the* movie image of Santa: by one of 1940s Hollywood's wilder lunacies, he was billed as support to O'Hara and Payne; his deserved Oscar thus came as Best Supporting Actor. The screenplay won an Oscar for Davies and Seaton. A tvm remake, broadcast in 1973, disappeared without trace. An homage to the movie is the rare vt of **batteries not included* (**1987**): *Miracle on 8th Street*. [JG]

2. US movie (**1994**). 20th Century-Fox. **Pr** John Hughes. **Exec pr** William S. Beasley, William Ryan. **Dir** Les Mayfield. **Screenplay** Hughes. **Based on 1**. **Starring** Richard Attenborough (Kriss Kringle), Dylan McDermott (Bryan Bedford), Elizabeth Perkins (Dorey Walker), Mara Wilson (Susan Walker). 111 mins. Colour.

In mood, this is a faithful remake of **1**, although the plot has been tweaked in order further to twist our heartstrings, and all has been relocated into what can probably best be described as a LAND-OF-FABLE 1990s NEW YORK. There are plotting flaws – it is not at all clear why New York State should be seeking to put Kringle behind bars – but the overall effect is delightful. Kringle gains his freedom because the judge is persuaded that, if the USA can believe so much in GOD that it states its trust on the dollar bill, then the court has no right to disbelieve in SANTA CLAUS. [JG]

MIRACLES An event that defies human understanding and is usually attributed to divine intervention. There is in fact a dissonance in the relationship between miracles and FANTASY. The performance of miracles and a belief in their divine origin is fundamental to religious faith, and this brings with it an acceptance of miracles as fact. But a fundamental premise of fantasy is that we are dealing with the impossible. Thus "miracle" is a meaningless term in a SECONDARY WORLD of MAGIC, where their performance is natural and therefore not miraculous. They can be perceived as miraculous only in our world, and thus tend to appear in works of either LOW FANTASY or SUPERNATURAL

FICTION. To work as fiction, miracles are best left unexplained, as in Brian MOORE's *Cold Heaven* (**1983**), where the miraculous events seem to form part of some GODGAME, or *The Miracle Boy* (**1927**) by Louis GOLDING, where the dead are restored to life (◊ RESURRECTION). The belief in miracles as core to faith was explored by Honoré de BALZAC in "Jesus-Christ en Flandre" ["Christ in Flanders"] (1831), where each individual's faith is tested by following Jesus out of a boat in a storm to walk across the sea. Such aspects of faith and belief are explored and challenged in *A Book of Miracles* (coll **1939**) by Ben HECHT. Because of their divine origin, all miracles should be beneficial; anything inexplicable and EVIL will be the work of the DEVIL and thus BLACK MAGIC, not a miracle. Miracles induce wonder and are akin to the MARVELLOUS. While they might suggest a sense of WRONGNESS, their performance results in a feeling of *rightness*, which may even have restored some BALANCE in an inherently evil world – as in MIRACLE ON 34TH STREET (**1947**, **1994**). Miracles or apparent miracles appear most frequently in religious fantasies (◊ CHRISTIAN FANTASY; RELIGION) and apocalyptic fiction (◊ APOCALYPSE), where the rise in miracles is a precursor to ARMAGEDDON. H.G. WELLS's "The Man who Could Work Miracles" (1898 *Illustrated London News*), filmed as *The* MAN WHO COULD WORK MIRACLES (**1936**), shows the folly of humanity attempting to operate with divine power. [MA]

MIRANDA UK movie (**1947**). ◊ SPLASH! (**1984**).

MIRBEAU, OCTAVE (1848-1917) French writer. ◊ CONTE CRUEL; DECADENCE.

MIRRLEES, HOPE (1887-1978) UK author whose single fantasy novel *Lud-in-the-Mist* (**1926**) is a minor classic. The people of the town of Lud have severed their links with a nearby FAERIE land redolent of unpredictability, arbitrary use of power and danger. In reaction there is an illegal traffic in fairy fruit, a matter so unspeakable that Lud's law courts hide it in elaborate euphemisms about contraband silk. This theme both echoes and argues with "Goblin Market" (1862) by Christina ROSSETTI, with the fruit's attraction generalized from a symbol of sexual craving into a broader yearning after unattainables – which SEHNSUCHT HM implies is the wellspring of all creativity.

The perceived WRONGNESS of the fruit and of Faerie is gradually dispelled by a detective subplot that transfers the taint to the principal fruit-smugglers, guilty of a long-ago murder. Meanwhile various children of Lud have been lured across the borders of Faerie, generally believed indistinguishable from death. Chanticleer, Mayor of Lud – having spent much of the book being comically bamboozled by smugglers – rises to the occasion by crossing the border, rescuing others' children from webs of illusion, and pursuing his own son via a terrifying leap into darkness. In due course there comes a renewal, with the borders opened and the ancient trade in magical fruit made legal: the final chapter notes in passing that "art was creeping back" to the mundane land. Nevertheless, for Chanticleer there is no end to those indefinable longings this side of death.

HM's writing, usually underrated, moves between gently crazy humour, poetic snatches, real menace and real poignancy. [DRL]

Other works: *The Book of the Bear: Being Twenty-One Tales Newly Translated from the Russian* (anth **1926**), FAIRYTALES, trans with HM's partner Jane Ellen Harrison (1850-1928), author of *Themis* (**1912**) and speculative anthropological studies.

MIRROR Mirrors have always been perceived as subtly magical. More particularly:

1. A mirrored face is uncannily lifelike and can easily be imagined as speaking autonomously; hence the talking magic mirror best known from the SNOW WHITE tale, a frequent fantasy prop which has often been spoofed – e.g. in Anthony ARMSTRONG's stories. (Breaking a mirror is bad luck since one shatters one's own image, with obvious sympathetic-MAGIC implications.)

2. The ability to reflect, instantly, whatever may be happening close at hand extends by natural analogy to the display of scenes far off in space or time: hence mirrors can be SCRYING devices, as with the Magic Mirror in DISNEY's SNOW WHITE AND THE SEVEN DWARFS (*1937*). Many traditional magic mirrors operate thus: the GOD Vulcan's showed past, present and future; the MOON King's mirror in LUCIAN's *True History* revealed events anywhere on Earth, as did MERLIN's in Edmund SPENSER's *The Faerie Queene* (**1590-96**). Cambuscan's mirror in Geoffrey CHAUCER's "The Squire's Tale" warned of national ill-fortune and the falsity of women. The mirror of Galadriel in J.R.R. TOLKIEN's *The Lord of the Rings* (**1954-5**) is a reflecting pool which displays potential futures. Avram DAVIDSON's *The Phoenix and the Mirror* (**1966**) invokes much learned ALCHEMY and ASTROLOGY for its preparation of a "virgin speculum" crafted in darkness, whose first fleeting image is a PORTENT.

3. A room's mirror reflection seems subtly awry owing to reversal, suggesting a different room in an OTHERWORLD behind the glass, with the mirror now a potential PORTAL through which, for example, Alice can climb in Lewis CARROLL's *Through the Looking-glass* (**1872**). Combining **2** and **3** implies portals to anywhere and anywhen, as in Stephen R. DONALDSON's **Mordant's Need**, where in accordance with a RATIONALIZED-FANTASY law the precise curvature of each mirror surface "tunes" it to a specific destination. Father Inire's mirrors and related numinous mirrors in Gene WOLFE's *The Book of the New Sun* (**1980-83**) allow both distant viewing and travel, and also allude to Jorge Luis BORGES's notes on fauna inhabiting mirrors in *The Book of Imaginary Beings* (**1967**). The mirror of the inset story in *Phantastes* (**1858**) by George MACDONALD shows the expected room, but with a new inhabitant compelled there by BONDAGE.

4. When mirror faces mirror the infinite corridor of repeated reflections is eerie in itself. Fritz LEIBER's "Midnight in the Mirror World" (**1964**) has a suicide's GHOST visible in the depths, moving closer by one reflection each night. The egregious savant de Selby in Flann O'BRIEN's *The Third Policeman* (**1967**) studies the sequence's far-off and therefore older images through a telescope, ultimately seeing himself aged 12. The VILLAIN of Terry PRATCHETT's *Witches Abroad* (**1991**) uses facing mirrors for her magic, feeding on her multiplied self but being spread thin along the line of images in a double metaphor of hubris and THINNING: "Mirrors give plenty, but they take away lots."

5. Mirrors can also work against magic. They may reflect SPELLS, or usefully *not* reflect the deadly component of Medusa's gaze. They offer images truer than human PERCEPTION, and do not show unrealities like VAMPIRES or GHOSTS. A disguising ILLUSION or GLAMOUR will probably fail when viewed in the mirror. [DRL]

MR ED US tv series (1960-66). Filmways/syndication and then CBS. **Pr** Herbert W. Browar, Howard Campbell.

Exec pr Arthur Lubin, Al Simon. **Dir** Rod Amateau, Lubin, John Rich, Ira Stewart. **Writers** William Burns and many others. **Created by** Lubin. **Graphic novelization** 7 issues of *Mr Ed* (1962-4) from Dell Publishing/Gold Key comics. **Starring** Alan Young (Wilbur Post). **Voice actor** Allan "Rocky" Lane (Mr Ed). 143 30min episodes. B/w.

The premise for this long-running sitcom is that a palomino horse is able to speak (◊ TALKING ANIMALS). The series was the brainchild of Lubin, who had earlier directed the FRANCIS MOVIES, about a talking mule. Architect Post moves from the city to a house in the country and finds a horse in his barn, left there by the prior owner. The horse, Mr Ed, can talk, but wishes to talk only to Wilbur. This causes trouble: people think Post is crazy. Mr Ed will, though, talk to others on the telephone: a call can be generated by an urge for take-out food or a desire to meet Clint Eastwood's horse.

ME had the distinction of being the first series to move from syndication to network tv. [BC]

MR MERLIN US tv series (1981-2). Columbia/CBS. **Pr** Joel Rogosin. **Exec pr** Larry Rosen, Larry Tucker. **Dir** John Astin, Bill Bixby and many others. **Writers** Tad Chehak, Tom Chehak and many others. **Novelization** *Mr Merlin – Episode 1* * (**1981**) and *Mr Merlin – Episode 2* * (**1981**) by William Rotsler. **Starring** Clark Brandon (Zachary Rogers), Barnard Hughes (Max Merlin), Elaine Joyce (Alexandra), Jonathan Prince (Leo Samuels). 21 30min episodes. Colour.

MERLIN is now living in modern San Francisco as garage owner Max Merlin. Told by WIZARDS even more powerful than himself that he has not been doing enough good deeds lately, he is ordered to find a SORCERER'S APPRENTICE. His choice is teenager Zachary. Zachary tries to be a good student, but one of his major interests is using the MAGIC he learns to attract women. Somehow, the pair manage a few rescues of those in distress. [BC]

MR TERRIFIC US tv series (1967). Universal/CBS. **Pr** David J. O'Connell. **Dir** Jack Arnold. **Writers** R.S. Allen and others. **Starring** Dick Gautier (Hal Walters), John McGiver (Barton J. Reed), Paul Smith (Harley Trent), Stephen Strimpell (Stanley Beamish/Mr Terrific). 13 30min episodes. Colour.

The ultra-secret Bureau of Special Projects finally succeeds in creating a pill that can impart superpowers; an unfortunate side-effect is that it makes people violently ill – all except Stanley Beamish, who works at a nearby gas station. Reluctantly Beamish takes on the identity of Mr Terrific, a costumed SUPERHERO. This comedy had the dubious distinction of debuting and being cancelled at the same time as another very similar effort, CAPTAIN NICE (1967). [BC]

MITCHISON, NAOMI (MARGARET) (1897-   ) Scottish writer who between 1913 and 1991 published novels, short stories, poetry, plays, essays, reviews, biography, memoir, and political and social polemic. NM's full bibliography lists over 100 books, and over 1000 short pieces, many for children. NM is a committed socialist and feminist. Much of her work is historical, sf or fantasy, its frequently subversive or didactic purposes disguised by displacement from the contemporary milieu. In her historical fictions, characters often believe in and practise MAGIC, but NM's purpose is normally one of ANTHROPOLOGY or ALLEGORY, and the fantastic elements are secondary to her characters and moral purpose.

Early work comprises mainly historical fiction set in Classical antiquity, telling of displaced or dispossessed characters involved in war or revolution. *The Corn King and the Spring Queen* (**1931**; vt *The Barbarian* 1961 US), NM's major historical fantasy, takes Corn King Tarrik and Spring Queen Erif Der from (fictional) Marob in Scythia to Sparta and Egypt during the 3rd century BC. NM's only major work of contemporary fiction, *We Have Been Warned* (**1935**), is set in the then near-future and addresses similar themes, but shorn of the protective veil of fantasy it was met with resistance and outrage.

The DREAM fantasy *Beyond this Limit* (**1935** chap) was done in collaboration with Wyndham LEWIS as illustrator, each responding to the other's work. In *The Bull Calves* (**1947**), set in Scotland in 1747, a good woman becomes a practising WITCH. In *The Big House* (**1950**), Su from "the big house" and fisherman's boy Winkie are companions on a TIME-TRAVEL adventure into the past with consequences for the present. *Travel Light* (**1952**) tells of Halla, reared by bears and DRAGONS, who goes to Micklegard and Marob, and converses with Valkyries (◊ NORDIC FANTASY). *To the Chapel Perilous* (**1955**) is a superb and fantastic SATIRE on the press, and a thought-provoking retelling of the GRAIL quest.

From the late 1950s NM made less use of fantastic elements in her historical fiction. Some contemporary children's stories deploy Indian and African backgrounds and FOLKTALES. Adult fiction includes short stories of all kinds, historical novels, and three sf novels, best-known being *Memoirs of a Spacewoman* (**1962**). [CM]

Other works (selective): *The Conquered* (**1923**); *The Fourth Pig* (coll **1936**); *Graeme and the Dragon* (**1954**); *Behold Your King* (**1957**); *Solution Three* (**1975**); *The Two Magicians* (**1978**) with Dick Mitchison; *Images of Africa* (coll **1980**); *Not by Bread Alone* (**1983**); *Beyond This Limit: Selected Shorter Fiction* (coll **1986**); *Early in Orcadia* (**1987**); *A Girl Must Live: Collected Stories and Poems* (coll **1990**).

Further reading: *Return to the Fairy Hill* (**1966**), one of several works of autobiography; *Naomi Mitchison: A Biography* (**1990**) by Jill Benton, including partial bibliography.

MODESITT, L(ELAND) E(XTON) Jr (1943-) US writer of sf novels who turned to fantasy with his popular **Recluce** series, so far comprising *The Magic of Recluce* (**1991**), *The Towers of the Sunset* (**1992**), *The Magic Engineer* (**1994**), *The Order War* (**1995**), *The Death of Chaos* (**1995**), and «*Fall of Angels*» (**1996**). The common setting is a world following RATIONALIZED-FANTASY rules which, as in Larry NIVEN's **Magic Goes Away** sequence, echo the implacable laws of physics. In LEM's systematization, Order and CHAOS are opposed but need not (though generally do) correspond to the usual GOOD AND EVIL dichotomy. Chaos WIZARDS command disruptive effects: firebolts, explosions. Order wizards, whose abilities initially seem confined to craftsmanship, HEALING and defence, can usefully manipulate the rules (◊ QUIBBLES). Thus *The Towers of the Sunset*, chronologically first in the series, exploits the notion that "ordering" water and wind gives offensive weapons of ice and storm, though at high cost to the wielder: hysterical blindness. *The Magic of Recluce* has an UGLY-DUCKLING hero who emerges as a natural Order master who single-handedly destroys the leading current Chaos wizard.

LEM's sf experience gives the later books an interesting turn by combining magic and technology (◊ TECHNOFANTASY). *The Magic Engineer*'s eponymous innovator is a Renaissance man who controversially develops steamships

with high-pressure systems and an outer armour of magically enhanced iron (◊ COLD IRON), harnessing the Chaos of Fire. Later, in *The Order War*, similar techniques allow more advanced weaponry – ultimately including an approximation to a military laser, which reshapes the world. (There is a sly indication that nuclear weapons would be feasible, but inadvisable.) In the Order-Chaos arms race, rigorous conservation laws apply: as with electric charge, one side's accumulation of "positive" magic increases the availability of "negative" magic to the other, and the final victory in *The Order War* part-cripples the resources of the nominal victors.

The magical system, with its planetary BALANCE, is a strength of these books; the writing, though, is often sloppy, with repeated phrases and plot structures, and a tiresome line in onomatopoeia. [DRL]

MOEBIUS Working name of influential French fantasy COMIC-strip artist Jean Giraud (1938-). M has a fine, eloquent line style and a remarkably fertile imagination, and is considered one of Europe's leading talents; his influence can be discerned in the work of younger fantasy and sf artists all over the world. His early work (as Jean Giraud, or Gir), mainly in the form of comic strips about the US West, itself showed the strong influence of COMICS artists like Milton Caniff (1907-1988) and Frank Robbins (1917-). In 1963 he began a collaboration with writer Jean-Michel Charlier (1924-1989), and together they created the Western series **Lieutenant Blueberry** (1963-70; graph coll **1965-90** 26 vols; trans first 4 vols **1977-9** UK; subsequent vols trans as *Lieutenant Blueberry* vols 1-4 **1991** US, *Marshall Blueberry* vols 1-2 **1990** US, *Young Blueberry* vols 1-3 **1989** US, and *Blueberry* vols 1-5 **1989-90** US).

In the early 1960s Moebius created a series of dark-humoured strips for the satirical monthlies *Hara Kiri* and *L'Echo des Savannes*, which he signed "Moebius", a sobriquet which he continues to use for all his fantasy and sf work. In the late 1960s he illustrated a series of sf books for the publisher Editions Opta and in 1975 he cofounded the magazine *Métal Hurlant* with another leading French sf artist, Philippe Druillet (1944-) and the writer Jean-Pierre Dionnet (1947-). This ground-breaking periodical inspired an English-language counterpart in the USA, HEAVY METAL, in which M's work featured prominently. This signalled the beginning of a long and particularly fertile period during which many of the themes and ideas which were to constitute what has been termed the "Moebius Universe" were first mooted. These elements were at first only tenuously interlinked, and some fascinating cross-feed from his work on **Blueberry** can be discerned, despite the historical accuracy of the one and the delightfully free improvisational quality of the other. M's range is very wide but visually remarkably consistent; and he creates stories and scenarios in many established genres of sf and fantasy plus a great number that are entirely unique, such as **Le Bandard Fou** (1975; trans as **The Horny Goof** in *Heavy Metal* February 1981).

Other creations for *Métal Hurlant* include **Le Garage Hermétique de Gerry Cornelius** (from 1975; trans as **The Airtight Garage of Jerry Cornelius** in *Heavy Metal* 1977-80), *Arzach* (1976; trans in *Heavy Metal* 1977), *The Long Tomorrow* (1976; trans in *Heavy Metal* 1977), which was scripted by moviemaker Dan O'Bannon (1946-), and the multi-part epic scripted by moviemaker Alejandro Jodorowski (1930-), **Les Aventures de John Difool**

["The Adventures of John Difool"] (1981-8): *L'Incal Noir* ["The Dark Incal"] (graph **1981**), *L'Incal Lumière* (graph **1982**; trans as "The Incal Light" in *Heavy Metal* 1982), *Ce Qui est en Bas* ["That Which is Below"] (1984; trans as "The Third Incal" in *Heavy Metal* 1984), *Ce Qui est en Haut* ["That Which is Above"] (graph **1985**) and *Le Cinquième Essence* ["The Fifth Essence"] *#1* (graph **1986**) and *#2* (graph **1988**). These latter stories have been published in collected form as *Incal #1* (graph coll **1988** UK/US), *#2* (graph coll **1988** UK/US) and *#3* (graph coll **1988** UK/US).

Tracing the publishing history of other Moebius material is made difficult by the fact that collections published in France and in the USA frequently have different contents. It is therefore expedient to list them separately. Books in French include *Gir 30x40* (graph coll **1974**), *Les Maîtres du Temps* ["The Time Masters"] (graph **1978**), *Tueur de Mondes* ["World Killer"] (graph coll **1980**), *Mémoire du Futur* ["Memory of the Future"] (graph coll **1983**), *Sur l'Étoile* ["Upon a Star"] (graph coll **1984**) and its sequel *Les Jardins d'Aedena* ["The Gardens of Aedena"] (graph coll **1986**) based on the philosophical concepts of Appel Guery, *Venise Celeste* ["Heavenly Venice"] (graph coll **1985**), *Made in L.A.* (graph coll **1988**) and *La Déesse* ["The Goddess"] (graph coll **1990**).

Collected works in English include *Moebius #1: Upon a Star* (graph coll **1986** US), *#2: Arzach and Other Fantasy Stories* (graph coll **1986** US), *#3: The Airtight Garage* (graph coll **1987** US), *#4: The Long Tomorrow and Other Science Fiction Stories* (graph coll **1988** US), *#5: The Gardens of Aedena* (graph coll **1988** US), *#6: Pharagonesia and Other Strange Stories* (graph coll **1988** US) and *#7: The Goddess* (graph coll **1989** US), plus *The Magic Crystal* (graph **1988** US), *Island of the Unicorn* (graph **1988** US) *Aurely's Secret* (graph **1989** US), *Chaos* (graph **1991** US) and others.

In the late 1980s Moebius moved to California and set up the company Starwatcher Graphics to publish his posters and fine-art pieces. He illustrated one two-episode story featuring Stan LEE's **Silver Surfer**, *Parable* (1988-9 US), and an ecological story for a special "Earth Day" issue of *Concrete* (1991 US). In the early 1990s several spinoff comic books featuring such Moebius creations as **The Airtight Garage** and **The Incal** appeared in the USA as collaborative ventures with other artists, further developing the Moebius Universe; they include **The Elsewhere Prince** (*#1-#6* 1990), **The Man from Ciguri** (1990-91), **The Onyx Overlord** (*#1-#4* 1992-3) and **Legends of Arzach**, a six-issue series of short stories, accompanied by colour prints of artwork by leading comics artists (1992-3), based on Moebius themes.

M has also been influential in the CINEMA, doing design work on Jodorowski's ill-fated *Dune* project, spacesuit designs for Ridley Scott's *Alien* (*1979*) and the creature for James Cameron's *The Abyss* (*1989*); his influence is evident also in Scott's *Blade Runner* (*1982*). He contributed designs for the feature film *Masters of the Universe* (1987) dir Gary Goddard. His credits in ANIMATED MOVIES include DISNEY's TRON (*1982*), and he designed *Les Maîtres du Temps* ["The Time Masters"] (*1982*) dir René Laloux, which was based on Stefan Wul's novel *L'Orphéline de Perdide* ["The Orphan from Perdide"] (*1958*). He was the leading designer on the animated feature LITTLE NEMO: ADVENTURES IN SLUMBERLAND (*1993*), based on Winsor MCCAY's newspaper strip.

A French postage stamp designed by and in honour of Moebius was issued in 1988. [RT]

Other works: *Moebius: Oeuvres Complètes #1: Le Bandard Fou, John Watercolor, Cauchemar Blanc* ["The Horny Goof, John Watercolor, White Nightmare"] (graph coll **1979**), *#2: Arzach, L'Homme est-il Bon?* ["Arzach, Is Man Good?"] (graph coll **1980**), *#3: Major Fatal* (graph coll **1980**), *#4: La Complainte de L'Homme-Programmé* ["The Programmed Man's Complaint"] (graph coll **1981**) and *#5: Le Désintégré Réintégré* ["The Disintegrated Reintegrated"]. Portfolios include *City of Fire* (**1985**) with Geoff Darrow, *Crystal Saga* (**1986**) and *Futurs Magiques* ["Magic Futures"] (**1987**).

MOLESWORTH, MARY (1836-1921) UK writer. ◊ CHILDREN'S FANTASY; Walter CRANE; GHOST STORIES; GHOST STORIES FOR CHILDREN; Roger Lancelyn GREEN; Charles KINGSLEY.

MOLIÈRE Working name of French dramatist and actor Jean-Baptiste Poquelin (1622-1673). ◊ COMMEDIA DELL'ARTE. [JC]

MOLIN, CHARLES Pseudonym of William MAYNE.

MOLYNEUX, BINGHAM T. [s] ◊ Kenneth MORRIS.

MOMADAY, N. SCOTT (1934-) US writer. ◊ BILLY THE KID.

MOMPESSON, SERGIUS [s] ◊ Kenneth MORRIS.

MONACO, RICHARD (1940-) US academic, composer, playwright, screenwriter, literary agent and novelist who first came to notice as a fantasist with his **Parsival** sequence of Arthurian novels: *Parsival; or, A Knight's Tale* (**1977**), *The Grail War* (**1979**), *The Final Quest* (**1980**) and *Blood and Dreams* (**1985**). Dark-hued and ironic, these retell the PERCEVAL tales from a disenchanted modern perspective. The **Leitus** novels – *Runes* (**1984**) and *Broken Stone* (**1985**) – are rather more conventional in their depiction of Romans versus druids in a LAND OF FABLE. *Journey to the Flame* (**1985**) is an adequately enjoyable pastiche of H. Rider HAGGARD (who features as a character); the plot concerns an attempt by pre-WWI German nationalists to find the lost CITY of Kôr, as described by Haggard in *She* (**1886**), hoping to exploit its occult powers. [DP]

Other works: *Unto the Beast* (**1987**), supernatural horror; *The Dracula Syndrome*, with Bill Burt (**1993**), nonfiction.

MONKEY 1. The TRICKSTER god of Chinese LEGEND, who begins life as a stone monkey and determinedly acquires tremendous powers of MAGIC, SHAPESHIFTING, INVULNERABILITY and IMMORTALITY. After upsetting the tranquillity of HEAVEN he is subdued by Buddha, and assists the pilgrim Hsüan-tsang or Tripitaka ("three baskets") in a QUEST for baskets of holy scriptures in India; obstacles include WIZARDS and DRAGONS. Wu Ch'êng-ên (*c*1505-*c*1580) used the legends and their Chinese stage representations as the basis for his 100-chapter vernacular classic of HUMOUR and ALLEGORY, *Hsi Yü Chi* ["Journey to the West"] (**1592**; cut trans by Arthur WALEY as *Monkey* 1942). Also of note is *Three Tales of Monkey* (coll **1967** US) trans Ruth Tooze (»PUSS-IN-BOOTS). A GRAPHIC-NOVEL adaptation is *The Ape* (graph **1986**) by Milo MANARA and Silverio Pisu. [DRL]

2. Japanese tv series (1979-81). NTV-Kokusai Heoi/BBC 2. **Dir** Yusuke Watanabe; English-language version (dubbed from Japanese screenplays trans David Weir) dir Michael Bakewell. **Based on** Wu Ch'êng-ên's original novel.

Low-budget successor to *The* WATER MARGIN, involving Monkey and his COMPANIONS in numerous martial-arts encounters. Despite the economy of the production, this was very fine, amid the humour and the action conveying a real power of legend. [JG/DRL]

MONKEY BUSINESS (ot *Darling, I am Growing Younger*)

US movie (*1952*). 20th Century-Fox. **Pr** Sol C. Siegel. **Dir** Howard Hawks. **Spfx** Ray Kellogg. **Screenplay** I.A.L. Diamond, Ben HECHT, Charles Lederer, Harry Segall. **Starring** Cary Grant (Dr Barnaby Fulton), Marilyn Monroe (Miss Laurel), Ginger Rogers (Edwina Fulton). 97 mins. B/w.

A team under myopic, middle-aged Barnaby Fulton is developing a rejuvenation drug. An experimental chimp serendipitously mixes the correct formula, and pours it into the water tank. Fulton, drinking from the tank, undergoes a form of IDENTITY EXCHANGE, becoming, in effect, an adolescent in an adult's body. After the effects have worn off, his wife Edwina makes the same mistake, with similar results. In this TECHNOFANTASY sex comedy, Fulton manages not to be seduced by either the adolescent Edwina or the bimboish secretary Miss Laurel; although filled with innuendo, *MB* dodges the full implications of its central fantasy. Rogers, 41 when the movie was made, proves expert at shedding 20 years; Grant, aged 48, is less successful. [JG]

MONKEYS ◊ APES; MONKEY; TRICKSTER.

MONOMYTH A term devised by Joseph Campbell (1904-1987) in *The Hero With a Thousand Faces* (**1949**) to designate the single "shape-shifting yet marvellously constant story" at the heart of the "mythological hero narrative". Campbell's simplest version is: "A hero ventures forth from the world of common clay into a region of supernatural wonder: fabulous forces are there encountered and a decisive victory is won: the hero comes back from this mysterious adventure with the power to bestow boons on his fellow man."

There is a loose but usable consonance between this three-part rendering of the monomyth and the model of FANTASY narrative employed in this encyclopedia, though of course the latter is meant to encompass many narratives which are *not* built around the central QUEST of a culture HERO. [JC]

MONSTER MOVIES In written fantasy, MONSTERS need not be huge; in the CINEMA they almost always are, unless occurring *en masse* – as swarms of mutant bees, shoals of mutant piranhas, or mutant whatevers. On occasion they can start small and end up huge – as in the **Alien** series of sf movies, *Arachnophobia* (**1990**), etc. – but generally they are vast from the outset. Even Frankenstein's monster (◊ FRANKENSTEIN MOVIES), although made from the parts of presumably normal-sized human beings, is almost always depicted as larger than any normal man.

DINOSAURS and APES are the most popular monsters, the former typified by the GODZILLA MOVIES – although *The Lost World* (**1924**) really started the trend – and the latter by the KING KONG MOVIES, which had spinoffs like *Mighty Joe Young* (**1949**). PREHISTORIC FANTASIES, like ONE MILLION BC and its various spinoffs, normally see humans battling, however implausibly, with giant creatures; *The* CLAN OF THE CAVE BEAR (*1985*) is a glorious exception. The taxonomy of the eponym of *The* CREATURE FROM THE BLACK LAGOON (*1954*) is unclear, although it is to be assumed that he (or she or it) is an amphibian.

More interesting are instances where humans *become* monsters. This happens to the protagonists of the various FLY movies but perhaps more importantly to the denizens of the WAINSCOT society at the heart of NIGHTBREED (*1990*). Vampires, werewolves and zombies (◊ VAMPIRE MOVIES; WEREWOLF MOVIES; ZOMBIE MOVIES) are all in their diverse ways humans monstrously transformed. The cinema has not generally dealt very well with such iconographic

characters: to take a single example, Lawrence Talbot, the werewolf protagonist of *The Wolf Man* (**1941**), within only a few years found himself in the debasing *Abbott and Costello Meet the Wolf Man* (**1948**). The general trend of Hollywood to discover an ICON and then swiftly degrade it is nowhere more evident than in the case of monster movies. [JG]

MONSTER OF MONSTERS vt of *Ghidora, The Three-Headed Monster* (*1965*). ◊ GODZILLA MOVIES.

MONSTERS In modern GENRE FANTASY, monsters are a writer's convenience (◊ PLOT COUPON) for placing yet another obstacle in the way of the hero's QUEST. J.R.R. TOLKIEN used them to particular good effect in *The Lord of the Rings* (**1954-55**), also creating new monsters, like the Balrog, the giant spider (◊ SPIDERS) Shelob, and the many-tentacled Watcher in the Water. More traditional monsters are often the purpose of a quest and form a fundamental part of folk tradition. The commonest monsters are DRAGONS and GIANTS, which feature heavily in LEGENDS, MYTHS, FOLKTALES and medieval ROMANCE. Diverse other monsters are treated separately in this encyclopedia under ANIMALS UNKNOWN TO SCIENCE, GODZILLA, GORGONS, IMAGINARY ANIMALS, MINOTAUR, MONSTER MOVIES, MYTHICAL CREATURES, SEA MONSTERS, SERPENTS, TROLLS, VAMPIRES, WEREWOLVES, WORMS and ZOMBIES. Most monsters instil dread and are killed to our hero's greater glory, though they may not be inherently evil; e.g., in BEOWULF, Grendel's mother may wreak havoc, but she is merely avenging the killing of her child. Similarly, KING KONG may through sheer size cause much destruction. The FRANKENSTEIN monster has not the capacity to be either good or evil: its "sin" is that its mere appearance causes revulsion (or, to read more deeply, that it has become the repository for its creator's well justified shame), and this leads to its performing evil acts.

Humans can be monsters in more subtle ways, where the monstrous is not apparent on the surface. Most monster-humans have become so through either POSSESSION or dabbling in the occult. Some monstrous women may be FEMMES FATALES. Historical examples of both who appear in HORROR are Gilles de Rais (1404-1440), the Marquis DE SADE and Elizabeth de Báthory (?   -1614).

A different definition of "monster" might illuminate. The word comes from the Latin *monstrum*, meaning something MARVELLOUS to be treated as a warning or PORTENT. Many of the Roman annalists noted the birth of freaks or monsters (animal or human) each year as if their occurrence was in some way related to the dread events of those years. The appearance of monsters thus indicates WRONGNESS, and is used in this form, alongside other cataclysms, as a warning in "The Call of Cthulhu" (1928 *WT*) by H.P. LOVECRAFT.

CHILDREN'S FANTASY has some of the best monsters because the authors can play more to a child's imagination. Excellent examples appear in the poems and stories of Lewis CARROLL, Edward LEAR and Terry JONES, the **Narnia** books by C.S. LEWIS, *The Phantom Tollbooth* (**1961**) by Norton JUSTER and *The Neverending Story* (**1979**) by Michael ENDE. R. CHETWYND-HAYES delights in creating monsters; a selection is in *The Monster Club* (coll **1976**).

ANTHOLOGIES include *Monsters Galore* (anth **1965**) ed Bernhardt J. Hurwood (1926-1987), *A Walk with the Beast* (anth **1969**) ed Charles M. Collins, *Monsters, Monsters, Monsters* (anth **1977**) ed Helen Hoke (1903-1990), *Bestiary!* (anth **1985**) and others in the series by Jack DANN and Gardner Dozois, and *The Monster Book of Monsters* (anth

1988) ed Michael O'Shaughnessy (1965-   ). Interesting studies include *The Book of Beasts* (**1954**) by T.H. WHITE, *Fictitious Beasts: A Bibliography* (**1961**) by Margaret W. Robinson, *The Book of Imaginary Beings* (**1967**; rev trans 1969) by Jorge Luis BORGES, and *The Magic Zoo* (**1979**) by Peter Costello (1946-   ).   [MA]

MONSTERS FROM THE UNKNOWN PLANET vt of *Mekagojira No Gyakushu* (**1975**). ◊ GODZILLA MOVIES.

MONSTER ZERO vt of *Kaiju Daisenso* (**1965**). ◊ GODZILLA MOVIES.

MONTELEONE, THOMAS F(RANCIS) (1946-   )
US author who has published nearly 100 short stories in various magazines and anthologies. His novels have ranged from explicit horror – *Night Train* (**1984**), *Lyrica* (**1987**) – to fantasy – *The Magnificent Gallery* (**1987**), *Dragonstar* (**1981**; rev in 3 vols **1983-9** with David Bischoff; see below) – to more recent mainstream thrillers with supernatural elements, like *The Blood of the Lamb* (**1992**), winner of the BRAM STOKER AWARD, *The Resurrectionist* (**1995**) and «Night of Broken Souls» (**1996**). He has also written for the stage and tv, including *Mister Magister* (**1981**) for *American Playhouse*, which won the Bronze Award at the International TV and Film Festival of New York and the Gabriel Award; his story "The Cutty Black Sow" (**1984**) was adapted for *Tales from the Darkside* in 1988. He began editing the **Borderlands** series of non-themed anthologies in 1990, and took its name for his own SMALL-PRESS imprint, specializing in hardcover novels and anthologies. TM's often controversial opinion column **The Mothers and Fathers Italian Association** has appeared in various small-press magazines.   [SJ]
Other works: *Seeds of Change* (**1975**); *The Time Connection* (**1976**); *The Time-Swept City* (**1977**); *The Secret Sea* (**1979**); *Night Things* (**1980**); *Guardian* (**1980**); *Dark Stars and Other Illuminations* (coll **1981**); *Ozymandias* (**1981**); the **Dragonstar** series with David Bischoff, being *Day of the Dragonstar* (**1983**), *Night of the Dragonstar* (**1985**) and *Dragonstar Destiny* (**1989**); *Traveler 8: Terminal Road* (**1986**); *Crooked House* (**1987**); *Fantasma* (**1989**).
As editor: *The Arts and Beyond* (anth **1977**); *R*A*M: Random Access Messages* (anth **1984**; vt *Microworlds* UK); *Borderlands* (anth **1990**), *#2* (anth **1991**), *#3* (anth **1992**) and *#4* (anth **1994**) with Elizabeth Monteleone.

MONTY PYTHON AND THE HOLY GRAIL UK movie (**1975**). Python/Michael White. **Pr** Mark Forstater. **Exec pr** John Goldstone. **Dir** Terry GILLIAM, Terry JONES. **Spfx** Julian Doyle, John Horton. **Screenplay** Graham Chapman, John Cleese, Gilliam, Eric Idle, Jones, Michael Palin, published as *Monty Python and the Holy Grail (Book)* (**1977**). **Starring** Chapman (Arthur/etc.), Cleese (Lancelot/Tim the Enchanter/etc.), Gilliam (Soothsayer/etc.), Idle (Sir Robin/etc.), Neil Innes (Robin's lead minstrel/etc.), Jones (Sir Bedevere/Prince Herbert/etc.), Palin (Sir Galahad/King of the Swamp Castle/etc.). 90 mins. Colour.
AD932, and ARTHUR recruits KNIGHTS for the ROUND TABLE (defeating, *inter alia*, the Black Knight), but decides against taking them to the all-singing, all-dancing CAMELOT. GOD appears in the sky to charge the company with a QUEST: to seek the GRAIL. This they undergo variously. Robin encounters a three-headed knight and flees in disarray. Galahad the Chaste is lured to Castle Anthrax and reluctantly rescued from the perils of its "but eight-score young blondes and brunettes, all between 16 and 19½". Arthur and Bedevere are told by a Soothsayer the only path to the Grail is via the Bridge of Death – a course that throws

them into the clutches of the Knights Who Say "Ni!", who demand a shrubbery as tribute. Lancelot, mistaking a situation, massacres a wedding party to save the wimpish Prince Herbert . . . And so the Pythonesque adventures continue until, with a sudden TIMESLIP, Arthur and the surviving knights are arrested by 20th-century police for causing an affray.
MPATHG is a movie of its age, and some of its HUMOUR seems puerile today; yet much still shines. It is that rare thing: a PARODY that outclasses its targets. Where the movie excels is in its superbly squalid portrayal of the Dark Ages, with the pretentious artificiality of the Round Table symbolized by the KNIGHTS, through lack of horses, employing pages to clatter coconut shells together. Many Arthurian LEGENDS are guyed, and not always affectionately. Whether such destructive SATIRE is praiseworthy is a qualm lost in laughter.   [JG]

MONTY PYTHON'S FLYING CIRCUS UK tv series (1969-70, 1972-4). BBC. **Pr** John Howard Davies, Ian MacNaughton. **Writers** Graham Chapman, John Cleese, Terry GILLIAM, Eric Idle, Terry JONES, Michael Palin. **Starring** Chapman, Cleese, Carol Cleveland, Idle, Jones, Palin. 45 30min episodes. Colour.
As a kind of TAPROOT TEXT for much UK tv and movie fantasy from the early 1970s on, *MPFC* is of great importance, though its contents tended to the Surrealist end of fantasy (◊ SURREALISM): the deconstructive energy of the show was inherently inimical to the internal narrative consistency that normally marks fantasy tales. Of the movies made by the Python team the most relevant fantasy titles are MONTY PYTHON AND THE HOLY GRAIL (*1975*) and MONTY PYTHON'S LIFE OF BRIAN (*1979*), although *And Now For Something Completely Different* (*1971*) and *Monty Python's The Meaning of Life* (*1983*) contain fantasticated sketches. Gilliam and Jones carried the spirit of this AFFINITY GROUP into fantasy CINEMA, and Jones has written several books of fantasy interest (◊◊ Michael FOREMAN; Brian FROUD). Palin and Jones wrote and appeared in the two series of *Ripping Yarns* (1977, 1979), many of whose parodic stories were fantasy or quasi-fantasy – e.g., "Across the Andes by Frog"; together they also wrote *Bert Fegg's Nasty Book for Boys and Girls* (**1974**; vt *Dr Fegg's Nasty Book of Knowledge* US; rev vt *Dr Fegg's Encyclopedia of All World Knowledge* 1985), as well as *Ripping Yarns* (coll **1978**) and *More Ripping Yarns* (coll **1980**). Palin has written several books, most nonfiction, but including *Small Harry and the Toothache Pills* (**1982**) and *The Mirrorstone: A Ghost Story with Holograms* (graph **1986**), a CHILDREN'S FANTASY illustrated by Alan LEE. Chapman died tragically young; his *A Liar's Autobiography* (**1980**) is fascinating. Cleese and Connie Booth (then married) created and starred in the nonfantasy *Fawlty Towers* (1975, 1979), and produced associated books. Idle wrote the scatalogical sf/fantasy novel *"Hello Sailor"* (**1975**) and *Pass the Butler* (**1982**), and, with Neil Innes, was the creator of *Rutland Weekend Television* (1975-6), a Pythonesque series that had a spinoff "documentary", *The Rutles*, parodying the career of The BEATLES; the series' associated book, by Idle, was *Rutland Dirty Weekend Book* (**1976**).   [JC/JG/DRL]
Monty Python books: *Monty Python's Big Red Book* (coll **1972**) and *The Brand New Monty Python Bok* (coll **1973**; vt *The Brand New Monty Python Papperbok* 1974) both ed Idle and assembled as *The Complete Works of Shakespeare and Monty Python: Volume One – Monty Python* (omni **1981**), with all Pythons variously involved; *Monty Python and the*

Holy Grail * (**1977**), movie tie; *Monty Python's Life of Brian; and Montypythonscrapbook* * (coll **1979**) ed Idle, containing the movie tie; *Monty Python's The Meaning of Life* * (**1983** US), movie tie; *The Complete Monty Python's Flying Circus: All the Words* (coll **1989** 2 vols).

MONTY PYTHON'S LIFE OF BRIAN UK movie (*1979*). Hand Made. **Pr** John Goldstone. **Exec pr** George Harrison, Denis O'Brien. **Dir** Terry JONES. **Design and animation** Terry GILLIAM. **Screenplay** Graham Chapman, John Cleese, Gilliam, Eric Idle, Jones, Michael Palin. **Starring** Chapman (Brian), Cleese (Reg/etc.), Gilliam (various), Idle (various), Jones (Brian's mother), Palin (Pilate/etc.), Gwen Taylor (Judith). 94 mins. Colour.

Brian Cohen, mistaken at birth by the Magi for the Messiah, grows up under the thumb of his mother. He escapes from her to join the ranks of the People's Front of Judea, carries out bungling operations for them, is seized by the Romans, escapes, is briefly carried away by a passing alien spaceship, is hailed by a credulous mob as the saviour, and is finally crucified.

Widely regarded as offensively blasphemous, *MPLOB* contains – beneath all the usual Python grotesqueries and surreal HUMOUR – some sober reflection on the nature of MESSIAHS (are they, as it were, products of nature or nurture?), on the arbitrary invention of RITUALS, selection of symbols of veneration, and accreditation of MIRACLES. CHRIST makes a brief appearance, and is not treated disrespectfully. The thrust of this fundamentally serious movie is expressed by Brian himself in a reluctant address to the multitude: "You don't need to follow me – you don't *need* to follow *anybody*. You've got to think for *yourselves*." Its most abiding image is of its finale, as ranked crucifixion victims sway together singing "Always Look on the Bright Side of Life", reminding us that human viciousness is, alas, no fantasy. [JG]

MOON In Classical mythology the Moon is associated with various forms of the GODDESS; in Teutonic and some other mythologies it is masculine. The association which links its fullness to madness (as in "lunacy") and lycanthropy is also parochial. The symbolism of the Moon is often derived by contrast with the attributes of the SUN and/or Earth. Some Classical writers wondered whether it might be the habitation of the souls of the dead and the source of ORACLES. It often represents the unattainable, the unlikely and the absurd, as in "crying for the Moon". All these meanings are amply reflected in fantasy fiction, especially in the more fanciful satirical lunar voyages (◊ SATIRE), whose tradition began with LUCIAN and is detailed in Marjorie Hope Nicolson's *Voyages to the Moon* (anth **1948**) and Roger Lancelyn GREEN's *Into Other Worlds* (anth **1958**).

The Man in the Moon, supposedly discernible in its visible features, is sometimes identified as Cain (◊ ACCURSED WANDERERS) and sometimes as Endymion (a lover of Selene or Diana who chose eternal sleep as his "reward"). The story of Endymion is recapitulated in numerous literary works, including "Endimion, the Man in the Moon" (**1588**) by John Lyly (c1554-1606), "Endymion" (**1842**) by William Aytoun (1813-1865) and "The New Endymion" (1879) by Julian HAWTHORNE.

In ARIOSTO's *Orlando Furioso* (**1516**) everything wasted on Earth – including misspent time, broken vows and unanswered prayers – is treasured on the Moon; this theme was recapitulated by several later writers. In Jacques Cazotte's *The Thousand-and-One Follies* (**1742**) the Moon's inhabitants are reportedly light-headed and light-minded.

Similarly playful devices are used in much modern fantasy, especially CHILDREN'S FANTASY; the tradition extends from *The Garden on the Moon* (**1895**) by Howard PYLE to *The Moon's Revenge* (**1987** chap) by Joan AIKEN. The symbolism of the Moon is explored far more earnestly and elaborately in Paul Auster's *Moon Palace* (**1986**).

The metaphorical link between the Moon and womanhood is emphasized by the menstrual cycle, giving rise to the SCHOLARLY FANTASY that women are somehow "enslaved" or forced to "pay tribute" to the Moon – a notion echoed in such works as "The Moon-Slave" (1901) by Barry PAIN and *Salome* (**1930**) by George S. VIERECK and Paul Eldridge. James Branch CABELL's Ettarre was held captive in a palace on the far side of the Moon but made her presence known via *The Music from Behind the Moon* (**1926**). Dion FORTUNE's *Moon Magic* (**1956**) was an early example of a now-fashionable occultism which combines lunar goddesses from various mythologies into a single overarching symbol of female nature – a notion cleverly redeveloped in such works as "The Woman who Loved the Moon" (1979) by Elizabeth LYNN and *Waking the Moon* (**1994**) by Elizabeth HAND.

The more sinister aspects of lunar mythology mostly fall within the domain of occult fiction, although tales like "The Moon-Stricken" (1899) by Bernard CAPES are of fantasy interest. The attempted creation of a sinister MESSIAH in Aleister CROWLEY's *Moonchild* (**1929**) is echoed in some other OCCULT FANTASIES, including "The Case of the Moonchild" (1945) by Margery LAWRENCE. [BS]

MOON, (SUSAN) ELIZABETH (NORRIS) (1945-   ) US writer whose experience of service in the Marine Corps provided some under-girding for her **Deed of Paksenarrion** series of otherworldly fantasies about a saintly female warrior (◊ MILITARY FANTASY): *Sheepfarmer's Daughter* (**1988**), *Divided Allegiance* (**1988**), *Oath of Gold* (**1989**), the prequel *Surrender None: The Legacy of Gird* (**1990**) and the follow-up *Liar's Oath* (**1992**); *The Deed of Paksenarrion* (all texts rev, omni **1992**) assembles the first three titles. EM has also written some sf, including Anne MCCAFFREY sharecrops (◊ SHARED WORLDS). [DP]

Other work: *Lunar Activity* (coll **1990**).

MOORCOCK, MICHAEL (JOHN) (1939-   ) UK writer and editor, based until the early 1990s in LONDON (where much of his most important work has been set), after which period he began to live part-time in Texas. In his early career, he wrote also as Bill Barclay, Michael Barrington (1 story in collaboration with Barrington J. Bayley), Edward P. Bradbury, James Colvin (a *New Worlds* housename) and Desmond Reid. He is the most important UK fantasy author of the 1960s and 1970s, and altogether the most significant UK author of SWORD AND SORCERY, a form he has both borrowed from and transformed. He is a central 20th-century exponent of URBAN FANTASY, and of both GASLIGHT ROMANCE and STEAMPUNK, and he can take primary responsibility for the TEMPORAL ADVENTURESS. He is an adroit and seminal manipulator of a concept of ALTERNATE REALITIES which treats the MULTIVERSE – a term he first used in 1965 – as an interweaving performance of worlds (◊ PLAYGROUND). The word "performance" is central: of all 20th-century fantasy writers of any popularity, MM is the most profoundly and multifariously theatrical. Almost all his work can be seen – either implicitly or, as in the **Jerry Cornelius** and related sequences, explicitly – as both reflecting and embodying the principles of the COMMEDIA DELL'ARTE. The Multiverse is a parade.

MM has been prolific since the late 1950s, although he began his writing life earlier by producing many adolescent fanzines, the first apparently being *Outlaw's Own* in about 1950. Because he has written copiously for nearly 50 years, because he constantly revises and retitles his texts, and because he habitually reshuffles the order in which those texts appear in omnibuses and collected editions, his bibliography is a nightmare. The 1990s have, for instance, seen two separate and markedly different collected editions of his central S&S series, plus associated texts, each omnibus sequence being granted the same overall title, **The Tale of the Eternal Champion** (see listing below for breakdown). Neither sequence is complete (e.g., both omit the **Jerry Cornelius** books, which have an indirect relationship to the **Eternal Champion** novels) and both are eccentric (each includes novels lacking an **Eternal Champion** figure).

SWORD AND SORCERY

MM's first SWORD-AND-SORCERY tales date from the 1950s, before the primitive notion of the "Ghost Worlds" had begun to evolve into the concept of the Multiverse. But a volume like *Sojan* (coll of linked stories **1977**), which assembles this early work, shows how early MM had begun to inject into routine action tales the characteristic sense of exploratory newness and of BELATEDNESS, of high-sounding awe and retrospective irony, that would mark his mature writings. From the first MM managed to convey his complex reaction to the genre in a confidingly *available* tone of voice. As a result, he almost singlehandedly created the UK brand of S&S. This accomplishment is now taken for granted, and MM's tone of voice has become the default manner in which S&S tales are properly told in the UK; younger writers must either sound like MM or work hard not to.

It was not until the 1960s, however, that MM began to evolve the complex relationship between world and HERO, between Multiverse and Eternal Champion, that became the overarching structure of all his purely genre work; he is still evolving that structure today.

In the terminology of this encyclopedia, the Multiverse can be described as an indefinitely (though not, it seems, infinitely) extendable array of ALTERNATE REALITIES, each one of which is a FANTASYLAND with a PALIMPSEST relationship to other fantasylands; at least one of these environments is a pure J.R.R. TOLKIEN-like SECONDARY WORLD, and others exhibit a LAND-OF-FABLE relationship to Europe. In the war between Order and CHAOS which supports most of MM's S&S plots, the Multiverse is both a fixed arena which has been stripped down for action (as the term "Fantasyland" implies) and a free, chaotic, primal swamp.

What binds the worlds together – in practical, storytelling terms – is the Eternal Champion. Each incarnation of the Eternal Champion – Erekosë, Elric, Kane of Old Mars, Hawkmoon, Corum and (less completely) Von Bek – finds himself in BONDAGE to the particular version of the Multiverse he has been brought into being to defend (or scour). Despite the seeming looseness of that Multiverse, therefore, many of MM's most popular characters are immured in lives and actions – however heroic – they desperately wish to escape. It is this serial Bondage of the Eternal Champion that gives substance to venues which would otherwise be subject to interminable fractionation.

This combination of bondage and multiversity is also an ironist's trick on his creations. Seen from above, the dozens of S&S tales which populate the Multiverse take on the air of a sequence of charades, as though they were COMMEDIA DELL'ARTE skits devoted not to love but to the joke of heroism; although there are no GODGAMES within these texts, the Multiverse as a whole can be seen as a Godgame of a manipulating author/ironist. There is a smell of greasepaint to the MM oeuvre; even the bondage of the Eternal Champion himself is peculiarly and deliberately arbitrary – most of the AVATARS are in any case granted frequent paroles, and manifest themselves in various worlds, under various names. The avatar masks they wear can, almost always, be shed. And, just as the Commedic instability of the Multiverse worlds constitutes a THINNING of the autonomous venue that most fantasy novels inhabit, so the Eternal Champion parodies the hero of the MONOMYTH, subjecting his generic model to a similarly constant thinning. Robert E. HOWARD's CONAN, though an immensely less sophisticated creation, is far more real to most readers than Elric, who is a deliberate PARODY of Conan. In the end, the effect can be nightmarish. Thinned worlds which are interminably interchangeable; thinned heroes entrapped in masquerade roles: it is not surprising that so much of MM's work generates an atmosphere of HORROR.

Tales exploiting the relationship between Multiverse and Eternal Champion proper make up only part of MM's oeuvre, but since the 1970s they have been quite ruthlessly interwoven with the full-fledged **Commedia** sequences featuring various versions of Jerry Cornelius, with MM's occasional sf, with his romances, and with his late nongenre works.

Erekosë "I am John Daker," confesses this first embodiment of the Eternal Champion in the prologue to *The Dragon in the Sword* (see below), "the victim of the whole world's dreams. I am Erekosë, Champion of Humanity, who slew the human race. I am . . . Elric Womanslayer, Hawkmoon, Corum and so many others – man, woman or androgyne. I have been them all." John Daker is a human translated to a standard Nordic Fantasyland, where he is known as Erekosë, and behaves valorously. He is the only avatar of the Eternal Champion to describe himself as such, or to name his siblings. He features mainly in *The Eternal Champion* (1962 *Science Fantasy*; exp **1970** US; rev 1978 US), *Phoenix in Obsidian* (**1970**; vt *The Silver Warriors* 1973 US), *The Quest for Tanelorn* (**1975**) – which brings various avatars together – *The Swords of Heaven, the Flowers of Hell* (graph **1979** US) illustrated by Howard CHAYKIN and *The Dragon in the Sword: Being the Third and Final Story in the History of John Daker* (**1986**; full text 1987 UK). *The Eternal Champion* (rev omni **1992**) excludes *The Quest for Tanelorn* and the GRAPHIC NOVEL.

Elric The tales featuring Elric of Melniboné, published over more than 30 years, constitute MM's first consequential work. By internal chronology, the sequence comprises *Elric of Melniboné* (**1972**; cut vt *The Dreaming City* 1972 US), *The Fortress of the Pearl* (**1989**), *The Sailor on the Seas of Fate* (fixup **1976**) – incorporating a rev of *The Jade Man's Eyes* (**1973** chap) – *The Weird of the White Wolf* (coll **1977** US) – incorporating stories from *The Stealer of Souls* (1961-2 *Science Fantasy*; coll **1963**) and *The Singing Citadel* (coll **1970**) – *The Sleeping Sorceress* (**1971**; vt *The Vanishing Tower* 1977 US), *The Revenge of the Rose: A Tale of the Albino Prince in the Years of his Wandering* (**1991**), *The Bane of the Black Sword* (1962 *Science Fantasy*; coll **1977** US) – incorporating the remaining stories from *The Stealer of Souls* and *The Singing Citadel* – and *Stormbringer* (1963-4 *Science Fantasy*;

cut **1965**; restored and rev 1977 US). Omnibuses of this material are: *The Elric Saga Part I* (omni **1984** US) containing *Elric of Melniboné*, *The Sailor on the Seas of Fate* and *The Weird of the White Wolf*; and *Part II* (omni **1984** US) containing *The Vanishing Tower*, *The Bane of the Black Sword* and *Stormbringer*. *Elric at the End of Time* (coll **1984**) assembles mostly earlier stories, including some from *Sojan*; *Elric: The Return to Melniboné* (graph **1973**) illustrated by Philippe Druillet (1944-   ) is a graphic novel; *Michael Moorcock's Elric: Tales of the White Wolf** (anth **1994** US) ed Edward E. Kramer and Richard Gilliam comprises stories by others set in the **Elric** universe.

Stormbringer, the first-written volume of the sequence, also terminates it, closing Elric's angst-ridden life as well; all subsequent volumes are prequels or interjections. The figure of Elric is, as noted, a direct parody of Conan; he is an albino weakling, introspective, haunted, treacherous, and the tool of his own SOUL-drinking SWORD Stormbringer, itself a parody of the normal S&S weapon, which may aspire to COMPANION status (if animate) but not normally to that of puppetmaster. Melniboné is – like HYPERBOREA – a land-of-fable vision of prehistoric Europe; Elric's treachery causes its total downfall. His humorlessness and unrelenting *Weltschmerz* may make him easy to mock, but he remains a remarkably vivid iconic figure.

Warrior of Mars The SCIENCE-FANTASY tales featuring Kane of Old Mars are PLANETARY ROMANCES which pastiche Egard Rice BURROUGHS with some skill; Kane's role as an Eternal Champion is an afterthought. The sequence comprises *Warriors of Mars* (**1965**; vt *The City of the Beast* 1970 US), *Blades of Mars* (**1965**; vt *The Lord of the Spiders* 1971 US) and *Barbarians of Mars* (**1965**; vt *Masters of the Pit* 1971), all assembled as *Warrior of Mars* (omni **1981** UK) and all first published as by Edward P. Bradbury.

Hawkmoon The two series featuring Dorian Hawkmoon, Duke of Coln, and the brusque but fatherly Count Brass are set in a land-of-fable version of a far-future Europe, and are more novelistically textured than any other **Eternal Champion** book, except the late **Von Bek** tales. Hawkmoon is an attractive figure, though MM's tendency to load him with PLOT COUPONS generates at times a sense of (perhaps deliberate) hilarity; his war against Granbretan has the virtue of being just. The **Runestaff** books are *The Jewel in the Skull* (**1967** US; rev 1977 US), *Sorcerer's Amulet* (**1968** US; vt *The Mad God's Amulet* 1969 UK), *Sword of the Dawn* (**1968** US; rev vt *The Sword of the Dawn* 1969 UK; rev 1977 US) and *The Secret of the Runestaff* (**1969** US; vt *The Runestaff* 1969 UK; rev 1977 US), all assembled as *The History of the Runestaff* (omni **1979** UK; rev vt *Hawkmoon* 1992). The **Count Brass** books are *Count Brass* (**1973**), *The Champion of Garathorm* (**1973**) and *The Quest for Tanelorn* (**1975**), all assembled as *The Chronicles of Castle Brass* (omni **1985** UK).

Corum Beginning in an Earth long before our present world has taken shape and featuring a hero who is the last of the ELF-like Vadhagh, this series can be seen as a fairly decorous assault upon J.R.R. TOLKIEN. Corum is much unlike Elric; metaphorically speaking, his back is to the reader, and he engages in his long war against Lord Arioch of Chaos with a certain elegance. There are two series. The **Swords** books are *The Knight of the Swords* (**1971**), *The Queen of the Swords* (**1971**) and *The King of the Swords* (**1971**), all assembled as *The Swords Trilogy* (omni **1977** US; vt *The Swords of Corum* 1986 UK; rev vt *Corum* 1992

UK). A second trilogy, the **Chronicles of Corum**, comprises *The Bull and the Spear* (**1973**), *The Oak and the Ram* (**1973**) and *The Sword and the Stallion* (**1974**), all assembled as *The Chronicles of Corum* (omni **1978** US).

Von Bek Graf Ulrich Von Bek is the first of his family, and the closest to a conventional Eternal Champion. *The War Hound and the World's Pain* (**1981** US) is a smoothly told but generically complex tale, combining elements of traditional SUPERNATURAL FICTION and MATTER-of-Britain fantasy, plus motifs out of the general CAULDRON OF STORY, including a WILD HUNT. Von Bek's family features in *The City in the Autumn Stars* (**1986**), and the series thereafter begins to concentrate upon the family as a kind of Commedia troupe; these two books are assembled with an added story as *Von Bek* (rev omni **1992**). Other titles, which extend the concept of the Eternal Champion to breaking point, include: *The Brothel in Rosenstrasse* (**1982**), a fantasy of sexual torment set in a very slightly displaced Europe; *Blood: A Southern Fantasy* (fixup **1995** US), which features variously spelled members of the **Von Bek** troupe in the US South in a tale whose main protagonists variously adventure through the Biloxi Fault, a rift in REALITY, plus its sequel, *Fabulous Harbours* (coll of linked stories **1995**), the epilogue to which also appeared as *The Birds of the Moon: A Travellers' Tale* (**1995** chap); and *Lunching with the Antichrist: A Family History: 1925-2015* (coll of linked stories **1995**).

COMMEDIA

Cornelius Just as Elric is a parody of Conan, so Jerry Cornelius is a direct parody of Elric, Elric turned inside out. Any sense that various AVATARS are parading before the footlights of the world, flaunting a succession of MASKS, must surely accord with MM's intentions. In his early appearances, Jerry Cornelius represents MM's first version of Harlequin ($\Diamond$ COMMEDIA DELL'ARTE): a portmanteau Pop 1960s ANTIHERO, an anarchic streetwise urban ragamuffin in James Bond gear, amorally deft at manipulating everything from women to the Multiverse itself. In his early adventures – during which the planet suffers various catastrophes – Jerry ranges from the present through the FAR FUTURE, randy (but loyal to his sister, Catherine Cornelius, who is a Columbine figure throughout the MM oeuvre), and evanescent. Jerry as Harlequin dominates the first two novels of the **Jerry Cornelius** sequence: *The Final Programme* (excerpts 1965-6 *New Worlds*; **1968** US; rev 1969 UK; rev 1977 US; rev 1979 UK), filmed as *The* FINAL PROGRAMME (*1973*; vt *The Last Days of Man on Earth* 1975 US), and *A Cure for Cancer* (1969 *New Worlds*; **1971**; rev 1977 US; rev 1979 UK). In the third and fourth volumes of the sequence – *The English Assassin* (**1972**; rev 1977 US; rev 1979 UK) and *The Condition of Muzak* (**1977**; rev 1977 US; further rev 1978 UK), which won the 1977 *Guardian* Fiction Prize – Jerry undergoes a transformation into Pierrot, and the latter novel (perhaps MM's finest) ends with the Cornelius figure stripped of his MASK, a failed rock singer in bondage to the mortal world. The omnibus which first assembled these four volumes was *The Cornelius Chronicles* (omni **1977** US; incorporating 1979 revs of individual titles vt 2 vols *The Cornelius Chronicles: Book One* 1988 UK and *Book Two* 1988 UK; vt *The Cornelius Quartet* 2 vols 1993). In *The Cornelius Chronicles, Volume II* (omni **1986** US) were assembled *The Lives and Times of Jerry Cornelius* (coll **1976**; exp 1987) and *The Entropy Tango: A Comic Romance* (fixup **1981**). In *The Cornelius Chronicles, Volume III* (omni **1987** US) were assembled *The Adventures of Una Persson and Catherine Cornelius in* *the Twentieth Century* (**1976**; cut vt *The Adventures of Una Persson and Catherine Cornelius* in omni 1980 US with *The Black Corridor* [see below]) and "The Alchemist's Question" (1984) from *The Opium General and Other Stories* (coll **1984**). *The Cornelius Calendar* (omni **1994**) assembles *The Adventures of Una Persson, The Entropy Tango,* "Gold Diggers of 1977" (1989) and "The Alchemist's Question" (1984).

The titles assembled in these subordinate omnibuses are fantasias upon the thematic material of the central quartet, but lack its cumulative intensity, though the Commedia dell'Arte effects remain central, if not as intensely conveyed in the absence of Jerry as Pierrot. They are perhaps most interesting for the space they grant for the development of the TEMPORAL ADVENTURESS, mainly in the guise of Una Persson, who toughly, sagely, resignedly, sexily and wittily traverses the venues of the Multiverse. She is as much a TRICKSTER as Jerry Cornelius in his early pomp, but is not tied to role. Further associated material appears in *The Nature of the Catastrophe* (anth **1971**; exp vt *The New Nature of the Catastrophe* 1993) ed MM and Langdon Jones, which contains stories and material by MM and other *New Worlds* writers who were invited to use **Cornelius** as, in effect, a SHARED WORLD; and in *The Great Rock'n'Roll Swindle* (**1980** chap in the format of a tabloid newspaper; rev vt "Gold Diggers of 1977" in *Casablanca* coll **1989**). *The Distant Suns* (1969 *The Illustrated Weekly of India*; **1975** chap) with Philip James (James CAWTHORN) has as its protagonist a Jerry Cornelius who bears no relation to the Jerry of the other books.

Dancers at the End of Time This sequence is set in a FAR-FUTURE venue rationalized in sf terms, and comprises an initial central trilogy: *An Alien Heat* (**1972**), *The Hollow Lands* (**1974** US) and *The End of All Songs* (**1976** US), all assembled as *The Dancers at the End of Time* (omni **1981**; rev 1991); this trilogy is accompanied by *Legends from the End of Time* (coll **1976** US) and *The Transformation of Miss Mavis Ming* (1976 *New Worlds* as "Constant Fire"; much exp **1976**; vt *A Messiah at the End of Time* 1978 US), both assembled as *Tales from the End of Time* (omni **1989** US; vt *Legends from the End of Time* 1993 UK). The DYING EARTH is of course a natural venue for Commedia dell'Arte routines. **Dancers at the End of Time**, though technically sf, sustainedly conveys a sense that genre distinctions are themselves part of the fantastic play.

Individual Commedia Titles *Gloriana, or the Unfulfill'd Queen: Being a Romance* (**1978**; rev 1993) is an ambiguous sexual parable, sanitized in the revision, set in a land-of-fable Elizabethan England. Most of the story takes place in Queen Gloriana's palace, a legacy from her father King Hern and an intricate EDIFICE. It is again and again described in terms that liken it to Gloriana's own mind, so that its murky underpassages have a double ominousness (which is not fully retained in the 1993 version). And indeed her nights are awful, for she is sexually unfulfill'd. References to Edmund SPENSER's *The Faerie Queene* (**1590-96**) and Mervyn PEAKE's **Gormenghast** sequence (**1946-59**) permeate the text; John DEE is in evidence as well, and many members of the **Cornelius** and **Eternal Champion** troupes.

Mother London (**1988**) is less fantasy than FABULATION; but its presentation of the MASQUE of the great city is complexly and movingly Commedic, and the multifaceted protagonist – three separate figures making up an insect-eye

self-portrait of MM – has all the theatrical jointedness of the true masker. Along with *The Condition of Muzak* and *Gloriana, Mother London* can stand as MM's most significant single creation. It is worth noting that all three novels are set in some or other version of LONDON.

NEW WORLDS

In the 1960s MM became editor of *New Worlds* magazine, a position he held, with some time off, from *#142* (May/June 1964) to its effective demise as a magazine with *#201* (March 1971). His editorship of *New Worlds* was far more significant for sf than for fantasy, as genre sf was in 1964 suffering profound turmoil, and MM's promoting of what became known as the New Wave caused grave upset and much excitement, especially in the USA. *New Worlds* published literate and in sf terms radically "modern" work by a number of writers – including Brian W. ALDISS, J.G. BALLARD, Samuel R. DELANY, Thomas M. DISCH, M. John HARRISON, John T. Sladek and Norman Spinrad – most of whom wrote fantasy and sf with equal facility. But in this context it was not their fantasy that aroused contention. After ceasing as a magazine, *New Worlds* continued as a series of ANTHOLOGIES until 1976, under the editorship (variously and in combination) of MM, Hilary Bailey (MM's wife 1962-78) and Charles Platt; another brief series in magazine format ran for several issues in 1978-9; a further anthology series, with MM's authorization, began in the 1990s with *New Worlds 1* (anth **1991**) ed David S. Garnett.

OTHER WORKS

In the 1980s MM began his only nongenre series, the **Colonel Pyat** sequence: *Byzantium Endures* (**1981**; cut 1981 US), *The Laughter of Carthage* (**1984**) and *Jerusalem Commands* (**1992**), with one further novel projected, «The Vengeance of Rome» (with a comma after "Carthage", the four titles read together as a sentence). These novels, which feature many characters from the **Cornelius** books, comprise an ambitious attempt to convey some sense of the charnel-house nature of the 20th century through the memoirs of Colonel Pyat, born in 1900, a liar, a Jewish antisemite, a betrayer of all he meets, an aviator – the dark underside of the Commedia parade. His careering course through history is projected to culminate at Auschwitz, though he himself survives to become another London manikin. The **Pyat** books represent a culmination of MM's long attempt to integrate popular fiction and Postmodernism, an argument and traversal made all the more interesting because of the large number of books through which it can be traced, and because MM has so frequently returned to early sequences (**Elric** in particular), transforming them in the process. MM was never easy to pigeonhole as a writer – an early text like *The Golden Barge* (written 1958; excerpt 1965 *New Worlds* as by William Barclay; **1979**) is as metaphysical as any of his mature experiments – and has come to be recognized as a major figure at the edge of sf and at the centre of fantasy, materially helping define all his chosen worlds.

In recent years MM has begun to write nonfiction statements, not dissimilar in tone to some of his *New Worlds* editorials, but now conceived autonomously. They include: a political pamphlet, *The Retreat from Liberty: The Erosion of Democracy in Today's Britain* (**1983** chap); *Letters from Hollywood* (**1986**) illustrated by Michael FOREMAN; an impatient and rather patchy examination of fantasy, *Wizardry and Wild Romance: A Study of Epic Fantasy* (**1987**), a chapter of which was based on *Epic Pooh* (**1978** chap), and which constitutes as a whole an apologia for his own ironized point of view; and *Fantasy: The 100 Best Books* (**1988**) with (in fact written almost entirely by) James CAWTHORN. He has also become involved in FEMINISM, and his advocacy of censorship-like controls over writing that is damaging to women has inspired some of the more recent revisions to his own work. At a comparatively young age, he remains the 20th century's central fantasist about fantasy.   [JC]

Other works: *Caribbean Crisis* (**1962** chap) with James Cawthorn, together writing as Desmond Reid; *The Sundered Worlds* (1962-3 *SF Adventures*; fixup **1965**; vt *The Blood Red Game* 1970; rev under first title 1992); *The Fireclown* (**1965**; vt *The Winds of Limbo* 1969 US); *The Twilight Man* (**1966**; vt *The Shores of Death* 1970); *The Deep Fix* (coll **1966**) as by James Colvin; *The LSD Dossier* (**1966**), a text drafted by Roger Harris and rewritten by MM, and its sequels, *Somewhere in the Night* (**1966** as by Bill Barclay; rev vt *the Chinese Agent* 1970 US as by MM) and *Printer's Devil* (**1966** as by Bill Barclay; rev vt *The Russian Intelligence* 1980 as by MM), comprising in rev form a series about a **Cornelius** analogue, **Jerry Cornell**; *The Wrecks of Time* (1965-66 *New Worlds* as by James Colvin; cut **1967** dos US as by MM; text restored vt *The Rituals of Infinity* 1971 UK); the **Oswald Bastable** STEAMPUNK sequence, comprising *The Warlord of the Air* (**1971** US), *The Land Leviathan* (**1974**) and *The Steel Tsar* (**1981**), all 3 assembled as *The Nomad of Time* (omni **1982**; rev vt *A Nomad of the Time Streams* 1993 UK); *The Black Corridor* (**1969** US) with Hilary Bailey (uncredited); *The Time Dweller* (coll **1969**); *The Ice Schooner* (1966-7 *Impulse*; **1969**; rev 1977 US; rev 1985 UK); 2 novels featuring **Karl Glogauer**, being *Behold the Man* (1966 *New Worlds*; exp **1969**) and *Breakfast in the Ruins* (**1972**), both assembled with an unconnected tale as *Behold the Man and Other Stories* (omni **1994**); the **Hawklords** sequence, comprising *The Time of the Hawklords* (**1976**) and *Queens of Deliria* (**1977**), the first with Michael Butterworth, the second by Butterworth alone though MM was credited without his authorization; *Moorcock's Book of Martyrs* (coll **1976**; vt *Dying for Tomorrow* 1978 US); *The Real Life Mr Newman* (1966 in *The Deep Fix*; **1979** chap); *My Experiences in the Third World War* (coll **1980**).

The Tale of the Eternal Champion: Each volume of the UK series, and those of the US series which appeared before the end of 1995, is listed below, in conjunction with the titles each assembles; for convenience, the sequences are listed by title only, without registering other bibliographical complexities, unless they are major. The UK series comprises: *The Tale of the Eternal Champion #1: Von Bek* (omni **1992**); *#2: The Eternal Champion* (omni **1992**), concerning **Erekosë**; *#3: Hawkmoon* (omni **1992**); *#4: Corum* (omni **1992**); *#5: Sailing to Utopia* (omni **1993**), which contains *The Ice Schooner, The Black Corridor* and *The Distant Suns*, none being **Eternal Champion** tales; *#6: A Nomad of the Time Streams* (omni **1993**), concerning **Oswald Bastable**; *#7: The Dancers at the End of Time* (omni **1993**); *#8: Elric of Melniboné* (omni **1993**); *#9: The New Nature of the Catastrophe* (coll **1993**); *#10: The Prince with the Silver Hand* (omni **1993**), concerning **Corum**; *#11: Legends from the End of Time* (omni **1993**); *#12: Stormbringer* (omni **1993**); *#13: Earl Aubec* (coll **1993**), which assembles miscellaneous material; *#14: Count Brass* (omni **1993**). The US series comprises *The Tale of the Eternal Champion #1: The Eternal Champion* (omni **1994** US), contents differing from *#2* above; *#2: Von Bek* (omni **1995** US); *#3: Hawkmoon*

(omni **1995** US); *#4: A Nomad of the Time Streams* (omni **1995**); *#5: Elric: Song of the Black Sword* (omni **1995**); 10 further volumes projected.

As editor: *The Best of New Worlds* (anth **1965**); *SF Reprise 1* (anth **1966**), *#2* (anth **1966**) and *#5* (anth **1967**), all assembled from issues of *New Worlds* without MM's consent; *Best SF Stories from New Worlds* (anth **1967**); *Best Stories from New Worlds 2* (anth **1968**; vt *Best SF Stories from New Worlds 2* 1969 US); *Best SF Stories from New Worlds 3* (anth **1968**); *The Traps of Time* (anth **1968**); *Best SF Stories from New Worlds 4* (anth **1969**); *The Inner Landscape* (anth **1969**), anon; *Best SF Stories from New Worlds 5* (anth **1969**); *Best SF Stories from New Worlds 6* (anth **1970**); *Best SF Stories from New Worlds 7* (anth **1971**); *New Worlds 1* (anth **1971**; vt *New Worlds Quarterly 1* 1971 US), *#2* (anth **1971**; vt *New Worlds Quarterly 2* 1971 US), *#3* (anth **1972**; vt *New Worlds Quarterly 3* 1972 US), *#4* (anth **1972**; vt *New Worlds Quarterly 4* 1972 US), *#5* (anth **1973**) and *#6* (anth **1973**; vt *New Worlds Quarterly 5* 1974 US), the last with Charles Platt; *Best SF Stories from New Worlds 8* (anth **1974**); *Before Armageddon* (anth **1975**); *England Invaded* (anth **1977**); *New Worlds: An Anthology* (anth **1983**).

Further reading: *The Tanelorn Archives: A Primary and Secondary Bibliography of the Works of Michael Moorcock, 1949-1979* (**1981**) by Richard Bilyeu; *The Entropy Exhibition: Michael Moorcock and the British "New Wave" in Science Fiction* (**1983**) by Colin GREENLAND; *Michael Moorcock: A Reader's Guide* (**1991** chap; rev 1992 chap) by John Davey (1962-   ); *Death is No Obstacle* (**1992**), book-length conversation between Greenland and MM about the latter's work.

MOORE, ALAN (1953-   ) UK writer whose output has included a number of groundbreaking GRAPHIC NOVELS and several long comic-book STORY CYCLES which have brought a new depth and complexity to established COMICS characters.

AM's first professional work was as Kurt Vile, under which pseudonym he wrote and drew two series for the weekly music paper *Sounds* (**Rosco Moscow** [1979-80] and **The Stars My Degradation** [1980-82]) and *Three Eyes McGurk and His Death Planet Commandos* in *Rip Off Comics #8* (1981 US). As Jill de Ray he wrote and drew **Maxwell the Magic Cat** for the *Northants Post* (1979-86). He began producing short **Future Shock** scripts under his own name for the UK sf weekly comic *2,000 A.D.*, some of which were reprinted as *Alan Moore's Shocking Futures* (graph coll **1986**) and *Alan Moore's Twisted Times* (graph coll **1986**). He also wrote stories for the UK MARVEL COMICS' *Dr Who Weekly*, *Dr Who Monthly*, *Star Wars* and *Captain Britain*. His first works of note were two long series, both started in the first issue of the UK anthology comic *Warrior* (1982-5): a Postmodernist treatment of the popular 1950s superhero CAPTAIN MARVEL entitled **Marvelman** – assembled with additional material as *Miracleman* (graph coll **1988** US), *The Red King Syndrome* (graph coll **1990** US) and *Olympus* (graph coll **1990** US) – and **V for Vendetta**, assembled with additional material as *V for Vendetta* (graph coll **1990** US), which featured an anarchist hero pitted against a fascist regime.

AM's several series for *2,000 A.D.* included **The Ballad of Halo Jones** (1984-6; graph coll **1986** 3 vols), **Skizz** (1983) and **D.R. and Quinch** (1983-5; as *D.R. and Quinch's Guide to Life* graph coll **1986**). His first work for US comic books was in *Saga of the Swamp Thing* in 1984. This was followed by the remarkable retro superhero GRAPHIC NOVEL *Watchmen* (1986-7; graph **1987** US; exp 1988 US) and *Batman: The Killing Joke* (graph **1988** US).

AM formed Mad Love (Publishing) Ltd in 1988 with Phyllis Moore and Debbie Delano, and this imprint produced the anti-homophobia anthology *AARGH! (Artists Against Rampant Government Homophobia)* (graph anth **1988**) and the currently moribund graphic novel «Big Numbers» (2 vols appeared in 1990). Other projects include: the Dickensian Jack the Ripper graphic novel *From Hell*, which began serialization in TABOO (in 1991), subsequently republished and continued in a planned 12 vols from Tundra; and the erotic *Lost Girls*, which also began serialization in *Taboo* (in 1992). [RT]

MOORE, BRIAN (1921-   ) Irish-born writer, a Canadian citizen resident in the USA from 1959, best-known for nongenre work like *Judith Hearne* (**1955** UK; vt *The Lonely Passion of Judith Hearne* 1956 US), the first novel now commonly ascribed to him, though he did in fact write some early detective thrillers under his own name and as by Bernard Mara and (later) Michael Bryan. Of his novels which incorporate some element of the fantastic, *Catholics* (**1972** Canada) is closest to sf. Of more direct fantasy interest are tales which equivocate with themes typical of SUPERNATURAL FICTION. In *Fergus* (**1970**) an expatriate Irish novelist is haunted by GHOSTS from his Irish past; in *The Mangan Inheritance* (**1979**), a similar challenge to his identity faces an Irish novelist who returns to Ireland. *The Great Victorian Collection* (**1975**) traces the consequences of the sudden incursion into this world of the eponymous collection, which appears suddenly in a California parking lot – apparently generated out of a DREAM experienced by the scholarly protagonist. *Cold Heaven* (**1983**), which initially has the air of a GHOST STORY, rapidly becomes a much more complicated exercise in what proves to be a GODGAME of sorts, imposed by GOD on the female protagonist. Eventually she is allowed to continue her life, and to leave her husband. [JC]

MOORE, C(ATHERINE) L(UCILLE) (1911-1987) US writer, a leading author of SCIENCE FANTASIES for WEIRD TALES before devoting herself to sf, frequently in collaboration with her husband, Henry KUTTNER. She was working as a bank secretary in Indianapolis when her first professional sale, "Shambleau", ran as the lead story in the November 1933 *WT*. This tale of an intergalactic soldier-of-fortune seduced and nearly destroyed by a medusa-like alien temptress (◊ GORGONS) blended imagery drawn from Classical MYTH with a strong undercurrent of eroticism. The story met with critical and popular acclaim, and catapulted CLM instantly into the ranks of *WT*'s foremost writers. She wrote almost exclusively for *WT* for the next three years and helped shape the legacy of its golden years. It was for almost two years not generally known that "C.L. Moore" was a woman.

"Shambleau" was the first of a dozen stories featuring **Northwest Smith**, whose adventures dominate the two collections *Shambleau and Others* (coll **1953**; with 3 of 7 stories cut vt *Shambleau* 1958; with 1 story cut vt *Shambleau* UK 1961) and *Northwest of Earth* (coll **1954**); they were collected as *Scarlet Dream* (coll **1981**; vt *Northwest Smith* 1982). Through **Smith**, CLM helped revamp the formulae of both space opera and HEROIC FANTASY. Smith's introspection and fallibility give him a more human dimension than his predecessors in heroic fantasy, and the depiction of his sexual

vulnerability represented a psychological maturity uncommon in the field. He encounters VAMPIRES in "Black Thirst" (1934) and "Scarlet Dream" (1934) and WEREWOLVES in "Werewoman" (1938), but CLM wreaks original variations on these familiar fantasy themes by linking them to earthly and alien mythologies. Smith's universe contrasts sharply with that of most heroic fantasy and sf of the era, insofar as it presents humanity as ignorant and ill equipped to comprehend alien worlds and species. Two **Smith** stories were collaborations: "Nymph of Darkness" (1935), with Forrest J Ackerman (1916-   ), and "Quest of the Starstone" (1937 WT), with Kuttner. The latter featured CLM's other series fantasy character, **Jirel of Joiry**, a strong-willed woman living in medieval France whom CLM had introduced in "The Black God's Kiss" (1934). The first SWORD-AND-SORCERY saga to feature a female protagonist, the **Jirel** stories bear a close resemblance in their plots and themes to the **Smith** tales. They were collected as *Jirel of Joiry* (coll of linked stories **1969**; vt *Black God's Shadow* 1977). Moore's other stories for fantasy and weird-fiction magazines include "Miracle in Three Dimensions" (1939 *Strange Stories*), "Fruit of Knowledge" (1940 *Unknown*), and "Daemon" (1946 *Famous Fantastic Mysteries*).

CLM published a handful of stories in sf magazines 1934-9, all redolent of the romantic spirit of her fantasy. Following her marriage to Kuttner in 1940, she worked almost exclusively in sf, usually in collaboration with him – under their own names and also under a score of pseudonyms, notably including Lewis Padgett and Lawrence O'Donnell. Although it is assumed on the basis of their earlier work that Kuttner was responsible for the meticulous plots and frequent comedy of their co-written efforts, CLM for their characterization and atmosphere, it is impossible to single out their individual contributions. The two became regular presences in *Astounding Science Fiction* during WWII. Key works include "Clash by Night" (**1943** as by O'Donnell) and its sequel *Fury* (**1947** as by O'Donnell; 1950; vt *Destination Infinity* 1958), both set in undersea cities on the planet Venus, and the TIME-TRAVEL story "Vintage Season" (**1946**). Four of the O'Donnell stories were collected under CLM's name as *Judgment Night* (coll **1952**; title novel only **1965**). Collections of stories from this period in which her hand is apparent include *Line to Tomorrow* (coll **1954** as by Padgett), *No Boundaries* (coll **1955** with Kuttner), *Clash by Night and Other Stories* (coll **1980** UK with Kuttner), and *Tomorrow and Tomorrow, and The Fairy Chessmen* (coll **1951** as by Padgett; 2nd short novel vt *Chessboard Planet* 1956 and vt *The Far Reality* 1966 UK as by Padgett). *The Best of C.L. Moore* (coll **1975**) ed Lester del Rey represents the most concerted effort to single out those stories which are primarily CLM's.

In the early postwar years CLM and Kuttner wrote a series of short novels strongly influenced by the fantasies of A. MERRITT, all of which but *Earth's Last Citadel* (**1943** *Argosy* as by HK and CLM; **1964** as by CLM and HK) are attributed to Kuttner or Padgett. However, *The Dark World* (**1946** *Startling Stories* as by Kuttner; **1965**), which was included with *Valley of the Flame* (1946 *Startling Stories* as by Keith Hammond; **1964** as by HK) and *Beyond Earth's Gates* (1949 as "The Portal in the Picture" by Kuttner; **1954** dos as by Padgett) in *The Startling Worlds of Henry Kuttner* (omni **1987**), portrays a character whose travel to a SECONDARY WORLD liberates the dark side of his personality, a plot complication consistent with CLM's deployment of psychologically conflicted characters in her earlier fiction.

Within a few years, both writers had switched to writing for the crime and mystery market. CLM's last solo sf novel was *Doomsday Morning* (**1957**). After Kuttner's death the following year, she concentrated on screenplays for television series including *Maverick* and *77 Sunset Strip*, and produced no more significant fantasy work.   [SD]

Other works: *The Time Axis* (1949 *Startling Stories* as by HK; **1965**); *The Mask of Circe* (1948 *Startling Stories* as by HK; **1971**); *A Gnome There Was* (coll 1950 as by Padgett); *Well of the Worlds* (1952 *Startling Stories*; 1953 as by Padgett; vt *The Well of the Worlds* 1965 as by HK); *Mutant* (fixup 1953 as by Padgett; 1954 UK as by Kuttner); *There Shall Be Darkness* (1954 chap Australia).

MORGAN, CHRIS (1946-   ) UK writer and editor who began to publish fiction with "Clown Fish and Anemone" for *Science Fiction Monthly* in 1975, and whose *The Shape of Futures Past: The Story of Prediction* (**1980**), based on an idea by Michael Scott ROHAN, takes a balanced view of the predictions made by sf and other writers before about 1945. His *Dark Fantasies* (anth **1989**) assembles a strong selection of original tales.   [JC]

Other works: *Fritz Leiber: A Bibliography 1934-1979* (**1979** chap); *Future Man* (**1980**); *Facts and Fallacies: A Book of Definitive Mistakes and Misguided Predictions* (**1981**) with David LANGFORD.

MORGAN LE FAY An ENCHANTRESS in the Arthurian LEGENDS who takes the primary role in opposing Arthur's actions and destroying him and his kingdom. She is an enigmatic figure who maintains her FAIRY origins throughout the legends, despite attempts to Christianize her.

Although by the time of Sir Thomas MALORY's *Le Morte Darthur* (**1485**) Morgan has become an evil character, seeking Arthur's death, when she was first incorporated into the cycle, in the *Vita Merlini* (?**1155**) by GEOFFREY OF MONMOUTH, she was portrayed as the head of a sisterhood who lived on the Isle of AVALON and had special healing powers. This image remains, somewhat enigmatically, at the end of *Le Morte Darthur* when she and her sisters take Arthur's dying body away to Avalon; because of her past actions, though, this final event takes on a more sinister aspect. Throughout the medieval Arthurian romances Morgan is testing and obstructing Arthur and his knights. The beheading-match challenge (◊ GAWAIN) in *Sir Gawain and the Green Knight* (14th century) is contrived by Morgan to test the resolve of the ROUND TABLE. In *Le Morte Darthur* she creates a false EXCALIBUR in the hope that her lover, Sir Accolon, will kill Arthur with the real one, but when that fails she steals the scabbard (which conveys the power of INVULNERABILITY).

In Malory, Morgan is the half-sister of Arthur, and the sister of Morgause (and thereby the aunt of Gawain, Gareth, Gaheris, Agravain and Mordred/ Modred). She married Urien, king of Rheged, and is the mother of Owain. These last two are historical kings who ruled at the end of the 6th century, which would make Urien's wife a contemporary of Myrddin Wyllt (◊ MERLIN). The character of Morgan has older Celtic origins, and is probably descended from the Mother GODDESS Modron who, as the mother of Mabon, appears in the earliest surviving Arthurian story, *Culhwch and Olwen* (11th century). She also earned the title the MORRIGAN, under which name she often appears in more recent Arthurian fiction.

Morgan's role as Queen of Avalon extends beyond the Arthurian canon into other STORY CYCLES. She appears in the anonymous 13th-century *chanson de geste* called *Ogier le Danois* ["Ogier the Dane"], where she falls in love with Ogier, a knight in the court of Charlemagne. When Ogier reaches 100 years of age she carries him off to Avalon and restores his youth. He stays there for 200 years before returning briefly to Earth to fight the Saracens. Ogier became Denmark's national hero; his story is told in *Holger Danske* (**1837**) by Bernhard Ingemann (1789-1862). William MORRIS retold the Avalon episode in *The Earthly Paradise* (**1868-70**), and Poul ANDERSON recreated the legend of Ogier and Morgan in *Three Hearts and Three Lions* (1953 *F&SF*; rev **1961**). In Italian folklore she became Fata Morgana (*fata* = fairy), retaining the sinister duality whereby she is both a dispenser of gifts and a ruthless WITCH. She is introduced thus in *Orlando Innamorato* (**1486**) by Matteo Boiardo (1434-1494). Morgan therefore took on the role of a FEMME FATALE, an identity she has retained into modern fiction.

T.H. WHITE brought Morgan into modern fiction in *The Sword in the Stone* (**1938**) and made her the central character in *The Witch in the Wood* (**1939**). A mystical interpretation was presented by Dion FORTUNE in *The Sea Priestess* (**1938**) and its sequel, *Moon Magic* (**1956**), which feature Miss Le Fay Morgan, a modern-day acolyte of the ancient goddess.

A more human presentation is made of Morgan in both *The Mists of Avalon* (**1982**) by Marion Zimmer BRADLEY, where she is the storyteller, and in the **Daughter of Tintagel** sequence by Fay SAMPSON, which follows Morgan's life. She also features in the children's books *On All Hollows' Eve* (**1984**) and *Out of the Dark World* (**1985**) by Grace CHETWIN. The image of Morgan as a saviour of the elder world reappears in *The Night of the Solstice* (**1987**) and its sequel *Heart of Valor* (**1992**) by L(isa) J. Smith, where she battles to save our world from the evils of the Wildworld. She is a central character in EXCALIBUR (*1981*).   [MA]

Further reading: *Herself* (**1992**), the last of Fay Sampson's **Daughter of Tintagel** series, ostensibly a novel but contains extensive discussion – as by Morgan – of the range of relevant literature and of changing ideas about her.

MORLEY, CHRISTOPHER (DARLINGTON) (1890-1957) US editor and writer, active as an author for nearly 50 years after the publication of his first book, *The Eighth Sin* (coll **1912**), which is poetry. His first novel of fantasy interest is *Where the Blue Begins* (**1922**; 1925 edn illus Arthur RACKHAM), a BEAST FABLE of considerable subtlety set in a Long Island and NEW YORK entirely populated by dogs; the narrative is absolutely deadpan, and the mild SATIRE involved is imparted without the slightest condescension. *Thunder on the Left* (**1925**) is a TIMESLIP tale whose direction is, unusually, forward in time. Its protagonist, aged 10, is sent as a "spy" into the future, where as an adult with a child's PERCEPTION of the new world he comes close to losing his innocence (◊ ET IN ARCADIA EGO); but he returns safely, cajoled into doing so by the GHOST of his dead sister. *The Trojan Horse* (**1937**) is another satire, placing a contemporary drama in 1185BC. Perhaps partly because of his wit and fluency, CM was often accused of unseriousness; he is much undervalued.   [JC]

Other works: *The Haunted Bookshop* (**1919**), a spy story set around the travelling bookshop featured in *Parnassus on Wheels* (**1917**), the two being CM's most famous titles; *The Arrow* (**1927** chap; exp vt *The Arrow and Two Other Stories* coll 1927 UK); *The Swiss Family Manhattan* (**1932**).

MÓRRÍGAN The Celtic GODDESS of war and slaughter, usually represented as the triple goddess Mórrígú (or Macha), Badb and Nemain, of which Mórrígú was the most powerful manifestation. The Mórrígan was usually represented by a hooded crow, the symbol later associated in FOLKLORE with the banshee, both images presaging DEATH. The triple identity of the Mórrígan is also seen in the Three Witches in SHAKESPEARE's *Macbeth* (performed 1606; **1623**). In Irish legend the Mórrígan appears in the story "Táin Bo Cuailnge" in the *Cúchulain Saga* of the **Ulster Cycle**, where she tries to win the love of CUCHULAIN: but when he spurns her she seeks his death. This story is retold by Morgan LLYWELYN in *Red Branch* (**1989**; vt *On Raven's Wing* 1990 UK) and by Rosemary SUTCLIFF in *The Hound of Ulster* (**1963**), where she appears as Maeve. This manifestation is also known as Mab or Medb, the Warrior Queen of Connacht, whose adventures are retold by James STEPHENS in *In the Land of Youth* (coll of linked stories **1924** UK). Maeve or Mab is sometimes known as the FAIRY QUEEN, which is more akin to the benign aspect of MORGAN LE FAY. Morgan is frequently referred to as the Mórrígan in recent Arthurian fiction (◊ ARTHUR). The Mórrígan, depicted as a wicked witch-like figure, is used with considerable effect by Alan GARNER in *The Weirdstone of Brisingamen* (**1960**) and by Pat O'SHEA in *The Hounds of the Mórrígan* (**1985**). [MA]

MORRESSY, JOHN (1930-) US writer and professor of English. He began his career as a mainstream novelist in 1966, switching to sf in the early 1970s. After a number of sf novels, and at the prompting of an editor, he decided to try his hand at GENRE FANTASY with two sequences: the dark-toned **Iron Angel** series, about three brothers on a magical QUEST – *Ironbrand* (**1980**), *Graymantle* (**1981**), *Kingsbane* (**1982**) and *The Time of the Annihilator* (**1985**) – and the determinedly lighthearted **Kedrigern** series, about a reluctant WIZARD and his family and friends – *A Voice for Princess* (**1986**), *The Questing of Kedrigern* (**1987**), *Kedrigern in Wanderland* (**1988**), *Kedrigern and the Charming Couple* (**1990**) and *A Remembrance for Kedrigern* (**1990**). Although proficient, JM's work in either mode has made no major impact. [DP]
Other work: *Other Stories* (coll **1983**).
MORRILL, ROWENA ◊ ROWENA.
MORRIS, JANET E(LLEN) (1946-) US writer who began publishing fiction with the ambitious sf **Silistra** sequence in 1977, and who has since published much sf and fantasy, mostly since 1982 in various forms of collaboration. Although her early sf possesses elements redolent of fantasy – the **Silistra** novels recall SWORD AND SORCERY, while the civilization in the **Dream Dancer** trilogy (**1980-82**) resembles that of Hellenistic Greece – JEM has written fantasy only in two SHARED-WORLD enterprises. The first is the **Tempus** series, based on the **Thieves' World** enterprise: *Beyond Sanctuary* * (**1985**), *Beyond the Veil* * (**1985**), *Beyond Wizardwall* * (**1986**), *Tempus* * (coll of linked stories **1987**), *City at the Edge of Time* * (**1988**) with Chris Morris, *Tempus Unbound* * (**1989**) with Chris Morris and *Storm Seed* * (**1990**) with Chris Morris. With C.J. CHERRYH, JEM created the **Heroes in Hell** enterprise: *Heroes in Hell* * (anth **1986**) with Cherryh, *Rebels in Hell* * (anth **1986**) with Cherryh, *Masters in Hell* * (anth **1987**), *Kings in Hell* * (**1987**) with Cherryh, *The Gates of Hell* * (fixup **1986**) with Cherryh, *Angels in Hell* * (anth **1987**), *Crusaders in Hell* * (anth **1987**), *War in Hell* * (anth **1988**), *Explorers in Hell* * (**1989**) with David DRAKE, *The Little Helliad* * (**1988**) with Chris Morris, *Prophets in Hell* * (anth **1989**). [GF]

Other works: *I, the Sun* (**1983**), historical novel.
MORRIS, (MARGARET) JEAN (1924-) UK writer who began her career with *Man and Two Gods* (**1953**), some plays and a series of detective novels as by Kenneth O'Hara, starting with *A View to a Death* (**1958**). As JM she began publishing YA fantasy novels with *The Path of the Dragons* (**1980**), which remains her best-known. Set in a richly conceived, technologically advanced ATLANTIS, it depicts – with a complexity reminiscent of the work of Ursula K. LE GUIN – the relation of the Atlantids to the DRAGONS who plough the skies, wise and inscrutable; and to the activities of men and gods in the throes of enacting – perhaps for the first time – the MYTHS which underlie the Greek PANTHEON. *Twist of Eight* (coll **1981**) contains REVISIONIST FANTASIES, including examinations of CINDERELLA and "True Thomas". *The Donkey's Crusade* (**1983**), set in the ostensibly Christianized context of a monkish QUEST for PRESTER JOHN, transforms the ASS into a wise COMPANION who is both TALKING ANIMAL and savant, and who guides the humans in his care into the LAND OF FABLE of the East. In *The Troy Game* (**1987**) a young man is sent on a quest by a MAGUS named Mennor (he resembles MERLIN) which ends – after a WILD-HUNT episode – in a mysterious EDIFICE in the heart of the "troy maze" (◊ LABYRINTH); the protagonist's RITE OF PASSAGE into wisdom is cogently presented. JM remains a central crafter of tales for her demanding market. [JC]
Other works: *Song Under the Water* (**1985**); *The Paper Canoe* (**1988**); *A New Magic* (**1990**) and its sequel, *A New Calling* (**1992**).
MORRIS, KENNETH (VENNOR) (1879-1937) UK writer born and raised in Wales; he was active in THEOSOPHY all his adult life, and was employed by the Universal Brotherhood and Theosophical Society in California 1908-30, there doing almost all of his creative work. Though he wrote three novels, 40 stories, a number of plays and considerable nonfiction, KM published only two books beyond Theosophical circles, and seemed indifferent to worldly success. This obscurity was deepened by his use of a wide range of pseudonyms for his short fiction, including C. ApArthur, Walshingham Arthur, Aubrey Tyndall Bloggsleigh, Floyd C. Egbert, Hank in Maggs, F. McHugh Hilman, Ambrosius Kesteven, Maurice Langran, Fortescue Lanyard, Venon Lloyd-Griffiths, Jefferson D. Malvern, Patton H. Miffkin, Bingham T. Molyneux, Sergius Mompesson, Evan Gregson Mortimer, Ephraim Soulsby Paton, Quintus Reynolds, Evan Snowdon, Wentworth Tompkins and Thomas J. Wildredge. His first novel appeared as by Cenydd Morus. It is only in the 1990s that his collected stories and his last novel have appeared.

Because of the range of names used, it is not easy to identify his first story, but Douglas A. Anderson, who edited *The Dragon Path: Collected Stories* (coll **1995**), thinks KM began publishing fiction with "Prince Lion of the Sure Hand: A Story for Children" for *The Crusader* in 1899. The stories assembled as *The Secret Mountain and Other Tales* (coll **1926**) were selected from work written after 1914, and generally promulgate Theosophical principles with unobtrusive and supple tact. The world of the senses can be understood as manifesting a stage of meaning, a phase in the progression towards a higher reality, but with full REALITY always immanent. Mortality is generally contrasted with IMMORTALITY as a matter of choice: mortals choose to serve, immortals choose transcendence. The settings run from the

Wales of CELTIC FANTASY, with which KM is most closely associated, through Greece, Rome, India and China. Sages and other LIMINAL BEINGS offer gnomic advice, and point the way upwards. The narratives tend to serenity.

He is best-remembered for the **Pwyll** sequence of Celtic fantasies: *The Fates of the Princes of Dyfed* (**1914**) as by Cenydd Morus and *Book of the Three Dragons* (**1930**), only the first two-thirds of the latter tale being so far published. That tale is initially a closely TWICE-TOLD (but ultimately very divergent) recasting of the first and third branches of the MABINOGION, and describes the long, slow, tortured coming to greatness and wisdom of Pwyll Pen Annwn, the mortal king of Dyfed and a HERO whose life is inextricably entangled with the GODS. The setting is in part a LAND-OF-FABLE Wales, in part FAERIE; the two realms CROSSHATCH constantly, and it is often unclear whether Pwyll is in mortal or immortal countries, whether he is in the Three Islands of the Mighty (◊ ARCHIPELAGO) or in the Country of the Immortals. Certainly the Wales he dominates is so interfused with MAGIC, and so tied in its weal to the fate of the heroes who rule it, that it may be considered a genuine LAND, almost as though it were a full SECONDARY WORLD.

Pwyll himself is radiant with bravery and compassion, but all too mortal. He is doomed to fall in love with and – after a year-long QUEST for a magic basket with which he will entrap the elf-like god who has claimed her – wed the goddess Rhiannon, one of the two versions of the GODDESS in this long tale; the other, Ceridwen, who wears Rhiannon's semblance at one point in order to test Pwyll, is "the foster-mother of the Immortals, and queen of all the green things in the world".

KM's style is "bardic", but without any descents into sentimentality or bathos; his control of long prose rhythms is perhaps unmatched in 20th-century fantasy literature. When KM leaves Celtic fantasy, his style becomes more liquid, faster, but dense. It is in this mode that he wrote his last novel, the posthumous *The Chalchiuhite Dragon: A Tale of Toltec Times* (**1992**); set in an Edenic pre-Columbian POLDER called Hutiznahuac, this tells of the coming to wisdom of the Toltec Topiltzin, who becomes a philosopher king after causing the assassination of Nopalitzin, whose son he adopts, and who turns out to be the REINCARNATION of the god Quetzalcoatl, who will bring even greater illumination to the land. KM's influence has been small, but he is central to the genre.   [JC]

Other works: *On Verse, "Free Verse", and the Dual Nature of Man* (**1924** chap); *Golden Threads in the Tapestry of History* (1915-16 *The Theosophical Path*; **1975**); *Through Dragon Eyes* (1915-16 *The Theosophical Path*; coll **1980** chap).

Further reading: *Lloyd Alexander, Evangeline Walton Ensley, Kenneth Morris: A Primary and Secondary Bibliography* (**1981**) by Robert H. Boyer and Kenneth J. Zahorski.

MORRIS, WILLIAM (1834-1896) UK author, poet, artist and designer, associated with the PRERAPHAELITES but best-remembered in general for furniture, wallpaper and fabric designs. Much of his many-sided working life was spent reacting against or retreating from a perceived debasement of Victorian popular taste by mass-production techniques in the wake of the Industrial Revolution. Outrage at the demolition and philistine restoration of historic buildings led him into politics – specifically, revolutionary Marxist socialism. Like most pioneer fantasists WM has been criticized as escapist, but in fact his lesser works are marked by insufficient ESCAPISM, by lack of full engagement with the imagined world. Thus his immense poetry cycles, looking far back to mythic times, are technically proficient but now seem curiously bloodless and remote. *The Defence of Guenevere and Other Poems* (coll **1858**) reworks some of the ARTHUR themes, though its best poems are snapshot accounts of tragic incidents in a brutal and materialist Middle Ages, while *The Earthly Paradise* (coll **1868-70** 3 vols) deals extensively in Greek MYTH – incorporating for example WM's *The Life and Death of Jason* (**1867**), which alone runs to an exhausting 10,000 lines. Still longer is *Sigurd the Volsung and the Fall of the Niblungs* (**1876**).

WM's first prose fantasy of any note is the short, confused "The Hollow Land" (1856 *Oxford and Cambridge Magazine*): an unjust knight enters an earthly paradise, and then departs in a striking disjunction that leaves him aged (clothes decayed, helmet full of earth and worms); he finally regains the LAND through devotion to pictorial art.

The first step towards the characteristic large-scale fantasies which have had such influence on the genre – and, indirectly, on sf – is *The House of the Wolfings* (**1889**). Here the setting is quasi-historical: a European Saxon community is resisting the decadent advances of late-Imperial Rome. The romantic-supernatural story contains a large admixture of verse. What later critics were to call "the Teutonic thing" or "the Northern thing" continued in *The Roots of the Mountains* (**1890**), another tale of a tribal community whose historical context is less definite.

The Story of the Glittering Plain, or The Land of Living Men (**1890**) is the first of WM's novels to be fantasy in something like the modern sense. Despite Nordic names and emblems the setting has an OTHERWORLD hue; elements of FAIRY-TALE morality are present. The Glittering Plain is a supposed UTOPIA, difficult to attain, whose inhabitants enjoy IMMORTALITY. Hallblithe, the hero, reaches it by sea with ominous ease and soon finds counterbalancing dystopian qualities: moral aridity and a deviously manipulative king. With this shift in PERCEPTION, the land transforms from goal to prison; the urge is to leave, which is not permitted and will also cancel any rejuvenation (◊◊ SHANGRI-LA). Although Hallblithe successfully returns to his homeland and wife-to-be, the subtle, continuing, unharrowed WRONGNESS of the Plain makes the story seem incomplete.

WM's prose style, based largely on Thomas MALORY's, was now well developed. An effect of period freshness is given by long streams of conjunction-linked descriptive phrases and clauses built from relatively simple words – though not all readers can tolerate the earlier speeches' high medieval syntax. Later this would be slightly but usefully toned down.

In *The Wood Beyond the World* (**1894**) a sea voyage again separates the more fantastic realms from the hero Walter's mundane home town, though the land of the Wood sends visions even there – of the land's witchy Mistress, her enslaved Maid, and a hideous, savagely energetic DWARF servitor. This is a summoning: when a storm blows his ship to unknown shores, Walter defies all advice and reason, abandons his fellows, and sets off through mountains and wastes to the Wood where he can meet the mysterious three . . . an unwelcome fourth being the Mistress's current paramour, whom she wishes Walter to replace and who like Walter has his eye on the Maid. After some mild titillation, the stage seems set for triangular games of love and power. But, perhaps too rapidly, Walter and the Maid escape with the aid of her own small magics, leaving the others variously dead. The happy ending is a distinct *non sequitur*, involving a hitherto unmentioned CITY which needs a king and has vowed to appoint the first outsider to arrive – i.e., Walter.

WM's acknowledged masterpiece is *The Well at the World's End* (**1896**), presenting a unified fantasy geography (which is a clear forerunner of J.R.R. TOLKIEN's kind of SECONDARY WORLD) rather than conventionally setting a DREAM- or magical land on unknown shores of the real world's sea. The lofty DICTION has been smoothed nearer to transparency, minor characters possess personalities and motivations aside from their roles in the story, and the previous two books' flat backdrop becomes a functioning medieval society.

A brief synopsis suggests fairytale simplicity: Ralph – youngest son of the minor king of Upmeads – sets out to win fame, after long effort achieves the ultimate goal of drinking from the youth-giving and life-prolonging Well at the World's End ($\Diamond$ FOUNTAIN OF YOUTH), and, with various evil lords overthrown, returns home with a worthy consort. There are many complications; for example, Ralph's idyll with another and much older woman (restored by the Well), whose own inset story of past BONDAGE to a WITCH is a recurring WM theme.

Meanwhile the landscape is unrolled with painterly delight and an endearing, rain-washed clarity. It is a long, varied haul from Upmeads to the furthest point where merchants care to travel, and then onward to the ultimate-sounding Utterness and Utterbol (whose Lord is a memorably unpleasant tyrant). But beyond again is the cloud-piercing mountain range called the Wall of the World, on whose far side lie numinous regions. Conventional MAGIC is shown as chancy and elusive; however, in the approaches to the "Ocean Sea" and the Well, the scenery itself carries an increasing magical charge. In the heart of the Thirsty Desert, with its thickening litter of corpses who lacked the proper TALISMAN, we find the venomous pool surrounding the leafless Dry Tree . . . which has been so often mentioned and invoked as a symbol in the earlier narrative that the name itself has gained power.

The homeward journey to Upmeads after the drink of life is something of a triumphal progress. Wrongness and injustice are everywhere corrected – or are found already corrected, Ralph's outward trip having often proved catalytic. Thus the Lord of Utterbol was dealt with by a bondsman Ralph acquired in battle and then abandoned for the QUEST of the Well. Finally Ralph is able to cash in on the very first favour owed him since his travels began, to aid a military action which redeems embattled Uplands itself. This closing of the circle produces a deeply childlike satisfaction; J.R.R. TOLKIEN followed WM's example in the post-climactic chapters of *LOTR*.

The final two WM fantasy novels were posthumously published. Each in its own way steps back from that mesmeric unfolding of invented landscape in *The Well at the World's End*. The ISLANDS of *The Water of the Wondrous Isles* (**1897**) lie on a great lake and provide "marvellous" tableaux and encounters rather than seeming part of any unified world, while *The Sundering Flood* (**1897**) is set in a rather small and closely mapped medieval realm whose "flood" is a mere river.

A central device in *The Water of the Wondrous Isles* is the Sending Boat, a SPIRIT-driven barge which, propitiated with blood and a spell, follows a set course over the lake. This is

the escape route taken by the heroine Birdalone, kidnapped and enslaved since childhood by a SHAPESHIFTING witch who punishes disobedience with TRANSFORMATION into dumb-animal form. The theme of young women's enforced servitude to witches seems something of a WM MAGGOT; it features twice in this book. The lake and its ARCHIPELAGO of five isles are indeed magical, but on the lake's far shore the story lapses into faintly ridiculous excesses of CHIVALRY. Three KNIGHTS whose ladies are imprisoned on one island have taken decisive rescue action by spending several presumed years constructing a lakeside castle, and the ladies' tokens which Birdalone brings from isle to castle lead to rhapsodies extending courtly love into something like clothes fetishism. (It is a refreshing complication that one knight switches his attentions to Birdalone.) There is an interestingly perverse episode where the knights, though falling slowly into the seductive toils of a Witch-Queen, repeatedly search her island for their women – who all along are cursed with INVISIBILITY and chained to pillars as impotent observers in the bad Queen's hall.

This book's fantastical isles were surely in C.S. LEWIS's mind during the writing of *The Voyage of the Dawn Treader* (**1952**). Repeated visits apparently lead to THINNING: the last Sending Boat journey finds several of the isles now mundanely populated (a puzzle, since the Water seems otherwise unnavigable), and the daemonic ferry fails at last with the witch's offstage death.

In *The Sundering Flood* the fantasy elements seem dispensable: the hero's and heroine's childhood dealings with DWARF folk (so-called, but actually more like brownies) are of little import, the hero could as plausibly have become a renowned soldier without his ageless warrior MENTOR and magic SWORD Board-cleaver, and so on. As a PLOT DEVICE, the supposedly impassable "flood" dividing the young lovers offends common sense: spears and even clothing can be thrown across it, yet no one thinks of ropes. In keeping with the author's socialist inclinations, the final struggle is of the "Small Crafts and the lesser commons" (aided by good knights) against a corrupt monarchy, which is abolished. It reads well despite structural longueurs, but is not a magical novel.

WM's principal fantasy heritage is the indefinitely extensible QUEST in which the LANDSCAPE itself plays a major character part; the sense of protracted journeying is buttressed by sheer length of narrative. He also gently subverted the contemporary tendency to idealize the "parfit gentil knight" – working back towards older root texts in which heroes can have ordinary human quirks and foibles in between the extreme poles of tragic flawedness and icy perfection. Thus, more than once and without particular moral alarm, a WM hero sleeps with another woman before attaining true marital bliss. Though this permissiveness does not quite extend to heroines, WM shows a basic honesty and uncoyness about SEX (a notable exception being the knightly woman-worship of *The Water of the Wondrous Isles*).

C.S. Lewis and J.R.R. Tolkien both acknowledged the influence of WM: Lewis wrote appreciatively, "No mountains in literature are as far away as distant mountains in Morris" – an effect of panoramic vastness to which Tolkien is frequently seen to aspire (one thinks also of Aslan's Country in Lewis). When the occasional oddities of WM's diction fade in the mind, that huge vista remains.

Also crucial to the work of his last years was his overseeing of the elaborate fine-art printing of the Kelmscott Press,

which issued a variety of medieval TAPROOT TEXTS. [DRL]
Other works: *A Dream of John Ball, and A King's Lesson*
(coll **1888**; *A Dream of John Ball* reissued alone **1915** US),
part verse, a medieval dream tale involving prophecies of the
future of socialism; *News from Nowhere, or An Epoch of Rest*
(**1890** US; rev **1891** UK), utopian socialist sf; *Child
Christopher* (**1895**); *The Collected Works of William Morris*
(**1910-15** 24 vols) ed May Morris (WM's daughter), supplemented by her *William Morris, Artist, Writer, Socialist*
(**1936** 2 vols); *The Letters of William Morris to his Family and
Friends* (**1950**) ed Philip Henderson.
As translator: *Aeneid* (**1875**); *Three Northern Love Stories*
(**1875**), from the Icelandic, with E. Magnusson; *Odyssey*
(**1887**).
Further reading: *Life of William Morris* (**1889**) by J.W.
Mackail; "William Morris" in *Rehabilitations and Other
Essays* (**1939**) by C.S. Lewis; *William Morris: Romantic to
Revolutionary* (**1955**) by E.P. Thompson; *William Morris:
His Life, Work and Friends* (**1967**) by Philip Henderson; *The
Work of William Morris* (**1967**) by Paul Thompson.

MORRISON, GRANT (1963-   ) Scottish writer of
numerous successful, critically acclaimed and sometimes
controversial COMICS. His plays *Red King Rising* (1989),
about Lewis CARROLL, and *Depravity* (1990), about Aleister
CROWLEY, were performed at the Edinburgh Fringe
Festival, the former winning the Fringe First and
Independent awards, the latter the *Evening News* Award.
    GM began his career in 1978 writing and drawing "Time
is a Four-Lettered Word" for the short-lived experimental
magazine *Near Myths*, for which he also produced two tales
about the **Jerry Cornelius**-like **Gideon Stargrave** (◊
Michael MOORCOCK). He worked for two years on a local
newspaper strip, **Captain Clyde**, while simultaneously
writing sf stories for D.C. Thomson's **Starblazer** pocket
library. He produced a text piece, "The Stalking" (*Batman
Annual 1986* **1985**), and a steady stream of stories for *2,000
A.D.*, *Spider-Man and Zoids* and *Dr Who Monthly* before his
first major success with **Zenith** (*2,000 A.D.* 1987-92; graph
part coll 5 vols **1988-90**), about a pop-star superhero in an
ALTERNATE WORLD where such superheroes were developed
during WORLD WAR II. Beginning as a relatively routine
story pitting superheroes against an occult Nazi menace
and H.P. LOVECRAFT horrors, the series became darker and
more apocalyptic, ending in a spectacular destruction of the
Earth.
    GM's work for *2,000 A.D.* brought him to the attention of
the US market, notably DC COMICS, for whom he revamped
Animal Man (*#1-#26* 1988-90; graph part coll **1991**), turning the hitherto unsuccessful character into a fresh,
environmentally friendly superhero, which allowed GM to
display his fine senses of humour and pathos. The stories
grew more abrasive and even outrageous as the saga progressed, ending with a sequence which introduced GM
himself as The Creator – Animal Man and others discovering their true BONDAGE as mere comic-strip heroes. GM also
applied his energies to **Doom Patrol** (*#19-#63* 1989-93;
graph part coll **1992**), slowly remoulding the group from a
bunch of misfits whose basic *raison d'être* had been to tackle
weird, often ridiculous composite MONSTERS into a spoof of
superhero comics, culminating in the introduction of
Doom Force (*#1* 1992), a PARODY of much of his own writing on the series.
    GM wrote **The New Adventures of Hitler** (first
episodes in the short-lived *Cut* 1989; full story in *Crisis*

1990), which led to the resignation of columnist Pat Kane
and the editor amid allegations of Nazi sympathies because
of its portrayal of a youthful Hitler seeking the GRAIL in
Liverpool, and **St Swithin's Day** (*Trident #1-#4* 1989-90;
graph coll **1990**), concerning an unemployed young man
who fantasizes about killing Margaret Thatcher. The same
year GM's GRAPHIC NOVEL *Arkham Asylum* (graph **1989**
US) was published, in which BATMAN is portrayed as a
potential psychotic, as much in need of treatment as the
weird villains he has fought.
    GM is considered one of the best REVISIONIST writers: his
other work includes the radical updating of the 1950s sf
hero **Dan Dare** as **Dare** (*Revolver* 1990-1, *Crisis* 1991;
graph coll **1991**), the popular tv series *The* AVENGERS in
Steed and Mrs Peel (*#1-#3* 1990-1), and *Kid Eternity* (*#1-#3*
1991). His *Sebastian O* (*#1-#3* 1993) was a Victorian thriller
of Wildean flamboyance and excesses, which qualities characterize much of GM's writing. His approach has always
been dark-edged, strongly imaginative and steeped in the
occult. Nowhere is this more evident than in *The Mystery
Play* (graph **1994** US), a multi-level murder mystery in
which SATAN kills GOD, as portrayed by actors during a theatrical performance.   [SH/RT]

MORROW, GEORGE (1869-1955) UK cartoonist and
illustrator, active from about 1890, and long (1906-54) associated with *Punch*. He was essentially a humorist, and his
strokes were broad. Relatively little of his work was fantasy, though he illustrated several books of interest,
including *The Flying Carpet* (anth **1926**) ed Cynthia
ASQUITH, *Swollen-Headed William* (**1914**) by E.V. Lucas
(1868-1938), *The Death of the Dragon* (coll **1934**) by J.B.
Morton, all three of E.A. WYKE-SMITH's fantasy novels and
a 1932 edition of Jonathan SWIFT's *Gulliver's Travels*.   [JC]

MORROW, JAMES (KENNETH) (1947-   ) US writer.
His early novels are sf, although they deploy motifs which
really belong to fantasy: *The Wine of Violence* (**1981**) features
a fluid which soaks up aggression; *The Continent of Lies* (**1984**)
involves a method of controlled dreaming; the plot of *This is
the Way the World Ends* is haunted by the dark spectres of the
"unadmitted" – people who would have been born if only
nuclear war had not destroyed the world. JM's first full-
blooded fantasy, *Only Begotten Daughter* (**1990**), is a sceptical
and bitterly sarcastic CHRISTIAN FANTASY about a female MES-
SIAH who has to contend with a suave and mendacious SATAN
as well as the excesses of near-future fundamentalists. The
Nebula AWARD-winning *City of Truth* (**1991**) is a fabular
account of a CITY in which truth-telling is compulsory, and its
more comfortable but no less perverse SHADOW where lying
is not merely permitted but *de rigueur*. *Towing Jehovah* (**1994**)
is an extended FABLE in which a disgraced and guilt-ridden
ship's captain is given a chance to redeem himself when
selected by the dying ANGELS to tow the dead body of GOD to
its final resting-place – a journey made difficult by the great
size of the corpse, the determination of the Vatican to keep
God's death a secret, and the dedication of a group of militant
atheists to thwart the mission.
    In parallel with these later works JM has issued various
randomly numbered elements in a series of **Bible Stories
for Adults**, which set out righteously and wrathfully to
revise the moral precepts of the original tales according to
modern priciples of liberalism and tolerance. The Nebula-
winning "Bible Stories for Adults #17: The Deluge" (1988)
is in *Swatting at the Cosmos* (coll **1990**), but the parables as a
whole still await assembly.

JM is by far the most significant contemporary writer of religious fantasies. [BS]

MORTIMER, EVAN GREGSON [s] ◊ Kenneth MORRIS.

MORTON, ANTHONY [s] ◊ Robert ARTHUR.

MORUS, CENYDD [s] ◊ Kenneth MORRIS.

MORWOOD, PETER Pseudonym of UK writer Robert Peter Smith (1956-   ), married to Diane DUANE and resident in the Republic of Ireland for many years. His competent but routine adventure fantasies fall into several series. The Japanese-mythology-tinged **Alban Saga** comprises *The Horse Lord* (**1983**), *The Demon Lord* (**1984**), *The Dragon Lord* (**1986**) and *The Warlord's Domain* (**1989**). The pseudo-Russian historical **Prince Ivan** trilogy comprises *Prince Ivan* (**1990**), *Firebird* (**1992**) and *The Golden Horde* (**1993**). The **Clan Wars** series reverts to the setting of the first sequence some five centuries on: *Greylady* (**1993**) and *Widowmaker* (**1994**). *Keeper of the City* * (**1989**), with Duane, is a contribution to the **Guardians of the Three** SHARED-WORLD series. PM's other works, most co-written by Duane, are sf NOVELIZATIONS and media spinoffs. [DP]

MOSURA TAI GOJIRA ot of *Godzilla Versus Mothra* (*1964*). ◊ GODZILLA MOVIES.

MOTHER GOOSE Nursery rhymes go far back into the past, but until fairly recently were not recorded. Occasional lines can be retrieved from the Middle Ages onwards, but whole verses and nursery-rhyme collections were not published until the early and middle 18th century, and then hesitantly, the first noteworthy collection being *Tommy Thumb's Pretty Song Book* (*c*1745). In the late 18th and early 19th centuries (especially with the Romantic reevaluation of children and childhood), however, children's literature became an important area of publication.

MG is first associated with nursery rhymes in *Mother Goose's Melody* issued by Newbery of London. Its date is uncertain; according to *The Oxford Dictionary of Nursery Rhymes* (**1952**) by Iona and Peter Opie there was a lost 1765 edition; a 1791 edition survives. MG herself derives ultimately from French literature, where Charles PERRAULT used the traditional term *contes de ma mère oye* to designate FAIRYTALES. The original French term is of shadowy origin.

In the UK MG was only one of several persons associated with nursery rhymes, but in the USA she became the prime dispenser, a position which she still retains. This is largely due to the Boston publisher Munroe and Francis, which included in its extensive line of juvenile books *Mother Goose's Melodies* (**1833**). The largest and most resourceful collection of classical rhymes to date, it was selected from various sources and illustrated by US wood-engravers. This book was reissued many times in slightly varying editions and remained in print well into the 1860s (there have been occasional later reissues). Since then countless editions of the rhymes have appeared, sometimes set to music, sometimes elaborately and beautifully illustrated, but almost all essentially and ultimately based on the early Boston editions.

As for the verses, they are a coloured patchbag of UK literature – with a few US additions. They have served various functions, sometimes as pure entertainment, sometimes as instruction, sometimes as both. They include counting rhymes ("One, Two, Buckle My Shoe"), dance games ("London Bridge Is Falling Down", "Oranges and Lemons"), finger and toe plays ("This Little Piggy Went to Market"), lullabies ("Rock-a-Bye Baby"), riddles ("Humpty Dumpty"), simple educational mnemonics ("A was an Apple Pie") and animal compassion rhymes ("I Love Little

Pussy"). Some are deliberate NONSENSE ("The Cat and the Fiddle"). Not all were originally children's verse. Some may have been magical in origin ("Jack Be Nimble", "Rain, Rain, Go Away", "Arthur O'Bower"); others are infantilized political squibs ("Little Jack Horner", "Bobby Shaftoe"); and yet others are decayed popular adult ballads ("Oh, Dear, What Can the Matter Be", "Lucy Locket Lost Her Pocket" – the last said to name two celebrated courtesans).

Most of the verses are traditional, but a few are of recorded authorship. "Mary Had a Little Lamb" (*Poems for Our Children* coll 1830) was written by Massachusetts poet Mrs Sarah Hale (1788-1879) and "There Was a Little Girl Who Had a Little Curl" (late 1850s) by Henry Wadsworth Longfellow (1807-1882). It has also been argued that some of the 18th-century collections were compiled by Oliver Goldsmith (1728-1774), who is known to have loved and recited nursery rhymes. A few of the verses have pan-European counterparts, like "London Bridge is Falling Down" and "Eenie Meenie Minee Mo".

Originally many of the verses accompanied things to do but, when nursery rhymes shifted from oral transmission to books, dancing, finger plays, music and action games dropped away. Today, MG is really a browsing book selectively read to children, for many of the verses are generally disregarded as obscure or uninteresting. A central core, however, remains very much alive as a source of wonderful images, subjects so idiotic as to be brilliant, and great (sometimes inadvert) NONSENSE.

Parodies and imitations of the verses are innumerable in almost every area of Western culture – advertising, politics, military life, personal satire, etc. Individual verses have also served as the heart or focal point of many larger works of literature, particularly in modern times. Lewis CARROLL's use of "Humpty Dumpty", "The Queen of Hearts" and "The Lion and the Unicorn" is well known. The mystery writer Agatha Christie (1890-1976) structured *A Pocketful of Rye* (**1953**) on "Sing a Song of Sixpence", and has metaphorically invoked MG elsewhere, as in *Crooked House* (**1949**). John Le Carré (real name David Cornwell; 1931-   ) echoed the children's rhyme in the title of his *Tinker, Tailor, Soldier, Spy* (**1974**). [EFB]

Further reading: *The Annotated Mother Goose* (**1962**) by William and Ceil Baring-Gould; *Mother Goose's Melodies, Facsimile of the Munroe and Francis (1833) Edition* (**1970**) ed E.F. BLEILER.

MOTHRA VERSUS GODZILLA vt of *Godzilla Versus Mothra* (*1964*). ◊ GODZILLA MOVIES.

MOTIFS In *The Folktale* (**1946**), Stith Thompson (1885-1976), defined the motif as "the smallest element in a tale having a power to persist in tradition". He then divided motifs into three categories: (a) "actors" in a tale, who may be described as characters defined by what they do or who are dictated by the STORY they are in; (b) ICONS and customs that are material to the continuing action; and (c) single actions which "have an independent existence", and which can therefore be *repeated*, these being the most important of the three. Gary K. Wolfe, in *Critical Terms for Science Fiction and Fantasy: A Glossary and Guide to Scholarship* (**1986**), defined motifs as "recurrent or signal narrative events or figures, especially in folklore". [JC]

MOUSE AND HIS CHILD, THE US ANIMATED MOVIE (*1977*). ◊ Russell HOBAN.

MOWGLI Central character of a series of stories in Rudyard KIPLING's **Jungle Books** and one of the key images of FERAL

CHILDREN. Mowgli is nowadays most familiar by courtesy of the animated DISNEY musical *The Jungle Book* (*1967*), which eliminates the original stories' darkly ironic commentary on the dubious benefits of "civilization". [BS]

MOZERT, ZOE (? –) US artist. ◊ OLIVIA.

MU ◊ LEMURIA.

MUJICA LAINEZ, MANUEL (1910-1984) Argentinian novelist who wrote the libretto (after his own novel) for the OPERA *Bomarzo* (**1967**) by Alberto Ginastera, and in whose *El unicornio* (**1965**; trans Mary Fitton as *The Wandering Unicorn* **1982** Canada), Melusine tells of her passion for a young 12th-century KNIGHT, whom she follows to JERUSALEM, before the tale ends in tears. The story is told with an aerated lightness from a 20th-century perspective by the immortal FAIRY, though it laments the THINNING of Europe into secular nation states. [JC]

MULE ◊ ASS.

MULISCH, HARRY (1927-) Dutch writer, mostly of nonfantasy. However, *De toekomst van gisteren* ["The Future of Yesterday"] (**1972**), a book-length essay explaining why HM decided not to write a projected book of that title in which HITLER WINS, shows HM's familiarity with the devices and themes of modern fantasy. *Hoogste tijd* (**1985**; trans Adrienne Dixon as *Last Call* **1987** UK) verges on FABULATION; its climactic chapter alludes repeatedly to Edgar Allan POE's *The Narrative of Arthur Gordon Pym* (**1838**). The three long stories in *Oude lucht* (coll **1977**; trans Adrienne Dixon in *New Writing and Writers #17-#19* 1980-82) are variously fantasicated. The title story (trans as "Antique Air") ends with the protagonist passing into an AFTERLIFE; "De Grens" ("The Boundary") is a Franz KAFKA-like tale of bureaucracy overtaking a man whose wife has died on the borderline between two jurisdictions; "Symmetrie" ("Symmetry") is a meditation on the spirit of the 19th-century scientific romance. [GF]

MULREADY, WILLIAM (1786-1863) UK illustrator. ◊ ILLUSTRATION.

MULTIPLE PERSONALITY Modern psychotherapy recognizes multiple-personality disorder as a form of mental illness, often confused in common parlance with schizophrenia. Although rare, it has an obvious literary appeal, and is frequently invoked as a device in murder mysteries, as in *The Florentine Dagger* (**1923**) by Ben HECHT and *Methinks the Lady* (**1945**) by Guy ENDORE. It is a particularly common device in the cinema, where the influence of the dramatized case-study *The Three Faces of Eve* (*1957*) and Alfred HITCHCOCK's *Psycho* (*1960*) was carried forward to implausible thrillers like *Color of Night* (*1994*). Some TIMESLIP stories have an MP element, after the fashion of *Charlotte Sometimes* (**1969**) by Penelope FARMER. A particularly extreme example is featured in "Schizoid Creator" (1953) by Clark Ashton SMITH. [BS]

See also: BEWITCHED (*1945*); Wilkie COLLINS; DOUBLES; EYES OF LAURA MARS (*1978*); IDENTITY EXCHANGE; INVISIBLE COMPANIONS; POSSESSION.

MULTIVERSE A term coined by Michael MOORCOCK (also, earlier and independently, by John Cowper POWYS) to describe a universe consisting of innumerable ALTERNATE WORLDS, all intersecting, laterally and (PALIMPSEST-fashion) vertically. Some of these parallel worlds operate according to sf premises; some – like the worlds in which various AVATARS of Moorcock's **Eternal Champion** series play out their linked destinies – operate in fantasy terms. Worlds governed by incompatible premises are not, however, barred

from one another, and in this sense the overall concept belongs more properly to fantasy than to sf; Moorcock himself treats his extremely large and varied oeuvre as though all its venues occupy niches in the one multiverse. Novels in which TEMPORAL ADVENTURESS figures appear – Moorcock and others have written them – normally allow these figures free access to various REALITIES within the multiverse. The "polycosmos" featured in John GRANT's work is analogous, but is depicted as having physical REALITY, while one character is capable of travelling outside its constituent universes to view the polycosmos "from above".

Moorcock is not much interested in PORTALS or other fantasy devices of the sort. In his works, shifts from one world to another, whether or not voluntary, are normally signalled by shifts in PERCEPTION. To perceive a new world in the multiverse is to inhabit it. [JC]

MUMMIES Mummies provide a particularly striking imaginative link between present and past, and they are frequently invoked as key images in TIMESLIP fantasies and tales of REINCARNATION. Early uses of the motif include the bizarre futuristic fantasy *The Mummy! A Tale of the Twenty-Second Century* (**1827**) by Jane Loudun (1807-1858) and the ironic "Some Words with a Mummy" (1845) by Edgar Allan POE. Théophile GAUTIER's *The Romance of a Mummy* (**1856**) and "The Mummy's Foot" (1863) are prime examples of the mummy's capacity to act as an imaginative stimulus. Reanimated mummies are, of course, a staple of cheap HORROR MOVIES, but tales displaying a more earnest fascination, including tales of revivification which are more sentimentally than horrifically inclined, enjoyed a considerable vogue at the end of the 19th century. Notable examples include *Pharaoh's Daughter* (**1889**) by Edgar Lee, *Iras: A Mystery* (**1896**) by Theo Douglas (Mrs H.D. Everett), *The Prince of Gravas* (**1898**) by Alfred C. Fleckenstein, *An Egyptian Coquette* (**1898**) by Clive Holland (1866-1959) and *The Mummy and Miss Nitocris* (fixup **1906**) by George Griffith (1857-1906). Griffith's *The Romance of Golden Star* (**1897**) features a rare South American mummy, while Nitocris is revived again in *Nile Gold* (**1929**) by John Knittel (1891-1970), the best of the Egyptian REVENANT tales. The romantic spirit of these works is echoed in Anne RICE's *The Mummy, or Ramses the Damned* (**1989**), but "Nofrit" (1947) by Eliot CRAWSHAY-WILLIAMS treats the theme with an ironic suspicion. The 1890s mummy boom was roundly sent up at the time by C.J. Cutcliffe Hyne (1866-1944) in "The Mummy of Thompson-Pratt" (1904), but it also extended to offbeat melodramas like "The Ring of Thoth" (1890) and "Lot No. 249" (1892) by Arthur Conan DOYLE and the confused and ultimately selfcontradictory *Pharos the Egyptian* (**1899**) by Guy Boothby (1867-1905). Sax ROHMER's mummy stories – of which the best is *The Brood of the Witch-Queen* (**1918**) – are thrillers of a similar stripe. Algernon BLACKWOOD's "The Nemesis of Fire" (1908), in which a mummy is linked to a fire ELEMENTAL, is more interesting than the work which presumably inspired it, Bram STOKER's incoherent *The Jewel of Seven Stars* (**1903**). Most recent works that feature mummies are horror, but *The Third Grave* (**1981**) by David Case and *Cities of the Dead* (**1988**) by Michael Paine are among those which also warrant consideration as fantasy. Two theme anthologies are *Mummy!* (anth **1980**) ed Bill Pronzini and *Mummy Stories* (anth **1990**) ed Martin H. GREENBERG. [BS]

MUNCHHAUSEN The figure of the fabulous Baron Munchhausen (or Munchausen), a traveller and recounter of

TALL TALES, has a bibliographic history as tangled and disreputable as the Baron's incredible accounts. The appearance of the anonymous *Baron Munchhausen's Narrative of his Marvellous Travels and Campaigns in Russia* (chap dated 1786 but **1785** UK; exp 1785; exp 1786; exp vt *Gulliver Revived* 1786; many other revs and exps) created the vogue for tales of the incredible and high-spirited German campaigner. Earlier versions of the first 17 tales had previously been published in two unsigned pieces in an improper Berlin periodical, *Vade Mecum fur Lustige Leute*, in 1781 and 1783. Hieronymus Karl Friederich, Freiherr von Munchhausen (1720-1797), a retired army captain who had long amused dinner guests with his straight-faced narration of fabulous supposed adventures, found his later years made miserable by this unsought notoriety. The author of the Oxford-published pamphlet (who probably also wrote the *Sea Adventures* in the 3rd edn) has been identified as Rudolf Erich RASPE, a brilliant German scientist and courtier who had fled in disgrace to England in 1775 after embezzling from collections of which he had been curator. Raspe had been acquainted with a kinsman of Munchhausen, and may have met the man.

By the time *Gulliver Revived*, announced as the third edition but now known to have been the fourth, was published, Raspe was no longer in control of his creation. Translations, further expansions, and SEQUELS BY OTHER HANDS followed for nearly a century, and Raspe's vigorous prose was slowly smothered by inferior accretions. This in no way impeded the work's popularity, and the 19th century saw editions illustrated by (among others) Thomas Rowlandson (1756-1827), George CRUIKSHANK and Gustave DORÉ. The tradition continues: a 1969 US edition is "profusely illustrated" by Ronald SEARLE. Munchhausen's tales have been the source of several movies, of which the most noteworthy is Terry GILLIAM's *The* ADVENTURES OF BARON MUNCHAUSEN (*1989*). [GF]

MUNDY, TALBOT Pseudonym of UK-born writer William Lancaster Gribbon (1879-1940), resident in the USA from 1909. His early life is obscure, though it is known that during the decade before 1909 he was a petty criminal, ivory poacher and confidence trickster in Africa, being twice imprisoned. Soon after his emigration to the USA he published his first story, "A Transaction in Diamonds" for *Adventure* in 1911. His first two novels, *Rung Ho!* (**1914**) and *King – of the Khyber Rifles* (**1916**), do not foreground any supernatural elements, though the latter features an AVATAR of the GODDESS, includes a trip through a LABYRINTH into the kind of UNDERGROUND world which features in many of his later books, is set in ORIENTAL-FANTASY venues and introduces characters like Athelstan King of the Secret Service. King features in *Caves of Terror* (1922 *Adventure* as "The Grey Mahatma"; **1924**), the first of the loosely linked sequence usually known as the **Jimgrim/Ramsden** series for its two main protagonists – James Schuyler Grim or Jimgrim, a US secret agent, and Jeff Ramsden, who generally serves as narrator: the book-length "Moses and Mrs Aintree" (1922 *Adventure*), *The Nine Unknown* (1923 *Adventure*; **1924**), *The Mystery of Khufu's Tomb* (1922 *Adventure* as "Khufu's Real Tomb"; **1933**), *The Devil's Guard* (1926 *Adventure* as "Ramsden"; **1926**; vt *Ramsden* 1926 UK) and *Jimgrim* (1930-1 *Adventure* as "King of the World"; **1931**; vt *Jimgrim Sahib* 1953).

TM was a follower of THEOSOPHY, and the **mgrim/Ramsden** sequence is based on the Theosophical

supposition that various occult sciences and manifestations come into this world as evidences of an ancient, ATLANTIS-derived wisdom which the world has forgotten, guarded in the present by cadres of SECRET MASTERS; men like Jimgrim can aspire to grasp this wisdom through various QUESTS in which their integrity and toughness strike a chord in the wise men who supervise us. This relationship between the mundane present and an occluded past is perhaps most strikingly visible in *The Nine Unknown*, in which the eponymous secret masters (who had seemed evil in *Caves of Terror*) are revealed as manifestations of the Good and an opposing cadre of nine Kali worshippers attempts to do wrong by generating confusions between themselves and the genuines sages.

In later volumes, like *The Devil's Guard* – constructed as a QUEST for Shambala ($\Diamond$ SHANGRI-LA) – Jimgrim begins to climb the ladder of understanding, and the sequence becomes increasingly mystical. *Jimgrim* introduces many sf devices into a FANTASY-OF-HISTORY frame, and climaxes in Jimgrim's world-saving CHRIST-like self-sacrifice.

Other novels which exhibit a similar sense of the relationship between the present and the past include: *Om: The Secret of Abhor Valley* (**1924**); *Black Light* (**1930**); *Full Moon* (**1935**; vt *There Was a Door* 1935 UK); and *The Thunder Dragon Gate* (**1937**), which is associational, though its sequel, *Old Ugly Face* (**1940**), returns to a Tibet conceived in Theosophical terms as a fount of supernatural wisdom from long ago.

Many of TM's novels – perhaps 25 further titles in all – are tightly constructed adventure tales, mostly set in the East, and only intermittently infused with anything more than a hint of the fantastic. TM's other series known to genre readers is **Tros of Samothrace**, various segments of which were published 1925-35 in *Adventure*. It comprises (in terms of internal chronology) *Tros of Samothrace* (**1934** UK; vt in 4 vols as *Tros* 1967, *Helma* 1967, *Helene* 1967, and *Liafail* 1967; new vt in 3 vols as *Lud of Lunden* 1976, *Avenging Liafail* 1976 and *The Praetor's Dungeon* 1976), *Queen Cleopatra* (**1929**) and *Purple Pirate* (**1935**). There is SWORD AND SORCERY in the long saga, during which Tros battles against Rome in Britain and Egypt, though he finally aids the conquerors before setting off on his last voyage, during which he hopes to travel to the ends of the earth.

TM told implausible tales with a clean-cut momentum, and with a seductive attention to plausible detail. [JC]

MUNN, H(AROLD) WARNER (1903-1981) US occasional writer, remembered best for his early stories in WEIRD TALES. His first, "The Werewolf of Ponkert" (1925 *WT*), arose from a comment by H.P. LOVECRAFT suggesting a story written from the WEREWOLF's viewpoint. HWM's resulting tale became the first of a series, **The Tales of the Master**. The series included a serial, "The Werewolf's Daughter" (1928 *WT*), and this and the initial story appeared as *The Werewolf of Ponkert* (fixup **1958**). HWM later reworked the other stories and added extensively to the series, most of these tales appearing initially in Robert WEINBERG's **Lost Fantasies** series, and then in book form as *Tales of the Werewolf Clan, #1: In the Tomb of the Bishop* (coll of linked stories **1979**) and *#2: The Master Goes Home* (coll of linked stories **1979**).

HWM's other main achievement for *WT*, before family responsibilities took him away from writing for almost 30 years, was *King of the World's Edge* (1939 *WT*; **1966**). This starts in the last days of ARTHUR, and follows the adventures of Myrdhinn ($ MERLIN), Gwalchmai ($ GAWAIN) and a Roman centurion, who leave Britain for new lands to the west, and find themselves in the kingdom of the Aztecs. The sequel, *The Ship from Atlantis* (**1967**), remained unpublished for 26 years; it follows the further adventures of Gwalchmai, who sets out for Rome but becomes lost in the Sargasso Sea and encounters a survivor from ATLANTIS. These two novels, later combined as *Merlin's Godson* (omni **1976**), are a precursor to HWM's magnum opus, *Merlin's Ring* (**1974**), which explores the Atlantean and Arthurian influences down through history to the time of JOAN OF ARC. HWM was fascinated by the Maid, and wrote an extensive narrative poem about her, *The Banner of Joan* (**1975**). Although essentially nonfantastic – other than in Joan's spirit-driven zeal ($ POSSESSION) – the poem may be seen as an epilogue to the **Merlin** sequence.

HWM's only other published novel was *The Lost Legion* (**1980**), a sister to *King of the World's Edge* but set 400 years earlier at the time of Caligula.

Throughout his life HWM was fascinated with the occult, and was often known as the Warlock of Tacoma (where he lived in Washington State). This knowledge pervades all of his fiction, particularly the later stories. After his retirement he produced a number of stories for SMALL-PRESS magazines, especially WEIRDBOOK. He developed a new sequence that sought to link Lovecraft's CTHULHU MYTHOS stories with a MACHEN-esque ancient race of Pictish FAIRIES. The published tales are "The Merlin Stone" (**1977**), "The Stairway to the Sea" (**1978**) and "The Wanderers of the Waters" (**1981**), all in *Weirdbook*.

Every birthday and Christmas 1974-80 HWM issued a booklet as a gift. These were usually poetry, but several are short fantasies: *The Affair of the Cuckolded Warlock* (**1975** chap), *What Dreams May Come* (**1978** chap), *In the Hulks* (**1979** chap), *The Transient* (**1979** chap) and *The Baby Dryad* (**1980** chap). [MA]

Other works: Poetry and reflections published as *Christmas Comes to Little Horse* (**1974** chap), *Twenty-Five Poems* (**1975** chap), *To All Amis* (**1976** chap), *Season Greetings with Spooky Stuff* (**1976** chap), *There Was a Man* (**1977** chap), *The Pioneers* (**1977** chap), *In Regard to the Opening of Doors* (**1979** chap), *Dawn Woman* (**1979** chap), *Fairy Gold* (**1979** chap), *Of Life and Love and Loneliness* (**1979** chap) and *The Book of Munn, or A Recipe for Roast Camel* (**1979**).

MUNSTER, GO HOME US movie (*1966*). $ *The* MUNSTERS (1964-6).

MUNSTERS, THE US tv series (1964-6). Universal/CBS. **Pr** Joe Connelly, Bob Mosher. **Dir** Norman Abbott and many others. **Created by** Al Burns, Chris Hayward. **Writers** Tom Adair and many others. **Novelization** *The Munsters* * (**1964**) by Morton Cooper. **Comics adaptation** *The Munsters* (16 issues 1965-8) from Gold Key Comics. **Starring** Yvonne DeCarlo (Lily Munster), Fred Gwynne (Herman Munster), Al Lewis (Grandpa), Beverly Owen (Marilyn Munster 1964), Butch Patrick (Eddie Munster), Pat Priest (Marilyn Munster 1964-6). 70 30min episodes. B/w.

Like *The* ADDAMS FAMILY (1964-6), which debuted the same year, this featured a strange family who thought they were normal and the rest of the world unusual; here the family members bear more than a passing resemblance to Universal's past HORROR-MOVIE stars. Herman, a hulking giant who resembles Universal's Frankenstein monster ($ FRANKENSTEIN MOVIES), is actually a kind and timid soul. His wife Lily appears just to have stepped out of a coffin. Her father, Grandpa, looks and acts like DRACULA ($ DRACULA MOVIES), often turning into a bat between magic SPELLS. The son, Eddie, is a juvenile wolfman. The only person in the household who looks normal is niece Marilyn; the rest of the family, in an amusing twist, assume she is plain because of the reactions of any putative boyfriend she brings home.

Home, at 1313 Mocking Bird Lane, Mockingbird Heights, is a decrepit Gothic mansion that seems to be the centre of a perpetual storm. Complete with a dungeon laboratory and secret passages, it is also the domicile of an unseen pet with fiery breath.

A theatrical movie, *Munster, Go Home* (*1966*) features the Munsters as heirs to a piece of property out West – a ghost town, of course. This outing had the tv cast but with Marilyn played by Debbie Watson. [BC]

See also: *The* MUNSTERS' REVENGE (*1981* tvm); *The* MUNSTERS TODAY (1988-9).

MUNSTERS' REVENGE, THE US movie (*1981* tvm). ABC. **Pr** Arthur Alsberg, Don Nelson. **Exec pr** Edward J. Montagne. **Dir** Don Weis. **Screenplay** Arthur Alsberg, Don Nelson. **Starring** Sid Caesar (Dr Diablo/ Emil Hornshymler), Yvonne DeCarlo (Lily Munster), Fred Gwynne (Herman Munster), Bob Hastings (Phantom of the Opera), Al Lewis (Grandpa), Jo McDonnell (Marilyn Munster), K.C. Martel (Eddie Munster), Howard Morris (Igor). 120 mins. Colour.

The cast of *The* MUNSTERS (1964-6) was reunited for this forgettable outing, which pitted them against the evil Dr Diablo and his henchman, Igor. Even a visit from their cousin, the PHANTOM OF THE OPERA, fails to liven things up. [BC]

MUNSTERS TODAY, THE US syndicated tv series (1988-9). Arthur Company. **Pr** Lloyd J. Schwartz. **Dir** Norm Abbott and many others. **Writers** Ed Haas, Norman Liebman. **Based on** *The* MUNSTERS (1964-6). **Starring** Jason Marsden (Eddie Munster), Lee Meriweather (Lily Munster), Howard Morton (Grandpa), John Schuck (Herman Muster), Hilary Van Dyke (Marilyn Munster). 26 60min episodes. Colour.

15 years after Grandpa accidentally froze the entire family, the Munsters thaw out for this sequel. The setting is the same as the original, but all attempts to update the humour failed. [BC]

MUPPET CHRISTMAS CAROL, THE US movie (*1992*). $ *A* CHRISTMAS CAROL.

MUPPET SHOW, THE US syndicated tv series (1976-81). ITC Entertainment. **Pr** Jack Burns, Jim HENSON. **Exec pr** David Lazer. **Dir** Philip Casson, Peter Harris. **Writers** Joseph A. Bailey, Burns, Henson, Don Hinkley, Jerry Juhl, Chris Langham, Marc London, David Odell, James Thurman. Colour. 120 30min episodes.

When Henson offered a series based on his Muppets to ABC, they quickly declined, seeing little possibility for a series seemingly aimed solely at children. Luckily Lord Lew Grade disagreed, and financed the venture, which became extremely popular with children and adults. Set backstage in a vaudeville-style theatre, the series stars Kermit the Frog as the beleaguered MC trying to get a new show together each week. Each episode centres on a human guest star and features a number of musical skits (e.g., Elton John singing "Crocodile Rock" with a band of crocodiles) and comedy variety acts. There are several recurring themes, such as "Pigs in Space", featuring Miss Piggy and other porcine performers as they explore the galaxy aboard the S.S. *Swinetrek*. From up in the theatre balcony, critics Statler and

Waldorf can be counted on for a stream of heckling, all of which Kermit and his performers pretty much ignore.

TMS was notable for its costumes and settings; many skits featured dozens of characters and elaborate spfx. The series enjoyed a long run in syndication, and spun off various movies and the animated series *The Muppet Babies* (CBS 1984-92). [BC]

See also: A CHRISTMAS CAROL.

MURNAU, F.W. Pseudonym of Westphalian movie director Friedrich Wilhelm Plumpe (1888-1931); he died young in a car crash. No prints of his earliest movies – including *Der Januskopf* (*1920*; vt *Dr Jekyll and Mr Hyde*; ◊ JEKYLL AND HYDE MOVIES) – seem to survive. His two masterpieces of fantasy interest are *Nosferatu* (*1922*; ◊ DRACULA MOVIES) and FAUST: EINE DEUTSCHE VOLKSSAGE (*1926*). FWM's great talent was his "eye": his extant silent movies all show a beautiful power of observation. With the coming of sound – he went to Hollywood in 1927 – his technique became less effective. [JG]

MURPHY, DENNIS JASPER Pseudonym of Charles R. MATURIN.

MURPHY, JILL (1949-) UK writer. ◊ CHILDREN'S FANTASY.

MURPHY, PAT(RICE ANNE) (1955-) US writer whose first acknowledged fiction is "Nightbird at the Window" (in *Chrysalis 5* anth **1979** ed Roy Torgeson). Most of her work combines speculative elements in ways that place them near the borders of rather than at the centres of genres. *The Shadow Hunter* (1980 *IASFM* as "Touch of the Bear"; exp **1982**) is an sf novel in which a Stone-Age man is abducted by a TIME-TRAVEL project into a deracinated future. *The Falling Woman* (**1986**) uses TIMESLIP – an archaeologist perceives (and eventually interacts with) the long-vanished inhabitants of the Mayan dig she is overseeing in Mexico – to tell a story of reconciliation and psychic HEALING. The novel won the Nebula AWARD. *The City, Not Long After* (as "Art in the War Zone" in *Universe 14* anth **1984** ed Robert Silverberg; exp **1988**) is ostensibly set in an implausibly genial post-holocaust SAN FRANCISCO (◊ HOLOCAUST AND AFTER), in which the gentle artisans who populate the city defend it against invading fundamentalists.

Besides the theme of temporal displacement, PM's stories frequently centre on physical TRANSFORMATIONS, often between human and animal. *Rachel in Love* (1987 *IASFM*; **1992** chap), about a chimpanzee whose mind has been imprinted with the personality of a dead teenaged girl, employs both themes; it won both the Nebula Award and the Theodore Sturgeon Memorial Award. "His Vegetable Wife" (1986 *Interzone*) offers a twist on the theme by John COLLIER. "Points of Departure" (1995 *F&SF*), not included in the collection of that name, proposes lycanthropy as a metaphor for the changes overcoming a rejected woman. "Going Through Changes" (1992 *F&SF*) involves virtual reality in a story that nevertheless quickly becomes fantasy, again involving biological transformation as a metaphor for female self-realization.

PM's early fiction was collected in *Points of Departure* (coll **1990**), which won the Philip K. Dick Memorial Award. She has published relatively little since. Her fiction is assured and intelligent. [GF]

Other works: *Letters from Home* (anth **1991** UK) with Pat Cadigan and Karen Joy Fowler, containing one story by each author.

MURPHY, SHIRLEY ROUSSEAU (1928-) US

painter, sculptor and writer of myth-tinged otherworldly YA fantasies, notably the **Children of Ynell** sequence – *The Ring of Fire* (**1977**), *The Wolf Bell* (**1979**), *The Castle of Hape* (**1980**), *Caves of Fire and Ice* (**1980**) and *The Joining of the Stone* (**1981**) – and the **Dragonbards** trilogy – *Nightpool* (**1985**), *The Ivory Lyre* (**1987**) and *The Dragonbards* (**1988**). SRM's work in these series shows the influence of C.S. LEWIS's **Narnia** novels. Her fantasy singletons include *The Grass Tower* (**1976**), *Silver Woven in My Hair* (**1977**), *Soonie and the Dragon* (**1979**), *Valentine for a Dragon* (**1984**), *Medallion of the Black Hound*, with Welch Suggs (**1989**), and a number of picture books for young children. *The Catswold Portal* (**1992**), an ANIMAL FANTASY, is her first novel to be aimed primarily at adults. [DP]

MURPHY, WARREN B(URTON) (1933-) US writer, married to Molly COCHRAN, with whom he frequently writes, and known mainly for the long **Destroyer** sequence of action-oriented sf/fantasy/thrillers featuring a contemporary AVATAR of Shiva the Destroyer whose TALENTS include the ability to fuse his corporal body with other forms of matter. These novels, many with WBM's initial collaborator, Richard Ben Sapir (1936-1987), begin with *Created, the Destroyer* (**1971**) with Sapir; nearly 100 titles had been published by 1995. A detailed author/title breakdown of the sequence can be found in R. Reginald's *Science Fiction and Fantasy Literature 1975-1991: A Bibliography* (**1992**). [JC]

See also: SFE.

MURRAY, FRIEDA A. (1948-) US writer. ◊ Roland J. GREEN.

MURRAY, MARGARET (1863-1963) UK cultural anthropologist and folklorist, author of *The Witch Cult in Western Europe* (**1921**). ◊ ANTHROPOLOGY.

MUSÄUS, JOHANN KARL (AUGUST) (1735-1787) German academic and writer who was among the first to collect together local FOLKTALES, some derived from oral tradition but others drawn from written sources. These were published as *Volksmärchen der Deutschen* ["German Folk Tales"] (**1782-7** 5 vols; cut trans [attributed to William BECKFORD] *Popular Tales of the Germans* **1791** UK; vt *Popular Tales* 1826 UK), and included: "Richilde", an early version of the SNOW WHITE story; "Die Nymphe des Brunnens" ["The Nymph of the Fountain"], which involves THREE WISHES but is otherwise similar to the CINDERELLA tale; and other stories involving ENCHANTMENT, TRANSFORMATIONS and GHOSTS. The stories were immensely popular and highly influential. "Die Entführung" ["The Abduction"] suggested "The Legend of the Bleeding Nun", which Matthew Gregory LEWIS, who knew JKM, incorporated into *The Monk* (**1796**), while one of the episodes in "Legenden der Rübezahl" (usually translated as "Elfin Freaks, or the Seven Legends of Number-Nip"), featuring a headless horseman, provided inspiration to Washington IRVING for "The Legend of Sleepy Hollow" (1820). Many of JKM's stories were translated anonymously or misattributed. He died relatively young, and his role as a popularizer of folktales was soon eclipsed by Ludwig TIECK, Johann August APEL, Friedrich de la Motte FOUQUÉ and the GRIMM BROTHERS, all of whose work he helped inspire. A good selection of JKM's work was included in the first volume of *German Romance* (anth **1827** UK) ed Thomas CARLYLE; this included the foundation myth *Libussa* (1782; chap **1844** UK) and the GHOST STORY "Dumb Love" [ot "Stumme Liebe"] (1782; vt "The Spectre-Barber" in *Tales of the Dead* anth **1813**). JKM was briefly rediscovered during the UK's fairytale bonanza in the 1840s, which saw translations as *Legends of Rubezahl and Other Tales* (coll trans William Hazlitt the Younger [1811-1893] **1844** UK) and *Select Popular Tales from the German of Musäus* (coll trans **1845** UK). [MA]

MUSES In Greek MYTH, the nine daughters of ZEUS and Mnemosyne (memory); goddesses of SONG and inspirers of the arts. They comprise Calliope (the leader, and muse of epic poetry), Clio (history), Erato (LOVE poetry), Euterpe (MUSIC and lyric verse), Melpomene (tragedy), Polyhymnia or Polymnia (sacred or sublime verse), Terpsichore (dance), Thalia (comedy) and Urania (astronomy). Fantasy rarely invokes them, though Terry PRATCHETT's **Discworld** has eight Muses of its own, led by "Cantaloupe". Max BEERBOHM ironically uses Clio as a PLOT DEVICE in *Zuleika Dobson* (**1911**), granting Beerbohm as "historian" the TALENT normally reserved for novelists, of reading characters' inner thoughts. Terpsichore features in the movie *Down to Earth* (**1947**), the follow-up to HERE COMES MR JORDAN (**1941**). John FOWLES's FABULATION *Mantissa* (**1982**) incarnates Erato as muse of the modern novel for a violently funny male-female confrontation featuring much SEX in its ALLEGORY of novelistic inspiration. Sex is also the mainspring of Neil GAIMAN's "Calliope" in *Dream Country* (graph coll **1991**), whose ANTIHERO writer holds Calliope in BONDAGE and acquires inspiration by repeatedly raping her. [DRL]

MUSIC 1. Fantasy in music Although FANTASY is a term normally restricted to *narrated* material – thus excluding delusions, dream imagery, fancies, and other forms which employ the mode of the FANTASTIC – it is worth noting that the term has complex ramifications for music history.

It was during the 16th century that independent instrumental music became fully established in Europe; ricercars and "fantasias" were common in Italy and Spain by the 1530s. A fantasia was defined in 1595 by the composer Thomas Morley as being what happens "when a musician taketh a point at his pleasure, and wresteth it and turneth it as he list, making either much or little of it according as shall seem best in his own conceit". The fantasia gradually evolved into the classical sonata form, and therefore lies at the core of the dominant form of Western music from about 1770. In the 19th century some Romantic composers endowed the term with a meaning closer to that in literature, as in Hector Berlioz's *Symphonie Fantastique* (1830) (see below; ◊◊ Marvin KAYE), while Robert Schumann's choice of the title *Fantasiestücke* for four sets of purely instrumental pieces (plus five other works with "fantasy" in their titles) was inspired by *Fantasiestücke* (coll **1814**), a collection of essays on music by E.T.A. HOFFMANN.

Ever since the first secular music dramas were composed at the end of the 16th century, composers have explored the whole of pre-20th-century fantasy literature in operas, cantatas, oratorios, songs and ballets, as well as symphonic poems and other instrumental works. Some themes have continued to fascinate composers throughout these four centuries, in particular the stories of ORPHEUS and FAUST. The extra-musical bases for these compositions range from word-for-word settings of existing texts or literal evocations of a scenario, via ones that have been paraphrased, abridged, distorted, parodied, excerpted or collaged, to those that were specially devised, sometimes entirely by the composer.

The repertory of operas whose plots contain elements of the fantastic is huge; and for obvious reasons a high

proportion of music containing narrative and voiced fantasy lies in this category. (◊ OPERA.)

Cantatas and oratorios – certainly in the 17th and 18th centuries, when staged works were often thought blasphemous and consequently banned – are frequently operas without the physical action. The earliest dramatic cantatas include Alessandro Stradella's *L'anima del purgatorio* ["The Spirit of Purgatory"] (*c*1670), which features Lucifer, an angel and tormented souls, his solo cantata *L'Arianna* (*c*1670), Giovanni Battista Bassani's collection of ten solo amorous cantatas, *Armonie delle Sirene* ["The Harmony of the Sirens"] (**1680**) and Marc-Antoine Charpentier's *Orphée descendant aux enfers* ["Orpheus's Descent into the Underworld"] (**1683**). Among the over 800 cantatas and similar works composed by Alessandro Scarlatti, about a dozen from between the early 1680s and 1716 deal with fantasy themes (mostly featuring Venus and/or Cupid), such as *Diana e Endimione* (*c*1680-85).

German and English composers turned to the medium slightly later, usually in more substantial compositions with several solo voices and choir; the only works by Johann Sebastian Bach that feature fantasy characters – with a subtitle that was common for operas, (*dramma per musica*) – are three of his secular cantatas: *Der zufriedengestellte Äolus* ["The Appeasement of Aeolus"] (**1725**) featuring Aeolus, Athena, Pomona and Zephyrus; *Der Streit zwischen Phoebus und Pan* ["The Contest of Phoebus and Pan"] (?**1729**), including also Momus, Mercury, Tmolus and Midas; and *Herkules auf dem Scheidewege* ["Hercules at the Crossroads"] (**1733**) with Hercules, Virtue, Pleasure and Mercury. Similar forces are used in: Handel's late cantata, *The Choice of Hercules* (**1750**), whose librettist, probably Thomas Morell, based the story indirectly on Xenophon's *Memorabilia* (after Prodicus of Ceos's *Horai*); Maurice Greene's dramatic pastoral *The Judgement of Hercules* (**1740**); John Stanley's ode *The Choice of Hercules* (*c*1750); and Stanley's *The Gay Nymph*, also composed around the middle of the century. Although the genre of the small-scale dramatic cantata faded in importance after the mid-18th century, later examples include several German cantatas on the PYGMALION story, Haydn's solo cantata *Arianna a Naxos* (?**1789**) and a much more substantial work with three solo voices and chorus by Thomas Linley, *Lyric Ode on the Fairies, Aerial Beings and Witches of Shakespeare* (**1776**), usually known as *The Shakespeare Ode*, based on the fantasy aspects of William SHAKESPEARE's *A Midsummer Night's Dream* (performed *c*1595; **1600**), *Macbeth* (performed *c*1606; **1623**) and *The Tempest* (performed *c*1611; **1623**).

Until well after the death of Handel in 1759, oratorios were often operas in disguise; early examples are therefore integrated into the OPERA entry. Sacred oratorios may depict supernatural subjects, but not in a fictional mode. Some later oratorios, like Edward Elgar's *The Dream of Gerontius* (**1900**) and Sir Michael TIPPETT's *The Mask of Time* (**1983**), occasionally put fantasy elements into visualizable form.

The region of SONG is of course enormous, and the subject matter of a huge number of songs might be fairly deemed fantastic. But with exceptions – like Franz Schubert's "Erlkönig" ["The Erl King"] – these fantastic elements fall more into the realm of image and dream than of narrative.

Ballet is a vexed question. Most traditional ballets are of course narrative, but they are not normally voiced; and very often the wraiths and deities who feature are in fact VISIONS experienced by the main cast. The origin of ballet in Greek dance and Roman PANTOMIME is of fantasy interest, but the interest is most cogently conveyed to fantasy audiences through verbal descriptions – often in terms which evoke RITUAL or REVEL – illustrative of an ongoing story. Some ballets are undoubtedly fantasy, and may be the main vehicle for some particular fantasy stories. There are very many examples from the 17th and 18th century of ballets, either independent or built into operas, in which GODS and GODDESSES disport themselves in movements that evoke ALLEGORY. Later individual titles of more direct interest – most including supernatural figures out of the repertory of early ROMANTICISM – include: Adolphe Adam's *Giselle* (**1841**); Friedrich Burgmüller's French ballet *La Péri* for which Théophile GAUTIER wrote the libretto; Jacques Offenbach's opera-ballet *Die Rheinnixen* (**1864**), which is concerned with watersprites, along with a large number of UNDINE ballets, including Adolphe Adam's *La fille du Danube* (**1836**), Cesare Pugni's *Ondine, ou La Naiade* (**1843**), Tchaikovsky's *Undina* (**1869**), Prokofiev's unperformed youthful *Undina* (**1904-07**) and Hans Werner Henze's *Undine* (**1958**); Tchaikovsky's *The Nutcracker* (**1891-2**), based on the E.T.A. HOFFMANN tale; Prokofiev's *Cinderella* (**1941-4**); and Leonard Bernstein's *Dybbuk* (**1954**). There are many more, Tchaikovsky being particularly prolific with ballets (aside from those already mentioned) like *Swan Lake* (**1877**) and *Sleeping Beauty* (**1890**), while much of Stravinsky's ballet work – like *The Firebird* (**1910**), *Petrushka* (**1911**) and *The Rite of Spring* (**1913**) – is of clear fantasy interest.

Many instrumental works have fantastic titles and/or are designed to evoke fantastic images in the listener – like Dukas's *The Sorcerer's Apprentice* (**1897**), based on the poem by GOETHE. However, these are not voiced and thus can scarcely be treated as narratives.

Fantasy musicals are very common, and have often been filmed; just a few examples of those discussed in this book are BRIGADOON (*1954*), MARY POPPINS (*1964*), FINIAN'S RAINBOW (*1968*), *The* ROCKY HORROR PICTURE SHOW (*1975*) and *The* LITTLE SHOP OF HORRORS (*1986*). [HD/JC]

2. Music in fantasy Many writers of fantasy also write music, usually SONGS, which are sometimes narrative. Professional composer/writers include Sarah Ash, John BRUNNER, Anthony BURGESS (under his real name, John Anthony Burgess Wilson), Samuel R. DELANY; E.T.A. HOFFMANN, Michael MOORCOCK and Somtow Sucharitkul (◊ S.P. SOMTOW). Most of these have performed their own works in passing, while others are better known as performers, often of their own compositions; they include Gael BAUDINO, Charles DE LINT, Ellen KUSHNER, Anne MCCAFFREY and Roger ZELAZNY.

The relationship between music and fantasy is in fact so profound and so pervasive that it might be easier to list texts with no relationship at all to music – whether superficial or deeply shaping – than to survey that relationship. In the TAPROOT TEXTS from which modern fantasy writers have taken themes and roles and plots, music and MAGIC are frequently associated (characters invoking magic almost invariably chant or sing or incant): the "music of the spheres" is a metaphor, sometimes taken literally in hermetic texts, of the sound of the Universe surceaselessly singing itself into being; on the mortal plane, as a form of rapture or intoxicant (◊ APOLLO; FAUST; ORPHEUS; PAN), music can be heard as an

embodiment (◊ AS ABOVE, SO BELOW) of higher orders of being which engenders profound changes in mortals, from RITUAL dances at the change of SEASONS to METAMORPHOSIS, as a sign that some TRICKSTER figure (like the PIED PIPER) is in the process of bedazzling the mundane world, or that a figure from this world (◊ TAM LIN) has been charmed out of reality. More mundanely, music is administered to HEROES in the midst of QUESTS; and music, as in the COMMEDIA DELL'ARTE, can be used to accompany and intensify action.

In modern fantasy, when they are performing their task, protagonists or COMPANIONS who are musicians or (more likely) BARDS or MINSTRELS tend to become LIMINAL BEINGS, and articulate in memorable form the relationship between different levels of being of the world. They put into a form – which may have the magic capacity to act as some kind of "song of power", as significantly in Greg BEAR's *Songs of Earth and Power* (omni **1994**), or which may constitute an act of PROPHECY – some version of the essential STORY being enacted, which may be memorized, or followed, or obeyed. This function may be relatively trivial – as in most GENRE FANTASY set in FANTASYLAND, where the straightforward entertainment value of SONG tends to be emphasized – or it may shape entire texts, an example of the latter being Peter S. BEAGLE's *The Innkeeper's Song* (**1993**), which reads (in part) as an excursus upon the eponymous ballad. Some significant novels – Thomas MANN's *Doctor Faustus* (**1947**) being the most obvious – have translated the Faustian pact (◊ FAUST) into an argument about music as POSSESSION, daimonic or secular. Very occasionally, as in Marvin KAYE's *Fantastique* (**1992**), which is constructed as a textual analogue of Hector Berlioz's *Symphonie Fantastique* (1830), a fantasy writer may take the actual shape of a composition and construct a structural paraphrase. Exercises of this sort are potentially of great interest.

In SUPERNATURAL FICTION, music tends to induce – for good or for ill – characters in this world to heed messages from, or actually to enter, a different plane; it is, in other words, a form of seduction or INITIATION. [JC]

MYERS, EDWARD (1950-   ) US writer whose only fiction of genre interest, the LOST-RACE **Mountain Trilogy** – *The Mountain Made of Light* (**1992**), *Fire and Ice* (**1992**) and *The Summit* (**1994**) – picks up MOTIFS and PLOT DEVICES from the period (*c*1900) when this category of fiction was in its prime, but applies a REVISIONIST-FANTASY analysis to the imperialism and racism endemic to his examples. So doing, he expands upon the lessons James HILTON (for instance) imparts through the world-well-lost tone of *Lost Horizon* (**1933**). In the first tale, a young anthropologist, psychically wounded by WORLD WAR I, comes across a lost civilization in a valley in the Andes. By the end of the trilogy, he and another white man, along with a citizen of Xirrixir, are undertaking a transcendental climb to the summit of the Mountain Made of Light, where a CITY rests in wait. The lessons are complex: the Xirrixirians retain a range of wisdom Westerners have lost; but the anthropologist and his white rival have an ability to see their lives in terms of a QUEST that drives the story onwards. [JC]

MYERS, JOHN MYERS (1906-1988) US writer and popular historian; he wrote historical novels and, later, numerous books about the US West. Although his early historical novels were well received – the best-known, *The Harp and the Blade* (**1941**), has some fantasy elements – JMM is remembered almost solely for *Silverlock* (**1949**), a RECURSIVE FANTASY that centres on a PICARESQUE voyage by a shipwrecked protagonist through the "Commonwealth" (of literature), where he encounters numerous characters and situations from world literature and MYTHOLOGY – the ASS of APULEIUS, BEOWULF, the GREEN KNIGHT, ROBIN HOOD, the DANTE's HELL, Friar John from RABELAIS, and many more. The novel is light and pleasant, rather in the manner of Christopher MORLEY.

Silverlock divides the two halves of JMM's career: before it he was almost exclusively an historical novelist; afterwards he devoted himself largely to popular history – although another fantasy was *The Moon's Fire-Eating Daughter* (**1981**), a picaresque caper, traversing TIME rather than geography, whose hapless hero (at the behest of APHRODITE) meets writers and makers throughout history.   [GF]

Further reading: *A Silverlock Companion* (**1988**) ed Fred Lerner (1945-   ), an exegesis of the novel, plus bibliography by James A. Crane.

MY SELF ◊ MAGAZINES.

MYSTERIOUS DR FU MANCHU, THE US movie (*1929*). ◊ FU MANCHU MOVIES.

MYSTERIOUS TRAVELER, THE 1. Weekly US radio show (1943-52). This mixed HORROR and suspense with crime and detection. It was narrated by the Mysterious Traveler, played by actor Maurice Tarplin. Many of the tales were written by pulp writer Robert ARTHUR. The Traveler, a phantom commuter on a train, introduced each show with the words: "This is the Mysterious Traveler inviting you to join me on another journey into the strange and terrifying. I hope you will enjoy the trip, that it will thrill you a little and chill you a little." His closing line was: "I take the same train every week, at the same time." Like other similar shows, TMT was killed by the advent of TELEVISION.   [SH/RT]

2. *The Mysterious Traveler Magazine*, US digest-size pulp anthology MAGAZINE (5 issues November 1951-September 1952; #5 retitled *The Mysterious Traveler Mystery Reader*), published by Grace Publishing Company and ed Robert ARTHUR. Based on **1**, this contained a mixture of mystery and horror. Despite reprinting stories by Ray BRADBURY, John Dickson CARR, Agatha Christie, August DERLETH, Sax ROHMER, Dorothy L. Sayers and Edgar Wallace, it was not a success.   [SH/RT]

3. US COMIC book entitled *Mysterious Traveler Comics*, published by Trans-World Publications, notable for an adaptation of Edgar Allen POE's *The Tell-Tale Heart*, with artwork and cover by Bob Powell. 1 issue (November 1948).   [SH/RT]

4. US comic book entitled *Tales of the Mysterious Traveler* (13 issues 1956-9) published by Charlton Comics and based on **1**, with each story introduced by the gaunt-faced phantom. The major contributor was artist/writer Steve DITKO. The title was revived briefly in 1985 but ran for only 2 issues (Vol 2 *#14-#15*); these also featured work by Ditko.   [SH/RT]

MYSTERY-STORY MAGAZINE ◊ HUTCHINSON'S MAGAZINES; MAGAZINES.

MYSTIC ◊ MAGAZINES.

MYTHAGO A term coined by Robert HOLDSTOCK and first used by him in "Mythago Wood" (1981 *F&SF*), a story which in expanded form as *Mythago Wood* (**1984**) begins the **Ryhope Wood** sequence. Mythagos – the word is put together from "myth" and "image" or, perhaps more plausibly, "imago" – are in Holdstock's words "heroic legendary characters from our inherited unconscious" (◊ JUNGIAN PSYCHOLOGY), metamorphs who wear (and are) MASKS of

our deep, constantly mutating MYTHS. In the sequence which gave them birth, they inhabit an almost impenetrable POLDER in rural England called Ryhope Wood. The wood's outskirts are relatively easy to near, but the further one penetrates INTO THE WOODS the larger it becomes (◊ LITTLE BIG), and, the closer one approaches the heartwood through LABYRINTHS and THRESHOLDS, the closer one comes to the utterly primordial mythagos (◊ TIME ABYSS) at the beginning of things: the point at which all the METAMORPHOSES or TRANSFORMATIONS or SHAPESHIFTINGS began, the point at which (it may be surmised) the inner psyche of the human being also started to take shape, the point from which (it may be) mythago and human began to walk separate paths.

Mythagos take various shapes in Holdstock's work: the wodewose or Wild Man, HEROES AND HEROINES from the MATTER of Britain, the GREEN MAN or ROBIN HOOD. Other writers have not used Holdstock's coining, but (as an example) Charles DE LINT in *Spiritwalk* (omni 1992) creates Native American manifestations of myth and story which might usefully have been called mythagos.   [JC]

MYTHICAL CREATURES These fall into several different categories. First are those whose origins are more or less lost in history – like UNICORNS and GORGONS. There are those born out of popular MYTH, like the MERMAIDS sailors "saw", like GIANTS, and also like WEREWOLVES and VAMPIRES. Then there are the creatures reported in TRAVELLERS' TALES like those of Sir John de MANDEVILLE, who recorded various bizarre MONSTERS; in this context, some real creatures were for a while regarded as mythical – for a long time Europe refused to believe in explorers' reports of the rhinoceros.

Finally, and in a quite separate category, are IMAGINARY ANIMALS. These are the creatures devised by writers of fantasy to populate their SECONDARY WORLDS.   [JG]

See also: ANIMALS UNKNOWN TO SCIENCE.

MYTH OF ORIGIN A CREATION MYTH provides an emotionally satisfying explanation for how the world and humanity came to be; an MOO pretends to explain some small facet of the world. The PROMETHEUS story is a MOO for the discovery of fire; the tale of Romulus and Remus mythologizes the founding of ROME. FOLKTALES contain many examples; e.g., the GRIMM BROTHERS' "The Straw, the Coal and the Bean": these objects' adventures conclude with the bean literally splitting with laughter and being sutured with black thread – so "all beans since then have a black seam". Rudyard KIPLING created several joyous MOOs for animals' shapes in *Just So Stories* (coll **1902**). Lord DUNSANY's "The Sword and the Idol" (1910) mythologizes the discovery of iron-smelting and Ernest BRAMAH's "The Story of Wan and the Remarkable Shrub" (1928) does the same for tea.

A story-form related to the MOO is the onomastic tale which purports to explain a place's or person's NAME; several such anecdotes appear in the MABINOGION, some lacking their punchlines. John JAMES's *Votan* (**1966**) is an MOO in another sense, an anti-myth reducing the Norse AESIR to RATIONALIZED FANTASY: squabbles and killings in the mundane village called Asgard are seen as UNDERLIERS from which the MYTH will grow.   [DRL]

MYTHOLOGY The term describes both the study of myths, particularly ancient myths, and the accumulation of a culture's myths, LEGENDS and FOLKTALES that becomes a nation's heritage. ◊ MYTHS.   [MA]

See also: MATTER.

MYTHOPOEIC AWARDS Awards presented by the

MYTHOPOEIC SOCIETY in fiction and scholarship categories; the award itself is a statuette of a seated lion. Eligibility for the fiction awards has customarily been book publication in the year preceding the award, though re-issues can be eligible. In the early years the scholarship award was occasionally given for a body of work, but eligibility was eventually limited to books published during the previous three years. In 1992 the fiction award was divided into an adult and a children's category, and a second scholarship award concerning general myth and fantasy was added to the first, which had always been limited to studies of work by the INKLINGS. The fiction awards, however, are given to works "in the spirit of the Inklings". [DB]

Mythopoeic Fantasy Award:
1971: *The Crystal Cave* by Mary STEWART
1972: *Red Moon and Black Mountain* by Joy CHANT
1973: *The Song of Rhiannon* by Evangeline WALTON
1974: *The Hollow Hills* by Mary Stewart
1975: *A Midsummer Tempest* by Poul ANDERSON
1981: *Unfinished Tales* by J.R.R. TOLKIEN
1982: *Little, Big* by John CROWLEY
1983: *The Firelings* by Carol KENDALL
1984: *When Voiha Wakes* by Joy Chant
1985: *Cards of Grief* by Jane YOLEN
1986: *Bridge of Birds* by Barry HUGHART
1987: *The Folk of the Air* by Peter S. BEAGLE
1988: *Seventh Son* by Orson Scott CARD
1989: *Unicorn Mountain* by Michael BISHOP
1990: *The Stress of Her Regard* by Tim POWERS
1991: *Thomas the Rhymer* by Ellen KUSHNER

Adult Literature:
1992: *A Woman of the Iron People* by Eleanor ARNASON
1993: *Briar Rose* by Jane Yolen
1994: *The Porcelain Dove* by Delia SHERMAN
1995: *Something Rich and Strange* by Patricia A. MCKILLIP

Children's Literature:
1992: *Haroun and the Sea of Stories* by Salman RUSHDIE
1993: *Knight's Wyrd* by Debra DOYLE and James D. Macdonald
1994: *The Kingdom of Kevin Malone* by Suzy McKee CHARNAS
1995: *Owl in Love* by Patrice Kindl

Scholarship Award (from 1992 award is for Inklings Studies):
1971: C.S. Kilby; Mary McDermott Shideler
1972: Walter Hooper
1973: *Master of Middle-Earth* by Paul H. Kocher
1974: *C.S. Lewis, Mere Christian* by Kathryn A. Lindskoog
1976: *Tolkien Criticism* by Richard C. West; *C.S. Lewis, An Annotated Checklist* by J.R. Christopher and Joan K. Ostling; *Charles W.S. Williams, A Checklist* by Lois Glenn
1981: Christopher Tolkien
1982: *The Inklings* by Humphrey Carpenter
1983: *Companion to Narnia* by Paul F. Ford
1984: *The Road to Middle-Earth* by T.A. Shippey
1985: *Reason and Imagination in C.S. Lewis* by Peter J. Schakel
1986: *Charles Williams, Poet of Theology* by Glen Cavaliero
1987: *J.R.R. Tolkien: Myth, Morality, and Religion* by Richard Purtill
1988: *C.S. Lewis* by Joe R. Christopher
1989: *The Return of the Shadow* ed Christopher Tolkien
1990: *The Annotated Hobbit* ed Douglas A. Anderson
1991: *Jack: C.S. Lewis and His Times* by George Sayer
1992: *Word and Story in C.S. Lewis* ed Peter J. Schakel and Charles A. Huttar

1993: *Planets in Peril* by David C. Downing
1994: *J.R.R. Tolkien: A Descriptive Bibliography* by Wayne G. Hammond with Douglas A. Anderson
1995: *C.S. Lewis in Context* by Doris T. Myers
Scholarship Award (Myth and Fantasy Studies):
1992: *The Victorian Fantasists* ed Kath Filmer
1993: *Strategies of Fantasy* by Brian ATTEBERY
1994: *Twentieth-Century Fantasists* ed Kath Filmer
1995: *Old Tales and New Truths* by James Roy King

MYTHOPOEIC SOCIETY US organization founded 1967 by Glen H. GoodKnight and devoted to fantasy, in particular the works of the INKLINGS. The MS has sponsored local branches and discussion groups across the USA. MS publications include the journal *Mythlore* (1969-current), the bulletin *Mythprint* (1968-current), and a sequence of fiction magazines: *Mythril* (1971-1980), *Mythellany* (1981-7), and *The Mythic Circle* (1987-current).

Annual Mythopoeic Conferences (Mythcons) began in 1970. The MYTHOPOEIC AWARDS were first presented at Mythcon 1971. [DB]

MYTHS A myth is something we like to believe in but know is false; it is thus a FABULATION. Alternatively, a myth may be almost anything once believed and subsequently proved false; myths are thereby close to FOLKLORE, although scientific refutation does not stop a myth remaining in the national consciousness – indeed, the very refutation of some myths leads to an alternate study that re-establishes the myth on a new scientific basis. Myths are closely allied to legend – a collection of linked LEGENDS may become a culture's mythology. Although the term "myth" is most often applied to ancient beliefs or events, it is frequently used to describe anything false and fanciful, and is often synonymous with FAIRYTALE. The word derives from the Greek *muthos*, meaning STORY. Myths are thus the basis of storytelling, and provide the TAPROOT TEXTS for many later tales. Some of the earliest fantasies are myths like the story of GILGAMESH and the Egyptian Book of the Dead, both dating from the second millennium BC, and the poems of HOMER and Hesiod from the 8th century BC.

Although all nations have their myths, FANTASY has been most influenced by the mythologies of the Greeks ($\lozenge$ GREEK AND LATIN CLASSICS), Celts ($\lozenge$ CELTIC FANTASY), Scandinavians ($\lozenge$ NORDIC FANTASY), Persians and Arabs ($\lozenge$ ARABIAN FANTASY), Chinese and Japanese ($\lozenge$ CHINOISERIE; ORIENTAL FANTASY) and, more recently, Native Americans and Australian Aborigines.

Collections of myths are frequently set in story form ($\lozenge$ REVISIONIST FANTASY), although early retellings were often aimed at children ($\lozenge$ CHILDREN'S FANTASY). Early redactors include Nathaniel HAWTHORNE with *A Wonder-Book for Boys and Girls* (coll **1851**) and *Tanglewood Tales* (coll **1853**), Charles KINGSLEY with *The Heroes* (coll **1856**), Thomas BULFINCH with *The Age of Fable* (coll **1855**) and Annie Keary (1825-1879), who produced *The Heroes of Asgard* (coll **1857**) with her sister Eliza. Similar compilations were made by Andrew LANG in *Tales of Troy and Greece* (coll **1907**) and by Richard Lancelyn GREEN in *Tales of the Greek Heroes* (coll **1958**), *Heroes of Greece and Troy* (coll **1960**) and *The Saga of Asgard* (coll **1960**; vt *Myths of the Norsemen* 1970), while Leon GARFIELD utilized myths in *The God Beneath the Sea* (**1970**) to bring the ancient world back to life. It was the fascination of William MORRIS for the ancient Greek and Nordic myths that turned his attention to producing fantastic fiction. Likewise Lord DUNSANY presented his earliest fantasies, *The Gods of Pegana* (coll **1905**), about a mythical PANTHEON. The use of the Greek myths in particular gave a degree of respectability to fantasy in the late Victorian and Edwardian periods, as shown in the works of Richard GAR-NETT in *The Twilight of the Gods* (coll **1888**; exp 1903) and Andrew Lang who, with H. Rider HAGGARD, retold the stories of Odysseus's final wanderings in *The World's Desire* (**1890**). This attitude is evident even as late as Eden PHILLPOTTS's stories like *The Miniature* (**1926**), though by then it was becoming fashionable to lampoon the myths, as in *The Night Life of the Gods* (**1931**) by Thorne SMITH. The Greek myths continue to be popular with writers. Some seek to rationalize them as straight history (◊ RATIONALIZED FANTASY). Notable in this area have been Robert Graves (1895-1985) – who retold the story of Jason and the Argonauts in *The Golden Fleece* (**1945**; vt *Jason*), and produced his own collations, *The Greek Myths* (**1955**) and *The Hebrew Myths* (**1964**), as well as utilizing the themes and images of ancient myths in many of his works – and Mary Renault (real name Mary Challans; 1905-1983) with her **Theseus** books, *The King Must Die* (**1958**) and *The Bull from the Sea* (**1962**). Further writers taking this approach include Edward Lucas White (1866-1934) with *Helen* (**1925**), John ERSKINE with *The Private Life of Helen of Troy* (**1925**), Edison Marshall (1894-1967) with *Earth Giant* (**1960**), about Hercules, Henry TREECE with his **Greek Trilogy** – *Jason* (**1961**), *Electra* (**1963**) and *Oedipus* (**1964**) – and Rosemary SUTCLIFF with *Black Ships Before Troy* (**1993**). Others accept the stories for what they are and treat the supernatural as part of everyday life, but still present the stories as about real people leading real lives; authors include S.P. SOMTOW with *The Shattered Horse* (**1986**), Marion Zimmer BRADLEY with *The Firebrand* (**1987**), Hilary Bailey

(1936-) with *Cassandra: Princess of Troy* (**1994**) and Patrick H. ADKINS with his **Titans** trilogy, which takes us back to the very dawn of the Greek myths. One of the most accomplished reworkers of myths was Thomas Burnett SWANN, almost all of whose books explored the THINNING of the elder world through over 4000 years, starting with *The Minikins of Yam* (**1976**), set in ancient Egypt. Roberto Calasso turned the Greek myths inside out for his radical reinterpretation in *Le Nozze di Cadmo e Armonia* (**1988**; trans Tim Parks as *The Marriage of Cadmus and Harmony* **1993** US).

Myths continue to be created (◊ URBAN LEGENDS), but in the form in which we usually regard them – as the earliest attempts by people to interpret and understand their world – they have a permanent freshness. [MA]

MY WORLD . . . AND WELCOME TO IT US tv series (1969-70, 1972). Sheldon Leonard Productions/NBC, CBS. **Pr** Danny Arnold. **Exec pr** Sheldon Leonard. **Dir** Arnold and many others. **Anim** David DePatie, Friz Freleng. **Writers** David Adler and many others. **Based on** the work of James THURBER. **Starring** Lisa Gerritsen (Lydia Monroe), Joan Hotchkis (Ellen Monroe), William Windom (John Monroe). 26 30min episodes. Colour.

Based on drawings and stories by Thurber, this series focused on the life of John Monroe, a cartoonist for a fictional Manhattan-based magazine. Intimidated by his wife, daughter, children, friends and work, Monroe retreats into a fantasy world of his own imagination, where he views himself as a HERO. Animated sequences were combined with live action to portray his OTHERWORLD. After a year on NBC, the series was brought back for a season of reruns on CBS. A 1959 pilot had previously failed to bring this Walter Mitty-like series to tv. [BC]

NABOKOV, VLADIMIR (1899-1977) Russian-born writer, translator and entomologist who lived most of his life in exile: before WWII in Europe; in the USA for many years from 1940; and from 1959 in Switzerland. From *c*1940 he wrote almost exclusively in English; in his later years, he spent much energy translating or supervising the translation into English of his earlier Russian works, and these versions, almost invariably revised, constitute the definitive texts. From the beginning of his career, VN imperiously created fictional worlds whose REALITIES were contingent upon play, upon linguistic artifice, upon the reader's success in decoding messages from the interior of the text; indeed, it has been suggested that much of VN's work from *Pnin* (**1957**) and "The Vane Sisters" (1959) to *Transparent Things* (**1972**) incorporates encrypted attempts at communication from dead characters, buried in the text, to the "live" world outside.

After some juvenile poetry, VN began publishing novels (all his Russian-language work is as by V. Sirin) with *Mashen'ka* (**1926** Germany; trans Michael Glenny and VN as *Mary* **1970** US), 30 years before coming to world-wide fame with the release of *Lolita* (**1955** France); an early, long-lost version of this novel was released much later as *The Enchanter* (**1987** US). Though all his work can be seen as articulated through the rhetorics of the FANTASTIC, little of it can be understood unequivocally as either sf or fantasy. There are automata in *Korol', Dama, Valet* (**1928** Germany; trans Dmitri Nabokov and VN as *King, Queen, Knave* **1968** US). *Zashchita Luzhina* (**1930** Germany; trans Michael Scammell and VN as *The Defense* **1964** US) is not fantasy, but the hallucinated intensity of its CHESS-master protagonist's projection of a chess defense into the world masterfully elides delusion and reality. *Soglyadatay* (**1930** France; trans Dmitri Nabokov and VN as *The Eye* **1965** US) is a POSTHUMOUS FANTASY in a sense much more explicit than that in which all his late novels (see above) may be readable as exhalations from the dead; and *Priglashenie na kasn'* (**1938** France; trans Dmitri Nabokov and VN as *Invitation to a Beheading* **1959** US), a political FABLE, conveys its protagonist into a posthumous world.

Izobretenie Val'sa (**1938** France; rev text trans Dmitri

Nabokov as *The Waltz Invention* **1966** US) is an sf play; and some of the pre-WWII plays assembled in *The Man from the USSR and Other Plays* (coll trans Dmitri Nabokov **1984** US) are fantastical. *Bend Sinister* (**1947**) is a dystopia; and *Pale Fire* (**1962** US) might be called an antifantasy, because the tale's extraordinary tragic intensity derives precisely from its denial of the solaces of an imposed STORY, its refusal to countenance a RURITANIAN reading of events. Yet it reads as a fantasy of PERCEPTION, the perception toyed with being that of the reader, to whom the truth of events is only slowly revealed. Collections containing tales of fantasy interest include: *Nabokov's Dozen* (coll **1958**); *Nabokov's Quartet* (coll **1966** US), which includes "Poseshchenie muzeya" (1939; trans as "The Visit to the Museum" 1963) and "The Vane Sisters"; *Nabokov's Congeries* (coll **1968** US), which includes "Lance" (1952); *Tyrants Destroyed and Other Stories* (coll trans Dmitri Nabokov and VN **1976** US); *Details of a Sunset and Other Stories* (coll **1976** US). This work is assembled as *The Stories of Vladimir Nabokov* (coll **1996** US).

Ada, or Ardor: A Family Chronicle (**1969** US) is set in a world called Anti-Terra – possibly the linguistic creation of the novel's protagonist, who passionately reveals and obscures with his "TWIN"; although VN consistently scorned all forms of psychology, there is some sense that Anti-Terra, in its wish-fulfilling liberality and lubricity, is a kind of SHADOW of the mundane world (◊◊ JUNGIAN PSYCHOLOGY). *Look at the Harlequins!* (**1974** US), though not technically a fantasy, renders the life of its autobiographical protagonist as a COMMEDIA DELL'ARTE parade. In the end, there are indeed GODGAMES in VN novels: each novel is itself the game in question. [JC]

NADIR, A.A. ◊ Achmed ABDULLAH.

NAKED LUNCH Canadian/UK movie (*1991*). First Independent. **Pr** Jeremy Thomas. **Dir** David Cronenberg (◊ *SFE*). **Creature fx** Chris Walas Inc. **Screenplay** Cronenberg. **Based on** *The Naked Lunch* (**1959**) by William S. BURROUGHS. **Starring** Judy Davis (Joan Frost/Joan Lee), Ian Holm (Tom Frost), Monique Mercure (Fadela), Julian Sands (Yves Cloquet), Roy Scheider (Dr Benway), Joseph Scorsiani (Kiki), Robert A. Silverman (Hans), Peter Weller

(William Lee). **Voice actor** Peter Boretski (creatures). 115 mins. Colour.

New York, 1953. Would-be writer William (Bill) Lee – an early pseudonym of Burroughs – and wife Joan are addicted to the pyrethrum he uses in his work as a bug exterminator. Police arrest him; they leave him alone to test the powder on a MONSTER bug, which tells him it is his case officer and he is a secret agent countering the efforts of "Interzone, a notorious free port on the North African coast, a haven for the mongrel scum of the Earth, an engorged parasite on the underbelly of the West". Smashing the bug, he tries to flee with Joan, but accidentally shoots her dead in a drunken re-enactment of William Tell's feat with the apple (Burroughs indeed killed his wife Joan this way). In a gay bar a Mugwump (a bizarre, semi-insectile beast the size of a man) gives him tickets to escape to Interzone – although the tickets, shown to a friend, look like the phial of cut pyrethrum given to him by genial GP Dr Benway. In Interzone he is befriended by Hans, who introduces him to rent-boy Kiki, who in turn introduces him to expatriate writers Tom and Joan Frost; she is a DOUBLE of Lee's dead wife. Interzone, we find, is co-extensive with REALITY; on rare occasion the camera allows us to see this, and even more rarely Lee sees it too. The rest of the plot is too complex for summary.

NL, full of SURREALISM, is widely regarded as being even more incoherent than the book on which it is based, perhaps because of its use of a wide array of fantasy tools with which mainstream movie viewers are unfamiliar. But the notion of writers being the tools of typewriters rather than the other way around is hammered home perhaps a little too firmly, and the typewriters' sexuality, and that of the giant bugs (who speak through dorsal vents that are like puckering anuses), reflects a little too crudely the sexual METAMORPHOSES in progress in Lee's mind. [JG]

Further reading: *Everything is Permitted: The Making of "Naked Lunch"* * (**1992**) ed Ira Silverberg.

NAMES The problem with names in fantasy is a subset of the more general problem of DICTION. Names have convincingly to derive from another time and place but also bear some sort of appropriate association.

One solution is to use words not usually used as names, but which bring with them verbal force – e.g., Mervyn PEAKE's **Gormenghast** trilogy has characters called Flay and Swelter – or to assemble names on a portmanteau principle – Peake, again, has characters called Prunesquallor and Steerpike, while the giants in Stephen R. DONALDSON's **Chronicles of Thomas Covenant the Unbeliever (1977-83)** have names like Saltheart Foamfollower.

Names which are simple statements of allegorical meaning – Sweet Rhyme and Pure Reason in Norton JUSTER's *The Phantom Tollbooth* (**1961**), for example – are appropriate to WONDERLANDS; names that tend in this sort of literalist direction are less appropriate in a FANTASYLAND context.

Invented languages bring their own problems. J.R.R. TOLKIEN was perhaps the only writer of fantasy with sufficient expertise to conceive both consistent languages and names deriving from them, but even he notoriously hits some dangerous false notes and crudities. He may not have intended the dark riders, the Nazgûl, to be read as a portmanteau of Nazi and ghoul, but that is inevitably how they have been read; his DARK LORD Sauron has no obvious connection with the reptilian, save those intrinsic to a DEVIL-figure in a CHRISTIAN FANTASY – the Old Serpent.

Tolkien used a mixture of procedures: many of the names in *The Lord of the Rings* (**1954-5**) are archaic names from Northern European languages – not entirely inappropriately, since his SECONDARY WORLD is clearly intended to be at least partly cognate with Europe. When he does use portmanteau names, it is by assembling syllables that have connotative force rather than by putting words together – Tom Bombadil for example. It is significant that he was prepared to abandon portmanteau names – Tinfang Warble is a notorious example – that on reflection and redrafting struck him as inherently silly; it is also significant that even Tolkien, who had thought long and hard about the issue, had trouble with names.

As with the more general case of diction as a whole, the only way to be sure-footed in the matter of fantasy names is to be linguistically sensitive. Jack VANCE's names draw eclectically from various sources of invention, but they feel right because they are integrated into his wayward style and odd touches of recondite learning. Terry PRATCHETT draws on the same battery of sources for fantasy names with equivalent skill, often engaging in elaborate jokes in the process – the Patrician, autocrat of Ankh-Morpork, has the family name Vetinarii, by analogy with the Florentine Medici (*vetinarii* = veterinary surgeons; *medici* = doctors).

It is generally assumed in fantasy that names have magic (◊ TRUE NAMES). [RK]

NANKAI NO DAIKETTO ot of *Ebirah, Terror of the Deep* (**1966**). ◊ GODZILLA MOVIES.

NARRENSCHIFF ◊ SHIP OF FOOLS.

NATHAN, ROBERT (GRUNTAL) (1894-1985) US writer who began publishing realistic fiction as early as 1915; his first novel, *Peter Kindred* (**1919**), appeared soon after. His success at realism was limited, however, and from his second novel, *Autumn* (**1921**), his 40 or so books tended almost invariably to be SUPERNATURAL FICTIONS, plus the occasional fantasy, all told in a consistently civilized, mild-mannered, moderately bittersweet tone. Given the soft but ironic focus of so much of his work, and his refusal to generate happy endings to romances which seemed to beg for them, the scale of his success over 50 years was notable.

Autumn, set in New England – plus its thematic sequel, *A Fiddler in Barly* (**1926**), and several other tales – constitutes an attempt to write a US rural idyll in PASTORAL mode. These tales are not, however, fantasies (though TALKING ANIMALS and other fantasy-like figures are sometimes present), and the pastoral element they display has little or no sense of EUCATASTROPHE. The main note they sound – perhaps the dominant note sounded by all his work – is of BELATEDNESS. Further titles of the same sort include *The Woodcutter's House* (**1927**) and *The Summer Meadows* (**1973**); others, like *The Orchid* (**1931**), attempt similar effects in urban settings, but 20th-century URBAN FANTASIES are very rarely successful in soft focus – and RN's certainly are not.

Several of RN's supernatural fictions play – again without conspicuous edge – on religious themes. *Jonah* (**1925**; vt *Son of Amittai* 1925 UK; vt *Jonah, or The Withering Vine* 1934 US) is based directly on the BIBLE, and is rather sharper than RN's later work. In *The Bishop's Wife* (**1928**) the archangel Michael (◊ ANGELS), while answering a call for help in building a new cathedral, becomes involved with the wife of a somewhat venal bishop, but has of course no sexual organs with which to consummate the relationship. In *There is Another Heaven* (**1929**), which plays on the POSTHUMOUS-FANTASY mode, a character from the previous novel finds that his parents in HEAVEN advocate an unpleasing sexual latitude, and a Jew who has converted to Christianity finds that this heaven is alien to him, and eventually re-fords the Jordan in a search for the heaven of his roots. Other titles involving humans and the traditional supernatural pantheon include: *Mr Whittle and the Morning Star* (**1947**), *The River Journey* (**1949**), *The Innocent Eve* (**1951**) – in which SATAN attempts to gain control of the A-bomb – *The Train in the Meadow* (**1953**) – another posthumous fantasy – *The Rancho of the Little Loves* (**1956**), *The Devil with Love* (**1963**) – a FAUST tale featuring a contemporary bargainer and a flustered Mephistopheles – and *Heaven and Hell and the Megas Factor* (**1975**), in which GOD and Satan band together to save humanity from the Moloch of technology.

But most of the novels for which RN is likely to be remembered are love stories in TIMESLIP or ALTERNATE-WORLD frames. The best-known is *Portrait of Jennie* (**1940**), filmed as PORTRAIT OF JENNIE (*1948*). A painter in the midst of an artistic crisis meets an 8-year-old girl named Jennie in Central Park, and sketches her. Over the next months, he meets her again and again, but each time she is years older. He becomes enamoured. They finally make love, but the next time he sees her she is dead in the ocean. In another world – or TIME, perhaps (as RN himelf claimed), exemplifying arguments by J.W. Dunne (1875-1949) about dissociative time states – she has been urging herself forward (as it were) so as to be able to come to him. The story (as with the movie) makes less sense in synopsis than it does while being read; in its poignance and the overbearingness of its premonitions of BELATEDNESS, it is probably RN's strongest single work. Other titles involving a very similar relationship between eros and the pathos of the TIMESLIP include *The Married Look* (**1950**), *So Love Returns* (**1958**), *The Wilderness-Stone* (**1960**) and *Mia* (**1970**).

The removed, gentle irony of RN's narrative voice does not vary greatly from novel to novel, but his stories are in fact quite widely varied in subject and venue: *The Puppet Master* (**1923**) comes close to HORROR when some PUPPETS come briefly alive; *One More Spring* (**1933**), though muted, has a CARNIVAL setting, and has been acknowledged by Peter S. BEAGLE as a direct influence upon his *A Fine and Private Place* (**1960**); *Road of Ages* (**1935**) is a political allegory about a new diaspora afflicting the Jews of Europe; in *The Enchanted Voyage* (**1936**) a sailboat carries its cast overland; *But Gently Day* (**1943**) rather confusedly conflates timeslip with a plot derived from Ambrose BIERCE's "An Occurrence at Owl Creek Bridge" (1892); *Sir Henry* (**1955**) incorporates traditional fantasy matters (a DRAGON, a sorcerer and a SHADOW self for its quixotic hero to joust with); in *The Fair* (**1964**), Arthurian matters CROSSHATCH with events in the contemporary world; and *The Elixir* (**1971**) engages its professor hero, haunted by Nimue ($ LADY OF THE LAKE), in a desert trek, during which he meets other characters crosshatched from the MATTER of Britain. [JC]
Other works: The **Tapiola** sequence, told from a dog's viewpoint, being *Journey of Tapiola* (**1938**) and *Tapiola's Brave Regiment* (**1941**), assembled as *The Adventures of Tapiola* (omni **1950**); *The Barly Fields* (omni **1938**), assembling *The Fiddler in Barly*, *The Woodcutter's House*, *The Bishop's Wife*, *There is Another Heaven* and *The Orchid*; *Winter in April* (**1938**); *They Went on Together* (**1941**); *The Sea-Gull Cry* (**1942**); *Long After Summer* (**1948**); *Nathan 3: The Seagull-Cry*; *The Innocent Eve*; *The River Journey* (omni

1952 UK); *The Color of Evening* (**1960**); *The Weans* (1956 *Harper's Magazine* as "Digging the Weans"; **1960** chap), RN's only sf title; *A Star in the Wind* (**1962**); *The Mallot Diaries* (**1965**), portraying a Neanderthal tribe in a lost-world enclave ($ LOST RACES); *Stonecliff* (**1967**).

NAVIGATOR, THE: A MEDIAEVAL ODYSSEY Australian/NZ movie (*1988*). Arenafilm/Film Investment Corporation of NZ. **Pr** John Maynard. **Exec pr** Gary Hannam. **Dir** Vincent Ward. **Spfx** Paul Nichola. **Screenplay** Geoff Chapple, Kely Lyons, Ward. **Starring** Noel Appleby (Ulf), Chris Haywood (Arno), Desmond Kelly (Smithy), Jay Lavea Laga'aia (Jay), Bill Le Marquand (Tom), Paul Livingston (Martin), Bruce Lyons (Connor), Hamish McFarlane (Griffin), Marshall Napier (Searle), Sarah Peirse (Linnet). 91 mins. Colour and b/w.

TN:AMO is set partly in 1348 (shown in b/w) and partly in 1988 (colour); the tale makes extensive use of flash-forwards, and some of flashbacks.

1348: young Griffin's prophetic DREAMS persuade the people of a Cumbrian mining village that they can be spared the Black Death if they send a mission to tunnel through the disc of the flat Earth to cast and erect a new copper spike on the spire of the cathedral they will find there. Led by Griffin's elder brother Connor, recently returned to the village from the outside world, the party emerges in a modern, industrialized NZ city, which they perceive alternately as the Celestial City and HELL. By chance they find a foundry on the verge of shutdown, where Smithy, Tom and Jay cast their cruciform spike. As they row across the harbour to the cathedral (they perceive an emergent submarine as Leviathan) Griffin has a further precognitive flash: one of them will fall to his death from the spire. In the event, it is Griffin himself; but as he falls he is returned to 14th-century Cumbria, where he has been telling his companions of his dream. However, it has not been "just a dream": Connor, it is found, survived the plague during his sojourn away from the village but has brought it back (there is a PIED-PIPER motif here); as Griffin foretold, only one will die, and that, as if he were a Year King ($ GOLDEN BOUGH), will be himself.

TN:AMO is one of a kind, and powerful. Derisory comparisons have been made by mainstream critics between some of its elements and MONTY PYTHON AND THE HOLY GRAIL (*1975*) (and similar notions were indeed later utilized also to comic effect in *Les* VISITEURS [*1993*]). Such mockery betrays some cultural illiteracy, for the moments of HUMOUR in this sophisticated, dual-levelled fantasy – it is as much an URBAN FANTASY as a medieval tale of WONDER – are skilfully deployed to point up the true nightmare in which the 14th-century party finds itself. [JG]

NECROMANCY Divination by consultation with the SPIRITS of the dead. Once considered morally neutral – in HOMER's *Odyssey* ODYSSEUS consults the shade of Tiresias, and in *Samuel* there is no condemnation of the woman who summons the shade of Samuel on Saul's behalf – the practice came to be associated with SORCERY to the extent that the word became almost synonymous with BLACK MAGIC; the gloss added to the latter passage by the Authorized Version thus refers to the woman as the WITCH of Endor. The term was so comprehensively blackened that the revival of a form of necromancy at the end of the 19th century required the invention of a new term: SPIRITUALISM. The movement's detractors were quick to point out the fudge, as Robert Hugh BENSON does in *The Necromancers* (**1909**). The bad repute into which the term has fallen has largely displaced its use into the field of HORROR, where it is employed loosely, but there is a substantial subgenre of spiritualist fantasy. Fantasies involving necromancy which do not fit neatly into that category include *The Soul of Lilith* (**1892**) by Marie CORELLI, *All Hallow's Eve* (**1945**) by Charles WILLIAMS and the tv series RANDALL & HOPKIRK (DECEASED) (1969-71). [BS]

NECRONOMICON ◊ BOOKS; CTHULHU MYTHOS; H.P. LOVECRAFT.

NEEDFUL THINGS US movie (*1993*). Castle Rock/New Line/Columbia. **Pr** Jack Cummins. **Exec pr** Peter Yates. **Dir** Fraser C. Heston. **Spfx** Garry Paller. **Mufx** Tibor Farkas. **Screenplay** W.D. Richter. **Based on** *Needful Things* (**1991**) by Stephen KING. **Starring** Bonnie Bedelia (Polly Chalmers), Ed Harris (Sheriff Alan Pangbourn), Ray McKinnon (Deputy Norris Ridgewick), Shane Meier (Brian Rusk), Amanda Plummer (Nettie Cobb), Max Von Sydow (Leland Gaunt), J.T. Walsh (Danforth "Buster" Keaton III). 116 mins. Colour.

King's original novel had the interesting premise that the TRICKSTER figure Leland Gaunt – probably the DEVIL, but with characteristics also of DEATH and the WANDERING JEW – opens a magic SHOP in the small Maine town Castle Rock. His stock consists of what each individual most wants; his price is part money but, more importantly, part a "favour" or "trick" they must perform for him. Initially minor, albeit malicious, these "tricks" have the effect of unwrapping the latent EVIL in each person's heart, so that soon the town is at war with itself in a frenzy of bloodshed. Only doughty Sheriff Pangbourn stands between Castle Rock and the abyss.

The novel was long and rambling. The movie, though following the plot fairly carefully, necessarily abridges matters, and this works much to its benefit. Most of the characterizations are stereotyped, but Von Sydow intriguingly plays Gaunt for the most part as a plausibly genial demon, thereby shocking us all the more when his ruthlessness surfaces. [JG]

NEIDERMAN, ANDREW (1940-) US writer. ◊ Virginia C. ANDREWS.

NERVAL, GÉRARD DE Pseudonym of Gérard Labrunie (1808-1855), French poet and journalist. He wrote in the waning days of French Romanticism, but his work anticipates the Symbolism that dominated the middle of the century. Neglected for nearly a century after his death (although Proust considered him a seminal figure), he has since the end of WWII become one of the most-studied French figures of the 19th century. His early collection, *Le main de gloire* (**1832**), consists of tales in the manner of E.T.A. HOFFMANN. The much later *Les Filles du Feu* (**1854**; transJames Whitall as *Daughters of Fire* 1922 US), his most famous volume of imaginative prose, contains "Sylvie" (trans anon **1887** chap US), widely regarded as his masterpiece: a semi-autobiographical romance of lost loves, it makes extremely sophisticated use of the DOPPELGÄNGER theme. *Aurelia, ou le rêve et la vie* (**1855**) is concerned with the "second life" of the DREAM state; it describes GN's late mental afflictions in terms of a descent into HELL. *Les Chimères* (omni with *Les Filles du Feu* **1854**; individual poems trans Andrew LANG 1872 and John Payne 1906) comprises a short sonnet sequence employing images from the TAROT, ALCHEMY and Egyptian and Greek MYTHOLOGY in an extremely dense verse. This last has in recent decades been translated in whole or in part many times. [GF]

Other works: *L'Alchimiste* (**1839**) with Alexandre DUMAS;

Scènes de la vie orientale (**1851**; trans as *The Women of Cairo: Scenes of Life in the Orient* **1929**), travel book containing some fantastic tales; "Fantaisie" (1832 *Annales romantiques*).

NESBIT, E(DITH) (1858-1924) UK writer of CHILDREN'S FANTASY, considered by many the first truly modern writer for children, and certainly very influential on the development of children's literature. She thought of herself first as a poet, but her husband's illness and the failure of his business soon after the birth of their first child in 1880 forced her to write whatever she could sell. Most of her earliest fiction (pre-1890) was written in collaboration with her husband, Hubert Bland (?1855-1914), under the joint pseudonym Fabian Bland. It was not until the publication of a series of stories about the **Bastable** children, based in part on her own childhood experiences, that EN found her niche. Collected in volume form as *The Story of the Treasure Seekers, Being the Adventures of the Bastable Children in Search of a Fortune* (coll **1899**), these established her as a major writer for children. But, although two more **Bastable** serial novels followed, she now turned to fantasy, writing a series of witty FAIRYTALES for *The Strand* and other papers, collected as *The Book of Dragons* (coll **1899**) and *Nine Unlikely Tales for Children* (coll **1901**). Her first fantasy novel, *Five Children and It* (1902 *The Strand* as "The Psammead or The Gifts"; **1902**), combined her two main strengths: the depiction of believable children and an inventive, humorous flair for fantasy. The children of the title discover "It" – a Psammead, or sand-fairy – in a sandpit during their summer holidays, and are delighted to learn it will grant them one WISH a day. Few of their wishes work out as they hope, although they do get their hearts' desire in the end. The formula EN established in this first book has been a fruitful tradition in fantasy ever since.

EN was self-admittedly influenced by the work of her contemporary F. ANSTEY, while others before her – including William Makepeace THACKERAY and Andrew LANG – had made humorous use of fairytale materials, but EN was the first to write fantasy in such a knowing, modern, non-pious way for children.

Two more books about the "five children" followed: in *The Phoenix and the Carpet* (1903-4 *The Strand*; **1904**) they have further adventures after discovering that their new carpet is a magic one, and in *The Story of the Amulet* (1905 *The Strand*; **1906**) they encounter the Psammead again: it becomes their guide on journeys through TIME in search of the missing half of a powerful Egyptian AMULET. These three novels have always been her most popular, and had the greatest influence on later generations of children's writers (of whom Edward EAGER acknowledged the debt most explicitly). C.S. LEWIS, Charles WILLIAMS and others have cited the lasting imaginative impact made on them by *The Story of the Amulet*.

EN's richest and most complex work, drawing again upon her own childhood, is *The Enchanted Castle* (1906-7 *The Strand*; **1907**), while her personal favourites were the two interlinked time fantasies (complete with Fabian socialist lessons) *The House of Arden* (**1908**) and *Harding's Luck* (**1909**). *The Magic City* (**1910**) developed out of EN's hobby of constructing miniature cities from household objects.

Despite her success with children's books, EN continued to churn out a varied body of work, including a number of romantic novels for adults. Only one, *Dormant* (**1911**; vt *Rose Royal* US), is fantasy, concerning a woman kept immune from ageing in suspended animation for over 50 years. EN also wrote HORROR stories, most of which are collected in *Grim Tales* (coll **1893**), *Something Wrong* (coll **1893**) and *Fear* (coll **1910**), the last comprising 5 stories from *Grim Tales* plus 6 newer stories. *In the Dark* (coll **1988**) ed Hugh Lamb (1947-) contains most of EN's horror stories. [LT]

Other works: *The Prophet's Mantle* (**1888**) as Fabian Bland; *The Secret of Kyriels* (**1899**); *The Wouldbegoods, Being the Further Adventures of the Treasure Seekers* (**1901**); *Thirteen Ways Home* (coll **1901**); *The Literary Sense* (coll **1903**); *The New Treasure Seekers* (**1904**); *The Railway Children* (**1906**); *The Incomplete Amorist* (**1906**); *Daphne in Fitzroy Street* (**1909**); *Salome and the Head* (**1909**; vt *The House with No Address*); *The Wonderful Garden* (**1911**); *The Magic World* (coll **1912**); *Wings and the Child, or The Building of Magic Cities* (**1913**), with photographs and diagrams for model-building; *Wet Magic* (1912-13 *The Strand*; **1913**); *The Lark* (**1922**); *To the Adventurous* (coll **1923**); *Five of Us, and Madeleine* (coll **1925**) with linking material by EN's daughter, Rosamund Sharp.

Further reading: *E. Nesbit: A Biography* (**1933**; rev 1966) by Doris Langley Moore; *A Woman of Passion: The Life of E. Nesbit 1858-1924* (**1987**) by Julia Briggs.

NESTED STORIES ◊ STORY CYCLE.

NEVERENDING STORY, THE A series of three (so far) movies.

1. The Neverending Story (vt *Die Unendliche Geschichte*) US/German movie (*1984*). Warner/Producers Sales Organization/Bavaria Studios/WDR/Neue Constantin Filmproduktion. **Pr** Bernd Eichinger, Dieter Geissler. **Exec pr** Mark Damon, John Hyde. **Dir** Wolfgang Petersen. **Spfx/vfx** Brian Johnson. **Screenplay** Petersen, Herman Weigel. **Based on** *The Neverending Story* (**1979**) by Michael ENDE. **Starring** Sydney Bromley (Engywook), Moses Gunn (Cairon), Noah Hathaway (Atreyu), Thomas Hill (Bookseller), Barret Oliver (Bastian), Alan Oppenheimer (Falkor's voice), Patricia Hayes (Urgl), Tami Stronach (Childlike Empress). 94 mins. Colour.

Young, bullied Bastian "borrows" from an antiquarian bookshop a book called *The Neverending Story*, which he takes to school and, cutting classes, reads in the attic. It tells of the SECONDARY WORLD of Fantastica (the world of human fantasies, we later learn), currently being devoured by the Nothing (human loss of hope and dreams); its Childlike Empress is dying for want of ... for want of no one knows what. But there is a warrior, Atreyu, who can battle the Nothing if anyone can; when he appears, he proves to be a (very feminine) boy of Bastian's age. Atreyu has assorted adventures with various expectable fantasy folk and MONSTERS before finally, Fantastica now almost entirely destroyed, the Childlike Empress herself tells him the only hope for the empire is that Bastian, the reader, be called in. His initially disbelieving intervention regenerates Fantastica.

TNS has two plots, one of which serves only as a basis for the other, which is the important one. The inner tale, of Atreyu's QUEST across Fantastica, is not of great interest: Fantastica is reduced to merely another FANTASYLAND. The *real* plot, framing and periodically interrupting the lesser one, concerns the interrelation between reader and STORY, and the way in which the story writes itself. This is pointed up early, when the Bookseller tells Bastian: "Your books are *safe*. By reading them, you get to become Tarzan, or Robinson Crusoe. ... Ah, but afterwards you get to be a little boy again." But *The Neverending Story* – the book Bastian filches – is not "safe": it is a *real* book. [JG]

2. The Neverending Story II: The Next Chapter US/German movie (*1990*). Warner/Soriba & Dehle. **Pr** Dieter Geissler. **Dir** George Miller. **Creature fx** Colin Arthur. **Screenplay** Karin Howard. **Starring** Jonathan Brandis, Clarissa Burt, Alexandra Johnes, Kenny Morrison. 89 mins. Colour.

This sees a return to Fantastica and a rehash of its predecessor's less interesting material. Lacking any proper plot, the movie is rather like a dull travelogue of FANTASYLAND. [JG]

3. The Neverending Story III German/US movie (*1994*). Warner/CineVoz Filmproduktion/Babelsberg. **Pr** Dieter Geissler, Tim Hampton. **Dir** Peter Macdonald. **Animatronics** Jim HENSON's Creature Workshop. **Screenplay** Jeff Lieberman. **Starring** Ryan Bellman, Jason Jack Black, Carole Finn, Melody Kay, James Richter. 95 mins. Colour.

Bastian goes back to Fantastica, now being torn apart under the attack of a vile Nastiness. He discovers from the Empress that, to help, he and his old COMPANIONS must battle with the Nasties in our REALITY. Complications include misuse of a MAGIC necklace that grants WISHES. In the end EVIL is thwarted and order restored. [JG]

See also: *The* DARK CRYSTAL (*1982*); *The* PRINCESS BRIDE (*1987*).

NEVER-NEVER LAND Also rendered as Never Never Land and Never Land, the OTHERWORLD which features in J.M. BARRIE's play *Peter Pan* (produced 1904; rev **1928**) and its prose sequels. PETER PAN lives there, along with his COMPANIONS and his foes. The pathos and melancholy of Barrie's overall concept infects his rendering of N-NL, for it is a venue which exudes both BELATEDNESS and a frail level of REALITY, like that of the FAIRY Tinkerbell; N-NL seems therefore constantly at risk from the mortal children who visit, for they are doomed to pass through TIME into the estrangements of adulthood. Time does not exist for N-NL's denizens, notably Peter Pan himself.

Fantasy otherworlds whose depiction is exaggerated and playful are sometimes called N-NLs – indeed, they are so-called because they actually contradict their readers' sense of reality. They are not uncommon in CHILDREN'S FANTASY – J.P. MARTIN's **Uncle** novels use such a setting.

The term "neverland" is sometimes used – as in *Neverland: Fabled Places and Fabulous Voyages of History and Legend* (**1976**) by Steven Frimmer (1928-) – to describe imagined and perhaps believed-in physical locations on this planet, like ATLANTIS, or "flyaway" islands like Brazil Rock, or the ARCHIPELAGO visited by Saint Brendan, or the kingdom of PRESTER JOHN. [JC]

NEW ADVENTURES OF WONDER WOMAN, THE Retitling of US tv series *The* ADVENTURES OF WONDER WOMAN (1976-9).

NEWBOLT, [Sir] HENRY (1862-1938) UK poet and writer, now best remembered for the patriotic poems assembled in *Admirals All and Other Verses* (coll **1897** chap; exp vt *The Island Race* 1898) and the nostalgic verse – including "Vitaï Lampada" – assembled in *Clifton Chapel and Other School Poems* (coll **1908**). Of his several novels, *Aladore* (**1914**) is a fantasy novel of some interest, set in a LAND-OF-FABLE medieval Europe and clearly indebted in style and tone to William MORRIS, but with a dreamlike chamber-music air of its own. Haunted by a vision (a child who wears his face, and who becomes, literally, his SHADOW), the protagonist undergoes a complex QUEST which initially leads him to attack a DARK TOWER at the heart of a CITY named Paladore, only to find that the battle he has joined is part of an immemorial AGON; later episodes, some redolent of CHRISTIAN FANTASY, reveal the existence of a GODGAME centred on the ultimate city of Aladore, where the protagonist finds his true love and his true self. [JC]

NEWLOVE, DONALD (1928-   ) US writer. His memoir of his years as an alcoholic and failed writer, *Those Drinking Days* (**1981**), remains his best-known work. *Eternal Life* (**1979**), although containing the autobiographical elements and hyperrealism that mark all his fiction, is a POSTHUMOUS FANTASY in which the protagonist tours the ASTRAL PLANE, meeting figures from history and his lost love. [GF]

Other works: *The Painter Gabriel* (**1970**); *Leo and Theodore* (**1972**) and *The Drunks* (**1974**), assembled as *Sweet Adversity* (rev omni **1978**); *Curanne Trueheart* (**1986**).

NEWMAN, KIM (1959-   ) UK writer, critic, broadcaster and former cabaret performer. KN worked extensively as a reviewer before publishing his first book, *Nightmare Movies* (**1984**; rev 1988), a critical history of HORROR MOVIES since 1968. It was followed by *Ghastly Beyond Belief* (anth **1985**) with Neil GAIMAN, assembling some of the worst and funniest SCIENCE-FICTION and FANTASY quotations.

Most of KN's fiction is heavily influenced by the media, and he often recursively (◊ RECURSIVE FANTASY) utilizes characters and situations from tv, cinema and literature for his short stories and novels. Much of his fiction can be thought of as existing in a MULTIVERSE, links being overlapping storylines and recurring characters – most notably the VAMPIRE heroine Geneviéve Dieudonné, private investigator Sally Rhodes and evil media mogul Derek Leech.

His début novel, *The Night Mayor* (**1989**), was an audacious blend of cyberpunk and hard-boiled-detective yarn in which a pair of professional Dreamers from the 21st century are sent into a criminal's *film noir* dreamscape. DREAMS were also used as major imagery in his next novel, *Bad Dreams* (**1990**), set in contemporary London and featuring a young US woman stalked by a psychic VAMPIRE. *Jago* (**1991**), KN's attempt to write an apocalyptic Stephen KING-type novel set in the UK's West Country, is perhaps KN's most ambitious novel to date, with its descriptions of REALITY warping around the members of a bizarre religious sect and its charismatic leader. *Anno Dracula* (1992 as "Red Reign"; exp **1992**; vt *Anno-Dracula* US), among his most popular and successful works, is set in an alternative 1888 where Count DRACULA is Queen Victoria's new consort in a world ruled by VAMPIRES. KN here skilfully interweaves famous (and not so famous) historical and fictional characters in a carefully constructed ALTERNATE WORLD. His next novel, *The Quorum* (**1994**), set against the background of London over the past 30 years, uses a Faustian PACT with the enigmatic Derek Leech as the basis for the story of three ambitious men and the boyhood friend whose happiness they sacrifice in return for their own success. With *The Bloody Red Baron* (**1995**), the author finally returned to the VAMPIRE world he had created for *Anno Dracula*: set during WWI in 1918, this focuses on the exploits of vampiric air ace Baron von Richthofen.

Some of KN's cleverest stories, such as "Famous Monsters" (**1988**) – Martians invade Hollywood – "The Original Dr Shade" (**1990**) – a legendary PULP hero returns in Margaret Thatcher's UK – "The Man who Collected Barker" (**1990**) – a fan's obsession with collecting – "The Big Fish" (**1993**) – Raymond Chandler meets H.P. LOVECRAFT – and "Out of the Night, When the Full Moon is

Bright" (**1994**) – Zorro recreated as a WEREWOLF in contemporary Los Angeles – are collected in *The Original Dr Shade and Other Stories* (coll **1994**) and *Famous Monsters* (coll **1995**). KN is also the author of **Back in the USSA**, a collaborative cycle of stories with Eugene Byrne set in an alternate world where the USA, rather than Russia, became a communist country, and the **Where the Bodies Are Buried** series, pastiching of contemporary horror movies and their sequels.

As Jack Yeovil he has written a number of ties set in the fantasy GAMES worlds of **Warhammer** and **Dark Future**, listed below, plus one original novel: *Orgy of the Blood Parasites* (**1994**; ot *Bloody Students*), a spoof of splatter HORROR.

KN's work is marked by pace, verve and imagination.[SJ]

Other works: *Wild West Movies* (**1990**), nonfiction.

As Jack Yeovil: *Drachenfels* * (**1989**); *Demon Download* * (**1990**); *Krokodil Tears* * (**1990**); *Beasts in Velvet* * (**1991**); *Comeback Tour* * (**1991**); *Genevieve Undead* * (**1993**); *Route 666* * (**1994**).

As editor: *Horror: 100 Best Books* (**1988**; rev 1992) with Stephen JONES, nonfiction; *In Dreams* (anth **1992**) with Paul J. McAuley; «The BFI Companion to Horror» (**1996**), nonfiction.

NEWMAN, ROBERT (HOWARD) (1909-1988) US writer, much involved in radio and tv work from 1936, and author of several nonfantastic adult novels – like *Identity Unknown* (**1945**) and *The Enchanter* (**1962**), the latter a thriller – before beginning to publish, normally for a YA audience, works of genre interest like *Corbie* (**1966**) and *The Boy who Could Fly* (**1967**). He remains best-known for the **Tertius** fantasy sequence – *Merlin's Mistake* (**1970**) and *The Testing of Tertius* (**1973**) – featuring young Tertius, upon whom MERLIN has bestowed the dangerous and seemingly unuseful gift of understanding and making use of future science, though Tertius is ignorant of matters like MAGIC. In the first volume, Merlin serves as an enigmatic TRICKSTER figure, guiding the course of a complex QUEST whose ostensible protagonist, the armiger Brian, is never in control of events; a ROBIN HOOD figure appears intermittently. In the second volume, Tertius helps ARTHUR cope with Attila the Hun. RN's style was polished and professional, though sometimes over-slick. [JC]

Other works: *The Shattered Stone* (**1975**); *Night Spell* (**1977**); *The Case of the Baker Street Irregulars* (**1978**; vt *A Puzzle for Sherlock Holmes* 1979 UK) (◊ SHERLOCK HOLMES).

NEWMAN, SHARAN (1949-   ) US author of adult and juvenile fantasy with a strong historical flavour. Her first book, *The Dagda's Harp* (**1977**), was a juvenile. The trilogy comprising *Guinevere* (**1981**), *The Chessboard Queen* (**1983**) and *Guinevere Evermore* (**1985**) gives a gently humorous account of the legends of ARTHUR, centred on the life and times of GUINEVERE: SN's iconoclastic and often irreverent reinterpretations of stock Arthurian characters and situations lend considerable freshness and novelty to the trilogy. [KLM]

Other works: *Death Comes as Epiphany* (**1993**); *The Devil's Door* (**1994**).

NEW MONTHLY MAGAZINE, THE ◊ MAGAZINES.

NEW NOVELISTS MAGAZINE ◊ MAGAZINES.

NEW, ORIGINAL WONDER WOMAN, THE US movie (*1975* tvm). Warner Bros./ABC. **Pr** Douglas S. Cramer. **Dir** Leonard J. Horn. **Screenplay** Stanley Ralph Ross. **Based on** COMIC-book characters created by Charles

Moulton. **Starring** Lynda Carter (Diana Prince/Wonder Woman), Cloris Leachman (Queen), John Randolph (General Philip Blankenship), Stella Stevens (Marcia), Lyle Waggoner (Major Steve Trevor). 90 mins. Colour.

This second pilot movie based on WONDER WOMAN – the first was WONDER WOMAN (*1974*) – returned to the original WORLD WAR II setting and look of the comic books. While flying over the Bermuda Triangle, US pilot Steve Trevor is forced to bail out over supposedly empty ocean. Luckily, he lands on the uncharted Paradise Island, the hidden home of the AMAZONS, and their Queen holds a contest to find a warrior to return with him to fight the Nazis. The winner takes the identity of Diana Prince, an aide to Trevor, and together they battle to stop a deadly Nazi plot.

This movie was far more successful than its predecessor, due largely to the casting of Lynda Carter in the lead role, for her skimpy costume helped draw the desired male audience. One of the more memorable effects was the transformation from Diana Prince to Wonder Woman – a rapid spinning, so fast that she blurred, returned the Amazon warrior to her true identity. The story continued the following year in the series *The* ADVENTURES OF WONDER WOMAN. [BC]

NEW ORLEANS US CITY located close to the Mississippi Delta and thus a natural THRESHOLD venue. It stands between the continent and the Gulf of Mexico; it is an outpost of the American way in an environment which is almost tropic; and it has been "notorious" for centuries as a venue for transgressive behaviour: smuggling, vice, jazz and miscegenation – over the years, French, Spanish, English, Indians and Afro-Caribbeans bred to create the Creole culture and tongue. It is an ideal setting for SUPERNATURAL FICTION and HORROR.

The first writer of any renown to make use of the venue was George Washington Cable (1844-1925), most of whose work is NO-based, including "Jean-ah Poquelin" (1875 *Scribner's*), a rationalized GHOST STORY in which various cultures and races come together. Some of Lafcadio HEARN's NO stories, as assembled in *Fantastics* (coll 1914), are also of interest; and some of the Gothic anecdotes assembled in *New Orleans Sketches* (coll 1954 Japan) by William Faulkner (1897-1962) have supernatural elements.

The most telling use of NO may be the *film-noir* vision of the city that dominates CAT PEOPLE (*1942*) and its various sequels; this may have influenced the choice of venue for *Angel Heart* (*1987*), a movie in which the original NEW YORK setting of William HJORTSBERG's *Falling Angel* (1978) is shifted south. But the most famous exploitation of NO is in the **Vampire Chronicles** and **Mayfair Witches** sequences by Anne RICE, herself born and resident in NO. The sequences have understandably fixed for a wide public the paradigm use of NO for DARK FANTASY; Valerie MARTIN's *A Recent Martyr* (1987), based on the **Cat People** movies, also makes effective use of the venue. Lucius SHEPARD's *Green Eyes* (1984), though not literally set there, could be described as occupying an NO hinterland. Terry PRATCHETT uses a thinly disguised version of the city as a FAIRYTALE backdrop in *Witches Abroad* (1991).

Many horror novels, especially those dealing with VOODOO, are either set in NO, or – like James HERBERT's *Portent* (1992) – feature characters whose transgressive ominousness is signalled by the the fact they come from the city. [JC]

NEW TWILIGHT ZONE, THE US syndicated tv series

(1985-8). MGM/UA. **Pr** Harvey Frand. **Exec pr** Philip DeGuere. **Dir** Gil Bettman, Bruce Bilson, Noel Black, Ben Bolt, Martha Coolidge, Wes CRAVEN, Joe Dante (◊ *SFE*), Robert Downey, Bill Duke, Don Carlos Dunaway, J.D. Feigelson, Ted Flicker, Rick Friedburg, William Friedkin, Kenneth Gilbert, John Hancock, Curtis Harrington, Sheldon Larry, Shelley Levinson, Paul Lynch, Bruce Malmuth, Bradford May, Jim McBride, Peter Medak, Sigmund Neafeld, B.W.L. Norton, Gerard Oswald, Alan Smithee, David Steinberg, Jeannot Szwarc, R.L. Thomas, Gus Trikonas, Paul Tucker, Tommy Lee Wallace, Claudia Weill. **Writers/based on stories by** Virginia Aldridge, Lynn Barker, Haskell Barkin, Steven Barnes (◊ *SFE*), Greg BEAR, Charles BEAUMONT, Steven Bochco, Ray BRADBURY, Alan BRENNERT, Michael Bryant, David Bennett Carren, Michael Cassutt, Paul Chitlik, Arthur C. Clarke (◊ *SFE*), Anne Collins, Ron Cobb, Robert Craig, James Crocker, Philip DeGuere, J.M. DeMatteis, Phyllis EISENSTEIN, Harlan ELLISON, Les Enroe, J.D. Feigelson, Jeremy Bertrand Finch, Charles E. Fritch, Joe Gannon, David Gerrold (◊ *SFE*), Ron Glass, Parke GODWIN, Gerrit Graham, Arthur Gray, Robert Hunter, Chris Hubbell, Steven KING, Anthony and Nancy Lawrence, Robin Lee, William M. Lee, Bryce Maritano, George R.R. MARTIN, Terry Matz, Patricia Messina, Robert R. McCammon, Gordon Mitchell, Rockne S. O'Bannon, Lan O'Kun, Rebecca Parr, Martin Pasko, Steven Rae, Edward Redlich, J. Michael REAVES, Carter Scholz, J. Neil Schulman, Rod SERLING, Sidney Sheldon, Robert SILVERBERG, Henry Slesar (◊ *SFE*), J. Michael Staczynski, Theodore STURGEON, Logan Swanson (i.e., Richard MATHESON), Donald Todd, Cal Willingham, William F. Wu, Roger ZELAZNY. **Based on** *The* TWILIGHT ZONE (1959-64). **Comics adaptation** *The New Twilight Zone* * (19 issues 1991-3) from Now Comics. **Anthology** *The New Twilight Zone* * (anth 1991) ed Alan BRENNERT and Martin H. GREENBERG. 80 60min and 30min episodes. Colour.

Hoping to cash in on the continued interest shown in reruns of *The* TWILIGHT ZONE, the producers launched the series with new stories and several remakes from the original series. Once again, the plots focused on tales with a twist, but unfortunately most were rather predictable (the same charge was levelled at the roughly contemporaneous AMAZING STORIES [1985-7]). Part of the blame for this evidently falls to CBS (the original networker), for the producers were vocal about interference; Harlan ELLISON, the story consultant, quit the show in frustration. Viewers were treated to episodes such as "Ye Gods", featuring a man struck by Cupid's arrow, "A Message from Charity", about an 18th-century girl accused of witchcraft when she sees the future, and a man playing poker with the DEVIL for his SOUL in "Dealer's Choice".

Despite an impressive line-up of writers, directors and actors, the series got off to a rocky start, perhaps due to the inevitable and justified comparisons with the original series. Critics were unanimous that episodes seemed slow and bloated, and a decision was made to cut the format from 60 to 30 mins to tighten up the stories. When all efforts to salvage the series for CBS failed and the network cancelled it, the producers kept it going in first-run syndication until they had enough episodes for future syndication. Despite their hopes, though, the series has not been seen since. [BC]

NEW WORLDS UK sf magazine. ◊ Michael MOORCOCK.

NEW YORK Images of the Statue of Liberty up to her neck

in the sands of the desert – like images of LONDON submerged under a cleansing lake – properly belong to post-HOLOCAUST sf (◊ SFE), not fantasy. A novel like *The Last American* (**1889**) by John Ames Mitchell (1845-1918), despite a storyline which evokes ARABIAN FANTASY, presents an image of NY as a CITY most easily understood in sf terms, and only with some difficulty in terms of fantasy. But NY (or at least Manhattan) does differ from URBAN-FANTASY venues like CAIRO, London, LOS ANGELES or even PARIS in that it can be perceived, in the mind's eye, from a distance, whole. In this, it projects something of the allure of the magic cities which can be found in SECONDARY-WORLD venues.

Manhattan (which is as much of NY as is dealt with in most texts) is highly foregrounded and highly dramatic. It is a city whose fate can be seen: a vast POLDER, with a circumambient THRESHOLD and PORTALS galore. In all of this, NY differs radically from London, the only city more frequently found in fantasy texts (apart, possibly, from NEW ORLEANS). London, though it contains worlds of drama, is very rarely itself the occasion of drama. Contrariwise, many fantasy novels deal with NY as an entity inside which, and to which and in terms of which, portentous events may occur. INSTAURATION FANTASIES like John CROWLEY's *Little, Big* (**1981**) or Mark HELPRIN's *Winter's Tale* (**1983**) plausibly key their intimations of change through a rendering of the great city drama of NY.

But NY does also serve – like London – as a circumambient venue for URBAN FANTASIES, though it is often the case that the high dramatic profile of NY seems to welcome stories whose implications are similarly foregrounded. Christopher MORLEY's BEAST-FABLE *Where the Blue Begins* (**1922**) is fittingly set in NY, as are E.B. WHITE's *Stuart Little* (**1945**), George SELDEN's *The Cricket in Times Square* (**1960**) and *Freddie the Pigeon: A Tale of the Secret Service* (**1972**) by Seymour Leichman (1933-); and when Scott BRADFIELD, in *Animal Planet* (**1995**), wishes to complexify the world within which his constantly mutating ALLEGORY works, he moves significant figures of his cast to Manhattan.

NY-based fantasy novels, often featuring elements of allegory or transfiguration, include: Stephen BOYETT's *Ariel* (**1983**); Angela CARTER's *The Passion of New Eve* (**1977**), whose vision of NY is significantly far more allegorized than the vision of London presented in her *Nights at the Circus* (**1984**); almost all of the novels of Jerome CHARYN; Tom DE HAVEN's *Funny Papers* (**1985**), which along with E.L. DOCTOROW's *The Waterworks* (**1994**) comes close to establishing a GASLIGHT-ROMANCE tradition for the city; Thomas M. DISCH's *On Wings of Song* (**1979** UK), an sf novel (◊ SFE) whose vision of NY is a haunted phantasmagoric; Diane DUANE's *So You Want to be a Wizard* (**1983**); the fables of contemporary violence assembled in Harlan ELLISON's *Deathbird Stories: A Pantheon of Modern Gods* (coll of linked stories **1975**); several of Jack FINNEY's TIMESLIP tales, including "The Third Level" (1952) and *Time and Again* (**1970**); Esther M. FRIESNER's **New York** sequence of CONTEMPORARY FANTASIES; Simon HAWKE's *The Wizard of 4th Street* (**1987**); William HJORTSBERG's *Falling Angel* (**1978**); *The Werewolf's Tale* (**1988**) by Richard Jaccoma (1943-); Tappan KING's and Viido Polikarpus's *Down Town: A Fantasy* (**1985**); and William KOTZWINKLE's *The Game of Thirty* (**1994**), which echoes *The Confessions* (**1962**) by Harry Mathews (1930-) (◊ SFE) in its use of the urban venue as LABYRINTH and GAME combined.

Movies with significant sequences set in NY include KING KONG (*1933*; *1976*), ROSEMARY'S BABY (*1968*), WOLFEN (*1981*), GHOSTBUSTERS (*1984*), SPLASH! (*1984*), HIGHLANDER (*1986*), Woody Allen's contribution to the anthology movie *New York Stories* (**1989**) and GREMLINS 2 (*1990*).

From the turn of the century, NY has also been used in COMICS as an immediately recognizable venue. Some of the better known examples are: Winsor MCCAY's *Little Nemo in Slumberland* (**1905-11**); SUPERMAN (**1938**-current), whose Metropolis may have originated as a portrait of Toronto but which soon became NY; *The Batman* (**1938**-current), whose Gotham City was always NY; most of the MARVEL COMICS interlinked set of SUPERHERO epics; and Alan MOORE's *Watchmen* (graph **1986-7**).

Often, in movies and comics produced by Americans, the unidentified city in which an urban fantasy is set is in fact recognizably NY: NY has become the default venue for both media. Had London not preceded it, NY might have become the default venue for written fantasy as well.   [JC]

NICHETTI, MAURIZIO (?1949-   ) Italian director, scriptwriter and actor, several of whose movies are of fantasy interest. After various jobs – including circus clown and comics gag-writer – MN co-wrote and starred in Bruno Bozzetto's ALLEGRO NON TROPPO (*1976*), a live-action/ANIMATED MOVIE; he took the mixture much further with VOLERE VOLARE (*1991*), which he co-wrote and co-directed with Guido Manuli. The latter is an archetypal TOON movie, ranking alongside WHO FRAMED ROGER RABBIT (*1988*) and COOL WORLD (*1992*). Also of note in a fantasy context is *The* ICICLE THIEF (*1989*) – which MN wrote, directed and starred in – a wistful, nostalgic PARODY of Cesare Zavattini's *Bicycle Thieves* (*1948*; ot *Ladri di Biciclette*; vt *Bicycle Thief* US) while at the same time a vicious SATIRE of the way tv stations can butcher movies through COMMERCIAL breaks, a matter over which other Italian directors have loudly complained. MN's work can tend to self-indulgence, but his imagination is rarely infertile.   [JG]

Other works written and directed: *Ratatplan* (*1979*); *Ho Fatto Splash* (*1980*); *Domani si Balla* (*1982*; vt *Tomorrow We Dance*).

NICHOLS, (JOHN) BEVERLEY (1898-1983) UK writer who, after a precocious 1920s career as a literary *enfant terrible*, became well known for gardening books like *Down the Garden Path* (**1932**). *Failures: Three Plays* (coll **1933**) contains an sf drama, "When the Crash Comes", in which the Communists take over Britain. BN is of fantasy interest for his CHILDREN'S FANTASIES, including *The Tree that Sat Down* (**1945**), about an enchanted FOREST and its denizens, *The Stream that Stood Still* (**1948**) and *The Mountain of Magic: A Romance for Children* (**1950**).   [JC]

NICHOLS, ROBERT (MALISE BOWYER) (1893-1944) UK academic, poet, dramatist and writer, known initially as a WWI poet because of the success of his first book, *Invocation: War Poems and Others* (coll **1915**). After WWI his poetic career faltered; *Fisbo, or The Looking-Glass Loaned* (**1934**) is a lumbering narrative SATIRE, insufficiently fantasticated to raise any spirits. His prose was somewhat livelier: *The Smile of the Sphinx* (**1920** chap), an ARABIAN FANTASY, appeared in revised form in *Romances of Idea, Volume One: Fantastica: Being the Smile of the Sphinx and Other Tales of Imagination* (coll **1923**), a large volume featuring the book-length "Golgotha & Co", which anticipates an apocalyptic WORLD WAR II. Through a heavily satirized LANDSCAPE the WANDERING JEW is discovered, playing the role of ANTICHRIST; and CHRIST is recrucified. No second volume of *Fantastica* appeared. [JC]

Other works: *Under the Yew* (**1928** chap), marginal fantasy; *Wings Over Europe: A Dramatic Extravaganza on a Pressing Theme* (**1929** US) with Maurice Browne (1881-1955), an sf play.

NICHOLS, RUTH (JOANNE) (1948-) Canadian writer whose fairly conventional but appealingly written otherworldy fantasies are intended for children or YA. *A Walk Out of the World* (**1969**), *The Marrow of the World* (**1972**), *Song of the Pearl* (**1976**) and *The Left-Handed Spirit* (**1978**), all written while she was still quite young, attracted praise and gained her Canada Council Fellowships and a Canadian Library Association award. Her later fiction has been infrequent, and mainly non-fantastic. [DP]

Other works: *The Burning of the Rose* (**1989**) and *What Dangers Deep* (**1992**), historical novels with slight fantasy elements.

NICHOLSON, JOSEPH SHIELD (1850-1927) UK economist and writer. The first of his three anonymous fantasy novels, *Thoth* (**1888**), is borderline SCIENCE FICTION, describing a technologically advanced but decadent culture coexistent with primitive but virile Periclean Athens. *A Dreamer of Dreams* (**1889**) is a moralistic fantasy whose protagonist's adventures in drug-assisted lucid dreaming lead him into a paranoid nightmare where he is blackmailed into a PACT WITH THE DEVIL. The most interesting is *Toxar* (**1890**), an extended *conte philosophique* in which an empathically gifted scholar-slave gives his masters the advice they demand, so that their warped ambitions may lead them to destruction. [BS]

NIGHTBREED (vt *Clive Barker's Nightbreed*) US movie (**1990**). J&M/Morgan Creek. **Pr** Gabriella Martinelli. **Exec pr** James G. Robinson, Joe Roth. **Dir** Clive BARKER. **Vfx/mufx** Image Animation. **Screenplay** Barker. **Based on** "Cabal" (in *Cabal* coll **1988** US) by Barker. **Starring** Anne Bobby (Lori), Doug Bradley (Lylesberg), Catherine Chevalier (Rachel), David Cronenberg (Dr Philip Decker), Charles Haid (Captain Eigerman), Oliver Parker (Pelonquin), Hugh Quarshie (Inspector Joyce), Kim and Nina Robertson (Babette), Hugh Ross (Narcisse), Bob Sessions (Pettine), Craig Sheffer (Boone), Malcolm Smith (Ashberry), Debora Weston (Sheryl Ann). 102 mins. Colour.

Seemingly respectable psychiatrist Decker is a SERIAL KILLER; he commits his barbarities wearing a grotesque MASK, so that they are, in effect, done by "someone else". His patient, Boone, has DREAMS depicting the crimes, and is framed by Decker. But Boone has dreams also of another place, seemingly nightmarish yet also welcoming: Midian. Fleeing the law, he finds this place: living UNDERGROUND is a WAINSCOT society of MONSTERS. "Killed" by the police at Decker's behest, he becomes a part of this community, made up of the surviving members of SHAPESHIFTER races otherwise long ago exterminated by mankind (as recorded in *Genesis*); he has always been, unknowing, one of them, the Nightbreed. Undead, they have INVULNERABILITY to much that might kill humans but are susceptible to, e.g., sunlight. Standard MONSTER-MOVIE plotting ensues – though this time the lynch mob is slaughtered by the monsters – until Boone, ritually given the new NAME Cabal, must lead the Nightbreed to a fresh refuge.

N, despite gore, is a straightforward exposition of a GENRE-FANTASY theme: the Nightbreed are a PARIAH ELITE, with Boone ignorantly passing as a member of humanity until witlessly undergoing a RITE OF PASSAGE (here, his "death"); once accepted among them he rapidly becomes their leader; chips down, mankind's mindless mob proves unable, despite vast weapons superiority, to prevail against the elite's abilities/sheer grit. [JG]

Further reading: *Clive Barker's Nightbreed: The Making of the Film* * (**1990**); *The Nightbreed Chronicles* * (coll **1990**) ed Stephen JONES.

NIGHTCOMERS, THE UK movie (*1971*, but copyright 1972). Avco Embassy/Kastner Ladd Kanter/Scimitar. **Pr** Michael Winner. **Dir** Winner. **Screenplay** Michael Hastings. **Based on** characters in "The Turn of the Screw" (1898) by Henry JAMES. **Novelization** *The Nightcomers* * (**1972**) by Hastings. **Starring** Harry Andrews (The Master), Stephanie Beacham (Margaret Jessel), Marlon Brando (Peter Quint), Christopher Ellis (Miles), Verna Harvey (Flora), Thora Hird (Mrs Grose), Anna Palk (Miss Giddens). 97 mins. Colour.

A prequel to the James novella, and hence to *The* INNOCENTS (*1961*), this shows the CHILDREN of the house corrupted both by the kinky sexual affair between Quint and Jessel, on whose activities they spy, and by the hypocrisy of all the adult characters. Quint is depicted as an Irish FERTILITY symbol – cruel and caring all at the same time – so that the "corruption" he spreads might be seen as an assault on "good" Christianity, as epitomized by the puritanical Grose and the outwardly prim Jessel (◊ RELIGION), yet the former woman shows no charity and the latter no chastity. The two mismatched lovers are murdered by the children, who believe this is the only way Quint and Jessel can remain at the house and be together. Its potential sensationalist thrills smothered by clumsiness of script (cod-historical) and direction (stodgy), this movie – itself nonfantastic – is best watched, if at all, as a curious appendage to *The Innocents*. [JG]

NIGHT CRY US digest MAGAZINE, 11 issues, quarterly, Winter 1984-Fall 1987, published by Montcalm Publishing, New York; *#1-#2* ed T.E.D. KLEIN, then ed Alan Rodgers (1959-).

NC began as a digest reprinting from from ROD SERLING'S THE TWILIGHT ZONE MAGAZINE, but when its newsstand sales outstripped those of its parent it became a magazine in its own right, concentrating on fiction plus a few associated author profiles, but eschewing the movie coverage and features of its parent. During its brief life *NC* became the focus of AMERICAN GOTHIC, especially URBAN FANTASY. It took the basic premise of Rod SERLING'S *The* TWILIGHT ZONE and set it in the modern CITY. Although never bending to the excesses of Splatterpunk, *NC* allowed the exponents of that subgenre to explore the terrors of our society in a controlled fashion. The result was some of the best of the early work of G.L. Raisor, David J. Schow (1955-), John Skipp (1957-) and Craig Spector (1958-), plus equally powerful works from Robert BLOCH, Ramsey CAMPBELL, Dennis ETCHISON, Dean KOONTZ, Richard Christian MATHESON and J.N. WILLIAMSON. The magazine's endeavours to disturb and dislocate were admirably enhanced by the artwork of J.K. Potter (1956-). *NC* relied wholly on newsstand sales; when in 1987 this dropped by over half the magazine was discontinued. *NC* stands out as the leading magazine of mid-1980s modern HORROR. [MA]

NIGHT JOURNEY The NJ constitutes a central moment in the RITE OF PASSAGE undertaken by fantasy protagonists.

The journey referred to is usually but not always depicted as a literal act of travel. The protagonist travels into a dark country, which may be the underside of the LAND or some interior territory occupied by his or her SHADOW. Here matters of significance to that protagonist's life are met, confronted and defeated (or, possibly, not defeated). RECOGNITION scenes attend or follow NJs. No protagonist who survives an NJ does so unaltered.

The term is used tellingly by P.C. HODGELL in her essay "The Night Journey in 20th Century English Fantasy" (*Riverside Quarterly* 1977), but she tends to restrict it to Christianizing work (◊ CHRISTIAN FANTASY) and to journeys involving engulfment, whether UNDERGROUND or inside beasts. The term is used in this encyclopedia in a broader sense: the main characteristics of an NJ are that it is awesome, to both protagonist and reader, and that it is instructive. Because the turning of the SEASONS is an apt symbol for the process of turning from ignorance into wisdom, NJs may be associated with solstices, as in Charles DICKENS's *A Christmas Carol* (1843) and Sir Michael Tippett's *The Midsummer Marriage* (1955) (◊ OPERA). The darkness of the NJ may be symbolic of the state of ignorance which the instruction is to relieve, but not necessarily. And darkness is not essential to an NJ: Sam and Frodo's time in the Marches of Mordor and the Dead Marshes is an NJ even though some of it takes place in bright sunlight.

NJs are undergone by OBSESSED SEEKERS and ACCURSED WANDERERS; a significant aspect of Hodgell's own heroine, Jamethiel, who has elements of both in her makeup, is that her predestined role as the champion of her people makes her intrinsically asocial and incapable of complete integration into any of the societies in which she moves. In the works of overtly Christian writers like C.S. LEWIS and Charles WILLIAMS, NJs are certain to culminate in Christianizing arrivals, but we should note that perhaps the most crucial of all NJs is not at all Christian, being that of the prophet Muhammad from the Dome of the Rock to the Empyrean.

The NJ may take place in a realm that is intrinsically one of shadow or delusion; Hodgell argues that William Hope HODGSON's Night Land and Borderland and David LINDSAY's Arcturus are both of this kind. Here the instruction is primarily that of the will to pierce through shadow and delusion, or to accept that there is nothing else; whether this acceptance is seen as bracingly tragic or merely as a failure of nerve depends on philosophic perspective. CHILDES – such as Stephen KING's Gunslinger – can be seen as taking an NJ of this kind, but usually only to the extent that they learn that a childe is indeed what they are. It is only because GENRE FANTASY is ultimately consoling that NJs generally have fortunate culminations.

Sometimes it is not the principal actor alone who undergoes NJs, nor do characters necessarily undergo just a single NJ: in J.R.R. TOLKIEN's *The Lord of the Rings* (1954-5) the Fellowship of the Ring are put repeatedly through NJs collectively and severally – whether in the Paths of the Dead or the Marches of Mordor.

NJs are rare in pure HEROIC FANTASY and SWORD AND SORCERY because the lively intelligence of the classic adventurer is so busy searching for something to purloin or someone to knife in the back. However, NJs are likely to occur whenever metaphysics starts to form a part of the TEMPLATE. Michael MOORCOCK's **Elric** is manipulated to make him the hero of Law and the victim of the SWORD Stormbringer, and

thus his entire life can be seen as an NJ. Even more relevantly, the extended NJ which forms the overall subject of Louise COOPER's **Indigo** sequence (**1988-93**) regularly provides the protagonist with one important moral insight per book as she learns to fight the DEMONS she has released into the world – demons based on aspects of herself.

The NJ is often a feature, perhaps the controlling image, of a *Bildungsroman*, whether that which is being made or forged is the protagonist of heroic fantasy or the chastened, sometimes metamorphosed viewpoint-figure of INSTAURATION FANTASY – though in the latter instance the NJ is likely to be subsidiary and largely a matter of passive endurance, as with Auberon in John CROWLEY's *Little, Big* (**1981**) and Sweeney in Elizabeth HAND's *Waking the Moon* (**1995**). [RK/JC]

NIGHT LIFE OF THE GODS, THE US movie (*1935*). ◊ Thorne SMITH.

NIGHTMARE BEFORE CHRISTMAS, THE (vt *Tim Burton's The Nightmare Before Christmas*) US stop-motion ANIMATED MOVIE (*1993*), with some traditional animation. Buena Vista/Touchstone/Burton-Di Novi. **Pr** Tim BURTON, Denise Di Novi. **Dir** Henry Selick. **Spfx** Gordon Baker, Dave Bossert, Miguel Domingo Cachuela, Chris Green. **Vfx** Pete Kozachik. **Digital fx** Walt DISNEY Feature Animation. **Anim sv** Eric Leighton. **Character fabricator sv** Bonita De Carlo. **Screenplay** Michael McDowell, Caroline Thompson. **Voice actors** Danny Elfman (Jack Skellington singing/Barrel/Clown with the Tear Away Face), William Hickey (Evil Scientist), Ed Ivory (Santa), Catherine O'Hara (Sally/Shock), Ken Page (Oogie Boogie), Paul Reubens (Lock), Chris Sarandon (Jack Skellington speaking), Glenn Shadix (Mayor). 76 mins. Colour.

The various festivals – Easter, HALLOWE'EN, Thanksgiving, CHRISTMAS, etc. – are run by the denizens of individual and mutually unaware towns, sited both in some OTHERWORLD and beneath respective TREES ringing a clearing in a FOREST whose location is enigmatic. Hallowe'en over, the leader of the ghoulish ELEMENTALS responsible for that festival, Jack Skellington, wanders into the forest and inadvertently stumbles via a treetrunk door into Christmas. He determines that this year the Hallowe'en folk, with himself as SANTA CLAUS, will also run Christmas – and better. After various disasters he LEARNS BETTER, and reinstates Santa in his rightful position.

TNBC is a visually riveting movie; it is improbable that any other stop-motion animated movie equals it for technical proficiency. The script, largely in verse and much of it narrated or sung, has a mesmeric, driving quality that is somehow light rather than oppressive; the underlying fantasy premise is fascinating. The voice acting and characterization are superb. And yet, as so often with Burton's work, one comes away with the feeling that it could have been *better*: for much of its duration it promises as climax something truly spectacular, but then it fails to deliver, as if Burton, nearing the end, were suddenly in a hurry to get the whole thing wrapped up and over with. Nevertheless, viewed as an artwork rather than as a story, *TNBC* is magnificent. [JG]

NIGHTMARE ON ELM STREET, A Series of six plus one fantasy-HORROR MOVIES, of widely varying quality though often with points of interest, concerning a murderous SPIRIT, Freddy Krueger – a slouching SERIAL KILLER with a hideously ravaged face and knife-fingered gloves – who has the power to kill people in their DREAMS; he is a SHAPESHIFTER and SOUL-eater. Like Dracula (◊ DRACULA

MOVIES) he is regularly destroyed at the end of one movie and revived, through some mechanism or another, at the start of the next.

1. *A Nightmare on Elm Street* US movie (**1984**). New Line Cinema/Media Home Entertainment/Smart Egg/Elm Street Venture. **Pr** Robert Shaye. **Exec pr** Stanley Dudelson, Joseph Wolf. **Dir** Wes CRAVEN. **Spfx** Jim Doyle, Theatrical Engines. **Mufx** David Miller. **Screenplay** Craven. **Starring** Ronee Blakley (Marge Thompson), Nick Corri (Rod Lane), Johnny Depp (Glen Lantz), Robert Englund (Freddy Krueger), Heather Langenkamp (Nancy Thompson), John Saxon (Lt Thompson), Amanda Wyss (Tina Gray). 91 mins. Colour.

Teenagers Tina, Don, Nancy and Glen slowly realize they are sharing horrific DREAMS of Freddy. Then, in one of Tina's dreams, Freddy attacks her, shredding her flesh – and the gory murder is simultaneously enacted in our waking REALITY. Boyfriend Don is arrested for the crime. Nancy, who has had narrow escapes in her own dreams, is convinced the nightmare figure is responsible, but before she can act it has hanged Don in his cell. Taken to a sleep laboratory, Nancy succeeds in dragging the murderer's hat into waking reality. Inside its brim is the name "Fred Krueger". Nancy's alcoholic mother confesses that, years ago, SERIAL KILLER Krueger was caught but released on a technicality, and that she and other parents burnt him alive. Nancy, by now many nights deliberately sleepless, ventures into the dream reality to haul Freddy into ours, where an accomplice can deal with him; but her accomplices let her down. Unknowing, she reifies Freddy, but at last destroys him (she thinks) when, believing he survives only because of the energy he leeches from her, withdraws that energy by refusing to perceive him (◊ PERCEPTION). In an ambiguous finale, Nancy steps into an OTHERWORLD where all seems happy and sunkissed, her friends and mother again alive; but, with an eye to sequels, Freddy strikes here as well ...

This is an extremely intelligent, well made HORROR MOVIE, achieving its effects less through the plentiful gore than through playing on our paranoia by blurring the line between dream and waking realities. [JG]

2. *A Nightmare on Elm Street Part 2: Freddy's Revenge* US movie (**1985**). New Line Cinema/Heron/Smart Egg/Elm Street Two Venture. **Pr** Robert Shaye. **Exec pr** Stephen Dieder, Stanley Dudelson. **Dir** Jack Sholder. **Vfx** Paul Boyington. **Mufx** Kevin Yagher. **Screenplay** David Chaskin. **Starring** Marshall Bell (Schneider), Robert Englund (Freddy), Clu Gulager (Mr Walsh), Hope Lange (Mrs Walsh), Kim Myers (Lisa Webber), Mark Patton (Jesse Walsh), Robert Rusler (Ron Grady), Sydney Walsh (Kerry). 84 mins. Colour.

The first sequel, set five years later in the old Thompson house at 1428 Elm Street, abandons the style and (occasional) subtlety of **1** in favour of a more straightforward tale of POSSESSION. The implausibility of the mundane aspects of the plot deprives the fantasy of all credibility, leaving the movie reliant solely on gore. [JG]

3. *A Nightmare on Elm Street Part 3: Dream Warriors* US movie (**1987**). New Line/Heron/Smart Egg. **Pr** Robert Shaye. **Exec pr** Wes CRAVEN, Stephen Diener. **Dir** Chuck Russell. **Spfx** Thomas Bellissimo, Doug Beswick Productions, Peter Chesney, Image Engineering, Bryan Moore. **Mufx** Greg Cannom, Mathew Mungel, Mark Shostrum, Kevin Yagher. **Vfx** Dream Quest Images, Hoyt Yeatman. **Anim fx** Jeff Burks. **Screenplay** Craven, Frank

Darabont, Russell, Bruce Wagner. **Starring** Patricia Arquette (Kristen Parker), Brooke Bundy (Elaine Parker), Kristen Clayton (Little Girl), Rodney Eastman (Joey), Robert Englund (Freddy), Larry Fishburne (Max), Ira Heiden (Will), Heather Langenkamp (Nancy Thompson), Nan Martin (Sister Mary Helena/Amanda Krueger), Priscilla Pointer (Dr Elizabeth Simms), Jennifer Rubin (Taryn), Ken Sagoes (Roland Kincaid), John Saxon (Lt Thompson), Penelope Sudrow (Jennifer), Craig Wasson (Dr Neil Gordon). 96 mins. Colour.

Defying the rule that series always deteriorate, this is substantially better and more imaginative than **2**. A group of teenagers suffering sleep-related mental deterioration are isolated in a psychiatric hospital; it proves they are the last of the children of the parents who burnt Freddy alive (◊ **1**). Nancy (from **1**) is brought in as a psychiatric assistant, and knows the truth: she finds psychiatrist Gordon open-minded but others less so. Freddy picks off the teenagers one by one, assailing them whenever they sleep. Enigmatic GHOST nun Sister Mary Helena appears repeatedly to Gordon, explaining that Freddy was her child through gang rape by criminally insane men; she tells him Freddy's remains must be buried if the unquiet SPIRIT is to be laid to rest. He sets about this with Nancy's estranged, now-alcoholic father – duelling a skeleton in a deserted car dump – while Nancy and the surviving teenagers plunge together into the DREAM reality to battle Freddy there; their advantage in that REALITY is that they can act in their (more powerful) dream personas. At last, of course, Freddy is defeated – for now.

The movie is stuffed with fantasy images – people are pulled into their own MIRROR reflections, a door in midair leads downwards to a HELL, a DOLL house encapsulates and represents the original house on Elm Street, etc. – as well as some from SCIENCE FICTION: a significant strand of TECH-NOFANTASY runs throughout. Despite Langenkamp's star billing, Arquette and Wasson make this movie theirs. The movie is something of an spfx-fest. [JG]

4. *A Nightmare on Elm Street Part 4: The Dream Master* US movie (**1988**). New Line/Heron/Smart Egg. **Pr** Robert Shaye, Rachel Talalay. **Exec pr** Stephen Diener, Sara Risher. **Dir** Renny Harlin. **Spfx** Image Engineering. **Vfx** Dream Quest Images. **Mufx** R. Christopher Biggs, Steve Johnson, Magical Media Industries, Screaming Mad George, Kevin Yagher. **Screenplay** Brian Helgeland, William KOTZWINKLE, Scott Pierce. **Starring** Brooke Bundy (Elaine Parker), Kristen Clayton (Little Girl), Robert Englund (Freddy), Danny Hassel (Dan Jordan), Tuesday Knight (Kristen Parker), Lisa Wilcox (Alice Johnson). 93 mins. Colour.

Freddy is, of course, not dead. Escaping his grave, he hunts down the survivors from **3**. Last to go is Kristen, who by now has confided all in friends Rick, Dan and notably the dreamy Alice. Although Kristen was the final child of the people who burnt Freddy alive, his appetite is not satiated, and Alice comes to realize that, through her dreams, she is "collecting" further teenagers to be killed by him. As each dies, Freddy takes their SOULS but Alice gains certain of their abilities. At last, with boyfriend Dan near death, she confronts Freddy in the dream REALITY with these powers, calling also on the mythological figure of the Dream Master – who guards the souls of sleepers – to help destroy him. Much of this movie is an incoherent excuse to create ever gorier spfx. Yet there are some good moments: a sequence in which Alice is sucked through a cinema screen to the world behind it is strikingly managed; and there is an excellent rendition of an anxiety DREAM. [JG]

5. *A Nightmare on Elm Street Part 5: The Dream Child* US movie (**1989**). New Line/Heron/Smart Egg. **Pr** Rupert Harvey, Robert Shaye. **Exec pr** Sara Risher, Jon Turtle. **Dir** Stephen Hopkins. **Spfx** Amorie G. Ellingson, Gary Sivertson. **Vfx** Alan Munro, Visual Concept Enginnering. **Mufx** David Miller. **Miniature fx** Jim Aupperle. **Screenplay** Leslie Bohem. **Starring** Erika Anderson (Greta), Robert Englund (Freddy), Danny Hassel (Dan Jordan), Whitby Hertford (Jacob), Kelly Jo Minter (Yvonne), Joe Seely (Mark), Lisa Wilcox (Alice Johnson). 89 mins. Colour.

Freddy ain't dead yet. Alice is pregnant by Dan, and Freddy is able to reinvade her life and slaughter a few teenagers (starting with Dan) *via* the foetus's dreaming mind. Alice has several encounters with her unborn child, who appears as a little boy called Jacob. There seems no way to stop Freddy except by, in either the DREAM or the mundane REALITY, discovering the bones of his dead mother Amanda so her SOUL may be released to take back her child. This is effected, and Alice's baby is born normally.

This movie is full of ideas, many interesting and most gruesome, but lacks any overall, driving idea. It is also full of spfx: there is insufficient space above for anything like a complete credits listing. One sequence exhibits both features at their best: the character Mark, an amateur illustrator, opens a COMIC book and discovers its story (about Freddy and the rest) is unfinished; he is drawn (pun intended) into the next empty frame, where he confronts Freddy in the dream reality; Mark transmutes into his own comic-book creation, The Phantom Prowler, and as such apparently defeats Freddy; but Freddy revives, converts the Prowler into a mere TOON, and rips him into paper tatters. [JG]

6. *Freddy's Dead: The Final Nightmare* US movie (**1991**). New Line. **Pr** Robert Shaye, Aron Warner. **Exec pr** Michael De Luca. **Dir** Rachel Talalay. **Vfx** Don Baker, Chandler Group, Dream Quest Images, John Scheele, Michael Shea. **Mufx** Magical Media Industries. **Screenplay** De Luca, Talalay. **Starring** Lezlie Deane (Tracy), Robert Englund (Freddy), Shon Greenblatt ("John"), Yaphet Kotto (Doc), Ricky Dean Logan (Carlos), Breckin Meyer (Spencer), Lisa Zane (Maggie). 90 mins. Colour.

Of course, Freddy is still "alive"; no explanation is offered. Ten years on, Springwood, Ohio, is devoid of children and teenagers, the adult residents living a curious CARNIVAL-like fantasy existence. The first half of this movie is standard for the series: a gory incoherence shot with some interesting ideas – as when a character becomes part of an arcade GAME, obeying its rules both within the game and in "real life". The second half more interestingly provides some sort of ratio-nale and conclusion to the series. Social worker Maggie is Freddy's child, adopted elsewhere when his crimes were discovered and through trauma amnesiac of her childhood; she, we discover, is the little girl (really named Catherine) who has so often acted as GUARDIAN OF THE THRESHOLD at the house on Elm Street. Travelling through an array of Freddy's memories, she discovers how childhood persecu-tion and physical abuse moulded the sadistic SERIAL KILLER; how at the time of his incineration the DREAM demons came to him, promising IMMORTALITY in the dream REALITY if he allowed them to possess him (◊ POSSESSION); and how his VENGEANCE on successive teenagers was sparked less by his

death than by the forcible removal of his daughter. He is indeed immortal (◊ IMMORTALITY) in the dream reality, but Maggie drags him into the mundane one and kills him. The second half justifies the movie. [JG]

7. *Wes Craven's New Nightmare* US movie (**1994**). Rank/New Line. **Pr** Marianne Maddalena. **Exec pr** Wes CRAVEN, Robert Shaye. **Dir** Craven. **Spfx** John C. Carlucci, Michael W. Menzel, Charles Schmitz and others. **Mechanical fx** Lou Carlucci. **Vfx** Flash Film Works, William Mesa. **Mufx** Howard Berger, Robert Kurtzman, David Miller Creations, Gregory Nicotero. **Screenplay** Craven. **Starring** Fran Bennett (Dr Heffner), Nick Corri (himself), Craven (himself), Robert Englund (Freddy/him-self), Miko Hughes (Dylan Porter), Tuesday Knight (herself), Heather Langenkamp (herself), Maddalena (her-self), Tracy Middendorf (Julie), David Newsom (Chase Porter), Sara Risher (herself), Sam Rubin (himself), John Saxon (himself), Shaye (himself). 112 mins. Colour.

Langenkamp (from **1** and **3**) is married to Porter, an spfx man among Craven's crew; they have a young son, Dylan. She is being tormented by phonecalls from a Freddy Krueger soundalike, by her terror of a series of earthquakes rocking the Los Angeles area, and by DREAMS relating to Freddy. Dylan is suffering from somnambulism – during which he watches **1** on an unplugged tv set and sings snatches of the series' recurrent nursery rhyme – and from fit-like attacks; he claims there is a man with claws at the foot of his bed who can be kept away only by the stuffed TOY Dinosaur Rex. Langenkamp is told Craven has been writing a fresh **NOES** script, and wishes her to play Nancy once more. She declines: now she has a child, HORROR MOVIES no longer seem such a good idea. Porter is killed in an apparent road accident; but his chest bears claw-marks. At his funeral Langenkamp has an HALLUCINATION (or a view into a diver-gent REALITY) that Freddy is in the coffin as well, and that both he and Porter are trying to drag Dylan with them. Actors from the series – notably Englund and Saxon – try to comfort her, but neither seems to believe her claim that Freddy is somehow emerging into "real life". Craven, how-ever, does: his new script, he says, portrays Freddy as an ancient EVIL being who can be at least temporarily captured by storytellers and forced to take on the shape the story-tellers decree; when the STORY is ended, however – as the **NOES** series has been – the SPIRIT is freed to act on the "real world" again. Craven believes that only through mak-ing one last movie, with Langenkamp as Nancy, can both the story and the reality of the being be terminated; she is the keeper of the PORTAL through which Freddy is trying to enter this world; Freddy is trying to bypass her through Dylan. Dylan is taken into hospital, where he is visited by babysitter Julie, whom Freddy dramatically slaughters. Dylan escapes hospital and, pursued by his mother, staggers home to find Dinosaur Rex. There, too, is Saxon; but he now has become Lt Thompson (from **1** and **3**), Nancy's father, and addresses her as his daughter. Langenkamp finds a trail of sleeping pills leading to Dylan's bed; recalling how the children at the end of *Hansel and Gretel*, his favourite FAIRYTALE, followed the trail of breadcrumbs home to safety, she swallows some of the pills and burrows to the foot of his bed, where she discovers the nightmare world. In that world she finds Craven's script, and reads from it that "There was no movie. There was only ... her ... life ..." Then she and Dylan battle Freddy directly, at last stuffing him into an oven (like the witch in *Hansel and Gretel*). As the flames consume him, he reverts (◊ SHAPESHIFTER) to his DEMON form and is plunged back to HELL.

This is an astonishing exercise in RECURSIVE FANTASY and an act of reverence to the power of STORY. It is also a contribution to the debate about the ill-effects of HORROR MOVIES on CHILDREN: its conclusion seems to be that, so long as there is resolution (as in *Hansel and Gretel*), no harm can ensue (a point it doesn't quite make). The most frightening of the series, this is probably the least gory; also to its credit, it portrays frankly the pain of bereavement, while parodying the shallowness of Hollywood society (Shaye, in particular, is either a brilliant self-parodist or did not realize what Craven was doing to him). There are reprises of scenes from **1**, notably in the butchery of Julie, which recalls that of **1**'s Tina; but they are more skilfully staged and timed, while Freddy himself has been redesigned to be less of a figure of fun. This is a noteworthy fantasy movie by any standard. [JG]

Novelizations: *The Nightmares on Elm Street, Parts 1, 2 and 3* * (**1987**) by Jeffrey Cooper; *The Nightmares on Elm Street, Parts 4 & 5* * (**1989**) by Joseph Locke (Ray Garton); *Wes Craven's New Nightmare* * (**1994**) by David Bergantino.

NIGHTMARES ◊ DREAMS.

NIGHT STALKER, THE US movie (**1971** tvm). ◊ KOLCHAK MOVIES.

NIGHT STRANGLER, THE US movie (**1972** tvm). ◊ KOLCHAK MOVIES.

NIGHT VISIONS ANTHOLOGY series from SMALL PRESS publisher Dark Harvest, most of whose contents are HORROR; volumes published several stories each by a small number of contributors. The series ran: *Night Visions 1* (anth **1984**; vt *Night Visions: In the Blood* 1988) ed Alan Ryan (1943-   ), including stories by Charles L. GRANT, Tanith LEE and Steve Rasnic Tem (1950-   ); *Night Visions 2* (anth **1985**) ed Charles L. Grant, with stories by Joseph Payne BRENNAN, David Morrell (1943-   ) and Karl Edward WAGNER; *Night Visions 3* (anth **1986**) ed George R.R. MARTIN, with stories by Clive BARKER, Ramsey CAMPBELL and Lisa TUTTLE; *Night Visions 4* (anth **1987**) ed Paul J. Mikol, with stories by Barker, Edward BRYANT, Dean R. KOONTZ and Robert R. McCammon (1952-   ); *Night Visions 5* (anth **1988**) ed Mikol, with stories by Stephen KING, Martin and Dan SIMMONS; *Night Visions 6* (anth **1988**) ed Mikol, with stories by Ray Garton, Sheri S. TEPPER and F. Paul WILSON; *Night Visions 7* (anth **1989**) ed Stanley Wiater (1953-   ), with stories by Gary Brandner (1933-   ), Richard Laymon (1947-   ), Witater and Chet WILLIAMSON; *Night Visions 8* (anth **1991**) ed Mikol, with stories by John Farris (1936-   ), Stephen Gallagher (1954-   ), Joe R. Landsdale (1951-   ) and McCammon; and *Night Visions 9* (anth **1991**) ed Mikol, with stories by Rick Hautala (1949-   ), James Kisner (1947-   ), Thomas Tessier (1947-   ) and Wilson. [JC]

NILES, DOUGLAS (?   -   ) US writer who has concentrated on ties, beginning with a tie (after Leiber's death) to Fritz LEIBER's **Fafhrd and the Grey Mouser** sequence, *Lankhmar, City of Adventure* * (**1985** chap) with Bruce Nesmith and Ken Rolston. DN has mainly contributed to the **Forgotten Realms** enterprise, beginning with *Lord of Doom: A DragonLance Adventure* * (**1986**), continuing the game-based **Moonshae** sequence, *Darkwalker on Moonshae* * (**1987**), *Black Wizards* * (**1988**), *Darkwell* * (**1989**) and *Prophet of Moonshae* * (**1992**); and going on with the **Maztica** sequence – *Ironhelm* * (**1990**), *Viperhand* * (**1990**) and

Feathered Dragon * (**1991**) – being fast-paced adventures in a world resembling pre-Columbian America. Other titles include: *Flint, the King* * (**1990**) with Mary KIRCHOFF, part of **DragonLance Preludes**; a contribution to the **Elven Nations** sequence, *The Kinslayer Wars* * (**1991**); two volumes in the **Druidhome** sequence, *The Coral Kingdom* * (**1992**) and *Druid Queen* * (**1993**); a contribution to the **Villains** sequence, *Emperor of Ansalon* * (**1993**); the first volume in the **Lost Histories** sequence, *The Kagonesti* * (**1995**); and two volumes in the **First Quest** sequence, *Pawns Prevail* * (**1995**) and *Suitors Duel* * (**1995**).

A Breach in the Watershed (**1995**), not a tie, is set in a moderately interesting tripartite FANTASYLAND divided into human, FAERIE and EVIL Darkblood regions. [JC]

NIMMO, JENNY (1944-   ) UK writer of fiction for children, notably the Welsh-flavoured fantasies *The Snow Spider* (**1986**), *Emlyn's Moon* (**1987**; vt *Orchard of the Crescent Moon* 1989 US) and *The Chestnut Soldier* (**1989**), assembled as *The Snow Spider Trilogy* (omni **1993**). These fine novels are set in the present day but based on the MABINOGION, somewhat after the fashion of Alan GARNER's *The Owl Service* (**1967**). The first two volumes won the Smarties Prize and the Welsh Arts Council Tir na n'Og Award respectively, and the series formed the basis of a trio of UK tv serials (1988-91) titled as per the books. In addition to many picture-books for younger children, JN has also written the novels *Ultramarine* (**1991**), *Rainbow and Mr Zed* (**1992**) and *Griffin's Castle* (**1994**). [DP]

NIMUË ◊ LADY OF THE LAKE.

NINIANE ◊ LADY OF THE LAKE; MERLIN.

NIÑO, ALEX (1940-   ) Filipino COMIC-book artist of remarkable versatility and originality whose pen-and-brush line style, uniquely personal colour sense and remarkably fertile imagination have subtly influenced many artists working in the US comics field. His technical virtuosity and innovative layouts, allied to a seemingly inexhaustible inventiveness, have led to his being described as an "artist's artist", with comic-book readers either loving or hating his work.

AN worked as a photographer and musician before approaching comic-book publishers as an illustrator. His first work of significance was the *Arabian Nights*-style *Calibot ng Persia* ["The Terror of Persia"] (Philipino Komiks 1965). A year later he wrote and drew *Gruaga - The Fifth Corner of the World* (Pioneer Komiks 1966), which included themes to which AN returned from time to time over the succeeding decade or so, including a short wordless story in *Weird Heroes* #6 (1977) and the portfolio *Dark Suns of Gruaga* (graph **1978**). He drew a wide range of fantasy stories for several publishers, his most ground-breaking achievement during this period being the serialized graphic novel *Mga Matang Nagliliyab* ["The Eyes that Glow in the Dark"] (*Alcala Komiks, c1970*), written by Marcelo B. Isidro.

AN began working for the US comics market in the early 1970s, drawing mainly short pieces for the DC COMICS horror titles *The House of Mystery, House of Secrets, Weird War,* etc., and for the Pendulum Press series of comic-book adaptations of classic literature, including H.G. WELLS's *The Time Machine* (graph **1973** US), *The Invisible Man* (graph **1974** US) and *The War of the Worlds* (graph **1974** US), Herman MELVILLE's *Moby-Dick* (graph **1974** US) and Alexandre DUMAS's *The Three Musketeers* (graph **1974** US). He also drew a few short stories featuring the Edgar Rice BURROUGHS character **Korak** for DC and some of the

Sunday pages of TARZAN for United Features Syndicate. He embarked upon his most creative and experimental period when he began illustrating US large-format b/w magazines with *Unknown Worlds of Science Fiction* (#3-#6 1975), for which he drew Michael MOORCOCK's *Behold the Man* and a remarkable adaptation of Harlan ELLISON's *Repent, Harlequin, Said the Tick Tock Man*. His experiments in style and page layout continued in his work for WARREN PUBLISHING, most significantly for the sf title *1984* (retitled *1994* in 1980), where his pages became more and more freely designed, the reader's eye being led expertly through a multitude of interlinked images across each spread. He began to play with visual narrative ideas, drawing stories in which the pages could be laid end to end to produce a single overall composition, and in one instance using the techniques of the wallpaper designer to achieve a single story, the pages of which could be laid alongside one another to create an endlessly repeating design. He experimented with many different styles, too, using at various times a frenzied pen hatching, a delicate brush line, a moody b/w on grey paper, or an exaggerated cartoon style.

AN's GRAPHIC NOVELS include an adaptation of Theodore STURGEON's *More than Human* (1953; graph 1978 US), *Space Clusters* (graph 1986 US) written by Arthur Byron Cover and *Tales of the One-Eyed Crow* (graph 1991 US) written by Dennis L. MCKIERNAN. He has also drawn two CONAN stories (in *Savage Sword of Conan #6*, 1978, and *#228*, 1994). [RT]

Other works: *Satan's Tears: The Art of Alex Niño* (graph 1977 US); *The Fantastic Worlds of Alex Niño* (graph 1975 US), portfolio.

NISBET, HUME (1849-1923) Scottish writer and artist, for part of his life resident in Australia. His travels in the Southern Hemisphere provided material for a run of adventure romances. Nine or so of his 46 novels contain fantasy elements. *The Jolly Roger* (1892) involves mass hypnotism and a hidden PIRATE island; *Valdemar the Viking* (1893) is an 11th-century ROMANCE almost certainly inspired by the works of H. Rider HAGGARD – a re-incarnated (◊ REINCARNATION) Greek is able to revive his lost love from the frozen vastness of the North Pole; *The Great Secret* (1895) reveals an ISLAND of the dead and the discovery of the Hesperides, while *The Empire Builders* (1900) reveals another Haggard-like LOST RACE in South Africa. These works are essentially imaginative potboilers. HN may be longer remembered for his short stories, of which the title story to *The Haunted Station* (coll 1894), a GHOST STORY set in the Australian outback, is his best. "The Demon Spell" is notable for its JACK THE RIPPER theme. Those collected in *Stories Weird and Wonderful* (coll 1900) utilize more traditional themes, with "The Old Portrait", about a VAMPIRE revived from a painting, being the most original. [MA]

Other works (selective): *Paths of the Dead* (1899), *The Revenge of Valerie* (1900), *The Divers* (1900), *A Crafty Foe* (1901), *A Colonial King* (1905).

NIVEN, LARRY Working name of US writer Laurence van Cott Niven (1938-), extensively honoured for his contributions to sf (◊ SFE). Occasional fantasies use the same imaginatively analytical approach to good RATIONALIZED-FANTASY effect. Thus "Convergent Series" (1967 *F&SF* as "The Long Night") amusingly employs a logical QUIBBLE to trap a DEMON in infinite mathematical regress. "Not Long Before the End" (1969 *F&SF*) launches the prehistoric **Magic Goes Away** sequence, with its premise that

mana (LN also spells it *manna*) which fuels MAGIC is an exhaustible natural resource: a wasteful SPELL designed to squander *mana* will drain all magic in the vicinity, destroying even "invincible" demons. THINNING is thus inevitable; even the spells which hold off the tectonic instability of ATLANTIS must fade. This notion has been used as a PLAYGROUND by other writers. In *The Magic Goes Away* (1978) the MOON is considered as an outside source of *mana*, the Midgard SERPENT proves to have degenerated (◊ DEBASEMENT) into a still world-circling but barely alive mountain range, and Earth's last major *mana*-focus is an uncontrollable GOD who threatens the END OF THE WORLD. LN wrote other more or less related **Magic Goes Away** stories, some assembled as *The Time of the Warlock* (coll 1984); SHARED WORLD additions from others appear in *The Magic May Return* * (anth 1981) and *More Magic* * (anth 1984).

The **Svetz** SCIENCE FANTASIES collected in *The Flight of the Horse* (coll 1973) assume that TIME TRAVEL is logically impossible and therefore pure fantasy – so a future sf time machine trawling the deep past for now-extinct horses, whales, etc., finds Pegasus, Leviathan (◊ SEA MONSTERS) ...

Inferno (1975) with Jerry Pournelle (1933-) is a RECURSIVE FANTASY revisiting the *Inferno* of DANTE ALIGHIERI's *The Divine Comedy* (*c*1304-21) and, with resolute tastelessness, extending Dante's own habit of allocating political enemies their place in HELL: *inter alia*, LN and Pournelle damn fanatical book collectors, environmentalists and Kurt Vonnegut (1922-). The story moves well, with the protagonist desperately rationalizing demonic horrors as TECHNOFANTASY constructs despite advice from his VIRGIL-like guide, Benito Mussolini (1883-1945). Hell is explained as "the violent ward for the theologically insane", its brutal sadism intended to shock SOULS awake from self-obsessed self-damnation (as C.S. LEWIS's arguably less compassionate God does not) and ready them for PURGATORY. By the time Mussolini escapes along the route taken by Dante, the protagonist has learned better (◊ LEARNS BETTER) and voluntarily returns to be the Virgil for others – a moral conclusion partly redeeming earlier crassness.

The **Dream Park** series – *Dream Park* (1981), *The Barsoom Project* (1989) and *Dream Park: The Voodoo Game* (1991 UK; vt *The California Voodoo Game* 1992), all with Steven Barnes (1952-) – is sf featuring fantasy role-playing GAMES of post-Disneyland realism, drawing on unusual MYTH systems: e.g., Melanesian and Eskimo. [DRL]

Other works: Much sf. ◊ SFE.

NODIER, CHARLES (1780-1844) Working name of French writer Jean Charles Emmanuel Nodier (1780-1844), whose early writings were aesthetically delicate FIN DE SIÈCLE romances and poems. Some of this work was read as subversive; one ode, *La Napoléone* (1802), landed him briefly in jail. His first work of genre interest – directly based on John POLIDORI's *The Vampyre: A Tale* (1819 chap) – is *Le Vampire* (1820) (◊ VAMPIRE), a play in which Polidori's Lord Ruthven (based on Lord BYRON) again appears and finally dies; this version was immediately adapted by James R. PLANCHÉ for the English stage as *The Vampire, or The Bride of the Isles* (1820). For some time it was suspected that CD had also written the anonymous *Lord Ruthwen ou les Vampires* (1820 2 vols), a follow-up to the Polidori tale that constitutes the first full-length vampire novel, but he was responsible only for the introduction; the book itself was written by Cyprien Bérard.

CN's later life was industrious and hermetic – he was the

founder of a reclusive "*cénacle*" whose members were or became influential in French Romanticism. CN's own fantasies are restricted to two titles, *Smarra, ou Les Démons de la nuit* ["Smarra, or The Demons of the Night"] (**1821**) and *Trilby, ou le lutin d'Argail* (**1822**; trans anon as *Trilby, the Fairy of Argyle* **1895** chap US); the two texts were retranslated by Judith Landry for *Smarra & Trilby* (omni **1993** UK). Inspired by Henry FUSELI's *The Nightmare* (1781), *Smarra* is a literary DREAM whose convolutions lead a Byronic CHILDE brigand downwards through WILL-O'-THE-WISP-ridden forests into ARABIAN-NIGHTMARE territory, where the eponymous vampire – a malign figure spun out of the dreamworld – sucks his blood. *Trilby*, which derives its Scottish geography and FOLKLORE resonance from Sir Walter SCOTT, is a tale of psychic BONDAGE in which a mortal woman is seduced and entrapped by an imp who takes on the semblance of a man; George DU MAURIER's *Trilby* (**1894**) makes no reference to this story.

It has been speculated that, sometime before 1813 – when relevant portions of the text in question were first published in France – CN wrote the initial, complicatedly supernatural sections that comprise the first 10 of the 66 chapters of *Manuscrit Trouvé à Saragosse* (full text **1989** France; trans Ian Maclean as *Manuscript Found at Saragossa* **1995** UK), a work whose traditional attribution to Jan POTOCKI seems, nevertheless, secure; certainly no plausible reason for CN's remaining quiet about the composition has been put forward. [JC]

Other works: *Trésor des Fèves et Leur des Pois* (trans anon as *Bean Flower and Pea Blossom* **1846** chap UK; new trans anon as *The Luck of the Bean-Rows* **1921** chap UK) and *Histoire du chien de Briquet* (trans as *The Woodcutter's Dog* **1922** chap UK), two literary FAIRYTALES in verse.

NOGGIN THE NOG, THE SAGA OF UK tv series (1959-65). ◊ *The* CLANGERS (1969-73).

NOLAN, WILLIAM F(RANCIS) (1928-) US writer and editor, active in various genres from the early 1950s, publishing his first story of genre interest, "The Joy of Living", for *If* in 1954; he scripted it for tv in 1971. In the early years of his career he was best known for his sf (◊ *SFE*), including *Logan's Run* (**1967**) along with its sequels, the movie (*1976*), and the tv series (1977; pilot plus 13 episodes). More recently he has concentrated on HORROR, with stories like those assembled in *Things Beyond Midnight* (coll **1984**), novels like *Helltrack* (**1990**) and tv scripts including *Trilogy of Terror II* (**1990**). *How to Write Horror Fiction* (**1991**) is an informed, casual guide.

Some of WFN's stories written with Charles BEAUMONT are DARK FANTASY, as are some later solo tales. His long-standing interest in the work of Ray BRADBURY produced *The Ray Bradbury Companion* (**1975**), which includes a bibliography, and *The Bradbury Chronicles: Stories in Honor of Ray Bradbury* (anth **1991**) with Martin H. GREENBERG. *The Work of Charles Beaumont* (**1986** chap; exp **1991** chap) is a bibliography. [JC]

NONSENSE Intentionally nonsensical fantasy is something of a UK tradition, thanks to Lewis CARROLL and Edward LEAR: Carroll's **Alice** books offer a kind of intellectual nonsense based on perverse READ-THE-SMALL-PRINT interpretations of logic and idiom, while Lear exploits disconcerting whimsies, non sequiturs, and unexplained nonce-words like "runcible". G.K. CHESTERTON argued in "A Defence of Nonsense" (in *The Defendant* coll **1901**) that this made Lear the superior fabulist, a view not universally shared. Reversing Carroll's dictum "Take care of the sense and the sounds will take care of themselves", the backbone provided by metre and rhyme makes nonsense verse more generally popular than pure nonsense prose (Lear's eccentric prose pieces, like "The Story of the Four Little Children Who Went Round the World" in *Nonsense Songs, Stories, Botany and Alphabets* [**1871**], are often forgotten). Carroll's JABBERWOCKY, its offspring *The Hunting of the Snark* (**1876**), and Lear's "The Dong with a Luminous Nose" (in *Laughable Lyrics* coll **1877**) have gained their own independent fame; Mervyn PEAKE also wrote some notable nonsense verse in (e.g.) *A Book of Nonsense* (coll **1972**). But nonsense in prose fantasy tends to come in fleeting snatches. When seeming gibberish is uttered by FOOLS, Tom o' Bedlams or (especially) ORACLES, it usually conceals bitter truth, as layers of meaning are concealed within the labyrinthine and Carrollian portmanteau-wordplay of James JOYCE's *Finnegans Wake* (**1939**). The absurd messages flung by the sinister Sunday to pursuing detectives in Chesterton's *The Man Who Was Thursday* (**1908**) seem crudely farcical, but hint that the pursuers are asking the wrong question of the Universe (for Sunday is Nature) and thus compelling silly answers. In James THURBER's *The Thirteen Clocks* (**1950**) the bad Duke's intention of slitting someone open from "guggle" to "zatch" uses nonsense words to cheerily grim effect. Diana Wynne JONES's *Fire and Hemlock* (**1985**) similarly employs a wildcard nonsense phrase, the "Obah Cypt", for a kind of PLOT COUPON whose nature emerges only late in the book. A seemingly meaningless quatrain imagined in the rattle of a TRAIN's wheels in Neil GAIMAN's *The Kindly Ones* (graph coll **1996**) contains a prophecy of the story's outcome. Nonsense in modern fantasy is often a MASK hiding important information. [DRL]

See also: ABSURDIST FANTASY; MAGIC WORDS; SATIRE; SURREALISM.

NORDIC FANTASY That body of FANTASY which draws its heart from the MYTHOLOGY of the Scandinavian and Teutonic races and incorporates the stories retold in the SAGAS. The Nordic LEGENDS formed the religion of the Saxons and Vikings who settled much of Western Europe in the 5th to 11th centuries. Along with Greek and Celtic mythology, the Nordic legends incorporate the biggest body of folk-memory in Western tradition. Their origins are later than Greek, Celtic or Hebrew mythology, probably dating to the early years of the first millennium AD. The Nordic myths suggest vast movements of peoples, as races settled in the northern latitudes. At their core are the adventures of the AESIR, led by ODIN, and the Vanir, who invade the Northlands and battle against the GIANTS or Jotunn, who live in Jottunheim. The Aesir befriend the DWARFS, who live in Alfheim. The Aesir eventually triumph as the more powerful race and settle in their city of Asgard. The Aesir, who became regarded as the GODS of the Nordic races, were eventually destroyed in RAGNAROK. The story of the Aesir is told in the Icelandic Eddas, of which the two most important are the *Elder Edda* (?**1090**), attributed to the Icelandic historian Saemund (11th century), but probably edited by him from other sources – the manuscript was rediscovered by Brynjulf Sveinsson, an Icelandic bishop, in 1643 – and the *Younger Edda* or *Prose Edda* (?**1220**) by Snorri Sturlason (1179-1241), another Icelandic historian. The latter, the more important of the two, was not rediscovered until 1625. It is from these Eddas that the *Volsunga Saga* (final form 13th century) is derived, early variants of which in turn structured the final version of the German *Nibelungenlied* (final shape 1210, but in embryonic form as early as AD960). Both lays feature the same HEROES, though with some slight changes of name – e.g., Sigurd in the *Volsunga Saga* becomes Siegfried, Gudrun becomes Kriemhild.

Because the Norsemen were strong on ancestor-worship, the early part of the SAGA is almost like a DYNASTIC FANTASY, but the main part of the story centres on the grandchildren of Volsung – the great-grandson of Odin – and in particular on Sigurd. After avenging the death of his father, Sigurd sets out with his tutor Regin, a WIZARD, to fight Fafnir (Regin's brother transformed into a DRAGON because he had murdered his father to gain his father's treasure). Sigurd defeats Fafnir and learns of the treachery of Regin, whom he also kills. He encounters Brynhild in an ensorcelled sleep in a castle guarded by flame (◊ SLEEPING BEAUTY); he wakes her and the two are betrothed. Later, though, Sigurd is given a potion of forgetfulness, and instead marries Gudrun, daughter of Queen Grimhild. Distraught, Brynhild seeks Sigurd's death, then kills herself. Sigurd's treasure is buried and lost. Gudrun places herself in exile but later also takes the potion of forgetfulness to rid herself of her memories. She later marries King Atli, the brother of Brynhild, who desires Sigurd's treasure. He summons help to his court but this comes in the shape of Gudrun's brothers, who are ill received. The battle that ensues kills most of the cast. The saga ends in the death of all the main players.

The Nordic and Teutonic myths were readily adopted by the German Romantics, particularly Baron de la Motte FOUQUÉ, who used elements in most of his stories, including the trilogy **Der Held des Nordens** ["The Hero of the North"] (omni **1810**), which inspired Richard WAGNER to write the **Ring Cycle** of OPERAS and Friedrich Hebbel (1813-1863) his dramatic trilogy *Die Nibelungen* (**1862**).

William MORRIS was totally smitten by "the Nordic thing", and studied Icelandic with Eiríkr Magnússon in order to translate the Sagas, travelling to Iceland in 1871 and 1873 to absorb the atmosphere. Although his early prose works, particularly *The Roots of the Mountains* (**1889**), use Nordic settings and characters, they do not draw heavily on the Nordic myths, and Morris moved further away in his later writings. The same is true of S. BARING-GOULD, who produced *Grettir the Outlaw* (**1889**) but did not explore the stories further. To this same period belongs *Eric Brighteyes* (**1891**) by H. Rider HAGGARD, but this is ostensibly a heroic adventure and not a fantasy.

The world of the Aesir, just before Ragnarok, was lightheartedly explored by L. Sprague DE CAMP and Fletcher PRATT in "The Roaring Trumpet" (1940 *Unknown*), later incorporated into *The Incomplete Enchanter* (fixup **1941**), introducing a strain of humour that has since been followed by Tom HOLT in *Expecting Someone Taller* (**1987**). A more serious treatment of the Norse gods was made by Harry Harrison (1925-   ) and Katherine MacLean (1925-   ) in the novella "Web of the Worlds" (1953 *Fantasy Fiction*; rev vt "Web of the Norns" 1958 *Science Fantasy*), in which the Norse FATES argue over the life of a young man. Poul ANDERSON has often returned to the Nordic legends – utilizing them in *The Broken Sword* (**1954**; rev 1971), *Hrolf Kraki's Saga* (**1973**) and the **Last Viking** sequence.

Other examples of fantasies derived from the Eddas and Sagas are: *Votan* (**1966**) by John JAMES; the **Odan** series by

Kenneth BULMER (as Manning Norvil); the **Vidar** sequence by Michael Jan FRIEDMAN; the **Bifrost Guardians** series by Mickey Zucker REICHERT; and the **Song** sequence by Thorarinn GUNNARSSON. A thorough treatment of the whole Nibelungenlied is *Rhinegold* (**1994**) by Stephan GRUNDY, which faithfully follows the legend but seeks to set it within a meaningful historic framework. Diana PAXSON has worked along similar lines with her **Wodan's Children** trilogy.

The Eddas and their offspring are not the only well known Nordic legends. Possibly better known in English-speaking countries is BEOWULF. Although of unknown authorship, the tale almost certainly originated in Denmark before the Saxon invasions of Britain, when it was further embellished, reaching its present form around the 8th century (the one extant manuscript dates from *c*1000). Three fantasy novels based on the tale are *Beowulf* (**1961**) by Rosemary SUTCLIFF, *Grendel* (1971) by John GARDNER and *The Tower of Beowulf* (**1995**) by Parke GODWIN.

The other main series of Teutonic legends revolve around the exploits of Dietrich of Bern, a hero based on the historical figure of Theodoric (?454-526), king of the Ostrogoths. Like ARTHUR and Sigurd, Dietrich has become larger-than-life, capable of superhuman feats, and doing battle against ogres, giants and dwarfs. Dietrich's adventures appear as early as the 7th-century *Hildebrandslied* ["Song of Hildebrand"] and he also figures in the *Nibelungenlied*, but his main STORY-CYCLE is that brought together in *Das Heldenbuch* ["The Book of Heroes"] (**1472**) by Kaspar von der Roen. Because Dietrich is expelled from his homeland by Ermenrich at the start of the story, and remains in exile for 30 years, he may be regarded as an early example of the ACCURSED WANDERER. His adventures were cast into FOLKTALE form in the Norwegian *Thidreksaga* ["Theodoric's Saga"] (?**1250**). Dietrich's exploits have not been the subject of REVISIONIST FANTASY as often as Sigurd's, but his story was admirably retold in *Epics and Romances of the Middle Ages* (trans **1883**) by Wilhelm Wägner. [MA]

NORFOLK, LAWRENCE (1963-   ) UK writer whose first novel, *Lempriere's Dictionary* (**1991**; cut 1992 US), an elaborate FANTASY OF HISTORY set in the early 19th century, depicts vast AUTOMATA and sinister immortals (◊ IMMORTALITY) under a phantasmagoric LONDON, murders that involve the re-enactment of classical myth, and a conspiracy that ties together the Siege of La Rochelle, the British East India Company and the origins of the French Revolution. The US edition is cut to a point which renders much of this unclear. LN's second novel, «*The Pope's Rhinoceros*» (1996), is a MAGIC-REALISM historical novel about a 16th-century QUEST by two dim men-at-arms for the eponymous beast and their own self-respect, a quest which briefly involves some use of African myth and magic. [RK]

NORMAN, JAY [s] ◊ Robert ARTHUR.

NORMAN, JOHN Pseudonym of US academic and writer John Frederick Lange Jr (1931-   ), whose debut novel *Tarnsman of Gor* (**1966**) was the first of a lengthy PLANETARY ROMANCE sequence. **Gor** is a world on the far side of our Sun, and its barbarous slave-owning culture is continually supplied with manpower (and, more significantly, *woman*power) shanghaied from Earth. The first book and its immediate sequels, *Outlaw of Gor* (**1967**) and *Priest-Kings of Gor* (**1968**) – all three assembled as *Gor Omnibus: The Chronicles of Counter Earth* (omni **1972** UK) – are passable exercises in the Edgar Rice BURROUGHS mode, as are #4 and #5, *Nomads of Gor* (**1969**) and *Assassin of Gor* (**1970**). However, later volumes degenerate into extremely sexist, sadomasochistic pornography involving the ritual humiliation of women, and as a result have caused widespread offence: *Raiders of Gor* (**1971**), *Captive of Gor* (**1972**), *Hunters of Gor* (**1974**), *Marauders of Gor* (**1975**), *Tribesmen of Gor* (**1976**), *Slave Girl of Gor* (**1977**), *Beasts of Gor* (**1978**), *Explorers of Gor* (**1979**), *Fighting Slave of Gor* (**1980**), *Guardsman of Gor* (**1981**), *Rogue of Gor* (**1981**), *Blood Brothers of Gor* (**1982**), *Savages of Gor* (**1982**), *Kajira of Gor* (**1983**), *Players of Gor* (**1984**), *Mercenaries of Gor* (**1985**), *Dancer of Gor* (**1985**), *Renegades of Gor* (**1986**), *Vagabonds of Gor* (**1987**) and *Magicians of Gor* (**1988**). Though deplored by critics, the series was popular, and its opening volumes inspired two obscure movies, *Gor* (*1987*) dir Fritz Kiersch and *Outlaw of Gor* (*1987*) dir John Cardos. In the early 1990s JN attempted to relaunch his by-then stalled career with a new series, the **Telnarian Histories** – *The Chieftain* (**1991**), *The Captain* (**1993**) and *The King* (**1993**) – but this seems to have caused much less of a stir. [DP] **Other works:** *Ghost Dance* (**1969**); *Imaginative Sex* (**1974**), nonfiction; *Time Slave* (**1975**).

NORNS ◊ FATES.

NORTH, ANDREW Pseudonym of Andre NORTON.

NORTON, ANDRE Working name of US writer and ex-librarian Alice Mary Norton (1912-), who has also written as Andrew North and Allen Weston. A prolific writer, initially of historical adventures, AN made her name as the author of the **Witch World** series which, along with other novels of the period, were adopted as SCIENCE FICTION but are essentially fantasy, as is much of her output. (Her sf output, mostly YA, is detailed in *SFE*.)

AN began writing in the 1930s with the nonfantasy *The Prince Commands* (**1934**), though at the same time she was seeking story sales in the sf field. Manuscripts accepted by William Crawford (1911-1984) for his semi-professional magazine *Marvel Tales* in 1934 had to wait nearly 40 years for their complete publication in *Garan the Eternal* (coll **1972**), although a few samples, such as "The People of the Crater" (1947 *Fantasy Book*), surfaced earlier. AN's first-published fantasy books were *Rogue Reynard* (**1947**) and *Huon of the Horn* (**1951**), both YA and based on medieval ROMANCEs (*Reynard the Fox* and *The Song of Roland* respectively). The lack of a good fantasy market caused her to focus on sf during the 1950s, but thanks to her editor at Ace Books, Donald A. WOLLHEIM, AN was able to return to the fantastic adventure with *Witch World* (**1963**). Although the story starts as sf it soon develops as a PLANETARY ROMANCE. Simon Tregarth passes through a PORTAL in a Cornish stone (the Siege Perilous) and finds himself in the land of Estcarp on the planet Witch World, where he helps the inhabitants against the invading Kolder. The natives are called witches because they have psi-powers (◊ TALENTS) aided by jewels. The popularity of the book resulted in a series, and in the later titles AN shifted away from the sf premise and developed instead the psychic/magical abilities of the witches and the world they live in, and the mysterious Old Ones who, we presume, were earlier inhabitants or colonists who developed the superior technology of the world and created the portals. The core series includes *Web of the Witch World* (**1964**), *Year of the Unicorn* (**1965**) – first three assembled as *Annals of the Witch World* (omni **1994**) – *Three Against the Witch World* (**1965**), *Warlock of the Witch World* (**1967**) and *Sorceress of the Witch*

World (**1968**). AN then developed a parallel series set in the lands of Arvon and High Halleck, which are across the sea from Estcarp; here the society is less matriarchal and the inhabitants have SHAPESHIFTING abilities. This series includes *Year of the Unicorn* (**1965**), *The Crystal Gryphon* (**1972**), *Spell of the Witch World* (coll **1972**), *The Jargoon Pard* (**1974**), *Gryphon in Glory* (**1981**) and, with A.C. Crispin (1950-), *Gryphon's Eyrie* (**1984**). Other stories in the series are rather more loosely adapted to the framework and fill in gaps in the world's history. The earliest setting is in *Horn Crown* (**1981**). The rest are *Trey of Swords* (**1977**), *Zarsthor's Bane* (**1978**), *Lore of the Witch World* (coll **1980**), *'Ware Hawk* (**1983**), *Were-Wrath* (**1984** chap), *Serpent's Tooth* (**1987** chap) and *The Gate of the Cat* (**1987**). **Witch World** has become a SHARED WORLD, with books either edited or codeveloped with AN. With P.M. Griffin (1947-), AN has written *Storms of Victory* (**1991**) and with Crispin *Songsmith* (**1992**). AN has compiled *Tales of the Witch World* (anth **1989**), *Tales of the Witch World II* (anth **1988**), *Four from Witch World* (anth **1989**), *Tales of the Witch World III* (anth **1990**), plus *Flight of Vengeance* (anth **1992**) in collaboration with Griffin and Mary Schaub, and *On Wings of Magic* (anth **1993**) with Patricia Mathews and Sasha Miller. A new subseries, **Secrets of the Witch World**, has been developed with Lyn McConchie, starting with *The Key of the Keplian* (**1995**). The series, while itself derivative – particularly of the works of Leigh BRACKETT – has also been immensely influential, most evidently in the **Darkover** series by Marion Zimmer BRADLEY and the **Diadem** books by Jo CLAYTON, and to some extent in the **Pern** stories by Anne MCCAFFREY. The stories improved once AN got into her stride – the books of the early 1970s are strongest on character and plot – yet, despite their popularity, have an increasing repetitiveness.

The same can be said of her **Magic** books, which are derivative of the CHILDREN'S FANTASY novels by Edith NESBIT, where everyday objects become magical TALISMANS able to whisk children into past times or worlds of legend. These TIME FANTASIES began with *Steel Magic* (**1965**; vt *Gray Magic* 1967), where knives and forks transported the children back to the age of ARTHUR. In *Octagon Magic* (**1967**) a Victorian house is a portal to the Civil War, and in *Fur Magic* (**1968**) an Indian medicine bag transports a girl back to the days of Indian legend. These three were assembled as *The Magic Books* (omni **1988**). Others in this loosely connected series are *Dragon Magic* (**1972**), in which a magic box transports four children back to each of their ancestral origins; *Lavender-Green Magic* (**1974**), where a maze in an English stately home transports children to Puritan England; and *Red Hart Magic* (**1976**), where the model of a 16th-century English inn is the portal to that period.

AN's other solo fantasies follow similar patterns, usually involving children or adults transported through time or into other dimensions. *Here Abide Monsters* (**1973**) takes two children into an alternate universe of AVALON. *Merlin's Mirror* (**1975**), another Arthurian fantasy, portrays MERLIN and ARTHUR as aliens with superscientific technology. Somewhat different is *Mark of the Cat* (**1992**) where AN develops Karen Kuykendall's cat paintings into a story about the CAT people. *The Hands of Llyr* (**1994**) tells of a couple's search for light in a land of darkness, and may almost be treated as an ALLEGORY. *Mirror of Destiny* (**1995**) uses a mirror to stop a war between mankind and FAERIE, and is perhaps the most refreshing of her novels.

Much of AN's later work is in collaboration. The main OTHERWORLD fantasy is the **Black Trillium** series, developed from a concept by literary agent Uwe Luserke. The first volume, *Black Trillium* (**1990**), is a collaboration with Marion Zimmer BRADLEY and Julian MAY, a modern FAIRY-TALE (with many TECHNOFANTASY and PLANETARY-ROMANCE elements) about three princesses seeking to regain their kingdom. The series was then continued by each writer individually, AN's novel being *Golden Trillium* (**1993**). With Mercedes LACKEY AN wrote the **Halfblood** series starting with *The Elvenbane* (**1991**) and *Elvenblood* (**1995**), set in a world where cruel ELVES do battle with men and shapeshifters. AN's less typical collaborations are with Phyllis Miller (1920-), beginning with the Nesbitesque *Seven Spells to Sunday* (**1979**) and including the admirable *House of Shadows* (**1984**). With Robert BLOCH she wrote *The Jekyll Legacy* (**1990**).

Although AN has concentrated on novels she has produced enough short fiction to fill the following volumes: *The Many Worlds of Andre Norton* (coll **1974**; vt *The Book of Andre Norton* 1975) ed Roger Elwood (1943-), *High Sorcery* (coll **1970**), *Perilous Dreams* (coll **1976**), *Moon Mirror* (coll **1988**), *Grand Master's Choice* (coll **1989**) and *Wizards' Worlds* (coll **1989**), the last three compiled with Ingrid Zierhut.

Despite frequent superficiality and repetitiveness, AN manages to create much memorable fiction. Its particular appeal has been in the development of female characters and the encouragement of other women writers – for this AN will have left a significant legacy. Her popularity and impact on the field has won her many honorific AWARDS, notably the Grand Master WORLD FANTASY AWARD and the GANDALF AWARD in 1977 and the Fritz Leiber Award in 1983.   [MA]

Other works: The **Moon Magic** or **Moon Singer** series, imitation **Witch World**, being *Moon of Three Rings* (**1966**), *Exiles of the Stars* (**1971**), *Flight in Yiktor* (**1986**), *Dare to Go A-Hunting* (**1990**). Singletons include *Dread Companion* (**1970**), *Forerunner Foray* (**1973**), *Wraiths of Time* (**1976**), *Quag Keep* * (**1978**) – based on the *Dungeons & Dragons* GAME – *Yurth Burden* (**1978**), which is another planetary romance, *Moon Called* (**1982**), *Wheel of Stars* (**1983**) and, with Phyllis Miller, *Ride the Green Dragon* (**1985**).

As editor: *Small Shadows Creep* (anth **1974**; cut 1974 UK), featuring children's ghosts; the **Magic in Ithkar** series with Roberts ADAMS, being *Magic in Ithkar* (anth **1985**), *#2* (anth **1985**), *#3* (anth **1986**) and *#4* (anth **1987**); and the **Catfantastic** series with Martin H. GREENBERG, being *Catfantastic* (anth **1989**), *II* (anth **1991**) and *III* (anth **1994**).

See also: Susan SHWARTZ.

NORTON, MARY (1903-1992) UK children's author and former actress, born Mary Pearson (married 1926), best-known for her series of stories about **The Borrowers** but also noted for her early work, which formed the basis for the DISNEY movie *Bedknobs and Broomsticks* (*1971*). Her first children's book, *The Magic Bed-Knob* (**1943** US), tells of Miss Price, a spinster who is learning to be a WITCH. After an accident with her broomstick, she transfers the SPELL to an old bedknob, which enables her and three children to go wherever they wish. This book and its sequel, *Bonfires and Broomsticks* (**1947**) – assembled as *Bedknob and Broomstick* (omni **1957**) – is much in the style of E. NESBIT, though MN is as much interested in the effects of MAGIC as in the adventures.

The Borrowers (**1952**) is one of the modern classics of CHILDREN'S FANTASY. The definitive WAINSCOT novel, it tells of a race of little people who live beneath the floors of old houses and survive by "borrowing". The Borrowers have become depleted over the centuries (◊ THINNING), and this story concentrates on one family, the Clocks (their home is beneath a grandfather clock), whose discovery means they have to flee and face the perils of the outside world. The later novels in the series trace their various adventures and fight for survival: *The Borrowers Afield* (**1955**), *The Borrowers Afloat* (**1959**), *The Borrowers Aloft* (**1961**) – the first four assembled as *The Borrowers Omnibus* (omni **1966**; vt *The Complete Adventures of the Borrowers* 1967 US) – and *The Borrowers Avenged* (**1982**). MN also wrote the related *Poor Stainless* (1966; **1971** chap; vt *The Last Borrowers' Story* 1994 chap). A movie – *The* BORROWERS (*1973* tvm) – has been based on the books, as was the BBC tv 6-part serial (1992), adapted by Richard Carpenter from the first two novels and starring Ian Holm as Pod, Penelope Wilton as Homily and Rebecca Callard as Arrietty.

MN's only other novel is *Are All the Giants Dead?* (**1975**), a delightful flight of fancy in which a rationalist studious boy finds himself in a world where all the characters of FAIRY-TALES live in bored retirement.   [MA]

NORVIL, MANNING Pseudonym of Kenneth BULMER.

NOSFERATU ◊ DRACULA; DRACULA MOVIES; VAMPIRES.

NOSFERATU German movie (*1922*). ◊ DRACULA MOVIES.

NOSFERATU THE VAMPYRE German/French movie (*1979*). ◊ DRACULA MOVIES.

NOSTRADAMUS French seer, born Michel de Nostredame (1503-1566) in Provence, France, where he spent most of his life, maintaining a medical practice and conducting his literary activities. Educated at Montpellier and apparently a competent physician, he began his prophetic writings (◊ PROPHECY) in the early 1540s with a series of short yearly almanacs in which he made predictions in verse. The new art of printing gratified a popular demand for supernaturally acquired "certainty", and such almanacs were a common literary production of the day.

It was with *Les Prophéties de Me. Michel Nostradamus* (**1555**), however, that Nostradamus became an author of contemporary reputation and a figure who has had an impact on later history and one taken seriously by many people even today. The *Prophéties* contain 353 quatrains like those of the almanacs, arranged in "centuries" of 100 verses, to which were added additional quatrains in 1557 and 1558, bringing the total up to 1040, plus eight posthumous verses. They constitute the largest body of prophetic verse prepared to that day, perhaps in all literature.

Although they evolved directly from the popular press, the quatrains were also linked to the strong stream of religious prophetic verse that continued from the Middle Ages, particularly the Sibylline Oracles (◊ SIBYL). Presenting a strange blending of classical erudition and popular folk material, they were fashioned according to the strict poetic techniques of the French Renaissance.

Nostradamus's quatrains are linguistically and poetically interesting in that they are often presented in a meta-language that effectively permits multiple interpretations of individual quatrains to fit varied circumstances. Grammatically, Nostradamus often negated time by using infinitives and past participles that could be read with different tenses and voices; he omitted copulas, used ambiguous prepositions and distributed modifiers. He made considerable use of phonetic homonyms (*cent*, *sang*; Po, Pau), spelling homonyms (*nef* – "ship" or "nave of a church"), puns, fractured words, anagrams (Rapis for Paris) and autantonyms (like Latin *inhabitabilis*, which may mean "inhabitable" or "uninhabitable"). Particularly useful were metaphoric words with multiple possible meanings; e.g., *sol* might be astrologically the sun, alchemically gold, politically the Emperor. In such cases he invoked symbol-sets current at the time: heraldic, alchemical, religious, contemporary political, astrological, etc. The result of this wordplay, which was often combined with considerable ingenuity, was a fantasy DICTION that probably has not been excelled for its purpose until modern times.

Nostradamus was writing about the world around him, in order to sell books. He thus covered court scandal, possible calamities in towns in southern France, religious topics of the Reformation, military adventures in Italy, rivalry between the Valois and Habsburgs, Turkish raids, crimes and social horrors of the day. Much of his verse is based on identifiable – albeit cryptically rendered – major events in the past, like the Imperial sack of Rome (1527) and the Battle of Preveza (1538). His attempts at datable future history are relatively few.

Maintaining a persona suitable for the verses, he liked to describe himself as a prophet inspired by watching the heavens and receiving a divine afflatus; he also suggested he received his prophetic voice from a family acquaintance with the supernatural, though there is no evidence that his ancestors, who were petty officials and merchants, had any concern with occult matters. In his quatrains he also described traditional procedures of ritual MAGIC. Oddly enough, he rejected the title "astrologer", although he did prepare horoscopes, some of which survive.

How much of this fitted the real Nostradamus and how much was a pose? The best answer is that Nostradamus, who was really a very bad prophet if his predictions are examined closely, was a man who liked to versify (i.e., was a minor poet) and consciously and cynically wrote cryptic verse for a market. In this he was successful, for he died a wealthy man. As for Nostradamus in his times, we know a little from recently published correspondence. His concealed religious sympathies (in an age of religious persecution) were with the Reformation, but he tried to remain on good terms with the Catholic Church. On one occasion, however, he had to leave Provence to avoid an interview with the Inquisition. Speculations in the older literature that he was secretly a practising Jew are incorrect, even though his ancestors, generations back, were mostly converts from Judaism.

During his lifetime Nostradamus received limited recognition. A trip to Paris to visit Queen Catherine de Médici, who was fascinated with the occult, was a failure, though Catherine seems to have maintained contact with him. He achieved considerable attention when one of his verses (I-35) seemed to prophesy the accidental death of King Henry II in a ceremonial joust in 1559. But in fact the wording of the verse was changed in later editions of the *Prophéties* – i.e., after the event – to fit the circumstances: in 1558 Nostradamus had predicted a long, glorious reign for Henry.

Two years before his death Nostradamus received recognition from Catherine and the young King Charles IX, who appointed him court physician. After his death his reputation increased, and his quatrains were frequently republished, even in fraudulently backdated editions with spurious material added for political purposes. The first English translation, a wretched piece of work with a very corrupt text and inaccurate renderings, appeared in 1672.

During the 20th century Nostradamus has been much in the public consciousness. The Nazis made use of his verses during WWII, and there are scores of books currently in print that utilize bad texts, mistranslations, and curious interpretations to discover, after the event, prophecies of just about every important calamity in recent history. In literature Nostradamus makes no significant appearance, although T.H. WHITE's rendering of a traditional anecdote about him in *The Maharaja and Other Stories* (1981) is pleasant. In addition to several tv programmes, including one hosted by Orson Welles (1915-1985), there have been a few movies – like *Nostradamus* (*1995*) – celebrating the Prophet of Provence, but none are of note.

An additional claim, made by L. Sprague DE CAMP in his heavily researched article "You Too Can Be a Nostradamus" (1942 *Esquire*) and cited by Bergen Evans in *The Spoor of Spooks* (**1951** UK), is that "there were many Nostradamuses – some 20 in all – the name having become a generic term for prophets".   [EFB]

Further reading: *Prophecies and Enigmas of Nostradamus* (**1979**) ed and trans Liberté E. Levert (E.F. BLEILER) establishes texts on a scholarly basis and offers correct translations that recognize 16th-century French, contemporary allusions and multivalences; *The Mask of Nostradamus* (**1990**) by James Randi provides a general, if somewhat rambling, account, incorporating recent scholarship and new insights on the man and his work. Bruce PENNINGTON's *Eschatus* (**1977**) is a book of FANTASY ART inspired by Nostradamus.

NOT AGAINST THE FLESH vt of VAMPYR (*1932*).

NOT AT NIGHT UK ANTHOLOGY series ed Christine Campbell Thomson (1897-1985) for Selwyn & Blount. It ran to 12 approximately annual vols, though subsequent paperback selections complicate the bibliography. The original series was *Not at Night* (anth **1925**), *More Not at Night* (anth **1926**), *You'll Need a Night Light* (anth **1927**), *Gruesome Cargoes* (anth **1928**), *By Daylight Only* (anth **1929**), *Switch on the Light* (anth **1931**), *At Dead of Night* (anth **1931**), *Grim Death* (anth **1932**), *Keep on the Light* (anth **1933**), *Terror By Night* (anth **1934**) and *Nightmare By Daylight* (anth **1936**). The final volume was a retrospective selection, *The Not at Night Omnibus* (anth **1937**). The first paperback selections, still ed Christine Campbell Thomson – not reprints of the original volumes – were *Not at Night* (anth **1960**) and *More Not at Night* (anth **1961**; vt *Never at Night* 1972). A third paperback selection was *Still Not at Night* (anth **1962**; vt *Only By Daylight* 1972). A US selection from the series was *Not at Night!* (anth **1928** US) ed Herbert Asbury.

NAN initially selected heavily from WEIRD TALES – although Thomson focused on gruesome, physical HORROR rather than supernatural or fantasy – but did give early UK appearances to Hugh B. CAVE, Mary Elizabeth COUNSELMAN, August W. DERLETH (who advised Thomson on some selections, and whose story "The Metronome" appeared in *Terror By Night* before its *WT* publication), Robert E. HOWARD, David H. KELLER, Frank Belknap LONG and H.P. LOVECRAFT. By *#4* Thomson was selecting a greater deal of original material plus reprints selected from UK sources – especially *Adventure-Story* and *Mystery-Story*

(◊ HUTCHINSON'S MAGAZINES) – although *WT* remained the source for over half the tales published in the series, so that **NAN** was more than once referred to as the UK edition of *WT*. Thomson published nine stories of her own (as Flavia Richardson) and a further nine by her husband, Oscar Cook (1888-1952), whose fictions reflected his background in the British East Indies. Most of these stories were commercial hackwork. Although **NAN** was popular, much of its content is routine and certainly not on a par with the parallel anthologies ed Cynthia ASQUITH and John GAWSWORTH. The series has been mined by later anthologists, especially by Herbert Van Thal for his PAN BOOK OF HORROR STORIES series. [MA]

NOVELIZATIONS This entry is primarily about movie novelizations: works of fiction based on screenplays or scenarios for cinematic feature movies or movie serials, and published in book form. There are other types of novelization, inspired by songs, stage plays, radio or tv scripts, COMICS, GAMES and so on; movie novelizations, as defined above, are therefore just one major subtype of a general phenomenon.

Movie novelizations were preceded by stage novelizations, examples of which may be found from early in the 19th century; those with fantasy themes range from the anonymous *The Bride of the Isles: A Tale Founded on the Popular Legend of the Vampire by Lord Byron* * (**1820** chap), taken from James Robinson PLANCHÉ's loose dramatization of John POLIDORI's *The Vampyre: A Tale* (**1819**), to *Peter and Wendy* (**1911**) by J.M. BARRIE, based on his own highly successful play.

The first movie novelizations arose partly as a result of US newspaper and magazine circulation wars, and they arrived in tandem with a new type of movie – the serial. The earliest US serial was the Edison company's *What Happened to Mary* (**1912**), an episodic melodrama which also ran in prose form in the *Ladies' World* magazine; the book version was by Robert Carlton Brown: *What Happened to Mary* * (**1913**). In 1913 popular writer Harold MacGrath was involved in the creation of a scenario for *The Adventures of Kathlyn*, the first serial to be produced by the Chicago-based Selig company. This was conceived in collaboration with the editor of the *Chicago Tribune* newspaper, the idea being that a version of the story would run in the paper simultaneously with the fortnightly release of episodes of the cinema serial. Like the earlier Edison project, it was a success, the paper's circulation was boosted, and MacGrath's novelization was subsequently published in book form: *The Adventures of Kathlyn* * (**1914**).

Within a decade, novelizations of more prestigious, big-budget feature movies were appearing. One of the first of fantasy note was *The Thief of Bagdad* * (**1924**), taken from *The* THIEF OF BAGDAD (**1924**). There was no novel or magazine story entitled "The Thief of Bagdad" for the publishers to put on the market, so they decided to commission one. They turned to Achmed ABDULLAH, known for his "eastern exoticism", and asked him to write a full-length narrative based on the movie's script.

Others from the 1920s which may be classed as fantasy include: *The Ten Commandments* * (**1924**) by Henry MacMahon, from Cecil B. de Mille's Biblical drama; *Faust* * (**?1927**) by Hayter Preston and Henry Savage, from F.W. MURNAU's FAUST: EINE DEUTSCHE VOLKSSAGE (**1926**); and *Noah's Ark* * (**1928**) by Arline de Haas, a Biblical fantasy from the movie dir Michael Curtiz. Such books were common by the mid-1920s, and it is interesting to note that the earliest novelizers were writers of some repute: MacGrath (as above), Arthur B. Reeve (1880-1936), E. Phillips Oppenheim (1866-1946), Louis Joseph Vance (1879-1933), Albert Payson Terhune (1872-1942), William Le Queux (1864-1927), Elinor Glyn (1864-1943) and Abdullah were all successful authors of their time – they had "name value". This situation changed by the later 1920s, as novelizations appeared in ever greater numbers from cheap hardcover publishers like Grosset & Dunlap in the USA and the Readers' Library in the UK. Professional novelizers arrived, low-profile hacks who had honed their skills in the movie fanzines. Among these writers were some who later did gain fame for work other than novelizations – for example Val Lewton (1904-1951), who was to become a notable "creative producer" of horror/fantasy movies in the Hollywood of the 1940s.

Fantasy novelizations of the 1930s and 1940s included *King Kong* * (**1932**) by Delos W. Lovelace (1894-1967) (◊ KING KONG MOVIES), *The Bride of Frankenstein* * (**1936**) by Michael Egremont (Michael HARRISON) (◊ FRANKENSTEIN MOVIES), the quasi-novelization *Man who Could Work Miracles* * (**1936**) by H.G. WELLS, based on his own short story (◊ *The* MAN WHO COULD WORK MIRACLES [**1936**]), *Dr Cyclops* * (**1940**) by Will Garth (believed to be Alexander Samalman; ◊ SFE), *A Guy Named Joe* * (**1946**) by James Cairns (◊ ALWAYS [**1989**]), *A Matter of Life and Death* * (**1946**) by Eric Warman (1904-1992) (◊ *A* MATTER OF LIFE AND DEATH [**1946**]), *The Chips Are Down* * (**1947**; trans **1951**) by Jean-Paul Sartre (1905-1980) – an interesting AFTERLIFE fantasy based on his own script for *Les jeux sont faits* (**1947**) dir Jean Delannoy – *Miracle on 34th Street* * (**1947**) by Valentine Davies (◊ MIRACLE ON 34TH STREET [**1947**]), *Miranda* * (**1947**) by Warwick Mannon (◊ SPLASH! [**1984**]), *Vice Versa* * (**1947**) by Warwick Mannon (◊ VICE VERSA [**1947**]) – an example of a "re-novelization", being taken from the script of a movie based on F. ANSTEY's original novel – and *It Happens Every Spring* * (**1949**) by Valentine Davies (1905-1961), based on his own script for *It Happens Every Spring* (**1949**) dir Lloyd Bacon.

There was a lull in the early 1950s, but similar books of the later 1950s and the 1960s included *Darby O'Gill and the Little People* * (**1959**) by Lawrence Edward Watkin (◊ DARBY O'GILL AND THE LITTLE PEOPLE [**1959**]), the quasi-novelization *Last Year at Marienbad* * (**1961**; trans Richard Howard **1962**) by Alain Robbe-Grillet (1922-   ), from his own script for *L'année dernière à Marienbad* (**1961**) dir Alain Resnais, *The Thief of Baghdad* * (**1961**) by Richard Wormser (1908-1977) (◊ *The* THIEF OF BAGDAD), *The Last Days of Sodom and Gomorrah* * (**1962**) by Wormser, *The Raven* * (**1963**) by Eunice Sudak, based on the Roger CORMAN movie remotely inspired by Edgar Allan POE, *Goodbye Charlie* * (**1964**) by Marvin H. Albert, from George Axelrod's stage play via the movie *Goodbye Charlie* (**1964**), and *Tarzan and the Valley of Gold* * (**1966**) by Fritz LEIBER, the only TARZAN sequel to be sanctioned officially by the Edgar Rice BURROUGHS estate (◊◊ TARZAN MOVIES). Most of these books were paperback originals which came and went with the initial release of the movie in question, rarely achieving reprints.

However, novelizations of hit movies, particularly those based on sf, fantasy and horror screenplays, were to become big business in the decade that followed. According to John Sutherland in *Bestsellers: Popular Fiction of the 1970s* (**1981**):

"A notable development of the mid-1970s was the emergence from the tie-in business of the 'novelization' as a superselling form of novel in its own right, thus reversing traditional ideas of text originality. For the first time, film spin-offs like *The Omen* headed American lists." Indeed, David Seltzer's *The Omen* * (**1976**), based on his own horror script (◊ *The* OMEN), has some claim to being, in Sutherland's words, "the bestselling novelization of all time. It had sold getting on for 5m. in the US, around 7m. worldwide by the late 1970s and made the #1 spot in America as a paperback. Since *The Omen* the previously despised novelization and its hack novelizer have acquired new dignity – at least in the eyes of the moneymen who run the film and publishing industries."

Other fantasy-movie novelizations of the 1970s included *Brother John* * (**1971**) by Leo P. Kelley (1928-   ), *Sinbad and the Eye of the Tiger* * (**1976**) by John Ryder Hall (real name William Rotsler; 1926-   ) (◊ SINBAD MOVIES), *The Slipper and the Rose* * (**1976**) by Bryan Forbes (1926-   ) (◊ CINDERELLA [*1950*]), *Jabberwocky* * (**1977**) by Ralph Hoover (◊ JABBERWOCKY [*1977*]), *The Last Wave* * (**1977**) by Petru Popescu (?1929-   ), *Heaven Can Wait* * (**1978**) by Leonore Fleischer (1932-   ) (◊ HEAVEN CAN WAIT [*1978*]), *Arabian Adventure* * (**1979**) by Keith Miles (1940-   ), and *Circle of Iron* * (**1979**) by Robert Weverka (1926-   ). The SWORD-AND-SORCERY movie fad of the early 1980s led to a glut of novelizations, among them *Hawk the Slayer* * (**1980**) by Terry Marcel and Harry Robertson, *Clash of the Titans* * (**1981**) by Alan Dean FOSTER (◊ CLASH OF THE TITANS [*1981*]), *Dragonslayer* * (**1981**) by Wayland DREW (◊ DRAGONSLAYER [*1981*]), *Time Bandits* * (**1981**) by Charles Alverson (1935-   ) (◊ TIME BANDITS [*1981*]), *Conan the Barbarian* * (**1982**) by L. Sprague DE CAMP and Lin CARTER (◊ CONAN MOVIES), *The Dark Crystal* * (**1982**) by A.C.H. Smith (1935-   ) (◊ *The* DARK CRYSTAL [*1982*]), *The Sword and the Sorcerer* * (**1982**) by Norman Winski (◊ *The* SWORD AND THE SORCERER [*1982*]), *Krull* * (**1983**) by Alan Dean FOSTER (◊ KRULL [*1983*]), *Conan the Destroyer* * (**1984**) by Robert JORDAN (◊ CONAN MOVIES), *Sword of the Valiant: The Legend of Sir Gawain and the Green Knight* * (**1984**) by Stephen Weeks and Henry Whittington, and *Ladyhawke* * (**1985**) by Joan D. VINGE (◊ LADYHAWKE [*1985*]). A rare example of a proper novelization of an ANIMATED MOVIE – although chapbooks for the very young proliferate – is *An American Tail: The Illustrated Story* * (**1986** chap) by Emily Perl Kingsley (1940-   ) (◊ *An* AMERICAN TAIL (*1986*).

Other subgenres of movie fantasy proved successful in novelized form, particularly the adventure-fantasies of Stephen SPIELBERG – *Raiders of the Lost Ark* * (**1981**) by Campbell Black (1944-   ), *Indiana Jones and the Temple of Doom* * (**1984**) by James Kahn (1947-   ) and *Indiana Jones and the Last Crusade* * (**1989**) by Rob MacGregor (◊ INDIANA JONES) – and a cycle of horror-comedies represented by such titles as *Ghostbusters* * (**1984**) by Larry Milne and *Ghostbusters II* * (**1989**) by Ed Naha (1950-   ) (◊ GHOSTBUSTERS) and *Gremlins* * (**1984**) by George Gipe (1933-1986) and *Gremlins 2: The New Batch* * (**1990**) by David F. Bischoff (1951-   ) (◊ GREMLINS [*1984*]). Gentler or more juvenile-oriented fantasy comedies also reached novelized form: *Splash!* * (**1984**) by Ian Don (◊ SPLASH! [*1984*]), *The Goonies* * (**1985**) by James Kahn (◊ *The* GOONIES [*1985*]), *One Magic Christmas* * (**1985**) by Martin Noble (1947-   ), *Return to Oz* * (**1985**) by Joan D. VINGE (◊

The WIZARD OF OZ), *Santa Claus: The Movie* * (**1985**) by Vinge (◊ SANTA CLAUS), *Young Sherlock Holmes* * (**1985**) by Alan Arnold (◊ SHERLOCK HOLMES), *Labyrinth* * (**1986**) by A.C.H. Smith (◊ LABYRINTH [*1986*]), *Harry and the Hendersons* * (**1987**; vt *Bigfoot and the Hendersons* 1987 UK) by Joyce Thompson (1948-   ), *Big!* * (**1988**) by B.B. and Neil W. Hiller (◊ BIG [*1988*]), *Biggles: The Movie* * (**1986**) by Larry Milne (◊ BIGGLES [*1986*]) and *Willow* * (**1988**) by Wayland DREW (◊ WILLOW [*1988*]), etc.

Since the late 1980s, fantasy-movie novelizations have been as numerous as ever, notable examples including *The Adventures of Baron Munchausen* * (**1989**) by Charles McKeown and Terry GILLIAM (◊ *The* ADVENTURES OF BARON MUNCHAUSEN [*1989*]), *Batman* * (**1989**) by Craig Shaw GARDNER (◊ BATMAN MOVIES), *Dick Tracy* * (**1990**) by Max Allan Collins (1948-   ) (◊ DICK TRACY [*1990*]), *Flatliners* * (**1990**) by Leonore Fleischer, *Teenage Mutant Ninja Turtles* * (**1990**) by Dave Morris (◊ TEENAGE MUTANT NINJA TURTLES), *The Addams Family* * (**1991**) by Elizabeth Faucher (◊ ADDAMS FAMILY MOVIES), *Bill & Ted's Bogus Journey* * (**1991**) by Robert Tine (◊ BILL & TED'S BOGUS JOURNEY [*1991*]), *The Fisher King* * (**1991**) by Leonore Fleischer (◊ *The* FISHER KING [*1991*]), *Hook* * (**1991**) by Terry BROOKS (◊ HOOK [*1991*]), *Buffy the Vampire Slayer* * (**1992**) by Richie Tankersley Cusick (1952-   ) (◊ BUFFY THE VAMPIRE SLAYER [*1992*]), *Last Action Hero* * (**1993**) by Robert Tine (◊ LAST ACTION HERO [*1993*]), *The Mask* * (**1994**) by Steve Perry (1947-   ) (◊ *The* MASK [*1994*]), *The Pagemaster* * (**1994**) by Todd Strasser (1950-   ) (◊ *The* PAGEMASTER [*1994*]) and *The Shadow* * (**1994**) by James Luceno. At the same time, tv novelizations and spinoffs have proliferated, taking their cues from such series as BEAUTY AND THE BEAST (1987-90), *Quantum Leap* (1989-94), TWIN PEAKS (1989-91), *The* ADVENTURES OF BATMAN AND ROBIN (1992-4), *The X-Files* (1993-current) and *The* HIGHLANDER (1992-current), among many others. Clearly, fantasy and horror novelizations are here to stay; even if despised by critics, the novelization is a form which cannot be ignored.   [DP]

NOVEL MAGAZINE, THE UK general-fiction PULP (393 issues April 1905-December 1937) which regularly published SUPERNATURAL FICTION and for 10 years (April 1913-January 1923) ran a monthly "uncanny story" feature. Among the mostly routine commercial fictions are many GHOST STORIES, especially during the post-WWI renaissance of SPIRITUALISM; also included were stories of precognition (◊ TALENTS), WEREWOLVES, pagan MAGIC and DREAM fantasies. Best-known stories from the series are: "Not on the Passenger List" (August 1915) by Barry PAIN, "The Haunted Chessmen" (March 1916) by E.R. Punshon (1872-1956) and "The Eighth Lamp" (July 1916) by Roy Vickers (1889-1965). Regular contributors included A.M. BURRAGE, Elliott O'Donnell (1872-1965) and Theo. DOUGLAS; the series later reprinted stories from the US pulps by Ray Cummings (1887-1957), Achmed ABDULLAH, Murray Leinster (1896-1975; ◊ SFE) and Charles B. Stilson (1880-1932). The overall executive editor of *NM* was Percy W. Everett (1870-1952), and for a period (1918-1922) the managing editor was E. Charles VIVIAN; after the latter's tenure ended the MAGAZINE switched primarily to romance and stories for women. Two selections were published as *Uncanny Stories* (anth **1916**) and *More Uncanny Stories* (anth **1918**) – assembled as *Ghost Stories and Other Queer Tales* (rev omni **1931**); all three volumes were probably under the editorial direction of Everett.   [MA]

NOW YOU SEE IT, NOW YOU DON'T US movie (*1972*). ◊ *The* SHAGGY DOG (*1959*).

NOYES, ALFRED (1880-1958) UK poet and academic, resident for substantial parts of his career in the USA. Highly regarded in his day for lengthy poems like *Drake: An English Epic* (**1906-8** 2 vols) and *The Torch-Bearers* (**1922-30** 3 vols), he was made a Companion of the British Empire in 1918 and received numerous honorary degrees. His main contribution to fantasy was the humorous *The Devil Takes a Holiday* (**1955**), in which SATAN arrives in California to condemn the use of atomic weapons. AN also wrote some fantastic short stories, gathered in *Walking Shadows: Sea Tales and Others* (coll **1918**) and *The Hidden Player* (coll **1924**), and a novella, *Beyond the Desert: A Tale of Death Valley* (**1920** chap US), but his principal efforts in fantasy are found in his many volumes of poetry and verse plays, especially *The Flower of Old Japan* (**1903**; rev vt *The Flower of Old Japan and Other Poems* 1907), *The Forest of Wild Thyme* (**1905**), *Sherwood, or Robin Hood and the Three Kings* (**1911**; rev vt *Robin Hood* 1926) and *Tales of the Mermaid Tavern* (**1913**). A Tennysonian post-Romantic, and later a convert to Roman Catholicism, AN became deeply unfashionable in the era of T.S. Eliot (1888-1965) and F.R. Leavis (1895-1978), but retains some admirers to this day. [DP]

Other works: *The Magic Casement: An Anthology of Fairy Poetry* (**1908**), edited; *Collected Poems* (**1910-20** 3 vols; exp 1927; exp 1947; exp 1963); *William Morris* (**1908**), *Tennyson* (**1932**) and *Voltaire* (**1936**), all nonfiction; *The Last Man* (**1940**; vt *No Other Man* US), scientific romance; *The Secret of Pooduck Island* (**1943** US), juvenile fantasy.

NUMBER OF THE BEAST ◊ GREAT BEAST; NUMEROLOGY.

NUMEROLOGY In its popular sense this tends to refer to various bastardizations of the *gematria*, a system of esoteric interpretation of Hebrew scriptures (◊ BIBLE; JEWISH RELIGIOUS LITERATURE) based on the swapping around of words which yield the same value when the numbers corresponding to their letters are added up. With creative choice of spelling and letter-values, any desired conclusion can be reached. 666, the Number of the GREAT BEAST, has been extracted from the names of Martin Luther and various popes; the evil company Lateinos and Romiith in Robert RANKIN's *East of Ealing* (**1984**) takes its name from words devised to "prove" the beastliness of the Catholic Church. FANTASIES OF HISTORY occasionally invoke such obsessive reasoning.

Of far greater fantasy interest are the magical associations of numbers. 1 is primal; 2 stands for many dyads like male and female, GOOD AND EVIL, order and CHAOS, YIN AND YANG; THREE is perhaps the most magic of all; 4 suggests the "corners" of the world and the ELEMENTS; a 5-pointed star is the heart of a pentacle; 6 recalls the Star of David as well as the magic symmetry of snowflakes; SEVEN has almost the numinosity of 3 (◊◊ SEVEN SAMURAI); 8 is rather unmagical

on Earth, and perhaps for that very reason is the basic magic number of Terry PRATCHETT's **Discworld**, whose spectrum has 8 colours; 9 is of course 3 times 3; 12 has biblical and CALENDAR resonances (the Apostles, the months); THIRTEEN reeks of bad luck. J.R.R. TOLKIEN's most effective short piece of verse, the RING-tallying epigraph of *The Lord of the Rings* (**1954-5**), makes its impact *via* the cumulative magic of 3, 7, 9 and 1. [DRL]

See also: QUINCUNX.

NURSERY RHYMES ◊ MOTHER GOOSE.

NUTTY PROFESSOR, THE US movie (*1963*). ◊ JEKYLL AND HYDE MOVIES.

NYCTALOPS ◊ MAGAZINES.

NYE, JODY LYNN (1957-   ) US writer whose first publications were episodes in the **Crossroads Adventure** series of ties; they include *Encyclopedia of Xanth* * (**1987**) and *Ghost of a Chance* * (**1988**), both tied to the **Xanth** sequence by Piers ANTHONY, and *Dragonharper* * (**1987**) and *Dragonfire* * (**1988**), both tied to the **Pern** sequence by Anne MCCAFFREY. With Anthony, JLN wrote *Piers Anthony's Visual Guide to Xanth* (**1989**); with McCaffrey her collaborations have been more extensive, including *The Dragonlover's Guide to Pern* * (**1989**), *The Death of Sleep* * (**1990**), which is set in McCaffrey's **Sassinak** world, and *Crisis on Doona* * (**1992**), also sf, set in the **Doona** sequence. Solo, JLN has published the **Mythology** sequence – *Mythology 101* (**1990**), *Mythology Abroad* (**1991**) and *Higher Mythology* (**1992**) – involving a WAINSCOT society of Little People (◊ CELTIC FANTASY) who live beneath a university library and have adventures with chosen humans. *Taylor's Ark* (**1993**) is a medical sf thriller, and *Medicine Show* (**1994**) roughly sequels it. [JC]

NYE, ROBERT (1939-   ) UK poet and novelist, a fellow of the Royal Society of Literature, and past-winner of the *Guardian* Fiction Prize, the Hawthornden Prize and other awards. His novels are principally historical, centred on actual or legendary individuals, and include such titles as *Falstaff* (**1976**) and *The Life and Death of My Lord Gilles de Rais* (**1990**). Of greatest fantasy interest are *Merlin* (**1978**) and *Faust* (**1980**) – bawdy, scatological, richly told, sometimes anachronistic reworkings of the traditional material (◊ MERLIN; FAUST). Some of RN's short stories – gathered in *Tales I Told My Mother* (coll 1969) and *The Facts of Life and Other Fictions* (coll 1983) – are fantastic. He has written many juvenile fantasies; e.g., *Taliesin* (**1966**), *Bee Hunter: Adventures of Beowulf* (**1968**; vt *Beowulf: A New Telling* US), *Wishing Gold* (**1970**) – these three are assembled as *Three Tales* (omni **1983**) – *Poor Pumpkin* (coll **1971**; vt *The Mathematical Princess and Other Stories* US), *Out of the World and Back Again* (coll **1977**) and *The Bird of the Golden Land* (**1980**). [DP]

NYOKA THE JUNGLE GIRL ◊ TARZAN.

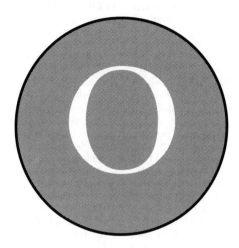

OAK The most revered of Britain's native TREES, held as sacred by many cultures including the druids, Germans and Greeks. It was believed to be sacred to the god of thunder, as it was often struck by lightning, and the Druids believed mistletoe grew on the oak after it had been struck. A symbol of virility and strength, it was also regarded as the tree of the dead and the home of dead spirits. ODIN hanged himself on an oak for nine days prior to his RESURRECTION, while Carlo COLLODI's PINOCCHIO is hanged from an oak before his rebirth as a REAL BOY. The oak can also bind spirits (◊ BONDAGE) – e.g., MERLIN was imprisoned in an oak by Viviane (◊ LADY OF THE LAKE) and Jennifer and Daniel were likewise treated in I MARRIED A WITCH (*1942*). (Ariel, in SHAKESPEARE's *The Tempest* [performed c1611; **1623**], was by contrast imprisoned by the witch Sycorax in a pine.) The death of an oak tree was a symbol of death in "The Old Oak Tree's Last Dream" (1857) by Hans Christian ANDERSEN. It is also maintained that the last leaf never falls from an oak tree and there are several Baltic FOLKTALES about PACTS WITH THE DEVIL where the deal hinges on the last leaf falling. Oaks were also capable of protecting from evil, and oak wood was often used in making gallows. Anything made from the wood of a gallow will protect against evil: see "The Stalls of Barchester Cathedral" (1910 *Contemporary Review*) by M.R. JAMES and "The Fatal Oak" (1947 in *Unholy Relics*) by M.P. Dare (1902-1962).

Oak groves were sacred meeting sites for the Druids. The oakmen were spirits of the oak (◊ DRYADS), DWARF-like beings who are guardians of the forest and its animals (the oak supports more wildlife than any other tree) – they appear in, e.g., Beatrix POTTER's *The Fairy Caravan* (**1929** US). A 13th-century oak at Windsor is associated with the spirit of HERNE THE HUNTER. The oak was believed to confer power, and was often used as the site for temporary parliaments. It is through the oak that the Hubelaires begin their rise to power in the **Tales of Cornwall** by David H. KELLER. An oak was probably the original of the stone in which the SWORD IN THE STONE was placed. ARTHUR thus becomes symbolic of the Oak King – see *In the Shadow of the Oak King* (**1991**) by Courtway Jones (real name John Alan Jones; 1923-   ). The Oak King is also associated with the

GREEN MAN and is often represented by a FOLIATE HEAD. Further references to this figure are in *Peace* (**1975**) by Gene WOLFE and *The Oak King's Daughter* (**1979** chap) by Charles DE LINT; De Lint utilizes the imagery of the FOREST in many of his books, and the oak is especially significant in *Into the Green* (**1993**), where a Merlin-like figure is trapped in a LITTLE-BIG oak. [MA]

OATES, JOYCE CAROL (1938-   ) US writer whose many novels and stories constitute an analysis of and panoramic perspective on the American Dream; she has more than once been likened to Honoré de BALZAC, whose feverish productivity she has more than matched (by 1996 she had published at least 65 full-length books); she was given a BRAM STOKER AWARD in 1994. Her work is exceedingly various, and touches on many genres, though even her most Gothic tales tend to offer nonfantastic explanations for the extremities they depict. Neither *Wonderland* (**1971**) nor *Son of the Morning* (**1978**), for instance, more than hints at the supernatural in dramatizing, in the first instance, a ghastly life haunted by DOUBLES, and, in the second, an evangelist's career which constitutes a savage PARODY of CHRIST's.

Yet JCO has written SUPERNATURAL FICTIONS of considerable interest. Three of her four large "genre" novels of the 1980s include elements of the fantastic in their omnivorous grasp. *Bellefleur* (**1980**), mostly set in the eponymous EDIFICE, incorporates a GHOST STORY, a CAT with TALENTS, a SHAPESHIFTER and a VAMPIRE into its panoramic array of symbols representative of the self-devouring underside of US life. In *A Bloodsmoor Romance* (**1982**), four sisters transgressively explore various byways of 19th-century Gothic fiction, undergoing METAMORPHOSIS, becoming spiritualists (◊ SPIRITUALISM) and, in one case, changing sex to become a scientist/MAGUS whose inventions include a time machine; in the end, each of the sisters survives her RITE OF PASSAGE, and is whole, though the world they occupy is grotesque. *Mysteries of Winterthurn* (**1984**) ironically contrasts the supernatural HORROR underlying a series of cases facing the detective Xavier Kilgarvan with the inadequate ratiocinations he brings to bear; in the end, he surrenders to the Gothic world and love comes to him.

Of JCO's later novels, *Zombie* (1995), which won the 1996 Bram Stoker Best Novel Award, is a SERIAL-KILLER horror tale about a man who tries to create a ZOMBIE without success. Most of her short fiction of interest – much of it constitutes a series of explorations in WEIRD-FICTION modes – is assembled in *Night-Side: Eighteen Tales* (coll 1977) and *Haunted: Tales of the Grotesque* (coll 1994). [JC]

Other works: *The Poisoned Kiss and Other Stories from the Portuguese* (coll 1975); *The Bingo Master* (in *Dark Forces* ed Kirby McCauley anth 1980; 1992 chap); *Demon and Other Tales* (coll 1996 chap).

O'BANNON, DAN (1946-) US moviemaker. ◊ MOEBIUS.

OBEAH ◊ VOODOO.

OBERON The king of the FAIRIES, best remembered from his appearance in SHAKESPEARE's *A Midsummer Night's Dream* (performed 1596; 1600). Set in the time of Theseus, it concerns Oberon's quarrel with Titania (◊ FAIRY QUEEN) over a young CHANGELING. Shakespeare's portrayal of Oberon has been highly influential, and has lent itself to many adaptations for OPERA, but it differs from the original depiction of Oberon in the medieval ROMANCE *Huon of Bordeaux* (15th century; trans Lord Berners ?1534 UK). Here he is an angelic-faced DWARF whom Huon encounters in a FOREST. He tells Huon that he is the son of Julius Caesar and MORGAN LE FAY. At his birth all good fairies gave him gifts, but a wicked fairy cursed him (◊ CURSES) – hence his diminutive stature. He helps Huon in his QUEST and makes Huon his successor as king of the fairies – a role recognized by Rudyard KIPLING in *Puck of Pook's Hill* (coll of linked stories 1906).

The **Huon** adventures are explored more by Christoph Wieland (1733-1813) in his verse ROMANCE *Oberon* (1780), where Oberon helps Huon in the impossible challenge set him by Charlemagne. Parallels are drawn with the *Midsummer Night's Dream* events, particularly in the confusion between lovers. The romance introduces strong Oriental elements (◊ ARABIAN FANTASY), which were popular at that time, and translates the story into the form of the FAIRYTALE. This romance formed the basis of Weber's opera *Oberon, König der Elfen* (1826), libretto by James R. PLANCHÉ – this libretto, plus a new one by Anthony BURGESS, was published as *Oberon Old and New* (1985).

The character of Oberon in the *Huon* romance is believed to have been adopted from Alberich, the king of the Dwarfs in the *Nibelungenlied* (◊ NORDIC FANTASY) who guarded the treasure of the Nibelungs. The name is sometimes rendered Auberon – the form used by John CROWLEY in *Little, Big* (1981), whose Auberon survives in a New York FAERIE-like POLDER.

Shakespeare's Oberon reappears in those REVISIONIST FANTASIES which seek to explore further the world of *A Midsummer Night's Dream*. This was well handled by Fletcher PRATT and L. Sprague DE CAMP in *Land of Unreason* (1941 *Unknown*; 1942), in which a US diplomat is transported to the land of faerie of Oberon and becomes the changeling. Poul ANDERSON's *A Midsummer Tempest* (1974) is set in an ALTERNATE WORLD where Shakespeare's plays are true and where Oberon and PUCK help King Charles I against Cromwell. In Roger ZELAZNY's **Amber** Oberon is the King of Amber who has mysteriously vanished; his appearance in *The Hand of Oberon* (1976) only deepens the intrigue. As with all the characters in this series, the names may suggest some relationship to their ARCHETYPE, but the characters do not reflect their true personality. [MA]

See also: *A* MIDSUMMER NIGHT'S DREAM (*1935*).

O'BRIEN, FLANN Working name, for fiction, of Irish writer, journalist and civil servant Brian O'Nolan or Briain ÓNualláin (1911-1966), also known for his humorous and polemical *Irish Times* newspaper column **Cruiskeen Lawn** (1940-66), written as Myles na Gopaleen (or gCopaleen). This HUMOUR was occasionally fantastic – with, for example, satirical TECHNOFANTASY notions about TRAINS – and is notorious for absurdist anecdotes about John KEATS and George Chapman (*c*1559-1634) that culminate in puns; a good sampler is *The Best of Myles: A Selection from "Cruiskeen Lawn"* (coll 1968 UK) ed Kevin O'Nolan, FO's brother.

FO's first novel was *At Swim-Two-Birds* (1939 UK), a multi-layered FABULATION of contemporary Irish life and CELTIC FANTASY which drew praise from James JOYCE and Dylan THOMAS. Its FRAME-STORY narrator is writing a novel about a would-be novelist, Trellis, whose characters resent their imposed roles (◊ BONDAGE) and go their own way whenever Trellis sleeps. Then Trellis impregnates his leading female character, and a farcical pooka (◊ PUCK) and Good FAIRY bargain for the child's SOUL. EVIL wins, at CARDS; the grown child is chosen to write a further STORY killing off Trellis and freeing his creations. But his too-elaborate narrative strategies are constantly interrupted, and eventually Trellis is saved when his own manuscript pages establishing the vengeful characters are accidentally burnt. The interruptions echo those inflicted earlier by modern Dubliners on lovingly pastiched recitations of Irish MYTH by FINN MAC COOL, contrasting with Finn's still earlier telling to a spellbound LAND-OF-FABLE audience: this THINNING of appreciation comments glumly on the modern MATTER of Ireland. Narrative layers can interpenetrate (◊ ARABIAN NIGHTMARE), as when the Pooka and Good Fairy encounter cowboys created by a Dublin PULP writer, and also the mythic King Sweeney whose CURSE of thinking himself a bird has been narrated by Finn Mac Cool. The contrasts of DICTION are finely handled throughout.

The Third Policeman (written 1939-40; 1967 UK) – inexplicably rejected by publishers in 1940 – is a grimly comic POSTHUMOUS FANTASY whose protagonist is a murderer who, unknowingly, has himself been murdered by his accomplice: the warped Ireland he inhabits is HELL, its unreality signalled by what Jorge Luis BORGES called "games with time and infinity". These centre on a police station whose geometry and perspective throb with WRONGNESS. One comic-sinister policeman's hobby is to create boxes within boxes that regress forever (◊ GREAT AND SMALL), a spear whose extreme sharpness extends far beyond the material point, and a new colour whose beholders suffer insanity; another is obsessed with a molecular interpretation of the Law of Contagion (◊ MAGIC) whereby bicycles absorb humanity from their riders (◊ ANIMATE/INANIMATE) and in return contaminate them with bicycle-ness. These policemen have interpreted cracks on a ceiling as a MAP, leading them to a timeless (◊ TIME IN FAERIE) UNDERGROUND boiler-room where space is mere repetition (to leave one room is to re-enter it or an identical one) and which mysteriously regulates the Universe. The more elusive third policeman has constructed his own station within the thickness of the walls of the murderee's house, and ominously has this man's face and voice (◊ DOPPELGÄNGER). Eventually the narrator finds his quondam accomplice, who dies of fright on seeing this GHOST. Both dead men come again to the police station, now only dimly recognized (◊ MEMORY

WIPE). Counterpointing this narrative, mock-learned foot-notes expound the bizarre natural philosophy of one de Selby, who theorizes, for example, that night is an ILLU-SION caused by accretions of black air. He is the MAGGOT of the protagonist, who murdered for money to publish a "de Selby Index"; FO, a Catholic, may be hinting at the theo-logical speculation that pursuing secular knowledge is itself sinful (◊◊ James BLISH).

FO pillaged parts of *The Third Policemen* – e.g., the bicycle theory – for *The Dalkey Archive* (**1964**), a romp in which mad scientist De Selby (now with a capital D) has learned to annihilate TIME by destroying the air's oxygen. By this means he matures whiskey with abnormal speed, calls up St Augustine, and threatens the END OF THE WORLD. Though very funny, the novel goes astray in a counterplot to distract De Selby by introducing him to the unexpectedly still living James JOYCE, who now wishes to become a Jesuit. A stage version was produced in Dublin as *When the Saints Go Cycling In* (1965).

FO was one of the finest of modern Irish fantasists.

[DRL]

Other works: *An Béal Bocht* (**1941**; trans Patrick C. Power as *The Poor Mouth* **1973** UK), part-fantastic SATIRE on pre-occupations of the Irish-language movement; *Faustus Kelly* (**1943**) as Myles na Gopaleen, a play featuring the DEVIL; *Rhapsody in Stephen's Green* (produced 1943; **1994** UK), satirical BEAST-FABLE play based on *The Insect Play* (**1921**) by the Brothers ČAPEK; *The Hard Life: An Exegesis of Squalor* (**1961** UK); *Stories and Plays* (coll **1973** UK); *Further Cuttings from Cruiskeen Lawn* (coll **1976** UK) ed Kevin O'Nolan; *The Various Lives of Keats and Chapman and The Brother* (coll **1976** UK) ed Benedict Kiely; *A Flann O'Brien Reader* (coll **1978** US) ed with commentary by Stephen Jones.

Further reading: *Myles: Portraits of Brian O'Nolan* (coll **1973**) ed Timothy O'Keeffe; *Flann O'Brien: An Illustrated Biography* (**1987**) by Peter Costello and Peter van de Kamp.

OBSESSED SEEKER One of the standard protagonists of GENRE FANTASY. OSs are often also ACCURSED WANDERERS, in which case the obsession and the CURSE may be identical, or the search may be for the curse's cure.

A major theme of HIGH FANTASY is the successful QUEST. The OS, by contrast, is involved in a search whose fictional point is that it will have no end; the search is a TEMPLATE rationale for wanderings that will take the character to a variety of locales and adventures. A classic sf example is **Dumarest** series (over 30 titles) by E.C. Tubb (1919-   ), the story of an unsuccessful search for Earth.

In practice, there is sometimes an overlap between OS fantasies and coupon fantasies. Louise COOPER's Indigo, in the **Indigo** series (**1988-93**), has a set number of demons to destroy, and thus is collecting PLOT COUPONS, but inasmuch as her search involves her in plots where her search for the demons is only one strand of her motivations or of the com-plex plots, she is an OS. She is also, of course, an accursed wanderer, in that her task is a punishment incurred for opening a PANDORA'S BOX which contained the demons.

[RK/JC]

OCAMPO, SILVINA (1903-   ) Argentine writer. ◊ Adolfo BIOY CASARES; Jorge Luis BORGES.

OCCULT DETECTIVES Sometimes referred to as Psychic Detectives or more loosely as Ghost Hunters, ODs are individuals with specialist, often arcane knowledge who seek to solve psychic phenomena; ODs are not necessarily

psychic in their own right, though a few, like George CHES-BRO's **Mongo**, have a range of TALENTS. They are more properly psychic researchers, though stories turn them into detectives to create tension and adventure. It is no coinci-dence that the rise in popularity of the OD paralleled the rise of the private-detective story in the wake of Arthur Conan DOYLE's SHERLOCK HOLMES. Although Doyle had considerable interest in OCCULTISM, Holmes never investi-gated an outright supernatural case. Stories involving ODs also came into popularity because of Victorian society's growing interest in the occult. The growth of SPIRITUALISM in the USA and the headline case (1848) of the Fox sis-ters – among the first spirit mediums – the remarkable demonstrations of Daniel Dunglas Home (1833-1886), the formation of the Theosophical Society (◊ THEOSOPHY) in 1875 and the Order of the GOLDEN DAWN in 1888, and above all the creation of the Society for Psychical Research (SPR) in 1882 all provided the fertile soil in which interest in stories about ODs grew.

Most of the early ODs were not spiritualists or psychic specialists or even priests, but doctors. The early cases focused on psychic manifestations as effects of the living, not necessarily the dead. This was the case of the stories written by Samuel Warren (1807-1877): **The Diary of a Late Physician** (1830-37 *Blackwood's*; selection as *Passages from the Diary of a Late Physician* **1831** US; rev 1832; rev 3 vols 1838), which are first-person narrations by a doctor inter-ested in the occult and macabre. More significant were the stories narrated by **Dr Martin Hesselius** in *In a Glass Darkly* (coll **1872**) by Joseph Sheridan LE FANU, although these really only provided a frame, the stories having orig-inally been written without such a device. Another doctor appears in *Stories from the Diary of a Doctor* (2 series **1894**; **1896**) by L.T. Meade (1854-1914) and Clifford Halifax, most of whose mysteries are bizarre if rationalized; with her other collaborator, Robert Eustace (real name Eustace Robert Barton; 1868-1943), Meade wrote *A Master of Mysteries* (1897 *Cassell's*; **1898**) featuring **John Bell**, a psy-chic investigator who unravels hoaxes. The Bell stories underscore a central feature of many OD stories. The mystery being investigated may be either genuinely super-natural or *appear* to be (◊ RATIONALIZED FANTASY). Both types are OD stories – indeed, the latter is more a reflection of the real thing – but for the purposes of this encyclopedia the emphasis is on stories involving the supernatural.

The first genuine OD in fiction, in that he is a specialist called in to investigate supernatural mysteries, is **Flaxman Low**, who appeared in two series of stories by E. & H. HERON presented as **Real Ghost Stories** and collected as *Ghosts* (1898-99 *Pearson's*; coll **1899**; cut vt [first series only] *Ghost Stories* 1916; full text vt *Flaxman Low, Psychic Detective* 1993). Low's investigations are modelled on those reported by the SPR, and are supported by photographs of the haunted locations, thus increasing the apparent veracity of the stories. Their success ushered in two decades of OD stories – a sort of Golden Age. Allen Upward (1863-1926) copied the Low format for his series **The Ghost Hunters** (1905-6 *Royal Magazine*), while *Pearson's* repeated the pre-sentation for **True Ghost Stories** by Jessie Adelaide Middleton in 1907. The next major leap, though, came with *John Silence, Physician Extraordinary* (coll **1908**; with new preface vt *John Silence* 1942) by Algernon BLACKWOOD. Because of the publisher's advertising campaign, which included some of the biggest posters then printed on hoardings and buses, the book became an overnight best-seller and established the OD as a genre in its own right. John Silence was another doctor who had undergone years of arcane training. Although they were presented as fiction, Blackwood had first planned the stories as a series of essays on psychic afflictions – this adds to their authenticity and conviction. When Blackwood declined to continue the stories, his publisher Eveleigh Nash turned to William Hope HODGSON, who created his **Thomas Carnacki** following the success of John Silence. Carnacki – not a doctor but a psychic researcher – has recourse to arcane manuscripts as well as modern scientific instruments. The stories were collected as *Carnacki the Ghost-Finder* (coll **1913**; exp 1947 US).

Over the next decade many of the popular magazines ran OD series, most of which have not appeared in book form. These included the long-running **Semi-Dual** series by J.U. GIESY, starting with "The Occult Detector" (1912 *Cavalier*) and ending with "The Ledger of Life" (1934 *Argosy*). Semi-Dual is a psychologist who can detect crime through psychic vibrations and uses ASTROLOGY as his main aid. His cases involve a wider cosmic scale rather than mere spirits, and feature battles against evil-doers in the mode of Sax ROHMER's **Fu Manchu**. Rohmer created his own psychic detective with **Moris Klaw**, who solves his mysteries by sleeping at the scene of the crime and dreaming the solution. Rohmer's stories began to appear in 1913 and were eventually collected as *The Dream-Detective* (coll **1920**; exp 1925 US). Other series from this period include: **Aylmer Vance** (a clairvoyant) by Alice and Claude Askew (both ? - 1917), which ran in *The Weekly Tale-Teller* during 1914; **Norton Vyse** by Mrs Champion de Crespigny (? -1935) in *The Premier Magazine* in 1919; and **Shiela Crerar**, the first woman OD, by Ella Scrymsour (1888-?) in *The Blue Magazine* in 1920. Luna Bartendale, another female OD, made her single appearance in *The Undying Monster* (**1922**), a better-than-average WEREWOLF tale by Jessie Douglas KERRUISH. **Dr Arnold Rhymer** was the psychic investigator created by Uel KEY in *The Broken Fang* (coll **1920**); although he is called a "specialist in spooks", most of the stories involve psychic espionage by the Germans during WORLD WAR I. Elliott O'Donnell (1872-1965) created ghosthunter **Damon Vane** in *The Novel Magazine* in 1922 as a way to recount his own psychic investigations. **Dr Taverner**, created by Dion FORTUNE in *The Secrets of Doctor Taverner* (coll **1926**), is more like Low and Silence in drawing upon his occult studies and experiences.

All of these stories are variants on a theme, and there were few new features to add. As a result the popularity of the OD genre had by the 1920s faded. Its main continuation was in the USA, with the stories of **Jules de Grandin** by Seabury QUINN; these ran in WEIRD TALES 1925-51. Not all of the stories involved supernatural events, and there was considerable emphasis upon sex and violence, especially in *The Devil's Bride* (1932 *WT*; **1976**), the longest of the cases and the most vicious. *WT* was also the home of **John Thunstone**, a more believable OD, created by Manly Wade WELLMAN. This series began in 1943 but the stories were not collected until *Lonely Vigils* (coll **1981**), after which Wellman added two more adventures, *What Dreams May Come* (**1983**) and *The School of Darkness* (**1985**). Wellman used authentic FOLKLORE and LEGENDS as background for his stories; they are thus among the most distinctive in the field.

E. Charles VIVIAN wrote a number of novels featuring Gregory George Gordon Green, known as **Gees**. Not all of these are SUPERNATURAL FICTION, and indeed Gees is introduced as an everyday investigator who, in later novels, becomes increasingly involved in the bizarre. *Grey Shapes* (**1937**), *Maker of Shadows* (**1938**), *The Ninth Life* (**1939**) and particularly *Her Ways Are Death* (**1941**) show Gees becoming more experienced in handling the occult, so that by the last novel he has become something of an adept.

The adventures of Margery LAWRENCE's **Miles Pennoyer**, who is modelled on John Silence, appear in *Number Seven Queer Street* (coll **1945**). The last Golden Age OD was **Lucius Leffing**, created by Joseph Payne BRENNAN. Although working in the modern USA, Leffing has a spiritual affinity with the Victorian Age and lives like a latter-day Sherlock Holmes. Most of his investigations are mundane because, by the time Brennan had created the character, the market for supernatural fiction had waned. These tales are published as *The Casebook of Lucius Leffing* (coll **1973**), *The Chronicles of Lucius Leffing* (coll **1977**), *The Adventures of Lucius Leffing* (coll **1990**) plus the short novel *Act of Providence* (**1979**) with Donald Grant (1927-).

Although an exorcist might be classified as an OD, he is not a detective in the normal sense, and the success of William Peter BLATTY's novel *The Exorcist* (**1971**) did not result in a revival of interest in OD stories – if anything it made them more of an anachronism. ODs since 1971 include: **Titus Crow** in books by Brian LUMLEY; **Francis St Clare** in many short stories and *The Psychic Detective* (**1993**) by R. CHETWYND-HAYES; **Ralph Tyler** by Mark Valentine (1959-), some of whose adventures were collected in *14 Bellchamber Towers* (coll **1987** chap); Jessica Amanda SALMONSON's **Penelope Pettiweather**, who appears in *Harmless Ghosts* (coll **1990** chap); and **Ernie Pine** in various stories by Rick Kennett (1956-), some of which have been collected as *The Reluctant Ghost-Hunter* (coll **1991** chap) – Kennett and A.F. Kidd (1953-) found that they had both independently written some Carnacki pastiches, and these were published as *No 472 Cheyne Walk* (coll **1992** chap). James HERBERT gave the medium a boost when he created the character of **David Ash**, a sceptical OD who finds real-life cases in *Haunted* (**1988**) and *The Ghosts of Sleath* (**1994**). Although ODs lack the commercial success they had a century ago they still retain a vibrant fascination for their ardent core of admirers.

Representative anthologies are *Horror Hunters* (anth **1971**) ed Roger Elwood and Vic Ghidalia; *The Supernatural Solution* (anth **1976**) ed Michel Parry; and *Supernatural Sleuths* (anth **1986**) ed Peter HAINING. [MA]
Further reading: "Fighters of Fear" by Mike ASHLEY in *Voices from Shadow* (anth **1994** chap) ed David Sutton.
See also: L. Adams BECK; Robert W. CHAMBERS; August DERLETH; Paul GALLICO; Marvin KAYE; KOLCHAK MOVIES; KOLCHAK: THE NIGHT STALKER; Evangeline WALTON.

OCCULT FANTASY The study of the occult is the pursuit of hidden or secret doctrine (the word comes from the Latin *occultere*, "to conceal"). The knowledge that is pursued is thus not *unknown*, but has been hidden from mankind as being *too dangerous to know*. The pursuit of the occult could thus be claimed to have started when ADAM AND EVE ate the Fruit of the Tree of Knowledge and understood that which was forbidden. Students of the occult may be individuals – famous ones from history include Pythagoras (?582-500BC), Apollonius of Tyana (1st century AD),

Michael Scot (?1175-1230), Albertus Magnus (?1206-1280), Roger Bacon (?1214-?1292), Christian ROSENCREUTZ, John DEE, Count Cagliostro (1743-1795) and, more recently, Aleister CROWLEY – or may be nameless members of a secret society or occult order, the studies of which would be related to one of the following: the CABBALA, Gnosticism (◊ GNOSTIC FANTASY), ROSICRUCIANISM, Hermeticism (◊ GOLDEN DAWN) and THEOSOPHY. Members of these organizations may aspire to higher levels of awareness or consciousness, and thus students of the occult are often classified as WIZARDS, adepts, warlocks or WITCHES. Stories related to the occult are thus often classified with BLACK MAGIC, SATANISM and WITCHCRAFT; while these have aspects in common with the occult, the terms are not synonymous.

True OFs are stories which explore the mysteries, often in pursuit of or being pursued by SECRET GUARDIANS or SECRET MASTERS, in the hope of a revelation. The aftermath of this is usually a conflict between GOOD AND EVIL which often serves as the climax of an OF.

OFs may be traced back to the Roman writers (◊ GREEK AND LATIN CLASSICS). The most complete surviving Latin novel, *Metamorphoses, or The Golden Ass* (?**150**AD) by Lucius APULEIUS has an occult basis. Although many medieval texts appeared on the subject of the occult, the subject was banned in Britain during the dominance of Puritanism and did not return until the rise of ARABIAN FANTASY provided an opportunity to blend the two interests. The two seminal works of OF both first appeared in French: *Le Diable Amoureux* (**1772**; trans as *The Devil in Love* 1793 UK) by Jacques CAZOTTE and *Vathek* (ot *An Arabian Tale* **1786** France) by William BECKFORD. In both, individuals enter PACTS WITH THE DEVIL to seek exceptional pleasures. Honoré de BALZAC continued the theme in France, with *Séraphita* (**1835**), while in the UK the primary Victorian exponent was Lord BULWER-LYTTON, particularly with *Zanoni* (**1842**) and *A Strange Story* (1861-2 *All the Year Round*; **1862**). Although Bulwer-Lytton established the occult novel in the UK it was not until the general rise of public interest in SPIRITUALISM and the occult sciences towards the late 19th century that OF became a genre in its own right. Much of this was due to the teachings of Helena BLAVATSKY, who established THEOSOPHY. Theosophy gave rise to a host of OFs, particularly the works of Marie CORELLI. At the same time Ignatius Donnelly (1831-1901) produced *Atlantis: The Antediluvian World* (**1882**), which brought the lost sciences of ATLANTIS (and by extension Mu, LEMURIA and other lost worlds) into the equation. The fascination for the LOST-RACE novels of H. Rider HAGGARD resulted in an upsurge of adventure fantasy, and this soon merged with the Theosophical Romance to produce a number of genuine OFs. The master of this form was Talbot MUNDY, whose *Om: The Secret of Abhor Valley* (**1924**) may be seen as the quintessential OF. Somewhat in the same mould were the works of Abraham MERRITT, although the occult was less intrusive. A similar mood pervades some of the novels of L. Adams BECK and Joan GRANT. Their influence lives on in the INDIANA JONES films, especially *Raiders of the Lost Ark* (**1981**).

The more traditional OF, related to black magic, was explored by Joris-Karl HUYSMANS in *Là-Bas* (**1891**; trans Keene Wallace as *Down There: Là-Bas* **1928** France). In this vein a number of UK writers, particularly those steeped in the teachings of Theosophy and the Golden Dawn, produced their own OFs, including Algernon BLACKWOOD, J.W. Brodie-Innes (1848-1923), Aleister CROWLEY, Dion FORTUNE, Arthur MACHEN and Sax ROHMER. Successors were Dennis WHEATLEY and even Elliott O'Donnell (1872-1965), with *The Dead Riders* (**1952**). It was the work of these writers, especially Crowley and Wheatley, that made the OF so closely associated with black magic. Several of them produced stories about OCCULT DETECTIVES, a subgenre that overlaps at some points with OF. Likewise, some of the antiquarian GHOST STORIES of M.R. JAMES and the JAMES GANG have elements of the OF, particularly James's "Casting the Runes" (1911). Karslake, in that story, like Oliver Haddo in *The Magician* (**1908**) by W. Somerset MAUGHAM, is based on Crowley.

H.P. LOVECRAFT used occultism as the basis for developing what became the CTHULHU MYTHOS, since the stories depend upon studies of ancient tomes, most notably *The Necronomicon* (◊ BOOKS). There is an argument that the **Cthulhu Mythos** is in its entirety an OF.

The emergence of GENRE FANTASY and the recommercialization of HORROR fiction during the past 30 years has blurred the distinction between OF and other subgenres. For instance, Lyndon HARDY's **Arcadia** sequence – particularly *Master of the Five Magics* (**1980**) – and the **Adept** series by Deborah Turner HARRIS and Katherine KURTZ both use occult studies as a basis for novels of HIGH FANTASY. Paramount among these are the works of Louise COOPER, especially the **Time Master** trilogy. Most fantasies featuring WIZARDS studying arcane lore might be regarded to have their bases in OF, but they have been overtaken by the wider popularity of SWORD AND SORCERY. The most distinctive OF in the fantasy genre during this period was the **Seedbearers** trilogy by Peter Valentine TIMLETT, which traces the downfall of Atlantis and the continuation of occult knowledge among the survivors.

Otherwise, OF remains closer to the realms of SUPERNATURAL FICTION, and particularly supernatural horror. James HERBERT uses Hitler's interest in the occult in *The Spear* (**1978**) while Mark FROST sweeps Arthur Conan DOYLE into a turbulent occult adventure in *The List of 7* (**1993**). A similar RECURSIVE FANTASY drawing upon the work of Arthur MACHEN is *The Devil's Maze* (**1983**) by Gerald Suster (1951-). Other modern fictions which have elements of OF include: *The Magus* (**1965**; rev 1977) by John FOWLES; *The Ceremonies* (**1984**) by T.E.D. KLEIN; *Son of the Endless Night* (**1985**) by John Farris (1936-); and the **Titus Crow** stories by Brian LUMLEY. [MA]

OCCULTISM A general term which describes the study of those "sciences" – ALCHEMY, certain forms of MAGIC, THEOSOPHY and the like – which may be defined as having a secret core of knowledge about matters themselves inherently esoteric. Students of occult sciences (like John DEE or, less convincingly, Aleister CROWLEY) might be better described as occultists. As the term describes a relationship between the exoteric mundane world and an esoteric body of knowledge by which that mundane world can secretly be understood or ruled, it is not often found in fantasy. It is, however, very common in SUPERNATURAL FICTION, where a similar relationship between the mundane and the Other tends to operate. It should be noted, however, that stories explicitly evoking the occult routinely fail as fiction in proportion to the extent of their authors' belief in the literal truth of the "science" being expounded. It might also be noted that occult knowledge and RITUAL – an example is the

BLACK MASS – may be presented by an author as either obscene or holy; rarely will an author be neutral. To write about O is to advocate O, or to denounce it. Stories featuring OCCULT DETECTIVES are similarly didactic in tone.

Authors whose works invoke occultism themes include E.F. BENSON, Charles BIRKIN, Algernon BLACKWOOD, Madame BLAVATSKY, Edward BULWER-LYTTON, Arthur Conan DOYLE, Nictzin DYALHIS, Dion FORTUNE, Ronald FRASER, Robert HICHENS, J.K. HUYSMANS, Margery LAWRENCE, Seabury QUINN, Sax ROHMER and Dennis WHEATLEY. There are many others. [JC]

See also: OCCULT FANTASY.

OCCULT SHORTS ◊ MAGAZINES.

OCEAN OF STORY An 11th-century Kashmiri court poet, Somadeva, assembled a vast collection of stories from a wide range of sources, some known, some forever obscure; many of the tales were of Indian origin, but many (it is likely) had come to the subcontinent from abroad. This gathering, known as the *Katha Sarit Sagara*, contains a wide range of adventures with supernatural elements – BEAST FABLES, VAMPIRE tales, etc. – and stands as a kind of encyclopedia of story types, a CAULDRON OF STORY. It first appeared in English under the original title (trans C.H. Tawney **1880-84** UK) but is now best-known in the version prepared by Norman Penzer from Tawney's translation, *The Ocean of Story* (10-vol anth **1924-8** UK), as ed Somadeva. In putting together his original compilation, Somadeva incorporated not only individual tales but at least two already existing STORY CYCLES: the *Panchatantra* ["Five Books"], in whose FRAME STORY a Brahmin, who must educate three princes in proper behaviour, does so by telling them a number of exemplary tales; and the *Vetalapanchavinsati* ["25 Tales of a Vampire"] (part trans Sir Richard Burton as *Vikram and the Vampire, or Tales of Hindu Devilry* **1870** UK), in which a friendly VAMPIRE spirit tells a king – who has gone to its cemetery to retrieve a corpse hanging from a tree, at the request of a beggar to whom he is morally indebted – a number of stories to prevent his returning to court, where he will be murdered. Not only, in other words, does *The Ocean of Story* preserve many classic tales, and comprise therefore a compost of tales from which much modern fantasy has grown – specifically shaping, for instance, Salman RUSHDIE's *Haroun and the Sea of Stories* (coll of linked stories **1990**) – it is also, along with *The Arabian Nights* (◊ ARABIAN FANTASY), a central example of the story cycle. Typically formed of large cycles, it also incorporates stories which turn out themselves to frame further sets of stories within stories (◊◊ ARABIAN NIGHTMARE).

More generally, the term OOS can be understood – and is so used in this encyclopedia – as referring to the current critical understanding that almost every traditional STORY exists in multiple versions; that it is exceedingly difficult to sort these versions into chaste stemmata (◊ STEMMA); and that we in the 20th century cannot lop our versions of traditional tales from the top of a linear tree but must instead cast our nets in an unfathomably complex ocean. [JC]

ODD COUPLES ◊ DUOS.

ODIN Norse GOD, combining two Teutonic deities: Tiwaz, the Sky Father, and Wotan/Woden, a war-god. Favoured by aristocrats and warriors, he has over 50 titles, including "All-Father", "the Hooded One" and "the Hanged God". Two ravens, Hugin and Munin (Thought and Memory), are his informers, His ring Draupnir spawns nine more rings on every ninth night. He flies on his eight-legged horse Sleipnir and goes among humans disguised. To gain wisdom, he traded one of his eyes for water from under the WORLD-TREE, Yggdrasil. RUNES appeared at his feet after he had wounded himself with his spear and hanged himself for nine days. At RAGNAROK, the Doom of the Gods, he faces the giant WOLF Fenris, who kills him.

He is a sacrificed god. His suffering is not for FERTILITY or REDEMPTION. Mounting a tree, self-inflicting pain to gain wisdom, and the sense of flying are characteristic experiences of SHAMANS. Odin is a battle-lord, but no great warrior. He casts his spear over armies to determine the outcome, advises on strategy and chooses who will form his army of the slain in VALHALLA. He is the forerunner of the cloaked, supernatural stranger, and Lord of the rings. [FS]

See also: NORDIC FANTASY.

ODLE, ALAN (ELSDEN) (1888-1948) UK artist and illustrator whose style was in the direct tradition of William Hogarth (1697-1764) but who displayed a disturbing element of SURREALISM in his subject matter. He is now recognized as one of the masters of grotesque ILLUSTRATION. Recurring themes in his work were dismemberment and transmuting plant and animal lifeforms. AO claimed that a major influence was the children's comic *Chatterbox*.

AO was art editor of *The Gypsy* (1915-16), a short-lived periodical in which several of his drawings were published. An alcoholic and consumptive, he was not expected to live long, but married the novelist Dorothy Richardson (1873-1957) in 1917 and lived an impoverished life with her in Cornwall for over 30 years. [RT]

ODYSSEUS By virtue of his starring role in HOMER's *Odyssey*, Odysseus (Ulysses in the Roman version) has become one of the key symbolic figures of subsequent fantasy: an ACCURSED WANDERER who eventually returns to save his wife Penelope from the suitors who have laid siege to his home. His story has been frequently retold and variously recast. His MYTH makes much of his TRICKSTER ingenuity, which secures the conquest of Troy before allowing him to outwit such adversaries as the Cyclops, the Lotus Eaters and CIRCE in the course of his much-interrupted voyage. Notable recapitulations and reconfigurations of the *Odyssey* include *Ulysses* (**1922**) by James JOYCE, *Penelope's Man* (**1927**) by John ERSKINE, *The Voyage Home* (**1958**) by Ernst Schnabel and *Space Chantey* (**1968**) by R.A. LAFFERTY. Tales based on particular episodes include "Letters from the Phaeacian Capital" (1914) by Oswald J. Couldrey and "Circe's Island" (1925) by Eden PHILLPOTTS. "On the Margin of the Odyssey" (1905) by Jules Lemaître (1853-1914) was the author's second addition to the myth, following the sarcastic "Nausicaa" (1894) – which also features Odysseus's son Telemachus, famous in France through being also the hero of the didactic mock-epic *Télémaque* (**1699**) by François Fénelon (1651-1715) – and Lemaître subsequently composed other "marginal" episodes featuring Helen and Penelope. Further adventures of Odysseus are featured in *The World's Desire* (**1890**) by H. Rider HAGGARD and Andrew LANG, "Death and Odysseus" (1915) by Lord DUNSANY, "Odysseus Goes Roving" (1929) by F. Britten Austin (1885-1941) and *Atlantis* (**1954**) by John Cowper POWYS. A recent characterization can be found in Marion Zimmer BRADLEY's *Iliad*-inspired novel *The Firebrand* (**1987**). [BS]

OFFUTT, ANDREW J(EFFERSON V.) (1934-   ) US writer who has often signed himself andrew j. offutt; he has also published as Andy Offutt and A.J. Offutt. His first published story was "And Gone Tomorrow" (*If*, 1954) as Andy Offutt. AJO's many novels in sf and other genres (◊ *SFE*), often under pseudonyms, have been executed vigorously but not always with great care.

Since the appearance of *Messenger of Zhuvastu* (1973), AJO has concentrated on fantasy, usually SWORD AND SORCERY, much of his work being collaborations, SHARED-WORLD efforts or ties.

After *My Lord Barbarian* (1977) most books were in series, including **Cormac Mac Art**, based on Robert E. HOWARD's character: *Sword of the Gaeel* * (1975), *The Undying Wizard* * (1976), *Sign of the Moonbow* * (1977), *The Mists of Doom* * (1977), *When Death Birds Fly* * (1980) with Keith TAYLOR and *The Tower of Death* * (1982) with Taylor. **War of the Wizards** comprises *Demon in the Mirror* (1978), *Eyes of Sarsis* (1980) and *Web of the Spider* (1981), all with Richard K. Lyon. **War of the Gods on Earth** comprises *The Iron Lords* (1979), *Shadows out of Hell* (1980) and *The Lady of the Snowmist* (1983). Outside series are *Chieftan of Andor* (1976, vt *Clansman of Andor* 1979 UK); and a CONAN parody, *The Black Sorcerer of the Black Castle* * (1976 chap), plus three **Conan** novels: *Conan and the Sorcerer* * (1978), *The Sword of Skelos* * (1979) and *Conan the Mercenary* * (1980). Other works are *King Dragon* (1980), *Shadowspawn* * (1987) – in the **Thieves' World** sequence – *Deathknight* (1990) and *The Shadow of Sorcery* (1993). AJO also edited the **Swords Against Darkness** series, comprising *Swords Against Darkness* (anth 1977; vt *Swords Against Darkness I* 1990), *II* (anth 1977), *III* (anth 1978), *IV* (anth 1979) and *V* (anth 1979). [JC/GF]

OGIER ◊ MORGAN LE FAY.

O'HAGAN, HOWARD (1902-1982) Canadian writer. ◊ CANADA.

O'HARA, KENNETH Pseudonym of Jean MORRIS.

"OH, GOD!" US movie (*1977*). Warner. **Pr** Jerry Weintraub. **Dir** Carl Reiner. **Screenplay** Larry Gelbart. **Based on** *Oh, God!* (1971) by Avery Corman (1935-   ). **Starring** Ralph Bellamy (Sam Raven), George Burns (God), John Denver (Jerry Landers), Teri Garr (Bobbie Landers), Paul Sorvino (Reverend Willie Williams). 104 mins. Colour.

GOD, in the form of a little old man in a baseball cap, appoints supermarket assistant manager Landers to bring to the world his new message: that humanity can work things out if it wants to. Obtusely, God makes Landers' task as difficult as possible, giving to no one else any unequivocal sign of His existence until finally, when Landers is facing a slander suit brought by a corrupt evangelist to whom he gave God's message ("Shut up"), He manifests in court. Even then, witnesses prefer to believe their "common sense" rather than the evidence of their eyes and ears.

It is hard to convey the full blandness of this movie. The platitudinous utterances of the Almighty seem designed to lure the atheist-humanist vote, rather than the theist one. The movie comes to life only with Sorvino's chillingly comedic performance as a profiteering evangelist – but even this represents a flaw: if the message of the movie is that we should reject fundamentalism and bigotry, its argument is nullified in that it presents no serious advocate of the opposite case, only a caricature.

The movie was sequelled by *Oh God! Book Two* (*1980*), which is a yet more anodyne rehash, and by OH GOD! YOU DEVIL (*1984*). [JG]

OH GOD! YOU DEVIL US movie (*1984*). Warner. **Pr**

Robert M. Sherman. **Exec pr** Irving Fein. **Dir** Paul Bogart. **Spfx** Ray Klein. **Screenplay** Andrew Bergman. **Starring** George Burns (God/Satan/Harry O. Tophet), Robert Desiderio (Billy Wayne), Roxanne Hart (Wendy Shelton), Eugene Roche (Charlie Gray), Ron Silver (Gary Frantz), Ted Wass (Bobby Shelton). 96 mins. Colour.

The third in the series that began with "OH, GOD!" (*1977*) at last discovers some waspishness in its retelling of the FAUST legend. Back in 1960, as young Shelton was suffering scarlet fever, his father invoked GOD to watch over the boy. Today Shelton is a struggling musician. He sells his SOUL to the DEVIL – God's counterpart and identical DOUBLE – in exchange for success; SATAN appears in the form of Harry O. Tophet, claiming to be agent for most of the big stars. The CONTRACT signed, Shelton discovers his life has been swapped with that of successful rock star Wayne (◊ IDENTITY EXCHANGE), whose own PACT WITH THE DEVIL has just expired. For a while Shelton/Wayne enjoys success and groupies, but then he yearns for wife Wendy, especially when he discovers that the ex-Wayne is now Bobby Shelton, and has always been, and that she is pregnant with his – not the ex-Wayne's – child. Shelton/Wayne goes in search of God, the only one who can extricate him from his contract, and discovers him in Las Vegas. There Satan and God play a game of poker (◊ CARDS) for his soul, God winning. Shelton regains his own life, wife, child and happiness.

This is quite impressive piece of fantasy, full of interesting touches and clever lines (HEAVEN, we are told, is half-empty these days: "They had to close one of the main dining-rooms"). There are some fine flashes of TECHNO-FANTASY. Clearly parallels can be drawn with *Phantom of the Paradise* (*1974*; ◊ PHANTOM OF THE OPERA), which is certainly a better movie; yet *OG!YD* has its own strengths, and should certainly not be dismissed as merely the last in a dying series. [JG]

OKRI, BEN (1959-   ) Nigerian-born UK writer, whose early novels and short stories were straightforward realist descriptions of African and occasionally UK life. His two MAGIC-REALISM novels of African smalltown life, *The Famished Road* (1991) and *Songs of Enchantment* (1992), deal with the dreamlike existence of a child whose early head-injury has put him in touch with spirits and with the dreams of others. Class-war politics in the town are dominated by a VAMPIRE enchantress bar-owner; the boy's parents survive enchantments, partly by telling each other and the boy traditional STORIES which echo the magic of their daily existence. These remarkable books combine a wide-eyed tone of voice with a sophisticated sense of place, constantly transgressing standard boundaries of tone; *The Famished Road* won the 1991 Booker Prize.

Astonishing the Gods (1995) is a fabulation about an invisible man (◊ INVISIBILITY) who discovers a CITY whose morally perfect inhabitants are invisible even to him. He journeys to its centre, and goes through an INITIATION which gradually render him perfect and invisible even to himself. This is less successful than the two African novels, partly because Okri is using the stock imagery of ordeals. [RK]

Other works: *Flowers and Shadows* (1980); *The Landscapes Within* (1981); *Incidents at the Shrine* (coll 1987); *Stars of the New Curfew* (coll 1988).

OLD GODS ◊ AESIR; ELDER GODS.

OLIPHANT, MRS Working name of UK writer Margaret Wilson Oliphant (1828-1897) for most of her 100 or so novels; she sometimes wrote anonymously. She began publishing work of fantasy interest with "A Christmas Tale" for *Blackwood's Magazine* in 1857 (vt "Witcherley Ways: A Christmas Tale" in *Tales from Blackwood* anth **1879**), a story which prefigures her **Stories of the Seen and Unseen**, the phrase she used as a surtitle on various occasions to mark her SUPERNATURAL FICTION. Like most of its successors, "A Christmas Tale" depicts family life as hollowed out by some central absence or deficiency. Her next story – "The Secret Chamber" (1878 *Blackwood's*, published initially as a CHRISTMAS tale and much exp vt *The Wizard's Son* **1883** US) – follows the same pattern: in the story a decent son obeys a feckless father's instructions to spend the night in a Secret Chamber, where the lad outwits the GHOST of an evil ancestor; in the novel, the son takes over a Scottish estate and manages to thwart the MALIGN SLEEPER his accession has awakened.

The Unseen, for MO, is the land beyond the living; and attempts to deal with the overwhelming poignance of death dominate her career, most significantly in *A Beleaguered City, Being a Narrative of Certain Recent Events in the City of Semur, in the Department of the Haute Bourgogne: A Story of the Seen and the Unseen* (**1879** US; exp dated 1880 but 1879 UK). The FRAME STORY is complex: various witnesses tell of the blaspheming of the citizens of Semur against the Lord, their eviction by the risen dead, the judgement then imposed, and finally the forgiveness. Some subsequent tales – like "The Open Door" (*Blackwood's* 1882) and *Old Lady Mary: A Story of the Seen and Unseen* (**1884** chap US) – have a similar impact, both stories resolving the obsessional returns of REVENANTS to the land of the living; but most of MO's further work suffers from a preoccupation with the AFTERLIFE, and her tales of the **Little Pilgrim** – assembled with other work in *A Little Pilgrim in the Unseen* (coll dated 1883 but **1882**) and *The Land of Darkness* (coll **1888**) – are deeply sentimental. Other stories of HEAVEN or the supernatural in general – like *The Lady's Walk* (**1883** US; much exp 1897 UK) and "The Land of Suspense" (*Blackwood's* 1897) – convey so desperate a sense of BELATEDNESS that they are hard to bear. Only once is there a genuine solace: the Stranger from the Little Pilgrim's Heaven, who visits Earth in *A Visitor and His Opinions: A Story of the Seen and Unseen* (1893 *Blackwood's*; **1902** chap US), finds much to admire in mortal life. Generally, however, MO did not; and a sense of thwarted longing impregnates her work. [JC]

Other works: *The Open Door and The Portrait: Two Stories of the Seen and Unseen* (coll **1885** US); *Two Stories of the Seen and Unseen: The Open Door; Old Lady Mary* (coll **1885**); *Stories of the Seen and Unseen* (coll **1889** US; with different contents **1902** UK); *"Dies Irae": The Story of a Spirit in Prison* (**1895** chap). Most of MO's best supernatural fiction has been re-sorted in *Selected Short Stories of the Supernatural* (coll **1985**) ed Margaret K. Gray and *A Beleaguered City and Other Stories* (coll **1988**) ed Merryn Williams.

Further reading: *Mrs Oliphant: "A Fiction to Herself": A Literary Life* (**1995**) by Elisabeth Jay.

OLIVIA (De BERNARDINIS) (1948-   ) US painter of light erotica, often with fantasy elements – e.g., giving people bird or butterfly wings, or the surface markings of an exotic animal such as a leopard or zebra. Her pictures usually consist of a single, languidly posed female against a white background, airbrush-painted with a delicate touch and a particular skill in rendering the texture of soft fur and feather, lace and human flesh.

Born in California, Olivia attended the School of Visual Arts in New York in 1967. From 1974, she secured a niche painting her erotic fantasy women for men's magazines. In 1977, with her husband Joel Beren, she set up the greetings-card business O Cards to publish her work, which has since appeared in calendars and on the covers of magazines such as HEAVY METAL.

A collection of Olivia's art is *Let Them Eat Cheesecake: The Art of Olivia* (graph **1993** US). [RT]

OLYMPUS A 3000m peak in Pieria designated in Greek MYTHOLOGY as the home of the gods, although the physical mountain ought perhaps to be reckoned a mere pointer to some loftier abode in the sky. Olympus is usually conceived as a place where much feasting and lovemaking went on, but fictional images of it are generally conceived with comic or satiric effect. It is the chief setting of "Ixion in Heaven" (1847) by Benjamin DISRAELI, *Olympian Nights* (**1902**) by John Kendrick BANGS, *Venus the Lonely Goddess* (**1949**) by John ERSKINE and *The Memoirs of Zeus* (**1963**) by Maurice Druon (1918-   ). It is more detailedly described in *O Men of Athens* (**1947**) by A.C. Malcolm. Olympus is depicted in various movies – e.g., *Clash of the Titans* (**1981**) – but generally not very interestingly. [BS]

OMAR KHAYYAM (vt *The Life, Loves and Adventures of Omar Khayyam*) US movie (**1956**). Paramount. **Pr** Frank Freeman Jr. **Dir** William Dieterle. **Spfx** John P. Fulton. **Screenplay** Barré Lyndon. **Starring** John Derek (Prince Malik), Raymond Massey (Shah), Debra Paget (Sharain), Michael Rennie (Hasani), Cornel Wilde (Omar). 101 mins. Colour.

An ARABIAN FANTASY more in the spirit of *The* THIEF OF BAGDAD (**1924**) and KISMET (**1930**) than *The Rubáiyát of Omar Khayyám* (12th century; trans **1859** anon by Edward Fitzgerald [1809-1883], revs as trans EF; rev 1868; rev 1872; rev 1879), although it is studded with quotations from the Fitzgerald version. Here the poet has an "aw, shucks" style and a habit of lurching into homespun cliché. Not only a poet but an expert on clocks and the CALENDAR, he is employed in all capacities by the Shah of Persia, with whose newest wife Omar is in love. Many buckles are swashed in a routine adventure.

This is a disappointing movie from Dieterle. One of the co-stars is indeed the same John Derek who would direct such movies as *Tarzan, the Ape Man* (**1981**; ◊ TARZAN MOVIES) and GHOSTS CAN'T DO IT (**1990**). [JG]

OMEN, THE A series of four movies: the **Damien** trilogy plus, much later, the loosely related **4** (which seemed intended to spawn a new cycle). The initial plan was a **Damien** tetralogy but, after the huge success of **1**, the relative box-office failure of **2** curtailed ambitions. The series, especially **1**, is mercilessly parodied (◊ PARODY) in *Good Omens* (**1990**) by Neil GAIMAN and Terry PRATCHETT. In addition to the movies, Gordon McGill, who novelized **3**, continued the book saga in *Omen IV: Armageddon 2000* (**1982**) and *Omen V: The Abomination* (**1985**).

1. *The Omen* US movie (**1976**). 20th Century-Fox. **Pr** Harvey Bernhard. **Exec pr** Mace Neufeld. **Dir** Richard Donner. **Spfx** John Richardson. **Screenplay** David Seltzer. **Novelization** *The Omen* * (**1976**) by Seltzer. **Starring** Martin Benson (Father Spiletto), Leo McKern (Carl Bugenhagen), Holly Palance (Holly), Gregory Peck (Robert Thorn), Lee Remick (Kathy Thorn), Harvey Stephens (Damien), Patrick Troughton (Father Brennan), David Warner (Keith Jennings), Billie Whitelaw (Mrs Baylock). 111 mins. Colour.

In Rome, at 6am on June 6 (i.e., 6/6/6; ☿ GREAT BEAST), Kathy gives birth to a stillborn child; husband Robert, US Ambassador to Italy, colludes with the hospital's Father Spiletto to replace it with the living child of a mother who died in childbirth at that same moment. At the CHANGELING Damien's fifth birthday, with Robert now Ambassador to the UK, a black dog appears and mesmerizes his nanny into hanging herself. Soon after, eccentric Irish priest Brennan comes to Robert to try to persuade him Damien is the ANTICHRIST, but is shown the door. A new nanny, Mrs Baylock, arrives without warning; she is in fact an emissary of SATAN. The tale becomes very complex, leading Robert to Megiddo, near Jerusalem, to consult with mystic/archaeologist Bugenhagen and eventually to conclude that Damien is indeed the GREAT BEAST. Using the set of SEVEN knives Bugenhagen gave him for the ritual killing of the child, Robert is poised to do the deed when shot dead by a policeman. At Robert's funeral, we find Damien hand-in-hand with the US President.

TO is a very classy HORROR MOVIE, riding on the coat-tails of The EXORCIST (*1973*), with excellent camerawork and spfx and some fine performances – notably by Peck (Charlton Heston had declined the part), Stephens and Warner – all disguising occasionally poor dialogue and several plotting infelicities. A highpoint is the sequence leading up to Brennan's impalement by a church's falling lightning-conductor: the howling WINDS and the flailing trees contribute to a scene of Nature attacking an individual that rivals the gathering of the birds in Alfred HITCHCOCK's *The BIRDS* (*1963*). [JG]

2. *Damien: Omen Two* (vt *Omen 2*) US movie (*1978*). 20th Century-Fox/Mace Neufeld. **Pr** Harvey Bernhard. **Dir** Don Taylor. **Spfx** Ira Anderson Jr. **Screenplay** Bernhard, Michael Hodges, Stanley Mann. **Novelization** *Damien: Omen II* * (**1978**) by Joseph Howard. **Starring** Alan Arbus (Pasarian), Lew Ayres (Bill Atherton), Lucas Donat (Mark Thorn), Robert Foxworth (Paul Buher), Lee Grant (Ann Thorn), Lance Henriksen (Sergeant Neff), William Holden (Richard Thorn), Leo McKern (Carl Bugenhagen), Nicholas Pryor (Charles Warren), Jonathan Scott-Taylor (Damien), Elizabeth Shepherd (Joan Hart), Sylvia Sidney (Marion), Meshach Taylor (Dr Kayne). 109 mins. Colour.

Immediately after the events of **1**, Bugenhagen, still at Megiddo, realizes Damien is depicted on a just-uncovered ancient mural; he tries to spread the news but is buried alive. Seven years later Damien lives with his Uncle Richard (his father's brother), Aunt Ann and cousin Mark, he and Mark attending Davidson Military Academy. Great-Aunt Marion knows Damien is evil, but a malign raven (in the first part of this movie ravens have the same symbolism as black dogs in **1**) induces a fatal coronary in her before she can take action. Various relics unearthed by Bugenhagen and his successors are being shipped from Israel to the Thorn Museum; among them, a statue of the Whore of Babylon astride the GREAT BEAST particularly fascinates Ann. With the antiquities comes reporter Joan Hart, who has discovered the truth about Damien; a raven pecks out her eyes and she stumbles into the path of a truck. Damien, ignorant of his true identity, has two demonic minders: at the academy Neff; elsewhere Thorn Industries VP Buher. Neff at last tells Damien the truth about himself. There are many deaths: Thorn Industries' Dr Kayne (Kane in the credits) discovers Damien has the blood of a jackal, and dies in a freak elevator accident; Cousin Mark discovers the truth, so

Damien fatally ruptures his brain; etc. Richard, too, deduces all: he plans to kill Damien with the ceremonial knives from **1**, but is knifed by Ann, who admits she has always been Damien's tool. Damien rewards her by destroying both surrogate parents in a boiler-room accident.

D:OT, a somewhat better movie than the synopsis suggests, is marred by a succession of good ideas not followed through: Buher and Neff correspond to two of the Four Horsemen of the APOCALYPSE (War and Famine), but there is no sign of the other two; Ann is a Whore of Babylon analogue, but seems to lack whorishness; the potentially interesting clash between Damien-the-nice-lad and Damien-the-Antichrist is heralded but never arrives. *D:OT* did poorly at the box office. [JG]

3. *The Final Conflict* (vt *Omen 3: The Final Conflict*) US movie (*1981*). 20th Century-Fox/Mace Neufeld. **Pr** Harvey Bernhard. **Exec pr** Richard Donner. **Dir** Graham Baker. **Spfx** Ian Wingrove. **Screenplay** Andrew Birkin. **Novelization** *Omen III: The Final Conflict* * (**1980**) by Gordon McGill. **Starring** Mason Adams (President), Rossano Brazzi (Father De Carlo), Don Gordon (Harvey Dean), Lisa Harrow (Kate Reynolds), Barnaby Holm (Peter), Sam Neill (Damien), Leueen Willoughby (Barbara Dean). 108 mins. Colour.

Now 32, head of Thorn Industries and fully dedicated to EVIL, Damien is busy engineering disasters so the company may profit from relief operations. Guided by the prophecies of the Apocryphal *Book of Hebron* that CHRIST will be reborn in the "Isle of Angels", he engineers also his ambassadorship to the UK. But others know of the imminent Second Coming: a rare conjunction of stars reprises the Star of Bethlehem, and from their alignment the precise place/instant of the MESSIAH's rebirth can be pinpointed; moreover, the knives of Negiddo have been recovered from the ashes of the Thorn Museum and are now with Father De Carlo, head of the monastery at Subiaco, Italy, where Father Spiletto has just died, confessing all. De Carlo and six fellow-monks, each with a knife (SEVEN Knives for Seven Brothers, as it were), travel to the UK to defend the Christ and kill the ANTICHRIST; most are disposed of pretty swiftly. Also disposed of by Damien's disciples are, Herod-like, all male babies born in the UK on the night of March 24. And so the plot meanders gorily until De Carlo confronts Damien in a ruined church. Damien kills De Carlo, but tv journalist Reynolds, whose son Damien has also killed, seizes the knife and slays him. As Damien dies, he admits defeat to a VISION of Jesus that appears over the altar: the Second Coming of Christ has been successfully achieved.

TFC is a clumsy movie, and Neill a heavy-handed Damien; the scientific illiteracy of the astronomy sequences is risible. Yet there are some merits. Addressing his massed disciples, Damien spells out a sort of MIRROR religion to Christianity, where pure EVIL has the virtue of pure GOOD, in which pain rather than love is the true beauty. And there are glimpses of the Antichrist as a TRICKSTER figure, bending the PERCEPTIONS of the soldiers of Good so that they unwittingly commit evil acts. [JG]

4. *Omen IV: The Awakening* US movie (*1991* tvm theatrically released outside USA). FNM/20th Century-Fox. **Pr** Harvey Bernhard. **Exec pr** Mace Neufeld. **Dir** Jorge Montesi, Dominique Othenin-Girard. **Spfx** Gary Paller. **Screenplay** Brian Taggert. **Starring** Duncan Fraser (Father James Mattson), Faye Grant (Karen York), Ann Hearn (Jo Thurson), Megan Leitch (Sister Yvonne),

Michael Lerner (Earl Knight), Andrea Mann (Lisa Roselli), Madison Mason (Dr Hastings), Asia Vieira (Delia), Michael Woods (Gene York). 97 mins. Colour.

A low-budget rerun of **1**, with the sexes comprehensively reversed. Some years after the end of **3** – whose Second Coming appears forgotten – Gene and Karen adopt a baby girl, Delia, from the St Francis Orphanage in Virginia. Young Sister Yvonne, who knows the truth of the child, has fears; her Mother Superior chastises her for them, but in the process has a heart attack; the same fate attends the priest who attempts baptism, while later the threatening father of Delia's kindergarten bully is decapitated in a freak car accident (shades of Jennings's fate in **1**). Gene's path towards political stardom is strangely eased. Delia grows up healthy but friendless to age 8, when she attains menarche: this is where the main action starts, but a promising setup dissolves into a mess of gory deaths and reverent nods towards "Alternative Wisdom". Delia proves to be the daughter of Damien (from **1-3**); moreover, thanks to the condition called fetus papyraceous, when born she carried the embryo of a twin brother, which Satanist GP Hastings later transplanted into Karen to engender younger brother Alexander. Karen tries to kill the CHILDREN, but Delia mentally forces her to turn the gun on herself.

A near-perfect performance by Vieira cannot dispel the tvm aura and the desperately contrived plotting, rich in clichés: the few occasions when crucifixes stay the right way up are causes for surprise. [JG]

OMENS These take many forms in fantasy: any phenomenon, from a rainbow to a rain of frogs, may be interpreted within the STORY context as a pointer to future events. Usually the importance will not be left to the reader's imagination: thus in Piers ANTHONY's *A Spell for Chameleon* (**1977**) the troubling omen of a chameleon seized by a hawk is finally interpreted by the hero as his own capture (by marriage) of the woman named Chameleon. Omens, despite their reflection in the word "ominous", may presage favourable as well as unfavourable events – unlike PORTENTS. [CB/DRL]

ONCE A HERO US tv series (1987). New World. **Pr** Ira Steven Behr, Kevin Inch, Paul Pompian. **Dir** Kevin Hooks and many others. **Created by** Dusty Kay. **Starring** Josh Blake (Woody Greely), Caitlin Clarke (Emma Greely), Robert Forster (Gumshoe), Dianne Kay (Rachel Kirk), Jeff Lester (Captain Justice/Brad Steele), Milo O'Shea (Abner Bevis). 90min pilot plus 6 1hr episodes (3 unaired). Colour.

The (fictional) COMIC-book SUPERHERO Captain Justice was once popular but is now largely forgotten. He comes to life when he realizes people are not reading his comics any more. Along with Gumshoe, a private detective, and his girlfriend, Kirk, Captain Justice crosses the "Forbidden Zone" that separates our world from the cartoon universe to try to salvage his life and career. Much to his surprise, he discovers he has no superpowers after the transition. Forced to take a secret identity, he poses as college professor Brad Steele, then sets out on a QUEST to revitalize Abner Bevis, the cartoonist who created the comic-book characters. *OAH* offers an intersting example of a live-action TOON struggling to survive in the mundane world. [BC]

ONCE AND FUTURE KING The phrase used – in *HIC IACET ARTHURUS, REX QUONDAM REX QUE FUTURUS* – by Sir Thomas MALORY in *Le Morte Darthur* (**1485**) to describe his own assertion that ARTHUR was not dead but only sleeping and would come again in Britain's hour of

need. This RESURRECTION-myth has been applied to a number of famous national heroes (◊ SLEEPER UNDER THE HILL). The title was used famously by T.H. WHITE. [MA]

ONE MILLION BC Two PREHISTORIC FANTASIES, **2** a remake of **1**.

1. *One Million BC* (vt *The Cave Dwellers*; vt *Man and his Mate* UK) US movie (**1940**). Hal Roach. **Pr** Hal Roach. **Dir** D.W. Griffith (uncredited), Hal Roach, Hal Roach Jr. **Spfx** Roy Seawright. **Screenplay** George Baker, Joseph Frickert, Grover Jones, Mickell Novak. **Starring** Nigel De Brulier (Peytow), Lon Chaney Jr (Akhoba), Mamo Clark (Nupondi), Jacqueline Dalya (Ataf), Edgar Edwards (Skakana), Mary Gale Fisher (Wandi), John Hubbard (Ohtao), Carole Landis (Loana), Victor Mature (Tumak), Inez Palange (Tohana). **Voice actors** Conrad Nagel (Narrator). 80 mins. B/w.

Lost mountaineers come across a cave where an archaeologist gives them his interpretation of a series of prehistoric wall paintings. According to this, Tumak, son of the chieftain of the brutish Rock People – who have yet to invent the spear but seem to have invented shaving – is driven out for picking a quarrel with his father, Akhoba. After an encounter with a rogue mastodon, he drifts downriver to be rescued by the pacifistic, more civilized Shell People, who teach him the art of sharing. But eventually he is driven from them, too, because of his greedy aggression; the beautiful Loana follows him. The two, after various scuffles with DINOSAURS and other MONSTERS, reach the Rock People once more, and teach them niceness. A volcanic eruption mars the idyll, but in adversity the two tribes discover cooperation and unite.

This is not a fine movie. Though it has a certain raw strength and surprisingly good spfx, it features some startlingly wooden acting. D.W. Griffith (1875-1948), originally involved, reportedly left the project because he felt the humans should use modern dialogue; in truth, this is a silent movie with a soundtrack. It may also be regarded as a sort of prehistoric TARZAN MOVIE; the UK vt was likely an attempt to cash in on the popularity of *Tarzan and His Mate* (**1934**). *OMB* gave rise not only to **2** but also – and in some ways more direct comparisons can be made – to *The* CLAN OF THE CAVE BEAR (**1985**). [JG]

2. *One Million Years BC* UK movie (**1966**). Associated British-Pathé/HAMMER/20th Century-Fox. **Pr** Michael Carreras. **Assoc pr** Aida Young. **Dir** Don Chaffey. **Spfx** George Blackwell. **Vfx** Ray HARRYHAUSEN. **Mufx** Wally Schneiderman. **Screenplay** Carreras. **Based on 1**. **Starring** Martine Beswick (Nupondi), Robert Brown (Akhoba), William Lyon Brown (Payto), Percy Herbert (Sakana), Richard James (Young Rock Man), Mayla Nappil (Tohana), John Richardson (Tumak), Lisa Thomas (Sura), Jean Waldon (Ahot), Raquel Welch (Loana). **Voice actor** Nicolette McKenzie (Loana). 100 mins. Colour and some sepia.

A faithful remake of the core tale of **1**, with somewhat less wooden acting and an enhanced sense of anachronism – not just in that (obviously) the extinction of the DINOSAURS occurred long before the emergence of *Homo* but also internally in the depiction of this prehistoric world: in particular, the women of the pacifistic, fair-haired shore-dwelling Shell People are curvaceously well fed, well manicured and very modern-seeming, with elegant hairstyles. As with the core of **1**, there is no dialogue aside from gibberish, grunts and names, plus a narrated start like that to a DISNEY True-Life Adventure short.

Hugely popular (that curvaceousness is only scantily covered), *OMYBC* sparked off a string of imitative PREHISTORIC FANTASIES, starting with *When Dinosaurs Ruled the Earth* (*1969*). By any objective measure the movie is nonsense, yet it is fun and has fine (and some less fine) spfx in the classic tradition. [JG]

O'NIELL, ROSE CECIL (1875-1944) US artist. ◊ ILLUSTRATION.

ONIONS, (GEORGE) OLIVER (1873-1961) UK writer who, after 1918, changed his name legally to George Oliver, but retained his given name for all his writing; he was married to the romantic novelist Berta Ruck (1878-1978). Outside the fantasy genre OO is best-known for his grim, unrelenting character novels, especially *In Accordance With the Evidence* (1912), a murder mystery, and *The Story of Ragged Robyn* (**1945**), which creates a nightmare landscape of the 17th-century Lincolnshire marshes and often challenges the reader to distinguish between REALITY and fantasy. Within the more orthodox fantasy field, however, OO is remembered mostly for "The Beckoning Fair One" (1911), considered by many to be among the best of all GHOST STORIES. It was inspired by OO hearing the sound of his wife combing her long hair, and becoming aware of the distinctiveness of sounds that might continue to haunt a house (◊ HAUNTED DWELLINGS). In the story it is only sounds and imagination that deliver the writer protagonist, Paul Oleron, into madness and starvation. OO himself, a no-nonsense Yorkshireman, had no truck with the supernatural, and this defiance of belief adds weight to the story's growing unease. It was included in OO's *Widdershins* (coll **1911**). Although he had written no earlier fantasies, his stories, particularly those in *Tales from a Far Riding* (coll **1902**) and *Draw in Your Stool* (coll **1909**), had shown an increasing tendency to explore the grimness of human existence. Once OO discovered the power of using the supernatural he was able to add the dimension of the inexplicable to enhance situational terrors and, often, the mental disintegration or TRANSMUTATION. "Benlian", about a sculptor who believes he can transfer his SPIRIT to that of his STATUE, and "Rooum", about a man living in fear of a spirit that constantly chases him, are the two best examples of this.

OO produced two further collections of ghost stories, *Ghosts in Daylight* (coll **1924**) and *The Painted Face* (coll **1929**). These were assembled with *Widdershins* and slightly revised as *The Collected Ghost Stories of Oliver Onions* (omni **1935**), from which was made the selection *Bells Rung Backwards* (coll **1953**). At their best, OO's ghost stories are powerfully charged explorations of psychical violation, their effects heightened by detailed character study and a preparedness to challenge the accepted.

OO's novels of the supernatural defy categorization. Two are rather slight, and could be classified as wish-fulfilment fantasies: *A Certain Man* (**1931**), about a man who acquires a MAGIC self-repairing overcoat that fits like a skin, and *A Shilling to Spend* (**1965**), about a self-perpetuating shilling. Both also contain messages concerning IMMORTALITY and survival, a theme turned on its head in *The Tower of Oblivion* (**1921**), about a man who is growing progressively younger. All of OO's works contain feelings of rejection and alienation, and none is more powerful in this respect than *The Hand of Kornelius Voyt* (**1939**), in which a boy comes under the psychic influence of the mesmeric Heinrich Opfer (◊ MESMERISM). As in *The Story of Ragged Robyn*, OO develops

an atmosphere of despair and of dislocation from reality. It is this ability that makes OO's stories among the most challenging and rewarding in SUPERNATURAL FICTION. [MA]
Other works: *The New Moon* (1918), about a future ideal UK state.

OPEN SESAME ◊ MAGIC WORDS.

OPERA The following listing is of operas, operettas, musicals and large-scale choral works that are based on myths, legends, folktales and so on or that contain fantasy elements. Space precludes the inclusion of detailed plot synopses or complete character lists. It is assumed that readers will be sufficiently familiar with such legends as ORPHEUS and Eurydice, PYGMALION, Acis and Galatea and FAUST, and with plays such as SHAKESPEARE's *A Midsummer Night's Dream*, to require no further explanation. Similarly, many titles are in themselves self-explanatory and are given with no additional gloss. Unless otherwise indicated, "after Aesop" refers to *The Fables* (◊ AESOPIAN FANTASY); "after Apollodorus" refers to the *Chronicle*, the verse history of Greece after the fall of Troy; "after APULEIUS" refers to *The Golden Ass*; "after ARIOSTO" refers to *Orlando furioso*; "after BOCCACCIO" refers to *The Decameron*; "after CERVANTES" refers to *Don Quixote*; "after EURIPIDES" refers to *Trojan Women*; "after GOETHE" refers to *Faust: A Tragedy*; "after Homer" refers to *The Odyssey*; "after MUSÄUS" refers to *Volksmärchen der Deutschen*; "after OVID" refers to *Metamorphoses*; "after PERRAULT* " refers to *Histoires et contes du temps passeé: Contes de ma mère l'oye*; "after Plutarch" refers to *Bioi Paralleloi* ["Parallel Lives"]; "after Tasso" refers to *Gerusalemme liberata* ["Jerusalem Delivered"] (**1580-81**) by Torquato Tasso; and "after VIRGIL" refers to the *Aeneid*.

Peri, Jacopo (Italy 1561-1633)
La Dafne ["Daphne"] (**1598**; rev 1598-1600). Libretto Ottavio Rinuccini, after OVID. Apollo, Venus, Cupid, Daphne, dragon, nymphs, Ovid.
L'Euridice ["Eurydice"] (**1600**). Libretto Ottavio Rinuccini, after *La favola d'Orfeo* by Angelo Poliziano, after OVID.
Caccini, Giulio (Italy *c*1550-1618)
Il rapimento di Cefalo ["The Apotheosis of Cephalus"] (**1600**). Libretto Gabriello Chiabrera, after OVID. Aurora, Apollo, Cupid, Diana (Cynthia), Mercury, Jupiter, Night, Sleep, Death, whale, crescent moon, flying chariot, gods and goddesses, signs of zodiac, Apollo, Muses, Poetry, winged horse (Pegasus), Fame, cities subject to Florence.
L'Euridice ["Eurydice"] (**1602**). Libretto Ottavio Rinuccini, after *La favola d'Orfeo* by Angelo Poliziano, after OVID.
Cavalieri, Emilio de' (Italy *c*1550-1602)
La rappresentatione di anima, e di corpo ["The Representation of the Soul and the Body"] (**1600**). Libretto Padre Agostino Manni with Dorisio Isorelli. Guardian Angel, Soul, Body, Time, Intellect, Counsel, Worldly Life, Pleasure, Deadly Sins, Prudence, Insight, angels, blessed souls, damned souls.
Monteverdi, Claudio (Italy 1567-1643)
La favola d'Orfeo ["The Story of Orpheus"] (**1607**). Libretto Alessandro Striggio, after *L'Euridice* by Ottavio Rinuccini, after *La favola d'Orfeo* by Angelo Poliziano, after OVID.
L'Arianna ["Ariadne"] (**1608**). Libretto Ottavio Rinuccini, after OVID. Venus, Cupid, Bacchus, Jupiter, Ariadne (transformed into a star constellation), Apollo.
Il ritorno d'Ulisse in patria ["The Return of Ulysses/Odysseus"] (**1640**). Libretto Giacomo Badoaro, after Homer. Neptune, Jupiter, Minerva, Juno, celestial spirits, naiads, nereids, sirens, Human Frailty, Time, Fortune, Cupid.

L'incoronazione di Poppea ["The Coronation of Poppea"] (**1643**). Libretto Giovanni Francesco Busenello, after *Annals et al.* by Tacitus. Athena, Mercury, Cupid, Venus, Fortune, Virtue, Cupid.

Gagliano, Marco da (Italy 1582-1643)
La Dafne ["Daphne"] (**1608**). Libretto Ottavio Rinuccini, after OVID. Apollo, Venus, Cupid, nymphs (including Daphne, transformed into a laurel tree), dragon, Ovid.
La Flora, ovvero Il natal de' fiori ["Flora, or The Birth of Flowers"] (**1628**). Libretto Andrea Salvadori. Zephyrus, Cupid, Venus, Mercury, Pluto, Apollo, Pan, nymphs, tritons, gods, Muses, Jealousy, storms, sea nymphs, wood nymphs, satyrs, cupids, winds, breezes, Jupiter. Tears of joy create flowers.

Landi, Stefano (Italy 1587-1639)
La morte d'Orfeo ["The Death of Orpheus"] (?**1619**). Libretto Francesco Pona (attrib), after Orphic myths. Eurydice's spirit denies knowing Orpheus, he becomes a demi-god.

Caccini, Francesca ["La Cecchina"] (Italy ?1587-c1640)
La liberazione di Ruggiero dall'isola d'Alcina ["The Rescue of Ruggiero from the Island of Alcina"] (**1626**). Libretto Ferdinando Saracinelli, after ARIOSTO. Sorceress (Alcina), siren, Neptune. Men on a magic island, transformed into plants and trees, are changed back.

Lawes, Henry (England 1596-1662)
Comus (**1634**). Libretto John MILTON. Comus (god of pleasure), spirit, nymph.

Cavalli, Pietro Francesco [Pier Francesco Caletti-Bruni] (Italy 1602-1676)
Le nozze di Teti e di Peleo ["The Marriage of Thetis and Peleus"] (**1639**). Libretto Orazio Persiani, after OVID. Thetis, Jupiter, Aeolus, Pluto, Mercury, Cupid, Juno, Hymen, Discord (Eris), Momus (Ridicule), gods, tritons, demons, nymphs.
Gli amori d'Apollo e di Dafne ["The Loves of Apollo and Daphne"] (**1640**). Libretto Giovanni Francesco Busenello, after OVID. Jupiter, Venus, Cupid, Apollo, Aurora, Tithonus, Pan, Peneius (river god), Daphne (transformed into a laurel tree), nymphs, faun, Somnus, Morpheus.
La Didone ["Dido"] (**1641**). Libretto Giovanni Francesco Busenello, after VIRGIL. Jupiter, Juno, Neptune, Mercury, Venus, Cupid, Aeolus, Fortune, ghosts, Iris.
L'Egisto (**1643**). Libretto not known. Night, Dawn, Venus, Cupid, Hours.
L'Ormindo ["Ormindo"] (**1644**). Libretto Giovanni Faustini. Cupid, Destiny, Harmony, winds, ghost.
Il Giasone ["Jason"] (**1649**). Libretto Giacinto Andrea Cicognini, after *Argonautica* by Apollonius of Rhodes. Cupid, Jupiter, Aeolus (Pluto), Zephyrus, Hercules, sorceress (Medea), demon, monsters, giant bull, golden fleece, magic ring, spirits, gods, demons, winds, Apollo, Cupid.
La Calisto ["Callisto"] (**1651**). Libretto Giovanni Faustini, after OVID. Jupiter, Mercury, Diana, Juno, Pan, Echo, satyrs, sylvan, nymph (Callisto, transformed into a bear, then into Ursa Minor), wood creatures, Furies, heavenly spirits, Nature, Eternity, Destiny.
L'Ercole amante ["Hercules in Love"] (**1662**). Libretto Francesco Buti, after OVID. Juno, Venus, Neptune, Mercury, Somnus, Hercules, Beauty, ghosts, Graces, nymphs, zephyrs, monsters, Diana (Cynthia), Hercules, Hebe, Female Purity.

Rossi, Luigi (Italy ?1597-1653)
L'Orfeo ["Orpheus"] (**1647**). Libretto Francesco Buti, after OVID.

Cesti, Pietro Antonio (Italy 1623-1669)
L'Orontea ["Orontea"] (**1649** or **1656**). Libretto Giovanni Filippo Apolloni, after Giacinto Andrea Cicognini. Cupid, Philosophy.
Il pomo d'oro ["The Golden Apple"] (**1668**). Libretto Francesco Sbarra, after EURIPIDES. Jupiter, Neptune, Pluto, Persephone, Bacchus, Mars, Apollo, Mercury, Juno, Athena, Venus, Diana, Hebe, Cupid, Hymen, Discord, Momus, Charon, Aeolus, winds (Auster, Eurus, Volturnus, Zephyrus), Graces, Furies, nymph (Oenone), Ganymede, dragons, Glory of Austria, Fire, winds, cupids, spirits of the air, Knights of the Earth, tritons, Ideal Beauty, Spain, Italy, Sardinia, Kingdom of Hungary, Kingdom of Bohemia, Austrian Crown Lands (Imperio), America, German Empire, gods and goddesses.

Locke, Matthew (England c1622-1677)
Cupid and Death (**1653**). Libretto James Shirley, after Aesop. Cupid, Mercury, Death, Despair, Nature, Elysium, Folly, Madness, Despair, satyr, spirits, apes. The arrows of Cupid and Death are mischievously interchanged.
The Tempest, or The Enchanted Island (**1667**). Libretto Thomas Shadwell, after William Davenant and John Dryden, after William SHAKESPEARE. Magician (PROSPERO), ARIEL, Neptune, Amphitrite, Oceanus, Aeolus, Tethys, Pride, Fraud, Rapine, Murder, devils, spirits, tritons, nereids.
Orpheus and Euridice (**1673**). Libretto Elkanah Settle, after OVID.
Psyche (**1675**). Libretto Thomas Shadwell, after *Psyché* by Thomas Corneille and Philippe Quinault, after Molière, after APULEIUS. Cupid, Venus, Apollo, Jupiter, Vulcan, Mars, Pan, Pluto, Persephone, Minos, Aeacus, Rhadamanthus, Bacchus, Zephyrus, river god, water nymphs, Furies, zephyrs, cyclops, ghosts, devils, Elysian lovers, Envy, Ambition, Power, Plenty, Peace, gods and goddesses, infernal spirits, devils, nymphs, dryads, sylvans, naiads, cupids, echoes. Jupiter makes Psyche immortal.

Legrenzi, Giovanni (Italy 1626-1690)
Il Giustino ["Justin"] (**1683**). Libretto Count Niccolò Beregan, after *Anekdota* by Procopius. Giant (Atlas), sea monsters.

Lully, Jean-Baptiste [Giovanni Battista Lulli] (Italy-France 1632-1687)
Les fêtes de l'Amour et de Bacchus ["The Festivities of Cupid and Bacchus"] (**1672**). Libretto Philippe Quinault, Isaac de Benserade and Président de Périgny, after Molière. Cupid, Bacchus.
Cadmus et Hermione ["Cadmus and Hermione"] (**1673**). Libretto Philippe Quinault, after OVID. Mars, Athena, Jupiter, Juno, Venus, Apollo, Cupid, Hymen, giants turned to stone, golden statues, dragon, Envy, serpent (Python). Sown dragon's teeth grow into warriors.
Alceste ou Le triomphe d'Alcide ["Alcestis, or The Triumph of Alcides"] (**1674**). Libretto Philippe Quinault, after *Alcestis* by EURIPIDES. Aeolus, Apollo, Thetis, Diana, Hercules, Charon, Pluto, Persephone, Alecto, tritons, Cerberus (including barking chorus), nymphs, naiads, sea sprites, ghosts, Muses, winds, flying creatures, Glory, Nymph of the Marne, Nymph of the Seine, Nymph of the Sea, Nymph of the Tuileries. Hercules (Alcides) brings Alcestis back from Hades.
Thésée ["Theseus"] (**1675**). Libretto Philippe Quinault, after OVID. Sorceress (Medea), Minerva, winged dragons pulling chariot, nymphs, cupids, demons, Venus, Mars, Ceres (Demeter), Bacchus.

Atys (**1676**.) Libretto Philippe Quinault, after *Fasti* by OVID. Cybele (fertility goddess), Somnus, Morpheus, Phantasy, Alecto (Fury), nymphs, Atys (transformed into a pine tree), zephyrs, spirits, good and bad dreams, Iris, Time, Flora (Spring), Melpomene.

Isis (**1677**). Libretto Philippe Quinault, after OVID. Mercury, Jupiter, Juno, Iris, Hebe, nymph Io (transformed into Egyptian goddess Isis), man changed into a bird, Argus (100-eyed watchman), Fates, nymphs, satyrs, naiads, Furies, shivering people, spirits, Fame, Neptune, Apollo, tritons, Muses, Arts, Pan, nymph Syrinx (transformed into reeds from which Pan made panpipes).

Psyché ["Psyche"] (**1678**). Libretto Thomas Corneille and Philippe Quinault, after Molière, after APULEIUS. Cupid, river god, water nymphs.

Bellérophon ["Bellerophon"] (**1679**). Libretto Thomas Corneille and Bernard le Bovier de Fontenelle, after Philippe Quinault, after *Theogony* by Hesiod. Apollo, Athena, Neptune, magician, monster (Chimera), winged horse (Pegasus), magicians, Apollo, Bacchus, Pan.

Proserpine ["Persephone"] (**1680**). Libretto Philippe Quinault, after OVID. Persephone, Ceres (Demeter), Pluto, Jupiter, Mercury, Alpheius, Arethusa, Discord, War, Peace, Discord, Peace, Abundance, Games, Pleasures, Victory, heroes.

Persée ["Perseus"] (**1682**). Libretto Philippe Quinault, after OVID. Gorgon (Medusa), monsters, gods.

Phaëton ou La volonté de briller ["Phaethon, or The Desire to Shine"] (**1683**). Libretto Philippe Quinault, after OVID. Apollo, earth goddess, Proteus (self-transformations into lion, tree, sea monster, fountain, flames), Isis, Jupiter, Triton, Autumn, Hours of the Day, flying chariot, ghosts, Furies, the Seasons, Hours of the Day, Astraia, Saturn.

Amadis de Gaule ["Amadis of Gaul"] (**1684**). Libretto Philippe Quinault, after *Amadis de Gaula* by García Rodríguez de Montalvo. Benevolent magician, malevolent magician, benevolent sorceress, malevolent sorceress, nymphs, ghosts, monster, demons, good and evil spirits, sorceress.

Roland (**1685**). Libretto Philippe Quinault, after ARIOSTO. Fairy, magic ring (conferring invisibility), magic fountain of love, cupids, sirens, river gods, sylvans, fairies, ghosts of dead heroes, Fairy King, Glory, Fame, Terror, genies.

Armide (et Renaud) ["Armida (and Rinaldo)"] (**1686**). Libretto Philippe Quinault, after Tasso. Sorceress (Armida), Hatred, naiad, demons, winged dragons pulling chariot, monsters, nymphs, Furies, demons, spirits, ghosts, Hatred and followers, Glory, Wisdom.

Acis et Galatée ["Acis and Galatea"] (**1686**). Libretto Jean Galbert de Campistron, after OVID.

Sartorio, Antonio (Italy 1630-1680)

L'Orfeo ["Orpheus"] (**1672**). Libretto Aurelio Aureli, after OVID.

Stradella, Alessandro (Italy 1639-1682)

San Giovanni Battista ["St John the Baptist"] (**1675**). Libretto Girardo Ansaldi, after the BIBLE. Pluto, Persephone, Betrayal, Vengeance, magician, St Paul, Modesty, Luxury. Story of Salome.

Charpentier, Marc-Antoine (France c1645-1704)

Les plaisirs de Versailles ["The Pleasures of Versailles"] (**1682**). Libretto anon. Music, Conversation, Comus (god of pleasure), *Le Jeu* "The Game", Pleasures.

Actéon ["Actaeon"] (?**1683**; rev 1685 as *Actéon changé en biche* ["Actaeon Changed into a Stag"]). Libretto anon, after OVID. Juno, Jupiter, Diana, Actaeon (transformed into a stag), nymphs.

La descente d'Orphée aux enfers ["The Descent of Orpheus to the Underworld"] (?**1685**). Libretto anon, after OVID.

La couronne de fleurs ["The Crown of Flowers"] (**1685**). Libretto anon, after *Le malade imaginaire* by Molière. Flora, Pan.

Les arts florissants ["The Flourishing Arts"] (**1686**). Libretto anon. Music, Poetry, Painting, Architecture, Peace, Discord, Furies.

David et Jonathas ["David and Jonathan"] (**1688**). Libretto Père François de Paule Bretonneau, after the Bible (I Samuel 28). Witch of Endor (*la Pythonisse*) summons up spirit of the prophet Samuel.

Médée ["Medea"] (**1693**). Libretto Thomas Corneille, after Pierre Corneille, after Seneca, after *Medea* by EURIPIDES. Sorceress (Medea), Jealousy, Vengeance, Cupid, demons, winged dragons pulling chariot, spirits, Victory, Glory, Bellona (war goddess).

Blow, John (England 1649-1708)

Venus and Adonis (?**1684**). Libretto anon, after William SHAKESPEARE, after OVID. Venus, Cupid, cupids, Graces.

Steffani, Agostino (Italy 1654-1728)

Niobe, regina di Tebe ["Niobe, Queen of Thebes"] (**1688**). Libretto Luigi Orlandi, after OVID. Magician, Apollo, Artemis, Tiresias, Niobe (turned to stone), SHAPESHIFTING monster, winged dragons, spirits, monsters. The walls of Thebes are destroyed by magic and restored by Zeus.

Purcell, Henry (England 1659-1695)

Dido and Aeneas (c**1683** or **1689**). Libretto Nahum Tate, after his *Brutus of Alba*, after VIRGIL. Sorceress, witches, spirit (as Mercury), Furies, will-o'-the-wisp, cupids, Phoebus, Venus, Spring, nereids, tritons, nymphs.

The Prophetess, or The History of Dioclesian (**1690**). Libretto Thomas Betterton, after John Fletcher and Philip Massinger. Silvanus, Flora, Pomona, river gods, heroes, Graces, Pleasures, fauns, nymphs, Furies, butterflies, Cupid, Bacchus.

King Arthur, or The British Worthy (**1691**). Libretto John Dryden. Good magician (MERLIN), evil magician, air spirit, earth spirit, Cupid, Cold Genius, sirens, nymphs, winged dragons pulling chariot, magic potion, magic forest, air spirits, cold spirits, sylvans, Aeolus, Britannia, Pan, Comus (god of pleasure), Venus, Honour, winds, tritons, nereids. (ARTHUR and Merlin are the only characters from the Arthurian legend.)

The Fairy Queen (**1692**). Libretto anon (?Thomas Betterton or Elkanah Settle), after *A Midsummer Night's Dream* by William SHAKESPEARE. Oberon, Titania, Puck, Juno, Hymen, fairies, nymph, elves, satyrs, monkeys, Night, Mystery, Secrecy, Sleep, Phoebus, the Seasons, swans, followers of Night, Green Men.

Timon of Athens (**1694**). Libretto Thomas Shadwell, after William SHAKESPEARE, after Plutarch, after LUCIAN. Timon, Cupid, Bacchus, nymphs.

The Indian Queen (**1695**). Libretto anon, after John Dryden and Sir Robert Howard. Magician, God of Dreams, spirits, Hymen, Cupid, Fame, Envy, followers of Envy.

Marais, Marin (France 1656-1728)

Ariane et Bacchus ["Ariadne and Bacchus"] (**1696**). Libretto Saint-Jean, after OVID. Bacchus, Ariadne (transformed into a star constellation).

Alcyone (**1706**). Libretto Antoine Houdar de la Motte, after OVID. Neptune, Somnus, magicians, sorceresses, zephyrs, dreams, sea gods, Pan, Apollo, Mount Tmolus, Muses, fauns, dryads, river gods, naiads.

Sémélé ["Semele"] (**1709**). Libretto Antoine Houdar de la Motte, after OVID. Jupiter, Juno, Somnus, Morpheus, Apollo, Cupid, Iris, Pan, Arcadia.

Campra, André (France 1660-1744)

L'Europe galante ["Amorous Europe"] (**1697**). Libretto Antoine Houdar de la Motte. Venus, Discord.

Le carnaval de Venise ["The Carnival of Venice"] (**1699**). Libretto Jean-François Regnard, after OVID.

Tancrède ["Tancredi"] (**1702**). Libretto Antoine Danchet, after Tasso. 4 magicians, dryads, wood nymph, enchanted forest, Peace, Vengeance, demons, naiads, fauns.

Les Muses ["The Muses"] (**1703**). Libretto Antoine Danchet. Muses.

Les fêtes vénitiennes ["Venetian Festivities"] (**1710**). Libretto Antoine Danchet. Cupid, Fortune, zephyrs, Carnival, Folly, Reason.

Idoménée ["Idomeneo"] (**1712**). Libretto Antoine Danchet, after *Idomenée* by Prosper de Crébillon père, after Greek legend. Venus, Aeolus, Neptune, Jealousy, Nemesis, sea monster.

Les amours des Vénus et de Mars ["The Loves of Venus and Mars"] (**1712**). Libretto Antoine Danchet.

Keiser, Reinhard (Germany 1674-1739)

Der geliebter Adonis ["The Beloved Adonis"] (**1697**). Libretto Christian Heinrich Postel, after OVID. Venus, Mars, Adonis (transformed into red roses).

Die sterbende Eurydice oder Orpheus erster Theil ["The Dying Eurydice, or Orpheus Part One"] (**1702**). Libretto Friedrich Christian Bressand, after OVID.

Die verwandelte Leyer des Orpheus oder Orpheus ander Theil ["The Transformed Lyre of Orpheus, or Orpheus Part Two"] (**1702**). Libretto Friedrich Christian Bressand, after OVID.

Eccles, John (UK 1668?-1735)

The Judgement of Paris (**1701**). Libretto William Congreve, after EURIPIDES. Juno, Athena, Venus, Mercury.

Semele (?**1707**; premiered 1972). Libretto William Congreve, after OVID. Jupiter, Juno, Somnus, Morpheus, Apollo, Cupid, Iris, Pan, Arcadia, zephyrs, cupids, nymphs.

Weldon, John (UK 1676-1736)

The Judgement of Paris (**1701**). Libretto William Congreve, after EURIPIDES. Juno, Athena, Venus, Mercury.

The Tempest, or The Enchanted Island (?**1712**). Libretto Thomas Shadwell, after William Davenant and John Dryden, after William SHAKESPEARE. Magician (PROSPERO), spirit (ARIEL), Amphitrite, Neptune, Aeolus, devils, spirits, winds.

Scarlatti, Domenico (Italy-Spain 1685-1757)

Tetide in Sciro ["Thetis on Scyros"] (**1712**). Libretto Carlo Sigismondo Capece, after OVID. Thetis, centaur (Chiron).

Amor d'un ombra e gelosia d'un aura ["Love of a Shadow and Jealousy of a Spirit"] (**1714**). Libretto Carlo Sigismondo Capece, after OVID. Nymph (Echo), breeze (Aura). Echo dies and is brought back to life.

Handel, George Frideric [Händel] (Germany-UK 1685-1759)

Agrippina (**1709**). Libretto Vincenzo Grimani, after *Annals* by Tacitus and *De vitae Caesarum* by Suetonius. Juno.

Rinaldo (**1711**; rev 1731). Libretto Giacomo Rossi, after Aaron Hill, after Tasso. Sorceress (Armida), good magician, sirens, spirits (transformed into furies), winged dragons (pulling chariot), monsters, magic wands, magic garden, mermaids, fairies.

Silla ["Sulla"] (?**1713**). Libretto Giacomo Rossi, probably

after Italian libretto, after Plutarch. God (unidentified, in a dream), furies, winged dragons (pulling chariot).

Teseo ["Theseus"] (**1713**). Libretto Nicola Francesco Haym, after *Thésée* by Philippe Quinault, after OVID. Sorceress (Medea), Minerva, spirits, winged dragons (pulling chariot), monsters.

Amadigi di Gaula ["Amadis of Gaul"] (**1715**). Libretto anon (?Nicola Francesco Haym or Giacomo Rossi), after *Amadis de Grèce* by Antoine Houdar de la Motte, after *Amadis de Gaula* by García Rodríguez de Montalvo. Sorceress, ghost, Furies, flying chariot, spirits.

Acis and Galatea (**1718**; rev 1732). Libretto John Gay *et al.*, after OVID.

Admeto, re di Tessaglia ["Admetus, King of Thessaly"] (**1727**). Libretto anon (?Nicola Francesco Haym or Paolo Antonio Rolli), after *L'Alceste* by Ortensio Mauro, after *Alcestis* by EURIPIDES. Alcestis, Hercules, voice of Apollo, underworld spirits.

Orlando ["Roland"] (**1733**). Libretto anon (?Nicola Francesco Haym), after Carlo Sigismondo Capece, after ARIOSTO. Magician (Zoroaster), spirits.

Arianna in Creta ["Ariadne in Crete"] (**1734**). Libretto anon (?Francis Colman), after *Arianna e Teseo* by Pietro Pariati, after Apollodorus. Athena, Somnus, minotaur, magic girdle.

Terpsicore ["Terpsichore"] (**1734**). Libretto anon (?Giacomo Rossi). Apollo, Erato, Muses, nymphs, Terpsichore.

Alcina (**1735**). Libretto anon, after *L'Isola di Alcina* by Antonio Marchi or Antonio Fanzaglia, after ARIOSTO. Sorceresses (Alcina and her sister), magic ring, magic island, underworld spirits. Men previously transformed into animals are changed back again.

Atalanta (**1736**). Libretto anon, after *Lacaccia in Etolia* by Belisario Valeriani, after OVID. Mercury, Graces, cupids.

Giustino ["Justin"] (**1737**). Libretto anon, after Pietro Pariati, after Count Nicolò Beregan, after *Anekdota* by Procopius. Fortune, unearthly voice, sea monster.

Saul (**1738**). Libretto attributed to Charles Jennens (?or Newburgh Hamilton), after the Bible (I Samuel 28). Witch of Endor summons up spirit of the prophet Samuel.

Semele (**1744**). Libretto attributed to Newburgh Hamilton, after William Congreve, after OVID. Jupiter, Juno, Somnus, Morpheus, Apollo, Cupid, Iris, Pan, Arcadia, zephyrs, cupids, nymphs.

Hercules (**1745**). Libretto Rev Thomas Broughton, after Giovanni Battista Guarini, after *Trachiniae* by Sophocles, after OVID. Hercules, nymphs, Furies.

Albinoni, Tommaso (Italy 1671-1750)

Il nascimento de l'Aurora ["The Birth of Aurora"] (?**1710**). Libretto anon. Apollo, Flora, zephyr.

Vivaldi, Antonio (Italy 1678-1741)

Ercole su'l Termodonte ["Hercules at the Thermodon"] (**1723**). Libretto Giacomo Francesco Bussani, after ?Apollodorus.

Il Giustino ["Justin"] (**1724**). Libretto Pietro Pariati, after Count Nicolò Beregan, after *Anedkota* by Procopius. Fortune, unearthly voice, sea monster. (See also George Frideric Handel 1737.)

Orlando ["Roland"] (**1727**). Libretto Grazio Braccioli, after ARIOSTO. Sorceress (Alcina), magic ring, magic island, winged horse.

La fida ninfa ["The Faithful Nymph"] (**1732**; rev 1737 as *Il giorno felice* ["The Happy Day"], but rev version lost). Libretto Scipione Maffei. Juno, Aeolus, nymph (Licoris).

Fux, Johann Joseph (Austria 1660-1741)

Dafne in lauro ["Daphne in the Laurel"] (**1714**). Libretto

Pietro Pariati, after OVID. Diana, Apollo, Cupid, Mercury, nymph (Daphne, transformed into a laurel tree).

Orfeo ed Euridice ["Orpheus and Eurydice"] (**1715**). Libretto Pietro Pariati, after OVID.

Angelica, vincitrice d'Alcina ["Angelica, Vanquisher of Alcina"] (**1716**). Libretto Pietro Pariati, after ARIOSTO. Sorceress (Alcina), Megaera (Fury), sea monster, Furies, spirits, ghosts.

Elisa (**1719**). Libretto Pietro Pariati, after VIRGIL. Venus, Cupid, Hymen. Elisa = Dido.

Psiche ["Psyche"] (**1720**). Libretto Apostolo Jener, after APULEIUS. Jupiter, Mercury, Venus, Cupid, Graces, cupids.

La corona d'Arianna ["The Crown of Ariadne"] (**1726**). Libretto Pietro Pariati, after OVID. Bacchus, Thetis, Ariadne (transformed into a star constellation).

Scarlatti, Alessandro (Italy 1660-1725)

Gli equivoci nel sembiante ["Mistaken Identities"] (**1679**). Libretto Domenico Filippo Contini. Nymphs.

Il Telemaco o sia L'isola di Circe ["Telemachus, or The Island of Circe"] (**1718**). Libretto Carlo Sigismondo Capece, after Homer. Sorceress (Circe), sailors (previously transformed into trees are changed back again), underworld spirits.

Caldara, Antonio (Italy-Austria 1670-1736)

Psiche ["Psyche"] (**1720**). See Johann Joseph Fux **1720**.

Achille in Sciro ["Achilles on Scyros"] (**1736**). Libretto Pietro Metastasio, after OVID. Thetis, centaur (Chiron), Glory, Love, Time.

Telemann, Georg Philipp (Germany 1681-1767)

Der geduldige Sokrates ["The Patience of Socrates"] (**1721**). Libretto Johann Ulrich von König trans of *La pazienza di Socrate* by Nicolò Minato. Cupid.

Orpheus, oder Die wunderbare Beständigkeit der Liebe ["Orpheus, or The Amazing Constancy of Love"] (**1726**). Libretto Telemann, after Friedrich Christian Bressand, after OVID.

Marcello, Benedetto (Italy 1686-1739)

Arianna ["Ariadne"] (?**1727**; premiered 1913). Libretto Vincenzo Cassani, after OVID. Bacchus, Ariadne (transformed into a star constellation), satyrs, fauns.

Pergolesi, Giovanni Battista (Italy 1710-1736)

La conversione e morte di San Guglielmo (d'Aquitania) ["The Conversion and Death of St William (of Aquitaine)"] (**1731**). Libretto Ignazio Mancini. Devil, angel.

Montéclair, Michel Pignolet de (France 1667-1737)

Jephté ["Jephthah"] (**1732**). Libretto Abbé Simon-Joseph Pellegrin, after the Bible. Includes Venus, Apollo.

Arne, Thomas (UK 1710-1778)

The Opera of Operas, or Tom Thumb the Great (**1733**). Libretto Elizabeth Haywood and William Hatchett, after *The Tragedy of Tragedies, or The Life and Death of Tom Thumb the Great* by Henry Fielding, after folktale. TOM THUMB, ghost.

Comus (**1738**). Libretto John Dalton, after John MILTON. Magician, Bacchus, Euphrosyne, Comus (god of pleasure), river goddess, nymphs, spirits, pastoral divinities, naiads.

Alfred (**1740**). Libretto James Thomson and David Mallet. Spirits, nymph.

The Judgement of Paris (**1742**). Libretto William Congreve, after EURIPIDES. Juno, Athena, Venus, Mercury, nymphs.

Lampe, John Frederick [Johann Friedrich] (Germany-UK 1703-1751)

The Opera of Operas, or Tom Thumb the Great (**1733**). Libretto Elizabeth Haywood, after *The Tragedy of Tragedies, or The Life and Death of Tom Thumb the Great* by Henry Fielding, after folktale. TOM THUMB, ghost.

The Dragon of Wantley (**1737**). Libretto Henry Carey, after English legends. Dragon, sea monster.

Rameau, Jean Philippe (France 1683-1764)

Hippolyte et Aricie ["Hippolytus and Aricia"] (**1733**). Libretto Abbé Simon-Joseph Pellegrin, after *Phèdre* by Jean Racine, after *Phaedra* by Seneca, after *Hippolytus* by EURIPIDES. Diana, Mercury, Neptune, Pluto, Fates, Tisiphone (Fury), sea monster, nymphs, underworld gods, Furies, zephyrs, Cupid, Jupiter.

Les victoires galantes ["The Amorous Victories"] (**1735**; rev 1736 as *Les Indes galantes* ["The Amorous Indies"]). Libretto Louis Fuzelier. Hebe, Cupid, Bellona (war goddess), cupids, Pleasures, sports.

Castor et Pollux ["Castor and Pollux"] (**1737**). Libretto Pierre-Joseph Bernard, after Greek legend. Jupiter, Cupid, Mercury, Hebe, a Pleasure, ghost, Castor and Pollux (transformed into stars), Arts, Pleasures, Celestial Pleasures, demons, blessed spirits, stars, Graces, Arts, Pleasures, cupids, monsters, planets, constellations, Jupiter, Mercury, Mars, Venus, Minerva, Hebe, Cupid.

Les fêtes d'Hébé, ou Les talents lyriques ["The Festivities of Hebe, or The Lyric Talents"] (**1739**). Libretto Antoine Gautier de Montdorge with Pierre-Joseph Bernard, Abbé Simon-Joseph Pellegrin, Alexandre Le Riche de la Pouplinière, after Greek legends. Hebe, Cupid, Momus (Ridicule), La Musique, Mercury, Terpsichore, Graces, nymphs of Terpsichore, water nymphs, fauns, sylvans.

Dardanus (**1739**). Libretto Charles-Antoine Le Clerc de la Bruère (rev 1744 with Abbé Simon-Joseph Pellegrin), after Greek legend. Magician, Venus, Neptune, monster, dreams, Pleasure, magic ring, magicians, dreams, cupids, Pleasures, Jealousy, Troubles, Suspicions, Venus, Cupid.

La princesse de Navarre ["The Princess of Navarre"] (**1745**). Libretto Voltaire. The Pyrenees collapse; a Temple of Love arises.

Platée ou Junon jalouse ["Plataea, or Juno Jealous"] (**1745**). Libretto Adrien-Joseph Le Valois d'Orville, after Jacques Autreau, after *Description of Greece* by Pausanias. Mercury, Juno, Jupiter, marsh nymph (Plataea), Momus (Ridicule), Folly, frogs, satyrs, naiads, maenads, nymphs, aquilons, dryads, Graces, Thespis (inventor of Comedy), Thalia (Comedy), Cupid.

Les fêtes de Polymnie ["The Festivities of Polyhymnia"] (**1745**). Libretto Louis de Cahusac, after Greek legend, historical fact, Middle Eastern legend. Polyhymnia (Lyric Poetry), Arts, Muses, Hercules, Hebe, fairy.

Le temple de la Gloire ["The Temple of Glory"] (**1745**). Libretto Voltaire. Bacchus (as a king), Envy.

Les fêtes de l'Hymen et de l'Amour, ou Les dieux d'Egypte ["The Festivities of Hymen and Cupid, or The Egyptian Gods"] (**1747**). Libretto Louis de Cahusac, after Egyptian legends. Osiris (vegetation god), Canopus (water god), Horus (god of the arts).

Zaïs (**1748**). Libretto Louis de Cahusac. Air spirit (Zaïs), Cupid, magic ring, spirits.

Pigmalion ["Pygmalion"] (**1748**). Libretto Ballot de Sovot, after Antoine Houdar de la Motte, after OVID.

Les surprises de l'Amour ["The Surprises of Cupid"] (**1748**). Libretto Pierre-Joseph Bernard, after Greek legend. Cupid, magic lyre, Anacreon.

Naïs (**1749**). Libretto Louis de Cahusac, after Greek legend. Neptune, sea nymph (Naïs), Tiresias, gods and goddesses, sea gods and goddesses, nymphs, Jupiter, Pluto, Flora, Titans, giants, zephyrs.

Zoroastre ["Zoroaster"] (**1749**). Libretto Louis de Cahusac. Magician, Vengeance, heavenly voice, spirits, demons.

La guirlande, ou Les fleurs enchantées ["The Garland, or The Magic Flowers"] (**1751**). Libretto Jean François Marmontel. Cupid.

Acanthe et Céphise, ou La sympathie ["Acanthus and Cephisa, or Empathy"] (**1751**). Libretto Jean François Marmontel. Fairy, genie, evil spirits, magic talisman.

La naissance d'Osiris, ou la fête Pamilie ["The Birth of Osiris, or The Pamylia Festival"] (**1754**). Libretto Louis de Cahusac. Jupiter, Osiris.

Les nymphs de Diane ["The Nymphs of Diana"] (rev as *Zéphyre* ["Zephirus"]) (written **1750s**; premiered 1967). Libretto anon, after Greek legend. Zephirus, nymph, breezes.

Les Paladins ["The Knights Errant"] (**1760**). Libretto attributed to Duplat de Monticourt, after *Le petit chien qui secoue de l'argent et des pierreries* by Jean de la Fontaine, after ARIOSTO. Fairy.

Abaris, ou Les Boréades ["Abaris, or The Boreads"] (**?1763**; premiered 1963). Libretto attributed to Louis de Cahusac, after Greek legend. Boreas (North Wind), Cupid, Polyhymnia (Muse), Apollo, nymph, Pleasures, Graces, Muses, Arts, boreads, winds, Pleasures, zephyrs, winds, hours, the Seasons, Talents, magic talisman.

Hasse, Johann Adolf (Germany 1699-1783)

Didone abbandonata ["Dido Abandoned"] (**1742**). Libretto Pietro Metastasio (1724), after VIRGIL. Neptune, ghost, Nereids.

Alcide al bivio ["Alcides at the Crossroads"] (**1760**). Libretto Pietro Metastasio, after *Memorabilia* by Xenophon, after *Horai* by Prodicus of Ceos. Hercules (Alcides), Iris, Pleasure, Virtue.

Egeria (**1764**). Libretto Pietro Metastasio, after ?OVID. Venus, Mars, Apollo, nymph (Egeria, transformed into a spring).

Jommelli, Niccolò (Italy 1714-1774)

Semiramide riconosciuta ["Semiramis Revealed"] (**1742**). Libretto Pietro Metastasio (1729), after Middle Eastern legend. Jupiter, Iris.

Didone abbandonata ["Dido Abandoned"] (**1747**). Libretto Pietro Metastasio (1724), after VIRGIL. Neptune, ghost, Nereids.

Fetonte ["Phaethon"] (**1753**). Libretto Mattia Verazi, after Philippe Quinault, after OVID. Thetis, Fortune, Proteus, Sun god, flying chariot, tritons, river gods.

Armida abbandonata ["Armida Abandoned"] (**1770**). Libretto Francesco Saverio de Rogatis, after Tasso. Sorceress (Armida), winged dragons pulling chariot, monsters.

Leclair, Jean Marie (l'Aîné) (France 1697-1764)

Scylla et Glaucus ["Scylla and Glaucus"] (**1746**). Libretto D'Albaret, after OVID. Sorceress (Circe), Hecate, nymph (Scylla, transformed into a dangerous rock in the sea), dryads, wood gods, nymphs, demons, Venus, Cupid, Spartan women (turned to stone).

Boyce, William (UK 1711-1779)

Peleus and Thetis (**1740** or **1747**). Libretto from *The Jew of Venice* by George Granville, Lord Lansdowne, after *The Merchant of Venice* by William SHAKESPEARE, after OVID, after *Prometheus Bound* by Aeschylus. Prometheus, Jupiter, Thetis, vulture.

The Secular Masque (**?1749**). Libretto John Dryden, after Greek legend. Janus, Chronos, Momus, Diana, Mars, Venus, nymphs.

The Tempest (**1757**). Libretto David Garrick, after William SHAKESPEARE. PROSPERO, ARIEL, Ceres (Demeter), Hymen.

Gluck, Christoph Willibald (Germany-Austria 1714-1787)

Le nozze d'Ercole e d'Ebe ["The Marriage of Hercules and Hebe"] (**1747**). Libretto anon, after Greek legend. Jupiter, Juno, Hercules, Hebe.

L'île de Merlin ou Le monde renversé ["Merlin's Island, or The World Turned Upside Down"] (**1758**). Libretto Louis Anseaume, after *Le Monde renversé* by Alain René Le Sage and d'Orneval. Magician (MERLIN), Pierrot, Scapino, magic island.

L'ivrogne corrigé, ou Le mariage du diable ["The Reformed Drunkard, or The Devil's Marriage"] (**1760**). Libretto Charles-Simon Favart, after Louis Anseaume and J.B. Lourdet de Sarterre. Pluto, furies, devils.

Tetide ["Thetis"] (**1760**). Libretto Giovanni Ambrogio Migliavacca, after Greek legend. Thetis, Mars, Apollo, Athena, Venus, Hymen.

Orfeo ed Euridice ["Orpheus and Eurydice"] (**1762**). Libretto Ranieri de' Calzabigi, after OVID.

Il Parnaso confuso ["Parnassus in Turmoil"] (**1765**). Libretto Pietro Metastasio. Apollo, Melpomene, Erato, Euterpe, Muses.

Il Telemaco o sia L'isola di Circe ["Telemachus, or The Island of Circe"] (**1765**). Libretto Marco Coltellini, after Carlo Sigismondo Capece, after Homer. Sorceress (Circe), sailors, underworld spirits. Sailors previously transformed into trees are changed back again.

Alceste ["Alcestis"] (**1767**). Libretto Ranieri de' Calzabigi, after *Alcestis* by EURIPIDES. Hercules, Thanatos, Apollo, underworld gods.

Paridi ed Elena ["Paris and Helen"] (**1770**). Libretto Ranieri de' Calzabigi, after Greek legend. Cupid, Athena.

Iphigénie en Aulide ["Iphigenia in Aulis"] (**1774**). Libretto François Le Blanc du Roullet, after *Iphigénie* by Jean Racine, after EURIPIDES. Diana.

Armide ["Armida"] (**1777**). Libretto Philippe Quinault, after Tasso. Sorceress (Armida), Fury of Hate, demons, naiad, monsters, nymphs, Furies, spirits.

Iphigénie en Tauride ["Iphigenia in Tauris"] (**1779**). Libretto Nicolas-François Guillard, after Claude Guimond de la Touche, after EURIPIDES. Diana, ghost, Furies (Eumenides), demons.

Echo et Narcisse ["Echo and Narcissus"] (**1779**). Libretto Baron Louis de Tschudi, after OVID. Cupid, nymphs, zephyrs, sylvans. The nymph Echo dies and is brought back to life.

Graun, Carl Heinrich (Germany, c1704-1759)

L'Armida (**1751**). Libretto Leopoldo de Villati, after Philippe Quinault, after Tasso. Sorceress (Armida), demons, nymphs.

L'Orfeo ["Orpheus"] (**1752**). Libretto Leopoldo de Villati, after M. du Boulier (?Boulai), after OVID.

Haydn, Franz Joseph (Austria 1732-1809)

Der krumme Teufel ["The Crooked Devil"] (**1751**). Libretto Joseph Kurz, after *Le diable boiteux* by F.C. Dancourt. Harlequin, devil, magician.

Asmodeus, der krumme Teufel (Der neue krumme Teufel) ["Asmodeus, The Crooked Devil (The New Crooked Devil)"] (**?1758**). Libretto Joseph Kurz. (Music lost; possibly a revised version of *Der krumme Teufel*.)

Hexenschabbas ["The Witches' Sabbath"] (**?1773**). Libretto anon. Opera lost.

Philemon und Baucis, oder Jupiters Reise auf die Erde ["Philemon and Baucis, or Jupiter's Journey to Earth"] (**1773**). Libretto anon, after Gottlieb Konrad Pfeffel, after OVID. Jupiter,

Mercury, nymphs, Hungarian Nation, Love of the Fatherland, Obedience, Devotion, Fidelity.

Il mondo della luna ["The World on the Moon"] (**1777**). Libretto Polisseno Fegajo Pastor, after Carlo Goldoni. A rich man, drugged by "magic potion", is convinced he is on the moon.

Die Feuersbrunst ["The Conflagration"] (**1777**). Libretto anon. Ghost.

La fedeltà premiata ["Fidelity Rewarded"] (**1781**). Libretto Haydn with anon, after *L'infedeltà fidele* by Giovanni Battista Lorenzi. Diana, wood nymph, sea monster, curse, nymphs, satyrs.

Orlando paladino ["Knight Roland"] (**1782**). Libretto Nunziato Porta, after Ariosto. Sorceress (Alcina), Charon, Orlando (temporarily turned to stone), nymphs, monsters.

Armida (**1784**). Libretto Nunziato Porta, after *Rinaldo* anon, after Giovanni Bertati, Jacopo Durandi and Francesco Saverio de Rogatis, after Tasso. Sorceress (Armida), nymphs, underworld spirits, Furies.

L'anima del filosofo, ossia Orfeo ed Euridice ["The Philosopher's Spirit, or Orpheus and Eurydice"] (**?1791**; premiered 1951). Libretto Carlo Francesco Badini, after OVID. Eurydice remains in Hades.

Rousseau, Jean-Jacques (Switzerland-France 1712-1778)
Le devin du village ["The Village Soothsayer"] (**1752**). Libretto Rousseau.

Mondonville, Jean-Joseph (France 1711-1772)
Titon et l'Aurore ["Tithonus and Aurora"] (**1753**). Libretto Abbé Voisenon and Abbé Savarre, after Greek legend. Cupid, Prometheus, Aeolus, Aurora (Dawn), nymph.

Traetta, Tommaso (Italy 1727-1779)
Ippolito ed Aricia ["Hippolytus and Aricia"] (**1759**). Libretto from *Hippolyte et Aricie* by Abbé Simon-Joseph Pellegrin, after *Phèdre* by Jean Racine, after *Phaedra* by Seneca, after *Hippolytus* by EURIPIDES. Diana, sea monster, Mercury, Neptune, Pluto, Fates, Tisiphone (Fury), nymphs, underworld gods, Furies, zephyrs, Cupid, Jupiter.

Armida (**1761**). Libretto Giacomo Durazzo and Giovanni Ambrogio Migliavacca, after *Armide* by Philippe Quinault, after Tasso. Sorceress (Armida), winged dragons (pulling chariot), Hatred, naiad, demons, monsters, nymphs, Furies, spirits, ghosts, Glory, Wisdom.

Ifigenia en Tauride ["Iphigenia in Tauris"] (**1763**). Libretto Marco Coltellini, after EURIPIDES. Artemis, ghosts, Furies (Eumenides), demons.

Bach, Johann Christian (Germany-UK 1735-1782)
Orione, ossia Diana vendicata ["Orion, or Diana Revenged"] (**1763**). Libretto Giovanni Gualberto Bottarelli, after Greek legend. Diana, Aurora (Dawn), Orion (transformed into a constellation).

Amadis de Gaule ["Amadis of Gaul"] (**1779**). Libretto Alphonse de Vismes, after Philippe Quinault, after *Amadis de Gaula* by García Rodríguez de Montalvo. Magician, demon, fairy, Righteousness, Valour, Discord, Lust, Envy, Hate, angels, ghosts, demons.

Mozart, Wolfgang Amadeus (Austria 1756-1791)
Apollo et Hyacinthus seu Hyacinthi Metamorphosis ["Apollo and Hyacinth, or The Metamorphosis of Hyacinth"] (**1767**). Libretto Pater Rufinus Widl, after Palaephatos or LUCIAN or OVID. Apollo, Hyacinthus (transformed into a flower), Zephyrus (transformed into a wind).

Ascanio in Alba ["Ascanius in Alba"] (**1771**). Libretto Abbate Giuseppe Parini, after VIRGIL. Venus, nymphs, faun, spirits, Genii, Graces.

Lucio Silla ["Lucius Sulla"] (**1772**). Libretto Giovanni de Gamerra and ?Pietro Metastasio, after Plutarch.

Il sogno di Scipione ["Scipio's Dream"] (**1772**). Libretto Pietro Metastasio (1735), after *De republica (Somnium Scipionis)* by Cicero. Fortune, Constancy, dead father, dead uncle, dead heroes, Heaven; Licenza. (See also recomposed version by Judith Weir **1991**.)

Idomeneo, re di Creta ["Idomeneus, King of Crete"] (**1781**). Libretto Abbate Giambattista Varesco, after Antoine Danchet, after *Idomenée* by Prosper de Crébillon *père*, after Greek legend. Neptune, sea monster.

Il dissoluto punito, o sia Il Don Giovanni ["The Rake Punished, or Don Juan"] (**1787**). Libretto Lorenzo da Ponte, after Giovanni Bertati, after *El burlador de Sevilla y convidado de pietra* by Tirso de Molina (attrib). Statue comes to life and drags DON JUAN to Hell.

Die Zauberflöte ["The Magic Flute"] (**1791**). Libretto Emanuel Schikaneder, after *Lulu, oder Die Zauberflöte* by A.J. Liebeskind, *Sethos* by Jean Terrasson *et al*. Queen of the Night, Genii, dragon, animals, magic flute, magic glockenspiel.

Mysliveek, Josef ["Venatorini"/"Il (divino) boemo"] (Czechoslovakia-Italy 1737-1781)
Il Bellerofonte ["Bellerophon"] (**1767**). Libretto G. Bonechi, after Philippe Quinault, after *Theogony* by Hesiod. Apollo, Athena, Neptune, magician, monster (Chimera), winged horse (Pegasus), magicians.

Paisiello, Giovanni (Italy 1740-1816)
L'osteria di Marechiaro ["The Marechiaro Inn"] (**?1769**). Libretto Francesco Cerlone. Goblin.

Don Chisciotte della Mancia ["Don Quixote of La Mancha"] (**1769**). Libretto Giovanni Battista Lorenzi, after CERVANTES. (See also recomposed version by Hans Werner Henze **1976**.)

Il credulo deluso ["The Credulous Deluded"] (**1774**). Libretto Carlo Goldoni from *Il mondo della luna* (**1750**). A rich man, drugged by "magic potion" is convinced he is on the moon. (Music apparently unrelated to **1783** setting of same libretto; see also *Il mondo della luna* ["The World on the Moon"] (**1777**) by Joseph Haydn.)

Il Socrate immaginario ["The Imaginary Socrates"] (**1775**). Libretto Ferdinando Galiani and Giovanni Battista Lorenzi.

Il mondo della luna ["The World on the Moon"] (**1783**). Libretto Carlo Goldoni. A rich man, drugged by "magic potion", is convinced he is on the moon. (See also *Il credulo deluso* **1774**.)

Fedra ["Phaedra"] (**1788**). Libretto Mario (?Luigi) Salvioni, after Carlo Innocenzo Frugoni, ?after *Phèdre* by Jean Racine, after *Phaedra* by Seneca, after *Hippolytus* by EURIPIDES. Mercury, Diana, Pluto, sea monster, underworld gods, wood deities.

Grétry, André-Modeste (Belgium-France 1741-1813)
Zémire et Azor ["Zemire and Azor"] (**1771**). Libretto Jean François Marmontel, after *La* BELLE ET LA BÊTE ["Beauty and the Beast"] by Mme J.M. LEPRINCE DE BEAUMONT, *Amour pour amour* by Pierre Claude Nivelle de la Chaussée. A prince, previously transformed into a monster, is changed back.

Céphale et Procris, ou L'amour conjugale ["Cephalus and Procris, or Married Love"] (**1773**). Libretto Jean François Marmontel, after OVID. Diana, Aurora [Dawn], Cronos.

Le jugement de Midas ["The Judgement of Midas"] (**1778**). Libretto Thomas d'Hèle and Louis Anseaume, after *Midas* by Kane O'Hara, after OVID. Apollo, Mercury, Pan. Midas receives ass's ears.

Amphitryon (**1786**). Libretto Michel Jean Sedaine, after Molière, after *Amphitruo* by Plautus. Jupiter, Mercury.

Salieri, Antonio (Italy-Austria 1750-1825)

Armida (**1771**). Libretto anon, after Tasso. Sorceress (Armida), winged dragons (pulling chariot), monsters, nymphs, Furies, demons.

Les Danaïdes ["The Danaïdes"] (**1784**). Libretto François Le Blanc du Roullet and Baron Louis de Tschudi, after Ranieri de' Calzabigi, after *Supplices* by Aeschylus. Demons, Furies, serpents.

La grotta di Trofonio ["Trofonio's Cave"] (**1785**). Libretto Giambattista Casti. Magician, magic cave (reverses people's characters), infernal spirits.

Falstaff, ossia Le tre burle ["Falstaff, or The Three Practical Jokes"] (**1799**). Libretto Carlo Prospero Defranceschi, after *The Merry Wives of Windsor* by William SHAKESPEARE. Herne the Hunter, Oberon, Titania, elves, spirits.

Sacchini, Antonio (Italy-UK-France 1730-1786)

Armida (? e Rinaldo) (**1772**; rev 1780 as *Renaud/Rinaldo*). Libretto Jean Joseph Lebout, after Giovanni di Gamorra, after Tasso. Sorceress (Armida), good magician, siren, mermaids, winged dragons (pulling chariot), magic garden.

Benda, Georg Anton [Jí Antonín] (Czechoslovakia-Germany 1722-1795)

Ariadne auf Naxos ["Ariadne on Naxos"] (**1775**). Libretto Johann Christian Brander, after OVID. Bacchus, Ariadne (transformed into a star constellation).

Medea (**1775**). Libretto Friedrich Wilhelm Gotter, after EURIPIDES. Sorceress, winged dragons (pulling chariot).

Righini, Vincenzo (Italy 1756-1812)

Il convitato di pietra, ossia Il dissoluto punito ["The Stone Guest, or The Rake Punished"] (**1777**). Libretto anon, after *El burlador de Sevilla y convidado de pietra* by Tirso de Molina (attrib). A statue comes to life and drags DON JUAN to Hell.

Alcide al bivio ["Alcides at the Crossroads"] (**1790**). Libretto Pietro Metastasio, after *Memorabilia* by Xenophon, after *Horai* by Prodicus of Ceos. Hercules ("Alcides"), nymphs, Genii.

Piccinni, Niccolò (Italy-France 1728-1800)

Roland (**1778**). Libretto Jean François Marmontel, after Philippe Quinault, after ARIOSTO. Fairy, magic ring (confers invisibility), enchanted forest, magic fountain of love.

Atys (**1780**). Libretto Jean François Marmontel, after Philippe Quinault, after *Fasti* by OVID. Cybele (Fertility Goddess), Somnus, Morpheus, Phantasy, Alecto (Fury), nymphs, spirits, good and bad dreams.

Iphigénie en Tauride ["Iphigenia in Tauris"] (**1781**). Libretto Alphonse de Congé Dubreuil, after Claude Guimond de la Touche, after EURIPIDES.

Didon ["Dido"] (**1783**). Libretto Jean François Marmontel, after VIRGIL.

Kraus, Joseph Martin (Germany-Sweden 1756-1792)

Proserpin ["Persephone"] (**1781**). Libretto Johan Henrik Kellgren, after OVID. Jupiter, Pluto, Persephone, Ceres (Demeter), ghost, underworld, gods, nymphs, hellish spirits.

Aeneas i Carthago (Dido och Aeneas) ["Aeneas in Carthage (Dido and Aeneas)"] (**1799**). Libretto Johan Henrik Kellgren, after King Gustav III, after *Didon* by Jean-Jacques Le Franc (Marquis de Pompignan). Venus, Juno, Jupiter, Honour, gods.

Benda, Friedrich (Germany 1745-1814)

Orpheus (**1785**). Libretto ?H. von L., after OVID.

Naumann, Johann Gottlieb (Germany 1741-1801)

Gustaf Wasa ["Gustav Vasa"] (**1786**). Libretto Johan Henrik Kellgren. Sweden's guardian angel, ghosts, Athena, Victory, Glory, Conscience, Hatred, Treachery, Revenge, ghosts, genii.

Orpheus og Eurydike ["Orpheus and Eurydice"] (**1786**). Libretto P.D. Biehl, after Ranieri de' Calzabigi, after OVID.

Gazzaniga, Giuseppe (Italy 1743-1818)

Don Giovanni Tenorio o sia Il convitato di pietra ["Don Juan Tenorio, or The Stone Guest"] (**1787**). Libretto Giovanni Bertati, after *El burlador de Sevilla y convidado de pietra* by Tirso de Molina (attrib). A statue comes to life, and demons drag DON JUAN to Hell.

Martín y Soler, Vicente (Spain-Russia 1754-1806)

L'arbore di Diana ["Diana's Arbour"] (**1787**). Libretto Lorenzo da Ponte, after Greek legend. Diana, Cupid, nymphs, magic tree, gods, goddesses. A shepherd is transformed into a tree and is changed back again; nymphs and shepherds are transformed into sheep and are changed back again.

Kunzen, Friedrich (Germany-Denmark 1761-1817)

Holger Danske ["Ogier the Dane"] (**1789**). Libretto Jens Baggesen, after *Oberon* by Christoph Martin Wieland, after *Huon de Bordeaux*, anon (13th century). Oberon, Titania, Puck, elf, sea nymphs (mermaids), magic horn, magic goblet, spirits.

Storace, Stephen (UK 1762-1796)

The Haunted Tower (**1789**). Libretto James Cobb. (Gothic story; partly lost.)

Wranitzky, Paul [Pavel Vranick?] (Bohemia-Austria 1756-1808)

Oberon, König der Elfen ["Oberon, King of the Elves"] (**1789**). Libretto Carl Ludwig Giesecke and Friederike Sophie Seyler, after *Oberon* by Christoph Martin Wieland, after *Huon de Bordeaux*, anon (13th century), *A Midsummer Night's Dream* by William SHAKESPEARE. Oberon, Titania, magic horn, magic goblet, nymphs, genii.

Fomin, Yevstigney (Russia 1761-1800)

Orfey i Evridika ["Orpheus and Eurydice"] (**1792**). Libretto Yakov Knyashnin, after OVID. Eurydice remains in Hades.

Haeffner, Johann Christian Friedrich (Germany-Sweden 1759-1833)

Alcides inträde i Världen ["Alcides' Entrance into the World"] (**1793**). Libretto A. N. Clewberg-Edelcrantz. Hercules ("Alcides").

Renaud ["Rinaldo"] (**1801**). Libretto N.B. Sparrschöld, after Leboeuf *et al.*, after Tasso.

Cherubini, Luigi (Italy-France 1760-1842)

Médée ["Medea"] (**1797**). Libretto François Benoît Hoffmann, after Pierre Corneille, after Seneca, after EURIPIDES. Sorceress (Medea), winged dragons (pulling chariot), Furies.

Anacréon, ou L'amour fugitif ["Anacreon, or Fugitive Love"] (**1803**). Libretto R. Mendouze. Venus, Cupid.

Pigmalione ["Pygmalion"] (**1809**). Libretto Stefano Vestris, after Antonio Simcone Sografi, after *Pygmalion* by Jean Jacques Rousseau, after OVID.

Ali Baba, ou Les quarante voleurs ["Ali Baba, or The Forty Thieves"] (**1833**). Libretto Eugène Scribe and Mélesville (Anne-Honoré Joseph Duveyrier), after *The Thousand and One Nights*.

Kauer, Ferdinand (Austria 1751-1831)

Das Donauweibchen ["The Danube Sprite"] (**1798**). Libretto Karl Friedrich Hensler. (Rewritten Russian version, *Lyesta*, *dnyeprovskaya rusalka* ["Lesta, the Dnieper Water Sprite"] **?1803** by Styepan Davïdov [?1777-1825].)

Weber, Carl Maria von (Germany 1786-1826)

Das Waldmädchen ["The Wood Nymph"] (**1800**; rev 1808-10 as *Silvana, oder Das Waldmädchen* ["Silvana, or The Wood Nymph"]). Libretto Karl von Steinberg (rev Franz Karl Hiemer), after German legend.

Rübezahl, oder Der Beherrscher der Geister ["Rübezahl, or The Ruler of the Spirits"] (**?1804-5**). Libretto J.G. Rhode, after MUSÄUS. Mountain spirit, magic sceptre, griffin, turnips (transformed into humans), spirits.

Der Freischütz ["The Free-Shooter"] (**1821**). Libretto Johann Friedrich Kind, after *Gespensterbuch (Der schwarze Jäger)* by Johann August Apel and Friedrich Laun. Devil (the Black Huntsman), magic bullet, ghost, evil spirits.

Euryanthe (**1823**). Libretto Helmine von Chézy, after *L'histoire du très noble et chevalereux prince Gérard, comte de Nevers, et de la très vertueuse et très chaste princesse Euriant de Savoye, sa mye,* anon (13th century). Serpent, ghost.

Oberon, or The Elf-King's Oath (**1826**). Libretto James Robertson Planché, after *Oberon* by Christoph Martin Wieland, after *Huon de Bordeaux,* anon (13th century), *A Midsummer Night's Dream* by William SHAKESPEARE. Oberon, Titania, Puck, elf, sea nymph (mermaid), magic horn, magic goblet, spirits.

Isouard, Nicolò (Malta-France 1775-1818)

Cendrillon ["Cinderella"] (**1810**). Libretto Charles Guillaume Etienne, *Cendrillon,* after PERRAULT. Magic rose but no fairy godmother.

Aladin, ou La lampe merveilleuse ["Aladdin, or The Wonderful Lamp"] (**1822**). Libretto Charles Guillaume Etienne, after *The Arabian Nights.* Magic lamp, genie, evil spirits.

Danzi, Franz (Germany 1763-1826)

Der Berggeist, oder Schicksal und Treue (Rübezahl) ["The Mountain Spirit, or Destiny and Fidelity (Rübezahl)"] (**1813**). Libretto ?Carl von Lohbauer, after MUSÄUS.

Mayr, Giovanni Simone [Johann Simon Mayer] (Germany-Italy 1763-1845)

Medea in Corinto ["Medea in Corinth"] (**1813**). Libretto Felice Romani, after *Medea* by EURIPIDES. Sorceress (Medea), winged dragons (pulling chariot), Furies.

Schubert, Franz (Austria 1797-1828)

Der Spiegelritter ["The Looking-Glass Knight"] (**?1811-13**; premiered 1949). Libretto August von Kotzebue. Good magician, evil magician, giant, dwarf, magic island, dragon, magic mirror.

Des Teufels Lustschloß ["The Devil's Pleasure Castle"]. (**?1813-15**; premiered 1879). Libretto August von Kotzebue. Evil magician, ghosts. An Amazon emerges from a tomb, statues come to life.

Sacontala ["Sacuntala"] (**?1820-21**; premiered 1971). Libretto Johann Philipp Naumann, after *Abhijnana-shakuntala* ["Sakuntala and the Ring of Recognition"] by Kalidasa (5th century). Demon, nymph, flying chariot, descending cloud, magic sword, magic shield, Heaven, heavenly voices, dryads, Genii, demons.

Donizetti, Gaetano (Italy 1797-1848)

Il Pimmalione ["Pygmalion"] (**?1816**; premiered 1960). Libretto anon, after Antonio Simcone Sografi, after *Pygmalion* by Jean Jacques Rousseau, after OVID.

Hoffmann, E(rnst) T(heodor Wilhelm) A(madeus) (Germany 1776-1822)

Undine (**1816**). Libretto by Friedrich de la Motte FOUQUÉ.

Spohr, Louis [Ludwig] (Germany 1784-1859)

Faust (**1816**). Libretto Joseph Karl Bernard, after *Fausts Leben, Taten und Höllenfahrt* by Friedrich Maximilian von Klinger.

Zemire und Azor ["Zemire and Azor"] (**1819**). Libretto Johann Jakob Ihlee, after *Zémire et Azor* by Jean François Marmontel, after *La BELLE ET LA BÊTE* ["Beauty and the Beast"] by Mme J.M. LEPRINCE DE BEAUMONT, *Amour pour amour* by Pierre Claude Nivelle de la Chaussée. Fairy, magic rose, magic mirror, spirits. Prince previously transformed into a monster is changed back.

Der Berggeist ["The Mountain Spirit"] (**1825**). Libretto Georg Döring, after ?MUSÄUS.

Hérold, Ferdinand (France 1791-1833)

La clochette, ou Le diable page ["The Bell, or The Devil Page"] (**1817**). Libretto Emmanuel Guillaume Marguerite Théaulon de Lambert, after *The Arabian Nights.* Story of Aladdin.

Zampa, ou La fiancée de marbre ["Zampa, or The Marble Bride"] (**1831**). Libretto Mélesville (Anne-Honoré Joseph Duveyrier). A female statue does not release pirate's ring placed on its finger, but drags him to death underwater.

Rossini, Gioachino [Gioacchino] (Italy 1792-1868)

Armida (**1817**). Libretto Giovanni Schmidt, after Tasso. Sorceress (Armida), demon (Astaroth), heavenly voice, magic wand, winged dragons (pulling chariot), demons, spirits, nymphs.

Semiramide ["Semiramis"] (**1823**). Libretto Gaetano Rossi, after *Sémiramis* by Voltaire, after Middle Eastern legend.

Boieldieu, François-Adrien (France 1775-1834)

Le petit chaperon rouge ["Little Red Riding Hood"] (**1818**). Libretto Emmanuel Guillaume Marguerite Théaulon de Lambert, after ?German folktale. A nobleman replaces the traditional wolf.

La dame blanche ["The White Lady"] (**1825**). Libretto Eugène Scribe, after *Guy Mannering, The Monastery* and *The Abbot* by Walter Scott.

Mercadente, Saverio (Italy 1795-1870)

L'apoteosi d'Ercole ["The Apotheosis of Hercules"] (**1819**). Libretto Giovanni Schmidt, after OVID. Jupiter, Hercules.

Amleto ["Hamlet"] (**1822**). Libretto Felice Romani, after Jean François Ducis, after William SHAKESPEARE.

Spontini, Gaspare (Italy-France 1774-1851)

Lalla Rookh (**1821**; rev 1822 as *Nurmahal, oder Das Rosenfest von Caschmir* ["Nurmahal, or The Rose Festival of Cashmir"]). Libretto Carl Alexander Herklots, after *Lalla Rookh* by Thomas Moore.

Alcidor (**1825**). Libretto Carl Alexander Herklots, after *Alcindor* by Rochon de Chabannes.

Kuhlau, Friedrich (Germany-Denmark 1786-1832)

Lulu (**1824**). Libretto Carl Christian Frederik Güntelberg, after *Lulu, oder Die Zauberflöte* by A.J. Liebeskind. Magician, witches, fairy, black elves, water elf, magic flute, magic ring (for transformation), magic rosebud, ghosts, elves.

Liszt, Franz [Ferenc] (Hungary-Germany 1811-1886)

Don Sanche, ou Le château d'amour ["Don Sanche, or The Castle of Love"] (**1825**). Libretto Emmanuel Guillaume Marguerite Théaulon de Lambert and de Rancé, after Claris de Florian.

Mendelssohn [-Bartholdy], Felix (Germany 1809-1847)

Die Hochzeit des Camacho ["Camacho's Wedding"] (**1827**). Libretto Friedrich Voigt (or Karl Klingemann), after CERVANTES. Cupid, ghost.

Loreley (**?1847**). Libretto Emanuel Geibel, after German legend. A human woman becomes a watersprite (Loreley).

Marschner, Heinrich (Germany 1795-1861)

Der Vampyr ["The Vampire"] (**1828**). Libretto Wilhelm August Wohlbrück, after *Les vampires* by Charles Nodier and P.F.A. Carmouche and A. de Jouffroy (trans as *Der Vampir oder die Totenbraut* by Heinrich Ludwig Ritter), after *The Vampyre* by John W. Polidori (attrib to Lord BYRON), after *Fragment of a Novel* (or *Augustus Darnell*) by Lord Byron.

Hans Heiling (**1833**). Libretto Eduard Devrient, after Karl Theodor Körner, after German legend.

Meyerbeer, Giacomo [Jakob Liebmann Meyer Beer] (Germany-France 1791-1864)

Robert le diable ["Robert the Devil"] (**1831**). Libretto Eugène Scribe, after Germain Delavigne. Devil (Robert's father), magic cypress branch, evil spirits. Tombstone statues of dead nuns come to life.

Wagner, Richard (Germany 1813-1883)

Die Feen ["The Fairies"] (?**1833-4**; premiered 1888). Libretto Wagner, after *La donna serpente* by Carlo Gozzi. Fairy King, magician, fairies (one temporarily turned to stone), bronze men, magic lyre, spirits.

Der fliegende Holländer ["The Flying Dutchman"] (**1843**). Libretto Wagner, after *Aus den Memoiren des Herrn von Schnabelewopski* by Heinrich Heine. Dutchman cursed to sail the seas forever in his ghost ship.

Tannhäuser und der Sängerkrieg auf dem Wartburg ["Tannhäuser and the Song Contest on the Wartburg"] (**1845**). Libretto Wager, after *Tannhäuser: eine Legende* by Heinrich Heine, *Der getreue Eckart und der Tannenhäuser* by Ludwig Tieck and *Der Wartburgkrieg*, anon (13th century), after historical fact and German legend.

Lohengrin (**1850**). Libretto Wagner, after *Lohengrin*, anon (c1290), after *Parzival* by Wolfram von Eschenbach, after historical fact and German legend. A prince, previously transformed into a swan, is changed back again.

Tristan und Isolde ["Tristan and Isolde"] (**1865**). Libretto Wagner, after *Tristan* by Gottfried von Strassburg, after *Tristran* by Thomas of Britain, after Tristram and Iseult legend.

Der Ring des Nibelungen ["The Ring of the Nibelungs"] (4 operas). Librettos Wagner, after *Nibelungenlied*, anon, *Thidreks Saga*, anon, *Volsunga Saga*, anon, *Poetic (Elder) Edda*, anon, *Prose (Younger) Edda* by Snorri Sturluson. *Das Rheingold* ["The Rhinegold"] (**1869**). Valhalla, Wotan, Fricka, Freia, Loge, Donner, Froh, Erda (Earth), giants (one transformed into a dragon), gnomes (Nibelungs), Rhine maidens, magic gold forged into magic ring and magic helmet, Nibelungs. *Die Walküre* ["The Valkyrie"] (**1870**). Wotan, Fricka, Loge, Brunnhilde, Valkyries. *Siegfried* (**1876**). Wotan, Erda, Brunnhilde, gnomes (Nibelungs), dragon, magic ring, magic helmet, magic sword, woodbird. *Götterdämmerung* ["The Twilight of the Gods"] (**1876**). Brunnhilde, Valkyre, gnome (Nibelung), Norns, Rhine maidens, winged horse, magic ring, magic helmet, magic potion.

Parzifal ["Parsifal"] (**1882**), Libretto Wagner, after *Parzival* by Wolfram von Eschenbach, after *Perceval le Gallois, ou Le conte de Graal* by Chrétien de Troyes, Welsh Arthurian legend of "Peredur Son of Efrawg" (11th century?) included in the MABINOGION. Witch, magician, Holy Grail, magic garden, holy spear, heavenly voices.

Auber, Daniel (France 1782-1871)

Le cheval de bronze ["The Bronze Horse"] (**1835**; rev 1857). Libretto Eugène Scribe. Siren on the planet Venus, winged bronze horse, magic armband, men are transformed into wooden pagodas and are changed back.

Le lac des fées ["The Fairy Lake"] (**1839**). Libretto Eugène Scribe and Mélesville (Anne-Honoré Joseph Duveyrier), after German ballad. Fairies are transformed from swans.

La part du diable, ou Carlo Broschi ["The Devil's Portion, or Carlo Broschi"] (**1843**). Libretto Eugène Scribe. Devil. (Carlo Broschi was original name of castrato singer Farinelli.)

Glinka, Mikhail (Russia 1804-1857)

Ruslan i Lyudmila ["Ruslan and Ludmila"] (**1842**). Libretto Glinka with Valerian Shirkov *et al.*, after Aleksandr Pushkin. Good magician, evil magician (dwarf Chernomor), witch, magic ring, magic sword, giant head, dwarves, nymphs, undines, enchanted maidens.

Pacini, Giovanni (Italy 1796-1867)

Medea (**1843**). Libretto Benedetto Castiglia, after EURIPIDES.

Dargomïzhsky, Aleksandr (Russia 1813-1869)

Torzhestvo Vakha ["The Triumph of Bacchus"] (?**1845**; premiered 1867). Libretto after Aleksandr Pushkin.

Rusalka (**1856**). Libretto Dargomïzhsky, after Aleksandr Pushkin. Dead girl becomes watersprite queen.

Kamyenniy gost ["The Stone Guest"] (**1872**). Libretto Aleksandr Pushkin, after *El burlador de Sevilla y convidado de pietra* by Tirso de Molina (attrib). A statue comes to life and drags DON JUAN to hell.

Lortzing, Albert (Germany 1801-1851)

Undine (**1845**). Libretto Lortzing with Philip Düringer, after Friedrich de la Motte FOUQUÉ.

Rolands Knappen, oder Das ersehnete Glück ["Roland's Miners, or The Desired Happiness"] (**1849**). Libretto Lortzing with G. Meisinger and K.[?Karl] Haffner, after MUSÄUS. Mountain Queen, magic gifts, gnomes, earth spirits.

Verdi, Giuseppe (Italy 1813-1901)

Giovanna d'Arco ["Joan of Arc"] (**1845**). Libretto Temistocle Solera, after *Die Jungfrau von Orleans* by Friedrich von Schiller. Good and evil spirits.

Macbeth (**1847**). Libretto Francesco Maria Piave with Andrea Maffei, after William SHAKESPEARE. Witches, ghost, apparitions, spirits, Hecate, devils.

Don Carlos (**1867**). Libretto François-Joseph Méry and Camille du Locle, after Friedrich von Schiller. Heavenly voice.

Falstaff (**1893**). Libretto Arrigo Boito, after *The Merry Wives of Windsor* by William SHAKESPEARE. Herne the Hunter, Fairy Queen, elves, spirits, fairies, witches.

Berlioz, Hector (France 1803-1869)

La damnation de Faust ["The Damnation of Faust"] (**1846**). Libretto Berlioz with Almire Gandonnière, after GOETHE.

Les Troyens ["The Trojans"] (?**1856-60**; premiered 1890). Libretto Berlioz, after VIRGIL. Mercury, ghosts, naiads, fauns, wood nymphs, satyrs.

Nicolai, Otto (Germany 1810-1849)

Die lustigen Weiber von Windsor ["The Merry Wives of Windsor"] (**1849**). Libretto Hermann Salomon Mosenthal, after William SHAKESPEARE. Herne the Hunter, Oberon, Titania, elves, spirits, insects.

Halévy, Fromental (France 1799-1862)

La tempestà ["The Tempest"] (**1850**). Libretto Pietro Giannone, after William SHAKESPEARE. Magician (PROSPERO), spirit (ARIEL).

Le juif errant ["The Wandering Jew"] (**1852**). Libretto Eugène Scribe and Jules-Henri Vernoy de Saint-Georges, after legend. Ahasuerus (but only in a single lifetime).

Ricci, Luigi (Italy 1805-1859) and Ricci, Federico (Italy 1809-1877)

Crispino e la comare, ossia Il medico e la morte ["Crispin and the Fairy, or The Doctor and Death"] (**1850**). Libretto Francesco Maria Piave, after Salvatore Fabbrichesi, after *Crispin médecin* by Noel Le Breton. Fairy godmother, vision of Hell.

Schumann, Robert (Germany 1810-1856)

Genoveva ["Genevieve"] (**1850**). Libretto Schumann with Robert Reinick, after *Das Leben und Tod der heiligen Genoveva* by Ludwig TIECK and *Genoveva* by Friedrich Hebbel. Sorceress, ghost, magic mirror, spirits.

Manfred (**1852**). Libretto Schumann after ?F.W. Suckow trans of play by Lord BYRON. Spirit (Manfred), Alpine fairy (Nemesis), Ariman (Prince of Earth and Air), Astarte (phantom lover), Fates, spirits, demons.

Szenen aus Goethes Faust ["Scenes from Goethe's Faust"] (?**1844-53**; premiered 1862). Libretto Schumann, after GOETHE.

Thomas, Ambroise (France 1811-1896)

Le songe d'une nuit d'été ["The Midsummer Night's Dream"] (**1850**). Libretto Joseph Bernard Rosier and Adolphe de Leuven. Not based on *A Midsummer Night's Dream* by William SHAKESPEARE. Queen Elizabeth I, William Shakespeare, various Shakespearean characters, including Sir John Falstaff.

Hamlet (**1868**). Libretto Jules Barbier and Michel Carré, after William SHAKESPEARE.

Adam, Adolphe (France 1803-1856)

La poupée de Nuremberg ["The Nuremberg Doll"] (**1852**). Libretto Adolphe de Leuven and A. de Beauplan, after *Der Sandmann* by E.T.A. HOFFMANN. Life-size female doll comes alive.

Flotow, Friedrich von (Germany 1812-1883)

Rübezahl (**1853**). Libretto Gustav Heinrich Gans zu Pulitz, after ?MUSÄUS.

Rubinstein, Anton (Russia 1829-1894)

Sibirskiye okhotniki ["The Siberian Huntsmen"] (**1854**). Libretto Andrei Sherebzov. Snow Queen, snow maidens.

Das verlorene Paradies ["Paradise Lost"] (**1858**). Libretto Arnold Schönbach, after John MILTON. Satan, angels, evil spirits.

Der Turm zu Babel ["The Tower of Babel"] (**1870**). Libretto Julius Rodenberg, after the Bible. Angels, evil spirits.

Dyemon ["The Demon"] (**1875**). Libretto Pavel Viskovatov and Apollon Maikov, after Mikhail Lyermontov. Demon, Angel of Goodness, demons, angels.

Nyeron ["Nero"] (**1879**). Libretto Rubinstein, after Jules Barbier. Ghosts.

Der Papagei ["The Parrot"] (**1884**). Libretto Hugo Wittmann, after Persian folktale. Magic mirror.

Balfe, Michael William (Ireland-UK 1808-1870)

Satanella, or The Power of Love (**1858**). Libretto Augustus Harris and Edmund Falconer (Edmund O'Rourke), after *Le diable amoureux* by Joseph Mazilier after Jacques Cazotte. Devil.

Offenbach, Jacques [Jakob] (Germany-France 1819-1880)

La chatte métamorphosée en femme ["The Cat Transformed into a Woman"] (**1858**). Libretto Eugène Scribe and Mélesville (Anne-Honoré Joseph Duveyrier).

Orphée aux enfers ["Orpheus in the Underworld"] (**1858**). Libretto Hector Crémieux and Ludovic Halévy, after OVID.

Daphnis et Chloé ["Daphnis and Chloe"] (**1860**). Libretto Clairville (Louis François Nicolaie) and Jules Cordier, after

Longus. Pan, Calisto, nymph.

Die Rheinnixen ["The Rhine Sprites"] (**1864**). Libretto Baron Alfred von Wolzogen trans of Charles Nuitter (Truinet) and Etienne Tréfeu. Watersprites, fairies, elves, goblins.

La belle Hélène ["Beautiful Helen"] (**1864**). Libretto Henri Meilhac and Ludovic Halévy, after *Helen* etc. by EURIPIDES. Juno, Minerva, Venus, Mercury.

Les contes d'Hoffmann ["The Tales of Hoffmann"] (**1881**). Libretto Jules Barbier, after Barbier and Michel Carré, after *Der Sandmann*, *Rath Krespel* and *Die Abenteuer der Silvester-Nacht* by E.T.A. HOFFMANN. Devil, mechanical woman, Muse of the poet Hoffmann.

Gounod, Charles (France 1818-1893)

Faust (**1859**). Libretto Jules Barbier and Michel Carré, after *Faust et Marguerite* by Carré, after GOETHE.

Philémon et Baucis ["Philemon and Baucis"] (**1860**). Libretto Jules Barbier and Michel Carré, after Jean de la Fontaine, after OVID. Jupiter, Vulcan.

Mireille (**1864**). Libretto Michel Carré, after *Mirèio* by Frédéric Mistral. Good witch, ghost, heavenly voice, river spirits.

Roméo et Juliette ["Romeo and Juliet"] (**1867**). Libretto Jules Barbier and Michel Carré, after William SHAKESPEARE.

Wallace, Vincent (Ireland 1812-1865)

Lurline (**1860**). Libretto Edward Fitzball, after German legend. Rhine nymph, Rhine maidens, magic harp.

David, Félicien (France 1810-1876)

Lalla Roukh ["Lalla Rookh"] (**1862**). Libretto Michel Carré and Lucas, after *Lalla Rookh* by Thomas Moore.

Bruch, Max (Germany 1838-1920)

Die Loreley ["The Loreley"] (**1863**). Libretto Emanuel Geibel, after German legend. A human woman becomes a watersprite (Loreley).

Moniuszko, Stanisław (Poland 1819-1872)

Strazny dwór ["The Haunted Manor"] (**1865**). Libretto Jan Ch'cifski.

Suppé, Franz von [Francesco de Suppé-Demelli] (Austria 1819-1895)

Die schöne Galathee ["Beautiful Galatea"] (**1865**). Libretto Poly Henrion (Leopold Kohl von Kohlenegg), after OVID. The statue of a young woman comes to life but is subsequently turned again to stone.

Boccaccio (**1879**). Libretto F. Zell (Camillo Walzel) and Richard Genée, after *Boccace, ou Le Décaméron* by J.F.A. Bayard, Adolphe de Leuven, Lhérie and de Beauplan, after BOCCACCIO and Boccaccio's life. Boccaccio, COMMEDIA DELL'ARTE characters, including Harlequin.

Millöcker, Karl (Austria 1842-1899)

Diana (**1867**). Libretto J. Braun, after ?*Die Freier* by Joseph von Eichendorff. Diana, Acteon, nymphs, animals.

Boito, Arrigo (Italy 1842-1918)

Mefistofele ["Mephistopheles"] (**1868**; rev 1860-67). Libretto Boito, after GOETHE.

Smetana, Bedřich (Czechoslovakia 1824-1884)

Dalibor (**1868**). Libretto Ervín Spindler, trans of text by Josef Wenzig. Ghost.

Tajemství ["The Secret"] (**1878**). Libretto Eliška Krásnohorská. Ghosts, spirits.

Čertova stûna ["The Devil's Wall"] (**1882**). Libretto Eliška Krásnohorská, after Czech legend. Devil, demons.

Cui, César (Russia 1835-1918)

Vil'yam Ratclif ["William Ratcliff"] (**1869**). Libretto Aleksey Pleshcheyev, after Heinrich Heine. Ghosts.

Snyezhniy bogatïr' ["The Snow Giant"] (**1906**; rev 1953 by D.

Sedych as *Ivan bogatïr'*). Libretto Marina Pol, after Russian fairytale. Dragon, swan maidens, enchanted maidens.

Krasnaya shapochka ["Red Riding Hood"] (**1911**; premiered 1922). Libretto Marina Pol, after PERRAULT.

Kot v sapogakh ["Puss in Boots"] (**1916**). Libretto Marina Pol, after PERRAULT.

Hervé [Florimond Ronger] (France 1825-1892)

Le petit Faust ["Little Faust"] (**1869**). Libretto Hector Crémieux and Adolphe Jaime, after GOETHE. FAUST is a school teacher in this parody of Charles Gounod's *Faust* 1859.

Rheinberger, Joseph (Germany 1839-1901)

Die sieben Raben ["The Seven Ravens"] (**1869**). Libretto Franz Bonn, ?after German fairytale. Seven brothers, previously transformed into ravens, are changed back again.

Tchaikovsky, Pyotr Ilyich (Russia 1840-1893)

Undina ["Undine"] (**1869**). Libretto Vladimir Sollogub, after Friedrich de la Motte FOUQUÉ.

Kuznyets Vakula ["Vakula the Smith"] (**1876**; rev 1885-6 as *Cherevichki* ["The Little Boots"]). Libretto Yakov Polonsky, after *Noch' pered Rozhdyestvom* by Nikolai Gogol. The plot is almost identical to the version of Gogol's story used in Nikolai Rimsky-Korsakov's *Noch' pered Rozhdyestvom* 1895.

Orlyeanskaya dyeva ["The Maid of Orleans/Joan of Arc"] (**1881**) Libretto Tchaikovsky, after *Die Jungfrau von Orleans* by Friedrich von Schiller, *Jeanne d'Arc* by Auguste Mermet, after Jules Barbier.

Pikovaya dama ["Pique Dame/The Queen of Spades"] (**1890**). Libretto Modyeste Tchaikovsky, after Aleksandr Pushkin.

Strauss, Johann the Younger (Austria 1825-1899)

Indigo und die vierzig Räuber ["Indigo and the Forty Thieves"] (**1871**; rev 1906 by Ernst Reiterer as *1000 Nacht*). Libretto M. Steiner, after *The Arabian Nights*.

Sullivan, Arthur (UK 1842-1900)

Thespis, or The Gods Grown Old (**1871**). Libretto W.S. Gilbert. Jupiter, Apollo, Mars, Diana, Mercury.

Iolanthe, or The Peer and the Peri (**1882**). Libretto W.S. Gilbert. Fairy Queen, Peri (spirit, Iolanthe), fairies.

Ruddigore, or The Witch's Curse (**1887**). Libretto W.S. Gilbert. The ghosts of 8 ancestors emerge from their portraits.

Cornelius, Peter (Germany 1824-1874)

Gunlöd (?**1866**; premiered 1891). Libretto Cornelius, after *Poetic (Elder) Edda*, anon, *Prose (Younger) Edda* by Snorri Sturluson. Odin, underworld goddess (Hela), giant, magic potion, spirits of light, underworld ghosts, giants.

Dvořák, Antonín (Czechoslovakia 1841-1904)

Vanda (**1876**). Libretto Václav Beneš-úmavsk? and František Zákrejs, after Julian Surzycki (attrib). Witch, ghosts.

Čert a Káča ["The Devil and Kate"] (**1899**). Libretto Adolf Wenig, after Czech folktale.

Rusalka (**1901**). Libretto Jaroslav Kvapil, after *Undine* by Friedrich de la Motte FOUQUÉ, *The Little Mermaid* by Hans Christian ANDERSEN and *Die versunkene Glocke* by Gerhart Hauptmann. Watersprite, Rusalka, transformed into a human, then into a will-o'-the-wisp.

Armida (**1904**). Libretto Jaroslav Vrchlick?, after Philippe Quinault, after Tasso. Sorceress (Armida), magician, sirens.

Ponchielli, Amilcare (Italy 1834-1886)

La gioconda ["The Joyful Girl"] (**1876**). Libretto Arrigo Boito (as Tobia Gorrio), after *Angelo, tyran de Padoue* by Victor Hugo.

Chabrier, Emmanuel (France 1841-1894)

Le Sabbat ["The Sabbath"] (**1877**). Libretto Armand Silvestre.

Massenet, Jules (France 1842-1912)

Le roi de Lahore ["The King of Lahore"] (**1877**). Libretto Louis Gallet, after *Voyage autour du monde* by Comte Roger de Beauvoir, after the *Mahabharata*.

Esclarmonde (**1889**). Libretto Alfred Blau and Louis de Gramont, after *Pathénopoeus de Blois*, anon (pre-1188). Magician (Byzantine King), sorceress (Princess Esclarmonde), magic island, magic sword.

Cendrillon ["Cinderella"] (**1899**). Libretto Henri Cain, after PERRAULT.

Grisélidis (**1901**). Libretto Armand Silvestre and Eugène Morand, after their play, after BOCCACCIO. Devil, female devil, St Agnes, spirits, heavenly voices.

Le jongleur de Notre-Dame ["The Juggler of Notre-Dame"] (**1902**). Libretto Maurice Léna, after *L'Etui de nacre* by Anatole France, after *Le tombeur de Notre-Dame* by Gaston Paris, after medieval mystery play.

Ariane ["Ariadne"] (**1906**). Libretto Catulle Mendès, partly after Apollodorus.

Bacchus (**1909**). Libretto Catulle Mendès, after *Râmayana* by Palmiki. Bacchus, monkeys, etc.

Don Quichotte ["Don Quixote"] (**1910**). Libretto Henri Cain, after *Le chevalier de la longue figure* by Jacques Le Lorrain, after CERVANTES.

Amadis (?**1889-90**; premiered 1922). Libretto Jules Claretie, after *Amadis de Gaula* by García Rodríguez de Montalvo. Fairies, magic jewels.

Panurge (**1913**). Libretto Maurice Boukay and Georges Spitzmuller, after *Pantagruel* by François RABELAIS.

Planquette, Robert (France 1848-1903)

Les cloches de Corneville ["The Bells of Corneville"] (**1877**). Libretto Clairville (Louis François Nicolaie) and Charles Gabet. Set in a haunted castle.

Rip (van Winkle) (**1882**). Libretto Philippe Gille and Henri Meilhac, after "Rip Van Winkle" by Washington IRVING. A man dreams that he is cursed to sleep for 20 years.

Catalani, Alfredo (Italy 1854-1893)

Elda (**1880**; rev 1889 as *Loreley*). Libretto Carlo d'Ormeville and Angelo Zanardini, after German legend. A woman becomes a watersprite (Loreley).

Rimsky-Korsakov, Nikolai (Russia 1844-1908)

Mayskaya noch' ["May Night"] (**1880**). Libretto Rimsky-Korsakov, after Nikolai Gogol.

Snyegurochka ["The Snow Maiden"] (**1882**). Libretto Rimsky-Korsakov, after Aleksandr Ostrovsky, after Russian folktale. Spring Fairy, Frost King, wood-spirit, Sun God, fairies, Shrovetide straw puppet, animals, birds, spring flowers.

Mlada (**1892**). Libretto Rimsky-Korsakov, after Stepan Gedeonov and Viktor Krylov. Good ghost (Mlada), underworld goddess, apparitions of Cleopatra, magician (Kashchei), black god (Chernobog), gods of heaven and hell, wood spirits, werewolves, ghosts, witches, shades, souls of heroes.

Noch' pered Rozhdyestvom ["Christmas Eve"] (**1895**). Libretto Rimsky-Korsakov, after Nikolai Gogol. Devil, witch, magician, winged horse, magicians, witches, good and evil spirits, ghosts, stars.

Sadko (**1898**). Libretto Rimsky-Korsakov with Vladimir Byel'sky and others, after 11th-century epic poem. Magicians, Sea King, Sea King's daughter (transformed into a river), apparition of a hero, mermaids (also in the form of swans), Sea Queen, gold and silver fish.

Skazka o tsare Saltanye, o synye yevo slavnom i moguchem bogatyrye Gvidone Saltanovich i o prekrasnoi tsaryevnye lyebyedi ["The Tale of Tsar Saltan, of his Son the Famous and

Mighty Hero Prince Guidon Saltanovich and of the Beautiful Swan Princess"] (**1900**). Libretto Vladimir Byel'sky, after "Swan Princess" by Aleksandr Pushkin. A prince is transformed into a bumble-bee and is changed back again.

Kashchey Byessmyertniy ["Kashchei the Immortal"] (**1902**). Libretto Rimsky-Korsakov, after Yevgyeniy Pyotrovsky. A magician's daughter transformed into a weeping willow.

Skazaniye o nyevidimom gradye Kityezhe i dyevye Fyevronii ["The Legend of the Invisible City of Kitezh and the Maiden Fevroniya"] (**1907**). Libretto Vladimir Byel'sky, after Russian legend. Kitezh avoids the Tartars by invisibility at the bottom of a lake; only its church bells are audible.

Zolotoy pyetushok ["The Golden Cockerel"] (**1909**). Libretto Vladimir Byel'sky, after Aleksandr Pushkin. A magic golden cockerel warns of danger.

Musorgsky, Modyest (Russia 1839-1881)
Sorochinskaya yarmarka ["Sorochinsky Fair"] (?**1874-81**; rev 1911). Libretto Musorgsky, after Nikolai Gogol. Black god (Chernobog), devils, witches (includes sabbath), dwarves. The opera was unfinished and has been revised and completed by several hands.

Delibes, Léo (France 1836-1891)
Lakmé (**1883**). Libretto Edmond Gondinet and Philippe Gille, after *Rarahu (Le mariage de Loti)* by Pierre Loti.

Puccini, Giacomo (Italy 1858-1924)
Le Villi ["The Willis"] (**1884**). Libretto Ferdinando Fontana, after *Les wilis* by Alphonse Karr, after legend.
Turandot (**1926**; completed 1925-6 by Franco Alfano). Libretto Giuseppe Adami and Renato Simoni, after Carlo Gozzi, after legend.

Reyer [Rey], Ernest (France 1823-1909)
Sigurd (**1884**). Libretto Camille du Locle and Alfred Blau, after *Nibelungenlied*.

Duparc, Henri (France 1848-1933)
Roussalka ["Rusalka"] (before **1885**). Libretto anon, after Aleksandr Pushkin.

Goldmark, Karl [Károly] (Hungary-Austria 1830-1915)
Merlin (**1886**; rev 1904). Libretto Siegfried Lipiner, after Arthurian legend (◊ ARTHUR). magician (MERLIN), fairy (Morgana ◊ MORGAN LE FAY), demon, magic veil, spirits.
Das Heimchen am Herd ["The Cricket on the Hearth"] (**1896**). Libretto Alfred Maria Willner, after Charles DICKENS.

Janáček, Leoš (Czechoslovakia 1854-1928)
Sárka (?**1887-8**; rev 1918 and 1924). Libretto Julius Zeyer, after his play, after *Dalimil's Chronicle*, anon (14th century). Magic weapons.
Výlety pánů Brouakovy ["The Excursions of Mr Brouaek"] (**1920**). Libretto Janáček with František Procházka and others, after novels by Svatopluk Čech. The drunken "hero" dreams of life on the moon and in the 15th century.
Příhody lišky Bystroušky ["The Cunning Little Vixen/Adventures of Vixen Sharp-Ears"] (**1924**). Libretto Janáček, after *Liska Bystrouska* (1920 *Lidove Noviny*; **1921**; trans Tatiana Firkusny, Maritza Morgan and Robert T. Jones as *The Cunning Little Vixen* 1985 US) by Rudolf Těsnohlídek (1882-1928) after drawings by Stanislav Lolek.
Vĭec Makropulos ["The Makropoulos Affair"] (**1926**). Libretto Janáček, after Karel ČAPEK. A singer's lifespan is much extended – to 337 years – by the elixir of life.

Lalo, Edouard (France 1823-1892)
Le roi d'Ys ["The King of Ys"] (**1888**). Libretto Edouard Blau, after Breton legend. A patron saint's statue comes to life.

Mahler, Gustav (Austria 1860-1911)

Rübezahl (**1890**; unfinished, now lost). Libretto Mahler, after ?MUSÄUS.

Delius, Frederick (UK-France 1862-1934)
Irmelin (?**1890-92**; premiered 1953). Libretto Delius, after ?legend and *The Princess and the Swineherd* by Hans Christian ANDERSEN.
The Magic Fountain (?**1894-5**; premiered 1977). Libretto Delius with Jutta Bell.
Koanga (**1904**). Libretto Charles F. Keary, after composer with Jutta Bell, after *The Grandissimes: A Story of Creole Life* by George Washington Cable. A voodoo priest's curse destroys a sugarcane plantation.

Leoncavallo, Ruggero (Italy 1857-1919)
Il Pagliacci ["The Clowns"] (**1892**). Libretto Leoncavallo. The play within the opera has COMMEDIA DEL'ARTE characters.
Edipo Re ["Oedipus the King"] (**1920**). Libretto Giovacchino Forzano, after *Oedipus Tyrannus* by Sophocles.

Albéniz, Isaac (Spain 1860-1909)
The Magic Opal (**1893**; retitled as *The Magic Ring*, rev 1894 as *La sortija*) (**1893**). Libretto Arthur Law.
Merlin (?**1898-1902**; premiered 1950). Libretto Francis Burdett Money-Coutts, after Arthurian legend (◊ ARTHUR).

Humperdinck, Engelbert (Germany 1854-1921)
Hänsel und Gretel ["Hansel and Gretel"] (**1893**). Libretto Adelheid Wette, after GRIMM BROTHERS. Children previously transformed into cakes are changed back again.
Dornröschen ["Sleeping Beauty"] (**1902**). Libretto E. Ebeling and B. Filhès, after "La Belle au bois dormant" by Charles PERRAULT.
Die Königskinder ["The Royal Children"] (**1910**). Libretto Ernst Rosmer (Else Bernstein-Porges), after her play.

Fibich, Zdeněk (Czechoslovakia 1850-1900)
Bouře ["The Tempest"] (**1895**). Libretto Jaroslav Vrchlick?, after *The Tempest* by William SHAKESPEARE. Magician (PROSPERO), spirit (ARIEL).

Mascagni, Pietro (Italy 1863-1945)
Guglielmo Ratcliff ["William Ratcliff"] (**1895**). Libretto Andrea Maffei, trans of play by Heinrich Heine.
Iris (**1898**). Libretto Luigi Illica.
Parisina (**1913**). Libretto Gabriele d'Annunzio, partly after Lord BYRON.
Il piccolo Marat ["The Little Marat"] (**1921**). Libretto Giovacchino Forzano and Giovanni Targioni-Tozzetti, after *Sous la Terreur* by Victor Martin.

Taneyev, Sergei (Russia 1856-1915)
Oresteya ["The Oresteia"] (**1895**). Libretto Alexei Venkstyern, after dramatic trilogy by Aeschylus.

d'Indy, Vincent (France 1851-1931)
Fervaal (**1897**). Libretto d'Indy, after "Axel" by Esaïas Tegner. Magic garden, heavenly voices.
L'étranger ["The Stranger"] (**1903**). Libretto d'Indy. Story not dissimilar to similar to Richard WAGNER's *Der fliegende Holländer* 1843.
La légende de Saint-Christophe ["The Legend of Saint Christopher"] (**1920**; rev 1922). Libretto d'Indy, after *Legenda aurea* by Archbishop Jacobus de Voragine (Jacopo da Varagine) (13th century). Satan, Queen of Pleasure, King of Gold, King of Heaven.

Foerster, Josef Bohuslav (Czechoslovakia 1859-1951)
Eva (**1899**). Libretto Foerster, after *Gazdina roba* ("The Farm Mistress") by Gabriela Preissov. Ghosts exist in Eva's imagination.

Lincke, Paul (Germany 1866-1946)
Frau Luna ["Mistress Moon/Castles in the Air"] (**1899**).

Libretto Heinz Bolton-Bäckers. A balloon flies to the moon.

Wagner, Siegfried (Germany 1869-1930)

Der Bärenhäuter ["The Man in the Bearskin"] (**1899**). Libretto S. Wagner. Devil, St Peter, Hell, demons, lost souls.

Der Kobold ["The Goblin"] (**1904**). Libretto S. Wagner. Hobgoblins, magic jewel.

Banadietrich (**1910**). Libretto S. Wagner. Devil, Death, watersprite, monstrous ghost, elves, spirits.

Sonnenflammen ["The Flames of the Sun"] (**1912**). Libretto S. Wagner with Peter Paul Pachl. Ghost.

An allem ist Hütchen schuld ["Hütchen is to Blame for Everything"] (**1917**). Libretto S. Wagner. Hütchen is a goblin.

Schwarzschwanenreich ["The Kingdom of the Black Swan"] (**1918**) Liberetto S. Wagner. Devil, magic lake.

Der Schmied von Marienburg ["The Blacksmith of Marienburg"] (**1923**). Libretto S. Wagner. Devil.

Der Friedensengel ["The Angel of Peace"] (**1926**). Libretto S. Wagner.

Das Flüchlein, das Jeder mitbekam ["The Curse that Affected Everyone"] (**?1928-30**; premiered 1984). Libretto S. Wagner. Princes previously transformed into swans are changed back again.

Fauré, Gabriel (France 1845-1924)

Prométhée ["Prometheus"] (**1900**; rev 1916). Libretto Jean Lorrain, after André Ferdinand Hérold, after *Erga* ["Works and Days"] by Hesiod, *Prometheus Bound* by Aeschylus.

Wolf-Ferrari, Ermanno (Italy-Germany 1876-1948)

Cenerentola (Aschenbrödel) ["Cinderella"] (**1900**). Libretto Maria Pezze-Pascolato (Julius Schweitzer), after GRIMM BROTHERS and Charles PERRAULT.

La veste di cielo (Das Himmelskleid) ["The Heavenly Clothes"] (**1927**). Libretto ?Wolf-Ferrari. Magician, moon fairy, magic donkey, heavenly clothing, moon court, air spirits, moon maidens, sun children.

Gli dei a Tebe (Der Kuckuck von Theben) ["The Gods at Thebes (The Cuckoo of Thebes)"] (**1943**). Libretto Ludwig Andersen and Mario Ghisalberti, after ?Apollodorus. The story of the conception of Hercules.

Magnard, Albéric (France 1865-1914)

Guercoeur (**?1897-1901**; premiered 1931). Libretto Magnard. A dead man returns to life on earth.

Pfitzner, Hans (Germany 1869-1949)

Die Rose vom Liebesgarten ["The Rose from the Garden of Love"] (**1901**). Libretto James Grun, after painting by Hans Thoma.

Das Christ-Elflein ["The Little Elf of Christ"] (**1906**). Libretto Pfitzner, after Ilse von Stach.

Palestrina (**1917**). Libretto Pfitzner, after life of composer Giovanni Palestrina (*c*1525-1594). Angelic voices, ghosts.

Das Herz ["The Heart"] (**1931**). Libretto Hans Mahner-Mons. Magician, ghost, demon (Asmodeus).

Strauss, Richard (Germany 1864-1949)

Feuersnot ["Fire Famine"] (**1901**). Libretto Ernst von Wolzogen, after Flemish legend "The Extinguished Fires of Oudenaarde".

Ariadne auf Naxos ["Ariadne on Naxos"] (**1912**). Libretto Hugo von Hofmannsthal, after *Le bourgeois gentilhomme* by Molière, after OVID.

Die Frau ohne Schatten ["The Woman without a Shadow"] (**1919**). Libretto Hugo von Hofmannsthal, after his short story. An empress without a shadow (previously transformed back to human form from a gazelle) buys one

from a woman. Emperor gradually turned to stone and is changed back again.

Die ägyptische Helena ["The Egyptian Helen"] (**1928**). Libretto Hugo von Hofmannsthal, after Greek legend.

Daphne (**1938**). Libretto Josef Gregor, after OVID.

Die Liebe der Danaë ["The Love of Danae"] (**?1938-40**; premiered 1952). Libretto Josef Gregor, after Hugo von Hofmannsthal, after Greek legend. Danae is transformed into a golden statue and changed back again.

Terrasse, Claude (France 1867-1923)

Les travaux d'Hercule ["The Labours of Hercules"] (**1901**). Libretto anon, after Apollodorus.

Cilèa, Francesco (Italy 1866-1950)

Adriana Lecouvreur ["Adrienne Lecouvreur"] (**1902**). Libretto Arturo Colautti, after Eugène Scribe and Ernest Legouvé, based on life of 18th-century tragic actress.

Nielsen, Carl (Denmark 1865-1931)

Saul og David ["Saul and David"] (**1902**). Libretto Einar Christiansen, after the Bible (I Samuel 28). Witch of Endor summons up spirit of the prophet Samuel.

Chausson, Ernest (France 1855-1899)

Le roi Arthus ["King Arthur"] (**1903**). Libretto Chausson, after Arthurian legend. Magician (MERLIN), winged maidens. ARTHUR ascends to heaven in a chariot.

Herbert, Victor (Ireland-US 1859-1924)

Babes in Toyland (**1903**). Libretto Glen MacDonough.

Sloane, A. Baldwin (US 1872-1926) and Tietjens, Paul (US 1877-1943)

The Wizard of Oz (**1903**). Libretto L. Frank BAUM, after *The Wonderful Wizard of Oz*. Witches, OZ.

Ganne, Louis (France 1862-1923)

Hans le joueur de flûte ["Hans the Flute Player"] (**1906**). Libretto Maurice Ordonneau. A magic flute.

Rakhmaninov, Sergei (Russia-US 1873-1943)

Francheska da Rimini ["Francesca of Rimini"] (**1906**). Libretto Modyest Tchaikovsky, after Aleksandr Pushkin, after *Inferno* by Dante. Ghosts of VIRGIL, Paolo and Francesca, Dante.

Prokofiev, Sergei (Russia 1891-1953)

Undina ["Undine"] (**1904-7**). Libretto Mariya Kilstett, after Friedrich de la Motte FOUQUÉ.

Lyubov' k tryom apel'sinam ["The Love for Three Oranges"] (**1921**). Libretto Prokofiev, after *Fiaba dell'amore delle tre melarancie* by Carlo Gozzi. Devil, witch (Fata Morgana), magician, princesses in oranges (one is later transformed into a rat and is changed back again), Ridiculous People, magic ribbon, devils, spirits, monsters, Tragedy, Comedy, Lyric Drama, Farce; includes COMMEDIA DELL'ARTE elements.

Ognyenniy angyel ["The Fiery Angel"] (**?1919-23**; premiered 1954). Libretto Prokofiev, after novel by Valery Bruysov.

German, Edward (UK 1862-1936)

Fallen Fairies, or The Wicked World (**1909**). Libretto W.S. Gilbert. Fairy Queen, fairies.

Monckton, Lionel (UK 1861-1924)

The Arcadians (**1909**). Libretto Martin Ambient and Alexander Mattock Thompson; lyric Arthur Harold Wimperis.

Bloch, Ernest (Switzerland-US 1888-1959)

Macbeth (**1910**). Libretto Edmond Fleg, after William SHAKESPEARE.

Granados, Enrique (Spain 1867-1916)

Liliana (**1911**). Libretto Apeles Mestres.

Busoni, Ferruccio (Italy-Germany 1866-1924)

Die Brautwahl ["The Bridal Choice"] (**1912**). Libretto

Busoni, after E.T.A. HOFFMANN. Revenant white magicians more than 300 years old.

Arlecchino oder Die Fenster ["Harlequin, or The Windows"] (**1917**). Libretto Busoni, after COMMEDIA DELL'ARTE characters.

Doktor Faust ["Doctor Faust"] (**1925**). Libretto Busoni, after 16th-century puppet play.

Debussy, Claude (France 1862-1918)

Le Diable dans le beffroi ["The Devil in the Belfry"] (**1912**). Libretto Debussy, after Edgar Allen POE.

La chute de la maison Usher ["The Fall of the House of Usher"] (**?1908-17**; premiered 1977). Libretto Debussy, after Charles Baudelaire trans of Edgar Allen POE.

Holbrooke, Josef (UK 1878-1958)

The Cauldron of Anwyn Libretto Thomas Evelyn Ellis (Lord Howard de Walden), after Welsh legends, including those known collectively as the MABINOGION. A trilogy consisting of *The Children of Don* (**1912**), *Dylan, Son of the Wave* (**1914**) and *Bronwen, Daughter of Llyr* (**1929**).

Schreker, Franz (Austria-Germany 1878-1934)

Der ferne Klang ["The Distant Sound"] (**1912**). Libretto Schreker. Young musician haunted by a "distant sound" realizes that it is true love.

Das Spielwerk und die Prinzessin ["The Carillon and the Princess"] (**1913**; rev 1915-16 as *Das Spielwerk*). Libretto Schreker. Magic carillon functions only in response to music, its playing revives dead violinist.

Der Schatzgräber ["The Treasure Digger"] (**1920**) Libretto Schreker. Magic lute and jewels.

Irrelohe (**1924**) Libretto Schreker. A family curse is broken by true love.

Der Schmied von Gent ["The Blacksmith of Ghent"] (**1932**). Libretto Schreker, after *Smetse Smee* by Charles de Coster. Lucifer, St Joseph, St Mary, St Peter, devils, angels.

Charpentier, Gustave (France 1860-1956)

Orphée ["Orpheus"] (**1913**). Libretto Charpentier, after OVID.

Boughton, Rutland (UK 1878-1960)

The Immortal Hour (**1914**). Libretto Boughton after play and poems by Fiona MacLeod (William Sharp). Lord of Shadow, Prince and Princess of the Land of the Ever-Young, spirit voices, spirits.

The Queen of Cornwall (**1924**). Libretto Boughton, after Thomas Hardy, after Tristram and Iseult legend.

Rabaud, Henri (France 1873-1949)

Mârouf, savetier du Caire ["Marouf, Cobbler of Cairo"] (**1914**). Libretto Lucien Népoty, after *The Arabian Nights*.

Ravel, Maurice (France 1875-1937)

La cloche engloutie ["The Sunken Bell"] (**1906-14**). Libretto André Ferdinand Hérold trans of *Die versunkene Glocke* by Gerhart Hauptmann.

L'enfant et les sortilèges ["The Child and the Magic Spells"] (**1925**). Libretto Colette.

Stravinsky, Igor (Russia-Switzerland-US 1882-1971)

Solovyei ["Le Rossignol"] (**1914**). Libretto Stravinsky with Stepan Mitusov, after "The Nightingale" by Hans Christian ANDERSEN.

Bayka/Renard ["The Fox"] (**1922**). Libretto C.F. Ramuz, after Russian folktales.

Oedipus Rex ["Oedipus the King"] (**1927**). Libretto Jean Daniélou, trans Jean COCTEAU, after *Oedipus Tyrannus* by Sophocles.

Perséphone ["Persephone"] (**1934**). Libretto André Gide, after his poem, after "Homeric Hymn", anon.

The Rake's Progress (**1951**). Libretto W.H. Auden and Chester Kallman, after engravings by William Hogarth.

The Flood (**1962**). Libretto Robert Craft, after Chester and York mystery plays.

Falla, Manuel de (Spain-Argentina 1876-1946)

El amor brujo ["Love the Magician"] (**1915**). Libretto Gregorio and María Martínez Sierra. Witch, will-o'-the-wisp.

Fuego fatuo ["Will-o'-the-wisp"] (**1919**). Libretto María Martínez Sierra.

Atlàntida ["Atlantis"] (**1928-46**; completed 1961 by Ernesto Halffter, rev 1976). Libretto de Falla, after Jacint Verdaguer. Christopher Columbus (as a boy), Hercules, giant (Atlas), Archangel, Pleiades (transformed into stars), dragon, voice of prophecy, Cyclops, Titans.

Pizzetti, Ildebrando (Italy 1880-1968)

Fedra ["Phaedra"] (**1915**). Libretto Gabriele d'Annunzio, after his play, after *Phaedra* by Seneca, after *Hippolytus* by EURIPIDES.

Holst, Gustav (UK 1874-1934)

Savitri (**1916**). Libretto Holst, after *Mahabharata*.

The Perfect Fool (**1923**). Libretto Holst. Magician, elixir of youth, spirits of earth, fire and water.

Stanford, Charles Villiers (Ireland-UK 1852-1924)

The Travelling Companion (**1916**; premiered 1925). Libretto Sir Henry Newbolt, after Hans Christian ANDERSEN. Wizard, goblin, flying princess (no shadow), corpse temporarily reincarnated, goblins, resuscitated skeletons.

Kalomíris, Manólis (Greece 1883-1962)

To Dakhtylídi tis Mánas ["The Mother's Ring"] (**1917**). Libretto Agnis Orfikós, after Yánnis Kambýsis. Destiny, nymphs, mountain nereid.

Braunfels, Walter (Germany 1882-1954)

Die Vögel ["The Birds"] (**1920**). Libretto Braunfels, after Aristophanes.

Galathea ["Galatea"] (**1930**). Libretto Braunfels, after marionette play by Silvia Baltus, after OVID.

Verkündigung ["Annunciation"] (**1948**). Libretto Jakob Hegner, from ?trans of *L'annonce faite à Marie* by Paul Claudel. A dead child is restored to life.

Paliashvili, Zachariy (Georgia 1871-1933)

Abesalom da Eteri ["Abesalom and Eteri"] (**1919**). Libretto Petr Mirianashvili, after Georgian legend. Magic necklace.

Korngold, Erich Wolfgang (Austria-US 1897-1957)

Die tote Stadt ["The Dead City"] (**1920**). Libretto Paul Schott [= composer and his father Julius], after *Bruges-la-morte* by Georges Rodenbach.

Das Wunder der Heliane ["The Miracle of Heliane"] (**1927**). Libretto Hans Müller, after *Die Heilige* by Hans Kaltneker. Heliane's magical powers restore a dead man to life, then, through the power of love, he is able to restore her.

Alfano, Franco (Italy 1875-1954)

La leggenda di Sakùntala ["The Legend of Sacuntala"] (**1921**; rev 1952 as *Sakùntala*). Libretto Alfano, after "Abhijnana-shakuntala" ["Sakuntala and the Ring of Recognition"] by Kalidasa (5th century). A curse can be lifted only by a ring, cloud of fire.

Hindemith, Paul (Germany-US 1895-1963)

Das Nusch-Nuschi ["The Nush-Nushi"] (**1921**). Libretto Franz Blei, after Burmese fairytale.

Cardillac (**1926**). Libretto Ferdinand Lion, after *Das Fräulein von Scuderi* by E.T.A. HOFFMANN.

Hin und Zurück ["There and Back"] (**1927**). Libretto Marcellus Schiffer. A wise man reverses time; a dead woman is restored to life.

Mathis der Maler ["Mathis the Painter"] (**1938**). Libretto Hindemith, after life and paintings of Matthias Grünewald. A recreation of the temptation of St Anthony.

Die Harmonie der Welt ["The Harmony of the Universe"] (**1957**). Libretto Hindemith, after life of Johannes Kepler. Characters from Kepler's life appear as heavenly bodies, creating the music of the spheres.

The Long Christmas Dinner (*Das lange Weihnachtsmahl*) (**1961**). Libretto Thornton Wilder, after own play. 90 Christmas dinners with four generations of a family are speeded up.

Schoeck, Othmar (Switzerland 1886-1957)

Das Wandbild ["The Portrait"] (**1921**). Libretto Ferruccio Busoni, after Chinese fairytale.

Venus (**1922**). Libretto Armin Rüeger, after *La Vénus d'Ille* by Prosper Mérimée. An unearthed statue of Venus temporarily comes to life.

Penthesilea (**1927**). Libretto Schoeck, after Heinrich von Kleist, after *Periegesis* by Pausanias.

Vom Fischer un syner Fru ["The Fisherman and his Wife"] (**1930**). Libretto Schoeck, after Philipp Otto Runge, ?after GRIMM BROTHERS. A flounder fulfils his wife's wishes until they become too exorbitant, then all are annulled.

Milhaud, Darius (France-US 1892-1974)

Les Euménides ["The Eumenides"] (**1917-22**; premiered 1949). Libretto Paul Claudel, from trans of play by Aeschylus.

Les malheurs d'Orphée ["The Sorrows of Orpheus"] (**1926**). Libretto Armand Lunel, after OVID. The story is modified to contemporary equivalents, with no fantasy elements but a fox, wolf, bear and wild boar.

L'enlèvement d'Europe ["The Abduction of Europa"] (**1927**). Libretto Henri Hoppenot, after Mopsus.

L'abandon d'Ariane ["Ariadne Abandoned"] (**1928**). Libretto Henri Hoppenot, after OVID. Ariadne is transformed into a constellation.

La délivrance de Thésée ["The Deliverance of Theseus"] (**1928**). Libretto Henri Hoppenot, after EURIPIDES.

Christophe Colomb ["Christopher Columbus"] (**1930**). Libretto Paul Claudel, after his play.

Médée ["Medea"] (**1939**). Libretto Madelaine Milhaud, after EURIPIDES.

Respighi, Ottorino (Italy 1879-1936)

La bella addormentata nel bosco ["Sleeping Beauty"] (**1922**; rev. 1966 by Gian Luca Tocchi]). Libretto Gian Bistolfi, after PERRAULT.

Belfagor (**1923**). Libretto Claudio Guastalla, after Ercole Luigi Morselli, after *Il demonio che prese moglie* (*Novella di Belfagor*) by Niccolò Machiavelli. Church bells ring by themselves.

La campana sommersa ["The Sunken Bell"] (**1927**). Libretto Claudio Guastalla, after *Die versunkene Glocke* by Gerhart Hauptmann.

Maria egiziaca ["St Mary of Egypt"] (**1932**). Libretto Claudio Guastalla, after *Le vite dei santi padri* by Domenico Cavalca. Voice of an angel, voice from the sea, angels.

La fiamma ["The Flame"] (**1934**). Libretto Claudio Guastalla, after *Anne Pedersdotter* by Hans Wiers-Jenssen. Women accused of witchcraft.

Vaughan Williams, Ralph (UK 1872-1958)

The Shepherds of the Delectable Mountains (**1922**). Libretto Vaughan Williams, after *The Pilgrim's Progress* by John Bunyan.

The Poisoned Kiss, or The Empress and the Necromancer (**1936**). Libretto Evelyn Sharp, after *The Poisoner Maid* by Richard Garnett, *Rappaccini's Daughter* by Nathaniel Hawthorne. Magician, sorceress, witches, hobgoblins, magic philosopher's stone, witches, goblins, good and evil spirits, forest creatures.

The Pilgrim's Progress (**1951**). Libretto Vaughan Williams, after John BUNYAN.

Burian, Emil František (Czechoslovakia 1904-1959)

Alladina a Palomid ["Alladina and Palomid"] (**1923**). Libretto anon, after *Alladines et Palomides* by Maurice Maeterlinck.

Novák, Vítûzslav (Czechoslovakia 1870-1949)

Lucerna ["The Lantern"] (**1923**; rev 1930). Libretto Hanuš Jelínek, after Alois Jirásek. Water goblins, spirits, fairies, elves.

Roussel, Albert (France 1869-1937)

La naissance de la lyre ["The Birth of the Lyre"] (**1923**). Libretto Théodore Reinach, after *Ichneutae* by Sophocles, "Homeric Hymn", anon.

Bantock, Granville (UK 1868-1946)

The Seal-Woman (**1924**). Libretto Marjory Kennedy-Fraser, after Hebridean legend.

Křenek, Ernst (Austria-US 1900-1991)

Die Zwingburg ["The Stronghold"] (**1924**). Libretto Franz Werfel. A statue of freedom kills its sculptor when the people reject freedom.

Orpheus und Eurydike ["Orpheus and Eurydice"] (**1926**). Libretto from play by Oskar Kokoshka, after OVID. Eurydice dies a second time through Orpheus' jealousy; as he is dying he rejects her ghost's entreaty for their reconciliation.

Das geheime Königreich ["The Secret Kingdom"] (**1928**). Libretto Křenek. A queen is transformed into a tree.

Leben des Orest ["The Life of Orestes"] (**1930**). Libretto Křenek, after *Orestes* by EURIPIDES.

Cefalo e Procri ["Cephalus and Procris"] (**1934**). Libretto Rinaldo Küfferle, after OVID.

Karl V ["Charles V"] (**1938**; rev 1954). Libretto Křenek, after life of Holy Roman Emperor.

Pallas Athene weint ["Pallas Athene Weeps"] (**1955**). Libretto Křenek, after life and times of Socrates.

Der goldene Bock ["The Golden Ram"] (**1964**). Libretto Křenek, after *Argonautica* by Apollonius of Rhodes. The tale of Jason and the Argonauts translated to 20th-century USA instead of the original Colchis.

Wellesz, Egon (Austria-UK 1885-1974)

Alkestis ["Alcestis"] (**1924**). Libretto Wellesz, after Hugo von Hofmannsthal, after EURIPIDES. Hercules brings Queen Alcestis back from Hades.

Die Bakchantinnen ["The Bacchantes"] (**1931**). Libretto Wellesz, after *The Bacchae* by EURIPIDES.

Malipiero, Gian Francesco (Italy 1882-1973)

L'Orfeide ["The Orpheid"] (**1925**) A trilogy including *#1 La morte delle maschere* ["The Death of the Masks"]; libretto Malipiero, after "La favola d'Orfeo" by Angelo Poliziano, after OVID; and *#3 Orfeo, ovvero L'ottava canzone* ["Orpheus, or the Eighth Song"]; libretto Malipiero, after OVID.

Merlino, mastro d'organi ["Merlin the Organ Master"] (**1934**). Libretto ?Malipiero. A magic organ kills its listeners.

Scott, Cyril (UK 1879-1970)

The Alchemist (**1925**) Libretto Scott, after Indian folktale. The story is similar to *Der Zauberlehrling* ["The Sorcerer's Apprentice"] by GOETHE, after "The Pathological Liar" by LUCIAN.

Sessions, Roger (US 1896-1985)

The Fall of the House of Usher (**1925**). Libretto anon, after Edgar Allen POE.

The Trial of Lucullus (**1947**). Libretto H.R. Hays, from *Das Verhör des Lukullus* by Betrolt Brecht. A dead Roman general fails in his trial for entry to the Elysian Fields.

d'Albert, Eugen [Eugène] (UK-Germany 1864-1932)
Der Golem ["The Golem"] (1926). Libretto Ferdinand Lion, after Jewish legend.

Szymanowski, Karol (Poland 1882-1937)
Król Roger ["King Roger"] (**1926**). Libretto Szymanowski with Jarosław Iwaszkiewicz, after composer's own novel *Ephebos*. Bacchus, dryads, satyrs.

Ibert, Jacques (France 1890-1962)
Angélique (**1927**). Libretto Nino (Michel Veber). Devil.

Persée et Andromède ["Perseus and Andromeda"] (**1929**). Libretto Nino (Michel Veber), after *Moralités légendaires* by Jules Laforgue, after OVID.

Rodgers, Richard (US 1902-1979)
A Connecticut Yankee (**1927**). Libretto Herbert Fields, lyric Lorenz Hart, after *A Connecticut Yankee at the Court of King Arthur* by Mark TWAIN. ♂♂ ARTHUR.

Toch, Ernst (Germany-US 1887-1964)
Die Prinzessin auf der Erbse ["The Princess and the Pea"] (**1927**). Libretto Benno Elkan, after Hans Christian ANDERSEN.

The Last Tale (**1960-62**). Libretto Melchior [Menyhért] Lengyel (trans Cornel Lengyel), after *The Arabian Nights*.

Weinberger, Jaromír (Czechoslovakia-US 1896-1967)
Švanda dudák ["Schwanda the Bagpiper"] (**1927**). Libretto Miloš Kareš with Max Brod, after "The Bagpiper of Straconice" by Josef Kajetán Tyl, after Czech folktale. Queen Iceheart is temporarily released from enchantment.

Reutter, Hermann (Germany 1900-1985)
Saul (**1928**). Libretto Reutter, after Alexander Lernet-Holenia, after the Bible (I Samuel 28). Witch of Endor summons up spirit of the prophet Samuel.

Dr Johannes Faust (**1936**). Libretto Ludwig Andersen, after 16th-century puppet play.

Odysseus (**1942**). Libretto Rudolf Bach, after Homer.

Don Juan und Faust ["Don Juan and Faust"] (**1950**). Libretto Ludwig Andersen, after Christian Dietrich Grabbe. A conflation of the DON JUAN and FAUST stories.

Hamlet (**1980**). Libretto Reutter, after William SHAKESPEARE.

Honegger, Arthur (Switzerland-France 1892-1955)
Amphion (**1929**). Libretto Paul Valéry, after OVID. Stones build Thebes by magic.

Jeanne (d'Arc) au bûcher ["Joan (of Arc) at the Stake"] (**1938**). Libretto Paul Claudel, after historical fact. Heavenly voices (Virgin Mary, St Margaret, St Catherine), Death, Stupidity, Pride, Avarice, Lust, Northern France (Heurtebise), Southern France (Mère aux Tonneaux), Porcus (pig), Asinus (ass), Pecus (sheep).

Martinů, Bohuslav (Czechoslovakia-France-US 1890-1959)
Trojí přání, aneb vrtkavosti Iivota (Trois souhaits, ou Les vicissitudes de la vie) ["The Three Wishes, or the Whims of Life"] (**?1929**; premiered 1971). Libretto Georges Ribemont-Dessaignes, after fairytale. A fairy grants three wishes.

Hry o Marii ["The Plays of Mary/The Miracle of Our Lady"] (**1935**). In 4 parts, of which 2 are of fantasy interest: Pt 2 *Mariken z Nimègue* ["Marie of Nijmegen"]; libretto Henri Ghéon and Vilém Závada, after 15th-century Flemish liturgical drama; Pt 4 *Sestra Paskalina* ["Sister Pascalina"]; libretto Martinů, after Julius Zeyer and folktales.

Dvakrát Alexandr (Alexandre bis) ["Alexander the Second"] (**?1937**. premiered 1964). Libretto André Wurmser. A speaking portrait.

Julietta: Snář ["Julietta, or the Key to Dreams"] (**1938**). Libretto Martinu, after *Juliette, ou La clé des songes* by Georges Neveux. A SURREALIST drama.

Ariadna (Ariane) ["Ariadne"] (**1961**). Libretto Martinů, after *Le voyage de Thésée* by Georges Neveux, after Apollodorus.

Shostakovich, Dmitri (Russia 1906-1975)
Nos ["The Nose"] (**1929**). Libretto Shostakovich with Aleksandr [?Arkadi] Preis, Georgi Yonin and Yevgyeniy Zamyatin, after Nikolai Gogol. A man's nose becomes a state councillor.

Lyedi Makbyet Mtsyenskogo uyezda ["Lady Macbeth of the Mtsensk District"] (**1934**; rev 1956-63 as *Katerina Izmaylova*, premiered 1962). Libretto Shostakovich with Aleksandr [?Arkadi] Preis, after *Katerina Izmaylova* by Nikolai Lyeskov, after *Macbeth* by William SHAKESPEARE.

Skazka o pope i rabotnike yego Balda ["The Tale of the Priest and his Servant Balda"] (**?1933-6**; premiered 1980). Libretto Sofia Khentova, after Aleksandr Pushkin. Devils.

Weill, Kurt (Germany-US 1900-1950)
Aufstieg und Fall der Stadt Mahagonny ["The Rise and Fall of the City of Mahagonny"] (**1930**). Libretto Bertolt Brecht. A metaphorical hell where everything is permitted.

Der Silbersee (Ein Wintermärchen) ["The Silver Lake (A Winter's Tale)"] (**1933**). Libretto Georg Kaiser. Winter suddenly turns to Spring, but the lake's ice surface is still sufficiently thick to walk on.

Die sieben Todsünden (der Kleinbürger) ["The Seven Deadly Sins (of the Petit-Bourgeois)"] (**1933**). Libretto Bertolt Brecht, after scenario by Edward James and Boris Kochno. Representations of Sloth, Pride, Anger, Gluttony, Lust, Covetousness, Envy.

Der Weg der Verheißung ["The Road of Promise"] (**1937**; rev 1936 as *The Eternal Road*). Libretto from *Der Weg der Verheißung* by Franz Werfel. The Witch of Endor summons up spirit of prophet Samuel.

Lady in the Dark (**1941**). Libretto Moss Hart, lyric Ira Gershwin.

One Touch of Venus (**1943**). Libretto S.J. Perelman and Ogden Nash, lyric Ogden Nash, after *The Tinted Venus* by F.J. ANSTEY, after *La Vénus d'Ille* by Prosper Mérimée. A 3000-year old statue of Venus comes to life and has supernatural powers.

Down in the Valley (**1948**). Libretto Arnold Sundgaard. Angels.

Love Life (**1948**). Libretto Alan Jay Lerner. A married couple's lives extended over 150 years.

Benjamin, Arthur (Australia-UK 1893-1960)
The Devil Take Her (**1931**). Libretto Alan Collard and John Gordon.

Bridge, Frank (UK 1879-1941)
The Christmas Rose (1931). Libretto Bridge, after Margaret Kemp-Welch and Constance Cotterell. Tears cause roses to grow in the snow in non-Biblical nativity story.

Casella, Alfredo (Italy 1883-1947)
La donna serpente ["The Snake Woman"] (**1932**). Libretto Cesare Lodovici, after Carlo Gozzi. The fairy of the title is transformed into an enormous snake and changed back again.

La favola d'Orfeo ["The Story of Orpheus"] (**1932**). Libretto Corrado Pavolini, after Angelo Poliziano, after OVID.

Hauer, Josef Matthias (Austria 1883-1959)
Die schwarze Spinne ["The Black Spider"] (**?1932**; premiered 1966). Libretto Hans Schlesinger, after Jeremias Gotthelf. A poisonous spider emerges from a woman's cheek.

Schulhoff, Erwin (Czechoslovakia 1894-1942)
Plameny ["The Flames"] (**1932**). Libretto Karel Josef Beneš, after *El burlador de Sevilla y convidado de pietra* by Tirso de Molina (attrib). Death, ghosts (including that of DON JUAN), Harlequin.

Elgar, Edward (UK 1857-1934)
The Spanish Lady (?1929-33; premiered 1994). Libretto Sir Barry Jackson, after *The Devil is an Ass* by Ben Jonson.

Hanson, Howard (US 1896-1981)
Merry Mount (**1933**). Libretto R.L. Stokes, after *The Maypole of Merry Mount* by Nathaniel HAWTHORNE.

Karel, Rudolf (Czechoslovakia 1880-1945)
Smrt kmotřiaka ["Godmother Death"] (**1933**). Libretto Stanislav Lom, after Czech folktale.
Tři (zlaté) vlasy dûda všvûda ["The Three (Golden) Hairs of Grandfather the Allknowing"] (**1948**). Libretto ?Karel with Zbynûk Vostřák.

Thomson, Virgil (US 1896-1989)
Four Saints in Three Acts (**1933**). Libretto Gertrude Stein.
The Mother of Us All (**1947**). Libretto Gertrude Stein, after life of US suffragette Susan B(rownwell) Anthony (1820-1906). Non-contemporaneous Americans, including Daniel Webster, John Adams, Andrew Johnson, Thaddeus Stevens, Ulysses S. Grant, statue/ghost (Susan B. Anthony), ghosts, composer and librettist (as narrators).
Lord Byron (**1972**). Libretto Jack Larson, includes texts by Lord BYRON, after life of the poet. Ghosts of poets, including Ben Jonson, John MILTON, Percy Bysshe Shelley.

Rocca, Lodovico (Italy 1895-1986)
Il Dibuk ["The Dybbuk"] (**1934**). Libretto Renato Simoni, after Shalom Anski, after Jewish legend. The possession by a dybbuk, which subsequently appears as a ghost.

Egk, Werner [Werner Mayer] (Germany 1901-1983)
Die Zaubergeige ["The Magic Fiddle"] (**1935**). Libretto Egk with Ludwig Andersen, after marionette play by Graf Franz von Pocci.
Peer Gynt (**1938**). Libretto Egk, after Henrik Ibsen.
Circe (**1948**; rev 1966 as *Siebzehn Tage und vier Minuten* ["Seventeen Days and Four Minutes"]). Libretto Egk, after *El mayor encanto, amor* by Pedro Calderón, after Homer.
Irische Legende ["Irish Legends"] (**1955**). Libretto Egk, after "The Countess Cathleen" by W.B. YEATS. The spirit of the damned FAUST, demons (tiger, vulture, owls, snake, hyena-people), angels, empty souls.
Die Verlobung in San Domingo ["The Betrothal in San Domingo"] (**1963**). Libretto Egk, after Heinrich von Kleist. Ghosts.

Lloyd, George (UK 1913-   )
Iernin (**1935**). Libretto William Lloyd. A fairy, previously turned to stone, is changed back again.

Enesco, Georges [George Enescu] (Romania-France 1881-1955)
Oedipe ["Oedipus"] (**1936**). Libretto Edmond Fleg, after *Oedipus Tyrannus* and *Oedipus Colonus* by Sophocles and *Seven Against Thebes* by Aeschylus.

Sutermeister, Heinrich (Switzerland 1910-   )
Die schwarze Spinne ["The Black Spider"] (**1936**). Libretto Albert Roesler, after Jeremias Gotthelf. A woman is transformed into a poisonous spider.
Die Zauberinsel ["The Magic Island"] (**1936**). Libretto Sutermeister, after *The Tempest* by William SHAKESPEARE. Magician (PROSPERO), spirit (ARIEL), nymphs, goblins, spirits, winds, Time, Twilight.
Niobe (**1946**). Libretto Peter Sutermeister, after OVID. Apollo,

Artemis, Niobe turned to stone, ghosts, voices of Nature.
Der rote Stiefel ["The Red Boot"] (**1951**). Libretto Sutermeister, after *Das kalte Herz* by Wilhelm Hauff. Good spirit, evil spirit.
Das Gespenst von Canterville ["The Canterville Ghost"] (**1964**) Libretto anon, after Oscar WILDE. Lively and mischievous ghost's despair at modern times and attitudes.
Der Flaschenteufel ["The Bottle Imp"] (**1971**) Libretto anon, after Robert Louis STEVENSON.

Ghedini, Giorgio Federico (Italy 1892-1965)
Maria d'Alessandria ["Maria of Alexandria"] (**1937**). Libretto Cesare Meano, after ?*Le vite dei santi padri* by Domenico Cavalca. Voice of dead man, heavenly voices.
Le baccanti ["The Bacchantes"] (**1948**). Libretto Tullio Pinelli, after *The Bacchae* by EURIPIDES. King Cadmus is transformed into a dragon.

Moore, Douglas (US 1893-1969)
The Headless Horseman (**1937**). Libretto Stephen Vincent BENÉT, after *The Legend of Sleepy Hollow* by Washington IRVING.
The Devil and Daniel Webster (**1939**). Libretto Stephen Vincent BENÉT, after his story. A jury is formed of damned souls from American history.

Françaix, Jean (France 1912-   )
Le diable boiteux ["The Limping Devil"] (**1938**). Libretto Françaix, after Alain René Le Sage. The devil (Asmodeus) is released from imprisonment in a bottle.

Orff, Carl (Germany 1895-1982)
Der Mond ["The Moon"] (**1939**). Libretto Orff, after GRIMM BROTHERS. Four young men steal the moon from a neighbouring country and hang it in a tree; when they die a quarter of the moon is buried in each grave; the dead in the underworld are woken by the light and celebrate so noisily that St Peter notices, hurls a thunderbolt and hangs the moon in the sky.
Die Bernauerin ["Agnes Bernauer"] (**1947**). Libretto Orff, after Bavarian ballad, after historical fact.
Antigonae ["Antigone"] (**1949**). Libretto from Friedrich Hölderlin trans of *Tiresias* by Sophocles.
Astutuli (eine bairische Komödie) ["Astutuli (A Bavarian Comedy)"] (**1953**). Libretto Orff, after medieval drama. Goblin (Goggolori), giant.
Comoedia de Christi resurrectione (Ein Osterspiel) ["The Comedy of the Resurrection of Christ (An Easter Play)"] (**1957**). Libretto Orff.
Oedipus der Tyrann ["Oedipus the King"] (**1959**). Libretto from Friedrich Hölderlin trans of *Oedipus* by Sophocles.
Ludus de nato infante mirificus (Ein Weihnachtsspiel) ["The Miraculous Play of the Infant Birth (A Christmas Play)"] (**1960**). Libretto Orff.
Prometheus. (**1968**). Libretto (in original Greek) from *Prometheus Bound* by Aeschylus.
De temporum fine comoedia (Das Spiel vom Ende der Zeiten) ["The Play of the End of Time"] (**1973**). Libretto Orff, after Greek, Latin, German sources.

Rosenberg, Hilding (Sweden 1892-1985)
Marionetter ["Marionettes"] (**1939**). Libretto K.A. Hagberg, trans of *Los intereses creados* by Jacinto Benavente. COMMEDIA DELL'ARTE characters.
Lycksalighetens ö ["The Isle of Bliss"] (**1945**). Libretto Per Daniel Amadeus Atterbom. West wind (Zephyrus), wind mother, star mother, Time, Fairy Queen, Mopsus, Genius of Youth, nymphs, nightingales.

Dallapiccola, Luigi (Italy 1904-1975)
Volo di notte ["Night Flight"] (**1940**). Libretto Dallapiccola,

after *Vol de nuit* by Antoine de Saint-Exupéry. Disembodied voice of warning.

Ulisse ["Ulysses/Odysseus"] (**1968**). Libretto Dallapiccola, after Homer via Dante, Alfred Lord Tennyson, James Joyce, Pier Paolo Pasolini. *The Odyssey* as metaphor for a journey of self-discovery.

Atterberg, Kurt (Sweden 1887-1974)

Aladdin (**1941**). Libretto Atterberg with Margareta Atterberg, trans of Bruno Hard-Warden and Ignaz Welleminsky, after *Thousand and One Nights*.

Stormen ["The Tempest"] (**1948**). Libretto Atterberg, after William SHAKESPEARE. Magician (PROSPERO), spirit (ARIEL), air spirits.

Britten, Benjamin (UK 1913-1976)

Paul Bunyan (**1941**). Libretto W.H. Auden, after North American legend. Voice of giant (Paul Bunyan), heron, beetle, squirrel, Moon, Wind, geese, trees.

Gloriana (**1953**). Libretto William Plomer, after *Elizabeth and Essex* by Lytton Strachey. Spirit of the Masque, phantom kings and queens; Time, Concord.

The Turn of the Screw (**1954**). Libretto Myfanwy Piper, after Henry JAMES.

Noye's Fludde (**1958**). Libretto from Chester mystery play.

A Midsummer Night's Dream (**1960**). Libretto Britten with Peter Pears, after William SHAKESPEARE.

Curlew River (**1964**). Libretto William Plomer, after 15th-century Noh play *Sumidagawa* by Juro Motomasa.

Owen Wingrave (**1971**). Libretto Myfanwy Piper, after Henry JAMES.

Death in Venice (**1973**). Libretto Myfanwy Piper, after *Der Tod in Venedig* by Thomas MANN.

Arlen, Harold [Hyman Arluck] (US 1905-1986)

The Wizard of Oz (**1942**). Libretto Frank Gabrielsen, after novel by L. Frank BAUM. Witches, land of OZ.

Martin, Frank (Switzerland 1890-1974)

Le vin herbé ["The Magic Potion"] (**1942**). Libretto Martin, after *Le roman de Tristan et Yseut* by Joseph Bédier, after Tristram and Iseult legend.

Der Sturm ["The Tempest"] (**1956**). Libretto Martin, after August Wilhelm von Schlegel trans of *The Tempest* by William SHAKESPEARE. Magician (PROSPERO), spirit (ARIEL), Iris, Ceres, Juno, nymphs, spirits.

Ullmann, Viktor (Austria 1898-*c*1944)

Der Tod dankt ab oder Der Kaiser von Atlantis (Der Kaiser von Atlantis oder die Todverweigerung) ["Death Abdicates, or the Emperor of Atlantis (The Emperor of Atlantis, or Death's Refusal)"] (**?1943-4**; premiered 1975). Libretto Peter Kien. Death renounces his power; Pierrot is a satire on Adolf Hitler and genocide.

Pijper, Willem (Holland 1894-1947)

Merlijn ["Merlin"] (**1939-46**; premiered 1952). Libretto S. Vestdijk, after Arthurian legend (◊ ARTHUR). Magician (MERLIN), etc.

Menotti, Gian Carlo (Italy-US-UK 1911-   )

The Medium (**1946**). Libretto Menotti. A fraudulent medium begins to question her rejection of the supernatural.

Poulenc, Francis (France 1899-1963)

Les mamelles de Tirésias ["The Breasts of Tiresias"] (**1947**). Libretto Poulenc, after Guillaume Apollinaire. Thérèse changes sex as Tirésias, her husband gives birth to more than 40,000 children.

Lane, Burton (US 1912-   )

Finian's Rainbow (**1947**). Libretto E.Y. Harburg and Fred Saidy, lyric Harburg. Magic crock of gold, leprechaun.

◊◊ FINIAN'S RAINBOW *1968*.

Loewe, Frederick (Austria-US 1904-1988)

Brigadoon (**1947**). Libretto Alan Jay LERNER. A "lost" Scottish village reappears for one day every century. ◊◊ BRIGADOON *1954*.

Camelot (**1960**). Libretto Alan Jay LERNER, after *The Once and Future King* by T.H. WHITE, after Arthurian legend (◊ ARTHUR). Magician (Merlyn; ◊ MERLIN), sprite (Nimuë), witch (MORGAN LE FAY). ◊◊ CAMELOT.

Bliss, Arthur (UK 1891-1975)

The Olympians (**1949**). Libretto J.B. PRIESTLEY, after legend. Greek gods regain their divinity once in every 100 years on midsummer night.

Tobias and the Angel (**1960**). Libretto Christopher Hassall, ?after the Book of Tobit (Apocrypha).

Burkhard, Willy (Switzerland 1900-1955)

Die schwarze Spinne ["The Black Spider"] (**1949**). Libretto Robert Faesi, after Georgette Boner, after Jeremias Gotthelf. A woman is transformed into poisonous spider.

Mohaupt, Richard (Germany-US-Germany 1904-1957)

Die Bremer Stadtmusikanten ["The City Musicians of Bremen"] (**1949**). Libretto Theo Phil, after German legend. A city is saved by animal musicians (cockerel, hen, cat, dog, donkey, bear).

Andriessen, Hendrik (Holland 1892-1981)

Philomena (**1950**). Libretto J. Engelman, after OVID.

Porter, Cole (1891-1964)

Out of This World (**1950**). Libretto Dwight Taylor and Reginald Lawrence, partly after *Amphitryon 38* by Jean Giraudoux, after Molière, after *Amphitruo* by Plautus; lyric Porter. Combines Greek gods and gangsters.

Aladdin (**1957**). Libretto S.J. Perelman, after *The Arabian Nights*.

Dan, Ikuma (Japan 1924-   )

Yuzuru [The Silver Heron] (**1951**). Setting of Junji Kinoshita, after Japanese legend.

Dessau, Paul (Germany [GDR] 1894-1979)

Das Verhör des Lukullus ["The Trial of Lucullus"] (**1951**; rev 1951 as *Die Verurteilung des Lukullus* ["The Condemnation of Lucullus"]). Libretto Bertolt Brecht. A dead Roman general fails in his trial for entry to Elysian Fields. (See also *The Trial of Lucullus* 1947 by Roger Sessions.)

Lanzelot ["Lancelot"] (**1969**). Libretto Heiner Müller and Ginka Tsholakova, after Hans Christian ANDERSEN and "The Dragon" by Yevgyeniy Schwarz. Dragon, cat, donkey, Hercules, Nemean lion, Lernean hydra, shark.

Einstein (**1974**). Libretto Karl Mickel, after sketches by Bertolt Brecht, after life of Albert Einstein. Galileo Galilei, Giordano Bruno, Leonardo da Vinci. A crocodile symbolizes imperialism.

Schaeffer, Pierre (France 1910-1995) and Henry, Pierre (France 1927-   )

Orphée 51 ou Toute la lyre ["Orpheus 51, or All the Lyre"] (**1951**; rev 1953 as *Orphée 53*). Libretto Pierre Schaeffer, after OVID.

Vlad, Roman (Romania-Italy 1919-   )

La storia di una mamma ["The Story of a Mother"] (**1951**). Libretto Gastone da Venezia, after Hans Christian ANDERSEN.

Liebermann, Rolf (Switzerland 1910-   )

Leonore 40/45 ["Leonora 40/45"] (**1952**). Libretto Heinrich Strobel. A guardian angel.

Die Schule der Frauen ["The School for Wives"] (**1955**). Libretto Heinrich Strobel, after *L'école des femmes* by Molière. Jean Baptiste Poquelin (Molière) intervenes several times in his own play to advise the actors.

Loesser, Frank (US 1910-1969)
Hans Christian Andersen (**1952**). Libretto Moss Hart, after life and stories of Hans Christian ANDERSEN.

Partch, Harry (US 1901-1974)
Oedipus – A Music-Dance Drama (**1952**). Libretto Partch, after *Oedipus Tyrannus* by Sophocles.
The Bewitched – A Dance Satire (**1957**). Libretto Partch.
Revelation in the Courthouse Park (**1961**). Libretto Partch, after *The Bacchae* by EURIPIDES.
Delusion of the Fury – A Ritual of Drama and Delusion (**1969**). Libretto Partch, after Japanese and African legends.

Styne, Jule [Jules Stein] (UK-US 1905-   )
Peter Pan (**1954**). Lyric Betty Comden and Adolph Green, after J.M. BARRIE.

Adler, Richard (US 1921-   ) and Ross, Jerry [Jerrold Rosenberg] (US 1926-1955)
Damn Yankees (**1955**). Libretto George Abbott, Douglass Wallopp and Richard Bissel, after *The Year the Yankees Lost the Pennant* by Wallopp. The Faust story transferred to baseball.

Foss, Lukas (Germany-US 1922-   )
Griffelkin (**1955**). Libretto A. Reed, after H. Foss, after GRIMM BROTHERS.

Tippett, Michael (UK 1905-   )
The Midsummer Marriage (**1955**). Libretto Tippett. Time is shifted at ruined temple to the Ancients.
King Priam (**1962**). Libretto Tippett, after *The Iliad* by Homer and EURIPIDES.
The Ice Break (**1977**). Libretto Tippett. A psychedelic messenger in the Paradise Garden.

Henze, Hans Werner (Germany-Italy 1926-   )
König Hirsch ["King Stag"] (1956; rev 1962 as *Il re cervo, oder Die Irrfahrten der Wahrheit* ["King Stag, or The Vagaries of Truth"]). Libretto Heinz von Cramer, after *Il re cervo* by Carlo Gozzi. A king is transformed into a stag by a magic word and is changed back. A woman has three mirror images.
Der junge Lord ["The Young Lord"] (**1965**). Libretto Ingeborg Bachmann, after the parable *Der Scheik von Alessandria und seine Sklaven* by Wilhelm Hauff. A monkey becomes human.
The Bassarids ["Die Bassariden"] (**1966**). Libretto W.H. Auden and Chester Kallman, after *The Bacchae* by EURIPIDES.
Don Chisciotte della Mancia ["Don Quixote of La Mancha"] (**1976**). Libretto Giuseppe di Leva, after Giovanni Battista Lorenzi, after CERVANTES. In old age, Don Quixote and Sancho Panza unhappily watch their younger selves and finally intervene.
Pollicino (**1980**). Libretto Giuseppe di Leva, after Carlo Collodi, GRIMM BROTHERS and Charles PERRAULT.
The English Cat ["Die englische Katze"] (**1983**). Libretto Edward Bond, after *Peines de coeur d'une chatte anglaise* by Honoré de Balzac. Cats, dogs, birds, fox, sheep, mouse, Moon, stars, Royal Society for the Protection of Rats.

Bäck, Sven-Erik (Sweden 1919-1994)
Tranfjädrarna ["The Crane Feathers"] (**1957**). Libretto Bertil Malmberg, after Junji Kinoshita, after Japanese legend. A bird-woman.

Fortner, Wolfgang (Germany 1907-1987)
Bluthochzeit ["Blood Wedding"] (**1957**). Libretto Fortner, after *Bodas de sangre* by Federico García Lorca. Death, moon.
In seinem Garten liebt Don Perlimplín Belisa ["The Love of Don Perlimplin and Belisa in his Garden"] (**1962**). Libretto Fortner, after *El amor de don Perlimplín con Belisa en su jardín* by Federico García Lorca. Goblins.

Tal, Josef [Joseph Gruenthal] (Germany-Israel 1910-   )
Sha'ul be En-Dor ["Saul at Ein Dor"] (**1957**). Libretto Rahel Vernon, after the Bible (I Samuel 28). Witch of Endor summons up spirit of the prophet Samuel.
Ashmedai ["Asmodeus"] (**1971**). Libretto Israel Eliraz, after the Talmud. Devil (Asmodeus).
Der Garten ["The Garden"] (**1988**). Libretto Israel Eliraz. Adam and Eve return to the Garden of Eden, after living for many years in "civilization".

Pauer, Jiří (Czechoslovakia 1919-   )
Ivanivř Slimejš ["The Garrulous Snail"] (**1958**). Libretto Míla Mellanov, after Joe Hloucha. Sea king's daughter, snail, monkey.
Červen Karkulka ["Little Red Riding Hood"] (**1960**). Libretto Míla Mellanov, after PERRAULT.

Searle, Humphrey (UK 1915-1982)
The Diary of a Madman (**1958**). Libretto anon, after Nikolai Gogol.
Hamlet (**1968**). Libretto Searle, after William SHAKESPEARE.

Glanville-Hicks, Peggy (Australia-US 1912-1990)
The Transposed Heads (**1954**). Libretto Glanville-Hicks, after *The Transposed Heads* by Thomas MANN.
The Glittering Gate (**1961**). Libretto Lord DUNSANY.
Nausicaa (**1961**). Libretto A. Reid and Robert Graves, after *Homer's Daughter* by Graves.

Johanson, Sven-Erik (Sweden 1919-   )
Borthytingarna ["The Changelings"] (**1959**). Libretto Johanson, after Helena Nyblom.
Sagan om ringen ["The Tale of the Rings"] (1972). Libretto Anderz Harning, after *The Lord of the Rings* by J.R.R. TOLKIEN.

Maconchy, Elizabeth (UK 1907-1994)
The Sofa (**1959**). Libretto Ursula Vaughan Williams, after *Le sopha* by Claude Prosper de Crébillon *fils*. A prince is transformed into a sofa but changed back.
The Birds (**1968**). Libretto anon, after Aristophanes.

Wagner-Régeny, Rudolf (Hungary-Germany [GDR] 1903-1969)
Prometheus (**1959**). Libretto Wagner-Régeny, after *Prometheus Bound* by Aeschylus.
Das Bergwerk zu Falun ["The Mine at Falun"] (**1961**). Libretto Wagner-Régeny, after Hugo von Hofmannsthal, after *Die Bergwerke zu Falun* by E.T.A. HOFFMANN.

Killmayer, Wilhelm (Germany 1927-   )
La buffonata ["Buffoonery"] (**1960**. Libretto Tankred Dorst, after COMMEDIA DELL'ARTE characters.
La tragedia di Orfeo ["The Tragedy of Orpheus"] (**1961**). Libretto anon, after *La favola d'Orfeo* by Angelo Poliziano, after OVID.
Yolimba oder die Grenzen der Magie ["Yolimba, or The Limits of Magic"] (**1964**). Libretto Killmayer with Tankred Dorst. Those killed by a magician's creation, Yolimba, are restored to life on the magician's death.

Suchoa, Eugen (Czechoslovakia [Slovakia] 1908-   )
Svätopluk ["Svatopluk"] (**1960**) Libretto Suchoa with Ivan Stodola and Jela Kraméryov. Magicians.

Castiglioni, Niccolò (Italy 1932-   )
Attraverso lo specchio ["Through the Looking Glass"] (**1961**). Libretto Alberto Ca' Zorzi Noventa, after *Alice in Wonderland* and *Through the Looking-glass* by Lewis CARROLL.

Jabberwocky (**1962**). Libretto anon, after Lewis CARROLL.
Klebe, Giselher (Germany 1925-   )
Alkmene ["Alcmena"] (**1961**). Libretto Klege, after *Amphitryon* by Heinrich von Kleist, after Molière, after *Amphitruo* by Plautus.
Das Märchen von der schönen Lilie ["The Tale of Beautiful Lily"] (**1969**). Libretto Klebe, after Johann Wolfgang GOETHE. When they are touched by Lilie, living people and animals die and dead ones come to life.
Maderna, Bruno (Italy-Germany 1920-1973)
Don Perlimplin (**1961**). Libretto Maderna, after *El amor de don Perlimplín con Belisa en su jardín* by Federico García Lorca. Goblins.
Hyperion (**1964**). Libretto Maderna with Virginio Puecher, after *Hyperion, oder der Eremit in Griechenland* by Friedrich Hölderlin, with phonemes by Hans G Helms.
Ward, Robert (US 1917-   )
The Crucible (**1961**). Libretto Bernard Stambler, after play by Arthur Miller about 1692 Salem witchcraft trials.
Zafred, Mario (Italy 1922-   )
Amleto ["Hamlet"] (**1961**). Libretto Zafred with Lilyan Zafred, after William SHAKESPEARE.
Moroi, Makoto (Japan 1930-   )
Yamamba ["A Mountain Witch"] (**1962**). Libretto anon, after Japanese legend.
Gyosha Paeton ["Phaethon, the Coachman"] (**1965**). Libretto anon, after OVID.
Cikker, Ján (Czechoslovakia [Slovakia] 1911-1989)
Veaer, noc a ráno (Mr Scrooge) ["Evening, Night and Morning"] (**1963**). Libretto Cikker with Ján Smrek, after *A Christmas Carol* by Charles DICKENS.
Zo Iivota hmyzu ["The Insect Play"] (**1987**). Libretto Cikker, after Karel and Josef ČAPEK.
Kelterborn, Rudolf (Switzerland 1931-   )
Die Erettung Thebens ["The Rescue of Thebes"] (**1963**). Libretto Kelterborn, after *Seven Against Thebes* by Aeschylus.
Kaiser Jovian ["Emperor Jovian"] (**1967**). Libretto Herbert Meier.
Ein Engel kommt nach Babylon ["An Angel Comes to Babylon"] (**1977**). Libretto Friedrich Dürrenmatt, after his play.
Ophelia (**1984**). Libretto Kelterborn with Herbert Meier, after *Hamlet* by William SHAKESPEARE.
de Leeuw, Ton (Holland 1926-   )
Alcestis (**1963**). Libretto de Leeuw, after *Hercules* by EURIPIDES.
De droom ["The Dream"] (**1965**). Libretto de Leeuw, from Haiku texts, after Chinese story. A magic pillow shows future life in a dream.
Argento, Dominick (US 1927-   )
The Masque of Angels (**1964**). Libretto John Olon-Scrymgeour. Cherubim, seraphim, powers, virtues.
The Voyage of Edgar Allen Poe (**1976**). Libretto Charles Nolte. Auguste Dupin, ghost ship; includes vocal setting of Edgar Allen POE's "Annabel Lee".
Bentzon, Niels Viggo (Denmark 1919-   )
Faust III (**1964**). Libretto Bentzon with Kjeld Kromann, after GOETHE, *Ulysses* by James Joyce and *The Trial* by Franz KAFKA. Mephistoples (including as NOSTRADAMUS), FAUST, Leopold Bloom, Josef K.
Die Automaten ["The Automata"] (**1974**). Libretto anon, after *Die Automate* by E.T.A. HOFFMANN.
Felciano, Richard (US 1930-   )
Sir Gawain and the Green Knight (**1964**). Libretto anon, after *Sir Gawain and the Green Knight*, anon (14th century) (◊ GAWAIN). ◊◊ ARTHUR.

Szokolay, Sándor (Hungary 1931-   )
Vérnász ["Blood Wedding"] (**1964**). Libretto Gyula Illyés, after *Bodas de sangre* by Federico García Lorca.
Hamlet (**1968**). Libretto Szokolay, after William SHAKESPEARE.
Wimberger, Gerhard (Austria 1923-   )
Dame Kobold ["The Lady Goblin"] (**1964**). Libretto Wimberger with Wolfgang Rinnert, after Hugo von Hofmannsthal, after *La dama duende* by Pedro Calderón.
Bennett, Richard Rodney (UK-US 1936-   )
The Mines of Sulphur (**1965**). Libretto Beverley Cross. Plague-carrying ghosts.
Blomdahl, Karl-Birger (Sweden 1916-1968)
Herr von Hancken ["Master von Hancken"] (**1965**). Libretto Erik Johan Lindegren, after Hjalmar Bergman. Death, devil, God.
Hanuš, Jan (Czechoslovakia [Czech Republic] 1915-   )
Pochode Prométheova ["The Torch of Prometheus"] (**1965**). Libretto ?Jaroslav Pokorn, after *Prometheus Bound* by Aeschylus. A modern Prometheus is an atomic scientist who defies his government.
Reimann, Aribert (Germany 1936-   )
Ein Traumspiel ["A Dream Play"] (**1965**). Libretto Carla Henius, after *Ett drömspel* by August STRINDBERG. The voice of god Indra.
Melusine (**1971**). Libretto Claus H. Henneberg, after Iwan Goll. Nature spirits, magic park.
Die Gespenstersonate ["The Ghost Sonata"] (**1984**). Libretto Reimann with Uwe Schendel, after *Spöksonaten* by August STRINDBERG. A mummy, the living dead.
Troades ["The Trojan Women"] (**1986**). Libretto Reimann and Gerd Albrecht, after *Die Troerinnen* by Franz Werfel, after EURIPIDES.
Sitsky, Larry (Russia-Australia 1934-   )
The Fall of the House of Usher (**1965**). Libretto Gwen Harwood, after Edgar Allen POE.
The Golem (?**1980**; premiered 1993). Libretto Gwen Harwood, after Jewish legend. A clay figure of a golem gradually comes to life.
Williamson, Malcolm (Australia-UK 1931-   )
The Happy Prince (**1965**). Libretto Williamson, after Oscar Wilde. Speaking statue of prince, swallow, angels.
The Growing Castle (**1968**). Libretto Williamson, after *Ett drömspel* ["A Dream Play"] by August STRINDBERG.
Lucky-Peter's Journey (**1969**). Libretto Edmund Tracey, after *Lycko-Pers resa* August STRINDBERG. Fairy, saints, gnome, birds, rats; starboys, land and sea, monsters, animals, trees, flowers, ocean waves.
Yun, Isang (Korea-Germany 1917-   )
Der Traum des Liu-Tung ["The Dream of Liu-Tung"] (**1965**). Libretto Hans Rudelberger and Winfried Bauernfeind, after drama by Ma Chi-yuan (14th century). An Immortal's dream spiritual pilgrimage lasts 18 years.
Die Witwe des Schmetterlings ["The Butterfly's Widow"] (**1965**). Libretto ?Hans Rudelberger and Winfried Bauernfeind. A dead man is transformed into a butterfly.
Geisterliebe ["Spirit Love"] (**1971**). Libretto Harald Kunz.
Sim Tjong (**1972**). Libretto Harald Kunz. Angel sent to earth, her heavenly mother, a dragon king, heavenly spirits.
Crosse, Gordon (UK 1937-   )
Purgatory (**1966**). Libretto Crosse, after W.B. YEATS. Ghosts.
Potter Thompson (**1975**). Libretto Alan GARNER. Two children from the past, knights sleeping in a cave until they are needed.

Marco, Tomás (Spain 1942-   )
Jabberwocky (**1966**). Libretto anon, after Lewis CARROLL.
Ginastera, Alberto (Argentina 1916-1983)
Bomarzo (**1967**). Libretto Manuel Mujica Láinez, after his novel. Stone giants come to life in a vision.
Lidholm, Ingvar (Sweden 1921-   )
Holländarn ["The Dutchman"] (**1967**). Libretto Herbert Grevenius, after dramatic fragment by August STRINDBERG.
Ett drömspel ["A Dream Play"] (**1992**). Libretto Lidholm, after August STRINDBERG. Damned souls, voice of god Indra.
Birtwistle, Harrison (UK 1934-   )
Punch and Judy (**1968**). Libretto Stephen Pruslin, after traditional puppet play.
Down by the Greenwood Side (**1969**). Libretto Michael Nyman, after mummer's play, *Ballad of the Cruel Mother*. Father Christmas, St George, Green Man
The Mask of Orpheus (**1986**). Libretto Peter Zinovieff, after OVID and other versions of the legend.
Yan Tan Tethera (**1986**). Libretto Tony Harrison. Devil (the Bad'un), sheep.
Gawain (**1991**; rev 1994). Libretto David Harsent, after *Sir Gawain and the Green Knight*, anon (14th century) (⇩ GAWAIN). ARTHUR, Green Knight.
The Second Mrs Kong (**1994**). Libretto Russell Hoban. Anubis, Eurydice, head of Orpheus, Vermeer, Pearl (in Vermeer's painting), Paganini, Kong, Death of Kong, Monstrous Messenger, Mirror, Doubt, Fear, Despair, Terror, the Dead.
Eaton, John (US 1935-   )
Heracles (**1968**). Libretto M. Fried, after Greek legend.
The Tempest (**1985**). Libretto Andrew Porter, after William SHAKESPEARE. Magician (PROSPERO), spirit (ARIEL).
Rota, Nino (Italy 1911-1979)
Aladino e la lampada magica ["Aladdin and the Magic Lamp"] (**1968**). Libretto Vinci Verginelli, after *The Arabian Nights*.
La visita meravigliosa ["The Miraculous Visit"] (**1970**). Libretto anon, after H.G. WELLS.
Kelemen, Milko (Yugoslavia/Croatia 1924-   )
Der Belagerungszustand ["The State of Siege"] (**1969**) Libretto Kelemen with Joachim Hess, after *La Peste* by Albert Camus.
Lorentzen, Bent (Denmark 1935-   )
Euridice ["Eurydice"] (**1969**). Libretto Lorentzen, after woodcuts by Palle Nielsen, after OVID.
Lutyens, Elizabeth (UK 1906-1983)
Isis and Osiris (**1969**). Libretto Lutyens, after Plutarch.
McCabe, John (UK 1939-   )
The Lion, the Witch and the Wardrobe (**1969**). Libretto Gerald Larner, after the Narnia series by C.S. LEWIS.
Penderecki, Krzysztof (Poland 1933-   )
Diably z Loudun (*Die Teufel von Loudun*) ["The Devils of Loudun"] (**1969**). Libretto, after John Whiting, after Aldous Huxley. Devilish possession; demon Asmodeus speaks through Prioress.
Raj utracony ["Paradise Lost"] (**1978**). Libretto Christopher Fry, after John MILTON. Milton (narrator), Adam, Eve, Satan, Beelzebub, Moloch, Belial, Mammon, Death, Sin, voice of God, archangels, angels, Hell, Garden of Eden.
Pousseur, Henri (Belgium 1929-   )
Votre Faust ["Your Faust"] (**1969**). Libretto Michel Butor, after various versions of FAUST story. Composer Henri (= Faust) is offered unlimited resources for his opera. The title refers to *Mon Faust* by Paul Valéry.

Berio, Luciano (Italy 1925-   )
Opera (**1970**). Libretto Berio with Furio Colombo, Umberto ECO and Susan Yankowitz, after Alessandro Striggio. A combination of elements of the Orpheus myth with 20th-century fact.
Un re in ascolto ["A King Listens"])**1984**). Libretto Berio, after Italo Calvino, after W.H. Auden, Friedrich Wilhelm Gotter and *The Tempest* by William SHAKESPEARE. PROSPERO is a theatrical impresario.
Bricusse, Leslie (UK 1931-   )
Scrooge (**1970**; rev 1992 for the stage) (**1970**). Libretto Bricusse with Ian Fraser and Herbert W. Spencer, after *A Christmas Carol* by Charles DICKENS.
Christou, Jani (Greece 1926-1970)
Oresteia (**1967-70**). Libretto anon, after dramatic trilogy by Aeschylus.
Rautavaara, Einojuhani (Finland 1928-   )
Apollon contra Marsyas ["Apollo versus Marsyas"] (**1970**). Libretto Bengt V. Wall, after Greek legend.
Schuller, Gunther (US 1925-   )
The Fisherman and His Wife (**1970**). Libretto John Updike, after GRIMM BROTHERS.
Kagel, Mauricio (Argentina-Germany 1931-   )
Staatstheater ["State Theatre"] (**1971**). Libretto Kagel. A parody of traditional operas, it includes mixture of recognizable existing costumes and props.
Mare Nostrum: Entdeckung, Befriedung und Konversion des Mittelmeerraumes durch einen Stamm aus Amazonien ["Our Sea: The Discovery, Liberation and Conversion of the Mediterranean Area by an Amazonian Tribe"] (**1975**) Libretto Kagel. A reversal of Columbus's discovery of the Americas.
Die Erschöpfung der Welt ["The Exhaustion of the World"] (**1980**). Libretto Kagel. A reversal of the Creation.
Aus Deutschland ["From Germany"] (**1981**). Libretto Kagel, after German Romantic poems and *Lieder*. Franz Schubert, Johann Wolfgang GOETHE, dogs, characters from well-known Lieder. Leiermann (and wife), Poetess, Night, Ship's Captain, Edward (and mother), suit of armour, grenadiers, Black King, Death and the Maiden, Rat Catcher, Hyperion, fortune-teller, nightingale, Music, youth, Mignon and DOPPELGÄNGER, rats.
La trahison orale (Der mündliche Verrat) ["Oral Treason"] (**1983**). Libretto Kagel, after *Les Evangiles du Diable, selon la croyance populaire* by Claude Seignolle. "A musical epic on the devil" (stories from oral tradition of encounters with the devil; exorcism, preventative incantation).
Pashchenko, Andrei (Russia 1883-1972)
Master i Margarita ["The Master and Margarita"] (**1971**). Libretto anon, after Mikhail BULGAKOV. Devil, female demon, cat demon.
Bussotti, Sylvano (Italy 1931-   )
Lorenzaccio (**1972**). Libretto Bussotti, after Alfred de Musset, *et al.* Alfred de Musset and George Sand transformed into Eros and Death.
Nottetempo (**1976**). Libretto Bussotti with Romano Amidei, after *Legenda aurea* (13th century) by Archbishop Jacobus de Voragine (Jacopo da Varagine), *Philoctetes* by Sophocles and life of Michelangelo. Michelangelo (= Philoctetes), ghosts from his past.
Davies, Peter Maxwell (UK 1934-   )
Taverner (**1972**). Libretto Davies, after life of English composer John Taverner (c1490-1545). Death, Antichrist, God the Father, Archangels, St John, Virgin Mary, demons.

The Martyrdom of St. Magnus (**1977**). Libretto Davies, after *Magnus* by George Mackay BROWN. ◊ TIME ABYSS; ANACHRONISM.

The Two Fiddlers (**1978**). Libretto Davies, after George Mackay BROWN. A fiddler loses 21 years, after one day with the trolls.

Cinderella (**1980**). Libretto Davies, after *Cendrillon* by Charles PERRAULT. Cat (instead of fairy).

The Lighthouse (**1980**). Libretto Davies, after historical fact. Ghosts.

Resurrection (**1988**). Libretto Davies. Zeus (or Hera), Pluto, Apollo, Antechrist, cats, demons.

Gruber, H.K. [Heinz Karl] (Austria 1943-   )
Gomorra (**1972**). Libretto Richard Bletschacher.
Gloria – A Pigtale (**1994**). Libretto Rudolf Herfurtner. The world inhabited by pigs.

Slonimsky, Sergei (Russia 1932-   )
Master i Margarita ["The Master and Margarita"] (?**1972-3**; premiered 1989). Libretto Yuri Dimitrin and V. Fialkovsky, after Mikhail BULGAKOV. Devil, female demon, cat demon.
Hamlet (**1990**). Libretto anon, after William SHAKESPEARE.

Kox, Hans (Holland 1930-   )
Dorian Gray (**1973**). Libretto Kox, after *The Picture of Dorian Gray* by Oscar WILDE.

Matthus, Siegfried (Germany [GDR] 1934-   )
Omphale (**1973**). Libretto Peter Hacks, after ?Apollodorus.

Nørgård, Per (Denmark 1932-   )
Gilgamesh (**1973**). Libretto Nírg†rd, after Sumerian legend.

Sciarrino, Salvatore (Italy 1947-   )
Amore e Psiche ["Cupid and Psyche"] (**1973**). Libretto A. Pes, after APULEIUS.
Lohengrin (**1983**). Libretto Sciarrino, after *Lohengrin fils de Parsifal* by Jules Laforgue, after historical fact and legend. Lohengrin flies to the moon on a cushion transformed into a swan.
Perseo e Andromeda ["Perseus and Andromeda"] (**1991**). Libretto Sciarrino, after *Moralités légendaires* by Jules Laforgue, after OVID.

Werle, Lars Johan (Sweden 1926-   )
Medusan och djävulen ["Medusa and the Devil"] (**1973**). Libretto Werle, after Elsa Grave.
Animalen ["The Animals"] (**1979**). Libretto Tage Danielsson. A congress is held by the animals to save mankind from self-destruction.

Knaifel, Aleksandr (Russia 1943-   )
The Canterville Ghost (**1974**). Libretto T.Kramarova, after Oscar WILDE. A lively and mischievous ghost's despair at modern times and attitudes.

Kunad, Rainer (Germany [GDR] 1936-   )
Sabellicus (**1974**). Libretto Kunad, after *The Tragicall History of Dr Faustus* by Christopher Marlowe by Christopher Marlowe *et al*. A Faustian pact the ruling feudal system rather than with the Devil. (Georgius Sabellicus was the first to adopt the FAUST name.)
Amphitryon (**1984**). Libretto Ingo Zimmermann, after ?Heinrich von Kleist, after Molière, after *Amphitruo* by Plautus.
Der Meister und Margarita ["The Master and Margarita"] (**1986**). Libretto Heinz Czechowski, after Mikhail BULGAKOV.

Musgrave, Thea (UK-US 1928-   )
The Voice of Ariadne (**1974**). Libretto Amalia Elguera, after *The Last of the Valerii* by Henry JAMES. A count is haunted by voice of Ariadne (on tape), after unearthing of plinth for her statue; the countess takes on her identity and becomes the missing statue.
A Christmas Carol (**1979**). Libretto Musgrave, after Charles DICKENS.
An Occurrence at Owl Creek Bridge (1982). Libretto ?Musgrave, after Ambrose BIERCE. ◊◊ POSTHUMOUS FANTASY.

Smalls, Charlie (US ? -   )
The Wiz (**1974**). Lyric Small, libretto William F. Brown, after *The Wonderful Wizard of Oz* by L. Frank BAUM. Teenagers put on their own performance of *The Wonderful Wizard of Oz*.

Bialas, Günter (Germany 1907-   )
Der gestiefelte Kater, oder Wie man das Spiel spielt ["Puss in Boots, or How to Play the Game"] (**1975**). Libretto Tankred Dorst, after Ludwig TIECK.

Zimmermann, Udo (Germany [GDR] 1943-   )
Der Schuhu und die fliegende Prinzessin ["The Owl and the Flying Princess"] (**1975**). Libretto Peter Hacks.

Andriessen, Louis (Holland 1939-   )
Orpheus (**1977**). Libretto Lodewijk de Boer, after OVID.

Balassa, Sándor (Hungary 1935-   )
Az ajtón kivül ["The Man Outside"] (**1977**). Libretto Géza Fodor, after *Draußen vor der Tür* by Wolfgang Borchert.

Ligeti, György (Hungary-Germany 1923-   )
Le grand macabre (**1978**). Libretto Ligeti with Michael Meschke, after *La balade du grand macabre* by Michel de Ghelderode.

Rihm, Wolfgang (Germany 1952-   )
Faust and Yorick ["Faust and Yorick"] (**1977**). Libretto Frithjof Haas, after *Faust et Yorick* by Jean Tardieu. FAUST seeks enlightenment in analysing Yorick's skull.
Jakob Lenz (**1979**). Libretto Michael Fröhling, after *Lenz* by Georg Büchner. A schizophrenic poet experiences imaginary voices and apparitions.
Die Hamletmaschine ["The Hamlet Machine"] (1987). Libretto Rihm, after Heiner Müller, after William SHAKESPEARE.
Oedipus (**1987**). Libretto Rihm, after *Oedipus Tyrannus* by Sophocles, *Oedipus, Reden des letzten Philosophen mit sich selbst* by Friedrich Nietzsche, *Oedipuskommentar* by Heiner Müller.

Bedford, David (UK 1937-   )
The Rime of the Ancient Mariner (1978). Libretto Bedford, after Samuel Taylor COLERIDGE.
The Death of Baldur (**1979**). Libretto Terry Bragg, after *Prose (Younger) Edda* by Snorri Sturluson.
Fridiof's Saga (**1980**). Libretto Terry Bragg, after *Prose (Younger) Edda* Snorri Sturluson.
The Ragnarok (**1983**). Libretto Terry Bragg, after *Poetic (Elder) Edda*, anon and *Prose (Younger) Edda* Snorri Sturluson.

Blackford, Richard (UK 1954-   )
Sir Gawain and the Green Knight (**1978**). Libretto John Emlyn Edwards, after *Sir Gawain and the Green Knight*, anon (14th century) (◊ GAWAIN). ARTHUR, giant (Green Knight), sorceress (MORGAN LE FAY).
Metamorphoses (**1983**). Libretto John Emlyn Edwards, after OVID. Ovid (narrator), Persephone (Proserpine), Ceres (Demeter), Arethusa, Phaethon, Phoebus, Jupiter, Mercury, tree nymph, flying chariot; several transformations.
Gawain and Ragnall (**1985**). Libretto Ian Barnett, after Arthurian legend (◊ ARTHUR), forest sprites.

Boyd, Anne (Australia (?UK 1946-   )
The Little Mermaid (**1978**). Libretto Robin Lee, after Hans Christian ANDERSEN.

Johnson, Robert Sherlaw (UK 1932-   )
The Lambton Worm (**1978**). Libretto Anne Ridler.
Bekku, Sadao (Japan 1922-   )
Aoi no Ue ["Princess Hollyhock"] (**1979**). Libretto Matsuko Suzuki, after two Noh plays, after *Genji monogatari* by Lady Murasaki.
Cowie, Edward (UK-?Australia 1943-   )
Commedia (**1979**). Libretto David Starsmeare, after COMMEDIA DELL'ARTE characters.
Hiller, Wilfried (Germany 1941-   )
Der Lindwurm und der Schmetterling oder Der seltsame Tauch (**1979**). Libretto Michael ENDE.
Niobe (**1979**). Libretto Hiller with Elisabeth Woska, after dramatic fragment "Niobe, Tantalus" by Aeschylus.
Tranquilla Trampeltreu, die beharrliche Schildkröte (**1981**). Libretto Michael ENDE.
Die Ballade von Norbert Nackendick oder Das nackte Nashorn (**1981**). Libretto Michael ENDE.
Die Fabel vom Elefanten Filemon Faltenreich oder die Fussballweltmeisterschaft der Fliegen (**1984**). Libretto Michael ENDE.
Der Goggolori ["The Goblin"] (**1985**). Libretto Michael ENDE, after folktale.
Die Jagd nach dem Schlarg (**1987**). Libretto Michael ENDE, after *The Hunting of the Snark* by Lewis CARROLL.
Das Traumfresserchen (?**1990**). Libretto Michael ENDE.
Der Rattenfänger (Ein Hamelner Totentanz) (?**1993**). Libretto ?Michael ENDE, ?after "The Pied Piper of Hamelin" by Robert BROWNING.
Glass, Philip (US 1937-   )
Satyagraha (**1980**). Libretto Glass with Constance de Jong, after *Bhagavad-Gita*. Krishna, Arjuna, Mahatma Gandhi, Leo Tolstoy, Rabindranath Tagore, Martin Luther King.
The Juniper Tree (**1985**). Libretto Arthur Yorinks, after GRIMM BROTHERS.
The Fall of the House of Usher (**1988**). Libretto Arthur Yorinks, after Edgar Allen POE.
Schat, Peter (Holland 1935-   )
Aap verslaat de knekelgeest ["Monkey Subdues the White-Bone Demon"] (**1980**). Libretto Schat, after 17th-century picture novel by Chao Hung-pen and Chien Hsiao-tai, after *Journey to the West* by Wu Cheng-en and Wang Hsing-pei. Magic, immortal, demon (self-transformations), evil fairy, monsters, monkey warriors.
Vivier, Claude (Canada 1948-1983)
Kopernikus (Rituel de Mort) ["Copernicus: Death Ritual"] (**1980**). Libretto Vivier. Experiences, after death, including sorceress, MERLIN, Queen of the Night, Tristan, Isolde, Master of the Waters, Mozart, Copernicus, Lewis CARROLL.
Lloyd Webber, Andrew (UK 1948-   )
Cats (**1981**). Libretto Trevor Nunn and Richard Stilgoe, after *Old Possum's Book of Practical Cats* by T.S. Eliot.
Starlight Express (**1984**). Libretto ?, lyric Richard Stilgoe.
The Phantom of the Opera (**1986**). Libretto Lloyd Webber with Richard Stilgoe, lyric Charles Hart and Stilgoe, after *Le Fantôme de l'Opéra* by Gaston LEROUX.
Stockhausen, Karlheinz (Germany 1928-   )
Licht (seven operas, from **1977** and continuing). Libretto Stockhausen, partly after *The Urantia Book*, anon. Principal protagonists: archangel Michael, Lucifer, Eve (each represented by a singer, an instrumentalist and a dancer). *Donnerstag aus Licht* (**1981**), *Samstag aus Licht* (**1984**), *Montag aus Licht* (**1988**), *Dienstag aus Licht* (**1993**), *Freitag aus Licht* (**1996**), «Mittwoch aus Licht» (**1993**; due for

completion in 1997) and «Sonntag aus Licht» (due for completion in 2002).
Burgon, Geoffrey (UK 1941-   )
Orpheus (**1982**). Libretto Peter Porter, after OVID.
Menken, Alan (US 1950-   )
Little Shop of Horrors (**1982**). Libretto Howard Ashman, after screenplay by Roger CORMAN and Charles Griffith. Filmed as *The* LITTLE SHOP OF HORRORS (*1986*).
Schwertsik, Kurt (Austria 1935-   )
Das Märchen von Fanferlieschen Schönefuáchen ["The Wondrous Tale of Fanferlizzy Sunnyfeet"] (**1983**). Libretto Karin and Thomas Körner, after Clemens von Brentano. A goat is transformed into a girl; a castle in the air is built in one night.
Bryars, Gavin (UK 1943-   )
Medea (**1984**). Libretto Robert Wilson (primarily in Greek, plus French and English), after EURIPIDES.
Knussen, Oliver (UK 1952-   )
Where the Wild Things Are (**1984**). Libretto Maurice SENDAK, after his book.
Higglety Pigglety Pop!, or There Must be More to Life (**1990**). Libretto Maurice SENDAK, after his book.
Nono, Luigi (Italy 1924-1990)
Prometeo (**1984**). Libretto ?Nono with Massimo Cacciari, after *Prometheus Bound* by Aeschylus.
Oliver, Stephen (UK 1950-1992)
Beauty and the Beast (**1984**). Libretto Oliver, based on Carlo COLLODI's translation of *La* BELLE ET LA BÊTE by Mme J.M. LEPRINCE DE BEAUMONT.
Shibata, Minao (Japan 1916-   )
El divino Orfeo ["The Divine Orpheus"] (**1984**). Libretto anon, after OVID.
Weir, Judith (UK 1954-   )
The Black Spider (**1984**). Libretto Weir, after *Die schwarze Spinne* by Jeremias Gotthelf. A poisonous spider emerges from woman's cheek.
Heaven Ablaze in his Breast (**1990**). Libretto Weir, after *Der Sandmann* by E.T.A. HOFFMANN.
The Vanishing Bridegroom (**1990**). Libretto Weir after J.F. Campbell's *Popular Tales of the West Highlands*, Pt 2 "The Disappearance" (a man loses 20 years in magic hill) and Pt 3 "The Stranger".
Scipio's Dream (**1991**). Libretto ?Weir, after Pietro Metastasio, after *De republica (Somnium Scipionis)* by Scipio. Fortune, Constancy, young Mozart, flying London taxi, inhabitants of Heaven working out in a gymnasium.
Blond Eckbert (**1994**). Libretto Weir, after Thomas CARLYLE's trans of *Der blonde Eckbert* by Ludwig TIECK.
Boehmer, Konrad (Germany-Holland 1941-   )
Dr Faustus (**1985**). Libretto Hugo Claus, after earliest FAUST stories. Homunculus created by Faust; apparition of Virgin Mary; a visit to future; but no Mephistopheles.
Bozay, Attila (Hungary 1939-   )
Csongor és Tünde ["Csongor and Tünde"] (**1985**). Libretto Bozay, after Mihály Vörösmarty.
Barker, Paul (UK 1956-   )
Phantastes (**1986**). Libretto Barker, after George MACDONALD.
Jolas, Betsy (US-France 1926-   )
Le Cyclope ["The Cyclops"] (**1986**). Libretto anon, after *Cyclops* by EURIPIDES.
Cage, John (US 1912-1992)
Europeras #1 and *#2* (**1987**), *Europeras #3* and *#4* (**1990**), *Europera #5* (**1991**). Scenario Cage. Arias and costumes collaged from classical operas.

Fénelon, Philippe (France 1952-   )
Le chevalier imaginaire ["The Imaginary Chevalier"] (?**1987**).
Libretto anon, after CERVANTES and *Die Wahrheit über
Sancho Pansa* by Franz KAFKA. In a deconstruction of
Cervantes' story, Don Quixote is imagined by Sancho Panza.

Finnissy, Michael (UK 1946-   )
The Undivine Comedy (**1988**). Libretto Finnissy, after *Nieboska
Komedia* by Zygmunt Krasiýski.

Turnage, Mark-Antony (UK 1960-   )
Greek (**1988**). Libretto Turnage with Jonathan Moore, after
Stephen Berkoff, after *Oedipus Tyrannus* by Sophocles.

Bergman, Erik (Finland 1911-   )
Der sjungande trädet ["The Singing Tree"] (**1989**). Libretto Bo
Carpelen, after "Prince Hatt in the Underworld", anon.

Casken, John (UK 1949-   )
The Golem (**1989**). Libretto Casken with Pierre Audi, based on
the legend of the golem. A clay figure of a golem gradually
comes to life.

Höller, York [Georg] (Germany 1944-   )
Der Meister und Margarita (**1989**). Libretto Höller, after *The
Master and Margarita* by Mikhail BULGAKOV.

Manzoni, Giacomo (Italy 1932-   )
Doktor Faustus (**1989**). Libretto Manzoni, after Thomas
MANN. FAUST is a composer.

Teitelbaum, Richard (US 1939-   )
Golems (**1989**). Libretto Teitelbaum, after Paul Wegener's
movie *Der Golem*, earlier golem stories and *Sefer Yezira*
["The Book of Creation"], anon (before 6th century). A
robot, as a modern golem, is activated by incantations of
magic formulae.

Townshend, Pete (UK 1945-   )
The Iron Man (**1989**). Libretto Townshend and David
Thacker, after "The Iron Man" by Ted Hughes.

Young, Douglas (UK 1947-   )
The Tailor of Gloucester (**1989**). Libretto Young with John
Michael Phillips, after Beatrix POTTER.

Bibalo, Antonio (Italy-Norway 1922-   )
Macbeth (**1990**). Libretto anon, after William SHAKESPEARE.

Döhl, Friedhelm (Germany 1936-   )
Medea (?**1990**). Libretto anon, after EURIPIDES.

Einem, Gottfried von (Austria 1918-   )
Tulifant (**1990**). Libretto Lotte Ingrisch, after ?*Tulifäntchen* by
Karl Immermann. The Earth, the Past, the Future, magic
golden horn, earthly and unearthly spirits.

Sondheim, Stephen (US 1930-   )
Into the Woods (**1990**). Libretto James Lapine, after fairytales,
lyric Sondheim. Most of the characters from "Little Red
Riding Hood", "Cinderella" (GRIMM BROTHERS version),
"Jack the Giant Killer", "Rapunzel", "The Goose that Lays
the Golden Egg", magic harp, voice of giant's wife,
"Sleeping Beauty", "Snow White".

Corigliano, John (US 1938-   )
The Ghosts of Versailles (**1991**). Libretto William M. Hoffman,
partly after *La mère coupable* by Pierre Beaumarchais.

Monk, Meredith (US 1943-   )
Atlas (**1991**). Libretto ?Monk. A spirit guide, quasi-Paradise.

Buller, John (UK 1927-   )
Bakxai ["The Bacchae"] (**1992**) Libretto ?Buller, (mostly in
Greek), after EURIPIDES.

Dusapin, Pascal (France 1955-   )
Medeamaterial (**1992**). Libretto Heiner Müller, after
?*Medea* by EURIPIDES. Medea in modern psychoanalytical
terms.

Hersant, Philippe (France 1948-   )

Le château des Carpathes (**1992**). Libretto Jorge Silva-Melo,
after Jules Verne.

Berkeley, Michael (UK 1948-   )
Baa Baa Black Sheep (**1993**). Libretto David Malouf, after *The
Jungle Book* by Rudyard Kipling and "Baa Baa Black Sheep".
Mowgli, Bagheera, Ka, Baloo, Akela, Grey Wolf, Punch,
Mrs Punch, wolves.

Harvey, Jonathan (UK 1939-   )
Inquest of Love (**1993**). Libretto Harvey with David Rudkin.
Life after death.

Xenakis, Iannis (Greece-France 1922-   )
Les Bacchantes ["The Bacchae"] (**1993**). Libretto anon, after
EURIPIDES.

Zender, Hans (Germany 1936-   )
Don Quijote de la Mancha ["Don Quixote of La Mancha"]
(**1993**). Libretto Zender, after CERVANTES. Devil, angel,
magician (MERLIN), ghost (oracle).

Butler, Martin (UK 1960-   )
Craig's Progress (**1994**). Libretto Stephen Pruslin.
Gingerbread witch, Santa Claus, gods, reindeer, planets,
comics superheroes.

Firsova, Elena (Russia-UK 1950-   )
The Nightingale and the Rose (**1994**). Libretto anon, after *The
Nightingale and the Rose* by Oscar WILDE and Christina
Rossetti.

Mason, Benedict (UK 1953-   )
Playing Away (**1994**). Libretto Howard Brenton. The FAUST
story transferred to football.

Moran, Robert (US 1937-   )
The Dracula Diary (**1994**). Libretto James Skofield, after VAM-
PIRE legend. (See also Philip Glass **1985**.)

Goehr, Alexander (UK 1932-   )
Arianna ["Ariadne"] (**1995**). Libretto Ottavio Rinuccini, after
OVID. Venus, Cupid, Bacchus, Jupiter, cupids. Ariadne and
Theseus are transformed into constellations.

Schnittke, Alfred (Russia 1934-   )
Historia von D. Johann Fausten ["The History of Dr. Johann
Faustus"] (**1995**). Libretto Jürg Morgener, after *Historia
vom D. Johann Fausten, dem weitbeschreyten Zauberer und
Schwartzkunstler* (Faustbuch). ◊ FAUST.   [HD]

Further reading: *The Complete Opera Book* by Gustav Kobbé
(**1918**; rev vt by Earl of Harewood as *Kobbé's Complete Opera
Book*; various editions; much cut as *The Pocket Kobbé's Opera
Book* by Earl of Harewood 1994); *L'opera; repertorio della
lirica dal 1597* (**1977**; trans as *Phaidon Book of the Opera: A
Survey of 780 Operas from 1597* 1979); *Handbuch der russis-
chen und sowjetischen Oper* (**1985**) by Sigrid Neef; *Gänzl's
Book of the Musical Theatre* (**1988**) by Kurt Gänzl and
Andrew Lamb; *Das groáe Handbuch der Oper* (**1991**; rev
1994) by Heinz Wagner; *The New Grove Dictionary of Opera*
(**1992** 4 vols) ed Stanley Sadie; *The Viking Opera Guide*
(**1993**; much cut vt *The Penguin Guide to Opera* 1995) ed
Amanda Holden; *International Dictionary of Opera* (**1993** 2
vols) ed C. Steven LaRue.

OPERATION MONSTERLAND vt of *Destroy All
Monsters* (**1968**). ◊ GODZILLA MOVIES.

ORACLES At the height of the Greek civilization, in the
5th and 6th centuries BC, many of the shrines to the GODS
had a priest (or more likely priestess) who was a channel for
divine advice. The shrines were in effect Greek churches,
where individuals came to consult their divinity. The
responses were always ambiguous, so that if events did not
turn out as predicted it was not an error of the GODS. The
most famous oracle was at the shrine to APOLLO at Delphi,

which was consulted from all over the Mediterranean world and thus became the centre of a world-wide intelligence network. Its priests had mastered the art of the ambiguous. By Roman times the oracles were no longer shrines so much as messages from the gods: the Romans based their PROPHECIES on the Sibylline Books (◊ SIBYL), but their and other cultures developed a whole industry of augurs, divinators and prophets who claimed PRECOGNITION and would utter an oracle for the relevant fee.

Oracles, in the widest sense, feature frequently in fantastic fiction, but in the strict form of ancient Greece they are clearly limited to FANTASIES OF HISTORY or to HEROIC FANTASY drawing upon historical sources. They are best represented in the classical fantasies of Thomas Burnett SWANN and similar stories evoking MYTHS and LEGENDS. The most famous Greek prophetess was Cassandra, daughter of King Priam of Troy: given the gift of prophecy by APOLLO after she refused to meet his demands, her CURSE was that no one would believe her. Cassandra's plight has fascinated many; she is the subject of *The Firebrand* (**1987**) by Marion Zimmer BRADLEY and *Cassandra: Princess of Troy* (**1994**) by Hilary Bailey (1936-   ).   [MA]

ORCHIDEENGARTEN, DER ◊ MAGAZINES.

ORDER ◊ CHAOS.

ORE, REBECCA Working name of US writer Rebecca B. Brown (1948-   ), who began publishing fiction with "Projectile Weapons and Wild Alien Water" (1986 *Amazing*); she had earlier published four poetry collections. She is best-known for the sf trilogy comprising *Becoming Alien* (**1988**), *Being Alien* (**1989**) and *Human to Human* (**1990**). *Slow Funeral* (**1994**) is a contemporary fantasy, set in a well realized US South. *Alien Bootlegger and Other Stories* (coll **1993**) collects her short fiction, which tends to sf.   [GF]

ORIENTAL DREAM vt of *Kismet* (**1944**). ◊ KISMET.

ORIENTAL FANTASY Any story set in the Far East and drawing upon indigenous beliefs, MYTHS, MAGIC or the supernatural is technically an OF, but a few more lines of definition can be drawn. Most OFs are set in a LAND-OF-FABLE Orient. Stories set in the contemporary East can be OFs if their veneer of REALITY is removed to reveal the fantastic, though this CHINOISERIE must be central to the story – otherwise they are merely Oriental thrillers. Initially OF and ARABIAN FANTASY were ill distinguished in Western eyes, especially as many stories in the *Arabian Nights* were drawn from more extensive lands than Arabia, including as far east as India. Most OFs are written by Western writers drawing upon Oriental imagery, myths and LEGENDS (◊ OCEAN OF STORY). Oriental writers may draw upon the same myths, and such stories are also OFs – although contemporary Oriental fantasists would no more tend to think of their works as OFs than would Europeans regard their fiction as "Occidental Fantasies".

Chinese literature flourished during the Tang Dynasty (AD618-907); a representative English-language selection of FAIRYTALES and SUPERNATURAL FICTIONS, most involving men or animals with TALENTS, is *The Dragon King's Daughter* (anth **1954**) ed/trans Yang Xianyi and Gladys Yang. Japanese scribes assembled two massive collections of CREATION MYTHS and hero tales in the 8th century as *Koji-Ki* and *Nihon-Gi*. The 9th-century *Yu Yang Ts Tsu* is a collection of Chinese FOLKTALES which includes the earliest known version of CINDERELLA. The best-known early OF is the 16th-century *Monkey* (trans Arthur Waley **1942** UK), compiled from nearly 1000 years of tradition by Wu

Chêng-ên (?1505-1580) (◊ MONKEY). Other Chinese myths, including many GHOST STORIES, were collected by P'U SUNG-LING in *Liao-Chai chih i* (**1679**; cut trans Herbert A. Giles as *Strange Stories from a Chinese Studio* 1908 UK; cut *Strange Stories from the Lodge of Leisures* 1913 UK), and Lafcadio HEARN retold many Chinese and Japanese legends in *Some Chinese Ghosts* (**1887** US), *In Ghostly Japan* (**1899** US) and *Kwaidan* (**1904** US), with a much wider Oriental selection in *Stray Leaves from Strange Literature* (**1884** US). Hearn's work formed the basis for several Japanese and Chinese movies, using spectacular spfx; the best are KWAIDAN (*1964* Japan) and *A Chinese Ghost Story* (*1987* Hong Kong), though the most remarkable of all Chinese fantasy movies is ZU: WARRIORS OF THE MAGIC MOUNTAIN (*1981*).

OFs became fashionable in the West after the success of the *Arabian Nights*, originally translated and published in France as *Les mille et une nuit* (**1704-17**) by Antoine Galland (1646-1715). Among the many similar collections assembled in its wake were three compiled by Thomas-Simon Gueullette (1683-1766), which used the same frame device. The individual titles were: *Les avantures merveilleuses du Mandarin Fum-Hoam; contes chinois* (coll **1723**; vt *Contes chinois, ou Les avantures merveilleuses du Mandarin Fum-Hoam* 1728; trans as *Chinese Tales, or The Wonderful Adventures of the Mandarin Fum-Hoam* **1725** UK; vt *The Transmigrations of the Mandarin Fum-Hoam, Chinese Tales* 1894 UK); *Les mille et un Quart-d'heure: Contes Tartares* (coll **1723**; trans as *One Thousand and One Quarters of Hours, being Tartarian Tales* ?**1726** UK; vt *Tartarian Tales; or a Thousand and One Quarters of Hours* 1759 UK); and *Les Sultanes de Guzarate, ou Les songes des hommes éveillés* (coll **1732**; vt *Les Mille et une soirées, Contes Mogols* 1765; trans as *Mogul Tales* **1736** UK). These were particularly influential in the UK and France, and for a while Arabian and Oriental stories became the fashion. Horace WALPOLE included a Chinese FAIRYTALE, "Mi Li", in *Hieroglyphic Tales* (coll **1785**), while Henry William Weber (1783-1818) compiled *Tales of the East* (anth **1812**), an immense three-decker set and the first fantasy ANTHOLOGY.

OFs drifted out of fashion during the height of GOTHIC FANTASY and the Romantic movement (◊ ROMANTICISM) – a rare example from the 19th century is "The Dragon-Fang Possessed by the Conjurer Piou-Lu" (1856 *Harper's*) by Fitz-James O'BRIEN – but interest returned with the rise in popularity of the fairytale and the wider exploration of folklorists (◊ FOLKTALES). By the turn of the century, with the success of *The Mikado* (1885) by W.S. GILBERT and Sir Arthur Sullivan and *Madam Butterfly* (1904) by Giacomo Puccini, Orientalism returned into fashion. With *The Wallet of Kai Lung* (coll of linked stories **1900**) Ernest BRAMAH launched the first of his long-running **Kai Lung** series set in a mythical China attuned more to Western minds than to Oriental tradition. The play *The Darling of the Gods* (**1902**) by David BELASCO and John Luther Long created a quasi-mythic Japan that so ensnared Lord DUNSANY that it turned him to writing his FABULATIONS, many with quasi-Oriental settings, starting with *The Gods of Pegana* (coll **1905**). Authors strongly influenced by Dunsany found themselves creating Oriental lands – including Kenneth MORRIS with stories collected as *The Secret Mountain* (coll **1926**) and more recently as *The Dragon Path* (coll **1995** US), H.P. LOVECRAFT in such tales as *The Cats of Ulthar* (1920 *The Tryout*; **1935** chap); Clark Ashton SMITH in "The

Willow Landscape" (1931 *Philippine Magazine*) and others; and Donald CORLEY in his collections *The House of Lost Identity* (coll **1927**) and *The Haunted Jester* (coll **1931**).

Most of these stories are set in a Land-of-Fable Orient. Other writers created fantasies set in the contemporary Orient, but peeling back the layers of REALITY. Best-known is Sax ROHMER, who appealed to the sinophobic mood of the West prevalent around the turn of the century (after the Boxer Rebellion of 1898) and created a Chinese villain in FU-MANCHU. Rohmer explored the medium further in *Tales of Chinatown* (coll **1922**) and *Tales of East and West* (coll **1932**; rev 1933 US). Stories showing the Chinese influence in LONDON and other cities (especially SAN FRANCISCO) are not necessarily OFs, but they can develop the elements of CHI-NOISERIE to create an almost mythical Oriental POLDER in the West. Like Rohmer, Thomas Burke (1886-1945) produced many such stories, collected as *Limehouse Nights* (coll **1916**), *Whispering Windows* (coll **1920**), *East of Mansion House* (coll **1926**) and *The Pleasantries of Old Quong* (coll **1931**). Stories of this type seldom age well, and it is only recently that a new generation of writers has been prepared to accept ethnic communities and explore the wealth of cultural diversity and knowledge they feature: a much-welcomed example was *Tea With the Black Dragon* (**1983**) by R.A. MACAVOY. The movie BIG TROUBLE IN LITTLE CHINA (*1986*) harvested the same territory.

Unfortunately, much OF of the first few decades of this century reflected the cultural stigmas and prejudices of the day (◊ MAGGOTS), when Orientals were regarded as the "Yellow Peril", a viewpoint which M.P. SHIEL had exploited, and it took work by writers with a closer affinity with the Orient to establish a balance. Their work began to emerge soon after WWI, and mostly in the US magazines. Leading the way was Achmed ABDULLAH, a Russian by birth but raised in India, who was able to infuse his fiction with authentic Oriental beliefs and create stories of mysticism and mystery for Western eyes. His best are collected as *Wings* (coll **1920**) and *Alien Souls* (coll **1922**). Starting with *The Ninth Vibration, and Other Stories* (coll **1922**) and *The Perfume of the Rainbow, and Other Stories* (coll **1923**), L. Adams BECK produced more romantic tales, mixing Oriental and Theosophical beliefs (◊ THEOSOPHY). Talbot MUNDY took this a stage further, creating action-packed Rohmer-like thrillers but utilizing Oriental LOST-WORLD landscapes; the best of these was *Om: The Secret of Abhor Valley* (**1924**). Mundy imitators, whose work occasionally showed greater originality, include H. BEDFORD-JONES, Harold LAMB and E. Hoffmann PRICE. Norvell W. PAGE, a highly prolific pulpster, wrote two OFs featuring PRESTER JOHN in *Flame Winds* (1939 *Unknown*; **1969**) and *Sons of the Bear God* (1939 *Unknown*; **1969**). A contemporary of theirs, but producing stories more in the vein of Bramah, was Frank OWEN, a popular contributor to WEIRD TALES, whose exotic, delicate and sometimes erotic work appeared steadily for over 30 years. Common to all of these authors was that their extensive travelling allowed them to bring experience and authenticity to their work. The same was true of Margaret YOURCENAR, whose stories in *Nouvelles Orientales* (coll **1938**; rev 1963; exp 1978; trans Alberto MANGUEL with the author as *Oriental Tales* 1985 UK/US) reflect an extensive knowledge of Oriental legends and culture. Similarly, Robert van Gulik (1910-1967), a Dutch ambassador to Japan, became fascinated with the traditional tales of the 7th-century Chinese magistrate Judge Dee. While on war

duties, he translated an anonymous 18th-century volume, *Dee Goong An*, into English as *Dee Goong An: Three Murder Cases of Judge Dee* (**1949** Japan; vt *Celebrated Cases of Judge Dee* 1976 US). Van Gulik then wrote new novels and stories about the character, most of them nonfantastic (though beautifully evocative); *The Haunted Monastery* (**1961** Malaya) comes closest to an OF.

Only a few of these authors continued to produce work into the 1950s – Price well into the 1980s, completing two excellent OFs towards the end of his life, *The Devil Wives of Li Fong* (**1979**) and *The Jade Enchantress* (**1982**) – but the interest in OFs faded after WWII and did not return in any strength until the revival of interest in fantasy had become established and settled down sufficiently to support more diversity. When it did return it displayed a greater richness and authenticity than before: not only were writers clearly in control of their subject matter but they were writing for a readership more widely attuned to the fantastic and generally educated beyond the perception of cultural stereotypes. As a result, the best OFs are those which have appeared since the end of the 1970s. These include: *The Rainbow Annals* (**1980**) and *Moonbird* (**1986**) by Grania DAVIS; the **Tomoe Gozen** series plus *Ou Lu Khen and the Beautiful Madwoman* (**1985**) by Jessica Amanda SALMONSON; the **Master Li** series by Barry HUGHART; the **Chia** series by Stephen MARLEY; and such singletons as *The Fairy of Ku-She* (**1988**) by M. Lucie CHIN, *Silk Road* (**1989**) by Jeanne LARSEN, *Imperial Lady* (**1989**) by Susan SHWARTZ and Andre NORTON, and *The Mask of the Sorcerer* (**1995**) by Darrell SCHWEITZER.

The OF has had an uneasy life, treading a fine line between the Western love of a mythical Orient and an inherent Western bigotry over a portion of the globe which existed almost entirely independent of the West for centuries. At its best it is exotic but unreal; at its worst it can serve only as a channel for a writer's or culture's prejudices. [MA]

ORLANDO UK movie (*1992*). Adventure Pictures (Orlando)/Lenfilm/Mikado/Rio/Sigma/British Screen. **Pr** Christopher Sheppard. **Dir** Sally Potter. **Spfx** Yuri Borovkov, Paul Corbould, Effects Associates, Viktor Okovitev. **Screenplay** Potter. **Based on** *Orlando: A Biography* (**1928**) by Virginia WOOLF. **Starring** Quentin Crisp (Elizabeth I), Jimmy Somerville (Singer/Angel), Tilda Swinton (Orlando), Charlotte Valandrey (Sasha), John Wood (Archduke Harry), Billy Zane (Shelmerdine). 93 mins. Colour.

1600, and the aged Elizabeth I adopts the Lord Orlando as a favourite, granting him a great estate in perpetuity on condition he does not grow old. Within a few years the Queen and Orlando's father are dead, and Orlando grimly affianced. In 1610, under James VI & I, a Muscovite Embassy is in London, and Orlando flouts convention by falling in love with the Ambassador's daughter, Sasha, but she spurns him. Cursing the perfidy of women, he embraces misogyny; then falls into a long sleep, presumably of revival, for already his agelessness is obvious. 1650 sees him as a poetry groupie, wishing for talent himself. In 1700 he is Ambassador to a Middle Eastern desert nation, pressed into service by the local Khan to help put down a rebellion. Appalled by the carnage, Orlando falls into another long sleep, this time wakening as a woman. Joining a refugee caravan she at last makes her way home, where she is received dubiously. In 1750 she finds herself the butt of that same misogyny she purveyed when male. Also, the law is moving to repossess her house, for as a woman she has no entitlement. Fleeing the offer of a marriage of convenience she runs through both the great house's LABYRINTH and the decades to find herself in 1850. There she abruptly encounters and as abruptly beds Shelmerdine, an adventurer and latter-day Tom Paine. As he departs (in a lush, Georgette Heyer bathos), she closes her eyes, opening them again to discover herself in WWI, heavily pregnant, stumbling through no-man's-land. She scrambles away to reach the later 20th century, where she tries to sell her autobiography and returns with her small daughter to the house, which is now – ornamental garden plants and all – under dustwraps. Moments later she is in the 1990s, where her daughter (still a child) videos her and the ANGEL who has come to sing to her that she is the union of male and female. "When she let go of the past," we are told, "she found her life was beginning."

O is both a subtle and surreal TIME FANTASY – visually and musically exquisite – and a continuing debate about the insubstantiality of physical form when set alongside the uniqueness of the self: "Same person," murmurs the TEMPORAL ADVENTURESS Orlando on wakening to find herself a woman; "No difference at all – just a different sex." This lack of any effect (beyond the superficial) in consequence of the sex-change subverts all IDENTITY-EXCHANGE tales, which assume the opposite. Sexual ambiguity pervades throughout: young Sasha, with whom the effeminate-seeming Lord Orlando falls in love, is a boyish woman; Elizabeth I is played by a man; all of the set-piece music is sung by males but in falsetto/countertenor; Shelmerdine and Orlando discuss how exchanging sexes would not change the roles they would wish to play in life, and it is she who both proposes marriage to him and seduces him, thereby playing what was certainly in 1850 the male role. The temporal surrealism is maintained through filmic techniques, with curious and not always logical cuts/fades between scenes, and with occasional anachronistic (but never incongruous-seeming) asides from Orlando to the camera. The whole is held together by an electric and luminously intelligent performance from Swinton, who is rarely off centre-screen.   [JG]

ORPHÉE (vt *Orpheus*) French movie (*1949*). André Paulvé/Films du Palais Royal. **Pr** Paulvé. **Dir** Jean COCTEAU. **Screenplay** Cocteau. **Based on** *Orphée* (*1927*) by Cocteau. **Starring** Maria Casarès (Death Princess), Marie Déa (Eurydice), Édouard Dermithe (Cégeste), Juliette Gréco (Aglaonice), Jean Marais (Orphée), François Périer (Heurtebise). 112 mins (usual version cut to 98 mins). B/w.

In France, soon after WWII, hugely popular poet Orphée is loathed by the trendy student Left Bank society. While in a café there he witnesses a brawl, at the end of which drunken young poet Jacques Cégeste is run over by a pair of mysterious motorcyclists. Cégeste's mysterious companion, known only as the Princess, commands that Orphée come with her and the corpse in her car (chauffeured by the ANGEL Heurtebise) to a gloomy château; there, partly witnessed by Orphée, she raises Cégeste from the dead and leads him, Heurtebise and the two motorcyclists through a MIRROR to the UNDERWORLD. Back home, Orphée becomes interested in enigmatic messages which he can pick up on the Princess's car radio; these he believes to be brilliant poetry, but when he sends them to a friend they are identified as unpublished fragments by the dead Cégeste, and rumours swell. Meanwhile the Princess – Orphée's personal DEATH – has taken to visiting him as he sleeps, for she is in love with him. Soon she has her motorcycling aides kill Eurydice. With Heurtebise's guidance, Orphée dons Death's forgotten gloves and probes through the mirror into the Underworld, seeking either Eurydice or the Death Princess – he knows not which. In the Underworld a tribunal reprimands Death and all her aides for their misbehaviour, and decrees that Orphée (who has now exchanged oaths of undying love with Death) and Eurydice may return to the land of the living provided he never claps eyes on her again. But one day, still compulsively jotting down the radio messages, he accidentally catches sight of her in the car's rear-view mirror. Before he can react, the lynch-mob arrives, and he is killed in the scuffling. He rejoins Death in the Underworld, but her love is greater than his, for, though she knows it means her own ultimate extinction, she sends him back through time to a point from which he and Eurydice may live happily ever after.

O is not an unambiguous movie; any attempt to produce a definitive interpretation is doomed. Rather, it provides a tangle of meanings and half-meanings from which the viewer is invited to choose. This ambiguity – whether wilful or profound – penetrates to every level; Marais, the archetypally heterosexual Orphée, was in fact Cocteau's lover, a fact which (deliberately) invests the screenplay with an additional layer of meaning. The pacing and cinematography are stunning, and the mood of the movie generally almost overpowering, although occasionally punctured by passages where flat narration replaces performance. Nevertheless, *O* is memorable.   [JG]

See also: *Le* TESTAMENT D'ORPHÉE (*1959*).

ORPHEUS A Thracian MINSTREL, son of the MUSE Calliope, whose playing was said to charm wild beasts, trees and even stones. He was the supposed founder of the enigmatic cult of Orphism which flourished in the 6th century BC. He travelled with Jason on the Argo (◊ GOLDEN FLEECE) and saved the ship's crew by drowning out the song of the SIRENS; when his wife Eurydice was bitten by a snake he went to HADES to reclaim her, charming the dead with his music and winning her a reprieve which he lost again by looking back too soon as he emerged from the UNDERWORLD.

Orpheus is a key symbolic figure handed down by Classical MYTHOLOGY to literary fantasy, especially in relation to the power of MUSIC. "Orpheus in Mayfair" (1909) by Maurice BARING describes the effect of Orpheus's playing on an unwary politician. His ill-fated journey into the Underworld has been frequently recast and reinterpreted; one of the earliest examples of such a reconfiguration is the 14th-century poem *Sir Orfeo*, which replays the story in the context of Anglo-Norman feudalism, and it became a favourite subject for OPERAS, including Monteverdi's *L'Orfeo* (*1607*), Gluck's *Orfeo ed Euridice* (1762) and Offenbach's *Orpheus in the Underworld* (1858). Another notable dramatic reconfiguration is Jean COCTEAU's *Orphée* (*1927*); Cocteau directed his own movie version, ORPHÉE (*1949*). Stories involving similar reworkings include "The Miracle" (1921) by John BERESFORD, which reverses the sex-roles, the third of "The Centenarist's Tales" in *The Salzburg Tales* (coll *1934*) by Christina Stead (1902-1983), *The Golden Age* (*1975*) by Constantine FitzGibbon (1919-1983) and *The Medusa Frequency* (*1987*) by Russell HOBAN. Sf transfigurations include *Wolfhead* (*1978*) by Charles L. Harness (1915-   ) and *Dinner at Deviant's Palace* (*1985*) by Tim POWERS.   [BS]

ORR, A. Working name of US writer Alice Ingram Orr Sprague (1950-   ), whose fantasy consists of the **World in Amber** sequence – *The World in Amber* (**1985**) and *In the Ice King's Palace* (**1986**) – about a world set literally in a drop of amber and centring on the MINSTREL/WIZARD Isme and his constant efforts to maintain BALANCE as discords strike. A third volume has for some time been anticipated.   [JC]

ORRERY A clockwork mechanism which represents the movement of the planets around the Sun. It was invented *c*1700 by George Graham (1673-1751), who presented one to Charles Boyle (1676-1731), fourth Earl of Orrery; by 1713 the device was known as an orrery. In fantasy – especially INSTAURATION FANTASIES like John CROWLEY's *Little, Big* (**1981**) and Elizabeth HAND's *Waking the Moon* (**1994**) – orreries are likely to represent the BALANCE of the world (◊◊ AS ABOVE SO BELOW), and their behaviour, certainly if disrupted, will tend to mark the promise or threat of transformative change. Even in SECONDARY-WORLD fantasies like Lisa A. Barnett's and Melissa SCOTT's *Point of Hopes* (**1995**) the presence of an orrery signals a probable threat to the balance of things.   [JC]

ORTIZ, JOSÉ Working name of Spanish COMICS artist José Ortiz Moya (1932-   ), whose energetic drawing style and prodigious output have made him one of the foremost talents in the field. His work has great vitality and he draws with precision in a deceptively informal fine pen line, freely augmented with a brush. He is an expert in creating dramatic atmospheres with a wide variety of textures and patterns, subtly tuned to the mood of each story he draws. These have included war stories, Westerns, children's FAIRYTALES, the goriest of horror pieces, humorous sf and historical and fantasy subjects, all of which he has tackled with consummate skill. He is said to be one of the fastest draughtsmen in the business.

JO's first work, at age 16, was for the Spanish comics publisher Maga. His early strips included **Balin**, **Pantera Nera** ["Black Panther"] and **Sigur el Vichingo** ["Sigur the Viking"]. In the early 1960s he worked for UK publishers through the agency Bardon Press Features, producing artwork for Fleetway's digest-size WWII comic books and for an sf series in colour for *Eagle* entitled **UFO Agent**. He also drew the newspaper strip **Carol Baker** for the *Daily Mirror* plus several stories for the teenage romance title *Valentine* and the children's fairytale magazine *Once Upon a Time*.

He created a kung fu action series entitled **Tse Khan** (1971-2) for the Spanish agency Selecciones Illustradas, *El Niño Salvaje* ["The Little Savage"] (**1973**), and embarked on a long series of true-story Westerns: **Los Grandes Mitos del Oest** ["Great Legends of the West"] (1973-4). He began working for WARREN PUBLISHING in 1974, drawing adaptations of stories by Edgar Allan POE and creating several long-running horror, sf and fantasy series including the very gory **Jackass** (*Eerie* 1974-5), the samurai fantasy **Skallywag** (*Eerie* 1976-7), **Moonshadow** (*Eerie* 1978) and **The Four Horsemen of the Apocalypse** (*Eerie* 1975). Meanwhile he created **Hombre** ["Man"], a gritty postholocaust series in colour for Norma Editorial, and drew several stories featuring TARZAN. He returned to the UK *Eagle* comic with the medieval sf saga **The Tower King** (1982) plus **The House of Daemon** (1982-3), **The Fifth Horseman** (1983-4) and several more.

The Spanish comics market was revitalized in the 1980s with the creation of fantasy comics like *Cimoc*, *Totem* and *Zona 84*, for which JO created the hard-boiled cop story

Morgan (1985; trans in *Aces* 1988 UK), **Jack the Ripper** (1989) and **Burton and Cyb** (1991), a humorous sf series. He produced two colour albums in the series published to mark the 500th anniversary of Columbus's discovery of America, **Relatos del Nuevo Mundo** ["Stories of the New World"] (1992).

JO currently (1996) works mainly for the Italian publisher Sergio Bonelli Editore, for which he draws the westerns **Tex** and **Ken Parker**.   [RT]

ORU KAIJU DAISHINGEKI ot of *Godzilla's Revenge* (**1969**). ◊ GODZILLA MOVIES.

ORWELL, GEORGE Working name of UK writer, journalist and polemicist Eric Arthur Blair (1903-1950). His sole work of pure fantasy is the short novel *Animal Farm: A Fairy Story* (**1945** chap). This uses the BEAST-FABLE format as a vehicle for a savagely ironical ALLEGORY of how socialist political ideals – which GO emphatically shared – became corrupted in the USSR. The animals of the eponymous farm revolt, expel their human master, and joyously outline a perfect society of comrades where "All animals are equal". But inexorably an administrative class emerges: the Pigs, who, led by "Napoleon" (Stalin), rapidly become a ruling class while still preaching equality. The master-stroke of political SATIRE is the debasement of the original SEVEN revolutionary Commandments to a single line: "All Animals Are Equal But Some Animals Are More Equal Than Others." More viscerally moving, though, is the closing scene of RECOGNITION in which the exploited animals see their porcine leaders – who have already taken to walking on two legs – entertaining other, human farmers, from whom they have become indistinguishable. *Animal Farm*'s publication was bitterly resented by pro-communists who lacked what GO called his "power of facing unpleasant facts"; but the fable endures and has remained in print ever since. The movie ANIMAL FARM (*1955*), while generally faithful to the novel, adds a final counter-revolution to EUCATASTROPHE.

GO also discussed the manipulation of PERCEPTION through language (◊◊ DICTION) in essays and in his famous dystopian sf novel *Nineteen Eighty-Four* (**1949**), whose "Newspeak" is designed to impose linguistic BONDAGE by making dissident ideas impossible to formulate. A piercing clarity of thought and expression characterizes virtually all GO's work.   [DRL]

Other works: *Orwell: The War Broadcasts* (coll **1985**) ed W.J. West includes GO's adaptation for BBC radio of Hans Christian ANDERSEN's "The Emperor's New Clothes" (broadcast 1943).

O'SHEA, PAT (1931-   ) Irish writer, resident in the UK since 1947, whose *The Hounds of the Mórrigan* (**1985**) is a successful CELTIC FANTASY written ostensibly for children (◊ CHILDREN'S FANTASY) but widely read by adults. The boy hero's discovery of the evil SERPENT Olc-Glas from Irish myth, bound (◊ BONDAGE) by St Patrick into a picture in a BOOK, precipitates a prolonged, colourful and breathlessly inventive chase across Ireland with the MORRIGAN, her two blackly comic WITCH aspects (making THREE; ◊ GODDESS) and her menacing but inept Hounds in pursuit. There is much CROSSHATCHING between modern Ireland and mythic TIR-NAN-OG, and TIMESLIP detours through earlier (e.g., Victorian) times. Numerous and varied TALKING ANIMALS – including CATS, BIRDS and even insects, among them a Napoleonic earwig – assist the fleeing boy and his sister, as do CUCHULAIN, Queen Maeve, the druid Cathbad, and such Irish GODS as the LOVE god Angus Óg and the all-powerful

Dagda. After dramatic obstacles, including a magic LABYRINTH, all ends well with the destruction of Olc-Glas and the Mórrígan's discomfited return to her role as queen of GHOSTS ... although the Dagda's reward of rainbows hardly compensates the children for their traditional MEMORY WIPE. *Finn Mac Cool and the Small Men of Deeds* (**1987**) is a TWICE-TOLD fable (◊ FINN MAC COOL). [DRL]

"OSSIAN" Quasi-pseudonym of Scottish writer James Macpherson (1736-1796). ◊ CELTIC FANTASY.

O'SULLIVAN, VINCENT (1868-1940) US-born writer long resident in Europe. He was a central figure of the English Decadent Movement (◊ DECADENCE), his early books of (mostly fantastic) verse – *Poems* (coll **1896**) and *The Houses of Sin* (**1897**) – being among the movement's key documents, as was his first story collection, *A Book of Bargains* (coll **1896**). The supernatural stories contained herein and in *A Dissertation upon Second Fiddles* (coll **1902**) are combined with several uncollected tales in *Master of Fallen Years: Complete Supernatural Stories of Vincent O'Sullivan* (coll **1995**) ed Jessica Amanda SALMONSON. They include some fine WEIRD FICTION; those of most fantasy interest are "The Monkey and Basil Holderness" (1895) and "Master of Fallen Years" (1921). [BS]

Other works: *Human Affairs* (coll **1907**).

OTHER, THE US movie (*1972*). ◊ Thomas TRYON.

OTHER CINDERELLA, THE US movie (*1977*). ◊ CINDERELLA (*1950*).

OTHERWORLD For many critics and readers, there is little or no distinction between an otherworld and a SECONDARY WORLD. The two terms are indeed often used interchangeably in this encyclopedia to designate autonomous worlds that are not *bound* to the mundane world (the lands of the dead from which GHOSTS make their forays, and other territories common to SUPERNATURAL FICTION, are therefore excluded) and which are *impossible* in terms of our normal understanding of the sciences and of history (the worlds described in PLANETARY ROMANCES are not, for instance, otherworlds or secondary worlds) and which are *self-coherent* in terms of STORY (that is, they exist in a time and place and REALITY in which stories can be told, believed and returned to – like the Middle-Earth created by J.R.R. TOLKIEN).

There are, however, two distinctions which may be made; each should be clear in context. (a) The term "otherworld" may refer to any sort of autonomous impossible world, including FAERIE and WONDERLANDS, while the secondary world is not normally thought of as being governed by the arbitrary rules that, for instance, operate the wonderlands of Lewis CARROLL. (b) A secondary world may or may not have some sort of connection to the mundane world through, for instance, the PORTALS and CROSSHATCHES which are so common in fantasy, but otherworlds certainly have some sort of connection with mundane reality. C.S. LEWIS's **Narnia** is a secondary world; but its protagonists are constantly toing and froing between one world and the other, and so it can also be described as an otherworld. [JC]

OTOMO, KATSUHIRO (1954-) Japanese COMICS illustrator and movie animator, a central figure in the world of MANGA. His work incorporates both sf (◊ SFE) and fantasy elements, though with an emphasis on the former in his most important title, the ongoing **Akira** sequence, which first appeared in serial form in 1982-6, and again from 1988; it has appeared in English from 1988. Several GRAPHIC NOVELS have been calved from the ongoing epic, and it inspired

a movie, AKIRA (*1987*), which KO wrote and directed. Of greater fantasy interest is the graphic novel, *Dohmu* ["A Dream of Childhood"] (1981; **1983**; trans **1992**), in which a SEVEN-SAMURAI band of children come together to fight off an older man with killing TALENTS. [JC]

OUTCOULT, R.F. (1863-1928) US comics pioneer. ◊ COMICS.

OUTLANDS ◊ MAGAZINES.

OUTLAW OF GOR Italian movie (*1987*). ◊ John NORMAN.

OUTLAWS US tv series (1987). CBS. **Starring** Christine Belford (Lt Maggie Randall), Patrick Houser (Billy Pike), William Lucking (Harland Pike), Charles Napier (Wolfson "Wolf" Lucas), Richard Roundtree (Isaiah "Ice" McAdams), Rod Taylor (Sheriff Jonathan Grail). Colour.

As Sheriff Grail chases four fleeing outlaws after a robbery in 1899, the five ride into a strange electrical storm and TIME TRAVEL to 1986 Houston. Grail convinces the gang to go straight once they realize where and when they are. Using the money from the robbery the five buy a ranch and open the Double Eagle Detective Agency. With weapons and techniques of the past, they set out to solve the crimes the Houston Police Department cannot handle. [BC]

OUT OF THE BLUE US tv series (1979). ABC. **Pr** William Bickley, Michael Warren. **Exec pr** Austin Kalish, Irma Kalish. **Dir** Peter Baldwin, Jeff Chambers, John Tracy. **Writers** Howard Albrecht, William Bickley, Laurie Gelman, Barry Kemp, Michael Warren. **Created by** Robert L. Boyett, Thomas L. Miller. **Starring** Olivia Barash (Laura Richards), Clark Brandon (Chris Richards), James Brogan (Angel Random), Dixie Carter (Marion MacNelmor), Hannah Dean (Gladys), Eileen Heckart (Boss Angel), Jason Keller (Jason Richards), Shabe Keller (Shane Richards), Tammy Lauren (Stacey Richards). 8 30min episodes. Colour.

When the Richards CHILDREN lose their parents in a plane crash, a guardian ANGEL called Random is assigned to watch over them. Random becomes their lodger and high-school teacher, revealing his secret only to the children. The short-lived sitcom centred on Random's efforts to help them through either good advice or judicious MAGIC. [BC]

OUTWARD BOUND US movie (*1930*). Warner/ Vitaphone. **Dir** Robert Milton. **Spfx** Fred Jackman. **Screenplay** J. Grubb Alexander. **Based on** *Outward Bound* (play 1925; novel **1929**) by Sutton VANE. **Starring** Helen Chandler, Dudley Diggs, Douglas Fairbanks Jr, Alec B. Francis, Leslie Howard, Montagu Love, Beryl Mercer, Allison Skipworth, Lyonel Watts (movie has no proper credits). 82 mins. B/w.

Stagebound, hammily acted version of a preachy play, of interest more for its concept of the AFTERLIFE than for its minimal narrative virtues. Thwarted London-suburb lovers Henry and Ann suicide and find that LIMBO is a nearly deserted liner where only they and the steward, Scrubby, realize all are dead. Gradually the secret leaks out as the SHIP OF FOOLS nears the Celestial City, which looks through the fog much like Manhattan. The passengers are interviewed and judged by a boarding Examiner – who has the subtlety and charisma of a tv evangelist. Henry and Ann, as suicides – like Scrubby – are sent back, Henry to be returned to life and Ann to remain forever shipbound; yet their love is great enough that Ann, too, regains life.

This POSTHUMOUS FANTASY was remade as *Between Two Worlds* (*1944*), set in WORLD WAR II, with Ann and Henry married to each other and the rest of the passengers having been killed in an air-raid *en route* to boarding a London-New York liner. The tale – especially the judgement sequence – is tiresomely extended, although Prior (here a hardbitten journalist) and Scrubby (Edmund Gwenn) contrive to hold the attention. In the finale, Ann is to be allowed to return to life, but prefers Limbo to separation from Henry; the reward for her virtue is that both are resuscitated. [JG]

OVID Generally used name for the Latin poet Publius Ovidius Naso (43BC-*c*AD17). Born in Sulmona, he early established himself among the Roman gentry, then turned single-mindedly and prolifically to writing; most of his copious work has survived.

His importance to fantasy rests chiefly on two works of his middle age. *Fasti* ["Festivals"] (*c*AD8; part rev *c*AD15) is either uncompleted or partly lost. We have six books, each surveying the Roman sacred CALENDAR for a single month. In using the elegiac metre, Ovid imitated his presumed Greek model, Callimachus's *Aitia* (◊ GREEK AND LATIN CLASSICS); but he retained much of the highly personal tone specific to Latin elegy, tending to vitiate the myths even in this, his most respectful treatment of them.

Metamorphoses (AD8), Ovid's longest work, also bears the form of a handbook, though this time in epic metre and with epic's greater distance. Not reverence, however: no other ancient epic, certainly, begins with an invocation to the poet's own soul. The poem's topic is METAMORPHOSIS, and (despite occasional bows to science or history) primarily the TRANSFORMATIONS of the characters of MYTH into birds, trees, etc. These stories are told so sweetly that the gradual revelation of the author's basic chill comes as an increasing shock; though LOVE is ever-present in the *Metamorphoses*, it very often means rape. The epic begins with the creation and ends with the Emperor Augustus, and remains consistently narrative; its first few digressions fit Callimachus's aesthetic unexceptionally. But the digressions multiply and intertwine, seemingly endlessly, until one can read only in constant awareness of the author's interventions. His open scepticism makes this a founding work of FANTASY.

Metamorphoses is, in any event, one of the most influential works in Western literature. It is at once the most comprehensive account of Latin MYTHOLOGY and a feat of fictional construction which taught innumerable later writers, among them CHAUCER and DANTE. In the shorter term, Lucius Annaeus SENECA and Publius Papinius STATIUS learned much from Ovid's style – the supplest in Latin literature, capable of rendering any mood or thought in its smooth flow – while turning his insouciant fictional voice toward now more familiar forms, the fictional mood of HORROR or the fictional world of GENRE FANTASY. [JB]

OWEN, FRANK (1893-1968) US author and editor, husband of the writer Ethel Owen. He wrote 10 mildly salacious novels in the 1930s as Roswell Williams, sometimes listed erroneously as his real name. FO is best-known for mannered tales of the Orient (◊ ORIENTAL FANTASY). *The House Mother* (**1929**) and *Rare Earth* (**1931**) are only marginally fantastic, and *The Scarlet Hill* (**1941**) has no supernatural element at all, but the fabular stories in *The Wind that Tramps the World: Splashes of Chinese Color* (coll **1929**) and *The Purple Sea: More Splashes of Chinese Color* (coll **1930**) are mostly fantasies of a delicately charming kind. Several, including the FEMME-FATALE story "The Tinkle of the Camel's Bell" (1928), had earlier appeared in WEIRD TALES, where FO had made his debut with "The

Man who Owned the World" (1923); others, including *Pale Pink Porcelain* (**1929** chap), were subsequently reprinted there. There is less fantasy in *Della Wu, Chinese Courtezan, and Other Oriental Love Tales* (coll **1931**) and *A Husband for Kutani* (coll **1938**), but the latter includes "Doctor Shen Fu", about a Chinese alchemist who possesses the ELIXIR OF LIFE. Shen Fu features also in "For Tomorrow We Die" (1942), "The Man who Amazed Fish" (1943) and "The Old Gentleman with the Scarlet Umbrella" (1951). Although FO continued publishing in *WT* and elsewhere until 1952, few stories appearing after 1938 or featuring non-Oriental settings have been reprinted; all but two of the stories in *The Porcelain Magician* (coll **1948**) are drawn from earlier collections.

FO co-wrote several children's collections with his wife, including *Coat Tales from the Pockets of the Happy Giant* (coll **1927**), *The Dream Hills of Happy Country* (coll **1928**), *Windblown Stories* (coll **1930**) and *The Blue Highway* (coll **1932**). Ethel Owen's solo works include *The Pumpkin People* (**1927**) and *Hallowe'en Tales and Games* (coll **1928**). [BS]

Other works: *Madonna of the Damned* (**1935**) as Roswell Williams; *Lovers of Lo Foh* (**1936**) as Williams.

As editor: *Murder for the Millions: A Harvest of Horror and Homicide* (anth **1946**); *Fireside Mystery Book* (anth **1947**); *Teen-age Mystery Stories* (anth **1948**).

OZ The first successful – and still the most famous – US fantasy OTHERWORLD, created by L. Frank BAUM in *The Wonderful Wizard of Oz* (**1900**; vt *The New Wizard of Oz* 1903). SEQUELS BY OTHER HANDS are numerous (see below). Oz is a rectangular, landlocked POLDER in the heart of the USA, and stands in vivid contrast to the Great Plains which presumably surround it, though the geography which connects this GARDEN to the circumambient desert is – albeit detailed by Baum – clearly imaginary, and turns anyone who trespasses upon it into dust (Dorothy arrives initially by air). To the north of Oz is Impassable Desert, to the east Shifting Sands, to the west Deadly Desert and to the south a Great Sandy Waste. Oz itself is divided into four colour-coded provinces (◊ COLOUR CODING): northern, forested Gillikin Country is purple; eastern, quaintly urban Munchkin Country is blue; western, windblown Winkie Country is yellow; and southern Quadling Country, inhabited by eccentrics, is red. At the heart of Oz is the Emerald City. The inhabitants of Oz are likewise colour-coded, their clothes and their skins reflecting their region of origin.

Oz is governed by Princess Ozma, though later volumes of Baum's sequence lay down some utopian principles for the organization of the land (◊ UTOPIAS); and Ozma – who is a classic UGLY DUCKLING (◊ HIDDEN MONARCH), having been "disguised" from birth as a poor boy named Tip and transformed back into a girl when it comes time for her to assume the throne (◊ GENDER DISGUISE) – rules in fact more through acclamation than by anything resembling hereditary right. The inhabitants of Oz are variously creatures of fantasy, though none of them reflect the European FAIRYTALE tradition from which J.R.R. TOLKIEN evolved his cast and which was commandeered by US GENRE-FANTASY writers from the late 1970s in their creation of stories set in FANTASYLAND. The creatures of Oz are sometimes generated by MAGIC rules (of a sort more commonly found in 19th-century fantasy than later), often through acts of animation (◊ ANIMATE/INANIMATE) that produce beings – like the Glass CAT, or the Scarecrow – whose physical characteristics directly reflect their natures. Magic itself – even the rationalized magic Baum clearly found comfortable – is banned, adding to the sense of Oz as a secular paradise. Though its surrounding desert is treated as impenetrable from the first volumes of the sequence, Baum clearly felt his polder needed a more thorough magical protection, and a later volume tells us that one of the four resident WITCHES – at Ozma's request – sets a series of SPELLS to make Oz invisible from the air and impossible to find by land.

Populous with eccentric beings whose greatest fulfilment is being themselves, rendered eternally safe from the wilderness of the world, Oz continues to represent what might be called the daydream of the MATTER of America. It is the dream that we will not spoil the territory through the act of discovering it. Oz can be found, by children; but it cannot be desecrated.

After Baum's death, the **Oz** sequence was continued by other hands, the first and most important of these being Ruth Plumly Thompson (1891-1976), who contributed *The Royal Book of Oz* * (**1921**) as by Baum, *Kabumpo in Oz* * (**1922**), *The Cowardly Lion of Oz* * (**1923**), *Grampa in Oz* * (**1924**), *The Lost King of Oz* * (**1925**), *The Hungry Tiger of Oz* * (**1926**), *The Gnome King of Oz* * (**1927**), *The Giant Horse of Oz* * (**1927**), *Jack Pumpkinhead of Oz* * (**1929**), *The Yellow Knight of Oz* * (**1930**), *Pirates in Oz* * (**1931**), *The Purple Prince of Oz* * (**1932**), *Ojo in Oz* * (**1933**), *Speedy in Oz* * (**1934**), *The Wishing Horse of Oz* * (**1935**), *Captain Salt in Oz* * (**1936**), *Handy Mandy in Oz* * (**1937**), *The Silver Princess in Oz* * (**1938**) and *Ozoplaning with the Wizard of Oz* * (**1939**); plus two late titles, *Yankee in Oz* * (**1972**) and *The Enchanted Island of Oz* * (**1976**). John Rea Neill, Baum's preferred illustrator from almost the beginning of the series, wrote *The Wonder City of Oz* * (**1940**), *The Scalawagons of Oz* * (**1941**), *Lucky Bucky in Oz* * **1942**) and *The Runaway in Oz* * (written 1943; **1995**). Jack Snow wrote *The Magical Mimics in Oz* * (**1946**), *The Shaggy Man of Oz* * (**1949**) and *Who's Who in Oz* * (**1954**). Further titles include: *The Hidden Valley of Oz* * (**1951**) by Rachel Cosgrove (1922-   ), who later wrote as E.L. Arch; *Merry-Go-Round in Oz* * (**1963**) and *The Forbidden Fountain of Oz* * (**1980**) by Eloise Jarvis McGraw (1915-   ) and Lauren McGraw Wagner; *A Barnstormer in Oz* * (**1982**) by Philip José FARMER; *Ozma and the Wayward Wand* * (**1985**) by Polly Berends (1939-   ); *Dorothy and the Seven-Leaf Clover* (**1985**) by Dorothy Haas; *Mister Tinker in Oz* * (**1985**) by James HOWE; *Dorothy and the Magic Belt* * (**1985**) by Susan Saunders (1945-   ); and *Return to Oz* * (**1985**) by Joan D. VINGE.

The influence of Oz as a model for US fantasy is vast but diffuse, though some earlier texts show its more overt effect; a typical example is William Rose BENÉT's *The Flying King of Kurio* (**1926**).   [JC]

OZICK, CYNTHIA (1928-   ) US novelist and critic, best-known for work published outside fantasy. Although she writes in the mainstream of literary modernism – her first novel, *Trust* (**1966**), takes to heart the solemn abjurations of Henry JAMES, and CO has written ironically but feelingly about her youthful devotion to the "mandarin" standards of "High Art" – she has shown no interest in genre fiction. A second influence in CO's work, the tradition of Yiddish literature (often based on fables, FOLKTALES and other fantastic sources) that ran from Sholem Aleichem to Isaac Bashevis SINGER, is evident in much of her short fiction, which frequently approaches the fantastic. The narrator of "The Pagan Rabbi" (*Hudson Review* 1966) reads the journal of a pious rabbi who fell in love with a DRYAD; as the rabbi was eventually a suicide, his account may be rationalized as delusional. "The Dock-Witch" (in *The Pagan Rabbi and Other Stories* coll **1971**) relates the tale of a man's obsessive affair with a woman who may be an UNDINE. Most of the stories in *Bloodshed and Three Novellas* (coll **1976**) and *Levitation: Five Fictions* (coll **1982**) similarly admit, without always confirming, some element of the supernatural: GOLEMS, the AFTERLIFE, magical powers. CO's short fictions are usually novellas; they are exactingly wrought, often with poised ambiguity (the influence of James can be inferred) regarding their fantasy content, or the reliability of their narrators; they have won numerous awards.   [GF]
Other works: *The Cannibal Galaxy* (**1983**), not sf; *The Messiah of Stockholm* (**1987**), not fantasy; *The Shawl* (coll of linked stories **1989** chap), two stories featuring a perhaps magical shawl. CO's nonfiction is assembled in *Art & Ardor: Essays* (coll **1983**), *Metaphor & Memory: Essays* (coll **1989**), and *What Henry James Knew and Other Essays on Writers* (coll **1993** UK).

PACTS WITH THE DEVIL One of the most popular themes in fantasy involves humans making deals with SATAN or his minions for wealth, IMMORTALITY or (borrowing from FAIRYTALE) the granting of a specified number of WISHES; the vigour of such stories is maintained by the ingenuity of the narrative twists invoked to bring the deals to unexpected conclusions. The idea of such pacts emerged from a medieval cautionary tale about a Bishop named Theophilus, but was then applied by theologians to the analysis of heresies and integrated into the abusive fantasies used to justify the persecution of heretics. All practitioners of WITCHCRAFT were assumed to have made such pacts, but theirs tended to be analogous to mere contracts of employment; it was the more grandiose CONTRACT made by FAUST that became the main prototype of literary pacts.

Diabolical pacts were a commonplace of Gothic HORROR, but the sealing of the pact and the devil's gloating return to claim his prize are there deployed as parentheses enclosing elaborate accounts of villainy and debauchery. In fantasy, by contrast, Satan and his agents tend to be portrayed as sharp operators whose cunningly worded contracts remain the focus of attention throughout while both parties engage in a contest of wits. The enigmatic Man in Gray who gives a bottomless purse for a shadow in Adalbert von CHAMISSO's *Peter Schlemihl* (**1814**) is an early prototype of the deceptive bargainer who achieves his end by subtle means. The dark sententiousness of *Melmoth the Wanderer* (**1820**) by Charles MATURIN (◊◊ GOTHIC FANTASY) gave way to the sarcastic SATIRE of Honoré de BALZAC's "Melmoth Reconciled" (1835), and there is a similarly gleeful irony in quasi-anecdotal fantasies like Washington IRVING's "The Devil and Tom Walker" (1824), the tale of Chips in Charles DICKENS's *The Uncommercial Traveller* (coll **1860**) and J. Sheridan LE FANU's "Sir Dominick's Bargain" (1872). A melodramatic element survived in such luridly flamboyant works as *The Necromancer* (**1857**) by G.W.M. Reynolds (1814-1879), *The Sorrows of Satan* (**1895**) by Marie CORELLI and *The Lord of the Dark Red Star* (**1903**) by Eugene Lee-Hamilton (1845-1907), but the main emphasis shifted to comedy, as in *The Gentleman in Black* (**1831**) by James DALTON, "An Episode in the Life of Mr Latimer" (1883) by

Walter Herries POLLOCK, "The Demon Pope" (1888) by Richard GARNETT, *A Deal with the Devil* (**1895**) by Eden PHILLPOTTS, *The Devil and the Inventor* (**1900**) by Austin Fryers, "Enoch Soames" (1919) by Max BEERBOHM, *The Devil in Woodford Wells* (**1946**) by Harold HOBSON and *The Missing Angel* (**1947**) by Erle Cox (1873-1950). In the best of the many 20th-century examples of the pact motif, however, comedy remains a frivolous gloss on more serious philosophical matters, as it is in Stephen Vincent BENÉT's "The Devil and Daniel Webster" (1937) – filmed as *The* DEVIL AND DANIEL WEBSTER (**1941**) – John COLLIER's several variations on the theme, Robert BLOCH's "That Hellbound Train" (1958), and such froth as *The* DEVIL AND MAX DEVLIN (**1981**) or the more serious *Phantom of the Paradise* (**1974**; ◊ PHANTOM OF THE OPERA). Only a few modern stories contrive to maintain an authentic sense of threat: *The Devil in Velvet* (**1951**) by John Dickson CARR, *The Devil and Mrs Devine* (**1974**) by Josephine LESLIE and *The Devil's Game* (**1980**) by Poul ANDERSON are only half-successful; Leon GARFIELD's surreal children's book *The Ghost Downstairs* (**1972**) does much better.

An excellent theme anthology is *Deals with the Devil* (anth **1958**) ed Basil Davenport (1905-1966). [BS]

PADGETT, LEWIS Pseudonym used by Henry KUTTNER and C.L. MOORE.

PAGE, NORVELL W(OOTEN) (1904-1961) US pulp-MAGAZINE writer who published a considerable amount of material as by Grant Stockbridge and a lesser amount as by Randolph Craig. His two best-known works as NWP are *Flame Winds* (1939 *Unknown*; **1969**) and *Sons of the Bear-God* (1939 *Unknown*; **1969**), SWORD-AND-SORCERY adventures concerning PRESTER JOHN. Under the Stockbridge house name NWP wrote about 90 short-novel-length adventures of the **Spider** – the alias of millionaire socialite Richard Wentworth, who regularly transformed himself into a cloaked-and-fanged avenger (◊ MAGAZINES). Those eventually published in book form include *Wings of the Black Death* (1933 *The Spider*; **1969**), *City of Flaming Shadows* (1934; **1970**), *Builders of the Black Empire* (1934; **1980**), *The City Destroyer* (1935; **1975**), *Hordes of the Red Butcher* (1935; **1975**), *Master of the Death Madness* (1935; **1980**), *Overlord of* the Damned (1935; **1980**), *Death Reign of the Vampire King* (1935; **1975**) and *Death and the Spider* (1942; **1975**) – in this last a supernatural personification of DEATH threatens the USA. As Craig, NWP wrote a couple of weird/fantastic thrillers featuring **Dr Skull**: *The City Condemned to Hell* (1939 *The Octopus*; **1975** chap) and *Satan's Incubators* (1939 *The Scorpion*; **1975** chap). An archetypal pulpster, whose work was more energetic and flamboyant than most others', NWP ceased writing fiction in 1943. [DP]

PAGEMASTER, THE US live-action/ANIMATED MOVIE (*1994*). 20th Century-Fox/Turner. **Pr** Paul Gertz, Michael R. Joyce, David Kirschner. **Dir** Joe Johnston. **Spfx** Bob Hill, Al Magliochetti, Philip Meador. **Vfx** Dennis Skotak, Robert Skotak, Richard T. Sullivan. **Anim dir** Maurice Hunt. **Screenplay** David Casci, Ernie Contreras, Kirschner. **Starring** Macaulay Culkin (Richard Tyler), Christopher Lloyd (Mr Dewey). **Voice actors** Culkin, Jim Cummings (Long John Silver), Whoopi Goldberg (Fantasy), Lloyd (Pagemaster), Leonard Nimoy (Jekyll/Hyde), Patrick Stewart (Adventure), Frank Welker (Horror). 75 mins. Colour.

A freak electric storm guides young, paranoid Richard to a vast LIBRARY, deserted aside from its eccentric librarian, Mr Dewey. Richard slips and knocks himself unconscious. A great painted dome in the library comes to life as a torrent of malignant, multi-hued paint, and he is transported into a TOON world. Seeking the library's exit, he is befriended by three books called Adventure, Fantasy and Horror, who agree to be the COMPANIONS on his QUEST if he will take them out with him. First, though, they have to negotiate the lands of HORROR (as epitomized by a haunted house [◊ HAUNTED DWELLINGS] occupied by JEKYLL AND HYDE), Adventure (populated by Captain Ahab from Herman MELVILLE's *Moby-Dick*, Long John Silver and his PIRATES from Robert Louis STEVENSON's *Treasure Island* [**1883**]) and Fantasy (where there is a brief skirmish with Lilliputians from Jonathan SWIFT's *Gulliver's Travels* [**1726**] and a much longer one with a DRAGON [from whose stomach Richard escapes with the aid of JACK's beanstalk], with some ARABIAN FANTASY swilled in). On return to the mundane world Richard is no longer slave to his fears.

TP is not shy of drawing on other sources; further debts are owed to Lewis CARROLL's *Alice in Wonderland* (**1865**), to PINOCCHIO (*1940*) and, of course, in terms of the overall notion, to *The Neverending Story* (**1983**) by Michael ENDE. Little is done with most of these borrowed ideas: the overall affect of *TP* is of a perfunctory touching of bases. The early animation – when a live-action Richard is being pursued among the library shelves by floods of paint – is impressive, but thereafter the animation defaults towards simplicity. *TP* apparently took three years to make, but it is hard to see why. [JG]

PAGET, FRANCIS EDWARD (1806-1882) UK writer. ◊ CHILDREN'S FANTASY; FAIRYTALE.

PAIN, BARRY (ERIC ODELL) (1864-1928) UK writer whose happiest creative period was the 1890s, though he remained active until the end of his life. His SUPERNATURAL FICTION – most famously the stories assembled in *Stories in the Dark* (coll **1901**) and *Stories in Grey* (coll **1911**) – is more assured than his fantasy proper; but the benefits of his sharp, bluff, edgy mind can be found throughout, although his humorous work – for which he was best-known during his lifetime – now grates. His first professional work, containing some pieces done at the end of the 1880s for *Granta*,

was assembled in *In a Canadian Canoe* (coll **1891**), which contains "The Celestial Grocery", whose narrator is taken to HEAVEN where the eponymous magic SHOP will sell him LOVE, fame and other things, including death. The most memorable tale in *Stories and Interludes* (coll **1891**), "The Glass of Supreme Moments", again deals with death, this time in corporeal form (◊ DEATH) – for it is She who opens for the protagonist a PORTAL that leads to a tower where the eponymous MIRROR opens his eyes to the future, which includes Her. Tales of this sort, in which the real world and the supernatural meet uneasily, appeared throughout BP's career. Sometimes highly effective, they too often withdraw the sanction of belief at their moments of highest intensity.

Most of BP's work is HORROR, with the supernatural seen as violating the natural world, but some stories are of more general interest. In "The Moon-Slave" (from *Stories in the Dark*) the female protagonist's love of dancing opens her to the deadly influences of Eros (always threatening in BP's work); after her betrothal banquet, she slips away into the wild and penetrates (◊ RECOGNITION) an abandoned maze (◊ LABYRINTH) where the MOON promises her MUSIC, though she must pay the price of becoming its slave. She returns monthly and dances there alone; but on the night before her wedding she finds herself *"no longer dancing alone"*. It is to be presumed that this ending titillated and horrified the readers of 1901, who tended to derive guilty arousal from any evocation of PAN. "Rose Rose" (in *Stories in Grey*) is an effective painter-and-model tale; significantly, however, the model turns into a GHOST and causes a suicide. Self-expression and punishment, in BP's only superficially jolly world, are usually married – inevitably so if a woman is involved.

Of lighter import are novels like: *The One Before* (**1902**), in which a mysterious jewel causes mild IDENTITY-EXCHANGE problems for a suburban cast; *Robinson Crusoe's Return* (**1906**; rev vt *The Return and Supperizing Reception of Robinson Crusoe of York, Parrot-Tamer* 1921 chap), whose protagonist has become immortal and finds the England of 1906 intolerably noisy; *The Shadow of the Unseen* (**1907**) with James Blyth, which dooms a young woman to death for consorting with the ghost of an ancestor WITCH; *An Exchange of Souls* (**1911**), which fluctuates between sf and horror in its attempts to get a fix on metempsychosis, as a dead scientist's SOUL gradually alters for the worse within the body of his fiancée. But *Going Home: Being the Fantastical Romance of the Girl with Angel Eyes and the Man who had Wings* (**1921**) is of greater interest. The eponymous duo, both afflicted by ET IN ARCADIA EGO "desiderium", meet and fall in love; in their mating they seem to acquire full ANGEL status, and they disappear into the sky on transcendental WINGS.

Unlocked from his era and his need to do commercial work to survive, BP could have been a writer of the first order. [JC]

Other works: *Three Fantasies* (coll **1904**); *The Diary of a Baby: Being a Free Record of the Unconscious Thought of Rosalys Ysolde Smith Aged One Year* (**1907**); *Stories Without Tears* (coll **1912**); *The New Gulliver and Other Stories* (coll **1913**), the title novella being sf; *One Kind and Another* (coll **1914**); *Collected Tales, Volume One* (coll **1916**), all previously published; *Short Stories of To-Day and Yesterday* (coll **1928**); *More Stories* (omni **1930**), volumes of interest here included being *In a Canadian Canoe* and *An Exchange of Souls*.

PALIMPSEST Originally, a palimpsest was any material – e.g., parchment, a slate – so treated that anything written upon it could be erased, and it could be written upon again. In later usage, the term came to describe any actual material – again e.g., parchment, though *not* a slate – known to have been written on more than once. In this encyclopedia we use the term in a sense analogous to its original meaning, to refer to what one might call a "writing world" – one which outlives the brief REALITIES inscribed, each after another, upon it. "Palimpsest" is a term, therefore, perhaps less useful for describing fantasy than SUPERNAT-URAL FICTION, many forms of which tend to postulate higher realities to which this world is subordinate, and which represent an occult lesson for those of us capable of perceiving it (◊ THEOSOPHY).

But there is a significant fantasy application for the term. The long tradition of serious and cod speculation based on gnostic doctrine (◊ GNOSTIC FANTASY) tends to postulate a central and more real world upon which innumerable SHADOW worlds are written, erased, and written again; and, although fantasies based on this sort of dualism tend to harbour fossilized supernatural-fiction motifs and assumptions, it seems wise to treat them as more than exercises in OCCULTISM. The novels of John CROWLEY – whose protagonists are typically caught in the shadows of the mundane world, and who search for the true STORY – are fantasies if for no other binding reason than that the truth (in a didactic sense) of the story is of radically less importance than the fact that it *is* a story, and an autonomous one.

Cosmologies – like the MULTIVERSE created by Michael MOORCOCK and Roger ZELAZNY's **Amber** – which treat huge arrays of mundane and fantasy realities as having been written in sand upon a central and fundamental reality can also be described as treating that substrate as a palimpsest. Though echoes of occult thinking haunt works of this sort, their overriding impulse is toward fantasy outcomes.

Both its origin and its usefulness in describing supernatural fictions give a sense that the term best describes vertical structures, stories in which new realities are written over – on top of – prior realities. When ALTERNATE REALI-TIES convene or portals open *horizontally* – so that there is a "geographical" or lateral mixing of kinds of world – then the term is less useful; in this encyclopedia the term CROSS-HATCH is used to signal lateral mixings. [JC]

PALL MALL MAGAZINE, THE ◊ MAGAZINES.

PALMER, THOMAS (1955-   ) US writer. His first novel, *The Transfer* (1983), was a thriller. In *Dream Science* (1990) a banking executive is suddenly plunged out of his mundane life into, apparently, the first of a series of ALTER-NATE WORLDS, from which he can on occasion – but only with a deep sense of WRONGNESS – return to our existence. These "other worlds" seem generated from his own private DREAMS and idealized memories, yet also have many of the characteristics of AFTERLIVES, as if this were an elaborate POSTHUMOUS FANTASY. But soon it becomes clear to him that they are not other worlds but different areas of REALITY, which is not the single LANDSCAPE we normally experience but a much more complicated, spiralling structure, with death being just one way – perceptual shift (◊ PERCEPTION) is another – whereby a PORTAL may be opened between one of these areas and the next. Further fantasies by TP are eagerly awaited. [JG]

PALWICK, SUSAN (1960-   ) US academic and writer. Her earliest published work is poetry; she won the American Academy of Poets Poetry Prize in 1981. SP first began publishing work of genre interest with "The Woman who Saved the World" for *IASFM* in 1985, and her short fiction has since been frequently anthologized. Many of her stories offer revisionist takes on FOLKTALES or canonical literature, such as "Ever After" (1987), which harshly treats the theme of the FAIRY GODMOTHER, or "Jo's Hair" (1995), which traces the history of the eponymous hairpiece following the end of Louisa May Allcott's *Little Women* (**1868**). She won the Rhysling Award for her poem "The Neighbor's Wife" (1985 *Amazing*). *Flying in Place* (**1992**), which won the WILLIAM L. CRAWFORD MEMORIAL AWARD, makes extremely effective use of the conventions of the GHOST STORY to tell a tale of familial sexual abuse. SP's erudition, ingenuity and imaginative sympathy far outweigh the occasional unsureness of her plotting, and mark her as a writer worth attending. [GF]

PAN In Greek mythology Pan was a horned and goat-legged Arcadian shepherd-god, similar in form to the SATYRS; he played seductive music on pipes made from the reeds into which the nymph Syrinx had been turned in order to escape his amorous advances. His shout had the ability to infect adversaries with the unreasoning terror of "panic". His identification with both the seductive and frightening aspects of "nature" made him a uniquely useful symbol in late-19th- and early-20th-century fantasy, his invocation as an ambiguous adversary in URBAN FANTASY mapping out the changing relationship of town and country. His utility was further increased by virtue of the fact that key aspects of his physical appearance had, by association with the Horned God (◊ DEVILS), been appropriated for incorporation into Christian images of the Devil, so that he became the chief substitute for SATAN in the sceptical tradition of UK Literary Satanism. This renewed image was quickly appropriated by lifestyle fantasists (◊ LIFESTYLE FANTASY) like Aleister CROWLEY, perhaps inspired by the example of Edgar JEPSON's *The Horned Shepherd* (**1904**) and *No. 19* (**1910**). Crowley's intellectual descendants often credited this allegedly suitable object of reverence with the title bestowed on him by Arthur MACHEN's classic melodrama "The Great God Pan" (1894), while Oscar WILDE's litany to the "Goat-foot god of Arcady", in "Pan" (1881), is echoed in such apologetic works as *The Goat-foot God* (**1936**) by Crowley's one-time disciple Dion FORTUNE as well as in tales in a more horrific vein like *The Devil and the Crusader* (**1909**) by Alice and Claude Askew.

Pan's symbolism of the wildness within and without is variously exemplified by such tales as "The Moon Slave" (1901) by Barry PAIN, "Pan" (1905) by James Huneker (1860-1921), "The Story of a Panic" (1911) by E.M. FORSTER, "The Music on the Hill" (1911) by SAKI, "The Man who Went Too Far" (1912) by E.F. BENSON, "The Golden Bough" (1934) by David H. KELLER, "Capra" (1951) by SARBAN and "The Forest of Unreason" (1961) by Robert F. YOUNG. Meeker and more benevolent versions appear in "Syrinx" (1891) by James Vila Blake, "Pan's Wand" (1903) by Richard GARNETT, "The Prayer of the Flowers" (1915) by Lord DUNSANY, "On the Knees of the Gods" (1940) by J. Allan Dunn (1872-1941) and "The Pipes of Pan" (1940) by Lester del Rey (1915-   ); he makes a cameo appearance in a visionary interlude in *The Wind in the Willows* (**1908**) by Kenneth GRAHAME. A more sinister Pan is featured in "The Master of Cotswold" (1944) by Nelson BOND.

The Plea of Pan (coll of linked stories **1901**) by Henry Nevinson (1856-1941) is a curious set of dialogues in which Pan and his followers reveal new depths of philosophical sophistication. These were carried forward into such works as *Pan and the Twins* (**1922**) by Eden PHILLPOTTS, in which Pan is a symbol of Epicurean joy in life whose worship is carefully contrasted with the asceticism of Christianity, and *The Oldest God* (**1926**) by Stephen MCKENNA, in which Pan plays a cleverly mischievous role in upsetting the attempts of a conference of Christians to come to terms with the problem of EVIL. Dunsany's *The Blessing of Pan* (**1927**) describes a revival of life-enhancing paganism caused by the music of Pan, which carries away a vicar as well as his village flock, according to an ambivalent pattern earlier laid out in "How Pan Came to Little Ingleton" (1926) by Margery LAWRENCE. In David H. Keller's *The Homunculus* (**1949**) Pan is the twin brother of LILITH and the archetypal male.   [BS]

PAN BOOK OF HORROR STORIES UK annual ANTHOLOGY series which ran to 30 volumes (**1959-89**), the first 25 vols ed Herbert van Thal (1904-1983) and the remainder ed Clarence Paget (1909-1991), after which the series was retitled **Dark Voices**.

Van Thal had long believed in the merits of an annual HORROR anthology. He interested the publisher Pan in *The Pan Book of Horror Stories* (anth **1959**; vt *The First Pan Book of Horror Stories* 1960). This volume, which rode on the back of the success of the HAMMER horror movies, was so successful that Pan agreed to an annual series, which thereafter missed only 1961. The initial anthologies were almost entirely of reprints, drawing on Victorian GHOST STORIES and tales from the 1920s and 1930s, especially from the CREEPS LIBRARY and NOT AT NIGHT series. These focused more on physical than supernatural horror, and this element was still the main content of the series when it switched to mostly original material, from *The Sixth Pan Book of Horror Stories* (anth **1965**) onwards. By the end of the 1960s the series had established a reputation for graphic violence that marginalized it from mainstream WEIRD FICTION, but which has since been seen as forerunning Splatterpunk. The series published some of the first stories by Basil Copper (1924-   ), Richard Davis (1945-   ), Alex Hamilton (1930-   ) and Tanith LEE, and revived the career of Charles BIRKIN. Its success not only launched many rival publications in the UK (e.g., FONTANA BOOK OF GREAT GHOST STORIES) but also raised new interest in the horror genre in the USA. The series in effect resurrected horror fiction as a publishing genre. Its highlight was the reprinting of stories by David Case (1937-   ), whose work had hitherto been overlooked.

From *The 21st Pan Book of Horror Stories* (anth **1980**), with the reprinting of stories by Stephen KING, the series began to receive wider acceptance among major horror writers, with stories appearing by Christopher FOWLER, Nicholas Royle (1963-   ), Alan Ryan (1943-   ) and Jessica Amanda SALMONSON. Yet it never rid itself of its image of graphic violence and added little by way of memorable fiction. Van Thal made a selection of his favourite stories from the first 3 vols, *Striking Terror!* (anth **1963**), while in the USA 3 vols appeared as *Selections from Pan Horror #3* (anth **1970** US), *#4* (anth **1970** US) and *#5* (anth **1970** US).

In 1989 it was agreed to give the series a facelift. Assisted by Stephen JONES, Clarence Paget assembled *Dark Voices: The Best from the Pan Book of Horror Stories* (anth **1990**) with personal tributes by many leading writers. The series then continued as **Dark Voices** under the joint editorship of Stephen Jones and David A. Sutton (1947-   ), with *#2* (anth **1990**), *#3* (anth **1991**), *#4* (anth **1992**), *#5* (anth **1993**) and *#6* (anth **1994**). Under Jones and Sutton the series more closely reflected modern horror, with stories (original and reprinted) by many of today's leading writers, including Ramsey CAMPBELL, Basil Copper, Stephen Gallagher (1954-   ), Robert HOLDSTOCK, Kathe Koja (1960-   ), Graham Masterton (1946-   ), Kim NEWMAN and David J. Schow (1955-   ). After *#6* the series was dropped by Pan and (retitled) taken up by Gollancz, starting with *Dark Terrors* (anth **1995**).   [MA]

Main series: *The Pan Book of Horror Stories* (anth **1959**; vt *The First Pan Book of Horror Stories* 1960); *Second* (anth **1960**), *Third* (anth **1962**), *Fourth* (anth **1963**), *Fifth* (anth **1964**), *Sixth* (anth **1965**), *Seventh* (anth **1966**), *Eighth* (anth **1967**), *Ninth* (anth **1968**), *Tenth* (anth **1969**), *Eleventh* (anth **1970**), *Twelfth* (anth **1971**), *13th* (anth **1972**), *14th* (anth **1973**), *15th* (anth **1974**), *16th* (anth **1975**), *17th* (anth **1976**), *18th* (anth **1977**), *19th* (anth **1978**), *20th* (anth **1979**), *21st* (anth **1980**), *22nd* (anth **1981**), *23rd* (anth **1982**), *24th* (anth **1983**), *25th* (anth **1984**), these all ed van Thal; *26th* (anth **1985**), *27th* (anth **1986**), *28th* (anth **1987**), *29th* (anth **1988**) and *30th* (anth **1989**), all ed Clarence Paget.

PANDORA ◊ GODDESS; PANDORA'S BOX.

PANDORA'S BOX The urn or jar which in Greek MYTH was unwisely opened by Pandora, the female gift of ZEUS to Epimetheus the Titan (brother of PROMETHEUS). This action – sheer curiosity rather than the breaking of an overt PROHIBITION – released afflictions like disease and old age on newly created mankind; the one consoling element was that the jar also contained hope.   [DRL]

PANTALONE ◊ COMMEDIA DELL'ARTE.

PANTHEONS Originally, a pantheon was a sacred building dedicated to the worship of all the GODS. It then came to designate a place or habitation where the gods assembled in a family group. It is now most usually used to describe the gods themselves, collectively, when so assembled. Classic examples found in fantasy include the Egyptian pantheon (◊ EGYPT), the closely linked Greek and Roman pantheons (◊ GREEK AND LATIN CLASSICS), the Hindu pantheon (◊ SANSKRIT CLASSICS), the gods and goddesses who dominate Irish, Welsh and Scottish mythology (◊ CELTIC FANTASY), the AESIR and their kin in the Nordic pantheon (◊ NORDIC FANTASY), and the Aztec pantheon. Authors of fictions not set in SECONDARY WORLDS sometimes use classic pantheons directly – like Roger ZELAZNY in *Lord of Light* (**1967**), which is based on the Hindu model, and Kenneth MORRIS, whose *The Chalchiuhite Dragon* (**1992**) is based on the Aztec. They also create pantheons out of whole cloth – like Zelazny in the **Amber** sequence, H.P. LOVECRAFT in the CTHULHU MYTHOS, Lord DUNSANY – whose *The Gods of Pegana* (coll of linked stories **1905**) is intentionally a kind of holy book establishing a pantheon and its traditions – and Clark Ashton SMITH, whose stories are full of gods consorting together. These pantheons tend to PARODY those which already exist.

Fantasies set in secondary worlds are normally fitted with a pantheon. Examples – from E.R. EDDISON and J.R.R. TOLKIEN through Philip José FARMER, Mary GENTLE, Tanith LEE, Michael MOORCOCK, Terry PRATCHETT and Jack VANCE – are innumerable. Pantheons are part of the furniture of much GENRE FANTASY.   [JC]

PANTOMIME The Latin *pantomimus* was a performer who re-enacted episodes from MYTHOLOGY solely through movement and gesture; he wore a variety of MASKS, and was accompanied by MUSIC, usually sung and played. In early-18th-century England, the term was less comprehensively understood to mean a tale told in dance; and the 18th-century harlequinade (◊ COMMEDIA DELL'ARTE), a dance performance in which Harlequin wins Columbine from Pantaloon, became known as the Pantomime before the end of the century. In the 19th century the harlequinade element, with its emphasis on dance, was relegated to the sidelines, and a fully acted-out, though usually burlesqued FAIRYTALE took centre stage. Popular subjects included CINDERELLA, Aladdin and Little Red Riding-Hood. This mature form soon became associated with CHRISTMAS, and incorporated safely detoothed elements of REVEL: cross-dressing (◊ GENDER DISGUISE), PARODY, etc. In the late 20th century, the Pantomime is a UK institution.

When used here, the term refers to the 19th-century extravaganza which influenced writers like Charles DICKENS (who influenced it in turn) and to its 20th-century incarnations, on the stage and elsewhere – like *The* MUPPET SHOW (1976-81). [JC]

PAPÉ, FRANK C(HEYNE) (1878-1972) Popular UK illustrator with a finely worked pen-and-ink style, using which he produced fully rendered pictures somewhat in the style of etchings. Although he had a redeeming sense of humour, his work has a dated look and lacks the strongly personal viewpoint of his contemporaries Arthur RACKHAM and Edmund DULAC. It nevertheless has undoubted charm, and was greatly appreciated by James Branch CABELL, whose novels *Jurgen* (1921) and *Figures of Earth* (1925) FP illustrated with particular success. He also illustrated fantasies by Anatole FRANCE. [RT]

PAPERHOUSE UK movie (*1988*). Vestron/Working Title/Tilby Rose. **Pr** Tim Bevan, Sarah Radclyffe. **Exec pr** Dan Ireland, M.J. Peckos. **Dir** Bernard Rose. **Spfx** Alan Whibley. **Screenplay** Matthew Jacobs. **Based on** *Marianne Dreams* (1958; vt *The Magic Drawing Pencil* 1960 US) by Catherine Storr (1913-   ). **Starring** Charlotte Burke (Anna Madden), Ben Cross (Mr Madden), Glenne Headly (Kate Madden), Gemma Jones (Dr Sarah Nichols), Sarah Newbold (Karen), Elliott Spiers (Marc). 92 mins. Colour.

Troublesome 11-year-old Anna has glandular fever, and during the consequent fainting fits – and then in DREAMS – finds herself in an OTHERWORLD, where stands, remote in the midst of a windswept Andrew Wyeth-like plain, the house she has been drawing in her sketchbook. In her waking hours she embellishes the drawings and thus embellishes the other REALITY, giving the house an interior and an occupant, Marc, a boy whose legs – which she omitted to draw – are paralysed. Anna finds that, in the mundane world, Marc is another patient of her own GP, Nichols; he suffers from muscular dystrophy. Anna tries through her drawings to aid him, but discovers that, once drawn, any line becomes unerasable; thus, when she tries to bring her father to the Otherworld as rescuer, her careless drawing instead creates that Otherworld's drunken, psychopathic VILLAIN – her incest fears incarnate. The two realities become progressively more interpenetrative. At last, in our reality, Marc dies; in the Otherworld, he escapes to an unknown destination by helicopter. Anna's dreams of the Otherworld end, but convalescing in Devon she discovers the lighthouse that stood near the "paperhouse", sees Marc's helicopter

and hears his voice, and knows that he is somewhere safe. The movie remains equivocal about its events: all may be a product of coincidence and Anna's fever dreams (◊ PERCEPTION).

On release *P* was treated by the critics as a HORROR MOVIE and, as such, considered inadequate. In fact, despite scary moments, it is a strikingly powerful piece of well worked-out fantasy. The children's performances make up for some less assured adult playing. With its superb evocation of childhood and its sharp ear for dialogue ("What's snogging like, then?" "Like kissing a vacuum cleaner."), *P* is a movie agleam with conviction. [JG]

PARACELSUS Pseudonym of Dr Theophrastus Bombastus von Hohenheim (*c*1493-1541), a physician involved in ALCHEMY; the pseudonym means "beyond Celsus", implying that he was a greater physician than the then-revered Celsus (*fl* 1st century AD), whose *De Medicina* was in 1478 one of the first medical texts to be printed and had a profound (and largely deleterious) effect on medieval medicine. The suggestion that the term "bombastic" derives from Paracelsus's real name is, according to the *OED*, erroneous. [JC/JG]

PARADISE ◊ HEAVEN.

PARALLEL WORLDS ◊ ALTERNATE WORLDS.

PARGETER, EDITH (MARY) (1913-1995) UK writer. Although she published extensively under her own name, her pseudonym Ellis Peters – attached to the medieval **Brother Cadfael** detective tales – became better known. *The City Lies Four-Square* (**1939**) is a touching account of the tribulations of a GHOST barred from HEAVEN, who seeks solace in mournful poetry and a platonic but emotionally charged friendship with the heroine. *By Firelight* (**1948**; vt *By This Strange Fire* US) is a harrowing TIMESLIP romance in which a more comprehensively alienated heroine receives intelligence from the past regarding the fate of a man framed for WITCHCRAFT by a lover; the novel includes an excellent recreation of an English witch-trial and generates a dramatic intensity of feeling. [BS]

PARIAH ELITE A group which, though despised and rejected by society, remembers and preserves the secret knowledge necessary to keep the world from ultimate THINNING. In other words, members of PEs are despised and rejected precisely for that which they *retain*: their knowledge of the Secret History of the World, their TALENTS, their memory of the ELDER GODS, their familiarity with the old TRUE NAMES and the real MAP of the territory (which allows them to escape the minions of the false king), their direct descent from the ELDER RACES, their memory of the way through the LABYRINTH, their access through PORTALS to the GOLDEN AGE ... But they sometimes do more than retain the past; it is always possible that the PE may be the SECRET MASTERS of the world (◊ FANTASIES OF HISTORY).

PEs can be found in all forms of fantasy and SUPERNATURAL FICTION. In the former, they are almost always viewed favourably, and are normally seen as preserving that which must be preserved, and rediscovered, if the LAND or world is to regain its health; the final RECOGNITION of them is likely to generate HEALING – a process typical of fully structured FANTASY. In supernatural fiction, on the other hand, the PE may well be seen as inimical. There can be no presumption that an association of WITCHES, or a secret society nested within the covert imperium of the Knights Templar, or a coven of immortal Tibetans is either good or evil.

The history of the Western World provides a convenient model for the notion of the PE through the fall of the Roman Empire, a collapse which impelled Europe into the

Dark Ages, a time when much that was valuable from the ancient world was preserved in enclaves around the periphery of Europe – or so it has been widely believed. Consequently, "invasive" groups – Jews, gypsies (the term comes from EGYPT), Knights Templar, beggars, JESTERS and other travelling players, Orientals, assassins, WITCHES, monks – were all varyingly likely to be thought of as secretly in control, or secretly planning to wrest power from the just rulers of the world. Sinister cabals, most of them reflecting the imagined agendas of outsider groups in real life, have always been a staple of supernatural fiction.

Before the 20th century, very few of these fictional cabals were seen as benevolent. More recently, they tend to be regarded as bearers of authentic wisdom, like the Little Brothers of the Apostles in Patrick HARPUR's *The Serpent's Circle* (**1985**), and are often depicted as victims of persecution by organized RELIGION, through inquisitors and witchfinders. A novel like *Knights of the Blood: Vampyr-SS* (**1993**) by Scott MacMillan (as "presented" by Katherine KURTZ) neatly encapsulates the shift: a group of Crusader knights who have suffered TRANSFORMATION into VAMPIRES, but refuse to live off human beings. Trouble comes to them in WORLD WAR II, when a Nazi officer attempts to make them into an Aryan cadre of the undead, and to rule the world through them.

Fantasy featuring PEs includes C.J. CHERRYH's **Morgaine** sequence, Michael de LARRABEITI's **Borribles** sequence, various novels by Barbara HAMBLY, Tanya HUFF's *Gate of Darkness, Circle of Light* (**1989**), Katherine KURTZ's **Deryni** sequence, Tanith LEE's **Birthgrave** sequence, Megan LINDHOLM's *Wizard of the Pigeons* (**1986**), Elizabeth SCARBOROUGH's **Songkiller Saga** and *Empire of the Eagle* (**1993**) by Susan M. SHWARTZ and Andre NORTON. [JC]

PARIS Paris has been for centuries, like LONDON, a CITY which has shaped the literary imaginations of those who inhabit it, and was itself in turn transformed, during the 19th century, by the architectural and city-planning dreams of those it had haunted. Therefore, though its origins lie as deep in time as London, Paris tends to manifest itself in URBAN FANTASIES as an intricate, interlocking LABYRINTH of EDIFICES constructed by men; it tends to be understood as a vast *exposition* rather than as an organic growth whose boundaries fade imperceptibly, as do London's, into WATER MARGINS. Paris is half a city that writers of fantasy can view, from without, as a focus of longing, and half a fever dream in stone – within which a figure like Quasimodo, in *The Hunchback of Notre-Dame* (**1831**) by Victor Hugo (1802-1885), or the failed GODGAME player in Gaston LEROUX's *The Phantom of the Opera* (**1910**) can caper and strain, but cannot escape (◊PHANTOM OF THE OPERA) .

As an urban-fantasy venue, Paris served as a backdrop – a backdrop with false exits and TROMPE L'OEIL effects – for 19th-century writers like Honoré de BALZAC and Eugène SUE. Their intricate vision of the city was central for UK writers like Charles DICKENS, whose *A Tale of Two Cities* (**1859**) reflects that vision; and it seems likely that the Paris of the Surrealists – the first relevant texts being Guillaume APOLLINAIRE's *The Heresiarch and Co* (coll **1910**) and *The Poet Assassinated* (**1916**) – is very closely shaped by their great 19th-century predecessors. Visions of Paris which combine Sue and SURREALISM include Lisa GOLDSTEIN's *The Dream Years* (**1985**), Macdonald HARRIS's *Herma* (**1981**) and *The Glowstone* (**1987**), William KOTZWINKLE's *Fata Morgana* (**1977**), Tanith LEE's **Secret Books of Paradys** sequence,

and the movie version (***1994***) of Anne RICE's *Interview with the Vampire* (**1976**). [JC]

PARIS QUI DORT French movie (***1923***). ◊ René CLAIR.

PARKINSON, DAN(IEL EDWARD) (1935-   ) US writer whose first novel, *Starsong* (**1988**), attractively fuses PREHISTORIC FANTASY and SCIENCE FANTASY in describing the return to Earth of a race of ELVES which emigrated in the Cretaceous Era. His other works are **Dragonlance** ties: *Dragonlance Heroes II: The Gates of Thorbardin* * (**1990**), #2 in the **Dragonlance Heroes II** sequence; and the **Dwarven Nations Trilogy** – #1: *The Covenant of the Forge* * (**1993**), #2: *Hammer and Axe* * (**1993**) and #3: *The Swordsheath Scroll* * (**1994**). [JC]

PARODY This term is often found in this encyclopedia in its customary sense, but it is also used to help describe the origins of fantasy as a genre, and to assist in defining an important structural pattern in fantasy as it has evolved.

1. The original Greek word meant a "subsidiary [or mock] song". Parody can be defined as an imitation of the work of an individual writer or group of writers, generally to mock, comically, that work, usually as SATIRE – but not always: close, devoted, analytical mimicry may be a form of wit; but it may also constitute a homage.

Authors of fantasy interest who have created parodies of earlier texts include ARISTOPHANES, LUCIAN, Geoffrey CHAUCER (in the "Tale of Sir Topas"), François RABELAIS, Miguel de CERVANTES, William Makepeace THACKERAY, Lewis CARROLL, Max BEERBOHM and Terry PRATCHETT ... and countless others.

2. From about the end of the 18th century, FANTASY began to take shape as a conscious genre, contrasting itself with a world of literature dominated by the mimetic novel and by Enlightenment concepts of the rule of Reason. Given the straitjacket tradition of the "real", and the powerful Romantic impulse to unshackle the self from the stultifying bondage of a constricting world, it is not surprising to find new, selfconscious, subversive fantasy texts responding to the world in terms of mockery and/or emulation. Mary SHELLEY's *Frankenstein* (**1818**), for instance, models itself in part upon (and savagely mocks) the 18th-century *Bildungsroman*, just as E.T.A. HOFFMANN's *The Life and Opinions of Kater Murr* (**1820-21**), which purports to be the autobiography of a CAT, is a direct parody of GOETHE's *The Apprenticeship of Wilhelm Meister* (**1796**).

3. The term "parody" is also used in this book to help understand an assumption about the nature of EVIL that lies at the heart of much CHRISTIAN FANTASY – especially the works of writers like C.S. LEWIS and J.R.R. TOLKIEN – and has shaped much subsequent fantasy. Christian fantasies tend to convey a sense that evil is a distortion of good, and has no creative capacity of its own – i.e., that evil is a *parody* of good, and that its irruption into a world is marked by manifestations of WRONGNESS. The DARK LORD in Tolkien (and numerous imitators) is a parody of the rightful monarch or lord, an ape of god; moreover, because he is incapable of true creation, any version of the LAND he rules will itself lack true FERTILITY (◊◊ FISHER KING; THINNING) and will thus likewise be a parody. Dark Lords and their domains are, in short, parasitic of true being. A modern fantasy which closes in EUCATASTROPHE and scenes of HEALING will need to represent a successful defeat of the deathly sovereignty of parody, and will tend to do so in passages in which parody shrivels away because the true and original STORY, full of being and armed against all mockery, has been recognized in the nick of time. [JC]

PARRISH, MAXFIELD Working name of Frederick Parrish (1870-1966), US fantasy illustrator and mural painter whose glowing colours, distinctive crisp-edged style and flat, stagelike compositions have influenced many modern fantasy illustrators. At the beginning of his working life he rejected his given name of Frederick in favour of his maternal grandmother's maiden surname, Maxfield. He worked in oils, rather in the manner of the Renaissance artists, drawing a meticulous outline design and laying on successive transparent glazes of colour. He was noted for a particularly rich, clear, almost luminous blue, known as the Maxfield Parrish Blue, which was built up in this manner. His knowledge of these traditional techniques was not, however, infallible, and shrinking of some paint layers with consequent cracking has occurred in various works.

MP studied under Howard PYLE and provided illustrations for magazines (e.g., *Scribner's*) and books of poems and fairytales. His best-known work includes illustrations for Kenneth Grahame's *Dream Days* (**1902**) and *The Golden Age* (**1899**), a series of advertisements for the Mazda Electric Company, and several murals on fairytale subjects for US hotels. However, the majority of his later paintings were produced specifically for reproduction in the form of fine-art prints or as calendars. [RT]

About the artist: *Maxfield Parrish* (graph **1973**) by Coy Ludwig.

PARRY, [Sir] EDWARD ABBOTT (1863-1943) UK writer and judge who signed at least some of his books His Honour Judge Edward Abbott Parry. His FAIRYTALES for older children – facetious, chummy, occasionally imaginative – were assembled in several volumes: the **Katawampus** sequence, being *Katawampus: Its Treatment and Cure* (**1895**) and *Butterscotia, or a Cheap Trip to Fairy Land* (**1896**); *The First Book of Krab: Christmas Stories for Young and Old* (coll **1897**); *The Scarlet Herring and Other Stories* (coll **1909**); and *Gamble Gold* (**1907**), illustrated by Harry FURNISS. He adapted for children Miguel de CERVANTES's novel as *Don Quixote of the Mancha* (**1900**). [JC]

PARSIVAL ◊ PERCEVAL; Richard WAGNER.

PASSING OF THE THIRD FLOOR BACK, THE UK movie (*1935*). Gaumont-British Picture Corp. **Dir** Berthold Viertel. **Screenplay** Michael Hogan, Alma Reville. **Based on** *The Passing of the Third Floor Back* (play **1907**) by Jerome K. JEROME. **Starring** Sara Allgood (Miss de Hooley), Frank Cellier (Wright), Mary Clare (Mrs Sharpe), Anna Lee (Vivian Tomkin), Beatrix Lehmann (Miss Kite), Jack Livesey (Larkcom), Catherine Nesbitt (Mrs Tomkin), René Ray (Stasia), Alexander Sarner (Gramophone Man), John Turnbull (Major George Tomkin), Conrad Veidt (Stranger), Ronald Ward (Chris Penny). 90 mins. B/w.

In a seedy boarding house everyone lives behind (metaphorical) MASKS donned to disguise their own inadequacies; all are compromising their SOULS in order to cope with "reality" ... all except Stasia the maid, on probation from the reformatory, and Wright the wealthy jerry-builder, who has effectively already lost his soul. Into their midst comes a new lodger, renting the third-floor room at the back. Through quiet suggestion he makes the rest draw their own virtues to the surface. But Wright remains impervious; seeing the effects the Stranger has achieved, he deliberately counters them, so that once more the lodgers plunge back into spite and meanness. As Wright is on the verge of seducing Stasia with the promise of "presents" he encounters a burglar and has a fatal heart attack. When the

lodgers discover, amid a hail of accusations, that neither Stasia nor the Stranger – both virtuous, and thus natural scapegoats – is guilty of the crime, they realize the error of their ways; decency and happiness once more prevail. Meanwhile the Stranger slips away unseen by all but Stasia, in whom he confides that he came in answer to her prayers.

TPOTTFB is generally discussed in terms of the Stranger being a manifestation of CHRIST, but it is more profitable to interpret him as an ANGEL, a view supported by two short, powerful scenes in which Wright adopts the role of the DEVIL, setting out explicitly the battle between the two of them for the lodgers' souls. Curiously, the movie's appeal was at the time judged to rest on Veidt's star qualities, but in fact he is of necessity cypher-like. [JG]

PASTORAL Although other antecedents can be identified, including some of the Psalms of the BIBLE, the *Idylls* of the Greek poet Theocritus (early 3rd century BC) are generally regarded as the first works of pastoral literature. In contrast to earlier poets who focused on the exploits of GODS and HEROES, Theocritus primarily (though, unlike his followers, not exclusively) celebrated the lives, problems and diversions of common rural people, especially shepherds. Early imitators of his work include VIRGIL in his *Eclogues* (or *Bucolics*) and, to an extent, in the didactic *Georgics*, and Longus (4th or 5th century AD) in *Daphnis and Chloe*. Later pastoral eclogues were written by DANTE ALIGHIERI, Francesco Petrarca Petrarch (1304-1374), BOCCACCIO, Edmund SPENSER (*The Shephearde's Calendar* [**1579**]), John MILTON (*Lycidas* [**1637**]) and Percy Bysshe SHELLEY (*Adonais* [**1821**]).

The events of pastoral poetry are normally realistic, although mild forms of MAGIC and certain rustic gods – like PAN or SATYRS – may appear. There also developed a form of narrative, the pastoral ROMANCE, examples of which include *Ameto* (**1341**) by Boccaccio, *Arcadia* (**1501**) by Jacopo Sannazaro (*c*1458-1530), *Arcadia* (final rev **1593**) by Sir Philip Sidney (1554-1586) and the sixth book of Spenser's *The Faerie Queene* (**1596**). Sannazaro's and Sidney's works helped to establish ARCADIA as a standard expression for an idealized rural setting. Beyond formal definition, almost any work which describes and praises the virtues of a simple agrarian life might be called pastoral, so that one could apply the term to, e.g., *The Compleat Angler* (**1653**) by Izaak Walton (1593-1683), the poems of Robert Frost (1874-1963) and, to move into fantasy proper, the novels of writers like William MORRIS, James STEPHENS, Clifford D. SIMAK and Thomas Burnett SWANN.

In a sense, however, fantasy in the modern sense is incompatible with the pastoral world, where there is no EVIL and no real conflict beyond comic misunderstandings and the like. Yet pastoral should not be seen as a projection of youthful innocence or childish delights; its inhabitants understand, and perhaps recall, evil perfectly well, but dwell in an environment where they have chosen – or managed – to *remove* evil. This is not a child's world but an old man's world; its characteristic figure is a wise old man, much like Walton's Piscator, who passes on his memories and knowledge to the younger men who will someday follow in his footsteps.

Pastoral is, therefore, not the *setting* of fantasy but rather the *goal* of fantasy, a world from which evil is finally banished, all tasks are accomplished, and necessary HEALING has occurred; and the "happy endings" of virtually all fantasies constitute a brief but powerful evocation of a newfound pastoral world. As a realm without dramatic conflict, it is not a place where a narrative striving only for drama can long remain, which is why many modern fantasies – particularly GENRE FANTASIES – often rush through their conclusions; but J.R.R. TOLKIEN's *The Lord of the Rings* (**1954-5**) interestingly lingers in this final stage (◊ EUCATASTROPHE), describing at length both the contented family life of Sam and the ongoing discontent of Frodo, and thus conveys the bittersweet spirit of true pastoral, the satisfaction of hard-earned achievements and the regret of dreams left unfulfilled. [GW]

PATHETIC FALLACY ◊ John RUSKIN; SEHNSUCHT.

PATON, EPHRAIM SOULSBY [s] ◊ Kenneth MORRIS.

PATON, [Sir] JOSEPH NOEL (1821-1901) UK artist ◊ ILLUSTRATION.

PATON WALSH, JILL (1939-) UK writer, mostly of YA novels. *A Chance Child* (**1978**) is a strong TIME-TRAVEL story of a boy who inadvertently goes back to the grim, industrialized 19th-century English Midlands. JPW has also written *Birdy and the Ghosties* (**1989** chap) for younger children. *Knowledge of Angels* (**1994** US), a historical novel, gained some notoriety when the UK edition, which had to be published by a press set up for the purpose by friends of the author, was nominated for the Booker Prize. [GF]

PATRICK Australian movie (*1978*). AIFC/Antony I. Ginnane/Filmways Australasia. **Pr** Richard Franklin, Antony I. Ginnane. **Exec pr** William Fayman. **Dir** Franklin. **Spfx** Conrad C. Rothmann. **Screenplay** Everett de Roche. **Starring** Bruce Barry (Brian Wright), Julia Blake (Matron Cassidy), Robert Helpmann (Dr Roget), Helen Hemingway (Paula Williams), Rod Mullinar (Ed Jacquard), Susan Penhaligon (Kathy Jacquard), Robert Thompson (Patrick). 115 mins. Colour.

Patrick lies in a coma in the seedy private hospital owned by Dr Roget. New nurse Kathy, separated from husband Ed, is given the task of supervising the human vegetable. Roget explains loosely that Patrick is being kept alive for the study of the grey area between life and death. But Kathy finds Patrick can communicate both through plosives and via the typewriter, which he can limitedly manipulate through psychokinesis (◊ TALENTS). Soon it is clear he is sexually obsessed by her, identifying her with his mother (whom he murdered); frightening accidents, some fatal, occur, despite attempts to kill Patrick. At last Kathy confronts him; he agrees to die if she will die with him. Under his mental control, she almost does so before Ed saves her.

P is a cult movie much disliked by mainstream critics, and certainly it has plotting, cinematographic and other flaws; Helpmann's hamming is embarrassing. But *P* has many strengths, not least the powerful image of the staring-eyed Patrick; and the script leaves us in doubt until satisfyingly late as to whether or not this is all going to be rationalized (◊ RATIONALIZED FANTASY) as the product of Kathy's own emotionally perplexed mind. Although the novel *Tetrasomy Two* (**1974**) by Oscar Rossiter (real name Vernon H. Skeels; 1918-) is nowhere credited, the similarity of situation and theme seems hardly coincidence. [JG]

PATRON Typically someone who supports and protects another. It was common from earliest days to look upon a GOD as a patron or protector; e.g., HERMES was upheld as the patron of shepherds, travellers and thieves. The Physicians of Kos formed a guild which they called the Asklepiadai, with Asklepios, the god of healing, as their patron. Guilds continue to take on patrons today, usually patron saints – St Pantaleon is the modern patron of physicians. Many

countries have their own patrons: England's St George brought with him the legend of George and the DRAGON. George allegedly became patron of England because ARTHUR flew the banner of St George. In popular tradition, however, Arthur himself is a more highly regarded patron as legend has it that he will return and save England in its moment of need. The PRERAPHAELITES chose another Arthurian link in adopting GALAHAD as their patron.

In fantasy patrons may take several forms. In FAIRYTALES they are usually FAIRY GODMOTHERS; in SLICK FANTASY they may be guardian ANGELS – the best-known example is the angel Clarence in IT'S A WONDERFUL LIFE (*1946*); in HIGH FANTASY they are normally WIZARDS – Gandalf serves as a patron to Bilbo and the DWARFS in *The Hobbit* (**1937**) and to Frodo and his COMPANIONS in *Lord of the Rings* (**1954-5**), in which role he may also be seen as a MENTOR. One of the best examples of patrons as protectors occurs in *Lammas Night* (**1983**) by Katherine KURTZ, where all of the UK's patrons, including the WITCHES, pool their powers to save the land during the Battle of Britain, as they had centuries before against the Spanish Armada. Patrons, like guardian angels, frequently work in secret. One of the objects of their work may be to protect a dispossessed prince or HIDDEN MONARCH, as with MERLIN and the young Arthur, Gandalf and Aragorn in *LOTR*, and Belgarath and Garion in David EDDINGS's **Belgariad** sequence. [MA]

See also: SECRET SHARER.

PATTRICK, WILLIAM Pseudonym of Peter HAINING.

PAVIĆ, MILORAD (1929-) Serbo-Croat writer who came to world attention with *Hazarski recnik* (**1988** in French trans; trans Christina Pribicevic-Zoric from the Serbo-Croat as *Dictionary of the Khazars: A Lexicon Novel in 100,000 Words* **1988** US), an elaborately playful mock-history of the IMAGINARY LAND of the Khazars, located somewhere among Turkey, Russia and the Slavic countries to the west; the novel is famous in part for being published in two editions, a "male" and a "female" version, which differ on only one page. At intervals there are hints that Khazar history has underwritten world history, but without follow-through. *Predeo Slikan Cajem* (**1990**; trans Christina Pribicevic-Zoric as *Landscape Painted with Tea* **1990** US) slips into playful SURREALISM in its textually foregrounded, Postmodernist rendition of a man's search for secrets his father left at Mount Athos in Greece; whiffs of the FANTASY OF HISTORY linger here as well. And *Unustrsnja strana vetra, ili, Roman o Heri i Leandru* (**1991**; trans Christina Pribicevic-Zoric as *The Inner Side of the Wind, or The Novel of Hero and Leander* **1993** US) – a text whose two versions are printed dos-a-dos so that both read towards the centre where the two lovers finally meet – tells the Hero and Leander tale in TWICE-TOLD fashion through the stories of a couple who respectively hail from the 18th and 20th centuries. [JC]

PAXSON, DIANA L(UCILE) (1943-) US writer who began publishing fantasy with the **Westria** novels, set in a post-HOLOCAUST California in which technology has been supplanted by MAGIC. *Lady of Light* (**1982**) and *Lady of Darkness* (**1983**) – assembled as *Lady of Light, Lady of Darkness* (omni **1990** UK; vt *The Mistress of the Jewels* 1991 US) – were followed by *Silverhair the Wanderer* (**1986**), *The Earthstone* (**1987**), *The Sea Star* (**1988**), *The Wind Crystal* (**1990**) and *The Jewel of Fire* (**1992**). **The Chronicles of Fionn Mac Cumhal** with Adrienne MARTINE-BARNES – *Master of Earth and Water* (**1993**), *The Shield*

Between the Worlds (**1994**) and *Sword of Fire and Shadow* (**1995**) – is CELTIC FANTASY (◊◊ FINN MAC COOL), as is *White Mare, Red Stallion* (**1986**). *The White Raven* (**1988**) is an Arthurian fantasy (◊◊ ARTHUR), while *The Serpent's Tooth* (**1991**) adapts William SHAKESPEARE's *King Lear*.[JCB/GF]
Other works: *Brisingamen* (**1984**) and *The Paradise Tree* (**1987**), fantasy adventures set in contemporary California; *The Wolf and the Raven* (**1993**) and *The Dragons on the Rhine* (**1995**), starting a series retelling the *Nibelungenlied*.

PAYE, ROBERT Pseudonym of Marjorie BOWEN.

PAYNE, ALAN [s] ◊ John W. JAKES.

PEACOCK, THOMAS LOVE (1785-1866) UK writer. ◊ ARTHUR; FLOOD; GHOST STORIES.

PEAKE, MERVYN (LAURENCE) (1911-1968) China-born UK artist and writer, much influenced by his first 12 years in China, where he lived in a missionary compound overlooking a territory as alien to his world as the land surrounding Gormenghast seems alien to the readers of his great trilogy; the ornate and forbidding EDIFICE of Gormenghast may reflect something of the Forbidden City of Peking. Indeed, MP's rendering of LANDSCAPE is so expressionist, so austere and daunting, that readers frequently assume the **Gormenghast** or **Titus Groan** sequence – *Titus Groan* (**1946**), *Gormenghast* (**1950**) and *Titus Alone* (**1959**; rev by Langdon Jones 1970; with additional material, as coll 1991 US), all three assembled as *The Titus Books* (omni **1983**; vt *The Gormenghast Trilogy* 1988 US) – must unequivocally be set in a SECONDARY WORLD, but MP's signals as to the nature of his venue are ambiguous throughout. The first volume seems to be set in an exaggerated (but not *impossible*) version of this world; the third volume is set in a surreal, dystopic WONDERLAND; and the second volume is indeed set in something closely resembling a secondary world.

The sequence ostensibly recounts the life of Titus Groan, 77th Earl of Groan, from birth through young manhood; but the infant Titus is hardly visible in the first volume, his dark SHADOW Steerpike dominates the second, and the somewhat faceless young heir only comes into his own in the surrealistic, sf-like third volume, after he has become an exile from his home and inheritance. A projected fourth volume may (or may not) have returned Titus to his domain, where some fitting RITE-OF-PASSAGE climax might have been mounted. The books – the first and third of which can be read alone without difficulty – are radically incomplete as a sequence.

Titus Groan is structured around the geography of the edifice of Gormenghast rather than around events as such – the main event being the birth of Titus. Traversing the vast aisles and abyssal LABYRINTHS of Gormenghast, the text introduces readers to the large cast, all of whom are defined in terms of their location in the hierarchical structure of the building. These characters include Lord Sepulchrave, a paradigm example of the KNIGHT OF THE DOLEFUL COUNTENANCE: withdrawn, withered, haunted by the RITUAL observances which dominate the operations of Gormenghast, and ultimately doomed (after his LIBRARY has burned down) to be eaten by owls. Other characters – like Sourdust and Barquentine, who control Ritual, Titus's simplemindedly sullen and obsessive sister Fuschsia, his awful and dimwitted Nannie Slagg, Flay, Swelter and Dr Prunesquallor – seem to be affects of the building itself. Though there is nothing literally fantastic in their lives, the ensemble of those lives is so improbable and so extreme as

to constitute a fantastic vision of the nature of the world.

In the first two books, the scullion Steerpike forces his *Realpolitik* way up the corrupt and vulnerable hierarchy of the Gormenghast world, until the half-grown Titus finally dispatches him as a great FLOOD threatens to demolish everything. (Gormenghast rather resembles the island of Sark – where MP spent much time – in profile; and when partially submerged is described as an ARCHIPELAGO.) The tale of Steerpike's rise and fall is as close to a plot as the trilogy contains, and is at points superbly conveyed. *Titus Alone* is written in a condensed fragmentary style which may reflect MP's fatal illness, which progressively crippled his faculties; but (in the restored version of 1970) the narrative clearly benefits from paring down, as Titus hurtles through a world evocative of Franz KAFKA's, and finally rediscovers his home domain, only to turn his back on it, in an abrupt final page. In an associated novella – *Boy in Darkness* (in *Sometime, Never* ed Kenyon Calthrop anth **1976**; **1976** chap) – Titus (here unnamed) undergoes BEAST-FABLE experiences with human/animal figures; typically of the form, these experiences constitute part of his education.

The trilogy is perhaps the most intensely visual fantasy ever written. It is like a GOTHIC FANTASY stripped of all plot accretions, of all intrusions of the supernatural, until only impacted ambience remains. It is something of a masterpiece, and is certainly *sui generis*.

MP's other full-length text, *Mr Pye* (**1953**), is governed by a SLICK-FANTASY plot, somewhat to its detriment. The eponymous hero comes to Sark, where his manipulation of the islanders for their own good causes the "Great Pal" to give him the slowly growing wings of an ANGEL; fighting back with deliberate wickednesses, Pye learns the impossibility of perfect BALANCE and finds himself sporting horns.

MP remains important for the central sequence. He is the most potent visionary the field has yet witnessed. [JC]
Other works (poetry): *Shapes & Sounds* (coll **1941** chap); *Rhymes Without Reason* (coll **1944** chap); *The Glassblowers* (coll **1950** chap); *The Rhyme of the Flying Bomb* (**1962** chap); *Poems and Drawings* (coll **1965**); *A Reverie of Bone and Other Poems* (coll **1967** chap); *Selected Poems* (coll **1972**); *A Book of Nonsense* (coll **1972**); *Twelve Poems 1939-1960* (coll **1975** chap).
Other works (miscellaneous): *Captain Slaughterboard Drops Anchor* (**1939** chap; rev 1945); *The Craft of the Lead Pencil* (**1946**); *Letters from a Lost Uncle* (**1948** chap); *Drawings by Mervyn Peake* (graph **1950**); *Figures of Speech* (graph **1954**); *The Drawings of Mervyn Peake* (graph **1974**), text by Hilary Spurling; *Mervyn Peake: Writings and Drawings* (coll **1974**), ed Maeve Gilmore (MP's widow) and Shelagh Johnson; *Peake's Progress: Selected Writings and Drawings* (coll **1978**; rev 1981), ed Maeve Gilmore; *Sketches from Bleak House* (graph **1983**).

PEARCE, (ANN) PHILIPPA (CHRISTIE) (1920-   )
UK writer of juvenile fiction; her earliest books were signed A. Philippa Pearce. She followed the nonfantastic *Minnow on the Say* (**1955**), illustrated by Edward ARDIZZONE, with the Carnegie Medal-winning *Tom's Midnight Garden* (**1958**), a classic TIMESLIP romance whose young hero meets a girl from Victorian times in a ghostly garden and learns several important lessons in the course of their problematic relationship. Although not a fantasy, *A Dog So Small* (**1962**) is a celebration of the power of the imagination. Some of PP's works for younger children are fantasies; more interesting material is found in her collections, especially the eerie tales

in *The Shadow-Cage and Other Tales of the Supernatural* (coll **1977**) and *Who's Afraid? and Other Strange Stories* (coll **1986**), which deftly adapt the tradition of M.R. JAMES for teenage readers. The title story of the former collection, about the effects of an enigmatic bottle upon its various custodians, is particularly notable.   [BS]

Other works: *The Strange Sunflower* (**1966**); *The Children of the House* (**1968**; vt *Children of Charlecote*) with Brian Fairfax-Lucy; *Squirrel Wife* (**1971**); *Beauty and the Beast* (**1972**); *What the Neighbours Did and Other Stories* (coll **1972**); *A Lion at School and Other Stories* (coll **1985**); *The Battle of Bubble and Squeak* (**1987**); *Emily's Own Elephant* (**1987** chap); *Old Belle's Summer Holiday* (**1989** chap); *The Toothball* (**1987**); *In the Middle of the Night* (**1991**); *Fresh* (**1992**); *Here Comes Tod* (**1992**).

PEARSON'S MAGAZINE ◊ MAGAZINES.

PEDROLINO ◊ COMMEDIA DELL'ARTE.

PEI, MARIO (ANDREW) (1901-1978) Italian-born US educator and writer, who long advocated the creation of a universal language; his popular studies, like *The Story of Language* (**1965**), were influential. In *The Sparrows of Paris* (**1958**) the eponymous international cabal of criminals and Communists – headed by a woman who can turn herself into a CAT – plans to transform Americans into WERE-WOLVES. *Tales of the Natural and Supernatural* (coll **1971**) collects stories of mild fantasy interest.   [JC]

PÉLADAN, JOSÉPHIN (1859-1916) French writer. ◊ DECADENCE.

PENALURICK, JAN [s] ◊ Charles DE LINT.

PENNINGTON, BRUCE (1944-   ) UK sf/fantasy artist and illustrator, with an epic imagination and a strong, colourful painting style. Although his representations of the human figure can sometimes be a little stiff, his work nevertheless has a distinctive clarity and richness of colour which he uses to telling effect.

BP from 1967 worked as a freelance book-cover illustrator, doing Westerns, historical novels and then a cover for Robert A. HEINLEIN's *Stranger in a Strange Land* (**1961**), after which he tended to specialize in sf and fantasy. His most important achievement was the apocalyptic *Eschatus* (graph **1977**), a sequence of paintings representing the prophecies of NOSTRADAMUS and the BIBLE interpreted as a vision of the future: a new Dark Age whose eventual downfall comes about with the arrival of a long-awaited Messiah who heralds a millennium of peace.   [RT]

Other works: *The Bruce Pennington Portfolio* (graph coll **1991**; cut vt *B.P.* **1994** chap).

Further reading: *Ultraterranium: The Paintings of Bruce Pennington* (graph coll **1991**).

PERCEPTION In many systematizations of MAGIC its effects are on characters' perceptions rather than the real world: its workings are ILLUSION. Anodos in George MAC-DONALD's *Phantastes* (**1858**), when influenced by a SHADOW, sees people and places as uglier and cruder than before, which may or may not be a true insight. A woman's GHOST in Charles WILLIAMS's *All Hallows Eve* (**1945**) wanders an erie LONDON which to her seems empty, for she cannot perceive the living. Characters' varying perceptions of the eponymous *something* in Fritz LEIBER's "A Bit of the Dark World" (1962) emphasize its inherent WRONGNESS. The villain of George Dunstan MARTIN's *The Soul Master* (**1984**) controls his slave army by POSSESSION, but lacks a full perception of the world: in an effective metaphor, his slaves are not only unable to see places within the resulting "gaps" but

cannot exist in them, seeming to teleport instantly across the unknown regions.

More interestingly, it could be argued that, if FANTASY (and debatably the literature of the FANTASTIC as a whole) has a purpose other than to entertain, it is to show readers *how to perceive*; an extension of the argument is that fantasy may try to alter readers' perception of REALITY. Of course, quack RELIGIONS (etc.) make similar attempts, but a major difference is that, while the latter attempt to convert people to their codified way of thinking, the best fantasy introduces its readers into a PLAYGROUND of rethought perception, where there are no restrictions other than those of the human imagination. In some modes of the fantastic – e.g., MAGIC REALISM and SURREALISM – the attempt to alter the reader's perception is overt, but most full-fantasy texts have at their core the urge to *change* the reader; that is, full fantasy is by definition a subversive literary form. This also in part explains why there is such a smooth continuum between written fantasy and FANTASY ART.

Other overt examples are particularly noteworthy in fantasy CINEMA: DICK TRACY (*1990*) forces us to perceive its world as if we were living in a COMIC strip; WINGS OF DESIRE (*1987*) makes us see our surroundings as an ANGEL might; *The* BRAVE LITTLE TOASTER (*1987*) makes us see electrical appliances as *people like us*, whereas human beings become potential GODS and thus mere appendages to the real world (the movie uses further perceptual devices: whenever a human looks at a gadget it is suddenly nothing more than a piece of machinery); Jan SVANKMAJER's FAUST (*1994*) toys with our perceptions to the extent that for hours after any viewing it is hard to cling on to the laws of daily logic. Countless other movies could be cited, but such effects are certainly not confined to the movies: for example, in M. John HARRISON's *A Storm of Wings* (**1980** US) we are forced to perceive events from alternating viewpoints, alien and human; in Alan GARNER's *The Owl Service* the relic of Blodeuwedd may be an owl or an array of flowers; in John BARTH's *The Last Voyage of Somebody the Sailor* (**1992**) it is left gloriously unclear as to which memory-string represents true history.

Further aspects of perception are important in fantasy. Three can be described together, when deployed, as "the Fantasy of Perception". The first, and probably the least interesting, is typified by the RATIONALIZED FANTASY, wherein the author knows what is going on but the reader does not – that is, the author plays narrative tricks to mislead our perception of events. An example of this being done quite well is the movie *The* BEAST WITH FIVE FINGERS (*1946*), where at the end it is revealed that we have been watching events – notably an amputated hand wreaking VENGEANCE – through the eyes of the insane murderer.

This example melds neatly into another variant of the fantasy of perception, in which not only the reader but the protagonist misperceives the STORY, so that there are sudden reversals of understanding. At its crassest, this is represented by the "I awoke and found that it had all been just a DREAM" tale, but there are many more sophisticated manipulations, as in John FOWLES's *The Magus* (**1965**), where a young man is the victim (as are we) of a carefully constructed GODGAME, and the movie ROSEMARY'S BABY (*1968*), which can be read as the tale of Rosemary's gross misperception of her predicament – she is being possessed (to use the term loosely) all right, but perhaps merely by her new sexual experiences rather than by the offspring of the DEVIL, with all else being a quasi-hallucinatory by-product. (Ira LEVIN's original *Rosemary's Baby* [**1967**] was less effective in this respect.) A similar reading is overwhelmingly possible for BARTON FINK (1990): in such terms, Barton is indeed transported into an alien landscape when he is brought to Hollywood, and his exaggerated perception of its alienness is a self-feeding cause, symptom and result of his mental decline. In William GOLDMAN's MAGIC (both book [**1978**] and movie [**1978**]) the ventriloquist dummy is independently intelligent because Corky perceives it as such – which is precisely why the tale is one of nightmare. In Terry GILLIAM's *The* FISHER KING (1991) the protagonist eventually finds the fantastication more rational than the rational. Gene WOLFE's *There Are Doors* (**1988**) *can* be read as an ALTERNATE-WORLD/ALTERNATE-REALITY fantasy, but it can equally well stand as a novel in which the narrator is constantly being deluded (or is deluding himself) about what is happening to him. In Donna TARTT's *The Secret History* (**1992**) events are so misinterpreted by the narrator that the past becomes unsure to both him and us. In Theodore STURGEON's *tour de force* of perceptual shifts, "To Here and the Easel" (1954), the reader soon sees that the frustrated artist in the mundane USA and the imprisoned KNIGHT in the setting of ARIOSTO's *Orlando Furioso* (**1516**) are the same protagonist, the fantasy obstacles being perceptual metaphors for a REALITY he cannot bear; the same interpretation of an apparent SECONDARY WORLD is indicated in William MAYNE's *A Game of Dark* (**1971**). The impact of "The Turn of the Screw" (1898) by Henry JAMES – filmed as *The* INNOCENTS (*1961*; vt *The Turn of the Screw*) – derives from the fact that we do not know if the governess's perception of GHOSTS is valid, or even if her belief that the now-dead couple's love affair psychologically scarred the children is justified.

All of the above examples play with our (and their protagonists') perception, but the Gilliam movie and the Wolfe novel take matters further: it becomes easier to believe the fantasy version of events than the rationalization (◊ RATIONALIZED FANTASY) that is on offer. Both the protagonists and ourselves experience a *perceptual shift* (again, this relates to the notion of fantasy as a deliberately perception-changing literature). Sometimes this can be right on the surface – for example, in HOOK (*1991*) the ageing PETER PAN has retained his realist perception throughout the voyage to Neverland and the discovery of all its strangenesses, so that when the Lost Boys invite him to join their feast he sees only empty plates; the turning-point of the movie comes when his perception shifts such that he can see the food – a moment of RECOGNITION. Perceptual shifts are the key to sf's conceptual breakthrough, when suddenly a new "truth" is revealed to both protagonist and reader; more profoundly, in – for example – a SLINGSHOT ENDING, no new "truth" is directly revealed, only that the old "truth" was a simplification or outright falsity, so that the interaction between reader and text is such that the reader's imagination spirals away into new worlds of perception that, in the end result, may have only a remote relationship to the original text. This need not occur at the end of the text – it need not be a sort of open-ended "twist in the tail". Overall, the most significant piece of fantastication in Gilliam's *The* ADVENTURES OF BARON MUNCHAUSEN (*1989*) is that it forces us into a whole string of perceptual shifts, notably concerning the venue in which events are actually occurring: On a stage? In "reality"? Inside the head of the Baron, a territory into which the child (with us in tow) can gain admission? Inside the head of the fascinated child?

A further (minor) issue of perception relates to critical analysis. For example, many of the novels of Sheri S. TEPPER – such as the series begun with *Grass* (**1989**) – can be read with equal validity as sf or as fantasy set in sf contexts. Such differences of perception are significant, however, mainly only to critics and to publishers' sales department: an intelligent author like Tepper has her own conceptual and philosophical agenda, so that categorization is unimportant, and her readers presumably have a similar attitude.

[JG/DRL]

See also: FANTASIES OF HISTORY; HALLUCINATION; INVISIBLE COMPANION; LITTLE BIG; PSYCHOLOGICAL THRILLER; TROMPE L'OEIL; Tsetvan TODOROV.

PERCEVAL In Arthurian LEGEND, a peasant boy who becomes a KNIGHT of ARTHUR's ROUND TABLE..In the early ROMANCES his role is central to the GRAIL; in later stories he takes something of a secondary seat to GALAHAD. Nevertheless, it is in the stories about Sir Perceval that the core of the Grail mythology is presented.

Perceval (sometimes rendered Percival) first appears in the unfinished *Perceval, ou Le Conte del Graal* (begun ?1182) by CHRÉTIEN DE TROYES, though Chrétien almost certainly drew the story from Celtic tradition and probably from the same source as the later Welsh text *Peredur* (?**1250**), which forms part of the MABINOGION. In both stories Peredur/Perceval is a lowly son of noble descent who works with farm animals and the soil. His mother has kept him ignorant of the ways of the world, but a chance encounter with a band of knights captures the boy's imagination and he eventually makes his way to Arthur's court where, by a series of challenges, he proves his worth. One of these challenges results in the boy witnessing a strange procession, variously described: the essence of all accounts is that a squire (or squires) carries a Bleeding Lance, followed by candlesticks or silver salvers, followed in turn by a vessel that purports to be the Grail. In Chrétien's *Perceval* this happens at the castle of the FISHER KING, and Perceval's failure to ask key questions about the nature of the Grail and whom it serves perpetuates the enchantment on the Fisher King, which enchantment results in the WASTE LAND. Perceval subsequently attains the Grail and even succeeds the Fisher King.

The Perceval story was developed in other medieval texts by writers who became fascinated by the Grail and with Chrétien's unfinished work. Most of these were French writers at the court of Count Philip of Flanders, and were writing in the period 1200-1250; they included Gauchier de Donaing, Manessier and Gerbert de Montreuil. Their work has been painstakingly edited by William Roach (1907-) in *The Continuations of the Old French "Perceval" of Chrétien de Troyes* (5 vols 1949-83 US). Roach earlier assembled another anonymous text known as *The Didot-Perceval* (**1200**; ed Roach **1941** US), named after the French owner of the manuscript. This version drew heavily on the now lost *Perceval* of the Burgundian poet Robert de Boron (? -1212). The other main text of this period, also an expansion of Chrétien's *Perceval*, is *Parzival* (*c*1200-1210), by Wolfram von Eschenbach (?1170-?1220), which was the main source of Richard WAGNER's *Parsifal* (1882). These texts add as much, if not more, to the ritual aspects of the Grail legend as they do to the Arthurian romance, and incorporate elements which have since become fundamental to Masonic and other traditions.

Although Perceval's role was diminished in later romances, his circumstance as a peasant-boy-turned-HERO has become a fundamental device in all fiction and a cliché of modern fantasy (especially GENRE FANTASY); examples include the **Prydain** series by Lloyd ALEXANDER, which drew directly upon the original legend, and the **Belgariad** by David EDDINGS. Perceval's story contains parallels with that of Arthur himself, and he certainly has many of the traits of the HIDDEN MONARCH – though his links with the Grail suggest a more fundamental Christian or even pre-Christian FERTILITY cult.

Perceval was used as a role-model of Christian virtue in *Sir Percival: A Story of the Past and the Present* (**1886**) by Joseph Henry Shorthouse (1834-1903), an ALLEGORY that challenged the new Victorian values. His story also appears in *Perronik the Fool* **1926**) by George Moore (1852-1933), though Moore drew his inspiration from a Breton folktale. In modern fantasy the most complete picture of Perceval's character and transcendence is *Percival and the Presence of God* (**1978**) by Jim Hunter (1939-   ); two other works exploring different facets of the character are *Firelord* (**1980**) by Parke GODWIN and the **Parsival** trilogy (**1977-80**) by Richard MONACO. Chrétien's original tale was filmed by Eric Rohmer (1920-   ) as *Perceval le Gallois* (*1978*).   [MA]
Further reading: *The Legend of Sir Perceval: Studies Upon its Origin, Development and Position in the Arthurian Cycle* coll (**1906-9**) by Jessie L. WESTON.

PERCY, THOMAS (1729-1811) UK antiquary. ◊ ROMANCE.
PEREDUR ◊ MABINOGION; PERCEVAL.
PERRAULT, CHARLES (1628-1703) French lawyer and civil servant who became the father of the FAIRYTALE. Enchanted by the FABLES of Jean de La Fontaine (1621-1695), CP penned some of his own in verse, though none were fantastic other than through their anthropomorphic use of animals. His first fantastic work was in the poem "Les souhaits ridicules" ["The Ridiculous Wishes"] (1693 *Le Mercure Galant*), which utilized the traditional THREE-WISHES story. This, with another poem, "Peau d'âne" ["Donkey-Skin"] (1693 *Le Mercure Galant*), was later published alongside an earlier nonfantastic poem as *Griseldis, Nouvelle, Avec le conte de Peau d'âne et celui des souhaits ridicules* (coll **1694**). CP now switched from verse to prose narrative. His *Histoires ou Contes du temps passé, avec des Moralitez* (coll **1697**; trans Guy Miège as *Histories, or Tales of Past Times* **1729** UK), which contained eight stories, had earlier existed as a five-story presentation manuscript dedicated to Princess Élizabeth-Charlotte, Louis XIV's niece, in 1695. The dedication was inscribed as by P. Darmancour, the name adopted by CP's son, Pierre Perrault (1678-1700), which gave rise to the belief that the son had contributed to and possibly even written the stories, a suspicion that has never totally been laid to rest. Of the stories in the published book, seven have since become standards, with CP's text serving as the primary source. The stories were "La Belle au Bois Dormant" ["The Beauty Sleeping in the Wood"] (earlier published in *Le Mercure Galant* February 1696) (◊ SLEEPING BEAUTY), "Le Petit Chaperon Rouge" ["Little Red Riding-Hood), "La Barbe Bleue" ["Bluebeard"], "Le Maître Chat ou le Chat Botte" (◊ PUSS-IN-BOOTS), "Les Fées" ["The Fairies"; trans vt "Diamonds and Toads"], "Cendrillon ou la Petite Pantoufle de Verre" (◊ CIN-DERELLA), "Riquet à la Houppe" ["Ricky with the Tuft"] and "Le Petit Poucet" (◊ TOM THUMB). All were drawn from oral tradition, and some had been evident in earlier collections of tales, though "Little Red Riding-Hood" may be original to CP; he probably derived it from a children's game. CP added further touches that were his own, most notably the glass slipper in "Cinderella", so that, although other folklorists collected this story from their own oral tradition – as in "Aschenputtel" by the Brothers GRIMM – it is CP's version that remains the definitive text. Because he also added short verses at the end of each story highlighting the moral of the tale, they rapidly became instructive as well as entertaining for children, so it can be said that CP first developed the fairytale as a children's literature, whereas his contemporary, Madame d'AULNOY, used the medium as a vehicle for SATIRE and thus more adult entertainment.

CP indirectly gave rise to another popular character. The frontispiece to the French edition of his fairytales depicted an old lady telling stories to a group of children; behind them, a plaque on the wall stated: "Contes de ma mère l'Oye." This was translated in the first English-language edition as "Mother Goose's Tales". Certainly by 1768 the title *Mother Goose's Tales* (coll **1768**) had supplanted CP's original, so giving rise to a new character in nursery rhymes (◊ MOTHER GOOSE).   [MA]
Other works and editions: *Recueil de divers ouvrages en prose et en vers* ["Collection of Diverse Works in Prose and in Verse"] (coll **1675**) which includes his earliest fables. Of the many editions of CP's fairytales, some have been beautifully illustrated; e.g., *Fairy Tales by Charles Perrault* (trans coll **1913** UK) illus Charles ROBINSON; *Old Time Stories told by Master Charles Perrault* (coll **1921** UK) trans A.E. Johnson, illus W. Heath ROBINSON (vt, but with illus by Gustave DORÉ from *Perrault's Fairy Tales* [trans coll **1912** UK] *Perrault's Fairy Tales* [trans coll **1969** US] ed E.F. BLEILER); *The Fairy Tales of Charles Perrault* (trans coll **1922** UK) illus Harry Clarke (1889-1931); *Tales of Passed Times* (trans coll **1922** UK) illus John Austen (1886-1948). Other revised translations of note include: *Perrault's Popular Tales* (coll **1888** UK) ed Andrew LANG; *The Fairy Tales of Master Perrault* (coll **1897**) trans Walter Ripman; *The Fairy Tales of Charles Perrault* (coll **1950** UK) trans Norman Denny; *The Fairy Tales of Charles Perrault* (coll **1957** UK) trans Geoffrey Brereton; *Perrault's Fairy Tales* (coll **1972** US) trans Sasha Moorsom; *The Fairy Tales of Charles Perrault* (coll **1977** UK) trans Angela CARTER. Jack ZIPES included all the tales in a new translation in his *Beauties, Beasts and Enchantment* (anth **1989** US).
Further reading: *Mémoires de ma vie* (cut **1759**; first full edition 1909; most recent trans *Memoirs of My Life* ed Jeanne Morgan Zarucchi **1989** US, with extensive introduction and notes) by CP; *Charles Perrault* (**1981** US) by Jacques Barchilon and Peter Flinders; *Perrault's Morals for Moderns* (**1985** US) by Jeanne Morgan.

PERRAULT, PIERRE French folklorist (1678-1700) ◊ Charles PERRAULT.

PERSIAN LITERATURE Modern fantasy's first and most important debt to PL is Zoroaster's cosmological dualism, which would be taken over by Jews in a weaker, monotheistic, version (◊ JEWISH RELIGIOUS LITERATURE); both forms of this dualism underlie most mythopoeic fantasy. The syncretistic religion founded by Mani (216-276) had a strikingly dramatic dualist account of cosmology, accessibly presented by Jes P. Asmussen in *Manichaean Literature* (anth **1975**). Another important aspect of Persian literature from an early date is its role in transmitting stories between East and West. To approach one such case, see Ramsay Wood's *Kalila and Dimna* (**1980**).

The closest analogue to fantasy in classical Persian literature is the elaborate mysticism prominent in Sufi writing, although Westerners rarely find that the poetry this led to reads like fantasy. The numerous classical works which do so read include: *Manteq at-Tair* (**1177**; trans Afkham Darbandi and Dick Davis as *The Conference of the Birds* **1984**), by Farid ud-Din Abu-Hamid Muhammad 'Attar (*c*1130-*c*1220), whose plot John CROWLEY borrowed in *Little, Big* (**1981**); the Persian national epic, *Shah-nama* (**1010**; much cut trans Reuben Levy as *The Epic of the Kings* **1967**) by Ferdowsi or Firdausi (*c*935-*c*1020); and *Haft Paykar* (**1197**; trans Julie Scott Meisami **1995**) by Abu Muhammad Ilyas ibn Yusuf ibn Zaki Mu'ayyad (*c*1141-*c*1209) writing as Nizami Ganjavi, a story from which became the plot of the various *Turandot* OPERAS.

In this century, Iranian modernists have on occasion adopted tools reminiscent of fantasy or FABULATION. Such novels include *Buf-i Kur* (**1937**; trans D.P. Costello as *The Blind Owl* **1957**) by Sadegh Hedayat (1903-1951) and *Sang-e Sabur* (trans Mohammed R. Ghanoonparvar as *The Patient Stone* **1979**; rev 1989) by Sadeq Chubak (1918-   ). Iranian drama has been much influenced by Absurdist and similar models, as in *Seh nemayeshname-ye 'arusaki* (**1963**; trans Gisele Kapuscinski as "Three Puppet Shows" in *Modern Persian Drama* anth **1987**) by Bahram Beyza'i (1938-   ). Other writers have begun using the tools of MAGIC REALISM; an example is "Sara" (1974; trans Farzin Yazdanfar in *A Walnut Sapling on Masih's Grave* anth **1993** ed John Green and Yazdanfar) by Sharnush Parsipur (1946-   ).

As for works closer to the Western fantasy genre, these do exist, but it has not proven possible to learn much about them. Massoud Khayyam, an aerospace engineer, has written *Qafas-i shatranj* ["The Chess Cage"], a "science-fiction fantasy". Meanwhile, although Arabic literature has historically included less of the fantastic, this has changed somewhat over the past three decades, and some of the contents of *Flights of Fantasy* (anth **1985**) ed Ceza Kassem and Malak Hashem can be profitably read as genre fantasy. And – like many Arabic writers – influenced by *Alf layla wa layla* ["One Thousand and One Nights"] (◊ ARABIAN FANTASY), the Turkish writer Güneli Gün has written *On the Road to Baghdad* (**1989**). There is every reason to believe that further research into modern fantasy in Iran and elsewhere in the region will be productive.

A good starting point for consideration of Persian literature is *Persian Literature* (anth **1988**) ed Ehsan Yarshater.   [JB]

PERSONA Swedish movie (*1966*). Svensk Filmindustri. **Pr** Ingmar BERGMAN. **Dir** Bergman. **Spfx** Evald Andersson. **Screenplay** Bergman, published as *Persona* (**1966**; trans in *Persona and Shame* **1972** US). **Starring** Bibi Andersson (Sister Alma), Gunnar Björnstrand (Vogler), Margaretha Krook (Dr Läkaren), Jörgen Lindström (Boy), Liv Ullmann (Elisabeth Vogler). 84 mins. B/w.

Almost certainly the most disturbing movie ever made about IDENTITY EXCHANGE, *P* has attracted countless interpretations rooted in psychoanalysis. Above all, we are never allowed to forget that *P* is a *movie*: it opens with the starting up of a projector and a jumbled sequence of images – some loathsome – from other movies; at a crux moment later on the projector apparently jams and the film is burnt out, with a further jumble of film clips before the main action recommences; close to *P*'s end we see for a moment the crew at work filming Ullmann; and the movie ends with the film coming free of the projector's sprockets.

That introductory jumble of filmic images may represent the suppressed images in the mind of either of the two main characters: Elisabeth, an actress who has suddenly become dumb/autistic; Alma, the nurse assigned to try to lure her back to speech during a long vacation in a remote beach cottage. Yet these events are set back from us by the use of something akin to a FRAME STORY: the first coherent minutes of *P* show a young boy waking in a mortuary (though whether this is a normal awakening or the emergence of his no-longer-incarnate SOUL is dubious) and projecting the overlain images of the two women's faces. As we find later, this sequence has a dual meaning: the women may be the products of the boy's PERCEPTION, or the consignment of the boy to a mental mortuary may be a product of Elisabeth's perception of the son she abhors.

The events at the cottage are superficially simple. Since Elisabeth is silent, even her facial reactions being minimal, Alma talks for both of them; as the relationship between them becomes close to that between lovers (or so Alma assumes), her talk becomes more revelatory and confessional, displaying an honesty Alma has hitherto allowed only in communicating with herself. But Alma reads one of Elisabeth's letters to her husband, and realizes Elisabeth has no love for her – indeed, regards her as a child, to be patronized. At this, Alma's love turns swiftly to hatred (the turning of this tide is marked by the jamming of the projector). Onto the *tabula rasa* of Elisabeth she projects motivations that will explain her own behaviour. Only one woman leaves the cottage: her physical appearance is Alma's, yet she may be Elisabeth, or perhaps Elisabeth/Alma ...

P is a powerful movie, and requires repeated viewing. The two central performances are spellbinding, especially Andersson's. This was Ullmann's first movie, although she already enjoyed a successful career on the Norwegian stage: Bergman chose her because of her physical resemblance to Andersson, though this is not especially pronounced – *P* gains from the fact that the facial differences of the two women become ignored: it is their *mental* congruence that makes them come to seem identical.   [JG]

PERUTZ, LEO (1884-1957) Austrian novelist and playwright; he left Austria for Israel after the 1938 Anschluss. Although he is not by instinct an sf writer, *Der Meister des juengsten Tages* (**1923**; trans Hedwig Singer as *The Master of the Day of Judgment* **1929** UK) can be understood as sf through the specificity of its premise that an ancient hallucinogen, when breathed by men of ambition, makes them see themselves so nakedly that they are forced to commit suicide; the eponymous wheat fungus featured in *Sanct Petri-Schnee* (**1933**; trans E.B.G. Stamper and F.M. Hodson as *The Virgin's Brand* **1934** UK; new trans Eric Mosbacher as *Saint Peter's Snow* **1990** UK) has the opposite effect on its human victims: it gives them faith.

Most of LP's work is distinctly fantasy or SUPERNATURAL FICTION, ironic, well crafted, engendering something of the same hectic doom-laden civility that so marks the work of, say, Karel ČAPEK. *Zwischen Neun und Neun* (**1918**; trans Lily Lore as *From Nine to Nine* **1926** US) is a POSTHUMOUS FANTASY in which – as in Ambrose BIERCE's "An Occurrence at Owl Creek Bridge" (1891) – the protagonist's long, complicated flight from a terrible event takes place, in reality, in an instant, at the point of death. In *Der Marques de Bolibar* (**1920**; trans Graham Rawson as *The Marquis de Bolibar* **1926** UK; new trans John Brownjohn as *The Marquis of*

Bolivar 1989 UK), the WANDERING JEW and the spirit of the eponymous marquis – which has taken POSSESSION of the narrator – narrowly defeat a German regiment fighting for Napoleon. Other fantasies include: *Die Geburt des Antichrist* ["The Birth of the Antichrist"] (**1921**); *Das Mangobaumwunder* ["The Miraculous Mango Tree"] (**1923**) with Paul Frank, in which a WIZARD has magically linked the fate of the Baron, his employer, to the eponymous tree; and *Nachts unter der steinernen Brücke* (**1953**; trans Eric Mosbacher as *By Night Under the Stone Bridge* 1989 UK), an atmospheric evocation of the medieval PRAGUE of Rabbi Loëw, creator of the first GOLEM. [JC]

PESTILENCE ◊ FOUR HORSEMEN.

PETER PAN The magical boy who can fly, who lives in NEVER-NEVER LAND, who is garbed in "the juices that ooze out of trees", and who refuses to grow up. He first appears in J.M. BARRIE's *The Little White Bird* (**1902**), in a section of that book later published as *Peter Pan in Kensington Gardens* (**1906**), illustrated by Arthur RACKHAM, where it is recounted that, as a baby, he became a friend (and victim) of FAIRIES, who bestowed unending childhood upon him (◊ TIME IN FAERIE). He is the hero of the play *Peter Pan, or The Boy who Would not Grow Up* (performed 1904; rev 1905; rev **1928**), reappearing as a pathos-choked REVENANT who cannot remember the passage of the SEASONS in *When Wendy Grew Up: An Afterthought* (produced 1908; **1957** chap) (◊◊ BELATEDNESS). He also features in Barrie's prose CHILDREN'S FANTASY *Peter and Wendy* (**1911**; vt *Peter Pan and Wendy* 1921; vt *Peter Pan* 1951). His first name was almost certainly taken from Peter Llewellyn Davies (1897-1960), one of several Davies brothers who obsessed the asexual, melancholic, haunted Barrie; his second name reflects the suburbanized PAN. A perpetual boy, he has intermittent longings, but is not a REAL BOY because ultimately he is doomed to changelessness.

The first movie version of the play was *Peter Pan* (**1924**); the most famous is PETER PAN (*1953*), an ANIMATED MOVIE from DISNEY; the most recent is HOOK (*1991*) from Steven SPIELBERG. The actress Mary Martin (?1913-1990) was known for playing the role on stage – PP was normally played by grown women – and she starred in the US tv version in 1955. Gilbert ADAIR's *Peter Pan and the Only Children* (**1987**) and Toby Forward's *Neverland* (**1989**) are among the SEQUELS BY OTHER HANDS.

Because of the surreal BELATEDNESS and fixity of his fate, the PP figure is more likely to serve as an UNDERLIER for HORROR than for fantasy. An example is "The Taking of Mr Billy" (1993) by Graham Masterton (1946-   ), which begins as a GASLIGHT ROMANCE but which unfolds as the tale of an unaging child-murderer named Piotr Pan. [JC]

Further reading: *Fifty Years of Peter Pan* (1954) by Roger Lancelyn GREEN; *The Peter Pan Chronicles: The Nearly One-Hundred-Year History of the Boy who Wouldn't Grow Up* (1993) by Bruce K. Hanson.

PETER PAN US ANIMATED MOVIE (*1953*). Disney. **Pr** Walt DISNEY. **Dir** Clyde Geronimi, Wilfred Jackson, Hamilton Luske. **Special processes** Ub IWERKS. **Based on** the various versions of PETER PAN by J.M. BARRIE. **Voice actors** Kathryn Beaumont (Wendy), Candy Candido (Indian Chief), Paul Collins (John), Hans Conried (Hook/Mr Darling), Bobby Driscoll (Peter), Tommy Luske (Michael), Bill Thompson (Smee/pirates). **Special credit** "With gratitude to the Hospital for Sick Children, Great Ormond Street, London." 77 mins. Colour.

Unlike the case with many other DISNEY animated features, the story in *PP* differs in detail but not in substance from the original play, the sole significant additional piece of fantastication being that, at movie's end, all the children travel back to London in the PIRATE ship, which is able to travel through the sky thanks to "pixie dust" supplied by Tinker Bell. As proof of her adventures, Wendy is able to point out to her parents the silhouette of the ship against the full Moon; Mr Darling sheepishly admits that when he was a boy he, too, once saw just such a flying ship ...

This version of the Barrie tale is of particular interest, however, in that it raised Tinker Bell from a blob of light to a cute, adult little female figure, a tradition continued in HOOK (*1991*). The figure has been very widely used by Disney but, more importantly, has become, independent of Peter Pan, a significant modern fantasy ICON. [JG]

See also: NEVER-NEVER LAND.

PETERS, ELLIS Pseudonym of Edith PARGETER.

PETE'S DRAGON US live-action/ANIMATED MOVIE (*1977*). Disney. **Pr** Jerome Courtland, Ron Miller. **Dir** Don Chaffey. **Spfx** Art Cruickshank, Danny Lee, Eustace Lycett. **Fx anim** Dorse A. Lanpher. **Anim art dir** Ken Anderson. **Anim dir** Don BLUTH. **Screenplay** Malcolm Marmorstein. **Starring** Cal Bartlett (Paul), Red Buttons (Hoagy), Jim Dale (Dr Terminus), Sean Marshall (Pete), Helen Reddy (Nora), Micky Rooney (Lampie), Charles Tyner (Merle), Shelley Winters (Lena Gogan). 129 mins (cut to 106 mins). Colour.

Sold (for complex reasons) into a Maine backwoods family, the Gogans, the boy Pete flees with his INVISIBLE COMPANION, a bumblingly genial (TOON) DRAGON, Elliott. Taken in by lighthouse keeper Lampie and daughter Nora, Pete settles down to being blamed for all the mishaps caused by the generally disbelieved-in Elliott's clumsiness. After various adventures, the INVISIBLE COMPANION is proven to exist and saves the day; he then flies off because another little boy needs him.

This ending has, of course, shades of MARY POPPINS (*1964*), and *PD*'s connection with its DISNEY predecessor was to make it much more disliked than might otherwise have been the case, for *PD* was originally launched by Disney's publicists with the promise that it was the new *Poppins* – a stratagem that had already failed disastrously for *Bedknobs and Broomsticks* (*1971*), another good live-action/animated movie. *PD*, in some ways a children's rehash of HARVEY (*1950*), does have problems with its muddled and somewhat clichéd script (even after desperate cutting and re-editing), but it has many good moments. [JG]

PET SEMATARY A series of two (so far) movies.

1. *Pet Sematary* US movie (*1989*). Paramount. **Pr** Richard P. Rubinstein. **Exec pr** Tim Zinnemann. **Dir** Mary Lambert. **Spfx** Image Engineering Inc. **Mufx** David Anderson, Lance Anderson. **Vfx** Fantasy II Film Effects Inc. **Screenplay** Stephen KING. **Based on** *Pet Sematary* (1983) by King. **Starring** Blaze Berdahl (Ellie Creed), Denise Crosby (Rachel Creed), Brad Greenquist (Victor Pascow), Fred Gwynne (Jud Crandall), Andrew Hubatsek (Zelda), Miko Hughes (Gage Creed), Dale Midkiff (Louis Creed). 103 mins. Colour.

Dr Louis Creed comes from Chicago to Ludlow, New England, with wife Rachel, daughter Ellie and toddler son Gage. Their new house is beside a rural road along which speed regular Orinco fuel lorries. Folksy neighbour Jud explains that so many pets have died on the road that local kids have for decades maintained the Pet "Sematary" in the woods nearby. On Louis's first day in the new job he tries unsuccessfully to save accident victim Pascow; the corpse revives for a few moments to warn him it will come to him. That night Pascow's GHOST leads him to the Sematary and points beyond, warning him never to go to "the place where the dead walk". Rachel and the kids go away for a few days, and Ellie's cat is killed on the road. Jud leads Louis beyond the Sematary to an ancient Indian burial place and tells him to bury the cat. A few hours later it returns to the house – alive, but vicious, smelly and glowing-eyed. The scenario is soon re-enacted with Gage, killed by an Orinco lorry. The risen Gage, with the cat as his familiar, slaughters Jud and then Rachel. Louis re-kills the cat and Gage but, incomprehensibly, buries Rachel in the Indian cemetery. She returns and kills him.

King's novel is one of his best but his script makes a rather trite movie. Yet there is some interest in *PS*, notably the depiction of Pascow's well wishing, albeit hardly amiable, ghost, and the use of the thundering Orinco lorries as a sort of Greek chorus. [JG]

2. Pet Sematary II US movie (*1992*). Paramount. **Pr** Ralph S. Singleton. **Dir** Mary Lambert. **Spfx** Peter M. Chesney, Design FX Company, Dean W. Miller. **Mufx** David Barton, Bart Mixon, Steve Johnson's XFX Inc. **Screenplay** Richard Outten. **Starring** Clancy Brown (Gus Gilbert), Anthony Edwards (Chase Matthews), Edward Furlong (Jeff Matthews), Jason McGuire (Drew Gilbert), Jared Rushton (Clyde Parker). 92 mins. Colour.

Some years after **1**, boy Jeff sees actress mother Renee accidentally electrocuted on set, and moves with estranged father Chase to Ludlow; school bully Clyde picks on Jeff, leading him to Sematary; fat Drew befriends Jeff; Drew's vile stepfather Gus, in fit of rage, kills Drew's dog Zowie, and the two lads bury it at Indian graveyard (with obvious consequences); body and cliché counts mount; nuff said. [JG]

PEYTON, RICHARD Pseudonym of Peter HAINING.

PHANTOM UK large digest MAGAZINE, 16 issues, monthly, April 1957-July 1958, published by Dalrow Publications, Bolton, Lancashire; ed Leslie Syddall.

Short-lived, *P* was at the time the only UK magazine devoted entirely to SUPERNATURAL FICTION. It concentrated on GHOST STORIES, but did cover other traditional supernatural themes, especially WEREWOLVES and WITCHCRAFT. From *#4*, through the influence of Forrest J. Ackerman (1916-) and August W. DERLETH, who served as agents providing the bulk of the material for the magazine, *P* began to contain more reprint material, mostly from WEIRD TALES. Of special interest is the final issue, which included new stories by Sophie Wenzel Ellis, Andre NORTON, Bob Olsen (1884-1956) and Idella P. Stone (1901-1982) among the reprints. The UK material was minor and mostly by local authors, though it included rare forays by R. Lionel Fanthorpe (1935-) and John S. Glasby (1928-) outside their Badger Books prison. [MA]

PHANTOM, THE MASKED AVENGER of newspaper COMIC strips and, to a lesser extent, comic books, with a skintight mauve costume, striped shorts and black boots, belt and eyemask, created in 1936 by Lee Falk (1905-). He is a man of mystery, known as The Ghost Who Walks, and is believed to be immortal, but the individual with whom the strip deals is the 21st to hold the title – which has been passed on from father to son since the 16th century, when the son of a seaman who had served as a cabin boy to

Christopher Columbus was washed up on a remote Bengal shore. After seeing his parents murdered by PIRATES he swore an oath on the skull of the murderer to devote his life to combating "all forms of piracy, greed and cruelty" in the jungles of Africa and remote cities of Asia. TP lives in a skull-shaped cave in the Deep Woods and his symbol, the Sign of the Skull, strikes terror into the hearts of wrong-doers. His wife is Olympic gold medallist diver, explorer, pilot, athlete and paramedic Diana Palmer, with whom he shared adventures for four decades before finally marrying in the 1980s. A grey wolf, Devil, is his faithful COMPANION.

TP was drawn at first by Ray Moore; when Moore suffered a grave injury in 1942 the art chores were passed on to Wilson McCoy, then via Bill Lignante to Seymore ("Sy") Barry (1928-), who drew the strip from 1963.

At first *TP* appeared only in daily newspapers; a Sunday strip was marketed by King Features from 1939. There have been several attempts to introduce *TP* into comic books by Harvey, Gold Key, King Features and Charlton Comics; artists have included Wallace WOOD, Jeff JONES and Al WILLIAMSON. [RT]

PHANTOM HITCH-HIKER ◊ URBAN LEGENDS.

PHANTOM OF HOLLYWOOD, THE US movie (*1974* tvm). ◊ PHANTOM OF THE OPERA.

PHANTOM OF THE MALL US movie (*1988*). ◊ PHANTOM OF THE OPERA.

PHANTOM OF THE OPERA At least 11 movies – plus a 1990 tv miniseries starring Burt Lancaster – have been based on *The Phantom of the Opera* (*1911*) by Gaston LEROUX, some more loosely than others. A number of books, aside from novelizations, have been based on an amalgam of Leroux's novel and one or more of the movie versions: they include *Phantom* (*1990*) by Susan Kay, which won the Romantic Novel of the Year Award, and, at the opposite end of the spectrum, *Maskerade* (*1995*) by Terry PRATCHETT. In fact, Leroux's original story is somewhat banal – as if he were making it up as he went along: the abiding importance of the POTO as a fantasy/horror ICON, as marked by the number of movie remakes and quasi-remakes – not to mention the 1986 musical – is almost entirely as a result of **1**.

1. The Phantom of the Opera US movie (*1925*). Universal/Jewel. **Pr** Carl Laemmle. **Dir** Robert Julian. **Photography** Milton Bridebecker, Virgil Miller. **Screenplay** Elliot J. Clawson, Raymond Schrock. **Starring** Arthur Edmund Carewe (Ledoux), Lon Chaney (Erik/Phantom), Snitz Edwards (Florine Papillon), Gibson Gowland (Simon), Norman Kerry (Raoul de Chagny), Mary Philbin (Christine Daae). 101 mins. B/w. Silent.

Christine, a young hopeful at the Paris Opéra, finds her path to stardom eased by a series of fortuitous deaths and resignations. The author of these crimes is the mysterious Phantom, occasionally seen in the shadowier parts of the building, furtively fleeing. As Paris worships at her feet she tells fiancé Raoul she can never marry him, for the OPERA is now her husband. Soon after, a voice speaks from the walls telling her that she must, indeed, think henceforth only of her career and of her "master". In due course, the now-masked Phantom seizes her not once but twice to his lair, five levels below the Opéra: the journey there, rich in fantasy motifs, is as through a series of PORTALS; she is escorted on a white horse down stairs and tunnels, then sculled across a lake of dark water. In his lair he tells her that his love for her is what she should regard, not the leering MASK he wears and the hideous face she reveals when she strikes the mask from him. In due course she is rescued by Raoul and the Phantom dies, but this is all of less importance than the portrayal of art (in this instance MUSIC) as the product of pain: the point is powerfully made; also powerful is the sequence in which, almost as self-PARODY, the Phantom brings a MASQUE to a standstill by appearing masked as the Red Death (◊ MASQUE OF THE RED DEATH; Edgar Allen POE). The Phantom himself can be taken as a symbol of the new ideas of music that became current from the early decades of this century, in particular the notion that, if one probes far enough beyond a veneer of ugliness, one can find beauty. Such evocations more than compensate for the movie's frequent creakiness, the dynamic between the two qualities being encapsulated in Chaney's performance, which is at the same time a crass caricature and a reification of the tormented soul. [JG]

2. *The Phantom of the Opera* US movie (**1930**). Universal. Most credits as **1**, but with Ernst Laemmle and Edward Sedgwick additionally credited for direction, Charles Van Enger for photography, and Frank McCormack and Tom Reed for screenplay. B/w, with some colour and tinted sequences. 89 mins.

Universal released this revised and truncated version of **1** to take advantage of the new technology. Most of the climactic scenes remain silent, and some of the dubbing is clumsy; the colour is restricted to the masque scene and some views of the Opéra. This version flopped. [JG]

3. *Phantom of the Opera* US movie (**1943**). Universal. **Pr** George Waggner. **Dir** Arthur Lubin. **Mufx** Jack Pierce. **Screenplay** Samuel Hoffenstein, Eric Taylor. **Starring** Nicki André (Mme Lorenzi), Edgar Barrier (Raoul Daubert), Nelson Eddy (Anatole Garron), Jane Farrar (Mme Biancarolli), Susanna Foster (Christine DuBois), Frank Puglia (Villeneuve), Claude Rains (Erique Claudin/Phantom). 92 mins. Colour.

Filled with the joys of sound and colour, Universal turned its remake of **1** (Columbia's *The Face Behind the Mask* [**1941**] can be viewed as a partial remake) into a musical extravaganza, complete with Nelson Eddy, and drowned most of the fantasy and PSYCHOLOGICAL-THRILLER elements in excruciatingly long, ostentatiously lavish operatic productions and some oafish humour. Paris Opéra understudy Christine is loved by leading baritone Garron and local police inspector Daubert – and also, though she does not know it, by modest veteran violinist Claudin, who has been secretly funding her career these past three years. When Claudin's fingers fail him and he is sacked, he tries to publish his piano concerto. A comedy of errors deceives him into justifiable homicide; the putative publisher's mistress throws etching acid in his face. Clad in cloak and MASK, leading a WAINSCOT existence in the sewers beneath the Opéra, Claudin – the "Phantom" – obsessively works and murders to further Christine's career. Various stratagems – including Liszt (played by Fritz LEIBER's father) performing Claudin's concerto – are used to lure the Phantom into the open, but not before more murder and the abduction of Christine to Claudin's subterranean home. The final unmasking lacks any shock, presumably through reluctance to lower the tone of this lush concoction by presenting anything more than tasteful disfigurement. The theme cried out for a gutter treatment – just what it got in **4**.

However, this version was extremely successful; Universal and Waggner capitalized on it the following year by producing, using the same sets, *The Climax* (**1944**), a sort of cross between the normal **Phantom** plot and George DU

MAURIER's *Trilby* (**1894**; ◊◊ SVENGALI): a murderous doctor, played by Boris Karloff, brings a young opera singer under his hypnotic control. [JG]

4. *The Phantom of the Opera* UK movie (**1962**). HAMMER/Universal. **Pr** Anthony Hinds. **Dir** Terence Fisher. **Mufx** Roy Ashton. **Screenplay** John Elder (Hinds). **Starring** Harold Goodwin (Bill), Michael Gough (Lord Ambrose D'Arcy), Herbert Lom (Professor L. Petrie/Phantom), Heather Sears (Christine Charles), Edward de Souza (Harry Hunter), Ian Wilson (Dwarf). **Voice actor** Pat Clark (Christine singing). 90 mins. Colour.

Caddish D'Arcy has become, to general astonishment, a fine composer these past few years. His new OPERA *Joan of Arc* (or *St Joan* – posters differ) has been bedevilled by bad luck and malicious pranks from the start, and its opening night is interrupted by a hanged stagehand swinging into view. The prima donna, Maria, quits, and producer Hunter hires ingénue Christine in her place. Alone in her dressing-room, she is addressed by a disembodied voice, warning of her D'Arcy. Later the voice speaks to Hunter and Christine together, a DWARF murders the theatre's ratcatcher, and the Phantom tries to lure Christine away. Christine has spurned D'Arcy's advances; he fires both her and Hunter. The two – romance blossoming – start probing the past of one Professor Petrie, a brilliant musician who, face accidentally covered in acid, threw himself in the river and presumably drowned; clearly D'Arcy stole the dead man's music. That night the murderous dwarf (the Igor to Petrie's Frankenstein) abducts Christine to the subterranean lair where his master plays sepulchral Bach organ music and teaches Christine properly to sing – using corporal punishment when necessary. A re-hired Hunter tracks Petrie to his watery lair, overpowers the dwarf, and elicits from Petrie the tale of how he was disfigured while burning copies of his music printed with D'Arcy's name as composer. By agreement Christine has more tuition under Petrie, and her opening night is a sensation; but then Petrie, D'Arcy's nemesis, in saving her from a falling chandelier is himself killed.

Cheap but superbly crafted, with an accurate eye and ear and a pleasing sense of GRAND GUIGNOL, this version (set in London) is like a breath of refreshingly fetid air after **3**; the wash of vigour affects even the music (by Edwin Astley). Led by Lom, most of the players give joyously hammy performances. Because of the inverted plot the effect is of a chillingly effective GHOST STORY (this is not, despite HORROR tropes, a HORROR MOVIE) transmuting into a RATIONALIZED FANTASY. The whole, despite much kitsch, is riveting. [JG]

5. *The Phantom of Hollywood* US movie (**1974** tvm). **Pr** Gene Levitt. **Exec pr** Burt Nodella. **Dir** Levitt. **Mufx** William Tuttle. **Screenplay** George Schenck, Robert Thom. **Based loosely on** Leroux's novel. **Starring** Skye Aubrey (Randy Cross), Jack Cassidy (Otto Bonner/Karl Bonner/Phantom) (movie lacks proper credits). 78 mins. Colour.

Decades ago Karl Bonner was being groomed for stardom, but then his face was accidentally mutilated. Since, he has lived in hiding beneath Worldwide Studios' Lot #2, shielded from detection by his studio-archivist brother Otto. But now Lot #2 is being sold off as real estate. Accordingly, masked as the Phantom and wielding a morningstar (later he switches to bow and arrow), Karl murders various bit characters – teenaged vandals, survey engineers and a security guard, before very publicly slaying his brother. He kidnaps Randy, daughter of the studio's boss, and attempts to use her as a bargaining counter; she sympathizes, but is rescued just before the final, climactic battle between the Phantom and the demolition bulldozers.

This curiously charming movie is made with the kind of care and affection not normally associated with a tvm. There are some impressive sequences: at the start, as the camera roams the empty, haunted windswept lot, there are intriguing intercuts from decades-old movies showing the thronged sets as they used to be; in the finale, when the Phantom announces that his world of fantasy is immortal, in contrast to the real-estate development, there is once again poignant intercutting, this time between his form plunging to its death and the bulldozers pulverizing the props that support that doomed world of fantasy. [JG]

5. *Phantom of the Paradise* US movie (*1974*). Harbor/Pressman-Williams/20th Century-Fox. **Pr** Edward R. Pressman. **Exec pr** Gustave Berne. **Dir** Brian DePalma. **Spfx** Greg Auer. **Mufx** John Chambers, Rolf Miller. **Choreographer** Harold Oblong. **Screenplay** DePalma. **Based very loosely on** Leroux's novel and on the FAUST legend. **Novelization** *Phantom of the Paradise* * (**1974**) by Bjarne Rostaing. **Starring** William Finley (Winslow Leach, the Phantom), Gerrit Graham (Beef), Jessica Harper (Phoenix), George Memmoli (Philbin), Paul Williams (Swan). 91 mins. Colour.

Hugely corrupt rock impresario and idol Swan (scribbled, his signature reads as SATAN), head of Death Records, steals the rock OPERA written by unknown Leach, based on the FAUST legend, and proposes to use it as the opening show for his grandiose new venue, the Paradise. When Leach attempts to rectify the wrong he is beaten up, framed for drug-dealing, sent to Sing Sing and subjected to total dental extraction; escaping Sing Sing to wreak vengeance on Death Records' hardware, he catches his head in a record-presser and is hideously mutilated, his voice wrecked. Dressed in cape and MASK, he starts haunting the Paradise, but soon strikes a deal with Swan: he will rewrite the opera to be sung by his true love, the singer Phoenix. The PACT WITH THE DEVIL produced by Swan is inches thick, binding for life, and must be signed in blood. But, as the Phantom works in secret, Swan reneges, and arranges for the piece to be performed by camp heavy-metal rocker Beef; on the rewrite's completion he has Leach immured. But Leach breaks free and assassinates Beef mid-opera; Phoenix takes over and is a wild success. That night, watching Swan and Phoenix make love, Leach tries to kill himself, but finds he cannot die; as Swan explains, the "lifetime" of the contract is his, not Leach's. Plans are laid for Swan and Phoenix to be married on-stage as the climax of the opera's next presentation, but secretly Swan arranges her public assassination – her perfection offends him. Leach, prowling Death Records' video library, discovers this plot; also that long ago Swan signed a contract with his own DEVIL (the notion of a sort of serial Devil is interesting) giving him youth and IMMORTALITY so long as the portrait (◊ PICTURES AND PORTRAITS) of him on the video should survive. Leach destroys the video and thwarts the assassin but – as Swan's hideousness is revealed to all – himself dies.

This is by far the most fantasticated and imaginative version of the tale; in addition to the CROSSHATCH elements noted there are references to the DRACULA and FRANKENSTEIN corpi and (obviously) to Oscar WILDE's *The Picture of Dorian Gray* (**1891**) (◊◊ *The* PICTURE OF DORIAN GRAY

[*1945*]), plus a glorious PARODY of the shower scene in *Psycho* (*1960*); the name of Swan's greasy sidekick, Philbin, echoes that of **1**'s co-star, Mary Philbin. The songs (by Williams) are surprisingly good – consistently better than in the in some ways comparable *The* ROCKY HORROR PICTURE SHOW (*1975*); in defiance of verisimilitude, even Beef's supposedly dreadful performances are pretty good. The direction is occasionally a little over-fussy in the trendy '70s fashion, with split-screen, montage effects and accelerated motion being deployed unnecessarily; but this is more than compensated for by its merits. *POTP*, though rarely scary, is frequently very funny; most of all, it races along with a frothy vitality, its inventiveness never flagging. [JG]

6. *Phantom of the Opera* US movie (*1983* tvm). Robert Halmi Inc. **Pr** Robert Halmi. **Exec pr** Robert Halmi Jr. **Dir** Robert Markowitz. **Spfx** Janos Kukoricza. **Phantom designed by** Stan Winston. **Screenplay** Sherman Yellen. **Starring** Diana Quick (Brigida Bianchi), Gellert Raksanyi (Lajos), Maximilian Schell (Sandor Korvin/Phantom), Jane Seymour (Elena Korvin/Maria Gianelli), Michael York (Michael Hartnell). 104 mins. Colour.

Beautiful but talentless Elena Korvin commits suicide after being humiliated on her first night as Marguerite in *Faust* because she refused to sleep with Baron Hunyadi, owner of the Budapest Opera. Her husband Sandor, the Opera's conductor, is disfigured while taking vengeance on some of the Baron's catspaws, but nursed back to health by mad ratcatcher Lajos in a chamber under the Opera House. Four years later Korvin sees the image of his dead wife in Maria Gianelli, an understudy in a new production of *Faust*, and starts giving her covert singing lessons ... and the standard plot is followed.

This version is sumptuously photographed and well scripted, but the screenplay's portrayal of Gianelli as a self-centred brat makes it impossible for a somewhat wooden Seymour to capture our sympathies. The notion of Gianelli as spiritual TWIN of the dead woman is dallied with but unexplored. Rather more is made of Korvin's MASK: when Gianelli strips it away to reveal his hideous face he declares that, too, to be a mask – "And what's behind it? Another mask? And another?" It is a rare moment of depth in a rather shallow movie. [JG]

7. *Phantom of the Opera* UK ANIMATED MOVIE (*1987* tvm). Emerald City/Taffner. **Pr** Al Guest, Jean Mathieson. **Dir** Guest, Mathieson. **Spfx anim** Julian Hynes. **Screenplay** Guest, Mathieson. **Voice actors** Aiden Grennell, Collette Proctor, Daniel Reardon, Jim Reid, Joseph Taylor (none individually credited). 48 mins. Colour.

A modest animated version – almost every line of the animation seems to have been decreed by the need for economy – yet highly enjoyable; for this is that *rara avis*, a version based directly on Leroux's work (albeit much simplified), and despite limitations a neatly scripted one.

8. *Phantom of the Mall: Eric's Revenge* US movie (*1989*). Fries Entertainment. **Pr** Thomas Fries. **Dir** Richard Friedman. **Mufx** Matthew Mungle. **Screenplay** Robert King, Tony Michelman, Scott J. Schneid, Frederick R. Ulrich. **Based very loosely on** Leroux's novel. **Starring** Morgan Fairchild (Mayor Karen Wilton), Derek Rydall (Eric Matthews/Phantom), Kari Whitman (Melody Austin). 92 mins. Colour.

Pretty Melody Austin is hired at the vast shopping mall built on the site where, a year ago, her boyfriend Eric and his parents were burnt alive by an arsonist. Eric, not dead after all, haunts the mall's ventilation system and, masked, slaughters various spear carriers. Melody is refreshingly unfazed when the MASK comes off, but has by now fallen for Another Man ... Though predictable, this gory movie is well enough made to be not as bad as it sounds. Not quite. [JG]

9. *Phantom of the Ritz* US movie (*1988*). Hancock Park. **Pr** Carol Marcus Plone. **Exec pr** Jerry Kutner. **Dir** Allen Plone. **Mufx** Dean Gates. **Screenplay** Tom Dempsey. **Based very loosely on** Leroux's novel. **Starring** Peter Bergman (Ed Blake), The Coasters (themselves), Russel Curry (Marcus), Joshua Sussman (Phantom), Deborah Van Valkenburgh (Nancy). 88 mins. Colour.

Inane semi-parodic variation in which a youth disfigured in a 1956 drag-race crash today lurks murderously in a derelict theatre restored and reopened by Blake as a rock venue. The Phantom falls for Blake's girlfriend Nancy, although deploring her taste in interior decor. Made on a short shoestring. [JG]

10. *The Phantom of the Opera* US movie (*1989*). 21st Century/Menaham Golan/Breton/Castle. **Pr** Harry Alan Towers. **Exec pr** Menaham Golan. **Dir** Dwight H. Little. **Mufx** Kevin Yagher. **Screenplay** Gerry O'Hara, Duke Sandefur. **Starring** Yehuda Efroni (Ratcatcher), Robert Englund (Erik Destler/Phantom/Foster), Terence Harvey (Hawking), Alex Hyde-White (Richard Dutton), Stephanie Lawrence (La Carlotta), Bill Nighy (Barton), Emma Rawson (Meg in London), Jill Schoelen (Christine Day), Molly Shannon (Meg in New York). 93 mins. Colour.

The Splatterpunk version had to come. Modern operatic hopeful Christine is hit by a sandbag while auditioning for the role of Marguerite in *Faust*, her chosen piece being from an unfinished opera by dead composer and SERIAL KILLER Destler. The blow throws her back to 1889, where she is understudy in London to the Marguerite of bitch diva La Carlotta. The 19th-century Christine is secretly instructed by a music tutor she never sees, and whom she believes is an ANGEL sent by her dead father; in fact it is Destler, who years ago made a PACT WITH THE DEVIL that his music would be loved forever and that he would have IMMORTAL-ITY and superphysical strength – but, in the small print ($\lozenge$ QUIBBLES), that he would be so hideous that none would ever love him for himself. Destler, dwelling in the sewers, mur-ders freely and very gorily (he is a sort of SPRING-HEELED JACK figure): none are safe, especially those who stand in Christine's way. At a MASQUE Destler (guised as DEATH) despatches La Carlotta (leaving her severed head in the soup) and abducts Christine to his cavern. After much more gore Christine sets all ablaze, seemingly destroying Destler; she recovers consciousness back in 20th-century New York. There is promptly taken up by *Faust*'s producer, Foster, whom she rapidly discovers is in fact the surviving Destler; this time she knifes him, dealing a death blow by destroying his sheet music. But yet, later, she passes a busker who seems all too familiar ...

This version owes much to **6** – notably: an attack by Destler in a Turkish bath; a plot against Christine by the London Opera's co-owner, involving a crooked opera critic who is murdered by the Phantom; the Phantom's ratcatcher sidekick; and the appearance of the Phantom's lair, with an antique organ and candles everywhere. The plot has the feel of one of John Dickson CARR's TIMESLIP detections, but the gore robs it of any fascination and the movie of any suspense. [JG]

PHANTOM OF THE PARADISE US movie (*1974*). $\lozenge$ PHANTOM OF THE OPERA.

PHANTOM OF THE RITZ US movie (*1988*). $\lozenge$ PHANTOM OF THE OPERA.

PHELPS, ELIZABETH STUART (1844-1911) US novel-ist and poet, also known as Elizabeth Stuart Phelps Ward or Mrs Herbert Dickinson Ward after her marriage in 1898. Her reputation rests on *The Gates Ajar* (**1868**), the story of a grieving sister whose brother died in the Civil War. An aunt makes her believe in the IMMORTALITY of the SOUL and the hope of an AFTERLIFE. The sequels, *Beyond the Gates* (**1883**), *The Gates Between* (**1887**) and *Within the Gates* (**1901**) are strong spiritualist books ($\lozenge$ SPIRITUALISM) that explore different aspects of the afterlife and its meaning to various individuals. Mark TWAIN wrote a satire of the first book, *Captain Stormfield's Visit to Heaven* (**1909**), but held back publication for over 40 years for fear it would upset people. ESP also wrote a number of GHOST STORIES. Some of the earliest, collected in *Men, Women and Ghosts* (coll **1869**) are humorous, although "Kentucky's Ghost", about a revengeful stowaway killed at sea, has much atmosphere. Her later stories are more melancholic and reflective; *Sealed Orders* (coll **1879**) and *The Empty House* (coll **1910**) contain the most effective of these, though not all are supernatural fiction. With her husband, Herbert D. Ward (1861-1932), she wrote *The Master of the Magicians* (**1890**). [MA]

PHILIP K. DICK AWARD The sf writer Philip K. Dick (1928-1992) ($\lozenge$ *SFE*) published many of his best novels as paperback originals. It was to honour him, and to recognize the fact that many of the best sf and fantasy novels continue to originate as paperbacks, that Thomas M. DISCH suggested the creation of an annual PKDA, with the active participation of Charles N. Brown (1937-), editor of *Locus*, and David G. HARTWELL. After a period, A.J. Budrys (1931-) and Hartwell succeeded Disch as administrators; other writers and editors have also helped run proceedings. The award – given through the Philadelphia SF Society – comes with a cash prize of $1000; the runner-up receives $500. The award is given each March for the best paperback novel published during the previous year. [JC]

Winners:

1983: *Software* by Rudy Rucker.
1984: *The Anubis Gates* by Tim POWERS.
1985: *Neuromancer* by William Gibson.
1986: *Dinner at Deviant's Palace* by Tim POWERS.
1987: *Homunculus* by James P. BLAYLOCK.
1988: *Strange Toys* by Patricia GEARY.
1989: *Four Hundred Billion Stars* by Paul J. McAuley and *Wetware* by Rudy Rucker.
1990: *Subterranean Gallery* by Richard Paul Russo.
1991: *Points of Departure* by Pat MURPHY.
1992: *King of Morning, Queen of Day* by Ian McDonald.
1993: *Through the Heart* by Richard Grant.
1994: *Growing Up Weightless* by John M. Ford and *Elvissey* by Jack Womack.
1995: *Mysterium* by Robert Charles Wilson.
1996: *Headcrash* by Bruce Bethke.

PHILLIPS, MICHAEL Pseudonym used by Charles BEAU-MONT, sometimes with William F. NOLAN.

PHILLPOTTS, EDEN (1862-1960) UK writer whose pro-lific and long literary career began with two brief comic fantasies in *The Idler*: "The Spectre's Dilemma" (1892) and "Friar Lawrence" (1892). Many of his early fantasy short stories were collected in *Fancy Free* (coll **1901**) and *Transit* *of the Red Dragon* (coll **1903**). His first full-length fantasy, *A Deal with the Devil* (**1895**), was a pastiche of F. ANSTEY in which the extra seven years earned by a PACT WITH THE DEVIL involve a dying man in accelerated rejuvenation.

My Laughing Philosopher (**1896**) records a series of imaginary conversations between a man and a bronze bust (◊ STATUES) on various matters of contemporary concern, foreshadowing the Epicurean sensibility and militant rationalism which were later to be extrapolated in a long series of extended FABLES, beginning with *The Girl and the Faun* (**1916**), a heartfelt tale of ill fated love. In *Evander* (**1919**) Bacchus and Apollo vie for the adoration of a woman. *Pan and the Twins* (**1922**) tracks a similar ideological conflict between two brothers, one a Christian and the other a follower of PAN; like Anatole FRANCE's saintly SATYRS, EP's Pan symbolizes an amiable Epicureanism preferable to ascetic Christianity. In *The Treasures of Typhon* (**1924**) a halfhearted disciple of Epicurus embarks on a QUEST for a magical plant, and is taught by the TREES (with which he can converse) to become more wholehearted. The fine novella "Circe's Island" (in *Circe's Island* coll **1925** with "The Girl and the Faun") describes a boy's quest to liberate his father from CIRCE's curse, aided by good advice from ODYSSEUS. *The Miniature* (**1926**) is a pessimistic commentary by the Classical GODS on the philosophical evolution of mankind, with much reflection on the psychological utility of RELIGION. *Arachne* (**1927**) is an effective recapitulation of the LEGEND, altering its conclusion to contrive a better moral. *Alycone (A Fairy Story)* (**1930**) chronicles the adventures of a third-rate poet.

Alongside his Classical fantasies EP wrote *The Lavender Dragon* (**1923**), a charming tale in which a benevolent DRAGON steals lonely humans to populate a UTOPIA whose ideals contrast sharply with the tyranny of feudalism, and *The Apes* (**1927**), an ironic ALLEGORY of evolution in which overconfident APES refuse to come to terms with the inevitability of their supersession. *The Owl of Athene* (**1936**) begins as a Classical fantasy in which a council of Olympians decides to put mankind to the test, but veers abruptly into the realms of scientific romance as the land surface of the Earth is invaded by giant crabs; all EP's subsequent *contes philosophiques* were framed as sf, although the nonsupernatural WITCHCRAFT featured in *The Hidden Hand* (coll of linked stories **1952**) involves an amiably subtle championship of rationalism. EP also published two notable CHILDREN'S FANTASIES, *The Flint Heart* (**1910**) and *Golden Island* (**1938**), and there are some briefer fables in *Thoughts in Prose and Verse* (coll **1924**). EP's extended fables were commercially unsuccessful, and he was eventually forced to abandon the form, but their occasional tendency to pomposity does not detract from their effectiveness or charm. [BS]

PHILOSOPHERS' STONE In ALCHEMY, a sought-after substance enabling TRANSMUTATION of base metals into gold – or, metaphorically, base SOULS into enlightened ones. It features in the PANDORA'S BOX chemistry sets of Diana Wynne JONES's *The Ogre Downstairs* (**1974**). Colin WILSON's *The Philosopher's Stone* (**1969**) uses the term in its metaphorical sense. [DRL]

PHOENIX A fabulous BIRD said to immolate itself upon its nest at regular intervals in order to be reborn and renewed. It was used by alchemists (◊ ALCHEMY) as a symbol of their endeavours, and hence was adopted as a shop-sign by early chemists. Notable fantasies featuring phoenixes include *The Phoenix and the Carpet* (**1904**) by E. NESBIT, the title novella of *The Man who Ate the Phoenix* (coll **1949**) by Lord DUN-

SANY and Roger Lancelyn GREEN's *From the World's End* (**1948**); the last was reprinted with Edmund Cooper's "The Firebird" (◊ *SFE*) in *Double Phoenix* (anth **1971**), intro Lin CARTER. [BS]

PHOENIX, THE US tv series (1981-2). ABC. **Pr** Leigh Vance. **Exec pr** Mark Carliner. **Dir** Reza S. Badiyi, Douglas Hickox. **Writers** Mark Carliner, David Guthrie, Anthony Lawrence, Nancy Lawrence, Leigh Vance. **Created by** Anthony and Nancy Lawrence. **Starring** Sheila Frazier (Mira), Richard Lynch (Justin Preminger), E.G. Marshall (Dr Ward Frazier), Judson Scott (Bennu of the Golden Light). 2hr pilot plus 5 1hr episodes. Colour.

An archaeological expedition in Peru finds a golden sarcophagus bearing the symbol of the PHOENIX; inside is Bennu, a traveller from the planet El DeBrande. He possesses strange powers derived from the Sun. He is not sure why he is on Earth, but discovers that another traveller, Mira, from his planet is here as well. Faintly remembering that he has been sent to change Earth's future, he sets off to find Mira. Each episode finds him involved with the lives of those he meets on his QUEST, and sees frequent use of his TALENTS to aid them. [BC]

PHYLOS THE THIBETAN Pseudonym of US writer Frederick Spencer Oliver (1866-1899), whose **Zailm Numinos** sequence – *A Dweller on Two Planets, or The Dividing of the Way* (**1905**) and *An Earth Dweller's Return* (**1940**) – presents the life of a dweller of ATLANTIS through various REINCARNATIONS, here and on Venus. [JC]

PICARESQUE A picaresque is, strictly, any prose fiction which tells the tale of a rogue-servant, or picaro. The picaro serves various masters; he travels hither and yon; his observations about his "betters" are a form of ESTATES SATIRE, though he has no commitment to the preservation of the established order (◊ THEODICY); he is seen as a survivor of war and its aftermath; in the end, he may find the home which – perhaps unconsciously – he had been seeking. The picaro is a human mote dislodged from a secure world, who witnesses a dissolute and/or dissolving social order, and who refastens himself to the normal world only at the end. Picaresques tend to be written, therefore, during periods of social upheaval or interregnum. The first examples known – the anonymous *Lazarillo de Tormes* (**1553**) and *Guzmán de Alfarache* (**1599-1604**) by Mateo Alemán (1547-?1614) – bracket the decline of Imperial Spain. *The Adventurous Simplicissimus* (**1669**) by Johann Grimmelshausen (?1625-1676) reflects the Thirty Years' War. Thomas MANN's *Confessions of Felix Krull* (**1954**), Gunter GRASS's *The Tin Drum* (**1959**), John BARTH's *The Sotweed Factor* (**1960**) and *The Painted Bird* (**1965**) by Jerzy Kosinski (1933-1991) are picaresque visions of our own century.

If the picaro's habit of finding safe haven in the end can be – remotely – deemed to resemble a QUEST, then many SWORD-AND-SORCERY tales can be seen to ape elements of the classic picaresque, and some of the **Fafhrd and Gray Mouser** tales by Fritz LEIBER inhabit a genuinely similar moral landscape; stories featuring CHILDE figures, like Steven KING's **Dark Tower** sequence, or the CONFIDENCE MAN, like Terry BISSON's *Talking Man* (**1986**), also deploy picaresque effects. DYING-EARTH tales – like Jack VANCE's **Cugel** sequence or Gene WOLFE's *The Book of the New Sun* (**1980-3**) – are picaresques, for the worlds there depicted are dioramas through which, faintly, satirical visions of the 20th century can be discerned.

As an adjective, the term can be fairly applied to any tale in which individual episodes dominate (or seem to dominate), though an element of SATIRE should probably be present for the term to convey any useful meaning. APULEIUS's *The Golden Ass* (*c165*) is in this double sense clearly picaresque, as are Angela CARTER's *The Infernal Desire Machines of Dr Hoffman* (**1972**), Barry HUGHART's *Bridge of Birds* (**1984**), Franz KAFKA's *America* (**1927**), William KOTZWINKLE's *Doctor Rat* (**1976**), Jeanne LARSEN's *Silk Road* (**1989**), John Myers MYERS's *Silverlock* (**1949**), David Henry WILSON's *The Coachman Rat* (**1985**) and Roger ZELAZNY's *Jack of Shadows* (**1971**). [JC]

PICNIC AT HANGING ROCK Australian movie (*1975*). HEF/South Australian Film Corporation/Australian Film Commission/Picnic Productions. **Pr** Hal McElroy, Jim McElroy. **Exec pr** John Graves, Patricia Lovell. **Dir** Peter Weir. **Screenplay** Cliff Green. **Based on** *Picnic at Hanging Rock* (**1967**) by Joan Lindsay (1896-1984). **Starring** Vivean Gray (Greta McCraw), Dominic Guard (Michael Fitzhubert), John Jarratt (Albert Crundall), Anne Lambert (Miranda), Helen Morse (Mlle de Portiers), Margaret Nelson (Sara), Rachel Roberts (Mrs Appleyard), Karen Robson (Irma Leopold), Christine Schuler (Edith Horton), Jane Vallis (Marion), Martin Vaughan (Ben Hussey). 115 mins. Colour.

On Valentine's Day 1900 the pupils of Appleyard College, Victoria, are sent on an expedition to picnic in the woods around Hanging Rock, a local crag; kept behind from the treat is orphan Sara, who has a crush on the beautiful Miranda. Oddly, the party's watches stop at noon. Four girls – Miranda, Marion, Irma and Edith – set off for the summit; the first three become infected by the mystical, acting as if in a trance, audaciously removing their boots and stockings. Edith, accompanying them on sufferance, becomes terrified by whatever is affecting the rest, and flees downslope, on her way seeing governess McCraw ascending stripped to her (plentiful) underwear; later she recalls also having seen a mysterious red cloud. The party returns to the College, having hunted McCraw and the three girls in vain. Also on the rock were visiting Englishman Michael with the Fitzhubert family driver Albert, another orphan; Michael, captivated on sight by Miranda, followed them uphill a small way. Police searches of the Rock reveal nothing; a week later the obsessed Michael goes with Albert on a private search, and eventually Albert discovers Irma, still alive but concussed and, although "intact", missing her corset – and, it soon proves, her memory of any relevant event. The mystery creates hysteria among the pupils, triggered into near-riot by the return of Irma, who is now obviously a woman rather than a girl, as if having gone through a RITE OF PASSAGE. Meanwhile, Appleyard – her school's future in ruins – victimizes the bereft Sara. At last Sara suicides, unknowing that her long-lost brother Albert is, in every sense, so near; he has a clairvoyant DREAM of her saying farewell to him. Michael, still seeing GHOSTS of Miranda, determines to leave the area; and soon Appleyard is found dead at the base of Hanging Rock, presumably having fallen while attempting to climb it – to beg forgiveness from Miranda for the death of Sara? The three missing girls are never found.

PAHR, in several senses a fantasy of PERCEPTION, is a movie whose ambience abides long in the memory: it has a slow beauty, a perfection that renders it the visual equivalent of MAGIC REALISM – a comparison enhanced by its subterfuge of presenting a fiction as if it were based on a true incident. *PAHR*'s linchpins are the personality and image of Miranda and of the Rock itself, its outcrops all too often seeming grim, ancient, secretive faces. [JG]

Further reading: *The Murders at Hanging Rock* (**1980**) by Yvonne Rousseau (1945-   ), a tongue-in-cheek analysis of the novel, also casts some light on the movie; *The Films of Peter Weir* (**1993**) by Don Shiach contains an extended discussion. (The novel's overtly fantastic final chapter – cut before publication – was published as *The Secret of Hanging Rock* [**1987** chap])

PICTURE OF DORIAN GRAY, THE US movie (*1913*). ◊ *The* PICTURE OF DORIAN GRAY (*1945*).

PICTURE OF DORIAN GRAY, THE US movie (*1945*). MGM. **Pr** Pandro S. Berman. **Dir** Albert Lewin. **Paintings of Gray** Ivan le Lorraine Albright, Henrique Medina. **Dir ph** Harry Stradling. **Screenplay** Lewin. **Based on** *The Picture of Dorian Gray* (**1891**) by Oscar WILDE. **Starring** Richard Fraser (James Vane), Lowell Gilmore (Basil Hallward), Hurd Hatfield (Dorian Gray), Angela Lansbury (Sybil Vane), Peter Lawford (David Stone), Morton Lowry (Adrian Singleton), Donna Reed (Gladys Hallward), George Sanders (Lord Henry Wotton), Douglas Walton (Allen Campbell). **Voice actor** Sanders (Narrator). 110 mins. B/w with four brief colour inserts.

London, 1886. While having his portrait (◊ PICTURES AND PORTRAITS) painted by Hallward, young Gray meets older roué Wotton and is almost corrupted by his cynical, epigram-laden worldliness. Wotton's comment that the portrait will remain young as Gray ages prompts the latter to say, in the presence of a STATUE of a divine Egyptian CAT, that he would trade his SOUL if matters could be the other way round. Soon Gray, exploring the low-spots of London, meets and woos chanteuse Sybil Vane; Wotton guides Gray in her seduction – after which Gray rejects her entirely. Next day he glances at his portrait and sees it has new lines of cruelty; full of remorse, he plans to wed Vane, but almost immediately hears she has suicided.

18 years of debauchery later, he is still youthful while his portrait is vile. When Hallward's niece Gladys seeks to marry him, however, he has enough decency to refuse her. Soon after, on the spur of the moment he shows Hallward what has become of the portrait; the painter, guessing the dreadful truth, is horrified; Gray kills him to keep the secret, and then, after all, woos Gladys. Just before their marriage Gray once more repents, and stabs the portrait through the heart. Through sympathetic MAGIC, this blow kills him, but as he dies he babbles a prayer begging GOD's forgiveness: the portrait becomes once more youthful while Gray's corpse takes on the hideous features.

This is a striking movie on many counts, not least a script packed with Wildean aphorisms. Almost as much happens off-screen as on, and there is virtually no visible violence; the characters, especially Gray, are presented with passionless exteriors – a notable exception is Sybil (for whose portrayal Lansbury won an Oscar nomination). Yet, probably because of these self-imposed disciplines, *TPODG* packs a considerable emotional punch, aided (despite some slightly cumbersome direction) by the excellent, *film noir*ish cinematography, which substitutes for spfx (and for which Stradling received an Oscar).

Comparisons should certainly be made between this and *Doctor Jekyll and Mister Hyde* (*1931*) (◊ JEKYLL AND HYDE MOVIES), since the two – and, of course, their respective source novels – deal with the very similar theme of respectability as a veneer disguising an individual's, if not humankind's, inherent component of EVIL. Moreover, both movies take it as axiomatic that good will *look* good, evil will *look* evil (although Wilde and *TPODG* play this both ways, in that Wotton's studied DECADENCE must be equated with evil – this is a work with not one but two VILLAINS, and it is hard to decide which is the more villainous). The moral quandary posed by this assumption appears to have occurred to neither author, as it has not to innumerable writers since. In the case of Robert Louis STEVENSON we might blame it on a Calvinist-style upbringing; Wilde has no such excuse – and certainly would have been offended had it been offered on his behalf. The respective moviemakers were, of course, simply aping their sources. In terms of the two movies as movies, the big difference is that, with the passage of 14 years, Lewin clearly felt that the increasing sophistication of his audience meant a movie no longer had to rely on sensationalism in order to tell a sensational tale. An evidence of the evolution of FANTASY in both its literary and, some decades later, its CINEMA modes is that, while Stevenson felt the necessity to dress up his notion in quasi-science – i.e., to make it a TECHNOFANTASY – Wilde (and *TPODG*) did not.

TPODG is the classic movie version of Wilde's novel but was far from the first – the earliest was probably a Danish version released in 1910. The most frequently quoted precursor is the 1913 US version released by the New York Motion Picture Company. The story was filmed in Denmark in 1913, in Russia and again the USA in 1915, in the UK in 1916, and in Germany and in Hungary in 1917. There were also a UK/Italian coproduction in 1969 (as *Il Dio Chiamato Dorian*; vt *The Picture of Dorian Gray*; vt *The Secret of Dorian Gray*), a straight US production in 1975, and a soft-porn US version, *Take Off* (*1978*), in which the central character, Daren Blue, has remained youthful through three decades of sexual conquest. There have been innumerable stage versions. Most recently *Dorian Gray: A Musical*, by the Rock Theatre Budapest, was given its Western premiere in London in 1995 (original Hungarian presentation date not known, through probably 1994). [JG]

PICTURES AND PORTRAITS There is an obvious sympathetic-MAGIC link between a likeness and its original, most famously exploited in Oscar WILDE's *The Picture of Dorian Gray* (**1891**) – where, by a process resembling Charles WILLIAMS's Substitution, the protagonist's increasing dissolution is reflected only in the portrait, not in his own face. Symbolic truth also emerges in Nathaniel HAWTHORNE's "Edward Randolph's Portrait" (in *Twice-Told Tales* coll **1837**), where the eponymous portrait radiates the guilt and remorse of its subject, who once betrayed the people of New England and suffered their CURSE; and in Charles Williams's *All Hallows' Eve* (**1945**), whose painter involuntarily depicts a sinister MAGUS and his audience as an imbecile preaching to things like beetles. In W.S. GILBERT's *Ruddigore* (**1887**) ancestors' portraited images are animated by their SPIRITS and step from their frames; Vigo the Carpathian's spirit likewise inhabits his portrait in *Ghostbusters II* (*1989*) (◊ GHOSTBUSTERS). Portraits may be PORTALS: that in James Branch CABELL's "The Delta of Radegonde" (1921) allows carnal access to the idealized Queen Radegonde painted 13 centuries earlier, and that of a Narnian ship in C.S. LEWIS's *The Voyage of the Dawn Treader* (**1952**) opens on the SECONDARY WORLD seas of

Narnia itself. Pictures which change can be SCRYING devices of a sort: M.R. JAMES's "The Mezzotint" (1904) shows a sinister figure approaching and leaving the depicted house, outlining a century-old episode of HORROR; the animated tapestry in Piers ANTHONY's *Castle Roogna* (**1979**) replays historical episodes from 800 years before; in Roald DAHL's *The Witches* (**1983**) a child is locked into a picture, where she is seen to grow old and die. The abstract picture in Susan COOPER's *Greenwitch* (**1974**) is a DISGUISE for painted SPELLS; the videotape picture of Swan in *The Phantom of the Paradise* (*1974*; ◊ PHANTOM OF THE OPERA) embodies his PACT WITH THE DEVIL. Photographs are often believed to capture SOULS – CINEMA technology is used precisely thus to trap the DEMON in Barbara HAMBLY's *Bride of the Rat God* (**1994**). Like MIRRORS, photographs may either show the unseen or fail to show the unreal: many a GHOST, VAMPIRE, etc., is inferred from a missing place in a group photograph, while the famous punchline of H.P. LOVECRAFT's "Pickman's Model" (1926) explains Pickman's inspiration for pictures of ghouls: "But by God, Eliot, *it was a photograph from life.*" The ancient picture of a "KNIGHT" in Gene WOLFE's *The Shadow of the Torturer* (**1980**) indicates a TIME ABYSS as we realize that it shows an astronaut on the MOON. [DRL]

PIED PIPER A figure whose origins are lost in legend, but whose story (◊ FOLKTALE) is best-known from the poem "The Pied Piper of Hamelin" (1842 in *Dramatic Lyrics*) by Robert BROWNING. 14th-century Hamelin is infested with rats (◊ MICE AND RATS), and the burghers are at a loss what to do. As if from nowhere (◊ ANSWERED PRAYERS) they are visited by a strange, quaintly dressed TRAVELLER or MINSTREL who looks a little like Harlequin (◊ COMMEDIA DELL'ARTE; JESTER); he claims he can rid the town of the rats. They welcome his offer and agree to pay him 1000 guilders. He begins to play his pipe and immediately the rats begin to follow him and perish in the River Weser, save one, who returns to Rat-Land to tell of the fate of his fellows, remarking on the promise of beauty and plenitude that was in the PP's MUSIC. The burghers refuse to pay the full fee, offering the PP 50 guilders. When he threatens them they pay no heed, so the PP plays a new tune and all the town's children follow him into the mountains, where a door admits all save one, a lame boy, who returns to tell the tale. The PP and the children are never seen again, although the poem ends with reference to a belief that there was a German-speaking community in TRANSYLVANIA who did not know of their origin. Browning based his poem on a version his father had written and on other historical accounts. The earliest recorded version of the story, according to Humphrey Carpenter and Mari Prichard in *The Oxford Companion to Children's Literature* (**1984**), was by the historian Heinrich von Herford, writing in 1450: a young man came to Hamelin in the year 1284 and lured all the children away with his pipe-playing. This may have its roots in the Children's Crusade of 1212. The link with the rats did not follow for some while. Other versions of the legend exist, the most common in story form being "The Ratcatcher", from *Affenschwanz* ["The Monkey's Tail"] (coll **1888**) by the French folklorist Charles Marelles, trans in *The Red Fairy Book* (anth **1890**) ed Andrew LANG; here the PP is dressed like a gypsy and plays the bagpipes, but the rest is the same.

The PP is sometimes perceived as a TRICKSTER, but he was only claiming his right. The CHILDREN believed that they were being taken to some SECRET GARDEN which could be accessed only via a door in the mountains (◊ PORTAL; THRESHOLD). The end result for Hamelin was a THING BOUGHT AT TOO HIGH A COST.

The tale's MOTIFS have been utilized in many stories. The whole episode is replayed in *Ashmadi* (**1985** Germany; English text *The Coachman Rat* 1987) by David Henry WILSON, from the perspective of the surviving rat, who is shown to betray his fellows in order to save mankind. The idea of music opening up a doorway to another land is used in *A Strange Land* (**1908**) by Felix Ryark. The trickster image is used by Mildred Clingerman in "The Gay Deceiver" (in *A Cupful of Space* coll **1961**), where the PP continues to seek revenge upon the descendants of the townsfolk, and again in "Devlin" (1953 *F&SF*) by W.B. Ready (1914   ), where the DEVIL plays a similar trick on an Irish marching band. In both "A Present from Brunswick" (1951 *F&SF* ot "Bargain from Brunswick") by John Wyndham (1903-1969) and "Eine Kleine Nachtmusik" (1965 *F&SF*) by Frederic BROWN and Carl Onspaugh, the PP's pipe comes to light and continues to have its effect. In "The Pied Piper Fights the Gestapo" (1942 *Fantastic Adventures*) Robert BLOCH portrays the PP using his skills against the Germans. In "A Distant Shrine" (1961 *Saturday Evening Post*) by William Sambrot (1920   ) the PP is discovered to have been a Martian, descendants of the lost children being found on Mars. The PP has been featured in a number of ANIMATED MOVIES, but the only live-action movies seem to have been *The Pied Piper* (**1971** UK) dir Jacques Demy, starring Donovan Leitch, Donald Pleasence, Michael Hordern, John Hurt and Roy Kinnear, and *The Pied Piper of Hamelin* (**1984** tvm US) dir Nicholas Meyer, starring Eric Idle. The PP is an UNDERLIER in other movies; e.g., *The* NAVIGATOR (**1988**). [MA]

PIEŃKOWSKI, JAN (MICHAL) (1936   ) Polish-born illustrator, in the UK from 1946, initially active in theatre design; he began to illustrate children's books with *Annie, Bridge and Charlie* (**1967**) by Jessie Gertrude Townsend, immediately following this with his first illustrations for a work of fantasy interest, *A Necklace of Raindrops* (coll **1968**) by Joan AIKEN. Further Aiken books he illustrated include *The Kingdom Under the Sea* (coll **1971**), *Tale of a One-Way Street* (coll **1978**), *Past Eight O'Clock* (coll **1986**) and *A Foot in the Grave* (coll **1989**). He has composed several illustrated books for younger children. His illustrations almost invariably present figures in black silhouette against vivid backgrounds. [JC]

PIERCE, MEREDITH ANN (1958   ) US writer of fantasies for older children. She began with *The Darkangel* (**1982**), which she wrote while still an undergraduate; an inventive fantasy involving a VAMPIRE figure, drawing its inspiration from a DREAM recounted in C.G. Jung's *Memories, Dreams, Reflections* (◊ JUNGIAN PSYCHOLOGY), this dramatizes EVIL as seductively beautiful with a vividness unusual in a YA novel. MAP published two sequels, *A Gathering of Gargoyles* (**1984**) and *The Pearl of the Soul of the World* (**1990**), all three being assembled as *The Darkangel Trilogy* (omni **1990**); the sequels did not materially enlarge MAP's accomplishment. The two extant volumes of the **Firebringer** trilogy – *Birth of the Firebringer* (**1985**) and *Dark Moon* (**1992**) – deal with good UNICORNS and evil wyrms (◊ WORM/WYRM). *The Woman who Loved Reindeer* (**1985**) is about a foundling babe whose appearance is the causal agent in allowing the protagonist to fulfil her destiny (◊ FATE). *Where the Wild Geese Go* (**1988** chap), a fantasy for

younger children, involves geese and a QUEST. The significance accorded hierarchical orders, destinies, conflicts between GOOD AND EVIL, and traditional fantasy elements in MAP's fiction give it a decidedly conservative cast; the imaginative force evident in her first novel suggests she may yet write something more original. [GF]

PIERCE, TAMORA (1954   ) US writer whose **Song of the Lioness** sequence for YA readers – *Alanna: The First Adventure* (**1983**), *In the Hand of the Goddess* (**1984**), *The Woman who Rides Like a Man* (**1986**; vt *The Girl Who Rides Like a Man* 1992 UK) and *Lioness Rampant* (**1988**) – offers an extended narrative based on GENDER DISGUISE. At the beginning, Alanna and her twin brother Alan change places, as she wishes to become not a WIZARD but a KNIGHT and he wishes the converse. Her FANTASYLAND adventures defending her liege lord against his wicked wizard brother occupy much of the remaining text. The image of a magic CAT – the "lioness rampant" which she becomes – guides her; and she receives the succour of the GODDESS. TP's second sequence, the **Immortals** series – *Wild Magic: The Immortals* (**1992**), *Wolf-Speaker* (**1994**), *The Emperor Mage* (**1994**) and *Realm of the Gods* (**1996**) – is set in the same world some time later, and concentrates on young Daine, a girl whose TALENT is conversing with animals, who act as her COMPANIONS in her adventures. [JC]

PIERROT ◊ COMMEDIA DELL'ARTE.

PIKE, CHRISTOPHER Pseudonym of US writer Kevin McFadden (?   -   ). Most of his novels, which have been immensely successful, are suspense or HORROR, although some qualify as fantasy; generally they are for the YA market. *Chain Letter 2: The Ancient Evi*l (**1992**; vt *Chain Letter 2* 1992 UK) is fantasy (although the volume to which it is a sequel is not) in that it reveals that the novels' protagonist was, in fact, possessed (◊ POSSESSION). Horror novels with fantasy elements include **The Last Vampire** series – *The Last Vampire* (**1994**), *#2: Black Blood* (**1994**) and *#3: Red Dice* (**1995**) – the **Remember Me** series – *Remember Me* (**1989**), *#2: The Return* (**1994**) and *#3: The Last Story* (**1995**) – *Witch* (**1990**) and *Spooksville #1: The Secret Path* (**1995**). *Scavenger Hunt* (**1989**) and *See You Later* (**1990**) are TIME-TRAVEL fantasies.

CP began publishing adult fiction with *Sati* (**1990**), whose eponymous heroine claims to be GOD. *The Season of Passage* (**1992**) and *The Cold One* (**1994**) are adult horror novels. [GF]

PILGRIM'S PROGRESS The abbreviated title by which John BUNYAN's 17th-century ALLEGORY is generally known. By virtue of being one of the landmarks of CHRISTIAN FANTASY – it was until a century ago among the most widely read texts in the English language – its imagery has exerted a considerable influence over the entire subgenre of QUEST fantasy. Although never directly imitated – C.S. LEWIS's disguised spiritual autobiography *The Pilgrim's Regress* (**1933**) is not the same thing at all – its existence as a key element of our cultural heritage both licenses and assists the decoding by which readers can (and are supposed to) discern the processes of intellectual maturation to which questing heroes are subject as they struggle against various perils and distractions. The lexicon of equivalencies by which it translates emotions into geography and architecture (the Slough of Despond, Doubting Castle, etc.) echoes throughout the subgenre, most resonantly in RITE-OF-PASSAGE parables. The state of grace symbolized by Bunyan's Celestial City is more often conceived nowadays in terms of secular moral enlightenment, attained by self-awareness and self-control, but this merely serves to remind us that allegory cuts both ways, and that signs and what they signify can sometimes change places. [BS]

PINI, RICHARD (ALAN) (1950-   ) US writer and illustrator who, with his wife Wendy PINI, created and controlled the various forms of ELFQUEST from its inception as a COMIC in 1978. Spinoffs in book form include: *Elfquest: The Novel!: Journey to Sorrow's End* (**1982**; vt *Elquest, the Novel* 1982; vt *Elfquest* 1984) with Wendy Pini; several GRAPHIC NOVELS (◊ Wendi PINI); and a series of anthologies: *The Blood of Ten Chiefs #1* * (anth **1986**) with Lynn ABBEY and Robert Lynn ASPRIN, *#2: Elfquest: Wolfsong* * (anth **1988**) with Asprin, *#3: Elfquest: Winds of Change* * (anth **1989**) and *#4: Against the Wind* * (anth **1990**), the latter two ed RP. With Wendy Pini he won a 1985 BALROG AWARD as Best Artist. [JC]

PINI, WENDY (1951-   ) US writer and illustrator who, with her husband Richard PINI, created and controlled the various forms of ELFQUEST from its inception as a COMIC in 1978. Richard Pini has been primarily responsible for text spinoffs; WP has been credited as being more responsible for GRAPHIC-NOVEL versions of the original comics. This **Elfquest** sequence includes *Elfquest #1* (graph **1981**), *#2* (graph **1982**), *#3* (graph **1984**) and *#4* (anth **1984**). The **Complete Elfquest** sequence of comics re-assembles colour versions of the original b/w strips, running from *The Complete Elfquest #1: Fire and Flight* (**1981**) through *#6: The Secret of Two-Edge* (**1989**). [JC]

PINKWATER, DANIEL M(ANUS) (1941-   ) US author and radio commentator. His books, mostly sf or CHILDREN'S FANTASY, have attracted a large audience of adult readers for their anarchic humour and subversive spirit. As with much children's fiction, genre categories prove largely inapplicable; no useful distinction is made by calling, e.g., *Guys from Space* (**1989** chap) sf and *Blue Moose* (**1975** chap) fantasy. Much of DMP's fiction has been treated as sf: admirers of his work read it without regard to genre. Most of DMP's books for young children are also illustrated by him.

Works that make prominent use of unequivocal fantasy elements include: *Wizard Crystal* (**1973**); *Magic Camera* (**1974**); the **Moose** trilogy, being *Blue Moose* (**1975** chap) as Manus Pinkwater, *Return of the Moose* (**1979** chap) and *The Moosepire* (**1986** chap); *I Was a Second Grade Werewolf* (**1983**); *Tooth-Gnasher Superflash* (**1981** chap); *Devil in the Drain* (**1984** chap); the **Magic Moscow** sequence, being *The Magic Moscow* (**1980** chap), *Attila the Pun* (**1981** chap) and *Slaves of Spiegel* (**1982** chap); *The Muffin Fiend* (**1986** chap); and *Wempires* (**1991** chap). This list scants DMP's novels for older children, most of which possess an sf rationale. *The Afterlife Diet* (**1995**), DMP's first novel for adults, is an acerbic AFTERLIFE fantasy. [GF]

PINOCCHIO In Carlo COLLODI's *Pinocchio* (**1883**), old Gepetto carves a PUPPET which turns out not only animate (◊ ANIMATE/INANIMATE) but a blockhead TRICKSTER. Pinocchio is naughty and self-destructive and, though he longs to be a REAL BOY, must LEARN BETTER before becoming flesh and blood. He is an UNDERLIER figure for novels like Jerome CHARYN's *Pinocchio's Nose* (**1983**) and Robert COOVER's *Pinocchio in Venice* (**1991**). He is now popular as the protagonist of DISNEY's PINOCCHIO (*1940*), where his anarchic naughtiness is instead represented as naive folly. [JC]

PINOCCHIO US ANIMATED MOVIE (*1940*). DISNEY. **Pr** Walt DISNEY. **Supervising dir** Hamilton Luske, Ben Sharpsteen. **Based on** *The Adventure of Pinocchio* (serialized from 1880; **1882**) by Carlo COLLODI. **Voice actors** Mel Blanc (Gideon), Walter Catlett (J. Worthington Foulfellow), Franki Darrow (Lampwick), Cliff Edwards (Jiminy Cricket), Dickie Jones (Pinocchio), Charles Judels (Coachman, Stromboli), Christian Rub (Geppetto), Evelyn Venable (Blue Fairy). 88 mins. Colour.

Geppetto makes clocks and automata (◊ TOYS), and one night he completes his new and most realistic marionette, Pinocchio. His prayer that night that the PUPPET might become a REAL BOY is heard by the Wishing Star, which sends the Blue Fairy to give the marionette life, though he is told that he can become a real boy only after he has discovered bravery, truth and unselfishness; she charges Jiminy Cricket, a cocky insectile character, to be Pinocchio's Conscience. Sent to school the next morning, Pinocchio is waylaid by TRICKSTER J. Worthington Foulfellow (aka Honest John), a fox, and his inept feline sidekick Gideon who, despite Jiminy's efforts, sell Pinocchio to evil puppeteer Stromboli, who imprisons him. However, the Blue Fairy reappears and, after the famous sequence in which Pinocchio's lies about his circumstances cause his nose to elongate, frees him. Homeward bound, Jiminy and Pinocchio are separated, and again the latter is intercepted by Foulfellow, now collecting little boys on behalf of the Coachman. Pursued by Jiminy, Pinocchio is taken by the Coachman to Pleasure Island, where hundreds of boys are seduced by decadent pleasures before being transformed into donkeys (◊ ASSES) for sale. Jiminy saves Pinocchio, but not before the marionette has become part-donkey. Reaching home, they discover their friends have been swallowed by the whale Monstro (◊ DOLPHINS AND WHALES). The two rescue them, but Pinocchio is apparently dead. However the "corpse" is filled with the Blue Fairy's radiance and comes alive: Pinocchio has earnt the status of being a real boy.

P is widely regarded as the best animated movie of all time. Jiminy Cricket has become a fantasy ICON of some note, as have the Blue Fairy and Pinocchio himself. Part of the reason for the technical brilliance was extensive use of a horizontal version of the new multiplane camera, which could "penetrate" layers of painted scenery to give a superb illusion of depth: the opening sequence, lasting mere seconds, had 12 planes and cost $25,000. The movie's total cost, $2,600,000, almost bankrupted the studio. Viewed today, the artistry of *P* is still impressive; no description of it can avoid superlatives. [JG]

PIPER, JOHN UK artist (1903-1992). ◊ Michael AYRTON.

PIRANDELLO, LUIGI (1867-1936) Italian writer and dramatist whose prolific career began with poetry – *Mal Giocondo* ["Unhappy Joy"] (coll **1889**) – and continued with a large amount of short fiction. Many of his *c*400 stories are set in Sicily: a grim sense of the tortures men and women put themselves through when sex and marriage are concerned expresses itself in CONTE-CRUEL mode. Some of the tales assembled in *A Character in Distress* (coll trans **1938** UK) are fantasy; but most of his fiction is set desolately in this world, as are all his novels, even *Si gira . . . (Quaderni di Serafino Gubbio Operatore)* (**1916**; trans C.K. Scott Moncrieff as *Shoot!: The Notebooks of Serafino Gubbio Cinematograph Operator* 1927 UK), an extremely early metaphysical exploration of the nature of movie REALITY.

LP remains most famous for his 44 plays, several of fantasy interest and most were first printed in an ongoing, frequently revised sequence of collected volumes published under the overall title **Maschere Nude** ["Naked Masks"] (**1919-49**). The greatest is probably *Sei personaggi in cerca d'autore* (**1921**; trans Edward Storer as *Six Characters in Search of an Author* in *Three Plays* coll **1922** UK), a seminal exploration of the relationship between art and REALITY – MASK and naked self – which cannot be understood in any simple terms as fantasy, though supernatural elements enter the fray. Six characters – who have some GHOST-like characteristics – invade rehearsals for an already written LP drama in search of an author, as they have been abandoned halfway through their STORY by the author who originally created them. The resulting tragicomic interplay between actors and characters becomes indissolubly confused, for reality is simultaneously evanescent and fixed in art; within the drama itself, the author never appears, the GODGAME is declined.

Further dramas of interest include: *All'uscita* (performed 1922; **1949**; trans Blanche Valentine Mitchell as *At the Exit* in *Pirandello's One-Act Plays* coll **1928** US), a POSTHUMOUS FANTASY; *Lazarus* (**1929**), in which a repressive puritan loses his faith through dying, experiencing posthumous nothingness, and being reborn; *Sogno, ma forse no* (1929 *La Lettura*; trans Samuel Putnam as *Dream, But Perhaps Not* 1930 *This Quarter*), in which DREAM and reality loop; *Quando si è qualcuno* (**1933**; trans Marta Abba as *When Someone is Somebody* in *The Mountain Giants and Other Plays* coll **1958** US), in which a famous writer is literally metamorphosed (◊ METAMORPHOSIS) by his fame into stone (◊ BONDAGE); *La favola del figlio cambiato* ["The Changeling"] (**1933**), a political SATIRE; and *I giganti dell montagna* (1931-2 var mags; **1938**; trans Marta Abba as title play in *The Mountain Giants and Other Plays*), a complex unfinished drama set in a mountain EDIFICE whose grotesque inhabitants interact with a troupe of travelling players, demanding that they perform something other than LP's own *The Changeling* for the mysterious, unseen, GIANTS who require entertainment and who tear apart those who do not please them. But the actors cannot fulfil the needs of the gods, and their chief is duly dismembered. [JC]

PIRANESI, GIOVANNI BATTISTA (1720-1778) Italian architect and artist, from 1740 in Rome, where he became best-known for two series of etchings. The *Vedute* (graph **1745**) comprises 135 etchings of Rome (◊ CITY; EDIFICE; URBAN FANTASY), depicting the great city in a fashion that inextricably mingles the ancient and the contemporary, the ruined and the populous, the vertical and the subterranean, all effects conveyed through a profoundly Romantic use of chiaroscuro. One of the frontispieces attached to a second sequence – "Two Roman Roads Flanked by Colossal Funerary Monuments" in *Le Antichità Romane* (graph **1756**) – almost literally transforms the dream of the city into a dream of edifice. Indeed, Horace WALPOLE described GBP's visions of Rome as "sublime dreams", and it is clear that their depictions of ruined (but animate) cities and edifices deeply affected the Gothic writers (◊ GOTHIC FANTASY) at the end of the 18th century. GBP's second great series, *Invenzioni Capric di Carceri* (graph **1745-50**; rev vt *Carceri d'Invenzione* 1761), a sequence of images of vast, imaginary prisons, only intensified his role as an architect of dreams. The close resemblance of GBP's work to the *Architectura* (graph 1593-4; rev 1598) of Wendel Dietterlin (1550-1599) may be coincidental; the affinity between the two artists is unmistakable.

A typical late demonstration of GBP's continuing relevance is the dustwrapper by Jon Blake to the UK edition of Stephen Laws's *Daemonic* (**1995**), which is based on GBP, and which depicts a LABYRINTH without exit. GBP's influence on fantasy (in distinction to HORROR) is indirect; but some of fantasy's darker edifices, like Mervyn PEAKE's Gormenghast, are Pirenesian; and illustrators like Mike WILKS – in *Pile: Petals from St Klaed's Computer* (**1979** chap) with Brian W. ALDISS – clearly hearken back to GBP, the first artist to envision the modern city as an edifice. [JC]

PIRATES Piracy is as old as marine trade and is still widely practised, but there was a heavily mythologized "Golden Age of Piracy" in the 16th and 17th centuries following the development of sailing ships capable of transoceanic navigation. Legions of pirates – some licensed as privateers – were active in the Caribbean and the Indian Ocean, but it was those who preyed on the ships which carried American bullion to Spain who became central to modern pirate mythology. The core of this mythology derives from *A General Historie of the Most Notorious Pirates* (**1724**) by "Captain Johnson", belatedly revealed as the novelist Daniel DEFOE; although many of the individuals covered therein were certainly real, the existence of others is dubious and much is invented. There is a sense, therefore, in which all fiction about pirates has a fantasy dimension – a fact appreciated in such nonsupernatural historical fantasias as *The Pyrates* (**1985**) by George MacDonald Fraser (1925-   ) as well as more wholehearted fantasies like "Typewriter in the Sky" (1940) by L. Ron HUBBARD and *There Were Two Pirates* (**1946**) by James Branch CABELL. Such is the essential romance of piracy that pirates are often treated with considerable sentimentality, as in Richard MIDDLETON's "The Ghost Ship" (1912), in which the "young" ghosts of an English village set forth in search of the adventure they never found in life. Robert Louis STEVENSON's Long John Silver remains an ARCHETYPE of ambiguity as well as providing a key image of the perfect pirate.

The most famous pirates in fantasy fiction are Captain Hook and his crew in J.M. BARRIE's *Peter Pan* (**1904**) and its many spinoffs. Their presence in NEVER-NEVER LAND is required as a key element of boyhood fantasy; it is for the same reason that the hero of William GOLDMAN's *The Princess Bride* (**1973**) has to put in a stint as the Dread Pirate Roberts. Other imaginary pirates of note include Scarcabomba in *The Green Lacquer Pavilion* (**1926**) by Helen BEAUCLERK, Captain Darkness in *The City of Frozen Fire* (**1950**) by Vaughan Wilkins (1890-1959), Black and Littlejack in *The Wonderful O* (**1958**) by James THURBER and Langoisse in *The Empire of Fear* (**1988**) by Brian STABLEFORD.

Pirates also found a natural home among the stock characters of SWORD AND SORCERY; Robert E. HOWARD's CONAN was often involved with them, most notably in "Queen of the Black Coast" (1934), which features the bloodthirsty female pirate Belit. Pirates figure prominently in the works of Edgar Rice BURROUGHS and the pastiches of Lin CARTER. A female pirate plays a major role in *The Eyes of Sarsis* (**1980**) by Andrew J. OFFUTT and Richard K. Lyon (1933-   ). Recent fantasies which make extravagant use of pirates in order to reflect more-or-less ironically on this tradition include *On Stranger Tides* (**1987**) by Tim POWERS, *Chase the Morning* (**1990**) by Michael Scott ROHAN and *Pussy, King of the Pirates* (**1995**) by Kathy Acker (1948-   ). [BS]

PIT AND THE PENDULUM, THE Two movies have been based on "The Pit and the Pendulum" (1843) by Edgar Allan POE.

1. US movie (*1961*). American-International. **Pr** Roger CORMAN. **Exec pr** Samuel Z. Arkoff, James H. Nicholson. **Dir** Corman. **Spfx** Pat Dinga. **Screenplay** Richard MATHESON. **Starring** Luana Anders (Catherine), Antony Carbone (Dr Leon), John Kerr (Francis Barnard), Vincent Price (Nicholas Medina), Barbara Steele (Elizabeth). 85 mins. Colour.

In this, the second of Corman's cycle of eight **Poe** movies, it is the mid-16th century. Hearing of the death of his sister Elizabeth, Francis comes to the forbidding castle of her widower, Don Nicholas Medina, to investigate. It seems she died of fear while exploring the grim torture chamber of Don Nicholas's father, Don Sebastian, where Nicholas as a boy saw Sebastian murder his wife Isabella's lover (Sebastian's brother Bartolome) before torturing the woman and then finally immuring her. Now Nicholas is tormented by fears that Elizabeth may have been buried alive – fears apparently confirmed when her coffin is opened and her corpse discovered disfigured by struggle. And her spirit walks the castle ...

In fact *TPATP* is a RATIONALIZED FANTASY. Elizabeth's death has been faked in a plot between her and the family physician Leon, her lover, to drive Nicholas insane. Finally they succeed: Nicholas believes himself to be his own father, and the adulterous pair to be Bartolome and Isabella; Leon slain, Bartolome's identification is switched to the hapless Francis, who is bound supine beneath the descending, razor-sharp pendulum. Nicholas's smoulderingly repressed sister Catherine and a servant save him in the nick of time; Elizabeth is inadvertently left immured in the torture chamber.

Aside from the pendulum sequence, the tale owes little to Poe's original, though its themes – notably the PREMATURE BURIAL – are Poe's, brought here as reprises from Corman's *The* HOUSE OF USHER (*1960*). [JG]

2. Italian/US movie (*1990*). Full Moon. **Pr** Albert Band. **Exec pr** Charles Band. **Dir** Stuart Gordon. **Spfx** Giovanni Corridori. **Mufx** Greg Cannom. **Screenplay** Dennis Paoli. **Starring** Jonathan Fuller (Antonio), Lance Henriksen (Torquemada), Oliver Reed (Cardinal), Rona de Ricci (Maria). 91 mins. Colour.

Bearing even less resemblance to Poe's original than does **1**, this tells of innocent baker Antonio and wife Maria, and their – mainly her – torment under the Spanish Inquisition in Toledo, 1492. It is a nasty bit of work, mixing sex with suffering. [JG]

PIXIES ◊ ELVES.

PLAGUE DOGS, THE UK/US ANIMATED MOVIE (*1982*). ◊ Richard ADAMS.

PLANCHÉ, JAMES ROBINSON (1796-1880) UK dramatist, author and translator/adaptor of at least 150 plays, PANTOMIMES and burlesques, many featuring FAIRIES and other MYTHICAL CREATURES; he also translated stories by the Comtesse d'AULNOY and Charles PERRAULT. His most famous single play is *The Vampyre, or The Bride of the Isles* (**1820**), an adaptation of Charles NODIER's *Le Vampire* (**1820**), itself closely based on John POLIDORI's *The Vampyre: A Tale* (**1819** chap). His libretto for Carl Maria Weber's OPERA *Oberon, or The Elf King's Oath* (**1826**), based indirectly on the 13th-century French epic *Huon de Bordeaux*, was reprinted by Anthony BURGESS in *Oberon Old and New* (anth **1985**). Other works of fantasy interest include: *Shere Afkun, the First Husband of Mourmahal: A Legend of Hindoostan* (**1823**), a poem; *Olympic Devils, or Orpheus and Eurydice* (performed 1831; **1836** chap), a play; *An Old Fairy Tale Told Anew* (**1865** chap) illus Richard DOYLE; and many of the dramas assembled as *The Extravagances of J.R. Planché, 1825-1871* (coll **1879** 5 vols). His work and creative methods are briefly but memorably pastiched in *The Lyre of Orpheus* (**1989**) by Robertson DAVIES. [JC]

PLANETARY ROMANCE Tales set on other planets are customarily classed as SCIENCE FICTION, and indeed most are; but there has long been one type, the planetary romance (a label which dates from the 1970s but refers to a much older phenomenon), which readers have felt belongs at least partly to fantasy, or rather to SCIENCE FANTASY. PRs are stories of adventure set almost entirely on the surface of some alien world, with an emphasis on swordplay (or similar), MONSTERS, telepathy (◊ TALENTS) or other under-explained "MAGIC", and near-human alien civilizations which often resemble those of Earth's pre-technological past (featuring royal dynasties, theocracies, etc.). The hero is usually from Earth, but the means of his or her "translation" to the far planet is often supernatural rather than technological, involving flying carpets, astral projection, angel-power and kindred devices. Spaceships are sometimes mentioned, but the complete lack of interest shown in the mechanics of space travel is one of the principal features distinguishing PR from space opera (which may be fantastic and illogical in its own ways); super-scientific spacecraft and other mighty machines are central to space opera, but rarely feature in planetary romance.

The planetary romance has its main origin in the LOST-RACE novels which flourished from the time of H. Rider HAGGARD's first African adventure stories in the 1880s. The conventions of these novels about present-day heroes stumbling upon primitive societies or backward civilizations in remote parts of the globe were crossed with those of the tale of interplanetary travel – e.g., *Across the Zodiac* (**1880**) by Percy Greg (1836-1889) and *A Plunge Into Space* (**1890**) by Robert Cromie (1856-1907) – and the resultant mix, usually stripped of any utopian/dystopian or sf ideative "content" evolved into the full-fledged romance of derring-do on an exotic planet. (The RURITANIAN romances of the 1890s were perhaps also influential.) Early examples of the new formula include *Mr Stranger's Sealed Packet* (**1889**) by Hugh MacColl, set on Mars (attained by antigravity machine); a pair of novels by Gustavus W. Pope, *Journey to Mars* (**1894**) and *Journey to Venus* (**1895**); and *Pharaoh's Broker* (**1899**) by Ellsworth Douglas (real name Elmer Dwiggins). The last of these shows the blending of genres very clearly: the hero discovers a Mars peopled by Ancient Egyptians, whose civilization has somehow arisen there by a process of "parallel evolution", and he is able to converse with them in Hebrew.

But the PR did not flourish until the opening decades of the new century. More harbingers of the form can be found in such novels as *Lieutenant Gullivar Jones: His Vacation* (**1905**; vt *Gulliver of Mars* 1964 US) by Edwin Lester ARNOLD, *Angilin: A Venite King* (**1907**) by A.L. Hallen and *A Trip to Mars* (**1909**) by Fenton Ash (real name Frank Atkins; 1840-1927), but the first classic of the type – the original, defining, or "template" work of planetary romance – is *A Princess of Mars* (1912 *All-Story*; **1917**) by Edgar Rice BURROUGHS. This opening episode in the adventures of **John Carter** on the red planet known to its inhabitants as Barsoom, has remained constantly in print and has inspired numerous imitations. The love story of

Carter and the evidently mammalian but egg-laying Martian princess was spun out in two sequel volumes, *The Gods of Mars* (1913 *All-Story*; **1918**) and *The Warlord of Mars* (1913-14 *All-Story*; **1919**), which in turn gave rise to a series of seven more books, featuring various protagonists in the same, highly variable, dream-like setting. Throughout, the "science" is absurd, the action brisk, the atmosphere beguiling: it all adds up to the ultimate escapist fantasy (◊ ESCAPISM).

Few of Burroughs's imitators had his flair, but most of them achieved some success in the pulp MAGAZINES of the interwar years. This was the heyday of the planetary romance, and its practitioners included: John U. GIESY, in *Palos of the Dog-Star Pack* (1918 *All-Story*; **1965**) and two sequels; Ralph Milne Farley (real name Roger Sherman Hoar; 1887-1963), in *The Radio Man* (1924 *Argosy*; **1948**) and two sequels; Otis Adelbert Kline (1891-1946), in *The Planet of Peril* (**1929**) and two sequels; Ray(mond King) Cummings (1887-1957) in *Tama of the Light Country* (1930 *Argosy*; **1965**) and one sequel; Otis Adelbert Kline again in *The Swordsman of Mars* (1933 *Argosy*; **1960**) and one sequel; Burroughs again, in a new self-imitative series, *Lost on Venus* (**1935**) and three sequels; and Robert E. HOWARD in his posthumous *Almuric* (1939 *Weird Tales*; **1964**). After WWII, the cycle of Burroughsian planetary romances began again, with such late-in-the-day pastiches as *Emperor of Mars* (**1950**) by John (Francis) Russell Fearn (1908-1960); *Warrior of Llarn* (**1964**) by Gardner F. FOX, *Warriors of Mars* (**1965**) by Edward P. Bradbury (Michael MOORCOCK), *Tarnsman of Gor* (**1966**) by John NORMAN, *The Goddess of Ganymede* (**1967**) by Mike Resnick (1942-   ), *Zanthar of the Many Worlds* (**1967**) by Robert Moore Williams (1907-1978), *Transit to Scorpio* (**1972**) by Alan Burt Akers (Kenneth BULMER) and *Jandar of Callisto* (**1972**) and *Under the Green Star* (**1972**) by Lin CARTER, all trailing various sequels.

The PR's most important line of evolution after Burroughs was in the sf MAGAZINES. Here, talented writers such as Leigh BRACKETT and her occasional collaborator Ray BRADBURY were to bring the form to its most romantic pitch. Brackett's *The Sword of Rhiannon* (1949 *Thrilling Wonder Stories*; **1953**) is representative of her stylish best, while Bradbury's famous *The Martian Chronicles* (fixup **1950**; vt *The Silver Locusts* UK) returns Burroughsian PR to its roots in true sf via a series of moral tales which are as much comments about life on Earth as about any imaginary planetary venue. A lighter, more humorous, version of PR also developed in the sf magazines of the period, reaching book form in such works as *Cosmic Manhunt* (**1954**) by L. Sprague DE CAMP, *Big Planet* (1952 *Startling Stories*; cut **1957**) by Jack VANCE and *The Green Odyssey* (**1957**) by Philip José FARMER. Many of Vance's subsequent novels have been PRs, as have been a number by de Camp and Farmer.

Thereafter, the history of PR is, by and large, the history of an sf form. In its most serious examples, like Ursula K. LE GUIN's *The Left Hand of Darkness* (**1969**) and Brian ALDISS's **Helliconia** trilogy (**1982-85**), the PR has mutated into sf, and is hence beyond the scope of this encyclopedia. Nevertheless, something of the old Burroughsian spirit of fantasy endures in a number of popular series, notably Marion Zimmer BRADLEY's **Darkover** and Anne MCCAFFREY's **Dragonriders of Pern**. These are sf at its most romantic and fantastic, though sometimes leavened with feminist themes (◊ GENDER); their popularity has moved numerous younger writers

(particularly women) to emulate them – thus ensuring a future for a century-old tradition.   [DP]

PLAYGROUND A term which acts as a complement – and as a conceptual opposite – to FANTASYLAND. Where a fantasyland can be regarded as an ecological framework within which a story can be set – too often, in the case of GENRE FANTASY, this setting is the sole fantasticated element of the tale – a playground is, rather, a set of related ideas or concepts which are open to the fantasy-creator to romp in. As an example, the field of quantum mechanics – and the concepts associated with it – has supplied a broad playground for sf writers, with fantasy writers only more recently joining in; among the latter have been Thomas PALMER with *Dream Science* (**1990**), Lisa TUTTLE with *Lost Futures* (**1992**), John GRANT with *The World* (**1992**) and, to a lesser extent but much earlier, Mark HELPRIN with *Winter's Tale* (**1983**) and possibly (depending on one's reading of the book as fantasy or sf) Gene WOLFE with *There Are Doors* (**1988**). Most of these books relate to the ALTERNATE-WORLDS aspect of the playground opened out by quantum mechanics, which aspect fantasy writers have chosen to interpret in terms of ALTERNATE REALITIES; such books – and movies, as in the case of Ralph BAKSHI's COOL WORLD (*1992*) – are, predictably, often CROSSHATCHES. Yet playgrounds need not have their origins outside the demesne that we would ordinarily consider fantasy, or at least LIFESTYLE FANTASY: countless writers, especially in the 19th century although the influence still lingers, discovered that THEOSOPHY presented a playground they could not resist, and almost the same could be said of FAERIE. In the earlier part of this century, Sir James FRAZER's *The Golden Bough* (**1890-1915**), with its "laws" of MAGIC, offered a playground in which romped writers like Fletcher PRATT and L. Sprague DE CAMP (the **Harold Shea** stories), Robert A. HEINLEIN ("Magic, Inc." [ot "The Devil Makes the Law" 1940]), Poul ANDERSON (*Operation Chaos* [coll of linked stories **1971**) and Randall GARRETT (the **Lord Darcy** series). Further examples of playgrounds could be cited, like the concept of the MULTIVERSE created (though he did not create the term itself) by Michael MOORCOCK.

An essential difference between a playground and a fantasyland is that, while the latter circumscribes – sets limits to what is allowed within the tale – the former permits, if not encourages, enlargement of itself.   [JG/DRL/JC]

See also: AGON; WONDERLAND.

PLAYING CARDS ◊ CARDS.

PLOOG, MIKE Working name of US artist and illustrator of COMIC books and ANIMATED MOVIES Michael Ploog (1940-   ); his bold, attractive line style is heavily influenced by that of Will EISNER. In 1969 MP began working for Filmation, doing character designs and layouts for BATMAN and SUPERMAN cartoons and the series *Autocat and Motormouse*. He went on to work for Eisner on *PS Magazine* and then to provide artwork for MARVEL COMICS' new monster comics such as *Werewolf by Night* (#1-#7 1972-3), *The Monster of Frankenstein* (#1-#6 1973-4), *Man Thing* (#5-#11 1974), *Planet of the Apes* (#1-#8 and others 1974-5) and *Kull the Destroyer* (#11-#15 1974). He co-created the Tolkienesque **Weirdworld** (in *Marvel Super Action #1* and *Marvel Premiere #38* 1976).

MP has done much movie work – with Ralph BAKSHI on WIZARDS (*1977*), LORD OF THE RINGS (*1978*) and *Heavy Metal: The Movie* (*1981*); doing storyboards and set designs for *Superman II* (*1980*), *Superman III* (*1983*) and *Supergirl*

(1984) (◊ SUPERMAN MOVIES); he was an assistant director on LITTLE SHOP OF HORRORS *(1986)* and a production designer on Michael Jackson's *Moonwalker* *(1988)*. He also worked on *Young Sherlock Holmes* *(1985)* (◊ SHERLOCK HOLMES) and *The* DARK CRYSTAL *(1982)*.

MP returned to comic-strip work again with an adaptation of Frank BAUM's *The Life and Adventures of Santa Claus* (graph **1991** France). His painted artwork for the card GAME *Guardians* is due for publication in 1996.   [RT]

PLOT COUPONS Term coined by UK critic Nick Lowe (1956-   ), who identified "collect-the-coupons" plotting as characteristic of uninventive FANTASYLAND narratives: the coupons are typically magical items (AMULETS, RINGS, SWORDS, etc.) all of which the characters must collect before, in Lowe's phrase, they can send off to the author for the ending. Scattered PCs are a too-convenient means of motivating a Cook's Tour (◊ PLOT DEVICES) of the MAP; fresh ones may be introduced *en route*, making the story indefinitely extensible. PCs are most pernicious when used to decouple cause and effect – e.g., when they grant their holder disproportionate power, or when a DARK LORD is defeated solely through the manipulation of PCs, leaving a sense of unearned HEALING. Lowe cites Lin CARTER's *The Black Star* (**1973**), whose eponymous jewel controls the outcome for no reason other than a whim of the GODS'; i.e., of the author's. Other PCs include the THREE arbitrary magic places which the protagonist must visit to regain himself – for a PC may also be a fulfilled task or RITUAL – in Fletcher PRATT's and L. Sprague DE CAMP's *Land of Unreason* (**1942**); the bull, spear, oak, ram, sword and stallion in Michael MOORCOCK's **The Chronicles of Corum**; the swords of Fred SABERHAGEN's **Swords** series, whose vast powers overshadow mere human characters; and the Destiny Stone in Robert VARDEMAN's and Victor MILAN's **The War of Powers** sequence, whose power to generate wild improbabilities and reversals of fortune allows the authors to manipulate their plot with less than the usual pretence of plausibility. Further examples abound. PCs are particularly prevalent, for obvious reasons, in multiple-choice GAMES.

But in J.R.R. TOLKIEN's *The Lord of the Rings* (**1954-5**), the One Ring is more than a PC: it has symbolic weight, its bearers suffer arduous NIGHT JOURNEYS, and painful battles must still be fought; and in Susan COOPER's *The Dark is Rising* (**1973**) the gathering of six Signs does not so much assemble a PC *deus ex machina* as form a kinetic RITE OF PASSAGE for the young protagonist. It is when QUEST objects dwindle from the stature of GRAIL or Ring to nothing more than narrative conveniences that the term PC may be pejoratively used.   [DRL]

See also: MCGUFFIN.

PLOT DEVICES Stock mechanisms used in fiction to initiate, complicate, advance, retard or terminate the plot. Some are common to all genres; the following selection naturally favours PDs prevalent in fantasy, cross-referring as appropriate to those having their own entries.

Amnesia Inflicting AMNESIA on a protagonist may be convenient as a pretext for exposition, as a way of making a sympathetic character into an OBSESSED SEEKER, or as a means of distancing a less sympathetic past. Wolff/Jadawin in Philip José FARMER's *The Maker of Universes* (**1965**) and Corwin in Roger ZELAZNY's *Nine Princes in Amber* (**1970**) are clearly gentler for their memory loss; but the protagonist of Tanith LEE's *The Birthgrave* (**1975**) is rendered more

dangerous, lacking knowledge of and control over the powers that go with her lost identity. A related "tidying-up" PD, common in CHILDREN'S FANTASY, is the concluding MEMORY WIPE.

Butchery The excessive violence of DARK LORDS is often crucial in motivating HEROES AND HEROINES against them. Often, the initiating violence is perpetrated against the protagonist's parents, as in Geoff RYMAN's *The Warrior who Carried Life* (**1985**) and Terry GOODKIND's *Wizard's First Rule* (**1994**). The atrocity is not always murder – in Michael MOORCOCK's *Stormbringer* (**1963**) Elric's wife suffers TRANSFORMATION into a MONSTER.

Clues In DETECTIVE/THRILLER FANTASY, clues have the same function as in the nonfantasy genres; readers can match their wits with the author, protagonist and unknown villain. In Barbara HAMBLY's *The Witches of Wenshar* (**1987**) the identity and motivation of the magical murderer can be deduced from probabilities. More generally, clues can dramatically foreshadow or prepare the ground for later revelation or RECOGNITION. In Tim POWERS's *The Anubis Gates* (**1983**) Doyle deduces from hearing someone whistling The BEATLES' "Yesterday" in Regency London that he is not alone in his TIME TRAVEL: this is both clue and Pistol Effect (*q.v.*).

Cook's Tour The traditional journey around the MAP of a FANTASYLAND, visiting every point of interest and perhaps collecting PLOT COUPONS; also, a device whereby an OBSESSED SEEKER or ACCURSED WANDERER is moved from location to location in a TEMPLATE series. In lands of ALLEGORY (◊◊ IMAGINARY LANDS) the tour reflects progress through intellectual or spiritual states.

Duel A frequent assumption is that right will prevail in a straightforward contest between two strong men. Accordingly, many fantasies climax in physical or magical single combat. At the end of Martha WELLS's *The Element of Fire* (**1993**) her battle-weary protagonist allows his enemy, the king's favourite, to assume him more weakened than he is and challenge him; righteousness overrules honesty on such occasions. The protagonist of Ellen KUSHNER's *Swordspoint* (**1987**) is unusual in being a professional duellist. Duels between wizards are rarely physical, and take the form of competitive METAMORPHOSES as in the movie *The Sword in the Stone* (**1963**; ◊ DISNEY; T.H. WHITE) – the duel between Dream and the DEMON Choronzon in Neil GAIMAN's *Preludes & Nocturnes* (graph coll **1991**) elegantly varies this traditional theme – or as struggles between SPELLS embodied as physical forces or objects, like the spfx duel in Roger CORMAN's *The Raven* (**1963**); or an alternating sequence of oddly named charms, one set of which eventually overwhelms the other: Jack VANCE's wizards incline to such duels, as does MARVEL COMICS's DOCTOR STRANGE.

Escape J.R.R. TOLKIEN argued that escape from prison is a prime image of fantasy, and that to object to its ESCAPISM is to reveal a secret tyrannous agenda. Certainly *LOTR* is full of escapes and rescues: they are small fixes of victory. Gandalf's escape from the UNDERWORLD and the Balrog effectively involves his death and RESURRECTION – for a Christian like Tolkien, all escapes are types of CHRIST's Resurrection, and all rescues reflect Christ's redemption of humankind.

Forgetting The decision to destroy an item of knowledge on behalf of humanity is a standard ending in much SUPERNATURAL FICTION, HORROR and DARK FANTASY. In Karel ČAPEK's *The Makropoulos Case* (**1922**) Emilia Marty decides that 300 years of life is enough and gives the IMMOR-TALITY-drug recipe to the girl Krista, who decides to burn it. Outside pure fantasy, forgetting still tends to mean the destruction of unique manuscripts and BOOKS – e.g., in Umberto ECO's *The Name of the Rose* (**1980**). This is a device whereby quite mundane fictions can ally themselves to FANTASIES OF HISTORY.

Impersonation A traditional device in RURITANIA, where DOUBLES of royal-family members are strangely easy to find. MAGIC impersonation can be sinisterly used, as when the evil fox in Susan COOPER's *The Grey King* (**1975**) takes on the forms of dogs.

Inns ◊ INNS.

Jealousy Emotions are powerful generators of plot, and those concerned with a sense of identity are perhaps peculiarly crucial to fantasy, in which identity (◊ HIDDEN MONARCHS; IDENTITY EXCHANGE; TRUE NAMES; SOUL) is always important. Villains become so because envious of what they are not: Barbara HAMBLY's Altiokis, in *The Ladies of Mandrigyn* (**1984**), resents wizardly apprenticeship and finds his own way to magical power; the heir to the throne, Elias, in Tad WILLIAMS's **Memory, Sorrow and Thorn** (**1988-93**), envies his brother Joshua's prowess and glamour and sets out to destroy him, thus wounding the LAND. Jealousy and envy are powerfully destructive temptations – Boromir in *LOTR* resents the likely replacement of his line of stewards by the HIDDEN MONARCH Aragorn, and attempts to steal the RING. All these characters implicitly fall short by seeking that to which they are not innately entitled, whereas UGLY DUCKLINGS effortlessly get that which is rightfully theirs.

Lies and Deceits A lie is both a highly metaphysical element in fantasy (since the nature of REALITY is always a concern) and one of its most simplistic generators of STORY; by definition it is a self-serving PARODY of the purity of Story. Lies are expected from SATAN, LOKI, DARK LORDS and TRICKSTERS. Virtuous tricksters often lie by economy with the truth: in Tim POWERS's *The Drawing of the Dark* (**1979**) Ambrosius does not so much deceive the protagonist as fail to explain things to him in enough detail.

Loopholes Fantasy tends to have sharply defined rules, with oaths, CONDITIONS, CONTRACTS, PROHIBITIONS, PROPHECIES, vows, WISHES, etc., possessing (through MAGIC) greater force than physical law. Loopholes are thus constantly sought. (◊◊ ANSWERED PRAYERS; QUIBBLES; READ THE SMALL PRINT.)

Mutilation Partly because ODIN gave up an eye for wisdom, fantasy protagonists are prone to losing body parts during NIGHT JOURNEYS or otherwise. Frodo loses a finger in *LOTR*; Fritz LEIBER's Fafhrd a hand; Barbara HAMBLY's Sun Wolf an eye; Tim POWERS's protagonists are peculiarly likely to lose bits.

Nazgûl In J.R.R. TOLKIEN's *LOTR* the Nazgûl, Sauron's mightiest henchmen, are – considering their powers – remarkably absent from the plot. Nazgûl typify menaces introduced more for ornament than use.

Oracles ◊ ORACLES.

Pistol Effect In Akira Kurosawa's film *Yojimbo* (**1961**) the Samurai hero witnesses the arrival of a villain – who produces and fires a six-shooter, revealing that we are not in generic samurai-time, as we had assumed. The Pistol Effect is any such sudden presentation of an object or concept which radically changes our perception of where we are in TIME, space or genre. Often it triggers the realization of a

TIME ABYSS. Examples abound – in Michael MOORCOCK's *The Runestaff* (**1969**) the ships of Granbretan have names including distorted versions of those of The BEATLES, shockingly indicating that this is our own world's future; the ancient painting (◊ PICTURES AND PORTRAITS) in Gene WOLFE's *The Shadow of the Torturer* (**1980**) proves to depict an astronaut on the MOON; in Barbara HAMBLY's *The Silicon Mage* (**1989**) the heroine's realization that a supposed godling is actually another interloper from elsewhere in the MULTIVERSE is signalled by her use of *pi* to communicate with what turns out to be an alien; the strangeness of Mary GENTLE's CITY in *Rats and Gargoyles* (**1990**) is indicated by the mention of a fifth compass point.

Plot Coupons ◊ MCGUFFIN; PLOT COUPONS.

Plot Voucher A wild-card PD issued to protagonists of GENRE FANTASY, early in their travels, often with an assurance that when the time comes they will realize its use. An inoffensive example is the magic tin whistle acquired by Fafhrd in Fritz LEIBER's *The Swords of Lankhmar* (**1968**).

Separation Principals, often part of a DUO or a SEVEN-SAMURAI or a DIRTY-DOZEN group, are scattered to collect PLOT COUPONS from different locations, or simply to give the reader a Cook's Tour (*q.v.*) of this particular FANTASY-LAND. Separation is so much a standard PD of large-scale GENRE FANTASY that we take it almost for granted.

Temptation Much GENRE FANTASY assumes a Christian Universe, where even the best of humans may fall into moral jeopardy – especially at the climax of their prolonged NIGHT JOURNEY, when they risk throwing away everything they have learned and earned. John BUNYAN's *The Pilgrim's Progress* (**1678**) is careful to point out a way to HELL even from the gates of HEAVEN; J.R.R. TOLKIEN's snobbery is interestingly evident as noble Frodo and gallant Boromir nearly succumb to the RING's temptation, while Samwise, though momentarily tempted, is in less moral danger because he knows his place. Often the temptation is merely to think wrong thoughts that will make the principal more vulnerable to the forces of evil – literally in Poul ANDERSON's *Three Hearts and Three Lions* (**1953**), when Holger's mildly carnal thought about a female COMPANION nullifies their magic protective circle and admits a GIANT. Under torture, in Tad WILLIAMS's **Memory, Sorrow and Thorn** (**1988-93**), Simon finds himself wrongly resenting his allies as having used him; such unjust self-pity in a potential MESSIAH would be a serious moral failing if continued. More interestingly, Covenant's temptations in Stephen R. DONALDSON's **Chronicles of Thomas Covenant** (**1977**) concern the improper use of power for good motives; Covenant has a personalized tempter rather than merely being self-betrayed. Personal tempters are rife in SLICK FANTASY, which relies so much on Faustian PACTS and magic SHOPS.

Walking In his *Inventing the Middle Ages* (**1992**), Norman Cantor remarks that one of the several services J.R.R. TOLKIEN and C.S. LEWIS did for medieval studies was to dramatize for lay readers the fact that travel on foot, or even on horseback if reasonable care is being taken of a single horse on a long journey, is prolonged, arduous and inconvenient. The walking pace of most fantasy journeys means that people move slowly through LANDSCAPES which often acquire moral meanings in the process; problems must be *solved* rather than merely moved away from at high speed.

Xenophobia Savage tribes, LOST RACES and intractable nonhumans routinely delay the progress of a Cook's Tour (*q.v.*) through unfounded hostility; conversely, J.R.R. TOLKIEN's orcs have promoted the xenophobic assumption that those who look unpleasant or merely different must be evil (◊◊ COLOUR-CODING) – a notion interestingly challenged by (for example) Tad WILLIAMS in **Memory, Sorrow and Thorn**. [RK/DRL]

Further reading: *The Tough Guide to Fantasyland* (**1996**) by Diana Wynne JONES.

PLUTO ◊ HADES; HELL.

POE, EDGAR ALLAN (1809-1849) US writer, poet and editor, one of the most important figures in the development of the US short story and a seminal influence on SUPERNATURAL FICTION, SCIENCE FICTION and detective fiction. Orphaned at age two, he was raised by a wealthy merchant, John Allan, of Richmond, Virginia. The family visited England in 1815; there EAP was educated until their return to the USA in 1820. His brief time at university showed his proclivity for languages, but his dissolute lifestyle soon plunged him into debt and drink. EAP argued with his foster-father and ran away from home. His first volume of poetry, *Tamerlane and Other Poems* (coll **1827** chap), privately published, was a financial failure. He spent time at West Point, but was expelled. Now cut off from John Allan, EAP found his poetry – two further volumes, *Al Aaraaf, Tamerlane, and Minor Poems* (coll **1829** chap) and *Poems* (coll **1831** chap), had appeared – while critically acclaimed, brought in no money, and he turned to fiction and essays; he also briefly edited four magazines, *Southern Literary Messenger* (December 1835-January 1837), *The Gentleman's Magazine* (July 1839-June 1840), *Graham's Magazine* (April 1841-May 1842) and *The Broadway Journal* (March 1845-January 1846). These contained many of his writings and allowed him an editorial platform for his favourite topics. He caused much interest by predicting – in "Charles Dickens" (*Saturday Evening Post* 1 May 1841) – the conclusion of Charles DICKENS's *Barnaby Rudge* (**1841**) after publication of the first three chapters. His essay, "The Philosophy of Composition" (1846 *Graham's*) is a remarkable assessment of the process of writing. Of the several collections of EAP's essays the most readily available is *Poems & Essays* (coll **1881** UK) ed Andrew LANG and the most complete is *Essays and Reviews* (**1984**) ed G.R. Thompson (1937-   ).

EAP's poetry was heavily influenced by the work of Samuel Taylor COLERIDGE and Lord BYRON, with a deep brooding melancholy pervaded by the darkness of the soul and intrusion of the ghosts of memory. This same mood influenced his early fiction, though this also showed his appreciation of the writings of E.T.A. HOFFMANN, Walter SCOTT and the German Romantics whose stories were appearing in BLACKWOOD'S MAGAZINE. EAP's first published story, "Metzengerstein" (1832 *Saturday Courier*) is heavily Gothic, with the usual trappings of revenge and FATE. It started a loosely connected series which EAP called **Tales of the Folio Club** (◊ CLUB STORY) in which 11 friends form a club with the entry qualifications of writing a short story. Two of these stories have become classics: "MS Found in a Bottle" (1833 *Baltimore Saturday Visitor*) – which won a newspaper competition and earned EAP national recognition – and "A Descent Into the Maelstrom" (1841 *Graham's*). Both qualify as TRAVELLER'S TALES. "MS Found in a Bottle" is a nautical horror story, and is largely a pilot for the much longer *The Narrative of Arthur Gordon Pym of Nantucket* (1837 *Southern Literary Messenger*; dated 1837 but **1838**; vt *Arthur Gordon Pym, or Shipwreck, Mutiny and Famine* 1841 UK; vt *The Wonderful Adventures of Arthur Gordon Pym* 1861 UK; reissued with continuation by Jules Verne as *The Mystery of Arthur Gordon Pym* **1960** UK). Neither version is overtly supernatural, although both contain ghost ships (◊ FLYING DUTCHMAN). More significantly, along with "A Descent Into the Maelstrom" – which EAP regarded as one of his best stories – they are each an ALLEGORY for a descent into madness through an awareness of the terrors of one's soul, a theme which is central to most of EAP's later fiction.

His particular torments were all related to death, and here *The Masque of the Red Death* (1842 *Graham's*; **1923** chap; vt *The Mask of the Red Death* 1969 chap) is central. Another allegory, it is a crescendo of mounting terror, in the form of a DANCE OF DEATH, as Prince Prospero endeavours to isolate himself from the plague but inevitably faces DEATH (◊◊ MASQUES). Around this story parade two other key themes. One is incarceration, especially PREMATURE BURIAL, as evident in *The Black Cat* (1843 *US Saturday Post*; **1914** chap), *The Tell-Tale Heart* (1843 *The Pioneer*; **1916** chap) – both PSYCHOLOGICAL THRILLERS of guilt – "The Premature Burial" (1844 *Dollar Newspaper*), and to some extent "The Oblong Box" (1844 *Godey's Lady's Book*). Incarceration is central to EAP's major CONTE CRUEL, "The Pit and the Pendulum" (1843 *The Gift*). The other theme is that of the return from the dead, which may be in the form of recovery from a trance or disease or may remain unexplained. Within this lie his trilogy of FEMME-FATALE stories, "Berenice" (1835 *Southern Literary Messenger*), which is also a VAMPIRE story, "Morella" (1835 *Southern Literary Messenger*) and the superior *Ligeia* (1838 *American Museum*; **1943** chap UK). In "The Facts in the Case of M. Valdemar" (1845 *Broadway Journal*; vt "The Facts of M. Valdemar's Case"; vt *Mesmerism, `In Articulo Mortis'* **1846** chap UK) EAP uses MESMERISM to perpetuate the life of the spirit after the death of the body. EAP's masterpiece in this territory, however, is *The Fall of the House of Usher* (1839 *Gentleman's Magazine*; **1903** chap). At the same time as he wrote about the doom of Roderick Usher, EAP was exploring the theme of linked SOULS and DOPPELGÄNGERS in "William Wilson" (1839 *Gentleman's Magazine*), but in *Usher* he takes the masterful step of linking Usher's SOUL with that of his house. In retrospect this is a natural conclusion of the GOTHIC FANTASY, with its emphasis on EDIFICE and torment, but EAP was first to forge that connection. Superficially the story deals with a brother whose sister, Madeline, is believed dead but who returns to life. Upon seeing her, Roderick dies and Madeline dies with him. At that moment the house falls. The haunting melancholy of death, particularly the death of a young woman, also fuelled EAP's poetry, especially "Lenore" (1831; rev 1843; rev 1845) and its remarkable sequel "The Raven" (1845 *American Review*) where death haunts the narrator in the form of a raven. EAP assembled his first volume of stories as *Tales of the Grotesque and Arabesque* (coll **1840** 2 vols).

Although EAP wrote other stories, including several humorous ones, it is his works of terror and anguish that made him famous and which had an incalculable effect on literature. It would be impossible to list all of the authors influenced by him, especially as such influence naturally becomes attenuated over the years. Of special significance, however, are Charles BAUDELAIRE, who translated EAP's work into French and championed it across Europe – EAP's fiction lends itself well to the French language, and EAP must be reckoned an important forefather of the literature of DECADENCE. DE MAUPASSANT was another literary heir. EAP's poetry was an acknowledged influence on Algernon Swinburne and Lord Tennyson, and his horror fiction influenced Robert Louis STEVENSON, M.P. SHIEL, W.C. MORROW, Ambrose BIERCE and H.P. LOVECRAFT (through whom his influence spread to many more, especially Thomas LIGOTTI). EAP's work shaped the development of the HORROR genre during the 19th century, particularly in the short-story form, and his tales featured regularly in ANTHOLOGIES of the day. It was a fascination for EAP's work that caused Jacob Henneberger (1890-1969) to launch WEIRD TALES in 1923. EAP's life, too, fascinates authors, and he appears as a character in many stories and novels. These include *The Man who Was Poe* (**1989**) by AVI, *The Hollow Earth* (**1990**) by Rudy Rucker (1946-   ), *Nevermore* (**1994**) by William HJORTSBERG, *The Bloody Red Baron* (**1996**) by Kim NEWMAN and *The Lighthouse at the End of the World* (**1995**) by Stephen Marlowe (1928-   ), which explores the last week of EAP's life. Sam Moskowitz (1920-   ) compiled a collection of such stories, including also rarities by EAP, in *The Man who Called Himself Poe* (**1969**; vt *A Man Called Poe* 1972 UK).

Only two other story collections were issued during EAP's lifetime. His emerging popularity was evident with the special printing of "The Murders in the Rue Morgue" (1841 *Graham's*), the first of his **Auguste Dupin** detective stories, along with "The Man who Was Used Up" in a booklet entitled *The Prose Romances of Edgar A. Poe* (coll **1843** chap). This item is the rarest of all of Poe's books. With the success of "The Raven", EAP was able to secure the New York publication of a collection of his poetry, *The Raven and Other Poems* (coll **1845**), and this was published as a companion to *Tales* (coll **1845**) – the two assembled as *Tales and The Raven* (omni **1846**) in a special hardcover edition. The contents for *Tales* had been selected by New York editor and literary specialist Evert Duyckinck (1816-1878); EAP was not happy with the choice.

After the death of his consumptive wife Virginia (1822-1847), EAP became even more dissolute and drunken. Nevertheless, the success of his books allowed EAP to return to his first love, poetry, and this brought him some relief. Also, he found love again, and this uplift in his spirit is reflected in his fiction – as in the **Ellison** sequence, "The Landscape Garden" (1842 *Snowden's Ladies Companion*) and "The Domain of Arnhem" (1847 *Columbian Magazine*) – and their spiritual sequel, "Landor's Cottage" (1849 *Flag of Our Union*), which are almost transcendental in their study of beauty and the creativity of wealth. It is interesting to speculate whether, had EAP lived longer, he would have moved away from horror fiction towards other fields.

Collections of EAP's fiction are too many and varied to list here. The first collected edition, *The Works of the Late Edgar Allan Poe* (coll **1850** 3 vols; **1856** vol 4) was ed Rufus Griswold (1815-1857), EAP's literary executor who, though trusted by EAP, ignored the author's own amendments to his stories and published original texts together with a malicious biography, which invented many of the myths still perpetuated about Poe. *The Complete Works of Edgar Allan Poe* (coll **1902** 17 vols) was ed James A. Harrison. The standard title of most EAP collections was first used on *Tales of Mystery and Imagination* (coll **1855** Canada); the most popular edition of this title, with variant contents, is *Tales of Mystery and Imagination* (coll **1919** UK), illustrated by

Harry CLARKE. Other books and collections of interest include: *The Raven* (**1883** UK chap), the last book to be illustrated by Gustav DORÉ; *The Poems of Edgar Allan Poe* (coll **1900** UK), illustrated by W. Heath ROBINSON; *The Bells and Other Poems* (coll **1912**), illustrated by Edmund DULAC; *The Works of Edgar Allan Poe* (coll **1927**); *The Book of Poe: Tales, Criticisms, Poems* (coll **1929**); *The Complete Tales and Poems of Edgar Allan Poe* (coll **1938**); and *The Portable Edgar Allan Poe* (coll **1945**). Peter HAINING has compiled *The Edgar Allan Poe Scrapbook* (coll **1977**) and *The Edgar Allan Poe Bedside Companion* (coll **1980**).   [MA]

Further reading: *Edgar Allan Poe: A Critical Biography* (**1941**) by Arthur H. Quinn; *The French Face of Edgar Allan Poe* (**1957**) by Patrick H. Quinn; *Edgar Allan Poe* (**1961**; rev 1977) by Vincent Buranelli; *Edgar Allan Poe: The Man Behind the Legend* (**1963**) by Edward WAGENKNECHT; *Edgar Allan Poe* (**1977**) by David Sinclair; *The Tell-Tale Heart: The Extraordinary Mr Poe* (**1978**) by Wolf Mankowitz; *The Life and Works of Edgar Allan Poe* (**1978**) by Julian Symons; *The Rationale of Deception in Poe* (**1979**) by David Ketterer; *A Psychology of Fear: The Nightmare Formula of Edgar Allan Poe* (**1980**) by David R. Saliba.

POGÁNY, WILLY Working name of versatile Hungarian illustrator William Andrew Pogány (1882-1955), who built a considerable reputation in the UK before transferring to the USA, where he achieved even greater success in many different fields of art. His illustration work was carried out in a number of different styles and techniques, including pen-and-ink, lithography and watercolour. His most successful medium was pen, with which he produced imaginative drawings of delightful delicacy and exotic decorativeness. His UK publisher, Harrap, was remarkably indulgent with WP, and published several volumes of his hand-lettered text, illustrated in a range of different media and styles, and printed on fine tinted papers. Examples of this are the three volumes of tales from Richard WAGNER – *Tannhäuser* (graph **1911**), *Parsifal* (graph **1912**) and *The Tale of Lohengrin* (graph **1913**) – Samuel Taylor COLERIDGE's *The Rime of the Ancient Mariner* (graph **1910**), and *The Rubáiyát of Omar Khaiyam* (**1909**). The effect is idiosyncratic and curious rather than artistically satisfying.

After 1915 WP settled in New York, where he widened his artistic activities to include mural painting and stage design and became an art director in Warner's First National Studios, Hollywood.   [RT]

POGLES' WOOD UK tv series (1966-7), early episodes titled *The Pogles*. ◊ *The* CLANGERS (1969-73).

POICTESME Still occasionally cited – like OZ – as an exemplary IMAGINARY LAND or LAND OF FABLE, this is the setting for much of James Branch CABELL's **Biography of the Life of Manuel**. Despite extensive links to mythical geography, it is supposedly a province of medieval France, longitude 4 degrees east, whose principal castles are Bellegarde and Storisende. Cabell and his frequent illustrator Frank C. PAPÉ both drew MAPS of Poictesme.   [DRL]

POLAND What is described as "fantasy" in Poland – the word has been borrowed directly from English – is a literature closely akin to the West's GENRE FANTASY: it is regarded as an offshoot of SCIENCE FICTION, differing mainly in that MAGIC rather than science is used in the writer's world-creation. Moreover, almost all Polish critics and writers have agreed that "fantasy" must be based on the Arthurian archetype – the MATTER of Britain – and this further divorces writers from the field, for such material is quite alien to Polish literature in general. In these terms, Polish 20th-century fantasy is really just an imitation of English-language fantasy, and there is very little of it. Outside this narrow definition, however – using the word FANTASY in the way it is generally used in this book – there is much of interest.

Polish fantasy was born in the early part of the 19th century as part of the Europe-wide surge in ROMANTICISM. Through the late 18th century the country had been progressively partitioned among Russia, Austria and Prussia, and an uprising had been crushed in 1794. (Further uprisings would be likewise put down in 1831 and 1863.) Works of Polish Romanticism – the popular literature of their day – usually had to be smuggled from abroad into a country that currently did not exist as such; there they were copied – usually by hand – and distributed under the shadow of severe penalties. Effectively, these were the first samizdats in modern literature.

Adam Mickiewicz (1798-1855) is a case in point. He was banished to Russia in 1824, and thereafter had a brilliant career in various parts of Europe. His first work of fantasy interest was a collection of poetry, *Ballady i romanse* ["Ballads and Romances"] (coll **1822**); it can be regarded as the wellspring of Polish Romanticism. The poems as a whole display quite elaborate folk stylistics, but are of interest primarily because they accept a complete coexistence of "this world" and "the other" – i.e., that REALITY has more than one level. Thus GHOSTS can return to give moral lessons to the living, people can undergo strange transformations, justice can be directly administered by the GODS, etc.

More important in the fantasy context is Mickiewicz's three-part verse drama *Dziady* ["Forefathers" or "Forefathers' Eve" – Dziady is a Byelorussian rite of ancestral-spirit worship] (Parts 2 and 4 **1823** Vilnius; Part 3 **1832** Paris; Part 1 [unfinished] posthumously in *Dziela tom 3* ["Works, Volume 3"] **1860** Paris). In Part 2 Gúslarz – who is a hybrid of priest, seer and witchdoctor – calls up ghosts for a group of peasants. The ghosts bemoan their sins and the peasants offer help. At last the ghosts vanish – except one, the Spectre, who refuses to listen to Gúslarz. In Part 4, on the night of Zaduszki (SAMHAIN), a hermit, Gustaw, suddenly appears in the parish house to tell the priest the story of his unhappy love. He stabs himself with a dagger, but is not hurt; then passionately defends the pagan folk-rites of Dziady against the Church. He then vanishes, leaving us with the enigma of whether he was alive or a ghost. Part 3, written after the suppression of the November Uprising of 1830-31, was a patriotic piece, but does contain some fantasticated material. In its prologue Gustaw is spoken to in DREAMS by satanic and angelic voices, which tell him of the future; he becomes a freedom-fighter, changing his name to Konrad. Later Konrad talks to God, trying to usurp part of God's power. In a final section, "The Vision of Father Peter" – heavily influenced by the APOCALYPSE – Konrad is absolved of his crime against God by a low priest, who goes on to formulate the "Messianic Theory": Poland is the "Christ of Nations" and, like CHRIST, must be tortured and murdered before rising to redeem the world.

Roughly contemporary with Mickiewicz was Juliusz Słowacki (1809-1849). His *Balladyna* (**1839**), a tragedy in five parts, is far from his best work but contains a veritable library of fantasy themes. The story is set in Poland's GOLDEN AGE, and has two strands, in both of which supernatural powers intervene. The first tells of the mythical King Popiel III, overthrown by a usurper, and of his crown, which is the TALISMAN guaranteeing his country's strength, the FERTILITY of its fields, etc. The second strand describes the fate of a nobleman, Kirkor, who is intent on Popiel's reinstatement.

A further contemporary was Zygmunt Krasiński (1812-1859). His works were extremely important in the political debates of the time; they also represent wonderful examples of the unrestricted imagination at work. In his drama *Irydion* ["Iridion"] (**1836**), for example, a Greek, Iridion, aided by the demonic Massynissa, mounts a bloody revolution against a decadent Ancient Rome. A Christian bishop thwarts his plans, persuading the Christians in Iridion's army to withdraw because the way to conquer is not through violence but through benign martyrdom. Iridion is sentenced to centuries of sleep, after which, by direct order from God, he rises and goes to Poland to resurrect the country through martyrdom.

The brutal crushing of the January Uprising of 1863 marked the end of Polish Romanticism; writers turned instead to Realism. Even so, succeeding decades saw the appearance of at least one major fantasy novel, *Ogniem i mieczem* (**1884**; trans as *With Fire and Sword* **1991** US) by the Nobel laureate Henryk Sienkiewicz (1846-1916). Superficially this is a realistic account of part of Poland's history; Sienkiewicz uses genuine 17th-century characters, settings, events and even language to reinforce that realism, and yet the heart of the book is fantasy. Sienkiewicz created a set of HEROES each highly specialized in his field – not unlike characters in a role-playing GAME. One is pure of heart and a superhumanly strong fighter; another has exquisite skill with the sabre; a third is wise beyond measure. These heroes, each the core of a set of COMPANIONS, engage in various QUESTS, both patriotic and private.

Poland was independent during the interbellum. Two authors of note emerged. Stanisław Ignacy Witkiewicz (1885-1939; ◊◊ *SFE*), with his philosophical and artistic theory of "pure form" and completely accidental action in the theatre, was the purest of Absurdist playwrights (◊ ABSURDIST FANTASY). Between 1893 (when he was aged 8) and 1934 he wrote over 30 plays; these, with their crashing mountains of carefully staged nonsense, are both tragic and refreshing. The other significant (in our context) writer of the period was Zofia Kossak-Szczucka (1890-1968), author of the large novel *Krzyżowcy* ["The Crusaders"] (**1935**). The power of this novel lies in Kossak-Szczucka's ability to capture the sense of wonder with which the medieval KNIGHTS at the centre of the story perceive the exotic civilization of Islam: ignorant of it, hating it, fighting it – the Crusaders are nevertheless captivated by it. The superior Arab civilization and science are, to them, indistinguishable from MAGIC. And magic of all kinds, both Christian and pagan, plays its part in the conquest of Jerusalem. At last, on the battlefield, the Polish knights are able to call up the Slavic Spectre to save the day. In return, she demands her usual fee: the life of the caller. If any indigenous novel could be said to have influenced modern Polish fantasy, this is it.

After WWII Polish literature was for decades devoid of fantasy, with sf being encouraged in its place – partly for political reasons, as elsewhere in Eastern Europe. The great tradition of Polish fantastication was forgotten. J.R.R. TOLKIEN's *The Lord of the Rings* (**1954-5**) was translated with much success, but had little effect on Polish writers.

Then, in the early 1980s, fantasy resurfaced – partly through young writers rebelling against the established sf "masters". The first hit was probably "Twierdza trzech studni" ["The Keep of Three Wells"] (1982) by Jarosław Grzedowicz (1965-   ). That same year saw the first of a flood of short stories from the prolific Jacek Piekara (1965-   ), followed by what was planned as a trilogy, **Imperium** ["Empire"], but of which only parts appeared: *Smoki Haldoru* ["The Dragons of Haldor"] (**1987**), the first volume, was followed by an omnibus containing this and *Przeklety tron* ["The Cursed Throne"] (omni **1989**). The collection *Stolin's Treasures* (coll **1990**) by Rafał A. Ziemkiewicz (1964-   ), containing two fantasy stories – "Hrebor Cudak" ["Hrebor the Odd"] and "Skarby Stolinów" ["Stolin's Treasure"] – was something of a *cause célèbre*. A fine sf writer, as evidenced by his highly emotional, fast-paced, politically based short stories, Ziemkiewicz was for a long while the voice of his generation; but his excursion into fantasy was less successful. His aim was to create fantasy based on the "Slavic archetype", but the result was a lame cross between *The Lord of the Rings* and *Krzyżowcy*.

The following year Zbigniew Nienacki (real name Zbigniew Nowicki; 1929-1994), an unsuccessful mainstream writer and very successful writer for YA, produced the trilogy *Dagome Iudex – Ja, Dago* ["I, Dago"] (**1989**), *Ja, Dago Piastun* ["I, Dago the Guardian"] (**1989**) and *Ja, Dago Wladca* ["I, Dago the Ruler"] (**1989**) – based fairly closely on Slavic MYTHS and RELIGION. Despite a plethora of SEX, the trilogy was a flop, and has had little influence on other writers.

The Polish fantasy scene today might, then, seem rather dismal, but two recently emergent young writers may change that. Feliks W. Kres (real name Witold Chmielecki; 1966-   ) – has published two collections and two novels – *Prawo sepów* ["The Rule of Vultures"] (linked coll **1991**), *Król bezmiarów* ["The King of the Endless Ocean"] (**1992**), *Strażniczka istnień* ["The Guardian of All Life"] (**1993**] and *Serce Gór* ["The Heart of the Mountains"] (linked coll **1994**) – as well as short stories in both the Polish sf/fantasy MAGAZINES, *Fantastyka* and *Fenix*. The typical hero of his well written, moody, broad-scope fantasies is hesitant and unsure, burdened by his past, and fighting not only with the outside world and magic but also with himself. Kres is not much interested in foreign fantasy, so his work may be expected to remain clear of cliché for a long while yet.

Andrzej Sapkowski (1948   ) was so hugely successful with his first short story, "Wiedźmin" ["He-Witch" or "Witchkiller" – a punning title] (1986), which won a story contest mounted by *Fantastyka*, that he decided to stick with its hero: a post-apocalypse, genetically altered supernatural killer of supernatural creatures – a kind of CONAN with a bad conscience. Sapkowski has now produced many stories as well as several books, often in his **Wiedźmin** series; his books are *Wiedźmin* (coll **1990**), *Miecz przesnaczenia* ["The Sword of Destiny"] (coll **1992**), *Ostatnie życzenie* ["The Last Wish of the Dying"] (coll **1993**), *Krew elfów* ["The Blood of Elves"] (**1994**), *Świat Króla Artura/Maladie* ["The World of King Arthur/*Maladie*"] (coll of 1 essay and 1 story **1995**) and *Oko Yrrhedesa* ["The Eyes of Yrrhedes"] (gamebook **1995**). The **Wiedźnin** stories are well told but poorly realized and derivative from foreign fantasy; Sapkowski's non-**Wiedźnin** material is, however, far more interesting and inventive, as exemplified by "Maladie", a beautiful variation on the Tristan and Isolde LEGEND.

As far as Polish movies are concerned, there is some animation but otherwise no fantasy-movie production to speak of.   [KS]

Further reading: *The Dedalus Book of Polish Fantasy* (trans coll **1996** UK) ed Wiesiek Powaga (1958-   ) features a very different range of Polish fantasy writers.

POLANSKI, ROMAN (1933-   ) French-born movie director, actor and screenplay writer, of Polish-Jewish stock. He has lived through interesting times: his mother was murdered by the Nazis in the concentration camp where he and his father were also incarcerated; his second wife, the actress Sharon Tate, was among those grotesquely murdered by Charles Manson's cult (as would have been RP himself had it not been that his airplane home was delayed); he fled from the USA to Europe in 1979, having spent some time in prison for statutory rape (it seems that he had no idea that the girl concerned was only 13). All of his work has been influenced by these dreadful experiences: even his comedies are affected by an acute sense of EVIL. His first – much-admired – feature was *Knife in the Water* (**1962**; ot *Noz w Wodzie*). In the UK RP made one of his strongest movies – *Repulsion* (**1965**), a PSYCHOLOGICAL THRILLER – as well as *Cul de Sac* (**1966**) and *The* FEARLESS VAMPIRE KILLERS, OR PARDON ME YOUR TEETH ARE IN MY NECK (**1967**). The latter made him Hollywood's darling, and he married Tate, one of its minor stars, in the movie's aftermath. In Hollywood RP made movies including ROSEMARY'S BABY (**1968**), one of the CINEMA's finest fantasies of PERCEPTION (based on the Ira LEVIN novel *Rosemary's Baby* [**1967**]) as well as *The Magic Christian* (**1969**) – based on the Terry Southern (1924-   ) novel – *Macbeth* (**1971**) – based on SHAKESPEARE – *Chinatown* (**1974**) and *The Tenant* (**1976**), the last a paranoid fantasy of merit. Since his exile from the USA, RP has made few movies, some of which are bad: of note are *Tess* (**1979**), based on Thomas Hardy's *Tess of the d'Urbervilles* (**1891**), and the Alfred HITCHCOCK homage *Frantic* (**1988**). At his best RP shows a fine understanding of his chosen medium – and he has contributed one of the major DARK-FANTASY movies – but overall his career must be described as patchy.   [JG]

Further reading: *Roman* (**1984**), autobiography.

POLDER Literally a polder – the word derives from Old Dutch – is a tract of low-lying land reclaimed from a body of water and generally surrounded by dykes; to ensure its continued existence, these dykes must be maintained. The protagonist of Gustav MEYRINK's *Das grüne Gesicht* (**1916**; trans as *The Green Face* **1992** UK) leaves Amsterdam after the end of WORLD WAR I for the safety of a polder from which he watches an apocalyptic WIND scour the rest of the world. This may be the only Dutch polder in fantasy.

Here we use the word analogously: polders are defined as enclaves of toughened REALITY, demarcated by boundaries (◊ THRESHOLDS) from the surrounding world. It is central to our definition of the polder that these boundaries are *maintained*; some significant figure within the tale almost certainly comprehends and has acted upon (in the backstory, or during the course of the ongoing plot) the need to maintain them. A polder, in other words, is an *active* MICROCOSM, armed against the potential WRONGNESS of that which surrounds it, an anachronism *consciously* opposed to wrong time. In fantasy terms, pacific enclaves become polders only when a liminal threshold must be passed to enter them, for only then are they defended.

Within the threshold that marks the outer boundary of a

polder, a PORTAL may exist; its presence may, indeed, explain the existence of the polder itself, and this points to the polder's central role in many CROSSHATCHES. The size of the polder may vary widely. At its smallest it may be co-extensive with a GARDEN (◊◊ PASTORAL); examples include: Tom Bombadil's realm within the malign Old Forest in J.R.R. TOLKIEN's *The Lord of the Rings* (**1954-5**); Medwyn's undiscoverable ARCADIAN valley, which dates from the time of the FLOOD, in Lloyd ALEXANDER's *The Book of Three* (**1964**); the Time Island which demurely protects the village of Barly in Robert NATHAN's *Autumn* (**1921**) and *The Fiddler in Barly* (**1926**); the secret valley in Tibet inhabited by The Masters in H.P. BLAVATSKY's *The Secret Doctrine* (**1888**), a location used also by James HILTON in *Lost Horizon* (**1933**) for SHANGRI-LA; and the poisonous EDEN and TIME ABYSS which the god Thasaidon, in Clark Ashton SMITH's "Xeethra" (1934), uses to embroil victims with irretrievably belated (◊ BELATEDNESS) former selves, of whom they become doomed REINCARNATIONS. Alternatively, the polder may be an extremely large and complex EDIFICE, the kind whose roots, like the banyan tree's, are extensive but linked to a single heart – an example is the House Absolute in Gene WOLFE's *The Book of the New Sun* (**1980-81**). Or the polder may occupy a secreted portion of a CITY, like Toontown in WHO FRAMED ROGER RABBIT (*1988*). At its largest, a polder may be a landlocked OTHERWORLD like OZ, or a RURITANIA.

Surrounding the polder is a world whose effects may – all unconsciously – be inimical: the secular nation state whose presence threatens any Ruritania; the winds of TIME which blow a BRIGADOON onwards; the state of CHAOS which threatens the land ruled by Earl Aubec in Michael MOORCOCK's **Count Brass** sequence; the vacuum of space which stresses the walls of any pocket universe; the THINNING which eats at the boundaries of FAERIE; the pressure of time which surrounds the wild WOOD with suburbs ...

Successful polders do not change. Polders change only when they are being devoured from without. [JC]

POLIDORI, JOHN (WILLIAM) (1795-1821) UK physician and writer, raised in Italy; he was Lord BYRON's physician for a brief period, which included the famous June evenings in 1816 when he, Byron, Percy Bysshe SHELLEY and Mary SHELLEY – inspired by "Family Portraits", the first tale in *Fantasmagoriana* (anth trans **1812**) by Jean Baptiste Benoît, which instructs a group of people to gather together and tell each other GHOST STORIES – told each other, in turn, four tales. Mary Shelley's tale became *Frankenstein, or The New Prometheus* (**1818**). Byron's was a VAMPIRE tale which, with the vampire elements excised, appeared as "Fragment of a Story" in *Mazeppo* (coll **1819**). JP's eventually became *Ernestus Berchtold, or The Modern Oedipus: A Tale* (**1819**), a Gothic fiction of very modest interest. But JP, having been dismissed by Byron in September 1816, then made use of his former employer's sketchy vampire tale, transforming it very substantially into *The Vampyre: A Tale* (1819 *New Monthly Magazine*; **1819** chap), both releases being published – whether or not with JP's sanction is uncertain – as by Byron. Byron repudiated the assertion of authorship, but the (correct) assumption that the Lord Ruthven depicted in the latter tale was him was soon universally accepted. The connection was all the more inevitable since JP had taken the name of his vampire antihero from *Glenarvon* (**1816**) by Lady Caroline Lamb (1785-1828), a *roman à clef* whose central villain, Clarence de

Ruthven, Lord Glenarvon, was a savage portrait of Byron, her ex-lover. So *The Vampyre* is important not only for its merits but because it fixed in place, for most of a century, the UNDERLIER motif of the Byronic vampire – the satanic, blanched, world-weary aristocrat whose eyes have a hypnotic effect, especially upon women, and in whom vampirism and seduction are part of the same process (◊◊ SUPERNATURAL FICTION). The languor of the Byronic vampire is a pose: for his energy is infernal. The story traces the growing awareness of the young romantic Aubrey that Ruthven, who fascinates him, is indeed a vampire; unfortunately for Aubrey, this discovery is made through the deaths of his fiancée, his sister and in due course himself. Imitations (◊ Charles NODIER; James Robinson PLANCHÉ) led eventually to Alexander DUMAS's *Le Vampire* (performed **1851**), which put a final gloss on the Byronic motif.

JP died young; recent research suggests that he may not have committed suicide. He was an uncle of Christina ROSSETTI and Dante Gabriel ROSSETTI. [JC]
Further reading: *Poor Polidori: A Critical Biography of the Author of "The Vampyre"* (**1991**) by D.L. MacDonald.
See also: FRANKENSTEIN MOVIES.

POLLACK, FREDERICK US writer (1946-   ). ◊ POSTHUMOUS FANTASY.

POLLACK, RACHEL (GRACE) (1945-   ) US writer, resident in the Netherlands 1973-90, who began publishing fiction with "Pandora's Bust" (1972 *New Worlds Quarterly #2*) as by Richard A. Pollack. Although her first two novels are sf (◊ SFE), they contain elements which carry the flavour of fantasy; and it is as fantasy that *Unquenchable Fire* (**1988** UK) and its sequel *Temporary Agency* (**1994** UK) ask to be read. Set in an alternate USA in which the forces of SHAMANISM actually work and a system of guilds, now ossified into bureaucracy, control these forces in the manner of a public utility, the novels vividly admix familiarity with strangeness. The former won the ARTHUR C. CLARKE AWARD. «Godmother Night» (1996) is a seemingly contemporary novel but is in fact a kind of sombre feminist FAIRYTALE. RP edited *Tarot Tales* (anth **1989** UK) with Caitlín Matthews, and designed a TAROT deck, "Shining Woman Tarot", which is sold with *Shining Woman Tarot Guide* (**1992** UK). [GF]

POLLOCK, WALTER HERRIES (1850-1926) UK author, who wrote a great deal in collaboration. With his mother, Lady Julia Pollock (?   -1899), and W.K. Clifford he compiled *The Little People and Other Tales* (coll **1874**). With Walter BESANT he wrote the novella "Sir Jocelyn's Cap" (1884-5) – which anticipated the method and tone of F. ANSTEY's fantasies – and *The Charm and Other Drawing-Room Plays* (coll **1896**), whose title piece is a marginal fantasy.

WHP collaborated with Andrew LANG on *He* (**1887**), a parody of H. Rider HAGGARD's *She* (**1887**) set in darkest London. It was published as "by the authors of *It, King Solomon's Wives* and *Bess*". (It was actually the US writer John De Morgan [1848-*c*1920] – who also published an anonymous volume called *He* (**1887**) – who wrote *It* [**1887**] and *Bess* [**1887**]; these were pastiches rather than parodies. One Henry Chartres Biron was the author of *King Solomon's Wives* [**1887**] by "Hyder Ragged". De Morgan's more earnest homage was *King Solomon's Treasures* [**1887**].)

WHP also wrote several stories with the US writer J. Brander Matthews (1852-1929), including the comic fantasy "Edged Tools" (1886). He wrote *The Were-Wolf: A Romantic Play in One Act* (**1898** chap) with Lilian Moubrey.

His solo short novel, about a family CURSE and a FEMME FATALE, "Lilith" (1874-5), was reprinted as the first item in *The Picture's Secret: A Story, to Which is Added An Episode in the Life of Mr Latimer* (coll **1883**), but appeared under its original title in *A Nine Men's Morrice: Stories Collected and Recollected* (coll **1889**). The second item in the former collection, which in the latter was re-separated into its two constituent parts – "Mr Morton's Butler" and "Lady Volant" – is WHP's best comic fantasy: a tale in which attempts to trick a young man into a PACT WITH THE DEVIL go awry in spite of his awesome naivety. *King Zub* (coll **1897**) reprints "Sir Jocelyn's Cap" and a few other fantasies: "The Phantasmatograph" (1899) remains unreprinted. [BS]

POLTERGEIST Series of three SUPERNATURAL-FICTION movies notable for their spfx and inventiveness, and for the performances of child actress Heather O'Rourke.

1. *Poltergeist* US movie (**1982**). MGM. **Pr** Frank Marshall, Steven SPIELBERG. **Dir** Tobe Hooper. **Spfx** Industrial Light & Magic. **Screenplay** Michael Grais, Spielberg, Mark Victor. **Novelization** *Poltergeist* * (**1982**) by James Kahn (◊ SFE). **Starring** Craig T. Nelson (Steve Freeling), Heather O'Rourke (Carol Anne Freeling), Oliver Robins (Robbie Freeling), Zelda Rubinstein (Tangina), Beatrice Straight (Dr Lesh), JoBeth Williams (Diane Freeling). 110 mins. Colour.

The Freeling family live in a housing estate erected – as they discover – on an old Native American burial place. Strangeness starts when 5-year-old Carol Anne hears the voices of "the tv people" in the white-noise static after closedown. Physical manifestations build until Carol Anne is snatched to another plane, although she can still be heard calling for her mother *via* the tv static. The Freelings bring in parapsychologist Lesh, who in turn summons dwarf medium Tangina. Carol Anne is rescued and the house seemingly cleared of evil SPIRITS – but only long enough for everyone to depart except Diane and the two younger children, who are relaxed and defenceless when ...

P, which is about a HAUNTING rather than about a POLTERGEIST event, has some good qualities, but is ever too eager to stifle its plot in a welter of (excellent) spfx. Despite references aplenty to fantasy ICONS (for example, the first main burst of activity features the tornado from *The* WIZARD OF OZ [1939]) and to various spiritualist and religious beliefs, somehow the movie fails to hold together as anything more than a spectacularly produced but rather humdrum entertainment.

Interestingly, at one point we see a snatch of an old movie on tv: *A Guy Named Joe* (**1944**), about a dead airman who, as a GHOST, observes his girlfriend's new romance. Clearly the tale fascinated Spielberg, for some years after *P* he directed its remake, ALWAYS (**1989**).

2. *Poltergeist II: The Other Side* US movie (**1986**). MGM/Freddie Fields/Victor-Grais. **Pr** Michael Grais, Mark Victor. **Exec pr** Fields. **Dir** Brian Gibson. **Vfx sv** Richard Edlund. **Conceptual artist** H.R. GIGER. **Screenplay** Grais, Victor. **Novelization** *Poltergeist II: The Other Side* * (**1986**) by James Kahn. **Starring** Julian Beck (Preacher Henry Kane), Geraldine Fitzgerald (Granma), Craig T. Nelson (Steve), Heather O'Rourke (Carol Anne), Oliver Robins (Robbie), Zelda Rubinstein (Tangina), Will Sampson (Taylor), JoBeth Williams (Diane). 90 mins. Colour.

Taking refuge with Diane's mother, Granma, the Freelings hope the nightmare is over. But one day Carol Anne is approached by a sinister stranger, Kane. Soon Granma dies, her SOUL giving a final farewell to Carol Anne *via* the latter's toy telephone. The next call on that phone, though, is not from Granma; the children's TOYS become animate, ectoplasmic swirls fill the air, and Carol Anne tearfully announces: "They're back!" Native American medium and SHAMAN Taylor is summoned to their aid by Tangina (from 1). Kane calls by the house, introduces himself, and uses various stratagems to try to be invited in, but is resisted. That night the HAUNTING begins. Taylor guards Carol Anne, explaining she is the one the SPIRITS want; later he tells Steve that Kane is a man filled by a DEMON, and that the corporate entity thinks its REALITY and this one is the same. Meantime Tangina is with Diane, showing her a photograph of Kane and inducing her to see clairvoyant (◊ TALENTS) VISIONS of what happened: Kane was leader of a sect that came West to found a UTOPIA; he hid them underground anticipating the END OF THE WORLD, refusing to let them leave when his PROPHECY went unfulfilled. He and his followers need Carol Anne, whom they regard as an ANGEL, because they tasted her lifeforce during her incursion (in 1) into the AFTERLIFE. There is some spectacular stuff hereafter before the demon is seen off, Carol Anne's ultimate rescuer being the soul of Granma.

Although its script is sometimes disjointed, *PII* is in many ways a more interesting movie than 1, *using* its material rather than merely splashing it noisily all over the screen. There is one moment that, with hindsight, has true pathos: speaking to Granma, Carol Anne says: "Don't want to grow up much ... Probably not much fun." Tragically, O'Rourke (1975-1988) was dead two years later. [JG]

3. *Poltergeist III* US movie (**1988**). MGM. **Pr** Barry Bernardi. **Exec pr** Gary Sherman. **Dir** Sherman. **Screenplay** Sherman, Brian Taggert. **Starring** Nancy Allen (Patricia), Lara Flynn Boyle (Donna), Nathan Davis (Kane), Richard Fire (Dr Seaton), Heather O'Rourke (Carol Anne), Zelda Rubinstein (Tangina), Tom Skerritt (Bruce). 97 mins. Colour.

The last in the series was made on a smaller budget: it has the feel of a tvm. This has tended to disguise its interest; unable to draw on exotic spfx, the makers instead drew on imagination, in particular on the imagery of MIRRORS – so that events seen in the mirror world (which is where, in this movie, the SPIRITS are presumed to reside) may be seen as disparate from those in "our" world. Carol Anne has been sent to live awhile with her Uncle Bruce, Aunt Patricia and cousin Donna in their apartment in a 98-storey Chicago city-in-a-tower-block, whose internal walls are almost everywhere fronted with mirrors (the setting is much like that of the second GREMLINS movie). Meddling child psychologist Dr Seaton, a caricature sceptic, inadvertently draws Preacher Kane (played by Nathan Davis, Julian Beck having died) back to Carol Anne. Much of the rest of the movie is a fairly sophisticated if sometimes derivative TECHNOFANTASY (including demonic cars *à la* CHRISTINE [*1983*]). Tangina sacrifices herself to Kane to save Carol Anne and the rest. [JG]

POLTERGEISTS Literally, noisy ghosts (from the German): mischievous household SPIRITS who traditionally make rapping noises, throw around furniture and small objects, etc. Real-world poltergeist reports very often involve homes containing disturbed adolescents, suggesting undisciplined TALENTS or (to more sceptical investigators) ingenious trickery. The poltergeist tradition is sufficiently widely accepted that a genuine and active specimen features in the otherwise nonfantastic detection *Buried for Pleasure* (**1948**) by Edmund Crispin (real name Bruce Montgomery; 1921-1978); typically, it is resistant to religious EXORCISM since the keynote is mischief and not EVIL. Poltergeist activity – a hurled plate – is the first sign of the wakening of mythic forces by the emotional turmoil of modern youngsters in Alan GARNER's *The Owl Service* (**1967**). Colin WILSON's *The Philosopher's Stone* (**1969**) includes a poltergeist episode whose ghostly phenomena result from a kind of morbid spiritual gestalt in a neurotic household, including frustrated adolescents: adopting the role of OCCULT DETECTIVE, the protagonist uses his own talents to feed positive "vibrations" into this vortex, dispersing it after a display of spectacularly amplified effects. The heroine of William MAYNE's *It* (**1977**) finds herself framed, as it were, in the "disturbed adolescent" template, since she has inadvertently acquired a WITCH's familiar whose undirected energies manifest as poltergeist behaviour, forcing her into uncharacteristic actions in order to rid herself of it. Terry PRATCHETT offers a roughly similar power-source for ghostly doings in *Reaper Man* (**1991**), where the retirement of DEATH produces a temporary backlog of uncollected SOULS whose aggregated energy comically "earths itself in random poltergeist activity".

In CINEMA, *The* EXORCIST (*1973*) is – like its parent novel *The Exorcist* (**1971**) by William Peter BLATTY – based on a supposed poltergeist event involving a child, though here the heart of the problem is POSSESSION. The POLTERGEIST movies utilize a female child more strategically, as innocent victim of HAUNTINGS and DEMONS, and lack the sense of ambiguous complicity appropriate to a youngster's role as source or lightning-rod in classic poltergeist cases. [DRL]

POOKA ◊ PUCK.

POOLE, JOSEPHINE Pseudonym of UK writer Jane Penelope Josephine Helyar (1933-   ), most of whose work is YA SUPERNATURAL FICTION – like *Moon Eyes* (**1965**), her first novel. In *The Visitor: A Story of Suspense* (**1972**) an adolescent boy discovers his tutor is in fact an ancient deity (◊ TIME ABYSS) who requires HUMAN SACRIFICE from his worshippers. Other books of interest include *Billy Buck* (**1972**), *The Loving Ghosts* (**1988**) – a GHOST STORY – *Angel* (**1989**), *This is Me Speaking* (**1990**), *Paul Loves Amy Loves Christo* (**1992**), *Scared to Death and other Ghostly Stories* (coll **1994**) and a TWICE-TOLD *Pinocchio* (**1994**).   [JC]

POPE, ELIZABETH MARIE (1917-1992) US writer whose first novel, *The Sherwood Ring* (**1958**), takes an orphan girl from her dying father's home in England to relatives in contemporary Orange County, New York, where she encounters Revolutionary War GHOSTS and TIME TRAVEL. It is, considering its sombre material, a strangely light-hearted and memorable romance. *The Perilous Gard* (**1974**), a Newbery Honor Book, is a darker-hued tale which retells TAM LIN.   [DGK]

POPE JOAN According to the earliest versions of the legend, Joan was a 9th-century Englishwoman reared in Germany. In male dress she accompanied a lover to Athens, where she studied extensively, then went to Rome, perhaps masquerading as a Benedictine monk. Being a person of intellect, learning and charm, she rose in office, eventually being elected pope and reigning for 2-3 years as Pope John. She then became pregnant by one of her household and, during a street procession, suffered a fatal miscarriage. According to some, her body was thrown into the Tiber;

according to others, she was buried at the spot. Later popes, when processing, avoided the place of her death, which was marked by a monument.

The origin of the legend, which first appeared in the 13th century, is not known, although it seems certain it has no historical basis; by the Reformation it was rejected except by religious polemicists. It was probably a folktale that emerged from a coalition of anecdotes about transvestite female religios, local folklore about Roman monuments, and a distorted memory of the early and middle 10th century, during the reigns of various Popes John – when for example Marozia (?   -938) scandalously controlled the papacy.

Despite the power of the theme, it has had few literary treatments. BOCCACCIO included a section on Joan in his *De claris mulieribus*. A lost English play, *Pope Joan*, was performed in 1591-2, and in 1689 *The Female Prelate, or The Life and Death of Pope Joan* by Elkanah Settle (1648-1724) was acted on the London stage. A less significant treatment occurs in *Pope Joan, or The Female Pontiff* (**1850-51**) by the sensational novelist G.W.M. Reynolds (1814-1879), where Joan is identified with the 9th-century John VII/VIII. The story, which concerns itself with a sentimentalized Joan before she entered the clergy, takes place in a Gothic-novel Spain and stresses sex and sadism, with plentiful harassment and tortures. Only in the last chapter does Joan appear as pope. To fit Victorian sensibilities she is not pregnant, but drops dead merely through being recognized and unmasked. Other novels include *The Woman who was Pope* (**1931**) by Clement Wood, *When Joan was Pope* (**1931**) by Richard Ince, *The Book of Joanna* (**1947**) by George Borodin (1903-   ) and *The Legend of Pope Joan* (**1983**) by Emily Hope. A collateral work, Marjorie BOWEN's supernatural *Black Magic* (**1909**), tells of one Ursula who became Pope Michael II. Much the outstanding fictional account of Pope Joan is *Papissa Joanna* (**1886**) by the Greek author Emmanuel Rhoïdes, which is essentially a very witty satire on the Greek Orthodox Church, Christianity, religion and politics. It has been translated into many languages – into English at least three times; of these translations *Pope Joan* (**1960**) by Lawrence Durrell (1912-1990) is the finest, though sometimes inaccurate.

There is a CARD game called Pope Joan, and Joan (as "La Papessa") often figures on one of the TAROT cards. Joan has been taken up by modern feminists; such works include Elizabeth Gould Davis's *The First Sex* (**1973**) and Elizabeth Goessmann's scholarly *Mulier Papa: Der Skandal eines weiblichen Papstes* (**1994**), which contains an elaborate historical apparatus and facsimile reproductions of many Renaissance texts. A UK movie, *Pope Joan* (*1972*; vt *The Devil's Impostor* US), is generally regarded as undistinguished.   [EFB]

Further reading: *The Female Pope* (**1988**) by Rosemary and Darroll Pardoe.

POPEYE US movie (*1980*). Paramount/DISNEY. **Pr** Robert Evans. **Exec pr** C.O. Erickson. **Dir** Robert Altman. **Spfx** Allen Hall. **Screenplay** Jules Feiffer. **Based on** the characters created by E.C. Segar. **Starring** Paul Dooley (Wimpy), Shelley Duvall (Olive Oyl), Wesley Ivan Hurt (Swee'Pea), Donald Moffat (Taxman), Paul L. Smith (Bluto), Ray Walston (Poopdeck Pappy), Robin Williams (Popeye). **Voice actor** Jack Mercer (Popeye in prologue animated by HANNA-BARBERA). 114 mins. Colour.

Searching the Seven Seas for his long-lost Pappy, the spinach-loathing Popeye comes to Sweethaven, a coastal community run as a tax-ridden dictatorship by the never-seen Commodore and his bullying sidekick Bluto. Popeye takes lodgings at the Oyls; the daughter, Olive, is engaged to Bluto. However, as she and the outsider go towards her engagement party they trip over baby Swee'Pea, abandoned like Popeye himself once was; their late arrival together is enough to convince a wrathful Bluto to break off the engagement. Swee'Pea is found to be precognitive (◊ TALENTS); almost immediately Wimpy steals the infant to sell to Bluto, who, now rebelling against his master, plans to use Swee'Pea's powers to track down the Commodore's treasure. Popeye, following, discovers the trussed-up, misanthropic old man is his father; together they and most of the cast pursue Bluto and the kidnapped Swee'Pea and Olive to Scab Island. Bluto proceeds to thrash Popeye; a giant octopus threatens infant and Olive; Pappy's treasure turns out to be mementoes of Popeye's infancy; and in a fit of final – and suicidal – sadism Bluto forces Popeye to swallow a can of spinach. Villain and octopus disposed of, Olive and Popeye declare undying love.

The decision to turn the COMICS and animated shorts into a full-length comedy musical directed by Altman was not, perhaps, a felicitous one. The attempts to make live actors behave physically as if animated figures have some success, generally where they rely on the actors' skills rather than on spfx.   [JG]

PORNOGRAPHIC FANTASY MOVIES A substantial portion of (soft) pornographic movies released fall into the fantasy category; to turn this the other way round, in any comprehensive listing of fantasy movies (◊ CINEMA), an astonishingly high percentage – about 10-15%, perhaps more – is pornography. These movies form, in effect, a separate subgenre, having little influence on, and being little influenced by, more orthodox fantasy movies. Many are released on sell-through video only, and almost all are reportedly – as fantasies – dire. Research is difficult in this field, in that few researchers are willing to sit through the stuff, but there are one or two notable exceptions to the general rule; for example, *Through the Looking Glass* (*1976*), in which a discarded middle-aged woman rediscovers her selfhood through being lured for SEX through a MIRROR by a DEMON who dwells in the mirrorworld, deploys some sophisticated fantasy. (Uncut versions of this movie are not generally available.)

The line between pornographic and "art" movies is a thin one, and many more widely available fantasy movies probably cross it. Yet there is a quite definite qualitative difference between these – even where their primary *raison d'être* is erotic – and the PFM proper; the two types of movie can on occasion approach a common ground, but they do so from different directions.   [JG]

PORTALS Very few fantasy texts lack them. They may be physical (doors, gates, tunnels, PICTURES, movie screens, MIRRORS, LABYRINTHS) or metaphorical; they may exist whenever a CROSSHATCH which mingles worlds, or a THRESHOLD which demarcates them, is sufficiently focused to be detected (◊ PERCEPTION), perhaps only by TALENTS; or, less commonly, they may be themselves transportable, in the form of AMULETS or RINGS or BOOKS – more often than not, a LITTLE-BIG relationship obtains between the outside and the "inside" of a portal. They may be located anywhere, from a nook or wardrobe or cranny to an EDIFICE or CITY, at whose heart may hum a thousand intersections. They may be signals of almost any significant transit point in the

typical GENRE FANTASY: transitions between this world and an OTHERWORLD or AFTERLIFE venue or ARABIAN NIGHTMARE; from one otherworld to another; from our time (via TIMESLIP or TIME TRAVEL) to another time; from this world to FAERIE; from one level of REALITY to another; from life into death; from a prior state of growth (◊ RITE OF PASSAGE) into empowered adulthood; from a prior state of being, via METAMORPHOSIS, into something rich and strange (◊ BONDAGE; WRONGNESS; THINNING); from AMNESIA through RECOGNITION into HEALING, whenever that central fantasy movement is dramatized or put in RITUAL form.

The distinction between a portal and a threshold may sometimes be blurred, but normally the term is used here to describe a liminal structure or aura, while a threshold is a sharp gradient between two places or conditions, a gradient which may define a BORDERLAND or POLDER. Though portals can of course be invisible, usually this invisibility is treated as somehow to be remarked upon (while any portal *may* be a PLOT DEVICE, all invisible portals almost certainly are). Visible or invisible, portals are likely to be warded – woven round with CONDITIONS and PROHIBITIONS – and to pass through a portal is likely to pass some kind of test, to gain a new level of understanding of power, to demonstrate oneself as a *chosen* one, whether through birth or actions or some other merit: in fantasy, it is very often the case that a character who finds a portal has in some sense been *found by* that portal. Portals are part of the grammar of significant STORY. Portals represent acts of selection and election.

In genre fantasy portals are generally passages from here to there; in DARK FANTASY, SUPERNATURAL FICTION and WEIRD FICTION, portals are more often seen as passages from the other world into this one.

Portals dominate the TAPROOT-TEXT tale of ORPHEUS, particularly Harrison Birtwistle's OVID-based *The Mask of Orpheus* (**1986**) (◊ OPERA). They initiate a high proportion of CHILDREN'S FANTASIES involving any form of transition from this world to another, from Lewis CARROLL's *Alice's Adventures in Wonderland* (**1865**) through E. NESBIT's *The Story of the Amulet* (**1906**) to C.S. LEWIS's *The Lion, the Witch and the Wardrobe* (**1950**). They are often used in PLANETARY ROMANCES as the opening through which the destined hero is called. They articulate three- or multi-dimensional worlds, like the wood at the heart of Robert HOLDSTOCK's **Ryhope Wood** sequence, most clearly in *The Hollowing* (**1993**). They shape large narrative sequences, like the four portals (the last constituting a SLINGSHOT ENDING) that divide the four volumes of Gene WOLFE's *The Book of the New Sun* (**1980-83**).

There is also a commonly found secondary use of the term. In sf, portals may be access points to forms of transit, usually instantaneous. In fantasy – particularly genre fantasies set in hard-wired FANTASYLANDS, and TECHNOFANTASIES or RATIONALIZED-FANTASY tales – portals may also perform this relatively mundane function, though in such texts the transition is usually effected by MAGIC.   [JC]

PORTENTS These tend to be OMENS with a darker, more literally ominous significance (◊◊ PROPHECY). If TREES bleed, or the sky is the wrong colour, or (as in J.R.R. TOLKIEN's *The Lord of the Rings* [**1954-5**]) the SUN itself is obscured, this will be a portent rather than an omen. Portents are usually explained in context, and have a specific PLOT-DEVICE function (if only to set a mood).   [CB/DRL]

PORTRAIT OF JENNIE (vt *Jennie* UK) US movie (*1948*).

Selznick/Vanguard. **Pr** David O. Selznick. **Dir** William Dieterle. **Spfx** Clarence Slifer. **Screenplay** Peter Berneis, Paul Osborn (introduction by Ben HECHT). **Based on** *Portrait of Jennie* (**1940**) by Robert NATHAN. **Starring** Ethel Barrymore (Miss Spinney), Joseph Cotten (Eben Adams), Lillian Gish (Mother Mary of Mercy), Jennifer Jones (Jennie Appleton), Cecil Kellaway (Matthews), Maude Simmons (Clara Morgan). 86 mins. B/w plus tinted sequence and brief Technicolor sequence.

Penniless Manhattan artist Adams meets in Central Park a child, Jennie, dressed in old-fashioned clothing and talking of long-gone locations as if they still existed. While with him she makes a WISH that he will wait for her while she grows up; then vanishes, leaving behind a scarf wrapped in a 1910 newspaper. Adams draws a sketch from memory, and immediately sells it to art dealers Matthews & Spinney, whose Miss Spinney soon becomes his champion. Shortly afterwards Adams again meets Jennie, now already a young woman: she is growing up fast to "catch" him. She fails to turn up for their next "date"; investigating her, Adams discovers Jennie's parents died decades ago, and next encounters Jennie herself weeping over their loss. Rapidly falling in love, he visits with her a ceremony at her convent school; and at last she (now a beautiful woman) sits for him in her studio. When months pass without her, Adams goes to the convent where he discovers from Mother Mary that Jennie died some years earlier when a tidal wave struck Land's End Point, Connecticut. Travelling there, he endeavours to cheat the past by saving her; but she refuses to comply and is lost to the waves – although leaving behind her scarf.

POJ has some crudities of realization and many flaws, including a long and irrelevant subplot, but holds much of interest: notably, it shows us Jennie's own inability properly to come to terms with her somewhat convoluted timeline, alternately being fully conscious of and quite ignorant of her rapid growth to adulthood, forgetting and remembering details, etc.; also, it explores the hypothesis that Jennie is a construct of Adams's own mind, desperately seeking a subject that will draw out his nascent artistic ability (◊ PERCEPTION; RATIONALIZED FANTASY). Its downbeat ending – very un-Hollywood – serves, because correct rather than glib, to render *POJ* a movie that abides in the memory – as does the reiterated theme of chill and winter. *POJ* is far more affecting than any of the early-1990s cycle of movies reprising the notion of love across the barrier of death – e.g., DEAD AGAIN (**1991**), GHOST (**1990**), GHOSTS CAN'T DO IT (**1990**) and TRULY MADLY DEEPLY (**1990**). [JG]

PORTRAITS ◊ PICTURES AND PORTRAITS.

POSSESSION The theme of demonic possession occurs only occasionally in GENRE FANTASY by comparison with HORROR or SUPERNATURAL FICTION. The reason is perhaps the difficulty of giving a protagonist who is possessed by a DEMON, and who is accordingly passive in many of his actions, any sort of that status as protagonist which is essential to emotional identification in an escapist genre. When possession occurs, it will accordingly be as a condition which the protagonist is trying to escape, as with Jason Blood in DC's **The Demon** comic, or as an attribute of villains, usually minor villains like the Ravers in Stephen R. DONALDSON's **Chronicles of Thomas Covenant the Unbeliever** (**1977-83**).

Actual possession by evil spirits has to be distinguished from mere subjection to their will; the holders of the various Rings in J.R.R TOLKIEN's **The Lord of the Rings** (**1954-55**) are in some cases eaten up by Sauron's influence, but not necessarily.

Possession is often indistinguishable in many of its attributes from the process of discovery that one is an AVATAR of a GOD or goddess, a demon or some other being. Angelica, the villainess of Elizabeth HAND's *Waking the Moon* (**1994**) chooses to become an avatar of the dark side of the GODDESS – she is perceived by the viewpoint character Sweeney as a woman possessed. Angelica's sacrificial castration of her lover Oliver frees him into an androgyny where he can be a vehicle for a gentler and more nurturing version of the Goddess – it is not entirely clear how much of the original Oliver survives. The protagonist of Jack WILLIAMSON's *Darker than you think* (**1948**) discovers that he is the dark Messiah of a primordial race of SHAPESHIFTERS; he experiences the early stages of this transformation as a somnambulism akin to possession.

Analogy with the sf tropes of symbiosis or parasitism with alien minds sometimes produces a version of demonic possession which allows the protagonist some degree of free will. His dependency on the SWORD Stormbringer, which is explicitly a demon, renders Elric, in Michael MOORCOCK's **Elric** stories (**1963**) in the position of constantly outwitting the sword's desire to corrupt him further by drinking, and passing on to him as the energy he needs for life, the souls not only of enemies but also of friends and lovers. Dorian Hawkmoon in Moorcock's **Hawkmoon** series (**1967-75**) is similarly half-controlled by the jewel that the evil scientists of Granbretan have implanted in his skull.

Such semi-possession is usually a source from which status as an ACCURSED WANDERER derives; the search to be free of it renders protagonists likely to become OBSESSED SEEKERS.
[RK]

See also: IDENTITY EXCHANGE.

POSTGATE, OLIVER (?   -   ) UK writer and tv screenwriter who with animator/director Peter Firmin (1928-) formed the production company Smallfilms, responsible for a string of cult children's series using either drawn or stop-motion animation or (often) a mixture. The first was *The Saga of Noggin the Nog* (1959-65), scripted by OP from a story by Firmin: it tells of a QUEST by a gentle Viking with his family and associates in their longship. *Ivor the Engine* (1962-4, 1976-7) anthropomorphized various railway features, the central character being a steam engine whose boiler is fired by a small DRAGON. *The Pogles* (1966-7), later retitled *Pogles' Wood*, intercut live action with stop-motion animation. The company's biggest success was almost certainly *The CLANGERS* (1969-73), featuring the little whistling inhabitants of a blue moon. *Bagpuss* (1974) was set in a junkshop, in which TOYS and other items would weekly be incanted into life (◊ ANIMATE/INANIMATE) in order to investigate and mend some new addition to the shop's stock. All the series – there were others – had considerable charm, the stories being told with a beautiful simplicity.
[JG]

POSTHUMOUS FANTASY As used over the past decade or so, this term has proved inadequate as a descriptive term; but a more fitting one seems difficult to find. Some critics use the term "afterlife fantasy" to designate any story whose protagonists are dead, but when they do so generally concentrate on the adventures of the protagonists in a HEAVEN of some sort. A PF is a tale which deals primarily with the RITE OF PASSAGE from death to a state of understanding. In the standard version of the PF the protagonist's SOUL generally begins its journey – several examples of the form are set on SHIPS and TRAINS – in a state of almost total ignorance, not usually even being aware that its mortal existence, within a human frame, has ended. The LANDSCAPE into which the soul moves may seem to be nothing more than his or her own native surroundings, but almost invariably a sense of haunted strangeness and WRONGNESS soon pervades, and within a very short period the landscape tends to take on the characteristics of a world or LABYRINTH which must be deciphered, while at the same time it is beginning to echo and grey out; the chiaroscuros and bureaucratic absurdities often found in URBAN FANTASIES now frequently start to oppress the soul. But throughout there is a growing pressure of meaningfulness in every sight, every sign. Eventually, the soul begins to recognize that the world it is now inhabiting constitutes a kind of diorama or theatre whose central STORY concerns the true meaning of the mortal life so recently departed. And, in due course, the soul begins to understand the Story; and the tale ends.

The PF is thus not really a fantasy at all but, because it posits a real world which has been departed and an AFTERLIFE which is superior or subsequent to that real world, it is a type of SUPERNATURAL FICTION. Tales whose focus is upon a relatively autonomous afterlife – such as the **Houseboat** sequence by John Kendrick BANGS and afterlife epics by authors like E.R. EDDISON and Philip José FARMER – are not PFs, in the terms of this narrow definition: they are fantasies whose SECONDARY-WORLD settings are posthumous; the relationship between that world and the world departed is of virtually no importance in the narrative structure of such tales. By contrast, in the PF proper that relationship is central: everything pivots upon the protagonist's act of understanding, an act that may be couched in terms that attain the consistency of ALLEGORY (in the sense intended by Northrop FRYE, who defines a text as allegorical "when the events of a narrative obviously and continuously refer to another simultaneous structure of events or ideas ..."). The posthumous world, in this kind of story, is always something else: it is the map of a soul.

DANTE's *The Divine Comedy*, not itself a PF, provided for future centuries much of the furniture of the mode. Before the 20th century, the form was uncommon; Charles KINGSLEY's *The Water-Babies: A Fairy-Tale for a Land-Baby* (1863) is one of relatively few significant examples. The tale that most famously presents the sensation of the PF – though technically it is a tale of nonsupernatural HORROR – is Ambrose BIERCE's "An Occurrence at Owl Creek Bridge" (1891). (The latter part of Nikos KAZANTZAKIS's *The Last Temptation of Christ* [1950; trans 1960] can be seen as a variant on this tale.) An earlier story which conveys some of the same effect is Pedro A. de ALARCÓN's *El amigo de la muerte* (1852; trans as *The Strange Friend of Tito Gil* 1890 US). The protagonist of Conrad AIKEN's most famous single tale, "Mr Arcularis" (1932) – later dramatized as *Mr Arcularis: A Play* (1957 chap) – hears his death in the throbbing of the engine of a great ocean liner; the text is ambiguous as to whether or not he has actually died, or is merely hearing death arrive. A similar ambiguity applies to the hero of William GOLDING's *Pincher Martin* (1956; vt *The Two Deaths of Christopher Martin* 1957 US), who clings tenaciously to a rock in the ocean after his ship has been sunk, refusing to die, rehearsing his life again and again.

PFs have been fairly common throughout the 20th century, though they have become less frequent in recent years. Texts which comprise or incorporate examples include: Margaret Allen Curtois's "The Land Without a Sun" (1907); H. Rider HAGGARD's *The Mahatma and the Hare* (1911); Coningsby W. DAWSON's *The Unknown Country* (1915 chap); Leo PERUTZ's *Zwischen Neun und Neun* (1918; trans Lily Lore as *From Nine to Nine* 1926 US); Par LAGERKVIST's *Det Eviga Leendet* (1920; trans as *The Eternal Smile* 1932 chap UK); Elmer Rice's *The Adding Machine* (1923); Rudyard KIPLING's "On the Gate", in *Debits and Credits* (coll 1926); Alan Sullivan's *The Days of their Youth* (1926); A.M. BURRAGE's "The Wrong Station" in *Some Ghost Stories* (coll 1927); Stephen King-Hall's *Posterity* (1927 chap), a play; Sutton VANE's *Outward Bound* (as play 1923; as novel 1929); Winifred Holtby's "The Man who Hated God" (1928); *The Bands of Orion* (1928) by Temple Lane; Wyndham LEWIS's *The Childermass* (1928), whose sequels – *Monstre Gai* (1955) and *Malign Fiesta* (1955) – are more properly fantasies of the afterworld; Rebecca WEST's *Harriet Hume: A London Fantasy* (1929); *As I Lay Dying* (1930) by William Faulkner (1897-1962); Vladimir NABOKOV's *Soglyadatay* (1930; trans Dimitri Nabokov as *The Eye* 1965 US); Lady Saltoun's *After* (1930); *Fiddlers' Green* (1931) by Albert Richard Wetjen (1900-1948); Michael Maurice's *Marooned* (1932); *The Wedding Garment* (1894) and *The Invisible Police* (1915 *Christian Endeavor World* as "The Great Crossing"; exp 1932) by Louis Pendleton (1861-1939); Claude HOUGHTON's *Julian Grant Loses his Way* (1933); *Hangman's Isle* (1933) by Glyn Griffith; *A Christmas Party* (1934) by Paul Bloomfield (1898-?); Marjorie LIVINGSTON's *The Future of Mr Purdue* (1935); Naomi MITCHISON's *Beyond this Limit* (1935 chap); *Intra Muros, or Within the Walls* (1936) by Rebecca Springer; Alexander LERNET-HOLENIA's *Baron Bagge* (1936; trans Jane B. Green in *Count Luna: Two Tales of the Real and Unreal* 1956 US); Rex WARNER's *Why Was I Killed?: A Dramatic Dialogue* (1943; vt *Return of the Traveller* 1944 US); *Huis-clos* (produced 1944; 1945; trans as *In Camera* in *The Flies and In Camera* 1946 and as *No Exit* in *No Exit and the Flies* 1947) by Jean-Paul Sartre (1905-1980); Philip Wylie's *Night Unto Night* (1944); Charles WILLIAMS's *All Hallows' Eve* (1945); Gerald KERSH's "In a Room Without Walls", in *Neither Man Nor Dog* (coll 1946); the movie *A MATTER OF LIFE AND DEATH* (1946) dir Michael Powell (1905-1990), novelized under the same title by Eric Warman (1904-1992); Lawrence Durrell's *Cefalû* (1947; vt *The Dark Labyrinth* 1958 US); *Message from a Stranger* (1948) by Marya Mannes (1904-1990); Almet JENKS's *The Huntsman at the Gate* (1952); Robert NATHAN's *The Train in the Meadow* (1953) and *The Summer Meadows* (1973); *The Investigator* (1956) by Reuben Ship; *The Bridge* (1957) by Pamela Frankau (1908-1967); *Memoirs of a Venus Lackey* (1968) by Derek Marlowe (1938-); Piero SCANZIANI's *Libro Biano* (1968; rev 1983; trans Linda Lappin as *The White Book* 1991 UK); Michael AYRTON's "Tenebroso", in *Fabrications* (coll 1972); Michael Frayn's *Sweet Dreams* (1973); Astrid LINDGREN's *Bröderna Lejonhjärta* (1973; trans Joan Tate as *The Brothers Lionheart* 1975 US); Larry NIVEN's and Jerry Pournelle's *Inferno* (1975); Clifford D. SIMAK's "The Ghost of a Model T" (1975); the movie *Alice, ou la Dernière Fugue* (1976; vt *Alice, or the Last Fugue*) dir Claude Chabrol (1930-); J.G. BALLARD's *The Unlimited Dream Company* (1979); *Dying, in Other Words* (1981) by Maggie Gee (1948-); *Lanark* (1981) by Alasdair Gray

(1934-); Peter CAREY's *Bliss* (**1981**); D.M. THOMAS's *The White Hotel* (**1981**); Susan COOPER's *Seaward* (**1983**), whose two protagonists, having fallen into the Country of Life and Death (◊◊ GODGAME), decide to return to the mortal world; Kathy Acker's *Don Quixote* (**1986**); *The Adventure* (**1986**), a book-length narrative poem by Frederick Pollack (**1946**-); Brian W. ALDISS's "North of the Abyss" (1989); Wolf MANKOWITZ's *Exquisite Cadaver* (**1990**); and Robertson DAVIES's *Murther and Walking Spirits* (**1991**).

Movie examples of PFs have appeared regularly over the years; unlike the case with written fiction, they show no particular decline in popularity during the 1990s. One of the earliest was OUTWARD BOUND (*1930*), based on the Sutton Vane novel noted above; it was remade as *Between Two Worlds* (*1944*). The definitive PF movie is probably HERE COMES MR JORDAN (*1941*), based on the play *Heaven Can Wait* by Harry Segall; confusingly, this has nothing to do with the movie HEAVEN CAN WAIT (*1943*), which is also a PF; but *Here Comes Mr Jordan* was remade as HEAVEN CAN WAIT (*1978*). A MATTER OF LIFE AND DEATH (*1946*) deals with similar material. *A Guy Named Joe* (*1944*) saw a dead airplane pilot return to oversee his girlfriend's establishment of a new relationship; the plot was reworked to become more of a true PF in ALWAYS (*1989*). The comedy BEETLEJUICE (*1988*) centred on a young couple who must discover that they now exist only as GHOSTS. GHOSTS CAN'T DO IT (*1990*) has some peripheral interest in the PF context. A large chunk of BILL & TED'S BOGUS JOURNEY (*1991*) is PF – and very well done – although the protagonists, by cheating DEATH, are able to return to the mortal world. DEFENDING YOUR LIFE (*1991*) is a recent example of the PF proper: the protagonist discovers himself in a LIMBO where he is judged. By far the most interesting movie PF of recent years has been the TECHNOFANTASY *The* BREAKTHROUGH (*1993*); here we are never permitted clear sight of a dead man's experience of the AFTERLIFE, but must witness his realization of his situation through the eyes of those still living, and the hi-tech machinery they use. This reversal of the normal viewpoint of the PF is debatably evident also in FIELD OF DREAMS (*1989*), and there are various SUPERNATURAL-FICTION movies – an example is PORTRAIT OF JENNIE (*1948*; vt *Jennie* UK) – which might be regarded as PFs seen from the wrong side of the "barrier".

Some further written examples stand out as remarkable. The protagonist of Flann O'BRIEN's *The Third Policeman* (**1967**) has murdered a man, and has himself been killed; caught in a surreal Ireland, he steals a bicycle and eventually scares his murderer to death, but fails to realize that (as is ultimately clear to the reader) he is himself dead and doomed to repeat his cyclical course for all eternity. The protagonist of Gene WOLFE's *Peace* (**1975**) either does not realize that he has been dead for many years, or does not reveal to us his knowledge; either way, his failure to come to proper terms with his mortal life is reflected in the fact that none of the stories embedded in his reminiscences ever reaches an end: there is no peace without an ending. The text of John CROWLEY's *Engine Summer* (**1979**) comprises the life-story of a narrator who has been dead for centuries, but who never realizes this fact upon being awakened in crystalline form to tell, again and again, his exemplary tale. *The Hereafter Gang* (**1991**) by Neal Barrett Jr very energetically describes its dead protagonist's coming to terms with himself in a Texas BORDERLAND town populous with ghosts, memories and auguries.

Afterlives (anth **1986**) ed Pamela Sargent contains some examples of PFs. [JC/JG]

POTIONS In GENRE FANTASY, potions are generally not simple medicines or poisons but SPELLS in liquid form, brewed by a WITCH or WIZARD: a potion of HEALING is understood to act not medically but magically. Other established potion effects include AMNESIA, TRANSFORMATION, sleep, involuntary LOVE-fixation, temporary INVISIBILITY or INVULNERABILITY, and general loss of free will (◊◊ DEBASEMENT). Lewis CARROLL's "DRINK ME" potion in *Alice's Adventures in Wonderland* (**1865**) reduces Alice to a height of ten inches; ASTERIX's tiny Gaulish village defies the might of Rome thanks to a potion of strength; the Yellow Adept in Piers ANTHONY's **Apprentice Adept** sequence is a potion specialist with a brew for every need. [CB/DRL]

See also: ELIXIR OF LIFE.

POTOCKI, [Count] JAN (HRABIA) (1761-1815) Polish military man and ethnologist whose first stories – which show the influences of ARABIAN FANTASY – appeared in the 1780s, embedded into his travel books. He is best-known for the *Manuscrit trouvé à Saragosse* ["The Manuscript Found at Saragossa"], a complex text whose publication history is so convoluted that doubts have been expressed as to JP's actual authorship of the full book, which is a CYCLE of nested stories told by various characters (some of whom are themselves characters in some of the stories being told by others) over a period of 66 nights. Such doubts seem to have been laid to rest, at least for the time being; but the story of the book's composition and publication is still complicated. JP (who wrote in French) is thought to have begun composing the book in about 1797, publishing parts of the whole over the next decades, and continuing to work on the manuscripts until his suicide.

The first version – *Manuscrit trouvé à Saragosse* (**1804** Russia and **1805** Russia; rev [day 12 and half of day 13 cut; day 14 added] vt *Les Dix Jours d'Alphonse van Worden* 1814 France; rev version trans Christine Donogher as *Tales from the Saragossa Manuscript: (Ten Days in the Life of Alphonse Van Worden)* 1990 UK) – was printed privately (JP's frequent practice) and deposited in a St Petersburg library; it contains the first 12½ days. Further scattered days (15, 18, 20, 26-29, 47-56) were published as *Avadoro, histoire espagnole* ["Avadoro, a Spanish Story"] (**1813** France) in 4 vols as by M.L.C.J.P (Monsieur Le Comte Jan Potocki). The full text appeared as *Rekopiz Znaleziony w Saragossie* (trans Edmund Chojecki from French manuscipts **1847** 6 vols Poland). The complete French manuscript for this latter version has never been found, though by the late 1980s – when a scholarly edition of *Manuscrit trouvé à Saragosse* (**1989**; trans Ian Maclean as *The Manuscript Found in Saragossa* **1995** UK) appeared – manuscript sources for all but days 21, 30, 46, 57-66 and the conclusion had come to light. A further translation of parts of the book is *The Saragossa Manuscript: A Collection of Weird Tales* (trans Elizabeth Abbott **1960** US), with further excerpts appearing as *The New Decameron: Further Tales from the Saragossa Manuscript* (trans **1966** US).

The first 10 days are the most conspicuously integrated, and describe a remarkable descent on the part of the protagonist – who falls asleep in various places, but always awakens beneath a scaffold in the embrace of two corpses – into an ARABIAN NIGHTMARE, each story he hears (and in which he participates in DREAM) dragging him deeper. Later episodes lighten the protagonist's burden, as he becomes more of a listener to tales within tales told by others, though he remains implicated in their outcomes. Some of the embedded tales are mundane, but most are SUPERNATURAL FICTIONS; DOUBLES and other haunted echoings appear and reappear throughout, and there are FANTASY-OF-HISTORY hints (never fully developed) of some secret knowledge that will change the listening protagonist's understanding of the world.

The full scholarly publication of the book has revealed a masterpiece. [JC/BS]

POTTER, BEATRIX (1866-1943) UK children's writer and illustrator whose art is focused on the acute manipulation of fantasy and reality.

Her stories, beginning with *The Tale of Peter Rabbit* (privately printed **1901**; commercially published 1902) belong in the nursery tradition of TALKING-ANIMAL stories. Peter is a young rabbit with a rabbit's taste for garden produce, yet he wears a blue coat and shoes; Mrs Tiggy-Winkle (a hedgehog) takes in laundry; Ribby (a cat) invites Duchess (a dog) to an elegant tea-party; etc. However, BP's animals largely behave true to type – although the party invitation is delivered by a postman and Mrs Rabbit shops at the baker's. There is a clarity of vision and an unsentimentality in her tales which is based upon accurate observation of both the physical forms and inner instincts of her subjects. She creates a genuinely novel invented world.

BP, an isolated child from a rigidly conventional middle-class Victorian background, turned to art and storytelling as a means of escape. Although several of her stories – including *Peter Rabbit* – were created with specific children in mind, she never "wrote down", and thus her tales are sometimes mini-epics with a potentially disturbing edge. Even the simplest, such as *The Story of a Fierce Bad Rabbit* (**1906** chap), can be savage little morality tales, though this darker undercurrent is lightened by her delicate art (often done from life). Many of the books written after her move to Sawrey in the Lake District (where she became a famer) use the real scenery and countryside and the actual fittings of her cottage. While her settings and plots are cosily domestic, the vocabulary is as challenging and ironic and the undercurrents as wild as in any novel for adults.

Beatrix Potter's seclusion and settings are used as the springboard for Bryan Talbot's graphic novel *The Tale of One Bad Rat* (graph **1995**). [AS]

Other titles: *The Tailor of Gloucester* (**1903** chap); *The Tale of Squirrel Nutkin* (**1903** chap); *The Tale of Benjamin Bunny* (**1904** chap); *The Tale of Two Bad Mice* (**1904** chap); *The Tale of Mrs Tiggy-Winkle* (**1905** chap); *The Pie and the Patty-Pan* (**1905** chap); *The Tale of Mr Jeremy Fisher* (**1906** chap); *The Story of a Fierce Bad Rabbit* (**1906** chap); *The Story of Miss Moppet* (**1906** chap); *The Tale of Tom Kitten* (**1907** chap); *The Tale of Jemima Puddle-Duck* (**1908** chap); *The Roly-Poly Pudding* (**1908** chap); *The Tale of the Flopsy Bunnies* (**1909** chap); *Ginger and Pickles* (**1909** chap); *The Tale of Mrs Tittlemouse* (**1910** chap); *The Tale of Timmy Tiptoes* (**1911** chap); *The Tale of Mr Tod* (**1912** chap); *The Tale of Pigling Bland* (**1913** chap); *Appley Dapply's Nursery Rhymes* (**1917** chap); *The Tale of Johnny Town-Mouse* (**1918** chap); *Cecily Parsley's Rhymes* (**1922** chap); *Peter Rabbit's Almanac* (**1929** chap); *The Fairy Caravan* (**1929** chap); *The Tale of Little Pig Robinson* (**1930** chap); *Sister Anne* (**1932** chap); *Wag-by-Wall* (**1944** chap); *The Faithful Dove* (**1955** chap).

Further reading: *The Tale of Beatrix Potter* (**1946**) by Margaret Lane.

POWER, CECIL Pseudonym of Grant ALLEN.

POWER, SUSAN MARY (1961-   ) US writer, of Sioux ancestry. Her first novel, *The Grass Dancer* (**1994**), is a family saga written in short vignettes that move around the century of White domination, demonstrating the power for GOOD AND EVIL of traditional ritual and MAGIC. What makes this something more like a GENRE FANTASY than MAGIC REALISM is that the MAGIC is exploited for plot purposes rather than left as backdrop. [RK]

POWERS ◊ TALENTS.

POWERS, TIM(OTHY) (1952-   ) US author who first came to serious attention with his third novel, *The Drawing of the Dark* (**1979**). Like most of his mature work, this is a FANTASY OF HISTORY: the first Siege of Vienna in the early 16th century turns out to be merely the symptom of a far greater struggle between light and darkness, between West and East. The ageing mercenary hero, Duffy, a version of the Eternal Champion (◊ Michael MOORCOCK) whose past AVATARS include both ARTHUR and Siegfried, is recruited by MERLIN to defend a tavern where the beer referred to in the title is made every century or so to renew the life of the FISHER KING, the SECRET MASTER of Western Europe. The plot is complicated by much emotional disaster – the personal relationships of TP's protagonists tend always to the chastened if not to the bleak – and by a variety of supernatural meddlers anxious to renew their own "lives" through a taste of the beer; in the casual description of these last, TP demonstrates the off-hand erudition that is one of his hallmarks.

The Anubis Gates (**1983**; rev 1984 UK) won the 1984 PHILIP K. DICK AWARD and has, perhaps justifiably, been the most admired of his books, partly because it is the least gloomy. Academic Brendan Doyle, is persuaded by a cancer-ridden millionaire to act as guide to TIME-travelling tourists, and soon finds himself marooned in Regency LONDON. His studies of the poet Ashbless have made him acquainted with a London of URBAN LEGENDS, and he is caught up with rival guilds of beggars, a theriomorph (◊ SHAPESHIFTING) and a group of vindictive Egyptian sorcerers; the nightmarish creations TP inflicts on Doyle include Horrabin the Clown, a memorable MAGUS whose vivisected MONSTERS and tortures make him the secret ruler of a hidden UNDERGROUND London. Doyle, already half-mad with grief over the death of his wife, is damaged by privation and then shifted into the muscular body of a treacherous former pupil; he is befriended by Lord BYRON and Samuel Taylor COLERIDGE, survives the massacre of the Mamelukes, and travels further back in time to Restoration England, where he prevents the legitimation of Monmouth.

In the post-HOLOCAUST *Dinner at Deviant's Palace* (**1985**) all the fantastic elements are superficially rationalized (◊ RATIONALIZED FANTASY) in sf terms as mutants or alien intruders; the alien villain, though, is clearly a DARK LORD, a fake MESSIAH, a VAMPIRE and the Lord of the Underworld. Like all of TP's protagonists, the musician hero Rivas is wounded, in his case in the hand. The same is true of John Shandagnac, the puppeteer hero of *On Stranger Tides* (**1987**); the wounds inflicted on TP's protagonists are always to a faculty which is part of their core, in both these cases to the hands crucial to their art. Though Doyle is starved, poisoned, shot at and eventually mutilated to the point of death, the crucial wound inflicted on him is the loss of his original body. Disguises that come close to loss of identity abound in TP; it is only belatedly that Doyle realizes that the "boy" Jackie is the woman Ashbless is destined to marry ($ GENDER DISGUISE), and John Shandagnac becomes the famous pirate chieftain Jack Shandy.

Shandagnac's wound makes it possible for him to defeat the PIRATE magus Blackbeard, another aspiring Dark Lord, by using the iron in blood to defuse magic. In *The Anubis Gates* various characters wear chains around their ankles as lightning conductors for a magic that has much in common with energy, while the mental powers of the alien VAMPIRE in *Dinner at Deviant's Palace* can be disrupted by the humming of particular sorts of MUSIC. Protagony in TP is closely bound up with acquiring not so much magic as the capacity to resist it and to refuse the temptations that it makes possible – Doyle is at one point offered the RESURRECTION of his dead wife.

There are other strands we can observe tracing their way through TP's work. As he matured as a writer, his heroines become steadily more important; Anna in *The Drawing of the Dark* is almost entirely peripheral, and Jackie in *The Anubis Gates* acquires protagony because her gender disguise makes her an honorary man. In *Dinner at Deviant's Palace* the ex-lover whom Rivas is trying to rescue proves to be little more than a MCGUFFIN, and the real heroine starts the book as an acolyte and dupe of the villain, while Beth in *On Stranger Tides* spends much of the time under the enchantments of her villainous father before finally awakening to make possible the transformative act of sacred marriage which enables the final defeat of Blackbeard. Josephine, the heroine of TP's next novel – *The Stress of Her Regard* (**1989**) – starts off as a vengeful madwoman, convinced her brother-in-law Crawford is responsible for the death of her sister and prepared to form alliances with the LAMIAS that are the true guilty parties, but makes a hard-won transformation into fully shared protagony. Far more spectacularly even than TP's male principals, she LEARNS BETTER through losing an eye; recovering her sanity is also a loss of identity, since she has in the process to abandon not only the delusion that she *is* her sister but also the delusion that her sister was a nice person.

The Stress of her Regard is by some way TP's darkest and best novel, since it deals in the loss of illusion and of hope, as well as in the death of children. When the poets Lord BYRON, Percy Bysshe SHELLEY and John KEATS make their break with the vampire silicon beings who have given them inspiration while destroying those they love, they lose their gifts and become in a real sense less themselves. Crawford, a dedicated doctor, has to help his friends to their deaths and finally perform a bizarre act of magical obstetrics, separating the realms of silicon and carbon for good – and in the process destroying the Holy Roman Empire of the Hapsburgs and ushering in the modern age.

RITUALS OF DESECRATION dominate TP's two next novels, which run their immediate predecessor close for bleakness. *Last Call* (**1992**; rev 1992), set in a Las Vegas turned into a WASTE LAND by the magical death of its FISHER KING, the gangster Bugsy Siegel, pits its one-eyed JACK protagonist Crane (whose name indicates that he is potentially the HIDDEN MONARCH) against his malignant MAGUS father; the father wishes to circumvent fate by tricking him, through magical poker games, into IDENTITY EXCHANGE. Crane is a driven gambler and drunk, haunted by the GHOST of his dead wife; he is also the FOOL perpetually failing to ask the right questions and give the right answers. This is, of all TP's novels, the one that comes closest to being an INSTAURATION FANTASY, in that it ends in positive transformation of Las Vegas as a whole rather than in the destruction of EVIL.

Expiration Date (**1995**; rev 1996) is even darker; its picture of contemporary Los Angeles is of a waste land which will never bloom except with mischief. TP's inventiveness with the reversal of tropes ($ REVISIONIST FANTASY) is at its most remarkable here. There is a malignant WAINSCOT society of those who are addicted to the eating of GHOSTS – which are usually the only marginally sentient psychic shells thrown off during death or under stress. TP gives us three principals – Sullivan, wounded by the suicide of his TWIN, and sought by his wicked STEPMOTHER as bait for the ghost of his father; Elizalde, a person who LEARNS BETTER, a psychiatrist who tried to use folk wisdom therapeutically and had the mortification of finding it works; and Kootie, a boy reared as a Theosophist ($ THEOSOPHY) MESSIAH who is desperately seeking normality after the murder of his parents, and who is possessed ($ POSSESSION) by the ghost of Thomas Alva Edison (1847-1931). The protagonist's wound in TP's work has always had its Oedipal aspect; here all three principals have their roots planted in both Freud and folklore. Notably, the main body of the novel ends with their self-creation as a family, as Powers relegates the complex double-dealing mayhem of the denouement to an epilogue; this structural trait has grown in his work, emphasizing the relative importances for him of, on the one hand, the emotional plot and, on the other, the mere gaudy incidents.

As always, TP is in this novel profligate in invention and in the relentless use of fantasy tropes: one of the villains is an amnesiac immortal ($ AMNESIA; IMMORTALITY), and the murder of Sullivan's father is yet another ritual of desecration. As so often in TP's writing, the stepmother DeLarava is herself a victim, stripped by her long-lived colleague of psychic energy at birth.

TP was one of the small AFFINITY GROUP that clustered around Philip K. Dick ($ SFE) in Dick's latter years; he is closely associated with James P. BLAYLOCK and K.W. Jeter (1950-   ), both as a friend and in their co-creation of what has come to be known as STEAMPUNK. Like Dick, TP has always combined endless ingenuity of plotting, witty and revealing dialogue and an eye for the presentation of detail with a strong moral sense; like Dick also, his plots have on occasion a ramshackle and improvisatory quality which is part of the reason for his constant revision of his texts. He is one of the most original and authoritative voices in the genre. [RK]

Other works: *The Skies Discrowned* (**1976** Canada as Timothy Powers; rev vt *Forsake the Sky* 1986 as TP), fantasy-tinged adventure sf; *Epitaph in Rust* (**1976** Canada as Timothy Powers; text restored vt *An Epitaph in Rust* 1989 as TP); *Night Moves* (**1986** chap); *The Way down the Hill* (**1986** chap).

Further reading: *A Checklist of Tim Powers* (**1991** chap) by Tom Joyce and Christopher P. Stephens.

POWYS, JOHN COWPER (1872-1963) UK writer, teacher and lecturer, who spent many years in the USA before settling in Wales in the late 1930s. He was the eldest of three literary brothers: Theodore ($ T.F. POWYS) and Llewelyn. JCP's work is the most philosophical and mystical. His earliest publications were poems, starting with *Odes and Other Poems* (coll **1896**), plus the strongly religious *Visions and Revisions* (coll **1915**) and the introspective essays

Suspended Judgements (coll **1916**). During this period JCP, who had been imbued by his father with an intensely theological outlook, was creating his own understanding of the world which began to materialize in a series of novels which used isolation and dislocation as MOTIFs in polarizing perspectives and creating a sense of *otherness*, not strong enough to be regarded as WRONGNESS but sufficient to be detectable. This became evident in *Wood and Stone* (**1915** US) in which existence in a UK West Country village is energized between the tension of Christianity of the resurrection (represented by the Holy Rood) and the Christianity of esoteric tradition (represented by the Holy GRAIL). The duality constrained Powys's writing for a decade until he was able to express his vision of GOOD AND EVIL in *Ducdame* (**1925** US) and *Wolf Solent* (**1929** US), both nonfantastic metaphysical novels which seek to explain the human condition in terms of a MULTIVERSE of conflict. *Wolf Solent* marked a transition in JCP's writing. Not only was it his first successful book, critically and financially, but it was the first in which he allowed the subject matter to explore itself rather than force it into shapes that even JCP could not fully understand. As a result JCP was able to channel his beliefs into more extensive mindscapes and this led to his first major fantasy, *A Glastonbury Romance* (**1932** US). As in *Wood and Stone*, he returns to the religious conflict of Christianity and paganism and the influence through time of the Holy Grail. This timeshift imagery continues in *Maiden Castle* (**1936** US). JCP then felt able to tackle two major historical novels set in ancient Wales – *Owen Glendower* (**1940** US), which is largely nonfantastic (bar the acceptance of the supernatural as part of life), and *Porius* (**1951**), set at the time of king ARTHUR and exploring the character of MERLIN.

During this transition JCP wrote a very different novel, *Morwyn, or The Vengeance of God* (**1937**), untypical in setting but typical in philosophy and expression. JCP utilizes the vision of DANTE's HELL to explore inhumanity. Hell is full of sadists, and JCP explores the difference between humanity's past cruelty in the demand for power and his current obsession with science, exemplified by vivisection. Ultimately JCP finds against scientific research. His hell features many notable historical figures: Tomas de Torquemada (1420-1498), the Marquis DE SADE, Merlin (again) and Taliesin (◊ MABINOGION). The narrator's journey through hell is also a RITE OF PASSAGE. He emerges wounded, like the FISHER KING, carrying the ills of the world.

JCP's final works are mostly no less remarkable in their strong philosophical vision. The best are *Atlantis* (**1954**), which tells of the last voyage of ODYSSEUS as he sails west and discovers ATLANTIS and America, and has a companion piece in *Homer and the Aether* (**1959**), a free rendition of *The Iliad* (◊ HOMER); and *The Brazen Head* (**1956**), about the search by Roger Bacon (*c*1214-1292) for the ultimate truth (◊ ALCHEMY). [MA]

Other works: *The Owl, the Duck, and – Miss Rowe! Miss Rowe!* (**1930** chap US) and *The Inmates* (**1952**) both novels of incarceration and madness; *Rabelais* (**1948**), nonfiction; *Up and Out* (coll **1952**), which contains two novellas "Up and Out" and "The Mountains of the Moon"; *Lucifer: A Poem* (**1956**); *All or Nothing* (**1960**); *Real Wraiths* (**1974** chap), *Two and Two* (**1974** chap), *Three Fantasies* (coll **1985**) ed Glen Cavaliero.

Further reading: *Autobiography* (**1934**); *The Saturnian Quest: A Chart of the Prose Works of John Cowper Powys*

(**1964**) by G. Wilson Knight; *John Cowper Powys: Old-Earth Man* (**1966**) by H.P. Collins; *The Powys Brothers* (**1967**) by Kenneth Hopkins; *Essays on John Cowper Powys* (anth **1972**) ed Belinda Humfrey; *John Cowper Powys* (**1973**) by Jeremy Hooker; *John Cowper Powys: Novelist* (**1973**) by Glen Cavaliero; *The Demon Within: A Study of John Cowper Powys's Novels* (**1973**) by John A. Brebner; *John Cowper Powys and the Magical Quest* (**1980**) by Morine Krissdottir; *John Cowper Powys in Search of a Landscape* (**1982**) by C.A. Coates; *The Brothers Powys* (**1983**) by Richard Perceval Graves; *The Ecstatic World of John Cowper Powys* (**1986**) by Harald W. Fawkner.

POWYS, LLEWELYN (1884-1939) UK writer. ◊ John Cowper POWYS; T.F. POWYS.

POWYS, T(HEODORE) F(RANCIS) (1875-1953) UK writer, brother of Llewelyn and John Cowper POWYS. Although TFP and John were brought up in the same strongly theological household, their approaches to fiction soon diversified, John's becoming deeply mystical while TFP's became more light-hearted and allegorical (◊ ALLEGORY). Their style seem as polarized as their individual interpretations of GOOD AND EVIL. TFP's religious background was evident from his first book, *An Interpretation of Genesis* (**1907**), and particularly in *Soliloquies of a Hermit* (**1916**), which established his personal view of religious experience and belief that EVIL was inherent in Man and could be controlled only if humankind opened itself to the moods of God. This philosophy runs through most of his fantasies, starting with "The Left Leg" (in *The Left Leg* 1928), set in a quiet UK West Country village where Farmer Mew represents EVIL incarnate and is on the point of possessing the village when GOD, in the form of Mr Jar, returns. This story is a pilot for TFP's best-known novel, *Mr Weston's Good Wine* (**1927**), where God and St Michael visit the village of Folly Down to test its inhabitants. Unlike John, who explored good and evil in a series of darkly philosophical studies, TFP was able to convey a similar concept in a humorous yet equally challenging way. His work is more typical of other SLICK-FANTASY novelists of the 1920s, like Martin D. ARMSTRONG and Lord DUNSANY. God in the form of Jar reappears in the later stories *The Key of the Field* (**1930** chap) and *The Only Penitent* (**1931** chap), where again he tests men to despair, like Job, to prove their worth in HEAVEN, and *Unclay* (**1931**), where the central figure is DEATH. Though not as enjoyable as *Mr Weston's Good Wine*, *Unclay* is perhaps more rewarding in its approach. TFP concludes his series of allegories with "The Two Thieves" (in *The Two Thieves* 1932), in which a man drinks the wines of the DEVIL and becomes the epitome of evil, until Jar returns to recover them. TFP wrote several collections of short fiction. *Fables* (coll **1929**; vt *No Painted Plumage* 1934) is a volume of surreal but ingenious ALLEGORIES. *The House of the Echo* (coll **1928**) and *The White Paternoster* (coll **1930**) contain the best of his more conventional SUPERNATURAL FICTIONS. [MA]

Other works: *Mr Tasker's Gods* (**1925**); *Mockery Gap* (**1925**); *The Only Penitent* (**1931**); *Bottle's Path and Other Stories* (coll **1946**); *God's Eyes A-Twinkle* (coll **1947**).

Further reading: *The Powys Brothers* (**1967**) by Kenneth Hopkins; *The Brothers Powys* (**1983**) by Richard Perceval Graves.

PRAED, MRS (ROSA CAMPBELL) (1851-1935) Australian-born novelist, *neé* Rosa Caroline Murray-Prior; in the UK after her marriage in the late 1880s. Many of her novels were romances set in her native land; once in the UK she developed a keen interest in OCCULTISM, then fashionable, and increasingly incorporated its themes into her novels. *Affinities: A Romance of Today* (**1885**) is a tale of the psychic domination of a vulnerable young woman by a decadent poet. H.P. BLAVATSKY's ideas, which play a marginal role in *Affinities*, provide the basis for the brief but wholehearted Theosophical fantasy (◊ THEOSOPHY) *The Brother of the Shadow: A Mystery of Today* (**1886**). *The Soul of Countess Adrian: A Romance* (**1891**) and *The Insane Root: A Romance of a Strange Country* (**1902**) are both tales of personality displacement, the latter being the more interesting by virtue of its intriguingly confused morality. *"As a Watch in the Night": A Drama of Waking and Dream, in Five Acts* (**1901**) is a VISIONARY FANTASY of problematic REINCARNATION which draws upon the same sources of inspiration as the allegedly documentary *Nyria* (**1904**; rev vt *The Soul of Nyria* 1931). *The Body of His Desire: A Romance of the Soul* (**1912**) and *The Mystery Woman* (**1913**) are feebler occult romances in the same vein. Some of the short stories in *Stubble Before the Wind* (coll **1908**) exhibit the same preoccupations as the novels. *Fugitive Anne* (**1902**) is a LOST-RACE novel set in Australia. [BS]

Other works: *The Ghost* (**1903**).

PRAGER, EMILY (1952-   ) US writer and humorist; she has worked as contributing editor and columnist for magazines such as *National Lampoon* and *Penthouse*. Her extravagant fictions often take the form of FABULATION: some of the tales in *A Visit to the Footbinder and Other Stories* (coll **1982**), notably the title story and "The Lincoln-Pruitt Anti-Rape Device: Memoirs of the Women's Combat Army in Vietnam", verge on alternate history (◊ ALTERNATE REALITIES). *Clea & Zeus Divorce* (1987) portrays the eponymous gods as a contemporary celebrity couple. [GF]

Other work: *Eve's Tattoo* (**1991**).

PRAGUE A CITY with continuing OCCULT associations, almost certainly because Rabbi Loëw (or Lowe) ben Bezalel (1512-1609) is supposed to have used his powers as a MAGUS and leader of a cabal of Jews to create a GOLEM, which henceforth could be imagined haunting the ghetto. It is this city that John DEE (in life and in fiction) visited in his search for hermetic truths, and which is the venue for tales like F. Marion CRAWFORD's *The Witch of Prague* (**1891**). The golem story itself has been recounted on innumerable occasions, most recently in *He, She and It* (**1991**; vt *Body of Glass* 1992 UK) by Marge Piercy (1936-   ). The central use of the Prague golem in a SUPERNATURAL FICTION is Gustav MEYRINK's *The Golem* (**1914**). Meyrink subsequently used Prague as a city hovering on the verge of dissolution, perhaps inspired in this by its geographical and political fragility after 1918, when it was no longer part of the Austro-Hungarian Empire. This city – hectic, doom-laden, Weimar-like, ornate and mysterious and courtly – underlies much of the fiction of Karel ČAPEK, Franz KAFKA and Leo PERUTZ. [JC]

PRANTERA, AMANDA (1942-   ) UK-born writer, in Italy since the age of 20, author of some fantasy and SUPERNATURAL FICTION, plus tales like *Strange Loop* (**1984**), *The Side of the Moon* (**1991**) and *Proto Zoë* (coll of linked stories 1992) which play with the UNCANNY. *The Cabalist* (**1985**) is set in a crepuscular, contemporary VENICE irradiated – as is usual in URBAN FANTASIES set here – by death. The protagonist is a MAGUS in a world no longer receptive to his art (◊ THINNING); he is haunted by a DEMON in the shape of a

child, who may (or may not) be the rightful inheritor of his wisdom. The frame story of *Conversations with Lord Byron on Perversion, 163 Years After his Lordship's Death* (**1987**) is TECHNOFANTASY, the conversations being manifested through a computer. *The Kingdom of Fanes* (**1995**) is set in FAERIE, where a complicated tale is told deadpan. [JC]

PRATCHETT, TERRY (1948-   ) UK writer whose **Discworld** fantasies (◊ HUMOUR) are phenomenal UK bestsellers; reportedly, 1% of all books sold in the UK are by TP, an astonishing share of the market for a single author. His first published story was "The Hades Business" (1963 *Science Fantasy*). His debut novel, *The Carpet People* (**1971** rev 1992), is a CHILDREN'S FANTASY set in a WAINSCOT world of tiny, warring creatures inhabiting the weave of a carpet, where a fallen sugar crystal is a vast natural resource and humans' crushing footsteps are perceived as the malice of a DARK LORD called Fray (◊ GREAT AND SMALL). *Strata* (**1981**), a parodic sf novel (◊ PARODY), plays with the idea of an artificial flat world with its own orbiting SUN, MOON and STARS painted on a Ptolemaic crystal sphere; this was the seed of the Discworld.

The **Discworld** comedies to date are *The Colour of Magic* (**1983**), *The Light Fantastic* (**1986**), *Equal Rites* (**1987**), *Mort* (**1987**), *Sourcery* (**1988**), *Wyrd Sisters* (**1988**), *Pyramids* (**1989**), *Guards! Guards!* (**1989**), *Eric* (**1990**) – whose heavily illustrated first edition gives equal credit to Josh KIRBY, regular artist of UK Discworld covers; this book has the pseudo-title *Faust* crossed out on the cover and *Eric* scrawled in – *Moving Pictures* (**1990**), *Reaper Man* (**1991**), *Witches Abroad* (**1991**), *Small Gods* (**1992**), *Lords and Ladies* (**1992**), *Men at Arms* (**1993**), *Soul Music* (**1994**), *Interesting Times* (**1994**), *Maskerade* (**1995**), and «*Feet of Clay*» (1996). Related short stories are "Troll Bridge" in *After the King: Stories in Honour of J.R.R. Tolkien* (**1992** US) ed Martin H. GREENBERG, and "Theatre of Cruelty" (1993 *W.H. Smith Bookcase*; exp rev 1993 US).

Discworld, echoing certain CREATION MYTHS, is supported by four elephants standing atop an immense turtle swimming endlessly through space. Naturally it has odd properties, such as a complicatedly silly CALENDAR, but after the initial books the structure is only fleetingly referred to. More importantly, this is a PLAYGROUND where REALITY is thin and fragile, and where "narrative causality" – the momentum of STORY – has shaping power; "morphic resonance" makes Discworld overtly a MIRROR of our world and its stories ... a distorting mirror.

The novels fall into rough thematic groups. First comes the **Rincewind** sequence, here named for the hilariously cowardly and inept WIZARD found fleeing in terror through book after book. Rincewind is present from the start in *The Colour of Magic*, which contains most of the series' overt PARODY (e.g., of Fritz LEIBER, H.P. LOVECRAFT and Anne MCCAFFREY). He remains reluctant escort to Discworld's first naive tourist, Twoflower (and his animated, psychopathic Luggage), in the romping continuation *The Light Fantastic* – which establishes the wizards' college Unseen University (◊ EDIFICE) in the great and malodorous CITY of Ankh-Morpork, while poking fun at ASTROLOGY, Druids, DWARFS, HEROIC FANTASY (via the nonagenarian hero Cohen the Barbarian; ◊ CONAN), magic SHOPS, SPELLS, TROLLS and other genre staples. *Sourcery* sees an increase in the potency of MAGIC luring Discworld's wizards into hubris and war: the APOCALYPSE and its FOUR HORSEMEN loom. A reluctant act of inept heroism saves the day but exiles

Rincewind to a Lovecraftian OTHERWORLD. In *Eric* he is summoned back by a callow would-be FAUST wishing to raise a DEMON: the ensuing rapid travelogue (shaped to provide Kirby with illustration opportunities) tours the Discworld's Aztec Empire, Trojan War, CREATION, END OF THE WORLD and HELL. In *Interesting Times*, TP's excursion into CHINOISERIE, both Rincewind and Cohen acquire a certain comic gravitas from their developed philosophies of, respectively, cowardice and barbarian conquest.

The next subgroup stars a WITCH with a whim of steel: **Granny Weatherwax**, who in *Equal Rites* attacks magical GENDER prejudice to force a talented girl's admission to the men-only faculty of Unseen University. *Wyrd Sisters* fills out Granny's tiny coven with the shameless old reprobate Nanny Ogg and the young, desperately earnest New Age witch Magrat: all are entangled in royal and thespian derring-do which repeatedly echoes and perverts William SHAKESPEARE's *Macbeth* (**1623**). *Witches Abroad* takes the trio to foreign parts for a clash of rival FAIRY GODMOTHERS over the proper outcome of CINDERELLA's story; FAIRYTALE references abound, incongruously mixed with VOODOO in the Discworld analogue of NEW ORLEANS. The title *Lords and Ladies* refers euphemistically to ELVES, here viciously sadistic invaders which only the returned coven (and a Morris-dancing team) can effectively oppose; the resonances are with *A Midsummer Night's Dream* (1600). *Maskerade* sees mayhem at the Ankh-Morpork OPERA House as Granny and Nanny wrestle with the narrative momentum of the Discworld's skewed re-enactment of PHANTOM OF THE OPERA. A distinctive character from the outset, Granny Weatherwax grows in stature as an indomitable figure who commands respect rather than liking, might easily have turned to BLACK MAGIC, and now determinedly preserves BALANCE: in *Maskerade* she impossibly catches a SWORD blade in her bare hand – but later, when time permits, restores balance by finally accepting the stroke and stitching up her delayed wound.

The **Death** stories hinge on Discworld's eternally grim and humourless straight-man DEATH, a self-confessed anthropomorphic personification manifesting in all books as a robed skeleton speaking in hollow capitals. Death has a soft spot for humanity, and a pathos stemming from inability to grasp human quirks like emotion. In *Mort* – the first Discworld novel with a successfully integrated plot – Death allows himself a holiday by hiring the eponymous apprentice, whose sympathy for the doomed proves disastrous. *Reaper Man* sees Death (temporarily) pensioned off by the auditors of REALITY and finally subject to TIME, while uncollected life-forces leave Ankh-Morpork plagued with POLTERGEISTS, ZOMBIES and worse. *Soul Music* has Death depressedly taking another holiday while his adoptive granddaughter is forced to take up his role ... but this novel principally concerns a manic craze for rock MUSIC which coincidentally afflicts Ankh-Morpork and environs, including a repeatedly implied pun about the star performer, who somehow looks Elvish.

TP's **City Guard** or **Night Watch** sequence (◊ DETECTIVE/THRILLER FANTASY; URBAN FANTASY) opens with the very successful *Guards! Guards!*, shot through with Chandleresque and police-procedural pastiche as the run-down, sleazy Ankh-Morpork Night Watch finds the city threatened by a killer – a DRAGON summoned in a scheme to overthrow the Patrician, the city's unlovable but efficient ruler, and instal a malleable king. *Men at Arms* features a

SERIAL KILLER with Discworld's only gun (for Ankh-Morpork is in the throes of Renaissance), and has much inventive detail about the city guilds. «Feet of Clay» confronts the Watch with a devious assassination plot involving GOLEMS. Notable continuing characters are Vimes, the Watch leader whose glum feeling for injustice and moody love of his city lead him to a kind of socialism (fanned by the irony of his being, uniquely, knighted for service to the Patrician, whom he hates), and Carrot, a Guardsman initially naive-seeming to the point of stupidity, who develops sharp wits and a sunnier love for Ankh-Morpork – of which he is, though refusing the throne, the HIDDEN MONARCH.

The rest are less overtly linked. *Pyramids* begins with a comic Ankh-Morpork Assassin's Guild practical examination and moves to Discworld's Ancient EGYPT, where the picture of a static and bankrupt society spending all its resources on pyramids and MUMMIES is more darkly humorous. *Moving Pictures*, perhaps TP's most hilariously shambolic novel, infects Discworld with the spirit of Hollywood as a magic-powered CINEMA industry rises at fast-forward speed, and falls – after countless real-world movie allusions and a climactic KING KONG travesty in which a 50ft woman scales Ankh-Morpork's tallest building while brandishing a terrified ape (who is the Librarian of Unseen University, another popular running character). *Small Gods*, the most grimly effective of the series, introduces a fundamentalist RELIGION enshrining one revealed "truth", to deny which is a punishable heresy: that Discworld is not flat but spherical. Echoing Galileo's "*Eppur si muove*", expiring heretics cry: "The Turtle moves." Even this theocracy's GOD, Om (now incarnated as a tortoise; ◊ BONDAGE; KENOSIS), is powerless against its inflexibility; the hero Brutha, a holy FOOL, undergoes a gruelling desert NIGHT JOURNEY before reaching EUCATASTROPHE. As TP has remarked, not everything in a **Discworld** book is intended as funny: a steadfast seriousness in matters of life and death underpins the humour of the finer novels.

Discworld narratives are increasingly shaped by the interplay of belief and Story. In *Moving Pictures* the movies compel belief, a source of power which threatens the world with MONSTERS; but the actor HERO realizes that, by the Hollywood logic of Story brought into operation through this belief, he can win by behaving not prudently but in wildly heroic, filmic style. In *Small Gods* Om has suffered THINNING through neglect, because his Church is following another Story of blood and conquest which pays scant lip-service to Om, and – until a climactic "miracle" makes many converts – Brutha (to whom Om becomes SECRET SHARER) is the sole believer. Granny Weatherwax, whose laser-willed belief in herself makes her unstoppable, repeatedly works to derail Stories whose impetus is being turned to bad ends. The city of Ankh-Morpork is haunted by the Story of the Return of the King: the Night Watch defeats several plans to depose the Patrician in favour of a ruler "qualified" by birth. Death himself often warns the newly dead that the Story of the AFTERLIFE is their own choice: what would they like to believe? Thus, in *Interesting Times*, an aged schoolmaster who had expected a dull afterworld shifts Stories to gatecrash VALHALLA.

Discworld's success has led to many spinoffs, the most substantial being *The Discworld Companion* * (**1994**) by TP and Stephen Briggs (1951-   ), a humorous concordance with much additional material. Briggs has adapted several of the novels for the stage. GRAPHIC-NOVEL versions are *The*

Colour of Magic (graph **1991**), *The Light Fantastic* (graph **1992**) and, most successfully, *Mort: A Discworld Big Comic* (graph **1994**), rewritten by TP for illustrator Graham Higgins (1953-   ). Two fold-out map books were devised by Briggs and TP and drawn by Stephen Player: *The Streets of Ankh-Morpork: Being a Concife and Possibly Even Accurate Mapp of the Great City of the Discworld* * (**1993**) and *The Discworld Mapp: Being the Onlie True & Mostlie Accurate Mappe of the Fantastyk & Magical Discworlde* * (**1995**). The *Unseen University Challenge* * (**1996**) by David LANGFORD is a **Discworld** quizbook.

TP's most ambitious non-**Discworld** novel is *Good Omens: The Nice and Accurate Prophecies of Agnes Nutter, Witch* (**1990**; rev 1990 US) with Neil GAIMAN, which spoofs the OMEN movies *via* a youthful ANTICHRIST who, with his cronies and dog (a hellhound), strongly recalls the **William** children's stories by Richmal Crompton (◊ John LAMBOURNE) – and who, despite the efforts of the FOUR HORSEMEN, HEAVEN and HELL, forestalls ARMAGEDDON. Additional diversions include the long-dead witch Agnes Nutter's book of devastatingly infallible PROPHECY, a small, inept army of Witchfinders, and the notion that the M25 motorway encircling London is an evilly emanating sigil. The YA **Johnny** books star a more normal but adventure-prone boy. *Only You Can Save Mankind* (**1992**) is a TECHNOFANTASY where the alien targets in a shoot-'em-up computer GAME require Johnny to accept their surrender and guard their retreat through the OTHERWORLD of "game space"; it was dramatized as a BBC radio serial by Bob Hescott in 1996. *Johnny and the Dead* (**1993** US; 1994 UK) entangles him with GHOSTS in a cemetery corruptly sold off for redevelopment (a topical UK allusion); «Johnny and the Bomb» (1996) involves TIME TRAVEL to WORLD WAR II.

Despite occasional patches of sloppiness ascribable to TP's high rate of production, the **Discworld** books are a remarkably fine set of fantasy comedies whose popular success derives not only from inventiveness with words and ideas but from understanding of how darker issues can, by contrast, intensify the comic highlights. Further **Discworld** novels are expected.   [DRL]
Other works: *The Dark Side of the Sun* (**1976**), sf; **The Bromeliad** or **Book of the Nomes** children's sf series, being *Truckers* (**1989**), *Diggers* (**1990**) and *Wings* (**1990**); *The* **Witches Trilogy** (omni **1994**), assembling *Equal Rites*, *Wyrd Sisters* and *Witches Abroad*; «Hogfather» (1996).
See also: RECURSIVE FANTASY; SATIRE.

PRATT, FLETCHER (1897-1956) US writer and historian, a prolific contributor to the Gernsback SCIENCE-FICTION pulp MAGAZINES, mostly as a translator. During the 1930s he produced much weighty nonfiction, semming to regard his fiction as light relief. He teamed up with L. Sprague DE CAMP to write comedies for UNKNOWN, most famously the **Incomplete Enchanter** or **Harold Shea** sequence, whose heroes visit a series of ALTERNATE WORLDS in which various ancient MYTHOLOGIES are reified and where they must pit their 20th-century wits against the naïve MAGIC of the indigenes (◊◊ RATIONALIZED FANTASY). *The Incomplete Enchanter* (1940 as "The Roaring Trumpet" and "The Mathematics of Magic"; fixup **1941**) conflates two novellas featuring the world of Norse mythology (in the run-up to RAGNAROK) and the world of Edmund SPENSER's *The Faerie Queene*. *The Castle of Iron* (1941; exp **1950**) involves Shea and his associates with the *dramatis personae* of ARIOSTO's *Orlando Furioso* (**1516**), whose close similarity to

Spenser's characters causes much confusion. *The Land of Unreason* (1941; exp **1942**), whose protagonist is a US tourist in the UK who is carried off by drunken leprechauns to OBERON's FAERIE, is more slapdash than its predecessors. The last novel FP and de Camp wrote for *Unknown* (although the magazine died before it appeared) was *The Carnelian Cube* (**1948**), in which an archaeologist wishes himself into a series of alternate worlds in the forlorn hope of finding one that will suit him. When FP and de Camp came together again after WWII they found a more congenial literary form in the anecdotal whimsies collected as *Tales from Gavagan's Bar* (coll **1953**; exp 1978); the TALL TALES (◊◊ CLUB STORIES) therein are a good deal slicker than the two slightly dispirited novellas which they belatedly added to the **Incomplete Enchanter** series, reprinted as *The Wall of Serpents* (1953-4 as "The Wall of Serpents" and "The Green Magician"; fixup **1960**; vt *The Enchanter Completed* 1980 UK). The former features the world of the KALEVALA, the latter the realm of Irish mythology. The first two volumes of the sequence were reissued as *The Compleat Enchanter: The Magical Misadventures of Harold Shea* (omni **1975**) and the third was added into *The Intrepid Enchanter* (omni **1988** UK; vt *The Complete Compleat Enchanter* 1989 US); the titles became even less appropriate when de Camp later added a further story to the sequence.

FP's post-war SCIENCE FICTION is earnest but not ambitious. Two stories of some fantasy interest are "The Wanderer's Return" (1951) and *The Undying Fire* (1953 as "The Conditioned Captain"; **1953**); the former is a transfiguration of the story of ODYSSEUS, the latter the story of Jason and the Argonauts (◊ GOLDEN FLEECE).

The two solo fantasy novels which FP wrote in the time before reteaming with de Camp are very different from his purely commercial work; they represent a determined and conscientious attempt to bring a new gravitas to SWORD AND SORCERY. *The Well of the Unicorn* (**1948** as by George U. Fletcher; 1967 as by FP) is a quasihistorical fantasy set in a world borrowed from Lord DUNSANY's play *King Argimenes and the Unknown Warrior* (**1911** chap); magic plays a marginal and largely symbolic role in a *Bildungsroman* tracking the growth to maturity of the son of a dispossessed landowner who turns rebel. Ahead of its time, it did not sell. *The Blue Star* (in *Witches Three* anth **1952**; 1969) – a more original and more impressive work – could not find a publisher until FP published it in one of a series of anthologies he was editing anonymously. The hero seduces a WITCH to win control of the eponymous TALISMAN on behalf of a revolutionary movement, thus initiating a long and difficult relationship; it is one of the finest HEROIC FANTASIES of its period, but FP concluded there was no point in attempting more fiction of this kind.   [BS]
PRATT, HUGO (1927-1995) UK artist. ◊ Milo MANARA.
PRATT, THEODORE (1901-1969) US writer who worked for the *New Yorker* and who was most prolific (as Timothy Brace) as an author of detections like *Murder Goes Fishing* (**1936**). His occasional excursions into the fantastic read as SLICK FANTASY. Best known is *Mr Limpet* (**1942**) – filmed as *The Incredible Mr Limpet* (**1964**) – which follows the adventures of a man who, granted his WISH to undergo METAMORPHOSIS into a fish, volunteers in WORLD WAR II to help defeat the Germans.   [JC]
Other works: *Mr Thurkle's Trolley* (**1947**); *Mr Atom* (**1969**).

PRECOGNITION The TALENT for seeing glimpses of the

future ($ PROPHECY); it differs from SCRYING in that mind-focusing devices like crystal balls are generally not required. Often precognition comes in brief, unexpected flashes which make sense only in hindsight. Thus the young hero of Diana Wynne JONES's *Power of Three* (1976), who has the "Gift of Sight Unasked", can give no reason for the dire warning he utters; it goes unheeded, like the prophecies of Cassandra in Greek MYTHOLOGY ($ ORACLES). In Randall GARRETT's *Too Many Magicians* (1967) a minimal precognitive talent – seeing only seconds into the future – makes its owner a deadly SWORD-fighter. [DRL]

PREEDY, GEORGE R. Pseudonym of Marjorie BOWEN.

PREHISTORIC FANTASY Most literary accounts of prehistoric life are best regarded as an extrapolation of historical fiction or as sf, but prehistoric episodes are sometimes included in VISIONARY FANTASIES and fantasies of REINCARNATION – *The Eternal Lover* (1925) by Edgar Rice BURROUGHS is both – and there is a notable subspecies of fantasy in which accounts of prehistoric events are juxtaposed or interwoven with accounts of much later events in order to display some kind of eternal recurrence; examples include *Chains* (1925) by Henry Barbusse (1874-1935), *Marden Fee* (1931) by Gerald BULLETT and *Red Shift* (1973) by Alan GARNER. Also of some fantasy interest are accounts of prehistory which allegorize the Biblical tale of ADAM AND EVE, some accounts of ATLANTIS and other LOST CONTINENTS, a few tales concerned with the remote origins of MAGIC and WITCHCRAFT – including those which dramatize the SCHOLARLY FANTASIES of anthropologists like James FRAZER – some accounts of ATAVISM and a few tales featuring DINOSAURS. Jean M. AUEL's series of PFs has been incredibly popular, as has Julian MAY's **Saga of Plioce Exile**, which might best be described as a prehistoric TECHNOFANTASY.

Fantasies which deal with entirely imaginary prehistories include Norman DOUGLAS's *In the Beginning* (1927) and S. Fowler Wright's *Dream* (1931). Several surreal PFs are featured in Italo CALVINO's *Cosmicomics* (coll 1965) and *T zero* (coll 1967). [BS]

PREISS, BYRON (CARY) (1953-   ) US book packager, anthologist, and author of three novels: *Guts* (1979) with C.J. Henderson, *Dragonworld* (1979) with J. Michael REAVES and *The Bat Family* (1984). Edited works include some of the **Weird Heroes** sequence, including *Weird Heroes #1* (anth 1975), *#2* (anth 1975), *#6* (anth 1977) and *#8* (anth 1978); the series as a whole was produced by his packaging firm, Byron Preiss Visual Publications Inc., or BPVP ($ SFE). Of greatest BPVP fantasy interest is the **Ultimate** series of theme anthologies. The first three – *The Ultimate Dracula* (anth 1991) with David Keller, Megan Miller and Martin H. GREENBERG (anon), *The Ultimate Frankenstein* (anth 1991) with John BETANCOURT, Keller and Miller and *The Ultimate Werewolf* (anth 1991) with Betancourt, Keller and Miller – comprise original stories assembled as 60-year anniversary celebrations of the original *Dracula* and *Frankenstein* movies ($ DRACULA MOVIES; FRANKENSTEIN MOVIES), and the 50th anniversary of *The Wolf Man* (1941). Most of the stories, unfortunately, eschew any connection with the movies or movie history. Other volumes in the sequence include *The Ultimate Dinosaur: Past*Present*Future* (anth 1992) with Robert SILVERBERG, *The Ultimate Zombie* (anth 1993) with Betancourt, *The Ultimate Witch* (anth 1993) with Betancourt, *The Ultimate Alien* (anth 1995) with Betancourt and Keith R.A.

DeCandido, and *The Ultimate Dragon* (anth 1995) with Betancourt and DeCandido. [JC]

PREMATURE BURIAL The fear of being accidentally buried alive is very real to some people. In some schizophrenics prone to catatonic fits the body can take on a *rigor mortis*-like state. It was a particular fear to Edgar Allan POE, who returned to the theme often, most notably in "The Premature Burial" (1844 *Dollar Newspaper*) and *The Fall of the House of Usher* (1839 *Gentleman's Magazine*; 1903 chap). The PLOT DEVICE became a convenience in sensational Gothic and Victorian fiction to explain apparent HAUNTINGS ($ RATIONALIZED FANTASY), REVENANTS or even VAMPIRES. The theme was used by several writers strongly influenced by Poe in their early works, especially M.P. SHIEL and H.P. LOVECRAFT. Poe's story leant itself to an extended movie adaptation as *Premature Burial* (1962), dir Roger CORMAN, scripted Charles BEAUMONT and Ray Russell (1924-   ); novelized as *Premature Burial* * (1962) by Max Hallan Danne. [MA]

PRERAPHAELITES The Preraphaelite/Pre-Raphaelite Brotherhood was founded informally by William Holman Hunt (1827-1910), John Everett MILLAIS and Dante Gabriel ROSSETTI c1849. Their intention was to signal their backwards-looking, medievalizing focus on painters prior to Raphael (1483-1520). By the early 16th century, they felt, European art had lost its youth, its sincerity, its high symbolic purpose, its clarity of intent. The works of Brotherhood members would reintroduce into the modern world – mainly through religious and sentimental portraits set against transcendentally glowing medieval LANDSCAPES – some of this lost purity. John RUSKIN agreed, and by 1852 or so Hunt, Millais, Rossetti and other like-minded artists became financially successful. With William MORRIS and Edward BURNE-JONES, Rossetti carried on with some form of the Brotherhood in Oxford.

The appeal of a doctrine which rejected the burgeoning and visibly unaesthetic 19th-century industrial world is clear. Its appeal to writers of fantasy – like Morris – is also clear; for a good Preraphaelite painting was like a PORTAL into an earlier, truer, cleaner, more meaningful world. [JC]

PRESCOT, DRAY Pseudonym of Kenneth BULMER.

PRESTER JOHN Legendary religious ruler of a Far Eastern Christian kingdom. Word of a distant Christian outpost began to filter back to the West from missionaries in the 12th century. A letter signed by PJ was supposed to have been received by the Holy Roman Emperor Frederick Barbarossa in 1165, though this has long been regarded as a hoax aimed at encouraging support for the Crusades by suggesting there would be help from the East. The Nestorian Christians who lived among the Mongols knew the ruler of this kingdom as Owanh-kohan, meaning Prince-Chief, but mistranslated it as John-Priest. PJ was supposedly defeated by Genghis Khan in 1202. The legend soon attached itself to other Christian myths, most especially that of the Holy GRAIL. PJ appears as the cousin of Lohengrin, and later as Guardian of the Grail in *Parzival* (?1200-1210) by Wolfram von Eschenbach (?1170-?1220) ($ PERCEVAL), a theme later embellished in *Der jüngere Titurel* (?1272) attributed to Albrecht von Scharfenberg (13th century). Although the Grail link with PJ has generally been forgotten, it was utilized Charles WILLIAMS in *War in Heaven* (1930). PJ's existence seemed to be authenticated by Marco Polo (1254-1324) in his wildly fabulated *Travels* (written 1298). Needless to say, that great fabricator of TRAVELLERS' TALES

John de MANDEVILLE (?   -1372) recounted his meeting with PJ in his own *Travels* (**1360**), making him a descendent of Ogier the Dane (◊ MORGAN LE FAY) and king of Teneduc in northern India. By the time PJ surfaced in *Orlando Furioso* (**1516**) by Lodovico ARIOSTO he had become the blind king of Ethiopia and the richest monarch in the world. It is in Ethiopia that John BUCHAN placed PJ in *Prester John* (**1910**). Frequent references by travellers to his continued existence made him seem immortal and some kind of accursed figure, like the WANDERING JEW. It is in this guise that Esther FRIESNER portrays him in *Yesterday We Saw Mermaids* (**1992**), where he appears as one of the Wise Men of the East who betrayed the birth of Christ to Herod and was thereafter cursed to rule a kingdom of magic beings forever. Friesner, rather like Jean MORRIS in *The Donkey's Crusade* (**1983**) and Peter S. BEAGLE in *The Folk of the Air* (**1986**), presents PJ's kingdom as a LAND OF FABLE weakened by THINNING and thus consigned to the world of MYTH. Norvell W. PAGE re-created PJ as a mighty hero and swordsman in *Flame Winds* (1939; **1967**) and its sequel *Sons of the Bear God* (1939; **1969**), where PJ (as Wan Tengri) is a gladiator who escapes from Alexandria and travels across Asia. Page, who wrote the first of these novels in a matter of weeks to meet a deadline, may have been encouraged to write about PJ by H. BEDFORD-JONES's use of the character in "The Singing Sands of Prester John" (1939), one of his **Trumpets from Oblivion** series. PJ has increasingly become an UNDERLIER. [MA]

PRICE, E(DGAR) HOFFMANN (1898-1988) US writer. He began to publish WEIRD FICTION – the genre for which he is remembered – with "Triangle with Variations" in *Droll Stories* in 1924. By the time he stopped writing for the pulp MAGAZINES in the 1950s he had published hundreds of stories, sometimes as Hamlin Daly and often drawing upon his Oriental and Near Eastern experiences for backgrounds. His best-known story from this period is probably "Through the Gates of the Silver Key" (*WT* 1934), with H.P. LOVECRAFT. Some of his fantastic fiction was later collected in *Strange Gateways* (coll **1967**) and *Far Lands, Other Days* (coll **1975**). In retirement EHP became annoyed at being remembered only as "one of the Lovecraft Circle", and in 1979 he resumed writing. In his final decade he wrote a Western, then two fantasies, *The Devil Wives of Li-Fong* (**1979**) and *The Jade Enchantress* (**1982**), both set in a capably described pre-modern China and describing the actions of the benign female SPIRITS who assist worthy men through various difficulties. EHP was an astrologer, a Theosophist, a practising Buddhist and a conservative Republican, and his late fantasies capably melded ideas from the first three fields. His final works were the **Operation** sequence, a progressively more chaotic 4-vol sf series. EHP may be remembered primarily for his vivid biographical sketches of his friends Robert E. HOWARD, Lovecraft and Clark Ashton SMITH. A volume of reminiscences and a late mystery remain unpublished. [RB]

PRIEST, CHRISTOPHER (McKENZIE) (1943-   ) UK writer whose work during the 1960s and 1970s was marketed as sf, although its sf metaphors were increasingly employed to subvert genre expectations. This subversion has since come to the fore. A significant turning point was *The Affirmation* (**1981**), a teasing FABULATION which plays disquietingly with issues of PERCEPTION and REALITY. Here the first-person narrator's record of "facts" soon proves shockingly untrue; his attempt to deal with his

unexceptional life by retelling it in fictionalized, metaphorical form becomes the tale of a man who must record his experiences in readiness for the AMNESIA which will be the price of IMMORTALITY – the two narrations, equally fictional, entwine in complex knots of STORY. *An Infinite Summer* (coll **1979**) includes stories sharing the **Dream Archipelago** setting featured in *The Affirmation*. *The Glamour* (**1984**; rev 1985 US; rev 1996) appears to be a CONTEMPORARY FANTASY featuring a kind of PARIAH ELITE whose literal, personal GLAMOUR confers not so much INVISIBILITY as unnoticeability. But this unseeable status carries metaphorical overtones of characters and events excised or edited from cinema, personal perception, memory, and life ... and from a story which ultimately revels in its own fictionality. Most recently *The Prestige* (**1995**) offers twin first-person narratives of Victorian stage MAGICIANS, bitter rivals, whose lives are both distorted and defined by the central deception in each of their trademark illusions. GASLIGHT ROMANCE shades into TECHNOFANTASY as one conjurer commissions Nikola Tesla (1856-1943) to duplicate and surpass his enemy's effect using not stage-magic but physics. DOUBLES, echoes and blurrings of identity already abound; a mishap with Tesla's device introduces a RATIONALIZED-FANTASY literalization of one rival's SHADOW. The modern-day FRAME STORY eventually brings the threads together on a note of GOTHIC FANTASY. This may be CP's finest novel; it won the James Tait Black Memorial Prize. [DRL]

See also: RECURSIVE FANTASY.

PRIESTLEY, J(OHN) B(OYNTON) (1894-1984) UK novelist and playwright whose once immensely popular novels have worn less well than his plays. Some of his work uses sf devices (◊ *SFE*); fantasy is not so much represented. In novels, the ABSURDIST movie universe in *Albert Comes Through* (**1933**) is relegated to RATIONALIZED FANTASY as a DREAM, and the genially humorous *The Thirty-First of June: A Tale of True Love, Enterprise and Progress, in the Arthurian and Ad-Atomic Ages* (**1961**) involves WIZARD-mediated PORTAL crossings between modern London and a medieval FANTASYLAND, the TRANSFORMATION of an ad exec into a DRAGON, and an ultimate CROSSHATCH linking of worlds with a view to future tourism. More seriously, JBP's interest in the TIME theories of J.W. Dunne (1875-1949) and P.D. Ouspensky (1878-1947) led to such "time plays" as: *Dangerous Corner* (**1932**), offering a timeline where opening a metaphorical PANDORA'S BOX produces a cascade of increasingly destructive revelations, and then an ALTERNATE REALITY where the dangerous sequence is averted by chance; *Time and the Conways* (**1937**), whose middle act is one character's transforming, Dunne-style vision of her imperfect future; and *I Have Been Here Before* (**1937**), pitting free will against Ouspensky's cycle of eternal return. *Two Time-Plays* (omni **1937**) and *Three Time Plays* (omni **1947**) assemble, respectively, the latter two and all three. *Johnson over Jordan* (**1939**) is a less successful AFTERLIFE fantasy whose late hero is questioned by spiritual examiners and relives parts of his life before leaving for regions unknown. [DRL]

Further reading: *The Amazing Theatre* (coll **1939**) by James Agate (1877-1947) reviews early performances of most cited plays.

PRINCESS AND THE GOBLIN, THE UK/Hungarian ANIMATED MOVIE (*1992*). Siriol/Pannonia/S4C/NHX. **Pr** Robin Lyons. **Exec pr** Marietta Dárdai, Steve Walsh. **Dir**

József Gémes **Spfx** Zsuzsanna Bulyiki, Piroska Martsa. **Anim dir** Les Orton. **Screenplay** Lyons. **Based on** *The Princess and the Goblin* (*1872*) by George MACDONALD. **Voice actors** Joss Ackland (King), Claire Bloom, Roy Kinnear, Paul Keating (Curdie singing), Sally Ann Marsh, Rik Mayall (Prince Froglip), Peggy Mount, Peter Murray, Victor Spinetti, Mollie Sugden (movie lacks proper credits). 111 minutes. Colour.

Led by their strident Queen, flu-ridden King and sadistic Prince Froglip, the GOBLINS plan to conquer the world of the Sun People (i.e., humanity) and enslave them UNDERGROUND; Froglip also hopes forcibly to marry Princess Irene so that, as heir to the human throne, he will have legal dominion over the Sun People. This plan is thwarted by miner's son Curdie and Irene, who was given by her "not-quite-a-GHOST" great-great-grandmother a ring and a ball of invisibly thin thread with the instruction that, if ever in trouble, she should follow the thread to "find your *own* MAGIC".

This was a conscious venture into DISNEY territory, and as such failed: while there is some pleasing work (notably in a final flood sequence, and in the SHAPESHIFTING of Irene's great-great-grandmother), it is lost in much that is clumsy. [JG]

PRINCESS BRIDE, THE US movie (*1987*). Act III Communications. **Pr** Rob Reiner, Andrew Scheinman. **Exec pr** Norman Lear. **Dir** Reiner. **Spfx** Nick Allder. **Animatronics** Rodger Shaw. **Swordmaster** Bob Anderson. **Screenplay** William GOLDMAN. **Based on** *The Princess Bride* (*1973*) by Goldman. **Starring** Andre the Giant (Fezzik), Billy Crystal (Miracle Max), Cary Elwes (Westley), Peter Falk (the Grandfather), Christopher Guest (Count Rugen), Mandy Patinkin (Inigo Montoya), Fred Savage (Grandson), Chris Sarandon (Prince Humperdinck), Wallace Shawn (Vizzini), Robin Wright (Buttercup). 98 mins. Colour.

In the FRAME STORY an ill boy's grandfather reads him *The Princess Bride*, full of swashbuckling swordplay and derring-do.

Farm lad Westley and sweet virginal Buttercup are in love. He goes off to seek fortune and is soon reported dead at the hands of Dread Pirate Roberts; she becomes engaged to oily Prince Humperdinck. Out riding, she is seized by DWARF Vizzini, swordsman Inigo and giant Fezzik; her murder is to be blamed on the neighbouring land of Guilder, so that Humperdinck may declare war. But Pirate Roberts – in fact, Westley, latest successor to the title – pursues, bests Inigo and Fezzik and tricks Vizzini into killing himself. Plot abounds until the goodies triumph.

There is plenty of STORY in *TPB* – by design. Goldman initially wrote his book as "just the good bits" extracted from a fictional BOOK, «S. Morgenstern's Classic Tale of True Love and High Adventure» – all padding excised. The structure whereby a fantasy tale is embedded in a reading of that tale is strongly reminiscent of *The* NEVERENDING STORY (*1984*); however, Goldman's book predated Michael ENDE's far more complex text by several years. *TPB*'s rejoicing use of cliché to drive its tale may have less lofty purpose than *The Neverending Story*'s, but it is difficult not to regard it as the better fantasy movie. [JG]

PRINCE VALIANT US movie (*1954*). 20th Century-Fox. **Pr** Robert L. Jacks. **Dir** Henry Hathaway. **Spfx** Ray Kellogg. **Mufx** Ben Nye. **Screenplay** Dudley Nichols. **Based on** the King Features comic strips by Harold R. FOSTER. **Starring** Brian Aherne (Arthur), Primo Carnera (Sligon), Sterling Hayden (Gawain), Janet Leigh (Aleta), Victor McLaglen (Boltar), James Mason (Sir Brack), Debra Paget (Ilene), Robert Wagner (Valiant). 100 mins. Colour.

The exiled Christian king of Scandia sends his son, Valiant, to become a KNIGHT of the ROUND TABLE, but ARTHUR explains knighthood is a higher honour than princehood, and must be earned. Apprenticed as squire to GAWAIN, Valiant soon suspects suave Sir Brack, bastard alternative claimant to the British throne – of being in fact the Black Knight, the traitor treating with the Scandian usurper Sligon – but can prove nothing. A tedious loveknot develops between Gawain, Brack and Valiant and the princesses Aleta and Ilene. Betrayed by Brack to the Vikings, Valiant – with Aleta – is transported to Sligon's castle in Scandia, where his parents are now also captive. Escaping, Valiant amid much swashbuckling aids the rebel Christian Vikings depose Sligon, whom Valiant kills with his father's sword. Back at CAMELOT, Valiant denounces Brack to Arthur and then kills the traitor in single combat, winning Aleta's hand and his own knighthood.

The plotting of this cod-Arthurian romance betrays its COMIC-strip origins, yet the tale is given the full Hollywood "epic" treatment. [JG]

PRINGLE, DAVID (WILLIAM) (1950-   ) Scottish editor and writer, long resident in England, who has mostly worked in sf (◊ *SFE*). He was editor of *Foundation* 1980-86, and with Malcolm Edwards (1949-   ) was one of the two prime founders of the sf magazine *Interzone*. By 1988 he had become its sole editor and publisher; he received a 1995 Hugo AWARD for the journal, and co-edited all 5 anthologies taken from the magazine: *Interzone: The First Anthology* (anth **1985**) with John CLUTE and Colin GREENLAND, *Interzone: The 2nd Anthology* (anth **1987**) with Clute and Simon Ounsley, *Interzone: The 3rd Anthology* (anth **1988**) with Clute and Ounsley, *Interzone: The 4th Anthology* (anth **1989**) with Clute and Ounsley, and *Interzone: The 5th Anthology* (anth **1991**) with Clute and Lee Montgomerie. As series editor for GW Books 1988-91 he commissioned (from 1990 with Neil Jones) several SHARED-WORLD fantasy and sf novels tied to GAMES like **Warhammer** and **Dark Future**; tied anthologies ed DP include *Ignorant Armies* * (anth **1989**), *Wolf Riders* * (anth **1989**) and *Red Thirst* * (anth **1990**) in the **Warhammer** series, *Route 666* * (anth **1990**) in the **Dark Future** series, and *Deathwing* * (anth **1990**) with Jones in the **Warhammer 40,000** series.

As a critic, DP was initially associated with J.G. BALLARD, whose works he promoted vigorously; the contents of Ballard's *A User's Guide to the Millennium: Essays and Reviews* (coll **1996**) were selected by DP. Critical studies include *J.G. Ballard: The First Twenty Years* (anth **1976** chap) ed with James Goddard, *Earth is the Alien Planet: J.G. Ballard's Four-Dimensional Nightmare* (**1979** chap US) and *J.G. Ballard: A Primary and Secondary Bibliography* (**1984** US). Subsequently, DP began to produce guides to various aspects of sf, fantasy and popular literature: *Science Fiction: 100 SF Authors* (**1978** chap), *Science Fiction: The 100 Best Novels: An English-Language Selection, 1949-1984* (**1985**), *Imaginary People: A Who's Who of Modern Fictional Characters* (**1987**; rev 1989; rev 1996); *Modern Fantasy: The Hundred Best Novels: An English-Language Selection, 1946-1987* (**1988**), *The Ultimate Guide to Science Fiction: An A-Z of SF Books* (**1990**; exp vt *The Ultimate Guide to Science Fiction, Second Edition: An A-Z of Science-Fiction Books by Title* 1995) with Ken Brown (uncredited), *St James Guide to Fantasy Writers* (**1996**) and *The Ultimate Encyclopedia of Science*

Fiction (**1996**). He also edited a posthumous collection of Theodore STURGEON stories, *A Touch of Sturgeon* (coll **1987** UK). He has contributed several entries to the first edition of the *SFE* and to this encyclopedia. [JC]

PRIZE GHOST STORIES ◊ GHOST STORIES (magazine).

PROENÇA, M. CAVALCANTI (1905-1996) Brazilian writer. ◊ BRAZIL.

PROHIBITIONS Judeo-Christian mythology has the transgression of a prohibition at its core – ADAM AND EVE disobey God's command and eat the fruit; Greek and Scandinavian mythology include transgression of prohibitions in their explanations of how evil came into the world; the theme occurs elsewhere. The idea that evil and pain are the consequences of some forbidden action is common enough that it perhaps suits a deep-seated human need; or it may simply be that the writers of wisdom literature are as committed as anyone else to the telling of STORY, one of the most fertile generators of story being the transgression of a CONDITION or a taboo. The breaking of a prohibition usually entails punishment; this punishment may be remitted, as in Christian doctrine. More often, the punishment will be rewritten in the course of the story as an ordeal or NIGHT JOURNEY; Psyche (◊ CUPID AND PSYCHE) looks at her sleeping lover despite his instructions, discovers him to be Cupid, and is cast out into the wilderness.

The consequence of breaking a prohibition often depends on the legitimacy of the authority of the person doing the forbidding. Bluebeard forbids his wife to open one particular door, relying on his authority as her husband; when she opens the door she discovers she was right to disobey him, since he was a murderer and thus had no rightful authority. In those versions of the legend in which she is rescued, the presumption remains that earlier wives who disobeyed him were not. It is only in such revisionist versions of the legend (◊ REVISIONIST FANTASY) as Paul Dukas's OPERA *Ariane et Barbe-Bleu* (**1907**) that her transgression has even the potential of saving anyone save herself.

The transgression of prohibitions is often closely linked to such themes as QUIBBLES, and the FOOL. In fantasy, as opposed to FAIRYTALE or LEGEND, the question is often one of knowing what the prohibition was in the first place and knowing how exactly one can be deemed to have transgressed it. The awakening of MALIGN SLEEPERS or other wanderings into prohibited realms is rarely announced by an explicit prohibition; normally the transgression lies in imprudence. Sometimes, on the other hand, the prohibition is entirely explicit and entirely unfair. The Endless, in Neil GAIMAN's **Sandman** graphic novels and particularly in *The Kindly Ones* (graph coll **1996**), are prohibited from directly harming each other or their kin; Dream is driven to his destruction by the Furies for finally taking the life of his decapitated yet living son Orpheus in a mercy killing. [RK]

PROMETHEUS In Greek MYTHOLOGY, the Titan who became the altruistic champion of humankind, first persuading ZEUS to accept lesser sacrifices and then stealing the fire of the GODS for human use; as punishment he was chained to a crag and an eagle was sent forth every day to peck out his constantly regenerated liver. He was also rendered in some accounts as the creator of humankind, moulding the first members of the race out of clay; it is for this reason that Mary SHELLEY's *Frankenstein* (**1818**) is subtitled "The Modern Prometheus". Given the obvious symbolism of his name (which means "forethought", while that of his dimmer brother Epimetheus signifies

"afterthought") it is not surprising that the "fire" which Prometheus stole has often been construed metaphorically, nor that he remains an important symbol in the vocabulary of modern fantasy. Percy Bysshe SHELLEY employed him in *Prometheus Unbound* (**1820**), and he is the central figure of John STERLING's "Cydon" (1829), Richard GARNETT's "The Twilight of the Gods" (1888) and Karel ČAPEK's "The Punishment of Prometheus" (in *Apocryphal Stories* coll **1943**). John UPDIKE's ALLEGORY *The Centaur* (**1963**) refers to a myth which appointed a wise centaur as Prometheus's tutor. Max BEERBOHM's "The Case of Prometheus" (1899) is far less reverent, but "Prometheus" (1983) by Alasdair Gray (1934-), whose protagonist is writing a new *Prometheus Unbound*, is more slyly ambivalent. [BS]

PROPHECY Prediction and prophecy are always of special significance in fantasy: if a foretelling is mentioned at all, it is almost certain to be accurate. Delphic ambiguity, QUIBBLES or mere obscurity of DICTION may cloud the issue, but a prophecy, OMEN or PORTENT will never simply be wrong or apply to a different LAND, HERO or century. It is a glimpse of perhaps inescapable FATE. Oedipus does not know of and thus cannot avoid his prophesied crimes of parricide and incest. But better-informed attempts to cheat Fate are frequently self-defeating, as with the story of the man who saw DEATH looking oddly at him, and fled to far-off Samarra (or Persepolis) – there to discover that Death had been surprised to see him in the other city when his next day's appointment with that man was in Samarra. Piers ANTHONY's **Apprentice Adept** sequence turns on a self-fulfilling prophecy that one sorcerer, the Blue Adept, will destroy the Red Adept; which Blue eventually does, solely in retaliation for Red's attempts to kill him and thus evade Fate.

Prophetic pronouncements may come from ORACLES, as in the Anthony story just cited; in gnomic verses, RIDDLES and FOLKLORE handed down since ancient times; in BOOKS of prophecy like that of Neil GAIMAN's and Terry PRATCHETT's *Good Omens* (**1990**); in DREAMS or VISIONS – often thought to be sent by the GODS; by SCRYING; by the casting of RUNES; by haruspication, the study of the entrails of a newly slaughtered SACRIFICE – chiefly practised by very ancient or EVIL prophets (it becomes black comedy in Roger ZELAZNY's *Creatures of Light and Darkness* [**1969**], where the disembowelee is a rival seer with his own interpretation of his entrails); by NECROMANCY; by use of clairvoyant or precognitive TALENT; by the study of STARS and horoscopes (◊ ASTROLOGY); by auspices and augurs; by TAROT, other CARDS, or the yarrow-stalks of the I Ching; by sortilege, opening a book at random and interpreting a line or passage as prophecy – as memorably done with the BIBLE in M.R. JAMES's "The Ash-Tree" (1904); or even by reading tea-leaves. NUMEROLOGY is very rarely employed, perhaps because it lacks numinous resonance and is recognized as too easily manipulable.

To preserve both free will and narrative suspense, prophecies tend to become wholly clear only in hindsight. The gods may warn by dropping hints, but they do not compel; the author reserves the right to be untrustworthy and invoke some QUIBBLE, so the coming-about of even a straightforward prediction can still be effective. There is much STORY satisfaction in threads of pleasingly fulfilled prophecy, such as are profusely woven throughout J.R.R. TOLKIEN's *The Lord of the Rings* (**1954-5**); these include prophecies turning on CONDITIONS, like the warning to

Legolas against approaching the sea. Fred SABERHAGEN's *The Broken Lands* (**1968**) effectively understates the detailed working-out of an initial prophecy in the climactic action, flattering readers who have traced the thread without heavy auctorial nudging. [CB/DRL]

See also: MERLIN; NOSTRADAMUS.

PROSE, FRANCINE (1947-   ) US writer. Her first novel, *Judah the Pious* (**1973**), is a magical QUEST tale which an elderly rabbi tells to the young king of Poland. *The Glorious Ones* (**1974**) dramatizes the fortunes of a 17th-century Italian COMMEDIA DELL'ARTE troupe who inhabit a world steeped in FATE, CURSES and miraculous events. As in nearly all of FP's work, these novels' fantastic nature owes more to their emphasis on theatricality, the creation of fictions and human receptivity to the otherworldly than to the question of whether supernatural events actually take place. *Marie Laveau* (**1977**) and *Animal Magnetism* (**1978**), both set in the 19th century, deal with subjects – respectively the famous VOODOO worker and MESMERISM – that inhabit both history and the literature of fantasy.

Household Saints (**1981**) concerns a post-WWII Italian-US family whose daughter abjures worldliness – and eventually the world – in her devotion to Saint Thérèse (1873-1897). *Hungry Hearts* (**1983**), set among Yiddish theatre in 1920s New York, resembles *The Glorious Ones* in its characters' tendency to interpret astonishing events in terms of the fantastic. *Bigfoot Dreams* (**1986**) hilariously limns the life of a staff writer for a supermarket tabloid whose stories seem to foretell actual events. *Primitive People* (**1992**) and *Hunters and Gatherers* (**1995**) are contemporary SATIRES.

FP's ironic but impassioned voice, her spare lyricism and her considerable HUMOUR have won her critical acclaim. [GF]

Other works: *Women and Children First* (coll **1988**); *The Peaceable Kingdom* (coll **1993**).

PROSPERO Duke of Milan who neglected his duties for his studies in OCCULTISM, and was deposed by his brother. So we understand from SHAKESPEARE's *The Tempest* (performed, *c*1611; **1623**) in which Prospero has been argued to represent the alchemist/mage John DEE. As mage, Prospero rules his island kingdom with CALIBAN and ARIEL and his daughter Miranda (◊ VIRGINITY) as subjects. Attracting his foes, he manipulates them towards a moment of revenge but at the last moment renounces his power, forgives his enemies and takes his place in the world.

Prospero is both type and example of MAGUS, exerting power through GLAMOUR and the control of supercelestial powers, and using threat and persuasion to cajole his subjects. He can be seen in John FOWLES's *The Magus* (**1966**) and Elizabeth WILLEY's *A Sorcerer and a Gentleman* (**1995**), and faintly in the many WIZARDS of HEROIC FANTASY. As emblem of the writer, he is an interesting foreshadow of metafiction, questioning the nature of persuasion through words. His power is that of the Grand Designer; he discourses – sometimes garrulously – but he ends by violently rejecting the power of "his book". It is this element which is focused on in Peter GREENAWAY's PROSPERO'S BOOKS (*1991*), which ironically is as much a paean of the image as it is of the written word. [AS]

PROSPERO'S BOOKS UK/French movie (*1991*). Allarts-Cinéa/Camera One-Penta/Canal+/Elsevier Vendex/Film Four/NHK/VPRO Television. **Pr** Kees Kasander. **Exec pr** Kasander, Denis Wigman. **Dir** Peter GREENAWAY. **Mufx**

Sjoerd Didden. **Screenplay** Greenaway, published with additional material as *Prospero's Books: A Film of Shakespeare's The Tempest* * (**1991**). **Based loosely on** *The Tempest* (performed *c*1611; **1623**) by William SHAKESPEARE. **Starring** Tom Bell (Antonio), Michael Clark (Caliban), John Gielgud (Prospero), Orphéo (Ariel), Isabelle Pasco (Miranda), Paul Russell (Ariel), Mark Rylance (Ferdinand), James Thierree (Ariel), Emil Wolk (Ariel). 120 mins. Colour.

This astonishing movie's base plot is that of *The Tempest*, but almost infinitely elaborated, convoluted and distorted, to the extent that any summary would be, like the movie itself, almost incomprehensible to those without a reasonable degree of familiarity with the play. It is strange that a movie which has storytelling as a primary preoccupation should itself not attempt to tell a STORY.

In jumbled TIME, PROSPERO writes and tells (i.e., creates) the events while himself being, along with a fourfold Ariel, their major participant; almost without exception he speaks the parts of every character, the spoken lines usually being underlain by the character's own voice as if events have been cast into an ALTERNATE REALITY that cán be experienced only via his sophisticated, magic-skewed PERCEPTION (or as if, possibly, they are entirely products of his [mis]perception). That said, Prospero is not the movie's central character: this role is reserved for his collection of BOOKS. These are lost, apocryphal or fictional, with titles like *An Atlas Belonging to Orpheus* (which MAPS, of course, HELL). They are also magical, although not grimoires. Prospero believes his powers derive from them, and his renunciation of MAGIC at the movie's end is allegorized by a shutting and then a destruction of them; but we, looking in on Prospero from outside (and, the movie hints, his jailers, for he is a part of the Story we all have in our mental/cultural handbaggage), can see that the books *are* the powers: the powers are not *his*, but *theirs*. They, not he, are the MAGUS.

Visually *PB* is numbing, overpowering. Exploiting to the full both the new techniques made available by advances in video technology and his own almost obsessive attention to detail, Greenaway packs the screen with more images, almost, than it can be expected to hold: in a sort of visual cacophony, images and scenes are piled one over the other, superimposed on or jutting into each other often three at a time, intermingling with or smothering each other. There is a richness of visual quotes: often one is arrested by discovering that suddenly there has formed on screen a representation of some noted painting (◊ FANTASY ART) – particularly from the Flemish School and the works of Sir Lawrence Alma-Tadema (1836-1912). Frequently scenes themselves are presented to us framed as PICTURES or as mirror reflections. In fact, MIRRORS are another key preoccupation: one of Prospero's books is *A Book of Mirrors*, and it is frequently implied that all the hosts of ELEMENTALS are mirrors of the human mind; Caliban in particular is a reflection of Prospero, for the MONSTER does not even mouth his lines but instead dances his role, his words being rendered by his source, Prospero. And, in the midst of all the dazzling visual effects, much of what lies at the heart of the movie is in the form of a *stage* production.

The result of this near-chaos is a deeply magical movie – that is, a movie concerned at its core with the portrayal of MAGIC. Watching *PB* is a bewildering, confusing, almost shocking experience ... which is exactly what we might expect an encounter with magic to be.

And then there is the matter of the final book, the 23rd,

the only one that refuses to be destroyed in Prospero's final renunciation. It is *A Book of Thirty-Five Plays* by Shakespeare, with space left for the 36th play, *The Tempest*, which Prospero at the last supplies. [JG]

PSI POWERS ◊ TALENTS.

PSYCHIC DETECTIVES ◊ OCCULT DETECTIVES.

PSYCHOLOGICAL THRILLER A subgenre of crime fiction which lurks somewhere in the infrared or ultraviolet of the spectrum of FANTASTIC literature, the PT concentrates on the disturbed (and usually but not always criminal) psyches of otherwise normal human beings, and the events that occur because of that disturbance. Many PTs do cross the line to become either fantasy or SUPERNATURAL FICTION, in the form of what could be called the Parapsychological Thriller; many RATIONALIZED FANTASIES are PTs, as are many GHOST STORIES and tales of SERIAL KILLERS; the distinction between some forms of HORROR and the PT is difficult to make (an example is Stephen KING's *Misery* [1987]); and the PT shares with the mainstream of FANTASY the urge to alter the reader's PERCEPTION, either because we see events and motivations via insanity or, if identifying with someone other than the mad person, find ourselves in a world of uncertain and shifting REALITIES because of the WRONGNESS of an environment in which there is, as it were, a wild card.

The PTs of most interest in the fantasy context are those which rest upon an ambiguity of perception. Wilkie COLLINS wrote several, though tended to provide a culminating rationalization which often destroyed the effect; but the classic example is "The Turn of the Screw" (1898) by Henry JAMES – filmed as *The* INNOCENTS (*1961*; vt *The Turn of the Screw*) – in which the truth of events is kept hidden from us. A fine 20th-century example is *The Other* (1971) by Thomas TRYON – filmed as *The Other* (*1972*) – in which we become only slowly aware that the narrating "good" twin may be guilty of the crimes he is blaming on his "bad", deceased brother, if that brother ever existed as other than an INVISIBLE COMPANION. In William GOLDMAN's *Magic* (1978) – filmed as MAGIC (*1978*) – it may be that a ventriloquist's dummy is EVIL or that the ventriloquist himself is crazy. Mark McShane's *Séance on a Wet Afternoon* (1961; vt *Séance* 1962 US) – filmed as *Séance on a Wet Afternoon* (*1964*) – depicts the mental disintegration of a fraudulent medium: convinced she has contacted her dead son, she becomes sure that her powers are real ... as, for all we know, they may be. Even comedies can be strongly affected by PT themes, as in *Topper Returns* (*1941*; ◊ TOPPER MOVIES) and the numerous rationalized HAUNTED-DWELLING movies epitomized by the various versions of *The Cat and the Canary* (*1927*, *1939*, *1979*, plus *Haunted Honeymoon 1986*). The list of primarily fantasy authors who have also published PTs is extensive. [JG]

See also: Virginia C. ANDREWS; *The* BEAST WITH FIVE FINGERS (*1946*); BEWITCHED (*1945*); Ramsey CAMPBELL; CANDYMAN; *The* CURSE OF THE CAT PEOPLE (*1944*); *The* HANDS OF ORLAC (*1960*); William F. HARVEY; David H. KELLER; Ira LEVIN; Edgar Allan POE; *The* WICKER MAN (*1973*).

PUCK A hobgoblin or nature spirit, called the pooka or phouka by the Irish and pwca by the Welsh, equated by the Scottish to the brownie (◊ ELVES) because of a helpful if mischievous, TRICKSTER nature. The earlier usage of the name emphasized the evil nature of the spirit (the name was at one time synonymous with the DEVIL). SHAKESPEARE used the word as a personal name in *A Midsummer Night's Dream* (performed 1596; **1600**), equating it with ROBIN GOODFELLOW, so that the two have since become synonymous, with Goodfellow's adventures grafted on to Puck's. Puck thus reappears in a number of stories which recreate Shakespeare's world, in particular *Land of Unreason* (1941 *Unknown*; **1942**) by Fletcher PRATT and L. Sprague DE CAMP and *A Midsummer Tempest* (**1974**) by Poul ANDERSON. Rudyard KIPLING uses the play as a medium to introduce Puck to two children in *Puck of Pook's Hill* (coll of linked stories **1906**) and its sequel *Rewards and Fairies* (coll of linked stories **1910**). Eleanor FARJEON uses a similar device in her books featuring **Martin Pippin**, starting with *Martin Pippin in the Apple-Orchard* (coll of linked stories **1921**).

Puck is something of a JACK-like figure and may be connected with Jack-in-the-Green, a trickster who is one manifestation of the GREEN MAN. Charles DE LINT introduces into *Moonheart* (**1984**) and the stories in *Spiritwalk* (coll **1992**) the spirit Pukwudji, who resembles a Native American Puck. [MA]

Further reading: *The Anatomy of Puck: An Examination of Fairy Beliefs Among Shakespeare's Contemporaries and Successors* (**1959**) by K.M. BRIGGS.

PULCINELLA ◊ COMMEDIA DELL'ARTE; GRAND GUIGNOL; PUPPETS.

PULLMAN, PHILIP (NICHOLAS) (1946-) UK writer, most significantly of YA fiction, though his first novel, *Galatea* (**1978**), was for adults, and his second – *Count Karlstein, or The Ride of the Demon Huntsman* (**1982**), which features a WILD HUNT conducted by a DEMON – was for children. His YA work tends to straddle genres; for instance, his sequence comprising *The Ruby in the Smoke* (**1985**), *The Shadow in the Plate* (**1987**; vt *The Shadow in the North* 1988 US) and *The Tiger in the Well* (**1990** US), set in a 19th-century LONDON, is told overall as a GASLIGHT ROMANCE, but the first volume is predominantly HORROR, the second could be considered STEAMPUNK, and the third is "realistic", though its villain is named Tzaddik, a figure out of the CABBALA. *Spring-Heeled Jack* (**1989**) treats SPRING-HEELED JACK as a kind of ROBIN HOOD who uses his TALENTS to help the poor.

The first volume of **His Dark Materials** – *Northern Lights* (**1995**; vt *The Golden Compass* 1996 US) – has been well received, gaining a 1996 Carnegie Award, the most prestigious UK award for children's literature. The projected trilogy, intended to reflect elements of John MILTON's *Paradise Lost* (**1667**), begins in an ALTERNATE-WORLD version of the 20th-century UK, marked by a differing physics and the fact that everyone has a COMPANION who seems to be a kind of DEMON. As each child passes through the RITE OF PASSAGE into adulthood, her/his demon, which has previously shapeshifted (◊ SHAPESHIFTERS) at will, freezes into a shape that MIRRORS its human's mature selfhood. The first volume, a complicated plot-driven QUEST, fuels expectations of a significant full tale. [JC]

Other works: *Frankenstein* (**1990**), a play adapted from Mary SHELLEY; *Sherlock Holmes and the Limehouse Horror* (**1992**); *The New Cut Gang: Thunderbolt's Waxwork* (**1994**); *The Tin Princess* (**1994**); *The Wonderful Story of Aladdin and the Enchanted Lamp* (**1994** chap).

PULPS A term used in two broad senses.

1. In 1896, US publisher Frank A Munsey (1854-1925) decided that his magazine *The Argosy* would henceforth print only fiction, for a wide audience. He decided to use the newly available cheap pulp paper, derived from wood pulp in a process invented by Friedrick Gottlob Keller (1816-1895) in 1844 but developed fully only towards the end of the century. *The Argosy* became the first pulp magazine, or "pulp". Between 1900 and WWII most genre-fiction magazines were pulps.

Pulp paper is thick, porous and friable, and smells; pulps tend to be in large format (usually *c*10in × 7in [25cm × 18cm]), with uncut edges, and they turn yellow within months. Their physical nature precluded their easy acceptance in the tidy US middle-class households in the years preceding 1939.

2. By association, pulp fiction is normally thought of as being lowbrow, action-oriented, pacy, brutal, sexist and ephemeral. There is some truth in this; but not much, and the use of the term to describe a particular type of fiction, regardless of the format of its original publication, has become less and less efficacious.

Pulp magazines are discussed at length in the article MAGAZINES. [JC]

PUNCH AND JUDY ◊ PUPPETS.

PUPPETS A particular subset of DOLLS which, controlled by inserted hands (glove puppets) or supporting strings or rods (marionettes), have been used for theatrical performances since antiquity. The classic UK puppet play is *Punch and Judy*, whose ANTIHERO is derived from the COMMEDIA DELL'ARTE's Pulcinella – adopted for English puppet plays *c*1662 and soon anglicized as Punchinello, then Punch. In this comic GRAND-GUIGNOL story the brutal TRICKSTER Punch gleefully kills off his infant child, his wife Judy and various other characters, including the DEVIL. The Punch and Judy showman in John MASEFIELD's *The Box of Delights* (**1935**) proves to be a disguised WIZARD; Neil GAIMAN's and Dave MCKEAN's *The Tragical Comedy or Comical Tragedy of Mr Punch* (graph **1994**) knowingly exploits the show's sinister aspects.

Puppets in fantasy are likely, like STATUES, to come alive; their BONDAGE is particularly oppressive since they are made to enact STORIES devised by others. Malign ventriloquist's dummies are notorious (◊ VENTRILOQUISM). Carlo COLLODI's *The Adventures of Pinocchio* (**1883**), adapted by DISNEY as PINOCCHIO (**1940**), sees wooden Pinocchio finally escape the bondage of puppethood to become a REAL BOY. G.K. CHESTERTON's play *The Surprise* (**1952**) contains an inset marionette-play which goes haywire when the puppets come alive, presenting in an ALLEGORY of GOD the Author's dismay at human misuse of free will. The situation is partly reversed in Diana Wynne JONES's *The Magicians of Caprona* (**1980**), where two children are magically shrunk to puppet-size (◊◊ GREAT AND SMALL) and compelled to play the parts of Punch and Judy. Severian's DREAM of fighting as a puppet in Gene WOLFE's *The Book of the New Sun* (**1980-83**) not only prefigures a later duel with a GIANT but hints at his life's manipulation by higher beings. *The Stress of Her Regard* (**1989**) by Tim POWERS has a grisly scene in which Percy Bysshe SHELLEY is compelled to use his dead daughter's body as a marionette. To become a puppet – as in popular metaphor – is generally DEBASEMENT. [DRL]

See also: Jim HENSON; *The* MUPPET SHOW; Jan SVANKMAJER.

PURCELL, HENRY (1659-1695) English composer. ◊ Michael AYRTON; OPERA.

PURGATORY Place of AFTERLIFE expiation where, according to Catholic doctrine, sinners are purged by a period of disciplinary suffering before admission to HEAVEN. It is often assumed as a "given" in CHRISTIAN FANTASY. DANTE's *Purgatorio* (◊◊ TAPROOT TEXTS) has the classic depiction: a conical mountain with successively higher circular ledges where various degrees of "refining fire" are joyfully submitted to – the 1955 Dorothy L. Sayers (1893-1957) translation adds useful MAPS. Larry NIVEN and Jerry Pournelle (1933-   ) annexe HELL as the "violent ward" of Purgatory in *Inferno* (**1976**), allowing even the damned a slim chance of Heaven. C.S. LEWIS's *The Great Divorce* (**1946**) considers a similar opportunity but focuses on purgatory as a transition rather than a place – a stripping-away of comforting spiritual baggage like selfishness or excessive sexual desire. Piers ANTHONY's **Incarnations of Immortality** sequence fuses purgatory with LIMBO as a painless but dull region where the celestial civil service toils eternally (◊ AS ABOVE, SO BELOW). [DRL]

PURPLE ROSE OF CAIRO, THE US movie (**1984**). Orion/Jack Rollins-Charles H. Joffe. **Pr** Robert Greenhut. **Exec pr** Joffe. **Dir** Woody Allen. **Screenplay** Allen. **Starring** Jeff Daniels (Tom Baxter/Gil Shepherd), Mia Farrow (Cecilia). 82 mins. Colour.

New Jersey in the Depression. Cecilia is sacked from her job and takes solace at the cinema, watching light romance *The Purple Rose of Cairo* over and over until she catches the eye of the movie's young romantic lead Baxter, quixotic archaeologist (a sort of Noel Coward version of INDIANA JONES). Falling in love with Cecilia, he climbs out of the movie REALITY and takes her away for what he promises will be endless romance and adventure, leaving the rest of the cast stuck squabbling behind the screen. But things cannot work out that easily: Baxter has no knowledge of the world beyond what has been written into his character; e.g., the prospect of serious lovemaking sans fade-out terrifies him. (Oddly, he seems unaffected by the transition from b/w to full colour.) Meanwhile, wherever the movie is showing, Baxters are forgetting their lines and attempting to leave the screen; producer Raoul Hirsch and the actor who played Baxter, Gil Shepherd, come to town to right the situation. Shepherd woos Cecilia, promising her a life with him in Hollywood, and at last she must choose between the idealized fictional character and the real human being. She chooses reality over fantasy, the human being over the romanticized hero ... and Shepherd, with Baxter gone and his problem solved, promptly ditches her.

TPROC is in many ways a slight work, although finally very moving; it functions better as a fantasy than as a comedy. Interesting is the fact that Allen declines to make this a RATIONALIZED FANTASY: there is no suggestion that the events might be merely Cecilia's escapist wish-fulfilment dreams. Given the option between making her life fictional and settling "sensibly" for the real, she makes the wrong choice – as she seems to realize, returning immediately to the cinema to involve herself in this week's new movie, presumably hoping for a second chance. Yet this juxtaposition is in itself simplistic, for Shepherd's Hollywood promises were surely always, like Baxter's wide-eyed plans, lures towards an impossible fantasy. The chilling message of *TPROC* is thus that, for the oppressed Cecilias of the world, there is no viable form of ESCAPISM on offer. [JG]

PUSS-IN-BOOTS One of the best-known BEAST-FABLES about an animal helping a human. The modern version comes from "Le Maître Chat ou le Chat Botté" ["The Master Cat or the Booted Cat"] (1697) by Charles PERRAULT, but did not originate with him, probably having an

Arabian source. An early Indian version has been translated as "The Monkey and Mr Janel Sinna" in *Three Tales of Monkey* (coll **1967** US) by Ruth Tooze (1892-1972) (◊◊ MONKEY). The tale was first written down in Europe in volume 2 of *Le Piacevoli Notti* ["The Delectable Nights"] (coll **1553**) by Gianfrancesco Straparola (?1480-1558), and was repeated with little variance as "Gagliuso" in *Lo Cunto de li Cunti* ["The Story of Stories"] (**1634**; vt *The Pentameron* 1674) by Giambattista Basile (1575-1632). Only in Perrault's story does the CAT wear boots, but otherwise the storyline is almost identical. A miller dies and his sons inherit in turn the mill, his ass and his cat. The youngest son believes he is most hard done-by with only the cat, but the cat through TRICKSTER skills, elevates him such that he marries the princess of the land. In Basile's version alone we learn that the Master is so pleased that he promises the cat that after it dies it will be preserved in a golden cage. To test this the cat feigns death and hears the master order that the cat be thrown from the window. At this the cat confronts the Master with his ingratitude and leaves.

The story is almost completely amoral, suggesting that fortune can be achieved through deceit and trickery – almost certainly how the general populace regarded the gentry in the Middle Ages. This made it less easily adaptable as a story for Victorian children, though it had become established as a PANTOMIME by the end of the 18th century. Surprisingly, Mrs CRAIK left the story unchanged in *The Fairy Book* (anth **1863**), but George CRUIKSHANK totally rewrote it for his *Fairy Library* (coll **1870**). In Cruikshank's version the cat reveals it was a man changed into a cat for not appreciating his good fortune, and that the miller had been a real Marquis. Laura Valentine (1814-1899) rewrote the beginning for *The Old Old Fairy Tales* (anth **1889**) to show that the son was not ungrateful but had been devoted to his father, by contrast with his two selfish brothers – shades of BEAUTY AND THE BEAST.

The equivocal nature of the story lends itself ideally to SATIRE, as adapted by Ludwig TIECK in his play *Der Gestiefelte Kater* ["The Booted Cat"] (**1797**), and provides a good basis for further interpretations. In *The Master Cat: The True and Unexpurgated Story of Puss in Boots* (as "From 'Puss in Boots'" in *A Chatto & Windus Almanack* anth **1926**; exp **1974**), David GARNETT develops Basile's version; he continues the cat's story after it leaves. Sylvia Townsend WARNER uses the same starting point for "The Castle of Carabas" (in *The Cat's Cradle Book* coll **1940** US). There is something of the theme in "The Cat King's Daughter" (in *The Cat King's Daughter* anth **1984** ed Fiona Waters) by Lloyd ALEXANDER, wherein a cat advises the princess how she might gain love and fortune, although this may owe as much to the LEGEND of Dick Whittington. The Whittington story, based on the historical Richard Whittington (?1358-1423), may have cross-fertilized with the tale developed by Straparola.

The story was taken to heart by Nicholas Stuart GRAY, who developed it into its most complete form, providing a life history and full set of feline adventures in *The Marvellous Story of Puss-in-Boots* (**1955**) and *The Further Adventures of Puss-in-Boots* (**1971**). [MA]

P'U SUNG-LING (1640-1715) Chinese writer whose *Liao-chai-chih-i* (coll **1766**; some later exps; trans H. Giles as *Strange Tales from a Chinese Studio* **1916** UK) assembles over 400 short stories, many SUPERNATURAL FICTION; BEAST-FABLES and tales of SHAPESHIFTER fairies are also frequent. [JC]

PYGMALION A king of Cyprus, a woman-hater and an artist. The story appears in *Metamorphoses* (written cAD1-8) by OVID. Pygmalion sculpts an ivory statue of APHRODITE and falls in LOVE with it. He then prays to the GODDESS for a wife as beautiful as this carven image, and She causes the statue to come alive. Pygmalion marries the living woman, Galatea.

The tale is important in various ways. Even more than DAEDALUS and Icarus, Pygmalion and Galatea have long served as UNDERLIER figures for stories depicting makers and their creations, most notably perhaps in the story of FRANKENSTEIN and his MONSTER. Just as significantly for the understanding of fantasy stories is the moment in the tale when inanimate stone cannot be told apart from animate human flesh (◊ ANIMATE/INANIMATE). This TROMPE L'OEIL instant is an exemplary model for the danger-filled, highly charged METAMORPHOSES which shape the outcomes of many fantasy texts. [JC]

PYLE, HOWARD (1853-1911) Very prolific and influential US painter, writer, illustrator and teacher, whose influence can be detected in the work of countless 20th-century illustrators; he is justifiably referred to as the father of US ILLUSTRATION. As a writer he had almost 200 texts to his credit, and as an illustrator over 3300 published works: around two for every week of his working life. More than half his written works were children's stories, his most famous works being an illustrated retelling of the ARTHUR legends – as *The Story of King Arthur and His Knights* (St Nicholas Magazine 1902-3; coll **1903**), *The Story of the Champions of the Round Table* (**1905**), *The Story of Sir Launcelot and his Companions* (**1907**) and *The Story of the Grail and the Passing of Arthur* (**1911**) – and the legends of ROBIN HOOD as *The Merry Adventures of Robin Hood* (**1893**), all of which are still (1996) in print. He wrote these in a self-conciously "medieval" style, but nonetheless they have considerable literary merit. Other important works include books on PIRATES and highwaymen, as well as many recounting LEGENDS, FOLKTALES and FAIRYTALES.

HP's illustrations show a wide variety of influences: his b/w book illustrations display indebtedness to the Arts and Crafts movement and the work of Walter CRANE, while his colour work, in oils, was executed very much in the traditional manner. It is in his painted work that he can be seen to have had the most profound effect on the development of US illustration, for although, owing to printing limitations, only his late paintings were reproduced in full colour, these comprise a very considerable body of work. His handling of colour and form was remarkable, and he had masterful control of composition, allowing him to paint heroic figures with a fine sense of the dramatic.

HP studied under Adolf van der Wielen (1843-1876) and at the Pennsylvania Academy of the Fine Arts, and published his first illustrated poem, "The Magic Pill", in *Scribner's Monthly* in 1876, drawn in the anthropomorphic style of Isidore Grandville (◊ ILLUSTRATION). In the same year he moved to New York and wrote and drew the first of a long series of illustrated fairytales, published in children's magazines and books. He also created a great number of painted illustrations for MAGAZINES such as *Scribner's*, *Harper's Weekly* and *Harper's Monthly*. In 1898 he returned to Pennsylvania and began teaching a selected group of students; two years later he opened the Howard Pyle School of Art at Chadds Ford. His fame and influence grew steadily: he exhibited in the World's Columbian Exposition in

Chicago (1893) and the 1895 World's Fair in Atlanta, he was listed among the top US illustrators in 1895, and in 1901 he was included in *A History of American Art* by S. Hartmann. A travelling exhibition of 111 examples of his work opened in Chicago in 1903. In 1905 he began to paint murals of military subjects, but his illustration output continued unabated, with 191 illustrations published in that year. He died on an extended trip to Italy to study the Italian masters, and was buried in Florence.

HP's legacy is vast and wide-ranging, and his influence is still evident in all areas of US illustration. [RT]

Further reading: *The Brandywine Tradition* (**1969**) by Henry C. Pitz; *Howard Pyle: Writer, Illustrator, Founder of the Brandywine School* (graph **1975**) by Pitz; *Howard Pyle* (graph **1975**), intro Rowland Elzea; *The Howard Pyle Studio: A History* (**1983**) by Howard Pyle Brokaw.

PYNCHON, THOMAS (RUGGLES) (1937-   ) US writer, all of whose works are to varying degrees FABULA-TIONS. *The Crying of Lot 49* (**1966**) is a cheerfully paranoid FANTASY OF HISTORY revolving around a secret, alternative postal system operating in the world's WAINSCOTS. Although often described as sf, *Gravity's Rainbow* (**1973**) contains numerous supernatural elements: characters communicate with the dead; an ANGEL appears; a bureaucratic AFTERLIFE is described. In *Vineland* (**1990**) a subplot features the Thanatoids, people who are literally deceased yet cannot proceed to the next phase of astral existence (◊ ASTRAL PLANE); they live in Thanatoid communities, and Pynchon describes with a straight face their acceptance in the contemporary world, which is an otherwise untransformed USA. This characteristic of Pynchon's last two novels – that certain scenes are *not set in the same universe* as others – is one of the most radical features of his fiction; it is nearly unique in modern literature. This lack of ontological fixity – scenes in *Gravity's Rainbow* drift into FANTASTIC venues (Happyville, Baby Bulb Heaven, the Raketen-Stadt) which cannot be explained as fantasies by the characters, and are never integrated with the rest of the novel in terms of narrative mimesis – may have had its origin in James JOYCE'S *Ulysses* (**1922**), where the Nighttown episode becomes a phantasmagoria that cannot be explained as a HALLUCINA-TION of any character or combination of characters, but seems – as Vladimir NABOKOV, one of TP's professors at Cornell, has proposed – to be something strikingly more free-form: "The book is itself dreaming and having visions." This unmooring of narrative from the novelistic universe of discourse is a fantastic journey beside which most GENRE FANTASY is home-bound. [GF]

Other works: *V.* (**1973**); *Slow Learner* (coll **1986**).

Q Pseudonym of Arthur QUILLER-COUCH.

QUANTUM THEORY ◊ ALTERNATE REALITIES; PLAY-GROUND.

QUATERMASS UK mini-series (1979) (video version vt *The Quatermass Conclusion* and cut to 102 mins). Euston Films/Thames TV. **Pr** Ted Childs. **Dir** Piers Haggard. **Writer** Nigel Kneale. **Starring** Simon MacCordindale (Joe Knapp), John Mills (Quatermass), Rebecca Saire (Hettie). 4 60min episodes. Colour.

There is an element of SUPERNATURAL FICTION in this tepid attempt to re-introduce Professor Bernard Quatermass, who again attempts to defend the fragile ISLAND realm of the UK from alien (and by definition unnatural) incursions. John Mills plays Quatermass as though a faded paterfamilias in a Chekhov play; the depiction of Stonehenge's mystic allure, which turns the head of lots of young hippies, also lacks bright hues. The aliens' plan is cunning: when the youngsters reach the great stone circle, the aliens will evaporate them. This proves moderately difficult to thwart. [JC]

QUATERMASS AND THE PIT Two screened TECHNO-FANTASIES, **1** being a TELEVISION serial, **2** being a movie based on **1**.

1. UK tv serial (1958-9). BBC. **Pr** Rudolph Cartier. **Spfx** Jack Kine, Bernard Wilkie. **Writer** Nigel Kneale. **Script published as** *Quatermass and the Pit* (**1960**) by Kneale. **Starring** Anthony Bushell (Colonel Breen), Christine Finn (Barbara Judd), Cec Linder (Dr Matthew Roney), André Morell (Professor Bernard Quatermass), John Stratton (Captain Potter), Brian Worth (James Fullalove). 6 35min episodes. B/w. Released as a video "tvm" (**1988**), 178 mins.

The TECHNOFANTASY *par excellence*. Excavations in LONDON under Hobbs Lane – the locale has a history of GHOSTS and POLTERGEISTS – reveal fossils of exceptionally early hominids and, alongside them, a five-million-year-old Martian spaceship. During the excavations, spectral manifestations are witnessed; the spaceship is engraved with cabbalistic signs. Historical records show that any ground disturbance in the area has triggered appearances of beings perceived as imps, GOBLINS and DEMONS. Coincidentally, chief archaeologist Roney has invented a prototype "opticencephalograph" – a sort of telepathy machine capable of projecting thoughts onto a tv screen. Through its use we discover that our racial memories of the DEVIL are in fact of the horned, insectile Martians; the RITUAL self-culling of that race gave rise to our legend of the WILD HUNT. The human species is a product of psychological tinkering by the Martians with our hominid ancestors: in effect, we are all still victims of demonic POSSESSION. Psychic disturbances radiate from the newly activated spaceship, and London falls into rioting chaos as the Wild Hunt is enacted afresh.

Kneale has a dour faith in human stupidity that leads the motivational aspects of the plot well beyond plausibility, and his fixed anthropocentrism is depressing (the Martians were EVIL because their social customs were alien to ours); there are some bizarre scientific gaffes. The production was obviously shoestring – props wobble. Yet the cumulative effect of *QATP* is exceptionally powerful, and its fiesta of fantasy ideas easily carries the interest through its several longueurs.

It is virtually inconceivable that Stephen KING was not strongly influenced by *QATP* in writing *The Tommyknockers* (**1987**). [JG]

2. (vt *Five Million Years to Earth* US) UK movie (**1967**). British-Pathe/Seven Arts-HAMMER/20th Century-Fox. **Pr** Anthony Nelson Keys. **Dir** Roy Ward Baker. **Spfx** Bowie Films. **Screenplay** Nigel Kneale. **Based on 1. Starring** James Donald (Roney), Julian Glover (Breen), Andrew Keir (Quatermass), Bryan Marshall (Potter), Barbara Shelley (Judd). 97 mins. Colour.

The same story as **1**, with frequent duplications of dialogue, but pared back and with slightly more effort taken to cover up the wilder scientific unlikelihoods. Did this movie stand alone its reputation would be higher; as it is – despite good performances and dramatic spfx – it suffers severely by comparison with **1**, almost certainly because **1**'s low-budget murkiness and slowness left more room for the viewer's imagination. [JG]

QUEEN MAB ◊ FAIRY QUEEN.

QUEEN OF AIR AND DARKNESS Phrase coined by A.E. Housman (1859-1936) in *Last Poems* (coll **1922**) to describe a seeming ENCHANTRESS or LAMIA – or perhaps merely a woman resented for her former sexual dominance – whom the male speaker feels justified in killing. Book 2 of T.H. WHITE's revised *The Once and Future King* (**1958**) uses the phrase as its title, referring to MORGAN LE FAY. Others apply it to the FAIRY QUEEN, as in Poul ANDERSON's "The Queen of Air and Darkness" (1971 *F&SF*), whose alien pseudo-FAERIE has sf roots (◊ RATIONALIZED FANTASY). [DRL]

QUESTING, FRANK (? -) Writer. ◊ DREAMTIME.

QUESTS Very broadly, quests come in two categories. There is the external quest, until recently engaged upon almost exclusively by men. Here the protagonist of a tale embarks upon a search – likely picking up PLOT COUPONS along the way – for something important to his survival or the survival of the LAND for which he is or will be responsible, travels beyond the fields we know into the place where he will be tested and found worthing of winning the prize, accomplishes this goal, returns home with the desired object, or partner, or knowledge. The earliest external quest tale in Western literature, HOMER's *Odyssey* (*c*850BC), is one of the first sustained narratives to have been conceived anywhere; and remains one of the finest. It underlies much of the quest literature of the Western world and much modern GENRE FANTASY.

Second, there is the internal quest, in which females may participate more equally. Here the protagonist, whose goal is (broadly) self-knowledge, embarks upon an internal search, engages upon a RITE OF PASSAGE and returns to the world as an integrated person, a MAGUS, or a SHAMAN, or ... DANTE's *The Divine Comedy* (*c*1330), John BUNYAN's *The Pilgrim's Progress* (**1678-9**) and Lewis CARROLL's *Alice's Adventures in Wonderland* (**1865**) are, at their various levels of seriousness, examples of the internal quest.

There are also quests which consciously join both elements, fantasies where full self-RECOGNITION combines with the gaining of an external goal in a tale whose various elements interweave, generating a sense of full STORY. In J.R.R. TOLKIEN's *The Lord of the Rings* (**1954-5**) Frodo's quest is double in this sense, and Stephen R. DONALDSON's **Chronicles of Thomas Covenant the Unbeliever** is a recent model for this attempt at integration.

Quests are sequential, suspenseful, event- and goal-oriented; they normally reach a conclusion (even the ACCURSED WANDERER's search for quietus is often rewarded, if only with death; and more than one tale of the WANDERING JEW and the FLYING DUTCHMAN provide ultimate solace); those who oppose the successful conclusion of a quest – from the KNIGHT OF THE DOLEFUL COUNTENANCE to the DARK LORD – can often be best understood as mere symbols of opposition; and quests require an identifiable protagonist – from UGLY DUCKLING to CHILDE to culture hero – plus, usually, an accumulating mass of COMPANIONS to strengthen and complicate the action. Quests are therefore basic to the telling of Story; as fantasy as a genre is inherently tied to Story, it is not surprising that almost all modern fantasy texts are built around, or incorporate, a quest.

Quests are less frequent in SUPERNATURAL FICTION, whose protagonists tend to react to transgressive circumstances rather than explore new territory, though the MAGUS figures often found in supernatural fiction and HORROR are often described as embarked upon foredoomed quests. [JC]

QUIBBLES Fantasy is full of pacts – especially PACTS WITH THE DEVIL – and other explicit or implicit CONTRACTS with entities, magical law or FATE (◊ CONDITIONS; CURSES; PROHIBITIONS; PROPHECIES). Human nature demands a loophole: a quibble allows reinterpretation of the contract, usually by strict adherence to the letter of the law (◊ READ THE SMALL PRINT). Nordic mythology anticipated William SHAKESPEARE's *The Merchant of Venice* (**1600**) when LOKI lost a bet and forfeited his head to the DWARF Brökk, but quibbled that his neck had not been wagered and must remain undamaged. In *Macbeth* (performed *c*1606; **1623**), a WITCH reassures that "none of woman born/Shall harm Macbeth"; but Macduff was born of a corpse. Frederic, in W.S. GILBERT's *The Pirates of Penzance* (**1879**), is indentured to PIRATES until his 21st birthday – but was born on 29 February, so must serve until aged 84. H.G. WELLS's obese Pyecraft in "The Truth about Pyecraft" (1903) suffers from quibbling euphemism: using a SPELL to lose weight, rather than fat, he becomes weightless. In James Branch CABELL's *The Music from Behind the Moon* (**1926**) the hero's lover is doomed to spend 725 years in BONDAGE, as written in the BOOK of the Norns of which "no man nor any god may alter any word" – but this allows the double quibble of inserting, after the 7 and using a quill pen from the DEVIL's wing, a decimal point. A prophecy in J.R.R. TOLKIEN's *The Lord of the Rings* (**1954-5**) says of the Nazgûl king, "not by the hand of man shall he fall": he is despatched by a woman and a hobbit. Som the Dead in Fred SABERHAGEN's *The Black Mountains* (**1971**), whose pact with DEATH has made him a walking dead man immune to any weapon, is destroyed by an innocently meant splash of healing POTION. In Terry PRATCHETT's *Moving Pictures* (**1990**) legends promise awful fates to any man opening the *Necrotelecomnicon* (◊ BOOKS); the actual opener suffers only mild migraine and eczema, since he is an orang-utan. Sometimes the quibble can void a contract through logical paradox: on the ISLAND of Barataria in Miguel de CERVANTES's *Don Quixote* (**1605-15**) those crossing a certain bridge must first state their business and are hanged if they lie – whereupon a traveller declares that his business is to die on the adjacent gallows, which paralyses the law. The WIZARD-kings of Norton JUSTER's *The Phantom Tollbooth* (**1961**) patch up their quarrel when informed that their determination to disagree about everything is flawed, for they agree about disagreeing. More surreally, James THURBER's *The Thirteen Clocks* (**1950**) offers an outrageous verbal quibble when a hand-touch fails to wake the frozen clocks: "If you can touch the clocks and never start them, then you can start the clocks and never touch them. That's logic ..."

Demons, FAIRIES and GODS are apt to invoke quibbles when granting WISHES (◊◊ ANSWERED PRAYERS); a classical example is ZEUS's gift to Tithonus of IMMORTALITY without any halt to normal aging.

Virtuous wishes can bring, as it were, easing of the clauses in a contract. GOETHE's FAUST is saved because, in the moment at which he utters the specific formula – *Verweile doch, du bist so schön!*/"Moment stay, thou art so fair" – it is in respect of an unselfish desire. William Butler YEATS's Countess Cathleen in his play *The Countess Cathleen* (**1892**; performed 1899) tries to sell her SOUL to buy food for starving peasants, but cannot be damned for an act of charity. Fredric BROWN's short-short "Millennium" sees SATAN's power destroyed by an ultimately unselfish wish.

In Gustav Holst's OPERA *Savitri* (**1916**) Death, who has taken Savitri's husband, offers her a consolatory boon and she asks for life, which he gives her. Life without her husband is meaningless, and so Death has perforce to return him. Here what is at stake is not just the logical implication of what is offered, nor the intelligence of the TRICKSTER heroine, but also a view of the Universe in which all is ILLUSION, and REALITY itself merely a quibble. [DRL/RK]

QUILLER-COUCH, [Sir] ARTHUR (THOMAS) (1863-1944) UK poet, novelist and critic, knighted in 1910 for his political services, and Professor of English Literature at Cambridge from 1912. He is perhaps best-known as editor of *The Oxford Book of English Verse* (1900). In his day he was more famous simply as Q. Born in Cornwall, descended from a notable local family – both his father, Richard Quiller-Couch (1816-1863) and grandfather, Jonathan Couch (1789-1870), were naturalists, doctors and antiquarians, well known for their collections of local LEGENDS – Q spent much of his early life there, and drew on his knowledge of legends and FOLKLORE in many of his stories. A few of his GHOST STORIES are still regularly reprinted, especially "The Roll-Call of the Reef" (1895 *The Idler*; in *Wandering Heath* coll **1895**) and "A Pair of Hands" (1898 *Cornish Magazine*; in *Old Fires and Profitable Ghosts* coll **1900**). These two volumes contain most of Q's SUPERNATURAL FICTION. He wrote many other ghost stories, but should be better known for his folkloristic fiction, which contains much local colour and atmosphere and, in keeping with many FOLKTALES, is often humorous. These are found in several collections, notably *Noughts and Crosses* (coll **1891**), "*I Saw Three Ships*" *and Other Winter's Tales* (coll **1892**), *The Laird's Luck* (coll **1901**), *The White Wolf and Other Fireside Tales* (coll **1902**) – which contains mostly nautical fantasies – *Two Sides of the Face* (coll **1903**) and *Shakespeare's Christmas* (coll **1905**). No specific collection of Q's super-

natural fiction exists, but *Selected Stories* (coll **1921**) and *Q's Mystery Stories* (coll **1937**) are representative. [MA]

QUINCUNX A pattern of five objects placed at the corner and centre of a square. The NUMEROLOGY of planting trees in such repeated patterns was a MAGGOT of Sir Thomas Browne (1605-1682), who discussed it in *The Garden of Cyrus, or The Quincunciall, Lozenge or Net-work Plantations of the Ancients* (**1658**). [DRL]

QUINET, EDGAR (1803-1875) French writer and politician. ◊ APOCALYPSE.

QUINN, SEABURY (GRANDIN) (1889-1969) US attorney, editor and writer, the majority of whose output appeared in the pulp MAGAZINES – beginning with "The Stone Image" (1919) in *The Thrill Book*. WEIRD TALES published more than 90 of his efforts, the majority of which were set in Harrisonville, N.J., and featured the courteous, dapper, vain, and lethal OCCULT DETECTIVE Jules de Grandin. SQ's only book-length **de Grandin** adventure is *The Devil's Bride* (*WT* 1932; **1976**), a mishmash involving a worldwide conspiracy of Yezidee adepts from Kurdistan, Leopardmen from Africa and twin DEVIL-worshipping Communists from Russia. Different in approach is *Roads* (*WT* 1938; **1938** chap; rev 1948), a sentimental WEIRD FICTION describing the fantastic origins of SANTA CLAUS. SQ wrote low-level commercial fantasy with occasional elements of HUMOUR and imagination but without originality of idea or execution. [RB]

Other works: *The Phantom-Fighter: The Memoirs of Jules de Grandin* (coll **1966**); *Is the Devil a Gentleman?* (**1970**); *The Adventures of Jules de Grandin* (coll **1976**); *The Casebook of Jules de Grandin* (coll **1976**); *The Hellfire Files of Jules de Grandin* (coll **1976**); *The Skeleton Closet of Jules de Grandin* (coll **1976**); *The Horror Chambers of Jules de Grandin* (coll **1977**).

RABELAIS, FRANÇOIS (?1494-1553) French scholar, physician, humanist and writer. Like many scholars of his time he took holy orders, but soon left the monastery behind. His central TAPROOT-TEXT importance for both sf and fantasy comes from the books now always known in English as *Gargantua and Pantagruel*, an omnibus assembled from *Les Horribles et espovantables faictz et prouesses du tre renommé Pantagruel* ["The Horrible and Astonishing Deeds of the Most Renowned Pantagruel"] (**1532**), which became Book II of the whole; *La Vie inestimable du grant Gargantua* ["The Inestimable Life of Gargantua"] (**1534**), which became Book I; *Le Tiers Livre des faictz et dictz heroiques du noble Pantagruel* ["The Third Book of Facts and Heroic Deeds of the Noble Pantagruel"] (**1546**), which became Book III; *Le Quart Livre* ["The Fourth Book"] (**1548**; exp **1552**), which became Book IV; and *Cinquième et dernier livre* ["The Fifth and Last Book"] (**1564**), which incorporates *L'Isle sonante* ["The Ringing Island"] (**1562**) and became Book V. The last of these, posthumous and cruder than the previous volumes, is generally thought to have been spatch-cocked together from fragments left by FR. Of the many translations the most vigorous and best-known early version is by Sir Thomas Urquhart (Books 1 and 2 **1653** UK, Book 3 **1693** UK) and Peter Le Motteux (Books 4 and 5 **1694** UK); the most successful contemporary version is by Burton Raffel (**1990** US).

The whole constitutes an immense and compendious comic anatomy of 15th-century France, the main targets of the SATIRE being the ineradicable contradictions ravaging late-medieval Christianity, though contemporary politics, the law and mores in general are also guyed. The most important single character other than the eponymous GIANTS is Panurge, a TRICKSTER figure, a plausible UNDERLIER source for the servant who dupes the master to ultimately run the show. The exuberance and linguistic inventiveness of *Gargantua and Pantagruel* have not, perhaps, had much influence on modern fantasy, though writers like Philip José FARMER and Jack VANCE both create worlds with something like FR's all-encompassing gusto, and certainly there are loud echoes of FR's exuberant voice in the works of John BARTH – most notably in *The Sot-Weed*

Factor (**1960**; rev 1967) – and Robert NYE, such as *Falstaff* (**1976**). Beyond that gusto, which is extremely difficult to mimic, FR may have influenced recent fantasy writers mainly by demonstrating the uses of exorbitance to illustrate sustained arguments about the way of the world. [JC]

RABKIN, ERIC S(TANLEY) (1946-) US academic, editor and writer who began publishing critical work in the field of the FANTASTIC with *The Fantastic in Literature* (**1976**), which presents his most sustained modelling thoughts on the structure of a genre like FANTASY. His basic premise is that a work that is fantastic manifests a clear reversal of the "ground rules" of reality, in order that a new set of ground rules can be established for – and within – the work in question. This new set of rules sets up a world which, though "dis-expected", remains internally consistent once understood.

Other works: by ESR include *Science Fiction: History, Science, Vision* (**1977**) with Robert Scholes (1929-) and *Arthur C. Clarke* (**1979** chap; rev 1980). He has edited two anthologies collecting sf and fantasy, *Fantastic Worlds: Myths, Tales & Stories* (anth **1979**) and *Science Fiction: A Historical Anthology* (anth **1983**), plus a number of critical anthologies, many of them with George E. Slusser (1939-) and most assembling selected papers from the annual Eaton Conference: *Bridges to Fantasy* (anth **1982**) with Slusser; *The End of the World* (anth **1983**) with Martin H. GREENBERG and Joseph D. Olander (1939-), *Co-Ordinates: Placing Science Fiction and Fantasy* (anth **1983**) with Slusser and Scholes, *No Place Else: Explorations in Utopian and Dystopian Fiction* (anth **1983**) with Greenberg and Olander; *Shadows of the Magic Lamp: Fantasy and Science Fiction in Film* (anth **1985**) with Slusser, *Hard Science Fiction* (anth **1986**) with Slusser, *Storm Warnings: Science Fiction Confronts the Future* (anth **1987**) with Slusser and Colin GREENLAND, *Intersections: Fantasy and Science Fiction* (anth **1987**) with Slusser, *Aliens: The Anthropology of Science Fiction* (anth **1987**) with Slusser, *Mindscapes: The Geographies of Imagined Worlds* (anth **1989**) with Slusser; *Styles of Creation: Aesthetic Technique and the Creation of Fictional Worlds* (anth **1992**) with Slusser and *Fights of Fancy: Armed Conflict in Science Fiction and Fantasy* (anth **1993**) with Slusser. [JC]

RACHILDE (1860-1953) French writer. ◊ DECADENCE.

RACKHAM, ARTHUR (1867-1939) UK illustrator, one of the most famous and influential artists of fantasy material of the early 20th century. He began publishing illustrations in his teens, his first professional work appearing in the magazine *Scraps* in 1884. His first illustrated book was Thomas Rhodes's *To the Other Side* (**1893**), which contains no element of the fantastic; soon, though, he began to specialize in fantasy, illustrating new stories – like S.J. Adair Fitzgerald's *The Zankiwank and the Bletherwitch* (**1896**) – as well as more traditional material like R.H. BARHAM's *The Ingoldsby Legends* (1898; exp 1907). His prime period began with a brilliant and haunting edition of Washington IRVING's *Rip Van Winkle* (1905), the first of a long series of versions of classic tales brought out at Christmas (◊ CHRISTMAS BOOKS). Others in the sequence included editions of J.M. BARRIE's *Peter Pan in Kensington Gardens* (1906; excerpts as *The Peter Pan Portfolio* 1912), Lewis CARROLL's *Alice's Adventures in Wonderland* (1907), William SHAKESPEARE's *A Midsummer Night's Dream* (1908), Friedrich FOUQUÉ's *Undine* (1909); Richard WAGNER's *The Rhinegold and the Valkyrie* (1910) and *Siegfried and the Twilight of the Gods* (1911), AESOP's *Fables* (1912), *Mother Goose: The Old Nursery Rhymes* (1913), Charles DICKENS's *A Christmas Carol* (1915), *The Allies' Fairy Book* (1916; vt *Fairy Tales from Many Lands* 1974), Thomas MALORY's *The Romance of King Arthur* (1917), Charles PERRAULT's *Cinderella* (1919) and *The Sleeping Beauty* (1920), and John MILTON's *Comus* (1921). The sequence continued, but AR's genius was not well suited to the post-WWI world, and with occasional exceptions – like Irving's *The Legend of Sleepy Hollow* (1928), Hans Christian ANDERSEN's *Fairy Tales* (1932) and Henrik Ibsen's *Peer Gynt* (1936) – the great illustrator slid out of the land of dreams.

That these dreams are uneasy may demonstrate AR's sensitivity to his subjects as well as the general disquiet underlying the Edwardian Indian Summer of his prime. Certainly his GOBLINS, ELVES and FAIRIES have a grotesque, uncannily fluid, menacing air; moreover, his backgrounds seem animate, constantly at the cusp of METAMORPHOSIS. His LANDSCAPE is Nordic, lowering, precipitate; the heart of his vision is not the GARDEN but the FOREST. Many artists of the period produced similar unease; but AR's draughtsmanship was superb, his range of techniques wide, and his use of influence – significant predecessors include William BLAKE, Richard DADD and Richard DOYLE – sagacious and synthesizing. Until WWI put paid to these profitable forebodings, he was supreme; *Arthur Rackham's Book of Pictures* (graph 1913) is a fine survey. After WWI there seemed, for the likes of AR, nothing left to nullify through the healing anticipations of art: the dreams had come true. [JC]

Other works (selective): Editions of *Two Old Ladies, Two Foolish Fairies and a Tom Cat, the Surprising Adventures of Tuppy and Tue* (1897) by Maggie Browne; *Gulliver's Travels* (1900; exp 1909) by Jonathan SWIFT; *Fairy Tales of the Brothers Grimm* (1900) followed by *Grimm's Fairy Tales* (1909; rev vt in 2 vols as *Hansel & Gretel and Other Tales* 1920 and *Snowdrop and Other Tales* 1920) and *Little Brother and Little Sister* (1917); *Puck of Pook's Hill* (coll of linked stories **1906** US) by Rudyard KIPLING, this being one of the rare times AR illustrated a first edition of importance; *Stories of King Arthur* (**1910**) by A.L. Haydon; *English Fairy Tales* (**1918**) by Flora Annie Steele; *Snickerty Nick and the Giant* (**1919** US) by Julia Ellsworth Ford; *Irish Fairy Tales* (**1920**) by James STEPHENS; *A Wonder Book* (1922) by Nathaniel

HAWTHORNE; *Where the Blue Begins* (**1925**) by Christopher MORLEY; *The Road to Fairyland* (**1926**) by Erica Fay; *The Tempest* (1926) and another edition of *A Midsummer Night's Dream* (1939) by William SHAKESPEARE; *The Lonesomest Doll* (**1928**) by Abbie Farwell Brown; *The Chimes* (1931) by Dickens; *The Night Before Christmas* (1931) by Clement C. Moore; *The King of the Golden River* (1932) by John RUSKIN; *Goblin Market* (1933) by Christina ROSSETTI; *The Arthur Rackham Fairy Book* (**1933**); *The Pied Piper of Hamelin* (1934) by Robert BROWNING; *Tales of Mystery and Imagination* (1935) by Edgar Allan POE; *The Wind in the Willows* (1940) by Kenneth GRAHAME. *Arthur Rackham* (graph **1974**) ed David Larkin is a useful compilation.

Further reading: *Arthur Rackham: The Wizard at Home* (**1914**) by Eleanor FARJEON.

RADCLIFFE, [Mrs] ANN (WARD) (1764-1823) UK novelist. Although her influential Gothic romances are full of delicious supernatural effects, only the last is in fact a SUPERNATURAL FICTION: in the others, things that seem supernatural invariably prove to have a natural explanation (◊ RATIONALIZED FANTASY). Those earlier novels are *The Castles of Athlin and Dunbayne: A Highland Story* (**1789**), *A Sicilian Romance* (**1790**), *The Romance of the Forest, Interspersed with some Pieces of Poetry* (**1791**), *The Mysteries of Udolpho: A Romance Interspersed with some Pieces of Poetry* (**1794**) and *The Italian, or The Confessional of the Black Penitents* (**1797**). Beyond their ingenious use of virtually animate LANDSCAPES – which evoke much terror in the breasts of the female protagonists of the first three titles – these tales are of significance in the development of fantasy for their concentration on great eerie Gothic castles which likewise seem animate, and which exhibit many additional features of the full-blown fantasy EDIFICE.

AR's last novel, *Gaston de Blondeville, or The Court of Henry III Keeping Festival in Ardennes* (written 1802; **1826**), moves past semblance to feature a genuine GHOST, who supports a claim that the king's favourite KNIGHT has murdered him (for the sake of a magic SWORD). The novel is also of interest in that it is set in the 13th century, and is thus an extremely early example of the historical novel. [JC]

RAEPER, WILLIAM (1959-   ) UK author whose fiction consists of a minor fantasy series – *A Witch in Time* (**1993**) and *Warrior of Light* (**1993**) – and of *The Troll and the Butterfly, and Other Stories* (coll **1987**). Of more general interest is his work on George MACDONALD: *Fantasy King: A Biography of George MacDonald* (**1987**; vt *George MacDonald* 1988) and *The Gold Thread: Essays on George MacDonald* (anth **1990**). The latter's essays attempt to redeem MacDonald from C.S. LEWIS's condescending vision of him as an idiot savant blessed with a sectarian Christian vision. [JC]

RAGGEDY ANN AND ANDY US ANIMATED MOVIE (**1977**). ◊ Richard WILLIAMS.

RAGNAROK The great battle described in the 10th-century Icelandic poem *Völuspá*, which provides a climax for the "Twilight of the Gods" in Norse MYTHOLOGY. The AESIR, attacked by the frost GIANTS and various other enemies, are all but obliterated – a fate presumably influenced by Christian ideas of the APOCALYPSE, probably also responsible for the appointment of BALDER as a MESSIAH figure. The seeming perversity of the idea is addressed in "The Riddle of Ragnarok" (1955) by Theodore STURGEON. Ragnarok is a particular favourite in the kind of pulp SCIENCE FANTASY which reinterprets myths in terms of sf's vocabulary of ideas, featuring in, e.g., *Exiles of Time* (1940; **1949**) by Nelson BOND and "A Yank at Valhalla" (1941; vt *The Monsters of Juntonheim* **1950**) by Edmond Hamilton (1904-1977). Hamilton also offered a straightforward fantasy account of the affair in "Twilight of the Gods" (1948). The hero of Fletcher PRATT's and L. Sprague DE CAMP's *The Incomplete Enchanter* (**1941**) escapes in the nick of time. *The Gods are Not Dead* (**1985**) by Michael J. FRIEDMAN is a "sequel" to the tale of Ragnarok. [BS]

RAIDERS OF THE LOST ARK (*1981*). ◊ INDIANA JONES.

RAMAL, WALTER Pseudonym of Walter DE LA MARE.

RAMSAY, JAY Pseudonym of Ramsey CAMPBELL.

RAMUZ, CHARLES-FERDINAND (1878-1947) Swiss-born French writer whose first works, like *Aline* (**1905**), were naturalistic tales in a somewhat affectedly simple DICTION, but whose later works – published in response to the apocalyptic experience of WWI – tended to fantasy as tinged with ALLEGORY, like his libretto for *The Soldier's Tale* (**1918**), put to music as a PANTOMIME-ballet by Igor Stravinsky (1882-1971), in which the eponymous returning soldier meets the DEVIL and (◊ READ THE SMALL PRINT) swaps his violin for a "magic" BOOK which will bring him fame. *Le Reigne de l'Esprit Mal* (**1917**; trans anon as *The Reign of the Evil One* **1922** US) is a CHRISTIAN FANTASY in which SATAN impersonates CHRIST, re-enacting his life in PARODY form and corrupting a small Swiss village; at the heart of the tale, a bleak saturnalian REVEL brings the dead to life, and elicits from Satan the statement that only the powers of Earth count, for we are now beyond GOOD AND EVIL. *Presence de la Mort* (**1922**; trans Allan Ross Macdougall and Alex Comfort as *The End of All Men* **1944** UK; rev trans vt *The Triumph of Death* 1946 UK) is sf. In *La Beauté sur la Terre* (**1927**; trans anon as *Beauty on Earth* **1929** UK) a village is maddened by the supernatural beauty of a servant girl. In *Derborence* (**1935**; trans S.F. Scott as *When the Mountain Fell* 1949 UK) the miraculous survival of a young man in an avalanche generates a widespread mythopoeic response. [JC]

RANDALL & HOPKIRK (DECEASED) (vt *My Partner the Ghost* US) UK tv series (1969-71). ITC. **Pr** Monty Berman. **Dir** Ray Austin, Roy Ward Baker, Cyril Frankel, Leslie Norman and others. **Writers** Austin, Mike Pratt, Ralph Smart, Tony Williamson, Ian Wilson and others. **Created by** Dennis Spooner. **Starring** Annette Andre (Jean Hopkirk), Kenneth Cope (Marty Hopkirk), Pratt (Jeff Randall). 26 60min episodes. Col.

A private detective has as his INVISIBLE COMPANION the GHOST of his dead partner. To complicate matters, the dead Hopkirk is jealous of his widow, with whom Randall has a developing relationship. *R&H(D)*, though hindered by cheap production, provided much good-humoured adventure and often considerable inventiveness. What intrigued about the series was that, in general, the stories worked well as detections. [JC/JG]

RANDOLPH, ELLEN Pseudonym of Marilyn ROSS.

RANKIN, ROBERT (FLEMING) (1949-   ) UK writer who first came to notice with the **Brentford** sequence of humorous fantasies: *The Antipope* (**1981**), *The Brentford Triangle* (**1982**) and *East of Ealing* (**1984**), assembled as *The Brentford Trilogy* (omni **1988**), plus *The Sprouts of Wrath* (**1988**). These are an eclectic mix of far-out notions and clichés freely drawn from sf, horror and fantasy, and amusingly grounded in the all-too-mundane London suburb of Brentford. RR's career suffered a mid-1980s hiatus, but was relaunched in the wake of the immense success of Terry PRATCHETT's comic fantasies – not that his work resembles Pratchett's very closely: RR is altogether wilder, more prepared to entertain ideas from the occult and from fringe beliefs of every kind, and at the same time more parochially "British" (with a touch of the Irish). The **Armageddon** trilogy – *Armageddon: The Musical* (**1990**), *They Came and Ate Us: Armageddon II, The B-Movie* (**1991**) and *The Suburban Book of the Dead: Armageddon III, The Remake* (**1992**) – features the Apocalypse and a time-travelling Elvis Presley (1935-1977) (◊◊ ICONS). The **Cornelius Murphy** trilogy – *The Book of Ultimate Truths* (**1993**), *Raiders of the Lost Car Park* (**1994**) and *The Greatest Show Off Earth* (**1994**) – plays with notions that resemble those of Charles Fort (1874-1932). Much more of the same is to be found in *The Most Amazing Man who Ever Lived* (**1995**), *The Garden of Unearthly Delights* (**1995**) and *A Dog Called Demolition* (**1996**), the last of which is blurbed, characteristically, as "a nightmare journey to hell and back, with only a brief stop at a Happy Eater to use the toilet". [DP]

RANSMAYR, CHRISTOPH (1954-   ) Austrian novelist whose *Die letzte Welt* (**1988**; trans John E. Woods as *The Last World: A Novel with an Ovidian Repertory* **1990** US) is set at the edge of the Ancient World, in a place of exile where the banished OVID – who did historically end his life in exile – generates a pattern of METAMORPHOSES through the power of STORY, eventually transforming the town into an arena where figures from his own works undergo various TRANSFORMATIONS. [JC]

RANSOME, ARTHUR (MITCHELL) (1884-1967) UK writer and editor, best-known for his late series of children's stories (none fantasy) which began with *Swallows and Amazons* (**1930**). His fantasy work precedes this series by several decades. *The Stone Lady: Ten Little Papers and Two Mad Stories* (coll **1905**) contains two or three stories – "Meddling with the Fairies" in particular – that emote a condescending "delicacy" typical of turn-of-the-century texts dealing with FAIRIES and other diminutive inhabitants of suburban gardens; and *Highways and Byways in Fairyland* (**1906**) is a geography of FAERIE, which is again seen in diminutive terms. The tales assembled in *The Hoofmarks of the Faun* (coll **1911**) are intermittently more significant, though they tend still to be crippled by sentimentality. In "The Hoofmarks of the Faun" (1908) itself a faun in northern exile falls in love with a young mortal lady, but his hoofs spook her. "Rolf Sigurdson" (1904) tells of long-ago days when DREAMS foretold reality, and of a man who goes to Yggdrasil (◊ WORLD-TREE) in a dream-state which curves in upon itself (◊ ARABIAN NIGHTMARE), but escapes. The protagonist of "The Ageing Faun", a tale which prefigures the work of Thomas Burnett SWANN, gives up IMMORTALITY to wed a woman, though warned by APOLLO that she'll die in childbirth and that he'll wither and perish likewise, having forgotten his motive for becoming mortal (this indeed comes to pass). *The Elixir of Life* (**1915**) is a HORROR novel: an aristocrat owes his immortality to the ELIXIR, which requires those who take it to murder to maintain its strength.

AR's first lasting success in fiction came with *Old Peter's Russian Tales* (coll **1916**), Russian-style FAIRYTALES told from the vantage of a FRAME STORY by Old Peter to his grandchildren. By now AR had mastered a DICTION which did not belittle his chosen subject matter. *The Soldier and Death: A Russian Folktale Told in English* (**1920** chap) retells a traditional tale with dignity.

AR was also active as a man of letters in the early years of the 20th century. *A History of Story-Telling: Studies in the Development of Narrative* (**1909**) is more than anecdotally useful in any history of the impulse to STORY. He edited (*c*1908) a series of selections from taletellers including Théophile GAUTIER, Nathaniel HAWTHORNE, E.T.A. HOFFMANN and Edgar Allan POE, and translated Rémy de GOURMONT's philosophical fantasy *Une nuit au Luxembourg* (**1906**; trans as *A Night in the Luxembourg* **1912** UK); his essay on that author in *Portraits and Speculations* (coll **1913**) is of interest. [JC]

Other works: *Aladdin and His Wonderful Lamp* (**1919** chap).

RAPTURE, THE US movie (*1991*). ◊ APOCALYPSE.

RASMUSSEN, ALIS A. (1958-   ) US writer. Most of her work has been sf, now as by Kate Elliott. Her first novel, *The Labyrinth Gate* (**1988**), was fantasy. The protagonists, newly wed and in possession of an odd TAROT pack, find that the elevator they have just descended is a PORTAL into an ALTERNATE-WORLD UK where MAGIC works and a matriarchy still rules, though it is THINNING. The GODDESS is evoked, and a TIME ABYSS or two. [JC]

Other works (sf): The **Highroad** sequence, being *A Passage of Stars* (**1990**), *Revolution's Shore* (**1990**) and *The Price of Ransom* (**1990**); the linked **Sword of Heaven** or **Jaran** sequence, as Elliott, being *Jaran* (**1992**), *An Earthly Crown* (**1993**), *His Conquering Sword* (**1993**) and *The Law of Becoming* (**1994**).

RASPE, RUDOLF ERICH (1737-1794) German professor and librarian who fled to England in 1775 after he was found to have been selling the precious gems and medals in his care. In England he set himself up as a mining expert and before long was found to be swindling his employer, Sir John Sinclair (1754-1835). RER fled again, this time to Ireland, where he died of a fever.

It is not surprising that RER should be interested in other rogues and scoundrels. He was probably friends with James MACPHERSON, who forged the **Ossian** poems. RER translated these and assembled similar Ossiana. He collected other TALL TALES, and thus met Baron von MUNCHHAUSEN. Drawing upon a variety of sources, RER serialized several stories in *Vademecum für lustige Leute* ["Handbook of Humorous People"] between 1781 and 1783 and then expanded these as *Baron Münchhausen's Narrative of his Marvellous Travels and Campaigns in Russia* (fixup **1785** chap UK; exp vt *The Surprising Travels and Adventures of Baron Münchhausen* 1792). Although RER's work set in train a fascination for exaggerated tales, it was not his version that became the most popular. His countryman, Gottfried Bürger (1747-1794) translated it back into German with considerable embellishment as *Wunderbare Reisen zu Wasser und zu Lande, Feldzüge und lustige Abenteuer des Freyherrn von Münchhausen* (**1786** chap UK; exp 1788) and it was this version, issued in English as *Singular Travels, Campaigns, Voyages and Sporting Adventures of Baron Munnikhouson, Commonly Pronounced Münchhausen, as He Relates them Over a Bottle When Surrounded by his Friends* (**1786** chap UK), that really attracted the public's attention. By the next edition, in late 1786, Munchhausen's adventures were being equated with those of Lemuel GULLIVER (◊◊ TRAVELLERS' TALES), and for a while the two names became inextricably linked. The stories have stayed in print ever since, a recent edition being *The Travels and Surprising Adventures of Baron Munchhausen* (coll **1993** US).

RER also wrote a gloriously atmospheric GOTHIC FANTASY, *Koenigsmark der Räuber, oder Der Schrecken aus Böhmen* (**1790** chap UK; trans by H.J. Sarratt as *Koenigsmark the Robber, or The Terror of Bohemia* 1801 chap UK). Recounted by Münchhausen, it tells of the leader of a company of brigands who seems so invincible that it is believed he is a warlock. The story includes many supernatural agencies, including a werewolf and a giant SPIDER. [MA]

RATIONALIZED FANTASY In this book we use this term in three linked and usually easily distinguishable senses.

1. In works such as *Too Many Magicians* (**1967**) by Randall GARRETT, stock fantasy elements are given a rationale that provides them with internal consistency and coherence. In such works the laws of MAGIC may be carefully codified, often through the elaborate systems of mysticism found in ritual ALCHEMY or in the CABBALA, or by using the "laws of magic" devised by Sir James FRAZER.

2. Many works appear purely fantastic but then are discovered, usually towards the end, through a foreshadowed or merely tacked-on shift of PERCEPTION, to be explicable in mundane terms (i.e., they are rationalized). Much of the Gothic fiction (◊ GOTHIC FANTASY) of the late 18th and early 19th centuries, perhaps most typically *The Mysteries of Udolpho* (**1794**) by Ann RADCLIFFE – parodied mercilessly in *Northanger Abbey* (**1818**) by Jane Austen (1775-1817) – was rationalized in this way, so that what appeared to be quasi-magical was in fact being stage-managed via elaborate machineries. In modern written fantasy, the explaining-away of the fantastic is comparatively rare, having for the most part been displaced to metafictions and PSYCHOLOGICAL THRILLERS like Theodore STURGEON's *Some of Your Blood* (**1961**). In the movies, this form of RF remained commoner until later – especially in CHILDREN'S FANTASY movies, where protagonists were very likely to wake and discover "it had all been a DREAM" – although the paradigm is an adult movie, *The* BEAST WITH FIVE FINGERS (*1946*). Unusually, the wizards in Dave DUNCAN's **The Seventh Sword** trilogy (**1988**) turn out to be masquerading technologists even though the world they inhabit is one in which the GODDESS and her magical servants regularly intervene.

3. Much contemporary fantasy turns out to be SCIENCE FANTASY, in that the explanations are managed not in terms of the *mundane* known but in terms of sf tropes. Thus in Julian MAY's **Saga of Pliocene Exile** (**1981-4**) the FAIRIES whom the adventurers meet in the past turn out to be aliens, the VAMPIRES in Barbara HAMBLY's *Those who Hunt the Night* (**1988**) have been changed by a virus with which it is possible to experiment in the laboratory, and so on. This sort of rationalized fantasy is often closely allied with, and intermingled with, REVISIONIST FANTASY and Slipstream. Perhaps the most common version of this is the trope whereby PARIAH ELITES of ELVES or Wiccans turn out to be, in our terms, mutants with psi powers (◊ TALENTS).

Sometimes, of course, the rationalization works the other way – genre fantasy has cheerfully naturalized standard sf icons by giving them a magical rationale or attaching their emotional weight to existing fantasy icons. For example, DRAGONS have acquired many of the characteristics of sf aliens in the course of being subjected to revisionist fantasy; Hambly's dragons in *Dragonsbane* (**1986**) have specifically arrived in the magical human world they now inhabit by winged flight from another planet.

Meaning **3** is often back-conflated with the other two, so that, for example, the petrification of a troll in Poul

ANDERSON's *Three Hearts and Three Lions* (**1961**) is a magical event with scientific consequences in the shape of the radioactivity released when carbon-based molecules change, by atomic fusion, into silicon – and so the CURSE associated with a dead troll's treasure is explained away as radiation sickness.

Much rationalized and quasi-rationalized fantasy takes place in ALTERNATE WORLDS; these are sometimes versions of the modern USA with magic substituted for technology, as in Robert HEINLEIN's "Magic Inc." (1950) and Poul ANDERSON's *Operation Chaos* (**1971**), and sometimes alternate histories in which historical process has favoured magic as much as technology, as in Orson Scott CARD's **Tales of Alvin Maker**, where an extended Cromwellian rule in England has concentrated the magically gifted as a breeding pool in the colonies to which they were deported.

In theory, RF which uses the real to explain away the fantastic is directly opposite to FANTASIES OF HISTORY, which provide fantastic explanations of the real; in practice, various works combine the two. Tim POWERS's *The Stress of Her Regard* (**1989**) at once provides a fantastic explanation of such phenomena as the tendency of Romantic poets to get TB, the persistence of the Austrian empire in Italy and the name and rituals of the Carbonari, yet offers that explanation – vampirism – in quasi-sf terms (silicon lifeforms).   [RK]

RATS ◊ MICE AND RATS.

RATS, THE US movie (*1982*). ◊ James HERBERT.

RAWN, MELANIE (ROBIN) (?1953-   ) US writer, most of whose work falls into two linked series, with a third and unconnected series underway.

The **Dragon Prince** series – *Dragon Prince* (**1988**), *Dragon Prince #2: Star Scroll* (**1989**) and *#3: Sunrunner's Fire* (**1990**) – features a standard semi-medieval FANTASYLAND divided into a number of small kingdoms, the dynastic squabbles of the ruling classes of which form most of the plot material. MR is a writer of DYNASTIC FANTASY in which the fantasy elements are for the most part local colour in a plot whose emotional resonances have as much to do with the sexual and financial shenanigans of soap opera as with fantasy. Nonetheless, the books feature the ability of some of the characters to wield magic fire as a weapon of enormous power and others to use a more generalized MAGIC, and include non-sentient DRAGONS who are used by the heroes to sniff for gold ore.

The **Dragonstar** series, set in the same fantasyland a generation later, comprises *Dragonstar #1: Stronghold* (**1990**), *#2: The Dragon Token* (**1992**) and *#3: Skybowl* (**1993**). Rohan, hero of the first series and now High King, faces complicated succession crises and the massive incursion of genocidally inclined barbarians from another continent. These barbarians, who travel in longships and have most of the standard attributes of Vikings, have the use of COLD IRON which renders them largely immune to the magic Rohan and his people rely on. MR allows the tone of this second sequence to become rather bleaker than expected.

The **Exiles** series, begun with *Exiles #1: The Ruins of Ambrai* (**1994**), deals with the aftermath of a magical catastrophe which has made certain sorts of magic workers a fairly standard PARIAH ELITE.   [RK]

Other works: *Quantum Leap: Knights of the Morningstar* * (**1994**), tv tie.

RAYMOND, ALEX(ANDER GILLESPIE) (1909-1956) Versatile and very influential US COMIC-strip creator and illustrator with a clear, bold, precise line style, among whose creations is one of the most abidingly popular and long-running newspaper strips in the world - FLASH GORDON. AR is, along with Harold R. FOSTER, one of the most celebrated comics artists of all time, and is quoted by almost every other creator in the field as a major influence. He utilized a pen and dry-brush technique, which when he started was more commonly used in magazine illustration; his technique was imitated so widely that, by the end of the 1930s, it was widespread in comics. He had a fertile imagination and great skill in delineating character.

AR worked on Wall Street until the crash of 1929, when he enrolled in the Grand Central School of Art. He became art assistant to Russ Westover on **Tilly the Toiler** in 1930, then worked with Lyman Young on **Tim Tyler's Luck**. He attempted to interest King Features Syndicate in an idea for a series about a group of scientsts who take a rocketship to another world; the idea was initially rejected, but then he reduced the number of scientists to one and added a pretty girl, and **Flash Gordon** was the result. At the same time he created two other features: **Secret Agent X9**, a police thriller scripted by Dashiell Hammett (1894-1961), and **Jungle Jim**, an exotic adventure strip. All three began publication in January 1934, but within less than two years the workload had become too great and responsibility for **Secret Agent X9** art was passed to Charles Flanders. AR continued to draw his other two features until 1944, when he joined the US Marine Corps, returning to strips in 1946 to create the detective series **Rip Kirby**. At the height of his fame he died in a car accident.   [RT]

RAYNER, OLIVE PRATT ◊ Grant ALLEN.

RAYNER, WILLIAM (1929-   ) UK writer in whose *Stag Boy* (**1972**) an ailing young English lad finds a horned helmet in a burial ground, and discovers, when he puts it on, that it transfers his consciousness into that of the great stag it had immemorially represented (◊ TOTEM). This MAGIC increasingly encloses the lad in the stag's mind; he is freed only after a violent scene in which the stag is killed by hunters. It is a powerful tale.   [JC]

READ, [Sir] HERBERT (EDWARD) (1893-1968) UK poet, critic and novelist whose *The Green Child: A Romance* (**1935**) can be understood either as utopian sf or as a fantasy in which a standard utopian vision – the initial South American society which the protagonist creates, then abandons – is literally absorbed into the UNDERWORLD realm where he finds surcease.   [JC]

See also: GREEN CHILD; UTOPIAS.

READ THE SMALL PRINT Fantasy CONTRACTS and PACTS WITH THE DEVIL tend to be reinterpreted through QUIBBLES. "Small print" issues are not so much quibbles as built-in consequences whose importance one contracting party fails to see. In Fredric BROWN's short "Nasty" (in *Nightmares and Geezenstacks* coll **1961**) an elderly lecher's WISH brings him enchanted swimming-trunks which restore youthful potency – *until* he removes them. John Dickson CARR's TIMESLIP detection *The Devil in Velvet* (**1951**) has a protagonist who believes himself immune to anger and accepts the CONDITION that losing his temper may lead to brief POSSESSION by a violent spirit. The DEVIL takes this clause very literally: following a momentary political annoyance, the hero "wakes" to find that in a few minutes of swordplay he has killed a man.   [DRL]

REAL BOY In Carlo COLLODI's *The Adventures of Pinocchio* (**1883**) and its DISNEY adaptation PINOCCHIO (*1940*), this is the reward of the PUPPET hero: to become human and "real". (A RECURSIVE-FANTASY echo features in DUCKTALES: THE MOVIE – TREASURE OF THE LOST LAMP [*1990*], where a transformed GENIE cries, "I'm a real boy!") PINOCCHIO's METAMORPHOSIS by the Blue FAIRY is itself the happy ending – but for Hans Christian ANDERSEN's Little MERMAID, her pain-wracked humanity seems a THING BOUGHT AT TOO HIGH A COST. More generalized RB transformations are undergone by the GOLEMS Grundy in Piers ANTHONY's *The Source of Magic* (1979), who becomes an elf, and Dorfl in Terry PRATCHETT's *Feet of Clay* (1996), who remains clay but gains speech and volition. Physical change follows from mental change: all these creatures discover that learning to care for human beings takes them halfway to RB status. Humanity comes with many responsibilities, and so the clockwork man Tik-Tok in L. Frank BAUM's OZ rejoices in *not* being an RB. [DRL]

REALITY "I've wrestled with reality for 35 years," says Elwood in HARVEY (*1950*), "and I'm happy, doctor: I finally won out over it." The appeal of FANTASY can be interpreted as lying in the dynamic relationship between its domain and what we would normally call reality. Fantasy can be an alternative to reality, can posit an ALTERNATE REALITY (or an array of them), can throw up "unrealized realities", can provide a putative explanation for today's reality (as in FANTASIES OF HISTORY, especially those dealing with secret societies/conspiracies), or can, in GENERIC FANTASY, create a SECONDARY WORLD or FANTASYLAND (effectively, a reality that is isolated from ours).

A major notion in fantasy is that our everyday reality is uncertain; through PERCEPTION or otherwise we may discover it to be only the *surface* of the true reality, which is infinitely more complicated than we think – a notion explored in, for example, Thomas PALMER's *Dream Science* (1990) and Lisa TUTTLE's *Lost Futures* (1992). Such novels are concerned with the *structure* of reality; a further major fantasy concern is with the relationship between objective and subjective realities. A trivial example of this is the type of RECURSIVE FANTASY in which fictional and historical characters interact – e.g., Kim NEWMAN's *Anno Dracula* (1992), where DRACULA, JACK THE RIPPER, JEKYLL AND HYDE *et al.* co-exist with Queen Victoria. More significant are works in which a created reality has or comes to have equal weight with our mundane one – COOL WORLD (*1992*), STAY TUNED (*1992*) and John GRANT's *The World* are examples. The "message" of such fantasies may be that all things are real, all possibilities reified; or, in corollary, that (as John Lennon put it) "nothing is real". [JG]

REALMS OF FANTASY US large-format slick MAGAZINE, bimonthly, October 1994-current, published by Sovereign Media, Herndon, VA; ed Shawna McCarthy (1954-   ).

A fantasy companion to *Science Fiction Age*, *ROF* has placed heavy emphasis on FANTASY ART, with full-colour interior illustrations by top artists including Bob Eggleton (1960-), Brian FROUD, James GURNEY and Don MAITZ. The fiction emphasizes HIGH FANTASY, though a blend of historical and mythographic LOW FANTASIES have also developed, especially drawing upon FOLKLORE roots. Authors include Louise COOPER, Neil GAIMAN, Tanith LEE, Connie WILLIS and Roger ZELAZNY, along with features by Terri WINDLING on folk roots and Gahan WILSON on books. [MA]

REAMY, TOM Working name of US writer and graphic designer Thomas Earl Reamy (1935-1977), whose first published story was "Twilla" in the September 1974 issue of *F&SF* and who won the 1976 Nebula AWARD for "San Diego Lightfoot Sue" (1975). Reamy's comparatively early death, within three years of his first publication, robbed US sf/fantasy of a significant talent – he also won the 1976 John Campbell AWARD. His only novel, *Blind Voices* (1978), published posthumously in a complete but not final draft, deals with the arrival and stay, in a small town in 1930s Kansas, of a travelling CIRCUS whose personnel include freaks; their position in relationship to the show's proprietor is one of BONDAGE from which the novel's slightly tentative and elegiac ending provides a measure of liberation. The sf pretext for their existence – genetic engineering – has little to do with the feel or plot of the novel, which can accordingly be regarded as RATIONALIZED FANTASY. It won the 1979 BALROG AWARD.

TR's short stories, collected as *San Diego Lightfoot Sue and Other Stories* (coll 1979), eclectically combine sf, HORROR and fantasy materials, often in the same story. "Insects in Amber", for example, has a disparate group of travellers, two of them telepaths meeting for the first time, trapped in a house by a MALIGN SLEEPER and helped by a GHOST. TR had a clear gift, only imperfectly realized at the time of his death, for taking the reader through several emotional extremes in the same tale to a culmination of cathartic HEALING. TR worked on the parodic *Flesh Gordon* (*1974*). [RK]

RE-ANIMATOR (vt *H.P. Lovecraft's Re-Animator*) US movie (*1985*). Empire. **Pr** Brian Yuzna. **Exec pr** Michael Avery, Bruce Curtis. **Dir** Stuart Gordon. **Spfx** Anthony Doublin, John Naulin. **Screenplay** Gordon, William J. Norris, Dennis Paoli. **Based very loosely on** "Herbert West, Re-Animator" by H.P. LOVECRAFT. **Starring** Bruce Abbott (Dan Cain), Jeffrey Combs (Herbert West), Barbara Crampton (Megan Halsey), David Gale (Dr Carl Hill), Robert Sampson (Dean Alan Halsey). 86 mins. Colour.

This black comedy is a PARODY of the FRANKENSTEIN MOVIES, transported to the campus of a modern US university – in fact, Miskatonic Medical School, Arkham, Mass. Herbert West has left Zurich University under something of a cloud, his experiments having apparently killed his tutor, the renowned Dr Gruber. Finding himself studying under Gruber's rival, Hill, he continues his experiments in trying to bring the dead back to life . . . starting with the cat of flatmate Cain, who is sleeping with Megan Halsey, daughter of the dean. Blood flies as Dean Halsey is killed and revived and then Hill, becoming murderous, proves almost unkillable. The finale sees a troop of vengeful ZOMBIES on the rampage.

Exceptionally tasteless and full of gratuitous gore and nudity, this cod TECHNOFANTASY has no redeeming characteristics whatsoever except that it is often – in large part thanks to Combs's po-faced portrayal of the obsessed West – genuinely funny. The sequel, *Re-Animator II* (*1989*), lacks this flair. [JG]

REAVES, J(AMES) MICHAEL (1950-   ) US writer, as well known for sf as for fantasy; he has also been a prolific producer of teleplays for children's Saturday-morning slots. JMR began publishing work of genre interest with "The Breath of Dragons" for *Clarion 3* (anth 1973) ed Robin Scott Wilson. *Dragonworld* (1979; rev 1983) with Byron PREISS is a YA adventure set in several FANTASYLAND kingdoms; the protagonist travels to the country of the DRAGONS to discover who has, in order to foment internecine strife, been murdering children. *Darkworld Detective* (coll of linked stories 1982) is comic SCIENCE FANTASY set on a Fantasyland planet, with a doltish detective and a missing

DARK LORD. The **Shattered World** sequence – *The Shattered World* (**1984**) and *The Burning Realm* (**1988**) – is set in a fantasized future, after Earth has been sundered into hundreds of planetoids whose intersecting orbits are governed by RUNES. The protagonist undergoes occasional METAMORPHOSIS into a bear and has an uneasy relationship with a manipulating sorcerer and more than one ENCHANTRESS; there are many dragons and other standard creatures. *Street Magic* (**1991**), again YA, is an URBAN FANTASY whose protagonist – a kind of HIDDEN MONARCH, unaware that he is a Lord of FAERIE – must discover his heritage and save his fellows. JMR is a smooth writer who has not yet stretched his gift. [JC]

Other works: *I, Alien* (**1978**); *Hellstar* (**1984**) with Steve PERRY; *Time Machine 3: Sword of the Samurai* * (**1984**) with Perry; *Dome* (**1987**) with Perry.

REBIRTH While REINCARNATION refers to a return to Earth in a fresh life, "rebirth" usually refers to a dramatic renewal of the living. The notion is often connected with religious revelation, as when zealous converts describe themselves as "born again"; it was discussed by C.G. Jung (1875-1961) as a product of the collective unconscious, manifested in such myth-images as the PHOENIX.

Orthodox fantasies of individual rebirth include *The God Within Him* (**1926**) by Robert HICHENS and *All or Nothing* (**1928**) by John BERESFORD, but images of quasitranscendental rebirth are more interesting. The idea is particularly common in fantasies which are assimilated to sf – notable examples are "Desertion" (1944) by Clifford D. SIMAK and *Nightwings* (**1969**) by Robert SILVERBERG – as well as earnest symbolic fantasies like *The Green Child* (**1935**) by Herbert READ and *The Passion of New Eve* (**1977**) by Angela CARTER. The notion of a rebirth of the whole world is featured in, again, fantasies on the borders of sf, like *Time's Dark Laughter* (**1982**) by James Kahn (1947-   ) and Gene WOLFE's *Book of the New Sun* (**1980-83**); it is a consistent feature of the work of some writers located in this grey area, including R.A. LAFFERTY in novels like *Fourth Mansions* (**1969**) and Storm CONSTANTINE in novels like *Sign of the Sacred* (**1993**). The rebirth of the entire Universe is a fairly common notion in sf: a fantasy example is John GRANT's *The World* (**1992**). [BS]

RECLUSE, THE US amateur MAGAZINE, quarto, 1 issue, 1927, published and ed W. Paul Cook (1880-1948).

TR is of interest because it was the first SMALL-PRESS magazine produced within the fantasy movement, in particular the H.P. LOVECRAFT circle of friends. Although the contents were not exclusively SUPERNATURAL FICTION, many were written by *WT* contributors. It included fiction and poetry by Frank Belknap LONG, H. Warner MUNN, Clark Ashton SMITH and Donald WANDREI, as well as the first appearance of Lovecraft's essay "Supernatural Horror in Literature". A facsimile was produced by Moshassuck Press in 1990.

Cook later assembled 4 issues of *The Ghost* (Spring 1943-July 1946), which contained mostly nonfiction, again concentrating on the Lovecraft circle. It was most noted for its series of reminiscences by E. Hoffmann PRICE. [MA]

RECOGNITION Stories have a habit of getting tied in knots, and then unfolding. First they entangle their protagonists, whose actions sometimes seem dictated by the needs of the STORY in which they have become engaged; then the light dawns, and the LABYRINTH becomes a path. For much mimetic or realistic literature, this process – inherent to the telling of Story – proves to be an embarrassment, a scandalous admission that fiction is an artifice. For most non-mimetic literature, the opposite is the case: the literatures of the FANTASTIC positively glory in the fact that they present, and embody, Story-shaped worlds.

Unsurprisingly, the two modes – realism and the fantastic – differ radically in the degree of prominence they grant to the central TRANSFORMATION scene which is defined – after Aristotle – as the moment of anagnorisis, or Recognition. For Aristotle, Recognition marks a fundamental shift in the process of a story from increasing ignorance to knowledge. As Terence Cave defines it in *Recognitions: A Study in Poetics* (**1988**): "It is the moment at which the characters understand their predicament fully for the first time. . . it makes the world (and the text) intelligible. Yet it is also a shift *into* the implausible: the secret unfolded lies beyond the realm of common experience; the truth discovered is 'marvellous' . . . the truth of fabulous myth of legend . . . The interest of recognition scenes in drama and narrative fiction is perhaps that, more than any other literary motif or element, they have the character of an old tale."

It is at this moment of Recognition that the inherent Story at the heart of most full fantasy texts is most visible, most "artificial", and most revelatory. At this moment in "the structurally complete fantasy tale" (Brian ATTEBERY's phrase) protagonists begin to understand what has been happening to them (he may have been an UGLY DUCKLING awaiting the moment he becomes king; she may have been re-enacting a CREATION MYTH in order that the LAND be reborn; they may discover what ARCHETYPE serves as an UNDERLIER figure and defines their fate; etc.). They understand, in other words, that they are in a Story; that, properly recognized (which is to say properly told), their lives have the coherence and significance of Story; that, in short, the Story has been telling *them*.

At this moment, characters might be thought of as gazing simultaneously into the past and into the future – backwards at the BONDAGE or AMNESIA of their beginnings (when their story was still leading them INTO THE WOODS), forwards through resolutions and (perhaps) METAMORPHOSIS at the MIRROR of the future which reflects their true being. It is a moment which may be signalled by TROMPE L'OEIL effects, when two REALITIES (the past and the future) dance in one moment or body, before TIME moves again, and they become incompossible. Story itself, at the moment of Recognition, seems to hold still in order to be recognized; and then – illuminated, as though the truth had made it free – begins to move. In fantasy texts this recuperative, inward-turning, heartlifting moment of Recognition is analogous to the moment of conceptual breakthrough (◊ *SFE*) which defines the essential thrust outwards towards increased knowing in an sf text. Conceptual breakthrough leads through the barrier to a realization of what the world is; Recognition is an acknowledgement that one has been there all the time.

The essential TAPROOT TEXTS which shaped the use of Recognition in fantasy are probably the late ROMANCES of SHAKESPEARE: *Pericles* (performed *c*1608; **1609**); *The Winter's Tale* (performed *c*1610; **1623**); *Cymbeline* (performed *c*1611; **1623**); and *The Tempest* (performed *c*1611; **1623**). Each of these plays is built with exquisite selfconscious "artifice" around convoluted, TIME-drenched stories which climax in moments or "motors" of Recognition "in which", Cave points out, "persons and their identities are displaced and recovered". Characters who are both identifiable figures and ICONS are like characters in a FAIRY-TALE; characters who are transparent to the Story that drives them are also central to fantasy. And almost every GODGAME tale, from *The Tempest* on, climaxes in scenes of Recognition.

A famous example of recuperative recognition is the moment in J.R.R. TOLKIEN's *The Lord of the Rings* (**1954-5**) when the eagles arrive during the last battle – an act prefigured in *The Hobbit* (**1937**) – and foreshadow the saving of the day, a foreshadowing recognized by all. A lesser-known example is in Peter S. BEAGLE's *The Last Unicorn* (**1968**), at the point when the saving transformations still hang in the balance, and the world stills utterly – then moves into renewal. A similar scene appears at the climax of Susan COOPER's *Seaward* (**1983**), when the girl protagonist decides not to follow her SELKIE nature and the boy protagonist decides not to pass on to the HEAVEN of TIR-NAN-OG. As with Beagle's characters, the moment takes place on the natural THRESHOLD of a beach, where the elements mix with TROMPE L'OEIL abandon.

The second half of Michael ENDE's *The Neverending Story* (**1979**) tells of its young protagonist's long search – after he has successfully saved Fantastica from the disbelief of the world – for his own self, which has become intrinsicate with the Land, but which he must find or he will never recover from the amnesia which is progressively thinning him (and Fantastica). Almost too late, he comes to a Goddess of vegetation, who feeds him and tells him he has gone "the long way around", being "'one of those who can't go back until they have found the fountain from which springs the Water of Life. And that's the most secret place in Fantastica. There's no simple way of getting there'. After a short silence she added: 'But every way that leads there is the right one.'" He finds the Water, within the jaws of a great WORM OUROBOROS and bathes, and recovers himself (and the Story he has been creating). The tale ends.

Almost all of the work of John CROWLEY can be seen as tied to moments of Recognition: the various protagonists of both *Little, Big* (**1981**) and the **Aegypt** sequence (**1987-94**) are obsessively concerned with the true nature of the Story that is telling them, for only when they have an inkling of that true nature can they know how to pass through the Labyrinth of climax; both *Little, Big* and **Aegypt** are enormously extended exercises in how to narrate the moment of cusp, the coital moment of the coming of Recognition; and to the extent they are examples of GNOSTIC FANTASY, they represent the passage through Recognition as a transition from darkness and sleep into true REALITY. Both Sharon SHINN's *The Shape-Changer's Wife* (**1995**) and William Browning SPENCER's *Zod Wallop* (**1995**) come to climaxes of Recognition. In Shinn's tale in particular the pattern is clear. Her protagonist's reluctant movement towards recognizing the eponymous wife's true nature – her name is Lilith, and she is a DRYAD who has been SHAPESHIFTED by a sorcerer into the bondage of human form, remaining so immured until the last pages of the novel – comes to a climax during the intricate moments when he recognizes his own nature as well, and prepares to assume the adult powers he had been eschewing.

In full-fantasy texts, Recognition marks the moment when the Story means itself. [JC]

RECURSIVE FANTASY 1. RF exploits existing fantasy settings or characters as its subject matter. But, since fantasy is an ancient mode of Story, full of retellings, TWICE-TOLD tales and recurring figures of MYTH (◊◊ ARCHETYPES; UNDERLIERS), it is useful to qualify this definition by requiring that RF deal with a *specific* former fiction. ARTHUR, for example, has boiled so long in the CAULDRON OF STORY as to become a universal ICON, detached from any single telling; the same has arguably happened in this century to DRACULA and SHERLOCK HOLMES. Fantasy forms related to RF include: fictional PARODY, such as Henry N. Beard's and Douglas C. Kenney's clownish reworking of J.R.R. TOLKIEN in *Bored of the Rings* (**1969**); pastiche SEQUELS BY OTHER HANDS; and REVISIONIST-FANTASY re-examinations of FAIRYTALES etc., as in Tanith LEE's *Red as Blood, or Tales from the Sisters Grimmer* (coll **1983**).

The flavour of true RF derives from intersecting levels of REALITY, with "real" protagonists encountering worlds and characters which are as "fictional" to them as to us. Examples include: Walter DE LA MARE's *Henry Brocken* (**1904**), whose hero meets various imaginary folk, including Jonathan SWIFT's Gulliver; John Myers MYERS's *Silverlock* (**1949**), whose Commonwealth of letters is an IMAGINARY LAND densely populated by fictions, from APULEIUS's Ass to the SHIP OF FOOLS to creations of Mark TWAIN and G.K. CHESTERTON; L. Sprague DE CAMP's and Fletcher PRATT's **Incomplete Enchanter** sequence, plunging modern Americans into stories which they know or think they know, including SPENSER's *The Faerie Queene* (**1590-6**) and ARIOSTO's *Orlando Furioso* (**1516**); Edward EAGER's *The Time Garden* (**1958**), whose author's debt to E. NESBIT is acknowledged in an encounter with some of her characters; the play *Rosencrantz and Guildenstern Are Dead* (**1967**) by Tom Stoppard (1937-), whose eponymous attendant lords are trapped in the machinery of SHAKESPEARE's *Hamlet* (**1603**), baffled and "real" when offstage, assured and "fictional" when speaking their lines; Larry NIVEN's and Jerry Pournelle's *Inferno* (**1976**), revisiting the HELL of DANTE ALIGHIERI; Marvin KAYE's *The Incredible Umbrella* (**1979**), which freewheels through a medley of W.S. GILBERT librettos, climaxing with Holmes and FRANKENSTEIN's monster in the 2D world of *Flatland* (**1884**) by Edwin A. Abbott (1839-1926); Robert A. HEINLEIN's *"The Number of the Beast"* (**1980**), which, though nominally sf, includes a visit to OZ; and Roger ZELAZNY's *The Changing Land* (**1981**) – whose EDIFICE setting is eventually transformed into that of William Hope HODGSON's *The House on the Borderland* (**1908**). Although John CROWLEY's *Little, Big* (**1981**) has no such overt appearance, its sheer density of allusions to Lewis CARROLL's work conveys a feeling of subliminal RF, or RF in disguise. A fine movie example is WHO FRAMED ROGER RABBIT? (**1988**), with its numerous TOON characters from past ANIMATED MOVIES.

2. RF's sense of tangled reality-levels may also be attained in metafictions without reference to past works. Celebrated examples include: *Six Characters In Search of an Author* (**1921**) by Luigi PIRANDELLO; Flann O'BRIEN's *At Swim-Two-Birds* (**1939**); Jorge Luis BORGES's "Tlön, Uqbar, Orbis Tertius" (1941), whose dizzying intellectual conspiracy proposes to relegate our real world to the status of fiction; and such images by M.C. ESCHER as the lithograph "Drawing Hands" (1948), where each of two pictured hands is engaged in drawing the other. William GOLDMAN's *The Princess Bride* (**1973**) dwells lovingly on the supposed editing of its eponymous fiction to present only the "good bits"; in Michael ENDE's *The Neverending Story* (**1979**) the child protagonist becomes literally absorbed into the Story of the

BOOK he is reading; the nostalgically evoked fantasy tales of Marshall France in Jonathan CARROLL's *The Land of Laughs* (**1980**) first leak characters into, and then prove to be scripts for, a small town's reality; the "real" and "fictional" diarists of Christopher PRIEST's *The Affirmation* (**1981**) create and reflect each other, like Escher's hands. Relevant movies include *The* PRINCESS BRIDE (*1987*), *The* NEVERENDING STORY (*1984*), and all those featuring characters' transitions between the worlds of movie and audience, using the screen as a PORTAL – like *The* PURPLE ROSE OF CAIRO (*1984*) and *The* LAST ACTION HERO (*1993*). [DRL]

RED BALLOON, THE (ot *Le Ballon Rouge*) French movie (*1955*). Films Montsouris. **Pr** Michel Pezin. **Dir** Albert Lamorisse. **Screenplay** Lamorisse. **Starring** Pascal Lamorisse (The Boy). 34 mins. Colour.

A simple tale of a little Parisian boy befriended by a red balloon seemingly possessed of free will; it even briefly dallies with a blue balloon. But eventually a gang of bullies seize it and burst it. On its "death" balloons all over Paris escape their owners and rally to the spot; they bear the boy aloft for a magical aerial journey to who knows where.

It is impossible to convey the extraordinary charm, poignancy and sense of MAGIC of *TRB*. It received La Palme d'Or at Cannes in 1956, and La Medaille d'Or du Cinéma Français in the same year; it was a huge international success, and is justly regarded not just as a classic of the fantasy CINEMA but also as probably the cinema's finest short. [JG]

REDEMPTION Salvation – deliverance from a previous state, usually EVIL, a frequent trope of CHRISTIAN FANTASY. Otherwise the term generally refers to a turning point of STORY, a point when the RITE OF PASSAGE or QUEST is resolved (◊ RECOGNITION), and when the protagonist *turns back* towards home, towards the face of GOD. A redeemed protagonist is the true protagonist at last. Also, it is frequently the case in fantasy – and in most other storytelling genres – that a semi-major character portrayed as or actually bad turns out to be good in the end, or does some final good act, thus redeeming her/himself either outside or within the bounds of the STORY. Although there are a few exceptions, it is a characteristic of HORROR that those who start off bad stay that way: there is no redemption. [JC/JG]

REDGROVE, PETER (WILLIAM) (1932-   ) UK writer, best-known as a poet, who wrote his first two novels in collaboration with his wife Penelope Shuttle (1947-   ). These were *The Terrors of Dr Treviles: A Romance* (**1974**) and *The Glass Cottage: A Nautical Romance* (**1976**), complex and rather overblown FABULATIONS which place their settings in a vivid metaphysical context somewhat after the fashion of John Cowper POWYS. PR's solo novels employ some sf motifs in developing highly eroticized OCCULT FANTASIES. *The God of Glass: A Morality* (**1979**) describes the career of a tainted MESSIAH. *The Sleep of the Great Hypnotist: The Life and Death and Life After Death of a Great Magician* (**1979**) concerns the attempted resurrection of a similarly megalomaniac character by his daughter, placed in BONDAGE by a mesmeric machine of his invention. *The Beekeepers: A Novel* (**1980**) and its sequel, *The Facilitators, or Mister Hole-in-the-Day*, also feature bizarre research projects conducted by characters of doubtful sanity in the attempt to obtain unholy power over their fellows and the prescribed order of existence. The tales in *The One who Set Out to Study Fear* (coll **1989**) – recast from originals by the Brothers GRIMM, with the ideological revisions

expectable in such work (◊ REVISIONIST FANTASY) – are not so conscientiously *avant garde*. [BS]

RED SONJA US movie (*1985*). MGM-United Artists/Famous Films/Dino De Laurentiis. **Pr** Christian Ferry. **Exec pr** A. Michael Lieberman. **Dir** Richard Fleischer. **Spfx** John Stirber, Universal City Studios. **Screenplay** Clive Exton, George Macdonald Fraser. **Based on** the creations of Robert E. HOWARD as further developed in the **Red Sonja** series by David C. SMITH and Richard Tierney. **Starring** Sandahl Bergman (Gedren), Brigitte Nielsen (Sonja), Arnold Schwarzenegger (Kalidor). 89 mins. Colour.

A thematic continuation of the series started with *Conan the Barbarian* (*1981*; ◊ CONAN MOVIES), this casts a woman, Sonja, in the CONAN role. Because Sonja has rejected wicked Queen Gedren's lesbian advances, the Queen slaughters her family. A SPIRIT appears to grant Sonja the strength that will enable her to exact her REVENGE. Assisted by Kalidor – an up-market Conan with less personality – Sonja eventually destroys Gedren and a TALISMAN. Morally dubious (Gedren's lesbianism is depicted as one of her EVIL attributes) and worse-acted than words can explain, *RS* is a great embarrassment. [JG]

REED, JEREMY (1951-   ) UK poet, critic and writer whose best poetry – unusually for a UK writer – has grown from the model of Surrealist poets like David Gascoyne (1916-   ), and whose novels consistently depend on implied IDENTITY EXCHANGES, episodes of POSSESSION, and DOUBLES; most of his work focuses on the nature of DECADENCE in a style which has reminded some critics of Ronald FIRBANK. In *Blue Rock* (**1987**) this thematic material is shaped into a SUPERNATURAL FICTION, one of whose protagonists is possessed by an animal spirit. Later novels often recount the lives of real 19th-century French figures through explicit (though fabulated) use of fantasy devices. *Red Eclipse* (**1989**) mixes occult psychic invasions with the imaginary journal of Charles BAUDELAIRE's black mistress. *Isidore: A Novel about the Comte de Lautréamont* (**1991**) features an interview with the dead novelist (◊ Comte de LAUTRÉAMONT; POSTHUMOUS FANTASY); in *When the Whip Comes Down: A Novel about de Sade* (**1992**) the Marquis DE SADE engages in TIMESLIP excursions to the 20th century; *Chasing Black Rainbows: A Novel about Antonin Artaud* (**1994**) continues the sequence, without moving into fantasy. *Diamond Nebula* (**1994**), set in the 23rd century, features a protagonist obsessed with Decadents of the later 20th century, including J.G. BALLARD. [JC]

REEVE, CLARA (1729-1807) UK writer of sentimental fiction, sometimes with a Gothic element. She is relevant to the development of fantasy for *The Champion of Virtue* (**1777**; vt *The Old English Baron* 1778), which contains a supernatural element – an informative GHOST – but more importantly introduces into the concept of the EDIFICE a central haunted and/or forbidden chamber, where (in this case) the hero must spend several nights of redemptive (◊ REDEMPTION). [JC]

REEVES, JAMES Working name of UK poet, writer and anthologist John Morris Reeves (1909-1978), brother of Joyce GARD, intensely prolific from the mid-1930s, when he began publishing anthologies like *The Modern Poet's World* (anth **1935**; rev vt *Poet's World* 1948), some of which – like the massive *The Golden Land: Stories, Poems, Songs New and Old* (anth **1958**) – were assembled for children. His copious verse began with *The Natural Need* (coll **1936**); some of this poetry, like *The Wandering Moon* (coll **1950**), was for children.

His prose for children can be divided into two categories. There are original stories like *Pigeons and Princesses* (**1956**) illus Edward ARDIZZONE and *The Strange Light* (**1964**), in which a young girl discovers a fragile POLDER inhabited by characters awaiting the chance to enter a BOOK. When an author conceives of them, they begin to shine; but if they are never imagined they fade, like J.M. BARRIE's Tinkerbell. And there are many volumes of TWICE-TOLD tales, beginning with *English Fables and Fairy Stories, Retold* (coll **1954**) and *The Exploits of Don Quixote, Retold* (**1959**) illus Ardizzone. JR is a quiet writer, but through the transparency of his style can be detected a sense of the constant wonderfulness within things and stories. [JC]

Other works:

For children: *Mulcaster Market: Three Plays for Young People* (coll **1951**); *The King who Took Sunshine* (**1954**), play; *Mulbridge Manor* (**1958**); *Titus in Trouble* (**1959**) illus Ardizzone; *Sailor Rumbelow and Britannia* (**1962**) illus Ardizzone; *The Pillar-Box Thieves* (**1965**); *Rhyming Will* (**1967**) illus Ardizzone; *Mr Horrox and the Gratch* (**1969**) illus Quentin BLAKE; *The Path of Gold* (**1972**); *The Lion that Flew* (**1974**) illus Ardizzone; *Clever Mouse* (**1976**); *Eggtime Stories* (coll **1978**); *The James Reeves Storybook* (coll **1978**) illus Ardizzone; *A Prince in Danger* (**1979**).

Twice-told tales: *Fables from Aesop, Retold* (coll **1961**); *Three Tales, Chosen from Traditional Sources* (coll **1965**) illus Ardizzone; *The Road to a Kingdom: Stories from the Old and New Testaments* (coll **1965**) illus Richard KENNEDY; *The Secret Shoemakers and Other Stories* (coll **1966**) illus Ardizzone; *The Cold Flame, Based on a Tale from the Collection of the Brothers Grimm* (**1967**) illus Charles KEEPING; *The Trojan Horse* (**1968**); *Heroes and Monsters: Legends of Ancient Greece Retold* (coll **1969/1971** 2 vols); *The Angel and the Donkey* (**1969**) illus Ardizzone; *Maildun the Voyager* (**1971**); *How the Moon Began* (**1971**) illus Ardizzone; *The Forbidden Forest and Other Stories* (coll **1973**) illus Raymond BRIGGS; *The Voyage of Odysseus: Homer's Odyssey Retold* (**1973**); *Two Greedy Bears* (**1974**); *Quest and Conquest: Pilgrim's Progress Retold* (**1976**); *Snow-White and Rose Red* (**1979**).

Anthologies (selective): *The Merry-Go-Round: A Collection of Rhymes and Poems for Children* (anth **1955**); *The Christmas Book* (anth **1968**) illus Briggs; *The Springtime Book: A Collection of Prose and Poetry* (anth **1976**); *The Autumn Book: A Collection of Prose and Poetry* (anth **1977**).

REEVES-STEVENS, (FRANCIS) GARFIELD (1953-) Canadian writer, better-known for HORROR and sf than for fantasy. His first novel, *Bloodshift* (**1981**), combines both genres in a complicated tale involving a contract killer and a renegade female VAMPIRE. *Dreamland* (**1985**), a TECHNOFANTASY with horror elements, is set in a Disneyland-like themepark which is also a BAD PLACE. *Nighteyes* (**1989** US) deals with UFOS. GR-S's outright fantasy is restricted to the **Chronicles of Galen Sword** sequence, done with Judith Reeves-Stevens: *Shifter* (**1990**) and *Nightfeeder* (**1991**), with sequels clearly necessary. Galen Sword is a rich young human in an URBAN-FANTASY version of NEW YORK who is obsessed by flashes of memory of his life in at least one other ALTERNATE REALITY; he sets off on a QUEST, with hired COMPANIONS, to penetrate the AMNESIA which blocks him from his proper home and role in the First World. The sophisticated use of fantasy devices in **Galen Sword** leads at times to expectations which may be

higher than the authors intend to meet; frustratingly, GR-S seems to lack the ambition his skill demands. [JC]

Other works: *Children of the Shroud* (**1987**); 3 **Star Trek** ties, all with Judith Reeves-Stevens – *Memory Prime* * (**1988** US), *Prime Directive* * (**1990** US) and *Federation* * (**1994** US); *The Making of Star Trek: Deep Space Nine* * (**1994** US), nonfiction; *Dark Matter* (**1990** US); *Alien Nation #1: The Day of Descent* * (**1993** US) with Judith Reeves-Stevens, tied to the 2nd (cancelled) series of the tv sf show.

REEVES-STEVENS, JUDITH (EVELYN) (?1953-) Canadian writer, all of whose works to date have been with Garfield REEVES-STEVENS. [JC]

REH: LONE STAR FICTIONEER ◊ CHACAL.

REICHERT, MICKEY ZUCKER (1962-) Working name of Miriam S. Zucker Reichert, who began publishing fantasy with *Godslayer* (**1987**), in which the warring gods of Norse MYTH (◊◊ NORDIC FANTASY) shanghai a US soldier from the Vietnam War for their own purposes. MZR soon became a prolific and successful writer of commercial fantasy, following *Godslayer* with four further volumes of what was soon called **The Bifrost Guardians**: *Shadow Climber* (**1988**), *Dragonrank Master* (**1989**), *Shadow's Realm* (**1990**) and *By Chaos Cursed* (**1991**). With **The Renshai Trilogy** – *The Last of the Renshai* (**1992**), *The Western Wizard* (**1992**), *Child of Thunder* (**1993**) – MZR began to write much larger books, which continue to make free use of Norse elements to tell an adventure with a large scope and cast. *The Legend of Nightfall* (**1993**) is a singleton, and *The Unknown Soldier* (**1994**) is sf. With *Beyond Ragnarok* (**1995**) MZR began a new series, **The Renshai Chronicle**, which is continued with *Prince of Demons* (**1996**). [GF]

REID, DESMOND [s] ◊ Michael MOORCOCK.

REID, FORREST (1876-1947) Irish writer, author of studies of two close friends, William Butler YEATS and Walter DE LA MARE. His most famous novel is *Peter Waring* (**1937**). Of genre interest are *Pender Among the Residents* (**1922**), a SUPERNATURAL FICTION about a house haunted by an unresolved scandal (◊ HAUNTED DWELLINGS), and *Demophon: A Traveller's Tale* (**1927**), which invokes the pagan GODS. [JC]

REID BANKS, LYNNE (1929-) UK writer who remains best-known for her first novel, *The L-Shaped Room* (**1960**), set in contemporary London; it was filmed in 1962. As a fantasy writer LRB has concentrated on YA work, though *The Farthest-Away Mountain* (**1976**), in which a girl and a frog who was once a prince must seek the eponymous domain to break a bad SPELL, is for a younger audience. The **Omri** sequence – *The Indian in the Cupboard* (**1980**), *The Return of the Indian* (**1986**), *The Secret of the Indian* (**1989**) and *The Mystery of the Cupboard* (**1993**) – focuses on a cupboard with a magic key. When a TOY figure – e.g., a toy Indian – is put inside the cupboard, it comes to life; or, rather, the key operates as a TIME-TRAVEL device which transports real persons into the 20th century, transforming them (◊ BONDAGE) into live toy-sized beings. Eventually Omri and his friends understand that the METAMORPHOSES to which they are subjecting real people – who tend at first to think they have died and entered an AFTERLIFE populated by enormous, fitful gods – are indeed literal imprisonments. Using the key, the mortal protagonists themselves travel into the past. The later volumes of the series are gratifyingly complex as adventures and as examinations – cogently shaped for a young audience – of the moral consequences of action. The series was the basis for the US movie *The Indian in the Cupboard* (*1995*).

LRB's most impressive single work, *Melusine: A Mystery* (**1988**), subjects an adolescent boy to an enriching RITE OF PASSAGE into sexual awareness when he and his family spend the summer in a French château inhabited by a surly teenage girl whose powers of METAMORPHOSIS are complexly interconnected with a shadowy pattern of enforced incest. [JC]

Other works: *The Adventure of King Midas* (**1974**); *I, Houdini: The Autobiography of a Self-Educated Hamster* (**1978**); *Maura's Angel* (**1984**); *The Fairy Rebel* (**1985**).

REINCARNATION The notion that the undying SOUL may be serially reincarnated is common to many myth-systems, but is particularly associated with Buddhism, which adds an element of moral judgement in asserting that the pattern of such reincarnations is determined by *karma*. Such ideas were enthusiastically taken up by writers of fantasies based in SPIRITUALISM and THEOSOPHY, which often take the form of "karmic romances" – whose troubled characters are deemed to be atoning for sins committed in former lives. Many karmic romances insist that true LOVE has the power to transcend TIME; the subgenre thus overlaps with the TIMESLIP romance. The most popular sources of reincarnated souls are ATLANTIS and ancient EGYPT.

Writers particularly devoted to the production of karmic romances include H. Rider HAGGARD – whose series begun with *She* (**1887**) might be reckoned the classic of the subgenre – Edwin Lester ARNOLD, Mrs Campbell PRAED, Dion FORTUNE, Shaw Desmond (1877-1960) and Joan GRANT. Notable further examples include *Ziska* (**1897**) by Marie CORELLI, *Cecilia* (**1902**) by F. Marion CRAWFORD, *A Son of Perdition* (**1912**) by Fergus HUME, *The Bridge of Time* (**1919**) by William Henry Warner, *Avernus* (**1924**) by Mary Bligh Bond (1895-), *When They Came Back* (**1938**) by Roy Devereux (real name Margaret R. Pember-Devereux), *I Live Again* (**1942**) by Warwick Deeping (1877-1950), *The Book of Ptath* (**1943**; **1947**) by A.E. VAN VOGT and *Alas, That Great City* (**1948**) by Francis Ashton (1904-). More refined karmic romances, less obsessed with erotic matters, include *The Star Rover* (**1915**) by Jack LONDON, *Julius Levallon* (**1916**) by Algernon BLACKWOOD, *The Man who was Born Again* (**1921**) by Paul Busson (1873-1924), *The Reincarnation of Peter Proud* (**1974**) by Max Ehrlich (1909-1983) and *Audrey Rose* (**1976**) by Frank De FELITTA (◊◊ AUDREY ROSE [**1977**]).

Perhaps surprisingly, stories in which humans are reincarnated as animals do not usually rationalize the move in terms of of karmic penalty; notable examples include "The Black Spaniel" (1905) by Robert HICHENS, "The Professor's Mare" (1913) by L.P. JACKS, "Don Juan in the Arena" (1934) by Christina Stead (1902-1983) and *Fluke* (**1977**) by James HERBERT. Don MARQUIS's *archy and mehitabel* (coll **1927**) stars as a poet reincarnated as a cockroach – the comic penalty for writing free verse; Lord DUNSANY's *My Talks with Dean Spanley* (**1936**) plays with animal-to-human reincarnation, featuring a Dean who dimly remembers being a spaniel. Reincarnation on other worlds was a particular fascination of the French astronomer Camille Flammarion (1842-1925); it is also featured in *Transmigration* (**1874**) by Mortimer Collins (1827-1876). *Contes philosophiques* considering various aspects of the notion include "Tulsah" (1896) by M.P. SHIEL, "The Choice: An Allegory of Blood and Tears" (1929) by Sydney Fowler Wright (1874-1965) and "The Vitanuls" (1967) by John BRUNNER. [BS]

See also: RESURRECTION.

RELIGION A term normally used here to designate a shared structure of belief, like Christianity or Wicca, rather than the fact that an individual person avows faith in a deity. Entries which refer to religious matters as part of human society include those dealing with ANTHROPOLOGY, CHRISTMAS, DEATH, DEMONS, FERTILITY, GODDESS, GODS and INITIATION. Entries which refer to religious matters in the context of fantasy include CHRISTIAN FANTASY, FANTASIES OF HISTORY, GNOSTIC FANTASY, HEAVEN, HELL, HISTORY IN FANTASY, INSTAURATION FANTASY, LIMBO, PARODY, PURGATORY and THINNING.

Non-Christian fantasies set in OTHERWORLDS frequently feature organized religions, usually depicted as corrupt, as in the background of Ursula K. LE GUIN's **Earthsea** sequence, or in the foreground of Terry PRATCHETT's *Small Gods* (**1992**). Individual faith is normally treated with greater sympathy. When they deal with religion, fantasies set in this world frequently focus upon faiths or belief structures under threat of THINNING from organized religion. As most fantasies dealt with in this book were written in Europe or the USA, the organized religion usually blamed for the fading away of the richness of the GOLDEN AGE is Christianity. The most sustained antireligious this-worldly fantasies of mourning are probably those of Thomas Burnett SWANN; but many fantasies set in recognizable versions of this world are unfriendly to establishment faiths. Often recently set tales like Esther FRIESNER's *Yesterday We Saw Mermaids* (**1991**) and Sara DALKEY's *Goa* (**1996**) also tend to treat forces like the Inquisition as external enemies. James Branch CABELL, Anatole FRANCE, Laurence HOUSMAN and others treat organized religion in terms of SATIRE. [JC]

RENAULT, MARY Pseudonym of UK historical novelist Mary Challans (1905-1983), long resident in South Africa. A few of her books are RATIONALISED FANTASIES based on LEGENDS. They include *The King Must Die* (**1958**) and *The Bull from the Sea* (**1962**), concerning Thesus and the MINOTAUR, with the former also bearing on the notion of the Year King (◊ GOLDEN BOUGH) – these two assembled as *The King Must Die/The Bull from the Sea* (omni **1992**) – and *The Persian Boy* (**1972**), which lightly fantasticates the life of Alexander the Great. These novels can be read as straight fantasies because of MR's ability to portray the world plausibly via her protagonists' PERCEPTION, which differs from a modern one. [JG]

REPOSSESSED US movie (*1990*). ◊ *The* EXORCIST.

RESURRECTION In fantasy the dead may return in many ways – as GHOSTS, VAMPIRES or ZOMBIES, or through REINCARNATION – but the simplest process of return is a straightforward revivification by some magical or miraculous agency. The resurrection of CHRIST is the foundation of Christian faith, holding out a similar promise for everyone else at the LAST JUDGEMENT. Fantasies dealing with New Testament MIRACLES or their aftermath include "Lazarus" (1906) by Leonid ANDREYEV, "Lazarus Returns" (1935) by Guy ENDORE and *This Above All* (**1933**) by M.P. SHIEL; Jesus returns to perform similar favours in various works, including *Gloria Victis* (**1897**) by J.A. Mitchell (1845-1918). A satirical deflation of an apocryphal tale is "The Miracle of the Great St Nicolas" (1909) by Anatole FRANCE.

Most tales of resurrection by magical means are gruesome HORROR after the fashion of one of the cautionary episodes in *The Magic Mirror* (coll **1866**) by William GILBERT, "The Monkey's Paw" (1902) by W.W. Jacobs (1863-1943), "The

Empire of the Necromancers" (1932) by Clark Ashton SMITH and *Pet Sematary* (**1983**) by Stephen KING. TECHNO-FANTASIES in which resurrection is accomplished, as in imitation of Mary SHELLEY's *Frankenstein* (**1818**), generally take an equally gloomy view; examples include "A Thousand Deaths" (1899) by Jack LONDON and *Frankenstein's Bride* (**1995**) by Hilary Bailey (1936-   ). Returnees from the dead often discover, like Tennyson's Enoch Arden (who was of course only *believed* dead), that life has progressed without them to a stage where no viable readmission is possible; examples include "The Man who Returned" (1934) by Edmond Hamilton (1904-1977). There is usually a penalty to be paid even in moral fantasies which sanction resurrection as a reward, as in *The Demon Lover* (**1927**) by Dion FORTUNE. Happier resurrections are, however, often featured in tales of revivified MUMMIES. Genuine "second chances" in life are rarely offered as gifts, but one exception is *The Strange Friend of Tito Gil* (**1852**) by Pedro de Alarcón (1833-1891).   [BS]

RETURN FROM WITCH MOUNTAIN US movie (*1978*). ◊ UFOS.

RETURN OF GODZILLA, THE vt of *Gigantis* (*1955*). ◊ GODZILLA MOVIES.

RETURN OF JAFAR, THE US ANIMATED MOVIE (*1994* tvm). Walt Disney Television Animation. **Pr** Tad Stones, Alan Zaslove. **Dir** Toby Shelton, Stones, Zaslove. **Screenplay** Kevin Campbell and 11 others. **Voice actors** Jason Alexander, Jeff Bennett, Val Bettin, Liz Calaway, Dan Castellaneta, Jim Cummings, Jonathan Freeman, Gilbert Gottfried, Brad Kane, Linda Larkin, B.J. Ward, Scott Weinger, Frank Welker (movie lacks proper credits). 66 mins. Colour.

This sequel to DISNEY's theatrical hit ALADDIN (*1992*) is, as befits a tvm, less spectacularly but not necessarily less imaginatively animated. The story is uncomplicated. Iago the parrot frees himself from the magic lamp but leaves Jafar (still a GENIE) to his fate. The BIRD ingratiates himself with Aladdin, who has yet formally to be betrothed to Jasmine. Abi Smal, a thief whose enmity Aladdin has earned, discovers the lamp and releases Jafar who, though still bound by the lamp's restrictions, seeks revenge on Aladdin. He fails and is destroyed. Aladdin, with Jasmine and others, determines to seek adventure in the wide world – thereby setting things up for a succeeding tv series. Some spectacular visuals are involved; the whole production has the same joyous flamboyance as its large-screen precursor.   [JG]

RETURN OF THE FLY US movie (*1959*). ◊ *The* FLY.

RETURN TO FANTASY ISLAND US movie (*1978* tvm) ◊ FANTASY ISLAND (*1977*).

RETURN TO OZ Two US movies (*1964/1985*) ◊ *The* WIZARD OF OZ.

RETURN TO 'SALEM'S LOT, A US movie (*1987*) ◊ 'SALEM'S LOT.

REVEL A sequence in a fantasy text when the proper ordering of the world and society is deliberately overturned (◊◊ TOPSY-TURVY), the lowly become kings and queens for a night, justice and mercy are mocked (and perhaps uncovered as shams), and creatures who dwell in the WAINSCOTS of the world come out to frolic and to parade. A revel is a time when all values and hierarchies are reversed; a time of sexual liberty, rudeness and unexpected mercy. The revel takes place in a psychic night, a night which has taken over from the day. A revel is therefore a PARODY of the normal world. The god of the revel is DIONYSUS.

Revels are almost always under the ultimate control of a MAGUS or monarch, or (for instance) the Lord of Misrule, who typically presides over the CHRISTMAS Saturnalia in honour of the turn of the SEASONS. Folk dramas, full of scabrous scenes of upsets and comic humiliations, were common in the Middle Ages, but became domesticated as they entered establishment culture; a tamed form of revel became a licensed part of formal theatrical performances, for instance, in the various "anti-masques" by Ben Jonson (1572-1637), first performed in the early 17th century (◊ MASQUE). And in what may still remain the central examples of how to compose revels while ultimately maintaining control over the CHAOS they induce – William SHAKESPEARE's *A Midsummer Night's Dream* (performed *c*1595; **1600**) and *The Tempest* (performed *c*1611; **1623**) – the reversals are righted in the end. In Peter GREENAWAY's PROSPERO'S BOOKS (*1991*), the subversive danger inherent in the unleashing of primordial forces is emphasized.

Subversion is a particular feature of the kind of SATIRE Mikhail Bakhtin described as carnivalesque (◊ CARNIVAL), and which was characteristic of works by writers like E.T.A. HOFFMANN, Edgar Allan POE and Nikolai GOGOL. In *Fantasy: The Literature of Subversion* (**1981**) Rosemary JACKSON agrees that the roots of the carnivalesque lie in the traditional Menippean satire, a form utilized in works like APULEIUS's *The Golden Ass* (*c*165), François RABELAIS's *Gargantua and Pantagruel* (**1532-64**), Jonathan SWIFT's *Gulliver's Travels* (**1726**) and Charles DICKENS's **Christmas Books**, particularly *The Haunted Man and the Ghost's Bargain* (**1848**) and, more importantly, "mundane" novels like *Our Mutual Friend* (**1865**). Understood as an acting out of the Menippean impulse, the revel can be seen as fantasy's way of incorporating transgression. (Revels in SUPERNATURAL FICTION and HORROR, where they are likely to be closely linked to the basic transgressive actions of the antagonists, tend to be described as events which seduce protagonists from normal life into something, as in the BLACK MASS, inherently EVIL.) The fantasy revel may loosen the stays of STORY, but normally so that, in the end, a new order can be recognized, and calm restored. A revel that does not end is hard to distinguish from APOCALYPSE.

There are revels in Robert W. CHAMBERS's *The King in Yellow* (coll **1895**); in much FIN DE SIÈCLE and Decadent writing (◊ DECADENCE), though the closest an Edwardian writer came to creating an upside-down world before WWI seems to have been the picnic of the under-beasts in Kenneth GRAHAME's *The Wind in the Willows* (**1908**). Two of Gustav MEYRINK's novels – *The Green Face* (**1916**) and *Walpurgisnacht* (**1917**) – are revels; and Thomas MANN's *The Magic Mountain* (**1924**) sublimates – as does much of his work – revel into elaborately taming discourse. Later fantasy texts dominated by the subversion of revel (there are not many) include: most DYING-EARTH tales, which are set in venues which could almost be defined as post-facto revel versions of our world; Charles G. FINNEY's *The Circus of Dr Lao* (**1935**) and the various tales, from authors like Ray BRADBURY and Theodore STURGEON, that examine the same ambivalent relation between nostalgia for a world well lost and subversion of the present; Elizabeth HAND's **Winterlong** sequence, which is as much sf as fantasy, the "antifantasies" that make up M. John HARRISON's **Viriconium** sequence; Wyndham LEWIS's *Malign Fiesta* (**1955**); almost any MAGIC-REALISM novel; and Michael MOORCOCK's *The Sound of Muzak* (**1977**).   [JC]

REVELATION OF ST JOHN ◊ APOCALYPSE.

REVENANTS GHOSTS who reappear after some time rather than immediately after death and in more material form. They are often presented as decomposing (though clothed). They are not ZOMBIES, as they retain an identity and purpose – possibly VENGEANCE. Like ghosts, revenants appear close to the place of death, though not necessarily in HAUNTED DWELLINGS. They are frequently depicted in graveyards or crypts, which is where they are most common in Victorian GHOST STORIES, but they can appear on a grander scale – a good example is in *The Beleaguered City* (**1879**) by Margaret OLIPHANT. Many of M.R. JAMES's ghosts are revenants; so is the thing that returns for "John Charrington's Wedding" (1891 *Temple Bar*) by E. NESBIT. Revenants are also common in ghost stories of the sea. The crew of the FLYING DUTCHMAN are revenants, and other examples occur in "Ringing the Changes" (1955) by Robert AICKMAN and *The Fog* (**1975**) by James HERBERT – also in *The FOG* (*1979*), seemingly unrelated. Further interesting movie examples are Clint Eastwood's *High Plains Drifter* (*1972*) and *Pale Rider* (*1985*). Revenants allow for more graphic descriptions than their incorporeal counterparts, and are thus frequently used by such exotic wordsmiths as Clark Ashton SMITH, Tanith LEE and Michael SHEA. [MA]

REVENGE ◊ VENGEANCE.

REVENGE OF THE STEPFORD WIVES US movie (*1980* tvm). ◊ *The* STEPFORD WIVES (*1974*).

REVISIONIST FANTASY Much of what is best in contemporary GENRE FANTASY derives from a conscientious attempt to make standard genre tropes over, to make the condition of fantasy new. For example, the thoroughgoing programme of revisionism that followed on from the "second-wave" FEMINISM of the 1960s was both a necessary piece of common human decency and a productive force, no matter how often its results became new clichés – matriarchal ELVES or herb-wise Wiccans persecuted by the patriarchs of the Inquisition, as in the works of Gael BAUDINO.

The expansion of FANTASY was a consequence of the counterculture of the 1960s, but that period's usually progressive values were largely antithetical to texts like J.R.R. TOLKIEN's *The Lord of the Rings* (**1954-5**). *LOTR* nonetheless came to be loved, imitated and in due course engaged with because it was read, not without good reason, as an attack on industrialization and militarism, and thus progressive in at least those respects, whatever its class, racial and sexual politics. It should not be thought that RF is essentially political: there are various examples of the mode, from Robin MCKINLEY's *Beauty* (**1978**) – revamping the BEAUTY AND THE BEAST LEGEND and Jane YOLEN's *Briar Rose* (**1992**), both of which are RFs with very serious intent, to the generally much more lighthearted revisionism of Terry PRATCHETT in such novels as *Maskerade* (**1995**), which cheerily regurgitates the PHANTOM OF THE OPERA tale. The McKinley and Yolen books could be regarded as having some form of political agenda, but what is important about them – and about the many Pratchett examples – is that they are creating something new in fantasy out of pre-existing materials.

Part of the essence of genre fiction is that it feeds constantly on itself (sometimes unconsciously); readers of genre fantasy want at least some of the time to be on familiar ground, to participate in the perpetuation of FANTASYLAND. In order to go on making things new while yet complying with the dictates of the market, many writers adopt the

strategy of expressing in their fiction their own dissidence from the assumed values of Fantasyland. This is a legitimate instinct.

In almost all revisionist genre fantasy there is a strong element of complicity with the thing disapproved of – when Tad WILLIAMS criticizes J.R.R. TOLKIEN for the racist and hierarchical MAGGOTS in *LOTR*, it is within a narrative framework and to some real extent among set-pieces and moralized landscapes derived from *LOTR*. In Williams's work this produces a useful creative tension between moral programme and imaginative sympathy; it is not always thus.

There is in much conscientiously RF an element of posturing, most apparent when that which is engaged with is the specific work of other writers. It ought to be that to engage in self-criticism would be less problematic; Ursula LE GUIN's *Tehanu* (**1990**), in which she takes issue with much of what she said in **The Earthsea Trilogy** (**1968-72**), proves otherwise – partly perhaps because of the strong implied criticism of readers for not seeing at the time with the clear moral vision Le Guin has herself now attained.

It is easier to try to delete the past than to learn, and teach, how to read it critically. When RF deals with traditional and folkloric material, particularly in CHILDREN'S FANTASY, it is at the risks of both patronizing the past and being criticized in similar terms in the future. The 19th-century attempt by the GRIMM BROTHERS to bowdlerize FAIRYTALES and LEGENDS as they collected them is now disliked. Later attempts to substitute mid-19th-century progressive values for "traditional" ones – satirized by Charles DICKENS in "Frauds on the Fairies" (1853 *Household Words*) – have dated badly, simply because their laudable core values were so often sidetracked into passing issues like temperance and dress reform. Many contemporary attempts – as in *Don't Bet On the Prince* (anth **1986**) ed Jack ZIPES – to recast fairytale material to suit modern notions of what values children ought to imbibe from the fantastic are equally dubious.

In a further but overlapping category – described in this encyclopedia as TWICE-TOLD – the retelling of stories, or elements of stories, is less a polemical revision than a way of meditating on STORY itself. For example, the tales in Angela CARTER's *The Bloody Chamber* (coll **1979**) tend to be both RF and twice-told, but the former more than the latter; by contrast, the Shakespearean echoes and parallels in her *Wise Children* (**1991**) and the transposed Westerns of her *American Dreams and Old World Wonders* (coll **1993**) are certainly twice-told – which is not to say that they do not score the odd polemical point as they go. The version of PUSS-IN-BOOTS in *The Bloody Chamber* is an RF in that it makes sensuous use of the cruelty implicit in the original tale, a twice-told tale inasmuch as it is about the tone of voice in which Puss narrates rather than about transforming the plot.

There are various strategies for revisionist treatment of fantasy ICONS. For example, Anne RICE's vampires are largely sympathetic and provided with some justification for their behaviour; Poppy Z. BRITE's, it is implied, should not be judged by human morality. [RK]

REYNARD THE FOX ◊ BEAST-FABLE.

REYNOLDS, QUINTUS [s] ◊ Kenneth MORRIS.

RHIANNON ◊ GODDESS; MABINOGION.

RICE, ANNE (O'BRIEN) (1941-) US writer of historical novels, erotica and fantasy, three genres which in her hands often blend into each other to a remarkable extent. Her first straightforward fantasy (debatably

SUPERNATURAL-FICTION) novels were the **Vampire Chronicles** – *Interview with the Vampire* (**1976**), *The Vampire Lestat* (**1985**), *The Queen of the Damned* (**1988**), *The Tale of the Body Thief* (**1992**) and *Memnoch the Devil* (**1995**). *Interview with the Vampire* was filmed as INTERVIEW WITH THE VAMPIRE (**1994**); there is also *The Vampire Lestat Graphic Novel* (graph **1991**).

AR was responsible, if not for creating, at least for popularizing and in many respects crystallizing the mythology of the REVISIONIST-FANTASY version of the VAMPIRE. *Interview with the Vampire*, a book which AR has explicitly linked to a period of alcoholic excess following the death of a daughter in childhood, is charged with an extraordinary and perverse eroticism in its description of the relationship between the new and reluctant vampire Louis, his begetter Lestat and the eternal child Claudia, whom Lestat creates as a vampire in order to trap Louis into a version of family life. The struggle for power between these three has elements of sexual ritual and family romance; Claudia's position, trapped in a child's body for decades of unlife, is a particularly poignant working-out of one aspect of the IMMORTALITY theme attached to vampirism. The settings – 19th-century NEW ORLEANS and PARIS – are gloomily atmospheric.

The FRAME STORY, in which Louis tells of his past to a contemporary journalist, is an interesting authenticating device in that Louis is giving an Awful Warning, but the journalist's immediate reaction is to try to become a vampire. Perhaps Rice's major piece of revisionism with the vampire legend was to reinvent the vampire as a sexual ICON for a period of polymorphous freedom.

Rice's free way with the vampire mythology continued its journey into the inventively rococo in *The Vampire Lestat* and its immediate sequel *The Queen of the Damned*. Lestat, himself now subject to revision once he becomes narrator, and revealed as a countrified aristocrat rather than the mere peasant boor mocked by Louis, acquires a new career as rock star and investigates the past of vampirism, tracing back the lines of begetting until he discovers transfigured, godlike and seemingly petrified Egyptian vampires and learns of vampirism's demonic origins. Lestat takes to restricting his depredations to criminals, a major revisionist trope, and we hear a certain amount of the bars where vampire groupies hang out in the hope that they will be visited there – another Rice invention from which other writers, notably Laurell K. HAMILTON in her **Anita Blake** sequence, have productively drawn.

The Tale of the Body Thief involves Lestat and his elderly lover, a member of the Talamasca order of SECRET GUARDIANS featured rather more extensively in the **Mayfair Witches** sequence (*see below*), in a complicated tale of IDENTITY EXCHANGE in which Lestat is briefly returned to human status. In *Memnoch the Devil*, Lestat finds himself taken up by GOD and the DEVIL and subjected to an extended revisionist history of humanity and its religions. As with Rice's other series, these books have become subject to the law of diminishing returns and repositories for whatever notions take her fancy; but it would be a mistake to let this reduce one's sense of the important contribution to the vampire mythos represented by the first two.

Less popular but probably more interesting has been her **Mayfair Witches** sequence, to date comprising *The Witching Hour* (**1990**), *Lasher* (**1993**) and *Taltos: Lives of the Mayfair Witches* (**1994**). The first is for much of its considerable length a cross between a DYNASTIC FANTASY and a straightforward SUPERNATURAL FICTION. The Mayfair family – a huge clan whose genealogical ramifications are much complicated by incestuous and quasi-incestuous, often pederast, couplings – are today based mainly in NEW ORLEANS, but their ancestry can be traced back to a WITCH burnt in the Scottish Highlands during the Witch Craze. Not all Mayfairs are possessed of special powers, but all subscribe to the system whereby in each generation there is one woman who will be their ruling witch. They are under investigation by a secret society of scholars, the Talamasca, whose motives are obscure. Many Mayfairs are aware that one of their mansions is haunted not merely by ancestral GHOSTS but by a weird and evil SPIRIT, Lasher. When the new leading witch, Rowan, weds a non-Mayfair (who proves much later in the saga to be a scion of a lost branch of the family), Lasher takes advantage of their union to bring himself into existence. Clearly nonhuman, he reaches adulthood in minutes and abducts his mother, upon whom he wishes to father a child, another "Taltos" like himself. This he succeeds in doing in *Lasher*, although eventually both he and the child are destroyed by Rowan and her husband. In *Taltos* – conceptually although not literally the most ambitious (the writing becomes progressively sloppier during the series) – much more is revealed to us, through the discovery of another of the Taltos species. These quasi-immortal (◊ IMMORTALITY) benevolent beings, gifted with an enhanced racial memory and older than humankind, came to a virtually uninhabited Britain when their insular LOST LAND erupted and vanished beneath the seas. There they built megaliths like Stonehenge before incursive humans began to slaughter them. The Taltos retreated to a remote Highland glen, but, unable to keep their secret forever, eventually passed themselves off as an unusually tall human race, the Picts, discovering they could on occasion breed with humans – usually disastrously, unless with "witches" (it is by now clear that AR is not using the term in its conventional sense) – and, successfully producing further Taltos, with the Little People (◊ FAERIE), a species perhaps even older than themselves. (King ARTHUR was maybe a pure or hybrid Taltos.) Modern "witches" are those humans who possess the genes so that they can, under correct circumstances, mate successfully with a Taltos to produce another, or achieve the same result between two humans if the pairing chances to be ideal. The seduction by a 13-year-old child-witch of Rowan's husband brings yet another Taltos into being. Clearly further sequels are in store.

AR's sole fantasy singleton is *The Mummy, or Ramses the Damned* (**1989**), which sees Ramses the Great (1304-1237BC) brought back to life in Edwardian England, where – and later in Cairo – he naïvely wreaks havoc. It has some interest but suffers from a tendency to descend into bodice-ripper mode. *Cry to Heaven* (**1990**), nonfantasy, is about 18th-century castrati and their erotic entanglements. Two singletons, both soft pornography as by Anne Rampling, are *Belinda* (**1986**), about a middle-aged man's affair with an underage girl, and *Exit to Eden* (**1985**) – filmed as *Exit to Eden* (**1995**) – about sexual slavery.

The topic of sexual slavery and the idea that fulfilment can be found only in the dominance of one partner by another, generally through sadistic acts, is taken emetically further in AR's much less soft-core pornographic fantasy series, the **Sleeping Beauty** trilogy, as by A.N. Roquelaure: *The Claiming of Sleeping Beauty* (**1983**), *Beauty's Punishment* (**1984**) and *Beauty's Release* (**1985**). The 15-year-old

SLEEPING BEAUTY is awoken not by a kiss but by rape and is taken by the rapist prince to a pornotopic FANTASYLAND where a Queen and her aristocrats subject high-born youths from neighbouring lands to humiliating and sadistic degradation on the basis that only such a regime can "train" the unfortunates – through perverting them absolutely – to rule their own lands well. Improbably, the men are forced to sport perpetual erections.

AR's protagonists are typically impossibly beautiful, dwelling in a world of fabulous glamour and vast riches; characterization is not her strength – her characters are those of a high-budget Hollywood movie, and as remote from both empathy and reality. This may be part of her popular appeal. The themes of pederasty (with either boys or girls and often as forcible rape) and sadistic sexual dominance – both of which could be regarded as a nonsupernatural form of vampirism – are in the Roquelaure novels made explicit; they are never far from the surface in much of the rest of AR's work. There is no doubt that AR is a compelling storyteller – though often the first 100 pages or so of a novel are heavy going – but it is concerning that her preoccupations should be thus. [RK/JG]

Further reading: *Prism of the Night: A Biography of Anne Rice* (**1991**) by Katherine Ramsland.

RICE, JAMES (1844-1882) UK writer. ◊ Walter BESANT.

RICHARD, MARK (1955-) US writer whose first book, *The Ice at the Bottom of the World* (coll **1989**), won the PEN Ernest Hemingway Award. MR's violently hallucinated, rather Gothic (◊ GOTHIC FANTASY) stories verge sometimes on FABULATION – especially with "Fishboy" (1989; exp vt *Fishboy: A Ghost's Story* **1993**), a story populated by freaks, GHOSTS and a monstrous narrator. Recalling Edgar Allan POE and Herman MELVILLE in its maritime Gothic, MR's novel offers itself as as much a demented inscape as a tale of the supernatural, although it resolves finally into a POSTHUMOUS FANTASY. [GF]

RICHARDS, SEAN Pseudonym of Peter HAINING.

RICHARDSON, MAURICE (LANE) (1907-1978) UK writer and journalist whose principal genre contribution is *The Exploits of Engelbrecht: Abstracted from The Chronicles of the Surrealist Sportsman's Club* (coll of linked stories **1950**). This, with enjoyable absurdist HUMOUR, involves Engelbrecht – a DWARF "Surrealist Boxer" – in such unlikely sports as WITCH-shooting, boxing literally against TIME in the form of a pugilistic grandfather clock, football on the MOON, fishing for water MONSTERS (Engelbrecht is bait) and a Surrealist CHESS-game leading to APOCALYPSE. *Fits and Starts* (coll **1979**) assembles fiction (including the amusing prose poem "Fiendish Nuptials: A Horroratorio", featuring DRACULA, the FRANKENSTEIN MONSTER, Edgar Allan POE, SHERLOCK HOLMES, etc.) and critical pieces on, *inter alia*, the MUNCHAUSEN tales and GHOST STORIES. [DRL]

RIDDELL, [Mrs] J.H. Married and most common working name of Irish writer (largely resident in England) Charlotte Elizabeth Lawson Cowan (1832-1906). She is regarded by many as the foremost Victorian writer of SUPERNATURAL FICTION, particularly GHOST STORIES, after J. Sheridan LE FANU, although these subjects account for perhaps one-tenth of her output. Most of her books deal with business and the City, exploring the drama and romance linked to finance and commerce. JHR herself was constantly fighting debt due to the bankruptcy of first her father, then her husband. It is possible she wrote under more pseudonyms than the three known – R.V. Sparling (under which appeared

her first book, *Zuriel's Grandchild* [**1856**]), Rainey Hawthorne and F.G. Trafford – and quite probably she had many stories published anonymously. Her most successful book was *George Geith of Fen Court* (**1864**), another City novel.

Four of JHR's novels deal with HAUNTED DWELLINGS. These are: *Fairy Water* (1873 *Routledge's Christmas Annual* chap; vt "The Haunted House at Latchford" in *Three Supernatural Novels of the Victorian Period* anth **1975** ed E.F. BLEILER); *The Uninhabited House* (1875 *Routledge's Christmas Annual* chap; in *Five Victorian Ghost Novels* (anth **1971**) ed Bleiler); *The Haunted River* (1877 *Routledge's Christmas Annual* chap; combined with *The Uninhabited House* as *The Haunted River and The Uninhabited House* omni **1883**); and *The Disappearance of Mr Jeremiah Redworth* (1878 *Routledge's Christmas Annual* chap). All rely as heavily on character as on effect, the EDIFICE of the haunted house being a prop for the resolution of some past crime (usually a murder or suicide). A fifth novel, *The Nun's Curse* (**1888**) provides a supernatural explanation for the problems that haunt a family estate.

JHR's short ghost stories are even more potent. She compiled her selection of the best in *Weird Stories* (coll **1882** [often wrongly cited as 1884]), again mostly haunted-house stories, though "Nut Bush Farm" includes a highly effective apparition in a field. JHR's other ghost stories are scattered through various collections, most notably *The Banshee's Warning* (coll **1894**); E.F. Bleiler gathered them together as *The Collected Ghost Stories of Mrs J.H. Riddell* (coll **1977** US), with bibliography and extensive introduction. [MA]

RIDDLES Riddling questions date back to antiquity, the two most famous being the SPHINX's "What goes on four legs in the morning, two legs in the afternoon, and three legs in the evening?" (whose answer, "Man", is subjected to intense analysis in Terry PRATCHETT's *Pyramids* [**1989**]) and Samson's riddle to the Philistines in the BIBLE: "Out of the eater came forth meat, and out of the strong came forth sweetness" (i.e., honey from bees nesting in a lion's carcase).

Anglo-Saxon riddles were assembled in the *Exeter Book* (*c*975-1000), and indeed the classic riddle resembles an Old English kenning – a poetic "cryptic clue" – rather than the pun-based riddles popularized in Victorian times. Some riddles are unfair: RUMPELSTILTSKIN's demand for his own name was intended as unanswerable; Lewis CARROLL's Mad Hatter in *Alice's Adventures in Wonderland* (**1865**) can offer no answer to his own "Why is a raven like a writing-desk?"; and in J.R.R. TOLKIEN's *The Hobbit* (**1937**) Bilbo's celebrated "What have I got in my pocket?" is an accidental subversion of the genuine riddle-game which it concludes. But when the Master Doorkeeper in Ursula K. LE GUIN's *A Wizard of Earthsea* (**1968**) requires, like Rumpelstiltskin, to be told his own guarded TRUE NAME, this tough-seeming riddle is a subtler test: the name may be had freely, if the supplicant WIZARD only thinks of asking for it. Poul ANDERSON sent up the riddle-game in *Three Hearts and Three Lions* (**1953**), whose hero foxes a formidable GIANT with children's catch-question riddles; Susan COOPER's *The Grey King* (**1975**) has a PLOT COUPON guarded by such riddling questions as "What is the shore that fears the sea?" (beech-wood, which is harmed by water); the Bridge of Death scene in MONTY PYTHON AND THE HOLY GRAIL (*1975*) again spoofs the riddle-game; Patricia MCKILLIP's *The Riddle-Master of Hed* (**1976**) and its successors pivot on unanswered riddles hiding deep secrets and PROPHECIES; Michael SWANWICK's *The Iron Dragon's Daughter* (**1993**) makes riddle-game exchanges the scoreboard in a TECHNOFANTASY duel of electronic weaponry. Generally, any GUARDIAN OF THE THRESHOLD is likely to require the answer to a riddle. [DRL]

RILEY, JUDITH MERKLE (1942-) US author of historical novels with a fantasy twist. *A Vision of Light* (**1989**) and *In Pursuit of the Green Lion* (**1991**) set the adventures of Margaret of Ashbury, a gentry-woman with psychic HEALING powers, against a meticulously researched background of 14th-century Europe. *The Oracle Glass* (**1994**) uses the intrigues and scandals of the court of Louis XIV to explore precognition, female roles and the nature of DISGUISE. Characterized by thorough research and a strong sense of period, all three display an underlying concern with the problems facing intelligent women in male-dominated societies. As Judith A. Merkle, JMR has published academic works in the field of political science, and the influence of this discipline pervades her fiction. [KLM]

RINGS Fantasy contains a plethora of rings: the combination of ornament, convenience of wear, mystic circle-shape and strategic position on a finger (allowing easy aiming, turning, rubbing, etc.) makes the ring a natural receptacle for MAGIC – although magic rings disconcertingly tend to be irremovable, and/or to slip off at their own whim. Legendarily, Solomon's ring made him all-knowing and able to speak with animals, and additionally bore the sigil with which he sealed up GENIES in jars; this ring is discussed in *Many Dimensions* (**1931**) by Charles WILLIAMS. INVISIBILITY is frequently conferred by rings; e.g., that of Gyges in classic MYTH, Agramante's in ARIOSTO's *Orlando Furioso* (**1516**), which also protects against all magic, the One Ring in J.R.R. TOLKIEN's *The Lord of the Rings* (**1954-5**) – which like the Ring of the Nibelungs (◊ Richard WAGNER) can also enslave others – Tolkien's SEVEN and Nine lesser rings, to wear which is BONDAGE, and the ring in E. NESBIT's *The Enchanted Castle* (**1907**), which secretly has whatever properties the holder says it has. WISH-granting rings have been endemic ever since Aladdin's accidental rubbing of his ring called up its GENIE slave. In Piers ANTHONY's *Castle Roogna* (**1979**) an ambiguous variant ring promises to grant WISHES and claims credit for their fulfilment through the hero's apparently unaided efforts. Yet other rings compel love in William Makepeace THACKERAY's *The Rose and the Ring* (**1855**), expand into a PORTAL for Liane the Wayfarer in Jack VANCE's *The Dying Earth* (**1950**), grant SUPERHERO powers in the *Green Lantern* COMIC and others, offer transport between worlds in C.S. LEWIS's *The Magician's Nephew* (**1955**), confer a TECHNOFANTASY almost-omnipotence in Michael MOORCOCK's **Dancers at the End of Time** sequence, and control a wide variety of magics in Roger ZELAZNY's *Knight of Shadows* (**1989**).

Notable non-magical rings include: that of Polycrates, thrown into the sea to propitiate Nemesis (◊ FATE) and dismayingly returned in the belly of a fish; the ring of Kings worn by Michael MOORCOCK's **Elric**; the ring (in fact armlet) of Erreth-Akbe in Ursula K. LE GUIN's *The Tombs of Atuan* (**1971**), which when made whole shows the lost RUNE of peace; and, arguably, the white gold ring of Stephen R. DONALDSON's **Thomas Covenant the Unbeliever**, a focus for the cataclysmic power of Covenant's unbelief. [DRL]
See also: AMULETS.

RIP VAN WINKLE ◊ Washington IRVING.

RITE OF PASSAGE Two categories of QUEST dominate fantasy writing. The external quest, like that of ODYSSEUS, is more commonly found; but the internal or ROP quest – in which the protagonist, often an UGLY DUCKLING, moves from childhood through PUBERTY (and/or MENARCHE) to adult empowerment, from AMNESIA to self-RECOGNITION, from ignorance to bliss – has become increasingly popular; and defines the central plot structure of many fantasies, and probably most of those written for YA readers.

The ROP may take many forms, though the underlying movement is from BONDAGE towards enablement, freedom, responsibility. A straightforward ROP tale will challenge its young protagonist with a dilemma during the solving of which s/he matures. Complexities and resonances begin to enrich the process when the ROP is conceived of as an INITIATION of the SOUL, a journey invoking a pattern of Departure, Absence and Return. This initiatory journey may be undertaken by a CHILDE protagonist without benefit of counsel; or by a person who wanders INTO THE WOODS of a GODGAME from which s/he returns wiser, with gifts, and perhaps married; or by a figure who, in order to LEARN BETTER, must undertake a NIGHT JOURNEY towards difficult self-realization; or by a person, of any age, whose psyche is internally disordered, and who enters a psychic UNDERWORLD where the ARCHETYPES of JUNGIAN PSYCHOLOGY (in particular the SHADOW) dance healing attendance; or by a HERO (e.g., GALAHAD), whose ROP is genuinely transcendental and leads towards a recoverable GRAIL.

The internal ROP may – in some literatures of the FANTASTIC – be embarked upon from within the psyche in such a manner that the experience can be understood as wholly subjective. In FANTASY, however, it is far more likely that the quest for selfhood will entail actual travel, actual confrontation, and actual return. One of the central strengths of fantasy lies precisely in this bias of the fantasy STORY towards *acting-out*. The ROP in fantasy becomes a RITUAL. It is part of the world, and it leads to a HEALING of both hero and world. [JC]

RITUAL A formal procedure of, usually, MAGIC or RELIGION. A common religious ritual in fantasy is EXORCISM; the major ritual of SATANISM is the BLACK MASS, a PARODY of the Catholic service. In BLACK MAGIC the purpose is generally conjuration of DEVILS; the slightest deviation from protocol will be deadly, as in E.R. EDDISON's *The Worm Ouroboros* (**1922**), James BLISH's *Black Easter* (**1968**) and John WHITBOURN's *A Dangerous Energy* (**1992**). Other rituals may be SPELLS involving the cooperation of several WIZARDS, like the Rite of AshkEnte in Terry PRATCHETT's **Discworld** sequence (which summons DEATH), or simple gestures of respect, like the ritual of purification required before reading the sacred BOOK in William MORRIS's *The Well at the World's End* (**1896**). INITIATION often takes a ritual form, as imposed on groups of would-be wizards in Roger ZELAZNY's *Madwand* (**1981**) or on individuals who must walk the Pattern (◊ LABYRINTH) in Zelazny's **Amber** sequence. Seemingly meaningless secular rituals are likely to be PLOT DEVICES encoding information in RIDDLE form, like the Musgrave Ritual deciphered by SHERLOCK HOLMES or the significant rote-games of CHILDREN in Barry HUGHART's **Master Li** books. Lost rituals, which must be rediscovered to save the LAND or enthrone the true monarch, occasionally feature as PLOT COUPONS. [DRL]
See also: HUMAN SACRIFICE; RITE OF PASSAGE; RITUALS OF DESECRATION.

RITUALS OF DESECRATION Blasphemous acts of deconsecration are common in HORROR and SUPERNATURAL FICTION, frequently appearing at the point when mundane protagonists must choose whether or not to be seduced by the blandishments of a predatory SATAN or VAMPIRE or WITCH or other proselytizer figure of evil. What they sense at this moment may be described as a linchpoint moment of WRONGNESS. In FANTASY texts, RODs are more likely to initiate a period of THINNING or at least to be a signal that thinning has started to occur, as in C.S. LEWIS's *The Last Battle* (**1956**). In the backstory to Steven R. DONALDSON's **Chronicles of Thomas Covenant the Unbeliever**, for instance, it is revealed that the good MAGUS Kevin has been tricked by the DARK LORD, Lord Foul, into participating in an ROD, in the belief that he either assist in desecrating the LAND or witness its total destruction. Almost all of M. John HARRISON's fantasies take place in the aftermath of some ROD, which is sometimes made explicit, but which can always be sensed. Tales of his in which the ROD is clearly present include: "The Incalling" (1978), an URBAN FANTASY set in a decayed LONDON whose protagonist apes a MAGIC rite which kills him, just as his surroundings are themselves thinning into entropic rigor mortis, and *The Course of the Heart* (**1992**), in which a man ritually rapes his daughter precisely to desecrate the act and her. [JC/DRL]

RIVERS The ordinary properties of rivers (especially *named* rivers), whether as territorial boundaries, defensive lines or trade/transport routes, tend to be echoed on a deeper level in fantasy. Here the boundary may be between GOOD AND EVIL; the defence may be against THINNING or incursion by a DARK LORD – Anduin, the Great River, is a significant line of defence in J.R.R. TOLKIEN's *The Lord of the Rings* (**1954-5**) – or by WITCHES, who traditionally cannot cross running water except in a boat made of eggshell. Rivers may be barometers of the LAND through which they flow, a diseased river indicating a general need for HEALING. Their age and continuity of flow indicate the ancient nature of the land's STORY (◊◊ TIME ABYSS). Subterranean rivers like Alph in XANADU (◊◊ Samuel Taylor COLERIDGE) have a special resonance, as does the Thames in LONDON and the Rhine in Richard WAGNER's **Ring** cycle.

The liquid in a fantasy river may have any property of a POTION. Notoriously, the river Lethe brings AMNESIA, and the name of the FOUNTAIN OF YOUTH is self-explanatory. An oddity of magical ecology in Piers ANTHONY's *A Spell for Chameleon* (**1977**) is a barren river wishing to restock itself: drinking its water brings TRANSFORMATION into a fish. What flows in some rivers may be not water but STARS, souls, the stuff of MAGIC, or – as with HELL's river Phlegethon – boiling blood. Rivers may contain tutelary GODS, as encountered in C.S. LEWIS's *Prince Caspian* (**1951**) and Gene WOLFE's *Solder of the Mist* (**1986**); other denizens may include DRAGONS, SELKIES, and nymphs or naiads who may or may not be perilous, like the Lorelei (◊ MERMAIDS; SIRENS). Travelling along or across any important river may mean a change of condition more significant than movement on the MAP: by crossing a mere brook, Alice in Lewis CARROLL's *Through the Looking-Glass* (**1872**) becomes a queen. [CB/DRL]

RIVKIN, J.F. Joint pseudonym of two unidentified authors, one of whom was apparently born in 1951. Of greatest interest is their **Silverglass** sequence – *Silverglass* (**1986**), *Web of Wind* (**1987**), *Witch of Rhostshyl* (**1989**) and *Mistress of Ambiguities* (**1991**) – featuring a female mercenary in SWORD-AND-SORCERY adventures in a brightly drawn FANTASYLAND as she guards the female WIZARD who has hired her. Other JFR works include *Runesword: The Dreamstone* *

(**1991**), and *Age of Dinosaurs #1: Tyrannosaurus Rex* (**1992**), which is sf. [JC]

"ROAD TO" MOVIES A series of inventive US musical comedies featuring Bing Crosby (1904-1977), Bob Hope (1903-) and Dorothy Lamour (1910-1996), their dynamic being the friendly rivalry of Crosby and Hope for Lamour's affections. In the series' heyday the movies were characterized by a zany SURREALISM, the intrusion of fantasy elements, and the recognition by the characters that they are not real but part of a STORY. Notable are the ARABIAN FANTASY *Road to Morocco* (**1942**) – which features a TALKING ANIMAL in the shape of a grumpy camel ("This is the screwiest picture I've ever been in") – and *Road to Rio* (**1947**), whose plot hinges on MESMERISM. *Road to Utopia* (**1945**) has nothing to do with UTOPIA – its concern is the Klondike – but does have dancing bears and talking fish. Others were *Road to Singapore* (**1940**), *Road to Zanzibar* (**1941**), *Road to Bali* (**1952**) – the only one in colour – and *Road to Hong Kong* (**1961**), a late, UK-produced and rather weary entry in which, although Lamour appears, her role is taken over by Joan Collins (1933-). [JG]

ROBBINS, TOD Working name of US writer Clarence Aaron Robbins (1888-1949), long resident in Europe and the UK. Much of his work is nonsupernatural HORROR, beginning with *Mysterious Martin* (**1912**; rev vt *The Master of Murder* 1933 UK), though a tinge of supernaturalism does adhere to this title through the mystery behind the eponymous author's BOOK, which seems to have the power to make those who read it into murderers. In *The Unholy Three* (**1917**; vt *The Three Freaks* 1934 UK) a macabre alliance is established among three CIRCUS performers: a strongman, a midget and a ventriloquist. The book was filmed as *The Unholy Three* (**1925**) dir Tod Browning and starring Lon Chaney, who also starred in the 1930 sound remake; TR's closely connected short story, "Spurs" (1926), was also used by Browning for his famous *Freaks* (**1932**). TR's short fiction was assembled as *Silent, White and Beautiful* (coll **1920**), *Who Wants a Green Bottle?* (coll **1926** UK), which shares some contents with the previous volume, and *In the Shadow* (coll of linked stories **1929**). [JC]
Other works: *Close Their Eyes Tenderly* (c**1940** Monaco).

ROBBINS, TOM Working name of US writer Thomas Eugene Robbins (1936-), whose novels, beginning with *Another Roadside Attraction* (**1971**), incorporate fantasy elements as aspects of his vision of a fable-choked USA. The pixillated mythopoesis of this first novel – in which the mummified body of CHRIST, guarded by Jesuits in the mountains of Washington State, focuses various loose-limbed QUEST episodes – already bestows a sense of BELATEDNESS upon the hippie consciousnesses at its heart. In this, TR's work resembles early stories of his near contemporary, William KOTZWINKLE.

Even Cowgirls Get the Blues (**1976**), *Still Life With Woodpecker* (**1980**) and *Half Asleep in Frog Pajamas* (**1994**) have relatively minor fantasy elements; but *Jitterbug Perfume* (**1984**) cannily invokes a wiseacre though fading PAN to preside over a fabulated paean to SEX and IMMORTALITY. *Skinny Legs and All* (**1990**) features five characters who have been brought to life after thousands of years by a Phoenician love GODDESS in the throes of sexual intercourse, as by-blows. Intersecting with their tales is an ongoing plot by a fundamentalist Christian preacher to bring about the END OF THE WORLD through the reconstruction of the Temple of Solomon in JERUSALEM.

TR's books lack bite for some; but for others their playful dance-like intricacies seem joyous. [JC]

ROBERSON, JENNIFER (1953-) US writer whose first novel, *Shapechangers* (**1984**), began an eight-volume series concerning the **Cheysuli**, a race possessing the ability to take animal form and to forge psychic bonds with creatures of that form ($ SHAPESHIFTERS). The series continues with *The Song of Homana* (**1985**), *Legacy of the Sword* (**1986**), *Track of the White Wolf* (**1987**), *A Pride of Princes* (**1988**), *Daughter of the Lion* (**1989**), *Flight of the Raven* (**1990**) and *A Tapestry of Lions* (**1992**). The **Sword** sequence – *Sword-Dancer* (**1986**), *Sword-Singer* (**1988**), *Sword-Maker* (**1989**) and *Sword-Breaker* (**1991**) – is SWORD AND SORCERY. [JCB] **Other works:** *Lady of the Forest* (**1992**), a bulky historical romance concerning ROBIN HOOD and Marian; «Lady of the Glen» (**1996**).

ROBERT DE BORON (? -1212) French chronicler. $ ARTHUR.

ROBERTS, [Sir] CHARLES G(EORGE) D(OUGLAS) (1860-1943) Canadian writer. $ CANADA.

ROBERTS, JOHN MADDOX (1947-) US writer, best-known in fantasy for his continuations of the CONAN stories originated by Robert E. HOWARD. As with all **Conan** continuations, the stories are formulaic, and JMR made it evident he was not planning to imitate Howard's style. JMR's Conan is essentially an adventurer who is less violent than Howard's original and uses his brain slightly more, without losing his barbarian origins. JMR's titles run *Conan the Valorous* * (**1985**), *Conan the Champion* * (**1987**), *Conan the Marauder* * (**1988**), *Conan the Bold* * (**1989**), *Conan the Rogue* * (**1991**), *Conan and the Treasure of the Python* * (**1993**), *Conan and the Manhunters* * (**1994**) and *Conan and the Amazon* * (**1995**). The majority of JMR's other work is either sf or mystery; his best mysteries are those in the **SPQR** historical detective series.

JMR's other fantasies are more properly SCIENCE FANTASIES. *King of the Wood* (**1983**) is set in an ALTERNATE REALITY where North America was inhabited from the 10th century by various European refugees, resulting in a Christian kingdom in the north and a pagan one in the south. The novel is set several centuries later. The **Stormlands** sequence is set on a once technologically advanced planet whose inhabitants have now sunk into barbarism after all bar precious metals have gone. The fantastic elements arise only through the apparent MAGIC of the TECHNOLOGY which is being rediscovered. Titles to date are *The Islander* (**1990**), *The Black Shields* (**1991**), *The Poisoned Lands* (**1992**), *The Steel Kings* (**1994**) and *Queens of Land and Sea* (**1994**). JMR's most intriguing work to date has been *Murder in Tarsis* * (**1996**), set in the **Dragonlance** SHARED WORLD but part of a new series of murder mysteries in a FANTASYLAND; JMR is able to exercise both his predilections in what may prove a rewarding series. [MA] **Other works:** The **SPQR** series, being *SPQR* (**1990**), *SPQR II: The Catiline Conspiracy* (**1991**), *The Sacrilege* (**1992**), *The Temple of the Muses* (**1992**), plus *Tödliche Saturnalien* ["Saturnalia"] (**1993** Germany) and *Tod Eines Centurio* ["Nobody Loves a Centurion"] (**1994** Germany), the last two as yet unpublished in English.

ROBERTS, KEITH (JOHN KINGSTON) (1935-) UK writer, illustrator and editor; he has also written as Alistair Bevan, John Kingston and David Stringer. Although he is closely associated with SCIENCE FICTION, many of his writings utilize settings and images of the fantastic and draw from the wellspring of British myth. He first appeared in SCIENCE FANTASY with "Anita" (1964), the start of the **Anita** series about a young female WITCH, collected as *Anita* (coll of linked stories **1970** US; with 1 extra story 1990 US). Deliberately humorous, the stories are at times rather cutesy. At this period KR appeared regularly in SCIENCE FANTASY. He illustrated many of the covers and became Associate Editor when the magazine reconstituted itself as *Impulse*, being Managing Editor for the final 5 issues. Some of KR's early stories blended science and the supernatural. The **Boulter** stories feature an inventor who encounters POLTERGEISTS in "Boulter's Canaries" (1965 *New Writings in SF*) and draws on the power of leys in "The Big Fans" (1977 *F&SF*). "The Scarlet Lady" (1966 *Impulse*) as Bevan is about a car haunted by previous victims. Both "Susan" (1965 *Science Fantasy*) and "The Pace that Kills" (1966 *Impulse*) feature psychic powers ($ TALENTS), and the first includes an example of the FEMME FATALE who is a key feature in many of KR's stories – he regards them as "primitive heroines" and wrote about them in *The Natural History of the P.H.* (**1988** chap): she is KR's female equivalent of MOORCOCK's ETERNAL CHAMPION. Some of these early stories were collected as *Machines and Men* (coll **1973**); others were later revived in *Winterwood and Other Hauntings* (coll **1989**), the limited edition also containing the bound-in booklet *The Event* (**1989** chap). *The Inner Wheel* (1965 *New Writings in SF*; much exp **1970**) has similarities in drawing on the theme of psychic powers to portray a gestalt superman.

KR's first success came with *Pavane* (coll of linked stories **1968**; rev with 1 extra story 1969 US), one of the classic ALTERNATE-WORLD books, set in a technologically backward Britain ruled by the Catholic Church – the Spanish Armada having defeated England. KR portrays a PASTORAL land where some of the old powers, in the form of DRYADS, still survive; although these are not key to the story, they provide an atmosphere of halted THINNING, and the book bears some comparison with those of Thomas Burnett SWANN. KR has repeated the approach in two other volumes: *The Chalk Giants* (coll of linked stories **1974**; cut 1975 US), a post-HOLOCAUST novel in which the UK is torn apart by a barbarism that is likened to the native power of the land; and *Kiteworld* (fixup **1985**), set in another quasimedieval world where the power of the Church holds sway. The theme of the primitive heroine comes sharply into focus in *Gráinne* (**1987**), which draws upon the Celtic legend ($ CELTIC FANTASY) of the GODDESS and explores her power in the modern age through the form of a mysterious young girl. These novels form KR's main corpus of fantastic fiction. They have affinities with the work of Robert HOLDSTOCK and Richard COWPER in the exploration of mythic figures. They find a natural conclusion in *The Road to Paradise* (**1988**), a contemporary TIMESLIP novel where the weight of Britain's past begins to haunt the heroine.

KR's other works are more in the tradition of science fiction. He has also written a historical novel, *The Boat of Fate* (**1971**), set toward the end of the Roman Empire. Some less traditional tales are interweaved into *The Passing of the Dragons* (coll 1977 US), *Ladies from Hell* (coll **1979**) and *The Lordly Ones* (coll **1986**). [MA] **Other works:** *The Furies* (1965 *Science Fantasy*; **1966**); *The Grain Kings* (coll **1976**); *Molly Zero* (**1980**); the **Kaeti** sequence, being *Kaeti and Company* (coll of linked stories **1986**), *Kaeti's Apocalypse* (**1986** chap) and *Kaeti on Tour* (coll **1992**); *A Heron Caught in Weeds* (coll **1987** chap), poetry;

Irish Encounters: A Short Travel (dated 1988 but **1989** chap).

ROBIN GOODFELLOW A hobgoblin or brownie (◊ ELVES) whose adventures were a popular element of FOLKLORE in medieval Britain. SHAKESPEARE equated him to PUCK in *A Midsummer Night's Dream* (performed 1596; **1600**), but popular tradition gives him a separate identity. The earliest surviving complete text of his adventures is the anonymous *Robin Goodfellow, His Mad Pranks and Merry Jests* (**1628** chap). RG is revealed as a half-FAIRY, the son of OBERON and a country girl. Although RG has the fairy love of trickery (◊ TRICKSTER) he has no supernatural powers. He runs away from home, has a dream of fairies and wakes to find Oberon has conferred on him fairy powers which he must use to help the needy. He thus enters upon a life of good-humoured trickery (◊ TRICKSTER), helping the needy at the expense of the pompous or lecherous – clear parallels with ROBIN HOOD. At length, having earned his RITE OF PASSAGE, RG is allowed to enter the land of FAERIE. His appearance in modern fantasy is usually as Puck. [MA]

ROBIN HOOD Britain's most famous folk-hero (◊ FOLKTALES); his adventures have long passed into tradition and almost form part of the MATTER of Britain. The connections between the real RH, the LEGENDS of his life and possible mythological origins are complicated and obscure. The basic story as perpetuated in many books, movies and tv series is nonfantastic. RH is shown as a dispossessed lord (usually treated as the Earl of Huntingdon), possibly a Saxon who fell foul of the Normans in the century or two after their conquest of England (the commonest period is during the reign of Richard I). He retreated to the woods, Sherwood Forest being the usual venue, where he gathered about him other hunted men (◊ COMPANIONS), the best-known being Little John, Friar Tuck, Allen A Dale, Will Scarlet(t) and Much the Miller. His sweetheart was Maid Marian. The outlaws stole from the rich but supported the needy (◊ ROBIN GOODFELLOW). In some stories RH eventually recovers his earldom. He was betrayed by the Prioress of Kirklees Abbey who poisoned him. As he lay dying he fired an arrow into the air, asking to be buried where the arrow fell. His traditional burial place is at Kirklees in Yorkshire.

This legend, with minor variations, has been the subject of many poems, ballads and books dating back to the 14th century, the oldest extant being *Robin Hood and the Monk* (**?1450**), although the most complete text is *A Gest of Robyn Hode* (written ?1400; **?1510**). The tale was given modern currency by Walter SCOTT in *Ivanhoe* (**1819**), and the adventures were developed in all their glory for children in *Robin Hood and Little John, or The Merry Men of Sherwood Forest* (**1840**) by Pierce Egan (1814-1880), which began the RH industry. Other books of associational interest include *The Merry Adventures of Robin Hood* (**1883**) by Howard PYLE, *Robin Hood* (**1927**) by E.C. VIVIAN, *The Chronicles of Robin Hood* (**1950**) by Rosemary SUTCLIFF and *The Adventures of Robin Hood* (**1956**) by Roger Lancelyn GREEN. (◊◊ *The* ADVENTURES OF ROBIN HOOD; 1955-9). As the UNDERLIER of the good-natured TRICKSTER rebel, RH also appears in *Silverlock* (**1949**) by John Myers MYERS and *The Last Unicorn* (**1968**) by Peter S. BEAGLE, while in *The Sword in the Stone* (**1938**; rev 1939 US) by T.H. WHITE RH (as Robin Wood) is one of the MENTORS of the young ARTHUR.

The more mystical and pagan associations with RH emerge through his connection with the May Day festivities, where his adventures form part of the elaborate Morris Dance routines. It was these celebrations, traditional from the 15th century, that brought together RH and Maid Marian, who is the personification of spring (◊ SEASONS; VIRGINITY) and who consorts with RH as the GREEN MAN. This shows RH more as a TRICKSTER, but also as a Lord of the Greenwood, protector of beasts and forest, akin to HERNE THE HUNTER but potentially greater, almost a Lord of FAERIE and also almost a HIDDEN MONARCH. RH thus becomes part of the MATTER of Britain, a MYTHAGO, with his forest as a POLDER of the old world. Plays for these festivities started to appear around 1475, although a much earlier play, *Jeu de Robin et Marion* (**?1283**) by Adam de la Halle (*c*1230-*c*1287), which appeared first in Norman France, linked the names if not the characters. Ben Jonson (1572-1637) sought to infuse the more rustic concepts, particularly the fey aspects of Maid Marian, into his unfinished play *The Sad Shepherd* (written 1637; first published in *The Workes of Benjamin Jonson* coll **1640**; vt *The Sad Shepherd, or A Tale of Robin Hood* **1783**). The image of RH as symbolic of Nature versus mankind, the forest versus the city, superstition versus law and order, with RH representing the old world threatened by THINNING, has considerable appeal, particularly at the eco-conscious end of the 20th century, and this was well developed in *The Death of Robin Hood* (**1981**) by Peter VANSITTART. The theme was continued with mystical appeal in the tv series ROBIN OF SHERWOOD (**1984-6**) and there was renewed interest among fantasy writers. Recent reconstructions of RH's life, predominantly nonfantastic but with underlying mythical motifs, include *The Outlaws of Sherwood* (**1988**) by Robin MCKINLEY, *Sherwood* (**1991**) and *Robin and the King* (**1993**) by Parke GODWIN and *The Lady of the Greenwood* (**1992**) by Jennifer ROBERSON, where Maid Marian is the central character. [MA]

Further reading: *Robin Hood* (**1982**) by J.C. Holt; *Robin Hood: Green Lord of the Wildwood* (**1993**) by John Matthews; *Robin Hood* (**1995**) by Graham Phillips.

ROBIN OF SHERWOOD UK tv series (1984-6). Goldcrest. **Pr** Esta Clarkham, Paul Knight. **Exec pr** Patrick Dromgoole. **Dir** Dennis Abey and many others. **Writers** Richard Carpenter, John Flanagan, Anthony Horowitz, Andrew McCullough. **Created by** Carpenter. **Starring** Robert Addie (Sir Guy de Gisburne), Jason Connery (Robin of Nottingham 1985-6), Philip David (Prince John), Nickolas Grace (Sheriff of Nottingham), Peter Llewellyn-Williams (Much), Clive Mantle (Little John), Richard O'Brien (Gulnar), Michael Praed (Robin of Loxley 1984-5), Phil Rose (Tuck), Mark Ryan (Nasir), Judi Trott (Marian). 20 60min and 3 120min episodes. Colour.

This retelling of ROBIN HOOD combined the familiar elements of the LEGEND with the additional twist of sorcery and black MAGIC. As the series begins the people of England are under the merciless rule of Norman dictators, epitomized by the evil Sheriff of Nottingham and his aide, Guy de Gisburne. To combat this menace, HERNE THE HUNTER, a forest SPIRIT, chooses the fugitive Robin of Loxley to serve him in his quest to help the oppressed. Dubbed "Robin Hood", the new hero faces a series of tests of strength and courage, many based on magic.

When Praed decided to leave after two seasons, Robin of Loxley was killed off and replaced by Robin of Nottingham. The battles against the sheriff continued apace, but the series lasted only one more year. An active fan following continues today, however, and there are fanzines, mailing lists and an annual convention.

Several novelizations were published: *Robin of Sherwood* * (**1984**) by Richard Carpenter, *Robin of Sherwood and the Hounds of Lucifer* * (**1985**) by Robin May and Carpenter, *Robin of Sherwood: The Hooded Man* * (**1986**) by Anthony Horowitz, and *Robin of Sherwood: Time of the Wolf* * (**1988**) by Carpenter; all were assembled as *The Complete Adventures of Robin of Sherwood* * (omni **1990**). [BC]

ROBINSON, CHARLES (1870-1937) Outstanding UK book illustrator with a clean, bright, decorative pen-line style showing distinct Art Nouveau influence; he also worked in a delicate line-and-watercolour style. His ILLUSTRATIONS have a sensitive ROMANTICISM allied to a strong sense of design and firmly controlled composition. Influenced by Walter CRANE, he had a genuine feeling for book design, and was a major contributor to the Arts and Crafts tradition of seeing the book as a work of art.

CR was one of three illustrator sons of London wood engraver Thomas Robinson (the other two were Thomas Heath Robinson [1869-1950] and William Heath ROBINSON). Unable to afford to accept the offer of a place at the Royal Academy Schools, he was apprenticed to a lithographer and attended evening classes at West London Art School and Heatherley's School of Art, working as a freelance illustrator in his father's studio in The Strand. His remarkable early success began with the publication of Robert Louis STEVENSON's *A Child's Garden of Verses* (**1895**), in which his drawings show a surprising maturity and sureness of touch. He was very prolific, illustrating well over 120 books and contributing to many leading MAGAZINES, including *The Graphic*, *Black and White* and *The Illustrated London News*. [RT]

ROBINSON, W(ILLIAM) HEATH (1872-1944) UK artist and illustrator – and brother of Charles ROBINSON – who from 1897 produced many finely composed, meticulously inked drawings and decorations for illustrated editions of fantasy by Hans Christian ANDERSEN, Miguel CERVANTES, Edgar Allan POE, William SHAKESPEARE, François RABELAIS, Walter DE LA MARE, Charles KINGSLEY and others. He also illustrated his own humorous CHILDREN'S FANTASIES, *The Adventures of Uncle Lubin* (**1902**) and *Bill the Minder* (**1912**): the drawings for Lubin's MUNCHHAUSEN-like exploits show WHR's fondness for cluttered masses of ramshackle paraphernalia. This foreshadowed the ridiculously overelaborate TECHNOFANTASY gadgets, usually held together with knotted string, which became the trademark of his HUMOUR – although his range was rather wider. Even WORLD WAR I became comic as WHR filled cartoon battlefields with gigantic vacuum cleaners to suck Germans out of dugouts, or mechanized belts of hot-water-bottles to warm kilted Highlanders' exposed flesh. He was the perfect illustrator for *The Incredible Adventures of Professor Branestawm* (**1933**) by Norman Hunter * (1899-1995). *Absurdities: A Book of Collected Drawings* (graph coll **1934**) is WHR's own large selection of his humorous artwork; *Railway Ribaldry* (graph coll **1935**) focuses on railways and TRAINS (◊◊ Rowland EMETT); *Inventions* (graph coll **1973**) assembles "gadget" drawings from WWI to WHR's death. The phrase "a Heath Robinson contraption" is still part of the language. [DRL]

Other works: *The Art of the Illustrator* (graph **1918**); *Flypapers* (graph coll **1919**); *Humours of Golf* (graph coll **1923**); *Let's Laugh* (graph coll **1939**; vt *Devices* 1977); *Heath Robinson at War* (graph coll **1942**); *The Penguin Heath Robinson* (graph coll **1966**) ed. R. Furneaux Jordan; *The W.*

Heath Robinson Illustrated Story Book (coll **1979**), children's stories 1916-21; *Great British Industries and Other Cartoons from The Sketch, 1906-1914* (graph coll **1985**). The unfinished «Uncle Lubin's Holiday» is included in *The Life and Art of W. Heath Robinson* (**1947**) by Langton Day.

With K.R.G. Browne: *How To Live in a Flat* (**1936**); *How to Be a Perfect Husband* (**1937**); *How to Make a Garden Grow* (**1938**); *How to Be a Motorist* (**1939**)

With H. Cecil Hunt: *How to Make the Best of Things* (**1940**); *How to Build a New World* (**1941**); *How to Run a Communal Home* (**1943**).

Further reading: *My Line of Life* (**1938**), autobiography; *The Illustrations of W. Heath Robinson: A Commentary and Bibliography* (**1983**) by Geoffrey C. Beare.

See also: Rube GOLDBERG.

ROCKING HORSE WINNER, THE UK movie (*1949*). ◊ D.H. LAWRENCE.

ROCK VIDEOS Short movies using popular songs as their sole soundtracks, issued by recording companies in order to sell records; as such they are, strictly speaking, COMMERCIALS, though their increasing sophistication also qualifies them as a distinct new artform. Basically there are three categories: performance videos, showing the artists performing in a concert setting; "mood" videos, interspersing performance scenes with various images to inspire the emotions suggested by the song; and narrative videos, which convey a story connected to the song. Videos in the latter two groups often draw upon or are relevant to fantasy.

Some videos clearly derive from actual fantasy texts: Tom Petty and the Heartbreakers' "Don't Come Around Here No More" (*1983*) features characters from Lewis CARROLL's **Alice** books; Blues Traveler's "Run Around" (*1995*) retells L. Frank BAUM's *The Wizard of Oz* (**1900**) with modern-dress analogues of Dorothy, Scarecrow, Tin Woodman and Cowardly Lion watching an impostor band as Toto pulls a curtain to reveal the real band playing behind them; and Michael Jackson's "Thriller" (*1983*) and "Leave Me Alone" (*1987*) pay homage to a number of horror and sf movies. A number of fantasy characters and creatures also appear: Enigma's "Return to Innocence" (*1994*) features backwards movie clips and images of a galloping UNICORN; Madonna's "Cherish" (*1989*) includes some well rendered mermen (◊ MERMAIDS AND MERMEN); The Red Hot Chili Peppers' "Soul to Squeeze" (*1993*) shows the lead singer as a SNAKE-haired GORGON; Paula Abdul's "Opposites Attract" (*1989*) shows the singer dancing with an anthropomorphic animated CAT; Elton John's "Blessed" (*1995*) presents a series of breathtaking fantasy landscapes while the singer floats as a head in a crystal ball; The Grateful Dead's "Touch of Grey" (*1987*) depicts the band as walking skeletons; and The Rolling Stones' "Love is Strong" (*1994*). Other memorable fantasy videos include: George Harrison's "Got My Mind Set on You" (*1987*), where the singer lounges on a chair while every object in the room around him comes to life; A-Ha's "Take on Me" (*1985*), where a woman is drawn into a COMIC book; Dire Straits' "Money for Nothing" (*1985*), where animated manual laborers enviously watch glamorous imaginary videos; and Sting's "If I Ever Lose my Faith" (*1993*), where many striking tableaux recreate the mood of a medieval ROMANCE, including one scene of Sting as CUCHULAIN, brandishing his sword at the sea. Two noteworthy videos with a definite fantasy narrative are Yes's "Owner of a Lonely Heart" (*1982*), an Orwellian nightmare about a man captured by anonymous oppressors who escapes by a METAMORPHOSIS into a bird, and Mike and the Mechanics' "All I Need is a Miracle" (*1987*), showing the improbable comic misadventures of a band manager.[GW]

ROCKWELL, NORMAN (1894-1978) US artist. ◊ ILLUSTRATION.

ROCKY HORROR PICTURE SHOW, THE UK movie (*1975*). 20th Century-Fox. **Pr** Michael White. **Exec pr** Lou Adler. **Dir** Jim Sharman. **Spfx** Colin Chilvers, Wally Weevers. **Screenplay** Richard O'Brien, Sharman. **Music and lyrics** O'Brien. **Incidental music** Richard Hartley. **Based on** The ROCKY HORROR SHOW (**1973**). **Starring** Jonathan Adams (Dr Everett V. Scott), Barry Bostwick (Brad Majors), Tim Curry (Dr Frank-N-Furter), Charles Gray (Criminologist), Peter Hinwood (Rocky Horror), Little Nell (Columbia), Meatloaf (Eddie), O'Brien (Riff Raff), Patricia Quinn (Magenta), Susan Sarandon (Janet Weiss). 101 mins. Colour.

The ultimate cult stage show put on screen to become the ultimate cult movie; performances of the original became excuses for wild fancy-dress parties among the audience, and the movie – after a shaky start – picked up where the stage show left off, so that even today showings of it are not so much screenings as events. The plot – insofar as it is relevant – sees straight couple Janet and Brad drop by an isolated castle where a convention of aliens from the planet Transsexual (in the galaxy Transylvania) is partying. Leader of them is Dr Frank-N-Furter, a camp, bisexual cross-dressing version of Mick Jagger, with a liberal dash of Elizabeth II. He has created for himself Rocky Horror, a muscle-bound sexual plaything. Various couplings – both homo and hetero – ensue, alongside cannibalism, murder and the seduction of Janet and Brad into the arms of full-blown DECADENCE. The net result is both a RECURSIVE homage to and PARODY of old sf, fantasy and HORROR MOVIES (especially RKO Radio Pictures), all linked by passages in which a narrator apes – right down to the last sneer – the 1950s and 1960s true-crime cinematic shorts presented by Edgar Lustgarten.

An interesting subplay involves references to Grant Wood's painting *American Gothic* (1930): O'Brien is first seen, in the distance, as a church servitor made up as and in the pose of the farmer with his pitchfork; towards movie's end, when O'Brien (as Riff Raff) shoots Frank-N-Furter and Rocky with a laser gun, that gun is trifurcate, in conscious mimicry of the pitchfork. There are other fleeting allusions to the painting, perhaps suggesting that the main plot is intended as Brad's or Janet's DREAM.

Mainstream critics have tended to dismiss *TRHPS* as unadulterated garbage; this is in large part true, but misses the point. It was sequelled by the inferior *Shock Treatment* (*1982*), in which a US town is transformed into a non-stop tv show; this flopped. [JG]

ROCKY HORROR SHOW, THE Rock musical (**1973**) by Richard O'Brien (1942-   ), produced in London, New York (1975) and Paris (1984). Along with the movie version, The ROCKY HORROR PICTURE SHOW (*1975*) (*which see for synopsis*) it has enjoyed enormous cult popularity. The central emphasis of its spoofing assault on conventions is TECHNO-FANTASY and HORROR. *Thing-Fish* (*1984*) by Frank Zappa (1940-1993) is a scabrous PARODY. [JC]

ROD SERLING'S NIGHT GALLERY US tv series (1971-3). Universal/NBC. **Pr** Jack Laird, William Sackheim. **Dir** John Badham, Leonard Nimoy, Steven SPIELBERG and many others. **Writers** Richard MATHESON, Rod SERLING and

many others. **Created by** Serling. **Ties** *Night Gallery* * (coll **1971**) by Serling; *Night Gallery 2* (coll **1972**) by Serling; *Rod Serling's Night Gallery Reader* * (anth **1987**) ed Martin H. GREENBERG, Carol Serling, Charles G. Waugh. **Starring** Serling (Host). **Guest stars** John Astin, John Carradine, Alex Cord, Patty Duke, Larry Hagman, Steve Lawrence, David McCallum, Burgess Meredith, Leslie Nielsen, Carl Reiner, Cesar Romero, Dean Stockwell, David Wayne, Fritz Weaver, Stuart Whitman and many others. Pilot plus 42 episodes mostly 1 hr, later 30 mins. Colour.

Several years after *The* TWILIGHT ZONE (1959-64) left the air, Serling returned with this anthology series set in a macabre art gallery. Each week he presented a number of stories – "Offered to you now, from *The Night Gallery*" – all introduced through the strange PICTURES hanging in the gallery. The series began with a 2hr pilot for NBC in 1969. It featured three stories: a man who has killed his rich uncle becomes obsessed with a painting of the family cemetery; a rich blind woman buys the eyes of a man in debt so she can see for 12 hours; an ex-Nazi concentration-camp officer is stalked by a former victim.

The series continued the same format, with two or three stories per episode. Twist endings were a trademark, and the early episodes succeeded in this quite well. As time went on, though, the story quality suffered and Serling himself, disenchanted, became less involved with the series.

While never achieving the success of *The Twilight Zone*, *RSNG* did produce some interesting episodes. In "A Question of Fear", for example, a man survives a night in a supposedly HAUNTED DWELLING but its owner convinces him that he has been poisoned and will become a giant slug-like creature; he kills himself, not realizing it was all a hoax. "Lindemann's Catch" features a MERMAID caught by a fisherman: he yearns for her to acquire human legs, but discovers this TRANSFORMATION involves her developing an amphibian head. [BC]

ROD SERLING'S THE TWILIGHT ZONE MAGAZINE US large-format semi-slick MAGAZINE, 60 issues April 1981-June 1989, monthly until December 1982, then bimonthly, published by Montcalm Publications, New York; ed T.E.D. KLEIN, April 1981-August 1985, Michael Blaine October 1985-October 1986, Tappan KING December 1986-June 1989.

RSTTZM endeavoured to capture the essence of Rod SERLING's tv series *The* TWILIGHT ZONE (1959-64) by exploring the intrusion of the supernatural into everyday life (◊ WRONGNESS) and lifting the lid off REALITY. There was an inevitable slant towards visual media, with movie and tv news and reviews, plus a programme-by-programme guide to *The Twilight Zone* (and later to *The Outer Limits* [1963-4] and ROD SERLING'S NIGHT GALLERY [1971-3]). However, fiction remained the core of the magazine, and under Klein's guidance *RSTTZM* developed into the premier supernatural-HORROR magazine of the early 1980s. Appearing at the height of Stephen KING's popularity, *RSTTZM* was able to attract a hungry readership. In addition to fiction by King, Klein presented similar fare by Ramsey CAMPBELL, Dennis ETCHISON, Gregory FROST, Charles L. GRANT and Richard Christian MATHESON. There was plenty that was also in keeping with the Serling approach, particularly by Parke GODWIN, George R.R. MARTIN, Joyce Carol OATES, Robert Sheckley (1928-   ), Lucius SHEPARD and Connie WILLIS. Then there were the slightly more offbeat stories, including material from Phyllis EISENSTEIN, Harlan ELLISON, Janet

Fox (1940-) and Tanith LEE. *RSTTZM* acquired the first fiction by Dan SIMMONS as a result of a story contest, and published first or early stories by Joe R. Lansdale (1951-), David J. Schow (1955-), Lewis Shiner (1950-) and John Skipp (1957-). Some of this fiction, which began to cohere during Blaine's editorship, was too strong for many readers, leading to a drop in circulation. King endeavoured to restore some of the original ambience, but overcompensated, putting in too much sf; also, the subtler fantasies he introduced jarred with the remaining shock-horror, and the original formula of *RSTTZM* had been shattered. The King issues do contain some of the best fantasies published by the magazine, especially those by James P. BLAYLOCK, Susan Casper (1947-), Elizabeth HAND – another *RSTTZM* discovery – Barbara Owens, Tim POWERS and William F. Wu (1951-). Overall, however, the mixture failed to gel with the readership, and circulation fell below viable levels.

RSTTZM was a key publication of the 1980s in the transformation of supernatural horror towards paranoid fantasy, though the magazine destroyed itself in the process. A retrospective large-format anthology selected from the magazine's first year was *Great Stories from Rod Serling's The Twilight Zone Magazine* (anth **1982**) ed Klein. *RSTTZM* began a spin-off digest magazine, NIGHT CRY, which reprinted from *RSTTZM*'s later issues, but it soon developed into a separate publication. [MA]

ROESSNER, MICHAELA Working name of US writer Michaela-Marie Roessner-Herman (1950-). Her first novel, *Walkabout Woman* (**1988**), concerns an aboriginal heroine's travails in recovering the DREAMTIME for her people; it won the John W. Campbell AWARD and the WILLIAM L. CRAWFORD MEMORIAL AWARD. *Vanishing Point* (**1993**) is sf; like her earlier novel it involves a charismatic young woman and other REALITIES (here rationalized through particle physics and ALTERNATE WORLDS) and suffers somewhat from a degree of moral smugness. [GF]

ROGERS, MARK (? -) US writer whose *The Dead* (**1989**) treats the APOCALYPSE in CHRISTIAN-FANTASY terms, with the LAST JUDGEMENT engendering scenes of horror: the dead arise again as ZOMBIES. In the end the protagonists are sent to an ALTERNATE-WORLD version of Earth for a new start. MR is not Mark E. ROGERS. [JC]

ROGERS, MARK E(ARL) (1952-) US writer and illustrator who began publishing work of genre interest with *The Bridle of Catzad-Dûm* (coll **1980** chap), and who became identified as a writer of comic fantasy (◊ HUMOUR) with his **Samurai Cat** sequence – *The Adventures of Samurai Cat* (**1984**; vt *Samurai Cat* 1984), *More Adventures of Samurai Cat* (**1986**), *Samurai Cat in the Real World* (**1989**), *The Sword of Samurai Cat* (**1991**) and *Samurai Cat Goes to the Movies* (**1994**) – which he also illustrated. The premise – that Miaowara Tomokato, a 16th-century samurai cat, quests through TIME and space to revenge his murdered warlord – serves as a pretext for textual and visual PARODIES of a wide range of GENRE FANTASY and fantasy CINEMA. MER's least predictable take-off point is, perhaps, Akira Kurosawa (1910-), some of whose movies – e.g., *Seven Samurai* (*1954*) in *Samurai Cat Goes to the Movies* – serve as slapstick venues. The **Zorachus** sequence – *Zorachus* (**1986**) and *The Nightmare of God* (**1988**) – is DARK FANTASY set in a decadent LAND. The **Blood of the Lamb** sequence – *The Expected One* (**1991**) and *The Devouring Void* (**1991**) – is marginally less grim. MER is not Mark ROGERS. [JC]

Other works: *The Runestone* (**1989** chap).

ROHAN, MICHAEL SCOTT (1951-) Scottish writer, son of a French-Mauritian father and a Scottish mother. He wrote sf before switching to fantasy in *The Ice King* (**1986**; vt *Burial Rites* 1987 US) with Allan SCOTT, writing together as Michael Scot. Although set in the present, the novel draws extensively on its authors' researches into Viking culture, which had earlier given rise to the nonfiction collaboration *The Hammer and the Cross* (**1981**). The same researches helped provide background material for MSR's **Winter of the World** trilogy: *The Anvil of Ice* (**1986**), *The Forge in the Forest* (**1987**) and *The Hammer of the Sun* (**1988**). In a world threatened by an Ice Age unleashed by powers inimical to humanity, an apprentice SMITH acquires various magical WEAPONS; his wanderings take him into more fertile lands and then across the ocean to a final conflict which will settle the destiny of humankind. The less earnest **Spiral** trilogy – *Chase the Morning* (**1990**), *The Gates of Noon* (**1992**) and *Cloud Castles* (**1993**) – establishes the real world as the "core" of a complex array of historical and legendary ALTERNATE REALITIES which intersect in such a way as to hurl the three central characters through a series of supernatural encounters, mostly involving eccentric juxtapositions of ancient and modern apparatus. The first two volumes employ exotic geographical settings, but the third returns to a politically unstable near-future Europe in order to construct a hectic modern GRAIL-quest and deal with the MATTER of Britain.

The Spell of Empire: The Horns of Tartarus (**1992**), another collaboration with Scott, is an amusing PICARESQUE set in an alternative Europe in which a Scandinavia-based Nibelung Empire confronts a Decadent Mediterranean-based Tyrrhennian Empire. The plot deploys three musketeers in an exuberantly playful fashion. *The Lord of Middle Air* (**1994**) is a historical fantasy starring the reputed 13th-century wizard Michael Scot, whom MSR claims as an ancestor of both himself and Sir Walter SCOTT. The Walter Scot who is the novel's hero reluctantly accepts his kinsman's aid in fighting a powerful sorcerer, venturing into FAERIE to obtain armaments adequate to his task.

MSR is an accomplished writer of action-adventure fantasies. He is refreshingly eclectic in his choice of materials and is capable of wringing new twists from traditional sources. [BS]

ROHMER, SAX Pseudonym of prolific thriller-writer Arthur Sarsfield Ward (1883-1959), best-known for a long series featuring FU MANCHU. Most of SR's imaginative fiction is OCCULT FANTASY or borderline sf, but some of it invites consideration as fantasy proper, especially that which draws on the folklore of the Near East. *The Quest of the Sacred Slipper* (**1914**) is his most substantial work in this vein. Other notable items can be found in *Tales of Secret Egypt* (coll **1918**), *The Haunting of Low Fennel* (coll **1920** UK) and *Tales of East and West* (coll **1932**; the 1933 US coll of this name is a selection from the UK coll and the previous item). A few of the OCCULT-DETECTIVE stories in *The Dream Detective* (coll **1920**) and some of the stories in *The Wrath of Fu Manchu and Other Stories* (coll **1973**) are also fantasies. *She Who Sleeps* (**1928**) is a rationalized romance of REINCARNATION involving a MUMMY; it is inferior to the melodramatic *Brood of the Witch Queen* (**1918**) and the rather slapdash *The Bat Flies Low* (**1935**). Another mummy is featured in *Seven Sins* (**1943**), the last of the **Gaston Max** series, none of which is fantasy – although *The Day the*

World Ended (**1933**) is marginal sf. The **Sumuru** series is of fantasy interest in that it features an exotic FEMME FATALE: *Nude in Mink* (**1950** US; vt *Sins of Sumuru* 1950 UK), *Sumuru* (**1951** US; vt *Slaves of Sumuru* 1952 UK), *The Fire Goddess* (**1952** US; vt *Virgin in Flames* **1953** UK), *Return of Sumuru* (**1954** US; vt *Sand and Satin* 1955 UK) and *Sinister Madonna* (**1956**).

The **Fu Manchu** series runs: *The Mystery of Dr Fu-Manchu* (fixup **1913**; vt *The Insidious Dr Fu-Manchu* 1913 US), *The Devil Doctor* (fixup **1916**; vt *The Return of Dr Fu-Manchu* 1916 US), *The Si-Fan Mysteries* (fixup **1917**; vt *The Hand of Fu-Manchu* 1917 US), *Daughter of Fu Manchu* (**1931**), *The Mask of Fu Manchu* (**1932**), *Fu Manchu's Bride* (**1933** US; vt *The Bride of Fu Manchu* 1933 UK), *The Trail of Fu Manchu* (**1934**), *President Fu Manchu* (**1936**), *The Drums of Fu Manchu* (**1938**), *The Island of Fu Manchu* (**1941**), *The Shadow of Fu Manchu* (**1948**), *Re-Enter Fu Manchu* (**1957** US; vt *Re-Enter Dr Fu Manchu* 1957 UK), *Emperor Fu Manchu* (**1959**) and the first four stories in *The Wrath of Fu Manchu and Other Stories*. The first three items are collected in *The Book of Fu-Manchu* (omni **1929** UK; the 1929 US edition includes also *The Golden Scorpion* [**1919**], a thriller). Cay Van Ash, co-author with Elizabeth Sax Rohmer of the SR biography *Master of Villainy* (**1972**) added the SHERLOCK HOLMES versus Fu Manchu novel *Ten Years Beyond Baker Street* (**1984**). [BS]

Other works: *The Sins of Séverac Bablon* (**1914**); *The Orchard of Tears* (**1918**); *The Green Eyes of Bast* (**1920**); *Bat Wing* (**1921**); *Fire-Tongue* (**1921**); *Tales of Chinatown* (coll **1922**); *Grey Face* (**1924**); *The Emperor of America* (**1929**); *The Sax Rohmer Omnibus* (omni **1938**); *Salute to Bazarada and Other Stories* (coll **1939**); *Wulfheim* (**1950**) as Michel Furey; *The Moon is Red* (**1954**); *The Secret of Holm Peel and Other Strange Stories* (coll **1970**).

ROLAND ◊ CHILDE; ROMANCE.

ROLT, L.T.C. (1910-1974) UK writer. ◊ GHOST STORIES.

ROMANCE Like the term FANTASY, "romance" refers to a bewildering number of forms of literary expression, and even to some things apparently unrelated to literature. Romance can mean a medieval poem, a paperback novel (with a woman and a castle on the cover), the "love interest" in a movie, or a commercial product such as champagne or exotic underwear. When the term is used to identify particular kinds of literature it is usually part of a compound – e.g., "chivalric romance" or "scientific romance" – and some critics use this trend as an excuse to bump "romance" up a level from genre to mode. Modern fantasy draws heavily upon many of the early forms of romance, and the two categories overlap considerably. The exact boundaries of romance matter less, though, than the historical and interpretive perspectives suggested by the term.

The earliest texts classed as romances are a few prose narratives written in Greek between the 1st and 3rd centuries AD; best-known is Longus's *Daphnis and Chloe* (*c*160AD). The stories generally involve couples separated by circumstance and eventually reunited, often by divine intervention. Their structure is episodic, their characters conventional, their action full of improbable adventures and their endings happy. Most contain supernatural characters or incidents. The structure is often quite complex, including stories-within-stories and elements of ALLEGORY. John J. Winkler credits these narratives with the invention of "romance" in the extraliterary sense: the idea of seeking out and falling in love with one's predestined partner.

A text bearing some relationship to these Greek romances is APULEIUS's *The Golden Ass* (2nd century AD), written in Latin but based on one or more Greek texts. *The Golden Ass* is more satirical than any of the Greek romances, but it shares with them the emphasis on fantastic adventures and an eventual reunion – not of lovers but of the hero Lucius and his human shape – brought about through magical means. Much of Apuleius's story is closer to farce or black comedy than to romance, but a lengthy embedded tale, the story of CUPID AND PSYCHE, follows the pattern of the lovers' quest. Another text that resembles the Greek romances in many respects is the Latin *Apollonius, Prince of Tyre* (3rd century AD).

The Classical narratives are romances only in hindsight: they were given the name to indicate their similarity to medieval romances. The term "romance" itself was first applied in the Middle Ages to any text translated into or composed in the vernacular languages rather than in Latin, but soon came to be applied to verse tales of chivalry and courtly love, such as CHRÉTIEN DE TROYES's *Lancelot* (*c*1177-81) and the *Lais* (before **1167**) of Marie de France, both from the 12th century. Medieval romances were based on traditional materials, such as the legends that had come to surround historical figures like Charlemagne and Alexander the Great or the court of ARTHUR. In retelling the familiar stories, the writer of romance felt free to embroider and embellish, often attaching incidents from one LEGEND to the hero of another. Medieval romances are characterized by several features that often reappear later, such as a QUEST, a knightly HERO, episodic structure, embedded stories and the intrusion of the supernatural into the more ordinary world of court and village.

Like the Greek romances, later medieval romances often shift towards either SATIRE or ALLEGORY. The most famous allegorical romance is *Romance of the Rose*. The first part, by Guillaume di Lorris (*c*1230), embodies in its narrative the various aspects of courtly love, while the second part, by Jean de Meun (*c*1275), shifts the focus to philosophy and misogyny.

Although the birthplace of the medieval romance was France, the fashion set in the French courts was eventually adopted by German, Italian, Scandinavian, English and Iberian writers. Among English poets, Geoffrey CHAUCER and John Gower (*c*1330-1408) both dabbled in it. Chaucer's *The Canterbury Tales* (coll *c*1387) includes a Classical tale told in romance style ("The Knight's Tale"), a version of one of the GAWAIN romances ("The Wife of Bath's Tale"), and a romance PARODY ("Chaucer's Tale of Sir Topas"). Gower demonstrates a similar range among the stories of his *Confessio Amantis* (coll **1390**), and also retells the Latin tale of Apollonius of Tyre, thereby providing an intermediary between the Classical romances and those of the Renaissance, for Gower's version became the source for SHAKESPEARE's *Pericles* (performed *c*1608; **1609**). Perhaps the greatest English romance is the anonymous 14th-century *Sir Gawain and the Green Knight* (◊ GAWAIN). One of the last major medieval romances is Sir Thomas MALORY's *Le Morte D'Arthur* (**1470**), in which elements from many earlier romances are combined into a single narrative cycle centring on the life of Arthur.

In the Renaissance, writers such as García de Montalvo with *Amadis of Gaul* (late 15th century), Sir Philip Sidney (1554-1586) with *Arcadia* (**1581**) and Robert Greene (1558-1592) with *Pandosto* (**1588**) followed Malory in constructing long prose narratives combining various motifs from earlier romances. Poets including Matteo Boiardo (1434-1494) with *Orlando Innamorato* (**1487**), Ludovico ARIOSTO with *Orlando Furioso* (**1532**), Torquato Tasso (1544-1594) with *Rinaldo* (**1562**) and Edmund SPENSER with *The Faerie Queene* (**1590-96**) so greatly expanded the scope and significance of the earlier romances that their works are often called romantic epics.

A group of late plays by Shakespeare constitute a subgenre of romance in themselves. Sometimes called tragicomedies, these set up tragic situations and then transform them into improbable but frequently breathtaking scenes of reconciliation and RECOGNITION. Among the Shakespearean romances, *Pericles* and *The Winter's Tale* (performed *c*1610; **1623**) are based on earlier romances – *Apollonius* and *Pandosto* respectively – while *Cymbeline* (performed *c*1611; **1623**) and *The Tempest* (performed *c*1611; **1623**) blend romance motifs with historical events.

18th- and 19th-century writers used the older forms of romance as justification for their own departures from the increasingly dominant realistic model of fiction. The scholarship of Sir Walter SCOTT, who wrote the original article on "Romance" for the *Encyclopedia Britannica* (**1824**), and Thomas Percy (1729-1811), editor of the *Reliques of Ancient English Poetry* (anth **1765**) – usually known simply as *Percy's Reliques* – reminded readers in the Age of Reason of an older tradition of literature based on the unreasonable and the extraordinary. Hence, when writers began to incorporate fantastic and improbable events into their fictions, they invoked the past with the words "Gothic" and "romance". Scott, Clara REEVE and Nathaniel HAWTHORNE are among those who not only wrote romances but offered critical defences of the form as an alternative to the realistic novel. In *The Progress of Romance* (**1785**) Reeve attempted to distinguish between the two without prejudice to the romance: "The Romance is an heroic fable, which treats of fabulous persons and things. – The Novel is a picture of real life and manners, and of the times in which it is written. The Romance in lofty and elevated language, describes what never happened nor is likely to happen. – The Novel gives a familiar relation of such things, as pass every day before our eyes, such as may happen to friend, or to ourselves . . ." Reeve's approach anticipates that of Northrop FRYE in *The Secular Scripture: A Study of the Structure of Romance* (**1976**).

The Gothic romances of Horace WALPOLE, Matthew LEWIS and Ann RADCLIFFE soon generated a number of widely divergent forms. These include the historical romances of Scott and James Fenimore Cooper (1789-1851) – from which developed the Western – the ambiguously supernatural US romances of Washington IRVING, Hawthorne and Herman MELVILLE, the detective stories of Edgar Allan POE, Wilkie COLLINS and Arthur Conan DOYLE, the pseudo-medieval romances of Wiliam MORRIS and Alfred Lord Tennyson (1809-1892) – which influenced J.R.R. TOLKIEN and C.S. LEWIS – the haunted love stories of Emily and Charlotte BRONTË, which gave birth to the woman's romance tradition of Daphne DU MAURIER and Mary STEWART, and the scientific romances of Mary SHELLEY, Poe, Edward BULWER-LYTTON, Edward Bellamy, Mark TWAIN, Jules Verne and H.G. WELLS, the forerunners of modern SCIENCE FICTION. Many of these varieties of romance continue in the form of mass-market formula fiction.

The Postmodernist version of romance seems to involve selfconscious elaboration of the popular formulae with a strong dose of metafiction. Indeed, it is often difficult to tell whether a particular story is a literary imitation of a popular genre or an attempt on the part of a gifted and ambitious genre writer to transcend the form. Writers who have produced works that might be considered Postmodern romance include Doris Lessing, Gene WOLFE, Umberto ECO, Angela CARTER, Ursula K. LE GUIN, A.S. BYATT, John Calvin Batchelor, Geoff RYMAN, Salman RUSHDIE and John CROWLEY.

It is difficult to make any statements that might cover all of this sprawling terrain. Perhaps the safest generalization about romance is that it is not realism. It does not aim primarily to reproduce the texture of ordinary life, though it may use realistic effects at various points in the story. The elements that dominate realistic fiction, such as the motivations and interactions of character, are frequently sketchy or stylized in a romance. Rather than presenting a coherent and motivated plot, romance emphasizes an inventive and open-ended story. This emphasis on STORY – on the pleasures of narrative – may be one of the reasons that romance has nearly always been deplored by critics looking for moral messages and high seriousness. It is also one of the reasons for the appeal to Postmodernist writers, who are often interested in the operations of narrative itself.   [BA]

ROMANTICISM A movement in the arts and philosophy whose reverberations profoundly affected intellectual, social and political life from the late 18th to the mid-19th century. It took the form of a rebellion against the rewards and supposed lessons of the Enlightenment, challenging the intellectual hegemony of science and reason and the social hegemony of tradition. It came to be seen as an extreme opposed to the orthodoxies of "Classicism", linked to if not straightforwardly reflective of the opposition between subjectivity and objectivity. Romanticism was correlated with a dramatic resurgence of interest in all matters psychological and supernatural, including FOLKLORE, MYTHOLOGY, DREAMS and TRANSCENDENCE. The spectacular rehabilitation of the imagination thus contrived was fundamental to the evolution of modern fantasy; the name forged a calculated link betwen the movement and the tradition of medieval ROMANCE which provides GENRE FANTASY with much of its imaginative apparatus. Some aesthetic theorists of the day attempted to account for the substitution of Classicism by Romanticism in terms of the supplementation of the idea of beauty with that of the sublime.

The French Romantic tradition was foreshadowed by the cult of sensibility and glorification of noble savagery associated with Jean-Jacques Rousseau (1712-1778), but the Romantic Movement *per se* began in Germany, where a number of writers began a defiant celebration of *Sturm und Drang* ("Storm and Stress") in the 1770s. Leading German Romantics included J.W. GOETHE, NOVALIS, Friedrich SCHILLER, Ludwig TIECK and Friedrich de la Motte FOUQUÉ. English Romanticism was foreshadowed by the invention of OSSIAN and by "graveyard poetry", before being theorized at the turn of the century by Nathan Drake, Samuel Taylor COLERIDGE and William BLAKE. Among its most significant early converts were Percy Bysshe SHELLEY, Lord BYRON and John KEATS. The central figures of the French Movement included Théophile GAUTIER, Victor HUGO and (via the translations of Charles BAUDELAIRE) Edgar Allan POE. The unique character of US imaginative literature is partly determined by the interaction of Romantic ideas imported from Europe with the mythology of the frontier (whose westward movement was determinedly obliterating a world strongly akin to the one whose loss European Romanticism was lamenting). The historical and the Gothic novel (◊ GOTHIC FANTASY) were both products of Romanticism, as were countless collections and imitations of FAIRYTALES. Although the Romantic movements of all the European nations declined after 1848, they were the parents of DECADENCE and Symbolism and the grandparents of SURREALISM and Expressionism.

Although GENRE FANTASY required a new period of rehabilitation in the 1960s, the tradition had never fallen into derelicton; Romanticism "declined" not because it had been superseded but because its essential message had become so widely taken for granted that it no longer required such passionately clamorous expression. [BS]

ROMULUS AND REMUS ◊ MYTH OF ORIGIN.

RONSON, MARK Pseudonym of Marc ALEXANDER.

ROOKE, LEON (1934-   ) US writer, resident in Canada since 1969. Long a short-story writer and playwright, LR reached a wider audience with *Shakespeare's Dog* (**1982**) an extravagant FABULATION narrated by the dog. In *The Magician in Love* (chap **1981**) the indeterminacy of locale and era pushes the story to the verge of fantasy, a tactic LR uses in much short fiction. The title of his sixth collection, *The Birth Control King of the Upper Volta* (coll **1982**), suggests the Absurdism that characterizes much of his other work. Some stories in *How I Saved the Province* (coll **1989**), are set in an Absurdist near future. [GF]
Other works: *Vault, A Story in Three Parts* (**1973** US); *A Good Baby* (**1989**); LR has published many short colls, which are drawn upon for *Sing Me No Songs, I'll Say You No Prayers: Selected Stories* (coll **1984** US); more recent colls include *A Bolt of White Cloth* (coll **1985**), *The Happiness of Others* (coll **1991**), *Who Do You Love?* (coll **1992**) and *Narciso allo Speccio* (coll **1995** Italy).

ROSEMARY'S BABY US movie (*1968*). Paramount/William Castle Enterprises. **Pr** William Castle. **Dir** Roman POLANSKI. **Spfx** Farciot Édouard. **Screenplay** Polanski. **Based on** *Rosemary's Baby* (**1967**) by Ira LEVIN. **Starring** Ralph Bellamy (Abraham Sapirstein), Sidney Blackmer (Roman Castevet), John Cassavetes (Guy Woodhouse), Maurice Evans (Hutch), Mia Farrow (Rosemary Woodhouse), Ruth Gordon (Minnie Castevet), Charles Grodin (Dr C.C. Hill). 137 mins. Colour.

Childlike Rosemary and unsuccessful actor husband Guy move into an apartment in an old block with an unsavoury history involving the WITCH Adrian Marcato. Their elderly neighbours, Minnie and Roman Castevet, soon become claustrophically close, although Guy relishes this; Roman tells him he is destined for greatness, and soon his successful rival for an important role is struck blind, so Guy gets the part. One night, after eating mousse presented by Minnie, Rosemary collapses; among her vivid DREAMS (or so they seem) is one of intercourse with the DEVIL. She awakes feeling raped and with scratches on her shoulders; Guy explains he had SEX with her as she slept, a rape which she accepts. Soon she finds herself pregnant; a lot later she finds out more about WITCHCRAFT, and deduces (falsely) that the witches want her baby for HUMAN SACRIFICE; at last, too, she realizes Guy is complicit (indeed, has sold her body in return for his own success), as is Sapirstein, her new obstetrician. Sapirstein and the rest of the coven drug her and she gives birth; later she is told the child is dead. But she hears through the partition a baby crying and breaks into the Castevets' apartment to find the full coven, plus the horrific

baby fathered on her by Satan as an evil MESSIAH. Soon, though, the maternal instinct triumphs and she accepts the child as hers.

RB is a long movie, but its time is not wasted and events are beautifully paced; despite its sensational material, it is oddly unsensationalist. Although everything is told from Rosemary's viewpoint, we are constantly ahead of her in understanding what is going on; yet this is handled so well that the situation does not seem artificial. Such restraint was not to be displayed by later movies on similar themes, like *The* EXORCIST (*1973*), *The* OMEN (*1976*) and the drab sequel *Look What's Happened to Rosemary's Baby* (*1973* tvm; vt *Rosemary's Baby 2*), which attempted to capitalize on the success of *RB* in a standard tale of the GREAT BEAST. In all of these, what you see is what you get, whereas *RB*'s strength is its delicious ambiguity; by design, it is left open to us to read the movie as an account of distorted PERCEPTION produced by Rosemary's neuroses. [JG]

ROSENBERG, JOEL (1954-   ) US writer who first began publishing fiction with "Like the Gentle Rains" for *IASFM* in 1982. JR is best-known for **The Guardians of the Flame** sequence, which began with the SWORD-AND-SORCERY fantasy *The Sleeping Dragon* (1983), a novel about role-playing gamers (◊ GAMES) drawn into their GAME-WORLD, but this element diminishes in importance in later volumes, which include *The Sword and the Chain* (**1984**) and *The Silver Crown* (**1985**) – these first three assembled as *Guardians of the Flame: The Warriors* (omni **1985**) – plus *The Heir Apparent* (**1987**) and *The Warrior Lives* (**1989**) – these two assembled as *Guardians of the Flame: The Heroes* (omni **1989**) – plus *The Road to Ehvenor* (**1991**) and *The Road Home* (**1995**). JR has also published several sf novels.

In recent years JR has inaugurated two new fantasy series: the **D'Shai** novels – *D'Shai* (**1991**) and *The Hour of the Octopus* (**1994**) – which are perhaps his best work, and the **Keeper of the Hidden Ways** series, which begins with *Keeper of the Hidden Ways: The Fire Duke* (**1995**). Both series benefit from being conceived in JR's maturity as a writer, and so enjoy a greater sophistication than those novels set in venues created at the outset of his career. [JCB/GF]

ROSENCREUTZ, CHRISTIAN Supposedly a German writer (1378-1484) whose works inspired – or who was himself directly responsible for – the founding of ROSICRUCIANISM. CR is credited with the authorship of at least part of the *Fama Fraternitas dess Loblichen Ordens des Rosenkreutzes* (**1614** chap), known in English as *The Fame of the Fraternity of the Meritorious Order of the Rosy Cross*, and of the *Confessio Fraternitas* (**1615**), two of the texts generally thought to have spurred the founding of the movement. The former describes CR's voyages to the Orient – the appropriate direction to take for a MAGUS seeking initiation and access to the secret history of the world (◊ FANTASIES OF HISTORY). CR's life is described, in terms indistinguishable from ROMANCE, in Johann Valentin Andreae's *Chymische Hochzeit* (**1616**), known in English under various titles but usually as *The Chemical Wedding of Christian Rosencreutz*. [JC]

ROSENDORFER, HERBERT (1934-   ) German judge and writer whose first novel, *Der Ruinenbaumeister* (**1969**; trans Mike Mitchell as *The Architect of Ruins* 1992 UK), is a complex tale, seemingly in the style of the ARABIAN NIGHT-MARE. Its protagonist descends through baroque convolutions of plot into the dream UNDERWORLD of the eponymous MAGUS, where he meets the WANDERING JEW and many other figures. Dark hints of German history bulge ominously into the knots of interrupted STORY-within-story that comprise the bulk of the text; but eventually the stories are concluded (which differentiates the text from the ultimate entrapment implied by the Arabian Nightmare mode), and the protagonist does make a kind of escape. *Deutsche Suite* (**1972**; trans Arnold Pomerans as *German Suite* **1979** UK) anatomizes post-Nazi German history through the life of a man understood to be an APE because his mother had slept with one. *Briefe in die deutsche Vergangenheit* (**1983**; trans Mike Mitchell as «Letters into the Chinese Past» 1996 UK) takes a satiric look at present Germany through the eyes of a 10th-century Chinese mandarin. *Stephanie, oder Das vorige Leben* (**1987**; trans Mike Mitchell as *Stephanie, or A Previous Existence* **1995** UK) is a compact, funny, moving TIMESLIP tale whose contemporary protagonist enters a life two centuries earlier, with disturbing consequences. Much of HR's other fantasy has not been translated; it includes *Grosses Solo für Anton* ["Big Solo for Anton"] (**1976**), whose protagonist becomes the last human alive after a nuclear HOLOCAUST. [JC]

ROSICRUCIANISM A secret doctrine apparently formulated in the early 17th century; the texts that announced the Rosicrucian Society's existence were probably inspired by the millennial anxieties that ransacked Europe during these decades of violent religious conflict. The immediate sources of Rosicrucian doctrine – all of which is esoteric – seems to be 16th-century Neoplatonism, and it has been suggested that the hermetic speculations in the CABBALA by John DEE may have been directly influential. Whatever the truth, the Rosicrucian Society was a brotherhood whose members affianced themselves to a tradition that SECRET MASTERS embodied the world's history and controlled it from behind the scenes. Their own MAGUS founder, Christian ROSENCREUTZ, is a typical Secret Master: he lives an abnormally long time; he travels into the East, where he is initiated into arcana and MAGIC; he returns with his gifts to Europe, where he sets them down in secret; after his death, he becomes a SLEEPER UNDER THE HILL; he is credited (long after his death) with the partial authorship of the *Fama Fraternitus dess Loblichen Ordens des Rosenkreutzes* (**1614** chap; trans **1652** UK), which tells his life story, and of the *Confessio Fraternitas* (**1615**), which purports to give an historical background, extending into Ancient Egypt, for the secret order.

More interestingly to general readers, Rosenkreutz is also the subject of *Chymische Hochzeit Christiani Rosenkreutz* (**1616**; trans as *The Hermetick Romance, or The Chymical Wedding* **1690** UK) by Johann Valentin Andreae, a ROMANCE ostensibly much concerned with ALCHEMY (Rosencreutz discovers the PHILOSOPHERS' STONE) but in fact a tale that would assuredly have TAPROOT-TEXT status were it better-known. Rosenkreutz is interrupted at Easter by an invitation to a royal wedding, and undertakes a perilous journey to the castle where the chymical marriage is held, though only after he has been put through various ordeals. Afterwards he penetrates the marriage chamber, where Venus (◊ GODDESS) reposes asleep; on discovery he is punished by being commanded to remain on guard over the secret chamber, which also contains books of wisdom recounting the secret history of the world. A plethora of occult imagery and the creation of at least one HOMUNCULUS adorn the central tale. Whether or not Andreae wrote not only this novel but also the two "nonfiction" founding texts of Rosicrucianism is moot – it seems on balance likely

he did. In later life, however, he abjured any relationship to the secret brotherhood he may have created out of whole cloth.

There is no real evidence that the Rosicrucian Order existed at all – as an actual group of men – during the 17th century. Nor is there much more evidence for the existence of the Order of the Gold and Rosy Cross, ostensibly founded on Rosicrucian principles in 1710. During the 18th century, however, various societies began to claim a Rosicrucian provenance, but none adhered in any significant fashion to the precepts of the original fabrication; and the picture is rendered far more complex by the competing and/or complementary activities and assertions of Freemasonry. It was not until the founding of the Hermetic Order of the GOLDEN DAWN in 1887 that the 17th-century material was given heed (though any claimed historical connection was spurious) and that rituals involving a SEVEN-sided tomb and a magically preserved Rosencreutz came into use. It is through the Golden Dawn connection that writers like Aleister CROWLEY and Arthur MACHEN may have come across and made use of certain fabulous images brought into being by Andreae.

The Invisible College in Mary GENTLE's *Rats and Gargoyles* (**1990**) is a direct reference to Rosicrucianism, as is Terry PRATCHETT's Unseen University. The 20th-century (Californian) Rosicrucian Society, whose ads were much enjoyed by readers of pulp MAGAZINES, seems at best a very remote descendant of the original. [JC]

ROSNY aîné, J.H. Working name of Belgian writer Joseph-Henri Boëx (1856-1940). ◊ ANTHROPOLOGY.

ROSS, CHARLES HENRY UK comics pioneer. ◊ COMICS.

ROSS, CLARISSA Pseudonym of Marilyn ROSS.

ROSS, DANA Pseudonym of Marilyn ROSS.

ROSS, MARILYN Best-known pseudonym of US writer William Edward Daniel Ross (1912-   ), who has written about 350 novels in various genres under this and at least 20 additional names, including Marilyn Carter, Rose Dana, Ruth Dorset, Ann Gilmer, Ellen Randolph, Clarissa Ross, Dana Ross, Olin Ross and Jane Rossiter – the last being the name attached to *Summer Season* (**1962**), his first novel. He is of genre interest mainly for the **Dark Shadows** sequence of novelizations tied to the tv series DARK SHADOWS (1966-71): *Dark Shadows* * (**1966**), *Victoria Winters* * (**1967**), *Strangers at Collins House* * (**1967**), *The Mystery at Collinwood* * (**1967**), *The Curse of Collinwood* * (**1968**), *Barnabas Collins* * (**1969**), *The Secret of Barnabas Collins* * (**1969**), *The Demon of Barnabas Collins* * (**1969**), *The Foe of Barnabas Collins* * (**1969**), *The Phantom and Barnabas Collins* * (**1969**), *Barnabas Collins Versus the Warlock* * (**1969**), *The Peril of Barnabas Collins* * (**1969**), *Barnabas Collins and the Mysterious Ghost* * (**1970**), *Barnabas Collins and Quentin's Dream* * (**1970**), *Barnabas Collins and the Gypsy Witch* * (**1970**), *House of Dark Shadows* * (**1970**) – which novelizes the movie version – *Barnabas, Quentin, and the Mummy's Curse* * (**1970**), *Barnabas, Quentin, and the Avenging Ghost* * (**1970**), *Barnabas, Quentin, and the Nightmare Assassin* * (**1970**), *Barnabas, Quentin, and the Crystal Coffin* * (**1970**), *Barnabas, Quentin, and the Witch's Curse* * (**1970**), *Barnabas, Quentin, and the Haunted Cave* * (**1970**), *Barnabas, Quentin, and the Frightened Bride* * (**1970**), *Barnabas, Quentin, and the Scorpio Curse* * (**1970**), *Barnabas, Quentin, and the Serpent* * (**1970**), *Barnabas, Quentin, and the Magic Potion* * (**1971**), *Barnabas, Quentin, and the Body Snatchers* * (**1971**), *Barnabas, Quentin, and Dr Jekyll's Son* * (**1971**), *Barnabas, Quentin, and the*

Grave Robbers * (**1971**), *Barnabas, Quentin, and the Sea Ghost* * (**1971**), *Barnabas, Quentin, and the Mad Magician* * (**1971**), *Barnabas, Quentin, and the Hidden Tomb* * (**1971**) and *Barnabas, Quentin, and the Vampire Beauty* * (**1972**).

Other works of interest include: *Lust Planet* (**1962**) as by Olin Ross; *Gemini in Darkness* (**1969**), *The Secret of the Pale Lover* (**1969**), *The Corridors of Fear* (**1971**), *Satan Whispers* (**1981**) and *Summer of the Shaman* (**1982**) as by Clarissa Ross; *The Vampire Contessa* (**1974**) and titles in the **Birthstone Gothic** collective sequence – *The Ghost and the Garnet* * (**1975**), *The Amethyst of Tears* * (**1975**) and *Shadow over Emerald Castle* * (**1975**) – all by Marilyn Ross.   [JC]

ROSS, OLIN Pseudonym of Marilyn ROSS.

ROSSETTI, CHRISTINA (GEORGINA) (1830-1894) Italian poet, born and raised in UK, sister of Dante Gabriel ROSSETTI and niece of John POLIDORI. Her grandfather published some of her poems privately as *Verses* (coll **1847**), and her work began to appear in *The Athenaeum* from 1848. Through her brother she became associated with the PRE-RAPHAELITES, and her verses appeared in the Brotherhood's magazine *The Germ* in 1850 as by Ellen Alleyne. Most of her work is religious, but some of her early poems are of a more imaginative nature. Her best-known is *Goblin Market and Other Poems* (coll **1862**). The title verse is a FAIRYTALE in which GOBLINS tempt two young sisters to eat fairy fruit. One succumbs and drifts into a delirium but the other refuses and tries to save her sister. CR denied the sexual undertones of the poem, but it has always retained notoriety. The title poem of *The Prince's Progress and Other Poems* (coll **1866**; reissed with above as *Goblin Market, The Prince's Progress and Other Poems* coll **1875**; vt *Poems* 1876 US) is an ALLEGORY in which a prince sets out to marry his bride but gives in to temptations *en route* and dallies too long so that his bride dies of despair. It was written as a reversal of the MOTIF of SLEEPING BEAUTY. *Sing-Song, a Nursery-Rhyme Book* (**1872**) was for children.

CR wrote several stories, their form and content influenced by the FAIRYTALES of Hans Christian ANDERSEN. They are more aggressive than her poetry, serving as a channel for darker emotions. "Hero" (1865 *Argosy*) was written in the mid-1850s during the wave of fairytale mania in the UK; it is a clever observation of the clash of cultures on the borders of FAERIE and of the METAMORPHOSIS that comes over the HERO in his QUEST for beauty and love. "Nick" (1857 *National Magazine*) is a violent interpretation of the THREE-WISHES motif in FABLE form. In *Speaking Likenesses* (**1874** chap) a child awakes to find that the others at her party have taken on the form that reflects their individual characteristics ($\lozenge$ TRANSFORMATION). There is also the lost "Folio Q. Case 2", written *c*1860, which CR destroyed after it failed to sell. Her brother William Michael Rossetti (1829-1919), who edited her *Complete Poetical Works* (**1904**), recalled that it was about a man cursed with no SHADOW, and regarded it as her best story. The most complete volume of CR's work currently available is *Poems and Prose* (coll **1994**) ed Jan Marsh, who has also written *Christina Rossetti: A Literary Biography* (**1994**).   [MA]

ROSSETTI, DANTE GABRIEL (CHARLES) (1828-1882) UK-born poet and painter of Italian parentage, brother of Christina ROSSETTI and William Michael Rossetti (1829-1919), nephew of John POLIDORI. His early friendships with Holman Hunt (1827-1910) and John Everett MILLAIS, all with a common interest in revolutionizing English art and bringing back the lost ROMANTICISM,

led to the formation of the PRERAPHAELITE Brotherhood in 1848. Their work changed the Victorian attitude to art completely. Supported by John RUSKIN and Lord TENNYSON and later joined by William MORRIS and others, the movement rapidly took hold despite initial intransigence among critics. DGR was perhaps the most influential of the group, less ardent than Morris and less productive than Hunt, but nevertheless produced a body of work that remains the epitome of Victorian Romanticism. His paintings, which drew on religious themes as well as Arthurian ($\lozenge$ ARTHUR) and Dantesque ($\lozenge$ DANTE ALIGHIERI) subjects, were spiritually and emotionally charged, evoking distant and otherworldly imagery. He endeavoured to bring that same sense of displacement to his poetry. "The Blessed Damozel" (1850 *The Germ*) is a romantic piece about a maiden from HEAVEN calling to her earthly love; *Sister Helen* (**1870** chap), an imitation medieval ballad, uses the theme of VENGEANCE through a waxen image; "The House of Life" (1881 in *Ballads and Sonnets*) is a sequence of mystical and arcane sonnets. Of all the Preraphaelites, DGR is perhaps the most representative and the most influential.   [MA]

ROSSITER, JANE Pseudonym of Marilyn ROSS.

ROSY CROSS The central visual SYMBOL of ROSICRUCIANISM, the image of a rose placed within the arms of a cross. It is a play on the name of the (probably fictitious) 17th-century German writer Christian ROSENCREUTZ.   [JC]

ROSZAK, THEODORE (1933-   ) US writer and cultural critic. His first book-length fiction was *Pontifex: A Revolutionary Entertainment for the Mind's Eye Theater* (**1975** UK); his second, *Bugs* (**1981**; vt *Bugs: A Novel of Terror in the Computer Age*), is a HORROR/TECHNOFANTASY/sf hybrid in which a telepathic child in fright causes literal bugs to infiltrate computer systems and then devour people. *Dreamwatcher* (**1985**) is a melodrama based on the premise that certain talented individuals have the TALENT to enter, interact with and sculpt other people's DREAMS. *Flicker* (**1991**) is a major novel, a complex and fascinating TECHNOFANTASY in which a movie buff, Gates, becomes fascinated by the legendary *film noir* director Max Castle, who vanished in the 1940s and whose movies have since become "lost classics". Gates revives the movies – including a portion of the unfinished *Heart of Darkness* (based on the Joseph CONRAD novel) done in conjunction with Orson Welles (1915-1985) – and finds that all have a disquiet that goes beyond what is on screen. At last he discovers that Castle was making use of the "flicker" – the moment between each frame – in order to subliminally propagandize for a nihilistic, corrupting secret society. Castle disappeared when he developed qualms; Gates, getting too close to the conspiracy ($\lozenge$ FANTASIES OF HISTORY), is likewise abducted by that society, discovering the aged Castle still alive on a remote ISLAND. Although fading towards its end, *Flicker* is a distinguished example of the fantasy of PARANOIA and possibly the finest extant fantasy of the CINEMA. The RECURSIVE FANTASY *The Memoirs of Elizabeth Frankenstein* (**1995**) shared the 1996 James Tiptree Jr AWARD. TR's novels are difficult to classify generically; this is probably why they are not better-known within the field.   [JG]

ROUND TABLE The RT of King ARTHUR was first mentioned in the *Roman de Brut* (**1155**) by the Norman monk Wace (?1110-?1175). Here it was a physical object designed to ensure equality between all of the KNIGHTS. Later stories gave various versions of its seating capacity, ranging from 13 up to a startling 1600. One of the seats, the Siege Perilous, remained unclaimed and awaited the purest of knights; any others who sought to sit there would die. At length GALAHAD claimed it.

Legends vary about the RT's origin, but the chief story is that it was made by MERLIN for Uther Pendragon and, after Uther's death, was inherited by Leodegrance, king of Cameliard, then came into Arthur's possession when he married Leodegrance's daughter GUINEVERE. Not all of Arthur's knights were also Knights of the RT: they had to earn that right through demonstrations of courage and valour – hence the many QUESTS. The principal Knights of the RT were Agravain, Bagdemagus, Bedivere, Bors, Dinadan, Gaheris, GALAHAD, GAWAIN, Lamorak, LANCELOT, Mordred (Modred), Pelleas, PERCEVAL, Sagremore, Tor, Tristan and Yvain.

The RT signified not only equality but unity. Hermetic interpretation of the RT was that it was an earthly representation of completeness, paralleling one on the spiritual plane. With Galahad's completion of the circle, the time was right for the Second Coming, presaged by the Holy GRAIL. The RT also represented the Order of Knights anointed by Arthur. This is what Edward III (reigned 1327-77) imitated in 1348 when he created the Noble Order of the Garter as the highest order of knighthood. [MA]

ROUNTREE, HARRY (1878-1950) Highly imaginative New Zealand-born book and magazine illustrator who worked in the UK from 1910. He was particularly adept at creating endearingly humorous drawings of animals and birds in a delicate pen-and-ink style with finely textured dense areas of shadow and contrasting untouched areas of light; he also worked in a broader watercolour style.

Born in Auckland, HR worked as a lithographer in a commercial art studio before moving to London at the age of 32. His first work in the UK was for *Little Folks*, where he showed the remarkable originality of his imagination in ARABIAN FANTASIES, and where his skill at HUMOUR first became evident. His work enjoyed great popularity for a while. He moved to St Ives, Cornwall, late in life, and died there in relative poverty; a commemorative plaque is displayed on the harbour jetty. [RT]

ROUSSEAU, JEAN-JACQUES (1712-1778) Swiss-born French philosopher and essayist. ◊ ANTHROPOLOGY.

ROVIN, JEFF Working name of US COMIC-book editor and writer Jeffrey Daniel Rovin (1951-   ), who has specialized from his first book of genre interest – *A Pictorial History of Science Fiction Films* (**1975**) – in the creation of reference works concerning various forms of fantastic literature. He has also written fiction of interest, including *The Madjan* (**1984**), *Re-Animator* * (**1987**) – novelizing RE-ANIMATOR (*1985*), itself loosely based on *Herbert West Reanimator* (*1977* chap) by H.P. LOVECRAFT – and *Starik* (**1988**), the latter with Sander Diamond (1942-   ). *Mortal Kombat* * (**1995**) is a tie based on a video game.

Of particular interest is *The Fantasy Almanac* (**1979**), an extensive illustrated glossary which focuses mainly on fictional characters, though some authors are discussed. Other works include *The Fabulous Fantasy Films* (**1977**), *The Encyclopedia of Superheroes* (**1985**), *The Encyclopedia of Super Villains* (**1987**), *The Encyclopedia of Monsters* (**1989**) and *The Illustrated Encyclopedia of Cartoon Animals* (**1991**). This last title, arranged alphabetically by fictional creature, encompasses examples from newspaper cartoons, COMICS and ANIMATED MOVIES. [JC]

Other works (nonfiction): *From Jules Verne to Star Trek: The Best of Science Fiction Movies and Television* (**1977**); *From*

the Land Beyond Beyond: The Films of Willis O'Brien and Ray Harryhausen (**1977**); a series of quiz books starting with *The Supernatural Movie Quizbook* (**1977**); *Mars!* (**1978**); *The Transgalactic Guide to Solar System M-17* (**1981**); *The Science Fiction Collector's Catalog* (**1982**).

ROWENA Working name of US fantasy/sf book-cover artist Rowena Morrill (1944-   ), whose paintings create cold, unsettling fantasy worlds with great precision of detail. She typically works in a combination of oils and acrylics, with a high-gloss glaze finish. Her training was in a soldiers' wives' art course on a US Army base. *The Fantastic Art of Rowena* (portfolio **1983**) offers a selection of her work. [RT]
Further reading: *Great Masters of Fantasy Art* (**1986**) by Eckart Sackmann.

ROWLANDSON, THOMAS (1756-1827) UK artist. ◊ COMICS.

ROWLEY, CHRISTOPHER (B) (1948-   ) US writer, best-known for his sf. His **Bazil Broketail** series – *Bazil Broketail* (**1992**), *A Sword for a Dragon* (**1993**), *Dragons of War* (**1994**) and *Battledragon* (**1995**) – attempts to duplicate his successful military-adventure sf in DRAGON fantasy. [GF]

RPG Popular contraction for "role-playing games". ◊ GAMES.

RUARK, ROBERT C(HESTER) (1915-1965) US author best-known for novels like *Something of Value* (**1955**) and *Uhuru* (**1962**), concerned with Kenya. Of fantasy interest is the short **Grenadine** sequence – *Grenadine Etching: Her Life and Loves* (**1947**) and *Grenadine's Spawn* (**1952**) – which recounts in a TALL-TALE idiom the life and devastating career of the eponymous WITCH, responsible for the cigarette and the invention of advertising, and of her rambunctious descendants. [JC]

RUBIÃO, MURILO (1916-1991) Brazilian writer of well crafted and delicate short stories, most collected in volumes like *O Ex-Mágico* ["The Ex-Magician"] (coll **1947**), *A Estrela Vermelha* ["The Red Star"; vt *A Casa do Girassol Vermelho* ["The House of the Red Sunflower"] (coll **1953**), *Os Dragões e outros contos* (coll **1965**; trans Thomas Colchie as *The Ex-Magician and Other Stories* **1979** US), *O Pirotécnico Zacarias* ["Zacarias, the Pyrotechnical"] (coll **1974**) and *O Convidado* ["The Guest"] (coll **1974**). He was not discovered by the critics until 1974. His stories display a strong influence of the BIBLE and ancient mythologies; he explores the Absurdism of daily life, with a mix of Ray BRADBURY's nostalgia and Franz KAFKA's pessimism. [BT]

RUBIMOR Pseudonym of US COMIC-strip artist Reuben Moriera (1922-   ). ◊ TARZAN.

RUFF, MATT (HEW THERON) (1965-   ) US writer who has published only the CONTEMPORARY-FANTASY college novel *Fool on the Hill* (**1988**), which he wrote in his early 20s. Its many overt influences include FOLKTALES, MAGIC REALISM, Thomas PYNCHON, J.R.R. TOLKEIN and debatably the movie KNIGHTRIDERS (*1981*). MR's style is surprisingly accomplished for such a young writer. A second novel was scheduled for 1996. [JG]

RULAH THE JUNGLE GODDESS ◊ TARZAN.

RUMPELSTILTSKIN FAIRYTALE collected by the GRIMM BROTHERS. A miller claims his daughter can spin GOLD from straw. Faced with death if she cannot perform, the girl accepts help from the eponymous DWARF, whose CONDITION is that she must hand over her firstborn if unable to discover the dwarf's NAME. Luckily he is overheard gloating aloud that no one knows his name is Rumpelstiltskin (there are many regional variations: Suffolk's Tom Tit Tot, Cornwall's DEMON Terrytop, Scotland's FAIRY Whuppity

Stoorie, etc.). Rumpelstiltskin himself plays an ominous part in Jonathan CARROLL's *Sleeping in Flame* (**1988**).   [DRL]

See also: RIDDLES; TRUE NAMES.

RUNES Originally the letters of the Norse alphabet, runes are often associated with MAGIC, PROPHECY and wisdom. The term generalizes to any invented alphabet of spiky straight-line segments, appropriate for scratching in stone. J.R.R. TOLKIEN associated such alphabets with DWARFS; M.R. JAMES's "Casting the Runes" (1911) turns on the sinister efficacy of a transferable CURSE thus written. By synecdoche, "rune" may mean a CHARM or SPELL written in runes: WIZARDS often protect their arcana with "runes of power" which are occult booby-traps. A Lost Rune of peace is sought and found in Ursula K. LE GUIN's *The Tombs of Atuan* (**1971**). Latterly, runes have been employed as a divinatory tool.   [CB/DRL]

RUPERT THE BEAR Enduring anthropomorphic UK newspaper comic strip featuring a bear cub or teddy bear with a white face, red jumper, yellow check trousers and matching check scarf. He achieved enormous popularity through publication in the *Daily Express* and in colour albums in the 1930s-50s, and still commands a considerable readership today. The strip's longevity owes much to the timeless quality of its PASTORAL setting (the village of Nutwood), its charming fantasy and FAIRYTALE themes, and its fascinating cast of characters which include quaintly dressed humanized animals, trolls, giants, elves and a wide variety of human types dressed in all kinds of exotic costume. Rupert's regular playmates are Bill Badger, Algy Pug, Edward Trunk the elephant and Podgy Pig, all of whom dress in the style of the 1920s. The stories draw upon most traditional sources and are recounted with great appreciation of a child's-eye view of the world, filled with magic and wonder, but with a loving mummy and daddy at home.

The Little Lost Bear began in the *Daily Express*, rather unobtrusively, in a single frame with two four-line verses underneath in November 1920. Its creator was Mary Tourtel (1874-1948), the wife of a night news editor and herself an already established children's illustrator. There was no consistent format for the feature in its early days: it could vary from one to four frames per day, but eventually settled to one, with the story narrated in verse underneath. It was an immediate success and reprints in book form were tremendously popular.

On Tourtel's retirement in 1935, the artwork and stories were undertaken by Alfred BESTALL, in whose hands the strip reached its greatest heights of success. Bestall's ability to depict both the idyllic rural surroundings of Nutwood and the fantastic lands to which Rupert is transported gave the feature a particular attractiveness. He established a format of two frames per day, displaying a remarkable talent for storytelling, and a new series of quarterly books (the **Rupert Adventure** series) was begun along with regular annual reprints (in colour from 1940), which during one period each attained sales figures in excess of one and a half million. The **Adventure** series continued until 1963, although Bestall did not write them and had assistance from Alex Cubie and Enid Ash on the drawing. His last **Rupert** frames were published on 22 July 1965; subsequent artists have included Cubie, Lucy Matthews and John Harrold. One of the most successful writers has been James Henderson.

Merchandising of Rupert has included jigsaw puzzles, china plates and mugs, soap, chocolate bars, paper tissues, ladies' underwear and GAMES of all kinds. A tv puppet series

was first broadcast in 1970; an animated ROCK VIDEO was made for Paul McCartney's "We All Stand Together"; this won a BAFTA. In the "Schoolkids' Issue" (1971) of *Oz* magazine a drawing of Rupert copulating resulted in celebrated legal proceedings.

Rupert is still (1996) published regularly in the *Daily Express*, and coloured reprints of old stories are serialized in the *Sunday Express Magazine*.   [RT]

Further reading: *Rupert: A Bear's Life* (**1985**) by George Perry.

RURITANIA The essential flavour of Ruritanian romance was captured by John BUCHAN in a passage at the end of *The House of the Four Winds* (**1935**). The hero is a Glasgow grocer who has fallen into adventure in the Mittel-European principality of Evallonia: "He had been a king, acclaimed by shouting mobs. He had kept a throne warm for a friend, and now he was vanishing into the darkness, an honourable fugitive, a willing exile. He was the first grocer in all history that had been a Pretender to a Crown. The clack of hooves on stone, the jingling of bits, the echo of falling water were like strong wine." Buchan's Evallonia stands towards the end of a popular tradition that began with Anthony Hope's Ruritania in the 1890s.

The Ruritanian romance, an outgrowth of the 19th-century historical novel and kissing-cousin to the LOST-RACE novel, consists of tales of love and adventure set in imaginary European countries, principalities or duchies, and usually involving UK or US "commoner" heroes who save the throne, defeat the villain, marry the princess, and so on. Robert Louis STEVENSON's *Prince Otto* (**1885**), about courtly intrigue in a vernal land called Grunewald, and A.C. Gunter's *Mr Barnes of New York* (**1887**), about a stalwart American caught up in a European vendetta, were precursors, but the fashion was truly initiated by Anthony Hope (real name Anthony Hope Hawkins; 1863-1933) in his bestselling *The Prisoner of Zenda* (**1894**), which set the style and gave the Ruritanian romance its name. (Two of Mark TWAIN's novels of the 1880s, *The Prince and the Pauper* [**1881**] and *A Connecticut Yankee in King Arthur's Court* [**1889**], may also have provided the subgenre with some of its basic motifs.) In the wake of Hope's book (soon adapted to the stage, and often filmed) Ruritanian romance enjoyed a vogue throughout the English-speaking world.

US examples of the type include Richard Harding Davis's *The Princess Aline* (**1895**) and *The King's Jackal* (**1899**), and Harold MacGrath's *Arms and the Woman* (**1899**) and *The Puppet Crown* (**1900**). In the UK, Anthony Hope was to provide several more – *The Heart of Princess Osra* (**1896**), *Rupert of Hentzau* (**1898**) and *Sophy of Kravonia* (**1906**) – and so did such imitators as H.B. Marriott-Watson (1863-1921), with *The Princess Xenia* (**1899**), John Oxenham (real name William Arthur Dunkerley; 1852-1941), with *A Princess of Vascovy* (**1900**), and even Winston S. Churchill (1874-1965), with *Savrola* (**1900**). Many of these novels were bestsellers, but all, excepting those of Hope himself, were to be eclipsed by the US success of *Graustark* (**1901**) by George Barr McCutcheon (1866-1928) and its sequels *Beverly of Graustark* (**1904**) and *The Prince of Graustark* (**1914**). The mythical Balkan realm of Graustark remains a household name in the USA, often confused with Ruritania (in *Imaginary Worlds: The Art of Fantasy* [**1973**] Lin CARTER refers to "*The Prisoner of Zenda* . . . set in an imaginary country called 'Graustark'").

Ruritanian romance was open to variation, and to spoof. It

became torrid love-romance in *Three Weeks* (**1907**) by Elinor Glyn (1864-1943), light comedy in R. ANDOM's *In Fear of a Throne* (**1911**) and outright farce in *The Prince and Betty* (**1912**) by P.G. Wodehouse (1881-1975). Other well known authors who dipped a toe into the Ruritanian puddle include Ford Madox FORD in *The New Humpty-Dumpty* (**1912**) as by Daniel Chaucer, Laurence HOUSMAN in *John of Jingalo: The Story of a Monarch in Difficulties* (**1912**), Frances Hodgson BURNETT in *The Lost Prince* (**1915**), George A. Birmingham in *King Tommy* (**1923**), Edgar Rice BURROUGHS in *The Mad King* (**1926**), Dornford Yates (real name Cecil William Mercer; 1885-1960) in *Blood Royal* (**1929**), Leslie Charteris (real name Leslie Charles Bowyer Yin; 1907-1993) in *The Last Hero* (**1930**; vt *The Saint Closes the Case*), E(dward) Phillips Oppenheim (1866-1946) in *Jeremiah and the Princess* (**1933**), J.B. PRIESTLEY and Gerald BULLETT in their collaborative *I'll Tell You Everything* (**1933**) and, as noted, Buchan.

The vogue had run its course by the mid-1930s, but it was to have an afterlife in children's fiction – and in fantasy. Andre NORTON's first published book has a splendidly Ruritanian title: *The Prince Commands, Being Sundry Adventures of Michael Karl, Sometime Crown Prince & Pretender to the Throne of Morvania* (**1934**). Similar motifs crop up in *Biggles Goes to War* (**1938**) by Captain W.E. Johns (1893-1968), in *There's No Escape* (**1950**) by Ian Serraillier (1912-1994) and in a whole series of excellent juveniles by Violet Needham (1876-1967), beginning with *The Black Riders* (**1939**) and *The Emerald Crown* (**1940**). A later juvenile series which bears the Ruritanian (or Graustarkian) stamp is Lloyd ALEXANDER's *Westmark* (**1981**) and sequels.

In adult fiction, there is more than a touch of Ruritania to be found in the humorous *The Mouse that Roared* (**1955**) by Leonard Wibberley (1915-1983) and sequels, in *Royal Flash* (**1970**) by George MacDonald Fraser (1925-), in William GOLDMAN's *The Princess Bride* (**1973**), in Avram DAVIDSON's *The Enquiries of Dr Eszterhazy* (**1975**), and even in Ursula LE GUIN's *Orsinian Tales* (**1976**) and *Malafrena* (**1979**). Simon HAWKE's sf *The Zenda Vendetta* (**1985**) and John Spurling's sequelizing *After Zenda* (**1995**) are direct homages to Hope. But it could be argued that the main influence of this tradition on modern fantasy has been a more diffuse one: in the FANTASYLANDS that have enjoyed such popularity since the 1960s, wherever we find castles and courts, princes and princesses, swordplay and horseback-riding, evil usurpers and brave "commoner" heroes, we also find something of the spirit of Ruritania. [DP]

RUSCH, KRISTINE KATHRYN (1960-) US writer and editor, editor of *F&SF* from 1991. Her first published story was "Sing" in *Aboriginal Science Fiction* in 1987; she won the 1990 John W. Campbell Memorial AWARD for Best New Writer. KKR published some two dozen stories before the appearance of her first novel; she has concentrated primarily on novels since then. Although much of her work is technically sf – in *The Gallery of His Dreams* (**1991** chap) the device by which the sufferings of US Civil War photographer Matthew B. Brady achieve a transcendental memorialization involve TIME TRAVEL – a good deal of it, such as "Skin Deep" (1988), with its plight of a member of a race of dwindling aliens passing for human in a farmer's family, or "Story Child" which tells of a child with HEALING abilities in a post-HOLOCAUST society – can be read as essentially fantasy retold in conventional sf terms.

The White Mists of Power (**1991**) is a fantasy, although most of KKR's immediately succeeding novels have been sf. With *The Fey: Sacrifice* (**1995** UK) and «The Changeling: The Second Book of the Fey» (1996) she has published the first volumes of a HIGH-FANTASY sequence. [GF]

Other works: *Facade* (**1993**); *Heart Readers* (**1993**); *Traitors* (**1993** UK); *The Big Game*, a **Star Trek: Deep Space Nine** tie with Dean Wesley Smith, both writing as Sandy Schofield (**1993**); *Sins of the Blood* (**1994**), a VAMPIRE novel; *Star Trek: Voyager: The Escape* (**1995**) with Smith.

RUSHDIE, (AHMED) SALMAN (1947-) Indian-born writer, in the UK from 1961. He was well-known before 1988, but has become unwillingly world-famous because *The Satanic Verses* (**1988**), a fabulation and SATIRE which made daringly irreverent comments about Islam, inspired a *fatwa*, or death "sentence", proclaimed against him by the Ayatolla Khomeini and the Islamic theocracy of Iran, but illegal in terms of international law (and offensive to civil societies throughout the world). He is a fabulist, an excavator of STORY, a writer of conglomerate tales where various genres – including passages of fantasy – intermingle in discords of great wit, forming together a medium for rendering experiences that literary realism could only inadequately express.

SR's first novel, *Grimus* (**1975**), is closer to fantasy throughout than any of his later books for adults, though its sf element includes references to extraterrestrial life, IMMORTALITY, with multidimensional adventures into various alternate universes. Its Amerindian protagonist, who is a protean blockhead TRICKSTER, ultimately in conflict with his SHADOW the satanic MAGUS Grimus, undertakes a QUEST through various REALITIES – some of them constructed on WONDERLAND lines – in search for his beloved sister. The tale is set primarily in the imaginary realm of Calf Island, the creation of a trinity of characters whose falling out has left it in Grimus's hands. The story is played out on several levels – in the interaction between the warring members of the trio and the trickster hero, Flapping Eagle; among the inhabitants of the island, who look to Grimus as a deity and are subject to paralysing bouts of "Dimension fever", or self-consciousness about their own triviality; and on the planet Thera, inhabited by sentient stone frogs named Gorfs who may have had a hand in the Island's creation simply by imagining the possibility of its existence.

In *Haroun and the Sea of Stories* (**1990**), which was written for YA readers and which won a 1992 MYTHOPOEIC SOCIETY award, young Haroun follows his storyteller father – who, after his wife leaves him, has found that there is no STORY left to tell – into a Wonderland-like OTHERWORLD, where he is eventually successful in retrieving the essence of Story from the black-hearted and the merciless who would attempt to shut the world down.

Other novels – they include *Midnight's Children* (**1980**), *Shame* (**1983**) and *The Moor's Last Sigh* (**1995**) – make copious use of fantasy MOTIFS, but within a context that allows no unadulterated fantasy understanding or reading to prevail, in a manner reminiscent of the most complex MAGIC REALIST tales of writers like Gabriel GARCÍA MÁRQUEZ. *Midnight's Children* and *Shame* are both attempts at the construction for the 20th century – and for readers whose cultural heritage is quilted with divergent traditions – of a MYTH OF ORIGIN for India in the first instance, Pakistan in the second.

Midnight's Children is SR's most sustained accomplishment.

Strongly autobiographical, it is the story of Saleem Sinai, one of 1001 children born at midnight on the day of India's independence in 1947; each of them – like the CHANGELING protagonist, switched at birth into a privileged existence – has TALENTS by virtue of which they prove capable of META-MORPHOSIS; of flight; of MAGIC. In all their intolerable complexity and uplift, they represent the birth and the future of their country. The totality of life in India which their extreme situations describe is enhanced further by Saleem's particular talent. He is linked telepathically to his 1000 coevals and, through a nightly "Midnight's Children Conference" inside his mind, he vicariously experiences their lives across the subcontinent, becomes a living embodiment of the vast new Indian state and the promise of its first free generation. After their powers have been understood by the authorities, they are given vasectomies.

SR endows all of the novel's characters with a mythic stature that also suggests the inescapability of past CYCLES of history and the inevitability of FATE: Shiva, whose privileged life Saleem has taken over, is an AVATAR of the destroyer GOD of ancient Hindu mythology. As in Günter GRASS's *The Tin Drum* (**1959**), whose style it calls to mind, there is no moment in the story that does not seem pregnant with a deeper meaning.

Like its predecessors, *The Satanic Verses* invokes fantasy, supernatural fiction, horror, magic realist tropes and mundane realism, interweaving this narrative mix into a CAULDRON OF STORY through which pixilated (but intensely representative) characters fly, swim, choke, survive. The novel is, of course, about RELIGION; it is also about the fate of exiles in a world – both this 20th-century one – both dominated by the exilic consciousness and surpassingly cruel to its spokespersons. It two protagonists, Gibreel Farishta and Saladin Chamcha, represent halves of a divided single personality. Actors skilled in the art of mimicry, they both miraculously survive a plunge of 29,000 feet into the English Channel following the terrorist bombing of an airplane with which the novel opens. Saladin finds himself endowed with satanic features, including a set of horns, and the ability involuntarily to assume the demonized image those around him project upon him. Gibreel, by contrast, boasts features of the archangel Gabriel, SR's coded representation of fundamentalist self-righteousness. An interpolated dream sequence involving Gibreel recapitulates events from the so-called "satanic verses", an episode recorded in Islamic oral tradition but not the Qu'ran, in which Shaitan (◊ SATAN) impersonates the archangel Gibreel to dupe Muhammad into believing false pronouncements from Allah. SR's contemporary rendering of the episode in a bordello setting, offered as a satirical critique of cultural DEBASEMENT, outraged fundamentalists through its profanation and resulted in the imposition of the *fatwa* on 14 February 1989.

Wizard of Oz (coll **1992** chap US) is a casual disquisition on L.Frank BAUM and Hollywood (◊ LOS ANGELES). Some of the stories assembled in *East, West* (coll **1994**), including "At the Auction of the Ruby Slippers", are fantasy.　　[SD/JC]

RUSKIN, JOHN (1819-1900) UK writer, poet and art critic. Although he wrote prodigiously, his impact on fantasy is more through his role as a critic and catalyst than as a direct contributor. His one major fantasy is *The King of the Golden River* (written 1841; dated 1851 but **1850**), illus Richard DOYLE. It tells of a young brother, Gluck, who is cruelly treated by his two elder brothers. Gluck is kind to an old

man, though his brothers throw the man out. Thereafter they are cursed, but Gluck survives to break the CURSE through his goodness. This was an early example of an English CHILDREN'S FANTASY. JR was a strong advocate of imaginative work for children, opposing the strict moral tales of the previous generation. He became a champion of the FAIRYTALE, writing an introduction to an 1868 edition of the GRIMM BROTHERS' *Fairy Tales*, and he encouraged the romantic exploration of art in all its forms. His ground-breaking essay "The Nature of Gothic" (in *The Stones of Venice* vol 2 **1853**) found in favour of the Gothic tradition and ushered in a revival of interest in Gothic art and architecture. His series of books called **Modern Painters** (**1843-60** 5 vols) is a sweeping assessment of art and architecture as an expression of nature, which gave a stamp of approval to the PRE-RAPHAELITES, particularly William MORRIS, on whom JR was a tremendous influence. He also supported and heavily influenced Kate GREENAWAY. Although JR antagonized many critics in his early years, his liberal attitude to art allowed ROMANTICISM to break free of its restrictive roots and encouraged the greater diversity of art to flourish from the 1850s on. He was certainly one of the influences that allowed FANTASY to become established.　　[MA]

Further reading: There are many books about JR. Of most interest are *Ruskin, the Great Victorian* (**1949**) by Derrick Leon; *John Ruskin: The Portrait of a Prophet* (**1949**) by Peter Quennell; *John Ruskin* (**1954**) by Joan Evans; and *The Wider Sea* (**1982**) by John Dixon Hunt.

RUSS, JOANNA (1937-　　) US writer. ◊ GENDER.

RUSSELL, (HENRY) KEN(NETH ALFRED) (1927-　　) UK movie director and screenwriter whose compulsion to shock has led to some very bad movies (◊ CINEMA) but also to a restless experimentation that has affected both written and movie fantasy at a quite profound level: he has shown that imaginative fictions should not be restricted by self-imposed boundaries, should tread where angels fear to. His early work was for the BBC arts documentary series *Monitor* (1958-65), for which he directed and sometimes wrote or co-wrote a series of near-movie-length dramatized biographies, notably of composers; these included *Elgar* (**1962** tvm) and *The Debussy Film* (**1965** tvm). The musical connection continued with *Isadora* (**1966** tvm), a much less "realistic" movie, this time about the dancer Isadora Duncan (1878-1927). His *Dante's Inferno* (**1967** tvm) for the BBC's *Omnibus* series is likewise well remembered, but thereafter KR has done little tv work, with the notable exception of *William and Dorothy* (**1978**) and *The Rime of the Ancient Mariner* (**1978**), respectively about the Wordsworths and Samuel Taylor COLERIDGE.

Most of this time KR had been working also for the big screen. *French Dressing* (**1964**), his first feature, is a parochial UK screwball comedy; his next, *Billion Dollar Brain* (**1967**), was based on the technothriller *Billion-Dollar Brain* (**1966**) by Len Deighton (1920-　　). These were comparatively straightforward, but thereafter KR's directoral style altered radically, so that he became a kind of poet of excess. *Women in Love* (**1969**), based on D.H. LAWRENCE's *Women in Love* (**1921**), was a shocker in its day because of its sexual content and nudity (mainstream cinema had already accepted female pudenda, but penises had been largely taboo). This was followed by *The Music Lovers* (**1970**), based on the life of P.I. Tchaikovsky (1840-1893) and seemingly trying to be more outrageous than its predecessor, with grandiloquent fantasy imagery blending into the fabulated biography. In *The* DEVILS (*1971*), based largely on *The Devils of Loudon* (**1952**) by Aldous Huxley, KR – while continuing to try to shock with sex and violence – explored issues of PERCEPTION: the main characters' perceptions of each other, either contrived (the state hates the priest, so the priest must be allied to SATAN) or unwitting (an abbess hates the priest because of her frustrated sexuality), are of great fantasy interest, but so too is KR's depiction of the past as, in L.P. HARTLEY's phrase, "a foreign country", one that we cannot correctly perceive through 20th-century eyes. Several flops followed – indeed, KR's career is one largely of flops – but his *Tommy* (*1975*), based on the "rock OPERA" *Tommy* (1969) by The Who (largely by Peter Townshend [1945-]), was a major success. *Gothic* (*1986*; ◊ FRANKENSTEIN MOVIES) presents an orgiastic version of that famous time at the Villa Diodati when Mary SHELLEY dreamt up *Frankenstein* (**1818**). *The Lair of the White Worm* (*1988*), based on Bram STOKER's *The Lair of the White Worm* (**1911**), has some pleasingly disgusting moments but is a mess. The conceit of *Salome's Last Dance* (*1988*) is that Oscar WILDE watches prostitutes enact a performance of his banned play *Salomé* (**1894**).   [JG]

Other works: *The Boy Friend* (*1971*), based on the musical by Sandy Wilson; *Savage Messiah* (*1972*), about the artists Henri Gaudier and Sophie Brzeska; *Mahler* (*1974*) and *Lisztomania* (*1975*), fabulated versions of the composers' lives; *Valentino* (*1977*), fabulation based on the life of Rudolf Valentino; *Altered States* (*1980*), horror/sf based on *Altered States* (**1978**) by Paddy Chayefsky (real name Sidney Aaron Chayefsky; 1923-1981); *Crimes of Passion* (*1984*), a PSYCHOLOGICAL THRILLER; *The Rainbow* (*1989*), based on D.H. Lawrence's *The Rainbow* (**1915**); *Whore* (*1991*). KR was one of the 10 directors involved in *Aria* (*1988*).

RUSSELL, W(ILLIAM) CLARK (1844-1911) US-born UK writer, most of whose work deals with the sea. *The Frozen Pirate* (**1887**) is an sf novel of interest, and some of the tales in *The Phantom Death and Other Stories* (coll **1895**) are SUPERNATURAL FICTIONS, though WCR has a tendency to rationalize incursions from beyond (◊ RATIONALIZED FANTASY). Of most fantasy interest is *The Death Ship: A Strange Story: An Account of a Cruise in "The Flying Dutchman"* (**1888**; vt *The Flying Dutchman* 1888 US), in which a young man is rescued from drowning by Vanderdecken's crew, travels aboard the spectral ship (for whose occupants TIME has stopped), and falls in love with an English girl who is also aboard; tragedy ensues, and the girl is shot trying to escape (◊ FLYING DUTCHMAN). In both this and Captain MARRYAT's *The Phantom Ship* (**1839**), the deaths of young women can be likened to Richard WAGNER's theme of the redemptive female death, as in his OPERA *The Flying Dutchman* (**1843**); in WCR's case, it may simply be an echo.   [JC]

RYMAN, GEOFF(REY CHARLES) (1951-) Canadian-born writer who moved to the US at age 11, and has

been resident in the UK since 1973. Although he has received a good deal of attention for his sf, his first novel, *The Warrior who Carried Life* (**1985**), was a fantasy employing numerous familiar elements – DRAGONS, the superstitious village milieu of medieval fantasy, the QUEST, a visit to the Land of the Dead – to write a deeply revisionist tale (◊ REVISIONIST FANTASY) of pacifism and redemption. *"Was. . ."* (**1992**; US vt *Was* 1992 US), GR's most accomplished book, is similarly revisionist in intent: the formal conceit of the novel (that the Dorothy of L. Frank BAUM's novel *The Wonderful Wizard of Oz* [**1900**] was based on a child whom Baum met during his brief time in Kansas) is elaborated by additional narrative strands, involving the life of Judy Garland (◊ *The* WIZARD OF OZ) and a young actor who is dying of AIDS; the work becomes a moving meditation upon the anguished nature of human sexuality (Dorothy is sexually abused by her uncle) and the ambiguous solace of fantasy. A dramatic version by Paul Edwards has been produced (Chicago 1996). GR himself has written/dir/performed in several sf plays based on works by other writers.   [GF]

Other works: *The Unconquered Country: A Life History* (1984 *Interzone*; exp **1986**), sf partaking of FABULATION; *The Child Garden, or A Low Comedy* (1988 *Interzone* as "Love Sickness"; exp **1989**), sf which won the ARTHUR C. CLARKE AWARD and the John W. Campbell Memorial AWARD; *Coming of Enkidu* (**1989** chap), which treats of the GILGAMESH story; *Unconquered Countries: Four Novellas* (coll **1994** US).

RYMER, JAMES MALCOLM (1814-1884) Scottish civil engineer and writer of penny dreadfuls, mostly anon, though sometimes as Malcolm J. Errym or Malcolm J. Merry. He was exceedingly prolific, and a true tally of his output will probably never be known. His work is often confused with that of his fellow contributor to Edward Lloyd's publishing house, Thomas Pecket Prest (1810-1859), but research seems to support JMR's authorship of the notorious *Varney the Vampire, or The Feast of Blood* (serially 1845-7; **1847**). It is possible that both Prest and JMR worked on the story during its 109 weekly parts. A long, episodic and puerile work, it was the most extensive VAMPIRE work before Bram STOKER's *Dracula* (**1897**). There is minimal plot. Sir Francis Varney has become a vampire (probably by a PACT WITH THE DEVIL) and proceeds to terrorize and suck the blood from weekly victims. JMR's works are all in the Gothic form (◊ GOTHIC FANTASY), all sensational, but most with rationalized supernaturalism (◊ RATIONALIZED FANTASY). Of his attributed works, *The Black Monk* (**1844**), also once ascribed to Prest, is an historical Gothic involving a formulaic haunted abbey in the time of Richard I. *The String of Pearls* (1846-8 *The People's Periodical*; vt *Sweeney Todd, the Demon Barber of Fleet Street* **1878**), which has sometimes been ascribed to JMR, is believed to be by Prest.   [MA]

SABBAT The assembly at which the WITCHES persecuted by the Inquisition were alleged to meet (◊ WITCHCRAFT). The mythology of the sabbat was largely constructed by Pierre de Lancre (*c*1550-1630), who extracted extraordinarily elaborate descriptions of such assemblies from his Basque informants (most of whom were children) and summarized his "findings" in *Tableau de l'Inconstance des Mauvais Anges et Demons* (**1612**), whose oft-reprinted and semipornographic frontispiece provided a paradigmatic visual representation of the monstrous acts supposedly practised at the sabbat. The sabbat is sometimes confused with the BLACK MASS, although the two notions have distinct histories and different implications. Literary descriptions of sabbats can be found in Harrison AINSWORTH's *The Lancashire Witches* (**1849**), *The Fiery Angel* (**1908**; trans **1930**) by Valeri Briussov (1873-1924), *The Last Devil* (**1927**) by Signe Toksvig (1891-?), *Melusine, or Devil Take Her* (**1936**) by Charlotte Haldane (1894-1969) and in Part I of *The Witches* (coll **1968**; trans **1969**) by Françoise Mallet-Joris. [BS]

SABERHAGEN, FRED (THOMAS) (1930-) US writer and editor best-known for his creation of a genuine sf MYTH with his **Berserker** stories of chilly, spacegoing machine intelligences bent on the destruction of all life (◊ SFE). In his SCIENCE-FANTASY **Empire of the East** trio – *The Broken Lands* (**1968**), *The Black Mountains* (**1971**) and *Changeling Earth* (**1973**; vt *Ardneh's World* 1988), assembled as *Empire of the East* (rev omni **1979**) – Earth has been saved from nuclear devastation by a quasi-mystical PERCEPTION-shift whereby energy sources "come alive" as ELEMENTALS and DEMONS, the ultimate demon resulting from the TRANSFORMATION of an actual nuclear fireball. MAGIC and PROPHECY become workable, following strict RATIONALIZED-FANTASY laws, and the traditional struggle of "West" and "East" becomes a SWORD-AND-SORCERY conflict – complicated by remnants of old technology, including a mechanical RESURRECTION system (◊◊ TECHNOFANTASY) and a sentient computer which is effectively the West's GOD. One entertaining episode features a technophile GENIE who will construct any required machine without regard for its feasibility (◊ ANSWERED PRAYERS). The whole succeeds very well as action-adventure; *#1* gains

force by echoing an episode from Indian MYTHOLOGY in which the god Indra binds himself by CONDITIONS which seemingly make it impossible for him to kill a certain demon – but there is a loophole (◊ QUIBBLES).

The more purely fantastic **Swords** sequels, set 2000 years later, are tenuously linked to the original by certain long-lived characters. The **Book of Swords** sequence comprises *The First Book of Swords* (**1983**), *Second* (**1983**) and *Third* (**1984**), assembled as *The Complete Book of Swords* (omni **1985**). Its premise is that, to amuse the gods, Vulcan creates 12 magical PLOT COUPONS: these named SWORDS have various powers which are scattered over Earth for humanity to fight about, and fight with. Vulcan's secret joke is that the swords can kill the gods, which duly occurs. All this was apparently devised by FS as a computer-GAME scenario: despite some ingenuity, the fictional result seems scrappy, leaving many loose ends and unused swords. The rambling **Book of Lost Swords** series tours the GAMEWORLD further in *The First Book of Lost Swords: Woundhealer's Story* (**1986**), *Second: Sightblinder's Story* (**1987**), *Third: Stonecutter's Story* (**1988**) – these assembled as *The Lost Swords: The First Triad* (omni **1988**) – *Fourth: Farslayer's Story* (**1989**), *Fifth: Coinspinner's Story* (**1989**), *Sixth: Mindsword's Story* (**1990**) – *#4-#6* assembled as *The Lost Swords: The Second Triad* (omni **1991**) – *Seventh: Wayfinder's Story* (**1992**) and *The Last Book of Swords: Shieldbreaker's Story* (**1994**) – these final two assembled as *The Lost Swords: Endgame* (omni **1994**). The series' recurring VILLAIN is a malign WIZARD who is anticlimactic after the battle against the gods; his temporary absence adds interest to *#3*'s lightweight DETECTIVE/THRILLER FANTASY.

The **Dracula** books begin as REVISIONIST FANTASY with the VAMPIRE Vlad Tepes or Drakulya arguing in *The Dracula Tape* (**1975**) that Bram STOKER unfairly maligned him; in *The Holmes-Dracula File* (**1978**) he helps SHERLOCK HOLMES save LONDON from a plague carried by the Giant Rat of Sumatra (◊ MICE AND RATS). This Dracula may drain the blood of deserving human villains, but normally contents himself with rats. His adventures continue rather tepidly in the modern USA, where he assists police and innocents against less benign vampires; his police contact imitates

Holmes in updating the stake-through-the-heart notion with wooden bullets. Further titles are *An Old Friend of the Family* (**1979**), *Thorn* (**1980**), *Dominion* (**1982**) – which features MERLIN and Nimue (◊ LADY OF THE LAKE) in modern NEW YORK – *A Matter of Taste* (**1990**), *A Question of Time* (**1992**) and *Seance for a Vampire* (**1994**), the last again featuring Holmes. *The Black Throne* (**1990**) with Roger ZELAZNY is another RECURSIVE FANTASY, this time featuring Edgar Allan POE. *Merlin's Bones* (**1995**) involves ARTHUR in a story whose action is set in both past and future.

FS's industry and craftsmanship are notable, but commercial pressures can extend his series too long after their original concepts have lost freshness.   [DRL]

Other works: *Bram Stoker's Dracula* * (**1992**) with James V. Hart; *An Armory of Swords* * (anth **1995**) ed FS, a SHARED-WORLD enterprise based on the **Swords** sequence; *Dancing Bears* (**1996**), SHAPESHIFTER fantasy.

SACRED GROVE ◊ GOLDEN BOUGH.

SACRED NUMBERS ◊ NUMEROLOGY; SEVEN; THREE.

SACRIFICE Fantasy's fierce morality of BALANCE and CONTRACTS insists that there are prices which must be paid. GODS and other higher beings may not want the offered goods or beasts, but require the sacrifice if only as a token of earnestness: ODIN sacrificed an eye for wisdom, and the BIBLE's *Leviticus* describes in great detail the various "burnt offerings" which will please the Lord. Animal sacrifices are routine in VOODOO, minor BLACK MAGIC and "primitive" RELIGIONS. Terry PRATCHETT has a rare comic example in *Mort* (**1987**), whose offering is chosen for its visibility to the myopic High Priest – a sacrificial elephant. Manuel in James Branch CABELL's *Figures of Earth* (**1921**) sacrifices his youthfulness to regain the woman whose life he had earlier spent to save his own. Further examples abound, but the greatest sacrificial drama lies in outright HUMAN SACRIFICE.

In a different sense, it is a common underlying theme of fantasy that something more abstract – LOVE, dignity, etc. – must be sacrificed for the fulfilment of the QUEST, or for the general good of the LAND, or for some other reason. The notion is that nothing of import can be attained without the cost of pain or loss.   [DRL]

See also: INITIATION; RITE OF PASSAGE.

SAGA The word "saga" is Norse for "tale", and sagas were essentially retellings of Scandinavian history and FOLKTALES in a narrative form. In giving them structure and continuity the sagamen, the Nordic equivalent of HOMER or SCHEHERAZADE, turned history into STORY, and made HEROES out of their ancestors. Most of the sagas are adventure tales, and are not fantastic. However, those derived from the Eddas (◊ NORDIC FANTASY) drew more heavily on MYTHS and LEGENDS. The original sagas were Nordic and Icelandic, and were written in the 11th-14th centuries. Most famous is *Burnt Njál's Saga* (11th century), about one of Iceland's most beloved and law-abiding countrymen. More in the realm of the supernatural is *Grettir's Saga* (11th century), an early form of HEROIC FANTASY. Grettir, a young hothead, is banished after killing someone in a quarrel. In exile he undergoes many adventures, mostly supernatural, including the episode often reprinted in the version best known as "The Sword of Glam" or "The Ghost of Glam" adapted by Andrew LANG in *The Book of Dreams and Ghosts* (coll **1897**). *Grettir's Saga* was translated in verse by William MORRIS and as a story by S. BARING-GOULD. Also famous is *Volsunga Saga* (final form 13th century). Poul ANDERSON retold another famous saga in *Hrolf Kraki's Saga*

(**1973**). The best-known sagaman is Snorri Sturluson (1179-1241), who wrote the *Heimskringla* ["Orb of the World"], telling the history of the Norwegian kings.

The word "saga" is often used for any extended STORY CYCLE or mythos, most notably the stories about ARTHUR, Charlemagne and Dietrich of Bern, but these are more properly treated as medieval ROMANCES. It is also used commercially to describe multi-volume GENRE FANTASIES. [MA]

SAGA OF NOGGIN THE NOG, THE UK tv series (1959-65). ◊ *The* CLANGERS (1969-73).

SAGE, JUNIPER Pseudonym of US writer Margaret Wise BROWN.

SAINT-AUBIN, HORACE DE Pseudonym of French writer Honoré de BALZAC.

ST JOHN, JOHN ALLEN (1872-1957) US illustrator. ◊ TARZAN.

SAKI Pseudonym of UK author and journalist Hector Hugh Munro (1870-1916), now best-remembered for his witty, barbed and epigrammatic short stories and CONTES CRUELS. Several of these are fantasy or SUPERNATURAL FICTION, often framed as CLUB STORIES, with a leaning to ANIMAL FANTASY – even nonfantasies casually loose hyenas, monkeys, WOLVES, etc., upon an exquisite English social scene. Relevant stories appear in all collections. *Reginald* (coll **1904**) spoofs FABLES with "Reginald on Besetting Sins: The Woman who Told the Truth", stressing the hazards of truth-telling. *Reginald in Russia* (coll **1910**) includes the ironically cruel WEREWOLF story "Gabriel-Ernest" and "The Soul of Laploshka", where a miser's GHOST will not rest until a two-franc debt is repaid by bestowing it on the deserving rich (◊ CONDITIONS). *The Chronicles of Clovis* (coll **1911**) has: "Tobermory", where the eponymous CAT who has seen too much scandal through country-house windows alarmingly becomes a TALKING ANIMAL; "Sredni Vashtar", whose ill child makes an oversized pet ferret his GOD and feels vindicated when it kills his oppressive guardian; "The Music on the Hill", where a sceptical modern woman disturbs an offering to PAN, who has her killed by a stag; and "Ministers of Grace", which replaces politicians by DOUBLES who, being rational ANGELS, run England even more disastrously. *Beasts and Super-Beasts* (coll **1914**) includes "Laura", whose mischievous heroine achieves REINCARNATION as a troublesome otter, and two celebrated tales-within-tales: "The Open Window", whose inset GHOST STORY is fabricated by a young girl to terrify a visitor, and "The Story-Teller", in whose invented NEVER-NEVER LAND the penalty of being odiously moral is to be eaten by a wolf – delighting child listeners and appalling their aunt. *The Toys of Peace* (coll **1919**) has: "The Wolves of Cernogratz", who like banshees howl only for the von Cernogratz dead, discomfiting their castle's parvenu owners; and "The Hedgehog", another unlikely eponymous GHOST. *The Square Egg* (coll **1924**) includes "The Infernal Parliament", a comic HELL where constituents are compelled to listen to all parliamentary speeches.   [DRL]

Other works: *The Complete Short Stories of Saki* (omni **1930**; exp 1948), including biographical memoir by Saki's sister Ethel M. Munro; *The Secret Sin of Septimus Brope, and Other Stories* (coll **1995** chap).

Further reading: *Saki: A Life of Hector Hugh Munro, With Six Short Stories Never Before Collected* (**1981**) by A.J. Langguth (1933-   ); "new" stories are associational only.

SALE, TIM (? -) ◊ Robert Lynn ASPRIN.

'SALEM'S LOT (vt *'Salem's Lot: The Movie*) US movie

(*1979* tvm) derived from a tv miniseries. Warner/Serendipity. **Pr** Richard Kobritz. **Exec pr** Stirling Silliphant. **Dir** Tobe Hooper. **Spfx** Frank Torro. **Mufx** Jack Young. **Screenplay** Paul Monash. **Starring** Lew Ayres (Jason Burke), Bonnie Bedelia (Susan Norton), Lance Kerwin (Mark Petrie), James Mason (Straker), Reggie Nalder (Barlow), David Soul (Ben Mears). 112 mins, cut for theatrical release from a 190 min tvm. Colour.

Novelist Mears returns to his Maine hometown, 'Salem's Lot, obsessed by the gloomy HAUNTED DWELLING on its outskirts, recently bought by another fresh arrival, suave Englishman Straker, who is opening a posh antique shop in partnership with the never-seen Barlow. Mears rapidly makes new acquaintants – notably pretty schoolmarm Susan Norton – and re-encounters old ones, such as school-teacher Jason Burke. All is soon not well in 'Salem's Lot: first to disappear is little boy Ralphie Glick. Ralphie returns from the dead to vampirize elder brother Danny, who in turn . . .

The corpse-count and associated POLTERGEIST activity accelerate, and it becomes clear that Straker and the old Marston house are at the focus of it all. Burke and Mears, with young horror buff Mark, investigate and become convinced that Straker shields and touts for an ancient VAMPIRE, the enigmatic Barlow. Few believe them, but eventually most of the surviving townsfolk are convinced, and wisely flee. At last Mears and Mark manage to kill Straker, stake Barlow and torch both house and town.

Though suffering many of the traditional failings of the tvm and with a few lurches due to cutting, 'SL is surprisingly effective, faithful in mood to King's novel. Where it succeeds brilliantly is in transferring many of the elements of the central DRACULA tale into the setting of neither remote Transylvania nor atmospheric 19th-century London but a typical late-20th-century US town: this melding of ancient and modern is what gives vibrancy to what might otherwise have been a humdrum movie. A sequel, *A Return to 'Salem's Lot* (*1987*), offers less interest but more gore. [JG]

SALMONSON, JESSICA AMANDA (1950-   ) US editor, poet and writer, born Jesse Amos Salmonson, who began publishing work of interest with several simultaneously released items for *The Literary Magazine of Fantasy and Terror* (1973), which she edited as Amos Salmonson; her fiction in the first issue included "A Great Experience" as by Patrick Lean and "Youngin" as by Josiah Kerr. Over its seven issues the journal published increasingly competent fiction and served as a forum for issues of FEMINISM; while editing it, JAS went through a sex and consequent name change, an experience which she recorded openly in the journal. At the same time she continued writing stories. Her first significant publication was *Amazons!* (anth **1979**), which won the 1980 WORLD FANTASY AWARD and – along with its sequel, *Amazons II* (anth **1982**) – presented a number of HEROIC-FANTASY tales featuring female warriors. Other anthologies with a similar remit, though not restricted to women warriors, include *Heroic Visions* (anth **1983**) and *Heroic Visions II* (anth **1986**). JAS's interest in AMAZONS led eventually to *The Encyclopedia of Amazons: Women Warriors from Antiquity to the Modern Era* (**1991**), which has over 1000 entries on figures real and fictional.

Much of JAS's work – including her first book, *Tragedy of the Moisty Morning* (**1978** chap), a novella – has appeared from SMALL-PRESS outlets, in which she has always taken a very active interest; most of her poetry has been released

through small firms. Her first novel sequence, however, **Tomoe Gozen** – *Tomoe Gozen* (**1981**), *The Golden Naginata* (**1982**) and *Thousand Shrine Warrior* (**1984**) – appeared from a trade house. Set in a LAND-OF-FABLE medieval Japan, it recreates in ORIENTAL-FANTASY terms the life of a woman Samurai: the exploits are not unusual, but the complexly delineated god- and demon-saturated "Naipon" in which they are set is evocative. *Ou Lu Khen and the Beautiful Madwoman* (**1985**), also set in the East (China this time), carries its two protagonists into a TIME ABYSS where an ancient EVIL lurks. *Anthony Shriek, or Lovers of Another Realm* (**1992**) is HORROR.

Full-length collections include *A Silver Thread of Madness* (coll **1989**), *John Collier and Fredric Brown Went Quarrelling Through my Head: Stories* (coll **1989**), *The Mysterious Doom and Other Ghostly Tales of the Pacific Northwest* (coll **1992**) – containing TWICE-TOLD Native American tales and some investigations into supernatural matters conducted by OCCULT DETECTIVE **Penelope Pettiweather**, on the same basis as the tales which were earlier assembled as *Harmless Ghosts: The Penelope Pettiweather Stories* (coll **1990** chap UK) – *Phantom Waters: Northwest Legends of Rivers, Lakes, and Shores* (coll **1995**) and *The Eleventh Jaguarandi and Other Mysterious Persons* (coll **1995**). But, although JAS remains prolific in her fiction, it may be that her editorial work has in recent years been more influential.

After *Tales by Moonlight* (anth **1983**) and *Tales by Moonlight II* (anth **1989**), both presenting original fantasies, she began to publish titles which reflected her strong interest in 19th- and early-20th-century writers of SUPERNATURAL FICTION, assembling *The Faded Garden: The Collected Ghost Stories of Hildegarde Hawthorne* (coll **1985**), by Nathaniel HAWTHORNE's grandaughter, *The Haunted Wherry and Other Rare Ghost Stories* (anth **1985**), *The Supernatural Tales of Fitz-James O'Brien #1: Macabre Tales* (coll **1988**) and *#2: Dream Stories and Fantasies* (coll **1988**) (◊ Fitz-James O'BRIEN), *What Did Miss Darrington See?: An Anthology of Feminist Supernatural Fiction* (anth **1989**) with an introduction by Rosemary JACKSON, *Wife or Spinster: Short Stories by 19th Century American Women* (anth **1991**) with Isabelle D. Waugh and Charles G. Waugh, *From Out of the Past: The Indiana Ghost Stories of Anna Nicholas* (coll **1991** chap UK), assembling the work of Anna Katherine Nicholas (1917-   ) and *Master of the Past: Complete Supernatural Stories of Vincent O'Sullivan* (coll **1994** UK), assembling scattered work of Vincent O'SULLIVAN. *Mr Monkey and Other Sumerian Fables* (anth **1995**), which stands somewhat aside from its predecessors, assembles Sumerian BEAST-FABLES (◊◊ MESOPOTAMIAN EPIC), which may be the earliest extant examples of the form. The scholarship in these collections and anthologies is thorough and secure, and the adventurousness is evident.

Throughout her work – some experimental – JAS constantly attempts to use the forms of fiction as weapons of discovery. [JC]

Other works: *The Swordswoman* (**1982**); *Young Tyrone: A Melodrama* (coll **1983** chap); *Ten Magnificent Peonies Presents SLIDE SHOW* (coll **1983** chap); *Hag's Tapestry* (coll **1986** chap UK); *Mystic Women: Their Ancient Tales and Legends Recounted by a Woman Inmate of the Calcutta Insane Asylum* (coll **1991**); *Bibliography* (**1992** chap dos); *Twenty-One Epic Novels* (coll **1995** chap), very short stories.

Other works (poetry): *The Black Crusader and Other Poems of Horror* (coll **1979** chap); *Moonstill Tulip Wine and Others* (coll **1979** chap); *Cheap Present* (**1980** chap); *On the Shores of Eternity* (coll **1981** chap); *Feigned Death and Other Sorceries* (coll **1983** chap); *In this Vent* (**1983** chap); *Innocent of Evil: Poems in Prose* (coll **1984** chap); *The Patient Child* (**1985** chap); *Fantasies in Black and White* (coll **1987** chap); *The Ghost Garden* (coll **1988** chap UK); *A Celestial Occurrence* (**1991** chap); *The Goddess Under Siege* (coll **1992** chap); *Sorceries and Sorrows (Early Poems)* (coll **1992** chap dos); *Songs of the Maenads* (coll **1993** chap); *The Horn of Tara* (coll **1995** chap); *Lake of the Devil: Poems of Morosity and Jest* (coll **1995** chap).

Other works (nonfiction): *Wisewomen and Boggy-Boos: A Dictionary of Lesbian Fairy Lore* (**1992**) with Jules Remedios Faye; *Miniature Vegetables* (**1994** chap).

SALOME The name traditionally given to the daughter of Herodias who – according to *Matthew* – pleased the tetrarch Herod with her dancing and claimed the head of John the Baptist as reward, thus securing herself the role of CHRISTIAN FANTASY's principal FEMME FATALE. She was sometimes called Herodias, and is confusingly renamed LILITH in "Lilith" (1890) by Jules Lemaître (1853-1914). She was adopted as a central symbol of DECADENCE, her image variously immortalized in two paintings by Gustave Moreau (1826-1898) and a series of illustrations by Aubrey BEARDSLEY. Her story is retold in an episode of *An Epic of Fine Women* (**1870**) by Arthur W. O'Shaughnessy (1844-1881), in *Salomé* (**1893** France; trans **1894**) by Oscar WILDE and in *Salome, Princess of Galilee* (**1951**) by Henry Denker (1912-). It is dramatically extended in *Salome the Wandering Jewess* (**1930**) by George VIERECK and Paul Eldridge (1888-1982), and reconfigured in "Salomé" (1992) by Brian STABLEFORD.

Movies of relevance include: the silent *Salome* (*1923* US) dir Charles Bryant and with set drawings by Beardsley, based on the Wilde play and starring Nazimova; *Salome* (*1953* US) dir William Dieterle, starring Rita Hayworth and Charles Laughton in a standard Hollywood biblical epic (here, though, Salome dances in an effort to *save* the life of John the Baptist); *Salome* (*1986* Italy) dir Claude D'Anna, a soft-porn version; and *Salome's Last Dance* (*1987* UK) dir Ken RUSSELL, which has Wilde as the audience to a performance of his play by a band of prostitutes. [BS/JG]

SALTEN, FELIX Pseudonym of Austrian dramatist, critic and writer Sigmund Salzmann (1869-1945), who moved to Switzerland after the 1938 Nazi Anschluss. His first reputation, pre-WWI, was as a drama critic; his novels of fantasy interest include *Der Hund von Florenz* (**1923**; trans Huntley Paterson as *The Hound of Florence* **1930** US, in which an artistic young man makes a WISH – that he could undergo METAMORPHOSIS into a dog and travel south to art-choked Italy – is granted, on alternate days. At novel's end, fully human again, he may become an artist; or die. FS remains best-known for the ANIMAL FANTASY *Bambi* (**1926**; trans Whittaker Chambers as *Bambi: A Life in the Woods* **1928** UK), done as an ANIMATED MOVIE by DISNEY as *Bambi* (*1942*). A baby deer grows up – learning how to cope with the terror generated by the weird TRICKSTER figure of a neighbouring human hunter – until he becomes a master of the woods. The less engrossing sequel was *Bambi's Children* (trans **1939** US). [JC]

SALTYKOV, MIKHAIL (1826-1889) Russian critic. ◊ AESOPIAN FANTASY.

SALVATORE, R(OBERT) A(NTHONY) (1959-) US writer, much of whose work consists of GAME-ties in

the **Forgotten Realms** franchise. These fall into various subseries: the **Icewind Dale** trilogy – *The Crystal Shard* * (**1988**), *Streams of Silver* * (**1989**) and *The Halfling's Gem* * (**1990**) – the **Dark Elf** trilogy – *Homeland* * (**1990**), *Exile* * (**1990**) and *Sojourn* * (**1991**) – the **Cleric** quintet – *Canticle* * (**1991**), *In Sylvan Shadows* * (**1992**), *Night Masks* * (**1992**), *The Fallen Fortress* * (**1993**) and *The Chaos Curse* * (**1994**) – and related books such as *The Legacy* * (**1992**), *Starless Night* * (**1993**) and *Siege of Darkness* * (**1994**). Singletons and other non-ties include *Echoes of the Fourth Magic* (**1990**), a Bermuda Triangle mystery-fantasy, *The Witch's Daughter* (**1991**), *The Sword of Bedwyr* (**1995**), an Arthurian fantasy (◊ ARTHUR) and the **Spearwielder's Tale** trilogy: *The Woods Out Back* (**1993**), *The Dragon's Dagger* (**1994**) and *Dragonslayer's Return* (**1995**). RAS is a fast writer who uses conventional motifs capably. [DP]

SAMHAIN Celtic festival held on the first day of November, and which HALLOWE'EN is the eve of. It marks the end of summer ("samhain" may mean literally that) and the beginning of winter. It is a festival during which PORTALS in the hollow hills known as The Sidhe open to a land of FAERIE, which the Celtic fairies known as Sidhe rule; the UNDERWORLD itself may yawn open, and the DEAD may become manifest. It has characteristics of the REVEL, and is likely to be found in tales that invoke CELTIC FANTASY. [DRL/JC]

SAMPSON, FAY (ELIZABETH) (1935-) UK writer of many books for children, including the fantasy singleton *The Chains of Sleep* (**1981**) and the **Pangur Bán** series: *Pangur Bán: The White Cat* (**1983**), *Finnglas of the Horses* (**1985**), *Finnglas and the Stones of Choosing* (**1986**), *Shape-Shifter: The Naming of Pangur Bán* (**1988**), *The Serpent of Senargad* (**1989**) and *The White Horse is Running* (**1990**). For adults, her most notable achievement has been the **Daughter of Tintagel** series of Arthurian fantasies (◊ ARTHUR): *Wise Woman's Telling* (**1989**), *White Nun's Telling* (**1989**), *Black Smith's Telling* (**1990**), *Taliesin's Telling* (**1991**) and *Herself* (**1992**), all assembled as *Daughter of Tintagel* (omni **1992**). *Star Dancer* (**1993**) is an elaborate historical fantasy based on Sumerian myth. [DP]

SANDMAN The name of four US COMIC-book characters, all associated in some manner with sleep and DREAMS.

1. Masked crimefighter (◊ MASKED AVENGER) with a green business suit and cape, a slouch hat, kid gloves and stylish white spats over his shiny black shoes; he wears a yellow gas-mask and carries a sleep-inducing gas gun of his own invention. He is the alter ego of millionaire playboy Wesley Dodds, whose adventures often came about through the capricious behaviour of his girlfriend and confidante Dian Belmont, a reformed safecracker. No origin story was ever published. He was created by Larry Dean (a pseudonym of writer Gardner Fox and artist Bert Christman) and first appeared in *New York World's Fair Comics* (April 1939), a special information book about the World's Fair with back-up comics features. He started as a regular feature in *Adventure Comics #40* (July 1939) and, despite his vow to bring "justice in a world of injustice", was in the early stories generally seen as an outcast, wanted by the police. He was a member of the original line-up of **The Justice Society of America**, reappearing, rejuvenated, when that superhero team surfaced again in the 1960s (in *Justice League of America #21*, 1963). Other artists – e.g., Ogden Whitney, Creig Flessel (1912-) and Chad Grothkopf (1914-) – also drew the character; it was Grothkopf, with DC editor Whit Ellsworth, who was responsible for a complete revamp of the character into a costumed SUPERHERO (◊ **2**) in 1941. The original gasmasked S was resurrected in *Sandman Mystery Theater* (1994-   ) with excellent scripts by Matt Wagner and Steven Seagle. Set once again in 1930s New York, in this latest version Wesley Dodds, traumatized by his father's death by mustard gas in WWI, becomes Sandman as a kind of exorcism.

2. Costumed SUPERHERO with sleep-inducing powers who wears a yellow tunic and tights, mauve shorts and hooded cape, initially drawn by Grothkopf (◊ **1**) and then Paul Norris (1914-   ). His first appearance was in *Adventure Comics #69* (1941), when he acquired a juvenile companion, Sandy Hawkins (aka Sandy the Golden Boy). From *#72* (1942) Joe Simon (1915-   ) and Jack Kirby took over writing and art, spinning action-filled fantasy plots dealing mainly with dreams and nightmares. This version suddenly replaced **1** in the **Justice Society** line-up without explanation in *All Star Comics #10* (1942). **2** last appeared in *Adventure Comics #102* (1946).

3. Costumed fantasy figure kitted out in yellow vest and tights with red trunks, boots, cape and mask; created by Simon and Kirby (who drew only one issue - their last collaboration). **3** featured in *The Sandman #1-#6* (1974-6). A denizen of "another dimension", he entered people's nightmares and fought the surreal beings responsible for them.

4. Hollow-eyed, sickly-looking "Keeper and Lord of Dreams", with tousled, early-morning hair and a long, black, figure-hiding gown, also known as Dream or Morpheus, created by Neil GAIMAN, who draws upon a wide range of literary and mythical sources for inspiration. In many stories he makes little more than a token appearance, and is something of a WANDERING-JEW figure; rarely a catalyst. Gaiman has a unique agreement with publishers DC COMICS which states that no other graphic writer shall use the character, and that this version of Sandman will cease when he tires of writing the series. *The Sandman Book of Dreams* (anth **1996**) ed Gaiman and Edward E. Kramer is an anthology of original text stories based on Gaiman's version.   [RT/DR]

SAN FRANCISCO A CITY which - unlike its vast URBAN-FANTASY sibling LOS ANGELES - can be seen from above, envisioned whole from the hills which surround it and it and give it the appearance of a POLDER. Also unlike LA, which cannot be seen and embraced as an entity, SF tends to be cherished by those who use it as a venue; and, because it is compact and has a correspondingly tellable history, SF is more usable for fantasy writers as a focus of STORY. The SF fire - which features in Beverly Sommers's *Time and Again* (**1987**) and others - provides a natural magnet for protagonists drawn into TIMESLIPS.

More broadly, SF's centrality as a city in the mundane world made it a natural focus for 19th-century writers of the fantastic. SF remains a natural theatre for tales like *Sarah Canary* (**1991**) by Karen Joy Fowler (1950-   ), *Virtual Light* (**1993**) by William Gibson (1948-   ) and *The City, Not Long After* (**1989**) by Pat MURPHY. Both Fowler and Murphy live in or near the city; other SF writers who use their home in their fiction include Lisa GOLDSTEIN, whose *A Mask for the General* (**1987**) and «Walking the Labyrinth» (1996) circle around the venue, and Jonathan Lethem (1964-   ), whose *Gun, With Occasional Music* (**1993**) and *Amnesia Moon* (**1995**) are both set in the city. Other fantasy tales set in SF - or which expicitly reflect its warm ambience - include James P. BLAYLOCK's *The Paper Grail* (**1991**),

the latter parts of Mercedes LACKEY's and Ellen GUON's **Bedlam's Bard** sequence, plus Lackey's *The Fire Rose* (**1995**), Fritz LEIBER's *Our Lady of Darkness* (**1977**), Michael J. REAVES's *Street Magic* (**1991**) and Zilpha Keatley SNYDER's *Black and Blue Magic* (**1966**).   [JC]

SANGRAAL, SANGRAIL or **SANGREAL** ◊ GRAIL.

SANJULIAN, MANUEL PEREZ (1941-   ) Prolific and acclaimed Spanish illustrator with a rich, subtle colour sense and an iconographic painting style. His work, mainly for book covers, has been widely published in the UK and USA. He favours a realistic, romantic style, but there is a considerable imaginative element in the incidental detail. His colour, too, is imaginative and exotic: he often uses rich, clear blues or greens in his skilful treatment of shadow areas. Some of the cover paintings he made for WARREN PUBLISHING (particularly those for VAMPIRELLA *#12-#16* and *#36* [1971-3], EERIE *#41* [1972] and *#66* [1975], and CREEPY *#46* and *#49* [1972]) have achieved iconic status of icons in fantasy painting, symbolizing a new approach to ILLUSTRATION in the genre. His composition is influenced by the work of Frank Brangwyn (1867-1956), Dean Cornwell (1892-1960) and E.A. Abbey (1852-1911). MPS works in gouache on illustration-board, producing a carefully rendered, closely referenced, full-size pencil drawing of his design before beginning each painting. MPS's first professional job was a cover for *The Diary of Anne Frank* (1964 Spanish edn). This was followed by illustrations for romantic-story magazines. He went on to produce a few digest-size historical and classics COMIC strips, possibly for the UK market, although he claims no one told him who they were for. He joined the agency Selecciones Illustradas, and his work has since appeared on movie posters and hundreds of book and magazine covers.   [RT]

Further reading: *Sanjulian (periodo 1970-1984)* (graph **1984**).

SANSKRIT LITERATURE Northern India's Classical language, Sanskrit - "that which is perfected" - was formalized in the 4th century BC. Its literature embodies a corpus extending from *c*1500BC through to at least AD1500; some scholars designate the cultural and thus literary "Golden Age" as the 4th-7th centuries AD. Content ranges from the sacred (Vedas) to what some consider the profane (*Kamasutra*), and style from relatively simple (*Buddhacarita*) to euphuistically ornate (*Vasavadatta*).

Although Sanskrit's literary inventory is replete with STORIES, many of which might qualify in a broad sense as "fantasies", an annotated listing of the texts in which storytelling *per se* appears would not be informative as to what the tradition itself esteems: *kavya*, or "literature as art", profoundly influenced writers in the subcontinent's vernaculars, and in the West inspired the likes of Jean de la Fontaine (1621-1695), GOETHE, Hans Christian ANDERSEN and Thomas MANN. Harold Bloom says that "All strong literary originality becomes canonical", and it is in Sanskrit *kavya* that strong literary originality is to be found. The following discussion is confined to the *kavya* tradition which most nearly approximates the Western concept of literature.

An acceptable starting point for the exploration of Sanskrit literature as art is the *Buddhacarita* (*c*1st century AD), Ashvaghosha's metrical life of the Buddha, trans by E.H. Johnston as *The Acts of the Buddha* (**1935**). Recounting the Buddha's journey to enlightenment, it has all the characteristics of *kavya*: poetic conceits, play of figurative language, and rich description. Unlike the Jatakas, which are popular scriptural tellings of the Buddha's various rebirths, this narrative is consciously crafted, with the *kavi* (poet) attending to both the manner and the matter of the telling.

The *Mahabharata*, attributed to one Vyasa, jams nearly every story in the Indian tradition into its some 100,000 two-line verses. It was revised and added to as it travelled from inception *c*400BC to completion *c*AD400, and hundreds of peripheral stories nestle within its grand narrative of an internecine war. For depictions of ALTERNATE WORLDS, MYTHICAL CREATURES and the supernatural, this is a TAPROOT TEXT. As it says itself: "What is not found here is found nowhere." The most accessible translation, though yet incomplete, is that begun by the late J.A.B. van Buitenen: *The Mahabharata* vols 1-3 (**1973-8**); a full version is that by P.C. Roy, *The Mahabharata* (**1919-35**). The much smaller and more artful *Ramayana* ("The Wanderings of Rama"), also taking shape over several centuries – 200BC-AD200 – is believed to be mostly the work of a single hand, Valmiki. Concerned with the exile of Rama, the abduction of Sita and the battle to reclaim her, this epic is revered in several Asian cultures. This immensely influential text is filled with the stuff of myth, legend and ethical precepts. A modern translation, *The Ramayana of Valmiki* (trans Robert Goldman, Sheldon Pollock, Rosalind Lefeber **1984-94**; 4 vols out of projected 6) is in progress. Among the several abridgements of the two epics, William Buck's are charming, illustrated, and accessible: *The Mahabharata* (**1973**) and *King Rama's Way* **1976**).

The *Pancatantra* ["The Five Books"] (*c*200BC, trans Arthur W. Ryder as *The Panchatantra* **1955**), a book of instruction in the path of wise conduct, is one of India's more renowned works and one of the earliest Indian texts to appear in the West, where, as translations of it proliferated, it became the template for many FABLE-style collections of tales. The main framing narrative is a slight one that concerns the education of a certain king's three "supremely blockheaded" sons. Each book then has its own frame, with the prose narrations linked by gnomic verses that serve to draw the reader deeper and deeper into a fantasy world consisting of stories-within-stories-within-stories.

Kalidasa (*c*4th century AD) is certainly the most famous of the Sanskrit poets, his masterpiece being the drama *Shakuntala*, the tale of a maiden wooed by a king and then, because of an enchantment, abandoned by him. Written in both prose and poetry, the play exemplifies the subtle and sophisticated aesthetic, *rasa*. Much has been written about the *rasa* theory, which, some commentators assert, applies to all Indian art, ancient and modern. Kalidasa's narrative poems *Kumarasambhava* (trans Hank Heifetz as *The Origin of the Young God* **1985**) and *Raghuvamsa* (trans K.N. Anantapadmanabhan **1973**) and his lyric fantasy *Meghaduta* (trans Leonard Nathan as *The Transport of Love* **1976**) possess the very best features of *kavya*. Kalidasa's complete extant works can be found in *Works of Kalidasa* (**1981-4**) ed and trans C.R. Devadhar.

The *Brhatkatha* ["Great Story"] is a work of *kavya* known only by reputation and recensions. One, Budhasvamin's *Brhatkathashlokasamgraha* ["A Collection in Verse of the Great Story"] (*c*8th century AD; trans Ram Prakash Poddar as *Budhasvamin's Brhatkatha Shlokasamgraha* **1986**), is incomplete, but what there is of it presents a coherent story enhanced by subsidiary tales. The other, Somadeva's *Brhatkathasaritsagara* ["The Ocean of the Rivers of the Great Story"] (*c*12th century AD), also known simply as the

Kathasaritsagara, is a huge collection of stories framed and intricately emboxed. A monumental 10-volume translation by C.H. Tawney as *The Ocean of Story* (**1925**) was ed N.M. Penzer, who meticulously traced literary and folklore references, Eastern and Western, and minutely detailed the scope of this extraordinary compendium of storytelling.

The *ne plus ultra* of *kavya* are the three prose fantasies of Dandin, Subandhu and Banabhatta, composed in the late 6th and early 7th centuries AD. Dandin's *Dashakumaracarita* (trans Arthur W. Ryder as *Tales of the Ten Princes* **1927**), a framed interweaving of related adventures in an easy prose, renders a fairly realistic depiction of street life. Subandhu's *Vasavadatta* (trans Louis H. Gray **1912**) is intended for the *rasika* or connoisseur and does not lend itself to a felicitous translation. The *kavya* tradition peaks with Banabhatta's *Kadambari* (trans Gwendolyn Layne **1991**), the story of the MOON god's descent into incarnation after incarnation and his love for the eponymous heroine. Among his contemporaries and in the vernaculars, Banabhatta was deemed the finest practitioner of prose Sanskrit.

As a perfected language, Sanskrit could only be gilded, never modified. *Kavis* eventually made the language itself a focus for what early Orientalists called "perverted ingenuity" – linguistic pyrotechnics. By the 12th century, manner finally completely overthrows matter so that two and even three stories are told simultaneously, depending on how the strings of compounds are broken up and which meanings are taken from the immense Sanskrit lexicon. Not much attention has been given to these curiosities, which are obviously difficult to translate, but they are in and of themselves examples of a kind of literature as FANTASY ART: what else does one call 30 stanzas which when read forwards tell Rama's story and, backwards, Krishna's? [GL]

SAN SOUCI, ROBERT D(ANIEL) (1946-   ) Canadian writer who has specialized mostly in HORROR novels – like *Emergence* (**1981** US), *Blood Offerings* (**1985** US) and *The Dreaming* (**1989**), which is a VAMPIRE tale – or on work for children like *The Enchanted Tapestry* (**1987** chap), a TWICE-TOLD tale from the Chinese, and *Young Merlin* (**1990** chap), which tells of MERLIN's early years. *Short & Shivery* (coll **1987**) and *Cut from the Same Cloth* (coll **1993**), the latter with Brian Pinkney, are sets of twice-told FOLKTALES. [JC]

SANTA CLAUS Now common name of St Nicholas (4th century), the patron saint of children, and also of Greece, Russia, pawnbrokers, apothecaries, perfumiers and sailors. Virtually nothing is known of his life: his biography by Methodius (?   -847) is a largely fictitious account of his MIRACLES.

The link between St Nicholas and Christmas developed in Germany where he is known as Kriss Kringle, derived from Christkindl, or the "Christ Child". This became merged with the Dutch reverence of Sintirklass, derived from St Nicholas, whose feast day (6 December) saw presents given to children – a tradition deriving from the Three Wise Men's bearing of gifts to CHRIST. Dutch settlers in New Amsterdam developed the story, linking it to the legend of a MAGICIAN who punished bad children and rewarded good ones. Washington IRVING reported this celebration in his quasifactual *History of New York* (**1809**). At this stage the festivity may still be regarded as pagan, but the acceptance of it as a Christian celebration came with the phenomenal success of the poem "An Account of a Visit from St Nicholas" (1823 *Troy Sentinel*; vt "'Twas the Night Before Christmas") by Clement Clarke Moore (1779-1863). This captured the public imagination, fused the worship of SC and the birth of Christ, and introduced the modern celebration of Christmas. It was Moore who created the concept of SC coming down the chimney, thus making him small and elfin. The sleigh drawn by reindeer, however, was probably already part of Dutch FOLKLORE drawn from Nordic legend, though its image was captured by an anonymous writer in the poem *The Children's Friend* (**1821** chap).

Before this mythologization of SC, the UK already had the legend of Father Christmas. Its pagan roots relate to the Roman festival of Saturn, the Roman equivalent of the Greek Kronos, the personification of TIME: he had a long beard and a garland of holly around his head. This image as the symbol for CHRISTMAS was sufficiently acceptable for Ben Jonson (1572-1637) to portray him in his play *Christmas, His Masque* (**1616**) as having a long beard and a crowned hat. Charles DICKENS drew upon this in *A Christmas Carol* (**1843**), where he created the Ghost of Christmas Present in the form of Father Christmas, a jovial, bearded giant with a fur-trimmed robe. The artist Thomas Nast (1840-1902) drew these threads together in his depictions (1863-6) of SC for *Harper's Weekly*, and these firmly embedded the figure of SC in US cultural iconography. SC soon became a figure in the children's stories and nursery rhymes of Christmas issues of newspapers and MAGAZINES. In "St Nicholas and the Gnome" (1878 in *Hannibal's Man*) by Leonard Kip (1826-1906) St Nicholas's joviality is restored by a gnome. In "Behind the White Brick" (1879 *St Nicholas Magazine*) by Frances Hodgson BURNETT a girl climbs up a chimney to a WONDERLAND where she meets various characters including SC.

First to write a book entirely about SC was L. Frank BAUM, with *The Life and Adventures of Santa Claus* (**1902**), a CHILDREN'S FANTASY which keeps most of the basic MOTIFS but develops a complete life history. SC is an abandoned child brought up by the ELVES of the FOREST of Burzee. He is given the name Neclaus, meaning "Necile's Little One", after the name of the wood-nymph who found him. When Neclaus grows to manhood he settles in Laughing Valley and spends his time making toys for children. He brings such delight to children that as he grows old the rulers of FAERIE agree he should be granted the Mantle of IMMORTALITY. Baum also wrote a short story, "A Kidnapped Santa Claus" (1904 *Delineator*), in which SC is kidnapped by DEMONS, and he introduced SC into OZ in *The Road to Oz* (**1909**).

In *Only Toys!* (**1903**) by F. ANSTEY SC helps two children continue to enjoy their toys by shrinking them down to toy-size. The children soon become ungrateful and SC punishes them by making the toys more human and aggressive. In the *Father Cristmas Letters* (written 1920s; **1976**) J.R.R. TOLKIEN provides a series of episodes in SC's life at the North Pole, while in the **Father Christmas** books Raymond BRIGGS depicts a grumpy SC who hates snow.

Few adult fantasies consider SC. The classic is *Roads* (1938 *WT*; **1938**) by Seabury QUINN. Klaus is a Norse gladiator in the court of King Herod who saves Christ during the Slaughter of the Innocents and is granted eternal life. Centuries later Klaus, accompanied by elves, flees the downfall of Rome to settle in VALHALLA. In "Proof Negative" (1956 *Science Fantasy*) John BRUNNER (as Trevor Staines) provides proof to a man about the existence of SC. A number of short stories featuring SC are in *Christmas Bestiary* (anth **1992**) ed Rosalind M. Greenberg and Martin

H. GREENBERG and in *Christmas Stars* (anth **1992**), *Christmas Forever* (anth **1993**) and *Christmas Magic* (anth **1994**), all ed David G. HARTWELL. SC is also spoofed in *Grailblazers* (**1994**) by Tom Holt, where he appears as a modern Woden (◊ ODIN) and must face Von Weinacht and the sinister reindeer Radulph.

SC has adapted patchily to the CINEMA. The classic is MIRACLE ON 34TH STREET (**1947**), in which an old man who is recruited as a store Santa claims to be the real SC. The movie was remade in 1994. Less successful was *Santa Claus: The Movie* (**1985**), but *One Magic Christmas* (**1985**), *The* NIGHTMARE BEFORE CHRISTMAS (**1993**) and *The Santa Clause* (**1995**) are worthy efforts. [MA]

SAPPHIRE AND STEEL UK tv series (1979-82) ITV/ATV Network Production. **Pr** Shaun O'Riordan. **Dir** Shaun O'Riordan, David Foster. **Writers** P.J. Hammond (all but the fifth "Adventure"), Don Houghton, Anthony Read. **Starring** David Cullings (Silver), Joanna Lumley (Sapphire), David McCallum (Steel). 34 30min episodes, divided into six "Adventures". Colour.

In this remarkable series the stability of the world is continually under threat from incursions of TIME. Sapphire and Steel are two agents who represent an unknown but certainly unearthly agency; in a mysterious sense they are personifications of the elements their names designate (in the lore of the CABBALA, "Sephirah" – the singular form of "Sephirot", the ten Divine Utterances which create the world – can be translated as meaning something like both "cipher" and "sapphire"), and they work to maintain the stability of REALITY whenever a disruption occurs in the flow of events. In the first "Adventure", the real world begins to distort itself according to the dictates of nursery rhymes (◊ STORY), causing parents to disappear from their children. In the second, GHOSTS of soldiers, trapped at the moment of their deaths, haunt a railway station (◊ TRAINS). Throughout the series, both characters and viewers are unable to trust their PERCEPTIONS of the world; and very frequently the former are haunted by DOPPELGÄNGERS themselves hoisted from various times. The storylines were slow, packed with indirection and musings, and the world unfolded was anything but secure, even at those moments when the agents proved successful for a span. [JC]

SARBAN Writing name of UK writer John William Wall (1910-1989), who was a career diplomat 1933-66. His writing career occupied less than half a decade, during which period he published the three short books that established his reputation as a subtle, literate teller of tales, conscious of the darker and less acceptable implications that underlie much popular literature. The title story of his first collection, *Ringstones, and Other Curious Tales* (coll **1951**; story alone vt *Ringstones* 1961 US), is told through a FRAME STORY, a device he used more than once to transform HORROR into FAIRYTALE. Two students worry about the contents of a manuscript one of them has received from Daphne Hazel, an old schoolfriend now hired to care for some children at a remote manor house, and whose experiences with the eldest child are deeply distressing: the lad indulges in sadomasochistic practices, casts a GLAMOUR upon her so that she cannot escape the region, may be an AVATAR of the ELDER GODS, and may almost have managed to lure her into FAERIE (◊ TAM LIN).

A vein of sexual implication runs through all of Sarban's work, and in a story like "The Khan" – which appears in his second collection, *The Doll Maker and Other Tales of the*

Uncanny (coll **1953**; cut to contain title story alone, vt *The Doll Maker* 1960 US) – hits a pitch of eloquent perversity not then commonly encountered. An Englishwoman in Persia, tied to a bad marriage, runs away INTO THE WOODS, where she comes across a welcoming castle; there she is bathed and perfumed and prepared for the coming of the Khan, who is in fact a bear.

Sarban's sole full-length novel, *The Sound of his Horn* (**1952**), balances between sf and fantasy. As an early HITLER-WINS tale it occupies an important position in sf history; but the strangely retrospective atmosphere of its telling constantly works against the claimed futurity of the central setting – a heavily forested, ornately rustic German countryside 100 years after WORLD WAR II has ended. Importantly, the central tale is surrounded, once again, by a frame story, which both casts the events the protagonist recounts into the narrative past and renders his horrific experiences in terms of myth-evoking STORY – with, moreover, a technically happy ending (for he survives to tell the tale). His arrival in the New Reich is classic fantasy: lost in WWII, he wanders INTO THE WOODS – a vast pine forest which mysteriously changes into a broadleaf LANDSCAPE under a transformed Moon – and finds himself trapped inside a magic fence surrounding the endlessly proliferating EDIFICE of the Reich Master Forester, who regularly arranges an ersatz WILD HUNT with human prey. Another diversion is the "hunting" of captive girls dressed as gamebirds, who are served, trussed, at the dinner-table. One of these sacrifices herself so that the protagonist may escape back into the world.

Whatever drove John Wall into his brief *floruit* as Sarban clearly did not last; no new stories surfaced after 1953.

[JC/CB]

SARGENT, CARL (? -) US writer, with Marc GAS-COIGNE, of ties in the SHARED-WORLD series of TECHNOFANTASIES, SHADOWRUN.

SATAN The Christian anti-GOD, otherwise known as the DEVIL. "Satan" is a Hebrew word meaning "adversary"; the term is used in the Old Testament as a trivial noun except in Job (where there is no textual implication that Satan is EVIL). Christian writers later co-opted the notion that there had been a war in HEAVEN as a result of which Satan, *alias* LUCIFER, had been consigned to HELL along with a host of rebel ANGELS, but had been given permission to return or send emissaries to Earth in order to tempt humans to sin; by way of example, Christian writers alleged that the SERPENT which tempted ADAM AND EVE in EDEN was Satan in disguise. The Christian Church became increasingly anxious about Satan's wiles, eventually deciding that many heretics and WITCHES had made PACTS WITH THE DEVIL; this conclusion licensed the use of torture and mass murder. In Christian art Satan is usually represented as a monstrous figue with a bestial face, horns (borrowed from the Horned God), a tail and sometimes cloven feet (borrowed from PAN) – an image much parodied on stage and in the cinema.

Satan remains the explicit or implicit adversary of much HORROR, although he usually remains a shadowy figure now that his appearance has been trivialized by too many jokes. His role in fantasy is complicated by the apologetic tradition, Literary Satanism, which sprang from William BLAKE's observation that John MILTON had been "of the Devil's party without knowing it" when he analysed Satan's character and motivation in *Paradise Lost* (**1667**). Percy Bysshe SHELLEY likewise refused to see Milton's Satan as a villain,

extolling him as a heroic rebel against tyranny, but substituted PROMETHEUS in his own cosmic fantasia *Prometheus Unbound* (**1820**); it was left to Charles BAUDELAIRE to issue "Les Litanies de Satan" in *Les Fleurs du Mal* (**1857**). Jules Michelet (1798-1874) amplified this prayer into a four-act ritual "Communion of Revolt" in his classic SCHOLARLY FANTASY *La sorcière* (**1862**), and Anatole FRANCE, after first employing SATYRS as stand-ins, brought full-blown Literary Satanism to the peak of its achievement in "L'humain tragédie" (1895; trans as *The Human Tragedy* **1917**) and *The Revolt of the Angels* (**1914**). In the latter, Satan – living quietly as a gardener – refuses to lead a new army of revolt on the grounds that the struggle against the tyranny of moral absolutism must be carried forward in the hearts and minds of men rather than on the battlefield.

Notable additions to the sceptical tradition of Literary Satanism include *The Memoirs of Satan* (**1932**) by William Gerhardie (1895-1977) and Brian Lunn (1893-?), *The Devil and the Doctor* (**1940**) by David H. KELLER, *Mister St John* (**1947**) by Raoul Fauré (1909-), "Talk of the Devil" (1948) by Ewan Butler, *The Innocent Eve* (**1951**) by Robert NATHAN and *The Master and Margarita* (written 1938; **1967**) by Mikhail BULGAKOV. The mischievous fables in *The Devil's Storybook* (coll of linked stories **1974**) and *The Devil's Other Storybook* (coll of linked stories **1987**) by Natalie BABBITT extend the tradition into CHILDREN'S FANTASY, as does *Satan: The Hiss and Tell Memoirs* (**1987**), by Jeremy Pascall (likely a pseudonym). A distinct note of sympathy is also evident in such devoutly anti-Satanic works as *The Sorrows of Satan* (**1895**) by Marie CORELLI and *The Devil Takes a Holiday* (**1955**) by Alfred NOYES. Sympathy of a rather different kind is displayed in fantasies which portray Satan as a pathetic no-hoper in severely reduced circumstances; notable examples are *The Devil, Poor Devil!* (**1934**) by Murray Constantine (Katharine BURDEKIN) and *The Return of Fursey* (**1948**) by Mervyn WALL, and there is something of the same ironic spirit in John UPDIKE's *The Witches of Eastwick* (**1984**), filmed as *The* WITCHES OF EASTWICK (*1987*). *The Day After Judgment* (**1972**) by James BLISH, the TECHNOFANTASY *Satan* (**1982**) by Jeremy Leven (1941-) and the **Incarnations of Immortality** sequence (**1983** onwards) by Piers ANTHONY all allow Satan to retain his adversarial role, but with various significant modifications.

In many modern SLICK-FANTASY tales of diabolical pacts and infernal comedies Satan becomes an urbane and witty social sophisticate, but he retains his adversarial role in some modern horror, especially stories dealing with demonic POSSESSION (although he is here usually represented by lesser minions) and the activities of Devil-worshippers. Some fundamentalist Christians allege that Satan is still at work in the world, aided by legions of active Satanists engaged in BLACK MAGIC, HUMAN SACRIFICE and ritual child abuse.

Appearances of Satan in the CINEMA are far too numerous to list. As polar opposites, each in their way an accurate portrayal, we can note Ridley Scott's LEGEND (*1985*), where Satan (in this SECONDARY WORLD called "Darkness") is a huge and hugely horned figure of slobbering lasciviousness, and Paul Bogart's *Oh God! You Devil!* (*1984*), in which Satan is the TWIN of God, both being amiably folksy elderly gents – and neither more ethically reliable than the other. [BS]

SATANIC RITES OF DRACULA, THE UK movie (*1973*). ◊ DRACULA MOVIES.

SATANISM A form of religious belief in which Christian values are turned upside down, SATAN is worshipped and God is reviled, RITUAL is upended (as in the BLACK MASS), and behaviour normally thought of as EVIL is promoted while "virtues" (like humility) are condemned. The origin of Satanism may lie as far back as Gnosticism (◊ GNOSTIC FANTASY), and certainly for centuries the religious establishments of the West tended to accuse those they wished to scapegoat (Cathars, WITCHES, etc.) of being Satanists. All the same, the 19th-century vogue for Satanism, and the behaviour of the lifestyle fantasists who thought of themselves as Satanists, seems literary in origin (◊ SATAN).

The term is of some use in the study of SUPERNATURAL FICTION. Authors (some of them advocates of or dabblers in DECADENCE) like Charles BAUDELAIRE, Marjorie BOWEN, Aleister CROWLEY, Gustave FLAUBERT, Anatole FRANCE, J.K. HUYSMANS and Oscar WILDE made ironic use of the reversal of values represented by the belief. The extent to which any individual writer actually ascribes to Satanism presumably varied. The wave of HORROR tales and movies following Ira LEVIN's *Rosemary's Baby* (**1967**), filmed as ROSEMARY'S BABY (*1968*), and William Peter BLATTY's *The Exorcist* (**1971**), filmed as *The* EXORCIST (*1973*), brought Satanism renewed popularity as a theme. [JC]

SATIRE A satire is a form of protest against the rotting of the world. It is normally written by a man (women satirists are less common) who is past his first youth, and who attacks the world from a conservative point of view. He may well create a model of the world which exaggerates the vices and follies of the real one, but however far from reality he may seem to stray he will not indulge in the Absurd or SURREALISM, or the flights of free fancy which might lead to the creation of a SECONDARY WORLD. The target of satire (◊◊ AESOPIAN FANTASY) is always the human condition *in situ*.

Though satirical moments are frequently found, full-blown satire is therefore somewhat rare in fantasy. APULEIUS's *The Golden Ass* (*c*165), for instance, contains satirical interludes, but the underlying RITE OF PASSAGE of the protagonist is not treated in terms of satire. However *Das Narrenschiff* (**1494**; trans Alexander Barclay as *The Ship of Fools* **1509** UK) by Sebastian Brant (1457-1521) *is* a satire (◊ SHIP OF FOOLS), and indeed loses all narrative impulsion through its author's need to focus on various examples of the folly of the world. In the 18th century, *Gulliver's Travels* (**1626**) – the savage masterpiece of Jonathan SWIFT, perhaps the greatest satirist yet born – is clearly a satire; as is much of Voltaire's work.

The 19th century is almost devoid of the form in fantasy, if one puts aside the special case of the WONDERLAND satires by Lewis CARROLL; but the 20th century has seen a revival, including Anatole FRANCE's *Penguin Island* (**1908**), James Branch CABELL's *Jurgen* (**1919**), John ERSKINE's *Adam and Eve* (**1927**), C.S. LEWIS's *The Screwtape Letters* (**1942**), George ORWELL's *Animal Farm* (**1945** chap) and *Nineteen Eighty-four* (**1949**), Michael AYRTON's *Tittivulus* (**1953**), most of the novels of Thomas BERGER – especially *Changing the Past* (**1989**) – Steve BAUER's *Satyrday* (**1980**), *An* AMERICAN TAIL (*1986*), *Lord Horror* (**1989**) and *Motherfucker* (**1996**) – both HORROR – by David Britton (1945-   ), Thomas M. DISCH's *The Priest* (**1994**) and Scott BRADFIELD's *Animal Planet* (**1995**). [JC]

See also: ESTATES SATIRE; SATYRS.

SATURNALIA ◊ REVEL.

SATYRS Lustful woodland sprits of Greek MYTHOLOGY, identical to the fauns of Roman mythology. Physically they resembled PAN in being horned and often having goatlike legs, but were more closely associated with DIONYSUS by virtue of seeming to be younger and more virile versions of the hoary Sileni, whose drunken, lecherous and boastful singular incarnation Silenus became the central figure of the "satyr plays" ancestral to the tradition of SATIRE.

The role played by satyrs and fauns in literary fantasy is broadly similar to Pan's in that they usually symbolize the reckless fecundity of Nature. Before Pan was taken up by UK Literary Satanists (◊ SATAN), Anatole FRANCE had used satyrs in likewise in "Amycus and Celestine" (1892) and "San Satiro" (1895). The Nicaraguan-born Spanish poet Ruben Dario (1867-1916) and the Portuguese poet Fernando Pessoa (1888-1935) deployed them in similar symbolic roles. Being much less powerful than Pan, they are sometimes seen in a sentimentally wistful light which can threaten to reduce them to cuteness; examples include "The Ageing Faun" (1912) by Arthur RANSOME, "The Inquisitive Satyr" (1914) by Oswald J. Couldrey, "The Faun" (1915) by C.C. Martindale and, most saccharinely, in a section of the DISNEY movie FANTASIA (*1940*). "The Curate's Friend" (1911) by E.M. FORSTER, *Mr Antiphilos, Satyr* (*1913*) by Rémy de GOURMONT and *The Girl and the Faun* (**1916**) by Eden PHILLPOTTS offer more robust and dignified images, as does *After the Afternoon* (**1941**) by Arthur MacArthur (1896-?   ), an eccentric sequel to "L'après-midi d'un faune" (1876) by Stéphane Mallarmé (1842-1898).

The central figures in "The Faun" (1894) by de Gourmont and "The Satyr" (1931) by Clark Ashton SMITH are straightforward symbols of lust, but the mute satyrs in *The Island of Captain Sparrow* (**1928**) by S. Fowler Wright (1874-1965) are curiously innocent, and the talkative one in *Satyrday* (**1980**) by Steven BAUER is sympathetically well rounded. The last surviving satyrs in a THINNING world play an educative role in *In the Beginning* (**1927**) by Norman DOUGLAS. Sympathetic satyrs play crucial roles in a few of Thomas Burnett SWANN's fantasies, most conspicuously in *Wolfwinter* (**1972**). [BS]

SAUNDERS, CHARLES R(OBERT) (1946-   ) US writer, long resident in Canada. Active in the fantasy SMALL PRESSES of the 1970s, he went on to write three novels in the **Imaro** series, SWORD-AND-SORCERY adventures with a black African hero: *Imaro* (**1981**), *The Quest for Cush* (**1984**) and *The Trail of Bohu* (**1985**). [DP]

Other work: *Robert E. Howard: Adventure Unlimited* (**1976** chap).

SAVAGE, FELICITY (ROSE) (1975-   ) Irish-born writer, now resident in the USA, whose first fantasy novel – published in two volumes as *Humility Garden: An Unfinished Biography* (**1995**) and *In Human Country* (**1996**) – is set in the southern hemisphere of a water-dominated world whose six small continents and adjacent ARCHIPELAGO have been governed for many centuries by a literal theocracy, for the ruling "Divinarch" is (or seems to be) a GOD. Humility Garden herself moves from being one of the apprentice "ghostiers" – a cadre of artists sanctioned to kill the young and beautiful at aesthetically appropriate moments, which moments they preserve through MAGIC and art for the delectation of the living – to intimacy with the gods. The narrative is rich and the SEX is intricate and believable. [JC]

SAVOY, THE ◊ MAGAZINES.

SAXTON, JOSEPHINE (MARY HOWARD) (1935-
) UK writer who began to publish work of genre
interest with "The Wall" for SCIENCE FANTASY in 1965, and
whose novels hover – or ricochet – between sf and fantasy.
"The Wall" itself is a good example of that hovering intensity: the tale, like much of her work, is set in a LANDSCAPE
which resembles a fantasy WASTE LAND while at the same
time serving as a kind of Magic Slate upon which the psyche
"writes" or creates SYMBOLS, and traverses in the traditional
QUEST for justice, or safety, or wholeness, or herself. The
eponymous wall traverses this typical landscape, hugely and
blankly. It is both abstract and crushingly physical.

Very early in her career, JS began to shape this landscape
and its inhabitants into a model for explorations into the LIT-
TLE BIG interiors of the collective unconscious (◊ JUNGIAN
PSYCHOLOGY). "The Consciousness Machine" (1968; rev in
*The Consciousness Machine; Jane Saint and the Backlash: The
Further Travails of Jane Saint* coll **1989**) is the first and perhaps most clearcut of these tale; its protagonist, interacting
with the eponymous TECHNOFANTASY machine, enters the
cave of the Jungian SHADOW, sorts the DREAMS which have
tortured her, and heals herself. A similar voyage of self-
discovery is central to JS's first novel, *The Hieros Gamos of
Sam and An Smith* (**1969** US), which "The Consciousness
Machine" was apparently written to elucidate (though it
appeared earlier). Other early novels – *Group Feast* (**1971**
US) and *Vector for Seven: The Weltanschaung [sic] of Mrs
Amelia Mortimer and Friends* (**1971** US) – explore the same
territory. The first, set in a seemingly infinite EDIFICE,
describes the stripping away of its protagonist's material
obsessions; in the second, seven COMPANIONS are caught in
BONDAGE to yet another landscape, until they sort themselves into a kind of unity.

JS returned some years later to the field with the **Jane
Saint** sequence – *The Travails of Jane Saint* (**1980**) and *Jane
Saint and the Backlash: The Further Travails of Jane Saint* (see
above) – in which the eponymous heroine traverses a landscape filled with emblematic figures, charged now with
burdens which are unpacked in feminist terms (◊ FEMI-
NISM). *The Queen of the States* (**1986**), perhaps her best work,
again applies a politicized feminist analytical wit to the
Surrealist (◊ SURREALISM) adventures of the protagonist,
who is, or dreams she is, being interrogated by aliens, or
maybe they are "simply" doctors – it all depends on which
REALITY she inhabits, and on how she can grasp its meaning.
The book is funny, distressed and acute. JS's short work is
normally most effective as remembered image, for she is a
crafter of ICONS which nag at her readers' memories and
dreams; they are assembled in *The Travails of Jane Saint and
Other Stories* (coll **1986**), *The Power of Time* (coll **1985**) and
Little Tours of Hell: Tall Tales of Food and Holidays (coll
1986). [JC]

SCANDINAVIA Literary and intellectual traditions among
the five Scandinavian countries (Denmark, Finland, Iceland,
Norway, Sweden) vary to a larger extent than is perhaps
obvious to the foreigner. While four of the languages are
derived from the same Old Norse roots, Finnish belongs to
the Finno-Hungarian group of languages. Similarly,
although politically the countries have often been linked or
involved in complex rituals of conquest, war and interdependence, the major foreign cultural influences have varied
during the last centuries. Finland was primarily influenced
by Russia, Sweden by Germany, and Denmark and Norway
to a greater extent by England. It is worth noting that, of the

five countries, Sweden in particular is inclined towards insu-
larity: while Swedish authors sell (both in translation and in
the original) in Norway and Denmark, the converse is
almost never true of Danish and Norwegian authors. Thus
the countries are here largely treated individually.

Iceland, obviously, has an immensely rich and rewarding
fantastic tradition in its *fornaldarsagor*, the pseudo-historical
SAGAS of giants, dwarfs, gods and heroes, in many cases
probably surviving orally since before Iceland's colonization
in the 9th century and written down in the 13th century.
These in turn inspired the *rímur* of the 14th
century – poetic renderings of the imaginary, fabled
pseudo-past of the Eddas. Peculiarly, however, this rich
mythological heritage – at least as far as its fantastic content
goes – has not much influenced modern Icelandic literature,
which for its nonrealist elements has largely turned to
MAGIC REALISM, often coloured by Icelandic tradition, but in
tone more reminiscent of South American than of Northern
European tradition. Instead, the mythology of the Eddas has
inspired primarily Anglo-Saxon fantasy authors, among
them William MORRIS, E.R. EDDISON, Fletcher PRATT, C.S.
LEWIS, J.R.R. TOLKIEN and Poul ANDERSON; indeed, a fairly
large portion of the sagas have been translated by fantasy
authors, Morris having translated the *Eyrbyggja* and the
Grettla and Eddison the *Egla*, while Anderson attempted
the considerably more risky task of reconstructing the lost
Hrolf Kraki's Saga from the extant fragments and para-
phrasings.

Denmark and Norway were united from the late Middle
Ages until 1814, when Norway was lost to Sweden due to
Denmark's unsuccessful alliance with Napoleonic France.
Thus the two countries share a common early literary tradition. Although large parts of the sagas stem originally
from Denmark and Norway, modern Danish/Norwegian
literature really begins with Ludvig Holberg (1684-1754), an
immensely prolific scholar and author of almost every kind
of literary work, many fantastic – e.g., his most famous satirical novel, *Nicolai Klimii iter Subterraneum* (**1741**; trans as
Journey of Niels Klim to the World Underground 1742; new
trans 1960), a lively utopian fantasy, comparable at its best
to Jonathan SWIFT.

In the early 19th century, the Romantic Movement (◊
ROMANCE) triumphed in Denmark. Its primary exponent,
Adam Oehlenschläger (1779-1850), inspired primarily by
the German Romantics, wrote a number of plays and narrative poems hugely influential in Denmark and clearly
fantastic in nature. To these belong his dramatic poem
Sankt Hansaften-Spil (**1802**; trans as *Midsummer Night's
Play*), the mythological legend *Vaulundur's Saga* (**1805**) and
the undiluted fantasy play *Aladdin* (**1805**). Among
Oehlenschläger's disciples, the most influential was Nicolai
Grundtvig (1783-1872), a poet, historian, educationalist and
bishop. Grundtvig not only contributed his own fantastic
and heroic Romantic poetry but also popularized Norse
MYTHOLOGY via his huge work on Nordic mythology and his
translations of numerous SAGAS, among them BEOWULF,
Saxo Grammaticus's *Gesta Danorum* and Snorri's
Heimskringla.

The next major Danish fantasist is the most famous: Hans
Christian ANDERSEN. The first four of his 168 fairytales
appeared in *Eventyr, fortalte for Børn* ["Fairytales, Told for
Children"] (coll **1835** chap): "The Tinderbox", "The
Princess on the Pea", "Little Claus and Big Claus" and
"Little Ida's Flowers". Later came Jens Peter Jacobsen

(1847-1885), whose poetry was strongly influenced by Edgar Allan POE and is evocative of a fantastic realm of supernatural horrors, strange romances and alien mythologies. The major work of Nobel laureate Johannes V. Jensen (1873-1949), **Den lange rejse** (**1908-22**; trans from 1922 as **The Long Journey**), is a 6-vol novel cycle detailing the long journey of humankind from prehistory (PREHISTORIC FANTASY) to the historical period, and incorporating ancient MYTHOLOGY, Icelandic sagas, BIBLE stories and pure imagination – as well as a wealth of scientific detail. Jensen also included a wealth of fantasies in his visionary **Myter** (**1906-44**; cut trans as *The Waving Rye*), which in its freeform combination of stories, myths and essays is probably his most original work.

Other major fantasy authors of the present century include Isak DINESEN (Karen Blixen), with her collections of Gothic mysteries and imaginative fantastic tales, and Martin A. Hansen (1909-1955), whose burlesque *Jonathans rejse* ["Jonathan's Journey"] (**1941**) tells of the righteous blacksmith Jonathan, who captures the DEVIL in a bottle and sets off to present his captive to the king, on his way travelling through many of Denmark's classical fairytale adventures. Frank Jaeger (1926-   ) is among the few Danes to have written a pure OTHERWORLD fantasy, *Iners*. Since Danish literature has always accepted the nonrealist, modern Danish prose can boast numerous examples of fantastic FABULATIONS, from those of Sven Holm (1940-   ) and Dorrit Willumsen through Ib Michael, Per Højholt – who has developed what he calls the "dead end" story inspired by Jorge Luis BORGES – and Svend Åge Madsen, whose numerous works are creating an ALTERNATE WORLD that comments on our own.

Currently the most important Danish fantasist is, arguably, Erwin Neutzsky-Wulff (1949-   ), a productive author of poetry, essays, sf and fantasy, among the latter being *Den treogtredivte marts* ["March 33rd"], *Faust* and *Verden* ["The World"], an immense novel trying to encompass the totality of human experience in an imaginative, experimental play on time and history.

In Norway, although Holberg played much the same part as a central literary percursor as in Denmark, the FANTASTIC did not become a similarly entrenched part of fiction, and Norwegian literature instead turned much more firmly in the direction of social realism. Nevertheless, occasional important works of fantasy have appeared, among them *Professor Umbrosius* by Sven Elvestad (1884-1934), where the history of the world is retold backwards and where the author offers occasional jubilant comments on how progress has now managed to replace such curses as democracy, schooling, writing and technology. Egil Rasmussen (1903-1964) published at least two novels which can be characterized as fairly pure Howardian SWORD AND SORCERY, *Legenden om Lovella* ["The Legend of Lovella"] and *En konge rider hjem* ["A King Rides Home"]; Peder' W. Cappelen (1931-   ) has used Norse and Icelandic mythology and sagas as the basis for short stories and a number of plays including *Tornerose* ["Sleeping Beauty"] and *Loke* ["Loki"].

The most-read fantasy author in Scandinavia today is the Norwegian Margit Sandemo. After a long career as a writer of popular stories and serials for weekly women's magazines, she started in 1980 an immense series of novels called **Sagan om isfolket** ["The Story of the Ice People"], totalling 45 volumes; in it she chronicles the Tengil family

through the centuries as they interact with historical events and with elves, fairies, goblins and magic.

On a different level, modern Norwegian fantasy can boast at least two authors of high accomplishment: Tor Åge Bringsvaerd (1939-   ) and Øyvind Myhre (1945-   ). Bringsvaerd's first major fantasy is the experimental *Syvsoverskens dystre frokost: En underholdningsroman på liv og død* ["The Sad Breakfast of the Late Sleeper: An Entertainment of Life and Death"] (**1976**), a Surrealist fragmental fantasy set in modern NEW YORK; the book remains one of the most impressive Modernist experiments in Norwegian literature. Bringsvaerd has later published more traditional mythologically anchored fantasy novels, including *Minotauros* (**1980**) and *Ker Shus* (**1983**).

Øyvind Myhre (1945-   ), on the other hand, is a wholly traditional storyteller, clearly influenced by the UK and US fantasy tradition. His five major fantasy novels all deal with the birth of the modern world, in the sense that Myhre writes of the break between the pagan and the Christian universe but also of the birth of intolerance, hierarchical systems and power structures. In *De sidste tider* ["The Last Days"] (**1976**) an individual who lacks destiny is born into a totally predestined world on its course towards destruction; *Kongen of gudene* ["The King and the Gods"] (**1979**) is set in 7th-century England and deals with a conflict between heathen and Christian; *Grønlandsfarerna* ["The Greenland Travellers"] (**1981**) tells of Torgils, who sets out on a mission to Christianize the world, but is visited by the wrath of the Aesir; it is an impressive portrayal of the last days of the saga period and of men torn between faith and tradition, comfort and belief. The most impressive of Myhre's novels may be *Makt* ["Power"] (**1983**), a historical fantasy of the monolith builders of Boyne Valley, Ireland. In a totally different vein, Myhre has written an impressive and original addition to the Lovecraftian CTHULHU MYTHOS, *Mørke over Dunwich* ["Darkness Over Dunwich"] (**1991**).

Finland was from the 13th century part of Sweden, but coveted by Russia; in 1809 Sweden lost control, and Finland was incorporated as an autonomous Grand Duchy within the Russian Empire. A Swedish-speaking minority remained as the ruling class, and Swedish remained the only official language until 1863; after Finland gained its independence in 1917, Swedish was granted equal status with Finnish as the official language. Today, roughly 6% of the inhabitants are Swedish-speaking.

The folklorist and philologist Elias Lönnroth (1802-1884) in a sense created the Finnish national literature by collecting folk poetry and myths among the Lapps, the Estonians and the Finnish tribes of Karelia and turning these into the coherent, lyrical "folk epic" KALEVALA (**1835**; exp 1849), which largely created the Finnish nationalist movement. However, apart from some works inspired directly by the *Kalevala* (perhaps most importantly those of the poet Eino Leino), Finnish literature has been fundamentally realist and inspired from Sweden, Germany and Russia; it remained for modern authors like Tove JANSSON and Irmelin Sandman Lilius (1936-   ) to provide the foundations for Finnish fantastic literature. Their importance can hardly be overstated; Jansson's **Moomin** stories and Sandman Lilius's many tales of the imaginary, magic CITY of **Tulavall** are major literary fantasy creations.

Finally, in Sweden, 19th-century Romance at last broke the dominance of Realism. The first important fantasist was

Per Daniel Amadeus Atterbom (1790-1855), whose unfinished fairy-play *Fågel Blå* ["Blue Bird"] (**1814**) presaged his masterpiece *Lycksalighetens ö* ["Island of Happiness"] (**1824-7**; rev 1855), an immense FAIRY poem of a journey outside TIME to a beautiful ISLAND in the sky. A later liberal Romantic, Carl Jonas Love Almqvist (1793-1866), created another masterpiece of Swedish epic literature, **Törnrosens bok** ["The Book of the Wild Rose"], a vast sequence of stories and plays published from 1832 onwards; his most important work was *Drottningens juvelsmycke* ["The Queen's Jewel Necklace"] (**1834**), a pseudohistorical novel about the murder of Gustav III but told largely around the mysterious, ambiguous figure of the androgynous heroine Tintomara.

Much different but no less important was Nobel laureate Selma Lagerlöf (1858-1940). In almost all her works the ASTRAL PLANE and natural MAGIC are included as self-evident ingredients, but only seldom do they occupy a central place. However, in *Herr Arnes penningar* (**1904**; trans as *Herr Arne's Hoard*) humans and SPIRITS cooperate to punish those guilty of murdering Arne; in *Nils Holgerssons underbara resa* (**1906-7**; trans as *The Wonderful Adventures of Nils*) she wrote a charming fantasy of a boy's travels through a magical Sweden on the back of a talking goose. In *Körkarlen* (**1912**; trans William Frederick Harvey as *Thy Soul Shall Bear Witness!* 1921 UK) she brilliantly portrays the emissary of Death who on a New Year's Eve comes driving his old cart into a small town to collect the spirits of the dead. Additionally, Lagerlöf was a master of the GHOST STORY; over 30 were collected by Sven Christer SWAHN in *Vägen mellan himmel och jord* ["The Road Between Heaven and Earth"] (coll **1992**).

Another Swedish Nobel laureate proved a first-class fantasist. Pär LAGERKVIST tended towards the fantastic and symbolic, from early stories like "Det eviga leendet" (1920; trans as "The Eternal Smile"), the novel and later play *Bödeln* (**1933**; trans as *The Hangman*), and in novels like *Sibyllan* (**1956**; trans as *The Sibyl*), set in ancient Delphi, and *Ahasverus död* (**1960**; trans as *The Death of Ahasuerus*), about the WANDERING JEW.

On a less exalted level, Sweden's primary contributions to current fantasy are CHILDREN'S FANTASY. Here Astrid LINDGREN is foremost, with her **Pippi Longstocking** stories plus the HIGH FANTASIES *Mio, min Mio* (**1954**; trans as *Mio, My Son*), *Bröderna Lejonhjärta* (**1973**; trans as *The Brothers Lionheart*) and *Ronja Rövardotter* (**1981**; trans as *Ronja, the Robber's Daughter*). A younger internationally successful author of subtle juvenile fantasies is Maria GRIPE, with, in particular, her **Shadows** series.

Other Swedes writing in the fantasy genre include Sam J. Lundwall (1941-) and Bertil Mårtensson (1945-), who has written the only truly successful Swedish high-fantasy trilogy, **Maktens vägar** ["The Ways of Power"] (**1979-83**). Peder Carlsson (1945-) has published two original, idiosyncratic philosophical fantasy novels, *Syns, syns inte* ["Now You See It, Now You Don't"] (**1976**) and *Enhörning på té* ["Did You Ever Have a Unicorn for Tea?"] (**1979**). Among younger authors, several writers have shown considerable originality in handling nonrealist themes, perhaps most impressively Anna-Karin Palm in *Faunen* ["The Faun"] (**1991**) and *Utanför bilden* ["Outside the Picture"] (**1992**). [J-HH]

SCANZIANI, PIERO (1908-) Italian writer whose most famous book, the nonfiction *Avventura dell'uomo* (**1957**; rev 1983; trans as *The Adventure of Man* **1991** UK), mixes ANTHROPOLOGY and a diffuse Gnosticism (◊ GNOSTIC FANTASY) in its attempt to articulate humanity's hidden nature in a clouded world. His fiction tends to convey a similar burden of message, though his second novel, *I cinque continenti* ["The Five Continents"] (**1942**; rev 1983), which includes a visit to a tiny ISLAND occupied by the last survivors of ATLANTIS, is lightly told. *Felix* (**1952**; rev 1980) places the biography of its main character within a FRAME STORY set in an arid AD2000. In *Libro bianco* (**1968**; rev 1983; trans Linda Lappin as *The White Book* **1991** UK) the newly dead protagonist is plunged into a surreal field of judgement (◊ POSTHUMOUS FANTASY); he eventually becomes an Adam (◊ ADAM AND EVE), but avoids making the same mistake this time round. *Entronauti* (**1969**; rev 1983; trans Linda Lappin as *The Entronauts* **1991** UK) depicts four journeys in search of the transcendental self; the protagonist is accompanied on these by an invisible archangel whose home is the Gnostic HEAVEN. [JC]

Further reading: *Piero Scanziani: A Man for Europe* (anth **1991**; trans Nicoletta Simbrorowski-Gill **1991** UK), ed anon.

SCARAMOUCHE ◊ COMMEDIA DELL'ARTE.

SCARBOROUGH, ELIZABETH ANN (1947-) US writer whose work has long been read as fantasy, though some of her more recent novels are CROSSHATCHES. She began with the **Argonia** sequence: *Song of Sorcery* (**1982**) and *The Unicorn Creed* (**1983**), both assembled as *Songs from the Seashell Archives #1* (omni **1987**), and *Bronwyn's Bane* (**1983**) and *The Christening Quest* (**1985**), both assembled as *Songs from the Seashell Archives #2* (omni **1988**). Each pair of tales, though set in the same amiable FANTASYLAND, features different protagonists. The first is a young WITCH who sets out on a QUEST to find her sister, encountering *en route* various figures from the repertory of GENRE FANTASY; eventually she succeeds, with the aid of various COMPANIONS, including a querulous UNICORN. The second, a princess under a CURSE (she cannot tell the truth), has similar adventures. Further comedies followed: *The Harem of Aman Akbar, or The Djinn Decanted* (**1984**), EAS's first singleton, is set in a LAND-OF-FABLE venue full of supernatural creatures from ARABIAN FANTASY; *The Drastic Dragon of Draco, Texas* (**1986**) invokes Native American material, the eponymous DRAGON being in fact a winged SERPENT who causes havoc; *The Goldcamp Vampire, or The Sanguinary Sourdough* (**1987**) is a SUPERNATURAL FICTION.

EAS finally turned – in *The Healer's War* (**1988**), which won the Nebula AWARD – to a theme which demanded a straight face. The protagonist of the book, a nurse in Vietnam, has some experiences which reflect EAS's own, and her depiction of the conflict combines MAGIC REALISM and reportage in a manner also found fitting by Bruce McAllister (1946-) and Lucius SHEPARD in their own Vietnam tales of nightmare – all three translate Southeast Asia into a hellish land of fable. The central action of EAS's book engages the protagonist with a wounded Vietnamese healer who recognizes in her a potential power akin to his own; he signals this by giving her an AMULET, that focuses her own TALENTS – which include telepathy and the healer's touch.

The female protagonist of *Nothing Sacred* (**1991**) and its sequel, *Last Refuge* (**1992**), is also haunted by the calamities afflicting this world. About a century hence she is immured in a Tibetan camp, which she discovers to be SHANGRI-LA, a

POLDER protective of her and of a Buddhist UTOPIA. She learns that the doctrine of REINCARNATION is literally true, and by the second volume is running QUESTS back into the world, where huge numbers of GHOSTS – victims of nuclear HOLOCAUST – await rebirth into flesh.

The **Songkiller Saga** – *Phantom Banjo* (**1991**), *Picking the Ballad's Bones* (**1991**) and *Strum Again?* (**1992**) – represents something of a descent. The forces of HELL decide that "Project Man" is doing moderately well, except that the human race continues to refuse to commit final *hara kiri*. The reason is folk music, which they set out to destroy. *The Godmother* (**1994**) is a partial recuperation: in contemporary Seattle a FAIRY GODMOTHER does her best to help out characters whose stories distortingly reflect familiar FAIRYTALES. [JC]

Other works: *The Powers that Be* (**1994**) with Anne MCCAFFREY, sf.

SCARFE, GERALD (1936-   ) UK illustrator, caricaturist and moviemaker who became famous with the political and culture SATIRE he published in the magazine *Private Eye* from 1961. So consummately grotesque is his draughtsmanship that mundane figures are routinely transformed into dreamlike MONSTERS whose most plausible venue is WONDERLAND. The ANIMATED MOVIE *A Long Drawn-Out Trip* (*1971*) is of interest; GS also did much design for *The Wall* (*1982*), a movie featuring Pink Floyd. His original work has been assembled in *Gerald Scarfe's People* (graph **1966**), *Indecent Exposure* (graph **1973**), *Expletive Deleted* (graph **1974**), *Gerald Scarfe* (graph **1982**), *Father Kissmass and Mother Claws* (graph **1985**), *Seven Deadly Sins* (graph **1987**), *Lines of Attack* (graph **1988**) and *Scarfeland* (graph **1989**). [JC]

SCARS OF DRACULA UK movie (*1970*). ◊ DRACULA MOVIES.

SCARTH Pretty, blonde and almost invariably naked female star of the UK daily newspaper SCIENCE-FANTASY strip *Scarth, A.D.2195*, created by writer Les Lilley and Spanish artist Luis Roca (1941-   ). The first full-frontal nude in UK newspaper strips, she appeared in *The Sun* 1969-72. Killed at the moment of her first appearance, she is resuscitated by 22nd-century medicine. Suffering AMNESIA and chronic nudity, she is taken up by an interplanetary fashion mogul. Her adventures are SATIRES on the fashion world of the Pop Art and Flower Power period. [RT]

SCAVONE, RUBENS TEIXEIRA (1925-   ) Brazilian writer. ◊ BRAZIL.

SCHEHERAZADE Sometimes called Shahrazad or Sheherezade, the narrator of the *Arabian Nights* (◊ ARABIAN FANTASY). King Shahryar, angered to discover his wife has been unfaithful, executes her and thereafter slays each new wife after the wedding night. Scheherazade, daughter of the vizier, cunningly – and with the help of her sister Dunyazad – contrives to begin a story on the wedding night that is unfinished at dawn. The king is sufficiently interested that he delays the execution to hear the end of the story. This continues for 1000 nights, after which Shahryar is so enamoured that he lets Scheherazade live.

This is one of the oldest forms of FRAME STORY, and has been used by other writers, such as Ernest BRAMAH in his **Kai Lung** tales. The stories inspired Nicolai Rimsky-Korsakov (1844-1908) to compose his symphonic suite *Shahrazad* (**1888**), which is used as a trigger for spiritual TRANSFORMATIONS in the movie SHADOW DANCING (*1988*). Scheherazade features in several of the stories and books by John BARTH, including "Dunyazadiad" (in *Chimera* coll of linked stories **1972**), where he uses a timewarp (◊ TIME) to sustain Scheherazade's storytelling, and notably in *The Last Voyage of Somebody the Sailor* (**1991**). In "The Thousand-and-Second Tale of Scheherazade" (1845 *Godey's Lady's Book*) Edgar Allan POE has her tell of many contemporary scientific wonders which the king finds too fanciful – he thus has her throttled. Larry NIVEN continued the story of Scheherazade in "The Tale of the Djinni and the Sisters" (in *Arabesques* anth **1988** ed Susan SHWARTZ). [MA]

SCHIFF, STUART D(AVID) (1946-   ) US editor and publisher. ◊ WHISPERS.

SCHIKANEDER, EMANUEL (JOHANN JOSEPH BAPTIST) (1751-1812) Austrian actor-manager, impresario and librettist of, among others, Mozart's *The Magic Flute* (**1791**). [MSR]

See also: OPERA; Sir Michael TIPPETT.

SCHMIDT, DENNIS A. (?   -   ) US writer of two sf series, the **Kensho** and **Questioner** sequences, and one fantasy sequence, the **Twilight of the Gods**: *Twilight of the Gods #1: The First Name* (**1985**), *#2: Groa's Other Eye* (**1986**) and *#3: Three Trumps Sounding* (**1987**). Out of the CAULDRON OF STORY from which writers of NORDIC FANTASY normally select individual strands, DAS has extracted almost every ICON and MOTIF. In a FANTASYLAND version of the world of Richard WAGNER's **Ring Cycle** of OPERAS, a large cast – also echoing the KALEVALA and sources as far removed as MESOPOTAMIAN EPIC – is complexly involved in SWORD-AND-SORCERY exploits. LOKI is female. It is perhaps unfortunate the sequence stopped at three. [JC]

SCHOLARLY FANTASY All worthwhile scholarship involves an element of fantasy. The leap from a heap of raw data to a theory is an act of imagination; that some venturesome theories may be rudely falsified by subsequent data need not diminish the achievement of their producers. Some SFs are produced by fools, some by honest misinterpreters, some by calculating charlatans, some by geniuses, and some by people who are fully conscious they are fantasizing. Discrimination between the categories on the grounds of sincere belief is for obvious reasons impracticable.

A particularly interesting set of SFs emerges from the human sciences (especially history), because understanding in those sciences is at least partly to do with trying to see things as other social actors see or saw them. Such acts of imaginative identification are intrinsically difficult to put to the proof, and such evidence as the past leaves behind is usually flexible enough to accommodate several different accounts of what people thought they were doing and why. When our interpretations of history have to come to terms with the fantasies which were entertained by the people of the past, the task of sorting out what was actually believed, to what extent and on what grounds becomes extremely difficult: we risk infection not only by the fantasies of the past but by all manner of new fantasies born of our attempts to theorize about the old ones. It is thus not surprising that many SFs once thought doomed have made spectacular comebacks, often in more complicated or dramatically transfigured forms, as the study of history has become more widespread and more thoughtful. Both deliberate and accidental SFs are more widespread now than ever before.

Where metaphysical matters are concerned, no evidence is or can be relevant to judgements of truth; thus treatises on theology – however unorthodox – need not be treated as SFs. Where the writings of Churchmen make empirical claims, however, they remain subject to the same judgement as any other kind of pseudoscience. Thus the set of accusations designed by the Inquisition to remove heretics from the moral community, in order that they might be tortured and killed, can be clearly seen as an SF, and one which has been of cardinal importance in the history of fantasy and OCCULT FANTASY. Another set of SFs which has had a major influence on literary fantasy is that dealing with ATLANTIS and other LOST LANDS AND CONTINENTS.

Fantasy inevitably makes avid use of SFs as PLAYGROUNDS, borrowing wherever convenient and rarely hesitating to invent. [BS]

Further reading: *The Natural History of Nonsense* (**1947**) by Bergen Evans; *In the Name of Science* (**1952**; exp vt *Fads and Fallacies in the Name of Science*) and *Science: Good, Bad and Bogus* (coll **1981**), both by Martin Gardner; *Can You Speak Venusian?: A Guide to the Independent Thinkers* (**1972**; rev 1976) by Patrick Moore; *Cults of Unreason* (**1973**) by Dr Christopher Evans; *The New Apocrypha* (**1973**) by John T. Sladek; *A Dictionary of Common Fallacies* (**1978**; exp 1980 2 vols) by Philip Ward; *A Directory of Discarded Ideas* (**1981**) by John GRANT; *Facts and Fallacies: A Book of Definitive Mistakes and Misguided Predictions* (**1981**) by David LANGFORD and Chris MORGAN; *Science and the Paranormal* (anth **1981**) ed G. Abell and B. Singer; *Ancient Astronauts, Cosmic Collisions, and Other Popular Theories about Man's Past* (**1984**) by William H. Stiebing Jr; *Pseudoscience and the Paranormal: A Critical Examination of the Evidence* (**1987**) by T. Hines.

SCHOONOVER, FRANK (1877-1972) US artist. ◊ ILLUSTRATION.

SCHWEITZER, DARRELL (CHARLES) (1952-   ) US writer, editor and critic, best-known in the latter category, especially in the SMALL-PRESS field, to which he has been a prolific contributor since 1970. He has also been an editorial assistant on ASIMOV'S SCIENCE FICTION MAGAZINE 1977-82, *Amazing Stories* 1982-86, and joint (later sole) editor on WEIRD TALES from 1977. He is currently editor of *WT*'s spiritual successor, WORLDS OF FANTASY AND HORROR.

DS works in many modes, but the overriding mood is of DARK FANTASY. His stories have echoes from writers like Lord DUNSANY and Clark Ashton SMITH. This was evident from his first professional sale "The Story of Obbok" (1973 *Whispers*). All of DS's work shows a delight in the use of language and creation of mood, sometimes at the expense of STORY. This melancholy is evident in the **Sir Julian** stories – about a medieval KNIGHT who is an ACCURSED WANDERER – collected as *We Are All Legends* (coll of linked stories **1981**). *The White Isle* (1975 *Weirdbook*; exp 1980 *Fantastic*; rev dated 1989 but **1990**) is about a prince who, like ORPHEUS, tries to bring his wife back from the Land of the Dead.

DS's first main body of work, the **Goddess** sequence, begins with "The Story of a Dadar" (**1982** *Amazing Stories*) and includes *The Shattered Goddess* (**1982**). It is set on a DYING EARTH after the death of the GODDESS, whose remains are kept as a holy relic. The story plots the rise of a WITCH who takes POSSESSION of a series of bodies in order to attain ultimately the body of the Goddess. She can be defeated only through the heroic SACRIFICE of one of her own creations. The novel has immense power in its climax and again sustains the forlorn mood of loss.

DS's other main series are the **Great River** stories, begun with "The Last of the Shadow Titans" (1985 *Amazing Stories*) and including the novel *The Mask of the Sorcerer* (as

"To Become a Sorcerer" 1991 *WT*; exp **1995** UK). The setting is the dawn of the world, when the GODS still interact with humankind. Evoking the mood of ancient EGYPT, the novel tells of a son's QUEST for his lost sister and his travels among the LANDSCAPES of death and legend. As before, DS explores the growing power of individuals and the inevitable sacrifice that leads to transcendence and loss of mortality.

DS's work does not lend itself well to HUMOUR, though his best attempts, the **Tom O'Bedlam** stories – begun with "Tom O'Bedlam's Night Out" (1977 *Fantastic*) are of note; some are collected in *Tom O'Bedlam's Night Out and Other Strange Excursions* (coll **1985**), which also includes selections from the **Goddess** and **Great River** series. *Transients and Other Disquieting Stories* (coll **1993**) contains generally more disturbing fantasies, including some of DS's modern horror tales. Most of DS's short fiction remains uncollected. [MA]

Other works: *The Meaning of Life and Other Awesome Cosmic Revelations* (coll **1988** chap); *Non Compos Mentis: An Affrontery of Limericks and Other Eldritch Metrical Terrors* (coll **1995** chap) verse.

Nonfiction: *Lovecraft in the Cinema* (**1975** chap); *The Dream-Quest of H.P. Lovecraft* (**1978** chap); *Conan's World and Robert E. Howard* (**1978** chap); *On Writing Science Fiction: The Editors Strike Back* (anth **1981**) with George Scithers and John M. Ford; *Constructing Scientifiction & Fantasy* (**1982** chap) with George Scithers and John Ashmead; *Pathways to Elfland: The Writings of Lord Dunsany* (**1989**); *Lord Dunsany: A Bibliography* (**1993**) with S.T. Joshi.

Interviews: *SF Voices* (coll **1976**); *Science Fiction Voices #1* (coll **1979** chap); *Science Fiction Voices #5* (coll **1981** chap); *Speaking of Horror* (coll **1994**).

As editor: *Essays Lovecraftian* (anth **1976**; rev vt *Discovering H.P. Lovecraft* anth 1987); *The Ghosts of the Heaviside Layer* (coll **1980** US) by Lord DUNSANY; *Exploring Fantasy Worlds* (anth **1985**); *Discovering Modern Horror Fiction* (anth **1985**); *Discovering Stephen King* (anth **1985**); *Discovering Modern Horror Fiction II* (anth **1988**); *Tales from the Spaceport Bar* (anth **1987**) and *Another Round at the Spaceport Bar* (anth **1989**) both with George Scithers (◊ TALL TALES); *Discovering Classic Horror I* (anth **1992**).

SCIENCE FANTASY In the 1950s Judith Merril (1923-   ) advocated the use of this term as consistent with her sense that SCIENCE FICTION could be understood as an aspect of FANTASY as a whole. In *Critical Terms for Science Fiction and Fantasy: A Glossary and Guide to Scholarship* (**1986**), Gary K. Wolfe (1946-   ) suggests that the term, defined as tightly as possible, "refers to a genre in which devices of fantasy are employed in a 'science-fictional' context (related to but distanced from the 'real world' by time, space, or dimension)".

Both HEROIC FANTASY and RATIONALIZED FANTASY texts have often been called SF, and both frequently underpin their worlds through explanations which draw on history (as in Robert E. HOWARD's **Conan** series) or science (as in the elaborate explanations of how MAGIC works in Poul ANDERSON's *Operation Chaos* [coll of linked stories **1971**]). In the works of authors like Leigh BRACKETT, Andre NORTON and Marion Zimmer BRADLEY – and many others – PLANETARY-ROMANCE venues can tend to take on FANTASYLAND airs very swiftly.

Perhaps the most useful application of the term is to DYING-EARTH tales – like those by Jack VANCE, Michael

MOORCOCK and Gene WOLFE – in which fantasy-like tales are told in venues not necessarily very well understood by the characters involved. It is here that SF and fantasy as a whole are closest: to invoke historical or scientific explanations for happenings that seem magical is likely to arouse a sense of TIME ABYSS, because these explanations will almost certainly refer back to wisdom or science that has been forgotten for aeons. In Dying-Earth tales history and science are like the stories which underlie most fantasy narratives, and when protagonists in such tales discover what makes their worlds tick, the effect is less that of conceptual breakthrough than of RECOGNITION. [JC]

See also: TECHNOFANTASY.

SCIENCE FANTASY UK digest MAGAZINE (#1-#3 large digest; pocketbook format from June/July 1964), 81 issues, Summer 1950-February 1966, quarterly until May 1954, then bimonthly to March 1965; title hyphenated for issues #1-#6; published by Nova Publications, London, until April 1964, thereafter by Roberts & Vinter, London; ed Walter Gillings (1912-1979) Summer 1950-Winter 1950/51, E.J. CARNELL Winter 1951/52-April 64, Kyril Bonfiglioli (1928-1985) June/July 1964-February 1966.

SF began as an sf companion to *New Worlds*, but once both came under the editorship of Carnell it began to paint on a wider canvas and, from 1956, its content became primarily fantasy. The best exponents were John BRUNNER – whose stories focused on the intrusion of supernatural forces into our own REALITY (◊ WRONGNESS), as in "The Kingdoms of the World" (1957) – and Kenneth BULMER, who produced a number of CROSSHATCH and SECONDARY-WORLD fantasies involving THRESHOLDS – such as "The Map Country" (1961) and the **Watkin's World** series beginning with "The Seventh Stair" (1961 as Frank Brandon). By 1961 *SF* was becoming more critically acclaimed than its companion, and it was nominated in 1962, 1963 and 1964 for the Hugo AWARD. The December 1960 issue was a special "Weird-Story" issue. *SF*'s peak came during 1961-4, when it published the first **Elric** stories by Michael MOORCOCK, the mythological and historical fantasies of Thomas Burnett SWANN – starting with "Where is the Bird of Fire?" in the April 1962 special "Fantasy" issue – and a number of challenging psychological fantasies by J.G. BALLARD, whose first story *SF* published. *SF* also issued two rare short stories by Mervyn PEAKE and the first story by Terry PRATCHETT.

When the magazine changed hands in 1964 it took a while to establish a new identity; only the writings of Keith ROBERTS, especially his **Anita** series, gave it any cohesion. After the February 1966 issue *SF* folded, but immediately re-emerged from the same publishers and still (briefly) under Bonfiglioli's editorship as *Impulse* (12 issues, March 1966-February 1967); after August 1966 it was retitled *SF Impulse*, with later issues ed Harry Harrison and Keith Roberts. *Impulse* was basically a continuation of *SF*, but the name-change and the reversion in numbering (to start with *Impulse #1*) confused retailers and distributors, and the magazine lost financially. *Impulse* was, nevertheless, a superior magazine, the highspot being Roberts's **Pavane** sequence; it also published early work by Chris Boyce (1943-   ), Christopher PRIEST and Brian STABLEFORD, and some wonderfully nightmarish fiction by Thomas M. DISCH. [MA]

SCIENCE FANTASY (US magazine) ◊ FANTASTIC.

SCIENCE FANTASY YEARBOOK (US magazine) ◊ FANTASTIC.

SCIENCE FICTION One of the genres of the FANTASTIC, a term which also encompasses FANTASY, SUPERNATURAL FICTION and supernatural HORROR. Readers, writers and critics are more secure in their understanding of the nature of sf than when they attempt to define fantasy, but the entry on Definitions of Science Fiction in the *SFE* and the entry on Science Fiction in *Critical Terms for Science Fiction and Fantasy* (1986) by Gary K. Wolfe (1946-   ) amply demonstrate that this security is only relative. In this encyclopedia the term science fiction (sf) normally refers to a genre separate from but overlapping with fantasy. By contrast with fantasy – which can be quickly defined as a body of self-coherent but impossible narratives – the label sf normally designates a text whose story is explicitly or implicitly extrapolated from scientific or historical premises. In other words, whether or not an sf story is plausible it can at least be *argued*. At its best sf contains a Sense of Wonder analogous to fantasy's RECOGNITION or SLINGSHOT ENDING. [JC]

SCIENCE FICTION CLASSICS ◊ FANTASTIC; FANTASTIC ADVENTURES.

SF IMPULSE ◊ SCIENCE FANTASY (magazine).

SCITHERS, GEORGE H(ARRY) (1929-   ) US writer and editor. ◊ John Gregory BETANCOURT; Darrell SCHWEITZER; TALL TALES.

SCLIAR, MOACYR (1937-   ) Brazilian writer and physician, born to a Jewish family. This background is reflected in his writing. His novels are usually mainstream; it is in his short stories that he has become associated with the FANTASTIC, through tales that can be read as FABLES or parables, often with a sardonic edge and a bitter humour about human nature. These stories are mostly collected in *O Carnaval dos Animais* (coll 1968; trans Eloah F. Giacorelli as *The Carnival of the Animals* 1985 US), *Histórias da terra trêmula* ["Tales of the Trembling Earth"] (coll 1976), *O olho enigmático* (coll 1986; trans as *The Enigmatic Eye* 1988 US) and *A orelha de Van Gogh* ["Van Gogh's Ear"] (coll 1989). *O centauro no jardim* (1980; trans Margaret A. Neves as *The Centaur in the Garden* 1984 US) tells of a Jewish family that flees the European pogroms for Brazil, where their centaur-like child is raised. [BT]

Other works: *The One-Man Army* (trans Eloah F. Giacomelli 1986 US), first novel; *Os Deuses de Raquel* (1978; trans Giacomelli as *The Gods of Raquel* 1986 US); *Max e os Felinos* (coll 1981; trans Giacomelli as *Max and the Cats* 1990 chap US); *The Strange Nation of Rafael Mendes* (trans Giacomelli 1988 US).

SCOT, MICHAEL Joint pseudonym of Michael Scott ROHAN and Allan SCOTT.

SCOTLAND, JAY [s] ◊ John W. JAKES.

SCOTT, ALLAN (JAMES JULIUS) (1952-   ) UK writer (half-Danish) whose first novel was *The Ice King* (1986 vt *Burial Rites* 1987 US) with Michael Scott ROHAN, published as by Michael Scot. This mingles NORDIC FANTASY with routine HORROR as a modern archaeological team's explorations of a sunken Viking ship off the Yorkshire coast release *draugar* – Icelandic undead, resembling both VAMPIRES and ZOMBIES, who can recruit new *draugar* from their kills and summon the FIMBULWINTER. AS's solo *The Dragon in the Stone* (1991) is a more distinguished novel, again exploring Nordic themes: light and savagely dark ELVES (who attack via POSSESSION), a PORTAL leading from modern Denmark through an OTHERWORLD and across TIME, a version of BEOWULF's fight with Grendel made darker by the fact that this Grendel is a needed GUARDIAN OF THE

THRESHOLD and was once human (◊ TRANSFORMATION), and the fatalistic sense of coming RAGNAROK. Eventually, at high cost, a complex knot in time satisfyingly emerges. *A Spell of Empire: The Horns of Tartarus* (**1992**) with Rohan, originally plotted as a GAME tie, offers an interesting but under-exploited ALTERNATE-WORLD setting with a German/Scandinavian/Hun "Nibelung Empire"; the action alternates between lighthearted swashbuckling and encounters with CHAOS-tainted MONSTERS; a sequel may follow.   [DRL]
Other works: *The Hammer and the Cross* (**1980**) with Rohan, nonfiction on the Christianization of the Vikings.

SCOTT, JEREMY Pseudonym of Kay DICK.

SCOTT, MELISSA (1960-   ) US writer who has been publishing sf since her first novel, *The Game Beyond* (**1984**), but who has written two fantasies of interest, both with Lisa A. Barnett (1958-   ). *The Armor of Light* (**1988**) is set in an ALTERNATE-WORLD 16th-century Scotland, whose king is threatened by a conspiracy of WITCHES; John DEE figures in the complex action. *Point of Hopes* (**1995**) occurs in a world with two suns and whose nature is shaped – and the destinies of whose inhabitants are governed – by the configurations of the skies as interpreted by ASTROLOGY, which in this world is a respectable science. MS and Barnett make clever use of their premise, positing a culture where occupations and other roles are determined by individuals' charts rather than by (for instance) their GENDER – by indirection, and without undue foregrounding, a sharp feminist lesson is conveyed. The plot itself is a somewhat cumbersome detection; the leisurely telling allows for much FANTASY-OF-MANNERS detailing of the workings of society.   [JC]

SCOTT, MICHAEL (1959-   ) Irish writer who has spe-cialized in both HORROR and fantasy since beginning to publish with *The Song of the Children of Lir* (**1983** chap; vt *The Children of Lir: An Irish Legend* 1986 chap UK), a YA CELTIC FANTASY in TWICE-TOLD mode, a tale also recounted in *Irish Folk and Fairy Tales* (coll **1984** UK), a volume of retold traditional material. Along with *Irish Folk and Fairy Tales #2* (coll **1984** UK) and *#3* (coll **1984** UK), this collec-tion was assembled as *Irish Folk & Fairy Tales Omnibus* (omni **1989** UK). MS's first sequence, the **Tales from the Land of Erin** – *A Bright Enchantment* (**1985** UK), *A Golden Dream* (**1985** UK) and *A Silver Wish* (**1985** UK) – carries on in the same manner, as do *The Last of the Fianna* (**1987** UK) and *The Quest of the Sons* (**1988** UK). The **Tales of the Bard** – *Magician's Law* (**1987** UK), *Demon's Law* (**1988** UK) and *Death's Law* (**1989** UK) – more daringly utilizes FOLK-LORE to enrich the story of the QUESTS of an Irish BARD, whose life follows a euhemerist course towards his eventual role as legend and embodiment of the old times before the New Religion threatened a universal THINNING.

MS's later work tends to horror, and includes *Banshee* (**1990** UK), *Image* (**1991** UK), *Reflection* (**1992** UK), *Imp* (**1993** UK), *Gemini Game* (**1993** UK) and *The Hallows* (**1995** UK). The last effectively marries genres, deploying horror tropes – a series of violent deaths decimates the Guardians responsible for the eponymous sacred artefacts from Ancient Britain – in a fantasy context.   [JC]
Other works: *Navigator: The Voyage of Saint Brendan* (**1988** UK) with Gloria Gaghan; *Wind Lord* (**1991**).

SCOTT, [Sir] WALTER (1771-1832) Scottish poet and novelist who grew up fascinated with the Scottish and English border ballads, particularly those involving the supernatural, which interest culminated in *The Minstrelsy of the Scottish Border* (coll **1802-3** 3 vols). His first substantial

original work, the narrative poem *The Lay of the Last Minstrel* (**1805** chap), is indebted to medieval ROMANCES; it involves the pranks of a GOBLIN page who steals the SPELL-book of the legendary WIZARD Michael Scot. WS wrote further narratives, mainly with historical settings (but including the forgettable Arthurian romance [◊ ARTHUR] *The Bridal of Triermain* [**1813** chap]), until his popularity was eclipsed by Lord BYRON, and he turned to writing historical novels. Although there are supernatural elements in several of these – e.g., WELAND SMITH appears in *Guy Mannering* (**1815**) – only *The Monastery* (**1820**) can fairly be called a fan-tasy, and that only through the character of the White Lady, a FAIRY who is half-banshee-protectress and half-PUCK. Supernatural elements elsewhere in WS's novels are usually rationalized.   [LH]

SCRAP BOOK, THE ◊ MAGAZINES.

SCRIBBLIES In January 1980 in Minneapolis a group of fans and writers – Stephen BRUST, Nathan Bucklin, Emma BULL, Kara DALKEY, Pamela DEAN, Will SHETTERLY and Patricia WREDE – formed themselves into a loose association, the "Interstate Writers' Workshop (IWW), better known as the Scribblies", on the model of the Industrial Workers of the World (IWW). Additionally (so the tale runs) the group called itself The Scribblies in remembrance of the Duke of Gloucester's reported comment to Edward Gibbon (1737-1794): "Another damned, thick, square book! Always scribble, scribble, scribble! Eh! Mr Gibbon?"

The Scribblies have never published a manifesto and have never claimed an agenda beyond mutual support and friend-ship; but for two reasons they do comprise an AFFINITY GROUP: under the general editorship of Bull and Shetterly, they created and wrote most of the **Liavek** SHARED-WORLD series of ANTHOLOGIES; and many of their most significant publications have been CONTEMPORARY FANTASY, frequently URBAN FANTASY, and arguably written under the influence of the early work of Peter S. BEAGLE (◊◊ FANTASY OF MANNERS).

The Scribblies' claim to be agenda-free has become diffi-cult to sustain since their creation of the loose movement called the Pre-Joycean Fellowship (PJF), as per the PRERAPHAELITE Brotherhood. Writers who designate them-selves (often on the title pages of their books) members of the PJF do in fact give their names to a recognizable posi-tion: an anti-elitist and anti-Modernist posture that so-called High Literature has lost its way, its humour and its natural audience. In 1996 the Scribblies remain active.[JC]

SCROOGE McDUCK US COMICS character, sometimes rendered as $crooge McDuck, created by Carl BARKS in "Christmas on Bear Mountain" (1947) for a one-shot DON-ALD DUCK comic. Scrooge, Donald's uncle, rapidly became more popular in the DISNEY comics than Donald himself; he also picked up along the way Donald's nephews Huey, Dewey and Louie as COMPANIONS as he embarked on ever more fantasticated adventures. Barks's **Scrooge** tales are now generally regarded as major achievements of 20th-century fantasy, although Disney was slow to realize the artist/writer's genius. (◊ Carl BARKS for bibliography.) The essence of Scrooge is that he is the richest duck in history, having hoarded billions which he is unwilling to spend – especially not the first dime he ever earned, which has become a kind of fetish to him. A frequent enemy is the sorceress Magica DeSpell, who has her eyes on that very dime.

Scrooge's first screen experience was in the animated tv educational featurette *Scrooge McDuck and Money* (1967).

His second was in *Mickey's Christmas Carol* (**1983**; ◊ *A* CHRISTMAS CAROL), where he played the part of Charles DICKENS's miser. He was the central character in the tv series *DuckTales* (from 1987) and its offshoot DUCKTALES: THE MOVIE – TREASURE OF THE LOST LAMP (**1990**). [JG]

SCROOGE MOVIES ◊ *A* CHRISTMAS CAROL.

SCRYING Properly the art of seeing distant or future images in a crystal ball, scrying has come to be a loose generic term for visual PROPHECY and farseeing using MIRRORS, fires, pools, smoke or any reflective surface. (Roger ZELAZNY's *Creatures of Light and Darkness* [**1969**] uses the term even more loosely, to describe haruspication – divination using entrails.) In GENRE FANTASY scrying is often no more than a rather inefficient, soundless equivalent of the sf "vidscreen"; more imaginatively employed, it becomes a vehicle not only for images but for personality and magical overtones. The showings of Denethor's *palantír* or scrying-stone in J.R.R. TOLKIEN's *The Lord of the Rings* (**1954-5**) are manipulated by the DARK LORD to emphasize his power and induce despair; John CROWLEY's *Aegypt* (**1987**) opens with a scrier's numinous VISION of ANGELS; the Magic Mirror in DISNEY's SNOW WHITE AND THE SEVEN DWARFS (**1937**) is probably today the most famous scrying focus of all. [CB/DRL]

SEAL MEN ◊ SELKIES.

SEA MONSTERS The fact that very large creatures like whales and giant squid (krakens) live UNDER THE SEA makes the deeps a logical home for MONSTERS. Greek MYTHOLOGY features not only the human-sized SIRENS but the many-headed Scylla (a former girl who has suffered METAMORPHOSIS and later menaces ODYSSEUS) and the vast SM from whom Perseus saves Andromeda – a battle closely echoed by Orlando's epic defeat of the Orc in ARIOSTO's *Orlando Furioso* (**1516**), in turn re-echoed in "To Here and the Easel" (1954) by Theodore STURGEON. The BIBLE speaks of the DRAGON in the sea and asks, "Canst thou draw out Leviathan with an hook?" – a feat which celebrated angler Izaak Walton (1593-1683) very nearly achieves in the eccentric "God's Hooks!" (1982) by Howard Waldrop (1946-   ). Leviathan is often identified as a whale, though Kipling's *Just So Stories* (coll **1902**) distinguishes between whales and far huger undersea creatures. Seeming ISLANDS may prove to be whales (or krakens) which plunge underwater when incautious visitors light a fire, as in one voyage of Sinbad (◊ ARABIAN FANTASY). The most famous of fictional whales is the eponymous albino of Herman MELVILLE's *Moby-Dick* (**1851**) – rivalled in popular appreciation by Monstro in DISNEY's PINOCCHIO (**1940**). A staple of TRAVELLERS' TALES is the sea SERPENT, depicted with great circumstantial realism in Rudyard KIPLING's "A Matter of Fact" (1892); one tries to crush the eponymous ship with its coils in C.S. LEWIS's *The Voyage of the Dawn Treader* (**1952**). The truly monstrous, mountain-sized undersea foes Abaia and Erebus in Gene WOLFE's *The Book of the New Sun* (**1980-83**) are never seen, despite encounters with the submarine female GIANTS who are Abaia's concubines. In Terry GILLIAM's TIME BANDITS (**1981**) a submarine monster proves to be a human-style GIANT, who wears the protagonists' new-gained ship as a hat. [DRL]

SÉANCE A method of contacting the spirit world through a human medium, who draws power from a group of individuals who link hands to maintain the chain. Mediums, previously called scryers (◊ SCRYING), came into prominence with the rise of interest in SPIRITUALISM during the second half of the 19th century. Spiritual manifestation may arise through ectoplasm. The séance itself is a potent image for spiritual contact and is sometimes the starting point in GHOST STORIES for contacting the dead, or unleashing a SPIRIT, which may result in POSSESSION – in fact during the séance the medium is possessed, though inactive, serving as a channel. In the most effective stories the séance unleashes a dangerous and powerful force, as in "Playing With Fire" (1900 *Strand*) by Arthur Conan DOYLE, "The Demon Spell" (1894) by Hume NISBET, "The Last Séance" (1926) by Agatha Christie (1890-1976), "The Gatecrasher" (1971) by R. CHETWYND-HAYES, and at the outset of *The List of 7* (**1993**) by Mark FROST. Mediums are often accused of charlatanism, and even before the end of the Victorian era the séance had become a cliché, subject to ridicule. But even this approach can be handled effectively by good authors as in "Mr Tilly's Séance" (1923 *Hutchinson's*) by E.F. BENSON and especially *Blithe Spirit* (**1941**) by Noel Coward (1899-1973). *Séance on a Wet Afternoon* (**1961**; vt *Séance* 1962 US) by Mark McShane (1930-   ), filmed as *Séance on a Wet Afternoon* (**1964**), is an effective PSYCHOLOGICAL THRILLER of the mental disintegration of a fraudulent medium who becomes convinced she has contacted her dead son. [MA]

SEARLE, RONALD (WILLIAM FORDHAM) (1920-   ) UK cartoonist with a biting satirical wit and an incisive line style. His work has an ebullient humour offset by a rather macabre elongation and distortion of the human form.

He studied at the Cambridge School of Art and served in WWII, being interned in a Japanese POW camp, where he produced some harrowing drawings, published as *Ronald Searle's Secret Sketchbook* (graph **1970**). After WWII he worked in London as a humorous illustrator and caricaturist for *Punch* and wrote and illustrated many humorous books, including the popular **St Trinians** series and James THURBER's *The 13 Clocks and The Wonderful O* (omni **1962** UK). In 1960 he settled in France, since when he has written and illustrated several books about Paris. His work is basically of comedy rather than fantasy interest. [RT]

SEASONS Most fantasy authors lived or live in the temperate zones. Most of their stories, whether set here or in some OTHERWORLD, naturally tend to replicate or exaggerate the climatic pattern of these zones where, from time immemorial, four seasons have been identified. For many fantasy writers, moreover, these four seasons are deeply significant for reasons that transcend geographical circumstance; and in non- or post-CHRISTIAN FANTASIES, which dominate the field, the four seasons are normally conceived as aspects of an everlasting CYCLE, one in which TIME returns upon itself, bringing gifts. A sense of the central importance of cycle in any understanding of Nature, history, humanity, religion and world – a sense basic to archaic religions in Mircea ELIADE's convincing description and analysis of faiths other than Judaism, Christianity and Islam – therefore permeates much modern fantasy, where any reference to the seasons is likely to carry with it a burden of significance, whether or not precisely articulated.

The iteration of the seasons into specific festivals is a most complex task, and most fantasy texts focus on only the most prominent of these days of high significance. A primal instinct of fantasy writers – which is to attempt to recover original or "true" archetypal (◊ ARCHETYPES) patterns – normally ensures that seasonal festivals (like Easter) are depicted and understood in ways which uncover and value what is deemed to be the true STORY. Easter (for instance) is more likely to be presented as the climax of the grave erotic dance between the GODDESS and her consort (◊◊ GOLDEN BOUGH) than as the climax of the essentially linear story of CHRIST's Passion. Seasonal festivals invoked by English-speaking writers of fantasy include Plough Sunday (and the other three agricultural days: Rogationtide, Lammas and the Harvest), Easter, May Day, MIDSUMMER NIGHT, HAL-LOWE'EN (also known as All Hallow's Eve or All Souls' Day; ◊◊ SAMHAIN), CHRISTMAS and New Year's Day.

When the seasons appear in a fantasy text without a specific iconic signification it is common for them to reflect – as a form of metaphysical pathos (◊ John RUSKIN) – the events taking place in the story. Any season, suitably described, can represent any stage in the unfolding of a full fantasy. A desolate Spring can represent THINNING; a fruc-tifying Winter can just as easily tell the reader that HEALING is underway. The normal impulse, however, is to exploit the usual vegetative cycle (a cycle which in any case subtends most archaic understanding of the meaning of the world) and to represent thinning through images of Autumn and Winter, RECOGNITION through moments (like the REVEL attendant upon New Year's Day) that invoke through images of METAMORPHOSIS the turning of the year, HEALING through Spring and PASTORAL – and the GENRE FANTASIES set in FANTASYLAND whose storylines are pastoral in the sense that they do not threaten the overall world with fur-ther change – through Summer. [JC]

SECONDARY BELIEF ◊ J.R.R. TOLKIEN.

SECONDARY WORLD Term coined by J.R.R. TOLKIEN in his seminal essay, "On Fairy Tales" (first delivered 1939; in *Essays for Charles Williams* anth **1947** ed anon C.S. LEWIS; exp in *Tree and Leaf* coll **1964**; rev 1988) for a particular kind of otherworld. An SW can be defined as an autonomous world or venue which is not bound to mundane reality (as are many of the domains common to SUPERNATURAL FIC-TION), which is impossible according to common sense and which is self-coherent as a venue for STORY (i.e., the rules by which its REALITY is defined can be learned by living them, and are not arbitrary like those of a WONDERLAND can be). All FANTASYLANDS are forms of the SW; the main distinction is that a fantasyland is a TEMPLATE venue inherently resistant to change, but no such repudiation of the possibility of METAMORPHOSIS, arguably essential to any definition of the fully structured fantasy, is implied by the use of the term SW. [JC]

SECOND SIGHT ◊ TALENTS.

SECRET ADVENTURES OF TOM THUMB, THE UK live-action/ANIMATED MOVIE (*1993*). Bolex Brothers/BBC Bristol/La Sept/Manga/Lumen. **Pr** Richard "Hutch" Hutchison. **Exec pr** Colin Rose. **Dir** Dave Borthwick. **Screenplay** Borthwick. **Starring** Deborah Collard (Tom's mother), Nick Upton (Tom's father). **Voice actors** Andrew Bailey, Marie Clifford, Tim Hands, Brett Lane, John Schofield, Peter Townsend, Upton, Helen Veysey, Paul Veysey (movie has no proper credits). 60 mins. Colour.

Among the most bizarre of all contributions to the cinema of TECHNOFANTASY. In an artificial-insemination plant a fly (the movie is obsessed with insects) falls into a flask; when the consequent baby is born he proves, to his parents' aston-ishment, to be minute, and so they name him Tom Thumb. Agents of a sinister genetic-engineering "laboratorium" raid the home and seize Tom. In the laboratorium, which is full of grotesques, Tom is subjected to painful tests. A mutant

reptilian frees him so that he may, at their behest, shut down the life-support systems of many of those grotesques. The two liberators escape down a waste-disposal chute, at whose other end they find a WAINSCOT tribe of tiny people who live by scavenging the toxic-waste discharge. These folk kill the mutant, but Tom is adopted by their militaris-tic leader, JACK, a giant-killer. After many adventures, Tom is reincarnated (◊ REINCARNATION) as a normal baby; but, as his father and mother coo over him, the father, looking up, sees their shared SHADOW on the wall forms a Christian nativity grouping. A group of flies forms a circle, creating a halo for the infant CHRIST.

The animation used is almost exclusively stop-motion; fascinatingly, all of the (quite extensive) live action is shot using jerky camera techniques to give the sensation that it, too, is stop-motion animated (so that live actors become, in effect, TOONS). There is virtually no dialogue, most com-munication being in the form of electronically distorted grunts and half-formed words – a stratagem that further emphasizes the impression that everything is animated. Because we are never allowed to forget that this is an ani-mation, technical inadequacies become strengths rather than failings. The quest for meaning in the tale is less rewarding; yet overall the movie succeeds brilliantly in its aim of unsettling. [JG]

See also: GREAT AND SMALL.

SECRET EMPIRE, THE US tv series (1979). Universal/ NBC. **Pr** Richard Milton, Paul Samuelson, B.W. Sandefur. **Exec pr** Kenneth Johnson. **Dir** Alan Crosland, Johnson, Joseph Pevney. **Writer** Johnson. **Created by** Johnson. **Starring** Pamela Brull (Princess Maya), Peter Breck (Jess Keller), Stephanie Kramer (Tara), Mark Lenard (Emperor Thorval), Diane Markoff (Princess Tara), David Opatoshu (Hator), Geoffrey Scott (Marshal Jim Donner), Carlene Watkins (Millie Thomas). **Voice actor** Brad Crandall (Narrator). 10 15min episodes. Colour.

Set in Arizona, 1880, this segment of the series CLIFFHANGERS begins with with US Marshal Donner dis-covering a society living in the UNDERGROUND city of Chimera, ruled by evil dictator Thorval. Aided by traitorous humans, Thorval plans to conquer Earth through mind control. Interestingly, all the surface scenes were filmed in b/w, the underground scenes in colour. Unfortunately, the series was cancelled before its outcome was revealed. [BC]

SECRET GARDEN Derived from *The Secret Garden* (**1911**) by Frances Hodgson BURNETT, the term refers to a private world which becomes something of a personal Paradise sur-rounded by oppressive reality. In Burnett's novel the place is a walled garden amid the bleak Yorkshire moors; the gar-den, imbued with the youthful zeal of the children, becomes a place of recovery and regeneration. An SG is thus any place of escape or retreat that provides a personal haven for the protagonist. The SG may be a WAINSCOT or POLDER in our own world, or in an OTHERWORLD or even elsewhere in TIME. It became a standard MOTIF in CHILDREN'S FANTASY, where most SGs are literally gardens – *Tom's Midnight Garden* (**1958**) by Philippa PEARCE, *The Time Garden* (**1958**) by Edward EAGER, *The Castle of Yew* (**1965** chap) by Lucy M. BOSTON and the garden at Chrestomanci Castle in *Charmed Life* (**1977**) by Diana Wynne JONES. It was in recognition of the SG as a key motif in children's fiction that Humphrey Carpenter (1946-   ) titled his study of the golden age of children's literature *Secret Gardens* (**1985**).

In ADULT FANTASY SGs are a similar retreat from everyday

life, though may not necessarily prove to be such a haven. Lionel Wallace's recollection of his enchanted garden has fatal results in "The Door in the Wall" (1906 *Daily Chronicle*) by H.G. WELLS. **Landover** in Terry BROOKS's *Magic Kingdom for Sale – Sold!* is almost an SG; the rooftops of **Gormenghast** become an SG for Steerpike in *Titus Groan* (**1946**) by Mervyn PEAKE, and one could argue that SHANGRI-LA is a remote and virtually inaccessible SG in *Lost Horizon* (**1933**) by James HILTON. [MA]

SECRET GUARDIANS SGs protect society or the world as a whole; they are, unusually, an idea as likely to appear in CONTEMPORARY FANTASY as in SWORD AND SORCERY. They are usually to be distinguished from the SECRET MASTERS of THEOSOPHY, whose influence is by precept and example and subtle spiritual influence rather than by active intervention. They are also to be distinguished from GODS, who are usually invulnerable if things go wrong; SGs are always in some degree of jeopardy themselves.

By definition, SGs are an element intrinsically associated with FANTASIES OF HISTORY, because in any context in which they occur they automatically make a statement about the nature of consensual reality. It is explicit that the order of WIZARDS in J.R.R. TOLKIEN's *The Lord of the Rings* (**1954-5**) are at one and the same time SGs and ANGELS, and have guarded Middle-Earth for millennia. The Benandanti in Elizabeth HAND's *Waking the Moon* (**1994**) have guarded humanity against the malignant aspects which they have confused with the true nature of the GODDESS since before recorded history.

SGs sometimes guard deliberately and sometimes by merely existing – in Central European Jewish legend, there are at any given time 12 just men, the Lamed Wufniks, whose virtue is the keystone of the Universe. SGs sometimes forget their purpose, as has happened with the Antaeus Brotherhood in Tim POWERS's *The Anubis Gates* (**1983**). They are not always especially competent - the League of the Scroll in Tad WILLIAMS's **Memory, Sorrow and Thorn** (**1988-93**) is manipulated by its DARK LORD opponent into potentially destructive decisions - and are sometimes almost as ruthless in their workings as the EVIL they oppose - the Diogenes Club in Kim NEWMAN's *Anno Dracula* (**1993**) sacrifice endless pawns in order to put a silver knife in the same place as the reluctantly vampirized Queen Victoria and make possible her liberatory suicide. Groups that believe themselves the authentic SGs of a society, but are not, such as the Children of Light in Robert JORDAN's **The Wheel of Time** sequence (**1990** onwards), are often a fertile source of DEBASEMENT and WRONGNESS in their world, whether or not they are being secretly manipulated by the dark lords or their allies.

SGs are often a HIDDEN MONARCH and his retinue – like Aragorn and the other rangers in *LOTR*. They are sometimes ACCURSED WANDERERS – e.g., Indigo in Louise COOPER's **Indigo** sequence (**1988-93**) travels from location to location destroying the demons she unwittingly awoke and freeing societies from their machinations. The culmination of an UGLY DUCKLING's or a BRAVE LITTLE TAILOR's career may be to be recruited to membership of a group of SGs.

The secrecy surrounding their work may make them a WAINSCOT group or even a PARIAH ELITE. They will often live, or deliberate, in a POLDER where they are safe and hidden. SEVEN-SAMURAI or DIRTY-DOZEN groupings will often be recruited by them, or discover that they have been working for them unknowingly for some time. SGs are the standard opponents of dark lords and potential dark lords. [RK]

SECRET HISTORY OF THE WORLD ◊ FANTASIES OF HISTORY.

SECRET LIFE OF WALTER MITTY, THE US movie (*1947*). RKO/Goldwyn. **Pr** Samuel Goldwyn. **Dir** Norman Z. McLeod. **Spfx** John Fulton. **Screenplay** Ken Englund, Everett Freeman. **Based on** "The Secret Life of Walter Mitty" (1932 *New Yorker*) by James THURBER. **Starring** Fay Bainter (Eunice Mitty), Boris Karloff (Dr Hugo Hollingshead), Danny Kaye (Walter Mitty), Virginia Mayo (Rosalind van Hoorn), Ann Rutherford (Gertrude Griswold), Konstantin Shayne (Peter van Hoorn/Krug). 108 mins. Colour.

Meek, mother-dominated pulp-magazine editor Mitty, to whom nothing ever seems to happen – his imminent marriage to Gertrude and her obnoxious poodle Queenie promises to bring more of the same – survives because of his DREAMS of fantastic adventure, triggered by everyday events; e.g., a headline about an RAF fighter ace makes Mitty likewise, shooting down Germans in between enlivening the officers' mess with his brilliant impersonation acts. At the end of each dream he walks, alone and full of pathos, back into the real world. In all of them the same beautiful woman appears in one role or another; then, in the mundane REALITY, Mitty encounters her (Rosalind) aboard his commuter train and plunges straight into a real-life jewel caper. *TSLOWM* travesties Thurber's story, where Mitty's utterly mundane life was counterpointed by his exotic fantasy existence. Considered as a Danny Kaye movie and in no other context, *TSLOWM* entertains for a while, but long outstays its welcome. [JG]

SECRET MASTERS A potent attraction of FANTASIES OF HISTORY is that they depict the seeming chaos of events as purposeful, often reflecting the guiding will of SMs. Typical "historical" choices of SM include: the ILLUMINATI; the Knights Templar, whose assumed omnipresent influence pervades Umberto ECO's *Foucault's Pendulum* (**1988**); followers of ROSICRUCIANISM; the German Vehmgericht or Holy Vehme, as in John WHITBOURN's *Popes and Phantoms* (**1993**); and Freemasons, like the decaying Masonic splinter-orders guarding PORTALS to the Universe-spanning LABYRINTH in Avram DAVIDSON's *Masters of the Maze* (**1965**). Although the SM concept is essentially paranoid, many examples are benevolent: the company of Logres in C.S. LEWIS's *That Hideous Strength* (**1945**), the Old Ones of Susan COOPER's **The Dark Is Rising** quintet, and the Ring of WITCHES in Diana Wynne JONES's *A Sudden Wild Magic* (**1992**). History may be seen as a by-product of conflicts between rival SMs, like the four animal-coded groups in R.A. LAFFERTY's *Fourth Mansions* (**1969**). Further examples include: the Noisy Bridge Rod and Gun Club which controls the USA in John CROWLEY's *Little, Big* (**1981**); the History Monks keeping Discworld events on course in Terry PRATCHETT's *Small Gods* (**1992**); and, according to a conspiracy theorist in some of Tom HOLT's comic fantasies, the British Milk Marketing Board. [DRL]

See also: PARIAH ELITE; SECRET GUARDIANS; WAINSCOT.

SECRET OF NIMH, THE US movie (*1982*). ◊ Don BLUTH.

SECRET SHARER A phrase adopted from "The Secret Sharer" (1910 *Harper's Magazine*) by Joseph CONRAD to describe that PLOT DEVICE where a character is either a split personality or has some other secret INVISIBLE COMPANION (◊◊ DOPPELGÄNGER; DOUBLES; IDENTITY EXCHANGE; SHADOW). Some examples, such as that of Joan of Arc (1412-1431), who heard "voices", may be regarded as either divine intervention or POSSESSION. Other cases – e.g., that of Bridie Murphy – may be seen as MULTIPLE PERSONALITIES or REINCARNATIONS. The device is often used in stories featuring CHILDREN who have an invisible playmate, of which DROP DEAD FRED (*1991*) is an extreme example. The device can be used for humour, as in HARVEY (*1950*), but is more often either sinister, as in "Cellmate" (1947 *WT*) by Theodore STURGEON, where a man is imprisoned with an unseen Siamese twin (◊ TWINS), or sympathetic, as in *Kehinde* (**1994**) by Buchi Emecheta (1944-   ), where the protagonist communes with the spirit of her stillborn twin sister, or "The Drawing Lesson" (1987) by Silvina Ocampo (◊ Adolfo BIOY CASARES), where the protagonist discusses her life with her younger self. The term was borrowed by Robert SILVERBERG for his sf novel *The Secret Sharer* (**1988**), in which a starship captain is possessed by a SPIRIT. DOLLS may become tangible projections of an SS, particularly in stories of VENTRILOQUISM. [MA]

SEERS ◊ PROPHECY.

SEGRELLES, VINCENTE (1940-   ) Spanish painter and illustrator whose book covers and eerie, exotic fantasy COMIC strips have received wide acclaim on at least three continents. He works on a large scale in oils; for his comic strips these paintings are varnished and cut to size, then mounted as individual frames in the sequence.

Nephew of famous painter/illustrator of fantasy and nightmare, José Segrelles Albert (1885-1969), VS painted from his youth. After studies at the Escuela de Apredidices de ENASA, he began his career as a graphic designer, working for advertising agencies. In 1970 he decided to dedicate himself solely to ILLUSTRATION, and contacted the publisher Ediciones AFHA, for whom he began to paint book covers. International recognition came in 1980 with the publication of *El Mercenario* (trans as *The Mercenary* in *Heavy Metal #56-#60* 1981-2; graph coll **1985** US). Further volumes in this series followed, comprising *La Formula* ["The Formula"], *Las Pruebas* ["The Trials"], *El Sacrificio* ["The Sacrifice"] and *La Forteleza* ["The Fortress"]. VS refers to these atmospheric stories, set in a misty, mountaintop world, full of ARABIAN-FANTASY motifs, as "fantasia logica" [logical fantasy], in which all aspects of the imaginary scenario are consistently envisaged. In 1989 he began a fruitful collaboration with Ray BRADBURY in *El Dragon* ["The Dragon"]. [RT]

Further reading: *Great Masters of Fantasy Art* (graph **1985** US) by Eckart Sackmann (though some biographical information incorrect).

SEHNSUCHT Term used by C.S. LEWIS to describe a yearning "desire for something that has never actually appeared in our experience". Lewis presented this version in "The Weight of Glory" (in *The Weight of Glory and Other Addresses* coll **1949** US) and went on to argue that *Sehnsucht* is not a longing that pertains to the natural world or the categories by which we respond – however deeply – to that world. It has nothing to do with beauty, or "sublimated eroticism"; it is not a synonym for the Pathetic Fallacy; it is not a projection. It is a longing for HEAVEN.

As a Christian, Lewis meant by "heaven" something like the presence or the immanence of God; and he claimed that *Sehnsucht* represents an actual calling of God to the human soul, a "wind of Joy" – a divine communication. The LAND-

SCAPE Lewis seems to visualize when recording his sense of *Sehnsucht* – a landscape often found in NORDIC FANTASY – is no more than dressing: *Sehnsucht* calls *through* the garb of landscape. Lewis would not, therefore, have used the term to describe sensations perceived in the mind's eye by a secular human reader when touched by a purely imaginary wind out of FAERIE, although that is the sense most often used in this encyclopedia.

As such, *Sehnsucht* is analogous to but actually contradicts any nostalgic sense of EDEN or of a lost GOLDEN AGE, a nostalgia most expressively rendered in the phrase ET IN ARCADIA EGO. [JC]

SELDEN, GEORGE Working name of US writer George Selden Thompson (1929-1989). His CHILDREN'S FANTASIES include *The Garden Under the Sea* (**1957**; vt *Oscar Lobster's Fair Exchange*) and *The Genie of Sutton Place* (**1973**), about a boy, a genie and a transformed dog; but he is most noted for his **Chester, Harry and Tucker** series, begun with *The Cricket in Times Square* (**1960**). This well loved ANIMAL FANTASY concerns the adventures of Chester Cricket, who is transported in a picnic hamper to New York, where he meets the streetwise Tucker Mouse and Harry Cat. The animals'-eye view of the great city is engagingly drawn. Affiliated books in the loose series, all well illustrated by Garth WILLIAMS, are *Tucker's Countryside* (**1969**), *Harry Cat's Pet Puppy* (**1974**), *Chester Cricket's Pigeon Ride* (**1981**), *Chester Cricket's New Home* (**1983**) and *Harry Kitten and Tucker Mouse* (**1986**). [DP]

SELKIES Legendary seal-people of Scottish FOLKTALE, also called silkies or the roane (Gaelic, "seal"). All large species of seal are supposedly selkies, capable of SHAPESHIFTING to their true forms as FAIRIES or humans, often by casting off their seal-skins (◊◊ SKINNED); they may interbreed with humans, producing chimerical offspring with webbed hands and feet. Examples of such human/seal shapeshifters appear in Larry NIVEN's "The Lion in His Attic" (1982), whose "lion" is a sealion, Susan COOPER's *Seaward* (**1983**) and Gordon R. DICKSON's *The Dragon Knight* (**1990**). [DRL]

SELLWOOD, A.V. (?   -   ) UK writer. ◊ Peter HAINING.

SENDAK, MAURICE (BERNARD) (1928-   ) US artist, illustrator and storyteller, of Polish descent. He began illustrating children's books with the US edition of *The Wonderful Farm* (1951 US) by Marcel AYMÉ, and has continued to produce his idiosyncratic ILLUSTRATIONS for new editions and translations of classic texts. His first picture book with his own text was *Kenny's Window* (graph **1956** chap), about a boy who has a DREAM in which he is set several impossible tasks. *Very Far Away* (graph **1957** chap) is a recognition of a child's need for an OTHERWORLD where it can escape unbearable reality. MS's reputation, however, took off with *Where the Wild Things Are* (graph **1963** chap). A boy, banished to his room for misbehaving, escapes to a self-created imaginary world, the world of the Wild Things, whose king he becomes. MS returned to the imaginary-world theme with *In the Night Kitchen* (graph **1970** chap), in which a young boy has adventures in an all-night bakery, and the more sinister *Outside Over There* (graph **1981** chap), derived from "The Goblins" by the GRIMM BROTHERS; the latter, about a young girl who must find her baby sister, stolen by GOBLINS, helped inspire LABYRINTH (*1986*). [MA]

Other works: *The Sign on Rosie's Door* (graph **1958** chap); *Hector Protector and As I Went Over the Water* (graph coll **1965** chap); *Higglety Pigglety Pop!, or There Must Be More to Life* (graph **1967** chap).

Other works illustrated: *Seven Tales* (coll **1959** US) by Hans Christian ANDERSEN, trans Eva Le Gallienne; *The Griffin and the Minor Canon* (graph **1963** chap) and *The Bee-Man of Orn* (graph **1964** chap) by Frank R. STOCKTON; *The Golden Key* (**1967**) and *The Light Princess* (**1969**) by George MACDONALD; and two books derived from Grimm, *The Juniper Tree and Other Tales* (coll **1973**) trans Lore Segal and Randall Jarrell, and *King Grisly-Beard* (**1973** chap) trans Edgar Taylor.

Further reading: *The Art of Maurice Sendak* (**1980**) by Selma G. Lanes; *Pipers at the Gates of Dawn: The Wisdom of Children's Literature* (**1983**) by Jonathan Cott.

SENECA, LUCIUS ANNAEUS (*c*4BC-AD65) Spanish-born Latin man of letters and Roman politician, ostensible author of the only 10 surviving Roman tragedies, none dated. Some are preoccupied with tyranny and with Stoic ethics; several offer killings and (in *Medea*'s case) MAGIC onstage; most are exaggerated in language and action and pointed in style. Some have reasonably coherent plots, but most are unified more by the outlook of HORROR, in which Seneca is the first known specialist and, though too little survives from his predecessors to judge properly, an immensely influential one. He taught the Renaissance tragedians of Western Europe much, including the use of GHOSTS and PORTENTS, the taste for blood, and the fascination with moral monsters. At his best – most sustainedly in *Hercules Furens*, *Thyestes* and *Medea* – Seneca remains powerfully disturbing to read today.

His compressed style, SHAKESPEARE's shadow and the Classicists' disdain combine to impede translation. *Four Tragedies and Octavia* (anth **1966**) trans E.F. Watling, *Three Tragedies* (coll **1986**) trans Frederick Ahl and *Seneca II* (anth **1995**) ed David R. Slavitt between them include all but *Agamemnon*, and offer several approaches to the task. [JB]

Other tragedies: *Troades* ["Trojan Women"]; *Phoenician Women*; *Phaedra*; *Oedipus*; *Hercules Oetaeus*, possibly spurious; *Octavia*, possibly spurious.

Other work: *Apocolocyntosis* (*c*54), an intermittently amusing POSTHUMOUS FANTASY.

See also: GREEK AND LATIN CLASSICS.

SENSE OF WONDER ◊ SCIENCE FICTION; SLINGSHOT ENDING; TIME ABYSS.

SENSIBLE MAN A recurring character type who (often as viewpoint character) comfortingly embodies rationality and common sense – but may therefore attract some irony when uncommon sense rules. Watson in the SHERLOCK HOLMES sequence is a determined SM, steadfast but unable to follow Holmes's leaps of genius. The SM hero of Hope MIRRLEES's *Lud-in-the-Mist* (**1926**) is guyed and disgraced by anarchic forces of FAERIE, but wins through by doggedly insisting on what is right, even to the brink of death. In DUO sequences like Fritz LEIBER's **Fafhrd and the Gray Mouser**, the SM role can alternate as one partner balances the other's feyness or excess; CLUB-STORY narrators like Lord DUNSANY's **Jorkens** affect a stolid SM persona as ballast for their outrageous TALL TALES. The eponymous SM in Joanna Russ's "The Man who Could not See Devils" (1970) has no PERCEPTION of and thus no vulnerability to omnipresent DEMONS and GHOSTS – a harbinger of THINNING. Stephen DONALDSON's Covenant in the **Chronicles of Thomas Covenant the Unbeliever** is a kind of PARODY of the SM, forced by the rules of his own existence into a shocking rejection of the SECONDARY WORLD's very reality. [DRL]

SEQUELS BY OTHER HANDS Fantasy STORIES normally convey a sense that surface tales are retellings of archetypal material. This sense of the TWICE-TOLD is, of course, much intensified whenever an easily identifiable nexus of story is being drawn upon; the network of interlocked MYTHS and LEGENDS and tales which come together in, for instance, the MATTER of Britain (◊◊ ARTHUR) is intricately interwoven, and many late versions of central material resemble SBOHs. However, the term is more sensibly used in a more restricted sense, one that also excludes, for example, SHARED-WORLD enterprises, stories "set in the universe of" some prior text, PARODIES, RECURSIVE FANTASIES and REVISIONIST FANTASIES. The SBOH proper continues a previous story, one almost certainly written by a named author. Some SBOHs are written after copyright has lapsed on the original work, but a sizeable proportion are written by agreement with the owner(s) of the original work.

Prequels by other hands – i.e., stories set before the beginnings of previously written tales – are less often found, but the same principles apply.

Parodies have been common since long before fantasy became an identifiable genre, and usually – like *He* (**1887**) by Andrew LANG and Walter Herries Pollock (1850-1926), which parodies H. Rider HAGGARD's *She* (**1887**) – repeat in transmogrified form the action of the original. The first SBOH of fantasy interest may be a continuation of Charles MATURIN's *Melmoth the Wanderer* (**1820**) by Honoré de BALZAC, "Melmoth Reconciled" (1835). Further examples (out of many): include Angelo Patri's *Pinocchio in America* * (**1928**), a sequel to Carlo COLLODI's *Pinocchio* (**1883**); T.H. WHITE's *Mistress Masham's Repose* * (**1946** US), adding to Jonathan SWIFT's *Gulliver's Travels* (**1726**), as does the **Antelope Company** sequence by Willis Hall (1929-   ), which novelizes his tv series, itself a sequel; *Eric Brighteyes 2: A Witch's Welcome* * (**1979**) by Mildred Downey BROXON (as Sigfridur Skaldaspillir), a sequel to Haggard's *Eric Brighteyes* (**1891**); *Ten Years Beyond Baker Street: Sherlock Holmes Matches his Wits with the Diabolical Dr Fu Manchu* * (**1984**) and *The Fires of Fu Manchu* * (**1987**) by Cay Van Ash (1918-1994), sequels to Sax ROHMER's FU MANCHU sequence (◊◊ SHERLOCK HOLMES); Gilbert ADAIR's *Alice Through the Needle's Eye: A Third Adventure for Lewis Carroll's "Alice"* * (**1984**) and *Peter Pan and the Only Children* * (**1987**), respectively sequelling Lewis CARROLL's **Alice** books and J.M. BARRIE's PETER PAN; *Neverland* (**1989**) by Toby Forward, another **Peter Pan** sequel, done for the Great Ormond Street Children's Hospital, London, the beneficiary of Barrie's own royalties and of part of the income from a further **Peter Pan** SBOH, Steven SPIELBERG's HOOK (*1991*); *Return to Shangri-La* * (**1987**) by Leslie Halliwell (1929-1989), a sequel to James HILTON's *Lost Horizon* (**1933**); Helen CRESSWELL's *The Return of the Psammead* * (**1992**), sequelling E. NESBIT's *Five Children and It* (**1902**); William HORWOOD's *The Willows in Winter* * (**1993**) and *Toad Triumphant* * (**1995**), sequels to Kenneth GRAHAME's *The Wind in the Willows* (**1908**); and Susan HILL's *Mrs De Winter* * (**1993**), a sequel to Daphne DU MAURIER's *Rebecca* (**1938**).

Tales which present the continuing adventures of a series character in further adventures, none necessarily connected to any previous plot, include additions to: Arthur Conan DOYLE's SHERLOCK HOLMES series; L. Frank BAUM's OZ books; the TARZAN series by Edgar Rice BURROUGHS; and the CONAN series by Robert E. HOWARD.

For many years, Philip José FARMER, in his **Wold Newton**

Family sequence, has been evolving an over-narrative through which many of the most familiar UNDERLIER figures of fantasy and sf are presented as members of the same family. Figures from Herman MELVILLE's *Moby-Dick* (**1851**) are linked to Tarzan, Sherlock Holmes, Doc Savage and others; and some of Farmer's texts may read as SBOHs. This seems, however, a special case, as is the family tree of movie figures constructed by David Thomson (1941-) in *Suspects* (**1985**). [JC]

SERGEANT PEPPER'S LONELY HEARTS CLUB BAND US ANIMATED MOVIE (*1978*). ◊ The BEATLES.

SERIAL KILLERS Real SKs represent a late-20th-century nadir of human stupidity, and thus are the antithesis of fantasy. Fictional SKs who are carriers of the toxins of HORROR have some small place here. Also, whether we like it or not, the image of the SK has gained iconographic status well outside horror, as in the PSYCHOLOGICAL THRILLER.

DRACULA is a serial killer, and most books or movies about VAMPIRES are about serial killing. The paradigm SK is of course JACK THE RIPPER, who is rarely a figure of fantasy interest, though some version of the man can be found in GASLIGHT ROMANCES like Kim NEWMAN's *Anno Dracula* (**1992**). JACK is a paradigm figure because unseen, indulgent in splatterpunk excesses of sadism, uncaught, has a rough TRICKSTER wit (as has Hannibal Lecter in *The Silence of the Lambs* [**1988**] by Thomas Harris [1940-]) and is chased fruitlessly by the law. Barbara HAMBLY's *Those Who Hunt the Night* (*1988*) is dependent on these characteristics. Other SK novels with some fantasy interest include James HERBERT's *Moon* (**1985**), Dean R. KOONTZ's *The Bad Place* (**1990**), Simon R. GREEN's *Shadows Fall* (**1994**) and Poppy Z. BRITE's *Exquisite Corpse* (**1996**). The *Hellblazer* comic (1988-) by Jamie DELANO and others, with **John Constantine** as created by Alan MOORE, features encounters with SKs.

Movies which make play with SKs are frequent. Those of fantasy interest include the KOLCHAK MOVIES, DON'T LOOK NOW (*1973*), EYES OF LAURA MARS (*1978*); CAT PEOPLE (*1982*), the NIGHTMARE ON ELM STREET series, LADY IN WHITE (*1988*), BARTON FINK (*1990*), NIGHTBREED (*1990*) and CANDYMAN (*1992*). [JC]

SERLING, ROD Working name of US screenwriter, tv producer and author Edward Rodman Serling (1924-1975), best-known for his tv work (from 1952), gaining early fame for *Requiem for a Heavyweight* (1956); filmed as *Requiem for a Heavyweight* (*1963*), and winning six Emmies over his career. His first series was perhaps his finest: The TWILIGHT ZONE (1959-64) anthology series, which gained three Hugo AWARDS (1960-62), deftly dramatized sf/fantasy scripts, often in a SLICK-FANTASY mode, by RS and from contributors like Charles BEAUMONT, Ray BRADBURY and Richard MATHESON. RS also hosted the show (a practice then common on US anthology series), and his lean and hungry telegenic mien soon became iconic. He was an emcee of the UNCANNY, and of THRESHOLDS. A second series, ROD SERLING'S NIGHT GALLERY (1970-73) was less successful.

RS was not a major writer of printed fiction. *The Season to be Wary* (coll **1967**), his only collection of independent stories, is not strong. His adaptations of scripts written for his two series are stronger, though he made little attempt to translate them into autonomous prose fictions. **Twilight Zone** work is assembled in *Stories from The Twilight Zone* * (coll **1960**), *More Stories from The Twilight Zone* * (coll **1961**), *New Stories from The Twilight Zone* * (coll **1962**), the latter two apparently ghostwritten by Walter B. Gibson

(1897-1985) from RS scripts – selections from all three were assembled as *From The Twilight Zone* * (coll **1962**), and all three in their entirety as *Stories from The Twilight Zone* * (omni **1986**). *Stories from the Twilight Zone* * (graph coll **1979**) is a COMICS version of the material. Gibson definitely ghostwrote two further collections: *Rod Serling's The Twilight Zone* * (coll **1963**) and *Rod Serling's Twilight Zone Revisited* * (coll **1964**), both assembled as *Rod Serling's Twilight Zone* * (omni **1984**). The **Night Gallery** work appears in *Night Gallery* * (coll **1971**) and *Night Gallery 2* * (coll **1972**).

RS's widow, Carol Serling (1929-), has edited three further spinoffs: *Rod Serling's Night Gallery Reader* * (anth **1987**) with Martin H. GREENBERG and Charles G. Waugh (1943-), *Return to the Twilight Zone* (anth **1994**) with Greenberg and *Adventures in the Twilight Zone* (anth **1995**). She also served as a founding consultant for ROD SERLING'S THE TWILIGHT ZONE MAGAZINE (1981-9). [JC]

Other works: *Rod Serling's Triple W: Witches, Warlocks and Werewolves* (anth **1963**) and *Rod Serling's Devils and Demons* (anth **1967**), both ghost-edited by Gordon R. DICKSON; *Rod Serling's Other Worlds* (anth **1978**), which may also have been ghost-edited.

See also: The NEW TWILIGHT ZONE (1985-8).

SERNINE, DANIEL Working name of Canadian writer Alain Lortie (1955-). ◊ CANADA.

SERPENTS Snakes and serpents have long been emblems of EVIL and stealth, thanks to their silence, sinister leglessness and venom; but the sloughing and regeneration of their skins can also signify renewal, as noted in George Bernard SHAW's *Back to Methuselah* (**1921**) and Avram DAVIDSON's *The Phoenix and the Mirror* (**1969**). In Norse MYTH, the Midgard serpent coils around the world (◊ WORM OUROBOROS) and symbolically around the WORLD-TREE Yggdrasil, whose roots (in some accounts) it gnaws. GORGONS have snakes for hair, and snakes are recurring agents of death in Greek myths: e.g., ORPHEUS's Eurydice died of a snake-bite. The Aztec PANTHEON features Quetzalcoatl the Feathered Serpent, who is incarnated in Kenneth MORRIS's *The Chalchiuhite Dragon* (**1992**); Australian Aboriginal tales of the DREAMTIME include the Rainbow Serpent. The serpent of EDEN (◊ SATAN), celebrated from the BIBLE and John MILTON's *Paradise Lost* (**1667**), reappears in many guises throughout CHRISTIAN FANTASY – for example, taking over a human body in C.S. LEWIS's *Perelandra* (**1943**); a woman suffers similar POSSESSION by a serpent ARCHETYPE in Charles WILLIAMS's *The Place of the Lion* (**1931**). Geoff RYMAN's REVISIONIST FANTASY treatment in *The Warrior who Carried Life* (**1985**) identifies the tempter serpent with Adam (◊ ADAM AND EVE), who was punished with this TRANSFORMATION. Woman/serpent SHAPESHIFTERS (◊ LAMIAS) are almost invariably malign, as in Bram STOKER's *The Lair of the White Worm* (**1911**) and C.S. Lewis's *The Silver Chair* (**1953**); some legends place LILITH in this category. The perceived WRONGNESS of a woman's being also a phallic snake is given a perverse twist in Philip José FARMER's *Blown* (**1969**), where an attached, symbiotic serpent-creature resides in one female character's vagina. It is the ultimate DEBASEMENT when Elric's wife is made into a serpent-bodied chimera in Michael MOORCOCK's *Stormbringer* (**1965**); Greg BEAR's *The Serpent Mage* (**1986**) features the eponymous transformed and debased WIZARD, inhabiting Loch Ness. The more naturalistic python, Kaa, of Rudyard KIPLING's *The Jungle Book* (coll **1894**) possesses the power of MESMERISM often ascribed to snakes. Huge serpents are common "encounter MONSTERS" in FANTASY-LAND and in GAMES; one appears in 7 FACES OF DR LAO (*1964*). That in Paula VOLSKY's *Illusion* (1991) is, unusually, a TECHNOFANTASY machine (◊ AUTOMATA).   [DRL]

See also: DRAGONS; SEA MONSTERS; WORM/WYRM.

SERVICE, PAMELA F. (1945-   ) US writer, primarily of YA fantasy and sf. As with much YA fiction, the distinction between sf and fantasy in PFS's work is haphazard: her first books, the **Winter** sequence – *Winter of Magic's Return* (1985) and *Tomorrow's Magic* (1987) – take place in a post-HOLOCAUST, but involve ARTHUR; while *Weirdos of the Universe, Unite!* (1992) offers figures from MYTHOLOGY battling an alien invasion. Other works that partake of fantasy include the GHOST STORY *When the Night Wind Howls* (1987), *The Reluctant God* (1988), *Vision Quest* (1989), *Wizard of Wind and Rock* (1990) and *Storm at the End of Time* (1995).   [GF]

SETH, VIKRAM (1952-   ) Indian writer. ◊ AESOPIAN FANTASY; BEAST-FABLE.

SETON, ERNEST (EVAN) THOMPSON (1886-1946). Canadian writer. ◊ CANADA.

SEUSS, DR Pseudonym of US writer Theodore Seuss Geisel (1904-1991), who also wrote as Theo. LeSieg. With dozens of his verse CHILDREN'S FANTASIES perpetually in print, he may be the most widely read fantasy writer of modern times. The ingredients of a typical DS book are lively doggerel verse, humorous drawings of innumerable IMAGINARY ANIMALS, and narratives which impart a clear but unobtrusive moral.

A few of his early books feature settings related to standard fantasy. In *To Think that I Saw It on Mulberry Street* (1937), the first, a boy imagines amazing sights to report to his father. In *The 500 Hats of Bartholomew Cubbins* (1938) a boy who tries to remove his MAGIC hat in front of a king problematically finds that another hat appears; this character reappears in *Bartholomew and the Oobleck* (1949), where the king's desire for something new to fall from the sky disastrously leads to a rain of green goo. In *The King's Stilts* (1939) a king's loss of his stilts almost ruins his kingdom. Other distinctive works include: *Horton Hatches the Egg* (1940) and *Horton Hears a Who* (1954), about a loyal elephant who helps, respectively, a vacationing mother bird and the minuscule inhabitants of a dust speck; *McElligot's Pool* (1947), about a boy who envisions SEA MONSTERS; *If I Ran the Zoo* (1950) and *If I Ran the Circus* (1956), where a boy wishes to recruit strange animals, like a ten-footed lion and an Elephant-Cat, to replace the standard zoo and circus offerings; *How the Grinch Stole Christmas* (1957), perhaps his most famous work, an enjoyable homage to Charles DICKENS's *A Christmas Carol* (1843) (a noteworthy animated tv version [1966] featured the voice of Boris Karloff); and *Dr Seuss's Sleep Book* (1962), which achieves a certain epic grandeur in describing how the many odd inhabitants of DS's universe go to sleep. A few later books drift into political issues: *The Lorax* (1971) is a fable about pollution; *Marvin K. Mooney, Will You Please Go Now!* (1972) was interpreted by many as an attack on Richard Nixon; and *The Butter Battle Book* (1984) inspired protests for implying that the competition between the USA and the USSR was equivalent to a dispute over which side of the bread should be buttered (recalling Jonathan SWIFT's Lilliputians's dispute over which end of the egg to open).

At first opposed by librarians and educators for his anarchic irreverence, DS was recruited for a famous series of **Beginner Books**, designed to teach children to read, beginning with *The Cat in the Hat* (1957), where a playful cat causes havoc in the home of two bored children; the character reappeared in *The Cat in the Hat Comes Back* (1958) and other books. DS wrote further books for the series. Many other children's authors have been inspired by DS's example. His list of other works is very long indeed.   [LL]

Further reading: *Dr Seuss from Then to Now: A Catalogue of the Retrospective Exhibition* (1986); *Dr Seuss*, by Ruth K. MacDonald (1988).

SEVEN Since the time of the ancient Babylonians and Greeks the number 7 has signified completeness. The Heavens consisted of 7 spheres (Sun, Moon, Mars, Venus, Mercury, Jupiter, Saturn). The Earth was created in 6 days and God rested on the 7th. To the alchemists there were 7 basic metals: tin, copper, iron, mercury, lead, silver, gold. The Seven Seas represented the totality of the oceans, and there were the Seven Deadly Sins and the Seven Ages of Man. The number was of great significance to cabbalist and hermetic philosophies (◊ CABBALA), particularly the Seven Seals of *Revelations* (◊ BIBLE). The occultist Eliphas Levi (real name Alphonse Louis Constant; 1810-1875) stated that "the number seven represents magical power in all its fullness; it is the mind reinforced by all elementary potencies, it is the soul served by Nature". Thus the use of 7 is often significant of power or perfection. Its magical import means the number arises more often in FANTASY than in SUPERNATURAL FICTION, other than OCCULT FANTASY, where *The List of 7* (1993) by Mark FROST is a recent example. It is well known from our childhood FAIRYTALES, particularly "Snow White and the Seven Dwarfs", 7-league boots, and the significance of the 7th son of a 7th son, this last used recently in *Seventh Son* (1987) by Orson Scott CARD. In the MABINOGION's story of *Culhwch and Olwen* (11th century) Culhwch is set 7 impossible tasks by the giant Ysbaddaden. This concept has been taken to its extreme in the **Seven Citadels** sequence by Geraldine HARRIS, where the novel's entire plot revolves around the impact of 7. In *At the Back of the North Wind* (1871) by George MACDONALD Diamond stays in the land of the North Wind for only 7 days, though it seems like as many years (◊ TIME IN FAERIE). In a similar vein Wintersland, in Joan AIKEN's play *Winterthing* (1972 chap US), disappears every 7th year. The symbolism and usage of 7 in fantasy fiction are frequent and often predictable.   [MA]

See also: 7 FACES OF DR LAO (*1964*); SEVEN SAMURAI; *The SEVENTH SEAL* (*1956*).

7 FACES OF DR LAO US movie (*1964*). MGM. **Pr** George Pal. **Dir** Pal. **Vfx** Paul B. Byrd, Wah Chang, Jim Danforth, Robert R. Hong, Ralph Rodine. **Mufx** William Tuttle. **Magic advisor** George L. Boston. **Screenplay** Charles BEAUMONT. **Based on** *The Circus of Dr Lao* (1935) by Charles G. FINNEY. **Starring** Barbara Eden (Angela Benedict), John Ericson (Ed Cunningham), Arthur O'Connell (Clint Stark), Tony Randall (Dr Lao/Apollonius of Tyana/Medusa/Merlin/Pan/Snowman). **Voice actor** Randall (Giant Serpent). 100 mins. Colour.

The small Midwest town of Abalone is on the skids. Cynical magnate Stark hopes to buy out the entire town, knowing the railroad will soon be built through it. Newcomer Cunningham, editor of the *Abalone Daily Star*, campaigns against him, but is supported by few except young widowed librarian Benedict – who brusquely rejects his romantic advances. To Abalone comes, alone on a jack-ass, the TRICKSTER Dr Lao, proclaiming his travelling CIRCUS. In fact, it is more of a CARNIVAL, with acts including Apollonius of Tyana, MERLIN, Medusa, PAN, the Abominable Snowman, a giant SERPENT and a catfish in a goldfish bowl; his claim that this is a Loch Ness Monster (◊ SEA MONSTERS) is widely derided. Dr Lao – ostensibly Chinese though displaying characteristics (and accents) of many races – is a sort of WANDERING-JEW figure: he is "7322 years old next autumn", but seems immortal (◊ IMMORTAL-ITY). He immediately pitches in on the side of the goodies; early on he befriends Benedict's young son Mike (the part goes uncredited – weirdly, since the role is central, and superbly played). On the first night Lao in his varying guises reveals unwelcome truths to various of the town's citizens. On the second (and last) night Lao displays to the assembled citizens the fall of a mythic city, which met its fate because of its citizens' greed. Abalone's citizens decide not to sell out to Stark, who thanks them for saving him from himself. Whenever you see the marvellous in life, Lao's voice-over reminds us in the finale, you become a part of the Circus of Dr Lao.

7FODL has many flaws, yet remains one of the core movies of fantastic CINEMA. Randall played at least seven parts and was unlucky not to receive an Oscar; one did go to Tuttle for the mufx. Nevertheless, 7FODL was largely a forgotten movie outside fantasy fandom until quite recently, when it became realized that its considerable strengths far outweighed its weaknesses. [JG]

SEVEN SAMURAI A term used here to designate a gathering of COMPANIONS, usually in a HEROIC-FANTASY venue, who have come together voluntarily in order to further a goal, and who frequently stay together after that goal has been accomplished. The number may not be SEVEN, but surprisingly many fantasy novels do in fact feature groups of seven, as in Lloyd ALEXANDER's *The Book of Three* (**1964**), by the end of which the protagonist, Taran, is accompanied by an ENCHANTRESS, a DWARF, a Gollum-figure, a BARD, a TALKING ANIMAL and the oracular pig. In «One for the Morning Glory» by John Barnes (1957-   ) the protagonist dithers over beginning to act until he has acquired six companions.

The essence of the SS grouping is that it is *voluntary*, that its goals are not simply self-concerned; unlike the case with a DIRTY DOZEN, there is no hierarchy (i.e., no foul-mouthed "sergeant" has to kick the group into shape); SS groups are not mercenaries, though their individual members may have been and may afterwards return to that trade. They have probably come together at the behest of a central figure, who himself or herself has a QUEST to obey, or a POLDER (perhaps a village) to defend; or they may unite to defend some other person – or perhaps the LAND itself – from a plight. Group members tend to display an assortment of TALENTS, among which military prowess ranks high (◊ MILITARY FANTASY); they may constitue a PARIAH ELITE, and in that guise defend the old ways, attempt to protect their land from, say, THINNING or the depredations of a DARK LORD; one of them will almost certainly be an UGLY DUCK-LING (possibly a HIDDEN MONARCH); one may be a MAGUS, also possibly in disguise; in recent decades, one may well be a woman, either in GENDER DISGUISE or, increasingly, not. SS groupings mimic chivalry.

There is no element of the supernatural or fantasy in *Shichi-nin no Samurai* (**1954**; vt *The Seven Samurai*),

probably the greatest movie by Akira Kurosawa (1910-   ), but the detail and sweep of his tale make it an almost inescapable model for any group of volunteers who fight for the good. The long sequences devoted to the actual recruitment of the seven have inspired opening chapters in many fantasy adventures; and the epic defence of the peasants' village – a task for which they have volunteered knowing they will probably not be paid, and may well die fulfilling – has served later moviemakers and fantasy writers as a model of sustained, organized bravery. [JC]

SEVENTH SEAL, THE (ot *Det Sjunde Inseglet*) Swedish movie (*1956*). Svensk Filmindustri. **Pr** Allan Ekelund. **Dir** Ingmar BERGMAN. **Screenplay** Bergman, published as *The Seventh Seal* * (**1957**; trans Lars Malmstrøm and David Kushner **1960** UK). **Starring** Bertil Anderberg (Raval), Bibi Andersson (Mia), Gunnar Björnstrand (Jöns), Bengt Ekerot (Death), Inga Gill (Lisa/Kunigunda), Maud Hansson (Witch), Inga Landgré (Karin), Gunnel Lindblom (The Woman), Nils Poppe (Jof), Erik Strandmark (Jonas Skat), Max von Sydow (Antonius Block). 95 mins. B/w.

The Christian KNIGHT Antonius Block returns with his atheistic squire Jöns after 10 years at the Crusades to find his homeland scourged by the Black Death and concomitant religious frenzy. The figure of DEATH comes to take him ("I have long walked beside you"), but Block persuades him to postpone his errand until after the two have played a game of CHESS, with Block being given his life should he win; the game both punctuates and *is* the next 24 hours, which is the scope of the movie. Block at one stage makes confession to a priest, whom he finally realizes is in fact Death, coaxing out of him his next chess move – but not before he has admitted his wish to die and his personal view of GOD ("We have to hew an image out of our fear [of death] and call it God"), and stated that his reason for begging the respite is so that he can commit a single meaningful act before his life ends.

The plot of *TSS* is not easily summarized. In its own terms it forms a perfect structure, with FATE performing its dance from the start to the rhythm of the chess game, while the tensions between the movie's opposites – between GOOD AND EVIL, between God and SATAN, between faith and atheism – are maintained and expertly manipulated through dialogue, action and performance alike (often surprisingly comic). There can be no disputing *TSS*'s visual power: it has been a trove for later moviemakers, including parodists (e.g., BILL & TED'S BOGUS JOURNEY [*1991*] and LAST ACTION HERO [*1993*]): particularly memorable are a procession of flagellants preaching the END OF THE WORLD and the final enactment of the DANCE OF DEATH (according to Bergman this latter was impromptu, with production crew filling in for those of the cast who had left for the day). *TSS* is arguably Bergman's masterpiece. [JG]

SEVERN, DAVID Pseudonym of UK writer David Storr Unwin (1918-   ), who wrote some adult work as David Unwin but has as DS concentrated on stories for children and YA readers, beginning with *Rick Afire!* (**1942**). Relatively little of his work is fantasy. *Dream Gold* (**1949**) tells of two boys whose identical DREAMS lead them to ISLANDS in the Pacific where they encounter PIRATES; *Drumbeats!* (**1953**) is a TIMESLIP tale set in Africa; *The Future Took Us* (**1957**) is sf; *The Girl in the Grove* (**1974**) is a GHOST STORY in which a young girl ghost creates a love triangle with two contemporary adolescents; in *The Wishing Bone* (**1977**) a TOY, grown to lifesize, forces captured children to make one WISH each, a task dauntingly difficult to accomplish without disaster. [JC]

SEX As a traditional engine of plot, sex has been central in the literatures of the West from the beginning. It is the direct cause of actions later taken, as in HOMER's *Iliad* (*c***850**BC) and many of the classic detective novels of the 20th century; it is the subject matter of the STORY, as in STORY CYCLES like Giovanni BOCCACCIO's *Decameron* (before **1353**), or in the love-romances of this century; it is the hidden substance that governs the surface shenanigans of the courtly ROMANCE, or the machinations that irradiate the MATTER behind MYTHS OF ORIGIN (◊◊ ARTHUR), or the seductions of SUPERNATURAL FICTION, where a sexual charge between the supernatural realm and this world is central, whether overt or hidden, justified or seen as miscegenation. The underlying language of HORROR, too, is suffused with images and gestures of sexual violation, and is the main carrier into the 20th century of the linkage between sex and the violating Other typical of GOTHIC FANTASY. However, though FANTASY also depends ultimately for its energy on sexual flux, the genre has traditionally avoided plots overtly based on sexual relations.

In recent years, of course, the politics of sex have affected fantasy, like all other popular genres. Before recent decades, any woman capable of acting a protagonist's role would likely be seen as appalling at the very least; more probably, she would be represented as a victim of POSSESSION, or an AVATAR of the dark side of the GODDESS (◊◊ LAMIA; SHE). Exceptions – like C.L. MOORE's tomboy **Shambleau**, whose PLANETARY-ROMANCE venue was typical of that category in providing a licence for female excesses – were rare. But autonomous females now proliferate; in modern SWORD AND SORCERY, for example, a mercenary is almost as likely to be a woman as a man (◊◊ TEMPORAL ADVENTURESS), and although most full-fantasy narratives still have male protagonists exceptions are numerous.

Overt depictions of the sexual act are relatively uncommon in GENRE FANTASY, though they are increasingly part of the *lingua franca* of supernatural fiction (notably concerning VAMPIRES) and horror. [JC]

See also: GENDER; GENDER DISGUISE.

SEX 2000 (vt *Cinderella 2000*) US movie (*1977*). ◊ CINDERELLA (*1950*).

SEYMOUR, MIRANDA (JANE) (1948-) US writer of some tales for younger children, and of several fantasies for YA and adult readers. They include: *Count Manfred* (**1976**); *Medea* (**1982**), which conflates three first-person accountings of the myth; *The Vampire of Verdonia* (**1986**), a VAMPIRE tale for YA readers; and *The Reluctant Devil: A Cautionary Tale* (**1990**), a wry SUPERNATURAL FICTION. A nonfiction study, *A Ring of Conspirators: Henry James and his Literary Circle* (**1988**), usefully locates Henry JAMES, Joseph CONRAD, Ford Madox FORD and others in the Edwardian context. [JC]

SHADE, THE CHANGING MAN Alien SUPERHERO of a short series of innovative, though somewhat chaotic, COMIC-book stories (*Shade, The Changing Man #1-#8*, 1977-8) conceived by Steve DITKO, with dialogue by Michael Fleischer. Rac Shade, a fugitive from the planet Meta dressed in red boots and a sleeveless red knicker-suit bedecked with gold gadgets, framed for treason by Metan criminals, steals the illegal Miraco vest (which endows him with awesome SHAPESHIFTING abilities) and flees to the Earth Zone, pursued by both the criminal element and

government agencies of the Meta Zone. Among these is his former fiancée, Mellu, who believes him responsible for the crippling of her parents. Her contact on Earth is Wizor, a red-bearded Metan with a mission to maintain a receiving station for Metans travelling from one dimension to the other via the Zero-Zone, a buffer dimension which contains the lethal Area of Madness. STCM showed Ditko at his most imaginative, introducing and discarding weird characters, all oozing metaphor and symbolism, with prodigal rapidity, and with Mellu in particular continually changing sides. The cast included psychologist Dr Sagan, the sinister Dr Lopak, the bizarre Form, the explosive Khaos, the mad Zokag – whose Em-rod was almost as powerful as Shade's M-vest – the megalomamiac Dr Z.Z. and SuDe, the Supreme Decider, a Metan crime-boss who was eventually to be revealed as Mellu's mother. Ditko is rumoured to have plotted the first 17 episodes and many plot threads remained unresolved when the series ceased at *#8*.

STCM was revived by writer Steve Milligan in a second series of *Shade, The Changing Man* (*#1-#70* 1990-96). Shade returns from the Area of Madness and enters the body of Troy Grenzer, a SERIAL KILLER on death row, escaping with the help of Kathy George, whose parents were Grenzer's last victims. Milligan used the series to attack consumerism among other US targets. In early issues, all that is mad and bad about the USA is embodied in a character known as The American Scream. Shade was reborn in *#33* (1993) under the Vertigo (for adults) imprint. The major artists were Chris Bachalo (1965-   ), Glynn Dillon, Sean Phillips (1965-   ) and Mark Buckingham (1966-   ). [SH/DR/RT]

SHADOW, THE ◊ MAGAZINES; MASKED AVENGER.

SHADOW DANCING US/Canadian movie (*1988*). Shapiro Glickenhaus. **Pr** Kay Bachman. **Exec pr** Don Haig, Robert Phillips. **Dir** Lewis Furey. **Spfx** Frank Carere. **Vfx** Light & Motion Corp. **Screenplay** Christine Foster. **Starring** James Kee (Paul), Gregory Osborne (Philip Crest), Christopher Plummer (Edmund Beaumont), Kay Tremblay (Sophie Beaumont), Nadine van der Velde (Jessica Blake/Liliane La Nuit). 100 mins. Colour.

An accident disables the lead female artiste of a modern-dance troupe, and mediocre but convenient tyro Jessica is taken on instead. Instantly she is "recognized" by the parrot now owned by theatre-owner Edmund but once owned by long-ago murdered dancer Liliane La Nuit, a Jessica lookalike whom he doomedly loved. On hearing Rimsky-Korsakov's SCHEHERAZADE Jessica undergoes the first of several, increasingly long IDENTITY EXCHANGES, in the process becoming a superb dancer. It is soon evident that she has become possessed by the spirit of dead Liliane, who destroyed the lives of most around her. Jessica has difficulty convincing boyfriend Paul that she is not having an affair with choreographer and troupe star Crest; although in fact she (i.e., Liliane) is. History seems set to repeat itself . . .

One of the cluster of movies in the late 1980s and early 1990s mixing GHOSTS/POSSESSION and sex – GHOST (*1990*) was the box-office peak of the trend – *SD* is a minor but not unrespectable offering, aided by some spectacularly staged modern dance (choreography by Timothy Spain; participation by members, including Osborne, of the National Ballet of Canada). The over-baroque plot doesn't quite make sense, but comes close enough to satisfy. [JG]

SHADOWRUN A SHARED-WORLD series of TECHNOFANTASY adventures owned by the FASA Corporation, which also owns the associated role-playing GAME. The sequence is set in the middle of the 21st century, in a world where technology and science CROSSHATCH with DRAGONS, druids, ELVES, MAGES, "Orks", TROLLS and other creatures standard to GENRE FANTASY. Titles include: *Shadowrun 1: Secrets of Power 1: Never Deal with a Dragon* * (**1990**), *#2: Choose Your Enemies Carefully* * (**1991**) and *#3: Find Your Own Truth* * (**1991**), all by Robert N. CHARRETTE; *#4: 2XS* * (**1992**) by Nigel FINDLEY; *#5: Changeling* * (**1992**) by Chris KUBASIK; *#6: Never Trust an Elf* * (**1992**) by Charrette; *#7: Into the Shadows* * (anth **1990**) ed Jordan K. Weisman; *#8: Streets of Blood* * (**1993**) by Carl SARGENT and Marc GASCOIGNE, which features a search for JACK THE RIPPER. After the **Streets of Power** subseries came *#9: Shadowplay* * (**1993**) by Findley, *#10: Night's Pawn* * (**1993**) by Tom Dowd, *#11: Striper Assassin* * (**1993**) by Nyx SMITH, *#12: Lone Wolf* * (**1994**) by Findley, *#13: Fade to Black* * (**1994**) by Smith, *#14: Nosferatu* * (**1994**) by Sargent and Gascoigne, *#15: Burning Bright* * (**1994**) by Dowd and *#16: Who Hunts the Hunter* * (**1995**) by Smith. [JC]

SHADOWS 1. In their literal, physical sense, shadows are both troubling – as temporary BAD PLACES in whose darkness vile things may lurk – and a reassuring indicator of human normality. Thus VAMPIRES and GHOSTS generally lack shadows, as does the boy Kay while unnaturally visiting the past (◊◊ TIMESLIP) in John MASEFIELD's *The Box of Delights* (**1935**). Conversely, in J.R.R. TOLKIEN's Middle-Earth, the INVISIBILITY of mortals who wear the RING is flawed by a betraying shadow. The regimented enemy horde in Gene WOLFE's *The Book of the New Sun* (**1980-83**), who speak only in preset slogans, are known as the Ascians or "men without shadows". A disturbing moment of WRONGNESS in James Branch CABELL's *Jurgen* (**1919**) comes when the hero sees that the shadow he casts is no longer his own; in the same book, MERLIN's shadow also acts with disconcerting independence.

Lord DUNSANY's *The Charwoman's Shadow* (**1926**) suggests that a lost shadow is part of the SOUL, whose restoration is transforming; similarly, the shadow-boy in Michael SWANWICK's *The Iron Dragon's Daughter* (**1993**) is a severed portion of the heroine's soul, whom she later re-absorbs. George MACDONALD's *Phantastes* (**1858**) features a disagreeable shadow which, once acquired, blights PERCEPTION of beauty and wonder; his "The Shadows" describes living shadows who are SECRET GUARDIANS of our spiritual well-being. PETER PAN's shadow serves as a PLOT DEVICE in J.M. BARRIE's play, but its easy detachability also indicates the gap between fey Peter and normal children. As Narnia dies in C.S. LEWIS's *The Last Battle* (**1956**), Aslan's gigantic shadow is the oblivion awaiting those who will not accept the light. The eponymous TRICKSTER of Roger ZELAZNY's *Jack of Shadows* (**1971**) has allegiance to neither light nor dark, but uniquely draws his power from shadow; in Zelazny's **Amber** sequence, the MULTIVERSE, including Earth, consists of mere shadows cast by the true REALITY of Amber (though this solipsism is undermined as the series progresses).

Wherever there is light, BALANCE requires there to be shadow – a truism which in fantasy has all the expected metaphorical ramifications. [DRL]

2. When it refers to a dark hidden side of the self, the term "shadow" is a close cousin to the term DOUBLE. Doubles are found throughout 18th-century GOTHIC FANTASY and 19th-century SUPERNATURAL FICTION, where they figure significantly in the work of authors like E.T.A. HOFFMANN,

James HOGG, Sheridan LE FANU, Robert Louis STEVENSON – whose *Strange Case of Dr Jekyll and Mr Hyde* (**1886**) is perhaps the most famous single tale to be constructed around the conflict between public and repressed self – and Oscar WILDE.

Though there is no fixed rule, the term "double" does tend to imply a malign or seductive relationship between a surface personality and a submerged aspect of that personality which *haunts* the surface self, threatening DEBASEMENT. Tales involving doubles frequently end in scenes of integration that make it clear that any marriage of the two halves of the self is likely to be fatal. This is consistent with a general tendency in supernatural fiction to treat the imploring side of the self as inherently obscene. As a threat to the daylight Victorian self, Hyde is typical: he is a kind of primordial force, an ape from humanity's bestial past; at the same time there is something filthily enticing about his manner, as though something female were welling up in his visage and slouch. Closely related is the ancient Egyptian notion of the Ka, which occasionally reappears in supernatural fictions like Dennis WHEATLEY's *The Ka of Gifford Hillary* (**1956**).

There can be no fixed rule, but in FANTASY double/shadow imagery is more likely to represent the separation of the self into discordant elements as a tragic circumstance that must somehow be transcended. The relationship between king and JESTER, or between the father of the GODS and the TRICKSTER, is a relationship between an official master and his or her shadow; and is almost always, in fantasy, accompanied by signals of potential HEALING. Even novels which end unhappily, like Pär LAGERKVIST's *The Dwarf* (**1944**) and Poul ANDERSON's *The Broken Sword* (**1954**), tend to present those warring opposites as an opportunity which may be seized, and which it would be a tragedy to miss.

The many tales featuring potential states of union between protagonists and shadows include: Edgar Rice BURROUGHS's **Tarzan** sequence, whose hero embodies an effortless marriage of "civilized" man and the 19th-century bugaboo apeman; Sir Henry NEWBOLT's *Aladore* (**1914**); Mervyn PEAKE's **Titus Groan** sequence, in which Titus and Steerpike orbit one another, half-selves forever disjunct; J.R.R. TOLKIEN's *The Lord of the Rings* (**1954-5**), where Frodo and Gollum are each other's shadow, just as Frodo and Samwise shadow one another, and where the DARK LORD is seen, definitively, as both fallen ANGEL and remorseless shadow PARODY of the Good; Philip José FARMER's **World of Tiers** sequence, in which the TRICKSTER Kickaha must ultimately fight his Trickster shadow; Stephen R. DONALDSON's **Chronicles of Thomas Covenant the Unbeliever**, which articulates the Tolkien schism more fiercely; Orson Scott CARD's **Alvin Maker** sequence, where Alvin is shadowed by his dark brother Calvin; Terry BISSON's *Talking Man* (**1986**); Nick BANTOCK's **Griffin & Sabine** sequence; Stephen KING's *The Dark Half* (**1989** UK), filmed as *The DARK HALF* (*1991*); Alan BRENNERT's *Time and Chance* (**1990**); Deborah GRABIEN's *Plainsong* (**1990**), in which CHRIST is a shadow of the WANDERING JEW; and Michael SWANWICK's *The Iron Dragon's Daughter* (**1993** UK), in which the "shadow-boy" is a split-off aspect of the protagonist's SOUL.

Shadows may also – in a usage entirely distinct from any application of the term "doubles" – be referred to in descriptions of REALITY structures underpinning entire imaginative worlds. Many tales, for instance, make use of the

Gnostic concept (◊ GNOSTIC FANTASY) of mundane reality as being an imperfect rendering of the Pleroma or celestial totality; in fantasy, any territory or world or LAND which depends upon some deeper reality for its existence can be understood as a shadow of that deeper reality. Both E.R. EDDISON's **Zimiamvia** and Roger ZELAZNY's **Amber** sequences establish a shadow/wholeness relationship between this world and the *realer* world where the action centres; and the relationship between Terra and Anti-Terra in Vladimir NABOKOV's *Ada* (**1969**) is one of shadow to substance, though it is never clear which version is the shadow.

It has often been suggested that the various stereotypical characters in the average ROMANCE are in fact aspects of one fully integrated being. This argument is clearly relevant to fantasy, for which the Romance form is a vital taproot (◊ TAPROOT TEXTS); and may help explain the complex and dynamic interactions between fantasy and JUNGIAN PSYCHOLOGY, in whose geography of the self the Shadow represents those aspects of the whole self which have been denied, and which must be re-integrated into the conscious personality if one wishes to become a mature and chivalrous adult. Writers like Ursula K. LE GUIN and Robert HOLD-STOCK have made constant use of the Jungian geography. Le Guin's essay "The Child and the Shadow" (1975) strongly argues the centrality of this understanding of the Shadow, in life and in fantasy both.   [JC]

SHADOWS US series of annual original ANTHOLOGIES ed Charles L. GRANT: *Shadows* (anth **1978**), *#2* (anth **1979**), *#3* (anth **1980**), *#4* (anth **1981**; vt *Shadows* 1987 UK), *#5* (anth **1982**), *#6* (anth **1983**), *#7* (anth **1984**), *#8* (anth **1985**), *#9* (anth **1986**), *#10* (anth **1987**), plus the retrospective *The Best of Shadows* (anth **1988**) and *Final Shadows* (anth **1991**). The series was dedicated to "stories of the unknown", seeking to resurrect the primitive fears residual within us, and so all the tales rely upon some supernatural intrusion into everyday normality (◊ SUPERNATURAL FICTION; WRONGNESS). Grant took the stance that SHADOWS are themselves ILLUSIONS, an interplay of light and dark, and used that theme to explore illusions in our mind and their relationships with the world about us. The result was an adult, cohesive series that published some of the best supernatural stories of the 1980s. *#1* won a 1979 WORLD FANTASY AWARD, as did its lead story, "Naples" by Avram DAVIDSON; a later World Fantasy Award-winner was "The Gorgon" by Tanith LEE, from *#5*. The series featured work by most of the leading writers in the field, including Ramsey CAMPBELL, Dennis ETCHISON, Stephen KING, William F. NOLAN, Alan Ryan (1943-   ), Jessica Amanda SALMONSON and Manly Wade WELLMAN.   [MA]

SHAGGY D.A., THE US movie (*1976*). ◊ *The* SHAGGY DOG (*1959*).

SHAGGY DOG, THE US movie (*1959*). Disney. **Assoc pr** Bill Walsh. **Dir** Charles Barton. **Screenplay** Lillie Hayward, Walsh. **Based loosely on** *Der Hund von Florenz* (**1923**) by Felix SALTEN. **Starring** Kevin Corcoran (Moochie Daniels), Tommy Kirk (Wilby Daniels), Fred MacMurray (Wilson Daniels), Alexander Scourby (Mikhail Andrassy), Roberta Shore (Franceska Andrassy). 104 mins. B/w.

Technology-crazy teenager Wilby, visiting a museum, is transformed by a magic RING into an Old English Sheepdog. Wilby's father Wilson is allergic to dogs; in his dog form the lad habitually frequents the home of his friend Franceska, of whose own pet he is near enough a DOUBLE.

While there he overhears Franceska's father plotting with spies, and he and brother Moochie resolve to snare them. In the resulting chase Franceska is thrown from a motorboat and saved by Wilby-as-dog. After the police have rounded up the spy-ring, it is Franceska's pet who is hailed as hero.

This leisurely movie – amusingly straplined, in line with the HORROR MOVIES of the day, "I Was a Teenage Boy!" – can be seen as the ancestor of a long string of generic DISNEY fantasy comedies. It was directly sequelled by *The Shaggy D.A.* (*1976*), which has Dean Jones as an adult Wilby, now running for the office of District Attorney despite the unfortunate reappearance of the MAGIC ring; and by *The Return of the Shaggy Dog* (*1987* tvm).   [JG]

SHAKESPEARE, WILLIAM (1564-1616) English poet and dramatist. Shakespearian venues, scenes, MOTIFS, ICONS and formal quotations – not to mention innumberable tags and scraps and catchphrases from the plays – have penetrated deeply into the matrix (or CAULDRON OF STORY) of Western literary and popular culture. In many of our acts of communication and storytelling WS underlies us, and we quote him often without knowing we do so. Many of his characters, too, are UNDERLIERS.

The universal influence of WS did not come about immediately. Although he remained well known throughout the 17th and into the 18th century, it was not until nearly 1800 that his works became unassailable lynchpins of literary tradition. The apotheosis of his work and life coincided roughly with the beginnings of FANTASY as a self-conscious genre. Fantasy has therefore been permeated by WS from its beginnings, both in English-speaking countries and on the European continent. The 20th century has seen much critical activity concerning WS, and this has resulted in a proper and necessary contextualizing of the works and humanizing of the man.

Such activity has had relatively little effect on the Cauldron of Story, but textual analysis, and a proliferation of editions of WS's works, make WS's bibliography difficult; here we give estimated year of first performance and year of first book publication; when relevant, the first folio – *Mr William Shakespeare's Comedies, Histories, and Tragedies* (coll **1623**) and the third folio (**1664**) – may be cited. No attempt is made to trace any further the history of the texts.

Only a few of the poems and plays are direct TAPROOT TEXTS. Two early narrative poems are of interest. *Venus and Adonis* (**1593**), taking its plot from OVID, describes the love of Venus for ADONIS, who spurns her in order to go hunting, where he is killed by a boar. In his *Shakespeare and the Goddess of Complete Being* (**1992**) Ted HUGHES argues that, through this late recension of the myth of the GODDESS and her consort, WS created part of a personal "Tragic Equation" out of which he generated the dynamic that governs his greatest plays. WS (Hughes suggests) combined this myth with the Roman tragedy of Lucrece, recounted by WS in *The Rape of Lucrece* (**1594**).

The first play of fantasy interest is *A Midsummer Night's Dream* (performed *c*1595; **1600**), a MIDSUMMER-NIGHT tale that CROSSHATCHES the world of ancient Greece with that of FAERIE. Various pairs, in LOVE with one another, travel INTO THE WOODS in an effort to escape the strictures of the mundane world; but PUCK dazzles their eyes with a MAGIC juice, causing them to fall in love with the wrong (or right) partners. Meanwhile, OBERON has tricked Titania (◊ FAIRY QUEEN) with the same SPELL, causing her to fall in love with the bewildered amateur actor Bottom, who has suffered

METAMORPHOSIS at Puck's hands and now has the head of an ASS. The REVEL continues sparklingly, the various protagonists undergo their NIGHT JOURNEY into a wiser morn ($\Diamond$ GODGAME), and all is sorted out. Poul ANDERSON's *A Midsummer Tempest* (**1974**) reflects the world of the play in an explicit and interesting fashion, as do Clemence DANE's *The Godson: A Fantasy* (**1964** chap), John CROWLEY's *Little, Big* (**1981**) and Neil GAIMAN's "A Midsummer Night's Dream" (**1990** *Sandman*); most deal (in different ways) with the questions of how humans would actually live in a world in which the lure of Faerie truly existed. *The Fairy Queen* (**1692**) by Henry Purcell (1659-1695), *Oberon* (**1826**) by Carl Maria von Weber (1786-1826) and *A Midsummer Night's Dream* (**1960**) by Benjamin Britten (1913-1976) are among the OPERAS based on the play. A fine movie version is *A* MIDSUMMER NIGHT'S DREAM (**1935**).

Neither *The History of Henry IV* (performed *c*1596; **1598**), nor its sequel, *Henry IV Part II* (performed *c*1598; **1600**), have direct fantasy content, but Falstaff has become an UNDERLIER in HEROIC FANTASY for the bluff, seemingly cowardly, heavy-drinking amorous older COMPANION who is much-beloved; the best novel about him is probably Robert NYE's *Falstaff* (**1976**). The INN in which Falstaff's life is mostly led, and which serves as a junction for others, is a central model for the inns so frequently found in GENRE FANTASY at those points when plots need to be hurried along.

There is a GHOST in *Hamlet* (performed *c*1600; **1603**; exp 1604) and there are WITCHES and PROPHECIES in *Macbeth* (performed *c*1606; **1623**), but neither play has had a huge effect on subsequent fantasy outside the realms of PARODY – as in Terry PRATCHETT's work, variously, and in the first episode of the UK tv series *The Black Adder* (1983). The first three of WS's late ROMANCES – *Pericles* (performed *c*1608; **1609**), *The Winter's Tale* (performed *c*1610; **1623**) and *Cymbeline* (performed *c*1611; **1623**) – incorporate at various points elements of fantasy, and many of the figures in these plays act with the profound ungovernable mysterious simplicity of figures in a FAIRYTALE. But they are of greatest interest as dramatic experiments. Each constitutes a manipulation of TIME, for each takes place over wide intervals. In each a HEALING occurs, reconciling LAND and family and marriage.

Though it does not have the same elongated narrative structure, *The Tempest* (performed *c*1611; **1623**) reflects these underlying strategies, compressing the long labyrinthine NIGHT JOURNEYS of the earlier plays into brief summations of various characters' past lives. The entire play occurs on an ISLAND governed by the MAGUS Prospero, Duke of Milan, exiled 12 years previously by usurpers. On his arrival, Prospero discovered ARIEL, a spirit of the air, and CALIBAN, a MONSTER born out of the chthonic earth who is the rightful owner of the place; the MAGUS has made both of them his servants. As the play opens, Prospero has used his MAGIC staff to call up a tempest which shipwrecks upon this POLDER various characters from his past. He then subjects them, and his daughter Miranda, to a complex GODGAME, bewitching their senses ($\Diamond$ PERCEPTION) as they stumble INTO THE WOODS in their search for clues as to who and where they are, and testing their natures. In the end, he judges and forgives all and, in scenes reminiscent of the LICENZA, permits his daughter ($\Diamond\Diamond$ VIRGINITY) to marry the young man she has fallen in LOVE with, releases Ariel and returns the island to the colonized Caliban. Prospero then

relinquishes his magic (in a scene which many commentators have taken as also representing WS's own farewell statement), and the play ends tranquilly. Many fantasy novels echo *The Tempest*'s structure, and many use characters closely modelled on Prospero and his two servants, significant examples including Rachel INGALLS's *Mrs Caliban* (**1982** UK) and Tad WILLIAMS's *Caliban* (**1994**). Many OPERAS as well, most inconsequential, have been based on the play. The most notable direct fantasy use of *The Tempest* is almost certainly Peter GREENAWAY's PROSPERO'S BOOKS (**1991**), though of course the sf movie *Forbidden Planet* (**1956**) remains more famous. *Shakespeare Stories* (anth **1982**) ed Giles Gordon (1940-   ) contains several **Tempest** fantasies.

Considering WS's supreme importance and the fact that his biography contains major lacunae permitting doubts (however crackpot) to be recurrently raised concerning his authorship, it is surprising how infrequently he has himself been a subject for fantasies. The most influential fictional portrayal of WS is probably Anthony BURGESS's *Nothing Like the Sun* (**1964**), which is not fantasy, though Burgess's later portrayals in "The Muse" (1968) and *Enderby's Dark Lady, or No End to Enderby* (1985) are. Works that portray WS as a radically transformative figure, like *Transformations* (fixup **1975**) by John Mella, are rare. A benign, rather wistful version of WS is brought forward in time for more or less tendentious purposes in William Dean HOWELLS's *The Seen and Unseen at Stratford-on-Avon* (1914) and *The Return of William Shakespeare* (1929) by Hugh Kingsmill (1889-1949); *Serinissima* (**1987**) by Erica Jong (1942-   ) is a slipstream fantasy employing WS, unusually, as a focus of erotic interest. An amusing portrayal of the young WS (seen through the tint of a fictitious minor novelist) is presented in John CROWLEY's *Aegypt* (**1987**). The most serious and complex portrayal of WS in fantasy is probably "Aweary of the Sun" (1994) by Gregory Feeley (1954-   ).

WS as the uncanny genius who transcends all literary bounds is a figure rarely evoked in GENRE FANTASY: the WS who appears (whether or not as an actual character) in most sf or fantasy represents either the emblem of canonical authority or else – as in Clifford D. SIMAK's *Shakespeare's Planet* (**1964**) and *The Goblin Reservation* (**1968**) – as a benign Fount of Narrative, a storyteller whose nature is essentially reassuring. Poul Anderson's *A Midsummer Tempest*, in which WS is the "Great Historian" whose every word is historical truth, also contains this element of domestication. Leon ROOKE's *Shakespeare's Dog* (**1983**), describes the squalor of WS's early life (which is implicitly likened to that of the dog who tells the tale) with a verbal energy which hints at something of the intensity of the man himself; perhaps significantly, Rooke's FABULATION is fantastic only in its use of narrator. [JC]

SHAMANISM The belief that the world is pervaded by good and evil spirits which must be ritualistically controlled or appeased. The shaman, combining the role of doctor, priest and sorcerer, controls access to the spirit world, supervises initiation into manhood, and maintains the myths and songs of his tribe. The word comes from the *saman* of the Tungus people of Siberia, but equivalent figures are found in many hunter-gatherer and pastoral cultures. In fantasy, the shaman may be good or evil, aiding the hero in his NIGHT JOURNEY, perhaps, or by contrast seeking to thwart or control him by magical means.

The way in which shamans gain access to the spiritual

world is much debated. Mircea ELIADE argued that it is achieved by meditation or fasting, or in trancelike states of ecstasy during ritualistic dancing or drumming. This view is more commonly reflected in fantasy than the findings of many cultural anthropologists that intoxicants, which Eliade regards as "mechanical and corrupt", often play an important role. However, there is persuasive evidence that human cultures have ritualistically used intoxicants since the Stone Age, and this use is still widespread: fly agaric toadstools by Siberian peoples, peyote by the Huichol Indians of Mexico, various snuffs by tribes of Amazonian Indians, etc. The messianic avocation by Aldous Huxley (1894-1963) and Dr Timothy Leary (1920-1996) of the use of psychoactive drugs to achieve enlightenment, particularly influential on the Beat Movement of writers – most notably seen in the nightmarish SATIRES of William S. BURROUGHS, was in part based upon anthropological studies of shamanistic rituals.

Notable appearances of shamans in fantasy include: in Scott BAKER's **Ashlu Cycle**, which turns on INITIATION; as the eponym of Terry BISSON's *Talking Man* (**1986**); in *The Clan of the Cave Bear* (*1985*); as an ambiguous female healer in Robertson DAVIES's *The Cunning Man* (**1985**); filling civil-service roles in the transformed USA of Rachel POLLACK's *Unquenchable Fire* (**1988**); in Rosemary SUTCLIFF's *Warrior Scarlet* (**1958**); and working the protagonists' TRANSFORMATION in John UPDIKE's *Brazil* (**1994**). But in FANTASYLAND, perhaps through unconscious association with "sham", a shaman is very often a magicless fraud imposing upon a gullible tribe – an assumption too easily made by Gene WOLFE's Severian in *The Sword of the Lictor* (**1982**), who discovers almost too late that a group of jungle sorcerers wields not only trickery but dangerous TALENTS.   [PJM/DRL]

See also: CHILDE; ODIN; SHADOWRUN; WORLD-TREE.

SHAMBALA/SHAMBHALA ◊ SHANGRI-LA.

SHANGRI-LA In James HILTON's *Lost Horizon* (**1933**), a lamasery in a valley or POLDER in the depths of Himalayan Tibet, which has long been hidden (◊ LOST RACES) from the outside world. In this sanctuary, a secret society of savants with much extended lifespans (◊ IMMORTALITY) have been contemplating recondite issues and living the kinds of lives mortals normally ascribe to the inhabitants of TIR-NAN-OG and other places beyond the bournes of this life. Hilton's depiction of this world may derive consciously from – and is certainly very similar to – Madame BLAVATSKY's description, in *The Secret Doctrine* (**1888**), of the home and activities of the Tibetan SECRET MASTERS who convey to such as her the message of THEOSOPHY. This relates in turn to the much more ancient notion of a Himalayan UTOPIA called Shambala or Shambhala. But the intensity of SEHNSUCHT conveyed through Hilton's depiction of S-L has just as important a source in the trauma of WORLD WAR I, for *Lost Horizon* is a novel written in the aftermath of what seemed to many a terminal apocalypse, a war which ended not war but civilization itself; and S-L is a "solution" to that sense of desolation. As such, it stands in radical contrast to another 20th-century polder constructed in an attempt to come to grips with WWI and to debate the nature of civilization; that polder is the MAGIC MOUNTAIN at the heart of Thomas MANN's *The Magic Mountain* (**1924**). While S-L is discovered years after 1918, WWI ends the Magic Mountain.

Although S-L, in Hilton's novel, is the name of the lamasery alone, the term soon became identified also with the Valley of the Blue Moon which surrounds it.   [JC]

SHANNA THE SHE DEVIL ◊ TARZAN.

SHAPESHIFTERS, SHAPESHIFTING The preferred terms in this encyclopedia for those who change shape (and for the act of thus changing shape) *repeatably* and *reversibly*, by innate MAGIC, TALENT or breeding. This is distinct from METAMORPHOSIS, whose tendency is to be radical, unique and permanent, and TRANSFORMATION, which is generally imposed by an outside magical agency, as with the FROG PRINCE.

Fantasy's best-known shapeshifters are the were-creatures or were-people (◊◊ THERIOMORPHY), the commonest being WEREWOLVES – a trope of HORROR because so often the shift is not voluntary but seen as a CURSE, a BONDAGE to the full MOON which brings on wolf-shape and/or a POSSESSION by a wolf's SPIRIT. The lighter fantasy favoured by UNKNOWN took a cheerier view of such shapeshifting and, typically ringing the changes, discussed a wider selection of target animals. Anthony BOUCHER's "The Compleat Werewolf" (1942) and Poul ANDERSON's "Operation Afreet" (1956) both accentuate the positive with heroes who find wolf-form useful when fighting, while in quieter times it offers a movie stunt-dog career; both consider other were-forms, including tigers – Anderson also offers a were-fennec or desert fox, while Boucher's broad HUMOUR expands from bears (Beorn in J.R.R. TOLKIEN's *The Hobbit* [**1937**] is seemingly a were-bear) to a were-ant and a were-diplodocus. Jack WILLIAMSON's altogether grimmer *Darker Than You Think* (**1948**) likewise enlarges the range of weredom with a pterosaur form. More traditional shapeshifters are the Chinese fox women, the Scots SELKIES or seal men, and VAMPIRES – whose subsidiary ability to shapeshift into bat form seems to have fed back into fiction after the vampire bat was so named in the 18th century. Shifting to SERPENT form is generally reserved for women of bad character (◊ LAMIA), like the witch in C.S. LEWIS's *The Silver Chair* (**1953**).

Further examples are too numerous to list. Variations include shapeshifters whose default form is beast rather than human – like the were-human wolf in Larry NIVEN's "What Good is a Glass Dagger?" (1972) and the CAT Greebo in Terry PRATCHETT's **Discworld** sequence, who after being briefly transformed to human form learns the knack and can repeat the effect by shapeshifting. The heroine of Piers ANTHONY's *A Spell for Chameleon* (**1977**) shapeshifts gradually and repeatedly between the poles of moronic beauty and highly intelligent ugliness; the hero decides this is a good thing since he will never become bored with her (◊ FEMINISM).

FAIRIES, WITCHES and WIZARDS, if they shapeshift at all, are not confined like were-creatures to a single alternative form; the shapeshifter villains of Patricia MCKILLIP's **Riddle-Master** trilogy can be pretty much what they wish. One traditional form of magical duel has the antagonists flashing from shape to shape in search of advantage: the solitary highlight of the movie *The Sword in the Stone* (**1963**; ◊ T.H. WHITE) is such a contest between MERLIN and "Madam Mim", who loses when polymath Merlin finally becomes a bacillus and infects her. Ursula K. LE GUIN, ever concerned with BALANCE, presents a darker side of shapeshifting in *A Wizard of Earthsea* (**1968**), where the temptation of another form can lead to loss of humanity as the animal body reshapes the SOUL within: her hero nearly loses himself when fleeing in hawk shape. Barbara HAMBLY offers a subtler version of this lure in *Dragonsbane* (**1986**), where becoming a DRAGON represents a clear improvement on being human. Voluntary shapeshifting offers broad vistas of wish-fulfilment, but may be a THING BOUGHT AT TOO HIGH A COST if it should become fixed as permanent METAMORPHOSIS. [DRL]

SHARED WORLDS Tales written by various hands but sharing a common setting (in fantasy, often a FANTASYLAND) are shared-world stories. They are distinguished from SEQUELS BY OTHER HANDS because individual stories in an SW venue often do not sequel one another and because there is not normally an original text from which later stories develop. In place of this original text, SW enterprises normally make use of a "bible", essentially an annotated set of rules laying down – on behalf of the owners and/or operators of the SW – the conditions governing the roles, actors, venues, storylines and potential implications of any story set in the SW.

The first modern SW enterprises were the CHRISTMAS annuals produced in the UK from *c*1860, and the first of fantasy interest was almost certainly *Mugby Junction* * (anth **1866** chap) ed Charles DICKENS, a self-contained Christmas issue of *All the Year Round*, in which several tales, including Dickens's own "No. 1 Branch Line. The Signalman" (◊ TRAINS), are told within the context of a complex FRAME STORY (◊◊ ANNUALS).

The most famous genre SWs are almost certainly those generated by the owners of **Star Trek** and **Dr Who**. The most famous HORROR example is probably H.P. LOVECRAFT's CTHULHU MYTHOS. Fantasy SWs include ventures connected with Marion Zimmer BRADLEY's **Darkover**, Philip José FARMER's **Riverworld**, Michael MOORCOCK's **Jerry Cornelius**, Andre NORTON's **Witch World**, **Thieves' World** ed Lynn ABBEY and Robert L. ASPRIN, **Borderland** ed Mark Alan Arnold and Terri WINDLING, *Shanadu* (anth **1953**) ed Robert E. Briney (1933-), Emma BULL's and Will SHETTERLY's **Liavek**, **Heroes in Hell** ed C.J. CHERRYH and Janet E. MORRIS, Bill Fawcett's **Guardians of the Three**, **Temps**, **The Weerde** and **Villains** – ed variously Neil GAIMAN, Mary GENTLE, Roz KAVENEY and Alex Stewart – George R.R. MARTIN's **Wild Cards**, Richard PINI's **Winds of Change**, *Witches Three* (anth **1952**) ed anon Fletcher PRATT, including Pratt's own *The Blue Star* (exp **1969**), and **Crafters** ed Christopher STASHEFF and Bill Fawcett.

There are also SW enterprises like **DragonLance**, which are owned by corporations and whose contents are tied to fantasy GAMES. [JC]

SHARP, MARGERY (1905-1991) UK writer known for decades as an author of nonfantasy books for adults, beginning with *Rhododendron Pie* (**1930**), in later years perhaps most famous for her **Miss Bianca** sequence of ANIMAL FANTASIES. The series runs: *The Rescuers* (**1959**), *Miss Bianca* (**1962**), *The Turret* (**1963**), *Miss Bianca in the Salt Mines* (**1966**), *Miss Bianca in the Orient* (**1970**), *Miss Bianca in the Antarctic* (**1971**), *Miss Bianca and the Bridesmaid* (**1972**), *Bernard the Brave* (**1976**), *Bernard into Battle* (**1978**) and *The Rescuers Down Under* * (**1991**). *Miss Bianca* and some other titles were memorably illustrated by Garth WILLIAMS. The books describe the WAINSCOT adventures of Miss Bianca, the sophisticated and inventive white mouse (◊ MICE AND RATS) who dominates the Mouse Prisoners' Aid Society, leading her COMPANIONS into various perilous situations in campaigns to rescue various prisoners from their various plights; her faithful helper Bernard dominates later volumes (◊ DUOS). *The Magical Cockatoo* (**1974**), an unconnected tale with a human protagonist, is noticeably thinner. Adventures of Miss Bianca have been filmed by DISNEY as *The Rescuers* (*1977*) and *The Rescuers Down Under* (*1990*). [JC]

SHAW, GEORGE BERNARD (1856-1950) Irish dramatist who spent most of his life in England; he won the Nobel Prize for Literature in 1925. Over a career lasting 70 years or more he wrote literary and musical criticism, numerous plays, fiction, polemic and much occasional writing. He was not by instinct a writer of fantasy, and the fiction and plays which deploy fantasy elements are at their core *arguments* about the world; for GBS, the FANTASTIC serves to illustrate or illuminate arguments, rarely serving as a self-contained narrative goal. All the same, there are unmistakable fantasy elements in his work. An AFTERLIFE interlude in HELL – dominated by a DON JUAN straight from the Mozart OPERA, who makes witty arguments about the nature of his new abode – enlivens *Man and Superman: A Comedy and a Philosophy* (**1903**), one of GBS's most sustained dramas; and *Androcles and the Lion* (performed 1913; in omni **1916**) and *Saint Joan* (**1923**) (◊ JOAN OF ARC) both have moments of fantasy. Most famously, *Back to Methuselah: A Metabiological Pentateuch* (**1921** US; rev 1921 UK; rev 1945) begins in EDEN and ends with a FAR-FUTURE dialogue featuring the GHOSTS of ADAM AND EVE, LILITH and the SERPENT. The late plays use both sf and fantasy MOTIFS in an Expressionist fashion which eschews plausibility but whose abandonment to cultural pessimism cuts very deep. They include: *The Apple Cart: A Political Extravaganza* (first English-language publication **1930**); *Too True to be Good: A Political Extravaganza* (performed 1932) and *On the Rocks: A Political Comedy* (performed 1933), both assembled in *Too True to be Good, Village Wooing & On the Rocks* (omni **1934**); *The Simpleton of the Unexpected Isles: A Vision of Judgment* (**1935**); *Geneva: A Fancied Page of History* (**1939**) and *Buoyant Billions* (**1948** Switzerland; with *Farfetched Fables* as omni **1950**).

Some of GBS's short fiction is more sustained fantasy. *The Adventures of the Black Girl in her Search for God* (**1932** chap) details the young woman's search for meaning, via interviews conducted with successive versions of GOD as presented in the BIBLE, and with sages from various periods of history; *Short Stories* (coll **1932**) includes "Aerial Football: The New Game" (1907), a POSTHUMOUS FANTASY; and *Short Stories, Scraps and Shavings* (omni **1934**), which includes revisions of both *Black Girl* and the earlier *Short Stories*, includes in "Don Giovanni Explains" a harkening back to the opera protagonist from *Man and Superman*. *The Black Girl in Search of God, and Some Lesser Tales* (coll **1946**) assembles similar material, revised.

GBS, financially independent for the final half-century of his career, could make constant revisions (often unsignalled) to reprints and resortings of his work. The bibliography of any GBS item, therefore, is likely to be complex; no attempt has been made here to trace the textual history of individual titles. [JC]

SHAYOL US slick-format small-press MAGAZINE, 7 issues, irregular, November 1977-1985, published by Flight Unlimited; ed Pat Cadigan (1953-).

One of the best-produced SMALL-PRESS magazines, beautifully designed and illustrated, this successor to CHACAL veered away from the association with Robert E. HOWARD to produce a wide range of FANTASY and SUPERNATURAL

FICTION. The emphasis remained on LOW FANTASY, but the very quality of this meant that sometimes it became a form of SLICK FANTASY, a trend especially noticeable in the works of Cadigan herself (her "Death From Exposure" won a 1979 BALROG AWARD), Michael BISHOP, Tom REAMY and Howard Waldrop (1946-   ). C.J. CHERRYH, Tanith LEE and Lisa TUTTLE were also represented. *S* featured artwork by Clyde Caldwell, Thomas Canty (1952-   ), Stephen Fabian (1930-   ), Hank Jankus (1929-1988), Tim Kirk (1947-   ) and Roger Stine (1952-1991). *S* won a 1981 WORLD FANTASY AWARD. [MA]

SHE The incarnation of the GODDESS in the **Ayesha** sequence by H. Rider HAGGARD (*which entry see for listing*), often referred to as "She" or "She-Who-Must-Be-Obeyed". Ayesha combines aspects of Aphrodite and Isis; she is imperious, ravishing, immortal and a LAMIA. Her sexual allure, which Haggard renders with a lack of horror unusual in a 19th-century writer of popular fiction, is central, and governs the plots of the **Ayesha** books. [JC]

SHE Sequence of three movies based on *She: A History of Adventure* (**1886**) by H. Rider HAGGARD.

1. *She* US movie (**1935**). Radio Pictures. **Pr** Merian C. Cooper. **Dir** Lancing C. Holden, Irving Pichel. **Vfx** Vernon Walker. **Screenplay** Dudley Nichols, Ruth Rose. **Starring** Nigel Bruce (Horace Holly), Helen Gahagan (She), Helen Mack (Tanya Dugdale), Randolph Scott (Leo Vincey). 89 mins. B/w.

An old Vincey family tale tells how the 15th-century John Vincey quested to the Arctic to find the FOUNTAIN OF YOUTH. Now Leo Vincey and manservant Holly follow that trail, picking up orphan Tanya on the way. Among the glaciers they are taken in by a cannibalistic LOST RACE; but these people are ruled by another, much more civilized people . . . who are in turn ruled by SHE. She lives behind a wall of flame and has been waiting centuries for the return of John Vincey, and immediately assumes Leo is John's REINCAR-NATION. Allured by her ethereal sensuality and her promise of IMMORTALITY, yet repelled by her cruelty, Leo opts instead for sensible, mundane Tanya. Arrogantly demonstrating in the immortality-granting Flame of Life how Tanya will wither and age, She inadvertently inflicts the full effects of the millennia upon herself, and dies.

Scott displays the sensitivity of a plank; Bruce previews the stalwart Watson he would play to Basil Rathbone's SHER-LOCK HOLMES; Mack is pretty and petite, but little else. Only Gahagan's performance conveys any sense of fantasy; she never made another movie. The screenplay, with its bizarrely gratuitous changes from the original, lumbers along; it is hardly helped by extended Busby Berkeley-style dance routines representing HUMAN-SACRIFICE rites. Yet *S* is not without importance: Frank CAPRA must have been influenced by it when he made LOST HORIZON (*1937*). [JG]

2. *She* UK movie (**1965**). Hammer/Associated British Picture Corporation/Warner-Pathé. **Pr** Michael Carreras. **Dir** Robert Day. **Spfx** George Blackwell, Bowie Films Ltd. **Mufx** Roy Ashton. **Screenplay** David T. Chantler. **Starring** Ursula Andress (Ayesha), Bernard Cribbins (Job), Peter Cushing (Major Horace L. Holly), Christopher Lee (Billali), Rosenda Monteros (Ustane), André Morell (Haumeid), John Richardson (Leo Vincey/Killikrates). 105 mins. Colour.

Three demobbed British Army servicemen – Vincey, Holly and Job – are *en route* home after WWI when Vincey is picked up in a Palestine bar by the improbably beautiful Ustane, and lured into the presence of Ayesha, a cruel ruler who has gained IMMORTALITY through bathing in the Secret Flame. For Vincey is the REINCARNATION of Killikrates, High Priest of Isis, the lover whom Ayesha slew millennia ago. After an arduous desert journey to Ayesha's lost-city civilization, the three men are accepted as guests while Ayesha woos Vincey. But Ustane, who has fallen in love with Vincey, follows the party, and his affections oscillate between her and Ayesha with bewildering swiftness. Ayesha eventually kills Ustane and turns Vincey into a catspaw, with the aim of giving him, too, eternal youth by having him bathe with her in the Flame. But her renewed immersion in the Flame reverses her immortality, and within minutes she ages until she is merely dust. Vincey is left to face eternity without her.

This was certainly one of HAMMER's more stylish productions, and was commercially very successful – leading to a sequel, **3**. Viewed today, it seems dated, with chanting painted natives seemingly plucked from a TARZAN MOVIE (Ayesha's own soldiers are, puzzlingly, kitted out as Roman legionaries). But its *brio* allows much to be forgiven. [JG]

3. *The Vengeance of She* UK movie (**1968**). 20th Century-Fox/Seven Arts-Hammer. **Pr** Aida Young. **Dir** Cliff Owen. **Spfx** Bowie Films Ltd. **Mufx** Michael Morris. **Screenplay** Peter O'Donnell. **Starring** Olinka Berova (Carol/Ayesha), Derek Godfrey (Men-Hari), Edward Judd (Dr Philip Smith), Daniele Noel (Sharna), John Richardson (Killikrates), Noel Willman (Za-Tor). 101 mins. Colour.

A modest sequel to **2**, reversing the sexes. Mini-skirted amnesiac Carol wanders vaguely towards North Africa, psychically guided (largely through DREAMS) and magically protected by magi in the service of Killikrates, who believes her the REINCARNATION of Ayesha. She gains the affection of psychiatrist Philip, who – after adventures – accompanies her to Killikrates's mountain fastness. There her mind is enslaved by scheming magus Men-Hari so that she becomes persuaded she is Ayesha, and prepares to pass through the Secret Flame and indeed become her *alter ego*. Killikrates' slaves rebel. Aided by slave-girl Sharna, who loves Killikrates, Philip confronts the pair just as Carol prepares to enter the Flame, and uses psychological suggestion to return her memory. When all is revealed as a plot by Men-Hari to attain IMMORTALITY for himself, Killikrates suicides by entering the Flame. Rebel leader Za-Tor calls down destruction upon all for having turned from white to black MAGIC – only Carol and Philip escape the disaster.

Berova, presumably chosen for her resemblance to Andress, is sweet but buxomly vacuous in the HAMMER style, and the hole left at the movie's core by her lack of presence remains unfilled. Yet there are some good moments, notably the RITUAL sequence in which Men-Hari calls up a DEMON to destroy the lone psychic who attempts to break the SPELL placed on Carol. [JG]

SHEA, MICHAEL (1946-   ) US writer of exotic fantasy and supernatural horror, not to be confused with the UK thriller writer Michael Shea (1938-   ). While MS's work may appear imitative, he is able to mix and blend styles and content to suit his subject matter. He first appeared in print with *A Quest for Simbilis* (**1974**), a continuation of Jack VANCE's **Cugel** tales, which had ended abruptly in *The Eyes of the Overworld* (coll of linked stories **1966**). Many critics believed MS was audacious to attempt this, but grudgingly admitted he had captured much of the style and flair of Vance and also had shown some originality. MS was quiet for a few years but, when he re-emerged with *Nifft the Lean* (coll of linked stories **1982**), he showed he had developed the exotic style of Vance (perhaps influenced by Clark Ashton SMITH) plus the ingenuity of Fritz LEIBER's **Gray Mouser** stories to produce an extravagant QUEST novel. It received a WORLD FANTASY AWARD. However, its companion piece, *In Yana, the Touch of Undying* (**1985**), about a vain opportunist's search for IMMORTALITY in a LAND OF FABLE, while equally inventive, is rather more selfindulgent. These novels are best appreciated in short doses. *The Color Out of Time* (**1984**; vt *The Colour Out of Time* 1986 UK) is MS's sequel to H.P. LOVECRAFT's "The Colour Out of Space" (1927), but here MS only borrows the setting and background, not attempting to pastiche Lovecraft's style. What he was first criticized for doing with Vance, MS was now criticized for not doing with Lovecraft.

MS's best work is his short fiction, especially the stories collected in *Polyphemus* (coll **1987**) where, although they betray the possible stylistic influence of Stephen KING, the stories are closer to MS's own voice. [MA]

Other works: *Fat Face* (**1987** chap), an addition to the CTHULHU MYTHOS; *I, Said the Fly* (1989 *The Omni Book of Science Fiction*; rev **1993** chap).

SHEARING, JOSEPH Pseudonym of Marjorie BOWEN.

SHEENA US movie (**1984**; vt *Sheena, Queen of the Jungle*). Columbia. **Pr** Paul Aratow. **Exec pr** Yoram Ben-Ami. **Dir** John Guillermin. **Screenplay** David Newman, Lorenzo Semple Jr, Leslie Stevens. **Based on** the COMICS by Will EISNER and Jerry Iger. **Starring** Tanya Roberts (Sheena), Trevor Thomas (Otwani), Elizabeth of Toro (Shaman). 115 mins. Colour.

Orphaned in Africa when her explorer parents – come in search of legendary healing sands – are killed in a rockfall, the infant Janet is told by the Shaman of the remote Zambouli tribe that her coming fulfils a PROPHECY: she is to be Sheena, Queen of the Jungle. When she is an adult she can talk with the animals and has limited telepathy (◊ TALENTS). A fairly standard adventure ensues, involving villainous US-educated local Prince Otwani, until Sheena leads the wild animals and the Zambouli to destroy the oppressors.

This is a glamour movie in the style of the notorious Bo/John Derek *Tarzan, the Ape Man* (**1981**), although it is much better and drifts less towards pornography. Indeed, this late JUNGLE MOVIE resonates more with the "mainstream" TARZAN MOVIES – as well as with SHE (**1965**) and ONE MILLION BC (**1966** version). The jungle footage is spectacularly beautiful; but *S* has an uncertain script and a desperate artificiality: battling through the jungle, Sheena is superbly manicured. [JG]

See also: SHEENA, QUEEN OF THE JUNGLE (1955).

SHEENA, QUEEN OF THE JUNGLE US COMICS series. ◊ Will EISNER; SHEENA (**1984**); SHEENA, QUEEN OF THE JUNGLE (1955).

SHEENA, QUEEN OF THE JUNGLE US syndicated tv series (1955). Nassour Studios. **Pr** Edward Nassour. **Exec pr** Don Sharpe, William Nassour. **Starring** Christian Drake (Bob), Irish McCalla (Sheena). 26 30min episodes. B/w.

With numerous TARZAN MOVIES over the years, it was inevitable that a female version would make it to the screen. McCalla brought her imposing 6ft 1in frame to the role of a white woman living in the jungles of Zambuli, an IMAGINARY LAND in Africa. Orphaned in a plane crash and raised by a local tribe, she vows to remain there to protect her friends, both human and animal. Dressed in a tiger skin, she is aided by her chimpanzee, Chim, and Bob, a local trader.

The production team later commented that the biggest problem wasn't filming in the wilds of Mexico, which stood in for Zambuli; rather, it was finding someone to double as McCalla in the often hazardous stunts. Unable to find a stuntwoman of suitable stature, they put a blonde wig on a stuntman and selected camera angles carefully. *S,QOTJ* was based on a COMIC-strip character created by Jerry Iger and Will EISNER. Anita Ekberg had originally been cast in the starring role but dropped out shortly before filming began. [BC]

See also: SHEENA (**1984**).

SHELLEY, MARY (WOLLSTONECRAFT) (1797-1851) UK writer, author of *Frankenstein, or The Modern Prometheus* (**1818**; rev 1831; vt *Frankenstein* 1897) cited by Brian W. ALDISS and others as the first SCIENCE-FICTION novel, although this is not a universal view (◊ TECHNOFANTASY). The novel is associated with the HORROR genre; though its structure derives from the form of GOTHIC FANTASY, it contains no supernatural elements. The creation of the MONSTER, left to the reader's imagination in the first edition, is described on a more rational scientific basis in the third edition. The origins of the novel are well documented. It grew from a challenge set by Lord BYRON and Percy Bysshe SHELLEY during their sojourn in Switzerland to see who could write the most frightening GHOST STORY. MS's idea came to her in a dream. The first edition had an unsigned preface by Percy Shelley, and many thought the novel to be his, disbelieving that MS, only 19 and a mere woman, could have produced such a work of profound horror.

In fact, she had a strong literary background. Her mother, the renowned pioneer of women's liberation, Mary Wollstonecraft (1759-1797), died in childbirth. Her father was the writer and publisher William GODWIN. Her stepmother was Mary Jane Godwin (*née* Clairmont, 1766-1841), who went into partnership with William in publishing children's books. MS's first published work was a poem, "Mounseer Nongtongpaw", written when she was 10, which she contributed to her father's **Juvenile Library** in 1808. She had contact during childhood with many important writers of the day, including Samuel Taylor COLERIDGE, Charles Lamb (1775-1834) and Percy Shelley, whom Mary married in 1816. After he drowned in 1822, MS returned to England, where she continued to write for another 30 years, though she never repeated the success of *Frankenstein*. Her best-known other novel is *The Last Man* (**1826**) where a plague destroys mankind and a survivor travels south from Europe like some Byronic ACCURSED WANDERER. The novel contains no supernatural elements, and reads more like a projection of the Byron-Shelley fraternity into the future, with MS using the novel as therapy for her own grief. Of her other novels, *Valperga, or The Life and Adventures of Castruccio, Prince of Lucca* (**1823**) is an historical Gothic romance set in 14th-century Italy, and *Falkner* (**1837**), her last completed novel and modelled on her father's *Caleb Williams* (1794), is a rather lacklustre romance.

MS did utilize the supernatural in her short fiction, much of which appeared in the prestigeous ANNUALS of the day, such as *The Keepsake*. All these stories have a melancholy air. The most effective is "The Mortal Immortal" (1833 *Keepsake*), in which a magician's assistant drinks half of an ELIXIR OF LIFE but regrets his IMMORTALITY and seeks death.

The story is a powerful expression of loss. Grief is also the basis for "The Invisible Girl" (1832 *Keepsake*), about a valley haunted by the GHOST of a girl who keeps vigil in the hope of the return of her lost sweetheart. Other stories deal, like *Frankenstein*, with the restoration of life, including "Roger Dodsworth: The Re-animated Englishman" (written 1826; 1863), which retells a popular hoax of the day about an Englishman found frozen in a glacier and resuscitated. "Transformation" (1830 *Keepsake*) tells of an IDENTITY EXCHANGE between a vindictive youth and a vengeful DWARF. The best of MS's short fiction was collected as *Tales and Stories* (coll 1891) ed Richard GARNETT, although the texts were altered; a more definitive edition is *Collected Tales and Stories* (coll 1976 US) ed Charles E. Robinson, which includes further fragments. [MA]

Further reading: Not all books about MS are reliable. Most relevant are: *Mary Shelley: A Biography* (1938) by R. Glynn Grylls; *Child of Light: A Reassessment of Mary Wollstonecraft Shelley* (1951) by Muriel Spark; *Mary Shelley: Author of Frankenstein* (1953) by Elizabeth Nitchie; *Mary Shelley* (1959) by Eileen Bigland; *Ariel Like a Harpy: Shelley, Mary and Frankenstein* (1972; vt *Mary Shelley's Frankenstein: Tracing the Myth* 1973 US) by Christopher Small; *Mary Shelley* (1972) by William A. Walling; *Moon in Eclipse: A Life of Mary Shelley* (1978) by Jane Dunn; *Mary Shelley* (1985) by Harold BLOOM. The most complete bibliography is *Mary Shelley: An Annotated Bibliography* (1975) by W.H. Lyles.

SHELLEY, PERCY BYSSHE (1792-1822) UK Romantic poet, with Lord BYRON a pariah in his lifetime for his atheism and domestic life, both born from his revolutionary convictions. Like many poets of his day (◊ ROMANTICISM), PBS employed mythological themes, and his long poems use figures from Greek MYTHOLOGY in extravagant narratives that give passionate form to his vision of life. *Prometheus Unbound* (1820), perhaps his greatest work, retells Aeschylus's tale of the Titan PROMETHEUS who brings fire to humanity in radically transformed terms; PBS's (largely misleading) use of mythological figures obscures the degree to which the work is deeply personal. Harold BLOOM notes that the audacity of PBS "gives us a vision of last things without the sanction of religious or mythological tradition. Blake does the same, but Blake is systematic where Shelley risks everything on one sustained imagining". This burst of imaginative invention places PBS's work in a realm beyond the traditional fantastic modes of FOLKTALE or myth. Reading John KEATS's verse narratives of battle among Greek gods, one is always conscious of being told of supernatural events, but PBS's strongest poems – *The Witch of Atlas* (written 1820; 1824 chap), *Adonais* (1821 chap), *Epipsychidion* (1821 chap), *The Triumph of Life* (written 1822; 1824) – are scarcely about character and event; they strain to break free of narrative entirely, and anticipate the following century's developments in verse.

PBS is a figure rarely evoked in prose fantasy. His presence is usually seen in citations of images from his poems that can be read as conventional sf or fantasy (the tableau in "Ozymandias" [1817]): he is, however, a notable figure in Tim POWERS's *The Stress of Her Regard* (1989) and in various FRANKENSTEIN MOVIES. Fantastic in a sense than no other poet in English was to that time (or, arguably, since), PBS is known in GENRE FANTASY almost exclusively as the husband of Mary SHELLEY. [GF]

Other works: *Queen Mab, A Philosophical Poem* (1813);

Alastor, or the Spirit of Solitude (1816); *The Masque of Anarchy* (written 1819; 1832).

SHELLEY, RICK Working name of US writer Richard Michael Shelley (1947-), who began his career with two fantasy series but has more recently concentrated on sf. The **Varayan Memoir** – *Song of the Hero* (1990), *The Hero of Varay* (1991) and *The Hero King* (1992) – follows the adventures of a young man from Earth who, on following his lost parents through a PORTAL into the first of a number of ALTERNATE WORLDS barring him from the final realm where the ELVES rule, discovers he is a HIDDEN MONARCH with heavy responsibilities. The QUEST which follows, dogged by PLOT DEVICES, is sharply recounted; later volumes incorporate some TECHNOFANTASY. The **Seven Towers** sequence – *The Wizard at Meq* (1994) and *The Wizard at Home* (1994) – is set in a LAND-OF-FABLE 12th-century England and features a WIZARD whose newly gained powers have offended some GODS. [JC]

Other works (sf): The **Spaceborne** sequence, being *Until Relieved* (1994), *Side Show* (1994) and *Jump Pay* (1995); *The Buchanan Campaign* (1995).

SHEPARD, E(RNEST) H(OWARD) (1879-1976) Prolific UK book illustrator with a fine pen-line style, most famous for his ILLUSTRATIONS for A.A. MILNE's **Winnie the Pooh** books. His drawings had great charm: they were strongly conceived but had the informal look of sketches, and during a long and successful working life he illustrated a great number of classics of English literature. Contemporary authors whom he illustrated, aside from Milne, included Kenneth GRAHAME, Frances Hodgson BURNETT and E.V. Lucas (1868-1938). His greatest talent was for depicting humanized animals, as in Milne's *Winnie the Pooh* (1926) and Grahame's *The Wind in the Willows* (1931 edn).

EHS studied at Heatherley's School of Art and at the Royal Academy Schools, and began work as a cartoonist for *Punch*. He became one of the best-loved children's illustrators of his day. [RT]

Further reading: *Drawn from Memory* (1957) and *Drawn from Life* (1961), both by EHS; *The Work of E.H. Shepard* (anth 1979) ed Rawle Knox.

SHEPARD, LUCIUS (TAYLOR) (1947-) US writer whose professional career began in the 1980s, but under whose name four stories and four articles were published 1952-5 in *Collins Magazine* (variously retitled *Collins, the Magazine to Grow Up With* and *Collins Young Elizabethan*), the first short story being the remarkably competent "Camp Greenville" (1953); in conversation with John CLUTE, LS (who would have been 6 in 1953) indicated that a family member had attached his name to these items. LS's first acknowledged work is a poem, *Cantata of Death, Weakmind & Generation* (1967 chap); his first adult prose of interest is the text to James Wolf's *Moon Flying* (portfolio 1978). He began to publish stories of genre interest with "The Taylorsville Reconstruction" for *Universe 13* (anth 1983) ed Terry CARR.

LS's work is mostly sf, HORROR or SUPERNATURAL FICTION, often focusing on psychic or literal POSSESSION. Straightforward fantasy is comparatively rare, the most striking example being the **Dragon Griaule** sequence – "The Man who Painted the Dragon Griaule" (1984 *F&SF*), *The Scalehunter's Beautiful Daughter* (1988) and *The Father of Stones* (dated 1988 but 1989) – set in an ALTERNATE WORLD "separated from this one by the thinnest margin of possibility" where DRAGONS exist. The Dragon Griaule – 750ft high and 6000ft long – dominates a huge valley and its human inhabitants; though he has been trapped through a WIZARD's spell for millennia in immobile BONDAGE, his hypnotic presence – "the cold tonnage of his brain" – seems to operate on those within or around him as a kind of psychic resonator, a form of tacit daemonic enthralment which intensifies the nature (and the limitations) of human experience. The scalehunter's daughter, escaping an attempted rape, enters Griaule's innards, where she finds a kind of WONDERLAND social order, which transforms her. The painter in the first story, which is set later (and is MAGIC REALISM), undertakes a commission to paint Griaule to death through the chemicals contained in the paints which are transfiguring the dragon into high art; but the painter himself, a KNIGHT OF THE DOLEFUL COUNTENANCE, dies this side of any transcendence. Other stories of interest are assembled in *The Jaguar Hunter* (coll **1987**; 1 story cut and 3 added 1988 UK; cut 1989 US), which won a WORLD FANTASY AWARD, *Nantucket Slayrides* (coll **1989**), with one of the three stories by Robert Frazier (1951-   ), *The Ends of the Earth* (coll **1991**), which also won a World Fantasy Award, and *Sports & Music* (coll **1994** chap).

Much of LS's work is set either in a LAND-OF-FABLE South America or literally there. His protagonists are often OBSESSED SEEKERS, as are those of the author he most often seems indebted to, Joseph CONRAD. *Kallimantan* (**1990** UK) evokes a specifically Conradian heart of darkness in describing the hegira through Borneo of a man obsessed with the native world and translated through a PORTAL into bleached transcendence. Both LS and Conrad are married to the geographical WATER MARGINS of the world, and both construct narratives in which obsession and BELATEDNESS come together in laments for unattainable understanding, and for moral closure which is rarely gained.

LS's other works are of indirect fantasy interest. His first novel, *Green Eyes* (**1984**), combines the scientific creation of ZOMBIES with a DARK-FANTASY climax in the Louisiana bayous near NEW ORLEANS. *Life During Wartime* (fixup **1987**) is also sf, a near-future tale involving drugs and alterered PERCEPTIONS set in a Vietnam-like Latin America; the first section is based on "]R & R" (1986), which won a Nebula AWARD. *The Golden* (**1993**), which won a LOCUS AWARD, portrays a complex, Realpolitik-ridden society of VAMPIRES; much of the book is set in a vast EDIFICE whose insterstices (and LIBRARY) are evocative of the work of Jorge Luis BORGES. *The Last Time* (**1995** chap) is horror.   [JC]

SHERLOCK HOLMES Arthur Conan DOYLE's Holmes/Watson partnership is an UNDERLIER for many DUOS in DETECTIVE/THRILLER FANTASY – Randall GARRETT's **Lord Darcy** and Master Sean are the obvious example, and Barry HUGHART's **Master Li** books pair a wilful genius with a stolid SENSIBLE MAN. The mythic stature of SH himself makes him a tempting RECURSIVE-FANTASY character to include in any URBAN FANTASY set in Victorian LONDON – and often elsewhere or elsewhen. August W. DERLETH published the **Solar Pons** series of pastiches, assembled as *The Adventures of Solar Pons* (coll **1945**; vt *Regarding Sherlock Holmes* 1974; vt *The Adventures of Solar Pons* 1975 UK); C.S. LEWIS's *The Magician's Nephew* (**1955**) uses SH's presence in Baker Street to date the story; *A Study in Terror ** (**1966**; vt *Sherlock Holmes vs Jack the Ripper* 1967 UK) by Paul W. Fairman (1916-1977) writing as Ellery Queen novelizes *A Study in Terror* (**1965**; vt *Fog*), where SH takes on the case of JACK THE RIPPER, a theme repeated

in *Murder by Decree* (**1978**), novelized by Robert Weverka as *Murder by Decree ** (**1979**); Nicholas Meyer (1945-   ) started an interesting recursive series with *The Seven-Per-Cent Solution* (**1974**); Philip José FARMER's *The Adventure of the Peerless Peer* (**1974**) pastiches the SH saga in the context of his **Wold Newton Family** saga; *Sherlock Holmes's War of the Worlds* (**1975**) by Manly Wade WELLMAN and Wade Wellman matches SH to H.G. WELLS, as does *Morlock Night* (**1979**) by K.W. Jeter (1950-   ); *The Giant Rat of Sumatra* (**1977**) by Richard L. Boyer becomes, disappointingly, a RATIONALIZED FANTASY, the "rat" proving to be a tapir; *Exit Sherlock Holmes* (**1977**) by Robert Lee Hall (1941-   ) has Moriarty as SH's *alter ego*, and involves TIME TRAVEL; Fred SABERHAGEN's *The Holmes-Dracula File* (**1978**) sees SH aided by a VAMPIRE against the Giant Rat of Sumatra; *Sherlock Holmes vs Dracula, or The Adventure of the Sanguinary Count* (**1978**) and *Dr Jekyll and Mr Holmes* (**1979**) by Loren D. Estleman (1952-   ) are of obvious associational interest; Michael KURLAND's pastiches to date are *The Infernal Device* (**1979**), *Death by Gaslight* (**1982**) and *A Study in Sorcery* (**1989**); *Time for Sherlock Holmes* (**1983**) by David Dvorkin (1943-   ) is another time-travel story, with SH having discovered the secret of eternal youth; the revenant SH of modern London in Robert RANKIN's *East of Ealing* (**1984**) is a comic travesty; *Ten Years Beyond Baker Street: Sherlock Holmes Matches his Wits with the Diabolical Dr Fu Manchu ** (**1984**) by Cay Van Ash (1918-1994) is explained by its subtitle; Esther FRIESNER's *Druid's Blood* (**1988**) has a version of SH saving a magical ALTERNATE-WORLD Victorian England; and Roger ZELAZNY's crowded *A Night in the Lonesome October* (**1993**) features him in a bit part. Further examples abound, notably Philip PULLMAN's *Sherlock Holmes and the Limehouse Horror* (**1992**). The ANIMAL-FANTASY homage *Basil of Baker Street* (**1974**) by Eve Titus became DISNEY's ANIMATED MOVIE *The Great Mouse Detective* (**1986**). SH movies of at least associational interest, either based on Doyle or recursive, include (aside from those mentioned above): *The Hound of the Baskervilles* (**1939**; **1958**; **1977**); *Sherlock Holmes and the Voice of Terror* (**1942**); *Sherlock Holmes and the Spider Woman* (**1944**); *The Scarlet Claw* (**1944**); *The Private Life of Sherlock Holmes* (**1970**) a GASLIGHT ROMANCE/TECHNOFANTASY which also involves the Loch Ness Moster; *They Might Be Giants* (**1971**), an excellent fantasy of PERCEPTION set in contemporary NEW YORK, where a latter-day SH calls up the Bleeker Street Irregulars; *Young Sherlock Holmes* (**1985**; vt *Young Sherlock Holmes and the Pyramid of Fear*) in which schoolboys Holmes and Watson battle a wicked cult and nasty MESMERISM; and *1994 Baker Street: Sherlock Holmes Returns* (**1994**). An anthology of recursive stories is *Sherlock Holmes Through Time and Space* (anth **1984**) ed Isaac ASIMOV, Martin GREENBERG and Charles G. Waugh.   [DRL/JG]

SHERMAN, DELIA Working name of US writer Cordelia Sherman (1951-   ), born in Japan but resident in the USA since childhood. She first appeared in print with "The Maid on the Shore" (*F&SF* 1987). Her first novel, *Through a Brazen Mirror* (**1989**), retells the traditional ballad "Famous Flower of Serving Men". When the sorceress mother of Fair Elinor has her husband and child murdered, Elinor dons GENDER DISGUISE and embarks on a career that results in her (him) becoming the king's chamberlain. The novel's artful multiplicity of style, the use of the ballad as a template and the emphasis on the individual's finding her/his place in the world unaided, as well as an ending which upsets every readerly expectation make the work a type-specimen of FANTASY OF MANNERS. It earned DS a nomination for the John W. Campbell Memorial AWARD.

The Porcelain Dove (**1993**), which won DS the MYTHOPOEIC AWARD, is unusually set in 18th-century France, with a smooth, unified and highly sophisticated style to match. A QUEST for the eponymous object lifts a CURSE and thereby creates a time POLDER where the characters are thereafter untouched by the outside world. But this is almost a subplot to the tale of a poor girl who becomes maid to an aristocratic wife and then part of the higher society; it is also a lesbian love story. DS clearly has more quietly subversive stories to tell. [DGK]

SHERMAN, FRANK DEMPSTER (1860-1916) US writer. ◊ John Kendrick BANGS.

SHERMAN, JOSEPHA (?   -   ) US writer, editor and folklorist. Her first two adult novels, arguably her strongest work, were *The Shining Falcon* (**1989**), which won the Compton Crook AWARD, and *The Horse of Flame* (**1990**); both are understated romantic fantasies that make use of Slavic and Russian FOLKLORE. *Child of Faerie, Child of Earth* (**1992**), *Windleaf* (**1993**) and *Gleaming Bright* (**1994**), all for children, draw on Slavic and Celtic folk traditions, while *A Strange and Ancient Name* (**1993**), *King's Son, Magic's Son* (**1994**) and *The Shattered Oath* (**1995**) are straightforward CELTIC FANTASIES. JS is a careful researcher, but her novels convey the flavour of fairytale or legend rather than historical fantasy. [JCB]

Other works: To the **Unicorn** SHARED-WORLD series JS has contributed *The Secret of the Unicorn Queen Book I: Swept Away* * (**1988**) and *Book V: The Dark Gods* * (**1989**); for **The Bard's Tale** GAME-tie series JS wrote *The Bard's Tale: Castle of Deception* * (**1992**) with Mercedes LACKEY and, solo, *The Bard's Tale: The Chaos Gate* * (**1994**); *A Sampler of Jewish-American Folklore* (anth **1992**); *Rachel the Clever and Other Jewish Folktales* (anth **1993**); *Once Upon a Galaxy* (**1994**), nonfiction about MYTHS behind contemporary legends; *Orphans of the Night* (anth **1995**), DARK FANTASY for children; «Lammas Night» (anth **1996**), mostly SWORD-AND-SORCERY stories based on a song lyric by Lackey.

SHERRELL, CARL (1929-1990) US commercial artist and writer whose sf and fantasy novels feature action routines and a certain philosophical poignance which do not always work together. Of greatest interest is the **Raum** sequence – *Raum* (**1977**) and *Skraelings* (**1987**) – about the gradual humanizing of a DEMON who, after being summoned from HELL to Earth by an unfortunately moral WIZARD, causes considerable mayhem in the first volume before meeting with ARTHUR and MERLIN, who teach him about life here, and falling in love with the lady Viviene. In the second volume he traces his kidnapped love through a Viking-dominated North, a trek which climaxes in North America ("skraeling" is a Viking term for the North American Indian). In *Arcane* (**1978**) a FOOL lives through a semi-divine life CYCLE according to a pattern governed by the TAROT. *The Space Prodigal* (**1981**) is sf and *The Curse* (**1989**) is horror. [JC]

SHETTERLY, WILL(IAM HOWARD) (1955-   ) US writer and publisher; he runs the SMALL PRESS Steeldragon Press. His first two novels, *Cats Have No Lord* (**1985**) and *Witchblood* (**1986**) made little impact, although *The Tangled Lands* (**1989**), set in a troubled FANTASYLAND created by a computer is more interesting. With his wife Emma BULL WS edited the **Liavek** series of anthologies – *Liavek* * (anth

1985), *The Players of Luck* * (anth **1986**), *Wizard's Row* * (anth **1987**), *Spells of Binding* * (anth **1988**) and *Festival Week* * (anth **1990**) – set in a venue that seems as determinedly charming as the **Thieves' World** series ed Robert L. ASPRIN is determinedly insalubrious.

WS's most impressive work to date is a YA novel in another perhaps too-pretty SHARED-WORLD venue: *Elsewhere* (**1991**) is set in the **Borderlands** universe created by Terri WINDLING. WS's story of teenaged runaways and punk ELVES shows an assurance his earlier fiction lacks. A sequel is *Nevernever* (**1993**). [GF]

Other works: *Double Feature* (coll **1994**) with Emma Bull.

SHIEL, M(ATTHEW) P(HIPPS) (1865-1947) UK writer of Irish-mulatto parentage, born (as Shiell) in Montserrat in the British West Indies. His SUPERNATURAL FICTION has a dedicated cult following, though he is better remembered for his SCIENCE FICTION and his unconventional detective and mystery fiction. His background, including his title as King of Redonda, an island in the Caribbean which his father claimed as his own, and his prolific output have made him of interest among collectors and bibliophiles. *The Works of M.P. Shiel: A Study in Bibliography* (**1948**; exp 1980 2 vols) by A. Reynolds Morse (1914-   ) and other books cited below give a guide through the maze of MPS's work. MPS was a highly competent linguist, and this gave him a remarkable if affected command of the English language. His florid but atmospheric style ideally suited the DECADENCE of the 1890s.

MPS started writing at age 12, but it was his discovery of the work of Edgar Allan POE in 1882 that really fired his imagination. His first sale was "The Doctor's Bee" (1889 *Rare Bits*). He translated stories for *The Strand Magazine* from 1891, but did not turn to writing full-time until the success of his first published book, *Prince Zaleski* (coll of linked stories **1895**), about a reclusive detective who solves crimes by logic and deduction from the depths of his archaic and exaggeratedly GOTHIC castle. MPS sustained this style through his short stories of this period, which include his best WEIRD FICTION, collected as *Shapes in the Fire* (coll **1896**) and *The Pale Ape and Other Pulses* (coll **1911**). Heavily influenced by Poe, overlain with images of decadent extravagance, these stories are unique in style and delivery and are among the most distinctive supernatural stories of the Victorian period. The best is "Vaila" (cut vt "The House of Sounds" in *The Pale Ape*), styled on "The Fall of the House of Usher", about an accursed house where a clock measures out its final days. Also from this period is "Huguenin's Wife" (1895 *Pall Mall*; rev in *The Pale Ape*), an effective story of REINCARNATION and transmigration. It shares with "Xélucha" (1896), "The Bride" (1902 *English Illustrated Magazine*) and others MPS's fascination for the FEMME FATALE returned from the grave. Although MPS continued to produce short stories for the next 40 years, little of his later work has been collected, although John GAWSWORTH included much in his ANTHOLOGIES. Some stories were reworked into episodic novels like *Here Comes the Lady* (coll of linked stories **1928**) and *The Invisible Voices* (fixup **1935**), mostly extensively revised and often for the poorer. Some of MPS's later stories were collaborations with (and sometimes uncredited revisions by) Gawsworth, Oswell Blakeston (1907-1985) and Edgar JEPSON. His most representative work is in *The Best Short Stories of M.P. Shiel* (coll **1948**) ed Gawsworth and *Xélucha and Others* (coll **1975**), the latter originally assembled by MPS for ARKHAM HOUSE in 1947.

As MPS's early writings were financially unsuccessful he turned to hackwork, including ghosting for others, especially Louis Tracy (1863-1928), with whom MPS sometimes collaborated on detective stories as Gordon Holmes. The extent of their work together has never been satisfactorily resolved, even though their styles were poles apart. Certainly MPS contributed extensively to "Tracy's" future-war novel *An American Emperor* (1896-97 *Pearson's Weekly*; **1897**), and this encouraged MPS's interest in such fiction. MPS's style remained ebullient but became more controlled for novels. These cover the whole range of commercial fiction, but the most significant is *The Purple Cloud* (1901 *Royal*; **1901**; rev 1929). Ostensibly sf – the first man to reach the North Pole returns to find all life on Earth has been destroyed by a poisonous gas released from volcanoes – the novel has the underlying theme of the BALANCE between GOOD AND EVIL, represented by the off-stage presence of supernatural agencies called Black and White. MPS's VILLAIN-turned-HERO explores the abandoned Earth like an ACCURSED WANDERER until at last he discovers another human. The movie *The World, the Flesh and the Devil* (*1959*) was based loosely on the book. MPS's interest in eternal wanderers and survivors from the grave reappears in *This Above All* (**1933**; vt *Above All Else* 1943), about the 20th-century lives and frustrations of those whom CHRIST raised from the dead and who have been rendered immortal (◊ IMMORTALITY); Christ himself survives as Raphael, living in a monastery in Tibet (◊◊ RELIGION; THEOSOPHY). These works link with MPS's interest in the theme of the Overman, which had emerged in his future-war novels *The Yellow Danger* (1898 *Short Stories* as "The Empress of the Earth"; **1898**) and *The Dragon* (1913 *Red Magazine* as "To Arms!"; **1913**; rev vt *The Yellow Peril* 1929), his political thriller *The Lord of the Sea* (**1901**; cut 1924 US), his short mystery stories featuring **Cummings King Monk**, his non-fantastic war novel *The Yellow Wave* (**1905**) and the mystical *How the Old Woman Got Home* (**1927**).

MPS was a mixed character. Essentially antireligious – his novel *The Last Miracle* (**1906**) seeks to discredit the Christian faith through the production of hoax MIRACLES – he still held strongly religious views, and believed scientific achievement would bring one closer to GOD. Although he could sometimes translate those views into his fiction, his emotions and linguistic pyrotechnics tended to dominate and obfuscate his message, leaving readers to draw what they wished from his fiction. The results are sometimes confused but seldom bland. [MA]

Other works: Mostly mysteries, but some of associational interest: *The Rajah's Sapphire* (**1896**) with W.T. Stead (1849-1912); *The Weird O' It* (1902 *Cassell's Saturday Journal* as "In Love's Whirlpool"; **1902**); *Unto the Third Generation* (**1903**); *The Isle of Lies* (**1908**); *Dr Krasinski's Secret* (**1929**), *The Black Box* (**1930**) and *The Young Men Are Coming!* (**1937**). The **Zaleski** and **Monk** stories were combined as *Prince Zaleski, and Cummings King Monk* (coll **1977**), with further discoveries in *The New King* (coll **1980**). Earlier unrevised and lesser-known material was assembled as *The Empress of the Earth, 1898; The Purple Cloud, 1901; Some Short Stories; Off-prints of the original editions* (omni **1979**).

Further reading: *The Quest for M.P. Shiel's Realm of Redonda* (**1979**) by A. Reynolds Morse; *Shiel in Diverse Hands: A Collection of Essays* (anth **1983**) ed Morse contains extensives studies and evaluations of MPS's work.

SHINING, THE UK movie (*1980*). Warner/Producer Circle. **Pr** Stanley Kubrick. **Exec pr** Jan Harlan. **Dir** Kubrick. **Screenplay** Diane Johnson, Kubrick. **Based on** *The Shining* (**1977**) by Stephen KING. **Starring** Scatman Crothers (Dick Halloran), Shelley Duvall (Wendy Torrance), Danny Lloyd (Danny Torrance), Barry Nelson (Ullman), Jack Nicholson (Jack Torrance), Philip Stone (Delbert Grady), Joe Turkel (Lloyd). 146 mins (usually cut to 119 mins). Colour.

Critically excoriated adaptation of King's SUPERNATURAL FICTION. Last year Delbert Grady, winter caretaker of the remote Outlook Hotel in Colorado, went berserk during the isolation, slaughtering his wife and daughters. This year struggling alcoholic writer Jack Torrance takes the job, hoping that same isolation will, through forcing him to dry out, help him overcome his writer's block. His psychic young son Danny voices his premonitions to himself through INVISIBLE COMPANION Tony; Tony predicts horror. As Jack and wife Wendy view the hotel, chef Halloran takes Danny aside and tells him he has recognized the boy's TALENT; he too has it, calling it the "Shining". After the Torrances are left alone, it is not long before the hotel's WRONGNESS makes itself felt. Jack becomes less and less rational, experiencing detailed HALLUCINATIONS and suffering nightmares (◊ DREAMS). The GHOSTS of Grady's TWIN daughters try to entice Danny to join them in the land of death, and later he is attacked by a mysterious naked woman in Room 237, the room he knows is the source of the EVIL. Terrified, the child psychically summons Halloran from his winter residence in Florida. This "betrayal" is communicated to Jack by Grady himself, taking the form of a waiter at a hallucinated 1920s party at which Jack is an honoured guest . . . The violence escalates as Jack succumbs to POSSESSION by the building, becoming progressively more infatuated by his own presumed cleverness (indeed, believing himself a JACK), until he is pursuing his terrified wife and son with a fire-axe . . .

It is a measure of the underlying strength of *TS* that it survives Nicholson's performance (apparently dictated by Kubrick), which seems modelled on the worst B-grade HORROR MOVIE, to stand as a powerful GHOST STORY. Much of the credit is down to the direction and the cinematography (by John Alcott), and to the truly chilling juxtaposition of such material with the brightly lit, sumptuously furnished, cheerily heated luxury hotel – which, as Wendy and Danny flit in terror through it, becomes an EDIFICE. Also successful is the intermixture with the supernatural fantasy of something quite different: some of the events we witness are, we know, products of the increasingly crazy Jack's PERCEPTION. The core of this movie's fright is that we don't know *which*. [JG]

SHINN, SHARON (RUTH) (1957-) US writer whose first novel, *The Shape-Changer's Wife* (**1995**), made an immediate impact through the grace and swiftness of its narrative, which unfolds the gradual realization of young Aubrey that not only the servants of the WIZARD who is teaching him SHAPESHIFTING but also the wizard's wife have all been shapeshifted from original nonhuman forms. The housekeeper is a SPIDER; Orion the handyman is a bear; and Lilith, with whom Aubrey falls in love, is a DRYAD. His slow and reluctant course towards a recognition of Lilith's true STORY – the painful BONDAGE into which she has been cast – is conveyed with genuine feeling throughout the text, and is an unusually clear example of the RECOGNITION of a passage into adulthood and power (◊◊ RITE OF PASSAGE).

[JC]

SHIP OF FOOLS This MOTIF dates back at least to medieval times, where a voyage in a ship – packed with every sort of person – is given as an ALLEGORY of the follies of society. The theme took significant literary shape with *Das Narrenschiff* (**1494**; trans Alexander Barclay as *The Ship of Fools* **1509** UK; best modern trans E.H. Zeydel **1944** US) by Sebastian Brant (1457-1521), a long narrative poem in which an SoF embarks for Narragonia ["Fool-Country"], which the passengers think will be a HEAVEN on Earth. They never reach their goal, as Brant's energy is almost entirely spent on SATIRE. Renaissance "Folly" literature – the most famous example being *Moriae encomium* (**1509**; trans J. Wilson as *The Praise of Folly* **1688** UK) by Desiderio Erasmus (1466-1536) – derives from Brant.

So pervasive was Brant's tale – and so well known were its ILLUSTRATIONS, long thought to have been by Albrecht Dürer – that any subsequent story set on a ship capable of carrying more than few passengers inevitably echoed the motif. More specific SoF narratives include Lewis CARROLL's *The Hunting of the Snark* (**1876**), Gerhart HAUPTMANN's *Atlantis* (**1912**), Robert Neumann's *Ship in the Night* (coll of linked stories **1932**), John BRUNNER's sf *Sanctuary in the Sky* (**1960** dos), *Ship of Fools* (**1962**) by Katherine Anne Porter (1890-1980) – an ALLEGORY of doom-bound humanity set in 1931 and intended to point symbolically towards WORLD WAR II – Gene WOLFE's *The Urth of the New Sun* (**1987**) and Iain SINCLAIR's *Radon Daughters* (**1994**). The original SOF, with an updated passenger list, is explicitly featured in John Myers MYERS's *Silverlock* (**1949**). [JC]

SHIVA In medieval and modern Hinduism, Vishnu and Shiva are both treated as supreme deity by their respective worshippers. More than any of the supreme GODS of the West – like ODIN or ZEUS – Shiva combines two contradictory aspects, each clearly articulated and separate, especially in early references. The "Shiva" aspect is the divine physician; the "Rudra" aspect is the great destroyer. The two aspects – even in the *Mahabharata*, where he is known mainly as Shiva – SHADOW one another. Shiva reconciles opposites; he is the ultimate ground upon which the BALANCE of the world is assured; he is also the Lord of the Dance, and depending on the nature of that dance he brings the world into being (◊ CREATION MYTHS) or ends it (◊ APOCALYPSE).

He is rarely specifically used in Western fantasy as an UNDERLIER, though in a general sense it is reasonable to assume that he provides a rough model for fictional gods who both raven and heal. The hero of Warren B. MURPHY's **Destroyer** sequence is based specifically upon Shiva the Destroyer; this aspect is also recreated in Roger ZELAZNY's *Lord of Light* (**1967**), with the help of TECHNOFANTASY weaponry. [JC]

SHOCK ◊ MAGAZINES.

SHOCK TREATMENT UK movie (**1982**). ◊ *The* ROCKY HORROR PICTURE SHOW (**1975**).

SHOP H.G. WELLS's "The Magic Shop" (in *Twelve Stories and a Dream* coll **1903**) established and gave its name to the useful PLOT DEVICE of a mysterious shop selling real MAGIC; Honoré de BALZAC had already used the device in *La Peau de chagrin* (**1831**). Often the wares are THINGS BOUGHT AT TOO HIGH A COST: a circus performer in "The Second Awakening of a Magician" (1930) by S.L. Dennis exchanges his SOUL for great strength, and accidentally crushes a loved woman with

a hug. Often the shop has vanished when one tries to return to it, though the second visit is the whole point of John COLLIER's "The Chaser". Terry PRATCHETT's *The Light Fantastic* (**1986**) suggests that wandering shops or "tabernae vagrantes" are under CURSES for giving bad service to WIZARDS. Further examples are numerous, including Lord DUNSANY's "The Bureau d'Echange de Maux" (in *Tales of Wonder* coll **1916**), whose vanishing shop allows you to swap one affliction for another that *seems* lesser. Others are in Theodore STURGEON's "Shottle Bop" (1941), Fritz LEIBER's "Bazaar of the Bizarre" (1963), whose shop sells rubbish disguised by ILLUSION, Diana Wynne JONES's *The Ogre Downstairs* (**1974**), Harlan ELLISON's "Shoppe Keeper" (1977), Piers ANTHONY's *On a Pale Horse* (**1983**) – where magic shops are routine in an ALTERNATE WORLD – and Peter DICKINSON's *A Box of Nothing* (**1985**), whose Nothing Shop sells precisely that. Still more unreliable, though perhaps less magical, are purchases at CARNIVAL stalls. On a larger scale, department stores house strange goings-on in John Collier's "Evening Primrose", where a WAINSCOT society emerges by night, and Thorne SMITH's *Rain in the Doorway* (**1933**), whose drunkenly anarchic OTHERWORLD store is an actualization of ESCAPISM during the US Depression. [DRL]

See also: ANSWERED PRAYERS; MANNEQUIN (**1987**); SLICK FANTASY.

SHULER, LINDA LAY (? -) US tv and radio producer and writer. Her ongoing **Time Circle Quartet** – to date *She Who Remembers* (**1987**) and *Voice of the Eagle* (**1992**) – depicts in fantasy terms the (Native American) Anasazi nation in the 13th century through the life of a woman, Kwani, whose ability to speak with the dead, and to foretell the future and to convey lore to the women of the tribe marks her as a kind of shaman (◊ SHAMANISM) linked to the MOON and the GODDESS. [JC]

SHWARTZ, SUSAN M(ARTHA) (1949-) US writer and businesswoman, known as much for sf as for fantasy. Her first story of genre interest, "The Fires of her Vengeance" in *The Keeper's Price* (anth **1979**) ed Marion Zimmer BRADLEY, is fantasy. Her first novel, *White Wing* (**1985**) with S.N. Lewitt, together as Gordon Kendall, is sf. The **Heirs to Byzantium** sequence – *Byzantium's Crown* (**1987**), *The Woman of Flowers* (**1987**) and *Queensblade* (**1988**) – begins as an ALTERNATE-WORLDS tale in which Antony and Cleopatra shifted the capital of the Roman Empire to Byzantium. In medieval times, Byzantium remains the centre of the world, which gradually evolves into a LAND-OF-FABLE Europe. MAGIC is introduced to the DYNASTIC-FANTASY storyline, which involves heirs to the throne, a power-mad wicked STEPMOTHER, druids with magic nous, and (in the final volume, which takes place after a savage THINNING of the world) the sacrifice of a brave woman to bring in the HEALING. There are some Arthurian echoes (◊ ARTHUR). Two unconnected stories – "Seven from Caer Sidi" (in *Invitation to Camelot* anth **1988** ed Parke GODWIN) and "The Count of the Saxon Shore" (in *Alternatives* anth **1989** ed Robert ADAMS and Pamela Crippen Adams) – also make use of the MATTER of Britain. *Silk Roads and Shadows* (**1988**), set in something closely resembling the historical Byzantium, is a RITE-OF-PASSAGE tale featuring the trials of a young woman (the Emperor's sister) who goes to China to steal the silkworms necessary to revitalize the Byzantine silk industry; some of the trials involve magic.

Imperial Lady: A Fantasy of Han China (**1989**) with Andre NORTON is a SUPERNATURAL FICTION set in a China whose land-of-fable elements are subdued, and sets its upright Chinese maiden forthright tasks of survival among the Huns to whom she has been sent by an unwilling Emperor (the tale is based on fact).

The Grail of Hearts (**1992**) returns to the Arthurian cycle, telling the GRAIL story from the viewpoint of Kundry, who is rather maligned by Richard WAGNER in *Parsifal* (**1883**), and who is revealed by SMS to be a version of the WANDERING JEW; here she is a kind of TEMPORAL ADVENTURESS, having timeslipped (◊ TIMESLIP) backwards into the arms of the FISHER KING, and having a complex, dark, spirited response to the events into whose heart she has been thrust. This is SMS's most interesting novel: it gives her a chance to exploit her extensive knowledge of history (she has a doctorate in medieval English literature) and to utilize to the full her usual narrative technique – allowing complexities and information to sidle inconspicuously into seemingly routinized venues. *Empire of the Eagle* (**1993**) with Norton is another historical fantasy, set at the edges of the pre-Christian Roman Empire and beyond, where magic still exists, as does a PARIAH ELITE of immortal sorcerers (◊ SORCERY). At her best, SMS's fantasy premises illuminate genuine arguments about the nature of history; there is an *interestingness* about her work. [JC]

Other works (sf): *Heritage of Flight* (fixup **1989**).

As editor: *Hecate's Children* (anth **1982**); *Habitats* (anth **1984**); *Moonsinger's Friends* (anth **1985**), in honour of Andre Norton; *Arabesques: More Tales of the Arabian Nights* (anth **1988**) and *Arabesques II* (anth **1989**) (◊ ARABIAN FANTASY); *Sisters in Fantasy* (anth **1995**) with Martin H. GREENBERG.

SIBYL A prophetess of the god APOLLO who formed a similar function to the ORACLES in predicting the future, usually in an enigmatic and ambiguous form. There were anything up to 12 sibyls in the ancient world, the most famous being Amalthaea at Kume (Cumae), near modern Naples. According to Livy (59BC-AD17), Amalthaea maintained all her prophecies in nine books which she offered to Tarquin the Proud, king of Rome (616BC-578BC). He refused to pay the price she asked, so Amalthaea burned three of the books and a year later offered the remaining six at the same price. Again he refused and again she burned three. A year later he bought the final three at the full price. These became known as the Sibylline Books, and remained Rome's primary source of consulting the GODS until destroyed in a fire in 83BC. *Wolfwinter* (**1972**) by Thomas Burnett SWANN is the story of the sibyl Erinna. [MA]

SIDHE ◊ FAIRIES.

SIEGEL, BARBARA (B.) (?1952-   ) US writer. ◊ Scott SIEGEL.

SIEGEL, SCOTT (WARREN) (1951-   ) US writer, most of whose work consists of ties contributed to various series, some SHARED-WORLD, some sf and some fantasy, generally for a YA audience and generally with his wife Barbara. For the **Dark Forces** sequence he wrote *The Companion* * (**1983**) and *Beat the Devil* * (**1983**). For the **Twistaplots** sequence he wrote *Ghost Riders of Goldspur* * (**1985**) with BS. All further titles are with BS. For the **Wizards, Warlocks and You** sequence they wrote *Wizards, Warlocks and You #6: Revenge of the Falcon Knight* * (**1985**), *#12: The Scarlet Shield of Shalimar* * (**1986**) and *#18: The Warrior Women of Weymouth* * (**1986**). For the **Dragonlance Preludes II** sequence they wrote *Dragonlance*

Preludes II #3: Tanis, the Shadow Years * (**1990**). For the **Ghostworld** sequence they wrote *Ghostworld #1: Beyond Terror* * (**1991**), *#2: Midnight Chill* * (**1991**), *#3: Dark Fire* * (**1992**) and *#4: Cold Dread* * (**1992**). [JC]

Other works (sf): In the **Junior Transformers** sequence, *Battle Drive* * (**1985** chap); in the **Which Way Books** sequence *The Champ of TV Wrestling* * (**1986**); a **Star Trek** tie, *Star Trek: Phaser Fight* * (**1986**); in the **G.I. Joe** sequence *G.I. Joe #6: Operation: Death Stone* * (**1986** chap), *#13: Operation: Snow Job* * (**1987** chap) and *#17: Operation: Sink or Swim* * (**1987** chap); in the **Firebrats** sequence, *Firebrats #1: The Burning Land* * (**1987**), *#2: Survivors* * (**1987**), *#3: Thunder Mountain* * (**1987**) and *#4: Shockwave* * (**1988**).

SIEGE PERILOUS ◊ ARTHUR; ROUND TABLE.

SIENKIEWICZ, BILL (1958-   ) US COMICS illustrator and writer whose early work has been compared to that of Neal ADAMS, combining highly polished and dynamic distortion with elements of photorealism. His GRAPHIC NOVELS with writer Frank MILLER include *Love and War* (graph **1986**) and *Elektra Assassin* (graph **1987**). He wrote and illustrated *Stray Toasters* (**1988-9**) for MARVEL COMICS and illustrated Alan MOORE's unfinished *Big Numbers* (2 issues 1990). [JC]

SIENKIEWICZ, HENRYK (1846-1916) Polish novelist. ◊ POLAND.

SILKE, JAMES R. (?   -   ) US author of the **Frank Frazetta's Death Dealer** sequence of ties: *Prisoner of the Horned Helmet* * (**1988**), *Lords of Destruction* * (**1989**), *Tooth and Claw* * (**1989**) and *Plague of Knives* * (**1990**). Written around Frank FRAZETTA's **Death Dealer** paintings, these are SWORD-AND-SORCERY epics notable for their violence. [JC]

SILKIES ◊ SELKIES.

SILVERBERG, ROBERT (1935-   ) US writer, author of more than 1000 stories since his first, "Gorgon Planet" for *Nebula Science Fiction* in 1954, and of an unknown number of novels (many under unrevealed pseudonyms in various genres) since *Revolt on Alpha C* (**1955**). He is central to the history of sf, especially for the stories and novels published 1967-76 (◊ SFE), but is not a comparatively significant author of fantasy. The **Majipoor** sequence – *Lord Valentine's Castle* (**1980**), *The Desert of Stolen Dreams* (**1981**), *Majipoor Chronicles* (coll of linked stories **1982**) and *Valentine Pontifex* (**1983**) – features extravagant (but coolly told) adventures in a PLANETARY-ROMANCE venue. *Gilgamesh the King* (**1984**) is a TWICE-TOLD rendering of GILGAMESH; the same character appears in "Gilgamesh in the Outback" (1986), which won a 1987 Hugo AWARD and which is incorporated into *To the Land of the Living* (fixup **1989**), an AFTERLIFE fantasy in which the dead are reborn into an Earth-like venue from which, after recycling their lives interminably, they tend to wish to escape; the rest of the fixup is spun off from the **Heroes in Hell** SHARED WORLD. The book sees Gilgamesh break through a PORTAL into present-day NEW YORK. [JC]

SIM, DAVE (1958-   ) US COMICS writer/artist. ◊ CEREBUS THE AARDVARK.

SIMAK, CLIFFORD D(ONALD) (1904-1988) US writer and journalist who in his long literary career chiefly produced sf (◊ SFE). Much of his fiction had – or yearned for – rustic settings, usually in the US Midwest, and was flavoured with nostalgia, gentle whimsy and a likeably old-fashioned moral sense; these qualities carried over to his fantasies, mostly late and minor works. Despite its title *The*

Werewolf Principle (**1967**) is sf, rationalizing SHAPESHIFTING in terms of alien psi TALENTS. *The Goblin Reservation* (**1968**) is a curious SCIENCE-FANTASY mixture, with supernatural beings like FAIRIES, a GHOST, GOBLINS and TROLLS now given scientific recognition in a FAR-FUTURE sf setting not without menace but heavily larded with whimsy. *Out of Their Minds* (**1969**) imagines FANTASYLAND as an sf reification of human belief: the thus-created DEVIL protests our shift of interest to UFOs, TOONS, etc., by shutting down Earth's modern technology, but is too easily thwarted by the reified Don Quixote. *Enchanted Pilgrimage* (**1975**) approaches straight fantasy; it is a generally routine QUEST with motley COMPANIONS through a MAGIC-ridden ALTERNATE WORLD whose odd inhabitants include an amorphous "Chaos Beast", reminiscent of CTHULHU-MYTHOS deities, that gives Caesarean birth to an apparent robot. Fantasy elements – WITCH, GIANT, magic SWORD, healing UNICORN horn – mix uneasily with sf hints of aliens and galactic society. Sf is further downplayed in the broadly similar *The Fellowship of the Talisman* (**1978**), whose supposedly feudal-English characters retain Midwestern folksiness of DICTION; a Midlands WASTE LAND is crossed and Earth's EVIL (alien evil against which even a DEMON assists) is dispelled by a token of CHRIST. *Where the Evil Dwells* (**1982**) again repeats the quest formula, with diminished energy. [DRL]

ŠIMÁNEK, JOSEF (1883-1959) Czech writer. ◊ CZECH REPUBLIC.

SIME, S(IDNEY) H(ERBERT) (?1865-1941) UK painter, illustrator and theatre designer; his early years are obscure. His work began to appear in the 1890s; his 1896 satirical caricatures of existence in the AFTERLIFE in the magazine *Pick-Me-Up* caused some controversy. Gradually shedding the influence of Aubrey BEARDSLEY, SHS evolved a style and an autonomy of content which meant his pictures did not so much illustrate the volumes in which they appeared as constitute a parallel vision of mythological LANDSCAPES, peopled by figures drawn with dreamlike intensity whose relation to the world was both intimate and remote, as in some Japanese watercolours. SHS's relationship to Lord DUNSANY is well known: his work appeared in most of Dunsany's early collections, many of the stories in which were inspired by SHS's images. Dunsany collections in which SHS collaborated include *The Gods of Pegana* (coll **1905**), *Time and the Gods* (coll **1906**), *The Sword of Welleran* (coll **1908**), *A Dreamer's Tales* (coll **1910**), *The Book of Wonder* (coll **1912**) and *Tales of Wonder* (coll **1916**). He composed frontispieces for some of the later novels, as well as for Arthur MACHEN's *The House of Souls* (coll **1906**) and *The Hill of Dreams* (**1907**) and William Hope HODGSON's *The Ghost Pirates* (**1909**).

Beasts that Might Have Been (1905 *The Sketch*; graph **1973**) prefigures SHS's most interesting later work, *Bogey Beasts: Jingles &c* (graph **1923**), a joke BESTIARY whose accompanying "jingles" were set to music by Josef Holbrooke (1878-1958). The beasts in both volumes are ludicrous, surreal, sometimes haunting.

SHS was not prolific, and in later years was something of a recluse. Two surveys by George Locke – *From an Ultimate Dim Thule: A Review of the Early Works of Sidney H. Sime* (**1973**) and *The Land of Dreams: A Review of the Work of Sidney H. Sime, 1905 to 1916* (**1975**) – provide a detailed coverage of his work, some of which is preserved in Worplesdon Memorial Hall, Surrey, where he spent his last 40 years. [JC]

SIMMONS, DAN (1948-   ) US writer, best-known for his horror and sf, for which he has been widely praised. *The Song of Kali* (**1985**), arguably HORROR, won a WORLD FANTASY AWARD. *Carrion Comfort* (1983 *Omni*; exp **1989**), a VAMPIRE novel, won a BRITISH FANTASY AWARD, a BRAM STOKER AWARD and a LOCUS AWARD. *Hyperion* (**1989**), which is sf, gained him a Hugo AWARD, and its sequel *The Fall of Hyperion* (**1990**) received a British Science Fiction Association Award. His first story was "The River Styx Runs Upstream" for *Rod Serling's The Twilight Zone Magazine* in 1982. Some of the stories he published over the next half dozen years – collected in *Prayers to Broken Stones* (coll **1990**) – are fantasy.

Much of DS's fiction CROSSHATCHES various genres, so that the horror novel *Children of the Night* (**1992**) offers an sf rationale, and the sf novel *The Hollow Man* (**1992**) has horror elements, including an invocation of DANTE'S inferno. The novellas in *Lovedeath: Five Tales of Love and Death* (coll **1993**) employ supernatural elements to dramatize variously horrific equations of Eros and Thanatos, placing them firmly in the company of DS's previous horror novels. *Fires of Eden* (**1994**) is dark fantasy/horror. *Endymion* (**1996**) is a more enigmatic work, and more difficult to classify than anything DS has done before. [GF/JG]

Other works: *Eyes I Dare Not Meet in Dreams*; *Phases of Gravity* (**1989**), a TIME-TRAVEL tale; *Banished Dreams* (**1990** chap); *Entropy's Bed at Midnight* (**1990** chap); *Going After the Rubber Chicken* (coll **1991** chap), speeches; *Summer of Night* (**1991** UK), DARK FANTASY; *Summer Sketches* (coll **1992**).

SIMMONS, JOHN (1823-1876) UK artist. ◊ ILLUSTRATION.

SIMPSON, N(ORMAN) F(REDERICK) (1919-   ) UK dramatist and novelist, most famous for *One Way Pendulum: A Farce in a New Dimension* (produced 1959; **1960**), which depicts a surreal WONDERLAND vision of English suburban life; characters include a young man who is teaching 500 talking-weight machines to sing the Hallelujah Chorus. *A Resounding Tinkle* (produced 1971) features Absurdist problems (◊ ABSURDIST FANTASY) with an elephant which is too large. NFS's only novel, *Harry Bleachbaker* (**1976**), describes attempts to rescue a man who has been drowning for months in the Mediterranean. [JC]

SINBAD ◊ ARABIAN FANTASY; SINBAD MOVIES.

SINBAD MOVIES Several movies have been based, usually loosely, on the classic ARABIAN FANTASY.

1. *Sinbad the Sailor* US movie (**1947**). RKO. **Pr** Stephen Ames. **Dir** Richard Wallace. **Spfx** Vernon L. Walker, Harold Wellman. **Screenplay** John Twist. **Starring** Douglas Fairbanks Jr (Sinbad), Maureen O'Hara (Shireen), Anthony Quinn (Emir), George Tobias (Abbu). 116 mins. Colour.

The 8th voyage. Sinbad and sidekick Abbu come across a dead ship. A MAP in the captain's quarters shows the route of Alexander the Great through the Sea of Oman to fabled Deriebah, where he left his treasure. Sinbad encounters mercenary-hearted siren Shireen, who is milking the tyrannous Emir of Daibul as the two plot to discover the route to Deriebah; despite herself, she loses her heart to the TRICKSTER adventurer . . . And so forth.

Fairbanks seeks to emulate the swashbucklery of his father in *The* THIEF OF BAGDAD (**1924**), and succeeds in producing a parade of supreme and ultimately tiresome camp, an effect underscored by a script of extreme floridity. Interestingly, *STS* largely eschews the fantastical: there is implausibility galore, but virtually nothing of the supernatural.   [JG]

2. *Son of Sinbad* US movie (**1955**). RKO. **Pr** Robert Sparks. **Dir** Ted Tetzlaff. **Screenplay** Jack Pollexfen, Aubrey Wisberg. **Starring** Mari Blanchard (Kristina), Sally Forrest (Ameer), Vincent Price (Omar Khayyám), Dale Robertson (Sinbad Jr), Lili St Cyr (Nerissa). 88 mins. Colour.

A spoof ARABIAN FANTASY in which the philanderous son of Sinbad the Sailor shares with Omar Khayyám adventures among elements from various *Arabian Nights* tales, with plenty of opportunities for harem women to appear in what at the time was naughtily little. The Forty Thieves are an AMAZON band, descendants of the men whom Ali Baba despatched. The magic lamp is a hypnotizing device which makes the daughter of an old scholar repeat the secret of Greek Fire. Price hams magnificently and some of the jokes are good, but the schoolboy sexism is depressing.   [JG]

3. *The Seventh Voyage of Sinbad* US movie (**1958**). Columbia/Morningside. **Pr** Charles H. Schneer. **Dir** Nathan Juran. **Spfx** Ray HARRYHAUSEN. **Screenplay** Kenneth Kolb. **Starring** Alfred Brown (Harufa), Richard Eyer (Boranni the Genie), Kathryn Grant (Parisa), Harold Kasket (Sultan of Chandra), Alec Mango (Caliph), Kerwin Mathews (Sinbad), Torin Thatcher (Sokurah). 89 mins. Colour.

Returning to Baghdad with Princess Parisa of Chandra, Sinbad's ship is blown off-course to the ISLAND of Colossa, where he and his men save devious sorcerer Sokurah from a Cyclops, although in the process losing Sokurah's magic lamp and its little-boy GENIE to the MONSTER. In Baghdad, they prepare for Sinbad's marriage to Parisa; but Sokurah, wanting his lamp, miniaturizes her, claiming the sole antidote is a POTION made from arcane ingredients available only on Colossa. Sinbad, best friend Harufa and tiny Parisa set off there with Sokurah, battling with rocs, two cyclopes and a DRAGON before finally having Parisa restored and causing the destruction of Sokurah. The genie, who has confessed to Parisa that his dream is to become a REAL BOY, is magically rendered such; and he gives the affianced pair the cyclopes' vast treasure as a wedding gift.

TSVOS is really just an excuse for Harryhausen's spfx; the movie proudly boasts itself the first ever in Dynamation, the process of stop-motion animation (◊ ANIMATED MOVIES) and mattework used to achieve the miniaturization of Parisa and the interplay of humans with MONSTERS. In fact, the significant advance was the adaptation of existing techniques to colour.

TSVOS was sequelled by **5**.   [JG]

4. *Captain Sindbad* (vt *Captain Sinbad*) US/German movie (**1963**). MGM/King Bros. **Pr** Frank King, Herman King. **Dir** Byron Haskin. **Spfx** Augie Lohman, Lee Zavitz. **Vfx** Tom Howard. **Screenplay** Harry Relis, Samuel B. West. **Starring** Pedro Armendariz (El Kerim), Heidi Bruhl (Jana), Abraham Sofaer (Galgo), Guy Williams (Sindbad). 88 mins. Colour.

Sindbad and Jana, Princess of Baristan, are in love; but the tyrannical *de facto* ruler of Baristan, El Kerim, wants her for himself. El Kerim is a tough foe: he is invincible because his heart no longer resides in his chest but at the top of a distant tower defended by a cordon of vicious MONSTERS (◊◊ KOSHCHEI). Occasionally assisted by tipsy court magician Galgo, Sindbad survives perils including – there is some cross-cultural leakage in this movie – the gladiatorial arena and a fight with the hydra to see El Kerim's heart destroyed

and win Jana. The spfx are, like the screenplay, distinctly ropy but carried through with such joyousness that it hardly matters; the acting is generally less convincing.   [JG]

5. *The Golden Voyage of Sinbad* UK movie (**1973**). Morningside/Columbia. **Pr** Ray HARRYHAUSEN, Charles H. Schneer. **Dir** Gordon Hessler. **Spfx** Harryhausen. **Screenplay** Brian Clemens. **Starring** Tom Baker (Prince Koura), Kurt Christian (Haroun), Takis Emmanuel (Achmed), John Phillip Law (Sinbad), Caroline Munro (Margiana), Martin Shaw (Rachid), Douglas Wilmer (Grand Vizier). 105 mins. Colour.

Sequel to **3** and forerunner of **8**. One of Sinbad's sailors shoots at a flying HOMUNCULUS, which drops an AMULET onto the ship. Donning the amulet, Sinbad sees VISIONS of a dancer with a single eye tattooed in her hand, and of a dark sorcerer. The amulet is in fact merely one-third of a PLOT COUPON: united, the three form a sea chart (◊ MAPS) and also, if presented to the Fountain of Destiny on LEMURIA, grant the bearer youth (◊ FOUNTAIN OF YOUTH), the Shield of Darkness (i.e., INVISIBILITY) and the Crown of Untold Riches. *En route* to gaining the last of these, Sinbad picks up the dancing girl, alternately thwarts and is thwarted by the sorcerer Koura, fights an idol of the GODDESS Kali (one sword in each of its six hands), escapes HUMAN SACRIFICE, sees the Griffin of Good killed by the one-eyed Centaur of EVIL (which Sinbad then slays), kills Koura and gains a righteous Vizier a sultanate. In short, the plot is hokum, but enjoyable hokum; the spfx vary between brilliant and risible. Two things stand out. Koura's performance of MAGIC (aided by the Demons of Darkness) tires, pains and prematurely ages him: it is pleasing, in this level of movie, to see magic coming with a price attached. And there is a very striking scene, where acting and spfx combine superbly, in which Koura activates, from a mandrake root and drops of his own blood, a new homunculus (to which, it proves, he is joined by bonds of sympathetic MAGIC).   [JG]

6. *Adventures of Sinbad* (vt *Sinbad the Sailor*) Japanese ANIMATED MOVIE (**1975**). Toei. **Pr** Hiroshi Okawa. **Dir** Taiji Yabushita. **Screenplay** Morio Kita, Osamu Tezuka. No other credits given. 82 mins. Colour.

Sinbad and his little friend Ali find an old man dying on the beach. He gives them a MAP to a faraway ISLAND where an underground cave is filled with jewels, guarded by DEMONS. With Ali's kitten, they stow away on a merchantman and reach a foreign land where a sinister vizier plots to marry the beautiful PRINCESS. Imprisoned, Sinbad catches the princess's eye and is released by her; her pet dove gives Ali a magic peacock feather which he rides like a magic carpet to freedom. Fleeing herself, she joins their QUEST; the vizier sends his winged minion, Hellbat (very like the character Fidget in DISNEY's *The Great Mouse Detective* [**1986**]), to spy on them, and pursues. After various adventures, all reach the enchanted island. At last, Sinbad gains not jewels but the love of the princess.

Although the animation is cheap, stylized and occasionally over-cute, it has humour, a fair deal of charm and some moments of great beauty.   [JG]

7. *The Adventures of Sinbad the Sailor* (ot *Pohady Tisice a Jedne Noci*; vt *Tales of 1001 Nights*) Czechoslovak ANIMATED MOVIE (**1975**). **Dir** Karel Zeman. 87 mins. Colour.

We have been unable to obtain a viewing copy of this movie, whose plot seems suspiciously similar to **6**'s. This may be a ghost title born of filmographic confusion.   [JG]

8. *Sinbad and the Eye of the Tiger* UK movie (**1977**).

Columbia/Andor. **Pr** Ray HARRYHAUSEN, Charles H. Schneer. **Dir** Sam Wanamaker. **Spfx** Harryhausen. **Screenplay** Beverley Cross. **Novelization** *Sinbad and the Eye of the Tiger* * (**1976**) by John Ryder Hall (William Rotsler; ◊ *SFE*). **Starring** Kurt Christian (Rafi), Taryn Power (Dione), Nadim Sawalha (Hassan), Jane Seymour (Farah), Damian Thomas (Kassim), Patrick Troughton (Melanthius), Patrick Wayne (Sinbad), Margaret Whiting (Zenobia). 113 mins. Colour.

The sequel to **5**. Sinbad arrives in port to seek marriage with Princess Farah. He discovers her father, the Caliph, has died, and that the coronation of her brother Kassim has been thwarted by Kassim's and Farah's wicked STEP-MOTHER, the WITCH Zenobia, who has by black MAGIC turned the rightful heir into a baboon in hopes of installing her oily son Rafi in his place. Sinbad realizes the only person who can help restore the *status quo* is the Greek alchemist Melanthius. With Farah and baboon in tow, Sinbad tracks down Melanthius and his beautiful daughter Dione. The alchemist tells them they must gain APOLLO's help at the Shrine of the Four Elements in far HYPERBOREA. There they duly venture, beset by various MONSTERS and pursued by a SHAPESHIFTING, POTION-swigging Zenobia with Rafi and her giant MINOTAUR-like automaton. With the help of a friendly troglodyte our heroes enter the shrine, pass the baboon through the light of Apollo to restore Kassim, have a stand-off with Zenobia and Rafi, and flee for their lives as the shrine crumbles about them.

The hokum as usual, but lacking the verve of **3** and **5**. The monsters are remarkably unconvincing; when a savage giant walrus waddles into the attack, no serious fantasy-lover can fail to rock with laughter. But this is not the only dated aspect of the movie: sexism is another. Although Whiting is permitted a certain villainous pantomime *brio*, Power and Seymour are allowed to be nothing but bimbos, often scantily clad (and capable of changing costumes mid-trek despite having no visible suitcases). Harryhausen and Schneer were to have a swansong in CLASH OF THE TITANS (*1981*), but in reality *SATEOTT* was the end of their line. [JG]

9. *The Adventures of Sinbad* UK ANIMATED MOVIE (*1979*). **Voice actor** Jon Pertwee (Narrator). 47 mins. Colour.

Evidently a minor production; we have not been able to obtain a viewing copy. A magic lamp has been stolen by the Old Man of the Sea; Sinbad regains it and thereby earns the love of a princess. [JG]

10. *Sinbad of the Seven Seas* Italian/US movie (*1989*). Cannon. **Pr** Enzo G. Castellari. **Dir** Castellari. **Vfx** Gruppo Memmo Milano Studio 4. **Spfx** Cataldo Galiano. **Screenplay** Tito Carpi, Castellari, Ian Danby, Egle Guarino. **Starring** Lou Ferrigno (Sinbad), Alessandra Martines (Alina), John Steiner (Jaffar), Roland Wybenga (Ali). 93 mins. Colour.

This chaotically scripted mishmash of themes from ARABIAN FANTASY, packed with *non sequiturs*, claims to be based on "The Thousand-and-Second Tale of Scheherazade" (1845) by Edgar Allan POE, but is not. To retrieve the magical GEMS that maintain prosperity in Basra – thereby restoring the Caliph, thwarting the usurping DEVIL-worshipping vizier Jaffar and uniting Prince Ali and Princess Alina – Sinbad takes the advice of an ORACLE and QUESTS with a crew including a Chinese Samurai (*sic*) to Skull Island (he defeats a rock MONSTER), to the Isle of the Warrior Women (he is seduced by the mind-VAMPIRE Queen of the AMAZONS) and to the Isle of the Dead (he battles with

GHOST warriors and – as the movie suddenly opts for humour – with the help of a martial-arts expert and her WIZARD father defeats gangs of flesh-eating ZOMBIES and a slime monster). Back in Basra, he bests a demonically conjured evil TWIN of himself. Some excellent sets and spfx are wasted in a mess of poor dubbing, appalling dialogue and hammy acting, not least from Ferrigno, better-known as tv's Incredible Hulk. [JG]

SINCLAIR, ANDREW (ANNANDALE) (1935-) UK writer of much fiction and nonfiction; he has also been a screenwriter and movie director. His major contribution to fantasy is the **Albion Triptych** – *Gog* (**1967**), *Magog* (**1972**) and *King Ludd* (**1988**) – a FABULATION about the MATTER of Britain which is half sentimental SATIRE and half mythopoesis. In the opening volume, a very tall man is washed up on a beach in Scotland, in 1945; he has lost his memory, but the names of the legendary GIANTS Gog and Magog are tattooed on his fists. His subsequent trek from Edinburgh to London is both a quest for personal identity and a mythological history of the British people. The second volume is more in the nature of mundane satire, but the third explores afresh the themes and images of the first book, retelling the story of the hero's entire life with many fantastic episodes set against a rich historical background. [DP/JC]

Other works: *Inkydoo, the Wild Boy* (**1976** chap); *The Facts in the Case of E.A. Poe* (**1979**); *The Sword and the Grail* (**1992** US), nonfiction.

SINCLAIR, CATHERINE (1800-1864) UK writer. ◊ CHILDREN'S FANTASY.

SINCLAIR, CLIVE (JOHN) (1948-) UK writer best-known for his first novel, *Bibliosexuality* (**1973**), which contains elements of FABULATION. *Hearts of Gold* (coll **1979**) and *Bedbugs* (coll **1982**) both contain some fantasy and SUPERNATURAL FICTION, including "Uncle Vlad", a VAMPIRE tale, in the first, and "Genesis", which features a fallen ANGEL, in the second. Generally, though, his stories make use of FABLE elements as an estranging device, and tend to hover along the peripheries of the fantastic. *Augustus Rex* (**1992**), however, steps vigorously over the line; narrated by Beelzebub, it tells how August STRINDBERG sells his SOUL (◊ PACTS WITH THE DEVIL) and is resurrected in 1961 in a LAND-OF-FABLE Scandinavia, which he soon transforms before slipping out of the Devil's clutches into genuine death. [JC]

SINCLAIR, UPTON (BEALL) (1878-1968) US writer who remains most famous for muckraking novels from early in his career, most notably *The Jungle* (**1905**), an exposé of conditions in the Chicago stockyards. His speculative works, mostly sf, are similarly designed to suggest ameliorative changes to the real world. In his prolific career, however, US did write some fantasy, most being also political SATIRE. *Prince Hagen: A Phantasy* (**1903**; rev vt *Prince Hagen: A Drama in Four Acts* performed 1909; **1921**) is a kind of spoof Bildungsroman, in which the eponymous heir to the throne of Nibelheim – an UNDERWORLD central to NORDIC FANTASY – comes to the surface where, as a Tammany Hall boss in turn-of-the-century NEW YORK, he threatens to corrupt the world. *Roman Holiday* (**1931**) is a TIMESLIP tale whose protagonist witnesses the political delusions of the Roman Republic. *The Gnomobile: A Gnice Gnew Gnarrative with Gnonsense, but Gnothing Gnaughty* (**1936**), filmed by DISNEY as *The Gnome-Mobile* (**1967**), describes the saving of a family of gnomes from destruction wreaked by the logging industry.

US also wrote several fantasies featuring MESSIAH figures, whose defeat by the modern world is described satirically. *They Call Me Carpenter* (**1922**) is delusional; in *Our Lady: A Story* (**1938**) Marya, mother of Jesus (◊ CHRIST), is brought by timeslip to the modern world; and *What Didymus Did* (**1954** UK; vt *It Happened to Didymus* 1958 US) argues that it is now too late for war to be stopped even through the intervention of an ANGEL and a man whom he has given the power to perform MIRACLES. [JC]

SIN-EATER One who magically takes on the sins of a dead person, usually by eating food placed on or passed over the deceased's chest, so that the deceased's soul may be delivered from PURGATORY. Sin-eating is found in many societies, both Christian and pagan, and was commonly practised up to the 19th century. The concept has been used in a number of horror novels, including Elizabeth Massie's *Sineater* (**1992**) and Ramsey CAMPBELL's *The Long Lost* (**1993**). *Rod Serling's Night Gallery: The Sins of the Father* (1972), with Richard Thomas as a young sin-eater, is based on a short story by Christianna Brand originally published in *The Fifth Pan Book of Horror Stories* (**1964**). [JF]

SINGER, ISAAC BASHEVIS (1904-1991) Polish writer who composed first in Hebrew, then solely in Yiddish; he was a US resident from 1935. He began publishing professionally in Warsaw magazines in 1925, and in early stories – as in late – translated into terms accessible to 20th-century audiences much Yiddish FOLKLORE as well as more obscure material from the CABBALA, the latter influence shaping much of his fiction into patterns in which the divine and the mundane rest their cases on each other (◊ AS ABOVE, SO BELOW). His first novel, *Satan in Goray* (**1935**; trans Jacob Sloan **1955**), describes the rise of a false MESSIAH in 17th-century Poland, and incorporates the harrowing POSSESSION of an entire Jewish community. Much of his work – though technically SUPERNATURAL FICTION, and populous with DEVILS and WITCHES and and GHOSTS – wholly embraces a worldview that sees the supernatural and the mundane as inextricably mixed, so that he must be understood as an author of fantasy. SATAN makes frequent appearances throughout IBS's work, and narrates several of the short stories.

Most of IBS's writing career was spent in the USA, and most of his work appeared in English translation before it came out in Yiddish. Collections include *Gimpel the Fool and Other Stories* (coll trans [in part by Saul Bellow] **1957**), *The Spinoza of Market Street* (coll trans **1961**), *Short Friday and Other Stories* (coll trans **1964**), *The Seance and Other Stories* (coll trans **1968**), *A Friend of Kafka* (coll trans **1970**), *A Crown of Feathers and Other Stories* (coll trans **1973**), *Passions and Other Stories* (coll trans **1975**), *Old Love* (coll trans **1979**), *Collected Stories* (coll trans **1982**), *The Image and Other Stories* (coll trans **1985**), *Gifts* (coll trans **1985**) and *The Death of Methuselah and Other Stories* (coll trans **1988**). Novels of interest include *The Magician of Lublin* (trans **1960**), *The Fools of Chelm and Their History* (trans **1973**), the latter set in a LAND-OF-FABLE Eastern European kingdom, and *The Penitent* (**1983**), a WORLD WAR II fantasy in which Hitler, the Devil and a false Messiah intertwine. In almost all cases, IBS collaborated in translating his work, some volumes having multiple translation credits, and the English texts have an authority independent of the manuscript originals.

IBS won the Nobel Literature Prize in 1978. [JC]

Other works (for children): *Zlateh the Goat and Other Stories* (coll trans **1966**) illus Maurice SENDAK; *Mazel and Shlimazel, or The Milk of a Lioness* (**1967**); *The Fearsome Inn* (**1967**); *When Schlemiel Went to Warsaw and Other Stories* (coll trans **1968**); *Joseph and Koza, or The Sacrifice to the Vistula* (trans **1970**); *Alone in the Wild Forest* (trans **1971**); *The Topsy-Turvy Emperor of China* (coll trans **1971**) illus William Pène du Bois; *A Tale of Three Wishes* (trans **1976**); *Naftali the Storyteller and His Horse, Sus, and Other Stories* (coll trans **1976**); *The Power of Light* (coll trans **1980**); *The Golem* (trans **1982**); *Stories for Children* (coll trans **1984**); *Meshugah* (trans **1994**).

SINGER, MARILYN (1948-) US writer of juvenile and YA fantasy, her first publications being the **Sam Spayed Series** for the former market: *The Fido Frame-Up* (**1983** chap), *A Nose for Trouble* (**1985** chap) and *Where There's a Will, There's a Wag* (**1987** chap). In her first YA novel, *Horsemaster* (**1985**), a young girl's DREAMS of a winged horse prefigure her discovery of a tapestry which turns out to be a PORTAL into a SECONDARY WORLD with a Middle Eastern coloration; here she must protect the tapestry on behalf of the next Horsemaster, who will rule the LAND. She herself, in this secondary world, is a princess. *Ghost Host* (**1987**) is a SUPERNATURAL FICTION. *Storm Rising* (**1989**) depicts both the doomed love of a male teenager who can see into the future (◊ TALENTS) and an older woman (she's 28), who is something of a GODDESS figure. *Charmed* (**1990**) carries its cast through ALTERNATE REALITIES – one of them a future Earth – in search of others who can join with them to fight off a world-threatening enemy known as Charmer; TALKING ANIMALS are involved. *California Demon* (**1992**) is for a somewhat younger audience – who might not catch the punning title – and involves the recapture of the eponymous escaped DEMON. [JC]

Other works: *Mitzi Meyer, Fearless Warrior Queen* (**1987**).

SINGERS ◊ MINSTRELS.

SIRENS In Greek MYTHOLOGY, chimerical BIRD-women who lived on the ISLAND of Anthemusa, luring sailors to their doom by means of seductive singing. ORPHEUS saved the *Argo*'s crew by drowning out their song; ODYSSEUS's crew filled their ears with wax while he was securely bound to the mast; after this latter failure the sirens drowned themselves. The Lorelei of the Rhine is a transplanted siren, and MERMAIDS (occasionally called sirens) are sometimes credited with similar proclivities. The sirens are an ARCHETYPE of the FEMME FATALE, and are featured in such roles as "The Song of the Sirens" (1919) by Edward L. White (1866-1934), E.M. FORSTER's "The Story of the Siren" (1928), Oliver ONIONS's "The Painted Face" (1929) and Lord DUNSANY's sarcastic **Jorkens** tale "The Grecian Singer" (1940). F. ANSTEY's "The Siren" (1884), contrariwise, is a sentimental tale of self-sacrificing LOVE. [BS]

SITWELL, SACHEVERELL (1897-1988) UK poet, critic and writer, the least-known of the Sitwell siblings when they were alive but in recent years increasingly deemed the most formidable of the three. Some of his narrative poetry, like *Dr Donne and Gargantua* (**1930**), is of fantasy interest; and *Poltergeists: An Introduction and Examination Followed by Chosen Instances* (**1940**), though it does not focus on POLTERGEISTS in fiction, is an eloquently presented compendium of instances. He remains of greatest interest for the **Entertainments of the Imagination** sequence of meditations: *Dance of the Quick and the Dead: An Entertainment of the Imagination* (**1936**), *Sacred & Profane Love* (**1940**), *Primitive Scenes and Festivals* (**1942**), *Splendours* *and Miseries* (**1943**), *The Hunters and the Hunted* (**1947**) and *Cupid and the Jacaranda* (**1952**). Neither fiction nor fantasy, the sequence nevertheless compellingly encompasses the fantasist's imaginative PLAYGROUND in the visual arts, music and fiction, doing so in a style reminiscent of Vernon LEE's nonfiction books about Italy. Though not formally part of the sequence, two further works – *Journey to the Ends of Time: Lost in the Dark Wood* (**1959**) and *For Want of the Golden City* (**1973**) – continue the elaborate traversal. An early volume of poems, *The Hundred and One Harlequins* (coll **1922**), makes the COMMEDIA DELL'ARTE connection explicit. [JC]

Further reading: *The World of Sacheverell Sitwell* (**1980**) by Denys Sutton.

666 ◊ DEVIL; GREAT BEAST.

SJUNDE INSEGLET, DET ot of *The* SEVENTH SEAL (*1956*).

SKALDASPILLIR, SIGFRIDUR Pseudonym of Mildred Downey BROXON.

SKELETON CREW ◊ MAGAZINES.

SKINNED There are two broad categories in fantasy where an act of flaying might be referred to.

1. The most famous in art history is probably the iconography (◊ ICONS) devoted to the musical contest between the Satyr Marsyas and APOLLO. Marsyas has been jinxed into hubris by the fact that his flute was made by Minerva, who then, however, cursed it. In his contest with Apollo, judged by the partial Muses, he is foredoomed to failure – as mortals tend to be when they compete with their deities – foredoomed to pay whatever penalty Apollo imposes for his pride at his skill with the flute. The penalty is to be skinned alive. Being flayed is thus what happens to mortals who contest with the gods.

2. The second broad use relates to the MAGIC belief that to wear the hide of a skinned animal or victim is to acquire its *numen* or strength. This is far from an exclusively Western idea: something very similar was important in the religion of the Aztec/Toltec people; a good example of this being explored in fantasy is Graham Watkins's *The Fire Within* (**1991**). One old CHRISTMAS ritual involves skin-wearing in order to gain (symbolically) "contact with the sanctity of the sacrificed victim", as Clement Miles puts it in *Christmas in Ritual and Tradition* (**1912**). The MYTHAGOES in Robert P. HOLDSTOCK's fantasies might well wear hides for this reason. Such transference of power probably underlies the wearing of skins – animal or human – by DARK LORDS or their minions, an example of this anthropologizing in a DARK-FANTASY context being the MAGUS Ralli-Faj – in A.A. ATTANASIO's «The Dark Shore» (**1996**) – who manifests himself as "a gray stick upon which [hangs] a wrinkled empty skin of brown leather flayed from a human body". Skin-wearing is a route to SHAPESHIFTING, as with "swan-maidens" who take on swan-form by donning a garment of feathers: examples appear in James Branch CABELL's *Figures of Earth* (**1921**) and Poul ANDERSON's *Three Hearts and Three Lions* (**1953**). (◊◊ SELKIES.)

This transference of the essence of the victim can also be seen in the use of a skinned hide as a DRUM: in Gustav MEYRINK's *Walpurgisnacht* (**1916**) the MIRROR character who had occultly inspired an apocalyptic WORLD WAR I revolution in PRAGUE afterwards has himself skinned with the instruction that his hide be made into a drum. When beaten, the drum will call the folk to arms. [JC]

SKINNER, MARTYN (? -   ) UK narrative poet whose *Sir Elfadore and Mabyna: A Poem in Four Cantos* (**1935**) deals with the fringes of the Arthurian cycle (◊ ARTHUR), and whose *Merlin. or The Return of Arthur* (**1951** chap) and *The Return of Arthur: A Poem of the Future* (**1955**) – both assembled and extended as *The Return of Arthur: A Poem of the Future* (**1966**) – explores the MATTER of Britain far more thoroughly, and in a fashion that attempts to wed fantasy and sf. In the first part Merlin wakes the ONCE AND FUTURE KING from his sleep under Avalon, along with some 20th-century heroes, and prepares them for the future; in the second part, the COMPANIONS take on the 1990s UK, which is threatened by Marxism; salvation comes in the full 1966 version. The epic is told in the rhyme-royal stanza form used by Lord BYRON in *Don Juan* (**1819-24**). *Old Rectory, or The Interview* (**1977**) is an sf poem, set in a post-HOLOCAUST plague-devastated Somerset. [JC]

SKIPPER, MERVYN (GARNHAM) (1886-1958) Australian journalist and writer whose **Meeting Pool** BEAST-FABLES – *The Meeting Pool: A Tale of Borneo* (coll **1929** UK) and *The White Man's Garden: A Tale of Borneo* (coll of linked stories **1930** UK) – presents with considerable energy a series of exemplary tales about proper relationships among the participating animals (as portrayed by the plants of the garden); and, more sadly, lessons in humanity's inevitable triumph over the prelapsarian life of Borneo. *The Fooling of King Alexander* (**1967** chap) is an illustrated version of a tale from the second **Meeting Pool** volume. Later MS became involved with the Montsalvat artists' colony near Melbourne, where he lived from 1935 until his death; of the 100 or so unpublished manuscripts held at Montsalvat, some may be fantasies. [JC]

SLAINE UK comics series, inspired by Celtic heroic myth and created by Pat MILLS, starring axe-wielding punk-haired Slaine Mac Roth, who first appeared in *2,000 A.D. #335* (**1983**), in which he first uttered his catchphrase: "Kiss my axe!" He inhabits TIR-NAN-OG. In his early appearances he wore a loincloth and broad belt and sported a gold torc at his throat; later he became a leather-clad punk, and later still he wore plaid trews and a sporran. The stories are told as though written retrospectively by Ukko, a vulgar, greedy, self-centred filthy-minded dwarf COMPANION. Most stories involve some conflict with the evil-smelling Slough Feg.

The early stories introduced Slaine and Ukko as a DUO of peripatetic fairground entertainers, but then Slaine is taken on as bodyguard to the drune lord Slough Throt, a quasireligious figure wearing antlers and a nondescript pile of stinking animal skins. Slaine gradually gains more respect and power until he becomes Sun King of the Sessir tribe. The first artist on the series was Michael McMahon, whose remarkable artwork, finished in rollerball pen, had a unique and distinctive clarity. Subsequent artists included Glenn Fabry, Massimo Bellardinelli, David Pugh and, most spectacularly, Simon BISLEY.

The great success of the collected editions of *Slaine: The Horned God* (*2,000 A.D. #626-#635, #650-#656, #662-#664* and *#688-#698*, 1988-91; graph coll **1989-91** 3 vols) has led to a multi-volume reprinting of the entire series beginning 1994. [RT]

SLEATOR, WILLIAM (WARNER III) (1945-   ) US writer of YA books, many sf or fantasy. *Blackbriar* (**1972**), WS's first book, is fantasy, as is *Among the Dolls* (**1975** chap), but he soon began to specialize in sf, which accounts for most of his oeuvre. *Fingers* (**1983**) is a supernatural thriller, while *The Spirit House* (**1991**) and its sequel, *Dangerous Wishes* (**1995**), make use of Thai gods and the theme of unwise bargains. All dramatize vividly the anxieties of children, for whom supernatural threats can be essentially intensifications of the terrors of a disconcerting world. [GF]

Other works: *Others See Us* (**1993**), a novel of telepathy.

SLEEPER UNDER THE HILL A theme common throughout much European LEGEND and FOLKLORE is that a past hero or king is only sleeping in a hall under a hill and at a time of national crisis will return to save the LAND. In Celtic and British myth the legend applies equally to Bran the Blessed, ARTHUR (though he is in AVALON, not under a hill) and MERLIN (trapped in a cave, according to one legend). The myth has also been claimed for Charlemagne (742-814), Frederick Barbarossa (1123-1190), his grandson Frederick II (1194-1250) and Ogier the Dane (◊ MORGAN LE FAY). Although superficially the myth may seem to draw upon the Christian belief of the RESURRECTION, it has deeper pagan origins in most cultures of the transmigration of souls through the AFTERLIFE to rebirth. The MABINOGION tells of a magic cauldron that brings soldiers back to life. This in turn has links to the holy GRAIL and the search for eternal life (◊ IMMORTALITY).

The commonest use of the SUTH device occurs in Arthurian fiction (◊ ONCE AND FUTURE KING). The revival of Arthur (often with Merlin, Morgan and some of the more famous Knights of the ROUND TABLE) is the starting point of *The Weirdstone of Brisingamen* (**1960**) by Alan GARNER, *Earthfasts* (**1966**) by William MAYNE, *The Sleepers* (**1968**) by Jane CURRY, *The King Awakes* (**1987**) by Janice ELLIOTT and *The Sleep of Stone* (**1991**) by Louise COOPER. Barbarossa returns in *Little, Big* (**1981**) by John CROWLEY, while in *Too Long a Sacrifice* (**1981**) Mildred Downey BROXON shows it is not only heroes who return, when 6th-century patriots re-emerge from their sleep to help in the Irish conflict. The SUTH MOTIF usually has strong religious connotations and may be linked to the legend of the Seven Sleepers of Ephesus, which tells of seven Christian youths who fled from the persecution of the Emperor Decius in the years AD249-251 and hid in a cave which was walled up. They re-emerged after nearly 200 years during the reign of Theodosius II. It also has alchemical connections (◊ ALCHEMY) as it is applied to Christian ROSENCREUTZ. Generally, though, the SUTH motif relates to a champion who will return and fight for a cause. It is thus not the same as the sleeper-awakes motif, as in "Rip Van Winkle" (**1819**) by Washington IRVING, or in many sf stories, where it is a PLOT DEVICE for moving into the future. Neither must it be confused with the MALIGN-SLEEPER motif, which is usually the antithesis of the SUTH. [MA]

SLEEPING BEAUTY First of Charles PERRAULT's FAIRY-TALES, and one of the best-known in the world. Perrault's "La Belle au bois dormant" ["The Beauty Sleeping in the Wood"] (**1696** *Le Mercure Galant*) has become the standard text, although the images and storyline amended for DISNEY's SLEEPING BEAUTY (*1959*) are probably better recalled. Perrault's story was first translated into English in *Histories, or Tales of Past Times* (coll **1729** UK) trans Robert Samber, and appeared in a separate booklet as *The Sleeping Beauty in the Wood* (**1764** UK chap).

In Perrault's version a daughter is born to a king and queen who have long been childless. At the christening all the FAIRIES in the land are invited as godmothers (◊ FAIRY GODMOTHER) save one, who is believed to be either dead or enchanted. But she arrives and, in her annoyance at not being invited, curses the babe, saying she will prick her hand on a spindle and die. However, a young fairy modifies the curse: the child will sleep for 100 years and then be woken by a king's son. The present king forbids any spindles in the castle, but 16 years later the princess, exploring a remote part of the castle, encounters an old lady who, knowing nothing of the decree, is working at her spindle. The princess pricks her hand and instantly falls into a deep sleep. The good fairy learns of the incident and arrives in a fiery chariot. So that the Princess will not be alone on waking she puts all the rest of the court to sleep except the king and queen, who leave. Within minutes the castle is encircled by a thick and impenetrable FOREST.

After 100 years the son of the current king sees the castle towers through the wood and learns about the ensorcelled princess. As he approaches the castle the trees part to let him through. He discovers the princess, kisses her, and she (and the court) awakens. They are married.

Perrault's story continues. They have two children, Aurore ("Morning") and Jour ("Day"). The prince inherits the throne, but his mother, portrayed as an ogress, hates her new daughter-in-law and the children and is determined to eat them. Although she orders each to be prepared for the pot, the cooks substitute a lamb, a goat and a hind. The ogress learns she has been deceived but, before she can kill the children herself, her son arrives home and she throws herself into the pot instead.

The basic concept of a princess ensorcelled within a protected castle, to be awakened after a stated time, is almost certainly based on a nature myth of the Sun-maiden forced to sleep through the Winter until re-awoken by the Spring. There is evidence of an SB-type story in the "Tale of Brynhild" in the *Volsunga Saga* (11th century; ◊ SAGA). The valkyrie Brynhild is imprisoned by ODIN in a castle to protect her from enforced marriage to a coward. She is touched with the thorn of sleep to save her beauty, and the castle is surrounded by a barrier of flame. Only Sigurd succeeds in penetrating the flames to waken her. The story has similarities to the GRAIL legend, particularly in the concept of an enchanted castle (◊ EDIFICE) which has been lain waste (◊ WASTE LAND) and is restored only by the efforts and triumphs of a pure and righteous prince (◊ PERCEVAL). In *The Uses of Enchantment* (**1976**), Bruno Bettelheim (1903-1990) more mundanely interprets the tale as a moral lesson to prepare young girls for the physical and sexual maturation of the body during adolescence.

A much more closely related tale is "Troylus and Zellandine" in the French ROMANCE *Perceforest* (14th century; printed **1528**). The baby princess Zellandine is cursed by the goddess Themis; the curse takes the form of a deep sleep after the princess begins to spin. She is later found by Prince Troylus, although he wakes her with not a kiss but a rape. It was almost certainly this story that was adapted by Giambattista Basile (1575-1632) in *The Pentamerone* (*Lo Cunto de li Cunti*) (**1634**) as "Sun, Moon, and Talia". Here Talia's fate is pre-ordained by three wise men, and it happens as predicted. The princess sleeps alone in the castle and, when discovered by the king, is raped by him. She, still sleeping, bears twins, who are cared for by the fairies though suckled at their mother's breast. One baby sucks on the mother's finger and draws out the sleep-inducing splinter. Talia wakens. The king returns to the castle and weds her. However, the king is already married; it is his wife (rather than mother) who takes on the ogress role in wishing to be rid of Talia and the two children.

The tale à la Perrault was picked up by the GRIMM BROTHERS and included in their "Dornröschen" ["Little Briar-Rose"] in *Kinder- und Hausmärchen* (coll **1812**; vol 2 **1814**; rev 1819, 1822). This is the story at its most basic and saccharine, and ends with the prince and princess happily married.

The story remains perennially popular. In addition to adaptations as PANTOMIME (first recorded performance 1806), ballet (Tchaikovsky's *The Sleeping Beauty* was first performed 1890) and Disney's animated feature, it has formed the basis of a number of modern REVISIONIST FANTASIES. One of the earliest was the nonfantastic "The Sleeping Beauty" by G.B. Stern (1890-1973) in *The Fairies Return* (anth **1934**), which translates the story into the English middle-class 1930s: a young girl is struck by a medical affliction; when the doctor restores her to health with a kiss he is struck off the medical register. Rather more fantastic is "Thorns" by Tanith LEE in *Young Winter's Tales 5* (anth **1974**) ed Marni Hodgkin, perhaps the best short-story version of the fairytale, in which the prince's kiss is literally the kiss of life.

Anne RICE went back to the earlier story of Sleeping Beauty's rape for her pornographic **Sleeping Beauty** series as by A.N. Roquelaure. *Beauty* (**1991**) by Sheri S. TEPPER conflates the SB tale with others, notably BEAUTY AND THE BEAST, while *Briar Rose* (**1992**) by Jane YOLEN brings the legend into war-torn Europe. Other treatments include *About the Sleeping Beauty* (**1975**) by P.L. TRAVERS, the **Wells of Ythan** sequence by Marc ALEXANDER, and *Marco Polo and the Sleeping Beauty* (**1988**) by Grania DAVIS and Avram DAVIDSON. [MA]

SLEEPING BEAUTY US ANIMATED MOVIE (**1959**). Disney. **Pr sv** Ken Peterson. **Sv dir** Clyde Geronimi. **Special processes** Ub IWERKS, Eustace Lycett. **Based on** the version of SLEEPING BEAUTY by Charles PERRAULT. **Voice actors** Barbara Jo Allen (Fauna), Eleanor Audley (Maleficent), Mary Costa (Princess Aurora/Briar Rose), Verna Felton (Flora), Taylor Holmes (Stefan), Barbara Luddy (Merryweather), Marvin Miller (Narrator), Bill Shirley (Prince Phillip). 75 mins. Colour.

14th-century King Stefan and his queen hold a christening party for daughter Aurora; to it come young Prince Phillip, to be betrothed to the infant, and the three good FAIRIES Flora, Fauna and Merryweather. The bad fairy Maleficent comes also, furious to be uninvited, and pronounces a CURSE on the infant: on the sunset of her 16th birthday Aurora will prick her finger on a spindle and die. Merryweather ameliorates the curse: Aurora will, instead, fall asleep until woken by true love's kiss. Stefan orders all the nation's spinning-wheels burnt; the three good fairies rear Aurora (as Briar Rose) in the remote forest, and agree to abandon MAGIC for the requisite 16 years. On her 16th birthday Aurora encounters Phillip, and the two fall in LOVE. The fairies use magic to aid preparations for her birthday party, and Maleficent's messenger raven detects this. Arriving at Stefan's castle, the duration of the curse almost outlived, Aurora is magically lured by Maleficent to a tower and pricks her finger. The three fairies put the court into the same deep sleep and free Phillip from Maleficent's clutches, arming him with the Shield of Virtue and the SWORD of Truth. Maleficent casts a FOREST of thorns

around the castle, but Phillip beats a way through. As a final throw, Maleficent METAMORPHOSES herself into a vast DRAGON and confronts Phillip; with the Sword of Truth he kills the dragon, and thus Maleficent and her EVIL. His kiss wakes Aurora and all the castle.

Astonishingly, this HIGH FANTASY was widely reviled on release: for its art, which deployed styles more commonly associated with painting; for its perceived imitation of SNOW WHITE AND THE SEVEN DWARFS (**1937**); and, bafflingly, because Maleficent wasn't thought frightening enough. *SB*, now seen as one of DISNEY's masterpieces, lost money on first release (it was the most expensive ANIMATED MOVIE to that time). Stung, Disney eschewed classic FAIRYTALES for over three decades, until BEAUTY AND THE BEAST (**1991**). [JG]

SLEEPING KING ◊ ONCE AND FUTURE KING; SLEEPER UNDER THE HILL.

SLEIGH, BARBARA (DE RIMER) (1906-1982) UK writer of novels and stories for children, many blending fantasy themes and elements into "real world" settings. BS is best known for the **Carbonel** series – *Carbonel* (**1955**), *The Kingdom of Carbonel* (**1958**), and *Carbonel and Calidor* (**1977**) – in which the child protagonists help the magical royal CAT Carbonel gain and hold his kingdom (in the rooftops of our world) against the opposition of feline rivals and human WITCHES. [KLM]
Other works: *The Patchwork Quilt* (**1956**); *The Singing Wreath and Other Stories* (coll **1957**); *The Seven Days* (**1958**); *No One Must Know* (**1962**); *North of Nowhere: Stories and Legends from Many Lands* (coll **1964**); *Jessamy* (**1967**); *Pen, Penny, Tuppence* (**1968**); *The Snowball* (**1969**); *West of Widdershins: A Gallimaufry of Stories Brewed in her own Cauldron* (coll **1971**); *The Smell of Privet* (**1971**), autobiography; *Stirabout Stories* (coll **1972**); *Ninety-Nine Dragons* (**1974**); *Funny Peculiar: An Anthology* (coll **1974**); *Charlie Chumbles* (**1977**); *Grimblegraw and the Wuthering Witch* (**1978**); *Winged Magic* (**1979**); *Broomsticks and Beasticles* (anth **1984**).

SLICK FANTASY The subgenre of fantasy, usually in short-story form, which deals with such matters as PACTS WITH THE DEVIL, THREE WISHES, IDENTITY EXCHANGE, ANSWERED PRAYERS, little SHOPS of the heart's desire, etc. These are described as "slick" partly because the driving engine of the story is elegant variation on a small number of set themes and partly because such tales were the sort of GENRE FANTASY most likely to get into slick MAGAZINES like the *Saturday Evening Post*. They also represent the type of fantasy most likely to be written by nongenre writers – e.g., P.G. Wodehouse's few excursions into the field. They overlap frequently with CLUB STORIES. John COLLIER, notably, and Lord DUNSANY often wrote such shorts. There are longer examples of the form – it might be argued that the novels of F. ANSTEY and Thorne SMITH fall within the category.

The mode can also overlap with HORROR; both Clive BARKER and Stephen KING have made use of SF tropes – the entire premise of King's *Needful Things* (**1991**) is a combination of the Shop trope with an implicit Faustian Pact. The lightness of touch and sardonic humour standard to SF do not obviate extremes of mayhem and emotional bleakness in the deployment of the standard themes.

SF can also overlap with HEROIC FANTASY; stories which deal with Delphic ambiguity or the working through of CURSES and PROHIBITIONS often tend in this direction. Specifically, the small subgenre of short stories dealing with the intrinsic and dangerous ambiguity of sexist language (◊ GENDER) – Tanith LEE's "Northern Chess" (1979) is a good example – will always tend to inhabit both subgenres. However, the values of heroic fantasy often overwhelm and refuse an SF trope that wanders in. In Fritz LEIBER's "Bazaar of the Bizarre" (1963) Fafhrd and the Gray Mouser find themselves in a Shop and manage to refuse its illusory blandishments. An SF usually works out quite simplistic moral implications to a very final end, whereas a TEMPLATE heroic-fantasy character will always be needed to fight another day. [RK]

SLINGSHOT ENDING The sf writer Kim Stanley Robinson (1952-) used this term when attempting to describe the typical ending of a Gene WOLFE tale. *The Book of the New Sun* (**1980-83**) and *There are Doors* (**1988**) both close as their protagonists begin to move towards a goal which has been anticipated from the beginning. But they move out of frame, out of the end of the book, and the story closes *as though* before its proper ending. But, though unexpecting readers might feel that the effect of the SE is of truncation, of not being told what should be told for proper completion, a true SE persuades its readers that the story has indeed been given – simply that they have to tell *themselves* the final outcome.

If the SE is not simply to be another device of the CONTE CRUEL or of HORROR in general, it is almost certainly necessary that the ending envisaged by the reader be a happy one. Thus the SE – though never common – is more often found in the literature of FANTASY (◊◊ STORY), where the happy ending tends to be built in, than elsewhere. The reasons for its infrequency are plain: an SE must be told in a fashion which surprises the reader but *also* compels ultimate assent, not an easy task; and it is a daring device, one that commercial publishers may resist. Examples of its recent use include Ursula K. LE GUIN's "In the Drought" (in *Xanadu 2* anth **1994** ed Jane YOLEN) and "The Rio Brain" (**1996** *Interzone*) by M. John HARRISON and Simon Ings (1965-), the latter tale being rejected – because of its "abrupt" ending – several times before eventual publication. The SE is more common, therefore, in the early, relatively uncommercial years of fantasy – when narrative risks could more easily be taken.

The last lines of John MILTON's *Paradise Lost* (**1667**; rev 1674) are an early example: "The World was all before them, where to choose / Thir place of rest, and Providence thir guide: / They hand in hand with wandring steps and slow / Through *Eden* took thir solitarie way." Two early-20th-century examples can be noted. In G.K. CHESTERTON's *The Man who Was Thursday: A Nightmare* (**1908**) the protagonist walks through glowing, suburban Saffron Park and catches sight of his friend's sister, whom (we guess, only after we have read the last sentence of the book) he will soon wed; that last sentence describes her (repeating the word "girl" in an unstated rendering of his suddenly heightened attention) as "the girl with the gold-red hair, cutting lilac before breakfast, with the great unconscious gravity of a girl". In Walter DE LA MARE's *The Three Mulla-Mulgars* (**1910**) the eponymous monkeys have been questing, from the beginning, for the land to which their father returned, fabled Assasimon. In their numerous adventures they have had no sight of this land. They are lost in the mountains. Suddenly Nod sees one of his brothers and another. "But not only these. For between them walked on high in a high, hairy cap, with a band of woven scarlet about

his loins, and a basket of honeycombs over his shoulder, a Mulgar of a presence and a strangeness, who was without doubt of the Kingdom of Assasimon." The book is over. The Mulgars begin.

Clearly, as in early volumes of *The Book of the New Sun*, the device can be used as a PORTAL into a further book, but the primary use of the SE, as in Wolfe, is to close the telling in a rush of wonder.

In sf, the SE can be used to convey the "Sense of Wonder"; the most famous example is from A.E. VAN VOGT's *The Weapon Makers* (fixup **1946**), which closes with a line that introduces a brand-new thought and a term not previously encountered in the book: "Here is the race that shall rule the sevagram." [JC]

SLIPPER AND THE ROSE, THE UK movie (**1976**). ◊ CINDERELLA (**1950**).

SLOVAKIA Slovakia lies on the cultural boundary between Eastern and Western Europe: its immediate neighbours are Ukraine to the east, POLAND to the north, Hungary to the south and AUSTRIA and the CZECH REPUBLIC to the west. In the early 11th century Slovakia fell to Hungary and after 1526, when Hungary was attached to Austria, it became a province of the Hapsburg Empire. Unlike the Czech lands, however, which were historically provinces of Austria, Slovakia remained part of the Hungarian sphere of influence within Austria-Hungary.

Following the break-up of Austria-Hungary in 1918, Slovakia was joined to the Czech provinces of Bohemia and Moravia to form the new state of Czechoslovakia, whose inhabitants included – besides Czechs and Slovaks – sizeable minorities of Germans, Hungarians, Gypsies and Jews; to this day Hungarians make up about 10% of Slovakia's population, and many notable Hungarian writers – including Mór Jókai (1825-1904), who was born in Upper Hungary – have had ties to what is now Slovakia.

Apart from a four-year break during WWII, Czechoslovakia survived as a political unit until, on 1 January 1993, it split into the two states of Slovakia and the Czech Republic. This entry concentrates on works in the Slovak language.

Slovak fantasy had its origins in the popular traditions of the ballad and FOLKTALE. Besides the characters common to all Slavic FOLKLORE, however, Slovakia has two dominant figures that are uniquely its own. First is the brigand Juraj Jánošík (1699-1713), who has been mythologized into a curious blend of outlaw and magician: he is said to have possessed a wide range of magical artefacts including a magic belt that gave its wearer great strength, a shirt that bestowed INVISIBILITY, and magic laces in his trousers. The second is the Hungarian noblewoman Erzsébet (or Elizabeth de) Báthory (1560-1614), notorious for her habit of bathing in the blood of murdered virgins (◊ VIRGINITY); she occurs as a bogeywoman in many of the more frightening Slovak folktales (and, of course, her story has contributed internationally to the corpus of VAMPIRE tales). Both figures were taken over in the 19th century by writers of Slovak ROMANTICISM. However, no Slovak writer of this period was a fantasist as such; the writers discussed below are, rather, mainstream Romantics who occasionally produced works with fantasy content.

Faustiáda ["The Faustiad"] (**1864**) by Jonáš Záborský (1812-1876), probably the most effective Slovak satirical novel (◊ SATIRE) of the 19th century, is a case in point. Part One takes place in HEAVEN, Part Two in HELL, and the finale in the invented town of Kocúrkov, world-famous for its enormous number of distilleries. FAUST comes to Kocúrkov to defeat the GIANT Puchor, who has inundated the whole country with cowsheds. He succeeds in blinding the giant and freeing the cattle, and Puchor is petrified in a fit of rage.

The poet Ján Botto (1829-1881) was another Romantic who made use of images from Slovak folklore. In his versified FAIRYTALE "Svetský víťaz" ["The Worldly Victor"] (1846) a HERO endowed with enormous strength and marked with a star on his forehead defeats first the DRAGONS who have ruled the world and then, with the help of 12 white eagles, the MONSTERS of the night who rule the Age of Darkness; after this he becomes Lord of the World. In "Smrt Jánošíkova" (1858; trans Ivan J. Kramoris as *The Death of Jánošík* 1944 US) Botto not only refers to the tale of Jánošík's magic belt but even has his hero marry the Queen of the FAIRIES after his execution.

František Švantner (1912-1950) was a somewhat later writer whose work frequently transgresses the boundary between realism and fantasy, in particular in the peculiarly half-conscious characters displayed by his protagonists. A good example could be almost any of the stories in *Malka* (coll 1942; title story trans Andrew Cincura as "Malka" in *An Anthology of Slovak Literature* [anth 1976 US] ed Cincura), in which elements of FATE and fantasy are often to the fore.

Another mid-20th-century writer of interest is the Modernist poet Rudolf Fábry (1915-1982), one of the founders of *nadrealismus* (a Slovak variant of SURREALISM). Although he was not a fantasist in the generic sense, his early collections – *Uťaté ruky* ["Severed Hands"] (coll 1935) and *Vodné hodiny, hodiny piesočné* ["Water Clocks, Clocks of Sand"] (coll 1938) – are a rich source of fantastic imagery.

The post-WWII years were slim for Slovak fantasists, as in the other Iron Curtain countries. Only in the 1960s did books with fantasy elements start to reappear, though again one cannot speak of any Slovak writer as a career fantasist; as before, mainstream writers made occasional forays into the fantastic. An extremely interesting example is Rudolf Sloboda (1938-   ), who developed a characteristic style of refined, short, witty fantasy story, of which the tales in *Uhorský rok* ["The Hungarian Year"] (coll 1968) are representative. These bizarre little texts might best be characterized as imitations of the fairytale.

MAGIC REALISM was introduced to Slovak literature by Peter Jaroš (1940-   ), most notably in *Tisícročná včela* ["The Thousand-Year-Old Bee"] (1979), successfully filmed by Juraj Jakubisko. The lives of his protagonists, the Pichanda family, are intertwined with the figures of popular folklore, and in several crucial scenes the heroes are visited by the mythical bee of the title.

Modern GENRE FANTASY began to gain significance in Slovakia only towards the end of the 1980s, when the Košice-based sf club 451°F (named in honour of Ray BRADBURY) began to publish fanzines, mostly under the editorship of Peter Sadovský. Although the club managed to bring together a number of promising young writers and critics, later attempts to reach a wider audience with the glossy, professionally printed fanzine *Legendy* ["Legends"] (1991) and its successor, the semi-prozine *Staré legendy* ["Old Legends"] (1992), were unsuccessful.

Due to the lack of domestic publishing outlets, the position of a new fantasy writer in Slovakia today is not encouraging. A type example is Jaroslav Lupečka (1958-

), a member of the 451°F circle. He won first prize in the annual Best Fantasy Competition in 1990 with "Stopou velkého Tora" ["In the Steps of the Great Thor"], but cannot find a Slovak publisher; a collection of his tales from this cycle is awaiting publication in the Czech Republic. (In this context it is interesting to note that the small Czech publisher Saga has recently begun to publish modern Slovak fantasy authors in translation, but giving the writers macho-sounding foreign pseudonyms!)

There can be little doubt that Slovak fantasy writers – and Slovak moviemakers – have had a lot of trouble making their voices heard since the political changes of 1989. Slovak fantasy readers still depend to a large extent on Czech publishers; indigenous fantasy authors – even prizewinners like Lupečka – have enormous difficulties in getting published. Indeed, in contrast to their Czech counterparts, Slovak publishers seem reluctant even to translate the bestselling UK and US authors; of them only Stephen KING has had a significant number of books published in Slovakia. One can only hope that, with time, the cultural thaw in Slovakia will reach the fantasy genre.

MOVIES

Slovak fantasy CINEMA has in recent decades fared somewhat better than the written form, although output has still been small. Of particular interest are two full-length ANIMATED MOVIES by the writer and caricaturist Viktor Kubal (1923-   ). *Zbojník Jurko* ["Jurko the Brigand"] (1976) and *Krvavá pani* ["The Bloody Lady"] (1980) respectively draw on the folktales about Jánošík and Báthory. Also interesting is *Sladký čas Kalimagdory* ["The Sweet Time of Kalimagdora"] (1968) dir Leopold Lahola (1918-1968), a Slovak adaptation of *Spáč ve zvěrokruhu* ["The Sleeper in the Zodiac"], the classic Czech fantasy novel by Jan Weiss (◊ CZECH REPUBLIC). Although he had previously written screenplays and directed several movies abroad, this was Lahola's first feature in Slovak. It describes the world of a group of people whose lives follow a seasonal cycle, as if they were plants.

Štefan Uher (1930-1993) directed a number of movies that drew on elements of folktale and fantasy. Perhaps the best is *Génius* ["Genius"] (1969), a tragicomedy about the role of sin in humanizing the DEVIL. But the most significant Slovak director of fantasy movies has undoubtedly been Juraj Jakubisko (1938-   ), almost all of whose movies contain elements of fantasy or MAGIC REALISM. Even in his early movies, like *Vtáčkovia, siroty a blázni* ["Little Birds, Orphans and Madmen"] (1969), he shows a profound interest in fantasy and metaphor, combining realistic and symbolic passages, visual ideas with moments of irrationality and ontological brutality, the comtemporary with the folkloristic – all conspiring as if to place the narrative of the movie outside space and TIME. A year later he began work on *Do videnia v pekle, priatelia* ["See You in Hell, My Friend"] (1990), a darkly comic burlesque constructed like a mosaic of ABSURDIST-FANTASY images, interlarded with SURREALISM. For political reasons this movie remained unfinished for two decades. All of Jakubisko's movies of the 1960s are characterized by this combination of a folkloristic style with an interest in modern artistic movements and ideas – a kind of Surrealism or Magic Realism mutated by the juxtaposition of Eastern Slovak folklore. Jakubisko fell out of favour in the 1970s and was unable to make further features; after his rehabilitation in the 1980s his first feature was *Tisícročná včela* (1983), adapted from the novel by Peter Jaroš; an expanded version was screened in four parts on tv. Jaroš's novel proved an excellent vehicle for Jakubisko's recurrent interest in the relation between REALITY and fantasy; if anything, the movie emphasizes the fantasy content of the original. It received the Czechoslovak Critics' Prize as movie of the year and did better box office than any other Slovak movie of the 1980s.

Jakubisko's next major movie, *Pehavý Max a strašidlá* ["Freckled Max and the Ghosts"], was less successful. Many critics dismissed it as a disjointed combination of the classical FAIRYTALE, a PARODY of HORROR MOVIES and half-baked Surrealism. It is nonetheless Jakubisko's only movie that belongs unequivocally to the fantasy genre, and is one of the better Slovak contributions to fantasy cinema.   [EN/CS]

SŁOWACKI, JULIUSZ (1809-1849) Polish poet. ◊ POLAND.

SMALL PRESSES The development of printing and computer technology – i.e., desk-top publishing (DTP) – has placed publishing within the technical and financial grasp of devotees. In the market, and in the perceptions of readers, SPs now seem more to resemble the earliest private presses than the amateur or fan-produced presses where the movement began.

A number of early private presses included fantasy among their publications, of which the most notable was the **Kelmscott Press**, produced by William MORRIS. Beautifully printed and illustrated, books from this press were intended to reissue and capture the magic of the medieval ROMANCE, including most of Morris's own fantasies. Its first book was Morris's *The Story of the Glittering Plain* (**1891**), but it also reprinted many of Caxton's earliest books, such as *The History of Reynard the Foxe* (**1481**; 1892), as well as other translations like *Sidonia the Sorceress* (**1849** UK; 1893) by William Meinhold (1797-1851), plus works by the poets Dante Gabriel ROSSETTI, John KEATS, Algernon Swinburne, Percy Bysshe SHELLEY and Samuel Taylor COLERIDGE. During the 1920s and 1930s there were many UK private presses producing specialist literary material, some of which was fantasy or supernatural fiction – Algernon BLACKWOOD, John COLLIER and T.F. POWYS, in particular, had work from such presses, and John GAWSWORTH ran his own Twyn Barlwun Press – but the real history of the specialist fantasy SP was in the USA.

The early history of the SP was dominated by the devotees of the work of H.P. LOVECRAFT and his circle. Lovecraft was himself a very active member of the amateur press movement, and it was this same dedication to artistic rather than commercial values that saw the first real fantasy SP, **Recluse Press**, run by W. Paul Cook (1880-1948). This began with a volume of poetry, *A Man from Genoa* (coll **1926** chap) by Frank Belknap LONG, but Cook is now best remembered for his attempts to publish the first hardcover book by Lovecraft, *The Shunned House* (**1928** chap). Cook's limited finances prevented him from completing the full binding of this book, and later bindings of Cook's original sheets have come from other publishers over the years.

Another pioneer of the SP was William Crawford (1911-1984) who, under the **Fantasy Pubs** imprint, produced *Men of Avalon/The White Sybil* (anth **1935** chap), two stories by David H. KELLER and Clark Ashton SMITH, plus the sf volume *Mars Mountain* (coll **1935**) by Eugene George Key (1907-1976), before establishing the **Visionary Publishing Company** to issue Lovecraft's *The Shadow Over Innsmouth* (**1936**).

After the death of Lovecraft several fans made attempts to bring aspects of his work into print, such as the *Notes & Commonplace Book* (**1938** chap) from **The Futile Press** of Clyde F. Beck (? -1985), but the real achievement came from August DERLETH and Donald WANDREI, who established ARKHAM HOUSE to publish the first genuine collection of Lovecraft's fiction, *The Outsider and Others* (coll **1939**).

After WWII a number of SPs appeared. Though they concentrated mostly on rescuing classic sf from the pulp MAGAZINES, several also published fantasy, most of it from UNKNOWN or WEIRD TALES. Lloyd A. ESHBACH's **Fantasy Press** (founded 1947), for instance, published mostly sf, but also *The Book of Ptath* (1943 *Unknown*; **1947**) by A.E. VAN VOGT and *Darker Than You Think* (1940 *Unknown*; **1949**) by Jack Williamson (1908-   ). **Prime Press**, whose editor was Oswald Train (1915-1988) – who later established his own **Oswald Train** imprint in 1968 – published more fantasy and SUPERNATURAL FICTION including *The Mislaid Charm* (1941 *Unknown*; **1947** chap) by Alexander M. Phillips (1907-   ), *Without Sorcery* (coll **1948**) by Theodore STURGEON, and three subsidized volumes by David H. KELLER: *The Homunculus* (**1949**), *The Eternal Conflict* (**1949**) and *The Lady Decides* (**1950**). **Shasta: Publishers**, run primarily by Melvin Korshak, is best-remembered for publishing *The Checklist of Fantastic Literature* (**1948**) by E.F. BLEILER, but also published *Slaves of Sleep* (1939 *Unknown*; **1948**) by L. Ron HUBBARD and *Kinsmen of the Dragon* (**1951**) by Stanley Mullen (1911-1973).

The most successful of the immediate post-WWII SPs was **Gnome Press**, run by David A. Kyle (1919-   ) and Martin L. Greenberg (1918-?1992), which consciously attempted to be a mass-market trade publisher. Its very name suggests an element of fantasy, and it did publish a much higher quota of fantasy than its sf brethren. Books of note include *The Porcelain Magician* (coll **1948**) by Frank OWEN; *The 31st of February* (coll **1949**) by Nelson S. Bond, *The Castle of Iron* (1941 *Unknown*; **1950**) by L. Sprague DE CAMP and Fletcher PRATT, *Conan the Conqueror* (1935-36 *WT* as "The Hour of the Dragon"; **1950**) by Robert E. HOWARD, followed by several Howard collections, *Typewriter in the Sky & Fear* (coll **1951**) by L. Ron HUBBARD, *Shambleau and Others* (coll **1953**) by C.L. MOORE and *Two Sought Adventure* (coll **1957**) by Fritz LEIBER.

With the rise of the paperback and the trade publishing of sf books, the postwar SPs faded away, Gnome Press being the last to go, in 1962. But during the 1960s fantasy SPs began to re-emerge. Donald M. Grant (1927-   ) had become an interesting though minor SP publisher with his **Grandon: Publishers** (founded 1949), some of whose titles were fantasy, including *Dwellers in the Mirage* (1932; 1950) by A. MERRITT and *The Werewolf of Ponkert* (coll **1958**) by H. Warner MUNN. But in the 1960s, after a hiatus, he came into his own with **Donald M. Grant, Publisher** (founded 1962). After releasing the revised edition of *A Golden Anniversary Bibliography of Edgar Rice Burroughs* (**1962**; rev 1964) by Henry Hardy Heins (1923-   ), he concentrated on Robert E. HOWARD, starting with *A Gent from Bear Creek* (**1937** UK; 1965) and including a definitive series of CONAN books. Grant also reprinted lesser known material from the pulp magazines, like *The Temple of Ten* (1921 *Adventure*; **1973**) by H. BEDFORD-JONES and W.C. Robertson, *The Bowl of Baal* (1916-17 *All Around*; **1975**) by Robert Ames Bennet (1870-1954) and *The Three Paladins* (1923 *Adventure*; **1977**) by Harold LAMB. Beginning with *The Dark Tower: The*

Gunslinger (**1982**), Grant was also among the first to produce SP editions of the works of Stephen KING.

Roy A. Squires (1920-1988) and Clyde Beck teamed in the 1960s to produce a poetry volume by Clark Ashton Smith, *The Hill of Dionysus* (coll **1962** chap); thereafter Squires continued on his own as **Roy A. Squires**, producing a series of hand-sewn card-bound chapbooks of extremely high quality. These featured mostly poetry, primarily by Smith but also by Ray BRADBURY, Fritz LEIBER and Robert E. HOWARD. At this same time Jack L. CHALKER was developing his **Mirage Press**. This focused initially on bibliographical works, including *The New H.P. Lovecraft Bibliography* (**1961** chap) by Chalker and *In Memoriam: Clark Ashton Smith* (**1963**) ed Chalker, but also *A Figment of a Dream* (**1962** chap) by David H. KELLER. Although nonfiction remained its core output, especially a series of studies and analyses of **Conan** and the first publication of *A Guide to Middle-Earth* (**1971**) by Robert Foster (1949-   ), Mirage did publish some less commercially oriented fiction, including *Dragons and Nightmares* (coll **1969**) by Robert BLOCH, *Is The Devil a Gentleman?* (coll **1970**) by Seabury QUINN and the poetry volume *Phantoms and Fancies* (coll **1972**) by L. Sprague DE CAMP.

SP publishing entered a new dimension in the 1970s, with sounder financing and better production techniques. Karl Edward WAGNER established **Carcosa** in 1973 to restore to print lost stories from the pulps, the same original mission of **Arkham House**. Wagner published only four books – *Worse Things Waiting* (coll **1973**) by Manly Wade WELLMAN, *Far Lands, Other Days* (coll **1975**) by E. Hoffmann PRICE, *Murgunstruum and Others* (coll **1977**) by Hugh B. CAVE and *Lonely Vigils* (coll **1981**) by Wellman – but Carcosa won the 1976 WORLD FANTASY AWARD, as did both the first Wellman and Cave books.

Since then the number of SPs has increased significantly, though most have been short-lived. There has also been a growth in specialist art books and the ever-continuing interest in the Lovecraftian/*WT* school, while other SPs concentrate on bibliographical and reference works, of which the two major operations have been **Starmount House** run by T.E. Dikty (1920-1991) and **Borgo Press** run by Robert Reginald (real name Michael Roy Burgess; 1948-   ). The following summarizes the primary SPs since 1973 which concentrated on fiction. All firms are US unless otherwise stated.

Ash-Tree Press (UK) Started by Christopher and Barbara Roden, who moved to Canada in 1996. Intended to rescue rare and forgotten GHOST STORIES plus new material in the classic ghost-story tradition. Began with a small anthology, *Lady Stanhope's Manuscript and Other Supernatural Tales* (anth **1994** chap) ed Roden, and moved to reprints, concentrating on M.R. JAMES and the JAMES GANG.

Axolotl Press Run by John C. Pelan. Concentrates on neatly produced chapbooks, like *Night Moves* (**1986** chap) by Tim POWERS, *Paper Dragons* (**1986** chap) by James P. BLAYLOCK and *Ascian in Rose* (**1987** chap) by Charles DE LINT.

Cheap Street Run by Jan Landau and George O'Nale. Produces small but beautifully published card-bound chapbooks, some with limited slip-cased editions. Started with *Ervool* (1944 *The Acolyte*; **1980** chap) by Fritz LEIBER. Others include *The Story of Pepita and Corindo* (**1982** chap) and other FAIRYTALES by Richard Cowper, *The Girl who Heard Dragons* (**1986** chap) by Anne MCCAFFREY, *Bibliomen: Twenty Characters Waiting for a Book* (coll **1984** chap) by Gene WOLFE.

Dark Harvest Run by Paul Mikol. Presents mostly works of SUPERNATURAL FICTION and HORROR, always with one eye toward the commercial market. Noted for its ANTHOLOGIES, including the NIGHT VISIONS series, it has also published titles like *Songs the Dead Men Sing* (coll **1983**) by George R.R. MARTIN and *Carrion Comfort* (**1989**) by Dan SIMMONS.

Fedogan & Bremer Run by Philip J. Rahman, this continues the **Arkham House** tradition, publishing works either by or inspired by the Lovecraft circle. Fantasy titles include *Colossus* (coll **1989**) by Donald WANDREI, *The Early Fears* (coll **1994**) by Robert BLOCH and *Time Burial* (coll **1995**) by Howard WANDREI.

Ghost Story Press (UK) Run by David and Kat Tibet with the editorial assistance of Richard DALBY; concentrates on reprinting rare GHOST STORIES, particularly those by the JAMES GANG, but has included some new material. Titles include *Tedious Brief Tales of Granta and Gramarye* (coll **1919**; exp 1993) by Arthur Gray (1852-1940), *Flaxman Low, Psychic Detective* (ot *Ghosts* coll **1899**; new intro 1993) by E. & H. HERON and *Master of Fallen Years* (coll **1995**), the complete supernatural stories of Vincent O'SULLIVAN ed Jessica Amanda SALMONSON.

Kerosina (UK) Run by a conglomerate which included as primary motivators James Goddard, Les Escott and Mike and Debby Moir. Escott later split to run his own **Morrigan** imprint (*q.v.*). Kerosina published both sf and fantasy by UK writers, concentrating on the work of Keith ROBERTS, whose titles included *Kaeti & Company* (coll of linked stories **1986**) with the chapbook *Kaeti's Apocalypse* (**1986** chap), *Grainne* (**1987**) with the chapbook *A Heron Caught in Weeds* (coll **1987** chap) and the singleton *The Natural History of the P.H.* (**1988** chap). Books by other authors included *The Jaguar Hunter* (coll **1988**) by Lucius SHEPARD and *The Days of March* (**1988**) by John BRUNNER.

Land of Enchantment Run by Christopher Zavisa, originally launched to promote the work of Berni WRIGHTSON with the portfolio *Dinosaurs* (graph **1977**) but which also produced *Cycle of the Werewolf* (**1983**) by Stephen KING and *Twilight Eyes* (**1985**) by Dean R. KOONTZ.

Mark V. Ziesing Imprint run by Mark Ziesing (1953-   ), who began by publishing avant garde material with his brother Michael (1946-   ) as **Ziesing Brothers**, including two books by Gene WOLFE, before establishing his own imprint. He has published *The Book of the Dead* (anth **1989**) ed John Skipp (1957-   ) and Craig Spector (1958-   ) and some rather more traditional fantasy and HORROR fiction, including *Beastmarks* (coll **1984**) by A.A. ATTANASIO, *The Silver Pillow* (**1987** chap) by Thomas M. DISCH, *The Scalehunter's Beautiful Daughter* (**1988**) by Lucis SHEPARD and *The Last Coin* (**1988**) by James P. BLAYLOCK.

Morrigan (UK) Run by Jim and Les Escott. Focuses on US authors of the FANTASTIC. Titles of interest include *East of Laughter* (**1988**) by R.A. LAFFERTY and *The Digging Leviathan* (**1988**), *Homunculus* (**1988**) and *The Magic Spectacles* (**1991**) by James P. BLAYLOCK.

Necronomicon Press Run primarily by Marc Michaud (1960-   ), assisted in its early days by S.T. JOSHI. Inaugurated to reprint rare items of Lovecraftiana, moving on to Lovecraftian scholarship but, since the mid-1980s, also issuing horror and supernatural fiction by divers authors, though keeping its roots in the WEIRD TALES school. It also issues the magazines *Lovecraft Studies*, *Studies in Weird Fiction*, CRYPT OF CTHULHU, *Cthulhu Codex* and *Necrofile: The Review of Horror Fiction*.

NESFA Press US speciality imprint of the New England Science Fiction Association. Although originally organized (1968) to publish the annual **Index to the Science Fiction Magazines** by Anthony Lewis (1941-   ), it began to publish an annual author collection for Boskone, the regional New England sf convention. Titles include *Unsilent Night* (coll **1981**) by Tanith LEE, *Storyteller* (coll **1992**) by Jane YOLEN, *Everard's Ride* (coll **1995**) by Diana Wynne JONES, *Ingathering: The Complete People Stories* (coll **1995**) by Zenna Henderson (1917-1983) and *The Silence of the Langford* (coll **1996**) by David LANGFORD.

Phantasia Started by Sid Altus and Alex Berman but run by Berman since 1984, publishing reprints, like *Wall of Serpents* (fixup 1960; **1978**) by L. Sprague DE CAMP and Fletcher PRATT, and limited editions like *The Magic Labyrinth* (**1980**) by Philip José FARMER. The press has specialized in books by C.J. CHERRYH, de Camp, Farmer and Alan Dean FOSTER.

Pulphouse Founded by Dean Wesley Smith (1950-   ). The ambitious production of sf, fantasy and horror in hardcover, chapbook and magazine formats finally overstretched financially; the firm closed in 1996. The most successful series was *Pulphouse: The Hardback Magazine* (12 issues; Fall 1988-Spring 1991), a quarterly magazine/anthology which alternated between sf, FANTASY and HORROR. At its peak in 1991-2 this was a significant publishing concern, operating on a commercial basis and moving beyond the SP arena. It also published a magazine format **Author's Choice Monthly** starting with *The Old Funny Stuff* (coll **1989**) by George Alec Effinger (1947-   ), and a chapbook **Short Story Paperback/Short Story Hardbacks** series starting with *Loser's Night* (**1991** chap) by Poul ANDERSON.

Scream/Press Run by Jeff Conner, concentrating on SUPERNATURAL FICTION. Titles include: *The Dark Country* (coll **1982**), *Red Dreams* (coll **1984**) and *The Blood Kiss* (coll **1988**) by Dennis ETCHISON; and *Cold Print* (coll **1985**) and *Scared Stiff* (coll **1987**) by Ramsey CAMPBELL. S/P's short-lived fantasy imprint **Dream/Press** published *Signs and Portents* (coll **1984**) by Chelsea Quinn YARBRO, *Collected Stories* (coll **1989**) by Richard MATHESON, etc.

Silver Scarab Imprint of Harry O. Morris (1949-   ), under which he published his magazine *Nyctalops* (19 issues May 1970-April 1991), dedicated to H.P. LOVECRAFT and Clark Ashton SMITH. Morris issued occasional books, including *Songs of a Dead Dreamer* (coll **1985**) by Thomas LIGOTTI.

Steeldragon Irregular short-lived publisher, unique in that it focused solely on genuine fantasy. Books included *To Reign in Hell* (**1984**) by Steven BRUST, *The Time of the Warlock* (**1984**) by Larry NIVEN and *Merlin's Booke* (coll of linked stories **1986**) by Jane YOLEN.

Underwood-Miller Run by Tim Underwood (1948-   ) and Chuck Miller (1952-   ), whose mainstay author was Jack VANCE. U-M began with the first hardcover edition of *The Dying Earth* (**1950**; rep 1976), issuing in the end over 30 Vance titles. U-M also published Roger ZELAZNY, starting with *The Bells of Shoredan* (1966; **1979** chap) and including *The Last Defender of Camelot* (coll **1981**). Other titles include *Leeson Park and Belsize Quare: Poems 1970-75* (coll **1983**) by Peter STRAUB, *Through the Ice* (**1989**) and *Balook* (**1991**) by Piers ANTHONY, and *Apocalypse* (**1989**) by Nancy SPRINGER. U-M produced author bibliographies of Vance, Philip K. Dick (1928-1982), L. Sprague DE CAMP and Zelazny. The team eventually split in 1993, but the two continue to produce books separately.

Weirdbook Press Begun in 1968 by W. Paul Ganley (1934-   ) with his magazine WEIRDBOOK, subsequently branching into books with *Hollow Faces, Merciless Moons* (coll **1977**) by William Scott Home (1940-   ) and *The Gothic Horror* (coll **1978**) by George T. Wetzel (1921-1983). From 1985, as **W. Paul Ganley: Publisher**, he has published books including *The House of Cthulhu and Other Tales of the Primal Land* (coll **1984**) by Brian LUMLEY and *Tom O' Bedlam's Night Out and Other Strange Excursions* (coll **1985**) by Darrell SCHWEITZER.

Whispers. ◊ WHISPERS.   [MA]

Further reading: *The Science-Fantasy Publishers: A Critical and Bibliographic History* (**1991**) by Jack L. CHALKER and Mark Owings (1945-   ).

SMEDS, DAVE (1955-   ) Legal name of US artist and writer who published his first story, "Dragon Touched", in *Dragons of Light* (anth **1980**) ed Orson Scott CARD. DS has published two volumes of **The War of the Dragons**: *The Sorcery Within* (**1985**) and *The Schemes of Dragons* (**1989**). His short stories have appeared widely; some have been announced as parts of longer works in progress.   [GF]

Other works: «*Worldly Pleasures*» (coll **1996**), erotic sf and fantasy.

SMITH, ARTHUR D(OUGLAS) HOWDEN (1887-1945) US writer of adventure stories for the pulp MAGAZINES. ADHS's fantastic series, beginning with ("The Forging" 1926 *Adventure*), involved the **Gray Maiden**, a blood-loving SWORD forged during the reign of the Pharaoh Thutmose III and quenched in the blood of one he loves and one he hates. The stories have weak characterizations but are thoroughly researched and conclude with copious bloodshed; six were collected in *Grey [sic] Maiden: The Story of a Sword through the Ages* (coll **1929**); additional stories remain uncollected. **Gray Maiden** stands as a direct ancestor to such works as Jaan KANGILASKI's **Seeking Sword** sequence and as a relative of other stories featuring uncanny WEAPONS – like Stormbringer in Michael MOORCOCK's **Elric** series.   [MA]

SMITH, BARRY ◊ Barry WINDSOR-SMITH.

SMITH, CLARK ASHTON (1893-1961) US writer. Primarily a poet – his best early work is in *The Star-Treader* (coll **1912**), *Ebony and Crystal: Poems in Verse and Prose* (coll **1923**) and *Sandalwood* (coll **1925**) – CAS wrote during one hectic period over 100 short stories and extended prose-poems, most published (some belatedly) in the pulp MAGAZINES – although he issued one slim collection, *The Double Shadow and Other Fantasies* (coll **1933** chap) himself. They were eventually assembled in a series of ARKHAM HOUSE volumes: *Out of Space and Time* (coll **1942**); *Lost Worlds* (coll **1944**); *Genius Loci* (coll **1948**); *The Abominations of Yondo* (coll **1960**); *Poems in Prose* (coll **1964**); *Tales of Science and Sorcery* (coll **1964**); and *Other Dimensions* (coll **1970**). Many of CAS's highly ornate and sometimes vividly erotic works were censored by magazine editors, but for some reason he never got around to correcting the book versions; many of the originals were destroyed by a fire but a few survived to be reconstructed for a series of Necronomicon Press booklets entitled **The Unexpurgated Clark Ashton Smith** ed Steve Behrends, whose six volumes are: *The Dweller in the Gulf* (cut 1933 as "The Dweller in Martian Depths"; **1987** chap); *Mother of Toads* (cut 1938; **1987** chap); *The Vaults of Yoh-Vombis* (cut 1932; **1988** chap); *The Monster of the Prophecy* (1932 cut; **1988** chap); *The Witchcraft of Ulua* (cut 1934; **1988**); and *Xeethra* (cut 1934;

1988 chap). The same publisher issued *Nostalgia of the Unknown: The Complete Prose Poetry* (coll **1988**) ed Marc and Susan Michaud, Steve Behrends and S.T. JOSHI. The remnants of CAS's fiction and working notes were assembled in *The Black Book of Clark Ashton Smith* (coll **1979**) and *Strange Shadows: The Uncollected Fiction and Essays of Clark Ashton Smith* (coll **1989**) ed Steve Behrends with Donald Sidney-Fryer and Rah Hoffman, which also explains how the novella *As it is Written* (**1982**) was mistakenly attributed to CAS and how the falseness of the attribution was discovered. (Some fragments "completed" by Lin CARTER and issued as "collaborations" are best ignored.)

CAS's most typical stories – which generally appeared in WEIRD TALES – constitute one of the most remarkable *oeuvres* in imaginative literature; they are in direct line of descent from French DECADENCE – CAS translated a good deal of Decadent and Parnassian poetry – and took the calculated exoticism of that movement to its logical limit, first in such poems as *The Hashish-Eater, or The Apocalypse of Evil* (1922; **1989** chap) and later in the final tales in his **Zothique** sequence. The fact that these stories found any medium of publication must be regarded as astonishing, but their subsequent acquisition of cult status assured their preservation, alongside the works of other members of the H.P. LOVECRAFT circle. CAS's highly ornamented prose is dedicated to the building of phantasmagoric dreamworlds more remote from human experience, and even from familiar mythology, than any described before. There is, however, nothing consolatory about his brand of ESCAPISM: his characters are usually led by the irresistible allure of the exotic to disappointment, damnation and doom. Although he shared the desperate *ennui* and intemperate *spleen* of the Decadents, CAS developed a worldview similar to that of Lovecraft, which drew eccentrically upon the imagery of science, involving both the awesomeness of the modern cosmic perspective and the detachment and clinicality of the scientific outlook. CAS used several IMAGINARY LANDS, including Averoigne, ATLANTIS and HYPERBOREA – shown to best effect in "The Seven Geases" (1934), in which a vainglorious magistrate is condemned to descend through a series of Tartarean realms to "the ultimate source of all miscreation and abomination" – before finding a perfect setting in the DYING-EARTH scenario of **Zothique**, with stories assembled initially as *Zothique* (coll **1970**) and in definitive form, printing fragments and using original manuscripts, as *Tales of Zothique* (coll **1995**) ed Will Murray with Steve Behrends. The sequence began with "The Empire of the Necromancers" (1932), a nightmarish extravaganza in which two necromancers conjure themselves an empire out of the dust of the ages and the reanimated corpses of the ancient dead (◊ BONDAGE). He brought the series to a dramatic apogee in the extravagant feasts of horror and bizarrerie of "Xeethra", "The Dark Eidolon" (1932) and "Necromancy in Naat" (1935). [BS]

Other works: *The Immortals of Mercury* (**1932** chap); *Hyperborea* (coll **1971**); *The Mortuary* (**1971** chap); *Xiccarph* (coll **1972**); *Sadastor* (1930; **1972** chap); *Poseidonis* (coll **1973**); *From the Crypts of Memory* (**1973** chap); *Prince Alcouz and the Magician* (**1977** chap); *The City of the Singing Flame* (coll **1981**); *The Monster of the Prophecy* (coll **1983**); *A Rendezvous in Averoigne* (coll **1988**).

Poetry (selective): *Odes and Sonnets* (coll **1918** chap); *Nero and Other Poems* (coll **1937** chap); *The Dark Château* (coll **1951** chap); *Selected Poems* (coll **1971**).

Nonfiction: *Planets and Dimensions: Collected Essays* (coll **1973** chap) ed Gary K. Wolfe; *The Devil's Notebook: Collected Epigrams and Pensées* (coll **1990** chap).

Further reading: *Emperor of Dreams: A Clark Ashton Smith Bibliography* (**1978**) by Donald Sidney-Fryer.

SMITH, DAVID C(LAUDE) (1952-) US writer who began publishing fantasy with the **Oron** series of SWORD-AND-SORCERY adventures: *Oron* (**1978**), *The Sorcerer's Shadow* (**1978**), *The Valley of Ogrum* (**1982**) and *Oron: The Ghost Army* (**1983**). The **Red Sonja** sequence – *Red Sonja #1: The Ring of Ikribu* * (**1981**) with HORROR poet and writer Richard Tierney (1936-), *#2: Demon Night* * (**1982**) with Tierney, *#3: When Hell Laughs* * (**1982**), *#4: Endithor's Daughter* * (**1982**) and *#5: Star of Doom* * (**1983**) – is based on a spinoff comic, *Red Sonja*, generated by Roy THOMAS from a minor character named Red Sonya of Rogatine, who appeared in a tale by Robert E. HOWARD. A third series, **Fall of the First World** – *Master of Evil* (**1983**), *Sorrowing Vengeance* (**1983**) and *The Passing of the Gods* (**1983**) – is set in a world which has an ARCHETYPE relationship to this one and features characters whose emblematic lives artfully prefigure the end of things. Other works include *For the Witch of the Mists* * (**1981**) with Tierney, another Robert E. Howard tie, this time to the **Bran Mak Morn** world. *The Fair Rules of Evil* (**1989**) and *The Eyes of Night* (**1991**) form a moderately effective occult horror sequence. [JC]

SMITH, DODIE Working name of Dorothy Gladys Smith (1896-1990), UK writer and popular playwright, now best-known for her CHILDREN'S FANTASY *The Hundred and One Dalmatians* (**1956**). In this charming and witty ANIMAL FANTASY, the resourceful Dalmatian dogs Pongo and Missis make a QUEST across modern England for their lost litter of pups, stolen by agents of the memorable aristocratic she-VILLAIN Cruella de Vil – who has amassed several score Dalmatian pups in darkest Suffolk, planning to have fur coats made of their skins. But England's dogs (in whose PERCEPTION humans are amiable pets) have a WAINSCOT society with its own country-wide communications via the "Twilight Barking" and "Midnight Barking". With aid from this network of scattered dogs and some helpful CATS, the entire pack of Dalmatians escapes to LONDON despite Cruella's malevolent pursuit. This story was adapted by DISNEY as the ANIMATED MOVIE *One Hundred and One Dalmatians* (**1961**); a live-action remake is in production. The novel's inferior sequel, *The Starlight Barking: More about The Hundred and One Dalmatians* (**1967**), is more extravagantly fantastic, with Sirius the Dog STAR placing an ENCHANTMENT on Earth which grants dogs special TALENTS – including flying – while all other species are cast asleep. Predictably, Pongo and the rest reject Sirius's offer of dog HEAVEN and loyally remain with humanity. It is for *The Hundred and One Dalmatians* that DS is fondly remembered. [DRL]

Other works (selective): *I Capture the Castle* (**1949**), associational; *The Midnight Kittens* (**1978**), another story for children.

Further reading: *Dear Dodie: Life of Dodie Smith* (**1996**) by Valerie Grove.

SMITH, JULIE DEAN (1960-) US writer whose **Caithan Crusade** sequence – *Call of Madness* (**1990**), *Mission of Magic* (**1991**), *The Sage of Sare* (**1992**) and *The Wizard King* (**1994**) – is more competent than the titles might hint. At the heart of the sequence, set in a FANTASY-LAND with CELTIC-FANTASY coloration, is a sometimes engrossing conflict between MAGIC and RELIGION. The female protagonist, whose TALENTS develop as she goes through a RITE OF PASSAGE into uneasy adulthood, is Princess of a LAND where magic causes madness and is in any case proscribed; her long fight to re-establish the proper training necessary to curb her talent, to find a husband, to defeat a wicked MAGUS, and to come to terms with her stubborn brother, all fit within GENRE-FANTASY formatting, but everything is done with panache and a happy skill.   [JC]

SMITH, JUNIUS B. (1883-1945) US writer. ◊ John Ulrich GIESY.

SMITH, NYX Working name of US writer N(athan) Y(ale) X(avier) Smith (?   -   ), author of ties for the SHARED-WORLD series of TECHNOFANTASIES, SHADOWRUN.   [JC]

SMITH, STEPHANIE A(NN) (1959-   ) US writer whose first novels, the **Snow-Eyes** sequence – *Snow-Eyes* (**1985**) and *The Boy who was Thrown Away* (**1987**) – are fantasy; *Other Nature* (**1995**) is sf. In *Snow-Eyes*, a young girl (Amarra) is taken from home to enter into a complex relationship with a goddess; similarly, in *The Boy who was Thrown Away* a young boy (Amant) is deprived of a normal childhood and must learn to control his ability as a SHAPESHIFTER while undergoing slavery, varieties of loss, the solace of an animal COMPANION, and lessons in MUSIC. He is mothered eventually by Amarra, and as a trained singer rescues his cousin from a bargain with DEATH, for she must die during each day, living only at night and only in the UNDERWORLD. His rescue of her constitutes a direct fantasia on the ORPHEUS myth. The venue is almost – but not quite – a LAND-OF-FABLE version of the CELTIC-FANTASY landscape. The telling is supple and undogmatic, with unexpected turns.   [JC]

Other work: *Conceived by Liberty: Maternal Figures and Nineteenth-Century American Literature* (**1995**), nonfiction.

SMITH, (JAMES) THORNE (Jr) (1893-1934) US writer, best-remembered for humorous novels, many turning on some fantastic PLOT DEVICE which thrusts the protagonist into grotesque predicaments, often via unwitting TRANSFORMATION, and leads him on a mild, mind-broadening version of the NIGHT JOURNEY. The **Topper** books, which led to the TOPPER MOVIES and the tv series TOPPER (1953-6), are *Topper: An Improbable Adventure* (**1926**; vt *The Jovial Ghosts: The Misadventures of Topper* 1933 UK) and *Topper Takes a Trip* (**1932**). These afflict the hapless hero Topper with the companionship of irresponsible GHOSTS who find death no obstacle to fast living. Characteristically, straitlaced Topper is principally tormented by the sexy female ghost, whose INVISIBILITY and intangibility are embarrassingly partial and intermittent. Cheerful bawdiness is a TS trademark, as is his Prohibition-era emphasis on heavy drinking as the prelude to unwise capers.

In *The Stray Lamb* (**1929**) the protagonist's incautious WISH to see humanity from outside is granted by a PAN figure, who transforms him into a succession of animals, culminating in a chimerical anthology creature (◊ IMAGINARY ANIMALS). *Turnabout* (**1931**) subjects a jaded husband and wife to IDENTITY EXCHANGE when a magic statue from EGYPT grants another unserious wish; after farcically experiencing pregnancy and childbirth the man in particular LEARNS BETTER. This was filmed as TURNABOUT (*1940*). *The Night Life of the Gods* (**1931**), particularly wild and shambolic in its comedy, has a protagonist who is irresponsible from the outset: he and his jealous mistress, a leprechaun's daughter, preside as lords of misrule over a riot of MAGIC and pseudoscientific petrification (e.g., of disagreeable relatives). This process is reversible and allows PYGMALION-like vivifying of Classical STATUES, loosing a partial PANTHEON from Graeco-Roman MYTHOLOGY on New York. Inevitably these include Bacchus and an armless APHRODITE. After copious misadventures the GODS weary of modern life's strictures – especially policemen – and willingly return to stone. They are joined by the protagonist and his lover, eternally embracing in stone. The movie is *The Night Life of the Gods* (**1935**).

Rain in the Doorway (**1933**) transports a harassed lawyer from the gloom of the Depression through a PORTAL to a TOPSY-TURVY department store run on Marx Brothers principles, featuring a Pornographic Department, permanently drunk owners, summary justice dealt out to awkward customers, and general anarchic wish-fulfilment. The protagonist's adventures are ultimately rationalized as insanity or skewed PERCEPTION; still, his inhibitions have "passed away forever", and TS is certain that this is a good thing. *Skin and Bones* (**1933**) returns to the TS formula of bizarre transformation when a chemical accident turns the hero, at unpredictable intervals, into a living skeleton. In *The Glorious Pool* (**1934**) a triangle of ageing protagonist, wife and mistress is complicated when a garden pool acquires FOUNTAIN-OF-YOUTH properties; the pool's nymph statue also comes alive to join the tangle of relationships.

Omnibus editions are: *The Thorne Smith 3-Decker* (omni **1936**), containing *The Stray Lamb*, *Turnabout* and *Rain in the Doorway*; *The Thorne Smith Triplets* (omni **1938**), containing *Topper Takes a Trip*, *The Night Life of the Gods* and the non-fantasy farce *The Bishop's Jaegers* (**1932**); and *The Thorne Smith Three-Bagger* (omni **1943**), containing *The Glorious Pool*, *Skin and Bones* and *Topper*.

Some of the uproariousness of TS's comedy has faded with time: his naughtiness inevitably seems less naughty today, and often comic effects are milked too assiduously through repetitive description and dialogue; but many passages remain very funny. His reiterated, heartfelt pleas against stifling conventions (middle-class propriety seen as BONDAGE) still sound a note of youthful charm. He should be read when young.   [DRL]

Other works: *Dream's End* (**1927**), a "serious" novel of HAUNTING and DREAM symbolism; *Lazy Bear Lane* (**1931**); *Did She Fall?* (**1936**), detective thriller; *The Passionate Witch* (**1941**), completed after TS's death by Norman H. MATSON and filmed as I MARRIED A WITCH (*1942*), and its sequel *Bats in the Belfry* (**1942**), by Matson alone.

See also: HUMOUR; SLICK FANTASY.

SMITHS These are generally figures of some mythic stature, harking back to the smith-GODS Vulcan in Greek MYTH, Ilmarinen in the KALEVALA, and WELAND SMITH. Other god-like smiths may be encountered, diminished by THINNING but still potent – like the thinned Weland of Rudyard KIPLING's *Puck of Pook's Hill* (coll **1906**), or Elof in Michael Scott ROHAN's **Winter of the World** trilogy, an AVATAR of Ilmarinen. Smiths' ability to handle and forge COLD IRON, besides making them unusually resistant to hostile MAGIC, goes with qualities of physical strength and honesty. As MENTORS they tend to attract UGLY DUCKLINGS who must learn to forge their own SWORDS as one RITE OF PASSAGE *en route* to leadership, like Taran in Lloyd ALEXANDER's *Taran Wanderer* (**1967**). Smiths know the LAND's old ways and hidden roads, and here function as LIMINAL BEINGS, as in Susan COOPER's *The Dark is Rising* (**1973**); the canny smith of Fay SAMPSON's *Black Smith's Telling* (**1990**) is a powerful force in the old RELIGION of the Horned God, though still no match for MORGAN LE FAY. DWARFS are traditionally excellent smiths. [CB/DRL]

SMOODIN, ROBERTA (1952-   ) US academic and novelist. Her first novel, *Ursa Major* (**1980**), sees a dancing, talking bear escape from its exhibitors and embark on an odyssey across the landscape of the 1970s USA; the book exhibits RS's talents as a fabulist. *Presto!* (**1982**) involves a struggling young MAGICIAN who possesses the genuine TALENT to make things vanish; his problems include intrigues with the Magicians' Union and a girlfriend who yearns to "disappear" via anorexia. RS's attraction to extremely figurative language and formal playfulness, as well as her use of the image of the conjurer as metaphor for the artist, recalls Vladimir NABOKOV, whose presence hovers also over *Imagining Ivanov* (**1985**), a nonfantasy about a young academic whose attempts to write the biography of a Nabokov-like émigré writer are undermined by the writer's mystifications. [GF]

Other works: *White Horse Cafe* (**1988**), contemporary novel.

SNAKES ◊ SERPENTS.

SNODGRASS, MELINDA M(ARILYN) (1951-   ) US writer who has been involved in **Star Trek** ventures as both writer of novelizations and consultant to the second tv series. Most of her work has been sf, but *Runespear* (**1987**) with Victor MILAN, which pits Nordic gods (◊ NORDIC FANTASY) vs Nazis in 1936, is fantasy, as is the historical *Queen's Gambit Declined* (**1989**). MMS edited *A Very Large Array: New Mexico Science Fiction and Fantasy* (anth **1987**). [GF]

SNOW, JACK (1907-1956) US writer. ◊ DOLLS.

SNOWDON, EVAN [s] ◊ Kenneth MORRIS.

SNOWMAN, THE UK ANIMATED MOVIE (*1982* tvm). ◊ Raymond BRIGGS.

SNOW QUEEN, THE Finnish movie (*1993*). Harollfilm. **Pr** Päivi Hartzell. **Dir** Hartzell. **Spfx** Lauri Pitkanen, Karl von Kugelgen. **Screenplay** Hartzell. **Based on** "The Snow Queen" (1846) by Hans Christian ANDERSEN. **Starring** Pirjo Bergström (The Beloved), Esko Hukkanen (The Fool), Sebastian Kaatrasalo (Kai), Tuula Nyman (The Witch), Saara Pakkasvirta (The Bandit Wife), Maria Pyykko (The Bandit Daughter), Elina Salo (The Sorceress of the North), Satu Silvo (The Snow Queen), Reijo Tuomi (Polar Bear Man), Juulia Ukkonen (The Princess), Outi Vainionkulma (Gerda), Paavo Westerberg (The Prince). 88 mins. Colour.

The children Kai and Gerda discover on the beach a music box which contains a tiny DOLL ballerina and three buttons, which latter become Kai's. From here on there is ambiguity as to whether what we are seeing is REALITY or a DREAM of Gerda's. Unknown to the children, long ago a green stone fell from the Earth into a well. If the Snow Queen could recover it and smash it with "the Black SWORD" she would own the world. But this she cannot do herself, for her hands are so cold they freeze the well's water every time she tries. So she one night recruits Kai to do the job for her, promising him that, if successful, he shall have the Crown of Darkness that will make him Ruler of the World. As they dash across the sky in a horse-drawn sleigh, she spies the MAGIC buttons and casts them out onto the land. The rest of the tale is of Gerda's QUEST to rescue Kai, with the discovered buttons acting as PLOT COUPONS. At last we are restored to the beach where Gerda and Kai

"again" discover the music box: this time they decide to rebury it in the sand.

TSQ is rich in fantasy and also in visual beauty; but the English-language version is appallingly dubbed. Silvo is magnificent as the Snow Queen – a torrent of ice-white hair dominates a performance that convinces one that here is a FEMME FATALE who is totally nonhuman – but Vainionkulma outshines even this. Had *TSQ* been made in Hollywood (or even been properly dubbed) it could be regarded as a classic of fantasy CINEMA – and, indeed, of filmed SWORD AND SORCERY, for that is what Hartzell has made Kaatrasalo's (small) part of the plot. [JG]

SNOW WHITE The story of Snow White and the Seven Dwarfs is among the most popular of all FAIRYTALES, particularly since it was chosen by Walt DISNEY for his first feature-length ANIMATED MOVIE, SNOW WHITE AND THE SEVEN DWARFS (*1937*). The tale derives from "Sneewitchen" ["Snowdrop"], in *Kinder- und Hausmärchen* (coll **1812**) by the GRIMM BROTHERS. It tells of a king who has a young daughter, Snowdrop. His wife dies and he marries a beautiful new queen, who becomes jealous of the beauty of her step-daughter, now seven. The queen asks her magic MIRROR who is the most beautiful in the land, and the mirror replies that it is Snowdrop. In her rage she orders that Snowdrop be taken INTO THE WOODS and killed, and only her heart – which the queen will eat – brought back. (Victorian translators soon bowdlerized this to have Snowdrop abandoned in the woods.) The soldier so charged, unable to kill Snowdrop, dupes the queen with a lamb's heart. Snowdrop chances upon the cottage of the seven DWARFS, who come home to find her asleep. The queen learns from her mirror that Snowdrop is still alive. She disguises herself as a pedlar and sells Snowdrop some lace which she ties so tight that Snowdrop collapses. The dwarfs revive her. The queen tries again, this time with a poisoned comb, but again the dwarfs revive Snowdrop. The third time, however, a poisoned apple succeeds. The dwarfs encase Snowdrop in a glass coffin where she remains for some years until a prince lifts her from the coffin and in so doing dislodges the poisoned apple. The queen dies of rage (in the original version she is forced to dance to her death in red-hot slippers) and Snowdrop marries the prince.

Like all fairytales the story has several forebears. In *Lo Cunto de li Cunti* ["The Story of Stories"] (**1634**; vt *The Pentameron* 1674) by Giambattista Basile (1575-1632) the story sometimes known as "The Young Slave" tells of a beautiful young girl, Lisa, who is poisoned by a comb, and is kept alive in a glass coffin, which grows with her. In "Richilde" (1782-87) by Johann Karl MUSÄUS, Richilde is a vain woman who has a magic mirror which tells her of her beauty. The mirror also informs her the handsomest man is Gombold, who is already married, but under Richilde's influence (◊ FEMME FATALE) Gombold divorces and marries Richilde. He already has a young daughter, Blanca, who grows up and becomes more beautiful than Richilde. The stepmother does all she can to be rid of the girl who, with the help of the court dwarfs, escapes.

SW was also used as a character by the Grimms in "Sneewitchen and Rosenrot" ["Snowdrop and Rose Red"] (1837), a story mostly by Wilhelm Grimm, who seems to have utilized elements from *Der Undankbare Zwerg* (**1818**) by Friedrich Kind (1768-1843) about an ungrateful dwarf. In the Grimm version, two sisters living with their mother are visited by a bear, whom they care for during the winter. In the spring the bear goes to protect his treasure from the DWARFS. The girls meet a dwarf and help him out of several predicaments, although he is never grateful. At length the bear returns, kills the dwarf and is revealed as a young prince who had been placed under a SPELL by the dwarf.

SW is a different girl in these stories, but always the name symbolizes purity and VIRGINITY. It is significant in "Snowdrop" that the girl is aged seven before her stepmother seeks to have her killed. SEVEN signifies completeness, and represents the first complete stage of childhood – it is relevant, too, that the girl is saved by seven dwarfs. The whole story of "Snowdrop" is replete with NUMEROLOGY references, implying that it may have some deeper alchemical ($\lozenge$ ALCHEMY) or GNOSTIC meaning. "Snowdrop and Rose Red", conversely, is probably drawn from a nature myth, with the bear signifiying winter, and the two sisters representing two roses who need to be cared for during the winter. In the summer the girls come forth and do battle with the dwarf, who has power over the SEASONS.

"Snowdrop" has formed the basis for REVISIONIST FANTASIES, including *Snow White* (**1967**) by Donald BARTHELME and *Suisan* (**1992**) by Phyllis AGINS. Tanith LEE has utilized the tale twice, once for a darker version, "Red as Blood" (1979 *F&SF*), and once translated into the modern day in "Snow-Drop" (in *Snow White, Blood Red* anth **1993** ed Ellen DATLOW and Terri WINDLING). Patricia C. WREDE reworked the other Snow White into Elizabethan England in *Snow White and Rose Red* (**1989**). Snow White is also retold for feminists by Róisín Sheerin in "Snow White" (in *Cinderella on the Ball* anth **1991** ed anon). [MA]

SNOW WHITE AND THE SEVEN DWARFS US ANIMATED MOVIE (*1937*). Disney. **Pr** Walt Disney. **Sv dir** David Hand. **Based on** the version in *Kinder- und Hausmärchen* (**1812-15**) by the GRIMM BROTHERS. **Voice actors** Roy Atwell (Doc), Adriana Caselotti (Snow White), Pinto Colvig (Grumpy, Sleepy), Billy Gilbert (Sneezy), Otis Harlan (Happy), Lucille LaVerne (Queen/Witch), Scotty Mattraw (Bashful), Moroni Olsen (Spirit of the Magic Mirror). 83 mins. Colour.

The wicked Queen asks her Magic MIRROR who is the fairest of them all, and to her fury it responds that there is one fairer than she . . . This retelling of the classic FAIRYTALE differs in no substantial way from its original; moreover, the movie has become such an ICON itself that any differences between it and the Grimms' version have now *become* part of the traditional tale.

SWATSD was the first feature-length ANIMATED MOVIE, and is thus a landmark in the history of the CINEMA. It was obviously a landmark, too, in the history of DISNEY, and would certainly have bankrupted the studio had it flopped; costing, in the end, $1,480,000, a huge sum in those days, it was before its release widely known in Hollywood as "Walt's Folly", a soubriquet that is one of many testimonies to the fact that the movie was most definitely Walt DISNEY's own project. New techniques were part of the expense (about $200 per foot, compared with $50-$70 per foot for the Disney shorts); also expensive were the deliberately created subliminal effects, including not only the colour of characters' clothing and eyes (selected to reflect personality) but even the textures of the colours used, which involved the development of new types of paints. Animals were imported to the Disney lot to serve as live-action reference for the animators; actors were used for the humans, SNOW WHITE herself being modelled by Marjorie Belcher, later better

known as the actress Marjorie Champion. All this was for a movie made at a time when no one knew if audiences would have the patience to sit through more than a few minutes of animation.

Fortunately for Disney, the critical and popular reception went beyond mere enthusiasm. Both at the time and in the decades since, however, some have objected to the Disnification of classic fairytales; and a few objecting voices were raised concerning the movie's scariness. Viewed today, however, *SWATSD* seems a fine film, its animation still impressive, its characterization generally excellent, its narrative drive compelling, and its sense of quintessential FANTASY – in particular through its juxtaposition of HUMOUR with imaginative transports – strong.

Live-action screen adaptations of the Grimm tale include *Snow White* (*1972*; East Germany; dir Gottfried Kolditz; starring Wolf-Dieter Panse, Doris Weikow; 70 mins; colour), *Snow White and the Three Stooges* (*1961* US; vt *Snow White and the Three Clowns*; dir Walter Lang; starring Carol Heiss, The Three Stooges; 107 mins; colour), *Snow White and the Seven Dwarfs* (*1987* US; dir Michael Berz; starring Sarah Patterson, Diana Rigg and many others; 85 mins; colour) and various soft-porn versions. [JG]

Further reading: *Walt Disney's "Snow White and the Seven Dwarfs"* (**1978**) ed Jack Solomon; *Walt Disney's "Snow White and the Seven Dwarfs" and the Making of the Classic Film* (**1987**) by Richard Holliss and Brian Sibley.

SNYDER, MIDORI (MADELEINE) (1954-   ) US writer who began publishing fantasy with "Demon" for *Bordertown* (anth **1986**) ed Terri WINDLING and Mark Alan Arnold, the second of the **Borderlands** SHARED-WORLD anthologies. Her first novel, *Soulstring* (**1987**), incorporates elements of the TWICE-TOLD (suitors must pass tests to win the hand of a sorcerer's daughter, and die if they fail) into a tangled but sometimes moving tale. The **Queen's Quarter** sequence – *New Moon* (**1989**), *Sadar's Keep* (**1990**) and *Beldan's Fire* (**1993**) – is interestingly dark (and more coherently told) HIGH FANTASY, based but not dependent on Celtic material ($\lozenge$ CELTIC FANTASY), depicting the centuries-long rule of immortal Zorah, who controls the ELEMENT of Fire, and whose tyranny is opposed by a complex cast. Much of the tale is set in a vast CITY whose labyrinthine ways are reminiscent of the LONDON of Charles DICKENS. *The Flight of Michael McBride* (**1994**) is also of interest, though MS's amiable style sometimes fails to convey the full rigours of the tale's mayhem. McBride, the young halfling son of a FAIRY mother and the mortal father who won her from a lord of FAERIE in a CHESS match, undertakes a long NIGHT JOURNEY through a US West CROSSHATCHED by deadly incursions of the Sidhe, during the course of which he finds himself locked into the form of a crow ($\lozenge$ TRANSFORMATION; BONDAGE) and succoured by the goddess Morrigu ($\lozenge$ MORGAN LE FAY). Eventually he comes to terms with the STORY told him (among other fables) by his mother, which he is involuntarily re-enacting; and he completes his RITE OF PASSAGE by winning back his own mortal bride from the frustrated fairy lord. [JC]

SNYDER, ZILPHA KEATLEY (1927-   ) US writer of children's novels, many being fantasies. *Black and Blue Magic* (**1966**) tells of a 12-year-old boy who can grow magic WINGS, the "black and blue" of the title refers to the fact that he is clumsy even when not trying to fly. *The Witches of Worm* (**1972**) involves a CAT that proves to be a WITCH's familiar, while the **Green-Sky Trilogy** – *Below the Root*

(1975), *And All Between* **(1976)** and *Until the Celebration* **(1977)** – deals rather solemnnly with a race of FAIRY-like creatures UNDERGROUND. ZKS has described herself as having an abiding penchant for the fantastic, which extends to her other works; some, such as *The Egypt Game* **(1967)** and *Eyes in the Fishbowl* **(1968)**, are ambiguous about the REALITY of their seeming fantasy content, while several nonfantasies have fantastical-sounding titles. Though not especially popular with young readers, ZKS has won numerous citations from the critical establishment, possibly because her work tends to be improving in nature, and slightly dull.   [GF]
Other works (selective): *The Headless Cupid* **(1971**; vt *A Witch in the Family* 1977 UK), borderline; *The Princess and the Giants* **(1973)**; *The Truth about Stone Hollow* **(1974)**; *Heirs of Darkness* **(1978)**; *Squeak Saves the Day and Other Tooley Stories* (coll **1988**), for younger children; *Song of the Gargoyle* **(1991)**; *The Trespassers* **(1995)**, a HORROR novel.

SOCIETY FOR CREATIVE ANACHRONISM (SCA) US-based organization which describes itself as a forum for the study and practice of the culture and technology of the medieval period (ending arbitrarily at AD1600), especially knightly battle (◊ KNIGHTS) and its accoutrements. Its publications include a quarterly magazine, *Tournaments Illuminated*. Most of the SCA's focus is on weekend camp gatherings at which members wear medieval clothing, adopt medieval names and sometimes personae, and fight "wars".

The SCA was founded in 1966, mostly by members of sf fandom, to which it retained a close connection for several years. Many fantasy authors have been active in the SCA, particularly in its early years, among them Poul ANDERSON, Randall GARRETT, Adrienne MARTINE-BARNES, Diana L. PAXSON and Paul Edwin ZIMMER. Younger authors of HEROIC FANTASY have also taken the opportunity for first-hand observation of medieval-style battles and technology. The **Westria** series by Paxson, set in a neo-medievalist post-HOLOCAUST California, is a particularly pure translation of SCA style into fantasy. RECURSIVE-FANTASY versions of the SCA itself appear in Peter S. BEAGLE's *The Folk of the Air* **(1986)**, Robert A. HEINLEIN's *"The Number of the Beast"* **(1980)** and the mystery *Murder at the War* **(1987)** by Mary Monica Pulver.   [DB]

SOLOVIEV, VLADIMIR (1853-1900) Russian philosopher. ◊ ANTICHRIST.

SOMADEVA Kashmiri poet (*fl*11th century). ◊ OCEAN OF STORY

SOMETHING WICKED THIS WAY COMES US movie (*1983*). DISNEY/Bryna/Buena Vista. **Pr** Peter Vincent Douglas. **Dir** Jack Clayton. **Vfx** Lee Dyer. **Screenplay** Ray BRADBURY, based on his *Something Wicked This Way Comes* **(1962)**. **Starring** Mary Grace Canfield (Miss Foley), Shawn Carson (Jim Nightshade), Royal Dano (Tom Fury), Richard Davalos (Mr Crosetti), Jack Dengel (Mr Tetley), Bruce M. Fischer (Mr Cooger), Pam Grier (Dust Witch), Vidal Peterson (Will Halloway), Jonathan Pryce (Mr Dark), Jason Robards (Charles Halloway), James Stacey (Ed). **Voice actor** Arthur Hill (Narrator). 95 mins. Colour.

Sometime in the October Country of Bradbury's youth, in the middle of the night, Dark's Pandemonium CARNIVAL comes to a rural US small town. Best friends Jim Nightshade and Will Halloway recognize WRONGNESS in the air, and their snooping reveals more: barber Crosetti is granted his erotic DREAMS, money-obsessed cigar-store owner Tetley wins $1000, schoolmarm Foley has her youth and beauty restored, one-legged Ed, once a football hero,

regains his lost leg . . . but all pay a terrible price. The two lads also see the carnival's owner, Mr Dark, run the carousel backwards to transform hefty barker Mr Cooger into a sinister child and torture lightning-rod seller Tom Fury in an electric chair to try to find when the storm is coming, for lightning will drive away the carnival's darkness and rain wash it clean. In short, the boys see too much: Mr Dark instructs the Dust WITCH to use her magic RING to send her spectral SPIDER army to round them up – unsuccessfully. Of course, no adult will believe the boys – until Mr Dark foolishly reveals too much to librarian Charles Halloway, who discovers that this October carnival has brought misery to the town before . . .

Bradbury's original DARK FANTASY was somewhat ponderous and formless, but had the advantage of his lyricism to generate brooding atmosphere; the movie, deprived of this, does its best with moody shots, lighting and (somewhat inadequately) music, but the end result, though laden with fantasy, falls short. The overall effect is of a helterskelter ride on a rainy day: breathtaking but cold, and you don't get anywhere.   [JG]

SOMTOW, S.P. Working name of Somtow Papinian Sucharitkul (1952-   ), Thai writer, composer and filmmaker. Much of SPS's early work, signed with his surname, was sf, including *Starship and Haiku* **(1981)**, a post-HOLOCAUST novel; The **Chronicles of the High Inquest** sequence, and the **Aquiliad** sequence, set in a western hemisphere under Roman influence. The SPS name has been used – although not exclusively – since 1985. In recent years, SPS has split his time between Thailand and the USA.

Perhaps the most vital of SPS's fantasies are *Riverrun* **(1991)** and *Forest of the Night* **(1992**; vt *Armorica* 1994 UK), the first two volumes of a projected trilogy, an ambitious exploration of REALITY in which SPS presents characters and viewpoints so fluid as to border on the chaotic. Theo Etchison and his family are travelling to Mexico when they are drawn into a reality-spanning war being fought by the DRAGON-children of the Darkling King, Strang, who effect changes on the Etchisons throughout the books. Different characters narrate sections, reflecting both varying perspectives and the shifting realities. What could have been a formless mess is – through unwavering auctorial control – a triumph, and one of the more original and overlooked series of the 1990s. The third volume, «Music of Madness», remains unpublished.

Also dealing with identity is *The Wizard's Apprentice* **(1993)**, an excellently rendered but minor juvenile in which a teenager, under a WIZARD's tutelage, reconciles his absentee parents' divorce and discovers a sense of self-worth. SPS has written two further juveniles: *The Fallen Country* **(1986)** and *Forgetting Places* **(1987)**.

SPS has also become a HORROR writer of some note. The first novel of his VAMPIRE trilogy, *Vampire Junction* **(1984)**, is often said to have prefigured the splatterpunk movement, while *Moon Dance* **(1989)**, an epic historical/WEREWOLF novel set in both the 1880s and 1960s, strays refreshingly from the conventional. SPS has become active as a moviemaker in recent years, as screenwriter, director and composer of the score for *The Laughing Dead* **(1989)**, a horror movie, and *Ill Met by Moonlight* (completed 1994; unreleased by late 1996), based on SHAKESPEARE's *A Midsummer Night's Dream*. The latter movie featured some genre writers in roles, including Edward BRYANT and Tim Sullivan.

Of associational interest is *Jasmine Nights* (**1994**), a highly episodic coming-of-age novel set in 1963 Thailand. This first appeared in serial form in a Thai newspaper.   [WKS/BM]
Other works: *Mallworld* (coll of linked stories **1981**); **The Chronicles of the High Inquest**, being *Light on the Sound* (**1982**; rev vt *The Dawning Shadow #1: Light on the Sound* 1986), *The Throne of Madness* (**1983**; rev vt *The Dawning Shadow #2: The Throne of Madness* 1986), *Utopia Hunters* (coll of linked stories **1984**) and *The Darkling Wind* (**1985**); the **Aquiliad** sequence, being *The Aquiliad* (**1983**; vt *The Aquiliad: Aquila in the New World* 1988), *#2: Aquila and the Iron Horse* (**1988**) and *#3: Aquila and the Sphinx* (**1988**); *The Shattered Horse* (**1986**), an ALTERNATE-WORLD Trojan-horse novel; *The Alien Swordmaster* * (**1985**) and *Symphony of Terror* * (**1988**), both *"V"* novelizations; the **Vampire** trilogy, being *Vampire Junction* (see above), *Valentine* (**1992**) and *Vanitas* (**1995**); *I Wake from a Dream of a Drowned Star City* (chap **1992**), sf novella.

SONG Like MUSIC in general, song is often deployed by modern fantasy writers – but often to very little effect: generally BARDS and MINSTRELS are, at least in GENRE FANTASY, merely stock characters. There are many examples, however, of song being a more important aspect of a text: one notable case is in J.R.R. TOLKIEN's *The Lord of the Rings* (**1954-5**), where the various songs act not as mere distractions but as a way of commenting on or carrying forward the main events of the text. Songs are given a similar function, although the technique is vastly different (for a start, the songs are generally existing pop/rock songs quoted in part), in many of the works of Stephen KING – especially in his FAR-FUTURE **Dark Tower** series, where what would be by now archaeologically old numbers by The BEATLES and others not only comment on the text but link it to our own age.
In MYTHOLOGY/LEGEND there are several examples of magical song, including the irresistibly alluring song of the SIRENS. Directly analogous is the Teutonic tale (◊ NORDIC FANTASY) of the Lorelei, a water nymph who inhabited the St Goar Rock in the River Rhine, and whose singing lured boatmen to their deaths; this legend was first set down by the German folklorist Clemens Brentano (1778-1842) (◊ FOLKLORE). Richard WAGNER took this further in the **Ring** cycle when he deployed the Rhinemaidens. William SHAKE-SPEARE mixed fantasy and song in most of his relevant plays.
Song, for obvious reasons, lies at the very core of tales based on ballads, like Ellen KUSHNER's *Thomas the Rhymer* (**1990**); the **Damiano** series by R.A. MACAVOY is likewise much concerned with the magical nature of song and minstrelsy. *The Kill Riff* (**1988**) by David J. Schow (1955-   ) mixes CONTEMPORARY FANTASY and HORROR in describing a man's obsession with a rock band and its songs. The title of Peter S. BEAGLE's *The Innkeeper's Song* (**1993**) speaks for itself; likewise John GRANT's HEROIC FANTASY *Albion* (**1991**) is framed as a song, and the latter author's "The Glad who Sang a Mermaid in from the Probability Sea" (1995) treats, in FAERIE terms, the Universe as a song-founded structure. But arguably the novel that comes closest to capturing the fantasticating, MAGIC-effecting power of song – although her *Star Dancer* (**1993**) has much to say on the subject as well – is Fay SAMPSON's *Taliesin's Telling* (**1991**): like other texts based on the legend of Taliesin (◊ MABINOGION), this is much concerned with the minstrel's ability to ensorcell crowds through the power of music and song, but Sampson uses the cadences of her prose to convey the trans-natural experience of being able to produce this effect.

Numerous examples could be produced of songs that tell fantasy tales, from traditional ballads (which are really FOLKTALES) onwards. In passing we can note Loreena McKennitt's setting, in *The Visit* (1991), of "The Lady of Shalott" by Alfred Lord Tennyson (1809-1892); also, many fantasy songs were produced by The BEATLES, Earth Opera, Forest and other groups – most especially The Incredible String Band – who flourished in the hippie era. Fantasy songs are reasonably common today in the (largely amateur) blend of folksong and sf/fantasy known as filk, although often these are intendedly humorous.   [JG]
See also: OPERA.

SONGWEAVER, CERIN [s] ◊ Charles DE LINT.

SORCERER'S APPRENTICE While his WIZARD master is absent, a boy tries his own version of a SPELL . . . which continues uncontrollably. This FAIRYTALE originated in a dialogue by LUCIAN; GOETHE's verse version, "Der Zauberlehrling", inspired the famous symphonic poem *L'Apprenti sorcier* (1897) by Paul Dukas (1865-1935). At least three movie adaptations exist, best-known being the Dukas sequence of DISNEY's FANTASIA (*1940*), in which MICKEY MOUSE plays the reckless apprentice who cannot stop his animated broom from fetching endless buckets of water – until rescued by the fatherly wizard Yen Sid ("Disney" backwards).   [DRL]

SORCERY Magic worked with evil intent. The sorcerer or black magician (◊ BLACK MAGIC) is a key figure in HEROIC FANTASY, to the extent that the PULP subgenre delimited by Robert E. HOWARD and his imitators was dubbed SWORD AND SORCERY. Although anthropologists sometimes differentiate between WITCHCRAFT and sorcery in respect of certain African tribes, the two terms are nearly synonymous in the Western tradition; the qualification is necessary because in literature as in legend male sorcerers employing RITUAL magic tend to be much more imposing figures than female witches, who are often imagined as disreputable hagwives. None of the Renaissance scholars who took an interest in ritual magic thought of themselves as sorcerers, but their critics often took a different view; textbooks of black magic were apocryphally attributed to Albertus Magnus (c1200-1280) and Cornelius Agrippa (1486-1535).
Sax ROHMER's study *The Romance of Sorcery* (**1914**) is a testament to the aesthetic fascination of the notion, but he was never able to capture its glamour in his fiction. Elliott O'Donnell's *The Sorcery Club* (**1912**) is further evidence that credulity is no advantage to the literary imagination. The names of the sorcerers of S&S are legion, and they constitute a remarkable compendium of linguistic exotica extending from Howard's Xaltotun and C.L. MOORE's (female) Jarisme to Roger ZELAZNY's Jelerak. Clark Ashton SMITH, whose sorcerers include Malygris, Pharpetron, Namirrha, Maal Dweb and Eibon, was particularly prolific in the field of eccentric nomenclature.   [BS]
See also: NECROMANCY; SORCERER'S APPRENTICE; WIZARDS.

SØRENSEN, VILLY (1929-   ) Danish writer and critic, much of whose short fiction – assembled as *Saere Historier* (coll **1953**; trans Maureen Neiiendam as *Strange Stories* **1956** UK; vt *Tiger in the Kitchen, and Other Strange Stories* 1957 US), *Ufarlige Historier* (coll **1955**; trans Paula Hostrup-Jessen as *Harmless Tales* **1991** US) and *Formynderfortaellinger* (coll **1964**; trans Hostrup-Jessen as *Tutelary Tales* **1988** US) – tends to use THRESHOLDS into the UNCANNY to represent the costs and abysses that attend modern humanity's incapacity to marry instinct and reason.

The tales frequently invoke the power of STORY – after the model of writers like E.T.A. HOFFMANN and Hans Christian ANDERSEN – to knit the world back together, problematically. VS's style has a childlike clarity, but much of his work has a disturbing affect. "A Tale of Glass", for instance – from the third collection – depicts with seeming simplicity the consequences of an optician's invention of a glass which transforms the PERCEPTION of those who peer through it so that the world looks good; the consequences are savage. *Ragnarok: En gudefortaelling* (**1982**; trans Hostrup-Jessen as *The Downfall of the Gods* **1989** US) unpacks the MATTER of RAGNAROK in a manner which removes from the tragedy all hints of ultimate redemption, irritating some scholars.

Of VS's prolific critical work, *Digtere of Daemoner* ["Writers and Demons"] (coll **1959**) contains an essay on Andersen, and *Kafkas digtning* ["The Works of Kafka"] (**1968**) is an important study. [JC]

SOUL The exact definition of this immaterial essence – or SPIRIT, or ASTRAL BODY – is a matter for RELIGION; fantasy normally assumes Cartesian dualism and the detachability of soul from body. Thus souls may travel the ASTRAL PLANE, or be swapped in IDENTITY EXCHANGE; POSSESSION imposes a stronger soul upon a weaker. In a fantasy world of TRANSFORMATION and METAMORPHOSIS, souls remain the vital key to identity. They may be stolen, as in Robert A. HEINLEIN's "The Unpleasant Profession of Jonathan Hoag" (1942), where the captured soul is bottled like a GENIE, or Fritz LEIBER's *Conjure Wife* (**1943**), where the loss is a dreadful DEBASEMENT. They may be devoured: by DEMONS in C.S. Lewis's "Screwtape Proposes a Toast" (1960), by a SWORD like Stormbringer in Michael MOORCOCK's **Elric** stories, by other forms of psychic VAMPIRE, or even by inhalation as a special "high" in Tim POWERS's *Expiration Date* (**1995**). WIZARDS and GIANTS in FAIRYTALES often detach their souls – usually symbolized by the heart (◊◊ KOSHCHEI) – to make themselves invulnerable, as in George MACDONALD's "The Giant's Heart" (1863), Fritz Leiber's "Adept's Gambit" (1947), Barry HUGHART's *Bridge of Birds* (**1984**) and many other works. Diana Wynne JONES puns effectively on the "heartlessness" of her Lothario wizard in *Howl's Moving Castle* (**1986**); her *The Lives of Christopher Chant* (**1988**) features a race kept subjugated by the ruler's confiscation of their souls. Creatures without souls may or may not envy this token of both mortality and survival after DEATH: J.R.R. TOLKIEN's ELVES call it the "Gift of Men" and also the "Doom of Men". Hans Christian ANDERSEN's Little MERMAID painfully acquires a soul, which is also the objective of the GOLEM in Piers ANTHONY's *The Source of Magic* (**1979**); soullessness is the heart of the horror in FRANKENSTEIN and ZOMBIE scenarios. In POSTHUMOUS FANTASY the soul generally does not, at least initially, realize the death of the body; realization leads to questions of an AFTERLIFE, with the soul conventionally despatched to LIMBO, HEAVEN, HELL or PURGATORY – perhaps still contactable via SPIRITUALISM, perhaps to be re-embodied at the LAST JUDGEMENT – or condemned to earthly BONDAGE as a GHOST. Outside CHRISTIAN FANTASY the destination may be HADES, the Happy Hunting Ground, VALHALLA, the KALEVALA's Tuonela, etc.; an alternative is the CYCLE of REINCARNATION until, at least in Buddhism, final escape from the wheel of karma into Nirvana. [DRL]

See also: TRUE NAME.

SOUSA, JOHN PHILIP (1854-1932) US bandmaster and composer of many famous marches. In his one novel, *The Fifth String* (**1902**), the mystical "fifth string" on a romantic violinist's instrument proves to be a literal manifestation of DEATH, who speaks compellingly through the MUSIC it makes. [JC]

SOUTHESK, THE EARL OF Title used by Scottish writer James Carnegie (1827-1905) for some purposes, though his first LOST-RACE novel, *Herminius: A Romance* (**1862**), was issued anonymously. The protagonist finds in North Wales a race of GIANTS who come from a land soon to be covered by the sea. In *Suomiria: A Fantasy* (**1899**) a lost race of miscegenate creatures, half-ape and half-goat, is discovered. [JC]

SOUTHEY, ROBERT (1774-1843) UK poet and writer who was thought for time the equal of his friends and colleagues Samuel Taylor COLERIDGE and William Wordsworth (1770-1850). He is of fantasy interest for some long narrative poems, disparagingly based on ARABIAN-FANTASY sources and now little read. They include *Thalaba the Destroyer* (**1800**), whose eponymous protagonist enters the UNDERWORLD and encounters various supernatural creatures and WIZARDS, whom he defeats roundly (he is an ALLEGORY of Christian virtue), and *The Curse of Kehama* (**1810**), in which the WANDERING JEW is discovered in India. Works of interest translated by RS include *Amadis of Gaul* (**1803**), compiled by Rodríguez Montalvo in 1508 from earlier sources, and *The Byrth, Lyf, and Actes of Kyng Arthur* (**1817**), an edition of Sir Thomas MALORY's *Le Morte Darthur*. In Volume 4 (**1837**) of *The Doctor* (coll **1834-47** 7 vols), which is a compendium of linked essays and tales and occasional pieces, RS put into print for the first time the "Story of the Three Bears"; although he did not claim original authorship, nor the other source was known, and he was long credited with it. Eventually an 1831 manuscript by the otherwise unknown Eleanor Mure came to light, giving a somewhat different but earlier version. (Goldilocks herself only slowly came into the world: in an 1849 text, RS's original old woman was a girl, who went through several name changes until 1904, when she was first called Goldilocks.) [JC]

SOUTO, MARCIAL (1947-   ) Spanish writer, editor and translator who lived for years in Montevideo and Buenos Aires. He has published two unique collections of fantastic tales, *Para bajar a un pozo de estrellas* ["Climbing Down Into a Well of Stars"] (**1983**) and *Trampas para pesadillas* ["Traps for Nightmares"] (**1988**). Brevity and a precision in his use of language, a quality uncharacteristic of Spanish, are the hallmarks of MS's highly original pieces. While in form some are classic short stories, the majority are poetic daydreams, almost philosophical perceptions bordering on the ineffable, asking the eternal question, "What if?" MS's work, frequently anthologized in Spanish and recently starting to appear in English ("The Man who Put Out the Sun" is in *Winter's Tales 11* anth **1995** UK), bears a stylistic resemblance to that of Jorge Luis BORGES only in that both writers are heavily influenced by their reading in English. MS has compiled several influential anthologies and edited two magazines, *El péndulo* (1979-90) and *Minotauro* (1983-7), in which much of the new fantasy writing of the River Plate appeared. He has also translated into Spanish a large part of the work of J.G. BALLARD as well as books by Ray BRADBURY, Samuel R. DELANY, Cordwainer Smith and Ambrose BIERCE. [NdG]

SOVEREIGN MAGAZINE, THE ◊ HUTCHINSON'S MAGAZINES.

SOYKA, OTTO (1882-1955) Austrian writer. ◊ AUSTRIA.

SPACEMAN AND KING ARTHUR, THE vt of *Unidentified Flying Oddball* (*1979*). ◊ *A* CONNECTICUT YANKEE.

SPACE TRAVEL (magazine) ◊ FANTASTIC ADVENTURES.

SPAIN Eight years after *Don Quixote* (vol 1 **1605**) had begun to kill off the KNIGHT-errant novel, the staple diet of the medieval imagination, Miguel de CERVANTES published his "Coloquio de los perros" (1613; trans C.A. Jones as "The Dogs' Colloquy" in *Exemplary Stories* coll **1972**), a dialogue between two dogs who, receiving the gift of speech for one night, tell each other their adventures against the dark, cruel background of a disintegrating society.

In 1790 José Cadalso (1741-1782) anticipated late-blossoming Spanish Romanticism with his *Noches lúgubres* ["Mournful Nights"], and almost half a century later Agustín Pérez Zaragoza (? -?), a self-proclaimed heir to Ann RADCLIFFE, published his *Galería fúnebre de espectros y sombras ensangrentadas* ["A Funeral Gallery of Ghosts and Bloodstained Shadows"] (coll **1831**), 21 short stories and three short novels about prodigies, marvellous events, itinerant corpses, bloodstained heads, and worse. Pedro Antonio de Alarcón (1833-1891), included in his *Narraciones inverosímiles* ["Implausible Stories"] (coll **1882**) one of the most famous fantasy short stories in the Spanish language – "La mujer alta" ["The Tall Woman"]. But the best Spanish fantasy writer of the 19th century was Gustavo Adolfo Bécquer (1836-1870), whose *Leyendas* ["Legends"] (coll **1871**) – 18 pieces about the faraway, exotic world of India, the atmosphere of the Middle Ages, and the relations between the living and the dead – combine humour and poetic fantasy.

In the 20th century Ramón del Valle Inclán (1866-1936), starting with his play *Luces de bohemia* ["Bohemian Lights"] (**1920**), put forward his special theory of the *esperpento* ["absurd"], which holds that Spain is a grotesque deformation of European civilization where the tragic meaning of life can be depicted only by means of a systematically deformed aesthetics. In his novels and plays he achieves this distortion by depriving his characters of the dimension of depth, comparing them with animals, PUPPETS or DOLLS.

Wenceslao Fernández Flórez (1884-1964) describes in *El secreto de Barba Azul* ["Bluebeard's Secret"] (**1923**) a far-fetched war between two imaginary countries, Surlandia and Westlavia, in which a notable strategist, General Mikrí, organizes a masterly retreat. His troops walk backwards around the world on the double, and so reconquer the starting-point. In *Las siete columnas* ["The Seven Pillars"] (**1926**), Flórez's most famous novel, the DEVIL withdraws the Seven Deadly Sins from circulation, thereby eliminating everything of interest and destroying civilization.

La princesa durmiente va a la escuela ["The Sleeping Princess Goes to School"] (written 1951; **1983**) by Gonzalo Torrente Ballester (1910-) freely mixes historical, fictional and mythological characters. In another of his novels, *La saga/fuga de J.B.* ["J.B.'s Saga/Escape"] (**1972**), past actions can be changed from the future, and history and legend can be altered at will.

One of Spain's most remarkable fantasy novels is *Industrias y andanzas de Alfanhuí* ["Industries and Deeds of Alfanhuí"] (**1951**) by Rafael Sánchez Ferlosio (1927-). Compared by critics with Lewis CARROLL's **Alice** books, PETER PAN and Lord DUNSANY's stories, it relates the magic adventures of a boy who, among other things, befriends a weathercock that comes down from the rooftop by night to catch lizards, learns how to extract red hues from the western sky, and at school writes in a strange alphabet of his own. A modern author of fantasy interest is Marcial SOUTO, whose work is better known in LATIN AMERICA.

Spain being a multilingual country, a good deal of its fantasy has been written in Galician and Catalan. Galician writer Ánxel Fole (1903-1986) has several fantasy pieces in *Álus do candil* ["By the the Light of the Oil Lamp"] (**1953**), *Contos da néboa* ["Tales of the Mist"] (**1973**) and *Historias que ninguén cré* ["Tales Nobody Believes In"] (**1987**). The novel *Cara a Times Square* ["Facing Times Square"] (**1980**) by Camilo Gonsar (1931-) describes an oneiric, Kafkaesque NEW YORK. Xosé Luís Méndez Ferrín (1938-) has gathered some of his best fantasies in *Percival e outras historias* ["Percival and Other Stories"] (**1958**), *O crepúsculo e as formigas* ["The Twilight and the Ants"] (**1961**) and *Eclipse e outras sombras* ["Eclipse and Other Shadows"] (**1971**). But Galicia's most important author of fantasy is Álvaro Cunqueiro (1911-1981), who makes a highly personal use of the Arthurian cycle (◊ ARTHUR), the magic of Brittany, and the Arabian Nights (◊ ARABIAN FANTASY), respectively, in *Merlín e familia e outras historias* ["Merlin and Family and Other Stories"] (**1955**), *As crónicas do sochantre* ["Chronicles of a Choirmaster"] (**1959**) and *Si o vello Sinbad volvese ás illas* ["If Old Sinbad Returned to the Islands"] (**1961**).

In Catalan, the tradition of fantasy goes all the way back to the 13th century, when Ramon Llull (1235-1315), a Majorcan poet, philosopher and mystic who wrote 243 books in Catalan, Arabic, Latin and Provençal, published his *Libre des maravelles* ["Book of Wonders"] (**1289**), about the spiritual and scientific pilgrimage of young Fèlix. The current master of fantasy in Catalan is Joan Perucho (1920-), whose books include *Amb la tècnica de Lovecraft* ["With Lovecraft's Technique"] (**1953**), *Llibre de cavalleries* ["Book of Knighthood"] (**1957**), the TIMESLIP story of a 20th-century young man mysteriously transported back to the Middle Ages, and *Les històries naturals* ["Natural Histories"] (**1960**), about a modern VAMPIRE in the Mediterranean landscape of Barcelona. Some of his shorter pieces are in *Roses, diables i somriures* ["Roses, Devils, and Smiles"] (coll **1965**), *Aparicions i fantasmes* ["Apparitions and Ghosts"] (coll **1968**) and *Històries apòcrifes* ["Apocryphal Tales"] (coll **1974**).

Two sf writers, Luis Vigil (1940-) and Domingo Santos (real name Pedro Domingo Mutiñó; 1941-), have published in Spanish the first three novels of a CONAN-like saga about a barbarian called Nomanor: *El mito de los harr* ["The Myth of the Harr"] (**1971**), *El bárbaro* ["The Barbarian"] (**1971**), and *La niebla dorada* ["The Golden Fog"] (**1974**). [MH/MS]

Further reading: *Historia natural de los cuentos de miedo* (**1974**) by Rafael Llopis; *Literatura fantástica de lengua española* (**1987**) by Antonio Risco; *El relato fantástico en España e Hispanoamérica* (anth **1991**) ed E. Morillas Ventura; *Anthropos #154-5* (March-April 1994), special issue about fantasy in Spanish.

SPARK, MURIEL (SARAH) (1918-) Scottish novelist, long resident in London, much of whose fiction – whether or not a particular title happens to be a SUPERNATURAL FICTION – engages in a variety of GODGAME manoeuvres. GOD Himself, or various TRICKSTER figures within the texts, or the implied author of those texts, constantly manipulate(s) the texture of REALITY, playing cardsharp games with TIME and narrative. It is, therefore, sometimes difficult to know if a supernatural (as opposed to a gamelike or allegorical) reading is intended.

MS's first novel, *The Comforters* (**1957**), however, is relatively easy to understand as supernatural. Its writer protagonist comes to believe – with reason – that someone or something, God or simply an Author, is "writing" her; in the end, the BOOK being written turns out to be *The Comforters*. Other characters appear and disappear according to whether or not they are "needed" by the ultimate Author of the text within which they are trapped. *Memento Mori* (**1959**) is also a supernatural fiction; its depiction of elderly people combines gerontological coldness and religious intensity, as a telephone voice – perhaps God's – foretells the deaths of various cast members. The protagonist of *The Bachelors* (**1960**) is a medium (◊ SPIRITUALISM) whose powers seem genuine. *The Ballad of Peckham Rye* (**1960**) is dominated by a figure who may well be the DEVIL. *The Hothouse by the East River* (**1973**) is a complex and rather cruel POSTHUMOUS FANTASY.

More routinely, several of MS's short stories – notably those in *The Go-Away Bird and Other Stories* (coll **1958**) – deal with the supernatural; *The Stories of Muriel Spark* (coll **1987**) contains these and some additional examples. *Child of Light: A Reassessment of Mary Wollstonecraft Shelley* (**1951**; vt *Mary Shelley: A Biography* 1987 US) is a competent study of Mary SHELLEY; it won a 1988 BRAM STOKER AWARD. [JC]

Other works: *Robinson* (**1958**), which makes nonfantastic use of Robinsonade motifs.

SPECTRE, THE Ultra-powerful SUPERHERO of US COMIC books. The godlike alter ego of murdered policeman Jim Corrigan, he has skull-pupilled eyes and wears a hooded green cloak and shorts over a grey skintight vest and leggings. At first his superpowers were almost infinite, including the ability to converse with GOD. Created by writer Jerry Siegel (1914-1996) and artist Bernard Bailey, he made his first appearance in *More Fun #52* (1940) in which his origin story told of Corrigan being encased in concrete and drowned, but denied eternal rest until he wiped out all crime on the planet. The evident difficulty of maintaining reader interest in so transcendentally powerful and invulnerable (◊ INVULNERABILITY) a character led to modifications in his abilities and in his relationship to Corrigan.

TS appeared in *More Fun* until #101 (1945), in *All Star #1-#23* (1940-45) and in *Showcase #60, #61* and *#64* (1960). A comic book named after him was *The Spectre* (#1-#10 1967-9). [RT]

SPEDDING (ALISON LOUISE) (1962-   ) UK writer, resident in Bolivia since 1989, who prefers to be known by her surname only (although she has published academic papers as Alison Spedding). Her fiction consists of one long work, **A Walk in the Dark**: *The Road and the Hills* (**1986**), *A Cloud Over Water* (**1988**) and *The Streets of the City* (**1988**). Set in an OTHERWORLD, it is a fantastication of the careers of Alexander the Great and his immediate successors. [DP]

SPELLS A spell is a consciously directed act of MAGIC which may take almost any form, depending on the laws of magic in operation. Commonly there will be a spoken element, ranging from a simple phrase or NAME to elaborately complex RITUAL incantations. Ingredients may need to be compounded into POTIONS or set on fire: in ALCHEMY and

BLACK MAGIC especially, the materials in the written recipe are likely to be coded symbols of something altogether different. Hand-gestures or passes may be required – called the "somatic element" by the scientific magic-investigators in L. Sprague DE CAMP's and Fletcher PRATT's **Incomplete Enchanter** series. With ingenuity, all these obvious components may be omitted: a subterfuge in Susan COOPER's *Greenwitch* (**1974**) is to paint the spell as abstract art, and in Diana Wynne JONES's *The Spellcoats* (**1979**) the binding STORY is woven into garments.

Much depends on the source of magic. GODS, SPIRITS or DEMONS will often be invoked by name – a protracted and impressive example being the demonic conjuration in James BLISH's *Black Easter* (**1968**). Michael MOORCOCK's Elric repeatedly calls on his demon-lord Arioch, and the cry "*Iä! Iä! Shub-Niggurath!*" is all too familiar in the CTHULHU MYTHOS. Magical fuel-supplies like the mana of Larry NIVEN's **Magic Goes Away** stories and David GEMMEL's eponymous **Sipstrassi** stones (which may be, in terms of RATIONALIZED FANTASY, an artificial intelligence with electromagnetic effectors) must be suitably directed; or the spellcaster's inner TALENT must be focused; or a victim of MESMERISM must be persuaded of the intended ILLUSION or GLAMOUR. In each case, special language may seem appropriate – e.g., the Latin used for incantations in Alan GARNER's *The Weirdstone of Brisingamen* (**1960**), the invented Black Speech of Sauron's binding-spell in J.R.R. TOLKIEN's *The Lord of the Rings* (**1954-5**), the Old Speech of true names in Ursula K. LE GUIN's *Earthsea*, and even pig-Latin in Poul ANDERSON's "Operation Afreet" (**1956**). The verse spell has also been traditional since, at least, the Cauldron Scene in SHAKESPEARE's *Macbeth*, and FANTASY-LAND magic is accompanied by much dire doggerel – most tirelessly in Piers ANTHONY's **Apprentice Adept** trilogy, whose hero's couplets worsen thanks to the CONDITION that each rhyme must be unique. Shea, in de Camp's and Pratt's *The Castle of Iron* (**1941**), finds pedestrian verse safest, since his spell adaptations of real poetry (Shakespeare, Percy Bysshe SHELLEY, Algernon Swinburne) give excessive results; the obscure CURSE in *Howl's Moving Castle* (**1986**) by Diana Wynne Jones is likewise borrowed, from John Donne's "Goe, and catch a falling star". Plausible verse cantrips tend towards extreme simplicity, like the Sending-Boat spell in William MORRIS's *The Water of the Wondrous Isles* (**1897**), Severian's childish guarding-spell in Gene WOLFE's *The Shadow of the Torturer* (**1980**), and the sung (◊ SONG) spell of INVISIBILITY in R.A. MACAVOY's *Damiano's Lute* (**1985**).

Many spells exhaust the caster (◊ BALANCE), like Gorice's arduous conjuration in *The Worm Ouroboros* (**1922**) by E.R. EDDISON. Jack VANCE's *The Dying Earth* (**1950**) has spells which must be painstakingly impressed on the mind (whose capacity is finite), and when cast are gone until re-learned; Vance also popularized the naming of a spell after its creator, as in "Phandaal's Gyrator", and both notions are influential in GAMES like *Dungeons & Dragons*. Terry PRATCHETT's **Discworld** books nod to this terminology – "Stacklady's Morphic Resonator", etc. – and adapt the idea of occupying mental space by proposing a Great Spell so terrifying that other spells are too frightened to share the WIZARD's mind; another menacingly sentient spell appears in Collin Webber's *Merlin and the Last Trump* (**1993**).

MAGIC WORDS are often brief spells: the ingenious Warlock in Niven's "Not Long Before the End" (**1969**)

prepares his master-spell in advance "like a telephone number already dialled but for one digit", to be activated by the single syllable "Four". [DRL]

See also: BOOKS.

SPENCE, (JAMES) LEWIS (THOMAS CHALMERS) (1874-1955) Scottish poet and author, a leading specialist on MYTHOLOGY, LEGENDS, FOLKLORE, MAGIC, OCCULTISM, ATLANTIS and ancient civilizations. He was well known as a fine reteller of great traditional legends and romances from all over the world – including those of Ancient EGYPT, Spain, Brittany, Mexico and Peru, Babylonia and Assyria (◊ MESOPOTAMIAN EPIC) and the North American Indians.

Only a small percentage of his writings were straight fiction. His best HORROR, SUPERNATURAL FICTION and fantasy (including several tales in Scottish dialect) were collected as *The Archer in the Arras* (coll **1932**). Among his uncollected fiction is a novella, "The Fellowship of the White Crane" (*Chambers Journal* 1926), based on Mexican arcane lore and describing a trek to the Caverna del Demonios. His monumental *Encyclopedia of Occultism* (**1920**) formed the basis of the heavily augmented and updated *Encyclopedia of Occultism and Parapsychology* (**1984-5** 3 vols) ed Leslie Shepard.

All LS's later books were devoted to the mythologies and arcane lore of the British Isles. In *The Magic Arts of Celtic Britain* (**1945**) the entire range of British-Celtic material connected with the occult was covered in depth. Other studies of importance include *British Fairy Origins* (**1946**), *The Fairy Tradition in Britain* (**1948**), *The Minor Traditions of British Mythology* (**1948**), *The History and Origins of Druidism* (**1949**) and *Second Sight* (**1951**). [RD]

Other works (poetry): *Le Roi d'Ys* (coll **1910**); *Songs Satanic and Celestial* (coll **1913**); *Plumes of Time* (coll **1926**); *Weirds and Vanities* (coll **1927**); *Collected Poems of Lewis Spence* (coll **1953**).

Other works (nonfiction): *A Dictionary of Mythology* (**1910**); *Mexico of the Mexicans* (**1917**); *The Gods of Mexico* (**1923**); *The Problem of Atlantis* (**1924**); *Atlantis in America* (**1925**); *The History of Atlantis* (**1926**); *The Magic and Mysteries of Mexico, or The Arcane Secrets and Occult Lore of the Ancient Mexicans and Maya* (**1930**); *The Problem of Lemuria* (**1932**); *Legendary London: Early London in Tradition and History* (**1937**); *The Mysteries of Britain* (**1928**); *The Mysteries of Egypt* (**1929**); *Boadicea* (**1937**); *The Occult Causes of the Present War* (**1940**); *Will Europe Follow Atlantis?* (**1942**); *The Occult Sciences in Atlantis* (**1943**); *The Religion of Ancient Mexico* (**1945**); *Myth and Ritual in Dance, Game and Rhyme* (**1947**).

SPENCER, WILLIAM BROWNING (1946-   ) US writer who began publishing fantasy with his first novel, *Maybe I'll Call Anna* (**1990**). *Résumé With Monsters* (**1995**) and *Zod Wallop* (**1995**) are both built around fictional BOOKS. The first is a SUPERNATURAL FICTION of sorts, though the representatives of the CTHULHU MYTHOS recreated by its protagonist in his imaginary book, «The Despicable Quest», flicker into "life" only near tale's end, when he masters his internal demons and finds love. The second, though couched as sf about reality-shifting drugs, works throughout as fantasy. The eponymous imaginary CHILDREN'S FANTASY exists in two versions, one astonishingly grim, the other rewritten so that there remains some hope for its cast. This is fortunate, as the imaginary «Zod Wallop» texts dictate their conflicting STORIES into the real world, where a number of protagonists and COMPANIONS jostle for *lebensraum* and search for the proper ending. This

ending takes place in a hotel transfigured into a fantasy EDIFICE, through the orifices of which an unravelling knot of TRANSFORMATIONS brings about a happy outcome. The tales assembled in *The Return of Count Electric and Other Stories* (coll **1993**) are surreal and intensely amusing, but almost never stray beyond the real world. [JC]

SPENSER, EDMUND (1552-1599) English Renaissance poet. ES's noteworthy shorter works include *The Shepheardes's Calender* (coll **1579**), a cycle of 12 PASTORAL poems about the lives of shepherds, and "Epithalamion" (1595), his delightful wedding poem; both include characters and imagery from MYTHOLOGY and FOLKLORE. But ES's greatest contribution to literature – and as model EPIC FANTASY and TAPROOT TEXT to fantasy – is his massive epic poem *The Faerie Queene* (Books I-III **1590**; Books IV-VI **1596**; with "Mutabilitie Cantos" added **1609**).

As originally planned, *The Faerie Queene* was to contain 12 books of 12 cantos, each book devoted to a KNIGHT representing a specific virtue adventuring through FAERIE. Uniting the books was to be the character of ARTHUR, who would appear at some crucial point to assist each knight and thus to display or acquire that knight's virtue. The poem would conclude with the marriage of Arthur, representing England, and the FAIRY QUEEN, representing Queen Elizabeth I, thus solving the problem of the Virgin Queen by symbolically marrying her to her country. However, ES completed only six books plus the two "Mutabilitie Cantoes", evidently written as an episode for a seventh book. The existing books visibly depart from the plan: Book IV, nominally about Cambel and Telamond, the Knights of Friendship, actually continues the adventures of Britomart, the female Knight of Chastity from Book III; and Book VI ends on a most unheroic note, as the Blatant Beast subdued by Calidore, the Knight of Courtesy, escapes to ravage the countryside once again. (Some speculate that ES abandoned the project because he became disillusioned, either by the world in general or by the physical and mental deterioration of Elizabeth.)

The Faerie Queene is clearly an ALLEGORY about the proper moral education of the ideal knight, but its meanings are by no means simple or straightforward, as ES immediately establishes: Book I begins with the Redcross Knight of Holiness rather handily subduing a monster named Error, only to fall victim to other, less obvious menaces. The lesson, for both knight and readers, is that events in Faerie will not always be easy to interpret.

Because of its extreme length (over 30,000 lines), most students read only Book I. This is unfortunate, since it is in some ways the most allegorical and least lively part of the whole. Book II, featuring Guyon, the Knight of Temperance, is much more memorable, with a remarkable final canto describing the Bower of Bliss. Britomart, heroine of Books III and IV, is a complex and well developed character, and many regret that she ends up marrying the rather stern and one-dimensional Artegall, Knight of Justice, in Book V. Book VI, the adventures of Calidore (modelled on Sir Philip Sidney [1554-1586]), is the most PASTORAL and arguably the most charming of all.

ES's poetry has two particular features of interest to fantasy readers. First, while ES wished to emulate the language of Geoffrey CHAUCER, he lacked the linguistic knowledge to reconstruct Middle English; instead, he devised his own mock-archaic DICTION, with many wilful misspellings and neologisms. Thus he was the first writer to create his own language to convey the distinct atmosphere of a fantasy world. Second, because ES worried that readers might have trouble understanding this language, he published *The Shephearde's Calendar* with extensive explanatory notes by "E.K." (presumably his friend Edmund Kirke, though some believe ES wrote them himself), and each canto of *The Faerie Queene* opens with a couplet summarizing its plot – making ES the first writer to publish annotated editions of his works.

Only a few modern works explicitly refer to ES. L. Sprague DE CAMP and Fletcher PRATT's **Incomplete Enchanter** series includes an interlude in ES's Faerie. Calidore and the Blatant Beast appear in John Myers MYERS's *Silverlock* (**1949**), and ES himself shows up in Myers's *The Moon's Fire-Eating Daughter* (**1981**). However, the indirect influence of *The Faerie Queene* can be felt in countless fantasy epics about noble heroes and heroines battling against evil menaces to achieve worthy goals. ES was the first to produce an epic that both incorporated countless mythological and folkloric traditions and exemplified the careful design and poetic quality of written literature. [GW]

SPIDER, THE ◊ MAGAZINES.

SPIDER-MAN US movie (*1977* tvm). CBS. **Pr** Edward J. Montagne. **Dir** E.W. Swackhamer. **Writer** Alvin Boretz. **Based on** the COMIC-book characters created by Stan LEE. **Starring** Ted Danson (Major Collins), Lisa Eilbacher (Judy Tyler), Nicholas Hammond (Peter Parker/Spider-Man), Hilly Hicks (Robbie Robertson), Michael Pataki (Captain Barbera), David White (J. Jonah Jameson). 120 mins. Colour.

This pilot to the series *The* AMAZING SPIDER-MAN (1978-9) finds Spider-Man battling an extortionist who is threatening to kill people using a mind-control device. [BC]

SPIDERS In Greek MYTHOLOGY, Arachne presumptuously challenges the GODDESS Athene/Minerva to a weaving competition and (having performed too well) suffers METAMORPHOSIS into a spider, giving the arachnids their name. Emphasis has since shifted from the beauty of the web to its debatably unprepossessing creator. Arachnophobia, the common fear of spiders, is routinely exploited in HORROR, an effective example being M.R. JAMES's "The Ash Tree" (1904) – whose unpleasantly oversized spiders were outdone by Lord DUNSANY's specimen "larger than a ram" in "The Fortress Unvanquishable, Save for Sacnoth" (1908). J.R.R. TOLKIEN established the definitive GENRE-FANTASY stereotype of giant spiders and webs in *The Hobbit* (**1937**) and *The Lord of the Rings* (**1954-5**); the latter's Shelob is memorably nasty. Similar creatures appear in Fritz LEIBER's "Bazaar of the Bizarre" (1963), Barry HUGHART's *Bridge of Birds* (**1984**), Mary BROWN's *The Unlikely Ones* (**1986**) and many other works. Such MONSTERS are tempting cases for rehabilitation: once communication is established, the huge spider of Piers ANTHONY's *Castle Roogna* (**1979**) becomes a valued COMPANION, and those in Colin WILSON's **Spider World** sequence prove at least worthy of respect.

Arachne-like woman/spider figures appear in James Branch CABELL's *Smirt* (**1934**) and the movie KISS OF THE SPIDER WOMAN (*1985*), the former emphasizing that some female spiders eat their mates; but the Spider Woman of Navajo myth is a benign WITCH living UNDERGROUND. Also benevolent is the eponymous spider of E.B. WHITE's *Charlotte's Web* (**1952**), spelling out "miraculous" messages with her web. ANANSI the Spider GOD is a TRICKSTER. In

COMICS, a normal-sized but radioactive spider famously bites teenage wimp Peter Parker, who gains spider-like TALENTS and becomes Spider-Man (◊ *The* AMAZING SPIDER-MAN; SPIDER-MAN (*1977* tvm); SUPERHEROES). [DRL]

SPIDER WOMAN ◊ GODDESS; KISS OF THE SPIDER WOMAN (*1985*); SPIDERS.

SPIELBERG, STEVEN (ALLAN) (1946-   ) US moviemaker who has made substantial contributions to the fantastic CINEMA. His first movie to receive attention was the 22min short *Amblin'* (*1969*), financed by a young entrepreneur called Dennis Hoffman on the basis that SS would direct for Hoffman, at some time during the subsequent 10 years, a feature movie for $25,000 plus a share of profits; this contract is currently (1996) the subject of a lawsuit. *Amblin'* – after which SS would in 1984 name his production company Amblin Entertainment – won the Atlanta Film Festival Award, gained widespread critical admiration, and was commercially released as a programme-filler with *Love Story* (*1970*); even before this, however, it earned him a seven-year contract with Universal/NBC. His first tv direction was an episode of ROD SERLING'S NIGHT GALLERY called *Eyes* (1969). Other shows for which he directed episodes included *Marcus Welby*, *The Name of the Game*, *The Psychiatrists* and *Columbo*. Next came three tv feature movies: *Duel* (*1971* tvm), scripted by Richard MATHESON, in which a young car-driver is threatened motivelessly by an unidentified truck-driver; *Something Evil* (*1972* tvm), which concerns POSSESSION and CURSES in a rural community; and *Savage* (*1972*; vt *The Savage Report*; vt *Watch Dog*), a detection. SS returned to tv over a decade later with AMAZING STORIES (1985-7).

His directorial debut on the big screen was *The Sugarland Express* (*1973*), a quirky road thriller which he co-wrote. It was followed by the hugely successful *Jaws* (*1975*), about a killer shark: this was the movie that put him on the map. *Close Encounters of the Third Kind* (*1977*; special edition *1980*), which he also scripted, is sf/TECHNOFANTASY about UFOs; overlong, it nevertheless has moments of great beauty, and was another box-office blockbuster. By contrast, *1941* (*1979*) – a screwball comedy – was a disaster, and people were ready to write SS off as a prodigy who had burnt himself out. Not so: two years later came *Raiders of the Lost Ark* (*1981*), the first of the INDIANA JONES movies, and he was back to his blockbusting status – even more so with the following year's offering, *E.T. – The Extra Terrestrial* (*1982*), an sf/technofantasy movie which broke all records. Though the tale is superficially only of a little alien stranded on Earth and befriended by children who must keep his existence a secret from the authorities, E.T. could as well be a stray from FAERIE, and certainly he performs MAGIC, as in the famous scene where the children's bicycles suddenly begin to fly; also, late in the movie, E.T. goes through a CHRIST-like cycle of death and RESURRECTION. *Indiana Jones and the Temple of Doom* (*1984*) and *Indiana Jones and the Last Crusade* (*1989*) completed the **Indiana Jones** sequence, although a fourth is reportedly on the stocks. *The Color Purple* (*1985*), based on the Alice Walker novel, is nonfantasy, as is *Empire of the Sun* (*1987*), based on the J.G. BALLARD novel. ALWAYS (*1989*) is a rather lacklustre POSTHUMOUS FANTASY, based on *A Guy Named Joe* (*1944*). HOOK (*1991*), a REVISIONIST FANTASY rooted in PETER PAN, was much disliked on release but has had more favourable reappraisals. *Schindler's List* (*1993*), based on *Schindler's Ark* (**1983**; vt *Schindler's List*) by Thomas Keneally (1935-   ) – which won the Booker

Prize – although dauntingly long and dealing with the grim subject of the Holocaust, was hugely successful. *Jurassic Park* (*1993*), based on *Jurassic Park* (*1990*) by Michael Crichton (1942-   ), is direly plotted and scripted, and not much better acted, but again broke box-office records because of its state-of-the-art spfx showing DINOSAURS which, according to the story, have been resurrected from fossil DNA.

SS has also been through the years a producer and executive producer. The movies concerned have been: *I Wanna Hold Your Hand* (*1978*), *Used Cars* (*1980*), *Continental Divide* (*1981*), POLTERGEIST (*1982*), *Twilight Zone – The Movie* (*1983*; ◊ *The* TWILIGHT ZONE [1959-64]), GREMLINS (*1984*) and its 1990 sequel, *The* GOONIES (*1985*), *Back to the Future* (*1985*) – about TIME TRAVEL, with sequels in 1989 and 1990 – *Young Sherlock Holmes* (*1985*; ◊ SHERLOCK HOLMES), *The Money Pit* (*1986*), *An* AMERICAN TAIL (*1986*) and its sequel in 1991, *InnerSpace* (*1987*) – a miniaturization sf story derivative of *Fantastic Voyage* (*1966*) – **batteries not included* (*1987*) – a dire movie, originally intended for tv, about little UFOS – WHO FRAMED ROGER RABBIT (*1988*), *Dad* (*1989*), *Joe Versus the Volcano* (*1990*), and *Arachnophobia* (*1990*), about big SPIDERS.

Unlike many other Hollywood creators, SS has never been afraid to surround himself with production/directorial talents who might have been seen as rivals; also, he has talent-spotted and groomed potential successors. Examples are Don BLUTH, Joe Dante, Richard Donner, Tobe Hooper, Kathleen Kennedy, David Kirschner, Barry Levinson, George Lucas, Frank Marshall and Robert ZEMECKIS – Kennedy and Marshall being particularly frequent collaborators.

SS has often been described as the Walt DISNEY of the 20th century's latter part, and the comparison is apt. While many of his "family" movies would – with their swearing and sexuality – have horrified Disney (but then so would have SPLASH! [*1984*]), they appeal to a market that is very similar, though displaced by a few decades. Also like Disney, he has been responsible more than anyone else of his generation for keeping fantasy thriving in the cinema: had it not been for SS's example, dozens of fantasy movies by *other* directors would never have been made. [JG]

SPIRIT OF THE BEEHIVE, THE (ot *El Espíritu de la Colmena*) Spanish movie (*1973*). Elías Querejeta. **Pr** Elías Querejeta. **Dir** Victor Erice. **Screenplay** Francisco J. Querejeta. **Starring** Teresa Gimpera (Teresa), Fernando Fernán Gómez (Fernando), Juan Margallo (Fugitive), Isabel Tellería (Isabel), Ana Torrent (Ana), José Villasante (Monster). 98 mins. Colour.

Hoyuelos (Segovia, Castile), 1940, with the Franco repression in full force. Young Ana's family is swathed in escapist fantasy (◊ ESCAPISM) of one form or another: her father Fernando is obsessed to exclusion with his beehives and the writing of a never-to-be-published *magnum opus* on the philosophy of apiary; her mother Teresa, seemingly much younger than Fernando, writes wistful letters to a probably nonexistent lover; elder sister Isabel has at least learnt to *control* her fantasies. The two children are much impressed by a village-hall screening of *Frankenstein* (*1931*; ◊ FRANKENSTEIN MOVIES). Afterwards Ana, who has identified strongly with the movie's Maria, asks why the MONSTER killed the little girl, and why the populace then killed the Monster. Isabel, to shut her up, replies that neither Maria nor the Monster are dead: she, Isabel, has met and spoken

with the Monster, which is a SPIRIT that rendered itself incarnate for the purposes of the movie. Isabel guides Ana to the derelict barn where, she claims, the Monster dwells – though he comes out only at night. Ana haunts the place by day, and when a criminal shelters there she believes him to be a different incarnation of the Spirit, bringing him food and articles of her father's clothing. The police kill the fugitive and Ana's activities are uncovered. She runs away from home. That night, while the hunt for her continues, she sees, in the water of the stream by which she kneels, her own reflected face becoming the Monster's – and then the Monster himself comes to comfort her. After her safe restoration to home, her family shed their fantasies; but Ana knows that the Monster is her friend and will come any time she calls him.

This is one of a group of movies that seek to explore the origins of FANTASY through the evocation of childhood, and is perhaps the most lovingly crafted of all; Erice's "eye" conjures the fantastical out of the mundane, in particular couching the somewhat bleak Spanish terrain such that it becomes filled with the resonances of a post-HOLOCAUST landscape. The two children, giving exceptional performances, enhance this effect of dual PERCEPTION – of seeing a REALITY that is not unique and absolute but instead layered, the relative importance of its individual layers being in a state of constant flux. [JG]

See also: CELIA (*1988*); *The* LORD OF THE FLIES; WHISTLE DOWN THE WIND (*1961*).

SPIRITS In the Western World, early philosophical theories of the nature of spirit conceived it as the vital principle – or breath of life – which both animated the body and mediated between the animal functions of the body and the SOUL. By the Middle Ages this dual function had been much complexified, but from about the turn of the 17th century spirit, conceived as literal vital principle, had become a more general metaphor, sometimes interchangeable with the soul, sometimes – as in the Holy Spirit – taken to designate a "person" of the Trinity. By the 19th century it had become common to use the term "spirits" to describe disembodied souls (◊ ASTRAL BODY), quite possibly occupying a "sphere" of their own.

The term is often encountered in HORROR and SUPERNATURAL FICTION, though less frequently in fantasy. It generally takes one of two broad senses: it describes disembodied vital principles which or who occupy regions unavailable to the senses, and which or who may or may not manifest the animate nature of the Universe; or it describes unhoused souls, who are also known as GHOSTS. These two broad senses may be linked: humans, for instance, may discover upon dying (◊ AFTERLIFE; POSTHUMOUS FANTASY) that they have entered (or returned to) a spirit world, which circumambiates and transcends our physical realm (◊ SPIRITUALISM). [JC]

SPIRITUALISM In philosophy the term refers to any system of thought that affirms the existence of an immaterial REALITY, but in modern parlance Spiritualism refers to a religious faith asserting that communication is possible between our world and the SOULS which have passed into an immaterial AFTERLIFE. Such communication is achieved via a medium, associated with a "control" in the spirit world, who may orchestrate various kinds of signals and ectoplasmic manifestations during a SÉANCE, which continues with the control summoning other spirits to answer questions posed by participants and deliver messages from the ASTRAL

PLANE. Spiritualism received its initial boost from the (fraudulent) "rappings" of the Fox sisters, which became a sensation in New York State in 1848, and the writings of their near neighbour Andrew Jackson Davis (1826-1910).

An entire subgenre of fiction deals with Spiritualism, much of it wholeheartedly credulous and some of it scathingly sceptical (and thus RATIONALIZED FANTASY). In most Spiritualist fantasies credulity and propagandistic intent succeed in effacing virtually all literary and imaginative interest; the few which contrive to retain a reasonable measure include *Urania* (**1889**) by Camille Flammarion, *The Land of Mist* (**1926**) by Arthur Conan DOYLE and *Time Must Have a Stop* (**1944**) by Aldous Huxley. Sceptical cautionary tales include *The Vasty Deep* (**1890**) by Stuart Cumberland (real name Charles Garner), *Vera the Medium* (**1908**) by Richard Harding Davis (1864-1916) and *Other Eyes Than Ours* (**1926**) by Ronald A. Knox (1888-1957). Modern fantasy tends to be sceptical and the medium has become a comic figure whose archetype is Madame Arcati in the play *Blithe Spirit* (**1941**) by Noel Coward (1899-1973), filmed as BLITHE SPIRIT (*1945*), although the medium Tangina in the POLTERGEIST movies has genuine power. A notable recent work in the farcical vein is *Strong Spirits* (**1994**) by Elisa de Carlo. [BS]

SPLASH! US movie (*1984*). Touchstone/Buena Vista. **Pr** Brian Grazer. **Exec pr** John Thomas Lenox. **Dir** Ron Howard. **Vfx** Mitch Suskin. **Screenplay** Lowell Ganz, Bruce Jay Friedman, Babaloo Mandel. **Starring** John Candy (Freddie), Tom Hanks (Alan), Daryl Hannah (Madison), Eugene Levy (Dr Walter Kornbluth). 110 mins. Colour.

Successful NEW YORK fruit wholesaler Alan recalls a childhood encounter with a MERMAID off Cape Cod, so travels there and indeed meets a mermaid. She tracks him to NEW YORK; their love is immediate and very physical (she is finned in water but legged on land). There are confusions as her naïveté confronts the CITY; she takes the name Madison from the avenue. Captured by obsessive marine zoologist Kornbluth, she is subjected to scientific study. Boorishly rejected by Alan, she physically deteriorates; it is Kornbluth who takes pity on her and, aided by Alan – who has seen the light – and Alan's rakehell brother Freddie, frees her. Alan opts for a life as a merman in The Mer-City with her rather than human life without.

Part-screwball, part-romantic SEX comedy, *S!* packs a reasonably erotic punch, yet with sufficient innocence to gain a PG rating. Certainly it was a departure for DISNEY, whose first movie under their new Touchstone byline this was. *Miranda* (*1947*) and its sequel *Mad About Men* (*1954*) have been claimed as *S!*'s forebears; later, parallels were to be drawn between *S!* and Disney's *The* LITTLE MERMAID (*1989*) – the initial encounter between man and mermaid has some similarities. *S!* was sequelled by the less ambitious *Splash, Too* (*1988* tvm) with different leading actors and director. Madison and Alan, now dwelling in the mer-city, have adventures after she dashes off to help a dolphin. [JG]

SPRING ◊ SEASONS.

SPRINGER, NANCY (1948   ) US writer, who began with the *The Books of Suns* (**1977**), which, revised, became the second volume of the HIGH-FANTASY **Vale** sequence: *The White Hart* (**1979**), *The Silver Sun* (1980), *The Sable Moon* (**1981**), *The Black Beast* (**1982**), *The Golden Swan* (**1983**) – the latter two assembled as *The Book of Vale* (omni **1983**) – and *Wings of Flame* (**1985**). NS's lyrical, sometimes florid style is also evident in *Chains of Gold* (**1986**) and the

Sea King trilogy: *Madbond* (**1987**), *Mindbond* (**1987**) and *Godbond* (**1988**). With *The Hex Witch of Seldom* (**1988**) and *Apocalypse* (**1989**) NS began to write CONTEMPORARY FANTASY, the well rendered rural Pennsylvanian settings and strongly portrayed female characters marking a distinct advance over her earlier work. She also began to write CHILDREN'S FANTASIES, including *Red Wizard* (**1990**) and *The Friendship Song* (**1992**). *Larque on the Wing* (**1994**), a contemporary novel dealing with GENDER changes and questions of sexual identity, and *Metal Angel* (**1994**), which features a bisexual ANGEL, incline more toward FABULATION than to GENRE FANTASY; the former won the James Tiptree AWARD. The novels' high spirits and eclectic story lines can be seen as a feminist counterpoint to the works of Tim POWERS and James BLAYLOCK. [JCB/GF]

Other works: *A Horse to Love* (**1987**) and *Not on a White Horse* (**1988**), both for children; *Chance & Other Gestures of the Hand of Fate* (coll **1987**), *Damnbanna* (**1992** chap) and *Stardark Songs* (chap **1994**), poetry; *The Blind God is Watching* (**1995** chap).

SPRING-HEELED JACK A legendary early-Victorian LONDON figure (◊◊ URBAN LEGENDS). He was (at first) a mysterious outlaw who attacked his victims while wearing a large helmet, a long black flowing cloak and tight-fitting white oilskin-like suit. His hands were claw-like, his eyes burned like coals, his ears were huge and his mouth spat blue flame. He was capable of jumping rooftops at a single bound. Guns didn't bother him. He made no noise at all. In early penny-dreadful appearances – like *Spring-Heeled Jack, the Terror of London* (*c*1870) – he was villainous, but he soon became in effect a MASKED AVENGER, defending the poor and helpless; this version stars in Philip PULLMAN's *Spring-Heeled Jack* (**1989**). In Tim POWERS's *The Anubis Gates* (**1983**) SHJ is far more menacing. [JC]

See also: JACK; JACK THE RIPPER.

SPUNDA, FRANZ (1890-1963) Austrian writer. ◊ AUSTRIA.

STABLEFORD, BRIAN M(ICHAEL) (1948-   ) UK writer, critic and academic, best-known for SCIENCE FICTION but equally adept in FANTASY, in which field he has developed a new reputation in the 1990s. BMS's early interest was in fantasy, and his liking for the works of Lord DUNSANY, Clark Ashton SMITH and Jack VANCE was evident in both his first professional sale, "Beyond Time's Aegis" (1965 *Science Fantasy* with Craig A. Mackintosh as Brian Craig), which he expanded in 1971 with other early unpublished stories into a novel, revised for publication as *Firefly* (fixup **1994** US); and his first published novel, *Cradle of the Sun* (**1969** dos US). Both are works of SCIENCE FANTASY set on a FAR-FUTURE Earth. In a similar vein, though far more aggressive, was his **Dies Irae** trilogy – *The Days of Glory* (**1971** US), *In the Kingdom of the Beasts* (**1971** US) and *Day of Wrath* (**1971** US) – which transposed HOMER's *Iliad* and *Odyssey* into a space-opera form of PLANETARY ROMANCE.

During the 1970s BMS concentrated mostly on a succession of hard-sf novels, and in the early 1980s he concentrated mostly on academic works. His only fantastic output during this period was a CHILDREN'S FANTASY, *The Last Days of the Edge of the World* (**1978**), set in the land of Caramorn, the closest to the World's Edge, where MAGIC is fading (◊ THINNING). A princess, hoping to delay her marriage to an unworthy prince, sets him three impossible questions to answer, the results of which translate Caramorn from a world of fantasy into a land of historical reality.

At the end of the 1980s BMS returned to fiction, this time with a greater emphasis on fantasy. As Brian Craig he contributed several novels and stories to the **Warhammer** fantasy series (◊ GAMES), including the **Orfeo** trilogy – *Zaragoz* * (**1989**), *Plague Daemon* * (**1990**) and *Storm Warriors* * (**1991**) – and the singleton *Ghost Dancers* * (**1991**). He also embarked upon a series of fantasies which drew upon our FOLKLORE nightmares. While the series was scientifically based, the presentation was supernatural HORROR. This started with *The Empire of Fear* (**1988**), set in an ALTERNATE WORLD where immortal VAMPIRES dominate. The novel remains basically sf – a scholar searches for the causes of vampirism in deepest Africa – but the book has all the atmosphere of a fantasy and is a close relative to the GASLIGHT ROMANCE. BMS continued this form in his **David Lydyard** trilogy – *The Werewolves of London* (**1990**), *The Angel of Pain* (**1991**) and *The Carnival of Destruction* (**1994**) – which spans the years from the time of Charles Darwin (1809-1892) to the end of WWII. Fallen ANGELS from the Earth's dawn reawaken in the Victorian era and channel their powers through mankind to achieve their APOCALYPSE. The vampire theme haunts two more books written at this time: *Young Blood* (**1992**), in which BMS contrasts the traditional image of the vampire with one created by modern mind-altering viruses, and *The Hunger and Ecstasy of Vampires* (1995 Interzone; exp **1996** US), a celebration of H.G. WELLS's *The Time Machine*, in which the Time Traveller travels to an alternate future where VAMPIRES rule.

BMS's **Genesys** sequence – *Serpent's Blood* (**1995**), *Salamander's Fire* (**1996**) and «Chimera's Cradle» (**1997**) – is a sophisticated PLANETARY ROMANCE with all the trappings of HEROIC FANTASY. BMS continues to play intellectual games with the ICONS of supernatural horror and fantasy but within the strict rules of scientific rationality.   [MA]

Other works (sf): *The Blind Worm* (**1970**); *To Challenge Chaos* (**1972** US); the **Hooded Swan** series, being *Halcyon Drift* (**1972** US), *Rhapsody in Black* (**1973** US), *Promised Land* (**1974** US), *The Paradise Game* (**1974** US), *The Fenris Device* (**1974** US) and *Swan Song* (**1975** US); *The Realms of Tartarus* (**1977** US; cut vt *The Face of Heaven* 1976 UK); *Man in a Cage* (**1976** US); *The Mind-Riders* (**1986** US); the **Daedalus Mission** series, being *The Florians* (**1976** US), *Critical Threshold* (**1977** US), *Wildeblood's Empire* (**1977** US), *The City of the Sun* (**1978** US), *Balance of Power* (**1979** US) and *The Paradox of the Sets* (**1979** US); *The Walking Shadow* (**1979**); *Optiman* (**1980** US; vt *War Games* 1981); *The Castaways of Tanagar* (**1981** US); the **Asgard** trilogy, being *Journey to the Centre* (**1982** US), *Invaders from the Centre* (**1990**) and *The Centre Cannot Hold* (**1990**); *The Gates of Eden* (**1983** US); *The Cosmic Perspective/Custer's Last Stand* (coll **1985** chap US); *Slumming in Voodooland* (**1991** chap US); *Sexual Chemistry* (coll **1991**); *The Innsmouth Heritage* (**1992** chap US), a sequel to H.P. LOVECRAFT's "Shadow Over Innsmouth" (1942); «Complications and Other Science Fiction Stories» (coll 1996 US); «Fables and Fantasies» (coll 1997 chap US).

As editor: *The Dedalus Book of Decadence (Moral Ruins)* (anth **1990**); *The Second Dedalus Book of Decadence: The Black Feast* (anth **1992**); *Tales of the Wandering Jew* (anth **1991**); *The Dedalus Book of British Fantasy: The 19th Century* (anth **1991**); *The Dedalus Book of Femmes Fatales* (anth **1992**).

STAINES, TREVOR Pseudonym of John BRUNNER.

STAIRWAY TO HEAVEN vt of *A* MATTER OF LIFE AND DEATH (*1946*).

STANTON, MARY (1947-   ) US writer whose two novels to date are the TALKING-ANIMAL fantasies *The Heavenly Horse from the Outermost West* (**1988**) and its sequel *Piper at the Gate* (**1989**; vt *Piper at the Gates of Dawn* UK).   [DP]

STARRETT, (CHARLES) VINCENT (EMERSON) (1886-1974) US literary figure and newspaperman, with a special interest in the fields of mystery and SUPERNATURAL FICTION; he is best-remembered for his writings about and pastiches of SHERLOCK HOLMES. His supernatural fiction is relatively undistinguished, although he was one of the better contributors to the earliest issues of WEIRD TALES in 1923. His weird fiction is collected as *Coffins for Two* (coll **1924**) and *The Quick and the Dead* (coll **1965**), the latter published by ARKHAM HOUSE. His greater legacy is through his work as a bibliophile and epistolarian. His extensive correspondences with writers, dealers and collecters, now housed at Yale University Library, contain a wealth of valuable historical material, particularly in relation to Ambrose BIERCE, Robert Louis STEVENSON, Arthur Conan DOYLE and especially Arthur MACHEN, whose work he championed in the USA – writing the monograph *Arthur Machen: A Novelist of Ecstacy and Sin* (1917 *Reedy's Mirror*; **1918** chap). Machen initially appreciated VS's interest and dedicated *The Secret Glory* (**1922**) to him. VS also published two collections of Machen's early work – *The Shining Pyramid* (coll **1923**) and *The Glorious Mystery* (coll **1924**) – with Machen's permission, but Machen forgot and accused VS of piracy. All was resolved, but not without acrimony. The correspondence about the affair was collected as *Starrett vs Machen: A Record of Discovery and Correspondence* (coll **1977** US) ed Michael Murphy. Further light is shed on VS's work in *Containing a Number of Things* (coll **1993** UK) ed R.B. Russell. More of VS's literary research is found in *Buried Caesars: Essays in Literary Appreciation* (coll **1923**).   [MA]

Other work: *Seaports in the Moon: A Fantasia on Romantic Themes* (**1928**).

STARS Chiefly of fantasy concern as a medium for OMENS and PORTENTS (◊◊ ASTROLOGY) – most famously the Star of Bethlehem in CHRISTIAN FANTASY. The Hyades cluster has a baleful reputation in the CTHULHU MYTHOS, thanks to mentions in (for example) Robert W. CHAMBERS's *The King in Yellow* (coll **1895**). Many stars of ill omen are actually PLANETS, like the Red Star that threatens Anne MCCAFFREY's Pern. Unfamiliar constellations may indicate to protagonists who have passed through a PORTAL that this is not Earth, or help distance a SECONDARY WORLD for the reader – as when J.R.R. TOLKIEN mentions "Remmirath, the Netted Stars" in *The Lord of the Rings* (**1954-5**). WIZARDS in Ursula K. LE GUIN's *A Wizard of Earthsea* (**1968**) must know the unchanging constellations which signal that they have entered the land of the dead. Fallen stars need not be figurative (◊ LUCIFER): two human-seeming characters in C.S. LEWIS's *The Voyage of the Dawn Treader* (**1952**) are former stars, one retired and one undergoing punishment. The SPIRIT of Sirius is likewise punished and condemned to live as a dog on Earth in Diana Wynne JONES's *Dogsbody* (**1975**); in her *Howl's Moving Castle* (**1986**) it is literally possible to catch a falling star which is a fire DEMON or ELEMENTAL. The huge eponymous mountain of Fritz LEIBER's "Stardock" (1965) was the launching-place for the world Nehwon's stars, of which one remains still docked: a jewel "big as the biggest oak tree". Unusually, Jack VANCE's "Morreion" (1973) features a magic-powered interstellar QUEST and includes a visit to a burnt-out star's surface.   [DRL]

STARTLING MYSTERY STORIES ◊ *The* MAGAZINE OF HORROR; MAGAZINES.

STASHEFF, CHRISTOPHER (1944-   ) US writer. His career began with and has remained largely dedicated to the **Rod Gallowglass** or **Warlock** PLANETARY-ROMANCE sequence. In order of internal chronology, the books are: *Escape Velocity* (1983), *The Warlock in Spite of Himself* (1969), CS's first book, and *King Kobold* (1969; rev vt *King Kobold Revived* 1984) – the first two assembled as *To the Magic Born* (omni 1986) and all three assembled as *Warlock to the Magic Born* (omni 1990 UK) – *The Warlock Unlocked* (1982) and *The Warlock Enraged* (1985) – both assembled with *King Kobold* as *The Warlock Enlarged* (omni 1986) and without it as *The Warlock Enlarged* (omni 1991 UK) – *The Warlock Wandering* (1986), *The Warlock is Missing* (1986) and *The Warlock Heretical* (1987) – the first two of these assembled as *The Warlock's Night Out* (omni 1988) and all three assembled as *The Warlock's Night Out* (omni 1991 UK) – *The Warlock Heretical* (1987), *The Warlock's Companion* (1988) and *The Warlock Insane* (1989) – all three assembled as *Odd Warlock Out* (omni 1989) – *The Warlock Rock* (1990), *Warlock and Son* (1991) and *M'Lady Witch* (1994). The sequence follows, with decreasing *joi de vivre*, the zany adventures of Rod Gallowglass and his clumsy robot sidekick who have found themselves on the planet Gramarye, where MAGIC works; they settle in and flourish. There is some TIME TRAVEL, and many creatures of FAERIE are comically rendered. Later volumes tend to preachiness on sexual matters. In the extremely similar **Rogue Wizard** series, featuring Gallowglass's son – *A Wizard in Bedlam* (1979), *A Wizard in Absentia* (1993), *A Wizard in Mind* (1995) and *A Wizard in War* (1995) – and in the likewise extremely similar **Wizard in Rhyme** series – *Her Majesty's Wizard* (1986), *The Oathbound Wizard* (1993), *The Witch Doctor* (1994) and *The Secular Wizard* (1994) – CS stuck to his last. With Bill Fawcett he edited a SHARED-WORLD series about the **Crafter** family of magicians – *The Crafters* (anth 1991) and *The Crafters 2: Blessings and Curses* (anth 1992) – and with L. Sprague DE CAMP he edited *The Enchanter Reborn* (anth 1992). *Sir Harold and the Monkey King* (1993 chap) is set in the same world as the last-named volume (the world of de Camp's and Fletcher PRATT's **Harold Shea** series). Most recently, CS has launched a new fantasy series, **The Star Stone**, commencing with *The Shaman* (1995) and *The Sage* (1996).   [JC/DP]
Other works: The **Starship Troupers** sf series, being *A Company of Stars* (1991), *We Open on Venus* (1994) and *A Slight Detour* (1994); *The Gods of War* (anth 1992); *Dragon's Eye* (anth 1994).

STATIUS, PUBLIUS PAPINIUS (cAD50-cAD95) Italian Latin epic poet. His *Thebaid* (cAD91; trans A.D. Melville 1992), modelled on VIRGIL's *Aeneid* (19BC), recalls its structure – travel and adventure followed by warfare. But PPS learned also from other writers: OVID's wit and flexibility, SENECA's horrific vision and Marcus Annaeus Lucanus's (◊ GREEK AND LATIN CLASSICS) political passion. His own voice was sentimental, descriptive and insistently subjective. As noted by J.H. Mozley and C.S. LEWIS, he prefigures the medieval, at once deeply emotional (notably for wives and children) and curiously intellectual in his coolly ordered treatment of battle and personification, and using such devices as DRAGONS and dark FORESTS. The unfinished *Achilleid* (cAD96; trans J.H. Mozley 1928) was even more romantic.

The *Thebaid* is modern fantasy's first true analogue. In it the GODS, even Jupiter (◊ ZEUS), are either helpless dolts or bloodthirsty fiends; though Theseus finally brings justice, only corpses and widows remain to receive that balm. It is a story of holocaust, and to write it he subjected his entire fictional world to the dictates of a consistent, intentional fantasy. He was the first Western writer to do so.   [JB]

STATUES The sympathetic-MAGIC link between a well made statue and the thing represented (whether a human, a GOD or an imagined ideal) lends itself to fantasy developments, usually ANIMATE/INANIMATE transitions. A statue may become animated, as in the tales of PYGMALION and DON JUAN – whose animations are inspired by, respectively, LOVE and VENGEANCE. In Ernest BRAMAH's "Kin Weng and the Miraculous Tusk" (in *Kai Lung Unrolls His Mat* coll 1928) the carved figure of a bird is so perfect that it comes alive and takes wing. Museum statues of Greek gods are animated in Thorne SMITH's *The Night Life of the Gods* (1931). NEW YORK gargoyles wake to vengeful life in Harlan ELLISON's "Bleeding Stones" (in *Deathbird Stories* coll 1975) and Parisian ones are friendly in DISNEY's *The Hunchback of Notre Dame* (1995); statues of MONSTERS atop a New York skyscraper come diabolically alive in GHOSTBUSTERS (1984), whose sequel animates the Statue of Liberty. MANNEQUIN (1987) sees a shop-window dummy animated *via* POSSESSION.

The reverse process – whereby living things turn into statuary – goes back to the MYTH of the GORGONS and their petrifying gaze, shared by mythical MONSTERS like the basilisk and cockatrice. It is a favourite form of BONDAGE imposed by evil magic-users: Mombi plans this fate for the boy Tip in L. Frank BAUM's *The Marvelous Land of Oz* (1904). However, a child turned to marble by an incautious WISH in E. NESBIT's *The Enchanted Castle* (1907) finds her (temporary) statuehood curiously idyllic. G.K. CHESTERTON's SCIENCE FANTASY "The Finger of Stone" (in *The Poet and the Lunatics* coll 1929) features water that causes rapid petrification. TROLLS traditionally become statues of themselves when touched by the SUN, as in J.R.R. TOLKIEN's *The Hobbit* (1937). In John COLLIER's "Evening Primrose" (1941) those threatening the secrecy of the WAINSCOT society, whose members pose as mannequins in a SHOP, are punished by being made over into "real" wax mannequins. The White WITCH of C.S. LEWIS's *The Lion, the Witch and the Wardrobe* (1950) routinely converts her opposition to statues. Robert HOLDSTOCK's "In the Valley of the Statues" (1979) hints at repeated, reversible statue/human TRANSFORMATIONS, and a mystic oneness of flesh and stone.

Even statues which remain statues may play a part in strange and/or symbolic relationships – one reason why the second of the BIBLE's Ten Commandments is a PROHIBITION against making any "graven image", particularly idols such as the Golden Calf. Legendarily, placing a RING on the finger of a statue of APHRODITE (as related in *The Anatomy of Melancholy* [1621] by Robert Burton [1577-1640]) is an unwitting CONTRACT of marriage which will clash dangerously with earthly relationships, as in F. ANSTEY's comic *The Tinted Venus* (1885); this theme is horrifically echoed in Tim POWERS's *The Stress of Her Regard* (1989). It is a magic statue that whimsically precipitates the IDENTITY EXCHANGE in Thorne SMITH's *Turnabout* (1931) (◊◊ TURNABOUT [1940]).   [DRL]
See also: FACE OF GLORY; GOG AND MAGOG; GOLEM.

STAY TUNED US movie (1992). Morgan Creek. **Pr** James G. Robinson. **Exec pr** Gary Barber, David Nicksay. **Dir**

Peter Hyams. **Spfx** George Erschbamer, John Thomas. **Mufx** Alex Gillis, Tom Woodruff Jr. **Vfx** Rhythm & Hues Inc. **Anim sv/character design** Chuck Jones. **Screenplay** Jim Jennewein, Tom S. Parker. **Starring** Pam Dawber (Helen Knable), Jeffrey Jones (Spike), Eugene Levy (Crowley), Heather McComb (Diane Knable), John Ritter (Roy Knable), David Tom (Darryl Knable). 87 mins. Colour.

High-flying executive Helen is on the verge of leaving husband Roy because of his addiction to tv. Kids Diane and Darryl, the latter an electronics whiz, leave their parents overnight to sort things out; instead Spike, a latter-day MEPHISTOPHELES, persuades Roy to sign the CONTRACT for an extravagant satellite system boasting 666 channels (◊ GREAT BEAST). Shortly thereafter Helen and Roy are sucked through the reception dish into Hellvision, the tv station catering to the DEVIL: mortals who survive 24 hours' channel-hopping between the scores of lethal shows on offer earn REDEMPTION, but almost always succumb, sacrificing their SOULS. Sometimes aided by sacked executive DEMON Crowley, the Knables endure 24 hours of programmes like *You Can't Win* (a quiz) and *Duane's Underworld*, and Roy is returned to earthly existence; Helen, however, is retained – she signed no contract and thus, according to Spike, is a trespasser (◊ QUIBBLES). Roy returns to rescue her and is whipped through a string of further parodied programmes before, aided by Darryl's electronics genius from "outside", vanquishing Spike.

Though the theme of people being dragged into tv shows (and video GAMES) has been overworked, this surprisingly funny TECHNOFANTASY misses few bases. A highlight is an excellent animated mid-section by Chuck Jones, with Helen and Roy as cartoon mice being pursued by a robot cat. There is also a fine PARODY of *film noir*. Some of the trailers and ads are inspired, notably those for *Three Men and Rosemary's Baby* and the prison drama *thirty-something-to-life*. Though Ritter is no Robin Williams, it is perhaps a pity, given Dawber's previous career, that a version of *Mork and Mindy* is missing from the list. [JG]

See also: REALITY.

STEADMAN, RALPH (1936-   ) UK illustrator with a wildly energetic pen-line and transparent colour style. His early work was heavily influenced by the Expressionist George Grosz (1893-1959) and by political cartoonist and caricaturist Gerald SCARFE, but he developed slowly into an illustrator of considerable power and originality. RS has worked as a caricaturist and cartoonist for *Punch*, *Private Eye*, *Rolling Stone* and *The New Statesman*. His illustrations for *Alice in Wonderland* (1967) and *Through the Looking Glass* (1972) won him the Francis Williams Illustrations Award in 1972. His standing as a major imaginative talent has steadily increased with the publication of his own books, including *Sigmund Freud* (graph **1979**), *I, Leonardo* (graph **1983**) and *The Big I AM* (graph **1988**). [RT]

STEAMPUNK A term applied more to SCIENCE FICTION than to fantasy, though some tales described as steampunk do cross genres. Steampunk stories are most commonly set in a romanticized, smoky, 19th-century LONDON, as are GASLIGHT ROMANCES. But the latter category focuses nostalgically on ICONS from the late years of that century and the early years of this – on DRACULA, JEKYLL AND HYDE, JACK THE RIPPER, SHERLOCK HOLMES and even TARZAN – and can normally be understood as combining SUPERNATURAL FICTION and RECURSIVE FANTASY, though some gaslight

romances can be read as FANTASIES OF HISTORY. Steampunk, on the other hand, can best be described as TECHNOFANTASY that is based, sometimes quite remotely, upon technological ANACHRONISM. Steampunk tales are thus often placed in an ALTERNATE WORLD, to allow their premised anachronisms full imaginative play.

As a marriage of URBAN FANTASY and the alternate-world tradition, steampunk can arguably be traced back to the influence of Charles DICKENS, whose vision of a labyrinthine, subaqueous London as moronic inferno underlies many later texts. Dickens's London, somewhat sanitized, also underlies the Babylon-on-the-Thames version of the great city created by authors like Robert Louis STEVENSON, Arthur Conan DOYLE, Bram STOKER and G.K. CHESTERTON in their fantasies – tales whose uneasy THEODICY underpins much contemporary gaslight romance. The two categories, steampunk and gaslight romance, point to two ways of rendering closely linked original material.

The term steampunk did not come into use until the late 1980s, and derives from the usage cyberpunk. Many examples of steampunk were, therefore, written before a word existed to describe them. Christopher PRIEST's *The Space Machine* (**1976**) combines steampunk and gaslight romance; his later *The Prestige* (**1995**) incorporates a strong steampunk subplot in the course of which Nikola Tesla (1856-1943) invents a matter transmitter. K.W. Jeter's *Morlock Night* (**1979**), a sequel to H.G. WELLS's *The Time Machine* (**1895**) which depicts Morlocks rampaging through the sewer system of Victorian London, may be considered the first genuine steampunk tale, while his later *Infernal Devices: A Mad Victorian Fantasy* (**1987**) is clearly written deliberately as steampunk. Further early texts include *The Crisis in Bulgaria, or Ibsen to the Rescue!* (graph **1956**) by Jocelyn Brooke (1908-1966), Herbert ROSENDORFER's *The Architect of Ruins* (**1969**), Harry Harrison's vigorous alternate history *A Transatlantic Tunnel, Hurrah!* (**1972**), John Mella's *Transformations* (fixup **1975**), Michael MOORCOCK's 1970s **Bastable** sequence (assembled as *A Nomad of the Time Streams* rev omni **1993**), "Black as the Pit, from Pole to Pole" (**1977**) by Stephen Utley (1948-   ) and Howard Waldrop (1946-   ), which chronicles the adventures of FRANKENSTEIN's Monster in a HOLLOW EARTH, and William KOTZWINKLE's *Fata Morgana* (**1977**).

Similarly early are the influential tales by the two authors who have become most identified with the term, which may have been invented to describe their work. James P. BLAYLOCK's steampunk novels include *The Digging Leviathan* (**1984**) and the later **St Ives** sequence: *Homunculus* (**1986**) and *Lord Kelvin's Machine* (**1992**). Tim POWERS's steampunk titles include *The Anubis Gates* (**1983**), *On Stranger Tides* (**1987**) and *The Stress of Her Regard* (**1989**). These are colourful, fast-paced SCIENCE FANTASIES involving anachronistic inventions and their eccentric inventors, occult WAINSCOT conspiracies, and UNDERGROUND criminal cabals. They share an essentially nostalgic vision of a crowded, hyperreal 19th-century London in which science, still mostly pursued by amateurs, has not lost its innocence to mechanized warfare, the HOLOCAUST and the atom bomb.

Other steampunk texts – most essentially sf – include *The Difference Engine* (**1990** UK) by William Gibson (1948-   ) and Bruce Sterling (1954-   ), *The Hollow Earth* (**1990**) by Rudy Rucker (1946-   ), in which Edgar Allan POE has adventures in a HOLLOW EARTH, Brian STABLEFORD's *The*

Werewolves of London (**1990**) and its sequels *The Angel of Pain* (**1991**) and *The Carnival of Destruction* (**1994**), closely argued metaphysical fantasies set in 19th-century London, though his *The Hunger and Ecstasy of Vampires* (**1996**) is gaslight romance, Lawrence NORFOLK's *Lempriere's Dictionary* (**1991**), *Anti-Ice* (**1993**) by Stephen Baxter (1957-   ), which ends the Crimean War with a Hiroshima-like armageddon – Baxter's *The Time Ships* (**1995**), more gaslight romance than steampunk, is a sequel to *The Time Machine* that is perhaps more faithful to the intentions of Wells than Jeter's *Morlock Nights* – Paul J. McAuley's *Pasquale's Angel* (**1994**), set in the Italy of LEONARDO DA VINCI, Colin GREENLAND's *Harm's Way* (**1994**), which expertly pastiches Dickens's style in a space-opera techno-fantasy in which the British Empire has extended its hegemony throughout the Solar System using sail-powered spaceships, Joan AIKEN's *Is* (**1992**) and *The Cockatrice Boys* (**1996** US), and *The Steampunk Trilogy* (coll **1995**) by Paul Di Filippo (1954-   ), collecting three tales which subvert 19th-century rationalism and much else with fantastic inventions and Lovecraftian apparitions: one describes substitution of a giant newt for Queen Victoria, another a romance between Emily Dickinson and Walt Whitman.

Not strictly steampunk, but echoing in gaslight-romance terms Steampunk's dense reworking of 19th-century London, Kim NEWMAN's *Anno Dracula* (**1992**) is an alternate history in which a triumphant Dracula forcibly marries Victoria and presides over a brutalized London in which the living and the undead uneasily mingle; the sequel, *The Bloody Red Baron* (**1995** US), is set in the battlefields of WORLD WAR I. The Dickensian squalor and child-labour of the DRAGON factory at the opening of Michael SWANWICK's *The Iron Dragon's Daughter* (**1993** UK) likewise echoes the spirit of steampunk's retro technofantasies.   [PJM/JC]

STEELE, V.M. Pseudonym of Dion FORTUNE.

STEINER, K. LESLIE Pseudonym of Samuel R. DELANY.

STEMMA Originally, a family tree. By transference, the term has come to designate "the tree of descent of a text" (*OED*). As Robert IRWIN argues in *The Arabian Nights: A Companion* (**1994**), Western Classical scholars have been much inclined to the deduction of stemmata, but the assumption is that there is in fact an original text to discover. This assumption has proven fruitful in Western literature, but (Irwin suggests) can be actively misleading in attempts to trace (say) *The Arabian Nights* (◊ ARABIAN FANTASY) or the *Katha Sarit Sagara* (◊ OCEAN OF STORY) back to an almost certainly nonexistent single source.   [JC]

STEPFORD WIVES, THE US movie (*1974*). Fadsin/Palomar/Columbia. **Pr** Edgar J. Scherick. **Exec pr** Gustave M. Berne. **Dir** Bryan Forbes. **Screenplay** William GOLDMAN. **Based on** *The Stepford Wives* (**1972**) by Ira LEVIN. **Starring** Peter Masterson (Walter Eberhart), Nanette Newman (Carol Van Sant), Patrick O'Neal (Dale "Diz" Coba), Paula Prentiss (Bobby Marco), Katharine Ross (Joanna Eberhart). 114 mins. Colour.

With their two standard children, lawyer Walter and photographer Joanna Eberhart move to dull town Stepford. He soon joins the mysterious Men's Association; she finds all the oddly beautiful wives are over-placid and obsessed with domestic trifles. It proves that the Men's Association, led by ex-Disneyland engineer Coba, is killing the women and replacing them with amazingly lifelike automata who are sexual dynamos and willing domestic slaves. The last we see is not Joanna but her replica, exchanging blandnesses

with the other "wives" against supermarket muzak.

TSW loses much – but not all – of the double-edged SATIRE of Levin's novel, aiming instead to suspend disbelief; its leisurely pace allows the development and climax the *feel* of plausibility, no mean feat given the subject matter. Further aiding this sense is the fact that this TECHNOFANTASY draws its themes not from SCIENCE FICTION but from traditional sources: the perfectly realized yet mindless automata are other-selves, DOPPELGÄNGERS, cuckoos in the nest, CHANGELINGS, GHOSTS . . . Despite the veneer of modernity, the horrors *TSW* evokes are horrors that already have aeons-old presence inside us: we believe in them because we always have.

TSW engendered two sequels. In *Revenge of the Stepford Wives* (*1980* tvm) journalist Sharon Glass comes to Stepford, digs out the truth, and causes the simulacra to rise up and, *à la* FRANKENSTEIN's MONSTER, destroy their creators; its climactic revenge scene has a chilly effectiveness, but in general there is a sense of existing material being reworked. *The Stepford Children* (*1987* tvm) ignores the 1980 tvm: the men of Stepford are now trying to complete their bliss by replacing also their children.   [JG]

STEPHENS, JAMES (1882-1950) Irish poet and prose writer, several of whose idiosyncratic books are fantasy: *The Crock of Gold* (**1912**), about two philosophers and sundry gods and leprechauns, and containing much lyrical blarney; *The Demi-Gods* (**1914**), about three ANGELS who visit Earth; and *Deirdre* (**1923**) – to modern tastes, perhaps JS's most satisfactorily focused novel – and its sequel *In the Land of Youth* (coll **1924**), both based on Irish mythology.   [DP]
Other works: *Irish Fairy Tales* (coll **1920**); *Etched in Moonlight* (coll **1928**).

STEPHENS, REED Pseudonym of Stephen R. DONALDSON.

STEPMOTHERS The FAIRYTALE and FOLKTALE cliché of the wicked stepmother is a PLOT DEVICE which serves to avoid the uncomfortable notion that biological mothers may also treat their children badly. The stories of CINDERELLA and SNOW WHITE offer classic examples. Modern CHILDREN'S FANTASY has rehabilitated step-parents with more thoughtful treatments of emotional stress in merged families, as in Zilpha Keatley SNYDER's *The Headless Cupid* (**1971**; vt *A Witch in the Family* 1977 UK) and Diana Wynne JONES's *The Ogre Downstairs* (**1974**).   [DRL]

STERLING, GEORGE (1869-1926) US writer. ◊ DECADENCE.

STERLING, JOHN (1806-1844) UK writer, an early co-proprietor of *Atheneum*, for which he wrote a number of remarkably vivid fantasies, including "Zamor" (1828), which describes a sobering VISION experienced by Alexander the Great, and "Cydon" (1829), an ALLEGORY about a youth driven to seek out the cave of PROMETHEUS. Eight more tales are embedded in the text of the otherwise naturalistic novel *Arthur Coningsby* (**1833** anon), five of which are reprinted, with other material, in Volume II of *Essays and Tales by John Sterling* (coll **1848** 2 vols) ed Julius Charles Hare. The other fantasy material includes a series which appeared in BLACKWOOD'S MAGAZINE as **Legendary Lore**, whose last item was the serial novel "The Onyx Ring" (1838-9); it is an earnest and deeply felt philosophical fantasy in which an unhappy man undergoes a series of IDENTITY EXCHANGES in search of the secret of happiness. The series began with the delicate prose poem "The Palace of Morgana" (1837); it also included "Land and Sea" (1838), a weird tale about a sea-sprite, and "A Chronicle of

England" (1840), a fine allegorical FAIRYTALE. Volume I of *Essays and Tales* includes a memoir of the author by Hare, whose concentration on JS's brief flirtation with religious faith (when he served as Hare's curate) so offended Thomas CARLYLE – the model for the most sympathetic character featured in "The Onyx Ring" – that Carlyle wrote his own biography, *The Life of John Sterling* (**1851**). It was to no avail: although JS was certainly the most significant UK pioneer of fantasy he was soon forgotten. [BS]

STEVENS, FRANCIS Pseudonym of US writer Mrs Gertrude Bennett, neé Barrows (1884-?1939). She wrote 12 stories for the pulp MAGAZINES 1916-20, beginning with "The Nightmare" (1917), but abruptly abandoned her writing career in 1920. *The Citadel of Fear* (1918 *Argosy*; **1970**) and *Claimed* (1920 *Argosy*; **1966**) are vivid weird novels with gaudy fantasy embellishments reminiscent of those employed in the work of A. MERRITT. *The Heads of Cerberus* (1919 *Thrill Book*; **1952**) is a fascinating dimensional fantasy which borders on sf in its description of a dystopian Philadelphia of the future. "Serapion" (1920 *Argosy*) is a novel of divided personality. "The Elf-Trap" (1919) is a notable short sentimental fantasy, perhaps her best work. Her other short stories were "Friend Island" (1918), the novelette "Behind the Curtain" (1918) and a tale of monstrous bacteria, "Unseen – Unfeared" (1919). Further novels were "Labyrinth" (1918) and "Avalon" (1919). Her final story, the slapdash LOST-RACE novella "Sunfire" (1923), was serialized in WEIRD TALES. [BS]

STEVENS, LAURENCE (1886-1960) US illustrator. ◊ Virgil FINLAY.

STEVENSON, ROBERT LOUIS (BALFOUR) (1850-1894) Scottish writer. His much-imitated short novel *Strange Case of Dr Jekyll and Mr Hyde* (**1888** chap); versions published after RLS's death usually add a prefatory *The*), a paradigmatic TECHNOFANTASY, was based on a nightmare; rumour alleges that his wife Fanny persuaded him to burn the first version and turn the story into a more strident moralistic fantasy in which the war between GOOD AND EVIL which rages in every human psyche would be clarified as well as made manifest (◊◊ JEKYLL AND HYDE; JEKYLL AND HYDE MOVIES). With his friend W.E. Henley (1849-1903) RLS had earlier written the melodramatic play *Deacon Brodie, or The Double Life* (**1880**; rev 1889) about a (real-life) criminal masquerader hanged in 1788.

Many of RLS's novellas and short stories, including the early works collected in *New Arabian Nights* (coll **1882** 2 vols), are baroque without being outrightly fantastic, but he did write several other moral fables with a horrific edge similar to that in *Jekyll and Hyde*. Most of the actual fantasies were assembled in *The Merry Men and Other Tales and Fables* (coll **1887**). The conscience-stricken protagonist of "Markheim" mistakes an ANGEL for SATAN. In *Will o' the Mill* (**1895** chap US) the private world of a selfish man is finally invaded by a not altogether hostile personification of DEATH. "The Merry Men" itself is a hallucinatory fantasy in which yet another Calvinist conscience does its brutal work; part of it is writen in Scottish dialect, as is the horrific "Thrawn Janet". "Olalla" is a more convoluted moral fantasy whose saintly heroine suffers for the sins of her animalistic relatives and cannot accept the mundane REDEMPTION offered by a wounded soldier. *Island Nights' Entertainments* (coll **1893**) includes *The Bottle Imp* (**1896** chap; vt *Kaëwe's Bottle* 1935 chap UK), RLS's version of the classic tale of the ultimate poisoned chalice reworked for the Samoans among

whom he had taken up residence, and "The Isle of Voices", an original fantasy likewise set in the South Seas. Much earlier RLS had written *Treasure Island* (1881-2 *Young Folks* as "The Sea Cook, or Treasure Island"; **1883**); while wholly lacking in supernatural elements, this adventure is so fantasticated that it is hard to regard it as outwith the fantasy genre.

All the above-mentioned stories are reprinted in *The Stories of Robert Louis Stevenson* (omni **1928** UK). The much slimmer *The Short Stories of Robert Louis Stevenson* (coll **1923** US) includes all the familiar fantasies and adds *The Waif Woman* (**1916** chap), but neither volume includes *Ticonderoga: A Legend of the West Highlands* (**1887** chap US) or *When the Devil was Well* (**1921** chap US). There are many other collections sampling RLS's work; one which assembles all his shorter works is *The Complete Shorter Fiction* (coll **1991**). [BS]

Other works (selective): *Thrawn Janet; Markheim: Two Tales* (coll **1903** chap); *Tales and Fantasies* (coll **1905**), which includes *The Misadventures of John Nicholson* (**1889** chap US) and the marginally weird *The Body-Snatcher* (1884; **1895** chap US); *Pan's Pipes* (**1910** chap); *Fables* (coll **1914**); *The Tales of Tusitala* (coll **1946**).

Further reading: *The Definitive Dr Jekyll and Mr Hyde Companion* (**1983**) by H.M. Geduld.

STEVEN SPIELBERG'S AMAZING STORIES ◊ AMAZING STORIES (1985-7).

STEVERMER, CAROLINE (J.) (1955-   ) US writer and newspaperwoman who began writing fantasy with *The Serpent's Egg* (**1988**). Set in a LAND-OF-FABLE 16th-century Europe, it is a DYNASTIC FANTASY about a conspiracy to counter a ruthlessly ambitious duke who has the queen's ear. Written in a plain but elegant style, it was a deft and very strong debut. *Sorcery & Cecilia* (**1988**) with Patricia WREDE is an epistolary Regency romance with MAGIC. It follows the conventions of its genre with satisfying faithfulness (each collaborator controlling a separate plot, the two eventually dovetailing), with the detailed and consistent magic adding an unusual *frisson*. *River Rats* (**1992**) received the ALA Best Books for Young Adults and the New York Public Library Books for the Teen Age. Its post-HOLOCAUST landscape seems to call for sf protocols, but, with its Mississippi riverboat which functions as a movable POLDER and the crew who embody a "rock'n'roll will save your life" philosophy that links the book closely to Terri WINDLING's **Borderlands** series (but without ELVES), it can profitably be read as fantasy. *A College of Magicks* (**1994**), set in a 1908 RURITANIA, concerns the Duchess of Galazon, sent to school to learn magic and who subsequently finds her place in the wider world. It combines *Bildungsroman*, dynastic fantasy, swashbuckling romance and an unusual postulation of the magical structure of the world. It is her most satisfying work to date.

In her tendency to let witty and ironic dialogue carry the narrative, in her use of Regency and Ruritania and rock'n'roll, CS has proved herself a practised hand at FANTASY OF MANNERS. [DGK]

Other works: *The Alchemist: Death of a Borgia* (**1981**) and *The Duke and the Veil* (**1981**), historical mysteries as by C.J. Stevermer.

STEWART, MARY (FLORENCE ELINOR) (1916-   ) UK writer of romantic thrillers since *Madam Will You Talk?* (**1955**), but best-known for her Arthurian books. The most significant are those in the **Merlin Trilogy**: *The Crystal*

Cave (**1970**; vt *Merlin of the Crystal Cave* 1991), *The Hollow Hills* (**1973**) and *The Last Enchantment* (**1979**), assembled as *Mary Stewart's Merlin Trilogy* (omni **1980** US). This was the first set of novels to explore the character of MERLIN in detail and set him in context in the Arthurian world. *The Crystal Cave* covers Merlin's youth and shows him trying to come to terms with his power of second sight (◊ TALENTS). The novel ends with the conception of ARTHUR. *The Hollow Hills* follows the same time-frame as T.H. WHITE's *The Sword in the Stone* (**1939**), but is an entirely different treatment of the subject. Merlin supervises Arthur's upbringing from afar and seeks to protect Britain by engineering Arthur's future. *The Last Enchantment*, the most powerful, considers Merlin's own fate. In all three MS draws from a common pool of legend but applies her own interpretation. Because the story is narrated by Merlin we learn much about his thoughts and motives, and he becomes a strong if sad character. The trilogy is one of the more convincing Arthurian works. MS returned to ARTHUR in *The Wicked Day* (**1983**), to telling the story of Mordred and his rebellion against Arthur; *The Prince and the Pilgrim* (**1995**), a sidebar work, takes a minor incident in Sir Thomas MALORY's *Morte Darthur* about King Mark's murder of his brother and makes it a QUEST for justice. The hero is the little-known Alexander (or Alisander) who is imprisoned by MORGAN LE FAY and must win his freedom.

In all her Arthurian novels MS has followed her own course, regardless of tradition, and developed a believable portrayal of events. The fantasy element is always slight, intended to flavour events, not dictate them.

Two of MS's romantic novels, both in the mood of Daphne DU MAURIER, feature sublinear supernatural events arising from psychic abilities (◊ TALENTS). *Touch Not the Cat* (**1976**) is a modern GOTHIC FANTASY in which a young woman makes a telepathic link with an unknown relative. *Thornyhold* (**1988**) is about a girl who inherits her aunt's house and begins to be influenced (POSSESSION is too strong a word) by whatever residuum of her aunt's presence still lingers.

MS has also written three novels for children. *The Little Broomstick* (**1971**), in which a child must triumph over a WITCH, is the least original. *Ludo and the Star Horse* (**1974**) is an instructional tour of the ZODIAC, with each of the signs portrayed in anthropomorphic form. *A Walk in Wolf Wood* (**1980**) is a TIMESLIP fantasy where two young children follow a strangely dressed man INTO THE WOODS and find themselves in the 14th century trying to help a WEREWOLF. [MA]

STEWART, SEAN (1965-   ) Canadian writer whose first novel, *Passion Play* (**1992**), was an sf detection. His second, *Nobody's Son* (**1993**), interestingly reverses the usual plot structure of the HIGH-FANTASY tale of individual growth (◊ RITE OF PASSAGE) by placing the normal climax in the first chapter. Young Shielder's Mark penetrates the dread Ghostwood (◊ INTO THE WOODS) from which no HERO has returned, successfully deals with the life-or-death QUIBBLES of a LIMINAL BEING (an old WITCH), is granted a necessary TALISMAN or two, breaks the SPELL that has locked the wood in a magic stasis (◊ TIME IN FAERIE), and returns to demand from the king that he grant – as promised – one WISH. Mark selects the king's youngest daughter as his bride; and the bulk of the tale warmly and engrossingly deals with the consequences of this choice. *Resurrection Man* (**1995**) is set in a complexly conceived ALTERNATE-WORLD USA, into which

MAGIC has begun to seep back through fissures in the texture of the previous mundane REALITY that have been opened, perhaps, by the Apocalyptic arousals of WORLD WAR II, when GOLEMS appeared in the death camps. The protagonist is an "angel" (◊ ANGELS) whose psychic powers lead him into the depths of family romance, an exploration which succinctly and potently comprises a portrait of a society profoundly in need of HEALING. SS is one of the fantasy writers of the 1990s who seems able to confirm the potential sharp relevance of the genre to modern life. [JC]

STOCKBRIDGE, GRANT [hn] ◊ Norvell W. PAGE.

STOCKTON FRANK R(ICHARD) (1834-1902) US writer and editor, a prolific author of short stories. Most of his fantasies were for children, including those in *Ting-a-Ling* (coll **1870**; vt *Ting-a-Ling Tales* UK), which contains three tales of the eponymous hero and the novella "The Magical Music". However, many of his quirky FABLES have a surreal edge and a sly suspicion of conventional moralism which are more fitted to adult consumption; the best are assembled in *The Bee-Man of Orn and Other Fanciful Tales* (coll **1887**). In *The Bee-Man of Orn* (**1964** chap) the transmogrified bee-man goes in search of his true self. In "The Griffin and the Minor Canon" a clergyman risks censure to befriend an ambivalent MONSTER. The museum in *The Queen's Museum* (**1915** chap) proves less educational than the queen intended until suitably enlivened. "Old Pipes and the Dryad" is a sentimental parable of rejuvenation. "The Banished King" puts a new slant on the Sphinx's fondness for RIDDLES. "The Philopena" is a bizarre tale in which the many interruptions to the unsmooth-running course of true LOVE include a Gryphoness and a Water Sprite. Three of these tales were reprinted with two of a similar ilk from *The Clocks of Rondaine and Other Stories* (coll **1892**) in *Fanciful Tales* (coll **1894**), all of whose contents were included in a more comprehensive selection, *The Queen's Museum and Other Fanciful Tales* (coll **1906**).

FRS was a pioneer of the comic GHOST STORY, best exemplified by "The Transferred Ghost" and "The Spectral Mortgage" in *The Lady, or the Tiger?, and Other Stories* (coll **1884**). The three stories in *Afield and Afloat* (coll **1900**) are weaker but still as good as anything by John Kendrick BANGS, who would carry the fledgling tradition forward. Some of FRS's fantasies for adults are offbeat comedies, including the title novella of *Amos Kilbright: His Adscititious Experiences, with Other Stories* (coll **1888**), but some are more contemplative, including "A Borrowed Month" in *The Christmas Wreck and Other Stories* (coll **1886**; vt *A Borrowed Month and Other Stories* 1887 UK) and "The Philosophy of Relative Existences" in *The Watchmaker's Wife and Other Stories* (coll **1893**; vt *The Shadrach and Other Stories* UK), which features shy ghosts from the future; the latter collection also includes the JEKYLL-AND-HYDE story "The Knife that Killed Po Hancy". "The Magic Egg", "The Bishop's Ghost and the Printer's Baby" and "Stephen Skarridge's Christmas" in *A Story-Teller's Pack* (coll **1897**) are offbeat moral fantasies with a hint of sarcasm, the last parodying Charles DICKENS's *A Christmas Carol* (**1843**). Some of the tales assembled in *The Science Fiction of Frank R. Stockton* (coll **1976**) ed Richard Gid Powers are fantasies. There are many other collections selected from FRS's works; *The Novels and Stories of Frank R. Stockton* (**1899-1904** 23 vols) is definitive . . . although it omits the works transcribed by the medium Etta de Camp on behalf of FRS's spirit, collected in *The Return of Frank R. Stockton* (coll **1913**)!

FRS's novels are less interesting; they include the LOST-RACE story *The Adventures of Captain Horn* (**1895**) and *The Vizier of the Two-Horned Alexander* (**1898**), a mildly facetious fantasy of IMMORTALITY. *The Stories of the Three Burglars* (**1889**) is presented as a novel but contains three interpolated tales, one of which is fantasy. *The Great Show in Kobol-Land* (**1891** chap) is an interesting novella for children – it is reprinted in *The Clocks of Rondaine and Other Stories* – whose hero and heroine ultimately reject the gift of the Cosmic Bean (the ultimate foodstuff) because they do not wish to be rulers of Lazyland. [BS]

Other works: *Roundabout Rambles in Lands of Fact and Fancy* (coll **1872**); *Tales out of School* (coll **1875**); *The Floating Prince and Other Fairy Tales* (coll **1881**); *A Chosen Few* (coll **1895**); *John Gayther's Garden and Stories Told Therein* (coll **1902**); *The Lost Dryad* (**1921** chap); *The Fairy Tales of Frank Stockton* (coll **1990**) ed Jack ZIPES.

STOKER, BRAM Working name of Irish theatrical manager and writer Abraham Stoker (1847-1912), in England from 1878, whose first and primary career was as manager (1878-1905) for the famed Victorian actor and impresario Sir Henry Irving (1838-1905). Some of this famed actor's roles – he appeared in *The Bells* by Leopold David Lewis (1828-1890), an 1871 adaptation of ERCKMANN-CHATRIAN's *Le Juif Polonais* (**1871**; trans as *The Polish Jew* **1871** UK), about a sinister hypnotist and murderer, and he played the FLYING DUTCHMAN in at least one production – may have had a formative influence on BS, who was profoundly influenced by Irving's mannered but domineering stage presence.

BS is known now almost exclusively for *Dracula* (**1897**), though he began his writing career much earlier, with "The Crystal Cup" (1872), which is HORROR, as was his first novel-length story, "The Primrose Path" (1875 *The Shamrock*). The FAIRYTALES assembled in *Under the Sunset* (coll **1882**) are sinister, though with touches of ALLEGORY that may have made them seem, to the Victorian mind, suitable for children. None of the works of these early years do more than hint mildly at what was to come.

Dracula – like Mary SHELLEY's *Frankenstein, or The Modern Prometheus* (**1818**), all Arthur Conan DOYLE's **Sherlock Holmes** books and Robert Louis STEVENSON's *Strange Case of Dr Jekyll and Mr Hyde* (**1886**) – is a text whose central figure has become an ICON of popular culture, and which is far better known through numerous movie adaptations (◊ DRACULA MOVIES) than it is in its own right (◊ DRACULA). Despite BS's stylistic infelicities, devotees of Dracula and VAMPIRE stories in general should perhaps return to the original text, for its merits are also evident. There have been many omnibus publications of the text, and several separate modern editions of value, including *The Annotated Dracula* (**1975** US) ed Leonard Wolf (1923-   ), *Dracula* (**1993**) ed Maurice Hindle (1944-   ), which is textually the most satisfactory modern edition, and *Dracula: Bram Stoker's Text of 1901* (**1994**).

As one of the most famous horror novels in existence, *Dracula* tells a tale which is now part of modern folklore. Jonathan Harker travels to Transylvania, where Count Dracula imprisons him briefly, arousing the deadly enmity of the Englishman. Most of the remaining story is set in LONDON, which the Count has invaded, primarily – in a manner definitive of the seductive impulsion that guides great SUPERNATURAL FICTION – through his taking the VIRGINITY and vampirizing of women, who then become erotic figures who menace the stability of things. Dracula is a paradigmatic version of the SHADOW who threatens the daylight world, and who must be expunged.

BS's further, consistently inferior, novels include: *The Mystery of the Sea* (**1902**), a supernatural fiction in which a man's TALENT (precognition or second sight) fails to enable him to regain a lost treasure; *The Jewel of Seven Stars* (**1907**; rev by another hand 1919), in which the ASTRAL BODY of a queen of ancient EGYPT attempts to reanimate her mummified body (◊ MUMMIES) by taking POSSESSION of a young girl; *The Lady of the Shroud* (**1909**), which features a rationalized female vampire, a touch of talents (precognition again), and a hero named Rupert who becomes king of his own RURITANIA; and *The Lair of the White Worm* (**1911**; full text ed Richard DALBY 1986), which confusedly evokes the downside of GODDESS imagery through the eponymous SHAPESHIFTER which possesses a glamorous lady (or vice versa), and attempts to infect a large cast with its venomous, accursed, erotic allure (◊ WORM/WYRM).

Some of BS's short fiction, however, attains something of the strength of *Dracula*. *Dracula's Guest and Other Weird Stories* (coll **1914**) includes the title story – an unpublished fragment from the novel – plus "The Judge's House" (1891 *Holly Leaves*), featuring an inimical HAUNTING, "The Secret of the Growing Gold" (1892 *Holly Leaves*), in which a dead woman's hair grows through living rock to proclaim her murder, and "The Squaw" (1893 *Holly Leaves*; vt "The Black Cat"), in which a CAT revenges itself on the human who has killed its kitten. *The Dualitists* (in *The Theatre Annual for 1887* anth **1886**; **1986** chap) is a supernatural fiction in which two children compose together a malevolent fantasy GAME. Some of BS's stories assembled in *Shades of Dracula* (coll **1982**) and *Midnight Tales* (coll **1990**), both ed Peter HAINING, had not previously appeared in book form. [JC]

STOKES, MANNING LEE (?   -   ) US writer, who wrote also as Bernice Ludwell, Lee Manning and, when in collaboration with Jerome Darwin Engel (1909-   ), Ford Worth. ◊ Roland J. GREEN.

STONIER, G(EORGE) W(ALTER) (1903-1985) Australian-born UK author, playwright and journalist; assistant literary editor of the *New Statesman and Nation* 1928-45. For this journal he also served as movie critic as "William Whitebait". *The Memoirs of a Ghost* (**1947**) is a POSTHUMOUS FANTASY narrated by a victim of the LONDON Blitz (◊◊ WORLD WAR II). Swept to and fro, while eagerly but futilely attempting to regain a life no longer his, this panic-stricken SPIRIT encounters other GHOSTS and steadily grows disillusioned with the AFTERLIFE. GWS recycled parts of the last chapter of the book into a short story, also titled "The Memoirs of a Ghost" (this time the protagonist is a traffic victim) for Cynthia ASQUITH's *The Second Ghost Book* (anth **1952**). [RD]

STORY 1. Any narrative which tells or implies a sequence of events, in any order which can be followed by hearers or readers, and which generates a sense that its meaning is conveyed through the actual telling, may be called a Story. A Story, in short, is a narrative discourse which is *told*.

One of the most useful distinctions between the FANTASTIC as a whole and FANTASY, considered as one of the literatures of the fantastic, is that fantasy texts are most easily understood as the telling of Story in this sense; other categories of the fantastic – some, like SURREALISM, of Modernist lineage – may well treat the telling of narrative as an act that warrants corrosive disparagement, deconstruction or dismantlement. SCIENCE FICTION, for instance, a genre of the Fantastic closely associated with fantasy, can certainly be treated in terms of Story but (to generalize wildly) may in fact be best understood through its presentation of themes, which can be analysed without regard to narrative sequence (that is, *what* is being described is primary, rather than *how* it is being told). An sf text can promulgate meaning through infodumps, striking examples, expositions of abstract arguments from science or history, arguments waged by talking heads in a vacuum, etc. A fantasy text almost invariably conveys its sense of things by conducting its protagonists (there are no significant fantasy texts without protagonists) to the end of their QUEST through sequences which hearers or readers understand as consecutive and essential moments in the telling of the tale.

Fantasy texts, in this understanding, can be characterized as always moving towards the unveiling of an irreducible substratum of Story, an essence sometimes obscure but ultimately omnipresent; the key events of a fantasy text are bound to each other, to the narrative world, and ideally to the tale's theme in a way that permits endless retellings ($\lozenge$ TWICE-TOLD), endless permutations of the narrative's unbound MOTIFS, and a sense of ending. In its purest form, the fantasy story resembles MYTH, which as C.S. Lewis has noted – in *An Experiment in Criticism* (**1961**) – retains its essential power despite the varying forms of its telling. The story of ORPHEUS, for instance, has survived endless permutations in literature, film, drama, and opera without substantial violation to its essential core.

It is partly this sense of familiarity – not only with a tale's possible earlier incarnations but with its relationship to the whole inherited body of taletelling – that produces the effect of *resonance* characteristic of so many fantasy stories, or narratives that are analogues of fantasy stories, like James JOYCE's *Ulysses* (**1922**) and John UPDIKE's *The Centaur* (**1963**), which retells the myth of Chiron. At the simplest level of language, such resonance is evoked by formulaic openings like the FAIRYTALE's "once upon a time"; at more complex levels it may be achieved through the presence of UNDERLIER figures (the TRICKSTER, the wise woman) or ritual events. All these devices signal to the reader – or hearer – of the tale that the events and characters described belong to the world of Story rather than to the world of represented actions which is implied by the novel of manners. As Brian Wicker states in *The Story-Shaped World* (**1975**): ". . . we may say that the characters in fairy tales [and their fantasy descendants] are 'good to think with'. Because they deal in final causes they explain thing which no amount of science based on 'association' can explain. The job of the fairytale is to show that Why? questions cannot be answered except in one way: by telling stories. The story does not contain the answer, it is the answer. The answer cannot be translated into factual, that is non-narrative form, for the answer is the narrative form."

It might seem an elementary act of critical apprehension to notice this central role of Story in the world, and particularly in the literature of fantasy. But 20th-century criticism has not much concentrated on Story (any more than it has paid attention to RECOGNITION, which may be defined as that narrative moment when Story knows itself), instead tending to devalue genres and individual works in any genre which are deemed to depend too deeply upon "primitive" devices such as storytelling. As Karl Kroeber (1926-   ) argues in *Retelling/Rereading: The Fate of Storytelling in Modern Times*

(**1992**), this bias against narrative extends back before E.M. FORSTER; but Forster's remarks on Story, in *Aspects of the Novel* (**1927**), though otiose, are sufficiently famous to quote: "The more we look at the story . . . the more we disentangle it from the finer growths that it supports, the less shall we find to admire. It runs like a backbone – or may I say a tape-worm, for its beginning and end are arbitrary. It is immensely old – goes back to neolithic times, perhaps to palaeolithic. Neanderthal man listened to stories, if one may judge by the shape of his skull . . . Yes – oh dear yes – the novel tells a story."

Forster's intellectual languor may be feigned, but the influence of this sort of thinking has generally been dire. Story is, after all, not only the most important mode, from time immemorial, chosen by humans for the conveyance of meaning, but is a primary technique ($\lozenge$ AESOPIAN FANTASY) for the inculcation of lessons inimical to the Thought Control Police who proliferate in this (as well as every previous) century, and who notoriously fear the anarchic, freeing power of the raw tale. "True moral education takes place," Philip PULLMAN suggests in "The Moral's in the Story" (1996 *Independent*), "whenever anyone, of whatever age, encounters a story with an open mind."

The devaluation of Story can be seen, in this light, as a profound abdication of critical responsibility. Stories may be popular, but they are potent, and their subversive potential was recognized from the first ($\lozenge$ LITERARY FAIRYTALES; WONDER TALE) by the 18th-century writers who began to shape fantasy as a genre. Fantasy may be hard to pin down – in part because no Story can be exactly paraphrased – but the analysis of fantasy texts in terms of thematic content, by critics better suited to some other endeavour, has largely contributed to the paucity of fantasy criticism worth remarking upon.

Given a blank slate and an even playing field, Stories shape the way the world is best understood. This may have something to do with remembered conventions, and with the inherent meaning-generating structure of the human brain; certainly it seems that, when humans are given the freedom to say anything they wish to say, they say Stories. And they embed these Stories in a Story-shaped world – a world which significantly does not resemble the "realistic" worlds enjoined by the mimetic tradition which dominated the writing of prose fiction in the Western world from the 18th century well into the 20th century. Arguably, at the end of the 20th century mimetic tradition increasingly fails to fulfil the most conservative expectations of how we can understand the nature of the world. More than perhaps ever before, human beings live (and perceive the meanings of their lives) in a maze of realities and illusions so multiplex and inchoate, that for many it is almost impossible to make sense of being alive. It could be that the late-century success of fantasy (and other genres of the fantastic) is partly due to these circumstances; and that we listen to stories at the *fin de millennium* in order to recuperate a sense that stories still exist. That we still can be told.   [JC/GW]

Mentions of Story in this Encyclopedia are frequent. Further references to the term may be found in the following entries (some have already been mentioned above); this list is not inclusive: AMNESIA; ANCESTRAL MEMORIES; ARABIAN FANTASY; ARABIAN NIGHTMARE; BARDS; BONDAGE; BOOKS; CAULDRON OF STORY; CINDERELLA; COMMEDIA DELL'ARTE; CYCLES; DEBASEMENT; DUOS; EDIFICE; ELDER RACES; FABULATION; FAIRYTALE; FANTASY; FANTASY ART; FOLKTALE; FRAME

STORY; GODDESS; GODGAME; GOOD AND EVIL; HEALING; INNS; INSTAURATION FANTASY; KNIGHT OF THE DOLEFUL COUNTENANCE; LANDSCAPE; LEARNS BETTER; LITERARY FAIRYTALES; MAGIC REALISM; MASKS; MATTER; METAMORPHOSIS; MOTIFS; MYTHS; OCEAN OF STORY; PALIMPSEST; PARODY; PERCEPTION; PLOT DEVICES; PORTALS; POSTHUMOUS FANTASY; QUESTS; RECOGNITION; REVEL; REVISIONIST FANTASY; RITE OF PASSAGE; ROMANCE; SAGA; William SHAKESPEARE; SLINGSHOT ENDING; STEMMA; TEMPLATE FANTASY; THEODICY; THEOSOPHY; THINNING; THRESHOLDS; TIME; TIME ABYSS; J.R.R. TOLKIEN; TOPOS; TWICE-TOLD; UNDERLIERS; WAINSCOTS; WONDER TALE; WRONGNESS.

2. Part of the definition of fantasy is that its protagonists tend to know they are in a Story of some sort, even if at first they do not know which one; at moments of RECOGNITION they find out just which Story it is that has, in some sense, *dictated* them. It would of course be injudiciously restrictive to claim that *all* fantasy texts convey a sense that their protagonists are under the control of an already-existing Story, and that sooner or later they come to an awareness of the fact; it is, however, the case that many fantasy texts are clearly and explicitly constructed so as to reveal the controlling presence of an underlying Story, and that the protagonists of many fantasy texts are explicitly aware they are acting out a tale.

Story as its own explicit subject matter is central in FOLKLORE and in the FAIRYTALE, and may reflect the underlying oral source of much of this traditional material. Stories told aloud seem inherently to call for an emphasis upon their already-told, their infinitely retellable nature; it is a natural enough impulse for these tales to tend to incorporate protagonists who are themselves aware of – and consciously live out, consciously heal themselves through acting out – the Stories that are telling them. In REVISIONIST FANTASY, this awareness of dictation, and a simultaneous revolt against the intended outcome, is clearly central as well.

In 20th-century fantasy, a more conspicuously self-conscious attitude towards Story becomes evident. As it comes to a close, the heroes of E.R. EDDISON's *The Worm Ouroboros* (**1922**) are at first dismayed that their Story may have ended (even though they have triumphed), and profoundly grateful when that Story begins again, on the last page of the book, and the CYCLE, blessedly, recommences. Most of the many characters in Peter S. BEAGLE's *The Last Unicorn* (**1968**) know very well they are in a "last unicorn" tale, and most of them know who they are in that tale – HERO, MAGUS, KNIGHT OF THE DOLEFUL COUNTENANCE. – the song which gives its name to Beagle's *The Innkeeper' Song* (**1993**) tells the Story inside, too. The stories told within John GARDNER's *In the Suicide Mountains* (**1977**) tell the Story of the book that contains them. The protagonist of Michael ENDE's *The Neverending Story* (**1979**) finds that the Fantastica he enters is a Story whose implications for his own RITE OF PASSAGE into responsibility he must understand, or he will never grow up. And the inhabitants of a small town in Jonathan CARROLL's *The Land of Laughs* (**1980**) are possessed by a deadly Story.

Into the 1980s and later, self-referentiality becomes very common, as in John CROWLEY's *Little, Big* (**1981**), Russell HOBAN's *The Medusa Frequency* (**1987**), Neil GAIMAN's "Parliament of Books" (1992 *Sandman 40*; in *Sandman: Fables and Reflections*, coll **1993**), Ursula LE GUIN's "The Poacher" (1993), Charles DE LINT's *Memory and Dream*

(**1994**) and A.S. BYATT's "The Story of the Eldest Princess" (1992). There are countless other examples. [JC]

STORY CYCLE A group of tales addressing a central theme or MATTER, and/or told within a shared over-narrative, usually in the shape of a FRAME STORY. So understood, the term has very nearly the same meaning as that ascribed to the term CYCLE (sense **1**), and is here used mainly to emphasize that the cycle being referred to has been written down as a series of narrative tales meant to be understood as fiction. [JC]

STOUT, WILLIAM (1949-) US fantasy artist and illustrator with a very wide range of artwork styles; he has worked mostly in CINEMA, but has many COMICS, album sleeves and illustrated books on DINOSAURS to his credit. Most of his work is in pen-line and transparent colour. His most obvious influences are Alphonse Mucha (1860-1939), Frank FRAZETTA, Al WILLIAMSON, Arthur RACKHAM and William Heath ROBINSON.

WS's first published work was a 1968 cover for COVEN 13. In 1971 he became an assistant to TARZAN artist Russ Manning (1929-1980); he collaborated with Harvey Kutzman on **Little Annie Fanny** in 1972, and in the mid-1970s produced several distinctive covers for "bootleg" record albums, including The Yardbirds' *More Golden Eggs* (1974), The Who's *Tales from the Who* and *Who's Zoo* (both 1974) and Wings's *Great Dane* (1976). He also drew stories and covers for several underground comics at this time.

WS's work in the entertainment industries has been considerable, his first being production design for a projected series of three **Buck Rogers** movies, eventually made as a tv series. Other movies include *More American Grafitti* (**1979**), *Conan the Barbarian* (**1981**; ◊ CONAN MOVIES), *Heavy Metal, The Movie* (**1981**), *Date with an Angel* (**1982**), *First Blood* (**1982**), *Conan the Destroyer* (**1984**), *Return of the Living Dead* (**1984**), RED SONJA (**1985**), *Invaders from Mars* (**1986**), *Masters of the Universe* (**1987**) and *Predator* (**1987**). He has done posters for *Amazon Women on the Moon* (**1987**) and MONTY PYTHON'S LIFE OF BRIAN (**1979**). He did design work on the 1984 New York stage production of *Sherlock's Last Case* and designed the Universal Studios Tours' live **Conan** show in 1987.

WS has had a lifelong interest in palaeontology and has been a member of the Society of Vertebrate Paleontology since the early 1980s; his work in this area has included the 1982 museum exhibition *Death of the Dinosaurs*, illustrations for *The New Dinosaur Dictionary* (**1976**) by Donald F. Glut (1944-), and *The Dinosaurs* (graph **1981**) with text by William Service, and was the key illustrator in Ray BRADBURY's *Dinosaur Tales* (coll **1983**). He collaborated with Byron PREISS on the juvenile *The Little Blue Dinosaur* (graph **1984**). [RT]

STRAND MAGAZINE, THE ◊ MAGAZINES.

STRANGE ADVENTURE OF DAVID GRAY, THE vt of ◊ VAMPYR (**1932**).

STRANGE CASE OF DR JEKYLL AND MISS OSBOURNE, THE French movie (**1981**). ◊ JEKYLL AND HYDE MOVIES.

STRANGE FANTASY ◊ FANTASTIC; FANTASTIC ADVENTURES; MAGAZINES.

STRANGE STORIES US pulp MAGAZINE, 13 issues, bimonthly, February 1939-February 1941, published by Better Publications, New York; ed Mort Weisinger (1915-1978).

A WEIRD-FICTION companion to *Thrilling Wonder Stories* and *Thrilling Mystery*, though less successful. Many of the same authors contributed, especially Robert BLOCH, Ralph Milne Farley (real name Roger Sherman Hoar; 1887-1963), Henry KUTTNER and Manly Wade WELLMAN, plus August W. DERLETH and Seabury QUINN from WEIRD TALES; between them Derleth, Bloch and Kuttner provided about 40 of the MAGAZINE's 148 stories. Its contents were similar to those of WEIRD TALES, though stronger on shocks and with less variety. *SS* paid only half a cent a word, so attracted authors' rejected or early stories and published little of merit. Exceptions are Kuttner's two **Prince Raynor** stories, David H. KELLER's "The Dead Woman" (a reprint), Derleth's "Logoda's Heads", C.L. MOORE's "Miracle in Three Dimensions" and Wellman's "Changeling", all of which appeared in the first four issues. *SS* was a failure by a publisher usually adept at exploiting markets. It folded when Weisinger left to edit *Superman* (◊ SUPERMAN). [MA]

STRANGEST STORIES EVER TOLD, THE ◊ FANTASTIC; FANTASTIC ADVENTURES; MAGAZINES.

STRANGE TALES US pulp MAGAZINE, 7 issues, bimonthly then quarterly, September 1931-January 1933, published by Clayton Magazines, New York; ed Harry Bates (1900-1981).

Short-lived companion to *Astounding Stories* and a close rival to WEIRD TALES in terms of the quality of its stories. Unlike *WT*, the emphasis was on action and adventure rather than unusual concepts. There were typical tales of BLACK MAGIC, MUMMIES, VAMPIRES and GHOSTS, though others showed more originality. Stories of merit included: "Wolves of Darkness" (January 1932) by Jack WILLIAMSON; "The Trap" (March 1932) by Henry S. WHITEHEAD; "The Thing that Walked on the Wind" (January 1933) by August W. DERLETH (◊◊ WENDIGO); and "Murgunstruum" (January 1933) by Hugh B. CAVE. Other popular contributors included Arthur J. Burks (1898-1974), Ray Cummings (1887-1957), Paul Ernst (1899-1985), Robert E. HOWARD, Gordon MacCreagh (1886-1953), Victor Rousseau (real name Victor Rousseau Emmanuel; 1879-1960) and Clark Ashton SMITH. Most of the stories from all but #7 were reprinted in *The* MAGAZINE OF HORROR and its companions. A facsimile anthology is *Strange Tales* (anth **1976**) ed William H. Desmond *et al.*

The title was also used for two post-WWII booklets (nd but February-March 1946) ed Walter Gillings (uncredited), reprinting fiction from US PULPS, especially *WT*. [MA]
See also: DOCTOR STRANGE.

STRANGE WORLDS ◊ EERIE.

STRAUB, PETER (1943-   ) US writer and poet whose first published genre work was *Julia* (**1975**; vt *Full Circle* 1977 UK), an effective GHOST STORY centring on a young woman whose daughter died in a tragic accident. Questions about her sanity and her husband's motives are played against apparent manifestations of the girl's cruel SPIRIT. PS followed with another tale of guilt and a malignant spirit, *If You Could See Me Now* (**1977**). He was working his way towards a definitive take on the ghostly tale, *Ghost Story* (**1979**). In this classic of the subgenre, a group of elderly men who meet regularly to exchange ghost stories become the targets of a SHAPESHIFTING creature that embodies the spirit of a woman whose death the men inadvertently caused in youth. It is both a tribute to the traditional ghost tales of Henry JAMES, Nathaniel HAWTHORNE *et al.* and an updating of such themes, featuring an intricate and varied contemporary Gothic plot. With this novel – filmed as GHOST

STORY (*1981*) – PS assumed a pre-eminent position in the HORROR genre, a role he embraced only briefly before moving further afield.

In *Shadowland* (**1980**; vt *Shadow Land* 1981 UK), a tale of MAGIC, ILLUSION and DARK FANTASY, the narrator recounts a bizarre summer of his adolescence spent at the home of an ageing magician who seeks a successor. PS's last venture into unadulterated supernatural horror was *Floating Dragon* (**1982**), which focuses on a small Connecticut community, juxtaposing the influences of an ancient, DRAGON-like entity with the effects of an accidentally released chemical weapon, DRG-16. A much-anticipated collaboration with Stephen KING resulted in *The Talisman* (**1984**), whose setting is split between the contemporary USA and the Territories, a dark parallel universe in which MAGIC holds sway and technology's advances stalled long ago. The 12-year-old protagonist Jack Sawyer (his name a homage to Mark TWAIN) befriends a WEREWOLF in the Territories and brings him back to the USA, where the two battle various adversaries in an epic QUEST whose ultimate aim is to save the life of Jack's mother. Elements of the fantastic and surreal are only hinted at in PS's **Blue Rose** trilogy, which begins with *Koko* (**1988**), a tale of four Vietnam vets seeking a killer, and which won the 1989 WORLD FANTASY AWARD despite minimal fantastic content. The remainder of the sequence consists of *Mystery* (**1990**), a slightly surreal novel of detection, and *The Throat* (**1993**), which won the 1994 BRAM STOKER AWARD. *The Hellfire Club* (**1996**) involves a SERIAL KILLER and reflects PS's continuing shift away from the fantastic.

PS's shorter fiction tends towards novellas. "The Ghost Village" (1992) won a World Fantasy Award. *Houses Without Doors* (coll **1990**) collects six long stories, including *Blue Rose* (**1985** chap) and the nightmarish *Mrs God* (exp **1990**), in which a professor is drawn to an English manor to research the works of his poetess grandmother, and finds himself caught in a vortex of hallucinatory images, including a preponderance of dead babies. [BM]
Other works: *Ishmael* (coll **1972** chap), *Open Air* (coll **1972** chap) and *Leeson Park and Belsize Square: Poems 1970-1975* (coll **1983**), verse; *Marriages* (**1973**); *The General's Wife* (**1982**), based on a previously unpublished excerpt from *Floating Dragon*; *Wild Animals* (omni **1984**), assembling *Julia*, *If You Could See Me Now* and the previously unpublished *Under Venus*; *Peter Straub's Ghosts* (coll **1995**).

STRIBLING, T(HOMAS) S(IGISMUND) (1881-1965) US author, most noted for his novels set in the US South, of which *The Store* (**1932**) won the Pulitzer Prize. TSS's fantasy, some verging on sf, tends to have been ignored except by enthusiasts. "The Green Splotches" (1920 *Adventure*), perhaps the best-known, is a South American adventure featuring aliens. "The Web of the Sun" (1922 *Adventure*) is a LOST-RACE novel set in a South American valley guarded by a giant SPIDER whose venom causes longevity (◊ IMMORTALITY). *East is East* (1922 *Argosy*; **1928**) is a routine adventure set in North Africa but involves a crystal which controls its own destiny and allows insight into the future (◊ SCRYING). "Christ in Chicago" (1926 *Adventure*) is an sf novel exploring the concept of eugenics but also involving faith-HEALING. "Mogglesby" (1930 *Adventure*), another African adventure, concerns the discovery of a tribe of intelligent APES. *These Bars of Flesh* (**1938**), a SATIRE on institutions, incorporates a research project into life after death. TSS wrote a series of very popular stories about the detective **Dr Poggioli**; all were unorthodox, but most were nonfantastic – although "The Governor of Cap Haitien" (1925 *Adventure*) involves VOODOO and "A Passage to Benares" (1926 *Adventure*) incorporates Oriental supernaturalism. These stories were collected as *Clues of the Caribbees* (coll **1929**) and *Best Dr Poggioli Detective Stories* (coll **1975**). [MA]

Further reading: "T.S. Stribling, Subliminal Science-Fictionist" by Sam Moskowitz in *Fantasy Commentator #40*, Winter 1989-90.

STRICKLAND, BRAD Working name of US writer William Bradley Strickland (1947-   ). His **Jeremy Moon** sequence – *Moon Dreams* (**1988**), *Nul's Quest* (**1989**) and *Wizard's Mole* (**1991**) – conveys its copywriter hero into a wittily constructed DREAM world, where he takes his stand. The juvenile *Dragon's Plunder* (**1992**) involves cut-throat PIRATES and a boy who can whistle up the wind. In addition to a number of sf novels, BS has completed several YA fantasy adventures begun by the late John BELLAIRS, including *The Ghost in the Mirror* (**1993**), *The Vengeance of the Witch-Finder* (**1993**) and *The Drum, the Doll, and the Zombie* (**1994**). [JC/DP]

Other works: *ShadowShow* (**1988**) and *Children of the Knife* (**1990**), both horror.

STRINDBERG, AUGUST (1849-1912) Swedish novelist and dramatist, intensely prolific (he wrote at least 70 plays) and controversial. His early work, both in fiction and in the theatre, tended to a heated naturalism in which increasingly savage passions (often described simplistically as misogyny) were let loose. From the 1890s on, the Expressionist violence of his writing allows a fantastic reading of ostensibly autobiographical sketches like *Inferno* (**1897**; trans Claud Field as *The Inferno* **1912** UK) and *Legenden* (**1898**; trans as *Legends* **1912** UK). A play like *Till Damaskus* (Parts I and II **1898**; Part III **1901-4**; trans as *To Damascus* **1933-5** UK) features a number of figures whose apparent identities dissolve into raging ARCHETYPE, and several of whom act as SHADOWS of one another: it is a fine example of the drama of JUNGIAN PSYCHOLOGY – though it was, of course, written long before Jung created his dramaturgy of the unconscious. Other plays of a similar intensity include *Ett drömspel* (**1902**; trans as *A Dream Play* **1912** UK; new trans Elizabeth Sprigge 1963 UK), which is dominated by a compassionate GODDESS figure, and *Spöksonaten* (**1907**; trans as *The Spook Sonata* **1916**; new trans Elizabeth Sprigge as *The Ghost Sonata* 1963 UK), an extraordinary presentation of a DREAM whose actors become terrifyingly real and archetypal simultaneously. [JC]

STROBL, KARL HANS (1877-1946) Moravian writer and editor. ◊ AUSTRIA.

STRUCK BY LIGHTNING US tv series (1979). Warner Bros./CBS. **Pr** Bob Ellison, John Thomas Lenox, Marvin Miller, Steve Pritzker. **Exec pr** Arthur Fellows, Terry Keegan. **Dir** Larry Shaw, Joel Zwick. **Writers** Lawrence J. Cohen, Fred Freeman, Bryan Joseph, Michael Russinow. **Created by** Keegan. **Starring** Jeff Cotler (Brian), Jack Elam (Frank), Bill Erwin (Glen Hillman), Jeffrey Kramer (Ted Stein), Millie Slavin (Nora), Richard Stahl (Walt Calvin). 3 30min episodes (plus 4 unaired). Colour.

Bringing the FRANKENSTEIN legend to modern times, this sitcom sees Ted Stein inherit the decrepit Brightwater Inn in Maine. He discovers that caretaker Frank was the creation of his great-great-grandfather – i.e., is the 230-year-old Creature. The show flopped. [BC]

STUART, FRANCIS Working name of Irish writer Henry Francis Montgomery Stuart (1902-   ), whose tumultuous career began in 1923 with a privately published volume of poems, *We Have Kept the Faith* (coll **1923** chap), was shadowed by his refusal to leave Nazi Germany – whose politics he disdained – in WWII, and continued into recent years. His first novel, *Women and God* (**1931** UK), gives off a distracted air of fantasy, though it is in fact mundane. His second, *Pigeon Irish* (**1932** UK), is sf. Early fantasies include *Try the Sky* (**1933** UK), in which a SHAMAN-like Native American woman – who creates three-dimensional models of a primordial RIVER resembling the Danube – takes flight with her lover into a surreal, desert Ireland, and *The Angel of Pity* (**1935** UK), in which an anticipated WORLD WAR II serves as venue for the passion of a female version of CHRIST. Late work includes *A Hole in the Head* (**1977**) and *Faillandia* (**1985**), a fantasticated satiric look at a corrupt near-future Ireland. [JC]

STURGEON, THEODORE (1918-1985) US writer, born Edward Hamilton Waldo, who changed his first name when he adopted his stepfather's surname. In the early part of his career he was more comfortable writing for John W. Campbell Jr's UNKNOWN – for which he wrote the haunted-ship story "Cargo" (1940) and the ironic *conte philosophique* "The Ultimate Egoist" (1941), among others – than *Astounding* (◊ SFE). Although he eventually made his name as a SCIENCE-FICTION writer, his primary interest always lay in parables of human maturation and transcendence, for which certain sf motifs merely served as convenient props. He wrote numerous HORROR stories, including the ultimate MONSTER story *"It"* (1940; **1948** chap) and the intense psychological study "Bianca's Hands" (1947). TS's WEIRD FICTION often portrays characters labouring under a mysterious compulsion which forces them to be cruel; examples include "Cellmate" (1947) and "The Perfect Host" (1948). His posthumously published account of his troubled relationship with his stepfather, *Argyll: A Memoir* (**1993** chap), helps explain the fascination of such themes, which achieved a more elaborate display in *The Dreaming Jewels* (**1950**; vt *The Synthetic Man*) and other sf tales of psionic persecution. The flipside of this coin is represented by heartfelt moral fables like "The Silken-Swift" (1953), in which a UNICORN plays out its traditional role as a moral indicator, applying its own criteria of female worthiness, and the hauntingly delicate "The Graveyard Reader" (1958). Most of TS's many story collections contain a few fantasies among the sf; the one with the highest proportion is *E Pluribus Unicorn* (coll **1953**).

Contending supernatural forces – which TS refused to characterize as good or EVIL – provide backgrounds for the novellas "One Foot and the Grave" (1949) and "Excalibur and the Atom" (1951), the latter contriving an uneasy alloy of Arthurian fantasy and hard-boiled detective fiction, but TS was not really interested in metaphysical contexts: he much preferred studying internal forces of compulsion and the behavioural strategies by which individuals might cope with them. Most of his stories in this vein lie, like the heartfelt "Need" (1960), on the sf/fantasy borderline, but the brilliant *Some of Your Blood* (**1961**) is a naturalistic study of a VAMPIRE who finds that his harmless method of assuaging his appetite is regarded with more intense revulsion than if he had stuck to traditional modes. Late in life, when TS was cursed by a decades-long writer's block through which he broke only briefly and ineffectively, he began constructing a metaphysical fantasy about a new MESSIAH, but the version of it published after his death as *Godbody* (**1986**) is an abruptly concluded fragment. [BS]

Other works: *Without Sorcery* (coll **1948**; cut vt *Not Without Sorcery* 1961); *A Way Home* (coll **1955**; cut 1956; cut vt *Thunder and Roses* 1957 UK); *Caviar* (coll **1955**); *A Touch of Strange* (coll **1958**; cut 1959); *Aliens 4* (coll **1959**); *Beyond* (coll **1960**); *Sturgeon in Orbit* (coll **1964**); *The Joyous Invasions* (coll **1965**); *Starshine* (coll **1966**); *Sturgeon is Alive and Well . . .* (coll **1971**); *The Worlds of Theodore Sturgeon* (coll **1972**); *To Here and the Easel* (coll **1973**); *Case and the Dreamer* (coll **1974**); *The Stars are the Styx* (coll **1979**); *The Golden Helix* (coll **1979**); *Alien Cargo* (coll **1987**); *A Touch of Sturgeon* (coll **1987**).

STYX ◊ RIVERS.

SUASSUNA, ARIANO (1927-   ) Brazilian poet and playwright. ◊ BRAZIL.

SUCCUBUS The female equivalent of the INCUBUS; i.e., a woman who visits men in their sleep to deliver temptations and nightmares. One Old English name for this demon was *hægge*; thus WITCHES became known as hags. The succubus was originally ugly, but in SUPERNATURAL FICTION became transformed into a temptress or FEMME FATALE with inevitable connections to the VAMPIRE legend (◊◊ LAMIA). Whereas incubi were unusual in Victorian fiction, succubi (or equivalent) became more frequent, almost certainly because of their erotic appeal to predominantly male writers and readers. The vampire in *Carmilla* (1871-2 *Dark Blue*; **1971** US) by J. Sheridan LE FANU is a version, as are the vampire girls in *Dracula* (**1897**) by Bram STOKER. Tanith LEE has explored various aspects of the motif in her stories, especially those collected in *Women as Demons* (coll **1989**) and *Nightshades* (coll **1993**). Perhaps the most sexually graphic fictional exploration is *The New Neighbor* (**1991**) by Ray Garton. [MA]

SUCHARITKUL, SOMTOW Real name of S.P. SOMTOW, under which some early books were first published.

SUE, EUGÈNE (1804-1857) French writer who achieved great but brief celebrity as a *feuilletonist* in the early 1840s, when radical periodicals fielded him as the chief rival of their royalist adversaries' champion storyteller, Alexandre DUMAS. His early melodramas of bloody piracy gave way to sweeping analyses of city life – paying particular attention to the criminal activities of rich and poor – like *Les Mystères de Paris* (1842-43; **1843**; trans J.D. Smith as *The Mysteries of Paris* **1844**). Another sprawling epic was *Le juif errant* (1844-5; trans D.M. Aird as *The Wandering Jew* **1845**), in which the descendants of a man who once aided the WANDERING JEW are summoned to Paris to receive the fortune which has been gathering interest for centuries. The supernatural elements are symbolic, after the fashion of *Ahasvérus* (**1833**) by Edgar Quinet (1803-1875); the Jew stands for dispossessed labourers and his consort Herodias for downtrodden womankind (Sue was a feminist of sorts as well as a radical socialist). [BS]

SULLIVAN, [Sir] ARTHUR (SEYMOUR) (1842-1900) UK composer. ◊ W.S. GILBERT; OPERA.

SUMMER ◊ SEASONS.

SUN As the most visible of all the heavenly bodies, and the one whose effects are most immediately felt by humans upon the planet, the Sun has always been a central actor in the great, meaningful show of the heavens. It is a central component in most CREATION MYTHS and in MYTHOLOGY in general. It is the central star in the ZODIAC around which the calculation of ASTROLOGY circulate; and is invoked in various forms of OCCULTISM. When thought of as taking godly form, it is usually male, and more often than not is either a SYMBOL of the father of the gods or may actually *be* GOD.

Modern fantasy – perhaps surprisingly – makes little use of the drama of the heavens which the Sun dominates (although in John GRANT's *Albion* [**1991**] the ruling elite worship the Sun as the creator god, and thus justify their oppression of the peasants, who worship no gods), concentrating rather on the much more assimilable MOON. Most GENRE FANTASIES set in FANTASYLANDS presume, without cosmological comment, a Sun very much like our own; in DYING-EARTH tales, its brightness is dimmed. The new star in Gene WOLFE's *The Book of the New Sun* (**1980-3**) is described, in sf terms, as a "white hole", though Severian's own association with the Sun god APOLLO weds sf and fantasy. Some writers with a metaphysical bent, like A.A. ATTANASIO in *The Dragon and the Unicorn* (**1994**), may treat the Sun as a generative (and conscious) principle of creation or destruction. Full-scale SECONDARY-WORLD fantasies, like J.R.R. TOLKIEN's *The Lord of the Rings* (**1954-5**), may also express some attitude toward the Sun, which in Tolkien's text is specifically female (and hence, perhaps, readily obscured as a PORTENT of woe to come to men) while the Moon is male; and some, like Jenny JONES's **Flight Over Fire** sequence, may treat the Sun as a being. But most often, in modern fantasy, the Sun is merely a source of heat and light.

Various metaphorical and literal uses of the Sun appear in *Chasing the Sun: A Journey Around the World in Verse* (anth **1992** chap) ed Sally Bacon. [JC]

See also: ICARUS.

SUNKEN LANDS ◊ ATLANTIS; FLOOD; LOST LANDS AND CONTINENTS.

SUPERBOY US syndicated tv series (1988-92). Viacom. **Pr** Robert Simmonds. **Exec pr** Alexander Salkind, Ilya Salkind. **Dir** Colin Chilvers and many others. **Writers** Elliot Anderson, David Gerrold (◊ SFE), Ilya Salkind and many others. **Based on** the COMIC-book characters created by Joe Shuster (1914-1992) and Jerry Siegel (1914-1996). **Starring** Jim Calvert (T.J. White 1988), Gerard Christopher (Clark Kent/Superboy 1989-92), Stacy Haiduk (Lana Lang), John Haymes Newton (Clark Kent/Superboy 1988), Sherman Howard (Lex Luthor 1989), Scott Wells (Lex Luthor 1988). 100 30min episodes. Colour.

The Salkinds, who had produced the SUPERMAN MOVIES, tried their hands one more time with this adventure series based on a younger version of the Man of Steel. The series, freely changing the comics' long-standing history of Superboy, began with Clark Kent enrolling in college as a journalism major. There he became friends with Lana Lang and T.J. White, the latter the son of Perry White of *Daily Planet* fame. The second season found a new actor playing Superboy (the first having suffered several run-ins with real police officers) and a new roommate, who often got Clark/Superboy involved in his questionable business dealings. Along the way Superboy had to deal with a host of problems that included MUMMIES, Metallo (a villain powered by kryptonite), Mr Mxyzptlk (an imp from the fifth dimension) and the apparent return from the dead of his parents from Krypton. The series shifted format again some years later when Clark and Lana left college and worked as interns at the Bureau for Extranormal Affairs; this brought them into contact with even stranger beings, including GHOSTS, WEREWOLVES, VAMPIRES and a parade of creatures from other dimensions. The series is probably most remembered today for Haiduk's performance as Lana.   [BC]

SUPERGIRL UK movie (*1984*). ◊ SUPERMAN MOVIES.

SUPERHEROES Fantasy characters with superhuman powers visually marked out from the rest of humanity by their fantastical apparel. First devised for exploitation in COMIC books, they are, although since featured in CINEMA and tv, still largely confined to that medium, their particular appeal for comics being their colourful and dramatically expressive costumes. Variations on the theme have proliferated to such an extent that the superhero has become virtually an ICON.

In the early 1930s, writer Jerome Siegel (1914-1996) and artist Joe Shuster (1914-1992) conceived SUPERMAN, the first superhero, loosely based on Philip Wylie's *Gladiator* (*1930*), as a newspaper strip, but were unable to interest anyone; the character did not see publication until *Action Comics #1* (1938). In this first 13-page story most of the defining characteristics of the superhero, which now look like clichés but were at the time highly original, were established. The story tells of an infant sent to Earth by rocketship from a doomed distant planet whose inhabitants have mental and physical abilities far superior to humans'. The child grows up to realize his superhuman powers and decides to dedicate them to the cause of justice. He keeps his unique abilities a secret from the world by leading a dual existence, using his powers only when wearing his distinctive costume. The feature caught the national imagination and *Action Comics*'s publisher, National (◊ DC COMICS), aggressively sued anyone attempting to publish any similar feature (◊ CAPTAIN MARVEL).

The characteristics which have come to define the typical superhero were thus established at the outset: the possession of one or several superhuman abilities; the use of these abilities against injustice and criminality; a dramatic costume expressing the unique powers; a perceived need to hide these powers from the general public, necessitating the establishment of an "ordinary" persona to hide the secret identity, so that the superhero becomes, in effect, a PARIAH ELITE; consequent difficulties in personal relationships, especially romantic ones. In the comic books there is a reluctance to attribute the special abilities to MAGIC: almost always there is a cod-science rationale; thus superheroes are generally TECHNOFANTASY figures.

Soon characters with superhuman strength, amazing physical skills and the ability to fly appeared in a great number of comic books. Timely Comics (later called MARVEL COMICS) introduced two durable superheroes: **The Sub-Mariner**, whose initial hatred of humans was transformed into a patriotic hatred of Nazis during WWII and who later took to crimefighting, and **The Human Torch**, an android with the ability to set himself and others ablaze (both started in *Marvel Comics #1*, 1939). A long list of superheroes appeared in the succeeding months – **The Shield, The Flash**, The SPECTRE, **Green Lantern, Captain America, Bulletman, Hourman**, etc. – plus superheroines, beginning in 1941 with WONDER WOMAN. In general the stories were simplistic tales: the superhero pitted his or her abilities against the VILLAINS, and won. This worked well during wartime, and superhero comics were enormously popular among members of the armed forces, but with the coming of peace readers turned more towards romance, Western, crime and horror comics. Superman and BATMAN were given a boost by the release of three cinema serials from Columbia Pictures: *Superman* (*1948*) and *Superman vs*

The Atom Man (*1950*) (◊ SUPERMAN MOVIES), and *Batman* (*1948*) (◊ BATMAN MOVIES), but few other superheroes survived this period.

The backlash against the perceived damaging influence of crime and horror comics in the early 1950s eventually created a second wave of popularity for the superhero. DC, who had continued throughout to publish **Superman, Batman** and **Wonder Woman**, began a revival with the creation of **The Legion of Superheroes** (◊ SUPERHERO TEAMS) in *Adventure Comics #247* (1958), the reader response to which encouraged them to revive many moribund superheroes. **The Green Arrow, The Flash, Hawkman, Aquaman** and **The Justice League** were resurrected and revamped. Superheroines, too, were given an overhaul, with **Wonder Woman** receiving a new origin story and makeover and **Supergirl** making her first appearance. Stan LEE at Marvel Comics enticed artist Jack Kirby away from DC, and together they began to create a line of superheroes which were to prove among the most popular comic-book characters of the 1960s and 1970s, beginning with **The Fantastic Four** (in *Fantasic Four #1*, 1961). **Spiderman, Thor** and others followed, all occasionally meeting up or finding themselves in conflict with one another in what became known as The Marvel Universe. The stories now featured monsters and costumed supervillains to give the plots more variety. The personalities of these super-individuals were developed and relationships between them were explored. But this process of humanizing the superheroes – of giving them personal problems – eventually backfired; sales again declined, and the early 1970s saw a doldrums. Attempts to maintain reader interest included many teamups and crossover stories, until DC and Marvel joined forces to produce a large-format comic book pitting their most popular superheroes against each other: *Superman versus The Amazing Spiderman* (1976). This was the first of several such cooperations. A further boost was given to many superheroes by the screening of tv shows and movies starring such long-established characters as **The Hulk, Wonder Woman, Spiderman** and **Superman**. Another means used to improve sales was to employ one of a number of particularly talented artists to breathe new life into failing characters: Neal ADAMS on **Batman** and **Green Lantern**, John Byrne on **The X-Men**, George Perez on **The Teen Titans** and Frank MILLER on **Daredevil**.

The late 1970s saw the beginning of changes in the distribution and retailing of comic books which affected their content. Street newsstands were closing and the chainstores often did not stock comics. In the early 1980s, new outlets began to appear in the form of specialist shops stocking comic books published by an increasing number of small, more creator-oriented publishing companies. A market for more mature and thought-provoking comic books emerged. Talented new writers and artists were re-examining and experimenting with the superhero mythos. Responding to the trend, DC, published two seminal GRAPHIC NOVELS: *Batman: The Dark Knight Returns* (#1-#4 1986; graph coll **1987**) by Frank Miller and *Watchmen* (#1-#12 1986-7; graph coll **1987**) by Alan MOORE. Both these works placed superheroes in the real world and examined the resultant ramifications. Since the publication of these two books, ideas associated with the superhero have continued to develop in comic books, if not in other media.

The modern comic-book superhero is not the universally applauded figure he once was: he is now the springboard for a complex array of subtle and profound ideas and a fertile visual image for the expression of and comment upon all aspects of the human condition.   [RT]

SUPERHERO TEAMS The notion of forming teams of SUPERHEROES was first used by Sheldon Mayer (1917-) and Gardner F. FOX when they created **The Justice Society of America** (later conveniently referred to as the JSA) in *All Star Comics #3* (1940). This group began with eight members, all well established characters in their own right: **Flash**, **Green Lantern**, **Hawkman**, **Hourman**, SANDMAN, **Dr Fate**, **The** SPECTRE and **The Atom**. The membership varied considerably – BATMAN, SUPERMAN, WONDER WOMAN and many others pitched in. The feature lasted until *All Star #57* (1951). With the second superhero boom in the 1960s, the notion was revived with **The Justice League of America** (the JLA) in *Brave and Bold #28* (1960), a similar team with some of the same members. These features gave Fox the opportunity to deal with troubled relationships between superpeople, which he handled expertly; he even established an annual team-up of teams after reviving the original JSA in 1963.

Imitators soon followed, many comics companies publishing books which starred two or more of their characters in the same story, though the membership of most such teams was fairly fluid. DC eventually used one of its titles, *Brave and Bold*, almost entirely for the purpose of teaming up various otherwise lone superheroes. This changed, however, with the creation of **The Legion of Superheroes** (*Adventure Comics #267* 1958), which featured a group of adolescents from the 30th century: Cosmic Boy, Lightning Lad and Saturn Girl.

A milestone in the history of STs was reached with the publication of MARVEL COMICS's *Fantastic Four #1* (1961), which featured a tightly knit team of supercharacters with "real" personality clashes and oddball dialogue. This formula was so successful (and Lee proved so adept at writing original dialogue for them) that other teams were soon formed, including **The Avengers** (in *The Avengers #1* 1963), another team-up of established characters, and **The X-Men** (in *The X-Men #1* 1963), a very original team of mutants; the latter, though it has undergone many changes, has remained popular ever since.

The ST remains a regular feature in comic books, but it is unusual for any team to last long, and although such teams remain a staple of children's animated tv series, they are less popular in the comic books of the 1990s, where standards of storytelling and characterization are higher. [RT]

See also: TEENAGE MUTANT NINJA TURTLES.

SUPERMAN The first and best-known SUPERHERO of US COMIC books. The character was created in the early 1930s by writer Jerry Siegel (1914-1996) and artist Joseph Shuster (1914-1992), and tagged "The Man of Steel". Based on ideas created by Philip WYLIE in his book *Gladiator* (1930), the feature was hawked around the newspaper syndicates for several years before bought by publisher Harry Donnenfield (for $130) for publication in *Action Comics #1* (1938). This story consisted of a brief introduction followed by a narrative made up of rearranged frames from the unsold newspaper version.

Born on the doomed planet Krypton, whose natives have mental and physical abilities far greater than the human, the infant Kal-El is launched towards Earth as a spectacular cataclysm destroys Krypton. He is adopted by a kindly couple, the Kents, who name him Clark. As he grows, he discovers that he can "hurdle skyscrapers, leap an eighth of a mile, run faster than a streamline train, and nothing less than a bursting shell can penetrate his skin!". His foster-parents advise him to hide his prodigious physical prowess; on their death he decides to "turn his titanic strength into channels that will benefit mankind. And so was created Superman, champion of the oppressed, the physical marvel who has sworn to devote his existence to those in need." As mild-mannered Clark Kent he works as a reporter for the *Daily Planet* newspaper, but dresses in a blue costume with red trunks, boots and cloak whenever he perceives a need for his remarkable abilities.

These abilities subsequently included flight and X-ray vision: indeed, he became so invulnerable that the stories became predictable and simplistic, and potentially fatal fragments of his home planet – green kryptonite – had to be introduced to provide plot variation. Various differently coloured kryptonites were added later.

From the outset, the character was enormously popular, with both daily and Sunday newspaper strips from 1939 and a radio show from 1941 (it was here that green kryptonite was first featured). The newspaper version was soon taken over by artist Wayne Boring (1916-1986), whose drawing was more accomplished than Shuster's, and the plots became more sophisticated than those in the comic books. This phenomenal success continued throughout the 1940s, but decline during the 1950s, and the newspaper feature was terminated in 1967. In the comic books, however, Superman remained popular. Shuster's rather crude drawing was quickly superseded: Curt Swan, Murphy Anderson (1926-), Neal ADAMS and others contributed much to the development and continued popularity of Superman, with scripts by talented writers like Alfred Bester (1913-1987), Edmond Hamilton (1904-1977), Henry KUTTNER and Manly Wade WELLMAN.

The difficulty of maintaining a high level of plot originality with so eminently invincible a character was a perennial problem. Super-VILLAINS whose powers matched Superman's had to be introduced; writer/editor Mort Weisinger (1915-1978) created the mad scientist Lex Luthor, the weird and surreal Mr Mxyzlptlk, and other eccentric adversaries. Striving to retain reader interest, the feature became increasingly implausible during the 1970s, leading ultimately to Frank MILLER's lampoon of Superman in his BATMAN graphic novel *The Dark Knight Returns* (graph **1986**). Publishers DC COMICS determined to rationalize Superman, doing so first in *Crisis on Infinite Earths* (*#1-#12* 1985-6) written by Marv Wolfman (1946-), and subsequently with a complete overhaul of the character by writer/artist John Byrne (1950-) beginning in *Adventures of Superman #424* (1987), in which finite limits were set upon Superman's powers and abilities: he could no longer TIME TRAVEL or go at the speed of light, nor survive in space longer than he could hold his breath. The ploy worked, largely through Byrne's originality and attractive drawing, and the announcement of Superman's engagement to long-time girlfriend Lois Lane (who finally twigged Clark's secret identity in 1990) was made much of in DC's publicity. This storyline did not, however, end as expected at the altar, but in the death of Superman (in *Man of Steel #18* 1993), with a further surge of publicity. A free memorial armband was included in *Superman Vol II #75* (1993) along with a clipping from the *Daily Planet*. All DC's superheroes went into mourning in a crossover storyline, "Death of a Friend". **Superman** titles were suspended until later that year, when "Reign of the Supermen" began in *Adventures of Superman #500*, culminating in the return of the "Last Son of Krypton" in *Superman Vol II #82* (1993). Superman, haunted by memories of his own death, obsessively pursues his killer, Doomsday, in the mini-series *Superman/Doomsday: Hunter/Prey (#1-#3* 1994).

DC has succeeded in retaining reader interest in Superman against what must be recognized as considerable odds. Despite the vast profits they had harvested, however, it was not until 1975 that they were finally persuaded to pay creators Siegel and Shuster (the latter by then blind and the former working as a clerk-typist) an annual stipend. The creators' names must now appear on every **Superman** story. [RT]

See also: *The* ADVENTURES OF LOIS AND CLARK (1993-current); *The* ADVENTURES OF SUPERMAN (1951-7); MAGIC WORDS; SUPERBOY (1988-92); SUPERHEROES; SUPERHERO TEAMS; SUPERMAN MOVIES; Roy THOMAS.

SUPERMAN MOVIES There have been numerous movies – and several tv series – based on the creations of Jerry Siegel (1914-1996) and Joe Shuster (1914-1992).

1. Series of 17 short ANIMATED MOVIES (*1941-43*). Paramount/Fleischer (*#1-#9*) and Paramount/Famous (*#10-#17*), by arrangement with Action Comics and Superman Magazines. **Pr** Max FLEISCHER. **Dir** Dave Fleischer except as noted. **Voice actors** Joan Alexander (Lois Lane), Clayton Collyer (Superman/Clark Kent). Each about 8$^{1}/_{2}$ mins. Colour.

The plots of these cartoons almost all share various similarities: Lois and Clark Kent are sent to investigate some threat or potential threat; they get separated; Lois's foolhardiness gets her into trouble; Clark says "This looks like a job for Superman" and transforms himself; as Superman he rescues Lois and saves the day; she gets the story and mocks Clark for his inefficiency or faintheartedness. Only in *#3* does Lois step out of her stereotype to help fight the menace; even then she must in due course be saved by the Man of Steel. The animation is quite good. The scripts rely on low-brow adventure rather than wit or imaginative invention. The series runs: *#1: Superman* (*1941*; vt *The Mad Scientist*); *#2: The Mechanical Monsters* (*1941*); *#3: Billion Dollar Limited* (*1942*); *#4: The Arctic Giant* (*1942*); *#5: The Bulleteers* (*1942*); *#6: The Magnetic Telescope* (*1942*); *#7: Electric Earthquake* (*1942*); *#8: Volcano* (*1942*); *#9: Terror on the Midway* (*1942*); *#10: Japoteurs* (*1942*) dir Seymour Kneitel; *#11: Showdown* (*1942*) dir Isidore Sparber; *#12: Eleventh Hour* (*1942*) dir Dan Gordon; *#13: Destruction, Inc.* (*1942*) dir Sparber; *#14: The Mummy Strikes* (*1943*) dir Sparber; *#15: Jungle Drums* (*1943*) dir Gordon; *#16: Underground World* (*1943*) dir Kneitel; *#17: Secret Agent* (*1943*) dir Kneitel.

2. Two serial movies and a feature were released between 1949 and 1951. These were:

Superman US serial movie (*1948*). Columbia. **Pr** Sam Katzman. **Dir** Spencer Gordon Bennet. **Screenplay** Lewis Clay, Royal K. Cole, Arthur Hoerl. **Starring** Kirk Alyn (Kent/Superman), Tommy Bond (Jimmy Olsen), Carol Forman (Spider Woman), Noel Neill (Lois Lane), Pierre Watkin (Perry White). 15 episodes. B/w.

Spider Woman seeks – aided by a ray and a piece of kryptonite – to destroy Superman and rule the world. Produced on a shoestring – visibly so – this serial was hugely successful.

Atom Man vs Superman US serial movie (*1950*). Columbia. **Pr** Sam Katzman. **Dir** Spencer Gordon Bennet. **Screenplay** David Mathews, George H. Plympton, Joseph Poland. **Starring** Kirk Alyn (Kent/Superman), Tommy Bond (Jimmy Olsen), Noel Neill (Lois Lane), Lyle Talbot (Lex Luthor), Pierre Watkin (Perry White). 15 episodes. B/w.

Luthor assails Superman, sacks Metropolis and abducts Lois. For Luthor this ends in tears. Despite the success of its predecessor, the budget on this movie was even tighter – and again it shows.

Superman and the Mole Men (vt *Superman and the Strange People*) US movie (*1951*). Lippert. **Pr** Robert Maxwell, Barney Sarecky. **Dir** Lee Sholem. **Screenplay** Maxwell (as Richard Fielding). **Starring** Phyllis Coates (Lois Lane), George Reeves (Kent/Superman). 67 mins. B/w.

The 1950s equivalent of a series pilot tvm (it was released theatrically but to promote the tv series *Superman*), this has little interest except for its anti-racist message. Superman saves the Mole Men from massacre by bigots. [JG]

3. *Superman: The Movie* US/UK movie (*1978*). Dovemead/International Film Production/Alexander & Ilya Salkind/Warner. **Pr** Pierre Spengler. **Exec pr** Ilya Salkind. **Dir** Richard Donner. **Spfx** Colin Chilvers. **Vfx** Roy Field. **Model fx** Derek Meddings, Brian Smithies. **Screenplay** Robert Benton, Norman Enfield, David Newman, Leslie Newman, Mario Puzo. **Novelization** *Superman: Last Son of Krypton* * (*1978*) by Elliot S. Magin. **Starring** Marlon Brando (Jor-El), Jackie Cooper (Perry White), Jeff East (Young Clark), Glenn Ford (Jonathan Kent), Gene Hackman (Lex Luthor), Margot Kidder (Lois Lane), Marc McClure (Jimmy Olsen), Christopher Reeve (Kent/Superman), Phyllis Thaxter (Martha Kent), Susannah York (Lara). 143 mins. Colour.

The planet Krypton is doomed. Jor-El and Lara send their baby son, along with a green crystal, to safety on Earth. He is adopted by the elderly, childless Kent couple, who call him Clark. At age 18 he is guided by the crystal to the Arctic, where it builds an ice palace and summons up Jor-El's personality to educate the boy for 12 long years. Afterwards, he takes a job in late-1970s Metropolis as a junior reporter on the *Daily Planet*, under editor Perry and alongside staffer Lois and cub photographer Jimmy. Although strictly forbidden by his father from interfering with human history, Clark impetuously transmutes into Superman to save Lois from a helicopter crash; he goes on a crimebusting spree. Evil genius Lex Luthor deflects two test-flying ICBMs to the San Andreas Fault as part of an estate scam. Superman thwarts him by flying into orbit, reversing the Earth's spin and thereby "turning back the clock" to a time before the missiles struck.

S, despite sf trappings, is a self-confessed fantasy, with Superman explicitly equating himself with PETER PAN, who "flew with children, Lois, in a FAIRYTALE". The variable spfx, produced by a huge team, are at their best excellent; Reeve plays his dual role with panache. Tucked away in cameo roles (although, irritatingly, given star billing) are such names as Trevor Howard and, in an introduction that cleverly trails **4**, Sarah Douglas and Terence Stamp; bit parts are played by Kirk Alyn and Noel Neill, who were Superman and Lois in the serial movies (◊ **2**). Lois is here rendered as a selfish, mercenary smartass, stupidly rash rather than pluckily intrepid. *S* is quality garbage, and blockbusted. [JG]

4. *Superman II* UK movie (**1980**). Dovemead/ International Film Production/Alexander & Ilya Salkind/ Warner. **Pr** Pierre Spengler. **Exec pr** Ilya Salkind. **Dir** Richard Donner (uncredited), Richard Lester. **Miniature fx** Derek Meddings. **Screenplay** David Newman, Leslie Newman, Mario Puzo. **Starring** Jackie Cooper (Perry White), Sarah Douglas (Ursa), Gene Hackman (Lex Luthor), Margot Kidder (Lois), Marc McClure (Jimmy Olsen), Jack O'Halloran (Kon), Christopher Reeve (Kent/Superman), Terence Stamp (Zod), Susannah York (Lara). 127 mins. Colour.

At the start of **3** three criminals – Zod, Ursa and Kon – were despatched from Krypton into space in a two-dimensional "cage" (the Phantom Zone). Now Superman, in despatching a terrorist bomb into space, shatters the "cage", allowing the trio to come to Earth, which they immediately decide to conquer. Lois, unmasking Superman's secret, accompanies him to his Arctic ice palace; there he is told by a projection of his mother, Lara, that if he wishes to live with Lois he must lose his superpowers and IMMORTALITY, an option he gladly embraces. Soon, however, he discovers he must reassume his powers to defeat Zod. The world is saved. When Lois complains of her emotional confusion knowing both Clark and Superman, he strips her of the relevant memories and reverts to being willing doormat Clark.

Partway through *SII* Donner, dir of **3**, quit, and Lester picked up the reins; it is difficult to identify who was responsible for what. *SII* has a more interesting premise than **3** and does not have to trace Superman's early history; it is marked by good performances from Reeve (as usual), Kidder (a refreshingly mellower Lois than in **3**), Stamp and Douglas. Nevertheless, it has longueurs absent from the earlier movie, and the spfx are significantly less impressive. Inappropriate (and implausible) slapstick sometimes destroys tension. The result lacks entirely the mythic power which **3** occasionally attains. [JG]

5. *Superman III* UK movie (**1983**). Dovemead/ Cantharus/Alexander & Ilya Salkind/Warner. **Pr** Pierre Spengler. **Exec pr** Ilya Salkind. **Dir** Richard Lester. **Spfx/miniatures** Colin Chilvers. **Optical/vfx sv** Roy Field. **Screenplay** David Newman, Leslie Newman. **Novelization** *Superman III* * (**1983**) by William KOTZWINKLE. **Starring** Jackie Cooper (Perry White), Margot Kidder (Lois Lane), Marc McClure (Jimmy Olsen), Annette O'Toole (Lana Lang), Richard Pryor (Gus Gorman), Christopher Reeve (Kent/Superman), Annie Ross (Vera Webster), Robert Vaughn (Ross Webster). 125 mins. Colour.

Played mainly for laughs, this self-parody sees computing idiot savant Gus being used by megalomaniac plutocrats Ross and Vera Webster to further schemes first to destroy the coffee industry of Colombia (a plan frustrated by Superman) and then to take over the world's oil supply. Meanwhile Clark has returned to Smallville for his high-school reunion and become entangled with one-time sweetheart Lana Lang, cameoed as a schoolgirl in **3** (Lois hardly appears in *SIII*). The Websters produce synthetic kryptonite, which changes his personality, so that, in effect, his evil TWIN emerges. All is righted in the end.

Although too comedic to be a satisfying movie, *SIII* has great fantasy interest. Aside from the evil-twin subplot, there are liberal doses of TECHNOFANTASY. Moreover, its tackling of a core problem of SUPERHERO adventures – that they are foregone conclusions unless the protagonists lose their INVULNERABILITY – is more imaginative and more fully

ramified than **4**'s similar attempt. However, despite **6** and **7**, this was really the end of the line for the series. [JG]

6. *Supergirl* UK movie (**1984**). Cantharus/Alexander Salkind. **Pr** Timothy Burrill. **Exec pr** Ilya Salkind. **Dir** Jeannot Szwarc. **Vfx** Roy Field, Derek Meddings. **Screenplay** David Odell. **Based on** the **Supergirl** COMICS. **Novelization** *Supergirl* * (**1984**) by Norma Fox Mazer. **Starring** Hart Bochner (Ethan), Peter Cook (Nigel), Faye Dunaway (Selena), Marc McClure (Jimmy Olsen), Peter O'Toole (Zaltar), Helen Slater (Kara/Linda Lee/Supergirl), Maureen Teefy (Lucy Lane), Brenda Vaccaro (Bianca). 124 mins. Colour.

Superman's cousin Kara lives in Argo City, whose savant is Zaltar. Through foolishness, Zaltar and Kara doom Argo City by losing its single power source, the Omegahedron, which hurtles Earthwards, falling into the hands of CARNI-VAL witch (indeed, self-styled WITCH of Endor) Selena, who uses it to confer on herself IMMORTALITY, and with slimy warlock Nigel and folksy fellow-witch Bianca plots world domination. Zaltar condemns himself to eternal penance in the extradimensional Phantom Zone, but Kara pursues the Omegahedron. On Earth she disguises herself as a high-school girl, Linda Lee, and finds herself rooming with Lucy Lane, younger sister of Lois and girlfriend of Jimmy Olsen. Selena, her powers of MAGIC growing steadily, conjures a love POTION to enslave hunky, inarticulate landscape gardener Ethan, but the SPELL misfires, and he falls instead for "Linda", in the process becoming a suave poet. After much more such stuff the USA is saved, the Omegahedron is recovered and Supergirl returns to Argo City, leaving a lovelorn Ethan.

Slater is an exceptionally charming Supergirl – the movie would have been lost without her – and the spfx are excellent. While there are some excellent – and excellently real-ized – flights of imagination, too often there is a tendency to default to an easy option. Supergirl's characterization is inconsistent: she is at one moment prepared to die fighting the good fight, the next wailing in childish fear. Dunaway suffers more from such inconsistency: her comic-cut role subverts her attempts to convey evil power and presence. Yet *S* is the most exuberantly fantasticated of the cycle begun with **3**: while it is less than the sum of its parts, not all of those parts are insignificant. [JG]

7. *Superman IV: The Quest for Peace* US movie (**1987**). Warner/Cannon/Golan-Globus. **Pr** Yoram Globus, Menahem Golan. **Exec pr** Michael J. Kagan. **Dir** Sidney J. Furie. **Vfx** Harrison Ellenshaw. **Screenplay** Lawrence Konner, Mark Rosenthal. **Novelization** *Superman IV* * (**1987**) by B.B. Hiller. **Starring** Jackie Cooper (Perry White), Gene Hackman (Lex Luthor), Mariel Hemingway (Lacy Warfield), Margot Kidder (Lois Lane), Marc McClure (Jimmy Olsen), Mark Pillow (Nuclear Man), Christopher Reeve (Kent/Superman), Sam Wanamaker (David Warfield). 89 mins. Colour.

International tension is high, and a schoolboy writes to Superman for a solution; Superman's response is to announce that he is now an Earthling rather than an alien and to filch all the world's nuclear missiles and herd them into the Sun. Meanwhile Lex Luthor clones from a strand of Superman's hair the evil Nuclear Man, superpowered in sunlight but not in shade. Much of *SIV* is taken up with battles between Nuclear Man and our hero (to whom it never occurs that he could end things by simply throwing a blanket over his adversary). All villains defeated, Superman rousingly broadcasts that the gift of world peace is not something for him or any single individual to bestow: it is up to all of us.

The story for *SIV* was largely generated by Reeve, who agreed (with a fresh production company) to make one last movie in the series if it had a "message". Sadly, this message is set in a clumsy context, and thus largely lost; in common with other Golan-Globus productions, there appear to be bits missing from the final cut, and the spfx are often poor. *SIV* is not a fine movie, but it is better than generally acknowledged. [JG]

SUPER MARIO BROS. US movie (*1993*). Lightmotive/ Allied/Cinergi. **Pr** Jake Eberts, Roland Joffé. **Dir** Annabel Jankel, Rocky Morton. **Vfx** Christopher Francis Woods. **Screenplay** Parker Bennet, Terry Runté, Ed Solomon. **Based on** concept and characters created by Shigeru Miyamoto and Takashi Tezuka, of Nintendo. **Novelization** *Super Mario Bros.* * (*1993*) by Todd Strasser (1950-   ). **Starring** Dennis Hopper (Koopa), Bob Hoskins (Mario Mario), John Leguizamo (Luigi Mario), Samantha Mathis (Daisy), Fiona Shaw (Lena). 104 mins. Colour.

A TECHNOFANTASY based on the best-selling console game. The meteorite that hit the Earth 65 million years ago did not so much extinguish the DINOSAURS as shift them into a parallel dimension, where they have evolved into humanoid form. 25 years ago tyrant Koopa usurped the throne, devolving the old king into a ubiquitous fungus. However, a loyal subject smuggled an egg of the king's through the interdimensional barrier and left it with a Brooklyn nunnery. The child of that egg is now archaeologist Daisy. But Koopa seeks this PRINCESS, for she bears the last fragment of the meteorite; if it is set in place with the rest (like the shard in *The* DARK CRYSTAL [*1982*]), the two REALITIES will be fused and he can conquer ours – besides, the tyrant seeks to make her Daisy Koopa. Two of his hoods drag her back to the ALTERNATE WORLD, and plumbers Mario and Luigi follow. Aided by the sentient fungus, they have a string of adventures reminiscent of a multiple-choice arcade game – collecting PLOT COUPONS and deliberating over options – all complicated by the desire of Koopa's rejected mistress Lena to use the rock-fragment for her own ends. As Lena struggles to fuse the two realities there are scenes reminiscent of the extraction of the Spike of Power in COOL WORLD (*1992*) before the villains are thwarted and the triumphant plumbers return to modern Brooklyn. Daisy, however, must be left in the reptilian dimension to sort out her kingdom . . . only then she appears in the plumbers' apartment, gun in hand, to set up an ending differing only in wording from that of *Back to the Future* (*1985*).

SMB seems a cynical exercise. [JG]

SUPERNATURAL CREATURES ◊ MYTHICAL CREATURES.

SUPERNATURAL FICTION Any story whose premises contradict the rules of the mundane world can be defined as supernatural fiction, but a definition so broad would logically incorporate all categories of FANTASY, all nonmundane HORROR, all TECHNOFANTASY and all SCIENCE FANTASY, and arguably all SCIENCE FICTION. It therefore makes sense to use the term more restrictively.

In SF the natural world is the base reality, and SFs take their argument from that base reality, even when they end by contradicting, transcending or teaching lessons to the base reality. SF is, therefore, more closely allied to science fiction than to fantasy. The supernatural world is *other* than the real world, and is generally seen as signalling WRONGNESS,

though in much OCCULT FANTASY, and in revisionist versions of (in particular) the VAMPIRE tale, the signals may be reversed, so that the supernatural world represents a higher rightness. The supernatural exists, therefore, in a contingent relationship to base reality, even though the relationship may be one of PARODY. This *contingency* of the supernatural element distinguishes the form from the central line of 20th-century fantasy: HIGH-FANTASY texts like J.R.R. TOLKIEN's *The Lord of the Rings* (**1954-5**) treat the SECONDARY WORLD as effectively autonomous; CROSSHATCH texts like Lord DUNSANY's *The King of Elfland's Daughter* (**1924**) similarly treat the OTHERWORLD as existing in its own right; and even tales set primarily in the real world, like Thorne SMITH's *The Night Life of the Gods* (**1931**), treat the fantastic as having a full, floating claim to independence.

Stories described as SFs – they begin historically with GOTHIC FANTASIES and include GHOST STORIES, stories of daemonic enthralment (◊ DIONYSUS; PAN), stories of the occult, WITCHCRAFT tales, stories involving satanic rites or PACTS WITH THE DEVIL, tales of POSSESSION, vampire tales, WEREWOLF tales (and some other tales of THERIOMORPHY) and much SLICK FANTASY – tend to *invade* or contradict the natural order to which they are ultimately bound. Usually narrated from a vantage-point situated in the real world, rather than from the vantage-point of the invading entity or influence, SFs generally reflect an initial disbelief in the incursion (or belief as a form of blasphemy), resistance to the violating supernatural element (or surrender to it, often sexual) or horror (or loathly wedlock). A sense of WRONGNESS pervades 20th-century fantasies when the secondary world or otherworld (or this world itself, in FANTASIES OF HISTORY and INSTAURATION FANTASIES) is threatened, often from within (◊◊ MALIGN SLEEPER, THINNING); in SF, on the other hand, wrongness accompanies threats to the real world from elsewhere, through incursions or wellings up of the supernatural. Everett F. BLEILER, in *The Guide to Supernatural Fiction* (**1983**), prefers the term "contranatural" to supernatural; it is a term which clearly conveys this essential quality of violation.

Unlike fantasy, SFs tend to pay relatively little attention to STORY, and certainly slight any conception of story as being a means of bringing attention to the MATTER of the world – hence, perhaps, Michael ENDE's use of a werewolf in *The Neverending Story* (**1979**) to tell the hero that beyond the protective realm of the secondary world he is a "neverending story" only in the sense that he is a lie. In fantasy, wrongness threatens the Story, and the successful telling (or finding, or recovery) of a Story almost invariably constitutes good news for the world (◊ HEALING); to tell the Story is to welcome the outcome. The opposite is generally the case in SFs, whose plots tend to expose dreaded (or wrongly longed-for) anomalies which are violating the world and which must be expunged, cast out or wed; if an SF changes the world permanently, it is likely to be by means of a showdown APOCALYPSE.

Owing perhaps to the problematic relationship that tends to exist between the real world and the invading supernatural element, SFs very frequently incorporate ongoing arguments meant to *explain* the invading element, often in a tone of elect knowingness. In many SFs – not always deliberately – this didacticism about the supernatural creates an atmosphere of doubt. This irresolution is sometimes a blessing – many of the best ghost stories turn on arguments, often never resolved, as to whether or not the ghost is real (i.e., they are fantasies of PERCEPTION) – but overall the ontological insecurity of SF as a mode can easily vitiate the pleasure of the tale, however delectable the insecurity may sometimes be. In fantasy, mysteries may abound – but almost invariably *within the frame* of the tale. We may not understand what we have read, but we continue to read within a fantasy frame (◊ FANTASY; Tzvetan TODOROV).

When the invasion of the mundane world comes together with explanations designed to justify or promulgate the principles underlying the invasion, SF tends to become a literature of *seduction*, at the level of either argument or action, or both. Like its close sibling, HORROR – and increasingly so the more its subject matter resembles pure horror – SF is a literature concerned with the body, with violations of the body, with conversions and immurements and seductions of the body. Compared with (until recently) most fantasy, there is a large amount of SEX in SF, a sense that the relationship between the supernatural and the mundane may best be understood in terms of the minglings of flesh. In fantasy wrongness can often be identified as a threat to STORY; in SFs wrongness can often be identified as seduction with evil intent.

At the end of the 18th century, and for much of the 19th, SF was astonishingly popular; in *The First Gothics* (**1987**) Frederick S. Frank (1935-   ) estimates that as many as 5000 Gothic novels appeared between Horace WALPOLE's *The Castle of Otranto* (**1764**) and Charles MATURIN's *Melmoth the Wanderer* (**1920**); of these, at least several hundred had supernatural elements. And, insofar as the Gothic novel serves as one of the central repositories of motif, location and plot for modern fantasy, SF itself can be seen as an essential incubator of the fully achieved 20th-century field – as the format through which MYTH, LEGEND, FOLKLORE, FAIRYTALES and the literary DREAM became available to the conscious fantasist.

As the 19th century progressed, SFs (mostly by this time ghost stories and tales of the occult) gradually began to separate itself from early works of fantasy by writers like Lewis CARROLL and George MACDONALD. Writers like Edgar Allan POE, Nathaniel HAWTHORNE, Edward BULWER-LYTTON, Sheridan LE FANU, E. NESBIT (in her tales for adults) and Arthur MACHEN were essentially writers of SFs. The most interesting 20th-century SFs have been defined – retroactively – as horror; but although that term may apply with some justice to the work of H.P. LOVECRAFT, it is more difficult to think of the ghost stories of M.R. JAMES, E.F. BENSON or Robert AICKMAN as readily understandable under that rubric. And, as the century progressed, it became increasingly difficult to give single labels to any writer of significance. Lovecraft himself must also be treated as a fantasy writer; and contemporary authors who have been justly associated with supernatural horror fiction – like Clive BARKER, Stephen KING and Peter STRAUB – are also creators of autonomous SECONDARY WORLDS.

Indeed, it might be argued that SF – like the genre sf that flourished from 1925 to 1965 or so – has lost over recent decades any secure platform in an agreed reality. The world at the end of the 20th century may well incorporate too many futures – too many intersecting versions of what is real – for a fiction to flourish which depends upon a dominating relationship between the "real" world and other worlds contingent upon it. Where SF (with the exception of the ghost story) once tended to look upon the past with

apprehension, it may be the case that BELATEDNESS now defines the form, and has become its solace.   [JC]

SUPERNATURAL STORIES ◊ MAGAZINES.

SUPER POWERS US tv series (1985). ◊ *The* ADVENTURES OF BATMAN AND ROBIN (1992-4).

SUPERPUP ◊ *The* ADVENTURES OF SUPERMAN (1951-7).

SUPERSTITION An irrational belief, often leading to a minor TABOO with an accompanying restorative RITUAL – for example, spilling salt is unlucky, but the bad luck may be averted by tossing a pinch over one's left shoulder. The inevitable twist in fantasy (and, more often, DARK FANTASY and HORROR) is for superstitions to be reinterpreted as rational attitudes. Libations (◊ SACRIFICE) will supposedly placate the GODS: in John BRUNNER's *The Traveler in Black* (coll **1971**) something unseen does indeed consume the poured-out wine before it reaches the floor. To the superstitious, numbers like THIRTEEN and 666 (◊ GREAT BEAST) are unlucky; in Terry PRATCHETT's *The Colour of Magic* (**1983**), saying "eight" may *in fact* invoke the unpleasant GOD Bel-Shamharoth whose number this is. Pessimistic superstitions about things going wrong – Murphy's Law, "whatever can go wrong, will" – are actualized in fantasy as literal GREMLINS, like Piers ANTHONY's WIZARD Murphy in *Castle Roogna* (**1979**), whose power is to make others' efforts go awry. Further examples abound. Superstition is the theme of Peter CROWTHER's **Narrow Houses** anthology sequence.   [DRL]

See also: MIRRORS; MOON; NUMEROLOGY; OMENS; PORTENTS.

SURREALISM Although the term may first have been coined by Guillaume APOLLINAIRE, Surrealism as an aesthetic movement dates from the *Manifeste du Surréalisme* ["Manifesto on Surrealism"] (**1924**; exp 1929) by André Breton (1896-1966), where a revolutionary art is promulgated which would eventually subsume the bourgeois rationalism of traditional realism. Influenced by the more nihilistic pre-WWI Dada movement, by Freudian theory and by individual artists like Giorgio di Chirico (1888-1978), Surrealism celebrated DREAM imagery and the putative "truth" of the unconscious, achieving its characteristic effects through the apparently irrational and unmotivated juxtaposition of realistic and fantastic images. Major painters included Salvador DALI, Max ERNST, René MAGRITTE and Yves Tanguy (1900-1955), with a more abstract version represented by Hans Arp (1887-1966) and Joan Miró (1893-1983).

Its deliberate randomness – which included "automatic" writing or drawing and "found" objects or poems – generally sets Surrealism apart from more traditional modes of fantasy, although the term itself is often appropriated to describe any number of works using incongruous image patterns, or mixing FANTASY with REALITY (as in MAGIC REALISM).

While a few novels – e.g., *The Eater of Darkness* (**1926**) by Robert M. Coates (1897-1973) – were written directly under the influence of Surrealist doctrine, the movement has generally survived better in poetry than in narrative, though it has had a substantial influence in the CINEMA (e.g., Luis BUÑUEL, Jean COCTEAU) and the theatre (e.g., Antonin Artaud [1896-1948], Eugene IONESCO). A number of sf and fantasy authors have used Surrealist patterns of imagery as a technique of estrangement: *City of the Iron Fish* (**1994**) by Simon Ings (1965-   ) is a recent example. Other authors – e.g., Lisa GOLDSTEIN in *The Dream Years* (**1985**) and Robert IRWIN in *Exquisite Corpse* (**1995**) – have incorporated aspects of the historical movement itself into works of fiction. [GW]

SURVIVOR, THE Australian movie (*1980*). ◊ James HERBERT.

SÜSKIND, PATRICK (1949-   ) German writer and playwright, best-known in English for *Das Parfum: die Geschichte eines Moerders* (**1985**; trans John E Woods as *Perfume: The Story of a Murderer* **1986** UK), a mordant tale of an 18th-century French orphan whose preternaturally acute sense of smell leads to obsession and serial murder (◊ SERIAL KILLERS); it won the WORLD FANTASY AWARD in 1987. *Die Taube* (**1987** chap; trans John E. Woods as *The Pigeon* **1988** US) tells of a man whose uneventful life is shaken to its foundations when he encounters a pigeon outside his hallway. Both works evoke the ineffable in the particulars of the phenomenal world, which PS dramatizes with grim relish. [GF]
Other works (selective): *Der Kontrabass* (**1984** chap; trans Michael Hofmann as *The Double Bass* 1987 UK), a play; *Die Geschichte von Herrn Sommer* (**1991**; trans John E Woods as *The Story of Mr Sommer* 1992 UK; vt *Mr Summer's Story* **1993** US).

SUSPENSE ◊ MAGAZINES.

SUTCLIFF, ROSEMARY (1920-1992) UK writer. The ability to create a realistic historical novel for children is in a sense one of the most testing challenges of the fantasist's art. RS's masterpieces of historical fiction are vivid re-creations rather than attempts to portray historical fact through story. While rarely straying beyond the boundaries of what *could* have happened in the later centuries of Roman rule in Britain and the succeeding Dark Ages, RS, like her mentor Rudyard KIPLING, set herself to describe history as part of a temporal tapestry. Thus *The Eagle of the Ninth* (**1954**), while containing at least one darkly numinous and certainly trans-real scene when its hero Marcus discovers the lost legion's missing standard in a British shrine, is told very much in the manner of a tale of the Imperial Northwest Frontier, with Marcus in the role of English subaltern and the Druid-inspired uprisings reminiscent of Indian struggles against the Raj. Suceeding novels, such as *The Silver Branch* (**1957**) and *The Lantern Bearers* (**1959**), portray Marcus's descendants, with the Romans and British developing elements of each other's culture and facing another wave of conquest and immigration from the Anglo-Saxons. We see the beginning of the MATTER of Britain in the latter novel and in the adult novel *Sword at Sunset* (**1964**) which depicts Artos, illegitimate nephew of Ambrosius the High King, as a warlord fighting the Saxon tribes to keep alight the memory of the Romano-British nation. In the events RS describes – especially the ambiguity of Artos's relationship with the incest-born Medraut and his use of the moon daisy, element of the White Goddess, to unite old faiths and new at the Battle of Badon – are both the ARTHUR of the later chroniclers and the seed of the later romances.

RS turned again and again to this period, revisiting the Romano-British "frontier" in *Mark of the Horse Lord* (**1965**) – a fierce study of espionage and assumed identity – and *Frontier Wolf* (**1980**). She retold Saxon and Irish legends such as *Beowulf* (**1961**; vt *Dragon Slayer* 1966) and *The Hound of Ulster* (**1963**), and returned to the Dark Ages in *The Shining Company* (**1990**) – based on the Welsh poem *The Gododdin* – and in retellings of the Arthur-story in *The Sword and the Circle* (**1981**) and *The Road to Camlann* (**1981**). She also wrote about Greece in *The Flowers of Adonis* (**1965**),

a study of the Athenian Alcibades which provides a multi-faceted picture of a charming but hollow genius. Apart from the **Marcus** sequence, though, perhaps her finest novel – and certainly the most akin to fantasy – is *Warrior Scarlet* (**1958**), in which Drem, a boy of a Bronze Age tribe, overcomes the disability of a withered arm to become a warrior. Within the limits of a book for children this is as powerful as possible a picture of a putative shamanistic society, with the sun-worshipping Golden People contrasted with the outcast Half People from whom they have wrested the LAND.

Apart from occasional suggestions of paranormal powers, RS remains a realistic writer, exploring the history of our here and now. But her imagination was powerful enough to create startling pictures of what could have been. [AS]
Other works: *The Armourer's House* (**1951**); *Brother Dusty-feet* (**1952**); *Simon* (**1953**); *Outcast* (**1955**); *Lady in Waiting* (**1957**); *The Bridge-Builders* (**1959**); *The Rider of the White Horse* (**1959**); *Knight's Fee* (**1960**); *Dawn Wind* (**1962**); *The Shield Ring* (**1962**); *Sword at Sunset* (**1964**); *The Chief's Daughter* (**1967**); *The High Deeds of Finn Mac Cool* (**1967**); *A Circlet of Oak Leaves* (**1968**); *The Witch's Brat* (**1970**); *Tristan and Iseult* (**1971**); *The Truce of the Games* (**1971**); *The Capricorn Bracelet* (**1973**); *The Changeling* (**1974**); *We Lived in Drumfyvie* (**1975**); *Blood Feud* (**1977**); *Shifting Sands* (**1977**); *Sun Horse, Moon Horse* (**1978**); *Song for a Dark Queen* (**1979**); *The Light beyond the Forest: The Quest for the Holy Grail* (**1980**).
Further reading: *Blue Remembered Hills* (**1983**), autobiography.
See also: CHILDREN'S FANTASY; GAWAIN AND THE GREEN KNIGHT; GHOST STORY (*1974*); GRAIL; MORRIGAN; MYTHS; NORDIC FANTASY; ROBIN HOOD.

SVANKMAJER, JAN (1934-   ) Czech director who uses a mixture of live action, animation and stop-motion animation (◊ ANIMATED MOVIES), puppetry and surrealist narrative techniques (◊ SURREALISM) to create movies that are peculiarly unsettling. His features to date are ALICE (*1988*; ot *Neco z Alenky*), based loosely on Lewis CARROLL'S **Alice** series, and FAUST (*1994*), based on diverse renditions of the FAUST legend. Like Andrei Tarkovsky (1932-1986), JS displays a deliberative fascination with his material that some find tedious, others infectious. [JG]

ŠVANTNER, FRANTIŠEK (1912-1950) Slovak writer. ◊ SLOVAKIA.

SVENGALI 1. *Svengali* US movie (*1931*). Warner. **Dir** Archie Mayo. **Screenplay** J. Grubb Alexander. **Based on** *Trilby* (**1894**) by George du Maurier (1834-1896). **Starring** John Barrymore (Svengali), Marian Marsh (Trilby). 81 mins. B/w.

This first movie version of du Maurier's classic (the plot is as in **2**) gives Barrymore plenty of opportunities to ham things up, much as Wolfit would; the movie is less vital when he is absent from the screen. The sets are excellent, and deft use is made of camera angles to create fantastication. Hollywood rarely paid much attention to what the European moviemakers were up to; this is an exception. Marsh is suitably distracting. [JG]

2. *Svengali* UK movie (*1954*). Renown/Alderdale. **Pr** George Minter. **Dir** Noel Langley. **Screenplay** Langley. **Starring** Derek Bond (Sandy, The Laird), Hubert Gregg (Durien), David Kossoff (Gecko), Terence Morgan (Billy), Hildegarde Neff (Trilby), Paul Rogers (Taffy), Harry Secombe (Barizel), Donald Wolfit (Svengali). **Voice actor**

Elizabeth Schwarzkopf (Trilby singing). 82 mins. Colour.

17-year-old Trilby becomes a model for the sculptor Durien, and makes friends with the three young English artists in the adjacent studio, falling in love with one of them, Billy; through them she also meets the street musicians Svengali and Gecko. Billy's family refuse to let him marry Trilby; in grief he hits Svengali, who pronounces a gypsy CURSE on him, and sure enough soon after Billy is crippled in a street accident. Svengali, who has hypnotized (◊ MESMERISM) Trilby once before to cure a headache, now assumes total dominance over her, telling her she shall become a great singer because, in effect, he will be doing the singing though the voice will come from her throat. As La Svengali she has a triumphant operatic career, but when she comes to London to sing at Covent Garden he has a heart attack mid-performance, and her operatic powers vanish abruptly. As he had before promised her, Svengali's death means that she too shall die – but Billy arrives on the scene just in time to lure her back to life with his love.

This imbalanced movie concentrates too much on the romance between Billy and Trilby at the expense of the story proper, and is also constrained by sexual primness. Wolfit hams his role to the rafters in an attempt to enliven to the proceedings, but Neff is so vapid that it is impossible to credit her as a heartbreaker. [JG]

3. *Svengali* US movie (*1983* tvm). Viacom. **Pr** Robert Halmi. **Dir** Anthony Harvey. **Screenplay** Frank Cucci. **Based on** a story by Sue Grafton, itself based loosely on du Maurier. **Starring** Jodie Foster (Zoe Alexander), Peter O'Toole (Anton J. Bosnyak). 100 mins. Colour.

An updating of the tale which ruthlessly defantasticates the original. Plucked from obscurity, singer Alexander achieves superstardom thanks to vocal coach Bosnyak, in whose absence she finds it impossible to sing. They become lovers but, when he cracks her obsession by deceiving her into thinking he was at a concert when he was not, she storms out. She attempts to return, but he explains his task is done: he has taught her to be an "intact" (i.e., autonomous) human being. The movie is affecting, but somewhat lacking in narrative impetus. [JG]

SWAHN, SVEN CHRISTER (1933-   ) Swedish author and critic. A prolific and versatile author of poetry, drama, short stories, novels, criticism and essays, he is without doubt the foremost living Swedish sf/fantasy author. Only two of his over 50 books have been translated into English. In fantasy, Swahn has written *Jag lovar dig* ["I Promise You"] (**1979**), a burlesque fantasy detailing a visit to a sunken city where a mythical BOOK of BLACK MAGIC has found its resting place. *Biskop Hattos torn* ["The Tower of Bishop Hatto"] (**1960**), a mythical story set in the 10th century, is based on the legend of the Rhine Rat Tower. *Stenjätten* ["The Stone Giant"] (**1965**) is an apocalyptic story centred on the STATUE of the giant Finn in Lund Cathedral; since Finn is probably the world's smallest giant, the entire story is humorously rendered as a miniature. *Jakten på Stora Sjörmen* ["The Hunt for the Great Sea Snake"] (**1974**) is a fantasy version of Herman MELVILLE's *Moby-Dick* (**1851**), with the whale swapped for a traditional SEA MONSTER; it is a psychologically complex and endlessly fascinating depiction of obsession and fate. *Havsporten* (**1975**; trans as *The Island Through the Gate*) tells of young Mikael's visit to the unknown ISLAND of Oberour, off Brittany, where he spends a magical, rapturous and fearful summer. *Skymningsgästerna* (**1977**; trans as *The Twilight*

Visitors **1980**) is a chillingly effective DOPPELGÄNGER story. *Stenbrottet* ["The Quarry"] (**1987**) is a quiet, wishful novel of love, set in the last few months before WWII, when Klas meets Lisa, who has the gift of MAGIC and can reach what she believes to be another world, but which Klas suspects is the realm of the dead. Finally, *Spöket på Myntgatan* ["The Coin Street Ghost"] (**1992**) is a ghost story set in 1710 and full of brilliantly realized and often hilarious historical detail.

Apart from these novels, SCS's major fantastic radio play *Kaspar Hauses fjärde dröm* ["The Fourth Dream of Kaspar Hauser"] should be mentioned. In addition, SCS has written a number of fantasy short stories, particularly GHOST STORIES, and some of his poetry is clearly of fantastic interest. He has dealt also with fantasy both as a critic and a translator. [J-HH]

Other works (selective): *Indianresan* ["The Red Indian Trip"] (**1957**), juvenile; *Tretton historier om spöken och annat* ["Thirteen Stories of Ghosts and Other Things"] (coll **1958**); *Mina kära döda* ["My Dear Dead Ones"] (coll **1977**); *Stanna alla klockor* ["Stop All Clocks"] (coll **1991**). [J-HH]

SWAMP THING Moss- and muck-encrusted vegetable monster star of US COMIC books, created by writer Len Wein and artist Bernie WRIGHTSON in *House of Secrets #92* (1971). Dr Alex Olsen is grotesquely mutated into ST after a disastrous explosion in his science laboratory, arranged by his jealous assistant, Damian Ridge, and returns to wreak vengeance. Wein rewrote and expanded upon this origin story for a new comic-book series (*Swamp Thing #1-#24* 1972-6), renaming the scientist Dr Alec Holland and describing the project on which he is working when the "accident" occurs as a "Biorestorative Formula". The first 10 issues of this series, also illustrated by Wrightson, made effective use of the contrast between the repulsive appearance of ST and his gentle, benign character; they were reprinted as *Roots of the Swamp Thing* (1986).

ST was resurrected for a new series, *Saga of the Swamp Thing* (1982-current), in which the character was used as a fairly uninteresting SUPERHERO, but was developed in depth and complexity by Alan MOORE (from #20 on). Subsequent writers have used ST as an ecological icon or a godlike figure representing the interests of the Earth's flora and fauna. Movie incarnations are *Swamp Thing* (*1982*) dir Wes Craven and *Return of the Swamp Thing* (*1989*; vt *Swamp Thing II*). [RT]

SWANN, THOMAS BURNETT (1928-1976) US poet, novelist and academic who taught English literature at Florida Atlantic University before turning to full-time writing in the late 1960s, after publishing some sentimental poetry, beginning with *Driftwood* (coll **1952** chap) for a vanity press; his academic studies included *Wonder and Whimsy: The Fantastic World of Christina Rossetti* (**1960**), *The Classical World of H.D.* (**1962**), *Ernest Dowson* (**1964**), *The Ungirt Runner: Charles Sorley, Poet of World War I* (**1965**) – who is homaged in *The Goat Without Horns* (see below) – and *A.A. Milne* (**1972**). As a critic he was personal and committed; most of his subjects were childhood loves, to whom he remained loyal.

Almost all his fiction – beginning with "Winged Victory" for *Fantastic Universe* in 1958 – fits into a single vision of the course of Western history, and can be seen as comprising a sustained meditation on the theme of THINNING, viewed through a reiterated central story in which the matriarchal, prelapsarian old order – represented by "Beasts", including MINOTAURS, fauns (◊ SATYRS), SIBYLS, DRYADS, HALFLINGS and occasional highly significant appeareances by the god PAN – is destroyed by the world-devouring patriarchy of the Achaeans, or Romans, or Christians. There are several venues – ancient Egypt, Crete, Rome, medieval Britain – but all have a similar LAND-OF-FABLE relationship to the mundane world, whose geography they rarely violate, and the general history of which is reinterpreted rather than ignored. Most of the novels describe RITES OF PASSAGE of children into ambivalent maturity; it is arguable that TBS saw adulthood and thinning as very similar conditions.

The order of publication of individual volumes of TBS's work is distinctly confusing. His series were all published in reverse chronological order, and the chronology of the overall meditation is likewise jumbled. It seems appropriate, therefore to follow the internal chronology compiled by Bob Roehm and published by Robert A. Collins in *Thomas Burnett Swann: A Brief Critical Biography and Annotated Bibliography* (**1979** chap).

The happiest of the novels is – naturally enough – the one set first. *The Minikins of Yam* (**1976**) follows the QUEST, a couple of millennia BC, of a young Pharaoh (in the company of a "minikin", a pert young girl-like figure typical of TBS's females) who must find out why his land has been thinning drastically. His father (in TBS's fiction a father was almost invariably a negative figure) has, it turns out, banished MAGIC, and with it the regenerative power of the Mother (◊ GODDESS). He reverses the edict, and EGYPT is saved.

Set in the mountains of Crete, the **Minotaur** sequence – *Cry Silver Bells* (**1977**), *The Forest of Forever* (**1971**) and TBS's first novel, *Day of the Minotaur* (1964-5 *Science Fantasy* as "The Blue Monkey"; **1966**) – is less idyllic. The three volumes are a litany of loss, as first the Cretans, then the Achaeans, relentlessly shrink the mountain POLDER of the folk, who include most of the fabulous creatures inhabiting the twilight regions of Classical MYTHOLOGY – centaurs, dryads and others of that ilk. In the end, Eunostos the Minotaur sets sail, with two children and other survivors, towards the Isles of the Blest. *Moondust* (**1968**) and *How Are the Mighty Fallen* (**1974**) are both set in Biblical Israel. In the first a CHANGELING has SEX in Jericho with an Israelite spy, and the walls tumble. The second tale occurs at the time of Saul and David. Jonathan is a GODDESS-worshipper and David's homosexual lover; he is soon killed and David is exiled.

With TBS's second sequence, the **Latium** series – *Queens Walk in the Dusk* (**1977**), *Green Phoenix* (**1972**) and *Lady of the Bees* (1962 *Science Fantasy* as "Where Is the Bird of Fire?"; exp **1976**) – the picture continues to darken, the polders to shrink. The first volume recounts the tragic story of Dido, who is betrayed by Aeneas, a forward-looking patriarch *in utero*. The viewpoint character of the remaining volumes is a dryad named Mellonia, who ages slowly, through heart-wrenching liaisons with short-lived mortals, into the time of Romulus and Remus, falling in love with the latter, who is halfling-like (and doomed). Two further novels – *Wolfwinter* (**1972**) and *The Weirwoods* (1965 *Science Fantasy*; **1967**) – are likewise set in Roman times. In the first, a sibyl recounts her adventures, climaxing in a pathos-ridden LOVE affair with an extremely short-lived faun; in the second, some humans learn lessons from the increasingly marginalized elder folk. The stories assembled in *The Dolphin and the Deep* (coll **1968**) also focus on this period, while those in *Where is the Bird of Fire?* (coll **1970**) range further ahead.

The Gods Abide (**1976**) constitutes something of a counterattack. It is set in the 4th century, when Christianity is beginning in earnest to root out previous Mediterranean faiths. But the Goddess attempts to save her followers from a GOD who (despite lamb's clothing) remains the Old Testament Yahweh at heart; and, although there is no real hope in the south, or even in Britain (where Christianity is beginning to take root), She creates a "Not-World" for her people, an ALTERNATE-REALITY polder where they will be forever safe.

TBS's remaining novels are set in the human world, into which occasional memories of the GOLDEN AGE intrude. They include: *The Tournament of Thorns* (fixup **1976**), set in the Middle Ages, *Will-o-the-Wisp* (1974 *Fantastic Stories*; dated 1976 but **1977** UK), which subjects the poet Robert Herrick (1591-1674) to experiences in the old world; *The Not-World* (**1975**), which introduces Elizabeth Barrett Browning (1806-1861) to the polder created by the Goddess in *The Gods Abide*; and *The Goat without Horns* (**1971**), which is set in the late 19th century, as close as TBS came to the present. The latter describes the love of a dolphin for Charlie – based on Charles Sorley (1895-1915) – a glowingly beautiful young man who arrives at a Caribbean ISLAND to tutor a young girl, falls in love with the mother, is saved from a were-shark by Gloomer the dolphin, and finally goes off with Gloomer to a secret enclave.

It is easy to mock TBS for sentimentality, for displacement of attention from adult sexuality to innocent relationships between boys and friendly older men, and for his sense that the best contrast to the reductions of history was a clambake of beasts in Arcadia. But the intensity of his sense of BELATEDNESS is at times overwhelming. [JC]

Further reading: *Thomas Burnett Swann: A Brief Critical Biography and Annotated Bibliography* (**1979** chap) by Robert A. Collins; "Thomas Burnett Swann" by John CLUTE in *Supernatural Fiction Writers* (anth **1985** 2 vols) ed E.F. BLEILER.

SWANWICK, MICHAEL (JENKINS) (1950-   ) US writer, most of whose work is sf, though some of his stories are fantasy, including "The Dragon Line" (1989 *Asimov's Science Fiction Magazine*), in which MERLIN, having survived into the modern world, fails in his attempt to save us from terminal pollution. The VAMPIRE featured in *In the Drift* (fixup **1985**) is in fact a mutant whose digestive tract is deficient.

MS's only novel of genuine fantasy interest is *The Iron Dragon's Daughter* (**1993** UK), a savage REVISIONIST FANTASY which can be understood as an exposé of FAERIE (or, more precisely, of readers' wish-fulfilment images of free-lunch OTHERWORLDS) as well as a remarkably ruthless deconstruction of the RITE-OF-PASSAGE structure of so many contemporary fantasy tales whose young female protagonists gain empowerment through MENARCHE. Passing disastrously from one REALITY to another, the CHANGELING anti-heroine undergoes four separate rites of passage, four versions of the central fantasy STORY whose protagonists escape BONDAGE through self-RECOGNITION, but she never truly comes to know herself, although her SHADOW, a kind of halfling, knows her well enough. The first section is set in a TECHNOFANTASY nightmare venue where war DRAGONS are manufactured at great human cost, and which she escapes only through the loss of a COMPANION's life. In the second – set in a cod CONTEMPORARY-FANTASY quasi-urban USA where one may experience TIME IN FAERIE at the very heart of the shopping mall – she is complicit in the ritual death by fire of the high-school "year Queen" (◊ GOLDEN BOUGH). In the third – set in a desolate CITY dominated by a university which gives degrees in ALCHEMY – she learns the true loss involved in using SEX to gain control over MAGIC. And in the fourth she has an ambiguous confrontation with the GODDESS who may rule the Universe through an EDIFICE which (conventionally) *is* a model of the Universe. She ends the book as a mortal woman in a mortal world: her reward is to be human.

There are few great anti-fantasies in the literature: *The Iron Dragon's Daughter*, more thoroughly perhaps than any predecessor, is one. It is a book which counts the costs of fantasy. [JC]

SWEET, DARRELL K. (1934-   ) US illustrator who began his career with covers for Ballantine Books, becoming the main cover artist for the firm after the inception of the Del Rey imprint. He has been associated, as a result, with Del Rey writers like Piers ANTHONY, Terry BROOKS, Jack L. CHALKER and Stephen R. DONALDSON. His cleanly lined, rounded, technically competent style – in which attractive females and insinuating MONSTERS feature in sharply characterized LANDSCAPES – has become one of the default representations of FANTASYLAND and its inhabitants for many readers. In recent years he has also done fantasy covers for Ace Books and Tor Books. [JC]

SWIFT, JONATHAN (1667-1745) Writer, satirist and cleric, of Yorkshire descent but born and raised in Ireland; most noted as the author of *Travels into Several Remote Nations of the World* (**1726** 2 vols; many subsequent editions, often cut; eventual vt *Gulliver's Travels* 1821). *Gulliver's Travels*, firmly in the tradition of TRAVELLERS' TALES, was a remarkable feat in the creation of imaginary worlds as a vehicle for SATIRE upon the political and religious establishments of the day. It works as both FANTASY and SCIENCE FICTION, for it requires the suspension of belief engendered by FANTASY while relying on the verisimilitude encouraged by scientific probity for Swift's satirical barbs to strike home. Its acceptance, and its later devolution into CHILDREN'S FANTASY, demonstrates that the fantastic elements were the most immediate, which is how the book is remembered today. Gulliver's experiences in Lilliput and Brobdingnag (among DWARFS and GIANTS) work almost on the level of FAIRYTALE, but the later travels to Laputa and the land of the Houynhnhnms are less popular because they lack the creation of "Sense of Wonder" in their intellectual endeavour to militate against, respectively, scientific and capitalist institutions. Nevertheless it is in the section on Laputa that JS allows the supernatural to intrude. On the island of Glubbdubdrib Gulliver encounters a community of sorcerers who can summon the spirits of the dead, allowing him to converse with Alexander, Julius Caesar, Aristotle and others. On the island of Luggnagg he meets the Struldbrugs, who are immortals (◊ IMMORTALITY), though they become senile in their 80s. *Gulliver's Travels* was immediately popular and spawned many SEQUELS BY OTHER HANDS, starting with the pseudonymous *Travels into Several Remote Nations of the World*, Vol. 3 (**1727**) as by Lemuel Gulliver, which took Gulliver back to Brobdingnag.

JS was a prolific essayist and pamphleteer. His early career, as secretary for Sir William Temple (1628-1699) for most of the period 1689-99, brought him into contact with the work of Charles PERRAULT, who had introduced an argument about the relative merits of ancient and modern

writers in a series of books starting in 1688. Temple extended the argument in his essay "Upon the Ancient and Modern Learning" (1690; in *Miscellania* **1692**), which JS would have transcribed, and this prompted JS to write the FABLE "An Account of a Battel Between the Antient and Modern Books in St James's Library" (written *c*1697; in *A Tale of a Tub* **1704**; vt *A Full and True Account of the Battel Fought last Friday between the Antient and the Modern Books in St James's Library* **1710** chap), usually known as "The Battle of the Books", in which the authors of renowned books take sides in a battle over the cause. JS may well have been the first writer in English to use the fairytale technique of Perrault in *A Tale of a Tub* (written 1696; **1704**) which has at its core a simple narrative of a father who has triplets and, upon his death, leaves them each a coat which will grow with them and last them all their lives, provided they don't change it. This FRAME STORY allowed JS to digress into a series of discourses in which he could attack the establishment. It was the publication of this work (albeit anonymous) that established JS's reputation.

JS used aspects of the fantastic in other SATIRES, such as *A Discourse Concerning the Mechanical Operation of the Spirit* (1704 in *A Tale of a Tub*; **1710** chap), where he challenged the growing interest in metaphysics, and most notably in the creation of the character Isaac Bickerstaff, who wrote a series of mock prophecies in *Predictions for the Ensuing Year* (**1708** chap), the first of his **Bickerstaff Papers**. [MA]

Further reading: *Gulliver's Travels, The Tale of a Tub, and The Battle of the Books* (coll **1919**); other compendia of interest are *A Tale of a Tub, The Battle of the Books and Other Satires* (coll **1909**; rev vt *A Tale of a Tub and other Satires* 1975) ed Kathleen Williams; *Jonathan Swift* (coll **1984**) ed Angus Ross and David Woolley.

SWINBURNE, ALGERNON (CHARLES) (1837-1909) UK poet. ◊ ARTHUR; LIFESTYLE FANTASY; SMALL PRESSES; SPELLS.

SWITCH US movie (*1991*). Columbia Tri-Star/Odyssey-Regency/Home Box Office/Cinema Plus/Beco. **Pr** Tony Adams. **Exec pr** Arnon Milchan, Patrick Wachsberger. **Dir** Blake Edwards. **Vfx** Michael Owens. **Screenplay** Edwards. **Starring** Ellen Barkin (Amanda Brooks), Bruce Martyn Payne (Devil), Jimmy Smits (Walter Stone). 103 mins. Colour.

Bullying, foul-mouthed, manipulative, chauvinist adman Steve is murdered by three ex-mistresses. In PURGATORY, he finds GOD undecided if he should go to HEAVEN or HELL, and is given a chance through REINCARNATION to avoid the latter if he can find just one female who likes him; the DEVIL intervenes, pointing out that Steve might brutalize yet another unsuspecting female into adoring him, and so Steve is reincarnated as bullying, foul-mouthed, manipulative Amanda. Blackmailing her way into Steve's old job, she finds herself surrounded by chauvinist males much like her old self, but her strong streak of homophobia bars her from taking up a lesbian relationship. The Devil offers her a job as a recruiter of SOULS or as mother of his child (referring to ROSEMARY'S BABY [*1968*]), but she rejects him. Instead, one drunken night she sleeps with Steve's best buddy Walter, becoming pregnant. She gives birth to a girl but dies – the baby being, of course, the one female who likes her.

S is a limp sex comedy, overburdened with profanity and leaving most of its plot's potentials unexplored: the most significant running gag is Amanda's difficulty in wearing high heels. [JG]

SWITHIN, ANTONY (?   -   ) UK writer, resident in Canada, whose published work to date consists of the four volumes of the **Perilous Quest for Lyonesse** series: *Princes of Sandastre* (**1990**), *The Lords of the Stoney Mountains* (**1991**), *The Winds of the Wastelands* (**1992**) and *The Nine Gods of Safaddne* (**1993**). Medieval historical adventures with fantasy ingredients, centred on an imaginary culture posited as existing on the real-world islet of Rockall (west of Ireland), they have more to do with ATLANTIS than with the traditional Arthurian Lyonesse. [DP]

SWORD AND SORCERESS US ANTHOLOGY series ed Marion Zimmer BRADLEY, 12 approximately annual vols to date: *Sword and Sorceress* (anth **1984**), *Sword and Sorceress II* (anth **1985**), *III* (anth **1986**), *IV* (anth **1987**), *V* (anth **1988**), *VI* (anth **1990**), *VII* (anth **1990**), *VIII* (anth **1991**), *IX* (anth **1992**), *X* (anth **1993**), *XI* (anth **1994**), *XII* (anth **1995**). The series concentrates on SWORD AND SORCERY and puts emphasis on female protagonists, without the stories necessarily being feminist (◊ FEMINISM) in theme or moral. Most contributors have been women. Bradley has used the anthologies to develop new writers, a crusade she has continued in MARION ZIMMER BRADLEY'S FANTASY MAGAZINE, which is similar to SAS in its emphasis on stories of wizardry and retribution. Regular contributors are C.J. CHERRYH, Phyllis Ann KARR, Mercedes LACKEY, Diana L. PAXSON, Jennifer Roberson (1953-   ), Laura Underwood (1954-   ) and Deborah Wheeler (1947-   ). [MA]

SWORD AND SORCERY In 1961 Michael MOORCOCK requested a term to describe the fantasy subgenre featuring muscular HEROES in violent conflict with a variety of VILLAINS, chiefly WIZARDS, WITCHES, evil SPIRITS and other creatures whose powers are – unlike the hero's – supernatural in origin. Fritz LEIBER suggested "Sword and Sorcery", and this term stuck. It is commonly considered synonymous with HEROIC FANTASY; both are aspects of ADVENTURER FANTASY, a term covering not only tales of the sort Moorcock specified but also stories whose protagonists need not be heroes. If adventurer fantasies are about men and women of all degrees (except the highest) attempting to survive, and make their livings, in both LANDS OF FABLE and FANTASY-LAND settings, then S&S might designate tales of heroes whose adventures occur in Lands of Fable (like CONAN's HYPERBOREA), while heroic fantasy might designate tales of heroes adventuring through SECONDARY WORLDS, like the Nehwon of Leiber's **Fahrd and the Gray Mouser** books. Such a distinction is over-finicky for practical use: this encyclopedia generally employs the time-honoured term S&S for adventurer fantasies fitting Moorcock's original specification. Some critics use the term dismissively or pejoratively; this book regards it as merely descriptive.

The true father of S&S may be an author no longer widely known for his stories of supernatural adventures: Alexandre DUMAS. As the most influential creator of swashbuckling historical adventure tales, Dumas provided a model not only for writers like Rafael Sabatini (1875-1950), who followed in his footsteps with works like *Scaramouche* (**1921**) – wherein swordsmanship (◊ SWORDS) was specifically valued – but also for PULP writers in every genre, including fantasy and WEIRD FICTION. Robert E. HOWARD, who was deeply versed in the ways and opportunities of pulp, was inevitably influenced by Dumas or Dumas's successors.

It is possible, therefore, to see Howard's **Conan** as a figure combining characteristics of two characters – D'Artagnan and Porthos – from Dumas's most famous single swashbuckler,

The Three Musketeers (**1844**; trans **1846** UK) with Auguste Maquet (anon). D'Artagnan is impulsive, intelligent, a superb athlete, successful with women, romantic, and totally involved in an exciting world – 17th-century France – which he has no wish to alter in any significant fashion; Porthos is vast, chthonic, almost supernatural in his physical strength and attributes, a trencherman, loyal to his friends, indifferent to the social order. Together, they make a very passable S&S hero. Dumas ages his heroes in sequels, and eventually sees them to their various deaths; but the Three Musketeers (plus One) remain, in the vision of our dreams, eternally young. Similarly, the hero (and from an early stage the heroine) of S&S does not age in the mind's eye; he or she fights on, for ever, in the dawn of a day we do not wish to end.

The venues of S&S are superficially various, but tend to default to a template geography which – whether a Land of Fable, a Fantasyland, ATLANTIS, a DYING EARTH, some PLANETARY-ROMANCE venue, etc. – boasts a range of usable features: swamps, mountains, rivers, INNS, villages, MINES, EDIFICES, walled CITIES. There will also be an emphasis on BORDERLANDS and WATER MARGINS: the typical S&S geography is almost necessarily designed to open gateways or PORTALS into what might negatively be called more of the same but what in competent hands may be understood as a constantly unrolling tapestry of LANDSCAPE. The geography of S&S is designed as an arena for heroes and heroines who awake each morning at the beginning of their lives; it is designed for more to happen.

Though written in a dense, highly literary, archaizing style that had little influence on the form, E.R. EDDISON's *The Worm Ouroboros* (**1922**) can stand as a model for this close interrelationship between the nature of the S&S hero and the design of the land through which s/he adventures. The book is shaped, as its title indicates, to return upon itself: once the heroes win the battle, the CYCLE of threat and AGON begins anew, to the manifest relief of everyone involved. There is no ageing in the book; it provides the clearest possible articulation of that underlying TEMPLATE structure of S&S which distinguishes it from the fantasy of writers like J.R.R. TOLKIEN. Another S&S precursor with the timeless quality of MYTH is *The Fortress Unvanquishable, Save for Sacnoth* (**1910** chap) by Lord DUNSANY.

But consensus S&S, being a genre perceived as born from PULP, begins with authors like Edgar Rice BURROUGHS – no matter that his **Barsoom** planetary-romance sequence, launched in 1912, is technically sf – and Robert E. Howard, who flourished in the 1920s and 1930s. S&S rapidly matured through the work of 1930s writers like Clark Ashton SMITH, whose **Zothique** tales, beginning with "The Empire of the Necromancers" (1932), launched the Dying Earth as a venue; C.L. MOORE, whose **Jirel of Joiry** stories introduced the first significant S&S heroine (◊ GENDER), Henry KUTTNER, whose *Elak of Atlantis* (1938-41 *WT*; **1985**) brought in ATLANTIS as an S&S venue, and Fritz Leiber, whose **Fahrd and the Gray Mouser** sequence features the best-loved S&S DUO of complementary characters, one arguably D'Artagnan-like and the other Porthos-like. Also important to the later spread and genre identity of S&S were the **Eric John Stark** stories by Leigh BRACKETT, beginning in 1949.

Four anthologies – *Swords and Sorcery* (anth **1963**), *The Spell of Seven* (anth **1965**), *The Fantastic Swordsmen* (anth **1967**) and *Warlocks and Warriors* (anth **1970**) – were assembled by L. Sprague DE CAMP from the extensive pulp literature of the 1930s and 1940s, and were crucial to the return to prominence of the subgenre from the 1960s onwards. Others imitated de Camp's example (◊ ANTHOLOGIES), including Lin CARTER with *Flashing Swords!* (anth **1973**) and its four successors, which promoted S&S through the 1970s to 1981.

In the 1960s Moorcock started publishing his **Elric** stories, and found the generation of such template heroic-fantasy sequences – many of real merit – a useful way of financing *New Worlds* magazine while he was its editor. These stories were seen by him as an explicit homage to those parts of the earlier pulp tradition that he valued, notably the works of Leiber . . . who was encouraged during Cele Goldsmith's 1958-65 editorship of FANTASTIC to produce more **Fafhrd and Gray Mouser** stories.

In developed S&S, protagonists may often act heroically, but are generally happy to do a deal – as in Michael SHEA's *Nifft the Lean* (**1982**), whose hero escapes HELL by bartering with a DEMON and swapping an unlovable COMPANION for the chance of freedom. The ultimate confrontation with the Lovecraftian MALIGN SLEEPERS of Barbara HAMBLY's **Darwath** trilogy (**1982-3**) leads not to their destruction but to persuading them to leave for a warmer world where they will be more comfortable; the war between, broadly, GOOD AND EVIL in Glen COOK's first **Black Company** trilogy (**1984-5**) culminates in an uneasy alliance against a far worse malign sleeper, and in the deposed leader of the forces of darkness going into exile as mistress of the virtuous narrator. Where epic fantasy celebrates heroic virtue, S&S prefers moderate virtue allied with good sense and a capacity to compromise.

Almost by definition, S&S is the version of heroic fantasy which lends itself most readily to HUMOUR. Many of Leiber's **Fafhrd and Gray Mouser** stories are outright comedies, and they retain a wry comic edge even when verging on HORROR. Alas, it is also true that such pseudo-comic botches as Craig Shaw GARDNER's **Ebenezum** trilogy, Dan MCGIRT's **Jason Cosmo** trio and John MORRESSY's **Kedrigern** stories belong here too. Terry PRATCHETT makes occasional references to epic fantasies in the freeflowing fantasy of his **Discworld** books, but it is to the standard matters of S&S that he returns most constantly, to wandering warriors like Cohen the Barbarian and his daughter Conina, to the guilds of assassins and thieves we already thought we knew from Leiber, to Rincewind the magician being constantly shifted to new locations, to GODS who play GODGAMES with the destiny of humanity and usually end up throwing the board at each other. A significant part of Pratchett's appeal is that he knows what is rich enough to PARODY – one reason for the popularity of his books being that they enable even the jaded S&S reader to re-experience the sheer enjoyment of early encounters with the form.

The various **Nevèryon** books of Samuel R. DELANY fit into S&S uneasily, but more readily than elsewhere; Delany is here pursuing various extra-literary and extra-generic agendas, but is also concerned to subject the subgenre to intense interrogation about its appeal and functioning. By taking a standard location like the Imperial CITY and trying to describe it as a working economy, Delany implicitly criticizes all S&S in which economics is a matter of there being merchants for mercenaries to rob or be hired by (◊◊ PATRONS). Analysing GENDER roles in the light of a rigorously conceived sexual politics, he manages to throw interesting light into dark corners and also to write some fascinating and scarifyingly kinky material. Generally, the **Nevèryon** books are a salutary corrective to the more clichéd aspects of S&S, and have had a quiet influence on it.

Others who have tapped the rich vein of S&S include: Robin BAILEY, whose **Frost** sequence is worked out in the light of FEMINISM; Marion Zimmer BRADLEY; Kenneth BULMER, with the **Dray Prescot** PLANETARY-ROMANCE series; Lin CARTER, whose passionate critical advocacy of fantasy was perhaps more effective than his actual fiction; L. Sprague DE CAMP; David DRAKE; Diane DUANE, with **Tale of the Five**; John GRANT, in his **Lone Wolf** ties; Robert HOLDSTOCK with his **Berserker** and (alternating with Angus WELLS) **Raven** sequences; John JAKES with **Brak the Barbarian**; Tanith LEE; Brian LUMLEY; John NORMAN, whose earlier and more readable **Gor** books are S&S planetary romances; Andrew J. OFFUTT; Fred SABERHAGEN, with the prolonged **Books of Lost Swords** series; Karl Edward WAGNER; and Roger ZELAZNY, in tales of his adventurer **Dilvish**. Since 1984, the SWORD AND SORCERESS anthology series ed Marion Zimmer Bradley has encouraged female S&S writers and protagonists, balancing the subgenre's traditional male bias. Movie S&S is generally of indifferent quality; examples include the CONAN MOVIES, DRAGONSLAYER (**1981**) – which is, unusually, a fine piece of work – RED SONJA (**1985**), portions of *The* SNOW QUEEN (**1993**), *The* SWORD AND THE SORCERER (**1982**) and WILLOW (**1988**). [JC/DRL/RK]

See also: MAGAZINES.

SWORD AND SORCERY (magazine) ◊ MAGAZINES.

SWORD & SORCERY ANNUAL ◊ FANTASTIC.

SWORD AND THE SORCERER, THE US movie (**1982**). Rank/Sorcerer Productions/Group One. **Pr** Brandon Chase, Marianne Chase. **Exec pr** Robert S. Bremson. **Dir** Albert Pyun. **Spfx sv** John Carter. **Screenplay** Tom Karnowski, Pyun, John Stuckmeyer. **Novelization** *The Sword and the Sorcerer* * (**1982**) by Norman Winski. **Starring** Kathleen Beller (Alana), Anna Bjorn (Elizabeth), Lee Horsley (Talon), Richard Lynch (Titus Cromwell), Simon MacCorkindale (Mikah), George Maharis (Machelli), Richard Moll (Xusia). 99 mins. Colour.

Charmless, gory SWORD AND SORCERY epic, intended as first in a series, set in a LAND OF FABLE. Nasty Cromwell, with the aid of raised-from-the-dead SORCERER Xusia, usurps the throne of King Richard's great kingdom, and subdues the surrounding kingdoms. Richard's youthful son Talon survives, and 11 years later is a mercenary renowned worldwide for saving kingdoms; distant relatives Mikah and Mikah's sister Alana plot rebellion. Alana is threatened with rape regularly and consequently knees a lot of groins. Sinister Machelli, Cromwell's chancellor, is in fact the SHAPESHIFTING Xusia, as is revealed to us when Alana knees him in the groin and he doesn't flinch. Blood flows as Talon, bearing his father's weird triple-bladed SWORD, backed by mercenaries with names like Darius, Philip and Captain Morgan, slays Cromwell and Xusia, gives Mikah the crown, screws Alana, and rejoins the carefree mercenary life, galloping off into the land of unmade sequels.

The plotting and dialogue seem drawn direct from role-playing gamebooks, and clichés proliferate; periodically carnage is presented as a dance, as if we should admire the sadistic aesthetics of mutilation. [JG]

SWORD IN THE STONE Arthurian motif first

introduced by Robert de Boron (?   -1212) in the early 13th century, repeated by Sir Thomas MALORY and other writers subsequently. After the death of King Uther, Britain was without a sole ruler. MERLIN placed a sword in a stone (or, in some versions, in a anvil set atop a stone) with the statement that it might be drawn only by the rightful new king. Contests took place, but no one could withdraw it until the young ARTHUR did, seeking a sword for his foster-brother Kay to use in a tournament. Kay at first claimed he had drawn the sword himself, but rapidly admitted Arthur had done so, and the feat was repeated for the benefit of the assembled company. Although Merlin proclaimed Arthur's kingship, the act of drawing the sword did not convince all of the leaders of the Britons, and Arthur had to face a series of battles to enforce his sovereignty. The SITS, although a symbol of his kingship, seems to have conferred no specific magical powers. It is not the same sword as EXCALIBUR, and after the kingmaking plays little part in the CYCLE. The phrase was adopted by T.H. WHITE as the title for the first volume of *The Once and Future King* (1958). Another sword in a stone appears in the cycle attached to the GRAIL: Arthur's knights see a sword in a stone floating down a river, bearing an inscription that only the best knight in the world might draw it. This proves to be GALAHAD.   [KLM]

SWORD IN THE STONE, THE US ANIMATED MOVIE (*1963*). ◊ ARTHUR; DISNEY; SWORD IN THE STONE; T.H. WHITE.

SWORDS Omnipresent default WEAPON of fantasy (hence SWORD AND SORCERY), the sword's phallic symbolism is obvious. The archetypal MAGIC sword is ARTHUR's EXCALIBUR or Caliburn, whose healing scabbard is an AMULET; in his **Life of Manuel**, James Branch CABELL named another magic blade Flamberge after the swords of Charlemagne and others; Board-cleaver in William MORRIS's *The Sundering Flood* (1897) carries the not uncommon CONDITION that once drawn it may not be resheathed without taking a life; Sacnoth in Lord DUNSANY's "The Fortress Unvanquishable, Save for Sacnoth" is more cunning in battle than its wielder; further examples abound. Notable broken swords include: Sigmund's, broken against ODIN's spear in Richard WAGNER's **The Ring of the Nibelungs** and remade by Siegfried; Narsil in J.R.R. TOLKIEN's *The Lord of the Rings* (1954-5), whose reforging as Anduril signals the emergence of its HIDDEN-MONARCH bearer, Aragorn; and the eponymous cursed blade of Poul ANDERSON's *The Broken Sword* (1954), broken by THOR, whose renewal stinks of WRONGNESS. Many swords embody SPIRITS (◊ BONDAGE): the SOUL-eating, strength-giving Stormbringer carried by Michael MOORCOCK's **Elric** is this doomed ANTIHERO's dark SHADOW; Glirendree in Larry NIVEN's "Not Long Before the End" (1969) is a transformed DEMON; the Living Blade in Diana Wynne JONES's *The Homeward Bounders* (1983) is a psychic projection (◊ TALENTS) that compensates for its owner's withered arm; the tediously chatty Kring in Terry PRATCHETT's *The Colour of Magic* (1983) expresses its desire to be a ploughshare. Stormbringer is the model for cursed swords (◊ CURSES) in GAMES, which can be let go only with great difficulty – or in extreme cases, as with Glirendree, not at all. Some swords are of unusual materials: G.K. CHESTERTON's eponymous *The Sword of Wood* (1928 chap) deals handily with an "enchanted" blade which defeats steel swords because magnetized; the Sword Called Llyr in Henry KUTTNER's *The Dark World* (1946) and Eirias in Susan COOPER's *Silver on the Tree* (1977) are both crystal,

enabling the former's concealment within a pane of glass; the reforged sword in Michael Scott ROHAN's *The Forge in the Forest* (1987) appears, in a sly TECHNOFANTASY hint, to be reinforced with carbon fibres. Some have odd properties: Orcrist and Glamdring in Tolkien's *The Hobbit* (1937) warn of nearby GOBLINS by glowing (presumably to the detriment of ambushes); that in *The Sword of Shannara* (1977) by Terry BROOKS merely compels truth, which alone destroys a less than resilient DARK LORD; that in Piers ANTHONY's *Wielding a Red Sword* (1986) is the empowering emblem of war; Need, the magic sword of Mercedes LACKEY's **Vows and Honour** sequence, must be carried by a woman and will not strike a woman, however inimical. Very many fantasy HEROES AND HEROINES routinely carry named swords, as Roland carried Durandal – a determined exception being tegeus-Cromis in M. John HARRISON's *The Pastel City* (1971), whose blade is unfashionably nameless. Examples of noteworthy swords and sword-bearers include: Oscar in Robert HEINLEIN's *Glory Road* (1963), waxing sentimental over his blade The Lady Vivamus; Fritz LEIBER's **Fafhrd and the Gray Mouser**, with Graywand and Scalpel (plus the Mouser's dagger Cat's Claw); the 12 named swords of Fred SABERHAGEN's **Swords** sequence, with their plethora of magical abilities; the eponymous swords of Tad WILLIAMS's **Memory, Sorrow and Thorn**; and Roger ZELAZNY's Corwin of **Amber** with Grayswandir. Of special note is Severian's lovingly maintained *Terminus Est* in Gene WOLFE's *The Book of the New Sun* (1980-83), an executioner's blade whose ingenious design (mercury flows in an internal channel to shift the centre of gravity) has led unwary readers to assume it magical.   [DRL]

See also: SWORD IN THE STONE; WELAND SMITH.

SYMBOLS AND SYMBOLISM In literary scholarship and criticism, any of a broad range of linguistic, cultural or personal signifiers suggesting extratextual meaning. Particularly in fantastic narratives, an early prevalence of one-to-one symbolic correspondences in MYTHS, FABLES and ALLEGORIES led to a widespread tendency to read any symbolic fantasy as allegory, and authors from George MACDONALD to J.R.R. TOLKIEN complained about the "allegorization" of their texts. In many fantasies, the use of symbols is more akin to what C.S. LEWIS called "sacramentalism", which is very nearly the opposite of allegory – "for the symbolist, it is we who are the allegory" (*Allegory of Love* 1936). Although the historical Symbolist movement in France, Russia, and England may have had little direct progeny in FANTASY literature, the sense of rebellion against social realism and the validation of personal and hermetic symbols almost certainly helped liberate fantasy from the stigma of apologue and broaden the possible range of its symbols.   [GW]

SYME, SIDNEY HERBERT (1867-1941) UK artist. ◊ ILLUSTRATION.

SYMONDS, JOHN (1914-   ) UK journalist and writer who began his career working for *Lilliput Magazine*. His first novel, *William Waste* (1947), is an ALLEGORY. Over the next decades he composed a large number of quiet, subversive, somewhat Surrealist examples of CHILDREN's FANTASY, beginning with *The Magic Currant Bun* (1952 chap US). Notable further titles include *Away to the Moon* (1956 chap US), *Lottie* (1957), describing the adventures of a DOLL and her dog in the 18th century, and *Dapple Grey: The Story of a Rocking-Horse* (1962), whose lost protagonist finds his way home. Little of JS's early adult fiction is fantasy, but in *The*

Guardian of the Threshold (**1980**), told in his usual spare and ironic manner, a MAGUS becomes involved with two Swedish aristocrats who have discovered the ELIXIR of life and are now immortal (◊ IMMORTALITY). Of his many plays, the second drama assembled in *The Lunatic Asylum is on Fire!*; *Zilpah* (coll **1982**) is a RURITANIA satire set in the future, *Oldcastle* (**1994**) features a contemporary magus in talks with the DEVIL, *Tower Above the Clouds: An Extravaganza* (**1994**) sees yet another magus spar with the spirit of MERLIN over his unjust immuring of his wife in the eponymous DARK TOWER, and over his incest with his daughter – echoes of SHAKESPEARE's *The Tempest* (performed *c* 1611; **1623**) being evident throughout – and *Lenin and the Tsar* (**1995** chap) is a kind of vaudeville in which various participants in the Russian Revolution, some dead, have their say. Two of JS's more recent novels – *Zélide* (**1984**) and *Sidony* (**1987**), the latter incorporating some TWICE-TOLD references to SLEEPING BEAUTY – have fantasy elements.

Much of JS's nonfiction has been devoted to figures of SUPERNATURAL-FICTION interest, including *Madam Blavatsky, Medium and Magician* (**1959**; vt *The Lady with the Magic Eyes* 1960 US; vt *In the Astral Light* 1965 UK) (◊ H.P. BLAVATSKY) and several books on Aleister CROWLEY: *The Great Beast: The Life of Aleister Crowley* (**1951**; rev 1971), *The Magic of Aleister Crowley* (**1958**) and *The King of the Shadow Realm: Aleister Crowley, His Life, and Magic* (**1989**). *The Medusa's Head, or Conversations Between Aleister Crowley and Adolf Hitler* (**1991**) is a novel, and with its companion-piece play, *The Trickster and the Devil* (**1992**), depicts Crowley as a TRICKSTER figure who flummoxes Hitler in the 1930s. JS has also edited several editions of Crowley's works, including *The Confessions of Aleister Crowley: An Autohagiography* (**1969**; rev 1979) with Kenneth Grant. [JC]

Other works (for children): *Travellers Three* (**1953** chap US); *The Isle of Cats* (**1955** chap; rev 1979); *Elfrida and the Pig* (**1959**); *The Story George Told Me* (**1963** chap); *Tom and Tabby* (**1964** chap US); *Grodge-Cat and the Window Cleaner* (**1965** chap US); *The Stuffed Dog* (**1967** chap); *Harold, The Story of a Friendship* (**1973** chap); *A Christmas Story* (**1977** chap).

SYMPATHY ◊ MAGIC.

SYRETT, NETTA Name used since childhood by UK writer Janet Syrett (?1870-1943), popular in her day for her FAIRYTALES and fairy plays. Her friend Grant ALLEN helped her place her first story, "That Dance at the Robson's" (1890 *Longman's*). She became associated with the John Lane set, particularly Henry Harland (1861-1905) and Aubrey BEARDSLEY, and was present at the conception of *The Yellow Book*. Lane published her first novel, *Nobody's Fault*

(1896), the first of over 40 books. Some are adult romances, a few with supernatural undertones. NS believed she was psychic (◊ TALENTS), particularly through vivid precognitive DREAMS. She used this theme in reverse in several books where her protagonists have VISIONS of past events: in *Barbara of the Thorn* (**1913**) a woman witnesses a past murder in Rome; in *The House that Was* (**1933**) a man relives events in an old house where his father committed suicide. *Angel Unawares* (**1936**) retold the childhood of her younger brother, onto whom she grafted her own psychic abilities.

NS began her fairytales with "Fairy-gold" (1896 *Temple Bar*), her first collection being *The Garden of Delight* (coll **1898**). GARDENS held a special fascination for NS, who used them as POLDERS of FAERIE where children might witness FAIRY activities, as in *Godmother's Garden* (**1918**). She edited the CHRISTMAS BOOK *The Dream Garden* (anth **1905**), with contributions from, among others, Hilaire Belloc, Fiona MACLEOD, E. NESBIT and Alfred NOYES. Widely travelled, NS loved to create fairytale worlds out of the CITIES she visited. The title story in *The Magic City and Other Fairy Tales* (coll **1903**) recreates landmarks in LONDON as places of magic, while *The Castle of Four Towers* (**1909**) transforms Siena into a fairy town. In *Magic London* (**1922**; rev 1933) a godmother's magic allows children to TIMESLIP to the past. NS's work was especially appreciated for its ability to create a presence of place with which children could identify. She wrote (and later produced at the Children's Theatre, which she helped establish in 1913) many fairy plays, including *Six Fairy Plays* (coll **1904**), *The Fairy Doll* (**1906**; also in *The Fairy Doll and Other Plays for Children* coll **1922** chap) and *Robin Goodfellow and Other Fairy Plays* (coll **1918**).

Some of NS's fairytales have an oppressive saccharinity, especially *Toby and the Odd Beasts* (**1921** chap) and *Rachel and the Seven Wonders* (**1921** chap), both in the **Royal Road Library** series, but at their best her stories share a close affinity with a child's imagination, making them true WONDER TALES. Her own joy in life and the delight of Nature comes through in her autobiography, *The Sheltering Tree* (**1939**). [MA]

Other works: *The Vanishing Princess* (**1910** chap) illus Charles ROBINSON; *The Endless Journey and Other Stories* (coll **1912**); *Stories from Medieval Romance* (coll **1913**); *Tinkelly Winkle* (**1923**); *The Magic Castle and Other Stories* (coll **1925**).

SZILAGYI, STEVE(N GLENN) (1952-) US painter, illustrator and writer whose *Photographing Fairies* (**1992**), set in 1920s England, makes RECURSIVE use of Arthur Conan DOYLE's belief in the Cottingley Fairies. Told in the first person by a gullible young US photographer, the novel unfolds its narrator's slow discovery that FAIRIES exist. [JC]

TABITHA US tv series (1977-8). Columbia/ABC. **Pr** William Asher, Robert Stambler, George Yanok. **Exec pr** Jerry Mayer. **Dir** Bruce Bilson and others. **Writers** Martin Donovan, Ed Jurist, Bernie Kahn, Mayer, Bernard Wolpert, Yanok. **Created by** Mayer. **Based on** BEWITCHED (1964-72). **Starring** David Ankrum (Adam Stevens), Lisa Hartman (Tabitha Stevens), Bruce Kimmel (Adam Stevens in the first pilot), Liberty Williams (Tabitha Stevens in the first pilot). 2 pilots plus 14 30min episodes. Colour.

This spinoff from *Bewitched* featured the adventures of Tabitha, seen in the earlier show as a young WITCH and now grown up and trying to make it in the working world. In the first pilot, Tabitha is an editorial assistant at a fashion magazine in San Francisco. The second pilot and subsequent series found her working as a production assistant at a Los Angeles tv station. Relatively little emphasis was put on Tabitha's magical capabilities, and the show degenerated into a standard sitcom. [BC]

TABOO Short-lived squarebound COMICS magazine (7 issues 1988-94) created by artists Steven Bissette and John Totleben (1958-) and published by Spiderbaby Graphics, with some surprisingly uninhibited contents. Early issues featured strips by Bissette, S. Clay Wilson, Paul Chadwick and MOEBIUS, but *T* was remarkable mainly because it featured the first publication of some works by Alan MOORE and Neil GAIMAN. The most important of these were the early episodes of Moore's Dickensian JACK THE RIPPER graphic novel *From Hell* and his erotica, *Lost Girls*. [RT]

TABOOS A term from ANTHROPOLOGY/comparative RELIGION, often used loosely to describe actions which are forbidden. However, the term originally described restrictions or bans relating to something sacred. That which is taboo is the sacred object or place or condition of being. Taboo actions are – by extension – actions which violate that sacred zone. Also spelled "tabu", "tapu" or "kapu", the term comes from the Polynesian.

In fantasy, the term is often applied with accuracy. Grant ALLEN's *The Great Taboo* (**1891**) treats its eponymous subject matter in an orthodox fashion; in Sheri S. TEPPER's *The Awakeners* (**1987**) a prohibition against the dead returning to the land of the living, or the living visiting the land of the

dead, corresponds to an underlying sense that contact between the two states is contaminating, a sense whose provenance extends very deep into human prehistory, and which informs LEGENDS like that of DON JUAN. In much of his work, an author like Philip José FARMER explores forbidden regions of the psyche.

It is in the loose sense that the term is used to describe subject matters which are, or have been at one time, forbidden to fantasy. SEX and RELIGION are both human meaning systems very frequently subject to censorship, perhaps because they are both potentially subversive. FANTASY, as a genre which began as a subversive assault on the "real" world (◊◊ AESOPIAN FANTASY), has been peculiarly vulnerable to restrictions; when it is also remembered that until recent decades the genre was normally conceived as essentially fit only for children, the tameness of much modern fantasy is unsurprising. In the last years of the 20th century, however, most censorship has weakened, except that exercised by publishers seeking to target particular markets. GENRE FANTASY, as a rule, therefore, avoids most "taboos". [JC]

See also: SUPERSTITION; WIZARDS.

TAKAHASHI, RUMIKO Japanese moviemaker (1957-). ◊ ANIME.

TAKEI, GEORGE (1939-) US actor and writer. ◊ Robert Lynn ASPRIN; *SFE*.

TAKE OFF US movie (**1978**). ◊ *The* PICTURE OF DORIAN GRAY (**1945**).

TALENTS The fantasy term for a range of innate, specialized abilities usually known as paranormal or psi powers. Both sf writers and real-world believers prefer technical Graeco-Latin names for talents: sf speaks of "telepathy" where a fantasy author might opt for "mindspeech". The most commonly deployed talents relate to extended PERCEPTION, alias ESP. Examples include: clairaudience, the hearing of distant or inaudible sounds; clairvoyance or "clear seeing", usually of far-off or hidden scenes (◊◊ SCRYING) – this term is sometimes used to cover all these perceptual talents; empathy, sensitivity to others' feelings and emotions – the eponymous seers of Kristine Kathryn RUSCH's *Heart Readers* (**1993**) detect the purity or otherwise of SOULS; PRECOGNITION, glimpsing the future; orientation or "bump of direction", a convenient PLOT DEVICE – Piers ANTHONY's *A Spell for Chameleon* (**1977**) extends the notion with a character who, on request, can point the way to anything; psychometry, the perception of inanimate objects' histories – used in Colin WILSON's *The Philosopher's Stone* (**1969**) to read the story of Mu (◊ LEMURIA) from a surviving artifact; and telepathy, the ability to read thoughts and/or project thoughts and influences into others' minds. This latter may merely facilitate communications in the form of "mindspeech" or, in Diane DUANE's *The Door Into Fire* (**1979**), "bespeaking"; or it may extend to total mental control on the order of POSSESSION, as in A.E. VAN VOGT's *The Book of Ptath* (**1947**). A further perceptual talent often deployed without comment in GENRE FANTASY is the "sixth-sense" ability of numerous characters to detect danger, EVIL, WRONGNESS or a generally "doom-laden" atmosphere.

The more physical talents include: bilocation, being in two places at the same time – an apparent ability of the evil Wither in C.S. LEWIS's *That Hideous Strength* (**1945**); HEALING by laying-on of hands; levitation, lifting oneself via telekinesis – which may extend to flight, as in Michael HARRISON's *Higher Things* (**1945**); pyrokinesis or firestarting, the ability to set things on fire – as possessed by Agni in Roger ZELAZNY's *Lord of Light* (**1967**) (◊◊ FIRESTARTER [*1984*]); telekinesis or psychokinesis, lifting or moving objects by mind alone – in Fritz LEIBER's "The Lords of Quarmall" (**1964**), the movement of pieces in a board-GAME reflects a telekinetic battle of wills; and teleportation, transporting objects instantly from place to place – transport often being of the talented person's own body, as in Phyllis EISENSTEIN's *Born to Exile* (**1978**), which features a whole clan of self-teleporters. (Talent is often hereditary in this way, as in Katherine KURTZ's **Deryni** and Roger ZELAZNY's **Amber** sequences.) Apportation is a subspecies of teleportation, usually associated with SPIRITUALISM, whereby mediums and others can summon small objects to themselves – as do the Ship's MIRROR-sails in Gene WOLFE's *The Urth of the New Sun* (**1987**).

POLTERGEIST phenomena are often interpreted as arising from uncontrolled talents emerging in adolescents undergoing MENARCHE. When the source of power is clearly external to the wielder (◊ MAGIC; MIRACLES; SPELLS), or the power is deployed by inherently magical beings like DEMONS, ELVES or GODS, the term "talent" is normally avoided. [DRL]

TALES FROM THE CRYPT Title of several horror COMIC books.

1. Published by EC COMICS, and formerly titled *Crypt of Terror*. As *TFTC* it ran from *#20* (**1950**) to *#46* (**1955**). It was remarkable for its high-quality writing and art and the goriness of some of the tales, which were used as examples in the Senate hearings investigating the unwholesomeness of horror comics.

2. Published by Gladstone Publishing (6 issues 1990-91), and reprinting several of the EC strips.

3. Featuring short horror tales and published (1992-current) by Russ Cochran. The title was also used on a giant-size one-off comic book by Cochran in 1992 and a b/w horror magazine published by Eerie Publications, of which only one issue (*#10* 1968) appeared. [RT]

TALES OF MAGIC AND MYSTERY ◊ MAGAZINES.

TALES OF THE FRIGHTENED ◊ MAGAZINES.

TALISMANS Objects with magical, apotropaic or luck-bringing powers, including AMULETS and CHARMS. They vary endlessly in form: the eponymous talisman of "The Monkey's Paw" (1902) by W.W. Jacobs (1863-1943) unpleasantly grants THREE WISHES; the EVIL Talisman of Set in Dennis WHEATLEY's *The Devil Rides Out* (**1934**) is the GOD's mummified phallus; the talisman of the Binder in Roger ZELAZNY's *Lord of Light* (**1967**) is a TECHNOFANTASY focus for the hero's TALENT; that in Clifford SIMAK's *The Fellowship of the Talisman* (**1978**) proves to be a holy manuscript about CHRIST. [DRL]

TALKING ANIMALS It is a rare ANIMAL FANTASY whose cast fails to "talk" in some way, usually through straightforward speech. More importantly, the term describes animals who can talk (or communicate analogously through TALENTS) to others of a different species, particularly to humans. Animals in BEAST-FABLES almost universally have the ability. TAs are found often in FOLKTALES and FAIRY-TALES. In LEGENDS, on the other hand, the talking animal is more likely to be a LIMINAL BEING who marks a THRESHOLD, who warns of a transgression, who must be conquered for the sake of the LAND, etc. Over the last century, the animal fantasy and the beast fable have tended to come together in tales where human and animal protagonists intermix; among many examples are the **Jungle Books** by Rudyard KIPLING, *The Wind in the Willows* (**1908**) by Kenneth GRAHAME, the **Dr Dolittle** sequence by Hugh LOFTING and the **Freddy the Pig** sequence by Walter R. BROOKS. The animals in tales of this sort tend – when not in communication with favoured humans – to inhabit POLDERS or WAINSCOTS of the human world, rather than to visit this world from an OTHERWORLD. The reason may be simple enough: it is in *this* world that they can be defined as animals that talk; in their own world (if they had one) they would simply be characters. In modern fantasies set in otherworlds it is normally the case that this protocol is observed, so that otherworld creatures who hauntingly resemble "animals" are treated as properly autonomous, though in HEROIC FANTASY they often take on faithful-COMPANION roles, and may poignantly die for the sake of the hero.

A secondary sense of the term is used by the makers of ANIMATED MOVIES to distinguish between (a) cartoons – like those featuring MICKEY MOUSE, BUGS BUNNY and DONALD DUCK – which rely on animals that have all the attributes of human beings except the physical form and (b) cartoons, certainly in the minority, which do not deploy such characters, although, as in the **Tom & Jerry** shorts, the protagonists may be animals. [JC/JG]

TALLIS, ROBYN Pseudonym of Debra DOYLE.

TALL TALES Tales where the personal recollection of events become exaggerated to unbelievable proportions, often leading to fantastic invention. Early TRAVELLERS' TALES are the forerunners of TTs, such as *Baron Münchhausen's Narrative of his Marvellous Travels and Campaigns in Russia* (fixup **1785** chap UK; exp vt *The Surprising Travels and Adventures of Baron Münchhausen* 1792) by Rudolf Erich RASPE. These established the criteria that have been often imitated, as in the **Joseph Jorkens** stories by Lord DUNSANY and the **Brigadier Ffellowes** adventures by Sterling Lanier (1927-   ). Such stories had their natural development in the FOLKLORE of the US frontier myth, featuring legendary heroes like the lumberjack Paul Bunyan, whose exploits were first recorded in *Paul Bunyan and His Big Blue Ox* (**1914** chap) by W.B. Laughead, and the cowboy Pecos Bill, whose adventures are included in *Tall Tales from Texas* (anth **1934**) ed Mody C. Boatright.

Similar TTs evolved around such historical characters as Daniel Boone (1734-1820), Davy Crockett (1786-1836) and Kit Carson (1809-1868), whose exploits passed into LEGEND and became exaggerated with every telling. This frontier bragging encouraged Mark TWAIN to write "The Celebrated Jumping Frog of Calavera County" (ot "Jim Smiley and His Jumping Frog" 1865 New York *Saturday Press*). Many regional collections of TTs exist – e.g., *Tall Tales of the Kentucky Mountains* (anth **1926**) ed Percy MacKaye and *We Always Lie to Strangers: Tall Tales from the Ozarks* (anth **1951**) ed Vance Randolph (1892-1980), both of which drew upon a rich vein of folk tradition that also inspired the **John the Balladeer** stories by Manly Wade WELLMAN.

The same attitude that encouraged TTs also engendered hoax stories. The most famous is the Moon Hoax by Richard Adams Locke (1800-1871), which first appeared in the *New York Sun* as a series of articles in 1834 (assembled as *The Moon Hoax* coll **1835** chap) and told of the discovery of flying beings on the Moon. Edgar Allan POE delighted in this and produced his own story, later retitled "The Balloon Hoax" (1844), also for the *Sun*, of a trans-Atlantic balloon crossing. George Horatio Derby (1823-1861), Bret Harte (1836-1902) and Mark TWAIN all perpetrated hoax stories.

TTs are often told as CLUB STORIES, and have a feeling of the TWICE-TOLD. Examples include the stories of A.J. ALAN, the **Murchison Morks** stories by Robert ARTHUR, the **Squaredeal Sam** stories by Nelson S. BOND, the **Papa Schimmelhorn** stories by Reginald Bretnor (1911-1992) in *The Schimmelhorn File* (coll **1979**), the **Hiram Holliday** stories by Paul GALLICO and the **Gavagan's Bar** stories by L. Sprague DE CAMP and Fletcher PRATT. The **Probability Zero** series of sf stories that ran in the magazine *Astounding* during WWII (and resurrected in *Analog*) is full of TTs. George H. Scithers (1929-   ) and Darrell SCHWEITZER have edited two volumes of such stories, *Tales from the Spaceport Bar* (anth **1987**) and *Another Round at the Spaceport Bar* (anth **1989**).

TTs are also known as Shaggy Dog stories, a phrase which itself has inspired countless tales and encouraged fantasists to create the inevitable inversion, the "Shaggy God" story. TTs can often become quite surreal (◊ SURREALISM) and are always meant as HUMOUR.   [MA]

TAM LIN Also called Tamlane, the subject of a Border ballad collected in Francis James Child's *English and Scottish Popular Ballads* (anth **1857-8**), and possibly dating back to the 14th century. Tam Lin is a young knight carried off by the FAIRIES who is fated to be their SEVEN-yearly sacrifice to HELL. He is rescued by Janet, whose child he has fathered, when she lies in wait for the Fairy Ride at HALLOWE'EN and takes hold of him, ignoring the fierce and terrible shapes into which he is transformed. As a narrative the ballad, like the similar *Thomas the Rhymer*, contains within it many of the most important "fairy" beliefs which – like the sinister FAIRY QUEEN, the TRANSFORMATIONS, the tithe to Hell and the amoral hedonism of the fairies – have become staples of modern fantasy. Perhaps the dangerous glamour of the story is best translated into modern idiom by Diana Wynne JONES in *Fire and Hemlock* (**1985**), but Ellen KUSHNER's historical fantasy *Thomas the Rhymer* (**1990**) also incorporates important elements of the story to good effect.   [AS]

TANNEN, MARY (1943-   ) US writer, originally of CHILDREN'S FANTASY. Her first adult novel, *Second Sight* (**1987**), deals with an urban psychic whose professional skills are complicated by a fitful but genuine TALENT. *After Roy* (**1989**), dealing with talking-chimp experiments (◊ APES) in a fictitious West African nation, contains what in other hands could be considered borderline sf elements, but MT's handling of her material – as with *Easy Keeper* (**1992**) and *Losing Edith* (**1995**) – resembles fantasy primarily in the reliance upon strong narrative drama and in climaxes that prove intrinsicate with the discovery of true kinship and location of the protagonist's place in his society.   [GF]
Other works (for children): *The Wizard Chidren of Finn* (**1981**) and *The Lost Legend of Finn* (**1982**); *Huntley Nutley and the Missing Link* (**1983**).

TAPROOT TEXTS Only in the last decades of the 18th century, when (at least in the West) a HORIZON OF EXPECTATIONS emerged among writers and readers, did a delimitable genre now called FANTASY appear. Before that there were writings which included the FANTASTIC – and such works can be described as taproot texts. To exemplify: The presence of ARIEL and of PROSPERO's staff in William SHAKESPEARE's *The Tempest* (performed *c*1611; **1623**) do not make that play a fantasy, according to this criterion; *The Tempest*, however defined generically, may contain elements of the fantastic, but these elements did not govern its audience's sense of its generic nature: it was, first and foremost, a play. On the other hand, GOETHE's *Faust, Part One* (**1808**) clearly reveals its author's consciousness that he is transforming a traditional story containing supernatural elements into a work mediated through – and in a telling sense defined by – those elements. For our purposes, *The Tempest* is best conceived as a TT and *Faust* as a fantasy.

The notion of the TT seems necessary – or at least desirable – for at least two reasons. The first is that a WATER MARGIN of not easily definable intentions marks what we may now read as an irreversible impulse towards fantasy over the last decades of the 18th century, and it seems advisable to have a blanket term available to use in order to distinguish relevant texts composed or written before those we can legitimately call fantasy. The second is that, because almost any form of tale written before the rise of the mimetic novel could be retroactively conceived as ur- or proto-fantasy, it seems highly convenient to apply to works from this OCEAN OF STORY a term – i.e., "taproot" – which emphasizes the *heightened* significance of the text mentioned. When we refer to a text as a TT, in other words, we describe one that contains a certain mix of ingredients and stands out for various reasons – not excepting quality.

The list of TTs, therefore, may be long, but it is by no means endless; and a clear degree of qualitative judgement will be apparent in any individual cataloguing. Beyond those already mentioned, some other texts seem to fit the taxonomical needs for which the term was devised.

Relevant texts from classical literature include HOMER's *Iliad* and *Odyssey* (composed by the 8th century BC); Hesiod's *Theogony* (composed 8th century BC), Aesop's *Fables* (composed before 560BC) (◊ AESOPIAN FANTASY); certain works of the Greek playwrights, like Aeschylus's *Prometheus Bound* (produced before 456BC) and Sophocles' *Oedipus Rex* (produced before 406BC); OVID's *Metamorphoses* (*c*AD1), Lucius APULEIUS's *The Golden Ass* (before AD155) and most of the surviving works of LUCIAN.

Relevant texts from the turn of the Renaissance onwards include DANTE's *The Divine Comedy* (before 1321), Giovanni BOCCACCIO's *Decameron* (before 1353), the various ROMANCES and epics that mass together around the MATTERS of Britain and France, including works like BEOWULF, *Sir Gawain and the Green Knight* (written *c*1370) (◊ GAWAIN) and Sir Thomas MALORY's *Le Morte Darthur* (**1485**) ed Thomas Caxton, some episodes of Geoffrey Chaucer's *The Canterbury Tales* (before 1400), Luigi Pulci's *The Greater Morgante* (**1470**; exp 1483), *Orlando Innamorato* (**1487**) by Matteo Maria Boiardo (1434-1494), Lodovico ARIOSTO's *Orlando Furioso* (**1516**), François RABELAIS's *Gargantua and Pantagruel* (**1532-64**), the *Nights* (**1550-53**) of Gianfrancesco Straparola, Luis de Camoes's *The Lusiads* (**1572**), Torquato Tasso's *Jerusalem Delivered* (**1581**), Edmund SPENSER's *The Faerie Queene* (**1590-96**), Christopher MARLOWE's *Dr Faustus* (written *c*1588), *A Midsummer Night's Dream* (performed *c*1595; **1600**) and other Shakespeare plays, Miguel de CERVANTES's *Don Quixote* (**1605-15**), the *Pentamerone* (**1634-6**) of Giambattista Basile, John MILTON's *Paradise Lost* (**1667**), John BUNYAN's *The Pilgrim's Progress* (**1678**) (◊◊ PILGRIM'S PROGRESS), Charles PERRAULT's *Tales of Mother Goose* (coll 1697), the various versions of *The Arabian Nights* (◊ ARABIAN FANTASY), Alexander Pope's *The Rape of the Lock* (**1714**) and Jonathan SWIFT's *Gulliver's Travels* (**1726**). The list could be considerably extended, but there is a distinction to be made: huge quantities of work can be treated as being of backdrop interest only; these titles cannot. [JC]

TARNMOOR, SALVATOR R. [s] ◊ Herman MELVILLE.

TAROT A pack of exotically symbolic CARDS, often credited with highly improbable ancient origins; it is extremely unlikely that Tarot cards existed before the development of pasteboard. Used in fantasy as a divinatory tool (◊ PROPHECY) and a generalized magical prop, such cards may bear little relation to Tarot packs sold here and now: often they are entirely invented – with fanciful new suits and trumps – and called Tarot merely for convenience. Italo CALVINO's *The Castle of Crossed Destinies* (**1969**) interestingly uses the "standard" Tarot as its medium of STORY. John CROWLEY's alternative "Least Trumps" in *Little, Big* (**1981**) offer some tantalizing new Tarot-card images. *Tarot Tales* (anth **1989**) ed Rachel POLLACK and Caitlín Matthews (?1952-   ) is a relevant anthology. [CB/DRL]

TARR, JUDITH (1955-   ) US writer whose first published work was **The Hound and the Falcon** trilogy: *The Isle of Glass* (**1985**), *The Golden Horn* (**1985**) and *The Hounds of God* (**1986**), assembled as *The Hound and the Falcon* (omni **1986**). A well realized medieval fantasy (despite its rationale that the human-seeming folk taken to be ELVES are the product of natural mutation), the trilogy shows JT's scrupulous historical accuracy and intelligent but sometimes florid characterization. These tendencies variously persisted over the next decade, as her work gained assurance and poise.

The **Avaryan** series – *The Hall of the Mountain King* (**1986**), *The Lady of Han-Gilen* (**1987**) and *A Fall of Princes* (**1988**), these three assembled as *Avaryan Rising* (omni **1988**), plus *Arrows of the Sun* (**1993**) and *Spear of Heaven* (**1994**) – though set on an invented planet is a DYNASTIC FANTASY involving mages and power politics. *A Wind in Cairo* (**1989**), a fantasy of medieval EGYPT, and *Ars Magica* (**1989**), about Pope Gerbert, are transitional works. They were followed by the expansive *Alamut* (**1989**), a historical fantasy set in the Kingdom of Jerusalem, and its sequel, *The Dagger and the Cross: A Novel of the Crusades* (**1991**). Set (very tangentially) in the world of **The Hound and the Falcon**, they are perhaps JT's best novels.

Lord of the Two Lands (**1993**) is an historical novel, with some fantastic elements, about Alexander the Great's conquest of Egypt; it has set the tone for much of JT's subsequent fiction, which has tended toward large historical fantasies or historical novels with minimal fantasy elements. These include *Throne of Isis* (**1994**), *Pillar of Fire* (**1995**) and *King and Goddess* (**1996**). [GF]

Other works: *His Majesty's Elephant* (**1993**), a children's novel about Charlemagne; *The Eagle's Daughter* (**1995**), a historical novel.

TARTT, DONNA (1964-   ) US writer whose *The Secret History* (**1992**) has slight fantasy interest in that its plot hinges on the successful attempt by a group of Classics students to recreate the requisite conditions for a bacchanal (◊ REVEL). During this they see the god DIONYSUS; the female among them is perceived, even by herself, as a deer, and is hunted; and they unwittingly tear to pieces an innocent bystander. That this is not merely a fantasy of PERCEPTION is indicated by the fact that, afterwards, one of their number displays the bite-marks of an unknown MONSTER. [JG]

TARZAN Hero of a long series of novels by Edgar Rice BURROUGHS, starting with *Tarzan of the Apes* (1912 *All-Story*; **1914**), and continued – often illicitly – in SEQUELS by OTHER HANDS, as in Fritz LEIBER's (authorized) novelization *Tarzan and the Valley of Gold* * (**1976**) and the REVISIONIST FANTASY *The Death of Tarzana Clayton* (**1985** chap) by Neville Farki. Orphaned as a baby in a LAND-OF-FABLE Africa, Tarzan is reared by APES and becomes King of the Jungle. It could be argued that the **Tarzan** stories are not fantasy at all (except in occasional detail), yet the Africa in which they are set is to all intents and purposes a SECONDARY WORLD, and Tarzan himself is able to converse with animals in their own languages; if the **Tarzan** stories are not fantasies, what *are* they?

Tarzan is UNDERLIER for many protagonists (not all male; ◊ SHEENA [*1984*]) of jungle stories and JUNGLE MOVIES – and has been, of course, himself the central figure of the many TARZAN MOVIES. He can also be claimed as underlier of many characters in SWORD AND SORCERY: he is the ultimate barbarian, entirely lacking academic education but instead educated by his closeness to Nature. [JG]

See also: ATLANTIS; A.A. ATTANASIO; AUSTRALIA; BOMBA MOVIES; James CAWTHORN; CITY; FANTASTIC ADVENTURES; Philip José FARMER; FERAL CHILDREN; Frank FRAZETTA; H. Rider HAGGARD; Tom HENIGHAN; IMAGINARY LANDS; Michael William KALUTA; Roy G. KRENKEL; LOST RACES; NOVELIZATIONS; ONE MILLION BC; SHADOWS; William STOUT; TREES; Joan D. VINGE.

Comics

The first adaptation of Tarzan into COMIC-strip form came about through the efforts of Joseph H. Neebe, a staff member of the Detroit advertising agency Campbell-Ewald. An admirer of Burroughs, he met the author in Los Angeles in 1928 and suggested an adaptation of *Tarzan of the Apes*. Neebe formed a subsidiary company, Famous Books and Plays Inc., to market the feature for publication in newspapers. He condensed the novel to 30,000 words and approached John Allen St John (1872-1957), the leading illustrator of ERB's work, to produce the drawings. St John declined, so Neebe turned to Harold R. FOSTER, a staff artist at Campbell Ewald. But Neebe, inexperienced in the field of comics syndication, was unable to sell the strip anywhere in the USA, so it first appeared in the UK weekly magazine *Tit Bits* (20 October 1928) – billed as "A serial story in pictures: 'Tarzan of the Apes' – your home picture-play!". Finally

Neebe sold the feature – in the form of 60 daily episodes of five frames per day, to 15 newspapers – through the Metropolitan Newspaper Service; the first US publication was on 7 January 1929. The complete story was published in book form by Grosset & Dunlap later the same year as *The Illustrated Tarzan Book* (graph **1929**). The daily newspaper series continued with *The Return of Tarzan* (from 17 June 1929), drawn by Rex Maxon (1892-1973), who went on to adapt further novels in the series. In 1930 United Features Syndicate bought Metropolitan Newspaper Service and Famous Books and Plays Inc., and it has retained syndication rights ever since.

By 1931 the popularity of the feature was deemed sufficient to justify a weekly Sunday page of previously unpublished narrative in colour. Maxon undertook both writing and drawing of this, in addition to the daily strip, but his work was plodding and uninspired. Foster took over the 12-frame Sunday page from 27 September 1931, while Maxon continued to draw the dailies (until August 1947).

Foster's version of Tarzan was initially clearly based on the actor Elmo Lincoln, the first movie Tarzan (◊ TARZAN MOVIES). Foster was not consistent: Tarzan's leopardskin knickers changed suddenly and inexplicably to the Weismuller-style loincloth in the middle of a conversation with a group of Chinese in 1938, and the leopard spots reappeared – equally mystifyingly – as Tarzan led a group of black warriors after a killer Tyrannosaur in 1945, but he proved a consummate draughtsman and storyteller, able to capture the spirit and flavour of Burroughs's writing. His work became enormously popular, and he was eventually offered the chance to create a feature of his own by the Hearst Corporation's King Features Syndicate. He thus abandoned **Tarzan** to create PRINCE VALIANT.

Foster was succeeded on 9 May 1937 by Burne HOGARTH, who had studied under St John. Hogarth's work was equally influential in the development of the US comic strip. His drawings of the jungle lord showed his admiration for Michelangelo's sculpture, and his jungle scenery, with its gnarled and twisted trees and rich foliage, showed the influence of Chinese painting. The storylines were provided by Donald Garden and displayed great literacy and imaginative breadth. Hogarth himself took over the scripting in 1943, and continued drawing **Tarzan** until 25 November 1947, when he abandoned it to create **Drago**, a short-lived adventure strip set in South America. He was replaced during this period by Rubimor (real name Reuben Moriera; 1922-). Hogarth returned with renewed enthusiasm in May 1948 and began to experiment with less formal page layouts and vigorous, dynamic frame compositions. His work now had a tremendous vitality that has rarely been matched.

Hogarth's last Sunday page was dated 20 August 1950; a conflict over the use of his work outside the USA (for which he was given neither credit nor payment) led to his refusing to renew his contract with United Features. He was succeeded 1950-54 by Bob Lubbers (1922-), then 1954-68 by John Celardo (1918-) and 1968-81 by Russ Manning (1929-1981). Manning was succeeded by Gil Kane (1926-), Mike Grell (1947-), Gray Morrow (real name Dwight Graydon Morrow; 1934-) and Joe Kubert (1926-). The dailies were drawn by Maxon until 1947, then by Dan Barry (1923-), whose work during the first few months was signed "Hogarth", although Burne Hogarth was not involved in their production. Other artists on the daily feature included Paul Reinman, Nick

Cardy (real name Nicolas Viscardy; 1920-), Lubbers, Celardo and Manning. In 1973 the daily feature ceased, and all subsequent publication was of reprinted earlier material.

Early comic books featuring Tarzan were collected reprints of the newspaper strips. These included *Tip Top Comics* (various issues 1936-61), *Comics on Parade* (1938) and *Sparkler Comics* (various issues 1941-55). The first comic books containing original material were *Tarzan and the Devil Ogre* and *Tarzan and the Tohr* (*Four-Color #134* and *#161*, both 1947), and the success of these led the following year to a monthly comic book entitled simply *Tarzan*, with art by Jesse Marsh (1907-1966), Russ Manning (1929-), Doug Wildey (1922-), Alberto GIOLITTI and others. This title continued regular publication by Dell/Gold Key until 1972, and enjoyed some impressive cover artwork by Moe Gollub, among others. DC COMICS then took over publication, publishing it until 1977 (with Kubert art), when it was taken up by MARVEL COMICS, who continued it until 1979, with art by John Buscema (1927-) and subsequently Sal Buscema (1936-). A four-issue book was published by Charlton Comics (who were under the misapprehension that copyright on the character had lapsed): *Jungle Tales of Tarzan* (**1964-5**), with art by Sam Glanzman (1924-). Malibu Comics picked up the character in 1992 with *Tarzan the Warrior*. In this offering Tarzan and Jane – now permanently young – have a number of hectic adventures with a group of aliens. Malibu followed this with the charming *Tarzan: Love, Lies and the Lost City* (3 issues 1992) and *Tarzan – The Beckoning* (7 issues 1993).

In the UK a number of reprint comic books have been published, the longest-running being *Tarzan Adventures* (1951-9) – reprinting the newspaper strips – which Michael MOORCOCK edited and provided material for during 1957-8. Others included *Tarzan of the Apes* (Top Sellers 1971-5), which reprinted the Dell material along with new stories, and *Tarzan Weekly* (Byblos 1977-8), which contained material unpublished in the USA supplied by a variety of international artists, including Manning, Mike PLOOG, Danny Bulandi (1946-), Alex NIÑO and José ORTIZ. New strips were also published in children's weeklies during the period when the various tv series were being broadcast.

Two original GRAPHIC-NOVEL adaptations by Hogarth were published: *Tarzan of the Apes* (graph **1972**) and *Jungle Tales of Tarzan* (graph **1976**). A hardcover collection of Foster's and Hogarth's work on the Sunday pages, projected to run to 18 volumes under the series title **Tarzan in Color**, is currently (1995) being published at the rate of one volume every three months.

A shamelessly plagiaristic pornographic digest-size comic, *Tarsan*, had a short run in Italy during 1981. Korak, son of Tarzan, has also starred in comics, but the characterization and storylines have been largely indistinguishable from **Tarzan** adventures proper.

Tarzan's comics success spawned many imitations, including **Kaanga, Jungle Lord** (1940-54), **Zago, Jungle Prince** (1948), **Jungle Jo** (1950), **Jo-Jo the Congo King** (1947), **Ka Zar, Lord of the Hidden Jungle** (1939-current) and **Thun-da, King of the Congo** (1952), which in turn led to an equally long list of scantily clad jungle ladies, including **Nyoka the Jungle Girl** (1942-57), **Sheena, Queen of the Jungle** (1939; ◊ Will EISNER), **Rulah the Jungle Goddess** (1948), **Zegra the Jungle Empress** (1949), **Shanna the She Devil** (1972 onwards) and **Jungle Lil** (1950). [RT]

Further reading: *The Edgar Rice Burroughs Library of Illustration* (graph **1975**) ed Russ Cochran.

Television

Aside from *Tarzan and the Trappers* (**1958** tvm), cobbled together from pilot episodes for an unmade tv series, and *Tarzan in Manhattan* (**1989** tvm), pilot for a stillborn tv series (◊ TARZAN MOVIES), Tarzan has had two primary tv appearances: *Tarzan* (57 episodes 1966-8), starring Ron Ely as the apeman (there was no Jane) – two episodes from which were stitched together as the theatrically released *Tarzan's Deadly Silence* (**1970** tvm) and a further two as the theatrically released *Tarzan's Jungle Rebellion* (**1970** tvm) – and *Tarzan* (50 episodes 1991-3), starring Wolf Larson as Tarzan and Lydie Denier as Jane; in both series the famous cry was produced using recordings from the Weissmuller movies. *Tarzan, Lord of the Jungle* (16 episodes 1976-7) was a children's animated series, with Tarzan voiced by Robert Ridgely; episodes were recycled in *The Tarzan/Lone Ranger/Zorro Adventure Hour* (1981-2) and, with new material, in *Tarzan and the Super Seven* (1978-80).   [JG]

TARZAN MOVIES The **Tarzan** JUNGLE MOVIES are of varying interest, and are treated accordingly.

1. *Tarzan of the Apes* US movie (**1918**). National Film Corporation of America/First National. **Pr** William Parsons. **Dir** Scott Sidney. **Screenplay** Lois Weber, Fred Miller. **Based on** *Tarzan of the Apes* (**1914**) by Edgar Rice BURROUGHS. **Starring** True Boardman (John Clayton, Earl of Greystoke), George B. French (Binns), Gordon Griffith (young Tarzan), Thomas Jefferson (Professor Porter), Colin Kenny (Cecil), Kathleen Kirkham (Alice, Lady Greystoke), Elmo Lincoln (Tarzan), Enid Markey (Jane Porter). *c*95 mins (8 reels). B/w. Silent.

There is a mutiny aboard the *Fuwalda*, bound for Africa, and a sailor called Binns saves Lord and Lady Greystoke from death at the hands of the crew. The trio are marooned on a jungle-girt shore; soon afterwards Binns is abducted by slavers. The Greystokes have a baby, but then die. The infant is reared by Kala the she-ape, who dubs him Tarzan. Years pass. Tarzan finds his parents' cabin and a knife, which weapon enables him to become top ape. Binns, free of the slavers, befriends Tarzan and teaches him to read; he then returns to England, where the incumbent Greystokes commit him to an asylum to shut him up about the true heir. However, a party is eventually sent to Africa, among them Jane Porter and her father, plus Tarzan's cousin Cecil. Tarzan secretly observes them and, when Cecil tries to seduce Jane, beats him up; he then rescues Jane from, in succession, a lion, a rapist and a hostile tribe. At last she agrees to remain in the jungle with him.

TOTA, a lavish production (it cost about $300,000, and was shot partly on location in Brazil) was a great hit, turning its production company and distributor into major commercial forces and making Elmo Lincoln (born Otto Elmo Linkenhelter) a star. However, Burroughs, earlier wildly enthusiastic about the bringing of his hero to the screen, seems to have cooled dramatically by the time of *TOTA*'s release, probably because the movie's depiction of a brutish, inarticulate giant was a long way from the civilized John Clayton of the written tales. Burroughs made unsuccessful efforts to halt the inevitable sequel, **2.**   [JG]

2. *The Romance of Tarzan* US movie (**1918**). National Film Corporation of America/First National. **Pr** William Parsons. **Dir** Wilfred Lucas. **Screenplay** Bess Meredyth. **Based on** *Tarzan of the Apes* (**1914**) by Burroughs. **Starring** True Boardman (John Clayton, Earl of Greystoke), George B. French (Binns), Gordon Griffith (young Tarzan), Thomas Jefferson (Professor Porter), Colin Kenny (Cecil), Kathleen Kirkham (Lady Greystoke), Elmo Lincoln (Tarzan), Cleo Madison (La Belle Odine), Enid Markey (Jane). *c*82 mins (7 reels). B/w. Silent.

This follows the early story of **1**, but diverges when the search party arrives from England. After Tarzan has saved the party from hostile natives, Cecil persuades the others he has seen Tarzan die, so they set sail. Eventually Tarzan makes his way to California, where he joins what passes for polite society near the Mexican border. By astonishing coincidence, a kidnapped Jane has also been brought here, and he rescues her. Foiling a murderous plot by Cecil and the loose but beautiful Odine (who succumbs to Tarzan's charms), Tarzan woos Jane; but Cecil has told her that Tarzan loves Odine, and she rejects her jungle lover, who returns disconsolately home. Odine confesses the truth to Jane who, with her father, sets forth once more for Africa.   [JG]

3. *The Revenge of Tarzan* US movie (**1920**). Numa/Goldwyn. **Pr** George M. Merrick. **Dir** Harry Revier. **Screenplay** Robert Saxmar. **Based on** *The Return of Tarzan* (**1915**) by Burroughs. **Starring** Franklin Coates (Paul D'Arnot), Armand Cortez (Nicholas Rokoff), Walter Miller (Ivan Paulovich), Gene Pollar (Tarzan), George Romain (Count de Coude), Karla Schramm (Jane), Estelle Taylor (Countess de Coude). *c*82 mins (7 reels). B/w. Silent.

Filmed as *The Return of Tarzan* but released cut from 9 reels to 7 and with the name changed, this represented a rapid decline even from **2**. It was the only movie of the new Tarzan, Gene Pollar (born Joe Pohler), a New York fireman. Tarzan, sojourning in civilization, tries to thwart a scam and earns the mortal enmity of Paulovich and Rokoff, who remain undefeated at movie's end – although at least Tarzan is reunited with Jane, who is serendipitously shipwrecked on the island on which Tarzan's enemies have marooned him.   [JG]

4. *The Son of Tarzan* US serial movie (**1920**). National Film Corporation of America. **Dir** Arthur J. Flaven, Harry Revier. **Screenplay** Roy Somerville. **Based on** *The Son of Tarzan* (**1917**) by Burroughs. **Starring** Eugene Burr (Ivan Paulovich), Mae Giraci (young Meriem), Gordon Griffith (young Korak/Jack), Manilla Martan (Meriem), Karla Schramm (Jane), Kamuela C. Searle (Korak), P. Dempsey Tabler (Tarzan). 15 episodes. B/w. Silent.

The real star of this tortuous tale is Jack, son of Tarzan and Jane, who have long been married and living as Lord and Lady Greystoke in civilization. Paulovich (from **3**) is still intent on revenge, especially when Jack disrupts Paulovich's exhibition in England of Akut, one of Tarzan's old ape friends. Akut and Jack abscond to Africa, where the lad rescues a girl-child, Meriem, from Arab slavers and becomes Korak. The two grow up together and fall in love. Discovering their son is still alive, Tarzan and Jane come to fetch him; both women are captured by and rescued from the slavers, whose sheik Korak eventually kills. Korak is saved from death at the stake by a friendly elephant, and, reunited, all set sail for home.

In reality, Searle was fatally injured during his "rescue" by the elephant, and Korak's last few scenes were filmed with a substitute, seen only from the rear. This tragedy significantly increased the serial's popularity. An edited feature-film version (*c*111 mins) was released in 1923 as *Jungle Trail of the Son of Tarzan*.   [JG]

5. *The Adventures of Tarzan* US serial movie (*1921*). Weiss Brothers-Numa/Great Western. **Dir** Robert F. Hill. **Screenplay** Hill, Lillian Valentine. **Based on** *The Return of Tarzan* (**1915**) by Burroughs. **Starring** Charles Islee (Professor Porter), Elmo Lincoln (Tarzan), Louise Lorraine (Jane), Frank Whitson (Nicholas Rokoff), Lillian Worth (La). 15 episodes, re-edited 1928 as 10 episodes. B/w. Silent.

The early story of Tarzan and his romance with Jane is rehashed; the rest of the serial is a nonstop collection of adventures in which Tarzan saves Jane from kidnappers, hostile natives, wild animals and more. The whole is fuelled by two MCGUFFINS: a nerve-gas formula stolen by the villainous Rokoff (from **3**) and the treasure of the jungle kingdom of Opar, the only MAP to which chances to be tattooed on Jane's shoulder. This was Lincoln's final appearance as Tarzan. [JG]

6. *Tarzan and the Golden Lion* US movie (*1927*). RC Pictures/FBO Gold Bond. **Exec pr** Joseph P. Kennedy. **Dir** J.P. McGowan. **Screenplay** William Wing. **Based on** *Tarzan and the Golden Lion* by Burroughs. **Starring** Dorothy Dunbar (Jane), Boris Karloff (Chief), Lui Yu-Ching (Weesimbo), Edna Murphy (Flora Hawkes), Fred Peters (Esteban Miranda), Jim Pierce (Tarzan). *c*75 mins (6 reels). B/w. Silent.

Tarzan's old friend Weesimbo tells him of the treasures of the lost City of Diamonds. Villainous Miranda, overhearing, abducts Weesimbo (plus Jane's niece Flora and her fiancé) and makes him act as guide. Tarzan pursues with his golden lion Jad-bal-ja, saves Flora from HUMAN SACRIFICE and returns home triumphant. Pierce was Burroughs's personal selection as Tarzan, but never played the role on screen again (although offered the part for **10**). In due course Pierce married Burroughs's daughter Joan, and the couple played Tarzan and Jane in a weekly radio series. [JG]

7. *Tarzan the Mighty* US serial movie (*1928*). Universal. **Pr** William Lord Wright. **Dir** Jack Nelson. **Screenplay** Ian McClosky Heath. **Based on** *Jungle Tales of Tarzan* (coll **1919**) by Burroughs. **Starring** Al Ferguson (Black John), Lorimer Johnston (Lord Greystoke), Natalie Kingston (Mary Trevor), Frank Merrill (Tarzan), Bobby Nelson (Bobby Trevor). 15 episodes. B/w. Silent.

Another rewrite of Tarzan's biography. Black John, of a villainous jungle "tribe" descended from PIRATES, discovers shipwrecked Mary Trevor and her little brother Bobby on the shore. Tarzan and Mary become interested in each other; Black John threatens Bobby will die unless Mary marries him. Tarzan rescues Mary, and she tells him some old documents he cannot read are proof he is the Greystoke heir. Black John obtains the documents and pretends to be first Tarzan and then Tarzan's uncle, Lord Greystoke – who has by chance arrived in search of the heir – in hope of gaining both girl and title. Leopards rip Black John to bits; Tarzan is offered the luxuries of England but determines to remain in the jungle with Mary as his loving "bride". [JG]

8. *Tarzan the Tiger* US serial movie (*1929*). **Pr** William Lord Wright. **Dir** Henry McRae. **Screenplay** Ian McClosky Heath. **Based on** *Tarzan and the Jewels of Opar* (**1918**) by Burroughs. **Starring** Mohammed Bey (Paul Panzer), Al Ferguson (Albert Werper), Natalie Kingston (Jane), Kithnou (La), Sheldon Lewis (Achmet Zek), Frank Merrill (Tarzan), Clive Morgan (Philip Annersley). 15 episodes. B/w. Silent.

Tarzan goes back to Opar (◊ **5**) to gather extra gold for the

Greystoke estates, but in an accident is hit on the head and loses his memory. Despite this, despite the villainous wiles of Zek, Werper and Annersley, and even despite the lustful grapplings of Queen La of Opar, Tarzan manages to recover not only sufficient jewels to pay off the estate debts but also to regain his memory, in the process twice saving Jane from a ghastly doom.

Kingston, who had in **7** played an alternative amour for Tarzan, was here cast as Jane: no wonder mighty man of jungle bit confused. A dubbed version of this, the last of the silent **Tarzans**, was released: it had sound effects, music and Tarzan's triumphal yell, but no dialogue. [JG]

9. *Tarzan the Ape Man* US movie (*1932*). MGM. **Pr** Bernard H. Hyman. **Dir** W.S. Van Dyke. **Screenplay** Cyril Hume, Ivor Novello. **Starring** Alfredo Codona (Weissmuller's body-double), Neil Hamilton (Harry Holt), Forrester Harvey (Beamish), Doris Lloyd (Mrs Cutten), Maureen O'Sullivan (Jane Parker), C. Aubrey Smith (James Parker), Johnny Weissmuller (Tarzan), Ivory Williams (Riano). 99 mins. B/w.

Jane comes to Africa to find her father. He, with partner Holt, is about to set out in search of the Elephant's Graveyard (and its ivory), reputed to lie beyond the TABOO-ridden Mutia Escarpment. She goes with them, and is soon captured as a curiosity by the wordless Tarzan, to whose sexuality she responds with innocent curiosity. She is recovered by Holt and her father, who kill an ape in the process; an angry Tarzan murders most of the black members of the expedition by way of revenge, but is then himself shot by Holt. The elephant Timba carries Jane to Tarzan, and she nurses him back to health, falling in love with him the while. Later the expedition is captured by a tribe of murderous dwarfs, and rescued by Tarzan and a herd of elephants. Jane, her father having died, determines to remain in the jungle as Tarzan's mate.

CINEMA's mythology persists in depicting this as a fine movie, but in fact it is not: it is little better than its precursors in the series, and the sexual electricity that made later Weissmuller/O'Sullivan **Tarzan** collaborations so memorable failed to spark here. Furthermore, it is unremittingly racist: the only way of dealing with darkies, it seems, is by use of the whip; indeed, the black members of the safari are less servants than slaves.

The movie had its genesis in *Trader Horn* (*1930*; ◊ JUNGLE MOVIES), which went way over budget and was almost scrapped. However, photographer Clyde de Vinna had shot more footage of African wildlife and African tribes than could be used in that movie, so MGM's Irving Thalberg decided another movie must be made to exploit this material; he struck a deal with Burroughs that, while his characters would be used, his plots would not. In the event Burroughs waived most of his rights concerning the use of his plots. [JG]

10. *Tarzan the Fearless* US serial and feature movie (*1933*). Wardour/Sol Lesser/Principal. **Pr** Lesser. **Dir** Robert F. Hill. **Screenplay** Walter Anthony, Basil Dickey, George Plympton, William Lord Wright. **Based on** an original story, possibly by Burroughs. **Starring** Mischa Auer (Eltar), Matthew Betz (Nick Moran), Symonia Boniface (Arab woman), Buster Crabbe (Tarzan), Frank Lackteen (Abdul), Philo McCullough (Jeff Herbert), E. Alyn Warren (Dr Brooks), Jacqueline Wells (Mary Brooks), Eddie Woods (Bob Hall). 12-episode serial, 85 mins feature (only feature survives). B/w.

Mary Brooks is part of a safari come to Africa to seek her father, a scientist who is a good friend of Tarzan's. She first encounters the apeman when she foolishly goes skinny-dipping in a croc-infested river (the apparent bareness of her buttocks possibly had consequences for **11**), and love ensues. Co-expeditioners Herbert and Moran have been offered £10,000 by the Greystoke estate if they can bring back proof Tarzan is dead, but give this up when they discover Dr Brooks has departed in search of a temple packed with treasure; also in the party are nice Bob Hall, who loves the gel, and sinister Arabs Abdul and his lovely sidekick, who kidnap Mary . . . There is much more of this sort: the movie is overloaded with plot, its genesis as a serial being very obvious in its need to have regular, increasingly arbitrary cliffhangers.

Existing prints of *TTF* are made up of the first 4 episodes of the serial, initially released together as an hour-long teaser, plus cobbled-on surviving bits from the other episodes. [JG]

11. *Tarzan and His Mate* US movie (*1934*). MGM. **Pr** Bernard H. Hyman. **Dir** Jack Conway (uncredited), Cedric Gibbons. **Screenplay** James Kevin McGuinness. **Starring** Paul Cavanagh (Martin Arlington), Alfredo Codona (Weissmuller's body-double), Nathan Curry (Saidi), Neil Hamilton (Harry Holt), Forrester Harvey (Beamish), Josephine McKim (O'Sullivan's body-double), Maureen O'Sullivan (Jane), Desmond Roberts (Henry Van Ness), William Stack (Tom Pierce), Johnny Weissmuller (Tarzan). 116 mins, cut to 95 mins for general release; restored 1991 video release 104 mins. B/w.

Holt (from **9**) is still optimistic that he can win Jane. He lures to Africa his sociopathic old friend Arlington, an ivory merchant keen to discover the Elephants' Graveyard; attempting to thwart them are Pierce and Van Ness. They meet Tarzan when he calls off a band of giant apes from slaughtering the safari's bearers; Arlington is immediately smitten by Jane. Tarzan and Jane live in a state of PASTORAL bliss, innocently sexual until the Westerners start to introduce foreign concepts like "modesty"; the BEAUTY AND THE BEAST theme is frequently underlined, as when Tarzan repeatedly saves Jane from wild animals, being an animal himself during these contests (most were to be recycled intact in later Weissmuller **Tarzan** movies). When Tarzan learns of the expedition's intended destination he declines to lead it further, regarding the pillaging of the Elephants' Graveyard as a desecration; Arlington, who has already murdered one of the bearers *pour encourager les autres*, promptly shoots and mortally wounds one of Tarzan's elephant friends, so that it leads them to the place. Tarzan heads a herd of elephants to stop the theft of the ivory; Arlington agrees to return it, but next day shoots Tarzan and persuades Jane her lover has been killed by a crocodile. But Tarzan is alive, as Jane soon discovers. By then, however, the expedition is being attacked by hostile natives and a vast herd (eh?) of lions. Tarzan arrives with his elephants to save Jane, but too late for the others.

This was a long movie for its day, and even longer at its premiere, before the Legion of Decency protested – possibly in part because Jacqueline Wells had bared much in **10** but in the main because, despite the adventures, *TAHM* is essentially a movie about SEX, depicting its innocence in surroundings uncorrupted by "morals": there is nothing remotely objectionable to the modern eye about Jane in her revealing garb or even, as we see her briefly (a body-double

was used), naked (a moment notably fictionalized in Theodor ROSZAK's *Flicker* [**1991**]); it is only when she is dressed in "acceptable" fashion costumes brought to Africa by the Englishmen that any hint of voyeurism intrudes. In short, *TAHM* deliberately argued that sex is a moral activity unless fetishized. Needless to say, the Legion of Decency saw only the offence of an underclad woman (Tarzan's even briefer customary costume escaped censure); a brief sequence in which Tarzan and Jane decide to head tree-wards for an (off-screen) bout of lovemaking served further to inflame the self-appointed guardians of morality – who won the argument, in the sense that none of the further MGM **Tarzan** movies dared to be ambitious.

TAHM cannot be described as a distinguished debate between innocence, on the one hand, and decorum, on the other – it is too marred by pointless adventures and bad spfx for that – yet it does represent an important landmark in the way the CINEMA was prepared to depict sexuality. It is certainly the best of the Weissmuller **Tarzan** movies. [JG]
See also: ONE MILLION BC (*1940*).

12. *The New Adventures of Tarzan* (vt *Tarzan's New Adventure*) US serial movie (*1935*). Burroughs/Tarzan Enterprises. **Pr** Ashton Dearholt. **Dir** Edward Kull. **Screenplay** Charles F. Royal. **Based on** original story by Burroughs. **Starring** Frank Baker (Major Martling), Herman Brix (aka Bruce Bennett; Tarzan), Don Costello (Raglan), Ashton Dearholt (Raglan), Harry Ernest (Gordon Hamilton), Ula Holt (Ula Vale), Dale Walsh (Alice Martling). 12 episodes. B/w.

With varying purposes, the cast set out from Africa to Guatemala in search of a statue stuffed with jewels: the Lost Goddess of the Dead City. After battles with the Mayans whose property the sacred object is, our heroes share the valuables among themselves.

This serial, in which Burroughs himself was directly involved (note production credits), was later edited as a feature, *The New Adventures of Tarzan* (*1935*). It was sequelled by **14**, also from Burroughs/Tarzan Enterprises. [JG]

13. *Tarzan Escapes* US movie (*1936*). MGM. **Assoc pr** Sam Zimbalist. **Dir** Richard Thorpe. **Screenplay** Cyril Hume. **Starring** John Buckler (Captain Fry), E.E. Clive (Masters), William Henry (Eric Parker), Benita Hume (Rita Parker), Darby Jones (Bomba), Herbert Mundin (Herbert Henry Rawlins), Maureen O'Sullivan (Jane), Johnny Weissmuller (Tarzan). 95 mins. B/w.

Jane's Uncle Peter has died, leaving her half his fortune. Her impecunious cousins – his other niece Rita and nephew Eric – come to Africa to persuade her to return briefly to England so she may claim her inheritance and give much of it to further Eric's studies. Unscrupulous big-game hunter Fry leads them on safari to the Mutia Escarpment, his plan being to capture the "great white gorilla" said to live there as King of the Apes and bring him back for public display in a specially constructed "durallium" cage ("I'll guarantee it'll hold anything"). Fry – who later tries to barter the lives of the rest with the murderous Hymandi tribe – eventually gets his comeuppance, and Tarzan and Jane are left in peace in their jungle idyll.

This version of *TE* was in fact the second one made, a completed previous movie having been scrapped because felt by MGM not to be exciting enough. It is marked by an increased domestication of Tarzan and especially Jane, who now sleeps fully clothed in a far more demure costume (◊ **11**); the pair have an elaborate tree house with cleverly constructed mod cons. (Amusingly, one erotic moment far more highly charged than anything that had gone before was allowed to slip through.) The Bomba here is the leader of Fry's native bearers, and is not to be confused with the hero of the BOMBA MOVIES. [JG]

14. *Tarzan and the Green Goddess* US movie (*1937*). Burroughs/Tarzan Enterprises. **Pr** Ashton Dearholt. **Dir** Edward Kull. **Spfx** Howard Anderson, Ray Mercer. **Screenplay** Charles F. Royal. **Based on** an original story by Burroughs. **Starring** Frank Baker (Major Martling), Herman Brix (Tarzan), Don Costello (Raglan), Ula Holt (Ula Vale). 72 mins. B/w.

This sequel to **12** was in fact directly derived from **12**, with some clever editing and a modicum of new footage being used to create a fresh story. The statue stolen by our heroes in **12** apparently contains, aside from the jewels, the formula for a super-explosive. After many adventures the baddies are slain and the goodies decorated, back in the UK, for their services to civilization. Although theoretically set in Guatemala the movie shows wildlife including lions and rhinoceroses. This was the end of Burroughs's dream of seeing his *own* version of Tarzan on screen. [JG]

15. *Tarzan's Revenge* US movie (*1938*). 20th Century-Fox/Principal. **Pr** Sol Lesser. **Dir** D. Ross Lederman. **Screenplay** Robert Lee Johnson, Jay Vann. **Based on** "a novel by Edgar Rice Burroughs", although it is difficult to establish which. **Starring** George Barbier (Roger Reed), C. Henry Gordon (Ben Aley Bey), Eleanor Holm (Eleanor Reed), Hedda Hopper (Penny Reed), John Lester Johnson (Koki), George Meeker (Nevin Potter), Corbet Morris (Jigger), Glenn Morris (Tarzan), Joe Sawyer (Olav). 70 mins. B/w.

Pretty, spoilt Eleanor Reed, her toping father Roger, her allergic mother Penny, her humourless, cowardly fiancé Nevin and their servant Jigger arrive near Toocompac, on the Luckbar River, for a hunting safari; they hire twitchy, drying-out alcoholic Olav as local guide, little realizing he is also in the pay of Sheikh Ben Aley Bey, ruler of a region further inland; earlier Bey has been captivated by Eleanor and he commands Olav to deliver her to his harem. Tarzan, also captivated by Eleanor and keen, too, to protect the wildlife, secretly dogs the safari, and saves the day. Eleanor decides to stay with Tarzan as the others return to Evanston, Illinois.

In many ways this is a much better movie than its approximate contemporaries in the Weissmuller series (although one is far more conscious of the fact that it is shot on set, not on location, and it lacks the dreadful recycled sequences. Yet it is largely affectless – in part because there is no O'Sullivan but perhaps in larger part because it is a story of white folk who encounter Tarzan in Africa rather than one about Tarzan's (and Jane's) interactions with white folk in his LAND-OF-FABLE home territory. [JG]

16. *Tarzan Finds a Son!* US movie (*1939*). MGM. **Pr** Sam Zimbalist. **Dir** Richard Thorpe. **Screenplay** Cyril Hume. **Starring** Laraine Day (Mrs Richard Lancing), Ian Hunter (Austin Lancing), Frieda Inescort (Mrs Austin Lancing), Morton Lowry (Richard Lancing), Maureen O'Sullivan (Jane), John Sheffield (Boy), Henry Stephenson (Sir Thomas Lancing), Johnny Weissmuller (Tarzan), Henry Wilcoxon (Sande). 95 mins. B/w.

Their baby son is the only survivor when the plane bearing Richard Lancing (favourite nephew of Neville, Lord Greystoke) and his wife crashes on the Mutia Escarpment.

The infant is rescued by chimps in a reprise of Tarzan's own earliest history; but then Cheeta brings the child to Tarzan and Jane, who adopt him – naming him, at Tarzan's insistence, "Boy". Five years later Lord Greystoke is dead, having left his fortune to his nephew Richard and Richard's heirs, to be held in trust for 20 years unless it can be proven Richard is dead. To the Mutia Escarpment comes a safari led by Lord Greystoke's brother Thomas and including Richard's cousin Austin and his wife (who hope to find all dead so they can inherit) plus local hunter-guide Sande. Invited to the tree-house for a grotesque parody of a suburban lunch party, they tell their story. Jane immediately sends Tarzan and Boy off on a swimming expedition so they will not see her lie as she claims the plane crash left no survivors. That swimming expedition – including underwater sequences with an elephant and a turtle – is one of the most charming and effective stretches in the entire **Tarzan** *oeuvre*; it ends with Tarzan saving Boy from tumbling over a huge waterfall. Sir Thomas, watching, recognizes features of Richard in Boy, and is all for doing the decent thing; the younger Lancings, abetted by Sande, decide to take Boy to England where, as guardians, they may exploit his fortune. The villains persuade Jane she owes it to Boy to let him take up his inheritance. She thus traps her mate in a ravine and sets off alongside Boy with the safari, promising to return. Sir Thomas is murdered by Austin, who then, despite Jane's warnings, leads the party into hostile Zambeli territory; they are captured and an orgy of HUMAN SACRIFICE starts. Jane is wounded aiding Boy escape to fetch Tarzan and, when Tarzan arrives at the head of a herd of elephants, seems dying. But his declaration of love revives her, and the family is reunited, the Lancings being peremptorily dismissed.

Although the persistent use of ludicrously speeded-up motion makes some action sequences hilarious rather than exciting, there is comparatively little use of matter recycled from earlier **Tarzan** movies, and overall this is a thoroughly enjoyable movie. It is also often affecting, thanks to the brilliance of O'Sullivan's performance as she struggles with the difficulties of explaining ethical complexities to one so determinedly simplistic as Tarzan; Weissmuller does well as her foil. This was Sheffield's first appearance: he played in seven further **Tarzan** movies (**17-23**) before going off to star in the BOMBA MOVIES. [JG]

17. *Tarzan's Secret Treasure* US movie (*1941*). MGM. **Pr** B.P. Fineman. **Dir** Richard Thorpe. **Spfx** Warren Newcombe. **Screenplay** Myles Connolly, Paul Gangelin. **Starring** Tom Conway (Medford), Philip Dorn (Vandermeer), Barry Fitzgerald (Dennis O'Doul), Cordell Hickman (Tumbo), Maureen O'Sullivan (Jane), Reginald Owen (Professor Elliott), John Sheffield (Boy), Johnny Weissmuller (Tarzan). 81 mins. B/w.

The first of the real Tarzan-by-numbers movies of the Weissmuller sequence. Life is yet more domesticated in the Tarzan household, with even an egg-timer and a refrigerator. Finding gold nuggets on the riverbed during a family swim, Boy is treated to a lecture by Jane and then Tarzan on the meaning of riches: gold is valueless, because you cannot eat it. Yet off Boy goes, leaving a note: "Gone to see Civilzashun. Back tomorrow." Instead he finds African boy Tumbo being threatened by a rhino, and saves his life. Tumbo's village is plague-ridden: Tumbo's mother dies as the two lads watch. The rest – including the famous Rubber Crocodile and O'Doul, a comic-cuts Irishman – is predictable.

Despite the dreary recycling, the Weissmuller sequence had kept up a reasonably high standard until now; this was the beginning of the slide. O'Sullivan, visibly ageing (**18** would be her last **Tarzan** movie), acts as if a redundancy notice has been served. There is a nastily racist moment when Cheeta, offered a photograph of a disc-lipped tribesman, kisses it passionately. Cordell Hickman's Tumbo, and his relationship with Boy, illumine an otherwise bleak landscape. [JG]

18. *Tarzan's New York Adventure* US movie (**1942**). MGM. **Pr** Frederick Stephani. **Dir** Richard Thorpe. **Spfx** Arnold Gillespie, Warren Newcombe. **Screenplay** Myles Connolly, William R. Lipman. **Starring** Charles Bickford (Buck Rand), Virginia Grey (Connie Beach), Russell Hicks (Judge Abbotson), Paul Kelly (Jimmy Shields), Cy Kendall (Ralph Sargent), Elmo Lincoln (Roustabout), Maureen O'Sullivan (Jane), John Sheffield (Boy), Johnny Weissmuller (Tarzan), Chill Wills (Manchester Mountford). 71 mins. B/w.

Nasty Rand and moderately nice Mountford arrive on the Mutia Escarpment in a plane piloted by very nice Shields, their aim to capture wild lions for Sargent's Circus, Long Island, USA. Tarzan gives them 24 hours to complete their mission and get out. They are leaving when Boy appears, impressing them by his tricks with three baby elephants. They plot to kidnap him, but Shields vetoes this; Mountford, too, is on Boy's side after Boy saves him from a lion. The Joconi attack. Boy calls Tarzan, but the Joconi set the shrub on fire. Believing Tarzan and Jane dead, the Westerners take Boy back to the USA. Tarzan, Jane and Cheeta pursue. The three have merry enfant-sauvage adventures in NEW YORK before rescuing Boy and, in court, being granted permanent custodianship of the lad.

Despite its fervent attempt to break the series mould, this is a flat movie. It was to be the last Weissmuller made for MGM: he moved with Sheffield to Sol Lesser at RKO, in the process losing O'Sullivan (who retired temporarily from the movies to tend a husband with typhus) and much of the series' scant remaining interest. Elmo Lincoln, who had played Tarzan in **1**, **2** and **5**, here had a bit part. This was a miserable swansong for O'Sullivan, an actress too fine for the role: she would not again attain cinematic fame until *Hannah and Her Sisters* (**1986**), where among her co-stars was her daughter Mia Farrow. [JG]

19. *Tarzan Triumphs* US movie (**1943**). RKO/Principal. **Pr** Sol Lesser. **Dir** William Thiele. **Screenplay** Roy Chanslor, Carroll Young. **Starring** Stanley Brown (Achmet), Pedro de Cordoba (Patriarch), Frances Gifford (Zandra), Stanley Ridges (Colonel von Reichart), Sig Rumann (Sergeant), Johnny Sheffield (Boy), Philip Van Zandt (Bausch), Johnny Weissmuller (Tarzan), Rex Williams (Corporal Reinhardt Schmidt). 78 mins. B/w.

A propaganda movie – anti both the Nazis and the US noninterventionist movement – but a surprisingly good addition to the *oeuvre*. Jane has gone to England to tend her sick mother (a close analogue to the real reason for O'Sullivan's absence; ◊ **18**). Boy, straying near the lost city of Palandria, is rescued from a scrape by Zandra, a woman of that city. Soon a Nazi plane lands paratroopers in Palandria; the radio operator, Schmidt, inadvertently falls as far afield as Tarzan's home, where he is nursed by Tarzan and Boy (and calls himself Shelton). He betrays their trust by trying to use his radio to call in a Nazi invasion force; when he attempts to kill Cheeta (who throughout *TT* is more of a TRICKSTER than ever before) he himself is killed by one of

her elephant friends. The Nazis swiftly enslave Palandria; Zandra flees when her brother Achmet is murdered, and seeks help from Tarzan. Boy is on her side, but Tarzan heeds the dictum that, if the Nazis do not directly harm him, he has no quarrel with them; at Boy's instigation, Zandra even dons Jane's costume to swim with Tarzan, hoping to seduce him to her way of thinking. However, when the Nazis come from Palandria to seize the radio and take Boy as well, Tarzan at last declares war.

The huge gap in *TT* is of course the absence of O'Sullivan; Gifford, physically much like her, does her best, but there can be no romance between her and Tarzan – mighty man of jungle not commit adultery. The shift from MGM to RKO thankfully meant the end (for a while) of recycled action scenes. As a final propagandistic joke, when German HQ hear Cheeta gabbling and screeching over the radio they assume it is a broadcast from Der Führer. [JG]

20. *Tarzan's Desert Mystery* US movie (**1943**). Sol Lesser/RKO. **Pr** Lesser. **Dir** William Thiele. **Screenplay** Edward T. Lowe, Carroll Young. **Starring** Lloyd Corrigan (Sheik), Nancy Kelly (Connie Bryce), Otto Kruger (Paul Hendrix/Heinrich), Robert Lowery (Selim), Joe Sawyer (Karl), Johnny Sheffield (Boy), Johnny Weissmuller (Tarzan). 70 mins. B/w.

Jane, still in England, sends a message begging Tarzan to take some of his jungle medicine to help fever-stricken Allied soldiers in Burma. With Boy and Cheeta, he crosses the desert to the patch of jungle where the relevant vines grow. *En route* he releases a wild stallion, Jana, from a gang of thugs. These thugs are in the entourage of Hendrix, alias Heinrich, a Nazi who has tricked the Sheik of nearby Birherari into accepting him as right-hand man. Elsewhere, Sheik Amir of Alakibra has asked peripatetic US MAGICIAN Bryce to convey to Selim, Birherari's Crown Prince, a note to the effect that Hendrix/Heinrich is a spy. Bryce, saved from a desert ambush by Tarzan and Boy, is escorted by them to Birherari. There Tarzan is immediately imprisoned for the theft of Jana. Selim receives the note from Bryce, but is immediately murdered by Heinrich; as he dies he gives it to Cheeta. Bryce, framed for the murder, is sentenced to hang. Tarzan bursts out of prison and, with Boy and Jana, saves Bryce; they go to the patch of jungle containing the medicinal vines – also found there are giant lizards (◊ DINOSAURS), man-eating plants and a bull-sized tarantula SPIDER, which almost has Boy before eating Heinrich instead. Back in Birherari our heroes are honoured.

The poorness of realization of this movie is bizarre: that a patch of jungle should exist in the middle of the desert is weird; that it should be populated by an elephant herd (which rescues Tarzan from a mantrap plant) beggars belief; that it should be further populated by prehistoric MONSTERS is, in context, almost acceptable. [JG]

21. *Tarzan and the Amazons* US movie (**1945**). RKO. **Pr** Sol Lesser. **Dir** Kurt Neumann. **Screenplay** John Jacoby, Marjorie L. Pfaelzer. **Starring** Don Douglas (Anders), Steven Geray (Brenner), Brenda Joyce (Jane), J.M. Kerrigan (Splivvers), Barton MacLane (Ballister), Shirley O'Hara (Athena), Maria Ouspenskaya (Queen), Johnny Sheffield (Boy), Henry Stephenson (Sir Guy Henderson), Johnny Weissmuller (Tarzan). 76 mins. B/w.

On the way to welcome back Jane (who has mysteriously, while away, morphed from brunette O'Sullivan to blonde Joyce), Tarzan and Boy save Athena from a couple of leopards and a black panther. They take her home to the a land "beyond the mountains", Palmyria, ruled matriarchally yet worshipping a male Sun GOD and his associated SERPENT and "ever-living TREE". A scientific expedition led by Henderson arrives on the same boat as Jane and spots Athena's dropped bracelet, which Cheeta has purloined. They determine, with the aid of murderous local agent Ballister, to seek Palmyria, a notion vetoed by Tarzan but approved by Boy, who guides them. In Palmyria they are promptly sentenced, as male intruders (Tarzan is exempt from this rule), to death – a sentence commuted to life imprisonment. Athena helps them escape, but Ballister and 'umorous Cockney sidekick Splivvers determine to loot the Sun God's sacred gold; Ballister murders both Athena and Henderson to still their dissent. Athena lives long enough to raise the alarm; the expedition flees with huge loss of life, Boy being recaptured and sentenced anew to death. Tarzan springs to the rescue, rushing surviving trekkers Ballister and Anders into a mudpit – where they drown – returning the stolen gold and persuading the Queen of Palmyria to free Boy.

The AMAZON-style society is quite well depicted, but *TATA* focuses more on action and mayhem. Nevertheless, in the context of **20** and **22**, *TATA* is not as poor a movie as it might have been, and established a trend in **Tarzan** movies of the plot hanging on an encounter with a weird LOST RACE. [JG]

22. *Tarzan and the Leopard Woman* US movie (*1946*). RKO. **Pr** Sol Lesser. **Dir** Kurt Neumann. **Screenplay** Carroll Young. **Starring** Acquanetta (Lea), Edgar Barrier (Dr Amir Lazar), Anthony Caruso (Mongo), Tommy Cook (Kimba), Dennis Hoey (District Commissioner), Brenda Joyce (Jane), Johnny Sheffield (Boy), Johnny Weissmuller (Tarzan). 72 mins. B/w.

The area around the town of Zambesi is plagued by adherents of a leopard cult (◊ CATS), whose High Priestess is the beautiful Lea. The British administrators – unaware that educated half-caste Dr Lazar is Lea's lover and a prime mover in the virulently anti-British cult – believe the killings are the work of real leopards, but Tarzan knows better. Lea's psychotic little brother Kimba vows to cut out a human heart as proof he is a warrior, and insinuates himself into the Tarzan family; he picks Jane as softest target. When a jungle convoy escorting a quartet of newly trained missionary teachers is massacred by the cult and Boy lands himself in danger, Tarzan uses jungle cunning to save goodies and kill baddies. But he, Jane and Boy are captured by the cult. Luckily Cheeta frees them all and the VILLAINS get a nasty comeuppance.

Weissmuller looks old, and possibly ill. Sheffield looks old, too: all semblance of the cute kid has gone. In a welter of Blacks-are-bad-or-at-the-very-least-untrustworthy, *TATLW* has few saving graces. [JG]

23. *Tarzan and the Huntress* US movie (*1947*). RKO. **Pr** Sol Lesser. **Dir** Kurt Neumann. **Screenplay** Jerry Gruskin, Rowland Leigh. **Starring** Ted Hecht (Ozira), Brenda Joyce (Jane), Barton MacLane (Paul Weir), Patricia Morison (Tanya Rawlins), Wallace Scott (Smitty), Johnny Sheffield (Boy), Charles Trowbridge (Farrod), John Warburton (Karl Marley), Johnny Weissmuller (Tarzan). 72 mins. B/w.

The huntress is Rawlins, who comes with Marley to collect live animal specimens for Western zoos; their local contact is corrupt trapper Weir (played by MacLane, a villain as recently as **21**). King Farrod permits exportation of only one

pair of each species; his ambitious nephew Ozira, an eye on the throne, is more tractable. Farrod is assassinated, but Tarzan manages in the end to save the animals: Marley, Weir and Ozira all meet ghastly ends; Rawlins, who has behaved reasonably correctly, is permitted to escape; the rightful heir ascends to Farrod's throne; the movie feels like one that lies towards the end of a series – which indeed it did, since Sheffield thankfully made this his swansong (◊ BOMBA MOVIES) and Weissmuller would have just one more outing. [JG]

24. *Tarzan and the Mermaids* US movie (*1948*). RKO. **Pr** Sol Lesser. **Assoc pr** Joe Noriega. **Dir** Robert Florey. **Screenplay** Carroll Young. **Starring** Linda Christian (Mara), Brenda Joyce (Jane), John Laurenz (Benji), Andrea Palma (Luana), Gustavo Rojo (Tiko), Fernando Wagner (Varga), Johnny Weissmuller (Tarzan), George Zucco (Palanth). 68 mins. B/w.

Earlier (**19-20**) O'Sullivan's departure was "explained": she was in England fulfilling various obligations; now the departure of Sheffield was similarly "explained". Another change was that long-time **Tarzan** musical sound-track composer Paul Sawtell was replaced by Dimitri Tiomkin (although the style is unaltered). Perhaps the biggest change of all was the disappearance of Kurt Neumann, who had been either associate producer or director, usually both, of the last four **Tarzan** movies (he reappeared for **29**). Indeed, even though veteran scripter Young was still at work, *TATM* has a different feel from its predecessors, beginning with a tourist-advertisement-like introduction to the lost land of the Aquaticans, denizens of the area where Tarzan's river debouches into the sea; they worship the "living" GOD Baloo, whom the virgin (◊ VIRGINITY) Mara is doomed to wed. She knows the "god" is a fake – the Germanic pearl thief Varga, who is in cahoots with High Priest Palanth – and anyway she loves Tiko, who has fled to the outside world. On the verge of the ritual marriage she too flees, only to be snared in one of Tarzan's fishing-nets. She is welcomed by the Tarzan family, but a passel of Aquatican males comes to retrieve her. Tarzan pursues; Jane is threatened; Varga is cast to his death from a high cliff; wrongs are righted.

Christian, despite being thespianically challenged, had the smoulder – which Joyce lacked – to have made a good Jane; it was not to be. An embarrassing newcomer was calypso-singing Benji, a family friend of and errand-runner for the Tarzans. Despite a great deal of dross in *TATM*, there is some splendid stuff – e.g., the camerawork and set design involved as Tarzan swims through the river's last subterranean stretches to the sea. It was fitting that Weissmuller's last **Tarzan** movie should be, at least visually, one of the best. [JG]

25. *Tarzan's Magic Fountain* US movie (*1949*). Sol Lesser/RKO. **Pr** Lesser. **Dir** Lee Sholem. **Screenplay** Harry Chandlee, Curt Siodmak. **Starring** Evelyn Ankers (Gloria James), Lex Barker (Tarzan), Albert Dekker (Trask), Charles Drake (Dodd), Brenda Joyce (Jane), Henry Kulky (Vredak), Elmo Lincoln (bit part), Alan Napier (Douglas Jessup). **Voice actor** Johnny Weissmuller (Tarzan's war cry; this was used in many further Tarzan movies). 74 mins. B/w.

The first Lex Barker **Tarzan**, and the last with Brenda Joyce as Jane; also the last **Tarzan** in which appeared Elmo Lincoln, who had played the lead role in **1**, **2** and **5**, plus a bit part in **18**.

The pilot Gloria James, who crashed in Africa 20 years ago, is not dead, as universally assumed, but has found refuge among the white LOST RACE who dwell in the secret Blue Valley, succoured by a FOUNTAIN OF YOUTH. Later, after restoration to the modern world, James returns to Africa with husband Jessup, both hoping to drink the fountain's waters and regain their youth; with them they bring crooked locals Trask and Dodd, who plan to steal the IMMORTALITY-conferring water for sale. After much adventuring, James and Jessup are accepted in the Blue Valley, whose secret is preserved; Trask and Dodd fall foul of Blue Valley warriors.

Barker played the part so much as Weissmuller had done that often one has to blink to realize it's a different man.

[JG]

26. *Tarzan and the Slave Girl* US movie (*1950*). Sol Lesser/RKO. **Pr** Lesser. **Dir** Lee Sholem. **Screenplay** Arnold Belgard, Hans Jacoby. **Starring** Robert Alda (Neil), Lex Barker (Tarzan), Vanessa Brown (Jane), Tony Caruso (Sengo), Denise Darcel (Lola), Hurd Hatfield (Prince), Mary Ellen Kay (Moana), Arthur Shields (Dr Campbell), Robert Warwick (High Priest). 74 mins. B/w.

Sengo and henchmen are abducting beautiful women from the jungle to improve the breeding stock of their plague-ridden lost city, Lyolia – whose civilization shares characteristics of Ancient Rome and Ancient Egypt, with citizens generally clad in Sherwood Forest garb. Among the abductees are Jane and Dr Campbell's nurse Lola. With Campbell (who has a serum for the plague) and Lola's boyfriend Neil, Tarzan pursues, arriving to find that vile Sengo has had the two spirited girls immured for showing resistance. Needless to say, all is sorted. Barker is very good in this movie, as is Darcel; Brown is a pretty but witless Jane.

[JG]

27. *Tarzan's Peril* (vt *Tarzan and the Jungle Queen*) US movie (*1951*). Sol Lesser/RKO. **Pr** Lesser. **Dir** Byron Haskin. **Screenplay** John Cousins, Samuel Newman, Francis Swann. **Starring** Edward Ashley (Connors), Lex Barker (Tarzan), Dorothy Dandridge (Melmendi), Douglas Fowley (Herbert Trask), Virginia Huston (Jane), George Macready (Radijeck), Alan Napier (Peters), Frederick O'Neal (Bulam). 79 mins. B/w.

Departing British Commissioner Peters and his replacement, Connors, witness the coronation of Queen Melmendi of the peaceful Ashuba and her indignant refusal to marry King Bulam of the warlike Yorango. Later the commissioners are shot dead by Radijeck, head of a trio of gunrunners including also Andrews and Trask. The baddies sell guns to Bulam, whose people attack the Ashuba, again demanding Melmendi as Bulam's bride, but now more forcefully. Tarzan, after rescuing both himself and a baby elephant, saves Melmendi and kills Bulam. Radijeck, having murdered Trask, tries to hold Jane as ransom for his life, but is killed by Tarzan. The first 10 minutes or so of this movie, filled with African drumming and dancing, are electric. Thereafter it becomes standard.

[JG]

28. *Tarzan's Savage Fury* US movie (*1952*). Sol Lesser/RKO. **Pr** Lesser. **Dir** Cyril Endfield. **Screenplay** Cyril Hume, Hans Jacoby, Shirley White. **Starring** Lex Barker (Tarzan), Tommy Carlton (Joey [Joseph Martin]), Dorothy Hart (Jane), Patric Knowles (Edwards), Charles Korvin (Rokoff). 80 mins. B/w.

Trekking in search of Tarzan and the fabled diamonds of the Wazuri, Sir Oliver Greystoke, Tarzan's cousin, is murdered by guide (and amateur MAGICIAN) Rokoff, who forces sidekick Edwards to assume Greystoke's identity to fool the jungle man. Meantime Tarzan has saved the life of and adopted the orphan Joey. Tarzan is suspicious of Rokoff and Edwards, but Jane is taken in and persuades Tarzan to help them. After a struggle with cannibals the major characters reach Wazuri territory, where the treachery is uncovered and Jane's life seems forfeit. Rokoff kills Edwards and almost kills Tarzan, but at last the apeman saves Jane and restores good relations with the Wazuri. Barker does his best amid a plethora of rerun footage from earlier movies. Hart was one of the best Janes – very reminiscent of O'Sullivan – but this was her only appearance.

[JG]

29. *Tarzan and the She-Devil* US movie (*1953*). Sol Lesser/RKO. **Pr** Lesser. **Dir** Kurt Neumann. **Screenplay** Karl Kamb, Carroll Young. **Starring** Lex Barker (Tarzan), Henry Brandon (M'Tara), Raymond Burr (Vargo), Tom Conway (Fidel), Michael Grainger (Philippe Lavar), Joyce MacKenzie (Jane), Monique Van Vooren (Lyra). 76 mins. B/w.

Barker's swansong as Tarzan and MacKenzie's sole appearance as Jane is a somewhat perfunctory movie, with much of the footage – notably the elephant sequences – being recycled from the JUNGLE MOVIE *Wild Cargo* (*1934*). Ivory hunters Vargo and Lavar decide to swindle their equally vile partners Lyra (the "She-Devil" of the title) and Fidel of a huge haul. The quartet enslave the men of the Lykopo tribe as bearers, but Tarzan frees them. Next they attempt to seize Jane as hostage so that Tarzan will call the elephants their way, but believe they have accidentally killed her. Likewise thinking Jane dead, Tarzan loses interest in life. On discovering her alive, however, he calls the elephants with a vengeance, so that all the baddies except Lyra (accidentally shot by Fidel) are trampled to death in the ensuing stampede.

[JG]

30. *Tarzan's Hidden Jungle* US movie (*1955*). Sol Lesser/RKO. **Pr** Lesser. **Dir** Harold Schuster. **Screenplay** William Lively. **Starring** Jack Elam (Burger), Charles Fredericks (De Groot), Ike Jones (Malenki), Vera Miles (Jill Hardy), Richard Reeves (Reeves), Gordon Scott (Tarzan), Peter Van Eyck (Dr Celliers). 73 mins. B/w.

The first of the Gordon Scott **Tarzans**, the last **Tarzan** distributed by RKO and the last filmed in b/w (except **33**, which was a tvm) is standard stuff, including lots of rerun wildlife footage from earlier Lesser **Tarzans**. A big-game safari wanders into Tarzan's area, then one of them, Reeves, goes into the territory of the Sukula tribe, where he is thrown into a lion pit for having breached the TABOO against killing animals. Led by De Groot and Burger, the hunters aim to use UN Dr Celliers as a shield to penetrate Sukula territory and round up the animals for massacre; but Celliers and his nurse, Hardy, have already been befriended by Tarzan, who thwarts the plot. This movie seems to belong to the previous decade.

Miles was wasted on a **Tarzan** movie but not on Tarzan: she and Scott married soon after.

[JG]

31. *Tarzan and the Lost Safari* US movie (*1957*). Solar. **Pr** John Croydon. **Dir** Bruce Humberstone. **Spfx** G. Blackwell, T. Howard. **Screenplay** Lillie Hayward, Montgomery Pittmann. **Starring** Peter Arne (Dick Penrod), Robert Beatty (Tusker Hawkins), George Coulouris (Carl Kraski), Yolande Donlan (Gamage Dean), Orlando Martins (Ogonooro), Betta St John (Diana Penrod), Gordon Scott (Tarzan), Wilfrid Hyde White (Doodles Fletcher). 84 mins. Colour.

The first **Tarzan** movie in colour. A plane containing rich adventurer Dick Penrod, his rebellious wife Diana, Dean, Fletcher and Kraski crashlands in the jungle territory of the Opar tribe, who "always try to take whites for sacrifice". Diana – whose job here is to look pretty, endanger herself and then shriek a lot – is almost immediately seized. Big-game hunter Hawkins, in league with the Opar – whose centuries-old ivory stockpile he covets – pretends to save her: his scheme is to deliver the other whites to the Opar but keep Diana for his lecherous self. Tarzan and Hawkins agree to take the party on foot to safety but, midway, Hawkins lets the Opar have them. Plenty of vile darkies are killed before Tarzan and Cheta (here so spelt) save the day. Some of the wildlife photography is fabulous, but too much of the movie was shot in the studio – there are incongruous echoes in several jungle episodes – and we have been here too often before: the novelty of colour is squandered on over-familiar scenes. [JG]

32. *Tarzan's Fight for Life* US movie (**1958**). Sol Lesser/MGM. **Pr** Lesser. **Dir** Bruce Humberstone. **Screenplay** Thomas Hal Phillips. **Starring** Eve Brent (Jane), James Edwards (Futa), Jill Jarmyn (Anne Sturdy), Henry Lauter (Dr Warwick), Carl Benton Reid (Dr Sturdy), Gordon Scott (Tarzan), Rickie Sorensen (Tantu), Woody Strode (Ramo). 86 mins. Colour.

For the first time Scott is given a Jane, in the form of Brent; she features also in **33**. The plot involves a war of intellect between a malicious witch-doctor, Futa, and three real (i.e., white) doctors trying to run a jungle hospital: Warwick, Sturdy and Sturdy's daughter Anne. The argument seems to have been won when Sturdy performs a successful appendectomy on Jane, but Futa has further tricks up his sleeve, in the end dying horribly when he drinks what he believes a cure-all but in fact a deadly poison. This is one of those movies that would somehow be better if it were worse – as it is, all it achieves is dullness. [JG]

33. *Tarzan and the Trappers* US movie (**1958** tvm). Sol Lesser. **Pr** Lesser. **Dir** Charles Haas, Sandy Howard. **Screenplay** Robert Leach, Frederick Schlick. **Starring** Lesley Bradley (Schroeder), Eve Brent (Jane), Sherman Crothers (Chief Tyana), Saul Gorse (Sikes), William Keene (Lapin), Maurice Marsac (René), Gordon Scott (Tarzan), Rickie Sorensen (Tantu). 74 mins. B/w.

Eventually released theatrically, this tvm was cobbled together from three pilot episodes produced for a tv series that never was. The story is, obviously, fitful, but involves Tarzan thwarting ruthless Schroeder and sidekick René – come to trap big game – and then Schroeder's brother Sikes, bent on revenge, and local trader Lapin, planning to steal the wealth of the lost city of Zarbo. Brent is adequate as Jane; Sorensen, as her and Tarzan's son Tantu, is fairly nauseating. There is considerable rerun wildlife footage from **30** (by Miki Carter) and elsewhere. This is the shoddiest of the **Tarzans** in terms of production; otherwise it is well up to the standard of some of Lesser's others. [JG]

34. *Tarzan's Greatest Adventure* US movie (**1959**). Sy Weintraub/Paramount. **Pr** Weintraub. **Dir** John Guillermin. **Screenplay** Berne Giler, Guillermin. **Starring** Sean Connery (O'Bannion), Scilla Gabel (Toni), Niall MacGinnis (Kriger), Al Mulock (Dino), Anthony Quayle (Slade), Gordon Scott (Tarzan), Sara Shane (Angie). 90 mins. Colour.

Regarded by fans as one of the best, if not *the* best, **Tarzan** movies, this succeeds because of a good script, good direction and, most especially, the pitting of Tarzan against a major villain (played by Anthony Quayle) as opposed to some of the caricatures he had confronted in earlier movies. This was Weintraub's first venture with the character – Lesser had finally given up – and the difference shows.

Tarzan and the murderous Slade are old enemies. Slade – plus criminal colleagues Dino, Kriger and O'Bannion and good-time girl Toni, carnally linked with Slade – plan to gut a diamond mine, and to this end murder a doctor and his staff in order to steal their dynamite. Tarzan and gutsy pilot Angie pursue the murderers, who proceed to murder each other with a view to increasing their share of the spoils. Last to go is of course Slade, whom Tarzan despatches himself. This movie is famous also for the sex-scene-that-never-was: featured in the posters, the entanglement of Tarzan with Angie was cut from the final edit. Tarzan would next meet a fine VILLAIN in **38**. [JG]

35. *Tarzan, the Ape Man* US movie (**1959**). MGM. **Pr** Al Zimbalist. **Dir** Joseph Newman. **Vfx** Robert R. Hoag, Lee LeBlanc. **Screenplay** Robert Hill. **Based on 9**. **Starring** Joanna Barnes (Jane), Cesare Danova (Harry Holt), Robert Douglas (Professor Parker), Denny Miller (Tarzan), Thomas Yangha (Riano). 82 mins. Colour.

A dire remake of **9**, using much of the same footage (tinted to give, supposedly, the effect of having been shot in colour) plus, to add interest, the famous Rubber Crocodile scene from **11**. Miller and Barnes tried to breathe life into their roles, but were scuppered by poor script and direction and by the decision of MGM to make this movie as cheaply as possible. [JG]

36. *Tarzan the Magnificent* (vt *Edgar Rice Burroughs' Tarzan the Magnificent*) US movie (**1960**). Sy Weintraub/Harvey Hayutin/Paramount. **Pr** Weintraub. **Dir** Robert Day. **Screenplay** Day, Berne Giler. **Starring** Earl Cameron (Tate), John Carradine (Abel Banton), Gary Cockrell (Johnny Banton), Lionel Jeffries (Ames), Ron MacDonnell (Ethan Banton), Jock Mahoney (Coy Banton), Betta St John (Fay Ames), Gordon Scott (Tarzan), Alexandra Stewart (Laurie), John Sullivan (Inspector Winters), Charles Tingwell (Dr Conway). 88 mins. Colour.

Tarzan captures Coy Banton – one of a murderous family band of robbers headed by Abel Banton – and proposes to take him to Kairobi and justice. In their first attempt to recover Coy, the Bantons strand a party of whites (Ames and wife Fay, Conway and girlfriend Laurie) up-country; with the help of Tate, a black, Tarzan treks with this party and Coy through rough country to Kairobi. Fay Ames takes a fancy to Coy, and several times betrays the party to the pursuing Bantons. By the time Kairobi is reached and Coy handed over for hanging, she and the remaining Bantons are dead. This is one of the most violent **Tarzans** but hardly magnificent except for the wildlife photography and the scenes among the Kikuyu and Masai. It was Scott's last movie in the part; ironically Mahoney, the psychopathic VILLAIN here, was to be Tarzan in the next two movies (**37** and **38**; ◊◊ **43**). [JG]

37. *Tarzan Goes to India* UK movie (**1962**; vt [dubbed in Hindustani] *Tarzan Mera Sathi* 1974). Sy Weintraub/MGM. **Pr** Weintraub. **Dir** John Guillermin. **Spfx** Roy Whybrow. **Screenplay** Robert Hardy Andrews, Guillermin. **Starring** Mark Dana (O'Hara), Leo Gordon (Bryce), Jai (Jai), Feroz Khan (Raju Kumar), Jock Mahoney (Tarzan), Murad (Maharaja), Simi (Kamara). 86 mins. Colour.

His old friend, the dying Maharaja, begs Tarzan to come to India because ruthless engineer O'Hara is about to flood a valley in order to create a power station, an operation that will require the translocation of thousands of people and, if they cannot somehow be guided out of the valley, the drowning of a herd of 300 elephants. Tarzan performs the feat with the help of Jai the Elephant Boy, Princess Kamara and Raju Kumar, but not before seeing the death of murderous ivory hunter Bryce. This could have been, with its ecological message and its charming stars, one of the all-time Tarzan greats; instead, for some reason, it ends up being just something of a muddle. [JG]

38. *Tarzan's Three Challenges* US movie (*1963*). Sy Weintraub/MGM. **Pr** Weintraub. **Dir** Robert Day. **Spfx** Cliff Richardson, Roy Whybrow. **Screenplay** Day, Berne Giler. **Starring** Earl Cameron (Mang), Christopher Carlos (Sechung), Ricky Der (Kashi), Robert Hu (Nari), Tsu Kobayashi (Cho-San), Jock Mahoney (Tarzan), Woody Strode (Khan/Tarim). 92 mins. Colour.

In an unnamed Oriental nation (in fact, the movie was shot in Thailand) the old ruler, Tarim, is dying. Tarzan is appointed to bring Tarim's successor, a young boy called Kashi, to the Crown City, where Kashi must undergo certain tests to show he is indeed the Chosen One – the REINCARNATION of all the previous rulers. Tarim's brother Khan wishes to install himself as ruler, to be succeeded by his son Nari, and sends assassins and spies to thwart Tarzan's expedition. All are fought off and Kashi passes the tests. Khan challenges the outcome, and Tarzan, on Kashi's behalf, ritually fights him to the death. This is a rather good movie, although Mahoney, as Tarzan, seems old and tired (he was apparently plagued by dysentery during filming). Strode is an impressively stylish villain. [JG]

39. *Tarzan and the Valley of Gold* US movie (*1966*). Sy Weintraub/American International. **Pr** Weintraub. **Dir** Robert Day. **Spfx** Ira Anderson, Ira Anderson Jr. **Screenplay** Clair Huffaker. **Starring** Mike Henry (Tarzan), Nancy Kovack (Sophia), Don Megowan (Mr Train), David Opatoshu (Vinero), Manuel Padilla Jr (Ramel), Francisco Riquerio (Manco). 90 mins. Colour.

A curious mixture – as were all the Mike Henry **Tarzans** – of James Bond and the jungle. Tarzan is summoned by an old friend to South America: a surviving Incan civilization, somewhere in the Andes, has lost the secret of its location to the international criminal Vinero, and it would be good if Tarzan could head off Vinero's men as they attempt to plunder the treasures of the entirely pacifistic Incans. Assisted by the leopard Bianco, the lion Major, the chimp Dinky and the small Incan boy Ramel – plus, in the later stages, Vinero's discarded mistress Sophia – he does exactly this. Vinero literally drowns in gold-dust; Tarzan barehandedly kills the villain's sadistic bodyguard Mr Train. Manco, the ruling Inca, accepts that sometimes a little violence is necessary if peace is to be retained. Tarzan and Sophia blast the tunnel leading to the Incan haven, so that with luck it will be centuries before it is rediscovered. While quite enjoyable, this movie seems much longer than its 90 mins. [JG]

40. *Tarzan and the Great River* US movie (*1967*). Paramount. **Pr** Sy Weintraub. **Dir** Robert Day. **Spfx** Ira Anderson. **Screenplay** Bob Barbash. **Starring** Paulo Grazindo (Professor), Mike Henry (Tarzan), Rafer Johnson (Barcuna), Diana Millay (Ann Phillips), Jan Murray (Sam Bishop), Manuel Padilla Jr (Pepe). 88 mins. Colour.

An above-par offering. Barcuna is cutting a swathe through the Amazon Basin with his revival of the enslaving Jaguar Cult. Tarzan some while ago gave his animal friends to a zoo there, and is now summoned by its curator, the Professor, to tackle Barcuna. Almost at once the Professor is murdered by cultists; Tarzan doffs his civilized clothing to enter the jungle with Cheeta and friendly lion Baron. *En route* he encounters riverboatman Bishop and his "first mate", cute boy Pepe, who cheerfully reprise – especially after being joined by medical missionary Phillips – *The African Queen* (*1951*). The Amazon is full of hippos and flamingoes, and Baron meets another lion in the South American jungle; but these howlers hardly matter – the fact that Dinky, the chimp playing Cheeta, was killed mid-movie for having attacked Henry seems to matter a lot more.

Doc Savage, Man of Bronze (*1975*) owes quite a lot to this movie. [JG]

41. *Tarzan and the Jungle Boy* US movie (*1968*). Sy Weintraub/Paramount. **Pr** Robert Day. **Dir** Robert Gordon. **Spfx** Gabriel Queiroz. **Screenplay** Steven Lord. **Starring** Steve Bond (Jukaro), Ronald Gans (Ken Matson), Alizia Gur (Myrna Claudel), Mike Henry (Tarzan), Edward Johnson (Buhara), Rafer Johnson (Nagambi). 90 mins. Colour.

Six years ago geologist Karl Brunik drowned in Africa, but his infant son Erik survived – much as Tarzan did years before – to become Jukaro the Jungle Boy, dwelling in the depths of Zegunda country. Journalists Claudel and Matson now fly in, a low-flying aerial photographer having caught a snap of Jukaro. Tarzan agrees to try to rescue the lad to civilization, a first difficulty being that the Zegunda law is that any non-Zegunda found on their territory is killed and a second being that Tarzan has earned the enmity of Nagambi, usurper king of Zegunda; a third is that Claudel decides to follow him, with Matson, into Zegunda country. In the end Nagambi is killed by his usurped brother Buharu, who thus recovers his throne; Jukaro decides, on Tarzan's advice, to return with Claudel to civilization. This is an entertaining exploit. [JG]

42. *Tarzan's Deadly Silence* US tvm released theatrically (*1970*). Sy Weintraub/National General. **Pr** Leon Benson. **Dir** Lawrence Dobkin, Robert L. Friend. **Spfx** Laurencio Cordero. **Screenplay** John Considine, Tim Considine, Lee Irwin, Jack A. Robinson. **Starring** Gregorio Acosta (Chico), Ron Ely (Tarzan), Jock Mahoney (Colonel), Michelle Nicols (Ruana), Manuel Padilla Jr (Jai), Woody Strode (Marshak). 88 mins. Colour.

Cobbled together from a two-parter in the tv series, this sees Tarzan defeat a neo-Nazi-style bully, The Colonel, who wants to set up a dictatorship in the jungle. Old faces reappeared in what was not a bad outing: Strode had been a villain in **32** and **38**; Mahoney had been a villain in **36**, then Tarzan in **37** and **38**; Padilla had starred as Tarzan's little-boy friend in **39** and **40**. There is a degree of sadism that might be excised from a modern tv version, as when The Colonel gratuitously whips a man to death. [JG]

43. *Tarzan's Jungle Rebellion* US tvm released theatrically (*1970*). Sy Weintraub/National General. **Pr** Leon Benson. **Dir** William Whitney. **Spfx** Laurencio Cordero. **Screenplay** Jackson Gillis. **Starring** Ron Ely (Tarzan), Jason Evers (Ramon), Sam Jaffe (Dr Singleton), Harry Lauter (Miller), William Marshall (Colonel Tatakombi), Manuel Padilla Jr (Jai), Ulla Stromstedt (Mary Singleton). 92 mins. Colour.

As with **42**, this was cobbled together from a two-part adventure in the tv series. The Singletons come to Africa in search (for academic purposes) of The Blue Stone of Heaven, and Tarzan agrees to help them find it. Tatakombi – a fascist – gets there first, and tries to use the jewel as a means of persuading the various tribes of the jungle to fall in under his dictatorship. Tarzan violently stops such plans in their tracks. Jaffe had decades earlier received recognition as Perrault in LOST HORIZON (*1937*).   [JG]

44. *Jungle Burger* (vt *La Honte de la Jungle*; vt *Shame of the Jungle*; vt *Tarzoon la Honte de la Jungle*; vt *Tarzoon the Shame of the Jungle*) Belgian-French ANIMATED MOVIE (*1975*). SND Valisa/Picha. **Pr** Boris Szulzinger. **Dir** Picha, Szulzinger. **Screenplay** Picha, Pierre Bartier; US screenplay Anne Beatts, Michael O'Donoghue. **Voice actors** John Belushi, Brian Doyle-Murray, Andrew Duncan, Christopher Guest, Bill Murray, Bob Perry, Emily Prager, Johnny Weissmuller Jr (Shame) (none but Weissmuller individually credited). 72 mins. Colour.

Extremely scatological and surprisingly clumsy – although energetically animated – PARODY of the **Tarzan**-movie *oeuvre*; the English-language version, for copyright reasons, is less overt about its origins. From her subterranean "spaceship", deep within what is either a cavern or the rectum of a recumbent giantess, many-breasted Queen Bazunga seeks world conquest, but first must hide her baldness by stealing someone's scalp. A detachment of her army – cloned mobile phalluses that fire explosive semen – seize June, mate of Shame, wimpish lord of the jungle. Shame rescues the nymphomaniacal June just before her scalp is cut off. In a minor subplot, a stereotyped **Tarzan**-movie safari arrives in the jungle from the West; three are devoured and the fourth finally becomes – as replacement for the destroyed Bazunga – a megalomaniac jungle queen set on world conquest. There is some flirtation with racism, and women seem regarded as mere fuck-objects.   [JG]

45. *Tarzan, the Ape Man* US movie (*1981*). MGM/Svengali. **Pr** Bo Derek. **Dir** John Derek. **Screenplay** Gary Goddard, Tom Rowe. **Based on 9**. **Starring** Bo Derek (Jane Parker), Richard Harris (James Parker), John Phillip Law (Harry Holt), Miles O'Keeffe (Tarzan), Maxime Philoe (Riano), Akushula Selayah (Nambia/Africa). 112 mins. Colour.

Orphaned adventuress Jane comes to Africa to find her father, batty Irish explorer Parker. He is mounting a trek to the fabled Elephants' Graveyard, and she goes too. On the Mutia Escarpment they discover the equally fabled Great Inland Sea; more to the point, they hear the night-time cries of legendary Tarzan, reputedly a 100ft-tall half-man half-ape. To this point *T,TAM* promises much. But then the archetypal Derek skinflick starts. Jane doffs all for a swim in the sea, is threatened by a lion that seems confused to find itself by the seaside, and is rescued by Tarzan. He crudely expresses amour, and is in turn driven off by Parker and expedition photographer Holt. Parker's mistress Africa is seized by hostile natives; Parker assumes Tarzan's guilt, and vows revenge. Tarzan abducts Jane; abandons her; saves her from a boa; is almost killed in the process; is nursed back to health by an elephant, chimps, an orang-utan and Jane; and has long but unconsummated byplay with the latter. The whites are all captured by natives and prepared for HUMAN SACRIFICE, which involves much manipulation of a naked Jane; Parker is slain by the monstrous tribal chief. But Tarzan, informed by Cheeta, comes to the rescue with a

herd of elephants, kills the chief, and rescues Holt and Jane, before going off into the jungle with the latter for a life of unbridled PASTORAL passion.

This is widely regarded as the direst of the **Tarzan** movies, but it has enough good bits (including some spectacular photography and moments of exquisite WRONGNESS) that, if cut by about 40 minutes, it would be highly regarded. As it is, it leaves a nasty taste: its intention seems to be to appeal to those who find eroticism in the sexual humiliation of women.   [JG]

46. *Greystoke: The Legend of Tarzan, Lord of the Apes* UK movie (*1984*). Warner/WEA. **Pr** Stanley S. Canter, Hugh Hudson. **Dir** Hudson. **Vfx** Albert J. Whitlock. **Mufx** Rick Baker. **Primate choreography** Peter Elliot. **Primate consultant** Roger Fouts. **Screenplay** Michael Austin, P.H. Vazak (Robert Towne). **Based on** *Tarzan of the Apes* (**1914**) by Burroughs. **Starring** John Alexander (White Eyes), Alisa Berk (Kala), Cheryl Campbell (Lady Alice Clayton), Elliot Cane (Silverbeard), James Fox (Lord Esker), Paul Geoffrey (Lord Jack Clayton), Ian Holm (Capitaine Phillippe D'Arnot), Christopher Lambert (John Clayton, Tarzan), Eric Langlois (young Tarzan), Alison Macrae (young Jane), Andie MacDowell (Jane Porter), Daniel Potts (young Tarzan), Ralph Richardson (6th Earl of Greystoke), John Wells (Sir Evelyn Blount). **Voice actor** Glenn Close (Jane Porter's voice). 130 mins. Colour.

In 1885 Jack and Alice Clayton go from Scotland to Africa as missionaries, are shipwrecked and build a rude home in the jungle. She soon gives birth to a boy but dies of malaria; the boy's father is almost immediately killed by one of the local colony of great APES, another of whom, Kala – who has just lost her own baby – seizes the infant and becomes his adoptive mother. The boy is reared as an ape, being an especially revered one when he discovers his father's knife and learns how to use it. When he is 12, pygmies kill Kala, and the boy retaliates by killing one of them. When he is full-grown, a UK zoological expedition arrives but, in response to their murderous ways, is massacred by the pygmies. One expedition member, D'Arnot, survives and is nursed back to health by the boy, who meanwhile kills the vicious ape-leader White Eyes to become bull of the colony. D'Arnot discovers the Claytons' treehouse, realizes the mysterious white savage is John Clayton, heir to the Greystoke title, and persuades him to quest through the jungle to reach, eventually, Britain. There Tarzan/John is at once accepted by his grandfather, the 6th Earl of Greystoke, as the true heir, and meets Greystoke's US ward, the lovely Jane. D'Arnot leaves young John with Greystoke; John and Jane soon fall in love, despite Jane's foppish suitor Lord Esker. Greystoke is killed in a domestic accident, and D'Arnot returns for the funeral. One day, at the British Museum, John discovers where the experimental apes are held and finds among them his adoptive father, Silverbeard, whom he releases; but a police marksman soon shoots Silverbeard, who dies in John's arms. This is the final straw for John so far as "civilization" is concerned: he returns to Africa and the beloved apes whose rightful leader he is.

Probably the most successful of the CINEMA's attempts during the 1980s and 1990s to breathe new life into the popular fantasy ICONS created during the earlier parts of the century, this movie had a tortuous history, being first written in the mid-1970s by Towne, who planned also to direct. For various reasons it was passed to Hudson, who had the screenplay substantially revised and then set out to create a masterpiece. Masterpiece this movie may not be, but it falls little short: there is an air of authenticity that transcends the star-studded cast, and the tale is genuinely involving; the photography is stunning, as are most of the key performances. The authenticity extends to the "choreography" of the apes and of John Clayton as a human reared among apes: this was derived not at random but from existing (and some fresh) studies of primate behaviour in the wild. In many ways *G* can be regarded as the first *real* **Tarzan** movie. It encourages one to reassess upwards the worth of Burroughs's creation. [JG]

47. *Tarzan in Manhattan* US movie (*1989* tvm). American First Run. **Pr** Charles Hairston. **Dir** Michael Schultz. **Screenplay** William Gough, Anna Sandor. **Starring** Kim Crosby (Jane Porter), Tony Curtis (Archimedes Porter), Joe Lara (Tarzan), Joe Seneca (Joseph), Jimmy Medina Taggert (Juan Lipschitz), Jan Michael Vincent (Brightmore). 100 mins. Colour.

This pilot for an unmade series is in effect a partial remake of *18*. Tarzan comes to NEW YORK to avenge the killing of the ape Kala and recover a kidnapped Cheeta. After various standard ENFANT-SAUVAGE miscomprehensions, he is befriended by cabdriver Jane Porter and, grudgingly, her private-eye father Archimedes. Cheeta is traced to the Brightmore Foundation, which is secretly experimenting in genetic engineering using apes for (fatal) trials. Brightmore himself tries with his thugs to kill Tarzan, but Tarzan kills him. Jane and Archimedes persuade Tarzan and Cheeta to stay on a while in Manhattan. [JG]

48. **Others** There has been a plethora of non-English-language **Tarzan** movies, most of them rarely seen in the West. They include: *Toofani Tarzan* Indian movie (*1937*), *Tarzan ki Beta* Indian movie (*1938*), *The Adventures of Chinese Tarzan* Singaporean 3-part movie (*1939-40*), *Toto Tarzan* Italian movie (*1950*), *Thozhan* Indian movie (*1959*), *Toofani Tarzan* Indian movie (*1962*), *Tarzan aur Gorilla* Indian movie (*1963*), *Rocket Tarzan* Indian movie (*1963*), *Tarzan aur Jadugar* Indian movie (*1963*), *Tarzan and Captain Kishore* Indian movie (*1964*), *Tarzan and Delilah* Indian movie (*1964*), *Tarzan aur Jalpari* Indian movie (*1964*), *Jungle Tales of Tarzan* West German movie (*1964*), *Tarzan and the Jewels of Opar* Jamaican movie (*1964*), *Tarzan and King Kong* Indian movie (*1965*), *Tarzan Comes to Delhi* Indian movie (*1965*), *Tarzan and the Circus* Indian movie (*1965*), *Tarzak Against the Leopardmen* Italian movie (*1965*), *Tarzan and Hercules* Indian movie (*1966*), *Tarzan ki Mehbooba* Indian movie (*1966*), *Tarzan aur Jadui Chirag* Indian movie (*1966*), *Tarzans Kampf mit dem Gorilla* West German short (*1968*), *Tarzan in Fairy Land* Indian movie (*1968*), *Tarzan .303* Indian movie (*1970*), *Tarzan of Bengal* Bangladeshi movie (*1976*), *Jane is Jane Forever* West German movie (*1978*), *La Infancia de Tarzan* Spanish movie (*1979*) and *Tarzan Andy* Italian movie (*1980*). Legal action by the Burroughs estate changed the titles of what were originally *Tazanova Smrt* Czechoslovakian movie (*1962*; vt *The Death of Tarzan* 1968 US), *Tarzan Chez les Coupeurs de Tête* Italian movie (*1963*), *Tarzan Roi de la Force Brutale* Italian movie (*1963*) and *Tarzan des Mers* USSR movie (*1963*). Parodies, aside from *44*, have included *Nature in the Wrong* US movie (*1932*), *Tarzan and Jane Regained Sort of . . .* US movie (*1964*) and the soft-porn *Tarzan and Boy* US movie (*c1955*), *Tarzan the Swinger* US movie (*1970*) and *Jungle Heat* US movie (*?1995*; vt *Dard'zan: The Humiliation of Jane* France). Two possibly ghost titles are «Tarzan's Beloved» Indian movie (1964) and «Tarzan and Cleopatra» Indian movie (1965).

Further reading: *Kings of the Jungle: An Illustrated Reference to "Tarzan" on Screen and Television* (*1994*) by David Fury provides sterling and generally extremely accurate coverage of all except *44* and *48*.

TAVERNS ◊ INNS.

TAYLOR, KEITH (JOHN) (1946-   ) Australian writer whose work is primarily CELTIC FANTASY. His best-known character is **Felimid mac Fal**, an Irish Bard, descended from both the Druids and the ancient FAIRY race of the Tuatha de Danann. With the aid of his harp, Felimid can create magic and control the forces of nature (◊ MUSIC). This is the main fantasy element in a series basically rooted in the historical reality of 6th-century Britain, at the time of the Saxon invasions. The series began with four short stories published in *Fantastic*, starting with "Fugitives in Winter" (1975), plus a fifth in the anthology *Swords Against Darkness II* (anth *1977*) ed Andrew J. OFFUTT; all bore the pseudonym Denis More. KT subsequently revised the stories as *Bard* (fixup *1981* US), the first of five novels about Felimid's wanderings in Britain and northern Europe. His Dark Age Britain is convincing and authentic, revealing a deep knowledge of the history and cultures of the period.

The later novels develop the character of Felimid but become rather more traditional in approach. In *Bard II* (*1984* US; vt *The First Long Ship* 1985 UK) Felimid encounters the female PIRATE Gudrun Blackhair, who has captured the magic ship *Ormungandr*, created by the DWARFS of Norse legend (◊ NORDIC FANTASY). Felimid falls in love with Gudrun and their adventures in the magical ship continue in *The Wild Sea* (*1986* US) and *Ravens' Gathering* (*1987* US). The final novel, *Felimid's Homecoming* (*1991* UK), brings Felimid back to Ireland to discover that much has changed since he left. Overall, the series is one of the best and most cohesive of Celtic fantasies, with well drawn characters and a sense of history. *The Wild Sea* won the Australian Ditmar AWARD.

After the **Bard** novels, KT turned to another series set in the early, turbulent days of Ireland. The **Danans** series tells of the attempt to unite two constantly warring tribes by arranging a marriage between the Queen of the Danans and the Chief of the Freths. An evil prince, however, has his own designs on Ireland and seeks to thwart the marriage. The series runs: *The Sorcerer's Sacred Isle* (*1989* US), *The Cauldron of Plenty* (*1989* US) and *Search for the Starblade* (*1989* US), the last a rather more formulaic QUEST novel.

KT's only solo novel to date is *Lances of Nengesdul* (*1982*), a PLANETARY ROMANCE in which a circus midget and acrobat is transported to a low-gravity planet and finds he is now superhuman. KT also wrote two novels with Andrew J. Offutt, continuing the **Cormac macArt** series by Robert E. HOWARD: *When Death Birds Fly* * (*1980* US) and *The Tower of Death* * (*1982* US). Among KT's short fiction, "Spirit Places" (1985 *in Faery!* ed Terri WINDLING), where spirits vie for rebirth in a new-born child, is a rare use of local Aboriginal MYTHS. [MA]

TAYLOR, ROGER (1938-   ) UK writer of EPIC FANTASY, originally influenced by hearing *The Lord of the Rings* (*1954-5*) by J.R.R. TOLKIEN dramatized on the BBC's Third Programme. The common thread in his writing is the thesis that the true warrior's way is achieved only by a person who uses the dark side of his nature for the protection of those weaker than himself. RT's first published work was

The Chronicles of Hawklan – *The Call of the Sword* (**1988**), *The Fall of Fyorlund* (**1989**), *The Waking of Orthlund* (**1989**) and *Into Narsindal* (**1990**) – a multilayered good-versus-evil saga. *Dream Finder* (**1991**) is set in a world based roughly on the city-states of Italy and again has enemies uniting against a common foe. His longest single work to date is **Nightfall**, written as one book but split into two by the publisher as *Farnor* (**1992**) and *Valderen* (**1993**), again utilizing RT's favourite device of a flawed hero forced to grow and accept his own powers and limitations before he can triumph over the enemy. *Whistler* (**1994**) takes RT into less familiar territory as his usual fantasy trappings are hung on a religious and political SATIRE (priest discovers power and forms his own fundamentalist cult to rule the world). *Ibryen* (**1995**) again has a flawed protagonist who must meet violence with violence to overcome the greater evil.   [JF]

TECHNOFANTASY In simplest terms, technofantasy is FANTASY that has scientific/technological trappings, or uses scientific/technological tools: it is distinguished from SCIENCE FICTION in that there is no attempt to justify such use in scientific or quasiscientific terms (sometimes there is a bit of gobbledygook, but both creator and audience know this for what it is) – although the closeness of the two forms may be seen if we regard Arthur C. Clarke's suggestion that "any sufficiently advanced technology is indistinguishable from magic" as a literary rather than a technological observation. More importantly, fantasy is not a genre whose subject matter should be confined to historic or fictional pasts – to GOLDEN AGES or LANDS OF FABLE, or to medieval-style SECONDARY WORLDS: fantasies can as well be set in the present or in futures as anywhere/anywhen else. Thus a robot capable of casting SPELLS that work is a creature of fantasy, not sf.

The ancestor of the technofantasy subgenre can be viewed as Mary SHELLEY's *Frankenstein, or The Modern Prometheus* (**1818**; rev **1831**), a novel which, while often described as the first true sf novel, in fact contains no science whatsoever outside vague allusions to the mysterious powers of electricity (the FRANKENSTEIN MOVIES, with their technological extravaganzas, may blind us to this) and is anyway much more concerned with a fantasy theme: the nature of GOOD AND EVIL. A like concern underlies Robert Louis STEVENSON's *Strange Case of Dr Jekyll and Mr Hyde* (**1888**) – a major difference between this and the similarly themed *The Portrait of Dorian Gray* (**1891**) by Oscar WILDE is that the former is a technofantasy and the latter defiantly not. H.G. WELLS's *The Invisible Man* (**1897**) is another novel which is often viewed as sf but which is arguably technofantasy (◊ INVISIBILITY).

Examples of technofantasy are too many to cite. Most have appeared during the past few decades, as technology has replaced GOD in the Western mind. Technofantasies see people fall into video games (as in Andrew M. Greeley's *God Game* [**1986**]) or tv programmes (as in STAY TUNED [**1992**]); spirits may communicate with the living via tv screens (as in the POLTERGEIST series and *Scrooged* [**1988**]; ◊ *A* CHRISTMAS CAROL); automobiles may be possessed (◊ POSSESSION) as in Stephen KING's *Christine* (**1983**) and especially *The Car* (**1977**), or somehow become sentient (◊ HERBIE MOVIES); such traditional entities as FAIRIES and FAERIE may be substituted by technology-derived but equally fantastic counterparts (as in *The* BRAVE LITTLE TOASTER [**1987**] and TOYS [**1992**]); technology may be used to create ZOMBIES (as in Ira LEVIN's *The Stepford Wives*

[**1972**]); a nuclear plant may adopt many of the attributes of deity (as in AKIRA [***1989***]); the DEVIL may be a machine (as in BILL & TED'S BOGUS JOURNEY [***1991***] and, in a different way, Jeremy Leven's *Satan* [**1982**]); a super-sophisticated robot may be capable of SHAPESHIFTING – a notion drawn from FOLKLORE and thus a core fantasy notion – as in *Terminator 2: Judgement Day* (**1991**); or . . . Such fantasy themes as the GOLEM, a fictitious artefact of an earlier science, can be approached from two such different directions as those in *He, She and It* (**1991**; vt *Body of Glass*) by Marge Piercy (1936-), which cobbles a fantasy notion onto purported sf, and Terry PRATCHETT's *Feet of Clay* (**1996**), which creates ridiculous quasi-science in order to justify his golems in a fantasy venue.

A particular technofantastic artefact spawned by the CINEMA is beginning to make its appearance in written fiction as well: the TOON, the animated character given a real existence in the human world. Also related to technofantasy are those fantasies couched in what are certainly sciencefictional terms, like Sheri S. TEPPER's *Raising the Stones* (**1990**), in which LEGEND and REALITY overlap, GODS existing in the real world but as a symbiotic fungus. Christopher STASHEFF has taken this PLANETARY-ROMANCE form of the subgenre to its extreme. And sf's cyberpunk subgenre is perhaps the ultimate expression of technofantasy: usually having no real relation to the potentials of current science, it allows its protagonists to indulge in such activities as flying through what are effectively FANTASYLANDS – where they QUEST, encounter and defeat the technofantastic equivalent of DRAGONS, and may even find LOVE.   [JG]

See also: SCIENCE FANTASY; STEAMPUNK.

TEENAGE MUTANT NINJA TURTLES Team of fantasy COMICS characters created by Kevin Eastman (1962-) and Peter Laird (1954-) in an independently published b/w comic book from 1984. They are four pizza-loving humanized turtle troubleshooters who live in a New York sewer and were transmuted through the action of radioactive mud. Intended at first as a parody of martial arts SUPERHERO TEAMS, they became enormously popular, making riches for their creators, who went on to create a very enlightened and forward-looking but ultimately ill-fated comic-book publishing company, Tundra.

A long-running animated tv children's series started in 1987.   [RT]

See also: TEENAGE MUTANT NINJA TURTLES MOVIES.

TEENAGE MUTANT NINJA TURTLES MOVIES The TEENAGE MUTANT NINJA TURTLES have featured in a number of movies.

1. *Teenage Mutant Hero Turtles: How it All Began* US movie (**1988** tvm). Murakami/Wolf/Swenson/Group W. **Sv pr** Fred Wolf. **Pr** Walt Kubiak. **Sv dir** Bill Wolf. **Anim dir** Chi Hyun Hwang, Soo An Kim, Dae Sick Moon. **Voice actors** Cam Clarke (Leonardo), Townsend Coleman (Michaelangelo), Barry Gordon (Donatello), Rob Paulsen (Raphael). 50 mins. Colour.

Using material drawn from the tv series, this, despite its title, gives origin story of the Turtles and their guru, the rat Splinter, for only the first few minutes. (The rest consists of some standard adventures.) Back in Japan, the Shredder was sneakily responsible for having Splinter thrown out of his martial-arts clan. In the USA, Splinter dwelt as a human in the sewers, having as pets some rats plus the baby turtles a child had accidentally dropped down a drain. A vengeful Shredder was responsible for pouring down a substance which mutated any creature into a cross between itself and the last animal it had been touching. The turtles – Donatello, Leonardo, Michaelangelo (*sic*) and Raphael – had just been in contact with Splinter, and Splinter had most recently handled a rat.   [JG]

2. *Teenage Mutant Ninja Turtles* US movie (*1990*). Golden Harvest/Limelight/Gary Propper. **Pr** David Chan, Kim Dawson, Simon Fields. **Exec pr** Raymond Chow. **Dir** Steve Barron. **Animatronics** Jim HENSON's Creature Shop. **Screenplay** Bobby Herbeck, Todd W. Langen. **Novelization** *Teenage Mutant Ninja Turtles* * (*1990*) by Dave Morris. **Starring** David Forman (Leonardo), Judith Hoag (April O'Neil), Elias Koteas (Casey Jones), Josh Pais (Raphael), James Saito (Shredder), Michelan Sisti (Michaelangelo), Leif Tilden (Donatello), Michael Turney (Danny Pennington). **Voice actors** Kevin Clash (Splinter), Corey Feldman (Donatello), David McCharen (Shredder), Michael McConnohie (Tok Chiu), Josh Pais (Raphael), Robbie Rist (Michaelangelo), Brian Tochi (Leonardo). 93 mins. Colour.

Petty crime is stalking the city, thanks to the organization of a vast subculture of teenage boys by the Shredder. Pitted against him are the Turtles and Splinter. Tv reporter O'Neil is drawn in because she suspects the true nature of the crime wave; a chance encounter adds amateur street vigilante Jones to the forces of virtue. In course of a complicated plot, it emerges that years ago the Shredder murdered Splinter's master.

TMNT is of interest primarily because it has far more integrity than anyone predicted – movies based on merchandised characters do not have a good track record. Despite the unlikeliness of the SUPERHEROES, the cast put in committed performances, responding to a script that takes its duties seriously, mixing SURREALISM and MAGIC REALISM – a subtext is that the weird is all around us, but that we usually fail to notice it – and recognizing that situational HUMOUR is at its best when allied to a driving narrative. Also of note is the quality of the animatronics.

TMNT was severely cut in the UK to excise the Turtles' use of *nunchaku*, regarded as likely to provoke infants to wanton acts of imitative violence.   [JG]

3. *Teenage Mutant Ninja Turtles II: The Secret of the Ooze* US movie (*1991*). 20th Century-Fox/Golden Harvest/Gary Propper. **Pr** David Chan, Kim Dawson, Thomas K. Gray. **Exec pr** Raymond Chow. **Dir** Michael Pressman. **Animatronics** Jim HENSON's Creature Shop. **Screenplay** Todd W. Langen. **Starring** Mark Caso (Leonardo), François Chau (Shredder), Kevin Clash (Splinter), Ernie Reyes Jr (Keno), Michelan Sisti (Michaelangelo), Leif Tilden (Donatello), Kenn Troum (Raphael), Paige Turco (April O'Neil), David Warner (Jordan Perry). **Voice actors** Adam Carl (Donatello), Laurie Faso (Raphael), David McCharen (Shredder), Robbie Rist (Michaelangelo), Brian Tochi (Leonardo). 87 mins. Colour.

This lacklustre sequel largely substitutes zany martial-arts brawls and dance routines for a plot. A research institute has discovered and is disposing of canisters of the radioactive slime responsible, 15 years ago, for mutating the Turtles. The Shredder, surviving his apparent demise at the end of **2**, captures one canister plus research scientist Perry, and forces the latter to create giant mutant versions of a wolf and a snapping-turtle. Aided by Perry and by martial-arts-devotee pizza-delivery boy Keno, the Turtles triumph over the villains, even the monster version of himself the

Shredder creates by drinking the ooze. Turco produces a surprisingly committed performance as a replacement for Hoag; Warner seems embarrassed by proceedings.   [JG]

4. *Teenage Mutant Ninja Turtles III: The Turtles are Back . . . In Time* US movie (*1992*). 20th Century-Fox/Golden Harvest/Clearwater/Gary Propper. **Pr** David Chan, Kim Dawson, Thomas K. Gray. **Exec pr** Raymond Chow. **Dir** Stuart Gillard. **Spfx** Joseph P. Mercurio. **Creature fx** Eric Allard, Rick Stratton. **Vfx** Richard Malzahn, Jeffrey A. Okun. **Screenplay** Gillard. **Starring** John Aylward (Niles), Mark Caso (Leonardo), David Fraser (Michaelangelo), Henry Hayashi (Kenshin), Matt Hill (Raphael), Elias Koteas (Casey Jones/Whit), Travis A. Moon (Yoshi), James Murray (Splinter), Jim Raposa (Donatello), Sab Shimono (Lord Noringa), Paige Turco (April O'Neil), Stuart Wilson (Walker), Vivian Wu (Mitsu). **Voice actors** Corey Feldman (Donatello), Tim Kelleher (Raphael), James Murray (Splinter), Robbie Rist (Michaelangelo), Brian Tochi (Leonardo). 96 mins. Colour.

Japan, 1603: the warlord Noringa is waging fierce war on rebels led by the beautiful Mitsu, loved by Noringa's son Kenshin. The warlord is armed by caddish English gun merchant Walker. New York City, today: April O'Neil rubs an antique Japanese sceptre and is instantly transported back to 1603; simultaneously Kenshin, handling the sceptre in his father's temple, is transported to 1992. The four turtles rush in pursuit of April, using the sceptre to exchange themselves with four of Noringa's guards. After various adventures the lovers are united, the villains are vanquished, and everyone is restored to their own times – although, in the most interesting part of the movie, Michaelangelo presents a very convincing argument for staying in 1603.

As a TIME-TRAVEL tale this is full of holes, but plenty of high spirits make it an enjoyable watch. It is seriously marred, however, by the fact that Walker and his relationship with his sidekick Niles seem lifted – right down to the accents – from Blackadder and Baldrick of the UK tv comedy series *Blackadder*. This was a significantly better movie than **3**; but, by the time it was released, Turtlemania was history.   [JG]

TEEN WOLF A series of two movies.

1. *Teen Wolf* US movie (*1985*). Atlantic. **Pr** Mark Levinson, Scott Rosenfelt. **Exec pr** Thomas Coleman, Michael Rosenblatt. **Dir** Rod Daniel. **Mufx** Tom Burman, Jefferson Dawn, Kyle Tucy. **Screenplay** Joseph Loeb III, Matthew Weisman. **Starring** Michael J. Fox (Scott Howard), James Hampton (Harold Howard), Susan Ursitti (Boof). 91 mins. Colour.

Highschool student Scott discovers to his initial horror that he is born of a line of WEREWOLVES. But when his lycanthropy suddenly becomes overt in the middle of a basketball game he finds the condition is not without advantages: his athletic superiority soon makes him the school hero as he brings the basketball team to the championship finals. The adulation, however, makes him insufferable; it takes the love of a good teenager (his girl-next-door childhood friend Boof) to make him realize that, if the idolization is directed towards the lupine MASK, it isn't worth anything, and that the only achievements that matter are those you earn yourself.

The moral is trite, but nicely handled – as is much of the rest of this amiable teen movie, despite suffering from a tendency towards blandness, as epitomized by Fox's somewhat androgynous charms: even as a wolf, he would earn the approbation of most mothers. (He was in his mid-20s when making *TW*.) Hampton and Ursitti give fine support performances. The glossiness and commercial success of this movie contrast bizarrely with its obvious precursor, *I Was a Teenage Werewolf* (*1957*). [JG]

2. *Teen Wolf Too* US movie (*1987*). Atlantic. **Pr** Kent Bateman. **Exec pr** Thomas Coleman, Michael Rosenblatt. **Dir** Christopher Leitch. **Mufx** John Logan, Michael Smithson. **Screenplay** R. Timothy Kring. **Starring** Jason Bateman (Todd Howard), Estee Chandler (Nicki). 94 mins. Colour.

Todd, cousin of **1**'s Scott, goes to Hamilton University to study biology, and is forced into the boxing team. Thereafter the plot is essentially that of **1** except that boxing replaces basketball, bullying is glorified, and much merriment is garnered from the infliction of brain damage; the pits are reached when the cast stage a hilarious fight hurling live frogs at each other. Some follow-ups are better than their originals, but not this one. [JG]

TELEKINESIS ◊ TALENTS.

TELEPATHY ◊ TALENTS.

TELEPORTATION ◊ TALENTS.

TEMPLATES A template series uses some repeated GENRE-FANTASY situation or PLOT DEVICE as a reliable generator of STORY in new settings and contexts. Much of Michael MOORCOCK's **Elric** sequence revolves template-fashion around the doomed prince's dependence on the SWORD Stormbringer which makes him a psychic VAMPIRE; he is separated from it, seeks herbal substitutes for its dangerous strength, encounters similar swords, regrets its recurring habit of killing his friends, etc. Some templates – like the problems of virtuous vampires, or streetwise/punk/rock'n'roll vampires and ELVES in US CITIES – have been exploited by multiple authors and have thus grown into subgenres. [DRL]

TEMPORAL ADVENTURESS A figure of late-20th-century fantasy. The roots of the TA can be traced back to the 19th century, where precursors can be detected in characters like H. Rider HAGGARD's SHE, an AVATAR of the GODDESS who travels through TIME by means of REINCARNATION/IMMORTALITY rather than TIMESLIP. The most important source for the TA is, however, probably the protagonist of Virginia WOOLF's *Orlando* (**1928**), who begins life in Elizabethan England as an androgynous-seeming young man, becomes a woman in Turkey, timeslips back to Augustan England, jumps a further century forward, and finally into 1928, effortlessly escaping, through these shifts and jumps, any attempts to define her role, her GENDER, her being. There is a slippery invulnerability about her passage which conveyed a dreamlike appeal to later writers, though it irritated Angela CARTER, whose own TA-like figures almost constantly pay for every somersault they make from sex to sex, from world to world. After *Orlando*, the central model for the TA may be the swashbuckling figure of Jirel, a SWORD-AND-SORCERY freelance who appears in various stories by C.L. MOORE, variously collected but most conveniently assembled in *Jirel of Joiry* (coll of linked stories **1969**; vt *Black God's Shadow* 1977). She supplies a necessary element of athleticism to the model.

All these influences are wedded together in two characters – Una Persson and Catherine Cornelius – who appear throughout Michael MOORCOCK's **Cornelius** sequence, most specifically in *The Adventures of Una Persson and Catherine Cornelius in the Twentieth Century* (**1976**), engaging in espionage, philosophy and debate, jumping from one aspect of the MULTIVERSE to another whenever cornered, always surviving. If there is a problem with the TA, it is perhaps that the feminist (◊ FEMINISM) message she promulgates can justify a certain smugness in her depiction. She tends to be all too smoothly immune to the anguish and death she herself is quite capable of creating, and of timeslipping out of reach of consequences. [JC]

TENNANT, EMMA (CHRISTINA) (1937-   ) UK writer whose first novel – *The Colour of Rain* (**1964**) as by Catherine Aydy – had little impact, perhaps through the inadequacy of its emulation of the style of her then father-in-law, Henry Green (1905-1973). Her first novel as ET, *The Time of the Crack* (**1973**; vt *The Crack* 1978), is understandable as an sf exploration of the metaphor of APOCALYPSE, though the eponymous crack – a mysterious faultline running through the bed of the Thames – is given no naturalistic explanation and is treated as the centrepiece in an Expressionist LANDSCAPE, a metaphysical barrier the book's female protagonists seek to cross (or transcend) in order to reach an imagined matriarchal ARCADIA on the "Other Side". ET's second acknowledged novel, *The Last of the Country House Murders* (**1974**), set in the future, is a clumsy parody of classic detective tales.

It is the novels of the next half-decade – though a few more recent books are also of fantasy interest – which established ET's reputation as a fabulist and in some eyes a significant Magic Realist (◊ MAGIC REALISM), who was often compared to Angela CARTER and writers like Gabriel GARCÍA MÁRQUEZ. In *Hotel de Dream* (**1976**) the DREAMS of various hotel residents, who find themselves sleeping compulsively, begin to interact with each other, and eventually corrupt the staid English city that surrounds the dreamers. *The Bad Sister* (**1978**) is the first of two fantasias on 19th-century Scottish novels in which DOUBLES – human or DEVIL – haunt defective humans; in this case the model is James HOGG's *Private Memoirs and Confessions of a Justified Sinner* (**1824**), and the sinner's Calvinism is replaced by a murderous FEMINISM. Robert Louis STEVENSON's *Strange Case of Dr Jekyll and Mr Hyde* (**1888**) similarly underlies *Two Women of London: The Strange Case of Ms Jekyll and Mrs Hyde* (**1989**), a tale whose Hyde figure murders a suspected rapist, and draws her respectable double downwards into madness. Both novels – along with *Faustine* (**1992**), which re-envisions FAUST – were assembled as *Travesties* (omni **1995**). In *Wild Nights* (**1979**) a Scottish childhood is recounted as a perilous, magic-filled ordeal. *Alice Fell* (**1980**) makes REVISIONIST-FANTASY play with a conflation of the myth of Persephone and Lewis CARROLL's *Alice's Adventures in Wonderland* (**1865**): the UNDERWORLD the protagonist falls into is the underside of LONDON.

Later work by ET has generally departed the fantastic, though several titles are of interest, including: *The Boggart* (**1980**) and *The Ghost Child* (**1984**), both CHILDREN'S FANTASIES; *The Magic Drum: An Excursion* (**1989**), a GHOST STORY; and *Sisters and Strangers: A Moral Tale* (**1990**), in which ADAM AND EVE, having survived to the present day, participate in a feminist vision of an ALTERNATE WORLD. [JC]

TENNESHAW, S.M. [hn] ◊ John W. JAKES.

TENNIEL, [Sir] JOHN (1820-1914) London-born UK artist and illustrator with a fine pen-line style derived from his early work as an engraver. He is most famous for his memorable ILLUSTRATIONS to Lewis CARROLL's **Alice** books. JT studied briefly at the Royal Academy, and despite being blinded in one eye in a fencing match with his father was a superb draughtsman. His illustrations to *Aesop's Fables* (1848) brought him to the attention of *Punch* magazine, for whom he worked for 50 years. His commission to illustrate *Alice in Wonderland* (**1865**) was not a happy one, since Carroll himself had hoped to provide the illustrations and was resentful about the publisher's rejection of his work. JT was reluctant to illustrate the sequel, but was eventually prevailed upon to do so – with great success. He was knighted in 1893. [RT]

TENNYSON, ALFRED LORD (1809-1892) UK poet. ◊ ARTHUR; RESURRECTION; ROMANCE; SONG.

TEOPHILO, DE RODOLPHO (1853-1922) Brasilian writer. ◊ BRAZIL.

TEPPER, SHERI S. (1929-   ) US writer whose extensive publications include sf, mystery (as by A.J. Orde, B.J. Oliphant), HORROR (under her own name and as E.E. Horlak) and poetry (under her then married name, Sheri S. Eberhart). Most of her work has been sf. Her first published fiction, the first of a trio of linked trilogies known collectively as the **True Game** series, is dramatized in terms of fantasy yet technically is sf: this readiness to conflate the conventions of distinct genres has characterized her work since, so that much of her writing is perhaps best described as TECHNOFANTASY.

In the rush of her first years as a published novelist, SST produced both of her two unqualified ventures into fantasy: the long novel *The Revenants* (**1984**) and the first volume of the **Marianne** trilogy, *Marianne, the Magus and the Manticore* (**1985**), assembled with *Marianne, the Madame and the Momentary Gods* (**1988**) and *Marianne, the Matchbox, and the Malachite Mouse* (**1989**) as *The Marianne Trilogy* (omni **1990** UK) – before the **True Game** sequence was fully published. Like the **True Game** novels (in which people with TALENTS are ranked in their hierarchical society rather like human CHESS pieces), the **Marianne** trilogy involves GAME-playing: the complex elements in SST's multivolume tales seem to require the organizing framework of a rule-bound pattern.

SST's subsequent books have tended to be more or less independent (if rather long) novels. *The Awakeners: Northshore* (**1987**) and *The Awakeners: Southshore* (**1987**) – the two volumes, written as one work, also published together as *The Awakeners* (**1987** UK) – are again written in the manner of fantasy although plainly set on another planet; SST was to continue combining sf and fantasy conventions as late as *A Plague of Angels* (**1993**). Only *Beauty* (**1991**; rev [preferred text] 1992 UK), which begins as a revisionist version of SLEEPING BEAUTY set in the 14th century (but soon is interrupted by a visitation of time travellers from a polluted 21st century), seems even for a while to be REVISIONIST FANTASY. Although her later work admixes fantasy elements, sometimes in dollops – *Grass* (**1989**), though set on an exotic planet settled by a galactic civilization, possesses various scene-setting elements from "medieval fantasy" – SST seems to have shifted her focus to sf. *Grass* was the first of a loose trilogy, the other two volumes being *Raising the Stones* (**1990**) and *Sideshow* (**1992**); though they cannot be classified as pure fantasy (although they are close to INSTAURATION FANTASIES), it is also very hard to regard them, despite their distant settings, as pure sf. The latter two are both concerned with the nature of godhood (◊ GODS): in the former the emphasis is on the relationship between mortals and gods, who are in this

instance manifest and "scientifically" rationalized; in the latter, the viewpoint shifts more to that of gods and their desire to go unworshipped, or even to divest themselves of godhood. Despite some longueurs, the books repay repeated reading, for SST is delving in very deep waters – and using fantasy techniques to probe some of the legitimate concerns of full fantasy.

Against SST's narrative fluency and exuberant invention must be weighed her tendency towards polished glibness and her occasional moment of stark improbability as she defaults into a stock PLOT DEVICE. Midway through the century's last decade, she seems characteristic of her time – it is a trait shared by such dissimilar writers as John CROWLEY and Dan SIMMONS – in combining genre conventions traditionally kept distinct; the storyteller becomes synthesist, subsuming diverse elements into audacious and unprecedented wholes. [GF/JCB]

Other works: The **True Game** sequence, being *King's Blood Four* (**1983**), *Necromancer Nine* (**1983**), *Wizard's Eleven* (**1984**) – these 3 assembled as *The True Game* (omni **1985** UK) – *The Song of Mavin Manyshaped* (**1985**), *The Flight of Mavin Manyshaped* (**1985**), *The Search of Mavin Manyshaped* (**1985**) – these 3, which come first in internal chronology, assembled as *The Chronicles of Mavin Manyshaped* (omni **1986** UK) – *Jinian Footseer* (**1985**), *Dervish Daughter* (**1985**), *Jinian Star-Eye* (**1986**) – these 3 assembled as *The End of the Game* (omni **1987**); *Blood Heritage* (**1986**) and its sequel, *The Bones* (**1987**), are superficially HORROR but read like CONTEMPORARY-FANTASY comedies; *The Gate to Women's Country* (**1988**), theoretically sf but reading like a fantasy, exploring topics of FEMINISM; *Still Life* (**1989** as by E.E. Horlak; 1989 UK as by SST), horror; *Shadow's End* (**1994**); *Gibbon's Decline and Fall* (**1996**), SCIENCE FANTASY.

TERMINATOR 2: JUDGEMENT DAY US movie (*1991*). ◊ TECHNOFANTASY.

TERROR AUSTRALIS ◊ MAGAZINES.

TERROR OF MECHAGODZILLA vt of *Mekagojira No Gyakushu* (*1975*). ◊ GODZILLA MOVIES.

TERROR TALES (magazine) ◊ MAGAZINES.

TESTAMENT D'ORPHÉE, LE, OU NE ME DEMANDEZ PAS POURQUOI (vt *The Testament of Orpheus*) French movie (*1959*). Editions Cinégraphique/Cinédis. **Pr** Jean Thuillier. **Dir** Jean COCTEAU. **Screenplay** Cocteau. **Spfx** Pierre Durin, Claude Pinoteau. **Starring** Yul Brynner, Maria Casarès (Death Princess), Cocteau (Himself), Henri Crémieux, Édouard Dermithe (Cégeste), Daniel Gélin, Jean-Pierre Léaud, Jean Marais, François Périer (Heurtebise) (movie lacks proper credits). 83 mins. B/w and colour.

A decade after ORPHÉE (*1949*), Cocteau returned to its themes with this, his last movie as director, and intendedly so; it is incomprehensible without prior viewing of *Orphée*, and little better with. Here Cocteau's identification of himself with Orphée/Cégeste, implicit in *Orphée*, is made explicit. He appears first as a man lost in time, like a GHOST; he engineers a means of RESURRECTION in the present. Thereafter he travels, with a resurrected Cégeste as guide, through a zone much like that of Death in the earlier movie, being brought before a tribunal where he is judged and found guilty by the Princess and Heurtebise of the charges of innocence and of repeatedly attempting (as a poet) to trespass in another world: his defence is that disobedience is a sacred duty – at least, for poets and children – and his sentence is being condemned to live. For their part, Heurtebise reveals, the two judges were sentenced for their crimes in the earlier movie to spend eternity condemning others. Still with Cégeste as MENTOR, Cocteau travels through a landscape of mythological figures – Anubis, Isolde, etc. – and sees parodies of his own values. After surviving a protracted delay at the mercy of a high-flown bureaucrat (Brynner), Cocteau is permitted an audience with Pallas Athene, who spears him for the impertinence of having once equated VIRGINITY with prostitution; among the mourners at his new false death are Charles Aznavour and Pablo Picasso (appearing as themselves). Resurrected once more, he eventually comes to a highway; two motorcyclists appear, but these are not the ANGELS of DEATH from *Orphée* but instead merely military police checking him out. To rescue him from their attentions, Cégeste appears to draw him back into the timeless zone.

Through this tangle, bathed in slightly self-important SURREALISM, various themes/preoccupations recur, notably the PHOENIX (a photograph of Orphée is reborn from flames), MASKS, the status of DREAM entities, the unreliability of TIME and the independence of artistic creations, which essentially make themselves yet may (like, to an extent, Cégeste) come to resent their progenitors. [JG]

TETSUJIN 28-GO Japanese tv series (1963 onwards). ◊ ANIME.

TETSUWAN ATOM Japanese tv series (1963 onwards). ◊ ANIME.

TEUTONIC FANTASY ◊ NORDIC FANTASY.

THACKERAY, WILLIAM MAKEPEACE (1811-1863) UK writer, author of much occasional journalism and SATIRE as well as novels like *Vanity Fair* (**1848**), whose sharp-eyed focus on the mundane world makes him an unlikely writer of fantasy. *John Bull and his Wonderful Lamp* (**1849** chap), as by Homunculus, is a satire of contemporary England within an ARABIAN-FANTASY frame. The last of his five CHRISTMAS BOOKS, *The Rose and the Ring, or The History of Prince Giglio and Prince Bulbo: A Fire Side Pantomime for Great and Small Children* (**1855**) as by M.A. Titmarsh, spoofs the FAIRYTALE, which by 1850 had become firmly designated a literature for children. The book includes a good FAIRY, several magic TALISMANS and an armamentarium of invincible magic WEAPONS; and all ends happily with the restoration of the rightful heir to the throne of a land much like England.

WMT was also highly esteemed as an illustrator, working regularly for *Punch* and illustrating many of his own works, including both *Vanity Fair* and *The Rose and the Ring*. Today, however, his illustrations seem stiff. [JC]
Other work (selective): *Bluebeard's Ghost* (**1843** chap).

THAT OBSCURE OBJECT OF DESIRE (ot *Cet Obscur Objet du Desir*) French movie (*1977*). ◊ Luis BUÑUEL.

THE BEATLES US animated tv series (1965-7). ◊ YELLOW SUBMARINE (*1968*).

THEMERSON, STEFAN (1910-1988) Polish-born scriptwriter, photographer and writer, active in his native land before moving to the UK prior to WWII, after which point he published increasingly in English. He was a member of the Collège de 'Pataphysique, and founded and ran the Gaberbocchus Press, which published Surrealist texts (◊ SURREALISM) with intelligence and vivacity. Of his fiction, *Professor Mmaa's Lecture* (**1953**) can be understood as satirical sf, but the book is structured in classic BEAST-FABLE format as a lecture, given by the eponymous termite academic, about the dangerous rise of mammals. In general his fiction utilizes – in a grab-bag fashion – various elements of

the FANTASTIC, distorting them through the dislocating lens of Semantic Poetry (a term he invented), paradox, absurdity and spoof. In ST's last two exuberant novels – *The Mystery of the Sardine* (**1986**) and *Hobson's Island* (**1988**) – the wordgames are both more complex and more vivid. [JC]
Other works: *Bayamus* (**1949**); *Wooff Wooff, or Who Killed Richard Wagner?* (**1951**); *Cardinal Pölätüo* (**1961**); *Tom Harris* (**1967**); *Special Branch (A Dialogue)* (**1972** chap); *Factor T; Followed by Beliefs, Tethered and Untethered, and the Pheromones of Fear* (coll **1972** chap); *General Piesc, or The Case of the Forgotten Mission* (**1976** chap).

THEODICY A word coined by Gottfried Leibnitz (1646-1716) to give a name to the doctrine that argued that a GOD who permitted EVIL to exist could be just. Basically, evil exists as a measure – in this best of all possible worlds – of good. The moment-to-moment and ultimate function of evil – a PARODY of good – is to make good visible. The tapestry of the world as it exists to our various perceptions – sunrise and sunset, the rich man in his castle and the poor man at his gate – is to be cherished, defended, maintained; it is a STORY whose warp and woof all need telling. God is, therefore, merciful.

It is easy to understand theodicy as a wisdom of winners, and there are certainly many tasteless moments, throughout world literature, in which humble folk are praised for doffing their caps to those who abuse them. Though the politics of fantasy writers are various, FANTASY as a genre – with its inherent bias towards stories focused upon a RECOGNITION of that which has always existed and is now gloriously restored – is peculiarly prone to bouts of thinking (and unthinking) theodicy. GENRE FANTASY in particular excessively valorizes hierarchy, ancient lineages, an unchallenged return of the SEASONS, a folk complexly colour-coded (◊ COLOUR CODING; ESTATES SATIRE) to social roles. Some FANTASIES OF HISTORY, and certainly much SUPERNATURAL FICTION – the sort, for instance, which takes doctrine and plot impulse from movements like THEOSOPHY – give a theodicy-governed high value to the back-story which generates the shape of the current world. [JC]

THEOSOPHY Literally, "knowledge of God". The Theosophical Society was an occult organization founded in 1875 by H.P. BLAVATSKY and two colleagues. Theosophy has a relationship to FANTASY similar to that which 19th-century SPIRITUALISM has to SUPERNATURAL FICTION: each was a PLAYGROUND – with some elements derived from earlier fiction – that could be easily entered (and transformed) by later writers of fiction. Transactions of this sort between realms of human behaviour and thought are common enough (◊ LIFESTYLE FANTASY). Because Blavatsky was probably a conscious charlatan, and certainly an opportunistic packrat when it came to assembling her cosmology, her own two main books – *Isis Revealed* (**1877**) and *The Secret Doctrine* (**1888**) – turn out to be anything but drily didactic; they are in fact enormous, entrancing honeypots of MYTH, FAIRYTALE, speculation, fabrication and tomfoolery – and so diffusely voluminous that it is difficult to know where their direct influence ceases and the spirit of the time takes over. But if there is any mystery about the speed with which fantasy in the late 19th century began to inhabit the kinds of environments J.R.R. TOLKIEN would later define as SECONDARY WORLDS, and to tell tales gripped by a sense of TIME ABYSS and irradiated by a movement towards EUCATASTROPHE, then it can be claimed that Theosophical cosmology played some role.

Just as important as the actual doctrines of Theosophy is the justifying narrative which accompanies their exposition. Blavatsky claimed to have been accorded the wisdom presented in *The Secret Doctrine* by the Hidden Masters or Secret Brothers (◊ SECRET MASTERS), who have resided since the beginning of things in a POLDER in the heart of Tibet; underneath their feet, in a LIBRARY secreted in an intricate network of UNDERWORLD caverns (◊ EDIFICE), is stored the occult knowledge of all the ages. The Masters' messages to Blavatsky constitute a secret history, which has been given to her (and to the elite which listens to her) as an explanation of the "inner government of the world", which is constituted in the form of a Great White Lodge of Hidden Masters. In its content, and by virtue of the framing devices which intensify the effect of that content, Theosophy is a sacred drama, a ROMANCE and a STORY. Those whose souls are sufficiently evolved to understand that drama know the tale is enacted in another place, beyond the THRESHOLD, in a longed-for Elsewhere, within a LAND exempt from secular accident. It cannot be suggested that Blavatsky consciously created a playground for HIGH FANTASY, or that any Theosophist consciously anticipated the use to which fantasy writers might put the Theosophical tendency, but it is clear that Theosophy paced along with figures like William MORRIS and Lord DUNSANY as they began to move away from the WONDERLANDS and LANDS OF FABLE of mid-19th-century fantasy towards the SECONDARY WORLD. Nor is it insignificant that Robert E. HOWARD both made use of the Theosophical canon of earlier worlds. Smith's **Zothique** sequence, moreover, places in a DYING-EARTH setting a highly ironized and decadent (◊ DECADENCE) revision of Theosophy's transcendental dream of future and higher stages of consciousness.

It is also noteworthy that John CROWLEY, whose *Little, Big* (**1981**) can be read as a *summae theologica* of modern fantasy, has one of the Bramble family describe the meaning of the book's title (i.e., that the inside is bigger than the outside [◊ LITTLE BIG]) in a lecture – concerning the nearly infinite interior worlds of FAERIE – that he delivers to the Theosophical Society.

Writers and others of interest involved in the Theosophical Society include Thomas Alva Edison (1847-1931), Mahatma Gandhi (1869-1948), Kenneth MORRIS, William Butler YEATS and Ella YOUNG. Edgar Rice BURROUGHS made some superficial use of Theosophy in creating his worlds. Many late lost-world tales – examples include James HILTON's *Lost Horizon* (**1933**) and various novels by Talbot MUNDY, in particular *Om: The Secret of Abhor Valley* (**1924**) – show the influence of Theosophy. Though she began in 1919 as a member of the Order of the GOLDEN DAWN, by 1923 Dion FORTUNE had moved to the Theosophical Society, a branch of which she spun off into her own Fraternity of the Inner Light, writing her later novels to illuminate this vision. Nor can the influence of Rudolf Steiner (1861-1925) – who began as a Theosophist then formed a breakaway group, the Anthroposophical Society, which espoused ancient sciences and postulated a spiritual world accessible primarily through a time- and space-transcending inner PERCEPTION – be ignored. Later writers of fantasy affected by Anthroposophy include Michael MOORCOCK, who attended a Rudolf Steiner school and whose TEMPORAL ADVENTURESS is an unmistakably Blavatskyan person.   [JC]

THERIOMORPHY Term derived from "theriomorphic" or "animal-shaped" (describing, e.g., the beast-headed GODS of EGYPT, or any representation of a deity in animal form); it is used by several fantasy critics and authors to describe the ability to change reversibly and repeatedly between human and animal form, as with WEREWOLVES. This encyclopedia generally prefers the more readily understood word SHAPESHIFTING.   [DRL]

See also: METAMORPHOSIS.

THIEF OF BAGDAD, THE Several movies – of which **1** and **2** are important – have been made of this ARABIAN FANTASY.

1. *The Thief of Bagdad* US movie (**1924**). United Artists. **Pr** Douglas Fairbanks. **Dir** Raoul Walsh. **Photography** Arthur Edeson. **Art dir** William Cameron Menzies. **Screenplay** Fairbanks (as Elton Thomas), Lotta Woods. **Novelization** *The Thief of Bagdad* * (**1924**) by Achmed ABDULLAH. **Starring** Snitz Edwards (Thief's "Evil Associate"), Fairbanks (Ahmed), Brandon Hurst (Caliph), Julanne Johnston (Princess), Sojin (Cham Shang the Great, Prince of the Mongols, King of Ho Sho, Governor of Wah Hoo and the Island of Wak), Anna May Wong (Mongol Slave). *c*155 mins. B/w, silent.

With his "Evil Associate", the thief Ahmed makes a rich living on the streets of BAGHDAD, and seems set to do even better when he steals the MAGIC rope belonging to a practitioner of the Indian Rope Trick. Using it, he breaks into the palace, where are stored the treasures brought by the Princess's three suitors; instead he falls in love with her, bringing away only her slipper. He forswears his criminal life and becomes a fourth suitor, as "Ahmed, Prince of the Isles, of the Seas, and of the Seven Palaces". Within the palace, the Princess's Mongol Slave is secretly in the pay of Cham Shang, who plans to take Baghdad either by marriage or by force. The Slave fakes a prediction that the successful suitor will be the first to touch the rose tree in the gardens; unfortunately, before Cham Shang gets there, Ahmed is thrown from his horse into the bush. The Caliph has Ahmed sentenced to death as an impostor, but he escapes. The Princess, now loving Ahmed, to whom she has given her RING, decrees that all her suitors have seven months in which to bring her a gift: the rarest gift will win her hand. Ahmed determines to enter this contest, and is directed by his imam to where there "is a silver chest that doth contain the greatest magic"; Ahmed sets off on his QUEST.

The Hermit of a defile in the Mountains of Dread Adventure tells him that many have sought the chest and none returned; yet offers help. He must first reach the Cavern of Enchanted Trees and touch the central tree with a TALISMAN the Hermit gives him. Ahmed survives the Valley of Fire, aided by the ASTRAL BODY of the Hermit. He kills a giant DRAGON (looking like a DINOSAUR) in the Valley of the Monsters before at last reaching the Enchanted Trees. The central TREE, on being touched, becomes briefly animate and gives him a chart to guide him to the Old Man of the Midnight Sea. The Old Man in turn directs him to the sea's bottom, where he finds a chest containing a star-shaped key, but guarded by a MONSTER and, more insidiously, by a troupe of alluring SIREN-like women. Returning to the surface, Ahmed is instructed by the Old Man to use the key to gain access to the Abode of the Wingèd Horse, which creature he rides to the Citadel of the Moon. Here the Hermit's astral body tells him the Magic

Chest is wrapped in a Cloak of INVISIBILITY. With both items he sets off for home on the horse.

Meantime the other three suitors have located their own gifts. The Persian Prince has bought a magic carpet from the bazaars of Shiraz; the Indian Prince has stolen the SCRYING crystal from the eye of "a forgotten [six-armed] idol near Kandahar", and the cruel Cham Shang has located in a secret shrine on the Island of Wak a magic apple that has the power to restore life to the dead; Cham Shang has also arranged that the Mongol Slave poison the Princess, so that he may save her life. The trio discover her condition using the crystal and fly aboard the carpet to the palace, where Cham Shang indeed cures her. Since she would have died had it not been for any one of the three gifts, the Princess – aware via the crystal ball that Ahmed is on his way – insists she must have time to decide. But Cham Shang orders his troops to seize the city by night. Ahmed arrives at dawn and, using his magic to create a vast horde of warriors, recaptures the city. Triumphant, he claims the Princess's hand. The stars in the night sky spell out for us the movie's moral: "Happiness Must Be Earned."

A milestone of early fantasy CINEMA, *TTOB* is packed with fantasy ideas; most are left undeveloped, but their effect *en masse* is impressive, and the movie thereby retains freshness and vitality today. There are other reasons why *TTOB* – despite much that is kitsch – still provides an astonishing visual experience. The play is less acted than danced: it is perhaps best approached as a ballet (Fairbanks was much influenced by the Diaghilev Ballet). The sets, designed by William Cameron Menzies (1896-1957; it was one of his first commissions), are sumptuously magnificent and highly imaginative; German Expressionist ideas played a role. The spfx are technologically limited, but are (mostly) cleverly deployed and serve their purpose well; again, there are so many of them that, despite frequent creakiness, the net effect is to convey the full sense of magic. Sojin is splendidly evil, Wong (real name Wong Lui Tsong; 1907-1961) – who later had a successful career in Hollywood (despite suffering racism) and more especially in the European cinema – produces a fascinating performance (quite upstaging Johnston whenever the two are together), and Fairbanks is Fairbanks.

TTOB was shot entirely on vast, specially constructed sets, and had a cast of thousands; it cost nearly $2,000,000 to make. Although warmly received by the critics, it was not as successful at the box office as predicted, probably because of its complexity. The best version available today, despite annoying tinting and an over-repetitive score by Carl Davis (based on Rimsky-Korsakov), is probably the one prepared in 1985 for Thames Television by Kevin Brownlow and David Gill, and subsequently released on video. [JG]

2. *The Thief of Baghdad* UK movie (*1940*). Korda/United Artists. **Pr** Alexander Korda. **Assoc pr** Zoltan Korda, William Cameron Menzies. **Dir** Ludwig Berger, Michael Powell, Tim Whelan (also Geoffrey Boothby, Charles David, Alexander Korda, Zoltan Korda, Menzies). **Spfx** Lawrence Butler, Tom Howard, John Mills. **Screenplay** Miles Malleson. **Starring** June Duprez (Princess), Rex Ingram (Djinn), John Justin (Ahmad), Malleson (Sultan), Sabu (Abu the Thief), Conrad Veidt (Jaffar). 106 mins. Colour.

Title apart, this is unrelated to **1**. The urchin TRICKSTER thief Abu is thrown into prison, where he finds Ahmad, deposed as King of BAGHDAD by the wicked vizier Jaffar. But

Abu has stolen the gaoler's key, and the two escape to Basra. There they glimpse the Princess – a capital crime – and Ahmad falls in love with her; when he steals into her garden and confronts her, the adoration is returned. Jaffar arrives and with a mechanical flying horse bribes the Sultan – an avid collector of automata ($\lozenge$ TOYS) – for the hand of the Princess. She flees (and, we learn, falls into slavery, being eventually bought by Jaffar). To silence them, Jaffar CURSES Ahmad to blindness and Abu to be a dog. About here the plot falls apart, spectacle and (mostly) dazzling spfx becoming the priorities. Notable among the further adventures of the restored pair are: a famous encounter between Abu and a giant djinn ($\lozenge$ GENIES), who eventually grants him THREE WISHES in return for freedom from his bottle; a fight with a giant SPIDER; and the use on the Princess by Jaffar of the Blue Rose of Forgetfulness, inhalation of whose fragrance causes AMNESIA – but not so total that she forgets Ahmad. At last, the lovers facing death, Abu is transported to the Land of Legend, where dwell those from the time before mankind lost its innocence. He is informed he will become the land's new king, but first he flies by stolen magic carpet to save his friends from execution – thereby fulfilling a PROPHECY that the city would be delivered from tyranny by a boy riding the clouds.

TTOB has sufficient spectacle and frequent enough flashes of wit that its deficiencies as a coherent piece of fantasy (and its dreary songs) can almost be ignored. Yet it lacks the magic of **1**. Its visual style, however, has influenced directors since, as well as the makers of various of the SINBAD MOVIES. In particular the DISNEY movie ALADDIN (*1992*) owes it a considerable debt: the modern movie's Jafar, Sultan and Genie are virtual carbon copies of their counterparts in *TTOB*, while there is much of Ahmad in the figure of Aladdin, and even traces of *TTOB*'s Abu in Aladdin's monkey sidekick Abu. [JG]

3. *The Thief of Baghdad* Italian/French movie (*1960*). Titanus/Lux. **Dir** Arthur Lubin. **Screenplay** Augusto Frassinetti, Filippo Sanjust, Bruno Vailati. **Starring** Georgia Moll (Princess), Steve Reeves (Thief). 90 mins. Colour.

A very minor version, repeating the essential ARABIAN-FANTASY tropes of **1** and **2**; this was Reeves's attempt to drop his image of being a mere mindless muscleman, but it was unsuccessful. [JG]

4. *The Thief of Baghdad* French/UK movie (*1978* tvm). Columbia/Palm/Victorine. **Pr** Aida Young. **Exec pr** Thomas M.C. Johnston. **Dir** Clive Donner. **Spfx** Allan Bryce, Ray Caple, Dick Hewill, Wladimir Ivanov, Louis Lapeyre, Zoran Perisic, John Stears. **Illusionist** Kovari. **Screenplay** Andrew Birkin, A.J. Carothers. **Starring** Kabir Bedi (Prince Taj), Daniel Emilfork (Genie), Frank Finlay (Abu Bakar), Ian Holm (Gatekeeper), Roddy McDowall (Hasan), Terence Stamp (Jaudur the Wazir), Pavla Ustinov (Princess Yasmine), Peter Ustinov (Caliph of Baghdad), Marina Vlady (Perizadah). 104 mins. Colour.

This version draws freely from **1** and **2** as well as other sources like *Captain Sindbad* (*1963*; $\lozenge$ SINBAD MOVIES). Prince Taj of Zakhar travels to BAGHDAD to woo Princess Yasmine, but in the desert his party is attacked by troops in the pay of his wicked wazir Jaudur. Eventually reaching Baghdad, ragged and penniless, Taj encounters the conjurer and thief Hasan, and the two become firm friends. In stolen garments, Taj presents his suit to the Caliph alongside the aged ruler of the Mongols and the fat Prince of

Kashmir. A fourth suitor arrives by magic carpet: Jaudur states Taj dead and proclaims himself King of Zakhar. Taj and Jaudur duel; Jaudur is unkillable because his SOUL is not in his body but secreted elsewhere. Adventures proliferate before all wrongs are righted.

McDowall makes an admirably fey thief and Stamp is excellent, but notable is Vlady (as Yasmine's handmaiden), who repeats the "flaw" of **1** by being much more interesting than the rather insipid Princess she serves. This is a modest version and the spfx creak, but the overall effect is exceptionally pleasing. [JG]

THINGS BOUGHT AT TOO HIGH A COST Short-term gains which have dire penalties in the long term. They are often a result of greed, and may be the outcome of treasures gained through THREE WISHES. The sampling of the fruit of the tree of knowledge by ADAM AND EVE – resulting in expulsion from EDEN – is the primary example. Anything gained as a result of a PACT WITH THE DEVIL will always be a TBATHAC, as exemplified in the FAUST legend and in such classic novels as *Vathek* (**1786**) by William BECKFORD and *Melmoth the Wanderer* (**1820**) by Charles MATURIN. Some TBATHACs may result in self-SACRIFICE to save others. [MA]

THINNING Fantasy tales can be described, in part, as fables of recovery. What is being regained may be (a) the primal STORY that the surface tale struggles to rearticulate, (b) the TRUE NAME, or home, of the protagonist, (c) the health of the LAND (◊◊ FISHER KING) through a process of HEALING, or indeed (d) the actual location of the land itself (◊ ARCADIA; OTHERWORLD; POLDER; TIME ABYSS). But, although it is true most fantasy stories *finish* – and tend to end in a EUCATA-STROPHE – it is also true that the happy endings of much fantasy derive from the notion that this is a *restoration*, that before the written story started there was a diminishment.

Even in HIGH FANTASY – which tends to be ringfenced from time's arrow – the SECONDARY WORLD is almost constantly under some threat of lessening, a threat frequently accompanied by mourning (◊ ET IN ARCADIA EGO) and/or a sense of WRONGNESS. In the structurally complete fantasy, thinning can be seen as a reduction of the healthy LAND to a PARODY of itself, and the thinning agent – ultimately, in most instances, the DARK LORD – can be seen as inflicting this damage upon the land out of envy.

The passing away of a higher and more intense REALITY provides a constant *leitmotif* in the immensely detailed mythology created by J.R.R. TOLKIEN. *The Lord of the Rings* (**1954-5**) comes at the end of aeons of slow loss. Within that text, local thinnings occur – those points when the elves are perceived as heading west, for instance, or when Frodo returns to the Shire to find it has been thinned into a secular WASTE LAND. Similarly, in *The Farthest Shore* (**1972**) by Ursula K. LE GUIN the exhaustion of the wells of MAGIC is a moving example of the thinning of an edenic secondary world (nor does Earthsea ever fully recover from the haemorrhage inflicted by the DARK LORD from the UNDERWORLD of the DEAD).

When the fantasy tale is not ringfenced beyond a THRESH-OLD difficult to pass, or beyond time, or within the autonomy of a secondary world or LAND, then thinning becomes more than a *leitmotif*: it becomes an explicit and central concern. In LOW FANTASY, CROSSHATCH fantasy, etc., rarely does the world provide venues unthreatened by one or more of a huge range of diminishings or dismissals of the old order: through the desiccations of the secular and of

technology; through the draining of MAGIC from the energy pools of creation; through the coming of DEATH into a world hitherto prelapsarian; through the crushing advent of *Homo sapiens*, which exterminates the old fauna and drives the inhabitants of FAERIE into WAINSCOTS or into PARIAH-ELITE roles; through the triumph of Christianity, which criminalizes worship of, e.g., the GODDESS; through the increase of entropy, which ages the world. Thinning is a sign of a loss of attention to the stories whose outcomes might save the heroes and the folk; it is a representation of the BONDAGE of the mortally real.

It would be rash to claim that any fantasy set in the real world must engage with thinning – indeed, tales dealing with the contemporary world, like Sean STEWART's *Resurrection Man* (**1995**), may well present a recongestion of REALITY – but fantasies set in history almost invariably deal with the loss of the old richness. Lisa GOLDSTEIN's *Strange Devices of the Sun and Moon* (**1993**) is entirely typical: "A great change is coming," prophesies a FAIRY in conclave: "This world and all we have known will pass away. Trees and stone, wind and rain, will be as naught. It will be a world of artifice, of vast gears interlocking in one enormous mechanism." Goldstein's fairy speaks for almost every character in every historical fantasy written in the 20th century.

In ALTERNATE-WORLD or MULTIVERSE stories, which often involve TIME TRAVEL, the weakening of the fabric of probability – which is generally a consequence of any tampering with TIME – can also be described as a process of thinning. When this process is treated with a dancer's disdain – as in the TEMPORAL-ADVENTURESS tales of Michael MOORCOCK and later UK writers – then the resulting story will, perhaps unwittingly, embody elements of consolation.

There are so many examples of thinning in fantasies set wholly or partially in the real world that it is perhaps unnecessary to mention any further examples. But some do stand out. The work of Thomas Burnett SWANN – almost all of whose novels deal with Christianity's slow elimination of a world crammed with pagan and supernatural beings – should be noted, if only for the violence of its lamenting; Swann's work constitutes a late-20th-century version of the sentimentalized PAN-worship engaged upon by the many Edwardian fantasists – including J.M. BARRIE, E.M. FORSTER, Kenneth GRAHAME, Arthur MACHEN, Barry PAIN and SAKI – who bemoaned the loss of childhood and the rise of suburbia. Many of the stories of Clifford D. SIMAK – an exemplar being "The Autumn Land" (**1971**) – treat the interface between the deracinating modern world and his favoured PASTORAL alternative in terms of patterns of thinning. A harsher and more complex rendering of similar material informs Peter VANSITTART's *The Death of Robin Hood* (**1981**), which carries the eponymous MYTHAGO figure from the Middle Ages to Nazi Germany. Esther M. FRIESNER's *Yesterday We Saw Mermaids* (**1992**) treats similar material from a perspective even later than Swann's. The type of RATIONALIZED FANTASY created for UNKNOWN almost invariably posits a "scientific" model for the workings of magic, which is therefore subjected to a constant threat of being drained; the fantasy tales of Larry NIVEN – in *The Magic Goes Away* (**1978**) and others – provide perhaps the clearest workings-out of the implications of treating magic as drainable. This model has been accepted very widely, certainly in GENRE FANTASIES featuring wizards who exhaust themselves casting SPELLS; more movingly, it can serve to represent the coming of age of hero and land in an extended paean like T.H. WHITE's *The Once and Future King* (**1958**) or its lighter-toned successor, the **Chronicles of Prydain** sequence (**1964-8**) by Lloyd ALEXANDER.

Thinning may be kept at bay, generally by diking it: physically through a polder of some sort, within which a toughened reality can be maintained through constant vigilance; promissorily through knowledge that somewhere a SLEEPER UNDER THE HILL awaits the call to restore to the world the savour of spring. [JC]

THIRTEEN The number most closely associated with bad luck. It represents the outsider, the dark side of life – there are 13 lunar months in a year, thus associating the number with the night, darkness and the power of the MOON. Its influence has become grafted onto Christianity because there were 13 at the Last Supper, one of whom betrayed CHRIST, though 13 at dinner has always been deemed unlucky – it also appears in Norse MYTHOLOGY when LOKI came uninvited to a banquet in VALHALLA, leading to the death of Baldur. Lord DUNSANY envokes this aspect in his GHOST STORY "Thirteen at Table" (1916 *Tales of Wonder*). The superstition associated with the number is very strong. In some skyscrapers there is no numbered 13th floor, thus the shock of its discovery in "The Thirteenth Floor" (1949 *WT*) by Frank Gruber (1904-1969). It becomes a step towards a personal HELL in "The Thirteenth Step" (in *The Fiend in You* anth **1962** ed Charles BEAUMONT) by Fritz LEIBER. A clock striking 13 is not only the ominous opening of *1984* (**1949**) by George ORWELL but invokes a PORTAL to another world in "Thirteen O'Clock" (1941 *Stirring Science Stories*) by C.M. Kornbluth (1923-1958) and *Tom's Midnight Garden* (**1958**) by Philippa PEARCE. Friday is sometimes considered an unlucky day (on a Friday Christ was crucified) so the combination of Friday and 13 is especially potent and may constitute the basis for Western culture's single most powerful superstition. It is used to establish a basis for HORROR fiction, most notably in the movie series that began with *Friday the Thirteenth* (**1980**), while its use as a starting point for humour is portrayed in FREAKY FRIDAY (**1976**), where the sudden IDENTITY EXCHANGE happens both on Friday 13th and to a 13-year-old. [MA]

THOMAS, D(ONALD) M(ICHAEL) (1935-) UK poet and novelist, born in Cornwall and occasionally a translator of Russian verse, two elements which echo through his own work. Before turning to prose DMT published several volumes of verse; his poetry has long made use of sf or fantasy themes, as in "The Head-Rape" (1968 *New Worlds*) and "Labyrinth" (1969 *New Worlds*). *The Devil and the Floral Dance* (**1978**) is a fantasy for young readers. *The Flute-Player* (**1979**), a parable of art and love in an imaginary Russia-like state, and *Birthstone* (**1980**; rev 1982), a fantasy of sexual roles set in Cornwall, featuring a protagonist whose personalities assume autonomous lives and begin to transform the world through their fantasies, show the influence of Sigmund Freud in their complex intertwinings of art, sexual love, and DEATH. *The White Hotel* (**1981**), very much the best of DMT's early prose, brings Freud onstage in the analysis (Freud's eventual monograph comprises part of the text, which is a composite in the manner that E.L. DOCTOROW calls "False Documents") of a hysterical woman whose visions of sexual obsession and mass violence eventually prove prophetic of the Final Solution; the book's final section is set in a disconcerting AFTERLIFE.

The **Russian Nights** quintet – *Ararat* (**1983**), *Swallow* (**1984**), *Sphinx* (**1986**), *Summit* (**1987**) and *Lying Together*

(**1990**) – is a set of "improvisational novels", according to the author (who first announced this as a quartet). They deal with a variety of subjects that dramatize themes familiar from DMT's earlier work: the art of translation, Soviet totalitarianism, early psychiatric documents and the mysteries of creation. Some of the tableaux – *Swallow*, perhaps the weakest, takes place at an international competition of poetic improvisations – are essentially fantastic. [JC/GF]

Other works: *Flying in to Love* (**1992**); *Pictures at an Exhibition* (**1993**); *Eating Pavlova* (**1994**).

THOMAS, DYLAN (MARLAIS) (1914-1953) Welsh poet, also playwright, prose writer and broadcaster; he died young of alcoholism. His work is of core interest to the literature of the FANTASTIC less because of its content than because of its PERCEPTION – a perception reflected in its DICTION. Most of his poems have to be unpacked from a welter of fantasticated imagery; as a single example, "Fern Hill" (1946 in *Deaths and Entrances* coll **1946**) is about childhood, but the imagery casts a glow of MAGIC REALISM (although the term was not then used) over everything. *Under Milk Wood* (**1954**), a radio drama developed under the title «Quite Early One Morning», was first broadcast in 1954; a version of the first part was published as "Llareggub" (1952 *Botteghe Oscure*), Llareggub being the name of the fictitious Welsh village in which various funny and quasi-Surrealist events transpire. DT's work was enormously influential on the New Apocalypse AFFINITY GROUP of poets. Some years after DT's death the US singer-songwriter Robert Allen Zimmerman (1941-) took the name Bob Dylan in homage. [JG]

Other works: *18 poems* (coll **1934**); *Twenty-five Poems* (coll **1936**); *The Map of Love* (coll **1939**), mixing poetry and prose; *Portrait of the Artist as a Young Dog* (coll **1940**), semi-autobiographical short stories; *The World I Breathe* (coll **1940**); *Collected Poems, 1934-52* (coll **1952**); *Adventures in the Skin Trade* (**1955**), unfinished novel; *A Prospect of the Sea* (coll **1955**), stories and essays. Other books and chapbooks have been derived from this material.

THOMAS, ROY (WILLIAM) (1940-) US writer of COMIC books. He co-created with Jerry Bails one of the first comics fanzines, *Alter Ego*, in 1961, including some pieces which were reprinted in the anthologies *All in Color for a Dime* (**1970**) and *The Comic-Book Book* (**1973**), both ed Richard A. Lupoff and Don Thompson. He was a high-school teacher in St Louis for four years, during which time he wrote scripts used in Charlton's *Son of Vulcan* (1966) and *Blue Beetle* (1966). An unsolicited script submission to National Comics (later called DC COMICS) resulted in a job offer that took him to New York as assistant editor to Mort Weisinger on *Superman*. Two weeks later he met Stan LEE, who invited him to write for MARVEL COMICS, and he transferred himself to that company as assistant editor. He wrote for a variety of Marvel titles, including *Doctor Strange*, *Spiderman* and *Sgt Fury*. It was in the last years of the 1960s that he hit his stride as a writer with **Sub-Mariner**, **The X-Men** and **The Avengers**, where he proved himself one of the finest comic-book writers of his time, consistently handling dozens of characters in a complicated plot sequence which culminated in the Kree-Skrull Wars conflict in 1972. This was drawn by Neal ADAMS and remains one of the highlights of 1970s comics. RT has also been praised for his exploration of SUPERHERO characters – building on the "superheroes with problems" theme developed by Lee – and his ability to endow even the most minor figures with character. On **The X-Men** he was less successful (even though he introduced Neal Adams as artist on the series), and it ceased at *#66* (1970).

RT's most outstanding series were those featuring Robert E. HOWARD's characters in *Conan the Barbarian* (115 issues by him) and *Savage Sword of Conan* (over 60 issues by him), for which he adapted many of the original novels in a multi-award winning run featuring artwork by Barry WINDSOR-SMITH and John Buscema (1927-   ), among others. It was RT who took **Red Sonja**, another Howard character, and transported her to the Hyborian Age in *Conan the Barbarian #23* (1972); she was featured in a comic book series of her own, drawn by Frank THORNE. To date RT has contributed at least 5000 pages of storytelling to **Conan**.

RT was instrumental in bringing **Star Wars** to Marvel, was responsible for co-creating **Wolverine** and **Morbius**, and helped develop **The New X-Men**. He became Marvel's editor-in-chief in 1972, when Lee became publisher, but resigned in 1974, although he continued to write for the company until 1980. He subsequently joined DC, where he handled numerous team-ups and created *Infinity Inc*. At the same time he was freelancing for smaller companies like First, for whom he adapted Michael MOORCOCK's **Elric** in a series of comics, including *Michael Moorcock's Elric of Melnibone* * (graph **1986**) with Michael T. Gilbert and P. Craig Russell. Another venture was *The DragonLance Saga Book Three* * (graph **1989**) with Tony de Zuniga, adapted from *Dragons of Winter Night* * (**1984**) by Margaret WEIS and Tracy HICKMAN.

RT continues to be one of the most successful and prolific writers of his generation, working on **Avengers West Coast** and **Doctor Strange**; he also adapted Frances Ford Coppola's *Bram Stoker's Dracula* (**1993**; ◊ DRACULA MOVIES) for comic-book publication by Topps as *Bram Stoker's Dracula* * (graph **1993**), with Mike Mignola and John Nyberg.   [SH/RT]

THOMAS, THOMAS T(HURSTON) (1948-   ) US writer. ◊ MASKS.

THOMAS THE RYMER ◊ TAM LIN.

THOMPSON, PAUL B. (1951-   ) US writer whose first novel, *Sundipper* (**1987**), is sf, but who has since concentrated on ties and other works for the GAMES firm TSR. These include *Red Sands* * (**1989**) with Tonya R. Carter, an ARABIAN FANTASY adventure, the first volume of the **DragonLance Preludes** sequence, *Darkness and Light* * (**1989**) with Carter, the first volume in the **DragonLance Preludes II** sequence, *Riverwind: The Plainsman* * (**1990**) with Carter, and two volumes in the **Elven Nations Trilogy** – *DragonLance: Firstborn* * (**1991**; vt *Elven Nations Trilogy #1: Firstborn* 1991 UK) and *DragonLance: The Qualinesti* * (**1991**), both with Carter.   [JC]
Other works: *Thorn and Needle* (**1992**).

THOMPSON, RUTH PLUMLY (1891-1976) US writer. ◊ L. Frank BAUM; OZ.

THORNE, FRANK (1930-   ) US COMICS artist with a loose, slick line style who has developed a distinctive line of feisty, sexy female comics characters. Early, FT drew newspaper strips in the style of Alex RAYMOND, including **Perry Mason** and **Dr. Guy Bennett**. He gradually developed his own clean, controlled pen-and-brush style, after doing comic-book versions of FLASH GORDON and **Jungle Jim** along with other work like **Twilight Zone** and **Boris Karloff**. A turning point came when he was selected as the

main artist on MARVEL COMICS's SWORD-AND-SORCERY *Red Sonja* (*#1-#11* 1977-9), which featured a character derived from Robert E. HOWARD's stories. His line style became loose and fluid and he went on to create another similar, but much more bawdy, barbarian woman: **Ghita of Alizarr**, which first appeared in WARREN PUBLISHING's *1984* magazine (*#7-#17* and *#16-#28* 1979-82); subsequent colour episodes have appeared in many countries. Other bawdy creations have included the sf strip **Lann** (syndicated worldwide since 1985) and the outrageous humour strip **Moonshine McJuggs** (from *c*1991).   [RT]

THORNLEY, RICHARD (1950-   ) UK writer. ◊ COYOTE; TRICKSTER.

THOUSAND AND ONE NIGHTS, THE ◊ ARABIAN FANTASY.

THREE Even more than SEVEN, 3 represents completeness – the beginning, middle and end; past, present and future; or the three dimensions of space – though with 3 the implications of perfection are stronger, as exemplified by the Holy Trinity. The Greeks expressed many aspects of life in forms of 3, such as the FATES, the Graces and the Furies. In stories of QUESTS the treasures sought sometimes fall into a pattern of 3 for completeness, as in *The Magic Three of Solatia* (**1974**) by Jane YOLEN. Children become acquainted with the completeness of 3 from its frequent use in FAIRY-TALES and nursery tales, like "Goldilocks and the Three Bears" (◊ Robert SOUTHEY), "The Three Billy-Goats Gruff" and "Three Blind Mice" (1609).

As in all SUPERSTITIONS, there is an ambivalence about 3. We may say "third time lucky" – an expression that can be traced back to a reference in *Sir Gawain and the Green Knight* (◊ GAWAIN) – but if things go wrong we also believe "accidents come in threes". In FANTASY and SUPERNATURAL FICTION the number often takes on this trickier or more sinister aspect, as in the THREE WISHES motif. The approach of allowing two people or events to set the scene and the third to resolve is a frequent PLOT DEVICE, as with the three Christmas ghosts that visit Scrooge in *A Christmas Carol* (**1843**) – it is the third who is the most sinister. Similar images are created in the GHOST STORIES "The Shadowy Third" (1916 *Scribner's*) by Ellen Glasgow (1874-1945), "The Third Time" (in *Powers of Darkness* coll **1934**) by Kenneth Ingram and "Three Gentlemen in Black" (1938 *WT*) by August DERLETH.

Fantasy's fascination with 3 is also exemplified by the proliferation of trilogies.   [MA]

THREE WISHES A popular PLOT DEVICE in fantasy, and common in oral tradition for centuries (particularly in ARABIAN FANTASY, with WISHES granted by GENIES freed from a bottle), but which was first written down as "The Three Wishes" (1757 *Le Magasin des Enfants*) by Jeanne-Marie LEPRINCE DE BEAUMONT. A FAIRY grants a man and his wife three wishes. The wife makes an accidental wish for a black pudding, and in cursing her for wasting the wish the husband wastes a second one. The third wish is needed to set things back as they started. The MOTIF usually highlights folly and greed. It was adapted to powerful effect by William Wymark Jacobs (1863-1943) in "The Monkey's Paw" (1902 *Harper's*) in which a TALISMAN grants the TW. The first wish for fortune results in the death of the couple's son. In her anguish the wife forces the husband to wish for the son to be alive again. The husband succeeds at the last minute, with the third wish, in stopping the consequent horror. The Irish novelist and folklorist William Carleton

(1794-1869) used a different version in "The Three Wishes" (*Traits and Stories of the Irish Peasantry* coll **1830**), in which a lazy good-for-nothing wastes his wishes on personal whims but then is able to use what he gained to outwit the DEVIL. Because it can often be used to convey a moral within a humorous context, the TW motif is often used in SLICK FANTASY, and it became a stock theme in stories in the 1950s and early 1960s in *The* MAGAZINE OF FANTASY AND SCIENCE FICTION and FANTASTIC, where many authors turned their hand to creating a new variant on the theme, among them Poul ANDERSON, Arthur Porges (1915-) and Robert F. YOUNG. [MA]

THRESHOLDS Thresholds may be physical, marking a gradient between two places or states of being, or metaphorical, marking some PERCEPTION of change. In both instances, they may contain and be focused by POR-TALS, but, while a portal always signals a threshold of some sort, a threshold does not require a focus. They may not even be meant to be liminal, or passable.

Physical thresholds normally form the spines of BORDER-LANDS, demarcating regions which borderlands join together; they announce the presence, or intrusion, of a CROSSHATCH; they constitute the perimeter of POLDERS; and, for those of peculiar TALENTS, they may comprise a MAP of the LAND.

Metaphorical thresholds are part of the sustaining warp and woof of a literature which addresses the "essence" of LANDSCAPE, and in which all transitions – anything which the author deems worthy of notice – are *ipso facto* significant. Thresholds tend to mark, therefore, jointures in plot and other points in the grammar of STORY; a whiff of difference (threshold moments are often described in terms resembling synaesthesia) may augur BONDAGE or WRONGNESS or THINNING; thresholds of anticipation, often conveyed as a welling up of the pressure of story, may circumambiate the moment of RECOGNITION in a full fantasy, the moment when characters and land pass the barrier into right knowing, METAMORPHOSIS, proper remembrance or HEALING.

The heart of all this is the sense that, in fantasy, where there is theoretically nothing that one cannot say (and therefore no excuse for saying something unmeant), thresholds are maps to the meaning of the text. [JC]

THRILL BOOK, THE US MAGAZINE, 16 issues, twice-monthly, 1 March-15 October 1919, published by Street & Smith, New York; ed Harold Hersey (1893-1956) March-June 1919 (8 issues) and Ronald Oliphant (1884-?) July-October 1919 (8 issues).

Often cited as the first sf magazine, *TTB* was chiefly an adventure magazine with a high quota of unusual or OCCULT FANTASY tapping into the post-WWI interest in SPIRITUALISM. The first eight issues were in the cheap dime-novel format which had earned Street & Smith its fortune; *TTB* shifted into PULP format for its last eight issues. In the latter format it clearly sought to imitate the highly popular *Adventure*, though its content of unusual stories was more in line with that of *The Black Cat* (◊ MAGAZINES). About half of its 120 or so stories are SUPERNATURAL FICTION. Most of these were traditional GHOST STORIES, though there were some POSTHUMOUS FANTASIES to acknowledge Spiritualism's popularity. Tales of LOST RACES, native MAGIC, MUMMIES and accursed jewels, common in the commercial magazines of the day, appeared. Few of the writers were significant, though H. BEDFORD-JONES, Greye La Spina (1880-1969), Murray Leinster (1896-1975), Seabury QUINN, Tod

ROBBINS and Francis STEVENS contributed. (Stevens's *The Heads of Cerberus* [1919; **1952**], set in an alternate future Philadelphia, is the only original story from *TTB* to have been reprinted.) Others of interest included: Clyde Broadwell's series **Tales of the Double Man**, about a spiritual TWIN; Harcourt Farmer's "When Brasset Forgot", about a man who can communicate with SPIDERS; and Robbins's "The Bibulous Baby", about life lived backwards. [MA]
Further reading: *The Annotated Index to The Thrill Book* (**1991**) by Richard Bleiler (1959-).

THRILLER, THE ◊ MAGAZINES.

THUMBELINA (vt *Don Bluth's Thumbelina*) Irish/US ANI-MATED MOVIE (**1994**). Warner Bros/Don Bluth Ltd/Don Bluth Ireland Ltd. **Pr** Don BLUTH, Gary Goldman, John Pomeroy. **Dir** Bluth, Goldman. **Screenplay** Bluth. **Based on** "Thumbelina" (1836) by Hans Christian ANDERSEN. **Voice actors** Jodi Benson (Thumbelina), Charo (Ma Toad), Carol Channing (Miss Fieldmouse), Gino Conforti (Jacquimo), Barbara Cook (Mother), June Foray (Queen Tabitha), Gilbert Gottfried (Berkeley Beetle), John Hurt (Mr Mole), Gary Imhoff (Prince Cornelius), Joe Lynch (Grundel), Kenneth Mars (King Colbert). 87 mins. Colour.

A solitary woman asks a good WITCH for a child, and is given a barleycorn seed. This flowers, and from the flower emerges a tiny adolescent girl, whom Mother christens Thumbelina. Thumbelina is lonely, because on the wrong scale for the rest of the world; she dreams of meeting and marrying the Prince of the FAIRIES, as depicted in her FAIRY-TALE books. Soon Cornelius, Prince of the Fairies, does indeed discover her, and the two fall in love, despite Thumbelina's lack of WINGS. The course of true LOVE, as usual, proves not untangled, but eventually they wed.

This somewhat underambitious movie tells its story effectively enough, but is marred by an overabundance of forgettable songs and some shoddy colour work: Thumbelina's hair is golden, orange and scarlet by turns, while her dress can be, from one moment to the next, white, mauve or blue. There are a few nice touches – e.g., in defiance of his parent's wishes Cornelius rides not a decorous butterfly but a loudly buzzing super-bee, equivalent to a Harley Davidson – but the overall standard is below what one expects from theatrically released animation. [JG]

THUN-DA, KING OF THE CONGO ◊ Frank FRAZETTA; TARZAN.

THURBER, JAMES (GROVER) (1894-1961) US writer and cartoonist, generally acknowledged as one of the 20th century's finest humorists. Many of his stories and distinctive cartoons show episodes in what he called the "War Between Men and Women" in his most celebrated piece, "The Secret Life of Walter Mitty" (1939; in *My World – And Welcome To It* coll **1942**) the would-be hero Mitty escapes a relentlessly unsympathetic wife in daydreams (◊ DREAMS) which parody fictional and CINEMA cliché ("The Commander's voice was like thin ice breaking") and whose miniaturist perfection was lost in the movie *The* SECRET LIFE OF WALTER MITTY (**1947**). *Fables for Our Time and Famous Poems Illustrated* (coll **1940**) and *Further Fables for Our Time* (coll **1956**) assemble humorously updated FAIRYTALES and sometimes bitterly wry FABLES (◊ ALLE-GORY) featuring a variety of TALKING ANIMALS, a UNICORN and subversive "morals" throughout. All JT's novels are short CHILDREN'S FANTASIES which adults can enjoy, and all show a love of words and language (◊ DICTION): alliteration, assonance, metrical prose with occasional rhymes. *The*

White Deer (**1945**) inventively elaborates the fairytale situation of king, princess and three suitors who must be tested – and also features a WITCH, a DWARF and a TRANSFORMATION of a princess into a deer, or vice versa. The exuberant *The Thirteen Clocks* (**1950**) has a memorably icy VILLAIN in a castle of frozen clocks (like the Mad Hatter, he has murdered TIME), an unseen clammy MONSTER made of "glup", a woman who weeps jewels, and a dotty TRICKSTER/MENTOR who keeps insisting in vain that he is no mere device (◊ PLOT DEVICES); it begs to be read aloud. Wordplay runs riot in the FABULATION *The Wonderful O* (**1955** chap), as monomaniac PIRATES excise the letter O from a gentle ISLAND's language and life ("Otto Ott, when asked his name, could only stutter") – until finally halted by the MAGIC WORD "freedom". [DRL]

Other works: *Many Moons* (**1943** chap) and *The Great Quillow* (**1944** chap), more fairytales; *The Thurber Carnival* (coll **1945**), a useful compendium of shorts and cartoons; *The 13 Clocks and The Wonderful O* (omni **1962** UK) illus Ronald SEARLE.

Further reading: *Remember Laughter: A Life of James Thurber* (**1994**) by Neil A. Grauer.

See also: HUMOUR; MY WORLD . . . AND WELCOME TO IT (1969-70, 1972).

THURSTON, E(RNEST) TEMPLE (1879-1933) UK writer of popular fiction whose *The Wandering Jew: A Play in Four Phases* (**1920**) carries the story of the WANDERING JEW to 1560, where he finds death at the hands of the Spanish Inquisition, and REDEMPTION at the thought of CHRIST; the play was novelized by his daughter, Emily Temple Thurston, as *The Wandering Jew* (**1934**). *Man in a Black Hat* (**1930**) pits a medical doctor against a Rosicrucian (ROSICRUCIANISM) who may be immortal, who can project himself through his ASTRAL BODY, and who is attempting to acquire a 16th-century BOOK full of MAGIC incantations; *The Rosicrucian* (coll **1930**) contains some SUPERNATURAL FICTION on similar subjects. [JC]

TICHEBURN, CHEVIOT Pseudonym of W. Harrison AINSWORTH.

TIECK, JOHANN LUDWIG (1773-1853) German writer, one of the primary figures of ROMANTICISM. Fascinated by medievalism and initially influenced by the Gothic movement, JLT produced a bloodthirsty HORROR novel, *Abdallah, oder das furchtbare Opfer* ["Abdallah, or The Dreadful Sacrifice"] (**1795**) in the style of William BECKFORD's *Vathek* (**1786**), but soon developed a more moralistic interpretation of FOLKTALES with *Der blonde Eckbert* (**1796**; as "Auburn Eckbert" in *Popular Tales and Romances of the Northern Nations* anth **1823** ed anon; vt "Fair-Haired Eckbert"; vt "Eckbert the Fair"), in which Eckbert's guilt for his misdeeds haunts him through ILLUSIONS. This story was treated by the Romantics as a model of the *novellen-märchen*, the type of tale developed by GOETHE where the supernatural was deployed to explain otherwise irrational events. It was adapted as an OPERA, *Blond Eckbert* (**1994**), by Judith Weir (1954-   ). "Auburn Eckbert" was included in JLT's *Volksmärchen* (omni **1797**), which also contained *Ritter Blaubart* ["Sir Bluebeard"] (1797) and *Der Gestiefelte Kater* ["Puss-in-Boots"] (1797), two ingeniously contrived plays within plays that utilized the fairytale motifs of Bluebeard and PUSS-IN-BOOTS for SATIRES on the literary and social scene. These established JLT's reputation. He continued to develop his own wonder tales, including: *Der Getreue Eckart* (**1799**; trans as "The Faithful Eckart and the Tannenhäuser"

in *The German Novelists* anth **1826** ed Thomas Roscoe; vt "Loyal Eckart"), about a valiant KNIGHT who sacrifices himself so that his SPIRIT can remain a guardian (◊ LIMINAL BEINGS); *Sehr wunderbare Historie von der Melusina* ["The Wonderful History of Melusina"] (**1800**), about a nobleman who marries Melusine but loses her when he discovers she is a MERMAID; *Der Runenberg* (**1804**; trans as "The Runenberg" in *German Romance* anth **1827** ed Thomas CARLYLE; vt "The Runic Mountain"), in which a LAMIA in a ruined castle lures a traveller first to riches and then to ruin; and *Die Elfen* (**1811**; trans as "Elfin-Land" in *Popular Tales and Romances of the Northern Nations* anth **1823** ed anon; vt "The Elves"), an early example of the INTO THE WOODS motif and of TIME IN FAERIE. JLT produced a new fairytale almost annually, many showing the power of ILLUSION over people. He collected these with earlier works as *Phantasus* (omni **1812-17** 3 vols; trans J.C. Hare as *Tales from the Phantasus* **1845** UK). He was one of the primary movers in the creation of SUPERNATURAL FICTION based on folk roots as distinct from GOTHIC FANTASY, which was developing in parallel. JLT wrote little in the second half of his life, preferring to translate and edit the works of others; in 1841 he became the reader to Friedrich Wilhelm IV of Prussia. [MA]

Other works: *Die verkehrte Welt* ["The Land of Upside Down"] (**1797**), a satire; *Anti-Faust, oder Geschichte eines dummen Teufels* ["Anti-Faust, or The Story of the Stupid Devil"], (**1801**), a spoof on FAUST.

Further reading: *Ludwig Tieck, the German Romanticist* (**1935**) by Edwin H. Zeydel; *Reality's Dark Dream: The Narrative Fiction of Ludwig Tieck* (**1979**) by William J. Lillyman.

TIM BURTON'S THE NIGHTMARE BEFORE CHRISTMAS vt of *The* NIGHTMARE BEFORE CHRISTMAS (*1993*).

TIME In the Dark Ages of the Western World, the clocking of the passage of Time was understood to be a qualitative endeavour, an effort to explain the significance of the moments, hours and days lived. Time was a form of significant shape, and was often expressed in terms of CYCLES. Today the concept of Time is radically different: Time can be measured linearly, and the more accurate the measurement the more Time is presumed to be understood (or, at the very least, commanded). It might be argued that the inherent idea of Time which operates in SCIENCE FICTION is the modern idea and that the concept of Time which governs fantasy is the medieval concept.

This sense that, for the medieval mind, Time enfolds the shape of events is analogous to the sense that, in fantasy texts, STORY expresses the shape of events. In both cases, what is significant is a form of *retelling*: the cycle of the SEASONS and the TWICE-TOLD structure of the fantasy story are different ways of expressing the conviction what counts, what is true, is to be *found again*. For the medieval mind, and for the fantasy writer or reader, to say "Once upon a time" is to seize the day. [JC]

See also: DREAMTIME; POLDER; TIME ABYSS; TIME FANTASIES; TIME IN FAERIE; TIMESLIPS; TIME TRAVEL.

TIME ABYSS Either a phenomenon or, more interestingly, a moment of PERCEPTION. As a perception it is closely analogous to the Sense of Wonder in SCIENCE FICTION, which may be defined as a shift in perspective so that the reader, having been made suddenly aware of the true scale of an event or venue, responds to the revelation with awe. The analogue in fantasy is the discovery by the reader that there is an immense gap between the time of the tale and the origin of whatever it is that has changed one's perspective on the world. The TA may be occasioned by almost anything: a RUNE, a NAME, a MEMORY, a LAND, an artefact, the WANDERING JEW . . . In Alan GARNER's *Elidor* (**1965**), when the CHILDE Roland discovers that his siblings have been profoundly immured – indeed, have fossilized – in a kind of amber for thousands of years, though only a short subjective time has passed for him, what is then experienced is a jolt of perception. Similarly, in Lloyd ALEXANDER's *The Book of Three* (**1964**), the discovery of the bones of a ship in a hidden valley becomes a TA when those bones turn out to be those of Noah's Ark. And J.R.R. TOLKIEN's *The Lord of the Rings* (**1954-5**) – once the immense backstory contained in *The Silmarillion* (**1977**) and other texts is understood – seems to hover at the very lip of of a profound TA.

There is a central moment in many fantasy narratives when the protagonist recognizes the STORY he has been living and remembers who he truly is. Depending on the nature of the story, and of the act of remembering, a TA may be experienced. Various protagonists in Robert HOLDSTOCK's **Ryhope Wood** sequence experience TAs of this sort, as they discover their own faces staring out at them from the heart of the woods and from the depths of time.

In fantasy, the TA almost always marks a gap between the present of the tale and some point deep in the past. Where an artefact or vista signals a gap between the tale's present and some point in the *future*, this is more likely to be perceived as an ANACHRONISM. [JC]

TIME BANDITS UK movie (**1981**). HandMade Films. **Pr** Terry GILLIAM. **Exec pr** George Harrison, Denis O'Brien. **Dir** Gilliam. **Spfx** John Bunker. **Screenplay** Gilliam, Michael Palin. **Novelization** *Time Bandits* * (**1980**) by Charles Alverson. **Starring** Kenny Baker (Fidgit), John Cleese (Robin Hood), Sean Connery (Agamemnon), David Daker (Kevin's Dad), Malcolm Dixon (Strutter), Shelley Duvall (Pansy), Mike Edmonds (Og), Sheila Fearn (Kevin's Mum), Katherine Helmond (Mrs Winston), Ian Holm (Napoleon), Michael Palin (Vincent), Jack Purvis (Wally), David Rappaport (Randall), Ralph Richardson (Supreme Being), Tiny Ross (Vermin), Peter Vaughan (Winston), David Warner (Evil Genius), Craig Warnock (Kevin), Jerold Wells (Benson). **Voice actor** Tony Jay (Supreme Being's projection). 113 mins. Colour.

Young Kevin, alienated from his tv-addicted, gadget-loving parents, is briefly visited in his bedroom one night by a horned horseman (foreshadowing the Red Knight in Gilliam's later *The* FISHER KING [**1991**]). Next night Kevin goes to bed armed with a torch and a Polaroid camera. Out of the wardrobe tumble six DWARFS (reminiscent more of Beachcomber's litigious red-bearded dwarfs than of SNOW WHITE's allies), led by Randall, in flight from the Supreme Being, whose MAP of the Universe they have stolen; the Universe was, apparently, a "botched job", and they were given the map so they could repair all the holes, but instead are using them for banditry. Dragging Kevin along (thus making a party of SEVEN), they flee the Supreme Being's wrathful projection to pass through a time-hole to 1796 and the midst of Napoleon's Italian campaign. Having robbed the Little Corporal, they dash to the Middle Ages, encountering ROBIN HOOD and his cutthroat band; again they flee, not knowing that the Universe's Evil Genius, trapped by the Supreme Being in the Fortress of Ultimate Darkness, is SCRYING on them. Desiring the map, he plants

in their minds, through brief POSSESSION, the notion of a QUEST to gain the greatest treasure of all . . . which will be found in his Fortress. But Kevin's mind is impervious to the Evil Genius. The Supreme Being's projection appears; two time-holes open, and Kevin, although by now an honorary dwarf, flees alone through one to land near 13th-century-BC Mycenae, just in time to save Agamemnon from a bull-headed warrior. But the dwarfs reappear to rob Agamemnon: they, Kevin and a large part of the court's jewellery vanish through a time-hole to the deck of the *Titanic*. The ship sinks and all seems lost for the group; but the Evil Genius intercedes, giving them the information that his Fortress lies in the Time of Legends (marked on the map) and drawing them through into that REALITY. Avoiding the clutches of an ocean-going ogre, Winston, they and the ogre's ship are carried ashore by a GIANT who uses the ship as a hat. Escaping again, they smash through an invisible barrier to reach the Fortress, where the Evil Genius and his henchmen appear to them in the guise of a tv gameshow's Quizmaster and attendants (who are also Kevin's Mum and Dad), trick them (◊ TRICKSTER) into handing over the map and cage them. The Evil Genius sets about plotting his recreation of the world, which will this time be dominated by hypertechnology. Kevin and the rest escape and recapture the map; in a symbolic version of the LAST BATTLE, the Evil Genius fights off Arthurian knights, Wild West cowboys, Trojan archers, a FLASH-GORDON-style fighter spaceship and a WORLD WAR II tank. Just in time, the Supreme Being arrives, destroys the Evil Genius (his own creation), sets the dwarfs to tidying up all the scattered chunks of EVIL, and tells them he planned they should have the map so that they would test his Universe-building handiwork, which they've done. The others abandon Kevin and go back to the task of creation, but one lump of Evil remains with him, unnoticed. The scene transmutes to his bedroom, where he is sleeping amid clouds of smoke: the house is on fire. He is saved. His parents find the smoking lump of Evil in their microwave; because Kevin warns them not to, they touch it, and are immediately incinerated. Kevin is left alone in the ashes of his former existence.

TB, perhaps the most significant achievement of the fantastic CINEMA, is an extremely complex and eventful CROSSHATCH using the images of boyhood dreams for its construction (Kevin reads a book on Greek heroes, there is a poster of Napoleon on his bedroom wall, some of the building-stones of the Fortress of Ultimate Darkness take the form of gigantic Lego blocks, the Evil Genius's technophilia reflects Kevin's parents' mindless obsession with gadgetry, etc.); *TB* tempts one to interpret it as a RATIONALIZED FANTASY – all is Kevin's jumbled DREAM – only to snatch any such possibility away again, with the emergence of the spare chunk of Evil in the mundane world and the fact that the lead fireman is Agamemnon (who seems to recognize him) in new guise. Some of the early sections of the movie are merely playful, reminding one of Gilliam's history with the MONTY PYTHON'S FLYING CIRCUS team; yet once the quest has been announced there is a deep vein of seriousness underlying all the slapstick and pyrotechnics. When Kevin is appalled by the Supreme Being's insouciance about the loss of life involved in testing the Universe, he is answered with callousness: "Why *do* we have to have Evil?" says Kevin. "Oh, I think it's something to do with free will," replies the Supreme Being with a dismissive wave of the hand. [JG]

TIME FANTASIES Stories in which time is shaped, stopped, saved, speeded up or travelled through are extremely common in fantasy. But time itself is perceived in significantly different ways in different eras and in different contexts; and the sense of time which permeates most fantasy differs as a whole from that which governs in general our sense of the nature of time in other literatures. To modern eyes, the medieval notion of TIME is itself fantastic, and any story – which means almost any fantasy – where time shapes itself around and gives significance to human and supernatural events will read like a TF. In this encyclopedia, when the term TF is used it may be assumed that a specific manipulation of time is being referred to. There is an additional assumption: that most stories in which time is manipulated are set, at least in part, in this world.

Both TIMESLIPS and TIME TRAVEL are common in fantasy. They are found particularly often in CHILDREN'S FANTASY. They tend to be caused by MAGIC, or by travel through PORTALS, or via DREAMS; almost all timeslip tales and most time-travel stories take place in this world. Tales featuring MALIGN SLEEPERS or ONCE AND FUTURE KING figures are not normally time fantasies, as it may normally be presumed that figures so described have actually slept through long periods during which time passes neutrally; but sometimes the chamber in which the DARK LORD or the HIDDEN MONARCH are sequestered is a tiny POLDER in which time has been stopped. Larger polders frequently encompass some distortion in time: time may be stopped; time may pass only at intervals, as in BRIGADOON; or, as in SLEEPING BEAUTY, may start again only once a SPELL has been broken; or time may pass more slowly (◊ TIME IN FAERIE). Time may also pass at differing rates between various OTHERWORLDS; the sense that time passes at the rate appropriate to the world it enfolds is, almost certainly, a default sense on the part of fantasy writers and readers.

Other kinds of tale which involve manipulations of time include: CYCLES, in which events repeat themselves like the SEASONS; MULTIVERSE tales, like those spun with very great interweaving complexity by Michael MOORCOCK; and time-trap tales in which characters, more narrowly, find themselves repeating their lives, as in Ken GRIMWOOD's *Replay* (**1986**), or perhaps – as in GROUNDHOG DAY (*1993*) – a single day. [JC]

TIME IN FAERIE Visitors to FAERIE find that TIME there is subjective, disengaged from real-world clocks and CALENDARS. Years of REVEL or BONDAGE may occupy one night or one instant of normal time; or years may pass outside during a perceived (◊ PERCEPTION) brief visit. The first situation suggests a time POLDER where in the extreme case no real time passes. Examples are: C.S. LEWIS's **Narnia** series, where a decades-long visit to the SECONDARY WORLD compresses into moments of Earthly time (and, conversely, an Earth year corresponds to many Narnian centuries); J.G. BALLARD's "The Garden of Time" (1962), whose MAGIC flowers can, while they last, be plucked to turn back oncoming THINNING and doom; and Michael Scott ROHAN's *The Lord of Middle Air* (**1994**), whose callow protagonist matures and fights long wars in Faerie, only to return at the instant he left. Non-Faerie variations may treat time as ILLUSION: Jurgen's entire year of adventures is cancelled and converted to DREAM in James Branch CABELL's *Jurgen* (**1919**); Jorge Luis BORGES's "The Secret Miracle" (**1943**) pauses time for the protagonist in a manner invisible to outsiders. *The Third Policeman* (**1967**) by Flann O'BRIEN has a region

called "eternity" where time continues but one does not age, nor (a special convenience) need to shave. Time polders are parodied in Michael SWANWICK's *The Iron Dragon's Daughter* (**1993** UK), where real-world time does not pass for visitors to an enchanted shopping mall.

The second and commoner form of Faerie time-slippage, where real time passes unnoticed, is likely to be harrowing. J.M. BARRIE's play *Mary Rose* (**1920**) recognizes this as the eponymous heroine returns unchanged to her 25-years-older husband and child; but BRIGADOON (*1954*) accentuates the positive by avoiding the painful issue of long-delayed return. James STEPHENS's *In the Land of Youth* (**1954**) toys with reasons for the magic world's differing time flow. Poul ANDERSON suggests a practical application in *Three Hearts and Three Lions* (**1953**), when ELVES plan to entertain the protagonist UNDERGROUND for a night which will be a century, preventing his interference in current events. Ursula K. LE GUIN's "The Dowry of the Angyar" (**1964**; vt "Semley's Necklace") adds a cruel TECHNO-FANTASY justification: here a night-long journey with GOBLIN-folk takes 16 years because, unknown to the victim, this is relativistic spaceflight. The stolen child Lilac, caught out of time in John Crowley's *Little, Big* (**1981**), is briefly glimpsed asleep on Father Time's lap.

Katharine BRIGGS's *A Dictionary of Fairies* (**1976**) contains an extensive listing of FAIRYTALE TIF examples.   [DRL]
See also: BELATEDNESS; CROSSHATCH; TIME FANTASIES; TIMESLIPS.

TIMESLIPS The facilities of memory and foresight are fundamental to conscious life and intelligence; their limitations are a prison through whose bars the imagination is constantly reaching. Stories of actual displacement in time (as opposed to time-distorting DREAMS) are of relatively recent provenance, but they have been produced in vast profusion during the last century. In spite of the paradoxes which stem from the idea, TIME TRAVEL was quickly adopted into SCIENCE FICTION. But abrupt and inexplicable timeslips are properly the stuff of fantasy.

Whether it is the individual consciousness which slips, taking up residence in another body or an earlier version of its own – as a sort of extension of the IDENTITY-EXCHANGE notion – or whether an object or a person is extracted from the timestream and thrown back again at a different point, literary timeslips are rarely random. They often link an ancestor with a descendant, and tales of time-crossed lovers are equally frequent; given that LOVE is so often represented in fiction as a kind of quasisupernatural force, it is perhaps unsurprising that it should occasionally be given the power to transcend time in order to secure a uniquely precious union (or, given that many such romances are distinctly bittersweet, to exaggerate the tragedy of its impossibility), and "timeslip romances" warrant consideration as a subgenre in their own right.

Tales of time dislocation associated with TIME IN FAERIE (or the kind of long sleep featured in Washington IRVING's "Rip van Winkle" [1819]) may be regarded as proto-timeslip fantasies, as can some early REINCARNATION romances and VISIONARY FANTASIES. Théophile GAUTIER's "Arria Marcella" (1852) was one of the earliest tales to blur the border between visionary fantasy and timeslip romance, but Mark TWAIN's *A Connecticut Yankee at King Arthur's Court* (**1889**) provided the most important precedent for modern timeslip stories. The theme was given a significant boost by the publication of J.W. Dunne's metaphysical treatise *An Experiment with Time* (**1937**), which helped inspire J.B. PRIESTLEY's "time plays", although J.M. BARRIE had already demonstrated that timeslips are among the easiest fantasy devices to contrive in the theatre. Barrie's harrowing *Mary Rose* (**1924**) is another landmark in the history of the motif, as are Arthur MACHEN's escapist fantasy *The Hill of Dreams* (**1907**), *Not in Our Stars* (**1923**) by Michael Maurice (1889-?   ) – an account of dislocated time-reversal – Christopher MORLEY's pessimistic study of the corrosions of maturation *Thunder on the Left* (**1925**) and Gerald BULLETT's sad tale of temporally mismatched partners, "Helen's Lovers" (1932).

Henry JAMES's unfinished *The Sense of the Past* (**1917**) would presumably have provided an important paradigm had he managed to complete it; the sentimental play *Berkeley Square* (**1928**) which John Balderston and J.C. Squire based on it is a poor substitute, although it was effectively filmed as *Berkeley Square* (**1933**) and as *I'll Never Forget You* (**1951**). Other notable tales of time-crossed lovers include *The Haunted Woman* (**1922**) by David LINDSAY, *Still She Wished for Company* (**1924**) by Margaret IRWIN, *Portrait of Jennie* (**1940**) by Robert NATHAN – filmed as PORTRAIT OF JENNIE (*1948*) – *The Twinkling of an Eye* (**1945**) by D.C.F. Harding, *By Firelight* (**1948**) by Edith PARGETER, *Time and Again* (**1970**) by Jack FINNEY, *Bid Time Return* (**1975**) by Richard MATHESON, *The Dream Years* (**1985**) by Lisa GOLDSTEIN and *Serenissima* (**1987**) by Erica Jong (1942-   ). Timeslip stories which allow characters a second chance to get their lives right (which they almost invariably waste) include Barrie's *Dear Brutus* (**1922**), *The Devil in Crystal* (**1944**) by Louis Marlow (1881-1966), *The Strange Life of Ivan Osokin* (**1947**) by P.D. Ouspensky (1878-1947), *Time Marches Sideways* (**1950**) by Ralph L. Finn (1912-   ) and *Changing the Past* (**1989**) by Thomas BERGER. Notable timeslip stories of more various inclination include *Before I Go Hence* (**1945**) by Frank BAKER, *The House on the Strand* (**1969**) by Daphne DU MAURIER, *The Mirror* (**1978**) by Marlys Millhiser (1938-   ) and *The Exile* (**1987**) by William KOTZWINKLE. John Dickson CARR failed to found a new subgenre of timeslip crime stories despite setting three notable precedents in *The Devil in Velvet* (**1951**), *Fear is the Same* (**1956**) as Carter Dickson and *Fire, Burn!* (**1957**).

In recent years timeslips have become a favourite didactic device in children's fiction, following such precedents in adult fantasy as *Friar's Lantern* (**1906**) by G.G. Coulton (1858-1947) – which assaulted G.K. CHESTERTON's and Hilaire Belloc's nostalgic affection for the medieval Church – *The Burning Ring* (**1927**) by Katharine BURDEKIN and *Night in No Time* (**1946**) by Eliot CRAWSHAY-WILLIAMS. They are deployed to particularly good effect in *A Traveller in Time* (**1939**) by Alison UTTLEY, *Tom's Midnight Garden* (**1958**) by Philippa PEARCE, *Charlotte Sometimes* (**1969**) by Penelope FARMER, *Beadbonny Ash* (**1973**) by Winifred Finlay (1910-1989) and *To Nowhere and Back* (**1975**) by Margaret J. ANDERSON. Timeslips have also recently become a cliché of MILITARY FANTASY, protagonists with specialist modern knowledge being projected into primitive contexts where they are able to commit more effective mayhem than the indigenous populations; Robert ADAMS's series begun with *Castaways in Time* (**1979**) is typical of the species.

Timeslips are numerous in the movies – too numerous to list. *Search for Grace* (*1994*) is an example of a straightforward treatment of the theme; DON'T LOOK NOW (*1973*) is altogether more enigmatic, since the timeslip may be either to the past, if the movie is read one way, or *have* been to the future, if read the other. The most famous "factual" occurrence was that experienced by Charlotte Moberly and Eleanor Jourdain in 1901 at Versailles: they for a while found themselves seemingly in the 18th century.   [BS]

TIME TRAVEL In SCIENCE FICTION rationalized TT into the past or future is common. In fantasy TT is usually embarked upon pastwards, and is sometimes difficult to distinguish from TIMESLIP. The main distinction seems to be that timeslip tales involve no actual corporeal travel through time: a human SOUL or SPIRIT may pass into the body of a human being from an earlier time, or, more usually, a person makes a mental venture (possibly involuntarily) into a past time and this intrusion has physical repercussions – as when lovers unite despite the decades separating them. In the former case, POSSESSION or IDENTITY EXCHANGE may be assumed. TT stories proper are those whose protagonists find themselves, body and soul, in a new venue, as in Mark TWAIN's *A Connecticut Yankee in King Arthur's Court* (**1889**); other examples, among many, are the **Magic** sequence by Andre NORTON, whose protagonists visit various eras with the aid of MAGIC, William MAYNE's *Earthfasts* (**1966**), in which an 18th-century lad travels accidentally to the present day, and Rudyard KIPLING's *Puck of Pook's Hill* (coll **1906**), where children experience history first-hand. TT which shifts characters backwards to a longed-for era is common; the stories Jack FINNEY assembled in *The Third Level* (coll **1959**) and *About Time* (coll **1986**) provide a definitive conspectus of the theme of TT as an affirmation of THEODICY.

A more complex kind of TT – it might indeed be thought of as a TIME FANTASY – appears in stories like Ursula K. LE GUIN's "April in Paris" (1962), where a number of characters from various eras are assembled at one point in space and time. Michael MOORCOCK's conjoined MULTIVERSE sequences offer similarly complexified versions of travel through time ($\lozenge\lozenge$ TEMPORAL ADVENTURESS).   [JC]

TIMLETT, PETER VALENTINE (1933-   ) UK writer of occult/New Age fantasy who was inspired to write fiction through what he learned by membership of the Society of Inner Light, the London-based occult group founded by Dion FORTUNE. His three published books form **The Seedbearers** trilogy; here PVT explores the idea of bearing the seed of knowledge and spiritual awareness up though the centuries. *The Seedbearers* (**1974**) is about the fall of ATLANTIS and the few who survive to pass the wisdom of the priests to the new world. It was followed by *The Power of the Serpent* (**1976**) and *The Twilight of the Serpent* (**1977**), the former about the building of Stonehenge, the latter about the conflict between the Culdees and the emergent Christian Church and the druids. After a long hiatus he returned to writing in the 1990s with a number of short stories.   [JF]

TIMLIN, WILLIAM M(ITCHESON) (1892-1943) UK-born artist and author, in South Africa from 1912. His reputation rests on one book, *The Ship that Sailed to Mars* (**1923**), which he sumptuously illustrated, and which was published using his own calligraphy. Despite its title, the story is wholly fantastic, modelled on FAIRYTALE. In his youth an Old Man believed in FAIRIES and used their skill and cunning to construct a ship in which they travelled through the wonders of space to Mars where they encountered a magical world. The storyline is minimal, the book's strength resting on its exotic descriptions and beautiful illustrations. At the time of his death WMT was working on a new volume, «The Building of a Fairy City»; this remains unpublished, though some of the artwork has been issued on postcards. [MA]

TINER, RON Working name of UK illustrator, COMICS artist and writer Ronald Charles Tickner (1940-   ), who began to draw sf, fantasy, sport, war and romance comics in the mid-1970s. His work has appeared in most UK children's comics, the Nigerian SUPERHERO title *Powerman* (1976-8) and the US comic book *Hellblazer* (1989-90); more recently he has concentrated on ILLUSTRATION. He has written *Figure Drawing Without a Model* (**1992**) and, with John GRANT, *The Encyclopedia of Fantasy and Science Fiction Art Techniques* (**1996**); his articles on illustration have appeared in various magazines. He co-ordinated and largely wrote the comics entries in the *SFE*, and has done the same for this encyclopedia, of which he is a Contributing Editor. [RT/JG]

TINTIN An adventurous young lad (also rendered Tin-Tin) who gets into various plights. He was created in 1929 by the Belgian COMICS artist Hergé (real name Georges Rémi [1907-1983]) for the children's supplement section of the newspaper *Le petit vingtième* ["The Little Twentieth"]; his first appearance in book form was in *Tintin in the Land of the Soviets* (graph **1930**), and his last was *Tin-Tin et les Picaros* ["Tintin and the Rogues"] (graph **1976**). In 1946 Tintin began to appear in a weekly magazine, along with other strips. Hergé, who had right-wing views, became controversial after WWII for alleged collaborationist activities, which included at least one **Tintin** adventure whose villain was a Jew. The series is undistinguished, but had a significant childhood effect on many fantasy writers working today. [JC]

Movies
Numerous **Tintin** tales, done in limited animation, have appeared on tv. From these have been derived several Belgian/French **Tintin** movies, most short and most with innumerable vts: *The Lake of Sharks* (**1972**; ot *Tintin et le Lac aux requins*), *Red Rackham's Treasure* (**1987**), *The Black Island* (**1987**), *The Calculus Affair* (**1987**), *The Crab with the Golden Claws* (**1987**), *The Secret of the Unicorn* (**1987**), *The Seven Crystal Balls* (**1987**) and *The Shooting Star* (**1987**); there may well be others. [JG]

TIPPETT, [Sir] MICHAEL (1905-   ) UK composer, of fantasy interest for a series of five OPERAS for which he also wrote the librettos. Of most intense interest is perhaps the first, *The Midsummer Marriage* (**1955**), a complex quasi-ALLEGORY – the libretto was separately published as *The Midsummer Marriage* (**1954** chap) – which combines some of the CROSSHATCH liquidity of William SHAKESPEARE's *A Midsummer Night's Dream* (performed *c*1595; **1600**) and the RITE-OF-PASSAGE structure of Mozart's *The Magic Flute* (**1791**; libretto by Emanuel SCHIKANEDER). *King Priam* (**1962**) is concerned with the Trojan War. *The Knot Garden* (**1970**) (◊ LABYRINTH) puts its protagonists through a metaphysical passage into self-recognition and forgiveness, echoing Shakespeare's *The Tempest* (performed *c*1611; **1623**). *The Ice Break* (**1977**) allegorizes the passage between the frozen BONDAGE of mortal selfhood into (once more) self-recognition. *New Year* (**1988**) presents a rite of the SEASONS. [JC]

TIR-NAN-OG The Celtic "Land of the Young" to which the old FAIRY folk retreated in the face of THINNING. This OTHERWORLD combines attributes of FAERIE, HEAVEN and the Earthly Paradise, and is often invoked in CELTIC FANTASY. Mortal visitors may suffer TIME-IN-FAERIE disjunctions, with centuries passing unheeded yet taking effect on departure (◊◊ SHANGRI-LA). Roger ZELAZNY borrowed the name in *Sign of the Unicorn* (**1975**) for an insubstantial kingdom which is literally in the sky. [DRL]

TITANIA ◊ FAIRY QUEEN.

TITHONUS ◊ IMMORTALITY; QUIBBLES.

TITMARSH, M(ICHAEL) A(NGELO) ◊ William Makepeace THACKERAY.

TODD, BARBARA EUPHAN (1890-1976) UK writer active as an author of CHILDREN'S FANTASY from *The 'Normous Saturday Fairy Book* (**1924**) with Marjory Royce and Moira Meighn. Most of her work is for younger readers; she remains of wider interest almost exclusively for the **Worzel Gummidge** sequence about the adventures of an animate scarecrow (◊ ANIMATE/INANIMATE): *Worzel Gummidge, or The Scarecrow of Scatterbrook* (**1936**), *Worzel Gummidge Again* (**1937**), parts of these two appearing as *Worzel Gummidge, The Scarecrow of Scatterbrook Farm* (**1947** US), plus *More About Worzel Gummidge* (**1938**), *Worzel Gummidge and Saucy Nancy* (**1947**), *Worzel Gummidge Takes a Holiday* (**1949**), *Worzel Gummidge and the Railway Scarecrows* (**1955**), *Worzel Gummidge at the Circus* (**1956**), *Worzel Gummidge and the Treasure Ship* (**1958**) and *Detective Worzel Gummidge* (**1963**). The three *Worzel Gummidge* tv series, starring Jon Pertwee (1919-1996), were first shown in 1979-81, 1987 and 1989. [JC]

TODOROV, TZVETAN (1939-   ) Bulgarian-born literary critic, from 1963 in France, where he soon established himself as one of the central figures of modern theoretical criticism, his work engaging with and sophisticating Russian formalism, French structuralism and the international dance of post-structuralism. He is of importance in the study of the FANTASTIC, though the impact of his definition of the fantastic is of less use in the study of FANTASY, where the MARVELLOUS is not a problematic to be solved, but a given. In *Introduction à la littérature fantastique* (**1970**; trans Richard Howard as *The Fantastic: A Structural Approach to a Literary Genre* **1973** US), TD defines the fantastic as something which occurs when an event is experienced "which cannot be explained by the laws of the . . . familiar world. The person who experiences the event must opt for one of two possible solutions: either he is the victim of the illusion of the senses, of a product of the imagination – and laws of the world then remain what they are; or else the event has indeed taken place, it is an integral part of reality – but then this reality is controlled by laws unknown to us . . . The fantastic occupies the duration of this uncertainty. Once we choose one answer or the other, we leave the fantastic for a neighbouring genre, the uncanny or the marvellous. The fantastic is that hesitation experienced by a person who knows only the laws of nature, confronting an apparently supernatural event." A congruent analysis has been evolved independently by John GRANT (◊ PERCEPTION) and others.

Authors whose works are cited in support of TT's thesis include E.T.A. HOFFMANN, Henry JAMES, Franz KAFKA, Gérard de NERVAL, Edgar Allan POE, Count Jan POTOCKI and VILLIERS DE L'ISLE ADAM. [JC]

TOLKIEN, CHRISTOPHER UK writer (1924-   ). ◊ J.R.R. TOLKIEN.

TOLKIEN, J(OHN) R(ONALD) R(EUEL) (1892-1973) UK writer and philologist, born in South Africa but resident in the UK from 1895. He specialized as a scholar in early forms of English, but is best-known as the 20th-century's single most important author of fantasy. After service in WWI he began work on *The Silmarillion*, and for the rest of his life continued to expand this "subcreation" (his term for the inventing of fantasy worlds) into a conscious MYTHOLOGY for England, as it was a culture (he thought) which lacked a true CREATION MYTH.

His academic career began in 1919. He became Professor of Anglo-Saxon at Oxford University in 1925, and was appointed Merton Professor of English at Oxford in 1945, a post he held until his retirement in 1959. It has been argued persuasively by Tom Shippey (1943-   ), in *The Road to Middle-Earth* (**1982**; rev 1992), that JRRT's profound grounding in philology did far more than provide a linguistic stew of real and imaginary languages out of which he dreamed his work, as though his tales were translated from a lost original, but in fact suggested to him a specific technique of world-making. Any argument of this sort must, of course, reckon with the central importance, for JRRT, of his own illustrations to his various works; many of these paintings and watercolours were composed prior to or in conjunction with the actual act of writing, and clearly represent an important inspiration for the tales. In *J.R.R. Tolkien: Artist & Illustrator* (**1995**) by Wayne G. Hammond and Christina Scull this material is presented and its importance argued for.

At Oxford before WWII JRRT formed a close literary association with Owen Barfield (1898-   ), C.S. LEWIS and Charles WILLIAMS; they met regularly, calling themselves The INKLINGS, and at their meetings read aloud drafts of fiction and other work, a habit facilitated by – and perhaps contributory to – their shared interest in narrative. They were all Christians (JRRT was Roman Catholic), and The Inklings AFFINITY GROUP is now thought of as a central forcing-house for 20th-century CHRISTIAN FANTASY. JRRT soon published *The Hobbit, or There and Back Again* (**1937**), a CHILDREN'S FANTASY set in the **Silmarillion** universe, and in Inklings sessions he now introduced draft portions of his masterwork, *The Lord of the Rings* (**1954-5**), the most influential fantasy novel ever written.

The Secondary World. Like almost everything JRRT brought into the light in his later years, the concept of the SECONDARY WORLD, which *LOTR* embodies in definitive form, had been evolving for decades, first being articulated in "On Fairy Tales", a 1939 lecture expanded for *Essays Presented to Charles Williams* (anth **1947**) ed anon C.S. Lewis, and further expanded for its appearance in *Tree and Leaf* (coll **1964**; rev 1988). The notion of the secondary world, as JRRT first defined and later embodied it, builds of course on the work of earlier writers: William MORRIS, Lord DUNSANY, possibly James Branch CABELL and certainly E.R. EDDISON, among others, had been creating partially autonomous fantasy worlds since before the turn of the century; but JRRT, through precept and example, gave final definitive legitimacy to the use of an internally coherent and autonomous LAND of FAERIE as a venue for the play of the human imagination. For the sf/fantasy writers who followed JRRT, this affirmation of autonomy was of very great importance. *LOTR* marked the end of apology. No longer was it necessary for fantasy writers to feel any lingering need to "normalize" their secondary worlds by framing them as TRAVELLERS' TALES, or DREAMS (entered via PORTALS) which prove exiguous at dawn, or TIMESLIP tales, or as BEAST-FABLES. Though each of these forms continues to be used, fantasy writers after about 1955 would invoke them as

a matter of aesthetic choice. JRRT gave fantasy a domain; it is of course another question as to whether, in recent years, his bequest has been properly honoured: countless purveyors of GENRE FANTASY have reduced the secondary world to the Identikit FANTASYLAND.

A fully imagined secondary world is, in theory, nothing more than a world which has been created by its teller, and which is governed by internally consistent rules to which the reader gives credence, and in terms of which anything can be believed – in which, as a random example, a "green sun will be credible", as JRRT puts it in "On Fairy-Stories" – as long as that which is believed in is *livable*. A world which operates according to the unlivable premises – however *arguable* they may be – of the typical WONDERLAND, or any fantasy environment constructed according to TOPSY-TURVY premises, is not a secondary world. JRRT generated the term during the decades of his work on the profoundly livable fantasy environment in which, eventually, the climactic action of *LOTR* was set. This environment – which we now think of as the paradigm secondary world – had three main characteristics: (a), as noted, it was livable, because the central LAND depicted by JRRT, Middle-Earth in the world of Arda, is imagined with such detail and solidity that it seems to breathe with the lives of its inhabitants; (b) it was legible, because it directly and profoundly expressed JRRT's deep attachment to the knowable LANDSCAPE of the Middle Ages, which is more real than our own REALITY, and brighter; and (c) it was fantasy because – as befits the stage upon which a transformative drama is being enacted – it was constantly in the throes of METAMORPHOSIS.

The geographical details of JRRT's secondary world have become a TEMPLATE which later writers have become accustomed to use as a *fixed* background against which all sorts of stories, very few of them full fantasies, can be told. But for JRRT, the detailed description of Middle-Earth, and the prefatory MAP which accompanies that description, do not constitute a template, because *LOTR* takes place after a long and profoundly transformative CREATION MYTH and cosmology and history of the world have been unfolded, climaxing at the point the trilogy begins. Middle-Earth – however much it may have been normalized into backdrop by others – was a fresh creation for JRRT, and it lies at the cusp of a final dramatic THINNING into the secular history of our own world. The Shire, which is the home of the Hobbits, is a classic POLDER; a variety of sites evoke a sense of TIME ABYSS.

Arda Almost every story and story-fragment pertinent to an understanding of the immensity of JRRT's mythology for England appeared only after his death, beginning with *The Silmarillion* (**1977**) and continuing with *Unfinished Tales of Númenor and Middle-Earth* (coll **1980**) ed Christopher Tolkien, JRRT's son. The main sequence of posthumous works has been published subsequently as **The History of Middle-Earth**, all ed Christopher Tolkien: *The History of Middle-Earth #1: The Book of Lost Tales 1* (coll **1983**), *#2: The Book of Lost Tales 2* (coll **1984**), *#3: The Lays of Beleriand* (coll **1985**), *#4: The Shaping of Middle-Earth* (coll **1986**), *#5: The Lost Road and Other Writings* (coll **1987**), *#6: The Return of the Shadow: The History of the Lord of the Rings 1* (coll **1988**), *#7: The Treason of Isengard: The History of the Lord of the Rings 2* (coll **1989**), *#8: The War of the Ring: The History of the Lord of the Rings 3* (coll **1990**), *#9: Sauron Defeated: The History of the Lord of the Rings 4* (coll **1992**), *#10: The Later Silmarillion 1: Morgoth's Ring* (**1993**) and *#11: The Later*

Silmarillion 2: The War of the Jewels (**1994**), plus a final volume, projected for 1996. The material assembled in these volumes is of variable interest as narrative, but can be summarized as follows.

Two things may be noted.

First, there are obvious similarities between JRRT's mythological history of the world and that propounded by H.P. BLAVATSKY in *The Secret Doctrine* (**1888**), the text which contains most of THEOSOPHY's long "back-story". These similarities include details of the initial cosmogony, the division of the long history of the world into Ages separated by cataclysms brought about by the EVIL of human beings (and other creatures), the fact that new species are introduced at the beginning of Ages, and the use of geographical regions like ATLANTIS. But it is likely that he and Blavatsky separately accessed the Western world's CAULDRON OF STORY (a term he uses in "On Fairy-Tales"). Moreover, the broad sweep of JRRT's narrative is retrospective, and mourns the passing away of the earlier Ages; Blavatsky resolutely "trumps" Darwin by extolling the rise to awareness (and therefore to a higher state) of her successive races.

Second, most of the details of JRRT's mythology have surfaced over the years since his death, are very much less known than the immediate story and background of *LOTR* itself, and are rather confusingly presented in the increasingly large number of volumes edited by his son. But the surface story of *LOTR* – like *The Hobbit* before it – is mediated through a network of allusions to earlier times, of which the events at the end of the Third Age constitute a thinned but intensely dramatic re-enactment. A brief resumé of the back-story is therefore likely to be useful, and also serves to emphasize the constant, central, defining TRANSFORMATIONS to which JRRT subjected his subcreation.

For convenience the story of Arda can be divided into eight parts, and these into two main divisions. The first five, though described in remarkable detail, remain comparatively distant in the depiction. Arda is created before TIME begins, and it is only with the Second Age, which begins 37,000 years before *LOTR*, that a chronology is set in place; the last of the five ends 30,000 years later. The second group of Ages, three in number – the Ages of the Sun – last something under 7000 years, and move into active back-story at the point Gollum finds the One RING, in an action which eventually governs the plot of *The Hobbit*. The eight ages are as follows:

1. The Prime Being or first principle brings about all of creation (◊◊ CREATION MYTHS) by conceiving of the GODS, who themselves sing Arda into existence (as in Theosophy, the Prime Being takes no further active role). Arda is a flat world, consisting of one huge continent, and is enclosed within spheres of light and aether. **2.** The gods (those who participate are known as the Valar, and their demigod partners – Gandalf is one – are known as the Maiar) continue their acts of creation, but one of the Valar (his name is Melkor; Sauron is his Maiar servitor) revolts. Although Melkor is defeated, Arda is no longer whole: the THINNING has begun, and the sense of loss so central to full fantasy – a sense which JRRT definitively embodies – begins to mount. **3.** The Age of Lamps, so-named because two sky-piercing magical lamps light the flat world, are nearly Edenic, until Melkor rises again from his fortress of Utumna, destroys the lamps, causes vast geological upheavals, and persuades the Valar to leave the heart of Arda and to establish themselves

westwards, in the Undying Lands. **4.** Melkor rules Middle-Earth in total darkness through the 10,000 years of the First Age of the Trees, so-named after two huge magical TREES constructed by the Valar to give themselves light and wisdom in the Undying Lands. DWARFS (whom JRRT habitually called "dwarves") come into being. **5.** In the Second Age of the Trees, the Valar bring STARS into the heavens, ELVES come into being, and eventually Melkor is defeated. But he arises yet again, destroys the magical Trees, "the Great Lights of the World", and steals the Silmaril jewels; *The Silmarillion*, which takes place in the next Age, is centrally concerned with the fate of these jewels, which contain the essence of light (another Theosophical trope) retained from the trees. Utter darkness falls. **6.** As the First Age of the Sun begins (7000 years before *LOTR*), the Valar kindle the SUN and the MOON, and humans come into being (as do the Hobbits, who are HALFLINGS, seemingly related to both humans and dwarfs); for 600 of these years, the Wars of Beleriand dominate Middle-Earth, until Melkor (also known as Morgoth) is finally driven from Arda, for good. But Beleriand, the Elven home, sinks beneath the waves. **7.** The Númenóreans, who are long-lived humans, found the island kingdom of Númenór, or Westernesse, or Atlantë (i.e., ATLANTIS), in the West, and prosper for 1000 years. But Sauron (Melkor's Maiar deputy) then creates the realm of Mordor in the Southeast, tricks some Elven-smiths into forging the Rings of Power, himself building the DARK TOWER of Barad-dûr and making the One Ring, as a result of which action he is the Lord of the Rings. He gives the Nine Rings of Mortal Men to nine corrupt kings, who are gradually transformed into the Nazgûl, SHADOW creatures as close to SUPERNATURAL FICTION in their deeds as JRRT could allow. Sauron then corrupts the Númenóreans, who go to war against the Valar, disastrously: Atlantë sinks; and in the ensuing APOCALYPSE Arda shrinks (or thins) into a mere globe, no longer connected to the ambient spheres. Sauron is overthrown but, though he loses the Ring, survives. **8.** The Third Age of the Sun culminates in the tale told in *LOTR*. Dissident races of humanity wage war or maintain uneasy peace for centuries; Sauron looks for the Ring. *LOTR* begins.

Afterwards, in the Fourth Age of the Sun, which JRRT does not describe, humanity rules Arda, which begins to revolve around the SUN.

Secondary Belief. To establish secondary belief in his secondary world – "secondary belief" being defined as the intense form of readerly acceptance required for proper belief in an autonomous subcreation – JRRT does two things.

First, he applies the principle that external descriptions or verifications of a secondary world, or the nature of any route into a secondary world, must be extrinsic to the reader's belief in that world. It is a principle which may have been articulated before, but JRRT was the first to embody it with full consistency. JRRT's profound influence on later generations of writers and readers derives, in part, from that unswerving application of principle.

Second, since the secondary world depicted in *The Hobbit* and *LOTR* have become a template for much of late-20th-century fantasy, it is necessary to remember that both novels were written *before* readers were accustomed to secondary worlds, and that JRRT's success in convincing early readers of the autonomy of his one comes from a deliberate application of techniques necessary to bring the vital secondary belief into being, techniques which almost *secretly* transform readers from secular appreciators of a text into something like parishioners. These techniques are various, but interlinked. A seemingly trival device can unpack the strategy as a whole. For instance, JRRT was (along with T.H. WHITE) probably the first fantasy author to mix DIC-TIONS within single passages, usually to contrast an archaic form of speech with the default language of the text at that stage; the result for JRRT (and White) is to open the surface narrative to a sense of TIME ABYSS, to a sense that a larger and older STORY lies under that surface, and defines it. The important thing is that both White and JRRT mix dictions deadpan, so that mixed passages *sound* as though they are "simply" part of the telling of the tale. (As has already been noted, much of *LOTR* was read aloud to gatherings of the Inklings, and a sense that the story is being *told* is never stronger than when dictions mix.)

For JRRT, this is not a seldom-used device: the deadpan mixing of radically different elements lies at the heart of his storytelling technique. In his work, the ordinary and the MARVELLOUS – or the "simple" event and the revelation that this present-day occurrence is a quote of profounder hap-penings from an immense back-story – inhabit the same overarching reality. In this fashion the ordinary and the marvellous are "heard" to validate each other, in a kind of utterly serious punning. Together they confirm the overar-ching reality they address and inhabit – which is the secondary world, and which becomes (for the first time in literature) utterly autonomous. This autonomy constituted a revolution.

Techniques of this sort enforce a trust in the reader that what is being told is a truth. As theorists have long empha-sized, there is no safety in metaphors in either sf or fantasy. What looks like comparison between two worlds more often than not turns out to be a literal description of the one world. A character like Bilbo Baggins in *The Hobbit* can gradually change into – or be revealed as always having been – a kind of HERO; in *LOTR* itself, slow growth-curves of understanding take on very much heavier implications. It is not simply that Strider turns out always to have been a HIDDEN MONARCH, that Gandalf is no mere WIZARD but something very much like a seraphim (♢ ANGEL), and that Tom Bombadil proves to be a kind of MYTHAGO, as ancient as Gandalf, and perhaps more mysterious; what is so deeply engaging for readers of *LOTR* is the sensation that they are being brought to *recognize* the "true" nature of the fictional characters, and that they are doing so within the implacable security of a technique that does not waver from the instil-lation of belief.

The pace of storytelling similarly works to unfold a sense that present events are PALIMPSESTS laid upon antique acts. What in *The Hobbit* is an exciting QUEST for a hoard guarded by a DRAGON, is re-understood in the context of *LOTR*, where we understand that the dragon is actually taking part in the fundamental drama that has riven Arda for aeons. In *LOTR*, a large cast of COMPANIONS (vol 1) undertake separate NIGHT JOURNEYS (vol 2) into a LAST BATTLE (vol 3) which re-enacts and completes that fundamental drama (and provides a convincing model for any definition of full fantasy). The effect is of a *whole tale*. The carefully achieved palimpsest of themes in these final pages is reminiscent of Richard WAG-NER's *Götterdämmerung* (**1867**), the last OPERA in his **Ring Cycle**, where an orgasmic confluence of *leitmotifs* marches that huge back-story into permanent death.

LOTR The storyline of *The Hobbit, or There and Back Again* (**1937**; rev 1951; rev 1966 US; further rev 1966; vt *The Annotated Hobbit* ed Douglas A. Anderson 1988 US; final corrected text 1995 UK) is familiar, and does not require extensive recounting. As already hinted, it tells the story of the Hobbit Bilbo Baggins, dragged by Gandalf into a QUEST with some companion dwarfs for a hoard guarded by the dragon Smaug, who has lurked in Erebor ever since (in the back-story) he banished the dwarfs from their underground kingdom. In a minor incident (darkened and given addi-tional emphasis in the 1951 version) Bilbo tricks a morally and physically decayed Hobbit named Gollum out of a RING. The story ends.

Decades later, *LOTR* begins; it is one extremely long sus-tained tale, initially published in three volumes – *The Fellowship of the Ring: Being the First Part of The Lord of the Rings* (**1954**; rev 1965 US), *The Two Towers: Being the Second Part of The Lord of the Rings* (**1954**; rev 1965 US) and *The Return of the King: Being the Third Part of The Lord of the Rings* (**1955**; rev 1965 US) – because of 1950s fears about attempting to market a FAIRYTALE over 1000 pages long; all three were assembled as *The Lord of the Rings* (omni **1968**; the edition of 1987 may be definitive). At the start, Gandalf begins to reveal to Frodo Baggins the implications behind his cousin Bilbo's long retention of what is in fact the One Ring of Power, lost by Sauron centuries earlier. The Fellowship of the Ring is founded to accompany Frodo on an immense trek to Mount Doom in Mordor, where the Ring was originally forged, and where it may be destroyed at last, ending the power of the dread Sauron, who is (we must remember) a kind of a PARODY of long-banished Melkor/Morgoth – so that his demise will mark the end of the story. That Sauron is a parody of the true DARK LORD befits JRRT's Christian sense of the nature of EVIL. The other actors in this final drama, though they too "pun" upon larger models, are treated much more gently – as witting or unwitting participants in a STORY which must be completed, a MONOMYTH in which they varyingly take on the role of the HERO (though Frodo endures the entire CYCLE), as AVATARS, or (as in the case of Gandalf and Tom Bombadil) simply themselves. But to be nobody but oneself in Middle-Earth – unless one is Sam Gamgee – is to be a walking TIME ABYSS.

Slowly, the long and tantalizingly incremental storyline of *LOTR* introduces Frodo and his companions – the most famous being the Sancho-Panza-like Sam Gamgee – to var-ious races (the elves and the dwarfs in particular) whose own role in Middle-Earth is simultaneously coming to a climax and to Tom Bombadil and Strider, whose true provenance and roles are only gradually revealed. Along the route the Fellowship of the Ring is haunted by the Nazgûl (or Ringwraiths) from the Second Age of the Sun, who ride Winged Beasts from the First Age before Time began, and who impart an element of HORROR into the tale; as befits denizens of the horror mode, they perish at the point at which the protagonists of *LOTR*, and Arda itself, pass through their ultimate ordeal and enter the EUCATA-STROPHE. That point is the precise moment at which the Ring is bitten off Frodo Baggins's finger by the insane Gollum, who then topples into the fiery abyss beneath Mount Doom. Weakened by its poison, and by the huge burden of implications it bears, Frodo himself had been unable to dispose of the Ring: JRRT's use of Gollum at this point is a coup of storytelling, and underlies the fact that, in his long scheme, *LOTR* is a dying fall, a tale of aftermath.

Describing Fantasy "On Fairy-Tales", which appears in slightly revised form as the title essay in *Tree and Leaf*, provides, among other riches, an extremely influential modelling of the structurally complete fantasy tale – that which is sometimes loosely referred to in this encyclopedia as "full fantasy". After some preliminary clearing of the way, JRRT begins by claiming that the "fairy story", by which we can understand him to mean fantasy, is a tale set in the ENCHANTMENT known as FAERIE, and which tells of marvels (◊ MARVELLOUS). He then describes four elements that are necessary to the FAIRYTALE, by which term he restricts himself to tales set in a SECONDARY WORLD. Oddly, these four elements – Fantasy, Recovery, Escape and Consolation – do not constitute a narrative analysis of fantasy, though they might seem to. Given JRRT's pervasive concern with narrative and his mastery of the techniques of telling a secondary-world tale, it is perhaps understandable that they have indeed been understood as representing phases of narrative . . . but, to repeat, they do not.

"Fantasy" incorporates JRRT's arguments about the nature of the secondary world, and if he had articulated his practices as a creator of secondary belief he might well have done so under this heading. About "Recovery" JRRT is not remarkably clear; and, when he indicates that the "recovery of freshness of vision" is only part of what he intends by the term, he does not go on to argue a case. It does seem, however, that he intends his audience to understand that the washed vision of Recovery returns us to a capacity to see things as we are meant (perhaps by STORY) to see them. For JRRT, "Escape" is not "the flight of the deserter" but the "escape of the prisoner". The prison – he makes clear by example – is the modern world; the secondary world may be imaginary, but it is at times preferable to think oneself inside such a world. It is here, perhaps, that JRRT comes closest to C.S. LEWIS's characteristic attitude to the modern world, and where he makes himself vulnerable to the charge that – because he condemns the "robot"-infested 20th century so vigorously – it follows that he values correspondingly the hierarchical world of *LOTR*, the Christian teleology underlining the cosmological tale it tells, the racism (which THEOSOPHY has also been accused of) inherent in the THEODICY-ridden vision of races ranked by degree which *LOTR* promulgates with such great insistence (and whose clones throng the FANTASYLANDS of his imitators). Finally, by "Consolation" JRRT means the EUCATASTROPHE – the happy event – which properly ends the fairytale, and which all-importantly defines the fairytale as something requiring an ending, a story which must be completed.

The underlying model of FANTASY in this encyclopedia has been developed in order to help describe the typical narrative movements of the form, and JRRT's four elements are not, therefore, referred to frequently, with the exception of the eucatastrophe associated with consolation. But "eucatastrophe" is, after all, a term which can clearly be used to describe a narrative movement. This said, it is obvious that any narrative understanding of fantasy espoused here must be consistent with the nature of *LOTR* – whether or not JRRT's own breakdown of elements is followed – because, if any model of fantasy fails to address the paradigm 20th-century fantasy text, then it is (by definition) a model which fails to describe fantasy.

Fantasy stories are defined here as stories which require completion, and which can be distinguished from their siblings (SUPERNATURAL FICTIONS and HORROR) by this requirement, as well as by more obvious differentiations. This basic movement of fantasy towards completion, which JRRT both propounds and exemplifies, could be described as a shift from a condition of BONDAGE into freedom, via a eucatastrophe (or, rarely, a tragic close) which has been *earned*. More usefully, and with direct reference to *LOTR*, this basic "sentence" of fantasy – which can be expressed in two words, "bondage loosens" – can be expanded into a longer utterance.

Towards the beginning of a paradigm fantasy text a sense of WRONGNESS will almost invariably be felt. The Hobbits' first sight of the Nazgûl in *LOTR* is a telling example of this intuition that the world is not what it should be, and that any return to the world that was lost may be profoundly taxing. The RECOGNITION of wrongness may be a recognition that the back-story that supports the secondary world has suddenly been brought to bear, or that a sudden dyslexia is corrupting the back-story (both of these recognitions occur in *LOTR*); or it may be a direct recognition of the THINNING of the world. Indeed, most early stages of most fantasy tales incorporate moments when wrongness is sensed, either simultaneously with or as a prelude to narrative sequences devoted to the description of thinning, which may be defined as a result of the loss of MAGIC, or of the slow death of the GODS, or of a transformation of the LAND into desert, or of an AMNESIA (the protagonist's, or the world's) about the true nature of the self or history or the secondary world, or of any of the consequences of the rule of a DARK LORD, whose diktats almost inevitably (as in *LOTR*) represent an estranging PARODY of true governance. *LOTR* as a whole represents not only the final phase of a thinning of Arda, a process of loss which began with the first destruction of EDEN at the beginning of the creation of the world, but other, local thinnings as well: several occur over the course of Frodo's long trek from the Shire to Mordor, and the Shire itself – when the victorious Hobbits finally come home again – has been thinned into a grotesque parody of the industrialization of the English Midlands.

In a full fantasy like *LOTR*, thinning cannot be the end of the matter. After the struggles of heroes, after protagonists have endured their NIGHT JOURNEYS, after the cause of the thinning has been identified, there must be a confrontation with the evil stasis that has bound the land into a rictus of its former edenic aliveness, and the self into BONDAGE to its SHADOW.

In this encyclopedia, a metaphorical term RECOGNITION is frequently used to describe (a) the moment at which – after penetrating the tangles and the LABYRINTHS and the unknowingness of the blinded – the protagonist finally gazes upon the heart of the thinned world, and recognizes the shape of his and its STORY; and (b) the sense of transition from stasis into what JRRT calls "consolation" but which we call HEALING. In *LOTR*, Frodo's several visions of Sauron in the form of an Eye, and his final tussle with Gollum, and many other moments in the complex tale, arguably represent (a). And (b) is signalled when the embattled Western armies stand exhausted at the cusp of defeat in their LAST BATTLE with the Dark Lord – to realize instinctively that they have passed through the valley of the shadow, when they hear the call that heralds the arrival of the LIMINAL BEINGS: "The eagles are coming!" A new world is suddenly open to view.

In *LOTR*, the healing is a complex process. For Arda itself, the expulsion of the Dark Lord constitutes the final moment in the long war that has defined the back-story. For humans, the establishment of a single kingdom of Gondor under Aragorn is a political healing, and will allow people to enter the Fourth Age (ours) in command of themselves. For the elves, the only true healing now is farewell; and for Frodo this is also the case. For the Hobbits in general, the harrowing of the Shire is healing enough. Because it both terminates a long mythology and initiates the diurnal secular world, *LOTR* is comedy and tragedy: an AGON of gods and angels which devastates the arena of Earth; and an ecology. Though the model of the fantasy novel used here does point to that complexity, *LOTR* munificently transcends the model.

Aftermath. JRRT's influence on fantasy and sf has been not merely profound but also demeaning. It is his work which has given licence to the fairies, elves, orcs, cuddly dwarfs, loquacious plants, singing barmen, etc., who inhabit FANTASYLAND, which itself constitutes a direct thinning of JRRT's constantly evolving secondary world. This trivialization is, perhaps, an inherent risk in a literature which is increasingly RECURSIVE in nature. But there are compensations. Over and above the value of his works themselves, the dialogue between JRRT and writers like Peter S. BEAGLE and Stephen R. DONALDSON (to name only two) has been immensely fruitful. And the books remain, untouched by the myriad borrowings. [JC]

Other works: *Songs for the Philologists* (coll **1936**) with E.V. Gordon and others; *Farmer Giles of Ham* (**1949** chap) and *Smith of Wootton Major* (**1967** chap), assembled as *Farmer Giles of Ham, The Adventures of Tom Bombadil* (omni **1975**; vt *Smith of Wootton Major and Farmer Giles of Ham* 1976 US); *The Adventures of Tom Bombadil and Other Verses from the Red Book* (coll **1962** chap); *The Tolkien Reader* (coll **1966**); *The Road Goes Ever On: A Song Cycle* (coll **1967**) with music by Michael Swann; *Bilbo's Last Song* (**1974** chap); *Tree and Leaf, Smith of Wootton Major, The Homecoming of Beorhtnoth* (omni **1975**); *The Father Christmas Letters* (coll **1976** chap); *Pictures by J.R.R. Tolkien* (graph **1979**; rev 1992); *Poems and Stories* (coll **1980**); *Mr Bliss* (**1982** chap).

Nonfiction (selective): *A Middle English Vocabulary* (**1922**), first of several works of varying interest, including an edition of *Sir Gawain and the Green Knight* (**1925**) with E.V. Gordon.

Further reading (selective): *J.R.R. Tolkien: A Biography* (**1977**) by Humphrey Carpenter, various atlases and concordances like *A Guide to Middle Earth* (**1971**) by Robert Foster, *The Tolkien Companion* (**1976**) by J.E.A. Tyler, and *Tolkien: The Illustrated Encyclopedia* (**1991**) by David Day. Among other biographical/critical works are *Tolkien and the Critics* (anth **1968**) ed Neil D. Isaacs and Rose A. Zimbardo, *Tolkien: A Look Behind the Lord of the Rings* (**1969**) by Lin CARTER, *Master of Middle Earth* (**1972**) by Paul H. Kocher, *Tolkien's World* (**1974**) by Randel Helms, *J.R.R. Tolkien: Architect of Middle-Earth* (**1976**) by Daniel Grotta-Kurska, *The Mythology of Middle-Earth* (**1977**) by Ruth S. Noel, *The Inklings* (**1979**) by Humphrey Carpenter, *J.R.R. Tolkien: This Far Land* (anth **1983**) ed Robert Giddings, *The Letters of J.R.R. Tolkien* (**1981**) ed Humphrey Carpenter with Christopher Tolkien, *The Road to Middle-Earth* (**1982**; rev 1992) by Tom Shippey and *The Comedy of the Fantastic: Ecological Perspectives on the Fantasy Novel* (**1985**) by Don D. Elgin.

TOLKIEN SOCIETY UK organization founded in 1969 by Vera CHAPMAN and dedicated to the furtherance of interest in J.R.R. TOLKIEN's life and works. Publications include an annual journal, *Mallorn*, a bimonthly newsletter, *Amon Hen*, and various special publications of value – e.g., *Leaves from the Tree: J.R.R. Tolkien's Shorter Fiction* (**1991**) by Tom Shippey (1943-) *et al*, the proceedings of a 1989 workshop, and most notably *Proceedings of the J.R.R. Tolkien Centenary Conference 1992* (**1995**), ed Patricia Reynolds and Glen H. GoodKnight, a special issue of *Mallorn* issued in collaboration with the MYTHOPOEIC SOCIETY. [DB]

TOLKIEN SOCIETY OF AMERICA US organization, not the first J.R.R. TOLKIEN fan club but the one most active during the boom years of Tolkien's popularity and thus the focus of a certain amount of media publicity. It was founded by Richard D. Plotz in February 1965, three months before the first US paperback of *The Lord of the Rings* (**1954-5**). Soon the TSOA was a nationwide group. Publication of *Tolkien Journal* and a bulletin on society events, *The Green Dragon*, was infrequent. After holding two Tolkien Conferences in 1968-9 (a third in 1970 was combined with the MYTHOPOEIC SOCIETY's Mythcon), the TSA was absorbed (1972) by the Mythopoeic Society, which published the final issue, *#15*, of *Tolkien Journal* that year. [DSB]

TOLSTOY, NIKOLAI (1935-) UK writer, a descendant of Leo Tolstoy (1828-1910). NS is known mainly for nonfiction on historical subjects, including a group-biography of his forebears, *The Tolstoys* (**1983**). His first novel, *The Founding of Evil Hold School* (**1968**), was a marginal comic fantasy in which it transpires that the headmaster of a minor public school has been raised as a child by African SERPENTS. Very different in tone and subject matter is NT's second fantasy (and major work of fiction to date), *The Coming of the King* (**1988**), the story of MERLIN, impressively researched and steeped in Welsh Dark-Age lore; sequels have been promised. [DP]

Other work: *The Quest for Merlin* (**1985**), nonfiction.

TOMLINE, F. Pseudonym of W.S. GILBERT.

TOMPKINS, WENTWORTH [s] ◊ Kenneth MORRIS.

TOM THUMB *The History of Tom Thumbe, the Little, for his Small Stature Surnamed, King Arthurs [sic] Dwarfe* (**1621** chap) by Richard Johnson (1573-?1659) was the first FAIRYTALE printed in England, though the story it recounts and modifies had long existed in oral form. In Johnson's version, one of King ARTHUR's rural councillors is unhappy because he lacks a son, and MERLIN causes his wife to give birth to a tiny child. Tom Thumb, named because of his smallness, attracts the protection of the FAIRIES, but his scampish behaviour gets him into all sorts of hot water. He ends his days as a tame TRICKSTER in the royal court. Subsequent retellings have been numerous – including a rendering, almost certainly from a parallel tradition, by the GRIMM BROTHERS, plus Hans Christian ANDERSEN's "Thumbelina", a female version – and vary the story widely. A real-life midget, Charles Stratton (1837-1883), called himself General Tom Thumb, and became famous. Movies based on the tale include TOM THUMB (*1958*), *The* SECRET ADVENTURES OF TOM THUMB (*1993*) and THUMBELINA (*1993*). [JC]

TOM THUMB UK live-action/ANIMATED MOVIE (*1958*). MGM/Galaxy. **Pr** George Pal. **Dir** Pal. **Spfx** Tom Howard. **Anim** Wah Chang, Gene Warren. **Screenplay** Ladislas Fodor. **Based on** the tale by the GRIMM BROTHERS. **Starring** Jessie Matthews (Jonathan's wife Anna), Bernard Miles (Honest Jonathan), Peter Sellers (Anthony), Russ Tamblyn (Tom Thumb), Terry-Thomas (Ivan), June Thorburn (Queen of the Forest), Alan Young (Woody). **Voice actors**

Stan Freberg, Dal McKennon. 98 mins. Colour.

A Grimms' FAIRYTALE – with several others admixed, plus strong flavourings of COLLODI's *Pinocchio* – rendered as a children's musical comedy/adventure might not seem a recipe for commercial or aesthetic success, yet this is one of the most famous children's movies ever made . . . or perhaps notorious, because many children were terrified by it. For the driving force of *TT* is the impotence of the very small, an impotence all too familiar to children: for example, a famous scene in which TOM THUMB struggles to avoid being crushed under hundreds of chaotically dancing feet at a fair must strike a chord in most infant hearts.

What marks *TT* out is Pal's sense of spectacle, apparent even in this very localized, unambitious tale. The prolific spfx (with much use of animation, almost exclusively stop-motion) are startling, and blend almost seamlessly into the live-action; each time one is led to accept the impossible, a new and even more lavish effect extends the range of one's credulity. The movie is far from flawless; for example, Tom's US brashness jars amid an almost exclusively English cast – though it may have been intended to impart a TRICK-STER spin to his characterization. Yet good performances from all back up the spfx. A nice RECURSIVE touch is that one of the books in Tom's nursery is a collected *Grimms' Fairy Tales*.

Alan Young, here the youthful romantic support, was later to become one of DISNEY's most significant voice actors, notably as SCROOGE MCDUCK. Dallas (here listed as "Dal") McKennon was another Disney stalwart, his many voice roles for the studio including Owl in SLEEPING BEAUTY (*1959*).   [JG]

TONARI NO TOTORO Japanese ANIMATED MOVIE (*1988*). ◊ ANIME.

TOONS From the beginnings of the CINEMA, live actors have been introduced into animated worlds: Walt DISNEY and Ub IWERKS produced the **Alice Comedies** (*1924-7*; ◊ ALICE IN WONDERLAND [*1951*]), Gene Kelly memorably danced with Tom and Jerry in *Anchors Aweigh* (*1946*) and the main characters of MARY POPPINS (*1964*) spent a while in an animated land. Until recently, when technological advances made the accomplishment easier, it was less frequent for animated characters to stray into the real world; one of the first successful forays occurred in some sections of the DISNEY compilation feature *The Three Caballeros* (*1944*).

To place animated characters – or Toons, as they are known in the paradigm movie of this sort, WHO FRAMED ROGER RABBIT (*1988*) – into the real world is to render them, despite or because of their loony anarchy and the skew-logic whereby they operate, profoundly vulnerable: as two-dimensional constructs in a three-dimensional world, they lack the solidity enjoyed by the denizens of that world, which naturally tends to be seen as implacable, lacking in free spirits, and ultimately victorious. If they are accepted at all by that world, it may be as an underclass. Toons tend also to be disadvantaged by their perceived naïvety: when Holli, in Ralph BAKSHI's COOL WORLD (*1992*), leads the explosive incursion of Doodles (as Toons are called in that movie) into our world, it is clear that, although our world may itself be destroyed or driven mad in the short term, the Doodles will ultimately be doomed because the simplicity of their rationales renders them inadequate to cope with the complexities of human intercourse; likewise Mark/The Phantom Prowler in *A Nightmare on*

Elm Street Part 5 (◊ *A* NIGHTMARE ON ELM STREET) naïvely believes he has destroyed Freddy Krueger but is then himself ripped to paper shreds.

Some interesting variations have been played on the theme. In LAST ACTION HERO (*1993*) Danny, the youthful protagonist, is thrust into a created ALTERNATE REALITY, that of his movie action hero: the other characters there react with incredulity when Danny is disturbed by the presence of a Toon cat operating alongside its firmly human police colleagues. In *The* MASK (*1994*) the transformed protagonist is a three-dimensional Toon in the human world (he owes much to, in particular, Tex AVERY's creations) and, against the conventional trend, is less rather than more vulnerable than the humans surrounding him. (The same observation might be made of the muppets (◊ *The* MUPPET SHOW [*1976-81*]), who – just – can likewise be regarded as three-dimensional Toons.) In Maurizio NICHETTI's VOLERE VOLARE (*1991*) the human protagonist becomes vulnerable through gaining unwanted Toon attributes, though those attributes themselves seem indestructible.

In written fiction, Toons – being a 20th-century creation – are most likely to appear in CONTEMPORARY FANTASIES, where they may represent in modern dress the various supernatural denizens of FAERIE whose existence, in countless CROSSHATCH tales, is threatened (◊ THINNING) by the rise of humanity; a story of this type is "Harry the Hare" (1971) by James P. Hemesath. It would be misleading, however, to regard them as exact TECHNOFANTASY equivalents of supernatural creatures, even though the analogy may be close; an important difference is that they are generally regarded by those around them – if regarded at all – as *artefacts* rather than independently originated beings, a quintessential difference which they themselves usually acknowledge: they have been *made*. A partial reversal of this scheme appears in John GRANT's *The World* (*1992*), in which a Toon rabbit is initially encountered as a quasi-independent mental construct, a part of the transtemporal cultural baggage littering a character's mind, but is later reproduced a thousandfold in artefact form as highly vulnerable holographic servitors. In Greg Snow's *Surface Tension* (*1991*; vt *That's All, Folks!* 1991 US) a human wakes up one morning to find himself transformed into a Toon, and is soon exploited by the media to satirical effect. Elsewhere Toons may be inhabitants of POLDERS: it can be argued that Toontown, in Gary K. Wolf's *Who Censored Roger Rabbit?* (*1981*) – as in the movie which it inspired – is exactly such a polder. And Toons are likely to fade, like those who take refuge in the eponymous waystation for fading DREAMS in *Shadows Fall* (*1994*) by Simon R. GREEN.   [JC/JG]

See also: ANIMATED MOVIES; DUNDERKLUMPEN! (*1974*); PUPPETS; TOYS.

TOPOS A characteristic pattern of action whose surface manifestations may be very various but whose underlying structure constitutes an essential element of STORY. A typical topos is the action of journeying INTO THE WOODS. Another is the exemplary transformation of a man into an ASS, with consequences comic and erotic and educational. A further is the search of the reluctant immortal (◊ ACCURSED WANDERER; FLYING DUTCHMAN; WANDERING JEW) for surcease.   [JC]

TOPPER US tv series (1953-6). CBS, ABC, NBC. **Pr** John W. Loveton, Bernard L. Schubert. **Dir** Richard L. Bare, Leslie Goodwins, James V. Kern, Lew Landers, Phillip Rapp. **Writers** Stephen Sondheim and many others. **Based**

on characters created by Thorne SMITH. **Comics adaptation** *Topper and Neil* * (1 issue 1957) from Dell Comics. **Starring** Leo G. Carroll (Cosmo Topper), Kathleen Freeman (Katy 1954-5), Thurston Hall (Mr Schuyler), Anne Jeffreys (Marion Kerby), Lee Patrick (Henrietta Topper), Edna Skinner (Maggie 1953-4), Robert Sterling (George Kerby). 78 30min episodes. B/w.

A follow-up to the TOPPER MOVIES, this features the further misadventures of Cosmo Topper, who has to contend with three unwelcome houseguests – the GHOSTS of Marian and George Kerby and their dog, a large St Bernard. The Kerbys, an irreverent and fun-loving couple, were killed in a skiing accident and, while still stuck on Earth, make it their posthumous mission to bring some fun into Topper's dull existence (◊ POSTHUMOUS FANTASY). Only Topper can see or hear them, and he eventually gives up trying to convince others. To make matters worse, their dog is an alcoholic and prone to drunken romps through the house.

The series enjoyed two successful years on CBS, then went on to re-runs on both ABC and NBC. The ghosts were played by real-life couple Anne Jeffreys and Robert Sterling, with Leo G. Carroll, best-known as Alexander Waverly in *The Man from U.N.C.L.E.* (◊ SFE), as Topper. [BC]

TOPPER MOVIES The amiable banker from Thorne SMITH's SLICK FANTASIES was the subject of three movies, a tv series (◊ TOPPER [1953-6]) and two tvms.

1. *Topper* US movie (**1937**). MGM. **Pr** Hal Roach. **Dir** Norman Z. McLeod. **Vfx** Roy Seawright. **Screenplay** Eric Hatch, Jack Jevne, Eddie Moran. **Based on** *Topper: An Improbable Adventure* (**1926**) by Smith. **Starring** Constance Bennett (Marion Kerby), Billie Burke (Clara Topper), Cary Grant (George Kerby), Roland Young (Cosmo Topper). 96 mins. B/w.

After he skids off the road at the site where irresponsible socialites Marion and George died in a car crash, their GHOSTS – capable of INVISIBILITY at will – "adopt" Topper, manager of the bank in which George was a major stockholder. Having got drunk with them, he is involved in a fight with some sailors, and is arrested. The press is full of the story of the disorderliness committed by him and a mysterious "babe" (i.e., Marion). His dominating, image-conscious wife Clara assumes social suicide, but to her astonishment the pillars of society court the Toppers, whom no one has before suspected of being interesting. Marion, peeved with George, lures Topper, who has long had a secret yen for her, to the swanky Seabreeze Hotel for a fling; there are riotous scenes – plus a reconciliation between Marion and George – before they depart, George driving maniacally. The car crashes in the same old spot, and Topper is concussed. On his recovery, Clara promises to be a new and delightful wife.

This comedy veers between wryness and shrillness, with wryness just winning; the same contest exists between Young's low-key restraint (as in *The* MAN WHO COULD WORK MIRACLES [**1946**]), which occasionally slips to just the right degree, and the staccato chatter of the Kerbys. The frequent materializations and dematerializations of the ghosts are among *T*'s many excellent, state-of-the-art spfx. [JG]

2. *Topper Takes a Trip* US movie (**1939**). Hal Roach/United Artists. **Pr** Roach. **Dir** Norman Z. McLeod. **Vfx** Roy Seawright. **Screenplay** Corey Ford, Jack Jevne, Eddie Moran. **Based on** *Topper Takes a Trip* (**1939**) by Smith. **Starring** Constance Bennett (Marion Kerby), Billie Burke (Clara Topper), Alexander D'Arcy (Baron), Cary

Grant (George Kerby, in extracts from 1), Alan Mowbray (Wilkins), Franklin Pangborn (Louis), Verree Teasdale (Mrs Parkhurst), Roland Young (Cosmo Topper). 85 mins. B/w.

As a result of events at the Seabreeze Hotel in 1, Clara has been egged by grim friend Mrs Parkhurst to divorce Topper; his testimony in court allows extensive extracts from 1 to be run as a recap. Her case dismissed, Clara is dragged by Parkhurst to the French Riviera; Marion returns from oblivion, without George but adopting the dog-GHOST Atlas, to help Topper recover his wife. In France, Parkhurst and oily hotel manager Louis try to fasten a gigolo Baron onto the rich Clara. Deprived of funds by his wife's lawyers, Topper goes to the casino, where an invisible Marion fiddles the roulette game for him; back at the hotel, she sneaks them into the suite shared by Parkhurst and Clara . . . and something of a bedroom farce (with added ghosts and a jail-break) ensues. Inevitably, the Toppers are joyously reconciled.

TTAT lacks some of the verve of 1, but makes up for it with some excellent comic routines. There is some desperate scripting to account for Cary Grant's absence – officially because he had become too expensive, but likely also because he had been comprehensively outshone in 1 by Bennett and Young. [JG]

3. *Topper Returns* US movie (**1941**). Hal Roach/United Artists. **Pr** Roach. **Dir** Roy Del Ruth. **Vfx** Roy Seawright. **Screenplay** Gordon Douglas, Jonathan Latimer, Paul Gerard Smith. **Starring** Eddie "Rochester" Anderson (Eddie), Joan Blondell (Gail Richards), Billie Burke (Clara Topper), Carole Landis (Ann Carrington), Dennis O'Keefe (Bob), Roland Young (Cosmo Topper). 87 mins. B/w.

Gail Richards and heiress Ann Carrington are travelling to the home of Ann's father, whom she has not seen since childhood. An assassin's bullet blows out the tyre of their taxi; when taxi-driver Bob goes for help, they hitch a lift with Topper and his chauffeur Eddie. That night in Carrington's gloomy castle the two women swap assigned bedrooms; a masked man stabs Gail to death in error, and her GHOST walks through the skies to Topper's house, where she blackmails him into coming back to the Carrington place with her to solve the murder. The ensuing hijinks comprise a broad PARODY of Haunted-House movies, PSYCHOLOGICAL THRILLERS and *film noir*, with elements of The PHANTOM OF THE OPERA (**1925**) thrown in. Quite unlike the earlier two **Topper** movies, it is one of a cluster of such spoofs, including *The Cat and the Canary* (**1939**), and is, even though its plot is not entirely resolved, one of the best. [JG]

4. *Topper Returns* US movie (**1973** tvm). NBC. **Pr** Walter Bien. **Exec pr** Arthur P. Jacobs. **Screenplay** AJ Carothers. **Dir** Hy Averback. **Starring** John Fink (George Kerby), Roddy McDowall (Cosmo Topper Jr), Reginald Owen (Jones), Stefanie Powers (Marion Kerby). 30 mins. Colour.

This unsuccessful pilot sequelled the original TOPPER tv series (1953-6). Cosmo Topper Jr has inherited all his uncle's worldly possessions – and three unworldly ones as well. The GHOSTS are as determined as ever to make at least one Topper join in their fun. [BC]

5. *Topper* US (**1979** tvm). Cosmo Productions/CBS. **Pr** Robert A. Papazian. **Exec pr** Kate Jackson, Andrew Stevens. **Screenplay** Maryann Kasiac, George Kirgo, Michael Schiff. **Dir** Charles S. Dubin. **Starring** Jackson (Marion Kerby), James Karen (Fred Korbell), Macon McCalman (Wilkins), Rue McClanahan (Clara

Topper), Andrew Stevens (George Kerby), Jack Warden (Cosmo Topper). 120 mins. Colour.

This further pilot for a **Topper** series followed several years later, this time starring Kate Jackson of *Charlie's Angels* fame and her then-husband Andrew Stevens. Perhaps one of the worst-received tvms of the year, this for some reason changed Topper from a banker to a lawyer. [BC]

TOPSY-TURVY An inversion of the world's order or the world's rules – in particular the upside-down logic which tends to obtain in arbitrary WONDERLANDS. Such reversals are most effective when contrasted with a rigid social order, which may explain their popularity in Victorian fantasies. Lewis CARROLL's Wonderlands present topsy-turviness on several levels; e.g., expected adult/child relations are subverted, with Alice striving towards common sense (◊◊ SENSIBLE MAN) while all Wonderland and Looking-Glass adults exhibit irrationality if not outright insanity; well known moralizing verses turn into amoral PARODY; and the games of ambiguity possible in the English language take precedence over the REALITY supposedly described. W.S. GILBERT's *Iolanthe* (**1885**) shows English peers displaced from their position at the top of the social order by FAIRIES, beings so far down the scale that they do not even exist; his ill-fated "magic lozenge" PLOT DEVICE disturbingly applies topsy-turviness to the SOUL, by turning people into what they have merely pretended to be. Another influential example is F. ANSTEY's inversion of the parent/child relationship via IDENTITY EXCHANGE in *Vice Versâ* (**1882**).

G.K. CHESTERTON's preferred form of topsy-turviness is a EUCATASTROPHE-like switch of PERCEPTION: in *The Man who Was Thursday* (**1908**) his most fanatical-seeming terrorist leaders prove to be the most dedicated policemen, and their devilish leader something close to GOD. Raymond BRIGGS's *Fungus the Bogeyman* (graph **1977**) presents an UNDERGROUND world where accepted views of the desirability of cleanliness *vs* filth are – comically – a systematic inversion of our own. Such skewed viewpoints lend themselves to SATIRE, a devastating example being Douglas Hofstadter's presentation in *Metamagical Themas* (coll **1985**) of a topsy-turvy ALTERNATE WORLD where English third-person pronouns indicate not GENDER but race – black or white. [DRL]

See also: CARNIVAL; ESTATES SATIRE; REVEL.

TOTEMS The adoption of a patron animal by a tribe or clan – most famously, among the North American Indians – is thought to bestow some of that animal's virtues by sympathetic MAGIC and also to offer protection, with part of the tribesman's SOUL being invested in the totem animal like a heart kept safe outside the body (◊◊ KOSHCHEI). Patron animals survive as emblems in heraldry: ARTHUR's device is often said to have been a bear. Fantasy sometimes invokes them in this simple, symbolic sense, as with the shepherd-folk in William MORRIS's *The Well at the World's End* (**1896**), whose totem (in whose name they may be rallied) is also the bear, as is that of Jean M. AUEL's central clan in *The Clan of the Cave Bear* (**1980**). The "Granbretan" villains of Michael MOORCOCK's **Hawkmoon** sequence wear various totemic MASKS; the movers and shakers in R.A. LAF-FERTY's *Fourth Mansions* (**1969**) have certain qualities of (respectively) toads, SERPENTS, eagles and badgers. But in settings where magic is real, the tendency is to take the tribe/animal relationship to its logical extreme of intimacy via SHAPESHIFTING. [DRL]

See also: SHAMANISM; SKINNED.

TO THE DEVIL A DAUGHTER (vt *To the Devil . . . A Daughter*) UK/West German movie (**1976**). HAMMER/Terra Filmkunst. **Pr** Roy Skeggs. **Dir** Peter Sykes. **Spfx** Les Bowie. **Mufx** Eric Allwright, George Blackler. **Screenplay** Chris Wicking, adapted by John Peacock. **Based on** *To the Devil – A Daughter* (**1953**) by Dennis WHEATLEY. **Starring** Anna Bentinck (Isabella Beddows), Honor Blackman (Anna Fountain), Denholm Elliott (Henry Beddows), Michael Goodliffe (George De Grass), Nastassja Kinski (Catherine Beddows), Christopher Lee (Father Michael Rayner), Eva Maria Meineke (Eveline De Grass), Anthony Valentine (David Kennedy), Richard Widmark (John Verney). 93 mins. Colour.

Movies like *The* EXORCIST (**1973**) had toppled Hammer from its throne as the "House of Horror", bringing new sophistication into the HORROR-MOVIE subgenre. This was Hammer's last-gasp attempt to strike back, trying, in conjunction with Terra Filmkunst, to recreate the success of the Wheatley adaptation *The* DEVIL RIDES OUT (**1968**). The result was a curious mixture of the Hammer tradition with more modern tropes concerning BLACK MAGIC and SATANISM, plus overt sexuality.

Renegade priest Rayner has turned to the dark side and seeks to incarnate Astaroth. His disciple Isabella Beddows agrees to bear the child who through various blood rituals will become Astaroth's AVATAR, knowing she herself will die in the process. 18 years later that child, Catherine, reared as a nun in Rayner's sham-Christian organization, The Children of Our Lord, is sought by Rayner for the second baptism on her birthday, on All Hallow's Eve (◊ HALLOWE'EN). Her natural father, Henry, enlists writer of bestselling occult books Verney to extricate her from her fate, even though he believes his PACT WITH THE DEVIL (reified as a metal quarter-MOON) will consume him in flames should he breach his original promise of silence. In the end, Verney succeeds.

Widely decried on release as muddled, the movie was a failure. In fact, although there *is* some muddle in the plotting, most of what the critics disliked was sophisticated storytelling. Of particular interest is the use by Rayner of sympathetic MAGIC to gain his ends; e.g., in a phone conversation with Henry he wraps a rope around his receiver, which is manifested as a SERPENT around Henry's wrist. Overall, this is quite an impressive piece of work, despite exploitational aspects (e.g., the sexy nude scene involving Kinski, who was only 16 at the time). [JG]

TOURTEL, MARY UK comics artist (1874-1930), creator of RUPERT THE BEAR. ◊◊ COMICS.

TOYS Children's toys are potent stimulators of fantasy; the juvenile imagination animates them, and actual "coming alive" seems a natural extension (◊ ANIMATE/INANIMATE). Sinister aspects of this awakening tend to involve DOLLS, which perhaps seem already too disturbingly human, although the rag doll in *The* NIGHTMARE BEFORE CHRISTMAS (**1993**) is, by contrast, a conscience. Animated toy soldiers are ever-popular, from H.G. WELLS's "The Magic Shop" (in *Twelve Stories and a Dream* coll **1903**) to Diana Wynne JONES's *Charmed Life* (**1977**); SAKI's nonfantastic "The Toys of Peace" (1914) satirically introduces worthy models of ballot boxes and politicians, which children rapidly adapt for bloodstained battle-scenes. In A.A. MILNE's **Winnie-the-Pooh** stories and John MASEFIELD's *The Midnight Folk* (**1927**) stuffed toys become full-fledged COMPANIONS, having equal status with children and TALKING

ANIMALS. Masefield's story interestingly breaks the tacit rule that such toys, being presumably animated by a child's belief, cannot interact with the adult world (◊◊ INVISIBLE COMPANION). In TOM THUMB (*1958*) the fact that Tom is an infant, even though he has a maturely adult (albeit minia-ture) body, is conveyed by the fact that only he can see his toys come alive. The toy status of the eponymous clockwork figures in Russell HOBAN's *The Mouse and His Child* (**1967**) is far more restricting, amounting to BONDAGE: their range of actions is limited and their machinery must be repeatedly rewound. The heroine of Neil GAIMAN's *A Game of You* (graph coll **1993**) comes to an important RECOGNITION that animal companions in "her" FANTASYLAND are fleshed-out memories of childhood toys. In the ANIMATED MOVIE *Toy Story* (**1995**), the toy spaceman Buzz Lightyear must learn to accept his toyhood and the impossibility of becoming a REAL BOY. This movie makes play with the easily imagined rivalry between toys competing for their owner's favours; Gene WOLFE takes this further in "The War Beneath the Tree" (1979), whose toys face a LAST BATTLE with the new intake supplanting them at CHRISTMAS.

Some toys are more than they seem. Eilonwy's glowing "bauble" in the **Chronicles of Prydain** sequence by Lloyd ALEXANDER holds potent MAGIC, and a brass walnut hidden among other playthings in Michael SWANWICK's *The Iron Dragon's Daughter* (**1993**) proves to be an important TECHNOFANTASY component. Randall GARRETT's *Too Many Magicians* (**1966**) features a magic educational toy whose built-in SPELL slowly fades unless unconsciously replenished by the child's developing TALENT. [DRL]

See also: CHILDREN'S FANTASY; SHOP.

TOYS US movie (*1992*). 20th Century-Fox/Baltimore Pictures. **Pr** Mark Johnson, Barry Levinson. **Dir** Levinson. **Spfx** Clayton Pinney. **Vfx** Mat Beck. **Screenplay** Valerie Curtin, Levinson. **Starring** Joan Cusack (Alsatia), Michael Gambon (Leland), LL Cool J (Patrick), Arthur Malet (Owens), Robin Williams (Leslie), Robin Wright (Gwen). 121 mins. Colour.

Dying toymaker Kenneth Zevo leaves his empire to his brother, discharged army General Leland Zevo, rather than to unpredictable son Leslie or daughter Alsatia. Leland mil-itarizes the factory, then secretly designs a whole range of deadly new weapons that will be only TOY-size and hence capable of penetrating enemy defences; the acme of his scheme is the employment of scores of children to operate weapons systems in the belief they are merely playing video GAMES. Discovering this ghastly truth, Alsatia, Leslie, his new girlfriend Gwen, factory supervisor Owens and Leland's son Patrick hit back. Leland turns his lethal toys on the five, and the entire factory becomes a battle zone: Leland's weaponry against a mass of traditional, harmless toys. Finally the main computer is smashed, deactivating all the war-toys except the most lethal, the Water Swine, which shoots Alsatia (who proves to be a robotic toy con-structed long ago by Kenneth to keep his son company, and who is thus repairable) and then Leland.

This TECHNOFANTASY was not much liked on release, with emphasis on its "too obvious" anti-war message. But that message was not unsophisticated. All the characters are really children: the difference is that Leland's childishness is potentially lethal. In a telling scene, Leland tries to sell his systems to three officials from Washington; we see much of this through the cameras of the X-ray surveillance team, and the crude images of the negotiators become the FOUR

HORSEMEN. Elsewhere, he visits a video arcade to try out the war-games and gets his kicks out of blasting the UN trucks rather than the enemy. The movie as a whole is couched in fantastic terms: outdoor scenes show an idealized landscape made up of swashes of primary colours; the Zevo mansion is a gigantic pop-up BOOK; the interior of the factory, spot-lessly clean and again in primary colours, is more like a playground than an industrial workplace (like the factory in WILLY WONKA AND THE CHOCOLATE FACTORY [1971], only more so), and has at its heart a miniaturized Manhattan through which the cast walk like GIANTS. *T* is remarkably powerful as both polemic and fantasy. [JG]

TRADER HORN US movie (*1930*). ◊ JUNGLE MOVIES; TARZAN MOVIES.

TRAINS The rise of the classic GHOST STORY and the con-struction of the first railroad networks came at about the same time in the UK, during the first half of the 19th cen-tury; and 20th-century readers of SUPERNATURAL FICTION may be forgiven the assumption that trains were from the first natural settings for stories featuring REVENANTS or mes-sengers bearing ill tidings. But this was not the case. For the first several decades trains were more likely to serve as emblems of the thrust of the future, and to feature in sf tales.

It was not until a generation had passed that the train began to become embedded into the "natural" world; the first significant story in which the supernatural has an effect upon the train – after "The 9:30 Up-Train" (1853) by S. BARING-GOULD – is probably Charles DICKENS's "No. 1 Branch Line. The Signalman", in *Mugby Junction: The Extra Christmas Number of All the Year Round* (anth **1866** chap), a CHRISTMAS BOOK which also includes "No. 5 Branch Line: The Engineer" by Amelia B. Edwards (1831-1892), in which a ghost prevents a train-wreck, thwarting a revenge. Edwards and Dickens collaborated the following year on another train-connected ghost story, "The Four-Fifteen Express" (1867); this, and almost all subsequent supernatural fiction set in trains or on railroads, tends to treat the train as a venue for the invasion or visitation of something prior, almost invariably a ghost. Several examples appear in *The Ghost Now Standing on Platform One* (anth **1990**) ed Peter HAINING (as Richard Peyton). Others are: the anonymous "The Parlor-Car Ghost" (1904), printed in various versions; Algernon BLACKWOOD's "Miss Slumbubble – and Claustrophobia" (1907); Bernard CAPES's "The Dark Compartment" (1915); "A Short Trip Home" (1927) by F. Scott FITZGERALD; "The Railway Carriage" (1931) by F. Tennyson Jesse (1889-1958); "A Journey by Train" (1935) by Henry L. Lawrence (1908-   ); and the second multi-episode "Adventure" in the tv series SAPPHIRE AND STEEL, in which a train station is haunted by soldiers killed in WORLD WAR I.

Fantasy stories set on trains, on the other hand, generally involve a voyage of some sort, and often treat the train as a method of transit between one world and another. Sometimes – as in Robert BLOCH's "That Hell-Bound Train" (1958) – the train itself is both the method of travel and (as it were) the destination; sometimes – as in "Lost in the Fog" (1919) by J.D. Beresford (1873-1947), A.M. BUR-RAGE's "The Wrong Station" (1927) and "Branch Line to Benceston" (1947) by Andrew Caldecott (1884-1951) – the train conveys its passenger through some sort of PORTAL into the otherworld: in the first instance, on the other side can be found a microcosm of Europe afflicted by WWI; in the second, it is a POSTHUMOUS-FANTASY venue; in the third, an ALTERNATE WORLD. Less explicitly, trains are frequently used to carry protagonists from familiar places to unknown regions, whether or not they are explicit OTHERWORLDS: the protagonist of Herbert ROSENDORFER's *The Architect of Ruins* (**1969**) begins his descent into the UNDERWORLD only after escaping from a surreal train voyage. Magical journeys (often in the middle of the night) also occur in Helen CRESSWELL's *The Night-Watchmen* (**1969**), Peter Collington's *The Midnight Circus* (graph **1992**) and *Escardy Gap* (**1996**) by Peter CROWTHER and James Lovegrove (**1965-**   ).

As the 20th century progressed, it became more and more likely that trains would come to represent a lost world of the past, and that fantasies incorporating travel by train would evoke such venues. The most famous of these tales is probably Jack FINNEY's "The Third Level" (1952), whose narrator discovers a third level beneath Grand Central Station in New York City; this level exists in 1894, where he longs to go. Rod SERLING's "A Stop at Willoughby" (1961) exploits a similar nostalgia, and Joan AIKEN's *The Cockatrice Boys* (in *Christmas Forever* anth **1993** ed David G. HARTWELL; exp **1996** US) uses its train setting as a solacing contrast to a monster-ridden contemporary England. In Richard KENNEDY's *The Boxcar at the Center of the Universe* (**1982** chap) a hobo tells of his search for the heart of the Universe, and in William KOTZWINKLE's "Boxcar Blues" (1989) hoboes escape DEATH by hopping freights into ALTERNATE REALITIES.

Trains (specifically engines) which are themselves animate are not common in adult fiction, a relatively rare example being Rudyard KIPLING's ".007" (1897). They are very commonly found in stories for younger children, where Little Engines That Could unceasingly proliferate.   [JC]

TRANSFORMATION Term used here for an inflicted METAMORPHOSIS. Metamorphosis is a magic and radical change in shape experienced, normally through an act of will. by its subject. When some external agent of change is involved, the term "transformation" is preferred. The FROG PRINCE is transformed by a WITCH; TAM LIN suffers transformation to a deer, SERPENT and hot iron without breaking his Janet's grip on him; MERLIN in T.H. WHITE's *The Sword in the Stone* (**1938**) transforms the young ARTHUR into various animals as part of his education. Trent in Piers ANTHONY's *A Spell for Chameleon* (**1977**) is known as the Transformer for his TALENT of changing any living creature into any other.   [DRL/JC]

See also: SHAPESHIFTERS; THERIOMORPHY; TRANSMUTATION.

TRANSMUTATION A term generally used here to refer to the transforming of inanimate matter into other forms of inanimate matter; e.g., the transmutation in ALCHEMY of base metals into gold. It is thus distinct from METAMORPHOSIS and TRANSFORMATION.   [JC]

TRANSYLVANIA A land which has variously formed part of Hungary and Rumania, and was for a time an independent principality. Its very name, meaning "across the forests" (◊ INTO THE WOODS) is evocative of strange and distant lands. It was to Transylvania that the PIED PIPER was reputed to have taken the children from Hamelin. The Báthory family, which ruled Transylvania in the 16th century, included the notorious Elisabeth de Báthory (1560-1614), who is recorded as having bathed in the blood of over 600 virgins in order to sustain her youthful beauty (◊ IMMORTALITY; VIRGINITY). Her story, filmed as *Countess Dracula* (**1970**) (◊ DRACULA MOVIES), added to the LEGENDS of WITCHCRAFT,

WEREWOLVES and VAMPIRES prevalent in that area, making it the ideal place for the castle of Count DRACULA in *Dracula* (**1897**) by Bram STOKER. It has formed the locale for many VAMPIRE stories and VAMPIRE MOVIES since.   [MA]

TRAUM DES ALLAN GREY, DER vt of ◊ VAMPYR (*1932*).

TRAUM DES DAVID GRAY, DER vt of ◊ VAMPYR (*1932*).

TRAVELLERS' TALES One of the oldest forms of fiction and the source of many TAPROOT TEXTS. TTs may involve QUESTS, but they are usually voyages of discovery, often to the ends of the Earth. TTs are usually presented as fact, and are always told in retrospect by the survivor(s) and usually have the status of TALL TALES; this distinguishes them from the FANTASTIC VOYAGE. TTs trace their roots back at least as far as the *Odyssey* (8th century BC) by HOMER, where ODYSSEUS relates his travels to King Alcinoüs. One of the first great Greek travellers who wrote of his adventures was Hecataeus of Miletos (?   -476BC), though his work (of which little survives) was apparently full of errors rather than deliberate fictions. Herodotus (?490-?425) severely criticized Hecataeus and, in his own *Historiai* (?430BC), established the basis for many future historical explorations. Some of these, such as the *Indika* (?400BC) by Ctesias, a Greek who was a physician to the Persian king Artaxerxes II (reigned 405-359), was the first book to be written entirely about India – though it is disputed whether Ctesias ever went there. It contains much invention and was upheld by LUCIAN of Samosata as a book of fiction. By the time of Lucian, in the mid-2nd century AD, TTs had become so prevalent that he was able to parody them in his *Verae historiae* ["True History"], which is an extravagant and amusing fiction taking adventurers beyond the Pillars of Hercules to the Isles of the Blest, the UNDERWORLD and the MOON. It has been suggested that Lucian wrote his story as a parody of *The Wonders Beyond Thule* by Antonius Diogenes. Although this book is regarded as one of the earliest of all Greek prose ROMANCES, the date of its composition is not known. The story is set *c*400BC, by its reference to historical characters, but Diogenes (who should not be confused with the Greek philosopher who lived ?400-?325), probably lived *fl*AD100. His work, which survives only in summary among the *Bibliotheke* of Photius (810-893), is a complex adventure story involving romantic attachments between various couples who travel throughout Europe, and eventually make their way to Thule (probably Iceland); two set forth on an intrepid journey to the North Pole, whence they come close to the Moon, described as a "land of purest light". Diogenes' work includes many references to supernatural wonders encountered on his characters' travels, including those caused by the magician Paapis, who curses his enemies with death during the day and life only at night (◊ VAMPIRES). A summary of this story included in *Collected Ancient Greek Novels* (anth **1989** US) ed Bryan P. Reardon with Lucian's *True History*.

The fascination for the TT did not pass with the fall of the Roman Empire. Hsuan Tsang (596-664) was a remarkable Chinese explorer who travelled throughout Asia and who in his memoirs, written in his last 20 years, told of the fierce DRAGONS that inhabited the mountains in remote western China.

Among the best-known TTs are those associated with St Brendan (?486-?575), as recorded in the anonymous *Navagatio Sancti Brendani Abbatis* ["The Voyage of St.

Brendan"] (pre-10th century), in which the Abbot travels to the Isles of the Blest in the Atlantic. The Celts delighted in *immrama*, or "voyage tales" of which the most famous, the *Immram Curaig Maíle Dúin* ["The Voyage of Maeldúin"], forms part of the *Book of the Dun Cow* (10th century, but written ?8th century). Maeldúin sets out with his companions to avenge the murder of his father but their boat is blown off course. They encounter a series of ISLANDS, each with its wonders and bizarre creatures – giant ants, giant horses, a beast that can turn its skin around, a CAT that turns into a ball of FIRE, sheep that change colour, shouting birds, undersea islands, a hermit cared for by the ANGELS, giant fountains, a bridge of crystal and much more. The Celts and the Vikings were great travellers, so it is no surprise that among their writings are many TTs (◊ SAGAS), including the voyages of MADOC and of Eirikr the Red to discover North America. Similarly, the Arab adventurers inspired the stories of Sinbad (◊ ARABIAN FANTASY). Books which reflect this period of TTs include *King of the World's Edge* (1939 *WT*; 1966) and its sequels by H. Warner MUNN.

With the real age of exploration, starting in the 13th century, TTs kept pace with the expanding map. In 1298 Marco Polo (1254-1324) recounted his *Travels* across Asia to China to a writer with whom he shared a prison cell; his veracity is now questioned. The biggest selling TT of this period was *The Travels of Sir John Mandeville* (written ?1357; 1360) (◊ Sir John MANDEVILLE) from whose book maps were drawn. Christopher Columbus (1451-1506) so believed Mandeville's book that when he discovered the West Indies he was convinced he had found Mandeville's Isles of Cathay. *Yesterday We Saw Mermaids* (1991) by Esther FRIESNER is almost a TT, set at the time of Columbus and exploring an alternate voyage.

The voyages of Columbus, Vasco da Gama and Ferdinand Magellan opened up the world in the 15th and 16th centuries and books of fantastic voyages appeared frequently. Such books as *Utopia* (Part 2 1516 in Latin; trans exp 1551) by Sir Thomas More (1478-1535), *Civitas Solis* (1623) by Tommaso Campanella (1568-1639), *The New Atlantis* (1629) by Francis Bacon (1561-1626) and *The Commonwealth of Oceana* (1656) by James Harrington (1611-1677) all use the popular form of the TT to establish the story's credentials, although the books themselves are UTOPIAS, and thus openly fictional. Imitations of the true TT returned to literature with the nonfantastic *Robinson Crusoe* (1719) by Daniel DEFOE and the inspirational *Travels into Several Remote Nations of the World* (1726 2 vols) by Jonathan SWIFT. Swift's account of the four voyages of Gulliver inspired many SEQUELS BY OTHER HANDS and brought the TT into modern literature. It was soon being rivalled by the exploits of Baron MÜNCHHAUSEN, as told by R.E. RASPE. Scores of Gulliver and Münchhausen imitations followed as the TT and fantastic-voyage forms gradually blended into proto-SCIENCE FICTION. Works which retain the element of fantasy include *Nicolai Klimii iter subterraneum* (1741 in Latin; trans as *A Journey to the World Under-Ground by Nicolas Klimius* 1742 UK) by Ludvig Holberg (1684-1754), *The Life and Adventures of Peter Wilkins* (1751) by Robert Paltock (1697-1767), *Symzonia* (1820) by the unidentified Adam Seaborn (possibly Nathaniel Ames [? -1835]) and *The Narrative of Arthur Gordon Pym of Nantucket* (*Southern Literary Messenger* 1838) by Edgar Allan POE. With Jules Verne (1828-1905), the TT and the fantastic voyage merged into what he called

Voyages Extraordinaire, the fantastic elements being dropped. TTs could not translate directly into genuine GENRE FANTASY, because their whole point was that they were presented as true. However, J.R.R. TOLKIEN's *The Hobbit* (1937) and *The Lord of the Rings* (1954-5) both retain some element of the TT: the first is presented as a record of his adventures as recorded by Bilbo Baggins and the second is purportedly retold by Sam Gamgee to his children. In this sense any first-person narrated fantasy quest may be treated as a TT. The form remained sublinear in stories of lost worlds, and the mood was brilliantly recaptured by Sterling Lanier (1927-) in his **Brigadier Ffellowes** stories, collected as *The Peculiar Exploits of Brigadier Ffellowes* (coll 1972) and *The Curious Quest of Brigadier Ffellowes* (coll 1986). The recrudescence of fantasy fiction in its broadest sense has provided a forum for TT to return. [MA]

TRAVERS, P(AMELA) L(YNDON) Pseudonym of Australian actress and writer Helen Lyndon Goff (1899-1996), resident in the UK from 1924, and who worked as PLT from her first appearance on the stage. She began publishing journalism in Sydney and continued her journlistic career in the UK until she began the **Mary Poppins** series of CHILDREN'S FANTASIES, for which she remains best-known: *Mary Poppins* (1934; rev 1981 US) and *Mary Poppins Comes Back* (1935), assembled as *Mary Poppins and Mary Poppins Comes Back* (omni 1937 US), plus *Mary Poppins Opens the Door* (1943 US), *Mary Poppins in the Park* (1952), *Mr Wiggs' Birthday Party, a Story from Mary Poppins* (1952 chap US), *The Magic Compass, a Story from Mary Poppins* (1953 chap US), *Mary Poppins from A to Z* (1962 US), *Mary Poppins in the Kitchen: A Cookery Book with a Story* * (1975 US) with Maruice Moore-Betty, *Mary Poppins in Cherry Tree Lane* (1982) and *Mary Poppins and the House Next Door* (1989 US). The DISNEY movie MARY POPPINS (1964) is based primarily on the first novel. Ostensibly nothing "more" than a nanny with TALENTS – abilities she uses to instruct and entertain her charges – Mary Poppins slips, at times, into a more profound guise. She could be described as a kind of psychopomp for English children in a period of historical transformation; while affirming certain values of an England that retained its Edwardian pomp and glow (the movie clearly sets its sentimentalized version of the tale before WWI), she also leads her children through a RITE OF PASSAGE into a more problematic world.

In *The Fox at the Manger* (1962 chap US) PLT created a sharp CHRISTIAN FANTASY whose animal protagonist – the fox – gives the Baby CHRIST the gift of cunning. *Friend Monkey* (1971 US) subjects a human family to the dubious assistance of Hanuman, the TRICKSTER monkey god, on his 1897 visit to England. *About the Sleeping Beauty* (coll 1975 US), illustrated by Charles KEEPING, contains an uplifting REVISIONIST-FANTASY version of the FAIRYTALE along with traditional renderings and an essay. [JC]

Other works: *Happy Ever After* (1940 chap US), a fable; *I Go by Sea, I Go by Land* (1941), associational; *In Search of the Hero: The Continuing Relevance of Myth and Fairy Tale* (1970 chap US), a lecture; *Two Pairs of Shoes: Folk Tales* (coll 1980 US), retold folktales; *What the Bee Knows: Reflections on Myth, Symbol and Story* (coll 1989 US), essays.

TREECE, HENRY (1911-1966) UK author who in the early 1940s began his career as anthologist, propagandist and writer, under the influence of Herbert READ, on behalf of poetry's New Apocalypse Movement, according to which the poet had a priest-like role to act as BARD. HT's poems were duly Celtic in tone ($ CELTIC FANTASY), but their fame is very dim. Within a few years, with *I Cannot Go Hunting Tomorrow: Short Stories* (coll **1946**), he had begun to shift into prose, and it was with stories, novels and plays – some for children – that he made his reputation. His range was considerable, and many of his historical novels invoke a fantasy glamour of presentation and significance without necessarily departing the mundane. His **Greek Trilogy** – *Jason* (**1961**), *Electra* (**1963**) and *Oedipus* (**1964**) – comprises tales of this sort; a mythological, mythopoeic glow suffuses them, but they are not fantasies.

Several of HT's historical novels deal with a quasi-historical 5th-century ARTHUR; although HT demythologizes certain aspects of the Arthurian cycle, MERLIN can still prophesy and other fantasy elements are mixed variously into some of the tales. "Princes of the Twilight" (in *The Haunted Garden* coll **1947**; the volume is mostly poetry) and *The Tragedy of Tristram* (broadcast 1950; in *The Exiles* coll of plays **1952**) introduce GHOSTS and other supernatural elements.

Legions of the Eagle (**1954**), *The Eagles Have Flown* (**1954**), *The Great Captains* (**1956**) and *The Green Man* (**1966**) all take place within the Arthurian world. Although clearly inclined to a euhemerist version of the mythos ($ EUHEMERISM) – at one point, the immensely powerful Artos the Bear thrusts a sword into an OAK log and, being the only one capable of freeing it, is acclaimed the ruler of Britain – an ambience of MYTH so powerfully suffuses them that the relatively few supernatural events in their pages seem fully appropriate. The king who is protagonist of *The Green Man* may only re-enact the SEASON myth of the GREEN MAN, but in doing so he seems preternaturally rooted in the tale.

Other titles of similar interest include: *The Golden Strangers* (**1956**; vt *The Invaders* 1957 US), a PREHISTORIC FANTASY, though once again with a severely restricted fantasy palette; *Red Queen, White Queen* (**1958**), about Boudicca; and *The Dream-Time* (**1967**). HT had a harsh imagination, but its icy clarity – its unequivocal pessimism about human virtue – makes him a writer very much of the times that came after his suicide. [JC]

Other works: *Ask for King Billy* (**1955**), borderline sf; *Viking's Dawn* (**1955**); *The Road to Miklagard* (**1957**); *The Return of Robinson Crusoe* (**1958**; vt *The Further Adventures of Robinson Crusoe* 1958 US); *The Golden One* (**1961**); *The Burning of Njal* (**1963**), a legend retold; *The Last of the Vikings* (**1964**; vt *The Last Viking* 1966 US); *The Windswept City* (**1967**); *Vinland the Good* (**1967**; vt *Westward to Vinland* 1967 US); *The Invaders: Three Stories* (coll **1972**).

Further reading: *Henry Treece* (**1969**) by Margery Fisher.

TREES Since humanity's primate ancestors were arboreal, trees are in a sense our original homes, which might explain their power over the human imagination. To be sure, entire FORESTS are often depicted as evil or dangerous, but images of individual trees are generally positive. Trees are regularly depicted as abodes, often hollowed-out havens of domesticity for the anthropomorphized forest creatures of CHILDREN'S FANTASY, as in Beatrix POTTER's **Peter Rabbit** books; one thinks also of the Swiss Family Robinson's treehouse and of TARZAN, happily swinging through the branches of trees and depicted (at least in some of the TARZAN MOVIES) as living in a treehouse. There have been many impressive CITIES in trees, ranging from Jules Verne's *The Village in the Treetops* (**1901**) to Revelwood in Stephen R. DONALDSON's **Thomas Covenant** series. More cosmically,

Norse mythology pictured the entire Universe as one enormous tree, Yggdrasil ($ WORLD-TREE); and central to the Christian CREATION MYTH is EDEN's Tree of the Knowledge of Good and Evil.

Trees have served also as characters. The features of a mature tree – its visible durability, stature and solidity – have undoubtedly served to establish the features of the imagined sentient tree: incredibly old, wise and powerful. In the original version of the CINDERELLA story it was a talking tree, not a FAIRY GODMOTHER, that helped the lass in her hour of need. This is that tradition that J.R.R. TOLKIEN drew upon in creating the Ents of *The Lord of the Rings* (**1954-5**): ancient, benevolent, knowledgeable, they lend dignity and substance to Frodo's battle against Mordor. Charming tree-people, who sometimes seem to change into human shape and dance, are found also in C.S. LEWIS's *Prince Caspian* (**1951**), though Lewis seems to be building more on the notion of Greek MYTHOLOGY that DRYADS spiritually inhabited trees. Trivialized with a smiling face, friendly trees are a staple of children's literature, often of the worst sort. While such trees are usually cast as males, another wise old tree, now envisioned as female, gave magical assistance and good advice to the heroine of DISNEY's *Pocahontas* (**1995**). In that movie and elsewhere we see a more modern aspect of the tree persona: defender of the natural world against the encroachments of civilization.

Yet sentient trees may also be viewed negatively. Because they usually cannot move or move only with difficulty, they might be seen, like Tolkien's Ents, as overly passive, unwilling to act in times of crisis. They might, like many very old creatures, become irritated by or hostile to the young, especially young humans who flaunt their mobility. Thus one frightening aspect of the dark and sinister forest is that the trees themselves might come to life and attack – which is exactly what Snow White imagines while fleeing through the woods in SNOW WHITE AND THE SEVEN DWARFS (*1937*). Also, Dorothy and her friends are attacked by mean-tempered trees in *The WIZARD OF OZ* (*1939*), and one can also mention the ambulatory tree which serves as a MONSTER in the movie *From Hell It Came* (**1957**). Further villainous movie trees occur in *The BLUE BIRD* (*1940*) and FERNGULLY: THE LAST RAINFOREST (*1992*). [GW]

TREMAYNE, PETER Pseudonym of UK writer and Celtic scholar Peter Berresford Ellis (1943-   ). Under his own name he has produced many books about Celtic history, culture and legend, starting with *Wales – A Nation Again* (**1968**). Of particular relevance to this encyclopedia are *A Dictionary of Irish Mythology* (**1987**) and *A Dictionary of Celtic Mythology* (**1992**). PT also used his own name for biographies of H. Rider HAGGARD, *H. Rider Haggard: A Voice from the Infinite* (**1978**), and Talbot MUNDY, *The Last Adventurer* (**1984**). He is currently working on a biography of E. Charles VIVIAN. He writes thrillers as Peter MacAlan.

As PT he has written over 30 books, almost all in the categories SUPERNATURAL FICTION, CELTIC FANTASY, SCIENCE FICTION and crime. When PT first turned to fiction there was one of the regular revivals of interest in the characters of DRACULA and FRANKENSTEIN. In *The Hound of Frankenstein* * (**1977**) the Baron hides on Bodmin Moor and turns again to his experiments, this time creating a dog. In his **Dracula Trilogy** PT went back to events before Bram STOKER's novel and traced three first-person accounts of their experiences with Dracula: *Dracula Unborn* * (**1977**; vt *Bloodright: Memoirs of Mircea, Son of Dracula* 1979 US), *The*

Revenge of Dracula * (**1978**) and *Dracula, My Love* *(**1980**), assembled as *Dracula Lives!* * (omni **1993**). PT experimented with other sequels (◊ SEQUELS BY OTHER HANDS) with *The Vengeance of She* * (**1978**), bringing the story of **Ayesha** into the modern day (◊ IMMORTALITY) and unconnected with the movie of the same name (◊ SHE); and the nonfantastic *The Return of Raffles* * (**1980**), featuring E.W. Hornung's rogue detective.

Not surprisingly, PT draws heavily upon Celtic myth and legend for many of his stories. The **Lan-Kern** series – *The Fires of Lan-Kern* (**1980**), *The Destroyers of Lan-Kern* (**1982**) and *The Buccaneers of Lan-Kern* (**1983**) – is based on legendary Cornwall, converting the myths into HEROIC FANTASY. *Raven of Destiny* (**1984**) is based on the historical figure of Bran MacMorgor, who invaded Greece in 279BC. Both *Ravenmoon* (**1988**; vt *Bloodmist* 1988 US) and *Island of Shadows* (**1991**) use Celtic legends, the first involving the Irish OTHERWORLD, the second telling the story of the female warrior Scáthach.

For a while in the 1980s PT threatened to rival Guy N. Smith (1939-   ) and James HERBERT for the number of novels in which people (or even humankind) is threatened by various monstrous creatures (◊ MONSTERS). These began with *The Ants* * (**1979**), a faithful sequel to "The Empire of the Ants" by H.G. WELLS. This was followed by *The Curse of Loch Ness* (**1979**), *Zombie!* (**1981**), *The Morgow Rises!* (**1982**), about a legendary Cornish SEA MONSTER, *Snowbeast* (**1983**), about a Scottish abominable snowman, *Kiss of the Cobra* (**1984**), about a Cobra cult in India and the working of a CURSE when a tomb is disturbed, *Swamp!* (**1985**), about a lake monster in the Florida everglades, *Nicor!* (**1986**), about a sea monster disturbed during oil-drilling off the coast of Venezuela, and *Trollnight* (**1987**) where an evil scientist has created a new race of TROLLS. Though these books are mostly action-based, PT was able through his fund of knowledge of the old legends to give them a degree of credibility that most similar books lack.

PT has written many short fictions. Those collected in *My Lady of Hy-Brasil* (coll **1987** US) draw on Celtic legend; those in *Aisling* (**1992**) contain more contemporary Irish horrors.

PT has also edited *Irish Masters of Fantasy* (anth **1979** Eire; vt *The Wondersmith and Other Macabre Tales* 1988 Eire) and compiled a volume of stories by William Hope HODGSON as *Masters of Terror 1: William Hope Hodgson* (coll **1977**).   [MA]

Other works: *Angelus!* (**1985**); the **Sister Fidelma** series of historical mysteries, featuring a 7th-century Irish advocate, being *Absolution by Murder* (**1994**), *Shroud for an Archbishop* (**1995**), *Suffer Little Children* (**1995**) and *The Subtle Serpent* (**1996**).

TREVOR, (LUCY) MERIOL (1919-   ) UK author. ◊ Raymond BRIGGS.

TRIAL, THE Two movies have been based on *Der Prozess* (**1925**; trans as *The Trial* 1937 UK) by Franz KAFKA.

1. *The Trial* (ot *Le Procès*) French/West German/Italian movie (**1962**). Paris Europa/Hisa/FI.C.IT. **Pr** Alexander Salkind. **Exec pr** Michael Salkind. **Dir** Orson Welles. **Screenplay** Welles. **Starring** Suzanne Flon (Miss Pittl), Arnoldo Foa (Inspector A), Elsa Martinelli (Hilda), Jeanne Moreau (Marika Bürstner), Anthony Perkins (Josef K), Romy Schneider (Leni), Akim Tamiroff (Bloch), Welles (Advocate Hastler). 118 mins. B/w.

It was probably a mistake to try to translate Kafka's novel,

most of whose "events" occur inside K's head, to the screen, where it would be judged as a *version* rather than as an independent movie. Welles sets out his stall almost at the start: after a spoken prologue (illustrated by scenes on pin-screen) telling the FABLE of the man who waits at the gate of justice, Welles announces that the novel is generally considered a DREAM or a nightmare; and it is in the form of an anxiety dream that the movie is couched. The plot is familiar: K wakes to find himself under arrest; he is subjected to a long series of PICARESQUE experiences and never does find out what he has been arrested *for*. The movie identifies clearly the legal system with the laws of God. Man's representative to God is incarnate (like CHRIST) in the person of K's supposed attorney, the bedridden Advocate Hastler who seems never to do anything but - like a dead MESSIAH - insists on being worshipped even when idle, absent, arbitrary or imbecile; the painter Titorelli, portraitist to the judges and therefore claimant of considerable influence over them, is in effect the gallery of saints, who promise much by way of intercession but, when the small print of their promises is read, can deliver little if anything. Like God's laws, those of K's country are never explained, and are probably inexplicable: this is the reason for the unstoppable trial's (i.e., life's) unfairness. Yet the movie to an extent undermines its own SATIRE by being so surreally dreamlike: within the EDIFICE that is the Hall of Justice abide almost all the other locales in which the various scenes are set – notably the vast office where K works (its tracts of synchronized typists reminiscent of scenes in Fritz Lang's *Metropolis* [**1926**]) and Hastler's littered, semi-devastated, mansion-grand home. As in any edifice, corridors may not lead to the same place twice. The parallel with Lewis CARROLL's *Alice in Wonderland* (**1865**) is recognized implicitly throughout, being made explicit in a brief sequence where a court guard emulates the White Rabbit.

A separate strand of the dream is distressingly misogynistic. Repeatedly K finds himself the focus of the erotic attentions (never, for one reason or another, consummated) of alluring women – in particular Leni, Hastler's nurse/mistress (who can be identified as the Church). These women are depicted as in some way debased: Miss Burstner, K's long-time neighbour, is an "exotic dancer"; K is only the latest in a long line of men seduced (or in his case nearseduced) by Leni, who preys sexually on arrested men, like a spiritual VAMPIRE. This strand peaks with K being pursued through catacombs by a horde of rapacious schoolgirls.

Visually *TT* is superb, with exquisite use of scale and shadow; the medium of b/w is exploited to the full. There are some stunning pieces of conceptualization, and the mood of the movie is hard to shake off.   [JG]

2. *The Trial* UK/Czech movie (**1992** tvm, also released theatrically). BBC/Europanda. **Pr** Louis Marks. **Exec pr** Reniero Compostella, Kobi Jaeger, Mark Shivas. **Dir** David Jones. **Screenplay** Harold Pinter. **Starring** Douglas Hodge (Inspector), Anthony Hopkins (The Priest), Michael Kitchen (Block), Kyle MacLachlan (Josef K), Alfred Molina (Titorelli), Catherine Neilson (Washer Woman), Jason Robards (Dr Huld), Juliet Stevenson (Fräulein Bürstner), Polly Walker (Leni). 120 mins. Colour.

This more literal version has little to add to **1**: it is less surreal, less melodramatic and less visually exciting; at the same time, because more tied to mundane (not dream) REALITY (partly through being more specific about extraneous details; e.g., K's crime is at least partly detailed to him), because the ALLEGORY is brought closer to the surface, and because its target appears to be more the interpreters of God than God Himself, it is a more *comprehensible* movie – and better as entertainment. Yet it does not live in the memory as **1** does – likely because it has forfeited the sense that K is the plaything of and throughout manipulated by unseen forces that are almost certainly *by definition* impossible to understand. This is palatable Kafka.   [JG]

TRICKSTER There always seems to be a trickster in the pack. Somewhere, embedded in the PANTHEON or *salon des refusées* of almost every MYTHOLOGY and RELIGION, a trickster figure can be found – in animal or human form, duping other gods and mortals; an ape of god who dupes himself; a SHAPESHIFTER who constantly undergoes METAMORPHOSIS; a polymorphically perverse blockhead (like PINOCCHIO), a scatological mocker, thief, culture hero; a raw, mutable first principle who manifests the inchoate birth of the world (◊ CREATION MYTHS) and who quite possibly, through the upwelling amorality of his being (◊ LOKI), helps bring the world to an end; a manifestation of raw appetite; a LIMINAL BEING who can never be fixed into place, except perhaps in the "advanced" religions of the world, where – like the early Jehovah of Judaism, or SATAN in the Christian tradition – he is constrained. Jehovah in the early books of the Old Testament is a tearaway; Satan is a trickster in BONDAGE.

Archaic mythologies – like those examined by Paul Radin (1883-1959) in *The Trickster: A Study in American Indian Mythology* (**1956**) – almost invariably incorporate CYCLES of tales dominated by primordial trickster figures, who usually appear in animal form: as a COYOTE in the lore of many southwestern Native American nations; as a rabbit or hare in northern cycles; as a raven along the Pacific Rim. As a SPIDER, sometimes under the name of ANANSI, trickster figures are found throughout Africa; as MONKEY he appears in Chinese LEGENDS; as ANUBIS (though his functions are various, and some of them are fixed) he appears in the mythology of EGYPT in the form of a man/jackal. Throughout these archaic tales, the trickster – inconsistently – does two things supremely well: he creates the world through TRANSFORMATIONS and through seedings; and he sticks his voracious head (or huge penis) into the workings of the world, sometimes deranging the gods and mortals he butts against, sometimes dismembering himself, sometimes destroying everything.

No trickster can, in the end, be trusted; it is perhaps a sign of their wisdom concerning the essential untrustworthiness of life that he has so central a place in archaic belief systems. Even in later mythologies – as with Loki in the Nordic mythologies and HERMES in Greek mythology – he functions as an untamed signal that the world is no secure haven. Significantly, tricksters like Loki or Hermes are closely related to more "reliable" gods like ODIN or ZEUS; and are often conceived as SHADOWS of their more straightforward, responsible kin. This sense of the unfixable but undeniable marriage of upper and lower – of light and dark, fixed and inchoate, ritual and anarchy, hierarchy and revelling, catechism and PARODY, ego and id – has perhaps not been satisfactorily addressed by the monotheisms characteristic of the religions of the Western World over the past two millennia; also, perhaps, the literatures of the FANTASTIC have so openly welcomed figures like the trickster out of a subversive need to undercut sermonizing versions of the meaning of life. In fantasy proper only CHRISTIAN FANTASY seems notably (and probably deliberately) chary of the

figure – J.R.R. TOLKIEN's *The Lord of the Rings* (**1954-5**) restricts itself to the sad figure of Gollum, and the closest C.S. LEWIS comes to creating one is Screwtape, in *The Screwtape Letters* (**1942**).

Animal tricksters in fantasy seem generally to derive from archaic lore. Reynard the Fox becomes the metamorph. Cat tricksters are a source for PUSS-IN-BOOTS. The archaic hare becomes Brer Rabbit (Joel Chandler HARRIS used FOLKLORE imported from Africa) and Brer Rabbit becomes BUGS BUNNY; Coyote becomes Wil E. Coyote (◊ Chuck JONES); the trickster as butthead becomes DONALD DUCK. US ANIMATED MOVIES in general tend to incorporate trickster metamorphoses and mockeries as a matter of course. Most animals depicted in cartoon form are SHAPESHIFTERS, and the cartoons in which they feature can normally be understood as BEAST-FABLES, a form immemorially conducive to mockery of the official version. Even DISNEY's notorious control over the punishment of anarchy slips slightly in the late ALADDIN (*1992*), whose eponymous trickster hero sometimes seems at the verge of talking back to the world.

Tricksters in human form thread through the literatures of the West. PROMETHEUS – who steals fire, as do many tricksters, but does so for the benefit of humanity – is an example of the trickster as culture hero. Shamans (◊ SHAMANISM) – like the eponymous hero of Terry BISSON's *Talking Man* (**1986**) – and MAGI – like PROSPERO, or Charles G. FINNEY's Dr Lao and the version of Aleister CROWLEY envisioned in John SYMONDS's *The Trickster and the Devil* (**1992**) – cannot be trusted to treat the mundane world with much respect. JACK-figures – from the Jack who tricks the Giant to JACK THE RIPPER to Fritz LEIBER's Gray Mouser – incorporate trickster elements. TOM THUMB is a trickster, and so are Oskar in Günter GRASS's *The Tin Drum* (**1959**) and the prankster Till Eulenspiegel ("Tyll Owlglass"), who has existed in German texts since *c*1483, first appearing in English in *A Merye Jest of a Man Called Howleglas* (*c***1560**). ROBIN HOOD is usually depicted as a trickster. As the Lord of Misrule, trickster figures emcee scenes of REVEL from behind their MASKS, as does the hero of The MASK (*1994*). As Harlequin (a figure who resembles HERMES, and who shapes Michael MOORCOCK's shapeshifting **Jerry Cornelius**), the trickster dominates the COMMEDIA DELL'ARTE. As the CONFIDENCE MAN (who also reflects the influence of Hermes) the trickster is UNDERLIER for 19th- and 20th-century tales that treat REALITY as problematic and morality as inherently inapposite in a shifting world. US TALL-TALE heroes – like Davy Crockett and Mike Fink, figures who underlie late fantasy texts like *Joyleg* (**1962**) by Avram DAVIDSON and Ward Moore (1903-1978) – are often tricksters. Over and above the huge number of tales featuring cunning rogues in Arab literature, ARABIAN FANTASY features figures like Ali Baba and Aladdin; in the Sufi tradition, the jokester Nasruddin is a trickster, and is so presented by Idries Shah (1924-   ) in *The Sufis* (**1964**).

Trickster gods feature in CELTIC FANTASY, NORDIC FANTASY, SLICK FANTASY, fantasy versions of various god families (◊◊ GREEK AND LATIN CLASSICS; MESOPOTAMIAN EPIC; SANSKRIT LITERATURE), where they are presented variously: as licenced FOOLS; as Machiavellian plotters; as co-creators of the universe; as grotesque or seductive shadows of the high gods; as emblems of the dangers and allures of FERTILITY, sometimes in the guise of PUCK; and so forth. The best-known fantasies to feature trickster gods who wear a human mask are probably Philip José FARMER's **World of Tiers**, where Kickaha is both god and man, and *Jack of Shadows* (**1971**) by Roger ZELAZNY, many of whose other books also star immortals with trickster miens. Tanith LEE's **Tales from the Flat Earth** also features in Chuz, the god known as Delusion's Master, a fully developed trickster, two-faced (like JANUS), and profoundly threatening to the BALANCE of the world.

When tricksters appear in HORROR – as in Stephen KING's *Needful Things* (**1991**) – they tend only to destroy, without passing through any change into a state which would allow them to impart or to manifest wisdom. In SUPERNATURAL FICTION, tricksters tend to disrupt the didactic flow – from supernatural realm to ignorant and seducible mundane world – typical of that genre; when they appear in fantasies, especially those set in OTHERWORLDS, they perform the full range of functions described in this entry, and in their volatile mutability are primary conveyors of wisdom in fantasy texts. [JC]

See also: BEETLEJUICE (*1988*); FOOL; PUCK.

TRILBY ◊ SVENGALI.

TRIPLE GODDESS ◊ GODDESS.

TROLL US movie (*1985*). ◊ GHOULIES (*1985*).

TROLLS MONSTERS of Scandinavian MYTH and NORDIC FANTASY; related Shetland myths call them trows. They have affinities with GIANTS (size, general malevolence, fondness for eating human flesh) and earth ELEMENTALS: they are associated with mountains and cold, and often turn to stone on exposure to daylight – as in J.R.R. TOLKIEN's *The Hobbit* (**1937**). Further famous trolls appear in *Peer Gynt* (**1867**) by Henrik Ibsen (1828-1906) and T.H. WHITE's comic-sinister "The Troll" (1935). Fantasy GAMES often base their trolls on the tough specimen in Poul ANDERSON's *Three Hearts and Three Lions* (**1953**), which regenerates even as it is hacked apart and must be burnt piecemeal. The REVISIONIST-FANTASY "trolls" in Tad WILLIAMS's **Memory, Sorrow and Thorn** are most un-troll-like and slightly resemble Tolkien's hobbits; the more traditional but nevertheless comic trolls of Terry PRATCHETT's **Discworld** (some of which do indeed turn to stone at dawn, only to revive at sunset) have done rather more to rehabilitate the creatures' image. Rose ESTES's **Troll** sequence features a society of child-kidnapping trolls beneath Chicago. [DRL]

TROMPE-L'OEIL This French term, meaning "deception of the eye", is used by art critics in discussing Illusionist paintings (◊ ILLUSION). Illusionism as a whole attempts to trick the beholder into "seeing" a painted surface as a solid object. A similar ambiguity can be established between two simultaneously present versions of an image: the eye may be deceived, which is a matter of perception, but the fact that the eye has been deceived does not affect the REALITY of the painted double scene. This is important for an understanding of FANTASY narratives.

Some of the most famous tales may be useful to take the case of E.T.A. HOFFMANN – whose work comes towards the beginning of FANTASY as a conscious literary form – "The Sandman" (1817), are profound studies of abnormal psychology, and their protagonists' visions of supernatural incursions into their lives are almost certainly intended by him to be taken as projections (◊ PERCEPTION) of diseased minds. The world they see is not, therefore, objectively there. On the other hand, in a Hoffmann LITERARY FAIRY-TALE like *The Golden Pot* (1814), the constant flow of TLO effects – the most vivid being the transformation of a door knocker into the FACE OF GLORY of a WITCH from the dawn

of time, and back again, so that the two realities become interchangeable – is precisely meant to describe the kind of objective and self-coherent world that fantasy narratives normally inhabit.

The most profound moments in many fantasy novels, moments when the meaning of the story tends to unfold itself, are often those moments when a METAMORPHOSIS of some sort occurs, or is about to occur. An example is when PAN, in Arthur MACHEN's *The Hill of Dreams* (**1907**), suddenly manifests himself after the protagonist has strayed INTO THE WOODS: "Green mosses were hair, and tresses were stark in grey lichen; a twisted root swelled into a limb." At such moments TLO effects are common, and serve as a convenient pointer to the mysterious heart of metamorphosis: the sense that it simultaneously contains both the being whose essence is changing and the being whose essence is taking shape. In fantasy texts TLO generally registers the fact that something or someone actually *is* one thing and *is* another. Among many examples are: the scarecrow in Nathaniel HAWTHORNE's "Feathertop" (1852), described as being perceived by others in passages simultaneously animate and inanimate (◊ ANIMATE/INANIMATE); in John COLLIER's "Evening Primrose" (1941) literal TLOs lead us into a discovery of the wainscot society inhabiting department stores; the protagonist of SARBAN's *The Sound of His Horn* (**1952**) observes the woman he loves kneeling to drink and sees her as a CAT; Gene WOLFE's *The Book of the New Sun* (**1980-83**) is full of TLO effects, including walls that are bulkheads and men with prostheses who are really cyborgs with flesh-bits; the transitions between this world and FAERIE in Wolfe's *Castleview* (**1990**) – as when a Cherokee jeep seems (rightly) to be, as well, a great Arthurian charger – are constructed around the effect; and in Peter S. BEAGLE's *The Innkeeper's Song* (**1993**) the man/woman woman/man warrior flickers from one state to another like a trick of the light. In movies, a noteworthy example is in Terry GILLIAM's *The* FISHER KING (*1991*), which depends on its chief protagonist's discovery that REALITY can be seen, simultaneously, as two quite different things; for example, the goblet that he must steal is both the GRAIL, at one level of reality, and just a goblet, at another. [JC]

TRON US live-action/ANIMATED MOVIE (*1982*). DISNEY/Lisberger-Kushner. **Pr** Donald Kushner. **Exec pr** Ron Miller. **Dir** Steven Lisberger. **Conceptual artists** Jean Giraud (◊ MOEBIUS), Peter Lloyd, Syd Mead. **Spfx** R.J. Spetter. **Vfx** Ellenshaw, Lisberger, John Scheele, Richard Taylor. **Computer fx** Taylor. **Effects anim** Lee Dyer. **Screenplay** Lisberger. **Novelization** *TRON* * (**1982**) by Brian Daley. **Starring** Bruce Boxleitner (Alan Bradley/TRON), Jeff Bridges (Kevin Flynn/CLU), Barnard Hughes (Walter Gibbs/DUMONT), Cindy Morgan (Lora/YORI), Dan Shor (RAM), David Warner (Ed Dillinger/SARK). 96 mins. Colour.

The ENCOM organization, headed by founder Gibbs, is run by Dillinger with the aid of the central computer's huge Master Control Program (MCP), which is ever growing through the theft of programs from elsewhere; MCP plans to take over the Pentagon and Kremlin while Dillinger plots to oust Gibbs. Coincidentally, Gibbs and assistant Lora are working within ENCOM on a system to encrypt and later decrypt objects and organisms using computer-linked lasers, the long-term goal being practicable matter transmission. Wacky genius Flynn devised ENCOM's vastly successful video-games programs, but saw his inventions stolen by Dillinger. ENCOM employee Bradley has devised a new debugging program, TRON, which MCP sees as a threat. Bradley, Lora and Flynn try to work their way into the computer. MCP activates the encryption laser to absorb Flynn into itself in the form of a program; there Flynn discovers a sort of video-game world where lesser programs are forced to engage in gladiatorial contests. The rest of the tale, set within the computer's net, sees Flynn, Tron and Lora's program YORI, aided by Gibbs's original "guardian" program DUMONT, outwit and destroy MCP and its chief lieutenant, SARK.

T, despite a clumsy script, is an extremely interesting piece of TECHNOFANTASY. The plot owes obvious debts to *The Wonderful Wizard of Oz* (**1900**) by L. Frank BAUM, *Journey to the Centre of the Earth* (**1864**) by Jules Verne (1828-1905) and even *Alice's Adventures in Wonderland* (**1865**) by Lewis CARROLL. The world within the main net is realized, through a mixture of live-action and computer animation, as if it were a SECONDARY WORLD. Individual programs are incarnated as the human beings who devised them – although there is religious dispute among the programs as to the existence or nonexistence of the Users (programmers and operators), regarded as GODS. There is a death-analogue, in that programs may be "derezzed" at the whim of MCP or through failure in any of the various games MCP makes them play. The programs display certain SHAPESHIFTING abilities, and Flynn, as an encrypted User rather than a piece of software, possesses some of the supernatural qualities of an AVATAR. The blurring between the two REALITIES is further enhanced through the depiction of the ENCOM building's interior as scrupulously clean, angular, over-automated and utterly inhuman; in a stunning final TROMPE L'OEIL the camera looks over nighttime LA and sees it as a board of flashing electronics. [JG]

TRUE NAME A convention of MAGIC is that knowing a TN gives power over the thus-named DEMON, DRAGON, WITCH, WIZARD or whatever. In the well known GRIMM BROTHERS tale, RUMPELSTILTSKIN must be named to be defeated. The TN is shorthand for deep understanding of the named thing's essence, identity or ACHILLES' HEEL, and is usually well guarded. The Horned King in Lloyd ALEXANDER's *The Book of Three* (**1964**) perishes when his TN is eventually divined by a pig who is an ORACLE; the LAMIA in Brian STABLEFORD's *The Last Days of the Edge of the World* (**1978**) clearly wants surcease but remains in BONDAGE until her TN has been laboriously called forth. When magic requires the TN to be spoken, an obvious QUIBBLE arises: Larry NIVEN's Warlock in "Not Long Before the End" (1969) has a TN impossible to pronounce, while the demon in Terry PRATCHETT's *Wyrd Systers* (**1988**) is strategically named WxrtHltl-jwlpklz. Ursula K. LE GUIN gives TNs proper dignity in *A Wizard of Earthsea* (**1968**), as the nouns of the True Speech of CREATION, which confer a self-limiting power – since a wizard desiring to enchant the whole sea must truly name its every reach, bay, cove, inlet, strait, shore . . . *ad infinitum*. The correspondence of TN and SOUL is emphasized in Le Guin's *The Farthest Shore* (**1973**), where giving up one's TN is the price asked for an ugly semblance of IMMORTALITY. [DRL]

See also: KALEVALA; MAGIC WORDS.

TRUE STRANGE STORIES ◊ GHOST STORIES (magazine).
TRUE TWILIGHT TALES ◊ GHOST STORIES (magazine).
TRULY MADLY DEEPLY UK movie (*1990* tvm, later released theatrically). Samuel Goldwyn/BBC Films. **Pr** Robert Cooper. **Exec pr** Mark Shivas. **Dir** Anthony Minghella. **Vfx** Ian Legg. **Screenplay** Minghella. **Starring** Michael Maloney (Mark), Bill Paterson (Sandy), Alan Rickman (Jamie), Christopher Różycki (Titus), Juliet Stevenson (Nina). 103 mins. Colour.

Superficially a GHOST STORY, this is primarily a movie about loss, and about coping with it. Translator Nina, recently bereaved of lover Jamie, is inconsolable. Yet, despite her self-absorption, friends rally round, each having a loss of their own to cope with. Then Jamie's GHOST comes back to live with her, their relationship picking up where it left off, and with refreshing lack of sentiment: they tease and grump about each other as if merely continuing a long-running and much-loved conversation. But, after he's fed her inward-focused grief for some while, things begin to sour between them. Her growing attraction towards chance-encountered amateur conjurer Mark is enough to make Nina realize she must end the relationship with Jamie, this time on her own terms rather than his.

It would be facile to regard *TMD* as a fantasy of PERCEPTION. Jamie's ghost is not merely a projection of Nina's grief: we see too much through his eyes for him to be other than an independent character. Yet *TMD*, released theatrically to cash in on the success of GHOST (*1990*), must have proved disappointing for admirers of the latter seeking further glitz: it abjures spfx except once – when Nina first meets Mark and he conjures a book into a pigeon, a feat that we are left to believe may be real MAGIC – relying instead on its script, some fine performances, perfectly measured timing, and the overall excellence of its direction. [JG]

TRYON, THOMAS (1926-1991) US actor and author, who turned to writing with *The Other* (**1971**), a bestseller which he adapted for the screen the following year. A "bad seed" tale, it presents a complex and nuanced portrait of a young boy so psychologically traumatized by the death of his TWIN brother that he imagines him still alive (◊ INVISIBLE COMPANION) and responsible for the increasingly evil acts of mischief the surviving twin perpetrates. *Harvest Home* (**1973**) is equally ambiguous in its approach to fantasy, evoking a mood of supernatural menace through its portrait of a rural New England town where pagan practices have survived into the present. Both novels are justly praised for their skilful narrative misdirection, which permits a variety of fantastic and nonfantastic interpretations of events until their revelatory climaxes.

Just when he seemed poised to be recognized as the preeminent contemporary literary horror novelist, TT shifted focus. His only other WEIRD FICTION, the posthumous *Night Magic* (**1995**), is set in the world of stage magic and concerns an illusionist whose skill at prestidigitation is based in Egyptian mysticism. The novel's overtly fantastic elements are uncharacteristic of TT's work in general, and may show the hand of Valerie MARTIN and John Cullen, who contributed to its final revision. [SD]

Other works: *Lady* (**1974**); *Crowned Heads* (coll **1976**); *All That Glitters* (coll **1986**); *The Night of the Moonbow* (**1988**); *Opal and Cupid* (**1990**), a juvenile; *The Wings of Morning* (**1990**).

TUCKER'S WITCH US tv series (1982-3). Leonard Hill Films/CBS. **Pr** William Bast, Steve Kline, John Thomas LeNnox. **Exec pr** Leonard Hill, Philip Mandelker. **Dir** Corey Allen, Rod Daniel, Randa Haines, Peter H. Hunt, Harvey S. Laidman, Harry Winer. **Writers** Bast and many others. **Starring** Barbara Barrie (Ellen Hobbes) Catherine

Hicks (Amanda Tucker), Tim Matheson (Rick Tucker), Bill Morey (Lt Sean Fisk), Alfre Woodard (Marcia Fulbright). Unaired pilot plus 12 60min episodes. Colour.

Amanda and Richard Tucker are PIs in Los Angeles; she is an apprentice WITCH, still working to master MAGIC. Even though not always able to control her SPELLS, Amanda keeps trying to use them to help solve cases, much to Richard's displeasure. Her failures often place them at odds with a police detective who wanted them to stay off his cases, showing that some clichés are hard to lose.　　[BC]

TURGENEV, IVAN (SERGEEVICH) (1818-1883) Russian novelist whose major novels – the last being *Otsy i deti* (**1862**; variously trans as *Fathers and Sons* and under other titles) – precede all but one of his SUPERNATURAL FICTIONS. Most of these tease at the THRESHOLD of the supernatural, and serve neatly to demonstrate Tsetvan TODOROV's theory of the FANTASTIC – that the fantastic occurs at a moment when it is profoundly uncertain whether or not events being described can be rationalized (◊ RATIONALIZED FANTASY). IT's first tale of interest, "Faust" (1856), is typical in that it hovers over the nature of the intervention – whether GHOST of her mother, or simply intuition – that causes a young woman to decide not to succumb to passion. Other similar stories include "A Strange Story" (1869), "Knock . . . Knock . . . Knock! . . ." (1871), "A Song of Triumphant Love" (1881) and "Klara Milich" (1882). "Father Alexei's Story" (1877) is HORROR.

There are three tales of more direct fantasy interest. In "Apparitions" (1864) – which has also been translated as "Phantoms", "Ghosts" and "Spectres" – a VAMPIRE-like female debilitatingly takes the protagonist on three TIMESLIP visitations to scenes that demonstrate the uselessness of all forms of human endeavour except, perhaps, LOVE. In "The Dog" (1866) the sound of supernatural canine claws induces a man to buy a real dog in time to save his life. In "The Dream" (1877) a son recognizes the father he has never met except in a recurring DREAM; unfortunately, the father is dead. The assorted "Poems in Prose" (1878-82) flicker delicately at the edge of the UNCANNY. *Dream Tales and Prose Poems* (coll trans Constance GARNETT **1897** UK). *Phantoms and Other Stories* (coll trans Isabel F. Hapgood **1904** US), *A Reckless Character and Other Stories* (coll trans Isabel F. Hapgood **1904** US) and *The Mysterious Tales of Ivan Turgenev* (coll trans Robert Dessaix **1979** Australia) are all useful, but none is complete; Dessaix's introduction is particularly helpful.　　[JC]

TURNABOUT US movie (*1940*). United Artists/Hal Roach. **Pr** Roach. **Dir** Roach. **Spfx** Roy Seawright. **Screenplay** Berne Giler, John McClain, Mickell Novak. **Based on** *Turnabout* (**1931**) by Thorne SMITH. **Starring** William Gargan (Joel Clare), John Hubbard (Tim Willows), Carole Landis (Sally Willows), Donald Meek (Henry), George Renavent (Mr Ram), Verree Teasdale (Laura Bannister). 83 mins. B/w.

In this first cinematic excursion into IDENTITY EXCHANGE, ex-showgirl Sally and her livewire advertising-executive husband Tim fantasize together about how much fun each would have in the other's role. Exotic STATUE Mr Ram overhears and, announcing this is the first time he has ever known the argumentative couple to agree on anything, grants their WISH. Swapping bodies (although not voices – laryngitis is blamed), they proceed to make a hash of each other's duties. Together the couple beg Mr Ram to be exchanged back, and he obliges.

Considered pretty daring in its time because of its recognition that the sexes were physically different (although the most obvious differences are ignored), *T* grates today. Hubbard's portrayal of Sally/Tim is simperingly patronizing; Landis's body-language as Tim/Sally is much better achieved, but the script's message seems to be that she is a more competent woman when inhabited by a man. The lip-sync of the dubbed voices is not always very good. The movie remains watchable for its host of lesser roles, with Teasdale and Meek outstanding, and for some inspired clowning.　　[JG]

See also: TURNABOUT (1979).

TURNABOUT US tv series (1979). Chertok Television/Universal/NBC. **Pr** Arnold Kayne, Michael Rhodes. **Exec pr** Sam Denoff. **Dir** Richard Crenna, William P. D'Angelo, Arnold Laven, Alex March. **Writers** Steven Bochco, Michael Rhodes. **Based on** *Turnabout* (**1931**) by Thorne SMITH. **Starring** Sharon Gless (Penny Alston), Bobbi Jordan (Judy Overmeyer), Bruce Kirby (Al Brennan), John Schuck (Sam Alston), James Sikking (Geoffrey St James), Richard Stahl (Jack Overmeyer). 13 30min episodes. Colour.

Sam and his wife Penny have become bored with their professions and each other, and are both convinced the other one has things better. Matters take a decidedly different turn when Penny buys a small statue from a gypsy, who tells her it has MAGIC powers. A casual comment about wishing they could trade places comes true, and they wake next morning in each other's bodies (◊ IDENTITY EXCHANGE). The episodes deal with the couple trying to pass as each other, and the bewilderment of their friends and neighbours over their "new" personalities. As expectable, there are lots of jokes about Sam having to deal with dressing like a woman and Penny's introduction to the men's locker room.　　[BC]

See also: TURNABOUT (*1940*).

TURN OF THE SCREW, THE vt of *The* INNOCENTS (*1961*).

TUROK, SON OF STONE ◊ Alberto GIOLITTI.

TUTTLE, LISA (1952-　　) US-born writer, resident in the UK since 1980. She worked as a journalist for five years on a daily newspaper in Austin and was an early member of the Clarion SF Writer's Workshop. She sold her first story in 1971 and won the John W. Campbell AWARD in 1974. Her first book, *Windhaven* (**1981**), was with George R.R. MARTIN. Much of LT's most powerful work has been in the fields of DARK FANTASY and HORROR, with such novels as *Familiar Spirit* (**1983**), *Gabriel: A Novel of Reincarnation* (**1987**), *Lost Futures* (**1992**) and *The Pillow Friend* (**1996**). Many of her best short stories – "The Horse Lord" (**1977**), "Treading the Maze" (**1981**), "Bug House" (**1980**), "Sun City" (**1980**) and "Flying to Byzantium" (**1985**) – are collected in *A Nest of Nightmares* (coll **1986**). Much of LT's fiction deals with fragile relationships combined with psychological or sexual TRANSFORMATION, often with a strongly feminist slant (◊ FEMINISM). LT edited the acclaimed horror anthology by female authors, *Skin of the Soul* (**1990**).　　[SJ]

Other works: *Catwitch* (**1983**), children's book with illustrator Una Woodruff; *Children's Literary Houses* (**1984**), children's book with Rosalind Ashe; *Angela's Rainbow* (**1983**), erotic fantasy; *Encyclopedia of Feminism* (**1986**), non-fiction; *A Spaceship Built of Stone and Other Stories* (coll **1987**); *Heroines: Women Inspired by Women* (**1988**), nonfiction; *Mike Harrison's Dreamlands* (**1990**), text for art book;

Memories of the Body: Tales of Desire and Transformation (coll **1992**); _Horrorscopes: Virgo, Snake Inside_ (**1995**), YA horror under hn "Maria Palmer"; _Panther in Argyll_ (**1996**), YA fantasy.

TUTUOLA, AMOS (1920-   ) Nigerian writer, of Yoruba origin. AT spent some time in missionary school, but stopped formal education early and worked in various manual jobs. His first English-language novel, _The Wild Hunter in the Bush of the Ghosts_ (written 1948; **1982**) was only accidentally rediscovered in the 1980s. His first published – and most popular – novel was _The Palm-Wine Drinkard_ (**1952**), about the wanderings of Drinkard to the Deads' Town, where his favourite, already-dead palm-wine tapster now dwells; there he meets dozens of supernatural creatures. A 7-year-old boy narrates _My Life in the Bush of Ghosts_ (**1954**), another novel about wandering, this time in a spirit world with towns populated by GHOSTS. In _Simbi and Satyr of the Dark Jungle_ (**1955**) AT changed style, for the first time organizing his adventures into chapters and narrating in the third person, but the novel lacks the freshness of its predecessors.

His later fantastic novels – _The Brave African Huntress_ (**1958**), _Feather Woman of the Jungle_ (**1962**), _The Witch-Herbalist of the Remote Town_ (**1981**) and _Pauper, Brawler and Slanderer_ (**1987**) – likewise lack his former vigour, but remain interesting examples of fantasy not based on European myths and legends. His only collection to contain mainly fantastic stories is _Ajaiyi and His Inherited Poverty_ (coll **1967**). All the above-cited works are written in an infectious form of basic English.   [JO]

TWAIN, MARK Pseudonym of Samuel Langhorne Clemens (1835-1910), US writer and humorist. Although much of MT's fiction strikes modern readers as fantastic, the determined rationalism that he brought to such works as "Extracts from Captain Stormfield's Visit to Heaven" (**1907**; vt _Captain Stormfield's Visit to Heaven_ **1909**), an AFTERLIFE fantasy and religious SATIRE but much concerned with astronomical matters (the protagonist rides a comet to a corporeal HEAVEN located outside the Solar System), and "Letters to the Earth" (written 1909; 1962), a satire on matters that Twain's audience would not regard as fantastic at all, means that he can better be understood as a writer of 19th-century SCIENCE FICTION (◊ _SFE_). Stories such as "A Ghost's Tale" (1888), in which the Cardiff GIANT makes a desultory appearance, and "The Canvasser's Tale" (1876), about a man who collects echoes, have more in common with the TALL TALE and the topical satire than with the fantastic.

Supernatural elements in Twain's short fiction tend to be employed as devices for specific effects, as in "A Horse's Tale" (1906), which is told in part by Buffalo Bill's horse, who sometimes speaks (◊ TALKING ANIMALS), although only to other horses. Like Nathaniel HAWTHORNE and Washington IRVING, MT wrote in an era when the US short story was still an offshoot of various folk and humorous traditions; most examples published before the 1880s can be (generously) defined as fantastic.

MT's longer works, even when dealing with such elements as TIME TRAVEL – like _A Connecticut Yankee in King Arthur's Court_ **1889**; vt _A Connecticut Yankee at the Court of King Arthur_) – Gothic themes and medieval settings, were essentially scientific. While _No. 44, The Mysterious Stranger: Being an Ancient Tale Found in a Jug, and Freely Translated from the Jug_ (written 1897-1908; in corrupt form as _The_

Mysterious Stranger **1916**; restored text in _Mark Twain's Mysterious Stranger Manuscripts_ **1969**; solo **1982** vt with subtitle added), with its superhuman protagonist – identified in the earliest draft as the "Young Satan" – seems an unequivocal fantasy, the character's full name in the final version, "Number 44, New Series 864,962", suggests a bureaucratic cosmology not unlike MT's other tales of Heaven and Earth; and while the species to which the protagonist belongs are forthrightly called "dream-sprites", MT's striking description of such creatures travelling through the Universe at "thought-speed" has greater affinities with the fiction of Olaf Stapledon (1886-1950) than with most spritely fantasies. Despite numerous points of correspondence with fantasy, MT cannot unanimously be claimed as a fantasist.   [GF]

See also: _A_ CONNECTICUT YANKEE.

TWEED, THOMAS F(REDERICK) (1890-1940) UK author of, among others, _Rinehard: A Melodrama of the Nineteen-Thirties_ (**1933**; vt _Gabriel Over the White House: A Novel of the Presidency_ 1933 US), filmed as GABRIEL OVER THE WHITE HOUSE (_1933_).   [JG]

TWELVE MONKEYS US movie (_1995_). ◊ Terry GILLIAM.

TWICE-TOLD The term "twice-told tale" comes from a line in William SHAKESPEARE's _King John_ (_c_1597): "Life is as tedious as a twice-told tale." It is almost certainly this sense that Nathanial HAWTHORNE intended to convey in his _Twice-Told Tales_ (coll **1837**), a volume which contains mostly SUPERNATURAL FICTIONS whose protagonists are locked into the BONDAGE of their nature and precisely "twice-told" in their conviction that nothing they can do can release them from punitive reiteration of that which damned their ancestors and which will damn them as well.

Here, however, we normally use the term to help characterize a FANTASY whose telling incorporates a clear _re_telling of the inherent STORY – very often of a FAIRYTALE or FOLKLORE or MYTH or LEGEND – _fore_grounding the existence of a previous version of the tale now being retold. It is clear (for instance) that the protagonist of David Henry WILSON's _The Coachman Rat_ (**1985**) is taking part in the tale of CINDERELLA, even though her name is never mentioned, and that the protagonist of Howard Waldrop's _A Dozen Tough Jobs_ (**1989**), which replays the Labours of Hercules in today's USA, is indeed a twice-told version of that legend. When a retelling also constitutes a substantive examination of the prior story – or of a corpus of stories – then the twice-told tale is also a REVISIONIST FANTASY.

Examples of both are extremely numerous. Collections are common in children's literature; adult twice-told tales, like the **Fairy Tale** sequence ed Terri WINDLING, clearly combine elements of the twice-told and revisionist fantasy. Two further adult examples are Kenneth MORRIS's _The Fates of the Princes of Dyfed_ (**1914**), a version of the first two Branches of the MABINOGION, and many of the stories assembled in Angela CARTER's _The Bloody Chamber and Other Stories_ (coll **1979**).   [JC]

TWILIGHT ZONE, THE US tv series (1959-64). Cayuga Productions/CBS. **Pr** William Froug, Herbert Hirschman, Buck Houghton. **Exec pr** Rod SERLING. **Dir** Justus Addiss and many others. **Writers/based on stories by** Charles BEAUMONT, Ambrose BIERCE, Jerome BIXBY, Ray BRADBURY, John COLLIER, Damon Knight, Richard MATHESON, Rod SERLING, Henry Slesar, Manly Wade WELLMAN and many others. **Comics adaptation** The Twilight Zone * (92 issues 1961-72) from Dell/Gold Key/Whitman Comics. **Starring**

Charles Aidman, Edward Andrews, Martin Balsam, Richard Basehart, Orson Bean, Shelly Berman, Ann Blyth, Charles Bronson, Carol Burnett, Sebastian Cabot, Art Carney, John Carradine, Jean Carson, James Coburn, Jackie Cooper, Wally Cox, Patricia Crowley, Bob Cummings, Susanne Cupito, James Daly, Richard Deacon, John Dehner, William Demerest, Andy Devine, Ivan Dixon, Howard Duff, Dan Duryea, Buddy Ebsen, Peter Falk, Anne Francis, Jonathan Harris, Richard Haydn, Pat Hingle, Earl Holliman, Sterling Holloway, Dennis Hopper, Dean Jagger, Buster Keaton, Phyllis Kirk, Jack Klugman, Martin Landau, Mary LaRoche, Cloris Leachman, Ida Lupino, Nancy Malone, Jean Marsh, Ross Martin, Strother Martin, Lee Marvin, Burgess Meredith, Kevin McCarthy, Doug McClure, Roddy McDowall, Gary Merrill, Vera Miles, Martin Milner, Elizabeth Montgomery, Agnes Moorehead, Howard Morris, Billy Mumy, Lois Nettleton, Julie Newmar, Barbara Nichols, Susan Oliver, J. Pat O'Malley, Suzy Parker, Nehemiah Persoff, Donald Pleasence, Robert Redford, Burt Reynolds, Don Rickles, Cliff Robertson, Mickey Rooney, Janice Rule, Albert Salmi, Telly Savalas, Robert Serling, Everett Sloane, Inger Stevens, Warren Stevens, Dean Stockwell, Harold J. Stone, Rod Taylor, Jack Warden, David Wayne, Dennis Weaver, Fritz Weaver, Jack Weston, James Whitmore, William Windom, Jonathan Winters, Ed Wynn, Keenan Wynn, Dick York, Gig Young. 134 30min episodes, 17 60min episodes. B/w.

A perennial winner on critics' lists of the best-ever tv series, *TWZ* was framed in the Fifth Dimension, where anything could happen and usually did. The brainchild of Serling, who contributed 89 of the stories aired, each episode usually had a twist ending. For example, in "Escape Clause" David Wayne makes a PACT WITH THE DEVIL to become immortal (◊ IMMORTALITY) and then kills someone for kicks, figuring he can escape death in the electric chair. He escapes, all right, but by being sentenced to life imprisonment, a decidedly unpleasant penalty for someone who will live forever: he should have READ THE SMALL PRINT.

TIME TRAVEL was a common theme, as in "A Stop at Willoughby", where a man flees into the past to escape the pressures of his busy life, and "The Odyssey of Flight 33", featuring an airliner somehow stuck in time. Other episodes are decidedly surreal, such as "A World of Difference", where an actor discovers his life has become a movie and he is no more than a character on a set. Others of the countless fantasy elements featured include a GENIE whose seemingly generous granting of WISHES belies his evil nature, a woman who discovers she is really a department-store mannequin, and an immortal college professor who can lecture reliably on history because he has actually lived through the events concerned.

What set this series apart from others of its ilk was the high quality of the stories and a relatively low dependence on spfx – indeed, many episodes had no spfx at all but relied instead on the power of imagination. In a medium that generally treats attention spans as an endangered species, *TWZ* proved a worthy exception. The series much later spawned the theatrical release *Twilight Zone: The Movie* (**1983**) and a syndicated series, *The* NEW TWILIGHT ZONE (1985-8). Both made use of stories first aired in the original series. *The Twilight Zone – Rod Serling's Lost Classics* (**1994** tvm) was an anthology movie based on unfinished stories by Serling.

[BC]

TWIN PEAKS US tv series/serial (1989-91). Propaganda Films/Lynch-Frost Productions/Worldvision/Spelling Entertainment. **Pr** Harley Peyton. **Exec pr** Mark FROST, David LYNCH. **Dir** Graeme Clifford, Caleb Deschanel, Duwayne Dunham, Uli Edel, James Foley, Frost, Lesli Linka Glatter, Stephen Gyllenhaal, Todd Holland, Tim Hunter, Diane Keaton, Lynch, Tina Rathborne, Jonathan Sanger. **Writers** Tricia Brock, Robert Engels, Frost, Scott Frost, Lynch, Peyton, Barry Pullman, Jerry Stahl. **Created by** Frost, Lynch. **Novelization** *The Secret Diary of Laura Palmer* * (**1990**) by Jennifer Lynch. **Starring** Mädchen Amick (Shelley Johnson), Dana Ashbrook (Bobby Briggs), Richard Beymer (Benjamin Horne), Lara Flynn Boyle (Donna Hayward), Joan Chen (Jocelyn Packard), Catherine Coulson (Log Lady), Eric Da Re (Leo Johnson), Sherilyn Fenn (Audrey Horne), Miguel Ferrer (Albert Rosenfeld), Warren Frost (Dr William Hayward), Piper Laurie (Catherine Martell), Sheryl Lee (Laura Palmer/Madeleine Ferguson), Peggy Lipton (Norma Jennings), Lynch (Chief Gordon Cole), Everett McGill (Ed Hurley), Kyle MacLachlan (Special Agent Dale Cooper), James Marshall (James Hurley), Chris Mulkey (Hank Jennings), Jack Nance (Pete Martell), Michael Ointkeen (Sheriff Harry S Truman), Wendy Robie (Nadine Hurley), Russ Tamblyn (Lawrence Jacoby), Ray Wise (Leland Palmer), Grace Zabriskie (Sara Palmer). 30 45min episodes. Colour.

In the small town Twin Peaks the body of the murdered Laura Palmer, highschool queen and assumedly pure lass, is discovered on the riverbank. FBI agent Cooper is called in, because this seems the further work of a SERIAL KILLER, although various of the local boys are suspected. In the event, it proves that the killer is Laura's father Leland, in his alternate personality as Bob – Bob is the DARK LORD of the series, which is full of the sense of WRONGNESS – but before that there is much wierdness and fantasy as the layers of the small town are stripped away.

Hypnotic and distracting, complex and cluttered, intelligent and trashy; with *TP*, David LYNCH and Mark FROST succeeded in making quintessential tv-noir. The series, building on the original full-length pilot, is part DARK FANTASY, part detective fiction, part soap opera, all compulsion. Hooked onto the simple premise of murder, *TP* develops far beyond expected boundaries to create a REALITY entirely its own. The extreme complexities of plot are matched only by the rich characterization and convoluted symbolism.

Where many tv shows highlight episodes taken in isolation out of their characters' fictional lives, *TP* builds up a textured web of reality, in which characters' ambitions and DREAMS, large or small, play equal weight, and fit together in continual overlapping patterns. From crime detection to teenage romance, from ecological concern to arson to cookery, the web is completed. In a place where "the owls are not what they seem", no action is necessarily trivial, and all actions have consequences. Amid its layers of supernatural power and elder wisdom, melodrama and violence, the show becomes, above all else, realistic, albeit on its own terms. The reality of *TP* is a product of its balanced blend of the mundane and the fantastical. Just as Agent Cooper brings his own eccentric beliefs and practices to the town, so the town by stages reveals its own eccentricities to him, and to the viewer. The star-crossed love of Ed Hurley and Norma Jennings remains external to the central concern – the murder of Laura Palmer, and the mystic power of the Black Lodge – yet it is neither marginal nor untouched by the strangeness of its context. Via the dominating, lurking, shadowy presence of Bob, through whom the destructive power of the Black Lodge is enacted, the themes of incest, domestic violence and sexual exploitation are lifted out of the mundane and given a new twist. In *TP* it becomes truly possible to hate the sin and love the sinner, and the helplessness of the oppressor becomes apparent. Leland Palmer is as much a victim as Laura – and Laura herself is as potentially destructive as Leland has been. Major Briggs, seemingly the most conservative of men, holds the key to some of the mysteries of the Black Lodge, and through his very nature moderates the series' careful balance of the natural and the supernatural. Human responsibility goes hand-in-hand with human weakness, and the answers are presented obliquely, implicitly. Agent Cooper's visions and dreams reprove his direct questions, yet present him with answers whose meaning must be hard sought-for. The creatures of Cooper's dreams – the servants and inhabitants of the Black and White Lodges – do indeed speak in tongues.

Assisted by a mesmeric soundtrack, and by careful, atmospheric cinematography, *TP* is rich in resonance and detail. Its plots belong in both the real world of the viewer and the fictive, fantasized world of the town Twin Peaks. Through its clever, intricate, blending of genres, it becomes one of the most powerful new realities within both fantasy and tv genres. The viewer is engaged: there is a sense that the realities of this world are continuing after the credits close.

A direct-to-video movie derived from the series was issued as *Twin Peaks* (*1989*): it contains material largely from episodes #1 and #3, plus some new footage, and is engrossing but, finally, incomprehensible. Lynch returned to his small town in TWIN PEAKS: FIRE WALK WITH ME (*1992*). [KLM/JG]

TWIN PEAKS: FIRE WALK WITH ME US movie (*1992*). Francis Bouygues/Ciby/Twin Peaks Productions. **Pr** Gregg Fienberg. **Exec pr** Mark FROST, David LYNCH. **Dir** Lynch. **Screenplay** Robert Engels, Lynch. **Based on** *The Secret Diary of Laura Palmer* * (*1990*) by Jennifer Lynch, itself based on the tv series TWIN PEAKS. **Starring** Mädchen Amick (Shelly Johnson), Michael J. Anderson (Man From Another Place), Dana Ashbrook (Bobby Briggs), Phoebe Augustine (Ronette Pulaski), David Bowie (Phillip Jeffries), Lenny von Dohlen (Harold Smith), Pamela Gidley (Teresa Banks), Chris Isaak (Chester Desmond), Moira Kelly (Donna Hayward), Sandra Kinder (Arlene), Sheryl Lee (Laura Palmer), Lynch (Chief Gordon Cole), Kyle MacLachlan (Special Agent Dale Cooper), James Marshall (James Hurley), Frank Silva (Bob), Harry Dean Stanton (Carl Rodd), Al Strobel (Philip Gerard, the One-Armed Man), Kiefer Sutherland (Sam Stanley), Ray Wise (Leland Palmer). 134 mins. Colour.

While the thrust of much GENRE FANTASY is to render everyday the fantastic, that of the *Twin Peaks* tv series was to fantasticate the (possibly) mundane; although *TP:FWWM* (shot in tv rather than wide-screen format) might thus be regarded as a piece of MAGIC REALISM, it is likely better treated as a fantasy of PERCEPTION or, taken more literally, as a self-parodic (◊ PARODY) BEAUTY AND THE BEAST fable, where Beauty is Laura Palmer and the Beast is both her father Leland and his bestial aspect, Bob (i.e., where the Beast lurks within the man rather than, as in the purer form of the fable, the man within the Beast). RECURSIVE elements abound, many traceable to old movies; the most profound subtextual reference is to the ORPHEUS myth (debatably via the Jean COCTEAU movie ORPHÉE [*1949*]). *TP:FWWM* is

also an exposition of Godel's Theorem, in that much of its axiomatic matter occurs off-screen.

Only the veneer of this very complicated movie can be summarized here; the discussion below should be read in conjunction with the entry on TWIN PEAKS. College girl Laura Palmer, Homecoming Queen, part-time prostitute and coke-addict, is both child/woman and virgin/whore, an incarnation of the GODDESS. She has been debauched since age 12 by a mysterious entity called Bob; that entity she now discovers is (at least, identifies with and perceives as) her father Leland. (She may have created Bob to "hide" a real incest or, conversely, grafted her father's image onto a "real" or subconsciously generated Bob.) That Bob has possessed her in more senses than one (◊ POSSESSION) is made explicit when she confesses part of her truth to friend Smith: for a moment she becomes a yellow-toothed VAMPIRE Bob, grunting the enigmatic words "Fire . . . Walk . . . With . . . Me". One day an old woman and a masked child present her with the picture of a room to hang on her bedroom wall; later she DREAMS herself into the picture where she finds both woman and child as well as the Red-Velvet Room in which wait Agent Cooper and the dwarfish Man From Another Place. In the mundane reality her life spirals downwards into sexual and other degradation. Forewarned of her doom by seeing an ANGEL vanish from a picture, she joins harlot Ronette in a shack for an orgy with two drunks. Leland arrives, batters one man insensible and whips the two bound girls to a disused railway carriage, where Gerard is too late to stop him beating the girls savagely, Laura to death: dying, she sees herself in a mirror as Bob, then watches her angel-self ascending to HEAVEN. At last she is in the Red-Velvet Room, where Leland and Bob are seen definitively as two separate individuals, and where Cooper stands with her as, laughing hysterically, she continues to watch her heaven-bound angel-self.

Laura is, of course, the core of the movie: from her first appearance she is rarely away from centre-screen. It is trendy to disparage Lee's performance, but in fact her portrayal of a multiply functioned role is magnificent. At one level she is the person who has lived forever behind MASKS which are now repeatedly cracking; she is both the college queen angel and the female DEVIL incarnate (explicitly in a sequence set in an INFERNO-like nightclub), albeit a Devil with vestiges of conscience, capable of switching in an instant between these polar-opposite roles (or, indeed, "twin peaks"); she is a child distraught when her father ticks her off for failing to wash her hands before meals, yet a duplicitous harlot trading on her "innocence"; she is both an idealized image of Woman and a real woman; she is a person onto whom roles are projected by others, rather than a personality in her own right. It seems evident that Lynch's purpose is to reconcile in his own mind all of these aspects of Woman, a task which he finally fails to perform. [JG]

TWINS Twins are treated relatively kindly in modern fantasy, but it was not always so. In almost every traditional society, TABOO enshrouds the birth of twins, and the fate which may be meted out to them (and their mothers). Twins evoke a sense of the sacred in peoples who lack a biological science which can assure them that simultaneous double births are natural: for good or for ill, twins are likely to caused by the gods.

Unsurprisingly, the twin gods who feature in numerous PANTHEONS tend to manifest a variety of aspects, not all of them easy to contemplate linked together. But there may be an underlying pattern, which may be described as communication, or passage (◊ RECOGNITION). Twin gods preside over rituals of FERTILITY; they found cities; they help (but also threaten) humanity with advice; they protect those who travel (but those who travel leave their homes, undertake great risks, and may cause change upon their return). Among the better-known twin gods are the Egyptian Osiris and Set, the Hindu Yama and Yami – whose incest is followed by their becoming rulers of the UNDERWORLD, a place where at least one twin may often expect to be enthroned or immured – the Hindu charioteers called the Asvins, the similar Greek charioteers Castor and Polydeuces, and their Roman equivalents, Castor and Pollux.

In myth and RELIGION and LEGEND, twins like Eve and LILITH, Esau and Jacob, Jesus and Thomas called Didymus, Zethus and Amphion, and Romulus and Remus are frequently distinguished from one another by the fact that one is human, the other supernatural. In this fashion, their compact is a form of communication between this world and another (◊ AS ABOVE, SO BELOW).

Examples of fantasies that have much to say about twins include: Charlotte BRONTË's *The Spell* (**1931** chap); Alexandre DUMAS's *The Corsican Brothers* (**1844**), in which Siamese twins are psychically linked, at a distance, and through extraordinary ordeals; David H. KELLER's *The Homunculus* (**1949**), in which PAN and LILITH are emblematical twins; H. Burgess DRAKE's *Hush-A-Bye Baby* (**1952**), in which unborn twins haunt their mother; Thomas TRYON's *The Other* (**1971**); many of the works of Vladimir NABOKOV, either explicitly or hidden within the text, but most specifically *Ada, or Ardor: A Family Chronicle* (**1969**); John DECHANCIE's **Castle Perilous** sequence, in which the legitimate ruler of the central EDIFICE and the Dark Lord who opposes him are twins; much of Steve ERICKSON's fiction; Raymond E. FEIST's *Faerie Tale* (**1988**), in which one twin is replaced by an ominous CHANGELING; Tamora PIERCE's *Lioness Rampant* (**1988**), in which twinning and GENDER DISGUISE intertwine; Elizabeth HAND's *Winterlong* (**1990**), where twins circle one another in moves that deliberately echo William SHAKESPEARE's *Twelfth Night* (performed *c*1602; **1623**); Victor KELLEHER's *Brother Night* (**1990** UK); Morgan LLYWELYN and Michael SCOTT's *Silverhand: The Arcana* (**1995**); and Tim POWERS's *Expiration Date* (**1995**). Movies and tv find the visual opportunities of twinning particularly compelling: examples include BEWITCHED (*1945*), I DREAM OF JEANNIE (1965-70), OH GOD! YOU DEVIL! (*1984*) – where GOD and SATAN are both played by George Burns – and *The* DARK HALF (*1991*).   [JC]

See also: DOPPELGÄNGERS; DARK LORD; DOUBLES; MIRRORS; PARODY; SHADOWS.

2AM ◊ MAGAZINES.

TWO FACES OF DR JEKYLL, THE UK movie (*1960*). ◊ JEKYLL AND HYDE MOVIES.

TWO WAGS ◊ John Kendrick BANGS.

UCHU SENKAN YAMOTO Japanese tv series (1974 onwards) and movie (*1977*). ◊ ANIME.

UFOS Unidentified flying objects – flying saucers – are, despite tabloid portrayals of sf fans, quite rare artefacts in sf stories outside a brief period during the 1950s when they were simply part of the cultural backdrop: the USA as a whole was in the middle of a "saucer flap" from which it has never entirely recovered. The reason that sf largely ignores ufos is not hard to appreciate: by its very nature the ufo, when viewed as an alien spacecraft, has no more scientific standing than a ghost or a goblin. Many modern researchers in the field believe that ufos are in fact psychological rather than physical phenomena, and have circumstantial evidence to back this up: we *need* ufos, is the claim, in the same way that we *need* GOD. In earlier ages mysterious lights in the sky were explained as WITCHES on broomsticks, while during the 19th century there were accounts of mysterious airships – as if people chose explanations that were always a step ahead of currently possible human technology (in the case of witches, of course, the technology was MAGIC). It is thus more rational, when discussing the deployment of ufos in fiction, to regard them as fantasy, rather than sf, devices.

Also, there is a very strong link between ufos and FAERIE. People who have given accounts of their meetings with ufonauts have produced descriptions markedly resembling those that folklore ascribes to FAIRIES of various kinds; those who claim to have been abducted in a ufo often complain of suffering a time distortion indistinguishable from TIME IN FAERIE, although frequently the imbalance is reversed (they were in the ufo for a long time but, on being returned to the mundane world, discovered that only moments had passed). It is not a surprise that a very famous US ufo case is universally known as that of the Hopkinsville Goblins.

Fantasy writers generally use ufos with caution. Bernard KING, in his **Chronicles of the Keeper** series, blends them with other devices of fantasy and supernatural fiction in order to derive a basic rationale for virtually *all* of the unknown. In *Out of Their Minds* (1969) Clifford D. SIMAK has a DEVIL protesting that humans have infested their subjective REALITY with such TECHNOFANTASY items rather than traditional figures like, say, himself. Garfield REEVES-STEVENS tackles the topic in his *Nighteyes* (**1989**). In Stephen KING's *The Tommyknockers* (**1987**) the occupants of a long-ago-crashed ufo are envisaged as sources of techno-magical EVIL. There have been a few notable screen treatments, especially *Escape to Witch Mountain* (*1974*) and its sequel, *Return from Witch Mountain* (*1978*) – in both of which it becomes apparent that, while the superficial explanation as to why two children have TALENTS is that they are lost members of a ufonaut culture, the subtext is that they are really denizens of a technofantasy Faerie who have become stranded in our world – and *The Flipside of Dominick Hide* (*1980* tvm), which is best read not literally (time machines [◊ TIME TRAVEL] look like flying saucers, and a time traveller returns to present-day London) but as a sort of ALLEGORY about innocents abroad, a notion linked with some of the ideas of C.G. Jung on the subject of ufos (◊ JUNGIAN PSYCHOLOGY). The most famous of all movie treatments of the theme, Steven SPIELBERG's *Close Encounters of the Third Kind* (*1977*), while generally treated as sf, is again much more easily interpreted as technofantasy: the moment of RECOGNITION as the vast spacecraft hoves into view is one of the most powerful in all CINEMA, as the GODS, who are also CHILDREN, arrive to offer humanity salvation. [JG]

UGLY DUCKLING Hans Christian ANDERSEN's "The Ugly Duckling" (1845) is a classic FABLE: an outsized, misfit "duckling" is rejected by other ducks – even his own supposed mother – but after a kind of NIGHT JOURNEY is greeted into the community of swans, since, now mature, he is indeed a handsome swan. The UD tale typifies the METAMORPHOSIS which comes naturally from within, given time – growing-up as a RITE OF PASSAGE (◊◊ MENARCHE). This maturing may be no more than a change in attitude or PERCEPTION: CINDERELLA realizes she may indeed be desirable, and the girl called Thing in Mary BROWN's *The Unlikely Ones* (**1986**) learns that the "ugliness" she hides behind her MASK is a DELUSION; gawky HIDDEN MONARCHS gain confidence, authority and *gravitas*; CHANGELINGS and outcasts need to rediscover their own folk, even if only a PARIAH ELITE; other young misfits grapple with emerging TALENT which may lead them to become full-blown WITCHES or WIZARDS. Swanhood comes in many forms. [DRL]

UGLY DUCKLING, THE UK movie (*1959*). ◊ JEKYLL AND HYDE MOVIES.

UHER, ŠTEFAN (1930-1993) Slovak moviemaker. ◊ SLOVAKIA.

ULYSSES ◊ ODYSSEUS.

UMANSKY, KAYE (?   -?   ) US writer. ◊ CHILDREN'S FANTASY.

UNCANNY Though often used to describe anything strange and unusual, the word strictly means something outside our knowledge, the same as "beyond our ken", and thus not necessarily supernatural. Tzvetan TODOROV's continuum of the FANTASTIC placed the uncanny at the most rationalized end of the scale (◊ RATIONALIZED FANTASY). However, because the uncanny is beyond our understanding it brings with it obvious connotations of fear, and the term is thus frequently used in relation to both HORROR and SUPERNATURAL FICTION. [MA]

UNCANNY TALES ◊ MAGAZINES.

UNDEAD ◊ VAMPIRES; ZOMBIES.

UNDERGROUND Underground regions where the dead go – like HELL – are normally described as UNDERWORLDS. In sf underground passages tend to lead downwards towards ancient artefacts, evidences of alien inhabitation, LOST-RACE domains or even the HOLLOW EARTH. In HORROR that which is underground has usually been there for a considerable time (◊ CTHULHU MYTHOS; ELDER RACES; MALIGN SLEEPER; TIME ABYSS), and its irruption into the daylight world is likely to constitute a desecration. In SUPERNATURAL FICTION underground caverns, hollows and tunnels may be places to take refuge and to recoup one's strength. In fantasy, what is underground may also explicitly MIRROR that which is open to the air: most fantasy EDIFICES show some vertical symmetry, with towers being echoed by underground structures – as with Jordan College (Oxford) in Philip PULLMAN's *Northern Lights* (**1995**). What is underground in fantasy may well be LITTLE BIG; and passages which expand, or continue for distances greater than seems possible, may also lead to PORTALS which deposit one in worlds of light. In fantasy, underground is the SHADOW of overground. [JC]

UNDERHILL, EVELYN (1875-1941) UK writer and mystic, briefly a member of A.E. WAITE's continuation of the GOLDEN DAWN. Most of EU's writings were on religious topics although she is best remembered for her books on mysticism: *Mysticism* (**1904**), *The Mystic Way* (**1913**) and *Practical Mysticism* (**1914**). EU also explored the subject in a series of novels. *The Grey World* (**1904**; vt *The Gray World* 1904 US) deals with the AFTERLIFE and REINCARNATION – a young boy dies and finds himself in the eponymous hinterland of death. His desire for life brings him back to the living but with memory of his experiences. *The Lost Word* (**1907**) and *The Column of Dust* (**1909**) both concern the study of ritual MAGIC and power over the spirit world through the use of sounds and SPELLS. The last book bears comparison with *The Human Chord* (**1910**) by Algernon BLACKWOOD. [MA]

UNDERLIERS In fantasy, many stories are built upon tales TWICE-TOLD (◊◊ REVISIONIST FANTASY) and upon characters who have long existed in the CAULDRON OF STORY. It is thus tempting to suggest that modern fantasy can be understood in terms of its working-out of primal ARCHETYPES (◊◊ ICONS) – although this is by no means accepted by all students of the field, and anyway such an interpretation makes the demand on many individual stories than may have been intended by their writers. Where a modern work makes literary play with the cauldron of story, the term "underlier" is useful.

The underlier GODS of the Greek and Roman pantheon (including in particular APOLLO, DIONYSUS, the TRICKSTER god HERMES, and ZEUS; ◊◊ GREEK AND LATIN CLASSICS) frequently shape our perceptions of contemporary figures; and fantasy writers, when they evoke various goddesses (like APHRODITE or Isis), very frequently intend readers to sense, within these figures, some hint of the GODDESS. Gods out of other pantheons are frequently used in CELTIC FANTASY and NORDIC FANTASY as underliers. The HEROES of MYTHOLOGY – GILGAMESH, ODYSSEUS, etc. – also serve. Characters of out LEGEND and FOLKLORE and FAIRYTALE figure again and again: ARTHUR, CINDERELLA, the FAIRY QUEEN, FAUST, the FLYING DUTCHMAN, the FROG PRINCE, OBERON, ROBIN HOOD, Rumpelstiltskin, SANTA CLAUS, SCHEHERAZADE, Sinbad, SLEEPING BEAUTY, SNOW WHITE, TAM LIN, the WANDERING JEW . . . the list is both loose and nearly endless.

Underlier figures who first came into existence through the works of individual authors are just as likely to be found, particularly in TAPROOT TEXTS. The VIRGIL who appears in DANTE's *Divine Comedy* (written *c*1320) is one, as are various characters created by William SHAKESPEARE – like ARIEL, CALIBAN, Falstaff, Hamlet and PROSPERO – the FRANKENSTEIN monster of Mary SHELLEY and many of the characters created by J.R.R. TOLKIEN. Again the list is enormous.

Types of character – from the scapegoat to the UGLY DUCKLING, from the types who animate the COMMEDIA DELL'ARTE to the COMPANION in HEROIC FANTASY, from unicorn to ELF – also serve to underlie fantasy tales. [JC]

UNDER THE SEA The undersea realm is, in a sense, an OTHERWORLD and the water's surface a PORTAL; the world below is populated by SEA MONSTERS and MERMAIDS, and littered with forgotten treasures, including the remains of sunken LOST LANDS like ATLANTIS and LEMURIA. Edgar Allan POE's "The City in the Sea" (1831) provides an influential image of deep-sea decay and THINNING. The chemicals of seawater are reflected in blood and amniotic fluid: SEX with undersea beings has a peculiar fascination and horror, as with H.P. LOVECRAFT's miscegenated Deep Ones in *The Shadow Over Innsmouth* (**1936**) and the more positive matings of Sterling Lanier's "The Kings of the Sea" (1968) (◊◊ SELKIES). The sea monster Abaia in Gene WOLFE's *The Book of the New Sun* (**1980-3**) is named for a mythical and decidedly phallic giant eel. When underwater CITIES and realms are still going concerns, the surrounding fluid provides a DREAM-like quality – as in the enchanted undersea episodes of John MASEFIELD's *The Midnight Folk* (**1927**) and Tanith LEE's *The Dragon Hoard* (**1971**), or the pursuit down a submerged stairway to Rebma, the pelagic DOUBLE of Amber, in Roger ZELAZNY's *Nine Princes in Amber* (**1970**). E. NESBIT's *Wet Magic* (**1913**) hinges on a magical undersea LIBRARY. [DRL]

See also: ISLANDS.

UNDERWORLD A term sometimes used to describe those parts of a CITY occupied by criminals or an underclass or – in fantasy texts – WAINSCOT societies which may not be human at all. More frequently what is being described is the land of the dead. It is usually thought to lie beneath the world, and is often called HADES, after the Greek god who rules it, or HELL; it may PARODY the world of the living, or work as a MIRROR (◊◊ AS ABOVE, SO BELOW), or serve as a venue for AFTERLIFE existences, some of them – as in John Kendrick BANGS's **Riverboat** sequence – perfectly pleasant. [JC]

UNDINE A water nymph belonging to one of the four classes of ELEMENTALS identified by PARACELSUS. One such is the central figure of Friedrich de la Motte FOUQUÉ's classic Romantic fantasy (◊ ROMANTICISM) *Undine* (**1811**); the use of the word as a proper name seems to have discouraged subsequent use of the generic term. The first cinematic treatment was *La Légende des Ondines* (**1910** France), followed by a number of US versions of Fouqué's tale: *Neptune's Daughter* (**1912**), *Undine* (**1912**) – a 35min short – and *Undine* (**1915**). *The Royal Ballet* (**1959** UK) features Margot Fonteyn dancing *Ondine* by Hans Werne Henze; *Sea Shadow* (**1965** US) is a 16mm ballet short based on Fouqué. [BS/JF]

UNICORN A noted MYTHICAL CREATURE, perhaps based on conflated TRAVELLERS' TALES of the rhinoceros and of the narwhal. The single horn has great powers of HEALING but is also a deadly weapon: one of the BRAVE LITTLE TAILOR's exploits was to trick a unicorn into fixing its horn in a tree when trying to impale him. More usually, VIRGINITY is required to catch the unicorn, though in Theodore STURGEON's "The Silken-Swift" (1953) the creature rejects a technical virgin to favour a raped woman who is a true innocent. Probably the finest fantasy to focus on a unicorn is Peter S. BEAGLE's *The Last Unicorn* (**1968**). Pier ANTHONY's **Apprentice Adept** sequence features an extensive herd hierarchy of musical unicorns who are also SHAPESHIFTERS. Relevant collections are *Unicorns!* (anth **1982**) and *Unicorns II* (anth **1992**) ed Jack DANN and Gardner Dozois (1947-   ), *The Unicorn Treasury* (anth **1988**) ed Bruce Coville and *Imortal Unicorn* (anth **1995**) ed Peter S. BEAGLE and Janet Berliner. [DRL]

UNKNOWN US pulp MAGAZINE, 39 issues, March 1939-October 1943, monthly until December 1940, then bimonthly, retitled *Unknown Worlds* from October 1941, published by Street & Smith, New York; ed John W. Campbell Jr (1910-1971).

Along with WEIRD TALES, this was one of the most influential of all fantasy magazines, and in content superior to its rival. *U* published more quality fantasy per issue than any other magazine, and reawakened public interest about the genre. Legend has it that the magazine was launched in order to publish *Sinister Barrier* (1939; **1943**; rev 1948) by Eric Frank Russell (1905-1978), but in fact Campbell had for a while considered the need for a magazine to cater for the good stories he was receiving at *Astounding Science-Fiction* which were insufficiently technocentric for his purposes. As with *Astounding*, he endeavoured to establish a core of writers for *U*; his two major contributors became L. Ron HUBBARD and L. Sprague DE CAMP. Hubbard produced 14 stories, of which eight were lead novels, including his classic PSYCHOLOGICAL THRILLER *Fear* (1940; **1957**). Whereas Hubbard usually cast his protagonists into DREAM worlds of their own making, de Camp and Fletcher PRATT – starting with "The Roaring Trumpet" (1940), later assembled with "The Mathematics of Magic" (1940) as *The Incomplete Enchanter* (fixup **1947**) – sent their hero **Harold Shea** into ALTERNATE WORLDS based on different national MYTHOLOGIES. In de Camp's and Pratt's worlds the fantasies were based on some external logic, whereas in Hubbard's they relied solely on internal logic.

Either way, Campbell sought to ensure the fantasy elements in *U* obeyed some set of laws, in effect treating the supernatural as another science (◊ RATIONALIZED FANTASY). He worked with Robert A. HEINLEIN to produce the extreme example of this in "The Devil Makes the Law" (1940; vt "Magic, Inc." in *Waldo and Magic, Inc.* coll **1950**), in which MAGIC is an inextricable part of life's bureaucracy. When Jack WILLIAMSON produced a WEREWOLF story, *Darker Than You Think* (1940; exp **1948**) he treated the subject scientifically, introducing a new evolutionary strain of humanity. The same attention to logic and law was made by Fritz LEIBER in his tale of modern WITCHCRAFT, *Conjure Wife* (1943; **1953**).

Some of the classic shorter fiction published in *U* included: "Trouble With Water" (1939) by H.L. GOLD; "The Cloak" (1939), a modern VAMPIRE story by Robert BLOCH; "Two Sought Adventure" (1939) by Leiber, his first appearance in print and the start of the **Fafhrd and Gray Mouser** series; "When It was Moonlight" (1940) by Manly Wade WELLMAN, a RECURSIVE FANTASY featuring Edgar Allan POE; "It" (1940) and "Yesterday was Monday" (1941), the latter a humorous WAINSCOT story (*U*'s stories were full of wainscots involving imps, gnomes, etc.), both by Theodore STURGEON; "They" (1941) by Heinlein; and "Smoke Ghost" (1941) by Leiber, arguably the first seriously modern GHOST story. There were many other excellent stories by Robert ARTHUR, Nelson S. BOND, Anthony BOUCHER, Fredric BROWN, Cleve Cartmill (1908-1964), Lester del Rey (1915-   ), Henry KUTTNER, Frank Belknap LONG, P. Schuyler Miller (1912-1974), Jane Rice and A.E. VAN VOGT.

U acquired for fantasy a respectability that all but its most exalted forms – e.g., the works of James Branch CABELL, Lord DUNSANY and Thorne SMITH – had hitherto lacked, and by treating it seriously gave it a new credibility. *U* established for some writers (especially de Camp and Hubbard) a milieu in which to shine, and boosted the careers of some writers (such as Leiber) who might otherwise have been relegated to lesser markets. It was through *U* that a commercial genre of fantasy was created, though it would be another 20 years before it established itself in the market.

U had to be sacrificed during WWII to ensure adequate paper supplies for *Astounding*. Interestingly, the UK reprint (abridged) edition of *U* rationed the material for longer and survived September 1939-Winter 1949 (41 issues). Plans to reissue *U* after WWII were aborted, although a retrospective ANTHOLOGY, *From Unknown Worlds* (anth **1948**; cut 1952 UK) ed Campbell, was issued to test the market. Further anthologies based on the magazine are *The Unknown* (anth **1963**) ed D.R. Bensen (1927-   ) and *The Unknown Five* (anth **1964**) ed Bensen, *Hell Hath Fury* (anth **1963**) ed George Hay (1922-   ), *Unknown* (anth **1988**) ed Stanley Schmidt (1944-   ), and *Unknown Worlds: Tales from Beyond* (anth **1988**) ed Schmidt and Martin H. GREENBERG. [MA]

Further reading: *The Annotated Guide to Unknown & Unknown Worlds* (**1991**) by Stefan DZIEMIANOWICZ.

UNKNOWN WORLDS (magazine) ◊ UNKNOWN.

UNSEEN PLAYMATE Robert Louis STEVENSON's preferred term for INVISIBLE COMPANION.

UNUSUAL STORIES ◊ MAGAZINES.

UPDIKE, JOHN (HOYER) (1932-   ) US writer, best-known for a large body of work outside fantasy, but who has veered into FABULATION, sometimes with an sf coloration. Although several of his novels are only superficially fantasies, *The Witches of Eastwick* (**1984**), filmed as *The* WITCHES OF EASTWICK (**1987**), a tale of three WITCHES in a small New England town and the diabolical person who beguiles them, is – however much it asks to be read as a sly ALLEGORY of newly empowered women and immemorially seductive men – sufficiently attentive to its narrative implications to be a true fantasy. *Brazil* (**1994**) retells the story of Tristan and Iseult (◊ ARTHUR) in modern Brazil in a magic-drenched manner deliberately evocative of MAGIC REALISM. [GF]

URBAN FANTASY A CITY is a *place*; urban fantasy is a *mode*. A city may be an ICON or a geography; the UF recounts an experience. A city may be seen from afar, and is generally seen clear; the UF is told from within, and, from the perspective of characters acting out their roles, it may be difficult to determine the extent and nature of the surrounding REALITY. UFs are normally texts where fantasy and the mundane world intersect and interweave throughout a tale which is significantly *about* a real city. There are many exceptions – Mary GENTLE's *Rats and Gargoyles* (**1990**), is set in a fantasy city "that is called the heart of the world" – but the general principle still holds: the city of a UF may be located in a SECONDARY WORLD, but in such a case it has been created not just as a backdrop but as an environment, as in Simon R. GREEN's **Hawk & Fisher** series.

TAPROOT TEXTS from which the UF evolved are not easy to find before the 18th century; perhaps the only example worth noting is the *Satyricon* by Petronius Arbiter (? -66), though some of the more mercilessly organized post-Renaissance utopian cities – like the circular example in Tommasso Campanella's *City of the Sun* (**1623**) – may have provided significant counterexamples to writers a few centuries later. Many of the stories assembled in *The Arabian Nights* (◊ ARABIAN FANTASY) are set in BAGHDAD and CAIRO, but use these cities only as backdrop.

It is reasonable to argue that UFs derive primarily from the notion of the EDIFICE, and edifices came into true literary existence only with *The Castle of Otranto* (**1765**) by Horace WALPOLE (◊ GOTHIC FANTASY). The headings under which Frederick S. FRANK anatomizes the form in *The First Gothics* (**1987**) also work to describe the early forms of UF: claustrophobic containment (◊ BONDAGE); subterranean pursuit; supernatural encroachment (◊ SUPERNATURAL FICTION); "extraordinary positions" and lethal predicaments; abeyance of rationality; possible victory of EVIL (◊ PARODY); supernatural gadgetry, contraptions, machinery, and demonic appliances; and "a constant vicissitude of interesting passions".

Early UFs tended to be described in terms beholden to the *Carceri d'Invenzione* (**1749-50**) and *Vedute* (**1745-78**) of Giovanni Battista PIRANESI – two sets of drawings in which urban scapes are seen in unmistakably theatrical terms. The first set depicts shadowy, illimitably complex imaginary prisons; the second confabulates ancient and modern Rome in images whose chiaroscuros are haunted and echoic. Piranesi was deeply influential in shaping the early 19th century's sense of the nightmare of the city, and remains important still, as demonstrated by explicit references in various texts (two instances being Russell HOBAN's *The Medusa Frequency* [**1987**], in which Piranesi is evoked to describe a Soho [◊◊ LONDON] that verges on the UNDERWORLD, and Lucius SHEPARD's *The Golden* [**1994**], where Castle Banat is seen in Piranesan terms). Piranesi's LABYRINTHS and abysses suggest pantheons of dark architects, innumerable troupes of players.

As it is a story centrally involving an edifice almost indistinguishable from the city whose heart it comprises, the novel which most clearly marks the relationship between edifice and UF is *Notre-Dame de Paris* (**1831**; trans Frederick Shoberl as *The Hunchback of Notre-Dame* **1833** UK) by Victor Hugo (1802-1885). This novel is not fantasy, but it was a seminal influence upon Eugène SUE, whose *Les mystères de Paris* (**1844**; trans anon as *The Mysteries of Paris* **1844** UK) is a full-fledged UF in everything but the absence of a specific supernatural element, though its central avenger hero – Rudolf von Gerolstein – is a superman figure whose roots in LEGEND lie very deep, for his ability to appear in any disguise makes him a kind of benign TRICKSTER. More importantly, the novel popularized the kind of multilayered episodic "Mysteries" plot that Charles DICKENS had already experimented with in novels like *Oliver Twist* (**1839**), which is set in London, and which Alexandre DUMAS would bring to perfection in *The Count of Monte Cristo* (**1846**), much of which takes place in PARIS.

Dickens and Sue both tended to imagine internal kingdoms within the city (made up of anything from criminal societies to collaborating lawyers) which operated as MICROCOSMS and PARODIES of the larger REALITY; whose inhabitants were described in terms as alienated, and as fascinated, as those used to describe the native nations of the Americas or Africa; and which were often physically interconnected, by tunnels and secret passages, with that larger world. These interior worlds, in works by both authors, tended to contain characters who were themselves SHADOWS of the makers and rulers of the ostensible world above (◊◊ AS ABOVE, SO BELOW), just as the subterranean chambers beneath the city shadowed the palaces which lorded its heights. Disguises were rife, as were melodrama and a fascinated grappling with the sense that whole segments of the city – whole populations – could be described according to their class, and were indeed frequently seen as fitting into extended panoramas illustrative of urban life. These panoramas enforce a sense that the work of both authors was clearly shaped by a central ongoing analogy between the city and the hierarchical world of the theatre, a sense that the city was not only a CAULDRON OF STORY but more specifically both setting for and participant in the telling of tales about a reality that could only be perceived as layered, interwoven, colour-coded (◊ COLOUR-CODING) and allegorical.

Much of Dickens's fantasy – most notably *A Christmas Carol* (**1843**) – was set in London, which with Paris has remained a central site for the UF mode, both being venues so irradiated with story and mystery that tales set there often have an air of the fantastic without in fact invoking the impossible. As the century passed, fantasy writers – from Robert Louis STEVENSON through G.K. CHESTERTON and beyond – made particularly intense use of London, as did the artist Gustave DORÉ; as the 20th century nears its end, both STEAMPUNK and GASLIGHT-ROMANCE texts tend to focus upon this city, possibly as a result of continued international fascination with the SHERLOCK HOLMES mythos, but they are by no means on their own: Peter ACKROYD's *Dan Leno and the Limehouse Golem* (**1994**; vt *The Trial of Elizabeth Cree: A Novel of the Limehouse Murders* 1995 US) and *The Great Fire of London* (**1982**) occupy the same territory, though not inherently fantasy (while at the same time not *not* fantasy), and Christopher FOWLER's *Roofworld* (**1988**) and Michael DE LARRABEITI's **Borribles** sequence are particularly powerful evocations of the WAINSCOTS that every Londoner knows exists but has seen, as it were, only out of the corner of an eye. Michael MOORCOCK, too, seems captivated by the locale, as in his FABULATION *Mother London* (**1988**). Paris is still an important UF locale, as it has been ever since Honoré de BALZAC's *La Peau de chagrin* (**1831**; trans as *Luck and Leather: A Parisian Romance* **1842** US; vt *The Wild Ass's Skin* **1888**; new trans Katharine Prescott Wormeley as *The Magic Skin* **1888** US); Gaston LEROUX's *The Phantom of the Opera* (**1910**) is steeped not just in the Opéra but in Paris itself. Other cities significant for the UF subgenre include Chicago – Martin H. GREENBERG's *Fantastic Chicago* (anth **1991**) is a typically specific anthology – Liverpool (primarily because of the work of Ramsey CAMPBELL), LOS ANGELES, MINNEAPOLIS, NEW YORK (most stunningly in Mark HELPRIN's *Winter's Tale* [**1983**], although HIGHLANDER [*1986*] also exploits its setting expertly and Thomas PYNCHON's *V* [*1963*] can be regarded as the paradigmatic "urban fabulation" – *Newer York: Stories of Science Fiction and Fantasy About the World's Greatest City* (anth **1991**) ed Lawrence WATT-EVANS is of interest), PRAGUE (whose environment hugely influenced Franz KAFKA and which of course was the setting for Gustav MEYRINK's *The Golem* [**1915**]), SAN FRANCISCO, Seattle (as in Megan LINDHOLM's *The Wizard of the Pigeons* [*1986*]), Toronto (a favourite venue for Charles DE LINT and other SCRIBBLIES), VENICE (the Daphne DU MAURIER story and the movie based on it, DON'T LOOK NOW [*1973*], are noteworthy) and Vienna (often used by Jonathan CARROLL).

As the 20th century has advanced, UF writers have intensified the model developed by Dickens and Sue. FANTASIES OF HISTORY are often set in cities, and CONTEMPORARY FANTASIES and INSTAURATION FANTASIES are also set more often than not within urban surrounds. Fantasies which may be describable under any or all of these three rubrics tend varyingly to share certain structural elements that make cities their natural venue: they tend to CROSSHATCH the mundane world with OTHERWORLDS, often locating within cities the PORTALS through which such intermixings are announced; they tend to emphasize the consanguinity not only of intersecting worlds, but of peoples, times (◊ TIMESLIP; TIME TRAVEL) and stories (◊◊ URBAN LEGENDS) as well; they tend – fairly enough – to treat the late 20th century as an essentially urban drama, so that conflicts within the city resonate throughout the worlds; and, like most fantasies, tend to try to achieve a sense of HEALING.

Imagined cities in which UFs are set are legion, from BATMAN's Gotham City (a much fantasticated New York) to the New Zealand city in *The* NAVIGATOR: A MEDIEVAL ODYSSEY (*1988*) to Fritz LEIBER's Lankhmar to Terry PRATCHETT's Ankh-Morpork, scene of the partly UF **City Guard/Night Watch** sequence. The mirroring cities of Gene WOLFE's *There Are Doors* (**1988**) are never identified, either in "this world" or "the other", but they are fully realized in a way that makes this one of the most profoundly urban of all UFs.

There is an increasing sense that writers may well be conceiving the typical inhabitant of the great cities as a kind of hunter-gatherer figure, one better able than suburbanites or farmers to cope with the crack-up of the immensely rigid world system created over the previous few thousand years. The urban venues of many cyberpunk novels, from William Gibson's *Neuromancer* (**1984**) on, seem in this light almost to read as Books of Instruction for survival in the new forest. [JC]

See also: Peter S. BEAGLE; Charles Brockden BROWN; Emma BULL; Peter CONRAD; *The* LAND OF FARAWAY (*1987*); David LYNCH; NIGHT CRY; PAN; J. Michael REEVES; STEAMPUNK; Jeanette WINTERSON; WOLF (*1994*); WOLFEN (*1981*).

URBAN LEGENDS Modern oral FOLKLORE – also spread via newspapers, photocopier graffiti and, increasingly, the Internet. The UL is typically a TALL TALE with a *frisson* of comeuppance or HORROR, related as having actually happened to a "friend of a friend" – usefully abbreviated to "foaf" by Rodney Dale in his *The Tumour in the Whale: A Collection of Modern Myths* (**1978**), whose title commemorates the WORLD WAR II UL of whale meat (offered as a substitute for beef) which proves to contain a live, pulsing tumour. ULs can feed into fantasy, and vice versa. Randall GARRETT builds on the belief that the image of a murderer can be photographed from the dead man's retina in "The Eyes Have It" (1964), a notion already used by Rudyard KIPLING; Harlan ELLISON makes creative use of the UL alligators infesting New York sewers, in his "Croatoan" (1975), as does Thomas PYNCHON in his FABULATION *V* (**1963**); Terry PRATCHETT's *Feet of Clay* (**1996**) spoofs ULs told against ethnic restaurants, when outraged DWARF customers discover the speciality dish, rat, to be disguised chicken.

Conversely, the title UL of *The Vanishing Hitchhiker: American Urban Legends and Their Meanings* (**1981**) by Jan Harold Brunvand (1933-   ), whose eponymous female GHOST disappears from a moving car, is so wearily reminiscent of innumerable GHOST STORIES that it cannot be told as fiction – only the spice of "foaf" pseudo-factuality gives it life. There are ULs of ghostly truckers driving immaterial vehicles along US Interstates, updating the FLYING DUTCHMAN; Steven SPIELBERG's *Duel* (*1971* tvm) owes something to this UL. The Greek MYTH of the poisoned shirt of Nessus resurfaced as a 1930s-40s UL involving a fatal dress, allegedly tainted with embalming fluid from a corpse and unscrupulously resold to a woman who died from wearing it. A classic case of a story published as fiction but retold as UL (often with indignant insistence on its truth) is Arthur MACHEN's "The Bowmen" (1914), which gave rise to the ANGELS OF MONS myth. [DRL]

Further reading: *The Natural History of Nonsense* (**1947**) by Bergen Evans; *The Book of Nasty Legends* (**1983**) by Paul Smith; *The Choking Doberman and Other "New" Urban Legends* (**1984**) and several further compilations by Jan Harold Brunvand.

See also: SPRING-HEELED JACK.

UROBOROS-SERPENT ◊ WORM OUROBOROS.

UTOPIAS Utopias have engaged a wide range of interests over many years, and so it is not surprising to find that attempts to make generic sense of the term have generated a considerable discord of usages. Richard Gerber's classic bibliographical study of the form may be entitled *Utopian Fantasy* (**1955**; exp 1973 US), but more recent scholars – like Brian ATTEBERY in "Fantasy as an Anti-Utopian Mode" (in *Reflections on the Fantastic* anth **1986** ed Michael R. Collings) – argue persuasively that, although fantasy and utopias are related, neither can plausibly be treated as a version of the other.

Set down as a fictional narrative or presentation, a utopia can be defined as a rendering in bodily form of an argument about society; but that argument, however extreme the results may be, must have some rational connection with the normal world. Even the *location* of a utopia, in space or time, constitutes an argument of connection or continuity. The utopias of previous centuries tended to be spatially displaced from normal societies but contemporaneous with them; those of the past century or so have tended to be set in the future, in an argument that *this* may possibly lead to *that*. These ideal societies tend to be profoundly *understandable* in terms of the rules that govern their creation, rules which are meant to present ideal solutions to the problems of how to organize human society (or, in the case of dystopias [sometimes called cacotopias], warnings against mistaken ideal solutions). Successful utopias have, therefore, little or no room for STORY.

The only fantasies that even superficially resemble the argued compact of a utopia are the rule-bound TOPSY-TURVY fantasies common to 19th-century writers like W.S. GILBERT and Lewis CARROLL. Certainly utopian societies exist in fantasy novels, especially those that feature one or more significant CITIES; but to say that utopias can be found within the overall structure of a fantasy tale does not mean that the fantasy *is* a utopian tale. The essential movement of the utopia is forwards, toward organized betterment of the world to come, or backwards to the present world, in order to illustrate arguable claims about the nature of this world; but the essential movement of fantasy is – as Attebery suggests – inwards, towards the HEALING of the land. Moreover, fantasy is impossible by nature; utopias are impossible only if they don't work.

Fantasies which incorporate (but ultimately dissolve) utopian worlds include L. Frank BAUM's **Oz** sequence (**1900-20**), Herbert READ's *The Green Child: A Romance* (**1935**), L. Sprague DE CAMP's "The Undesired Princess" (1942 *Unknown*; in *The Undesired Princess* coll **1951**) and Robert Graves's *Watch the North Wind Rise* (**1949** US; vt *Seven Days in New Crete* 1949 UK). Utopias – or, more frequently, dystopias – told in FABLE form through animal viewpoints are best understood as BEAST-FABLES or AESOPIAN FANTASIES; George ORWELL's *Animal Farm: A Fairy Tale* (**1945** chap) and Richard ADAMS's *Watership Down* (**1972**) exemplify. [JC]

UTTLEY, ALISON Working name of UK children's writer Alice Jane Uttley (1884-1976). She was the second woman honours graduate of Manchester University, gaining a BSc in physics in 1909. Her first book, written despite the disapproval of her husband, was *The Squirrel, The Hare and the Little Grey Rabbit* (**1929**). After his death she supported herself and her only son by her writing, publishing over 30 **Little Grey Rabbit** tales and over 100 novels and collections in all, many using as background the Victorian farming community in Derbyshire where she grew up. Her most famous and successful novel for older children, *A Traveller in Time* (**1939**), is a fantasy based again on her rural childhood, concerning a local plot to murder Mary Queen of Scots in 1569 and a girl from the 20th century who uncontrollably TIMESLIPS back 400 years to witness the tragedy. It reflects Uttley's interest in dreams, also discussed in her autobiography, *The Stuff of Dreams* (**1953**). [JF]

VACE, GEOFFREY [s] ◊ Hugh B. CAVE.

VALHALLA ODIN's feasting-hall in Asgard. Its walls are spears, its roof golden shields. Valkyries swoop on the battlefield choosing who shall die. Half go to Freya, GODDESS of LOVE. Women follow their men into battle or kill themselves to join them. The rest of the slain go to Odin's Valhalla to form his army, the Einheriar. Daily, they ride out to fight. At night their wounds are healed. They drink mead from a goat which feeds on Yggdrasil (◊ WORLD-TREE) and eat the same boar every night. Men dying in bed ask to be marked with a spear, to gain Valhalla and avoid the goddess Hel (◊ HELL). [FS]

See also: AESIR; NORDIC FANTASY.

VALLEJO, BORIS (1941-) Illustrator of exotic/erotic fantasy subjects with a style heavily influenced by that of Frank FRAZETTA, although readily distinguishable, being more languid in conception and displaying a more selfconsciously sensual elegance. BV works from photographic reference in oil colours over an acrylic underpainting on gessoed illustration board. His compositions are in the form of posed tableaux, typically featuring idealized nude or near-nude male and female figures.

BV was born in Peru and studied medicine before switching to art. In 1964 he emigrated to the USA and settled in New York, producing covers for WARREN PUBLISHING and for MARVEL COMICS's *Savage Sword of Conan*. He went on to produce cover paintings for fantasy books, including reissues of the **Gor** novels by John NORMAN (7 titles 1976) and the Ballantine reissue of the novels of Edgar Rice BURROUGHS (24 titles 1977), plus a very popular series of picture calendars. He has written two books about his painting techniques – *The Fantastic Art of Boris Vallejo* (graph **1978**) and *Boris Vallejo: Fantasy Art Techniques* (graph **1983**) – and has published several books of his art with poems and prose fiction by his wife, Doris, including *Mirage* (graph **1982**), *Enchantment* (graph **1984**) and *Ladies: Retold Tales of Goddesses and Heroines* (graph **1994**). [RT]

VAMP US movie (*1986*). New World/Balcor. **Pr** Donald P. Borchers. **Dir** Richard Wenk. **Spfx** Chris Chesney, Peter Chesney, Tom Chesney, Jarn Heil, Image Engineering Inc, Joseph Viscosil. **Mufx** Greg Cannom, Pamela S. Westmore.

Screenplay Wenk. **Starring** Sandy Baron (Vic), Billy Drago (Snow), Grace Jones (Katrina), Chris Makepeace (Keith), Deedee Pfeiffer (Amaretto/Alison), Robert Rusler (AJ), Gedde Watanabe (Duncan). 94 mins. Colour.

College boys Keith, AJ and Duncan seek a stripper for the frat-house party. They go to the wrong side of the tracks, where there are more VAMPIRES than humans on the streets. Pursued by the punk vampire gang led by Snow, they reach the After Dark Club, where AJ tries to hire the most exotic stripper of all, Katrina; but the club's staff and artistes are vampires. None of the non-vampires survive except Keith and his forgotten-now-rediscovered childhood friend Alison.

V seems uncertain whether to be a skinflick or a comedy-HORROR MOVIE, and locked in this dichotomy leaves a somewhat nasty taste in the mouth – although, primarily because of Jones, it has a huge cult following. Far better comedy-vampire movies released about the same time were FRIGHT NIGHT (*1985*) and The LOST BOYS (*1987*). [JG]

VAMPIRELLA Slant-eyed, black-haired, voluptuous beauty from the planet Draculon, where blood, rather than water, is the life-sustaining liquid. She wears a revealing red swimsuit with tiny wing collars and a gold bat motif across her crotch, plus calf-length black boots. She was created by Forrest J. Ackerman and artist Frank FRAZETTA, with costume design by Trina Robbins, and first appeared in the comics magazine *Vampirella* in 1969. Her early adventures, in which she comes to Earth to prey on humans, were written, with tongue firmly in cheek, by Ackerman and drawn by Tom Sutton (1937-). In 1971 the artwork was taken over by the Spanish artist José (Pepé) Gonzalez (1943-): in his hands Vampirella became a cuddly sex object, and her popularity increased dramatically among young male readers. Scripts were now written by Archie Goodwin (1937-) and featured the vampire-hunter Adam Van Helsing (a descendant of Bram STOKER's character in *Dracula* [**1897**]), with whom Vampirella eventually developed a romance. Later scripts were by John Cochran, T. Casey Brennan and others, with art occasionally by Gonzalo Mayo and Rudy Nebres. Some of the cover artwork featuring Vampirella, painted by Manuel SANJULIAN, Jordi

Penalva, and Enrich, are virtually ICONS of FANTASY ART. The magazine ceased publication, however, with other Warren magazines in 1983.

Vampirella, however, remained in the popular memory, and was revived in 1991 with *Vampirella: Morning in America (#1-#4)* and a continuing series of new stories began in 1992. Novelizations by Ron GOULART have been *Bloodstalk* * (**1975**), *On Alien Wings* * (**1975**), *Deadwalk* * (**1976**), *Blood Wedding* * (**1976**), *Deathgame* * (**1976**) and *Snakegod* * (**1976**). [RT]

VAMPIRE MOVIES By the early 1900s, the fledgling CINEMA had adapted the term "vamp" or "vampire" to describe such FEMME FATALES of the silent screen as Theda Bara (1890-1955). This resulted in a number of titles which, while sounding like vampire movies, had in fact nothing to do with VAMPIRES. The first movie version of Bram STOKER's *Dracula* (**1897**) might possibly have been a Hungarian feature called *Drakula* (**1921**), but it has been lost. The German actor Max Schreck (1879-1936) became the screen's first vampire when he stalked through F.W. MURNAU's *Nosferatu* (**1922**), an unauthorized 1922 adaptation of Stoker's book (◊ DRACULA MOVIES). Hollywood was not far behind: Lon Chaney Sr (1883-1930) portrayed a bug-eyed, top-hatted vampire in MGM's murder mystery *London After Midnight* (**1927**; vt *The Hypnotist* UK), also now seemingly lost. When Universal decided to film *Dracula* in the early 1930s, its first choice of an actor to portray the Count was Chaney, who would have been reunited with his *London After Midnight* director, Tod Browning (1882-1962). Unfortunately, before the project could be realized, Chaney died from throat cancer. Over the following months, various actors were considered or announced for the role. The studio finally chose Hungarian Bela Lugosi (real name Bela Ferenc Dezso Blasko; 1882-1956) who had played the part on stage since 1927. A superior Spanish version was filmed simultaneously, starring Carlos Villarias. Universal's *Dracula* (**1930**) firmly established the vampire in the public's consciousness, spawned the first great cycle of US HORROR MOVIES and typecast its star for life. The following year, Carl DREYER's VAMPYR (**1932**) was loosely inspired by J. Sheridan LE FANU's story "Carmilla" (1871). When Browning decided to remake *London After Midnight* as *Mark of the Vampire* (**1935**) at MGM he cast Lugosi as an actor disguised as a vampire. Meanwhile, Universal waited six years before making *Dracula's Daughter* (**1936**), in which only Edward Van Sloan (as Van Helsing) returned from the original movie. Although never the actor his father had been, Lon Chaney Jr (real name Creighton Chaney; 1906-1973) carved something of a niche for himself at Universal, playing most of the studio's major monsters. It was almost inevitable that he would be cast as the Count in *Son of Dracula* (**1943**). Universal combined many of its classic creatures in *House of Frankenstein* (**1944**) and *House of Dracula* (**1945**), both featuring Shakespearean actor John Carradine (real name Richmond Reed Carradine; 1906-1988) as a suave count. Lugosi played another Dracula-like vampire in Columbia's entertaining *The Return of the Vampire* (**1943**), and finally returned to recreate his most famous role on screen one last time alongside the slapstick antics of comedians Bud Abbott (real name William Abbott; 1895-1974) and Lou Costello (real name Louis Cristillo; 1906-1959) in *Abbott and Costello Meet Frankenstein* (**1948**) (◊ FRANKENSTEIN MOVIES). Val Lewton's underrated study of vampire mythology, *Isle of*

the Dead (**1945**), benefited from the presence of Boris Karloff (real name William Henry Pratt; 1887-1969) but Hollywood more usually restricted its bloodsuckers to such B-movie mediocrity as *Dead Men Walk* (**1943**), *The Vampire's Ghost* (**1945**) and *Valley of the Zombies* (**1946**). With the sf boom of the 1950s, US bloodsuckers took on distinctly more alien forms in such movies as *The Thing from Another World* (**1951**), *Blood of Dracula* (**1957**; vt *Blood is My Heritage* UK), *Not of This Earth* (**1957**) and *The Vampire* (**1957**; vt *The Mark of the Vampire*). In *The Return of Dracula* (**1957**; vt *The Fantastic Disappearing Man* UK) the count was transported to suburban southern California, while *Curse of the Undead* (**1959**) featured a vampire gunslinger in an incongruous Wild West setting. Mexican actor Germán Robles starred as vampire Count Lavud in both *The Vampire* (**1957**; ot *El Vampiro*) and *The Vampire's Coffin* (**1957**; ot *El Ataud del Vampiro*). Lugosi travelled to the UK to appear in the comedy *Mother Riley Meets the Vampire* (**1952**; vt *My Son the Vampire* US). He died in 1956 and was buried in his Dracula cape and tuxedo. Also in the UK, Hammer Films decided to team its star duo of Peter Cushing (1913-1995) and Christopher Lee (1922-) in *Dracula* (**1958**; vt *Horror of Dracula* US). It was a huge box-office hit, and the studio followed up with eight sequels (◊ DRACULA MOVIES), only two of which do not have Lee as the Count. While Hammer expanded the mythology with *The Kiss of the Vampire* (**1962**; vt *Kiss of Evil*), Lee himself was also not adverse to appearing in European vampire movies like *Uncle Was a Vampire* (**1959**; ot *Tempi duri per i Vampiri*; vt *Hard Times for Vampires* UK), *Hercules in the Haunted World* (**1961**; ot *Ercole al centro della Terra*; vt *Hercules in the Center of the Earth* US), *Crypt of Horror* (**1963**; ot *La Cripta e L'Incubo*; vt *Terror in the Crypt* US) and *The Blood Demon* (**1967**; ot *Die Schlangengrube und das Pendel*; vt *The Torture Chamber of Dr Sadism* US). Italian director Mario Bava used variations on the vampire theme for three of his best movies, *Black Sunday* (**1960**; ot *La Maschera del Demonio*; vt *Revenge of the Vampire* UK), *Black Sabbath* (**1963**; ot *I Tri Volti della Paura*) and *Planet of the Vampires* (**1965**; ot *Terrore nello Spazio*). Roman POLANSKI's dark comedy *The FEARLESS VAMPIRE KILLERS, OR PARDON ME, YOUR TEETH ARE IN MY NECK* (**1967**) was an inspired homage to both the mythology and Hammer, while Carradine recreated his role of a top-hatted Count in *Billy the Kid versus Dracula* (**1965**). During the 1970s, many vampire films added sex and nudity to the horror, from Hammer's lesbian bloodsuckers to the films of European directors Jess (Jesús) Franco and Jean Rollin. Some of the more interesting variations on the theme turned up in *Count Yorga, Vampire* (**1970**), *Daughters of Darkness* (**1971**; ot *La Rouge aux Levres*), *Blacula* (**1972**), *Captain Kronos Vampire Hunter* (**1972**), *Dracula's Dog* (**1977**, vt *Zoltan . . . Hound of Dracula* UK) and *Nosferatu the Vampyre* (**1979**). With the growth in the video market, producers aimed such movies as *The HUNGER* (**1983**), FRIGHT NIGHT (**1985**), VAMP (**1986**), *The LOST BOYS* (**1987**), Kathryn Bigelow's revisionist *Near Dark* (**1987**) and *Vampire's Kiss* (**1988**) at a more sophisticated teenage audience. VAMPIRES IN HAVANA (**1985**) is an oddball – and very interesting – Cuban ANIMATED MOVIE. The release of Francis Ford Coppola's $40-million *Bram Stoker's Dracula* (**1992**) led to a resurgence of interest in movie vampires. However, for every *Def by Temptation* (**1990**), *The Reflecting Skin* (**1990**), *Cronos* (**1992**), *Nadja* (**1995**) or *From Dusk Til Dawn* (**1996**) there was a *Red Blooded American Girl* (**1990**),

BUFFY THE VAMPIRE SLAYER (*1992*), *Innocent Blood* (*1992*), *Tale of a Vampire* (*1992*) and *Vampire in Brooklyn* (*1995*). Neil Jordan's big-budget version of Anne RICE's novel INTERVIEW WITH THE VAMPIRE (*1994*) boasted such stars as Tom Cruise, Brad Pitt and Antonio Banderas, while Mel Brooks belatedly spoofed Coppola's movie with *Dracula, Dead and Loving It* (*1995*). On tv, vampires have regularly appeared in such series as *The* MUNSTERS (1964-6), *Dark Shadows* (1966-71 and 1990-91), *The Munsters Today* (1988-91), DRACULA – THE SERIES (1990), *Little Dracula* (1991), *Forever Knight* (1992-6) and even a yuppie OPERA, called THE VAMPYRE: A SOAP OPERA (*1992* tvm). [SJ]

Further reading: *The Dracula Book* (1975) by Donald F. Glut; *The Seal of Dracula* (1975) by Barrie Pattison; *The Vampire Film* (1975) by James Ursini and Alain Silver; *Hollywood Gothic* (1990) by David J. Skal; *Dracula: The Vampire Legend on Film* (1992) by Robert Marrero; *The Illustrated Vampire Movie Guide* (1993) by Stephen JONES.

VAMPIRE OF VENICE Italian movie (*1988*). ◊ DRACULA MOVIES.

VAMPIRES Literary vampires have roots in two distinct folkloristic traditions: the Greek notion of the seductive LAMIA and the various Eastern European superstitions regarding cannibalistically inclined reanimated corpses. Most of the latter resemble the revenants of George Romero's *Night of the Living Dead* (*1968*) far more than the aristocratic DRACULA of Bram STOKER's *Dracula* (*1897*), which cleverly combined elements from John POLIDORI's caricature of Lord BYRON in *The Vampyre* (1819) and J. Sheridan LE FANU's account of the lamiaesque "Carmilla" (1872) with selected foQoristic trappings. Dracula became the archetype of the vampire of occult fiction, but that figure retained considerable fantasy interest by virtue of the sensuality which it inherited, partly from lamia stories like Théophile GAUTIER's "La morte amoureuse" (1836) and partly from the awesomely ambivalent Byronic charisma possessed by Polidori's and Stoker's villains. No matter how hard writers of HORROR tried to confine vampires to straightforwardly monstrous roles, this charismatic quality could not be entirely suppressed, and it became increasingly flagrant in the CINEMA as standards of censorship were relaxed (◊ VAMPIRE MOVIES). In addition, the vampire's qualified IMMORTALITY became increasingly unconvincing as a form of damnation, and such tentative experiments as Jane GASKELL's *The Shiny Narrow Grin* (1964) were followed a decade later by a spectacularly sudden *bouleversement* of attitude in *Les vampires d'Alfama* (1975; trans as *The Vampires of Alfama* 1976) by Pierre Kast (1920-1984), *The Dracula Tape* (1975) by Fred SABERHAGEN and *Interview with the Vampire* (1976) by Anne RICE, which together proved to be the forerunners of, arguably, the most prolific outburst of activity in the history of imaginative literature.

Rice's bestselling series became the flagship of a huge fleet of studies in exotic Existentialism in which the practical problems and moral predicaments of vampires came under intense scrutiny. Although some of these stories are so comprehensively rationalized as to be sf, the vast majority belong more to fantasy than SUPERNATURAL FICTION; such horror as they generate is deployed as perverse glamour rather than fearful repulsion. LILITH is sometimes invoked as a symbolic ancestor-figure of the separate species, gifted with problematic immortality, to which most modern vampires are deemed to belong. The most significant examples of this REVISIONIST FANTASY include the **Varkela** series

(1979-83) by Susan Petrey (1945-1980), Chelsea Quinn YARBRO's series begun with *Hotel Transylvania* (1978), Suzy McKee CHARNAS's *The Vampire Tapestry* (1980), Geoffrey FARRINGTON's *The Revenants* (1983), Barbara HAMBLY's *Immortal Blood* (1988), Nancy A. COLLINS's *Sunglasses After Dark* (1989), Freda WARRINGTON's *A Taste of Blood Wine* (1992) and its sequel, Brian LUMLEY's **Vampire World** sequence, Storm CONSTANTINE's *Burying the Shadow* (1992), Poppy Z. BRITE's *Lost Souls* (1992), Lucius SHEPARD's *The Golden* (1993) and Tom Holland's *The Vampyre: Being the True Pilgrimage of George Gordon, Sixth Lord Byron* (1995). Many of the stories in Ellen DATLOW's anthologies *Blood is Not Enough* (anth 1989) and *A Whisper of Blood* (anth 1991) consciously explore the erotic implications of the vampire motif, while those in *Love in Vein* (anth 1994) ed Brite with Martin H. GREENBERG take such explorations to calculatedly perverse extremes. Less earnest exercises in revisionism – whose status as fantasy is even more assured – include *The Partaker* (1980) by R. CHETWYND-HAYES, *Dracula's Diary* (1982) by Michael Geare (1919-) and Michael Corby, and *Suckers* (1993) by Anne Billson. [BS]

See also: TRANSYLVANIA.

VAMPIRES IN HAVANA (ot *¡Vampiros en La Habana!*; vt *The Vampires of Havana*) Cuban ANIMATED MOVIE (*1985*). Studio ICAIC/TV Española/Durniok. **Pr** Paco Prats. **Dir** Juan Padrón. **Screenplay** Padrón. **Voice actors** Irela Bravo, Carlos González, Frank González, Mirella Guillot, Manuel Marin, Padrón, Carmen Solar (movie lacks proper credits). 80 mins. Colour.

It is 1933. Since the VAMPIRES of the world organized themselves under DRACULA in 1870, two camps have sprung up – one in Chicago under Johnny Terrori, himself in hock to the sinister Al Tapone and running blood-raids rather like Prohibition-era crimes, the other in Düsseldorf. Banished years ago from Düsseldorf were the Count's son, Werner Amadeus von Dracula, and Werner's infant nephew, Joseph Emmanuel; now living in Havana, the former has perfected his formula that allows vampires to go out during daytime – indeed, Joseph (now generally known as Pepito or Pepe) has been reared on the stuff, and has no idea he is the HIDDEN MONARCH of the Dracula clan. The Chicago mob descend on Havana (which currently suffers under the dictatorship of General Bachado), intent on suppressing the formula, because their company Vampire Beaches Inc. makes money persuading vampires to holiday at indoor pseudo-beaches; the European mob arrive to steal the formula, planning to sell the chemical worldwide as "Vampisol". In the end, Pepe is found to have been cured of vampirism by the formula, the Bachado regime falls and the two vampire mobs get their comeuppances.

The deceptive crudity of the animation adds considerably to the charm of this unflaggingly entertaining movie, which was first released subtitled in English in 1987. A herd of minor characters (e.g., a vampire who can metamorphose [◊ METAMORPHOSIS] into a WOLF but alas finds himself the lust-object of every dog in Havana) keep the action rolling along merrily. [JG]

VAMPYR (vt *Castle of Doom* US; vt *Not Against the Flesh*; vt *The Strange Adventure of David Gray*; vt *Der Traum des Allan Grey*; vt *Der Traum des David Grey*; vt *Vampyr, ou l'Étrange Aventure de David Gray*) German/French movie (*1932*). Tobis Klangfilm/Carl Dreyer. **Pr** Carl Th. DREYER. **Dir** Dreyer. **Screenplay** Dreyer, Christen Jul. **Based on** "Carmilla" (1871-2) by J. Sheridan LE FANU. **Starring** N.

Babanini, Albert Bras, Henriette Gerard (Marguerite Chopin), Jan Hieronimko, Rena Mandel (Giséle), Sybille Schmitz (Léone), Maurice Schutz, Julian West (Grey/Gray). 83 mins; most extant versions about 68 mins. B/w.

Beautiful, moody, surreal, atmospheric and largely incomprehensible VAMPIRE-MOVIE classic. A young Englishman, Allan Grey or David Gray, arrives in an inn in a French village. There he falls asleep and presumably DREAMS the rest of the movie. The plot is extremely complex and fails to hang together: Dreyer was aiming less at narrative tension than at the creation of a shift of PERCEPTION in his audience; in an explanation, he noted how one's perceptions of a room (the lighting, etc.) change instantly if one is told there is a corpse hidden behind the door, and that he wanted to create the same effect on film. At the time, the Danish movie industry was in recession. He obtained private funding for *V* from Baron Nicholas de Gunzberg in exchange for de Gunzberg starring (as "Julian West"); oddly, the baron's performance is excellent.   [JG]

VAMPYR, THE: A SOAP OPERA UK movie (*1992* tvm; originally a 5-episode series). BBC/Arts & Entertainment Network. **Pr** Janet Street-Porter. **Dir** Nigel Finch. **Spfx** Mitch Mitchell. **Vfx** Steve Bowman. **Libretto** Charles Hart. **Music** Heinrich Marschner (1795-1861). **Based on** Marschner's opera *Der Vampyr* (1827). **Starring** Omar Ebrahim (Ripley), Colenton Freeman (George), Fiona O'Neill (Miranda Davenant), Philip Salmon (Alex), Roberto Salvatori (James Berkeley), Sally-Ann Shepherdson (Emma), Richard Van Allan (Sir Hugo Davenant), Willemijn Van Gent (Ginny), Winston (High Priestess of Satan), Sarah Jane Wright (Susie). **Voice actor** Robert Stephens (Narrator). **Musicians** BBC Philharmonic, Britten Singers. **Conductor** David Parry. *c*115 mins. Colour.

London, 1793: a VAMPIRE is chased to his apparent death. 200 years later, as a side-effect of London's Docklands development, the vampire, Ripley, is woken from his sleep and rapidly rises to be a cutthroat tycoon. Told by SATAN's emissary that, in order to gain a further year of life, he must kill three young women on successive nights, he seduces and vampirizes three yuppies: Ginny, a fashion model; Emma, an executive; and, almost, heiress Miranda, the lover of his sidekick Alex. Miranda remaining faithful to Alex, Ripley blackmails her father into allowing an instant marriage. But Alex and George (Ripley's chauffeur) both realize much of the truth and rush separately to stop the marriage and destroy the vampire.

This attempt to produce something that is simultaneously OPERA and tits'n'terror HORROR MOVIE is stylish, and interestingly mixes naturalistic backgrounds with stylized ones; the cast perform the fairly simple story more than adequately. The METAMORPHOSES are prettily handled. But too often the incongruity between opera and reality becomes ridiculous – as during the various graphic sex scenes, when the partners, far from breathless, sing complicated duets. The analogy between vampires and yuppie property developers is stated but left undeveloped. A clever, fascinating curio.   [JG]

VAN ASTEN, GAIL (?   -   ) US writer whose first novel, *The Blind Knight* (1988), pits a medieval descendent of Uther Pendragon (◊ ARTHUR) against a world disinclined to treat him as a likely candidate for knighthood – he is a blind albino. But his QUEST for meaning and knighthood,

with the MAGIC aid of a WIZARD's daughter, proves successful. The **Roland** sequence – *Charlemagne's Champion* (**1990**) and *The Dark Sword's Lover* (**1990**) – is likewise set in a LAND-OF-FABLE Europe, and deals with the MATTER of France, complexly intermixing history and LEGEND.   [JC]

VANCE, JACK Working name of US writer John Holbrook Vance (1916-   ), whose contributions to sf are extensive (◊ *SFE*) and who early in his career established the DYING-EARTH subgenre of FAR-FUTURE fantasy with *The Dying Earth* (coll of linked stories **1950**). The remote setting owes something to Clark Ashton SMITH's **Zothique** stories; JV's achievement was stylistic, with an ironic narrative tone appropriate to an exhausted world of melancholy and dying falls, haunted by strange MONSTERS (erbs, gids, deodands), where the dimmed SUN wobbled precariously through a dark-blue sky and might at any moment go out. The generally amoral characters' barbed formality of DICTION, distantly reminiscent of Ernest BRAMAH, is expertly turned to dramatic or comic effect; in a famous passage an augur (◊ PROPHECY) states his fees: "For the twenty terces I phrase the answer in clear and actionable language; for ten I use the language of cant, which occasionally admits of ambiguity; for five, I speak a parable which you must interpret as you will; and for one terce, I babble in an unknown tongue." Though MAGIC is omnipresent and the implied underlying science forgotten, even the known SPELLS are far fewer than of yore (◊ THINNING). The episodes are flimsily plotted but replete with striking incidents and images: WIZARDS create flawed life or shrink rivals to manikin size (◊ GREAT AND SMALL) to imprison them in a minimalist LABYRINTH with a miniature DRAGON; the BEAUTY AND THE BEAST theme is effectively reworked; a cruel freebooter over-fond of torture attempts a theft in defiance of all warnings and finds that his private PORTAL to an OTHERWORLD offers no escape from the dread guardian "Chun the Unavoidable"; a CITY is divided into colour-coded (◊ COLOUR-CODING) factions with Greys unable to see wearers of Green and vice versa; a QUEST for knowledge leads to the ancient Museum of Man, where the defeat of an invading DEMON leaves the protagonist's yearning unsatisfied in a LIBRARY containing all knowledge but no index.

Further **Dying Earth** volumes, though still fragmentary, are more robustly PICARESQUE, with TRICKSTER protagonists – notably Cugel the Clever. In *The Eyes of the Overworld* (coll of linked stories **1966**) Cugel swaggers and bluffs his way through adventures involving the eponymous lenses (which provide all-senses ILLUSION so that mud huts become palaces and rags exquisite garb), a demon-controlling AMULET, an awesome lake-dwelling GIANT, a wizard's binding conjuration of the entire Universe into a jellylike mass which Cugel hungrily eats, TIME TRAVEL, and numerous deceptions; repeatedly Cugel overreaches and comes to grief. His story continues in *The Bagful of Dreams* (**1979** chap) and *The Seventeen Virgins* (**1974** *F&SF*; **1979** chap), both incorporated into *Cugel's Saga* (fixup **1983**). Michael SHEA's *A Quest of Simbilis* * (**1974**) is an "alternative" Cugel sequel to *The Eyes of the Overworld*, written with JV's permission. *Rhialto the Marvelous* (coll of linked stories **1984**) stars Rhialto, one of several squabbling magicians whose power derives from unreliable imps called sandestins; they are threatened by a WITCH unfairly employing transsexuality spells and by a betrayed comrade retrieved from a dead STAR at, literally, the edge of the Universe; there is a fine journey in a magically powered spacegoing palace. (This story appeared as *Morreion: A Tale of the Dying Earth* in *Flashing Swords #1* anth **1973** ed Lin CARTER; **1979** chap.) A sense of TIME ABYSS is well conveyed when a precious object falls into the sea and Rhialto proceeds to a future aeon in which the seabed will be dry.

Uncommonly for fantasy, JV's coolly exotic style is distinctive enough to PARODY – as in "The Star Sneak" (1974 *F&SF*) by Larry Tritten. His gift for appropriate character- and place-NAMES is remarkable, as is his ability to weave unusual words into seamless prose; "deodand", an object forfeited to the crown for having caused a human death, is a wittily apt name for a murderous MONSTER. JV applied this increasingly polished style to sf as well as fantasy, which, coupled with his interest in constrained or obsessed societies (◊ ANTHROPOLOGY) rather than hard-sf trappings, tends to blur the genre lines. Thus *Big Planet* (1952 *Startling Stories*; cut **1957**; further cut 1958; text restored 1978) updated PLANETARY ROMANCE from the naiveties of Edgar Rice BUR-ROUGHS to sf respectability without loss of wonder; but Big Planet is also the setting of *Showboat World* (**1975**; vt *The Magnificent Showboats of the Lower Vissel River, Lune XXIII South, Big Planet* 1983), whose picaresque adventures and quarrels of captains on a great RIVER have a flavour of fantasy and of Cugel. A fine but notionally sf example of JV's anthropological creativeness is "The Moon Moth" (1961), where etiquette requires wearing the correct MASK and accompanying one's conversation with MUSIC. In *The Dragon Masters* (**1963** dos) – a Hugo AWARD winner – the feudal society is post-technological, with DRAGONS and GIANTS the products of genetic engineering; *The Last Castle* (**1967** dos), which won JV another Hugo and a Nebula, uses lords, castles and feudalism to emphasize Decadence in a story whose underlying rationale is again sf. *Green Magic* (1936 *F&SF*; **1979** chap) contains a memorable TECHNO-FANTASY device as an earthy GOLEM with a tv camera eye is sent to explore the elfin otherworld of the eponymous magic – whose secrets, eventually learned by the human wizard, may well be THINGS BOUGHT AT TOO HIGH A COST.

A late fantasy trilogy is **Lyonesse**, set two generations before the time of ARTHUR (and featuring a supposed UNDERLIER of the ROUND TABLE) in the 10 kingdoms of the "Elder Isles" west of France and the English Channel's mouth – now sunken lands. The component books are *Suldrun's Garden* (**1983**; rev 1983; vt *Lyonesse: Book 1: Suldrun's Garden* 1984 UK), *Lyonesse II: The Green Pearl* (**1985**; rev vt *The Green Pearl* 1986) and *Lyonesse III: Madouc* (**1989**; vt *Madouc* 1990); this last won a WORLD FANTASY AWARD. Against a background of appealing political intrigue and slightly unsatisfactory war (JV seems to lack enthusiasm for battle scenes), **Lyonesse** features fine episodes of magic, casual cruelty, vengeance, whimsy and gorgeous landscape in a yet more polished version of the **Dying Earth** manner. Further fantastic elements include FAIRIES (one character's age is anomalous owing to a fairy-mound visit; ◊ TIME IN FAERIE), fairy CURSES of bad luck, prophecies from a magic MIRROR, a CHANGELING, visits to magic and demonic other-worlds, journeys hedged about with CONDITIONS and PROHIBITIONS, spells that meddle with TIME, SHAPESHIFTING and transformations, a three-headed GIANT, the GRAIL, and an offstage MALIGN SLEEPER undersea, whose ultimate brief wakening drowns the legendary city of Ys.

JV's current projected works are reportedly sf, but his influence on fantasy has been immense, enduring and grate-fully acknowledged – by, for example, Gene WOLFE, whose

The Book of the New Sun (**1980-83**) is the most significant re-articulation to date of the DYING-EARTH theme. JV received the World Fantasy Award for Lifetime Achievement in 1984.   [DRL]

Other works: *Eight Fantasms and Magics* (coll **1969**; with 2 stories cut vt *Fantasms and Magics* 1978 UK); *Green Magic: The Fantasy Realms of Jack Vance* (coll **1979**); *Rhialto the Marvelous* * (coll **1984**; with one story by another hand added anth 1985).

VANDE VELDE, VIVIAN (1951-   ) US writer of CHIL-DREN'S FANTASY. *Once Upon a Test* (coll **1984** chap) contains three humorous fairytales in which princes and princesses break convention. VVV repeated this approach in the slightly more adult *Tales from the Brothers Grimm and the Sisters Weird* (coll **1995**), a collection REVISIONIST FAN-TASIES: Though they lack the bite of Tanith LEE or the ring of the MARVELLOUS created by Jane YOLEN, they are enjoy-able excursions. *A Hidden Magic* (**1985**) involves a young princess lost in the woods who becomes involved in a SOR-CERER's battle against a WITCH. *A Well-Timed Enchantment* (**1990**) is a TIMESLIP fantasy in which a girl and her cat retrieve an old watch from the past. In *User Unfriendly* (**1991**) some young boys find themselves trapped in a mal-functioning fantasy GAME. *Dragon's Bait* (**1992**) involves a young girl, wrongly accused of WITCHCRAFT, who befriends the DRAGON she was sacrificed to. *Companions of the Night* (**1995**) is a VAMPIRE romance emphasizing the attraction of the vampire motif for adolescents. In all these books VVV is able to pump new life and energy into traditional themes.
                                                                            [MA]

VANE, (VANE HUNT) SUTTON (1888-1963) UK poet and playwright, more widely published in the USA than the UK. His classic POSTHUMOUS FANTASY *Outward Bound* (play **1924**; novel **1929**) is set aboard a cruise ship whose passen-gers gradually realize they are *en route* to the Seat of Judgement – or is it all a DREAM shared by two collaborators in a suicide pact? It was filmed twice, as OUTWARD BOUND (***1930***) and *Between Two Worlds* (***1944***).   [BS]

VAN GULIK, ROBERT (1910-1967) Dutch writer. ◊ ORI-ENTAL FANTASY.

VAN LUSTBADER, ERIC (1946-   ) US writer, married to sf and fantasy editor Victoria Schochet; much of his later fiction has been published as by Eric Lustbader. Although his first novel, *The Sunset Warrior* (**1977**) was sf, EVL almost immediately began shifting his focus to fantasy, beginning with the subsequent volumes of the **Sunset Warrior** trilogy: *Shallows of Night* (**1978**) and *Dai-San* (**1978**). *Beneath an Opal Moon* (**1980**) is set in the same uni-verse, which has not been treated with much concern for its ostensible sf nature for some time. The Oriental SWORD-AND-SORCERY flavour of these novels is intensified in EVL's later work, all published outside genre fiction and marketed (successfully) as bestselling action-adventure. His subse-quent fictions have been contemporary martial-arts action novels of eastern intrigue; many contain supernatural ele-ments, though none centrally. Those novels concerning **Nicholas Linnear** – they include *The Ninja* (**1980**),*The Miko* (**1984**), *White Ninja* (**1990**), *The Kaisho* (**1993**), *Floating City* (**1994**) and *Second Skin* (**1995**) – tend to involve MAGIC powers.   [GF]

Other works: The **China Maroc** sequence, being *Jian* (**1985**) and *Shan* (**1986**); *Black Heart* (**1983**); *Zero* (**1987**); *French Kiss* (**1989**); *Angel Eyes* (**1991**); *Black Blade* (**1993**).

VANSITTART, PETER (1920-   ) UK writer whose

novels in various genres, beginning with the sf dystopia *I Am the World: A Romance* (**1942**), have over the decades become an important set of narrative visions of the interrelations between the data of history and the MYTHS that generate STORY. Most of his best novels are historical fictions, and his fantasies, too, are interwoven with the history of Europe. *The Story Teller* (**1968**) is typical: its protagonist has lived for over 500 years, and the phases of his life correspond to the phases of European history he witnesses and helps to shape, as well as to the movement of the SEASONS. Two further novels – *The Death of Robin Hood* (1981) and *Parsifal* (**1988**) – similarly examine history and myth through iconic figures whose IMMORTALITY is a metaphor of the immortality of Story. The recurring ROBIN HOOD figure in the former, MYTHAGO-like, recurs as a living image in the imaginations of men and women in various eras; and the protagonist of the latter is both an actor in the MATTER of Britain and the victim of Richard WAGNER's 19th-century TRANSFORMATION of the hero into a creature fit for contemporary Germany. The earlier *Lancelot* (1978) likewise ironizes the Arthurian cycle, but without the same savagery. Late novels, like *A Safe Conduct* (**1995**), constantly press at the edge of the supernatural; the 15th-century figures in this tale of the Children's Crusade believe in MAGIC, and act their beliefs out: it is constantly moot whether their behaviour is delusory.

Two collections of TWICE-TOLD tales – *The Dark Tower: Tales from the Past* (coll **1965**) and *The Shadow Land: More Stories from the Past* (coll **1967**) – convey more directly, for YA readers, the same sense that Story and history are intimately wed. In his denseness of language, and in the dangerousness of his narrative visions, PV can be understood as a significant precursor of mythopoeic fantasy writers like Paul HAZEL and Robert HOLDSTOCK.   [JC]

VAN VOGT, A(LFRED) E(LTON) (1912-   ) Canadian-born writer, a central figure of SCIENCE FICTION's Golden Age; much of his significant work was written before he moved to the USA in late 1944. The dreamlike momentum of the plots and rationales of his best sf are reminiscent of the fantasy of anxiety, but most of his work is certainly sf. Of the stories assembled in *Out of the Unknown* (coll **1948**) with E. Mayne Hull (1905-1975), "The Sea Thing" (1940 *Unknown*), "The Witch" (1943 *Unknown*) and "The Ghost" (1943 *Unknown*), all by AEVV alone, are RATIONALIZED FANTASIES in the UNKNOWN mode; AEVV's single fantasy novel, *The Book of Ptath* (1943 *Unknown*; exp **1947**; vt *Two Hundred Million A.D.* 1964), also attempts to ground its plot in something like science. The GOD Ptath – whose power derives MYTHAGO-like from the collective will of the folk of Earth – has (after millions of years of godhood) subjected himself to BONDAGE within a series of mortal forms (i.e., to successive REINCARNATIONS) so that he may relearn what it means to be human. Ptath awakens in a state of AMNESIA in the FAR FUTURE and, both confused and aided by memories of his last mortal persona, Holroyd from 1944, gradually regains knowledge of his true selfhood as Ptath, combats the evil goddess Ineznia, and blazes forth in his full glory as the god who will reunite the planet.

Beyond its hypnotic, wind-burned pace, the tale is perhaps most memorable as an early form of what, a few decades later, might have been described as a full-blown PLANETARY ROMANCE set on a DYING EARTH: the billions of citizens of Gonwonlane (and the lands with which it is at war) inhabit endless, volcano-riven LANDSCAPES, engage in SWORD-AND-SORCERY exploits made necessary because the Earth is metal-poor and high technologies are now impossible, and are ruled by gods who derive their power through the transference of mental energies. But AEVV is not much interested in depicting social complexities or conveying a sense of elegy, and in the end *The Book of Ptath* stands alone.

[JC]

VARDEMAN, ROBERT E(DWARD) (1947-   ) US writer of huge productivity in various genres. His moderately inventive, action-packed but ultimately routine fantasy adventures began with the **War of Powers** series, all with Victor MILAN: *The Sundered Realm* (**1980**), *The City in the Glacier* (**1980**), *The Destiny Stone* (**1980**) – these first three assembled as *The War of Powers* (omni **1984** UK) – and *The Fallen Ones* (**1982**), *In the Shadow of Omizantrim* (**1982**) and *Demon of the Dark Ones* (**1982**) – these latter assembled as *The War of Powers II: Istu Awakened* (omni **1985** UK). His solo work includes the **Cenotaph Road** series: *Cenotaph Road* (**1983**), *The Sorcerer's Skull* (**1983**), *World of Mazes* (**1983**), *Iron Tongue* (**1984**), *Fire and Fog* (**1984**) and *Pillar of Night* (**1984**). Another collaborative series is the **Swords of Raemllyn**, all with Geo. W. Proctor (1946-   ): *To Demons Bound* (**1985**), *A Yoke of Magic* (**1985**), *Blood Fountain* (**1985**) – the first three assembled as *Swords of Raemllyn: Book 1* (omni **1992** UK) – *Death's Acolyte* (**1986**), *The Beasts of the Mist* (**1986**), *For Crown and Kingdom* (**1987**) – the second three assembled as *Swords of Raemllyn: Book 2* (omni **1992** UK) – and *Swords of Raemllyn: Book 3* (coll **1995** UK), which assembles three book-length stories not published separately: "Blade of the Conqueror", "The Tombs of A'bre" and "The Jewels of Life". More solo work is: the **Jade Demons** series – *The Quaking Lands* (**1985**), *The Frozen Waves* (**1985**), *The Crystal Clouds* (**1985**) and *The White Fire* (**1986**), assembled as *The Jade Demons Quartet* (omni **1987** UK); the **Keys to Paradise** series, initially published as *The Keys to Paradise* as by REV (**1986** UK), then in separate volumes as by Daniel Moran – *The Flame Key* (**1987**), *The Skeleton Lord's Key* (**1987**) and *The Key of Ice and Steel* (**1988**); and the **Demon Crown** series – *The Glass Warrior* (**1989**), *Phantoms of the Wind* (**1989**) and *A Symphony of Storms* (**1990**), assembled as *The Demon Crown Trilogy* (omni **1990** UK). REV has also written much sf and crime fiction.[DP]

Other works (horror): *The Screaming Knife* (**1990**); *A Resonance of Blood* (**1992**); *Death Channels* (**1992**); *The Accursed* (coll **1994**).

VEIGA, JOSÉ J. (1915-   ) Brazilian writer, closely associated with MAGIC REALISM and the fantastic. Also a journalist, he worked in London for the BBC 1945-9; back in Brazil he became a translator for the Brazilian edition of *Reader's Digest*. His first book was *Os Cavalinhos de Platiplanto* ["The Little Horses of Platiplanto"] (coll **1959**), whose stories are mostly about rural life. His first novel, an immediate success, was *A Hora dos Ruminantes* (**1966**; trans Pamela G. Bird as *The Three Trials of Manirema* 1970 US), in which a small village is taken over by strange men before being invaded by dogs and then cattle. It was followed by *A Máquina Extraviada* (coll **1967**; trans Bird as *The Misplaced Machine and Other Stories* 1970 US), *Sombras de Reis Barbudos* ["Shadows of Bearded Kings"] (**1972**) and *Os Pecados da Tribo* ["The Sins of the Tribe"] (**1976**). Veiga's stories are a clever mix of rural-life naturalism and the Kafkaesque, in which many critics see allegories of Brazil under military rule. *A Casca da Serpente* ["The Serpent's Skin"] (**1989**) features Brazilian historical figures like

Antonio Conselheiro (a messianic 19th-century leader), the poet Sousândrade and the Russian anarchist Piotr Kropotkin. [BT]

VENGEANCE A potent PLOT DEVICE which motivates many stories, or provides a suitable comeuppance, like the savage revenge of the cheated PIED PIPER. In Greek MYTHOLOGY, the THREE Furies, Tisiphone, Alecto and Megaera – euphemistically, the Eumenides or Kindly Ones – are the GODS' official instruments of vengeance for unpaid blood debts: they feature in the play *The Family Reunion* (**1939**) by T.S. Eliot (1888-1963) and in Neil GAIMAN's «The Kindly Ones» (graph coll 1996). GHOSTS very frequently seek to be avenged on their murderers, sometimes through human intermediaries, as in SHAKESPEARE's *Hamlet* (**1603**) and Terry PRATCHETT's Shakespearean *Wyrd Sisters* (**1988**). Vengeance-driven protagonists command instant sympathy if their grievance is real – e.g., the tortured and mutilated Cara in Geoff RYMAN's *The Warrior who Carried Life* (**1985**). Roger ZELAZNY's ambiguous heroes often have this motivation: Corwin in *Nine Princes in Amber* (**1970**) wants revenge for the infliction upon him of exile, AMNESIA and later blindness; the eponymous TRICKSTER in *Jack of Shadows* (**1971**) is casually executed, and, reborn, embarks on a destructive vengeance which upsets the world; Zelazny's **Dilvish** has a similar grievance. The PICARESQUE adventures of Jack VANCE's still less sympathetic ANTIHERO **Cugel** are strung on a thread of planned revenge against the Laughing Magician whom he tried to rob, and who punished him perhaps excessively. [DRL]

VENGEANCE OF SHE, THE UK movie (*1968*). ◊ SHE.

VENICE A CITY of stone and water which is now slowly drowning. For writers of fantasy and SUPERNATURAL FICTION, Venice is thus an emblem of transition, of DEATH, of METAMORPHOSIS, a place haunted by REVENANTS and LIMINAL BEINGS. It is a city whose past and whose present condition cannot be ignored; unlike LONDON or NEW YORK, it cannot be treated as a level playing field upon which to set a tale. Venice is always a *character* in any story set there.

In earlier centuries, this was not the case. William SHAKESPEARE's *The Merchant of Venice* (produced *c*1596; **1600**) has no "Venetian" redolence. But by the time Wilkie COLLINS published *The Haunted Hotel: A Mystery of Modern Venice* (**1878**) and Vernon LEE began to write the SUPERNATURAL FICTIONS assembled in *Hauntings* (coll **1890**) and later volumes, Venice had become usable for writers as a city with more past than future, a natural focus for tales in which the past – in the form of transgressive beckonings on the part of supernatural creatures – calls upon the present or draws victims across the THRESHOLD into another state. The most famous use of Venice as a theatre for transgressive change is probably Thomas MANN's *Death in Venice* (**1913**).

Other works set partly or wholly in Venice include: *Gestures* (**1986**) by H.S. Bhabra (1955-); Guy BOOTHBY's *Farewell Nikola* (**1900**); Jerome CHARYN's *Pinocchio in Venice* (**1991**); Daphne DU MAURIER's "Don't Look Now" (**1971**), filmed as DON'T LOOK NOW (*1973*); Italo CALVINO's *Invisible Cities* (**1972**), in which all the fantastical cities described by Marco Polo are reflections of Venice; *Reincarnation in Venice* (**1979**) by Max Ehrlich (1909-1983); Steve ERICKSON's *Days Between Stations* (**1985**); MacDonald HARRIS's *Pandora's Galley* (**1979**); William GOLDMAN's *The Silent Gondoliers* (**1983**); Amanda PRANTERA's *The Cabalist* (**1985**); Muriel SPARK's *Territorial Rights* (**1979**); *The Stone Virgin* (**1985**) by Barry Unsworth

(1930-); Jeanette WINTERSON's *The Passion* (**1987**); and Elinor WYLIE's *The Venetian Glass Nephew* (**1925**). [JC]

VENTRILOQUISM This art has been an entertainment since the 15th century. It is likely that priests and prophetesses used it as a means of presenting the voices of the GODS at ORACLES, and even more so that various mediums performed the same trick at SÉANCES. Ventriloquism was introduced into GOTHIC FANTASY by Charles Brockden BROWN in *Wieland, or The Transformation* (**1798**) in order to rationalize the apparent supernatural events (◊ RATIONALIZED FANTASY), a PLOT DEVICE he explored further in the unfinished sequel *Memoirs of Carwin, the Biloquist* (written 1798; fragment **1803**).

More recent HORROR fiction has moved away from the practice of ventriloquism to explore the relationship between the ventriloquist and his dummy (◊ DOLLS), with its potential for split personalities and IDENTITY EXCHANGE. The best of these stories leave it uncertain as to whether the dummy's growing existence is supernatural or merely in the PERCEPTION of the ventriloquist, which makes them a subset of the GHOST STORY (◊◊ DOPPELGÄNGERS). Probably the best-known story on this theme is *Magic* (**1976**) by William GOLDMAN – filmed as MAGIC (*1978*) – but equally effective are "The Extraordinarily Horrible Dummy" (in *Penguin Parade #6* anth **1939**) by Gerald KERSH and "Farewell Performance" (in *The Clock Strikes Twelve* col **1939**) by H. Russell WAKEFIELD, while John COLLIER lampoons the theme in "Spring Fever" (in *Fancies and Goodnights* coll **1951**), where the dummy falls in love. [MA]

VENTURE SCIENCE FICTION ◊ *The* MAGAZINE OF FANTASY AND SCIENCE FICTION.

VENUS ◊ APHRODITE.

VERBEEK, GUSTAVE (1867-1937) US comics pioneer. ◊ COMICS.

VERGIL ◊ VIRGIL.

VERLAINE, PAUL (1844-1896) French writer. ◊ DECADENCE.

VESS, CHARLES (1951-) US fantasy illustrator and COMICS artist with a fine, delicate line style and a romantic imagination, strongly influenced by the English fairytale illustrators William Heath ROBINSON and Arthur RACKHAM. His work has a subtle Victorian quality; this makes it all the more unusual that he has become a leading artist in US comic books, where subtlety and delicacy are rare qualities. He works in coloured inks, using them in transparent glazes to achieve a clear, atmospheric, sometimes almost luminous effect.

CV has taught at the William King Regional Art Center in Abington, Virginia (1980-82), and at the Parson School of Design in New York (1992). His early comics work was on **Spiderman**, **Thor**, etc., but his reputation rests mainly on his more fanciful comics in EPIC ILLUSTRATED and some more recent pieces with writer Neil GAIMAN – on SANDMAN and the GRAPHIC NOVEL *StarDust* (graph **1993**). CV also wrote and drew the **Spiderman** story "Spirits of the Earth" (1991), inspired by a visit to the Orkney Islands. [RT]

VICE VERSA Two movies have been based (the latter very loosely) on *Vice Versâ, or A Lesson to Fathers* (**1882**) by F. ANSTEY.

1. UK movie (*1947*). Rank/Two Cities/General. **Pr** George H. Brown, Peter Ustinov. **Dir** Ustinov. **Spfx** Henry Harris. **Mufx** Geoffrey Rodway. **Screenplay** Ustinov. **Starring** Petula Clark (Dulcie), David Hutcheson (Paradine), James Robertson Justice (Dr Grimstone), Roger Livesey

(Paul Bultitude), Anthony Newley (Dick Bultitude), Kay Walsh (Fanny Verlayne), Joan Young (Alice). 111 mins. B/w.

Dastardly Marmaduke Paradine plunders the eye of the Hyena God from an Indian temple and on his return to England, believing the stone cursed, gives it to his cousin – stuffy, parsimonious bereaved stockbroker Paul Bultitude. While bundling off son Dick to hellish school Grimstone's, Paul foolishly yearns for his own schooldays, the happiest of his life, and the stone grants his wish, transforming him into a replica of Dick. Father and son discover the stone will grant only one WISH per person; Dick petulantly wishes he could take over his father's role. As his father, Dick copes with the flirty affections of gold-seeking chorus girl Fanny, kindles the passions of maid Alice, avoids being killed in a duel over Fanny by the militaristic Earl of Gosport, and is persuaded by Paradine to invest the family fortune in the – presumed crazy – development of the horseless carriage. Meantime, Paul is discovering the beastly realities of life under the thumb of vicious headmaster Grimstone while fending off the youthful enthusiasms of the martinet's daughter Dulcie. Dick's younger brother is persuaded to wish for a return to the *status quo ante*, the villains are routed, and father and son become firmest friends. The whole is told from the vantage of later years by Paul (now wed to Alice) on the day of Dick's marriage to Dulcie.

Although the humour – and for that matter the spfx – tend to trundle rather than scamper, *VV* has great richnesses of comedy, including several classic scenes and some fine character performances. It remains the default movie presentation of the IDENTITY-EXCHANGE theme. [JG]

2. US movie (*1988*). Columbia/Clement/La Frenais. **Pr** Dick Clement, Ian La Frenais. **Exec pr** Alan Ladd Jr. **Dir** Brian Gilbert. **Vfx** Louis Schwartzberg. **Spfx** Dennis Dion. **Screenplay** Clement, La Frenais. **Starring** Corinne Bohrer (Sam), Gloria Gifford (Marcie), Jane Kaczmarek (Robyn Seymour), Swoozie Kurtz (Tina), William Prince (Mr Avery), David Proval (Turk), Judge Reinhold (Marshall Seymour), Fred Savage (Charlie Seymour). 98 mins. Colour.

Executive Chicago yuppie and divorcé Marshall accidentally comes into possession of a mystical ornamental skull from Thailand. Before he can return it to smugglers Tina and Turk, he and 11-year-old son Charlie each make a nonce-WISH that they could exchange lives, and an IDENTITY EXCHANGE occurs. Charlie has to cope with boardroom battles and backstabbing at the store while Marshall must cope with schoolroom bullies and a martinet teacher; both must cope with their relationships to Marshall's ex-wife Robyn, his improbably beautiful girlfriend Sam, and the efforts of the smugglers to regain the skull. All turns out well, of course. Though slight, the movie is exceptionally good-humoured, sometimes very funny and constantly charming, with Savage particularly excellent as a man in a boy's body. [JG]

VIDEOS, ROCK ◊ ROCK VIDEOS.

VIERECK, GEORGE (SYLVESTER) (1884-1962) US journalist, best-known as a writer for his collaborations with Paul Eldridge (1888-1982). He was one of the literary "Bohemians" who attempted to import into the USA the aesthetic ideals of European DECADENCE. *The House of the Vampire* (**1907**) is a homoerotic tale of psychic vampirism (◊ VAMPIRES) seemingly inspired by Oscar WILDE. With Eldridge, GV wrote a trilogy of erotic fantasies begun with *My First Two Thousand Years: The Autobiography of the Wandering Jew* (**1928**), which transformed the traditionally

miserable figure of the WANDERING JEW into an urbane opportunist whose war against "the Great God Ennui" takes the form of a QUEST for the secret of "unendurable pleasure indefinitely prolonged". In pursuit of this end the wanderer frequently meets his female counterpart, whose parallel tale is told in *Salome, The Wandering Jewess* (**1930**). This SALOME's quest is the liberation of women, but she is perennially defeated by biological circumstance and sets out in the end to engineer a new hermaphroditic human species. *The Invincible Adam* (**1932**) tells the story of Kotikokura, who has evolved from protohumanity to acquire civilized charm and intelligence but is still subject to the urgings of the "rib" (a penile bone), which other men have lost but he retains. The trilogy's pretensions to psychological depth are cleverly parodied in *The Memoirs of Satan* (**1932**) by William Gerhardie (1895-1977) and Brian Lunn (1893-?   ). GV's peculiar combination of prurience and eccentric FEMINISM is given freer but more lighthearted rein in his solo novel *Gloria* (**1952**), a comedy which proposes that all the great lovers of history and legend were, in fact, woefully inept. [BS]

VILLAINS This term is primarily used for the more routine bad guys in GENRE FANTASY, HEROIC FANTASY, LOW FANTASY, etc. Mere villains lack the stature of ANTIHEROES or of major forces of EVIL like DARK LORDS, DEMONS and SATAN. They tend to have peculiarly human flaws: Gandolf of Utterbol in William MORRIS's *The Well at the World's End* (**1896**) indulges his sadism so thoughtlessly and indiscriminately that his own followers heap honours on his killer; Altiokis in Barbara HAMBLY's *The Ladies of Mandrigyn* (**1984**) is fatally small-minded; the Supreme Grand Master in Terry PRATCHETT's *Guards! Guards!* (**1989**) bullies his acolytes but cowers abjectly before the freed DRAGON which they have summoned. SWORD AND SORCERY in particular, owing to its open-ended nature, requires a steady supply of WIZARD and WITCH villains to be defeated by CONAN and his literary descendants; the relentless publishing schedules of COMICS similarly demand a flow of plausible opposition for SUPERHEROES. [DRL]

VILLIERS DE L'ISLE-ADAM (JEAN-MARIE-MATHIAS-PHILIPPE-AUGUSTE), COMTE DE (1838-1889) French writer who first gained fame for two volumes of short stories: *Contes Cruels* (coll **1883**; trans Robert Baldick as *Cruel Tales* **1963** UK) and *Nouveaux Contes cruels* ["New Cruel Tales"] (coll **1888**), some of the contents of which, along with stories from the first volume, appeared as *Sardonic Tales* (coll trans Hamish Miles **1927** US). The term CONTE CRUEL is taken from these volumes, which subvert the fable-like content and moral suasion of the traditional French "conte" through a GRAND GUIGNOL-like exorbitance of cruelty – derived in VDLA's case through Charles BAUDELAIRE's translations of Edgar Allan POE. *Claire Lenoir* (in *Tribulat Bonhomet* coll **1887**; trans Arthur Symons **1925** US) is a HORROR tale of posthumous POSSESSION, in which the philistine Dr Tribulat Bonhomet causes mayhem while understanding nothing of the cruelty of the Universe.

Axel (**1872** *La Renaissance Littéraire et Artisticque*; **1885-6** *Jeune France*; rev **1890**; trans H.P.R. Finberg **1925** UK; new trans June Guicharnaud **1970** US) is a play whose dithyrambic expansiveness precludes stage-presentation. The eponymous Rosicrucian count and MAGUS – victim of an ethic which entails renunciation and death over any attempt to press through life – inhabits an impregnable EDIFICE safe from the savageries of the world. His famous declaration – "Vivre? Les serviteurs feront cela pour nous ["Live? The servants will do that for us]" – and the final love-death ceremony he enters into with his inamorata make him a central example of the KNIGHT OF THE DOLEFUL COUNTENANCE, in which guise he has influenced generations of fantasy writers, in particular those inclined to the creation of DYING-EARTH tales.

Much of VDLA's work – including *Prolégomenès* ["Prolegomena"] (**1862**), the first part of an unfinished occult novel – was published privately, and has not yet been made available in translation. [JC]
Other work: *L'Ève future* (**1886**; trans Marilyn Gaddis Rose as *The Eve of the Future* 1981 US; new trans Robert M. Adams as *Tomorrow's Eve* 1982 US), sf.

VINES, (WALTER) SHERARD (1890-?   ) UK academic and writer. In his sharp and exuberant satirical fantasy *Return, Belphegor!* (**1932**), SATAN – under threat of extinction by unbelief – dispatches a DEMON to start a new wave of religious persecutions, culminating in the burning of heretics.
[BS]

VINGE, JOAN (CAROL) D(ENNISON) (1948-   ) US writer, almost all of whose work has been sf, albeit with a fantasy feel and often drawing on MYTHOLOGY or FAIRYTALE for underpinning. Her fantasy output is almost entirely confined to movie ties: *Tarzan, King of the Apes* * (**1983**) (◊ TARZAN MOVIES), *Return to Oz: A Novel* * (**1985**) (◊ The WIZARD OF OZ), *Ladyhawke* * (**1985**) (◊ LADYHAWKE [*1985*]), *Santa Claus, the Movie: A Novel* * (**1985**) and the juvenile *Santa Claus, the Movie Storybook* * (**1985**), and *Willow* * (**1988**) (◊ WILLOW [*1988*]). JDV is a fine stylist; it is a pity that, in fantasy, she seems to have set her sights low. [JG]

VIPONT, ELFRIDA (1902-   ). UK writer. ◊ Raymond BRIGGS.

VIRGIL Conventional English name for Italian Latin poet Publius Vergilius Maro (70BC-19BC), who led an apparently quiet life, distinguished after 39BC by the indirect patronage of the (then future) Emperor Augustus, whose reign has ever since been symbolized, foremost, by Virgil's poetry.

Later generations fathered on him the poems now called the *Appendix Virgiliana*: one or two very brief works among these may represent his juvenilia, though not the two fantastic works, the short epic *Ciris* and the curiously effective mock epic *Culex*, which deals with the death and afterlife of a wasp. The first work we know to be Virgil's is the ten *Eclogues* (**42BC-39BC**; coll 39BC). These pastoral poems, though modelled on Theocritus's (◊ GREEK AND LATIN CLASSICS), narrow the latter's range to the specific, implausible but not fantastic land of Arcadia, the basis of the Western European PASTORAL tradition. Only one (the sixth) tells of anything fantastic, and it is more a precis of mythical history than a tale. (The fourth has become most famous, as an alleged prophecy of CHRIST.)

These poems won him entrée to the Augustan circle, where he found his deeper role, as a public, committed poet. During a decade of civil war he wrote the *Georgics* (**29BC**; vt *Bucolics*), a didactic poem about the peaceable conduct of farming. Often misinterpreted as a programme for the actual revival of Italian agriculture, the work propounds an ideal close to the poet's, and Augustus's, heart. It contains one short, confusedly told and profoundly strange MYTH towards its close. The *Georgics* has been called the best poem in Latin; it is the last its author completed.

He then accepted the task later imperial poets routinely shunned, that of celebrating the emperor in epic. The surprising result, posthumously published, was the *Aeneid* (*c*19BC). This tells not of Augustus's wars but of his legendary ancestor, Aeneas. Its first half describes Aeneas's homeseeking (as opposed to the homecoming of HOMER's ODYSSEUS), its second his war to found a city (as opposed to Homer's Achilleus's war to sack one). Few readers find the second part lives up to the first, but the early episodes of the sack of Troy, Aeneas' romance with Dido, and his visit to the UNDERWORLD, as well as the concluding nightmare of his surrender to battle rage, are all moving; the style is consistently clear and noble; and the epic is by common consent the greatest in Latin. The *Aeneid* is not always accepted as fantasy today, though fantasy is in fact what Virgil's contemporaries expected of epic. Virgil never admits to providing it (as do his successors OVID and STATIUS), and the second half of his work, generally realistic, overshadows the frequent passages of wonder in the first. Moreover, his GODS are too civilized and decorous for the usual excitements of myth.

However, it is precisely in its least fantastic aspect that the *Aeneid* fundamentally shaped later fantasy. In its characters' consistent political concern, in its narrative elevation of tone and content, in its tragic outlook on both LOVE and war, and especially in its HERO's piety and struggles with his fated purpose (these last being elements fully original to it), the *Aeneid* prefigured the greater part of Western epic, whose writers usually had read it. If Ovid gave the fantasy tradition its voice, Virgil gave it the words it has most loved to say.

He has also figured as a character in much fiction, most notably DANTE ALIGHIERI's *Inferno* and *Purgatorio*. [JB]
Further reading: For a full discussion of Virgil's role in medieval literature see *Virgilio nel medio evo* (**1872**; rev 1896; trans E.F.M. Benecke as *Vergil in the Middle Ages* 1895; rev 1908; rev Giorgio Pasquali 1937) by Domenico Comparetti. For the romantic tales which bear little relation to the historical Virgil but which, in turn, underlie Avram DAVIDSON's **Vergil Magus** stories, a more sympathetic account is *Virgil the Necromancer* (**1934**) by John Webster Spargo. Also of interest is the poem "Virgil the Sorcerer" (1924) by Robert Graves.

VIRGINITY The notion of female virginity as special and magical has spun off a variety of MYTHS and SUPERSTITIONS, duly incorporated into fantasy. Virginity supposedly preserves and enhances an inner power which is peculiarly female. Britomart in Edmund SPENSER's *The Faerie Queene* (**1590-96**) taps this power in her career as woman KNIGHT, perhaps echoing the real-world figure of JOAN OF ARC as well as Elizabeth I. Thus, too, virgins are notoriously able to capture UNICORNS; frustrated adolescent girls are believed to be the power-source for POLTERGEISTS; and WITCHES' abilities often depend on virginity. In both history and fiction, princes and kings have been reluctant to marry other than virgins in case a bastard might inherit; the most famous fictional case is probably that in SHAKESPEARE's *The Tempest*, where the importance of the virginity of PROSPERO's daughter Miranda is spelt out – if not a virgin, she may not marry Frederick. Poul ANDERSON's *Operation Chaos* (**1971**) suggests that the THREE traditional phases of womanhood, Maiden, Mother and Crone, each have their own mode of MAGIC: loss of virginity means temporary incapacity while Mother-style magic is learned. Virgin status tends to be less significant for men, though it is required of one WIZARD in John BRUNNER's *The Traveler in Black* (**1971**) and is the saving of a lad in E.H. VISIAK's *Medusa* (**1929**). Christianity generally insists on the perpetual virginity not only of the Madonna (◊ GODDESS) but of CHRIST: hence the extreme outrage occasioned by portrayals of Christ figures who engage in SEX in MONTY PYTHON'S LIFE OF BRIAN (*1979*) and *The* LAST TEMPTATION OF CHRIST (*1988*) (◊ Nikos KAZANTZAKIS).

"Unspoilt" virgins are thought to enhance the magical effect of HUMAN SACRIFICE, and are the favoured tribute to DRAGONS, as in the Greek myth of Perseus and Andromeda and its numerous fantasy progeny – e.g., DRAGONSLAYER (*1981*). Math in the MABINOGION may rest his feet only in a virgin's lap. The passing of virginity is of course a significant RITE OF PASSAGE; that it should occur unwillingly, under conditions of DEBASEMENT, is the central WRONGNESS of the BLACK MASS. In Fletcher PRATT's *The Blue Star* (*1952*) the transition is the woman's INITIATION into WITCHCRAFT and also forms a permanent bond, the "great marriage", with her sexual partner. A virgin's first sexual arousal initiates her cycle of SHAPESHIFTING in CAT PEOPLE (*1942* and *1982*).

Virginity and chastity are generally regarded as pleasing to GODS as well as men – hence celibate priests, vestal virgins, and the use of "maid" as an honorific (could Maid Marian's long association with ROBIN HOOD really have been sexless?). As usual, double standards tend to apply. According to Sir James FRAZER, a Year King is expected to provide sexual satisfaction to numerous wives. Queens, however, are enjoined to chastity; GUINEVERE's adultery is seen as centrally damaging to the LAND. Michael MOORCOCK provides a REVISIONIST-FANTASY reworking of this theme in *Gloriana, or the Unfulfill'd Queen* (*1978*), whose eponymous queen experiences much joyless sex before she and her Land of Albion are healed when at last she achieves orgasm.

One tradition has it that the blood of virgins is of special quality. They are the preferred prey of VAMPIRES in many tales. Although the historical details are murky, Elisabeth de Bathóry – a vampire in a different sense – is believed to have had her servants murder some 600 virgins so that she could bathe in their blood and thus gain eternal youth (◊ TRANSYLVANIA). [DRL]

See also: FEMINISM; GENDER; SNOW WHITE; *The* WICKER MAN (*1973*).

VIRGIN MARY ◊ GODDESS.

VISIAK, E.H. Working name of UK poet, critic and writer Edward Harold Physick (1878-1972), whose best work shadows sf, HORROR and fantasy modes, employing speculative metaphysics in a manner similar to the fiction of his friend David LINDSAY. In stories of this sort, the consensual world is *argued* – sometimes in passages of considerable length – as being an expression of the DREAM world, or world of ARCHETYPES. Such texts do not easily fit into template definitions of the various genres of the FANTASTIC. EHV's first novel, *The Haunted Island* (*1910*), is clearly fantasy, and engagingly deploys GHOSTS and MAGIC in a tale of PIRATES set on a mysterious ISLAND in the 17th century. *Medusa: A Story of Mystery, and Ecstasy, & Strange Horror* (*1929*) is as hard to categorize as Lindsay's *A Voyage to Arcturus* (*1920*). The tale moves gradually, in a slow crescendo, from its beginnings in a normal-seeming 19th-century England through adventures at sea and finally into a literal pit of fantasy – a vast circular hole occupied by the eponymous SEA MONSTER which eats sexually aware men alive. The protagonist is a young boy who remains sexually innocent (◊ VIRGINITY), though haunted by other guilts: he survives while his COMPANIONS perish.

None of EHV's remaining work conveys a similar intensity. "The Shadow" – a book-length FLYING-DUTCHMAN tale (in *Crimes, Creeps and Thrills* anth **1936** ed anon John GAWSWORTH) – is overburdened with SHADOWS who emblematically represent aspects of a pirate, apparently long-dead but clearly some sort of REVENANT in search of redemption. Late in life, EHV wrote an essay for *The Strange Genius of David Lindsay* (anth **1970**). [JC]

VISIONARY FANTASY Despite the efforts of Sigmund Freud (1856-1939) and many others we still have no well founded theory which might allow us to extract meanings from actual DREAMS – or, indeed, to confirm that there are any consistent meanings to be extracted. Literary dreams are very different; if a character's dream had no relevance to the pattern of meanings contained in a story's plot there would be no point in describing it. Sometimes the connection is frivolous – the move by which a fanciful and ultimately incoherent plot is excused by declaring in conclusion that "it was all a dream" has been so overused as to be no longer acceptable (◊ RATIONALIZED FANTASY) – but in many other cases the connections are more complicated, as when dreams are credited with divinatory and precognitive powers (◊ PRECOGNITION).

Meaningful literary dreams often serve a merely mechanical function within a plot, but there is a significant subgenre of VFs in which a natural or drug-assisted dream is used to provide an imaginary space within which a fiction-within-the-fiction can be displayed or enacted. Some of these fictions-within-fictions belong to other genres, but the most interesting are fantasies crammed so full of meaning as to be supersaturated. These may be moralistic exercises, as in the CHRISTMAS BOOKS of Charles DICKENS; metaphysical rhapsodies, as in *A Night in the Luxembourg* (**1906**) by Rémy de GOURMONT; or satirical extravaganzas, as in *The Cream of the Jest* (**1917**) by James Branch CABELL. An important subcategory of VFs deals directly with the personal politics of ESCAPISM, developing theses regarding the utilities of fantasy and the hazards of its overindulgence; an interesting spectrum of modified apologetic arguments can be found in *A Dreamer of Dreams* (**1889**) by Joseph Shield NICHOLSON, *Peter Ibbetson* (**1891**) by George du Maurier (1834-1896), *The Hill of Dreams* (**1907**) by Arthur MACHEN, *Fantastic Traveller* (**1931**) by Maude Meagher (?1895-?1977), *Smirt* (**1934**) by Cabell, "Typewriter in the Sky" (**1940**) by L. Ron HUBBARD, *The Dream Quest of Unknown Kadath* (**1943**) by H.P. LOVECRAFT, *Marianne Dreams* (**1958**) by Catherine Storr (1913-) and *The Neverending Story* (**1979**) by Michael ENDE. [BS]

VISIONS Apparitions or seemings that are perceived only by the chosen recipient generally carry some numinous charge and are seen while awake – as distinct from DREAMS. William MORRIS's *The Well at the World's End* (**1896**) makes this distinction when the hero has nighttime visions of loved women: he may have been asleep, making it only a dream. GODS and ANGELS habitually send visions of themselves to mortals: the Virgin Mary (◊ GODDESS) appears thus to King Alfred in G.K. CHESTERTON's *The Ballad of the White Horse* (**1911**). Some visions employ a stripped-down symbolism or iconography, as when Sauron in J.R.R. TOLKIEN's *The Lord of the Rings* (**1954-5**) is repeatedly seen as only a single red eye. Receptiveness to visions may indicate sanctity in a context of RELIGION – though the DEVIL too can play this game; the hallucinatory temptation of St Anthony has inspired several painters, including Salvador DALI and Max ERNST. Alternatively, vision-proneness may be regarded as a

VISITEURS, LES French movie (*1993*). Gaumont/France 3/Alpilles/Amigo/Canal Plus/Languedoc Roussillon. **Pr** Alain Terzian. **Dir** Jean-Marc Poiré. **Spfx** Buf Compagne, Duboi, Jean-Marc Mouligné. **Mufx** Jacques Gastineau. **Screenplay** Christian Clavier, Poiré. **Starring** Christian Bujeau (Jean-Pierre), Marie-Anne Chazel (Ginette), Clavier (Jacquouille la Fripouille/Jacquart), Tara Gano (Witch of Malcombe), Valérie Lemercier (Frénégonde de Pouille/ Comtesse Béatrice), Jean Reno (Godefroy, comte de Montmirail), Pierre Vial (Eusebius/M. Ferdinand Eusèbe). 107 mins. Colour.

In 1122 Godefroy is made comte de Montmirail, and goes to woo the fair Frénégonde. En route, his party captures the WITCH of Malcombe, whom he proposes to burn; but she makes him see Frénégonde's father as a bear, which Godefroy kills. His life ruined, Godefroy consults the WIZARD Eusebius, who prepares a POTION that will transport Godefroy back through TIME to the moment before the deed, which he may then avert. But instead Godefroy and vassal Jacquouille are transported into the future. Just before their departure the remorseful Eusebius tells them he will leave the antidote in the château dungeon for them to find in whatever future age they may arrive. In the 1990s, after much miscomprehension, they eventually enter the household of Béatrice (today's comtesse de Montmirail), her yuppie dentist husband Jean-Pierre and their terrified/mortified children. The old château de Montmirail is now a hotel owned by the camp Jacquart, who proves a descendant of Jacquouille; there is displayed the very RING that Godefroy bears on his finger, the time paradox almost creating a rift in time's fabric. Godefroy and Jacquouille open the secret entrance to the dungeon and discover that Eusebius' grimoire has rotted with age; near it, though, is a scribbled telephone number, which leads them to the wizard's descendant Eusèbe and the recipe. Jacquouille has rediscovered a stash of jewels he stole in the 12th century and decides to stay in the 20th with strumpetish Ginette; he sends Jacquart in his place with Godefroy to the 12th century.

This TIME-TRAVEL fantasy is a madcap romp, and often hilarious; it contains elements of contemporary social SATIRE (although strictly *en passant*), of the bawdy humour of the UK comic Benny Hill (very popular in France), of Monty Python (most overtly MONTY PYTHON AND THE HOLY GRAIL [*1975*]) and of numerous SLICK FANTASIES based on the comic misunderstandings endured when historical characters are magicked into the modern age; an obvious precursor was the UK tv series CATWEAZLE (1970-71). For all this lack of apparent originality, it had an immediate freshness, as if Clavier (like Lemercier a Parisian *café théâtre* star) and Poiré had reinvented the concepts and scenarios. *LV* was colossally popular in France, breaking all box-office records for a home-produced movie and outgrossing even Steven SPIELBERG's contemporaneously released *Jurassic Park* (*1993*). [JG]

VIVIAN, E(VELYN) C(HARLES) (1882-1947) UK writer and editor, born Charles Henry Cannell, who changed his name but subsequently used his original cognomen as pseudonym on a few Oriental adventure stories, including *The Guardian of the Cup* (1930), sometimes erroneously recorded as a fantasy. His naturalistic novels occasionally contain fantasy elements, but his main contributions to the genre are his LOST-RACE stories and those volumes in the **Gees** series of detective stories – written as Jack Mann – which confront the protagonist with supernatural adversaries (◊ OCCULT DETECTIVES).

In *City of Wonder* (1922), explorers of the legendary CITY of Kir-Asa encounter sinister ghostly guardians left over from the Theosophists' (◊◊ THEOSOPHY) version of LEMURIA. *Fields of Sleep* (1923) sees a last remnant of the Babylonian empire entrapped by an unbreakable addiction to the scent of a flower; its sequel, *People of the Darkness* (1924), features an UNDERWORLD inhabited by descendants of an unhuman race which once lived in ATLANTIS. *The Lady of the Terraces* (1925) involves survivors of a pre-Incan civilization whose heyday is described in a tale narrated by the protagonist of the earlier novel, *A King There Was* (1926). *Woman Dominant* (1929) is another romance of South American exploration, describing a tribe in which the women have reduced their menfolk to docile passivity by means of a drug. Although sometimes rather slapdash in execution, ECV's are among the more interesting 20th-century lost-race stories.

Although *Gees' First Case* (1936) was perfectly straightforward, the second in the series, *Grey Shapes* (1937), involved Gees with WEREWOLVES who are relics of an ancient race, memory of which is preserved in the mythology of the Sidhe (◊ FAIRIES). *Nightmare Farm* (1937) confronts Gees with the supernatural beings who once guarded Kir-Asa (making the novel a sequel of sorts to *City of Wonder*). After one more mundane adventure Gees faced a much sterner challenge from another magically adept survivor of the Azilian race first featured in *Grey Shapes*, whose background is more extensively described. The introspective and sombre tone of this novel was carried over into *The Ninth Life* (1939), a FEMME-FATALE story about a survivor from Ancient Egypt; this soon abandons the pretence of being a detective story and becomes far more interesting. The lovelorn Gees, unable to see that he ought to get it together with his omnicompetent secretary, again does little detecting in *Her Ways are Death* (1939), in which he fails to save another *femme fatale* – who is likewise doomed, in spite of supernatural assistance from Thor, by virtue of being under the sway of evil deities. The syncretic mythological system underlying the mysteries is given its fullest elaboration here, seemingly carrying far more conviction than would be required of a mere literary device. The last book in the series, *The Glass Too Many* (1940), re-uses a motif from *Maker of Shadows*, setting it in a conventional country-house murder mystery. ECV had earlier written as Mann a series of boys' adventure stories, one of which – *Coulson Goes South* (1933) – briefly involves its clean-cut hero with some trivial magical shenanigans. [BS]

Other works: *Passion-Fruit* (1912), associational; *Dead Man's Chest* (1934), associational.

See also: HUTCHINSON'S MAGAZINES; NOVEL MAGAZINE.

VIVIANE ◊ LADY OF THE LAKE.

VOID ◊ MAGAZINES.

VOLCANO MONSTER, THE vt of *Gigantis* (*1955*). ◊ GODZILLA MOVIES.

VOLERE VOLARE Italian animated/live-action movie (*1991*). Metro/Bambu/Pentafilm/Omni. **Pr** Ernesto di Sarro. **Exec pr** Silvio Berlusconi, Mario & Vittorio Cecchi Gori. **Dir** Guido Manuli, Maurizio NICHETTI. **Spfx** Movie Engineering Natali. **Screenplay** Manuli, Nichetti. **Starring** Angela Finocchiaro (Martina), Nichetti (Maurizio), Patrizio

Roversi (Patrizio), Mariella Valentini (Loredana). 96 mins. Colour.

Martina is not quite a call girl: she caters for her clients' erotic fantasies (e.g., allowing her nude form to be covered in chocolate by a kinky chef), but apparently without intercourse. Shy, incompetent Maurizio is, with half-brother Patrizio, in the business of putting sound effects onto foreign movies: he does old ANIMATED MOVIES while Patrizio does blue movies. Martina, badly guided by semi-prostitute friend Loredana, seeks the ideal man, but bumps into Maurizio, who inadvertently accompanies her to fulfil a couple of her "commissions". Her clients like him, and her business falls off when she is unable to bring him again. She suggests a partnership, but as they negotiate he discovers his hands have become those of a TOON, and are now independent of his body and will. When, some while later, he and Martina return to her apartment, his hands strip him; but then the rest of his body becomes a toon's, and he flees. Returning, his body bandaged (as per the INVISIBLE MAN), he confesses all. She realizes she has found her ideal man – the one who doesn't really exist. And so to bed.

In terms of both animation and fantastication, *VV* is a less ambitious movie than WHO FRAMED ROGER RABBIT (*1988*) and COOL WORLD (*1992*), the two other principal animated/live-action movies of the period. Yet it has its individual interest in that, rather than focus on the gulf between toondom and humanness, it displays by example how negligibly narrow that gulf is; for example, even before Maurizio's transformation it is clear he is already more of a toon figure than most toons are – Martina later, on first seeing the toon Maurizio, remarks as she eyes his trembling nakedness, "Well, you haven't changed all that much." Except, paradoxically, that he is in many ways less vulnerable – and evidently more sexually attractive – as a toon. [JG]

VOLSKY, PAULA (? -) US writer whose debut novel, *The Curse of the Witch-Queen* (**1982**), comprises a genial tangle of multiple overlapping CURSES, including one which compels its victim to overeat. The **Sorcerer's Lady** trilogy begins as URBAN FANTASY in a CITY recalling 17th-century VENICE. *The Sorcerer's Lady* (**1986**) sees the heroine's arranged marriage to a WIZARD, whom she comes to LOVE despite his arrogance and hubris, which lead to his death. In *The Sorcerer's Heir* (**1988**) she hides with their son amid UNDERGROUND allies through whom the growing son obsessively plans revenge. The heroine hesitates rather ineffectually until others eventually act to save the city in *The Sorcerer's Curse* (**1989**). This sequence is occasionally reminiscent of Jack VANCE, to whom more obvious homage is paid in the engagingly good-humoured *The Luck of Relian Kru* (**1987**). The hapless Kru is cursed with comic ill-luck stemming from his untrained wizard TALENT: pursued by a suave assassin, he takes refuge with a TRICKSTER wizard who coerces him into hazardous QUESTS. An interesting system of MAGIC emerges, involving complex finger exercises developing into literal prestidigitation by reaching through OTHERWORLD dimensions. All ends well.

Illusion (**1991** UK) marks the rough bicentennial of the French Revolution, whose events it painstakingly mirrors. Magic, which like the aristocracy who can wield it has suffered THINNING, consists of either ILLUSION or the animation of TECHNOFANTASY machines known as "Sentients". These form metaphors for aspects of the Terror: a flame-belching serpent that terrorizes the streets,

an insectile spymaster whose flying workers replace informers, and a death machine resembling an automated Iron Maiden which bloodily represents the guillotine. These props are effective enough to make *Illusion* more than merely a disguised historical romance. *The Wolf of Winter* (**1993**) has an atmospherically bleak sense of Dark Ages chill, as a monarch's youngest brother, initially commanding sympathy as a bullied weakling, is aided by an unreliable MENTOR to find strength through NECROMANCY and associated drugs. He proceeds, Macbeth-like, to murder his way to the throne. Here necromancy is the forcing of one's will on earthbound SPIRITS of the dead: every graveyard holds a potential army. The remote LIBRARY where the usurper's niece has been hidden proves to harbour "white necromancers" wishing to free all such spirits through EXORCISM; magical war follows, and the villain's predestined end holds some pathos. PV has a knack for interesting variants of magic. [DRL]

VON AUE, HARTMANN (?1160-?1215) German chronicler. ◊ ARTHUR.

VON ESCHENBACH, WOLFRAM (?1170-?1220) German chronicler. ◊ ARTHUR; PERCEVAL.

VON STRASSBURG, GOTTFRIED (?   -?1210) German chronicler. ◊ ARTHUR.

VOODOO The religious folk cult of the West Indies, especially in Haiti, and the Creoles of the Southern USA. The term comes from the *vodun*, or snake-god worship of Benin (formerly Dahomey) in West Africa. It focuses heavily on the worship of the *loa*, spirits of local gods and ancestors, who often demand ritual SACRIFICE. During a cult service a *loa* may possess (◊ POSSESSION) an individual. This may extend to the dead. ZOMBIES are the most outward manifestation of voodoo, though in fact is a minor element of the religion. Early misconceptions of native African religions led to many words and phrases still associated with pagan worship and WITCHCRAFT, especially mumbo-jumbo, hoodoo (a variant of voodoo popularized by Lafcadio HEARN) and the placing of hexes (◊ CURSES).

Voodoo was much misunderstood by White Americans and popular journalism, especially that by Hearn and George W. Cable (1844-1925), the latter in *Old Creole Days* (coll **1879**), served to perpetuate these beliefs among colourful interpretations. The best SUPERNATURAL FICTIONS incorporating voodoo have been by those writers who have directly witnessed and experienced it, including Grant ALLEN, Hugh B. CAVE and Henry S. WHITEHEAD. William B. Seabrook (1886-1945) narrated his experiences in *The Magic Island* (**1929**), episodes from which are often reprinted in ANTHOLOGIES. Dennis WHEATLEY sensationalizes the more popular concepts in *Strange Conflict* (**1941**), but Patricia GEARY offers a more intelligent usage in *Strange Toys* (**1987**). In *Night Boat* (**1980**) Robert R. McCammon (1952-) used a voodoo curse placed on a U-boat to convert the whole crew into the undead.

Most relevant movies have concentrated on the zombie theme (◊ ZOMBIE MOVIES), usually without finesse. However, *The Ghost Breakers* (**1940**), starring Bob Hope and Paulette Goddard, was especially effective, and *I Walked With a Zombie* (**1943**), dir Jacques Tourneur (1904-1977) and scripted Curt Siodmak (1902-), treated the theme sympathetically.

Voodoo! (anth **1980**) ed Bill Pronzini (1943-) is one of the few ANTHOLOGIES to go beyond the zombie theme. [MA]

WAGENKNECHT, EDWARD (CHARLES) (1900-   )
US academic and anthologist, one of the earliest to champion SUPERNATURAL FICTION in the literary mainstream. In addition to numerous studies of US literature and CINEMA, EW edited a number of early anthologies, many presenting works of fantasy long unavailable (or never available) to a general audience: *Six Novels of the Supernatural* (anth **1944**), *The Fireside Book of Ghost Stories* (anth **1948**) and *Murder by Gaslight: Victorian Tales* (anth **1949**). EW wrote the introduction to (and probably assembled) Walter DE LA MARE's posthumous *Eight Tales* (coll **1971**) and also edited *The Supernaturalism of New England* by John Greenleaf Whittier (coll **1969**), *The Letters of James Branch Cabell* (coll **1975**) and an anthology of critical studies, *Seven Masters of Supernatural Fiction* (anth **1991**). His own critical studies include: *Utopia Americana* (**1929** chap), a study of L. Frank BAUM; *Edgar Allan Poe: The Man Behind the Legend* (**1963**); and volumes, usually introductory, on Geoffrey CHAUCER, Washington IRVING, John MILTON, Sir Walter SCOTT, William SHAKESPEARE, Mark TWAIN and others.   [GF]

WAGNER, KARL EDWARD (1945-1994) US writer and editor who wrote several SWORD-AND-SORCERY novels before working almost exclusively in the HORROR genre. KEW's first novel, *Darkness Weaves with Many Shades* (**1970**; rev vt *Darkness Weaves* 1978) serves to introduce his ANTIHERO Kane, an immortal warrior clearly in the tradition of Robert E. HOWARD's CONAN, but in whom can also be seen aspects of Melmoth and other GOTHIC-FANTASY archetypes. Kane is further distinguished from his barbarian brethren by his intelligence, introspection and humour. Kane's adventures continue in *Death Angel's Shadow* (**1973**), *Bloodstone* (**1975**), *Dark Crusade* (**1976**) and several stories, which are collected in *Night Winds* (coll **1978**) and *The Book of Kane* (coll **1985**). Over the course of these, KEW gradually amplifies his grim protagonist's characteristics while developing many intriguing secondary characters and continuing to provide some welcome twists. The **Kane** books helped reinvigorate the S&S subgenre, providing some interesting variations to a category that had previously been moribund. The series is marked by a propensity toward supernatural adversaries, a tendency that seems to portend KEW's later move to horror.

KEW also worked more directly with Howard's legacy, writing a sequel to the **Bran Mak Morn** series, *Legion from the Shadows* * (**1976**), plus the **Conan** novel *The Road of Kings* * (**1979**) and the screenplay for an unproduced CONAN MOVIE, and re-editing several **Conan** titles for Berkley.

With David DRAKE, KEW cowrote *Killer* (**1985**), a historical fantasy set in Imperial Rome and involving a contest between a beast hunter and a supernatural creature. KEW's short horror fiction is extremely accomplished, although marked by an increasingly dark, brooding tone in later years. Horrors are hidden beneath the surface in both ".220 Swift" and "Where the Summer Ends"; in the former, an updating of Arthur MACHEN's "Novel of the Black Seal", a LOST RACE of degenerate "little people" lead a subterranean existence in the Appalachians, while in the latter a prolifically invasive vine known as *kudzu*, imported from Japan, threatens to overgrow areas of the US South and conceals the presence of nasty creatures beneath its green carpet. KEW won a WORLD FANTASY AWARD in 1983 for his VAMPIRE/REINCARNATION novella "Beyond Any Measure", and won three BRITISH FANTASY AWARDS for his short fiction, plus a Special British Fantasy Award for editing (1983).

As editor, KEW was responsible for 15 volumes of DAW's **Year's Best Horror** series, which included some fantasy, and for the three-volume **Echoes of Valor** series, which reprinted S&S tales from the PULP era. He also briefly tried his hand at publishing, acting as a partner in the firm Carcosa House (winner of a 1975 World Fantasy Award for Special Award), which during the 1970s published single-author collections of pulp-era material by Manly Wade WELLMAN, E. Hoffman PRICE and Hugh B. CAVE.   [BM]

Other works: *Sign of the Salamander* (**1975** chap); *In a Lonely Place* (coll **1983**); *Why Not You and I?* (coll **1987**); *Unthreatened by the Morning Light* (coll **1989** chap).

As editor: The Year's Best Horror anthologies *#8* (anth **1980**), *#9* (anth **1981**), *#10* (anth **1982**), *#11* (anth **1983**), *#12* (anth **1984**), *#13* (anth **1985**), *#14* (anth **1986**), *#15* (anth **1987**), *#16* (anth **1988**), *#17* (anth **1989**), *#18* (anth **1990**), *#19* (anth **1991**), *#20* (anth **1992**), *#21* (anth **1993**) and *#22* (anth **1994**), assembled as *HorrorStory, Volume 3*

(1992) containing #8 and #9, HorrorStory, Volume 4 (1990) containing #10, #11 and #12, HorrorStory, Volume 5 (1989) containing #13, #14 and #15; the **Echoes of Valor** anthologies, being *Echoes of Valor* (anth 1987), *II* (anth 1989) and *III* (anth 1991); *Intensive Scare* (anth 1990).

WAGNER, RICHARD (1813-1883) German composer, the most widely significant theorist and OPERA composer/librettist of the 19th century. His concept of *Gesamtkunstmusik* – through which the interdependence of music, words and stagecraft was argued as being as necessary to any art capable (as he declared his own operas were) of fully expressing the soul of a nation – had a revolutionary effect on the world of music, and a perhaps less happy effect on the world in general. The idea that opera could express the soul of a nation was dependent upon the equally radical belief that nations had souls; the history of the 20th century has amply demonstrated the perniciousness of the concept.

Ten of RW's 13 completed operas are fantasies. *The Fairies* (**1833**), from Carlo GOZZI's *La Donna Serpente*, is set in FAERIE. *The Flying Dutchman* (**1843**) is a classic FLYING-DUTCHMAN story. *Tannhäuser* (**1845**) and *Lohengrin* (**1850**) are dramatizations of German LEGENDS. The **Ring** cycle – *Das Rheingold* (**1853**), *Die Walküre* (**1856**), *Siegfried* (**1856-71**) and *Götterdämmerung* (**1869-74**) – incorporates much Teutonic mythology. *Tristan and Isolde* (**1859**) expands a sidebar to the ARTHUR cycle into an extraordinary love story, involving a famous love POTION. *Parsifal* (**1882**) is also set within the Arthurian cycle (◊ PERCEVAL).

The long dramatic poems which RW set to music are highly readable in their own right. [JC]

WAIN, LOUIS (1860-1939) UK artist. ◊ ILLUSTRATION.

WAINSCOTS It has always been an assumption that behind the wainscots one may find invisible societies of animals – MICE AND RATS, etc. It follows that invisible or undetected societies living in the interstices of the dominant world – normally but not necessarily human – can be called "wainscot societies".

These societies – "wainscots" for short – are sometimes made up of humans indistinguishable from normal humans except for where they live; in fantasy wainscots comprising normal humans are uncommon, though they sometimes feature in sf texts like William Tenn's *Of Men and Monsters* (**1968**). The rooftop London gangs (◊ LONDON) featured in Christopher FOWLER's *Roofworld* (**1988**) form a wainscot, as do the eponymous "rats" in Stephen Elboz's *The House of Rats* (**1991**) – the "rats" in question being humans who, evicted from their village, have created "a home between the walls and beneath the floors" of a vast, doomed EDIFICE, and the street people of the tv series *Neverwhere* (1996) written and novelized by Neil GAIMAN. The labyrinthine nature of the CITY in most URBAN FANTASY evokes a landscape of intersecting wainscots – though in the dance of plots typical of urban fantasy it is often difficult to determine who is hiding from whom. But certainly the society ruled by Horrabin the beggar-king in Tim POWERS's *The Anubis Gates* (**1983**) is a wainscot, as is the empoldered underground society featured in Phyllis EISENSTEIN's "Subworld" (**1983**) (◊ NEW YORK). HORROR texts are often riddled with cellar wainscots, which may contain humans who have degenerated; horror texts tend also to feature invasions from the wainscot, although this is not inevitable: in Clive BARKER's NIGHTBREED (*1990*) the horror occurs when the outside (human) world attempts to exterminate the wainscot.

More usually the wainscots that appear in FANTASY texts relate to the dominant world in three main fashions: (a) They may be distinguished from the dominant world specifically through the nature of their inhabitants, who may differ from "normal" humans in size (usually by being tiny), in scarcity, by species (◊ ANIMAL FANTASY) or other basic fantasy criteria – they may be dead, or invisible or immortal; they may be SHAPESHIFTERS; they may be DOUBLES of those who live in the open. (b) They may *seem* visible to the world, but in fact conduct their true lives in private – it is here that the notion of the wainscot feeds into the notion of the secret organization which covertly governs the world (◊ FANTASIES OF HISTORY). (c) The wainscot may constitute an OTHERWORLD which is CROSSHATCHED with ours, in which case the assumption of human or "normal" dominance may become problematical; the inhabitants of this crosshatched otherworld may possibly be spellbound humans, but are more likely to be one or other of the species of FAERIE, as in Lisa GOLDSTEIN's *Strange Devices of the Sun and Moon* (**1993**). Frequently in stories of the latter sort an ANIMATE/INANIMATE dynamic will operate, and it will be found that features of the world that seem inert turn out to be alive, like (in the Goldstein novel) the treestump that turns out to be a man pointing the way to Faerie.

Examples are very numerous. The Borrowers in Mary NORTON's **Borrowers** sequence – filmed in part as *The* BORROWERS (*1973* tvm) – are small, as are Miss Bianca and the other mice in Margery SHARP's **Rescuers** sequence – filmed in part by DISNEY as *The Rescuers* (**1977**) and *The Rescuers Down Under* (**1990**) – the Wildkeepers in Nigel GRIMSHAW's **Wildkeepers** sequence, the Smalls in Charles DE LINT's *The Little Country* (**1991**), the eponymous clan of rag DOLLS in the **Mennyms** sequence by Sylvia WAUGH, the rat inhabitants of Lankhmar Below in Fritz LEIBER's *The Swords of Lankhmar* (**1968**), and the Nomes in Terry PRATCHETT's **Nomes** sequence. *Mistress Masham's Repose* (**1946** US) by T.H. WHITE features a wainscot community of Lilliputians, brought back to England by Gulliver. The protagonist of Raymond BRIGGS's *Fungus the Bogeyman* (graph **1977**) is a member of a somewhat loathly wainscot. There are two separate hidden nations in Diana Wynne JONES's *Power of Three* (**1976**), one underwater, one underground. The department-store dwellers in John COLLIER's "Evening Primrose" (**1941**) depend on the PERCEPTION that they are in fact mannequins. The antagonists who haunt Christopher PRIEST's *The Glamour* (**1984**) are effectively invisible, also through a control of perception. The *tesh* in Ian MacDonald's "Some Strange Desire" (1993) are shapeshifters. The Shadow Boy in Michael SWANWICK's *The Iron Dragon's Daughter* (**1993** UK) doubles the protagonist.

The family of WITCHES in one of Ray BRADBURY's early story sequences – including "The Traveller" (1946), "Homecoming" (1946), "Uncle Einar" (1947) and "The April Witch" (1952) – constitute a wainscot of the sort whose inhabitants inhabit the world in disguise, but conduct their genuine lives invisibly, as do the ramified and hierarchical society of WIZARDS in Diane DUANE's **Wizard** sequence, the concourse of all those born in a caul who periodically rescue the world in Guy Gavriel KAY's *Tigana* (**1990**), the Crafters in the **Crafters** SHARED-WORLD anthology sequence ed Christopher STASHEFF and Bill Fawcett, and the VAMPIRES in Dan SIMMONS's *Carrion Comfort* (**1989**). Zenna Henderson made a career almost solely out of a long series of wainscot tales, **The People**, where the human-like beings are extraterrestrials stranded on Earth – but they could as well be fairies, since their superhuman abilities are really just TALENTS flimsily clad as sf.

ELVES and brownies and other creatures – as in "The Gnurrs Come from the Voodvork Out" (1950) by Reginald Bretnor – frequently impact upon the human world from wainscot havens. Fables in which brownies (or hobs, etc.) secretly interact with favoured humans are extremely numerous; modern examples include William MAYNE's *Hob and the Goblins* (1993) and Barbara HAMBLY's "The Little Tailor and the Elves" (1994). More significant today is the complex interface where two worlds – the normal world and the otherworld – crosshatch; when Smoky Barnable, in John CROWLEY's *Little, Big* (1981), travels from New York to Edgewood he traverses a THRESHOLD into a crosshatched world inhabited by denizens of Faerie as well as characters seemingly shaped from the FAIRYTALES written by the man who will become his father-in-law. One world – in novels of this sort – is generally invisible to the other, except at points of crisis. In Midori SNYDER's *The Flight of Michael McBride* (1994), for instance, it is only when the protagonist needs help to escape that his world becomes suddenly animate, and he glimpses the true crosshatch complexity of things.

Insofar as they are designed to survive, and are frequently described in terms that emphasize the vigilance necessary to maintain them, many wainscots are so similar to POLDERS that the two terms can very often be used to describe the same setting. It is not surprising, therefore, that in the structure of the fantasy STORY both wainscot and polder tend to be found at those points where THINNING is being experienced in the world or by protagonists.   [JC]

WAITE, A(RTHUR) E(DWARD) (1857-1942) US-born UK writer, a foremost explainer of modern mysticism during the FIN-DE-SIÈCLE occult revival which influenced the fantasy of writers from Arthur MACHEN to H.P. LOVECRAFT and Charles WILLIAMS. AEW was a leader of and ritual-writer for mystical orders such as the Hermetic Order of the GOLDEN DAWN and the Fellowship of the Rosy Cross (◊ ROSICRUCIANISM), which latter he founded. His study of the 17th-century Rosicrucian movement, *The Brotherhood of the Rosy Cross* (1924), is a significant historical work. *The Secret Doctrine in Israel* (1913) is a study of the CABBALA. AEW was also a poet and a writer of Victorian LITERARY FAIRYTALES – e.g., *Prince Starbeam* (1889) and *The Golden Stairs* (1893). His memoir, *Shadows of Life and Thought* (1938), tells much about the literary side of modern mysticism.   [DMH]

WAKEFIELD, H(ERBERT) RUSSELL (1888-1964) UK writer, one-time private secretary to Lord Northcliffe, noted for his GHOST STORIES, some of which rank alongside those of M.R. JAMES, with whom he is sometimes compared. Although most of his stories are formulaic they are well crafted and frequently atmospheric, and often feature vengeful ghosts (◊ VENGEANCE). HRW was first inspired to write by an experience he had at a reputedly haunted house in 1917; this resulted in "The Red Lodge", in his first volume *They Return at Evening* (coll 1928). Other collections are *Old Man's Beard* (coll 1929; vt *Others Who Returned* 1929 US), *Imagine a Man in a Box* (coll 1931), *Ghost Stories* (coll 1932), *A Ghostly Company* (coll 1935), *The Clock Strikes Twelve* (coll 1940; cut vt *Stories from The Clock Strikes Twelve* 1961 US) and *Strayers from Sheol* (coll 1961), this last published by ARKHAM HOUSE. Richard DALBY compiled a volume of HRW's stories – including the later uncollected ones – as *The Best Ghost Stories of H. Russell Wakefield* (coll 1978).   [MA]

WALEY, ARTHUR (DAVID) Pseudonym of UK scholar, critic, translator and writer Arthur David Schloss (1889-1966), whose renderings of Japanese and Chinese literature – beginning with *170 Chinese Poems* (anth trans 1918) – were of central importance in bringing much great poetry and fiction to the West. Of greatest importance to fantasy is his version of *Hsi Yü Chi* ["Journey to the West"] (*c*1550) by Wu Ch'êng-ên (*c*1505-*c*1580), which he rendered in his partial translation as *Monkey* (trans 1942), after the TRICKSTER demigod (◊ MONKEY) who accompanies the priest Hsüan-tsang on a marvel-filled journey to India in search of the Tripitaka or "three baskets" containing Buddhist scriptures in scroll form. In *The Real Tripitaka and Other Pieces* (coll 1952) AW provides historical background for the pilgrimage described in *Monkey*; the rest of the volume incorporates *The Lady who Loved Insects* (1928 chap), along with other original and adapted ORIENTAL FANTASIES, including "Mrs White", about a SHAPESHIFTING white python who becomes enamoured of a mortal.   [JC]

WALKER, WENDY (1951-   ) US writer whose first collection, *The Sea-Rabbit, or The Artist of Life* (coll 1988) applies an alchemy of FEMINISM to the conventions of fantasy and fable. Her elaborate *The Secret Service* (1992) recounts the efforts of English spies to undo a conspiracy against the young King of England in an ALTERNATE-WORLD 19th-century. The spies have the power to shapeshift (◊ SHAPESHIFTERS). Walker's brilliant touch is in the sensual passages mapping changes in the spies' PERCEPTIONS of the world as they become a goblet, a rose or a statue. *Stories out of Omarie* (coll 1995) collects tales based on the lays of Marie de France (◊ ROMANCE): they are exotic and medieval in tone but Postmodern in their reinterpretation of the role of women and the powers of the storyteller.   [HW]

WALL, MERVYN (1908-   ) Irish writer, initially active as a playwright, secretary to the Irish Arts Council 1957-74. His first novel, *The Unfortunate Fursey* (1946), is a polished comedy in which various haunters plaguing an Irish monastery attach themselves to a luckless lay brother, who is thrown out into the wicked world; magic powers are foisted upon him by a WITCH, but SATAN takes a liking to him. More lighthearted than Anatole FRANCE's exercises in Literary Satanism, but just as sentimental, it is perhaps the finest work in that vein written in English. In *The Return of Fursey* (1948) the harassed hero recklessly sells his SOUL to the only real friend he has ever had but subsequently finds that SATAN is in direly reduced circumstances. The two novels were assembled as *The Complete Fursey* (omni 1985). *The Garden of Echoes, A Fable for Children and Grown-Ups* (1982; 1988), whose young heroines thwart a plan to assassinate SANTA CLAUS, first appeared in a special double issue, entirely devoted to MW's work, of the US-based *The Journal of Irish Literature*.   [BS]

Other works: *A Flutter of Wings* (coll 1974).

WALLOP, DOUGLASS Working name of US writer John Douglass Wallop III (1920-1985), whose *The Year the Yankees Lost the Pennant* (1954), was made into the movie *Damn Yankees* (*1958*) (◊ BASEBALL); it tells how a PACT WITH THE DEVIL leads to what was in 1954 a remarkable playoff defeat for the New York Yankees. *What Has Four Wheels and Flies?: A Tale* (1959) is a BEAST-FABLE in which a pack of dogs gets into trouble with a CAR that more or less drives (and flies) itself.   [JC]

Other works: *The Mermaid in the Swimming Pool* (1968).

WALPOLE, HORACE (1717-1797) UK writer and man of

letters, son of the prime minister Sir Robert Walpole (1676-1745) and (from 1791) the 4th Earl of Orford. HW's impact on GOTHIC FANTASY was immense. In 1747 he bought the house of Strawberry Hill, near Twickenham; over the next 20 years he developed it into an ornate Gothic castle, thus establishing an interest in that artform. He extended this to the Gothic novel with *The Castle of Otranto* (dated **1765** but 1764), the seminal work of Gothic fantasy. HW's love of hoaxes led him to issue this as a translation by William Marshal from the Italian of Onuphrio Muralto, purportedly published in Naples in 1529. HW claimed as his theme the natural retribution arising from "the sins of the fathers". The book was immensely popular and set in motion not just the Gothic novel but the whole genre of SUPERNATURAL FICTION.

HW did not attempt to repeat this success, though he came close with *The Mysterious Mother* (**1768**), a non-supernatural novel which follows through the consequences of discovering the family curse of incest. It was a shocking theme for its day. HW printed it from his own Strawberry Hill press, which he had established in 1757, the first private press of any importance in England. It was later issued in a larger print run by the firm of Graham & Dodlsey in 1781. The book may have appalled the public, but it set the trend for the more provocative works of Gothic fantasy, especially *Vathek* (**1786**) by William BECKFORD and *The Monk* (**1796**) by Matthew Gregory LEWIS.

Most of HW's other works, including those he published through his press, were books and catalogues on art and literature. The rarest volume is *Hieroglyphic Tales* (coll **1785**; rev with addition of a previously uncollected story 1982 ed Kenneth W. Gross), printed in an edition of only six copies and only more widely available when published in HW's collected *Works* (omni **1798**). This is a group of FAIRYTALES, clearly commentaries upon HW's political and social experiences, drawing from ARABIAN FANTASY, CELTIC FANTASY and ORIENTAL FANTASY. They show the influence of stories in translation by Madame d'AULNOY and particularly of Jonathan SWIFT's *Gulliver's Travels* (**1726**).

HW's love of the extravagant led him to coin the word "serendipity" in 1754, which he concocted from the Arabian fantasy "The Three Princes of Serendip", in which the princes are always making chance discoveries. A spoof was *An Account of the Giants Lately Discovered* (**1766** chap), which purported to tell of GIANTS discovered in Patagonia during a recent voyage by Admiral John Byron (1723-1786). [MA]

WALPURGISNACHT ◊ REVEL.

WALTON, EVANGELINE Working name of US writer Evangeline Walton Ensley (1907-1996), best known for her quartet of books based on the four branches of the MABINOGION. The first of these, *The Virgin and the Swine* (**1936**; vt *The Island of the Mighty* 1970) is the most accomplished, deftly blending the ancient Welsh LEGENDS to ensure a mixture of myth and didacticism which gives the tales their strength. After the reprinting of this volume in the **Ballantine Adult Fantasy** series under the editorship of Lin CARTER, her second book in the series came to light and was published as *The Children of Llyr* (**1971**), and she completed the quartet with *The Song of Rhiannon* (**1972**) and *The Prince of Annwn* (**1974**).

A sickly child, EW turned to writing from an early age. Stories written in her twenties finally saw print as "Above Ker-Is" (in *The Fantastic Imagination II* anth **1978** ed Robert H. Boyer and Kenneth J. Zahorski) and "The Mistress of

Kaer-Mor" (in *The Phoenix Tree* anth **1980** ed Boyer and Zahorski), both of which show her interest in legends and FOLKTALES. EW also wrote *Witch House* (**1945**; rev 1950 UK), an atmospheric HAUNTED-DWELLING tale.

EW worked on several historical novels. *The Cross and the Sword* (1956; *Son of Darkness* 1957 UK) is set at the time of the Viking invasion of Britain, and has minimal fantasy content. Her series about Theseus was put to the side because of work by Mary RENAULT, and only one volume finally appeared, *The Sword is Forged* (**1983**). She received a Special WORLD FANTASY AWARD in 1985 and a Lifetime Achievement Award in 1989. [MA]

WANDERING JEW From the first written version of the tale – in a Latin chronicle set down in Bologna in 1223 – the WJ appears more frequently in the literatures of Europe than almost any other single figure. He is the central ACCURSED WANDERER of Europe, and, though the pattern of his story has profound archetypal roots (the LEGEND of the WILD HUNT – with its hints of Christianity's victory over the worship of ODIN – is a close cousin), he is so potent a figure that he exists independently of any source.

The basic STORY is straightforward. At some point on the road to Calvary, weary from carrying the cross on which he will be crucified, CHRIST rests at a door for an instant; in some versions he asks the owner of the house – most commonly a cobbler from Jerusalem named Ahasuerus – for some water. But Ahasuerus rebuffs Christ angrily, saying "Get off! Away with you!" or "Go where you belong!" Christ responds either "Truly I go away, but tarry thou until I come again" or "I sure will reste, but thou shalte walke" (in the version quoted in Thomas Percy's *Reliques of Ancient English Poetry* [anth **1765**]). In either wording the message is clear. Ahasuerus – or Cartaphilus, or Buttadeus, or Isaac Laquedem – is condemned to wander the Earth without surcease until the Second Coming. Sometimes his IMMORTALITY takes the form of eternal middle age; sometimes he ages from early manhood to senescence, then falls into a stupor only to reawaken as a young man. In early recountings, Ahasuerus is seen as haunted, haggard, irreproachably and constantly remorseful; in later versions he remains haunted but is more and more frequently depicted as a man made wise by experience. He is occasionally allowed to die – like his less resonant seafaring latecomer analogue, the FLYING DUTCHMAN.

There is no real mystery behind the widespread appeal of the WJ. IMMORTALITY itself – though for the most part conscientiously disparaged in all but the most recent versions of the legend – is of course an intriguing subject, and much longed for, at least in the imagination of readers. But immortality needs a story; the initiating scene that begins the tale of Ahasuerus is dramatic and detailed, and his consequent fate is to remain embedded in history. This interminable BONDAGE to the world gives Ahasuerus unlimited access to experience, bestowing upon him a far more relevant worldly wisdom than is granted to Cain. Moreover, not only does Ahasuerus, by remaining eternally alive, sanction Christian belief in the Second Coming – an event he validates simply by continuing to await it – but, even more resonantly and romantically, the subtext of his tale is that of the triumph of Christianity over the elder faiths: his long durance is a *parading* of the defeated foe.

The main reason for the popularity of the tale may be the simplest explanation of all. Unlike that of FAUST, DON JUAN, HERNE THE HUNTER or the KNIGHT OF THE DOLEFUL

COUNTENANCE (or any other iconic figure from European cultures), the tale of the WJ is one of dramatic irony, a parade which passes in *secret* (though the reader is often given that secret, or guesses it, long before the rest of the cast is allowed to know). Because he is immortal, and because he is cursed, Ahasuerus never lives in the open; in almost any rendition of the tale, his identity must always be discovered, or confessed. Almost every narrative in which he appears – putting to one side the relatively late category of the first-person reminiscence – plays on secrecy, and gains suspense through delaying the moment of revelation. The WJ is an engine of story.

After the 13th-century manuscripts – the most notable being the *Chronica Majora* of Matthew Paris (*c*1200-1259) – the first significant appearance of the legend in print comes in a 1602 pamphlet which purports to recount various sightings of the WJ during the previous century. This pamphlet was variously expanded, modified, and translated into various languages, ultimately inspiring texts like *The Wandering Jew Telling Fortunes to Englishmen* (**1640**); and by the middle of the 18th century the basic story was profoundly familiar. By around 1770 in Germany, the Sturm und Drang ["Storm and Stress"] Movement had begun violently to promulgate images of unmitigable passion and openness, of rebellious vitality as embodied in great men; and the story of the WJ – sometimes recounted through imagery perhaps more appropriate to the story of PROMETHEUS – provided an extremely useful TEMPLATE. *Der ewige Jude* (**1783**) by Christian Schubart (1739-1791) was the first text fully to exploit the image of the WJ as a larger-than-life wanderer; its translation into English in 1801 was influential upon Percy Bysshe SHELLEY and others. Other German Romantics who composed versions include Nikolaus Lenau (1802-1850), Clemens Brentano (1778-1842) and Adalbert von Chamisso (1781-1838). In England, the WJ appears in Matthew LEWIS's *The Monk* (**1796**), in *St Leon* (**1799**) by William Godwin (1756-1836), in George Croly's *Salathiel* (**1828**), in *The Undying One* (**1830**) by Caroline Norton (1808-1877), in George MACDONALD's *Thomas Wingfold, Curate* (**1876**) and in Philip Norton's *Sub Sole* (**1890**). In France, Ahasuerus features in Jan POTOCKI's *Manuscrit Trouvé à Saragosse* (**1804** and later), in Edgar Quinet's *Ahasvérus* (**1833**) – an extremely complex narrative poem in which the WJ, accompanied by a fallen ANGEL who loves him, passes down the centuries as the central figure of a PARIAH ELITE – most famously in Eugène SUE's *The Wandering Jew* (**1844-5**) and in Alexandre DUMAS's *Isaak Lakadam* (**1853**), which remained unfinished. Other 19th-century versions of the story include Hans Christian ANDERSEN's *Ahasverus* (**1844**), an epic poem which remains untranslated.

Only a few of the innumerable 20th-century versions need mentioning. The figure is mocked by Guillaume APOLLINAIRE in *L'Heresiarch et cie* (coll **1910**; trans as *The Heresiarch and Co* **1965** US). In Gustav MEYRINK's *Das grune Gesicht* (**1916**; trans as *The Green Face* **1992** UK) the WJ is the eponymous MASK-like visage which prefigures the end of the world. An extremely popular play, *The Wandering Jew* (**1920**) by E. Temple THURSTON, carries the WJ through world history until the 16th century, at which point he is allowed a redemptive death; it was filmed as *The Wandering Jew* (*1933*), with Thurston's own *The Wandering Jew* * (**1934**) novelizing the play. In Robert NICHOLS's "Golgotha & Co", which appears in *Fantastica* (coll **1923**), the WJ is a defiant industrialist who successfully disparages the point of any Second Coming. George VIERECK and Paul Eldridge presented the story in *My First Two Thousand Years: The Autobiography of the Wandering Jew* (**1928**; cut 1956) as a revisionist mockery in which the WJ searches happily for unending pleasure. Evelyn Waugh provides a walk-on part in *Helena* (**1950**). In Pär LAGERKVIST's *Sibyllan* (**1956**; trans as *The Sybil* **1958** UK) and *Ahasverus' död* (**1960**; trans as *The Death of Ahasuerus* **1962** UK), Ahasuerus is dourly defiant to the end. There are further appearances in Walter M. Miller Jr's *A Canticle for Liebowitz* (**1960**), John Boyd's *The Last Starship from Earth* (**1968**), Herbert ROSENDORFER's *Der Ruinenbaumeister* (**1969**; trans as *The Architect of Ruins* **1992** UK), *The Wandering Jew* (**1981**; trans 1984 US) by Stefan Heym (1913-   ), Raymond E. FEIST's **Riftwar** sequence – where he is recorded as the father of MERLIN – *Snail* (**1984**) by Richard Miller (1925-   ), David LANGFORD's and John GRANT's *Earthdoom* (**1987**), centrally in *The Wandering Jew* (**1987** chap) by Michelene WANDOR, in Deborah GRABIEN's *Plainsong* (**1990**), where he fathers a replacement Christ, and in Wolf MANKOWITZ's *A Night with Casanova* (**1991**). He is linked to the FISHER KING in Susan SHWARTZ's *The Grail of Hearts* (**1992**).

Studies include *The Wandering Jew* (**1881**) by Moncure Daniel Conway (1832-1907), George Y. Anderson's *The Legend of the Wandering Jew* (**1965**), and Brian STABLEFORD's introduction to his anthology of stories, *Tales of the Wandering Jew* (anth **1991**). [JC]

WANDOR, MICHELENE (1940-   ) UK writer and dramatist, often concerned in her fiction and nonfiction with issues of FEMINISM, most interestingly perhaps in *Look Back in Gender* (**1987**), a study of UK theatre. *Guests in the Body* (coll **1986**) contains some SUPERNATURAL FICTION. Some plays are of interest, including *Penthesilia* (**1977** chap), based on the play (**1808**) by Heinrich von Kleist (1777-1811), and *The Wandering Jew* (**1987** chap), with Mike Alfreds (1934-   ), based on the novel by Eugène SUE – the published text being a cut version of the 5hr original production. [JC]

WANDREI, DONALD (1908-1987) US writer and editor, founder with August DERLETH in 1939 of ARKHAM HOUSE, the SMALL PRESS initially launched to publish the work of H.P. LOVECRAFT. DW resigned from the firm after WWII, was briefly active in it following Derleth's death in 1971, and was later involved in bitter litigation with AH's new owners.

DW's first published work was poetry, and many believe this was his best work. Like his fiction, DW's poetry output was diverse but displayed a predisposition towards the cosmic. His poems are collected in *Ecstasy and Other Poems* (coll **1928** chap) and *Dark Odyssey* (coll **1931** chap); *Poems for Midnight* (coll **1965**) is largely drawn from the earlier two volumes, and *The Collected Poems* (coll **1988** chap) contains all known poems.

DW wrote approximately 50 tales of sf, fantasy and horror, frequently crosshatching. In particular, sf trappings can be found in many of his tales, such as "Giant Plasm" (**1939** *WT*), in which the survivors of a sinking ocean liner reach an uncharted ISLAND where they encounter a large animated and aggressive grey "block" – and also the remains of an alien spacesuit. In this and many of DW's other stories, there is no tidy explanation or convenient rationale. His more traditional fantasies include "The Painted Mirror" (**1937** *Esquire*), later televised in ROD SERLING'S NIGHT GALLERY (1971-3), which concerns a boy who finds a magic

MIRROR that acts as a PORTAL to another world. "Don't Dream" (1939), DW's only story to appear in UNKNOWN, features a protagonist whose imaginings are seemingly being transformed into real-world events. Much of DW's work seems to exhibit a vague dissatisfaction with or even repulsion from the world at large. His short fiction is collected in *The Eye and the Finger* (coll **1944**) and *Strange Harvest* (coll **1965**), both with covers by DW's brother Howard WANDREI. DW's only novel, *The Web of Easter Island* (**1948**), is a powerful CTHULHU MYTHOS tale in which the unearthing of a strange statuette leads to the discovery of an impending invasion by an ancient alien race.

DW was reclusive and eccentric later in life; in 1984, he was awarded a WORLD FANTASY AWARD for Life Achievement, but declined it. [BM]

Other works: *Colossus: The Collected Science Fiction of Donald Wandrei* (coll **1989**)

As editor: 3 vols of **Lovecraft's Selected Letters** – *1911-1924* (coll **1965**), *1925-1929* (coll **1968**) and *1929-1931* (coll **1971**) – with Derleth.

WANDREI, HOWARD (1909-1956) US illustrator and writer whose work spanned fantasy, sf, horror and mystery. The majority of HW's fiction was for PULP magazines, and he was overshadowed by his brother Donald WANDREI, whose fiction appeared in some of the same publications. HW is generally better remembered for his artwork: his illustrations are notable for their intricate, layered styling, with an "ouroborus" sense of endlessly interconnecting elements. The combination of his exquisitely detailed line-work with often garish colour schemes resulted in a bizarre, arresting effect.

Under his own name and as Robert Coley and H.W. Guernsey he produced nearly 200 stories. The best of his fantasy exhibits a strong vein of humour and a tendency for romantic subplots. In "The Hexer" (1939 *Unknown*) a bizarre little man wields the magical power to give a literal interpretation to people's personality traits – a "nosy" reporter suddenly sports a huge nose, etc. In "The Monacle" (1939) a woman inadvertently utters a SPELL that revives Ardanth, a sorcerer from ancient Thebes, and then spends much of the ensuing story whimsically seeking romantic involvement with him.

Although several collections of HW's fiction were announced over the years by ARKHAM HOUSE, none appeared. In fact HW was largely, and unjustly, forgotten until the publication of *Time Burial* (coll **1995**), the first of at least three projected collections. [BM]

WANGERIN, WALTER Jr (1944-) US writer of various religious texts for young children, with titles like *God, I've Gotta Talk to You* (**1974**) and *My First Bible Book About Jesus* (**1983**), He came to prominence in fantasy with his widely praised BEAST-FABLE *The Book of the Dun Cow* (**1978**) and its downbeat sequel *The Book of Sorrows* (**1985**). These concern Chauntecleer the rooster and his barnyard friends (some of the characters' names are borrowed from CHAUCER's *Canterbury Tales*) and a terrible threat to their world – the whole clearly designed as a religious ALLEGORY. Also of fantasy interest is *The Crying for a Vision* (**1995**). [DP]

Other works (for children): *Elisabeth and the Water Troll* (**1991**); *Branta and the Golden Stone* (**1993**).

WAR ◊ MILITARY FANTASY.

WARBOROUGH, MARTIN LEACH ◊ Grant ALLEN.

WARD, CHRISTOPHER (LONGSTRETH) (1868-1943) US writer whose *Twisted Tales* (coll **1924**) contains various PARODIES, one being of H.G. WELLS's *The Dream* (**1924**). He is better known for his book-length parody of David GARNETT's *Lady Into Fox* (**1922** chap), told in a DICTION which if anything improves on Garnett's own rendering of 18th-century modes. Its title gives the gist of the thing: *Gentleman Into Goose: Being the Exact and True Account of Mr Timothy Teapot Gent., of Puddleditch, in Dorset, that was Changed to a great Grey Gander at the wish of his Wife. How, though a Gander, he did wear Breeches and Smoak a Pipe. How he near lost his life to his Dog Tyger. You have, also an Account of his Gallantries with a Goose, very Diverting to Read, with many other Surprizing Adventures, full of Wonder and Merriment, and a Full Relation of the Manner of his Sad Dismal End. Worthy to be had in all Families for a Warning to Wives and by all Batchelors intending Marriage* (**1924** chap). [JC]

WARD, ELIZABETH STUART PHELPS ◊ Elizabeth Stuart PHELPS.

WARNER, MARINA (SARAH) (1946-) UK writer and critic, author of several influential studies of the mythic representations that govern our perceptions of the world, including *Alone of All Her Sex: The Myth and Cult of the Virgin Mary* (**1976**), *Monuments and Maidens: The Allegory of the Female Form* (**1985**) and *Managing Monsters: Six Myths of Our Time* (coll **1994**). Of specific fantasy interest is *From the Beast to the Blonde: On Fairy Tales and their Tellers* (**1994**), which concentrates on demystifying the relationship between the tales themselves and their (mainly female) tellers. Her *Wonder Tales: Six Stories of Enchantment* (anth trans Gilbert ADAIR, John Ashbery, Ranjit Bolt, A.S. BYATT and Terence Cave **1994**) assembles literary fairytales from late-17th-century France (◊ WONDER TALES), presenting them as coded exercises in subversion – an argument consistent with that of the editors of this encyclopedia that FANTASY was born subversive but has suffered periodic normalizations.

MW's fictions examine similar themes, though generally in nonfantastic terms; but they always convey a sense of the underlying potency of STORY in the shaping of individual lives and the cultures that enwrap those lives. Several of the stories in *The Mermaids in the Basement* (coll **1993**) are TWICE-TOLD versions of FAIRYTALES like "The Princess and the Frog Prince", or retellings of the lives of mythic figures like the Queen of Sheba. [JC]

WARNER, REX (1905-1986) UK writer and translator whose early novels can be treated as Absurdist sf or as fantasy, the ambivalence tending to give an effect of the UNCANNY; they are often cited as examples of the influence of Franz KAFKA. In his first adult novel, *The Wild Goose Chase* (**1937**), three brothers chase a goose until they arrive in a surreal country, where they participate in an ambiguous revolution. *The Aerodrome: A Love Story* (**1941**) is a political ALLEGORY set in an abstracted England. *Why Was I Killed?* (**1943**; vt *Return of the Traveler* 1944 US), a POSTHUMOUS FANTASY, begins to evince the diminution of writerly zest that, within a few more years, put an effective end to RW's career as a novelist of interest. He became noted instead for his translations of Aeschylus and Euripides and others; his later novels are historical. [JC]

WARNER, SYLVIA TOWNSEND (1893-1978) UK writer who began her literary career in 1922 as an editor of scholarly series of books of Tudor Church Music. Her first novel, *Lolly Willowes, or The Loving Huntsman* (**1926**), is a subtly told SUPERNATURAL FICTION whose eponymous protagonist, stifled by her spinsterly life, leaves London for the country, where she continues to feel the presence of an enticing otherness, to which she gives her allegiance ($\lozenge$ PACTS WITH THE DEVIL). She then meets the DEVIL, who is a gamekeeper or "Loving Huntsman", and affirms her relationship to him and her escape from a male-dominated world in identifiably feminist terms ($\lozenge$ GENDER). Next she hits the road, disappearing from our ken.

The Cat's-Cradle Book (coll **1940** US) assembles a series of REVISIONIST stories, some based directly on traditional FAIRYTALES and all contained within a FRAME STORY which introduces them as part of a huge range of tales told by CATS to children, human and feline. The stories are recounted with succinct grace; most are singularly ruthless. "The Castle of Carabas" treats the descendants of the miller's son who betrayed PUSS-IN-BOOTS as a succession of KNIGHTS OF THE DOLEFUL COUNTENANCE who swoon at the sight of a cat, so great is their guilt and attendant AMNESIA, for they know not the reason for their BONDAGE, and refuse to learn. *The Kingdoms of Elfin* (coll of linked stories **1977**) sets 17 tales in a variety of matriarchal kingdoms inhabited by FAIRIES generally ignorant of the human world, and normally unperceivable by humans, though STW posits a wide range of relationships between her SECONDARY WORLDS and the Earth of human history. Some of the kingdoms seem to be POLDERS, some CROSSHATCHES, some WAINSCOTS. When fairies do intersect with humans, they sometimes re-enact ($\lozenge$ TWICE-TOLD) various of the crueller fairytales, as in "Foxcastle" and "The One and the Other", where humans are taken from normal life. The overall effect is of an axiom-sharp, dance-like assemblage of moral lessons, couched with an almost impersonal joy. *Selected Stories of Sylvia Townsend Warner* (coll **1988**) contains further stories of interest. STW's life of T.H. WHITE – *T.H. White: A Biography* (**1967**) – is exemplary. [JC]

WAR OF THE MONSTERS vt of *Gojira Tai Gaigan* (*1972*). $\lozenge$ GODZILLA MOVIES.

WARREN PUBLISHING Innovative US publisher of fantasy, sf and horror magazines.

James Warren (1931-   ) studied at the University of Pennsylvania and the Philadelphia Museum School of Art, and worked as an assistant advertising manager. Inspired by the example of *Playboy* entrepreneur Hugh Hefner (1926-   ), he formed his own company, Jay Publishing, and produced *After Hours* (4 issues 1957), a cheesecake and fiction magazine. In #4 was a feature entitled "Scream-o-Scope is Here", made up of photographs of movie monsters with humorous captions by Forrest J. Ackerman (1916-   ). The remarkable populariy of this feature (JW received more than 300 fan letters) inspired Jay Publishing's next project, *Fantastic Monsters of Filmland* (1957), whose publication happened to coincide with the release to tv of 52 classic HORROR MOVIES by Universal. Following the format of the tv show, in which the movies were interspersed with light comedy, the project was a great success and Ackerman and JW were able to get finance, from Kable News, to begin publishing a regular magazine of this type. *Famous Monsters of Filmland* had cover paintings depicting monster portraits by Albert Nutzell and later by Basil Gogos. A very important factor in the magazine's long success was JW's dogged insistence on maintaining a high quality of both picture and text.

In 1965, he began publishing COMIC-strip magazines using the proven format popularized by *The* MYSTERIOUS TRAVELER and other such radio shows of the 1940s and 1950s: short horror tales introduced and rounded off by a

"horror host". The first of these magazines was CREEPY, which featured some remarkable full-colour covers by Frank FRAZETTA and internal b/w art by many leading comics artists. This was followed by the very similar EERIE later the same year, and by a short-lived horror war title, *Blazing Combat*. A long-running success was VAMPIRELLA (from 1969), followed in 1978 by the highly original adult sf magazine *1984* (renamed *1994* from Feb 1980). Other projects included *The Rook* (1979-82) and *The Goblin* (1982), both capitalizing on popular features that first appeared in other WP titles.

In his continual search for quality features, JW approached European and Philippine comics creators, including José ORTIZ, Luis BERMEJO, Esteban MAROTO, José Gonzalez (1943-   ), Victor de la Fuente (1927-   ), Alex NIÑO and Paul Gillon, and reprinted many high-quality European comics series. These proved too sophisticated for the average US comic-book reader. Despite some remarkable cover paintings by some of the world's finest fantasy artists (Manuel SANJULIAN, Patrick WOODROFFE, Ian MILLER, etc.), sales figures were anyway low when agreement was reached with Harlan ELLISON to publish an adaptation of "A Boy and his Dog" (1969). Ellison then withdrew his permission, so JW commissioned Niño to draw a rather similar tale written by Bill DuBay (as Alabaster Redzone) featuring a young woman and a dog-like mutant monster, and published it as *Mondo Megillah* (in *1984 #3* 1978). Ellison went legal. At this time, sales were also being badly affected by the introduction of the new quality colour sf/fantasy comics magazines HEAVY METAL and EPIC ILLUSTRATED.

WP ceased in 1983 and all its properties were sold.

[DR/RT]

WARRINGTON, FREDA (1956-   ) UK author of DARK-FANTASY novels with an emphasis on strong females. All her works have romantic overtones, usually with a love story at the core. Her SWORD-AND-SORCERY **Blackbird** quartet – *A Blackbird in Silver* (**1986**), *A Blackbird in Darkness* (**1986**), *A Blackbird in Amber* (**1987**) and *A Blackbird in Twilight* (**1988**), plus the associated *Darker than the Storm* (**1991**), which, though not a direct sequel, uses some of the characters and the locations of the previous books – is set in an ALTERNATE WORLD like that of Michael MOORCOCK's **Eternal Champion**. *The Rainbow Gate* (**1989**) is CONTEMPORARY FANTASY, while *Sorrow's Light* (**1993**) is again S&S. FW's historical VAMPIRE trilogy – *A Taste of Blood Wine* (**1992**), *A Dance in Blood Velvet* (**1994**) and *The Dark Blood of Poppies* (**1995**) – showcases her distinctive blend of lushness, romance and dark shadows. *Dark Cathedral* (**1996**), a contemporary dark fantasy, discusses the conflicts of WITCHCRAFT and Christianity. A sequel, «Pagan Moon», is due. [JF]

WASTE LAND In the GRAIL story, the LAND of the FISHER KING becomes waste, reflecting the state of the king himself. In CHRÉTIEN DE TROYES' *Perceval* (after 1182) this occurs because PERCEVAL has failed to ask the correct questions in the Grail Castle. (Elsewhere the wasting of the land is consequent upon the DOLOROUS STROKE dealt to the Fisher King.) Once the correct questions are asked, however, the land is restored. Like many of the lands in Arthurian legend, its location is imprecise. The symbolism of the physical (and sexual) incapacity of a king being reflected in the barrenness of his land has been claimed as evidence for the survival of pagan motifs in the Grail legend, tied in, perhaps, to ancient FERTILITY rituals. This interpretation has tended to obscure the equally valid one that the waste itself represents the spiritual barrenness of the king, who has failed in his duty as keeper of the Grail.

The image of the WL has enjoyed mixed popularity among modern fantasy writers, many of whom relegate it to a minor role, possibly fighting shy of the overtly Christian elements. T.S. Eliot (1888-1965) used the image to explore spirituality in an overtly Christian context; the movie EXCALIBUR (*1981*) tied it to Arthur's political failures after his discovery of GUINEVERE's adultery (◊ VIRGINITY). The most dramatic modern reconstruction is probably that of Richard MONACO, who in his **Parsival** sequence presents the wasting of the land in terms of savage warfare, greed and violence. [KLM]

Further reading: *The Grail: From Celtic Myth to Christian Symbol* (*1963*) by R.S. Loomis; *The Evolution of the Grail Legend* (*1968*) by D.D.R. Owen.

WATER BABIES, THE UK/Polish live-action/ANIMATED MOVIE (*1978*). Ariadne/Studio Miniatur Filmowych/ Productions Associates/Pethurst. **Pr** Peter Shaw. **Dir** Lionel Jeffries. **Screenplay** Michael Robson. **Based on** *The Water-Babies* (*1863*) by Charles KINGSLEY. **Starring** Bernard Cribbins (Masterman), Samantha Gates (Ellie), Joan Greenwood (Lady Harriet), James Mason (Grimes), Tommy Pender (Tom), David Tomlinson (Sir John), Billie Whitelaw (Mrs Doasyouwouldbedoneby/Mrs Tripp). **Voice actors** Cribbins (Electric Eel), Mason (Killer Shark), Pender (Tom), plus Olive Gregg, David Jason, Lance Percival, Jon Pertwee, Una Stubbs. 93 mins. Colour.

York, 1850. Young Tom, brutalized by his master, Grimes the chimneysweep, and Grimes's henchman Masterman, is taken by them to sweep the chimneys of Harthover Hall, home of Sir John and Lady Harriet; there he discovers that the housekeeper, Mrs Tripp, is identical with an old peasant woman they've passed *en route* and also with Mrs Doasyouwouldbedoneby, a faked "bodiless head" whom Tom saw in a sideshow. Tom is accused of theft, flees and – with dog Toby – falls into the river; from here on events are in animation underwater and in live-action above. Tom is taught to swim by a salmon and an otter, who tell him that if he wants to return to surface life he must find the Water Babies, who live in mid-ocean. He has various adventures while undertaking this QUEST. Returning to real life, Tom is immediately grabbed by Grimes and Masterman, who take him back to Harthover Hall to burgle it. They are caught thanks to Tom's pluck; he is exonerated, then adopted by the family to the joy of daughter Ellie.

The live-action bits of *TWB* have much to do with Kingsley's novel, the animated sections less. Although the two children turn in fine performances, the live-action portions seem halfhearted, while the animation is characterized by a fatal cuteness and repeated choruses of a song called "High Cockalorum". [JG]

WATER MARGIN, The Japanese tv series (1975). NTV Tokyo. **Dir** Toshio Masuda, Michael Bakewell (for English dubbing). **Writer** David Weir (English adaptation). **Starring** Hajine Hana (Wu Sung), Yoshiyo Matuso (Hsiao Lan), Atsuo Nakamura (Lin Chung), Kei Sato (Kao Chia), Sanae Tschida (Hu San-niang). 10 45min episodes. Colour.

Based on the 14th-century Chinese novel *Shui Hu Chuan* (trans Pearl Buck as *All Men Are Brothers* 1933 US 2 vols), this is credited to an otherwise unknown author, Shih Nai-an, whose 70-chapter recension of traditional tales is a masterpiece of adventure. A band of 108 reborn COMPANIONS gather in the unmapped and essentially undefinable borderland regions surrounding the central empire, Lian Shan Po; from these "water margins" the companions wage guerrilla warfare against the corrupt central government. The series translates them into supernatural figures whose TALENTS are various and spectacular. The empire becomes a vast beleaguered POLDER protected by bad MAGIC and the emperor a kind of DARK LORD. Episodes were exuberant and eventful. [JC]

See also: WATER MARGINS.

WATER MARGINS A term taken from the tv series *The WATER MARGIN*, where it describes the unmapped and ultimately unmappable regions which surround a central empire, a vast POLDER whose rulers attempt to stave off, by the use of MAGIC and treachery, various revolutionary incursions from the HEROES who inhabit the unknown regions. WMs surround a central LAND or reality, and fade indefinitely into the distance, beyond the edges of any MAP. Fantasies set in SECONDARY WORLDS are commonly supplied with maps whose edges are not BORDERLANDS but WMs.

FANTASY itself has several times been described as a fuzzy set – as a grouping defined not by boundaries but by central examples (◊ Brian ATTEBERY). In the "darkness" surrounding this fuzzy set can be found various regions – impossible to fix exactly – into which the genre called fantasy does not really extend. Stories told for children too young to distinguish reliably between what is real and what is "fantasy" are an example, as are much FOLKLORE, many MYTHS, many TAPROOT TEXTS, much SCIENCE FANTASY and SCIENCE FICTION, the further reaches of SUPERNATURAL FICTION and HORROR, and MAGIC REALISM. The list of texts which marginally escape easy definition as fantasy, even with the remit of the fuzzy set, is very extensive.

Such texts can be described as inhabiting the "water margins". The term is used as a tool of perspective, from within the point of view of the fuzzy set of fantasy as a field, and not prescriptively. Of course, from the perspective of other fields of fiction, fantasy may itself legitimately be deemed a water margin. [JC]

WATERSHIP DOWN UK ANIMATED MOVIE (*1978*). Nepenthe. **Pr** Martin Rosen. **Dir** John Hubley, Rosen. **Anim dir** Tony Guy. **Screenplay** Rosen. **Based on** *Watership Down* (*1972*) by Richard ADAMS. **Novelization** *The Watership Down Film Picture Book* * (*1978*) by Adams. **Voice actors** Joss Ackland (Black Rabbit), Richard Briers (Fiver), Michael Graham-Cox (Bigwig), Michael Hordern (Frith/Narrator), John Hurt (Hazel). 92 mins. Colour.

In the beginning there was the great god Frith, the Sun, who made the world and its creatures; to all except the rabbit was given the wish to kill the rabbit, whose gifts were instead speed and TRICKSTER cunning. Thus goes rabbit MYTHOLOGY, which, illuminated by stylized animation, prefaces *WD*. In the present, the animation switching to a more pastel treatment, the rabbit Hazel attends his psychic UGLY-DUCKLING brother Fiver's forecasts of doom for the warren; they and Bigwig lead a small band off to find somewhere new – the Watership Down of Fiver's dreams. Long after, Watership Down reached and settled, an aged Hazel gladly accepts the summons of the rabbitish DEATH-figure, the Black Rabbit, to join the warren of the dead.

WD makes a surprisingly good fist of presenting Adams's bestselling fantasy on screen – especially since (a) this was Rosen's first attempt at an animated movie (in 1982 he used a similar team of voice actors for his animated version of Adams's *The Plague Dogs* [**1977**]), and (b) he took over the direction midway, John Hubley having resigned angrily. The animation is sometimes mediocre, but generally good and sometimes more; the rabbit mythology/cosmogony is fascinating, and the figure of the Black Rabbit, done in prehistoric-cave-painting style with a simplified MASK for a face, has power. However, *WD* lacks impact: the contraction of Adams's epic novel to a 92min movie renders the plot formless-seeming; the cast-list is overlong for any hope of audience identification; and *WD*, despite graphic incidents along the way, is devoid of overall drama. [JG]

WATSON, IAN (1943-   ) UK writer who began to publish work of genre interest with "Roof Garden Under Saturn" for *New Worlds* in 1969, and who has established himself as one of the most important UK sf writers of his generation. His first six novels – beginning with what remains his best-known single title, *The Embedding* (**1973**) – are all sf. It is only with *The Gardens of Delight* (**1980**) that he began to write book-length fantasy, though even this text is given an sf frame. Searching for a lost colony, a starship lands on a planet which has been transfigured into the LANDSCAPE created by Hieronimus BOSCH in his painting, "The Garden of Earthly Delights" (*c*1500). This landscape, and the METAMORPHOSIS-ridden figures who inhabit it, form an ALLEGORY depicting various fates potentially destined for the human SOUL after death. This AFTERLIFE arena is transformed into a spiritual testing ground, its inhabitants (the colonists originally searched for) being occupied in self-metamorphosing attempts to gain enlightenment. The GODGAME nature of the book is strengthened when it turns out that the central MAGUS controlling this STORY is a man named Knossos (a play on *gnosis*, just as John FOWLES's Conchis plays on "conscious"). Underlying hints that Bosch was versed in ALCHEMY and that his painting depicts an alchemical striving for enlightenment (◊◊ ROSICRUCIANISM) are incorporated, along with, perhaps, a grain of salt. Ultimately, there is an sf rationale for the transformations and the striving; but the multiplex QUEST nature of the tale and its circumambient tone make it a fantasy of importance.

The **Black Current** sequence – *The Book of the River* (fixup **1984**), *The Book of the Stars* (**1984**) and *The Book of Being* (**1985**), assembled as *The Books of the Black Current* (omni **1986** US) – is also fantasy, though with some markedly convoluted hints of an sf frame. The protagonist, Yaleen, lives on a planet bisected by a great RIVER which turns out to be a vast soul-collecting WORM, itself in conflict with an interstellar God-figure; Yaleen, who dies more than once in the sequence, travels posthumously to Earth, comes back to her home planet in the body of a baby, and assists the Worm in its conflict with the overweening GOD figure; in the end, REALITY proves exceedingly shiftable, and the sequence – a scintillating intellectual farrago – closes abruptly.

A metaphysical bias is apparent throughout IW's work, and several novels subject their casts and venues to such abrupt and absolute TRANSFORMATIONS and transcendences that fantasy and sf explanations may dizzyingly interweave. The **Black Current** is perhaps the most glittering example; but other late IW novels display much of the same dancelike ingenuity. They include: *Queenmagic, Kingmagic* (**1986**), set initially in a world governed by the rules of CHESS, though the cast soon proceed to venues where other rules govern proceedings; *Meat* (**1988**), a HORROR novel, which depicts a

world fighting back against those who refuse to eat meat; and *The Fire Worm* (1986 *Interzone* as "Jingling Geordie's Hole"; much exp **1988**), also horror, where past-life experiences which evoke Ramon Lull (?1232-1315) are conflated with the conjuration of the eponymous alchemical worm. The **Book of MANA** sequence – *Lucky's Harvest* (**1993**) and *The Fallen Moon* (**1994**) – combines elements from the KALEVALA, a hugely complicated set of metamorphosis-prone DYNASTIC FANTASY interrelationships between humans and aliens on the planet Kaleva and space opera. The planet's moon houses SHADOWS of the cast, and seems to be gestating a new REALITY where the same cast, variously transfigured, may continue much the same tale; and the sentient asteroid which transported a load of human colonists to Kaleva did so (it emerges) for a price: to be told a proper STORY. As in other IW novels, TRICKSTERS abound, including the "demon fastboy" Jack Pakkenk (◊◊ JACK).

IW's short fiction, much of which wickedly mixes genres in the same fashion as his longer works, has been assembled in *The Very Slow Time Machine* (coll **1979**), *Sunstroke* (coll **1982**), *Slow Birds* (coll **1985**), *The Book of Ian Watson* (coll **1985** US), *Evil Water* (coll **1987**), *Salvage Rites* (**1989**), *Stalin's Teardrops* (coll **1991**) and *The Coming of Vertumnus* (coll **1994**). Two anthologies ed IW – *Pictures at an Exhibition* (anth **1981**) and *Changes: Stories of Metamorphoses, Both Psychological and Physical* (anth **1983** US), the latter with Michael BISHOP – contain fantasy. [JC]

WATT-EVANS, LAWRENCE Working name of US writer Lawrence Watt Evans (1954-   ), who began publishing work of genre interest with "Paranoid Fantasy #1" in 1975 for *American Atheist*, as Lawrence Evans; he created his hyphenated surname in 1979 on finding there was another writer named Lawrence Evans. He began publishing full-length work with his first sequence, the **Lords of Dus** series – *The Lure of the Basilisk* (**1980**), *The Seven Altars of Dûsaara* (**1981**), *The Sword of Bheleu* (**1982**) and *The Book of Silence* (**1984**) – which gave the impression that it would retell the 12 labours of Hercules as a series of inspired PLOT COUPONS; but the series terminated before its SWORD-AND-SORCERY hero, mighty-thewed Garth, completed a full Herculean quota. LWE's second series, the **Legend of Ethshar** – *The Misenchanted Sword* (**1985**), *With a Single Spell* (**1987**), *The Unwilling Warlord* (**1989**), *Taking Flight* (**1993**) and *The Spell of the Black Dagger* (**1994**) – is more sustained. The first novel gradually makes clear that the apparently routine FANTASYLAND over whose fate various foes interminably clash is actually a genuine LAND held in BONDAGE; this original touch of BELATEDNESS, however, does not always sufficiently counteract the GENRE-FANTASY plotting which carries various unwilling heroes.

The **War Surplus** series – *The Cyborg and the Sorcerers* (**1982**) and *The Wizard and the War Machine* (**1987**) – again laces S&S with harsher material, this time elements of the military-sf genre, along with a cyborg protagonist. *Split Heirs* (**1993**) with Esther FRIESNER is a spoof featuring crazed TWINS, a DRAGON and other devices. *Out of this World* (**1994**) begins a new fantasy series, again genre-crossing. LWE remains an ingenious, potentially major writer, but tends to pull his punches. [JC]

Other works: *The Chromosomal Code* (**1984**); *Shining Steel* (**1986**); *Denner's Wreck* (**1988**); *Nightside City* (**1989**); *Nightmare People* (**1990**), horror; *Newer York: Stories of Science Fiction and Fantasy About the World's Greatest City* (anth **1991**), ed; *The Rebirth of Wonder* (coll **1992**); *Crosstime Traffic* (**1992**).

WAUGH, SYLVIA (?1935-   ) UK writer who began with the **Mennyms** YA sequence – *The Mennyms* (**1993**), *Mennyms in the Wilderness* (**1994**) and *Mennyms Under Siege* (**1995**), with further volumes projected – about a WAINSCOT family of rag DOLLS who come to life after the death of their human maker, and who because they are lifesize are able to live inconspicuously in a suburban home in the north of England. [JC]

WAYLAND SMITH ◊ WELAND SMITH.

WAY OF THE DRAGON, THE vt of KUNG FU: THE NEXT GENERATION (*1987* tvm).

WEAPONS The conventional weaponry of fantasy generally defaults to what might be found in a medieval or Renaissance armoury. Accurate descriptions and terminology can lend conviction to stories with ALTERNATE-WORLD historical settings (◊◊ HISTORY IN FANTASY; MILITARY FANTASY). Gene WOLFE cleverly assimilates sf energy weapons into the fantastic setting of *The Book of the New Sun* (**1980-83**) by assigning them ancient names like "contus" or "korseke", and there is a certain sf thrill in devices like Greek fire – featured in *Son of Sinbad* (*1955*) (◊ SINBAD MOVIES) – or the just-plausible 30ft solar reflector used to incinerate invading ships in Terry PRATCHETT's *Small Gods* (**1992**) or the special gunpowder in Roger ZELAZNY's *The Guns of Avalon* (**1972**), which unusually makes firearms operable in FANTASYLAND. However, to be of specific fantasy interest a weapon should exploit MAGIC.

Magic SWORDS are endemic, and any object may be enchanted in almost any way, notably AMULETS, armour, RINGS, TALISMANS, etc.; in J.R.R. TOLKIEN's *The Lord of the Rings* (**1954-5**) the battering-ram Grond used on the gates of Minas Tirith carries "spells of ruin". WIZARDS are apt to hurl fireballs and gouts of raw magic; as with enchanted objects, there are innumerable varieties of offensive SPELLS. TECHNOFANTASY weaponry is more usefully anchored in reality: the confusion projector of Randall GARRETT's *Too Many Magicians* (**1967**) has the brass-bound solidity of Victorian scientific apparatus; a gun in Piers ANTHONY's *Blue Adept* (**1981**) fires a bullet which will lethally animate within the hero's flesh; the devastating eponymous weapon of Michael Scott ROHAN's *The Hammer of the Sun* (**1988**) achieves, in effect, a magically triggered nuclear explosion; and conventional iron projectiles in L.E. MODESITT's *The Order War* (**1995**) are infused with essential "order", causing CHAOS wizards to explode when hit. Many weapons are similarly aimed at ACHILLES' HEELS: anything of COLD IRON is anathema to ELVES; Norse MYTH relates that only LOKI's improvised spear of mistletoe could kill Baldur; in OZ, water destroys the Wicked WITCH of the West; and garlic and holy water hurt VAMPIRES – in Kristine Kathryn RUSCH's *The Fey: The Sacrifice* (**1995**) priests have problems of conscience with profane use of their religion's sacramental water as the sole effective weapon against invaders. Among many other examples is the effect kryptonite weapons have on SUPERMAN. [DRL]

WEAVER, MICHAEL D. (1961-   ) US writer whose *Mercedes Nights* (**1987**) and *My Father, Immortal* (**1989**), are sf but who also wrote the **Wolf-Dreams** sequence: *Wolf-Dreams* (**1987**), *Nightreaver* (**1988**) and *Bloodfang* (**1989**), all assembled as *Wolf-Dreams* (omni **1989** UK). Set in a LAND-OF-FABLE Europe and North America, the sequence features the adventures of a lesbian WEREWOLF swordsperson – who undergoes METAMORPHOSIS into a great white wolf whenever enraged – in various SWORD-AND-SORCERY adventures, with many complications. [JC]

WEBBER, COLLIN (1939-   ) UK writer whose first novel, *Merlin and the Last Trump* (**1993**), was a runner-up in the BBC/Gollancz competition won by John WHITBOURN. The romp involves MERLIN and one of ARTHUR's KNIGHTS in TIME TRAVEL to modern LONDON and beyond, into an sf future and the abode of a DEMON planning humanity's destruction in a LAST BATTLE (with 80% success, souring the often genial HUMOUR); PLOT DEVICES include WISHES and a sentient, murderous SPELL. *Ribwash* (**1994**) is a sequel. [DRL]

WEBER, HENRY WILLIAM (1783-1818) UK anthologist. ◊ ANTHOLOGIES.

WEINBERG, ROBERT (1946-   ) US author, editor, bookseller and collector who has worked extensively in the sf, fantasy, horror, mystery and Western fields. A prolific writer, RW sold his first story in 1967 and has since worked as a freelance newspaper journalist, contributed hundreds of articles to books and magazines and edited over 120 volumes, including various anthologies with Martin H. GREENBERG and Stefan R. DZIEMIANOWICZ. Also well known for his nonfiction, RW won the WORLD FANTASY AWARD for *The Weird Tales Story* (anth **1977**; exp of *WT50* **1974**) and again for his exhaustive *Biographical Dictionary of Science Fiction Artists* (**1988**). His *A Logical Magician* (**1994**; vt *A Modern Magician* UK) and its sequel *A Calculated Magic* (**1995**), involving a mathematician in modern Chicago who finds himself working for MERLIN, are humorous fantasy adventures in the tradition of UNKNOWN and the stories of L. Sprague DE CAMP. RW has written several horror novels and a number of fantasy GAME ties. His contribution to later incarnations of WEIRD TALES has been important. [SJ]

Other works: *The Hero Pulp Index* (**1971**), nonfiction; *The Annotated Guide to Robert E. Howard* (**1978**), nonfiction; *The Devil's Auction* (**1988**); *The Black Lodge* (**1990**); *The Armageddon Box* (**1990**); *The Dead Man's Kiss* (**1992**); *The Louis L'Amour Companion* (**1992**), nonfiction; *Vampire Diary* (**1995**); *Blood War* (**1995**); *Unholy Allies* (**1995**); *The Unbeholden* (**1996**); *The Road to Hell* (**1996**).

As editor: *Far Below* (anth **1974**); *Famous Pulp Classics #1* (anth **1974**); *The Man Behind Doc Savage* (anth **1974**), nonfiction; *Famous Fantastic Classics #1* (anth **1975**); *Famous Fantastic Classics #2* (anth **1975**); *The Adventures of Jules de Grandin* (coll **1976**) by Seabury QUINN; *The Casebook of Jules de Grandin* (coll **1976**) by Quinn; *The Skeleton Closet of Jules de Grandin* (coll **1976**) by Quinn; *The Hellfire Files of Jules de Grandin* (coll **1976**) by Quinn; *Lost Fantasies #4* (anth **1976**); *The Horror Chambers of Jules de Grandin* (coll **1977**) by Quinn; *Lost Fantasies #5* (anth **1977**); *Lost Fantasies #6* (anth **1977**); *Weird Tales: 32 Unearthed Terrors* (anth **1988**) with Dziemianowicz and Greenberg; *Lovecraft's Legacy* (anth **1990**) with Greenberg; *Rivals of Weird Tales* (anth **1990**), *Famous Fantastic Mysteries* (anth **1991**), *Weird Vampire Tales* (anth **1992**), *A Taste for Blood* (anth **1992**), *The Mists from Beyond* (anth **1993**), *Nursery Crimes* (anth 1993), *100 Ghastly Little Ghost Stories* (anth **1993**), *To Sleep, Perchance to Dream . . . Nightmare* (anth **1993**), *100 Creepy Little Creatures* (anth **1994**), *100 Wild Little Weird Tales* (anth **1994**) and *Between Time and Terror* (anth **1995**) all with Dziemianowicz and Greenberg; *Great Writers and Kids Write Spooky Stories* (anth **1995**) with Greenberg and Jill Morgan; *100 Vicious Little Vampires* (anth **1995**) and *100*

Wicked Little Witches (anth **1995**) with Dziemianowicz and Greenberg; *Miskatonic* (anth **1996**) with Greenberg; *Rivals of Dracula* (anth **1996**), *Virtuous Vampires* (anth **1996**), *100 Astonishing Little Aliens* (anth **1996**), *100 Tiny Little Terrors* (anth **1996**) and *365 Scary Stories* (anth **1997**) with Dziemianowicz and Greenberg.

WEIR, PETER (1944-   ) Australian movie director and occasionally screenwriter, some of whose early work is of distinct fantasy interest. *Homesdale* (**1971**), a short, is an Absurdist fiction set in what may be either a lunatic asylum or a country club. *The Cars that Ate Paris* (**1974**), which he co-wrote, his first feature, is a striking piece of paranoid TECHNOFANTASY: a pair of outsiders come to a remote Australian town, Paris, and slowly discover that the town's economy is based entirely on scavenging from passers-through, whose cars are deliberately wrecked by local yobs driving cars souped-up to represent MONSTERS. It was with PICNIC AT HANGING ROCK (**1975**), however, that PW really hit his stride: the cinematic equivalent of MAGIC REALISM, it is paradoxically so powerful a work of fantasy because one is never quite sure whether or not it *is* fantasy, the ambiguity being redolent of full fantasy while at the same time all could be rationalized. *The Last Wave* (**1977**), also co-written, depicts a white Australian lawyer who has precognitive DREAMS which relate both to a second FLOOD and to the Aboriginal DREAMTIME; in waking life he is persuaded by Aborigines that the APOCALYPSE is indeed nigh – and the movie's ending shows what is either the "Last Wave" or his precognitive PERCEPTION of the oncoming END OF THE WORLD. *The Plumber* (**1979** tvm), which PW also wrote, is a fringe URBAN-FANTASY exercise in paranoia. *Gallipoli* (**1981**) is a powerful movie about WWI. *The Year of Living Dangerously* (**1982**) is a journalist thriller set in Indonesia. *Witness* (**1985**), PW's first Hollywood movie, is an exceptional thriller, and also directly pertinent to FANTASY, even though there is nothing of magic or the supernatural: a hard-boiled US cop enters an Amish community which is, in effect, an OTHERWORLD. In most Hollywood outpourings such a figure would introduce the pacifists to the merits of American-as-apple-pie violence; here, though – although at last he must resort to violence to beat off destroyers – it is the "otherworld" which changes *him*.

PW has veered away from the fantastic. It is hoped he will return. [JG]

Other works: *The Mosquito Coast* (**1986**); *Dead Poets Society* (**1989**); *Green Card* (**1991**).

WEIRD AND OCCULT LIBRARY ◊ MAGAZINES.

WEIRDBOOK US large-format small-press MAGAZINE, irregular (usually annual), April 1968-current; published and ed W. Paul Ganley (1934-   ), Buffalo, New York.

W is basically an amateur magazine of the old school, produced more out of love for magazine production than for any possible commercial or artistic achievement. The dedication with which the magazine is compiled has long been recognized, resulting in *W* being twice (1987, 1992) recipient of the WORLD FANTASY AWARD. It has always striven to follow in the footsteps of WEIRD TALES, but initially without pretensions to match quality. Its early issues were filled mostly with short-short HORROR stories, many by aspiring writers, although the presence of Joseph Payne BRENNAN and H. Warner MUNN, plus previously unpublished work by Robert E. HOWARD, forged links with *WT*.

W hit its stride by *#9* (1975) in terms of both quality fiction and production values, and since then has successfully replicated the formula of *WT* in its mixture of unusual stories, covering the complete range of fantasy and SUPERNATURAL FICTION. It has attracted prose and poetry from such writers as L. Sprague DE CAMP, Dennis ETCHISON, Joe R. Lansdale (1951-   ), Tanith LEE, Brian LUMLEY, Gerald W. Page (1939-   ), Jessica Amanda SALMONSON, Darrell SCHWEITZER and J.N. WILLIAMSON, many of whom are regulars. It has also featured an attractive series of covers by Stephen Fabian (1930-   ). Both the 10th anniversary issue (*#13* 1977) and the 20th (a double issue, *#23/#24* 1988) were also printed in hardcover. For some years Ganley published a companion magazine, *Eerie Country* (9 issues 1976-82), generally of slightly lesser quality; he also issued a selection, *Weirdbook Sampler* (anth **1987** chap). Since 1977 he has run Weirdbook Press (in 1985 renamed W. Paul Ganley: Publisher) on a rather more formal basis, releasing one or two separate books per year, mostly CTHULHU MYTHOS-based books by Brian LUMLEY and fantasies by Darrell SCHWEITZER. [MA]

WEIRD FICTION Term used loosely to describe FANTASY, SUPERNATURAL FICTION and HORROR tales embodying transgressive material: tales where motifs of THINNING and the UNCANNY predominate, and where subject matters like OCCULTISM or SATANISM may be central, and DOPPEL-GÄNGERS thrive. The magazine WEIRD TALES is rich in WF. Authors associated with WF include Grant ALLEN, Scott BAKER, Charles BEAUMONT, E.F. BENSON, Ambrose BIERCE, William Peter BLATTY, Robert BLOCH, Joseph Payne BRENNAN, Ramsey CAMPBELL, Jonathan CARROLL, August DERLETH, Dennis ETCHISON, Stefan GRABINSKI, William Hope HODGSON, Stephen KING, T.E.D. KLEIN, Dean R. KOONTZ, Sheridan LE FANU, Thomas LIGOTTI, Bentley LITTLE, Frank Belknap LONG, H.P. LOVECRAFT, Brian LUMLEY, Arthur MACHEN, Richard MATHESON, Anne RICE, James Malcolm RYMER, SARBAN, Dan SIMMONS, Peter STRAUB, Bram STOKER, Karl Edward WAGNER, Donald WANDREI, Chet WILLIAMSON, J.N. WILLIAMSON and F. Paul WILSON. [JC]

WEIRD MYSTERY (magazine) ◊ FANTASTIC; FANTASTIC ADVENTURES.

WEIRD SHORTS ◊ MAGAZINES.

WEIRD STORY MAGAZINE ◊ MAGAZINES.

WEIRD TALES US pulp MAGAZINE, large-format May 1923-May/July 1924; digest September 1953-September 1954, 279 issues, March 1923-September 1954, monthly March 1923-December 1939 except occasional combined issues, including May/June/July 1924, then gap until November 1924; then again monthly except combined issues February/March 1931, April/May 1931, June/July 1931, August/September 1936 and June/July 1939; bimonthly from January 1940; published by Rural Publications, Chicago, March 1923-May/July 1924, Popular Fiction Publishing Co., Chicago, November 1924-October 1938, Short Stories Inc., New York, November 1938-September 1954; ed Edwin Baird (1886-1957) March 1923-April 1924, Otis Adelbert Kline (1891-1946) May/July 1924, Farnsworth WRIGHT November 1924-December 1939, Dorothy McIlwraith (1891-1976) January 1940-September 1954.

WT is *the* legendary PULP fantasy magazine, second only to UNKNOWN in significance and influence. The magazine's start was inauspicious: its editor, Baird, had been employed to compile the companion *Detective Tales* and had no interest in HORROR; he thus filled *WT* with unimaginative traditional GHOST STORIES, macabre villainy, etc. Admittedly

WT discovered during this period some writers who would be among its most popular and influential – e.g., H.P. LOVE-CRAFT, Frank OWEN, Seabury QUINN and Clark Ashton SMITH (at that stage only for his poetry) – but the magazine remained generally uninspiring, and was unprofitable. However, its publisher, Jacob C. Henneberger (1890-1969), remained convinced of its importance, and after the bumper May/July 1924 issue, which contained the controversial necrophiliac story "The Loved Dead" by C.M. Eddy (1896-1967) (revised by Lovecraft) and was thus reputedly withdrawn from sale, he sold the more lucrative *Detective Tales* in order to concentrate on *WT*.

The new editor, Farnsworth Wright, although erratic and idiosyncratic in his selections, nevertheless led the magazine into its Golden Age, the 1930s. *WT* used the subtitle *The Unique Magazine*, which remained singularly appropriate, as Wright was always prepared to publish unusual and offbeat stories which would rarely have appeared elsewhere – at least, not until *WT*'s popularity inspired such rivals as STRANGE TALES and then STRANGE STORIES.

The most influential contributor at this time was undoubtedly Lovecraft. Although Wright did not himself always like Lovecraft's fiction, its cumulative power had a striking effect on readers and writers alike, particularly after the appearance of "The Call of Cthulhu" (February 1928), which began the CTHULHU MYTHOS. Other writers, mostly friends of Lovecraft's and regular contributors in their own right, began to add to the sequence, prominent among them being Frank Belknap LONG, August W. DERLETH, E. Hoffmann PRICE and Donald WANDREI, and by the mid-1930s Lovecraft could also number among his adherents Robert BLOCH and Henry KUTTNER; his circle included also Clark Ashton Smith and Robert E. HOWARD, although neither were Lovecraftian writers. In 1928 Smith began a prolific sequence of baroque and exotic HIGH-FANTASY tales, many set in either the distant past or the FAR FUTURE, where MAGIC and NECROMANCY took the place of science. Howard likewise looked to primeval days, though more to explore native strength pitted against evil wizardry. His sequence of stories, which culminated in the adventures of CONAN, were among the first of the SWORD-AND-SORCERY subgenre.

Outside the Lovecraft Circle, other writers carved their own niches. Seabury Quinn, *WT*'s most prolific contributor (and for a period its most popular), produced an extremely long series of OCCULT-DETECTIVE stories featuring **Jules de Grandin**; Edmond Hamilton (1904-1977) was a regular supplier of sf, though by the 1930s he was using *WT* as a medium in which to explore more fanciful ideas; Frank Owen produced many romantic pseudo-ORIENTAL FAN-TASIES; and C.L. MOORE contributed her semi-erotic fantasy adventures of **Jirel of Joiry** and **Northwest Smith**. Other regulars included Paul Ernst (1899-1985), David H. KELLER, the one-plot writer Bassett Morgan (real name Grace Jones; 1885-1974) – whose stories usually featured people's brains transplanted into animals – Greye La Spina (1880-1969), H. Warner MUNN, Henry S. WHITEHEAD, Hugh B. CAVE, Nictzin DYALHIS – whose tales of witchcraft and sorcery typify the individualism of the magazine and its contributors – and the UK writers Arlton Eadie (real name Leopold Eady; 1886-1935) and G.G. Pendarves (real name Gladys Trenery; 1885-1938).

WT's heyday was represented not only by its writers. The work of its artists had an immense effect on the magazine's character, particularly the bold and often erotic covers by Margaret Brundage (1900-1976), which caused the magazine to be banned in some countries, and the colourful action-orientated covers by J. Allen St John (1872-1957), who also designed what became the magazine's most lasting logo. Although C.C. Senf (1879-1948) and Hugh Rankin (1879-1957) were also regular cover and interior illustrators, they have not had the lasting impact of Brundage or of *WT*'s leading artists of the late 1930s and the 1940s: Virgil FINLAY, Hannes BOK, Boris Dolgov, Matt Fox (1906-   ) and Lee Brown Coye (1907-1981).

By the late 1930s the magazine's character had begun to change. Howard and Lovecraft had died, Smith had almost stopped writing, Quinn's **Jules de Grandin** was becoming stale, and Wright, who suffered from Parkinson's disease, eventually became too ill to edit. The magazine was sold to a new publisher (Short Stories, Inc.) and the editorial offices moved to New York where, once Wright retired, Dorothy McIlwraith – editor of the existing *Short Stories* – took over. The McIlwraith issues are usually regarded as inferior to Wright's; in fact, though seldom attaining Wright's high-points, they also omitted the lows. The magazine's mood changed, presenting more modern, psychological fiction, and excluding S&S. Bloch, Derleth, Hamilton and Quinn remained regular contributors, and were now joined by Ray BRADBURY, Fredric BROWN, Mary Elizabeth COUNSELMAN, Gardner F. FOX, Carl Jacobi (1908-   ), Harold Lawlor (1910-?1993), Fritz LEIBER, Theodore STURGEON and Manly Wade WELLMAN, whose **John Thunstone** series brought a more realistic native background to the occult-detective theme. Among the last major new contributors to the magazine were Joseph Payne BRENNAN and Richard MATHESON.

LATER INCARNATIONS

Although *WT* failed to survive the general pulp-magazine decline of the mid-1950s – a switch to digest format was to no avail – the memory lingered, and *WT* acquired quasi-mythical status. Occasional SMALL-PRESS magazines were launched in an attempt to continue the tradition – among them Joseph Payne Brennan's *Macabre*, Paul Ganley's WEIRDBOOK, Stuart Schiff's WHISPERS and Stephen Jones's and David A. Sutton's FANTASY TALES. The rights to the title had been acquired by Leo Margulies (1900-1975), who frequently planned to revive the magazine. In the interim he assembled (with the help of Sam Moskowitz [1920-   ]) four anthologies selected entirely from the magazine: *The Unexpected* (anth **1961**), *The Ghoul Keepers* (anth **1961**), *Weird Tales* (anth **1964**) and *Worlds of Weird* (anth **1965**). He eventually reissued *WT*, via Renown Publications, in mock-pulp format in 1973, under the editorship of Moskowitz. It had four quarterly issues (Summer 1973-Summer 1974), relying heavily on reprints from earlier issues and from turn-of-the-century magazines; it was most noted for Moskowitz's serialized study of the works of William Hope HODGSON.

When Margulies died in 1975 the rights were acquired by Robert WEINBERG, who published his own tributes to the magazine. He licensed the title to Lin CARTER, who sold the idea of a *WT* series to Zebra Books, New York. The new *WT* was issued in paperback format, ostensibly as an ANTHOLOGY series (4 issues Spring 1981-Summer 1983). Carter sought to revive the magic of the Wright years by reprinting lesser-known material, acquiring new stories from old-time contributors, and resurrecting previously unpublished material by former contributors. Although lovingly edited, this incarnation was fatally marred by Carter's failure to see that the best material available was that by new writers.

After the contract with Carter was terminated, Weinberg licensed the title to a California publisher, Bellerophon Network, owned by Brian Forbes. Amid much rumour and speculation about the editorship, two minor and poorly distributed issues, ed Gil Lamont, appeared (Fall 1984, Winter 1985). Underfinanced, the magazine once more drew on reprints and lesser material from the archives of Forrest J. Ackerman (1916-   ) – who had also been mooted as editor. Like Carter before him, Lamont was unable to capture the true essence of the magazine.

The most recent incarnation was instigated in 1987 by George H. Scithers (1929-   ) through his Terminus Publishing Co., Philadelphia. Assisted by Darrell SCHWEITZER and John Betancourt (1963-   ) – the latter for only the first two years – Scithers produced a magazine considerably more faithful to the original. Initially published in mock-pulp format, the first issue (#290 Spring 1988) was remarkably evocative of the Wright WT, with artist George Barr (1937-   ) successfully pastiching the illustrations of St John, Finlay, Rankin and Bok. The magazine maintained a quarterly schedule for some years. Almost every issue carried an author and an artist feature: #290 (Spring 1988) Gene WOLFE/George Barr; #291 (Summer 1988) Tanith LEE/Stephen Fabian (1930-   ); #292 (Fall 1988) Keith TAYLOR/Carl Lundgren (1947-   ); #293 (Winter 1989) Avram DAVIDSON/Hank Jankus (1929-1988); #294 (Fall 1989) Karl Edward WAGNER/J.K. Potter (1956-   ); #295 (Winter 1989/90) Brian LUMLEY/Vincent DI FATE; #296 (Spring 1990) David J. Schow (1955-   )/Janet Aulisio (1952-   ); #297 (Summer 1990) Nancy SPRINGER/Kelly FREAS; #298 (Fall 1990) Chet WILLIAMSON/no featured artist; #299 (Winter 1990/91) Jonathan CARROLL/Thomas Kidd (1955-   ); #300 (Spring 1991) Robert BLOCH/Gahan WILSON; #301 Ramsey CAMPBELL/Bob Walters (1949-   ); #302 (Fall 1991) William F. NOLAN/Bob Eggleton (1960-   ); #303 (Winter 1991/92) Thomas LIGOTTI/no featured artist; #304 (Spring 1992) John BRUNNER/Jill Bauman (1942-   ).

The editors did not fall into the trap of their three predecessors – i.e., trying to recapture the magazine's past through using writers of the past. Instead they encouraged new writers to explore the diversity of WEIRD FICTION in all its forms, as originally managed by Wright. Scithers and Schweitzer received a Special WORLD FANTASY AWARD in 1992 for their achievement. However, the cost of production and distribution forced the magazine to change to a larger format in 1992, and this deprived it of some of the WT "aura". The author/artist issues continued, but the magazine was now less regular: #305 (Winter 1992/93) F. Paul WILSON/Bob Eggleton; #306 (Spring 1993) Nina Kiriki HOFFMAN/Nicholas Jainschigg (1961-   ); #307 (Summer 1993) Ian WATSON/no featured artist; #308 (Spring 1994), Tanith LEE/Phil Parks.

After #308 the licence to use the title was terminated. The publishers continued the magazine under a new title, WORLDS OF FANTASY AND HORROR: the numbering began afresh with #1 (Summer 1994), and the format and style were the same. Although the spirit of WT remains, the loss of the title took with it some of the soul. The field awaits the next development in the afterlife of WT.

The influence of the original magazine remains palpable,

not only through the continuing legacy of authors like Lovecraft, Howard, Bloch and Bradbury but in the desire of writers and publishers to recapture its aura and essence. Somewhere in the imagination reservoir of all US (and many non-US) GENRE-FANTASY and HORROR writers is part of the spirit of WT.

REPRINT EDITIONS AND ANTHOLOGIES
WT has seen many variant reprint editions and has been a rich source for anthologists. It began early in the UK in 1925 with the first of the NOT AT NIGHT series ed Christine Campbell Thomson. Her first three volumes drew almost wholly from WT, to the extent that the series was regarded as a UK edition; later titles used more original material, but in some cases these were merely stories being published in the UK in advance of their WT appearance. During 1942 Gerald G. Swan (◊ MAGAZINES) released three unnumbered and cut editions of the September 1940, November 1940 and January 1941 issues. After WWII Thorpe & Porter issued a more regular and more complete UK edition, starting with the July 1949 edition (UK release November 1949) and then running complete reprints of the November 1949 to May 1954 US editions (though omitting July 1953 and swapping the order of the March 1953 and May 1953 issues).

A separate Canadian printing began in 1935 (in order to control the covers), but a completely separate Canadian edition began from the January 42 US edition (dated May 1942 for Canada). These versions had new covers by Canadian artists and, from the September 1942 Canadian issue, began to include some stories in advance of their US publication and to change some author by-lines – creating new pseudonyms (e.g., H.P. Lovecraft became J.H. Brownlow). The Canadian series ran May 1942-November 1951 (58 issues), the equivalent of the US January 1942-November 1951 issues, but omitting the US November 1944 and January 1948. From March 1948 the Canadian edition was identical and concurrent with the US one.

The first US ANTHOLOGY to draw wholly from WT was The Moon Terror (anth 1927) ed anon Farnsworth Wright as a subscription bonus; the selection was weak and the anthology took years to sell out. More serious attention came with a US selection from the UK Not at Night series, Not at Night! (anth 1928) ed Herbert Asbury (1891-1963), followed by Beware After Dark (anth 1929) ed T. Everett Harré (1884-1948) and, most significantly, Creeps by Night (anth 1931) ed Dashiell Hammett (1894-1961) and The Other Worlds (anth 1941) ed Philip D. Stong (1899-1957). All these titles, drawing heavily though not exclusively from WT, gave the WT authors a wider profile. Anthologies selecting solely from WT include, in addition to the Margulies ones already cited: Far Below and Other Horrors (anth 1974) ed Robert Weinberg; the 9-vol Lost Fantasies series: #1 The Bride of Osiris (coll 1975) by Otis Adelbert Kline, #2 Loot of the Vampire (coll 1975) by Thorp McClusky, #3 The Gargoyle (coll 1975) by Greye La Spina, Lost Fantasies #4 (anth 1976), #5 (anth 1977), #6 (anth 1977), #7 Dreadful Sleep (1938; 1977) by Jack Williamson, #8 The Lake of Life (anth 1978) and #9 The Sin Eaters (anth 1979), all ed Weinberg – Weird Tales (anth 1976; cut 2 vols vt Weird Tales 1978 and More Weird Tales 1978) ed Peter HAINING, which in the hardcover edition is in facsimile from the original magazine; Weird Legacies (anth 1977) ed Mike ASHLEY; Weird Tales: The Magazine that Never Dies (anth 1988) ed Marvin KAYE; Weird Tales: 32 Unearthed Terrors

(anth **1988**) ed Stefan R. DZIEMIANOWICZ, Weinberg and Martin H. GREENBERG; *The Eighth Green Man and Other Strange Folk* (anth **1989**) ed Weinberg; *100 Wild Little Weird Tales* (anth **1994**) ed Weinberg, Dziemianowicz and Greenberg. Despite its title, *Best of Weird Tales* (anth **1995**) ed John BETANCOURT selects solely from the Terminus years. [MA]
Further reading: *WT50* (anth **1974**; exp vt *The Weird Tales Story* 1977) ed Weinberg, a tribute; *The Collector's Guide to Weird Tales* (**1985**) by Fred Cook and Sheldon R. Jaffery.

WEIRD TERROR TALES ◊ *The* MAGAZINE OF HORROR; MAGAZINES.

WEIS, MARGARET (EDITH) (1948-   ) US writer, known mainly for her novels and short stories in collaboration with role-playing-GAME designer Tracy HICKMAN. The earlier of these fall into various game-tie sequences: the **DragonLance Chronicles** – *Dragons of Autumn Twilight* * (**1984**), *Dragons of Winter Night* * (**1985**) and *Dragons of Spring Dawning* * (**1985**) – assembled as *DragonLance Chronicles* * (omni **1988**); the **DragonLance Legends** – *Time of the Twins* * (**1986**), *War of the Twins* * (**1986**) and *Test of the Twins* * (**1986**) – assembled as *DragonLance Legends* * (omni **1988**); and associated volumes such as *DragonLance Adventures* * (coll **1987**), *The Magic of Krynn* * (anth **1987**), *Kender, Gully Dwarves, and Gnomes* * (anth **1987**), *Love and War* * (anth **1987**) – the last three titles gathered in *DragonLance Tales* * (omni **1991**) – *The Reign of Istar* * (anth **1992**), *The Cataclysm* * (anth **1992**), *The War of the Lance* * (anth **1992**), *The Dragons of Krynn* * (**1994**) and *DragonLance: The Second Generation* * (coll **1994**). The considerable commercial success of the early **DragonLance** books encouraged MW and her collaborator to venture out of the realm of game ties and to essay a similar type of light fantasy adventure in series format, beginning with the **Darksword** trilogy – *Forging the Darksword* (**1988**), *Doom of the Darksword* (**1988**) and *Triumph of the Darksword* (**1988**) – continuing with the **Rose of the Prophet** trilogy – *The Will of the Wanderer* (**1989**), *The Paladin of the Night* (**1989**) and *The Prophet of Akhran* (**1989**) – and concluding with the lengthier **Death Gate Cycle** – *Dragon Wing* (**1990**), *Elven Star* (**1990**), *Fire Sea* (**1991**), *Serpent Mage* (**1992**), *The Hand of Chaos* (**1993**), *Into the Labyrinth* (**1993**) and *The Seventh Gate* (**1994**). Without Hickman, MW next tried her hand at a form of fantasy-flavoured space opera (in the vein of the **Star Wars** movies) in the **Star of the Guardians** tetralogy – *The Lost King* (**1990**), *King's Test* (**1991**), *King's Sacrifice* (**1991**) and *Ghost Legion* (**1993**) – and, with the aid of a new collaborator, Don Perrin, its follow-up **Knights of the Black Earth** series, commencing with *The Knights of the Black Earth* (**1995**) and *Robot Blues* (**1996**). [DP]
Other works: *Endless Catacombs*, as Margaret Baldwin Weis (**1984**); *Riddle of the Griffin*, as Susan Lawson, with Roger E. Moore (**1985**); *The Art of Dungeons & Dragons*, as Margaret Baldwin Weis (**1985**), nonfiction; *Leaves from the Inn of the Last Home: The Complete Krynn Source Book*, with Hickman and Mary L. Kirchoff (**1987**), nonfiction; *A Dragon Lover's Treasury of the Fantastic* (anth **1994**).

WEISMAN, JORDAN K. (?   -   ) US writer of a TIE in the SHARED-WORLD series of TECHNOFANTASIES, SHADOWRUN. [JC]

WEISS, JAN (1892-1972) Czech writer. ◊ CZECH REPUBLIC; SLOVAKIA.

WELAND SMITH (Norse Völund, German Wieland, English Wayland Smith) Supernatural SMITH of Teutonic MYTHOLOGY.

Weland married a swan maiden. He was captured by King Nithud, who took his SWORD, gave his RING to his daughter Bodvild and hamstrung him. He was imprisoned on an ISLAND and made to work at his forge. In revenge, Weland killed the king's two young sons when they visited him secretly. He sent their father drinking-vessels made from their skulls, their mother jewels from their eyes, their sister breast-ornaments of their teeth. Bodvild brought him her ring to mend; he raped her, then escaped on WINGS he had made.

He has features of both TRICKSTER and shaman (◊ SHAMANISM). His story preserves a memory of ironworking as a mysterious craft brought from overseas. In England, Wayland's Smithy is an ancient burial chamber; legend says that if metal and payment are left outside, the finished article will appear in a day or two, provided the client makes no attempt to watch.

Forging a weapon for a HERO is potent magic. Weland is known as a wise ELF-lord. He made a magic sword for the AESIR and the armour in which BEOWULF fought Grendel. The fairy blacksmith's mystery is symbolized in ARTHUR's swords: Arthur drew his sword of kingship from a black stone, and Excalibur passed from existence when plunged into water. [FS]

WELLES, (GEORGE) ORSON (1915-1985) US actor and director. ◊ Archibald MACLEISH; *The* TRIAL (*1962*).

WELLMAN, MANLY WADE (1903-1986) US writer, born in Angola, brother of the writer Paul I. Wellman (1898-1966). MWW worked in several genres, including sf and Westerns, and won awards for crime fiction and "true crime" writing, but his HORROR and fantasy – often hybridizing the two – includes some highly distinctive material that might be reckoned his best. He was a regular contributor to WEIRD TALES, making his debut there with "Back to the Beast" (*WT* 1927); some of his *WT* work appeared as by Gans T. Field, including a series chronicling the exploits of OCCULT DETECTIVE **Judge Pursuivant**. The **Pursuivant** series was collected along with the similar **John Thunstone** series and one other story in *Lonely Vigils* (coll **1981**). **Thunstone** novels are *What Dreams May Come* (**1983**) and *The School of Darkness* (**1985**).

The cream of MWW's other weird and fantasy work for the pulps, including some as by Levi Crow, were previously assembled in *Worse Things Waiting* (coll **1973**). Among the best of these are several stories set during the US Civil War, including "The Valley was Still" (1939) and "Fearful Rock" (1939); these were early manifestations of his keen interest in devising supernatural Americana specifically adapted to the history and geography of the USA. A collection stressing this aspect of his work is *The Valley So Low: Southern Mountain Tales* (coll **1987**). "Frogfather" (1945) and "Sin's Doorway" (1946) introduced the character who was to be the mainstay of MWW's later career, **John the Balladeer** or **Silver John**, a modern wandering minstrel who encounters all manner of supernatural phenomena in the hills of North Carolina. A more sophisticated series of **Silver John** tales from *F&SF* was assembled as *Who Fears the Devil?* (coll of linked stories **1963**; exp vt *John the Balladeer* 1988). The **Silver John** novels have not the delicacy of the best short stories, but are interesting by virtue of their elaboration of a curious syncretic mythology based in an imaginary US prehistory; they are *The Old Gods Waken* (**1979**), *After Dark*

(1980), *The Lost and the Lurking* (1981), *The Hanging Stones* (1982) and *The Voice of the Mountain* (1985).

Although usually considered sf, *Twice in Time* (1939 *Startling Stories*; cut 1957; full text plus "The Timeless Tomorrow" [1947] coll 1988) is an unrationalized TIMESLIP romance; it is written more carefully, and seemingly with more feeling, than the rest of MWW's rather *gauche* work for the sf pulps. [BS]

Other works: *Romance in Black* (1938 *WT* as "The Black Drama"; 1946 chap UK) as by Gans T. Field; *The Beyonders* (1977); *Cahena: A Dream of the Past* (1986).

WELLS, ANGUS (1943-   ) UK writer and former publisher's editor who, since 1976, has been a prolific producer of genre fiction, mostly Westerns, and mostly pseudonymously. Apart from an sf novel, *Star Maidens* * (1977), as by Ian Evans, his first fantasy was the **Raven** series plotted with Robert HOLDSTOCK. AW wrote the first novel, *Raven, Swordmistress of Chaos* (1978), with Holdstock, plus two others in the series, *The Frozen God* (1978) and *A Time of Dying* (1979), solo. The books are SWORD-AND-SORCERY with more sex and violence than previously common in that genre. Much later came the **Book of the Kingdoms** series – *The Wrath of Ashar* (1988), *The Usurper* (1989) and *The Way Beneath* (1990) – routine HEROIC FANTASY about the god Ashar whose efforts to rule the world are thwarted by the GODDESS Kyrie, who uses humans as her minions. Most of the standard PLOT COUPONS and PLOT DEVICES are here, with the subplot that the hero, Kedryn, is blinded at the end of the first novel and must first regain his sight by a visit to the UNDERWORLD before his QUEST against Ashar can continue. The **Godwars** trilogy – *Forbidden Magic* (1991), *Dark Magic* (1992 US) and *Wild Magic* (1993 US) – is less predictable. The hero is seeking to stop a WIZARD from awakening a MALIGN-SLEEPER god, but is tricked into supporting the wrong wizard so that things get considerably worse before they get better. The last two books become rather more formulaic, but AW's use of MAGIC is exciting and convincing. *Lords of the Sky* (1994) is AW's most introspective novel to date. Its premise is simple but ingenious. In a SECONDARY WORLD, one land is being invaded by airships. A student wonders whether he can invoke the DRAGONS, long believed extinct, to help combat the invaders. The novel is a thoughtful study of the ethical issues raised among a normally peaceful race forced to defend their land. AW's latest series is the **Exiles Saga**, starting with *Exile's Children* (1995) and «Exile's Challenge» (1996). [MA]

WELLS, H(ERBERT) G(EORGE) (1886-1946) UK writer often regarded as the father of modern SCIENCE FICTION. Because he is so closely associated with that genre it is easy to assume that all of his imaginative fiction is sf, but HGW wrote a number of stories using supernatural or fantastic MOTIFS, only sometimes seeking to rationalize the events in scientific terms. The first group of purely SUPERNATURAL FICTION includes: "The Temptation of Harringay" (1895 *St James's Gazette*), about an artist tempted by a DEMON who comes alive from one of his paintings; "The Moth" (ot "A Moth – Genus Novo" 1895 *Pall Mall Gazette*) about an entomologist who believes he has discovered a new genus of moth until he learns no one else can see it (thereafter he believes it is the spirit of an old colleague with whom he has feuded – like similar stories by Sheridan LE FANU and Guy DE MAUPASSANT it may be interpreted either as a psychological GHOST STORY or a story of madness); "Pollock and the Porroh Man" (1895 *New Budget*), about a man who

meddles with an African shaman and is thereafter haunted until he commits suicide; "The Red Room" (1896 *The Idler*; vt "The Ghost of Fear"), about a room haunted not by a GHOST but by the residuum of fear; "The Apple" (1896 *The Idler*), a FABLE about the fear of knowledge in which a man rejects a fruit believed to be from the TREE of knowledge in EDEN. "The Man who Could Work Miracles" (1898 *Illustrated London News*), which can be seen as a companion piece to "The Apple" as it explores the consequences of a meek little man suddenly finding he has absolute power, which inevitably leads to total destruction – filmed as THE MAN WHO COULD WORK MIRACLES (*1936*; script by HGW published as *The Man who Could Work Miracles* * 1936 chap); "The Stolen Body" (1898 *The Strand*), involving experiments in astral travel resulting in a body suffering a spiritual POSSESSION; "Mr Skelmersdale in Fairyland" (1901 *The Strand*), about a man who thinks he has visited FAERIE in a dream but, on realizing his experience was truth, seeks unsuccessfully to return (this story has companion pieces in "The Door in the Wall" [1906 *Daily Chronicle*], where a man looks for his SECRET GARDEN to find it only in death, and "The Beautiful Suit" [1909 *Collier's Weekly*; ot "A Moonlight Fable"], in which a man finds perfection in a new suit – all are parables of the search for eternal bliss); "The Story of the Inexperienced Ghost" (1902 *The Strand*), a humorous ghost story in which a man helps a ghost to return to the land of the dead; "The Truth about Pyecraft" (1903 *The Strand*), about an Indian SPELL which makes a man weightless; and "The Magic Shop" (1903 *The Strand*), about a SHOP where all tricks are genuine MAGIC.

The second group of stories of fantasy interest attempt a scientific explanation. "The Flowering of the Strange Orchid" (1894 *Pall Mall Budget*) is about a new species of plant which proves to be vampiric. "In the Avu Observatory" (1894 *Pall Mall Budget*) and "Aepyronis Island" (1894 *Pall Mall Budget*) concern the discovery of ANIMALS UNKNOWN TO SCIENCE. "Under the Knife" (1896 *New Review*) tells of a patient who, during an operation, has either a HALLUCINATION or an astral journey (◊ ASTRAL PLANE). "The Plattner Story" (1896 *New Review*) is about a chemistry teacher blown into another dimension where he sees what he believes are the SOULS of the dead – here HGW seeks to rationalize the growing interest in psychic research. "The Story of the Late Mr Elvesham" (1896 *The Idler*) uses a tablet to effect an IDENTITY EXCHANGE. "A Story of the Stone Age" (1897 *The Idler*) may be regarded as a PREHISTORIC FANTASY. "The Presence by the Fire" (1897 *Penny Illustrated*) is a rationalized ghost story. "Mr Marshall's Doppelganger" (1897 *Gentlewoman*) considers two interpretations of a DOPPELGÄNGER. "A Vision of Judgment" (1899 *Butterfly*) considers divine retribution leading to REINCARNATION on a planet orbiting Sirius. "The Empire of the Ants" (1905 *The Strand*) tells of a new species of ant from South America which threatens to dominate the world. "The Strange Story of Brownlow's Newspaper" (1931 *The Strand*) is a TIMESLIP tale.

Both sets of stories demonstrate that HGW explored as many themes of fantasy and the supernatural as he did the sciences. No single collection contains all of them, although *The Short Stories of H.G. Wells* (coll 1927; vt *The Famous Short Stories of H.G. Wells* 1938 US; vt *The Complete Short Stories of H.G. Wells* 1965) contains most. Shorter collections of merit are *Thirty Strange Stories* (coll 1897), *Twelve Stories and a Dream* (coll 1903), *Tales of Wonder* (coll 1923), *The*

Valley of Spiders (coll **1964**) and *The Inexperienced Ghost* (coll **1965** US).

HGW's excursions into the fantastic were less frequent at novel length, and usually less successful. *The Wonderful Visit* (**1895**) tells of an ANGEL who comes to Earth from another dimension. In *The Sea Lady* (**1902**) a MERMAID appears to a family off the south coast of England. *The Dream* (**1924**) might be regarded as a timeslip story, but the initial future setting is really only a storyframe in which to add perspective to a mundane novel of 20th-century life. *The Croquet Player* (**1936**) is more successful, perhaps because of its cynicism: a doctor reveals his belief that an area of marshland is being haunted by the evil spirits of primitive men whose remains have recently been found, but the doctor's psychiatrist explains that all is hallucination. This novel and *The Camford Visitation* (**1937**), in which a disembodied alien visits a centre of learning to exhort the academics on the nature of their studies, are typical of HGW's later attitude toward humankind's inability to come to terms with the horrors of the social and political situation. HGW continued these barbed attacks to the end of his life. *The Happy Turning* (**1945**), in which he seems to be preparing to meet his maker, includes two discussions with CHRIST about the nature of Christianity. These final works emphasize the bitter nature HGW had developed, so different from the excitement of his early fantasies. [MA]

WELLS, JOHN JAY ◊ Marion Zimmer BRADLEY; Juanita COULSON.

WELLS, MARTHA (? -) US author of *The Element of Fire* (**1993**), which is a well told bit of GENRE FANTASY set in a fairly standard FANTASYLAND, and of *City of Bones* (**1995**), an ARABIAN FANTASY. [JG]

WENDELESSEN [s] ◊ Charles DE LINT.

WENDIGO In Native American tradition – especially among the Algonquins – an evil SPIRIT; it is also called Weedigo, Weetigo, Whittico, Windago, Windigo, Witiko, etc., and is variously portrayed as a giant skeleton of ice or, more so in SUPERNATURAL FICTION than in LEGEND, as a spirit of the WINDS. Native Americans believed you could be possessed by a Wendigo (◊ POSSESSION) and thereby turned into one, a psychosis usually manifested by way of devouring. Its main attribute is that it eats people and can transfer that affliction to any it bites – the parallels with European myths concerning VAMPIRES are obvious. The Wendigo has been likened to the call-of-the-wild personified – perhaps most recently by Margaret Atwood in *Strange Things: The Malevolent North in Canadian Literature* (**1995**), where she describes it as the embodiment of the spirit of the Canadian North.

The primary supernatural story is "The Wendigo" (in *The Lost Valley* coll **1910**) by Algernon BLACKWOOD, in which a member of a hunting party in northern Canada is abducted by the Wendigo and spiritually drained. Blackwood thus converted the Wendigo from a physical man-eater to a form of psychic or spiritual vampire. This aspect was further developed by August DERLETH, who linked the Wendigo to his CTHULHU MYTHOS writings in "The Thing that Walked on the Wind" (1933 *Strange Tales*) and "Ithaqua" (1941 *Strange Stories*). This has been further explored by Brian LUMLEY in "Born of the Winds" (1975 *F&SF*), which is more true to the legend than Lumley's later use of the creature in his **Titus Crow** novels, where the Wendigo more closely resembles an interplanetary Abominable Snowman. Elements of the legend occur also in *The Wabeno Feast*

(**1973**) by Wayland DREW, and it has been used as a motif in stories by Thomas Easton (1944-) and Thomas F. MONTELEONE. The direct-to-video movie *Wendigo* (**1978**) was based very loosely on Blackwood's story, although the spirit evidently drained any lifeforce from the moviemakers. A detailed anthology of items historical, fictional and psychological is *Windigo: An Anthology of Fact and Fantastic Fiction* (anth **1982**) ed John Robert COLOMBO. [MA]

WENZELL, ALBERT BECK (1864-1917) US artist. ◊ ILLUSTRATION.

WERE-CREATURES ◊ SHAPESHIFTERS.

WEREWOLF MOVIES The lycanthrope (◊ WEREWOLVES) made its CINEMA debut in the two-reel drama *The Werewolf* (**1913**), quickly followed by *The White Wolf* (**1914**), both based on Native American legends. Although there were four silent films entitled *The Wolf Man* (**1915**, **1918**, **1924** and **1924**), none actually dealt with supernatural lycanthropy. Following the success of *Dracula* (**1930**) and *Frankenstein* (**1931**), Universal began looking around for another horror hit. French-born writer/director Robert Florey developed a treatment called *The Wolf Man*, possibly to star Boris Karloff. The idea went through several rewrites, and by the time it reached the screen as *Werewolf of London* (**1935**) it was dir Stuart Walker and featured Broadway star Henry Hull as a botanist who transformed into a satanic-looking werewolf (created by the studio's great make-up artist Jack Pierce). With his portrayal of the doomed lycanthrope Lawrence Talbot in *The Wolf Man* (**1941**) Lon Chaney Jr finally created a classic character of his own and helped launch the second great cycle of HORROR MOVIES. The movie became the studio's biggest moneymaker of the season, and Universal continued Talbot's quest for a cure in *Frankenstein Meets the Wolf Man* (**1943**) and the multi-monster marathons *House of Frankenstein* (**1944**), *House of Dracula* (**1945**) and *Abbott and Costello Meet Frankenstein* (**1948**). The phenomenon of *The Wolf Man* could not be ignored by other studios, who quickly churned out their own variations with *The Mad Monster* (**1942**), *The Undying Monster* (**1942**, vt *The Hammond Mystery* UK), CAT PEOPLE (**1942**), *The Ape Man* (**1943**), *Cry of the Werewolf* (**1944**), *The Return of the Vampire* (**1944**), *She-Wolf of London* (**1946**; vt *The Curse of the Allenbys* UK), *The Catman of Paris* (**1946**) and *The Creeper* (**1948**). The hirsute star of *The Werewolf* (**1956**) was accidentally created by an injection of wolf's blood, and it was a mad doctor who combined his experimental serum with hypnotic regression to transform a troubled tearaway into American International Pictures' *I Was a Teenage Werewolf* (**1957**). In the confusing *Daughter of Dr Jekyll* (**1957**) (◊ JEKYLL AND HYDE MOVIES) it was not Gloria Talbott's titular heroine but a kindly old doctor who was revealed as the scientifically created werewolf. At least the FEMME FATALES of *Cult of the Cobra* (**1955**) and CAT GIRL (**1957**) both had supernatural explanations for their transformations. Although based on the 1933 novel *The Werewolf of Paris* by Guy Endore, the setting of Hammer's *The Curse of the Werewolf* (**1960**) was changed to Spain. Lon Chaney Jr travelled to Mexico to recreate two of his most famous roles, as a werewolf and a mummy, in *Face of the Screaming Werewolf* (**1960**; ot *La Casa del Terror*). The ageing star also encountered another lycanthrope in the equally dire *House of the Black Death* (**1966**), and other cheap-looking werewolves were featured in *Werewolf in a Girls' Dormitory* (**1961**; ot *Lycanthropus*; vt *I Married a Werewolf* UK), *Devil Wolf of Shadow Mountain* (**1964**), *Dr Terror's House of*

Horrors (**1965**), *Dr Terror's Gallery of Horrors* (**1966**) and *The Mummy and the Curse of the Jackals* (**1969**). Spanish actor Paul Naschy (Jacinto Molina) first portrayed the doomed El Hombre Lobo, Waldemar Daninsky, in *Frankenstein's Bloody Terror* (**1967**; ot *La Marca del Hombre*; vt *Hell's Creatures* UK), and he recreated the character through various incarnations in *Nights of the Werewolf* (**1968**; ot *Las Noches del Hombre Lobo*), *Dracula vs Frankenstein* (**1969**; ot *El Hombre Que Vino del Ummo*; vt *Assignment Terror* US), *The Fury of the Wolfman* (**1970**; ot *La Furia del Hombre Lobo*), *The Werewolf vs the Vampire Woman* (**1970**; ot *La Noche de Walpurgis*; vt *Shadow of the Werewolf* UK), *Dr Jekyll and the Werewolf* (**1971**; ot *Doctor Jekyll y el Hombre Lobo*), *Curse of the Devil* (**1973**; ot *El Retorno de Walpurgis*), *Night of the Howling Beast* (**1975**; ot *La Maldición de la Bestia*; vt *The Werewolf and the Yeti* UK), *The Craving* (**1980**; ot *El Retorno del Hombre Lobo*), *La Bestia y la Espada Mágica* (**1983**) and *Lycantropus* (**1996**). A coven of Satanists transformed Hell's Angels into *Werewolves on Wheels* (**1971**), the President's press aide became *The Werewolf of Washington* (**1973**), and no one believed the protagonist of *The Boy who Cried Werewolf* (**1974**). Peter Cushing starred in both *The Beast Must Die* (**1974**), a horror whodunnit in which the audience had to guess the identity of the werewolf, and *Legend of the Werewolf* (**1974**). Director Joe Dante's smart and scary *The* HOWLING (**1980**) was infinitely superior to the six direct-to-video sequels that followed. Another movie that successfully combined lycanthropes and laughs was *An* AMERICAN WERE-WOLF IN LONDON (**1981**) dir John Landis, who also turned Michael Jackson into a werewolf for the extended ROCK VIDEO *Thriller* (**1983**). WOLFEN (**1981**) and *Silver Bullet* (**1985**) were based on bestselling books by Whitley Strieber and Stephen KING respectively, and *The* COMPANY OF WOLVES (**1984**) used Angela CARTER's revisionist FAIRY-TALES for its inspiration. TEEN WOLF (**1985**) was a surprise hit. Jack Nicholson played a Manhattan book editor who transformed in the big-budget WOLF (**1994**), while *Night Stalkers* (**1995**) was made for just $600 with an all-deaf cast. *An American Werewolf in Paris*, a long-awaited sequel to *An American Werewolf in London*, was announced in 1996. Werewolves have also been featured in such tvms as *Moon of the Wolf* (**1972**), *Scream of the Wolf* (**1974**), the French *Hugues de Loup* (**1974**), *The Werewolf of Woodstock* (**1975**), *Death Moon* (**1978**) and *Full Eclipse* (**1993**), while series containing lycanthropic leads include *The* MUNSTERS (1964-6), *The* MUNSTERS TODAY (1988-91), *Werewolf* (1987-8), *She-Wolf of London* (1990-91) and the Hanna-Barbera cartoon *Fangface* (1978). [SJ]

Further reading: *Classic Movie Monsters* (**1978**) by Donald F. Glut; *Il Cinema dei Licantropi* (**1987**) by Riccardo Esposito; *The Illustrated Werewolf Movie Guide* (**1996**) by Stephen JONES.

WEREWOLVES European legends of WOLF-human SHAPESHIFTING – also known as lycanthropy – have been recorded since the 11th century or earlier, and perhaps arise from a far older tradition of wearing wolfskin in tribes whose TOTEM was the wolf. Classical werewolves take on wolf form at full MOON – involuntarily and often unwillingly (◊ BONDAGE) – are highly averse to (or, in some versions, have an affinity for) *Aconitum* plants, known as wolfbane; and like other supernatural MONSTERS may be harmed only by silver WEAPONS. *The Book of Were-wolves* (**1865**) by Sabine BARING-GOULD is a useful summary of the FOLKTALES. Werewolves have become stock ingredients in the

CAULDRON OF STORY, and now routinely appear in GENRE FANTASY. (The Red Riding-Hood FAIRYTALE may seem an UNDERLIER of the werewolf theme, but the reverse is apparently true.) Although Rudolf Erich RASPE's *Koenigsmark the Robber* (**1790**) is a TALL-TALE exception, earlier fictional werewolves tend to inhabit SUPERNATURAL FICTIONS or HORROR thrillers, examples being Johann August APEL's "The Boarwolf" (1812), Jessie Douglas KERRUISH's influential *The Undying Monster* (**1922**), Dion FORTUNE's *The Demon Lover* (**1927**) and Guy ENDORE's *The Werewolf of Paris* (**1933**). The last of these did much to concretize later fantasy's view of lycanthropy, also heavily influenced by WEREWOLF MOVIES. SAKI's "Gabriel-Ernest" (1909) typically thrusts a werewolf into English country society. James Branch CABELL contrasts the simple EVIL of werewolfhood with the worse hypocrisy of an inquisitorial bishop in *The White Robe* (**1928**). One exhibit in Charles G. FINNEY's *The Circus of Dr Lao* (**1935**) is a werewolf.

In reaction to the old tradition of werewolves as inherently evil – as still implied in C.S. LEWIS's *Prince Caspian* (**1951**) – there have been several sympathetic REVISIONIST-FANTASY treatments. H. Warner MUNN's "The Werewolf of Ponkert" (1925 *WT*) moved the lycanthrope away from simple monstrosity by making it the viewpoint character. Jack WILLIAMSON's notable *Darker Than You Think* (**1948**) pits were-creatures against the human race, presenting shapeshifting so enticingly that it seems reasonable for the hero to join the other side. The werewolf protagonists of Anthony BOUCHER's "The Compleat Werewolf" (1942) and Poul ANDERSON's *Operation Chaos* (**1971**) are wholly likeable and wholly on the side of good; the inept Polacek in L. Sprague DE CAMP's and Fletcher PRATT's *The Castle of Iron* (**1941**), and the female Angua of the City Watch in Terry PRATCHETT's **Discworld**, are – though sorely tempted when in wolf shape – too decent to prey on humans. C.L. MOORE had earlier considered the female aspect in "Werewoman" (1938). The obvious link between menstruation, the MOON and the werewolf change is made in Peter BEAGLE's comic *Lila, the Werewolf* (**1969** chap) and, far more darkly, in Alan MOORE's **Swamp Thing** episode "The Curse" (in *Swamp Thing Volume 5* graph coll **1988**).

The popularity of the werewolf theme continues. Further examples include: Angela CARTER's "The Company of Wolves" (1979), revitalizing the Red Riding-Hood story; Suzy McKee CHARNAS's first-person werewolf narration "Boobs" (1989); *Wilderness* (**1991**) by Dennis Danvers, a love story about an unwilling female werewolf and her man; Bernard KING's *Vargr-Moon* (**1986**); Stephen KING's routine *Cycle of the Werewolf* (**1983**) and also, with Peter STRAUB, *The Talisman* (**1984**) – whose werewolf becomes a good COMPANION for a QUEST; Tanith LEE's *Lycanthia, or The Children of Wolves* (**1981** US) and *Heart-Beast* (**1992**); S.P. SOMTOW's vivid *Moon Dance* (**1989**); E.C. VIVIAN's *Grey Shapes* (**1937**); Michael D. WEAVER's **Wolf-Dreams** SWORD-AND-SORCERY sequence starring a lesbian werewolf swordsperson; and novels by Chelsea Quinn YARBRO. Examples of werewolf theme anthologies are: *Werewolf: Horror Stories of the Man-Beast* (anth **1987**) ed Peter HAINING; *Werewolves* (anth **1988**) ed Debra DOYLE, Martin H. GREENBERG and Jane YOLEN; *The Ultimate Werewolf* (anth **1991**) ed John BETANCOURT, Byron PREISS, David Keller and Megan Miller; *The Mammoth Book of Werewolves* (anth **1994**) ed Stephen JONES; and *Tomorrow Bites* (anth **1995**) ed Greg COX and T.K.F. Weiskop. [DRL]

See also: TRANSYLVANIA.

WES CRAVEN'S NEW NIGHTMARE US movie (*1994*). ◊ *A* NIGHTMARE ON ELM STREET.

WEST, JOHN [s] ◊ Robert ARTHUR.

WEST, [Dame] REBECCA Pseudonym of UK writer Cecily Isabel Fairfield Andrews (1892-1983), a lover of H.G. WELLS and a formidable cultural critic and author of striking fictions from 1911; her pseudonym is the name of the heroine of the play *Rosmersholm* (**1886**) by Henrik Ibsen (1828-1906), a role she played on stage. She is of fantasy interest for *Harriet Hume: A London Fantasy* (**1929**), a POSTHUMOUS FANTASY whose telepathically sensitive heroine cannot tolerate her lover's mortal faults while they are both alive, but once they are dead finds him an ideal companion; they then become deeply intimate with LONDON in the course of exploring their new state. [JC]

WESTALL, ROBERT (ATKINSON) (1929-1993) UK writer, primarily of children's fiction, mostly GHOST STORIES FOR CHILDREN. RW uses Northumbrian settings, characters and dialect in a number of his stories, starting with *The Wind Eye* (**1976**), a TIMESLIP fantasy where a family comes under the HEALING influence of St Cuthbert and *The Watch House* (**1977**), about a haunted lifeboat museum. *The Devil on the Road* (**1978**) is another timeslip fantasy: a young biker returns to the 17th century to save a WITCH from hanging. RW's other SUPERNATURAL-FICTION novels are *The Scarecrows* (**1981**), *Ghost Abbey* (**1988**), *Blitzcat* (**1989**), *If Cats Could Fly. . .?* (**1990**), *The Promise* (**1990**), *Yaxley's Cat* (**1991**) and *The Wheatstone Pond* (**1993**).

RW broke the mould a few times, although all his novels draw on images of horror. *Futuretrack 5* (**1983**), a rebellion against conservatism, is set in a divided future society; *The Cats of Seroster* (**1984**) is a medieval fantasy where cats endeavour to restore the power of their ancient guardian; *Urn Burial* (**1987**) depicts an ongoing war between dogs and cats on a cosmic scale; and in *Gulf* (**1992**) a boy experiences the horrors of the Gulf War through a telepathic link with an Iraqi soldier.

RW was also an excellent short-story writer. The stories allowed him more opportunity to focus atmosphere and showed him as a capable imitator of M.R. JAMES, especially in the long title story to *The Stones of Muncaster Cathedral* (coll **1991**; title story alone **1993** chap US) about a steeplejack whose work on a cathedral unleashes an evil SPIRIT. RW's short stories have been published in *Break of Dark* (coll **1982**), *The Haunting of Chas McGill and Other Stories* (coll **1983**), *The Other* (**1985** chap), *Rachel and the Angel, and Other Stories* (coll **1986**), *Ghosts and Journeys* (coll **1988**), *The Call, and Other Stories* (coll **1989**), *A Walk on the Wild Side* (coll **1989**), *The Christmas Ghost* (**1992** chap) – included with *The Christmas Cat* (**1991** chap) in *Christmas Spirit* (omni **1994**) – and *Fearful Lovers, and Other Stories* (coll **1992**; vt *In Camera and Other Stories* 1993 US). RW's only volume of GHOST STORIES intended chiefly for adults is *Antique Dust* (coll **1989**), drawing on his experiences as an antique dealer and art collector. The best of his supernatural stories, including some previously unpublished, were collected as *Demons and Shadows* (coll **1993** US) and *Shades of Darkness* (coll **1994** US).

Like Leon GARFIELD, RW gave little consideration to the traditional proprieties of children's fiction. He was a teacher for nearly 30 years and knew the psychology of adolescents. He littered his books with profanities and images of growing sexual awareness, treating children like adults. He pulled no punches in his development of horrific situations, and many stories are genuinely frightening. He knew the terrors of war and used those images as warnings in his fiction. His work includes some of the best children's ghost stories of the 20th century. [MA]

Other works (for younger children): *Rosalie* (**1987**); *The Creature in the Dark* (**1988**); *Old Man on a Horse* (**1989**).

As editor: *Ghost Stories* (anth **1988**).

WESTON, ALLEN Pseudonym of Andre NORTON.

WESTON, JESSIE L(AIDLAW) (1850-1928) Influential and prolific UK editor, translator and scholar of medieval literature. JLW's works have had a profound effect upon modern Arthurian criticism and literature. In a series of monographs – *The Legend of Sir Gawain* (**1897**), *The Legend of Sir Lancelot du Lac* (**1901**), *The Legend of Sir Perceval* (**1906-9** 2 vols), *The Quest for the Holy Grail* (**1913**) and *From Ritual to Romance* (**1920**) – JLW sought to disinter pre-Christian motifs and ideas from the extant medieval texts. Influenced by the work of Sir James FRAZER, she identified a number of universal mythological themes running throughout the ARTHUR corpus, including vegetation rites (birth, rebirth and the connection of king and land), otherworldly mistresses, wives and lovers, and quasi-matrilineal kinship patterns. These she paralleled with Classical and other European MYTHOLOGIES, attributing their appearance in the Arthurian context to a survival of memories of pagan ideologies and practices. Although now largely abandoned by the modern academic world, JLW's work has had far-reaching effects upon modern writers of Arthurian fiction, notably Marion Zimmer BRADLEY, Gillian BRADSHAW and Bernard Cornwell. JLW's other works included a series of translations of medieval Arthurian romances not represented in the work of Sir Thomas MALORY: these remain useful and accessible. [KLM]

Translations (selective): *Morien* (trans **1901**); *Sir Gawain at the Grail Castle* (trans **1903**); *Sir Gawain and the Lady of Lys* (trans **1907**) (◊ GAWAIN).

Other work: *The Apple Mystery in Arthurian Romance* (**1925**).

WHALE, JAMES (1896-1957) UK-born Hollywood movie director, who began his career as a cartoonist, then took to the theatre. His London stage version of *Journey's End* (performed 1928; **1929**) by R.C. Sherriff (1896-1975) having been a major success, he came to Hollywood to film it – as *Journey's End* (*1930*). His first big hit, though, was *Frankenstein* (**1931**; ◊ FRANKENSTEIN MOVIES), one of the great classics of "Hollywood Gothic"; it was sequelled by *The Bride of Frankenstein* (*1935*), which was even better. Before that came *The Old Dark House* (*1932*), the precursor of the comic HAUNTED-DWELLING movie *à la* the various versions of *The Cat and the Canary*; it was based on J.B. PRIESTLEY's *Benighted* (**1927**; vt *The Old Dark House* 1928 US). *The* INVISIBLE MAN (*1933*), based on the H.G. WELLS novel, is a well judged mix of humour and the bizarre (◊◊ INVISIBILITY). *Remember Last Night* (*1935*) is a comedy murder mystery of peripheral fantasy interest. Thereafter JW shifted his attentions elsewhere, making a mixture of major movies – e.g., *Showboat* – and minor ones, his standards slowly declining. In his fantasy movies JW's great talent was for creating atmosphere through lighting and angles; often this disguises poor scripts. [JG]

Other works (selective): *The Man in the Iron Mask* (*1939*); *Green Hell* (*1940*), a JUNGLE MOVIE set in South America.

WHALES ◊ SEA MONSTERS.

WHALE-TUMOUR STORIES ◊ URBAN LEGENDS.
WHAM, TOM (? -) US author. ◊ Rose ESTES.
WHARTON, EDITH (NEWBOLD) (1862-1937) US writer whose many novels sharply examine US society, the focus of her most successful work – like *The Age of Innocence* (**1920**) – being on the oppositions between private passions and the world. She wrote about 12 stories of genre interest, all SUPERNATURAL FICTION, almost all involving GHOSTS. The first – "The Fulness of Life" (1893), a minor POSTHUMOUS FANTASY – does not appear in later collections. The remainder – first assembled (along with other work) in *Tales of Men and Ghosts* (coll **1910**), *Xingu and Other Stories* (coll **1916**) and *Here and Beyond* (coll **1926**) – are assembled in *Ghosts* (coll **1937**; vt *The Ghost Stories of Edith Wharton* 1973). "Afterward" (1910) is set in a country house where ghosts are never recognized as such until afterwards, which is too late. "The Eyes" (1910) is a superb study of sexual ambivalence, ending in a self-confrontation that some critics have likened to the revelations exploded in the climax of Oscar WILDE's *The Picture of Dorian Gray* (**1891**). In "Pomegranate Seed" (1931) a widower who has remarried receives letters from his demanding dead wife: the last letter beckons him presumably to the UNDERWORLD. "All Souls'" (1937) marries virtuoso descriptions of snow and isolation to a chillingly evoked rendering of the theft of a day. Subtle and stalwart, EW is a central 20th-century practitioner of the GHOST STORY. [JC]

WHARTON, WILLIAM Pseudonym of US painter and writer Albert DuAime (1927-), long resident in Paris; most of his novels are FABULATIONS whose MAGIC-REALISM techniques transform autobiography – almost all expand upon episodes in his own life – into contemporary myth. *Birdy* (**1979** US) translates its young protagonist's obsession with BIRDS and FLYING into a parable of the life of the artist. *A Midnight Clear* (**1982** US) presents the Christmastide experiences of a squad of US soldiers in WORLD WAR II as a revelatory rite. *Tidings* (**1987** US) is also set at CHRISTMAS, this time in rural 1980s France; a US expatriate philosopher obsessively gathers his widespread family together to celebrate the SEASON, an event made possible by a conflation of solstitial MIRACLES. *Franky Furbo* (**1989** US) depicts a talking fox which has come back from the future to possess the body of a human. [JC]

WHEATLEY, DENNIS (YEATS) (1897-1977) Prolific and once bestselling UK author of thrillers, historical romances and a loosely linked occult sequence, the **Black Magic** stories. These add *frissons* of BLACK MAGIC, BLACK MASSES and SATANISM to the basic adventure-thriller TEMPLATE. The first and liveliest is *The Devil Rides Out* (**1935**; juvenile edn cut Alison Sage 1987), featuring a SABBAT, a set-piece scene in a pentacle subjected to magical assaults culminating with the mounted ANGEL of Death, a Satanist who lacks a SHADOW, a MCGUFFIN-hunt for the evil TALISMAN of Set, and a *deus ex machina* finale; this was filmed as *The* DEVIL RIDES OUT (*1968*). A sequel, *Strange Conflict* (**1941**) uneasily extends WORLD WAR II espionage to the ASTRAL PLANE, throwing in VOODOO, ZOMBIES and PAN for good measure. *The Haunting of Toby Jugg* (**1948**) makes grisly play with the DEMON in SPIDER form which terrifies the crippled hero, but loses its force in anti-Communist polemic: DW's asides on race and politics are frequently embarrassing. *To the Devil – A Daughter* (**1953**) features an innocent virgin (◊ VIRGINITY) who at night undergoes fearsome POSSESSION, making her sophisticated and interested in

SEX; she has been PACT-bound to the DEVIL as preliminary to suffering HUMAN SACRIFICE, which will transfer her SOUL to a HOMUNCULUS. It is typical of DW's PLOT DEVICES that a malign altar should be fortuitously struck by lightning: "God had intervened." The movie is TO THE DEVIL A DAUGHTER (*1976*).

Other relevant novels are: *The Ka of Gifford Hillary* (**1956**), whose "murdered" protagonist's eponymous ASTRAL BODY struggles for VENGEANCE; *The Satanist* (**1960**); *They Used Dark Forces* (**1964**), mixing elements of MAGIC and ASTROLOGY (and the inevitable Black Mass) into an over-long WWII thriller; *Unholy Crusade* (**1967**); *The White Witch of the South Seas* (**1968**), with more voodoo; *Gateway to Hell* (**1970**); and *The Irish Witch* (**1973**). Additionally, DW produced some novels of LOST LANDS, shading between sf and fantasy: *The Fabulous Valley* (**1934**); *They Found Atlantis* (**1936**), disclosing an inhabited ATLANTIS; *Uncharted Seas* (**1938**), set in a Sargasso replete with SEA MONSTERS – this was filmed as *The Lost Continent* (*1968*) – and the Antarctic *The Man who Missed The War* (**1945**), assembled with the previous two as *Worlds Far from Here* (omni **1952**). *Star of Ill-Omen* (**1952**) presents routine UFOS.

A modest power of storytelling partly redeems DW's narrative longueurs, clichéd thought and repetitious pursuits, confrontations and escapes. The **Black Magic** books' spice of wickedness once held particular appeal for adolescents who found them daringly "adult", but HORROR's excesses have long since passed them by. [DRL]

Other works (selective): *Gunmen, Gallants and Ghosts* (coll **1943**); *The Devil Rides Out and Gateway To Hell* (omni **1992**).

As editor: *A Century of Horror Stories* (anth **1935**; cut vt *Quiver of Horror* 1965); *Uncanny Tales 1* (anth **1974**); *Uncanny Tales 2* (anth **1974**).

WHELAN, MICHAEL (1950-) Remarkably popular award-winning US artist and illustrator whose attractively coloured, somewhat languid paintings have consistently attracted the highest accolades. MW began painting covers in 1974, his most successful early works being for reissues of the Edgar Rice BURROUGHS **Barsoom** series and Michael MOORCOCK's **Elric** books. He exhibited his original paintings in New York, and his book *Wonderworks* (graph **1979**) enhanced his reputation. He was awarded the Hugo AWARD as Best Professional Artist every year 1980-86 and again in 1988, 1989 and 1991, after which he withdrew from the contest. Other volumes of his work are *Michael Whelan's Works of Wonder* (graph **1987**) – for which he won an extra Hugo – and *The Art of Michael Whelan* (graph **1993**). [RT]

WHEN THE WIND BLOWS UK ANIMATED MOVIE (*1987*). ◊ Raymond BRIGGS.

WHEN THINGS WERE ROTTEN US tv series (1975). Paramount/ABC. **Pr** Norman Steinberg. **Exec pr** Mel Brooks. **Dir** Bruce Bilson, Marty Feldman, Peter H. Hunt, Jerry Paris, Coby Ruskin, Joshua Shelley. **Writers** John Boni, Brooks, Norman Stiles and many others. **Created by** Boni, Brooks, Stiles. **Starring** Richard Dimitri (Bertram/Renaldo), Dick Gautier (Robin Hood), Jane A. Johnston (Princess Isabelle), Bernie Kopell (Allan-A-Dale), Henry Polic II (Lord Hubert, Sheriff of Nottingham), Ron Rifkin (Prince John), Misty Rowe (Maid Marian), David Sabin (Little John), Dick Van Patten (Friar Tuck). 13 30min episodes. Colour.

A Brooks-style wacky version of the ROBIN HOOD legend featuring a dimwitted Robin and equally dimwitted Merry

Men, plus a dumb-blonde Marian. The villains were likewise caricatures. Brooks's *Robin Hood: Men in Tights* (*1993*) might be regarded as a natural follow-up. [BC]

WHISPERS 1. US SMALL-PRESS digest MAGAZINE, irregular, 16 issues (though, because of 8 double issues, numbering to *#23/#24*), July 1973-October 1987, published and ed Stuart D. Schiff (1946-   ).

W was originally planned as a continuation of August W. DERLETH's *The Arkham Collector* (10 issues Summer 1967-Summer 1971; ◊ ARKHAM HOUSE) and was thus to be called *Whispers from Arkham*, but legal considerations required the curtailed title. The intention was to provide news of forthcoming titles from Arkham House and other small presses – a service that *W* continued to provide, although its news was often dated – but from the outset space was always made for fiction, articles and poetry. The fiction was initially mainly by H.P. LOVECRAFT disciples – the opening story, "House of Cthulhu" by Brian LUMLEY, being an indication of direction. By *#2* the magazine's ambience had already clearly widened to encompass the broader WEIRD TALES school of authors – with stories from Fritz LEIBER and Henry Hasse (1913-1977) – and by the next issue *W* had begun to establish its own identity, with fiction from writers of a new generation – establishing, in a way, a modern equivalent of the *WT* school. *#3* (1974) featured Karl Edward WAGNER's "Sticks", based on an incident that had happened to artist Lee Brown Coye (1907-1981), who provided the cover. The story won the BRITISH FANTASY AWARD; *W* itself went on to win the first WORLD FANTASY AWARD in 1975. The emphasis on new work by *WT* writers never faded – later issues featured Hugh B. CAVE, Carl Jacobi (1908-   ) and Manly Wade WELLMAN – but *W* increasingly became a core publication for new and even established writers at a time when markets for fantasy and SUPERNATURAL FICTION were poor. Contributors included Robert AICKMAN, Ramsey CAMPBELL, David Campton (1924-   ), Dennis ETCHISON, Charles L. GRANT, R.A. LAFFERTY, Richard Christian MATHESON, William F. NOLAN and Ray Russell (1924-   ). *W* also featured good cover art by Tim Kirk (1947-   ), Stephen Fabian (1930-   ), Frank Utpatel (1905-1980) and John Stewart (1942-   ).

The pressure of Schiff's studies and his professional work as a dentist restricted the time he could spend on *W*; as issues became variously delayed, he converted them to double issues, which provided better value. These were eagerly awaited, as they tended to focus on a particular author. The first of these (*#11/#12* October 1978), on Wellman, was also the first issue of *W* to be professionally typeset; the next (*#13/#14* October 1979), on Leiber, had *W*'s first wraparound colour cover. Other special issues were *#15/#16* (March 1982), on Campbell, *#17/#18* (August 1982), on Stephen KING, and *#19/#20* (October 1983), on Whitley Strieber. *W*'s fiction, while remaining true to the diversity and colour of the *WT* tradition, also helped succour the roots of the new school of horror that emerged in the wake of Stephen King's success. The magazine won the World Fantasy Award again in 1977, 1983 and 1985; "The Bone Wizard" by Alan Ryan (1943-   ) (from *#21/#22*) shared the 1985 World Fantasy Award. The success of the magazine gave rise to an anthology series (◊ 2).

In 1975 Schiff founded Whispers Press, through which he issued hardbound copies of the magazine: *Volume I* (*1975*) (*#1-#4*); *II* (*1978*) (*#5-#8*) and *IV* (*1979*) (*#13-#16*) – there was no hardbound *III*, which was simply the magazine issue

itself. The last 5 double issues – *#15/#16*, *#17/#18*, *#19/#20*, *#21/#22* and *#23/#24* – were also released in hardcover editions. In addition Whispers Press issued *A Winter Wish and Other Poems* (coll **1977**) by Lovecraft, *Rime Isle* (**1977**) by Leiber, *Strange Eons* (**1978**) by Robert BLOCH, *Heroes and Horrors* (coll **1978**) by Leiber, *The Scallion Stone* (coll **1980**) by Basil A. Smith (1908-1969) – GHOST STORIES in the M.R. JAMES tradition – *Psycho II* (**1982**) by Bloch, *Foundation's Edge* (**1982**) by Isaac ASIMOV and *The Tomb* (**1984**) by F. Paul WILSON. Unfortunately, production costs became too great, and both Whispers Press and *W* itself died. [MA]

2. An ANTHOLOGY series grew out of *Whispers* magazine: *Whispers* (anth **1977**), *Whispers II* (anth **1979**), *III* (anth **1981**), *IV* (anth **1983**), *V* (anth **1985**) and *VI* (anth **1987**), all ed Stuart David Schiff. The first volume consisted primarily of reprints from the magazine, but later volumes each contained only 5-6 such reprints, the remainder of their contents being new. The series tended to publish darker and more sinister material than the magazine. "Elle est Trois (La Mort)" by Tanith LEE, from *Whispers IV*, won the 1984 WORLD FANTASY AWARD. *The Best of Whispers* (anth **1994**) was selected from the earlier anthologies and from the unpublished «Whispers VII». [MA]

WHISPERS PRESS ◊ WHISPERS.

WHISTLE DOWN THE WIND UK movie (*1961*). Rank/Allied/Beaver. **Pr** Richard Attenborough. **Dir** Bryan Forbes. **Screenplay** Willis Hall, Keith Waterhouse. **Based on** *Whistle Down the Wind: A Modern Fable* (**1958**) by Mary Hayley Bell. **Starring** Alan Barnes (Charles Bostock), Alan Bates (Blakey), Norman Bird (Eddie), Roy Holder (Jackie), Diane Holgate (Nan Bostock), Bernard Lee (Bostock), Hayley Mills (Kathy Bostock), Elsie Wagstaff (Auntie Dorothy). 99 mins. B/w.

In the north of England three farm children – Kathy, Nan and Charles – come through misunderstandings to believe that Blakey, a murderer hiding in the barn, is in fact the returned MESSIAH. Anxious to protect him from a second martyrdom, they conceal his existence from the adult world, but news spreads among the local children, who at last throng to see him being removed by officialdom to a likely death. The fantasy that the children have erected around his presence is, however, merely confirmed by this final event.

This astonishingly affecting movie is made so by the performances of the three children; in most movies Holgate's portrayal of the second oldest child would be regarded as exceptional, but she is almost eclipsed by Mills and, most particularly, Barnes, as the youngest. This is one of a grouping of movies that depict the generation of powerful fantasies by young minds, hence giving us a glimpse at the origins of RELIGION and MYTHOLOGY; others include *The* LORD OF THE FLIES (*1963*), *The* SPIRIT OF THE BEEHIVE (*1973*) and CELIA (*1988*). [JG]

WHITBOURN, JOHN (1958-   ) UK writer whose first novel, *A Dangerous Energy* (**1992**), won the 1991 First Fantasy Novel Competition jointly organized by BBC Radio 4's *Bookshelf* and the publishers Gollancz. The novel is set in an ALTERNATE-WORLD UK, ingeniously outlined through a history examination paper, where Royalists won the Civil War and the stern guardianship of the Catholic Church has – as in Keith ROBERTS's *Pavane* (**1968**) – retarded technology; this Church controls and polices MAGIC. The selfish, amoral ANTIHERO Oakley is magically initiated by ELVES, enters the Church, and proves brutally effective in "crusades" against England's heretical

Protestant "Levellers". He advances himself by misusing DEMONS, source of all major magical power; these and their conjuration exude an effective stench of WRONGNESS, as in James BLISH's *Black Easter* (**1968**). Oakley's climactic experiment in demonological research, though horrifying and hugely destructive, leads him to a futile dead end. He dies quietly in bed, to learn (like C.S. LEWIS's damned) that his lifetime's efforts have achieved only his utter severance from GOD. *Popes and Phantoms* (**1993**) is a blackly witty PICARESQUE whose protagonist Admiral Slovo plays games with already skewed Renaissance history as agent of SECRET MASTERS, encountering REVENANTS, Borgias, Machiavelli, Michelangelo, Henry VII, Martin Luther, SATAN, etc. *To Build Jerusalem* (**1995**) returns to the England of *A Dangerous Energy*, where a new demon from beyond the hierarchy explored by Oakley (who briefly appears) capriciously aids the Levellers and has abducted the King and his court to a demonic OTHERWORLD. This is investigated by an elite Vatican emissary and assassin who, reversing the spiritual progress of Oakley and Slovo, is already dehumanized at the outset but during the action – skirmishes, conjurations, PORTAL-crossings and battle against MONSTERS – moves some way towards ordinary empathy. JW writes well, with dry wit. [DRL]

Other works: The **Binscombe Tales**, being *Binscombe Tales* (coll **1989** chap) and *Rollover Night* (coll **1990** chap), *A Binscombe Tale for Christmas* (**1994** chap and *A Binscombe Tale for Summer* (**1996** chap), GHOST STORIES/SUPERNATURAL FICTIONS; *Popes & Phantoms* (coll **1992** chap), two episodes from the novel.

WHITE, BABBINGTON [s] ◊ Mary E. BRADDON.

WHITE, E(LWYN) B(ROOKS) (1899-1985) US humorist and essayist whose longtime contributions to *The New Yorker* included the **Talk of the Town** column. His best-known fantasies are for children. *Stuart Little* (**1945**) is about a tiny child, the size of a mouse, who, after various adventures, goes in search of a bird whose life he had previously saved. The story is amusing, but ends abruptly, as if EBW tired of it or intended a sequel. *Charlotte's Web* (**1952**) explores further friendships between animals when a SPIDER saves a pig's life through the messages she weaves in her web. *The Trumpet of the Swan* (**1970**) is a FABLE about how a mute swan learns to trumpet and becomes a celebrity. [MA]

WHITE, MEL (1949-   ) US artist. ◊ Robert Lynn ASPRIN.

WHITE, T(ERENCE) H(ANBURY) (1906-1964) Indian-born UK writer, in the UK from 1911, whose overwhelming nostalgia for a lost land of England expresses itself most vividly in his two best-known works, *Farewell Victoria* (**1933**) and the superlative tragicomic fantasia *The Once and Future King* (**1958**) (◊◊ ONCE AND FUTURE KING; SLEEPER UNDER THE HILL), which made him famous late in life. Despite the fame and fortune that attended THW's closing years, that life was a tragic one. His parents separated, violently, when he was five. He was brought up by relatives, educated at a public school notable for cruelty, and was homosexual but could not admit it. He flourished in his craft only as a young man – a period between the publication of his first book, *Loved Helen and Other Poems* (coll **1927** chap), when he was 21, and the end of WWII, by which point almost everything for which he is now remembered had either been written in draft form or published. The decades after the death of a beloved dog in November 1944, when THW was 38, were desolate.

Two early THW titles are of interest. *Earth Stopped, or Mr Marx's Sporting Tour* (**1934**) builds up to the onslaught of a world-destroying HOLOCAUST; its sequel, *Gone to Ground: A Novel* (coll of linked stories **1935**), comprises a series of tales – mostly SUPERNATURAL FICTIONS – told to each other by eight survivors of the holocaust hiding in a cave complete with a well-stocked bar (◊ CLUB STORY). Without any source being cited – all hints of the FRAME STORY were carefully excised, and individual titles were supplied for each item – all the supernatural tales assembled in *The Maharajah, and Other Stories* (coll **1981**) ed Kurth Sprague were abstracted from *Gone to Ground*. Tales which appear in both volumes include: "The Spaniel Earl", in which a young man convinced he is a dog is finally mated to a girl who thinks she is a bitch; "The Troll", which is HORROR, but with a detached comic irony which distances any real horror (the proposed TROLL-victim's delight at finding himself sane entirely outweighs the prospect of being the thing's next meal); "The Point of Thirty Miles", which details the death of a SHAPESHIFTING wolf; and "The Black Rabbit", in which a boy poacher is caught by a keeper who turns out to be PAN, and whose perorations close the volume, wiping the slate clean. Tales which appear only in *Gone to Ground* each comprise a single chapter, none titled; typical of these is Chapter 12, in which a Greek-speaking MERMAID is discovered swimming up a local English stream, totally lost.

THW is now remembered for almost none of his early work (he had published 10 books by 1935) except *Farewell, Victoria*; but he remains of central importance for *The Once and Future King* (omni **1958**), a fantasia on Sir Thomas MALORY's *Le Morte Darthur* (**1485**), retelling the ARTHUR cycle as a profound meditation on the MATTER of Britain. *The Once and Future King* is THW's masterwork, and is one of the central *reactive* fantasy texts of the century.

The late publication date of *The Once and Future King* is deceptive. A version of the sequence, which incorporated revised versions of three previously published novels, had been prepared as earlier as 1941, and was rejected by THW's publishers because the final section – ultimately published as *The Book of Merlyn* (**1977** US) – argued for pacifism at a bad time to make such arguments. THW did not try to publish the sequence again until almost 15 years had passed. The 1958 version comprises the three earlier novels, substantially cut and recast; plus a fourth section, "The Candle in the Wind", a version of which had been ready in 1941 but which was now revised so as to replace *The Book of Merlyn*, and to terminate the sequence properly. The previously published novels, which in their revised form make up the bulk of the final version, are: *The Sword in the Stone* (**1938**; rev 1939 US), which was made into a philistine ANIMATED MOVIE by DISNEY in 1963; *The Witch in the Wood* (**1939** US), the weakest part of the sequence, and retitled "The Queen of Air and Darkness" in the recasting (which was extremely thorough); and *The Ill-Made Knight* (**1940** US). *The Once and Future King* was adapted by Alan Jay LERNER as *Camelot: A New Musical* (performed 1960; **1961**), with music by Frederick Loewe (1901-1988); the musical was filmed as *Camelot* (**1967**). Both stage and movie versions are watery.

There is a critical consensus that *The Once and Future King* is greater than the sum of its parts, though readers who restrict themselves to that volume as an aesthetic whole do forfeit something of the extraordinary freshness and easy amplitude of the original *The Sword in the Stone*; this has frequently been reprinted in its original form, though not necessarily with THW's illustrations. Alone, *The Sword in the Stone* stands as one of the finest CHILDREN'S FANTASIES of the 20th century. The opening chapters take place in the dawn of the world, even though the worm of history – the inexorable rhythm of a STORY which must be told – can soon be felt imparting an ET IN ARCADIA EGO shadow to the goings-on of the comic cast, all of whom seem to have been brought together in order to give young Wart an education – there is even an anachronistic ROBIN HOOD among them. We may soon guess that Wart is a HIDDEN MONARCH, for the title itself (obviously) recalls a key moment in the story of Arthur, and certainly there can be no doubt of his identity once Merlyn (◊ MERLIN), who lives backwards in TIME, becomes his tutor; but as a solitary tale *The Sword in the Stone* is primarily concerned with childhood (◊ CHILDREN) and a MAGIC-imbued education, and only secondarily with the man to come. In that education Merlyn METAMORPHOSES Wart into the shape of various animals and (Merlyn hopes) is thereby given wisdom from the behaviour of animals, who act in obedience to their nature.

When read as the first of several integrated parts, the innocence of these pages is shadowed very much more obviously by the weight of what we know is about to follow. As a whole, *The Once and Future King* is a lament: for the loss of childhood; for the loss of the vision of governance by virtue of which Britain may enjoy the brief GOLDEN AGE of Arthur's reign; and for the THINNING of a land which can only decreasingly sustain either innocence or virtue. The litany is unremitting. The relationship between THW's own text and Malory is that of the yearning epilogue to the real Story, which can no longer be told straight. The manner of telling of the tale itself moves from the nostalgic timelessness of children's fantasy (*The Sword in the Stone* is, as noted, a classic of that literature, but *The Ill-Made Knight*, focusing on Malory on LANCELOT's adultery with GUINEVERE, is not a children's book) towards the ultimate, secular, adult rhythms of "The Candle in the Wind". And – most remarkably of all – as the tale moves from Wart's childhood to King Arthur's old age, the Britain in which it is set itself moves forward through time from the Dark Ages to Malory's own 15th century: the last sentence of the book (beginning, "The cannons of his adversary were thundering in the tattered morning . . .") propels the dying Arthur into the unpitying light of a much later, narrower day.

It was, perhaps, just as well that *The Book of Merlyn* never became attached to *The Once and Future King*. Just before the LAST BATTLE, Merlyn reappears to Arthur, reintroduces him to the animals of *The Sword in the Stone*, metamorphoses him again to learn some new lessons, returns him to the cruel dawn and Mordred. The rhythm and tragic intensity of *The Once and Future King*, as published, are here flouted.

The overall plot of *The Once and Future King* is a simplification of Malory, and brings into sharper focus the essential tragic situation – Arthur's illegitimate birth, which leads to his begetting Mordred upon his half-sister, MORGAN LE FAY, the Queen of Air and Darkness (◊ ENCHANTRESS) – describing it as "the tragedy, the Aristotelian and comprehensive tragedy, of sin coming home to roost . . . and perhaps it may have been due to her, but it seems, in tragedy, that innocence is not enough". The GRAIL stories, which make up a large part of Malory, are almost completely ignored, being recounted to Arthur by defeated knights on their return to the ROUND TABLE, thus neatly defining the central Christian

contribution to the MATTER in terms of BELATEDNESS. In the end *The Once and Future King* is by far the most serious – and very clearly the most successful – attempt to translate the Matter of Britain into a tale that speaks to the concerns of this century. Its title (but little else) makes some claim that King Arthur's life is part of a CYCLE, that his slumber presages a return and an ultimate EUCATASTROPHE. But far more harshly than J.R.R. TOLKIEN, who began *The Lord of the Rings* (**1954-5**) at the same time THW began *The Once and Future King*, THW closes the door on any real solace. His Britain is a LAND OF FABLE, not a SECONDARY WORLD, and is tied to the wheel of secular time, which allows no returns.

Mistress Masham's Repose (**1946** US) tells how a WAINSCOT body of Lilliputians, transported to England by Gulliver, have survived in the capacious grounds of the vast estate of Malplaquet for 200 years until a young girl almost destroys them by treating them as pets. The protagonist of *The Elephant and the Kangaroo* (**1947** US) is a mocking self-portrait of the author; after being warned by the Archangel Michael of the imminence of a second FLOOD, he proclaims himself a new Noah in a hilariously pixilated Eire (where THW spent the WWII years). Both titles were written before 1945. *The Master* (**1957**) is an sf juvenile, notable mainly for its portrait of the Merlyn-like Master, 157 years old, who significantly resembles the public personality THW had constructed for himself. [JC]

Other works: *The Goshawk* (**1951**), about training a hawk; *The Book of Beasts, Being a Translation from a Latin Bestiary of the Twelfth Century* (**1954**), a BESTIARY copiously and wittily footnoted by THW; *The White/Garnett Letters* (**1968**) ed David GARNETT; *Letters to a Friend: The Correspondence Between T.H. White and I. J. Potts* (**1982**) ed François Gallix.

Further reading: *T.H. White: A Biography* (**1967**) by Sylvia Townsend WARNER.

WHITE, TIM (1952-) Prolific UK artist and illustrator of fantasy/sf subjects, with a finely detailed, fresh-coloured painting style. His outstanding ability to depict convincing alien environments has marked him out as one of the most successful and talented of modern UK illustrators in the genre. His work shows some influence from the Surrealists (particularly René MAGRITTE) and also from Maxfield PARRISH.

TW worked for an advertising agency before turning freelance as a fantasy illustrator. An impressive collection of his work is *The Science Fiction and Fantasy World of Tim White* (graph **1981**). [RT]

WHITEHEAD, HENRY S(AINT CLAIR) (1882-1932) Significant member of the H.P. LOVECRAFT circle, and in 1925-9 Episcopalian Archdeacon in the Virgin Islands, whose folklore and legends subsequently appeared in much of his fiction, in particular the stories featuring **Gerald Canevin**, an *alter ego* for HSW, whose family name was Caer-n'-avon. HSW's writing career began in 1905. The first of his 25 stories in WEIRD TALES was the slight and sentimental "Tea Leaves" (*WT* 1924), in which prophecies read from tea leaves bring a woman wealth and love. "Jumbee" (*WT* 1926) is darker, featuring death portents and were-dogs (◊ SHAPESHIFTERS). "The People of Pan" (*WT* 1929) describes a lost race of Greek PAN-worshippers. "The Passing of a God" (*WT* 1930) questions the nature of divinity: a cancerous tumour, acclaimed as a GOD by VOODOO worshippers, proves indeed a malignant divinity. Several of HSW's finest supernatural stories appeared in *Adventure*, his vividly described Virgin Islands backgrounds overcoming that MAGAZINE's no-fantasy policy. HSW's most convincing fantastic fiction is unambiguous and straightforward, written from a moral perspective that leaves no doubt that the supernatural exists. The majority of his fantastic stories were collected in two early ARKHAM HOUSE volumes, *Jumbee and Other Uncanny Stories* (coll **1944**; vt in 2 vols as *Jumbee and Other Voodoo Tales* and *The Black Beast and Other Uncanny Tales* 1976 UK) and *West India Lights* (coll **1946**). [RB]

WHITTEMORE, EDWARD (?1933-1995) US writer who, after *Quin's Shanghai Circus* (**1974**), a novel with some fabulistic elements, produced in the **Jerusalem Quartet** – *Sinai Tapestry* (**1979**), *Jerusalem Poker* (**1980**), *Nile Shadows* (**1983**) and *Jericho Mosaic* (**1986**) – an ambitious though occasionally incoherent FANTASY OF HISTORY. EW's style is almost damagingly clear, as it sometimes reduces the sense necessary to his project – that layers underlie layers, almost to infinity. As a literary fantasy, the quartet might seem to occupy a space between the **Alexandria Quartet** by Lawrence Durrell (1912-1990), Thomas PYNCHON's *V* (**1963**) and *The Recognitions* (**1955**) by William Gaddis (1920-   ). In the first volume, three characters begin to shape the century: Strongbow, a KNIGHT OF THE DOLEFUL COUNTENANCE who fails to pass on his legacy of knowledge and power (he more or less owns the Ottoman Empire) to his son, Stern, protagonist in later volumes; Wallenstein, who discovers the true BIBLE, which disqualifies Western Civilization, and decides to forge the old one to replace it; and Haj Harun, who may be immortal ($\lozenge$ IMMORTALITY), who is possibly an eternal dupe, or possibly is playing a GODGAME with the 20th century, especially in the second volume, which is dominated by a poker game, the winner to gain Jerusalem. As the sequence progresses, clarity fades and jumbles; and the tale ends disconsolately. [JC]

WHO FRAMED ROGER RABBIT US live-action/ANIMATED MOVIE (*1988*). Warner/Touchstone/Amblin. **Pr** Frank Marshall, Robert Watts. **Dir** Robert ZEMECKIS. **Spfx** Industrial Light & Magic. **Anim dir** Richard WILLIAMS. **Screenplay** Jeffrey Price, Peter S. Seaman. **Based on** *Who Censored Roger Rabbit?* (**1981**) by Gary K. Wolf. **Novelization** *Who Framed Roger Rabbit* * (**1988**) by Martin Noble (1947-   ). **Starring** Joanna Cassidy (Dolores), Bob Hoskins (Eddie Valiant), Stubby Kaye (Marvin Acme), Christopher Lloyd (Judge Doom), Alan Tilvern (R.K. Maroon). **Voice actors** Mel Blanc (Bugs Bunny/Daffy Duck/Porky Pig/Sylvester/Tweety Bird), Charles Fleischer (Roger Rabbit/Benny the Cab/Greasy/Psycho), Lou Hirsch (Baby Herman), Amy Irving (Jessica Rabbit's singing voice), Kathleen Turner (uncredited; Jessica Rabbit's speaking voice), Richard Williams (Droopy Dog). 103 mins. Colour.

Private eye Valiant has turned to drink since the murder of his brother by a TOON – one of the animated characters who co-exist with humans in 1947 Tinseltown. Another Toon, Roger Rabbit – co-star in Maroon Cartoons' **Baby Herman** series of animated shorts – is fluffing his lines because concerned his wife Jessica has become involved with the owner of Toontown, Marvin Acme. Valiant, a confirmed Toon-hater, reluctantly accepts from studio boss R.K. Maroon the commission to dog Jessica and capture her indiscretions on film, thereby hopefully shocking Roger out of his anxiety. Valiant takes the photographs, but then Acme is murdered and Roger is chief suspect. Judge Doom, sadistic arbiter of all matters Toonish, pronounces Roger guilty without trial and declares the rabbit will be, on

apprehension, dissolved in a fluid called The Dip. More devious crookery is afoot: Acme's will – reportedly leaving Toontown to the Toons – has gone missing. Valiant is informed by girlfriend Dolores that a combine called Cloverleaf Industries is buying up everything in the hope of turning LOS ANGELES into a freeway-dominated waste; if Acme's will is not found by midnight, Toontown too will fall to Cloverleaf. Aided by a sentient taxi (Benny the Cab), Valiant pursues Jessica into Toontown where he encounters a cornucopia of historical animated characters – MICKEY MOUSE, BUGS BUNNY, Droopy Dog (voiced by Richard WILLIAMS) etc. – and is shot at by Judge Doom, revealed as the true villain behind Cloverleaf. Captured by Doom and his weasel/bent-cop sycophants the Toon Patrol, Valiant and Jessica prepare for death. Doom's plan is mass murder – to spray Toontown with The Dip, erasing it from the face of the Earth. Valiant uses TRICKSTER means to sabotage this plot and save Toontown and our heroes.

WFRR is a landmark in the history of the ANIMATED MOVIE – a glance at the credits list above is evidence of that – and may even be regarded as a culmination of the second phase of that genre (if the first is taken as having ended with the integration of colour and sound). The interaction between live and animated characters is masterfully handled; Williams's team, using largely traditional animation techniques, achieved an effect so commonplace in live-action movies as usually to go unnoticed, but rare in animated movies, that of allowing the camera to roam rather than be fixed, and thus the principal animated characters have a realism and solidity to match their live counterparts. But technical considerations should not outdazzle the fact that *WFRR* is among the CINEMA's most significant achievements in the sphere of fantasy. The heart of this INSTAURATION FANTASY is an exploration of the relationship between two worlds, one based firmly in REALITY (the mundane events in the movie are based on a genuine 1940s corruption scandal) and the other belonging to MYTH and MAGIC: we can make a direct analogy, reading the TOONS as FAIRIES in TECHNOFANTASY guise (and, by definition, capable of such fairy-like feats as SHAPESHIFTING), between the plot of *WFRR* (beneath its convoluted, Chandleresque surface story) and the situations depicted in classic FAIRYTALES. [JG]

WICKER MAN $\lozenge$ GREEN MAN; *The* WICKER MAN (*1973*).

WICKER MAN, THE UK movie (*1973*). British Lion/Summerisle. **Pr** Peter Snell. **Dir** Robin Hardy. **Screenplay** Anthony Shaffer. **Novelization** *The Wicker Man* * (**1978**) by Hardy and Shaffer. **Starring** Diane Cilento (Miss Rose), Geraldine Cowper (Rowan Morrison), Britt Ekland (Willow MacGregor), Lindsay Kemp (Alder MacGregor), Christopher Lee (Lord Summerisle), Jennifer Martin (Myrtle Morrison), Ingrid Pitt (Registrar), Irene Sunters (May Morrison), Edward Woodward (Sgt Neil Howie). 102 mins. Colour.

Curious cult movie depicting the clash between Christian churchianity and the FERTILITY religions; interestingly, Christianity (embodied in the protagonist) is incapable of comprehending the mental set of its rival, yet paganism is shown as able to embrace (or at least identify with) Christian beliefs. Prissy God-fearing police sergeant and SENSIBLE MAN Howie comes to remote Summerisle, in the Hebrides, to investigate anonymous report of a missing girl, Rowan. There he finds widespread "depravity". Howie discovers Rowan was the Harvest Queen last year, when

Summerisle's normally prolific fruit-crops failed; research-ing, he concludes she was sacrificed and SKINNED in a fertility RITUAL. On Mayday, Howie infiltrates the ceremonies; Rowan is still alive, seemingly groomed for HUMAN SACRIFICE. But she is not the intended sacrifice: all has been a GODGAME designed to lure Howie to come here of his own will, as a virgin (◊ VIRGINITY) and FOOL, and as king for the day. Still praying to the Christ, Howie is set aflame alongside other livestock and vegetable offerings inside the towering "Wicker Man".

Certainly among CINEMA's most interesting and literate PSYCHOLOGICAL THRILLERS, TWM has some quaintnesses, some comic moments, some considerable suspense, and an omnipresent strangeness that cannot be dissipated even by the ham Scottishness of various of the cast. Although overtly nonfantasy, it serves almost as a directory of DARK FANTASY. At the close, Howie, clad in a virgin shift, personi-fies the martyred CHRIST; but GOD has forsaken him. [JG]

WILDE, OSCAR (FINGAL O'FLAHERTIE WILLS) (1854-1900) Anglo-Irish writer, most famous for the plays he wrote 1891-5; almost all his fantasies were written in the preceding phase of his career. Much of his early poetry – collected in *Poems* (coll **1881**) – is redolent of fantasy; his most significant fantasy poem, *The Sphinx* (**1894** chap), came later. *The Canterville Ghost* (1887; **1906** chap) – several times filmed (◊ *The* CANTERVILLE GHOST) – is a melancholy comedy lamenting the Americanization of English culture while acknowledging the irresistible force of US pragma-tism; the tradition-bound GHOST cannot prevail against the positive thinking of his mansion's new tenants. The tale was reprinted in *Lord Arthur Savile's Crime and Other Stories* (coll **1891**), whose title story (1887; **1890** chap) employs OW's sparkling wit as a sarcastic gloss on the attempts to cheat destiny made by its hapless protagonist.

The CHILDREN'S FANTASIES in *The Happy Prince and Other Stories* (coll **1888**) are embellished with a subversively cyni-cal irony. In *The Nightingale and the Rose* (**1911** chap US) the BIRD who sacrifices her life in order that a student may have the red rose demanded of him by the girl he loves turns out to have shed her life-blood for nothing. The four stories in *A House of Pomegranates* (coll **1891**) extrapolate this trend to its extreme. *The Young King* (1888; *c*1895 chap US) refuses the regalia of his office after discovering the hardships which his people endure in paying for his coronation, but no one respects his decision. In *The Birthday of the Infanta* (1889; *c*1895 chap US) a similarly ostentatious display of callous wealth is the background to a harrowing tale of disillusion. *The Fisherman and his Soul* (**1895** chap US) deftly combines the motifs of Hans Christian ANDERSEN's "The Little Mermaid" and "The Shadow": the fisherman's rejected SOUL returns periodically to tempt him with visions of a world full of exotic promise, but when he gives in he finds the soul irredeemably spoiled. *The Star-Child* (*c*1895 chap US) tracks the tribulations of an infant betrayed by delusions of grandeur. These short fictions have been assembled in many subsequent collections for children; they were heavily influ-enced by the French tradition of sophisticated and morally subversive *contes*, which OW adopted as enthusiastically as he embraced the theories and mannerisms of French DECA-DENCE, ironically displayed in *The Picture of Dorian Gray* (**1891**). Dorian's adoption of a Joris-Karl HUYSMANS-inspired LIFESTYLE FANTASY while his portrait suffers the displaced legacy of his selfindulgence ultimately comes to nothing because he cannot control his reaction when he is

irresistibly drawn to meet his tainted DOUBLE face-to-face. Looked at from another viewpoint, the novel owes a great deal to Robert Louis STEVENSON's TECHNOFANTASY *Strange Case of Dr Jekyll and Mr Hyde* (**1896**), in that it portrays an attempt to dissociate the good from the EVIL within a single individual, although OW focused on *superficial* GOOD AND EVIL (the picture, after all, being itself morally neutral). The most noteworthy cinematic version of the tale has been *The* PICTURE OF DORIAN GRAY (***1945***).

OW made his own contribution to French Decadence in the play *Salomé, drame en un acte* (**1893** chap France; trans Lord Alfred Douglas as *Salome: A Tragedy in One Act* **1894** chap), which Marcel Schwob and Pierre LOUŸS helped him polish; it was first produced in Paris in 1896, after being refused a licence for English production in 1892. OW also produced a cycle of six *Poems in Prose* (1893-4; coll **1905** chap France), of which the last and longest is a curiously plaintive CHRISTIAN FANTASY. The last items are omitted from the version of *The Collected Works of Oscar Wilde* (coll **1908**) ed Robert Ross, but are included in *The Complete Works of Oscar Wilde* (coll **1931**).

Following his imprisonment after the collapse of his libel case against the Marquess of Queensberry (who accused him of "posing as a Somdomite" [*sic*]) and subsequent pros-ecution for gross indecency, OW wrote nothing except *The Ballad of Reading Gaol* (**1898**). The abortion of his brilliant career was a tragedy without parallel in English letters. The best of his fantasies are heartfelt parables which seek to explain how human folly, vanity, greed and infidelity serve as the roots of all evil and misery – which is, of course, the highest and truest purpose of fantasy. [BS]

WILDER, ALAN [s] ◊ John W. JAKES.

WILDER, CHERRY Pseudonym of New Zealand-born writer Cherry Barbara Grimm, née Lockett (1930-   ), in Australia 1954-76 and then in Germany. After publishing short fiction and poetry she turned to sf and fantasy, and chose the name Wilder. Her first four novels are sf, but she produced a major work of fantasy in her **Rulers of Hylor** trilogy: *A Princess of the Chameln* (**1984** US), *Yorath the Wolf* (**1984** US) and *The Summer's King* (**1986** US). At first, this has a deceptively conventional appearance – the young ruler of a peaceful, magical country is threatened by enemies of great power and ruthlessness – but the language and char-acterization mark the narrative as superior to most fantasies of its sort. However, CW's most significant achievements may lie in her subtle short stories, many on the borderline between sf and fantasy. A too-rare sampling in book form is *Dealers in Light and Darkness* (coll **1995** US). [DP]

Other work: *Cruel Designs* (**1988** UK), horror.

WILDER, THORNTON (NIVEN) (1897-1975) US nov-elist and playwright best-known for work outside fantasy; he three times won the Pulitzer Prize. However, TW did con-sistently utilize devices and themes (historical drama, GHOSTS, the persistence of paganism in the modern world, the OCEAN OF STORY) more common to fantastic literature than to the fiction of literary Modernism, whose critics gen-erally embraced his work. Although the non-naturalistic elements of his plays – including the most famous, *Our Town, New York* (**1938**) – seem to possess some affinity with the "theatre within the theatre" of Luigi Pirandello (1867-1936), the decided homeliness of TW's supernatural figures (they include the voices of the dead, ANGELS and biblical fig-ures) has more in common with the 19th-century US short story than with European Modernism. Although the debt of

The Skin of Our Teeth (**1942**) to James JOYCE's *Finnegans Wake* (**1939**) has been remarked, TW's use of fabulism in theatre preceded his acquaintance with that book (on which he was an authority), as his one-act play "Pullman Hiawatha" (**1931**) shows. TW's fictions show an abiding interest in the interpenetration of the antique and the modern, from his first novel, *The Cabala* (**1926**), which viewed contemporary Rome in pagan terms, to his last, *Theophilus North* (**1973**), whose protagonist likens 1920s Newport, RI, to the seven cities of Troy. Despite his familiarity (and sympathy) with the impulses of European Modernism, TW's congeniality toward fantasy springs from older sources. [GF]

Other works (selective): *The Bridge of San Luis Rey* (**1927**); *The Woman of Andros* (**1930**); *Heaven's My Destination* (**1935**); *The Ides of March* (**1948**); *The Eighth Day* (**1967**).

Plays: *The Angel that Troubled the Waters* (coll **1928**); *The Long Christmas Dinner* (coll **1931**); *The Merchant of Yonkers* (**1938**; rev vt *The Matchmaker* 1954 UK).

WILD HUNT Legendary supernatural hunt which pursues and tears SOULS. Alternative names include the Cornish "DEVIL's Dandy Dogs", Dartmoor's "Wish Hounds" or "Yell Hounds" and the Welsh "Cwn Annwn" or hounds of HELL (these signal rather than cause death). The WH takes to the skies as the "Gabriel Hounds", "Gabriel Ratchets" or simply the "Host", tireless and inescapable as FATE. QUATERMASS AND THE PIT gives the WH a TECHNOFANTASY rationale rooted in race memory. In straight fantasy it is generally linked with the stag-antlered HERNE THE HUNTER, versions of whom lead the WH in tales such as: Alan GARNER's *The Moon of Gomrath* (**1963**), whose hunters' sheer love of bloodshed horrifies those they save; Penelope LIVELY's *The Wild Hunt of Hagworthy* (**1971**), where the spirit of the hunt infects young morris dancers; Susan COOPER's *The Dark is Rising* (**1973**), which effectively invokes the WH to disperse the forces of EVIL; and Diana Wynne JONES's *Dogsbody* (**1975**), whose hunt-Master, a sad but powerful UNDERWORLD figure, repeatedly offers himself as his own dogs' prey. The WH is perhaps too frequently used as a PLOT DEVICE lacking the appropriate sense of danger: the warning that this is an uncontrollable manifestation of Old, Wild or Earth MAGIC is more often issued than followed through; one notable exception is Guy Gavriel KAY's *The Wandering Fire* (**1986**), where the WH is summoned and impartially begins to slaughter both foes and allies. Further novels featuring the WH include Jean MORRIS's *The Troy Game* (**1987**), Charles DE LINT's *Greenmantle* (**1988**), Brian STABLEFORD's *Plague Warriors* * (**1991**), Tom DEITZ's *Dreamseeker's Road* (**1995**) and Jane YOLEN's *The Wild Hunt* (**1995**). [DRL]

See also: SARBAN.

WILDREDGE, THOMAS J. [s] ◊ Kenneth MORRIS.

WILKS, MIKE Working name of UK illustrator and writer Michael Thomas Wilks (1947-   ). He began to publish drawings of fantasy interest with his highly intricate, architecturally imaginative portrayal of the eponymous EDIFICE or CITY featured in Brian W. ALDISS's long narrative poem *Pile: Petals from St Klaed's Computer* (graph **1979**). Pile itself – conceived by MW in terms evocative of the work of M.C. ESCHER and Giovanni Battista PIRANESI – is a claustrophobic expression of urban BONDAGE. *The Weather Works* (graph **1983**), a narrative poem and which MW both wrote and illustrated, again describes a kind of edifice, in this case the works where weather is made. *The Ultimate Alphabet* (graph **1986**; augmented with explanatory text vt *The Annotated*

Ultimate Alphabet 1988) – comprises 26 compound illustrations, each devoted to objects whose name begins with a particular letter (the one devoted to the letter S alone contains 1229 separate items). *The Ultimate Noah's Ark* (graph **1993**) consists of a single painting – presented whole and then broken into 16 details – containing 353 paired animals, variously placed, plus one singleton which readers are asked to find. MW's work combines an astonishing grasp of detail plus an exuberance of imagination. [JC]

Other works: *In Granny's Garden* (**1980** chap), with narrative poem by Sarah Harrison (1946-   ), for children.

WILLARD, NANCY (MARGARET) (1936-   ) US writer. Most of her fiction and poetry is for children, but she has written two fantasy novels for adults. In the vividly sentimental *Things Invisible to See* (**1984**) a young soldier serving in WORLD WAR II makes a bargain with DEATH which ties his fate to a BASEBALL game in which his team must play Death's team of all-time greats. In *Sister Water* (**1993**) a man hired to look after the ailing mother of a young widow obtains modest (but charmingly inept) supernatural assistance when falsely accused of homicide.

NW's essays on writing, collected in *Telling Time: Angels, Ancestors and Stories* (coll **1993**), insist on the necessity of writing from experience while accepting that the supernatural plays a vital role in that aspect of experience which consists of private fantasies and shared stories; essays partly on fantasy include "Danny Weinstein's Magic Book", which uses parables to justify the writing of parables, and "How Poetry Came into the World and Why God Doesn't Write It", which is part-ALLEGORY and part-anthology. Some of the short stories in *Lively Anatomy of God* (coll **1968** chap), *Childhood of the Magician* (coll **1973** chap) and *Angel in the Parlor: Five Stories and Eight Essays* (coll **1982**) have slight fantasy elements, but fantasy plays a more prominent role in NW's children's stories, which include the **Anatole** series – *Sailing to Cythera and Other Anatole Stories* (coll **1985** chap), *The Island of the Grass King: The Further Adventures of Anatole* (coll **1985** chap) and *Uncle Terrible: More Adventures of Anatole* (coll **1985**) – and the novel *Firebrat* (**1988**). Agents of good, including ANGELS and ancestral SPIRITS are very much more in evidence in NW's work than agents of evil, often serving in a relatively humble capacity; in *The High Rise Glorious Skittle Skat Roarious Sky Pie Angel Food Cake* (**1990** chap) angels help bake a cake. *A Visit to William Blake's Inn: Poems for Innocent and Experienced Travelers* (coll **1981**) was a Newbery Medal winner. [BS]

Other works: *The Well-Mannered Balloon* (**1976** chap); *Simple Pictures are Best* (**1977** chap); *Stranger's Bread* (**1977** chap); *Highest Hit* (**1978**); *Papa's Panda* (**1979**); *The Marzipan Moon* (**1981** chap); *The Nightgown of the Sullen Moon* (**1983** chap); *Night Story* (**1986** chap); *The Mountains of Quilt* (**1987** chap); *East of the Sun and West of the Moon: A Play* (**1989** chap); *A Nancy Willard Reader: Selected Poetry and Prose* (coll **1991**); *Beauty and the Beast* (**1992** chap); *The Sorcerer's Apprentice* (**1993** chap); *Starlit Somersault Downhill* (**1993**); *Among Angels* (**1994**) with Jane YOLEN.

Poetry: *Water Walker* (coll **1989**); *Pish, Posh, Said Hieronymus Bosch* (coll **1991** chap).

WILLEY, ELIZABETH (1960-   ) US writer whose ongoing **Prospero** sequence – *The Well-Favored Man: The Tale of the Sorcerer's Nephew* (**1994**), *A Sorcerer and a Gentleman* (**1995**) and *The Price of Blood and Honor* (**1996**) – constitutes an unusually complex DYNASTIC FANTASY set in a plurality of worlds, though mainly in the land of Landuc, where generation-long conflicts attend Prospero's killing of his father, King Panurgus. The names of protagonists and places give a sense of rich fabric to the long tale – though at points one also senses that EW does not always have her MULTIVERSE under full control – and variously archaic DICTIONS are used with astuteness to signal the age of various featured characters (some extremely old). In the second volume, exiled to his ISLAND, PROSPERO has created out of the beasts and fishes a new race of humans, who go to war on his behalf for control of Landuc and of the well which sorts the various worlds while also having to do with the granting of something like IMMORTALITY to favoured individuals. [JC]

See also: FANTASY OF MANNERS.

WILLIAM L. CRAWFORD MEMORIAL AWARD

Generally referred to simply as the Crawford Award, this was named for William L(evi) Crawford (1911-1984), a seminal SMALL-PRESS publisher and editor active from around 1933 until the 1970s. Selected by a panel of judges, the AWARD is given each spring for the best first fantasy novel published during the previous 18 months and is presented under the auspices of the INTERNATIONAL ASSOCIATION FOR THE FANTASTIC IN THE ARTS at its annual conference. The prize, $100, is sponsored by Andre NORTON. [JC]

Winners:.

1985: Charles DE LINT, *Moonheart*.
1986: Nancy WILLARD, *Things Invisible to See*.
1987: Judith TARR, **The Hound and the Falcon** trilogy.
1988: Elizabeth Marshall Thomas, *Reindeer Moon*.
1989: Michael ROESSNER, *Walkabout Woman*.
1990: Jeanne LARSEN, *Silk Road*.
1991: Michael Scott ROHAN, **The Winter of the World** trilogy.
1992: Greer Ilene GILMAN, *Moonwise*.
1993: Susan PALWICK, *Flying in Place*.
1994: Judith KATZ, *Running Fiercely Toward a High Thin Sound*.
1995: Jonathan Lethem, *Gun, With Occasional Music*.
1996: Sharon SHINN, *The Shape-Changer's Wife*.

WILLIAMS, CHARLES (WALTER STANSBY) (1886-

1945) UK writer, editor and poet, and a key member with C.S. LEWIS and J.R.R. TOLKIEN of the INKLINGS. His novels are of considerable importance and influence in CHRISTIAN FANTASY. Often they are theological thrillers in which powerfully numinous scenes and descriptions sit a little uneasily with the DETECTIVE/THRILLER FANTASY elements; for example, *War in Heaven* (1930) uses the true GRAIL as a sought-after MCGUFFIN while Scotland Yard complains that this is "an infernally religious case". Memorable episodes include the villains' magical attempt to destroy the "Graal" by enforced THINNING, and a graphically horrid BLACK-MAGIC booby-trap. In *The Place of the Lion* (1931) mundane REALITY is invaded by living Platonic ARCHETYPES – including the LION, SERPENT and PHOENIX – into whom lesser creatures and people are absorbed: the world's butterflies vanish into the one Butterfly and a malicious woman's affinity with the Serpent leads to POSSESSION. The archetypes' metaphysical threat is countered by acceptance and by giving them their TRUE NAMES, as in EDEN. *Many Dimensions* (1931) leans towards sf as the MAGIC Stone from Solomon's crown is blasphemously threatened by commercial exploitation of its powers, including teleportation (◊ TALENTS), miraculous multiplication into endless copies, and strange versions of TIME TRAVEL where visiting the past leads to a

CYCLE of BONDAGE, while a forward trip brings subtle metaphysical WRONGNESS. Ultimately a young woman (who, like others in CW's canon, platonically adores a much older MENTOR), rather than commanding the Stone, places her will at its disposal and becomes the vehicle for HEALING – reuniting the copies; she then dies. *The Greater Trumps* (1932) introduces the "original" TAROT pack (CW knew its reviser A.E. WAITE through the Order of the GOLDEN DAWN) and a set of golden images corresponding to the CARDS, all magically dancing except for the FOOL – whose stillness, however, appears to a seer's eyes as motion so rapid as to be omnipresence. In a struggle over their ownership, the cards are misused to generate a killing storm which threatens the END OF THE WORLD. *Descent into Hell* (1937) is perhaps the most grimly powerful and successful of CW's novels, despite a confusing plethora of uncanny events. It articulates his doctrine of "Substitution" or "Substituted Love" – which he and his mystic circle, the Companions of the Co-inherence, attempted to practise in reality – as another old and wise mentor volunteers to accept for himself the anguish suffered by a woman who dreads meeting her DOPPELGÄNGER; she is later able to give the same aid via TIMESLIP to a martyr burned centuries before. Other features are a suicide's pathetic GHOST, the traversing of hallucinated landscapes in an ALLEGORY of a woman's approach to death, and the steady, horrific progress towards damnation of a scholar who through acts of pettiness gains the nasty love of a SUCCUBUS and who continues downward into empty nightmare. *All Hallows' Eve* (1945) traces the downfall of a black MAGUS whom an artist cannot help painting as an idiot preaching to a congregation of insects. This "Simon Leclerc" plans the SACRIFICE of his own daughter, who is saved through Substitution by a dead woman who is CW's best-drawn female character and whose new SPIRIT-life in post-WWII LONDON – where the living crowds do not register on her PERCEPTION – is shown with bleak intensity. Simon's magical abilities grow cruder and feebler with misuse (◊ DEBASEMENT) as he abandons pure force of will to incarcerate the inconvenient ghost (and her companion) in a GOLEM, and is eventually reduced to pins stuck in images, before his final confrontation by his own DOUBLES.

CW's verse, in *Taleissin through Logres* (coll **1938**) and *The Region of the Summer Stars* (coll **1944**), deals extensively with ARTHUR and the MATTER of Britain, often using quirky and personal religious imagery: representations of EVIL, for example, include Islam and the headless Emperor of "P'o'lu" on the other side of the world, whose minions have not hands but tentacles as a symbol of evil's imprecision. Even C.S. LEWIS's appreciative commentary in *Arthurian Torso: Containing the Posthumous Fragment of the Figure of Arthur by Charles Williams and A Commentary on the Arthurian Poems of Charles Williams by C.S. Lewis* (**1948**) shows occasional signs of perplexity. FEMINISM recoils (though Lewis does not) from CW's poetic endorsement of the view that women are simultaneously divine and unfit for church service owing to their menstrual curse.

Despite the novels' occasionally stilted dialogue, and exposition which can sometimes seem over-didactic, CW's remarkable gift for conveying both numinosity and spiritual corruption has retained a devoted cult audience. [DRL]

Other works: *Shadows of Ecstasy* (**1933**), first-written and weakest of the novels; *Witchcraft* (**1941**), nonfiction; *The Figure of Beatrice* (**1943**), on DANTE ALIGHIERI; *The Image of*

the City and Other Essays (coll **1958**) ed and with critical introduction by Anne Ridler.

Further reading: *An Introduction to Charles Williams* (**1959**) by Alice Mary Hadfield.

WILLIAMS, GARTH (1912-1996) US writer and illustrator, in Mexico from 1961, who remains best-known for his illustrations for E.B. WHITE's two wry BEAST-FABLES, *Stuart Little* (**1945**) and *Charlotte's Web* (**1952**). GW's style, which wove intimacy and a sense of the MARVELLOUS, was ideal for this kind of story. He conveyed a similar WAINSCOT ambience in the two main series he illustrated: the **Borrowers** sequence by Margery SHARP, GW's illustrations beginning with the second volume, *Miss Bianca* (**1952**) and including *The Turret* (**1963**) and *Miss Bianca in the Salt Mines* (**1966**), and George SELDEN's **Chester, Harry and Tucker** sequence. Other work included *Bedtime for Frances* (**1960** chap), Russell HOBAN's first book of fiction, plus non-genre books by Laura Ingalls Wilder (1867-1957) and Charlotte Zolotow (1915-   ). His own CHILDREN'S FANTASIES were directed towards younger children. The best-known is *The Rabbits' Wedding* (**1958** chap), which offended US segregationists by describing the marriage between a black and a white rabbit. [JC]

WILLIAMS, KENT (1962-   ) US illustrator with a delicate, loose, watercolour style. His early work appeared in EPIC ILLUSTRATED, HEAVY METAL and other magazines. He provided cover illustrations for Penguin Books and Bantam Books. His work has regularly been exhibited in the Society of Illustrators' annual exhibition and his cover painting for Iris Murdoch's *The Good Apprentice* (**1987**) was awarded the society's Silver Medal. He has produced art for a number of comic books, the most successful of which is *Blood: A Tale* (#1-#4, 1987-8), written by J.M. DeMatteis, in which the calmness of some of his charmingly painted nude figures is in stark contrast to the unpleasantness of the story. [RT]

WILLIAMS, MARK [s] ◊ Robert ARTHUR.

WILLIAMS, MICHAEL (LEON) (1952   ) US writer and poet known initially for his contributions to the **DragonLance Heroes** series of GAME-ties – *Weasel's Luck** (**1989**), *Galen Beknighted** (**1990**) and *The Oath and the Measure** (**1991**) – mainly concerning the adventures of the cynical hero Galen Pathwarden. His poems appeared in various **DragonLance** titles ed Margaret WEIS and Tracy HICKMAN, and his short stories for anthologies edited by the same duo. A non-tie trilogy, **From Thief to King** – *A Sorcerer's Apprentice* (**1990**), *A Forest Lord* (**1991**) and *The Balance of Power* (**1992**) – concerns the education into magic and spiritual kingship of another hero, Brennart of Maraven. A religious subtext underlies these and other works. Further **DragonLance** ties have been written in collaboration with Teri Williams, including *Before the Mask** (**1993**) and *The Dark Queen** (**1994**), but MW's most ambitious novel to date is *Arcady* (**1996**), which bases its fantasy world on the religious mythology of William BLAKE. MW should not be confused with the sf novelist Michael Lindsay Williams (1940-   ). [DP]

WILLIAMS, RICHARD (1933-   ) Canadian-born UK animator and animation director, widely regarded – notably among other animators – as the finest living exponent of his art. But business seems not to be his forte, for much of his talent has been expended on minor projects, including title sequences (e.g., *What's New Pussycat?* [**1965**], *Casino Royale* [**1967**]) and tv commercials – some of which have won awards. Such for-the-money excursions have been done as a

decades-sustained attempt to subsidise his lifework, «The Thief and the Cobbler», a movie that has not been and may never be released: in the 1990s, as RW's finances yet again became parlous, the movie-so-far was bought by DISNEY on the grounds that it would be finished by that studio, but (according to unconfirmed reports) the standard was so high that the project was shelved indefinitely.

RW is best-known for WHO FRAMED ROGER RABBIT (**1988**), for which he was Animation Director. For this he and his team used traditional animation techniques – RW is possibly the last major animator to resist modern technology – to create the finest TOON illusion yet seen. (He was also involved in the two Disney **Roger Rabbit** shorts that followed: *Tummy Trouble* [**1989**] and *Rollercoaster Rabbit* [**1990**].) But there have been other major works. *Charles Dickens's A Christmas Carol, Being a Ghost Story of Christmas* (**1971** tvm), although only short and made for tv, is extremely impressive (◊ *A* CHRISTMAS CAROL), as is the full-length theatrical feature *Raggedy Ann and Andy* (**1977**), based on stories by Johnny Gruelle (1880-1938), which mixes some live action with the animation: TOYS come alive to QUEST in rescue of an abducted DOLL. It is very much to be hoped that RW will re-emerge to demonstrate yet again to all makers of ANIMATED MOVIES the standards they should be seeking to attain. [JG]

WILLIAMS, ROSWELL Pseudonym of Frank OWEN.

WILLIAMS, TAD Working name of US writer Robert Paul Williams (1957-   ). TW's first novel was *Tailchaser's Song* (**1985**) an ANIMAL FANTASY featuring CATS, but his principal contribution to the genre has been the HIGH-FANTASY **Memory, Sorrow and Thorn** trilogy: *The Dragonbone Chair* (**1988**), *Stone of Farewell* (**1990**) and *To Green Angel Tower* (**1993**; vt in 2 vols *Siege* **1994** UK and *Storm* **1994** UK). This is explicitly a REVISIONIST FANTASY, intended as a criticism of the alleged implicit racism in J.R.R. TOLKIEN's *The Lord of the Rings* (**1954-5**) and more generally of the assumption throughout GENRE FANTASY that GOOD AND EVIL can be simply allocated along national or species boundaries; there is a DARK LORD in the trilogy, and he has his bodyguard of undead, but his particularly horrible scheme of DEBASEMENT in response to genuine wrongs. Neither humanity nor its elven rivals, the Sithi, is native to Osten Ard; it is explicit that both arrived in historical time, the ELVES ahead of various waves of human incomers, each of them semi-cognate with nations of Earth – the usual quasi-Vikings, Slavic plains folk, etc. – and that a genocidal war, largely won by humanity, ensued.

Much of the novel is centred on the Hayholt, castle of PRESTER JOHN (nothing is made of this coincidence of names at a literal level), who is High King and dragon-slayer, whose death after a long life of apparently benevolent rule opens the way to the intrigues of the undead Ineluki and his human pawn, John's heir Elias. The Hayholt, also formerly Ineluki's stronghold against humanity, is an EDIFICE full of tunnels, forges and LABYRINTHS; it is a haunted palace of the mind, symbolic of the recesses in which humanity has hidden its bad conscience about the Sithi, as well as the location for the climactic NIGHT JOURNEY undergone by the UGLY DUCKLING and HIDDEN MONARCH Simon in the final stage of his heroic attainment.

Much of the trilogy is a tour of the WATER MARGINS of Osten Ard undertaken by Simon and his various allies – Elias's virtuous daughter Miriamele, her uncle Josua and various members of the League of the Scroll, a secret order of scholars and MAGES. The collection of artefacts, notably the three SWORDS which give their name to the trilogy, is not the usual PLOT-COUPON collection but a search for knowledge deliberately suppressed – TW makes effective use of the TIME ABYSS as symbol for the guilt of attempted genocide. The swords are taken severally to the Hayholt in an attempt to defeat Ineluki. The interpretation the League of the Scroll places on this event is disastrously wrong; this is one of several points in the trilogy in which TW sardonically reverses *LOTR*. Where suffering in *LOTR* hardens the will of the virtuous against the wicked, though the actual act of destruction is done inadvertently by the already corrupt, in TW's work Simon's crucifixion on a waterwheel creates in him an empathy and compassion which enable him to refuse Ineluki from the moral high ground, just as Miriamele's rape by a minor villain provides the sympathetic analogy which makes her capable of killing her father to free him from POSSESSION by Ineluki – TW's enlightened moral relativism extends to the creation of an entirely virtuous parricide.

At the same time, TW owes much to Tolkien, in his marshalling of multiple secondary protagonists; his critique of Tolkien's absolutist morality is as much a killing of the father as the gradual uncovering of the truth about Prester John, who turns out not to have been quite the benevolent old patriarch he seemed. Of all the bestselling high-fantasy sequences of recent years, this is perhaps the one in which the author's intelligence is most effectively embodied in plot, character and vividly described action and backdrops. TW's only real weakness, though it is at times a damaging one, is a tendency to sentimentalized psychologizing, which comes to the fore most in his dialogue.

TW's other published works are two novellas – *Child of an Ancient City* (**1992** chap) with Nina Kiriki HOFFMAN, an interesting exercise in cross-cultural empathy which places a traditional encounter with a VAMPIRE in a specifically Islamic ARABIAN-FANTASY storytelling context, and *Caliban's Hour* (**1994**), another REVISIONIST FANTASY in which SHAKESPEARE's Caliban's getting of language and attempted rape of Miranda are complexly and interestingly deconstructed.
[RK]

WILLIAMSON, AL (1931-   ) US fantasy/sf COMIC-book illustrator with a fertile imagination and a firm, decorative line style. AW studied under Burne HOGARTH at the School of Visual Arts; his first published work appeared in *Heroic #51* (1948). He worked for several publishers including ACG and Eastern before joining EC COMICS in 1952, often working with Frank FRAZETTA and Roy KRENKEL. His strong, slick, naturalistic draughtsmanship quickly matured and he continued with EC until the company failed in 1955, when he moved to Atlas to work on war, romance, adventure and horror tales. His association with characters created by Alex RAYMOND began when he was taken on as an assistant by John Prentice (1920-   ) to work on the daily newspaper strip *Rip Kirby*. He drew several stories for WARREN PUBLISHING's new horror titles EERIE and CREEPY, and then drew three books of anther Raymond creation, *Flash Gordon* (#1, #4, #5 1966-7). In #4 he drew a 5-page backup story featuring a third Raymond character, *Secret Agent X9*, as a result of which he was offered the opportunity to take over the syndicated newspaper strip of the same name (subsequently retitled *Secret Agent Corrigan*): he drew it for 13 years.

AW then began drawing comic-book adaptations of movies, starting with George Lucas's *The Empire Strikes Back* (**1980**) which led to his taking on the **Star Wars** newspaper strip, where his flair for creating imaginative PLANETARY-ROMANCE scenarios was fully exploited: he continued drawing this until 1986. He also adapted Dino de Laurentis's *Flash Gordon* (**1980**) (◊ FLASH GORDON MOVIES), Ridley Scott's *Blade Runner* (**1982**) and Lucas's *Return of the Jedi* (**1983**). His **Star Wars** newspaper strips are presently (1996) being published in collected editions as *Classic Star Wars*. In the 1990s he is returning once again to **Flash Gordon**.
[RT]

Further reading: *Al Williamson* (graph **1983**) by James Van Hise.

WILLIAMSON, CHET Working name of US writer Chester Carlton Williamson (1948-   ), who began publishing work of genre interest in the mid-1980s. CW is a versatile writer with a gift for exceptional characterization, and typically infuses his work with a variety of subthemes. "Blue Notes" (1989), concerning a debt-ridden, drug-addicted saxophonist, proposes that, just as there are "blue notes" to be found in the aural areas between defined notes, there is a similar "twilight zone" between life and death. "Confessions of St James" (1989) is a penetrating tale of a pastor with a taste for human flesh who finds justification for his dark urges in the ingesting of the Host during the communion rite. In "The Bookman" (1991) a book collector attending an estate sale finds that the house's former occupant was compelled to collect books in an attempt to satiate an invisible presence with a bibliophilic bent. In the disturbing "Coventry Carol" (1994) the GHOST of a miscarried foetus haunts the distraught parents.

Much in the style of Shirley JACKSON's *The Haunting of Hill House* (**1959**), CW's debut novel *Soulstorm* (**1986**) features five people trapped in a mountaintop mansion, where they must face first the ghosts of every person who has ever been damned, and later an immensely evil force. An array of spirits is also at the heart of *Ash Wednesday* (**1987**), a study of the ravages of guilt, in which all of the people who ever died in a small Pennsylvania town suddenly return as ghosts to confront the town's residents. *Reign* (**1990**) is an interesting, if sometimes sluggish, attempt to blend novel and stage play, most notably by switching to a script format during climactic scenes: a veteran theatre star who has had an extremely successful career playing a single character finds himself and his troupe visited by the murderous embodiment of that character. *Second Chance* (**1994**) is a TIME-TRAVEL fantasy.
[BM]

Other works: *Lowland Rider* (**1988**); *McKain's Dilemma* (**1988**); *Dreamthorp* (**1989**); *The House Of Fear* (**1989** chap); *Ravenloft: Mordenheim* * (**1994**), game tie; *Hell: A Cyberpunk Thriller* * (**1995**), sf game tie.

WILLIAMSON, JACK Working name of US writer John Stewart Williamson (1908-   ), who has concentrated for most of the 70 years of his active career on sf. He won a 1976 Nebula Grand Master award; he was given a 1973 Pilgrim Award for his work in sf scholarship; but he was also given a 1994 WORLD FANTASY award for Life Achievement. But fantasy constitutes a small part of his large oeuvre; novels like *Dragon's Island* (**1951**) or *Demon Moon* (**1994**) are, despite their titles, genuine sf. *Darker Than You Think* (1940 *Unknown*; exp **1948**), JW's most single famous non-sf tale, hovers between rationalized explanations and pure fantasy. A race of SHAPESHIFTERS, *homo Lycanthropus*, has long lived secretly alongside *homo sapiens*; but as the tale opens is threatened with exposure. The protagonist – who turns out to be the Child of Night, a HIDDEN MONARCH figure destined to rule the WEREWOLF race – learns of his fate by falling in love with a woman shapeshifter, a relationship depicted with a tortured (and still haunting) erotic frankness unusual in genre literature of the 1940s.

In his early career, JW was much influenced by the style of Abraham MERRITT, and a novel like *Golden Blood* (1933 *Weird Tales*; rev **1964**) combines SWORD AND SORCERY and romanticized sf in a manner indebted to that writer. *The Reign of Wizardry* (1940 *Unknown*; rev **1964**; again rev 1979) similarly transforms Minoan Crete into an arena for exorbitant adventures. In general, JW's impulse was to rationalize tales whose fantasy trappings are ultimately deceptive. [JC]

Other works: *Lady in Danger* (1934 *Weird Tales* as "Wizard's Isle"; **1945** chap UK), with a tale by E. Hoffman PRICE added; *Dreadful Sleep* (1938 *Weird Tales*; **1977** chap)).

WILLIAMSON, J.N. Working name of US writer and editor Gerald Neal Williamson (1932-   ). An extremely prolific author, JNW rode the wave of the 1980s HORROR boom, his published output cresting and breaking virtually in synch with the market as a whole. Among his novels displaying a strong fantasy inclination are *Playmates* (**1982**), in which a child's INVISIBLE COMPANIONS are revealed to be Irish DEMONS whose attacks are preceded by a banshee's wail, and *The Dentist* (**1983**), in which the title character's quest for IMMORTALITY hinges upon a magical TALISMAN. A variety of GHOSTS are at the heart of *Horror House* (**1981**), *Ghost Mansion* (**1981**) and its sequel *Horror Mansion* (**1982**), *Ghost* (**1984**), *The Longest Night* (**1985**), *Dead to the World* (**1988**), and *Shadows of Death* (**1989**). An immortal female VAMPIRE is the evil protagonist of the **Lamia Zacharius** series – *Death-Coach* (**1981**), *Death-Angel* (**1981**), *Death-School* (**1982**), and *Death-Doctor* (**1982**) – and vampires also figure in *Bloodlines* (**1994**). Evil cults are the gist of *The Black School* (**1989**), *Hell Storm* (**1991**), and *The Monastery* (**1992**). TALENTS are key in the **Martin Ruben** series, concerning a parapsychologist – *The Ritual* (**1979**), *Premonition* (**1981**) and *Brotherkind* (**1982**) – and *The Evil One* (**1982**) and *Babel's Children* (**1984**). [BM]

Other works: *The Houngan* (**1980**; vt *Profits*, 1984); *The Banished* (**1981**); *Queen of Hell* (**1981**); *The Tulpa* (**1981**); *Extraterrestrial* (**1982**) as Julian Shock; *The Offspring* (**1984**); *Wards of Armageddon* (**1986**) with John Maclay, sf; *Evil Offspring* (**1987**); *Noonspell* (**1987**); *The Night Seasons* (**1991**); *The Naked Flesh of Feeling* (coll **1991** chap); *The Book of Websters* (**1993**); *Don't Take Away the Night* (**1993**); *The Fifth Season* (coll **1994** chap).

As editor: The **Masques** anthologies, being *Masques* (anth **1984**), *II* (anth **1987**), *III* (anth **1989**; vt *Fleshcreepers* 1990 UK); *IV* (anth **1991**), the first 2 volumes being showcased in *The Best of Masques* (anth **1988**).

Nonfiction: *The New Devil's Dictionary: Creepy Cliches and Sinister Synonyms* (coll **1985**); *How to Write Tales of Horror, Fantasy, and Science Fiction* (anth **1987**).

WILLIAMSON, PHILIP G(EORGE) (1955-   ) UK writer whose first novels were two dark contemporary SATIRES, *The Great Pervader* (**1985**) and *Dark Night* (**1989**), both as Philip First; under that name he also assembled a collection of similar stories, *Paper Thin and Other Stories* (coll **1987**). As an author of fantasy, PGW has concentrated exclusively upon various books in the overall **Firstworld** sequence. The **Firstworld Chronicles** – *Dinbig of Khimmur*

(**1990**), *The Legend of Shadd's Torment* (**1993**) and *From Enchantery* (**1993**) – is a sustained narrative, almost always focused upon Dinbig of Khimmur, a WIZARD, adventurer and spy. The further **Firstworld** volumes – *Moonblood* (**1993**), *Heart of Shadows* (**1994**) and *Citadel* (**1995**) – share some of the same characters and venues, but are set earlier. There is an underlying sense of seriousness of purpose. [JC]

WILLIS, CONNIE Working name of US writer Constance Elaine Trimmer Willis (1945-   ), known primarily for her sf. She began with "Santa Titicaca" in *Worlds of Fantasy* in 1971, but didn't start appearing regularly in the genre until the early 1980s. In the interim, she published many stories in confessions magazines.

CW's fictions have a number of common elements. The first is her use of TIME-TRAVEL, presumably acquired from her love of Robert A. HEINLEIN's work – and the second is the use of misunderstanding and misdirection, which harks back to the screwball comedies of 1920s and 1930s Hollywood. Her use of comic elements is muted somewhat in her more ambitious solo novels, *Lincoln's Dreams* (**1987**) and *Doomsday Book* (**1992**). The former is a heartfelt contemporary novel in which a Civil War researcher befriends a young woman haunted by the GHOST of Robert E. Lee. Here, through judicious use of exposition on the Civil War and Lee's horse, Traveller, Willis creates a present-day immediacy and a lasting resonance. Even more harrowing is *Doomsday Book*, properly sf, which shares some characters and settings with "Fire Watch" (1982), an earlier TIME-TRAVEL tale set in LONDON during the Blitz. It concerns a young woman sent back to 14th-century England, where the Black Plague reigns. The attention to detail is enormous, but CW does not shirk the moral or emotional consequences of the setting.

Much of CW's shorter work is collected in *Fire Watch* (**1984**) and *Impossible Things* (**1994**), the latter's title being a reference to Lewis CARROLL. *Fire Watch* in particular is noteworthy for the variety of its contents, which range from the poignant "A Letter from the Clearys" (1992) – a post-HOLOCAUST story – to the comic "Blued Moon" (1994). Stories in *Impossible Things* which are not explicitly sf include "Winter's Tale" (1988) – an alternate history of SHAKESPEARE's retirement and homecoming – and "Jack" (1991), a story of pluck and vampirism (◊ VAMPIRES). Of CW's uncollected stories, two are of fantasy interest: *Distress Call* (1981; with essay "On Ghost Stories" **1991** chap) and "Substitution Trick" (1985), about Houdini.

CW has won six Nebula and five Hugo AWARDS. [WKS]

Other works: *Water Witch* (**1982**) and *Light Raid* (**1989**), lightweight sf with Cynthia Felice; *Uncharted Territory* (**1994**), *Remake* (**1995**) and *Bellwether* (**1996**), all lightweight comic sf.

WILLO THE WISP UK animated tv series (1981-2). BBC. **Created by** Nicholas Spargo, with characters by D.N. and P.M. Spargo. **Voice actors** Kenneth Williams (all voices). 24 5min episodes. Colour.

Although designed for the pre-suppertime children's slot, this rapidly became a cult programme for parents. The LIMINAL BEING Willo the Wisp overlooks Doyley Woods, Oxfordshire, UK, where various strangenesses occur. Regular characters are Arthur (a Cockney caterpillar), Mavis Cruet (a FAIRY best described in her presence as "plumpish" rather than with the "f" word), Evil Edna (a WITCH who resembles a tv set with legs, and who loathes everyone) and The Beast (a handsome prince in bestial

BONDAGE). The animation was both rudimentary and brilliant. *WTW* was rarely uninventive and always very funny; a typical side-character was The Rabbit From Watership Down-Under. [JG]

WILLOW US movie (*1988*). MGM/Lucasfilm/Imagine Entertainment. **Pr** Nigel Wooll. **Exec pr** George Lucas. **Dir** Ron Howard. **Spfx** John Richardson. **Screenplay** Bob Dolman. **Novelizations** *Willow* * (**1988**) by Wayland Drew and *Willow* * (**1988**) by Joan D. VINGE. **Graphic novelization** *Marvel Comics Presents Willow, The Illustrated Version* * (graph **1988**) by Jo Duffy. **Starring** Warwick Davis (Willow), Patricia Hayes (Raziel), Val Kilmer (Madmartigan), Jean Marsh (Bavmorda), Gavan O'Herlihy (Airk), Julie Peters (Kiaya), Pat Roach (Kael), Joanne Whalley (Sorsha). 126 mins. Colour.

"It is a time of dread. Seers have foretold the birth of a child who will bring about the downfall of the powerful Queen Bavmorda. Seizing all pregnant women in the realm, the evil Queen vows to destroy the child when it is born." When a girl-child bearing "the mark" appears, Princess Sorsha informs her mother, who prepares to have the baby slain (echoes of the infant CHRIST); but the midwife has fled with her charge, and succeeds in setting it adrift on the river on a raft of reeds (like the infant Moses). The child is discovered by the family of farmer Willow, who dwells in a village of DWARFS. Various derivative adventures occur before right is restored; e.g., the Queen of the Fairies resembles the Blue Fairy in PINOCCHIO (*1940*) and her attendants resemble Tinker Bell in PETER PAN (*1953*). The script is subservient to a plethora of action sequences and (often excellent) spfx, so what might have been a good piece of SWORD AND SORCERY comes across more as a high-pitched, eventually rather monotonous mishmash of event: *W*'s climax lacks any emotional intensity because there have already been so many foreshocks. [JG]
Further reading: *The Willow Source Book* * (**1988**) by Allan Varney.

WILLY WONKA AND THE CHOCOLATE FACTORY US movie (*1971*). Warner Bros/Wolper. **Pr** Stan Margulies, David L. Wolper. **Dir** Mel Stuart. **Spfx** Logan R. Frazee. **Screenplay** Roald DAHL. **Based on** *Charlie and the Chocolate Factory* (**1964**) by Dahl. **Starring** Jack Albertson (Grandpa Joe), Michael Bollner (Augustus), Julie Dawn Cole (Veruca), Dodo Denney (Mrs Teevee), Roy Kinnear (Mr Salt), Gunter Meibner (Slugworth/Wilkinson), Denise Nickerson (Violet), Peter Ostrum (Charlie Bucket), Ursula Reit (Mrs Gloop), Leonard Stone (Mr Beauregarde), Paris Themmen (Mike), Gene Wilder (Mr Wonka). 98 mins. Colour.

Years ago the Wonka chocolate factory closed its doors to outsiders. Now there is a grand competition: five gold tickets will be contained in Wonka bars, bringing the winners a tour round the factory (with the relative of their choice) and a lifetime supply of chocolate. Young Charlie, living in abject poverty with his mother and four bedridden grandparents, is the last of the five, and his Grandpa Joe is elected to go with him. Charlie refuses money from Wonka's industrial rival Slugworth to bring away the secret of the Wonka Everlasting Gobstopper. Mr Wonka himself, though human, proves a sinister TRICKSTER character, and during the tour allows the four other children to be punished (possibly fatally) by their various vices. Even Charlie is not flawless in behaviour, but he honestly returns the Everlasting Gobstopper he has been given rather than sell it to Slugworth; all proves to have been a trial (and

"Slugworth" a Wonka *agent provocateur*) to determine which child would be virtuous enough to inherit the factory.

WWATCF has been much disliked as a family comedy musical: the songs are weak, there are few jokes, and the morals are trite and clumsy. But sitting amid all the tacky trappings is an interestingly chilly fantasy: it is effectively a FAIRYTALE, with Wonka, as the guide to FAERIE, neither benevolent nor actively malevolent, merely nonhuman and whimsically disinterested in the fates of those humans who refuse to treat with MAGIC on its own terms. Distrust me, he seems to be saying, when I seem to offer you Paradise for free and tell you there is no need to read the CONTRACT carefully: no one is entitled to a place here unless they have earned it, and Paradise deals severely with trespassers. [JG]

WILSON, COLIN (HENRY) (1931-   ) UK writer, critic, student of crime and the paranormal; his personal version of Existentialism was initially aired in his first book, *The Outsider* (**1956**). CW is one of the few writers who have interestingly subverted CTHULHU MYTHOS themes to their own ends, beginning with the curious SCIENCE FANTASY *The Mind Parasites* (**1967**). This book's science is best not examined; it borrows H.P. LOVECRAFT's trappings (August DERLETH features as a character) only to transcend them with the RECOGNITION that the eponymous parasites, who like malign psychic VAMPIRES drain human energy and creativity, are renegade fragments of the human unconscious – mental cancers. The metaphor is further illuminated in the discursive *The Philosopher's Stone* (**1969**), which suggests that only lazy habits of thought stand between us and a Nietzschean transcendence conferring IMMORTALITY and paranormal TALENTS. Here CW's REVISIONIST-FANTASY examination of the Cthulhu Mythos and its MALIGN SLEEPERS (through psychic probing across a TIME ABYSS) redefines them as victims: though still powerful and menacing, they have been cast into coma through over-reaching in their own development – giving humanity a chance to push onward and meet these ELDER GODS as mental equals when they wake again. *The Return of the Lloigor* (in *Tales of the Cthulhu Mythos*, anth **1969** US; rev **1974** chap UK) is a more conventional **Cthulhu Mythos** HORROR story with a Welsh setting, echoing the quasi-documentary treatment of Lovecraft's own "The Call of Cthulhu" (1928) and also invoking Arthur MACHEN.

CW's commentaries on fantasy and SUPERNATURAL FICTION appear in: *The Strength to Dream: Literature and the Imagination* (**1962**), covering many relevant topics including GHOST STORIES, E.T.A. HOFFMAN, H.P. LOVECRAFT and J.R.R. TOLKIEN; *The Strange Genius of David Lindsay* (**1970**; with CW's material only, cut vt *The Haunted Man* 1979 US) with E.H. VISIAK and J.B. Pick, on David LINDSAY; and *Tree by Tolkien* (**1973** chap).

The **Gerard Sorme** sequence, beginning with CW's first novel, *Ritual in the Dark* (**1960**), and continuing with *Man Without a Shadow: The Diary of an Existentialist* (**1963**; vt *The Sex Diary of Gerard Sorme* 1963 US) develops fantasy elements in *The God of the Labyrinth* (**1970**; vt *The Hedonists* 1971 US). This features an old erotic cult called the Sect of the Phoenix, copious and rebarbative SEX leading to friendly POSSESSION by a long-dead rake, and intimations of psychic abilities like ASTRAL-BODY travel resulting from abnormally prolonged orgasm – CW is here adapting conventions of pornography (as earlier he used the **Cthulhu Mythos**) as new scaffolding for his dream of transcendence. His mystical agenda and sometimes uncritical-seeming enthusiasm for the paranormal emerge even in detections like *The Schoolgirl Murder Case* (**1974**), whose police investigator consults an OCCULT DETECTIVE specializing in VISIONS, and its sequel *The Janus Murder Case* (**1984**).

A late SCIENCE-FANTASY venture is the **Spider World** series, set in Earth's FAR FUTURE: *Spider World: The Tower* (**1987**; vt in 3 vols as *Spider World 1: The Desert* 1988 US, *Spider World 2: The Tower* 1989 US and *Spider World 3: The Fortress* 1989 US), *Spider World: The Delta* (**1987**) and *Spider World: The Magician* (**1992**). The initial handling of sf tropes is routine: giant SPIDERS have enslaved the remnants of humanity, and there is rebellion. But *The Delta* stresses CW's habitual urging of mental self-improvement, here mediated by a vegetative soul recalling the Life Force of *Man and Superman* (**1903**) and *Back to Methuselah* (**1921**) by Bernard Shaw (1856-1950) – about whom CW has also written in *Bernard Shaw: A Reassessment* (**1969**).   [DRL]

Other works (selective): *The Space Vampires* (**1976** vt *Lifeforce* 1985 US), an sf VAMPIRE homage to A.E. VAN VOGT's story "Asylum" (1942); *Dark Dimensions: A Celebration of the Occult* (anth **1978** US), ed.

WILSON, DAVID HENRY (1937-   ) UK dramatist, translator and author who has spent considerable time in Germany, where he published his first adult novel in his own German translation, *Ashmadi* (**1985** Germany; original English text as *The Coachman Rat* 1987 UK), a remarkable BEAST-FABLE which ingeniously interrogates more than one FAIRYTALE (◊ REVISIONIST FANTASY). Much of his fiction for younger children is intriguingly fanciful.

The Coachman Rat is narrated by its protagonist, who has been born as a rat (◊ MICE AND RATS) in the WAINSCOT interstices of a small medieval town, and who longs to associate with humans, despite family warnings that the human race rules the world because it is murderous. He soon arranges to be captured by Amadea, a young girl who turns out to be a CINDERELLA figure, and is further trapped into her STORY, being transformed into a coachman by the "Woman of Light" who serves as the FAIRY GODMOTHER. At midnight, after the ball, though he returns to rat form he continues to understand human speech – and not the squeaks of his own kind. After PICARESQUE adventures, during which a scientist attempts to dissuade him from belief in MAGIC, he is taken over by a humane student; but both are tricked by a revolutionary agitator, with the result that Robert (as the rat is now known) inadvertently betrays Amadea, who is burnt at the stake. Transformed at this point by the Woman of Light back into a coachman, Robert enacts a savage revenge upon the agitator, who has become a corrupt tyrant. But his inhuman savagery, and the plague which now devastates both rat and human worlds, induces him to star in another STORY, the legend immortalized by Robert BROWNING in "The Pied Piper of Hamelin" (1842). He drowns his fellow rats to save the human world; and, himself stricken with plague, prepares to die.   [JC]

Other works (for younger children): *Elephants Don't Sit on Cars* (**1977**); *The Fastest Gun Alive* (**1978**); *Getting Rich with Jeremy James* (**1979**); *Beside the Sea with Jeremy James* (**1980**); *How to Stop a Train with One Finger* (**1984**); *Superdog* (**1984**); *Do Goldfish Play the Violin?* (**1985**); *There's a Wolf in My Pudding* (**1986**); *Superdog the Hero* (**1986**); *Yucky Ducky* (**1988**); *Superdog in Trouble* (**1988**); *Gander of the Yard* (**1989**); *Little Billy and the Wump* (**1990**); *Gideon Gander Solves the World's Greatest Mysteries* (**1993**); *Please Keep Off the Dinosaur* (**1993**).

WILSON, F(RANCIS) PAUL (1946-   ) US writer who began publishing work of genre interest in the early 1970s. FPW's career has gradually shifted first from sf to HORROR/DARK FANTASY, and more recently to medical thrillers. He began his move to horror with *The Keep* (**1981**), a memorable tale of supernatural EVIL set in a Transylvanian castle occupied by Nazi forces during WORLD WAR II. The Nazi interlopers accidentally awaken an ancient creature, whose initial attacks seem to indicate the presence of a VAMPIRE but whose true nature is quite different, and who is ultimately opposed by an enigmatic supernatural adversary; it was filmed as *The* KEEP (*1983*). *The Keep*, the first novel in the **Adversary** sequence, was followed by *Reborn* (**1990**), in which the protagonist inherits a fortune from an elderly scientist he has never met. His subsequent search for the reasons behind the bequest leads to the discovery of a cloning experiment and the reemergence of the evil creature from *The Keep*. Wilson enhances the scope of the story in *Reprisal* (**1991**), set 20 years later as the creature's power continues to grow, and *Nightworld* (**1992**).

FPW's fondness for PULP-style action and adventure is perhaps most clearly reflected in *The Tomb* (**1984**), in which a shipload of monstrous creatures is transported from India to the USA to fulfil an old CURSE. FPW's rebellious anti-hero, Repairman Jack, is called upon to save the day, a role he has since reprised in several short stories. *Black Wind* (**1988**) is another novel of WWII, this time centred in the Pacific theatre, where a mystical order of Japanese monks, the Kakureta Kao, seeks to harness the power of the Black Wind, a power that would guarantee a Japanese victory. This is one of FPW's longest and most intricate works, yet it maintains an action-oriented approach. His recent move to medical thrillers is presaged by *The Touch* (**1986**), in which the conveyance of a magical and seemingly glorious power of HEALING leads to dire consequences. A related story, "Dat-tay-vao", based on a Vietnam flashback scene cut from the novel, can be found in *Soft and Others* (coll **1989**). Horror elements can still be found in FPW's most recent novels – *The Select* (**1993** UK as by Colin Andrews; 1994 as by FPW) and *Implant* (**1994** UK as by Colin Andrews; 1995 as by FPW) – but these clearly herald a new direction.

Regardless of genre or subject matter, FPW's work is well crafted.   [BM]

Other works: *Healer* (1972 *Analog* as "Pard"; exp **1976**); *Wheels within Wheels: A Novel of the LaNague Federation* (1971 *Analog*; exp **1978**), winner of the first Prometheus AWARD; *The Tery* (1973 *Fiction 4* as "He Shall Be John"; exp **1979** chap dos; further exp as coll **1990**); *An Enemy of the State* (**1980**); *Black Wind* (**1988**); *Dydeetown World* (fixup **1989**); *Ad Statum Perspicuum* (coll **1990**); *Midnight Mass* (**1990** chap), a VAMPIRE novella; *Pelts* (**1990** chap); *Sibs* (**1991**; vt *Sister Night* 1993 UK); *Buckets* (**1991** chap); *The Barrens* (**1992**), an excellent Lovecraftian novella; *Freak Show* (anth **1992**); *The LaNague Chronicles* (omni **1992**), collecting *Healer*, *Wheels within Wheels*, and *An Enemy of the State*; *Diagnosis: Terminal* (anth **1996**).

WILSON, GAHAN (1930-   ) US cartoonist, illustrator and writer; following a first appearance in *Amazing Stories* (1954), his artwork became widely known through slick magazines like *Playboy*, *Collier's* and later *The New Yorker*. In 1964 he began his long association with *The* MAGAZINE OF FANTASY AND SCIENCE FICTION, contributing numerous full-page cartoons – mostly macabre – plus movie reviews. His first collection was *Gahan Wilson's Graveside Manner* (graph coll **1965**); others include *The Man in the Cannibal Pot* (graph coll **1967**), *I Paint What I See* (graph coll **1971**), *Playboy's Gahan Wilson* (graph coll **1973**; same title for completely different graph coll **1980**), *Gahan Wilson's Cracked Cosmos* (graph coll **1975**); *Weird World of Gahan Wilson* (graph coll **1975**), "*. . . and then we'll get him!*" (graph coll **1978**), *Is Nothing Sacred?* (graph coll **1982**), *Gahan Wilson's America* (graph coll **1985**) *Still Weird: A Look Back – And Forward* (graph coll **1994**) and «*Weirder Yet*» (graph coll 1996). GW draws in distinctively sprawling lines which encourage distortions and grotesques, a good visual match for his habitual black HUMOUR (inevitably compared with that of Charles ADDAMS); funerals, DEATH, DEVILS, oozing MONSTERS, VAMPIRES and seedy versions of HEAVEN and HELL regularly appear, as do sinister CHILDREN. *Nuts* (graph coll **1979**) assembles his eponymous *National Lampoon* cartoon strips on vividly remembered fantasies and horrors of US childhood. GW's fiction includes a part-graphic story in Harlan ELLISON's *Again, Dangerous Visions* (anth **1972**): its title is an unpronounceable blob, which proves carnivorous. *Gahan Wilson's Diner* (*1973*) is an ANIMATED MOVIE. *Eddy Deco's Last Caper: An Illustrated Mystery* (1987) is farcical DETECTIVE/THRILLER FANTASY with integral illustrations; another spoof mystery, *Everybody's Favorite Duck* (**1988**) also has fantasy elements, with FU MANCHU and SHERLOCK HOLMES characters. GW received a special WORLD FANTASY AWARD in 1981 (earlier, he had designed this trophy), and the BRAM STOKER AWARD for life achievement in 1982.

[DRL/SD]

Other works: *Harry, the Fat Bear Spy* (**1973**); *The Bang Bang Family* (**1974**); *Harry and the Sea Serpent* (**1976**); *Harry and the Snow Melting Ray* (**1978**); *Spooky Stories for a Dark and Stormy Night* (coll **1994**) – all CHILDREN'S FANTASY.

As editor: *Gahan Wilson's Favorite Tales of Horror* (anth **1976**); *First World Fantasy Awards* (anth **1977**).

WILSON, J. ARBUTHNOT Pseudonym of Grant ALLEN.

WINCH, JOHN Pseudonym of Marjorie BOWEN.

WIND IN THE WILLOWS, THE Various movies have been based on *The Wind in the Willows* (1908) by Kenneth GRAHAME.

1. *The Adventures of Ichabod and Mr Toad* (vt as two separate items *The Wind in the Willows* and *The Legend of Sleepy Hollow*) US movie (*1949*). DISNEY. **Pr** Ben Sharpsteen. **Dir** James Algar, Clyde Geronimi, Jack Kinney. **Special processes** Ub IWERKS. **Based on** Grahame's book and Washington IRVING's "The Legend of Sleepy Hollow" (1819-20). **Voice actors** Claude Allister (Rat), Eric Blore (J. Thaddeus Toad), Colin Campbell (Mole), Campbell Grant (Angus MacBadger), Alec Harford (Winkie), J. Pat O'Malley (Cyril Proudbottom), plus Bing Crosby (narrator of *Ichabod* section) and Basil Rathbone (narrator of *Toad* section). 68 mins. Colour.

The *TWITW* section of this movie is a vigorous but necessarily sketchy adaptation. MacBadger, Toad's financial advisor, calls in Rat and Mole to ask help in curtailing Toad's expensively destructive mania for gypsy-style caravanning. The next they know Toad is in court for car theft, and is jailed on false charges. Disguised as Toad's grandmother, Toad's horse Cyril brings the prisoner a washerwoman costume, using which Toad escapes and steals a train (◊ GENDER DISGUISE); the police pursue in a second train (providing an animated homage to the Keystone Cops, but in English uniform). The Weasels take over Toad Hall, but are cast out by the heroic quartet. This

is not classic Disney work, but it is possibly the best movie version of the tale so far. [JG]

2. *The Wind in the Willows* UK stop-motion ANIMATED MOVIE (*1983* tvm). Cosgrove-Hall/Thames TV. **Pr** Brian Cosgrove, Mark Hall. **Exec pr** John Hambley. **Dir** Hall. **Screenplay** Rosemary Anne Sisson. **Voice actors** Ian Carmichael (Water Rat), Michael Hordern (Badger), David Jason (Toad), Richard Pearson (Mole), Beryl Reid (Judge). 78 mins. Colour.

A faithful adaptation, if anything bolstering the original's reactionary viewpoint (with the Weasels voiced as sniggering working-class oicks) and its antifantasy (with the imaginative being reduced to cosy, plodding domesticity). In this lethargic version, all is done with ponderous tastefulness: even the menacing of Mole in the Wild Wood is blandly unfrightening. The animation is technically good and the models and sets are meticulously made, but all is somehow bloodless. [JG]

3. *The Wind in the Willows* Japanese/US ANIMATED MOVIE (*1983* tvm). **Dir** Jules Bass, Arthur Rankin Jr. **Voice actors** Eddie Bracken, José Ferrer, Paul Frees, Roddy McDowall, Charles Nelson Reilly (movie lacks proper credits). 97 mins. Colour.

A dutiful version, with some nice moments of animation among the rest. [JG]

WINDLING, TERRI (1958-   ) US editor, illustrator and book publisher whose passion for FAIRYTALE contributed to a recent revival and reassassment of the field. TW's influence has been most noticeable as an editor. She became an editor at Ace Books in 1979 and produced four ANTHOLOGIES, the **Elsewhere** trilogy, *Elsewhere* (anth **1981**), *II* (anth **1982**) and *III* (anth **1984**), all with Mark Alan Arnold, and *Faery!* (anth **1985**). These showed her wide knowledge of FANTASY beyond standard GENRE-FANTASY boundaries and helped broaden the understanding of a field that had become moribund during the 1970s. She developed the **Ace Fantasy** line of books, which showcased the early work of Steven R. BOYETT, Emma BULL, M. Lucie CHIN and Charles DE LINT, and enticed Jane YOLEN into writing novel-length work for adults. TW also launched the **MagicQuest** YA series, mostly reprint, bringing into paperback for the first time work by Paul R. FISHER, Elizabeth Marie POPE, and Ruth NICHOLS, alongside work by Diana Wynne JONES, Tanith LEE and Yolen, and publishing the first work of Patricia C. WREDE. Before leaving Ace TW established the **Fairy Tales** series, starting with Steve BRUST's *The Sun, the Moon and the Stars* (**1987**). During this period she also edited, with Mark Alan Arnold, the **Bordertown** SHARED-WORLD series, including the anthologies *Borderland* (anth **1986**), *Bordertown* (anth **1986**) and *Life on the Border* (anth **1991**), the last edited solo. These are set on the border between the human world and FAERIE and include punk ELVES.

In 1987 JY went freelance, serving as Consulting Editor to Tor Books and developing their fantasy line. She also started, with Ellen DATLOW, the **Year's Best Fantasy and Horror** series (anth **1988**-current) with TW selecting the fantasy and Datlow the horror. The series runs *The Year's Best Fantasy: First Annual Collection* (anth **1988**; vt *Demons and Dreams: The Best Fantasy and Horror 1* 1989 UK), *The Year's Best Fantasy: Second Annual Collection* (anth **1989**; vt *Demons and Dreams: The Best Fantasy and Horror 2* 1990 UK), with a permanent title switch to *The Year's Best Fantasy and Horror: Third Annual Collection* (anth **1990**), *Fourth* (anth **1991**), *Fifth* (anth **1992**), *Sixth* (anth **1993**), *Seventh* (anth **1994**), *Eighth* (anth **1995**) and *Ninth* (anth **1996**). TW's reading for these volumes is extensive, encompassing the widest possible range of new fantasy, and her selection is the most comprehensive available. Also with Datlow, TW has edited a series of modern fairytales – mostly REVISIONIST FANTASY with a strong element of FEMINISM – *Snow White, Blood Red* (anth **1993**), *Black Thorn, White Rose* (anth **1994**) and *Ruby Slippers, Golden Tears* (anth **1995**). Solo, TW compiled *The Armless Maiden and Other Tales for Childhood's Survivors* (anth **1995**), intended to draw attention to the plight of abused children. The volume includes TW's first published fiction, "The Green Children", drawn from FOLKLORE. TW's first novel, «The Wood Wife» (1996), uses a well known folktale as a basis for CONTEMPORARY FANTASY.

TW's impact on fantasy since the 1980s has helped establish a wider awareness and understanding of the field.[MA]

WINDSOR-SMITH, BARRY (1949-   ) London-born artist and illustrator of US COMIC books, with a strong, decorative line style. He worked as Barry Smith until 1983. He began his career with a series of back-cover pin-ups of SUPERHEROES for the UK MARVEL-COMICS reprint weeklies *Fantastic* and *Terrific* in 1967. He travelled to the US and began working for Marvel on *X Men* (#53 1969), but was forced to return to the UK in 1968 when his visitor's visa expired. He continued to work for US comic books, however, drawing *Daredevil* (#*50*-#*52* 1969), *Nick Fury, Agent of SHIELD* (#*12* 1969) and *The Avengers* (#*66*-#*67* 1969 and #*98*-#*100* 1972) and some short horror tales. A turning point came when he was chosen as the artist to draw Marvel's adaptation of Robert E. HOWARD's CONAN stories. Here his extraordinary narrative talent became increasingly evident, reaching a peak with the particularly inventive and skilful *Song of Red Sonja* (*Conan #24* 1973). For his work on this series he was awarded two Academy of Comic Book Art Awards (1971, 1972).

In 1974 BWS left the comic-book industry to found the Gorblimey Press to publish his portfolios and prints. In 1975, he joined Jeff JONES, Mike KALUTA and Bernie WRIGHTSON at The Studio – premises the artists shared and the subject of a coloured art book, *The Studio* (graph **1979**) with text by "J.S.". This was a period of experimentation, with only a few short comic strip pieces being published in EPIC ILLUSTRATED #16 (1983).

BWS returned to comics with *Machine Man #1-#4* (1984-5), and has since worked on many mainstream comic-book characters, including *The Uncanny X-Men* (intermittently 1984-7), *Daredevil* (#*236* 1986), *Wolverine* (1991), *Archer and Armstrong* (1992-3), *Eternal Warrior* (1993) and many others. His own creation, *Rune*, featuring an ancient, cannibalistic power-seeking alien, began publication in 1994.   [SH/RT]

WINGS Wings enable flight, and in fantasy are usually symbols of freedom, although a strong cautionary note was introduced into the dream of human flight by the story of ICARUS. Wings come in three distinct versions whose symbolism is markedly different. Feathered wings like those of BIRDS confer a certain nobility upon chimeras thus endowed, especially ANGELS (and their analogues) and winged horses. Batlike and pterosaurian wings are among the standard accoutrements of DEMONS and the more ominous kinds of VAMPIRES. Insectile wings are almost exclusively restricted to the effete and calculatedly quaint kind of FAIRIES beloved of Victorian artists, signifying

frailty. The symbolic roles of various chimeras, therefore, depend to some extent on what kind of wings they have (◊ ANIMALS UNKNOWN TO SCIENCE), although the partial redemption of DRAGONS from their monstrous role by modern writers has helped change the implication of pterosaurian wings.

The fate of Icarus is analysed in *Story for Icarus* (**1958**) by Ernst Schnabel. Sentimental fantasies about bird-winged humans who find solace in elevation include *Going Home* (**1921**) by Barry PAIN, *An Alien from Heaven* (**1929**) by Nathalia Crane and "He That Hath Wings" (1938) by Edmond Hamilton (1904-1977). The winged women in *Angel Island* (**1914**) by Inez Haynes Gillmore (1873-1970) and the protagonist of Nancy SPRINGER's *Metal Angel* (**1994**), on the other hand, have their wings clipped in order to humble them. The wings grown by the protagonist of Mervyn PEAKE's *Mr Pye* (**1953**) are an ironically mixed blessing. A notable TRAVELLER'S TALE featuring winged humans is *The Life and Adventures of Peter Wilkins* (**1751**) by Robert S. Paltock (1697-1767). Bat-winged humans are much rarer, although one is featured in "The Garden of Fear" (1934) by Robert E. HOWARD, and they occasionally crop up in ironic sf stories like *Childhood's End* (**1953**) by Arthur C. Clarke. Humanoid aliens with insectile wings are ironically featured in "Magnanthropus" (1961) and its sequel by Manly Banister (1914-1986).   [BS]

WINGS OF DESIRE (ot *Der Himmel über Berlin*) French/West German movie (*198***7**). Argos. **Pr** Anatole Dauman, Wim Wenders. **Dir** Wenders. **Screenplay** Peter Handke, Wenders. **Starring** Solveig Dommartin (Marion), Peter Falk (Himself), Bruno Ganz (Damiel), Otto Sander (Cassiel). 128 mins. B/w and colour.

Two angels, Damiel and Cassiel, attend Berlin, feeding on human thoughts and emotions as they have done since humanity first appeared on the site. Their IMMORTALITY has its cost: they cannot experience life directly – only leached of its colours, tastes and emotions. Yet Damiel falls in love with a beautiful trapeze artiste, Marion, and his growing passion for her finally succeeds in transforming him into a mortal man. Also in Berlin is Peter Falk, visiting to appear in a movie being made there; unlike everyone else except some children, he can directly sense the presence of the invisible angels – because, as he eventually confesses to the mortalized Damiel, he was once an angel himself, but took mortal form 30 years ago. Drifting through Berlin's nightlife, the mortal Damiel at last encounters Marion, and her love for him is instant, as if she had always known him.

Slow-moving, largely in b/w (the way the angels see the world) and with dialogue in English and (subtitled) German and French, *WOD* might seem at first daunting, yet it rapidly becomes absorbing and very convincing; the suspension of disbelief is attained not through spfx but by skilful nuances of acting (others do not walk *through* the angels but always manage narrowly, and quite naturally, to avoid bumping into them), by the alternations between b/w and colour, and by the rhythmic background patterning of people's thoughts (as heard by the angels), which rises to the heights of an abstracted, timeless-seeming chorale when the pair visit a public library, where the musings of the readers are overlaid on the susurration of ideas present in the books. Like all the best fantasies, it leads us to a shift in PERCEPTION: the angels' plane of existence becomes natural to us, while our own mortal world of sensation becomes a fantastic and infinitely desirable venue.   [JG]

WINTERSON, JEANETTE (1959-   ) UK writer whose second novel, *Boating for Beginners* (**1985**), is a REVISIONIST FANTASY making experimental and feminist (◊ GENDER) play with the story of Noah. *The Passion* (**1987**), largely set in a fabulated VENICE, surrounds the historical figure of Napoleon with a bevy of COMPANIONS, some with supernatural powers; in the second strand of this complex tale a woman with webbed feet falls in love with Napoleon's cook, but both eventually succumb to the terminal labyrinths of her native Venice. *Sexing the Cherry* (**1989**) engages more complexly with the matter and manner of fantasy, and constitutes an orthodox revisionist fantasy – in a mode reminiscent of Angela CARTER – through sections which examine, by retelling them, various FAIRYTALES, including "Rapunzel" and "The Twelve Dancing Princesses". At the heart of complex spirals of narrative lies an URBAN-FANTASY version of 17th-century LONDON and a central character who – though technically male – seems at moments as convincing a TEMPORAL ADVENTURESS as any descendant of Virginia WOOLF's Orlando. *Sexing the Cherry* is one of the most significant texts on the WATER MARGINS of fantasy.   [JC]

WISHES The simplest and superficially most attractive form of MAGIC, whereby desires are fulfilled without complexities of RITUALS or SPELLS – hence the pejorative phrase "wish-fulfilment fantasy". Traditionally, wishes are granted by a DEMON, FAIRY, GENIE or TALISMAN – most often a RING – and generally go awry (◊ THREE WISHES). Occasionally wishing may be seen as a TALENT, perhaps not fully subject to conscious control, as in "Oddy and Id" (1950) by Alfred Bester (1913-1987) and *Merlin and the Last Trump* (**1993**) by Collin WEBBER. Wishes are endlessly subject to QUIBBLES and READ-THE-SMALL-PRINT traps: e.g., requests for beauty in Anthony BOUCHER's "Nellthu" (1955 *F&SF*) and for restored vitality in Larry NIVEN's "The Wishing Game" (1987) are mockingly granted without the expected accompaniment of renewed youth. Because wishes and their pitfalls are well known through their popularity in SLICK FANTASY, there is a quasi-RECURSIVE-FANTASY sense in which characters are prepared for the challenge – like Moffett in John CROWLEY's *Aegypt* (**1987**), for whom the proper use of wishes has been a lifelong MAGGOT. When the normal restriction to only three WISHES is lifted, other limitations tend to be substituted: E. NESBIT's *Five Children and It* (**1902**) and *The Enchanted Castle* (**1907**) stipulate time limits after which wishes simply wear off, while in Ursula K. LE GUIN's sf *The Lathe of Heaven* (**1971**) the inadequacy of the human imagination means that each easy wished-for solution to the world's ills makes matters worse; it becomes necessary to stop. The Absurdist ploy of repeatedly wishing for more wishes is often tried in HUMOUR, from Anthony ARMSTRONG's FAIRYTALE travesty "The Prince's Birthday Present" (1932) to Tom HOLT's *Djinn Rummy* (**1995**).   [DRL]

See also: ANSWERED PRAYERS.

WITCHCRAFT Illicit magic; the term is not clearly distinguishable from "sorcery", though the word "witch" is nowadays usually associated with women while sorcery (i.e., MAGIC which apes priestly ritual) tends to be seen as a male prerogative. Some writers use "warlock" to designate male witches, although this usage has no proper etymological warrant; during the witch-persecutions of the 15th-17th centuries the word "witch" was applied to both sexes.

Western Europe inherited a series of witch-images from Classical myth and literature, including FEMMES FATALES like CIRCE and Medea and hagwives like Erichtho (in Lucan's *Pharsalia*), OVID's Dipsas and Horace's Canidia. The alleged power of witches is mocked in the depiction of Pamphile in LUCIAN's *The Golden Ass*. Witches in Teutonic mythology (◊ NORDIC FANTASY) are all hagwives and hedge-riders, usually busy mixing potions in their cauldrons; this was the image inherited by the witches in SHAKESPEARE's *Macbeth* (performed *c*1606; **1623**), although the invocation of Hecate links these to the Classical tradition. A similar Hecate-led crew appears in *The Witch* (*c*1620) by Thomas Middleton (1580-1627), one of several English witch-plays produced in the wake of the trial (1612) of the Lancashire Witches, whose fictional representations include the play *The Lancashire Witches* (produced 1681) by Thomas Shadwell (*c*1642-1692) and the novels *The Lancashire Witches* (**1848**) by W. Harrison AINSWORTH and *Mist Over Pendle* (**1951**) by Robert Neill.

Although witch-persecution was never as fierce in Britain as on the European mainland, several notable British writers – including Daniel DEFOE and Sir Walter SCOTT – produced popular "nonfiction" books full of witch-anecdotes which provided sources for later writers. The fact that only one significant witch-trial ever took placed in the USA (at Salem, 1692) did not diminish that country's interest in the least. The tale is retold in numerous literary works, ranging from the wholly credulous *Yesterday Never Dies* (**1941**) by Esther Barstow Hammond to the scathingly sceptical play *The Crucible* (**1953**) by Arthur Miller (1915-   ); and the incident inspired Nathaniel HAWTHORNE's keen interest in witchcraft, displayed in "Young Goodman Brown" (1835) and *The Scarlet Letter* (**1850**). The only other witchcraft trial to have attracted such copious attention is that in 1634 of the French priest Urbain Grandier, featured in *Dreams of Roses and Fire* (**1949** Sweden) by Eyvind Johnson (1900-1976) and in the documentary *The Devils of Loudun* (**1952**) by Aldous Huxley (1894-1963), the latter filmed by Ken RUSSELL as *The* DEVILS (*1970*). Other novels based (usually very loosely) on actual cases include *Sidonia the Sorceress* (**1848** Germany) by Wilhelm Meinhold (1797-1851), *The Blue Firedrake* (**1892**) by Thomas Wright (1859-1936), *The Devil's Mistress* (**1910**) by J.W. Brodie-Innes (1848-1923) and the third part of *The Witches* (fixup **1969** France) by Françoise Mallet-Joris; only the last is as effective as the purely exemplary trial described in *By Firelight* (**1948**) by Edith PARGETER, and Meinhold's *Sidonia* is far inferior to his wholly imaginary account, *The Amber Witch* (**1843** Germany).

The image of the witch underwent a dramatic overhaul following the publication of the carefully calculated SCHOLARLY FANTASY *La sorcière* (**1862**) by Jules Michelet (1798-1874), an essay in strident anticlericalism that used the Church's persecution of witches as proof of its tyranny and irrationality, arguing that by charging innocent women – many of them midwives and practitioners of folk medicine – with making PACTS WITH THE DEVIL and worshipping SATAN at SABBATS the Inquisition had actually justified (and perhaps created) Satanism as a form of rebellion. This thesis was modified by Charles Godfrey Leland (1824-1903) in his literary hoax *Aradia* (**1899**) to the claim that the persecuted witches had actually been pagan worshippers of a Mother GODDESS – a thesis recapitulated in *The Horned Shepherd* (**1904**) by Edgar JEPSON and much elaborated in *The Witch Cult in Western Europe* (**1921**) by Margaret Murray (1863-1963). This notion has had a great impact on modern GENRE FANTASY, whose witch-images have mostly been modelled according to some version of the thesis. Novels attempting to represent the ANTHROPOLOGY of witchcraft more realistically are very rare, although *The Witch in the Cave* (**1986**) by Martin H. Brice makes a noteworthy attempt. Cautionary tales warning against the dangers of witch-hunting are more common; they include *Talk of the Devil* (**1956**) by Frank BAKER and *The Devil on the Road* (**1978**) by Robert WESTALL.

The idea of witchcraft surviving unnoticed in a modern setting is commonplace, and is sometimes ambivalently invoked in URBAN FANTASY as a problematic disruption of normality. Stories of this kind which combine comedy and sentimentality in varying proportions include *Living Alone* (**1919**) by Stella BENSON, *Lolly Willowes, or The Loving Huntsman* (**1926**) by Sylvia Townsend WARNER, *The Passionate Witch* (**1941**) by Thorne SMITH and Norman MATSON – filmed as I MARRIED A WITCH (*1942*) – the play *Bell, Book and Candle* (**1956**) by John Van Druten and *A Likeness to Voices* (**1963**) by Mary Savage. More earnest uses of the same notion are mostly best considered under the heading of OCCULT FANTASY.

Most of the many relevant theme anthologies include far more occult fiction than fantasy, but some recent ones tip the balance the other way; one notable example is *Hecate's Caldron* (anth **1982**) ed Susan SHWARTZ. [BS]

WITCHCRAFT & SORCERY ◊ COVEN 13/WITCHCRAFT & SORCERY.

WITCHES Practitioners of WITCHCRAFT (or, to stretch a point, SHAMANISM), either male or female, although in fiction usually female. Such characters turn up quite rarely today in SECONDARY-WORLD fantasies – although there is a spectacular example in *Conan the Barbarian* (*1981*; ◊ CONAN MOVIES), while of course L. Frank BAUM deployed GOOD AND EVIL witches in his POLDER, OZ – but they are slightly more common in OCCULT FANTASY, HORROR and CONTEMPORARY FANTASY. The image of the hag riding a broomstick is subverted in Sheri S. TEPPER's quasi-horror novels *Blood Heritage* (**1986**) and *The Bones* (**1987**), where a clutch of witches is presented as if straight from a sherry party, full of homely wisdom, and in Keith ROBERTS's **Anita** series, whose eponym is the kind of lass one might expect to find dancing in a disco. Witches of a more traditional sort feature in the movie HOCUS POCUS (*1993*), although only to be sent up, and witchcraft is likewise a target for humour in several of Terry PRATCHETT's **Discworld** novels, such as *Witches Abroad* (**1991**). It is rare, in fact, for witches to be treated very seriously in modern fantasy: the movie I MARRIED A WITCH (*1942*), based on *The Passionate Witch* (**1941**) by Thorne SMITH and Norman MATSON, is typical in regarding its central young witch as both mirthful and infernally sexy. Roald DAHL's *The Witches* (**1983**), filmed as *The* WITCHES (*1989*), provides a very notable exception: its witches are intended to terrify, especially since they may be encountered on any street-corner. [JG]

WITCHES, THE UK movie (*1966*). ◊ *The* WITCHES (*1989*).

WITCHES, THE US movie (*1989*). Warner Bros/Lorimar/Henson. **Pr** Mark Shivas. **Exec pr** Jim HENSON. **Dir** Nicolas Roeg. **Animatronics** Jim Henson's Creature Shop. **Screenplay** Allan Scott. **Based on** *The Witches* (**1983**) by Roald DAHL. **Starring** Rowan Atkinson (Mr Stringer), Brenda Blethyn (Mrs Jenkins), Jasen Fisher (Luke), Jane Horrocks (Irvine), Anjelica Huston (Eva Ernst), Bill

Paterson (Herbert Jenkins), Charlie Potter (Bruno Jenkins), Mai Zetterling (Grandma/Helga). 91 mins. Colour.

There are child-hating WITCHES everywhere – Luke's Norwegian Grandma tells him – detectable only by the purplish tinge in their eyes; in fact they are bald and hideous, but wear wigs and masks to hide this, and have no toes so always wear plain, sensible shoes. Grandma's childhood friend Erika was seized by a witch and only later discovered locked into a painting, where she stayed trapped as she grew old and died; Grandma herself, in her youth, searched the world for the Grand High Witch, but never found her.

When Luke's parents are killed in an accident, he goes to live with Grandma in England, and eventually, to holiday at the Hotel Excelsior. Here they find a Royal Society for the Prevention of Cruelty to Children convention. As Luke discovers, hiding in the conference room, this is a front for a witches' convention, and the Grand High Witch herself is present in the guise of Eva Ernst. Her plan is that the witches should open hundreds of sweetshops, doctoring the sweets with her magic POTION, Formula 86, which turns people into mice; she demonstrates on Luke's new friend, glutton Bruno. Then Luke is caught and likewise turned into a mouse. With the help of Bruno and Grandma he succeeds in lacing the soup at the convention banquet with purloined Formula 86, and as mice the witches meet their nemesis. Ernst's much-abused secretary, Miss Irvine, turns Luke back into a boy. He and Grandma, with Ernst's US address book, plan to continue the hunt . . .

Although modest, *TW* is surprisingly enjoyable. However, the portrayal of witches as looking just like plainly dressed women, living alone or together, as propounded in both the movie and, even more so, Dahl's book, has led to some examples of reclusive women being tormented by vindictive children.

TW has no connection with *The Witches* (**1966**), which is rather good HAMMER fare about a new schoolmistress in an English village discovering that the locals are dabbling in ritual MAGIC and DEVIL-worship. [JG]

WITCHES OF EASTWICK, THE US movie (*1987*). Warner/Guber-Peters/Kennedy-Miller. **Pr** Neil Canton, Peter Guber, Jon Peters. **Exec pr** Rob Cohen, Don Devlin. **Dir** George Miller. **Spfx** Industrial Light & Magic. **Mufx** Rob Bottin. **Screenplay** Michael Cristofer. **Based on** *The Witches of Eastwick* (**1984**) by John UPDIKE. **Starring** Veronica Cartwright (Felicia), Cher (Alex), Jack Nicholson (Daryl Van Horne), Michelle Pfeiffer (Sukie), Susan Sarandon (Jane). 118 mins. Colour.

A sex comedy that occasionally pretends to greater profundities, *TWOE* centres on widowed brunette Alex, divorced redhead Jane and abandoned blonde Sukie, three apparently upright members of Eastwick's small community. Wishing a "foreign prince" would come to town to captivate them, they are seemingly rewarded by the arrival of (apparently) the Prince of Darkness himself, in the guise of Van Horne ("just your average horny little DEVIL"). He seduces (and, we later find, impregnates) all three – or perhaps, instead, he helps them unbutton their own repressions – and performs MIRACLES for their delight. But when he causes starchy pillar of the community Felicia Alden to be murdered the three "WITCHES" shun Van Horne and finally concoct their own MAGIC, mutilating a DOLL of him until, after a last transformation into the GREAT BEAST, he is driven away. Yet, 18 months later, he returns as a tv image in an attempt to corrupt their baby sons.

There is much informed use of fantasy motifs in *TWOE*, and within the movie resonances are quietly sounded between images from SUPERSTITION and real-world events in the story. Alex's clay earth-mother sculptures find echoes in Van Horne's views of women as repositories of FERTILITY (the movie is extremely rich in fertility symbolism throughout) and of witches in particular as women of power who render men impotent with fear – although Van Horne treats them as TOYS, an equation underscored when the "witches" thwart him and he behaves like a temperamental child. Asked at one point who he really *is*, he responds simply: "Anybody you want me to be" – a good enough definition of SATAN.   [JG]

WITCH'S TALES, THE ◊ MAGAZINES.

WITKIEWICZ, STANISŁAW IGNACY (1885-1939) Polish novelist and playwright. ◊ POLAND.

WIZ, THE US movie (*1978*). ◊ *The* WIZARD OF OZ.

WIZARD OF MARS, THE US movie (*1964*). ◊ *The* WIZARD OF OZ.

WIZARD OF OZ, THE Many movies have been based on the OZ series by L. Frank BAUM. The listing below is far from complete.

1. *The Wizard of Oz* US movie (*1939*). MGM. **Pr** Mervyn LeRoy. **Dir** Victor Fleming (colour), King Vidor (sepia). **Spfx** Arnold Gillespie. **Screenplay** Noel Langley, Florence Ryerson, Edgar Allan Wolfe. **Based primarily on** *The Wonderful Wizard of Oz* (*1900*). **Starring** Ray Bolger (Hunk/Scarecrow), Billie Burke (Glinda, the Witch of the North), Judy Garland (Dorothy), Jack Haley (Hickory/Tin Man), Margaret Hamilton (Miss Gulch/Wicked Witch of the West), Bert Lahr (Zeke/Cowardly Lion), Frank Morgan (Professor Marvel/Wizard of Oz). 102 mins. Sepia and colour.

The CINEMA's classic PANTOMIME. Orphan Dorothy, living with her Auntie Em and Uncle Henry, is persecuted by vile neighbour Miss Gulch over the activities of dog Toto, and runs away. Fortune-teller Professor Marvel uses his tricks to make her return home, but as she gets there a tornado blows up, and a bit of flying debris knocks her unconscious. In her DREAM the house is blown away, landing in both Technicolor (up to now the movie is in sepia) and OZ, and *on* the Wicked WITCH of the East, killing her – to the delight of the dwarfish Munchkins and the beautiful Glinda, the (good) Witch of the North. But Dorothy, worried for Auntie Em, wants to return to Kansas. Follow the Yellow Brick Road, advises Glinda, to reach the Emerald City, where the WIZARD of Oz will aid you. En route Dorothy encounters and befriends three zanies resembling her aunt's and uncle's three farmhands: the Scarecrow, who has no brain, the Tin Man, who has no heart, and the Lion, who has no courage. The three join her QUEST, each in search of his missing attribute, to the fury of the Wicked Witch of the West, SCRYING on them; she resembles Miss Gulch and wants Dorothy dead, if only for the MAGIC Ruby Slippers she wears, taken from the feet of the Wicked Witch of the East (and, like Hans ANDERSEN's Red Shoes, irremovable before death). EVIL is in due course confounded, and the wizard is shown to be a charlatan: magicless, he has been using TECHNOFANTASY devices to create his imposing persona. He gives the three zanies tokens of the attributes they have discovered in themselves, and offers to take Dorothy back to Kansas by balloon, but – as in a typical anxiety dream – the craft sails without her. However, Glinda appears and tells her that the way home is to tap her heels three times and say

"There's no place like home" – and sure enough it works.

One of the most famous movies of all time, debatably responsible for the continued popularity of Baum's book, the fountainhead of a MYTHOLOGY of its own (as in *"Was . . ."* [*1992*] by Geoff RYMAN, and LABYRINTH [*1986*]). It seems pointless to criticize any particular aspect of *TWOO* since it is the totality of the movie that impresses, rather than any of its component parts. Several of the songs (by Harold Arlen and E.Y. Harburg) have become standards, and numerous lines from them and from the screenplay are now everyday quotations.

This was not the first time the **Oz** stories had been brought to the screen: silents were *Dorothy and the Scarecrow of Oz* (*1910*), *The Land of Oz* (*1910*), *The Patchwork Girl of Oz* (*1914*), *His Majesty, The Scarecrow of Oz* (*1914*), which was produced by Baum himself, *The Magic Cloak of Oz* (*1914*) and *The Ragged Girl of Oz* (*1919*).   [JG]
Further reading: *The Wizard of Oz* (*1989*) by John Fricke, Jay Scarfone and William Stillman; *The Wizard of Oz* (*1992* chap) by Salman RUSHDIE.

2. *Return to Oz* US ANIMATED MOVIE (*1964*). Rankin-Bass/Videocraft/Crawley. **Pr** Jules Bass, Arthur Rankin. **Voice actors** Carl Banas (Dandy Lion/Wizard), Susan Conway (Dorothy), Pegi Loder (Glinda/Wicked Witch), Larry Mann (Tin Man), Alfie Scopp (Socrates the Strawman). *c*50 mins. Colour.

Dorothy returns to help her trio of friends, who have lost the attributes they gained in **1**.   [JG]
3. *The Wizard of Mars* (vt *Alien Massacre*; vt *Horror of the Red Planet*; vt *Horrors of the Red Planet*) US movie (*1964*). **Dir** David L. Hewitt. **Starring** Eve Bernhardt, John Carradine, Roger Gentry, Vic McGee, Jerry Rannow (movie lacks proper credits). 81 mins. Colour.

Obscure low-budget sf revamping of the story in which three men and a woman are crew of the first mission to Mars, crashland, and in their search for oxygen come to the ruins of an ancient civilization, ruled by a solitary survivor – who is the analogue of the wizard.   [JG]
4. *Journey Back to Oz* US ANIMATED MOVIE (copyright 1971 [often listed as 1964 – presumably in confusion with **2**], released 1974). Filmation/EBA/Warner. **Pr** Norm Prescott, Lou Scheimer. **Dir** Hal Sutherland. **Screenplay** Bernard Evslin, Fred Ladd, Prescott. **Based on** *The Marvelous Land of Oz* (*1904*). **Voice actors** Milton Berle (Cowardly Lion), Herschel Bernardi (Woodenhead Stallion III), Paul Lynde (Pumpkinhead), Ethel Merman (Mombi), Liza Minelli (Dorothy), Mickey Rooney (Scarecrow), Rise Stevens (Glinda), Danny Thomas (Tin Man). 90 mins. Colour.

This direct sequel to **1** features rather stylized animation – too many of the characters are too strongly reminiscent of those in other animated movies – done in crudish colours and frequently with static backgrounds; the pacing is erratic due to an excess of songs. The voice track is more interesting: Hamilton, Wicked Witch of the West in **1**, has a cameo role as Aunt Em, while Minelli, Garland's daughter, provides an uncanny replication of Dorothy's spoken (though not singing) voice. General standards, however, are those of a tvm; voicing distinctions go to some of the minor parts, supplied by Mel Blanc, Dallas McKennon and Larry Storch. Another surprise name in the small print is that of Don BLUTH as a layout artist.

In 1978 this movie was transformed into a tvm with the addition of live-action sections featuring Bill Cosby.   [JG]
5. *The Wiz* US movie (*1978*). Universal/Motown. **Pr**

Robert Cohen. **Dir** Sidney Lumet. **Spfx** Al Griswold, Albert Whitlock. **Screenplay** Joel Schumacher. **Based on** stage musical by William Brown and Charlie Smith. **Starring** Lena Horne (Glinda), Michael Jackson (Scarecrow), Mabel King (Evillene, the Wicked Witch of the West), Richard Pryor (The Wiz), Diana Ross (Dorothy), Ted Ross (Cowardly Lion), Nipsy Russell (Tin Man). 134 mins. Colour.

An all-singing, all-dancing, all-black version of the tale, set in NEW YORK amid contemporary pop music. Harlem teacher is frightened of the big wide world. As she puts out the garbage one snowy night, her dog Toto scampers away and she chases him; both are sucked into a nightmare version of the CITY by a tornado. The Munchkins thank her on her arrival: Evermean, the Wicked Witch of the East, had ensorcelled them into being mere graffiti, but now that Dorothy has accidentally killed the witch they can become independent beings again. The URBAN FANTASY continues – for example, the Cowardly Lion has been hiding in one of the stone lions outside the New York Public Library, The Wiz is an unsuccessful politician, and the Wicked Witch of the West runs a sweatshop. [JG]

6. *The Wizard of Oz* Japanese-US ANIMATED MOVIE (*1982*). Toho. **Pr** John Danylkiw. **Exec pr** Alan L. Gleitsman. **Dir** Danylkiw. **Based on** *The Wizard of Oz*. **Voice actors** Lorne Greene (Wizard of Oz), Elizabeth Hanna (Wicked Witch of the West), Aileen Quinn (Dorothy), John Stocker (Tin Man), Wendy Thatcher (Glinda), Billy Van (Scarecrow), Thick Wilson (Cowardly Lion). 78 mins. Colour.

This modest version is not a straight remake of **1** but appears to be derived independently from Baum's novel. Its plot is a pared-down version of **1**'s but with various novel elements. Most are of no great profit and some – e.g., too much time is spent underscoring the bonding between the central group of COMPANIONS – are tiresome. One point of interest is that the Wizard (who resembles Greene, the voice actor, facially) appears to each of the crew in a different guise (a big disembodied head for Dorothy, an ANGEL for the Scarecrow, a rhino for the Tin Man and a "great ball of fire" for the Lion) – i.e., he is a fake SHAPESHIFTER. The animation is limited, but passable; some of the backgrounds are excellent watercolours. [JG]

7. *Return to Oz* US movie (*1985*). DISNEY/Silver Screen Partners II/Oz Productions Ltd/Buena Vista. **Pr** Paul Maslansky. **Exec pr** Gary Kurtz. **Dir** Walter Murch. **Spfx** Ian Wingrove. **Vfx** Peter Krook. **Claymation dir** Will Vinton. **Screenplay** Gill Dennis, Murch. **Based on** *The Land of Oz* and *Ozma of Oz*. **Starring** Fairuza Balk (Dorothy), Jean Marsh (Nurse Wilson/Mombi), Emma Ridley (Princess Ozma), Nicol Williamson (Dr Worley/Nome King). **Voice actors** Sean Barrett (Tik Tok), Denise Bryer (Bellina), Lyle Conway (Gump), Brian Henson (Jack Pumpkinhead). 110 mins. Colour.

Months after **1**, Dorothy is sleeping little, and often babbling about the magical country she has visited. Convinced she is mad, Aunt Em takes her to Dr Worley for primitive (this is 1898) EST. Supervised by sinister Nurse Wilson, Dorothy is about to be treated when a thunderstorm shorts everything out; she and a mysterious nameless girl flee the clinic but are swept away by floods. Next morning Dorothy awakes to find herself in Oz, which is in disrepair. With friends Tik Tok, a mechanical man proud of his lifelessness, and Jack Pumpkinhead, a scarecrow, she destroys the Nome

King – who is vulnerable to eggs – and uses the MAGIC of the Ruby Slippers to restore the Emerald City's *status quo*, also freeing the rightful ruler, Princess Ozma – the little girl at the clinic. Back in Kansas, Dorothy is discovered by Toto sleeping beside the river, having survived drowning. Later, sometimes, Ozma appears to Dorothy in her MIRROR.

RTO was poorly received: it lacks the songs and lightheartedness that people expected, and the whole production has a certain murky muddiness about it. Its real fault, however, was that despite appearances it is not a CHILDREN'S FANTASY: instead, it is a fairly straightforward GENRE FANTASY, complete with various PLOT COUPONS, and there are several strong points of reminiscence between it and *The* NEVERENDING STORY (*1984*), although it is a betterconstructed movie. Balk makes an exquisite Dorothy. *RTO* deserves better than the almost total obscurity into which it vanished soon after release: it has pleasing ambition. [JG]

8. *The Wonderful Wizard of Oz* US ANIMATED MOVIE (*1987*). **Voice actor** Margot Kidder (Narrator). 93 mins. Colour.

Animated remake of **1**. We have been unable to find out anything about this movie except that it exists. A similar lack of information pertains to *The Wonderful Land of Oz* (*1969*), *Oz* (*1976*) and *Twentieth-Century Oz* (*1978*); it is possible that at least one of these may in fact be a documentary about Australia. [JG]

WIZARDS In GENRE FANTASY, a wizard is most often a male, human practitioner of MAGIC. Several hierarchies have been proposed: Lyndon HARDY's *Master of the Five Magics* (**1980**) suggests ascending ranks of thaumaturgist, alchemist, magician, sorcerer and wizard. Other magical titles include: adept, enchanter, hedge-wizard, mage, MAGUS, necromancer (◊ NECROMANCY), shaman (◊ SHAMANISM) and warlock. Alchemists are more often viewed as protoscientists (◊ ALCHEMY). This encyclopedia's preferred term for stage conjurers is MAGICIAN; where magic is unacceptable or TABOO, wizards may pretend they are stage magicians using sleight of hand, as in G.K. CHESTERTON's *Magic* (**1913**) and Stephen BOWKETT's *Spellbinder* (**1985**) (◊◊ ZATARA). Conversely, the eponym of L. Frank BAUM's *The Wonderful Wizard of Oz* (**1900**) is a fake wizard hiding behind "magical" stage effects.

Questions of GENDER and FEMINISM arise: are WITCHES simply female wizards (as warlocks are male witches, and ENCHANTRESSES female enchanters), or are WITCHCRAFT and wizardry profoundly different? The question is variously answered. In Ursula K. LE GUIN's **Earthsea** books, the Roke Island school of magic admits no women, and women's magic is proverbially "weak" and "wicked"; Terry PRATCHETT's *Equal Rites* (**1987**) turns on whether a girl may be admitted to male-only Unseen University for wizardly training, since her magic does not operate in the witch mode expected of women; but Diana Wynne JONES's *Witch Week* (**1982**) simply designates its world's magic users, whatever their sex, as witches, and Diane DUANE's *So You Want to be A Wizard* (**1983**) comfortably accommodates both female and male wizards.

Learning wizardry is apt to involve uncomfortable INITIATIONS and/or RITES OF PASSAGE, like the ordeal by poison through which wizard-to-be Sun Wolf passes in Barbara HAMBLY's *The Ladies of Mandrigyn* (**1984**). For younger trainee wizards, the process is partly a metaphor for accepting adult power and responsibility: in L.E. MODESITT's **Recluce** books, despised youngsters tend to mature with improbable speed and pit themselves single-handed against leading CHAOS wizards.

Careers for major wizards, besides instructing new recruits, are not numerous. Those who manipulate the essential stuff of magic which differentiates FANTASY from mundane narrative should be deeply entangled in the STORY; at the simplest they will directly attack EVIL, like DOCTOR STRANGE in COMICS, or may function as the representative of evil. They may be kings, like cruel Gorice XII in E.R. EDDISON's *The Worm Ouroboros* (**1922**) or the various kings of Piers ANTHONY's **Xanth** (where powerful TALENT is required of all monarchs). Far more frequently they are MENTORS, like MERLIN, or Gandalf in J.R.R. TOLKIEN's *The Lord of the Rings* (**1954-5**); or form a PARIAH ELITE, as in Hambly's *The Rainbow Abyss* (**1991**). Wizards who practise BLACK MAGIC and make PACTS WITH THE DEVIL generally hope for wealth, LOVE, VENGEANCE or – like the initiates in C.S. LEWIS's *That Hideous Strength* (**1945**) – secular power, and will be withered by their activities. Unusually, the slightly withered Theron Ware of James BLISH's *Black Easter* (**1968**) employs black magic to seek scientific knowledge. In Diana Wynne Jones's **Chrestomanci** sequence, the eponymous enchanter works for an ALTERNATE-WORLD English government to prevent exploitation of ordinary folk *via* magic; in the society of Randall GARRETT's **Lord Darcy** stories, wizards replace our world's scientists, with the Dr Watson character Master Sean o Lochlainn being a forensic sorcerer.

Historically, noted scholars often attracted LEGENDS of wizardhood: examples include VIRGIL, Roger Bacon (◊ James BLISH), Michael Scot (◊ Michael Scott ROHAN; Sir Walter SCOTT), and the original of FAUST. [DRL]

See also: BARDS; SMITHS; SORCERER'S APPRENTICE.

WIZARDS US ANIMATED MOVIE, with some live-action and archive footage (**1977**). 20th Century-Fox/Bakshi Productions. **Pr** Ralph BAKSHI. **Dir** Bakshi. **Screenplay** Bakshi. **Voice actors** James Connell (President), Steve Gravers (Blackwolf), Bob Holt (Avatar), David Proval (Peace), Richard Romanus (Weehawk), Jesse Welles (Elinore). 80 mins. Colour.

Two million years after nuclear war devastated the world, the bad (radioactive) areas are populated by mutants while the denizens of FAERIE have regained their rightful heritage in the good areas. Delia, Queen of the Fairies, gives birth to TWINS who, both destined to be WIZARDS, are polar opposites: Avatar is GOOD and Blackwolf EVIL. On Delia's death, years later, Blackwolf assumes he will inherit, but the brothers battle and he loses, fleeing to the heartlands of the mutants. There he revives the lost arts of technology and starts a war, which he loses, against the peace- and MAGIC-loving ELVES and FAIRIES. A second time he calls up an army of beings from HELL, and is this time aided by his discovery of a secret weapon, a movie projector (described as a "dream machine"), with which he can project images of Hitler and the Third Reich to inspire his hordes and terrify his foes. Avatar, now an aged-hippy-type wizard, undertakes with COMPANIONS a QUEST to destroy the projector and thus spare the world a renewed HOLOCAUST. After various adventures – when already the WILD HUNT is over and the hordes of Hell are winning the LAST BATTLE – the brothers face up for a magical confrontation. The projector is destroyed, the spectres from Hell evaporate, and the world is saved.

This short movie is full of riches; the plot is slightly ramshackle but driven well enough by the power of its ALLEGORY and the occasional savagery of its SATIRE; it has much HUMOUR, too, with nice touches of PARODY (and a visual quote from Winsor MCCAY's *Gertie* [**1914**]). The analogies with WORLD WAR II are extended as far as showing divisive enmity among the various forces on the side of GOOD; the image of the swastika is everywhere; DEMONS are garbed as SS officers; and in the finale Blackwolf is identified overtly as a second Hitler. Yet the overall impression of *W* is of beauty. Clearly shot on a limited budget, it makes a strength out of limited-animation techniques. Much effort has been put into the backgrounds; the hand of Ian MILLER, one of the artists involved, is distinctly recognizable. In memory *W* tends to be overshadowed by Bakshi's offering only a year later, the incoherent, rambling LORD OF THE RINGS (**1978**); in truth *W* is, on a modest scale, a highly significant piece of cinematic fantasy (and of TECHNOFANTASY) – not to be equalled in its conceptual ambitions by Bakshi until COOL WORLD (**1992**). [JG]

WIZARDS AND WARRIORS US tv series (1983). Warner Bros./CBS. **Pr** Robert Earll, S. Bryan Hickox, Bill Richmond. **Exec pr** Judith Allison, Don Reo. **Dir** Bill Bixby, Richard Colla, Kevin Connor, James Frawley, Paul Krasny. **Writers** Allison, Reo, Richmond. **Created by** Reo. **Starring** Randi Brooks (Witch Bethel), Jeff Conaway (Prince Erik Greystone), Julia Duffy (Princess Ariel Baaldorf), Tim Dunigan (Geoffrey Blackpool), Tom Hill (King Edwin Baaldorf), Phyllis Katz (Cassandra), Jay Kerr (Justin Greystone), Walter Olkewicz (Marko), Julia Payne (Queen Lattinia Baaldorf), Duncan Regehr (Prince Dirk Blackpool), Clive Revill (Wizard Vector), Ian Wolfe (Wizard Traquil). 8 1hr episodes. Colour.

This SWORD-AND-SORCERY adventure series is set in a GOLDEN AGE. The kingdom Camarand is imperilled by Prince Dirk Blackpool, ruler of a nearby land, who uses black MAGIC and military force in his efforts to invade. Blackpool is constantly thwarted by the good Prince Erik Greystone, who is engaged to the king's daughter.

Each episode featured some sort of magical element. Many were conjured up by Vector, an evil WIZARD in the service of Blackpool. Unknown to Blackpool, though, was Vector's secret desire to seize the throne himself, a factor that made some of his SPELLS less than reliable. Throughout, Greystone could be counted on to thwart Blackpool's and Vector's plans, and those of the other wizards, witches and demons encountered along the way. [BC]

WODEN ◊ ODIN.

WODEWOSE ◊ Robert HOLDSTOCK; MYTHAGO.

WOLF Wolves typify dangerous wildness. The Little Red Riding-Hood FAIRYTALE provides a classic example; the creatures' mere presence in the ALTERNATE-WORLD Britain of Joan AIKEN's *The Wolves of Willoughby Chase* (**1962**) is redolent of WRONGNESS. In DANTE's *Inferno* the wolf symbolizes betrayal, with the "sins of the wolf" being punished in the inmost circles of HELL; in the Norse myth of RAGNAROK, the wolf Fenrir (Fenris) will ultimately eat the MOON. Children suckled or raised by wolves are likely to become HEROES, like Romulus and Remus (◊ MYTH OF ORIGIN) and Rudyard KIPLING's Mowgli. Wolves, often under the name "Wargs", appear as servants of EVIL throughout J.R.R. TOLKIEN's sagas of Middle-Earth. Good wolves are rare, though a TALKING-ANIMAL specimen features in Gordon R. DICKSON's *The Dragon and the George* (**1976**). [DRL]

See also: WEREWOLVES.

WOLF US movie (*1994*). Columbia. **Pr** Douglas Wick. **Exec pr** Robert Greenhut, Neil Machlis. **Dir** Mike Nichols. **Spfx** Stan Parks, Daniel A. Sudick. **Vfx** Eric Brevig, Scott Farrar, John Nelson. **Mufx** Rick Baker. **Screenplay** Harrison, Wesley Strick. **Starring** Jack Nicholson (Will Randall), Michelle Pfeiffer (Laura Alden), Christopher Plummer (Raymond Alden), Om Puri (Dr Vijay Alezais). 122 mins. Colour.

Driving back from Vermont to New York in a storm, publishing editor Randall knocks over a wolf in the road; it proves not to be dead, and bites him before fleeing. Back at the office, amid corporate backstabbing, he finds he has enhanced senses. His hyperacute sense of scent reveals that his wife is conducting an affair, and he ditches her, soon finding himself drawn in by Laura, the spoilt, rebellious daughter of the plutocrat who has taken over Randall's publishing company. Yet at nights, as the Moon grows full, Randall finds himself taking half-remembered nocturnal excursions, during which he is capable of lupine athletic feats – and lupine savagery.

This is an elegant attempt to graft the European WEREWOLF legend onto US URBAN FANTASY – an attempt that, despite *W*'s cool reception, succeeds well. Nicholson shows occasional signs of trying to lurch into the hammishness that came close to destroying *The* SHINING (*1980*), but generally the transformation scenes are extremely effective, drawing on body language, slow-motion photography and only a modicum of mufx. Pfeiffer for once acts the part of a woman rather than an overgrown girl. The gory ending steers clear of splatter-movie corniness because of the comparative restraint of what has gone before. Even Randall's *de rigueur* consultation with a scholarly werewolf expert (Alezais), a risible element of most such movies, attains a certain dignity through skilled handling of the characters' motivations.

The sf/horror/fantasy genre publishers Tor Books are specially credited for assistance: it is hard to believe that the office cauldron of Randall's company is based on Tor. [JG]

WOLFE, GENE (RODMAN) (1931-   ) US writer, one of the central figures of 20th-century sf and an important author of fantasy. *The Book of the New Sun* (**1980-83**) is a difficult text which has a pervading fantasy "tone", though its underpinning is sf; it may nevertheless be treated as a tale depicting that which the secular world cannot offer (i.e., as a CHRISTIAN FANTASY) if the protagonist Severian's true identity is deemed to be that of CHRIST reborn. The Universe described in *The Urth of the New Sun* (**1987** UK) could be read literally as that depicted in the CABBALA. The meaning of the sequence as a whole is intricately layered, and not easily amenable to genre downloading; but in the end it does seem primarily to conduct itself in terms of SCIENCE FICTION. That said, it is interesting how often in this encyclopedia *The Book of the New Sun* turns up as a cited example of fantasy: it is an important CROSSHATCH. Some of the short volumes published separately – and presented as constituting tales from *The Book of the Wonders of Urth and Sky*, one of the fictional BOOKS central to the main sequence – are fantasies. They include *The Boy who Hooked the Sun* (**1985** chap) and *Empires of Foliage and Flower* (**1987** chap). The latter in particular is a tale of strong interest, featuring a GOD-like Father Thyme (◊ CYCLES; SEASONS) who travels westward over the world, always older; if he turns eastwards, he becomes younger. The girl he takes with him undergoes a complex RITE OF PASSAGE, but is eventually returned to childhood.

GW began publishing work of genre interest with "The Dead Man" for *Sir* in 1965, and concentrated on sf for the first decade of his career. His first fantasy novel, *Peace* (**1975**), may be his finest. Intricately told through nests of interconnecting STORIES – none, significantly, is ever concluded – and seemingly from the viewpoint of a man in late middle age reflecting with some amiability on his life in a small town somewhere in the US Midwest, it is in fact a POSTHUMOUS FANTASY whose implications are complexly appalling. Dennis Weer, the narrator, who when his narrative begins has been dead for many years, turns out to have been at the epicentre of an unknown (but not small) number of deaths, and may have actually murdered several of the novel's characters, whose lives are anyway truncated by the failure of any of the embedded stories actually to end. The elegy evoked by the novel has, therefore, a singularly uneasy affect; and any peace obtainable by the protagonist is very much that of the grave.

The atmosphere of *The Devil in a Forest* (**1976**) is consistent with fantasy, and echoes of the story of King Wenceslas generate a sense of the TWICE-TOLD throughout, but there are no supernatural elements. Though echoes of L. Frank BAUM's *The Wonderful Wizard of Oz* (**1900**) permeate *Free Live Free* (**1984**; rev 1985 UK), the tale is perhaps best understood as an sf exercise in the intricacies of TIME paradoxes. *Soldier of the Mist* (**1986**) and its sequel *Soldier of Arete* (**1989**) make up the open-ended (and perhaps incomplete) **Latro** sequence, set in pre-Classic Greece. It is narrated by a warrior who has been punished – by a goddess or *the* GODDESS – for a sin which he cannot remember, and which the reader cannot decipher. His CURSE is to have his memories wiped at the end of every day or waking period; his response is to write down everything he can remember at the end of each day – these scripts make up the text of the two novels – and to use what he has written as an aide mémoire. Eventually he is given some godly assistance, and seems on the verge of creating a universe of memory by treating all the events he lives through – and therefore the text of the novels – as a theatre of memory.

The protagonist of *There Are Doors* (**1988**) is again haunted by the Goddess, travelling between ALTERNATE REALITIES in his search for her, visiting increasingly bleak versions of the same city (◊ URBAN FANTASY) which constitute a series of RECOGNITIONS that he must travel through; in the end, it seems he is given a chance to emigrate, finally, in her direction. (The novel can instead be read as a fantasy of PERCEPTION, with the alternate realities being born from the mind of the protagonist.) *Castleview* (**1990**), set in present-day Illinois (◊ CONTEMPORARY FANTASY), unfolds a complex CROSSHATCH with a FAERIE occupied by MORGAN LE FAY, who is in search of an ARTHUR to end the THINNING of Faerie and to continue the great STORY. Neither of these last fantasies has attracted the attention given to *The Book of the New Sun* or to GW's more recent sequence, **The Book of the Long Sun** (see below), but each constitutes a significant contribution to modern fantasy. As in sf, his influence on other writers is pervasive, but hard to pinpoint. The clearest influence – in fantasy and in sf both – is almost certainly the model of the first-person confessional memoir presented in *The Book of the New Sun*; recent fantasy novels which unmistakeably show the effects of this model include *Assassin's Apprentice* (**1995** UK) by Megan LINDHOLM (as Robin Hobb) and *The Mask of the Sorcerer* (**1995**) by Darrell SCHWEITZER.

Several of GW's best-known stories – including *The Hero as Werwolf* (in *The New Improved Sun* anth **1975** ed Thomas M. DISCH and Charles Naylor; **1991** chap) and "The Detective of Dreams" (1980) – are fantasy, as are some of the tales incorporated into *Bibliomen: Twenty Characters Waiting for a Book* (coll **1984** chap; rev vt *Bibliomen: Twenty-Two Characters in Search of a Book* 1995 chap). Other fantasy titles include *At the Point of Capricorn* (**1984** chap), *The Arimaspian Legacy* (**1987** chap), *Slow Children at Play* (**1989** chap) and *The Old Woman whose Rolling Pin is the Sun* (**1991** chap).

In fantasy, as in sf, GW's originality lies in the sense that – more thoroughly than almost any of his contemporaries – he is in the process of finishing the stories he tells. His work signals the late maturity of the genres he graces. [JC]

Other works (sf): *Operation ARES* (**1970**); *The Fifth Head of Cerberus* (fixup **1972**); *The Island of Doctor Death and Other Stories and Other Stories* (coll **1980**), not to be confused with *The Death of Doctor Island* (in *Universe 3* anth **1973** ed Terry CARR; **1990** chap dos); *The Book of the New Sun*, being 4 main titles making up one novel, *The Shadow of the Torturer* (**1980**) and *The Claw of the Conciliator* (**1981**), both assembled as *The Book of the New Sun, Volumes I and II* (omni **1983** UK; vt *Shadow and Claw* 1994 US), and *The Sword of the Lictor* (**1982**) and *The Citadel of the Autarch* (**1982**), both assembled as *The Book of the New Sun, Volumes III and IV* (omni **1985** UK; vt *Sword and Citadel* 1994 US), plus *The Castle of the Otter* (coll dated 1982 but **1983**); *Gene Wolfe's Book of Days* (coll **1981**); *The Wolfe Archipelago* (coll **1983**); *Plan[e]t Engineering* (coll **1984**); *Storeys from the Old Hotel* (coll **1988** UK); *For Rosemary* (coll **1988** chap UK), poetry; *Endangered Species* (coll **1989**); *Seven American Nights* (in *Orbit 20* anth **1978** ed Damon KNIGHT); *Pandora by Holly Hollander* (**1990**), associational; *Letters Home* (coll **1991**), nonfiction; *Castle of Days* (coll/omni **1992**), containing *Gene Wolfe's Book of Days*, *The Castle of the Otter* and additional matter; *The Young Wolfe* (coll **1993**); *Orbital Thoughts* (coll **1993** chap); **The Book of the Long Sun** series, so far comprising *Nightside the Long Sun* (**1993**), *Lake of the Long Sun* (**1994**), *Caldé of the Long Sun* (**1994**) and «Exodus from the Long Sun» (1996).

WOLFEN US movie (*1981*). Orion/Warner. **Pr** Rupert Hitzig. **Exec pr** Alan King. **Dir** Michael Wadleigh. **Spfx** Conrad Brink, Ronnie Ottesen. **Vfx** Robert Blalack. **Mufx** Carl Fullerton. **Screenplay** David Eyre, Wadleigh. **Based on** *The Wolfen* (1978) by Whitley Strieber (1945-   ). **Starring** Albert Finney (Dewey Wilson), Gregory Hines (Woody Whittington), Tom Noonan (Ferguson), Edward James Olmos (Eddie Holt), Dick O'Neill (Police Chief Warren), Diane Venora (Rebecca Neff). 115 mins. Colour.

For 20,000 years the Native Americans and the wolves, or Wolfen, co-existed, organized societally along similar lines; but then the White Man came and virtually exterminated both. The cleverest Wolfen concealed themselves in the slums of the White Man's cities, where they survived in small family units by scavenging garbage and preying on derelicts. Now the territory of one NEW YORK Wolfen family is threatened by a housing project, and they begin to kill humans to defend it. Law officers Wilson, Neff and Whittington are put on the case by Warren, and slowly, incredulously, they uncover the truth. In a final confrontation Wilson destroys a model of the housing project, and the watching Wolfen vanish into thin air – although they still roam somewhere out there in the CITY.

Strieber's novel was an attempt to draw sf out of URBAN LEGEND: the Wolfen are a cryptozoological species. The movie – a much more stylish affair than the novel – is almost throughout a demystifying exercise: these are very clever wolves, but wolves nonetheless. Yet the final, enigmatic vanishing throws everything wide open in a classic SLING-SHOT ENDING: in what becomes with hindsight an URBAN FANTASY, are the Wolfen conjurations of Wilson's delusions, fuelled by alcohol and Native-American legends, or are they supernatural creatures? [JG]

WOLF MAN, THE US movie (*1940*). ◊ MONSTER MOVIES; WEREWOLVES.

WOLLHEIM, DONALD A(LLEN) (1914-1990) US publisher, editor and occasional writer. In the first two roles he had a considerable influence on the development and marketing of commercial SCIENCE FICTION and GENRE FANTASY, particularly as editor at Ace Books 1952-71, and then with his own DAW Books from 1972. It was DAW who helped establish HIGH FANTASY as a marketable commodity in the late 1960s, when he published an unauthorized edition of TOLKIEN's *The Lord of the Rings* in June 1965. (He was within his rights: the book had not been lodged for copyright in the USA.) Ballantine Books, about to publish the authorized and revised edition, took legal action, and DAW eventually made payment and withdrew the Ace editions, but not before the publicity had drawn considerable attention to them. DAW rapidly developed a quota of fantasy at Ace Books, and even more extensively at DAW Books, in particular the works of Marion Zimmer BRADLEY, Lin CARTER, Jo CLAYTON, C.J. CHERRYH, Tanith LEE, Brian LUMLEY, Michael MOORCOCK, John NORMAN, Andre NORTON, Jennifer ROBERSON, Michael SHEA and Thomas Burnett SWANN. Many of these works fall into the category of PLANETARY ROMANCE or SWORD-AND-SORCERY. DAW also inaugurated the series the **Year's Best Horror Stories** in 1972, edited in turn by Richard Davis (1945-   ), Gerald W. Page (1939-   ) and Karl Edward WAGNER, and the **Year's Best Fantasy Stories** edited first by Lin Carter and then by Art Saha (1923-   ).

DAW's involvement with SUPERNATURAL FICTION and FANTASY dates back to his days as an sf fan when he corresponded with H.P. LOVECRAFT. He was already producing minor fan broadsheets, but in 1935 he took over a club magazine and retitled it *The Phantagraph*, which he then edited on-and-off for 11 years (1935-46). In its heyday this was the leading amateur magazine of supernatural and fantasy fiction, including stories, poems and articles by many of the leading genre writers of the day (Lovecraft, Robert E. HOWARD, Robert BLOCH, Abraham MERRITT) and many that would establish themselves over the decade (James BLISH, Cyril M. Kornbluth [1923-1958], Henry KUTTNER, Robert W. Lowndes and Emil Petaja). DAW later assembled a volume of mementoes from the magazine, *Operation: Phantasy* (anth **1967** chap), and published a semi-professional magazine of supernatural fiction, *Fanciful Tales of Space and Time* (1 issue October 1936). This gave DAW some editorial experience, which allowed him to take on the suicidally underfinanced professional magazines, *Stirring Science Stories* (4 issues February 1941-March 1942) and *Cosmic Stories* (3 issues March-July 1941). *Stirring* had its own separate fantasy section with the running head *Stirring Fantasy Fiction*, which published some fine stories by David H. KELLER, Kornbluth and Clark Ashton SMITH.

When these MAGAZINES folded, DAW moved to Ace

Magazines, where he edited a variety of pulps. He also edited the first sf paperback anthology, *The Pocket Book of Science Fiction* (anth **1943**). At the end of 1946 DAW contacted Avon Books, who were establishing themselves as a paperback publisher, and this resulted in the first of what became a long series of digest-sized anthologies, the **Avon Fantasy Reader**. The first *Avon Fantasy Reader* appeared in February 1947 and the series ran for 18 volumes in total (◊ AVON FANTASY READER). Its success encouraged other publishers to enter the digest field and contributed heavily to the emergence of genre sf and fantasy. In addition to various magazines for Avon, DAW also compiled the first all-original anthology of fantasy and sf, *The Girl with the Hungry Eyes and Other Stories* (anth **1949**), featuring stories about various FEMMES FATALES, plus *The Fox Woman and Other Stories* (coll **1949**), the only volume of A. Merritt's short fiction. 20 years later George Ernsberger, the then editor at Avon Books, worked with DAW to compile two retrospective selections from the series, *The Avon Fantasy Reader* (anth **1969**) and *The 2nd Avon Fantasy Reader* (anth **1969**).

DAW returned to Ace Books in 1952, establishing its sf imprint and introducing the famous **Ace Double** dos-a-dos format. For Ace he edited over 20 anthologies, of which those with a supernatural or fantasy content are *The Macabre Reader* (anth **1959**), *More Macabre* (anth **1961**) and *Swordsmen in the Sky* (anth **1963**), a volume of PLANETARY ROMANCES. He also compiled for another publisher *Terror in the Modern Vein* (anth **1955**; cut in 2 vols as *Terror in the Modern Vein* 1961 UK and *More Terror in the Modern Vein* 1961 UK), one of the first anthologies to highlight the modern approach to HORROR. In all of DAW's anthologies there is a wide appreciation of the field beyond the genre magazines.

DAW's own fiction is mostly sf, although he contributed some effective supernatural horror stories to his own magazines, to F&SF and to MAGAZINE OF HORROR, often under the pen names David Grinnell, Millard Verne Gordon and Martin Pearson. Some are in *Two Dozen Dragon's Eggs* (coll **1969**), *The Men from Ariel* (coll **1982**) and *Up There and Other Strange Directions* (coll **1988**). DAW's own reflections upon his career are in *The Universe Makers* (**1971**). [MA]

WOLVES ◊ WOLF.

WONDERLAND The paradigm is the UNDERGROUND world in Lewis CARROLL's *Alice's Adventures in Wonderland* (**1864**), closely followed by the MIRROR in his *Through the Looking-glass* (**1871**). Wonderlands are worlds based on logical rules. Alice's Wonderland operates according to precepts which represent a rigorous working-out of various propositions carried to a point of absurdity. If Carroll's domain feels at times nightmarishly flimsy, it does so because its REALITY is tied to propositions. A Wonderland may therefore at any point be *refuted*. Indeed, that is precisely how Alice escapes in the end: by drawing attention to the rules which fabricate the world, at which point the dream palaces collapse into a pack of cards.

Because the original Wonderland was generated by rules, it has proved remarkably easy to generate PARODY Wonderlands through the application of modified sets of rules. Parodies and SATIRES abound, and AESOPIAN FANTASIES are not infrequent. Among the many parodies of *Alice in Wonderland* are John Kendrick Bangs's *Alice in Blunderland* (**1907**), Huang Chun-Sin's *Alice in Manialand* (**1959** chap), Max Kester's and James Dyrenforth's *Adolf in Blunderland* (**1939**), Sam J. Lundwall's *Alice's World* (**1971**),

Hercules Molloy's *Oedipus in Disneyland: Queen Victoria's Reincarnation as Superman* (**1972**) and Horace Wyatt's *Malice in Kultureland* (**1914**), while Emma TENNANT's *Alice Fell* (**1980**) treats Wonderland as an askew gloss upon the UNDERWORLD with which Persephone must cope. Gilbert ADAIR's *Alice Through the Needle's Eye* (**1984**), a homage as much as a continuation, creates a Wonderland of its own (◊ SEQUELS BY OTHER HANDS).

Fantasies built around QUESTS – especially if their authors have a satirical bent – may readily incorporate Wonderland societies, rule-bound POLDERS visited by protagonists in their search for truth, and soon escaped from. The inhabitants of the colour-coded (◊ COLOUR-CODING) land of the Quirks, in Nancy KRESS's *The Prince of Morning Bells* (**1981**), determine reality according to their Model of Forces, and literally cannot see phenomena which violate the rules. The multicoloured LIMINAL BEINGS who guide the protagonist to his quest's end in Salman RUSHDIE's *Haroun and the Sea of Stories* (**1990**) are Wonderland dwellers, as are the CHESS figures who inhabit the IMAGINARY LAND of Antipodes in *The Chess Garden, or The Twilight Letters of Gustav Uyterhoeven* (**1995**) by Brooks Hansen (1965-   ). The Dream-generated pocket worlds in Jonathan Lethem's *Amnesia Moon* (**1995**) are, like Kress's, colour-coded, and reality within each enclosure is determined *a priori* by a central dreamer, whose oneiric diktat constitutes a set of arbitrary rules for his victims. [JC]

WONDER TALE In her introduction to *Wonder Tales: Six Stories of Enchantment* (anth **1994**), Marina WARNER argues for the use of "wonder tales" as an umbrella term to cover FOLKTALES, FAIRYTALES, MÄRCHEN and other categories of story that celebrate the world as magical. Some of these categories are oral; others, like the literary fairytales assembled in *Wonder Tales* itself are highly selfconscious types of written literature.

Making the best of a bad distinction, Jack ZIPES, in his introduction to *Spells of Enchantment: The Wondrous Fairy Tales of Western Culture* (anth **1991**; vt *The Penguin Book of Western Fairy Tales* 1993 UK), suggests that the term WT best describes earlier, oral forms of the tale of enchantment, and that, while "many oral wonder tales had been concerned with the humanization of natural forces, the literary fairytale, beginning with CUPID AND PSYCHE, shifted the emphasis more toward the civilization of the protagonist", and toward his or her making a life in a social world. Zipes's choice of subtitle does, however, give the game away. [JC]

WONDER WOMAN Dark-haired US COMIC-book superheroine dressed in star-spangled blue shorts and a low-cut, strapless red top with a gold eagle motif, red high-heeled boots and a gold tiara. She also wears gold, bullet-deflecting bracelets which, if chained together by a man, become "bracelets of submission", placing her under his control. She was created by psychologist William Moulton Marston – the inventor of the polygraph – under the pseudonym Charles Moulton, and, along with BATMAN and SUPERMAN, is one of DC COMICS' most enduring characters. Her first appearance was in *All Star Comics #8* (December/January 1941-2) where the beginning of her origin story (completed in *Sensation Comics #1* 1942) is set on Paradise Island, a lost world of heroic immortal women, ruled by the wise and courageous Queen Hippolyte. They see the future threatened by world war and cause US army officer Steve Trevor to crashland on their island, so that one of their number can return with him to "help fight the forces of hate and oppression". Queen Hippolyte's daughter, Diana, falls in love with Trevor and is chosen to accompany him. With an invisible robot plane and a golden lasso that gives her control over anyone she ensnares with it, she battles the Nazi menace. She assumes the identity of plain, bespectacled Diana Prince as a cover and is aided by a group of adolescent girls – the Holliday Girls.

Marston used the character to express and explore ideas about male and female relationships, psychological theories and philosophical notions. WW's list of adversaries included some colourful villains and villainesses, including Gestapo agent Paula von Gunther, Hypnota the Great – a woman posing as a man, who mentally enslaves her own sister – and Dr Psycho, a psychopathic madman who hypnotizes his fiancée to gain her assistance in his plan to dominate and control all women. It was themes such as this that led Frederick Wertham (1895-1981) to cite WW as "one of the most harmful" comic books in his *Seduction of the Innocent* (**1953**).

WW was drawn with a touch of whimsy by H.G. Peter (? -1948), who took on a team of assistants to help with the art, which he continued to produce until 1950, when he was abruptly fired; he died soon afterwards. Marston wrote the stories until his death in 1947, after which, written by Robert Kanigher (1915-) and drawn by Ross Adru (1925-) and Mike Esposito, WW became a more routine superheroine adventure comic book, with more emphasis on her gadgets and her invisible plane.

In 1968 (*Wonder Woman #179*) Mike Sekowski took over both script and art. He re-clad WW in a trendy jumpsuit, killed off Steve Trevor and introduced an ancient Chinese mystic, I Ching. WW became a katate-chopping feminist icon. This phase was short-lived: she returned to being the costumed superheroine in *Wonder Woman #204* (1973). Other efforts were made to breathe new life into the character in the 1980s, most notably with writer Kurt Busiek's (1960-) and artist Trina Robbins's attempt to return to Marston's conception (in *The Legend of Wonder Woman* 1986). A new series of WW comic books began in 1987 with art by George Perez (1954-).

The feminist magazine *Ms* published a book-length vintage collection with a perceptive introduction by Gloria Steinem and an "interpretive essay" by feminist psychologist Dr Phyllis Chester entitled *Wonder Woman* (graph **1972**). [RT]

See also: *The* ADVENTURES OF WONDER WOMAN (1976-9); *The* NEW, ORIGINAL WONDER WOMAN (*1975* tvm); WONDER WOMAN (*1974* tvm).

WONDER WOMAN US movie (*1974* tvm). ABC/Warner. **Pr** John G. Stephens. **Exec pr** John D.F. Black. **Dir** Vincent McEveety. **Screenplay** Black. **Starring** Cathy Lee Crosby (Diana Prince/Wonder Woman), Kaz Garas (Steve Trevor), Ricardo Montalban (Abner Smith), Andrew Prine (George Calvin). *c*75 mins. Colour.

Sent from an all-woman island community of immortals to persuade the outside world that it should recognize women's sensitivity over the crassness of men, Diana becomes secretary to Trevor (who knows much of her secret) at the War Department, taking the name Diana Prince. She recovers stolen lists of spies from international arch-crook Smith and his psychopathic sidekick Calvin. The plot is complicated by Angela, another woman from the island, who has come into our world intent on crime. The

fantasy elements in this cheaply made movie almost always occur off-screen; WW's superpowers are restricted to keen vision, skill in martial arts and strongness. Aside from the fantasy premise, this is barely distinguishable from other trendy spy/crime capers of the time. [JG]

WONGAR, B(IRIMBIR) Pseudonym of Serbian-born writer Sreten Bozic (1932-), from 1960 in Australia, where he spent several years living with the Aborigines, eventually taking an Aboriginal name for his writing, all of which reflects and meditates upon their life. His first novel, *The Trackers* (**1975**), makes explicit metaphorical use of his own situation, being the tale of a white man who awakens one morning with black skin, encounters Aborigines, identifies with them, and begins to understand the "darkness" of their fate. The **Nuclear Trilogy** – *Walg* (**1983**), *Karan* (**1985**) and *Gabo Djara* (**1987**) – carries on the theme of METAMORPHOSIS. In the first volume an Aboriginal woman escapes the white world and gives birth to her child in the true LAND. In the second a male Aborigine similarly escapes, and is transformed into a TREE. The third recasts the whole history of loss through the aeon-spanning story of the eponymous ant, who manifests the land. [JC]

Other works: *Aboriginal Myths* (anth **1972**) as Streten Bozic with Alan Marshall (1902-1984); *The Track to Bralgu* (coll **1978**); *Babaru* (coll **1982**); *Marngit* (coll **1992**).

WOOD, BRIDGET (1947-) UK writer whose first novels – beginning with *Mark of the Fox* (**1982**) – were historical. She started producing work of genre interest with the **Lute** sequence of SUPERNATURAL FICTIONS – *The Minstrel's Lute* (**1987**) and *Satanic Lute* (**1987**). More recently, as Frances Gordon, she has published a VAMPIRE novel, *Blood Ritual* (**1994**). The **Wolfking** series – *Wolfking* (**1991**), *The Lost Prince* (**1992**) and *Rebel Angel* (**1993**) in the main sequence, plus *Sorceress* (**1994**) – is of direct fantasy interest. The first three feature one or another kind of TIMESLIP by means of which characters from the future travel into a LAND OF FABLE something like 3rd-century Ireland, whose rightful rulers, members of the dynasty founded by Cormac mac Airt, must constantly strive both to regain power wrested from them by forces of Darkness and to maintain their bloodline, as they are part WOLF. Elements of DARK FANTASY – primarily through an emphasis on imaginative and/or invasive SEX – intermingle with touches of SCIENCE FANTASY, for there seems to be an inhabited asteroid in the works. *Sorceress*, set rather later, is a tale more central to fantasy as normally conceived, dealing as it does with the THINNING of pagan Ireland under the onslaught of Christianity.

BW's later work shows a growing subtlety. [JC]

WOOD, WALLACE (1927-1981) Popular US COMIC-book artist with a tight, firm, brush-line style and a special flair for drawing beautiful women and space opera. WW studied at the Minneapolis School of Art and at Burne HOGARTH's Cartoonists' and Illustrators' School, then entered comics as a letterer before turning to artwork. His most significant early work involved three weeks of Sunday newspaper strips for Will EISNER's *Spirit* feature. He worked for several comic-book publishers before joining EC COMICS at the time when Gaines and Feldstein were introducing their **New Trend** line of horror and sf books. He also drew copiously for *Mad* magazine, packing his pages with a multitude of sight gags and idiosyncratic details: his work appeared in all of the first 12 issues. He drew spot illustrations for MAGAZINES like *Galaxy* and worked for all the major comic-book companies. However, he alternated these periods of intensive, round-the-clock working with equally intense periods of heavy drinking, with the result that both his health and his work suffered. He nevertheless continued to work for MARVEL COMICS on **The Avengers** and **Daredevil** and created and drew *T.H.U.N.D.E.R. Agents* (1965) for Tower.

WW created the "alternative comic book" *Witzend* in which he did several experimental pieces including the **Wizard King** saga and *The Adventures of Sally Forth*, a fantasy version of *Playboy*'s **Little Annie Fanny**.

In his final years he worked sporadically as an inker for DC and drew stories for a pornographic comic book. He committed suicide.   [RT]

WOODCOTT, KEITH Pseudonym of John BRUNNER.

WOODFORDE, CHRISTOPHER (1907-1962) UK writer. ◊ GHOST STORIES FOR CHILDREN; JAMES GANG.

WOODROFFE, PATRICK (1940-   ) UK artist, designer, etcher, painter and illustrator of fantasy and sf subjects with a meticulous, obsessive hatch-and-stipple line style and a finely detailed, richly coloured painting style. His work is fanciful and often poetic, with elements of the surreal, and with images of pubescent girl dolls in lace underwear a recurring motif. His influences include Salvador DALÍ and Hieronymous BOSCH, and his early work shows his affinity with the 1960s illustrator Alan ALDRIDGE. He is also a typographer.

PW became a freelance artist and illustrator in 1972. He has produced hundreds of book and magazine covers and record-sleeve designs. Books of his work include: *Mythopoeikon* (graph **1976**; cut vt *P.W.* 1994); *The Adventures of Tinker, the Hole-Eating Duck* (graph **1979**), juvenile; *Pentateuch* (graph **1980**; rev vt *The Second Earth: The Pentateuch Retold* 1987) with text by Dave Greenslade; *Hallelujah Anyway* (coll **1984**), poetry; *Micky's New Home* (graph **?1985**), juvenile; *The Dorbott of Vacuo, or How to Live with the Fluxus Quo* (graph **1987**); and *Pastures in the Sky* (graph **1993**).   [RT]
Further reading: *A Closer Look* (graph **1986**) by PW examines his techniques.

WOODS ◊ FORESTS.

WOOLF, (ADELINE) VIRGINIA (1882-1941) UK author and critic. Most of her works lie outside the fantasy genre. A member of the Bloomsbury Group, she was a leading exponent of stream-of-consciousness writing. Her exploration of inner life brought a new slant to interpretations of fictive reality, and has had a significant influence on modern literary fiction. Among her novels, only *Orlando* (**1928**) is fantasy. It recounts the career through five centuries of Orlando, born a man in the reign of Elizabeth I. Falling into a series of coma-like sleeps, Orlando journeys forward through TIME, transcending the limitations of mortality. Through the character of Orlando, VW confronts the transient nature of human mores and customs and the fluidity of human sexuality. As a youth, Orlando is beloved by the aged Queen Elizabeth, and loves a reckless, heartless Russian princess: experience does not automatically confer either comprehension or wisdom, and human relations are often thwarted by misunderstanding, ignorance and confusion. Transformed during one sleep-period into a woman, Orlando chafes against socially imposed constructions of femininity and female behaviour. Vigorous, unusual and provocative, *Orlando* is a landmark book in the borderland of the fantasy genre. It was filmed as ORLANDO (*1992*).   [KLM]
Further reading: *Virginia Woolf: A Biography* (1974) by Quentin Bell.

WOOLLEY, PERSIA (1935-   ) US author of Arthurian fiction. *Child of the Northern Spring* (**1987**), *Queen of the Summer Stars* (**1990**), and *Guinevere: The Legend in Autumn* (**1991**) follow the first-person narrative of GUINEVERE from childhood to old age. PW attempts to amalgamate as many as possible of the surviving medieval Arthurian tales into her story: the result is a breathless jumble of characters, cultural tropes and themes. These blend uncomfortably with pervasive and intrusive late-20th-century ideas of feminism, morality and political responsibility, lending an uneven tone to the whole.   [KLM]

WORLD FANTASY AWARD These awards have been presented annually at the World Fantasy Convention since 1975. Two places on the ballot in each category are reserved for nominations by members of both the current convention and the preceding year's; further nominations are supplied by a panel of five selected judges; the final voting is by members of the current convention. The trophy itself was designed by Gahan WILSON.   [JG/DRL]

Novel

1975: *The Forgotten Beasts of Eld* by Patricia A. MCKILLIP.
1976: *Bid Time Return* by Richard MATHESON.
1977: *Doctor Rat* by William KOTZWINKLE.
1978: *Our Lady of Darkness* by Fritz LEIBER.
1979: *Gloriana* by Michael MOORCOCK.
1980: *Watchtower* by Elizabeth A. LYNN..
1981: *The Shadow of the Torturer* by Gene WOLFE.
1982: *Little, Big* by John CROWLEY.
1983: *Nifft the Lean* by Michael SHEA.
1984: *The Dragon Waiting* by John M. FORD.
1985: *Mythago Wood* by Robert P. HOLDSTOCK and *The Bridge of Birds* by Barry M. HUGHART.
1986: *Song of Kali* by Dan SIMMONS.
1987: *Perfume* by Patrick SUSKIND.
1988: *Replay* by Ken GRIMWOOD.
1989: *Koko* by Peter STRAUB.
1990: *Madouc* by Jack VANCE.
1991: *Thomas the Rhymer* by Ellen KUSHNER and *Only Begotten Daughter* by James MORROW.
1992: *Boy's Life* by Robert R. McCammon (1952-   ).
1993: *Last Call* by Tim POWERS.
1994: *Glimpses* by Lewis SHINER (1950-   ).
1995: *Towing Jehovah* by James Morrow.

Lifetime Achievement

1975: Robert BLOCH.
1976: Fritz LEIBER.
1977: Ray BRADBURY.
1978: Frank Belknap LONG.
1979: Jorge Luis BORGES.
1980: Manly Wade WELLMAN.
1981: C.L. MOORE.
1982: Italo CALVINO.
1983: Roald DAHL.
1984: L. Sprague DE CAMP, Richard MATHESON, E. Hoffman PRICE, Jack VANCE, Donald WANDREI (this year the award was voted by previous recipients).
1985: Theodore STURGEON.
1986: Avram DAVIDSON.
1987: Jack FINNEY.
1988: E.F. BLEILER.
1989: Evangeline WALTON.
1990: R.A. LAFFERTY.
1991: Ray Russell (1924-   ).
1992: Edd Cartier (1914-   ).

1993: Harlan ELLISON.
1994: Jack WILLIAMSON.

Best Anthology/Collection
(From 1988 this became two awards, one for anthologies and the other for collections.)
1975: *Worse Things Waiting* by Manly Wade WELLMAN.
1976: *The Enquiries of Dr Esterhazy* by Avram DAVIDSON.
1977: *Frights: New Stories of Suspense and Supernatural Terror* ed Kirby McCauley (1941-   ).
1978: *Murgunstrumm and Others* by Hugh B. CAVE.
1979: *Shadows* ed Charles L. GRANT.
1980: *Amazons!* ed Jessica Amanda SALMONSON.
1981: *Dark Forces* ed Kirby McCauley.
1982: *Elsewhere* ed Terri WINDLING and Mark Alan Arnold.
1983: *Nightmare Seasons* by Charles L. Grant.
1984: *High Spirits* by Robertson DAVIES.
1985: *Clive Barker's Books of Blood #1-#3* by Clive BARKER.
1986: *Imaginary Lands* ed Robin MCKINLEY.
1987: *Tales of the Quintana Roo* by James M. Tiptree Jr (◊ SFE).
1988: *The Architecture of Fear* ed Kathryn Cramer and Peter D. Pautz, *Dark Descent* ed David G. HARTWELL and *The Jaguar Hunter* by Lucius SHEPARD.
1989: *The Year's Best Fantasy: First Annual Collection* ed Ellen DATLOW and Terri Windling, *Angry Candy* by Harlan ELLISON and *Stories from the Old Hotel* by Gene WOLFE.
1990: *The Year's Best Fantasy: Second Annual Collection* ed Ellen Datlow and Terri Windling and *Collected Stories* by Richard MATHESON.
1991: *Best New Horror* ed Stephen JONES and Ramsey CAMPBELL and *The Start of the End of It All and Other Stories* by Carol EMSHWILLER.
1992: *The Year's Best Fantasy and Horror: Fourth Annual Collection* ed Ellen Datlow and Terri Windling and *The Ends of the Earth* by Lucius Shepard.
1993: *Metahorror* ed Dennis ETCHISON and *The Sons of Noah and Other Stories* by Jack Cady (1932-   ).
1994: *Full Spectrum 4* ed Lou Aronica, Betsy Mitchell, Amy Stout and *Alone With the Horrors* by Ramsey CAMPBELL.
1995: *Little Deaths* ed Ellen DATLOW and *The Calvin Coolidge Home for Dead Comedians and A Conflagration Artist* by Bradley Denton (1958-   ).

Best Short Fiction
(This award was divided from 1982, as indicated by the separate listings below.).
1975: "Pages from a Young Girl's Diary" by Robert AICKMAN.
1976: "Belsen Express" by Fritz LEIBER.
1977: "There's a Long, Long Trail A-Winding" by Russell Kirk (1918-1994).
1978: "The Chimney" by Ramsey CAMPBELL.
1979: "Naples" by Avram DAVIDSON.
1980: "The Woman Who Loved the Moon" by Elizabeth A. LYNN and "Mackintosh Willy" by Ramsey Campbell.
1981: "The Ugly Chicken" by Howard Waldrop.

Best Novella
1982: "The Fire When it Comes" by Parke GODWIN.
1983: "Confess the Seasons" by Charles L. GRANT and "Beyond All Measure" by Karl Edward Wagner.
1984: "Black Air" by Kim Stanley ROBINSON.
1985: "The Unconquered Country" by Geoff RYMAN.
1986: "Naldelman's God" by T.E.D. KLEIN.
1987: "Hatrack River" by Orson Scott CARD.
1988: "Buffalo Gals Won't You Come Out Tonight" by

Ursula K. LE GUIN.
1989: "The Skin Trade" by George R.R. MARTIN.
1990: "Great Work of Time" by John CROWLEY.
1991: "Bones" by Pat MURPHY.
1992: "The Ragthorn" by Robert P. HOLDSTOCK and Garry KILWORTH.
1993: "The Ghost Village" by Peter STRAUB.
1994: "Under the Crust" by Terry Lamsley.
1995: "Last Summer at Mars Hills" by Elizabeth HAND.

Best Short Story
1982: "The Dark Country" by Dennis ETCHISON.
1983: "The Gorgon" by Tanith LEE.
1984: "Elle Est Troi (La Mort)" by Tanith Lee.
1985: "Still Life with Scorpion" by Scott BAKER.
1986: "Paper Dragons" by James P. BLAYLOCK.
1987: "Red Light" by David J. Schow (1955-   ).
1988: "Friends' Best Man" by Jonathan CARROLL.
1989: "Winter Solstice, Camelot Station" by John M. FORD.
1990: "The Illusionist" by Steven MILLHAUSER.
1991: "A Midsummer Night's Dream" by Neil GAIMAN and Charles VESS.
1992: "The Somewhere Doors" by Fred Chappell.
1993: "This Year's Class Picture" by Dan SIMMONS and "Graves" by Joe Haldeman (◊ SFE).
1994: "The Lodger" by Fred Chappell.
1995: "The Man in the Black Suit" by Stephen KING.

Best Artist
1975: Lee Brown Coye.
1976: Frank FRAZETTA.
1977: Roger DEAN.
1978: Lee Brown Coye.
1979: Alicia Austin and Dale Enzenbacher.
1980: Don MAITZ.
1981: Michael WHELAN.
1982: Michael Whelan.
1983: Michael Whelan.
1984: Stephen Gervais.
1985: Edward GOREY.
1986: Thomas Canty, Jeff Jones.
1987: Robert Gould.
1988: J.K. Potter.
1989: Edward Gorey.
1990: Tom Canty.
1991: Dave MCKEAN.
1992: Tim Hildebrandt.
1993: James Gurney.
1994: Alan Clarke, J.K. Potter.
1995: Jacek Yerka.

Special Awards
1975: Betty and Ian Ballantine (◊ BALLANTINE ADULT FANTASY), Stuart D. Schiff (◊ WHISPERS).
1976: Donald Grant (1927-   ), Carcosa Press.
1977: Alternate World Recordings, *Whispers* ed Stuart D. Schiff.
1978: E.F. BLEILER, Robert WEINBERG.
1979: Edward L. Ferman (◊ SFE), Donald H. Tuck (*SFE*).
1980: Donald M. GRANT, *Fantasy Newsletter* ed Paul Allen.
1981: Donald A. WOLLHEIM, *Shayol* ed Pat Cadigan and Arnold Fenner.
1982: Edward L. Ferman, *Fantasy Newsletter* ed Robert Collins.
1983: Donald M. Grant, *Whispers* ed Stuart D. Schiff.
1984: Ian and Betty Ballantine, Joy Chant and George

Sharp, *Fantasy Tales* ed Stephen JONES and Daivd A. Sutton.
1985: Chris Van Allsburg for *The Mysteries of Harris Burdick*, Stuart D. Schiff for *Whispers*.
1986: Pay LoBrutto, Douglas E. Winter.
1987: Jane YOLEN for *Favorite Tales from Around the World*, Jeff Connor for Scream Press, W. Paul Ganley for Weirdbook.
1988: David G. HARTWELL for Arbor/Tor Anthologies, David B. Silva for *The Horror Show*, Robert and Nancy Garcia, *American Fantasy*.
1989: Robert WEINBERG, Terri WINDLING, plus Kristine Kathryn RUSCH and Dean Wesley Smith for *Pulphouse*.
1991: Arnold Fenner, Richard Chizmarfor for "Cemetery Dance".
1992: George Scithers and Darrell SCHWEITZER for WEIRD TALES, W. Paul Ganley for WEIRDBOOK.
1993: Jeanne Cavelos, Doug and Tomi Lewis.
1994: Underwood-Miller, Marc Michaud for Necronomicon Press (◊ SMALL PRESSES).
1995: Ellen DATLOW, Bryan Cholfin for Broken Mirrors Press.

Convention Awards
(These were issued for only a few years.).
1978: Glenn Lord.
1979: Kirby McCauley.
1980: Stephen KING.
1981: Gahan WILSON.
1982: Joseph Payne Brennan and Roy KRENKEL.
1983: ARKHAM HOUSE.
1984: Donald M. GRANT.
1985: Evangeline WALTON.
1986: Donald A. WOLLHEIM.
1987: Andre NORTON.

WORLDS OF FANTASY US digest MAGAZINE, 4 issues 1968-71, published by Galaxy Publishing, New York (*#1*) and thereafter by Universal Publishing, New York; *#1-#2* ed Lester Del Rey, *#3-#4* ed Ejler Jakobsson (1911-1986).

The first issue of *WOF* was an experimental one-shot following the increasing success of the fantasy-oriented issues of *If* (by then known as *Worlds of If*) and the book-publishing success of HEROIC FANTASY. The cover boldly proclaimed the names L. Sprague DE CAMP, Robert E. HOWARD and J.R.R. TOLKIEN. The emphasis in the longer fiction was on SWORD AND SORCERY; the shorter fiction emulated the UNKNOWN-style approach to CONTEMPORARY FANTASY. The success of *#1* resulted in the magazine being established on a quarterly basis under its new publisher, following the sale of its companion magazine *Galaxy*. *WOF* sought to continue in the same vein, but the production was shoddier and the fiction variable. The strength lay in the long lead novels, which included *The Tombs of Atuan* (Win 1970; **1971**) by Ursula K. LE GUIN and *Reality Doll* (Spring 1971; vt *Destiny Doll* **1971**) by Clifford D. SIMAK. The short fiction ran the range from psychic VAMPIRES and PACTS WITH THE DEVIL to wish-fulfilment and MAGIC. Contributors included Robert SILVERBERG, Frederik Pohl, James Tiptree Jr, Robert BLOCH and Michael BISHOP, with two guest editorials by Theodore STURGEON, exploring the nature of fantasy. Suffering from a low budget and weak distribution, *WOF* never realized its poential. [MA]

WOF should not be confused with the earlier UK magazine *Worlds of Fantasy* (14 issues [Summer] 1950-[Summer] 1954), a mediocre juvenile sf magazine. [MA]

WORLDS OF FANTASY AND HORROR US SMALL-PRESS, large-format MAGAZINE (Summer 1994-current), irregular. An extension of WEIRD TALES, it has continued the concept of featuring individual authors in each issue, starting with *#2* (Spring 1995), which focused on Charles DE LINT. [MA]

WORLD-TREE Nordic mythology conceived of the world as having at its centre the ash tree Yggdrasil, whose branches stretched over HEAVEN and Earth. It had three roots, one passing into the realm of the GODS (Asgard), one into the realm of the Frost giants (Jotunheim), and one into the underworld (Niflheim). Near each root lay a fountain, the one in Heaven being the fountain of Urd, tended by the Norns, from which Yggdrasil was watered daily. Near the root in Jotunheim was the fountain of Mimir, the source of wisdom: ODIN paid an eye to drink from it. And near the root in Niflheim was the fountain Hvergelmir, source of RIVERS. The gods had an assembly place beside the tree where they met daily to dispense justice, and the goat Heidrunn, who browsed in the branches of the tree, provided milk for Odin's warriors. In the topmost branches perched an eagle, who warned the gods of attacks by GIANTS. The tree itself was subject to constant attacks: stags roamed the branches eating green shoots, while the SERPENT Nidhögg gnawed at the underworld root. Yggdrasil thus linked the worlds of gods and men and giants, as well as the underworld. Similar ideas are found in the MYTHOLOGIES of North and Central Asia, Finland and Siberia, and seem to have links to SHAMANISM, in particular to shamanic journeying. Odin himself is said to have acquired mystic knowledge while hanging for nine days upon the tree – an image used by Guy Gavriel KAY in his **Fionavar Tapestry** series. [KLM]

WORLD WAR I 1. Before WWI Western civilization had sustained itself, without fundamental collapse, through centuries of war and change; but the Great War, as it was called, marked – at least in the perception of its survivors – the beginning of the end. Later events, however terrible – like WORLD WAR II – only seemed to confirm the terrible message of the moral exhaustion of a way of life which had dominated the world for several centuries.

This sense of APOCALYPSE and termination may have been exaggerated, but WWI was unquestionably thus perceived, even from before its actual (though widely predicted) outbreak. Kenneth GRAHAME, whose *The Wind in the Willows* (**1908**) has been read as a fully conscious retreat from the looming conflict, stopped writing entirely when WWI began. Edgar Rice BURROUGHS's creation in 1912 of the PLANETARY ROMANCE may have been so irresistible to his many early readers because it offered an escapist venue (◊ ESCAPISM). Gustav MEYRINK, in *The Green Face* (**1916**), foresaw a world ravaged by "Spectres, monstrous yet without form and only discernible through the devastation they wrought . . . But there was another phantom, still more horrible, that had long since caught the foul stench of a decaying civilization in its gaping nostrils and now raised its snake-wreathed countenance from the abyss where it had lain, to mock humanity with the realization that the juggernaut they had driven the last four years in the belief it would clear the world for a new generation of free men was a treadmill in which they were trapped for all time. [Trans Mike Mitchell **1992**.]" And the narrator of "Lost in the Fog", from *Nineteen Impressions* (coll **1918**) by J.D. Beresford (1873-1947), after going astray on the wrong TRAIN, finds himself in a village which is a MICROCOSM of Europe, and in which WWI is raging interminably.

Introducing a 1987 reprint of John BUCHAN's *These for Remembrance: Memoirs of Six Friends Killed in the Great War* (**1919** chap), Peter VANSITTART suggests that WWI opened an abyss into "a great emptiness" for Buchan; that his old world, rich in the felt "roundness of being", had been transmogrified into a flattened, disrupted post-WWI world vulnerable to all sorts of barbarism, and that his late fiction is a series of tales of aftermath. As William Butler YEATS put it in "The Second Coming" (1921): "The centre cannot hold." Thomas MANN's *The Magic Mountain* (**1924**) (◊ MAGIC MOUNTAIN), which takes place over a seven-year period directly preceding WWI, can be seen as an intricate and massive monument to a civilization whose centre was about to collapse. Gerhart HAUPTMANN's *Till Eulenspiegel* (**1928**) features an aviator who, disillusioned with the post-WWI world, manages to escape it entirely. A late recension of some of these themes is "The Great Loves" by Dan SIMMONS, from *LoveDeath* (coll **1993**).

For many of those who lived through WWI and became writers of fantasy, the long, golden Edwardian sunset seemed an ET IN ARCADIA EGO vision of a lost world, a kind of POLDER which could be occupied in dream. Fantasy authors who began to write during WWI – many being in active service – include E.R. EDDISON, Hugh LOFTING, J.R.R. TOLKIEN and E.A. WYKE-SMITH. None of these authors, however, made direct use of WWI, not even in a fantasticated context: it may be that their refusal of its implications was too profound for their imaginations to pick at the wound. Nor was WWI itself of much use as a venue to fantasy writers like James HILTON who were born slightly later than those just cited. As with some of the protagonists of Buchan's later novels, Hilton's HEROES are hollowed-out dwellers in the aftermath; and their escape is to SHANGRI-LA or to occult understandings or to OTHERWORLDS. Similarly, the soldier protagonist of A. MERRITT's *Three Lines of Old French* (1919 *All-Story*; **1937** chap) seems to escape into something like a medieval paradise, though he awakes back in this world.

2. Most large wars are full of theoretical crux points which, in the hands of the writer of ALTERNATE-WORLD fantasies or sf, can generate different outcomes to the combat. In *Contro-passato prossimo* (**1975**; trans Hugh Shankland as *Past Conditional: A Retrospective Hypothesis* **1989** UK) by Guido Morselli (1912-1973) and *The Carnival of Destruction* (**1994**) by Brian STABLEFORD Germany wins. In *Ostrv Krym* (**1981**; trans anon as *The Island of Crimea* **1984**) by Vassily Aksyonov (1932-) the Bolshevik revolution fails. BIGGLES (*1986*) offers a TIMESLIP as explanation of why Germany lost. Generally, though, TIME-TRAVEL fantasies set partly in or around WWI, like Robert A. HEINLEIN's *Time Enough for Love* (**1970**) and Jerry Yulsman's *Elleander Morning* (**1984**), tend to use the conflict as background, as a STORY which cannot be changed. Alternate-world tales in which WWI never took place or is transmuted beyond recognition are rare: they include Michael MOORCOCK's **Nomad of Time** sequence and Kim NEWMAN's *The Bloody Red Baron* (**1995** US), the latter a VAMPIRE tale which, if anything, darkens the reality.

Fantasies set *in* WWI are very much more numerous. The most famous probably remains Arthur MACHEN's "The Bowmen" (1914), in which he created the legend of the ANGELS OF MONS; later stories in the same mode include

The Broken Soldier and the Maid of France (**1919** chap) by Henry Van Dyke (1852-1933). Machen himself, in *The Terror* (**1917**), created a fuller legend of the war, in which the animals – at first mistaken for Germans – revolt against humanity's corrupt and destructive rule. William Faulkner (1897-1962), in *A Fable* (**1954**), introduced an element of CHRISTIAN FANTASY into the tale of a saintly rebel who engineers a brief armistice, and whose body magically disappears after his execution.

GHOST STORIES set in WWI include: Achmed ABDULLAH's "To be Accounted For" and "Renunciation" (both 1920); "Secret Service" (1922) by F. Britten Austin (1885-1941), one of several tales in which a German spy, either a literal ghost or an astral projection, is featured, Buchan's "Dr Lartius" from *The Runagates Club* (coll **1928**) being another; several stories assembled in Lord DUNSANY's *Tales of War* (coll **1918**); "Kitchener at Archangel" (1935) by Stephen Graham (1884-1975); and "The Unbolted Door" (1931) by Marie Belloc Lowndes (1868-1947). Others titles of interest are: *Living Alone* (**1919**) by Stella BENSON; *The Wind Between the Worlds* (**1920**) by Alice Brown (1857-1948); H. Burgess DRAKE's *The Remedy* (**1925**); James Francis DWYER's *Evelyn: Something More than a Story* (**1929**), whose eponymous frail widow communicates with the WWI dead, to her profit; "The Men of Avalon" (1935) by David H. KELLER, in which MERLIN bucks up Lord Kitchener; *Fly Away Peter* (**1982**) by David Malouf (1934-); *The Deer-Smellers of Haunted Mountain* (**1921**) by John J. Meyer (1873-1948), an unusual example of a WWI tale featuring a secret cadre of Germans still bent on world conquest; and *The Bridge of Time* (**1919**) by William Henry Warner, a TIME-TRAVEL tale in which an ancient Egyptian sorcerer sends a young prince forward in time to WWI, where he meets his long-lost love. [JC]

WORLD WAR II Hitler's "Final Solution" was an obscenity it is hard, mentally, to cope with (although similar obscenities have been perpetrated since). Because atrocities make good copy, WWII has been a popular focus for all sorts of fiction, fantasy included. More kindly one can propose that another feature of WWII made it particularly usable by fiction writers. Both World Wars were tragic, but WWI was *remembered* as an unmitigated tragedy, a grinding apocalyptic process whose outcome was always foreseeable, even though some of the details (like the USA's entry into the conflict) might have been unexpected at the time. WWII, on the other hand, has been remembered as a melodrama, full of strange and uncanny ups and downs, with terrifying new weapons galore, feats of derring-do on a daily basis, and protagonists who were not only MONSTERS in real life but also, in fictional terms, highly effective ICONS of villainy. Despite the attempts of propagandists on both sides, no wholly evil figure emerges from WWI to occupy the world's imagination, no one of a viciousness so unmitigated that it seems almost supernatural; Hitler, on the other hand, has all the lineaments of a DARK LORD, and the Reich he hoped to found was a PARODY of the true LAND.

It is hard, on any other ground, to explain the fascination of the HITLER-WINS story. Either in the traditional form of the future-war novel that utters dreadful warnings or (after 1945) in the form of ALTERNATE-WORLD tales in which Germany prevails, the Hitler-Wins tale has been popular for well over half a century. The best are probably SARBAN's *The Sound of his Horn* (**1952**) and *The Man in the High Castle* (**1962**) by Philip K. Dick (1928-1992); the most popular recently has been *Fatherland* (**1992**) by Robert Harris (**1957-**   ). In a cognate vein are tales in which occult or other supernatural forces are brought to bear – either by the Nazis (Himmler was a devout believer in the occult) or by forces in opposition to them – or in which characters return or TIMESLIP to a Nazi-controlled Europe, often in a HORROR frame, with the Nazis often seen as vampiric or as in the clutches of a Dark Lord. Examples include: Dale Estey's *A Lost Tale* (**1980**; vt *Fortress Island* 1981), in which Druids protect the Isle of Man; Romain Gary's *The Dance of Genghis Cohn* (**1967**); Katherine KURTZ's *Lammas Night* (**1983**) and portions of her **Adept** sequence with Deborah Turner HARRIS; *The Ring Master* (**1987**) by David Gurr (1936-   ); Barbara HAMBLY's *The Magicians of Night* (**1992**); William KOTZWINKLE's *The Exile* (**1987**), whose protagonist timeslips into the body of a German black-marketeer, eventually to be trapped there after torture has driven the owner's mind forward in time and into the protagonist's own body; Frederick MACLEISH's *Prince Ombra* (**1982**); Scott MacMillan's *Knights of the Blood: Vampyr-SS* (**1993**); *Runespear* (**1987**) by Victor MILAN and Melinda SNODGRASS; *The Incredible Mr Limpet* (**1942**) by Theodore PRATT; Jon Ruddy's *The Bargain* (**1990**), in which Hitler does a deal with DRACULA; Lois Tilton's *Darkness on the Ice* (**1993**); several novels by Dennis WHEATLEY, including *The Man who Missed the War* (**1945**) and *They Used Dark Forces* (**1964**); Connie WILLIS's "Fire Watch" (1982); and F. Paul WILSON's **Adversary Cycle**, the first volume of which was filmed as *The* KEEP (*1983*). Of the relatively few relevant alternate-world tales that do not focus on a Nazi victory, the most interesting is probably the **WorldWar** sequence (**1994-**current) by Harry Turtledove (1949-   ).

Very occasionally the "Final Solution" has been approached by writers of the fantastic attempting to make sense of the event through a use of the tools of FABLE, as in Jane YOLEN's *Briar Rose* (**1992**) and Georges Perec's *W, or The Memory of Childhood* (**1975**). But there is a chariness about the use of overt fantasy to describe horrors which are too easily diminished through a genre generally designed for ESCAPISM. Novels like *The Birthday King* (**1962**) by Gabriel Fielding (1916-   ), *The Erl King* (**1970**) by Michel Tournier (1924-   ), Herbert ROSENDORFER's *German Suite* (**1972**), Thomas PYNCHON's *Gravity's Rainbow* (**1973**), Michael MOORCOCK's **Pyat** sequence and Martin AMIS's *Time's Arrow* (**1991**) are, fittingly, not fantasies – although the last can be read as such. But the black world they evoke has some of the characteristics of fantasy: a heightened sense, perhaps, of the horror of BONDAGE. Richard Condon's one true tragedy, *An Infinity of Mirrors* (**1964**), reads like a FANTASY OF HISTORY caught in the bondage of the real WWII. Both *The Key to the Great Gate* (**1947**) by Hinko Gottlieb (1886-1948) and *Slaughterhouse Five* (**1969**) by Kurt Vonnegut (1922-   ) confabulate arguments about the nature of space and time in order to pry their protagonists loose from that bondage; in these two novels, an intense pathos underlies the "comedy" of escape. Tales like Poul ANDERSON's *Three Hearts and Three Lions* (**1961**) and L. Sprague DE CAMP's and Fletcher PRATT's *Land of Unreason* (**1942**) catapult their heroes out of the nightmare into a SECONDARY WORLD where acts of derring-do are more possible, and more palatable.

Less centrally, WWII has generated some GHOST STORIES, like *Death Into Life* (**1946**) by Olaf Stapledon (1886-1950), whose dead pilot learns about the larger nature of the

Universe, Robert NATHAN's *But Gently Day* (**1943**), whose dead pilot has POSTHUMOUS-FANTASY experiences in the American Civil War, *On a Dark Night* (**1949**) by Anthony West (1914-1987), whose protagonist experiences after death a series of HELLS, *The Send-Off* (**1973**) by Christopher Leach (1925-   ) and *The Shepherd* (**1975**) by Frederick Forsyth (1938-   ), both the latter combining BELATEDNESS and nostalgia as they involve the return of long-dead soldiers who evince no awareness of the passage of time.

The greatest fantasy to make use of WWII is probably Thomas MANN's *Doctor Faustus* (**1947**), in which both the composer Leverkühn and Germany itself are seen, by analogy, to have made similar PACTS WITH THE DEVIL.   [JC]

WORM OUROBOROS The world-circling SERPENT which devours its own tail; this closed loop symbolizes eternity. In E.R. EDDISON's *The Worm Ouroboros* (**1922**) it represents eternal return, as the GODS permit the STORY to loop back on itself. A miniature Ouroboros fills a circular moat in Piers ANTHONY's *The Source of Magic* (**1979**).   [DRL]

WORM/WYRM The term used most frequently in NORDIC FANTASY for the DRAGON (the Nordic for dragon is *wyrm*). Nordic dragons usually lack the ability to fly, and are thus closely allied to the SERPENT, especially the sea-serpent. The name "worm" suggests a loathly beast, more disgusting and less fearsome than the dragon, which has an honourable ancestry. Fafnir, the dragon in the *Volsunga Saga*, is often referred to as a worm. In some stories, as in the FOLKTALE of "The Lambton Worm" (14th century), the worm can reunite if cut in two. In that story a young squire, the heir of Lambton, catches a large worm (or eft) while fishing, but throws it back. Years later, when he returns from the Crusades, he finds the worm has grown to alarming proportions and is menacing the countryside. He wears an armour of spikes in order to trap it. This story bears some comparison with *Farmer Giles of Ham* (**1949** chap) by J.R.R. TOLKIEN. Robert E. HOWARD used the term for his Nordic fantasy "The Valley of the Worm" (1934 *WT*). In *The Lair of the White Worm* (**1911**) Bram STOKER describes a disgusting worm that exists in a cave but has a symbiotic relationship with a woman. David H. KELLER used similar imagery for his horror story "The Worm" (1929 *Amazing Stories*). More recent examples include *The Fire Worm* (**1988**) by Ian WATSON and *Inside the Worm* (**1993**) by Robert Swindells (1939-   ).   [MA]

WORRELL, KATHLEEN L. (?   -?   ) US writer whose *The Red Serape* (coll ?1924) contains several fantastic stories making use of such supernatural themes as an immortal FEMME FATALE and a man reformed by a cross that mysteriously appears his forehead. KLW was a capable and sometimes surprising writer.   [RB]

WOTAN ◊ ODIN.

WRATH OF THE GODS Series of graphic stories set in ancient Greece, written by Michael MOORCOCK and published from January 1963 as a centrespread in the UK children's weekly comic *Boys' World*, continuing until just after its amalgamation with *Eagle* in October 1964. The feature had excellent colour artwork from Ron EMBLETON (for the first 23 issues) and John M. BURNS. The stories recounted the adventures of an impetuous young man, Arion, and his various encounters with the minor gods and magical figures of the ancient Greek myths. Compared with the usual standard of comics writing at the time they were well paced and inventive, but are more memorable for the groundbreaking artwork.   [RT]

WREDE, PATRICIA C(OLLINS) (1953-) US writer best-known for CHILDREN'S FANTASY. Most of PCW's books fit into one of two series. **Lyra** is an OTHERWORLD which humans share with FAIRIES, called the Shee, and FOREST folk, called the Wyrd. The stories usually feature female lead characters seeking to solve some personal anguish in a warm and pleasant world that is being threatened by the evil Shadow Born. They bear some comparison with TOLKIEN's work, but more in mood than in presentation. The series so far runs *Shadow Magic* (**1982**), *Daughter of Witches* (**1983**), *The Harp of Imach Thyssel* (**1985**), *Caught in Crystal* (**1987**) and *The Raven Ring* (**1994**). PCW's other series is set in the **Enchanted Forest**. The first, *Talking to Dragons* (**1985**; rev 1993), is a delightful HIGH FANTASY with an Alice-in-Wonderland mood. Daystar, a young boy, lives in a cottage on the edge of the Enchanted Forest, which is a POLDER of enchantment. He witnesses his mother melt down a WIZARD. She tells him he is now old enough to enter the Enchanted Forest and must not return until he is ready. She gives him a SWORD which Daystar later learns is the Sword of the sleeping king, lost since Daystar's birth. Once he is in the FOREST, events happen fast as Daystar struggles to understand who he is and why he's there. Although the story is a RITE OF PASSAGE, his actions soon appear to be just a small part of a much vaster tapestry, particularly the role of the DRAGONS in the war of the wizards. PCW then went to the back-story of the Enchanted Forest in *Dealing With Dragons* (**1990**; vt *Dragonsbane* 1993 UK), *Searching for Dragons* (**1991**; vt *Dragon Search* 1994 UK) and *Calling on Dragons* (**1993**). All four were assembled as *The Enchanted Forest Chronicles* (omni **1996**).

Snow White and Rose Red (**1989**) is a REVISIONIST FANTASY setting the original FAIRYTALE in Elizabethan England at the time of John DEE and *Mairelon the Magician* (**1991**) is another FANTASY OF HISTORY, set this time in Regency England.

PCW's stories may be formulaic, but she brings to them vitality and enthusiasm. [MA]

Other works: *The Seven Towers* (**1984**); *Sorcery and Cecelia* (**1988**) with Caroline STEVERMER; *Book of Enchantments* (**1996**).

WRIGHT, FARNSWORTH (1888-1940) US writer and editor, best-known as editor of WEIRD TALES for 179 issues (November 1924-March 1940). He allowed relatively free rein to his writers, although he could be both discouraging and encouraging with equal lack of logic. This particularly applied to H.P. LOVECRAFT, whose best works emerged in the early 1930s at longer lengths which FW discouraged, preferring shorter fiction. Nevertheless FW developed *WT* from a relatively routine horror pulp MAGAZINE to create what has become a legend. His wide tastes allowed for an extravagance of fiction, whether the development of SWORD-AND-SORCERY under Robert E. HOWARD, the cosmic fiction of Lovecraft, the OCCULT-DETECTIVE stories of Seabury QUINN, the CHINOISERIES of E. Hoffmann PRICE and Frank OWEN, the terror tales of Paul Ernst (1899-1985) or the space operas and pandimensional adventures of Edmond Hamilton (1904-1977) and Nictzin DYALHIS. He also bravely decorated the magazine with the risqué ILLUSTRATIONS of Margaret Brundage (1900-1976), and encouraged and developed the work of Hannes BOK and Virgil FINLAY.

FW's mother taught music and inspired in FW his zeal for the Classics and for art; he loved poetry and encouraged its appearance in *WT*, one of the few pulp markets for verse. His first job was as a reporter, but he was drafted into the US Army in 1917 and served in the infantry in WWI. His health deteriorated in the 1920s when he contracted Parkinson's disease, which prematurely aged him. Although FW produced a few stories, they are unmemorable. His poetry is more delicate, but he limited its appearance. He anonymously compiled an anthology of *WT* stories as a subscription bonus: *The Moon Terror* (anth **1927**), unfortunately representative of the worst of *WT*'s early years. He edited *WT*'s short-lived companion magazine *Oriental Stories* (14 issues 1930-34; retitled *The Magic Carpet Magazine* from 1933) and attempted to launch **Wright's Shakespeare Library** in 1935 with a pulp-format edition of *A Midsummer Night's Dream*. Despite the illustrations by Virgil Finlay, the book flopped.

FW's nephew, David Wright O'Brien (1918-1944), was killed during WWII after a brief but prolific period as a contributor to the Ziff-Davis pulp magazines, including FANTASTIC ADVENTURES, to which he contributed many humorous fantasies. [MA]

Further reading: "Farnsworth Wright" by E. Hoffmann PRICE in *The Ghost #2* (July 1944).

WRIGHTSON, BERNI (1948-) Prolific US fantasy COMIC-book artist with a unique talent for macabre HORROR. His art is distinguished by a confident line and skilful use of stark black, and by forceful composition. Despite a predilection for grotesque mutation and decay, his work still often manages to retain a humane quality that renders it endearing rather than repulsive.

BW's early work included **Nightmaster** (*DC Showcase #83-#84* 1969), which he did with Jeff JONES and Mike KALUTA, and covers and short stories for *House of Secrets* and *House of Mystery*. One of these was "The SWAMP THING" (*House of Secrets #92* 1971), again a collaboration with Jones and Kaluta; this became a series the following year, winning BW several awards from the Academy of Comic Book Arts. He also collaborated with Vaughn Bodé on some cartoon erotica for *Swank* magazine under the title **Purple Pictography** (1971), and published a collection of his own horror stories entitled *Badtime Stories* (1971). After *Swamp Thing* (#1-#10 1972-4) BW drew several horror stories for WARREN PUBLISHING's EERIE and CREEPY magazines along with many spot illustrations, then occupied a warehouse in New York with Jeff Jones, Mike Kaluta and Barry WINDSOR-SMITH, where all four became involved in the production of personal portfolios and other projects. A book about this period of experimentation is *The Studio* (graph **1979**), which seems to be by the four artists (although "JS" is credited for the introduction). It was during this period that BW began producing a remarkable series of illustrations for Mary SHELLEY's *Frankenstein* (**1818**), eventually published with the original text as *Mary Wollstonecraft Shelley's Frankenstein* (graph **1983**; vt *Bernie Wrightson's Frankenstein* 1984), coinciding with the publication of another BW-illustrated volume, Stephen KING's *Cycle of the Werewolf* (**1983**). BW began doing design work for a movie provisionally titled *Traveler*, but it was eventually abandoned. He returned to comic books with *Batman: The Cult* (#1-#4 1988; graph coll **1989**), in which his drawing style took an unexpected new direction, being more spontaneous and less polished.

BW has maintained a consistent personal direction in his work while remaining a firm favourite with comics fans. The number of comic books he has drawn is relatively small but his work remains very popular, and each new project is eagerly welcomed – a unique achievement in a field with such fickle enthusiasms. [RT]

Other works: *The Bernie Wrightson Exhibition* (graph **1977**); *Apparitions* (portfolio **1978**); *The Mutants* (graph coll **1980**).

Further reading: *Berni Wrightson: A Look Back* (graph **1980**) ed Christopher Zavisa.

WRIGHTSON, (ALICE) PATRICIA (1921–) Australian writer whose subtle and detailed portrayals of her native land began with *The Crooked Snake* (**1955**), about a children's secret society which saves some endangered species. The central LANDSCAPE depicted in her work is that of Australia, then and now, and the basic story which she tells and retells is of the THINNING of the old MAGIC land with the invasion of white settlers. This does not preclude sympathetic treatment of life in Sydney, in nonfantasies like *I Own the Racecourse!* (**1968** UK; vt *A Racecourse for Andy* 1968 US), but her central examinations of the nature of "ownership" – and of the mutual incomprehension of Aborigines and whites when "ownership" is at issue – give her best work a sharp, elegiac flavour.

Her first novel of genre interest, *Down to Earth* (**1965** US), is sf. *An Older Kind of Magic* (**1972** UK) serves as a kind of prelude to later fantasies: three urban children save a public GARDEN from a developer with the assistance of a variety of WAINSCOT spirits native to the enclosure and to the local Aboriginal DREAMTIME. In *The Nargun and the Stars* (**1973** UK) a displaced white boy, Simon, finds himself in conflict with the eponymous rock spirit, who is embodied in stone and whose short-term defeat at the hands of the sympathetic child cannot be permanent – for stone will last, but Simon will soon be a "whisper in the dark". A similar conflict breaks out in *A Little Fear* (**1983** UK) between a dogged elderly woman and another unhuman spirit from the heart of the earlier country.

PW's central work of fantasy is **Wirrun** trilogy: *The Ice is Coming* (**1977** US), *The Dark Bright Water* (**1978** US) and *Behind the Wind*(**1981** UK; vt *Journey Behind the Wind* 1981 US), assembled as *The Book of Wirrun* (omni **1985**). The central conflict – between ancient ice spirits and encroached-upon humanity – is here rendered in more subtle terms, because the human protagonist, Wirrun, is himself an Aborigine. In the first volume his alliance with the most ancient Nargun of all, to oppose the advent of the new ice age, is necessary to maintain a balance in the world between GOOD AND EVIL and to fight CHAOS. The sequels change the adversaries but not the war.

Later novels include: *Night Outside* (**1985** US), about elderly street-people; *Moon-Dark* (**1987** UK), which once again argues for Aboriginal/white cooperation to cope with the long trials laid upon the land, and in which a war among the animals (described from the viewpoint of a TALKING ANIMAL) is caused by human demands upon their habitat; and *Balyet* (**1989** UK), which reiterates similar material in the form of a GHOST STORY. [JC]

Other works: *The Old, Old Ngarang* (coll **1989** UK).

WRONGNESS A term most vividly associated with SUPERNATURAL FICTION and HORROR, but inherent to fantasy as well. Supernatural fictions are stories in which the real world is impinged upon or violated by supernatural elements; they thus include GOTHIC FANTASY, GHOST STORIES, OCCULT FANTASY, WITCHCRAFT tales, stories involving satanic rites or PACTS WITH THE DEVIL, tales of POSSESSION, VAMPIRE tales and stories of WEREWOLVES.

The central moments in many supernatural fictions take place in an EDIFICE or BAD PLACE that somehow *does not add up*, and whose architectural disproportions generate – as in Edgar Allan POE's "The Fall of the House of Usher" (1841), or in much of H.P. LOVECRAFT's CTHULHU MYTHOS – a sense that a PORTAL into moral abyss – a lesion or wrongness which must eventually be purged – has invaded. Consequently, though central to supernatural fiction, wrongness's intrusion is not generally sensed as signalling a profound or inherent illness in the circumambient world. It comes from abroad, and in most instances – even after it has dug itself in – it can be deported.

It is otherwise with "pure" FANTASY. A central premise of the SECONDARY WORLD or OTHERWORLD is that it cannot ultimately be understood in terms of real versus wrong, mundane versus contingent. This is not to claim that fantasies *exclude* supernatural fiction or horror – for the movement of a fantasy tends to co-opt any passing elements of these two genres of the fantastic. Fantasy REALITIES, in other words, *incorporate* the dichotomies that define and generate supernatural fictions. A sense of wrongness, therefore, when it bears in upon the protagonist of a fantasy text, generally signals not a threat from abroad but the apprehension of some profound change in the essence of things, though perhaps initially signalled in terms that evoke supernatural fiction. The sense of wrongness, in fantasy, is a recognition that the world is – or is about to become – no longer right, that the world has been subject to, or soon will be subject to, a process of THINNING.

Wrongness and thinning are two essential moments in the "grammar" of fantasy; but in the flow of STORY are distinct, even though they frequently manifest together. A textbook example is the moment, late in Lisa GOLDSTEIN's *Summer King, Winter Fool* (**1994**), when Taja – a potential poet-mage – overhears an incompetent poet attempting to call the god of summer. "She [Taja] felt the invocation as a terrible wrongness, a sickness that invaded her soul"; and within a few sentences the wrongness has awakened a bad MAGIC which turns the bad poet and his companions to stone – a change that could be described as the ultimate thinning-down of being, short of death.

A sensation of wrongness often warns of – or accompanies – a thinning of FAERIE or of MAGIC in general; it frequently prefigures or accompanies the departure of the folk, or the creation of some sort of POLDER, which may well have been constructed precisely to counteract a world gone wrong. Examples are common.

Wrongness may signal the TRANSFORMATION of the LAND that ensues when a DARK LORD triumphs or threatens to triumph, turning the old world into a PARODY of its prior being; METAMORPHOSES of this sort variously threaten Middle-Earth – most movingly the Hobbits' Shire, which subsequently requires harrowing – in J.R.R. TOLKIEN's *The Lord of the Rings* (**1954-5**), and in much subsequent HIGH FANTASY.

Wrongness also warns of TIME distortions – which may include the uncanny frisson generated by TIME IN FAERIE, when years may pass in a night or a night may be discovered to have lasted for years, or the analogous melancholy felt when "normal" time resumes or begins, a paradigm example being the moment in Richard WAGNER's **Ring Cycle** (**1851-67**) when the GODS suddenly age. Or it can be something ostensibly rather simpler. The sense of visceral shock felt in Lisa A. Barnett's and Melissa SCOTT's *Point of Hopes* (**1995**) when all the clocks in the city go suddenly awry, when even the central magists' ORRERY fails any longer to turn with the heavens, is so intense because the world of the novel is literally governed by the stars, by ASTROLOGY; when the clocks fail to work, the world goes wrong.

It can also mark a state of BONDAGE, the unnatural freezing of reality generated when a METAMORPHOSIS goes wrong, or cannot happen, or is imposed as a punishment. Ariel's interminable scream, when he is locked into the tree, is a scream of bondage. Though that scream is not heard but simply recollected in William SHAKESPEARE's *The Tempest* (performed *c*1611; **1623**), it epitomizes wrongness. [JC] **See also:** ARABIAN NIGHTMARE; BALANCE; CROSSHATCH; ELDER GODS; HAUNTED DWELLINGS; HORROR; LABYRINTHS; MALIGN SLEEPER; TALENTS.

WURTS, JANNY (1953-   ) US writer and artist, married to Don MAITZ. Her first novel, *Sorcerer's Legacy* (**1982**; rev 1989), is in traditional FANTASYLAND vein. The **Cycle of Fire** trilogy – *Stormwarden* (**1984**), *Keeper of the Keys* (**1988**) and *Shadowfane* (**1988**) – mixes sf with MAGIC. The **Empire** trilogy, with Raymond E. FEIST – *Daughter of the Empire* (**1987**), *Servant of the Empire* (**1990**) and *Mistress of the Empire* (**1992**) – is set in a fantasticated Byzantine Empire. *The Master of Whitestorm* (**1992**) is another FANTASYLAND singleton. The **Wars of Light and Shadows** – *The Curse of the Mistwraith* (**1993**), *The Ships of Merior* (**1994**) and *Warhost of Vastmark* (**1995**) – concerns the adventures of two half-brothers, one controlling the power of light, the other the power of shadow. *That Way Lies Camelot* (coll **1994**) mixes sf and fantasy stories, some originally written for shared-world anthologies. JW has produced effective cover paintings for several of her own books. [DP]

WYATT, HORACE (1876-1954) US writer. ◊ WONDERLAND.

WYETH, N(EWELL) C(ONVERS) (1882-1945) Prolific and influential US painter and illustrator, mostly of historical subjects, with a rich-coloured, monumental style; he was strongly influenced by his teacher, Howard PYLE. The human figures in his paintings are, as was NCW himself, enormously strong and vigorous. His compositions are simple and dramatic. He worked in oils for the first half of his life, but later developed a working process that involved the use of egg tempera. His methods involved several stages in the progress of a painting, and it was his unusual energy and sureness of touch which allowed him to produce such a vast quantity of work. At the time of his death in a railway-crossing accident, he had become one of the USA's best-loved illustrators.

NCW studied under Pyle at Chadds Ford, and became Pyle's most faithful disciple, painting very much in the same manner and choosing similar subject matter; he continued the teaching tradition there after Pyle's death in 1911. NCW produced over 3000 ILLUSTRATIONS, numerous huge murals and a great many still-lifes, landscape paintings and pictures of the Old West. He transmitted his enthusiasm for the arts to his children: one son was Andrew Wyeth (1917-   ) and a grandson, James Wyeth (1946-   ), is another highly respected painter. [RT]

WYKE-SMITH, E(DWARD) A(UGUSTINE) (1871-1935) UK writer; born Smith, he later took on the fuller name by deedpoll. He began writing fantasy tales for his children as an apparent antidote to the experience of WORLD WAR I. *Bill of the Bustingforths* (**1921**), like all his fantasy work, makes lighthearted REVISIONIST-FANTASY play with old material, introducing its young protagonists into a FAERIE, one of whose inhabitants is a DWARF who once helped Little Red Riding-Hood carve up the wolf; like his other CHILDREN'S FANTASIES it is illustrated by George MORROW. *The Last of the Barons* (dated 1922 but **1921**) has some fantasy elements, and *The Second Chance* (**1923**) is an adult tale involving sf-like rejuvenation. EAW-S remains known mainly for his last tale, *The Marvellous Land of Snergs* (**1927**), a fantasy which influenced J.R.R. TOLKIEN's *The Hobbit* (1937), as Tolkien acknowledged. The Land of Snergs is a POLDER, hidden from larger, harsher folk. In this polder the FLYING DUTCHMAN and various other figures live in contented retirement. [JC]

WYLIE, ELINOR (HOYT) (1885-1928) US poet and writer, married to William Rose BENÉT. She repudiated her first work – *Incidental Numbers* (coll **1912** chap), a volume of poems – and is of fantasy interest primarily for two late novels. In *The Venetian Glass Nephew* (**1925**) an 18th-century cardinal has a glassblower craft for him a live nephew made out of glass (◊ ANIMATE/INANIMATE); when the nephew falls in love with a full-blooded woman, disaster seems ominously likely, until she agrees in the end to be transformed through MAGIC into porcelain. The proceedings are observed with care by Carlo GOZZI, author of many famous Venetian FAIRYTALES in theatrical forms influenced by the COMMEDIA DELL'ARTE. *The Orphan Angel* (**1926**; vt *Mortal Image* 1927 UK) features Percy Bysshe SHELLEY, who has not died but – transported to the USA in the good ship *Witch of the West* – travels across the new continent on a QUEST for a mysterious female. [JC]

WYLIE, JONATHAN Joint pseudonym of married UK writers Mark Smith (1952-   ) and Julia Smith (1955-   ). While working for a UK publishing house, they established themselves as competent writers of GENRE FANTASY with the **Servants of Ark** trilogy – *The First Named* (**1987**), *The Centre of the Circle* (**1987**) and *The Mage-Born Child* (**1988**) – and the **Unbalanced Earth** trilogy – *Dreams of Stone* (**1989**), *The Lightless Kingdom* (**1989**) and *The Age of Chaos* (**1989**). Turning freelance, they produced *Dream-Weaver* (**1991**) and *Shadow-Maze* (**1992**; 1994 in US as by Mark and Julia Smith), which showed an increase in originality and density of imagination. These were followed by the **Island and Empire** trilogy – *Dark Fire* (**1993**), *Echoes of Flame* (**1994**) and *The Last Augury* (**1994**) – and two further singletons, *Other Lands* (**1995**) and *Across the Flame* (**1996**). With their liking for young protagonists who win through, the JW team has proven itself more than capable of delivering the customary escapist pleasures to a youthful readership. [DP]

WYRM ◊ WORM/WYRM.

XANADU The name of Shang-tu, the Mongolian summer capital historically founded by Kublai Khan in 1256, was anglicized in the 17th century as Xamdu or Xaindu and then eclipsed forever by S.T. COLERIDGE's euphonious variation in "Kubla Khan" (1816). This poem established Xanadu as a gorgeously unreachable IMAGINARY LAND containing Alph, the sacred RIVER. Coleridge's vision of the Khan's "pleasure-dome" features in L. Sprague DE CAMP's and Fletcher PRATT's *The Castle of Iron* (**1950**), where events in Xanadu suffer a monotonous eternal return because the poem is unfinished, and in Greg BEAR's *The Infinity Concerto* (**1984**), where the poem is completed as a SPELL and its idyllic scene destroyed. **Xanadu (1992**-current 3 vols) is a HEROIC-FANTASY anthology series ed Jane YOLEN. [DRL] **Further reading:** *The Road to Xanadu* (**1927**) by John Livingstone Lowes (1867-1945).

YANKEE IN KING ARTHUR'S COURT, A vt of *A Connecticut Yankee in King Arthur's Court* (**1949**). ◊ A CONNECTICUT YANKEE.

YANKEE MAGAZINE ◊ MAGAZINES.

YANKEE WEIRD SHORTS ◊ MAGAZINES.

YARBRO, CHELSEA QUINN (1942-   ) US writer whose work of genre interest began with "The Posture of Prophecy" for *If* in 1969, and whose early work – like *Time of the Fourth Horseman* (**1976**), *Cautionary Tales* (coll **1978**) and *False Dawn* (in *Strange Bedfellows* anth **1973** ed Thomas N. Scoria; exp **1978**) – was generally sf. Since then she has concentrated on HORROR and SUPERNATURAL FICTION, and is now best-known for the **Saint-German** VAMPIRE tales, the main sequence of which is *Hôtel Transylvania: A Novel of Forbidden Love* (**1978**), *The Palace* (**1978**), *Blood Games* (**1980**), *Path of the Eclipse* (**1981**), *Tempting Fate* (**1982**), *The Saint-Germain Chronicles* (coll of linked stories **1983**), *Out of the House of Life* (**1990**), *The Spider Glass* (**1991** chap), *Darker Jewels* (**1993**), *Better in the Dark* (**1993**), *The Vampire Stories of Chelsea Quinn Yarbro* (coll **1994**) and *Mansions of Darkness: A Novel of Saint-Germain* (**1996**). A second sequence, featuring **Atta Olivia Clemens**, Saint-Germain's lover is *A Flame in Byzantium* (**1987**), *Crusader's Torch* (**1988**) and *A Candle for D'Artagnan* (**1989**).

The 4000-year-old Saint-Germain – loosely based on the historical Count de Saint-Germain (?   -?1784) – is perhaps the most extreme example to have become popular of the vampire as hero. He is saturnine and very attractive to women (like Lord BYRON), but his powers are limited to his apparent IMMORTALITY and his paranormal strength; his vampiric kiss – described gently by CQY – is conceived of as a way of passing on the gift of life; his technical impotence (he is without semen), plus the fact that once transfigured into vampires his women can no longer sleep with him, generate a sometimes sentimental poignance. CQY likes Saint-Germain thoroughly, and, because of his wisdom, his learning and his gift of life, so (it seems) should we. Certainly most of the elements of horror in the sequence are generated by humans, not by the smooth swashbuckler MAGUS.

The sequence has not been published in order of internal chronology, and individual volumes tend to be relatively autonomous. The first tale is set in 18th-century France, and establishes the identity between the real and the fictional Saint-Germain; the second is set in 15th-century Florence; the third in Nero's Rome; the fourth in 13th-century Asia (featuring Genghis Khan); the fifth in 20th-century Austria; the sixth in ancient EGYPT; the seventh in 16th-century Russia; and so forth. The stories are romantic and packed with historical verisimilitude. The **Atta Olivia Clemens** tales are darker and more violent, perhaps because, as a woman, Clemens is less in control of events.

Other works of fantasy interest by CQY include: *Ariosto: Ariosto Furioso, a Romance for an Alternate Renaissance* (**1980**), partly set in a world derived from ARIOSTO's own *Orlando Furioso* (1516-32) and partly in an ALTERNATE-WORLD version of Renaissance Italy; *The Godforsaken* (**1983**), a WEREWOLF tale of the Spanish Inquisition; *Beastnights* (**1989**), a werewolf tale of contemporary SAN FRANCISCO; and *A Mortal Glamor* (**1985**), set in 14th-century Avignon, where a convent is invested by the DEVIL.

CQY is a committed believer in OCCULTISM and a professional TAROT reader. Unsurprisingly, her supernatural fiction is notable for a sense of advocacy – a sense that the strangenesses being described would do the world good, if only the world would listen. [JC]

Other works: The **Ogilvie, Tallant & Moon** detective series, with some fantasy elements, being *Ogilvie, Tallant & Moon* (**1976**; vt *Bad Medicine* 1990), *Music When Sweet Voices Die* (**1979**; vt *False Notes* 1991), *Poison Fruit* (**1991**) and *Cat's Claw* (**1992**); the **Michael** series of occult quasifictional tracts, comprising *Messages from Michael on the Nature of the Evolution of the Human Soul* (**1979**) and *More Messages from Michael* (**1986**); *Dead & Buried* * (**1980**), a movie tie; *Bloodgames* (**1980**); *Sins of Omission* (**1980**); *On Saint Hubert's Thing* (**1982** chap); *CQY* (**1982** chap); *Hyacinths* (**1983**), sf; *Nomads* * (**1984**), a movie tie; *Signs & Portents* (coll **1984**); *Locadio's Apprentice* (**1984**) and *Four Horses for Tishtry* (**1985**), both associational; *To the High Redoubt* (**1985**); *A Baroque Fable* (**1986**), humorous fantasy; *Floating Illusions* (**1986**); *Firecode* (**1987**); *Taji's Syndrome* (**1988**), an sf medical horror novel; *The Law in Charity* (**1989**), a

Western; *Crown of Empire* (**1994**), **#4** in the **Crisis of Empire** sequence created by David DRAKE.
Other work (as Vanessa Pryor): *A Taste of Wine* (**1982**).
YEAR KING ◊ GOLDEN BOUGH.
YEATS, WILLIAM BUTLER (1865-1939) Irish poet, dramatist and mystic, recipient of the Nobel Prize for Literature in 1923. WBY was a close associate of George Russell (◊ AE), William MORRIS, George Moore (1852-1933), Douglas Hyde (1860-1949), J.M. Synge (1871-1909), Katharine Tynan (1861-1931), Lord DUNSANY and Lady Augusta Gregory (1852-1932). He is of most interest for two strands of his work: his studies of FOLKLORE and of OCCULTISM. The two are intertwined to a large extent, particularly in his poetry, stories and sketches. His folklore researches were part of an exploration of his national heritage and identity, engendering a firm understanding of his roots which he believed was fundamental to the Celtic revival. His study of Irish legends and tales, which he did with Russell and Hyde, emerged in *Fairy and Folk Tales of the Irish Peasantry* (anth **1888**; vt *Irish Fairy and Folk Tales* 1893; exp 1895 US; vt *Irish Folk Stories and Fairy Tales* 1957 US). This is a series of studies, anecdotes, poems and tales drawn from a variety of sources, encompassing the essential feyness of the Irish heritage. He also assembled a less detailed volume for children, *Irish Fairy Tales* (anth **1892**; assembled with above as *Fairy and Folk Tales of Ireland* omni **1977**). WBY continued this revival of the Celtic heritage in *Stories from Carleton* (coll **1889**), a selection of material from the folklorist William Carleton (1794-1869) and *Representative Irish Tales* (anth **1891**; cut vt *Irish Tales* ?1892 US), selecting from other Irish writers. However, he soon rebelled against the clinical collation of folktales and yearned for his own expression of SEHNSUCHT.

This had already emerged in his poetry, starting with *The Wanderings of Oisin and Other Poems* (coll **1889**), the title poem dealing with Ireland's legendary bard Oisin, who travels on a fairy steed and spends TIME IN FAERIE. His sensitive retelling of legends continued in *Poems* (coll **1895**; rev 1899), *The Wind Among the Reeds* (coll **1899**) and *In the Seven Woods* (coll **1903**), which very effectively combine WBY's mystical leanings with his interpretation of the legends, and his verse-play *The Shadowy Waters* (**1900**; rev 1906), which utilizes Irish mythology for the setting of a visionary QUEST.

Folktales also inspired his dramas. The first, "The Countess Kathleen", was published as part of *The Countess Kathleen and Various Legends and Lyrics* (**1892**; rev vt *The Countess Cathleen* 1912). It tells of a noble lady who, during a famine, makes a PACT WITH THE DEVIL so that her people may have food. Because of her sacrifice God forgives her. When this was first performed in 1899 many found it blasphemous.

WBY joined a local Hermetic Society in 1886, studied THEOSOPHY, and met Madame BLAVATSKY (he was expelled from the Theosophical Society because of the ardour of his studies); he also joined the GOLDEN DAWN in March 1890 – he would become an adept and later the Imperator of the London Temple. These studies resulted in a strange early volume of two short novels, *John Sherman and Dhoya* (coll **1891**), which juxtaposed occult and mythological perspectives in order to contrast the moods of the poet and the sorcerer.

These two strands came together in a series of books that were later combined as *Mythologies* (omni **1959**; cut vt *The Secret Rose and Other Stories* 1982). The bibliography of these books is complicated and the following does not pretend completeness. The first of the volumes was *The Celtic Twilight* (coll 1893; exp 1902), a miscellany of tales, FABLES and sketches showing the relationship between humankind and the FAIRIES. The stories in *The Secret Rose* (coll **1897**) are more mystical, using WBY's knowledge of ROSICRUCIANISM to interpret some of the legends of Ireland. Several of the stories involve the thoughts and adventures of **Red Hanrahan**. WBY was unsatisfied with these: Lady Gregory helped him revise them as *Stories of Red Hanrahan* (coll **1905**).

Of particular interest in *The Secret Rose* is the story "Rosa Alchemica" (1896 *The Savoy*), an enlightening study of ALCHEMY, which introduced the character of **Michael Robartes**, an adept modelled on Macgregor Mathers (real name Samuel Liddell Mathers; 1854-1918), one of the founders of the Golden Dawn, whom WBY would use in his later occult stories, plays and essays collected as *Michael Robartes and the Dancer* (coll **1921**) and *Stories of Michael Robartes and his Friends* (coll **1932**). Other occult writings include *The Tables of the Law; The Adoration of the Magi* (coll **1897**) and *Per Amica Silentia Lunae* (coll **1918**).

In addition to these, WBY produced an outspoken volume of miscellaneous writings about mysticism and magic, *Ideas of Good and Evil* (coll **1903**), plus *A Vision* (coll **1925**; exp 1937), partly based on apparent automatic writing by his wife Georgie Hyde-Lees (1892-1968).

These volumes contain the bulk of WBY's fictional and semifictional occult and folkloristic writings, although all his work (certainly prior to 1905, when he left the Golden Dawn) was influenced by these passions. Unlike Arthur MACHEN and Algernon BLACKWOOD, who were able to merge their occult beliefs into a narrative, WBY's feelings were too strong to contain in simple stories. As a consequence his work is not as familiar to devotees of fantastic fiction as it should be. [MA]
Other works: *Mosada* (**1886** chap), offprint of poem from *Dublin University Review*; *Selections from the Writings of Lord Dunsany* (coll **1912**), ed; *The Poems of W.B. Yeats* (omni **1949**); *The Poems* (**1990**; rev 1994) ed Daniel Albright, the most complete collection. WBY also ed the occasional publications *Beltaine* (3 issues May 1899-April 1900), *Samhain* (7 issues October 1901-November 1908) and *The Arrow* (5 issues October 1906-August 1909).
Further reading: WBY's own autobiographical writings, collected as *Autobiographies* (omni **1955**). Of the many books about WBY the most relevant are *Yeats: The Man and the Masks* (**1948**) by Richard Ellmann, *The Unicorn: William Butler Yeats' Search for Reality* (**1954**) by Virginia Moore, *William Butler Yeats* (**1971**) by Denis Donoghue, *Yeats's Golden Dawn* (**1974**) by George Mills Harper, *W.B. Yeats and Irish Folklore* (**1980**) by Mary Helen Thuente and *A Bibliography of the Writings of W.B. Yeats* (**1951**; 3rd rev Russell K. Alspach 1968) by Allan Wade.
YELLOW BOOK, THE ◊ Aubrey BEARDSLEY; MAGAZINES; Oscar WILDE.
YELLOW BRICK ROAD ◊ OZ.
YELLOW SUBMARINE UK ANIMATED MOVIE (*1968*). Apple/King Features/United Artists. **Pr** Al Brodax. **Dir** George Dunning. **Screenplay** Brodax, Jack Mendelsohn, Lee Minoff, Erich Segal. **Voice actors** Paul Angelis (Ringo/Head Blue Meanie), Peter Batten (George), John Clive (John), Dick Emery (Lord Mayor/Max/Jeremy

Hillary Booby, the Nowhere Man), Geoffrey Hughes (Paul), Lance Percival (Old Fred). 87 mins. Colour.

A surreal FANTASYLAND called Pepperland comes under the tyranny of the Blue Meanies, deploying a monstrous flying creature (The Glove) and a variety of missiles, including bright green apples, that leach colour from the land and its people, petrifying the latter. Old Fred takes the ancestral Yellow Submarine in search of help, finds The Beatles in Liverpool, and brings them back through various adventures to Pepperland; *en route* they pick up the Nowhere Man as an ally. In Pepperland they impersonate Sergeant Pepper's Lonely Hearts Club Band, restore colour to the domain, and beat the Blue Meanies into submission with a barrage of hippy sentiments and Beatles numbers.

Psychedelic animation to a backing of old Beatles songs might seem a grim prospect, but YS is surprisingly enjoyable, with a constant stream of visual and conceptual fantasy notions tripping from the screen. The influences of Hieronymus BOSCH and Salvador DALI, though much diluted, are evident. The "Lucy in the Sky with Diamonds" sequence, done in a much different, far more Impressionistic style than the rest (although in the same livid colours), is striking. The movie had its genesis in the popular US animated tv series *The Beatles* (1965-7), produced by Brodax. [JG]

YEP, LAURENCE (MICHAEL) (1948-   ) US writer, most of whose books have been juveniles, including some sf and the highly successful *Dragonwings* (1975), a nonfantasy story about Chinese-Americans. *Child of the Owl* (1977) and several other mainstream novels are also about the Chinese experience in the USA. A fantasy series, the **Shimmer and Thorn** sequence – *Dragon of the Lost Sea* (1982), *Dragon Steel* (1985), *Dragon Cauldron* (1991) and *Dragon War* (1992) – deals engagingly with the magical adventures of a SHAPESHIFTING dragon called Shimmer and an orphan boy called Thorn. The narratives are much enlivened by elements drawn from Chinese lore and legend. *The Rainbow People* (coll 1989) and *Tongues of Jade* (coll 1991) assemble juvenile stories rewritten from Chinese-US FOLKTALES. [JC/DP]

Other works: *The Curse of the Squirrel* (1987); *The Shell Woman and the King: A Chinese Folktale* (1993 chap); *The Junior Thunder Lord* (1993 chap); *Butterfly Boy* (1993 chap); *The Man who Tricked a Ghost* (1993 chap); *Tiger Woman* (1994 chap); *The Boy who Swallowed Snakes* (coll 1994); *The Ghost Fox* (1994 chap), all for young children.

YGGDRASIL ◊ WORLD-TREE.

YIN AND YANG In ancient Chinese belief, the complementary principles of life, with Yin corresponding to the Earth (female, dark, passive) and Yang to HEAVEN (male, light, active). The preferred complex BALANCE is reflected in the yin-yang symbol, where dark and light flow into each other and each has a blob of the opposite colour at its heart. Ursula K. LE GUIN's **Earthsea** series celebrates this needed interplay of light and dark, life and death, a dance on which the world depends; but R.A. LAFFERTY, disliking such implied compromise between GOOD AND EVIL, made bladed yin-yangs the weapon of the Enemy in *Aurelia* (1982). [DRL]

YOLEN, JANE (HYATT) (1939-   ) US writer and editor, best-known for her children's books, especially CHILDREN'S FANTASY, of which she also became a publisher with her own YA imprint, Jane Yolen Books, for Harcourt Brace (1990-96). She has written or compiled over 130 books since *Pirates in Petticoats* (1963), nonfiction about

women pirates, and *See This Little Line* (graph 1963), a picture book in rhyme, both for children. Most of her work is fantasy, but it includes volumes of poetry, music books, picture-books, new and revisionist FAIRYTALES, YA novels and studies of children's literature.

Much of JY's work is short fiction, and some of her novels are in episodic form. *The Magic Three of Solatia* (1974), her earliest children's book also to appeal to adults, is a sequence of four novellas tracing the impact of three magic buttons (◊ TALISMANS) on a group of people and how the power must be faced and controlled. JY has mastered the fairytale form like no other modern writer, capturing the essence of folk tradition and presenting it for modern readers. Such stories have been collected as *The Girl who Cried Flowers and Other Tales* (coll 1974 chap), *The Moon Ribbon and Other Tales* (coll 1977 chap), *The Hundredth Dove and Other Tales* (coll 1977 chap), *Dream Weaver* (coll 1979 chap), *Neptune Rising: Songs and Tales of the Undersea Folk* (coll 1982), *Tales of Wonder* (coll 1983), *The Whitethorn Wood and Other Magicks* (coll 1984 chap), *Dragonfield and Other Stories* (anth 1985); *The Faery Flag* (coll 1989), *Storyteller* (coll 1992), *Here There Be Dragons* (coll 1993), *Here There Be Unicorns* (coll 1994), *Here There Be Witches* (coll 1995) and *Here There Be Angels* (coll 1996). Individual short stories, excluding those intended for very young children, have also been published as *The Girl who Loved the Wind* (1972 chap), *The Boy who Had Wings* (1974 chap), *The Lady and the Merman* (1979 chap) and *The Sword and the Stone* (1985 F&SF; 1991 chap). JY has also retold the original fairytale of SLEEPING BEAUTY as *The Sleeping Beauty* (1986 chap) and humorously twisted it as *Sleeping Ugly* (1981 chap), and has retold the ballad of TAM LIN in *Tam Lin* (1990 chap). JY's closest rivals – Diana Wynne JONES, Tanith LEE and Patricia MCKILLIP – are, because first and foremost fantasists, seldom able to approach the TWICE-TOLD form of STORY that JY has made her own.

JY's longer works take various forms. *The Mermaid's Three Wisdoms* (1978) is about a deaf girl and a wordless MERMAID who become friends through sign language. *The Wild Hunt* (1995) retells the WILD-HUNT legend but in an ALTERNATE WORLD. *Briar Rose* (1992) is a powerful retelling of the SLEEPING-BEAUTY tale, transposed to WORLD WAR II, where a young Jewish girl is gassed but is revived by the kiss of life. The HOLOCAUST was also the setting for the TIMESLIP novel *The Devil's Arithmetic* (1988).

With the **Great Alta** sequence – *Sister Light, Sister Dark* (1988) and *White Jenna* (1989), assembled as *The Book of Great Alta* (omni 1990) – JY's fiction took on mythic proportions. Again she chose an episodic format to build varying perspectives of the events. Jenna is a young girl who begins to live the prophecy that states she will become the MESSIAH. The book has much in common with the **Dune** series by Frank Herbert (1920-1986), but JY uses her narrative skills to allow the reader to question the real status of Jenna and to contrast the shadowy layers of GOOD AND EVIL.

JY sometimes uses other planets as settings for her stories, but these are only tokenly sf. The **Pit Dragon** trilogy – *Dragon's Blood* (1982), *Heart's Blood* (1984) and *A Sending of Dragons* (1987) – is set on a penal planet where DRAGONS are essential to the economy. *Cards of Grief* (1984), also set on an alien world, is a homage to the storytelling art and its place in the development of culture and civilization.

Merlin's Booke (coll of linked stories 1986) brings together an idiosyncratic collection of poems and stories about

MERLIN from his youth to his old age, each reflecting a different aspect of his life. JY returned to the world of King ARTHUR in *The Dragon's Boy* (**1990**), the children's book *Merlin and the Dragons* (**1995** chap) – retelling the story of Merlin and the fate of Vortigern – and the YA anthology *Camelot* (anth **1995**). She also wrote the **Young Merlin** trilogy, «Passager» (1996), «Hobby» (1996) and «Merlin» (1997); while she retains the core of the Merlin myth, her treatment of events is often far from orthodox.

JY is one of the leading storytellers of the 20th century. [MA]

Other works (excluding titles for the very young): *The Witch who Wasn't* (**1964**); *Gwinellen, the Princess who Could Not Sleep* (**1965** chap); *The Emperor and the Kite* (**1968** chap); *The Wizard of Washington Square* (**1969**); *The Seventh Mandarin* (**1970**); *The Transfigured Hart* (**1975**); *The Boy who Spoke Chimp* (**1981**); *Brothers of the Wind* (**1981**); *Children of the Wolf* (**1984**); *Wizard's Hall* (**1991**); very many others.

As editor: *Zoo 2000* (anth **1973**); *Shape Shifters* (anth **1978**); *Dragons & Dreams* (anth **1986**) and *Spaceships and Spells* (anth **1987**), these two with Martin H. GREENBERG and Charles G. Waugh; *Favourite Folktales from Around the World* (anth **1986**); *2041 ad* (anth **1991**); and, with Greenberg, *Werewolves* (anth **1988**), *Things That Go Bump in the Night* (anth **1989**), *Vampires* (anth **1991**), the **Xanadu** series – *Xanadu* (anth **1993**), *#2* (anth **1994**) and *#3* (anth **1995**) – and *The Haunted House* (anth **1995** chap).

Nonfiction: *The Wizard Islands* (**1973**); *Writing Books for Children* (**1973**; rev 1983); *Touch Magic: Fantasy, Faerie and Folklore in the Literature of Childhood* (**1981**); *Guide to Writing for Children* (**1989**).

YOUNG, ELLA (1867-1956) US writer deeply involved in the US version of THEOSOPHY, and friend of Kenneth MORRIS. In her own writing she focused on CELTIC FANTASY. *The Wonder Smith and his Son* (coll **1927**) and *The Tangle-Coated Horse and Other Tales: Episodes from the Fionn Saga* (coll **1929**) are TWICE-TOLD tales based on Irish material. *The Unicorn with Silver Shoes* (**1932**) illus Robert LAWSON is an original tale, though resembling both Morris and James STEPHENS in its telling of the trip of an Irish hero to the AFTERLIFE, where he meets Angus, the laughing god and others. [JC]

YOUNG, ROBERT F(RANKLIN) (1915-1986) US writer, better remembered for his SCIENCE FICTION despite a regular production of short fantasy stories for over 30 years after "The Black Deep Thou Wingest" (1953 *Startling Stories*). Like many of his best stories, this had a strong sentimental flavour, almost of BELATEDNESS. He used this to strong effect in his short fiction, where he was able to demonstrate the effect of THINNING, whether on this world, as in "The Forest of Unreason" (1961 *Fantastic*), "Neither Stairs Nor Door" (1963 *Fantastic*) and "The House that Time Forgot" (1963 *Fantastic*), or a colonized planet, as in "To Fell a Tree" (1959 *F&SF*; exp as *The Last Yggdrasil* 1982), an early example of ecological sf. RFY's fascination for GIANTS is found in "Goddess in Granite" (1957 *F&SF*), "The Giantess" (1973 *F&SF*) and the Gulliver-like "The Journal of Nathaniel Worth" (1978 *Fantastic*).

RFY could also be humorous, often satirical, particularly in a number of REVISIONIST FANTASIES that he produced based on old LEGENDS and FAIRYTALES. These include "Santa Clause" (1959 *F&SF*), "There Was an Old Woman who Lived in a Shoe" (1962 *F&SF*), "A Knyght Ther Was" (1963 *Analog*), "The Quest of the Holy Grille" (1964 *Amazing*; exp trans as *La Quête de la Sainte Grille* **1975** France), "Peeping Tommy" (1965 *Galaxy*), "Rumpelstiltskinski" (1965 *Amazing*), and the ARABIAN FANTASY "City of Brass" (1965 *Amazing*; exp as *The Vizier's Second Daughter* **1985**). Right until his last story, "The Giant, the Colleen, and the Twenty-One Cows" (1987 *F&SF*), RFY continued to transplant FOLKTALES – in this case JACK and the beanstalk – to new worlds.

Most of RFY's best sf was collected as *The Worlds of Robert F. Young* (coll **1965**) and *A Glass of Stars* (coll **1968**), but his best fantasies are uncollected. [MA]

Other works: *Starfinder* (fixup **1980**); *Eridahn* (**1964** *If* as "When Time was New"; exp **1983**), a TIME-TRAVEL adventure in prehistory.

YOUNG AGAIN US movie (*1986* tvm). Sharmhill/DISNEY. **Pr** Steven H. Stern. **Dir** Stern. **Screenplay** Barbara Hall. **Starring** Jack Gilford (Angel), Keanu Reeves (Mick Riley at 17), Jessica Steen (Tracy Gordon), Robert Urich (Mick Riley at 40), Lindsay Wagner (Laura Gordon). 85 mins. Colour.

Thanks to a WISH foolishly expressed in front of an ANGEL, 40-year-old Riley becomes his 17-year-old self, but in the present. As a teenager he falls in love all over again with the girl, Laura, he used to date, who is now in her late 30s. A second wish makes him 40 again, and now he can win her heart. This is by no means a bad IDENTITY-EXCHANGE movie, yet eventually is conquered by its own blandness. [JG]

YOUNG CONNECTICUT YANKEE IN KING ARTHUR'S COURT, A US movie (*1995*). ◊ *A* CONNECTICUT YANKEE.

YOUNG FRANKENSTEIN US movie (*1974*). ◊ FRANKENSTEIN MOVIES.

YOUNG SHERLOCK HOLMES AND THE PYRAMID OF FEAR (vt *Young Sherlock Holmes*) US movie (*1985*). ◊ SHERLOCK HOLMES; Steven SPIELBERG.

YOURCENAR, MARGUERITE Pseudonym of French writer Marguerite de Crayencour (1903-1987), who remains best-known in English-speaking lands for her *Mémoires d'Hadrien* (**1951**; trans Grace Frick with MY as *Memoirs of Hadrian* **1954** US), an historical meditation. MY had a long career. *Le Jardin des chimères* ["The Garden of Chimaeras"] (**1921**) is a dialogue in verse on the subject of ICARUS. *Feux* (coll **1936**; rev 1968; trans as *Fires* **1981** US) assembles prose poems which meditate, through abstract narratives featuring Greek characters, on the relationship between myth and love. *Nouvelles orientales* (coll **1938**; rev 1963; exp 1978; trans Alberto Manguel with MY as *Oriental Tales* **1985** UK) is a set of ORIENTAL FANTASIES which deals sophisticatedly of similar material. In *L'Oeuvre au noir* (**1968**; trans as *The Abyss* **1976** US), set in Renaissance Italy, a FAUST-like MAGUS attempts through ALCHEMY to gain an occult understanding and control over the cosmos. [JC]

ZÁBORSKÝ, JONÁŠ (1812-1876) Slovak writer. ◊ SLOVAKIA.
ZAGO, JUNGLE PRINCE ◊ TARZAN.
ZATANNA ◊ ZATARA.
ZATARA Caped, top-hatted and tuxedoed MAGICIAN of US COMIC books, Giovanni Zatara was created by writer Gardner Fox (1911-1986) and artist Fred Guardiner (1913-) in *Action Comics #1* (1938): he was one of the dozens of tuxedoed magicians in swirling capes to combat supernatural villains in US comic books in response to the great popularity of stage illusionists and the success of Lee Falk's *Mandrake the Magician*. Originally a routine vaudeville circuit conjurer whose tricks were so old (learned from a book he had received as a child) that he was regularly booed off the stage, Zatara's fortunes changed soon after his 19th birthday, when he found one of the notebooks of his ancestor, Leonardo da Vinci, which was written in reverse to preserve its secrecy. Reading aloud, Zatara found that he was no mere illusionist but a magician of awesome power, and from that moment began to perform truly mystical deeds by saying everything backwards: thus, to rid himself of flying snakes, he merely had to pronounce "Og yawa uoy gniylf sekans!". Artists on the feature included Joe Kubert (1926-). Over the next 12 years Zatara battled demons, evil mummies, zombies and, towards the end of his days, when crime stories were selling better, killers and gangsters. He took his last bow in *World's Finest #51* (1950).

It was later revealed, however, that on one of his many trips to ATLANTIS, he had fathered a daughter, Zatanna, who was created by Fox and artist Murphy Anderson (1926-) in *Hawkman #4* (1964) and introduced to the DC COMICS universe by a useful PLOT DEVICE: she was seeking her missing father. Half Atlantean (on her mother's side) she looked more like a conjurer's assistant than a magician in her own right. Her quest to find her father continued through various titles (including *Atom*, *Green Lantern* and *Justice League of America*), and she finally located him in *Justice League of America #51* (1967). She was eventually to become a member of the JLA (in *#161*, 1978), leaving in 1986 (*#257*).

Giovanni Zatara was killed in *Swamp Thing #50* (1987) when Earth's greatest magicians were gathered together to face a threat to the supernatural side of DC's universe. He also appeared in the late 1980s in *Young All-Stars*, which featured adventures from the 1940s, when he had lived on a large estate, Shadowcrest, with his faithful servant, Tong.

Zatanna remains a minor DC character with only occasional appearances in recent years, notably in the second series of *Spectre* (1987-9), a *Zatanna Special* (1987) drawn by Gray Morrow, *The Books of Magic #2* (1990) by Neil GAIMAN and Scott Hampton, and *Zatanna: Come Together* (*#1-#4*, 1993) by Lee Marrs and Esteban MAROTO. In DC's current continuity, Zatanna is retired from magic and resides in San Francisco. [SH/RT]
ZEGRA THE JUNGLE EMPRESS ◊ TARZAN.
ZELAZNY, ROGER (JOSEPH) (1937-1995) US writer who in 1962 gained an MA from Columbia University in Elizabethan and Jacobean drama. An intense interest in the dark intricacies and metaphysical daring displayed by the dramatists in whom he specialized infused all RZ's best work, giving even to some of his more routine output (in later years) a sense that he could always increase the density and pressure of his storytelling had it suited him to do so.

After "Mister Fuller's Revolt" for *Literary Cavalcade* in 1954 and various other nonprofessional appearances RZ began publishing sf with "Passion Play" for *Amazing* in 1962. He will almost certainly remain most admired for the sf – some as by Harrison Denmark – published in the first half-decade of his career. *Four for Tomorrow* (coll **1967**; vt *A Rose for Ecclesiastes* 1969 UK) and *The Doors of His Face, the Lamps of His Mouth, and Other Stories* (coll **1971**) assemble stories which remain central to the genre; the title story of the second volume, also published as *The Doors of His Face, the Lamps of His Mouth* (1965 *F&SF*; **1991** chap), won a Nebula for Best Novelette. His first novels – *This Immortal* (1965 *F&SF* as ". . . And Call me Conrad"; exp **1966**), which won a 1966 Hugo AWARD, *He Who Shapes* (**1989** dos) (1965 *Amazing*; **1989** dos), first published in book form as *The Dream Master* (1965 *Amazing* as "He Who Shapes"; exp **1966**), which won a 1966 Nebula Award as Best Novella, and *Lord of Light* (**1967**), which won a 1968 Hugo – are texts which help define 1960s sf.

Some of this material can be partially understood in fantasy terms. "A Rose for Ecclesiastes" (1963) and *The Doors of his Face* are set in PLANETARY-ROMANCE versions of, respectively, Mars annd Venus; the latter, by incorporating a pattern of references to Herman MELVILLE's *Moby-Dick* (**1851**), makes its action most clearly understandable as an acting-out of something approaching ARCHETYPE. Similarly, the UNDERLIER figure shaping the feats of the IMMORTAL hero of *This Immortal* is Herakles, though Conrad Nomikos is also a TRICKSTER, and maybe a GOD; he is the first representative of the kind of HERO RZ was almost obsessively taken with – the self-mocking, immortal, romantic, profoundly sagacious, dusk-ridden jokester Hero of a Thousand Faces (◊ MONOMYTH), an emblem of BELATEDNESS (and self-pity), but also the exuberant bearer of human tidings to the gods (and vice versa). *Lord of Light* – though underpinned as sf – reads for much of its course as full-blown fantasy: only after a while do we learn that a planet has been colonized by the Hindu crew of an exploring human starship, and that long before the crew used technology to make themselves into a PANTHEON of Hindu gods, thereafter ruling the world, its human colonists (the ship's passengers) and its natives. TECHNOFANTASY may ultimately be subverted by rational explanation in this tale, but the aura of MAGIC does not fade. *Creatures of Light and Darkness* (**1969**) less convincingly places the pantheon of ancient EGYPT in a similar position of power over space and time.

Even in his seminal early sf, the pattern of revelation in RZ's work tends not to conceptual breakthrough but to retroactive moments of RECOGNITION; the move of the RZ text is backwards to MYTH, and readers normally meet his most common protagonists at a point when their immortality conveys a sense of TIME ABYSS. The RZ fantasy which exemplifies this pattern most clearly – and the sequence he is most widely known for – is the **Chronicles of Amber**, which breaks into two linked series. The first, built around the figure of Corwin, is *Nine Princes in Amber* (**1970**), *The Guns of Avalon* (**1972**), *Sign of the Unicorn* (**1975**), *The Hand of Oberon* (**1976**) and *The Courts of Chaos* (**1978**), all five assembled as *The Chronicles of Amber* (omni **1979** 2 vols). The second, focusing on Corwin's son Merlin, is *Trumps of Doom* (**1985**), *Blood of Amber* (**1986**), *Sign of Chaos* (**1987**), *Knight of Shadows* (**1989**) and *Prince of Chaos* (**1991**). Two further related texts are *A Rhapsody in Amber* (coll **1981** chap) and *Roger Zelazny's Visual Guide to Castle Amber* (**1988**), the latter with Neil Randall. Amber itself is a central REALITY whose ontological substance is denser than the innumerable "Shadow" realities – one of which contains our world and its history – it underlies and which are cast off like spume. Similarly, Corwin and his many siblings – whose names indicate their UNDERLIER roles – are more real than mortals, or the GODS of any Shadow realm. The STORY of our world is, therefore, a shadowy semblance of the true story. There may be a touch of Gnostic thought here, though it is scanted; more centrally, RZ makes considerable play with JUNGIAN PSYCHOLOGY, and his characters duly treat various SHADOW worlds as though they were dioramas of the Collective Unconscious, which they are privileged to explore and manipulate. But, as the sequence continues, it becomes clear that Amber itself is not an ultimate reality; it shares a YIN-AND-YANG relationship with the courts of CHAOS, a relationship which generates a sense that whatever ultimate reality exists moves in CYCLES.

The sequence begins with Corwin, who seems a human afflicted with AMNESIA but who is (we learn) one of the sons of Oberon, the supposedly dead king of Amber, and –

in the TAROT-based symbology of the sequence – a FOOL. Aspects of DYNASTIC FANTASY – sometimes overcomplex – percolate throughout the tale as the various siblings vie for the throne. So destructive are these conflicts by the second volume that Amber is seen as a WASTE LAND and Corwin's QUEST for self-identity and empowerment also becomes a quest for the GRAIL; images of HEALING proliferate from this point. By the end of the fifth volume, Corwin – by understanding that Chaos underlies Amber in the same way that Amber underlies the phenomenal Universe – begins to be able to bestow something like peace upon the shape of the LAND.

The second sequence is much less organized and is fuller of PLOT DEVICES typical of DYNASTIC FANTASY; it may not conclude with the last published volume, which is absent-minded and digressive.

Two further sequences are of mild interest. The **Wizard World** series – *Changeling* (**1980**) and *Madwand* (**1981**), assembled as *Wizard World* (omni **1989**) – parallels **Amber** through a protagonist who is exiled on Earth and must valorously gain back his role and his world. The **Dilvish** series – *The Bells of Shoredan* (1966 *Fantastic*; **1979** chap), *The Changing Land* (**1981**) and *Dilvish, the Damned* (coll of linked stories **1982**) – features another exiled protagonist, who escapes his initial BONDAGE and engages in SWORD-AND-SORCERY adventures *en route* to VENGEANCE and reinstatement.

Some of RZ's earlier singletons are of strong fantasy interest. *Jack of Shadows* (**1971**) is set on a non-rotating world whose light side is understandable in sf terms but whose dark side is run by MAGIC. The JACK hero undergoes a variety of PICARESQUE adventures in his QUEST for his true name, his revenge and his empowerment. Some fantasy stories are included in *The Last Defender of Camelot* (coll **1980**; exp 1981), *Unicorn Variations* (coll **1983**) – which won a 1984 BALROG AWARD – *Frost and Fire* (coll **1989**) and *Gone to Earth* (coll dated 1991 but **1992**). Late fantasies include *The Black Throne* (**1990**) with Fred SABERHAGEN, a RECURSIVE fantasy in which Edgar Allan POE and his DOUBLE switch ALTERNATE WORLDS, with some lighthearted swashbuckling as a consequence. *The Mask of Loki* (**1990**) with Thomas T. Thomas (1948-   ) is SCIENCE FANTASY. The **Azzie Elbub** sequence with Robert Sheckley (1928-   ) – *Bring Me the Head of Prince Charming* (**1991**), *If at Faust You Don't Succeed* (**1993**) and *A Farce to be Reckoned With* (**1995**) – centres on AGONS in which the forces of GOOD AND EVIL contest the moral running of the world. *A Night in the Lonesome October* (**1993**) quirkily and succinctly recounts a GASLIGHT ROMANCE featuring all the usual late-Victorian suspects (DRACULA, SHERLOCK HOLMES, JACK THE RIPPER, etc.) but through the voice of a TALKING ANIMAL.

RZ's last books were more powerful than many of those published in the 1970s and 1980s: he may have wished to concentrate once again. His death was premature.   [JC]
Other works (mostly sf): *Damnation Alley* (**1969**); *Isle of the Dead* (**1969**) and *To Die in Italbar* (**1973**), both featuring **Francis Sandow**; *Today We Choose Faces* (**1973**); *Poems* (coll **1974** chap); *Doorways in the Sand* (**1976**); *Bridge of Ashes* (**1976**); *Deus Irae* (**1976**) with Philip K. Dick (1928-1992); *The Illustrated Zelazny* (graph coll **1978**; rev vt *The Authorized Illustrated Book of Roger Zelazny* 1979); *Roadmarks* (**1979**); *For a Breath I Tarry* (1966 NW; **1980** chap); *When Pussywillows Last in the Catyard Bloomed* (coll **1980** chap), poetry; *Coils* (**1980**) with Saberhagen; *Today We Choose*

Faces/Bridge of Ashes (omni **1981**); *To Spin is Miracle Cat* (coll **1981**), poems; *Eye of Cat* (**1982**), which though sf powerfully evokes the classic NIGHT JOURNEY of Native American SHAMANISM; *A Dark Traveling* (**1987**), a fantasy juvenile; *The Graveyard Heart* (1964 *Amazing*; **1990** chap dos); *Home is the Hangman* (1975 *Analog*; **1990** chap dos); *Flare* (**1992**) with Thomas; *Way Up High* (**1992** chap); *Here There be Dragons* (**1992** chap); *Wilderness* (**1994**) with Gerald Hausman (1945-), associational.

As editor: *Nebula Award Stories Three* (anth **1968**); *Warriors of Blood and Dream* (anth **1995**) with Martin H. GREENBERG; *Forever After* (anth **1995**); *Wheel of Fortune* (anth **1995**) with Greenberg; *The Williamson Effect* (anth **1996**).

Further reading: *Roger Zelazny: A Primary and Secondary Bibliography* (**1980**) by Joseph L. Sanders; *Roger Zelazny* (**1993**) by Jane Lindskold.

ZEMECKIS, ROBERT (1952-) US moviemaker, originally talent-spotted by Steven SPIELBERG, with whom he has since often worked: the two have similar directorial styles and work with similar material – although it is unlikely that Spielberg would have been able to cope with WHO FRAMED ROGER RABBIT (**1988**), RZ's masterpiece, which investigates levels of fantasy to which the older director has not yet aspired. Of fantasy interest are: *Romancing the Stone* (**1984**), which resembles an INDIANA JONES movie (RZ was not responsible for the pallid sequel, *Jewel of the Nile* [**1985**]), *Back to the Future* (**1985**), which he wrote, plus its sequels *#II* (**1989**) and *#III* (**1990**), all three being TIME-TRAVEL adventures; and DEATH BECOMES HER (**1992**), a comedy, with state-of-the-art spfx, about the price of physical IMMORTALITY. [JG]

ZEUS The supreme GOD of the Greek PANTHEON, ruler of OLYMPUS, called Jupiter or Jove by the Romans. This last name links him phonetically with the Judeo-Christian All-Father. Percy Bysshe SHELLEY preferred to use "Jupiter" as the name of the tyrant god in *Prometheus Unbound* (**1820**), while Benjamin DISRAELI used "Jove" in the political ALLEGORY "Ixion in Heaven" (1847). Zeus was the parent of many demigods as well as many gods, frequently assuming exotic disguises for the purpose of seducing human women; his consequent reputation as a determined lecher is treated with humorous contempt in several comic fantasies, of which the most profound is John ERSKINE's *Venus the Lonely Goddess* (**1949**). He is also an absurd comic figure in *Olympian Nights* (**1902**) by John Kendrick BANGS. His symbolic presence is highly significant – in an assortment of ways – in *The Miniature* (**1926**) by Eden PHILLPOTTS, *Mistress of Mistresses* (**1935**) and its prequels by E.R. EDDISON and the series by Ronald FRASER which includes *Jupiter in the Chair* (**1958**). Time has mellowed him considerably in *Mr Kronion* (**1949**) by Susan Alice KERBY, but his life-story is respectfully reinterpreted in *The Memoirs of Zeus* (**1963**) by Maurice Druon (1918-). [BS]

ZEYER, JULIUS (1841-1901) Czech writer. ◊ CZECH REPUBLIC.

ZIMMER, PAUL EDWIN (1943-) US writer, brother of Marion Zimmer BRADLEY, with whom he wrote his first novels, a pair of sf adventures. A fantasy series by PEZ alone, **Dark Border** – *The Lost Prince* (**1982**), *King Chondos' Ride* (**1982**) and *A Gathering of Heroes* (**1987**) – about two brothers tussling for their inheritance in a magical kingdom, and a pendant novel, *Ingulf the Mad* (**1989**), are enlivened by the author's detailed knowledge of swordsmanship. His fantasy singletons are *Woman of the*

Elfmounds (dated 1979 but **1980**) and *Blood of the Colyn Muir* (**1988**), with Jon DeCles. [JC/DP]

ZIPES, JACK (DAVID) (1937-) US professor of German and internationally acknowledged authority on FAIRYTALES. JZ has worked extensively not only as storyteller but in the development of children's theatre, in the USA, Canada and Germany. His early research was into German ROMANTICISM, resulting in his first book *The Great Refusal: Studies of the Romantic Hero in German and American Literature* (**1970** Germany). He then extended his research into the fairytale, from its FOLKTALE roots through its development as the LITERARY FAIRYTALE and its adaptation for SATIRE, didacticism, commercialism, FEMINISM and REVISIONIST FANTASY. This led to *Breaking the Magic Spell: Radical Theories of Folk & Fairy Tales* (coll **1979**), a volume of essays exploring the evolution of the genre. JZ's conclusion was that the fairytale had become desensitized and over-commercialized, partly because of the work of Walt DISNEY and partly because of later sf and fantasy CINEMA, which utilized the images of the fairytale without the underlying messages. JZ subsequently produced a series of studies and ANTHOLOGIES to explore further the different aspects of the fairytale and present them in context, so that modern readers could appreciate the full development of the form. His major works are *Fairy Tales and the Art of Subversion: The Classic Genre for Children and the Process of Civilization* (**1983**) and *Fairy Tale as Myth/Myth as Fairy Tale* (coll **1994**). His study of the GRIMM BROTHERS, *The Brothers Grimm* (**1988**), goes beyond their lives and works to explore their influence on German literature. He has also published his own translation of their stories in *The Complete Fairy Tales of the Brothers Grimm* (coll **1987** US), and has continued to explore the German passion for the fairytale in *Fairy Tales and Fables from the Weimar Days* (anth **1989**; vt *Utopian Tales from Weimar* 1990 UK), which examines its subversive political use in the 1920s, and *The Fairy Tales of Hermann Hesse* (coll **1995**), which shows HESSE's use of the fairytale to reflect the mood of modern ROMANTICISM.

JZ's anthologies began with a detailed analysis of the Little Red Riding-Hood story, *The Trials and Tribulations of Little Red Riding Hood* (anth **1983**; rev 1993), with 35 versions of the tale from the Grimm Brothers to Angela CARTER. *Don't Bet on the Prince* (anth **1986**) considered the contemporary feminist use of the fairytale. *Victorian Fairy Tales: The Revolt of the Fairies and Elves* (anth **1987**) explores the moral and didactic role of the fairytale for the Victorians. In *Beauties, Beasts and Enchantments* (anth **1989**) JZ turns to classic French fairytales, showing their satirical nature and exploring the transition from folktale to literary fairytale. *Spells of Enchantment* (anth **1991**; vt *The Penguin Book of Western Fairy Tales* 1993 UK) is a massive compilation tracing the history of the WONDER TALE from the 2nd century to today. *The Outspoken Princess and the Gentle Knight* (anth **1994**) looks at the modern fairytale. [MA]

ZODIAC The imagined circular belt in the sky against which the SUN and planets are seen to move; Western ASTROLOGY divides it into 12 "signs", which have acquired some symbolic force: Aries the Ram, Taurus the Bull, Gemini the TWINS, Cancer the crab, Leo the lion, Virgo the virgin (◊ VIRGINITY), Libra the scales, Scorpio the scorpion, Sagittarius the archer, Capricorn the goat, Aquarius the water-carrier and Pisces the fishes. John CROWLEY's *Little, Big* (**1980**) has an effective moment of WRONGNESS when the painted zodiac in the dome of NEW YORK's Grand Central

Station reverses so that its signs run at last in the correct direction. Rudyard KIPLING's strange fable "The Children of the Zodiac" (1891) brings the signs to life (omitting Capricorn, since Gemini is taken as two) as GODS, of whom only the death-gods like Cancer are truly immortal. [DRL]

ZOLTAN, HOUND OF DRACULA (vt *Dracula's Dog*; vt *Zoltan, Hound of Hell*) US movie (**1977**). ◊ DRACULA MOVIES.

ZOMBIE MOVIES Ever since Bela Lugosi's "Murder" Legendre ordered the revived corpses of his plantation workers to shamble across the screen in *White Zombie* (**1932**) it has been the CINEMA which has most influenced our perception of the walking undead. The same producers failed to repeat their success with *Revolt of the Zombies* (**1936**), but Bob Hope's comedic encounter with a zombie in *The Ghost Breakers* (**1940**) set the tone for such poverty-row movies as *King of the Zombies* (**1941**), *Revenge of the Zombies* (**1943**; vt *The Corpse Vanished* UK), *Voodoo Man* (**1944**) and *Zombies on Broadway* (**1945** vt *Loonies on Broadway* UK). Val Lewton's *I Walked With a Zombie* (**1943**) was an atmospheric reworking of Charlotte BRONTË's *Jane Eyre* (**1847**). Dean Martin and Jerry Lewis remade the Hope vehicle as *Scared Stiff* (**1953**), but in *Creature with the Atom Brain* (**1955**), *The Gamma People* (**1955**), *Teenage Zombies* (**1957**), *Invisible Invaders* (**1959**) or the infamous *Plan 9 from Outer Space* (**1959**) the walking undead of the 1950s were more likely to be scientifically created than the result of supernatural sorcery, as in *Voodoo Island* (**1957**; vt *Silent Death* UK), *The Zombies of Mora Tau* (**1957**; vt *The Dead that Walk* UK) and *Night of the Ghouls* (**1959**). Barbara Steele was menaced by revived Black Plague victims in the Italian *Terror Creatures from the Grave* (**1965**; ot *Cinque Tombe per un Medium*) and a Cornish graveyard gave up its dead in Hammer's *The Plague of the Zombies* (**1966**). But it was director George Romero's low budget *Night of the Living Dead* (**1968**), about an alien contamination which transformed the newly dead into cannibals, that became a cult hit; it spawned two direct sequels, *Dawn of the Dead* (**1978**; vt *Zombies* UK) and *Day of the Dead* (**1985**), and inspired numerous imitations. While continental European cinema added explicit gore to the mix in such films as *Bracula – The Terror of the Living Dead* (**1972**; ot *La Orgia de los Muertos*; vt *The Hanging Woman* UK), *Horror Rises from the Tomb* (**1972**; ot *El Espanto Surge de la Tumba*), *Vengeance of the Zombies* (**1972**; ot *La Rebellion de las Muertas*), *The Living Dead at the Manchester Morgue* (**1974**; ot *No Profanar el Sueno de los Muertos*; vt *Don't Open the Window* US), *Zombie Holocaust* (**1979**; ot *La Regina dei Cannibali*; vt *Doctor Butcher MD* [*Medical Deviate*] US), *Zombie Lake* (**1980**; ot *Le Lac des Morts Vivants*), *Zombie 3* (**1980**; ot *Le Notti del Terrore*) and countless others, such movies as *Children Shouldn't Play with Dead Things* (**1972**), *Neither the Sea nor the Sand* (**1972**), *Messiah of Evil* (**1973**), *Sugar Hill* (**1974**; vt *Voodoo Girl* UK), *Shock Waves* (**1976**; vt *Almost Human* UK) and *Dead & Buried* (**1981**) at least attempted to do something different with the genre. Spanish director Amando de Ossorio chronicled the exploits of the blind undead Templarios Knights in *Tombs of the Blind Dead* (**1971**; ot *La Noche del Terror Ciego*), *Return of the Evil Dead* (**1973**; ot *El Ataque de los Muertos sin Ojos*), *Horror of the Zombies* (**1974**; ot *El Buque Maldito*) and *Night of the Seagulls* (**1975**; ot *La Noche de las Gaviotas*), while Italian director Lucio Fulci carved out his own niche with *Zombie Flesh-Eaters* (**1979**; ot *Zombi 2*; vt *Zombie* US), *City of the Living Dead* (**1980**; ot *La Paura nella Citta dei Morti*; vt *Gates of Hell* US) and *The*

Beyond (**1981**; ot *E tu Vivrai nel Terrore . . . L'Aldila*; vt *Seven Doors to Death* US). Writer/director Sam Raimi's *The Evil Dead* (**1982**) reinvented the mythology for a new generation, and he returned with a superior sequel/remake, *Evil Dead II* (**1987**) and a silly third installment, *Army of Darkness* (**1992**). Michael Jackson led the walking dead in an exuberant dance routine for the extended music video *Thriller* (**1983**) (◊ ROCK VIDEOS), and the zombies that supposedly inspired Romero's 1968 movie formed the basis of a trio of belated spinoffs: *The Return of the Living Dead* (**1984**), *Return of the Living Dead Part II* (**1987**) and *Return of the Living Dead 3* (**1993**). Not to be outdone, Romero himself produced a colour remake of his original as *Night of the Living Dead* (**1990**) dir Tom Savini. Wes Craven's *The Serpent and the Rainbow* (**1987**) was a serious attempt to adapt Wade Davis' nonfiction book about voodoo in Haiti, but more typical of this period was such exploitation fare as *I Was a Teenage Zombie* (**1986**), *Redneck Zombies* (**1986**), *The Video Dead* (**1987**), *Chopper Chicks in Zombietown* (**1991**) and *Braindead* (**1992**; vt *Dead Alive* US), which took the zombie theme to new extremes of absurdity. At least *Weekend at Bernie's II* (**1992**), in which the titular character was periodically revived by the spell of a voodoo priestess, was *supposed* to be a comedy. [SJ]

Further reading: *Classic Movie Monsters* (**1978**) by Donald F. Glut; *Zombie: The Living Dead* (**1976**) by Rose London; *The Dead that Walk* (**1986**) by Leslie Halliwell.

ZOMBIES A term rarely found in fantasy, though often in HORROR and SUPERNATURAL FICTION. A zombie is a dead person who has been stimulated or reanimated into a kind of undead existence, almost invariably by means of VOODOO ritual. A zombie – in its traditional manifestation – is a powerful, mute, slavelike being which does not seem to suffer pain. Zombies are difficult to make interesting in prose – *Walking Dead* (**1977**) by Peter DICKINSON being a notable exception – but are vividly present in countless ZOMBIE MOVIES. To fill its pages, *The Mammoth Book of Zombies* (anth **1993**) ed Stephen JONES extends the term to cover REVENANTS. The threat of zombification can be a useful PLOT DEVICE, as in Dennis WHEATLEY's *Strange Conflict* (**1941**); such fear of DEBASEMENT is exploited in Barbara HAMBLY's *The Ladies of Mandrigyn* (**1984**), whose zombie-like "nuuwa" are humans whose brains have been partially eaten. Novels with TECHNOFANTASY rationales for zombies include Lucius SHEPARD's *Green Eyes* (**1984**), where the reanimation involves not only graveyard earth but tailored bacteria, and Mark FROST's *The List of 7* (**1993**). REVISIONIST-FANTASY zombies generally follow Shepard's in having intelligence and volition, thus raising civil-rights issues; examples include Piers ANTHONY's **Xanth** (where zombies are usually but not necessarily mentally impaired, owing to brain decay) and Terry PRATCHETT's **Discworld** – notably *Reaper Man* (**1991**) and *Witches Abroad* (**1991**). [JC/DRL]

See also: Hugh B. CAVE.

ZU: WARRIORS FROM THE MAGIC MOUNTAIN (vt *Zu: Warriors of the Magic Mountain*; vt *Zu: Warriors from the Mystic Mountain*) Hong Kong movie (**1982**). Film Workshop. **Pr** Raymond Chow. **Dir** Tsui Hark. **Spfx** Robert Blalock, John Scheele, Arnie Wong. **Starring** Yuen Biao (Ti Ming-chi), Adam Cheng Siu-Chow (Master Ting), Brigitte Lin Ching-Hsia (The Countess), Samo Hung Kam-Bo (Red Zu Warrior/Reverend Longbrows). 93 mins (HK cut)/98 mins (US cut). Colour.

A seminal historical-fantasy movie by an influential

director, *Zu* deals with adult themes of war and personal responsibility against a background drawn from Chinese mythology. Notable for its spectacular mountainscapes and imaginative though low-budget spfx – which achieved much with silk streamers and coloured light – it was a landmark in the HK fantasy genre and has achieved cult status in the West. The slightly shambolic storyline moves through elaborately choreographed set-pieces of MAGIC and martial arts as a young soldier in haunted 2nd-century China gains a prideful MENTOR (who later turns to EVIL) and a COMPANION in a QUEST for spiritual SWORDS which alone can defeat the "Blood MONSTER", temporarily held in check by WIZARD Longbrows and his sky MIRROR. A heavily emphasized moral is that, to win over Evil, agents of Good must act in perfect unison. Yuen Biao, Samo Hung and Brigitte Lin continue to be leading fantasy-action stars.   [KLM/DRL]